Handbook of MICROBIOLOGICAL MEDIA Fourth Edition

Ronald M. Atlas

ASM PRESS
Washington, D.C.

CRC Press
Taylor & Francis Group
Boca Raton London New York

CRC Press is an imprint of the
Taylor & Francis Group, an **informa** business

CRC Press
Taylor & Francis Group
6000 Broken Sound Parkway NW, Suite 300
Boca Raton, FL 33487-2742

Printed in the United States of America on acid-free paper
10 9 8 7 6 5 4 3 2 1

International Standard Book Number: 978-1-4398-0406-3 (Hardback)

Library of Congress Cataloging-in-Publication Data

Atlas, Ronald M., 1946-
Handbook of microbiological media / Ronald M. Atlas. -- 4th ed.
p. ; cm.
Includes bibliographical references and index.
ISBN 978-1-4398-0406-3 (hardcover : alk. paper)
1. Microbiology--Cultures and culture media--Handbooks, manuals, etc. I. Title.
[DNLM: 1. Culture Media--Handbooks. 2. Culture Media--Laboratory Manuals. 3. Microbiological Techniques--Handbooks. 4. Microbiological Techniques--Laboratory Manuals. QW 39 A881h 2010]

QR66.3.A85 2010
579.028'2--dc22 2009047096

Visit the Taylor & Francis Web site at
http://www.taylorandfrancis.com

and the CRC Press Web site at
http://www.crcpress.com

Table of Contents

Overview

The fourth edition of the *Handbook of Microbiological Media* includes the formulations and descriptions of 7,080 media used for cultivating microorganisms—more than 1500 more than in the previous edition. These include both classic and modern media used for the identification, cultivation, and maintenance of diverse bacteria, archaea, and fungi. Some of these microbiological media are produced by major suppliers of dehydrated media—including Oxoid, HiMedia, and BD Diagnostics (Difco, BBL, and GIBCO). These include all the media normally used in the clinical microbiology diagnostic laboratory and for the routine examination of food and water. Other media described in the fourth edition of the *Handbook of Microbiological Media* are used to cultivate specific strains of bacteria, archaea, fungi, and protists, including many anaerobes and extremophiles. The fourth edition of the *Handbook of Microbiological Media* includes the media needed to cultivate the numerous microorganisms currently available from the world's global bioresource centers (BRCs).

The breadth of culture media in this comprehensive resource is enormous and has greatly expanded in recent years with the exploration of extreme habitats and the use of molecular methods to identify new lineages of bacteria and archaea. The media also represent significant advances in the ability to use chromogenic substrates to identify specific species and strains of bacteria, e.g., *E. coli* O157 and methicillin resistant *Staphylococcus aureus* (MRSA). These media are extremely useful for clinical diagnostics and for the protection of the food supply from pathogenic microorganisms. Additionally many culture media are now available that are free of animal components. Plant based media eliminate possible contamination with prions which is important for production of vaccines and pharmaceuticals.

Organization

The media described in the *Handbook of Microbiological Media* are organized alphabetically. Synonyms for media are listed. The description of each medium includes its name(s), composition, instructions for preparation, commercial sources, safety cautions where needed, and uses.

Names of Media

Media often have numerous names. For the most part the fourth edition of the *Handbook of Microbiological Media* retains the original names assigned in the literature. In some cases media with identical compositions produced by different companies have different names. For example, Trypticase™ Soy Agar produced as a BBL product of BD Diagnostic Systems, Tryptone Soy Agar produced by Oxoid Unipath, and Tryptic Soy Agar produced as a Difco product of BD Diagnostic Systems have identical compositions. Many media also are known by acronyms. TSA, for example, is the common acronym for Trypticase™ Soy Agar. The fourth edition of the *Handbook of Microbiological Media* gives the various synonymous names and directs the reader to see the entry where the information about that medium is given. In cases where modifications to a medium yield a new medium, such media generally are listed with the original medium name, followed by the term modified—for example, TSA, Modified, rather than Modified TSA. Media that do not have formal names are listed according to the organism grown on that medium—for example, *Bacillus stearothermophilus* Broth.

Composition of Media

Media for the cultivation of microorganisms contain the substances necessary to support the growth of microorganisms. Due to the diversity of microorganisms and their diverse metabolic pathways, there are numerous media. Even slight differences in the composition of a medium can result in dramatically different growth characteristics of microorganisms.

When methods for culturing microorganisms were first developed in the nineteenth century, largely by Robert Koch and his colleagues, animal and plant tissues were principally used as sources of nutrients used to support microbial growth. One of the major discoveries of Fanny Hesse in Koch's laboratory was that agar could be used to form solidified culture media on which microorganisms would grow. Extracts of plants and animal tissues were prepared as broths or mixed with agar to form a variety of culture media. Virtually any plant, animal, or animal organ was considered for use in preparing media. Infusions were prepared from beef heart, calf brains, and beef liver, as a few examples. These classic infusions still form the primary components of many media that are widely used today, such as Brain Heart Infusion Agar and Liver Broth. In the last few years attempts have been made to reformulate many media containing animal tissue extracts with plant materials. This has added greatly to the number of media that are commercially available.

The composition section of each medium describes the ingredients that make up the medium, their amounts, and the pH. It lists those ingredients in order of decreasing amount. Solids are listed first showing the weights to be added, followed by liquids showing the volumes to be included in the medium.

The composition uses generic terms where these are applicable. For example, pancreatic digest of casein is marketed by various manufacturers as trypticase, tryptone, and other commercial product names. While there may well be differences between these products, such differences are undefined. Variations also occur between batches of products produced as digests of animal tissues.

Media for the cultivation of microorganisms have a source of carbon for incorporation into biomass. For autotrophs, the carbon source most often is carbon dioxide, which may be supplied as bicarbonate within the medium. Carbohydrates, such as glucose, or other organic compounds, such as acetate, various lipids, proteins, hydrocarbons, and other organic compounds, are included in media as sources of carbon for heterotrophs. These carbon sources may also serve as the supply of energy. Other compounds, such as ammonium ions, nitrite ions, elemental sulfur, and reduced iron, may be used as the sources of energy for the cultivation of autotrophs. Nitrogen also is required for microbial growth. It may be supplied as inorganic nitrogen compounds for the cultivation of some microorganisms but more commonly is supplied as proteins, peptones, or amino acids. Phosphates and metals, such as magnesium and iron, are also necessary components of microbiological media. Phosphates may also serve as buffers to maintain the pH of the medium within the growth tolerance limits of the microorganism being cultivated. Various additional growth factors may also be included in the media.

Agars

Agar is the most common solidifying agent used in microbiological media. Agar is a polysaccharide extract from marine algae. It melts at 84°C and solidifies at 38°C. Agar concentrations of 15.0g/L typically are used to form solid media. Lower concentrations of 7.5–10.0g/L are used to produce soft agars or semisolid media. Below are some agars used as solidifying agents in various media.

Agar Bacteriological (Agar No. 1)

An agar with low calcium and magnesium.

Agar, Bacto

A purified agar with reduced pigmented compounds, salts, and extraneous matter.

Agar, BiTek™
A

Agar prepared as a special technical grade.

Agar, Flake

A technical-grade agar.

Agar, Grade A

A select-grade agar containing minerals.

Agar, Granulated

A high-grade granulated agar that has been filtered, decolorized, and purified.

Agarose

A low-sulfate neutral gelling fraction of agar that is a complex galactose polysaccharide of near neutral charge.

Agar, Purified

A very high-grade agar that has been filtered, decolorized, and purified by washing and extraction of refined agars. It has reduced mineral content.

Agar Technical (Agar No. 3)

A technical-grade agar.

Ionagar

A purified agar.

Noble Agar

An agar that has been extensively washed and is essentially free of impurities.

Purified Agar

An agar that has been extensively washed and extracted with water and organic solvent. .

Peptones

Many complex media, that is, media in which not all the specific chemical components are known, contain peptones as the source of nitrogen. Peptones are hydrolyzed proteins formed by enzymatic or acidic digestion. Casein most often is used as the protein substrate for forming peptones, but other substances, such as soybean meal, also are commonly employed.

Below is a list of some of the peptones that are used as ingredients in various media.

Acidase™ Peptone

A hydrochloric acid hydrolysate of casein. It has a nitrogen content of 8% and is deficient in cystine and tryptophan.

Bacto Casitone

A pancreatic digest of casein.

Bacto Peptamin

A peptic digest of animal tissues.

Bacto Peptone

An enzymatic digest of animal tissues. It has a high concentration of low molecular weight peptones and amino acids.

Bacto Proteose Peptone

An enzymatic digest of animal tissues. It has a high concentration of high molecular weight peptones.

Bacto Soytone

A enzymatic hydrolysate of soybean meal.

Bacto Tryptone

A pancreatic digest of casein.

Bacto Triplets

An enzymatic hydrolysate containing numerous peptides, including those of higher molecular weights.

Biosate™ Peptone

A hydrolysate of plant and animal proteins.

Casein Hydrolysate

A hydrolysate of casein prepared with hydrochloric acid digestion under pressure and neutralized with sodium hydroxide. It contains total nitrogen of 7.6% and NaCl of 28.3%.

Gelatone

A pancreatic digest of gelatin.

Gelysate™ Peptone

A pancreatic digest of gelatin deficient in cystine and tryptophan and which has a low carbohydrate content.

HiVeg Peptone

An enzymic hydrolysate of vegetable proteins that gives comparable growth promoting properties as animal origin peptone.

HiVeg Peptone No. 1

An enzymic hydrolysate of specially selected vegetable proteins that gives comparable growth promoting properties as animal origin peptone No. 1.

HiVeg Peptone No. 2

An enzymic hydrolysate of vegetable proteins that gives comparable growth promoting properties as gelatin peptone.

HiVeg Peptone No. 3

An enzymic hydrolysate of vegetable proteins that gives comparable growth promoting properties as proteose peptone.

HiVeg Peptone No. 4

An enzymic hydrolysate of vegetable proteins that gives comparable growth promoting properties as mycological peptone.

HiVeg Peptone No. 5

An enzymic hydrolysate of vegetable proteins that gives comparable growth promoting properties as bio peptone.

HiVeg Peptone B

An enzymic hydrolysate of vegetable proteins that gives comparable growth promoting properties as proteose peptone.

HiVeg Peptone C

An enzymic hydrolysate of vegetable proteins that gives comparable growth promoting properties as animal origin peptone A.

HiVeg Special Peptone

An enzymic hydrolysate of vegetable proteins that gives comparable growth promoting properties as neopeptone and animal special peptone.

Lactoalbumin Hydrolysate

A pancreatic digest of lactoalbumin, a milk whey protein. It has high levels of amino acids. It contains total nitrogen of 11.9% and NaCl of 1.4%.

Liver Digest Neutralized

A papaic digest of liver that contains total nitrogen of 11.0% and NaCl of 1.6%.

Mycological Peptone

A peptone that contains total nitrogen of 9.5% and NaCl of 1.1%.

Myosate™ Peptone

A pancreatic digest of heart muscle.

Neopeptone

An enzymatic digest of protein.

Peptone Bacteriological Neutralized

A mixed pancreatic and papaic digest of animal tissues. It contains total nitrogen of 14.0% and NaCl of 1.6%.

Peptone P

A peptic digest of fresh meat that has a high sulfur content and contains total nitrogen of 11.12% and NaCl of 9.3%.

Peptonized Milk

A pancreatic digest of high-grade skim milk powder. It has a high carbohydrate and calcium concentration. It contains total nitrogen of 5.3% and NaCl of 1.6%.

Phytone™ Peptone

A papaic digest of soybean meal. It has a high vitamin and a high carbohydrate content.

Polypeptone™ Peptone

A mixture of peptones composed of equal parts of pancreatic digest of casein and peptic digest of animal tissue.

Proteose Peptone

A specialized peptone prepared from a mixture of peptones that contains a wide variety of high molecular weight peptides. It contains total nitrogen of 12.7% and NaCl of 8.0%.

Proteose Peptone No. 2

An enzymatic digest of animal tissues with a high concentration of high molecular weight peptones.

Proteose Peptone No. 3

An enzymatic digest of animal tissues. It has a high concentration of high molecular weight peptones.

Soy Peptone (Soya Peptone)

A papaic digest of soybean meal with a high carbohydrate concentration. It contains total nitrogen of 8.7% and NaCl of 0.4%.

Soytone

A papaic digest of soybean meal.

Special Peptone

A mixture of peptones, including meat, plant, and yeast digests. It contains a wide variety of peptides, nucleotides, and minerals. It contains total nitrogen of 11.7% and NaCl of 3.5%.

Thiotone™ E Peptone

An enzymatic digest of animal tissue.

Trypticase™ Peptone

A pancreatic digest of casein. It has a very low carbohydrate content and a relatively high tryptophan content.

Tryptone

A pancreatic digest of casein. It contains total nitrogen of 12.7% and NaCl of 0.4%.

Tryptone T

A pancreatic digest of casein with lower levels of calcium, magnesium, and iron than tryptone. It contains total nitrogen of 11.7% and NaCl of 4.9%.

Tryptose

An enzymatic hydrolysate containing high molecular weight peptides. It contains total nitrogen of 12.2% and NaCl of 5.7%.

Meat and Plant Extracts

Meat and plant infusions are aqueous extracts that are commonly used as sources of nutrients for the cultivation of microorganisms. Such infusions contain amino acids and low molecular weight peptides, carbohydrates, vitamins, minerals, and trace metals. Extracts of animal tissues contain relatively high concentrations of water-soluble protein components and glycogen. Extracts of plant tissues contain relatively high concentrations of carbohydrates.

With regard to infusions, many media list as an ingredient infusion from beef heart or another animal tissue. This ingredient is prepared by boiling a given amount of the animal tissue (e.g., 500.0g), and then using the liquid or, more commonly, drying the broth and using the solids from the infusion. The actual weight of the dry solids extracted from the hot water used to create the infusion varies, and so the ingredient typically is simply listed as 500.0g beef heart infusion, although the actual weight of solids recovered from the infusion and used in the medium is far less. Brain heart infusion is prepared from calf brains and beef heart.

Below is a list of some of the meat and plant extracts that are used as ingredients in various media.

Bacto Beef

A desiccated powder of lean beef.

Bacto Beef Extract

An extract of beef (paste).

Bacto Beef Extract Desiccated

An extract of desiccated beef.

Bacto Beef Heart for Infusion

A desiccated powder of beef heart.

Bacto Liver

A desiccated powder of beef liver.

HiVeg Acid Hydrolysate

An acid hydrolysate of vegetable proteins suitable for use in culture media requiring amino acid mixture with growth promotional characteristics matches with casein acid hydrolysate.

HiVeg Acid Hydrolysate No.1

An acid hydrolysate of vegetable proteins with growth enhancing performance similar to casein acid hydrolysate.

HiVeg Extract

An extract of vegetable proteins with growth enhancing performance similar to beef extract.

HiVeg Extract No. 1

An extract of vegetable proteins with growth enhancing performance similar to meat extract.

HiVeg Extract No. 2

An extract of vegetable proteins with growth enhancing performance similar to liver extract.

HiVeg Hydrolysate

An enzymatic hydrolysate of vegetable proteins with growth enhancing performance similar to casein enzyme hydrolysate (tryptone).

HiVeg Hydrolysate No. 1

An enzymatic hydrolysate of vegetable proteins with growth enhancing performance similar to milk proteins (tryptose).

HiVeg Hydrolysate No. 2

An enzymatic hydrolysate of vegetable proteins with growth enhancing performance similar to liver hydrolysate.

HiVeg Hydrolysate No. 3

An enzymatic hydrolysate of vegetable proteins with growth enhancing performance similar to peptonized milk.

HiVeg Hydrolysate No. 4

An enzymatic hydrolysate of vegetable proteins rich in amino acids.

HiVeg Hydrolysate No. 6

A hydrolysate of vegetable proteins with growth enhancing performance similar to casein enzyme hydrolysate Type-II.

HiVeg Hydrolysate C

An enzymatic hydrolysate of plant extract with growth enhancing performance similar to casein enzyme hydrolysate.

HiVeg Infusion

An infusion from vegetables with growth enhancement similar to heart infusion.

HiVeg Infusion No. 1

An infusion from vegetables with growth enhancement similar to liver infusion.

HiVeg Infusion Powder

An infusion from vegetables with growth enhancement similar to brain heart infusion.

HiVeg Special Infusion
An infusion from vegetables with growth enhancement similar to brain heart infusion.

Lab-Lemco
A meat extract powder.

Liver Desiccated
Dehydrated ox livers.

Malt Extract
A water-soluble extract from germinated grain dried by low-temperature evaporation. It has a high carbohydrate content. It contains total nitrogen of 1.1% and NaCl of 0.1%.

Growth Factors

Many microorganisms have specific growth factor requirements that must be included in media for their successful cultivation. Vitamins, amino acids, fatty acids, trace metals, and blood components often must be added to media. In some cases, specific defined components are used to meet the growth factor requirements. Incorporation of growth factors is used to enrich, that is, to increase the numbers of particular species of microorganisms. Most often, mixtures of growth factors are used in microbiological media. Acid hydrolysates of casein commonly are used as sources of amino acids. Extracts of yeast cells also are employed as sources of amino acids and vitamins for the cultivation of microorganisms. Many media, particularly those employed in the clinical laboratory, contain blood or blood components that serve as essential nutrients for fastidious microorganisms. X factor (heme) and V factor (nicotinamide adenine dinucleotide) often are supplied by adding hemoglobin, IsoVitaleX, and/or Supplement VX. Below is a list of some of the growth factors that are used as ingredients in various media.

Bacto Casamino Acids
A mixture of amino acids formed by acid hydrolysis of casein.

Bacto Vitamin-Free Casamino Acids
A mixture of amino acids formed by acid hydrolysis of casein that is free of vitamins.

Bovine Albumin
Bovine albumin fraction V 0.2% in 0.85% saline solution.

Bovine Blood, Citrated
Calf blood washed and treated with sodium citrate as an anticoagulant.

Bovine Blood, Defibrinated
Calf blood treated to denature fibrinogen without causing cell lysis.

Campylobacter Growth Supplement
Sodium pyruvate, sodium metabisulfite, and $FeSO_4$.

Castenholz Salts
Agar, $NaNO_3$, Na_2HPO_4, KNO_3, nitrilotriacetic acid, $MgSO_4 \cdot 7H_2O$, $CaSO_4 \cdot 2H_2O$, NaCl, $FeCl_3$, $MnSO_4$, H_3BO_3, $ZnSO_4$, $CoCl_2 \cdot 6H_2O$, Na_2MoO_4, $CuSO_4$, and H_2SO_4.

CM1 Broth Powder
Lab Lemco beef extract, yeast extract, peptone, NaCl.

CVA Enrichment
Glucose, L-cysteine·HCl·H_2O, vitamin B_{12}, L-glutamine, L-cystine·2HCl, adenine, nicotinamide adenine dinucleotide, cocarboxylase, guanine·HCl, $Fe(NO_3)_3$, *p*-aminobenzoic acid, and thiamine·HCl.

Cysteine Sulfide Reducing Agent
L-Cysteine·HCl·H_2O and $Na_2S \cdot 9H_2O$.

Dubos Medium Albumin
Albumin fraction V, glucose, and saline solution.

Dubos Oleic Albumin Complex
Alkalinized oleic acid, albumin fraction V, and saline solution.

Egg Yolk Emulsion
Chicken egg yolks and whole chicken egg.

Egg Yolk Emulsion, 50%
Chicken egg yolks, whole chicken egg, and saline solution.

EY Tellurite Enrichment
Egg yolk suspension with potassium tellurite.

Fildes Enrichment
A peptic digest of sheep or horse blood that is a rich source of growth factors, including hemin and nicotinamide adenine dinucleotide.

Fresh Yeast Extract Solution
Live, pressed, starch-free, hydrolyzed Baker's yeast.

Hemin Solution
Hemin and NaOH.

Hemoglobin
Dried bovine hemoglobin. Used to provide hemin required by many fastidious microorganisms.

Hemoglobin Solution 2%
Provides hemin required by many fastidious microorganisms.

HiVeg Yeast Hydrolysate
A dried extract from autolysing yeast cells (*Saccharomyces*) that is rich in vitamins and other nutritive substances such as free amino acids.

Hoagland Trace Elements Solution, Modified
H_3BO_3, $MnCl_2 \cdot 4H_2O$, $AlCl_3$, $CoCl_2$, $CuCl_2$, KI, $NiCl_2$, $ZnCl_2$, $BaCl_2$, Na_2MoO_4, $SeCl_4$, $SnCl_2 \cdot 2H_2O$, $NaVO_3 \cdot H_2O$, KBr, and LiCl.

Horse Blood, Citrated
Horse blood washed and treated with sodium citrate used as an anticoagulant.

Horse Blood, Defibrinated
Horse blood treated to denature fibrinogen without causing cell lysis.

Horse Blood, Hemolysed
Horse blood treated to lyse cells.

Horse Blood, Oxalated
Horse blood treated with potassium oxalate as an anticoagulant.

Horse Serum
Horse blood is allowed to clot at 2°–8°C so that the serum separates; the serum is filter sterilized. Serum usually is inactivated by heating to 56°C for 30 min to eliminate lipases that would cause degradation of lipids and inactivation of complement.

Hutner's Mineral Base
$MgSO_4 \cdot 7H_2O$, $CaCl_2 \cdot 2H_2O$, $FeSO_4 \cdot 7H_2O$, $Na_2B_4O_7 \cdot 10H_2O$, $(NH_4)_2MoO_4$, $FeSO_4 \cdot 7H_2O$, $MnSO_4 \cdot 7H_2O$, $Co(NO_3)_2 \cdot 6H_2O$, $ZnSO_4 \cdot 7H_2O$, $CuSO_4 \cdot 5H_2O$, EDTA, and nitrilotriacetic acid.

IsoVitaleX® Enrichment
Glucose, L-cysteine·HCl, L-glutamine, L-cystine, adenine, nicotinamide adenine dinucleotide, vitamin B_{12}, thiamine pyrophosphate, guanine·HCl, $Fe(NO_3)_3 \cdot 6H_2O$, *p*-aminobenzoic acid, and thiamine·HCl.

Legionella Agar Enrichment
L-Cysteine and ferric pyrophosphate.

Legionella BCYE Growth Supplement

ACES buffer/KOH, ferric pyrophosphate, L-cysteine-HCl, and α-ketoglutarate. For the enrichment of *Legionella* species.

Leptospira Enrichment

Lyophilized pooled rabbit serum containing hemoglobin that provides long-chain fatty acids and B vitamins for growth of *Leptospira* species.

Metals "44"

$ZnSO_4{\cdot}7H_2O$, $FeSO_4{\cdot}7H_2O$, $MnSO_4{\cdot}7H_2O$, $Na_2B_4O_7{\cdot}10H_2O$, $CuSO_4{\cdot}5H_2O$, $Co(NO_3)_2{\cdot}6H_2O$, and EDTA.

Middlebrook ADC Enrichment

NaCl, bovine albumin fraction V, glucose, and catalase. The albumin binds free fatty acids that may be toxic to mycobacteria.

Middlebrook OADC Enrichment

NaCl, bovine albumin, glucose, oleic acid, and catalase. The albumin binds free fatty acids that may be toxic to mycobacteria; the enrichment provides oleic acid used by *Mycobacterium tuberculosis* for growth.

Mycoplasma Enrichment without Penicillin

Horse serum, fresh autolysate of yeast-yeast extract, and thallium acetate. Provides cholesterol and nucleic acids for growth of *Mycoplasma* species. The thallium selectively inhibits other microorganisms.

Mycoplasma Supplement

Yeast extract and horse serum.

Nitsch's Trace Elements

$MnSO_4$, H_3BO_3, $ZnSO_4$, Na_2MoO_4, $CuSO_4$, $CoCl_2{\cdot}6H_2O$, and H_2SO_4.

Oleic Albumin Complex

NaCl, bovine albumin fraction V, and oleic acid. The albumin binds free fatty acids that may be toxic to mycobacteria, and the enrichment provides oleic acid that is used by *Mycobacterium tuberculosis* for growth.

PPLO Serum Fraction

Serum fraction A.

Rabbit Blood, Citrated

Rabbit blood washed and treated with sodium citrate as an anticoagulant.

Rabbit Blood, Defibrinated

Rabbit blood treated to denature fibrinogen without causing cell lysis.

RPF Supplement

Fibrinogen, rabbit plasma, trypsin inhibitor, and potassium tellurite. For the selection and nutrient supplementation of *Staphylococcus aureus*.

Sheep Blood, Citrated

Sheep blood washed and treated with sodium citrate as an anticoagulant.

Sheep Blood, Defibrinated

Sheep blood treated to denature fibrinogen without causing cell lysis.

SLA Trace Elements

$FeCl_2{\cdot}4H_2O$, H_3BO_3, $CoCl_2{\cdot}6H_2O$, $ZnCl_2$, $Na_2MoO_4{\cdot}2H_2O$, $MnCl_2{\cdot}4H_2O$, $NiCl_2{\cdot}6H_2O$, $CuCl_2{\cdot}2H_2O$, and $Na_2SeO_3{\cdot}5H_2O$.

Soil Extract

African Violet soil and Na_2CO_3.

Supplement A

Yeast concentrate with Crystal Violet.

Supplement B

Yeast concentrate, glutamine, coenzyme, cocarboxylase, hematin, and growth factors.

Supplement C

Yeast concentrate.

Supplement VX

Essential growth factors V and X.

Trace Elements Mixture

Ethylenediamine tetraacetic acid (EDTA), $ZnSO_4{\cdot}7H_2O$, $CaCl_2$, $MnCl_2{\cdot}4H_2O$, $FeSO_4{\cdot}7H_2O$, $(NH_4)_6Mo_7O_{24}{\cdot}4H_2O$, $CoCl_2{\cdot}6H_2O$, and $CuSO_4{\cdot}5H_2O$.

Trace Elements Solution HO-LE

H_3BO_3, $MnCl_2{\cdot}4H_2O$, sodium tartrate, $FeSO_4{\cdot}7H_2O$, $ZnCl_2$, $CoCl_2{\cdot}6H_2O$, $CuCl_2{\cdot}2H_2O$, and $Na_2MoO_4{\cdot}2H_2O$.

Trace Elements Solution SL-6

H_3BO_3, $CoCl_2{\cdot}6H_2O$, $ZnSO_4{\cdot}7H_2O$, $MnCl_2{\cdot}4H_2O$, $NiCl_2{\cdot}6H_2O$, $Na_2MoO_4{\cdot}H_2O$, and $CuCl_2{\cdot}2H_2O$.

Trace Elements Solution SL-7

$FeCl_2{\cdot}4H_2O$, $CoCl_2{\cdot}6H_2O$, $MnCl_2{\cdot}4H_2O$, $ZnCl_2$, H_3BO_3, $Na_2MoO_4{\cdot}2H_2O$, $NiCl_2{\cdot}6H_2O$, $CuCl_2{\cdot}2H_2O$, and HCl.

Trace Elements Solution SL-8

Disodium EDTA, $FeCl_2{\cdot}4H_2O$, $CoCl_2{\cdot}6H_2O$, $MnCl_2{\cdot}4H_2O$, $NiCl_2{\cdot}6H_20$, $ZnCl_2$, H_3BO_3, $NaMoO_4{\cdot}2H_2O$, and $CuCl_2{\cdot}2H_2O$.

Trace Elements Solution SL-10

$FeCl_2{\cdot}4H_2O$, $CoCl_2{\cdot}6H_2O$, $MnCl_2{\cdot}4H_2O$, $NiCl_2{\cdot}6H_2O$, H_3BO_3, $ZnCl_2$, $Na_2MoO_4{\cdot}2H_2O$, $CuCl_2{\cdot}2H_2O$, and HCl (25% solution).

Trace Metals A-5 Mix

$ZnSO_4{\cdot}7H_2O$, $Co(NO_3)_2{\cdot}6H_2O$ $Na_2MoO_4{\cdot}2H_2O$, $CuSO_4{\cdot}5H_2O$, H_3BO_3, and $MnCl_2{\cdot}4H_2O$.

VA Vitamin Solution

Nicotinamide, thiamine·HCl, *p*-aminobenzoic acid, biotin, calcium pantothenate, pyridoxine·2HCl, and cyanocobalamin.

Vitamin K_1 Solution

Vitamin K_1 and ethanol.

Vitox Supplement

Glucose, L-cysteine·HCl, L-glutamine, L-cystine, adenine sulfate, nicotinamide adenine dinucleotide, cocarboxylase, guanine·HCl, $Fe(NO_3)_3{\cdot}6H_2O$, *p*-aminobenzoic acid, vitamin B_{12}, and thiamine·HCl.

Wolfe's Mineral Solution

$MgSO_4{\cdot}7H_2O$, nitriloacetic acid, NaCl, $MnSO_4{\cdot}H_2O$, $FeSO_4{\cdot}7H_2O$, $CoCl_2{\cdot}6H_2O$, $CaCl_2$, $ZnSO_4{\cdot}7H_2O$, $CuSO_4{\cdot}5H_2O$, $AlK(SO_4)_2{\cdot}12H_2O$, $Na_2MoO_4{\cdot}2H_2O$, and H_3BO_3.

Wolfe's Vitamin Solution

Pyridoxine·HCl, thiamine·HCl, riboflavin, nicotinic acid, calcium pantothenate, *p*-aminobenzoic acid, thioctic acid, biotin, folic acid, and cyanocobalamin.

Yeast Autolysate Growth Supplement

Yeast autolysate fractions, glucose, and $NaHCO_3$.

Yeast Dialysate

Active, dried yeast.

Yeast Extract

A water-soluble extract of autolyzed yeast cells.

Yeast Extract Powder

A dried extract obtained from yeast cells (*Saccharomyces*) prepared under controlled conditions that retains its vitamin content and other nutritive values such as free amino acids.

Yeastolate

A water-soluble fraction of autolyzed yeast cells rich in vitamin B complex.

Selective Components

Many media contain selective components that inhibit the growth of nontarget microorganisms. Selective media are especially useful in the isolation of specific microorganisms from mixed populations. In many media for the study of microorganisms in nature, compounds are included in the media as sole sources of carbon or nitrogen so that only a few types of microorganisms can grow. Selective toxic compounds are also frequently used to select for the cultivation of particular microbial species. The isolation of a pathogen from a stool specimen, for example, where there is a high abundance of nonpathogenic normal microbiota, requires selective media. Often, antimicrobics or other selectively toxic compounds are incorporated into media to suppress the growth of the background microbiota while permitting the cultivation of the target organism of interest. Bile salts, selenite, tetrathionate, tellurite, azide, phenylethanol, sodium lauryl sulfate, high sodium chloride concentrations, and various dyes—such as eosin, Crystal Violet, and Methylene Blue—are used as selective toxic chemicals. Antimicrobial agents used to suppress specific types of microorganisms include ampicillin, chloramphenicol, colistin, cycloheximide, gentamicin, kanamycin, nalidixic acid, sulfadiazine, and vancomycin. Various combinations of antimicrobics are effective in suppressing classes of microorganisms, such as enteric bacteria.

Below are some of the selective agents, principally antimicrobic mixtures used for the selective isolation of pathogens.

Ampicillin Selective Supplement

Ampicillin. Used in media for the selection of *Aeromonas hydrophila.*

Anaerobe Selective Supplement GN

Hemin, menadione, sodium succinate, nalidixic acid, and vancomycin. For the selection of Gram-negative anaerobes.

Anaerobe Selective Supplement NS

Hemin, menadione, sodium pyruvate, and nalidixic acid. For the selection of non-sporulating anaerobes.

Bacillus cereus Selective Supplement

Polymyxin B. For the selection of *Bacillus cereus.*

Bordetella Selective Supplement

Cephalexin. For the selection of *Bordetella* species.

Brucella Selective Supplement

Polymyxin B, bacitracin, cycloheximide, nalidixic acid, nystatin, and vancomycin. For the selection of *Brucella* species.

Campylobacter Selective Supplement Blaser-Wang

Vancomycin, polymyxin B, trimethoprim, amphotericin B, cephalothin. For the selection of *Campylobacter* species.

Campylobacter Selective Supplement Butzler

Bacitracin, cycloheximide, colistin sulfate, sodium cephazolin, and novobiocin. For the selection of *Campylobacter* species.

Campylobacter Selective Supplement Preston

Polymyxin B, rifampicin, trimethoprim, and cycloheximide. For the selection of *Campylobacter* species.

Campylobacter Selective Supplement Skirrow

Vancomycin, trimethoprim, and polymyxin B. For the selection of *Campylobacter* species.

CCDA Selective Supplement

Cefoperazone and amphotericin B. For the selection of *Campylobacter* species.

Cefoperazone Selective Supplement

Cefoperazone. For the selection of *Campylobacter* species.

CFC Selective Supplement

Cetrimide, fucidin, and cephaloridine. For the selection of pseudomonads.

Chapman Tellurite Solution

Potassium tellurite 1% solution.

Chloramphenicol Selective Supplement

Chloramphenicol. For the selection of yeasts and filamentous fungi.

Clostridium difficile Selective Supplement

D-Cycloserine and cefoxitin. For the selection of *Clostridium difficile.*

CN Inhibitor

Cesulodin and novobiocin. It inhibits enteric Gram-negative microorganisms.

CNV Antimicrobic

Colistin sulfate, nystatin, and vancomycin.

CNVT Antimicrobic

Colistin sulfate, nystatin, vancomycin, and trimethoprim lactate.

Colbeck's Egg Broth

Egg emulsion and saline solution.

Fraser Supplement

Ferric ammonium sulfate, nalidixic acid, and acriflavin hydrochloride. For the selection of *Listeria* species.

Gardnerella vaginalis Selective Supplement

Gentamicin sulfate, nalidixic acid, and amphotericin B. For the selection of *Gardnerella vaginalis.*

GC Selective Supplement

Yeast autolysate, glucose, Na_2HCO_3, vancomycin, colistin methane sulfonate, nystatin, and trimethoprim. For the selection of *Neisseria* species.

Helicobacter pylori Selective Supplement Dent

Vancomycin, trimethoprim, cefulodin, and amphotericin B. For the selection of *Helicobacter pylori.*

Kanamycin Sulfate Selective Supplement

Kanamycin sulfate. For the selection of enterococci.

LCAT Selective Supplement

Lincomycin, colistin sulfate, amphotericin B, and trimethoprim. For the selection of *Neisseria* species.

Legionella BMPA Selective Supplement

Cefamandole, polymyxin B, and anisomycin. For the selection of *Legionella* species.

Legionella GVPC Selective Supplement

Glycine, vancomycin hydrochloride, polymixin B sulfate, and cycloheximide. For the selection of *Legionella* species.

Legionella MWY Selective Supplement

Glycine, polymyxin B, anisomycin, vancomycin, Bromthymol B, and Bromcresol Purple. For the selection of *Legionella* species.

Listeria Primary Selective Enrichment Supplement

Nalidixic acid and acriflavin. For the selection of *Listeria* species.

Listeria Selective Enrichment Supplement

Nalidixic acid, cycloheximide, and acriflavin. For the selection of *Listeria* species.

Listeria Selective Supplement MOX

Colistin and moxalactam. For the selection of *Listeria monocytogenes*.

Listeria Selective Supplement Oxford

Cycloheximide, colistin sulfate, acriflavin, cefotetan, and fosfomycin. For the selection of *Listeria* species.

Modified Oxford Antimicrobic Supplement

Moxalactam and colistin sulfate.

MSRV Selective Supplement

Novobiocin. For the selection of *Salmonella*.

Mycoplasma Supplement G

Horse serum, yeast extract, thallous acetate, and penicillin. For the selection of *Mycoplasma* species.

Mycoplasma Supplement P

Horse serum, yeast extract, thallous acetate, glucose, Phenol Red, Methylene Blue, penicillin, and *Mycoplasma* broth base. For the selection of *Mycoplasma* species.

Mycoplasma Supplement S

Yeast extract, horse serum, thallium acetate, and penicillin.

Oxford Antimicrobic Supplement

Cycloheximide, colistin, acriflavin, cefotetan, and fosfomycin.

Oxgall

Dehydrated fresh bile. For the selection of bile-tolerant bacteria.

Oxytetracycline GYE Supplement

Oxytetracycline in a buffer. For the selection of yeasts and filamentous fungi.

PALCAM Selective Supplement

Polymyxin B, acriflavin hydrochloride, and ceftazidime. For the selection of *Listeria monocytogenes*.

Perfringens OPSP Selective Supplement A

Sodium sulfadiazine. For the selection of *Clostridium perfringens*.

Perfringens SFP Selective Supplement A

Kanamycin sulfate and polymyxin B. For the selection of *Clostridium perfringens*.

Perfringens TSC Selective Supplement A

D-Cycloserine. For the selection of *Clostridium perfringens*.

Sodium Desoxycholate

Sodium salt of desoxycholic acid.

Sodium Taurocholate

Sodium salt of conjugated bile acid—75% sodium taurocholate and 25% bile salts. For the selection of bile-tolerant bacteria.

STAA Selective Supplement

Streptomycin sulfate, cycloheximide, and thallous acetate. For the selection of *Brochothrix thermosphacta*.

Staph/Strep Selective Supplement

Nalidixic acid and colistin sulfate. For the selection of *Staphylococcus* species and *Streptococcus* species.

Streptococcus Selective Supplement COA

Colistin sulfate and oxolinic acid. For the selection of *Streptococcus* species.

Sulfamandelate Supplement

Sodium sulfacetamide and sodium mandelate. For the selection of *Salmonella* species.

Tellurite Solution

A solution containing potassium tellurite. Inhibits Gram-negative and most Gram-positive microorganisms. For the isolation of *Corynebacterium* species, *Streptococcus* species, *Listeria* species, and *Candida albicans*.

Tinsdale Supplement

Serum, potassium tellurite, and sodium thiosulfate. For the selection of *Corynebacterium diphtheriae*.

V C A Inhibitor

Vancomycin, colistin, anisomycin, and trimethoprim. Inhibits most Gram-negative and Gram-positive bacteria and yeasts. For the isolation of *Neisseria* species.

V C A T Inhibitor

Vancomycin, colistin, anisomycin, and trimethoprim lactate. Inhibits most Gram-negative and Gram-positive bacteria and yeasts. For the isolation of *Neisseria* species.

V C N Inhibitor

Colistin, vancomycin, and nystatin. Inhibits most Gram-negative and Gram-positive bacteria and yeasts. For the isolation of *Neisseria* species.

V C N T Inhibitor

Colistin, vancomycin, nystatin, and trimethoprim lactate. Inhibits most Gram-negative and Gram-positive bacteria and yeasts. For the isolation of *Neisseria* species.

Yersinia Selective Supplement

Cefsulodin, irgasan, and novobiocin. For the selection of *Yersinia enterocolitica*.

Differential Components

The differentiation of many microorganisms is based upon the production of acid from various carbohydrates and other carbon sources or the decarboxylation of amino acids. Some media include indicators, particularly of pH, that permit the visual detection of changes in pH resulting from such metabolic reactions. A number of new media also include chromogenic dyes that change color when specific enzymatic reactions occur. Some of these have been developed based upon molecular biology determinations of specific genes that are useful for the differentiation of bacterial taxa.

Below is a list of some commonly used pH indicators.

pH Indicator	pH Range	Acid Color	Alkaline Color
m-Cresol Purple	0.5–2.5	Red	Yellow
Thymol Blue	1.2–2.8	Red	Yellow
Bromphenol Blue	3.0–4.6	Yellow	Blue
Bromcresol Green	3.8–5.4	Yellow	Blue
Chlorcresol Green	4.0–5.6	Yellow	Blue
Methyl Red	4.2–6.3	Red	Yellow
Chlorphenol Red	5.0–6.6	Yellow	Red
Bromcresol Purple	5.2–6.8	Yellow	Purple
Bromthymol Blue	6.0–7.6	Yellow	Blue
Phenol Red	6.8–8.4	Yellow	Red
Cresol Red	7.2–8.8	Yellow	Red
m-Cresol Purple	7.4–9.0	Yellow	Purple
Thymol Blue	8.0–9.6	Yellow	Blue
Cresolphthalein	8.2–9.8	Colorless	Red
Phenolphthalein	8.3–10.0	Colorless	Red

pH Buffers

Maintaining the pH of media usually is accomplished by the inclusion of suitable buffers. Because microorganisms grow optimally only within certain limits of a pH range, the pH generally is maintained within a few tenths of a pH unit.

For the phosphage buffers, the pH is established by using varying volumes of equimolar concentrations of Na_2HPO_4 and NaH_2PO_4.

pH	Na_2HPO_4 (mL)	NaH_2PO_4 (mL)
5.4	3.0	97.0
5.6	5.0	95.0
5.8	7.8	92.2
6.0	12.0	88.0
6.2	18.5	81.5
6.4	26.5	73.5
6.6	37.5	62.5
6.8	50.0	50.0
7.0	61.1	38.9
7.2	71.5	28.5
7.4	80.4	19.6
7.6	86.8	13.2
7.8	91.4	8.6
8.0	94.5	5.5

Trademarks

The names of some media, components of media, and other terms are registered trademarks. The trademarked items referred to in the *Handbook of Microbiological Media* are listed below.

American Type Culture Collection® and ATCC® are trademarks of the American Type Culture Collection.

Bacto®, BiTek®, and Difco® are trademarks of Difco Laboratories (registered trademarks owned by Becton Dickinson and Company).

Oxoid® and Lab–Lemco® are trademarks of Unipath Ltd.

HiVeg® and HiChrome® are registered trademarks of HiMedia Laboratories Pvt. Limited, India.

CHROMagar® registered trademark of CHROMagar Microbiology diagnostics

CandiSelect 4® registered trademark of BioRad

Acidase®, BBL®, Biosate®, CTA Medium®, DTA Medium®, DCLS Agar®, Desoxycholate®, Desoxycholate Agar®, Desoxycholate Citrate Agar®, Enterococcosel®, Eugonagar®, Eugonbroth®, GC-Lect®, Gelysate®, IsoVitaleX®, Mycobactosel®, Mycophil®, Mycosel®, Myosate®, Phytone®, Polypeptone®, Salmon®-β-D-GAL, Selenite-F Enrichment®, Thiotone®, Trichosel®, Triton®, Trypticase®, TSA II®, and TSI Agar® are trademarks of Becton Dickinson and Co.

Preparation of Media

The ingredients in a medium are usually dissolved, and the medium is then sterilized. When agar is used as a solidifying agent, the medium must be heated gently, usually to boiling, to dissolve the agar. In some cases where interactions of components, such as metals, would cause precipitates, solutions must be prepared and occasionally sterilized separately before mixing the various solutions to prepare the complete medium. The pH often is adjusted prior to sterilization, but in some cases sterile acid or base is used to adjust the pH of the medium following sterilization. Many media are sterilized by exposure to elevated temperatures. The most common method is to autoclave the medium. Different sterilization procedures are employed when heat-labile compounds are included in the formulation of the medium.

Tyndallization

Exposure to steam at 100°C for 30 min will kill vegetative bacterial cells but not endospores. Such exposure can be achieved using flowing steam in an Arnold sterilizer. By allowing the medium to cool and incubate under conditions where endospore germination will occur and by repeating the 100°C–30 min exposure on three successive days, the medium can be sterilized because all the endospores will have germinated and the heat exposure will have killed all the vegetative cells. This process of repetitive exposure to 100°C is called tyndallization, after its discoverer, John Tyndall.

Inspissation

Inspissation is a heat exposure method that is employed with high-protein materials, such as egg-containing media, that cannot withstand the high temperatures used in autoclaving. This process causes coagulation of the protein without greatly altering its chemical properties. Several different protocols can be followed for inspissation. Using an Arnold sterilizer or a specialized inspissator, the medium is exposed to 75°–80°C for 2 hr on each of three successive days. Inspissation using an autoclave employs exposure to 85°–90°C for 10 min achieved by having a mixture of air and steam in the chamber, followed by a 15 min exposure during which the temperature is raised to 121°C using only steam under pressure in the chamber; the temperature then is slowly lowered to less than 60°C.

Autoclaving

Autoclaving uses exposure to steam, generally under pressure, to kill microorganisms. Exposure for 15 min to steam at 15 psi—121°C is most commonly used. Such exposure kills vegetative bacterial cells and bacterial endospores. However, some substances do not tolerate such exposures, and lower temperatures and different exposure times are sometimes employed. Media containing carbohydrates often are sterilized at 116°–118°C in order to prevent the decomposition of the carbohydrate and the formation of toxic compounds that would inhibit microbial growth. Below is a list of pressure–temperature relationships.

Pressure—psi	Temperature—°C
0	100.0
1	101.9
2	103.6
3	105.3
4	106.9
5	108.4
6	109.8
7	111.3
8	112.6
9	113.9
10	115.2
11	116.4
12	117.6
13	118.8
14	119.9
15	121.0
16	122.0
17	123.0
18	124.0
19	125.0
20	126.0
21	126.9
22	127.8
23	128.7
24	129.6
25	130.4

Filtration

Filtration is commonly used to sterilize media containing heat-labile compounds. Liquid media are passed through sintered glass or membranes, typically made of cellulose acetate or nitrocellulose, with small pore sizes. A membrane with a pore size of 0.2mm will trap bacterial cells and, therefore, sometimes is called a bacteriological filter. By preventing the passage of microorganisms, filtration renders fluids free of bacteria and eukaryotic microorganisms, that is, free of living organisms, and hence sterile. Many carbohydrate solutions, antibiotic solutions, and vitamin solutions are filter sterilized and added to media that have been cooled to temperatures below 50°C.

Caution about Hazardous Components

Some media contain components that are toxic or carcinogenic. Appropriate safety precautions must be taken when using media with such components. Basic fuchsin and acid fuchsin are carcinogens, and caution must be used in handling media with these compounds to avoid dangerous exposure that could lead to the development of malignancies. Thallium salts, sodium azide, sodium biselenite, and cyanide are among the toxic components found in some media. These compounds are poisonous, and steps must be taken to avoid ingestion, inhalation, or skin contact. Azides also react with many metals, especially copper, to form explosive metal azides. The disposal of azides must avoid contact with copper or achieve sufficient dilution to avoid the formation of such hazardous explosive compounds. Cycloheximide is toxic. Avoid skin contact or aerosol formation and inhalation. Media with sulfur-containing compounds may result in the formation of hydrogen sulfide, which is a toxic gas. Care must be used to ensure proper ventilation. Media with human blood or human blood components must be handled with great caution to avoid exposure to human immunodeficiency virus and other pathogens that contaminate some blood supplies. Proper handling and disposal procedures must be followed with blood-containing as well as other media that are used to cultivate microorganisms.

Uses of Media

The *Handbook of Microbiological Media* contains all the media used to cultivate bacteria, archaea, fungi, and protists of the American Type Culture Collection, the media used to cultivate bacteria, archaea, and fungi of the Deutsche Sammlung von Mikroorganismen (DSM), the media used to cultivate bacteria, archaea, and fungi of the Japanese Collection of Microorganisms (JCM), the media used to cultivate bacteria of the British National Culture Collections of Industrial and Marine Bacteria, the media used to cultivate bacteria of the Spanish Culture Collection of Microorganisms, the media used to cultivate bacteria of the Belgian Culture Collection of Microorganisms (BCCM), the media used to cultivate bacteria of the Finnish VTM Culture Collection of Microorganisms, the media used to cultivate bacteria of the Russian Culture Collection of Microorganisms, and the media used for the testing of waters, wastewaters, and foods—including those recommended by the USEPA and FDA for the standard methods examination of water and food.

Sources of Media

The *Handbook of Microbiological Media* includes the media produced by major suppliers of dehydrated media, including Oxoid Unipath, HiMedia, and BD Diagnostic Systems which supplies Difco and BBL products. There also are a number of suppliers of these media that service different regions. Some of these suppliers also can provide prepared media. This is especially useful for some laboratories that do not have the personnel to oversee the quality assurance needed to prepare media. Quality assurance is a critical part of media preparation.

References

Below is a list of references that can be consulted for further information about media used for the isolation, cultivation, and differentiation of microorganisms.

AOAC International. Best Practices in Microbiological Methodology. 2006. http://www.fda.gov/Food/ScienceResearch/LaboratoryMethods/ucm124900.htm

Baird, R. M. and Lee, W. H. 1995. Media used in the detection and enumeration of *Staphylococcus aureus*. *International Journal of Food Microbiology* 26(1):15–24.

Basu S., Pal, A. and Desai, P. K. 2005. Quality control of culture media in a microbiology laboratory. *Indian Journal of Medical Microbiology* 23(3):159–163.

BD Diagnostic. Difco & BBL Manual: Dehydrated Culture Media and Reagents for Microbiology. 2003. Becton, Dickinson and Co., Sparks, MD. http://www.bd.com/ds/technicalCenter/inserts/difcoBblManual.asp

Beuchat, L. R. 1993. Selective media for detecting and enumerating foodborne yeasts. *International Journal of Food Microbiology* 19(1):1–14.

Blood, R. M. and Curtis, G. D. 1995. Media for 'total' Enterobacteriaceae, coliforms and Escherichia coli. *International Journal of Food Microbiology* 26(1):93–115.

Bridson, E.Y., ed. *The Oxoid Manual.* 1998. Unipath Ltd. Basingstoke, Hampshire, England. http://www.oxoid.com/UK/blue/catbrowse/catbrowse.asp

Busse, M. 1995. Media for *Salmonella. International Journal of Food Microbiology.* 26(1):117–131.

Clesceri, L.S., Greenberg, A.E., and Eaton, A.D. 2005. *Standard Methods for the Examination of Water and Wastewater.* American Public Health Association Publications, Washington, DC. http://www.standardmethods.org/store/

Clinical and Laboratory Standards Institute. 2004. Quality Assurance for Commercially Prepared Microbiological Culture Media. Standard M22-A3. Clinical and Laboratory Standards Institute, Wayne, PA.

Corry, J. E. L., Curtis, G D. W., and Baird, R. M. 2003. *Handbook of Culture Media for Food Microbiology,* 2nd ed. Elsevier, Amsterdam.

Curtis, G. D. and Lee, W. H. 1995. Culture media and methods for the isolation of *Listeria monocytogenes*. *International Journal of Food Microbiology* 26(1):1–13.

de Boer, E. 1992. Isolation of *Yersinia enterocolitica* from foods. *International Journal of Food Microbiology* 17(2):75–84.

Domig, K. J., Mayer, H. K., and Kneifel, W. 2003. Methods used for the isolation, enumeration, characterisation and identification of *Enterococcus* spp. 1. Media for isolation and enumeration. *International Journal of Food Microbiology* 88(2-3):147–164.

Donovan, T. J. and van Netten, P. 1995. Culture media for the isolation and enumeration of pathogenic *Vibrio* species in foods and environmental samples. *International Journal of Food Microbiology* 26(1):77–91.

Downes, F. and Ito, K. 2001. *Compendium of Methods for the Microbiological Examination of Foods.* American Public Health Association, Washington, D.C.

Ertola, R.J., Giulietti, A. M., and Castillo, F. J. 1995. Design, formulation, and optimization of media. *Bioprocess Technology* 21:89–137.

Falkow, S., Rosenberg, E., Schleifer, K.-H., Stackebrandt, E., Dworkin, M. (Eds.) 2007. *The Prokaryotes,* 3rd ed., Vols. 1–7 , Springer, NY.

Finegold, S.M. and Martin, W. J. 1990. *Diagnostic Microbiology.* Mosby Co., St. Louis, MO.

Forbes, B.A.,Sahm, D.F., and Weissfeld, A.S. 2007. *Bailey and Scott's Diagnostic Microbiology,* 12th Ed. Mosby Ltd., St. Louis, MO.

Froud, S. J. 1999. The development, benefits and disadvantages of serum-free media. *Developments in Biological Standardization* 99:157–166.

HiMedia. 2006. *The HiVeg Manual.* HiMedia Laboratories Pvt. Limited. Mumbai, India.

HiMedia. 2009. *The HiMedia Manual.* HiMedia Laboratories Pvt. Limited. Mumbai, India.

Holzapfel, W. H. 1992. Culture media for non-sporulating gram-positive food spoilage bacteria. I*nternational Journal of Food Microbiology.* 17(2):113–133.

Jayme, D. W., and Greenwold, D. J. 1991. Media selection and design: wise choices and common mistakes. *Bio/Technology* 9(8):716–721.

Jousimies-Somer, H., Summanen, P. E., Citron, D. M., Baron, E. J., Wexler, H. M., and Finegold, S. M. 2002. *Anaerobic Bacteriology Manual*, 6th ed. Star Publishing Co., Belmont, CA.

Manafi, M. 1996. Fluorogenic and chromogenic enzyme substrates in culture media and identification tests. *International Journal of Food Microbiology* 31(1-3):45–58.

Murray, P. R., Volume Editors E. J. Baron, Jorgensen, J. H., Landry, M. L., and Pfaller, M. A. 2007. *Manual of Clinical Microbiology,* 9th ed. ASM Press, Washington, DC.

Odds, FC. 1991. Sabouraud's agar. *Journal of Medical & Veterinary Mycology* 29(6):355–359.

Persing, D. H., Tenover, F. C., Tang, Y-W. et al. 2003. *Molecular Microbiology: Diagnostic Principles and Practice.* ASM Press, Washington, DC.

Peterson, L R. 1997. Effect of media on transport and recovery of anaerobic bacteria. *Clinical Infectious Diseases* 25 Suppl 2:S134–136.

Starliper, C. E. 2008. General and specialized media routinely employed for primary isolation of bacterial pathogens of fishes. *Journal of Wildlife Diseases.* 44(1):121–132.

Stoakes, L., Reyes, R., Daniel, J., Lennox, G., John, M. A., Lannigan, R., and Hussain, J. 2006. Prospective comparison of a new chromagen medium, MRSASelect, to CHROMagar MRSA and mannitol-salt medium with oxacillin or cefoxitin for detection of methicillin-resistant *Staphylococcus aureus*. *Journal of Clinical Microbiology* 44:637–639.

Truant, A. L. 2002. *Manual of Commercial Methods in Clinical Microbiology.* ASM Press, Washington, DC.

U.S. Food and Drug Administration. Bacteriological Analytical Manual. 2000. http://www.fda.gov/Food/ScienceResearch/LaboratoryMethods/BacteriologicalAnalyticalManualBAM/default.htm

van Netten, P., and Kramer, J. M. 1992. Media for the detection and enumeration of *Bacillus cereus* in foods: a review. *International Journal of Food Microbiology* 17(2):85–99.

Vimont, A., Vernozy-Rozand, C., and Delignette-Muller, M. L. 2006. Isolation of *E. coli* O157:H7 and non-O157 STEC in different matrices: review of the most commonly used enrichment protocols. *Letters in Applied Microbiology* 42(2):102–108.

Winn, Jr., W. C., Allen, S. D., Janda, W. M., Koneman, E. W., Schreckenberger, P. C., Procop, G. W., and Woods, G.L., eds. 2005. *Color Atlas and Textbook of Diagnostic Microbiology,* 6th ed. J. B. Lippincott Co., Philadelphia, PA.

Web Resources

Below is a list of Web sites that provide information about media and microbial cultures.

American Type Culture Collection (ATCC), a global biological resource.
http://www.atcc.org/catalogs/catalogs.html

Bacteria/Culture Media Protocols
http://www.protocol-online.org/prot/Microbiology/Bacteria/Culture_Media___Plates/index.html

BD (Becton, Dickinson and Company)
http://www.bd.com/

Belgian Coordinated Collections of Microorganisms / LMBP Plasmid Collection, Ghent University, Department of Molecular Biology
http://wdcm.nig.ac.jp/CCINFO/CCINFO.xml?643

Czechoslovokian Collection of Microorganisms (CCM).
http://www.sci.muni.cz/ccm/index.html

Finnish Culture Collection, Valtion Teknillinen Tutkimuskeskus (VTT).
http://culturecollection.vtt.fi/

German Resource Center for Biological Materrial. (German Collection of Microorganisms and Cell Cultures) (DSMZ).
http://www.dsmz.de/

Gibco Invitrogen Cell Culture Products
http://www.invitrogen.com/site/us/en/home/Applications/Cell-Culture.html?cid=invggl123000000000095s&

Hardy Diagnostics
http://www.hardydiagnostics.com/?gclid=CMifuc62tJsCFR7yDAodZlWRQg

HiMedia.
http://www.himedialabs.com/

Japanese Collection of Microorganisms and Microbial Cultures.
http://www.jcm.riken.go.jp

Netherlands Centraalbureau voor Schimmelcultures (CBS).
http://www.cbs.knaw.nl/collection/AboutCollections.aspx

Oxoid Ltd.
http://www.oxoid.com/uk/blue/index.asp

Spanish Collection of Microorganisms (Colección Española de Cultivos Tipo Catalogo de Cepas). http://www.cect.org/english/index.htm

United Kingdom National Collection of Yeast Cultures.
http://www.ifr.bbsrc.ac.uk/ncyc

United Kingdom National Culture Collection Microbiological Resources. http://www.ukncc.co.uk

U. S. Food and Drug Administration FDA Bacteriological Analytical Manual Online (BAM)
http://www.fda.gov/Food/ScienceResearch/LaboratoryMethods/BacteriologicalAnalyticalManualBAM/default.htm

U. S. Environmental Protection Agency. Microbiological Methods.
http://www.epa.gov/nerlcwww/online.htm

World Federation of Culture Collections.
http://www. wfcc.info

A Medium, 5X

Composition per liter:

K_2HPO_4	52.5g
KH_2PO_4	22.5g
$(NH_4)_2SO_4$	5.0g
Sodium citrate·$2H_2O$	2.5g
Carbon source solution	10.0mL
$MgSO_4$·$7H_2O$ solution	1.0mL

pH 7.0 ± 0.2 at 25°C

Carbon Source Solution:

Composition per 100.0mL:

Glycerol or sucrose	20.0g

Preparation of Carbon Source Solution: Add glycerol or glucose to distilled/deionized water and bring volume to 100.0mL. Mix thoroughly. Filter sterilize.

$MgSO_4$·$7H_2O$ Solution:

Composition per 100.0mL:

$MgSO_4$·$7H_2O$	24.65g

Preparation of $MgSO_4$·$7H_2O$ Solution: Add $MgSO_4$·$7H_2O$ to distilled/deionized water and bring volume to 100.0mL. Mix thoroughly. Filter sterilize.

Preparation of Medium: Add components, except carbon source solution and $MgSO_4$·$7H_2O$ solution, to distilled/deionized water and bring volume to 1.0L. Mix thoroughly. Gently heat and bring to boiling. Autoclave for 15 min at 15 psi pressure–121°C. Cool to 45°–50°C. To prepare medium for use (1×), aseptically dilute 200.0mL of 5× stock solution with 789.0mL of sterile distilled/deionized water. Aseptically add 10.0mL of sterile carbon source solution and 1.0mL of sterile $MgSO_4$·$7H_2O$ solution. Mix thoroughly. Aseptically distribute into sterile tubes or flasks.

Use: For the cultivation of *Escherichia coli.*

A 1 Broth

Composition per liter:

Pancreatic digest of casein	20.0g
Lactose	5.0g
NaCl	5.0g
Salicin	0.5g
Triton™ X-100	1.0mL

pH 6.9 ± 0.1 at 25°C

Source: This medium is available as a premixed powder from BD Diagnostic Systems.

Preparation of Medium: Add components to distilled/deionized water and bring volume to 1.0L. Mix thoroughly. Gently heat and bring to boiling. Distribute into test tubes containing an inverted Durham tube. Autoclave for 10 min at 15 psi pressure–121°C.

Use: For the detection of fecal coliforms in foods, treated wastewater, and seawater by a most-probable-number (MPN) method. Multiple dilutions of samples (3, 5, or 10 replicates per dilution) are added to tubes containing A 1 broth. After incubation, test tubes with gas accumulation in the Durham tubes are scored positive and those with no gas as negative. A MPN table is consulted to determine the most probable number of fecal coliforms.

A-1 HiVeg Broth

Composition per liter:

Plant hydrolysate	20.0g
Lactose	5.0g
NaCl	5.0g
Polyethylene glycol p-isoactylphenyl ether (Triton™ X-100)	1.0g
Salicin	0.5g

pH 6.9 ± 0.1 at 25°C

Source: This medium is available as a premixed powder from HiMedia.

Preparation of Medium: Add components to distilled/deionized water and bring volume to 1.0L. Mix thoroughly. Gently heat and bring to boiling. Distribute into test tubes containing an inverted Durham tube. Autoclave for 10 min at 15 psi pressure–121°C.

Use: For the detection of fecal coliforms in foods, treated wastewater, and seawater by a most-probable-number (MPN) method. Multiple dilutions of samples (3, 5, or 10 replicates per dilution) are added to tubes containing A 1 broth. After incubation, test tubes with gas accumulation in the Durham tubes are scored positive and those with no gas as negative. A MPN table is consulted to determine the most probable number of fecal coliforms.

A-1 Medium (BAM M1)

Composition per liter:

Pancreatic digest of casein	20.0g
Lactose	5.0g
NaCl	5.0g
Salicin	0.5g
Triton™ X-100	1.0mL

pH 6.9 ± 0.1 at 25°C

Source: This medium is available as a premixed powder from BD Diagnostic Systems.

Preparation of Medium: Add components to distilled/deionized water and bring volume to 1.0L. Mix thoroughly. Adjust pH to 6.9. Distribute into test tubes containing an inverted Durham tube. Medium may be cloudy prior to autoclaving. Autoclave for 10 min at 15 psi pressure–121°C.

Use: For the detection of fecal coliforms in foods and waters by a most-probable-number (MPN) method. Multiple dilutions of samples (3, 5, 10 replicates per dilution) are added to tubes containing A-1 medium. After incubation, test tubes with gas accumulation in the Durham tubes are scored positive and those with no gas as negative. A MPN table is consulted to determine the most probable number of fecal coliforms.

A1 Medium (DSMZ Medium 1054)

Composition per liter:

Agar	20.0g
Starch	10.0g
Yeast extract	4.0g
Bacto peptone	2.0g
Seawater (Biomaris) (natural or artificial)	1.0L

pH 7.3 ± 0.2 at 25°C

Preparation of Medium: Add components to seawater and bring volume to 1.0L. Mix thoroughly. Gently heat and bring to boiling. Distribute into tubes or flasks. Autoclave for 15 min at 15 psi pressure–121°C. Pour into sterile Petri dishes or leave in tubes.

Use: For the cultivation and maintenance of *Saccharomonospora saliphila.*

A 1 Minimal Medium

Composition per liter:

L-Asparagine	5.0g
$(NH_4)_2SO_4$	5.0g
Sodium pyruvate	5.0g
$MgSO_4 \cdot 7H_2O$	2.0g
Spermadine·3HCl	0.125g
L-Asparagine	0.1g
L-Isoleucine	0.1g
L-Methionine	0.1g
L-Phenylalanine	0.1g
L-Valine	0.1g
L-Leucine	0.05g
KH_2PO_4	0.013g
$FeCl_3 \cdot 6H_2O$	2.7mg
$CaCl_2$	1.1mg
Cyanocobalamin	1.0mg
Tris(hydroxymethyl)aminomethane buffer (0.01*M* solution, pH 7.6)	1.0L

pH 7.6 ± 0.2 at 25°C

Preparation of Medium: Add solid components to 1.0L of Tris buffer. Mix thoroughly. Filter sterilize. Aseptically distribute into tubes or flasks.

Use: For the cultivation of *Myxococcus xanthus*.

A 3 Agar

Composition per 202.4mL:

Agar base	140.0mL
Supplement solution	62.4mL

pH 6.0 ± 0.2 at 25°C

Agar Base:

Composition per liter:

Pancreatic digest of casein	17.0g
Ionagar No. 2	7.5g
NaCl	5.0g
Papaic digest of soybean meal	3.0g
K_2HPO_4	2.5g
Glucose	2.5g

Source: Ionagar No. 2 is available from Oxoid Unipath.

Preparation of Agar Base: Add components, except agar, to distilled/deionized water and bring volume to 1.0L. Adjust pH to 5.5. Add agar. Mix thoroughly. Gently heat and bring to boiling. Distribute into screw-capped bottles in 140.0mL volumes. Autoclave for 15 min at 15 psi pressure–121°C. Cool to 45°–50°C.

Supplement Solution:

Composition per 62.4mL:

Horse serum-urea solution	40.0mL
Fresh yeast extract solution	20.0mL
Penicillin solution	2.0mL
Phenol Red solution	0.4mL

Preparation of Supplement Solution: Aseptically combine components. Mix thoroughly.

Horse Serum-Urea Solution:

Composition per 40.0mL:

Urea	0.2g
Horse serum, unheated	40.0mL

Preparation of Horse Serum-Urea Solution: Add urea to 40.0mL of horse serum. Mix thoroughly. Filter sterilize.

Fresh Yeast Extract Solution:

Composition:

Baker's yeast, live, pressed, starch-free	25.0g

Preparation of Fresh Yeast Extract Solution: Add the live Baker's yeast to 100.0mL of distilled/deionized water. Autoclave for 90 min at 15 psi pressure–121°C. Allow to stand. Remove supernatant solution. Adjust pH to 6.6–6.8. Filter sterilize.

Penicillin Solution:

Composition per 10.0mL:

Penicillin G	1,000,000U

Preparation of Penicillin Solution: Add penicillin to distilled/deionized water and bring volume to 10.0mL. Mix thoroughly. Filter sterilize.

Phenol Red Solution:

Composition per 10.0mL:

Phenol Red	0.1g

Preparation of Phenol Red Solution: Add Phenol Red to distilled/deionized water and bring volume to 10.0mL. Mix thoroughly. Filter sterilize.

Preparation of Medium: Aseptically combine 140.0mL of cooled, sterile agar base and 62.4mL of sterile supplement solution. Mix thoroughly. Pour into sterile Petri dishes or distribute into sterile tubes.

Use: For the cultivation of *Ureaplasma urealyticum* from urine. Also used for the cultivation of other *Ureaplasma* species.

A 3B Agar

Composition per 101.5mL:

Agar base	80.0mL
Supplement solution	21.5mL

pH 6.0 ± 0.2 at 25°C

Agar Base:

Composition per liter:

Pancreatic digest of casein	17.0g
Ionagar No. 2	7.5g
NaCl	5.0g
Papaic digest of soybean meal	3.0g
K_2HPO_4	2.5g
Glucose	2.5g

Source: Ionagar No. 2 is available from Oxoid Unipath.

Preparation of Agar Base: Add components, except agar, to distilled/deionized water and bring volume to 1.0L. Adjust pH to 5.5. Add agar. Mix thoroughly. Gently heat and bring to boiling. Distribute into screw-capped bottles in 80.0mL volumes. Autoclave for 15 min at 15 psi pressure–121°C. Cool to 45°–50°C.

Supplement Solution:

Composition per 21.5mL:

Horse serum-urea solution	20.0mL
Penicillin solution	1.0mL
L-Cysteine·HCl·H_2O solution	0.5mL

Preparation of Supplement Solution: Aseptically combine components. Mix thoroughly.

Horse Serum-Urea Solution:

Composition per 40.0mL:

Urea	0.2g
Horse serum, unheated	40.0mL

Preparation of Horse Serum-Urea Solution: Add urea to 40.0mL of horse serum. Mix thoroughly. Filter sterilize.

Penicillin Solution:

Composition per 10.0mL:

Penicillin G	1,000,000U

Preparation of Penicillin Solution: Add penicillin to distilled/deionized water and bring volume to 10.0mL. Mix thoroughly. Filter sterilize.

L-Cysteine·HCl·H_2O Solution:

Composition per 10.0mL:

L-Cysteine·HCl·H_2O	0.2g

Preparation of L-Cysteine·HCl·H_2O Solution: Add L-cysteine·HCl·H_2O to distilled/deionized water and bring volume to 10.0mL. Mix thoroughly. Filter sterilize.

Preparation of Medium: Aseptically combine 80.0mL of cooled, sterile agar base and 21.5mL of sterile supplement solution. Mix thoroughly.

Use: For the cultivation of *Ureaplasma urealyticum* from urine. Also used for the cultivation of other *Ureaplasma* species.

A 7 Agar (Shepard's Differential Agar)

Composition per 205.7mL:

Agar base	160.0mL
Supplement solution	45.7mL

pH 6.0 ± 0.2 at 25°C

Agar Base:

Composition per 165.0mL:

Pancreatic digest of casein	2.72g
Agar	2.1g
NaCl	0.8g
Papaic digest of soybean meal	0.48g
K_2HPO_4	0.4g
Glucose	0.4g
$MnSO_4$·H_2O	0.15g

Preparation of Agar Base: Add components, except agar, to distilled/deionized water and bring volume to 165.0mL. Adjust pH to 5.5. Add agar. Mix thoroughly. Autoclave for 15 min at 15 psi pressure–121°C. Cool to 45°–50°C.

Supplement Solution:

Composition per 45.72mL:

Horse serum, unheated	40.0mL
Fresh yeast extract solution	2.0mL
Penicillin solution	2.0mL
CVA enrichment	1.0mL
L-Cysteine·HCl·H_2O solution	0.5mL
Urea solution	0.22mL

Preparation of Supplement Solution: Aseptically combine components. Mix thoroughly.

Fresh Yeast Extract Solution:

Composition per 100.0mL:

Baker's yeast, live, pressed, starch-free	25.0g

Preparation of Fresh Yeast Extract Solution: Add the live Baker's yeast to 100.0mL of distilled/deionized water. Autoclave for 90 min at 15 psi pressure–121°C. Allow to stand. Remove supernatant solution. Adjust pH to 6.6–6.8. Filter sterilize.

Penicillin Solution:

Composition per 10.0mL:

Penicillin G	1,000,000U

Preparation of Penicillin Solution: Add penicillin to distilled/deionized water and bring volume to 10.0mL. Mix thoroughly. Filter sterilize.

CVA Enrichment:

Composition per liter:

Glucose	100.0g
L-Cysteine·HCl·H_2O	25.9g
L-Glutamine	10.0g
L-Cystine·2HCl	1.0g
Adenine	1.0g
Nicotinamide adenine dinucleotide	0.25g
Cocarboxylase	0.1g
Guanine·HCl	0.03g
$Fe(NO_3)_3$	0.02g
Vitamin B_{12}	0.01g
p-Aminobenzoic acid	0.013g
Thiamine·HCl	3.0mg

Preparation of CVA Enrichment: Add components to distilled/deionized water and bring volume to 1.0L. Mix thoroughly. Filter sterilize.

L-Cysteine·HCl·H_2O Solution:

Composition per 10.0mL:

L-Cysteine·HCl·H_2O	0.4g

Preparation of L-Cysteine·HCl·H_2O Solution: Add L-cysteine·HCl·H_2O solution to distilled/deionized water and bring volume to 10.0mL. Mix thoroughly. Filter sterilize.

Urea Solution:

Composition per 10.0mL:

Urea, ultrapure	1.0g

Preparation of Urea Solution: Add urea to distilled/deionized water and bring volume to 10.0mL. Mix thoroughly. Filter sterilize.

Preparation of Medium: Aseptically combine 160.0mL of cooled, sterile agar base and 45.9mL of sterile supplement solution. Mix thoroughly. Pour into sterile Petri dishes or distribute into sterile tubes.

Use: For the cultivation and differentiation of *Ureaplasma urealyticum* from urine based on its ability to produce ammonia from urea. Bacteria that produce ammonia appear as golden to dark brown colonies. Also used for the cultivation of other *Ureaplasma* species.

A 7 Agar, Modified

Composition per 205.7mL:

Agar base	160.0mL
Supplement solution	45.7mL

pH 6.0 ± 0.2 at 25°C

Agar Base:

Composition per 165.0mL:

Agar	10.0g
Pancreatic digest of casein	2.72g
NaCl	0.8g
Papaic digest of soybean meal	0.48g
K_2HPO_4	0.4g
Glucose	0.4g
$MnSO_4$·H_2O	0.15g

Preparation of Agar Base: Add components, except agar, to distilled/deionized water and bring volume to 165.0mL. Adjust pH to 5.5. Add agar. Mix thoroughly. Autoclave for 15 min at 15 psi pressure–121°C. Cool to 45°–50°C.

Supplement Solution:
Composition per 45.72mL:

Horse serum, unheated	40.0mL
Fresh yeast extract solution	2.0mL
Penicillin solution	2.0mL
CVA enrichment	1.0mL
L-Cysteine·HCl·H_2O solution	0.5mL
Urea solution	0.22mL

Preparation of Supplement Solution: Aseptically combine components. Mix thoroughly.

Fresh Yeast Extract Solution:
Composition per 100.0mL:

Baker's yeast, live, pressed, starch-free	25.0g

Preparation of Fresh Yeast Extract Solution: Add the live Baker's yeast to 100.0mL of distilled/deionized water. Autoclave for 90 min at 15 psi pressure–121°C. Allow to stand. Remove supernatant solution. Adjust pH to 6.6–6.8. Filter sterilize.

Penicillin Solution:
Composition per 10.0mL:

Penicillin G	1,000,000U

Preparation of Penicillin Solution: Add penicillin to distilled/deionized water and bring volume to 10.0mL. Mix thoroughly. Filter sterilize.

CVA Enrichment:
Composition per liter:

Glucose	100.0g
L-Cysteine·HCl·H_2O	25.9g
L-Glutamine	10.0g
L-Cystine·2HCl	1.0g
Adenine	1.0g
Nicotinamide adenine dinucleotide	0.25g
Cocarboxylase	0.1g
Guanine·HCl	0.03g
$Fe(NO_3)_3$	0.02g
p-Aminobenzoic acid	0.013g
Vitamin B_{12}	0.01g
Thiamine·HCl	3.0mg

Preparation of CVA Enrichment: Add components to distilled/deionized water and bring volume to 1.0L. Mix thoroughly. Filter sterilize.

L-Cysteine·HCl·H_2O Solution:
Composition per 10.0mL:

L-Cysteine·HCl·H_2O	0.4g

Preparation of L-Cysteine·HCl·H_2O Solution: Add L-cysteine·HCl·H_2O solution to distilled/deionized water and bring volume to 10.0mL. Mix thoroughly. Filter sterilize.

Urea Solution:
Composition per 10.0mL:

Urea, ultrapure	1.0g

Preparation of Urea Solution: Add urea to distilled/deionized water and bring volume to 10.0mL. Mix thoroughly. Filter sterilize.

Preparation of Medium: Aseptically combine 160.0mL of cooled, sterile agar base and 45.9mL of sterile supplement solution. Mix thoroughly. Pour into sterile Petri dishes or distribute into sterile tubes.

Use: For the cultivation and differentiation of *Ureaplasma urealyticum* from urine based on its ability to produce ammonia from urea. Bacteria that produce ammonia appear as golden to dark brown colonies. Also used for the cultivation of other *Ureaplasma* species.

A 7B Agar

Composition per 205.7mL:

Agar base	160.0mL
Supplement solution	45.7mL

pH 6.0 ± 0.2 at 25°C

Agar Base:
Composition per 165.0mL:

Pancreatic digest of casein	2.72g
Agar	2.1g
NaCl	0.8g
Papaic digest of soybean meal	0.48g
K_2HPO_4	0.4g
Glucose	0.4g
Putrescine·2HCl	0.33g
$MnSO_4·H_2O$	0.15g

Preparation of Agar Base: Add components, except agar, to distilled/deionized water and bring volume to 165.0mL. Adjust pH to 5.5. Add agar. Mix thoroughly. Autoclave for 15 min at 15 psi pressure–121°C. Cool to 45°–50°C.

Supplement Solution:
Composition per 45.72mL:

Horse serum, unheated	40.0mL
Fresh yeast extract solution	2.0mL
Penicillin solution	2.0mL
CVA enrichment	1.0mL
L-Cysteine·HCl·H_2O solution	0.5mL
Urea solution	0.22mL

Preparation of Supplement Solution: Aseptically combine components. Mix thoroughly.

Fresh Yeast Extract Solution:
Composition per 100.0mL:

Baker's yeast, live, pressed, starch-free	25.0g

Preparation of Fresh Yeast Extract Solution: Add the live Baker's yeast to 100.0mL of distilled/deionized water. Autoclave for 90 min at 15 psi pressure–121°C. Allow to stand. Remove supernatant solution. Adjust pH to 6.6–6.8. Filter sterilize.

Penicillin Solution:
Composition per 10.0mL:

Penicillin G	1,000,000U

Preparation of Penicillin Solution: Add penicillin to distilled/deionized water and bring volume to 10.0mL. Mix thoroughly. Filter sterilize.

CVA Enrichment:
Composition per liter:

Glucose	100.0g
L-Cysteine·HCl·H_2O	25.9g
L-Glutamine	10.0g
L-Cystine·2HCl	1.0g
Adenine	1.0g

Nicotinamide adenine dinucleotide 0.25g
Cocarboxylase 0.1g
Guanine·HCl 0.03g
$Fe(NO_3)_3$ 0.02g
p-Aminobenzoic acid 0.013g
Vitamin B_{12} 0.01g
Thiamine·HCl 3.0mg

Preparation of CVA Enrichment: Add components to distilled/deionized water and bring volume to 1.0L. Mix thoroughly. Filter sterilize.

L-Cysteine·HCl·H_2O Solution:
Composition per 10.0mL:
L-Cysteine·HCl·H_2O 0.4g

Preparation of L-Cysteine·HCl·H_2O Solution: Add L-cysteine·HCl·H_2O solution to distilled/deionized water and bring volume to 10.0mL. Mix thoroughly. Filter sterilize.

Urea Solution:
Composition per 10.0mL:
Urea, ultrapure 1.0g

Preparation of Urea Solution: Add urea to distilled/deionized water and bring volume to 10.0mL. Mix thoroughly. Filter sterilize.

Preparation of Medium: Aseptically combine 160.0mL of cooled, sterile agar base and 45.9mL of sterile supplement solution. Mix thoroughly. Pour into sterile Petri dishes or distribute into sterile tubes.

Use: For the cultivation and differentiation of *Ureaplasma urealyticum* from urine based on its ability to produce ammonia from urea. Bacteria that produce ammonia appear as golden to dark brown colonies. Also used for the cultivation of other *Ureaplasma* species.

A 8B Agar

Composition per 84.6mL:
Agar base 80.0mL
Supplement solution 4.6mL
pH 6.0 ± 0.2 at 25°C

Agar Base:
Composition per 165.0mL:
Pancreatic digest of casein 2.72g
Agar 2.1g
NaCl 0.8g
Papaic digest of soybean meal 0.48g
K_2HPO_4 .... 0.4g
Glucose 0.4g
$MnSO_4 \cdot H_2O$ 0.15g
$CaCl_2 \cdot 2H_2O$ 0.03g
Putrescine·2HCl 34.0mg

Preparation of Agar Base: Add components, except agar, to distilled/deionized water and bring volume to 165.0mL. Adjust pH to 5.5. Add agar. Mix thoroughly. Autoclave for 15 min at 15 psi pressure–121°C. Cool to 45°–50°C.

Supplement Solution:
Composition per 4.6mL:
Horse serum, unheated 1.0mL
Fresh yeast extract solution 1.0mL
Penicillin solution 1.0mL
Urea solution 1.0mL
L-Cysteine·HCl·H_2O solution 0.5mL
GHL tripeptide solution 0.1mL

Preparation of Supplement Solution: Aseptically combine components. Mix thoroughly.

Fresh Yeast Extract Solution:
Composition per 100.0mL:
Baker's yeast, live, pressed, starch-free 25.0g

Preparation of Fresh Yeast Extract Solution: Add the live Baker's yeast to 100.0mL of distilled/deionized water. Autoclave for 90 min at 15 psi pressure–121°C. Allow to stand. Remove supernatant solution. Adjust pH to 6.6–6.8. Filter sterilize.

Penicillin Solution:
Composition per 10.0mL:
Penicillin G 1,000,000U

Preparation of Penicillin Solution: Add penicillin to distilled/deionized water and bring volume to 10.0mL. Mix thoroughly. Filter sterilize.

GHL Tripeptide Solution:
Composition per 10.0mL:
GHL (Glycyl-L-histidyl-L-lysine acetate) tripeptide 0.2g

Preparation of GHL Tripeptide Solution: Add GHL (Glycyl-L-histidyl-L-lysine acetate) tripeptide to distilled/deionized water and bring volume to 10.0mL. Mix thoroughly. Filter sterilize.

L-Cysteine·HCl·H_2O Solution:
Composition per 10.0mL:
L-Cysteine·HCl·H_2O 0.4g

Preparation of L-Cysteine·HCl·H_2O Solution: Add L-cysteine·HCl·H_2O solution to distilled/deionized water and bring volume to 10.0mL. Mix thoroughly. Filter sterilize.

Urea Solution:
Composition per 10.0mL:
Urea, ultrapure 1.0g

Preparation of Urea Solution: Add urea to distilled/deionized water and bring volume to 10.0mL. Mix thoroughly. Filter sterilize.

Preparation of Medium: Aseptically combine 80.0mL of cooled, sterile agar base and 4.6mL of sterile supplement solution. Mix thoroughly. Pour into sterile Petri dishes or distribute into sterile tubes.

Use: For the cultivation of *Ureaplasma urealyticum* from urine. Also used for the cultivation of other *Ureaplasma* species.

AAHC Medium for YAK Clones

Composition per liter:
Glucose 20.0g
Acid hydrolysate of casein 10.0g
Yeast nitrogen base without amino acids 6.7g
Adenine hemi-sulfate 40.0mg

Preparation of Medium: Add glucose, yeast nitrogen base without amino acids, and adenine hemi-sulfate to distilled/deionized water and bring volume to 100.0mL. Mix thoroughly. Filter sterilize. Add acid hydrolysate of casein to distilled/deionized water and bring volume to 900.0mL. Mix thoroughly. Gently heat and bring to boiling. Autoclave for 15 min at 15 psi pressure–121°C. Cool to room temperature. Aseptically combine the two sterile solutions. Mix thoroughly. Aseptically distribute into sterile tubes or flasks.

Use: For the cultivation and maintenance of *Saccharomyces cerevisiae.*

AAM Medium (DSMZ Medium 57)

Composition per liter:

Casamino acids	7.0g
Yeast extract	7.0g
Tryptone	5.0g
Meat extract	5.0g
Na-acetate	2.5g
$MgSO_4 \cdot 7H_2O$	200.0mg
$MnSO_4 \cdot 2H_2O$	50.0mg
Tween™ 80	1.0mL

pH 5.4 ± 0.2 at 25°C

Preparation of Medium: Add components to distilled/deionized water and bring volume to 1.0L. Mix thoroughly. Adjust pH to 5.4. Distribute into tubes or flasks. Autoclave for 15 min at 15 psi pressure–121°C.

Use: For the cultivation and maintenance of *Lactobacillus* spp.

AATCC Bacteriostasis Agar (American Association of Textile Chemists and Colorists Bacteriostasis Agar)

Composition per liter:

Agar	15.0g
Peptone	10.0g
Beef extract	5.0g
NaCl	5.0g

pH 7.2 ± 0.2 at 25°C

Preparation of Medium: Add components to distilled/deionized water and bring volume to 1.0L. Mix thoroughly. Gently heat and bring to boiling. Autoclave for 15 min at 15 psi pressure–121°C. Pour into sterile Petri dishes or distribute into sterile tubes.

Use: For the maintenance of cultures of *Escherichia coli* and *Staphylococcus aureus.* For the detection of antibacterial activity of fabrics. Test cultures of *Escherichia coli* or *Staphylococcus aureus* are inoculated onto an agar plate and a sample of sterile fabric is placed on the surface. Lack of bacterial growth indicates the fabric has antibacterial activity.

AATCC Bacteriostasis Agar
See: **FDA Agar**

AATCC Bacteriostasis Broth
See: **FDA Broth**

AATCC Bacteriostasis HiVeg Agar

Composition per liter:

Agar	15.0g
Plant peptone	10.0g
Plant extract	5.0g
NaCl	5.0g

pH 7.2 ± 0.2 at 25°C

Source: This medium is available as a premixed powder from HiMedia.

Preparation of Medium: Add components to distilled/deionized water and bring volume to 1.0L. Mix thoroughly. Gently heat and bring to boiling. Autoclave for 15 min at 15 psi pressure–121°C. Pour into sterile Petri dishes or distribute into sterile tubes.

Use: For for the detection of antibacterial activity of fabrics.

AATCC Bacteriostasis HiVeg Broth

Composition per liter:

Plant peptone	10.0g
Plant extract	5.0g
NaCl	5.0g

pH 6.8 ± 0.2 at 25°C

Source: This medium is available as a premixed powder from HiMedia.

Preparation of Medium: Add components to distilled/deionized water and bring volume to 1.0L. Mix thoroughly. Gently heat and bring to boiling. Distribute into tubes or flasks. Autoclave for 15 min at 15 psi pressure–121°C.

Use: For for the detection of antibacterial testing of antiseptics and disinfectants.

AATCC Mineral Salts Iron Agar (American Association of Textile Chemists and Colorists Mineral Salts Iron Agar)

Composition per liter:

Agar	20.0g
$(NH_4)_2NO_3$	3.0g
KH_2PO_4	2.5g
K_2HPO_4	2.0g
$MgSO_4 \cdot 7H_2O$	0.2g
$FeSO_4 \cdot 7H_2O$	0.1g

pH 5.6 ± 0.2 at 25°C

Preparation of Medium: Add components to distilled/deionized water and bring volume to 1.0L. Mix thoroughly. Gently heat and bring to boiling. Distribute into tubes or flasks. Autoclave for 15 min at 15 psi pressure–121°C.

Use: For testing the resistance of textiles to fungi that cause mildew and rot. Also used to test the effectiveness of fungicides used on textiles for preventing the growth of fungi. Cultures of *Chaetomium globosum* or *Aspergillus niger* are inoculated onto the plate and a sample of fabric is placed on top. Lack of growth of these fungi on the textile is indicative of resistance to mildew.

Abeyta-Hunt Bark Agar

Composition per 1016.0mL:

Beef heart, infusion from	500.0g
Agar	15.0g
Tryptose	10.0g
NaCl	5.0g
Yeast extract	2.0g
Horse blood, lysed	50.0mL
Cefoperazone solution	4.0mL
Rifampicin solution	4.0mL
Amphotericin B solution	4.0mL
Ferrous sulfate pyruvate metabisulfite solution	4.0mL

pH 7.4 ± 0.2 at 25°C

Amphotericin B Solution:
Composition per 100.0mL:

Amphotericin B	0.05g

Preparation of Amphotericin B Solution: Add amphotericin B to distilled/deionized water and bring volume to 100.0mL. Mix thoroughly. Filter sterilize.

Cefoperazone Solution:
Composition per 100.0mL:

Sodium cefoperazone	0.08g

Preparation of Cefoperazone Solution: Add sodium cefoperazone to distilled/deionized water and bring volume to 100.0mL. Mix thoroughly. Filter sterilize.

Rifampicin Solution:
Composition per 100.0mL

Rifampicin	0.25g

Preparation of Rifampicin Solution: Add rifampicin to 70.0mL ethanol. Mix thoroughly. Add distilled/deionized water to bring volume to 100.0mL. Mix thoroughly. Filter sterilize.

Ferrous Sulfate, Pyruvate, Metabisulfite Solution:
Composition per 100.0mL:

$FeSO_4$	6.25g
Na-pyruvate	6.25g
Na-metabisulfite	6.25g

Preparation of Ferrous Sulfate, Pyruvate, Metabisulfite Solution: Add Na-pyruvate to 20mL distilled/deionized water. Mix thoroughly. Add Na-metabisulfite and $FeSO_4$. Bring volume to 100.0mL. Mix thoroughly. Filter sterilize.

Preparation of Medium: Add components, except cefoperazone solution, amphotericin B solution, rifampicin solution, ferrous sulfate pyruvate metabisulfide solution, and horse blood, to distilled/deionized water and bring volume to 950.0mL. Mix thoroughly. Gently heat and bring to boiling. Autoclave for 15 min at 15 psi pressure–121°C. Cool to 50°C. Aseptically add 4.0mL cefoperazone solution, 4.0mL amphotericin B solution, 4.0mL rifampicin solution, 4.0mL ferrous sulfate pyruvate metabisulfide solution, and 50.0mL lysed horse blood. Mix thoroughly. Pour into sterile Petri dishes or distribute into sterile tubes.

Use: For the isolation and cultivation of *Campylobacter* spp.

ABY Agar
(Acid Bismuth Yeast Agar)

Composition per liter:

Agar	20.0g
Glucose	20.0g
$Bi_2(SO_3)_2$	8.0g
$(NH_4)_2SO_4$	3.0g
KH_2PO_4	3.0g
$MgSO_4 \cdot 7H_2O$	0.25g
$CaCl_2 \cdot 2H_2O$	0.25g
Biotin	10.0μg

pH 7.2 ± 0.2 at 25°C

Preparation of Medium: Add components to distilled/deionized water and bring volume to 1.0L. Mix thoroughly. Gently heat and bring to boiling. Distribute into tubes or flasks. Autoclave for 15 min at 15 psi pressure–121°C. Cool tubes in a slanted position.

Use: For the selective isolation and differentiation of *Candida albicans* from other *Candida* species. *Candida albicans* and *Candida tropicalis* colonies appear as smooth, brownish-black round colonies. Other *Candida* species are differentially pigmented or produce diffusible pigments. Usually used in conjunction with BiGGY agar to differentiate further *Candida*; on BiGGY agar, *Candida albicans* appears as brown to black colonies with no pigment diffusion and no sheen, whereas *Candida tropicalis* appears as dark brown colonies with black centers, black pigment diffusion, and a sheen.

AC Agar
(AC Medium)

Composition per liter:

Proteose peptone No. 3	20.0g
Glucose	5.0g
Beef extract	3.0g
Yeast extract	3.0g
Malt extract	3.0g
Ascorbic acid	0.2g
Agar	1.0g

pH 7.2 ± 0.2 at 25°C

Source: This medium is available as a premixed powder from BD Diagnostic Systems.

Preparation of Medium: Add components to distilled/deionized water and bring volume to 1.0L. Mix thoroughly. Gently heat and bring to boiling. Distribute into tubes or flasks. Autoclave for 15 min at 15 psi pressure–121°C.

Use: For the cultivation and isolation of anaerobes, microaerophiles, and aerobes. Recommended for the sterility testing of solutions and other materials not containing mercurial preservatives.

AC Broth

Composition per liter:

Proteose peptone No. 3	20.0g
Glucose	5.0g
Beef extract	3.0g
Yeast extract	3.0g
Malt extract	3.0g
Ascorbic acid	0.2g

pH 7.2 ± 0.2 at 25°C

Source: This medium is available as a premixed powder from BD Diagnostic Systems.

Preparation of Medium: Add components to distilled/deionized water and bring volume to 1.0L. Mix thoroughly. Gently heat and bring to boiling. Distribute into tubes or flasks. Autoclave for 15 min at 15 psi pressure–121°C.

Use: For the cultivation and isolation of a wide variety of microorganisms, including anaerobes, microaerophiles, and aerobes. Recommended for the sterility testing of solutions and other materials not containing mercurial preservatives.

AC HiVeg Agar

Composition per liter:

Plant peptone No. 3	20.0g
Glucose	5.0g
Plant extract	3.0g
Yeast extract	3.0g
Malt extract	3.0g
Ascorbic acid	0.2g
Agar	1.0g

pH 7.2 ± 0.2 at 25°C

Source: This medium is available as a premixed powder from HiMedia.

Preparation of Medium: Add components to distilled/deionized water and bring volume to 1.0L. Mix thoroughly. Gently heat and bring to boiling. Distribute into tubes or flasks or bottles. Autoclave for 15 min at 15 psi pressure–121°C.

Use: For the cultivation and isolation of anaerobes, microaerophiles, and aerobes. Recommended for the sterility testing of solutions and other materials not containing mercurial preservatives.

AC HiVeg Broth

Composition per liter:

Plant peptone No. 3	20.0g
Glucose	5.0g
Plant extract	3.0g
Yeast extract	3.0g
Malt extract	3.0g
Ascorbic acid	0.2g

pH 7.2 ± 0.2 at 25°C

Source: This medium is available as a premixed powder from HiMedia.

Preparation of Medium: Add components to distilled/deionized water and bring volume to 1.0L. Mix thoroughly. Gently heat and bring to boiling. Distribute into tubes or flasks. Autoclave for 15 min at 15 psi pressure–121°C.

Use: For the cultivation and isolation of a wide variety of microorganisms, including anaerobes, microaerophiles, and aerobes. Recommended for the sterility testing of solutions and other materials not containing mercurial preservatives.

AC Medium

See: **AC Agar**

Acanthamoeba Medium

Composition per liter:

Proteose peptone	15.0g
Glucose	15.0g
KH_2PO_4	0.3g
L-Methionine	14.9mg
Thiamine	1.0mg
Biotin	0.2mg
Vitamin B_{12}	1.0μg
Salt solution	1.0mL

pH 5.5 ± 0.2 at 25°C

Salt Solution:

Composition per 100.0mL:

$MgSO_4 \cdot 7H_2O$	2.46g
$CaCl_2 \cdot 2H_2O$	0.15g
$FeCl_3$	0.02g

Preparation of Salt Solution: Add components to distilled/deionized water and bring volume to 100.0mL. Mix thoroughly.

Preparation of Medium: Add components to distilled/deionized water and bring volume to 1.0L. Mix thoroughly. Adjust pH to 5.5. Filter through Whatman paper to remove particles. Distribute into screw-capped tubes or flasks. Autoclave for 15 min at 15 psi pressure–121°C.

Use: For the cultivation of *Acanthamoeba* species.

ACB90 Medium (DSMZ Medium 298h)

Composition per liter:

NaCl	1.0g
KCl	0.5g
$MgCl_2 \cdot 6H_2O$	0.4g
NH_4Cl	0.25g
KH_2PO_4	0.2g
$CaCl_2 \cdot 2H_2O$	0.15g
Resazurin	1.0mg
$NaHCO_3$ solution	10.0mL
Butanediol solution	10.0mL
$Na_2S \cdot 9H_2O$ solution	10.0mL
Vitamin solution	10.0mL
Seven vitamin solution	10.0mL
Glucose solution	10.0mL
Trace elements solution SL-10	1.0mL

pH 7.2 ± 0.2 at 25°C

$Na_2S \cdot 9H_2O$ Solution:

Composition per 10.0mL:

$Na_2S \cdot 9H_2O$	0.36g

Preparation of $Na_2S \cdot 9H_2O$ Solution: Add $Na_2S \cdot 9H_2O$ to distilled/deionized water and bring volume to 10.0mL. Mix thoroughly. Autoclave under 100% N_2 for 15 min at 15 psi pressure–121°C. Cool to room temperature.

$NaHCO_3$ Solution:

Composition per 10.0mL:

$NaHCO_3$	2.5g

Preparation of $NaHCO_3$ Solution: Add $NaHCO_3$ to distilled/deionized water and bring volume to 10.0mL. Mix thoroughly. Sparge with 80% N_2 + 20% CO_2. Filter sterilize.

Butanediol Solution:

Composition per 10.0mL:

2,3 butanediol	0.9g

Preparation of Butanediol Solution: Add butanediol to distilled/deionized water and bring volume to 10.0mL. Mix thoroughly. Sparge with 100% N_2. Filter sterilize.

Trace Elements Solution SL-10:

Composition per liter:

$FeCl_2 \cdot 4H_2O$	1.5g
$CoCl_2 \cdot 6H_2O$	190.0mg
$MnCl_2 \cdot 4H_2O$	100.0mg
$ZnCl_2$	70.0mg
$Na_2MoO_4 \cdot 2H_2O$	36.0mg
$NiCl_2 \cdot 6H_2O$	24.0mg
H_3BO_3	6.0mg
$CuCl_2 \cdot 2H_2O$	2.0mg
HCl (25% solution)	10.0mL

Preparation of Trace Elements Solution SL-10: Add $FeCl_2 \cdot 4H_2O$ to 10.0mL of HCl solution. Mix thoroughly. Add distilled/deionized water and bring volume to 1.0L. Add remaining components. Mix thoroughly. Sparge with 80% N_2 + 20% CO_2. Autoclave for 15 min at 15 psi pressure–121°C.

Vitamin Solution:

Composition per liter:

Pyridoxine-HCl	10.0mg
Thiamine-HCl·$2H_2O$	5.0mg
Riboflavin	5.0mg

Nicotinic acid	5.0mg
D-Ca-pantothenate	5.0mg
p-Aminobenzoic acid	5.0mg
Lipoic acid	5.0mg
Biotin	2.0mg
Folic acid	2.0mg
Vitamin B_{12}	0.1mg

Preparation of Vitamin Solution: Add components to distilled/deionized water and bring volume to 1.0L. Mix thoroughly. Sparge with 80% H_2 + 20% CO_2. Filter sterilize.

Seven Vitamin Solution:
Composition per liter:

Pyridoxine hydrochloride	300.0mg
Thiamine-HCl·$2H_2O$	200.0mg
Nicotinic acid	200.0mg
Vitamin B_{12}	100.0mg
Calcium pantothenate	100.0mg
p-Aminobenzoic acid	80.0mg
D(+)-Biotin	20.0mg

Preparation of Seven Vitamin Solution: Add components to distilled/deionized water and bring volume to 1.0L. Sparge with 100% N_2. Mix thoroughly. Filter sterilize.

Cellobiose Solution:
Composition per 10.0mL:

Cellobiose	0.7g

Preparation of Cellobiose Solution: Add cellobiose to distilled/deionized water and bring volume to 10.0mL. Mix thoroughly. Sparge with 100% N_2. Filter sterilize.

Preparation of Medium: Prepare and dispense medium under 80% N_2 + 20% CO_2 gas atmosphere. Add components, except $NaHCO_3$ solution, butanediol solution, $Na_2S·9H_2O$ solution, vitamin solution, seven vitamin solution, cellobiose solution, and trace elements solution SL-10, to distilled/deionized water and bring volume to 939.0mL. Mix thoroughly. Adjust pH to 7.2. Sparge with 80% N_2 + 20% CO_2. Autoclave for 15 min at 15 psi pressure–121°C. Aseptically and anaerobically add 10.0mL $NaHCO_3$ solution, 10.0mL butanediol solution, 10.0mL $Na_2S·9H_2O$ solution, 10.0mL vitamin solution, 10.0mL seven vitamin solution, 10.0mL cellobiose solution, and 1.0mL trace elements solution SL-10. Mix thoroughly. Aseptically and anaerobically distribute into sterile tubes or bottles. After inoculation, flush and repressurize the gas head space of culture bottles with sterile 80% N_2 + 20% CO_2 to 1 bar overpressure.

Use: For the cultivation of an unclassified bacterium (*Brevigemma cellulytica*) DSM 11249.

ACC Medium

Composition per liter:

Proteose peptone	20.0g
Agar	12.0g
Glycerol	1.5g
K_2SO_4	1.5g
$MgSO_4·7H_2O$	1.5g
Antibiotic solution	10.0mL

pH 7.2 ± 0.2 at 25°C.

Antibiotic Solution:
Composition per 10.0mL:

Cycloheximide	0.075g
Ampicillin	0.05g
Chloramphenicol	0.0125g

Preparation of Antibiotic Solution: Add components to distilled/deionized water and bring volume to 10.0mL. Mix thoroughly. Filter sterilize.

Caution: Cycloheximide is toxic. Avoid skin contact or aerosol formation and inhalation.

Preparation of Medium: Add components, except antibiotic solution, to distilled/deionized water and bring volume to 990.0mL. Mix thoroughly. Gently heat and bring to boiling. Autoclave for 15 min at 15 psi pressure–121°C. Cool to 45°–50°C. Aseptically add sterile antibiotic solution. Mix thoroughly. Pour into sterile Petri dishes or distribute into sterile tubes.

Use: For the selective isolation and cultivation of fluorescent *Pseudomonas* species.

ACE Medium

Composition per liter:

NaCl	18.0g
$MgCl_2·6H_2O$	6.0g
$NaHCO_3$	2.0g
$MgSO_4·7H_2O$	1.0g
Yeast extract	1.0g
L-Cysteine·HCl	0.5g
KCl	0.335g
NH_4Cl	0.25g
$CaCl_2·2H_2O$	0.14g
K_2HPO_4	0.14g
$Fe(NH_4)_2(SO_4)_2·6H_2O$	2.0mg
Resazurin	1.0mg
Glucose solution	20.0mL
Wolfe's vitamin solution	10.0mL

pH 7.6 ± 0.2 at 25°C

Glucose Solution:
Composition per 50.0mL:

D-Glucose	10.0g

Preparation of Glucose Solution: Add glucose to distilled/deionized water and bring volume to 50.0mL. Mix thoroughly. Sparge with 100% N_2. Autoclave for 15 min at 15 psi pressure–121°C.

Wolfe's Vitamin Solution:
Composition per liter:

Pyridoxine·HCl	10.0mg
p-Aminobenzoic acid	5.0mg
Lipoic acid	5.0mg
Nicotinic acid	5.0mg
Riboflavin	5.0mg
Thiamine·HCl	5.0mg
Calcium DL-pantothenate	5.0mg
Biotin	2.0mg
Folic acid	2.0mg
Vitamin B_{12}	0.1mg

Preparation of Wolfe's Vitamin Solution: Add components to distilled/deionized water and bring volume to 1.0L. Mix thoroughly. Filter sterilize.

Preparation of Medium: Add components, except L-cysteine·HCl, $NaHCO_3$, and glucose solution, to distilled/deionized water and bring volume to 980.0mL. Mix thoroughly. Adjust pH to 7.6. Gently heat and bring to boiling. Continue boiling for 3 min. Cool to room temperature while sparging with O_2-free 100% N_2. Add L-cysteine·HCl and $NaHCO_3$. Anaerobically distribute 9.8mL volumes into anaerobic tubes. Autoclave for 15 min at 15 psi pressure–121°C. Aseptically and anaerobically add 0.2mL of sterile glucose solution to each tube. Mix thoroughly.

Use: For the cultivation of unclassified bacterium DSM 6211 and unclassified bacterium DSM 6226.

Acetamide Agar

Composition per liter:

Agar	15.0g
Acetamide	10.0g
NaCl	5.0g
K_2HPO_4	1.0g
$NH_4H_2PO_4$	1.0g
$MgSO_4 \cdot 7H_2O$	0.2g
Bromthymol Blue	0.08g

pH 6.9 ± 0.2 at 25°C

Preparation of Medium: Add components to distilled/deionized water and bring volume to 1.0L. Mix thoroughly. Gently heat and bring to boiling. Adjust pH. Distribute into tubes or flasks. Autoclave for 15 min at 15 psi pressure–121°C. Cool tubes in a slanted position to produce a long slant.

Use: For the differentiation of nonfermentative Gram-negative bacteria, especially *Pseudomonas aeruginosa.* Can be used as a confirmatory test for water analysis. Bacteria that deamidate acetamide turn the medium blue.

Acetamide Agar

Composition per liter:

Agar	15.0g
Acetamide	10.0g
NaCl	5.0g
K_2HPO_4	1.39g
KH_2PO_4	0.73g
$MgSO_4 \cdot 7H_2O$	0.5g
Phenol Red	0.012g

pH 6.9 ± 0.2 at 25°C

Source: This medium is available as a premixed powder from BD Diagnostic Systems.

Preparation of Medium: Add components to distilled/deionized water and bring volume to 1.0L. Mix thoroughly. Gently heat and bring to boiling. Adjust pH. Distribute into tubes or flasks. Autoclave for 15 min at 15 psi pressure–121°C. Cool tubes in a slanted position to produce a long slant.

Use: For the differentiation of nonfermentative Gram-negative bacteria, especially *Pseudomonas aeruginosa.* Can be used as a confirmatory test for water analysis. Bacteria that deamidate acetamide turn the medium red.

Acetamide Broth

Composition per liter:

Acetamide	10.0g
NaCl	5.0g
K_2HPO_4	1.39g
KH_2PO_4	0.73g
$MgSO_4 \cdot 7H_2O$	0.5g
Phenol Red	0.012g

pH 6.9 ± 0.2 at 25°C

Preparation of Medium: Add components to distilled/deionized water and bring volume to 1.0L. Mix thoroughly. Adjust pH. Autoclave for 15 min at 15 psi pressure–121°C.

Use: For the differentiation of nonfermentative Gram-negative bacteria, especially *Pseudomonas aeruginosa.* Can be used as a confirmatory test for water analysis. Bacteria that deamidate acetamide turn the broth purplish red.

Acetamide Broth

Composition per liter:

Acetamide	2.0g
KH_2PO_4	1.0g
NaCl	0.2g
$MgSO_4$, anhydrous	0.2g
$Na_2MoO_4 \cdot 2H_2O$	5.0mg
$FeSO_4$	0.5mg

pH 7.0 ± 0.2 at 25°C

Source: This medium is available from HiMedia.

Preparation of Medium: Add components, except acetamide, to distilled/deionized water and bring volume to 1.0L. Mix thoroughly. Add acetamide. Adjust pH to 7.0. Autoclave for 15 min at 15 psi pressure–121°C.

Use: For the differentiation of nonfermentative Gram-negative bacteria, especially *Pseudomonas aeruginosa.*

Acetamide Cetrimide Glycerol Mannitol Selective Medium

Composition per liter:

Agar	15.0g
K_2SO_4	10.0g
D-Mannitol	5.0g
$MgCl_2 \cdot 6H_2O$	1.4g
Cetrimide	0.3g
Peptone	0.2g
Acetamide solution	100.0mL
Glycerol	5.0mL

pH 7.0 ± 0.2 at 25°C

Acetamide Solution:

Composition per 100.0mL:

Acetamide	10.0g
Phenol Red	0.012g

Preparation of Acetamide Solution: Add components to distilled/deionized water and bring volume to 100.0mL. Mix thoroughly. Filter sterilize.

Preparation of Medium: Add components, except acetamide solution, to distilled/deionized water and bring volume to 900.0mL. Mix thoroughly. Adjust pH to 7.0. Gently heat and bring to boiling. Autoclave for 20 min at 15 psi pressure–121°C. Cool to 45°–50°C. Aseptically add sterile acetamide solution. Mix thoroughly. Pour into sterile Petri dishes.

Use: For the cultivation of *Pseudomonas aeruginosa, Pseudomonas fluorescens, Pseudomonas putida, Pseudomonas alcaligenes, Pseudomonas cepacia,* and *Pseudomonas pseudoalcaligenes.*

Acetamide Medium

Composition per liter:

Agar, noble	20.0g
Glucose	20.0g

KH_2PO_4	15.0g
$CsCl_2$ solution	12.5mL
Acetamide solution	10.0mL
$CaCl_2 \cdot 2H_2O$ solution	4.1mL
$MgSO_4 \cdot 7H_2O$ solution	2.4mL
Trace elements solution	1.0mL

$CsCl_2$ Solution:
Composition per 100.0mL:

$CsCl_2$	16.84g

Preparation of $CsCl_2$ Solution: Add $CsCl_2$ to distilled/deionized water and bring volume to 100.0mL. Mix thoroughly. Autoclave for 15 min at 15 psi pressure–121°C.

Acetamide Solution:
Composition per 100.0mL:

Acetamide	5.91g

Preparation of Acetamide Solution: Add acetamide to distilled/deionized water and bring volume to 100.0mL. Mix thoroughly. Autoclave for 15 min at 15 psi pressure–121°C.

$CaCl_2 \cdot 2H_2O$ Solution:
Composition per 100.0mL:

$CaCl_2 \cdot 2H_2O$	14.7g

Preparation of $CaCl_2 \cdot 2H_2O$ Solution: Add $CaCl_2 \cdot 2H_2O$ to distilled/deionized water and bring volume to 100.0mL. Mix thoroughly. Autoclave for 15 min at 15 psi pressure–121°C.

$MgSO_4 \cdot 7H_2O$ Solution:
Composition per 100.0mL:

$MgSO_4 \cdot 7H_2O$	24.65g

Preparation of $MgSO_4 \cdot 7H_2O$ Solution: Add $MgSO_4 \cdot 7H_2O$ to distilled/deionized water and bring volume to 100.0mL. Mix thoroughly. Autoclave for 15 min at 15 psi pressure–121°C.

Trace Elements Solution:
Composition per liter:

$FeSO_4 \cdot 7H_2O$	5.0g
$CoCl_2 \cdot 6H_2O$	3.7g
$MnSO_4 \cdot 1H_2O$	1.6g
$ZnSO_4 \cdot 7H_2O$	1.4g

Preparation of Trace Elements Solution: Add components to distilled/deionized water and bring volume to 1.0L. Mix thoroughly.

Preparation of Medium: Add components, except $CsCl_2$ solution, acetamide solution, $CaCl_2 \cdot 2H_2O$ solution, and $MgSO_4 \cdot 7H_2O$ solution, to distilled/deionized water and bring volume to 971.0mL. Mix thoroughly. Gently heat and bring to boiling. Autoclave for 15 min at 15 psi pressure–121°C. Cool to 50°–55°C. Aseptically add 12.5mL of sterile $CsCl_2$ solution, 10.0mL of sterile acetamide solution, 4.1mL of sterile $CaCl_2 \cdot 2H_2O$ solution, and 2.4mL of sterile $MgSO_4 \cdot 7H_2O$ solution. Mix thoroughly. Pour into sterile Petri dishes or distribute into sterile tubes.

Use: For the cultivation and maintenance of *Trichoderma longibrachiatum.*

Acetamide Medium (BAM M2)

Composition per liter:

Stock basal solution	400.0mL
Stock acetamide solution	100.0mL

pH 6.9 ± 0.2 at 25°C

Stock Basal Solution:
Composition per 400.0mL:

Agar	0.5g
KH_2PO_4 solution, 0.5*M*	14.0mL
K_2HPO_4 solution, 0.5*M*	6.0mL
PR-CV solution	1.0mL

Preparation of Stock Basal Solution: Add components, except PR-CV solution, to distilled/deionized water and bring volume to 400.0mL. Mix thoroughly. Gently heat and bring to boiling with agitation to dissolve agar. Add 1.0mL PR-CV solution.

PR-CV Solution:
Composition per 200.0mL:

Phenol Red	2.0g
Crystal Violet	0.2g

Preparation of PR-CV Solution: Add components to distilled/deionized water and bring volume to 200.0mL. Mix thoroughly. Add 5*N* NaOH while stirring until components are dissolved.

Stock Acetamide Solution:
Composition per 100.0mL:

Acetamide	1.0g

Preparation of Stock Acetamide Solution: Add acetamide to distilled/deionized water and bring volume to 100.0mL. Mix thoroughly. Store over methylene chloride in a screw-capped container. Can be stored indefinitely at room temperature.

Preparation of Medium: Combine 400.0mL stock basal solution and 100.0mL stock acetamide solution. Mix thoroughly. Distribute into tubes or flasks. Steam for 10 min at 100°C. Cool.

Use: For the differentiation of nonfermentative Gram-negative bacteria, especially *Pseudomonas aeruginosa,* e.g., in milk.

Acetamide Nutrient Broth

Composition per liter:

Acetamide	2.0g
KH_2PO_4	1.0g
NaCl	0.2g
$MgSO_4$, anhydrous	0.158g
$Na_2MoO_4 \cdot 2H_2O$	5.0mg
$FeSO_4$	0.5mg

pH 7.0 ± 0.2 at 25°C

Source: This medium is available from HiMedia.

Preparation of Medium: Add components, except acetamide, to distilled/deionized water and bring volume to 1.0L. Mix thoroughly. Add acetamide. Adjust pH to 7.0. Autoclave for 15 min at 15 psi pressure–121°C.

Use: For the differentiation of nonfermentative Gram-negative bacteria, especially *Pseudomonas aeruginosa.*

Acetate Agar

Composition per liter:

Agar	20.0g
Glucose	10.0g
Peptic digest of animal tissue	5.0g
Meat extract	5.0g
Yeast extract	5.0g
Sodium acetate	27.22g
Tween™ 80	0.5mL

pH 5.4 ± 0.2 at 25°C

Source: This medium is available from HiMedia.

Preparation of Medium: Add components to distilled/deionized water and bring volume to 900.0mL. Mix thoroughly. Gently heat and bring to boiling. Adjust pH to 5.4. Distribute into tubes or flasks. Autoclave for 15 min at 15 psi pressure–121°C. Pour into Petri dishes or leave in tubes.

Use: For the isolation and cultivation of *Leuconostoc* species and *Pediococcus* species.

Acetate Agar

Composition per liter:

Meat extract	50.0g
Glucose	10.0g
Peptone	5.0g
Yeast extract	5.0g
Sodium acetate buffer	100.0mL
Tween™ 80	0.5mL

pH 5.4 ± 0.2 at 25°C

Sodium Acetate Buffer:

Composition per liter:

Sodium acetate·3H_2O	272.2g

Preparation of Sodium Acetate Buffer: Add sodium acetate to distilled/deionized water and bring volume to 1.0L. Mix thoroughly. Adjust pH to 5.4 with glacial acetic acid. Filter sterilize.

Preparation of Medium: Add components, except sodium acetate buffer, to distilled/deionized water and bring volume to 900.0mL. Mix thoroughly. Gently heat and bring to boiling. Adjust pH to 5.4. Autoclave for 15 min at 15 psi pressure–121°C. Cool to 45°–50°C. Aseptically add 100.0mL of sterile sodium acetate buffer. Mix thoroughly. Aseptically distribute into sterile tubes or flasks.

Use: For the isolation and cultivation of *Leuconostoc* species and *Pediococcus* species.

Acetate Agar

Composition per liter:

Agar	15.0g
Yeast extract	2.0g
Sodium acetate	1.0g
Pancreatic digest of casein	1.0g

pH 7.4 ± 0.2 at 25°C

Preparation of Medium: Add components to distilled/deionized water and bring volume to 1.0L. Mix thoroughly. Gently heat and bring to boiling. Distribute into tubes or flasks. Autoclave for 15 min at 15 psi pressure–121°C. Pour into sterile Petri dishes or leave in tubes.

Use: For the cultivation and maintenance of *Caryophanon latum.*

Acetate Agar (BAM M3)

Composition per liter:

Agar	20.0g
NaCl	5.0g
Sodium acetate	2.0g
KH_2PO_4	1.0g
$(NH_4)_2PO_4$	1.0g
$MgSO_4$	0.2g
Bromthymol Blue	0.08g

pH 6.7 ± 0.2 at 25°C

Preparation of Medium: Add components, except $MgSO_4$, to distilled/deionized water and bring volume to 1.0L. Mix thoroughly. Gently heat and bring to boiling. Adjust pH to 6.7. Add 0.2g $MgSO_4$. Mix thoroughly. Autoclave for 15 min at 15 psi pressure–121°C. Cool to 45°–50°C. Aseptically distribute into tubes or flasks. For slants distribute aliquots into tubes prior to autoclaving. After autoclaving allow tubes to cool in inclined position to obtain a 5cm slant.

Use: For the cultivation of *Leuconostoc* species.

Acetate Differential Agar (Sodium Acetate Agar) (Simmons' Citrate Agar, Modified)

Composition per liter:

Agar	20.0g
NaCl	5.0g
Sodium acetate	2.0g
$(NH_4)H_2PO_4$	1.0g
K_2HPO_4	1.0g
$MgSO_4·7H_2O$	0.2g
Bromthymol Blue	0.08g

pH 6.8 ± 0.2 at 25°C

Source: This medium is available as a premixed powder from BD Diagnostic Systems.

Preparation of Medium: Add components to cold distilled/deionized water and bring volume to 1.0L. Mix thoroughly. Gently heat and bring to boiling. Distribute into tubes to produce a 1 cm butt and 30 cm slant. Autoclave for 15 min at 15 psi pressure–121°C. Cool tubes in a slanted position.

Use: For the differentiation of *Shigella* species from *Escherichia coli* and also for the differentiation of nonfermenting Gram-negative bacteria. Bacteria that can utilize acetate as the sole carbon source turn the medium blue.

Acetate HiVeg Agar

Composition per liter:

Sodium acetate·3H_2O	27.22g
Agar	20.0g
Glucose	10.0g
Yeast extract	5.0g
Plant peptone	5.0g
Plant extract No.1	5.0g
Tween™ 80	0.5mL

pH 5.4 ± 0.2 at 25°C

Source: This medium is available as a premixed powder from HiMedia.

Preparation of Medium: Add components to distilled/deionized water and bring volume to 1.0L. Mix thoroughly. Gently heat and bring to boiling. Autoclave for 15 min at 15 psi pressure–121°C. Cool to 45°–50°C. Aseptically adjust pH to 5.4. Pour into sterile Petri dishes or distribute into sterile tubes.

Use: For the isolation and cultivation of *Leuconostoc* species and *Pediococcus* species.

Acetic Acid Agar

Composition per liter:

Agar	25.0g
Sodium acetate, anhydrous	4.5g
Yeast extract	1.0g

Wort solution ... 500.0mL
Sucrose solution ... 500.0mL

Wort Solution:

Composition per 500.0mL:

Malt extract ... 55.0g

Preparation of Wort Solution: Add malt extract to distilled/deionized water and bring volume to 500.0mL. Mix thoroughly.

Sucrose Solution:

Composition per 500.0mL:

Sucrose ... 50.0g

Preparation of Sucrose Solution: Add sucrose to distilled/deionized water and bring volume to 500.0mL. Mix thoroughly.

Preparation of Medium: Combine components. Mix thoroughly. Gently heat and bring to boiling. Distribute into tubes or flasks. Autoclave for 15 min at 15 psi pressure–121°C. Pour into sterile Petri dishes or leave in tubes.

Use: For the cultivation and maintenance of *Tetragenococcus halophilus.*

Acetic Acid Bacterium Medium (DSMZ Medium 989)

Composition per liter:

Agar	15.0g
Peptone	5.0g
Yeast extract	5.0g
Glucose	5.0g
$MgSO_4·7H_2O$	1.0g

pH 6.6–7.0 at 25°C

Preparation of Medium: Add components to distilled/deionized water and bring volume to 1.0L. Mix thoroughly. Gently heat and bring to boiling. Distribute into tubes or flasks. Autoclave for 15 min at 15 psi pressure–121°C. Pour into sterile Petri dishes or leave in tubes.

Use: For the cultivation and maintenance of acetic acid bacteria.

Acetic Acid Broth

Composition per liter:

Sodium acetate, anhydrous	4.5g
Yeast extract	1.0g
Wort solution	500.0mL
Sucrose solution	500.0mL

Wort Solution:

Composition per 500.0mL:

Malt extract ... 55.0g

Preparation of Wort Solution: Add malt extract to distilled/deionized water and bring volume to 500.0mL. Mix thoroughly.

Sucrose Solution:

Composition per 500.0mL:

Sucrose ... 50.0g

Preparation of Sucrose Solution: Add sucrose to distilled/deionized water and bring volume to 500.0mL. Mix thoroughly.

Preparation of Medium: Combine components. Mix thoroughly. Distribute into tubes or flasks. Autoclave for 15 min at 15 psi pressure–121°C.

Use: For the cultivation of *Tetragenococcus halophilus.*

Acetitomaculum Medium

Composition per liter:

NaCl	7.0g
Glucose	5.0g
$NaHCO_3$	4.0g
Yeast extract	1.5g
$MgCl_2·6H_2O$	1.2g
KCl	0.5g
NH_4Cl	0.3g
$CaCl_2·2H_2O$	0.15g
Na_2SO_4	0.1g
$NiCl_2·6H_2O$	1.0mg
Resazurin	0.5mg
$Na_2WO_4·2H_2O$	0.1mg
Ascorbic acid	50.0μg
Choline chloride	50.0μg
D-myo-Inositol	50.0μg
Nicotinamide	50.0μg
Glucose solution	50.0mL
Trace elements solution	10.0mL
Vitamin solution	10.0mL
Reducing agent solution	10.0mL

pH 7.0 ± 0.2 at 25°C

Glucose Solution:

Composition per 50.0mL:

D-Glucose ... 5.0g

Preparation of Glucose Solution: Add glucose to distilled/deionized water and bring volume to 50.0mL. Mix thoroughly. Dispense solution anaerobically under 100% N_2 gas. Autoclave for 15 min at 15 psi pressure–121°C.

Trace Elements Solution:

Composition per liter:

$MgSO_4·7H_2O$	3.0g
Nitrilotriacetic acid	1.5g
$CaCl_2·2H_2O$	.1.0g
NaCl	1.0g
$MnSO_4·2H_2O$	0.5 g
$CoSO_4·7H_2O$	0.18 g
$ZnSO_4·7H_2O$	0.18 g
$FeSO_4·7H_2O$	0.1g
$NiCl_2·6H_2O$	0.025 g
$KAl(SO_4)_2·12H_2O$	0.02g
$CuSO_4·5H_2O$	0.01g
H_3BO_3	0.01g
$Na_2MoO_4·2H_2O$	0.01g
$Na_2SeO_3·5H_2O$	0.3 mg

Preparation of Trace Elements Solution: Add nitrilotriacetic acid to 500.0mL of distilled/deionized water. Dissolve by adjusting pH to 6.5 with KOH. Add distilled/deionized water to 1.0L. Add remaining components. Mix thoroughly.

Vitamin Solution:

Composition per liter:

Pyridoxine·HCl	10.0mg
Calcium DL-pantothenate	5.0mg
Lipoic acid	5.0mg
Nicotinic acid	5.0mg
p-Aminobenzoic acid	5.0mg
Riboflavin	5.0mg
Thiamine·HCl	5.0mg
Biotin	2.0mg

Folic acid 2.0mg
Vitamin B_{12} 0.1mg

Preparation of Vitamin Solution: Add components to distilled/deionized water and bring volume to 1.0L. Mix thoroughly.

Reducing Agent Solution:
Composition per 110.0mL:

L-Cysteine·HCl·H_2O 2.5g
$Na_2S·9H_2O$ 2.5g

Preparation of Reducing Agent Solution: Add 110.0mL of distilled/deionized water to a 250.0mL flask. Boil under N_2 gas for 1 min. Cool to room temperature. Add L-cysteine·HCl·H_2O and dissolve. Adjust to pH 9 with 5*N* NaOH. Add washed $Na_2S·9H_2O$ and dissolve. Distribute in 10.0mL volumes into tubes. Autoclave for 10 min at 15 psi pressure–121°C.

Preparation of Medium: Add components, except $NaHCO_3$, glucose solution, and reducing agent solution, to distilled/deionized water and bring volume to 920.0mL. Gently heat and bring to boiling. Continue boiling for 3 min. Cool to room temperature under 80% N_2 + 20% CO_2. Add solid $NaHCO_3$ and bring pH to 6.8–7.0 by gassing. Distribute anaerobically under 80% N_2 + 20% CO_2. Autoclave for 15 min at 15 psi pressure–121°C. Prior to inoculation of cultures, aseptically and anaerobically add 0.1mL of sterile reducing agent solution and 0.5mL of sterile glucose solution to each tube containing 9.4mL of sterile basal medium.

Use: For the cultivation and maintenance of *Acetitomaculum ruminus*.

Acetitomaculum ruminus Medium (LMG Medium 224)

Composition per liter:

NaCl 7.0g
$NaHCO_3$ 4.0g
Yeast extract 1.5g
$MgCl_2·6H_2O$ 1.2g
KCl 0.5g
NH_4Cl 0.3g
KH_2PO_4 .... 0.2g
$CaCl_2·2H_2O$ 0.15g
Na_2SO_4 .... 0.1g
Resazurin 0.5mg
Ascorbic acid 50.0μg
Myo-inositol 50.0μg
Niacinamide 50.0μg
Choline chloride 50.0μg
Glucose solution 10.0mL
L-Cysteine·HCl·H_2O solution 10.0mL
$Na_2S·9H_2O$ solution 10.0mL
Wolfe's mineral solution 10.0mL
Wolfe's vitamin solution 10.0mL

pH 6.9 ± 0.2 at 25°C

Glucose Solution:
Composition per 10.0mL:

Glucose 5.0g

Preparation of Glucose Solution: Add glucose to distilled/deionized water and bring volume to 10.0mL. Mix thoroughly. Sparge with 100% N_2. Filter sterilize.

L-Cysteine·HCl·H_2O Solution:
Composition per 100.0mL:

L-Cysteine·HCl·H_2O 3.0g

Preparation of L-Cysteine·HCl·H_2O Solution: Bring 100.0mL of distilled/deionized water to boiling. Cool to room temperature while sparging with 100% N_2. Add L-cysteine·HCl·H_2O to the 100.0mL of anaerobic water. Distribute into serum bottles fitted with butyl rubber stoppers and aluminum seals. Do not grease stoppers. Autoclave for 20 min at 15 psi pressure–121°C.

$Na_2S·9H_2O$ Solution:
Composition per 100.0mL:

NaOH .... 1 pellet
$Na_2S·9H_2O$ 3.0g

Preparation of $Na_2S·9H_2O$ Solution: Bring 100.0mL of distilled/deionized water to boiling. Cool to room temperature while sparging with 100% N_2. Dissolve 1 pellet of NaOH in the anaerobic water. Weigh out a little more than 3.0g of $Na_2S·9H_2O$. Briefly rinse the crystals in distilled/deionized water. Dry the crystals by blotting on paper towels or filter paper. Weigh out 2.5g of washed $Na_2S·9H_2O$ crystals. Add to the 100.0mL of anaerobic NaOH solution. Distribute into serum bottles fitted with butyl rubber stoppers and aluminum seals. Do not grease stoppers. Pressurize to 60kPa with 100% N_2. Autoclave for 15 min at 15 psi pressure–121°C. Store at room temperature in an anaerobic chamber.

Wolfe's Mineral Solution:
Composition per liter:

$MgSO_4·7H_2O$ 3.0g
Nitrilotriacetic acid 1.5g
NaCl 1.0g
$MnSO_4·2H_2O$ 0.5g
$CoCl_2·6H_2O$ 0.1g
$ZnSO_4·7H_2O$ 0.1g
$CaCl_2·2H_2O$ 0.1g
$FeSO_4·7H_2O$ 0.1g
$NiCl_2·6H_2O$ 0.025g
$KAl(SO_4)_2·12H_2O$ 0.02g
$CuSO_4·5H_2O$ 0.01g
H_3BO_3 .... 0.01g
$Na_2MoO_4·2H_2O$ 0.01g
$Na_2SeO_3·5H_2O$ 0.3mg

Preparation of Wolfe's Mineral Solution: Add nitrilotriacetic acid to 500.0mL of distilled/deionized water. Adjust pH to 6.5 with KOH. Add remaining components. Add distilled/deionized water to 1.0L. Adjust pH to 6.8.

Wolfe's Vitamin Solution:
Composition per liter:

Pyridoxine·HCl 10.0mg
p-Aminobenzoic acid 5.0mg
Lipoic acid 5.0mg
Nicotinic acid 5.0mg
Riboflavin 5.0mg
Thiamine·HCl 5.0mg
Calcium DL-pantothenate 5.0mg
Biotin 2.0mg
Folic acid 2.0mg
Vitamin B_{12} 0.1mg

Preparation of Wolfe's Vitamin Solution: Add components to distilled/deionized water and bring volume to 1.0L. Mix thoroughly. Filter sterilize.

Preparation of Medium: Prepare and dispense medium under 80% N_2 + 20% CO_2. Add components, except bicarbonate, glucose solution, L-cysteine·HCl·H_2O solution, and $Na_2S·9H_2O$ solution, to distilled/deionized water and bring volume to 970.0mL. Mix thoroughly. Add solid bicarbonate and equilibrate pH to 6.8–7 by gassing. Gently heat and bring to boiling. Continue boiling for 3 min. Cool to room temperature while sparging with 80% N_2 + 20% CO_2. Anaerobically distribute 9.7mL volumes into anaerobic tubes. Autoclave for 20 min at 15 psi pressure–121°C. Aseptically and anaerobically add 0.1mL of sterile glucose, 0.1mL of sterile L-cysteine·HCl·H_2O solution, and 0.1mL of sterile $Na_2S·9H_2O$ solution to each tube. Mix thoroughly.

Use: For the cultivation of *Acetitomaculum ruminus.*

Acetivibrio cellulolyticus Medium

Composition per 1170.0mL:

Cellobiose or cellulose (MN 300, Whatman CF II, Kleenex tissue paper, or HCl-treated cotton) 3.0g
$NaHCO_3$ 2.0g
L-Cysteine·HCl 0.25g
$Na_2S·9H_2O$ 0.25g
$FeSO_4·7H_2O$ 0.02g
Resazurin 0.001g
Mineral solution 1 .... 75.0mL
Mineral solution 2 .... 75.0mL
Cellobiose solution 50.0mL
Trace elements solution 10.0mL
Vitamin solution 10.0mL
Reducing agent solution 10.0mL

pH 7.2 ± 0.2 at 25°C

Mineral Solution 1:

Composition per liter:

K_2HPO_4 .... 3.9g

Preparation of Mineral Solution 1: Add K_2HPO_4 to distilled/deionized water and bring volume to 1.0L. Mix thoroughly.

Mineral Solution 2:

Composition per liter:

$(NH_4)_2SO_4$ 6.0g
K_2HPO_4 .... 2.4g
$MgSO_4·7H_2O$ 1.2g
$CaCl_2·2H_2O$ 0.72g
NaCl 0.59g

Preparation of Mineral Solution 2: Add components to distilled/deionized water and bring volume to 1.0L. Mix thoroughly.

Cellobiose Solution:

Composition per 50.0mL:

D-Cellobiose 5.0g

Preparation of Cellobiose Solution: Add cellobiose to distilled/deionized water and bring volume to 50.0mL. Mix thoroughly. Sparge under 100% N_2 gas for 3 min. Filter sterilize. Store under N_2 gas.

Trace Elements Solution:

Composition per liter:

$MgSO_4·7H_2O$ 3.0g
Nitrilotriacetic acid 1.5g
$CaCl_2·2H_2O$ 1.0g
NaCl 1.0g
$MnSO_4·2H_2O$ 0.5g
$CoSO_4·7H_2O$ 0.18g
$ZnSO_4·7H_2O$ 0.18g
$FeSO_4·7H_2O$ 0.1g
$NiCl_2·6H_2O$ 0.025g
$KAl(SO_4)_2·12H_2O$ 0.02g
$CuSO_4·5H_2O$ 0.01g
H_3BO_3 .... 0.01g
$Na_2MoO_4·2H_2O$ 0.01g
$Na_2SeO_3·5H_2O$ 0.3mg

Preparation of Trace Elements Solution: Add nitrilotriacetic acid to 500.0mL of distilled/deionized water. Dissolve by adjusting pH to 6.5 with KOH. Add distilled/deionized water to 1.0L. Add remaining components. Mix thoroughly. Adjust pH to 7.0 with 1*N* KOH.

Vitamin Solution:

Composition per liter:

Pyridoxine·HCl 10.0mg
Calcium DL-pantothenate 5.0mg
Lipoic acid 5.0mg
Nicotinic acid 5.0mg
p-Aminobenzoic acid 5.0mg
Riboflavin 5.0mg
Thiamine·HCl 5.0mg
Biotin 2.0mg
Folic acid 2.0mg
Vitamin B_{12} 0.1mg

Preparation of Vitamin Solution: Add components to distilled/deionized water and bring volume to 1.0L. Mix thoroughly.

Reducing Agent Solution:

Composition per 110.0mL:

L-Cysteine·HCl·H_2O 2.5g
$Na_2S·9H_2O$ 2.5g

Preparation of Reducing Agent Solution: Add 110.0mL of distilled/deionized water to a 250.0mL flask. Boil under N_2 gas for 1 min. Cool to room temperature. Add L-cysteine·HCl·H_2O and dissolve. Adjust to pH 9 with 5*N* NaOH. Add washed $Na_2S·9H_2O$ and dissolve. Distribute under N_2 gas in 10.0mL volumes into tubes. Autoclave for 10 min at 15 psi pressure–121°C.

Preparation of Medium: Add components, except cellobiose solution and reducing agent solution, to distilled/deionized water and bring volume to 940.0mL. Gently heat and bring to boiling. Continue boiling for 3 min. Cool to room temperature under 80% N_2 + 20% CO_2. Adjust pH to 7.6 by gassing. Distribute anaerobically under 80% N_2 + 20% CO_2. Autoclave for 15 min at 15 psi pressure–121°C. After autoclaving, the pH of the medium will be 7.2. Prior to inoculation of cultures, aseptically and anaerobically add 0.1mL of sterile reducing agent solution and 0.5mL of sterile cellobiose solution to each tube containing 9.4mL of sterile basal medium.

Use: For the cultivation and maintenance of *Acetivibrio cellulolyticus.*

Acetivibrio Desulfovibrio Medium (LMG Medium 105)

Composition per liter:

Solution A 869.0mL
Solution C 100.0mL
Solution D 10.0mL
Solution E 10.0mL
Solution F 10.0mL
Solution B 1.0mL

pH 7.7 ± 0.2 at 25°C

Solution A:

Composition per 869.0mL:

Na_2SO_4 .... 3.0g
NaCl 1.0g

KCl	0.5g
$MgCl_2·6H_2O$	0.4g
NH_4Cl	0.3g
KH_2PO_4	0.2g
$CaCl_2·2H_2O$	0.15g

Preparation of Solution A: Add components to distilled/deionized water and bring volume to 869.0mL. Mix thoroughly. Prepare and autoclave part A under 80% N_2 + 20% CO_2. Autoclave for 15 min at 15 psi pressure–121°C. Cool to room temperature.

Solution B:

Composition per liter:

$FeCl_2·4H_2O$	1.5g
$CoCl_2·6H_2O$	0.19g
$MnCl_2·4H_2O$	0.1g
$ZnCl_2$	0.07g
H_3BO_3	0.06g
$Na_2MoO_4·2H_2O$	0.04g
$NiCl_2·6H_2O$	0.02g
$CuCl_2·2H_2O$	0.02g
HCl, 25%	10.0mL

Preparation of Solution B: Add the $FeCl_2·4H_2O$ to the HCl. Add distilled/deionized water and bring volume to 1.0L. Add remaining components. Mix thoroughly. Autoclave under 100% N_2 for 15 min at 15 psi pressure–121°C. Cool to room temperature.

Solution C:

Composition per 100.0mL:

$NaHCO_3$	5.0g

Preparation of Solution C: Add the $NaHCO_3$ to distilled/deionized water and bring volume to 100.0mL. Mix thoroughly. Filter sterilize. Gas with 80% N_2 + 20% CO_2 to remove residual O_2.

Solution D:

Composition per 10.0mL:

Sodium butyrate	0.7g
Sodium caproate	0.3g
Sodium octanoate	0.15g

Preparation of Solution D: Add components to distilled/deionized water and bring volume to 10.0mL. Mix thoroughly. Autoclave under 100% N_2 for 15 min at 15 psi pressure–121°C. Cool to room temperature.

Solution E:

Composition per 10.0mL:

Yeast extract	1.0g
Thiamine·HCl	100.0μg
p-Aminobenzoic acid	40.0μg
D(+)-Biotin	10.0μg

Preparation of Solution E: Add components to distilled/deionized water and bring volume to 10.0mL. Mix thoroughly. Autoclave under 100% N_2 for 15 min at 15 psi pressure–121°C. Cool to room temperature.

Solution F:

Composition per 10.0mL:

$Na_2S·9H_2O$	0.4g

Preparation of Solution F: Add $Na_2S·9H_2O$ to distilled/deionized water and bring volume to 10.0mL. Mix thoroughly. Autoclave under 100% N_2 for 15 min at 15 psi pressure–121°C. Cool to room temperature.

Preparation of Medium: To 869.0mL of sterile cooled Solution A, aseptically add the remaining sterile solutions in the following order: solution B, solution C, solution D, solution E, and solution F. Mix thoroughly. Adjust pH to 7.7. Anaerobically distribute under 90% N_2 + 10% CO_2 into sterile tubes or flasks.

Use: For the cultivation of *Acetivibrio ethanolgignens* and *Desulfovibrio sapovorans.*

Acetobacter Agar

Composition per liter:

Agar	15.0g
Yeast extract	5.0g
$(NH_4)_2SO_4$	3.3g
KH_2PO_4	1.0g
$MgSO_4$	0.25g
Vitamin solution	200.0mL
Glucose solution	15.0mL
Trace elements solution	1.0mL

Trace Elements Solution:

Composition per 100.0mL:

$CaCl_2·2H_2O$	1.457g
$FeSO_4·7H_2O$	0.366g
$ZnSO_4·7H_2O$	0.178g
$MnSO_4·H_2O$	0.101g
$Na_2MoO_4·2H_2O$	23.4mg
$CuSO_4·5H_2O$	7.8mg

Preparation of Trace Elements Solution: Add components to distilled/deionized water and bring volume to 100.0mL. Mix thoroughly. Filter sterilize.

Vitamin Solution:

Composition per 100.0mL:

m-Inositol	200.0mg
Calcium DL-pantothenate	40.0mg
Nicotinic acid	40.0mg
Pyrdoxine·HCl	40.0mg
Thiamine·HCl	40.0mg
p-Aminobenzoic acid	20.0mg
Riboflavin	20.0mg
Biotin	0.2mg
Folic acid	0.2mg

Preparation of Vitamin Solution: Add components to distilled/deionized water and bring volume to 100.0mL. Mix thoroughly. Filter sterilize.

Glucose Solution:

Composition per 100.0mL:

Glucose	40.0g

Preparation of Glucose Solution: Add components to distilled/deionized water and bring volume to 100.0mL. Mix thoroughly. Filter sterilize.

Preparation of Medium: Add components, except vitamin solution, glucose solution, and trace elements solution, to distilled/deionized water and bring volume to 784.0mL. Mix thoroughly. Gently heat and bring to boiling. Autoclave for 15 min at 15 psi pressure–121°C. Cool to 50°–55°C. Aseptically add 200.0mL of sterile vitamin solution, 15.0mL of sterile glucose solution, and 1.0mL of sterile trace elements solution. Mix thoroughly. Pour into sterile Petri dishes or distribute into sterile tubes.

Use: For the cultivation and maintenance of *Acetobacter xylinum.*

Acetobacter Agar

Composition per liter:

Glucose	50.0g
$CaCO_3$	30.0g
Agar	15.0g
Yeast extract	10.0g

Preparation of Medium: Add components to tap water and bring volume to 1.0L. Mix thoroughly. Gently heat and bring to boiling. Distribute into tubes or flasks. Autoclave for 15 min at 15 psi pressure–121°C. Pour into sterile Petri dishes or leave in tubes. Cool rapidly.

Use: For the cultivation and maintenance of *Acetobacter aceti, Acetobacter diazotrophicus, Acetobacter hansenii, Acetobacter liquefaciens, Acidomonas methanolica, Frateuria aurantia, Gluconobacter cerinus,* and *Gluconobacter oxydans.*

Acetobacter Agar (Glucose)

Composition per liter:

Agar	15.0g
$CaCO_3$	10.0g
Yeast extract	10.0g
Glucose	3.0g

pH 7.4 ± 0.1 at 25°C

Source: This medium is available as a premixed powder from Sigma-Aldrich.

Preparation of Medium: Add components to tap water and bring volume to 1.0L. Mix thoroughly to ensure that $CaCO_3$ is evenly distrubted. Gently heat and bring to boiling. Distribute into tubes or flasks. Autoclave for 15 min at 15 psi pressure–121°C. Pour into sterile Petri dishes or leave in tubes. Cool rapidly.

Use: For the cultivation and maintenance of glucose positive *Acetobacter* species.

Acetobacter Broth

Composition per liter:

Yeast extract	5.0g
$(NH_4)_2SO_4$	3.3g
KH_2PO_4	1.0g
$MgSO_4$	0.25g
Vitamin solution	200.0mL
Glucose solution	15.0mL
Trace elements solution	1.0mL

Trace Elements Solution:

Composition per 100.0mL:

$CaCl_2 \cdot 2H_2O$	1.457g
$FeSO_4 \cdot 7H_2O$	0.366g
$ZnSO_4 \cdot 7H_2O$	0.178g
$MnSO_4 \cdot H_2O$	0.101g
$Na_2MoO_4 \cdot 2H_2O$	23.4mg
$CuSO_4 \cdot 5H_2O$	7.8mg

Preparation of Trace Elements Solution: Add components to distilled/deionized water and bring volume to 100.0mL. Mix thoroughly. Filter sterilize.

Vitamin Solution:

Composition per 100.0mL:

m-Inositol	200.0mg
Calcium DL-pantothenate	40.0mg
Nicotinic acid	40.0mg
Pyrdoxine·HCl	40.0mg
Thiamine·HCl	40.0mg
p-Aminobenzoic acid	20.0mg
Riboflavin	20.0mg
Biotin	0.2mg
Folic acid	0.2mg

Preparation of Vitamin Solution: Add components to distilled/deionized water and bring volume to 100.0mL. Mix thoroughly. Filter sterilize.

Glucose Solution:

Composition per 100.0mL:

Glucose	40.0g

Preparation of Glucose Solution: Add glucose to distilled/deionized water and bring volume to 100.0mL. Mix thoroughly. Filter sterilize.

Preparation of Medium: Add components, except vitamin solution, glucose solution, and trace elements solution, to distilled/deionized water and bring volume to 784.0mL. Mix thoroughly. Autoclave for 15 min at 15 psi pressure–121°C. Cool to 50°–55°C. Aseptically add 200.0mL of sterile vitamin solution, 15.0mL of sterile glucose solution, and 1.0mL of sterile trace elments solution. Mix thoroughly. Aseptically distribute into sterile tubes or flasks.

Use: For the cultivation of *Acetobacter xylinum.*

Acetobacter diazotrophicus Agar

Composition per liter:

Glucose	50.0g
$CaCO_3$	30.0g
Agar	25.0g
Yeast extract	10.0g

pH 5.5 ± 0.2 at 25°C

Preparation of Medium: Add components to distilled/deionized water and bring volume to 1.0L. Mix thoroughly to evenly distribute $CaCO_3$. Bring pH to 5.5. Gently heat and bring to boiling. Distribute into tubes or flasks. Autoclave for 15 min at 15 psi pressure–121°C. Cool rapidly to 50°–55°C. Pour into sterile Petri dishes or leave in tubes.

Use: For the cultivation and maintenance of *Acetobacter diazotrophicus.*

Acetobacter europaeus Medium

Composition per liter:

Glucose	5.0g
Peptone	3.0g
Yeast extract	2.0g
Acetic acid	40.0mL
Ethanol	30.0mL

Preparation of Acetic Acid: Filter sterilize 40.0mL of acetic acid using a teflon filter.

Preparation of Ethanol: Filter sterilize 30.0mL of ethanol using a teflon filter.

Preparation of Medium: Add components, except acetic acid and ethanol, to distilled/deionized water and bring volume to 930.0mL. Mix thoroughly. Autoclave for 15 min at 15 psi pressure–121°C. Aseptically add 40.0mL of sterile acetic acid and 30.0mL of sterile ethanol. Mix thoroughly. Aseptically distribute into sterile tubes or flasks.

Use: For the cultivation of *Acetobacter europaeus.*

Acetobacter/Gluconobacter Agar

Composition per liter:

Glucose	100.0g
Agar	25.0g
$CaCO_3$	20.0g
Yeast extract	10.0g

pH 7.5 ± 0.2 at 25°C

Preparation of Medium: Add components to distilled/deionized water and bring volume to 1.0L. Mix thoroughly. Gently heat and bring to boiling. Distribute into tubes or flasks. Autoclave for 15 min at 15 psi pressure–121°C. Pour into sterile Petri dishes or leave in tubes.

Use: For the cultivation and maintenance of *Acetobacter aceti, Acetobacter liquefaciens, Acetobacter pasteurianus, Acetobacter xylinum, Frateuria aurantia,* and *Gluconobacter oxydans.*

Acetobacter HiVeg Agar with Plant Extract

Composition per liter:

Glucose	20.0g
Agar	20.0g
$CaCO_3$	10.0g
Plant hydrolysate	5.0g
Plant extract No. 2	2.0g

pH 7.4 ± 0.1 at 25°C

Source: This medium is available as a premixed powder from HiMedia.

Preparation of Medium: Add components to tap water and bring volume to 1.0L. Mix thoroughly to ensure that $CaCO_3$ is evenly distributed. Gently heat and bring to boiling. Distribute into tubes or flasks. Autoclave for 15 min at 15 psi pressure–121°C. Pour into sterile Petri dishes or leave in tubes. Cool rapidly.

Use: For the cultivation and maintenance of glucose positive *Acetobacter* species.

Acetobacter Medium

Composition per liter:

Agar	15.0g
Autolyzed yeast	10.0g
$CaCO_3$	10.0g
Glucose	3.0g

pH 7.0 ± 0.2 at 25°C

Preparation of Medium: Add components to distilled/deionized water and bring volume to 1.0L. Mix thoroughly. Gently heat and bring to boiling. Distribute into tubes to produce a 1 cm butt and 30cm slant. Autoclave for 15 min at 15 psi pressure–121°C. Agitate tubes to mix $CaCO_3$. Cool tubes rapidly in a slanted position to keep the $CaCO_3$ in suspension.

Use: For the cultivation and maintenance of *Acetobacter* species and *Gluconobacter* species.

Acetobacter peroxydans Medium

Composition per liter:

Agar	15.0g
Malt extract	15.0g
Yeast extract	5.0g
Ethanol (50% solution)	60.0mL

Ethanol Solution:

Composition per 100.0mL:

Ethanol (50% solution)	100.0mL

Preparation of Ethanol Solution: Filter sterilize.

Preparation of Medium: Add components, except ethanol solution, to distilled/deionized water and bring volume to 940.0mL. Mix thoroughly. Autoclave for 15 min at 15 psi pressure–121°C. Cool to 25°C. Add 60.0mL of sterile ethanol solution. Mix thoroughly. Aseptically distribute into sterile tubes or flasks.

Use: For the cultivation and maintenance of *Acetobacter peroxydans* and *Acetobacter pasteurianus.*

Acetobacter xylinum Medium

Composition per liter:

Glucose	20.0g
Peptone	5.0g
Yeast extract	5.0g
Na_2HPO_4	2.7g
Citric acid	1.5g

Preparation of Medium: Add components to distilled/deionized water and bring volume to 1.0L. Mix thoroughly. Gently heat and bring to boiling. Distribute into tubes or flasks. Autoclave for 15 min at 15 psi pressure–121°C.

Use: For the cultivation and maintenance of *Acetobacter xylinum.*

Acetobacter xylinum Medium

Composition per liter:

Glucose	50.0g
Yeast extract	5.0g

Preparation of Medium: Add components to distilled/deionized water and bring volume to 1.0L. Mix thoroughly. Distribute into tubes or flasks. Autoclave for 15 min at 15 psi pressure–121°C.

Use: For the cultivation of *Acetobacter xylinum.*

Acetobacterium Autotrophic Medium (DSMZ Medium 135)

Composition per liter:

$NaHCO_3$	10.0g
Yeast extract	2.0g
NH_4Cl	1.0g
K_2HPO_4	0.45g
KH_2PO_4	0.33g
$MgSO_4 \cdot 7H_2O$	0.1g
Resazurin	1.0mg
Fructose solution	25.0mL
Trace elements solution	20.0mL
Vitamin solution	20.0mL
Cysteine solution	10.0mL
$Na_2S \cdot 9H_2O$ solution	10.0mL

pH 8.2 ± 0.2 at 25°C

$Na_2S \cdot 9H_2O$ Solution:

Composition per 10.0mL:

$Na_2S \cdot 9H_2O$	0.5g

Preparation of $Na_2S \cdot 9H_2O$ Solution: Add $Na_2S \cdot 9H_2O$ to distilled/deionized water and bring volume to 10.0mL. Sparge with N_2. Autoclave for 15 min at 15 psi pressure–121°C. Cool to 25°C. Store anaerobically.

Cysteine Solution:

Composition per 10.0mL:

L-Cysteine·HCl·H_2O	0.5g

Preparation of Cysteine Solution: Add L-cysteine·HCl·H_2O to distilled/deionized water and bring volume to 10.0mL. Mix thoroughly. Sparge with 100% N_2. Autoclave for 15 min at 15 psi pressure–121°C.

Fructose Solution:

Composition per 25.0mL:

Fructose	10.0g

Preparation of Fructose Solution: Add fructose to distilled/deionized water and bring volume to 25.0mL. Mix thoroughly. Sparge with 100% N_2. Autoclave for 15 min at 15 psi pressure–121°C. Cool to room temperature.

Trace Elements Solution:

Composition per liter:

$MgSO_4 \cdot 7H_2O$	3.0g
Nitrilotriacetic acid	1.5g
NaCl	1.0g
$MnSO_4 \cdot 2H_2O$	0.5g
$CoSO_4 \cdot 7H_2O$	0.18g
$ZnSO_4 \cdot 7H_2O$	0.18g
$CaCl_2 \cdot 2H_2O$	0.1g
$FeSO_4 \cdot 7H_2O$	0.1g
$NiCl_2 \cdot 6H_2O$	0.025g
$KAl(SO_4)_2 \cdot 12H_2O$	0.02g
H_3BO_3	0.01g
$Na_2MoO_4 \cdot 4H_2O$	0.01g
$CuSO_4 \cdot 5H_2O$	0.01g
$Na_2SeO_3 \cdot 5H_2O$	0.3mg

Preparation of Trace Elements Solution: Add nitrilotriacetic acid to 500.0mL of distilled/deionized water. Dissolve by adjusting pH to 6.5 with KOH. Add remaining components. Add distilled/deionized water to 1.0L. Mix thoroughly.

Vitamin Solution:

Composition per liter:

Pyridoxine-HCl	10.0mg
Thiamine-HCl·$2H_2O$	5.0mg
Riboflavin	5.0mg
Nicotinic acid	5.0mg
D-Ca-pantothenate	5.0mg
p-Aminobenzoic acid	5.0mg
Lipoic acid	5.0mg
Biotin	2.0mg
Folic acid	2.0mg
Vitamin B_{12}	0.1mg

Preparation of Vitamin Solution: Add components to distilled/deionized water and bring volume to 1.0L. Mix thoroughly. Sparge with 80% H_2 + 20% CO_2. Filter sterilize.

Preparation of Medium: Prepare and dispense medium under 80% H_2 + 20% CO_2 gas atmosphere. Add components, except $NaHCO_3$, Na_2CO_3 solution, $Na_2S \cdot 9H_2O$ solution, vitamin solution, and cysteine solution, to distilled/deionized water and bring volume to 935.0mL. Mix thoroughly. Gently heat and bring to boiling. Boil for 5 min. Cool while sparging with 80% H_2 + 20% CO_2. Add 10.0g solid $NaHCO_3$. Equilibrate with 80% N_2 + 20% CO_2 until pH is approximately 7.4. Distribute into bottles. Autoclave under 80% H_2 + 20% CO_2 for 15 min at 15 psi pressure–121°C. Aseptically and anaerobically add approximately 0.25mL sterile Na_2CO_3 solution to each 10.0mL of medium so that pH is adjusted to 8.2. For every 10.0mL of medium inject 0.1mL $Na_2S \cdot 9H_2O$ solution, 0.2mL vitamin solution, and 0.1mL cysteine solution. Incubate under 80% H_2 + 20% CO_2 gas atmosphere.

Use: For the cultivation and heterotrophic growth of *Acetobacterium* spp.

Acetobacterium carbinolicum Medium

Composition per 1011.2mL:

$NaHCO_3$	4.5g
Na_2SO_4	2.84g
NaCl	1.17g
Yeast extract	1.0g
$MgCl_2 \cdot 6H_2O$	0.4g
KCl	0.3g
NH_4Cl	0.27g
KH_2PO_4	0.2g
$CaCl_2 \cdot 2H_2O$	0.15g
Resazurin	0.5mg
Reducing agent solution	10.0mL
Ethanol solution	1.2mL
Trace elements solution	1.0mL
Vitamin solution	1.0mL

pH 7.0–7.2 at 25°C

Trace Elements Solution:

Composition per liter:

$FeCl_2 \cdot 4H_2O$	1.5g
$CoCl_2 \cdot 6H_2O$	120.0mg
$MnCl_2 \cdot 4H_2O$	100.0mg
$ZnCl_2$	68.0mg
H_3BO_3	62.0mg
$Na_2MoO_4 \cdot 2H_2O$	24.0mg
$NiCl_2 \cdot 6H_2O$	24.0mg
$CuCl_2 \cdot 2H_2O$	17.0mg
HCl (0.05*M* solution)	1000.0mL

Preparation of Trace Elements Solution: Add components one at a time to 1.0L of 0.05*M* HCl solution. Mix thoroughly.

Vitamin Solution:

Composition per 100.0mL

Thiamine·HCl	10.0mg
p-Aminobenzoic acid	4.0mg
D(+)-Biotin	1.0mg

Preparation of Vitamin Solution: Add components to distilled/deionized water and bring volume to 100.0mL. Mix thoroughly. Filter sterilize. Flush with 80% N_2 + 20% CO_2.

Reducing Agent Solution:

Composition per 10.0mL:

$Na_2S \cdot 9H_2O$	0.36g

Preparation of Reducing Agent Solution: Add $Na_2S \cdot 9H_2O$ to distilled/deionized water and bring volume to 10.0mL. Mix thoroughly. Autoclave for 10 min at 15 psi pressure–121°C.

Ethanol Solution:

Composition per 10.0mL:

Ethanol (95% solution)	10.0mL

Preparation of Ethanol Solution: Filter sterilize. Sparge with N_2 gas for 1 min.

Preparation of Medium: Add components, except $NaHCO_3$, ethanol, and reducing agent solution, to distilled/deionized water and bring volume to 1.0L. Mix thoroughly. Gently heat and bring to boiling. Boil for a few minutes. Allow to cool to room temperature under 80% N_2 + 20% CO_2. Add the $NaHCO_3$ and adjust pH to 6.9–7.1. Distribute into tubes or flasks under 80% N_2 + 20% CO_2. Autoclave for 15 min at 15 psi pressure–121°C. Cool to room temperature. Before inoculation, add sterile anaerobic Na_2CO_3 (0.25mL of 5% Na_2CO_3 per 10.0mL of medium) to bring the pH to 8.2. Add sterile ethanol and reducing agent solution.

Use: For the cultivation and maintenance of *Acetobacterium malicum* and *Acetobacterium carbinolicum.*

Acetobacterium dehalogenans Medium (DSMZ Medium 787)

Composition per liter:

$NaHCO_3$	10.0g
Yeast extract	2.0g
NH_4Cl	1.0g
K_2HPO_4	0.45g
KH_2PO_4	0.33g
$MgSO_4 \cdot 7H_2O$	0.1g
Resazurin	1.0mg
$NaHCO_3$ solution	30.0mL
Na_2CO_3 solution	20.0mL
Trace elements solution	20.0mL
Vitamin solution	20.0mL
Na-syringate soltuion	10.0mL
Cysteine solution	10.0mL

pH 7.4 ± 0.2 at 25°C

$NaHCO_3$ Solution:

Composition per 100.0mL:

$NaHCO_3$	10.0g

Preparation of $NaHCO_3$ Solution: Add $NaHCO_3$ to distilled/deionized water and bring volume to 100.0mL. Mix thoroughly. Sparge with 80% N_2 + 20% CO_2. Filter sterilize.

Cysteine Solution:

Composition per 10.0mL:

L-Cysteine·HCl·H_2O	0.3g

Preparation of Cysteine Solution: Add L-cysteine·HCl·H_2O to distilled/deionized water and bring volume to 10.0mL. Mix thoroughly. Sparge with 100% N_2. Autoclave for 15 min at 15 psi pressure–121°C. Cool to 25°C.

Na-syringate Solution:

Composition per 20.0mL:

Na-syringate	1.2g

Preparation of Na-syringate Solution: Add Na-syringate to distilled/deionized water and bring volume to 20.0mL. Mix thoroughly. Sparge with 100% N_2. Filter sterilize.

Trace Elements Solution:

Composition per liter:

$MgSO_4 \cdot 7H_2O$	3.0g
Nitrilotriacetic acid	1.5g
NaCl	1.0g
$MnSO_4 \cdot 2H_2O$	0.5g
$CoSO_4 \cdot 7H_2O$	0.18g
$ZnSO_4 \cdot 7H_2O$	0.18g
$CaCl_2 \cdot 2H_2O$	0.1g
$FeSO_4 \cdot 7H_2O$	0.1g
$NiCl_2 \cdot 6H_2O$	0.025g
$KAl(SO_4)_2 \cdot 12H_2O$	0.02g
H_3BO_3	0.01g
$Na_2MoO_4 \cdot 4H_2O$	0.01g
$CuSO_4 \cdot 5H_2O$	0.01g
$Na_2SeO_3 \cdot 5H_2O$	0.3mg

Preparation of Trace Elements Solution: Add nitrilotriacetic acid to 500.0mL of distilled/deionized water. Dissolve by adjusting pH to 6.5 with KOH. Add remaining components. Add distilled/deionized water to 1.0L. Mix thoroughly.

Vitamin Solution:

Composition per liter:

Pyridoxine-HCl	10.0mg
Thiamine-HCl·$2H_2O$	5.0mg
Riboflavin	5.0mg
Nicotinic acid	5.0mg
D-Ca-pantothenate	5.0mg
p-Aminobenzoic acid	5.0mg
Lipoic acid	5.0mg
Biotin	2.0mg
Folic acid	2.0mg
Vitamin B_{12}	0.1mg

Preparation of Vitamin Solution: Add components to distilled/deionized water and bring volume to 1.0L. Mix thoroughly. Sparge with 80% N_2 + 20% CO_2. Filter sterilize.

Preparation of Medium: Prepare and dispense medium under 80% N_2 + 20% CO_2 gas atmosphere. Add components, except $NaHCO_3$ solution, Na_2CO_3 solution, vitamin solution, Na-syringate solution, and cysteine solution, to distilled/deionized water and bring volume to 900.0mL. Mix thoroughly. Gently heat and bring to boiling. Boil for 5 min. Cool while sparging with 80% N_2 + 20% CO_2. Distribute 9.0mL aliquots into serum bottles. Autoclave under 80% N_2 + 20% CO_2 for 15 min at 15 psi pressure–121°C. Aseptically and anaerobically add approximately 0.20mL sterile Na_2CO_3 solution to each 9.0mL of medium so that pH is adjusted to 7.4. For every 9.0mL of medium inject 1.0mL $NaHCO_3$ solution, 0.15mL Na-syringate solution, 0.2mL vitamin solution, and 0.17mL cysteine solution.

Use: For the cultivation of *Acetobacterium dehalogenans.*

Acetobacterium Heterotrophic Medium (DSMZ Medium 135)

Composition per liter:

$NaHCO_3$	10.0g
Yeast extract	2.0g
NH_4Cl	1.0g
K_2HPO_4	0.45g
KH_2PO_4	0.33g
$MgSO_4 \cdot 7H_2O$	0.1g
Resazurin	1.0mg
Na_2CO_3 solution	25.0mL
Trace elements solution	20.0mL
Vitamin solution	20.0mL
Fructose solution	10.0mL
Cysteine solution	10.0mL
$Na_2S \cdot 9H_2O$ solution	10.0mL

pH 8.2 ± 0.2 at 25°C

$Na_2S \cdot 9H_2O$ Solution:

Composition per 10.0mL:

$Na_2S \cdot 9H_2O$	0.5g

Preparation of $Na_2S \cdot 9H_2O$ Solution: Add $Na_2S \cdot 9H_2O$ to distilled/deionized water and bring volume to 10.0mL. Sparge with N_2.

Autoclave for 15 min at 15 psi pressure–121°C. Cool to 25°C. Store anaerobically.

Cysteine Solution:

Composition per 10.0mL:

L-Cysteine·HCl·H_2O	0.5g

Preparation of Cysteine Solution: Add L-cysteine·HCl·H_2O to distilled/deionized water and bring volume to 10.0mL. Mix thoroughly. Sparge with 100% N_2. Autoclave for 15 min at 15 psi pressure–121°C.

Fructose Solution:

Composition per 10.0mL:

Fructose	5.0g

Preparation of Fructose Solution: Add fructose to distilled/deionized water and bring volume to 10.0mL. Mix thoroughly. Sparge with 100% N_2. Autoclave for 15 min at 15 psi pressure–121°C. Cool to room temperature.

Trace Elements Solution:

Composition per liter:

$MgSO_4·7H_2O$	3.0g
Nitrilotriacetic acid	1.5g
NaCl	1.0g
$MnSO_4·2H_2O$	0.5g
$CoSO_4·7H_2O$	0.18g
$ZnSO_4·7H_2O$	0.18g
$CaCl_2·2H_2O$	0.1g
$FeSO_4·7H_2O$	0.1g
$NiCl_2·6H_2O$	0.025g
$KAl(SO_4)_2·12H_2O$	0.02g
H_3BO_3	0.01g
$Na_2MoO_4·4H_2O$	0.01g
$CuSO_4·5H_2O$	0.01g
$Na_2SeO_3·5H_2O$	0.3mg

Preparation of Trace Elements Solution: Add nitrilotriacetic acid to 500.0mL of distilled/deionized water. Dissolve by adjusting pH to 6.5 with KOH. Add remaining components. Add distilled/deionized water to 1.0L. Mix thoroughly.

Vitamin Solution:

Composition per liter:

Pyridoxine-HCl	10.0mg
Thiamine-HCl·$2H_2O$	5.0mg
Riboflavin	5.0mg
Nicotinic acid	5.0mg
D-Ca-pantothenate	5.0mg
p-Aminobenzoic acid	5.0mg
Lipoic acid	5.0mg
Biotin	2.0mg
Folic acid	2.0mg
Vitamin B_{12}	0.1mg

Preparation of Vitamin Solution: Add components to distilled/deionized water and bring volume to 1.0L. Mix thoroughly. Sparge with 80% H_2 + 20% CO_2. Filter sterilize.

Preparation of Medium: Prepare and dispense medium under 80% N_2 + 20% CO_2 gas atmosphere. Add components, except $NaHCO_3$, Na_2CO_3 solution, $Na_2S·9H_2O$ solution, vitamin solution, fructose soltuion, and cysteine solution, to distilled/deionized water and bring volume to 925.0mL. Mix thoroughly. Gently heat and bring to boiling. Boil for 5 min. Cool while sparging with 80% N_2 + 20% CO_2. Add 10.0g solid $NaHCO_3$. Equilibrate with 80% N_2 + 20% CO_2 until pH is approximately 7.4. Distribute into bottles. Autoclave under 80% N_2 + 20% CO_2 for 15 min at 15 psi pressure–121°C. Aseptically and anaerobically add approximately 0.25mL sterile Na_2CO_3 solution to each 10.0mL of medium so that pH is adjusted to 8.2. For every 10.0mL of medium inject 0.1mL $Na_2S·9H_2O$ solution, 0.1mL fructose solution, 0.2mL vitamin solution, and 0.1mL cysteine solution.

Use: For the cultivation and heterotrophic growth of *Acetobacterium* spp.

Acetobacterium Medium

Composition per 1060.0mL:

$NaHCO_3$	10.0g
Yeast extract	2.0g
NH_4Cl	1.0g
K_2HPO_4	0.45g
KH_2PO_4	0.33g
$MgSO_4·7H_2O$	0.1g
Resazurin	1.0mg
Fructose solution	50.0mL
Trace elements solution	20.0mL
Vitamin solution	20.0mL
Reducing agent solution	10.0mL

pH 8.2 ± 0.2 at 25°C

Fructose Solution:

Composition per 50.0mL:

Fructose	10.0g

Preparation of Fructose Solution: Add fructose to distilled/deionized water and bring volume to 50.0mL. Mix thoroughly. Sparge under 100% N_2 gas for 3 min. Filter sterilize. Store under N_2 gas.

Trace Elements Solution:

Composition per liter:

$MgSO_4·7H_2O$	3.0g
Nitrilotriacetic acid	1.5g
$CaCl_2·2H_2O$	1.0g
NaCl	1.0g
$MnSO_4·2H_2O$	0.5g
$CoSO_4·7H_2O$	0.18g
$ZnSO_4·7H_2O$	0.18g
$FeSO_4·7H_2O$	0.1g
$NiCl_2·6H_2O$	0.025g
$KAl(SO_4)_2·12H_2O$	0.02g
$CuSO_4·5H_2O$	0.01g
H_3BO_3	0.01g
$Na_2MoO_4·2H_2O$	0.01g
$Na_2SeO_3·5H_2O$	0.3mg

Preparation of Trace Elements Solution: Add nitrilotriacetic acid to 500.0mL of distilled/deionized water. Dissolve by adjusting pH to 6.5 with KOH. Add distilled/deionized water to 1.0L. Add remaining components. Mix thoroughly.

Vitamin Solution:

Composition per liter:

Pyridoxine·HCl	10.0mg
Calcium DL-pantothenate	5.0mg
Lipoic acid	5.0mg
Nicotinic acid	5.0mg
p-Aminobenzoic acid	5.0mg
Riboflavin	5.0mg
Thiamine·HCl	5.0mg
Biotin	2.0mg

Folic acid....2.0mg
Vitamin B_{12}....0.1mg

Preparation of Vitamin Solution: Add components to distilled/deionized water and bring volume to 1.0L. Mix thoroughly.

Reducing Agent Solution:
Composition per 10.0mL:
L-Cysteine·HCl·H_2O....0.5g
$Na_2S·9H_2O$....0.5g

Preparation of Reducing Agent Solution: Add 10.0mL of distilled/deionized water to a test tube. Gently heat and bring to boiling. Boil under N_2 gas for 1 min. Cool to room temperature. Add L-cysteine·HCl·H_2O and dissolve. Adjust pH to 9 with 5*N* NaOH. Add washed $Na_2S·9H_2O$ and dissolve. Autoclave for 10 min at 15 psi pressure–121°C.

Preparation of Medium: Add components, except $NaHCO_3$, fructose solution, and reducing agent solution, to distilled/deionized water and bring volume to 1.0L. Mix thoroughly. Gently heat and bring to boiling. Boil for a few minutes. Allow to cool to room temperature under 80% N_2 + 20% CO_2. Add the $NaHCO_3$ and adjust pH to 7.4. Distribute into tubes or flasks under 80% N_2 + 20% CO_2. Autoclave for 15 min at 15 psi pressure–121°C. Cool to room temperature. Before inoculation, add sterile anaerobic Na_2CO_3 (0.25mL of 5% Na_2CO_3 per 10.0mL of medium) to bring the pH to 8.2. Add sterile fructose solution and reducing agent solution.

Use: For the cultivation and maintenance of *Acetobacterium* species.

Acetobacterium Medium

Composition per 1060.0mL:
Yeast extract....2.0g
NH_4Cl....1.0g
K_2HPO_4....0.45g
$NaHCO_3$....1.0g
KH_2PO_4....0.33g
$MgSO_4·7H_2O$....0.1g
Resazurin....1.0mg
Fructose solution....50.0mL
Trace elements solution....20.0mL
Vitamin solution....20.0mL
Reducing agent solution....10.0mL
Metal solution....1.0mL
pH 6.5 ± 0.2 at 25°C

Fructose Solution:
Composition per 50.0mL:
Fructose....10.0g

Preparation of Fructose Solution: Add fructose to distilled/deionized water and bring volume to 50.0mL. Mix thoroughly. Sparge under 100% N_2 gas for 3 min. Filter sterilize. Store under N_2 gas.

Trace Elements Solution:
Composition per liter:
$MgSO_4·7H_2O$....3.0g
Nitrilotriacetic acid....1.5g
$CaCl_2·2H_2O$....1.0g
NaCl....1.0g
$MnSO_4·2H_2O$....0.5g
$CoSO_4·7H_2O$....0.18g
$ZnSO_4·7H_2O$....0.18g
$FeSO_4·7H_2O$....0.1g
$NiCl_2·6H_2O$....0.025g
$KAl(SO_4)_2·12H_2O$....0.02g
$CuSO_4·5H_2O$....0.01g
H_3BO_3....0.01g
$Na_2MoO_4·2H_2O$....0.01g
$Na_2SeO_3·5H_2O$....0.3mg

Preparation of Trace Elements Solution: Add nitrilotriacetic acid to 500.0mL of distilled/deionized water. Dissolve by adjusting pH to 6.5 with KOH. Add distilled/deionized water to 1.0L. Add remaining components. Mix thoroughly.

Vitamin Solution:
Composition per liter:
Pyridoxine·HCl....10.0mg
Calcium DL-pantothenate....5.0mg
Lipoic acid....5.0mg
Nicotinic acid....5.0mg
p-Aminobenzoic acid....5.0mg
Riboflavin....5.0mg
Thiamine·HCl....5.0mg
Biotin....2.0mg
Folic acid....2.0mg
Vitamin B_{12}....0.1mg

Preparation of Vitamin Solution: Add components to distilled/deionized water and bring volume to 1.0L. Mix thoroughly.

Reducing Agent Solution:
Composition per 10.0mL:
L-Cysteine·HCl·H_2O....0.5g
$Na_2S·9H_2O$....0.5g

Preparation of Metal Solution: Add 10.0mL of distilled/deionized water to a test tube. Gently heat and bring to boiling. Boil under N_2 gas for 1 min. Cool to room temperature. Add L-cysteine·HCl·H_2O and dissolve. Adjust pH to 9 with 5*N* NaOH. Add washed $Na_2S·9H_2O$ and dissolve. Autoclave for 10 min at 15 psi pressure–121°C.

Metal Solution:
Composition per liter:
$Na_2SeO_3·5H_2O$....3.0mg
$Na_2WO_4·2H_2O$....0.5g

Preparation of Reducing Agent Solution: Add 10.0mL of distilled/deionized water to a test tube. Gently heat and bring to boiling. Boil under N_2 gas for 1 min. Cool to room temperature. Add L-cysteine·HCl·H_2O and dissolve. Adjust pH to 9 with 5*N* NaOH. Add washed $Na_2S·9H_2O$ and dissolve. Autoclave for 10 min at 15 psi pressure–121°C.

Preparation of Medium: Add components, except $NaHCO_3$, fructose solution, and reducing agent solution, to distilled/deionized water and bring volume to 1.0L. Mix thoroughly. Gently heat and bring to boiling. Boil for a few minutes. Allow to cool to room temperature under 80% N_2 + 20% CO_2. Add the $NaHCO_3$ and adjust pH to 6.5. Distribute into tubes or flasks under 80% N_2 + 20% CO_2. Autoclave for 15 min at 15 psi pressure–121°C. Cool to room temperature. Before inoculation, add sterile fructose solution and reducing agent solution.

Use: For the cultivation and maintenance of *Clostridium thermoautotrophicum.*

Acetobacterium Medium

Composition per 1001.0mL:
Solution A....870.0mL
Solution C....100.0mL
Solution D....10.0mL

Solution E (Vitamin solution)	10.0mL
Solution F	10.0mL
Solution B (Trace elements solution SL-10)	1.0mL

pH 7.1–7.4 at 25°C

Solution A:

Composition per 870.0mL:

Na_2SO_4	3.0g
NaCl	1.0g
Yeast extract	0.5g
KCl	0.5g
$MgCl_2 \cdot 6H_2O$	0.4g
NH_4Cl	0.3g
KH_2PO_4	0.2g
$CaCl_2 \cdot 2H_2O$	0.15g
Resazurin	1.0mg

Preparation of Solution A: Add components to distilled/deionized water and bring volume to 870.0mL. Mix thoroughly. Gently heat and bring to boiling. Continue boiling for 3–4 min. Allow to cool to room temperature while gassing under 80% N_2 + 20% CO_2. Continue gassing until pH reaches below 6.0. Seal the flask under 80% N_2 + 20% CO_2. Autoclave for 15 min at 15 psi pressure–121°C.

Solution B (Trace Elements Solution SL-10):

Composition per liter:

$FeCl_2 \cdot 4H_2O$	1.5g
$CoCl_2 \cdot 6H_2O$	190.0mg
$MnCl_2 \cdot 4H_2O$	100.0mg
$ZnCl_2$	70.0mg
$Na_2MoO_4 \cdot 2H_2O$	36.0mg
$NiCl_2 \cdot 6H_2O$	24.0mg
H_3BO_3	6.0mg
$CuCl_2 \cdot 2H_2O$	2.0mg
HCl (25% solution)	10.0mL

Preparation of Solution B (Trace Elements Solution SL-10): Add $FeCl_2 \cdot 4H_2O$ to 10.0mL of HCl solution. Mix thoroughly. Add distilled/deionized water and bring volume to 1.0L. Add remaining components. Mix thoroughly. Gas under 100% N_2. Autoclave for 15 min at 15 psi pressure–121°C.

Solution C:

Composition per 100.0mL:

$NaHCO_3$	5.0g

Preparation of Solution C: Add $NaHCO_3$ to distilled/deionized water and bring volume to 100.0mL. Mix thoroughly. Filter sterilize. Gas under 80% N_2 + 20% CO_2.

Solution D:

Composition per 10.0mL:

Methoxyacetate	0.9g

Preparation of Solution D: Add methoxyacetate to distilled/deionized water and bring volume to 10.0mL. Mix thoroughly. Gas under 100% N_2. Autoclave for 15 min at 15 psi pressure–121°C.

Solution E (Vitamin Solution):

Composition per liter:

Pyridoxine·HCl	10.0mg
Calcium DL-pantothenate	5.0mg
Lipoic acid	5.0mg
Nicotinic acid	5.0mg
p-Aminobenzoic acid	5.0mg
Riboflavin	5.0mg
Thiamine·HCl	5.0mg
Biotin	2.0mg
Folic acid	2.0mg
Vitamin B_{12}	0.1mg

Preparation of Solution E (Vitamin Solution): Add components to distilled/deionized water and bring volume to 1.0L. Mix thoroughly. Gas under 100% N_2. Autoclave for 15 min at 15 psi pressure–121°C.

Solution F:

Composition per 10.0mL:

$Na_2S \cdot 9H_2O$	0.4g

Preparation of Solution F: Add $Na_2S \cdot 9H_2O$ to distilled/deionized water and bring volume to 10.0mL. Mix thoroughly. Gas under 100% N_2. Autoclave for 15 min at 15 psi pressure–121°C.

Preparation of Medium: Aseptically and anaerobically combine solution A with solution B, solution C, solution D, solution E, and solution F, in that order. Mix thoroughly. Anaerobically distribute into sterile tubes or flasks under 80% N_2 + 20% CO_2.

Use: For the cultivation and maintenance of *Acetobacterium* species.

Acetobacterium Medium (ATCC Medium 1612)

Composition per liter:

Fructose	10.0g
$NaHCO_3$	10.0g
Yeast extract	2.0g
NH_4Cl	1.0g
L-Cysteine·HCl·H_2O	0.5g
$Na_2S \cdot 9H_2O$	0.5g
K_2HPO_4	0.45g
KH_2PO_4	0.33g
$MgSO_4 \cdot 7H_2O$	0.1g
Resazurin	1.0mg
Wolfe's mineral solution	20.0mL
Wolfe's vitamin solution	20.0mL

pH 7.4 ± 0.2 at 25°C

Wolfe's Mineral Solution:

Composition per liter:

$MgSO_4 \cdot 7H_2O$	3.0g
Nitrilotriacetic acid	1.5g
NaCl	1.0g
$MnSO_4 \cdot H_2O$	0.5g
$FeSO_4 \cdot 7H_2O$	0.1g
$CoCl_2 \cdot 6H_2O$	0.1g
$CaCl_2$	0.1g
$ZnSO_4 \cdot 7H_2O$	0.1g
$CuSO_4 \cdot 5H_2O$	0.01g
$AlK(SO_4)_2 \cdot 12H_2O$	0.01g
H_3BO_3	0.01g
$Na_2MoO_4 \cdot 2H_2O$	0.01g

Preparation of Wolfe's Mineral Solution: Add nitrilotriacetic acid to 500.0mL of distilled/deionized water. Dissolve by adjusting pH to 6.5 with KOH. Add distilled/deionized water to 1.0L. Add remaining components.

Wolfe's Vitamin Solution:

Composition per liter:

Pyridoxine·HCl	10.0mg
Thiamine·HCl	5.0mg
Riboflavin	5.0mg

Nicotinic acid	5.0mg
Calcium pantothenate	5.0mg
p-Aminobenzoic acid	5.0mg
Thioctic acid	5.0mg
Biotin	2.0mg
Folic acid	2.0mg
Cyanocobalamin	100.0μg

Preparation of Wolfe's Vitamin Solution: Add components to distilled/deionized water and bring volume to 1.0L. Mix thoroughly.

Preparation of Medium: Add all components, except fructose, and bring volume to 1.0L with distilled/deionized water. Mix thoroughly. Equilibrate to pH 7.4 by gassing with 80% N_2 + 20% CO_2. Distribute into test tubes. Autoclave for 15 min at 15 psi pressure–121°C. Add sterile anaerobic Na_2CO_3 (0.25mL of 5% Na_2CO_3 per 10.0mL of medium) to bring the pH to 8.2. Add sterile fructose solution to give a final concentration of 1%. If autotrophic growth is desired omit fructose and gas with 80% H_2 + 20% CO_2.

Use: For the cultivation and maintenance of *Acetobacterium* species, *Clostridium aceticum*, and other bacteria that can ferment fructose to acetic acid.

Acetobacterium Medium (ATCC Medium 1019)

Composition per liter:

$NaHCO_3$	3.0g
Yeast extract	1.0g
NH_4Cl	1.0g
KH_2PO_4	0.4g
K_2HPO_4	0.4g
$MgSO_4 \cdot 7H_2O$	0.1g
Fructose (20% solution)	25.0mL
Wolfe's vitamin solution	10.0mL
Wolfe's mineral solution	10.0mL
Resazurin (0.01% solution)	1.0mL

pH 6.7 ± 0.2 at 25°C

Wolfe's Vitamin Solution:

Composition per liter:

Pyridoxine·HCl	10.0mg
Thiamine·HCl	5.0mg
Riboflavin	5.0mg
Nicotinic acid	5.0mg
Calcium pantothenate	5.0mg
p-Aminobenzoic acid	5.0mg
Thioctic acid	5.0mg
Biotin	2.0mg
Folic acid	2.0mg
Cyanocobalamin	100.0μg

Preparation of Wolfe's Vitamin Solution: Add components to distilled/deionized water and bring volume to 1.0L. Mix thoroughly.

Wolfe's Mineral Solution:

Composition per liter:

$MgSO_4 \cdot 7H_2O$	3.0g
Nitrilotriacetic acid	1.5g
$MnSO_4 \cdot H_2O$	0.5g
NaCl	1.0g
$FeSO_4 \cdot 7H_2O$	0.1g
$CoCl_2 \cdot 6H_2O$	0.1g
$CaCl_2$	0.1g
$ZnSO_4 \cdot 7H_2O$	0.1g
$CuSO_4 \cdot 5H_2O$	0.01g
$AlK(SO_4)_2 \cdot 12H_2O$	0.01g
H_3BO_3	0.01g
$Na_2MoO_4 \cdot 2H_2O$	0.01g

Preparation of Wolfe's Mineral Solution: Add nitrilotriacetic acid to 500.0mL of distilled/deionized water. Dissolve by adjusting pH to 6.5 with KOH. Add distilled/deionized water to 1.0L. Add remaining components.

Preparation of Medium: Add all components, except fructose, to distilled/deionized water and bring volume to 975.0mL. Boil to remove dissolved O_2. Add 40.0mL of a solution containing 1.25% L-cysteine·HCl·H_2O and 1.25% $Na_2S \cdot 9H_2O$. Autoclave for 15 min at 15 psi pressure–121°C. Immediately gas with 90% N_2 + 10% CO_2 to maintain anaerobiosis until cooled to 50°C. Add 25.0mL of a filter-sterilized 20% fructose solution. If necessary, adjust pH to 6.7. Aseptically distribute into tubes under anaerobic conditions. Cap with rubber stoppers.

Use: For the cultivation and maintenance of *Acetobacterium* species.

Acetobacterium Medium (DSMZ 614)

Composition per liter:

$MgCl_2 \cdot 6H_2O$	0.52g
Yeast extract	0.5g
KCl	0.33g
KH_2PO_4	0.33g
NH_4Cl	0.33g
$CaCl_2 \cdot 2H_2O$	0.22g
Resazurin	1.0mg
Fructose solution	100.0mL
Wolfe's mineral solution	10.0mL
Wolfe's vitamin solution	10.0mL
$NaHCO_3$ solution	10.0mL
$Na_2S \cdot 9H_2O$ solution	10.0mL

pH 6.8–7.0 at 25°C

Fructose Solution:

Composition per 100.0mL:

D-Fructose	10.0g

Preparation of Fructose Solution: Add fructose to distilled/deionized water and bring volume to 100.0mL. Mix thoroughly. Sparge with 100% N_2. Autoclave for 15 min at 15 psi pressure–121°C.

Wolfe's Vitamin Solution:

Composition per liter:

Pyridoxine·HCl	10.0mg
p-Aminobenzoic acid	5.0mg
Lipoic acid	5.0mg
Nicotinic acid	5.0mg
Riboflavin	5.0mg
Thiamine·HCl	5.0mg
Calcium DL-pantothenate	5.0mg
Biotin	2.0mg
Folic acid	2.0mg
Vitamin B_{12}	0.1mg

Preparation of Wolfe's Vitamin Solution: Add components to distilled/deionized water and bring volume to 1.0L. Mix thoroughly.

Wolfe's Mineral Solution:
Composition per liter:

$MgSO_4 \cdot 7H_2O$	3.0g
Nitrilotriacetic acid	1.5g
NaCl	1.0g
$MnSO_4 \cdot 2H_2O$	0.5g
$CoCl_2 \cdot 6H_2O$	0.1g
$ZnSO_4 \cdot 7H_2O$	0.1g
$CaCl_2 \cdot 2H_2O$	0.1g
$FeSO_4 \cdot 7H_2O$	0.1g
$NiCl_2 \cdot 6H_2O$	0.025g
$KAl(SO_4)_2 \cdot 12H_2O$	0.02g
$CuSO_4 \cdot 5H_2O$	0.01g
H_3BO_3	0.01g
$Na_2MoO_4 \cdot 2H_2O$	0.01g
$Na_2SeO_3 \cdot 5H_2O$	0.3mg

Preparation of Wolfe's Mineral Solution: Add nitrilotriacetic acid to 500.0mL of distilled/deionized water. Adjust pH to 6.5 with KOH. Add remaining components. Add distilled/deionized water to 1.0L. Adjust pH to 6.8.

$NaHCO_3$ Solution:
Composition per 10.0mL:

$NaHCO_3$	1.0g

Preparation of $NaHCO_3$ Solution: Add $NaHCO_3$ to distilled/deionized water and bring volume to 10.0mL. Mix thoroughly. Sparge with 80% N_2 + 20% CO_2. Autoclave for 15 min at 15 psi pressure–121°C.

$Na_2S \cdot 9H_2O$ Solution:
Composition per 10.0mL:

$Na_2S \cdot 9H_2O$	0.7g

Preparation of $Na_2S \cdot 9H_2O$ Solution: Add $Na_2S \cdot 9H_2O$ to distilled/deionized water and bring volume to 10.0mL. Mix thoroughly. Sparge with 100% N_2. Autoclave for 15 min at 15 psi pressure–121°C. Before use, neutralize to pH 7.0 with sterile HCl.

Preparation of Medium: Prepare and dispense medium under 100% N_2. Add components, except fructose solution, $NaHCO_3$ solution, and $Na_2S \cdot 9H_2O$ solution, to distilled/deionized water and bring volume to 880.0mL. Mix thoroughly. Adjust pH to 6.8–7.0. Sparge with 100% N_2. Autoclave for 15 min at 15 psi pressure–121°C. Aseptically and anaerobically add 100.0mL of sterile fructose solution, 10.0mL of sterile $NaHCO_3$ solution, and 10.0mL of sterile $Na_2S \cdot 9H_2O$ solution. Mix thoroughly. Aseptically and anaerobically distribute into sterile tubes or bottles.

Use: For the cultivation of *Acetobacterium bakii* and *Acetobacterium paludosum.*

Acetobacterium 2 Medium

Composition per 1050.0mL:

NaCl	20.0g
$NaHCO_3$	10.0g
Yeast extract	2.0g
Ethylene glycol	1.25g
NH_4Cl	1.0g
K_2HPO_4	0.45g
KH_2PO_4	0.33g
$MgSO_4 \cdot 7H_2O$	0.1g
Resazurin	1.0mg
Ethylene glycol solution	20.0mL
Trace elements solution	20.0mL
Vitamin solution	20.0mL
Reducing agent solution	10.0mL

Trace Elements Solution:
Composition per liter:

$MgSO_4 \cdot 7H_2O$	3.0g
Nitrilotriacetic acid	1.5g
$CaCl_2 \cdot 2H_2O$	1.0g
NaCl	1.0g
$MnSO_4 \cdot 2H_2O$	0.5g
$CoSO_4 \cdot 7H_2O$	0.18g
$ZnSO_4 \cdot 7H_2O$	0.18g
$FeSO_4 \cdot 7H_2O$	0.1g
$NiCl_2 \cdot 6H_2O$	0.025g
$KAl(SO_4)_2 \cdot 12H_2O$	0.02g
$CuSO_4 \cdot 5H_2O$	0.01g
H_3BO_3	0.01g
$Na_2MoO_4 \cdot 2H_2O$	0.01g
$Na_2SeO_3 \cdot 5H_2O$	0.3mg

Preparation of Trace Elements Solution: Add nitrilotriacetic acid to 500.0mL of distilled/deionized water. Dissolve by adjusting pH to 6.5 with KOH. Add distilled/deionized water to 1.0L. Add remaining components. Mix thoroughly.

Vitamin Solution:
Composition per liter:

Pyridoxine·HCl	10.0mg
Calcium DL-pantothenate	5.0mg
Lipoic acid	5.0mg
Nicotinic acid	5.0mg
p-Aminobenzoic acid	5.0mg
Riboflavin	5.0mg
Thiamine·HCl	5.0mg
Biotin	2.0mg
Folic acid	2.0mg
Vitamin B_{12}	0.1mg

Preparation of Vitamin Solution: Add components to distilled/deionized water and bring volume to 1.0L. Mix thoroughly.

Reducing Agent Solution:
Composition per 10.0mL:

L-Cysteine·HCl·H_2O	0.5g
$Na_2S \cdot 9H_2O$	0.5g

Preparation of Reducing Agent Solution: Add 10.0mL of distilled/deionized water to a test tube. Gently heat and bring to boiling. Boil under N_2 gas for 1 min. Cool to room temperature. Add L-cysteine·HCl·H_2O and dissolve. Adjust pH to 9 with 5*N* NaOH. Add washed $Na_2S \cdot 9H_2O$ and dissolve. Autoclave for 10 min at 15 psi pressure–121°C.

Ethylene Glycol Solution:
Composition per 20.0mL:

Ethylene glycol	5.0g

Preparation of Ethylene Glycol Solution: Add ethylene glycol to distilled/deionized water and bring volume to 20.0mL. Mix thoroughly. Sparge under 100% N_2 gas for 3 min. Filter sterilize. Store under N_2 gas.

Preparation of Medium: Add components, except $NaHCO_3$, ethylene glycol solution, and reducing agent solution, to distilled/deionized water and bring volume to 1.0L. Mix thoroughly. Gently heat and bring to boiling. Boil for a few minutes. Allow to cool to room temperature under 80% N_2 + 20% CO_2. Add the $NaHCO_3$ and adjust pH to

7.4. Distribute into tubes or flasks under 80% N_2 + 20% CO_2. Autoclave for 15 min at 15 psi pressure–121°C. Cool to room temperature. Before inoculation, add sterile anaerobic Na_2CO_3 (0.25mL of 5% Na_2CO_3 per 10.0mL of medium) to bring the pH to 8.2. Add sterile ethylene glycol solution and reducing agent solution.

Use: For the cultivation and maintenance of *Acetobacterium woodii.*

Acetobacterium sp. KoMAc1 Medium (DSMZ Medium 478)

Composition per 1001.0mL:

Solution A	870.0mL
Solution C	100.0mL
Solution D	10.0mL
Solution E (Vitamin solution)	10.0mL
Solution F	10.0mL
Solution B (Trace element solution SL-10)	1.0mL

pH 7.1–7.4 at 25°C

Solution A:

Composition per 870.0mL:

Na_2SO_4	3.0g
NaCl	1.0g
Yeast extract	0.5g
KCl	0.5g
$MgCl_2 \cdot 6H_2O$	0.4g
NH_4Cl	0.3g
KH_2PO_4	0.2g
$CaCl_2 \cdot 2H_2O$	0.15g
Resazurin	1.0mg

Preparation of Solution A: Add components to distilled/deionized water and bring volume to 870.0mL. Mix thoroughly.

Solution B (Trace Elements Solution SL-10):

Composition per liter:

$FeCl_2 \cdot 4H_2O$	1.5g
$CoCl_2 \cdot 6H_2O$	190.0mg
$MnCl_2 \cdot 4H_2O$	100.0mg
$ZnCl_2$	70.0mg
$Na_2MoO_4 \cdot 2H_2O$	36.0mg
$NiCl_2 \cdot 6H_2O$	24.0mg
H_3BO_3	6.0mg
$CuCl_2 \cdot 2H_2O$	2.0mg
HCl (25% solution)	10.0mL

Preparation of Solution B (Trace Elements Solution SL-10): Add $FeCl_2 \cdot 4H_2O$ to 10.0mL of HCl solution. Mix thoroughly. Add distilled/deionized water and bring volume to 1.0L. Add remaining components. Mix thoroughly. Sparge with 100% N_2. Autoclave for 15 min at 15 psi pressure–121°C.

Solution C:

Composition per 100.0mL:

$NaHCO_3$	5.0g

Preparation of Solution C: Add $NaHCO_3$ to distilled/deionized water and bring volume to 100.0mL Mix thoroughly. Filter sterilize. Flush with 80% N_2 + 20% CO_2 to remove dissolved oxygen.

Solution D:

Composition per 10.0mL:

Methoxyacetate	0.9g

Preparation of Solution D: Add methoxyacetate to distilled/deionized water and bring volume to 10.0mL. Mix thoroughly. Sparge with 100% N_2. Autoclave for 15 min at 15 psi pressure–121°C.

Solution E (Vitamin Solution):

Composition per liter:

Pyridoxine-HCl	10.0mg
Thiamine-HCl·$2H_2O$	5.0mg
Riboflavin	5.0mg
Nicotinic acid	5.0mg
D-Ca-pantothenate	5.0mg
p-Aminobenzoic acid	5.0mg
Lipoic acid	5.0mg
Biotin	2.0mg
Folic acid	2.0mg
Vitamin B_{12}	0.10mg

Solution E (Vitamin Solution): Add components to distilled/deionized water and bring volume to 1.0L. Mix thoroughly. Sparge with 100% N_2. Autoclave for 15 min at 15 psi pressure–121°C.

Solution F:

Composition per 10.0mL:

$Na_2S \cdot 9H_2O$	0.4g

Preparation of Solution F: Add $Na_2S \cdot 9H_2O$ to distilled/deionized water and bring volume to 10.0mL. Mix thoroughly. Sparge with 100% N_2. Autoclave for 15 min at 15 psi pressure–121°C.

Preparation of Medium: Gently heat solution A and bring to boiling. Boil solution A for a few minutes. Cool to room temperature. Gas with 80% N_2 + 20% CO_2 gas mixture to reach a pH below 6. Autoclave for 15 min at 15 psi pressure–121°C. Cool to room temperature. Sequentially add 1.0mL solution B, 100.0mL solution C, 10.0mL solution D, 10.0mL solution E, and 10.0mL solution F. Distribute anaerobically under 80% N_2 + 20% CO_2 into appropriate vessels. Addition of 10–20mg sodium dithionite per liter from a 5% (w/v) solution, freshly prepared under N_2 and filter-sterilized, may stimulate growth.

Use: For the cultivation of *Acetobacterium* sp.

Acetobacterium sp. Medium (DSMZ Medium 614)

Composition per liter:

$MgCl_2 \cdot 6H_2O$	0.52g
Yeast extract	0.5g
KCl	0.33g
NH_4Cl	0.33g
KH_2PO_4	0.33g
$CaCl_2 \cdot 2H_2O$	0.22g
Resazurin	1.0mg
Trace elements solution	10.0mL
Vitamin solution	10.0mL
$NaHCO_3$ solution	10.0mL
Substrate solution	10.0mL
$Na_2S \cdot 9H_2O$ solution	10.0mL

pH 7.5 ± 0.2 at 25°C

Substrate Solution:

Composition per 10.0mL:

Na-lactate	4.0g

Preparation of Substrate Solution: Add Na-lactate to distilled/deionized water and bring volume to 10.0mL. Mix thoroughly. Sparge with 10% N_2. Filter sterilize.

$NaHCO_3$ Solution:

Composition per 10.0mL:

$NaHCO_3$	1.0g

Preparation of $NaHCO_3$ Solution: Add $NaHCO_3$ to distilled/deionized water and bring volume to 10.0mL. Mix thoroughly. Sparge with 80% N_2 + 20% CO_2. Filter sterilize.

$Na_2S \cdot 9H_2O$ Solution:

Composition per 10.0mL:

$Na_2S \cdot 9H_2O$	0.7g

Preparation of $Na_2S \cdot 9H_2O$ Solution: Add $Na_2S \cdot 9H_2O$ to distilled/deionized water and bring volume to 10.0mL. Sparge with N_2. Autoclave for 15 min at 15 psi pressure–121°C. Cool to 25°C. Store anaerobically. Before use, neutralize sodium sulfide with sterile hydrochloric acid.

Vitamin Solution:

Composition per liter:

Pyridoxine-HCl	10.0mg
Thiamine-HCl·$2H_2O$	5.0mg
Riboflavin	5.0mg
Nicotinic acid	5.0mg
D-Ca-pantothenate	5.0mg
p-Aminobenzoic acid	5.0mg
Lipoic acid	5.0mg
Biotin	2.0mg
Folic acid	2.0mg
Vitamin B_{12}	0.1mg

Preparation of Vitamin Solution: Add components to distilled/deionized water and bring volume to 1.0L. Mix thoroughly. Sparge with 80% H_2 + 20% CO_2. Filter sterilize.

Trace Elements Solution:

Composition per liter:

$MgSO_4 \cdot 7H_2O$	3.0g
Nitrilotriacetic acid	1.5g
NaCl	1.0g
$MnSO_4 \cdot 2H_2O$	0.5g
$CoSO_4 \cdot 7H_2O$	0.18g
$ZnSO_4 \cdot 7H_2O$	0.18g
$CaCl_2 \cdot 2H_2O$	0.1g
$FeSO_4 \cdot 7H_2O$	0.1g
$NiCl_2 \cdot 6H_2O$	0.025g
$KAl(SO_4)_2 \cdot 12H_2O$	0.02g
H_3BO_3	0.01g
$Na_2MoO_4 \cdot 4H_2O$	0.01g
$CuSO_4 \cdot 5H_2O$	0.01g
$Na_2SeO_3 \cdot 5H_2O$	0.3mg

Preparation of Trace Elements Solution: Add nitrilotriacetic acid to 500.0mL of distilled/deionized water. Dissolve by adjusting pH to 6.5 with KOH. Add remaining components. Add distilled/deionized water to 1.0L. Mix thoroughly.

Preparation of Medium: Prepare and dispense medium under 100% N_2 gas atmosphere. Add components, except $NaHCO_3$ solution, substrate solution, $Na_2S \cdot 9H_2O$ solution, and vitamin solution, to distilled/deionized water and bring volume to 960.0mL. Mix thoroughly. Adjust pH to 7.5. Sparge with 100% N_2. Autoclave for 15 min at 15 psi pressure–121°C. Aseptically and anaerobically add 10.0mL $NaHCO_3$ solution, 10.0mL substrate solution, 10.0mL $Na_2S \cdot 9H_2O$ solution, and 10.0mL vitamin solution. Mix thoroughly. Aseptically and anaerobically distribute into sterile tubes or bottles.

Use: For the cultivation of *Acetobacterium fimetarium* DSM 8238.

Acetobacterium sp. Medium (DSMZ Medium 614)

Composition per liter:

$MgCl_2 \cdot 6H_2O$	0.52g
Yeast extract	0.5g
KCl	0.33g
NH_4Cl	0.33g
KH_2PO_4	0.33g
$CaCl_2 \cdot 2H_2O$	0.22g
Resazurin	1.0mg
Trace elements solution	10.0mL
Vitamin solution	10.0mL
$NaHCO_3$ solution	10.0mL
Substrate solution	10.0mL
$Na_2S \cdot 9H_2O$ solution	10.0mL

pH 6.9 ± 0.2 at 25°C

Substrate Solution:

Composition per 10.0mL:

Fructose	10.0g

Preparation of Substrate Solution: Add fructose to distilled/deionized water and bring volume to 10.0mL. Mix thoroughly. Sparge with 10% N_2. Filter sterilize.

$NaHCO_3$ Solution:

Composition per 10.0mL:

$NaHCO_3$	1.0g

Preparation of $NaHCO_3$ Solution: Add $NaHCO_3$ to distilled/deionized water and bring volume to 10.0mL. Mix thoroughly. Sparge with 80% N_2 + 20% CO_2. Filter sterilize.

$Na_2S \cdot 9H_2O$ Solution:

Composition per 10.0mL:

$Na_2S \cdot 9H_2O$	0.7g

Preparation of $Na_2S \cdot 9H_2O$ Solution: Add $Na_2S \cdot 9H_2O$ to distilled/deionized water and bring volume to 10.0mL. Sparge with N_2. Autoclave for 15 min at 15 psi pressure–121°C. Cool to 25°C. Store anaerobically. Before use, neutralize sodium sulfide with sterile hydrochloric acid.

Vitamin Solution:

Composition per liter:

Pyridoxine-HCl	10.0mg
Thiamine-HCl·$2H_2O$	5.0mg
Riboflavin	5.0mg
Nicotinic acid	5.0mg
D-Ca-pantothenate	5.0mg
p-Aminobenzoic acid	5.0mg
Lipoic acid	5.0mg
Biotin	2.0mg
Folic acid	2.0mg
Vitamin B_{12}	0.1mg

Preparation of Vitamin Solution: Add components to distilled/deionized water and bring volume to 1.0L. Mix thoroughly. Sparge with 80% H_2 + 20% CO_2. Filter sterilize.

Trace Elements Solution:

Composition per liter:

$MgSO_4 \cdot 7H_2O$	3.0g
Nitrilotriacetic acid	1.5g
NaCl	1.0g
$MnSO_4 \cdot 2H_2O$	0.5g
$CoSO_4 \cdot 7H_2O$	0.18g

$ZnSO_4 \cdot 7H_2O$	0.18g
$CaCl_2 \cdot 2H_2O$	0.1g
$FeSO_4 \cdot 7H_2O$	0.1g
$NiCl_2 \cdot 6H_2O$	0.025g
$KAl(SO_4)_2 \cdot 12H_2O$	0.02g
H_3BO_3	0.01g
$Na_2MoO_4 \cdot 4H_2O$	0.01g
$CuSO_4 \cdot 5H_2O$	0.01g
$Na_2SeO_3 \cdot 5H_2O$	0.3mg

Preparation of Trace Elements Solution: Add nitrilotriacetic acid to 500.0mL of distilled/deionized water. Dissolve by adjusting pH to 6.5 with KOH. Add remaining components. Add distilled/deionized water to 1.0L. Mix thoroughly.

Preparation of Medium: Prepare and dispense medium under 100% N_2 gas atmosphere. Add components, except $NaHCO_3$ solution, substrate solution, $Na_2S \cdot 9H_2O$ solution, and vitamin solution, to distilled/deionized water and bring volume to 960.0mL. Mix thoroughly. Adjust pH to 6.8–7.0. Sparge with 100% N_2. Autoclave for 15 min at 15 psi pressure–121°C. Aseptically and anaerobically add 10.0mL $NaHCO_3$ solution, 10.0mL substrate solution, 10.0mL $Na_2S \cdot 9H_2O$ solution, and 10.0mL vitamin solution. Mix thoroughly. Aseptically and anaerobically distribute into sterile tubes or bottles.

Use: For the cultivation of *Acetobacterium fimetarium* DSM 8237 and *Acetobacterium fimetarium* DSM 8239.

Acetobacterium tundrae Medium (DSMZ Medium 900)

Yeast extract	1.0g
$MgCl_2 \cdot 6H_2O$	0.52g
KCl	0.33g
NH_4Cl	0.33g
KH_2PO_4	0.33g
$CaCl_2 \cdot 2H_2O$	0.3g
Resazurin	0.5mg
Fructose solution	30.0mL
$NaHCO_3$ solution	20.0mL
L-Cysteine solution	10.0mL
$Na_2S \cdot 9H_2O$ solution	10.0mL
Trace mineral solution SL-10	1.0mL
Seven vitamin solution	1.0mL

pH 7.0 ± 0.2 at 25°C

L-Cysteine Solution:
Composition per 10.0mL:

L-Cysteine·HCl·H_2O	0.3g

Preparation of L-Cysteine Solution: Add L-cysteine·HCl·H_2O to distilled/deionized water and bring volume to 10.0mL. Mix thoroughly. Sparge with 100% N_2. Autoclave for 15 min at 15 psi pressure–121°C.

$NaHCO_3$ Solution:
Composition per 20.0mL:

$NaHCO_3$	5.0g

Preparation of $NaHCO_3$ Solution: Add $NaHCO_3$ to distilled/deionized water and bring volume to 20.0mL. Mix thoroughly. Sparge with 80% N_2 + 20% CO_2. Filter sterilize.

$Na_2S \cdot 9H_2O$ Solution:
Composition per 10.0mL:

$Na_2S \cdot 9H_2O$	0.3g

Preparation of $Na_2S \cdot 9H_2O$ Solution: Add $Na_2S \cdot 9H_2O$ to distilled/deionized water and bring volume to 10.0mL. Mix thoroughly. Autoclave under 100% N_2 for 15 min at 15 psi pressure–121°C. Cool to room temperature.

Fructose Solution:
Composition per 30.0mL:

Glucose	5.0g

Preparation of Fructose Solution: Add glucose to distilled/deionized water and bring volume to 30.0mL. Mix thoroughly. Sparge with 100% N_2. Filter sterilize.

Seven Vitamin Solution:
Composition per liter:

Pyridoxine hydrochloride	300.0mg
Thiamine-HCl·$2H_2O$	200.0mg
Nicotinic acid	200.0mg
Vitamin B_{12}	100.0mg
Calcium pantothenate	100.0mg
p-Aminobenzoic acid	80.0mg
D(+)-Biotin	20.0mg

Preparation of Seven Vitamin Solution: Add components to distilled/deionized water and bring volume to 1.0L. Sparge with 100% N_2. Mix thoroughly. Filter sterilize.

Trace Elements Solution SL-10:
Composition per liter:

$FeCl_2 \cdot 4H_2O$	1.5g
$CoCl_2 \cdot 6H_2O$	190.0mg
$MnCl_2 \cdot 4H_2O$	100.0mg
$ZnCl_2$	70.0mg
$Na_2MoO_4 \cdot 2H_2O$	36.0mg
$NiCl_2 \cdot 6H_2O$	24.0mg
H_3BO_3	6.0mg
$CuCl_2 \cdot 2H_2O$	2.0mg
HCl (25% solution)	10.0mL

Preparation of Trace Elements Solution SL-10: Add $FeCl_2 \cdot 4H_2O$ to 10.0mL of HCl solution. Mix thoroughly. Add distilled/deionized water and bring volume to 1.0L. Add remaining components. Mix thoroughly. Sparge with 80% N_2 + 20% CO_2. Autoclave for 15 min at 15 psi pressure–121°C.

Preparation of Medium: Prepare and dispense medium under 80% N_2 + 20% CO_2 gas atmosphere. Add components, except seven vitamin solution, $NaHCO_3$ solution, fructose solution, L-cysteine-solution, and $Na_2S \cdot 9H_2O$ solution, to distilled/deionized water and bring volume to 929.0mL. Mix thoroughly. Sparge with 80% N_2 + 20% CO_2. Distribute into anaerobe tubes or bottles. Autoclave for 15 min at 15 psi pressure–121°C. Aseptically and anaerobically add, per liter of medium, 1.0mL seven vitamin solution, 20.0mL $NaHCO_3$ 30.0mL fructose solution, 10.0mL L-cysteine-HCl·H_2O solution, and 10.0mL $Na_2S \cdot 9H_2O$ solution. Mix thoroughly. The final pH should be 7.0.

Use: For the cultivation of *Acetobacterium tundrae.*

Acetobacteroides glycinophilus Medium

Composition per 1020.0mL:

Na_2HPO_4	5.8g
KH_2PO_4	3.0g
NH_4Cl	1.0g
$MgCl_2 \cdot 6H_2O$	0.2g
Resazurin	1.0mg
$CaCl_2 \cdot 2H_2O$	0.13g

Trace elements solution	10.0mL
Vitamin solution	5.0mL
Yeast extract solution	5.0mL
Glycine solution	5.0mL
$NaHCO_3$ solution	5.0mL
$Na_2S·9H_2O$ solution	5.0mL

pH 7.2–7.4 at 25°C

Trace Elements Solution:

Composition per liter:

Nitrilotriacetic acid	2.8g
NaCl	1.0g
$FeCl_3·4H_2O$	0.2g
$CoCl_2·6H_2O$	0.17g
$CaCl_2·2H_2O$	0.1g
$MnCl_2·4H_2O$	0.1g
$ZnCl_2$	0.1g
$NiCl_2·6H_2O$	0.026g
$CuCl_2$	0.02g
H_3BO_3	0.01g
$Na_2MoO·2H_2O$	0.01g
$Na_2SeO·5H_2O$	0.02g

Preparation of Trace Elements Solution: Add nitrilotriacetic acid to 500.0mL of distilled/deionized water. Dissolve by adjusting pH to 6.5 with KOH. Add distilled/deionized water to 1.0L. Add remaining components. Mix thoroughly.

Vitamin Solution:

Composition per liter:

Pyridoxine·HCl	10.0mg
Calcium DL-pantothenate	5.0mg
Lipoic acid	5.0mg
Nicotinic acid	5.0mg
p-Aminobenzoic acid	5.0mg
Riboflavin	5.0mg
Thiamine·HCl	5.0mg
Biotin	2.0mg
Folic acid	2.0mg
Vitamin B_{12}	0.1mg

Preparation of Vitamin Solution: Add components to distilled/deionized water and bring volume to 1.0L. Mix thoroughly.

Yeast Extract Solution:

Composition per 5.0mL:

Yeast extract	0.5g

Preparation of Yeast Extract Solution: Add yeast extract to distilled/deionized water and bring volume to 5.0mL. Mix thoroughly. Sparge under 100% N_2 gas for 3 min. Autoclave for 15 min at 15 psi pressure–121°C. Store under N_2 gas.

Glycine Solution:

Composition per 5.0mL:

Glycine	1.5g

Preparation of Glycine Solution: Add glycine to distilled/deionized water and bring volume to 5.0mL. Mix thoroughly. Sparge under 100% N_2 gas for 3 min. Autoclave for 15 min at 15 psi pressure–121°C. Store under N_2 gas.

$NaHCO_3$ Solution:

Composition per 5.0mL:

$NaHCO_3$	0.5g

Preparation of $NaHCO_3$ Solution: Add $NaHCO_3$ to distilled/deionized water and bring volume to 5.0mL. Mix thoroughly. Sparge under 100% N_2 gas for 3 min. Autoclave for 15 min at 15 psi pressure–121°C. Store under N_2 gas.

$Na_2S·9H_2O$ Solution:

Composition per 5.0mL:

$Na_2S·9H_2O$	0.5g

Preparation of $Na_2S·9H_2O$ Solution: Add $Na_2S·9H_2O$ to distilled/deionized water and bring volume to 5.0mL. Mix thoroughly. Sparge under 100% N_2 gas for 3 min. Autoclave for 15 min at 15 psi pressure–121°C. Store under N_2 gas.

Preparation of Medium: Add components, except yeast extract solution, glycine solution, $NaHCO_3$ solution, and $Na_2S·9H_2O$ solution, to distilled/deionized water and bring volume to 1.0L. Adjust pH to 7.2–7.4 with NaOH. Mix thoroughly. Gently heat and bring to boiling. Boil for a few minutes. Allow to cool to room temperature under 100% N_2. Distribute into tubes or flasks under 100% N_2. Autoclave for 15 min at 15 psi pressure–121°C. Cool to room temperature. Before inoculation, aseptically and anaerobically add yeast extract solution, glycine solution, $NaHCO_3$ solution, and $Na_2S·9H_2O$ solution.

Use: For the cultivation and maintenance of *Acetobacteroides glycinophilus*.

Acetogen Medium

Composition per 421.8mL:

$NaHCO_3$	2.4g
NH_4Cl	0.2g
Yeast extract	0.2g
Stock salts solution #1	40.0mL
Potassium phosphate buffer	20.0mL
Clarified rumen fluid	20.0mL
Stock salts solution #2	4.0mL
Trace minerals solution	4.0mL
Vitamin solution	4.0mL
Reducing agent solution	4.0mL
Tungstate solution	0.4mL
Resazurin (0.1% solution)	0.4mL

Potassium Phosphate Buffer:

Composition per 830.0mL:

K_2HPO_4	15.68g
KH_2PO_4	4.72g

Preparation of Potassium Phosphate Buffer: Dissolve K_2HPO_4 in 600.0mL of distilled/deionized water and KH_2PO_4 in 230.0mL of distilled/deionized water. Mix the two solutions together and use.

Stock Salts Solution #1:

Composition per liter:

KCl	1.6g
NaCl	1.4g
$MgSO_4·7H_2O$	0.2g

Preparation of Stock Salts Solution #1: Add components to distilled/deionized water and bring volume to 1.0L. Mix thoroughly.

Stock Salts Solution #2:

Composition per liter:

$CaCl_2·2H_2O$	0.1g

Preparation of Stock Salts Solution #2: Add components to distilled/deionized water and bring volume to 1.0L. Mix thoroughly.

Trace Minerals Solution:
Composition per liter:

Nitrilotriacetic acid	1.5g
$MgSO_4 \cdot 7H_2O$	3.0g
$MnSO_4 \cdot H_2O$	0.5g
NaCl	1.0g
$NiCl_2 \cdot 6H_2O$	0.1g
$FeSO_4 \cdot 7H_2O$	0.1g
$CoCl_2 \cdot 6H_2O$	0.1g
$CaCl_2$	0.1g
$ZnSO_4 \cdot 7H_2O$	0.1g
$Na_2SeO_3 \cdot 5H_2O$	0.01g
$CuSO_4 \cdot 5H_2O$	0.01g
$AlK(SO_4)_2 \cdot 12H_2O$	0.01g
H_3BO_3	0.01g
$Na_2MoO_4 \cdot 2H_2O$	0.01g

Preparation of Trace Minerals Solution: Add nitrilotriacetic acid to 500.0mL of distilled/deionized water. Dissolve by adjusting pH to 6.5 with KOH. Bring volume to 1.0L with distilled/deionized water. Add remaining components. Mix thoroughly.

Vitamin Solution:
Composition per liter:

Pyridoxine·HCl	10.0mg
Ascorbic acid	5.0mg
Calcium pantothenate	5.0mg
Choline chloride	5.0mg
Lipoic acid	5.0mg
i-Inositol	5.0mg
Niacinamide	5.0mg
Nicotinic acid	5.0mg
p-Aminobenzoic acid	5.0mg
Pyridoxal·HCl	5.0mg
Riboflavin	5.0mg
Thiamine·HCl	5.0mg
Biotin	2.0mg
Folic acid	2.0mg
Vitamin B_{12}	0.1mg

Preparation of Vitamin Solution: Add components to distilled/deionized water and bring volume to 1.0L. Mix thoroughly. Store frozen.

Tungstate Solution:
Composition per liter:

$Na_2WO_4 \cdot 2H_2O$	99.0mg

Preparation of Tungstate Solution: Add components to distilled/deionized water and bring volume to 1.0L. Mix thoroughly.

Reducing Agent Solution:
Composition per 110.0mL:

L-Cysteine·HCl·H_2O	2.5g
$Na_2S \cdot 9H_2O$	2.5g

Preparation of Reducing Agent Solution: Add 110.0mL of distilled/deionized water to a 250.0mL round-bottomed flask. Boil under N_2 gas for 1 min. Cool to room temperature. Add L-cysteine·HCl and dissolve. Adjust to pH 9 with 5*N* NaOH. Add washed $Na_2S \cdot 9H_2O$ and dissolve. Distribute in amounts needed into tubes or flasks. Autoclave for 10 min at 15 psi pressure–121°C.

Preparation of Medium: Add components, except $NaHCO_3$ and reducing agent, to distilled/deionized water and bring volume to 417.8mL. Mix thoroughly. Gently heat and bring to boiling under 80% N_2 + 20% CO_2. Cool to 45°–50°C. Add $NaHCO_3$ and reducing agent. Distribute into tubes or flasks under 80% N_2 + 20% CO_2. Autoclave for 15 min at 15 psi pressure–121°C. After inoculation, exchange headspace with 80% H_2 + 20% CO_2.

Use: For the cultivation and maintenance of acetogenic anaerobes such as some *Clostridium* species.

Acetogenium Medium

Composition per liter:

$Na_2HPO_4 \cdot 12H_2O$	6.1g
$NaH_2PO_4 \cdot H_2O$	4.5g
L-Cysteine·HCl	0.5g
$Na_2S \cdot 9H_2O$	0.5g
NaCl	0.45g
NH_4Cl	0.31g
K_2HPO_4	0.22g
KH_2PO_4	0.22g
$(NH_4)_2SO_4$	0.22g
$MgSO_4 \cdot 7H_2O$	0.09g
$CaCl_2 \cdot 2H_2O$	6.0mg
$FeSO_4 \cdot 7H_2O$	2.0mg
Resazurin	1.0mg
Trace elements solution	10.0mL

pH 6.5 ± 0.2 at 25°C

Trace Elements Solution:
Composition per liter:

$MgSO_4 \cdot 7H_2O$	3.0g
Nitrilotriacetic acid	1.5 g
$CaCl_2 \cdot 2H_2O$	1.0g
NaCl	1.0g
$MnSO_4 \cdot 2H_2O$	0.5g
$CoSO_4 \cdot 7H_2O$	0.18g
$ZnSO_4 \cdot 7H_2O$	0.18g
$FeSO_4 \cdot 7H_2O$	0.1g
$NiCl_2 \cdot 6H_2O$	0.025g
$KAl(SO_4)_2 \cdot 12H_2O$	0.02g
$CuSO_4 \cdot 5H_2O$	0.01g
H_3BO_3	0.01g
$Na_2MoO_4 \cdot 2H_2O$	0.01g
$Na_2SeO_3 \cdot 5H_2O$	0.3mg

Preparation of Trace Elements Solution: Add nitrilotriacetic acid to 500.0mL of distilled/deionized water. Dissolve by adjusting pH to 6.5 with KOH. Add distilled/deionized water to 1.0L. Add remaining components. Mix thoroughly. Adjust pH to 7.0 with KOH.

Preparation of Medium: Add components to distilled/deionized water and bring volume to 1.0L. Adjust pH to 7.2–7.4 with NaOH. Mix thoroughly. Gently heat and bring to boiling. Boil for a few minutes. Allow to cool to room temperature under 80% H_2 + 20% CO_2. Distribute into tubes or flasks under 80% H_2 + 20% CO_2. Autoclave for 15 min at 15 psi pressure–121°C.

Use: For the cultivation and maintenance of *Acetogenium kivui.*

Acetohalobium Medium

Composition per 1025.0mL:

NaCl	150.0g
$MgCl_2 \cdot 6H_2O$	4.0g
$CaCl_2 \cdot 2H_2O$	0.33g
KCl	0.33g
KH_2PO_4	0.33g
NH_4Cl	0.33g

Resazurin	1.0mg
Trace elements solution	10.0mL
Vitamin solution	10.0mL
Trimethylamine·HCl solution	10.0mL
Yeast extract solution	5.0mL
$NaHCO_3$ solution	5.0mL
$Na_2S·9H_2O$ solution	5.0mL

pH 7.6 ± 0.2 at 25°C

Trace Elements Solution:
Composition per liter:

$MgSO_4·7H_2O$	3.0g
Nitrilotriacetic acid	1.5 g
$CaCl_2·2H_2O$	1.0g
NaCl	1.0g
$MnSO_4·2H_2O$	0.5g
$CoSO_4·7H_2O$	0.18g
$ZnSO_4·7H_2O$	0.18g
$FeSO_4·7H_2O$	0.1g
$NiCl_2·6H_2O$	0.025g
$KAl(SO_4)_2·12H_2O$	0.02g
$CuSO_4·5H_2O$	0.01g
H_3BO_3	0.01g
$Na_2MoO_4·2H_2O$	0.01g
$Na_2SeO_3·5H_2O$	0.3mg

Preparation of Trace Elements Solution: Add nitrilotriacetic acid to 500.0mL of distilled/deionized water. Dissolve by adjusting pH to 6.5 with KOH. Add distilled/deionized water to 1.0L. Add remaining components. Mix thoroughly. Adjust pH to 7.0 with KOH.

Vitamin Solution:
Composition per liter:

Pyridoxine·HCl	10.0mg
Calcium DL-pantothenate	5.0mg
Lipoic acid	5.0mg
Nicotinic acid	5.0mg
p-Aminobenzoic acid	5.0mg
Riboflavin	5.0mg
Thiamine·HCl	5.0mg
Biotin	2.0mg
Folic acid	2.0mg
Vitamin B_{12}	0.1mg

Preparation of Vitamin Solution: Add components to distilled/deionized water and bring volume to 1.0L. Mix thoroughly. Filter sterilize.

Trimethylamine·HCl Solution:
Composition per 10.0mL:

Trimethylamine·HCl	2.4g

Preparation of Trimethylamine·HCl Solution: Add trimethylamine·HCl to distilled/deionized water and bring volume to 10.0mL. Mix thoroughly. Sparge under 100% N_2 gas for 3 min. Autoclave for 15 min at 15 psi pressure–121°C. Store under N_2 gas. 4.5g of glycine betaine may be used in place of trimethylamine·HCl.

Yeast Extract Solution:
Composition per 5.0mL:

Yeast extract	0.05g

Preparation of Yeast Extract Solution: Add yeast extract to distilled/deionized water and bring volume to 5.0mL. Mix thoroughly. Sparge under 100% N_2 gas for 3 min. Autoclave for 15 min at 15 psi pressure–121°C. Store under N_2 gas.

$NaHCO_3$ Solution:
Composition per 5.0mL:

$NaHCO_3$	0.5g

Preparation of $NaHCO_3$ Solution: Add $NaHCO_3$ to distilled/deionized water and bring volume to 5.0mL. Mix thoroughly. Sparge under 100% N_2 gas for 3 min. Autoclave for 15 min at 15 psi pressure–121°C. Store under N_2 gas.

$Na_2S·9H_2O$ Solution:
Composition per 5.0mL:

$Na_2S·9H_2O$	0.5g

Preparation of $Na_2S·9H_2O$ Solution: Add $Na_2S·9H_2O$ to distilled/deionized water and bring volume to 5.0mL. Mix thoroughly. Sparge under 100% N_2 gas for 3 min. Autoclave for 15 min at 15 psi pressure–121°C. Store under N_2 gas.

Preparation of Medium: Add components, except vitamin solution, trimethylamine·HCl solution, yeast extract solution, $NaHCO_3$ solution, and $Na_2S·9H_2O$ solution, to distilled/deionized water and bring volume to 1.0L. Adjust pH to 7.2–7.4 with NaOH. Mix thoroughly. Gently heat and bring to boiling. Boil for a few minutes. Allow to cool to room temperature under 100% N_2. Distribute into tubes or flasks under 100% N_2. Autoclave for 15 min at 15 psi pressure–121°C. Cool to room temperature. Before inoculation, aseptically and anaerobically add vitamin solution, trimethylamine·HCl solution, yeast extract solution, $NaHCO_3$ solution, and $Na_2S·9H_2O$ solution. If necessary, adjust pH to 7.6 with sterile anaerobic Na_2CO_3 solution.

Use: For the cultivation and maintenance of *Acetohalobium arabaticum.*

Acetohalobium Medium

Composition per liter:

NaCl	150.0g
$MgCl_2·6H_2O$	4.0g
$NaHCO_3$	4.0g
Trimethylamine·HCl	2.4g
$Na_2S·9H_2O$	0.5g
$CaCl_2·2H_2O$	0.33g
KCl	0.33g
KH_2PO_4	0.33g
NH_4Cl	0.33g
Yeast extract	0.05g
Resazurin	1.0mg
Wolfe's mineral solution	10.0mL
Wolfe's vitamin solution	10.0mL

pH 7.8 ± 0.2 at 25°C

Wolfe's Mineral Solution:
Composition per liter:

$MgSO_4·7H_2O$	3.0g
Nitrilotriacetic acid	1.5g
NaCl	1.0g
$MnSO_4·2H_2O$	0.5g
$CoCl_2·6H_2O$	0.1g
$ZnSO_4·7H_2O$	0.1g
$CaCl_2·2H_2O$	0.1g
$FeSO_4·7H_2O$	0.1g
$NiCl_2·6H_2O$	0.025g
$KAl(SO_4)_2·12H_2O$	0.02g
$CuSO_4·5H_2O$	0.01g
H_3BO_3	0.01g
$Na_2MoO_4·2H_2O$	0.01g
$Na_2SeO_3·5H_2O$	0.3mg

Preparation of Wolfe's Mineral Solution: Add nitrilotriacetic acid to 500.0mL of distilled/deionized water. Adjust pH to 6.5 with KOH. Add remaining components one at a time. Add distilled/deionized water to 1.0L. Adjust pH to 6.8.

Wolfe's Vitamin Solution:

Composition per liter:

Pyridoxine·HCl	10.0mg
p-Aminobenzoic acid	5.0mg
Lipoic acid	5.0mg
Nicotinic acid	5.0mg
Riboflavin	5.0mg
Thiamine·HCl	5.0mg
Calcium DL-pantothenate	5.0mg
Biotin	2.0mg
Folic acid	2.0mg
Vitamin B_{12}	0.1mg

Preparation of Wolfe's Vitamin Solution: Add components to distilled/deionized water and bring volume to 1.0L. Mix thoroughly.

Preparation of Medium: Prepare and dispense medium under 80% N_2 + 20% CO_2. Add components, except $NaHCO_3$ and $Na_2S·9H_2O$, to distilled/deionized water and bring volume to 1.0L. Mix thoroughly. Gently heat and bring to boiling for 3 min. Cool to room temperature while sparging with 80% N_2 + 20% CO_2. Add $NaHCO_3$. Mix thoroughly. Continue sparging for 5 min. Add $Na_2S·9H_2O$. Mix thoroughly. Adjust pH to 7.8 with Na_2CO_3. Anaerobically distribute into tubes. Autoclave for 15 min at 15 psi pressure–121°C.

Use: For the cultivation of *Acetohalobium arabaticum.*

Acetohalobium Medium

Composition per liter:

NaCl	150.0g
$MgCl_2·6H_2O$	4.0g
$NaHCO_3$	4.0g
Glycinebetaine	4.0g
$Na_2S·9H_2O$	0.5g
$CaCl_2·2H_2O$	0.33g
KCl	0.33g
KH_2PO_4	0.33g
NH_4Cl	0.33g
Yeast extract	0.05g
Resazurin	1.0mg
Wolfe's mineral solution	10.0mL
Wolfe's vitamin solution	10.0mL

pH 7.8 ± 0.2 at 25°C

Wolfe's Mineral Solution:

Composition per liter:

$MgSO_4·7H_2O$	3.0g
Nitrilotriacetic acid	1.5g
NaCl	1.0g
$MnSO_4·2H_2O$	0.5g
$CoCl_2·6H_2O$	0.1g
$ZnSO_4·7H_2O$	0.1g
$CaCl_2·2H_2O$	0.1g
$FeSO_4·7H_2O$	0.1g
$NiCl_2·6H_2O$	0.025g
$KAl(SO_4)_2·12H_2O$	0.02g
$CuSO_4·5H_2O$	0.01g
H_3BO_3	0.01g
$Na_2MoO_4·2H_2O$	0.01g
$Na_2SeO_3·5H_2O$	0.3mg

Preparation of Wolfe's Mineral Solution: Add nitrilotriacetic acid to 500.0mL of distilled/deionized water. Adjust pH to 6.5 with KOH. Add remaining components one at a time. Add distilled/deionized water to 1.0L. Adjust pH to 6.8.

Wolfe's Vitamin Solution:

Composition per liter:

Pyridoxine·HCl	10.0mg
p-Aminobenzoic acid	5.0mg
Lipoic acid	5.0mg
Nicotinic acid	5.0mg
Riboflavin	5.0mg
Thiamine·HCl	5.0mg
Calcium DL-pantothenate	5.0mg
Biotin	2.0mg
Folic acid	2.0mg
Vitamin B_{12}	0.1mg

Preparation of Wolfe's Vitamin Solution: Add components to distilled/deionized water and bring volume to 1.0L. Mix thoroughly.

Preparation of Medium: Prepare and dispense medium under 80% N_2 + 20% CO_2. Add components, except $NaHCO_3$ and $Na_2S·9H_2O$, to distilled/deionized water and bring volume to 1.0L. Mix thoroughly. Gently heat and bring to boiling. Continue boiling for 3 min. Cool to room temperature while sparging with 80% N_2 + 20% CO_2. Add $NaHCO_3$. Mix thoroughly. Continue sparging for 5 min. Add $Na_2S·9H_2O$. Mix thoroughly. Adjust pH to 7.8 with Na_2CO_3. Anaerobically distribute into tubes. Autoclave for 15 min at 15 psi pressure–121°C.

Use: For the cultivation of *Acetohalobium arabaticum.*

Acetomicrobium faecalis Medium

Composition per 1010.0mL:

$NaHCO_3$	6.0g
Sodium acetate	5.0g
Glucose	4.0g
Pancreatic digest of casein	2.0g
Yeast extract	2.0g
L-Cysteine·HCl	0.5g
NaCl	0.5g
K_2HPO_4	0.225g
KH_2PO_4	0.225g
$(NH_4)_2SO_4$	0.225g
$MgSO_4·7H_2O$	0.1g
$CaCl_2·2H_2O$	0.07g
Resazurin	1.0mg
Trace elements solution	10.0mL
Vitamin solution	10.0mL

Trace Elements Solution:

Composition per liter:

$MgSO_4·7H_2O$	3.0g
Nitrilotriacetic acid	1.5g
$CaCl_2·2H_2O$	1.0g
NaCl	1.0g
$MnSO_4·2H_2O$	0.5g
$CoSO_4·7H_2O$	0.18g
$ZnSO_4·7H_2O$	0.18g
$FeSO_4·7H_2O$	0.1g
$NiCl_2·6H_2O$	0.025g
$KAl(SO_4)_2·12H_2O$	0.02g
$CuSO_4·5H_2O$	0.01g

H_3BO_3	0.01g
$Na_2MoO_4 \cdot 2H_2O$	0.01g
$Na_2SeO_3 \cdot 5H_2O$	0.3mg

Preparation of Trace Elements Solution: Add nitrilotriacetic acid to 500.0mL of distilled/deionized water. Dissolve by adjusting pH to 6.5 with KOH. Add distilled/deionized water to 1.0L. Add remaining components. Mix thoroughly. Adjust pH to 7.0 with KOH.

Vitamin Solution:
Composition per liter:

Pyridoxine·HCl	10.0mg
Calcium DL-pantothenate	5.0mg
Lipoic acid	5.0mg
Nicotinic acid	5.0mg
p-Aminobenzoic acid	5.0mg
Riboflavin	5.0mg
Thiamine·HCl	5.0mg
Biotin	2.0mg
Folic acid	2.0mg
Vitamin B_{12}	0.1mg

Preparation of Vitamin Solution: Add components to distilled/deionized water and bring volume to 1.0L. Mix thoroughly. Filter sterilize.

Preparation of Medium: Add components, except vitamin solution, to distilled/deionized water and bring volume to 1.0L. Mix thoroughly. Gently heat and bring to boiling. Boil for a few minutes. Allow to cool to room temperature under 80% N_2 + 20% CO_2. Distribute into tubes or flasks under 80% N_2 + 20% CO_2. Autoclave for 15 min at 15 psi pressure–121°C. Cool to room temperature. Before inoculation, aseptically and anaerobically add vitamin solution.

Use: For the cultivation and maintenance of *Acetomicrobium faecalis*.

Acetomicrobium flavidum Agar (LMG Medium 71)

Composition per liter:

Agar	15.0g
Pancreatic digest of casein	10.0g
Lab-Lemco beef extract	3.0g
Yeast extract	3.0g
Glucose	2.0g
L-Cysteine hydrochloride	0.5g
Salt solution	40.0mL
Tween™ 80	1.0mL

pH 7.2 ± 0.2 at 25°C

Salt solution:
Composition per liter:

$NaHCO_3$	10.0g
K_2HPO_4	1.0g
KH_2PO_4	1.0g
$CaCl_2 \cdot 2H_2O$	0.26g
$MgSO_4 \cdot 7H_2O$	0.2g

Preparation of Salt Solution: Add components to 1.0L of distilled/deionized water. Mix thoroughly.

Preparation of Medium: Add components to distilled/deionized water and bring volume to 1.0L. Mix thoroughly. Gently heat and bring to boiling. Distribute into tubes or flasks. Autoclave for 15 min at 15 psi pressure–121°C. Pour into sterile Petri dishes or leave in tubes.

Use: For the cultivation of *Acetomicrobium flavidum*.

Acetomicrobium flavidum Agar

Composition per liter:

Agar	15.0g
Casitone	10.0g
Beef extract	3.0g
Yeast extract	3.0g
Glucose	2.0g
L-Cysteine·HCl	0.5g
Salt solution	40.0mL
Tween™ 80	1.0mL

pH 7.2 ± 0.2 at 25°C

Salt Solution:
Composition per liter:

$NaHCO_3$	10.0g
K_2HPO_4	1.0g
KH_2PO_4	1.0g
$CaCl_2 \cdot 2H_2O$	0.26g
$MgSO_4 \cdot 7H_2O$	0.2g

Preparation of Salt Solution: Add components to distilled/deionized water and bring volume to 1.0L. Mix thoroughly. Adjust pH to 7.2.

Preparation of Medium: Add components to distilled/deionized water and bring volume to 1.0L. Mix thoroughly. Gently heat and bring to boiling. Distribute into tubes or flasks. Autoclave for 15 min at 15 psi pressure–121°C. Pour into sterile Petri dishes or leave in tubes.

Use: For the cultivation and maintenance of *Acetomicrobium flavidum.*

Acetomicrobium flavidum Broth (LMG Medium 71)

Composition per liter:

Pancreatic digest of casein	10.0g
Lab-Lemco beef extract	3.0g
Yeast extract	3.0g
Glucose	2.0g
L-Cysteine hydrochloride	0.5g
Salt solution	40.0mL
Tween™ 80	1.0mL

pH 7.2 ± 0.2 at 25°C

Salt solution:
Composition per liter:

$NaHCO_3$	10.0g
K_2HPO_4	1.0g
KH_2PO_4	1.0g
$CaCl_2 \cdot 2H_2O$	0.26g
$MgSO_4 \cdot 7H_2O$	0.2g

Preparation of Salt Solution: Add components to 1.0L of distilled/deionized water. Mix thoroughly.

Preparation of Medium: Add components to distilled/deionized water and bring volume to 1.0L. Mix thoroughly. Distribute into tubes or flasks. Autoclave for 15 min at 15 psi pressure–121°C.

Use: For the cultivation of *Acetomicrobium flavidum.*

Acetomicrobium flavidum Broth

Composition per liter:

Casitone	10.0g
Beef extract	3.0g
Yeast extract	3.0g
Glucose	2.0g

L-Cysteine·HCl 0.5g
Salt solution 40.0mL
Tween™ 80 .... 1.0mL

pH 7.2 ± 0.2 at 25°C

Salt Solution:

Composition per liter:

$NaHCO_3$ 10.0g
K_2HPO_4 .... 1.0g
KH_2PO_4 .... 1.0g
$CaCl_2 \cdot 2H_2O$ 0.26g
$MgSO_4 \cdot 7H_2O$ 0.2g

Preparation of Salt Solution: Add components to distilled/deionized water and bring volume to 1.0L. Mix thoroughly. Adjust pH to 7.2.

Preparation of Medium: Add components to distilled/deionized water and bring volume to 1.0L. Mix thoroughly. Distribute into tubes or flasks. Autoclave for 15 min at 15 psi pressure–121°C.

Use: For the cultivation of *Acetomicrobium flavidum.*

Acetonema Medium

Composition per liter:

Betaine·H_2O 6.7g
$NaHCO_3$ 4.0g
NaCl 2.25g
Yeast extract 2.0g
Pancreatic digest of casein 2.0g
NH_4Cl 0.5g
$MgSO_4 \cdot 7H_2O$ 0.5g
K_2HPO_4 .... 0.348g
$CaCl_2 \cdot 2H_2O$ 0.25g
KH_2PO_4 .... 0.227g
$FeSO_4 \cdot 7H_2O$ 2.0mg
Resazurin 1.0mg
$NaHSeO_3$ 26.3μg
Glucose solution 50.0mL
Vitamin solution 10.0mL
Reducing agent solution 10.0mL
Trace elements solution SL-10 .... 1.0mL

pH 7.0 ± 0.2 at 25°C

Vitamin Solution:

Composition per liter:

Pyridoxine·HCl 10.0mg
Calcium DL-pantothenate 5.0mg
Lipoic acid 5.0mg
Nicotinic acid 5.0mg
p-Aminobenzoic acid 5.0mg
Riboflavin 5.0mg
Thiamine·HCl 5.0mg
Biotin 2.0mg
Folic acid 2.0mg
Vitamin B_{12} 0.1mg

Preparation of Vitamin Solution: Add components to distilled/deionized water and bring volume to 1.0L. Mix thoroughly.

Glucose Solution:

Composition per 50.0mL:

D-Glucose 5.0g

Preparation of Glucose Solution: Add glucose to distilled/deionized water and bring volume to 50.0mL. Mix thoroughly. Dispense solution anaerobically under 100% N_2 gas. Autoclave for 15 min at 15 psi pressure–121°C.

Reducing Agent Solution:

Composition per 10.0mL:

Dithiothreitol 0.154g

Preparation of Reducing Agent Solution: Add 10.0mL of distilled/deionized water to a flask. Boil under N_2 gas for 1 minute. Cool to room temperature. Add dithiothreitol and dissolve. Autoclave for 10 min at 15 psi pressure–121°C.

Trace Elements Solution SL-10:

Composition per liter:

$FeCl_2 \cdot 4H_2O$ 1.5g
$CoCl_2 \cdot 6H_2O$ 190.0mg
$MnCl_2 \cdot 4H_2O$ 100.0mg
$ZnCl_2$ 70.0mg
$Na_2MoO_4 \cdot 2H_2O$ 36.0mg
$NiCl_2 \cdot 6H_2O$ 24.0mg
H_3BO_3 .... 6.0mg
$CuCl_2 \cdot 2H_2O$ 2.0mg
HCl (25% solution) 10.0mL

Preparation of Trace Elements Solution SL-10: Add $FeCl_2 \cdot 4H_2O$ to 10.0mL of HCl solution. Mix thoroughly. Add distilled/deionized water and bring volume to 1.0L. Add remaining components. Mix thoroughly.

Preparation of Medium: Add components, except $NaHCO_3$, glucose solution, and reducing agent solution, to distilled/deionized water and bring volume to 940.0mL. Gently heat and bring to boiling. Continue boiling for 3 min. Cool to room temperature under 80% N_2 + 20% CO_2. Add $NaHCO_3$ and bring pH to 7.0 by gassing. Distribute anaerobically under 80% N_2 + 20% CO_2 into tubes or flasks. Autoclave for 15 min at 15 psi pressure–121°C. Prior to inoculation of cultures, aseptically and anaerobically add 0.1mL of sterile reducing agent solution and 0.5mL of sterile glucose solution to each tube containing 9.4mL of sterile basal medium.

Use: For the cultivation and maintenance of *Acetonema longum* and *Clostridium mayombei.*

Acetylglucosamine Medium (*N*-Acetylglucosamine Medium)

Composition per liter:

N-Acetylglucosamine 20.0g
Beef extract 10.0g
Peptone 10.0g
Yeast extract 5.0g
K_2HPO_4 .... 2.0g
Triammonium citrate 2.0g
$MgSO_4 \cdot 7H_2O$ 0.2g
$MnSO_4 \cdot 4H_2O$ 0.05g
Tween™ 80 .... 1.0mL

pH 6.2 ± 0.4 at 25°C

Preparation of Medium: Add components to distilled/deionized water and bring volume to 1.0L. Mix thoroughly. Adjust pH to 6.2. Distribute into tubes or flasks. Autoclave for 15 min at 15 psi pressure–121°C.

Use: For the cultivation of bacteria that can utilize *N*-acetylglucosamine.

Acholeplasma Medium (ATCC Medium 1039)

Composition per liter:

Papaic digest of soybean meal 10.0g
Agar 3.0g

PPLO broth without Crystal Violet	900.0mL
Fresh yeast extract solution	100.0mL

pH 7.8 ± 0.2 at 25°C

PPLO Broth without Crystal Violet:

Composition per 900.0mL:

Beef heart, infusion from	225.0g
Peptone	9.0g
NaCl	4.5g

Source: PPLO broth without Crystal Violet is available as a premixed powder from BD Diagnostic Systems.

Preparation of PPLO Broth without Crystal Violet: Add components to distilled/deionized water and bring volume to 900.0mL. Mix thoroughly. Autoclave for 15 min at 15 psi pressure–121°C. Cool to room temperature.

Fresh Yeast Extract Solution:

Composition per 100.0mL:

Baker's yeast, live, pressed, starch-free	25.0g

Preparation of Fresh Yeast Extract Solution: Add the live Baker's yeast to 100.0mL of distilled/deionized water. Mix thoroughly. Autoclave for 90 min at 15 psi pressure–121°C. Allow to stand. Remove supernatant solution. Adjust pH to 6.6–6.8.

Preparation of Medium: Add components to distilled/deionized water and bring volume to 1.0L. Mix thoroughly. Gently heat and bring to boiling. Distribute into test tubes or flasks. Autoclave for 10 min at 15 psi pressure–121°C.

Use: For the cultivation and maintenance of *Acholeplasma* species.

Acholeplasma Medium (ATCC Medium 1215)

Composition per 1020.0mL:

PPLO broth without Crystal Violet	700.0mL
Fetal bovine serum, heat inactivated	100.0mL
Fresh yeast extract solution	100.0mL
Tween™-glucose-BSA solution	100.0mL
Phenol Red (0.1% solution)	20.0mL

PPLO Broth without Crystal Violet:

Composition per 700.0mL:

Beef heart, infusion from	175.0g
Peptone	7.0g
NaCl	3.5g

Source: PPLO broth without Crystal Violet is available as a premixed powder from BD Diagnostic Systems.

Preparation of PPLO Broth without Crystal Violet: Add components to distilled/deionized water and bring volume to 700.0mL. Autoclave for 15 min at 15 psi pressure–121°C. Cool to room temperature.

Fresh Yeast Extract Solution:

Composition per 100.0mL:

Baker's yeast, live, pressed, starch-free	25.0g

Preparation of Fresh Yeast Extract Solution: Add the live Baker's yeast to 100.0mL of distilled/deionized water. Autoclave for 90 min at 15 psi pressure–121°C. Allow to stand. Remove supernatant solution. Adjust pH to 6.6–6.8.

Tween™-Glucose-BSA Solution:

Glucose	2.0g
Tween™ 80	0.1g
Bovine serum albumin, fraction V (1% solution)	100.0mL

Preparation of Tween™-Glucose-BSA Solution: Add glucose and Tween™ 80 to 100.0mL of bovine serum albumin solution and mix thoroughly. Filter sterilize solution through a 0.2μm membrane filter.

Preparation of Medium: Aseptically mix components. Distribute into sterile tubes or flasks.

Use: For the cultivation and maintenance of *Acholeplasma* species.

Achromobacter Choline Medium

Composition per liter:

NaCl	30.0g
Agar	18.0g
Choline chloride	5.0g
K_2HPO_4	1.0g
$MgSO_4 \cdot 7H_2O$	0.5g
$FeSO_4 \cdot 7H_2O$	0.01g

Preparation of Medium: Add components to distilled/deionized water and bring volume to 1.0L. Mix well and warm gently until dissolved. Autoclave for 15 min at 15 psi pressure–121°C. Pour into sterile Petri dishes.

Use: For the cultivation and maintenance of *Achromobacter cholinophagum* and other bacteria that can utilize choline as a carbon source.

Achromobacter Choline Medium, Modified

Composition per liter:

NaCl	30.0g
Agar	15.0g
Choline chloride	5.0g
K_2HPO_4	1.0g
$MgSO_4 \cdot 7H_2O$	1.0g
$FeSO_4 \cdot 7H_2O$	0.018g

pH 7.4 ± 0.2 at 25°C

Preparation of Medium: Add agar, $MgSO_4 \cdot 7H_2O$, and $FeSO_4 \cdot 7H_2O$ to 500.0mL distilled/deionized water. Mix thoroughly. Bring volume to 1.0L with distilled/deionized water. Gently heat and bring to boiling. Add choline chloride. Mix thoroughly. Distribute into tubes or flasks. Autoclave for 15 min at 15 psi pressure–121°C. Pour into sterile Petri dishes or leave in tubes.

Use: For the cultivation and maintenance of *Achromobacter cholinophagum*.

Achromobacter Medium (ATCC Medium 457)

Composition per liter:

K_2HPO_4	7.32g
Ammonium tartrate	4.6g
KH_2PO_4	1.09g
$MgSO_4 \cdot 7H_2O$	0.04g
$FeSO_4 \cdot 7H_2O$	0.04g
$CaCl_2 \cdot 2H_2O$	0.014g
$MgSO_4 \cdot 7H_2O$	0.002g

pH 7.5 ± 0.2 at 25°C

Preparation of Medium: Add components to distilled/deionized water and bring volume to 1.0L. Mix well and warm gently until dissolved. Distribute into test tubes or flasks. Autoclave for 15 min at 15 psi pressure–121°C.

Use: For the cultivation and maintenance of *Achromobacter* species and *Alcaligenes* species.

Achromobacter Medium (ATCC Medium 589)

Composition per liter:

Agar	20.0g
K_2HPO_4	7.0g
Methionine	5.0g
KH_2PO_4	2.0g
$(NH_4)_2SO_4$	1.0g
Sodium citrate	0.4g
$MgSO_4 \cdot 7H_2O$	0.1g

Preparation of Medium: Add components to distilled/deionized water and bring volume to 1.0L. Mix thoroughly. Gently heat and bring to boiling. Autoclave for 15 min at 15 psi pressure–121°C. Pour into sterile Petri dishes.

Use: For the cultivation and maintenance of *Achromobacter* species.

Achromobacter pestifer Medium

Composition per liter:

Agar	15.0g
Yeast extract	12.5g
Beef extract	10.0g
Peptone	10.0g
NaCl	5.0g

pH 7.2 ± 0.2 at 25°C

Preparation of Medium: Add components to distilled/deionized water and bring volume to1.0L. Mix thoroughly. Gently heat and bring to boiling. Distribute into tubes or flasks. Autoclave for 15 min at 15 psi pressure–121°C.

Use: For the cultivation and maintenance of *Achromobacter pestifer.*

Acid Bismuth Yeast Agar

See: **ABY Agar**

Acid Broth

Composition per liter:

Glucose	5.0g
Proteose peptone	5.0g
Yeast extract	5.0g
K_2HPO_4	4.0g

pH 5.0 ± 0.2 at 25°C

Preparation of Medium: Add components to distilled/deionized water and bring volume to 1.0L. Mix thoroughly. Distribute into tubes or flasks. Autoclave for 15 min at 15 psi pressure–121°C.

Use: For the isolation of bacteria from canned foods.

Acid Broth

Composition per liter:

Invert sugar	10.0g
Peptic digest of animal tissue	10.0g
Yeast extract	7.5g

pH 4.0 ± 0.2 at 25°C

Source: This medium is available as a premixed powder from HiMedia.

Preparation of Medium: Add components to distilled/deionized water and bring volume to 1.0L. Mix thoroughly. Distribute into tubes or flasks. Autoclave for 15 min at 15 psi pressure–121°C.

Use: For the isolation of bacteria from canned foods.

Acid Egg Medium

Composition per 1640.0mL:

Potato starch	30.0g
KH_2PO_4	12.3g
Malachite Green	0.4g
$MgSO_4 \cdot 7H_2O$	0.3g
Penicillin G	100,000IU
Fresh egg mixture	1.0L
Glycerol	12.0mL

Source: This medium is available as a prepared medium from Oxoid Unipath.

Preparation of Medium: Add components to 1.0L of fresh egg mixture. Mix thoroughly. Gently heat and bring to boiling. Bring volume to 1640.0mL with distilled/deionized water. Distribute into tubes or flasks. Autoclave for 15 min at 15 psi pressure–121°C with tubes in an upright position.

Use: For the cultivation and maintenance of *Mycobacterium tuberculosis*.

Acid Glucose Salts Medium

Composition per liter:

Glucose	5.0g
$MgSO_4 \cdot 7H_2O$	0.5g
$(NH_4)_2SO_4$	0.15g
KH_2PO_4	0.1g
KCl	50.0mg
$Ca(NO_3)_2$	10.0mg

pH 3.0 ± 0.2 at 25°C

Preparation of Medium: Add components to distilled/deionized water and bring volume to 1.0L. Mix thoroughly. Distribute into tubes or flasks. Autoclave for 15 min at 15 psi pressure–121°C.

Use: For the cultivation of *Thiobacillus organoparus.*

Acid HiVeg Broth

Sucrose	10.0g
Plant peptone	10.0g
Yeast extract	7.5g

pH 4.0 ± 0.2 at 25°C

Source: This medium is available as a premixed powder from HiMedia.

Preparation of Medium: Add components to distilled/deionized water and bring volume to 1.0L. Mix thoroughly. Distribute into tubes or flasks. Autoclave for 15 min at 15 psi pressure–121°C. Aseptically adjust pH to 4.0.

Use: For the isolation of acid tolerant bacteria from canned foods.

Acid Products Test Broth

Composition per liter:

Invert sugar	10.0g
Peptone	10.0g
Yeast extract	7.5g

pH 4.0 ± 0.2 at 25°C

Preparation of Medium: Add components to distilled/deionized water and bring volume to 1.0L. Mix thoroughly. Gently heat while stirring and bring to boiling. Cool to 25°C. Adjust pH to 4.0 with 25% tartaric acid solution. Distribute into screw-capped flasks in 300.0mL volumes. Autoclave for 15 min at 15 psi pressure–121°C.

Use: For the cultivation of acid tolerant microorganisms from foods. For the sterility testing of canned foods.

Acid Rhodospirillaceae Medium

Composition per 1050.0 mL:

Ammonium acetate	1.5g
KH_2PO_4	0.5g
$MgSO_4.7H_2O$	0.4g
NaCl	0.4g
NH_4Cl	0.4g
Disodium succinate	0.25g
Yeast extract	0.2g
$CaCl_2{\cdot}2H_2O$	0.05g
Ferric citrate solution	5.0mL
Trace elements solution SL-6	1.0mL
Vitamin B_{12} solution	0.4mL
Neutralized sulfide solution	variable

pH 5.7 ± 0.2 at 25°C

Ferric Citrate Solution:

Composition per 10.0mL:

Ferric citrate	10.0mg

Preparation of Ferric Citrate Solution: Add ferric citrate to distilled/deionized water and bring volume to 10.0mL. Mix thoroughly. Sparge under 100% N_2 gas for 3 min. Autoclave for 15 min at 15 psi pressure–121°C. Store under N_2 gas.

Trace Elements Solution SL-6:

Composition per liter:

$MnCl_2{\cdot}4H_2O$	0.5g
H_3BO_3	0.3g
$CoCl_2{\cdot}6H_2O$	0.2g
$ZnSO_4{\cdot}7H_2O$	0.1g
$Na_2MoO_4{\cdot}2H_2O$	0.03g
$NiCl_2{\cdot}6H_2O$	0.02g
$CuCl_2{\cdot}2H_2O$	0.01g

Preparation of Trace Elements Solution SL-6: Add components to distilled/deionized water and bring volume to 1.0L. Mix thoroughly.

Vitamin B_{12} Solution:

Composition per 100.0mL:

Vitamin B_{12}	10.0mg

Preparation of Vitamin B_{12} Solution: Add vitamin B_{12} to distilled/deionized water and bring volume to 100.0mL. Mix thoroughly. Sparge under 100% N_2 gas for 3 min. Autoclave for 15 min at 15 psi pressure–121°C. Store under N_2 gas.

Neutralized Sulfide Solution:

Composition per 100.0mL:

$Na_2S{\cdot}9H_2O$	1.5g

Preparation of Neutralized Sulfide Solution: Add $Na_2S{\cdot}9H_2O$ to distilled/deionized water in a 250.0mL screw-capped bottle fitted with a butyl rubber septum and bring volume to 100.0mL. Add a magnetic stir bar. Mix thoroughly. Sparge under 100% N_2 gas for 3 min. Autoclave for 15 min at 15 psi pressure–121°C. Cool to room temperature. Adjust pH to about 7.3 with sterile 2*M* H_2SO_4. Do not open the bottle to add H_2SO_4; use a sterile syringe. Stir the solution continuously to avoid precipitation of elemental sulfur. The final solution should be clear and yellow in color.

Preparation of Medium: Add components, except neutralized sulfide solution, to distilled/deionized water and bring volume to 1050.0mL. Mix thoroughly. Gently heat and bring to boiling. Boil for 3–4 min under a stream of 100% N_2. Distribute 45.0mL of the prepared medium into 50.0mL screw-capped tubes that have been flushed with 100% N_2. Autoclave for 15 min at 15 psi pressure–121°C. Cool to room temperature. Before inoculation, aseptically and anaerobically add 0.25–0.50mL of neutralized sulfide solution.

Use: For the cultivation and maintenance of members of the family Rhodospirillaceae, including *Rhodomicrobium vannielii* and *Rhodopseudomonas acidophila.*

Acid Tomato Broth

Composition per liter:

Glucose	10.0g
Peptone	10.0g
Yeast extract	5.0g
$MgSO_4{\cdot}7H_2O$	0.2g
$MnSO_4{\cdot}4H_2O$	0.05g
Tomato juice	250.0mL
L-Cysteine solution	0.5mL

L-Cysteine Solution:

Composition per 10.0mL:

L-Cysteine	0.1g

Preparation of L-Cysteine Solution: Add 0.1g of L-cysteine to distilled/deionized water and bring volume to 10.0mL. Mix thoroughly. Filter sterilize.

Preparation of Medium: Add components, except L-cysteine solution, to distilled/deionized water and bring volume to 999.5mL. Mix thoroughly. Adjust pH to 4.8. Autoclave for 15 min at 15 psi pressure–121°C. Aseptically add 0.5mL of sterile L-cysteine solution. Mix thoroughly. Aseptically distribute into sterile tubes or flasks.

Use: For the cultivation of a variety of fungi.

Acidaminobacter Medium

Composition per liter:

$NaHCO_3$	2.0g
Glycine	1.5g
NaCl	1.2g
KCl	0.4g
$MgCl_2{\cdot}6H_2O$	0.4g
$Na_2S{\cdot}9H_2O$	0.3g
KH_2PO_4	0.2g
Na_2SO_4	0.2g
Yeast extract	0.2g
$CaCl_2{\cdot}2H_2O$	0.15g
Resazurin	1.0mg
$Na_2SeO_3{\cdot}5H_2O$	30.0μg
Vitamin solution	10.0mL
$NaHCO_3$ solution	5.0mL
$Na_2S{\cdot}9H_2O$ solution	5.0mL
Trace elements solution SL-10	1.0mL

pH 7.4 ± 0.2 at 25°C

Trace Elements Solution SL-10:

Composition per liter:

$FeCl_2{\cdot}4H_2O$	1.5g
$CoCl_2{\cdot}6H_2O$	190.0mg

$MnCl_2 \cdot 4H_2O$	100.0mg
$ZnCl_2$	70.0mg
$Na_2MoO_4 \cdot 2H_2O$	36.0mg
$NiCl_2 \cdot 6H_2O$	24.0mg
H_3BO_3	6.0mg
$CuCl_2 \cdot 2H_2O$	2.0mg
HCl (25% solution)	10.0mL

Preparation of Trace Elements Solution SL-10: Add $FeCl_2 \cdot 4H_2O$ to 10.0mL of HCl solution. Mix thoroughly. Add distilled/deionized water and bring volume to 1.0L. Add remaining components. Mix thoroughly.

Vitamin Solution:

Composition per liter:

Pyridoxine·HCl	10.0mg
Calcium DL-pantothenate	5.0mg
Lipoic acid	5.0mg
Nicotinic acid	5.0mg
p-Aminobenzoic acid	5.0mg
Riboflavin	5.0mg
Thiamine·HCl	5.0mg
Biotin	2.0mg
Folic acid	2.0mg
Vitamin B_{12}	0.1mg

Preparation of Vitamin Solution: Add components to distilled/deionized water and bring volume to 1.0L. Mix thoroughly. Filter sterilize.

$NaHCO_3$ Solution:

Composition per 5.0mL:

$NaHCO_3$	0.5g

Preparation of $NaHCO_3$ Solution: Add $NaHCO_3$ to distilled/deionized water and bring volume to 5.0mL. Mix thoroughly. Sparge under 100% N_2 gas for 3 min. Autoclave for 15 min at 15 psi pressure–121°C. Store under N_2 gas.

$Na_2S \cdot 9H_2O$ Solution:

Composition per 5.0mL:

$Na_2S \cdot 9H_2O$	0.5g

Preparation of $Na_2S \cdot 9H_2O$ Solution: Add $Na_2S \cdot 9H_2O$ to distilled/deionized water and bring volume to 5.0mL. Mix thoroughly. Sparge under 100% N_2 gas for 3 min. Autoclave for 15 min at 15 psi pressure–121°C. Store under N_2 gas.

Preparation of Medium: Add components, except vitamin solution, $NaHCO_3$ solution, and $Na_2S \cdot 9H_2O$ solution, to distilled/deionized water and bring volume to 1.0L. Mix thoroughly. Gently heat and bring to boiling. Boil for a few minutes. Allow to cool to room temperature under 100% N_2. Distribute into tubes or flasks under 100% N_2. Autoclave for 15 min at 15 psi pressure–121°C. Cool to room temperature. Before inoculation, aseptically and anaerobically add vitamin solution, $NaHCO_3$ solution, and $Na_2S \cdot 9H_2O$ solution. Mix thoroughly. Check that final pH is 7.4.

Use: For the cultivation and maintenance of *Acidaminobacter hydrogenoformans.*

Acidaminococcus fermentans Medium

Composition per liter:

Casamino acids	10.0g
Glucose	5.0g
Pancreatic digest of casein	5.0g
Yeast extract	5.0g
Sodium glutamate	4.0g
KH_2PO_4	2.0g
Arginine	1.0g
Glycine	1.0g
L-Cysteine·HCl	0.5g
DL-Tryptophan	0.1g
Tween™ 80	0.5mL

pH 7.0 ± 0.2 at 25°C

Preparation of Medium: Add components to distilled/deionized water and bring volume to 1.0L. Mix thoroughly. Adjust pH to 7.0. Distribute into tubes or flasks. Autoclave for 15 min at 15 psi pressure–121°C.

Use: For the cultivation and maintenance of *Acidaminococcus fermentans.*

Acidaminococcus Medium VR

Composition per liter:

Acid-hydrolyzed casein (vitamin and salt free)	20.0g
Glucose	5.0g
L-Cysteine·HCl·H_2O	0.35g
DL-Tryptophan	0.1g
Guanine	0.01g
Uracil	0.01g
Hypoxanthine	0.01g
Pyridoxal	1.0mg
Calcium pantothenate	1.0mg
Thiamine	50.0μg
Niacin	50.0μg
Riboflavin	50.0μg
p-Aminobenzoic acid	10.0μg
Biotin	2.0μg
Folic acid	1.0μg
Vitamin B_{12}	1.0μg
VR salts A	30.0mL
VR salts B	4.0mL

pH 7.0 ± 0.2 at 25°C

VR Salts A:

Composition per 500.0mL:

Na_2HPO_4	37.5g
KH_2PO_4	12.5g

Preparation of VR Salts A: Add components to distilled/deionized water and bring volume to 500.0mL. Mix thoroughly.

VR Salts B:

Composition per liter:

$MgSO_4 \cdot 7H_2O$	24.0g
$CaCl_2 \cdot 2H_2O$	0.5g
$FeSO_4 \cdot 7H_2O$	0.5g
$ZnSO_4 \cdot 7H_2O$	0.25g
$MnSO_4 \cdot H_2O$	0.25g
$CoCl_2 \cdot 6H_2O$	0.25g
$VSO_4 \cdot 7H_2O$	0.25g
$Na_2MoO_4 \cdot 2H_2O$	0.25g
$CuSO_4 \cdot 5H_2O$	0.125g

Preparation of VR Salts B: Add components to distilled/deionized water and bring volume to 700.0mL. Add 2.0mL of concentrated HCl and heat until dissolved. Add 5.0g of nitrilotriacetic acid to 300.0mL distilled/deionized water. Adjust pH with 10*N* NaOH to 7.0. Stir vigorously and slowly add the nitrilotriacetic acid solution to the larger volume of salt solution until dissolved. Add distilled/deionized water and bring volume to 1.0L. Filter through paper. Store in a cool place.

Preparation of Medium: Filter sterilize vitamins as separate solution. Add aseptically to sterile basal medium. If necessary, adjust pH with solid K_2CO_3 to 7.0. Prepare and distribute medium anaerobically using Hungate techniques with 100% N_2 gas.

Use: For the cultivation and maintenance of *Acidaminococcus fermentans*.

Acidianus brierleyi Medium

Composition per liter:

Sulfur flowers	10.0g
$(NH_4)_2SO_4$	3.0g
$K_2HPO_4 \cdot 3H_2O$	0.5g
$MgSO_4 \cdot 7H_2O$	0.5g
KCl	0.1g
$Ca(NO_3)_2$	0.01g
Yeast extract solution	10.0mL

pH 1.5–2.5 at 25°C

Yeast Extract Solution:

Composition per 10.0mL:

Yeast extract	0.2g

Preparation of Yeast Extract Solution: Add yeast extract to distilled/deionized water and bring volume to 10.0mL. Mix thoroughly. Autoclave for 15 min at 15 psi pressure–121°C.

Preparation of Medium: Add components, except sulfur flowers and yeast extract solution, to distilled/deionized water and bring volume to 990.0mL. Mix thoroughly. Gently heat and bring to boiling. Autoclave for 15 min at 15 psi pressure–121°C. Sulfur flowers are sterilized separately by steaming for 3 hr on 3 consecutive days. Aseptically combine the basal solution, sterile sulfur flowers, and sterile yeast extract solution. Adjust pH to 1.5–2.5 with 6*N* H_2SO_4.

Use: For the cultivation and maintenance of *Acidianus brierleyi.*

Acidianus infernus Medium

Composition per liter:

$(NH_4)_2SO_4$	1.3g
Yeast extract	1.0g
Sulfur flowers	1.0g
KH_2PO_4	0.28g
$MgSO_4 \cdot 7H_2O$	0.25g
$CaCl_2 \cdot 2H_2O$	0.07g
$FeCl_3 \cdot 6H_2O$	0.02g
$Na_2B_4O_7 \cdot 10H_2O$	4.5mg
$MnCl_2 \cdot 4H_2O$	1.8mg
$ZnSO_4 \cdot 7H_2O$	0.22mg
$CuCl_2 \cdot 2H_2O$	0.05mg
$Na_2MoO_4 \cdot 2H_2O$	0.03mg
$VOSO_4 \cdot 2H_2O$	0.03mg
$CoSO_4$	0.01mg

pH 2.5 ± 0.2 at 25°C

Preparation of Medium: Add components to distilled/deionized water and bring volume to 1.0L. Mix thoroughly. Adjust pH to 2.5 with 10*N* H_2SO_4. Distribute into tubes or flasks. Autoclave for 15 min at 15 psi pressure–121°C.

Use: For the aerobic cultivation and maintenance of *Acidianus infernus, Acidianus brierleyi,* and *Desulfurolobus ambivalens.*

Acidianus infernus Medium

Composition per liter:

$(NH_4)_2SO_4$	1.3g
Sulfur flowers	1.0g
KH_2PO_4	0.28g
$MgSO_4 \cdot 7H_2O$	0.25g
$CaCl_2 \cdot 2H_2O$	0.07g
$FeCl_3 \cdot 6H_2O$	0.02g
$Na_2B_4O_7 \cdot 10H_2O$	4.5mg
$MnCl_2 \cdot 4H_2O$	1.8mg
Resazurin	1.0mg
$ZnSO_4 \cdot 7H_2O$	0.22mg
$CuCl_2 \cdot 2H_2O$	0.05mg
$Na_2MoO_4 \cdot 2H_2O$	0.03mg
$VOSO_4 \cdot 2H_2O$	0.03mg
$CoSO_4$	0.01mg
Yeast extract solution	10.0mL

pH 2.5 ± 0.2 at 25°C

Yeast Extract Solution:

Composition per 10.0mL:

Yeast extract	0.5g

Preparation of Yeast Extract Solution: Add yeast extract to distilled/deionized water and bring volume to 10.0mL. Mix thoroughly. Autoclave for 15 min at 15 psi pressure–121°C.

Preparation of Medium: Add components, except sulfur flowers and yeast extract solution, to distilled/deionized water and bring volume to 990.0mL. Mix thoroughly. Gently heat and bring to boiling. Allow to cool under 80% N_2 + 20% CO_2. Autoclave for 15 min at 15 psi pressure–121°C. Sulfur flowers are sterilized separately by steaming for 3 hr on 3 consecutive days. Aseptically and anaerobically combine the basal solution, sterile sulfur flowers, and sterile yeast extract solution. Adjust pH to 2.5 with 6*N* H_2SO_4. Pressurize the culture bottles to 100kPa with 80% N_2 + 20% CO_2.

Use: For the anaerobic cultivation and maintenance of *Acidianus infernus, Acidianus brierleyi,* and *Desulfurolobus ambivalens.*

Acidianus infernus Medium

Composition per liter:

Sulfur flowers	5.0g
$(NH_4)_2SO_4$	1.3g
Yeast extract	1.0g
KH_2PO_4	0.28g
$MgSO_4 \cdot 7H_2O$	0.25g
$CaCl_2 \cdot 2H_2O$	0.07g
$FeCl_3 \cdot 6H_2O$	0.02g
$Na_2B_4O_7 \cdot 10H_2O$	4.5mg
$MnCl_2 \cdot 4H_2O$	1.8mg
$ZnSO_4 \cdot 7H_2O$	0.22mg
$CuCl_2 \cdot 2H_2O$	0.05mg
$Na_2MoO_4 \cdot 2H_2O$	0.03mg
$VOSO_4 \cdot 2H_2O$	0.03mg
$CoSO_4$	0.01mg

pH 2.0–2.5 at 25°C

Preparation of Medium: Add components to distilled/deionized water and bring volume to 1.0L. Mix thoroughly. Adjust pH to 2.0–2.5 with 10*N* H_2SO_4. Distribute into tubes or flasks. Autoclave for 15 min at 15 psi pressure–121°C.

Use: For the aerobic cultivation and maintenance of *Acidianus brierleyi, Acidianus infernus,* and *Desulfurolobus ambivalens.*

Acidianus infernus Medium

Composition per liter:

Sulfur flowers	5.0g
$(NH_4)_2SO_4$	1.3g
KH_2PO_4	0.28g
$MgSO_4·7H_2O$	0.25g
$CaCl_2·2H_2O$	0.07g
$FeCl_3·6H_2O$	0.02g
$Na_2B_4O_7·10H_2O$	4.5mg
$MnCl_2·4H_2O$	1.8mg
Resazurin	1.0mg
$ZnSO_4·7H_2O$	0.22mg
$CuCl_2·2H_2O$	0.05mg
$Na_2MoO_4·2H_2O$	0.03mg
$VOSO_4·2H_2O$	0.03mg
$CoSO_4$	0.01mg
Yeast extract solution	10.0mL

pH 2.5 ± 0.2 at 25°C

Yeast Extract Solution:

Composition per 10.0mL:

Yeast extract	0.2g

Preparation of Yeast Extract Solution: Add yeast extract to distilled/deionized water and bring volume to 10.0mL. Mix thoroughly. Autoclave for 15 min at 15 psi pressure–121°C.

Preparation of Medium: Add components, except sulfur flowers and yeast extract solution, to distilled/deionized water and bring volume to 990.0mL. Mix thoroughly. Gently heat and bring to boiling. Allow to cool under 80% N_2 + 20% CO_2. Autoclave for 15 min at 15 psi pressure–121°C. Sulfur flowers are sterilized separately by steaming for 3 hr on 3 consecutive days. Aseptically and anaerobically combine the basal solution, sterile sulfur flowers, and sterile yeast extract solution. Adjust pH to 2.5 with 6*N* H_2SO_4. Pressurize the culture bottles to 200kPa with 80% N_2 + 20% CO_2.

Use: For the anaerobic cultivation and maintenance of *Acidianus brierleyi, Acidianus infernus,* and *Desulfurolobus ambivalens.*

Acidianus infernus Medium

Composition per liter:

$(NH_4)_2SO_4$	1.3g
Sulfur flowers	1.0g
KH_2PO_4	0.28g
$MgSO_4·7H_2O$	0.25g
$CaCl_2·2H_2O$	0.07g
$FeCl_3·6H_2O$	0.02g
Yeast extract	0.02g
$Na_2B_4O_7·10H_2O$	4.5mg
$MnCl_2·4H_2O$	1.8mg
$ZnSO_4·7H_2O$	0.22mg
$CuCl_2·2H_2O$	0.05mg
$Na_2MoO_4·2H_2O$	0.03mg
$VOSO_4·2H_2O$	0.03mg
$CoSO_4$	0.01mg

pH 2.5 ± 0.2 at 25°C

Preparation of Medium: Add components to distilled/deionized water and bring volume to 1.0L. Mix thoroughly. Adjust pH to 2.5 with 10*N* H_2SO_4. Distribute into tubes or flasks. Autoclave for 15 min at 15 psi pressure–121°C.

Use: For the aerobic cultivation and maintenance of *Desulfurolobus ambivalens, Acidianus brierleyi, and Acidianus infernus.*

Acidianus infernus Medium

Composition per liter:

$(NH_4)_2SO_4$	1.3g
Sulfur flowers	1.0g
KH_2PO_4	0.28g
$MgSO_4·7H_2O$	0.25g
$CaCl_2·2H_2O$	0.07g
$FeCl_3·6H_2O$	0.02g
$Na_2B_4O_7·10H_2O$	4.5mg
$MnCl_2·4H_2O$	1.8mg
Resazurin	1.0mg
$ZnSO_4·7H_2O$	0.22mg
$CuCl_2·2H_2O$	0.05mg
$Na_2MoO_4·2H_2O$	0.03mg
$VOSO_4·2H_2O$	0.03mg
$CoSO_4$	0.01mg
Yeast extract solution	10.0mL

pH 2.5 ± 0.2 at 25°C

Yeast Extract Solution:

Composition per 10.0mL:

Yeast extract	2.0mg

Preparation of Yeast Extract Solution: Add yeast extract to distilled/deionized water and bring volume to 10.0mL. Mix thoroughly. Autoclave for 15 min at 15 psi pressure–121°C.

Preparation of Medium: Add components, except sulfur flowers and yeast extract solution, to distilled/deionized water and bring volume to 990.0mL. Mix thoroughly. Gently heat and bring to boiling. Allow to cool under 80% N_2 + 20% CO_2. Autoclave for 15 min at 15 psi pressure–121°C. Sulfur flowers are sterilized separately by steaming for 3 hr on 3 consecutive days. Aseptically and anaerobically combine the basal solution, sterile sulfur flowers, and sterile yeast extract solution. Adjust pH to 2.5 with 6*N* H_2SO_4. Pressurize the culture bottles to 100kPa with 80% N_2 + 20% CO_2.

Use: For the anaerobic cultivation and maintenance of *Acidianus brierleyi, Acidianus infernus,* and *Desulfurolobus ambivalens.*

Acidicaldus Medium (DSMZ Medium 1038)

Composition per liter:

$MgSO_4·7H_2O$	0.5g
$(NH_4)_2SO_4$	0.45g
KCl	0.05g
KH_2PO_4	0.05g
$Ca(NO_3)_2·4H_2O$	14.0mg
Glucose solution	10.0mL
Yeast extract solution	10.0mL

pH 2.5 ± 0.2 at 25°C

Glucose Solution:

Composition per 10.0mL:

Glucose	1.0g

Preparation of Glucose Solution: Add components to distilled/deionized water and bring volume to 10.0mL. Mix thoroughly. Autoclave for 15 min at 15 psi pressure–121°C. Cool to room temperature.

Yeast Extract Solution:

Composition per 10.0mL:

Yeast extract	0.2g

Preparation of Yeast Extract Solution: Add components to distilled/deionized water and bring volume to 10.0mL. Mix thoroughly.

Autoclave for 15 min at 15 psi pressure–121°C. Cool to room temperature.

Preparation of Medium: Add components to distilled/deionized water and bring volume to 980.0mL. Mix thoroughly. Gently heat and bring to boiling. Distribute into tubes or flasks. Autoclave for 15 min at 15 psi pressure–121°C. Cool to room temperature. Adjust pH to 2.5. Aseptically add 10.0mL sterile glucose solution and 10.0mL sterile yeast extract solution. Mix thoroughly. Aseptically distribute into sterile tubes or flasks.

Use: For the cultivation and maintenance of *Acidicaldus organivorans.*

Acidic Rhodospirillaceae Medium

Composition per 1006.0mL:

Disodium succinate	1.0g
KH_2PO_4	0.5g
$MgSO_4·7H_2O$	0.4g
NaCl	0.4g
NH_4Cl	0.4g
Yeast extract	0.2g
$CaCl_2·H_2O$	50.0mg
Ferric citrate solution	5.0mL
Trace elements solution	1.0mL

Ferric Citrate Solution:

Composition per 100.0mL:

Ferric citrate	0.1g

Preparation of Ferric Citrate Solution: Add ferric citrate to distilled/deionized water and bring volume to 100.0mL. Mix thoroughly. Filter sterilize.

Trace Elements Solution:

Composition per liter:

H_3BO_3	0.3g
$CoCl_2·6H_2O$	0.2g
$ZnSO_4·7H_2O$	0.1g
$MnCl_2·4H_2O$	0.03g
$Na_2MoO_4·2H_2O$	0.03g
$NiCl_2·6H_2O$	0.02g
$CuCl_2·2H_2O$	0.01g

Preparation of Trace Elements Solution: Add components to distilled/deionized water and bring volume to 1.0L. Mix thoroughly. Filter sterilize.

Preparation of Medium: Add components, except ferric citrate solution and trace elements solution, to distilled/deionized water and bring volume to 1.0L. Mix thoroughly. Autoclave for 15 min at 15 psi pressure–121°C. Aseptically add 5.0mL of sterile ferric citrate solution and 1.0mL of sterile trace elements solution. Mix thoroughly. Aseptically distribute into sterile tubes or flasks.

Use: For the cultivation of *Rhodopseudomonas acidophila.*

Acidic Tomato Medium for *Leuconostoc*

Composition per liter:

Agar	15.0g
Glucose	10.0g
Peptone	10.0g
Yeast extract	5.0g
$MgSO_4·7H_2O$	0.2g
$MnSO_4·4H_2O$	0.05g
Tomato juice	250.0mL

pH 4.8 ± 0.2 at 25°C

Preparation of Medium: Add solid components to 750.0mL of distilled/deionized water. Add tomato juice. Mix well and warm gently until dissolved. Autoclave for 15 min at 15 psi pressure–121°C. Pour into sterile Petri dishes.

Use: For the cultivation and maintenance of *Leuconostoc oenos* and other *Leuconostoc* species.

Acidified Potato Dextrose Agar (APDA)

Composition per liter:

Glucose	20.0g
Agar	15.0g
Potatoes, infusion from	500.0mL
Lactic acid solution	5.0mL

pH 5.6 ± 0.2 at 25°C

Potato Infusion:

Composition per 500.0mL:

Potatoes	300.0g

Preparation of Potato Infusion: Peel and dice potatoes. Add 500.0mL of distilled/deionized water. Gently heat and bring to boiling. Continue boiling for 30 min. Filter through cheesecloth. Reserve filtrate.

Lactic Acid Solution:

Composition per 10.0mL:

Lactic acid	2.5g

Preparation of Lactic Acid Solution: Add lactic acid to distilled/deionized water and bring volume to 10.0mL. Mix thoroughly. Filter sterilize.

Preparation of Medium: Add components, except lactic acid solution, s to distilled/deionized water and bring volume to 1.0L. Mix thoroughly. Gently heat and bring to boiling. Distribute into tubes or flasks. Autoclave for 20 min at 15 psi pressure–121°C. Cool to 50°C. Add 5.0mL of lactic acid solution. Mix thoroughly. Pour into sterile Petri dishes or leave in tubes.

Use: For the isolation, cultivation, and identification of oak wilt fungi.

Acidimicrobium Medium

Composition per liter:

$MgSO_4·7H_2O$	0.5g
$(NH_4)_2SO_4$	0.4g
K_2HPO_4	0.2g
KCl	0.1g
$FeSO_4·7H_2O$	10.0mg
Yeast extract solution	20.0mL

pH 2.0 ± 0.2 at 25°C

Yeast Extract Solution:

Composition per 20.0mL:

Yeast extract	10.0g

Preparation of Yeast Extract Solution: Add yeast extract to distilled/deionized water and bring volume to 20.0mL. Mix thoroughly. Autoclave for 15 min at 15 psi pressure–121°C.

Preparation of Medium: Add components, except yeast extract solution, to distilled/deionized water and bring volume to 980.0mL. Mix thoroughly. Adjust pH to 2.0 with H_2SO_4. Autoclave for 15 min at 15 psi pressure–121°C. Aseptically add 20.0mL of sterile yeast extract solution. Mix thoroughly. Aseptically distribute into sterile tubes or flasks.

Use: For the heterotrophic cultivation of *Sulfobacillus acidophilus.*

Acidimicrobium Medium

Composition per liter:

$FeSO_4 \cdot 7H_2O$	13.9g
$MgSO_4 \cdot 7H_2O$	0.5g
$(NH_4)_2SO_4$	0.4g
K_2HPO_4	0.2g
KCl	0.1g

pH 1.7 ± 0.2 at 25°C

Preparation of Medium: Add components to distilled/deionized water and bring volume to 1.0L. Mix thoroughly. Adjust pH to 1.7 with H_2SO_4. Distribute into tubes or flasks. Autoclave for 15 min at 15 psi pressure–121°C.

Use: For the autotrophic cultivation of *Sulfobacillus acidophilus.*

Acidiphilium Medium

Composition per liter:

$(NH_4)_2SO_4$	2.0g
K_2HPO_4	0.5g
$MgSO_4 \cdot 7H_2O$	0.5g
KCl	0.1g
Glucose solution	10.0mL
Yeast extract solution	10.0mL

pH 3.0 ± 0.2 at 25°C

Glucose Solution:

Composition per 10.0mL:

D-Glucose	1.0g

Preparation of Glucose Solution: Add glucose to distilled/deionized water and bring volume to 10.0mL. Mix thoroughly. Filter sterilize.

Yeast Extract Solution:

Composition per 10.0mL:

Yeast extract	0.3g

Preparation of Yeast Extract Solution: Add yeast extract to distilled/deionized water and bring volume to 10.0mL. Mix thoroughly. Autoclave for 15 min at 15 psi pressure–121°C.

Preparation of Medium: Add components, except glucose solution and yeast extract solution, to distilled/deionized water and bring volume to 1080.0mL. Mix thoroughly. Gently heat and bring to boiling. Adjust pH to 3.0 using 1*N* H_2SO_4. Autoclave for 15 min at 15 psi pressure–121°C. Cool to room temperature. Before inoculation, aseptically add glucose solution and yeast extract solution. Mix thoroughly.

Use: For the cultivation and maintenance of *Acidiphilium cryptum.*

Acidobacterium Medium

Composition per liter:

$(NH_4)_2SO_4$	2.0g
Glucose	1.0g
K_2HPO_4	0.5g
$MgSO_4 \cdot 7H_2O$	0.5g
KCl	0.1g
Yeast extract	0.1g

pH 3.5 ± 0.2 at 25°C

Preparation of Medium: Add components to distilled/deionized water and bring volume to 1.0L. Mix thoroughly. Adjust pH to 3.5 with H_2SO_4. Distribute into tubes or flasks. Autoclave for 15 min at 15 psi pressure–121°C.

Use: For the cultivation of *Acidobacterium capsulatum.*

Acidolobus aceticus Medium (DSMZ Medium 901)

Composition per 1055mL:

Sulfur, powdered	10.0g
NH_4Cl	0.33g
KCl	0.33g
KH_2PO_4	0.33g
$MgCl_2 \cdot 6H_2O$	0.33g
$CaCl_2 \cdot 2H_2O$	0.33g
Resazurin	0.5mg
Yeast extract solution	30.0mL
$Na_2S \cdot 9H_2O$ solution	15.0mL
Vitamin solution	10.0mL
Trace elements solution SL-10	1.0mL

pH 3.5–3.8 at 25°C

Trace Elements Solution SL-10:

Composition per liter:

$FeCl_2 \cdot 4H_2O$	1.5g
$CoCl_2 \cdot 6H_2O$	190.0mg
$MnCl_2 \cdot 4H_2O$	100.0mg
$ZnCl_2$	70.0mg
$Na_2MoO_4 \cdot 2H_2O$	36.0mg
$NiCl_2 \cdot 6H_2O$	24.0mg
H_3BO_3	6.0mg
$CuCl_2 \cdot 2H_2O$	2.0mg
HCl (25% solution)	10.0mL

Preparation of Trace Elements Solution SL-10: Add $FeCl_2 \cdot 4H_2O$ to 10.0mL of HCl solution. Mix thoroughly. Add distilled/deionized water and bring volume to 1.0L. Add remaining components. Mix thoroughly. Sparge with 80% N_2 + 20% CO_2. Filter sterilize.

Vitamin Solution:

Composition per liter:

Pyridoxine-HCl	10.0mg
Thiamine-HCl·$2H_2O$	5.0mg
Riboflavin	5.0mg
Nicotinic acid	5.0mg
D-Ca-pantothenate	5.0mg
p-Aminobenzoic acid	5.0mg
Lipoic acid	5.0mg
Biotin	2.0mg
Folic acid	2.0mg
Vitamin B_{12}	0.1mg

Preparation of Vitamin Solution: Add components to distilled/deionized water and bring volume to 1.0L. Mix thoroughly. Sparge with 80% H_2 + 20% CO_2. Filter sterilize.

$Na_2S \cdot 9H_2O$ Solution:

Composition per 20.0mL:

$Na_2S \cdot 9H_2O$	0.6g

Preparation of $Na_2S \cdot 9H_2O$ Solution: Add $Na_2S \cdot 9H_2O$ to distilled/deionized water and bring volume to 20.0mL. Mix thoroughly. Autoclave under 100% N_2 for 15 min at 15 psi pressure–121°C. Cool to room temperature. Neturalize to pH 7.0 with HCl.

Yeast Extract Solution:

Composition per 30.0mL:

Yeast extract	3.0g

Preparation of Yeast Extract Solution: Add yeast extract to distilled/deionized water and bring volume to 30.0mL. Mix thoroughly.

Sparge with 100% N_2. Autoclave under 100% N_2 for 15 min at 15 psi pressure–121°C. Cool to room temperature.

Preparation of Medium: Prepare and dispense medium under 100% CO_2. Add components, except vitamin solution, $Na_2S{\cdot}9H_2O$ solution, and yeast extract solution, to distilled/deionized water and bring volume to 1.0L. Mix thoroughly. Adjust pH to 3.5 with H_2SO_4. Distribute to anaerobe tubes or bottles. Heat to 90°C for 1 hr on each of 3 successive days. Aseptically and anaerobically add, per liter of medium, 10.0mL sterile vitamin solution, 15.0mL of sterile $Na_2S{\cdot}9H_2O$ solution, and 30.0mL sterile yeast extract solution. Mix thoroughly. The final pH should be 3.5–3.8.

Use: For the cultivation of *Acidilobus aceticus.*

Acidomonas Agar

Composition per liter:

Solution A	500.0mL
Solution B	500.0mL

Solution A:

Composition per 500.0mL:

Glucose	10.0g
Peptone	5.0g
Malt extract	3.0g
Yeast extract	3.0g

Preparation of Solution A: Add components to distilled/deionized water and bring volume to 1.0L. Mix thoroughly. Adjust pH to 4.0. Autoclave for 15 min at 15 psi pressure–121°C. Cool to 50°–55°C.

Solution B:

Composition per 500.0mL:

Agar	20.0g

Preparation of Solution B: Add 20.0g of agar to distilled/deionized water and bring volume to 500.0mL. Mix thoroughly. Gently heat and bring to boiling. Autoclave for 15 min at 15 psi pressure–121°C. Cool to 50°–55°C.

Preparation of Medium: Aseptically mix 500.0mL of solution A with 500.0mL of solution B. Pour into sterile Petri dishes or leave in tubes.

Use: For the cultivation and maintenance of *Acidomonas methanolica.*

Acidomonas methanolica Agar

Composition per liter:

Solution A	500.0mL
Solution B	500.0mL

pH 4.0 ± 0.2 at 25°C

Solution A:

Composition per 500.0mL:

Glucose	20.0g
Yeast extract	5.0g
$(NH_4)_2SO_4$	3.0g
KH_2PO_4	1.0g
$MgSO_4{\cdot}7H_2O$	0.7g
NaCl	0.5g
$Ca(NO_3)_2{\cdot}4H_2O$	0.4g
$K_2HPO_4{\cdot}3H_2O$	0.16g

Preparation of Solution A: Add components to distilled/deionized water and bring volume to 500.0mL. Mix thoroughly. Autoclave for 15 min at 15 psi pressure–121°C. Cool to 50°–55°C.

Solution B:

Composition per 500.0mL:

Agar	20.0g

Preparation of Solution B: Add agar to distilled/deionized water and bring volume to 500.0mL. Mix thoroughly. Gently heat and bring to boiling. Autoclave for 15 min at 15 psi pressure–121°C. Cool to 50°–55°C.

Preparation of Medium: Aseptically mix 500.0mL of solution A and 500.0mL of solution B. Mix thoroughly. Aseptically adjust pH to 4.0. Pour into sterile Petri dishes or distribute into sterile tubes.

Use: For the cultivation and maintenance of *Acidomonas methanolica.*

Acidophilic *Bacillus stearothermophilus* Agar

Composition per liter:

Part B	600.0mL
Part A	400.0mL

pH 5.0 ± 0.2 at 25°C

Part A:

Composition per 400.0mL:

Soluble starch	10.0g
Pancreatic digest of casein	5.0g
Yeast extract	5.0g
KH_2PO_4	1.0g
$CaCl_2{\cdot}2H_2O$	0.5g
$MnCl_2{\cdot}4H_2O$	0.5g

Preparation of Part A: Add components to distilled/deionized water and bring volume to 400.0mL. Mix thoroughly. Gently heat and bring to boiling. Adjust pH to 4.7. Autoclave for 15 min at 15 psi pressure–121°C. Cool to 50°C.

Part B:

Composition per 600.0mL:

Agar	20.0g

Preparation of Part B: Add agar to distilled/deionized water and bring volume to 600.0mL. Autoclave for 15 min at 15 psi pressure–121°C. Cool to 50°C.

Preparation of Medium: Aseptically combine solution A and solution B. Mix thoroughly. Adjust pH to 5.0. Pour into sterile Petri dishes.

Use: For the cultivation and maintenance of *Bacillus stearothermophilus* and other acidophilic *Bacillus* species.

Acidophilic *Bacillus stearothermophilus* Broth

Composition per liter:

Soluble starch	10.0g
Pancreatic digest of casein	5.0g
Yeast extract	5.0g
KH_2PO_4	1.0g
$CaCl_2{\cdot}2H_2O$	0.5g
$MnCl_2{\cdot}4H_2O$	0.5g

pH 5.0 ± 0.2 at 25°C

Preparation of Medium: Dissolve all components in 1.0L of distilled/deionized water. Mix thoroughly. Gently heat and bring to boiling. Adjust to pH 5.0. Autoclave for 15 min at 15 psi pressure–121°C. Precipitate will dissolve after cooling and mixing.

Use: For the cultivation and maintenance of *Bacillus stearothermophilus* and other acidophilic *Bacillus* species.

Acidophilium Agar

Composition per liter:

Solution A	500.0mL
Solution B	500.0mL

pH 3.5 ± 0.2 at 25°C

Solution A:

Composition per 500.0mL:

Agar	12.0g

Preparation of Solution A: Add agar to distilled/deionized water and bring volume to 500.0mL. Mix thoroughly. Gently heat and bring to boiling. Autoclave for 15 min at 15 psi pressure–121°C. Cool to 50°C.

Solution B:

Composition per 500.0mL:

Mannitol	1.0g
$MgSO_4 \cdot 7H_2O$	0.5g
$(NH_4)_2SO_4$	0.1g
Tryptone soya broth	0.1g
KCl	50.0mg
KH_2PO_4	50.0mg
$Ca(NO_3)_2$	10.0mg

Preparation of Solution B: Add components to distilled/deionized water and bring volume to 1.0L. Mix thoroughly. Bring pH to 3.5. Autoclave for 15 min at 15 psi pressure–121°C. Cool to 50°C.

Preparation of Medium: Aseptically combine 500.0mL of solution A with 500.0mL of solution B. Mix thoroughly. Pour into sterile Petri dishes or aseptically distribute into sterile tubes.

Use: For the cultivation and maintenance of *Acidiphilium cryptum* and other *Acidiphilium* species.

Aciduliprofundum Medium (DSMZ Medium 1083)

Composition per liter:

NaCl	30.0g
$MgSO_4 \cdot 7H_2O$	3.5g
$MgCl_2 \cdot 6H_2O$	2.75g
$CaCl_2 \cdot 2H_2O$	0.38g
KCl	0.33g
NaBr	0.05g
$(NH_4)_2SO_4$	0.10g
KH_2PO_4	0.28g
Wolfe's mineral elixir	1.0mL
Resazurin	0.5mg
Sodium citrate	2.94g
Yeast extract	1.0g
Tryptone	1.0g
Sulfur, powdered	10.0g
$Na_2S \cdot 9H_2O$ solution	10.0mL

pH 4.5 ± 0.2 at 25°C

Wolfe's Mineral Elixir:

Composition per liter:

$MgSO_4 \cdot 7H_2O$	30.0g
NaCl	10.0g
$MnSO_4 \cdot 2H_2O$	5.0g
$(NH_4)_2NiSO_4 \cdot 6H_2O$	2.8g
$CoCl_2 \cdot 6H_2O$	1.8g
$ZnSO_4 \cdot 7H_2O$	1.8g
$FeSO_4 \cdot 7H_2O$	1.0g
$CaCl_2 \cdot 2H_2O$	1.0g
$KAl(SO_4)_2 \cdot 12H_2O$	0.18g
$CuSO_4 \cdot 5H_2O$	0.1g
H_3BO_3	0.1g
$Na_2MoO_4 \cdot 2H_2O$	0.1g
Na_2SeO_4	0.1g
$Na_2WO_4 \cdot 2H_2O$	0.1g

Preparation of Wolfe's Mineral Elixir: Adjust pH of 1.0L of distilled/deionized water to 1.0 with dilute H_2SO_4. Add remaining components one at a time. Mix throughly to dissolve.

$Na_2S \cdot 9H_2O$ Solution:

Composition per 10.0mL:

$Na_2S \cdot 9H_2O$	0.5g

Preparation of $Na_2S \cdot 9H_2O$ Solution: Add $Na_2S \cdot 9H_2O$ to distilled/deionized water and bring volume to 10.0mL. Mix thoroughly. Autoclave under 100% N_2 for 15 min at 15 psi pressure–121°C. Cool to room temperature. Adjust pH to 4.5.

Preparation of Medium: Add components, except $Na_2S \cdot 9H_2O$ solution and sulfur, to distilled/deionized water and bring volume to 990.0mL. Mix thoroughly. Gently heat and bring to boiling. Boil for 1 min. Cool to room temperature while sparging with 80% N_2 + 20% CO_2. Distribute into serum bottles containing the sulfur. Distribute under 80% N_2 + 20% CO_2, e.g., 20mL into 120mL serum bottles. Autoclave for 60 min at 3 psi pressure–105°C. Cool to 25°C. Aseptically inject $Na_2S \cdot 9H_2O$ solution, 0.2mL per 20mL medium. Mix thoroughly. Adjust pH to 4.5.

Use: For the cultivation of *Aciduliprofundum* sp.

Actidione® Agar (Cycloheximide Agar)

Composition per liter:

Glucose	50.0g
Agar	15.0g
Pancreatic digest of casein	5.0g
Yeast extract	4.0g
KH_2PO_4	0.55g
KCl	0.425g
$CaCl_2 \cdot 2H_2O$	0.125g
$MgSO_4 \cdot 7H_2O$	0.125g
Bromocresol Green	22.0mg
Actidione® (cycloheximide)	10.0mg
$FeCl_3$	2.5mg

pH 5.5 ± 0.2 at 25°C

Source: Actidione® Agar is available as a prepared medium from Oxoid Unipath.

Caution: Cycloheximide is toxic. Avoid skin contact or aerosol formation and inhalation.

Preparation of Medium: Add components to distilled/deionized water and bring volume to 1.0L. Mix thoroughly. Gently heat and bring to boiling. Distribute into tubes or flasks. Autoclave for 15 min at 15 psi pressure–121°C. Pour into sterile Petri dishes or leave in tubes.

Use: For the enumeration and detection of bacteria in specimens containing large numbers of yeasts and molds.

Actidione HiVeg Agar with Actidione®

Composition per liter:

Glucose	50.0g
Agar	15.0g

Plant hydrolysate	5.0g
Yeast extract	4.0g
KH_2PO_4	0.55g
KCl	0.425g
$CaCl_2 \cdot 2H_2O$	0.125g
$MgSO_4 \cdot 7H_2O$	0.125g
Bromocresol Green	22.0mg
Actidione® (cycloheximide)	10.0mg
$MnSO_4 \cdot 4H_2O$	2.5mg
$FeCl_3$	2.5mg

pH 5.5 ± 0.2 at 25°C

Source: This medium is available as a premixed powder from HiMedia.

Caution: Cycloheximide is toxic. Avoid skin contact or aerosol formation and inhalation.

Preparation of Medium: Add components to distilled/deionized water and bring volume to 1.0L. Mix thoroughly. Gently heat and bring to boiling. Distribute into tubes or flasks. Autoclave for 15 min at 15 psi pressure–121°C. Pour into sterile Petri dishes or leave in tubes.

Use: For the enumeration and detection of bacteria in specimens containing large numbers of yeasts and molds.

Actidione HiVeg Agar Base with Actidione®

Composition per liter:

Glucose	50.0g
Agar	15.0g
Plant hydrolysate	5.0g
Yeast extract	4.0g
KH_2PO_4	0.55g
KCl	0.425g
$CaCl_2 \cdot 2H_2O$	0.125g
$MgSO_4 \cdot 7H_2O$	0.125g
Bromocresol Green	22.0mg
$MnSO_4 \cdot 4H_2O$	2.5mg
$FeCl_3$	2.5mg
Cycloheximide solution	10.0mL

pH 5.5 ± 0.2 at 25°C

Source: This medium, without actidione (cycloheximide), is available as a premixed powder from HiMedia.

Cycloheximide Solution:

Composition per 10.0mL:

Cycloheximide	0.025g

Preparation of Cycloheximide Solution: Add cycloheximide to distilled/deionized water and bring volume to 10.0mL. Mix thoroughly. Filter sterilize.

Caution: Cycloheximide is toxic. Avoid skin contact or aerosol formation and inhalation.

Preparation of Medium: Add components, except cycloheximide solution, to distilled/deionized water and bring volume to 990.0mL. Mix thoroughly. Gently heat and bring to boiling. Distribute into tubes or flasks. Autoclave for 15 min at 15 psi pressure–121°C. Cool to 50°C. Aseptically add 10.0mL cycloheximide solution. Pour into sterile Petri dishes or leave in tubes.

Use: For the enumeration and detection of bacteria in specimens containing large numbers of yeasts and molds.

Actidione HiVeg Agar Base without Actidione® with Antibiotics

Composition per liter:

Glucose	50.0g
Agar	15.0g
Plant hydrolysate	5.0g
Yeast extract	4.0g
KH_2PO_4	0.55g
KCl	0.425g
$CaCl_2 \cdot 2H_2O$	0.125g
$MgSO_4 \cdot 7H_2O$	0.125g
Bromocresol Green	22.0mg
$MnSO_4 \cdot 4H_2O$	2.5mg
$FeCl_3$	2.5mg
Antibiotic solution	10.0mL

pH 5.5 ± 0.2 at 25°C

Antibiotic Solution:

Composition per 10.0mL:

Ampicillin or streptomycin	20.0mg

Source: This medium is available as a premixed powder from HiMedia.

Preparation of Antibiotic Solution: Add components to distilled/deionized water and bring volume to 10.0mL. Mix thoroughly. Filter sterilize.

Preparation of Medium: Add components, except antibiotic solution, to distilled/deionized water and bring volume to 990.0mL. Mix thoroughly. Gently heat and bring to boiling. Distribute into tubes or flasks. Autoclave for 15 min at 15 psi pressure–121°C. Cool to 50°C. Aseptically add 10.0mL antibiotic solution. Pour into sterile Petri dishes or leave in tubes.

Use: For the selective isolation of dermatophytes.

Actinobacillus lignieresii Medium

Composition per 1010.0mL:

Agar	10.0g
Hartley's digest broth	900.0mL
Filde's enrichment	100.0mL
Antibiotic solution	10.0mL

pH 7.5 ± 0.2 at 25°C

Hartley's Digest Broth:

Composition per 10.0L:

Ox heart	3000.0g
Pancreatin	50.0g
Na_2CO_3, anhydrous (0.8% solution)	5.0L
HCl, concentrated	80.0mL

Preparation of Hartley's Digest Broth: Finely mince the ox heart. Add the meat to 5.0L of distilled/deionized water. Gently heat and bring to 80°C. Add Na_2CO_3 solution. Cool to 45°C. Add pancreatin and maintain at 45°C for 4 hr while stirring. Add the HCl and steam at 100°C for 30 min. Cool to room temperature. Adjust pH to 8.0 with 1*N* NaOH. Gently heat and bring to boiling. Continue boiling for 25 min. Filter while hot through Whatman #1 filter paper. Cool to room temperature. Adjust pH to 7.5.

Filde's Enrichment Solution:

Composition per 206.0mL:

Pepsin	1.0g
NaCl (0.85% solution)	150.0mL

Sheep blood, defibrinated 50.0mL
HCl 6.0mL

Source: Filde's enrichment solution is available as a premixed powder from BD Diagnostic Systems and Oxoid Unipath.

Preparation of Filde's Enrichment Solution: Combine components. Mix thoroughly. Incubate at 56°C for 4 hr. Bring pH to 7.0 with 20% NaOH. Adjust pH to 7.2 with HCl. Do not autoclave. Add 0.25 mL of chloroform and store at 4°C. Before use, heat to 56°C to remove chloroform.

Antibiotic Solution:

Composition per 10.0mL:

Oleandomycin phosphate.. 0.02g
Neomycin sulfate 1.5mg

Preparation of Antibiotic Solution: Add components to distilled/deionized water and bring volume to 10.0mL. Mix thoroughly. Filter sterilize.

Preparation of Medium: Add agar to 900.0mL of Hartley's digest broth. Mix thoroughly. Gently heat and bring to boiling. Autoclave for 15 min at 15 psi pressure–121°C. Cool to 45°–50°C. Aseptically add 100.0mL of Filde's enrichment solution and 10.0mL of antibiotic solution. Mix thoroughly. Pour into sterile Petri dishes or distribute into sterile tubes.

Use: For the isolation and cultivation of *Actinobacillus lignieresii.*

Actinobolin Medium

Composition per liter:

Milk, peptonized 15.0g
Glucose 10.0g
Yeast extract... 5.0g
KH_2PO_4 .... 2.0g
Sorbitan monooleate complex 1.0g
Tomato juice...100.0mL
Actinobolin 1.0mg/mL

Preparation of Medium: Add components to distilled/deionized water and bring volume to 1.0L. Mix thoroughly. Gently heat and bring to boiling. Continue boiling for 2–3 min. Distribute into tubes or flasks. Autoclave for 15 min at 15 psi pressure–121°C.

Use: For the cultivation of *Enterococcus durans.*

Actinomyces Agar

Composition per liter:

Agar 20.0g
K_2HPO_4.. 13.0g
Heart muscle, solids from infusion.. 10.0g
Peptic digest of animal tissue.. 10.0g
Glucose 5.0g
Yeast extract.. 5.0g
NaCl 5.0g
Pancreatic digest of casein.. 4.0g
KH_2PO_4.. 2.0g
$(NH_4)_2SO_4$.. 1.0g
L-Cysteine·HCl·H_2O.. 1.0g
Soluble starch... 1.0g
$MgSO_4 \cdot 7H_2O$ 0.2g
$CaCl_2 \cdot 2H_2O$.. 0.01g

pH 6.9 ± 0.2 at 25°C

Preparation of Medium: Add components to distilled/deionized water and bring volume to 1.0L. If a semisolid medium is desired, add 7.0g of agar instead of 20.0g. Mix thoroughly. Gently heat and bring to boiling. Distribute into tubes or flasks. Autoclave for 10 min at 15 psi pressure–121°C. Pour into sterile Petri dishes or leave in tubes.

Use: For the maintenance or cultivation of a variety of anaerobic bacteria, including *Actinomyces* species, *Eubacterium* species, *Fusobacterium* species, *Propionibacterium* species, and others.

Actinomyces Broth

Composition per liter:

K_2HPO_4... 13.0g
Heart muscle, solids from infusion... 10.0g
Peptic digest of animal tissue 10.0g
Glucose 5.0g
Yeast extract... 5.0g
NaCl... 5.0g
Pancreatic digest of casein... 4.0g
KH_2PO_4... 2.0g
$(NH_4)_2SO_4$ 1.0g
L-Cysteine·HCl·H_2O... 1.0g
Soluble starch.. 1.0g
$MgSO_4 \cdot 7H_2O$ 0.2g
$CaCl_2 \cdot 2H_2O$ 0.01g

pH 6.9 ± 0.2 at 25°C

Source: This medium is available as a premixed powder from BD Diagnostic Systems.

Preparation of Medium: Add components to distilled/deionized water and bring volume to 1.0L. Mix thoroughly. Distribute into tubes or flasks. Autoclave for 10 min at 15 psi pressure–121°C.

Use: For the maintenance or cultivation of a variety of anaerobic bacteria including *Actinomyces* species, *Eubacterium* species, *Fusobacterium* species, *Propionibacterium* species, and others.

Actinomyces Broth

Composition per liter:

Beef heart, infusion from.. 500.0g
KH_2PO_4... 15.0g
Peptic digest of animal tissue 10.0g
Glucose 5.0g
Yeast extract... 5.0g
NaCl... 5.0g
Pancreatic digest of casein... 4.0g
KH_2PO_4... 2.0g
$(NH_4)_2SO_4$ 1.0g
L-Cysteine·HCl·H_2O... 1.0g
Soluble starch.. 1.0g
$MgSO_4 \cdot 7H_2O$ 0.2g
$CaCl_2 \cdot 2H_2O$ 0.02g

pH 7.2 ± 0.2 at 25°C

Source: This medium is available as a premixed powder from BD Diagnostic Systems.

Preparation of Medium: Add components to distilled/deionized water and bring volume to 1.0L. Mix thoroughly. Distribute into tubes or flasks. Autoclave for 10 min at 15 psi pressure–121°C.

Use: For the maintenance or cultivation of a variety of anaerobic bacteria, including *Actinomyces* species, *Eubacterium* species, *Fusobacterium* species, *Propionibacterium* species, and others.

Actinomyces Broth (DSMZ Medium 1029)

Composition per liter:

Pancreatic digest of casein	17.0g
KH_2PO_4	15.0g
Yeast extract	10.0g
Glucose	5.0g
NaCl	5.0g
Heart muscle, solids from infusion	2.0g
$(NH_4)_2SO_4$	1.0g
L-Cysteine·HCl·H_2O	1.0g
Soluble starch	1.0g
$MgSO_4 \cdot 7H_2O$	0.2g
$CaCl_2 \cdot 2H_2O$	0.01g

pH 6.9 ± 0.2 at 25°C

Preparation of Medium: Add components to distilled/deionized water and bring volume to 1.0L. Mix thoroughly. Gently heat and bring to boiling. Autoclave for 15 min at 15 psi pressure–121°C.

Use: For the maintenance or cultivation of a variety of *Actinomyces* species and other anaerobic bacteria.

Actinomyces HiVeg Agar

Composition per liter:

Agar	20.0g
KH_2PO_4	15.0g
Plant hydrolysate No. 1	10.0g
Plant special influsion	10.0g
Yeast extract	5.0g
Glucose	5.0g
NaCl	5.0g
Plant hydrolysate	4.0g
KH_2PO_4	2.0g
$(NH_4)_2SO_4$	1.0g
L-Cysteine·HCl·H_2O	1.0g
Soluble starch	1.0g
$MgSO_4 \cdot 7H_2O$	0.2g
$CaCl_2 \cdot 2H_2O$	0.02g

pH 6.9 ± 0.2 at 25°C

Source: This medium is available as a premixed powder from HiMedia.

Preparation of Medium: Add components to distilled/deionized water and bring volume to 1.0L. Mix thoroughly. Gently heat and bring to boiling. Autoclave for 15 min at 15 psi pressure–121°C. Pour into sterile Petri dishes or distribute into sterile tubes.

Use: For the maintenance or cultivation of a variety of anaerobic bacteria, including *Actinomyces* species, *Eubacterium* species, *Fusobacterium* species, *Propionibacterium* species, and others.

Actinomyces HiVeg Broth

Composition per liter:

KH_2PO_4	15.0g
Plant hydrolysate No. 1	10.0g
Plant special influsion	10.0g
Yeast extract	5.0g
Glucose	5.0g
NaCl	5.0g
Plant hydrolysate	4.0g
KH_2PO_4	2.0g
$(NH_4)_2SO_4$	1.0g
L-Cysteine·HCl·H_2O	1.0g
Soluble starch	1.0g
$MgSO_4 \cdot 7H_2O$	0.2g
$CaCl_2 \cdot 2H_2O$	0.02g

pH 7.2 ± 0.2 at 25°C

Source: This medium is available as a premixed powder from HiMedia.

Preparation of Medium: Add components to distilled/deionized water and bring volume to 1.0L. Mix thoroughly. Distribute into tubes or flasks. Autoclave for 10 min at 15 psi pressure–121°C.

Use: For the maintenance or cultivation of a variety of anaerobic bacteria, including *Actinomyces* species, *Eubacterium* species, *Fusobacterium* species, *Propionibacterium* species, and others.

Actinomyces humiferus Medium

Composition per liter:

Pancreatic digest of casein	17.0g
NaCl	5.0g
Pancreatic digest of soybean meal	3.0g
K_2HPO_4	2.5g
Glucose	2.5g
Horse blood	50.0mL

Preparation of Medium: Add components to distilled/deionized water and bring volume to 1.0L. Mix thoroughly. Distribute into tubes or flasks. Autoclave for 15 min at 15 psi pressure–121°C.

Use: For the cultivation of *Actinomyces humiferus.*

Actinomyces Isolation Agar

Composition per liter:

Agar	15.0g
Glycerol	5.0g
Sodium propionate	4.0g
Sodium caseinate	2.0g
K_2HPO_4	0.5g
Asparagine	0.1g
$MgSO_4 \cdot 7H_2O$	0.1g
$FeSO_4 \cdot 7H_2O$	0.001g

Preparation of Medium: Add components to distilled/deionized water and bring volume to 1.0L. Mix thoroughly. Gently heat and bring to boiling. Distribute into tubes or flasks. Autoclave for 15 min at 15 psi pressure–121°C. Pour into sterile Petri dishes or leave in tubes.

Use: For the isolation and cultivation of *Actinomyces* species.

Actinomycete Growth Medium

Composition per liter:

Succinic acid	1.18g
L-Glutamine	0.29g
$CaCl_2 \cdot 2H_2O$	0.2g
KH_2PO_4	0.2g
$MgSO_4 \cdot 7H_2O$	0.2g
NaCl	0.1g
m-Inositol	0.09g
Ferric EDTA	0.037g
$MnSO_4 \cdot H_2O$	4.5mg
H_3BO_3	1.5mg
$ZnSO_4 \cdot 7H_2O$	1.5mg
Nicotonic acid	0.5mg
Pyridoxine-HCl	0.5mg

Thiamine-HCl	0.1mg
$CuSO_4 \cdot 5H_2O$	0.04mg
$Na_2MoO_4 \cdot 2H_2O$	0.025mg

pH 6.4 ± 0.2 at 25°C

Preparation of Medium: Add components to distilled/deionized water and bring volume to 1.0L. Mix thoroughly. Distribute into tubes or flasks. Autoclave for 15 min at 15 psi pressure–121°C.

Use: For the cultivation of actinomycetes.

Actinomycete Isolation Agar

Composition per liter:

Agar	15.0g
Glycerol	5.0g
Sodium propionate	4.0g
Sodium caseinate	2.0g
K_2HPO_4	0.5g
Asparagine	0.1g
$MgSO_4 \cdot 7H_2O$	0.1g
$FeSO_4 \cdot 7H_2O$	1.0mg

pH 8.1± 0.2 at 25°C

Source: This medium is available as a premixed powder from BD Diagnostic Systems.

Preparation of Medium: Add components, except glycerol, to distilled/deionized water and bring volume to 1.0L. Mix thoroughly. Gently heat and bring to boiling. Add 5.0g of glycerol. Distribute into tubes or flasks. Autoclave for 15 min at 15 psi pressure–121°C.

Use: For the isolation and cultivation of aerobic *Actinomyces* from soil and water.

Actinomycete Isolation HiVeg Agar

Composition per liter:

Agar	15.0g
Sodium propionate	4.0g
Plant protein	2.0g
L-Asparagine	0.1g
K_2HPO_4	0.5g
$MgSO_4 \cdot 7H_2O$	0.1g
$FeSO_4$	1.0mg
Glycerol	5.0mL

pH 8.1 ± 0.2 at 25°C

Source: This medium, without glycerol, is available as a premixed powder from HiMedia.

Preparation of Medium: Add components to distilled/deionized water and bring volume to 1.0L. Mix thoroughly. Gently heat and bring to boiling. Autoclave for 15 min at 15 psi pressure–121°C. Pour into sterile Petri dishes or distribute into sterile tubes.

Use: For the isolation and propagation of actinomycetes from soil and water.

Actinoplanes Medium

Composition per liter:

Oatmeal, baby cereal	60.0g
Yeast	2.5g
K_2HPO_4	1.0g
KCl	0.5g
$MgSO_4 \cdot 7H_2O$	0.5g
$FeSO_4 \cdot 7H_2O$	0.01g

Preparation of Medium: Add components to distilled/deionized water and bring volume to 1.0L. Mix thoroughly. Distribute into tubes or flasks. Autoclave for 15 min at 15 psi pressure–121°C.

Use: For the cultivation and maintenance of *Actinoplanes* species.

Actinopolyspora Medium

Composition per liter:

Agar	20.0g
Maltose	10.0g
N-Z-amine A	2.0g
Yeast extract	1.0g
Beef extract	1.0g

pH 7.3 ± 0.2 at 25°C

Preparation of Medium: Add components to distilled/deionized water and bring volume to 1.0L. Mix thoroughly. Gently heat and bring to boiling. Distribute into tubes or flasks. Autoclave for 15 min at 15 psi pressure–121°C. Pour into sterile Petri dishes or leave in tubes.

Use: For the cultivation and maintenance of *Actinopolyspora thermovinacea*.

Activated Carbon Medium (DSMZ Medium 811)

Composition per liter:

Agar	15.0g
$Na_2HPO_4 \cdot 12H_2O$	9.0g
Activated carbon	5.0g
KH_2PO_4	1.5g
NH_4Cl	1.5g
$MgSO_4 \cdot 7H_2O$	0.2g
$CaCl_2 \cdot 2H_2O$	20.0mg
NH_4-Fe-III-Citrate	1.2mg
Trace elements solution TS2	1.0mL

pH 7.5 ± 0.1 at 25°C

Trace Elements Solution TS2:

Composition per liter:

$Na_2MoO_4 \cdot 4H_2O$	900.0mg
H_3BO_3	300.0mg
$CoCl_2 \cdot 6H_2O$	200.0mg
$ZnSO_4 \cdot 7H_2O$	100.0mg
$MnCl_2 \cdot 4H_2O$	30.0mg
$NiCl_2 \cdot 6H_2O$	20.0mg
Na_2SeO_3	20.0mg
$CuCl_2 \cdot 2H_2O$	10.0mg

Trace Elements Solution TS2: Add components to distilled/deionized water and bring volume to 1.0L. Mix thoroughly.

Preparation of Medium: Add components, except activated carbon, to distilled/deionized water and bring volume to 900.0mL. Mix thoroughly. After all components are dissolved add activated carbon. Bring volume to 1.0L with distilled/deionized water. Adjust pH to 7.5. Distribute into tubes or flasks. Autoclave for 15 min at 15 psi pressure–121°C. Pour into Petri dishes or leave in tubes.

Use: For the cultivation of *Streptomyces thermoautotrophicus*.

Adams Agar

Composition per liter:

Agar	20.0g
Sodium acetate	2.3g
Glucose	0.4g

pH 7.2 ± 0.2 at 25°C

Source: This medium is available as a premixed powder from HiMedia.

Preparation of Medium: Add components to distilled/deionized water and bring volume to 1.0L. Mix thoroughly. Gently heat and bring to boiling. Autoclave for 15 min at 5–8 psi pressure–108°–112°C. Pour into sterile Petri dishes or distribute into sterile tubes. For tubes allow to solidify in a slanted position.

Use: For the examination of sporulation in yeasts.

AE Medium (4a/3e) (*Gluconobacter* Medium) (LMG Medium 269)

Composition per liter:

Agar	15.0g
Glucose	10.0g
Peptone	3.0g
Yeast extract	2.0g
Acetic acid, glacial	40.0mL
Ethanol, 96%	30.0mL

Preparation of Medium: Add components, except acetic acid and ethanol, to 930.0mL distilled/deionized water. Mix thoroughly. Gently heat and bring to boiling. Autoclave for 15 min at 15 psi pressure–121°C. Cool to 45°–50°C. Aseptically add 40.0mL filter sterilized acetic acid and 30.0mL filter sterilized 96% ethanol. Mix thoroughly. Pour into sterile Petri dishes or distribute into sterile tubes.

Use: For the cultivation of *Gluconacetobacter entanii* and other *Gluconacetobacter* spp.

A_2E_6 Medium

Composition per liter:

NaCl	23.48g
$MgCl_2 \cdot 6H_2O$	10.63g
Na_2SO_4	3.92g
Glucose	3.0g
Tris buffer	3.0g
Glutamic acid	1.5g
$CaCl_2$, anhydrous	1.11g
KCl	0.66g
$NaHCO_3$	0.19g
Sodium glycerophosphate	0.15g
KBr	0.1g
$(NH_4)_2SO_4$	0.05g
$SrCl_2 \cdot 6H_2O$	0.04g
H_3BO_3	0.03g
$FeCl_3 \cdot 6H_2O$	0.01g
K_2HPO_4	0.01g
Metal mixture	3.0mL
Vitamin solution	1.0mL

pH 6.4–6.6 at 25°C

Metal Mixture:

Composition per 100.0mL:

EDTA	1.0g
H_3BO_3	1.0g
$MnCl_2 \cdot 4H_2O$	0.15g
$FeCl_3 \cdot 6H_2O$	0.05g
$ZnCl_2$	0.01g
$CoCl_2 \cdot 6H_2O$	0.005g

Preparation of Metal Mixture: Add components to distilled/deionized water and bring volume to 100.0L. Mix thoroughly.

Vitamin Solution:

Composition per liter:

Thiamine	1.0g
Biotin	0.003g

Preparation of Vitamin Solution: Add components to distilled/deionized water and bring volume to 1.0L. Mix thoroughly. Filter sterilize.

Preparation of Medium: Add components, except vitamin solution, to distilled/deionized water and bring volume to 999.0mL. Mix thoroughly. Adjust pH to 6.4–6.6. Autoclave for 15 min at 15 psi pressure–121°C. Aseptically add 1.0mL of sterile vitamin solution. Mix thoroughly. Aseptically distribute into sterile screw-capped tubes or flasks.

Use: For the cultivation of *Crypthecodinium cohnii.*

AE Sporulation Medium, Modified

Composition per 1079.2mL:

Polypeptone™	10.0g
Yeast extract	10.0g
Na_2HPO_4	4.36g
Ammonium acetate	1.5g
KH_2PO_4	0.25g
$MgSO_4 \cdot 7H_2O$	0.2g
Raffinose solution	39.6mL
Na_2CO_3 solution	13.2mL
$CoCl_2 \cdot 6H_2O$ solution	13.2mL
Sodium ascorbate solution	13.2mL

pH 7.8 ± 0.1 at 25°C

Raffinose Solution:

Composition per 100.0mL:

Raffinose	10.0g

Preparation of Raffinose Solution: Add raffinose to distilled/deionized water and bring volume to 100.0mL. Mix thoroughly. Filter sterilize.

Na_2CO_3 Solution:

Composition per 100.0mL:

Na_2CO_3	7.0g

Preparation of Na_2CO_3 Solution: Add Na_2CO_3 to distilled/deionized water and bring volume to 100.0mL. Mix thoroughly. Filter sterilize.

$CoCl_2 \cdot 6H_2O$ Solution:

Composition per 100.0mL:

$CoCl_2 \cdot 6H_2O$	0.32g

Preparation of $CoCl_2 \cdot 6H_2O$ Solution: Add $CoCl_2 \cdot 6H_2O$ to distilled/deionized water and bring volume to 100.0mL. Mix thoroughly. Filter sterilize.

Sodium Ascorbate Solution:

Composition per 100.0mL:

Sodium ascorbate	1.5g

Preparation of Sodium Ascorbate Solution: Add sodium ascorbate to distilled/deionized water and bring volume to 100.0mL. Mix thoroughly. Filter sterilize. Use freshly prepared solution.

Preparation of Medium: Add components—except raffinose solution, Na_2CO_3 solution, $CoCl_2 \cdot 6H_2O$ solution, and sodium ascorbate solution—to distilled/deionized water and bring volume to 1.0L. Mix thoroughly. Adjust pH to 7.5 using 2*M* sodium carbonate solution. Distribute into tubes in 15.0mL volumes. Autoclave for 15 min at 15 psi pressure–121°C. Aseptically add 0.6mL of sterile raffinose solution, 0.2mL of sterile Na_2CO_3 solution, and 0.2mL of sterile $CoCl_2 \cdot 6H_2O$ solution to each tube. Mix thoroughly. Prior to inoculation, steam medium for 10 min. Cool to 25°C. Aseptically add 0.2mL of sterile sodium ascorbate solution to each tube.

Use: For the cultivation and sporulation of *Clostridium perfringens.*

Aero Pseudo Selective Agar

Composition per liter:

Agar	12.0g
Starch, soluble	20.0g
Sodium glutamate	2.0g
KH_2PO_4	2.0g
$MgSO_4 \cdot 7H_2O$	0.5g
Phenol Red	0.36g

pH 7.2 ± 0.2 at 25°C

Source: This medium is available from HiMedia.

Selective Supplement Solution:

Composition per 10.0mL:

Pimaricin	0.01g
Penicillin G	100,000 units

Preparation of Selective Supplement Solution: Add components to distilled/deionized water and bring volume to 10.0mL. Mix thoroughly. Filter sterilize.

Preparation of Medium: Add components, except selective supplement solution, to distilled/deionized water and bring volume to 990.0mL. Mix thoroughly. Autoclave for 15 min at 15 psi pressure–121°C. Cool to 50°C. Aseptically add selective supplement solution. Mix thoroughly. Pour into Petri dishes or aseptically distribute into sterile tubes.

Use: For the selective cultivation of *Pseudomonas* spp. and *Aeromonas* spp. For the detection of *Aeromonas* and *Pseudomonas* in foods, water, and food processing equipment.

Aerobic Low Peptone Basal Medium
See: ALP Basal Medium

Aeromonas Differential Agar (Dextrin Fuchsin Sulfite Agar)

Composition per liter:

Dextrin	15.0g
Agar	13.0g
Pancreatic digest of casein	10.0g
Na_2HPO_4	7.75g
NaCl	5.0g
Beef extract	3.0g
Na_2SO_3	1.6g
Acid Fuchsin solution	50.0mL

pH 7.5 ± 0.2 at 25°C

Acid Fuchsin Solution:

Composition per 50.0mL:

Acid Fuchsin	0.25g
Aqueous dioxan, 5%	50.0mL

Preparation of Acid Fuchsin Solution: Add Acid Fuchsin to 50.0mL of 5% aqueous dioxan. Mix well to dissolve.

Caution: Acid Fuchsin is a potential carcinogen and care must be taken to avoid inhalation of the powdered dye and contamination of the skin.

Preparation of Medium: Add components to distilled/deionized water and bring volume to 1.0L. Mix thoroughly. Gently heat while stirring and bring to boiling. Distribute into tubes or flasks. Autoclave for 15 min at 15 psi pressure–121°C. Pour into sterile Petri dishes or leave in tubes.

Use: For the isolation and differentiation of *Aeromonas* species from other Gram-negative rods such as *Pseudomonas* and Enterobacteriaceae. Specimens with low numbers of *Aeromonas* may first be enriched by growth in starch broth for 4–9 days. After 24 hrs of growth on this agar, colonies are sprayed with Nadi reagent (1% solution of *N,N,N′,N′*-tetramethyl-*p*-phenylene-diammonium dichloride). A positive Nadi reaction (dextrin degradation) is indicated by a purple color at the periphery of the colony. Dextrin fermentation is also indicated by red colonies. *Aeromonas* species appear as large, convex, dark red colonies with a purple periphery.

Aeromonas hydrophila Medium

Composition per liter:

Inositol	10.0g
Pancreatic digest of casein	10.0g
L-Ornithine·HCl	5.0g
Proteose peptone	5.0g
Agar	3.0g
Yeast extract	3.0g
Mannitol	1.0g
Ferric ammonium citrate	0.5g
$Na_2S_2O_3 \cdot 5H_2O$	0.4g
Bromcresol Purple	0.02g

pH 6.7 ± 0.2 at 25°C

Preparation of Medium: Add components to distilled/deionized water and bring volume to 1.0L. Mix thoroughly. Gently heat until dissolved. Adjust pH to 6.7. Distribute into tubes in 5.0mL volumes. Autoclave for 12 min at 15 psi pressure–121°C.

Use: For the isolation and cultivation of *Aeromonas hydrophila.*

Aeromonas Isolation Medium

Composition per liter:

Agar	12.5g
$Na_2S_2O_3$	10.67g
Special peptone	5.0g
NaCl	5.0g
Xylose	3.75g
L-Lysine·HCl	3.5g
Yeast extract	3.0g
Sorbitol	3.0g
Bile salts	3.0g
Inositol	2.5g
L-Arginine·HCl	2.0g
Lactose	1.5g
Ferric ammonium citrate	0.8g
Bromthymol Blue	0.04g
Thymol Blue	0.04g
Ampicillin solution	2.5mL

pH 8.0 ± 0.1 at 25°C

Source: This medium without ampicillin is available from Sigma Aldrich.

Ampicillin Solution:

Composition per 5.0mL:

Ampicillin	10.0mg

Preparation of Ampicillin Solution: Add ampicillin to distilled/deionized water and bring volume to 5.0mL. Mix thoroughly. Filter sterilize.

Preparation of Medium: Add components, except ampicillin solution, to distilled/deionized water and bring volume to 1.0L. Mix thoroughly. Gently heat and bring to boiling. Do not autoclave. Cool to 50°C. Aseptically add 2.5mL of ampicillin solution. Pour into sterile Petri dishes.

Use: For the isolation and selective differentiation of *Aeromonas hydrophila* and other *Aeromonas* species from clinical specimens and foods. *Aeromonas* species appear as small (0.5–1.5mm), dark green colonies with darker centers.

Aeromonas Isolation HiVeg Medium

Composition per liter:

Agar	12.5g
$Na_2S_2O_3$	10.67g
Plant special peptone	5.0g
NaCl	5.0g
Xylose	3.75g
L-Lysine·HCl	3.5g
Yeast extract	3.0g
Sorbitol	3.0g
Synthetic detergent	3.0g
Inositol	2.5g
L-Arginine·HCl	2.0g
Lactose	1.5g
Ferric ammonium citrate	0.8g
Bromthymol Blue	0.04g
Thymol Blue	0.04g
Ampicillin solution	2.5mL

pH 8.0 ± 0.1 at 25°C

Source: This medium without ampicillin is available from HiMedia.

Ampicillin Solution:

Composition per 5.0mL:

Ampicillin	10.0mg

Preparation of Ampicillin Solution: Add ampicillin to distilled/deionized water and bring volume to 5.0mL. Mix thoroughly. Filter sterilize.

Source: Ampicillin supplement solution is also available from HiMedia.

Preparation of Medium: Add components, except ampicillin solution, to distilled/deionized water and bring volume to 1.0L. Mix thoroughly. Gently heat and bring to boiling. Do not autoclave. Cool to 50°C. Aseptically add 2.5mL of ampicillin solution. Pour into sterile Petri dishes.

Use: For the isolation and selective differentiation of *Aeromonas hydrophila* and other *Aeromonas* species from clinical specimens and foods. *Aeromonas* species appear as small (0.5–1.5mm), dark green colonies with darker centers.

Aeromonas Medium (Ryan's *Aeromonas* Medium)

Composition per liter:

Agar	12.5g
$Na_2S_2O_3$	10.67g
Proteose peptone	5.0g
NaCl	5.0g
Xylose	3.75g
L-Lysine·HCl	3.5g
Yeast extract	3.0g
Sorbitol	3.0g
Bile salts No. 3	3.0g
Inositol	2.5g
L-Arginine·HCl	2.0g
Lactose	1.5g
Ferric ammonium citrate	0.8g
Bromthymol Blue	0.04g
Thymol Blue	0.04g

pH 8.0 ± 0.1 at 25°C

Source: This medium is available as a dehydrated powder from Oxoid Unipath.

Preparation of Medium: Add components to distilled/deionized water and bring volume to 1.0L. Mix thoroughly. Gently heat and bring to boiling. Do not autoclave. Cool to 50°C and aseptically add 5.0mg of ampicillin. Pour into sterile Petri dishes.

Use: For the isolation and selective differentiation of *Aeromonas hydrophila* and other *Aeromonas* species from clinical and nonclinical specimens. *Aeromonas* species appear as small (0.5–1.5mm), dark green colonies with darker centers.

Aeropyrum JXT Medium (DSMZ Medium 820)

Composition per liter:

Yeast extract	1.0g
Trypticase™ peptone	1.0g
$Na_2S_2O_3 \cdot 5H_2O$	1.0g
Seawater	1000.0mL

pH 7.1 ± 0.2 at 25°C

Artificial Seawater:

Composition per liter:

NaCl	23.477g
$MgCl_2 \cdot 6H_2O$	4.981g
Na_2SO_4	3.917g
$CaCl_2$	1.12g
KCl	664.0mg
$NaHCO_3$	192.0mg
H_3BO_3	26.0mg
$SrCl_2$	24.0mg
KBr	6.0mg
NaF	3.0mg

Preparation of Artificial Seawater: Add components to distilled/deionized water and bring volume to 1.0L. Mix thoroughly. Filter sterilize.

Preparation of Medium: Add components to 1.0L artificial seawater (filtered natural seawater can be used instead of artificial seawater). Mix thoroughly. Adjust pH to 7.0–7.2. Distribute into tubes or flasks. Autoclave for 15 min at 15 psi pressure–121°C.

Use: For the cultivation of *Aeropyrum pernix*.

AFPA
(*Aspergillus flavus/parasiticus* Agar)

Composition per liter:

Yeast extract	20.0g
Agar	15.0g
Peptone	10.0g
Ferric ammonium citrate	0.5g
DChloramphenicol	100.0mg
ichloran (Botran®)	2.0mg

pH 6.3 ± 0.2 at 25°C

Source: This medium is available as a dehydrated powder from Oxoid Unipath.

Preparation of Medium: Add components, except chlroamphenicol, to distilled/deionized water and bring volume to 1.0L. Mix thoroughly. Gently heat while stirring and bring to boiling. Add 100.0mg of chloramphenicol. Autoclave for 15 min at 15 psi pressure–121°C. Pour into sterile Petri dishes.

Use: For the selective isolation and enumeration of *Aspergillus flavus* and *Aspergillus parasiticus*. Colonies of these fungi appear with dark yellow-orange color on the reverse side.

AG Medium
(DSMZ Medium 955)

Composition per liter:

$CaCO_3$	7.0g
Peptone	5.0g
Yeast extract	5.0g
Malt extract	2.0g
Glycerol	1.5g
Glucose	1.0g

pH 6.8 ± 0.2 at 25°C

Preparation of Medium: Add components to distilled/deionized water and bring volume to 1.0L. Mix thoroughly. Distribute into tubes or flasks. Autoclave for 15 min at 15 psi pressure–121°C.

Use: For the cultivation of *Kozakia baliensis*.

Agar Medium A
See: **Antibiotic Medium 1**

Agar Medium C
See: **Antibiotic Medium 4**

Agar Medium for Differential Enumeration of Lactic Streptococci

Composition per 1170.0mL:

Agar	15.0g
Carboxymethylcellulose	15.0g
Calcium citrate	10.0g
Pancreatic digest of casein	5.0g
Yeast extract	5.0g
L-Arginine·HCl	5.0g
Casamino acids	2.5g
K_2HPO_4	1.25g
Calcium carbonate solution	100.0mL
Nonfat milk solution	50.0mL
Bromcresol Purple solution	20.0mL

pH 5.9 ± 0.2 at 25°C

Calcium Carbonate Solution:

Composition per 100.0mL:

$CaCO_3$	3.0g

Preparation of Calcium Carbonate Solution: Add $CaCO_3$ to distilled/deionized water and bring volume to 100.0mL. Mix thoroughly. Autoclave for 15 min at 15 psi pressure–121°C.

Nonfat Milk Solution:

Composition per 100.0mL:

Nonfat milk	11.0g

Preparation of Nonfat Milk Solution: Add nonfat milk to distilled/deionized water and bring volume to 100.0mL. Mix thoroughly. Autoclave for 15 min at 15 psi pressure–121°C.

Bromcresol Purple Solution:

Composition per 20.0mL:

Bromcresol Purple	0.02g

Preparation of Bromcresol Purple Solution: Add Bromcresol Purple to distilled/deionized water and bring volume to 20.0mL. Mix thoroughly. Filter sterilize.

Preparation of Medium: Add agar to 500.0mL of distilled/deionized water. Gently heat and bring to boiling. In a separate flask, add carboxymethylcellulose and calcium citrate to 500.0mL of distilled/deionized water. Gently heat while stirring until a white, turbid suspension is formed. Combine the two solutions. Add the pancreatic digest of casein, yeast extract, K_2HPO_4, casamino acids, and arginine. Mix thoroughly. Gently heat and bring to boiling. Adjust pH to 5.6 with 6*N* HCl. Distribute into screw-capped bottles in 100.0mL volumes. Autoclave for 15 min at 15 psi pressure–121°C. Cool to 45°–50°C. Immediately prior to pouring plates, aseptically add 5.0mL of sterile nonfat milk solution, 10.0mL of sterile calcium carbonate solution, and 2.0mL of sterile Bromcresol Purple solution to each screw-capped bottle. Mix thoroughly. The pH should be 5.9. Pour into cold, sterile Petri dishes.

Use: For the cultivation, differentiation, and enumeration of *Lactobacillus lactis, Lactobacillus lactis* subspecies *cremoris,* and *Lactobacillus lactis* subspecies *diacetylactis*. Lactose-fermenting bacteria such as *Lactobacillus lactis* subspecies *cremoris* appear as yellow colonies. Arginine-utilizing bacteria such as *Lactobacillus lactis* and *Lactobacillus lactis* subspecies *diacetylactis* appear as purple colonies. Citrate-utilizing bacteria such as *Lactobacillus lactis* subspecies *diacetylactis* appear as colonies surrounded by a clear zone.

Agar Medium P
(PM Indicator Agar)

Composition per liter:

Agar	15.0g
Glucose	5.25g
Peptone	5.0g
Beef extract	3.0g
Pancreatic digest of casein	1.7g
Tween™ 80	1.0g
NaCl	0.5g
Papaic digest of soybean meal	0.3g
K_2HPO_4	0.25g
Bromcresol Purple	0.06g

pH 7.8 ± 0.2 at 25°C

Preparation of Medium: Add components to distilled/deionized water and bring volume to 1.0L. Mix thoroughly. Gently heat and bring to boiling. Distribute into tubes or flasks. Autoclave for 15 min at 15 psi pressure–121°C. Pour into sterile Petri dishes or leave in tubes.

Use: For the cultivation of *Bacillus stearothermophilus* for the detection of penicillin in milk.

AGRE 1964
See: **Medium for Thermophilic Actinomycetes**

Agrobacterium Agar

Composition per liter:

Agar	15.0g
Mannitol	8.0g
NaCl	5.0g
Yeast extract	5.0g
$(NH_4)_2SO_4$	2.0g
Casamino acids	0.5g

pH 6.6 ± 0.2 at 25°C

Preparation of Medium: Add components to distilled/deionized water and bring volume to 1.0L. Mix thoroughly. Adjust pH to 6.6. Gently heat and bring to boiling. Distribute into tubes or flasks. Autoclave for 15 min at 15 psi pressure–121°C. Pour into sterile Petri dishes or leave in tubes.

Use: For the cultivation and maintenance of *Agrobacterium rhizogenes* and *Agrobacterium tumefaciens.*

Agrobacterium Agar

Composition per liter:

Agar	15.0g
Glucose	10.0g
Yeast extract	10.0g
$(NH_4)_2SO_4$	1.0g
KH_2PO_4	0.25g

Preparation of Medium: Add components to tap water and bring volume to 1.0L. Mix thoroughly. Gently heat and bring to boiling. Distribute into tubes or flasks. Autoclave for 15 min at 15 psi pressure–121°C. Pour into sterile Petri dishes or leave in tubes.

Use: For the cultivation and maintenance of *Agrobacterium azotophilum, Agrobacterium radiobacter, Agrobacterium rhizogenes, Agrobacterium rubi, Agrobacterium tumefaciens,* and *Agrobacterium vitis.*

Agrobacterium Agar with Biotin

Composition per liter:

Agar	15.0g
Mannitol	8.0g
NaCl	5.0g
Yeast extract	5.0g
$(NH_4)_2SO_4$	2.0g
Casamino acids	0.5g
Biotin solution	0.1mL

pH 6.6 ± 0.2 at 25°C

Biotin Solution:

Composition per 10.0mL:

Biotin	0.2mg

Preparation of Biotin Solution: Add biotin to distilled/deionized water and bring volume to 10.0mL. Mix thoroughly. Filter sterilize.

Preparation of Medium: Add components, except biotin solution, to distilled/deionized water and bring volume to 999.9mL. Mix thoroughly. Gently heat and bring to boiling. Adjust pH to 6.6. Autoclave for 15 min at 15 psi pressure–121°C. Cool to 50°–55°C. Aseptically add 0.1mL of sterile biotin solution. Mix thoroughly. Pour into sterile Petri dishes or distribute into sterile tubes.

Use: For the cultivation and maintenance of slow-growing *Agrobacterium rhizogenes* and *Agrobacterium tumefaciens.*

Agrobacterium Mannitol Medium

Composition per liter:

Mannitol	10.0g
L-Glutamate	2.0g
KH_2PO_4	0.5g
Yeast extract	0.3g
$MgSO_4 \cdot 7H_2O$	0.2g
NaCl	0.2g

pH 7.0 ± 0.2 at 25°C

Preparation of Medium: Add components to distilled/deionized water and bring volume to 1.0L. Mix thoroughly. Adjust pH to 7.0. Autoclave for 15 min at 15 psi pressure–121°C.

Use: For the cultivation of *Agrobacterium rhizogenes.*

Agrobacterium Medium

Composition per liter:

Agar	18.0g
Erythritol	5.0g
$NaNO_3$	2.5g
$CaCl_2$	0.2g
$MgSO_4 \cdot 7H_2O$	0.2g
NaCl	0.2g
KH_2PO_4	0.1g
Ferric EDTA	1.3mg
Biotin	2μg
Supplement	10.0mL

pH 7.0 ± 0.2 at 25°C

Supplement:

Composition per liter:

Cycloheximide	0.25g
Bacitracin	0.1g
Na_2SeO_3	0.1g
Tyrothricin	1.0mg

Preparation of Supplement: Add components to distilled/deionized water and bring volume to 10.0mL. Mix thoroughly. Filter sterilize.

Caution: Cycloheximide is toxic. Avoid skin contact or aerosol formation and inhalation.

Preparation of Medium: Add components, except supplement, to distilled/deionized water and bring volume to 990.0mL. Mix thoroughly. Adjust pH to 7.0 with 1*N* NaOH. Gently heat and bring to boiling. Autoclave for 15 min at 15 psi pressure–121°C. Cool to 45°–50°C. Aseptically add sterile supplement. Mix thoroughly. Pour into sterile Petri dishes or distribute into sterile tubes.

Use: For the selective isolation and cultivation of *Agrobacterium* species biotype 2.

Agrobacterium Medium

Composition per liter:

Agar	20.0g
Mannitol	10.0g
$NaNO_3$	4.0g
$MgCl_2$	2.0g

Calcium propionate	1.2g
$Mg_3(PO_4)_2$	0.2g
$MgSO_4$	0.1g
$MgCO_3$	0.075g
$NaHCO_3$	0.075g
Supplement	100.0mL

pH 7.1 ± 0.2 at 25°C

Supplement:

Composition per 100.0mL:

Berberine	0.275g
Cycloheximide	0.2g
Bacitracin	0.1g
Na_2SeO_3	0.1g
Penicillin G	0.06g
Streptomycin sulfate	0.03g
Tyrothricin	1.0mg

Preparation of Supplement: Add components to distilled/deionized water and bring volume to 100.0mL. Mix thoroughly. Filter sterilize.

Caution: Cycloheximide is toxic. Avoid skin contact or aerosol formation and inhalation.

Preparation of Medium: Add components, except supplement, to distilled/deionized water and bring volume to 900.0mL. Mix thoroughly. Gently heat and bring to boiling. Autoclave for 15 min at 15 psi pressure–121°C. Cool to 45°–50°C. Aseptically add 100.0mL of sterile supplement. Mix thoroughly. Pour into sterile Petri dishes or distribute into sterile tubes.

Use: For the selective isolation and cultivation of *Agrobacterium* species.

Agrobacterium Medium

Composition per liter:

Agar	12.0g
Lactose	5.0g
Na_2HPO_4	1.8g
KNO_3	1.0g
$MgSO_4 \cdot 7H_2O$	0.1g
Supplement	100.0mL

pH 6.8 ± 0.2 at 25°C

Supplement:

Composition per 100.0mL:

$MnSO_4 \cdot 4H_2O$	3.35g
Ferric EDTA	2.5mg

Preparation of Supplement: Add components to distilled/deionized water and bring volume to 100.0mL. Mix thoroughly. Filter sterilize.

Preparation of Medium: Add components, except supplement, to distilled/deionized water and bring volume to 900.0mL. Mix thoroughly. Gently heat and bring to boiling. Autoclave for 1 min at 25 psi pressure–130°C. Cool to 45°–50°C. Aseptically add sterile supplement. Mix thoroughly. Pour into sterile Petri dishes or distribute into sterile tubes.

Use: For the selective isolation and cultivation of *Agrobacterium* species.

Agrobacterium Medium D1

Composition per liter:

Agar	15.0g
Mannitol	15.0g
LiCl	6.0g
$NaNO_3$	5.0g
K_2HPO_4	2.0g
$MgSO_4 \cdot 7H_2O$	0.2g
Bromthymol Blue	0.1g
$Ca(NO_3)_2 \cdot 4H_2O$	0.02g

pH 7.2 ± 0.2 at 25°C

Preparation of Medium: Add components to distilled/deionized water and bring volume to 1.0L. Mix thoroughly. Gently heat and bring to boiling. Distribute into tubes or flasks. Autoclave for 15 min at 15 psi pressure–121°C. Cool to 45°–50°C. Adjust pH to 7.2. Pour into sterile Petri dishes or leave in tubes.

Use: For the selective isolation and cultivation of *Agrobacterium* species.

Agrobacterium tumefaciens Modified Roy and Sasser Medium for Grapevine Strains

Composition per 1020mL:

Agar	15.0g
Adoniitol	4.0g
H_3BO_3	1.0g
K_2HPO_4	0.9g
KH_2PO_4	0.7g
NaCl	0.2g
$MgSO_4 \cdot 7H_2O$	0.2g
Yeast extract	0.14g
Cycloheximide solution	10.0mL
Triphenyl tetrazolium chloride solution	1.0mL
D-Cycloserine solution	1.0mL
Trimethoprim solution	1.0mL

Cycloheximide Solution:

Composition per 10.0mL:

Cycloheximide	0.025g

Preparation of Cycloheximide Solution: Add cycloheximide to distilled/deionized water and bring volume to 10.0mL. Mix thoroughly. Filter sterilize.

Caution: Cycloheximide is toxic. Avoid skin contact or aerosol formation and inhalation.

Triphenyl Tetrazolium Chloride Solution:

Composition per 10.0mL:

Triphenyltetrazolium chloride	0.8g

Preparation of Triphenyl Tetrazolium Chloride Solution: Add triphenyltetrazolium chloride to distilled/deionized water and bring volume to 10.0mL. Mix thoroughly. Filter sterilize.

D-Cycloserine Solution:

Composition per 10.0mL:

D-Cycloserine	0.2g

Preparation of D-Cycloserine Solution: Add D-cycloserine to distilled/deionized water and bring volume to 10.0mL. Mix thoroughly. Filter sterilize.

Trimethoprim Solution:

Composition per 10.0mL:

Trimethoprim	0.025g

Preparation of Trimethoprim Solution: Add trimethoprim to distilled/deionized water and bring volume to 10.0mL. Add a drop of dilute HCl. Mix thoroughly. Gently heat while mixing until dissolved. Filter sterilize.

Preparation of Medium: Add components, except cycloheximide solution and triphenyl tetrazolium chloride solution, D-cycloserine solution, and trimethoprim solution to distilled/deionized water and bring volume to 1.0L. Mix thoroughly. Adjust pH to 7.2. Distribute 100.0mL into flasks. Gently heat and bring to boiling. Autoclave for 15 min at 15 psi pressure–121°C. Cool to 50°C. Aseptically add, per 100.0mL of medium, 0.1 mL sterile triphenyl tetrazolium chloride solution, 0.1mL sterile D-cycloserine solution, 0.1mL sterile trimethoprim solution, and 1.0mL cycloheximide solution. Mix thoroughly. Aseptically pour into sterile Petri dishes.

Use: For the selective cultivation of *Agrobacterium tumefaciens* biovar 3.

Agrobacterium tumefaciens Selective Medium

Composition per 1020mL:

Agar	15.0g
L(–)Arabitol	3.04g
K_2HPO_4	1.04g
KH_2PO_4	0.54g
Sodium taurocholate	0.29g
$MgSO_4 \cdot 7H_2O$	0.25g
NH_4NO_3	0.16g
Cycloheximide solution	10.0mL
Selenite solution	10.0mL
Crystal Violet (0.1% solution)	2.0mL

Selenite Solution:

Composition per 10.0mL:

NaOH	0.5g
$Na_2SeO_3 \cdot 5H_2O$	0.1g

Preparation of Selenite Solution: Add components to distilled/deionized water and bring volume to 1.0L. Mix thoroughly. Filter sterilize.

Cycloheximide Solution:

Composition per 10.0mL:

Cycloheximide	0.02g

Preparation of Cycloheximide Solution: Add cycloheximide to distilled/deionized water and bring volume to 10.0mL. Mix thoroughly. Filter sterilize.

Caution: Cycloheximide is toxic. Avoid skin contact or aerosol formation and inhalation.

Preparation of Medium: Add components, except cycloheximide solution and selenite solution, to distilled/deionized water and bring volume to 1.0L. Mix thoroughly. Distribute 100.0mL into flasks. Gently heat and bring to boiling. Autoclave for 15 min at 15 psi pressure–121°C. Cool to 50°C. Aseptically add, per 100.0mL of medium, 1.0mL sterile selenite solution and 1.0mL cycloheximide solution. Mix thoroughly. Aseptically pour into sterile Petri dishes.

Use: For the selective cultivation of *Agrobacterium tumefaciens* biovar 1.

Agrobacterium tumefaciens Selective Medium

Composition per 1020mL:

Agar	15.0g
Erythritol	3.05g
K_2HPO_4	1.04g
KH_2PO_4	0.54g
Sodium taurocholate	0.29g
$MgSO_4 \cdot 7H_2O$	0.25g
NH_4NO_3	0.16g
Cycloheximide solution	10.0mL
Selenite solution	10.0mL
Malachite Green (0.1% solution)	5.0mL
Yeast extract (1% solution)	1.0mL

Selenite Solution:

Composition per 10.0mL:

NaOH	0.5g
$Na_2SeO_3 \cdot 5H_2O$	0.1g

Preparation of Selenite Solution: Add components to distilled/deionized water and bring volume to 1.0L. Mix thoroughly. Filter sterilize.

Cycloheximide Solution:

Composition per 10.0mL:

Cycloheximide	0.02g

Preparation of Cycloheximide Solution: Add cycloheximide to distilled/deionized water and bring volume to 10.0mL. Mix thoroughly. Filter sterilize.

Caution: Cycloheximide is toxic. Avoid skin contact or aerosol formation and inhalation.

Preparation of Medium: Add components, except cycloheximide solution and selenite solution, to distilled/deionized water and bring volume to 1.0L. Mix thoroughly. Distribute 100.0mL into flasks. Gently heat and bring to boiling. Autoclave for 15 min at 15 psi pressure–121°C. Cool to 50°C. Aseptically add, per 100.0mL of medium, 1.0mL sterile selenite solution and 1.0mL cycloheximide solution. Mix thoroughly. Aseptically pour into sterile Petri dishes.

Use: For the selective cultivation of *Agrobacterium tumefaciens* biovar 2.

AGS

See: **Arginine Glucose Slant**

AH5 Medium

Composition per 205.9mL:

Agar base	160.0mL
Supplement solution	45.9mL

pH 6.0 ± 0.2 at 25°C

Agar Base:

Composition per 165.0mL:

Pancreatic digest of casein	2.72g
Agar	2.1g
NaCl	0.8g
Papaic digest of soybean meal	0.48g
K_2HPO_4	0.4g
Glucose	0.4g

Preparation of Agar Base: Add components, except agar, to distilled/deionized water and bring volume to 165.0mL. Adjust pH to 5.5. Add agar. Mix thoroughly. Autoclave for 15 min at 15 psi pressure–121°C. Cool to 45°–50°C.

Supplement Solution:

Composition per 45.9mL:

Horse serum, unheated	40.0mL
Fresh yeast extract solution	2.0mL
Penicillin solution	2.0mL
CVA enrichment	1.0mL
L-Cysteine·HCl·H_2O solution	0.5mL
Urea solution	0.4mL

Preparation of Supplement Solution: Aseptically combine components. Mix thoroughly.

Fresh Yeast Extract Solution:
Composition per 100.0mL:
Baker's yeast, live, pressed, starch-free 25.0g

Preparation of Fresh Yeast Extract Solution: Add the live Baker's yeast to 100.0mL of distilled/deionized water. Autoclave for 90 min at 15 psi pressure–121°C. Allow to stand. Remove supernatant solution. Adjust pH to 6.6–6.8.

Penicillin Solution:
Composition per 10.0mL:
Penicillin G 1,000,000U

Preparation of Penicillin Solution: Add penicillin to distilled/deionized water and bring volume to 10.0mL. Mix thoroughly. Filter sterilize.

CVA Enrichment:
Composition per liter:
Glucose 100.0g
L-Cysteine·HCl·H_2O 25.9g
L-Glutamine 10.0g
L-Cystine·2HCl 1.0g
Adenine 1.0g
Nicotinamide adenine dinucleotide 0.25g
Cocarboxylase 0.1g
Guanine·HCl 0.03g
$Fe(NO_3)_3$ 0.02g
p-Aminobenzoic acid 0.013g
Vitamin B_{12} 0.01g
Thiamine·HCl 3.0mg

Preparation of CVA Enrichment: Add components to distilled/deionized water and bring volume to 1.0L. Mix thoroughly. Filter sterilize.

L-Cysteine·HCl·H_2O Solution:
Composition per 10.0mL:
L-Cysteine·HCl·H_2O 0.4g

Preparation of L-Cysteine·HCl·H_2O Solution: Add L-cysteine·HCl·H_2O solution to distilled/deionized water and bring volume to 10.0mL. Mix thoroughly. Filter sterilize.

Urea Solution:
Composition per 10.0mL:
Urea 1.0g

Preparation of Urea Solution: Add urea to distilled/deionized water and bring volume to 10.0mL. Mix thoroughly. Filter sterilize.

Preparation of Medium: Aseptically combine cooled, sterile components. Mix thoroughly. Pour into sterile Petri dishes or distribute into sterile tubes.

Use: For the cultivation of *Ureaplasma urealyticum* from urine and exudates and for the cultivation of other *Ureaplasma* species.

AJYE Medium
See: **Apple Juice Yeast Extract Medium**

AK Agar No. 2 (Sporulating Agar)

Composition per liter:
Agar 15.0g
Pancreatic digest of gelatin 6.0g
Pancreatic digest of casein 4.0g
Yeast extract 3.0g
Beef extract 1.5g
Glucose 1.0g
$MnSO_4 \cdot 7H_2O$ 0.3g
pH 6.6 ± 0.2 at 25°C

Source: This medium is available as a premixed powder from BD Diagnostic Systems.

Preparation of Medium: Add components to distilled/deionized water and bring volume to 1.0L. Mix thoroughly. Gently heat while stirring and bring to boiling. Distribute into tubes or flasks. Autoclave for 20 min at 15 psi pressure–121°C. Make sure medium is dissolved before autoclaving.

Use: For the preparation of spore suspensions used to detect antibiotic residues in milk and dairy products.

AKI Medium

Composition per liter:
Peptone 15.0g
NaCl 5.0g
Yeast extract 4.0g
Sodium bicarbonate solution 30.0mL
pH 7.2 ± 0.2 at 25°C

Sodium Bicarbonate Solution:
Composition per 100.0mL:
$NaHCO_3$ 10.0g

Preparation of Sodium Bicarbonate Solution: Add sodium bicarbonate to distilled/deionized water and bring volume to 100.0mL. Mix thoroughly. Filter sterilize. Use freshly prepared solution.

Preparation of Medium: Add components, except sodium bicarbonate solution, to distilled/deionized water and bring volume to 970.0mL. Mix thoroughly. Autoclave for 15 min at 15 psi pressure–121°C. Cool to 45°–50°C. Aseptically add sterile sodium bicarbonate solution. Mix thoroughly. Aseptically distribute into sterile tubes or flasks. Prepare medium freshly.

Use: For the cultivation of *Vibrio cholerae* and other *Vibrio* species.

Albumin Fatty Acid Broth, *Leptospira* Medium
See: **Bovine Albumin Tween™ 80 Medium, Ellinghausen and McCullough, Modified**

Albumin Fatty Acid Semisolid Medium, Modified
See: **Bovine Albumin Tween™ 80 Semisolid Medium, Ellinghausen and McCullough, Modified**

Alcal Mannose Medium

Composition per liter:
K_2HPO_4 .... 15.1g
KH_2PO_4 .... 5.6g
Mannose 1.0g
Yeast extract 1.0g
Casamino acids 0.5g
$MgSO_4 \cdot 7H_2O$ 0.4g
$CaCl_2 \cdot 2H_2O$ 50.0mg
$FeSO_4 \cdot 7H_2O$ 10.0mg

Preparation of Medium: Add components to distilled/deionized water and bring volume to 1.0L. Mix thoroughly. Distribute into tubes or flasks. Autoclave for 15 min at 15 psi pressure–121°C.

Use: For the cultivation of *Bacillus circulans*.

Alcaligenes Agar

Composition per liter:

Agar	10.0g
Peptone	5.0g
Ammonium lactate	3.0g
Meat extract	3.0g
Ferric citrate	0.2g

pH 7.0 ± 0.2 at 25°C

Preparation of Medium: Add ferric citrate to distilled/deionized water and bring volume to 100.0mL. In a separate flask, add remaining components to distilled/deionized water and bring volume to 900.0mL. Mix thoroughly. Adjust pH to 7.0. Steam the two solutions for 20 min on three consecutive days. Aseptically combine the two solutions. Pour into sterile Petri dishes or distribute into sterile tubes.

Use: For the cultivation of *Alcaligenes* species.

Alcaligenes Medium

Composition per liter:

Peptone	5.0g
Beef extract	3.0g
Ferric citrate	0.2g
Ammonium lactate solution	3.0mL

pH 7.0 ± 0.2 at 25°C

Ammonium Lactate Solution:

Composition per 100.0mL:

Lactic acid	60.0g

Preparation of Ammonium Lactate Solution: Dissolve lactic acid in 100.0mL of distilled/deionized water. Neutralize with NH_4OH to pH 7.0.

Preparation of Medium: Add peptone, beef extract, and ammonium lactate to distilled/deionized water and bring volume to 1.0L. Mix thoroughly. Gently heat and bring to boiling. Autoclave for 15 min at 15 psi pressure–121°C. Add ferric citrate aseptically. Mix thoroughly. Aseptically distribute into tubes or flasks.

Use: For the cultivation of *Alcaligenes tolerans*.

Alcaligenes Medium

Composition per liter:

Tris	6.06g
NaCl	4.68g
KCl	1.49g
NH_4Cl	1.07g
Na_2SO_4	0.43g
$Na_2HPO_4 \cdot 12H_2O$	0.23g
$MgCl_2 \cdot 6H_2O$	0.2g
$CaCl_2 \cdot 2H_2O$	0.03g
Ferric ammonium citrate	0.005g
Sodium succinate solution	10.0mL
$CuSO_4$ solution	2.5mL
Trace elements solution SL-7	1.0mL

Sodium Succinate Solution:

Composition per 100.0mL:

Sodium succinate	40.0g

Preparation of Sodium Succinate Solution: Add sodium succinate to distilled/deionized water and bring volume to 100.0mL. Mix thoroughly. Filter sterilize.

$CuSO_4$ Solution:

Composition per 100.0mL:

$CuSO_4$	16.0g

Preparation of $CuSO_4$ Solution: Add $CuSO_4$ to distilled/deionized water and bring volume to 100.0mL. Mix thoroughly. Filter sterilize.

Trace Elements Solution SL-7:

Composition per 1001.0mL:

$CoCl_2 \cdot 6H_2O$	200.0mg
$MnCl_2 \cdot 4H_2O$	100.0mg
$ZnCl_2$	70.0mg
H_3BO_3	60.0mg
$Na_2MoO_4 \cdot 2H_2O$	40.0mg
$CuCl_2 \cdot 2H_2O$	20.0mg
$NiCl_2 \cdot 6H_2O$	20.0mg
HCl (25%)	1.0mL

Preparation of Trace Elements Solution SL-7: Add components to distilled/deionized water and bring volume to 1.0L. Mix thoroughly.

Preparation of Medium: Add components, except $CuSO_4$ solution and sodium succinate solution, to distilled/deionized water and bring volume to 987.5mL. Mix thoroughly. Autoclave for 15 min at 15 psi pressure–121°C. Aseptically add 10.0mL of sterile $CuSO_4$ solution and 2.5mL of sterile sodium succinate solution. Mix thoroughly. Aseptically distribute into sterile tubes or flasks.

Use: For the cultivation of *Alcaligenes* species.

Alcaligenes N5 Medium

Composition per liter:

Sodium succinate$\cdot 2H_2O$	5.0g
KH_2PO_4	0.75g
NH_4Cl	0.67g
K_2HPO_4	0.61g
$MgSO_4 \cdot 7H_2O$	0.2g
$CaCl_2 \cdot 2H_2O$	0.03g
$MnCl_2 \cdot 4H_2O$	3.0mg
$FeCl_3$	2.4mg
$Na_2MoO_4 \cdot 2H_2O$	1.0mg

Preparation of Medium: Add components to distilled/deionized water and bring volume to 1.0L. Mix thoroughly. Gently heat while stirring and bring to boiling. Distribute into tubes or flasks. Autoclave for 15 min at 15 psi pressure–121°C.

Use: For the cultivation and maintenance of *Alcaligenes faecalis*.

Alcaligenes NA YE Medium (*Alcaligenes* Nutrient Agar Yeast Extract Medium)

Composition per liter:

Agar	15.0g
Pancreatic digest of gelatin	5.0g
Yeast extract	5.0g
Beef extract	3.0g

pH 7.0 ± 0.2 at 25°C

Preparation of Medium: Add components to distilled/deionized water and bring volume to 1.0L. Mix thoroughly. Gently heat while stirring and bring to boiling. Distribute into tubes or flasks. Autoclave

for 15 min at 15 psi pressure–121°C. Pour into sterile Petri dishes or leave in tubes.

Use: For the cultivation and maintenance of *Alcaligenes* species.

Alcaligenes NB YE Agar (*Alcaligenes* Nutrient Broth Yeast Extract Agar)

Composition per liter:

Agar	15.0g
Pancreatic digest of gelatin	5.0g
Yeast extract	5.0g
Beef extract	3.0g

Preparation of Medium: Add components to distilled/deionized water and bring volume to 1.0L. Mix thoroughly. Gently heat while stirring and bring to boiling. Distribute into tubes or flasks. Autoclave for 15 min at 15 psi pressure–121°C. Pour into sterile Petri dishes or leave in tubes.

Use: For the cultivation and maintenance of *Alcaligenes faecalis*.

Alcaligenes NB YE Broth (*Alcaligenes* Nutrient Broth Yeast Extract Broth)

Composition per liter:

Pancreatic digest of gelatin	5.0g
Yeast extract	5.0g
Beef extract	3.0g

Preparation of Medium: Add components to distilled/deionized water and bring volume to 1.0L. Mix thoroughly. Gently heat while stirring and bring to boiling. Distribute into tubes or flasks. Autoclave for 15 min at 15 psi pressure–121°C.

Use: For the cultivation of *Alcaligenes faecalis*.

Alcaligenes NB YE Medium (*Alcaligenes* Nutrient Broth Yeast Extract Medium)

Composition per liter:

Pancreatic digest of gelatin	5.0g
Yeast extract	5.0g
Beef extract	3.0g

pH 7.0 ± 0.2 at 25°C

Preparation of Medium: Add components to distilled/deionized water and bring volume to 1.0L. Mix thoroughly. Distribute into tubes or flasks. Autoclave for 15 min at 15 psi pressure–121°C.

Use: For the cultivation and maintenance of *Alcaligenes* species.

Alcaligenes Nutrient Agar Yeast Extract Medium

See: ***Alcaligenes*** **NA YE Medium**

Alcaligenes Nutrient Broth Yeast Extract Agar

See: ***Alcaligenes*** **NB YE Agar**

Alcaligenes Nutrient Broth Yeast Extract Broth

See: ***Alcaligenes*** **NB YE Broth**

Alcaligenes Nutrient Broth Yeast Extract Medium

See: ***Alcaligenes*** **NB YE Medium**

Alcaligenes xylosoxydans Medium with Benzoate

Composition per liter:

Solution A	500.0mL
Solution B	500.0mL

pH 7.4 ± 0.2 at 25°C

Solution A:

Composition per 500.0mL

K_2HPO_4	0.65g
KH_2PO_4	0.19g

Preparation of Solution A: Add components to distilled/deionized water and bring volume to 500.0mL. Mix thoroughly. Autoclave for 15 min at 15 psi pressure–121°C.

Solution B:

Composition per 500.0mL:

Sodium glutamate	4.0g
$NaNO_3$	0.5g
$MgSO_4 \cdot 7H_2O$	0.1g
Trace elements solution SL-4	2.0mL

Preparation of Solution B: Add components to distilled/deionized water and bring volume to 500.0mL. Mix thoroughly. Autoclave for 15 min at 15 psi pressure–121°C.

Trace Elements Solution SL-4:

Composition per liter:

EDTA	0.5g
$FeSO_4 \cdot 7H_2O$	0.2g
Trace elements solution SL-6	100.0mL

Trace Elements Solution SL-6:

Composition per liter:

$MnCl_2 \cdot 4H_2O$	0.5g
H_3BO_3	0.3g
$CoCl_2 \cdot 6H_2O$	0.2g
$ZnSO_4 \cdot 7H_2O$	0.1g
$Na_2MoO_4 \cdot 2H_2O$	0.03g
$NiCl_2 \cdot 6H_2O$	0.02g
$CuCl_2 \cdot 2H_2O$	0.01g

Preparation of Trace Elements Solution SL-6: Add components to distilled/deionized water and bring volume to 1.0L. Mix thoroughly.

Preparation of Trace Elements Solution SL-4: Add components to distilled/deionized water and bring volume to 1.0L. Mix thoroughly.

Preparation of Medium: Aseptically combine solution A and solution B. Mix thoroughly. Adjust pH to 7.4. Distribute into sterile tubes or flasks.

Use: For the cultivation and maintenance of *Alcaligenes xylosoxydans*.

Alcaliphilic *Amphibacillus* Strains Medium (DSMZ Medium 931)

Composition per liter:

Na_2CO_3	63.6g
$NaHCO_3$	50.4g
KH_2PO_4	0.2g
$MgCl_2$	0.1g
NH_4Cl	0.5g
KCl	0.2g
Resazurin	0.01g
Sucrose solution	50.0mL
$Na_2S \cdot 9H_2O$ solution	10.0mL

Yeast extract solution 10.0mL
Vitamin solution 10.0mL
Trace elements solution 1.0mL

pH 9.5-10.0 at 25°C

Sucrose Solution:

Composition per 50.0mL:

Sucrose 5.0g

Preparation of Sucrose Solution: Add sucrose to distilled/deionized water and bring volume to 50.0mL. Mix thoroughly. Sparge with 100% N_2. Autoclave for 15 min at 15 psi pressure–121°C. Cool to room temperature.

Yeast Extract Solution:

Composition per 10.0mL:

Yeast extract 0.2g

Preparation of Yeast Extract Solution: Add yeast extract to distilled/deionized water and bring volume to 10.0mL. Mix thoroughly. Sparge with 100% N_2. Autoclave under 100% N_2 for 15 min at 15 psi pressure–121°C. Cool to room temperature.

$Na_2S \cdot 9H_2O$ Solution:

Composition per 10.0mL:

$Na_2S \cdot 9H_2O$ 0.7g

Preparation of $Na_2S \cdot 9H_2O$ Solution: Add $Na_2S \cdot 9H_2O$ to distilled/deionized water and bring volume to 10.0mL. Mix thoroughly. Autoclave under 100% N_2 for 15 min at 15 psi pressure–121°C. Cool to room temperature.

Trace Elements Solution:

Composition per liter:

$MgSO_4 \cdot 7H_2O$ 3.0g
Nitrilotriacetic acid 1.5g
NaCl 1.0g
$MnSO_4 \cdot 2H_2O$ 0.5g
$CoSO_4 \cdot 7H_2O$ 0.18g
$ZnSO_4 \cdot 7H_2O$ 0.18g
$CaCl_2 \cdot 2H_2O$ 0.1g
$FeSO_4 \cdot 7H_2O$ 0.1g
$NiCl_2 \cdot 6H_2O$ 0.025g
$KAl(SO_4)_2 \cdot 12H_2O$ 0.02g
H_3BO_3 .... 0.01g
$Na_2MoO_4 \cdot 4H_2O$ 0.01g
$CuSO_4 \cdot 5H_2O$ 0.01g
$Na_2SeO_3 \cdot 5H_2O$ 0.3mg

Preparation of Trace Elements Solution: Add nitrilotriacetic acid to 500.0mL of distilled/deionized water. Dissolve by adjusting pH to 6.5 with KOH. Add remaining components. Add distilled/deionized water to 1.0L. Mix thoroughly.

Vitamin Solution:

Composition per liter:

Pyridoxine-HCl 10.0mg
Thiamine-HCl·$2H_2O$ 5.0mg
Riboflavin 5.0mg
Nicotinic acid 5.0mg
D-Ca-pantothenate 5.0mg
p-Aminobenzoic acid 5.0mg
Lipoic acid 5.0mg
Biotin 2.0mg
Folic acid 2.0mg
Vitamin B_{12} 0.1mg

Preparation of Vitamin Solution: Add components to distilled/deionized water and bring volume to 1.0L. Mix thoroughly. Sparge with 80% H_2 + 20% CO_2. Filter sterilize.

Preparation of Medium: Prepare and dispense medium under 100% N_2 gas atmosphere. Add components, except $NaHCO_3$, NH_4Cl, Na_2CO_3, sucrose solution, $Na_2S \cdot 9H_2O$ solution, yeast extract solution, and vitamin solution, to distilled/deionized water and bring volume to 920.0mL. Mix thoroughly. Gently heat and bring to boiling. Boil for 5 min. Cool to room temperature while sparging with 100% N_2. Add solid $NaHCO_3$, NH_4Cl, and Na_2CO_3. Mix thoroughly. Distribute into anaerobe tubes or bottles. Autoclave for 15 min at 15 psi pressure–121°C. Aseptically and anaerobically add per liter of medium 50.0mL sucrose solution, 10.0mL yeast extract solution, 10.0mL $Na_2S \cdot 9H_2O$ solution, and 10.0mL vitamin solution. The final pH should be 9.5–10.0.

Use: For the cultivation of *Amphibacillus fermentum* and *Amphibacillus tropicus.*

Alcanivorax borkumensis Medium (DSMZ Medium 809)

Composition per liter:

NaCl 23.0g
Sodium pyruvate 10.0g
$MgCl_2 \cdot 2H_2O$ 6.16g
$MgSO_4 \cdot 7H_2O$ 5.8g
$NaNO_3$ 5.0g
$CaCl_2 \cdot 2H_2O$ 1.47g
$Na_2HPO_4 \cdot 7H_2O$ 0.89g
$FeSO_4 \cdot 7H_2O$ 0.03g

pH 7.0–7.5 at 25°C

Preparation of Medium: Add components to distilled/deionized water and bring volume to 1.0L. Mix thoroughly. Distribute into tubes or flasks. Autoclave for 15 min at 15 psi pressure–121°C.

Use: For the cultivation of *Alcanivorax borkumensis.*

Algae Culture Broth

Composition per liter:

$NaNO_3$ 1.0g
$MgSO_4 \cdot 7H_2O$ 0.513g
NH_4Cl 0.5g
K_2HPO_4 .... 0.25g
$CaCl_2 \cdot 2H_2O$ 0.058g
$FeCl_3$ 3.0mg

pH 7.4 ± 0.2 at 25°C

Source: This medium is available from HiMedia.

Preparation of Medium: Add components to distilled/deionized water and bring volume to 1.0L. Mix thoroughly. Gently heat while stirring and bring to boiling. Distribute into tubes or flasks. Autoclave for 15 min at 15 psi pressure–121°C.

Use: For the isolation and cultivation of algae.

Algal Proteose Agar

Composition per liter:

Agar 15.0g
Proteose peptone 1.0g
Bristol's solution 1.0L

Bristol's Solution:

Composition per 1000.1mL:

$NaNO_3$ solution	10.0g
KH_2PO_4 solution	7.0g
K_2HPO_4 solution	3.0g
$MgSO_4 \cdot 7H_2O$ solution	3.0g
$CaCl_2$ solution	1.0g
NaCl solution	1.0g
$FeCl_3$ solution	0.1mL

$NaNO_3$ Solution:

Composition per 400.0mL:

$NaNO_3$	10.0g

Preparation of $NaNO_3$ Solution: Add $NaNO_3$ to distilled/deionized water and bring volume to 400.0mL. Mix thoroughly.

$CaCl_2$ Solution:

Composition per 400.0mL:

$CaCl_2$	1.0g

Preparation of $CaCl_2$ Solution: Add $CaCl_2$ to distilled/deionized water and bring volume to 400.0mL. Mix thoroughly.

$MgSO_4 \cdot 7H_2O$ Solution:

Composition per 400.0mL:

$MgSO_4 \cdot 7H_2O$	3.0g

Preparation of $MgSO_4 \cdot 7H_2O$ Solution: Add $MgSO_4 \cdot 7H_2O$ to distilled/deionized water and bring volume to 400.0mL. Mix thoroughly.

K_2HPO_4 Solution:

Composition per 400.0mL:

K_2HPO_4	3.0g

Preparation of K_2HPO_4 Solution: Add K_2HPO_4 to distilled/deionized water and bring volume to 400.0mL. Mix thoroughly.

KH_2PO_4 Solution:

Composition per 400.0mL:

KH_2PO_4	7.0g

Preparation of KH_2PO_4 Solution: Add KH_2PO_4 to distilled/deionized water and bring volume to 400.0mL. Mix thoroughly.

NaCl Solution:

Composition per 400.0mL:

NaCl	1.0g

Preparation of NaCl Solution: Add NaCl to distilled/deionized water and bring volume to 400.0mL. Mix thoroughly.

$FeCl_3$ Solution:

Composition per 100.0mL:

$FeCl_3$	1.0g

Preparation of $FeCl_3$ Solution: Add $FeCl_3$ to distilled/deionized water and bring volume to 100.0mL. Mix thoroughly.

Preparation of Bristol's Solution: Add 10.0mL of $NaNO_3$ solution, 10.0mL of $CaCl_2$ solution, 10.0mL of $MgSO_4 \cdot 7H_2O$ solution, 10.0mL of $NaNO_3$ solution, 10.0mL of K_2HPO_4 solution, 10.0mL of KH_2PO_4 solution, and 10.0mL of NaCl solution to distilled/deionized water and bring volume to 1.0L. Mix thoroughly. Add 0.1mL of $FeCl_3$ solution. Mix thoroughly.

Preparation of Medium: Add proteose peptone and agar to 1.0L of Bristol's solution. Mix thoroughly. Gently heat and bring to boiling. Distribute into tubes or flasks. Autoclave for 15 min at 15 psi pressure–121°C. Pour into sterile Petri dishes or leave in tubes.

Use: For the cultivation of *Ankistrodesmus angustus*, *Ankistrodesmus braunii*, *Botrydium becherianum*, *Botrydium cystosum*, *Botrydium stoloniferum*, *Bracteacoccus grandis*, *Bumilleria sicula*, *Characium polymorphum*, *Chlamydomonas* species, *Chlorella* species, *Chlorosphaera klebsii*, *Coelastrum proboscideum*, *Crucigenia apiculata*, *Dictyochloris fragrans*, *Dictyosphaerium ehrenbergianum*, *Dictyosphaerium pulchellum*, *Elakatothrix viridis*, *Haematococcus lacustris*, *Interfilum paradoxum*, *Klebsormidium subtilissimum*, *Lobomonas piriformis*, *Mesotaenium caldariorum*, *Mischococcus sphaerocephalus*, *Monodus subterraneus*, *Muriella aurantiaca*, *Muriella decolor*, *Nephrochlamys subsolitaria*, *Nephrodiella brevis*, *Oocystis* species, *Ophiocytium majus*, *Pediastrum tetras*, *Polyedriella helvetica*, *Protosiphon botryoides*, *Scenedesmus armatus*, *Scenedesmus communis*, *Scenedesmus obliquus*, *Tetracystis disociata*, *Tribonema aequale*, *Ulothrix gigas*, *Vitreochlamys incisa*, and *Vischeria punctata*.

Alginate Utilization Medium

Composition per liter:

Solution B	500.0mL
Solution A	400.0mL
Solution C	100.0mL

Solution A:

Composition per 400.0mL:

Marine salts	38.0g

Preparation of Solution A: Add marine salts to distilled/deionized water and bring volume to 400.0mL. Mix thoroughly. Autoclave for 15 min at 15 psi pressure–121°C.

Solution B:

Composition per 500.0mL:

Agar	20.0g
Sodium alginate	10.0g

Preparation of Solution B: Add components to distilled/deionized water and bring volume to 500.0mL. Mix thoroughly. Autoclave for 15 min at 15 psi pressure–121°C.

Solution C:

Composition per 100.0mL:

Tris·HCl buffer	0.067g
$NaNO_3$	0.047g
Ferric EDTA	66.5mg
Sodium glycerophosphate	6.67mg
Thiamine·HCl	67.0μg
Vitamin B_{12}	1.3μg
Biotin	0.67μg

Preparation of Solution C: Add components to distilled/deionized water and bring volume to 100.0mL. Mix thoroughly. Filter sterilize.

Preparation of Medium: Aseptically combine solutions A, B, and C. For liquid medium, omit agar from solution B.

Use: For the cultivation of microorganisms that can utilize alginate as a carbon source. Growth on alginate (production of alginase) is a diagnostic test used in the differentiation of *Vibrio* species.

Alicyclobacillus acidoterrestris Agar

Composition per 1001.0mL:

Solution A	500.0mL
Solution C	500.0mL
Solution B	1.0mL

pH 4.0 ± 0.2 at 25°C

Solution A:

Composition per 500.0mL:

Glucose	5.0g
KH_2PO_4	3.0g
Yeast extract	2.0g
$MgSO_4 \cdot 7H_2O$	0.5g
$CaCl_2 \cdot 7H_2O$	0.25g
$(NH_4)_2SO_4$	0.2g

Preparation of Solution A: Add components to distilled/deionized water and bring volume to 500.0mL. Mix thoroughly. Adjust pH to 4.0. Distribute into tubes or flasks. Autoclave for 15 min at 15 psi pressure–121°C. Cool to 50°–55°C.

Solution B:

Composition per liter:

$MnCl_2 \cdot 4H_2O$	0.5g
H_3BO_3	0.3g
$CoCl_2 \cdot 6H_2O$	0.2g
$ZnSO_4 \cdot 7H_2O$	0.1g
$Na_2MoO_4 \cdot 2H_2O$	0.03g
$NiCl_2 \cdot 6H_2O$	0.02g
$CuCl_2 \cdot 2H_2O$	0.01g

Preparation of Solution B: Add components to distilled/deionized water and bring volume to 1.0L. Mix thoroughly. Filter sterilize. Warm to 50°–55°C.

Solution C:

Composition per 500.0mL:

Agar	15.0g

Preparation of Solution C: Add agar to distilled/deionized water and bring volume to 500.0mL. Mix thoroughly. Gently heat and bring to boiling. Autoclave for 15 min at 15 psi pressure–121°C. Cool to 50°–55°C.

Preparation of Medium: Aseptically combine 500.0mL of solution A, 1.0mL of solution B, and 500.0mL of solution C. Mix thoroughly. Pour into sterile Petri dishes or distribute into sterile tubes.

Use: For the cultivation and maintenance of *Alicyclobacillus acidoterrestris.*

Alicyclobacillus acidoterrestris Broth

Composition per 1001.0mL:

Solution A	1.0L
Solution B	1.0mL

pH 4.0 ± 0.2 at 25°C

Solution A:

Composition per liter:

Glucose	5.0g
KH_2PO_4	3.0g
Yeast extract	2.0g
$MgSO_4 \cdot 7H_2O$	0.5g
$CaCl_2 \cdot 7H_2O$	0.25g
$(NH_4)_2SO_4$	0.2g

Preparation of Solution A: Add components to distilled/deionized water and bring volume to 1.0L. Mix thoroughly. Adjust pH to 4.0. Distribute into tubes or flasks. Autoclave for 15 min at 15 psi pressure–121°C.

Solution B:

Composition per liter:

$MnCl_2 \cdot 4H_2O$	0.5g
H_3BO_3	0.3g
$CoCl_2 \cdot 6H_2O$	0.2g
$ZnSO_4 \cdot 7H_2O$	0.1g
$Na_2MoO_4 \cdot 2H_2O$	0.03g
$NiCl_2 \cdot 6H_2O$	0.02g
$CuCl_2 \cdot 2H_2O$	0.01g

Preparation of Solution B: Add components to distilled/deionized water and bring volume to 1.0L. Mix thoroughly. Filter sterilize.

Preparation of Medium: Aseptically combine 1.0L of solution A with 1.0mL of solution B. Mix thoroughly. Aseptically distribute into sterile tubes or flasks.

Use: For the cultivation of *Alicyclobacillus acidoterrestris.*

Alicyclobacillus acidoterrestris Medium (LMG Medium 141)

Composition per liter:

Agar	30.0g
Glucose	5.0g
K_2HPO_4	3.0g
Yeast extract	2.0g
$MgSO_4 \cdot 7H_2O$	0.5g
$CaCl_2 \cdot 2H_2O$	0.25g
$(NH_4)_2SO_4$	0.2g
Agar solution	500.0mL
Trace elements solution	1.0mL

pH 4.0 ± 0.2 at 25°C

Trace Elements Solution:

Composition per liter:

$CaCl_2 \cdot 2H_2O$	0.66g
$Na_2MoO_4 \cdot 2H_2O$	0.3g
$ZnSO_4 \cdot 7H_2O$	0.18g
$CoCl_2 \cdot 6H_2O$	0.18g
$CuSO_4 \cdot 5H_2O$	0.16g
$MnSO_4 \cdot 4H_2O$	0.15g
H_3BO_3	0.1g

Preparation of Trace Elements Solution: Add components to distilled/deionized water and bring volume to 1.0L. Mix thoroughly.

Agar Solution:

Composition per 500.0mL:

Agar	30.0g

Preparation of Agar Solution: Add agar to distilled/deionized water and bring volume to 500.0mL. Mix thoroughly. Gently heat and bring to boiling. Autoclave for 15 min at 15 psi pressure–121°C. Cool to 45°–50°C.

Preparation of Medium: Add components, except agar solution, to distilled/deionized water and bring volume to 500.0mL. Mix thoroughly. Adjust pH to 4.0 with H_2SO_4. Autoclave for 15 min at 15 psi pressure–121°C. Cool to 45°–50°C. Aseptically add 500.0mL agar solution. Mix thoroughly. Aseptically pour into sterile Petri dishes or distribute into sterile tubes.

Use: For the cultivation of *Alicyclobacillus acidoterrestris.*

Alicyclobacillus Agar (DSMZ Medium 402)

Composition per liter:

Glucose	5.0g
KH_2PO_4	3.0g
Yeast extract	2.0g
$MgSO_4 \cdot 7H_2O$	0.5g

$CaCl_2·2H_2O$ 0.25g
$(NH_4)_2 SO_4$ 0.2g
Agar solution 500.0mL
Trace elements solution SL-6 .... 1.0mL
pH 4.0 ± 0.2 at 25°C

Trace Elements Solution SL-6:

Composition per liter:

$MnCl_2·4H_2O$ 0.5g
H_3BO_3 .... 0.3g
$CoCl_2·6H_2O$ 0.2g
$ZnSO_4·7H_2O$ 0.1g
$Na_2MoO_4·2H_2O$ 0.03g
$NiCl_2·6H_2O$ 0.02g
$CuCl_2·2H_2O$ 0.01g

Preparation of Trace Elements Solution SL-6: Add components to distilled/deionized water and bring volume to 1.0L. Mix thoroughly. Autoclave for 15 min at 15 psi pressure–121°C. Cool to room temperature.

Agar Solution:

Composition per 500mL:

Agar 15.0g

Preparation of Agar Solution: Add agar to distilled/deionized water and bring volume to 500.0mL. Mix thoroughly. Gently heat and bring to boiling. Autoclave for 15 min at 15 psi pressure–121°C. Cool to 45°C.

Preparation of Medium: Add components, except trace elements solution SL-6 and agar solution, to distilled/deionized water and bring volume to 499.0mL. Mix thoroughly. Adjust pH to 4.0. Autoclave for 15 min at 15 psi pressure–121°C. Cool to 45°C. Aseptically add 1.0mL of sterile trace elements solution SL-6 and 500.0mL agar solution. Mix thoroughly. Pour into sterile Petri dishes or aseptically distribute into sterile tubes or flasks.

Use: For the cultivation and maintenance of *Alicyclobacillus* spp., *Bacillus* sp., *and Bacillus naganoensis.*

Alicyclobacillus cycloheptanicus Agar (LMG Medium 174)

Composition per 1001.0mL:

Solution A 500.0mL
Agar solution 500.0mL
Trace elements solution SL-6 .... 1.0mL
pH 4.0 ± 0.2 at 25°C

Solution A:

Composition per 500.0mL:

Yeast extract 5.0g
Glucose 5.0g
K_2HPO_4 .... 3.0g
$MgSO_4·7H_2O$ 0.5g
$CaCl_2·2H_2O$ 0.25g
$(NH_4)_2SO_4$ 0.2g

Preparation of Solution A: Add components to distilled/deionized water and bring volume to 500.0mL. Mix thoroughly. Adjust to pH 4.0. Autoclave for 15 min at 15 psi pressure–121°C. Cool to 45°–50°C.

Trace Elements Solution SL-6:

Composition per liter:

H_3BO_3 .... 0.3g
$CoCl_2·6H_2O$ 0.2g
$ZnSO_4·7H_2O$ 0.1g
$MnCl_2·4H_2O$ 0.03g
$Na_2MoO_4·H_2O$ 0.03g
$NiCl_2·6H_2O$ 0.02g
$CuCl_2·2H_2O$ 0.01g

Preparation of Trace Elements Solution SL-6: Add components to distilled/deionized water and bring volume to 1.0L. Mix thoroughly. Adjust pH to 3.4. Filter sterilize.

Agar Solution:

Composition per 500.0mL:

Agar 30.0g

Preparation of Agar Solution: Add agar to distilled/deionized water and bring volume to 500.0mL. Mix thoroughly. Gently heat and bring to boiling. Autoclave for 15 min at 15 psi pressure–121°C. Cool to 45°–50°C.

Preparation of Medium: Aseptically combine 500.0mL solution A, 500.0mL sterile agar solution, and 1.0mL sterile trace elements solution SL-6. Mix thoroughly. Aseptically pour into sterile Petri dishes or distribute into sterile tubes.

Use: For the cultivation of *Alicyclobacillus cycloheptanicus.*

Alicyclobacillus cycloheptanicus Medium (LMG Medium 174)

Composition per 1001.0mL:

Solution A 1.0L
Trace elements solution SL-6 .... 1.0mL
pH 4.0 ± 0.2 at 25°C

Solution A:

Composition per liter:

Yeast extract 5.0g
Glucose 5.0g
K_2HPO_4 .... 3.0g
$MgSO_4·7H_2O$ 0.5g
$CaCl_2·2H_2O$ 0.25g
$(NH_4)_2SO_4$ 0.2g

Preparation of Solution A: Add components to distilled/deionized water and bring volume to 1.0L. Mix thoroughly.

Trace Elements Solution SL-6:

Composition per liter:

H_3BO_3 .... 0.3g
$CoCl_2·6H_2O$ 0.2g
$ZnSO_4·7H_2O$ 0.1g
$MnCl_2·4H_2O$ 0.03g
$Na_2MoO_4·H_2O$ 0.03g
$NiCl_2·6H_2O$ 0.02g
$CuCl_2·2H_2O$ 0.01g

Preparation of Trace Elements Solution SL-6: Add components to distilled/deionized water and bring volume to 1.0L. Mix thoroughly. Adjust pH to 3.4.

Preparation of Medium: Add 1.0mL trace elements solution SL-6 to 1.0L of solution A. Mix thoroughly. Adjust pH to 4.0. Distribute to tubes or flasks. Autoclave for 15 min at 15 psi pressure–121°C.

Use: For the cultivation of *Alicyclobacillus cycloheptanicus.*

Alicyclobacillus ferrooxydans Medium (DSMZ Medium 1201)

Composition per liter:

$MgSO_4·7H_2O$ 0.5g
$(NH_4)_2SO_4$ 0.4g

K_2HPO_4	0.2g
Yeast extract	0.2g
$K_2S_4O_6$	0.15g
KCl	0.1g
$MnSO_4 \cdot H_2O$	0.01g
Iron sulfate solution	70.0mL

pH 1.8–2.5 at 25°C

Iron Sulfate Solution:

Composition per 100.0mL:

$FeSO_4 \cdot 7H_2O$	20.0g

Preparation of Iron Sulfate Solution: Add components to 0.2*N* $H_2SO_4$and bring volume with distilled/deionized water to 100.0mL. Mix thoroughly. Filter sterilize.

Preparation of Medium: Add components, except iron sulfate solution, to distilled/deionized water and bring volume to 970.0mL. Mix thoroughly. Gently heat and bring to boiling. Autoclave for 15 min at 15 psi pressure–121°C. Cool to 50°C. Adjust pH to approximately 2.2. Aseptically add 70.0mL of iron sulfate solution. Mix thoroughly. The pH should be 1.8–2.5.

Use: For the maintenance or cultivation of *Alicyclobacillus ferrooxydans*.

Alicyclobacillus Medium (DSMZ Medium 402)

Composition per liter:

Glucose	5.0g
KH_2PO_4	3.0g
Yeast extract	2.0g
$MgSO_4 \cdot 7H_2O$	0.5g
$CaCl_2 \cdot 2H_2O$	0.25g
$(NH_4)_2\ SO_4$	0.2g
Trace elements solution SL-6	1.0mL

pH 4.0 ± 0.2 at 25°C

Trace Elements Solution SL-6:

Composition per liter:

$MnCl_2 \cdot 4H_2O$	0.5g
H_3BO_3	0.3g
$CoCl_2 \cdot 6H_2O$	0.2g
$ZnSO_4 \cdot 7H_2O$	0.1g
$Na_2MoO_4 \cdot 2H_2O$	0.03g
$NiCl_2 \cdot 6H_2O$	0.02g
$CuCl_2 \cdot 2H_2O$	0.01g

Preparation of Trace Elements Solution SL-6: Add components to distilled/deionized water and bring volume to 1.0L. Mix thoroughly. Autoclave for 15 min at 15 psi pressure–121°C.

Preparation of Medium: Add components, except trace elements solution SL-6, to distilled/deionized water and bring volume to 999.0mL. Mix thoroughly. Adjust pH to 4.0. Autoclave for 15 min at 15 psi pressure–121°C. Cool to 25°C. Aseptically add 1.0mL of sterile trace elements solution SL-6. Mix thoroughly. Aseptically distribute into sterile tubes or flasks.

Use: For the cultivation and maintenance of *Alicyclobacillus* spp., *Bacillus* sp., *and Bacillus naganoensis*.

Alicyclobacillus Medium (DSMZ Medium 402)

Composition per liter:

Yeast extract	5.0g
Glucose	5.0g
KH_2PO_4	3.0g
$MgSO_4 \cdot 7H_2O$	0.5g
$CaCl_2 \cdot 2H_2O$	0.25g
$(NH_4)_2\ SO_4$	0.2g
Trace elements solution SL-6	1.0mL

pH 4.0 ± 0.2 at 25°C

Trace Elements Solution SL-6:

Composition per liter:

$MnCl_2 \cdot 4H_2O$	0.5g
H_3BO_3	0.3g
$CoCl_2 \cdot 6H_2O$	0.2g
$ZnSO_4 \cdot 7H_2O$	0.1g
$Na_2MoO_4 \cdot 2H_2O$	0.03g
$NiCl_2 \cdot 6H_2O$	0.02g
$CuCl_2 \cdot 2H_2O$	0.01g

Preparation of Trace Elements Solution SL-6: Add components to distilled/deionized water and bring volume to 1.0L. Mix thoroughly. Autoclave for 15 min at 15 psi pressure–121°C.

Preparation of Medium: Add components, except trace elements solution SL-6, to distilled/deionized water and bring volume to 999.0mL. Mix thoroughly. Adjust pH to 4.0. Autoclave for 15 min at 15 psi pressure–121°C. Cool to 25°C. Aseptically add 1.0mL of sterile trace elements solution SL-6. Mix thoroughly. Aseptically distribute into sterile tubes or flasks.

Use: For the cultivation and maintenance of *Alicyclobacillus cycloheptanicus*.

Alkalibacterium olivapovliticus Agar (DSMZ Medium 923)

Composition per liter:

Yeast extract	5.0g
Na glutamate	1.0g
$(NH_4)_2SO_4$	1.0g
K_2HPO_4	0.15g
$MgSO_4 \cdot 7H_2O$	0.025g
Agar solution	400.0mL
Na_2CO_3 solution	100.0mL

pH 9.5 ± 0.2 at 25°C

Agar Solution:

Composition per 400.0mL:

Agar	20.0g

Preparation of Agar Solution: Add agar to distilled/deionized water and bring volume to 400.0mL. Mix thoroughly. Gently heat and bring to boiling. Autoclave for 15 min at 15 psi pressure–121°C. Cool to 55°C.

Na_2CO_3 Solution:

Composition per 100.0mL:

Na_2CO_3	10.0g

Preparation of Na_2CO_3 Solution: Add $NaHCO_3$ to distilled/deionized water and bring volume to 100.0mL. Mix thoroughly. Autoclave for 15 min at 15 psi pressure–121°C. Cool to 55°C.

Preparation of Medium: Add components, except agar solution and Na_2CO_3 solution, to distilled/deionized water and bring volume to 500.0mL. Mix thoroughly. Autoclave for 15 min at 15 psi pressure–121°C. Cool to 55°C. Aseptically add 100.0mL sterile Na_2CO_3 solution. Mix thoroughly. Aseptically add 400.0mL sterile agar solution. Mix thoroughly. Pour into Petri dishes or aseptically distribute into sterile tubes.

Use: For the cultivation of *Bacillus sp.* and *Alkalibacterium olivapovliticus (Alkalibacterium olivoapovliticus).*

Alkalibacterium olivapovliticus Medium (DSMZ Medium 923)

Composition per liter:

Yeast extract	5.0g
Na glutamate	1.0g
$(NH_4)_2SO_4$	1.0g
K_2HPO_4	0.15g
$MgSO_4 \cdot 7H_2O$	0.025g
Na_2CO_3 solution	100.0mL

pH 9.5 ± 0.2 at 25°C

Na_2CO_3 Solution:

Composition per 100.0mL:

Na_2CO_3	10.0g

Preparation of Na_2CO_3 Solution: Add Na_2CO_3 to distilled/deionized water and bring volume to 100.0mL. Mix thoroughly. Autoclave for 15 min at 15 psi pressure–121°C. Cool to 25°C.

Preparation of Medium: Add components, except Na_2CO_3 solution, to distilled/deionized water and bring volume to 900.0mL. Mix thoroughly. Autoclave for 15 min at 15 psi pressure–121°C. Cool to 25°C. Aseptically add 100.0mL sterile Na_2CO_3 solution. Mix thoroughly. Aseptically distribute into sterile tubes or flasks.

Use: For the cultivation of *Bacillus sp.* and *Alkalibacterium olivapovliticus (Alkalibacterium olivoapovliticus).*

Alkaliflexus Medium (DSMZ Medium 1175)

Composition per liter:

NH_4Cl	0.2g
$MgCl_2 \cdot 6H_2O$	0.05g
KH_2PO_4	0.2g
$Na_2S \cdot 9H_2O$ solution	100.0mL
Yeast extract	100.0mL
Cellobiose solution	50.0mL
Na_2CO_3 solution	50.0mL
$NaHCO_3$ solution	50.0mL

pH 10.0 ± 0.2 at 25°C

Yeast Extract Solution:

Composition per 100.0mL:

Yeast extract	0.2g

Preparation of Yeast Extract Solution: Add components to distilled/deionized water and bring volume to 100.0mL. Mix thoroughly. Sparge with N_2. Autoclave for 15 min at 15 psi pressure–121°C. Cool to room temperature.

Cellobiose Solution:

Composition per 50.0mL:

Cellulobiose	3.0g

Preparation of Cellobiose Solution: Add components to distilled/deionized water and bring volume to 50.0mL. Mix thoroughly. Sparge with N_2. Autoclave for 15 min at 15 psi pressure–121°C. Cool to room temperature.

Na_2CO_3 Solution:

Composition per 50.0mL:

Na_2CO_3	7.4g

Preparation of Na_2CO_3 Solution: Add components to distilled/deionized water and bring volume to 50.0mL. Mix thoroughly. Sparge with N_2. Autoclave for 15 min at 15 psi pressure–121°C. Cool to room temperature.

$NaHCO_3$ Solution:

Composition per 50.0mL:

$NaHCO_3$	18.5g

Preparation of $NaHCO_3$ Solution: Add components to distilled/deionized water and bring volume to 50.0mL. Mix thoroughly. Sparge with N_2. Autoclave for 15 min at 15 psi pressure–121°C. Cool to room temperature.

$Na_2S \cdot 9H_2O$ Solution:

Composition per 100.0mL:

$Na_2S \cdot 9H_2O$	0.5g

Preparation of $Na_2S \cdot 9H_2O$ Solution: Add $Na_2S \cdot 9H_2O$ to distilled/deionized water and bring volume to 100.0mL. Mix thoroughly. Autoclave under 100% N_2 for 15 min at 15 psi pressure–121°C. Cool to room temperature.

Preparation of Medium: Add components, except $Na_2S \cdot 9H_2O$ solution, cellobiose solution, and yeast extract solution, to distilled/deionized water and bring volume to 750.0mL. Mix thoroughly. Gently heat and bring to boiling. Boil for several minutes. Cool to room temperature while sparging with N_2. Add the Na_2CO_3 solution and the $NaHCO_3$ solution. The pH should be 10.0. Distribute into serum bottles or Hungate tubes. Seal the tubes under N_2. Autoclave for 15 min at 15 psi pressure–121°C. Cool to 25°C. Aseptically inject $Na_2S \cdot 9H_2O$, cellobiose, and yeast extract solutions to give concentrations of 10%, 5%, and 10%, respectively.

Use: For the maintenance or cultivation of *Alkaliflexus* spp.

Alkaline *Bacillus* Medium

Composition per liter:

Agar	15.0g
Peptone	10.0g
Glucose	10.0g
Yeast extract	5.0g
K_2HPO_4	1.0g
Na_2CO_3 solution	100.0mL

pH 8.5–11.0 at 25°C

Na_2CO_3 Solution:

Composition per 100.0mL:

Na_2CO_3	10.0g

Preparation of Na_2CO_3 Solution: Add Na_2CO_3 to distilled/deionized water and bring volume to 100.0mL. Mix thoroughly. Filter sterilize.

Preparation of Medium: Add components, except Na_2CO_3 solution, to distilled/deionized water and bring volume to 900.0mL. Gently heat while stirring and bring to boiling. Autoclave for 15 min at 10 psi pressure–115°C. Cool to 45°–50°C. Aseptically add sterile Na_2CO_3 solution. Mix thoroughly. Pour into sterile Petri dishes or distribute into sterile tubes.

Use: For the cultivation and maintenance of alkalophilic microorganisms such as *Bacillus alcalophilus*, *Bacillus circulans,* and other *Bacillus* species.

Alkaline Cellulose Agar

Composition per liter:

Solution A	900.0mL
Solution B	100.0mL

Solution A:

Composition per 900.0mL:

Agar	15.0g
Cellulose powder MN 300	15.0g
NH_4NO_3	2.0g
K_2HPO_4	1.0g
Peptone	1.0g
Yeast extract	0.5g
$CaCl_2$	0.4g
$MgSO_4 \cdot 7H_2O$	0.4g

Preparation of Solution A: Add components to distilled/deionized water and bring volume to 900.0mL. Mix thoroughly. Adjust pH to 7.0 with 1*N* HCl. Gently heat and bring to boiling. Autoclave for 15 min at 15 psi pressure–121°C. Cool to 50°–55°C.

Solution B:

Composition per 100.0mL:

Na_2CO_3	0.5g

Preparation of Solution B: Add 0.5g of Na_2CO_3 to distilled/deionized water and bring volume to 100.0mL. Mix thoroughly. Adjust pH to 9.4 with 6% $NaHCO_3$ solution. Autoclave for 15 min at 15 psi pressure–121°C. Cool to 50°–55°C.

Preparation of Medium: Aseptically combine 900.0mL of solution A with 100.0mL of solution B. Mix thoroughly. Pour into sterile Petri dishes or leave in tubes.

Use: For the cultivation and maintenance of cellulose-utilizing bacteria.

Alkaline HiVeg Peptone Water

Composition per liter:

NaCl	10.0g
Plant peptone	10.0g

pH 8.4 ± 0.2 at 25°C

Source: This medium is available as a premixed powder from HiMedia.

Preparation of Medium: Add components to distilled/deionized water and bring volume to 1.0L. Mix thoroughly. Adjust pH to 8.5. Distribute into tubes or flasks. Autoclave for 10 min at 15 psi pressure–121°C.

Use: For the enrichment of *Vibrio* species from foods.

Alkaline Nutrient Agar

Composition per liter:

Agar	15.0g
Peptone	5.0g
Beef extract	3.0g

pH 9.5–10.0 at 25°C

Preparation of Medium: Add components to distilled/deionized water and bring volume to 1.0L. Mix thoroughly. Gently heat and bring to boiling. Distribute into tubes or flasks. Autoclave for 15 min at 15 psi pressure–121°C. Aseptically adjust to pH 9.5–10.0 with sterile 9% Na_2CO_3 solution. Pour into sterile Petri dishes or leave in tubes.

Use: For the cultivation and maintenance of *Bacillus alcalophilus* and *Bacillus* species.

Alkaline Nutrient Agar

Composition per liter:

Agar	20.0g
Pancreatic digest of gelatin	5.0g
Beef extract	3.0g

pH 10.0 ± 0.2 at 25°C

Preparation of Medium: Add components to distilled/deionized water and bring volume to 1.0L. Mix thoroughly. Gently heat and bring to boiling. Distribute into tubes or flasks. Autoclave for 15 min at 15 psi pressure–121°C. Aseptically adjust pH with 10% sterilie $Na_2S_2O_3$ solution. Mix thoroughly. Pour into sterile Petri dishes or leave in tubes.

Use: For the cultivation and maintenance of *Bacillus alcalophilus.*

Alkaline Nutrient Agar

Composition per liter:

Agar	15.0g
Peptone	5.0g
NaCl	5.0g
Yeast extract	2.0g
Beef extract	1.0g
Sodium sesquicarbonate solution	100.0mL

pH 9.7 ± 0.2 at 25°C

Sodium Sesquicarbonate Solution:

Composition per 100.0mL:

Na_2CO_3, anhydrous	10.6g
$NaHCO_3$	8.42g

Preparation of Sodium Sesquicarbonate Solution: Add components to distilled/deionized water and bring volume to 100.0mL. Mix thoroughly. Filter sterilize. Warm to 50°–55°C.

Preparation of Medium: Add components, except sodium sesquicarbonate solution, to distilled/deionized water and bring volume to 900.0mL. Mix thoroughly. Gently heat and bring to boiling. Autoclave for 15 min at 15 psi pressure–121°C. Cool to 50°–55°C. Aseptically add sterile sodium sesquicarbonate solution. Mix thoroughly. Adjust pH to 9.7. Pour into sterile Petri dishes or distribute into sterile tubes.

Use: For the cultivation of alkaliniphilic bacteria, including *Bacillus alcalophilus, Bacillus cohnii,* and other *Bacillus* species.

Alkaline Peptone Agar

Composition per liter:

NaCl	20.0g
Agar	15.0g
Peptone	10.0g

pH 8.5 ± 0.2 at 25°C

Preparation of Medium: Add components to distilled/deionized water and bring volume to 1.0L. Mix thoroughly. Gently heat and bring to boiling. Adjust pH to 8.5. Distribute into tubes. Autoclave for 15 min at 15 psi pressure–121°C. Allow tubes to cool in a slanted position.

Use: For the cultivation of *Vibrio cholerae* and other *Vibrio* species.

Alkaline Peptone Salt Broth (APS Broth)

Composition per liter:

NaCl	30.0g
Peptone	10.0g

Preparation of Medium: Add components to distilled/deionized water and bring volume to 1.0L. Mix thoroughly. Adjust pH to 8.5. Distribute into tubes in 10.0mL volumes. Autoclave for 10 min at 15 psi pressure–121°C.

Use: For the cultivation of *Vibrio cholerae* and other *Vibrio* species from foods.

Alkaline Peptone Water

Composition per liter:

NaCl	10.0g
Peptone	10.0g

pH 8.5 ± 0.2 at 25°C

Preparation of Medium: Add components to distilled/deionized water and bring volume to 1.0L. Mix thoroughly. Adjust pH to 8.5. Distribute into tubes or flasks. Autoclave for 10 min at 15 psi pressure–121°C.

Use: For the cultivation and transport of *Vibrio cholerae* and other *Vibrio* species from foods.

Alkaline Peptone Water

Composition per liter:

Peptone	10.0g
NaCl	5.0g

pH 9.0 ± 0.2 at 25°C

Preparation of Medium: Add components to distilled/deionized water and bring volume to 1.0L. Mix thoroughly. Adjust pH to 9.0. Distribute into tubes or flasks. Autoclave for 20 min at 15 psi pressure–121°C.

Use: For the cultivation of a variety of alkalophilic microorganisms, especially *Vibrio* species.

Alkaline Peptone Water

Composition per liter:

NaCl	30.0g
Peptone	20.0g

pH 8.4 ± 0.2 at 25°C

Source: This medium is available from HiMedia.

Preparation of Medium: Add components to distilled/deionized water and bring volume to 1.0L. Mix thoroughly. Adjust pH to 8.4. Distribute into tubes or flasks. Autoclave for 20 min at 15 psi pressure–121°C.

Use: For the cultivation of a variety of alkalophilic microorganisms.

Alkaline Polypectate Agar

Composition per liter:

Agar	16.0g
Na_2CO_3	10.0g
Peptone	6.0g
Sodium polypectate	5.0g
Yeast extract	3.0g
K_2HPO_4	1.0g
$MgSO_4 \cdot 7H_2O$	0.2g
$MnSO_4$	40.0mg

pH 10.0 ± 0.2 at 25°C

Preparation of Medium: Add components to distilled/deionized water and bring volume to 1.0L. Mix thoroughly. Bring pH to 10.0. Gently heat and bring to boiling. Distribute into tubes or flasks. Autoclave for 15 min at 15 psi pressure–121°C. Pour into sterile Petri dishes or leave in tubes.

Use: For the cultivation of pectinolytic bacteria.

Alkaline Starch Agar

Composition per liter:

Starch	20.0g
Agar	16.0g
Na_2CO_3	10.0g
Peptone	6.0g
Yeast extract	3.0g
K_2HPO_4	1.0g
$MgSO_4 \cdot 7H_2O$	0.2g
$MnSO_4$	40.0mg

pH 9.7 ± 0.2 at 25°C

Preparation of Medium: Add components to distilled/deionized water and bring volume to 1.0L. Mix thoroughly. Bring pH to 9.7. Gently heat and bring to boiling. Distribute into tubes or flasks. Autoclave for 15 min at 15 psi pressure–121°C. Pour into sterile Petri dishes or leave in tubes.

Use: For the cultivation of alkiliniphilic starch-utilizing bacteria.

Alkaline Xylan Agar

Composition per liter:

Agar	15.0g
Larchwood xylan	10.0g
Polypeptone™	5.0g
Yeast extract	5.0g
K_2HPO_4	0.45g
$MgSO_4 \cdot 7H_2O$	0.2g

pH 10.0 ± 0.2 at 25°C

Preparation of Medium: Add components to distilled/deionized water and bring volume to 1.0L. Mix thoroughly. Gently heat and bring to boiling. Distribute into tubes or flasks. Autoclave for 15 min at 15 psi pressure–121°C. Aseptically adjust pH with 10% sterilie $Na_2S_2O_3$ solution. Mix thoroughly. Pour into sterile Petri dishes or leave in tubes.

Use: For the cultivation and maintenance of *Bacillus* species.

Alkaline Xylan Broth

Composition per liter:

Xylan	10.0g
Yeast extract	3.0g
NH_4NO_3	2.0g
K_2HPO_4	1.0g
Polypepton	300.0mg
$MgSO_4 \cdot 7H_2O$	200.0mg
$CaCl_2 \cdot 2H_2O$	100.0mg
$FeSO_4 \cdot 7H_2O$	5.0mg
$MnSO_4 \cdot 7H_2O$	5.0mg
Resazurin	1.0mg

pH 10.0 ± 0.2 at 25°C

Preparation of Medium: Add components to distilled/deionized water and bring volume to 1.0L. Mix thoroughly. Autoclave for 15 min at 15 psi pressure–121°C. Aseptically adjust pH with 10% sterile $Na_2S_2O_3$ solution. Mix thoroughly. Aseptically distribute into sterile tubes or flasks.

Use: For the cultivation of *Amphibacillus xylanus*.

Alkaline Xylan Medium

Composition per 1001.0mL:

Xylan	10.0g
Yeast extract	3.0g
NH_4NO_3	2.0g
K_2HPO_4	1.0g
Polypeptone™	0.3g
$MgSO_4 \cdot 7H_2O$	0.2g
$CaCl_2 \cdot 2H_2O$	0.1g
$FeSO_4 \cdot 7H_2O$	5.0mg
$MnSO_4 \cdot 7H_2O$	5.0mg
Resazurin	1.0mg
Titanium citrate solution	1.0mL
Na_2CO_3 (10% solution)	variable

pH 10.0 ± 0.2 at 25°C

Titanium Citrate Solution:

Composition per 50.0mL:

$TiCl_2$	0.75g
Trisodium citrate	2.58g

Preparation of Titanium Citrate Solution: Add components to approximately 30.0mL of distilled/deionized water. Mix thoroughly. Adjust pH to 7.0 with Na_2CO_3. Bring volume to 50.0mL with distilled/deionized water.

Preparation of Medium: Prepare and dispense medium under 100% N_2. Add components, except titanium citrate solution, to distilled/deionized water and bring volume to 1.0L. Mix thoroughly. Gently heat and bring to boiling. Continue boiling for 3 min. Cool to room temperature while sparging with 100% N_2. Anaerobically distribute into anaerobic tubes. Autoclave for 15 min at 15 psi pressure–121°C. Cool to room temperature. Adjust pH to 10.0 with filter-sterilized 10% Na_2CO_3 solution. Immediately prior to inoculation, reduce medium by adding 1.0mL of titanium citrate solution per liter of medium.

Use: For the cultivation of *Amphibacillus xylanus.*

Alkaline Yeast Extract Malt Medium

Composition per liter:

Malt extract	10.0g
Yeast extract	4.0g
Glucose	4.0g
Na_2CO_3 (10% solution)	100.0mL

pH 8.5–11.0 at 25°C

Preparation of Medium: Add components, except Na_2CO_3, to distilled/deionized water and bring volume to 900.0mL. Mix thoroughly. Gently heat and bring to boiling. Autoclave for 15 min at 15 psi pressure–121°C. Separately sterilize a 10% Na_2CO_3 solution and aseptically add 100.0mL. Adjust pH, if necessary, to 8.5–11.0.

Use: For the cultivation of *Nocardiopsis dassonvillei.*

Alkaliphilic *Halomonas* Medium (DSMZ Medium 1034)

Composition per liter:

Yeast extract	10.0g
Sodium citrate	3.0g
$MgSO_4 \cdot 7H_2O$	1.0g
Solution C	100.0mL
Solution D	10.0mL
Solution A	1.0mL
Solution B	1.0mL

pH 9.0 ± 0.2 at 25°C

Solution A:

Composition per 10.0mL:

$MnCl_2 \cdot 4H_2O$	3.6mg

Preparation of Solution A: Add components to distilled/deionized water and bring volume to 10.0mL. Mix thoroughly. Autoclave for 15 min at 15 psi pressure–121°C. Cool to room temperature.

Solution B:

Composition per 10.0mL:

$FeSO_4 \cdot 7H_2O$	0.5g

Preparation of Solution B: Add components to distilled/deionized water and bring volume to 10.0mL. Mix thoroughly. Autoclave for 15 min at 15 psi pressure–121°C. Cool to room temperature.

Solution C:

Composition per 100.0mL:

NaCl	100.0g

Preparation of Solution C: Add components to distilled/deionized water and bring volume to 100.0mL. Mix thoroughly. Autoclave for 15 min at 15 psi pressure–121°C. Cool to room temperature.

Solution D:

Composition per 10.0mL:

Na_2CO_3	3.0g

Preparation of Solution D: Add components to distilled/deionized water and bring volume to 10.0mL. Mix thoroughly. Autoclave for 15 min at 15 psi pressure–121°C. Cool to room temperature.

Preparation of Medium: Add components, except solutions A, B, C, and D, to distilled/deionized water and bring volume to 890.0mL. Mix thoroughly. Gently heat and bring to boiling. Autoclave for 15 min at 15 psi pressure–121°C. Cool to 25°C. Aseptically add solutions A, B, C, and D. Mix thoroughly. The pH should be 9.0.

Use: For the maintenance or cultivation of alkaliphilic *Halomonas* spp.

Alkaliphilic Methanogen Medium

Composition per liter:

NaCl	15.0g
$NaHCO_3$	10.0g
Methanol	5.0g
Na_2CO_3	4.0g
$Na_2S \cdot 9H_2O$	1.0g
NH_4Cl	0.5g
Yeast extract	0.5g
KH_2PO_4	0.3g
$NiCl_2 \cdot 6H_2O$	2.0mg
Resazurin	0.5mg
Wolfe's mineral solution	10.0mL
Selenite/tungstate solution	1.0mL

pH 9.2–9.4 at 25°C

Wolfe's Mineral Solution:

Composition per liter:

$MgSO_4 \cdot 7H_2O$	3.0g
Nitrilotriacetic acid	1.5g
NaCl	1.0g
$MnSO_4 \cdot 2H_2O$	0.5g
$CoCl_2 \cdot 6H_2O$	0.1g
$ZnSO_4 \cdot 7H_2O$	0.1g

$CaCl_2 \cdot 2H_2O$ 0.1g
$FeSO_4 \cdot 7H_2O$ 0.1g
$NiCl_2 \cdot 6H_2O$ 0.025g
$KAl(SO_4)_2 \cdot 12H_2O$ 0.02g
$CuSO_4 \cdot 5H_2O$ 0.01g
H_3BO_3 .... 0.01g
$Na_2MoO_4 \cdot 2H_2O$ 0.01g
$Na_2SeO_3 \cdot 5H_2O$ 0.3mg

Preparation of Wolfe's Mineral Solution: Add nitrilotriacetic acid to 500.0mL of distilled/deionized water. Adjust pH to 6.5 with KOH. Add remaining components. Add distilled/deionized water to 1.0L. Adjust pH to 6.8.

Selenite/Tungstate Solution:
Composition per liter:
NaOH 0.5g
$Na_2WO_4 \cdot 2H_2O$ 4.0mg
$Na_2SeO_3 \cdot 5H_2O$ 3.0mg

Preparation of Selenite/Tungstate Solution: Add components to distilled/deionized water and bring volume to 1.0L. Mix thoroughly. Sparge with 100% N_2. Autoclave for 15 min at 15 psi pressure–121°C.

Preparation of Medium: Prepare and dispense medium under 100% N_2. Add components, except $NaHCO_3$ and $Na_2S \cdot 9H_2O$, to distilled/deionized water and bring volume to 1.0L. Mix thoroughly. Gently heat and bring to boiling. Continue boiling for 5 min. Cool to room temperature while sparging with 100% N_2. Add $NaHCO_3$ and $Na_2S \cdot 9H_2O$. Mix thoroughly. Anaerobically distribute into tubes. Autoclave for 15 min at 15 psi pressure–121°C. Adjust pH to 9.2–9.4.

Use: For the cultivation of *Methanohalophilus zhilinae.*

Alkaliphilic Spirochete Medium

Composition per 1011.0mL:
Na_2CO_3 .... 10.0g
NaCl 10.0g
NH_4Cl 1.0g
K_2HPO_4 .... 0.2g
KCl 0.2g
Yeast extract 0.5g
$NaHCO_3$ solution 50.0mL
Sucrose solution 20.0mL
$Na_2S \cdot 9H_2O$ solution 10.0mL
Wolfe's vitamin solution 10.0mL
Trace elements solution SL-6 .... 1.0mL

pH 9.7 ± 0.2 at 25°C

$NaHCO_3$ Solution:
Composition per 50.0mL:
$NaHCO_3$ 15.0g

Preparation of $NaHCO_3$ Solution: Add $NaHCO_3$ to distilled/deionized water and bring volume to 50.0mL. Mix thoroughly. Filter sterilize. Sparge with 100% N_2.

Sucrose Solution:
Composition per 20.0mL:
Sucrose 5.0g

Preparation of Sucrose Solution: Add sucrose to distilled/deionized water and bring volume to 20.0mL. Mix thoroughly. Sparge with 100% N_2. Autoclave for 15 min at 15 psi pressure–121°C.

$Na_2S \cdot 9H_2O$ Solution:
Composition per 10.0mL:
$Na_2S \cdot 9H_2O$ 1.0g

Preparation of $Na_2S \cdot 9H_2O$ Solution: Add $Na_2S \cdot 9H_2O$ to distilled/deionized water and bring volume to 10.0mL. Mix thoroughly. Sparge with 100% N_2. Autoclave for 15 min at 15 psi pressure–121°C. Before use, neutralize to pH 7.0 with sterile HCl.

Wolfe's Vitamin Solution:
Composition per liter:
Pyridoxine·HCl 10.0mg
p-Aminobenzoic acid 5.0mg
Lipoic acid 5.0mg
Nicotinic acid 5.0mg
Riboflavin 5.0mg
Thiamine·HCl 5.0mg
Calcium DL-pantothenate 5.0mg
Biotin 2.0mg
Folic acid 2.0mg
Vitamin B_{12} 0.1mg

Preparation of Wolfe's Vitamin Solution: Add components to distilled/deionized water and bring volume to 1.0L. Mix thoroughly. Filter sterilize. Sparge with 100% N_2.

Trace Elements Solution SL-6:
Composition per liter:
$MnCl_2 \cdot 4H_2O$ 0.5g
H_3BO_3 .... 0.3g
$CoCl_2 \cdot 6H_2O$ 0.2g
$ZnSO_4 \cdot 7H_2O$ 0.1g
$Na_2MoO_4 \cdot 2H_2O$ 0.03g
$NiCl_2 \cdot 6H_2O$ 0.02g
$CuCl_2 \cdot 2H_2O$ 0.01g

Preparation of Trace Elements Solution SL-6: Add components to distilled/deionized water and bring volume to 1.0L. Mix thoroughly.

Preparation of Medium: Prepare and dispense medium under 100% N_2. Add components, except $NaHCO_3$ solution, sucrose solution, $Na_2S \cdot 9H_2O$ solution, and Wolfe's vitamin solution, to distilled/deionized water and bring volume to 910.0mL. Mix thoroughly. Adjust pH to 9.7 with 6*N* NaOH (about 15.0mL). Gently heat and bring to boiling. Cool to room temperature while sparging with 100% N_2. Autoclave for 15 min at 15 psi pressure–121°C. Aseptically and anaerobically add 50.0mL of sterile $NaHCO_3$ solution, 20.0mL of sterile sucrose solution, 10.0mL of sterile $Na_2S \cdot 9H_2O$ solution, and 10.0mL of sterile Wolfe's vitamin solution. Mix thoroughly. Aseptically distribute into sterile tubes or flasks.

Use: For the cultivation of *Spirochaeta africana, Spirochaeta alkalica,* and *Spirochaeta asiatica.*

Alkaliphilic Sulphur Respiring Strains Medium (DSMZ Medium 925)

Composition per liter:
Mineral base 997.0mL
KSCN solution 10.0mL
Trace elements solution 2.0mL
Magnesium chloride solution 1.0mL

pH 10.0 ± 0.2 at 25°C

Mineral Base:
Composition per liter:
Na_2CO_3 .... 20.0g
$NaHCO_3$ 10.0g

NaCl 5.0g
K_2HPO_4 .... 1.0g

Preparation of Mineral Base: Add components to distilled/deionized water and bring volume to 1.0L. Mix thoroughly. Sparge with 100% N_2. Autoclave for 20 min at 110°C. The pH should be about 10.0.

Trace Elements Solution:
Composition per liter:

H_3BO_3 .... 300.0mg
$CoCl_2·6H_2O$ 200.0mg
$ZnSO_4·7H_2O$ 100.0mg
$MnCl_2·4H_2O$ 30.0mg
$Na_2MoO_4·4H_2O$ 30.0mg
$NiCl_2·6H_2O$ 20.0mg
$CuCl_2·2H_2O$ 10.0mg
EDTA 5.0mg
$FeSO_4·7H_2O$ 2.0mg

Preparation of Trace Elements Solution: Add components to distilled/deionized water and bring volume to 1.0L. Adjust pH to 3.0 with HCl. Mix thoroughly. Sparge with 100% N_2. Autoclave for 15 min at 15 psi pressure–121°C.

Magnesium Chloride Solution:
Composition per 10.0mL:

$MgCl_2·6H_2O$ 2.0g

Preparation of Magnesium Chloride Solution: Add $MgCl_2·6H_2O$ to distilled/deionized water and bring volume to 10.0mL. Mix thoroughly. A white colloid will dissolve after mixing. Autoclave for 15 min at 15 psi pressure–121°C.

KSCN Solution:
Composition per 10.0mL:

KSCN 1.5g

Preparation of KSCN Solution: Add KSCN to distilled/deionized water and bring volume to 10.0mL. Mix thoroughly. Autoclave for 15 min at 15 psi pressure–121°C. Cool to room temperature.

Preparation of Medium: Aseptically add 10.0mL sterile KSCN solution, 2.0mL sterile trace elements solution, and 1.0mL sterile magnesium chloride solution to 987.0mL sterile mineral base. Aseptically distribute to sterile tubes, flasks, or bottles.

Use: For the cultivation of *Thialkalivibrio paradoxus* DSM 13531 and *Thialkalivibrio thiocyanoxidans* DSM 13532.

Alkaliphilic Sulphur Respiring Strains Medium (DSMZ Medium 925)

Composition per liter:

Mineral base 967.0mL
Thiosulfate solution 20.0mL
Ammonium chloride solution 10.0mL
Trace elements solution 2.0mL
Magnesium chloride solution 1.0mL

pH 10.0 ± 0.2 at 25°C

Mineral Base:
Composition per liter:

Na_2CO_3 .... 20.0g
$NaHCO_3$ 10.0g
NaCl 5.0g
K_2HPO_4 .... 1.0g

Preparation of Mineral Base: Add components to distilled/deionized water and bring volume to 1.0L. Mix thoroughly. Sparge with 100% N_2. Autoclave for 20 min at 110°C. The pH should be about 10.0.

Trace Elements Solution:
Composition per liter:

H_3BO_3 .... 300.0mg
$CoCl_2·6H_2O$ 200.0mg
$ZnSO_4·7H_2O$ 100.0mg
$MnCl_2·4H_2O$ 30.0mg
$Na_2MoO_4·4H_2O$ 30.0mg
$NiCl_2·6H_2O$ 20.0mg
$CuCl_2·2H_2O$ 10.0mg
EDTA 5.0mg
$FeSO_4·7H_2O$ 2.0mg

Preparation of Trace Elements Solution: Add components to distilled/deionized water and bring volume to 1.0L. Adjust pH to 3.0 with HCl. Mix thoroughly. Sparge with 100% N_2. Autoclave for 15 min at 15 psi pressure–121°C.

Magnesium Chloride Solution:
Composition per 10.0mL:

$MgCl_2·6H_2O$ 2.0g

Preparation of Magnesium Chloride Solution: Add $MgCl_2·6H_2O$ to distilled/deionized water and bring volume to 10.0mL. Mix thoroughly. A white colloid will dissolve after mixing. Autoclave for 15 min at 15 psi pressure–121°C.

Thiosulfate Solution:
Composition per liter:

$Na_2S_2O_3·5H_2O$ 9.92g

Preparation of Thiosulfate Solution: Add $Na_2S_2O_3·5H_2O$ to distilled/deionized water and bring volume to 1.0L. Mix thoroughly. Autoclave for 15 min at 15 psi pressure–121°C. Cool to room temperature.

Ammonium Chloride Solution:
Composition per 10.0mL:

NH_4Cl 0.27g

Preparation of Ammonium Chloride Solution: Add NH_4Cl to distilled/deionized water and bring volume to 10.0mL. Mix thoroughly. Autoclave for 15 min at 15 psi pressure–121°C. Cool to room temperature.

Preparation of Medium: Aseptically add 20.0mL sterile thiosulfate solution, 10.0mL sterile ammonium chloride solution, 2.0mL sterile trace elements solution, and 1.0mL sterile magnesium chloride solution to 967.0mL sterile mineral base. Aseptically distribute to sterile tubes, flasks, or bottles.

Use: For the cultivation of *Thialkalivibrio thiocyanoxidans* DSM 13541.

Alkaliphilic Sulphur Respiring Strains Medium (DSMZ Medium 925)

Composition per liter:

Mineral base 967.0mL
Thiosulfate solution 20.0mL
Potassium nitrate solution 10.0mL
Trace elements solution 2.0mL
Magnesium chloride solution 1.0mL

pH 10.0 ± 0.2 at 25°C

Mineral Base:

Composition per liter:

Na_2CO_3 .... 20.0g
$NaHCO_3$ 10.0g
NaCl 5.0g
K_2HPO_4 .... 1.0g

Preparation of Mineral Base: Add components to distilled/deionized water and bring volume to 1.0L. Mix thoroughly. Sparge with 100% N_2. Autoclave for 20 min at 110°C. The pH should be about 10.0.

Trace Elements Solution:

Composition per liter:

H_3BO_3 .... 300.0mg
$CoCl_2 \cdot 6H_2O$ 200.0mg
$ZnSO_4 \cdot 7H_2O$ 100.0mg
$MnCl_2 \cdot 4H_2O$ 30.0mg
$Na_2MoO_4 \cdot 4H_2O$ 30.0mg
$NiCl_2 \cdot 6H_2O$ 20.0mg
$CuCl_2 \cdot 2H_2O$ 10.0mg
EDTA 5.0mg
$FeSO_4 \cdot 7H_2O$ 2.0mg

Preparation of Trace Elements Solution: Add components to distilled/deionized water and bring volume to 1.0L. Adjust pH to 3.0 with HCl. Mix thoroughly. Sparge with 100% N_2. Autoclave for 15 min at 15 psi pressure–121°C.

Magnesium Chloride Solution:

Composition per 10.0mL:

$MgCl_2 \cdot 6H_2O$ 2.0g

Preparation of Magnesium Chloride Solution: Add $MgCl_2 \cdot 6H_2O$ to distilled/deionized water and bring volume to 10.0mL. Mix thoroughly. A white colloid will dissolve after mixing. Autoclave for 15 min at 15 psi pressure–121°C.

Thiosulfate Solution:

Composition per liter:

$Na_2S_2O_3 \cdot 5H_2O$ 9.92g

Preparation of Thiosulfate Solution: Add $Na_2S_2O_3 \cdot 5H_2O$ to distilled/deionized water and bring volume to 1.0L. Mix thoroughly. Autoclave for 15 min at 15 psi pressure–121°C. Cool to room temperature.

Potassium Nitrate Solution:

Composition per 100.0mL:

KNO_3 .... 10.1g

Preparation of Potassium Nitrate Solution: Add KNO_3 to distilled/deionized water and bring volume to 100.0mL. Mix thoroughly. Autoclave for 15 min at 15 psi pressure–121°C. Cool to room temperature.

Preparation of Medium: Aseptically add 20.0mL sterile thiosulfate solution, 10.0mL sterile potassium nitrate solution, 2.0mL sterile trace elements solution, and 1.0mL sterile magnesium chloride solution to 967.0mL sterile mineral base. Aseptically distribute to sterile tubes, flasks, or bottles.

Use: For the cultivation of *Thialkalivibrio paradoxus* DSM 13542.

Alkaliphilic Sulphur Respiring Strains Medium (DSMZ Medium 925)

Composition per liter:

Mineral base 967.0mL
Thiosulfate solution 20.0mL
KSCN solution 10.0mL
Trace elements solution 2.0mL
Magnesium chloride solution 1.0mL

pH 10.0 ± 0.2 at 25°C

Mineral Base:

Composition per liter:

Na_2CO_3 .... 20.0g
$NaHCO_3$ 10.0g
NaCl 5.0g
K_2HPO_4 .... 1.0g

Preparation of Mineral Base: Add components to distilled/deionized water and bring volume to 1.0L. Mix thoroughly. Sparge with 100% N_2. Autoclave for 20 min at 110°C. The pH should be about 10.0.

Trace Elements Solution:

Composition per liter:

H_3BO_3 .... 300.0mg
$CoCl_2 \cdot 6H_2O$ 200.0mg
$ZnSO_4 \cdot 7H_2O$ 100.0mg
$MnCl_2 \cdot 4H_2O$ 30.0mg
$Na_2MoO_4 \cdot 4H_2O$ 30.0mg
$NiCl_2 \cdot 6H_2O$ 20.0mg
$CuCl_2 \cdot 2H_2O$ 10.0mg
EDTA 5.0mg
$FeSO_4 \cdot 7H_2O$ 2.0mg

Preparation of Trace Elements Solution: Add components to distilled/deionized water and bring volume to 1.0L. Adjust pH to 3.0 with HCl. Mix thoroughly. Sparge with 100% N_2. Autoclave for 15 min at 15 psi pressure–121°C.

Magnesium Chloride Solution:

Composition per 10.0mL:

$MgCl_2 \cdot 6H_2O$ 2.0g

Preparation of Magnesium Chloride Solution: Add $MgCl_2 \cdot 6H_2O$ to distilled/deionized water and bring volume to 10.0mL. Mix thoroughly. A white colloid will dissolve after mixing. Autoclave for 15 min at 15 psi pressure–121°C.

Thiosulfate Solution:

Composition per liter:

$Na_2S_2O_3 \cdot 5H_2O$ 9.92g

Preparation of Thiosulfate Solution: Add $Na_2S_2O_3 \cdot 5H_2O$ to distilled/deionized water and bring volume to 1.0L. Mix thoroughly. Autoclave for 15 min at 15 psi pressure–121°C. Cool to room temperature.

KSCN Solution:

Composition per 10.0mL:

KSCN 0.5g

Preparation of KSCN Solution: Add KSCN to distilled/deionized water and bring volume to 10.0mL. Mix thoroughly. Autoclave for 15 min at 15 psi pressure–121°C. Cool to room temperature.

Preparation of Medium: Aseptically add 20.0mL sterile thiosulfate solution, 10.0mL sterile KSCN solution, 2.0mL sterile trace elements solution, and 1.0mL sterile magnesium chloride solution to 967.0mL sterile mineral base. Aseptically distribute to sterile tubes, flasks, or bottles.

Use: For the cultivation of *Thialkalivibrio* sp. DSM 13533.

Alkaliphilic Sulphur Respiring Strains Medium (DSMZ Medium 925)

Composition per liter:

Mineral base	967.5mL
Thiosulfate solution	20.0mL
KSCN solution	10.0mL
Trace elements solution	2.0mL
Magnesium chloride solution	0.5mL

pH 10.0 ± 0.2 at 25°C

Mineral Base:
Composition per liter:

Na_2CO_3	20.0g
$NaHCO_3$	10.0g
NaCl	5.0g
K_2HPO_4	1.0g

Preparation of Mineral Base: Add components to distilled/deionized water and bring volume to 1.0L. Mix thoroughly. Sparge with 100% N_2. Autoclave for 20 min at 110°C. The pH should be about 10.0.

Trace Elements Solution:
Composition per liter:

H_3BO_3	300.0mg
$CoCl_2 \cdot 6H_2O$	200.0mg
$ZnSO_4 \cdot 7H_2O$	100.0mg
$MnCl_2 \cdot 4H_2O$	30.0mg
$Na_2MoO_4 \cdot 4H_2O$	30.0mg
$NiCl_2 \cdot 6H_2O$	20.0mg
$CuCl_2 \cdot 2H_2O$	10.0mg
EDTA	5.0mg
$FeSO_4 \cdot 7H_2O$	2.0mg

Preparation of Trace Elements Solution: Add components to distilled/deionized water and bring volume to 1.0L. Adjust pH to 3.0 with HCl. Mix thoroughly. Sparge with 100% N_2. Autoclave for 15 min at 15 psi pressure–121°C.

Magnesium Chloride Solution:
Composition per 10.0mL:

$MgCl_2 \cdot 6H_2O$	2.0g

Preparation of Magnesium Chloride Solution: Add $MgCl_2 \cdot 6H_2O$ to distilled/deionized water and bring volume to 10.0mL. Mix thoroughly. A white colloid will dissolve after mixing. Autoclave for 15 min at 15 psi pressure–121°C.

Thiosulfate Solution:
Composition per liter:

$Na_2S_2O_3 \cdot 5H_2O$	9.92g

Preparation of Thiosulfate Solution: Add $Na_2S_2O_3 \cdot 5H_2O$ to distilled/deionized water and bring volume to 1.0L. Mix thoroughly. Autoclave for 15 min at 15 psi pressure–121°C. Cool to room temperature.

KSCN Solution:
Composition per 10.0mL:

KSCN	0.5g

Preparation of KSCN Solution: Add KSCN to distilled/deionized water and bring volume to 10.0mL. Mix thoroughly. Autoclave for 15 min at 15 psi pressure–121°C. Cool to room temperature.

Preparation of Medium: Aseptically add 20.0mL sterile thiosulfate solution, 10.0mL sterile KSCN solution, 2.0mL sterile trace elements solution, and 0.5mL sterile magnesium chloride solution to 967.5mL sterile mineral base. Aseptically distribute to sterile tubes, flasks, or bottles.

Use: For the cultivation of *Thioalkalivibrio versutus* DSM 13738, *Thioalkalivibrio versutus* DSM 13741, and *Thioalkalivibrio denitrificans.*

Alkaliphilic Sulphur Respiring Strains Medium (DSMZ Medium 925)

Composition per liter:

Mineral base	947.5mL
Thiosulfate solution	40.0mL
KSCN solution	10.0mL
Trace elements solution	2.0mL
Magnesium chloride solution	0.5mL

pH 10.0 ± 0.2 at 25°C

Mineral Base:
Composition per liter:

Na_2CO_3	20.0g
$NaHCO_3$	10.0g
NaCl	5.0g
K_2HPO_4	1.0g

Preparation of Mineral Base: Add components to distilled/deionized water and bring volume to 1.0L. Mix thoroughly. Sparge with 100% N_2. Autoclave for 20 min at 110°C. The pH should be about 10.0.

Trace Elements Solution:
Composition per liter:

H_3BO_3	300.0mg
$CoCl_2 \cdot 6H_2O$	200.0mg
$ZnSO_4 \cdot 7H_2O$	100.0mg
$MnCl_2 \cdot 4H_2O$	30.0mg
$Na_2MoO_4 \cdot 4H_2O$	30.0mg
$NiCl_2 \cdot 6H_2O$	20.0mg
$CuCl_2 \cdot 2H_2O$	10.0mg
EDTA	5.0mg
$FeSO_4 \cdot 7H_2O$	2.0mg

Preparation of Trace Elements Solution: Add components to distilled/deionized water and bring volume to 1.0L. Adjust pH to 3.0 with HCl. Mix thoroughly. Sparge with 100% N_2. Autoclave for 15 min at 15 psi pressure–121°C.

Magnesium Chloride Solution:
Composition per 10.0mL:

$MgCl_2 \cdot 6H_2O$	2.0g

Preparation of Magnesium Chloride Solution: Add $MgCl_2 \cdot 6H_2O$ to distilled/deionized water and bring volume to 10.0mL. Mix thoroughly. A white colloid will dissolve after mixing. Autoclave for 15 min at 15 psi pressure–121°C.

Thiosulfate Solution:
Composition per liter:

$Na_2S_2O_3 \cdot 5H_2O$	9.92g

Preparation of Thiosulfate Solution: Add $Na_2S_2O_3 \cdot 5H_2O$ to distilled/deionized water and bring volume to 1.0L. Mix thoroughly. Autoclave for 15 min at 15 psi pressure–121°C. Cool to room temperature.

KSCN Solution:
Composition per 10.0mL:

KSCN	0.5g

Preparation of KSCN Solution: Add KSCN to distilled/deionized water and bring volume to 10.0mL. Mix thoroughly. Autoclave for 15 min at 15 psi pressure–121°C. Cool to room temperature.

Preparation of Medium: Aseptically add 40.0mL sterile thiosulfate solution, 10.0mL sterile KSCN solution, 2.0mL sterile trace elements solution, and 0.5mL sterile magnesium chloride solution to 947.5mL sterile mineral base. Aseptically distribute to sterile tubes, flasks, or bottles.

Use: For the cultivation of *Thialkalimicrobium aerophilum (Thioalkalimicrobium aerophilum)* DSM 13739 and *Thialkalimicrobium sibiricum (Thioalkalimicrobium sibericum)* DSM 13740.

Alkaliphilic *Thermococcus* Medium (DSMZ Medium 926)

Composition per 1082.0mL:

Base solution	1000.0mL
Glycine solution	50.0mL
Yeast extract solution	20.0mL
Polysulfide solution	12.0mL

Base Solution:

Composition per 2000.0mL:

NaCl	27.7g
$MgSO_4 \cdot 7H_2O$	7.0g
$MgCl_2 \cdot 6H_2O$	5.5g
K_2HPO_4	1.0g
KCl	0.65g
$NaHCO_3$	0.32g
NaBr	0.1g
H_3BO_3	0.03g
KI	15.0mg
$CaCl_2 \cdot 2H_2O$	0.05mg
Trace elements solution	20.0mL

Preparation of Base Solution: Sparge 2.0L of distilled/deionized water with 100% N_2 to remove O_2. Add components to 2000.0mL of O_2-free distilled/deionized water. Mix thoroughly. Sparge with 100% N_2. Autoclave for 15 min at 15 psi pressure–121°C. Cool to room temperature. Do not adjust the pH.

Trace Elements Solution:

Composition per liter:

$MgSO_4 \cdot 7H_2O$	3.0g
Nitrilotriacetic acid	1.5g
NaCl	1.0g
$MnSO_4 \cdot 2H_2O$	0.5g
$CoSO_4 \cdot 7H_2O$	0.18g
$ZnSO_4 \cdot 7H_2O$	0.18g
$CaCl_2 \cdot 2H_2O$	0.1g
$FeSO_4 \cdot 7H_2O$	0.1g
$NiCl_2 \cdot 6H_2O$	0.025g
$KAl(SO_4)_2 \cdot 12H_2O$	0.02g
H_3BO_3	0.01g
$Na_2MoO_4 \cdot 4H_2O$	0.01g
$CuSO_4 \cdot 5H_2O$	0.01g
$Na_2SeO_3 \cdot 5H_2O$	0.3mg

Preparation of Trace Elements Solution: Add nitrilotriacetic acid to 500.0mL of distilled/deionized water. Dissolve by adjusting pH to 6.5 with KOH. Add remaining components. Add distilled/deionized water to 1.0L. Mix thoroughly.

Polysulfide Solution:

Composition per 100.0mL:

$Na_2S \cdot 9H_2O$	1.2g
Sulfur	0.16g

Preparation of Polysulfide Solution: Sparge 100.0mL distilled/deionized water with 100% N_2. Add $Na_2S \cdot 9H_2O$. Mix thoroughly. Add sulfur. The solution will be dark yellow. Sparge with 100% N_2. Autoclave for 15 min at 15 psi pressure–121°C. Cool to room temperature.

Yeast Extract Solution:

Composition per 100.0mL:

Yeast extract	10.0g

Preparation of Yeast Extract Solution: Add yeast extract to distilled/deionized water and bring volume to 100.0mL. Mix thoroughly. Sparge with 100% N_2. Autoclave for 15 min at 15 psi pressure–121°C. Cool to room temperature.

Glycine Solution:

Composition per 100.0mL:

Glycine	15.0g

Preparation of Glycine Solution: Add glycine to distilled/deionized water and bring volume to 100.0mL. Mix thoroughly. Sparge with 100% N_2. Autoclave for 15 min at 15 psi pressure–121°C. Cool to room temperature.

Preparation of Medium: Aseptically and anaerobically add 50.0mL sterile glycine solution, 20.0mL sterile yeast extract solution, and 12.0mL sterile polysulfide solution to 1.0L sterile base solution. Mix thoroughly. Aseptically and anaerobically distribute to sterile tubes or bottles. Do not adjust pH.

Use: For the cultivation of *Thermococcus alcaliphilus* DSM 10322.

Alkaliphilic *Thermococcus* Medium (DSMZ Medium 926)

Composition per 1082.0mL:

Base solution	1000.0mL
Casamino acids solution	20.0mL
Polysulfide solution	8.0mL

Base Solution:

Composition per 2000.0mL:

NaCl	27.7g
$MgSO_4 \cdot 7H_2O$	7.0g
$MgCl_2 \cdot 6H_2O$	5.5g
K_2HPO_4	1.0g
KCl	0.65g
$NaHCO_3$	0.32g
NaBr	0.1g
H_3BO_3	0.03g
KJ	15.0mg
$CaCl_2 \cdot 2H_2O$	0.05mg
Trace elements solution	20.0mL

Preparation of Base Solution: Sparge 2.0L of distilled/deionized water with 100% N_2 to remove O_2. Add components to 2000.0mL of O_2-free distilled/deionized water. Mix thoroughly. Sparge with 100% N_2. Autoclave for 15 min at 15 psi pressure–121°C. Cool to room temperature. Do not adjust the pH.

Trace Elements Solution:

Composition per liter:

$MgSO_4 \cdot 7H_2O$	3.0g
Nitrilotriacetic acid	1.5g

NaCl	1.0g
$MnSO_4·2H_2O$	0.5g
$CoSO_4·7H_2O$	0.18g
$ZnSO_4·7H_2O$	0.18g
$CaCl_2·2H_2O$	0.1g
$FeSO_4·7H_2O$	0.1g
$NiCl_2·6H_2O$	0.025g
$KAl(SO_4)_2·12H_2O$	0.02g
H_3BO_3	0.01g
$Na_2MoO_4·4H_2O$	0.01g
$CuSO_4·5H_2O$	0.01g
$Na_2SeO_3·5H_2O$	0.3mg

Preparation of Trace Elements Solution: Add nitrilotriacetic acid to 500.0mL of distilled/deionized water. Dissolve by adjusting pH to 6.5 with KOH. Add remaining components. Add distilled/deionized water to 1.0L. Mix thoroughly.

Polysulfide Solution:
Composition per 100.0mL:

$Na_2S·9H_2O$	1.2g
Sulfur	0.16g

Preparation of Polysulfide Solution: Sparge 100.0mL distilled/deionized water with 100% N_2. Add $Na_2S·9H_2O$. Mix thoroughly. Add sulfur. The solution will be dark yellow. Sparge with 100% N_2. Autoclave for 15 min at 15 psi pressure–121°C. Cool to room temperature.

Casamino Acids Solution:
Composition per 100.0mL:

Casamino acids	10.0g

Preparation of Casamino Acids Solution: Add casamino acids to distilled/deionized water and bring volume to 100.0mL. Mix thoroughly. Sparge with 100% N_2. Autoclave for 15 min at 15 psi pressure–121°C. Cool to room temperature.

Preparation of Medium: Aseptically and anaerobically add 20.0mL sterile casamino acids solution, and 12.0mL sterile polysulfide solution to 1.0L sterile base solution. Mix thoroughly. Aseptically and anaerobically distribute to sterile tubes or bottles. There may be precipitation of material and the medium will turn pale yellow due to the addition of the polysulfide. The color will disappear as the strain grows. Do not adjust pH.

Use: For the cultivation of *Thermococcus acidaminovorans* DSM 11096.

Alkalophile Medium

Composition per liter:

Agar	15.0g
Peptone	5.0g
NaCl	5.0g
Yeast extract	2.0g
Beef extract	1.0g
Sodium sesquicarbonate solution	15.0mL

pH 9.5 ± 0.2 at 25°C

Sodium Sesquicarbonate Solution:
Composition per 100.0mL:

Sodium sesquicarbonate	9.0g

Preparation of Sodium Sesquicarbonate Solution: Add sodium sesquicarbonate to distilled/deionized water and bring volume to 100.0mL. Mix thoroughly. Filter sterilize.

Preparation of Medium: Add components, except sodium sesquicarbonate solution, to distilled/deionized water and bring volume to 985.0mL. Mix thoroughly. Gently heat and bring to boiling. Autoclave for 15 min at 15 psi pressure–121°C. Cool to 50°–55°C. Aseptically add 15.0mL of filter-sterilized sodium sesquicarbonate solution to adjust pH to 9.5. Mix thoroughly. Pour into sterile Petri dishes or distribute into sterile tubes.

Use: For the cultivation of *Bacillus alcalophilus, Bacillus circulans, Bacillus submarinus*, and other *Bacillus* species.

Alkalophilic Halophile Agar

Composition per liter:

Solution A	500.0mL
Solution B	500.0mL

pH 9.5 ± 1.0 at 25°C

Solution A:
Composition per 500.0mL:

NaCl	200.0g
$Na_2CO_3·10H_2O$	50.0g

Preparation of Solution A: Add components to distilled/deionized water and bring volume to 500.0mL. Mix thoroughly. Autoclave for 15 min at 15 psi pressure–121°C. Cool to 50°–55°C.

Solution B:
Composition per liter:

Agar	20.0g
Yeast extract	10.0g
Casamino acids	7.5g
Trisodium citrate	3.0g
KCl	2.0g
$MgSO_4·7H_2O$	1.0g
$FeSO_4·7H_2O$	50.0mg
$MnCl_2·4H_2O$	0.36mg

Preparation of Solution B: Add components to distilled/deionized water and bring volume to 500.0mL. Mix thoroughly. Autoclave for 15 min at 15 psi pressure–121°C. Cool to 50°–55°C.

Preparation of Medium: Aseptically combine 500.0mL of solution A with 500.0mL of solution B. Mix thoroughly. Bring pH to 9.5. Pour into sterile Petri dishes or distribute into sterile tubes.

Use: For the cultivation and maintenance of *Natronobacterium gregoryi, Natronobacterium magadii, Natronobacterium pharaonis, Natronobacterium vacuolata,* and *Natronococcus occultus.*

Alkalophilic Halophile Broth

Composition per liter:

Solution A	500.0mL
Solution B	500.0mL

pH 9.5 ± 1.0 at 25°C

Solution A:
Composition per 500.0mL:

NaCl	200.0g
$Na_2CO_3·10H_2O$	50.0g

Preparation of Solution A: Add components to distilled/deionized water and bring volume to 500.0mL. Mix thoroughly. Autoclave for 15 min at 15 psi pressure–121°C.

Solution B:
Composition per liter:

Yeast extract	10.0g
Casamino acids	7.5g
Trisodium citrate	3.0g

KCl 2.0g
$MgSO_4 \cdot 7H_2O$ 1.0g
$FeSO_4 \cdot 7H_2O$ 50.0mg
$MnCl_2 \cdot 4H_2O$ 0.36mg

Preparation of Solution B: Add components to distilled/deionized water and bring volume to 500.0mL. Mix thoroughly. Autoclave for 15 min at 15 psi pressure–121°C.

Preparation of Medium: Aseptically combine 500.0mL of solution A with 500.0mL of solution B. Mix thoroughly. Bring pH to 9.5. Aseptically distribute into sterile tubes or flasks.

Use: For the cultivation of *Natronobacterium gregoryi, Natronobacterium magadii, Natronobacterium pharaonis, Natronobacterium vacuolata,* and *Natronococcus occultus.*

Alkvisco Medium

Composition per liter:
Agar 15.0g
Beef extract 10.0g
Peptone 10.0g
NaCl 5.0g
Acrylonitrile 0.5g
KCN 10.0mg
pH 6.5–8.0 at 25°C

Caution: Cyanide is toxic. Acrylonitrile is a carcinogen; use appropriate precautions.

Preparation of Medium: Add components, except acrylonitrile, to distilled/deionized water and bring volume to 980.0mL. Mix thoroughly. Gently heat and bring to boiling. Autoclave for 10 min at 15 psi pressure–121°C. Add acrylonitrile to 20.0mL of distilled/deionized water and filter sterilize. Add aseptically to the sterile basal medium.

Use: For the cultivation and maintenance of *Bacillus subtilis* and *Corynebacterium* species.

Allantoin Agar

Composition per liter:
Agar 15.0g
$Na_2HPO_4 \cdot 12H_2O$ 9.0g
NaCl 5.0g
KH_2PO_4 .... 1.5g
Meat extract 1.0g
Yeast extract 1.0g
$MgSO_4 \cdot 7H_2O$ 0.2g
$MnCl_2 \cdot 4H_2O$ 20.0mg
$CaCl_2$ 1.2mg
Glucose-allantoin solution 100.0mL

Glucose-Allantoin Solution:
Composition per 100.0mL:
Glucose 5.0g
Allantoin 1.0g

Preparation of Glucose-Allantoin Solution: Add components to distilled/deionized water and bring volume to 100.0mL. Mix thoroughly. Filter sterilize. Warm to 50°C.

Preparation of Medium: Add components, except glucose-allantoin solution, to distilled/deionized water and bring volume to 900.0mL. Mix thoroughly. Gently heat and bring to boiling.Autoclave for 15 min at 15 psi pressure–121°C. Cool to 50°–55°C. Aseptically add 100.0mL of sterile glucose-allantoin solution. Mix thoroughly. Pour into sterile Petri dishes or distribute into sterile tubes.

Use: For the cultivation and maintenance of *Bacillus* species.

Allantoin Broth

Composition per liter:
$Na_2HPO_4 \cdot 12H_2O$ 9.0g
NaCl 5.0g
KH_2PO_4 .... 1.5g
Meat extract 1.0g
Yeast extract 1.0g
$MgSO_4 \cdot 7H_2O$ 0.2g
$MnCl_2 \cdot 4H_2O$ 20.0mg
$CaCl_2$ 1.2mg
Glucose-allantoin solution 100.0mL

Glucose-Allantoin Solution:
Composition per 100.0mL:
Glucose 5.0g
Allantoin 1.0g

Preparation of Glucose-Allantoin Solution: Add components to distilled/deionized water and bring volume to 100.0mL. Mix thoroughly. Filter sterilize.

Preparation of Medium: Add components, except glucose-allantoin solution, to distilled/deionized water and bring volume to 900.0mL. Mix thoroughly. Autoclave for 15 min at 15 psi pressure–121°C. Aseptically add 100.0mL of sterile glucose-allantoin solution. Mix thoroughly. Aseptically distribute into sterile tubes or flasks.

Use: For the cultivation and maintenance of *Bacillus* species.

Allantoin Mineral Medium

Composition per liter:
Allantoin 20.0g
Agar 15.0g
K_2HPO_4 .... 0.8g
$MgSO_4 \cdot 7H_2O$ 0.5g
KH_2PO_4 .... 0.2g
$CaCl_2 \cdot 2H_2O$ 0.05g
$FeSO_4 \cdot 7H_2O$ 0.01g
$MnSO_4 \cdot H_2O$ 1.0mg

Preparation of Medium: Add components to distilled/deionized water and bring volume to 1.0L. Mix thoroughly. Gently heat and bring to boiling. Distribute into tubes or flasks. Autoclave for 15 min at 15 psi pressure–121°C. Pour into sterile Petri dishes or leave in tubes.

Use: For the cultivation and maintenance of *Bacillus fastidiosus, Bacillus* species, *Mycobacterium vaccae,* and *Saccharopolyspora rectivirgula.*

Allen and Arnon Medium with Nitrate

Composition per 1000.25mL:
Noble agar 10.0g
KNO_3 .... 0.253g
$NaNO_3$ 0.212g
Solution A 25.0mL
Solution B 6.25mL

Solution A:
Composition per 2.0L:
$MgSO_4 \cdot 7H_2O$ (4% solution) 500.0mL
$CaCl_2 \cdot 2H_2O$ (1.2% solution) 500.0mL
NaCl (3.8% solution) 500.0mL
Microelements stock solution 500.0mL

Preparation of Solution A: Prepare individual solutions and combine.

Microelements Stock Solution:

Composition per 1090.0mL:

H_3BO_3	572.0mg
$MnCl_2·4H_2O$	360.0mg
$ZnSO_4·7H_2O$	44.0mg
MoO_3	36.0mg
$CuSO_4·5H_2O$	15.8mg
$CoCl_2·6H_2O$	8.0mg
NH_4VO_3	4.6mg
A & A FeEDTA solution	160.0mL

Preparation of Microelements Stock Solution: Add components to distilled/deionized water and bring volume to 1090.0mL. Mix well.

A & A FeEDTA Solution:

Composition per 550.0mL:

Disodium EDTA·$2H_2O$	20.4g
$FeSO_4·7H_2O$	13.7g
KOH	5.2g

Preparation of A & A FeEDTA Solution: Dissolve 5.2g of KOH in 186.0mL of distilled/deionized water. Add 20.4g of disodium EDTA·$2H_2O$. Add 13.7g of FeSO4·7H2O to 364.0mL of distilled/deionized water. Combine the EDTA solution with the $FeSO_4$ solution. Sparge solution with filtered air until color changes. The pH is about 3.5.

Solution B:

Composition per 500.0mL:

K_2HPO_4	28.0g

Preparation of Solution B: Add K_2HPO_4 to distilled/deionized water and bring volume to 500.0mL.

Preparation of Medium: Add agar, KNO_3, and $NaNO_3$ to distilled/deionized water and bring volume to 969.0mL. Mix thoroughly. Gently heat and bring to boiling. Add 25.0mL of solution A. Autoclave for 15 min at 15 psi pressure–121°C. Add 6.25mL of solution B aseptically after sterilization.

Use: For the cultivation and maintenance of *Anabaena* species and *Nostoc* species.

Allisonella Medium (DSMZ Medium 1006)

Composition per liter:

DL-Histidine	7.8g
Na_2CO_3	4.0g
Yeast extract	4.0g
Trypticase	1.0g
$(NH_4)_2SO_4$	480.0mg
NaCl	480.0mg
K_2HPO_4	292.0mg
KH_2PO_4	292.0mg
$MgSO_4·7H_2O$	100.0mg
$CaCl_2·2H_2O$	64.0mg
Resazurin	1.0mg
Cysteine solution	10.0mL
Volatile fatty acid mixture	3.1mL

pH 6.0 ± 0.2 at 25°C

Cysteine Solution:

Composition per 10.0mL:

L-Cysteine·HCl·H_2O	0.5g

Preparation of Cysteine Solution: Add L-cysteine·HCl·H_2O to distilled/deionized water and bring volume to 10.0mL. Mix thoroughly. Sparge with 100% N_2. Autoclave for 15 min at 15 psi pressure–121°C.

Volatile Fatty Acid Mixture:

Composition per 7.75mL:

Acetic acid	4.25mL
Propionic acid	1.50mL
Butyric acid	1.0mL
DL-2-Methyl butyric acid	0.25mL
iso-Butyric acid	0.25mL
iso-Valeric acid	0.25mL
n-Valeric acid	0.25mL

Preparation of Volatile Fatty Acid Mixture: Combine components. Mix thoroughly.

Preparation of Medium: Add components, except cysteine solution, volatile fatty acid mixture, and bicarbonate, to distilled/deionized water and bring volume to 1.0L. Mix thoroughly. Gently heat and bring to boiling. Boil for several minutes. Cool to room temperature while sparging with 100% CO_2. Add the solid Na_2CO_3 and 3.1 mL volatile fatty acid mixture and 10.0mL cysteine solution. Adjust the pH to 6.0. Distribute into serum bottles or Hungate tubes under 100% CO_2. Autoclave for 15 min at 15 psi pressure–121°C.

Use: For the maintenance or cultivation of *Allisonella* spp.

Almond Curd Agar

Composition per 4–6 servings:

Almonds	0.5 pound
Agar	0.5 cup
Sweetened water	2.0 cups

Sweetened Water:

Composition per 2.0 cups:

Sucrose	0.5 cup

Preparation of Sweetened Water: Bring 2.0 cups of tap water to boiling. Add sucrose. Mix thoroughly until dissolved. Cool to 4°C.

Preparation of Medium: Blanch and skin almonds. Add 0.25 pound in blender with 2.0 cups of tap water. Blend to a paste. Filter through two layers of cheesecloth. Squeeze cheesecloth to obtain all liquid. Discard solids. Repeat process with remaining 0.25 pound of almonds and 2.0 cups of tap water. In a separate container add agar to 3.0 cups of water. Gently heat and bring to boiling. Boil while stirring continuously until agar dissolves and begins to thicken. Add almond filtrate. Cook for 1 min. Pour into a rectangular pan. Cool to room temperature. Bring curd to 4°C. Prior to utilization, cut into cubes or diamond shapes. Place approximately 0.25 cup sweetened water into a container and add almond curd cubes.

Use: For the refreshment and culinary enjoyment of microbiologists.

ALOA Medium (Agar *Listeria* Ottavani & Agosti) (BAM M10a)

Composition per liter:

Agar	18.0g
Peptone	18.0g
LiCl	10.0g
Yeast extract	10.0g
Tryptone	6.0g
NaCl	5.0g

Na_2HPO_4 .... 2.5g
Na-pyruvate 2.0g
Glucose 2.0g
Mg-glycerophosphate 1.0g
$MgSO_4$ 0.5g
5-Bromo4-chloro-indolyl-β-D-glucopyranoside 0.05g
Phosphatidylinositol solution 50.0mL
Nalidixic acid solution 5.0mL
Ceftazidime solution 5.0mL
Cycloheximide solution 5.0mL
Polymyxin B solution 5.0mL

pH 7.2 ± 0.2 at 25°C

Nalidixic Acid Solution:
Composition per 5.0mL:
Nalidixic acid 0.02g

Preparation of Nalidixic Acid Solution: Add nalidixic acid to distilled/deionized water and bring volume to 5.0mL. Mix thoroughly. Filter sterilize.

Ceftazidime Solution:
Composition per 5.0mL:
Ceftazidime 0.02g

Preparation of Ceftazidime Solution: Add ceftazidime to distilled/deionized water and bring volume to 5.0mL. Mix thoroughly. Filter sterilize.

Cycloheximide Solution:
Composition per 5.0mL:
Cycloheximide 0.05g
Ethanol 2.5mL

Preparation of Cycloheximide Solution: Add cycloheximide to 2.5mL of ethanol. Mix thoroughly. Bring volume to 5.0mL with distilled/deionized water. Filter sterilize.

Caution: Cycloheximide is toxic. Avoid skin contact or aerosol formation and inhalation.

Polymyxin B Solution:
Composition per 5.0mL:
Polymyxin B 76700U

Preparation of Polymyxin B Solution: Add polymyxin B to distilled/deionized water and bring volume to 5.0mL. Mix thoroughly. Filter sterilize.

Phosphatidylinositol Solution:
Composition per 50.0mL:
L-α-phosphatidylinositol 2.0g

Preparation of Phosphatidylinositol Solution: Add L-α-phosphotidylinositol to cold distilled/deionized water and bring volume to 50.0mL. Stir for 30 min so a homogeneous suspension is obtained. Autoclave for 15 min at 15 psi pressure–121°C. Cool to 48–50°C.

Preparation of Medium: Add components, except phosphatidylinositol solution, nalidixic acid solution, cetazidime solution, cycloheximide solution, and polymyxin B solution, to distilled/deionized water and bring volume to 930.0mL. Mix thoroughly. Adjust the pH to 7.2. Gently heat and bring to boiling. Autoclave for 15 min at 15 psi pressure–121°C. Cool to 45°–50°C. Aseptically add 50.0mL sterile phosphatidylinositol solution, 5.0mL sterile nalidixic acid solution, 5.0mL sterile cetazidime solution, 5.0mL sterile cycloheximide solution, and 5.0mL sterile polymyxin B solution. Mix thoroughly. Pour into Petri dishes or distribute into sterile tubes.

Use: For the isolaltion and cultivation of *Listeria* spp.

ALOA Medium (Agar *Listeria* Ottavani & Agosti) (BAM M10a)

Composition per liter:
Agar 18.0g
Peptone 18.0g
LiCl 10.0g
Yeast extract 10.0g
Tryptone 6.0g
NaCl 5.0g
Na_2HPO_4 .... 2.5g
Na-pyruvate 2.0g
Glucose 2.0g
Mg-glycerophosphate 1.0g
$MgSO_4$ 0.5g
5-Bromo4-chloro-indolyl-β-D-glucopyranoside 0.05g
Phosphatidylinositol solution 50.0mL
Amphotericin B solution 10.0mL
Nalidixic acid solution 5.0mL
Ceftazidime solution 5.0mL
Polymyxin B solution 5.0mL

pH 7.2 ± 0.2 at 25°C

Nalidixic Acid Solution:
Composition per 5.0mL:
Nalidixic acid 0.02g

Preparation of Nalidixic Acid Solution: Add nalidixic acid to distilled/deionized water and bring volume to 5.0mL. Mix thoroughly. Filter sterilize.

Ceftazidime Solution:
Composition per 5.0mL:
Ceftazidime 0.02g

Preparation of Ceftazidime Solution: Add ceftazidime to distilled/deionized water and bring volume to 5.0mL. Mix thoroughly. Filter sterilize.

Amphotericin B Solution:
Composition per 10.0mL:
Amphotericin B 0.01g
Dimethylforamide 7.5mL
HCL, 1*M* 2.5mL

Preparation of Amphotericin B Solution: Add amphotericin B to 2.5mL of 1*M* HCl. Mix thoroughly. Add 7.5 mL of dimethlyformamide. Mix thoroughly. Filter sterilize.

Polymyxin B Solution:
Composition per 5.0mL:
Polymyxin B 76700U

Preparation of Polymyxin B Solution: Add polymyxin B to distilled/deionized water and bring volume to 5.0mL. Mix thoroughly. Filter sterilize.

Phosphatidylinositol Solution:
Composition per 50.0mL:
L-α-phosphatidylinositol 2.0g

Preparation of Phosphatidylinositol Solution: Add L-α-phosphotidylinositol to cold distilled/deionized water and bring volume to 50.0mL. Stir for 30 min so a homogeneous suspension is obtained. Autoclave for 15 min at 15 psi pressure–121°C. Cool to 48–50°C.

Preparation of Medium: Add components, except phosphatidylinositol solution, nalidixic acid solution, cetazidime solution, amphotericin B solution, and polymyxin B solution, to distilled/deionized water and bring volume to 920.0mL. Mix thoroughly. Adjust the pH to 7.2. Gently heat and bring to boiling. Autoclave for 15 min at 15 psi pressure–121°C. Cool to 45°–50°C. Aseptically add 50.0mL sterile phosphatidylinositol solution, 5.0mL sterile nalidixic acid solution, 5.0mL sterile cetazidime solution, 10.0mL sterile amphotericin B solution, and 5.0mL sterile polymyxin B solution. Mix thoroughly. Pour into Petri dishes or distribute into sterile tubes.

Use: For the isolaltion and cultivation of *Listeria* spp. For the isolation and cultivation of *Literia* spp. according to ISO standard 11290.

ALP Basal Medium (Aerobic Low Peptone Basal Medium)

Composition per liter:

Agar	15.0g
$(NH_4)_2SO_4$	1.0g
Pancreatic digest of casein	0.5g
Yeast extract	0.5g
$MgSO_4 \cdot 7H_2O$	0.2g
KCl	0.2g
Phenol Red	0.02g
Substrate solution	50.0mL

pH 7.8 ± 0.2 at 25°C

Substrate Solution:

Composition per 50.0mL:

Substrate	0.1g

Preparation of Substrate Solution: Add substrate to distilled/deionized water and bring volume to 50.0mL. Use sugars, carbohydrates, *n*-butanol, other alcohols, or any acidogenic carbon source. Mix thoroughly. Filter sterilize.

Preparation of Medium: Add components, except substrate solution, to distilled/deionized water and bring volume to 950.0mL. Mix thoroughly. Gently heat and bring to boiling. Adjust pH to 7.8. Distribute into screw-capped tubes in 3.0mL volumes. Autoclave for 15 min at 15 psi pressure–121°C. Cool to 45°–50°C. Aseptically add 0.15mL of sterile substrate solution to each tube. Mix thoroughly. Allow tubes to cool in a slanted position.

Use: For the cultivation and differentiation of microorganisms based on their ability to utilize a variety of carbon sources such as carbohydrates, alcohols, and other acidogenic substrates.

ALP Basal Medium Low pH (Aerobic Low Peptone Basal Medium)

Composition per liter:

Agar	15.0g
$(NH_4)_2SO_4$	1.0g
Pancreatic digest of casein	0.5g
Yeast extract	0.5g
Glucose	0.2g
$MgSO_4 \cdot 7H_2O$	0.2g
KCl	0.2g
Phenol Red	0.02g
Substrate solution	50.0mL

pH 6.5 ± 0.2 at 25°C

Substrate Solution:

Composition per 50.0mL:

Substrate	0.1g

Preparation of Substrate Solution: Add substrate to distilled/deionized water and bring volume to 50.0mL. Use gelatin, aliphatic acids, or any alkalogenic carbon source. Mix thoroughly. Filter sterilize.

Preparation of Medium: Add components, except substrate solution, to distilled/deionized water and bring volume to 950.0mL. Mix thoroughly. Gently heat and bring to boiling. Adjust pH to 6.5. Distribute into screw-capped tubes in 3.0mL volumes. Autoclave for 15 min at 15 psi pressure–121°C. Cool to 45°–50°C. Aseptically add 0.15mL of sterile substrate solution to each tube. Mix thoroughly. Allow tubes to cool in a slanted position.

Use: For the cultivation and differentiation of microorganisms based on their ability to utilize a variety of carbon sources such as gelatin, aliphatic acids, and other alkalophilic substrates.

Alternative Thioglycollate Medium (NIH Thioglycollate Broth)

Composition per liter:

Casein enzymatic hydrolysate	15.0g
Glucose	5.5g
Yeast extract	5.0g
NaCl	2.5g
L-Cystine	0.5g
Sodium thioglycollate	0.5g

pH 7.1 ± 0.1 at 25°C

Source: This medium is available from HiMedia.

Preparation of Medium: Add components to distilled/deionized water and bring volume to 1.0L. Mix thoroughly. Distribute into tubes or flasks. Autoclave for 10 min at 15 psi pressure–121°C.

Use: For the sterility testing of certain biological products that are turbid or viscous.

Alternative Thioglycollate Medium

Composition per liter:

Plant hydrolysate	15.0g
Glucose	5.5g
Yeast extract	5.0g
NaCl	2.5g
L-Cystine	0.5g
Sodium thioglycollate	0.5g

pH 7.1 ± 0.1 at 25°C

Source: This medium is available from HiMedia.

Preparation of Medium: Add components to distilled/deionized water and bring volume to 1.0L. Mix thoroughly. Distribute into tubes or flasks. Autoclave for 10 min at 15 psi pressure–121°C.

Use: For the sterility testing of certain biological products that are turbid or viscous.

Alteromonas denitrificans Medium

Composition per liter:

Peptone	0.5g
Pancreatic digest of casein	0.5g
Yeast extract	0.5g
Aged seawater	800.0mL

Preparation of Medium: Add components, except aged seawater, to tap water and bring volume to 200.0mL. Mix thoroughly. Autoclave for 15 min at 15 psi pressure–121°C. Aseptically add 800.0mL of filter-sterilized aged seawater. Mix thoroughly. Aseptically distribute into sterile tubes or flasks.

Use: For the cultivation of *Alteromonas denitrificans*.

Alteromonas Medium (LMG Medium 28)

Composition per 1012mL:

Agar base	1.0L
Methanol	10.0mL
Solution A	1.0mL
Solution B	1.0mL

pH 7.0± 0.2 at 25°C

Agar Base:

Composition per liter:

NaCl	20.0g
Agar	15.0g
$(NH_4)_2SO_4$	2.0g
K_2HPO_4	2.0g
KH_2PO_4	1.0g
$MgSO_4·7H_2O$	0.3g

Preparation of Agar Base: Add components to 1.0L of distilled/deionized water. Mix thoroughly.

Solution A:

Composition per 100.0mL:

$MnSO_4·2H_2O$	76mg
$FeSO_4.7H_2O$	28mg
$CuSO_4·5H_2O$	25mg
$Na_2MoO_4·2H_2O$	24mg
$CoCl_2·6H_2O$	24mg
$CaCl_2·2H_2O$	15mg
$ZnSO_4·7H_2O$	0.14mg

Preparation of Solution A: Add components to 100.0mL of distilled/deionized water. Mix thoroughly.

Solution B:

Composition per 100.0mL:

Vitamin B_{12}	0.1mg

Preparation of Solution B: Add Vitamin B_{12} to 100.0mL of distilled/deionized water. Mix thoroughly. Filter sterilize.

Preparation of Medium: Add 1.0mL Solution A to 1.0L Agar Base. Mix thoroughly. Adjust pH to 7.0. Gently heat and bring to boiling. Autoclave for 15 min at 15 psi pressure–121°C. Cool to 45°–50°C. Aseptically add 1.0mL sterile Solution B and 10.0mL filter-sterilized methanol. Mix thoroughly. Pour into sterile Petri dishes or distribute into sterile tubes.

Use: For the cultivation of *Alteromonas* spp.

AMB Agar

Composition per liter:

Agar	15.0g
Starch, soluble	5.0g
Pancreatic digest of casein	2.5g
$MgSO_4·7H_2O$	0.5g
K_2HPO_4	0.25g

Preparation of Medium: Add components to distilled/deionized water and bring volume to 1.0L. Mix thoroughly. Gently heat and bring to boiling. Distribute into tubes or flasks. Autoclave for 15 min at 15 psi pressure–121°C. Pour into sterile Petri dishes or leave in tubes.

Use: For the cultivation of myxobacteria.

AMB Broth

Composition per liter:

Starch, soluble	5.0g
Pancreatic digest of casein	2.5g
$MgSO_4·7H_2O$	0.5g
K_2HPO_4	0.25g

Preparation of Medium: Add components to distilled/deionized water and bring volume to 1.0L. Mix thoroughly. Distribute into tubes or flasks. Autoclave for 15 min at 15 psi pressure–121°C.

Use: For the cultivation of myxobacteria.

AMB Medium

Composition per liter:

Meat infusion	25.0g
K_2HPO_4	15.0g
Glucose	5.0g
Yeast extract	5.0g
Pancreatic digest of casein	4.0g
L-Cysteine	1.0g
$(NH_4)_2SO_4$	1.0g
Starch, soluble	1.0g
$MgSO_4$	0.2g
$CaCl_2$	0.01g

pH 6.9 ± 0.2 at 25°C

Preparation of Medium: Add components to distilled/deionized water and bring volume to 1.0L. Mix thoroughly. Gently heat and bring to boiling. Distribute into tubes or flasks. Autoclave for 15 min at 15 psi pressure–121°C.

Use: For the cultivation and maintenance of *Eubacterium alactolyticum, Eubacterium budayl, Eubacterium moniliforme,* and *Eubacterium tortuosum.*

American Association of Textile Chemists and Colorists Bacteriostasis Agar

See: **AATCC Bacteriostasis Agar**

American Association of Textile Chemists and Colorists Bacteriostasis Broth

See: **FDA Broth**

American Association of Textile Chemists and Colorists Mineral Salts Iron Agar

See: **AATCC Mineral Salts Iron Agar**

American Society for Testing and Materials Nutrient Salts Agar

See: **ASTM Nutrient Salts Agar**

American Trudeau Society Medium

See: **ATS Medium**

AMH (DSMZ Medium 1110)

Composition per liter:

NaCl	20.0g
Sulfur, powdered	5.0g

$MgCl_2 \cdot 6H_2O$	3.0g
KCl	0.5g
NH_4Cl	0.25g
KH_2PO_4	0.2g
$CaCl_2 \cdot 2H_2O$	0.15g
Resazurin	0.5mg
Vitamin solution	10.0mL
$Na_2S \cdot 9H_2O$ solution	10.0mL
$NaHCO_3$ solution	10.0mL
Trace elements solution SL-10	1.0mL
Selenite-tungstate solution	1.0mL

pH 7.0 ± 0.2 at 25°C

$NaHCO_3$ Solution:
Composition per 10.0mL:

$NaHCO_3$	2.5g

Preparation of $NaHCO_3$ Solution: Add $NaHCO_3$ to distilled/deionized water and bring volume to 10.0mL. Mix thoroughly. Sparge with 20% CO_2 + 80% H_2. Autoclave for 15 min at 15 psi pressure–121°C. Cool to room temperature.

$Na_2S \cdot 9H_2O$ Solution:
Composition per 10.0mL:

$Na_2S \cdot 9H_2O$	0.5g

Preparation of $Na_2S \cdot 9H_2O$ Solution: Add $Na_2S \cdot 9H_2O$ to distilled/deionized water and bring volume to 10.0mL. Mix thoroughly. Autoclave under 100% N_2 for 15 min at 15 psi pressure–121°C. Cool to room temperature.

Vitamin Solution:
Composition per liter:

Pyridoxine-HCl	10.0mg
Thiamine-HCl·$2H_2O$	5.0mg
Riboflavin	5.0mg
Nicotinic acid	5.0mg
D-Ca-pantothenate	5.0mg
p-Aminobenzoic acid	5.0mg
Lipoic acid	5.0mg
Biotin	2.0mg
Folic acid	2.0mg
Vitamin B_{12}	0.1mg

Preparation of Vitamin Solution: Add components to distilled/deionized water and bring volume to 1.0L. Mix thoroughly. Sparge with 80% H_2 + 20% CO_2. Filter sterilize.

Selenite/Tungstate Solution:
Composition per liter:

NaOH	0.5g
$Na_2WO_4 \cdot 2H_2O$	4.0mg
$Na_2SeO_3 \cdot 5H_2O$	3.0mg

Preparation of Selenite/Tungstate Solution: Add components to distilled/deionized water and bring volume to 1.0L. Mix thoroughly. Sparge with 100% N_2. Autoclave for 15 min at 15 psi pressure–121°C.

Trace Elements Solution SL-10:
Composition per liter:

$FeCl_2 \cdot 4H_2O$	1.5g
$CoCl_2 \cdot 6H_2O$	190.0mg
$MnCl_2 \cdot 4H_2O$	100.0mg
$ZnCl_2$	70.0mg
$Na_2MoO_4 \cdot 2H_2O$	36.0mg
$NiCl_2 \cdot 6H_2O$	24.0mg
H_3BO_3	6.0mg
$CuCl_2 \cdot 2H_2O$	2.0mg
HCl (25% solution)	10.0mL

Preparation of Trace Elements Solution SL-10: Add $FeCl_2 \cdot 4H_2O$ to 10.0mL of HCl solution. Mix thoroughly. Add distilled/deionized water and bring volume to 1.0L. Add remaining components. Mix thoroughly. Sparge with 100% N_2. Autoclave for 15 min at 15 psi pressure–121°C.

Preparation of Medium: Add components, except bicarbonate solution, sulfide solution, sulfur, and vitamin solution, to distilled/deionized water and bring volume to 970.0mL. Mix thoroughly. Gently heat and bring to boiling. Boil for several minutes. Cool to room temperature while sparging with 80% H_2 + 20% CO_2. Add the bicarbonate solution, sulfide solution, and vitamin solution. Distribute into scew-capped tubes or bottles. Autoclave for 15 min at 15 psi pressure–121°C. Place the sulfur in screw-capped tubes or bottles. Autoclave for 15 min at 8 psi pressure–112°C. Before use, aseptically and anoxically layer the sulfur onto the surface of sterile liquid basal medium. Adjust the final pH to 7.0. After inoculation, pressurize culture vials to 1 bar overpressure with 80% H_2 and 20% CO_2 gas mixture.

Use: For the maintenance or cultivation of *Nautilia profundicola.*

Amies Modified Transport Medium with Charcoal

Composition per liter:

Charcoal	10.0g
Agar	4.0g
NaCl	3.0g
Na_2HPO_4	1.15g
Sodium thioglycolate	1.0g
KCl	0.2g
KH_2PO_4	0.2g
$CaCl_2 \cdot 2H_2O$	0.1g
$MgCl_2 \cdot 6H_2O$	0.1g

pH 7.2 ± 0.2 at 25°C

Source: This medium is available as a premixed powder from BD Diagnostic Systems.

Preparation of Medium: Add components to distilled/deionized water and bring volume to 1.0L. Mix thoroughly. Gently heat and bring to boiling. Distribute into flasks or tubes. Autoclave for 20 min at 15 psi pressure–121°C. While cooling, turn tubes to uniformly suspend charcoal.

Use: For the transport of swab specimens to prolong the survival of microorganisms, especially *Neisseria gonorrhoeae,* between collection and culturing. Addition of charcoal to this medium neutralizes metabolic products that may be toxic to *Neisseria gonorrhoeae.*

Amies Modified Transport Medium with Charcoal

Composition per liter:

Charcoal	10.0g
NaCl	8.0g
Agar	3.6g
Na_2HPO_4	1.15g
Sodium thioglycolate	1.0g
KCl	0.2g
$CaCl_2 \cdot 2H_2O$	0.1g
$MgCl_2 \cdot 6H_2O$	0.1g
KH_2PO_4	0.2g

pH 7.2 ± 0.2 at 25°C

Source: This medium is available as a premixed powder from BD Diagnostic Systems and Oxoid Unipath.

Preparation of Medium: Add components to distilled/deionized water and bring volume to 1.0L. Mix thoroughly. Gently heat and bring to boiling. Distribute into flasks or tubes. Autoclave for 20 min at 15 psi pressure–121°C. While cooling, turn tubes to uniformly suspend charcoal.

Use: For the transport of swab specimens to prolong the survival of microorganisms, especially *Neisseria gonorrhoeae,* between collection and culturing. Addition of charcoal to this medium neutralizes metabolic products that may be toxic to *Neisseria gonorrhoeae*.

Amies Transport Medium without Charcoal

Composition per liter:

Agar	4.0g
NaCl	3.0g
Na_2HPO_4	1.15g
Sodium thioglycolate	1.0g
KCl	0.2g
$CaCl_2 \cdot 2H_2O$	0.1g
$MgCl_2 \cdot 6H_2O$	0.1g
KH_2PO_4	0.2g

pH 7.2 ± 0.2 at 25°C

Source: This medium is available as a premixed powder from BD Diagnostic Systems.

Preparation of Medium: Add components to distilled/deionized water and bring volume to 1.0L. Mix thoroughly. Gently heat and bring to boiling. Distribute into flasks or tubes. Autoclave for 20 min at 15 psi pressure–121°C.

Use: For the transport of swab specimens to prolong the survival of microorganisms, especially *Neisseria gonorrhoeae,* between collection and culturing.

Amies Transport Medium without Charcoal

Composition per liter:

NaCl	8.0g
Agar	3.6g
Na_2HPO_4	1.15g
Sodium thioglycolate	1.0g
KCl	0.2g
$CaCl_2 \cdot 2H_2O$	0.1g
$MgCl_2 \cdot 6H_2O$	0.1g
KH_2PO_4	0.2g

pH 7.2 ± 0.2 at 25°C

Source: This medium is available as a premixed powder from BD Diagnostic Systems.

Preparation of Medium: Add components to distilled/deionized water and bring volume to 1.0L. Mix thoroughly. Gently heat and bring to boiling. Distribute into flasks or tubes. Autoclave for 20 min at 15 psi pressure–121°C.

Use: For the transport of swab specimens to prolong the survival of microorganisms, especially *Neisseria gonorrhoeae,* between collection and culturing.

Aminiphilus Medium (DSMZ Medium 1082)

Composition per liter:

Trypticase peptone	10.0g
Casamino acids	10.0g
NaCl	1.0g
KCl	0.5g
$MgCl_2 \cdot 6H_2O$	0.4g
NH_4Cl	0.3g
KH_2PO_4	0.2g
$CaCl_2 \cdot 2H_2O$	0.15g
Resazurin	0.5mg
$NaHCO_3$ solution	10.0mL
Vitamin solution	10.0mL
$Na_2S \cdot 9H_2O$ solution	10.0mL
Trace elements solution SL-10	1.0mL

pH 7.1 ± 0.2 at 25°C

$NaHCO_3$ Solution:

Composition per 10.0mL:

$NaHCO_3$	2.5g

Preparation of $NaHCO_3$ Solution: Add $NaHCO_3$ to distilled/deionized water and bring volume to 10.0mL. Mix thoroughly. Sparge with 20% CO_2 + 80% H_2. Autoclave for 15 min at 15 psi pressure–121°C. Cool to room temperature.

$Na_2S \cdot 9H_2O$ Solution:

Composition per 100.0mL:

$Na_2S \cdot 9H_2O$	0.6g

Preparation of $Na_2S \cdot 9H_2O$ Solution: Add $Na_2S \cdot 9H_2O$ to distilled/deionized water and bring volume to 100.0mL. Mix thoroughly. Autoclave under 100% N_2 for 15 min at 15 psi pressure–121°C. Cool to room temperature.

Vitamin Solution:

Composition per liter:

Pyridoxine-HCl	10.0mg
Thiamine-HCl·2H$_2$O	5.0mg
Riboflavin	5.0mg
Nicotinic acid	5.0mg
D-Ca-pantothenate	5.0mg
p-Aminobenzoic acid	5.0mg
Lipoic acid	5.0mg
Biotin	2.0mg
Folic acid	2.0mg
Vitamin B_{12}	0.1mg

Preparation of Vitamin Solution: Add components to distilled/deionized water and bring volume to 1.0L. Mix thoroughly. Sparge with 80% H_2 + 20% CO_2. Filter sterilize.

Trace Elements Solution SL-10:

Composition per liter:

$FeCl_2 \cdot 4H_2O$	1.5g
$CoCl_2 \cdot 6H_2O$	190.0mg
$MnCl_2 \cdot 4H_2O$	100.0mg
$ZnCl_2$	70.0mg
$Na_2MoO_4 \cdot 2H_2O$	36.0mg
$NiCl_2 \cdot 6H_2O$	24.0mg
H_3BO_3	6.0mg
$CuCl_2 \cdot 2H_2O$	2.0mg
HCl (25% solution)	10.0mL

Preparation of Trace Elements Solution SL-10: Add $FeCl_2 \cdot 4H_2O$ to 10.0mL of HCl solution. Mix thoroughly. Add distilled/deionized water and bring volume to 1.0L. Add remaining components. Mix thoroughly. Sparge with 100% N_2. Autoclave for 15 min at 15 psi pressure–121°C.

Preparation of Medium: Add components, except $Na_2S \cdot 9H_2O$ solution, bicarbonate solution, and vitamin solution, to distilled/deionized water and bring volume to 970.0mL. Mix thoroughly. Gently heat and bring to boiling. Boil for several minutes. Cool to room temperature while sparging with 20% CO_2 + 80% N_2. Dispense into tubes or flasks. Autoclave for 15 min at 15 psi pressure–121°C. Cool to 25°C. Aseptically and anoxically add the $Na_2S \cdot 9H_2O$ solution, bicarbonate solution, and vitamin solution. The pH should be 7.1.

Use: For the maintenance or cultivation of *Aminiphilus* spp.

Amino Acid Assay Medium

Composition per liter:

Glucose	50.0g
Sodium acetate	40.0g
NH_4Cl	6.0g
KH_2PO_4	1.2g
K_2HPO_4	1.2g
Asparagine	0.8g
L-Glutamic acid	0.6g
Pyridoxamine·HCl	0.6g
Pyridoxal·HCl	0.6g
DL-Valine	0.5g
L-Lysine·HCl	0.5g
DL-Isoleucine	0.5g
DL-Leucine	0.5g
L-Arginine·HCl	0.5g
DL-Threonine	0.4g
$MgSO_4 \cdot 7H_2O$	0.4g
DL-Alanine	0.4g
DL-Phenylalanine	0.2g
L-Tyrosine	0.2g
Lysine	0.2g
L-Aspartic acid	0.2g
DL-Methionine	0.2g
L-Proline	0.2g
L-Histidine·HCl	0.12g
DL-Serine	0.1g
L-Cystine	0.1g
DL-Tryptophan	0.08g
$MnSO_4 \cdot 7H_2O$	0.04g
$FeSO_4$	0.02g
NaCl	0.02g
Adenine sulfate	0.02g
Guanine·HCl	0.02g
Uracil	0.02g
Xanthine	0.02g
Nicotinic acid	2.0mg
Pyridoxine·HCl	2.0mg
Thiamine·HCl	1.0mg
Calcium pantothenate	1.0mg
Riboflavin	1.0mg
p-Aminobenzoic acid	0.2mg
Folic acid	0.02mg
Biotin	2.0μg

pH 6.7 ± 0.2 at 25°C

Source: This medium is available, without cystine, lysine, or methionine, as a premixed powder from BD Diagnostic Systems.

Preparation of Medium: Add components to 1.0L of distilled water, omitting the specific amino acid to be assayed for in the procedure. Heat to boiling for 2–3 min. Distribute into tubes. Autoclave for 10 min at 15 psi pressure–121°C.

Use: For the microbiological assay for amino acids. *Pediococcus acidilactici* ATCC 8042 and *Enterococcus hirae* ATCC 8043 are used as test microorganisms.

Amino-butyric Acid Medium

Composition per liter:

Agar	15.0g
DL-Amino-butyric acid	10.0g
K_2HPO_4	7.0g
Glucose	5.0g
KH_2PO_4	3.0g
$(NH_4)_2SO_4$	1.0g
$MgSO_4 \cdot 7H_2O$	0.5g

pH 7.0 ± 0.2 at 25°C

Preparation of Medium: Add components to distilled/deionized water and bring volume to 1.0L. Mix thoroughly. Gently heat and bring to boiling. Autoclave for 15 min at 15 psi pressure–121°C. Pour into sterile Petri dishes.

Use: For the cultivation and maintenance of *Serratia marcescens* and other microorganisms that can utilize amino-butyric acid as a carbon source.

Ammonifex Medium

Composition per 1010.0mL:

NaCl	1.0g
$Na_2S \cdot 9H_2O$	0.5g
$K_2HPO_4 \cdot 3H_2O$	0.3g
KH_2PO_4	0.22g
$(NH_4)_2SO_4$	0.22g
$NaHCO_3$	0.2g
$MgSO_4 \cdot 7H_2O$	0.09g
$CaCl_2 \cdot 2H_2O$	0.06g
$FeSO_4 \cdot 7H_2O$	2.0mg
$NiCl_2 \cdot 6H_2O$	0.2mg
Resazurin	0.5mg
KNO_3 solution	10.0mL
Selenite/tungstate solution	3.0mL
Wolfe's mineral solution	1.0mL

pH 5.4 ± 0.2 at 25°C

Selenite/Tungstate Solution:

Composition per liter:

NaOH	0.5g
$Na_2WO_4 \cdot 2H_2O$	4.0mg
$Na_2SeO_3 \cdot 5H_2O$	3.0mg

Preparation of Selenite/Tungstate Solution: Add components to distilled/deionized water and bring volume to 1.0L. Mix thoroughly. Sparge with 100% N_2. Autoclave for 15 min at 15 psi pressure–121°C.

Wolfe's Mineral Solution:

Composition per liter:

$MgSO_4 \cdot 7H_2O$	3.0g
Nitrilotriacetic acid	1.5g
NaCl	1.0g
$MnSO_4 \cdot 2H_2O$	0.5g
$CoCl_2 \cdot 6H_2O$	0.1g
$ZnSO_4 \cdot 7H_2O$	0.1g
$CaCl_2 \cdot 2H_2O$	0.1g
$FeSO_4 \cdot 7H_2O$	0.1g
$NiCl_2 \cdot 6H_2O$	0.025g
$KAl(SO_4)_2 \cdot 12H_2O$	0.02g
$CuSO_4 \cdot 5H_2O$	0.01g

H_3BO_3 .... 0.01g
$Na_2MoO_4·2H_2O$ 0.01g
$Na_2SeO_3·5H_2O$ 0.3mg

Preparation of Wolfe's Mineral Solution: Add nitrilotriacetic acid to 500.0mL of distilled/deionized water. Adjust pH to 6.5 with KOH. Add remaining components. Add distilled/deionized water to 1.0L. Adjust pH to 6.8.

KNO_3 Solution:
Composition per 100.0mL:
KNO_3 .... 10.0g

Preparation of KNO_3 Solution: Add KNO_3 to distilled/deionized water and bring volume to 100.0mL. Mix thoroughly. Sparge with 100% N_2. Autoclave for 15 min at 15 psi pressure–121°C.

Preparation of Medium: Prepare and dispense medium under 100% N_2. Add components, except $Na_2S·9H_2O$ and KNO_3 solution, to distilled/deionized water and bring volume to 1.0L. Mix thoroughly. Add $Na_2S·9H_2O$. Mix thoroughly. Adjust pH to 6.5 with 20% HCl. Sparge with 100% N_2 for 30 min. Dispense anaerobically into tubes in 10.0mL aliquots. Evacuate headspace under vacuum. Pressurize tubes to 200 kPa with 80% H_2 + 20% CO_2 gas mixture. Autoclave for 15 min at 15 psi pressure–121°C. Adjust pH to 5.4. Prior to use, aseptically and anaerobically add 0.1mL of sterile KNO_3 to each tube.

Use: For the cultivation of *Ammonifex* species.

Ammonium Mineral Salts Agar
See: **AMS Agar**

Ammonium Mineral Salts Agar without Methanol
See: **AMS Agar without Methanol**

Ammonium Phosphate Agar

Composition per liter:
Agar 15.0g
Glucose 10.0g
$(NH_4)_3PO_4$ 1.0g
KCl 0.2g
$MgSO_4·7H_2O$ 0.2g
Bromocresol Purple 0.05g
pH 7.0 ± 0.2 at 25°C

Source: This medium is available from HiMedia.

Preparation of Medium: Add components to distilled/deionized water and bring volume to 1.0L. Mix thoroughly. Gently heat and bring to boiling. Autoclave for 15 min at 15 psi pressure–121°C. Pour into sterile Petri dishes.

Use: For the cultivation of microorganisms that use ammonium phosphate as a source of nitrogen. For the differentiation of micrococci from staphylococci.

Ammonium Yeast Extract Medium
See: ***Bacillus pasteurii*** **NH_4 YE Medium**

AMO.1 Medium

Composition per 1012.0mL:
Yeast extract 10.0g
NaCl 5.8g
N-Methylhydantoin 5.64g
$NaHCO_3$ 4.5g
L-Serine 2.0g
L-Threonine 2.0g
Pancreatic digest of casein 0.5g
L-Cysteine 0.5g
$MgCl_2·6H_2O$ 0.4g
KCl 0.3g
NH_4Cl 0.27g
KH_2PO_4 .... 0.2g
$CaCl_2·2H_2O$ 0.15g
Wolfe's vitamin solution 10.0mL
$NaHSeO_3$ solution 1.0mL
Trace elements solution SL-10 .... 1.0mL
pH 8.3 ± 0.2 at 25°C

Wolfe's Vitamin Solution:
Composition per liter:
Pyridoxine·HCl 10.0mg
p-Aminobenzoic acid 5.0mg
Lipoic acid 5.0mg
Nicotinic acid 5.0mg
Riboflavin 5.0mg
Thiamine·HCl 5.0mg
Calcium DL-pantothenate 5.0mg
Biotin 2.0mg
Folic acid 2.0mg
Vitamin B_{12} 0.1mg

Preparation of Wolfe's Vitamin Solution: Add components to distilled/deionized water and bring volume to 1.0L. Mix thoroughly.

Na_2SeO_3 Solution:
Composition per liter:
$Na_2SeO_3·5H_2O$ 0.2mg

Preparation of Na_2SeO_3 Solution: Add $Na_2SeO_3·5H_2O$ to distilled/deionized water and bring volume to 1.0L. Mix thoroughly. Sparge with 100% N_2. Autoclave for 15 min at 15 psi pressure–121°C.

Trace Elements Solution SL-10:
Composition per liter:
$FeCl_2·4H_2O$ 1.5g
$CoCl_2·6H_2O$ 190.0mg
$MnCl_2·4H_2O$ 100.0mg
$ZnCl_2$ 70.0mg
$Na_2MoO_4·2H_2O$ 36.0mg
$NiCl_2·6H_2O$ 24.0mg
H_3BO_3 .... 6.0mg
$CuCl_2·2H_2O$ 2.0mg
HCl (25% solution) 10.0mL

Preparation of Trace Elements Solution SL-10: Add $FeCl_2·4H_2O$ to 10.0mL of HCl solution. Mix thoroughly. Add distilled/deionized water and bring volume to 1.0L. Add remaining components. Mix thoroughly. Sparge with 100% N_2. Autoclave for 15 min at 15 psi pressure–121°C.

Preparation of Medium: Prepare anaerobically. Add components, except $NaHCO_3$ and L-cysteine, to distilled/deionized water and bring volume to 1.0L. Sparge with 100% N_2 for 30 min. Adjust pH to 8.3 with 10*N* NaOH. Add $NaHCO_3$ and L-cysteine (solid substances). Mix thoroughly. Sparge with 80% N_2 + 20% CO_2 mixture. Flush the headspace of the medium vessel with the 80% N_2 + 20% CO_2 mixture. Autoclave for 15 min at 15 psi pressure–121°C. Sparge with 80% N_2 + 20% CO_2 mixture for 30 min. Aseptically and anaerobically distribute into sterile tubes or bottles.

Use: For the cultivation of *Tissierella creatinini*.

Amoebobacter Medium

Composition per 4990.0mL:

Solution A	4000.0mL
Solution B	860.0mL
Solution E	100.0mL
Solution F	20.0mL
Solution C (Vitamin B_{12} solution)	5.0mL
Solution D	5.0mL

pH 7.3 ± 0.2 at 25°C

Solution A:

Composition per 4000.0mL:

$MgSO_4$	2.5g
KH_2PO_4	1.7g
NH_4Cl	1.7g
KCl	1.7g
$CaCl_2·2H_2O$	1.25g
$Na_2S_2O_3·5H_2O$	0.25g
Sodium acetate·$3H_2O$	0.14g

Preparation of Solution A: Add components to distilled/deionized water and bring volume to 4.0L. Mix thoroughly. Adjust pH to 6.0. Dispense into a 5L flask with four openings at the top (two openings are in a central silicon rubber stopper and two openings are gas-tight screw caps). Add a teflon-coated magnetic stir bar to the flask. Autoclave for 45 min at 15 psi pressure–121°C. Cool to room temperature under 100% N_2 at 0.05–0.1 atm pressure (use a manometer to measure low pressure).

Solution B:

Composition per 860.0mL:

Distilled/deionized water	860.0mL

Preparation of Solution B: Add 860.0mL of distilled/deionized water to a cotton-stoppered flask. Autoclave for 20 min at 15 psi pressure–121°C. Cool to room temperature under 100% N_2 in an anaerobic jar.

Solution C (Vitamin B_{12} Solution):

Composition per 5.0mL:

Vitamin B_{12}	1.0mg

Preparation of Solution C (Vitamin B_{12} Solution): Add vitamin B_{12} to distilled/deionized water and bring volume to 5.0mL. Mix thoroughly. Filter sterilize.

Solution D:

Composition per liter:

Disodium ethylendiamine-tetraacetate (Disodium EDTA)	3.0g
$FeSO_4·7H_2O$	1.1g
H_3BO_3	0.3g
$CoCl_2·6H_2O$	0.19g
$MnCl_2·2H_2O$	50.0mg
$ZnCl_2$	42.0mg
$NiCl_2·6H_2O$	24.0mg
$Na_2MoO_4·2H_2O$	18.0mg
$CuCl_2·2H_2O$	2.0mg

Preparation of Solution D: Add components to distilled/deionized water and bring volume to 1.0L. Mix thoroughly. Autoclave for 15 min at 15 psi pressure–121°C.

Solution E:

Composition per 100.0mL:

$NaHCO_3$	7.5g

Preparation of Solution E: Add $NaHCO_3$ to distilled/deionized water and bring volume to 100.0mL. Mix thoroughly. Sparge with 100% CO_2 until saturated. Filter sterilize under 100% CO_2 into a sterile, gas-tight 100.0mL screw-capped bottle.

Solution F:

Composition per 100.0mL:

$Na_2S·9H_2O$	10.0g

Preparation of Solution F: Add $Na_2S·9H_2O$ to distilled/deionized water and bring volume to 100.0mL. Mix thoroughly. Dispense into a screw-capped bottle. Sparge with 100% N_2 for 3–4 min. Autoclave for 15 min at 15 psi pressure–121°C.

Preparation of Medium: Saturate cooled solution A under 100% CO_2 at 0.05–0.1 atm pressure for 30 min with magnetic stirring. Add 860.0mL of solution B, 5.0mL of solution C, 5.0mL of solution D, 100.0mL of solution E, and 20.0mL of solution F through one of the screw-cappped openings under 95% N_2 and 5% CO_2 with magnetic stirring. Adjust pH to 7.3 with sterile 2*M* HCl or sterile 2*M* Na_2CO_3 solution. Aseptically and anaerobically distribute the medium through the medium outlet tube into sterile 100.0mL bottles under 95% N_2 + 5% CO_2 at 0.05–0.1 atm pressure. Leave a small gas bubble in each bottle to accommodate pressure changes. After 24 hr, the iron in the medium will precipitate out of solution as black flocs.

Use: For the cultivation and maintenance of *Amphibacillus xylanus, Amoebobacter pedioformis,* and *Amoebobacter purpureus.*

Amphibacillus Medium

Composition per liter:

Agar	15.0g
Glucose	10.0g
Yeast extract	3.0g
NH_4NO_3	2.0g
K_2HPO_4	1.0g
Polypeptone™ (pancreatic digest of casein and peptic digest of animal tissue)	0.3g
$MgSO_4·7H_2O$	0.2g
$CaCl_2·2H_2O$	0.1g
$FeSO_4·7H_2O$	5.0mg
$MnSO_4·7H_2O$	5.0mg

pH 9.7 ± 0.2 at 25°C

Sodium Sesquicarbonate Solution:

Composition per 100.0mL:

Na_2CO_3, anhydrous	10.6g
$NaHCO_3$	8.42g

Preparation of Sodium Sesquicarbonate Solution: Add components to distilled/deionized water and bring volume to 100.0mL. Mix thoroughly. Filter sterilize. Warm to 50°–55°C.

Preparation of Medium: Add components, except sodium sesquicarbonate solution, to distilled/deionized water and bring volume to 900.0mL. Mix thoroughly. Gently heat and bring to boiling. Autoclave for 15 min at 15 psi pressure–121°C. Cool to 50°–55°C. Aseptically add sterile sodium sesquicarbonate solution. Mix thoroughly. Adjust pH to 9.7. Pour into sterile Petri dishes or distribute into sterile tubes.

Use: For the cultivation and maintenance of *Amphibacillus xylanus.*

Ampicillin Dextrin Agar

Composition per liter:

Agar	15.0g
Dextrin	10.0g

NaCl	3.0g
Yeast extract	2.0g
KCl	2.0g
$MgSO_4 \cdot 7H_2O$	0.2g
$FeCl_3 \cdot 4H_2O$	0.1g
Selective supplement solution	10.0mL

pH 8.0 ± 0.2 at 25°C

Source: This medium is available from HiMedia.

Selective Supplement Solution:
Composition per 10.0mL:

Sodium deoxycholate	100.0mg
Ampicillin	10.0mg

Preparation of Selective Supplement Solution: Add components to distilled/deionized water and bring volume to 10.0mL. Mix thoroughly. Filter sterilize.

Preparation of Medium: Add components, except selective supplement solution and kanamycin solution, to distilled/deionized water and bring volume to 990.0mL. Mix thoroughly. Autoclave for 15 min at 15 psi pressure–121°C. Aseptically add 10.0mL of sterile selective supplement solution. Mix thoroughly. Pour into Petri dishes or aseptically distribute into sterile tubes.

Use: For the selective isolation and differentiation of *Aeromonas* spp. from water samples.

Ampicillin Dextrin Agar with Vancomycin (ADA-V)

Composition per liter:

Agar	13.0g
Dextrin	11.4g
Tryptose	5.0g
NaCl	3.0g
KCl	2.0g
Yeast extract	2.0g
$MgSO_4 \cdot 7H_2O$	1.0g
Bromothymol Blue	0.08g
$FeCl_3 \cdot 6H_2O$	0.06g
Sodium deoxycholate	1.0mg
Ampicillin solution	10.0mL
Vancomycin solution	10.0mL

pH 8.0 ± 0.2 at 25°C

Ampicillin Solution:
Composition per 10.0mL:

Ampicillin	10.0mg

Preparation of Ampicillin Solution: Add components to distilled/deionized water and bring volume to 10.0mL. Mix thoroughly. Filter sterilize.

Vancomycin Solution:
Composition per 10.0mL:

Vancomycin	2.0mg

Preparation of Vancomycin Solution: Add components to distilled/deionized water and bring volume to 10.0mL. Mix thoroughly. Filter sterilize.

Preparation of Medium: Add components, except sodium deoxycholate, ampicillin solution, and vancomycin solution, to distilled/deionized water and bring volume to 980.0mL. Mix thoroughly. Adjust pH to 8.0. Add sodium deoxycholate. Autoclave for 15 min at 15 psi pressure–121°C. Cool to 50°C. Aseptically add ampicillin and vancomycin solutions. Mix thoroughly. Pour into Petri dishes or aseptically distribute into sterile tubes.

Use: For the selective isolation and differentiation of *Aeromonas* spp. from water samples.

Ampicillin Kanamycin Nutrient Agar

Composition per liter:

Agar	15.0g
Peptone	5.0g
NaCl	5.0g
Yeast extract	2.0g
Beef extract	1.0g
Ampicillin solution	10.0mL
Kanamycin solution	10.0mL

Ampicillin Solution:
Composition per 10.0mL:

Ampicillin	50.0mg

Preparation of Ampicillin Solution: Add ampicillin to distilled/deionized water and bring volume to 10.0mL. Mix thoroughly. Filter sterilize.

Kanamycin Solution:
Composition per 10.0mL:

Kanamycin	25.0mg

Preparation of Kanamycin Solution: Add kanamycin to distilled/deionized water and bring volume to 10.0mL. Mix thoroughly. Filter sterilize.

Preparation of Medium: Add components, except ampicllin solution and kanamycin solution, to distilled/deionized water and bring volume to 980.0mL. Mix thoroughly. Autoclave for 15 min at 15 psi pressure–121°C. Aseptically add 10.0mL of sterile ampicillin solution and 10.0mL of sterile kanamycin solution. Mix thoroughly. Aseptically distribute into sterile tubes or flasks.

Use: For the cultivation of fungi and various antibiotic-resistant bacteria, including *Escherichia coli.*

Ampicillin L Broth Medium

Composition per liter:

Pancreatic digest of casein	10.0g
NaCl	5.0g
Yeast extract	5.0g
Glucose	1.0g
Ampicillin solution	10.0mL

pH 7.0 ± 0.2 at 25°C

Ampicillin Solution:
Composition per 10.0mL:

Ampicillin	50.0mg

Preparation of Ampicillin Solution: Add ampicillin to distilled/deionized water and bring volume to 10.0mL. Mix thoroughly. Filter sterilize.

Preparation of Medium: Add components, except ampicillin solution, to distilled/deionized water and bring volume to 990.0mL. Mix thoroughly. Bring pH to 7.0. Autoclave for 15 min at 15 psi pressure–121°C. Aseptically add 10.0mL of sterile ampicillin solution. Mix thoroughly. Aseptically distribute into sterile tubes or flasks.

Use: For the cultivation of *Escherichia coli.*

Ampicillin TY Salt Medium

Composition per liter:

NaCl	10.0g
Pancreatic digest of casein	10.0g
Yeast extract	5.0g
Ampicillin solution	10.0mL

pH 7.0 ± 0.2 at 25°C

Ampicillin Solution:

Composition per 10.0mL:

Ampicillin	50.0mg

Preparation of Ampicillin Solution: Add ampicillin to distilled/deionized water and bring volume to 10.0mL. Mix thoroughly. Filter sterilize.

Preparation of Medium: Add components, except ampicillin solution, to distilled/deionized water and bring volume to 990.0mL. Mix thoroughly. Bring pH to 7.0. Autoclave for 15 min at 15 psi pressure–121°C. Aseptically add 10.0mL of sterile ampicillin solution. Mix thoroughly. Aseptically distribute into sterile tubes or flasks.

Use: For the cultivation of various antibiotic resistant bacteria, including *Escherichia coli.*

AMS Agar (Ammonium Mineral Salts Agar)

Composition per liter:

Agar	15.0g
$MgSO_4 \cdot 7H_2O$	1.0g
K_2HPO_4	0.7g
KH_2PO_4	0.54g
NH_4Cl	0.5g
$CaCl_2 \cdot 2H_2O$	0.2g
$FeSO_4 \cdot 7H_2O$	4.0mg
H_3BO_4	0.3mg
$CoCl_2 \cdot 6H_2O$	0.2mg
$ZnSO_4 \cdot 7H_2O$	0.1mg
$Na_2MoO_4 \cdot 2H_2O$	0.06mg
$MnCl_2 \cdot 4H_2O$	0.03mg
$NiCl_2 \cdot 6H_2O$	0.02mg
$CuCl_2 \cdot 2H_2O$	0.01mg

pH 6.8 ± 0.2 at 25°C

Preparation of Medium: Add components to distilled/deionized water and bring volume to 1.0L. Mix thoroughly. Gently heat and bring to boiling. Autoclave for 15 min at 15 psi pressure–121°C. Add sterile methanol to a concentration of 0.5% aseptically to cooled basal medium.

Use: For the cultivation and maintenance of bacteria that can utilize methanol as a carbon source, such as *Methylobacterium* species, *Methylomonas* species, and *Methylophilus* species.

AMS Agar without Methanol (Ammonium Mineral Salts Agar without Methanol)

Composition per liter:

Agar	15.0g
$MgSO_4 \cdot 7H_2O$	1.0g
K_2HPO_4	0.7g
KH_2PO_4	0.54g
NH_4Cl	0.5g
$CaCl_2 \cdot 2H_2O$	0.2g
$FeSO_4 \cdot 7H_2O$	4.0mg
H_3BO_4	0.3mg
$CoCl_2 \cdot 6H_2O$	0.2mg
$ZnSO_4 \cdot 7H_2O$	0.1mg
$Na_2MoO_4 \cdot 2H_2O$	0.06mg
$MnCl_2 \cdot 4H_2O$	0.03mg
$NiCl_2 \cdot 6H_2O$	0.02mg
$CuCl_2 \cdot 2H_2O$	0.01mg

pH 6.8 ± 0.2 at 25°C

Preparation of Medium: Add components to distilled/deionized water and bring volume to 1.0L. Mix thoroughly. Gently heat and bring to boiling. Autoclave for 15 min at 15 psi pressure–121°C.

Use: For the cultivation and maintenance of *Methylosinus trichosporium* and other methane-oxidizing bacteria. Cultures are grown under an atmosphere of 50% methane.

AMS Medium

Composition per liter:

NaCl	26.0g
$MgSO_4 \cdot 7H_2O$	12.0g
Peptone	5.0g
Beef extract	3.0g
$CaCl_2 \cdot 2H_2O$	1.5g
KCl	0.7g

Preparation of Medium: Add components to distilled/deionized water and bring volume to 1.0L. Mix thoroughly. Gently heat and bring to boiling. Distribute into tubes or flasks. Autoclave for 15 min at 15 psi pressure–121°C.

Use: For the cultivation of *Alteromonas espejiana.*

AMS Medium, Modified

Composition per liter:

NaCl	24.0g
Proteose peptone	10.0g
$MgSO_4 \cdot 7H_2O$	7.0g
$MgCl_2 \cdot 6H_2O$	5.3g
Yeast extract	3.0g
KCl	0.7g

pH 7.0 ± 0.2 at 25°C

Preparation of Medium: Add components to distilled/deionized water and bring volume to 1.0L. Mix thoroughly. Gently heat and bring to boiling. Distribute into tubes or flasks. Autoclave for 15 min at 15 psi pressure–121°C.

Use: For the cultivation of *Alteromonas haloplanktis, Alteromonas macleodii, Alteromonas nigrifaciens, Alteromonas rubra, Cytophaga lytica, Cytophaga marinoflava,* and *Pseudomonas elongata.*

Amygdalin Medium

Composition per liter:

Peptone	10.0g
Beef extract	5.0g
NaCl	5.0g
Agar	3.0g
Amygdalin solution	200.0mL
Bromthymol Blue (0.05% solution)	5.0mL

pH 7.0 ± 0.2 at 25°C

Amygdalin Solution:

Composition per 200.0mL:

Amygdalin	10.0g

Preparation of Amygdalin Solution: Add amygdalin to distilled/deionized water and bring volume to 200.0mL. Mix thoroughly. Filter sterilize.

Preparation of Medium: Add components, except amygdalin solution, to distilled/deionized water and bring volume to 800.0mL. Mix thoroughly. Gently heat and bring to boiling. Adjust pH to 7.0. Autoclave for 20 min at 15 psi pressure–121°C. Cool to 45°–50°C. Aseptically add sterile amygdalin solution. Mix thoroughly. Aseptically distribute into sterile tubes with cotton plugs. Allow tubes to cool in a slanted position, forming a short slant.

Use: For the cultivation and differentiation of *Serratia* species based on their ability to produce acid and HCN from amygdalin.

AN1 Medium

Composition per liter:

Pancreatic digest of casein	10.0g
Sulfur, powdered	8.0g
NaCl	2.5g
K_2HPO_4	1.5g
Disodium thioglycolate solution	1.0g
Resazurin	1.0mg

pH 7.3 ± 0.2 at 25°C

Disodium Thioglycolate Solution:
Composition per 10.0mL:

Disodium thioglycolate	1.0g

Preparation of Disodium Thioglycolate Solution: Add disodium thioglycolate to distilled/deionized water and bring volume to 10.0mL. Mix thoroughly. Filter sterilize.

Preparation of Medium: Add components, except powdered sulfur and disodium thioglycolate solution, to distilled/deionized water and bring volume to 990.0mL. Mix thoroughly. Gently heat and bring to boiling. Allow to cool under 100% N_2. Autoclave for 15 min at 15 psi pressure–121°C. Powdered sulfur is sterilized separately by steaming for 3 hr on 3 consecutive days. Aseptically and anaerobically combine the basal solution, sterile powdered sulfur, and sterile disodium thioglycolate solution under 100% N_2.

Use: For the cultivation of *Thermococcus* species.

Anacker-Ordal Agar (DSMZ Medium 1039)

Composition per liter:

Agar	11.0g
Pancreatic digest of casein	0.5g
Yeast extract	0.5g
Sodium acetate	0.2g
Meat extract	0.2g

pH 7.2 ± 0.1 at 25°C

Preparation of Medium: Add components to distilled/deionized water and bring volume to 1.0L. Mix thoroughly. Gently heat and bring to boiling. Distribute into tubes or flasks. Autoclave for 15 min at 15 psi pressure–121°C. Pour into sterile Petri dishes or leave in tubes.

Use: For the cultivation and maintenance of *Flexibacter* spp.

Anacker and Ordal Medium

Composition per liter:

Agar	10.0g
Pancreatic digest of casein	0.5g
Yeast extract	0.5g
Sodium acetate	0.2g
Beef extract	0.2g

pH 7.3 ± 0.1 at 25°C

Preparation of Medium: Add components to distilled/deionized water and bring volume to 1.0L. Mix thoroughly. Gently heat and bring to boiling. Distribute into tubes or flasks. Autoclave for 15 min at 15 psi pressure–121°C. Pour into sterile Petri dishes or leave in tubes.

Use: For the cultivation and maintenance of *Flexibacter columnaris*.

Anacker and Ordal Medium, Enriched

Composition per liter:

Agar	10.0g
Pancreatic digest of casein	5.0g
Yeast extract	0.5g
Sodium acetate	0.2g
Beef extract	0.2g

pH 7.3 ± 0.1 at 25°C

Preparation of Medium: Add components to distilled/deionized water and bring volume to 1.0L. Mix thoroughly. Gently heat and bring to boiling. Distribute into tubes or flasks. Autoclave for 15 min at 15 psi pressure–121°C. Pour into sterile Petri dishes or leave in tubes.

Use: For the cultivation and maintenance of *Flexibacter psychrophilus*.

Anaerobacter Medium

Composition per liter:

Yeast extract	1.0g
$CaCl_2 \cdot 2H_2O$	0.33g
KCl	0.33g
KH_2PO_4	0.33g
$MgCl_2 \cdot 6H_2O$	0.33g
NH_4Cl	0.33g
Resazurin	1.0mg
$NaHCO_3$ solution	30.0mL
Sucrose solution	10.0mL
L-Cysteine·HCl solution	10.0mL
$Na_2S \cdot 9H_2O$ solution	10.0mL
Trace elements solution SL-10	1.0mL

pH 7.0 ± 0.2 at 25°C

$NaHCO_3$ Solution:
Composition per 30.0mL:

$NaHCO_3$	1.5g

Preparation of $NaHCO_3$ Solution: Add $NaHCO_3$ to distilled/deionized water and bring volume to 30.0mL. Mix thoroughly. Sparge under 80% N_2 + 20% CO_2 for 3 min. Autoclave for 15 min at 15 psi pressure–121°C.

Sucrose Solution:
Composition per 100.0mL:

Sucrose	20.0g

Preparation of Sucrose Solution: Add sucrose to distilled/deionized water and bring volume to 100.0mL. Mix thoroughly. Sparge with 100% N_2. Autoclave under 100% N_2 for 15 min at 15 psi pressure–121°C.

L-Cysteine·HCl Solution:
Composition per 10.0mL:

L-Cysteine·HCl	0.3g

Preparation of L-Cysteine·HCl Solution: Add L-cysteine·HCl to distilled/deionized water and bring volume to 10.0mL. Mix thoroughly. Sparge with 100% N_2. Autoclave under 100% N_2 for 15 min at 15 psi pressure–121°C.

$Na_2S \cdot 9H_2O$ Solution:

Composition per 10.0mL:

$Na_2S \cdot 9H_2O$	0.3g

Preparation of $Na_2S \cdot 9H_2O$ Solution: Add $Na_2S \cdot 9H_2O$ to distilled/deionized water and bring volume to 10.0mL. Mix thoroughly. Sparge with 100% N_2. Autoclave under 100% N_2 for 15 min at 15 psi pressure–121°C.

Trace Elements Solution SL-10:

Composition per liter:

$FeCl_2 \cdot 4H_2O$	1.5g
$CoCl_2 \cdot 6H_2O$	190.0mg
$MnCl_2 \cdot 4H_2O$	100.0mg
$ZnCl_2$	70.0mg
$Na_2MoO_4 \cdot 2H_2O$	36.0mg
$NiCl_2 \cdot 6H_2O$	24.0mg
H_3BO_3	6.0mg
$CuCl_2 \cdot 2H_2O$	2.0mg
HCl (25% solution)	10.0mL

Preparation of Trace Elements Solution SL-10: Add $FeCl_2 \cdot 4H_2O$ to 10.0mL of HCl solution. Mix thoroughly. Add distilled/deionized water and bring volume to 1.0L. Add remaining components. Mix thoroughly.

Preparation of Medium: Add components, except $NaHCO_3$ solution, sucrose solution, L-cysteine·HCl solution, and $Na_2S \cdot 9H_2O$ solution, to distilled/deionized water and bring volume to 940.0mL. Mix thoroughly. Sparge under 100% N_2 for 3–4 min. Autoclave under 100% N_2 for 15 min at 15 psi pressure–121°C. Aseptically and anaerobically add the sterile $NaHCO_3$ solution, sucrose solution, L-cysteine·HCl solution, and $Na_2S \cdot 9H_2O$ solution. Mix thoroughly. Final pH of the medium should be 7.0.

Use: For the cultivation and maintenance of *Anaerobacter polyendosporus.*

Anaerobaculum thermoterrenum Medium (DSMZ Medium 104a)

Composition per liter:

Yeast extract	10.0g
NaCl	8.0g
Trypticase™ peptone	5.0g
Peptone	5.0g
Beef extract	5.0g
Glucose	5.0g
K_2HPO_4	2.0g
L-Cysteine·HCl	0.5g
Resazurin	1.0mg
$Na_2S \cdot 9H_2O$ solution	50.0mL
Salt solution	40.0mL
Glucose solution	10.0mL

pH 7.0± 0.2 at 25°C

Salt Solution:

Composition per liter:

$NaHCO_3$	10.0g
NaCl	2.0g
K_2HPO_4	1.0g
KH_2PO_4	1.0g
$MgSO_4 \cdot 7H_2O$	0.5g
$CaCl_2 \cdot 2H_2O$	0.25g

Preparation of Salt Solution: Add components to distilled/deionized water and bring volume to 1.0L. Mix thoroughly.

$Na_2S \cdot 9H_2O$ Solution:

Composition per 100.0mL:

$Na_2S \cdot 9H_2O$	5.0g

Preparation of $Na_2S \cdot 9H_2O$ Solution: Add $Na_2S \cdot 9H_2O$ to distilled/deionized water and bring volume to 100.0mL. Sparge with N_2. Autoclave for 15 min at 15 psi pressure–121°C. Cool to 25°C. Store anaerobically.

Glucose Solution:

Composition per 100.0mL:

D-Glucose	10.0g

Preparation of Glucose Solution: Add D-glucose to distilled/deionized water and bring volume to 100.0mL. Mix thoroughly. Sparge with N_2. Autoclave for 15 min at 15 psi pressure–121°C. Cool to 25°C. Store anaerobically.

Preparation of Medium: Add components, except L-cysteine·HCl, glucose solution, and $Na_2S \cdot 9H_2O$ solution, to distilled/deionized water and bring volume to 940.0mL. Mix thoroughly. Gently heat and bring to boiling. Sparge with CO_2. Add L-cysteine·HCl. Autoclave for 15 min at 15 psi pressure–121°C. Cool to 25°C. Aseptically add 50.0mL sterile $Na_2S \cdot 9H_2O$ solution and 10.0mL sterile glucose solution. Mix thoroughly. Adjust pH to 7.0 with 8*N* NaOH. Distribute into sterile tubes or flasks under anaerobic N_2.

Use: For the cultivation and maintenance of *Anaerobaculum thermoterrenum.*

Anaerobe Agar (LMG Medium 41)

Composition per liter:

Agar Base	800.0mL
Solution A	100.0mL
Solution B	100.0mL

pH 6.9 ± 0.2 at 25°C

Agar Base:

Composition per 800.0mL:

Agar	30.5g
Tryptone	5.0g
Yeast extract	5.0g
$(NH_4)_2SO_4$	0.5g
Sodium thioglycolate	0.5g
$MgSO_4 \cdot 7H_2O$	0.1g
$Fe(NH_4)_2(SO_4)_2 \cdot 6H_2O$	55.0mg
$Na_2MoO_4 \cdot 2H_2O$	2.4 mg
$Na_2SeO_3 \cdot 5H_2O$	0.23 mg

Preparation of Agar Base: Add components to distilled/deionized water and bring volume to 800.0mL. Mix thoroughly. Gently heat and bring to boiling. Autoclave for 15 min at 15 psi pressure–121°C. Cool to 45°–50°C.

Solution A:

Composition per 100.0mL:

Glucose	18.0g
K_2HPO_4	7.0g
KH_2PO_4	5.5g

Preparation of Solution A: Add components to 100.0mL of distilled/deionized water. Mix thoroughly. Filter sterilize.

Solution B:

Composition per 100.0mL:

$NaHCO_3$	10.0g

Preparation of Solution B: Add $NaHCO_3$ to 100.0mL of distilled/deionized water. Mix thoroughly. Filter sterilize.

Preparation of Medium: Aseptically add solutions A and B to the agar base. Adjust pH to 6.9. Mix thoroughly. Pour into sterile Petri dishes or distribute into sterile tubes. Incubate anaerobically under 100% CO_2 gas atmosphere.

Use: For the cultivation of anaerobic bacteria.

Anaerobe Agar

Composition per 1001.5mL:

Agar	20.0g
Pancreatic digest of casein	17.0g
NaCl	5.0g
Yeast extract	5.0g
Papaic digest of soybean meal	3.0g
K_2HPO_4	2.5g
Glucose	2.5g
L-Cystine solution	5.0mL
Vitamin K_1 solution	1.0mL
Hemin solution	0.5mL

pH 7.5 ± 0.2 at 25°C

L-Cystine Solution:

Composition per 5.0mL:

L-Cystine	0.4g
NaOH (1*N* solution)	5.0mL

Preparation of L-Cystine Solution: Add L-cystine to 5.0mL of NaOH solution. Mix thoroughly.

Vitamin K_1 Solution:

Composition per 100.0mL:

Vitamin K_1	1.0g
Ethanol	99.0mL

Preparation of Vitamin K_1 Solution: Add vitamin K_1 to 99.0mL of absolute ethanol. Mix thoroughly. Filter sterilize.

Hemin Solution:

Composition per 100.0mL:

Hemin	1.0g
NaOH (1*N* solution)	20.0mL

Preparation of Hemin Solution: Add hemin to 20.0mL of 1*N* NaOH solution. Mix thoroughly. Bring volume to 100.0mL with distilled/deionized water. Autoclave for 15 min at 15 psi pressure–121°C. Cool to 45°–50°C.

Preparation of Medium: Add components, except vitamin K_1 solution and hemin solution, to distilled/deionized water and bring volume to 1.0L. Mix thoroughly. Gently heat while stirring and bring to boiling. Adjust pH to 7.5. Autoclave for 15 min at 15 psi pressure–121°C. Cool to 45°–50°C. Aseptically add 1.0mL of sterile vitamin K_1 solution and 0.5mL of sterile hemin solution. Mix thoroughly. Pour into sterile Petri dishes or distribute into sterile tubes.

Use: For the cultivation of anaerobes from cosmetic products.

Anaerobe Agar (BAM M11)

Composition per 1001.5mL:

Agar	15.0g
Pancreatic digest of casein	15.0g
Pancreatic digest of soybean meal	5.0g
NaCl	5.0g
L-Cysteine·HCl·H_2O solution	5.0mL
Vitamin K_1 solution	1.0mL
Hemin solution	0.5mL

pH 7.0 ± 0.2 at 25°C

L-Cysteine·HCl·H_2O Solution:

Composition per 5.0mL:

L-Cysteine·HCl·H_2O	0.4g
NaOH (1*N* solution)	5.0mL

Preparation of L-Cysteine·HCl·H_2O Solution: Add L-cysteine·HCl·H_2O to 5.0mL 1*N* NaOH. Mix thoroughly. Filter sterilize.

Hemin Solution:

Composition per 100.0mL:

Hemin	1.0g

Preparation of Hemin Solution: Add hemin to distilled/deionized water and bring volume to 100.0mL. Mix thoroughly. Autoclave for 15 min at 15 psi pressure–121°C. Cool. Refrigerate at 4°C for storage.

Vitamin K_1 Solution:

Composition per 100.0mL:

Vitamin K_1	1.0g
Ethanol, absolute	20.0mL

Preparation of Vitamin K_1 Solution: Add vitamin K_1 to 100.0mL of 95% ethanol. Mix thoroughly. Solution may require 2–3 days with intermittent shaking to completely dissolve. Filter sterilize. Refrigerate at 4°C for storage.

Preparation of Medium: Add components, except hemin solution and vitamin K_1 solution, to distilled/deionized water and bring volume to 1.0L. Mix thoroughly. Gently heat and bring to boiling. Adjust pH to 7.0 ± 0.2 at 25°C. Autoclave for 15 min at 15 psi pressure–121°C. Cool to 45°–50°C. Aseptically add 0.5mL of sterile hemin solution and 1.0mL of sterile vitamin K_1 solution. Mix thoroughly. Pour into sterile Petri dishes or distribute into sterile tubes. Reduce medium for 24h by incubation in an anaerobic glove box or GasPak jar prior to use.

Use: For the cultivation of anaerobic bacteria such as *Brucella* and *Clostridium* spp.

Anaerobe Medium (LMG Medium 41)

Composition per liter:

Agar Base	800.0mL
Solution A	100.0mL
Solution B	100.0mL

pH 6.9 ± 0.2 at 25°C

Agar Base:

Composition per 800.0mL:

Tryptone	5.0g
Yeast extract	5.0g
$(NH_4)_2SO_4$	0.5g
Sodium thioglycolate	0.5g
Agar	0.5g
$MgSO_4·7H_2O$	0.1g

$Fe(NH_4)_2(SO_4)_2 \cdot 6H_2O$ 55.0mg
$Na_2MoO_4 \cdot 2H_2O$ .... 2.4 mg
$Na_2SeO_3 \cdot 5H_2O$ .... 0.23 mg

Preparation of Agar Base: Add components to distilled/deionized water and bring volume to 800.0mL. Mix thoroughly. Gently heat and bring to boiling. Autoclave for 15 min at 15 psi pressure–121°C. Cool to 45°–50°C.

Solution A:
Composition per 100.0mL:
Glucose 18.0g
K_2HPO_4 .... 7.0g
KH_2PO_4 .... 5.5g

Preparation of Solution A: Add components to 100.0mL of distilled/deionized water. Mix thoroughly. Filter sterilize.

Solution B:
Composition per 100.0mL:
$NaHCO_3$ 10.0g

Preparation of Solution B: Add components to 100.0mL of distilled/deionized water. Mix thoroughly. Filter sterilize.

Preparation of Medium: Aseptically add solutions A and B to the sterile agar base. Adjust pH to 6.9. Mix thoroughly. Distribute into sterile tubes. Incubate anaerobically under 100% CO_2 gas atmosphere.

Use: For the cultivation of anaerobic bacteria.

Anaerobe Medium No. 1

Composition per 1011.0mL:
Beef extract 10.0g
Peptone 10.0g
Glucose 5.0g
Yeast extract 5.0g
NH_4Cl 1.0g
K_2HPO_4 .... 0.45g
KH_2PO_4 .... 0.33g
$MgSO_4 \cdot 7H_2O$ 0.1g
L-Cysteine·HCl·H_2O solution 10.0mL
$Na_2S \cdot 9H_2O$ solution 10.0mL
Resazurin (0.1% solution) 1.0mL
pH 7.5 ± 0.2 at 25°C

L-Cysteine·HCl·H_2O Solution:
Composition per 10.0mL:
L-Cysteine·HCl·H_2O 0.5g

Preparation of L-Cysteine·HCl·H_2O Solution: Add L-cysteine·HCl·H_2O to distilled/deionized water and bring volume to 10.0mL. Mix thoroughly. Gas under 100% N_2. Autoclave for 15 min at 15 psi pressure–121°C.

$Na_2S \cdot 9H_2O$ Solution:
Composition per 10.0mL:
$Na_2S \cdot 9H_2O$ 0.5g

Preparation of $Na_2S \cdot 9H_2O$ Solution: Add $Na_2S \cdot 9H_2O$ to distilled/deionized water and bring volume to 10.0mL. Mix thoroughly. Gas under 100% N_2. Autoclave for 15 min at 15 psi pressure–121°C.

Preparation of Medium: Add components, except L-cysteine·HCl·H_2O solution and $Na_2S \cdot 9H_2O$ solution, to distilled/deionized water and bring volume to 980.0mL. Mix thoroughly. Autoclave for 15 min at 15 psi pressure–121°C. Aseptically add 10.0mL of sterile L-cysteine·HCl·H_2O solution and 10.0mL of sterile $Na_2S \cdot 9H_2O$ solution. Mix thoroughly. Aseptically distribute into sterile tubes or flasks.

Use: For the cultivation of *Acetobacterium woodii* and *Acetobacterium wieringae.*

Anaerobic Acetoin Medium

Composition per 1006.0mL:
Solution A 916.0mL
Solution B 70.0mL
Solution C 10.0mL
Solution D 10.0mL
pH 7.2 ± 0.2 at 25°C

Solution A:
Composition per 916.0mL:
Acetoin 1.5g
Pancreatic digest of casein 1.0g
Resazurin 1.0mg
Mineral solution 50.0mL
Rumen fluid, clarified 50.0mL
Vitamin solution 5.0mL
Trace elements solution SL-10 .... 1.0mL

Mineral Solution:
Composition per liter:
Nitrilotriacetic acid 12.5g
NaCl 1.0g
$FeCl_3 \cdot 4H_2O$ 0.2g
$MnCl_2 \cdot 4H_2O$ 0.1g
$CaCl_2 \cdot 2H_2O$ 0.1g
$ZnCl_2$ 0.1g
$CuCl_2$ 0.02g
Na_2SeO_3 .... 0.02g
$CoCl_2 \cdot 6H_2O$ 0.017g
H_3BO_3 .... 0.01g
$Na_2MoO_4 \cdot 2H_2O$ 0.01g

Preparation of Mineral Solution: Add nitrilotriacetic acid to 500.0mL of distilled/deionized water. Adjust pH to 6.5 with KOH. Add remaining components. Add distilled/deionized water to 1.0L. Mix thoroughly.

Vitamin Solution:
Composition per liter:
Pyridoxine·HCl 6.2mg
Nicotinic acid 2.5mg
Thiamine·HCl 1.25mg
p-Aminobenzoic acid 1.25mg
Pantothenic acid 0.62mg
Biotin 0.25mg

Preparation of Vitamin Solution: Add components to distilled/deionized water and bring volume to 1.0L. Mix thoroughly.

Trace Elements Solution SL-10:
Composition per liter:
$FeCl_2 \cdot 4H_2O$ 1.5g
$CoCl_2 \cdot 6H_2O$ 190.0mg
$MnCl_2 \cdot 4H_2O$ 100.0mg
$ZnCl_2$ 70.0mg
$Na_2MoO_4 \cdot 2H_2O$ 36.0mg
$NiCl_2 \cdot 6H_2O$ 24.0mg
H_3BO_3 .... 6.0mg
$CuCl_2 \cdot 2H_2O$ 2.0mg
HCl (25% solution) 10.0mL

Preparation of Trace Elements Solution SL-10: Add $FeCl_2 \cdot 4H_2O$ to 10.0mL of HCl solution. Mix thoroughly. Add distilled/deionized

water and bring volume to 1.0L. Add remaining components. Mix thoroughly.

Preparation of Solution A: Add components to distilled/deionized water and bring volume to 916.0mL. Adjust pH to 7.2. Gently heat and bring to boiling. Continue boiling for a few minutes. Allow to cool to room temperature under 80% N_2 + 20% CO_2. Distribute into bottles under 80% N_2 + 20% CO_2. Autoclave for 15 min at 15 psi pressure–121°C.

Solution B:
Composition per 70.0mL:

$NaHCO_3$	3.5g

Preparation of Solution B: Add $NaHCO_3$ to distilled/deionized water and bring volume to 70.0mL. Mix thoroughly. Filter sterilize. Sparge with 80% N_2 + 20% CO_2 for 15 min.

Solution C:
Composition per 10.0mL:

L-Cysteine·HCl	0.3g

Preparation of Solution C: Add L-cysteine·HCl to distilled/deionized water and bring volume to 10.0mL. Mix thoroughly. Sparge with 100% N_2 for 3–4 min. Autoclave under 100% N_2 for 15 min at 15 psi pressure–121°C.

Solution D:
Composition per 10.0mL:

$Na_2S{\cdot}9H_2O$	0.3g

Preparation of Solution D: Add $Na_2S{\cdot}9H_2O$ to distilled/deionized water and bring volume to 10.0mL. Mix thoroughly. Sparge with 100% N_2 for 3–4 min. Autoclave under 100% N_2 for 15 min at 15 psi pressure–121°C.

Preparation of Medium: To 916.0mL of sterile solution A add 70.0mL of sterile solution B, 10.0mL of sterile solution C, and 10.0mL of sterile solution D. Mix thoroughly.

Use: For the cultivation of anaerobic bacteria that can metabolize acetoin.

Anaerobic Agar

Composition per liter:

Pancreatic digest of casein	20.0g
Agar	15.0g
NaCl	5.0g
Sodium thioglycolate	2.0g
Sodium formaldehyde sulfoxylate	1.0g

pH 7.2 ± 0.2 at 25°C

Preparation of Medium: Add components to distilled/deionized water and bring volume to 1.0L. Mix thoroughly. Gently heat and bring to boiling. Adjust pH to 7.2. Distribute into tubes until medium is 3 inches deep. Autoclave for 20 min at 15 psi pressure–121°C.

Use: For the anaerobic cultivation of *Bacillus* species and *Sporolactobacillus* species.

Anaerobic Agar

Composition per liter:

Pancreatic digest of casein	20.0g
Agar	15.0g
Yeast extract	15.0g
NaCl	5.0g
Sodium thioglycolate	2.0g
Sodium formaldehyde sulfoxylate	1.0g

pH 7.2 ± 0.2 at 25°C

Preparation of Medium: Add components to distilled/deionized water and bring volume to 1.0L. Mix thoroughly. Gently heat and bring to boiling. Adjust pH to 7.2. Distribute into tubes until medium is 3 inches deep. Autoclave for 20 min at 15 psi pressure–121°C.

Use: For the anaerobic cultivation of *Bacillus* species, especially *Bacillus larvae*, *Bacillus popilliae*, and *Bacillus lentimorbus*.

Anaerobic Agar

Composition per liter:

Agar	20.0g
Pancreatic digest of casein	20.0g
Glucose	10.0g
NaCl	5.0g
Sodium thioglycolate	2.0g
Sodium formaldehyde sulfoxylate	1.0g
Methylene Blue	2.0mg

pH 7.2 ± 0.2 at 25°C

Source: This medium is available as a premixed powder from HiMedia and BD Diagnostic Systems.

Preparation of Medium: Add components to distilled/deionized water and bring volume to 1.0L. Mix thoroughly. Gently heat and bring to boiling. Adjust pH to 7.2. Distribute into tubes until medium is 3 inches deep. Autoclave for 15 min at 15 psi pressure–121°C.

Use: For the cultivation of a variety of anaerobic microorganisms, especially *Clostridium* species.

Anaerobic Agar

Composition per liter:

Pancreatic digest of casein	17.5g
Agar	15.0g
Glucose	10.0g
Papaic digest of soybean meal	2.5g
NaCl	2.5g
Sodium thioglycolate	2.0g
Sodium formaldehyde sulfoxylate	1.0g
L-Cystine	0.4g
Methylene Blue	2.0mg

pH 7.2 ± 0.2 at 25°C

Source: This medium is available as a premixed powder from BD Diagnostic Systems.

Preparation of Medium: Add components to distilled/deionized water and bring volume to 1.0L. Mix thoroughly. Gently heat and bring to boiling. Autoclave for 15 min at 15 psi pressure–121°C. Use with Brewer anaerobic Petri dishes or in tubes or ordinary plates and incubate in anaerobic jars.

Use: For the cultivation of *Clostridium* species and for anaerobic microorganisms.

Anaerobic Agar, Brewer

Composition per liter:

Agar	15.0g
Proteose peptone	10.0g
Pancreatic digest of casein	5.0g
Yeast extract	5.0g
NaCl	2.5g
Sodium thioglycolate	2.0g

Sodium formaldehyde sulfoxylate 1.0g
Resazurin 2.0mg
pH 7.2 ± 0.2 at 25°C

Source: This medium is available as a premixed powder from HiMedia.

Preparation of Medium: Add components to distilled/deionized water and bring volume to 1.0L. Mix thoroughly. Gently heat and bring to boiling. Adjust pH to 7.2. Distribute into tubes or flasks. Autoclave for 15 min at 15 psi pressure–121°C. Pour into Petri dishes or leave in tubes.

Use: For the cultivation of a variety of anaerobic microorganisms, especially *Clostridium* species.

Anaerobic Agar without Dextrose

Composition per liter:
Pancreatic digest of casein 17.5g
Agar 15.0g
NaCl 2.5g
Sodium thioglycolate 2.0g
Sodium formaldehyde sulfoxylate 1.0g
Methylene Blue 2.0mg
pH 7.2 ± 0.2 at 25°C

Source: This medium is available as a premixed powder from HiMedia.

Preparation of Medium: Add components to distilled/deionized water and bring volume to 1.0L. Mix thoroughly. Gently heat and bring to boiling. Adjust pH to 7.2. Distribute into tubes or flasks. Autoclave for 15 min at 15 psi pressure–121°C. Pour into Petri dishes or leave in tubes.

Use: For the cultivation of a variety of anaerobic microorganisms, especially *Clostridium* species.

Anaerobic Basal Agar with Blood

Composition per liter:
Peptic digest of animal tissue 16.0g
Agar 12.0g
Yeast extract 7.0g
NaCl 5.0g
Starch 1.0g
Glucose 1.0g
Sodium pyruvate 1.0g
L-Arginine 1.0g
Sodium succinate 0.5g
$Fe_4(P_2O_7)\cdot H_20$ 0.5g
$NaHCO_3$ 0.4g
L-Cysteine HCl 0.25g
Dithiothreitol 0.25g
Hemin 5.0mg
Vitamin K 5.0mg
Horse blood, defibrinated 100.0mL
pH 7.0 ± 0.2 at 25°C

Source: This medium is available from HiMedia.

Preparation of Medium: Add components, except horse blood, to distilled/deionized water and bring volume to 900.0mL. Mix thoroughly. Adjust pH to 7.0. Autoclave for 15 min at 15 psi pressure–121°C. Cool to 50°–55°C. Aseptically add 100.0mL of sterile horse blood. Mix thoroughly. Pour into sterile Petri dishes or leave in tubes.

Use: For the cultivation of anaerobic microorganisms, especially *Bacteroides* species and other fastidious anaerobes.

Anaerobic Blood Agar Base with Blood and Neomycin

Composition per liter:
Casein enzymic hydrolysate 14.5g
Agar 14.0g
Papaic digest of soybean meal 5.0g
NaCl 5.0g
Growth factors 1.5g
Sheep blood, sterile defibrinated 50.0mL
Selective supplement solution 10.0mL
pH 7.3 ± 0.2 at 25°C

Source: This medium is available from HiMedia.

Selective Supplement Solution:
Composition per 10.0mL:
Neomycin sulfate 30.0mg

Preparation of Selective Supplement Solution: Add neomycin sulfate to distilled/deionized water and bring volume to 10.0mL. Mix thoroughly. Filter sterilize.

Preparation of Medium: Add components, except sheep blood and selective supplement, to distilled/deionized water and bring volume to 940.0mL. Mix thoroughly. Adjust pH to 7.3. Autoclave for 15 min at 15 psi pressure–121°C. Cool to 50°–55°C. Aseptically add 50.0mL of sterile sheep blood and 10.0mL of selective supplement solution. Mix thoroughly. Pour into sterile Petri dishes or leave in tubes.

Use: For the isolation and cultivation of Group A and Group B streptococci from throat cultures and other clinical samples.

Anaerobic Broth

Composition per liter:
Pancreatic digest of casein 17.5g
Glucose 10.0g
NaCl 2.5g
Papaic digest of soybean meal 2.5g
Sodium thioglycolate 2.0g
Sodium formaldehyde sulfoxylate 1.0g
L-Cystine 0.4g
Methylene Blue 2.0mg
pH 7.2 ± 0.2 at 25°C

Preparation of Medium: Add components to distilled/deionized water and bring volume to 1.0L. Mix thoroughly. Gently heat and bring to boiling. Distribute into tubes or flasks. Autoclave for 15 min at 15 psi pressure–121°C.

Use: For the cultivation of a variety of anaerobic and microaerophilic microorganisms.

Anaerobic Cellulolytic Medium

Composition per liter:
NH_4Cl 1.0g
Cellobiose 1.0g
Yeast extract 1.0g
$MgSO_4$ 0.5g
KCl 0.5g
L-Cysteine·HCl·H_2O 0.5g
K_2HPO_4 .... 0.4g
Resazurin 1.0mg
Wolfe's mineral solution 20.0mL
Na_2CO_3 solution 10.0mL
$Na_2S\cdot 9H_2O$ solution 10.0mL
pH 6.9 ± 0.1 at 25°C

Wolfe's Mineral Solution:
Composition per liter

$MgSO_4 \cdot 7H_2O$	3.0g
Nitrilotriacetic acid	1.5g
NaCl	1.0g
$MnSO_4 \cdot H_2O$	0.5g
$FeSO_4 \cdot 7H_2O$	0.1g
$CoCl_2 \cdot 6H_2O$	0.1g
$CaCl_2$	0.1g
$ZnSO_4 \cdot 7H_2O$	0.1g
$CuSO_4 \cdot 5H_2O$	0.01g
$AlK(SO_4)_2 \cdot 12H_2O$	0.01g
H_3BO_3	0.01g
$Na_2MoO_4 \cdot 2H_2O$	0.01g

Preparation of Wolfe's Mineral Solution: Add nitrilotriacetic acid to approximately 500.0mL of distilled/deionized water and adjust pH to 6.5 with KOH to dissolve. Bring volume to 1.0L with distilled/deionized water. Add other compounds. Mix thoroughly.

Na_2CO_3 Solution:
Composition per 100.0mL:

Na_2CO_3	10.0g

Preparation of Na_2CO_3 Solution: Add Na_2CO_3 to distilled/deionized water and bring volume to 100.0mL. Mix thoroughly. Filter sterilize.

$Na_2S \cdot 9H_2O$ Solution:
Composition per 100.0mL:

$Na_2S \cdot 9H_2O$	15.0g

Preparation of $Na_2S \cdot 9H_2O$ Solution: Add $Na_2S \cdot 9H_2O$ to distilled/deionized water and bring volume to 100.0mL. Mix thoroughly. Filter sterilize.

Preparation of Medium: Add components, except Na_2CO_3 solution and $Na_2S \cdot 9H_2O$ solution, to distilled/deionized water and bring volume to 980.0mL. Boil medium under 80% N_2 + 10% CO_2 + 10% H_2 until medium is colorless. Cool and distribute anaerobically into test tubes in 10.0mL volumes using 80% N_2 + 10% CO_2 + 10% H_2. Stopper the tubes anaerobically. Autoclave for 15 min at 15 psi pressure–121°C. Cool to room temperature. Aseptically add 10.0mL of sterile Na_2CO_3 solution and 10.0mL of sterile $Na_2S \cdot 9H_2O$ solution to each tube. Mix thoroughly.

Use: For the cultivation and maintenance of microorganisms that can utilize cellobiose as sole carbon source, such as *Clostridium cellulovorans*.

Anaerobic Cholesterol Medium (DSMZ Medium 858)

Composition per 1010mL:

Cholesterol	0.5g
Solution A	900.0mL
Solution B	50.0mL
Solution F ($NaHCO_3$ solution)	50.0mL
Solution D (Vitamin solution)	5.0mL
Solution C (Trace elements solution SL-10)	2.5mL
Solution E (Selenite tungstate solution)	2.5mL

pH 7.0 ± 0.2 at 25°C

Solution A:
Composition per 200mL:

$NaNO_3$	0.16g
$MgSO_4 \cdot 7H_2O$	0.10g
NH_4Cl	0.10g
$CaCl_2$	0.02g

Preparation of Solution A: Add components to distilled/deionized water and bring volume to 200.0mL. Sparge with 100% N_2. Mix thoroughly.

Solution B:
Composition per 100.0mL:

KH_2PO_4	1.0g

Preparation of Solution B: Add KH_2PO_4 to distilled/deionized water and bring volume to 100.0mL. Sparge with 80% N_2 + 20% CO_2. Mix thoroughly. Autoclave for 15 min at 15 psi pressure–121°C.

Solution C (Trace Elements Solution SL–10):
Composition per liter:

$FeCl_2 \cdot 4H_2O$	1.5g
$CoCl_2 \cdot 6H_2O$	190.0mg
$MnCl_2 \cdot 4H_2O$	100.0mg
$ZnCl_2$	70.0mg
$Na_2MoO_4 \cdot 2H_2O$	36.0mg
$NiCl_2 \cdot 6H_2O$	24.0mg
H_3BO_3	6.0mg
$CuCl_2 \cdot 2H_2O$	2.0mg
HCl (25% solution)	10.0mL

Preparation of Solution C (Trace Elements Solution SL–10): Add $FeCl_2 \cdot 4H_2O$ to 10.0mL of HCl solution. Mix thoroughly. Add distilled/deionized water and bring volume to 1.0L. Add remaining components. Mix thoroughly. Sparge with 80% N_2 + 20% CO_2. Autoclave for 15 min at 15 psi pressure–121°C.

Solution D (Vitamin Solution):
Composition per liter:

Vitamin B_{12}	50.0mg
Pantothenic acid	50.0mg
Riboflavin	50.0mg
Alpha-lipoic acid	50.0mg
p-Aminobenzoic acid	50.0mg
Thiamine-HCl·$2H_2O$	50.0mg
Nicotinic acid	25.0mg
Nicotine amide	25.0mg
Biotin	20.0mg
Folic acid	20.0mg
Pyridoxamine-HCl	10.0mg

Preparation of Solution D (Vitamin Solution): Add components to distilled/deionized water and bring volume to 1.0L. Sparge with 100% N_2. Mix thoroughly. Filter sterilize.

Solution E (Selenite Tungstate Solution):
Composition per liter:

NaOH	0.5g
$Na_2WO_4 \cdot 2H_2O$	4.0mg
$Na_2SeO_3 \cdot 5H_2O$	3.0mg

Preparation of Solution E (Selenite Tungstate Solution): Add components to distilled/deionized water and bring volume to 1.0L. Mix thoroughly. Sparge with 100% N_2. Filter sterilize.

Solution F ($NaHCO_3$ Solution):
Composition per 100.0mL:

$NaHCO_3$	8.4g

Preparation of Solution F ($NaHCO_3$ Solution): Add $NaHCO_3$ to distilled/deionized water and bring volume to 100.0mL. Mix thoroughly. Sparge with 80% N_2 + 20% CO_2. Autoclave for 15 min at 15 psi pressure–121°C.

Preparation of Medium: Distribute solution A into anaerobic tubes or bottles. Distribute appropriate amounts of cholesterol, 5.0mg cholesterol per 10.0mL solution A. Sparge with 80% N_2 + 20% CO_2. Autoclave for 15 min at 15 psi pressure–121°C. Cool to room temperature. Add to 9.0mL solution A: 500µl sterile solution B, 500µl sterile solution F, 25µl sterile solution C, 25µl sterile solution E, and 50µl sterile solution D. Adjust pH to 7.0.

Use: For the cultivation of unclassified bacterium DSM 12783 and *Sterolibacterium denitrificans* DSM 13999 = ATCC BAA-354.

Anaerobic Citrate Medium

Composition per liter:

Ferric citrate	17.0g
Sodium acetate	6.8g
$NaHCO_3$	2.5g
NH_4Cl	1.5g
$NaH_2PO_4 \cdot H_2O$	0.6g
KCl	0.1g
Trace elements solution	10.0mL
Wolfe's vitamin solution	10.0mL

pH 7.0 ± 0.2 at 25°C

Trace Elements Solution:

Composition per liter:

Na_2WO_4	25.0mg
$NiCl_2 \cdot 6H_2O$	24.0mg
Wolfe's mineral solution	1.0L

Wolfe's Mineral Solution:

Composition per liter:

$MgSO_4 \cdot 7H_2O$	3.0g
Nitrilotriacetic acid	1.5g
NaCl	1.0g
$MnSO_4 \cdot H_2O$	0.5g
$CaCl_2$	0.1g
$CoCl_2 \cdot 6H_2O$	0.1g
$FeSO_4 \cdot 7H_2O$	0.1g
$ZnSO_4 \cdot 7H_2O$	0.1g
$AlK(SO_4)_2 \cdot 12H_2O$	0.01g
$CuSO_4 \cdot 5H_2O$	0.01g
H_3BO_3	0.01g
$Na_2MoO_4 \cdot 2H_2O$	0.01g

Preparation of Wolfe's Mineral Solution: Add nitrilotriacetic acid to approximately 500.0mL of water and adjust to pH 6.5 with KOH to dissolve the compound. Bring volume to 1.0L with remaining water and add remaining compounds one at a time.

Preparation of Trace Elements Solution: Combine components. Mix thoroughly.

Wolfe's Vitamin Solution:

Composition per liter:

Pyridoxine·HCl	10.0mg
p-Aminobenzoic acid	5.0mg
Lipoic acid	5.0mg
Nicotinic acid	5.0mg
Riboflavin	5.0mg
Thiamine·HCl	5.0mg
Calcium DL-pantothenate	5.0mg
Biotin	2.0mg
Folic acid	2.0mg
Vitamin B_{12}	0.1mg

Preparation of Wolfe's Vitamin Solution: Add components to distilled/deionized water and bring volume to 1.0L. Mix thoroughly.

Preparation of Medium: Prepare and dispense medium under 80% N_2 + 20% CO_2. Add ferric citrate to 500.0mL of boiling distilled/deionized water. Mix thoroughly. Cool to room temperature while sparging with 80% N_2 + 20% CO_2. Add remaining components. Add distilled/deionized water and bring volume to 1.0L. Mix thoroughly. Adjust pH to 7.0 with NaOH. Continue sparging with 80% N_2 + 20% CO_2. Anaerobically distribute into tubes. Autoclave for 15 min at 15 psi pressure–121°C.

Use: For the cultivation of *Geobacter metallireducens.*

Anaerobic CNA Agar (Anaerobic Colistin Nalidixic Acid Agar)

Composition per liter:

Agar	13.0g
Pancreatic digest of casein	12.0g
Peptic digest of animal tissue	5.0g
NaCl	5.0g
Yeast extract	3.0g
Beef extract	3.0g
Cornstarch	1.0g
Glucose	1.0g
L-Cysteine·HCl·H_2O	0.5g
Vitamin K_1	10.0mg
Hemin	10.0mg
Colistin	10.0mg
Nalidixic acid	10.0mg
Sheep blood, defibrinated	50.0mL

Source: This medium is available as a premixed powder from BD Diagnostic Systems.

Preparation of Medium: Add components, except sheep blood, to distilled/deionized water and bring volume to 950.0mL. Mix thoroughly. Gently heat and bring to boiling. Autoclave for 15 min at 15 psi pressure–121°C. Cool to 45°–50°C. Aseptically add 50.0mL of sterile, defibrinated sheep blood. Mix thoroughly. Pour into sterile Petri dishes.

Use: For the selective isolation of anaerobic streptococci.

Anaerobic CNA Agar Base with Blood

Composition per liter:

Agar	13.5g
Casein enzymic hydrolysate	12.0g
Peptic digest of animal tissue	5.0g
NaCl	5.0g
Yeast extract	3.0g
Beef extract	3.0g
Corn starch	1.0g
Glucose	1.0g
L-Cystine hydrochloride	0.5g
Dithiothreitol (DTE)	0.1g
Vitamin K_1	0.01g
Hemin	0.01g
Colistin	0.01g
Nalidixic acid	0.01g
Sheep blood, sterile defibrinated	50.0mL

pH 7.0 ± 0.2 at 25°C

Source: This medium is available from HiMedia.

Preparation of Medium: Add components, except sheep blood, to distilled/deionized water and bring volume to 950.0mL. Mix thoroughly. Adjust pH to 7.0. Autoclave for 15 min at 15 psi pressure–121°C. Cool to 50°–55°C. Aseptically add 50.0mL of sterile sheep blood. Mix thoroughly. Pour into sterile Petri dishes or leave in tubes.

Use: For the selective isolation of anaerobic streptococci.

Anaerobic Egg Yolk Agar

Composition per 1080.0mL:

Agar	20.0g
Proteose peptone	20.0g
NaCl	5.0g
Pancreatic digest of casein	5.0g
Yeast extract	5.0g
Egg yolk emulsion, 50%	80.0mL

pH 7.0 ± 0.2 at 25°C

Egg Yolk Emulsion, 50%:
Composition per 100.0mL:

Chicken egg yolks	11
Whole chicken egg	1
NaCl (0.9% solution)	50.0mL

Preparation of Egg Yolk Emulsion, 50%: Soak whole eggs with 1:100 dilution of saturated mercuric chloride solution for 1 min. Crack eggs and separate yolks from whites. Mix egg yolks with 1 chicken egg. Beat to form emulsion. Measure 50.0mL of egg yolk emulsion and add to 50.0mL of 0.9% NaCl solution. Mix thoroughly. Filter sterilize. Warm to 45°–50°C.

Preparation of Medium: Add components, except egg yolk emulsion, 50%, to distilled/deionized water and bring volume to 1.0L. Mix thoroughly. Gently heat and bring to boiling. Autoclave for 15 min at 15 psi pressure–121°C. Cool to 45°–50°C. Aseptically add 80.0mL of sterile egg yolk emulsion, 50%. Mix thoroughly. Pour into sterile Petri dishes or distribute into sterile tubes. Allow plates to dry at 35°C for 24 hr.

Use: For the cultivation of *Clostridium* species. For the cultivation of *Yersinia enterocolitica.*

Anaerobic Egg Yolk Agar

Composition per liter:

Agar	20.0g
Proteose peptone	20.0g
Pancreatic digest of casein	5.0g
NaCl	5.0g
Yeast extract	5.0g
Egg yolk emulsion, 50%	20.0mL

pH 7.0 ± 0.2 at 25°C

Egg Yolk Emulsion, 50%:
Composition per 100.0mL:

Chicken egg yolks	2
NaCl (0.9% solution)	10.0mL

Preparation of Egg Yolk Emulsion, 50%: Soak eggs with 1:100 dilution of saturated mercuric chloride solution for 1 min. Crack eggs and separate yolks from whites. Beat to form emulsion. Measure 10.0mL of egg yolk emulsion and add to 10.0mL of 0.9% NaCl solution. Mix thoroughly. Filter sterilize. Warm to 45°–50°C.

Preparation of Medium: Add components, except egg yolk emulsion, to distilled/deionized water and bring volume to 980.0mL. Mix thoroughly. Gently heat and bring to boiling. Autoclave for 15 min at 15 psi pressure–121°C. Cool to 45°–50°C. Aseptically add sterile egg yolk emulsion. Mix thoroughly. Pour into sterile Petri dishes. Allow plates to dry at 35°C for 24 hr.

Use: For the cultivation of *Yersinia enterocolitica.*

Anaerobic Egg Yolk Agar (BAM M12)

Composition per liter:

Agar	20.0g
Proteose peptone	20.0g
Pancreatic digest of casein	5.0g
NaCl	5.0g
Yeast extract	5.0g
Egg yolk emulsion, 50%	80.0mL

pH 7.0 ± 0.2 at 25°C

Egg Yolk Emulsion, 50%:
Composition per 80.0mL:

Chicken egg yolks	2 or more
NaCl (0.85% solution)	40.0mL

Preparation of Egg Yolk Emulsion, 50%: Wash fresh eggs with stiff brush and drain. Soak eggs in 70% ethanol for 1 hour. Crack eggs aseptically and separate yolks from whites. Drain contents of yolk sacs into sterile stoppered graduate cylinder and discard sacs. Measure 40.0mL of egg yolk emulsion and add 40.0mL of 0.85% NaCl solution. Mix thoroughly by inverting graduate cylinder. Warm to 45°–50°C.

Preparation of Medium: Add components, except egg yolk emulsion, to distilled/deionized water and bring volume to 1.0L. Mix thoroughly. Gently heat and bring to boiling. Autoclave for 15 min at 15 psi pressure–121°C. Cool to 45°–50°C. Aseptically add 80.0mL sterile egg yolk emulsion. Mix thoroughly. Pour into sterile Petri dishes. Allow plates to dry at ambient temperature for 2–3 days or at 35°C for 24 hr. Check plates for contamination before use.

Use: For the cultivation of *Yersinia enterocolitica.*

Anaerobic Egg Yolk Base with Egg Yolk Emulsion

Composition per liter:

Agar	20.0g
Proteose peptone	20.0g
Casein enzymatic hydrolysate	5.0g
Yeast extract	5.0g
NaCl	5.0g
Egg yolk emulsion	80.0mL

pH 7.0 ± 0.2 at 25°C

Source: This medium is available from HiMedia.

Egg Yolk Emulsion:
Composition per liter:

Egg yolks	30.0mL
NaCl, 0.9% solution	70.0mL

Preparation of Egg Yolk Emulsion: Soak eggs with 1:100 dilution of saturated mercuric chloride solution for 1 min. Crack 11 eggs and separate yolks from whites. Mix egg yolks. Measure 30.0mL of egg yolk emulsion and add to 70.0mL of 0.9% sterile NaCl solution. Mix thoroughly. Warm to 45°–50°C.

Preparation of Medium: Add components, except egg yolk emulsion, to distilled/deionized water and bring volume to 920.0mL. Mix thoroughly. Gently heat and bring to boiling. Autoclave for 15 min at 15 psi pressure–121°C. Cool to 45°–50°C. Aseptically add sterile egg yolk emulsion. Mix thoroughly. Pour into sterile Petri dishes.

Use: For the cultivation of *Clostridium perfringens* from foods.

Anaerobic D-Gluconate Medium

Composition per liter:

Agar	15.0g
Pancreatic digest of casein	10.0g
Yeast extract	5.0g
D-Gluconate	4.0g
$MgSO_4 \cdot 7H_2O$	2.5g
$(NH_4)_2SO_4$	1.4g
L-Cysteine·HCl·H_2O	1.0g
$CaCl_2 \cdot 2H_2O$	0.15g
$FeSO_4 \cdot 7H_2O$	0.02g
Resazurin	1.0mg
$NaHCO_3$ solution	10.0mL

pH 7.1 ± 0.2 at 25°C

$NaHCO_3$ Solution:

Composition per 100.0mL:

$NaHCO_3$	10.0g

Preparation of $NaHCO_3$ Solution: Add $NaHCO_3$ to distilled/deionized water and bring volume to 100.0mL. Mix thoroughly. Filter sterilize.

Preparation of Medium: Add components, except $NaHCO_3$ solution, to distilled/deionized water and bring volume to 990.0mL. Prepare anaerobically under 100% N_2. Autoclave for 15 min at 15 psi pressure–121°C. Aseptically add 10.0mL of the sterile $NaHCO_3$ solution. Mix thoroughly. Adjust pH to 7.1.

Use: For the cultivation and maintenance of microorganisms that can utilize D-gluconate as a carbon source, such as *Bacteroides pectinophilus*.

Anaerobic Glucuronic Acid Medium

Composition per liter:

Agar	15.0g
Pancreatic digest of casein	10.0g
Yeast extract	5.0g
Glucuronic acid	4.0g
$MgSO_4 \cdot 7H_2O$	2.5g
$(NH_4)_2SO_4$	1.4g
L-Cysteine·HCl·H_2O	1.0g
$CaCl_2 \cdot 2H_2O$	0.15g
$FeSO_4 \cdot 7H_2O$	0.02g
Resazurin	1.0mg
$NaHCO_3$ solution	10.0mL

pH 7.1 ± 0.2 at 25°C

$NaHCO_3$ Solution:

Composition per 100.0mL:

$NaHCO_3$	10.0g

Preparation of $NaHCO_3$ Solution: Add $NaHCO_3$ to distilled/deionized water and bring volume to 100.0mL. Mix thoroughly. Filter sterilize.

Preparation of Medium: Add components, except $NaHCO_3$ solution, to distilled/deionized water and bring volume to 990.0mL. Prepare anaerobically under 100% N_2. Autoclave for 15 min at 15 psi pressure–121°C. Aseptically add 10.0mL of the sterile $NaHCO_3$ solution. Mix thoroughly. Adjust pH to 7.1.

Use: For the cultivation and maintenance of microorganisms that can utilize D-glucuronate as a carbon source, such as *Bacteroides galacturonicus*.

Anaerobic HiVeg Agar

Composition per liter:

Agar	20.0g
Plant hydrolysate	20.0g
Glucose	10.0g
NaCl	5.0g
Sodium thioglycolate	2.0g
Sodium formaldehyde sulfoxylate	1.0g
Methylene Blue	2.0mg

pH 7.2 ± 0.2 at 25°C

Source: This medium is available as a premixed powder from HiMedia.

Preparation of Medium: Add components to distilled/deionized water and bring volume to 1.0L. Mix thoroughly. Gently heat and bring to boiling. Adjust pH to 7.2. Distribute into tubes until medium is 3 inches deep. Autoclave for 15 min at 15 psi pressure–121°C.

Use: For the cultivation of a variety of anaerobic microorganisms, especially *Clostridium* species.

Anaerobic HiVeg Agar (Brewer)

Composition per liter:

Agar	15.0g
Glucose	10.0g
Plant petone No. 3	10.0g
Plant hydrolysate	5.0g
NaCl	5.0g
Yeast extract	5.0g
Sodium thioglycolate	2.0g
Sodium formaldehyde sulfoxylate	1.0g
Resazurin	2.0mg

pH 7.2 ± 0.2 at 25°C

Source: This medium is available as a premixed powder from HiMedia.

Preparation of Medium: Add components to distilled/deionized water and bring volume to 1.0L. Mix thoroughly. Gently heat and bring to boiling. Adjust pH to 7.2. Distribute into tubes until medium is 3 inches deep. Autoclave for 15 min at 15 psi pressure–121°C.

Use: For the cultivation of a variety of anaerobic microorganisms, especially *Clostridium* species.

Anaerobic HiVeg Agar Base with Egg Yolk Emulsion

Composition per liter:

Agar	20.0g
Plant petone No. 3	20.0g
Plant hydrolysate	5.0g
NaCl	5.0g
Yeast extract	5.0g
Egg yolk emulsion	100.0mL

pH 7.2 ± 0.2 at 25°C

Source: This medium, without egg yolk emulsion, is available as a premixed powder from HiMedia.

Egg Yolk Emulsion:

Composition per liter:

Egg yolks	30.0mL
NaCl, 0.9% solution	70.0mL

Preparation of Egg Yolk Emulsion: Soak eggs with 1:100 dilution of saturated mercuric chloride solution for 1 min. Crack 11 eggs and separate yolks from whites. Mix egg yolks. Measure 30.0mL of egg yolk emulsion and add to 70.0mL of 0.9% sterile NaCl solution. Mix thoroughly. Warm to 45°–50°C.

Preparation of Medium: Add components, except egg yolk emulsion, to distilled/deionized water and bring volume to 900.0mL. Mix thoroughly. Gently heat and bring to boiling. Autoclave for 15 min at 15 psi pressure–121°C. Cool to 45°–50°C. Aseptically add sterile egg yolk emulsion. Mix thoroughly. Pour into sterile Petri dishes.

Use: For the cultivation of *Clostridium perfringens* from foods.

Anaerobic HiVeg Agar without Dextrose

Composition per liter:

Plant hydrolysate	17.5g
Agar	15.0g
NaCl	2.5g
Sodium thioglycolate	2.0g
Sodium formaldehyde sulfoxylate	1.0g
Methylene Blue	2.0mg

pH 7.2 ± 0.2 at 25°C

Source: This medium is available as a premixed powder from HiMedia.

Preparation of Medium: Add components to distilled/deionized water and bring volume to 1.0L. Mix thoroughly. Gently heat and bring to boiling. Distribute into tubes or flasks. Autoclave for 15 min at 15 psi pressure–121°C. Pour into sterile Petri dishes or leave in tubes.

Use: For the cultivation of a variety of anaerobic microorganisms. With added blood for the detection of hemolytic activity of clostridia, streptococci, and other anaerobic bacteria. With added carbohydrate for fermentation studies.

Anaerobic HiVeg Agar without Dextrose and Eh Indicator

Composition per liter:

Plant hydrolysate	20.0g
Agar	15.0g
NaCl	5.0g
Sodium thioglycolate	2.0g
Sodium formaldehyde sulfoxylate	1.0g

pH 7.2 ± 0.2 at 25°C

Source: This medium is available as a premixed powder from HiMedia.

Preparation of Medium: Add components to distilled/deionized water and bring volume to 1.0L. Mix thoroughly. Gently heat and bring to boiling. Distribute into tubes or flasks. Autoclave for 15 min at 15 psi pressure–121°C. Pour into sterile Petri dishes or leave in tubes.

Use: For the cultivation of a variety of anaerobic microorganisms. With added blood for the detection of hemolytic activity of clostridia, streptococci, and other anaerobic bacteria.

Anaerobic LKV Blood Agar

Composition per liter:

Agar	15.0g
Pancreatic digest of casein	13.0g
Peptic digest of animal tissue	10.0g
NaCl	5.0g
Yeast extract	2.0g
Glucose	1.0g
$NaHSO_3$	0.1g
Sheep blood, laked	50.0mL
Antibiotic solution	10.0mL
Hemin solution	1.0mL
Vitamin K_1 solution	1.0mL

pH 7.1–7.8 at 25°C

Source: This medium is available as a premixed powder from BD Diagnostic Systems.

Antibiotic Solution:

Composition per 10.0mL:

Kanamycin	0.075g
Vancomycin	7.5mg

Preparation of Antibiotic Solution: Add components to distilled/deionized water and bring volume to 10.0mL. Mix thoroughly. Filter sterilize.

Vitamin K_1 Solution:

Composition per 100.0mL:

Vitamin K_1	0.1g
Ethanol	99.0mL

Preparation of Vitamin K_1 Solution: Add vitamin K_1 to 99.0mL of absolute ethanol. Mix thoroughly.

Hemin Solution:

Composition per 100.0mL:

Hemin	0.01g
NaOH (1*N* solution)	20.0mL

Preparation of Hemin Solution: Add hemin to 20.0mL of 1*N* NaOH solution. Mix thoroughly. Bring volume to 100.0mL with distilled/deionized water.

Preparation of Medium: Add components—except sheep blood, antibiotic solution, and vitamin K_1 solution—to distilled/deionized water and bring volume to 939.0mL. Mix thoroughly. Gently heat and bring to boiling. Autoclave for 15 min at 15 psi pressure–121°C. Cool to 45°–50°C. Aseptically add 50.0mL of sterile sheep blood, 10.0mL of sterile antibiotic solution, and 1.0mL of sterile vitamin K_1 solution. Mix thoroughly. Pour into sterile Petri dishes or distribute into sterile tubes.

Use: For the isolation and cultivation of anaerobic Gram-negative microorganisms, especially *Bacteroides* species.

Anaerobic Oxalate Medium

Composition per 1011.0mL:

Solution A	870.0mL
Solution C	100.0mL
Solution D	20.0mL
Solution E (Vitamin solution)	10.0mL
Solution F	10.0mL
Solution B (Trace elements solution SL-10)	1.0mL

pH 7.1–7.4 at 25°C

Solution A:
Composition per 870.0mL:

Na_2SO_4	3.0g
NaCl	1.0g
KCl	0.5g
$MgCl_2 \cdot 6H_2O$	0.4g
NH_4Cl	0.3g
KH_2PO_4	0.2g
$CaCl_2 \cdot 2H_2O$	0.15g
Resazurin	1.0mg

Preparation of Solution A: Add components to distilled/deionized water and bring volume to 870.0mL. Mix thoroughly. Gently heat and bring to boiling. Continue boiling for 3–4 min. Allow to cool to room temperature while gassing under 80% N_2 + 20% CO_2. Continue gassing until pH reaches below 6.0. Seal the flask under 80% N_2 + 20% CO_2. Autoclave for 15 min at 15 psi pressure–121°C.

Solution B (Trace Elements Solution SL-10):
Composition per liter:

$FeCl_2 \cdot 4H_2O$	1.5g
$CoCl_2 \cdot 6H_2O$	190.0mg
$MnCl_2 \cdot 4H_2O$	100.0mg
$ZnCl_2$	70.0mg
$Na_2MoO_4 \cdot 2H_2O$	36.0mg
$NiCl_2 \cdot 6H_2O$	24.0mg
H_3BO_3	6.0mg
$CuCl_2 \cdot 2H_2O$	2.0mg
HCl (25% solution)	10.0mL

Preparation of Solution B: Add $FeCl_2 \cdot 4H_2O$ to 10.0mL of HCl solution. Mix thoroughly. Add distilled/deionized water and bring volume to 1.0L. Add remaining components. Mix thoroughly. Gas under 100% N_2. Autoclave for 15 min at 15 psi pressure–121°C.

Solution C:
Composition per 100.0mL:

$NaHCO_3$	5.0g

Preparation of Solution C: Add $NaHCO_3$ to distilled/deionized water and bring volume to 100.0mL. Mix thoroughly. Filter sterilize. Gas under 80% N_2 + 20% CO_2.

Solution D:
Composition per 20.0mL:

Ammonium oxalate	3.0g
Yeast extract	1.0g
Sodium acetate	0.41g

Preparation of Solution D: Add components to distilled/deionized water and bring volume to 20.0mL. Mix thoroughly. Gas under 100% N_2. Autoclave for 15 min at 15 psi pressure–121°C.

Solution E (Vitamin Solution):
Composition per liter:

Pyridoxine·HCl	10.0mg
Calcium DL-pantothenate	5.0mg
Lipoic acid	5.0mg
Nicotinic acid	5.0mg
p-Aminobenzoic acid	5.0mg
Riboflavin	5.0mg
Thiamine·HCl	5.0mg
Biotin	2.0mg
Folic acid	2.0mg
Vitamin B_{12}	0.1mg

Preparation of Solution E (Vitamin Solution): Add components to distilled/deionized water and bring volume to 1.0L. Mix thoroughly. Gas under 100% N_2. Autoclave for 15 min at 15 psi pressure–121°C.

Solution F:
Composition per 10.0mL:

$Na_2S \cdot 9H_2O$	0.4g

Preparation of Solution F: Add $Na_2S \cdot 9H_2O$ to distilled/deionized water and bring volume to 10.0mL. Mix thoroughly. Gas under 100% N_2. Autoclave for 15 min at 15 psi pressure–121°C.

Preparation of Medium: Aseptically and anaerobically combine solution A with solution B, solution C, solution D, solution E, and solution F, in that order. Mix thoroughly. Anaerobically distribute into sterile tubes or flasks under 80% N_2 + 20% CO_2.

Use: For the cultivation of *Clostridium oxalicum* and *Oxalobacter vibrioformis.*

Anaerobic Thioglycollate Medium Base with Serum

Composition per liter:

Casein enzymic hydrolysate	17.0g
Meat extract	7.5g
D-Glucose	6.0g
Liver hydrolysate	3.0g
Papaic digest of soybean meal	3.0g
NaCl	2.5g
Agar	0.7g
Sodium thioglycollate	0.5g
L-Cysteine	0.25g
Na_2SO_3	0.1g
Serum, sterile	100.0mL

pH 7.3 ± 0.2 at 25°C

Source: This medium is available from HiMedia.

Preparation of Medium: Add components, except serum, to distilled/deionized water and bring volume to 900.0mL. Mix thoroughly. Adjust pH to 7.0. Autoclave for 15 min at 15 psi pressure–121°C. Cool to 50°–55°C. Aseptically add 100.0mL of sterile serum. Mix thoroughly. Aseptically distribute into sterile tubes.

Use: For the selective isolation of anaerobic bacteria.

Anaerobic Trypticase™ Soy Agar with Calf Blood (ATCC Medium 1664)

Composition per liter:

Pancreatic digest of casein	15.0g
Agar	15.0g
Papaic digest of soybean meal	5.0g
NaCl	5.0g
Calf blood, defibrinated	100.0mL

pH 7.3 ± 0.2 at 25°C

Preparation of Medium: Add components, except calf blood, to distilled/deionized water and bring volume to 900.0mL. Mix thoroughly. Prepare medium anaerobically with 80% N_2 + 10% CO_2 + 10% H_2. Gently heat while stirring and bring to boiling for 1 min. Autoclave for 15 min at 15 psi pressure–121°C. Do not overheat. Cool to 45°–50°C. Aseptically add 100.0mL sterile, defibrinated calf blood. Pour into sterile Petri dishes.

Use: For the isolation and cultivation of fastidious as well as nonfastidious microorganisms. For the differentiation of *Haemophilus* species.

Anaerobic Tryptone Soya Agar

Composition per liter

Agar	20.0g
Casein enzymatic hydrolysate	15.0g
Papaic digest of soybean meal	5.0g
NaCl	5.0g
Yeast extract	5.0g
L-Cysteine	0.4g
Hemin	5.0mg
Vitamin K_1	10.0mg

pH 7.5 ± 0.2 at 25°C

Source: This medium is available as a premixed powder from HiMedia.

Preparation of Medium: Add components to distilled/deionized water and bring volume to 1.0L. Mix thoroughly. Gently heat and bring to boiling. Distribute into tubes or flasks. Autoclave for 15 min at 15 psi pressure–121°C. Pour into sterile Petri dishes or leave in tubes.

Use: For the detection of anaerobic bacteria in cosmetics such as talcum powder.

Anaerobic TVLS Medium

Composition per liter:

Pancreatic digest of casein	17.0g
Beef extract	7.5g
Glucose	6.0g
Enzymatic hydrolysate of soybean meal	3.0g
Liver hydrolysate	3.0g
NaCl	2.5g
Na_2SO_3	0.7g
Sodium thioglycolate	0.5g
L-Cysteine·HCl·H_2O	0.25g
Agar	0.1g
Bovine serum	100.0mL

pH 7.3 ± 0.2 at 25°C

Preparation of Medium: Add components, except bovine serum, to distilled/deionized water and bring volume to 900.0mL. Mix thoroughly. Gently heat and bring to boiling. Autoclave for 15 min at 15 psi pressure–121°C. Cool to 45°–50°C. Aseptically add 100.0mL of bovine serum. Distribute into sterile tubes.

Use: For the isolation and cultivation of anaerobic microorganisms.

Anaerobiospirillum thomasii Medium (DSMZ Medium 800)

Composition per 1070mL:

Pancreatic digest of casein	10.0g
Gelatin peptone	10.0g
NaCl	5.0g
Yeast extract	5.0g
Glucose	1.0g
L-Arginine	1.0g
Sodium pyruvate	1.0g
Hemin	5.0mg
Menadione	0.5mg
Fildes enrichment solution	100.0mL
$NaHCO_3$ solution	50.0mL
$Na_2S·9H_2O$ solution	10.0mL
Cysteine solution	10.0mL

pH 6.9 ± 0.2 at 25°C

Fildes Enrichment Solution:

Composition per 206.0mL:

Pepsin	1.0g
NaCl (0.85% solution)	150.0mL
Sheep blood, defibrinated	50.0mL
HCl	6.0mL

Source: Fildes enrichment solution is available as a premixed powder from BD Diagnostic Systems and Oxoid Unipath.

Preparation of Fildes Enrichment Solution: Combine components. Mix thoroughly. Incubate at 56°C for 4 hr. Bring pH to 7.0 with 20% NaOH. Adjust pH to 7.2 with HCl. Do not autoclave. Add 0.25 mL of chloroform and store at 4°C. Before use, heat to 56°C to remove chloroform.

$Na_2S·9H_2O$ Solution:

Composition per 10.0mL:

$Na_2S·9H_2O$	0.3g

Preparation of $Na_2S·9H_2O$ Solution: Add $Na_2S·9H_2O$ to distilled/deionized water and bring volume to 10.0mL. Mix thoroughly. Autoclave under 100% N_2 for 15 min at 15 psi pressure–121°C. Cool to room temperature.

Cysteine Solution:

Composition per 10.0mL:

L-Cysteine·HCl·H_2O	0.3g

Preparation of Cysteine Solution: Add L-cysteine·HCl·H_2O to distilled/deionized water and bring volume to 10.0mL. Mix thoroughly. Sparge with 100% N_2. Autoclave for 15 min at 15 psi pressure–121°C. Cool to room temperature.

$NaHCO_3$ Solution:

Composition per 100.0mL:

$NaHCO_3$	5.0g

Preparation of $NaHCO_3$ Solution: Add $NaHCO_3$ to distilled/deionized water and bring volume to 100.0mL. Mix thoroughly. Sparge with 80% N_2 + 20% CO_2. Filter sterilize.

Preparation of Medium: Prepare and dispense medium under 80% N_2 + 20% CO_2 gas atmosphere. Add components, except Fildes enrichment solution, cysteine solution, $NaHCO_3$ solution, and $Na_2S·9H_2O$ solution, to distilled/deionized water and bring volume to 1.0L. Mix thoroughly. Distribute into tubes or flasks. Autoclave for 15 min at 15 psi pressure–121°C. Cool while sparging with 80% N_2 + 20% CO_2. Aseptically and anaerobically add 100.0mL Fildes enrichment solution, 10.0mL cysteine solution, 50.0mL $NaHCO_3$ solution, and 10.0mL $Na_2S·9H_2O$ solution. Aseptically and anaerobically distribute to sterile tubes or bottles.

Use: For the cultivation of *Anaerobiospirillum thomasii.*

Anaerobranca gottschalkii Medium (DSMZ Medium 895)

Composition 1070mL:

NaCl	10.0g
$(NH_4)_2SO_4$	1.0g
K_2HPO_4	0.5g
L-Cysteine	0.5g
NH_4Cl	0.4g
Yeast extract	0.25g
Tryptone	0.25g
$Na_2S_2O_3·5H_2O$	0.1g
$MgSO_4·7H_2O$	0.1g

$CaCl_2 \cdot 2H_2O$ 0.05g
$FeSO_4 \cdot 7H_2O$ 2.0mg
Resazurin 0.5mg
Na_2CO_3solution.......... 50.0mL
Soluble starch solution.......... 20.0mL
Trace elements solution 10.0mL
Vitamin solution, 10 fold conc. 1.0mL
pH 9.4 ± 0.2 at 25°C

Trace Elements Solution:
Composition per liter:
$MgSO_4 \cdot 7H_2O$ 3.0g
Nitrilotriacetic acid 1.5g
NaCl 1.0g
$MnSO_4 \cdot 2H_2O$ 0.5g
$CoSO_4 \cdot 7H_2O$ 0.18g
$ZnSO_4 \cdot 7H_2O$ 0.18g
$CaCl_2 \cdot 2H_2O$ 0.1g
$FeSO_4 \cdot 7H_2O$ 0.1g
$NiCl_2 \cdot 6H_2O$ 0.025g
$KAl(SO_4)_2 \cdot 12H_2O$ 0.02g
H_3BO_3 .... 0.01g
$Na_2MoO_4 \cdot 4H_2O$ 0.01g
$CuSO_4 \cdot 5H_2O$ 0.01g
$Na_2SeO_3 \cdot 5H_2O$ 0.3mg

Preparation of Trace Elements Solution: Add nitrilotriacetic acid to 500.0mL of distilled/deionized water. Dissolve by adjusting pH to 6.5 with KOH. Add remaining components. Add distilled/deionized water to 1.0L. Mix thoroughly.

Vitamin Solution:
Composition per 100.0mL:
Pyridoxine-HCl 10.0mg
Thiamine-HCl·$2H_2O$ 5.0mg
Riboflavin 5.0mg
Nicotinic acid 5.0mg
D-Ca-pantothenate 5.0mg
p-Aminobenzoic acid 5.0mg
Lipoic acid 5.0mg
Biotin 2.0mg
Folic acid 2.0mg
Vitamin B_{12} 0.1mg

Preparation of Vitamin Solution: Add components to distilled/deionized water and bring volume to 100.0mL. Mix thoroughly. Sparge with 80% H_2 + 20% CO_2. Filter sterilize.

Na_2CO_3 Solution:
Composition per 100.0mL:
Na_2CO_3 .... 5.0g

Preparation of Na_2CO_3 Solution: Add Na_2CO_3 to distilled/deionized water and bring volume to 100.0mL. Mix thoroughly. Sparge with 100% N_2. Autoclave for 15 min at 15 psi pressure–121°C.

Soluble Starch Solution:
Composition per 50.0mL:
Starch, soluble 5.0g

Preparation of Soluble Starch Solution: Add starch to distilled/deionized water and bring volume to 50.0mL. Mix thoroughly. Sparge with 100% N_2. Autoclave for 15 min at 15 psi pressure–121°C.

Preparation of Medium: Add components, except starch solution, Na_2CO_3 solution, and L-cysteine, to distilled/deionized water and bring volume to 1.0L. Mix thoroughly. Gently heat and bring to boiling. Cool to 25°C while sparging with 100% N_2. Add 0.5g L-cysteine. Mix thoroughly. Distribute to anaerobe tubes or bottles. Autoclave for 15 min at 15 psi pressure–121°C. Cool to 25°C. Aseptically and anaerobically add, per liter of medium, 20.0mL sterile starch solution, and 50.0mL sterile Na_2CO_3 solution. Final pH is 9.3–9.5.

Use: For the cultivation of *Anaerobranca gottschalkii.*

Anaerobranca Medium

Composition per 1015.0mL:
Yeast extract 5.0g
$Na_2HPO_4 \cdot 2H_2O$ 3.9g
Sodium fumarate 1.5g
KCl 0.5g
KH_2PO_4 .... 0.5g
L-Cysteine·HCl·H_2O 0.125g
$Na_2S \cdot 9H_2O$ 0.125g
Wolfe's vitamin solution 10.0mL
Wolfe's mineral solution 5.0mL
pH 8.5 ± 0.2 at 25°C

Wolfe's Vitamin Solution:
Composition per liter:
Pyridoxine·HCl 10.0mg
p-Aminobenzoic acid 5.0mg
Lipoic acid 5.0mg
Nicotinic acid 5.0mg
Riboflavin 5.0mg
Thiamine·HCl 5.0mg
Calcium DL-pantothenate 5.0mg
Biotin 2.0mg
Folic acid 2.0mg
Vitamin B_{12} 0.1mg

Preparation of Wolfe's Vitamin Solution: Add components to distilled/deionized water and bring volume to 1.0L. Mix thoroughly. Filter sterilize.

Wolfe's Mineral Solution:
Composition per liter:
$MgSO_4 \cdot 7H_2O$ 3.0g
Nitrilotriacetic acid 1.5g
NaCl 1.0g
$MnSO_4 \cdot 2H_2O$ 0.5g
$CoCl_2 \cdot 6H_2O$ 0.1g
$ZnSO_4 \cdot 7H_2O$ 0.1g
$CaCl_2 \cdot 2H_2O$ 0.1g
$FeSO_4 \cdot 7H_2O$ 0.1g
$NiCl_2 \cdot 6H_2O$ 0.025g
$KAl(SO_4)_2 \cdot 12H_2O$ 0.02g
$CuSO_4 \cdot 5H_2O$ 0.01g
H_3BO_3 .... 0.01g
$Na_2MoO_4 \cdot 2H_2O$ 0.01g
$Na_2SeO_3 \cdot 5H_2O$ 0.3mg

Preparation of Wolfe's Mineral Solution: Add nitrilotriacetic acid to 500.0mL of distilled/deionized water. Adjust pH to 6.5 with KOH. Add remaining components. Add distilled/deionized water to 1.0L. Adjust pH to 6.8.

Preparation of Medium: Prepare and dispense medium under 100% N_2. Add components, except Wolfe's vitamin solution, to distilled/deionized water and bring volume to 990.0mL. Mix thoroughly. Adjust pH to 8.5. Sparge with 100% N_2. Autoclave for 15 min at 15 psi pressure–121°C. Aseptically and anaerobically add 10.0mL of sterile

Wolfe's vitamin solution. Mix thoroughly. Aseptically and anaerobically distribute into sterile tubes or bottles.

Use: For the cultivation of *Anaerobranca horikoshii.*

Anaerocellum Medium

Composition per liter:

Cellobiose or starch	5.0g
$NaHCO_3$	1.5g
$Na_2S·9H_2O$	0.5g
Yeast extract	0.5g
$CaCl_2·2H_2O$	0.33g
KCl	0.33g
KH_2PO_4	0.33g
$MgCl_2·6H_2O$	0.33g
NH_4Cl	0.33g
Resazurin	0.5mg
$NaHCO_3$ solution	100.0mL
Cellobiose or starch solution	50.0mL
Vitamin solution	10.0mL
$Na_2S·9H_2O$ solution	10.0mL
Trace elements solution SL-10	1.0mL

pH 7.1–7.3 at 25°C

$NaHCO_3$ Solution:

Composition per 100.0mL:

$NaHCO_3$	5.0g

Preparation of $NaHCO_3$ Solution: Add $NaHCO_3$ to distilled/deionized water and bring volume to 100.0mL. Mix thoroughly. Filter sterilize. Gas under 80% N_2 + 20% CO_2.

Cellobiose or Starch Solution:

Composition per 50.0mL:

Cellobiose or starch	5.0g

Preparation of Cellobiose or Starch Solution: Add cellobiose or starch to distilled/deionized water and bring volume to 50.0mL. Mix thoroughly. Filter sterilize. Gas under 100% N_2.

$Na_2S·9H_2O$ Solution:

Composition per 10.0mL:

$Na_2S·9H_2O$	0.5g

Preparation of $Na_2S·9H_2O$ Solution: Add $Na_2S·9H_2O$ to distilled/deionized water and bring volume to 10.0mL. Mix thoroughly. Gas under 100% N_2. Autoclave for 15 min at 15 psi pressure–121°C.

Vitamin Solution:

Composition per liter:

Pyridoxine·HCl	10.0mg
Calcium DL-pantothenate	5.0mg
Lipoic acid	5.0mg
Nicotinic acid	5.0mg
p-Aminobenzoic acid	5.0mg
Riboflavin	5.0mg
Thiamine·HCl	5.0mg
Biotin	2.0mg
Folic acid	2.0mg
Vitamin B_{12}	0.1mg

Preparation of Vitamin Solution: Add components to distilled/deionized water and bring volume to 1.0L. Mix thoroughly. Gas under 100% N_2. Autoclave for 15 min at 15 psi pressure–121°C.

Trace Elements Solution SL-10:

Composition per liter:

$FeCl_2·4H_2O$	1.5g
$CoCl_2·6H_2O$	190.0mg
$MnCl_2·4H_2O$	100.0mg
$ZnCl_2$	70.0mg
$Na_2MoO_4·2H_2O$	36.0mg
$NiCl_2·6H_2O$	24.0mg
H_3BO_3	6.0mg
$CuCl_2·2H_2O$	2.0mg
HCl (25% solution)	10.0mL

Preparation of Trace Elements Solution SL-10: Add $FeCl_2·4H_2O$ to 10.0mL of HCl solution. Mix thoroughly. Add distilled/deionized water and bring volume to 1.0L. Add remaining components. Mix thoroughly.

Preparation of Medium: Add components, except $NaHCO_3$ solution, cellobiose or starch solution, and $Na_2S·9H_2O$ solution, to distilled/deionized water and bring volume to 830.0mL. Mix thoroughly. Gently heat and bring to boiling. Continue boiling for 3–4 min. Allow to cool to room temperature under 80% N_2 + 20% CO_2. Distribute into bottles under 80% N_2 + 20% CO_2. Autoclave for 15 min at 15 psi pressure–121°C. Aseptically and anaerobically add sterile $NaHCO_3$ solution, sterile cellobiose or starch solution, and sterile $Na_2S·9H_2O$ solution. Mix thoroughly.

Use: For the cultivation and maintenance of *Anaerocellum thermophilum* and *Dictyoglomus turgidus.*

Anaerocellum Medium

Composition per liter:

$NaHCO_3$	1.5g
$Na_2S·9H_2O$	0.5g
$CaCl_2·2H_2O$	0.33g
KCl	0.33g
KH_2PO_4	0.33g
$MgCl_2·6H_2O$	0.33g
NH_4Cl	0.33g
Yeast extract	0.2g
Resazurin	0.5mg
$NaHCO_3$ solution	100.0mL
Cellobiose or starch solution	50.0mL
Vitamin solution	10.0mL
$Na_2S·9H_2O$ solution	10.0mL
Trace elements solution SL-10	1.0mL

pH 7.1–7.3 at 25°C

$NaHCO_3$ Solution:

Composition per 100.0mL:

$NaHCO_3$	5.0g

Preparation of $NaHCO_3$ Solution: Add $NaHCO_3$ to distilled/deionized water and bring volume to 100.0mL. Mix thoroughly. Filter sterilize. Gas under 80% N_2 + 20% CO_2.

Cellobiose or Starch Solution:

Composition per 50.0mL:

Cellobiose or starch	5.0g

Preparation of Cellobiose or Starch Solution: Add cellobiose or starch to distilled/deionized water and bring volume to 50.0mL. Mix thoroughly. Filter sterilize. Gas under 100% N_2.

$Na_2S·9H_2O$ Solution:

Composition per 10.0mL:

$Na_2S·9H_2O$	0.5g

Preparation of $Na_2S{\cdot}9H_2O$ Solution: Add $Na_2S{\cdot}9H_2O$ to distilled/deionized water and bring volume to 10.0mL. Mix thoroughly. Gas under 100% N_2. Autoclave for 15 min at 15 psi pressure–121°C.

Vitamin Solution:

Composition per liter:

Pyridoxine·HCl	10.0mg
Calcium DL-pantothenate	5.0mg
Lipoic acid	5.0mg
Nicotinic acid	5.0mg
p-Aminobenzoic acid	5.0mg
Riboflavin	5.0mg
Thiamine·HCl	5.0mg
Biotin	2.0mg
Folic acid	2.0mg
Vitamin B_{12}	0.1mg

Preparation of Vitamin Solution: Add components to distilled/deionized water and bring volume to 1.0L. Mix thoroughly. Gas under 100% N_2. Autoclave for 15 min at 15 psi pressure–121°C.

Trace Elements Solution SL-10:

Composition per liter:

$FeCl_2{\cdot}4H_2O$	1.5g
$CoCl_2{\cdot}6H_2O$	190.0mg
$MnCl_2{\cdot}4H_2O$	100.0mg
$ZnCl_2$	70.0mg
$Na_2MoO_4{\cdot}2H_2O$	36.0mg
$NiCl_2{\cdot}6H_2O$	24.0mg
H_3BO_3	6.0mg
$CuCl_2{\cdot}2H_2O$	2.0mg
HCl (25% solution)	10.0mL

Preparation of Trace Elements Solution SL-10: Add $FeCl_2{\cdot}4H_2O$ to 10.0mL of HCl solution. Mix thoroughly. Add distilled/deionized water and bring volume to 1.0L. Add remaining components. Mix thoroughly.

Preparation of Medium: Add components, except $NaHCO_3$ solution, cellobiose or starch solution, and $Na_2S{\cdot}9H_2O$ solution, to distilled/deionized water and bring volume to 830.0mL. Mix thoroughly. Gently heat and bring to boiling. Continue boiling for 3–4 min. Allow to cool to room temperature under 80% N_2 + 20% CO_2. Distribute into bottles under 80% N_2 + 20% CO_2. Autoclave for 15 min at 15 psi pressure–121°C. Aseptically and anaerobically add sterile $NaHCO_3$ solution, sterile cellobiose or starch solution, and sterile $Na_2S{\cdot}9H_2O$ solution. Mix thoroughly.

Use: For the cultivation and maintenance of *Anaerocellum thermophilum* and *Dictyoglomus turgidus*.

Anaerofilum Medium

Composition per liter:

$NaHCO_3$	4.0g
Sodium formate	2.0g
Sodium acetate	1.0g
Yeast extract	1.0g
L-Cysteine·HCl	0.5g
KH_2PO_4	0.5g
$Na_2S{\cdot}9H_2O$	0.5g
$MgSO_4{\cdot}7H_2O$	0.4g
NaCl	0.4g
NH_4Cl	0.4g
$CaCl_2{\cdot}2H_2O$	0.05g
$FeSO_4{\cdot}7H_2O$	2.0mg
Resazurin	1.0mg
Glucose solution	20.0mL
Fatty acid mixture	20.0mL
Trace elements solution SL-10	1.0mL

pH 6.7 ± 0.2 at 25°C

Fatty Acid Mixture:

Composition per 20.0mL:

α-Methylbutyric acid	0.5g
Isobutyric acid	0.5g
Isovaleric acid	0.5g
Valeric acid	0.5g

Preparation of Fatty Acid Mixture: Add components to distilled/deionized water and bring volume to 20.0mL. Mix thoroughly. Adjust pH to 7.5 with concentrated NaOH.

Glucose Solution:

Composition per 20.0mL:

D-Glucose	50.0g

Preparation of Glucose Solution: Add glucose to distilled/deionized water and bring volume to 20.0mL. Mix thoroughly. Sparge with 100% N_2. Autoclave for 15 min at 15 psi pressure–121°C.

Trace Elements Solution SL-10:

Composition per liter:

$FeCl_2{\cdot}4H_2O$	1.5g
$CoCl_2{\cdot}6H_2O$	190.0mg
$MnCl_2{\cdot}4H_2O$	100.0mg
$ZnCl_2$	70.0mg
$Na_2MoO_4{\cdot}2H_2O$	36.0mg
$NiCl_2{\cdot}6H_2O$	24.0mg
H_3BO_3	6.0mg
$CuCl_2{\cdot}2H_2O$	2.0mg
HCl (25% solution)	10.0mL

Preparation of Trace Elements Solution SL-10: Add $FeCl_2{\cdot}4H_2O$ to 10.0mL of HCl solution. Mix thoroughly. Add distilled/deionized water and bring volume to 1.0L. Add remaining components. Mix thoroughly.

Preparation of Medium: Prepare and dispense medium anaerobically under 80% H_2 + 20% CO_2. Add components, except glucose solution, to distilled/deionized water and bring volume to 980.0mL. Mix thoroughly. Adjust pH to 6.7. Sparge with 80% H_2 + 20% CO_2. Autoclave for 15 min at 15 psi pressure–121°C. Aseptically and anaerobically add 20.0mL of sterile glucose solution. Aseptically and anaerobically distribute into sterile tubes or bottles.

Use: For the cultivation of *Anaerofilum agile* and *Anaerofilum pentosovorans*.

Anaerolinea Medium (DSMZ Medium 1004)

Composition per liter:

$NaHCO_3$	2.5g
Yeast extract	2.3g
NH_4Cl	0.54g
$MgCl_2{\cdot}6H_2O$	0.2g
$CaCl_2{\cdot}2H_2O$	0.15g
KH_2PO_4	0.14g
Resazurin	1.0mg
Glucose solution	10.0mL
L-Cysteine solution	10.0mL
Vitamin solution	10.0mL
$Na_2S{\cdot}9H_2O$ solution	10.0mL

Selenite tungstate solution 1.0mL
Trace elements solution SL-11 .... 1.0mL
pH 7.0 ± 0.1 at 25°C

$Na_2S{\cdot}9H_2O$ Solution:
Composition per 100.0mL:
$Na_2S{\cdot}9H_2O$ 0.25g

Preparation of $Na_2S{\cdot}9H_2O$ Solution: Add $Na_2S{\cdot}9H_2O$ to distilled/deionized water and bring volume to 100.0mL. Mix thoroughly. Autoclave under 100% N_2 for 15 min at 15 psi pressure–121°C. Cool to room temperature.

Vitamin Solution:
Composition per liter:
Pyridoxine-HCl 10.0mg
Thiamine-HCl·$2H_2O$ 5.0mg
Riboflavin 5.0mg
Nicotinic acid 5.0mg
D-Ca-pantothenate 5.0mg
p-Aminobenzoic acid 5.0mg
Lipoic acid 5.0mg
Biotin 2.0mg
Folic acid 2.0mg
Vitamin B_{12} 0.1mg

Preparation of Vitamin Solution: Add components to distilled/deionized water and bring volume to 1.0L. Mix thoroughly. Sparge with 80% H_2 + 20% CO_2. Filter sterilize.

Glucose Solution:
Composition per 10.0mL:
Glucose 2.2g

Preparation of Glucose Solution: Add glucose to distilled/deionized water and bring volume to 10.0mL. Mix thoroughly. Sparge with 100% N_2. Filter sterilize.

L-Cysteine Solution:
Composition per 10.0mL:
L-Cysteine·HCl·H_2O 0.25g

Preparation of L-Cysteine Solution: Add L-cysteine·HCl·H_2O to distilled/deionized water and bring volume to 10.0mL. Mix thoroughly. Sparge with 100% N_2. Autoclave for 15 min at 15 psi pressure–121°C.

Trace Elements Solution SL-11:
Composition per liter:
$FeCl_2{\cdot}4H_2O$ 1.5g
$CoCl_2{\cdot}6H_2O$ 190.0mg
$MnCl_2{\cdot}4H_2O$ 100.0mg
$ZnCl_2$ 70.0mg
$Na_2MoO_4{\cdot}H_2O$ 36.0mg
$NiCl_2{\cdot}6H_2O$ 24.0mg
H_3BO_3 .... 6.0mg
Na_2-EDTA 5.2g
$CuCl_2{\cdot}2H_2O$ 2.0mg

Preparation of Trace Elements Solution SL-11: Add components to distilled/deionized water and bring volume to 1.0L. Mix thoroughly. Adjust pH to 6.0.

Selenite/Tungstate Solution:
Composition per liter:
NaOH 0.5g
$Na_2WO_4{\cdot}2H_2O$ 4.0mg
$Na_2SeO_3{\cdot}5H_2O$ 3.0mg

Preparation of Selenite/Tungstate Solution: Add components to distilled/deionized water and bring volume to 1.0L. Mix thoroughly. Sparge with 100% N_2. Autoclave for 15 min at 15 psi pressure–121°C.

Preparation of Medium: Add components, except vitamin solution, $NaHCO_3$, L-cysteine solution, $Na_2S{\cdot}9H_2O$ solution and glucose solution, to distilled/deionized water and bring volume to 960.0mL. Mix thoroughly. Gently heat and bring to boiling. Cool to room temperature while sparging with 20% CO_2 + 80% N_2. Add solid bicarbonate. Mix thoroughly. Adjust pH to 7.0. Dispense into tubes or bottles. Autoclave for 15 min at 15 psi pressure–121°C. Cool to 25°C under 20% CO_2 + 80% N_2. Aseptically and anaerobically add sterile glucose, L-cysteine, vitamin, and $Na_2S{\cdot}9H_2O$ solutions. The final pH should be 7.0.

Use: For the cultivation and maintenance of *Anaerolinea* spp.

Anaerolinea Medium with Sucrose (DSMZ Medium 1004)

Composition per liter:
$NaHCO_3$ 2.5g
NH_4Cl 0.54g
$MgCl_2{\cdot}6H_2O$ 0.2g
$CaCl_2{\cdot}2H_2O$ 0.15g
KH_2PO_4 .... 0.14g
Yeast extract 0.1g
Resazurin 1.0mg
Sucrose solution 20.0mL
L-Cysteine solution 10.0mL
Vitamin solution 10.0mL
$Na_2S{\cdot}9H_2O$ solution 10.0mL
Selenite/tungstate solution 1.0mL
Trace elements solution SL-11 .... 1.0mL
pH 7.0 ± 0.2 at 25°C

Sucrose Solution:
Composition per 20.0mL:
Sucrose 7.2g

Preparation of Sucrose Solution: Add sucrose to distilled/deionized water and bring volume to 20.0mL. Mix thoroughly. Sparge with 100% N_2. Filter sterilize.

$Na_2S{\cdot}9H_2O$ Solution:
Composition per 100.0mL:
$Na_2S{\cdot}9H_2O$ 0.25g

Preparation of $Na_2S{\cdot}9H_2O$ Solution: Add $Na_2S{\cdot}9H_2O$ to distilled/deionized water and bring volume to 100.0mL. Mix thoroughly. Autoclave under 100% N_2 for 15 min at 15 psi pressure–121°C. Cool to room temperature.

Vitamin Solution:
Composition per liter:
Pyridoxine-HCl 10.0mg
Thiamine-HCl·$2H_2O$ 5.0mg
Riboflavin 5.0mg
Nicotinic acid 5.0mg
D-Ca-pantothenate 5.0mg
p-Aminobenzoic acid 5.0mg
Lipoic acid 5.0mg
Biotin 2.0mg
Folic acid 2.0mg
Vitamin B_{12} 0.1mg

Preparation of Vitamin Solution: Add components to distilled/deionized water and bring volume to 1.0L. Mix thoroughly. Sparge with 80% H_2 + 20% CO_2. Filter sterilize.

L-Cysteine Solution:
Composition per 10.0mL:

L-Cysteine·HCl·H_2O	0.25g

Preparation of L-Cysteine Solution: Add L-cysteine·HCl·H_2O to distilled/deionized water and bring volume to 10.0mL. Mix thoroughly. Sparge with 100% N_2. Autoclave for 15 min at 15 psi pressure–121°C.

Trace Elements Solution SL-11:
Composition per liter:

$FeCl_2·4H_2O$	1.5g
$CoCl_2·6H_2O$	190.0mg
$MnCl_2·4H_2O$	100.0mg
$ZnCl_2$	70.0mg
$Na_2MoO_4·H_2O$	36.0mg
$NiCl_2·6H_2O$	24.0mg
H_3BO_3	6.0mg
Na_2-EDTA	5.2g
$CuCl_2·2H_2O$	2.0mg

Preparation of Trace Elements Solution SL-11: Add components to distilled/deionized water and bring volume to 1.0L. Mix thoroughly. Adjust pH to 6.0.

Selenite/Tungstate Solution:
Composition per liter:

NaOH	0.5g
$Na_2WO_4·2H_2O$	4.0mg
$Na_2SeO_3·5H_2O$	3.0mg

Preparation of Selenite/Tungstate Solution: Add components to distilled/deionized water and bring volume to 1.0L. Mix thoroughly. Sparge with 100% N_2. Autoclave for 15 min at 15 psi pressure–121°C.

Preparation of Medium: Add components, except vitamin solution, $NaHCO_3$, sucrose solution, L-cysteine solution, and $Na_2S·9H_2O$ solution, to distilled/deionized water and bring volume to 950.0mL. Mix thoroughly. Gently heat and bring to boiling. Cool to room temperature while sparging with 20% CO_2 + 80% N_2. Add solid bicarbonate. Mix thoroughly. Adjust pH to 7.0. Dispense into tubes or bottles. Autoclave for 15 min at 15 psi pressure–121°C. Cool to 25°C under 20% CO_2 + 80% N_2. Aseptically and anaerobically add sterile sucrose, L-cysteine, vitamin, and $Na_2S·9H_2O$ solutions. The final pH should be 7.0.

Use: For the cultivation and maintenance of *Anaerolinea thermolimosa, Bellilinea caldistulae,* and *Levilinea saccharolytica.*

Anaerolinea Medium without Glucose (DSMZ Medium 1004)

Composition per liter:

$NaHCO_3$	2.5g
NH_4Cl	0.54g
$MgCl_2·6H_2O$	0.2g
$CaCl_2·2H_2O$	0.15g
KH_2PO_4	0.14g
Yeast extract	0.1g
Resazurin	1.0mg
L-Cysteine solution	10.0mL
Vitamin solution	10.0mL
$Na_2S·9H_2O$ solution	10.0mL
Selenite/tungstate solution	1.0mL
Trace elements solution SL-11	1.0mL

pH 7.0 ± 0.1 at 25°C

$Na_2S·9H_2O$ Solution:
Composition per 100.0mL:

$Na_2S·9H_2O$	0.25g

Preparation of $Na_2S·9H_2O$ Solution: Add $Na_2S·9H_2O$ to distilled/deionized water and bring volume to 100.0mL. Mix thoroughly. Autoclave under 100% N_2 for 15 min at 15 psi pressure–121°C. Cool to room temperature.

Vitamin Solution:
Composition per liter:

Pyridoxine-HCl	10.0mg
Thiamine-HCl·$2H_2O$	5.0mg
Riboflavin	5.0mg
Nicotinic acid	5.0mg
D-Ca-pantothenate	5.0mg
p-Aminobenzoic acid	5.0mg
Lipoic acid	5.0mg
Biotin	2.0mg
Folic acid	2.0mg
Vitamin B_{12}	0.1mg

Preparation of Vitamin Solution: Add components to distilled/deionized water and bring volume to 1.0L. Mix thoroughly. Sparge with 80% H_2 + 20% CO_2. Filter sterilize.

L-Cysteine Solution:
Composition per 10.0mL:

L-Cysteine·HCl·H_2O	0.25g

Preparation of L-Cysteine Solution: Add L-cysteine·HCl·H_2O to distilled/deionized water and bring volume to 10.0mL. Mix thoroughly. Sparge with 100% N_2. Autoclave for 15 min at 15 psi pressure–121°C.

Trace Elements Solution SL-11:
Composition per liter:

$FeCl_2·4H_2O$	1.5g
$CoCl_2·6H_2O$	190.0mg
$MnCl_2·4H_2O$	100.0mg
$ZnCl_2$	70.0mg
$Na_2MoO_4·H_2O$	36.0mg
$NiCl_2·6H_2O$	24.0mg
H_3BO_3	6.0mg
Na_2-EDTA	5.2g
$CuCl_2·2H_2O$	2.0mg

Preparation of Trace Elements Solution SL-11: Add components to distilled/deionized water and bring volume to 1.0L. Mix thoroughly. Adjust pH to 6.0.

Selenite/Tungstate Solution:
Composition per liter:

NaOH	0.5g
$Na_2WO_4·2H_2O$	4.0mg
$Na_2SeO_3·5H_2O$	3.0mg

Preparation of Selenite/Tungstate Solution: Add components to distilled/deionized water and bring volume to 1.0L. Mix thoroughly. Sparge with 100% N_2. Autoclave for 15 min at 15 psi pressure–121°C.

Preparation of Medium: Add components, except vitamin solution, $NaHCO_3$, L-cysteine solution, and $Na_2S·9H_2O$ solution , to distilled/deionized water and bring volume to 970.0mL. Mix thoroughly.

Gently heat and bring to boiling. Cool to room temperature while sparging with 20% CO_2 + 80% N_2. Add solid bicarbonate. Mix thoroughly. Adjust pH to 7.0. Dispense into tubes or bottles. Autoclave for 15 min at 15 psi pressure–121°C. Cool to 25°C under 20% CO_2 + 80% N_2. Aseptically and anaerobically add sterile L-cysteine, vitamin, and $Na_2S \cdot 9H_2O$ solutions. The final pH should be 7.0.

Use: For the cultivation and maintenance of *Leptolinea tardivitalis*.

Anaeromyxobacter Medium (DSMZ Medium 1200)

Composition per liter:

Solution A	900.0mL
Solution B	90.0mL
Solution C	18.0mL
Solution D	18.0mL
Solution F	18.0mL
Solution E	4.0mL

pH 7.2 ± 0.2 at 25°C

Solution A:

Composition per 900.0mL:

NaCl	1.0g
$MgCl_2 \cdot 6H_2O$	0.5g
Sodium acetate	0.4g
NH_4Cl	0.3g
KCl	0.3g
KH_2PO_4	0.2g
$CaCl_2 \cdot 2H_2O$	15.0mg
Resazurin	1.0mg
Selenite/tungstate solution	2.0mL
Trace element solution SL-10B	1.0mL

Selenite/Tungstate Solution:

Composition per liter:

NaOH	0.5g
$Na_2WO_4 \cdot 2H_2O$	4.0mg
$Na_2SeO_3 \cdot 5H_2O$	3.0mg

Preparation of Selenite/Tungstate Solution: Add components to distilled/deionized water and bring volume to 1.0L. Mix thoroughly. Sparge with 100% N_2. Autoclave for 15 min at 15 psi pressure–121°C.

Trace Elements Solution SL-10B:

Composition per liter:

$FeCl_2 \cdot 4H_2O$	1.5g
H_3BO_3	300.0mg
$CoCl_2 \cdot 6H_2O$	190.0mg
$MnCl_2 \cdot 4H_2O$	100.0mg
$ZnCl_2$	70.0mg
$Na_2MoO_4 \cdot 2H_2O$	36.0mg
$NiCl_2 \cdot 6H_2O$	24.0mg
$CuCl_2 \cdot 2H_2O$	2.0mg
HCl (25% solution)	10.0mL

Preparation of Trace Elements Solution SL-10B: Add $FeCl_2 \cdot 4H_2O$ to 10.0mL of HCl solution. Mix thoroughly. Add distilled/deionized water and bring volume to 1.0L. Add remaining components. Mix thoroughly.

Preparation of Solution A: Add components to distilled/deionized water and bring volume to 900.0mL. Mix thoroughly. Sparge with 20% CO_2 + 80% N_2. Autoclave for 15 min at 15 psi pressure–121°C. Cool to room temperature.

Solution B:

Composition per 100.0mL:

$NaHCO_3$	5.0g

Preparation of Solution B: Add $NaHCO_3$ to distilled/deionized water and bring volume to 100.0mL. Mix thoroughly. Sparge with 20% CO_2 + 80% H_2. Autoclave for 15 min at 15 psi pressure–121°C. Cool to room temperature.

Solution C:

Composition per 50.0mL:

DL-Dithiothreitol	385.0mg

Preparation of Solution C: Add DL-dithiothreitol to distilled/deionized water and bring volume to 50.0mL. Mix thoroughly. Sparge with 100% N_2. Filter sterilize.

Solution D:

Composition per 50.0mL:

L-Cysteine·HCl·H_2O	37.5mg
$Na_2S \cdot 9H_2O$ solution	10.0mL

Preparation of Solution D: Add L-cysteine·HCl·H_2O to distilled/deionized water and bring volume to 40.0mL. Mix thoroughly. Add 10.0mL $Na_2S \cdot 9H_2O$ solution. Sparge with 20% CO_2 + 80% H_2. Autoclave for 15 min at 15 psi pressure–121°C. Cool to room temperature.

$Na_2S \cdot 9H_2O$ Solution:

Composition per 10.0mL:

$Na_2S \cdot 9H_2O$	40.0mg

Preparation of $Na_2S \cdot 9H_2O$ Solution: Add $Na_2S \cdot 9H_2O$ to distilled/deionized water and bring volume to 10.0mL. Mix thoroughly.

Solution E:

Composition per 50.0mL:

Fumarate	4.0g

Preparation of Solution E: Add fumarate to distilled/deionized water and bring volume to 50.0mL. Mix thoroughly. Sparge with 100% N_2. Filter sterilize.

Solution F (Vitamin Solution):

Composition per liter:

Pyridoxine-HCl	10.0mg
Thiamine-HCl·$2H_2O$	5.0mg
Riboflavin	5.0mg
Nicotinic acid	5.0mg
D-Ca-pantothenate	5.0mg
p-Aminobenzoic acid	5.0mg
Lipoic acid	5.0mg
Biotin	2.0mg
Folic acid	2.0mg
Vitamin B_{12}	0.1mg

Preparation of Solution F (Vitamin Solution): Add components to distilled/deionized water and bring volume to 1.0L. Mix thoroughly. Sparge with 80% H_2 + 20% CO_2. Filter sterilize.

Preparation of Medium: Aseptically and anoxically add 1.0mL solution B per 10 mL solution A. Adjust pH to 7.2. Add 0.2mL solution C, 0.2mL solution D, 0.2mL solution F, and 0.05mL solution E, each per 10mL solution A.

Use: For the cultivation and maintenance of *Anaeromyxobacter* spp.

Anaerospirillum Medium

Composition per liter:

Polypeptone™	10.0g
Glucose	10.0g

Yeast extract	5.0g
Na_2CO_3	3.0g
NaCl	2.0g
K_2HPO_4	1.0g
$MgSO_4·7H_2O$	0.2g

pH 6.5 ± 0.2 at 25°C

Preparation of Medium: Add components to distilled/deionized water and bring volume to 1.0L. Mix thoroughly. Gently heat and bring to boiling. Autoclave for 15 min at 15 psi pressure–121°C. Distribute into tubes or flasks using anaerobic techniques and 100% CO_2 as gas phase.

Use: For the cultivation of *Anaerospirillum succiniciproducens.*

Anaerovibrio burkinabensis Medium

Composition per 1011.0mL:

Solution A	870.0mL
Solution C	100.0mL
Solution D	10.0mL
Solution E (Vitamin solution)	10.0mL
Solution F	10.0mL
Solution G	10.0mL
Solution B (Trace elements solution SL-10)	1.0mL

pH 6.8–7.2 at 25°C

Solution A:

Composition per 870.0mL:

Na_2SO_4	3.0g
NaCl	1.0g
KCl	0.5g
$MgCl_2·6H_2O$	0.4g
NH_4Cl	0.3g
KH_2PO_4	0.2g
$CaCl_2·2H_2O$	0.15g
Resazurin	1.0mg

Preparation of Solution A: Add components to distilled/deionized water and bring volume to 870.0mL. Mix thoroughly. Gently heat and bring to boiling. Continue boiling for 3–4 min. Allow to cool to room temperature while gassing under 80% N_2 + 20% CO_2. Continue gassing until pH reaches below 6.0. Seal the flask under 80% N_2 + 20% CO_2. Autoclave for 15 min at 15 psi pressure–121°C.

Solution B (Trace Elements Solution SL-10):

Composition per liter:

$FeCl_2·4H_2O$	1.5g
$CoCl_2·6H_2O$	190.0mg
$MnCl_2·4H_2O$	100.0mg
$ZnCl_2$	70.0mg
$Na_2MoO_4·2H_2O$	36.0mg
$NiCl_2·6H_2O$	24.0mg
H_3BO_3	6.0mg
$CuCl_2·2H_2O$	2.0mg
HCl (25% solution)	10.0mL

Preparation of Solution B (Trace Elements Solution SL-10): Add $FeCl_2·4H_2O$ to 10.0mL of HCl solution. Mix thoroughly. Add distilled/deionized water and bring volume to 1.0L. Add remaining components. Mix thoroughly. Gas under 100% N_2. Autoclave for 15 min at 15 psi pressure–121°C.

Solution C:

Composition per 100.0mL:

$NaHCO_3$	5.0g

Preparation of Solution C: Add $NaHCO_3$ to distilled/deionized water and bring volume to 100.0mL. Mix thoroughly. Filter sterilize. Gas under 80% N_2 + 20% CO_2.

Solution D:

Composition per 10.0mL:

Sodium lactate	2.5g

Preparation of Solution D: Add sodium lactate· to distilled/deionized water and bring volume to 10.0mL. Mix thoroughly. Gas under 100% N_2. Autoclave for 15 min at 15 psi pressure–121°C.

Solution E (Vitamin Solution):

Composition per liter:

Pyridoxine·HCl	10.0mg
Calcium DL-pantothenate	5.0mg
Lipoic acid	5.0mg
Nicotinic acid	5.0mg
p-Aminobenzoic acid	5.0mg
Riboflavin	5.0mg
Thiamine·HCl	5.0mg
Biotin	2.0mg
Folic acid	2.0mg
Vitamin B_{12}	0.1mg

Preparation of Solution E (Vitamin Solution): Add components to distilled/deionized water and bring volume to 1.0L. Mix thoroughly. Gas under 100% N_2. Autoclave for 15 min at 15 psi pressure–121°C.

Solution F:

Composition per 10.0mL:

$Na_2S·9H_2O$	0.4g

Preparation of Solution F: Add $Na_2S·9H_2O$ to distilled/deionized water and bring volume to 10.0mL. Mix thoroughly. Gas under 100% N_2. Autoclave for 15 min at 15 psi pressure–121°C.

Solution G:

Composition per 10.0mL:

Yeast extract	1.0g

Preparation of Solution F: Add yeast extract to distilled/deionized water and bring volume to 10.0mL. Mix thoroughly. Gas under 100% N_2. Autoclave for 15 min at 15 psi pressure–121°C.

Preparation of Medium: Aseptically and anaerobically combine solution A with solution B, solution C, solution D, solution E, solution F, and solution G, in that order. Mix thoroughly. Anaerobically distribute into sterile tubes or flasks under 80% N_2 + 20% CO_2.

Use: For the cultivation and maintenance of *Anaerovibrio burkinabensis.*

Ancalomicrobium adetum Medium

Composition per liter:

Ammonium sulfate	0.25g
Glucose	0.25g
Na_2HPO_4	71.0mg
Modified Hutner's basal salts	20.0mL
Vitamin solution	10.0mL

Modified Hutner's Basal Salts:

Composition per liter:

$MgSO_4·7H_2O$	29.7g
Nitrilotriacetic acid	10.0g
$CaCl_2·2H_2O$	3.34g
$FeSO_4·7H_2O$	99.0mg

Ammonium molybdate 9.25mg
Metals "44" 50.0mL

Preparation of Modified Hutner's Basal Salts: Dissolve the nitrilotracetic acid first and neutralize the solution with KOH. Add other components and adjust the pH to 7.2 with KOH or H_2SO_4. There may be a slight precipitate. Store at 5°C.

Metals "44"
Composition per 100.0mL:

$ZnSO_4 \cdot 7H_2O$ 1.1g
$FeSO_4 \cdot 7H_2O$ 0.5g
$CuSO_4 \cdot 5H_2O$ 0.04g
EDTA 0.25g
$MnSO_4 \cdot 7H_2O$ 0.154g
$Co(NO_3)_2 \cdot 6H_2O$ 0.025g
$Na_2B_4O_7 \cdot 10H_2O$ 0.018g

Preparation of Metals "44": Add components to distilled/deionized water and bring volume to 100.0mL. Mix thoroughly. Autoclave for 15 min at 15 psi pressure–121°C. Add aseptically to sterile modified Hutner's basal salts solution.

Vitamin Solution:
Composition per liter:

Thiamine HCl 5.0g
Calcium DL-pantothenate 5.0mg
Nicotinamide 5.0mg
Riboflavin 5.0mg
Biotin 2.0mg
Folic acid 2.0mg
Vitamin B_{12} 0.1mg

Preparation of Vitamin Solution: Add components to distilled/deionized water and bring volume to 1.0L. Mix thoroughly. Filter sterilize.

Preparation of Medium: Add components, except modified Hutner's basal salts solution and vitamin solution, to distilled/deionized water and bring volume to 970.0mL. Mix thoroughly. Autoclave for 15 min at 15 psi pressure–121°C. Aseptically add 20.0mL of modified Hutner's basal salts solution and 10.0mL of sterile vitamin solution. Mix thoroughly. Aseptically distribute into sterile tubes or flasks.

Use: For the cultivation of *Ancalomicrobium adetum.*

Ancalomicrobium Medium

Composition per liter:

Glucose 0.25g
$(NH_4)_2SO_4$ 0.25g
Na_2HPO_4 .... 0.071g
Hutner's basal salts solution 20.0mL
Vitamin solution 10.0mL

pH 7.0 ± 0.2 at 25°C

Hutner's Basal Salts Solution:
Composition per liter:

$MgSO_4 \cdot 7H_2O$ 29.7g
Nitrilotriacetic acid 10.0g
$CaCl_2 \cdot 2H_2O$ 3.335g
$FeSO_4 \cdot 7H_2O$ 99.0mg
$(NH_4)_6MoO_7O_{24} \cdot 4H_2O$ 9.25mg
"Metals 44" 50.0mL

"Metals 44":
Composition per 100.0mL:

$ZnSO_4 \cdot 7H_2O$ 1.095g
$FeSO_4 \cdot 7H_2O$ 0.5g
Sodium EDTA 0.25g
$MnSO_4 \cdot H2O$ 0.154g
$CuSO_4 \cdot 5H_2O$ 39.2mg
$Co(NO_3)_2 \cdot 6H_2O$ 24.8mg
$Na_2B_4O_7 \cdot 10H_2O$ 17.7mg

Preparation of Metals "44": Add sodium EDTA to distilled/deionized water and bring volume to 90.0mL. Mix thoroughly. Add a few drops of concentrated H_2SO_4 to retard precipitation of heavy metal ions. Add remaining components. Mix thoroughly. Bring volume to 100.0mL with distilled/deionized water.

Preparation of Hutner's Basal Salts Solution: Add nitrilotriacetic acid to 500.0mL of distilled/deionized water. Adjust pH to 6.5 with KOH. Add remaining components. Add distilled/deionized water to 1.0L. Adjust pH to 6.8. Filter sterilize.

Vitamin Solution:
Composition per liter:

Calcium DL-pantothenate 5.0mg
Nicotinamide 5.0mg
Riboflavin 5.0mg
Thiamine·HCl 5.0mg
Biotin 2.0mg
Folic acid 2.0mg
Cyanocobalamin 0.1mg

Preparation of Vitamin Solution: Add components to distilled/deionized water and bring volume to 1.0L. Mix thoroughly. Filter sterilize.

Preparation of Medium: Add components, except Hutner's basal salts solution and vitamin solution, to distilled/deionized water and bring volume to 970.0mL. Mix thoroughly. Adjust pH to 7.0. Autoclave for 15 min at 15 psi pressure–121°C. Aseptically add 20.0mL of sterile Hutner's basal salts solution and 10.0mL of sterile vitamin solution. Mix thoroughly. Aseptically distribute into sterile tubes or flasks.

Use: For the cultivation of *Ancalomicrobium adetum.*

Ancylobacter/Spirosoma Agar

Composition per liter:

Agar 20.0g
Glucose 1.0g
Peptone 1.0g
Yeast extract 1.0g

pH 6.8 ± 0.2 at 25°C

Preparation of Medium: Add components to distilled/deionized water and bring volume to 1.0L. Mix thoroughly. Adjust pH to 6.8. Gently heat and bring to boiling. Distribute into tubes or flasks. Autoclave for 15 min at 15 psi pressure–121°C. Pour into sterile Petri dishes or leave in tubes.

Use: For the cultivation of *Ancylobacter aquaticus, Ancylobacter* species, *Aquaspirillum metamorphum, Aquaspirillum serpens, Flectobacillus major, Methylobacterium mesophilicum, Runella slithyformis, Shewanella putrefaciens,* and *Spirosoma linguale.*

Ancylobacter Spirosoma Medium (DSMZ Medium 7)

Composition per liter:

Agar 15.0g
Glucose 1.0g

Peptone	1.0g
Yeast extract	1.0g

pH 7.1 ± 0.2 at 25°C

Preparation of Medium: Add components to distilled/deionized water and bring volume to 1.0L. Mix thoroughly. Gently heat and bring to boiling. Distribute into tubes or flasks. Autoclave for 15 min at 15 psi pressure–121°C. Pour into sterile Petri dishes or leave in tubes.

Use: For the cultivation and maintenance of *Spirosoma linguale.*

Andersen's Pork Pea Agar

Composition per 1685.0mL:

Agar	16.0g
Peptone	5.0g
Pancreatic digest of casein	1.6g
K_2HPO_4	1.25g
Soluble starch	1.0g
Sodium thioglycolate	0.5g
Pork infusion	800.0mL
Thioglycolate agar	660.0mL
Pea infusion	200.0mL
$NaHCO_3$ solution	25.0mL

pH 7.2 ± 0.2 at 25°C

Pork Infusion:

Composition per liter:

Pork, fresh lean ground	454.0g

Preparation of Pork Infusion: Add ground pork to distilled/deionized water and bring volume to 1.0L. Autoclave for 60 min at 0 psi pressure–100°C. Filter through two layers of cheesecloth. Cool to 4°C. Skim fat from surface. Warm to 25°C. Centrifuge at 5000 rpm for 10 min. Discard pellet.

Pea Infusion:

Composition per 450.0mL:

Green peas, fresh or frozen	454.0g
Diatomaceous earth (celite)	10.0g

Preparation of Pea Infusion: Add green peas to 450.0mL of distilled/deionized water. Blend until smooth. Autoclave for 60 min at 0 psi pressure–100°C. Centrifuge at 5000 rpm for 10 min. Discard pellet. Clarify supernatant solution with diatomaceous earth (celite). Filter through Whatman #4 filter paper. Use filtrate solution.

Thioglycolate Agar:

Composition per liter:

Agar	20.75g
Pancreatic digest of casein	15.0g
Glucose	5.5g
Yeast extract	5.0g
NaCl	2.5g
L-Cystine	0.5g
Sodium thioglycolate	0.5g
Resazurin	1.0mg

pH 7.1 ± 0.2 at 25°C

Preparation of Thioglycolate Agar: Add components to distilled/deionized water and bring volume to 1.0L. Mix thoroughly. Gently heat and bring to boiling. Autoclave for 15 min at 15 psi pressure–121°C. Cool to 45°–50°C.

$NaHCO_3$ Solution:

Composition per 100.0mL:

$NaHCO_3$	5.0g

Preparation of $NaHCO_3$ Solution: Add $NaHCO_3$ to distilled/deionized water and bring volume to 100.0mL. Mix thoroughly. Filter sterilize.

Preparation of Medium: Combine components, except $NaHCO_3$ solution and thioglycolate agar. Mix thoroughly. Adjust pH to 7.2. Autoclave for 5 min at 15 psi pressure–121°C. While medium is still hot, add 25.0g of celite. Filter through Whatman #4 filter paper with suction. Autoclave for 12 min at 15 psi pressure–121°C. Cool to 45°–50°C. Aseptically add 25.0mL of sterile $NaHCO_3$ solution. Mix thoroughly. Pour into sterile Petri dishes in 15.0mL volumes. Allow agar to solidify. Cover agar with 10.0mL of sterile, cooled thioglycolate agar.

Use: For the cultivation of mesophilic *Clostridium* species. For the recovery of endospores from foods following heat treatments.

Anderson's Marine Agar

Composition per liter:

Agar	15.0g
Peptone	2.5g
Yeast extract	2.5g
$FePO_4$	0.1g
Filtered, aged seawater	750.0mL

pH 7.4–7.6 at 25°C

Preparation of Medium: Add components to distilled/deionized water and bring volume to 1.0L. Mix thoroughly. Gently heat and bring to boiling. Adjust pH to 7.4–7.6. Distribute into tubes or flasks. Autoclave for 15 min at 15 psi pressure–121°C. Pour into sterile Petri dishes or leave in tubes.

Use: For the cultivation and maintenance of *Vibrio* species.

Anderson's Marine Broth

Composition per liter:

Peptone	2.5g
Yeast extract	2.5g
$FePO_4$	0.1g
Filtered, aged seawater	750.0mL

pH 7.4–7.6 at 25°C

Preparation of Medium: Add components to distilled/deionized water and bring volume to 1.0L. Mix thoroughly. Adjust pH to 7.4–7.6. Distribute into tubes or flasks. Autoclave for 15 min at 15 psi pressure–121°C.

Use: For the cultivation of *Vibrio* species.

Anderson's Marine Medium

Composition per liter:

Peptone	2.5g
Yeast extract	2.5g
$FePO_4$	0.1g
Filtered, aged seawater	750.0mL

pH 7.4 ± 0.2 at 25°C

Preparation of Medium: Add components, except seawater , to distilled/deionized water and bring volume to 250.0mL. Mix thoroughly. Autoclave for 15 min at 15 psi pressure–121°C. Aseptically add 750.0mL of sterile aged seawater. Mix thoroughly. Bring pH to 7.4. Aseptically distribute into sterile tubes or flasks.

Use: For the cultivation of *Flavobacterium* species, *Micrococcus* species, *Planococcus* species, *Pseudomonas fluorescens*, and *Vibrio* species.

Andrade HiVeg Peptone Water

Composition per liter:

Plant peptone	10.0g
NaCl	5.0g
Andrade indicator	0.1g

pH 7.4 ± 0.2 at 25°C

Source: This medium is available as a premixed powder from HiMedia.

Preparation of Medium: Add components to distilled/deionized water and bring volume to 1.0L. Mix thoroughly. Gently heat and bring to boiling. Distribute into tubes or flasks. Autoclave for 15 min at 15 psi pressure–121°C. Pour into sterile Petri dishes or leave in tubes.

Caution: Acid Fuchsin in Andrade indicator is a potential carcinogen and care must be taken to avoid inhalation of the powdered dye and contact with the skin.

Use: For the determination of carbohydrate fermentation reactions of microorganisms, particularly members of the Enterobacteriaceae. A specific carbohydrate is added to the medium to test the fermentation of that carbohydrate. A Durham tube is used to collect gas produced during the fermentation reaction. Acid production is indicated by a pink color.

Andrade Peptone Water

Composition per liter:

Peptic digest of animal tissue	10.0g
NaCl	5.0g
Andrade indicator	0.1g

pH 7.4 ± 0.2 at 25°C

Source: This medium is available as a premixed powder from HiMedia.

Preparation of Medium: Add components to distilled/deionized water and bring volume to 1.0L. Mix thoroughly. Gently heat and bring to boiling. Distribute into tubes or flasks. Autoclave for 15 min at 15 psi pressure–121°C. Pour into sterile Petri dishes or leave in tubes.

Caution: Acid Fuchsin in Andrade indicator is a potential carcinogen and care must be taken to avoid inhalation of the powdered dye and contact with the skin.

Use: For the determination of carbohydrate fermentation reactions of microorganisms, particularly members of the Enterobacteriaceae. A specific carbohydrate is added to the medium to test the fermentation of that carbohydrate. A Durham tube is used to collect gas produced during the fermentation reaction. Acid production is indicated by a pink color.

Andrade Peptone Water with HiVeg Extract No. 1

Composition per liter:

Plant peptone	10.0g
NaCl	5.0g
Plant extract No. 1	3.0g
Andrade indicator	0.1g

pH 7.4 ± 0.2 at 25°C

Source: This medium is available as a premixed powder from HiMedia.

Preparation of Medium: Add components to distilled/deionized water and bring volume to 1.0L. Mix thoroughly. Gently heat and bring to boiling. Distribute into tubes or flasks. Autoclave for 15 min at 15 psi pressure–121°C. Pour into sterile Petri dishes or leave in tubes.

Caution: Acid Fuchsin in Andrade indicator is a potential carcinogen and care must be taken to avoid inhalation of the powdered dye and contact with the skin.

Use: With added carbohydrates, for the determination of carbohydrate fermentation reactions of microorganisms, particularly members of the Enterobacteriaceae. A specific carbohydrate is added to the medium to test the fermentation of that carbohydrate. A Durham tube is used to collect gas produced during the fermentation reaction. Acid production is indicated by a pink color.

Andrade Peptone Water with Meat Extract

Composition per liter:

Peptic digest of animal tissue	10.0g
NaCl	5.0g
Meat extract	3.0g
Andrade indicator	0.1g

pH 7.4 ± 0.2 at 25°C

Source: This medium is available as a premixed powder from HiMedia.

Preparation of Medium: Add components to distilled/deionized water and bring volume to 1.0L. Mix thoroughly. Gently heat and bring to boiling. Distribute into tubes or flasks. Autoclave for 15 min at 15 psi pressure–121°C. Pour into sterile Petri dishes or leave in tubes.

Caution: Acid Fuchsin in Andrade indicator is a potential carcinogen and care must be taken to avoid inhalation of the powdered dye and contact with the skin.

Use: With added carbohydrates, for the determination of carbohydrate fermentation reactions of microorganisms, particularly members of the Enterobacteriaceae. A specific carbohydrate is added to the medium to test the fermentation of that carbohydrate. A Durham tube is used to collect gas produced during the fermentation reaction. Acid production is indicated by a pink color.

Andrade's Broth

Composition per liter:

Pancreatic digest of gelatin	10.0g
NaCl	5.0g
Beef extract	3.0g
Andrade's indicator	10.0mL
Carbohydrate solution	50.0mL

pH 7.4 ± 0.2 at 25°C

Source: This medium is available as a prepared medium from BD Diagnostic Systems, in tubes containing adonitol, arabinose, cellobiose, dulcitol, fructose, galactose, glucose, inositol, lactose, maltose, mannitol, raffinose, rhamnose, salicin, sorbitol, sucrose, trehalose, or xylose.

Andrade's Indicator

Composition per 100.0mL:

NaOH (1*N* solution)	16.0mL
Acid Fuchsin	0.1g

Preparation of Andrade's Indicator: Add Acid Fuchsin to NaOH solution and bring volume to 100.0mL with distilled/deionized water.

Carbohydrate Solution:

Composition per 100.0mL:

Carbohydrate	10.0g

Preparation of Carbohydrate Solution: Add carbohydrate to distilled/deionized water and bring volume to 100.0mL. Adonitol, arabinose, cellobiose, dulcitol, fructose, galactose, glucose, inositol, lactose, maltose, mannitol, raffinose, rhamnose, salicin, sorbitol, sucrose, trehalose, xylose, or other carbohydrates may be used. Mix thoroughly. Filter sterilize.

Preparation of Medium: Add components, except carbohydrate solution, to distilled/deionized water and bring volume to 1.0L. Mix thoroughly. Gently heat and bring to boiling. Distribute in 10.0mL volumes into test tubes containing inverted Durham tubes. Autoclave for 15 min at 15 psi pressure–121°C. Cool to 25°C. Add 0.5mL of sterile carbohydrate solution to each tube.

Caution: Acid Fuchsin is a potential carcinogen and care must be taken to avoid inhalation of the powdered dye and contact with the skin.

Use: For the determination of carbohydrate fermentation reactions of microorganisms, particularly members of the Enterobacteriaceae. A Durham tube is used to collect gas produced during the fermentation reaction. Acid production is indicated by a pink color.

Andrade's Carbohydrate Broth and Indicator (BAM M13)

Composition per liter:

Pancreatic digest of gelatin	10.0g
NaCl	10.0g
Beef extract	3.0g
Carbohydrate solution	100.0mL
Andrade's indicator	10.0mL

pH 7.2 ± 0.2 at 25°C

Source: This medium is available as a prepared medium from BBL Microbiology Systems, in tubes containing adonitol, arabinose, cellobiose, glucose, dulcitol, fructose, galactose, inositol, lactose, maltose, mannitol, raffinose, rhamnose, salicin, sorbitol, sucrose, trehalose, or xylose.

Andrade's Indicator
Composition per 26.0mL:

NaOH (1*N* solution)	16.0mL
Acid Fuchsin	0.21g

Preparation of Andrade's Indicator: Add Acid Fuchsin to NaOH solution and bring volume to 26.0mL with distilled/deionized water.

Carbohydrate Solution:
Composition per 100.0mL:

Carbohydrate	5.0–10.0g

Preparation of Carbohydrate Solution: Add carbohydrate to distilled/deionized water and bring volume to 100.0mL. For glucose, lactose, sucrose, and mannitol, add 10.0g to distilled/deionized water and bring volume to 100.0mL. For dulcitol, salicin, and other carbohydrates, add 5.0g to distilled/deionized water and bring volume to 100.0mL. Mix thoroughly. Filter sterilize.

Preparation of Medium: Add components, except carbohydrate solution, to distilled/deionized water and bring volume to 1.0L. Mix thoroughly. Gently heat and bring to boiling. Cool. Aseptically add 100mL of sterile carbohydrate solution to 900mL of sterile medium. Mix thoroughly. Aseptically distribute into tubes or flasks. Alternately, prior to autoclaving, distribute 9.0mL volumes into test tubes containing inverted Durham tubes. Autoclave for 15 min at 15 psi pressure–121°C. Cool to 25°C. Add 1.0mL of sterile carbohydrate solution to each tube.

Caution: Acid Fuchsin is a potential carcinogen and care must be taken to avoid inhalation of the powdered dye and contact with the skin.

Use: For the determination of carbohydrate fermentation reactions of microorganisms, particularly members of the Enterobacteriaceae. A Durham tube is used to collect gas produced during the fermentation reaction. Acid production is indicated by a pink color.

Anisoin Minimal Medium

Composition per liter:

$KH_2PO_4 \cdot 3H_2O$	3.8g
K_2HPO_4	2.1g
NH_4Cl	2.0g
$MgSO_4 \cdot 7H_2O$	0.3g
Anisoin	0.136g
NaCl	0.1g
Trace elements solution	1.0mL

pH 6.7 ± 0.2 at 25°C

Trace Elements Solution:
Composition per liter:

$Fe_2(SO_4)_3 \cdot H_2O$	0.6g
$CoSO_4 \cdot 7H_2O$	0.2g
$CuSO_4 \cdot 5H_2O$	0.2g
$MnSO_4 \cdot H_2O$	0.2g
$ZnSO_4 \cdot 7H_2O$	0.2g

Preparation of Trace Elements Solution: Add components to distilled/deionized water and bring volume to 1.0L. Mix thoroughly.

Preparation of Medium: Add components to distilled/deionized water and bring volume to 1.0L. Mix thoroughly. Distribute into tubes or flasks. Autoclave for 15 min at 15 psi pressure–121°C.

Use: For the cultivation of *Pseudomonas fluorescens.*

ANO2 Fungus II

See: ***Neocallimastix*** **Medium**

Anoxybacillus amylolyticus Medium (DSMZ Medium 1046)

Composition per liter:

Yeast extract	6.0g
NaCl	6.0g

pH 5.6 ± 0.2 at 25°C

Preparation of Medium: Add components to tap water and bring volume to 1.0L. Mix thoroughly. Gently heat and bring to boiling. Distribute into tubes or flasks. Autoclave for 15 min at 15 psi pressure–121°C.

Use: For the cultivation and maintenance of *Anoxybacillus amylolyticus.*

Anoxybacillus Medium (DSMZ Medium 898)

Composition per liter:

$NaHCO_3$	10.0g
NaCl	5.0g
Na_2CO_3	2.76g
NH_4Cl	1.0g
Yeast extract	0.5g
KH_2PO_4	0.2g
KCl	0.2g
$MgCl_2 \cdot 6H_2O$	0.1g
Resazurin	0.5mg
Glucose solution	50.0mL

Vitamin solution	10.0mL
$Na_2S·9H_2O$ solution	10.0mL
Trace elements solution SL-10	1.0mL

pH 9.5–9.7 at 25°C

$Na_2S·9H_2O$ Solution:

Composition per 10.0mL:

$Na_2S·9H_2O$	0.5g

Preparation of $Na_2S·9H_2O$ Solution: Add $Na_2S·9H_2O$ to distilled/deionized water and bring volume to 10.0mL. Mix thoroughly. Sparge with 100% N_2. Autoclave for 15 min at 15 psi pressure–121°C. Before use, neutralize to pH 7.0 with sterile HCl.

Glucose Solution:

Composition per 50.0mL:

Glucose	5.0g

Preparation of Glucose Solution: Add glucose to distilled/deionized water and bring volume to 50.0mL. Mix thoroughly. Sparge with 100% N_2. Filter sterilize.

Vitamin Solution:

Composition per liter:

Pyridoxine-HCl	10.0mg
Thiamine-HCl·$2H_2O$	5.0mg
Riboflavin	5.0mg
Nicotinic acid	5.0mg
D-Ca-pantothenate	5.0mg
p-Aminobenzoic acid	5.0mg
Lipoic acid	5.0mg
Biotin	2.0mg
Folic acid	2.0mg
Vitamin B_{12}	0.1mg

Preparation of Vitamin Solution: Add components to distilled/deionized water and bring volume to 1.0L. Mix thoroughly. Sparge with 80% H_2 + 20% CO_2. Filter sterilize.

Trace Elements Solution SL-10:

Composition per liter:

$FeCl_2·4H_2O$	1.5g
$CoCl_2·6H_2O$	190.0mg
$MnCl_2·4H_2O$	100.0mg
$ZnCl_2$	70.0mg
$Na_2MoO_4·2H_2O$	36.0mg
$NiCl_2·6H_2O$	24.0mg
H_3BO_3	6.0mg
$CuCl_2·2H_2O$	2.0mg
HCl (25% solution)	10.0mL

Preparation of Trace Elements Solution SL-10: Add $FeCl_2·4H_2O$ to 10.0mL of HCl solution. Mix thoroughly. Add distilled/deionized water and bring volume to 1.0L. Add remaining components. Mix thoroughly. Sparge with 100% N_2. Autoclave for 15 min at 15 psi pressure–121°C.

Preparation of Medium: Prepare and dispense medium under 100% N_2. Add components, except glucose solution, $Na_2S·9H_2O$ solution, and vitamin solution, to distilled/deionized water and bring volume to 930.0mL. Mix thoroughly. Sparge with 100% N_2 for 30–60 min. Adjust pH to 8.0–8.5 with NaOH. Autoclave for 15 min at 15 psi pressure–121°C. Aseptically and anaerobically add 50.0mL sterile glucose solution, 10.0mL sterile $Na_2S·9H_2O$ solution, and 10.0mL sterile vitamin solution. Mix thoroughly. The final pH should be 9.5–9.7. Aseptically and anaerobically under 100% N_2 distribute into sterile tubes or bottles.

Use: For the cultivation of *Anoxybacillus pushchinoensis (Anoxybacillus pushchinensis).*

Anoxynatronum Medium (DSMZ Medium 1187)

Composition per liter:

KCl	0.2g
NH_4Cl	0.5g
K_2HPO_4	0.2g
$MgCl_2·6H_2O$	0.1g
$NaHCO_3$ solution	50.0mL
Na_2CO_3 solution	50.0mL
Glucose solution	50.0mL
$Na_2S·9H_2O$ solution	10.0mL
Yeast extract solution	10.0mL
Trace elements solution SL-10	1.0mL
Vitamin solution	1.0mL

pH 9.0 ± 0.2 at 25°C

Na_2CO_3 Solution:

Composition per 50.0mL:

Na_2CO_3	25.0g

Preparation of Na_2CO_3 Solution: Add Na_2CO_3 to distilled/deionized water and bring volume to 50.0mL. Mix thoroughly. Sparge with N_2. Autoclave for 15 min at 15 psi pressure–121°C. Cool to room temperature.

$NaHCO_3$ Solution:

Composition per 50.0mL:

$NaHCO_3$	25.0g

Preparation of $NaHCO_3$ Solution: Add $NaHCO_3$ to distilled/deionized water and bring volume to 50.0mL. Mix thoroughly. Sparge with N_2. Autoclave for 15 min at 15 psi pressure–121°C. Cool to room temperature.

Yeast Extract Solution:

Composition per 10.0mL:

Yeast extract	0.2g

Preparation of Yeast Extract Solution: Add yeast extract to distilled/deionized water and bring volume to 10.0mL. Mix thoroughly. Autoclave for 15 min at 15 psi pressure–121°C. Cool to room temperature.

Glucose Solution:

Composition per 50.0mL:

Glucose	5.0g

Preparation of Glucose Solution: Add glucose to distilled/deionized water and bring volume to 50.0mL. Mix thoroughly. Sparge with 100% N_2. Filter sterilize.

Vitamin Solution:

Composition per liter:

Pyridoxine-HCl	10.0mg
Thiamine-HCl·$2H_2O$	5.0mg
Riboflavin	5.0mg
Nicotinic acid	5.0mg
D-Ca-pantothenate	5.0mg
p-Aminobenzoic acid	5.0mg
Lipoic acid	5.0mg
Biotin	2.0mg
Folic acid	2.0mg
Vitamin B_{12}	0.1mg

Preparation of Vitamin Solution: Add components to distilled/deionized water and bring volume to 1.0L. Mix thoroughly. Sparge with 80% H_2 + 20% CO_2. Filter sterilize.

$Na_2S{\cdot}9H_2O$ Solution:
Composition per 10.0mL:

$Na_2S{\cdot}9H_2O$	0.7g

Preparation of $Na_2S{\cdot}9H_2O$ Solution: Add $Na_2S{\cdot}9H_2O$ to distilled/deionized water and bring volume to 10.0mL. Mix thoroughly. Autoclave under 100% N_2 for 15 min at 15 psi pressure–121°C. Cool to room temperature.

Trace Elements Solution SL-10:
Composition per liter:

$FeCl_2{\cdot}4H_2O$	1.5g
$CoCl_2{\cdot}6H_2O$	190.0mg
$MnCl_2{\cdot}4H_2O$	100.0mg
$ZnCl_2$	70.0mg
$Na_2MoO_4{\cdot}2H_2O$	36.0mg
$NiCl_2{\cdot}6H_2O$	24.0mg
H_3BO_3	6.0mg
$CuCl_2{\cdot}2H_2O$	2.0mg
HCl (25% solution)	10.0mL

Preparation of Trace Elements Solution SL-10: Add $FeCl_2{\cdot}4H_2O$ to 10.0mL of HCl solution. Mix thoroughly. Add distilled/deionized water and bring volume to 1.0L. Add remaining components. Mix thoroughly. Sparge with 100% N_2. Autoclave for 15 min at 15 psi pressure–121°C.

Preparation of Medium: Add components, except vitamin solution, yeast extract solution, glucose solution, $NaHCO_3$ solution, Na_3CO_3 solution, and $Na_2S{\cdot}9H_2O$ solution, to distilled/deionized water and bring volume to 830.0mL. Mix thoroughly. Gently heat and bring to boiling. Cool to room temperature while sparging with 100% N_2. Dispense into tubes or bottles. Autoclave for 15 min at 15 psi pressure–121°C. Cool to 25°C under 100% N_2. Aseptically and anaerobically add sterile vitamin solution, yeast extract solution, glucose solution, $NaHCO_3$ solution, Na_3CO_3 solution, and $Na_2S{\cdot}9H_2O$ solution. The final pH should be 9.0.

Use: For the cultivation and maintenance of *Anoxynatronum* spp.

Anthracis Chromogenic Agar

Composition per liter:
Proprietary.

Source: This medium is available as a premixed powder from BIOSYNTH International, Inc.

Preparation of Medium: Per manufacturer's directions.

Use: For the rapid identification and isolation of *Bacillus anthracis* based on the detection of phosphatidylcholine-specific phospholipase C activity by 5-bromo–4-chloro–3-indoxyl-cholinphosphate hydrolysis. The medium incorporates chromogenic substrates for detecting specific enzyme activities in *Bacillus anthracis, B. cereus,* and *B. thuringiensis*. The enzymes targeted by the chromogenic medium are not present in other *Bacillus* species, allowing for specific isolation of these three *Bacillus* species. Inclusion of inhibitory compounds into the medium prevents the growth of environmental contaminants. The use of proprietary chromogenic substrates, X-IP and X-CP, allows for the differentiation of *Bacillus anthracis* from near-neighbors *B. cereus* and *B. thuringiensis.* Cream to pale teal-blue colored of *Bacillus anthracis* after 20–24h, teal-blue colonies of *Bacillus anthracis* after 36–48 h at 35–37°C. Dark teal-blue colonies of *Bacillus cereus/Bacillus thuringiensis* after 20–24h at 35–37°C.

Anthranilic Acid Medium, Revised

Composition per 1040.0mL:

Na_2HPO_4	6.0g
KH_2PO_4	3.0g
NH_4Cl	1.0g
NaCl	0.5g
Glucose solution	25.0mL
$CaCl_2$ solution	10.0mL
$MgSO_4$ solution	10.0mL
Anthranilic acid solution	5.0mL

Glucose Solution:
Composition per 100.0mL:

D-Glucose	20.0g

Preparation of Glucose Solution: Add D-glucose to distilled/deionized water and bring volume to 100.0mL. Mix thoroughly. Filter sterilize.

$CaCl_2$ Solution:
Composition per 100.0mL:

$CaCl_2{\cdot}2H_2O$	0.147g

Preparation of $CaCl_2$ Solution: Add $CaCl_2$ to distilled/deionized water and bring volume to 100.0mL. Mix thoroughly. Filter sterilize.

$MgSO_4$ Solution:
Composition per 10.0mL:

$MgSO_4{\cdot}7H_2O$	2.47g

Preparation of $MgSO_4$ Solution: Add $MgSO_4{\cdot}7H_2O$ to distilled/deionized water and bring volume to 100.0mL. Mix thoroughly. Filter sterilize.

Anthranilic Acid Solution:
Composition per 100.0mL:

Anthranilic acid	1.0g
Ethanol (95% solution)	100.0mL

Preparation of Anthranilic Acid Solution: Add anthranilic acid to 100.0mL of ethanol. Mix thoroughly. Filter sterilize.

Preparation of Medium: Add components, except glucose solution, $CaCl_2$ solution, $MgSO_4$ solution, and anthranilic acid solution, to distilled/deionized water and bring volume to 990.0mL. Mix thoroughly. Gently heat and bring to boiling. Autoclave for 15 min at 15 psi pressure–121°C. Cool to 45°–50°C. Aseptically add 25.0mL of sterile glucose solution, 10.0mL of sterile $CaCl_2$ solution, 10.0mL of sterile $MgSO_4$ solution, and 5.0mL of sterile anthranilic acid solution. Mix thoroughly. Pour into sterile Petri dishes or distribute into sterile tubes.

Use: For the cultivation of *Escherichia coli.*

Antibiotic Assay Medium No. 1 (Seed Agar)

Composition per liter:

Agar	15.0g
Peptone	6.0g
Casein enzymatic hydrolysate	4.0g
Yeast extract	3.0g
Beef extract	1.5g
Glucose	1.0g

pH 6.6 ± 0.2 at 25°C

Source: This medium is available as a premixed powder from HiMedia.

Preparation of Medium: Add components to distilled/deionized water and bring volume to 1.0L. Mix thoroughly. Gently heat and bring to boiling. Distribute into tubes or flasks. Autoclave for 15 min at 15 psi pressure–121°C. Pour into sterile Petri dishes or leave in tubes.

Use: For antibiotic assay testing. Widely employed as seed agar in the preparation of plates for microbiological agar diffusion antibiotic assays.

Antibiotic Assay Medium No. 2 (Base Agar)

Composition per liter:

Agar 15.0g
Peptone 6.0g
Yeast extract 3.0g
Beef extract 1.5g

pH 6.6 ± 0.2 at 25°C

Source: This medium is available as a premixed powder from HiMedia.

Preparation of Medium: Add components to distilled/deionized water and bring volume to 1.0L. Mix thoroughly. Gently heat and bring to boiling. Distribute into tubes or flasks. Autoclave for 15 min at 15 psi pressure–121°C. Pour into sterile Petri dishes or leave in tubes.

Use: For antibiotic assay testing. For use as a base layer in antibiotic assay testing. Especially useful for the plate assay of bacitracin and penicillin G.

Antibiotic Assay Medium No. 3 (Assay Broth)

Composition per liter:

Peptone 5.0g
K_2HPO_4 .... 3.68g
NaCl 3.5g
Beef extract 1.5g
Yeast extract 1.5g
KH_2PO_4 .... 1.32g
Glucose 1.0g

pH 7.0 ± 0.2 at 25°C

Source: This medium is available as a premixed powder from HiMedia.

Preparation of Medium: Add components to distilled/deionized water and bring volume to 1.0L. Mix thoroughly. Gently heat and bring to boiling. Distribute into tubes or flasks. Autoclave for 15 min at 15 psi pressure–121°C.

Use: For antibiotic assay testing. Used for the serial dilution assay of penicillins and other antibiotics. Used in the turbidimetric assay of penicillin and tetracycline with *Staphylococcus aureus*.

Antibiotic Assay Medium No. 4 (Yeast Beef Agar)

Composition per liter:

Agar 15.0g
Peptone 6.0g
Yeast extract 3.0g
Beef extract 1.5g
Glucose 1.0g

pH 6.6 ± 0.2 at 25°C

Source: This medium is available as a premixed powder from HiMedia.

Preparation of Medium: Add components to distilled/deionized water and bring volume to 1.0L. Mix thoroughly. Gently heat and bring to boiling. Distribute into tubes or flasks. Autoclave for 15 min at 15 psi pressure–121°C. Pour into sterile Petri dishes or leave in tubes.

Use: For antibiotic assay testing.

Antibiotic Assay Medium No. 5 (Streptomycin Assay Agar with Yeast Extract)

Composition per liter:

Agar 15.0g
Peptone 6.0g
Yeast extract 3.0g
Beef extract 1.5g

pH 7.9 ± 0.2 at 25°C

Source: This medium is available as a premixed powder from HiMedia.

Preparation of Medium: Add components to distilled/deionized water and bring volume to 1.0L. Mix thoroughly. Gently heat and bring to boiling. Distribute into tubes or flasks. Autoclave for 15 min at 15 psi pressure–121°C. Pour into sterile Petri dishes or leave in tubes.

Use: For antibiotic assay testing. For the streptomycin assay using the cylinder plate technique and *Bacillus subtilis* as test organism.

Antibiotic Assay Medium No. 6

Composition per liter:

Casein enzymatic hydrolysate 17.0g
NaCl 5.0g
Papaic digest of soybean meal 3.0g
Glucose 2.5g
K_2HPO_4 .... 2.5g
$MnSO_4 \cdot H_2O$ 0.03g

pH 7.0 ± 0.1 at 25°C

Source: This medium is available as a premixed powder from HiMedia.

Preparation of Medium: Add components to distilled/deionized water and bring volume to 1.0L. Mix thoroughly. Gently heat and bring to boiling. Distribute into tubes or flasks. Autoclave for 15 min at 15 psi pressure–121°C. Pour into sterile Petri dishes.

Use: For antibiotic assay testing. For inoculum development and spore induction of *Bacillus subtilis* for antibiotic assays.

Antibiotic Assay Medium No. 8 (Base Agar with low pH)

Composition per liter:

Agar 15.0g
Peptone 6.0g
Yeast extract 3.0g
Beef extract 1.5g

pH 5.9 ± 0.1 at 25°C

Source: This medium is available as a premixed powder from HiMedia.

Preparation of Medium: Add components to distilled/deionized water and bring volume to 1.0L. Mix thoroughly. Gently heat and bring to boiling. Distribute into tubes or flasks. Autoclave for 15 min at 15 psi pressure–121°C. Pour into sterile Petri dishes.

Use: For antibiotic assay testing. For use as the base agar and the seed agar in the plate assay of tetracycline. For use as the seed agar in the plate assay of vancomycin, mitomycin, and mithramycin.

Antibiotic Assay Medium No. 9 (Polymyxin Base Agar)

Composition per liter:

Agar	20.0g
Casein enzymatic hydrolysate	17.0g
NaCl	5.0g
Papaic digest of soybean meal	3.0g
K_2HPO_4	2.5g
Glucose	2.5g

pH 7.2 ± 0.1 at 25°C

Source: This medium is available as a premixed powder from HiMedia.

Preparation of Medium: Add components to distilled/deionized water and bring volume to 1.0L. Mix thoroughly. Gently heat and bring to boiling. Distribute into tubes or flasks. Autoclave for 15 min at 15 psi pressure–121°C. Pour into sterile Petri dishes.

Use: For antibiotic assay testing. For base agar for the plate assay of carbenicillin, colistimethate, and polymyxin B.

Antibiotic Assay Medium No. 10 (Polymyxin Seed Agar)

Composition per liter:

Casein enzymatic hydrolysate	17.0g
Agar	12.0g
NaCl	5.0g
Papaic digest of soybean meal	3.0g
K_2HPO_4	2.5g
Glucose	2.5g

pH 7.2 ± 0.2 at 25°C

Source: This medium, without polysorbate 80, is available as a premixed powder from HiMedia.

Preparation of Medium: Add components to distilled/deionized water and bring volume to 1.0L. Mix thoroughly. Gently heat and bring to boiling. Distribute into tubes or flasks. Autoclave for 15 min at 15 psi pressure–121°C. Pour into sterile Petri dishes.

Use: For antibiotic assay testing. For seed agar for the plate assay of carbenicillin, colistimethate, and polymyxin B.

Antibiotic Assay Medium No. 11 (Neomycin, Erythromycin Assay Agar)

Composition per liter:

Agar	15.0g
Peptone	6.0g
Casein enzymatic hydrolysate	4.0g
Yeast extract	3.0g
Beef extract	1.5g
Glucose	1.0g

pH 8.3 ± 0.2 at 25°C

Source: This medium, without polysorbate 80, is available as a premixed powder from HiMedia.

Preparation of Medium: Add components to distilled/deionized water and bring volume to 1.0L. Mix thoroughly. Gently heat and bring to boiling. Distribute into tubes or flasks. Autoclave for 15 min at 15 psi pressure–121°C. Pour into sterile Petri dishes.

Use: For antibiotic assay testing. For analyzing the neomycin content in pharmaceutical peparations.

Antibiotic Assay Medium No. 12 (Nystatin Assay Agar)

Composition per liter:

Agar	25.0g
Peptone	10.0g
Glucose	10.0g
NaCl	10.0g
Yeast extract	5.0g
Beef extract	2.5g

pH 6.0 ± 0.1 at 25°C

Source: This medium is available as a premixed powder from HiMedia.

Preparation of Medium: Add components to distilled/deionized water and bring volume to 1.0L. Mix thoroughly. Gently heat and bring to boiling. Distribute into tubes or flasks. Autoclave for 15 min at 15 psi pressure–121°C. Pour into sterile Petri dishes.

Use: For antibiotic assay effectiveness testing. For the assay of antifungal antibiotics like amphotericin and nystatin.

Antibiotic Assay Medium No. 13

Composition per liter:

Glucose	20.0g
Peptone	10.0g

pH 5.6 ± 0.2 at 25°C

Source: This medium is available as a premixed powder from HiMedia.

Preparation of Medium: Add components to distilled/deionized water and bring volume to 1.0L. Mix thoroughly. Gently heat and bring to boiling. Distribute into tubes or flasks. Autoclave for 15 min at 15 psi pressure–121°C.

Use: For testing the effectivness of antibiotics on yeast and molds.

Antibiotic Assay Medium No. 19

Agar	23.5g
Glucose	10.0g
NaCl	10.0g
Peptone	9.4g
Yeast extract	4.7g
Beef extract	2.4g

pH 6.1 ± 0.2 at 25°C

Source: This medium is available as a premixed powder from HiMedia.

Preparation of Medium: Add components to distilled/deionized water and bring volume to 1.0L. Mix thoroughly. Gently heat and bring to boiling. Distribute into tubes or flasks. Autoclave for 15 min at 15 psi pressure–121°C. Pour into sterile Petri dishes.

Use: For assaying the mycostatic activity of pharmaceutical preparations. For seed agar for the plate assay to test the effectiveness of nystatin, amphotericin B, and natamycin.

Antibiotic Assay Medium No. 20 (Yeast Beef Broth)

Composition per liter:

Peptone	15.0g
Glucose	11.0g
Yeast extract	6.5g
K_2HPO_4	3.68g
NaCl	3.5g
Beef extract	1.5g
KH_2PO_4	1.32g

pH 6.6 ± 0.2 at 25°C

Source: This medium is available as a premixed powder from HiMedia.

Preparation of Medium: Add components to distilled/deionized water and bring volume to 1.0L. Mix thoroughly. Gently heat and bring to boiling. Distribute into tubes or flasks. Autoclave for 15 min at 15 psi pressure–121°C.

Use: For assaying the mycostatic activity of pharmaceutical preparations.

Antibiotic Assay Medium No. 32

Composition per liter:

Agar	15.0g
Peptone	6.0g
Casein enzymatic hydrolysate	4.0g
Yeast extract	3.0g
Beef extract	1.5g
Glucose	1.0g
$MnSO_4 \cdot 4H_2O$	0.3g

pH 6.6 ± 0.2 at 25°C

Source: This medium is available as a premixed powder from HiMedia.

Preparation of Medium: Add components to distilled/deionized water and bring volume to 1.0L. Mix thoroughly. Gently heat and bring to boiling. Distribute into tubes or flasks. Autoclave for 15 min at 15 psi pressure–121°C. Pour into sterile Petri dishes.

Use: For preparing inoculum of *Bacillus subtilis* ATCC 6633 during assay of dihydrostreptomycin and vancomycin.

Antibiotic Assay Medium No. 34

Composition per liter:

Peptone	10.0g
Beef extract	10.0g
Glycerol	10.0g
NaCl	3.0g

pH 7.0 ± 0.2 at 25°C

Source: This medium is available as a premixed powder from HiMedia.

Preparation of Medium: Add components to distilled/deionized water and bring volume to 1.0L. Mix thoroughly. Gently heat and bring to boiling. Distribute into tubes or flasks. Autoclave for 15 min at 15 psi pressure–121°C. Pour into sterile Petri dishes.

Use: For antibiotic assay effectiveness testing of bleomycin using *Mycobacterium smegmatis* ATCC 607.

Antibiotic Assay Medium No. 35

Composition per liter:

Agar	17.0
Peptone	10.0g
Beef extract	10.0g
NaCl	3.0g

pH 7.0 ± 0.2 at 25°C

Source: This medium is available as a premixed powder from HiMedia.

Preparation of Medium: Add components to distilled/deionized water and bring volume to 1.0L. Mix thoroughly. Gently heat and bring to boiling. Distribute into tubes or flasks. Autoclave for 15 min at 15 psi pressure–121°C. Pour into sterile Petri dishes.

Use: For antibiotic assay effectiveness testing of bleomycin using *Mycobacterium smegmatis* ATCC 607.

Antibiotic Assay Medium No. 36

Composition per liter:

Agar	15.0g
Casein enzymatic hydrolysate	15.0g
NaCl	5.0g
Papaic digest of soybean meal	5.0g

pH 7.3 ± 0.2 at 25°C

Source: This medium is available as a premixed powder from HiMedia.

Preparation of Medium: Add components to distilled/deionized water and bring volume to 1.0L. Mix thoroughly. Gently heat and bring to boiling. Distribute into tubes or flasks. Autoclave for 15 min at 15 psi pressure–121°C. Pour into sterile Petri dishes.

Use: A general purpose medium for cultivating a wide variety of fastidious microorganisms.

Antibiotic Assay Medium No. 37

Composition per liter:

Casein enzymatic hydrolysate	17.0g
NaCl	5.0g
Papaic digest of soybean meal	3.0g
Glucose	2.5g
K_2HPO_4	2.5g

pH 7.3 ± 0.2 at 25°C

Source: This medium is available as a premixed powder from HiMedia.

Preparation of Medium: Add components to distilled/deionized water and bring volume to 1.0L. Mix thoroughly. Distribute into tubes or flasks. Autoclave for 10 min at 15 psi pressure–121°C.

Use: A general purpose medium for cultivating a wide variety of fastidious microorganisms.

Antibiotic Assay Medium No. 38

Composition per liter:

Agar	15.0g
Peptone	15.0g
Glucose	5.5g
Papaic digest of soybean meal	5.0g

NaCl	4.0g
L-Cysteine·HCl·H_2O	0.7g
Na_2SO_3	0.2g

pH 7.0 ± 0.2 at 25°C

Source: This medium is available as a premixed powder from HiMedia.

Preparation of Medium: Add components to distilled/deionized water and bring volume to 1.0L. Mix thoroughly. Gently heat and bring to boiling. Distribute into tubes or flasks. Autoclave for 15 min at 15 psi pressure–121°C. Pour into sterile Petri dishes.

Use: For microbiological assay of ticarcillin using *Pseudomonas aeruginosa* ATCC 29336.

Antibiotic Assay Medium No. 39

Composition per liter:

Peptone	5.0g
K_2HPO_4	3.68g
NaCl	3.5g
Beef extract	1.5g
Yeast extract	1.5g
KH_2PO_4	1.32g
Glucose	1.0g

pH 7.9 ± 0.2 at 25°C

Source: This medium is available as a premixed powder from HiMedia.

Preparation of Medium: Add components to distilled/deionized water and bring volume to 1.0L. Mix thoroughly. Gently heat and bring to boiling. Distribute into tubes or flasks. Autoclave for 15 min at 15 psi pressure–121°C.

Use: For the microbiological assay of neomycin and streptomycin using *Klebsiella pneumoniae* ATCC 10031 as the test organism.

Antibiotic Assay Medium No. 40

Composition per liter:

Yeast extract	20.0g
Agar	10.0g
Glucose	10.0g
Casein enzymatic hydrolysate	2.5g
Peptone	2.5g
KH_2PO_4	2.0g
Tween™ 80	0.1g

pH 6.7 ± 0.2 at 25°C

Source: This medium is available as a premixed powder from HiMedia.

Preparation of Medium: Add components to distilled/deionized water and bring volume to 1.0L. Mix thoroughly. Gently heat and bring to boiling. Distribute into tubes or flasks. Autoclave for 15 min at 15 psi pressure–121°C. Pour into sterile Petri dishes.

Use: For the microbiological assay of Thiostrepton using *Streptococcus faecium* ATCC 10541.

Antibiotic Assay Medium No. 41

Composition per liter:

Glucose	20.0g
Sodium citrate	10.0g
Casein enzymatic hydrolysate	9.0g
Yeast extract	5.0g
K_2HPO_4	1.0g
KH_2PO_4	1.0g

pH 6.8 ± 0.2 at 25°C

Source: This medium is available as a premixed powder from HiMedia.

Preparation of Medium: Add components to distilled/deionized water and bring volume to 1.0L. Mix thoroughly. Distribute into tubes or flasks. Autoclave for 10 min at 15 psi pressure–121°C.

Use: For the microbiological assay of thiostrepton using *Streptococcus faecium* ATCC 10541.

Antibiotic Assay Medium B

Composition per liter:

Casein enzymatic hydrolysate	17.0g
Agar	15.0g
Glucose	5.0g
NaCl	5.0g
Papaic digest of soybean meal	3.0g
K_2HPO_4	2.5g

pH 7.3 ± 0.2 at 25°C

Preparation of Medium: Add components to distilled/deionized water and bring volume to 1.0L. Mix thoroughly. Gently heat and bring to boiling. Distribute into tubes or flasks. Autoclave for 15 min at 15 psi pressure–121°C. Pour into sterile Petri dishes.

Use: For the microbiological assay of colistimethate using *Bordetella bronchiseptica* or *Escherichia coli.*.

Antibiotic Assay Medium C

Composition per liter:

Peptone	6.0g
K_2HPO_4	3.68g
NaCl	3.5g
Yeast extract	3.0g
Beef extract	1.5g
KH_2PO_4	1.32g
Glucose	1.0g

pH 7.0 ± 0.2 at 25°C

Preparation of Medium: Add components to distilled/deionized water and bring volume to 1.0L. Mix thoroughly. Distribute into tubes or flasks. Autoclave for 10 min at 15 psi pressure–121°C.

Use: For the microbiological assay of Rifampin using *Escherichia coli*, Colistimethate using *Escherichia coli*, erythromycin, framycetin, gentamicin, gramicidin, kanamycin, neomycyin, and vancomycin using *Staphylococcus aureus*, and gramicin using *Enterococcus hirae*.

Antibiotic Assay Medium D

Composition per liter:

Casein peptone	5.0g
K_2HPO_4	3.68g
NaCl	3.5g
KNO_3	2.0g
Heart extract	1.5g
Yeast extract	1.5g
KH_2PO_4	1.32g

pH 7.0 ± 0.2 at 25°C

Preparation of Medium: Add components to distilled/deionized water and bring volume to 1.0L. Mix thoroughly. Distribute into tubes or flasks. Autoclave for 10 min at 15 psi pressure–121°C.

Use: For the microbiological assay of erythromycin using *Klebsiella pneumoniae*.

Antibiotic Assay Medium E

Composition per liter:

$Na_2HPO_4 \cdot 12H_2O$	26.9g
Agar	10.0g
Peptone	5.0g
Meat extract	3.0g

pH 7.9 ± 0.2 at 25°C

Preparation of Medium: Add components to distilled/deionized water and bring volume to 1.0L. Mix thoroughly. Gently heat and bring to boiling. Distribute into tubes or flasks. Autoclave for 15 min at 15 psi pressure–121°C. Pour into sterile Petri dishes.

Use: For the microbiological assay of framycetin using *Bacillus subtilus*.

Antibiotic Assay Medium F

Composition per liter:

Agar	23.5g
Glucose	10.0g
NaCl	10.0g
Peptone	9.4g
Yeast extract	4.7g
Beef extract	2.4g

pH 6.0 ± 0.2 at 25°C

Preparation of Medium: Add components to distilled/deionized water and bring volume to 1.0L. Mix thoroughly. Gently heat and bring to boiling. Distribute into tubes or flasks. Autoclave for 15 min at 15 psi pressure–121°C. Pour into sterile Petri dishes.

Use: For the microbiological assay of nystatin using *Saccharomyces cerevisiae* or *Candida tropicalis*.

Antibiotic Assay Medium G

Composition per liter:

Agar	15.0g
Meat extract	10.0g
Peptone	10.0g
NaCl	3.0g

pH 7.0 ± 0.2 at 25°C

Preparation of Medium: Add components to distilled/deionized water and bring volume to 1.0L. Mix thoroughly. Gently heat and bring to boiling. Distribute into tubes or flasks. Autoclave for 15 min at 15 psi pressure–121°C. Pour into sterile Petri dishes.

Use: For the microbiological assay of bleomycin using *Mycobacterium smegmatis*.

Antibiotic Assay Medium H

Composition per liter:

D-Glucose	10.0g
Casein enzymatic hydrolysate	6.0g
Yeast extract	2.0g

pH 8.0 ± 0.2 at 25°C

Preparation of Medium: Add components to distilled/deionized water and bring volume to 1.0L. Mix thoroughly. Distribute into tubes or flasks. Autoclave for 10 min at 15 psi pressure–121°C.

Use: For the microbiological assay of apramycin using *Salmonella cholerasuis*.

Antibiotic Assay Medium L-AODC

Composition per liter:

Agar	15.0g
Glucose, anhydrous	10.0g
Yeast extract	2.5g
K_2HPO_4	0.69g
KH_2PO_4	0.45g

pH 6.0 ± 0.2 at 25°C

Preparation of Medium: Add components to distilled/deionized water and bring volume to 1.0L. Mix thoroughly. Gently heat and bring to boiling. Distribute into tubes or flasks. Autoclave for 15 min at 15 psi pressure–121°C. Pour into sterile Petri dishes.

Use: For the microbiological assay of monensin using *Bacillus subtilus*.

Antibiotic Assay Medium M-AODC

Composition per liter:

Agar	20.0g
Glucose	10.0g
Yeast extract	2.5g
K_2HPO_4	0.69g
KH_2PO_4	0.45g

pH 6.0 ± 0.2 at 25°C

Preparation of Medium: Add components to distilled/deionized water and bring volume to 1.0L. Mix thoroughly. Gently heat and bring to boiling. Distribute into tubes or flasks. Autoclave for 15 min at 15 psi pressure–121°C. Pour into sterile Petri dishes.

Use: For the microbiological assay of monensin using *Bacillus subtilus*.

Antibiotic HiVeg Assay Medium No. 1 (Antibiotic HiVeg Assay Medium - A) (Seed HiVeg Agar)

Composition per liter:

Agar	15.0g
Plant peptone	6.0g
Plant hydrolysate	4.0g
Yeast extract	3.0g
Plant extract	1.5g
Glucose	1.0g

pH 6.6 ± 0.2 at 25°C

Source: This medium is available as a premixed powder from HiMedia.

Preparation of Medium: Add components to distilled/deionized water and bring volume to 1.0L. Mix thoroughly. Gently heat and bring to boiling. Distribute into tubes or flasks. Autoclave for 15 min at 15 psi pressure–121°C. Pour into sterile Petri dishes or leave in tubes.

Use: For antibiotic assay testing. Widely employed as seed agar in the preparation of plates for microbiological agar diffusion antibiotic assays.

Antibiotic HiVeg Assay Medium No. 2 (Antibiotic HiVeg Assay Medium - B) (Seed HiVeg Agar)

Composition per liter:

Agar	15.0g
Plant peptone	6.0g

Yeast extract 3.0g
Plant extract 1.5g

pH 6.6 ± 0.2 at 25°C

Source: This medium is available as a premixed powder from HiMedia.

Preparation of Medium: Add components to distilled/deionized water and bring volume to 1.0L. Mix thoroughly. Gently heat and bring to boiling. Distribute into tubes or flasks. Autoclave for 15 min at 15 psi pressure–121°C. Pour into sterile Petri dishes or leave in tubes.

Use: For antibiotic assay testing. For use as a base layer in antibiotic assay testing. Especially useful for the plate assay of bacitracin and penicillin G.

Antibiotic HiVeg Assay Medium No. 3 (Antibiotic HiVeg Assay Medium - C)

Composition per liter:

Plant peptone 5.0g
K_2HPO_4 .... 3.68g
NaCl 3.5g
Yeast extract 1.5g
Plant extract 1.5g
KH_2PO_4 .... 1.32g
Glucose 1.0g

pH 7.0 ± 0.2 at 25°C

Source: This medium is available as a premixed powder from HiMedia.

Preparation of Medium: Add components to distilled/deionized water and bring volume to 1.0L. Mix thoroughly. Gently heat and bring to boiling. Distribute into tubes or flasks. Autoclave for 15 min at 15 psi pressure–121°C.

Use: For antibiotic assay testing. Used for the serial dilution assay of penicillins and other antibiotics. Used in the turbidimetric assay of penicillin and tetracycline with *Staphylococcus aureus*.

Antibiotic HiVeg Assay Medium No. 4 (Yeast Beef HiVeg Agar)

Composition per liter:

Agar 15.0g
Plant peptone 6.0g
Yeast extract 3.0g
Plant extract 1.5g
Glucose 1.0g

pH 6.6 ± 0.2 at 25°C

Source: This medium is available as a premixed powder from HiMedia.

Preparation of Medium: Add components to distilled/deionized water and bring volume to 1.0L. Mix thoroughly. Gently heat and bring to boiling. Distribute into tubes or flasks. Autoclave for 15 min at 15 psi pressure–121°C. Pour into sterile Petri dishes or leave in tubes.

Use: For antibiotic assay testing.

Antibiotic HiVeg Assay Medium No. 5 (Streptomycin HiVeg Agar with Yeast Extract) (Antibiotic HiVeg Assay Medium - E)

Composition per liter:

Agar 15.0g
Plant peptone 6.0g
Yeast extract 3.0g
Plant extract 1.5g

pH 7.9 ± 0.2 at 25°C

Source: This medium is available as a premixed powder from HiMedia.

Preparation of Medium: Add components to distilled/deionized water and bring volume to 1.0L. Mix thoroughly. Gently heat and bring to boiling. Distribute into tubes or flasks. Autoclave for 15 min at 15 psi pressure–121°C. Pour into sterile Petri dishes or leave in tubes.

Use: For antibiotic assay testing. For the streptomycin assay using the cylinder plate technique and *Bacillus subtilis* as test organism.

Antibiotic HiVeg Assay Medium No. 6

Composition per liter:

Plant hydrolysate 17.0g
NaCl 5.0g
Papaic digest of soybean meal 3.0g
Glucose 2.5g
K_2HPO_4 .... 2.5g
$MnSO_4 \cdot H_2O$ 0.03g

pH 7.0 ± 0.1 at 25°C

Source: This medium is available as a premixed powder from HiMedia.

Preparation of Medium: Add components to distilled/deionized water and bring volume to 1.0L. Mix thoroughly. Gently heat and bring to boiling. Distribute into tubes or flasks. Autoclave for 15 min at 15 psi pressure–121°C. Pour into sterile Petri dishes.

Use: For antibiotic assay testing. For inoculum development and spore induction of *Bacillus subtilis* for antibiotic assays.

Antibiotic HiVeg Assay Medium No. 8 (Base HiVeg Agar w/ low pH) (Antibiotic HiVeg Assay Medium F)

Composition per liter:

Agar 15.0g
Plant peptone 6.0g
Yeast extract 3.0g
Plant extract 1.5g

pH 5.9 ± 0.1 at 25°C

Source: This medium is available as a premixed powder from HiMedia.

Preparation of Medium: Add components to distilled/deionized water and bring volume to 1.0L. Mix thoroughly. Gently heat and bring to boiling. Distribute into tubes or flasks. Autoclave for 15 min at 15 psi pressure–121°C. Pour into sterile Petri dishes.

Use: For antibiotic assay testing. For use as the base agar and the seed agar in the plate assay of tetracycline. For use as the seed agar in the plate assay of vancomycin, mitomycin, and mithramycin.

Antibiotic HiVeg Assay Medium No. 9 (Polymyxin HiVeg Base Agar)

Composition per liter:

Agar 20.0g
Plant hydrolysate 17.0g
NaCl 5.0g
Papaic digest of soybean meal 3.0g

K_2HPO_4	2.5g
Glucose	2.5g

pH 7.2 ± 0.1 at 25°C

Source: This medium is available as a premixed powder from HiMedia.

Preparation of Medium: Add components to distilled/deionized water and bring volume to 1.0L. Mix thoroughly. Gently heat and bring to boiling. Distribute into tubes or flasks. Autoclave for 15 min at 15 psi pressure–121°C. Pour into sterile Petri dishes.

Use: For antibiotic assay testing. For base agar for the plate assay of carbenicillin, colistimethate, and polymyxin B.

Antibiotic HiVeg Assay Medium No. 10 (Polymyxin Seed HiVeg Agar) (Antibiotic HiVeg Assay Medium H)

Composition per liter:

Plant hydrolysate	17.0g
Agar	12.0g
Polysorbate 80	10.0g
NaCl	5.0g
Papaic digest of soybean meal	3.0g
K_2HPO_4	2.5g
Glucose	2.5g

pH 7.2 ± 0.2 at 25°C

Source: This medium, without polysorbate 80, is available as a premixed powder from HiMedia.

Preparation of Medium: Add components to distilled/deionized water and bring volume to 1.0L. Mix thoroughly. Gently heat and bring to boiling. Distribute into tubes or flasks. Autoclave for 15 min at 15 psi pressure–121°C. Pour into sterile Petri dishes.

Use: For antibiotic assay testing. For seed agar for the plate assay of carbenicillin, colistimethate, and polymyxin B.

Antibiotic HiVeg Assay Medium No. 11 (Neomycin, Erythromycin HiVeg Assay Agar)

Composition per liter:

Agar	15.0g
Plant peptone	6.0g
Plant hydrolysate	4.0g
Yeast extract	3.0g
Plant extract	1.5g
Glucose	1.0g

pH 8.3 ± 0.2 at 25°C

Source: This medium, without polysorbate 80, is available as a premixed powder from HiMedia.

Preparation of Medium: Add components to distilled/deionized water and bring volume to 1.0L. Mix thoroughly. Gently heat and bring to boiling. Distribute into tubes or flasks. Autoclave for 15 min at 15 psi pressure–121°C. Pour into sterile Petri dishes.

Use: For antibiotic assay testing. For analyzing the neomycin content in pharmaceutical preparations.

Antibiotic HiVeg Assay Medium No. 12 (Nystatin HiVeg Assay Agar)

Composition per liter:

Agar	25.0g
Plant peptone	10.0g
Glucose	10.0g
NaCl	10.0g
Yeast extract	5.0g
Plant extract	2.5g

pH 6.0 ± 0.1 at 25°C

Source: This medium is available as a premixed powder from HiMedia.

Preparation of Medium: Add components to distilled/deionized water and bring volume to 1.0L. Mix thoroughly. Gently heat and bring to boiling. Distribute into tubes or flasks. Autoclave for 15 min at 15 psi pressure–121°C. Pour into sterile Petri dishes.

Use: For antibiotic assay effectiveness testing. For the assay of antifungal antibiotics like amphotericin and nystatin.

Antibiotic HiVeg Assay Medium No. 13

Composition per liter:

Glucose	20.0g
Plant peptone	10.0g

pH 5.6 ± 0.2 at 25°C

Source: This medium is available as a premixed powder from HiMedia.

Preparation of Medium: Add components to distilled/deionized water and bring volume to 1.0L. Mix thoroughly. Gently heat and bring to boiling. Distribute into tubes or flasks. Autoclave for 15 min at 15 psi pressure–121°C.

Use: For testing the effectivness of antibiotics on yeast and molds.

Antibiotic HiVeg Assay Medium No. 19 (Antibiotic HiVeg Assay Medium G)

Agar	23.5g
Glucose	10.0g
NaCl	10.0g
Plant peptone	9.4g
Yeast extract	4.7g
Plant extract	2.4g

pH 6.1 ± 0.2 at 25°C

Source: This medium is available as a premixed powder from HiMedia.

Preparation of Medium: Add components to distilled/deionized water and bring volume to 1.0L. Mix thoroughly. Gently heat and bring to boiling. Distribute into tubes or flasks. Autoclave for 15 min at 15 psi pressure–121°C. Pour into sterile Petri dishes.

Use: For assaying the mycostatic activity of pharmaceutical preparations. For seed agar for the plate assay to test the effectiveness of nystatin, amphotericin B, and natamycin.

Antibiotic HiVeg Assay Medium No. 20 (Yeast Beef HiVeg Broth)

Composition per liter:

Plant peptone	15.0g
Glucose	11.0g
Yeast extract	6.5g
K_2HPO_4	3.68g
NaCl	3.5g
Plant extract	1.5g
KH_2PO_4	1.32g

pH 6.6 ± 0.2 at 25°C

Source: This medium is available as a premixed powder from HiMedia.

Preparation of Medium: Add components to distilled/deionized water and bring volume to 1.0L. Mix thoroughly. Gently heat and bring to boiling. Distribute into tubes or flasks. Autoclave for 15 min at 15 psi pressure–121°C.

Use: For assaying the mycostatic activity of pharmaceutical preparations.

Antibiotic HiVeg Assay Medium No. 32

Composition per liter:

Agar	15.0g
Plant peptone	6.0g
Plant hydrolysate	4.0g
Yeast extract	3.0g
Plant extract	1.5g
Glucose	1.0g
$MnSO_4 \cdot 4H_2O$	0.3g

pH 6.6 ± 0.2 at 25°C

Source: This medium is available as a premixed powder from HiMedia.

Preparation of Medium: Add components to distilled/deionized water and bring volume to 1.0L. Mix thoroughly. Gently heat and bring to boiling. Distribute into tubes or flasks. Autoclave for 15 min at 15 psi pressure–121°C. Pour into sterile Petri dishes.

Use: For preparing inoculum of *Bacillus subtilis* ATCC 6633 during assay of dihydrostreptomycin and vancomycin.

Antibiotic HiVeg Assay Medium No. 35 (Antibiotic HiVeg Assay Medium - I)

Composition per liter:

Agar	17.0g
Plant peptone	10.0g
Plant extract	10.0g
Glycerol	10.0g
NaCl	3.0g

pH 7.0 ± 0.2 at 25°C

Source: This medium is available as a premixed powder from HiMedia.

Preparation of Medium: Add components to distilled/deionized water and bring volume to 1.0L. Mix thoroughly. Gently heat and bring to boiling. Distribute into tubes or flasks. Autoclave for 15 min at 15 psi pressure–121°C. Pour into sterile Petri dishes.

Use: For antibiotic assay effectiveness testing of bleomycin using *Mycobacterium smegmatis* ATCC 607.

Antibiotic HiVeg Assay Medium No. 36 (Antibiotic HiVeg Assay Medium - J)

Composition per liter:

Agar	15.0g
Plant hydrolysate	15.0g
NaCl	5.0g
Papaic digest of soybean meal	5.0g

pH 7.3 ± 0.2 at 25°C

Source: This medium is available as a premixed powder from HiMedia.

Preparation of Medium: Add components to distilled/deionized water and bring volume to 1.0L. Mix thoroughly. Gently heat and bring to boiling. Distribute into tubes or flasks. Autoclave for 15 min at 15 psi pressure–121°C. Pour into sterile Petri dishes.

Use: A general purpose medium for cultivating a wide variety of fastidious microorganisms.

Antibiotic HiVeg Assay Medium No. 37

Composition per liter:

Plant hydrolysate	15.0g
NaCl	5.0g
Papaic digest of soybean meal	5.0g

pH 7.3 ± 0.2 at 25°C

Source: This medium is available as a premixed powder from HiMedia.

Preparation of Medium: Add components to distilled/deionized water and bring volume to 1.0L. Mix thoroughly. Distribute into tubes or flasks. Autoclave for 10 min at 15 psi pressure–121°C.

Use: A general purpose medium for cultivating a wide variety of fastidious microorganisms.

Antibiotic HiVeg Assay Medium No. 38

Composition per liter:

Agar	15.0g
Plant peptone	15.0g
Glucose	5.5g
Papaic digest of soybean meal	5.0g
NaCl	4.0g
L-Cysteine·HCl·H_2O	0.7g
Na_2SO_3	0.2g

pH 7.2 ± 0.2 at 25°C

Source: This medium is available as a premixed powder from HiMedia.

Preparation of Medium: Add components to distilled/deionized water and bring volume to 1.0L. Mix thoroughly. Gently heat and bring to boiling. Distribute into tubes or flasks. Autoclave for 15 min at 15 psi pressure–121°C. Pour into sterile Petri dishes.

Use: For microbiological assay of ticarcillin using *Pseudomonas aeruginosa* ATCC 29336.

Antibiotic HiVeg Assay Medium No. 39

Composition per liter:

Plant peptone	5.0g
K_2HPO_4	3.68g
NaCl	3.5g
Plant extract	1.5g
Yeast extract	1.5g
KH_2PO_4	1.32g
Glucose	1.0g

pH 7.9 ± 0.2 at 25°C

Source: This medium is available as a premixed powder from HiMedia.

Preparation of Medium: Add components to distilled/deionized water and bring volume to 1.0L. Mix thoroughly. Gently heat and bring to boiling. Distribute into tubes or flasks. Autoclave for 15 min at 15 psi pressure–121°C.

Use: For the microbiological assay of neomycin and streptomycin using *Klebsiella pneumoniae* ATCC 10031 as the test organism.

Antibiotic HiVeg Assay Medium No. 40

Composition per liter:

Yeast extract	20.0g
Agar	10.0g
Glucose	10.0g
Plant hydrolysate	2.5g
Plant peptone	2.5g
KH_2PO_4	2.0g
Tween™ 80	0.1g

pH 6.7 ± 0.2 at 25°C

Source: This medium is available as a premixed powder from HiMedia.

Preparation of Medium: Add components to distilled/deionized water and bring volume to 1.0L. Mix thoroughly. Gently heat and bring to boiling. Distribute into tubes or flasks. Autoclave for 15 min at 15 psi pressure–121°C. Pour into sterile Petri dishes.

Use: For the microbiological assay of thiostrepton using *Streptococcus faecium* ATCC 10541.

Antibiotic HiVeg Assay Medium No. 41

Composition per liter:

Glucose	20.0g
Sodium citrate	10.0g
Plant hydrolysate	9.0g
Yeast extract	5.0g
K_2HPO_4	1.0g
KH_2PO_4	1.0g

pH 6.8 ± 0.2 at 25°C

Source: This medium is available as a premixed powder from HiMedia.

Preparation of Medium: Add components to distilled/deionized water and bring volume to 1.0L. Mix thoroughly. Distribute into tubes or flasks. Autoclave for 10 min at 15 psi pressure–121°C.

Use: For the microbiological assay of thiostrepton using *Streptococcus faecium* ATCC 10541.

Antibiotic Medium 1 (Penassay Seed Agar) (Seed Agar)/(Agar Medium A)

Composition per liter:

Agar	15.0g
Pancreatic digest of gelatin	6.0g
Pancreatic digest of casein	4.0g
Yeast extract	3.0g
Beef extract	1.5g
Glucose	1.0g

pH 6.6 ± 0.1 at 25°C

Source: This medium is available as a premixed powder from BD Diagnostic Systems.

Preparation of Medium: Add components to distilled/deionized water and bring volume to 1.0L. Mix thoroughly. Gently heat and bring to boiling. Distribute into tubes or flasks. Autoclave for 15 min at 15 psi pressure–121°C. Pour into sterile Petri dishes or leave in tubes.

Use: For antibiotic assay testing, detection of antibiotics in milk, and determination of the antimicrobial effectiveness of antibiotics.

Antibiotic Medium 1 with Tetracycline

Composition per liter:

Agar	15.0g
Pancreatic digest of gelatin	6.0g
Pancreatic digest of casein	4.0g
Yeast extract	3.0g
Beef extract	1.5g
Glucose	1.0g
Tetracycline solution	10.0mL

pH 6.6 ± 0.1 at 25°C

Tetracycline Solution:

Composition per 10.0mL:

Tetracycline	0.02g

Preparation of Tetracycline Solution: Add tetracycline to distilled/deionized water and bring volume to 10.0mL. Mix thoroughly. Filter sterilize.

Preparation of Medium: Add components, except tetracycline solution, to distilled/deionized water and bring volume to 1.0L. Mix thoroughly. Gently heat and bring to boiling. Autoclave for 15 min at 15 psi pressure–121°C. Cool to 45°–50°C. Aseptically add sterile tetracycline solution. Mix thoroughly. Pour into sterile Petri dishes or distribute into sterile tubes.

Use: For the selective cultivation and maintenance of *Salmonella choleraesuis*.

Antibiotic Medium 1 with Tetracycline, Streptomycin, and Chloramphenicol

Composition per liter:

Agar	15.0g
Pancreatic digest of gelatin	6.0g
Pancreatic digest of casein	4.0g
Yeast extract	3.0g
Beef extract	1.5g
Glucose	1.0g
Antibiotic solution	10.0mL

pH 6.6 ± 0.1 at 25°C

Antibiotic Solution:

Composition per 10.0mL:

Tetracycline	0.02g
Streptomycin	0.02g
Chloramphenicol	0.02g

Preparation of Antibiotic Solution: Add components to distilled/deionized water and bring volume to 10.0mL. Mix thoroughly. Filter sterilize.

Preparation of Medium: Add components, except antibiotic solution, to distilled/deionized water and bring volume to 1.0L. Mix thoroughly. Gently heat and bring to boiling. Autoclave for 15 min at 15 psi pressure–121°C. Cool to 45°–50°C. Aseptically add sterile antibiotic solution. Mix thoroughly. Pour into sterile Petri dishes or distribute into sterile tubes.

Use: For the selective cultivation and maintenance of *Salmonella choleraesuis*.

Antibiotic Medium 2 (Base Agar) (Penassay Base Agar)

Composition per liter:

Agar	15.0g
Pancreatic digest of gelatin	6.0g
Yeast extract	3.0g
Beef extract	1.5g

pH 6.6 ± 0.1 at 25°C

Source: This medium is available as a premixed powder from BD Diagnostic Systems and Oxoid Unipath.

Preparation of Medium: Add components to distilled/deionized water and bring volume to 1.0L. Mix thoroughly. Gently heat and bring to boiling. Distribute into tubes or flasks. Autoclave for 15 min at 15 psi pressure–121°C. Pour into sterile Petri dishes.

Use: For use as a base layer in antibiotic assay testing. Especially useful for the plate assay of bacitracin and penicillin G.

Antibiotic Medium 3 (Penassay Broth)

Composition per liter:

Pancreatic digest of gelatin	5.0g
NaCl	3.5g
Yeast extract	1.5g
Beef extract	1.5g
Glucose	1.0g
K_2HPO_4	3.68g
KH_2PO_4	1.32g

pH 7.0 ± 0.05 at 25°C

Source: This medium is available as a premixed powder from BD Diagnostic Systems and Oxoid Unipath.

Preparation of Medium: Add components to distilled/deionized water and bring volume to 1.0L. Mix thoroughly. Gently heat and bring to boiling. Distribute into tubes or flasks. Autoclave for 15 min at 15 psi pressure–121°C.

Use: For antibiotic assay testing. Used for the serial dilution assay of penicillins and other antibiotics. Used in the turbidimetric assay of penicillin and tetracycline with *Staphylococcus aureus*. For the cultivation and maintenance of *Bacillus subtilis, Salmonella choleraesuis,* and *Staphylococcus aureus.* For the cloning of plasmids in *Streptococcus mutans.*

Antibiotic Medium 3 Plus

Composition per liter:

Agar	15.0g
Peptone	5.0g
K_2HPO_4	3.68g
NaCl	3.5g
Yeast extract	2.5g
Glucose	1.75g
Beef extract	1.5g
KH_2PO_4	1.32g

pH 7.0 ± 0.05 at 25°C

Preparation of Medium: Add components to distilled/deionized water and bring volume to 1.0L. Mix thoroughly. Gently heat and bring to boiling. Distribute into tubes or flasks. Autoclave for 15 min at 15 psi pressure–121°C. Pour into sterile Petri dishes or leave in tubes.

Use: For antibiotic assay testing and for the cultivation of *Escherichia coli.*

Antibiotic Medium 4 (Yeast Beef Agar) (Agar Medium C)

Composition per liter:

Agar	15.0g
Pancreatic digest of gelatin	6.0g
Yeast extract	3.0g
Beef extract	1.5g
Glucose	1.0g

pH 6.6 ± 0.05 at 25°C

Source: This medium is available as a premixed powder from BD Diagnostic Systems.

Preparation of Medium: Add components to distilled/deionized water and bring volume to 1.0L. Mix thoroughly. Gently heat and bring to boiling. Distribute into tubes or flasks. Autoclave for 15 min at 15 psi pressure–121°C. Pour into sterile Petri dishes or leave in tubes.

Use: For antibiotic assay testing.

Antibiotic Medium 5 (Streptomycin Assay Agar with Yeast Extract)

Composition per liter:

Agar	15.0g
Pancreatic digest of gelatin	6.0g
Yeast extract	3.0g
Beef extract	1.5g

pH 7.9 ± 0.1 at 25°C

Source: This medium is available as a premixed powder from BD Diagnostic Systems and Oxoid Unipath.

Preparation of Medium: Add components to distilled/deionized water and bring volume to 1.0L. Mix thoroughly. Gently heat and bring to boiling. Distribute into tubes or flasks. Autoclave for 15 min at 15 psi pressure–121°C. Pour into sterile Petri dishes.

Use: For antibiotic assay testing. For the streptomycin assay using the cylinder plate technique and *Bacillus subtilis* as test organism.

Antibiotic Medium 6

Composition per liter:

Pancreatic digest of casein	17.0g
NaCl	5.0g
Papaic digest of soybean meal	3.0g
Glucose	2.5g
K_2HPO_4	2.5g
$MnSO_4 \cdot H_2O$	0.03g

pH 7.0 ± 0.1 at 25°C

Source: This medium is available as a premixed powder from BD Diagnostic Systems.

Preparation of Medium: Add components to distilled/deionized water and bring volume to 1.0L. Mix thoroughly. Gently heat and bring to boiling. Distribute into tubes or flasks. Autoclave for 15 min at 15 psi pressure–121°C. Pour into sterile Petri dishes.

Use: For antibiotic assay testing.

Antibiotic Medium 7

Composition per liter:

Agar	15.0g
Pancreatic digest of gelatin	6.0g

Yeast extract	3.0g
Beef extract	1.5g

pH 7.0 ± 0.1 at 25°C

Preparation of Medium: Add components to distilled/deionized water and bring volume to 1.0L. Mix thoroughly. Gently heat and bring to boiling. Adjust pH to 7.0. Distribute into tubes or flasks. Autoclave for 15 min at 15 psi pressure–121°C. Pour into sterile Petri dishes.

Use: For use as a base layer in antibiotic assay testing. Especially useful for the plate assay of bacitracin and penicillin G.

Antibiotic Medium 8 (Base Agar with Low pH)

Composition per liter:

Agar	15.0g
Pancreatic digest of gelatin	6.0g
Yeast extract	3.0g
Beef extract	1.5g

pH 5.9 ± 0.1 at 25°C

Source: This medium is available as a premixed powder from BD Diagnostic Systems.

Preparation of Medium: Add components to distilled/deionized water and bring volume to 1.0L. Mix thoroughly. Gently heat and bring to boiling. Distribute into tubes or flasks. Autoclave for 15 min at 15 psi pressure–121°C. Pour into sterile Petri dishes.

Use: For antibiotic assay testing. For use as the base agar and the seed agar in the plate assay of tetracycline. For use as the seed agar in the plate assay of vancomycin, mitomycin, and mithramycin.

Antibiotic Medium 9 (Polymyxin Base Agar)

Composition per liter:

Agar	20.0g
Pancreatic digest of casein	17.0g
NaCl	5.0g
Papaic digest of soybean meal	3.0g
K_2HPO_4	2.5g
Glucose	2.5g

pH 7.2 ± 0.1 at 25°C

Source: This medium is available as a premixed powder from BD Diagnostic Systems.

Preparation of Medium: Add components to distilled/deionized water and bring volume to 1.0L. Mix thoroughly. Gently heat and bring to boiling. Distribute into tubes or flasks. Autoclave for 15 min at 15 psi pressure–121°C. Pour into sterile Petri dishes.

Use: For antibiotic assay testing. For base agar for the plate assay of carbenicillin, colistimethate, and polymyxin B.

Antibiotic Medium 10 (Polymyxin Seed Agar)

Composition per liter:

Pancreatic digest of casein	17.0g
Agar	12.0g
Polysorbate 80	10.0g
NaCl	5.0g
Papaic digest of soybean meal	3.0g
K_2HPO_4	2.5g
Glucose	2.5g

pH 7.3 ± 0.2 at 25°C

Source: This medium is available as a premixed powder from BD Diagnostic Systems.

Preparation of Medium: Add components to distilled/deionized water and bring volume to 1.0L. Mix thoroughly. Gently heat and bring to boiling. Distribute into tubes or flasks. Autoclave for 15 min at 15 psi pressure–121°C. Pour into sterile Petri dishes.

Use: For antibiotic assay testing. For seed agar for the plate assay of carbenicillin, colistimethate, and polymyxin B.

Antibiotic Medium 11 (Neomycin Assay Agar)

Composition per liter:

Agar	15.0g
Pancreatic digest of gelatin	6.0g
Pancreatic digest of casein	4.0g
Yeast extract	3.0g
Beef extract	1.5g
Glucose	1.0g

pH 8.0 ± 0.1 at 25°C

Source: This medium is available as a premixed powder from BD Diagnostic Systems and Oxoid Unipath.

Preparation of Medium: Add components to distilled/deionized water and bring volume to 1.0L. Mix thoroughly. Gently heat and bring to boiling. Distribute into tubes or flasks. Autoclave for 15 min at 15 psi pressure–121°C. Pour into sterile Petri dishes.

Use: For antibiotic assay testing. For base agar and seed agar for the plate assay to test the effectiveness of neomycin sulfate, amoxicillin, ampicillin, clindamycin, cyclacillin, erythromycin, gentamycin, neomycin, oleandomycin, and sisomycin.

Antibiotic Medium 12

Composition per liter:

Agar	25.0g
Peptone	10.0g
Glucose	10.0g
NaCl	10.0g
Yeast extract	5.0g
Beef extract	2.5g

pH 6.0 ± 0.1 at 25°C

Source: This medium is available as a premixed powder from BD Diagnostic Systems.

Preparation of Medium: Add components to distilled/deionized water and bring volume to 1.0L. Mix thoroughly. Gently heat and bring to boiling. Distribute into tubes or flasks. Autoclave for 15 min at 15 psi pressure–121°C. Pour into sterile Petri dishes.

Use: For antibiotic assay effectiveness testing.

Antibiotic Medium 13 (Sabouraud Liquid Broth, Modified) (Fluid Sabouraud Medium)

Composition per liter:

Glucose	20.0g
Pancreatic digest of casein	5.0g
Peptic digest of animal tissue	5.0g

pH 5.7 ± 0.1 at 25°C

Source: This medium is available as a premixed powder from BD Diagnostic Systems.

Preparation of Medium: Add components to distilled/deionized water and bring volume to 1.0L. Mix thoroughly. Gently heat and bring to boiling. Distribute into tubes or flasks. Autoclave for 15 min at 15 psi pressure–121°C. Pour into sterile Petri dishes.

Use: For testing the effectivness of antibiotics on yeast and molds.

Antibiotic Medium 19 (Nystatin Assay Agar)

Composition per liter:

Agar	23.5g
Glucose	10.0g
NaCl	10.0g
Pancreatic digest of gelatin	9.4g
Yeast extract	4.7g
Beef extract	2.4g

pH 6.1 ± 0.2 at 25°C

Source: This medium is available as a premixed powder from BD Diagnostic Systems.

Preparation of Medium: Add components to distilled/deionized water and bring volume to 1.0L. Mix thoroughly. Gently heat and bring to boiling. Distribute into tubes or flasks. Autoclave for 15 min at 15 psi pressure–121°C. Pour into sterile Petri dishes.

Use: For assaying the mycostatic activity of pharmaceutical preparations. For seed agar for the plate assay to test the effectiveness of nystatin, amphotericin B, and natamycin.

Antibiotic Medium 20

Composition per liter:

Glucose	11.0g
Pancreatic digest of casein	10.0g
Yeast extract	6.5g
Pancreatic digest of gelatin	5.0g
K_2HPO_4	3.68g
NaCl	3.5g
Beef extract	1.5g
KH_2PO_4	1.32g

pH 6.6 ± 0.2 at 25°C

Preparation of Medium: Add components to distilled/deionized water and bring volume to 1.0L. Mix thoroughly. Gently heat and bring to boiling. Distribute into tubes or flasks. Autoclave for 15 min at 15 psi pressure–121°C. Pour into sterile Petri dishes.

Use: For assaying the mycostatic activity of pharmaceutical preparations.

Antibiotic Medium 21

Composition per liter:

Glucose	11.0g
Pancreatic digest of gelatin	5.0g
K_2HPO_4	3.68g
NaCl	3.5g
Yeast extract	1.5g
Beef extract	1.5g
KH_2PO_4	1.32g

pH 6.6 ± 0.2 at 25°C

Preparation of Medium: Add components to distilled/deionized water and bring volume to 1.0L. Mix thoroughly. Gently heat and bring to boiling. Distribute into tubes or flasks. Autoclave for 15 min at 15 psi pressure–121°C. Pour into sterile Petri dishes.

Use: For assaying the mycostatic activity of pharmaceutical preparations.

Antibiotic Sulfonamide Sensitivity Test Agar (ASS Agar)

Composition per liter:

Agar	12.0g
Proteose peptone	10.0g
Beef extract	10.0g
NaCl	3.0g
Glucose	2.0g
Na_2HPO_4	2.0g
Sodium acetate	1.0g
Adenine	0.01g
Guanine	0.01g
Uracil	0.01g
Xanthine	0.01g

pH 7.4 ± 0.2 at 25°C

Preparation of Medium: Add components to distilled/deionized water and bring volume to 1.0L. Mix thoroughly. Gently heat and bring to boiling. Distribute into tubes or flasks. Autoclave for 15 min at 15 psi pressure–121°C. Pour into sterile Petri dishes or leave in tubes.

Use: For testing the antimicrobial effectiveness of antibiotics and sulfonamides. For detecting the presence of antimicrobial substances in milk, urine, and other fluids.

Antifungal Assay Agar

Composition per liter:

Glucose	50.0g
Agar	15.0g
Sodium citrate	4.5g
Pancreatic digest of casein	4.0g
Citric acid	1.0g
K_2HPO_4	0.55g
KCl	0.425g
$CaCl_2 \cdot 2H_2O$	0.125g
$MgSO_4 \cdot 7H_2O$	0.125g
Inositol	0.025g
$MnSO_4 \cdot 4H_2O$	2.5mg
Niacin	2.5mg
Caclium pantothenate	2.5mg
$FeCl_3$	2.5mg
Pyridoxine hydrochloride	0.25mg
Thiamine	0.25mg
Biotin	0.008mg

pH 5.5 ± 0.2 at 25°C

Source: This medium is available as a premixed powder from Sigma Aldrich.

Preparation of Medium: Add components to distilled/deionized water and bring volume to 1.0L. Mix thoroughly. Gently heat and bring to boiling. Distribute into tubes or flasks. Autoclave for 15 min at 15 psi pressure–121°C. Pour into sterile Petri dishes.

Use: For assaying antifungal activity of pharmaceutical products and other materials by cylinder plate or disc method.

Antifungal Assay HiVeg Agar

Composition per liter:

Glucose	50.0g
Agar	15.0g

Sodium citrate	4.5g
Plant hydrolysate	4.0g
Citric acid	1.0g
K_2HPO_4	0.55g
KCl	0.425g
$CaCl_2·2H_2O$	0.125g
$MgSO_4·7H_2O$	0.125g
Inositol	0.025g
$MnSO_4·4H_2O$	2.5mg
Niacin	2.5mg
Calcium pantothenate	2.5mg
$FeCl_3$	2.5mg
Pyridoxine hydrochloride	0.25mg
Thiamine	0.25mg
Biotin	0.008mg

pH 5.5 ± 0.2 at 25°C

Source: This medium is available as a premixed powder from HiMedia.

Preparation of Medium: Add components to distilled/deionized water and bring volume to 1.0L. Mix thoroughly. Gently heat and bring to boiling. Distribute into tubes or flasks. Autoclave for 15 min at 15 psi pressure–121°C. Pour into sterile Petri dishes.

Use: For assaying antifungal activity of pharmaceutical products and other materials by cylinder plate or disc method.

Antimicrobial Inhibitor Test Agar pH 6.0

Composition per liter:

Agar	13.0g
NaCl	5.0g
Tryptone	3.5g
Meat extract	3.5g
Bacillus subtilis spore suspension	1.0mL

pH 6.0 ± 0.2 at 25°C

Preparation of Medium: Add components, except *Bacillus subtilis* spore suspension, to distilled/deionized water and bring volume to 990.0mL. Mix thoroughly. Adjust pH to 6.0. Gently heat and bring to boiling. Distribute into tubes or flasks. Autoclave for 15 min at 15 psi pressure–121°C. Cool to 50°C. Add *Bacillus subtilis* spore suspension. Mix thoroughly. Pour into sterile Petri dishes.

Use: For residual analysis of antimicrobial components in meat and organ samples, using *Bacillus subtilis* ATCC 6633 as test organism.

Antimicrobial Inhibitor Test Agar pH 7.2

Composition per liter:

Agar	13.0g
Peptone	7.0g
NaCl	5.0g
$Na_3PO_4·12H_2O$	0.8g
Selective supplement solution	10.0mL
Bacillus subtilis spore suspension	1.0mL

pH 7.2 ± 0.2 at 25°C

Selective Supplement Solution:
Composition per 10.0mL:

Trimethoprim	5.0mg

Preparation of Selective Supplement Solution: Add trimethoprim to distilled/deionized water and bring volume to 10.0mL. Mix thoroughly. Filter sterilize.

Preparation of Medium: Add components, except selective supplement solution and *Bacillus subtilis* spore suspension, to distilled/deionized water and bring volume to 1.0L. Mix thoroughly. Adjust pH to 7.2. Gently heat and bring to boiling. Distribute into tubes or flasks. Autoclave for 15 min at 15 psi pressure–121°C. Cool to 50°C. Add selective supplement solution and *Bacillus subtilis* spore suspension. Mix thoroughly. Pour into sterile Petri dishes.

Use: For residual analysis of antimicrobial components in meat and organ samples, using *Bacillus subtilis* ATCC 6633 as test organism.

Antimicrobial Inhibitor Test Agar pH 8.0

Composition per liter:

Tryptone	3.5g
Meat extract	3.5g
NaCl	5.0g
$Na_3PO_4·12H_2O$	2.5g
Agar	13.0g
Micrococcus luteus suspension	10.0mL
Bacillus subtilis spore suspension	1.0mL

pH 8.0 ± 0.2 at 25°C

Preparation of Medium: Add components, except *Bacillus subtilis* spore suspension, to distilled/deionized water and bring volume to 990.0mL. Mix thoroughly. Adjust pH to 8.0. Gently heat and bring to boiling. Distribute into tubes or flasks. Autoclave for 15 min at 15 psi pressure–121°C. Cool to 50°C. Add *Bacillus subtilis* spore suspension and *Micrococcus luteus* (10^4 CFU per mL) suspension. Mix thoroughly. Pour into sterile Petri dishes.

Use: For residual analysis of antimicrobial components in meat and organ samples, using *Bacillus subtilis* ATCC 6633 and *Micrococcus luteus* ATCC 9341 as test organisms.

Antimycin Medium

Composition per liter:

Yeast extract	20.0g
Peptone	10.0g
Glycerol	30.0mL
Antimycin solution	10.0mL

pH 5.5 ± 0.2 at 25°C

Antimycin Solution:
Composition per 10.0mL:

Antimycin	1.0mg

Preparation of Antimycin Solution: Add antimycin to distilled/deionized water and bring volume to 10.0mL. Mix thoroughly. Filter sterilize.

Preparation of Medium: Add components, except antimycin solution, to distilled/deionized water and bring volume to 990.0mL. Mix thoroughly. Gently heat and bring to boiling. Adjust pH to 5.5 with HCl. Autoclave for 15 min at 15 psi pressure–121°C. Cool to 50°C. Aseptically add 10.0mL of sterile antimycin solution. Mix thoroughly. Aseptically distribute into sterile tubes.

Use: For the cultivation and maintenance of *Candida utilis*.

Antimycotic Sensitivity Test Agar

Composition per liter:

Agar	25.0g
Glucose	20.0g
Pancreatic digest of casein	19.0g
Sodium citrate	10.0g

Yeast extract 10.0g
Na_2HPO_4 .... 1.0g

pH 6.0 ± 0.2 at 25°C

Source: This medium is available as a premixed powder from HiMedia.

Preparation of Medium: Add components to distilled/deionized water and bring volume to 1.0L. Mix thoroughly. Gently heat and bring to boiling. Distribute into tubes or flasks. Autoclave for 15 min at 15 psi pressure–121°C. Pour into sterile Petri dishes.

Use: For for testing antimycotic sensitivity by the diffusion method.

AO Agar

Composition per liter:

Agar 11.0g
Sodium acetate 0.5g
Pancreatic digest of casein 0.5g
Yeast extract 0.5g
Beef extract 0.2g

pH 7.2 ± 0.2 at 25°C

Preparation of Medium: Add components to distilled/deionized water and bring volume to 1.0L. Mix thoroughly. Gently heat and bring to boiling. Distribute into tubes or flasks. Autoclave for 15 min at 15 psi pressure–121°C. Pour into sterile Petri dishes or leave in tubes.

Use: For the isolation and cultivation of *Cytophaga* species, *Herpetosiphon* species, *Saprospira* species, and *Flexithrix* species.

AO Agar

Composition per liter:

Agar 4.0g
Sodium acetate 0.5g
Pancreatic digest of casein 0.5g
Yeast extract 0.5g
Beef extract 0.2g

pH 7.2 ± 0.2 at 25°C

Preparation of Medium: Add components to distilled/deionized water and bring volume to 1.0L. Mix thoroughly. Gently heat and bring to boiling. Distribute into tubes or flasks. Autoclave for 15 min at 15 psi pressure–121°C. Pour into sterile Petri dishes or leave in tubes.

Use: For the maintenance of *Cytophaga* species, *Herpetosiphon* species, *Saprospira* species, and *Flexithrix* species.

AOAC Letheen Broth (Association of Official Analytical Chemists Letheen Broth)

Composition per liter:

Peptic digest of animal tissue 10.0g
Polysorbate 80 .... 5.0g
NaCl 5.0g
Beef extract 5.0g
Lecithin 0.7g

pH 7.0 ± 0.2 at 25°C

Source: This medium is available as a premixed powder from BD Diagnostic Systems.

Preparation of Medium: Add components to distilled/deionized water and bring volume to 1.0L. Mix thoroughly. Gently heat and bring to boiling. Distribute into tubes in 10.0mL volumes. Autoclave for 15 min at 15 psi pressure–121°C.

Use: For the determination of phenol coefficients of disinfectant products containing cationic surface-active materials. Use according to *Official Methods of Analysis* of the *Association of Official Analytical Chemists* (AOAC).

Aolpha Medium

Composition per 1041.0mL:

NaCl 100.0g
Agar 15.0g
$MgSO_4 \cdot 7H_2O$ 9.5g
$MgCl_2 \cdot 6H_2O$ 5.0g
KCl 5.0g
Peptone 5.0g
Yeast extract 1.0g
$CaCl_2 \cdot 2H_2O$ 0.2g
$(NH_4)_2SO_4$ 0.1g
KNO_3 .... 0.1g
Metals solution 20.0mL
Phosphate solution 20.0mL
Vitamin solution 1.0mL

pH 7.0 ± 0.2 at 25°C

Metals Solution:

Composition per liter:

$MgSO_4 \cdot 7H_2O$ 29.7g
Nitrilotriacetic acid 10.0g
$CaCl_2 \cdot 2H_2O$ 3.3g
$FeSO_4 \cdot 7H_2O$ 99.0mg
$Na_2MoO_4 \cdot 2H_2O$ 12.7mg
Metals "44" 50.0mL

Preparation of Metals Solution: Solubilize nitrilotriacetic acid with KOH. Dissolve remaining ingredients. Adjust pH to 7.2 with KOH or H_2SO_4. Autoclave for 15 min at 15 psi pressure–121°C. Add aseptically to sterile basal medium.

Metals "44":

Composition per 100.0mL:

$ZnSO_4 \cdot 7H_2O$ 1.1g
$FeSO_4 \cdot 7H_2O$ 0.5g
EDTA 0.25g
$MnSO_4 \cdot 7H_2O$ 0.154g
$CuSO_4 \cdot 5H_2O$ 0.04g
$Co(NO_3)_2 \cdot 6H_2O$ 0.025g
$Na_2B_4O_7 \cdot 10H_2O$ 0.018g

Preparation of Metals "44": Add components to distilled/deionized water and bring volume to 100.0mL. Mix thoroughly. Autoclave for 15 min at 15 psi pressure–121°C. Add aseptically to sterile basal medium.

Phosphate Solution:

Composition per liter:

K_2HPO_4 .... 2.5g
KH_2PO_4 .... 2.5g

Preparation of Phosphate Solution: Add components to distilled/deionized water and bring volume to 1.0L. Mix thoroughly. Autoclave for 15 min at 15 psi pressure–121°C. Add aseptically to sterile basal medium.

Vitamin Solution:

Composition per liter:

Pyridoxine·HCl 10.0mg
Calcium pantothenate 5.0mg

Nicotinamide	5.0mg
Riboflavin	5.0mg
Thiamine·HCl	5.0mg
Biotin	2.0mg
Folic acid	2.0mg
Cyanocobalamin	0.1mg

Preparation of Vitamin Solution: Add components to distilled/deionized water and bring volume to 1.0L. Mix thoroughly. Filter sterilize and add aseptically to sterile basal medium.

Preparation of Medium: Add components—except Metals "44", phosphate solution, and vitamin solution—to distilled/deionized water and bring volume to 1.0L. Mix thoroughly. Gently heat and bring to boiling. Adjust pH of basal medium to 7.0. Autoclave for 15 min at 15 psi pressure–121°C. Cool to 50°C and aseptically add the Metals "44", phosphate, and vitamin solutions.

Use: For the cultivation and maintenance of *Halomonas meridiana* and other *Halomonas* species.

Aphanomyces Synthetic Medium

Composition per liter:

D-Glucose	5.0g
KH_2PO_4	2.0g
L-Asparagine	0.75g
$MgCl_2$	0.05g
$FeCl_3$	5.0mg
$MnCl_2$	5.0mg
$ZnCl_2$	5.0mg
L-Methionine	0.02mg

pH 5.5 ± 0.2 at 25°C

Preparation of Medium: Add components to distilled/deionized water and bring volume to 1.0L. Mix thoroughly. Autoclave for 15 min at 15 psi pressure–121°C. Adjust pH to 5.5. Aseptically distribute into sterile tubes or flasks.

Use: For the cultivation of *Aphanomyces* species.

Aplanobacterium Medium

Composition per liter:

Agar	20.0g
Glucose	10.0g
Peptone	5.0g
Yeast extract	5.0g

pH 7.2 ± 0.2 at 25°C

Preparation of Medium: Add components to distilled/deionized water and bring volume to 1.0L. Mix thoroughly. Gently heat and bring to boiling. Autoclave for 15 min at 15 psi pressure–121°C. Pour into sterile Petri dishes or leave in tubes.

Use: For the cultivation and maintenance of *Xanthomonas* species.

Apple Juice Yeast Extract Medium (AJYE Medium)

Composition per 1200.0mL:

Agar	30.0g
Yeast extract	10.0g
Apple juice	1.0L

pH 4.8 ± 0.2 at 25°C

Preparation of Medium: Add yeast extract to 1.0L of apple juice. Mix thoroughly. Adjust pH to 4.8. Autoclave for 10 min at 9 psi pressure–114°C. Cool to 45°–50°C. In a separate flask, add agar to 200.0mL of distilled/deionized water and bring volume to 1.0L. Mix thoroughly. Gently heat and bring to boiling. Autoclave for 15 min at 15 psi pressure–121°C. Cool to 45°–50°C. Aseptically combine the sterile apple juice solution with the sterile agar solution. Mix thoroughly. Pour into sterile Petri dishes.

Use: For the cultivation of *Zymomonas* species.

APRY Agar

Composition per liter:

Fructose	30.0g
NaCl	25.0g
Glucose	20.0g
Agar	15.0g
Casein enzymatic hydrolysate	10.0g
Yeast extract	2.5g
Peptic digest of animal tissue	5.0g
Acetic acid, glacial	5.0mL
Selective supplement solution	5.0mL
Potassium sorbate solution	1.0mL

pH 6.0 ± 0.2 at 25°C

Source: This medium is available from HiMedia.

Potassium Sorbate Solution:
Composition per 10.0mL:

Potassium sorbate	1.0g

Preparation of Potassium Sorbate Solution: Add components to distilled/deionized water and bring volume to 10.0mL. Mix thoroughly. Filter sterilize.

Selective Supplement Solution:
Composition per 10.0mL:

Chlortetracycline	50.0mg

Preparation of Selective Supplement Solution: Add chlortetracycline to distilled/deionized water and bring volume to 10.0mL. Mix thoroughly. Filter sterilize.

Preparation of Medium: Add components, except acetic acid, selective supplement solution, and potassium sorbate solution, to distilled/deionized water and bring volume to 990.0mL. Mix thoroughly. Autoclave for 15 min at 15 psi pressure–121°C. Cool to 50°C. Aseptically add acetic acid, selective supplement solution, and potassium sorbate solution. Mix thoroughly. Pour into Petri dishes or aseptically distribute into sterile tubes.

Use: For the cultivation of acid resistant yeasts, including *Zygosaccharomyces bailii* and *Zygosaccharomyces rouxii* in salads, sauces, and dressings.

APRY Agar Base with Acetic Acid and Sorbate

Composition per liter:

Fructose	30.0g
NaCl	25.0g
Glucose	20.0g
Agar	15.0g
Pancreatic digest of casein	10.0g
Peptic digest of animal tissue	5.0g
Yeast extract	2.5g
Acetic acid (conc.)	5.0mL
Potassium sorbate (10%)	1.0mL

pH 5.5 ± 0.2 at 25°C

Source: This medium without acetic acid and potassium sorbate is available as a premixed powder from HiMedia.

Preparation of Medium: Add components to distilled/deionized water and bring volume to 1.0L. Mix thoroughly. Gently heat and bring to boiling. Autoclave for 15 min at 15 psi pressure–121°C. Cool to 45–50°C. Aseptically add 5.0mL sterile acetic acid and 1.0mL potassium sorbate solution. Mix thoroughly. Pour into sterile Petri dishes or distribute into sterile tubes.

Use: For the detection and isolation of acid resistant yeasts, *Zygosaccharomyces bailii* and *Zygosaccharomyces rouxii,* in salads, sauces, and dressings.

APRY Broth Base with Chloramphenicol

Composition per liter:

Glucose	30.0g
Fructose	20.0g
Pancreatic digest of casein	15.0g
Polysorbate 80	10.0g
Peptic digest of animal tissue	5.0g
Yeast extract	2.5g
Choramphenicol solution	2.0ml

pH 6.5 ± 0.2 at 25°C

Source: This medium without chloramphenicol is available as a premixed powder from HiMedia. Chloramphenicol supplement is available from HiMedia.

Chloramphenicol Solution:
Composition per 2.0mL:

Chloramphenicol	50.0mg
Ethanol	2.0mL

Preparation of Chloramphenicol Solution: Add chloramphenicol to ethanol and bring volume to 2.0mL. Mix thoroughly. Filter sterilize.

Preparation of Medium: Add components to distilled/deionized water and bring volume to 1.0L. Mix thoroughly. Gently heat and bring to boiling. Autoclave for 15 min at 15 psi pressure–121°C. Cool to 45–50°C. Aseptically add 2.0mL chloramphenicol soultion. Mix thoroughly. Distribute into sterile tubes or flasks.

Use: For the detection and isolation of acid resistant yeasts, *Zygosaccharomyces bailii* and *Zygosaccharomyces rouxii,* in salads, sauces, and dressings.

APRY Broth

Composition per liter:

Glucose	30.0g
Fructose	20.0g
Casein enzymatic hydrolysate	15.0g
Yeast extract	2.5g
Peptic digest of animal tissue	5.0g
Polysorbate 80	10.0mL
Acetic acid, glacial	5.0mL
Selective supplement solution	5.0mL
Potassium sorbate solution	1.0mL

pH 6.0 ± 0.2 at 25°C

Source: This medium is available from HiMedia.

Potassium Sorbate Solution:
Composition per 10.0mL:

Potassium sorbate	1.0g

Preparation of Potassium Sorbate Solution: Add components to distilled/deionized water and bring volume to 10.0mL. Mix thoroughly. Filter sterilize.

Selective Supplement Solution:
Composition per 10.0mL:

Chlortetracycline	50.0mg
Chloramphenicol	50.0mg

Preparation of Selective Supplement Solution: Add components to distilled/deionized water and bring volume to 10.0mL. Mix thoroughly. Filter sterilize.

Preparation of Medium: Add components, except acetic acid, selective supplement solution, and potassium sorbate solution, to distilled/deionized water and bring volume to 990.0mL. Mix thoroughly. Autoclave for 15 min at 15 psi pressure–121°C. Cool to 50°C. Aseptically add acetic acid, selective supplement solution, and potassium sorbate solution. Mix thoroughly. Aseptically distribute into sterile tubes or flasks.

Use: For the cultivation of acid resistant yeasts, including *Zygosaccharomyces bailii* and *Zygosaccharomyces rouxii* n salads, sauces, and dressings.

APT Agar

Composition per liter:

Agar	15.0g
Pancreatic digest of casein	12.5g
Glucose	10.0g
Yeast extract	7.5g
NaCl	5.0g
K_2HPO_4	5.0g
Sodium citrate	5.0g
Na_2CO_3	1.25g
$MgSO_4 \cdot 7H_2O$	0.8g
Polysorbate 80	0.2g
$MnCl_2 \cdot 4H_2O$	0.14g
$FeSO_4 \cdot 7H_2O$	0.04g
Thiamine·HCl	1.0mg

pH 6.7 ± 0.2 at 25°C

Source: This medium is available as a premixed powder from BD Diagnostic Systems.

Preparation of Medium: Add components to distilled/deionized water and bring volume to 1.0L. Mix thoroughly. Gently heat and bring to boiling. Distribute into tubes or flasks. Autoclave for 15 min at 13 psi—118°–121°C. Pour into sterile Petri dishes or leave in tubes.

Use: For the cultivation and enumeration of bacteria, especially heterofermentative lactobacilli, from meat and other foods. For the cultivation of streptococci.

APT Agar

Composition per liter:

Agar	13.5g
Pancreatic digest of casein	10.0g
Glucose	10.0g
Yeast extract	7.5g
NaCl	5.0g
KH_2PO_4	5.0g
Sodium citrate	5.0g
Na_2CO_3	1.25g
$MgSO_4 \cdot 7H_2O$	0.8g

Polysorbate 80	0.2g
$MnCl_2 \cdot 4H_2O$	0.14g
$FeSO_4 \cdot 7H_2O$	0.04g

pH 6.7 ± 0.2 at 25°C

Source: This medium is available as a premixed powder from BD Diagnostic Systems.

Preparation of Medium: Add components to distilled/deionized water and bring volume to 1.0L. Mix thoroughly. Gently heat and bring to boiling. Distribute into tubes or flasks. Autoclave for 15 min at 13 psi—118°–121°C. Pour into sterile Petri dishes or leave in tubes.

Use: For the cultivation and enumeration of bacteria, especially heterofermentative lactobacilli, from meat and other foods. For the cultivation of streptococci.

APT Broth

Composition per liter:

Pancreatic digest of casein	12.5g
Glucose	10.0g
Yeast extract	7.5g
NaCl	5.0g
K_2HPO_4	5.0g
Sodium citrate	5.0g
Na_2CO_3	1.25g
$MgSO_4 \cdot 7H_2O$	0.8g
Polysorbate 80	0.2g
$MnCl_2 \cdot 4H_2O$	0.14g
$FeSO_4 \cdot 7H_2O$	0.04g
Thiamine·HCl	1.0mg

pH 7.7 ± 0.2 at 25°C

Source: This medium is available as a premixed powder from BD Diagnostic Systems.

Preparation of Medium: Add components to distilled/deionized water and bring volume to 1.0L. Mix thoroughly. Gently heat and bring to boiling. Distribute into tubes or flasks. Autoclave for 15 min at 15 psi—118°–121°C.

Use: For the cultivation of lactic acid bacteria. For the cultivation of heterofermentative lactobacilli from meat and other foods. The American Public Health Association recommends adding 100.0μg/L of thiamine to this medium.

APT Broth

Composition per liter:

Pancreatic digest of casein	10.0g
Glucose	10.0g
Yeast extract	7.5g
NaCl	5.0g
KH_2PO_4	5.0g
Sodium citrate	5.0g
Na_2CO_3	1.25g
$MgSO_4 \cdot 7H_2O$	0.8g
Polysorbate 80	0.2g
$MnCl_2 \cdot 4H_2O$	0.14g
$FeSO_4 \cdot 7H_2O$	0.04g

pH 7.7 ± 0.2 at 25°C

Source: This medium is available as a premixed powder from BD Diagnostic Systems.

Preparation of Medium: Add components to distilled/deionized water and bring volume to 1.0L. Mix thoroughly. Gently heat and bring to boiling. Distribute into tubes or flasks. Autoclave for 15 min at 15 psi—118°–121°C.

Use: For the cultivation of lactic acid bacteria. For the cultivation of heterofermentative lactobacilli from meat and other foods. The American Public Health Association recommends adding 100.0μg/L of thiamine to this medium.

APT HiVeg Agar

Composition per liter:

Agar	15.0g
Plant hydrolysate	12.5g
Glucose	10.0g
Yeast extract	7.5g
NaCl	5.0g
KH_2PO_4	5.0g
Sodium citrate	5.0g
Na_2CO_3	1.25g
$MgSO_4 \cdot 7H_2O$	0.8g
Polysorbate 80	0.2g
$MnCl_2 \cdot 4H_2O$	0.14g
$FeSO_4 \cdot 7H_2O$	0.04g

pH 6.7 ± 0.2 at 25°C

Source: This medium is available as a premixed powder from HiMedia.

Preparation of Medium: Add components to distilled/deionized water and bring volume to 1.0L. Mix thoroughly. Gently heat and bring to boiling. Distribute into tubes or flasks. Autoclave for 15 min at 15 psi–121°C. Pour into sterile Petri dishes or leave in tubes.

Use: For the cultivation and enumeration of bacteria, especially heterofermentative lactobacilli, from meat and other foods. For the cultivation of streptococci.

APT HiVeg Broth

Composition per liter:

Plant hydrolysate	12.5g
Glucose	10.0g
Yeast extract	7.5g
NaCl	5.0g
K_2HPO_4	5.0g
Sodium citrate	5.0g
Na_2CO_3	1.25g
$MgSO_4 \cdot 7H_2O$	0.8g
Polysorbate 80	0.2g
$MnCl_2 \cdot 4H_2O$	0.14g
$FeSO_4 \cdot 7H_2O$	0.04g
Thiamine·HCl	1.0mg

pH 6.7 ± 0.2 at 25°C

Source: This medium is available as a premixed powder from HiMedia.

Preparation of Medium: Add components to distilled/deionized water and bring volume to 1.0L. Mix thoroughly. Gently heat and bring to boiling. Distribute into tubes or flasks. Autoclave for 15 min at 15 psi–121°C.

Use: For the cultivation of lactic acid bacteria. For the cultivation of heterofermentative lactobacilli from meat and other foods.

Aquabacter spiritensis Medium (LMG Medium 225)

Sodium succinate 2.0g
Yeast extract 1.0g
KH_2PO_4 .... 1.0g
Peptone 0.4g
NH_4Cl 0.2g
NaCl 0.2g
$MgSO_4 \cdot 7H_2O$ 0.2g
$CaCl_2 \cdot 2H_2O$ 10.0mg
Ferric citrate 5.0mg
Vitamin solution 20.0mL
Trace elements solution SL-6 .... 1.0mL
pH 7.0 ± 0.2 at 25°C

Trace Elements Solution SL-6 :
Composition per liter:
H_3BO_3 .... 0.3g
$CoCl_2 \cdot 6H_2O$ 0.2g
$ZnSO_4 \cdot 7H_2O$ 0.1g
$MnCl_2 \cdot 4H_2O$ 0.03g
$Na_2MoO_4 \cdot H_2O$ 0.03g
$NiCl_2 \cdot 6H_2O$ 0.02g
$CuCl_2 \cdot 2H_2O$ 0.01g

Preparation of Trace Elements Solution SL-6 : Add components to distilled/deionized water and bring volume to 1.0L. Mix thoroughly. Adjust pH to 3.4.

Vitamin Solution:
Composition per liter:
Calcium DL-pantothenate 5.0mg
Riboflavin 5.0mg
Thiamine·HCl 5.0mg
Biotin 2.0mg
Folic acid 2.0mg
Vitamin B_{12} 0.1mg

Preparation of Vitamin Solution: Add components to distilled/deionized water and bring volume to 1.0L. Mix thoroughly. Filter sterilize.

Preparation of Medium: Add components, except vitamin solution, to 980.0mL distilled/deionized water. Mix thoroughly. Autoclave for 15 min at 15 psi pressure–121°C. Cool to 25°C. Aseptically add 20.0mL sterile vitamin solution. Mix thoroughly. Aseptically distribute to sterile tubes or flasks.

Use: For the cultivation of *Aquabacter spiritensis*.

Aquaspirillum Autotrophic Agar

Composition per liter:
Noble agar 15.0g
$Na_2HPO_4 \cdot 12H_2O$ 9.0g
KH_2PO_4 .... 1.5g
NH_4Cl 1.0g
$MgSO_4 \cdot 7H_2O$ 0.2g
$CaCl_2 \cdot 2H_2O$ 0.01g
Ferric ammonium citrate 5.0mg
$NaHCO_3$ solution 10.0mL
Trace elements solution 3.0mL
pH 7.1 ± 0.2 at 25°C

$NaHCO_3$ Solution:
Composition per 10.0mL:
$NaHCO_3$ 0.5g

Preparation of $NaHCO_3$ Solution: Add the $NaHCO_3$ to distilled/deionized water and bring volume to 10.0mL. Mix thoroughly. Filter sterilize.

Trace Elements Solution:
Composition per liter:
H_3BO_3 .... 30.0mg
$CoCl_2 \cdot 6H_2O$ 20.0mg
$ZnSO_4 \cdot 7H_2O$ 10.0mg
$MnCl_2 \cdot 4H_2O$ 3.0mg
$Na_2MoO_4 \cdot 2H_2O$ 3.0mg
$NiCl_2 \cdot 6H_2O$ 2.0mg
$CuCl_2 \cdot 2H_2O$ 1.0mg

Preparation of Trace Elements Solution: Add components to distilled/deionized water and bring volume to 1.0L. Mix thoroughly.

Preparation of Medium: Add components, except $NaHCO_3$ solution, to double-distilled water and bring volume to 990.0mL. Mix thoroughly. Gently heat and bring to boiling. Autoclave for 15 min at 15 psi pressure–121°C. Cool to 45°–50°C. Aseptically add sterile $NaHCO_3$ solution. Mix thoroughly. Pour into sterile Petri dishes or distribute into sterile tubes. For autotrophic growth, incubate under 85% H_2 + 10% CO_2 + 5% O_2.

Use: For the autotrophic cultivation and maintenance of *Aquaspirillum autotrophicum*.

Aquaspirillum Autotrophic Broth

Composition per liter:
$Na_2HPO_4 \cdot 12H_2O$ 9.0g
KH_2PO_4 .... 1.5g
NH_4Cl 1.0g
$MgSO_4 \cdot 7H_2O$ 0.2g
$CaCl_2 \cdot 2H_2O$ 0.01g
Ferric ammonium citrate 5.0mg
$NaHCO_3$ solution 10.0mL
Trace elements solution 3.0mL
pH 7.1 ± 0.2 at 25°C

$NaHCO_3$ Solution:
Composition per 10.0mL:
$NaHCO_3$ 0.5g

Preparation of $NaHCO_3$ Solution: Add the $NaHCO_3$ to distilled/deionized water and bring volume to 10.0mL. Mix thoroughly. Filter sterilize.

Trace Elements Solution:
Composition per liter:
H_3BO_3 .... 30.0mg
$CoCl_2 \cdot 6H_2O$ 20.0mg
$ZnSO_4 \cdot 7H_2O$ 10.0mg
$MnCl_2 \cdot 4H_2O$ 3.0mg
$Na_2MoO_4 \cdot 2H_2O$ 3.0mg
$NiCl_2 \cdot 6H_2O$ 2.0mg
$CuCl_2 \cdot 2H_2O$ 1.0mg

Preparation of Trace Elements Solution: Add components to distilled/deionized water and bring volume to 1.0L. Mix thoroughly.

Preparation of Medium: Add components, except $NaHCO_3$ solution, to double-distilled water and bring volume to 990.0mL. Mix thoroughly. Gently heat and bring to boiling. Autoclave for 15 min at 15 psi pressure–121°C. Cool to 45°–50°C. Aseptically add sterile $NaHCO_3$ solution. Mix thoroughly. Aseptically distribute into sterile tubes or flasks. To grow autotrophically, incubate under 85% H_2 + 10% CO_2 + 5% O_2.

Use: For the autotrophic cultivation of *Aquaspirillum autotrophicum*.

Aquaspirillum Heterotrophic Agar

Composition per liter:

Noble agar	15.0g
$Na_2HPO_4 \cdot 12H_2O$	9.0g
KH_2PO_4	1.5g
NH_4Cl	1.0g
Sodium succinate	1.0g
$MgSO_4 \cdot 7H_2O$	0.2g
$CaCl_2 \cdot 2H_2O$	0.01g
Ferric ammonium citrate	5.0mg
Trace elements solution	3.0mL

pH 7.1 ± 0.2 at 25°C

Trace Elements Solution:

Composition per liter:

H_3BO_3	30.0mg
$CoCl_2 \cdot 6H_2O$	20.0mg
$ZnSO_4 \cdot 7H_2O$	10.0mg
$MnCl_2 \cdot 4H_2O$	3.0mg
$Na_2MoO_4 \cdot 2H_2O$	3.0mg
$NiCl_2 \cdot 6H_2O$	2.0mg
$CuCl_2 \cdot 2H_2O$	1.0mg

Preparation of Trace Elements Solution: Add components to distilled/deionized water and bring volume to 1.0L. Mix thoroughly.

Preparation of Medium: Add components to double-distilled water and bring volume to 1.0L. Mix thoroughly. Gently heat and bring to boiling. Autoclave for 15 min at 15 psi pressure–121°C. Pour into sterile Petri dishes or distribute into sterile tubes.

Use: For the heterotrophic cultivation and maintenance of *Aquaspirillum autotrophicum*.

Aquaspirillum Heterotrophic Broth

Composition per liter:

$Na_2HPO_4 \cdot 12H_2O$	9.0g
KH_2PO_4	1.5g
NH_4Cl	1.0g
Sodium succinate	1.0g
$MgSO_4 \cdot 7H_2O$	0.2g
$CaCl_2 \cdot 2H_2O$	0.01g
Ferric ammonium citrate	5.0mg
Trace elements solution	3.0mL

pH 7.1 ± 0.2 at 25°C

Trace Elements Solution:

Composition per liter:

H_3BO_3	30.0mg
$CoCl_2 \cdot 6H_2O$	20.0mg
$ZnSO_4 \cdot 7H_2O$	10.0mg
$MnCl_2 \cdot 4H_2O$	3.0mg
$Na_2MoO_4 \cdot 2H_2O$	3.0mg
$NiCl_2 \cdot 6H_2O$	2.0mg
$CuCl_2 \cdot 2H_2O$	1.0mg

Preparation of Trace Elements Solution: Add components to distilled/deionized water and bring volume to 1.0L. Mix thoroughly.

Preparation of Medium: Add components to double-distilled water and bring volume to 1.0L. Mix thoroughly. Gently heat and bring to boiling. Autoclave for 15 min at 15 psi pressure–121°C. Aseptically distribute into sterile tubes or flasks.

Use: For the heterotrophic cultivation of *Aquaspirillum autotrophicum*.

Aquaspirillum Medium (DSMZ Medium 888)

Composition per 1026mL:

Casamino acids	1.5g
$(NH_4)_2SO_4$	1.0g
$MgSO_4 \cdot 7H_2O$	1.0g
Agar	0.5g
$CaCl_2 \cdot 6H_2O$	30.0mg
$Na_2H_2PO_4$	10.0mg
Sodium succinate solution	10.0mL
Thiosulfate solution	10.0mL
Standard vitamin solution	5.0mL
Trace elements solution SL-10	1.0mL

pH 7.5 ± 0.2 at 25°C

Standard Vitamin Solution:

Composition per 100.0mL:

Thiamine-HCl·$2H_2O$	50.0mg
Nicotinic acid	50.0mg
Pyridoxine-HCl	50.0mg
Ca-pantothenate	50.0mg
Riboflavin	10.0mg
Vitamin B_{12}	1.0mg
Folic acid	0.2mg
Biotin	0.1mg

Preparation of Standard Vitamin Solution: Add components to distilled/deionized water and bring volume to 100.0mL. Mix thoroughly. Filter sterilize.

Trace Elements Solution SL-10:

Composition per liter:

$FeCl_2 \cdot 4H_2O$	1.5g
H_3BO_3	300.0mg
$CoCl_2 \cdot 6H_2O$	190.0mg
$MnCl_2 \cdot 4H_2O$	100.0mg
$ZnCl_2$	70.0mg
$Na_2MoO_4 \cdot 2H_2O$	36.0mg
$NiCl_2 \cdot 6H_2O$	24.0mg
$CuCl_2 \cdot 2H_2O$	2.0mg
HCl (25% solution)	7.7mL

Preparation of Trace Elements Solution SL-10: Add $FeCl_2 \cdot 4H_2O$ to 10.0mL of HCl solution. Mix thoroughly. Add distilled/deionized water and bring volume to 1.0L. Add remaining components. Mix thoroughly. Sparge with 100% N_2. Autoclave for 15 min at 15 psi pressure–121°C. Cool to room temperature.

Thiosulfate Solution:

Composition per 10.0mL:

$Na_2S_2O_3 \cdot 5H_2O$	1.0g

Preparation of Thiosulfate Solution: Add $Na_2S_2O_3 \cdot 5H_2O$ to distilled/deionized water and bring volume to 10.0mL. Mix thoroughly. Autoclave for 15 min at 15 psi pressure–121°C. Cool to room temperature.

Sodium Succinate Solution:

Composition per 10.0mL:

Sodium succinate	1.0g

Preparation of Sodium Succinate Solution: Add sodium succinate to distilled/deionized water and bring volume to 10.0mL. Mix thoroughly. Autoclave for 15 min at 15 psi pressure–121°C. Cool to room temperature.

Preparation of Medium: Add components, except sodium succinate solution, thiosulfate solution, standard vitamin solution, and trace elements solution SL-10, to distilled/deionized water and bring volume to 1.0L. Mix thoroughly. Autoclave for 15 min at 15 psi pressure–121°C. Cool to room temperature. Aseptically add 10.0mL sterile sodium succinate solution, 10.0mL sterile thiosulfate solution, 5.0mL sterile standard vitamin solution, and 1.0mL sterile trace elements solution SL-10. Mix thoroughly. Adjust pH to 7.5. Aseptically distribute into sterile tubes or flasks.

Use: For the cultivation of *Aquaspirillum* spp.

Aquincola Medium (DSMZ Medium 1178)

Composition per liter:

Yeast extract	1.0g
Peptone	1.0g
Fructose	0.5g

pH 7.0 ± 0.2 at 25°C

Preparation of Medium: Add components to tap water and bring volume to 1.0L. Mix thoroughly. Gently heat and bring to boiling. Distribute into tubes or flasks. Autoclave for 15 min at 15 psi pressure–121°C.

Use: For the cultivation and maintenance of *Aquincola* spp.

Aquisalimonas Agar (DSMZ Medium 1182)

Composition per liter:

Solution A	500.0mL
Solution B	500.0mL

pH 9.5 ± 0.2 at 25°C

Solution A:
Composition per 500.0mL:

Agar	20.0g
Glucose	10.0g
Peptone	5.0g
Yeast extract	5.0g
K_2HPO_4	1.0g
$MgCl_2 \cdot 6H_2O$	0.2g

Preparation of Solution A: Add components to distilled/deionized water and bring volume to 500.0mL. Mix thoroughly. Autoclave for 15 min at 15 psi pressure–121°C. Cool to 50°C.

Solution B:
Composition per 500.0mL:

NaCl	80.0g
Na_2CO_3	20.0g

Preparation of Solution B: Add components to distilled/deionized water and bring volume to 500.0mL. Mix thoroughly. Autoclave for 15 min at 15 psi pressure–121°C. Cool to 50°C.

Preparation of Medium: Aseptically combine 500.0mL solution A and 500.0mL solution B. The final pH should be 9.5. Pour into Petri dishes or distribute into sterile tubes.

Use: For the cultivation and maintenance of *Aquisalimonas* spp.

Aquisalimonas Medium (DSMZ Medium 1182)

Composition per liter:

Solution A	500.0mL
Solution B	500.0mL

pH 9.5 ± 0.2 at 25°C

Solution A:
Composition per 500.0mL:

Glucose	10.0g
Peptone	5.0g
Yeast extract	5.0g
K_2HPO_4	1.0g
$MgCl_2 \cdot 6H_2O$	0.2g

Preparation of Solution A: Add components to distilled/deionized water and bring volume to 500.0mL. Mix thoroughly. Autoclave for 15 min at 15 psi pressure–121°C. Cool to room temperature.

Solution B:
Composition per 500.0mL:

NaCl	80.0g
Na_2CO_3	20.0g

Preparation of Solution B: Add components to distilled/deionized water and bring volume to 500.0mL. Mix thoroughly. Autoclave for 15 min at 15 psi pressure–121°C. Cool to room temperature.

Preparation of Medium: Aseptically combine 500.0mL solution A and 500.0mL solution B. The final pH should be 9.5.

Use: For the cultivation and maintenance of *Aquisalimonas* spp.

Arabinose Agar Base with Selective Supplement

Composition per liter:

Peptone, special	23.0g
Agar	15.0g
Arabinose	10.0g
NaCl	5.0g
Corn starch	1.0g
Phenol Red	0.1g
Selective supplement solution	10.0mL

pH 7.8 ± 0.2 at 25°C

Source: This medium is available from HiMedia.

Selective Supplement Solution:
Composition per 10.0mL:

Thallium acetate	0.2g
Nalidixic acid	25.0mg

Preparation of Selective Supplement Solution: Add components to distilled/deionized water and bring volume to 10.0mL. Mix thoroughly. Filter sterilize.

Preparation of Medium: Add components, except selective supplement solution, to distilled/deionized water and bring volume to 990.0mL. Mix thoroughly. Adjust pH to 7.8. Gently heat and bring to boiling. Distribute into tubes or flasks. Gently heat and bring to boil. Do not autoclave. Cool to 50°C. Add selective supplement solution. Mix thoroughly. Pour into sterile Petri dishes.

Use: For selective isolation of *Enterococcus faecium* from feces, sewage, and water supplies.

ARC51 Medium (DSMZ Medium 1098)

Composition per liter:

Sea salts, Sigma	35.0g
Sodium acetate	1.6g
Mercaptoethanesulfonic acid (coenzyme M)	0.14g
NH_4Cl	0.1g
Yeast extract	0.1g
K_2HPO_4	0.05g

Resazurin	0.5mg
$NaHCO_3$ solution	10.0mL
$Na_2S\cdot 9H_2O$ solution	10.0mL
$Na_2S_2O_3$ solution	10.0mL
Wolfe's mineral elixer	1.0mL
Seven vitamin solution	1.0mL

pH 6.5 ± 0.2 at 25°C

Seven Vitamin Solution:

Composition per liter:

Pyridoxine hydrochloride	300.0mg
Thiamine-HCl·$2H_2O$	200.0mg
Nicotinic acid	200.0mg
Vitamin B_{12}	100.0mg
Calcium pantothenate	100.0mg
p-Aminobenzoic acid	80.0mg
D(+)-Biotin	20.0mg

Preparation of Seven Vitamin Solution: Add components to distilled/deionized water and bring volume to 1.0L. Sparge with 100% N_2. Mix thoroughly. Filter sterilize.

Wolfe's Mineral Elixir:

Composition per liter:

$MgSO_4\cdot 7H_2O$	30.0g
NaCl	10.0g
$MnSO_4\cdot 2H_2O$	5.0g
$(NH_4)_2NiSO_4\cdot 6H_2O$	2.8g
$CoCl_2\cdot 6H_2O$	1.8g
$ZnSO_4\cdot 7H_2O$	1.8g
$FeSO_4\cdot 7H_2O$	1.0g
$CaCl_2\cdot 2H_2O$	1.0g
$KAl(SO_4)_2\cdot 12H_2O$	0.18g
$CuSO_4\cdot 5H_2O$	0.1g
H_3BO_3	0.1g
$Na_2MoO_4\cdot 2H_2O$	0.1g
Na_2SeO_4	0.1g
$Na_2WO_4\cdot 2H_2O$	0.1g

Preparation of Wolfe's Mineral Elixir: Adjust pH of 1.0L of distilled/deionized water to 1.0 with dilute H_2SO_4. Add remaining components one at a time. Mix thoroughly to dissolve.

$Na_2S\cdot 9H_2O$ Solution:

Composition per 10.0mL:

$Na_2S\cdot 9H_2O$	0.5g

Preparation of $Na_2S\cdot 9H_2O$ Solution: Add $Na_2S\cdot 9H_2O$ to distilled/deionized water and bring volume to 10.0mL. Mix thoroughly. Autoclave under 100% N_2 for 15 min at 15 psi pressure–121°C. Cool to room temperature.

$Na_2S_2O_3$ Solution:

Composition per 10.0mL:

$Na_2S_2O_3\cdot 5H_2O$	2.5g

Preparation of $NaHCO_3$ Solution: Add $Na_2S_2O_3\cdot 5H_2O$ to distilled/deionized water and bring volume to 10.0mL. Mix thoroughly. Sparge with N_2. Autoclave for 15 min at 15 psi pressure–121°C. Cool to room temperature.

$NaHCO_3$ Solution:

Composition per 50.0mL:

$NaHCO_3$	2.0g

Preparation of $NaHCO_3$ Solution: Add $NaHCO_3$ to distilled/deionized water and bring volume to 50.0mL. Mix thoroughly. Sparge with N_2. Autoclave for 15 min at 15 psi pressure–121°C. Cool to room temperature.

Preparation of Medium: Add components, except vitamin solution, bicarbonate solution, thiosulfate solution, and $Na_2S\cdot 9H_2O$ solution, to distilled/deionized water and bring volume to 970.0mL. Mix thoroughly. Gently heat and bring to boiling. Boil for 3 min. Cool to room temperature while sparging with 20% CO_2 + 80% N_2. Dispense into balch tubes or serum bottles under atmosphere of 20% CO_2 + 80% N_2. Autoclave for 15 min at 15 psi pressure–121°C. Cool to 25°C under an atmosphere of 20% CO_2 + 80% N_2. Aseptically and anaerobically add sterile vitamin solution, bicarbonate solution, thiosulfate solution, and $Na_2S\cdot 9H_2O$ solution. Adjust final pH to 6.5. After inoculation adjust overpressure to 1.5 bar with 20% CO_2 + 80% N_2.

Use: For the cultivation and maintenance of *Archaeoglobus infectus.*

Archaeoglobus Medium

Composition per liter:

NaCl	18.0g
$NaHCO_3$	5.0g
$MgCl_2\cdot 6H_2O$	4.0g
$MgSO_4\cdot 7H_2O$	3.45g
Sodium L-lactate	1.5g
Yeast extract	0.5g
KCl	0.34g
NH_4Cl	0.25g
$CaCl_2\cdot 2H_2O$	0.14g
K_2HPO_4	0.14g
$Fe(NH_4)_2(SO_4)_2\cdot 7H_2O$	2.0mg
Resazurin	1.0mg
$Na_2S\cdot 9H_2O$ solution	25.0mL
Trace elements solution	10.0mL

pH 6.9 ± 0.2 at 25°C

Trace Elements Solution:

Composition per liter:

$MgSO_4\cdot 7H_2O$	3.0g
Nitrilotriacetic acid	1.5g
NaCl	1.0g
$MnSO_4\cdot 2H_2O$	0.5g
$CoSO_4\cdot 7H_2O$	0.18g
$ZnSO_4\cdot 7H_2O$	0.18g
$FeSO_4\cdot 7H_2O$	0.1g
$CaCl_2\cdot 2H_2O$	0.1g
$NiCl_2\cdot 6H_2O$	0.025g
$KAl(SO_4)_2\cdot 12H_2O$	0.02g
$CuSO_4\cdot 5H_2O$	0.01g
H_3BO_3	0.01g
$Na_2MoO_4\cdot 2H_2O$	0.01g
$Na_2SeO_3\cdot 5H_2O$	0.3mg

$Na_2S\cdot 9H_2O$ Solution:

Composition per 50.0mL:

$Na_2S\cdot 9H_2O$	1.0g

Preparation of $Na_2S\cdot 9H_2O$ Solution: Prepare and dispense solution anaerobically under 80% N_2 + 20% CO_2. Add $Na_2S\cdot 9H_2O$ to distilled/deionized water and bring volume to 50.0mL. Mix thoroughly. Adjust pH to 7.0. Autoclave for 15 min at 15 psi pressure–121°C.

Preparation of Trace Elements Solution: Add nitrilotriacetic acid to approximately 500.0mL of distilled/deionized water. Dissolve by adding KOH and adjust pH to 6.5. Add remaining components.

Bring volume to 1.0L with additional distilled/deionized water. Adjust pH to 7.0 with KOH.

Preparation of Medium: Add components, except $NaHCO_3$ and $Na_2S{\cdot}9H_2O$ solution, to distilled/deionized water and bring volume to 1.0L. Mix well and heat to boiling for a few minutes. Cool rapidly to room temperature while gassing with 80% N_2 + 20% CO_2. Add $NaHCO_3$ and adjust pH to 6.9. Distribute anaerobically under 80% N_2 + 20% CO_2 and pressurize sealed containers up to 2 bar pressure. Autoclave for 15 min at 15 psi pressure–121°C. Prior to inoculation of cultures, add 0.25mL of sterile $Na_2S{\cdot}9H_2O$ solution to each tube containing 9.75mL of sterile basal medium.

Use: For the cultivation and maintenance of *Archaeoglobus fulgidus*.

Archaeoglobus profundus Medium

Composition per liter:

NaCl	18.0g
$MgCl_2{\cdot}6H_2O$	4.0g
$MgSO_4{\cdot}7H_2O$	3.45g
Na_2SO_4	2.7g
Sodium acetate	1.0g
$NaHCO_3$	1.0g
Yeast extract	0.5g
KCl	0.34g
NH_4Cl	0.25g
$CaCl_2{\cdot}2H_2O$	0.14g
K_2HPO_4	0.14g
$Fe(NH_4)_2(SO_4)_2{\cdot}7H_2O$	2.0mg
Resazurin	1.0mg
$Na_2S{\cdot}9H_2O$ solution	25.0mL
Trace elements solution	10.0mL

pH 6.5 ± 0.2 at 25°C

Trace Elements Solution:

Composition per liter:

$MgSO_4{\cdot}7H_2O$	3.0g
Nitrilotriacetic acid	1.5g
NaCl	1.0g
$MnSO_4{\cdot}2H_2O$	0.5g
$CoSO_4{\cdot}7H_2O$	0.18g
$ZnSO_4{\cdot}7H_2O$	0.18g
$FeSO_4{\cdot}7H_2O$	0.1g
$CaCl_2{\cdot}2H_2O$	0.1g
$NiCl_2{\cdot}6H_2O$	0.025g
$KAl(SO_4)_2{\cdot}12H_2O$	0.02g
$CuSO_4{\cdot}5H_2O$	0.01g
H_3BO_3	0.01g
$Na_2MoO_4{\cdot}2H_2O$	0.01g
$Na_2SeO_3{\cdot}5H_2O$	0.3mg

$Na_2S{\cdot}9H_2O$ Solution:

Composition per 50.0mL:

$Na_2S{\cdot}9H_2O$	1.0g

Preparation of Trace Elements Solution: Add nitrilotriacetic acid to approximately 500.0mL of distilled/deionized water. Dissolve by adding KOH and adjust pH to 6.5. Add remaining components. Bring volume to 1.0L with additional distilled/deionized water. Adjust pH to 7.0 with KOH.

Preparation of $Na_2S{\cdot}9H_2O$ Solution: Prepare and dispense solution anaerobically under 80% N_2 + 20% CO_2. Add $Na_2S{\cdot}9H_2O$ to distilled/deionized water and bring volume to 50.0mL. Mix thoroughly. Adjust pH to 7.0. Autoclave for 15 min at 15 psi pressure–121°C.

Preparation of Medium: Add components, except $NaHCO_3$ and $Na_2S{\cdot}9H_2O$ solution, to distilled/deionized water and bring volume to 1.0L. Mix well and heat to boiling for a few minutes . Cool rapidly to room temperature while gassing with 80% N_2 + 20% CO_2. Add $NaHCO_3$ and adjust pH to 6.9. Distribute anaerobically under 80% N_2 + 20% CO_2 and pressurize sealed containers up to 2 bar pressure. Autoclave for 15 min at 15 psi pressure–121°C. Prior to inoculation of cultures, add 0.25mL of sterile $Na_2S{\cdot}9H_2O$ solution to each tube containing 9.75mL of sterile basal medium.

Use: For the cultivation and maintenance of *Archaeoglobus profundus*.

Archaeoglobus veneficus Medium (DSMZ Medium 796)

Composition per liter:

NaCl	18.0g
$MgCl_2{\cdot}6H_2O$	7.15g
$NaHCO_3$	5.0g
KCl	0.33g
NH_4Cl	0.25g
$K_2HPO_4{\cdot}3H_2O$	0.18g
$CaCl_2{\cdot}2H_2O$	0.14g
$Fe(NH_4)_2(SO_4)_2{\cdot}6H_2O$	2.0mg
Resazurin	0.5mg
Na_2SO_3 solution	20.0mL
$Na_2S{\cdot}9H_2O$ solution	10.0mL
Na-acetate solution	4.0mL
Trace elements solution	1.0mL

pH 1.0 ± 0.2 at 25°C

$Na_2S{\cdot}9H_2O$ Solution:

Composition per 20.0mL:

$Na_2S{\cdot}9H_2O$	0.5g

Preparation of $Na_2S{\cdot}9H_2O$ Solution: Add $Na_2S{\cdot}9H_2O$ to distilled/deionized water and bring volume to 20.0mL. Sparge with N_2. Autoclave for 15 min at 15 psi pressure–121°C. Cool to 25°C. Store anaerobically.

Na_2SO_3 Solution:

Composition per 10.0mL:

Na_2SO_3	0.5g

Preparation of Na_2SO_3 Solution: Add Na_2SO_3 to distilled/deionized water and bring volume to 10.0mL. Mix thoroughly. Sparge with 100% N_2. Filter sterilize.

Na-acetate Solution:

Composition per 10.0mL:

Na-acetate	2.5g

Preparation of Na-acetate Solution: Add Na-acetate to distilled/deionized water and bring volume to 10.0mL. Mix thoroughly. Sparge with 100% N_2. Filter sterilize.

Trace Elements Solution:

Composition per liter:

NaCl	5.0g
$MnCl_2{\cdot}4H_2O$	2.9g
$(NH_4)_2Ni(SO_4)_2$	1.0g
$FeSO_4{\cdot}7H_2O$	0.5g
$CoCl_2{\cdot}6H_2O$	0.5g
$CaCl_2{\cdot}2H_2O$	0.5g
$ZnSO_4{\cdot}7H_2$	0.5g
$CuSO_4{\cdot}5H_2O$	0.05g

H_3BO_3	0.05g
$KAl(SO_4)_2 \cdot 12H_2O$	0.05g
$Na_2MoO_4 \cdot 4H_2O$	0.05g
$Na_2WO_4 \cdot 2H_2O$	0.05g
$Na_2SeO_3 \cdot 5H_2O$	0.05g

Preparation of Trace Elements olution: Add components to distilled/deionized water and bring volume to 1.0L. Mix thoroughly. Sparge with 100% N_2.

Preparation of Medium: Prepare and dispense medium under 80% N_2 + 20% CO_2 gas mixture. Add components, except Na-acetate solution, $Na_2S \cdot 9H_2O$ solution, and $NaHCO_3$, to 986.0mL distilled/deionized water. Mix thoroughly. Gently heat and bring to boiling. Boil for 5 min. Cool to 25°C while sparging with 80% N_2 + 20% CO_2. Add the solid $NaHCO_3$. Equilibrate with the 80% N_2 + 20% CO_2 gas. Adjust pH to 7.0 with HCl. Add 10.0mL $Na_2S \cdot 9H_2O$ solution. Distribute into serum bottles under 80% N_2 + 20% CO_2 gas; 25.0mL medium in 100mL serum bottles. Adjust pH to 1.0. Autoclave for 15 min at 15 psi pressure–121°C. Cool to 25°C. Aseptically and anaerobically inject 0.1mL sterile Na-acetate solution and 0.5mL Na_2SO_3 solution per 25.0mL medium. Mix thoroughly. After inoculation pressurize to 1 bar atmosphere with 80% H_2 + 20% CO_2 gas.

Use: For the cultivation of *Archaeoglobus veneficus.*

Archangium violaceum Medium

Composition per liter:

Monosodium glutamate	1.0g
L-Leucine	0.5g
L-Tyrosine	0.5g
L-Isoleucine	0.3g
L-Proline	0.25g
$MgSO_4 \cdot 7H_2O$	0.2g
L-Lysine	0.15g
L-Arginine	0.1g
L-Asparagine	0.1g
L-Serine	0.1g
L-Threonine	0.1g
L-Valine	0.1g
L-Alanine	0.05g
L-Glycine	0.05g
L-Histidine	0.05g
L-Methionine	0.05g
$Ca_3(PO_4)_2$	0.02g
KCl	0.02g
Tris(hydroxymethyl)aminomethane buffer (0.02m solution, pH 7.5)	1.0L

pH 7.5 ± 0.2 at 25°C

Preparation of Medium: Add solid components to 1.0L of Tris buffer. Mix thoroughly. Filter sterilize. Aseptically distribute into tubes or flasks.

Use: For the cultivation of *Archangium violaceum.*

Arcobacter Broth Base with Selective Supplement

Composition per liter:

Peptone	18.0g
NaCl	5.0g
Yeast extract	1.0g
Selective supplement solution	10.0mL

pH 7.2 ± 0.2 at 25°C

Source: This medium is available from HiMedia.

Selective Supplement Solution:

Composition per 10.0mL:

Cefoperazone	16.0mg
Amphotericin B	5.0mg

Preparation of Selective Supplement Solution: Add components to distilled/deionized water and bring volume to 10.0mL. Mix thoroughly. Filter sterilize.

Preparation of Medium: Add components, except selective supplement solution, to distilled/deionized water and bring volume to 990.0mL. Mix thoroughly. Autoclave for 15 min at 15 psi pressure–121°C. Cool to 50°C. Aseptically add selective supplement solution. Mix thoroughly. Pour into Petri dishes or aseptically distribute into sterile tubes.

Use: For the enrichment and cultivation of *Arcobacter* spp.

Arcobacter Medium

Composition per liter:

Agar	15.0g
Special peptone	10.0g
Beef extract	5.0g
NaCl	5.0g
Yeast extract	5.0g
Sodium glutamate	2.0g
Sodium succinate	2.0g
$MgCl_2 \cdot H_2O$	1.0g
Horse blood, defibrinated	50.0mL

pH 7.0 ± 0.2 at 25°C

Source: Special peptone is available from Oxoid Unipath.

Preparation of Medium: Add components, except horse blood, to distilled/deionized water and bring volume to 950.0mL. Mix thoroughly. Gently heat and bring to boiling. Adjust pH to 7.0. Autoclave for 15 min at 15 psi pressure–121°C. Cool to 50°–55°C. Aseptically add 50.0mL of sterile horse blood. Mix thoroughly. Pour into sterile Petri dishes or distribute into sterile tubes.

Use: For the cultivation and maintenance of *Arcobacter* species.

Arcobacter nitrofigilis Agar (LMG Medium 86)

Composition per liter:

Agar	15.0g
NaCl	15.0g
Lab-Lemco beef extract	10.0g
Special peptones, Oxoid	10.0g

pH 7.1 ± 0.2 at 25°C

Preparation of Medium: Add components to distilled/deionized water and bring volume to 1.0L. Mix thoroughly. Gently heat and bring to boiling. Distribute into tubes or flasks. Autoclave for 15 min at 15 psi pressure–121°C. Pour into sterile Petri dishes or leave in tubes.

Use: For the cultivation and maintenance of *Arcobacter nitrofigilis.*

Arenavirus Plaquing Medium

Composition per liter:

Eagle's basal medium	1.0L
Agarose solution	1.0L
Fetal calf serum, inactivated	100.0mL

pH 7.0 ± 0.2 at 25°C

Eagle's Basal Medium:
Composition per liter:

HEPES (*N*-2-Hydroxyethylpiperazine-*N'*-2-ethanesulfonic acid) buffer	9.53g
NaCl	6.8g
$NaHCO_3$	2.2g
Glucose	1.0g
KCl	0.4g
$CaCl_2 \cdot 2H_2O$	0.2g
NaH_2PO_4	0.125g
$MgSO_4 \cdot 7H_2O$	0.1g
L-Isoleucine	0.026g
L-Leucine	0.026g
L-Lysine	0.026g
L-Threonine	0.024g
L-Valine	0.0235g
L-Tyrosine	0.018g
L-Arginine	0.0174g
L-Phenylalanine	0.0165g
L-Cystine	0.012g
L-Histidine	8.0mg
L-Methionine	7.5mg
L-Tryptophan	4.0mg
Inositol	1.8mg
Biotin	1.0mg
Calcium pantothenate	1.0mg
Choline chloride	1.0mg
Folic acid	1.0mg
Nicotinamide	1.0mg
Pyridoxal·HCl	1.0mg
Thiamine·HCl	1.0mg
Riboflavin	0.1mg

pH 7.2–7.4 at 25°C

Source: This medium is available as a premixed powder from BD Diagnostic Systems.

Preparation of Eagle's Basal Medium: Add components to distilled/deionized water and bring volume to 1.0L. Mix thoroughly. Adjust pH to 7.0 with NaOH. Filter sterilize.

Agarose Solution:
Composition per liter:

Agarose	20.0g

Preparation of Agarose Solution: Add agarose to distilled/deionized water and bring volume to 1.0L. Mix thoroughly. Autoclave for 15 min at 15 psi pressure–121°C. Cool to 45°–50°C.

Preparation of Medium: To 1.0L of sterile Eagle's basal medium, aseptically add 1.0L of sterile, cooled agarose solution and 100.0mL of fetal calf serum. Mix thoroughly. Pour into sterile Petri dishes.

Use: For the cultivation of animal tissue culture cells used for the growth of arenaviruses.

Arginine Assay Medium
See: **Amino Acid Assay Medium**

Arginine Broth (BAM M44)

Composition per liter:

L-Arginine	5.0g
Peptone or gelysate peptone	5.0g
Yeast extract	3.0g
Glucose	1.0g
Bromcresol Purple	0.02g

pH 6.5 ± 0.2 at 25°C

Preparation of Medium: Add components to distilled/deionized water and bring volume to 1.0L. Mix thoroughly. Adjust pH so that it will be 6.5 ± 0.2 after sterilization. Distribute into 16 × 150mm screw-capped tubes in 5.0mL volumes. Autoclave medium with loosely capped tubes for 10 min at 15 psi pressure–121°C. Screw the caps on tightly for storage and after inoculation.

Use: For the cultivation and differentiation of bacteria based on their ability to decarboxylate the amino acid arginine. Bacteria that decarboxylate arginine turn the medium turbid purple.

Arginine Broth with Sodium Chloride (BAM M44)

Composition per liter:

L-Arginine	5.0g
Peptone or gelysate peptone	5.0g
Yeast extract	3.0g
Glucose	1.0g
Bromcresol Purple	0.02g

pH 6.5 ± 0.2 at 25°C

Preparation of Medium: Add components to distilled/deionized water and bring volume to 1.0L. Mix thoroughly. Adjust pH so that it will be 6.5 ± 0.2 after sterilization. Distribute into 16 × 150mm screw-capped tubes in 5.0mL volumes. Autoclave medium with loosely capped tubes for 10 min at 15 psi pressure–121°C. Screw the caps on tightly for storage and after inoculation.

Use: For the cultivation and differentiation of *Vibrio* spp. based on their ability to decarboxylate the amino acid arginine. Bacteria that decarboxylate arginine turn the medium turbid purple.

Arginine Dihydrolase Broth

Composition per liter:

L-Arginine	10.0g
NaCl	5.0g
Agar	3.0g
Peptone	1.0g
K_2HPO_4	0.3g
Bromcresol Purple	0.016g

pH 6.0 ± 0.2 at 25°C

Source: This medium is available as a premixed powder from HiMedia.

Preparation of Medium: Add components to distilled/deionized water and bring volume to 1.0L. Mix thoroughly. Adjust pH so that it will be 6.5 ± 0.2 after sterilization. Distribute into 16 × 150mm screw-capped tubes in 5.0mL volumes. Autoclave medium with loosely capped tubes for 10 min at 15 psi pressure–121°C. Screw the caps on tightly for storage and after inoculation.

Use: For the cultivation and differentiation of bacteria based on their ability to decarboxylate the amino acid arginine. Bacteria that decarboxylate arginine turn the medium turbid purple.

Arginine Dihydrolase HiVeg Broth

Composition per liter:

L-Arginine	10.0g
NaCl	5.0g
Agar	3.0g

Plant peptone.......1.0g
K_2HPO_4.......0.3g
Bromcresol Purple.......0.016g

pH 6.0 ± 0.2 at 25°C

Source: This medium is available as a premixed powder from HiMedia.

Preparation of Medium: Add components to distilled/deionized water and bring volume to 1.0L. Mix thoroughly. Adjust pH so that it will be 6.5 ± 0.2 after sterilization. Distribute into 16 × 150mm screw-capped tubes in 5.0mL volumes. Autoclave medium with loosely capped tubes for 10 min at 15 psi pressure–121°C. Screw the caps on tightly for storage and after inoculation.

Use: For the cultivation and differentiation of bacteria based on their ability to decarboxylate the amino acid arginine. Bacteria that decarboxylate arginine turn the medium turbid purple.

Arginine Dihydrolase Medium, Modified

Composition per liter:

L-Arginine.......5.0g
Yeast extract.......3.0g
Glucose.......1.0g
Bromcresol Purple.......0.015g

pH 6.8 ± 0.2 at 25°C

Source: This medium is available as a premixed powder from HiMedia.

Preparation of Medium: Add components to distilled/deionized water and bring volume to 1.0L. Mix thoroughly. Adjust pH so that it will be 6.5 ± 0.2 after sterilization. Distribute into 16 × 150mm screw-capped tubes in 5.0mL volumes. Autoclave medium with loosely capped tubes for 10 min at 15 psi pressure–121°C. Screw the caps on tightly for storage and after inoculation.

Use: For the cultivation and differentiation of bacteria based on their ability to decarboxylate the amino acid arginine. For the confirmation of *Enterobacter sakazakii* from milk and dairy products.

Arginine Glucose Slants (AGS)

Composition per liter:

NaCl.......20.0g
Agar.......13.5g
Pancreatic digest of casein.......10.0g
L-Arginine·HCl.......5.0g
Peptone.......5.0g
Yeast extract.......3.0g
Glucose.......1.0g
Ferric ammonium citrate.......0.5g
$Na_2S_2O_3 \cdot 5H_2O$.......0.3g
Bromcresol Purple.......0.02g

pH 6.8–7.0 at 25°C

Preparation of Medium: Add components to distilled/deionized water and bring volume to 1.0L. Mix thoroughly. Gently heat and bring to boiling. Distribute into tubes. Autoclave for 12 min at 15 psi pressure–121°C. Allow tubes to cool in a slanted position.

Use: For the cultivation and differentation of *Vibrio* species.

Arhodomonas Medium (DSMZ Medium 941)

Composition per liter:

NaCl.......150.0g
NH_4Cl.......1.0g
Na-acetate.......1.0g
$MgSO_4 \cdot 7H_2O$.......0.2g
KCl.......0.1g
KH_2PO_4.......0.1g
Peptone.......0.1g
$CaCl_2 \cdot 2H_2O$.......0.04g
Trace elements solution SL-7.......1.0mL
Vitamin solution, concentrated.......1.0mL

pH 7.2 ± 0.2 at 25°C

Trace Elements Solution SL-7:
Composition per liter:

$FeCl_2 \cdot 7H_2O$.......1.5g
$CoCl_2 \cdot 6H_2O$.......190.0mg
$MnCl_2 \cdot 4H_2O$.......100.0mg
$ZnCl_2$.......70.0mg
$Na_2MoO_4 \cdot 2H_2O$.......36.0mg
$NiCl_2 \cdot 6H_2O$.......24.0mg
H_3BO_3.......62.0mg
$CuCl_2 \cdot 2H_2O$.......17.0mg
HCl (25% solution).......6.5mL

Preparation of Trace Elements Solution SL-7: Add $FeCl_2 \cdot 7H_2O$ to 10.0mL of HCl solution. Mix thoroughly. Add distilled/deionized water and bring volume to 1.0L. Add remaining components. Mix thoroughly. Sparge with 80% N_2 + 20% CO_2. Autoclave for 15 min at 15 psi pressure–121°C.

Vitamin Solution, Concentrated:
Composition per 100.0mL:

Pyridoxine-HCl.......10.0mg
Thiamine-HCl·$2H_2O$.......5.0mg
Riboflavin.......5.0mg
Nicotinic acid.......5.0mg
D-Ca-pantothenate.......5.0mg
p-Aminobenzoic acid.......5.0mg
Lipoic acid.......5.0mg
Biotin.......2.0mg
Folic acid.......2.0mg
Vitamin B_{12}.......0.1mg

Preparation of Vitamin Solution, Concentrated: Add components to distilled/deionized water and bring volume to 100.0mL. Mix thoroughly. Filter sterilize.

Preparation of Medium: Add components to distilled/deionized water and bring volume to 1.0L. Mix thoroughly. Distribute into tubes or flasks. Autoclave for 15 min at 15 psi pressure–121°C.

Use: For the cultivation of *Arhodomonas aquaeolei.*

Armstrong *Fusarium* Medium

Composition per liter:

Glucose.......20.0g
$Ca(NO_3)_2 \cdot 4H_2O$.......8.4g
KH_2PO_4.......1.09g
KCl.......0.22g
$FeCl_3$.......0.2μg
$MnSO_4$.......0.2μg
$ZnSO_4$.......0.2μg

Preparation of Medium: Add components to distilled/deionized water and bring volume to 1.0L. Mix thoroughly. Filter sterilize.

Use: For the cultivation of *Fusarium* species.

Arthrobacter Broth

Composition per liter:

Glucose	10.0g
Yeast extract	7.0g
K_2HPO_4	1.0g
KNO_3	0.5g
$MgSO_4·7H_2O$	0.2g
$CaCl_2·2H_2O$	0.1g
$FeCl_3·6H_2O$	10.0mg

Preparation of Medium: Add components to tap water and bring volume to 1.0L. Mix thoroughly. Gently heat and bring to boiling. Distribute into tubes or flasks. Autoclave for 15 min at 15 psi pressure–121°C. Pour into sterile Petri dishes or leave in tubes.

Use: For the cultivation and maintenance of *Arthrobacter atrocyaneus, Arthrobacter aurescens, Arthrobacter crystallopoietes, Arthrobacter globiformis, Arthrobacter histidinolovorans,* and *Arthrobacter oxydans.*

Arthrobacter Medium

Composition per liter:

Mannitol	10.0g
K_2HPO_4	1.77g
KH_2PO_4	0.68g
$MgSO_4·7H_2O$	0.2g
NaCl	0.14g
$CaCl_2$	0.132g
Yeast extract	0.08g
H_3BO_3	2.9mg
$FeSO_4·7H_2O$	2.5mg
$Na_2MoO_4·2H_2O$	2.5mg
$CoSO_4·7H_2O$	1.2mg
$ZnSO_4·7H_2O$	1.2mg
$CuSO_4·5H_2O$	0.1mg
$MnCl_2·4H_2O$	0.09mg

pH 7.0 ± 0.2 at 25°C

Preparation of Medium: Add components to distilled/deionized water and bring volume to 1.0L. Mix thoroughly. Gently heat and bring to boiling. Distribute into tubes or flasks. Autoclave for 15 min at 15 psi pressure–121°C.

Use: For the cultivation and maintenance of *Arthrobacter* species.

Arthrobacter Medium

Composition per liter:

Agar	10.0g
Casein	1.0g
Glucose	1.0g
K_2HPO_4	1.0g
Yeast extract	0.7g
$MgSO_4·7H_2O$	0.25g
$(NH_4)_2SO_4$	0.25g

pH 6.9–7.0 at 25°C

Preparation of Medium: Add components to tap water and bring volume to 1.0L. Mix thoroughly. Gently heat and bring to boiling. Autoclave for 15 min at 15 psi pressure–121°C. Pour into sterile Petri dishes.

Use: For the isolation, cultivation, and enumeration of *Arthrobacter* species from soil.

Arthrobacter Medium

Composition per liter:

Agar	15.0g
Peptone	10.0g
Yeast extract	10.0g
K_2HPO_4	2.0g
Rhodotorulic acid (δ-*N*-acetyl-L-ornithine) or desferal	20.0μg

pH 7.4 ± 0.2 at 25°C

Preparation of Medium: Add components to distilled/deionized water and bring volume to 1.0L. Mix thoroughly. Gently heat and bring to boiling. Distribute into tubes or flasks. Autoclave for 15 min at 15 psi pressure–121°C. Pour into sterile Petri dishes or leave in tubes.

Use: For the cultivation and maintenance of *Aureobacterium flavescens.*

Arthrobacter YCWD

Composition per liter:

Pancreatic digest of casein	10.0g
Yeast extract	1.0g

pH 7.2 ± 0.2 at 25°C

Preparation of Medium: Add components to distilled/deionized water and bring volume to 1.0L. Mix thoroughly. Gently heat and bring to boiling. Distribute into tubes or flasks. Autoclave for 15 min at 15 psi pressure–121°C.

Use: For the cultivation and maintenance of *Arthrobacter* species.

Artificial Deep Lake Medium

Composition per liter:

NaCl	180.0g
$MgCl_2·6H_2O$	75.0g
Noble agar	15.0g
Sodium succinate	10.0g
$MgSO_4·7H_2O$	7.4g
KCl	7.4g
$CaCl_2·2H_2O$	1.0g
Yeast extract	1.0g
Vitamin solution	10.0mL

pH 7.4 ± 0.2 at 25°C

Vitamin Solution:

Composition per liter:

Biotin	30.0mg
Cyanocobalamin	20.0mg
Thiamine·HCl	10.0mg

Preparation of Vitamin Solution: Add components to distilled/deionized water and bring volume to 1.0L. Mix thoroughly. Filter sterilize and add aseptically to sterile basal medium.

Preparation of Medium: Add components, except vitamin solution, to distilled/deionized water and bring volume to 990.0mL. Mix thoroughly. Gently heat and bring to boiling. Adjust medium to pH 7.4. Autoclave for 15 min at 15 psi pressure–121°C. Cool to 50°C. Aseptically add 10.0mL of vitamin solution. Pour into sterile Petri dishes or leave in tubes.

Use: For the cultivation and maintenance of *Halobacterium lacusprofundi*.

Artificial Organic Lake Medium
See: ***Halomonas subglaciescola*** **Medium**

Artificial Organic Lake Peptone Medium

Composition per 1001.0mL:

NaCl	30.0g
$MgSO_4·7H_2O$	9.5g
KCl	5.0g
Peptone	5.0g
Yeast extract	1.0g
$CaCl_2·2H_2O$	0.2g
KNO_3	0.1g
$(NH_4)_2SO_4$	0.1g
Modified Hutner's basal salts	20.0mL
Phosphate supplement	20.0mL
Artificial organic lake vitamin solution	1.0mL

pH 7.3 ± 0.2 at 25°C

Modified Hutner's Basal Salts:

Composition per liter:

$MgSO_4·7H_2O$	29.7g
Nitrilotriacetic acid	10.0g
$CaCl_2·2H_2O$	3.34g
$FeSO_4·7H_2O$	99.0mg
Ammonium molybdate	9.25mg
Metals "44"	50.0mL

Preparation of Modified Hutner's Basal Salts: Dissolve the nitrilotracetic acid first and neutralize the solution with KOH. Add the other components and adjust the pH to 7.2 with KOH or H_2SO_4. There may be a slight precipitate. Store at 5°C.

Metals "44"

Composition per liter:

$ZnSO_4·7H_2O$	1.1g
$FeSO_4·7H_2O$	0.5g
$CuSO_4·5H_2O$	0.04g
EDTA	0.25g
$MnSO_4·7H_2O$	0.154g
$Co(NO_3)_2·6H_2O$	0.025g
$Na_2B_4O_7·10H_2O$	0.018g

Preparation of Metals "44": Add components to distilled/deionized water and bring volume to 100.0mL. Mix thoroughly. Autoclave for 15 min at 15 psi pressure–121°C. Add aseptically to sterile modified Hutner's basal salts solution.

Phosphate Supplement:

Composition per liter:

K_2HPO_4	2.5g
KH_2PO_4	2.5g

Preparation of Phosphate Supplement: Add components to distilled/deionized water and bring volume to 20.0mL. Mix thoroughly. Autoclave for 15 min at 15 psi pressure–121°C.

Artificial Organic Lake Vitamin Solution:

Composition per liter:

Pyridoxine·HCl	10.0mg
Calcium DL-pantothenate	5.0mg
Nicotinamide	5.0mg
Riboflavin	5.0mg
Thiamine·HCl	5.0mg
Biotin	2.0mg
Folic acid	2.0mg
Cyanocobalamin	0.1mg

Preparation of Artificial Organic Lake Vitamin Solution: Add components to distilled/deionized water and bring volume to 1.0L. Mix thoroughly. Filter sterilize. Store at 5°C.

Preparation of Medium: Add components, except modified Hutner's basal salts solution, phosphate supplement solution, and vitamin solution, to distilled/deionized water and bring volume to 959.0mL. Mix thoroughly. Bring pH to 7.3. Autoclave for 15 min at 15 psi pressure–121°C. Cool to 50°C. Aseptically add 20.0mL of sterile modified Hutner's basal salts solution, 20.0mL of sterile phosphate supplement solution, and 1.0mL of sterile artificial organic lake vitamin solution. Mix thoroughly. Aseptically distribute into sterile tubes or flasks.

Use: For the cultivation of *Halomonas meridiana*.

Artificial Seawater Medium
See: **ASW Medium**

Artificial Seawater Medium (DSMZ Medium 1010)

Composition per liter:

NaCl	26.4g
$MgSO_4·7H_2O$	6.8g
$MgCl_2·6H_2O$	5.7g
$CaCl_2·2H_2O$	1.47g
KCl	0.66g
Resazurin	0.5g
K_2HPO_4	0.2g
KBr	0.09g
$Na_2S·9H_2O$ solution	10.0mL
Vitamin solution	10.0mL
Sodium lactate solution	10.0mL
Ammonium chloride solution	10.0mL
Bicarbonate solution	10.0mL
Dihydrogen phosphate solution	10.0mL
Seven vitamin solution	1.0mL
Selenite/tungstate solution	1.0mL
Trace elements solution SL-10	1.0mL

pH 7.3 ± 0.2 at 25°C

Trace Elements Solution SL-10:

Composition per liter:

$FeCl_2·4H_2O$	1.5g
$CoCl_2·6H_2O$	190.0mg
$MnCl_2·4H_2O$	100.0mg
$ZnCl_2$	70.0mg
$Na_2MoO_4·2H_2O$	36.0mg
$NiCl_2·6H_2O$	24.0mg
H_3BO_3	6.0mg
$CuCl_2·2H_2O$	2.0mg
HCl (25% solution)	10.0mL

Preparation of Trace Elements Solution SL-10: Add $FeCl_2·4H_2O$ to 10.0mL of HCl solution. Mix thoroughly. Add distilled/deionized water and bring volume to 1.0L. Add remaining components. Mix thoroughly. Sparge with 100% N_2. Autoclave for 15 min at 15 psi pressure–121°C.

Selenite/Tungstate Solution:

Composition per liter:

NaOH	0.5g
$Na_2WO_4 \cdot 2H_2O$	4.0mg
$Na_2SeO_3 \cdot 5H_2O$	3.0mg

Preparation of Selenite/Tungstate Solution: Add components to distilled/deionized water and bring volume to 1.0L. Mix thoroughly. Sparge with 100% N_2. Autoclave for 15 min at 15 psi pressure–121°C.

$Na_2S \cdot 9H_2O$ Solution:

Composition per 10.0mL:

$Na_2S \cdot 9H_2O$	0.5g

Preparation of $Na_2S \cdot 9H_2O$ Solution: Add $Na_2S \cdot 9H_2O$ to distilled/deionized water and bring volume to 10.0mL. Mix thoroughly.

Vitamin Solution:

Composition per liter:

Pyridoxine-HCl	10.0mg
Thiamine-HCl·$2H_2O$	5.0mg
Riboflavin	5.0mg
Nicotinic acid	5.0mg
D-Ca-pantothenate	5.0mg
p-Aminobenzoic acid	5.0mg
Lipoic acid	5.0mg
Biotin	2.0mg
Folic acid	2.0mg
Vitamin B_{12}	0.1mg

Preparation of Vitamin Solution: Add components to distilled/deionized water and bring volume to 1.0L. Mix thoroughly. Sparge with 80% H_2 + 20% CO_2. Filter sterilize.

Seven Vitamin Solution:

Composition per liter:

Pyridoxine hydrochloride	300.0mg
Thiamine-HCl·$2H_2O$	200.0mg
Nicotinic acid	200.0mg
Vitamin B_{12}	100.0mg
Calcium pantothenate	100.0mg
p-Aminobenzoic acid	80.0mg
D(+)-Biotin	20.0mg

Preparation of Seven Vitamin Solution: Add components to distilled/deionized water and bring volume to 1.0L. Sparge with 100% N_2. Mix thoroughly. Filter sterilize.

Dihydrogen Phosphate Solution:

Composition per 10.0mL:

KH_2PO_4	0.2g

Preparation of Dihydrogen Phosphate Solution: Add KH_2PO_4 to distilled/deionized water and bring volume to 10.0L. Sparge with 100% N_2. Mix thoroughly. Filter sterilize.

Sodium Lactate Solution:

Composition per 10.0mL:

Sodium lactate	2.3g

Preparation of Sodium Lactate Solution: Add sodium lactate to distilled/deionized water and bring volume to 10.0L. Sparge with 100% N_2. Mix thoroughly. Filter sterilize.

Bicarbonate Solution:

Composition per 10.0mL:

NH_4Cl	0.25g

Preparation of Bicarbonate Solution: Add NH_4Cl to distilled/deionized water and bring volume to 10.0L. Sparge with 100% N_2. Mix thoroughly. Filter sterilize.

Ammonium Chloride Solution:

Composition per 10.0mL:

$NaHCO_3$	2.5g

Preparation of Ammonium Chloride Solution: Add $NaHCO_3$ to distilled/deionized water and bring volume to 10.0L. Sparge with 100% N_2. Mix thoroughly. Filter sterilize.

Preparation of Medium: Add components, except dihydrogen phosphate, ammonium chloride, bicarbonate, lactate, vitamin, and sulfide solutions, to distilled/deionized water and bring volume to 940.0mL. Mix thoroughly. Gently heat and bring to boiling. Boil for 3 min. Cool to room temperature while sparging with 20% CO_2 + 80% N_2. Dispense into tubes or bottles under atmosphere of 20% CO_2 + 80% N_2. Autoclave for 15 min at 15 psi pressure–121°C. Cool to 25°C under an atmosphere of 20% CO_2 + 80% N_2. Aseptically and anaerobically add sterile dihydrogen phosphate, ammonium chloride, bicarbonate, lactate, vitamin, and sulfide solutions. Adjust final pH to 7.3.

Use: For the cultivation and maintenance of *Desulfobacterium corrodens*.

Artificial Seawater Medium with Propionate (DSMZ Medium 1010)

Composition per liter:

NaCl	26.4g
$MgSO_4 \cdot 7H_2O$	6.8g
$MgCl_2 \cdot 6H_2O$	5.7g
$CaCl_2 \cdot 2H_2O$	1.47g
KCl	0.66g
Resazurin	0.5g
K_2HPO_4	0.2g
KBr	0.09g
$Na_2S \cdot 9H_2O$ solution	10.0mL
Vitamin solution	10.0mL
Phenylpropionate solution	10.0mL
Ammonium chloride solution	10.0mL
Bicarbonate solution	10.0mL
Dihydrogen phosphate solution	10.0mL
Seven vitamin solution	1.0mL
Selenite/tungstate solution	1.0mL
Trace elements solution SL-10	1.0mL

Trace Elements Solution SL-10:

Composition per liter:

$FeCl_2 \cdot 4H_2O$	1.5g
$CoCl_2 \cdot 6H_2O$	190.0mg
$MnCl_2 \cdot 4H_2O$	100.0mg
$ZnCl_2$	70.0mg
$Na_2MoO_4 \cdot 2H_2O$	36.0mg
$NiCl_2 \cdot 6H_2O$	24.0mg
H_3BO_3	6.0mg
$CuCl_2 \cdot 2H_2O$	2.0mg
HCl (25% solution)	10.0mL

Preparation of Trace Elements Solution SL-10: Add $FeCl_2 \cdot 4H_2O$ to 10.0mL of HCl solution. Mix thoroughly. Add distilled/deionized water and bring volume to 1.0L. Add remaining components. Mix thoroughly. Sparge with 100% N_2. Autoclave for 15 min at 15 psi pressure–121°C.

Selenite/Tungstate Solution:
Composition per liter:

NaOH	0.5g
$Na_2WO_4 \cdot 2H_2O$	4.0mg
$Na_2SeO_3 \cdot 5H_2O$	3.0mg

Preparation of Selenite/Tungstate Solution: Add components to distilled/deionized water and bring volume to 1.0L. Mix thoroughly. Sparge with 100% N_2. Autoclave for 15 min at 15 psi pressure–121°C.

$Na_2S \cdot 9H_2O$ Solution:
Composition per 10.0mL:

$Na_2S \cdot 9H_2O$	0.5g

Preparation of $Na_2S \cdot 9H_2O$ Solution: Add $Na_2S \cdot 9H_2O$ to distilled/deionized water and bring volume to 10.0mL. Mix thoroughly.

Vitamin Solution:
Composition per liter:

Pyridoxine-HCl	10.0mg
Thiamine-HCl·$2H_2O$	5.0mg
Riboflavin	5.0mg
Nicotinic acid	5.0mg
D-Ca-pantothenate	5.0mg
p-Aminobenzoic acid	5.0mg
Lipoic acid	5.0mg
Biotin	2.0mg
Folic acid	2.0mg
Vitamin B_{12}	0.1mg

Preparation of Vitamin Solution: Add components to distilled/deionized water and bring volume to 1.0L. Mix thoroughly. Sparge with 80% H_2 + 20% CO_2. Filter sterilize.

Seven Vitamin Solution:
Composition per liter:

Pyridoxine hydrochloride	300.0mg
Thiamine-HCl·$2H_2O$	200.0mg
Nicotinic acid	200.0mg
Vitamin B_{12}	100.0mg
Calcium pantothenate	100.0mg
p-Aminobenzoic acid	80.0mg
D(+)-Biotin	20.0mg

Preparation of Seven Vitamin Solution: Add components to distilled/deionized water and bring volume to 1.0L. Sparge with 100% N_2. Mix thoroughly. Filter sterilize.

Dihydrogen Phosphate Solution:
Composition per 10.0mL:

KH_2PO_4	0.2g

Preparation of Dihydrogen Phosphate Solution: Add KH_2PO_4 to distilled/deionized water and bring volume to 10.0L. Sparge with 100% N_2. Mix thoroughly. Filter sterilize.

Phenylpropionate Solution:
Composition per 10.0mL:

3-Phenylpropionate	0.45g

Preparation of Phenylpropionate Solution: Add 3-phenylpropionate to distilled/deionized water and bring volume to 10.0L. Sparge with 100% N_2. Mix thoroughly. Filter sterilize.

Bicarbonate Solution:
Composition per 10.0mL:

$NaHCO_3$	0.25g

Preparation of Bicarbonate Solution: Add $NaHCO_3$ to distilled/deionized water and bring volume to 10.0L. Sparge with 100% N_2. Mix thoroughly. Filter sterilize.

Ammonium Chloride Solution:
Composition per 10.0mL:

NH_4Cl	2.5g

Preparation of Ammonium Chloride Solution: Add NH_4Cl to distilled/deionized water and bring volume to 10.0L. Sparge with 100% N_2. Mix thoroughly. Filter sterilize.

Preparation of Medium: Add components, except dihydrogenphosphate, ammonium chloride, bicarbonate, phenylprionate, vitamin, and sulfide solutions, to distilled/deionized water and bring volume to 940.0mL. Mix thoroughly. Gently heat and bring to boiling. Boil for 3 min. Cool to room temperature while sparging with 20% CO_2 + 80% N_2. Dispense into tubes or bottles under an atmosphere of 20% CO_2 + 80% N_2. Autoclave for 15 min at 15 psi pressure–121°C. Cool to 25°C under an atmosphere of 20% CO_2 + 80% N_2. Aseptically and anaerobically add sterile dihydrogen phosphate, ammonium chloride, bicarbonate, phenylprionate, vitamin, and sulfide solutions. Adjust final pH to 7.3.

Use: For the cultivation and maintenance of an unidentified bacterium from Guaymas basin sediment.

Arylsulfatase Agar (Wayne Sulfatase Agar)

Composition per liter:

Agar	15.0g
Na_2HPO_4	2.5g
L-Asparagine	1.0g
KH_2PO_4	1.0g
K_2HPO_4	1.0g
Trisodium phenolphthalein sulfate	0.65g
Pancreatic digest of casein	0.5g
Ferric ammonium citrate	0.05g
$MgSO_4 \cdot 7H_2O$	0.01g
$CaCl_2 \cdot 2H_2O$	0.5mg
$ZnSO_4 \cdot 7H_2O$	0.1mg
$CuSO_4$	0.1mg
Glycerol	10.0mL

pH 7.0 ± 0.2 at 25°C

Source: This medium is available as a premixed powder from BD Diagnostic Systems.

Preparation of Medium: Add glycerol to approximately 800.0mL of distilled/deionized water. Mix thoroughly. Add remaining components and bring volume to 1.0L with distilled/deionized water. Mix thoroughly. Gently heat and bring to boiling. Distribute into tubes. Autoclave for 15 min at 15 psi pressure–121°C. Cool tubes in an upright position.

Use: For the biochemical differentiation of species of *Mycobacterium*. Inoculate tubes with *Mycobacterium* cultures and incubate aerobically at 35°C for 3–14 days. Add 0.5–1.0mL of 2*N* Na_2CO_3 to each tube and observe color change within 30 min. Development of a pink color is indicative of *Mycobacterium fortuitum* or *Mycobacterium chelonae*. *Mycobacterium tuberculosis* gives a negative reaction.

Ascospore Agar

Composition per liter:

Agar	30.0g
Potassium acetate	10.0g

Yeast extract	2.5g
Glucose	1.0g

pH 6.4 ± 0.2 at 25°C

Source: This medium is available from HiMedia.

Preparation of Medium: Add components to distilled/deionized water and bring volume to 1.0L. Mix thoroughly. Gently heat and bring to boiling. Distribute into tubes or flasks. Autoclave for 15 min at 15 psi pressure–121°C. Pour into sterile Petri dishes or leave in tubes.

Use: For the enrichment of ascosporogenous yeasts and their production of ascospores.

Ashby's Glucose Agar

Composition per liter:

Glucose	20.0g
Agar	15.0g
$CaCO_3$	5.0g
K_2HPO_4	0.2g
$MgSO_4 \cdot 7H_2O$	0.2g
NaCl	0.2g
K_2SO_4	0.1g

pH 7.4 ± 0.2 at 25°C

Source: This medium is available from HiMedia.

Preparation of Medium: Add components to distilled/deionized water and bring volume to 1.0L. Mix thoroughly. Distribute into tubes or flasks. Autoclave for 15 min at 15 psi pressure–121°C. Pour into Petri dishes or leave in tubes.

Use: For the cultivation of *Azotobacter* spp. from soil.

Ashby's Mannitol Agar

Composition per liter:

Mannitol	20.0g
Agar	15.0g
$CaCO_3$	5.0g
K_2HPO_4	0.2g
$MgSO_4 \cdot 7H_2O$	0.2g
NaCl	0.2g
K_2SO_4	0.1g

pH 7.4 ± 0.2 at 25°C

Source: This medium is available from HiMedia.

Preparation of Medium: Add components to distilled/deionized water and bring volume to 1.0L. Mix thoroughly. Distribute into tubes or flasks. Autoclave for 15 min at 15 psi pressure–121°C. Pour into Petri dishes or leave in tubes.

Use: For the cultivation of *Azotobacter* spp. from soil.

Ashby's Nitrogen-Free Agar

Composition per liter:

Agar	15.0g
Mannitol	15.0g
$CaCl_2 \cdot 2H_2O$	0.2g
K_2HPO_4	0.2g
$MgSO_4 \cdot 7H_2O$	0.2g
MoO_3 (10% solution)	0.1mL
$FeCl_3$ (10% solution	0.05mL

pH 7.2 ± 0.2 at 25°C

Preparation of Medium: Add components to distilled/deionized water and bring volume to 1.0L. Mix thoroughly. Gently heat and bring to boiling. Distribute into tubes or flasks. Autoclave for 15 min at 15 psi pressure–121°C. Pour into sterile Petri dishes or leave in tubes.

Use: For the isolation and cultivation of bacteria, such as *Azotobacter* species and cyanobacteria, that can utilize atmospheric N_2 as sole nitrogen source.

Ashdown's Medium

Composition per liter:

Pancreatic digest of casein	14.5g
Agar	14.0g
Papaic digest of soybean meal	5.0g
NaCl	5.0g
Neutral Red	50.0mg
Crystal Violet	5.0mg
Gentamicin	4.0mg
Glycerol	40.0mL

pH 7.3 ± 0.2 at 25°C

Preparation of Medium: Add components to distilled/deionized water and bring volume to 1.0L. Autoclave for 15 min at 15 psi pressure–121°C. Pour into sterile Petri dishes or leave iin tubes.

Use: For the isolation and cultivation of *Burkholderia pseudomallei* from clinical specimens.

ASLA Agar

Composition per liter:

Sodium lactate	20.0g
Davis agar	10.0g
$(NH_4)_2SO_4$	3.0g
Na_2HPO_4	1.2g
L-Cysteine·HCl	0.5g
$MgSO_4 \cdot 7H_2O$	0.2g
$MnSO_4 \cdot 4H_2O$	0.05g
$FeSO_4 \cdot 7H_2O$	0.04g
Vitamin solution	10.0mL

pH 6.5 ± 0.2 at 25°C

Vitamin Solution:

Composition per liter:

Biotin	0.1g
Calcium pantothenate	0.1g
p-Aminobenzoic acid	0.1g
Thiamine	0.1g

Preparation of Vitamin Solution: Add components to distilled/deionized water and bring volume to 1.0L. Mix thoroughly. Filter sterilize. Store solution at –20°C.

Preparation of Medium: Add components, except vitamin solution, to distilled/deionized water and bring volume to 990.0mL. Mix thoroughly. Gently heat and bring to boiling. Autoclave for 15 min at 15 psi pressure–121°C. Cool to 45°–50°C. Aseptically add 10.0mL of sterile vitamin solution. Mix thoroughly. Pour into sterile Petri dishes or distribute into sterile tubes.

Use: For the selective isolation and cultivation of most *Propionibacterium* species from foods.

ASM Medium

Composition per 1001.2mL:

Na_2HPO_4	866.0mg
$NH_4 \cdot Cl$	535.0mg
KH_2PO_4	531.0mg
K_2SO_4	174.0mg

$MgSO_4 \cdot 7H_2O$ 37.0mg
$CaCl_2 \cdot 2H_2O$ 7.35mg
Trace elements solution 1.0mL
$FeSO_4$ solution 0.2mL

Trace Elements Solution:
Composition per liter:
$ZnSO_4 \cdot 7H_2O$ 288.0mg
$MnSO_4 \cdot 4H_2O$ 224.0mg
$CuSO_4 \cdot 5H_2O$ 125.0mg
KI 83.0mg
H_3BO_3 .... 61.8mg
$Na_2MoO_4 \cdot 2H_2O$ 48.4mg
$CoCl_2 \cdot 6H_2O$ 47.6mg
H_2SO_4, 1*M* 1.0mL

Preparation of Trace Elements Solution: Add components to distilled/deionized water and bring volume to 1.0L. Mix thoroughly. Autoclave for 15 min at 15 psi pressure–121°C.

$FeSO_4$ Solution:
Composition per liter:
$FeSO_4 \cdot 7H_2O$ 278.0mg

Preparation of $FeSO_4$ Solution: Add $FeSO_4 \cdot 7H_2O$ to distilled/deionized water and bring volume to 10.0mL. Mix thoroughly. Filter sterilize.

Preparation of Medium: Add components, except trace elements solution and $FeSO_4$ solution, to distilled/deionized water and bring volume to 1.0L. Mix thoroughly. Autoclave for 15 min at 15 psi pressure–121°C. Aseptically add 1.0mL of trace elements solution and 0.2mL of sterile $FeSO_4$ solution. Mix thoroughly. Aseptically distribute into sterile tubes or flasks.

Use: For the cultivation of *Mycobacterium* species.

ASN-III Agar

Composition per liter:
NaCl 25.0g
$MgSO_4 \cdot 7H_2O$ 3.5g
$MgCl_2 \cdot 6H_2O$ 2.0g
$NaNO_3$ 0.75g
$K_2HPO_4 \cdot 3H_2O$ 0.75g
$CaCl_2 \cdot 2H_2O$ 0.5g
KCl 0.5g
Na_2CO_3 .... 0.02g
Citric acid 3.0mg
Ferric ammonium citrate 3.0mg
Magnesium EDTA 0.5mg
Vitamin B_{12} 10.0μg
Agar solution 100.0mL
A-5 trace metals 1.0mL
pH 7.3 ± 0.2 at 25°C

Agar Solution:
Composition per 100.0mL:
Noble agar 10.0g

Preparation of Agar Solution: Add agar to glass-distilled water and bring volume to 100.0mL. Mix thoroughly. Gently heat and bring to boiling. Autoclave for 15 min at 15 psi pressure–121°C. Cool to 45°–50°C.

A-5 Trace Metals:
Composition per liter:
H_3BO_3 .... 2.86g
$MnCl_2 \cdot 4H_2O$ 1.81g
$ZnSO_4 \cdot 7H_2O$ 0.222g
$CuSO_4 \cdot 5H_2O$ 0.079g
$Co(NO_3)_2 \cdot 6H_2O$ 0.049g
$Na_2MoO_4 \cdot 2H_2O$ 0.039g

Preparation of A-5 Trace Metals: Add components to distilled/deionized water and bring volume to 1.0L. Mix thoroughly.

Preparation of Medium: Add components, except agar solution, to glass-distilled water and bring volume to 900.0mL. Mix well and heat gently until dissolved. Filter sterilize. Warm to 45°–50°C. Aseptically add agar solution. Mix thoroughly. Pour into sterile Petri dishes or distribute into sterile tubes.

Use: For the cultivation of *Xenococcus* species. For the isolation of cyanobacteria from marine habitats.

ASN-III Broth

Composition per liter:
NaCl 25.0g
$MgSO_4 \cdot 7H_2O$ 3.5g
$MgCl_2 \cdot 6H_2O$ 2.0g
$NaNO_3$ 0.75g
$K_2HPO_4 \cdot 3H_2O$ 0.75g
$CaCl_2 \cdot 2H_2O$ 0.5g
KCl 0.5g
Na_2CO_3 .... 0.02g
Citric acid 3.0mg
Ferric ammonium citrate 3.0mg
Magnesium EDTA 0.5mg
Vitamin B_{12} 10.0μg
A-5 trace metals 1.0mL
pH 7.3 ± 0.2 at 25°C

A-5 Trace Metals:
Composition per liter:
H_3BO_3 .... 2.86g
$MnCl_2 \cdot 4H_2O$ 1.81g
$ZnSO_4 \cdot 7H_2O$ 0.222g
$CuSO_4 \cdot 5H_2O$ 0.079g
$Co(NO_3)_2 \cdot 6H_2O$ 0.049g
$Na_2MoO_4 \cdot 2H_2O$ 0.039g

Preparation of A-5 Trace Metals: Add components to distilled/deionized water and bring volume to 1.0L. Mix thoroughly.

Preparation of Medium: Add components to glass-distilled water and bring volume to 1.0L. Mix well and heat gently until dissolved. Filter sterilize.

Use: For the cultivation of *Xenococcus* species. For the isolation of cyanobacteria from marine habitats.

ASP-2 Medium

Composition per liter:
NaCl 18.0g
$MgSO_4 \cdot 7H_2O$ 5.0g
KCl 0.6g
$NaNO_3$ 0.05g
Trace elements solution 10.0mL
Tris buffer solution 4.0mL
$CaCl_2 \cdot 2H_2O$ solution 2.8mL
$Na_2SiO_3 \cdot 9H_2O$ solution 1.5mL
Vitamin solution 1.0mL

K_2HPO_4 solution 0.5mL
Vitamin B_{12} solution 0.1mL

pH 7.6 ± 0.2 at 25°C

$CaCl_2 \cdot 2H_2O$ Solution:
Composition per 100.0mL:
$CaCl_2 \cdot 2H_2O$ 13.0g

Preparation of $CaCl_2 \cdot 2H_2O$ Solution: Add $CaCl_2 \cdot 2H_2O$ to distilled/deionized water and bring volume to 100.0mL. Mix thoroughly.

K_2HPO_4 Solution:
Composition per 100.0mL:
K_2HPO_4 .... 1.0g

Preparation of K_2HPO_4 Solution: Add K_2HPO_4 to distilled/deionized water and bring volume to 100.0mL. Mix thoroughly.

Tris Buffer Solution:
Composition per 100.0mL:
Tris[hydroxymethyl]aminomethane buffer 25.0g

Preparation of Tris Buffer Solution: Add 25.0g of tris to 65.0mL of distilled/deionized water. Titrate to pH 7.6–7.7 with concentrated HCl. Bring to 100.0mL with distilled/deionized water. Recheck pH after 12 hr.

Vitamin B_{12} Solution:
Composition per 100.0mL:
Vitamin B_{12} 2.0mg

Preparation of VitaminB$_{12}$ Solution: Add vitamin B_{12} to distilled/deionized water and bring volume to 100.0mL. Mix thoroughly.

Trace Elements Solution:
Composition per liter:
EDTA 3.0g
$MnCl_2 \cdot 4H_2O$ 432.0mg
$FeCl_3 \cdot 6H_2O$ 384.0mg
H_3BO_3 .... 342.0mg
$ZnCl_2$ 31.5mg
$CoCl_2 \cdot 6H_2O$ 2.0mg
CuCl or $CuCl_2$ 0.25mg

Preparation of Trace Elements Solution: Add components to distilled/deionized water and bring volume to 1.0L. Mix thoroughly.

Vitamin Solution:
Composition per 100.0mL:
Inositol 500.0mg
Thymine 300.0mg
Thiamine·HCl 50.0mg
Calcium D-(+)-pantothenate 10.0mg
Nicotinic acid 10.0mg
PABA 1.0mg
Folic acid 0.2mg
Biotin 0.1mg

Preparation of Vitamin Solution: Add components to distilled/deionized water and bring volume to 100.0mL. Mix thoroughly.

$Na_2SiO_3 \cdot 9H_2O$ Solution:
Composition per 100.0mL:
$Na_2SiO_3 \cdot 9H_2O$ 10.0g

Preparation of $Na_2SiO_3 \cdot 9H_2O$ Solution: Add $Na_2SiO_3 \cdot 9H_2O$ to distilled/deionized water and bring volume to 100.0mL. Mix thoroughly.

Preparation of Medium: Add components to distilled/deionized water and bring volume to 1.0L. Mix thoroughly. Filter sterilize. Aseptically distribute into sterile, screw-capped tubes or flasks.

Use: For the cultivation of *Chlamydomonas* species.

Asparaginate Glycerol Agar

Composition per liter:
Agar 15.0g
Sodium asparaginate 1.0g
K_2HPO_4 .... 1.0g
Glycerol 10.0mL

pH 7.0 ± 0.2 at 25°C

Preparation of Medium: Add components to distilled/deionized water and bring volume to 1.0L. Mix thoroughly. Gently heat and bring to boiling. Distribute into tubes or flasks. Autoclave for 15 min at 15 psi pressure–121°C. Pour into sterile Petri dishes or leave in tubes.

Use: For the cultivation and maintenance of *Nocardia transvalensis*.

Asparagine Broth

Composition per liter:
DL-Asparagine 30.0g
K_2HPO_4 .... 1.0g
$MgSO_4 \cdot 7H_2O$ 0.5g

pH 6.9–7.2 at 25°C

Preparation of Medium: Add components to distilled/deionized water and bring volume to 1.0L. Mix well until dissolved. Adjust pH to between 6.9 and 7.2. Distribute into tubes or flasks. Autoclave for 15 min at 15 psi pressure–121°C.

Use: For a presumptive test medium in the differentiation of nonfermentative Gram-negative bacteria, especially *Pseudomonas aeruginosa*. For use in the multiple tube technique in the microbiological analysis of recreational waters.

Asparagine Broth (Coccidioidin and Histoplasmin Broth)

Composition per liter:
Glucose 10.0g
L-Asparagine 7.0g
NH_4Cl 7.0g
$MgSO_4 \cdot 7H_2O$ 1.5g
K_2HPO_4 .... 1.31g
Sodium citrate 0.9g
Ferric citrate 0.3g
Glycerol 25.0mL

pH 6.8 ± 0.2 at 25°C

Source: This medium is available from HiMedia.

Preparation of Medium: Add glycerol to distilled/deionized water and bring volume to 1.0L. Mix thoroughly. Add remaining components. Mix thoroughly. Autoclave for 15 min at 15 psi pressure–121°C.

Use: For the preparation of the cultivation of *Coccidioides* and *Histoplama* for the production of coccidoidin and histoplasmin antigens for immunologic work.

Asparagine Gelatin Lactate Medium Base with Lactate

Composition per liter:
Gelatin 150.0g
Lactate 5.0g
Asparagine 1.0g
K_2HPO_4 .... 0.5g

$MgSO_4 \cdot 7H_2O$ 1.0g
$FeNH_4(SO_4)_2 \cdot 12H_2O$ 1.0mg
pH 7.0 ± 0.2 at 25°C

Source: This medium is available from HiMedia.

Preparation of Medium: Add components to distilled/deionized water and bring volume to 1.0L. Mix well until dissolved. Adjust pH to between 6.9 and 7.2. Distribute into tubes or flasks. Autoclave for 15 min at 10 psi pressure–115°C. Pour into Petri dishes or leave in tubes

Use: For the isolutation of sulfur bacteria such as *Desulfovibrio* spp.

Asparagine Nitrate Medium

Composition per liter:
Agar 15.0g
Sodium citrate 8.5g
KNO_3 .... 1.0g
L-Asparagine 1.0g
KH_2PO_4 .... 1.0g
$MgSO_4 \cdot 7H_2O$ 1.0g
$CaCl_2 \cdot 2H_2O$ 0.2g
$FeCl_3$ 0.1mg
pH 7.2 ± 0.2 at 25°C

Source: This medium is available from HiMedia.

Preparation of Medium: Add components to distilled/deionized water and bring volume to1.0L. Mix thoroughly. Autoclave for 15 min at 15 psi pressure–121°C. Pour into Petri dishes or leave in tubes.

Use: For the isolation and cultivation of denitrifying bacteria.

Asparagine Proline Broth

Composition per liter:
K_2SO_4 .... 10.0g
DL-Asparagine 2.0g
L-Proline 1.0g
K_2HPO_4 .... 1.0g
$MgSO_4 \cdot 7H_2O$ 0.5g
Ethanol 25.0mL
pH 7.2 ± 0.2 at 25°C

Source: This medium is available from HiMedia.

Preparation of Medium: Add components to distilled/deionized water and bring volume to 1.0mL. Mix thoroughly. Distribute into screw cap tubes or bottles. Close the caps almost completely. Autoclave for 15 min at 15 psi pressure–121°C. Quickly seal the caps to prevent evaporation of ethanol.

Use: For the enrichment and cultivation of *Pseudomonas aeruginosa.*

Aspergillus Differential Medium

Composition per liter:
Agar 15.0g
Pancreatic digest of casein 15.0g
Yeast extract 10.0g
Ferric citrate 0.5g

Preparation of Medium: Add components to distilled/deionized water and bring volume to 1.0L. Mix thoroughly. Gently heat and bring to boiling. Distribute into tubes in 7.0mL volumes. Autoclave for 15 min at 15 psi pressure–121°C. Allow tubes to cool in a slanted position.

Use: For the cultivation and differentiation of *Aspergillus flavus. Aspergillus flavus* appears as bright orange colonies.

Aspergillus Differentiation Medium Base with Chloramphenicol

Composition per liter:
Yeast extract 20.0g
Agar 15.0g
Peptic digest of animal tissue 10.0g
Ferric ammonium citrate 0.5g
Chloramphenicol 0.1g
Dichloran 2.0mg
pH 6.3 ± 0.2 at 25°C

Source: This medium is available from HiMedia.

Preparation of Medium: Add components to distilled/deionized water and bring volume to 1.0L. Mix thoroughly. Gently heat and bring to boiling. Autoclave for 15 min at 15 psi pressure–121°C. Pour into Petri dishes or leave in tubes.

Use: For the detction of aflatoxin producing *Aspergillus* spp. from foods.

Aspergillus flavus/parasiticus Agar Base

See: **AFPA Base**

Aspergillus Medium

Composition per liter:
Agar 15.0g
$NaNO_3$ 6.0g
Casamino acids 1.0g
Peptone 1.0g
Yeast extract 1.0g
Adenine 0.15g
Vitamin solution 10.0mL
pH 6.0 ± 0.2 at 25°C

Vitamin Solution:
Composition per 100.0mL:
Biotin 0.01g
Nicotinic acid 0.01g
p-Aminobenzoic acid 0.01g
Pyridoxine·HCl 0.01g
Riboflavin 0.01g
Thiamine·HCl 0.01g

Preparation of Vitamin Solution: Add components to distilled/deionized water and bring volume to 100.0mL. Mix thoroughly. Autoclave for 10 min at 15 psi pressure–121°C.

Preparation of Medium: Add components, except vitamin solution, to distilled/deionized water and bring volume to 990.0mL. Mix thoroughly. Gently heat and bring to boiling. Adjust pH to 6.0. Autoclave for 15 min at 15 psi pressure–121°C. Cool to 50°C. Aseptically add 10.0mL of sterile vitamin solution. Mix thoroughly. Pour into sterile Petri dishes or distribute into sterile tubes.

Use: For the cultivation and maintenance of *Aspergillus amstelodami, Aspergillus awamori, Aspergillus flavus,* and *Aspergillus nidulans.*

Aspergillus nidulans Minimal Medium

Composition per 950.0mL:
Solution A 500.0mL
Solution B 250.0mL
Solution C 200.0mL
pH 6.5 ± 0.2 at 25°C

Solution A:

Composition per 500.0mL:

$NaNO_3$	6.0g
KCl	1.52g
KH_2PO_4	1.52g

Preparation of Solution A: Add components to distilled/deionized water and bring volume to 500.0mL. Mix thoroughly. Adjust pH to 6.5 with 2*N* NaOH. Autoclave for 15 min at 15 psi pressure–121°C. Cool to 50°–55°C.

Solution B:

Composition per 250.0mL:

Agar	15.0g
$MgSO_4 \cdot 7H_2O$	0.52g
$FeSO_4 \cdot 7H_2O$	1.0μg
$ZnSO_4 \cdot 7H_2O$	1.0μg

Preparation of Solution B: Add components to distilled/deionized water and bring volume to 250.0mL. Mix thoroughly. Gently heat and bring to boiling. Autoclave for 15 min at 15 psi pressure–121°C. Cool to 50°–55°C.

Solution C:

Composition per 200.0mL:

Glucose	10.0g

Preparation of Solution C: Add glucose to distilled/deionized water and bring volume to 200.0mL. Mix thoroughly. Autoclave for 15 min at 15 psi pressure–121°C. Cool to 50°–55°C.

Preparation of Medium: Aseptically combine sterile solution A, sterile solution B, and sterile solution C. Mix thoroughly. Pour into sterile Petri dishes or distribute into sterile tubes.

Use: For the cultivation and maintenance of *Aspergillus nidulans*.

Aspergillus Test Medium

Composition per liter:

Agar	20.0g
Malt extract	20.0g
Peptone	1.0g
Glucose solution	100.0mL
Supplement solution	100.0mL

pH 6.5 ± 0.2 at 25°C

Glucose Solution:

Composition per 100.0mL:

Glucose	20.0g

Preparation of Glucose Solution: Add glucose to distilled/deionized water and bring volume to 100.0mL. Mix thoroughly. Filter sterilize. Warm to 50°C.

Supplement Solution:

Composition per 100.0mL:

Uridine	2.44g
Arginine	200.0mg
Methionine	50.0mg
Riboflavin	2.5mg
Nicotinic acid	2.0mg
p-Aminobenzoic acid	1.0mg
Pyrdoxine·HCl	0.05mg
Biotin	0.02mg

Preparation of Supplement Solution: Add components to distilled/deionized water and bring volume to 100.0mL. Mix thoroughly. Filter sterilize. Warm to 50°C.

Preparation of Medium: Add components, except glucose solution and supplement solution, to distilled/deionized water and bring volume to 800.0mL. Mix thoroughly. Adjust pH to 6.5. Gently heat and bring to boiling. Autoclave for 15 min at 15 psi pressure–121°C. Cool to 50°–55°C. Aseptically add 100.0mL of sterile glucose solution and 100.0mL of sterile supplement solution. Mix thoroughly. Pour into sterile Petri dishes or distribute into sterile tubes.

Use: For the cultivation and maintenance of *Emericella (Aspergillus) nidulans*.

ASS Agar

See: **Antibiotic Sulfonamide Sensitivity Test Agar**

Association of Official Analytical Chemists Letheen Broth *See:* AOAC Letheen Broth

Asticcacaulis Medium

Composition per liter:

Agar	15.0g
Pancreatic digest of casein	0.5g
Yeast extract	0.5g
Sodium acetate	0.2g

pH 6.8 ± 0.2 at 25°C

Preparation of Medium: Add components to distilled/deionized water and bring volume to 1.0L. Mix thoroughly. Gently heat and bring to boiling. Distribute into tubes or flasks. Autoclave for 15 min at 15 psi pressure–121°C. Pour into sterile Petri dishes or leave in tubes.

Use: For the isolation and cultivation of *Asticcacaulis* species.

ASTM Nutrient Salts Agar (American Society for Testing and Materials Nutrient Salts Agar)

Composition per liter:

Agar	15.0g
KH_2PO_4	0.7g
K_2HPO_4	0.7g
$MgSO_4 \cdot 7H_2O$	0.7g
NH_4NO_3	1.0g
NaCl	5.0mg
$FeSO_4 \cdot 7H_2O$	2.0mg
$ZnSO_4$	2.0mg
$MnSO_4 \cdot H_2O$	1.0mg

pH 6.5 ± 0.2 at 25°C

Preparation of Medium: Add components to tap water and bring volume to 1.0L. Mix thoroughly. Gently heat and bring to boiling. Distribute into tubes or flasks. Autoclave for 15 min at 15 psi pressure–121°C. Pour into sterile Petri dishes or leave in tubes.

Use: For determination of the susceptibility of plastics to fungal degradation.

ASW Medium (Artificial Seawater Medium)

Composition per liter:

NaCl	27.0g
Agar	15.0g
$MgSO_4 \cdot 7H_2O$	6.6g
$MgCl_2 \cdot 6H_2O$	5.6g
$CaCl_2 \cdot 2H_2O$	1.5g

KNO_3	1.0g
KH_2PO_4	0.07g
$NaHCO_3$	0.04g
Tris-HCl buffer (1.0*M*, pH 7.6)	20.0mL
Chelated iron solution	1.0mL
Trace metal solution	1.0mL

Trace Metal Solution:
Composition per 100.0mL:

H_3BO_3	60.0mg
$MnCl_2·4H_2O$	40.0mg
$(NH_4)_6Mo_7O_{24}·4H_2O$	37.0mg
$CuCl_2·2H_2O$	4.0mg
$ZnCl_2$	4.0mg
$CoCl_2·6H_2O$	1.5mg

Preparation of Trace Metal Solution: Add components to distilled/deionized water and bring volume to 100.0mL. Mix thoroughly.

Chelated Iron Solution:
Composition per 100.0mL:

$FeCl_3·4H_2O$	240.0mg
EDTA	14.6g

Preparation of Chelated Iron Solution: Add EDTA to distilled/deionized water and bring volume to 100.0mL. Mix thoroughly. Adjust pH to 7.6. Add $FeCl_3·4H_2O$. Mix thoroughly.

Preparation of Medium: Add components to distilled/deionized water and bring volume to 1.0L. Mix thoroughly. Gently heat and bring to boiling. Distribute into tubes or flasks. Autoclave for 15 min at 15 psi pressure–121°C. Pour into sterile Petri dishes or leave in tubes.

Use: For the cultivation of *Porphyridium purpureum.*

ATB Acid Tomato Broth

Composition per liter:

Glucose	10.0g
Peptone	10.0g
Yeast extract	5.0g
$MgSO_4·7H_2O$	0.2g
$MgSO_4·4H_2O$	0.05g
Tomato juice	250.0mL

Preparation of Medium: Add components to distilled/deionized water and bring volume to 1.0L. Mix thoroughly. Distribute into tubes or flasks. Autoclave for 15 min at 15 psi pressure–121°C.

Use: For the cultivation of *Lactobacillus fructivorans, Lactobacillus homohiochii, Lactobacillus kefiranofaciens,* and *Leuconostoc oenos.*

Atlas Oil Agar

Composition per liter:

Bushnell-Haas agar	990.0mL
Oil	10.0mL

pH 7.0 ± 0.2 at 25°C

Bushnell-Haas Agar:
Composition per 990.0mL:

Agar	15.0g
KH_2PO_4	1.0g
K_2HPO_4	1.0g
NH_4NO_3	1.0g
$MgSO_4·7H_2O$	0.2g
$FeCl_3$	0.05g
$CaCl_2·2H_2O$	0.02g

Preparation of Bushnell-Haas Agar: Add components to distilled/deionized water and bring volume to 990.0mL. Mix thoroughly. Gently heat and bring to boiling. Autoclave for 15 min at 15 psi pressure–121°C. Cool to 60°C.

Preparation of Medium: Filter sterilize oil. Aseptically add 10.0mL of sterile oil to 990.0mL of cooled, sterile Bushnell-Haas agar. Put mixture into a sterile blender container. Blend on low speed to minimize the incorporation of air into the medium. Pour into sterile Petri dishes.

Use: For the cultivation and enumeration of hydrocarbon-utilizing bacteria by direct plating of water and sediment samples.

AT5N Medium

Composition per liter:

$CaCO_3$	10.0g
$(NH_4)_2SO_4$	1.5g
K_2HPO_4	0.5g
$MgSO_4$	50.0mg
$KHCO_3$	30.0mg
$CaCl_2·2H_2O$	20.0mg

Preparation of Medium: Add components to tap water and bring volume to 1.0L. Mix thoroughly. Gently heat and bring to boiling. Distribute into tubes or flasks. Autoclave for 15 min at 15 psi pressure–121°C.

Use: For the cultivation of bacteria that oxidize ammonia, especially those from wastewater.

Atopobium/Olsenella Medium (LMG Medium 152)

Composition per liter:

Yeast extract	10.0g
Peptone	5.0g
Casitone	5.0g
Glucose	5.0g
$(NH_4)_2SO_4$	0.5g
L-Cysteine·HCl	0.5g
Resazurin	1.0mg
Mineral solution	40.0mL
Fatty acid mixture	3.1mL
Tween™ 80	2.0mL
Hemin solution	0.5mL
Vitamin K_1	0.2mL

pH 6.9 ± 0.2 at 25°C

Mineral Solution:
Composition per liter:

$NaHCO_3$	10.0g
NaCl	2.0g
K_2HPO_4	1.0g
KH_2PO_4	1.0g
$MgSO_4·7H_2O$	0.48g
$CaCl_2·2H_2O$	0.3g

Preparation of Mineral Solution: Add components to distilled/deionized water and bring volume to 1.0L. Mix thoroughly.

Fatty Acid Mixture:
Composition per 31.0mL:

Acetic acid	17.0mL
Propionic acid	6.0mL
n-Butyric acid	4.0mL

n-Valeric acid	1.0mL
iso-Valeric acid	1.0mL
iso-Butyric acid	1.0mL
DL-2-Methylbutyric acid	1.0mL

Preparation of Fatty Acid Mixture: Combine components. Mix thoroughly. Adjust pH to 7.5 with concentrated NaOH.

Hemin Solution:
Composition per 1.0mL:

Hemin	5.0mg
NaOH (1*N* solution)	1.0mL

Preparation of Hemin Solution: Add hemin to 1.0mL of NaOH solution. Mix thoroughly.

Preparation of Medium: Add components, except L-cysteine·HCl, hemin solution, and fatty acid mixture, to distilled/deionized water and bring volume to 1.0L. Mix thoroughly. Gently heat and bring to boiling. Continue boiling for 5 min. Cool to room temperature while sparging with 100% CO_2. Add L-cysteine·HCl, hemin solution, and fatty acid mixture. Adjust pH to 6.9 with 8*N* NaOH while continuing to sparge with 100% CO_2. After pH has been reached, sparge with 100% N_2. Anaerobically distribute into tubes or flasks. Autoclave for 15 min at 15 psi pressure–121°C.

Use: For the cultivation and maintenance of *Atopobium rimae* and *Olsenella uli.*

ATS Medium (American Trudeau Society Medium)

Composition per liter:

Potato	20.0g
Malachite Green	0.2g
Egg yolk emulsion	500.0mL
Glycerol	10.0mL

pH 6.5–7.0 at 25°C

Source: This medium is available as a prepared medium from BD Diagnostic Systems.

Egg Yolk Emulsion:
Composition:

Chicken egg yolks	11
Whole chicken egg	1

Preparation of Egg Yolk Emulsion: Soak eggs with 1:100 dilution of saturated mercuric chloride solution for 1 min. Crack eggs and separate yolks from whites. Mix egg yolks with 1 chicken egg.

Preparation of Medium: Add components to distilled/deionized water and bring volume to 1.0L. Distribute into tubes. Autoclave for 15 min at 15 psi pressure–121°C in a slanted position.

Use: For the isolation and cultivation of *Mycobacterium* species other than *Mycobacterium leprae*. Especially useful for the detection of *Mycobacterium tuberculosis* from clinical specimens such as cerebrospinal fluid, pleural fluid, and tissues.

Aureobacterium Agar

Composition per liter:

Agar	20.0g
Casamino acids	10.0g
Yeast extract	2.0g
$MgSO_4 \cdot 7H_2O$	1.0g

pH 7.0 ± 0.2 at 25°C

Preparation of Medium: Add components to distilled/deionized water and bring volume to 1.0L. Mix thoroughly. Adjust pH to 7.0. Distribute into tubes or flasks. Autoclave for 15 min at 15 psi pressure–121°C. Pour into sterile Petri dishes or leave in tubes.

Use: For the cultivation of *Aureobacterium arabinogalactanolyticum, Aureobacterium esteraromaticum, Aureobacterium keratanolyticum, Aureobacterium schleiferi, Aureobacterium terrae,* and *Aureobacterium trichothecenolyticum.*

Aureobacterium Agar

Composition per liter:

Agar	20.0g
Polypeptone™	10.0g
Yeast extract	2.0g
$MgSO_4 \cdot 7H_2O$	1.0g

pH 7.0 ± 0.2 at 25°C

Preparation of Medium: Add components to distilled/deionized water and bring volume to 1.0L. Mix thoroughly. Adjust pH to 7.0. Distribute into tubes or flasks. Autoclave for 15 min at 15 psi pressure–121°C. Pour into sterile Petri dishes or leave in tubes.

Use: For thc cultivation of *Aureobacterium* species.

Aureobacterium terregens Medium

Composition per liter:

Casamino acids	2.0g
K_2HPO_4	2.0g
Diammonium citrate	1.0g
Glucose	1.0g
Yeast extract	1.0g
$MgSO_4 \cdot 7H_2O$	0.5g
$FeCl_3 \cdot 6H_2O$	10.0mg
Acetylacetone solution	1.0mL

pH 7.0 ± 0.2 at 25°C

Acetylacetone Solution:
Composition per 100.0mL:

Acetylacetone	10.0g
Ethanol (95% solution)	100.0mL

Preparation of Acetylacetone Solution: Add acetylacetone to 100.0mL of ethanol. Mix thoroughly. Filter sterilize.

Preparation of Medium: Add components, except acetylacetone solution, to distilled/deionized water and bring volume to 990.0mL. Mix thoroughly. Autoclave for 15 min at 15 psi pressure–121°C. Cool to room temperature. Aseptically add 1.0mL of acetylacetone solution. Mix thoroughly. Aseptically distribute into sterile tubes or flasks.

Use: For the cultivation and maintenance of *Aureobacterium terregens.*

Aureomycin® Rose Bengal Glucose Peptone Agar

Composition per liter:

Agar	20.0g
Glucose	10.0g
Peptone	5.0g
KH_2PO_4	1.0g
$MgSO_4 \cdot 7H_2O$	0.5g
Rose Bengal	0.035g
Aureomycin solution	200.0mL

pH 5.4 ± 0.2 at 25°C

Aureomycin Solution:
Composition per 200.0mL:

Aureomycin·HCl 0.07g

Preparation of Aureomycin Solution: Add aureomycin·HCl to distilled/deionized water and bring volume to 200.0mL. Mix thoroughly. Filter sterilize.

Preparation of Medium: Add components, except aureomycin solution, to distilled/deionized water and bring volume to 800.0mL. Mix thoroughly. Gently heat and bring to boiling. Autoclave for 15 min at 15 psi pressure–121°C. Cool to 45°–50°C. Aseptically add 200.0mL of sterile aureomycin solution. Mix thoroughly. Pour into sterile Petri dishes or distribute into sterile tubes.

Use: For the cultivation and enumeration of fungi isolated from sewage and polluted waters.

Autotrophic *Nitrobacter* Medium (DSMZ Medium 756c)

Composition per liter:

$NaNO_2$ 2.0g
Stock solution 100.0mL
Trace elements solution 1.0mL

pH 7.5 ± 0.2 at 25°C

Stock Solution:
Composition per liter:

NaCl 5.0g
KH_2PO_4 .... 1.5g
$MgSO_4 \cdot 7H_2O$ 0.5g
$CaCO_3$ 0.07g

Preparation of Stock Solution: Add components to distilled/deionized water and bring volume to 1.0L. Mix thoroughly.

Trace Elements Solution:
Composition per liter:

$FeSO_4 \cdot 7H_2O$ 97.3mg
H_3BO_3 .... 49.4mg
$ZnSO_4 \cdot 7H_2O$ 43.1mg
$(NH_4)_6Mo_7O_{24} \cdot 4H_2O$ 37.1mg
$MnSO_4 \cdot 2H_2O$ 33.8mg
$CuSO_4 \cdot 5H_2O$ 25.0mg

Preparation of Trace Elements Solution: Add components to distilled/deionized water and bring volume to 1.0L. Mix thoroughly.

Preparation of Medium: Add components to distilled/deionized water and bring volume to 1.0L. Mix thoroughly. Adjust pH to 8.6. Distribute into tubes or flasks. Autoclave for 15 min at 15 psi pressure–121°C. Allow to stand for 2–3 days so that pH adjusts itself to 7.4–7.6.

Use: For the cultivation of *Nitrobacter winogradskyi.*

Autotrophic *Nitrobacter* Medium (LMG Medium 247)

Composition per liter:

$NaNO_2$ 2.0g
Stock solution 100.0mL
Trace elements solution 1.0mL

pH 7.5 ± 0.2 at 25°C

Stock Solution:
Composition per liter:

NaCl 5.0g
KH_2PO_4 .... 1.5g
$MgSO_4 \cdot 7H_2O$ 0.5g
$CaCO_3$ 0.07g

Preparation of Stock Solution: Add components to distilled/deionized water and bring volume to 1.0L. Mix thoroughly.

Trace Elements Solution:
Composition per liter:

$(NH_4)Mo7O_2$ 437.10mg
$FeSO_4 \cdot 7H_2O$ 97.30mg
$ZnSO_4 \cdot 7H_2O$ 43.10mg
H_3BO_3 .... 39.40mg
$MnSO_4 \cdot H_2O$ 33.80mg
$CuSO_2 \cdot 5H_2O$ 25.00mg

Preparation of Trace Elements Solution: Add components to distilled/deionized water and bring volume to 1.0L. Mix thoroughly.

Preparation of Medium: Add components to distilled/deionized water and bring volume to 1.0L. Mix thoroughly. Adjust pH to 8.6 with NaOH. Distribute into tubes or flasks. Autoclave for 15 min at 15 psi pressure–121°C. Allow the medium to stand for 2–3 days so that the pH can adjust itself to pH 7.4–7.6.

Use: For the cultivation of autotrophic *Nitrobacter* spp.

Auxanographic Agar Medium
See: **Carbon Assimilation Medium**

AUY

Composition per liter:

Yeast extract 0.5g

Preparation of Medium: Add yeast extract to distilled/deionized water and bring volume to 1.0L. Mix thoroughly. Filter through Whatman #1 filter paper. Distribute 15.0mL into 20 × 125mm screw-capped tubes. Autoclave for 15 min at 15 psi pressure–121°C.

Use: For the cultivation of *Tokophrya infusionum.*

AV Agar with Vitamins

Composition per liter:

Agar 15.0g
Glucose 1.0g
Glycerol 1.0g
L-Arginine 0.3g
K_2HPO_4 .... 0.3g
NaCl 0.3g
$MgSO_4 \cdot 7H_2O$ 0.2g
Vitamin solution 100.0mL
Trace salts solution 1.0mL

Vitamin Solution:
Composition per 100.0mL:

p-Aminobenzoic acid 0.5mg
Calcium pantothenate 0.5mg
HCl 0.5mg
Inositol 0.5mg
Niacin 0.5mg
Pyridoxine 0.5mg
Riboflavin 0.5mg
Thiamine·HCl 0.5mg
Biotin 0.25mg

Preparation of Vitamin Solution: Add components to distilled/deionized water and bring volume to 1.0L. Mix thoroughly. Filter sterilize.

Trace Salts Solution:

Composition per liter:

$FeSO_4 \cdot 7H_2O$	10.0g
$CuSO_4 \cdot 5H_2O$	1.0g
$MnSO_4 \cdot 7H_2O$	1.0g
$ZnSO_4 \cdot 7H_2O$	1.0g

Preparation of Trace Salts Solution: Add components to distilled/deionized water and bring volume to 1.0L. Mix thoroughly.

Preparation of Medium: Add components, except vitamin solution, to distilled/deionized water and bring volume to 900.0mL. Mix thoroughly. Gently heat and bring to boiling. Autoclave for 15 min at 15 psi pressure–121°C. Cool to 45°–50°C. Aseptically add 100.0mL of sterile vitamin solution. Mix thoroughly. Pour into sterile Petri dishes or distribute into sterile tubes.

Use: For the isolation and cultivation of *Actinomadura* species, *Actinopolyspora* species, *Excellospora* species, and *Microspora* species.

16AV Medium (DSMZ Medium 298f)

Composition per liter:

NaCl	1.0g
KCl	0.5g
$MgCl_2 \cdot 6H_2O$	0.4g
NH_4Cl	0.25g
KH_2PO_4	0.2g
$CaCl_2 \cdot 2H_2O$	0.15g
Resazurin	1.0mg
$NaHCO_3$ solution	10.0mL
Butanediol solution	10.0mL
$Na_2S \cdot 9H_2O$ solution	10.0mL
Yeast extract solution	10.0mL
Galactose solution	10.0mL
Trace elements solution SL-10	1.0mL

pH 7.2 ± 0.2 at 25°C

$Na_2S \cdot 9H_2O$ Solution:

Composition per 10.0mL:

$Na_2S \cdot 9H_2O$	0.36g

Preparation of $Na_2S \cdot 9H_2O$ Solution: Add $Na_2S \cdot 9H_2O$ to distilled/deionized water and bring volume to 10.0mL. Mix thoroughly. Autoclave under 100% N_2 for 15 min at 15 psi pressure–121°C. Cool to room temperature.

$NaHCO_3$ Solution:

Composition per 10.0mL:

$NaHCO_3$	2.5g

Preparation of $NaHCO_3$ Solution: Add $NaHCO_3$ to distilled/deionized water and bring volume to 10.0mL. Mix thoroughly. Sparge with 80% N_2 + 20% CO_2. Filter sterilize.

Butanediol Solution:

Composition per 10.0mL:

2,3 butanediol	0.9g

Preparation of Butanediol Solution: Add butanediol to distilled/deionized water and bring volume to 10.0mL. Mix thoroughly. Sparge with 100% N_2. Filter sterilize.

Galactose Solution:

Composition per 10.0mL:

Galactose	2.0g

Preparation of Galactose Solution: Add galactose to distilled/deionized water and bring volume to 10.0mL. Mix thoroughly. Sparge with 100% N_2. Filter sterilize.

Trace Elements Solution SL-10:

Composition per liter:

$FeCl_2 \cdot 4H_2O$	1.5g
$CoCl_2 \cdot 6H_2O$	190.0mg
$MnCl_2 \cdot 4H_2O$	100.0mg
$ZnCl_2$	70.0mg
$Na_2MoO_4 \cdot 2H_2O$	36.0mg
$NiCl_2 \cdot 6H_2O$	24.0mg
H_3BO_3	6.0mg
$CuCl_2 \cdot 2H_2O$	2.0mg
HCl (25% solution)	10.0mL

Preparation of Trace Elements Solution SL-10: Add $FeCl_2 \cdot 4H_2O$ to 10.0mL of HCl solution. Mix thoroughly. Add distilled/deionized water and bring volume to 1.0L. Add remaining components. Mix thoroughly. Sparge with 80% N_2 + 20% CO_2. Autoclave for 15 min at 15 psi pressure–121°C.

Yeast Extract Solution:

Composition per 10.0mL:

Yeast extract	1.0g

Preparation of Yeast Extract Solution: Add yeast extract to distilled/deionized water and bring volume to 10.0mL. Mix thoroughly. Sparge with 100% N_2. Filter sterilize.

Preparation of Medium: Prepare and dispense medium under 80% N_2 + 20% CO_2 gas atmosphere. Add components, except $NaHCO_3$ solution, butanediol solution, $Na_2S \cdot 9H_2O$ solution, galactose solution, yeast extract solution, and trace elements solution SL-10, to distilled/deionized water and bring volume to 949.0mL. Mix thoroughly. Adjust pH to 7.2. Sparge with 80% N_2 + 20% CO_2. Autoclave for 15 min at 15 psi pressure–121°C. Aseptically and anaerobically add 10.0mL $NaHCO_3$ solution, 10.0mL butanediol solution, 10.0mL $Na_2S \cdot 9H_2O$ solution, 10.0mL galactose soltuion, 10.0mL yeast extract solution, and 1.0mL trace elements solution SL-10. Mix thoroughly. Aseptically and anaerobically distribute into sterile tubes or bottles. After inoculation, flush and repressurize the gas head space of culture bottles with sterile 80% N_2 + 20% CO_2 to 1 bar overpressure.

Use: For the cultivation of unclassified bacterium DSM 8385.

Avian *Mycoplasma* Agar

Composition per liter:

Agar, not inhibitory to mycoplasmas	10.0g
PPLO broth without Crystal Violet	700.0mL
Swine or horse serum, heat inactivated at 56°C for 30 min.	150.0mL
Fresh yeast extract solution	100.0mL
Phenol Red solution	20.0mL
Glucose solution	10.0mL
Arginine solution	10.0mL
NAD solution	10.0mL

PPLO Broth without Crystal Violet:

Composition per 700.0mL:

Beef heart, infusion from	175.0g
Peptone	7.0g
NaCl	3.5g

Source: PPLO broth without Crystal Violet is available as a premixed powder from BD Diagnostic Systems.

Preparation of PPLO Broth without Crystal Violet: Add components to distilled/deionized water and bring volume to 700.0mL. Autoclave for 15 min at 15 psi pressure–121°C. Cool to 25°C. Beef heart for infusion may be substituted; 100.0g of beef heart for infusion is equivalent to 500.0g of fresh heart tissue.

Fresh Yeast Extract Solution:

Composition per 100.0mL:

Baker's yeast, live, pressed, starch-free 25.0g

Preparation of Fresh Yeast Extract Solution: Add the live Baker's yeast to 100.0mL of distilled/deionized water. Autoclave for 90 min at 15 psi pressure–121°C. Allow to stand. Remove supernatant solution. Adjust pH to 6.6–6.8.

Phenol Red Solution:

Composition per 20.0mL:

Phenol Red 0.02g

Preparation of Phenol Red Solution: Add Phenol Red to distilled/deionized water and bring volume to 20.0mL. Mix thoroughly. Filter sterilize.

Glucose Solution:

Composition per 10.0mL:

Glucose 1.0g

Preparation of Glucose Solution: Add glucose to distilled/deionized water and bring volume to 10.0mL. Mix thoroughly. Filter sterilize.

Arginine Solution:

Composition per 10.0mL:

Arginine 1.0g

Preparation of Arginine Solution: Add arginine to distilled/deionized water and bring volume to 10.0mL. Mix thoroughly. Filter sterilize.

NAD Solution:

Composition per 10.0mL:

NAD 0.1g

Preparation of NAD Solution: Add NAD to distilled/deionized water and bring volume to 10.0mL. Mix thoroughly. Filter sterilize.

Preparation of Medium: Add 10.0g of agar to 700.0mL of PPLO broth without Crystal Violet. Gently heat to boiling with frequent mixing. Autoclave for 15 min at 15 psi pressure–121°C. Cool to 50°–55°C. Warm other components to 50°–55°C using a water bath. Aseptically combine all components. Mix thoroughly. Pour into sterile Petri dishes or sterile tubes.

Use: For the cultivation and maintenance of *Mycoplasma* species.

Avian *Mycoplasma* Broth

Composition per liter:

PPLO broth without Crystal Violet 700.0mL
Swine or horse serum, heat inactivated at 56°C for 30 min. 150.0mL
Fresh yeast extract solution 100.0mL
Phenol Red solution 20.0mL
Glucose solution 10.0mL
Arginine solution 10.0mL
NAD solution 10.0mL

PPLO Broth without Crystal Violet:

Composition per 700.0mL:

Beef heart, infusion from 175.0g
Peptone 7.0g
NaCl 3.5g

Source: PPLO broth without Crystal Violet is available as a premixed powder from BD Diagnostic Systems.

Preparation of PPLO Broth without Crystal Violet: Add components to distilled/deionized water and bring volume to 700.0mL. Autoclave for 15 min at 15 psi pressure–121°C. Cool to 25°C. Beef heart for infusion may be substituted; 100.0g of beef heart for infusion is equivalent to 500.0g of fresh heart tissue.

Fresh Yeast Extract Solution:

Composition per 100.0mL:

Baker's yeast, live, pressed, starch-free 25.0g

Preparation of Fresh Yeast Extract Solution: Add the live Baker's yeast to 100.0mL of distilled/deionized water. Autoclave for 90 min at 15 psi pressure–121°C. Allow to stand. Remove supernatant solution. Adjust pH to 6.6–6.8.

Phenol Red Solution:

Composition per 20.0mL:

Phenol Red 0.02g

Preparation of Phenol Red Solution: Add Phenol Red to distilled/deionized water and bring volume to 20.0mL. Mix thoroughly. Filter sterilize.

Glucose Solution:

Composition per 10.0mL:

Glucose 1.0g

Preparation of Glucose Solution: Add glucose to distilled/deionized water and bring volume to 10.0mL. Mix thoroughly. Filter sterilize.

Arginine Solution:

Composition per 10.0mL:

Arginine 1.0g

Preparation of Arginine Solution: Add arginine to distilled/deionized water and bring volume to 10.0mL. Mix thoroughly. Filter sterilize.

NAD Solution:

Composition per 10.0mL:

NAD 0.1g

Preparation of NAD Solution: Add NAD to distilled/deionized water and bring volume to 10.0mL. Mix thoroughly. Filter sterilize.

Preparation of Medium: Aseptically combine components. Distribute into sterile tubes or flasks.

Use: For the cultivation and maintenance of *Mycoplasma* species.

Axenic *Dimastigella* Medium

Composition per liter:

Sonneborn's base *Paramecium* medium 980.0mL
Vitamin solution 10.0mL
Heat-killed bacterial suspension 10.0mL

Sonneborn's Base *Paramecium* Medium:

Composition per liter:

Rye grass cerophyll 2.5g
Na_2HPO_4 .... 0.5g

Preparation of Sonneborn's Base *Paramecium* Medium: Add cerophyll to distilled/deionized water and bring volume to 1.0L. Mix thoroughly. Gently heat and bring to boil. Boil for 5 min. Filter through Whatman #1 filter paper. Add 0.5g of Na_2HPO_4. Bring volume to 1.0L with distilled/deionized water. Mix thoroughly. Distribute 10.0mL volumes into tubes. Autoclave for 15 min at 15 psi pressure–121°C.

Source: Cerophyll can be obtained from Ward's Natural Science Establishment, Inc. Dairy Goat Nutrition distributes Grass Media Culture, which is equivalent. Cereal Leaf Product from Sigma Chemical is similar to cerophyll.

Vitamin Solution:

Composition per 100.0mL:

Calcium D-(+)-pantothenate	0.05g
Nicotinamide	0.05g
Pyridoxal·HCl	0.05g
Riboflavin	0.05g
Pyridoxamine·HCl	0.025g
Folic acid	0.025g
Thiamine·HCl	0.15g
Biotin	0.0125mg
DL-Thioctic acid	0.5mL

Preparation of Vitamin Solution: Add components to distilled/deionized water and bring volume to 100.0mL. Mix thoroughly. Filter sterilize. For long-term storage, preserve under nitrogen at –20°C.

Heat-Killed Bacterial Suspension:

Composition per 100.0mL:

Heat-killed *Klebsiella pneumoniae*	10^{12} cells

Preparation of Heat-Killed Bacterial Suspension: Inoculate a loopful of *Klebsiella pneumoniae* subsp. *pneumoniae* ATCC 27889 into 5.0mL of nutrient broth. Incubate at 35°C overnight. Transfer 0.5mL aliquots of nutrient broth with bacterial suspension to each of ten 1.0L Erlenmeyer flasks, each containing 250.0mL of nutrient broth. Incubate cultures at 35°C for 24 hr. Aseptically transfer bacterial suspensions to 500.0mL sterilized screw-capped centrifuge bottles. Fill bottles with a maximum of 400.0mL. Centrifuge in a refrigerated centrifuge at 5000 rpm for 10 min. Decant supernatant and resuspend pellets in Page's balanced salt solution. Pool all suspensions in a single bottle. Centrifuge in a refrigerated centrifuge at 5000 rpm for 10 min. Discard supernatant and resuspend pellet in Page's balanced salt solution. Final volume of cell suspension should be approximately 400.0mL. Decant supernatant and resuspend pellets in Page's balanced salt solution. Centrifuge in a refrigerated centrifuge at 5000 rpm for 10 min. Discard supernatant and resuspend pellet in Page's balanced salt solution. Decant supernatant and resuspend pellets in Page's balanced salt solution. Final volume of cell suspension should be approximately 400.0mL. Centrifuge in a refrigerated centrifuge at 5000 rpm for 10 min. Decant supernatant and resuspend pellets in Page's balanced salt solution. Final volume this time should only be 100.0mL. Agitate to ensure that cells are thoroughly suspended. Transfer to a 125.0mL screw-capped serum bottle and bring volume to 100.0mL with Page's balanced salt solution. Serially dilute the suspension to a dilution of 10^{-9}. Plate 0.1mL aliquots in triplicate from the 10^{-7} to 10^{-9} dilution tubes. Place the aliquots in the center of 100.0mm Petri plates containing nutrient agar and spread evenly over the surfaces with a sterile glass rod. Incubate plates at 35°C overnight. Place the 125.0mL screw-capped serum bottle containing the bacterial suspension in 100.0mL of Page's balanced salt solution into a 60°C water bath. Make sure that the liquid level of the water bath is above that of the suspension in the bottle. At 10-min intervals, swirl the bottle. Incubate for a total of 30 min. Allow the bottle to cool to room temperature. This treatment should kill all bacterial cells. Determine bacterial cell concentration from the serial dilution plates. Adjust the concentration of the heat-killed bacteria to 10^{10} cells per mL. As a check that the cells are not viable, add 3 drops of the cell suspension prepared in step 10 to the edge of a 100.0mm Petri plate containing nutrient agar. Hold the plate vertically to allow the drops to move to the opposite edge. Incubate plate at 35°C for 48 hr.

Nutrient Broth:

Composition per liter:

Pancreatic digest of gelatin	5.0g
Beef extract	3.0g

Preparation of Nutrient Broth: Add components to distilled/deionized water and bring volume to 1.0L. Mix thoroughly. Distribute into tubes or flasks. Autoclave for 15 min at 15 psi pressure–121°C.

Page's Balanced Salt Solution:

Composition per liter:

Solution 1	500.0mL
Solution 2	500.0mL

Solution 1:

Composition per 500.0mL:

Na_2HPO_4	2.84g
KH_2PO_4	2.72g

Preparation of Solution 1: Add components to distilled/deionized water and bring volume to 1.0L. Mix thoroughly. Autoclave for 15 min at 15 psi pressure–121°C. Cool to 25°C.

Solution 2:

Composition per 500.0mL:

$MgSO_4·7H_2O$	8.0mg
$CaCl_2·2H_2O$	8.0mg
NaCl	0.24g

Preparation of Solution 2: Add components to distilled/deionized water and bring volume to 1.0L. Mix thoroughly. Autoclave for 15 min at 15 psi pressure–121°C. Cool to 25°C.

Preparation of Page's Balanced Salt Solution: Aseptically combine 500.0mL of solution 1 with 500.0mL of solution 2.

Nutrient Agar:

Composition per liter:

Agar	15.0g
Pancreatic digest of gelatin	5.0g
Beef extract	3.0g

Preparation of Nutrient Agar: Add components to distilled/deionized water and bring volume to 1.0L. Mix thoroughly. Gently heat and bring to boiling. Distribute into tubes or flasks. Autoclave for 15 min at 15 psi pressure–121°C. Pour into sterile Petri dishes.

Preparation of Medium: Aseptically add 10.0mL of the vitamin solution to 980.0mL of Sonneborn's base *Paramecium* medium. Mix thoroughly. Autoclave for 15 min at 15 psi pressure–121°C. Cool to 25°C. Aseptically distribute 10.0mL aliquots into T-25 tissue culture flasks. Add 0.1mL of the heat-killed bacterial suspension to each flask. Inoculate immediately with *Dimastigella* species.

Use: For the cultivation of *Dimastigella trypaniformis* and other *Dimastigella* species.

Ayers and Johnson Agar (Stock Culture Agar)

Composition per liter:

Beef heart, infusion from	500.0g
Proteose peptone	10.0g

Gelatin 10.0g
Agar 7.5g
Casein, purified 5.0g
Na_2HPO_4 .... 4.0g
Sodium citrate 3.0g
Glucose 0.5g
pH 7.50 ± 0.2 at 25°C

Preparation of Medium: Add components to distilled/deionized water and bring volume to 1.0L. Mix thoroughly. Autoclave for 15 min at 15 psi pressure–121°C.

Use: For the maintenance of cultures of streptococci and other microorganisms.

Azide Agar

See: ***Enterococcus*** **Agar**

Azide Blood Agar

Composition per liter:
Agar 15.0g
Pancreatic digest of casein 5.0g
Peptic digest of animal tissue 5.0g
NaCl 5.0g
Beef extract 3.0g
NaN_3 .... 0.2g
Sheep blood, defibrinated 50.0mL
pH 7.2 ± 0.2 at 25°C

Source: This medium is available as a premixed powder from BD Diagnostic Systems and Oxoid Unipath.

Caution: Sodium azide is toxic. Azides also react with metals and disposal must be highly diluted.

Preparation of Medium: Add components, except sheep blood, to distilled/deionized water and bring volume to 950.0mL. Mix thoroughly. Gently heat and bring to boiling. Autoclave for 15 min at 15 psi pressure–121°C. Cool to 45–50°C. Aseptically add 50.0mL of sterile defibrinated sheep blood. Pour into sterile Petri dishes or distribute into sterile tubes. Allow tubes to cool in a slanted position.

Use: For the isolation and differentiation of streptococci and staphylococci from specimens containing mixed flora and from nonclinical specimens such as water and sewage.

Azide Blood Agar Base with Blood

Composition per liter:
Agar 15.0g
Peptone, special 10.0g
NaCl 5.0g
Beef 3.0g
NaN_3 .... 0.2g
Sheep blood, defibrinated 50.0mL
pH 7.2 ± 0.2 at 25°C

Source: This medium without sheep blood is available as a premixed powder from HiMedia.

Caution: Sodium azide is toxic. Azides also react with metals and disposal must be highly diluted.

Preparation of Medium: Add components, except sheep blood, to distilled/deionized water and bring volume to 950.0mL. Mix thoroughly. Gently heat and bring to boiling. Autoclave for 15 min at 15 psi pressure–121°C. Pour into sterile Petri dishes or leave in tubes.

Use: For the isolation and differentiation of streptococci and staphylococci from specimens containing mixed flora and from nonclinical specimens such as water and sewage.

Azide Blood Agar Base, HiVeg with Blood

Composition per liter:
Agar 15.0g
Plant special peptone 10.0g
NaCl 5.0g
Plant extract 3.0g
NaN_3 .... 0.2g
Sheep blood, defibrinated 50.0mL
pH 7.2 ± 0.2 at 25°C

Source: This medium without sheep blood is available as a premixed powder from HiMedia.

Caution: Sodium azide is toxic. Azides also react with metals and disposal must be highly diluted.

Preparation of Medium: Add components, except sheep blood, to distilled/deionized water and bring volume to 950.0mL. Mix thoroughly. Gently heat and bring to boiling. Autoclave for 15 min at 15 psi pressure–121°C. Pour into sterile Petri dishes or leave in tubes.

Use: For the isolation and differentiation of streptococci and staphylococci from specimens containing mixed flora and from nonclinical specimens such as water and sewage.

Azide Blood Agar with Crystal Violet (Packer's Agar)

Composition per liter:
Agar 15.0g
Pancreatic digest of casein 5.0g
Peptic digest of animal tissue 5.0g
NaCl 5.0g
Beef extract 3.0g
NaN_3 .... 0.9g
Crystal Violet 2.0mg
Sheep blood, defibrinated 50.0mL
pH 7.2 ± 0.2 at 25°C

Caution: Sodium azide is toxic. Azides also react with metals and disposal must be highly diluted.

Preparation of Medium: Add components, except sheep blood, to distilled/deionized water and bring volume to 950.0mL. Mix thoroughly. Gently heat and bring to boiling. Autoclave for 15 min at 15 psi pressure–121°C. Cool to 45°–50°C. Aseptically add 50.0mL of sterile defibrinated sheep blood. Pour into sterile Petri dishes or distribute into sterile tubes. Allow tubes to cool in a slanted position.

Use: For the isolation and enumeration of fecal streptococci from nonclinical specimens such as water and food. Also used for the isolation of *Streptococcus pneumoniae* and *Erysipelothrix rhusiopathiae*.

Azide Broth (Azide Glucose Broth) (Azide Dextrose Broth)

Composition per liter:
Pancreatic digest of casein 15.0g
Glucose 7.5g
NaCl 7.5g

Beef extract 4.5g
NaN_3 .... 0.2g

pH 7.2 ± 0.2 at 25°C

Source: This medium is available as a premixed powder from BD Diagnostic Systems.

Caution: Sodium azide is toxic. Azides also react with metals and disposal must be highly diluted.

Preparation of Medium: Add components to distilled/deionized water and bring volume to 1.0L. Mix thoroughly. Gently heat and bring to boiling. Distribute into tubes or flasks. Autoclave for 15 min at 15 psi pressure–121°C. Prepare double-strength broth for samples larger than 1.0mL.

Use: For the detection and enrichment of fecal streptococci in water and sewage. Also used in the multiple-tube technique as a presumptive test for the presence of fecal streptococci.

Azide Broth, Rothe (Azide Glucose Broth, Rothe) (Azide Dextrose Broth, Rothe)

Composition per liter:

Peptone 20.0g
Glucose 5.0g
NaCl 5.0g
K_2HPO_4 .... 2.7g
KH_2PO_4 .... 2.7g
NaN_3 .... 0.2g

pH 6.8 ± 0.2 at 25°C

Source: This medium is available as a premixed powder from Oxoid Unipath.

Caution: Sodium azide is toxic. Azides also react with metals and disposal must be highly diluted.

Preparation of Medium: Add components to distilled/deionized water and bring volume to 1.0L. Mix thoroughly. Gently heat and bring to boiling. Distribute into tubes or flasks. Autoclave for 15 min at 15 psi pressure–121°C. Prepare double-strength broth for samples larger than 1.0mL.

Use: For the detection of enterococci in water and sewage.

Azide Citrate Broth

Composition per liter:

Pancreatic digest of casein 20.0g
Sodium citrate 10.0g
Yeast extract 5.0g
Glucose 5.0g
NaCl 5.0g
K_2HPO_4 .... 4.0g
KH_2PO_4 .... 1.5g
NaN_3 .... 0.25g

pH 7.0 ± 0.2 at 25°C

Caution: Sodium azide is toxic. Azides also react with metals and disposal must be highly diluted.

Preparation of Medium: Add components to distilled/deionized water and bring volume to 1.0L. Mix thoroughly. Gently heat and bring to boiling. Distribute into tubes or flasks. Autoclave for 15 min at 15 psi pressure–118°C. Prepare double-strength broth for samples larger than 1.0mL.

Use: For the detection and enrichment of fecal streptococci in water and sewage.

Azide Dextrose Broth
See: **Azide Broth**

Azide Dextrose Broth, Rothe
See: **Azide Broth, Rothe**

Azide Dextrose HiVeg Broth

Composition per liter:

Plant special peptone 15.0g
Glucose 7.5g
NaCl 7.5g
Plant extract 4.5g
NaN_3 .... 0.2g

pH 7.2 ± 0.2 at 25°C

Source: This medium is available as a premixed powder from HiMedia.

Caution: Sodium azide is toxic. Azides also react with metals and disposal must be highly diluted.

Preparation of Medium: Add components to distilled/deionized water and bring volume to 1.0L. Mix thoroughly. Gently heat and bring to boiling. Autoclave for 15 min at 12 psi pressure–118°C.

Use: For the detection and enrichment of fecal streptococci in water and sewage. Also used in the multiple-tube technique as a presumptive test for the presence of fecal streptococci in water, sewage, food, and other materials suspected of sewage contamination.

Azide Glucose Broth
See: **Azide Broth**

Azide Glucose Broth, Rothe
See: **Azide Broth, Rothe**

Azide Medium

Composition per liter:

Peptone 10.0g
K_2HPO_4 .... 5.0g
Glucose 5.0g
NaCl 5.0g
Yeast extract 3.0g
KH_2PO_4 .... 2.0g
NaN_3 .... 0.25g
Bromcresol Purple solution 2.0mL

pH 7.2 ± 0.2 at 25°C

Bromcresol Purple Solution:

Composition per 10.0mL:

Bromcresol Purple 0.16g
Ethanol 10.0mL

Preparation of Bromcresol Purple Solution: Add Bromcresol Purple to ethanol and bring volume to 10.0mL. Mix thoroughly.

Caution: Sodium azide is toxic. Azides also react with metals and disposal must be highly diluted.

Preparation of Medium: Add components to distilled/deionized water and bring volume to 1.0L. Mix thoroughly. Distribute into tubes or flasks. Autoclave for 15 min at 15 psi pressure–121°C.

Use: For the cultivation of *Streptococcus* species and *Staphylococcus* species from clinical and nonclinical specimens.

Azoarcus Medium (LMG Medium 202)

Composition per liter:

Solution A	750.0mL
Phosphate buffer solution	250.0mL

pH 6.8 ± 0.2 at 25°C

Solution A:

Composition per 750.0mL:

Malic acid	5.0g
KOH	4.5g
$MgSO_4 \cdot 7H_2O$	0.2g
NaCl	0.1g
$CaCl_2$	20.0mg
$MnSO_4 \cdot H_2O$	10.0mg
$Na_2MoO_4 \cdot 2H_2O$	2.0mg
Ferric EDTA solution	10.0mL

Ferric EDTA Solution:

Composition per liter:

Ferric EDTA	0.066g

Preparation of Ferric EDTA Solution: Add ferric EDTA to distilled/deionized water and bring volume to 10.0mL. Mix thoroughly.

Preparation of Solution A: Add 5.0g malic acid to 500.0mL distilled/deionized water. Adjust pH to 7.0 with KOH (approximate amount of 4.5g). Add other components. Mix thoroughly. Bring volume to 750.0mL. Adjust pH to 6.8. Autoclave for 15 min at 15 psi pressure–121°C. Cool to 25°C.

Phosphate Buffer Solution:

Composition per liter:

$Na_2HPO_4 \cdot 2H_2O$	5.8g
KH_2PO_4	4.5g

Preparation of Phosphate Buffer Solution: Add components to distilled/deionized water and bring volume to 1.0L. Mix thoroughly. Adjust pH to 6.8. Autoclave for 15 min at 15 psi pressure–121°C. Cool to 25°C.

Preparation of Medium: Aseptically combine 750.0mL sterile solution A with 250.0mL sterile phosphate buffer solution. Aseptically distribute to sterile tubes or flasks.

Use: For the cultivation of *Azoarcus indigens*.

Azoarcus VM Medium (LMG Medium 252)

Composition per liter:

Agar	15.0g
Beef extract	3.0g
DL-malic acid	2.5g
KOH	2.5g
KH_2PO_4	1.5g
NaCl	1.1g
K_2HPO_4	1.0g
Yeast extract	1.0g
$MgSO_4 \cdot 7H_2O$	0.2g
$CaCl_2$	0.2g
Fe EDTA	66.0mg
$MnSO_4 \cdot H_2O$	10.0mg
$Na_2MoO_4 \cdot 2H_2O$	2.0mg
Biotin	0.1mg
NH_4Cl	0.5mg

pH 6.8 ± 0.2 at 25°C

Preparation of Medium: Add components to distilled/deionized water and bring volume to 1.0L. Mix thoroughly. Adjust pH to 6.8. Gently heat and bring to boiling. Distribute into tubes or flasks. Autoclave for 15 min at 15 psi pressure–121°C. Pour into sterile Petri dishes or leave in tubes.

Use: For the cultivation and maintenance of *Azospira oryzae* and *Azonexus fungiphilus*.

Azorhizobium caulinodans Agar (LMG 119)

Composition per liter:

Agar	15.0g
Beef extract	5.0g
Peptone	5.0g
Sucrose	5.0g
Yeast extract	1.0g
$MgSO_4$	0.24g

Preparation of Medium: Add components to distilled/deionized water and bring volume to 1.0L. Mix thoroughly. Gently heat and bring to boiling. Distribute into tubes or flasks. Autoclave for 15 min at 15 psi pressure–121°C. Pour into sterile Petri dishes or leave in tubes.

Use: For the cultivation and maintenance of *Azorhizobium caulinodans*.

Azorhizophilus paspali Agar

Composition per liter:

Agar	20.0g
Sucrose	20.0g
$Na_2MoO_4 \cdot 2H_2O$	0.5g
$MgSO_4 \cdot 7H_2O$	0.2g
KH_2PO_4	0.15g
K_2HPO_4	0.05g
$FeCl_3$	0.01g

pH 6.9 ± 0.2 at 25°C

Preparation of Medium: Add components to distilled/deionized water and bring volume to 1.0L. Mix thoroughly. Gently heat and bring to boiling. Distribute into tubes or flasks. Autoclave for 15 min at 15 psi pressure–121°C. Pour into sterile Petri dishes or leave in tubes.

Use: For the cultivation and maintenance of *Azorhizophilus paspali*.

Azospirillum amazonense Medium (LGI Medium)

Composition per liter:

Sucrose	5.0g
Agar	1.75g
KH_2PO_4	0.6g
K_2HPO_4	0.2g
$MgSO_4 \cdot 7H_2O$	0.2g
$CaCl_2 \cdot 2H_2O$	0.02g
$FeCl_3$	0.01g
$Na_2MoO_4 \cdot 2H_2O$	2.0mg
Bromthymol Blue solution	5.0mL

pH 6.0 ± 0.2 at 25°C

Bromthymol Blue Solution:

Composition per 100.0mL:

Bromthymol Blue 0.5g

Preparation of Bromthymol Blue Solution: Add Bromthymol Blue to 100.0mL of 0.2*N* KOH. Mix thoroughly.

Preparation of Medium: Add components to distilled/deionized water and bring volume to 1.0L. Mix thoroughly. Gently heat and bring to boiling. Distribute into tubes or flasks. Autoclave for 15 min at 15 psi pressure–121°C. Pour into sterile Petri dishes or leave in tubes.

Use: For the cultivation and maintenance of *Azospirillum amazonense*.

Azospirillum lipoferum Agar Medium

Composition per liter:

Glucose 20.0g
Agar 15.0g
K_2HPO_4 .... 0.8g
$MgSO_4 \cdot 7H_2O$ 0.5g
KH_2PO_4 .... 0.2g
$FeCl_3 \cdot 6H_2O$ 0.1g
Yeast extract 0.1g
$CaCl_2 \cdot 2H_2O$ 0.02g
$Na_2MoO_4 \cdot 2H_2O$ 0.02g

pH 6.9 ± 0.2 at 25°C

Preparation of Medium: Add components to distilled/deionized water and bring volume to 1.0L. Mix thoroughly. Gently heat and bring to boiling. Distribute into tubes or flasks. Autoclave for 15 min at 15 psi pressure–121°C. Pour into sterile Petri dishes or leave in tubes.

Use: For the cultivation of *Azospirillum lipoferum*.

Azospirillum lipoferum Agar Medium

Composition per liter:

Agar 15.0g
Calcium malate 10.0g
K_2HPO_4 .... 0.8g
$MgSO_4 \cdot 7H_2O$ 0.5g
KH_2PO_4 .... 0.2g
$FeCl_3 \cdot 6H_2O$ 0.1g
Yeast extract 0.1g
$CaCl_2 \cdot 2H_2O$ 0.02g
$Na_2MoO_4 \cdot 2H_2O$ 0.02g

pH 6.9 ± 0.2 at 25°C

Preparation of Medium: Add components to distilled/deionized water and bring volume to 1.0L. Mix thoroughly. Gently heat and bring to boiling. Distribute into tubes or flasks. Autoclave for 15 min at 15 psi pressure–121°C. Pour into sterile Petri dishes or leave in tubes.

Use: For the cultivation of *Azospirillum lipoferum*.

Azospirillum lipoferum Medium

Composition per liter:

Calcium malate 10.0g
K_2HPO_4 .... 1.0g
$MgSO_4 \cdot 7H_2O$ 0.5g
$CaCl_2 \cdot 2H_2O$ 0.02g

pH 6.5 ± 0.2 at 25°C

Preparation of Medium: Add components to distilled/deionized water and bring volume to 1.0L. Mix thoroughly. Distribute into tubes or flasks. Autoclave for 15 min at 15 psi pressure–121°C.

Use: For the isolation and cultivation of *Azospirillum lipoferum*.

Azospirillum Medium

Composition per liter:

Sodium malate 5.0g
Agar 1.75g
KH_2PO_4 .... 0.4g
$MgSO_4 \cdot 7H_2O$ 0.2g
K_2HPO_4 .... 0.1g
NaCl 0.1g
$CaCl_2 \cdot 2H_2O$ 0.02g
$FeCl_3$ 0.01g
$Na_2MoO_4 \cdot 2H_2O$ 2.0mg
Bromthymol Blue solution 5.0mL

pH 6.8 ± 0.2 at 25°C

Bromthymol Blue Solution:

Composition per 10.0mL:

Bromthymol Blue 0.5g
Ethanol 10.0mL

Preparation of Bromthymol Blue Solution: Add Bromthymol Blue to 10.0mL of ethanol. Mix thoroughly.

Preparation of Medium: Add components to distilled/deionized water and bring volume to 1.0L. Mix thoroughly. Distribute into tubes or flasks. Autoclave for 15 min at 15 psi pressure–121°C.

Use: For the cultivation of *Azospirillum* species isolated from roots.

Azospirillum Medium

Composition per 950.0mL:

$MnSO_4 \cdot H_2O$ 2.0g
$(NH_4)_2SO_4$ 1.0g
K_2HPO_4 .... 0.25g
$MgSO_4 \cdot 7H_2O$ 0.2g
NaCl 0.1g
Yeast extract 0.05g
$CaCl_2 \cdot 2H_2O$ 0.02g
$FeSO_4 \cdot 7H_2O$ 0.01g
Bromthymol Blue 25.0mg
$Na_2MoO_4 \cdot 2H_2O$ 1.0mg
Biotin 0.1mg
Glucose solution 25.0mL
Sodium malate solution 25.0mL
Bromthymol Blue solution 5.0mL

pH 7.1 ± 0.2 at 25°C

Glucose Solution:

Composition per 100.0mL:

D-Glucose 20.0g

Preparation of Glucose Solution: Add glucose to distilled/deionized water and bring volume to 100.0mL. Mix thoroughly. Filter sterilize.

Sodium Malate Solution:

Composition per 100.0mL:

Sodium malate 20.0g

Preparation of Sodium Malate Solution: Add sodium malate to distilled/deionized water and bring volume to 100.0mL. Mix thoroughly. Filter sterilize.

Bromthymol Blue Solution:

Composition per 100.0mL:

Bromthymol Blue 0.5g

Preparation of Bromthymol Blue Solution: Add Bromthymol Blue to 100.0mL of 0.2*N* KOH. Mix thoroughly.

Preparation of Medium: Add components, except glucose solution and sodium malate solution, to distilled/deionized water and bring volume to 950.0mL. Mix thoroughly. Autoclave for 15 min at 15 psi pressure–121°C. Cool to room temperature. Aseptically add 25.0mL of sterile glucose solution and 25.0mL of sterile sodium malate solution. Mix thoroughly. Aseptically distribute into sterile tubes or flasks.

Use: For the cultivation and maintenance of *Azospirillum* species.

Azospirillum Medium with 0.17% Agar

Composition per liter:

Malic acid	5.0g
Agar	1.75
K_2HPO_4	0.5g
$FeSO_4 \cdot 7H_2O$	0.5g
$MgSO_4 \cdot 7H_2O$	0.2g
NaCl	0.1g
$CaCl_2 \cdot 2H_2O$	0.02g
$MnSO_4 \cdot H_2O$	0.01g
$Na_2MoO_4 \cdot 2H_2O$	2.0mg
Bromthymol Blue	2.0mg
Potassium hydroxide solution	50.0mL

pH 6.8 ± 0.2 at 25°C

Source: This medium is available from HiMedia.

Potassium Hydroxide Solution:

Composition per 50.0mL:

KOH	4.0g

Preparation of Potassium Hydroxide Solution: Add KOH to distilled/deionized water and bring volume to 50.0mL. Mix thoroughly. Filter sterilize.

Preparation of Medium: Add components, except potassium hydroxide solution, to distilled/deionized water and bring volume to 950.0mL. Mix thoroughly. Autoclave for 15 min at 15 psi pressure–121°C. Cool to 50°C. Aseptically add potassium hydroxide solution,. Mix thoroughly. Pour into Petri dishes or aseptically distribute into sterile tubes.

Use: For the enrichment and cultivation of *Azospirillum* spp.

Azotobacter Agar

Composition per liter:

Agar	15.0g
Sucrose	10.0g
$MgSO_4 \cdot 7H_2O$	0.2g
KH_2PO_4	0.15g
K_2HPO_4	0.05g
$CaCl_2$	0.02g
Na_2MoO_4	0.002g
$FeCl_3$	1.0μg

Preparation of Medium: Add components to distilled/deionized water and bring volume to 1.0L. Mix thoroughly. Gently heat and bring to boiling. Distribute into tubes or flasks. Autoclave for 15 min at 15 psi pressure–121°C. Pour into sterile Petri dishes or leave in tubes.

Use: For the cultivation and maintenance of *Azorhizophilus paspali.*

Azotobacter Agar (Glucose)

Composition per liter:

Agar	15.0g
Glucose	10.0g
Soil extract	5.0g
K_2HPO_4	1.0g
$MgSO_4 \cdot 7H_2O$	0.2g
NaCl	0.2g
$FeSO_4$	5.0mg

pH 7.6 ± 0.2 at 25°C

Source: This medium is available from HiMedia.

Preparation of Medium: Add components to distilled/deionized water and bring volume to 1.0L. Mix thoroughly. Gently heat and bring to boiling. Distribute into tubes or flasks. Autoclave for 15 min at 15 psi pressure–121°C. Pour into sterile Petri dishes or leave in tubes.

Use: For the isolation and cultivation of glucose positive *Azotobacter* species from soil.

Azotobacter Agar (Mannitol)

Composition per liter:

Agar	15.0g
Mannitol	20.0g
Soil extract	5.0g
K_2HPO_4	1.0g
$MgSO_4 \cdot 7H_2O$	0.2g
NaCl	0.2g
$FeSO_4$	1.0mg

pH 7.6 ± 0.2 at 25°C

Source: This medium is available from HiMedia.

Preparation of Medium: Add components to distilled/deionized water and bring volume to 1.0L. Mix thoroughly. Gently heat and bring to boiling. Distribute into tubes or flasks. Autoclave for 15 min at 15 psi pressure–121°C. Pour into sterile Petri dishes or leave in tubes.

Use: For the isolation and cultivation of mannitol positive *Azotobacter* species from soil.

Azotobacter Agar, Modified I

Composition per liter:

Agar	15.0g
Sucrose	10.0g
Glucose	10.0g
$MgSO_4 \cdot 7H_2O$	0.2g
KH_2PO_4	0.15g
$CaSO_4 \cdot 2H_2O$	0.1g
K_2HPO_4	0.05g
$CaCl_2$	0.02g
Na_2MoO_4	2.0mg
$FeCl_3$	1.0mg
$Na_2MoO_4 \cdot 2H_2O$	1.0mg

pH 7.2 ± 0.2 at 25°C

Preparation of Medium: Add components to distilled/deionized water and bring volume to 1.0L. Mix thoroughly. Gently heat and bring to boiling. Adjust pH to 7.2. Distribute into tubes or flasks. Autoclave for 15 min at 15 psi pressure–121°C. Pour into sterile Petri dishes or leave in tubes.

Use: For the cultivation and maintenance of *Azotobacter* species.

Azotobacter Agar, Modified II

Composition per liter:

Sucrose	20.0g
Agar	15.0g
KH_2PO_4	0.15g
$MgSO_4 \cdot 7H_2O$	0.2g
K_2HPO_4	0.05g
$CaCl_2$	0.02g
Na_2MoO_4	2.0mg
$FeCl_3$	1.0mg
$Na_2MoO_4 \cdot 2H_2O$	1.0mg

pH 6.2 ± 0.2 at 25°C

Preparation of Medium: Add components to distilled/deionized water and bring volume to 1.0L. Mix thoroughly. Gently heat and bring to boiling. Adjust pH to 6.2. Distribute into tubes or flasks. Autoclave for 15 min at 15 psi pressure–121°C. Pour into sterile Petri dishes or leave in tubes.

Use: For the cultivation and maintenance of *Azotobacter* species and *Beijerinckia derxii.*

Azotobacter Basal Agar

Composition per liter:

Agar	15.0g
K_2HPO_4	1.0g
$MgSO_4 \cdot 7H_2O$	0.2g
NaCl	0.2g
$FeSO_4 \cdot 7H_2O$	5.0mg
Soil extract	100.0mL

pH 7.2 ± 0.2 at 25°C

Soil Extract:

Composition per 200.0mL:

African Violet soil	0.5g
Na_2CO_3	0.5g

Preparation of Soil Extract: Add components to tap water and bring volume to 200.0mL. Autoclave for 60 min at 15 psi pressure–121°C. Filter through Whatman filter paper.

Preparation of Medium: Add components, including filtered soil extract, to tap water and bring volume to 1.0L. Mix thoroughly. Gently heat and bring to boiling. Distribute into tubes or flasks. Autoclave for 15 min at 15 psi pressure–121°C. Pour into sterile Petri dishes or leave in tubes.

Use: For the cultivation of a variety of bacteria, including *Azomonas* species, *Azotobacter* species, and others when a carbon source is added.

Azotobacter Basal Broth

Composition per liter:

K_2HPO_4	1.0g
$MgSO_4 \cdot 7H_2O$	0.2g
NaCl	0.2g
$FeSO_4 \cdot 7H_2O$	5.0mg
Soil extract	100.0mL

pH 7.2 ± 0.2 at 25°C

Soil Extract:

Composition per 200.0mL:

African Violet soil	0.5g
Na_2CO_3	0.5g

Preparation of Soil Extract: Add components to tap water and bring volume to 200.0mL. Autoclave for 60 min at 15 psi pressure–121°C. Filter through Whatman filter paper.

Preparation of Medium: Add components, including filtered soil extract, to tap water and bring volume to 1.0L. Mix thoroughly. Distribute into tubes or flasks. Autoclave for 15 min at 15 psi pressure–121°C.

Use: For the cultivation of a variety of bacteria, including *Azomonas* species, *Azotobacter* species, and others when a carbon source is added.

Azotobacter Broth

Composition per liter:

Sucrose	10.0g
$MgSO_4 \cdot 7H_2O$	0.2g
KH_2PO_4	0.15g
K_2HPO_4	0.05g
$CaCl_2$	0.02g
Na_2MoO_4	0.002g
$FeCl_3$	1.0μg

Preparation of Medium: Add components to distilled/deionized water and bring volume to 1.0L. Mix thoroughly. Distribute into tubes or flasks. Autoclave for 15 min at 15 psi pressure–121°C.

Use: For the cultivation of *Azorhizophilus paspali.*

Azotobacter Broth

Composition per liter:

Glucose	10.0g
$CaCO_3$	5.0g
K_2HPO_4	0.9g
$CaCl_2 \cdot 2H_2O$	0.1g
$MgSO_4 \cdot 7H_2O$	0.1g
KH_2PO_4	0.1g
$FeSO_4 \cdot 7H_2O$	10.0mg
$Na_2MoO_4 \cdot 2H_2O$	5.0mg

pH 7.3 ± 0.2 at 25°C

Preparation of Medium: Add components to distilled/deionized water and bring volume to 1.0L. Mix thoroughly. Gently heat and bring to boiling. Distribute into tubes or flasks. Autoclave for 15 min at 15 psi pressure–121°C. Pour into sterile Petri dishes or leave in tubes.

Use: For the cultivation and maintenance of *Azotobacter beijerinckii, Azotobacter chroococcum, Azotobacter vinelandii,* and *Derxia gummosa.*

Azotobacter Broth, Modified I

Composition per liter:

Sucrose	10.0g
Glucose	10.0g
$MgSO_4 \cdot 7H_2O$	0.2g
KH_2PO_4	0.15g
$CaSO_4 \cdot 2H_2O$	0.1g
K_2HPO_4	0.05g
$CaCl_2$	0.02g
Na_2MoO_4	2.0mg
$FeCl_3$	1.0mg
$Na_2MoO_4 \cdot 2H_2O$	1.0mg

pH 7.2 ± 0.2 at 25°C

Preparation of Medium: Add components to distilled/deionized water and bring volume to 1.0L. Mix thoroughly. Gently heat and bring

to boiling. Adjust pH to 7.2. Distribute into tubes or flasks. Autoclave for 15 min at 15 psi pressure–121°C.

Use: For the cultivation of *Azotobacter* species.

Azotobacter Broth, Modified II

Composition per liter:

Sucrose	20.0g
KH_2PO_4	0.15g
$MgSO_4 \cdot 7H_2O$	0.2g
K_2HPO_4	0.05g
$CaCl_2$	0.02g
Na_2MoO_4	2.0mg
$FeCl_3$	1.0mg
$Na_2MoO_4 \cdot 2H_2O$	1.0mg

pH 6.2 ± 0.2 at 25°C

Preparation of Medium: Add components to distilled/deionized water and bring volume to 1.0L. Mix thoroughly. Gently heat and bring to boiling. Adjust pH to 6.2. Distribute into tubes or flasks. Autoclave for 15 min at 15 psi pressure–121°C.

Use: For the cultivation of *Azotobacter* species and *Beijerinckia derxii.*

Azotobacter chroococcum Agar

Composition per liter:

Agar	20.0g
$CaCO_3$	20.0g
Glucose	20.0g
K_2HPO_4	0.8g
$MgSO_4 \cdot 7H_2O$	0.5g
KH_2PO_4	0.2g
$FeCl_3 \cdot 6H_2O$	0.1g
$Na_2MoO_4 \cdot 2H_2O$	0.05g

pH 7.4–7.6 at 25°C

Preparation of Medium: Add components to distilled/deionized water and bring volume to 1.0L. Mix thoroughly. Gently heat and bring to boiling. Distribute into tubes or flasks. Autoclave for 15 min at 15 psi pressure–121°C. Pour into sterile Petri dishes or leave in tubes.

Use: For the cultivation and maintenance of *Azotobacter chroococcum.*

Azotobacter chroococcum Agar

Composition per liter:

Agar	20.0g
Glucose	20.0g
K_2HPO_4	0.8g
$MgSO_4 \cdot 7H_2O$	0.5g
KH_2PO_4	0.2g
$FeCl_3 \cdot 6H_2O$	0.1g
$CaCl_2 \cdot 2H_2O$	0.05g
$Na_2MoO_4 \cdot 2H_2O$	0.05g

pH 7.4–7.6 ± 0.2 at 25°C

Preparation of Medium: Add components to distilled/deionized water and bring volume to 1.0L. Mix thoroughly. Gently heat and bring to boiling. Distribute into tubes or flasks. Autoclave for 15 min at 15 psi pressure–121°C. Pour into sterile Petri dishes or leave in tubes.

Use: For the cultivation and maintenance of *Azotobacter chroococcum.*

Azotobacter chroococcum Medium

Composition per liter:

$CaCO_3$	20.0g
Glucose	20.0g
K_2HPO_4	1.0g
$MgSO_4 \cdot 7H_2O$	0.5g

Preparation of Medium: Add components to distilled/deionized water and bring volume to 1.0L. Mix thoroughly. Distribute into tubes or flasks. Autoclave for 15 min at 15 psi pressure–121°C.

Use: For the cultivation of *Azotobacter chroococcum.*

Azotobacter Medium

Composition per liter:

Agar	15.0g
$CaCO_3$	5.0g
K_2HPO_4	0.9g
$CaCl_2 \cdot 2H_2O$	0.1g
KH_2PO_4	0.1g
$MgSO_4 \cdot 7H_2O$	0.1g
$FeSO_4 \cdot 7H_2O$	0.01g
$Na_2MoO_4 \cdot 2H_2O$	5.0mg
Glucose solution	25.0mL
Mannitol solution	25.0mL

pH 7.3 ± 0.2 at 25°C

Glucose Solution:

Composition per 25.0mL:

D-Glucose	5.0g

Preparation of Glucose Solution: Add glucose to distilled/deionized water and bring volume to 25.0mL. Mix thoroughly. Filter sterilize. Warm to 50°–55°C.

Mannitol Solution:

Composition per 25.0mL:

Mannitol	5.0g

Preparation of Mannitol Solution: Add mannitol to distilled/deionized water and bring volume to 25.0mL. Mix thoroughly. Filter sterilize. Warm to 50°–55°C.

Preparation of Medium: Add components, except glucose solution and mannitol solution, to distilled/deionized water and bring volume to 950.0mL. Mix thoroughly. Autoclave for 15 min at 15 psi pressure–121°C. Cool to 50°–55°C. Aseptically add 25.0mL of sterile glucose solution and 25.0mL of sterile mannitol solution. Mix thoroughly. Pour into sterile Petri dishes or distribute into sterile tubes.

Use: For the cultivation and maintenance of *Azotobacter* species.

Azotobacter Medium (ATCC Medium 14)

Composition per liter:

Sucrose	20.0g
Agar	15.0g
K_2HPO_4	0.8g
Yeast extract	0.5g
KH_2PO_4	0.2g
$MgSO_4 \cdot 7H_2O$	0.2g
$CaSO_4 \cdot 2H_2O$	0.1g
$FeCl_3$	1.0mg
$Na_2MoO_4 \cdot 2H_2O$	1.0mg

pH 7.2 ± 0.2 at 25°C

Preparation of Medium: Add components to distilled/deionized water and bring volume to 1.0L. Mix thoroughly. Gently heat and bring to boiling. Distribute into tubes or flasks. Autoclave for 15 min at 15 psi pressure–121°C. Pour into sterile Petri dishes or leave in tubes.

Use: For the cultivation of a variety of bacteria, including *Azomonas* species, *Azotobacter* species, *Beijerinckia derxii, Pseudomonas azotocolligans,* and *Rhodococcus erythropolis*.

Azotobacter Medium (ATCC Medium 240)

Composition per liter:

Agar	15.0g
$MgSO_4 \cdot 7H_2O$	0.2g
KH_2PO_4	0.15g
K_2HPO_4	0.05g
$CaCl_2$	0.02g
$Na_2MoO_4 \cdot 2H_2O$	2.0mg
$FeCl_3$	1.0mg

pH 7.2 ± 0.2 at 25°C

Preparation of Medium: Add components to distilled/deionized water and bring volume to 1.0L. Mix thoroughly. Gently heat and bring to boiling. Distribute into tubes or flasks. Autoclave for 15 min at 15 psi pressure–121°C. Pour agar medium into sterile Petri dishes or leave in tubes.

Use: For the cultivation and maintenance of a variety of bacteria, including *Azotobacter* species.

Azotobacter Medium (ATCC Medium 1771)

Composition per liter:

Agar	15.0g
Glucose	10.0g
KH_2PO_4	0.22g
$CaSO_4 \cdot 2H_2O$	0.1g
$MgSO_4 \cdot 7H_2O$	0.098g
NaCl	0.058g
K_2HPO_4	0.058g
$FeSO_4 \cdot 7H_2O$	5.0mg
$Na_2MoO_4 \cdot 2H_2O$	0.2mg

pH 7.2 ± 0.2 at 25°C

Preparation of Medium: Add components to distilled/deionized water and bring volume to 1.0L. Mix thoroughly. Gently heat and bring to boiling. Distribute into tubes or flasks. Autoclave for 15 min at 15 psi pressure–121°C. Pour into sterile Petri dishes or leave in tubes.

Use: For the cultivation and maintenance of a variety of bacteria, including *Azotobacter* species.

Azotobacter paspali Medium

Composition per liter:

Agar	20.0g
Sucrose	20.0g
$CaCO_3$	1.0g
$MgSO_4 \cdot 7H_2O$	0.2g
KH_2PO_4	0.15g
K_2HPO_4	0.05g
$CaCl_2$	0.02g
$Na_2MoO_4 \cdot 2H_20$	2.0mg
Bromthymol Blue solution	10.0mL
$FeCl_3$ (10% solution)	0.1mL

pH 7.0 ± 0.2 at 25°C

Bromthymol Blue Solution:
Composition per 10.0mL:

Bromthymol Blue	0.5g
Ethanol	10.0mL

Preparation of Bromthymol Blue Solution: Add Bromthymol Blue to 10.0mL of ethanol. Mix thoroughly.

Preparation of Medium: Add components to distilled/deionized water and bring volume to 1.0L. Mix thoroughly. Gently heat and bring to boiling. Distribute into tubes or flasks. Autoclave for 15 min at 15 psi pressure–121°C. Pour into sterile Petri dishes or leave in tubes.

Use: For the cultivation and maintenance of *Azotobacter paspali.*

Azotobacter Supplement (ATCC Medium 11)

Composition per liter:

Agar	15.0g
K_2HPO_4	1.0g
$MgSO_4 \cdot 7H_2O$	0.2g
NaCl	0.2g
$FeSO_4 \cdot 7H_2O$	5.0mg
Soil extract	100.0mL
Glucose solution	100.0mL

pH 7.6 ± 0.2 at 25°C

Soil Extract:
Composition per 200.0mL:

African Violet soil	0.5g
Na_2CO_3	0.5g

Preparation of Soil Extract: Add components to tap water and bring volume to 200.0mL. Autoclave for 60 min at 15 psi pressure–121°C. Filter through Whatman filter paper.

Glucose Solution:
Composition per 100.0mL:

Glucose	20.0g

Preparation of Glucose Solution: Add glucose to distilled/deionized water and bring volume to 100.0mL. Mix thoroughly. Filter sterilize.

Preparation of Medium: Add components, except glucose solution, to tap water and bring volume to 900.0mL. Mix thoroughly. Adjust pH to 7.6. Autoclave for 15 min at 15 psi pressure–121°C. Cool to 50°–55°C. Aseptically add 100.0mL of sterile glucose solution. Mix thoroughly. Pour into sterile Petri dishes or leave in tubes.

Use: For the cultivation of *Azomonas agilis* and *Azotobacter chroococcum.*

Azotobacter Supplement (ATCC Medium 12)

Composition per liter:

Agar	15.0g
K_2HPO_4	1.0g
$MgSO_4 \cdot 7H_2O$	0.2g
NaCl	0.2g
$FeSO_4 \cdot 7H_2O$	5.0mg

Soil extract	100.0mL
Mannitol solution	100.0mL

pH 7.6 ± 0.2 at 25°C

Soil Extract:

Composition per 200.0mL:

African Violet soil	0.5g
Na_2CO_3	0.5g

Preparation of Soil Extract: Add components to distilled/deionized water and bring volume to 200.0mL. Autoclave for 60 min at 15 psi pressure–121°C. Filter through Whatman filter paper.

Mannitol Solution:

Composition per 100.0mL:

Mannitol	20.0g

Preparation of Mannitol Solution: Add mannitol to distilled/deionized water and bring volume to 100.0mL. Mix thoroughly. Filter sterilize.

Preparation of Medium: Add components, except mannitol solution, to tap water and bring volume to 900.0mL. Mix thoroughly. Adjust pH to 7.6. Autoclave for 15 min at 15 psi pressure–121°C. Cool to 50°–55°C. Aseptically add 100.0mL of sterile mannitol solution. Mix thoroughly. Pour into sterile Petri dishes or leave in tubes.

Use: For the cultivation of *Azotobacter* species and *Azomonas* species.

Azotobacter Supplement (ATCC Medium 13)

Composition per liter:

Agar	15.0g
K_2HPO_4	1.0g
$MgSO_4·7H_2O$	0.2g
NaCl	0.2g
$FeSO_4·7H_2O$	5.0mg
Soil extract	100.0mL
Glucose solution	100.0mL

pH 6.0 ± 0.2 at 25°C

Soil Extract:

Composition per 200.0mL:

African Violet soil	0.5g
Na_2CO_3	0.5g

Preparation of Soil Extract: Add components to tap water and bring volume to 200.0mL. Autoclave for 60 min at 15 psi pressure–121°C. Filter through Whatman filter paper.

Glucose Solution:

Composition per 100.0mL:

Glucose	20.0g

Preparation of Glucose Solution: Add glucose to distilled/deionized water and bring volume to 100.0mL. Mix thoroughly. Filter sterilize.

Preparation of Medium: Add components, except glucose solution, to tap water and bring volume to 900.0mL. Mix thoroughly. Adjust pH to 6.0. Autoclave for 15 min at 15 psi pressure–121°C. Cool to 50°–55°C. Aseptically add sterile glucose solution. Mix thoroughly. Pour into sterile Petri dishes or leave in tubes.

Use: For the cultivation of *Beijerinckia* species.

Azotobacter Supplement (ATCC Medium 15)

Composition per liter:

Agar	15.0g
K_2HPO_4	1.0g
$MgSO_4·7H_2O$	0.2g
NaCl	0.2g
$FeSO_4·7H_2O$	5.0mg
Soil extract	100.0mL
Mannitol solution	100.0mL

pH 6.0 ± 0.2 at 25°C

Soil Extract:

Composition per 200.0mL:

African Violet soil	0.5g
Na_2CO_3	0.5g

Preparation of Soil Extract: Add components to distilled/deionized water and bring volume to 200.0mL. Autoclave for 60 min at 15 psi pressure–121°C. Filter through Whatman filter paper.

Mannitol Solution:

Composition per 100.0mL:

Mannitol	20.0g

Preparation of Mannitol Solution: Add mannitol to distilled/deionized water and bring volume to 100.0mL. Mix thoroughly. Filter sterilize.

Preparation of Medium: Add components, except mannitol solution, to tap water and bring volume to 900.0mL. Mix thoroughly. Adjust pH to 6.0. Autoclave for 15 min at 15 psi pressure–121°C. Cool to 50°–55°C. Aseptically add 100.0mL of sterile mannitol solution. Mix thoroughly. Pour into sterile Petri dishes or leave in tubes.

Use: For the cultivation of *Azomonas macrocytogenes*.

Azotobacter vinelandii Medium

Composition per liter:

Sodium benzoate	1.0g
K_2HPO_4	0.5g
Mannitol	0.5g

Preparation of Medium: Add components to distilled/deionized water and bring volume to 1.0L. Mix thoroughly. Distribute into tubes or flasks. Autoclave for 15 min at 15 psi pressure–121°C.

Use: For the cultivation of *Azotobacter vinelandii* from water samples.

Azotobacter vinelandii Medium

Composition per liter:

Sodium benzoate	1.0g
K_2HPO_4	0.5g
Ethanol	1.0mL

Preparation of Medium: Add components, except ethanol, to distilled/deionized water and bring volume to 999.0mL. Mix thoroughly. Autoclave for 15 min at 15 psi pressure–121°C. Cool to 45°–50°C. Aseptically add 1.0mL of filter-sterilized ethanol. Mix thoroughly. Aseptically distribute into sterile tubes or flasks.

Use: For the cultivation of *Azotobacter vinelandii* from soil.

B Broth (Medium for *Ureaplasma*)

Composition per 100.25mL:

Yeast extract	0.1g
GHL (Glycyl-L–histidyl-L–lysine)	2.0μg

PPLO broth without Crystal Violet....50.0mL
Horse serum, not inactivated....10.0mL
Bromthymol Blue (0.4% solution)....1.0mL
Urea solution....0.25mL

pH 6.0 ± 0.2 at 25°C

B Broth (Medium for *Ureaplasma*)

Composition per 100.25mL:

Yeast extract....0.1g
GHL (Glycyl-L–histidyl-L–lysine)....2.0μg
PPLO broth without Crystal Violet....50.0mL
Horse serum, not inactivated....10.0mL
Bromthymol Blue (0.4% solution)....1.0mL
Urea solution....0.25mL

pH 6.0 ± 0.2 at 25°C

PPLO Broth without Crystal Violet:

Composition per 50.0mL:

Beef heart, infusion from....1.62g
Peptone....0.32g
NaCl....0.16g

Source: PPLO broth without Crystal Violet is available as a premixed powder from BD Diagnostic Systems.

Preparation of PPLO Broth without Crystal Violet: Add components to distilled/deionized water and bring volume to 50.0mL. Mix thoroughly.

Urea Solution:

Composition per 10.0mL:

Urea....1.0g

Preparation of Urea Solution: Add urea to distilled/deionized water and bring volume to 10.0mL. Mix thoroughly. Filter sterilize.

Preparation of Medium: Add components—except GHL, urea solution, and horse serum—to double glass-distilled water and bring volume to 90.0mL. Mix thoroughly. Gently heat and bring to boiling. Autoclave for 15 min at 15 psi pressure–121°C. Cool to 50°–55°C. To 90.0mL of the sterile medium, aseptically add 2.0μg of GHL, 10.0mL of horse serum, and 0.25mL of sterile urea solution. Mix thoroughly. Aseptically distribute into tubes or flasks.

Use: For the cultivation and maintenance of *Ureaplasma urealyticum* and other *Ureaplasma* species.

B/1t 7 A Medium

Composition per liter:

Agar....20.0g
K_2HPO_4....7.0g
KH_2PO_4....3.0g
Glucose....2.0g
$(NH_4)_2SO_4$....1.0g
$MgSO_4·7H_2O$....0.1g
$CaCl_2·2H_2O$....0.01g
Indole....0.01g
$FeSO_4·7H_2O$....0.5mg

Preparation of Medium: Add components to distilled/deionized water and bring volume to 1.0L. Mix thoroughly. Gently heat and bring to boiling. Distribute into tubes or flasks. Autoclave for 15 min at 15 psi pressure–121°C. Pour into sterile Petri dishes or leave in tubes.

Use: For the cultivation and maintenance of *Escherichia coli* and other bacteria.

B_{12} Assay HiVeg Medium (Vitamin B_{12} Assay HiVeg Medium)

Composition per liter:

Glucose....40.0g
Sodium acetate....20.0g
Plant hydrolysate....15.0g
Ascorbic acid....4.0g
Polysorbate 80....2.0g
K_2HPO_4....1.0g
KH_2PO_4....1.0g
DL-Tryptophan....0.4g
$MgSO_4·7H_2O$....0.4g
L-Cysteine....0.4g
Asparagine....0.2g
Adenine sulfate....0.02g
$FeSO_4$....0.02g
Guanine hydrochloride....0.02g
$MnSO_4$....0.02g
NaCl....0.02g
Uracil....0.02g
Xanthine....0.02g
Pyridoxal·HCl....4.0mg
Pyridoxine·HCl....4.0mg
Niacin....2.0mg
p-Aminobenzoic acid....2.0mg
Riboflavin....1.0mg
Thymine·HCl....1.0mg
Calcium pantothenate....1.0mg
Pyridoxamine·HCl....0.8mg
Folic acid....0.2mg
Biotin....0.01mg

pH 6.0 ± 0.2 at 25°C

Source: This medium is available as a premixed powder from HiMedia.

Preparation of Medium: Add components to distilled/deionized water and bring volume to 1.0L. Mix thoroughly. Distribute into tubes or flasks. Autoclave for 15 min at 15 psi pressure–121°C.

Use: For the determination of the vitamin B_{12} content of pharmaceutical products and other materials. *Lactobacillus leischmanii* ATCC 7830 is used as a test organism. A standard curve can be generated by adding known concentrations of cyanocobalamin and measuring the growth response turbidimetrically at 530 nm.

B_{12} Assay Medium

Composition per liter:

Glucose....20.5g
Lactose....20.0g
Amino acids, vitamin-free casamino acids....15.0g
Sodium acetate....10.0g
K_2HPO_4....2.5g
Polysorbate 80....2.0g
Ascorbic acid....1.0g
L-Arginine....0.5g
L-Histidine....0.25g
L-Phenylalanine....0.25g
L-Valine....0.25g
L-Asparagine....0.2g
$MgSO_4·7H_2O$....0.2g
Mercaptoacetic acid....0.13g

Calcium pantothenate	0.1g
L-Tryptophan	0.1g
$MnSO_4$	0.08g
Adenine	0.04g
Guanine	0.04g
Thymine	0.04g
Uracil	0.04g
$(NH_4)_2SO_4 \cdot FeSO_4 \cdot 6H_2O$	0.03g
KCN	5.0mg
Pyridoxal·HCl	1.0mg
Niacin	1.0mg
Riboflavin	1.0mg
Thiamine·HCl	0.5mg
p-Aminobenzoic acid	0.5mg
Folic acid	0.05mg

pH 6.0 ± 0.2 at 25°C

Source: This medium is available as a premixed powder from BD Diagnostic Systems.

Caution: Cyanide is toxic.

Preparation of Medium: Add components to distilled/deionized water and bring volume to 1.0L. Mix thoroughly. Gently heat and bring to boiling. Continue boiling for 2–3 min. Allow precipitate to settle out. Distribute supernatant into tubes in 5.0mL volumes. Add standard solution or test solutions to each tube. Adjust the volume of each tube to 10.0mL with distilled/deionized water. Autoclave for 15 min at 15 psi pressure–121°C.

Use: For the determination of the vitamin B_{12} content of pharmaceutical products and other materials. *Lactobacillus leischmanii* ATCC 7830 is used as a test organism. A standard curve can be generated by adding known concentrations of cyanocobalamin and measuring the growth response turbidimetrically at 530 nm.

B_{12} Culture Agar, USP

Composition per liter:

Agar	15.0g
Glucose	10.0g
Proteose peptone No. 3	7.5g
Yeast extract	7.5g
KH_2PO_4	2.0g
Polysorbate 80	0.1g
Tomato juice	100.0mL

pH 6.8 ± 0.1 at 25°C

Source: This medium is available as a premixed powder from BD Diagnostic Systems.

Preparation of Medium: Add components to distilled/deionized water and bring volume to 1.0L. Mix thoroughly. Gently heat and bring to boiling. Distribute into tubes in 10.0mL volumes. Autoclave for 15 min at 15 psi pressure–121°C. Cool tubes in an upright position.

Use: For the cultivation and maintenance of *Lactobacillus leischmannii* ATCC 7830 to be used as the test organism in the Vitamin B_{12} assay according to the USP.

B_{12} Inoculum Broth, USP

Composition per liter:

Glucose	10.0g
Proteose peptone No. 3	7.5g
Yeast extract	7.5g
K_2HPO_4	2.0g
Polysorbate 80	0.1g
Tomato juice	100.0mL

pH 6.8 ± 0.1 at 25°C

Source: This medium is available as a premixed powder from BD Diagnostic Systems.

Preparation of Medium: Add components to distilled/deionized water and bring volume to 1.0L. Mix thoroughly. Gently heat and bring to boiling. Distribute into tubes in 10.0mL volumes. Autoclave for 15 min at 15 psi pressure–121°C.

Use: For the preparation of inoculum cultures of *Lactobacillus leischmanii* ATCC 7830, which is used as the test organism in the vitamin B_{12} assay according to the USP.

B_{12} Medium

See: **Vitamin B_{12} Medium**

B_{12} Medium (DSMZ Medium 236)

Composition per liter:

Agar	15.0g
Casein hydrolysate	6.0g
K_2HPO_4	0.2g
$MgSO_4 \cdot 7H_2O$	0.2g
Asparagine	0.15g
Vitamin B_{12}	40.0μg
$FeSO_4 \cdot 7H_2O$	trace
Glycerol	2.0mL

pH 7.0 ± 0.2 at 25°C

Preparation of Medium: Add components to distilled/deionized water and bring volume to 1.0L. Mix thoroughly. Gently heat and bring to boiling. Distribute into tubes or flasks. Autoclave for 15 min at 15 psi pressure–121°C. Pour into sterile Petri dishes or leave in tubes.

Use: For the cultivation and maintenance of *Escherichia coli.*

B_{12} Nutrient Agar

See: **Vitamin B_{12} Nutrient Agar**

BA Medium

See: **BA Medium with Cellulose**

BA Medium with Cellobiose

Composition per liter:

$NaHCO_3$	2.6g
NH_4Cl	1.0g
Yeast extract	0.75g
$K_2HPO_4 \cdot 3H_2O$	0.4g
$MgCl_2 \cdot 6H_2O$	0.1g
NaCl	0.1g
$CaCl_2 \cdot 2H_2O$	0.05g
Resazurin	0.5mg
Cellobiose solution	50.0mL
$Na_2S \cdot 9H_2O$ solution	10.0mL
Wolfe's mineral solution	10.0mL
Wolfe's vitamin solution	10.0mL

pH 6.9–7.0 at 25°C

Cellobiose Solution:

Composition per 50.0mL:

Cellobiose	4.0g

Preparation of Cellobiose Solution: Add cellobiose to distilled/deionized water and bring volume to 50.0mL. Mix thoroughly. Filter sterilize.

$Na_2S \cdot 9H_2O$ Solution:

Composition per 10.0mL:

$Na_2S \cdot 9H_2O$	0.25g

Preparation of $Na_2S \cdot 9H_2O$ Solution: Add $Na_2S \cdot 9H_2O$ to distilled/deionized water and bring volume to 10.0mL. Mix thoroughly. Sparge with 100% N_2. Autoclave for 15 min at 15 psi pressure–121°C. Before use, neutralize to pH 7.0 with sterile HCl.

Wolfe's Mineral Solution:

Composition per liter:

$MgSO_4 \cdot 7H_2O$	3.0g
Nitrilotriacetic acid	1.5g
NaCl	1.0g
$MnSO_4 \cdot 2H_2O$	0.5g
$CoCl_2 \cdot 6H_2O$	0.1g
$ZnSO_4 \cdot 7H_2O$	0.1g
$CaCl_2 \cdot 2H_2O$	0.1g
$FeSO_4 \cdot 7H_2O$	0.1g
$NiCl_2 \cdot 6H_2O$	0.025g
$KAl(SO_4)_2 \cdot 12H_2O$	0.02g
$CuSO_4 \cdot 5H_2O$	0.01g
H_3BO_3	0.01g
$Na_2MoO_4 \cdot 2H_2O$	0.01g
$Na_2SeO_3 \cdot 5H_2O$	0.3mg

Preparation of Wolfe's Mineral Solution: Add nitrilotriacetic acid to 500.0mL of distilled/deionized water. Adjust pH to 6.5 with KOH. Add remaining components. Add distilled/deionized water to 1.0L. Adjust pH to 6.8.

Wolfe's Vitamin Solution:

Composition per liter:

Pyridoxine·HCl	10.0mg
p-Aminobenzoic acid	5.0mg
Lipoic acid	5.0mg
Nicotinic acid	5.0mg
Riboflavin	5.0mg
Thiamine·HCl	5.0mg
Calcium DL-pantothenate	5.0mg
Biotin	2.0mg
Folic acid	2.0mg
Vitamin B_{12}	0.1mg

Preparation of Wolfe's Vitamin Solution: Add components to distilled/deionized water and bring volume to 1.0L. Mix thoroughly. Filter sterilize.

Preparation of Medium: Prepare and dispense medium under 80% N_2 + 20% CO_2 gas mixture. Add components, except cellobiose solution, $Na_2S \cdot 9H_2O$ solution, Wolfe's mineral solution, and Wolfe's vitamin solution, to distilled/deionized water and bring volume to 920.0mL. Mix thoroughly. Sparge with 80% N_2 + 20% CO_2 gas mixture. Autoclave for 15 min at 15 psi pressure–121°C. Aseptically and anaerobically add 50.0mL of sterile cellobiose solution, 10.0mL of sterile Wolfe's mineral solutionn, 10.0mL of sterile Wolfe's vitamin solution, and 10.0mL of sterile $Na_2S \cdot 9H_2O$ solution. Mix thoroughly. Aseptically and anaerobically distribute into sterile tubes or bottles.

Use: For the cultivation of *Caldicellulosiruptor lactoaceticus.*

BA Medium with Cellulose (DSMZ Medium 671)

Composition per liter:

$NaHCO_3$	2.6g
Cellulose	2.0g
NH_4Cl	1.0g
Yeast extract	0.75g
$K_2HPO_4 \cdot 3H_2O$	0.4g
$MgCl_2 \cdot 6H_2O$	0.1g
NaCl	0.1g
$CaCl_2 \cdot 2H_2O$	0.05g
Resazurin	0.5mg
$Na_2S \cdot 9H_2O$ solution	10.0mL
Wolfe's mineral solution	10.0mL
Wolfe's vitamin solution	10.0mL

pH 6.9–7.0 at 25°C

$Na_2S \cdot 9H_2O$ Solution:

Composition per 10.0mL:

$Na_2S \cdot 9H_2O$	0.25g

Preparation of $Na_2S \cdot 9H_2O$ Solution: Add $Na_2S \cdot 9H_2O$ to distilled/deionized water and bring volume to 10.0mL. Mix thoroughly. Sparge with 100% N_2. Autoclave for 15 min at 15 psi pressure–121°C. Before use, neutralize to pH 7.0 with sterile HCl.

Wolfe's Mineral Solution:

Composition per liter:

$MgSO_4 \cdot 7H_2O$	3.0g
Nitrilotriacetic acid	1.5g
NaCl	1.0g
$MnSO_4 \cdot 2H_2O$	0.5g
$CoCl_2 \cdot 6H_2O$	0.1g
$ZnSO_4 \cdot 7H_2O$	0.1g
$CaCl_2 \cdot 2H_2O$	0.1g
$FeSO_4 \cdot 7H_2O$	0.1g
$NiCl_2 \cdot 6H_2O$	0.025g
$KAl(SO_4)_2 \cdot 12H_2O$	0.02g
$CuSO_4 \cdot 5H_2O$	0.01g
H_3BO_3	0.01g
$Na_2MoO_4 \cdot 2H_2O$	0.01g
$Na_2SeO_3 \cdot 5H_2O$	0.3mg

Preparation of Wolfe's Mineral Solution: Add nitrilotriacetic acid to 500.0mL of distilled/deionized water. Adjust pH to 6.5 with KOH. Add remaining components. Add distilled/deionized water to 1.0L. Adjust pH to 6.8.

Wolfe's Vitamin Solution:

Composition per liter:

Pyridoxine·HCl	10.0mg
p-Aminobenzoic acid	5.0mg
Lipoic acid	5.0mg
Nicotinic acid	5.0mg
Riboflavin	5.0mg
Thiamine·HCl	5.0mg
Calcium DL-pantothenate	5.0mg
Biotin	2.0mg
Folic acid	2.0mg
Vitamin B_{12}	0.1mg

Preparation of Wolfe's Vitamin Solution: Add components to distilled/deionized water and bring volume to 1.0L. Mix thoroughly. Filter sterilize.

Preparation of Medium: Prepare and dispense medium under 80% N_2 + 20% CO_2 gas mixture. Add components, except $Na_2S \cdot 9H_2O$ solution, Wolfe's mineral solution, and Wolfe's vitamin solution, to distilled/deionized water and bring volume to 970.0mL. Mix thoroughly. Sparge with 80% N_2 + 20% CO_2 gas mixture. Autoclave for 15 min at 15 psi pressure–121°C. Aseptically and anaerobically add 10.0mL of sterile Wolfe's mineral solution, 10.0mL of sterile Wolfe's vitamin solution, and 10.0mL of sterile $Na_2S \cdot 9H_2O$ solution. Mix thoroughly. Aseptically and anaerobically distribute into sterile tubes or bottles.

Use: For the cultivation of *Caldicellulosiruptor lactoaceticus* and *Caldicellulosiruptor kristjanssonii.*

Baar's Medium for Sulfate Reducers

Composition per liter:

Sodium lactate	3.5g
$MgSO_4 \cdot 7H_2O$	2.0g
K_2HPO_4	1.0g
$CaSO_4$	1.0g
NH_4Cl	0.5g
Ferrous ammonium sulfate solution	10.0mL
Yeast extract solution	10.0mL

pH 7.5 ± 0.2 at 25°C

Ferrous Ammonium Sulfate Solution:

Composition per 10.0mL:

$Fe(NH_4)_2(SO_4)_2$	0.5g

Preparation of Ferrous Ammonium Sulfate Solution: Add $Fe(NH_4)_2(SO_4)_2$ to distilled/deionized water and bring volume to 10.0mL. Mix thoroughly. Autoclave for 15 min at 15 psi pressure–121°C.

Yeast Extract Solution:

Composition per 10.0mL:

Yeast extract	1.0g

Preparation of Yeast Extract Solution: Add yeast extract to distilled/deionized water and bring volume to 10.0mL. Mix thoroughly. Autoclave for 15 min at 15 psi pressure–121°C.

Preparation of Medium: Add components, except ferrous ammonium sulfate solution and yeast extract solution, to tap water and bring volume to 980.0mL. Mix thoroughly. Gently heat and bring to boiling. Autoclave for 15 min at 15 psi pressure–121°C. Cool to 45°–50°C. Aseptically add 10.0mL of sterile ferrous ammonium sulfate solution and sterile yeast extract solution. Aseptically distribute into tubes or flasks.

Use: For the cultivation and maintenance of *Desulfotomaculum nigrificans*.

Baar's Medium for Sulfate Reducers, Modified

Composition per 1020.0mL:

Component I	400.0mL
Component III	400.0mL
Component II	200.0mL
Ferrous ammonium sulfate solution	20.0mL

pH 7.5 ± 0.2 at 25°C

Component I:

Composition per 400.0mL:

Sodium citrate	5.0g
$MgSO_4$	2.0g
$CaSO_4$	1.0g
NH_4Cl	1.0g

Preparation of Component I: Add components to distilled/deionized water and bring volume to 400.0mL. Mix thoroughly. Adjust pH to 7.5. Autoclave for 15 min at 15 psi pressure–121°C.

Component II:

Composition per 200.0mL:

K_2HPO_4	0.5g

Preparation of Component II: Add K_2HPO_4 to distilled/deionized water and bring volume to 200.0mL. Mix thoroughly. Adjust pH to 7.5. Autoclave for 15 min at 15 psi pressure–121°C.

Component III:

Composition per 400.0mL:

Sodium lactate	3.5g
Yeast extract	1.0g

Preparation of Component III: Add components to distilled/deionized water and bring volume to 400.0mL. Mix thoroughly. Adjust pH to 7.5. Autoclave for 15 min at 15 psi pressure–121°C.

Ferrous Ammonium Sulfate Solution:

Composition per 20.0mL:

$Fe(NH_4)_2(SO_4)_2$	1.0g

Preparation of Ferrous Ammonium Sulfate Solution: Add $Fe(NH_4)_2(SO_4)_2$ to distilled/deionized water and bring volume to 20.0mL. Mix thoroughly. Filter sterilize.

Preparation of Medium: Aseptically combine component I, component II, and component III. Mix thoroughly. Distribute 5.0mL volumes into tubes under 97% N_2 + 3% H_2. Add medium to tubes while still warm to exclude as much O_2 as possible. Aseptically add 0.1mL of sterile ferrous ammonium sulfate solution to 5.0mL of medium immediately prior to inoculation.

Use: For the cultivation and maintenance of *Desulfovibrio, Desulfobulbus, Desulfotomaculum,* and *Thermodesulfobacterium* species.

Baar's Medium for Sulfate Reducers, Modified with 2.5% Sodium Chloride

Composition per 1020.0mL:

Component I	400.0mL
Component III	400.0mL
Component II	200.0mL
Ferrous ammonium sulfate solution	20.0mL

pH 7.5 ± 0.2 at 25°C

Component I:

Composition per 400.0mL:

NaCl	25.0g
Sodium citrate	5.0g
$MgSO_4$	2.0g
$CaSO_4$	1.0g
NH_4Cl	1.0g

Preparation of Component I: Add components to distilled/deionized water and bring volume to 400.0mL. Mix thoroughly. Adjust pH to 7.5. Autoclave for 15 min at 15 psi pressure–121°C.

Component II:

Composition per 200.0mL:

K_2HPO_4	0.5g

Preparation of Component II: Add K_2HPO_4 to distilled/deionized water and bring volume to 200.0mL. Mix thoroughly. Adjust pH to 7.5. Autoclave for 15 min at 15 psi pressure–121°C.

Component III:

Composition per 400.0mL:

Sodium lactate	3.5g
Yeast extract	1.0g

Preparation of Component III: Add components to distilled/deionized water and bring volume to 400.0mL. Mix thoroughly. Adjust pH to 7.5. Autoclave for 15 min at 15 psi pressure–121°C.

Ferrous Ammonium Sulfate Solution:

Composition per 20.0mL:

$Fe(NH_4)_2(SO_4)_2$	1.0g

Preparation of Ferrous Ammonium Sulfate Solution: Add $Fe(NH_4)_2(SO_4)_2$ to distilled/deionized water and bring volume to 20.0mL. Mix thoroughly. Filter sterilize.

Preparation of Medium: Aseptically combine component I, component II, and component III. Mix thoroughly. Distribute 5.0mL volumes into tubes under 97% N_2 + 3% H_2. Add medium to tubes while still warm to exclude as much O_2 as possible. Aseptically add 0.1mL of sterile ferrous ammonium sulfate solution to 5.0mL of medium immediately prior to inoculation.

Use: For the cultivation of *Desulfovibrio africanus* and other *Desulfovibrio* species that prefer 2.5% NaCl.

Bacillus acidocaldarius Agar

Composition per liter:

Solution A	500.0mL
Solution B	500.0mL

pH 3.0–4.0 at 25°C

Solution A:

Composition per 500.0mL:

KH_2PO_4	3.0g
Yeast extract	1.0g
Glucose	1.0g
$MgSO_4 \cdot 7H_2O$	0.5g
$CaCl_2 \cdot 2H_2O$	0.25g
$(NH_4)_2SO_4$	0.2g

Preparation of Solution A: Add components to distilled/deionized water and bring volume to 500.0mL. Adjust pH to 3.0–4.0. Mix thoroughly. Autoclave for 10 min at 15 psi pressure–121°C. Cool to 50°–55°C.

Solution B:

Composition per 500.0mL:

Agar	30.0g

Preparation of Solution B: Add agar to distilled/deionized water and bring volume to 500.0mL. Mix thoroughly. Gently heat and bring to boiling. Autoclave for 15 min at 15 psi pressure–121°C. Cool to 50°–55°C.

Preparation of Medium: Aseptically mix 500.0mL of solution A and 500.0mL of solution B. Mix thoroughly. Aseptically adjust pH to 3.0–4.0. Pour into sterile Petri dishes or distribute into sterile tubes.

Use: For the cultivation and maintenance of *Bacillus acidocaldarius*.

Bacillus acidocaldarius Agar

Composition per liter:

Solution A	500.0mL
Solution B	500.0mL

pH 3.5 ± 0.5 at 25°C

Solution A:

Composition per 500.0mL:

Yeast extract	1.0g
KH_2PO_4	0.6g
$MgSO_4 \cdot 7H_2O$	0.5g
$CaCl_2 \cdot 2H_2O$	0.25g
$(NH_4)_2 \cdot SO_4$	0.2g

Preparation of Solution A: Add components to distilled/deionized water and bring volume to 500.0mL. Mix thoroughly. Bring pH to 3.5. Autoclave for 15 min at 15 psi pressure–121°C. Cool to 50°C.

Solution B:

Composition per 500.0mL:

Agar	20.0g
Glucose	1.0g

Preparation of Solution B: Add components to distilled/deionized water and bring volume to 500.0mL. Mix thoroughly. Autoclave for 15 min at 15 psi pressure–121°C. Cool to 50°C.

Preparation of Medium: Aseptically combine 500.0mL of solution A with 500.0mL of solution B. Mix thoroughly. Pour into sterile Petri dishes or aseptically distribute into sterile tubes.

Use: For the cultivation and maintenance of *Alicyclobacillus acidocaldarius*.

Bacillus acidoterrestris Agar

Composition per 1001.0mL:

Solution A	500.0mL
Solution C	500.0mL
Solution B (Trace elements solution SL-6)	1.0mL

pH 4.0 ± 0.2 at 25°C

Solution A:

Composition per 500.0mL:

Glucose	5.0g
KH_2PO_4	3.0g
Yeast extract	2.0g
$MgSO_4 \cdot 7H_2O$	0.5g
$CaCl_2 \cdot 2H_2O$	0.25g
$(NH_4)_2SO_4$	0.2g

Preparation of Solution A: Add components to distilled/deionized water and bring volume to 500.0mL. Mix thoroughly. Adjust pH to 4.0. Autoclave for 15 min at 15 psi pressure–121°C.

Solution C:

Composition per 500.0mL:

Agar	15.0g

Preparation of Solution C: Add agar to distilled/deionized water and bring volume to 500.0mL. Gently heat and bring to boiling. Autoclave for 15 min at 15 psi pressure–121°C. Cool to 50°–55°C.

Solution B (Trace Elements Solution SL-6):

Composition per liter:

$MnCl_2 \cdot 4H_2O$	0.5g
H_3BO_3	0.3g
$CoCl_2 \cdot 6H_2O$	0.2g
$ZnSO_4 \cdot 7H_2O$	0.1g
$Na_2MoO_4 \cdot 2H_2O$	0.03g
$NiCl_2 \cdot 6H_2O$	0.02g
$CuCl_2 \cdot 2H_2O$	0.01g

Preparation of Solution B (Trace Elements Solution SL-6): Add components to distilled/deionized water and bring volume to 1.0L. Mix thoroughly. Autoclave for 15 min at 15 psi pressure–121°C.

Preparation of Medium: Aseptically combine 500.0mL of sterile solution A, 500.0mL of sterile solution C, and 1.0mL of sterile solution B. Mix thoroughly. Pour into sterile Petri dishes or distribute into sterile tubes.

Use: For the cultivation and maintenance of *Bacillus acidoterrestris, Alicyclobacillus acidoterrestris,* and *Alicyclobacillus cycloheptanicus.*

Bacillus acidoterrestris Broth

Composition per 1001.0mL:

Solution A	1.0L
Solution B (Trace elements solution SL-6)	1.0mL

pH 4.0 ± 0.2 at 25°C

Solution A:

Composition per liter:

Glucose	5.0g
KH_2PO_4	3.0g
Yeast extract	2.0g
$MgSO_4 \cdot 7H_2O$	0.5g
$CaCl_2 \cdot 2H_2O$	0.25g
$(NH_4)_2SO_4$	0.2g

Preparation of Solution A: Add components to distilled/deionized water and bring volume to 1.0L. Mix thoroughly. Adjust pH to 4.0. Autoclave for 15 min at 15 psi pressure–121°C.

Solution B (Trace Elements Solution SL-6):

Composition per liter:

$MnCl_2 \cdot 4H_2O$	0.5g
H_3BO_3	0.3g
$CoCl_2 \cdot 6H_2O$	0.2g
$ZnSO_4 \cdot 7H_2O$	0.1g
$Na_2MoO_4 \cdot 2H_2O$	0.03g
$NiCl_2 \cdot 6H_2O$	0.02g
$CuCl_2 \cdot 2H_2O$	0.01g

Preparation of Solution B (Trace Elements Solution SL-6): Add components to distilled/deionized water and bring volume to 1.0L. Mix thoroughly. Autoclave for 15 min at 15 psi pressure–121°C.

Preparation of Medium: Aseptically combine 1.0L of sterile solution A and 1.0mL of sterile solution B. Mix thoroughly. Aseptically distribute into sterile tubes or flasks.

Use: For the cultivation and maintenance of *Bacillus acidoterrestris, Alicyclobacillus acidoterrestris,* and *Alicyclobacillus cycloheptanicus.*

Bacillus Agar

Composition per liter:

Agar	20.0g
$(NH_4)_2SO_4$	1.3g
Glucose	1.0g
Yeast extract	1.0g
KH_2PO_4	0.37g
$MgSO_4 \cdot 7H_2O$	0.25g
$CaCl_2 \cdot 2H_2O$	0.07g
$FeCl_3$	0.02g

pH 4.0 ± 0.2 at 25°C

Preparation of Medium: Add components to distilled/deionized water and bring volume to 500.0mL. Mix thoroughly. Gently heat and bring to boiling. Adjust pH to 3.5. Prepare a separate agar solution by adding 20.0g/500.0mL of distilled/deionized water. Autoclave solutions separately for 15 min at 15 psi pressure–121°C. Cool to 50°–55°C. Aseptically combine both solutions. This procedure avoids acid hydrolysis of the agar. Pour into sterile Petri dishes or leave in tubes.

Use: For the cultivation of acidophilic *Bacillus* species such as *Bacillus acidocaldarius.*

Bacillus Agar, Modified

Composition per liter:

Agar	20.0g
Glucose	1.0g
Yeast extract	1.0g
KH_2PO_4	0.6g
$MgSO_4 \cdot 7H_2O$	0.5g
$CaCl_2 \cdot 2H_2O$	0.25g
$(NH_4)_2SO_4$	0.2g

pH 3.0–4.0 at 25°C

Preparation of Medium: Add components, except agar and glucose, to distilled/deionized water and bring volume to 500.0mL. Mix thoroughly. Gently heat and bring to boiling. Adjust pH to 3.5. Prepare a separate agar and glucose solution by adding 20.0g of agar and 1.0g of glucose to 500.0mL of distilled/deionized water. Autoclave solutions separately for 15 min at 15 psi pressure–121°C. Cool to 50°–55°C. Aseptically combine both solutions. This procedure avoids acid hydrolysis of the agar. Pour into sterile Petri dishes or leave in tubes.

Use: For the cultivation of acidophilic *Bacillus* species such as *Bacillus acidocaldarius.*

Bacillus Agar, 1/4 Strength

Composition per liter:

Agar	18.0g
Yeast extract	2.5g
Pancreatic digest of casein	1.0g

pH 7.2 ± 0.2 at 25°C

Preparation of Medium: Add components to distilled/deionized water and bring volume to 1.0L. Mix thoroughly. Gently heat and bring to boiling. Distribute into tubes or flasks. Autoclave for 15 min at 15 psi pressure–121°C. Pour into sterile Petri dishes or leave in tubes.

Use: For the cultivation and maintenance of *Bacillus megaterium.*

Bacillus benzoevorans Agar

Composition per liter:

Agar	15.0g
Yeast extract	6.0g
Peptone	5.0g
NaCl	5.0g
$Na_2HPO_4 \cdot 12H_2O$	3.6g
Sodium benzoate	2.0g
Beef extract	1.0g
KH_2PO_4	0.98g
NH_4Cl	0.5g
$MgSO_4 \cdot 7H_2O$	0.03g
Trace elements solution	0.2mL

pH 7.0–7.2 at 25°C

Trace Elements Solution:

Composition per 100.0mL:

$FeSO_4 \cdot 7H_2O$	0.1g
$MnCl_2 \cdot 4H_2O$	0.1g
$ZnSO_4 \cdot 7H_2O$	0.1g

Preparation of Trace Elements Solution: Add components to distilled/deionized water and bring volume to 100.0mL. Mix thoroughly.

Preparation of Medium: Add components to distilled/deionized water and bring volume to 1.0L. Mix thoroughly. Gently heat and bring to boiling. Distribute into tubes or flasks. Autoclave for 15 min at 15 psi pressure–121°C. Pour into sterile Petri dishes or leave in tubes.

Use: For the cultivation and maintenance of *Bacillus benzoevorans.*

Bacillus benzoevorans Agar

Composition per liter:

Modified Palleroni and Doudoroff mineral base medium	450.0mL
Enriched *Cytophaga* agar	450.0mL
Sodium benzoate solution	100.0mL

pH 7.0 ± 0.2 at 25°C

Modified Palleroni and Doudoroff Mineral Base Medium:
Composition per 500.0mL:

Agar	15.0g
$Na_2HPO_4 \cdot 12H_2O$	6.0g
KH_2PO_4	2.4g
$NH_4 \cdot Cl$	1.0g
$MgSO_4 \cdot 7H_2O$	0.5g
$CaCl_2 \cdot 6H_2O$	0.01g
$FeCl_3 \cdot 6H_2O$	0.01g

Preparation of Modified Palleroni and Doudoroff Mineral Base Medium: Add components to distilled/deionized water and bring volume to 450.0mL. Mix thoroughly. Adjust pH to 7.2. Gently heat and bring to boiling. Autoclave for 15 min at 15 psi pressure–121°C. Cool to 50°C.

Enriched *Cytophaga* Agar:
Composition per 500.0mL:

Agar	15.0g
Pancreatic digest of casein	0.5g
Beef extract	0.5g
Yeast extract	0.5g
Sodium acetate	0.2g

Preparation of Enriched *Cytophaga* Agar: Add components to distilled/deionized water and bring volume to 450.0mL. Mix thoroughly. Adjust pH to 6.8. Gently heat and bring to boiling. Autoclave for 15 min at 15 psi pressure–121°C. Cool to 50°C.

Sodium Benzoate Solution:
Composition per 100.0mL:

Sodium benzoate	5.0g

Preparation of Sodium Benzoate Solution: Add sodium benzoate to distilled/deionized water and bring volume to 100.0mL. Mix thoroughly. Filter sterilize.

Preparation of Medium: Aseptically combine 450.0mL of modified Palleroni and Doudoroff mineral base medium, 450.0mL of enriched *Cytophaga* agar, and 100.0mL sodium benzoate solution. Mix thoroughly. Pour into sterile Petri dishes or aseptically distribute into sterile tubes.

Use: For the cultivation of *Bacillus benzoevorans*.

Bacillus Broth

Composition per liter:

$(NH_4)_2SO_4$	1.3g
Glucose	1.0g
Yeast extract	1.0g
KH_2PO_4	0.37g
$MgSO_4 \cdot 7H_2O$	0.25g
$CaCl_2 \cdot 2H_2O$	0.07g
$FeCl_3$	0.02g

pH 4.0 ± 0.2 at 25°C

Preparation of Medium: Add components to distilled/deionized water and bring volume to 1.0L. Mix thoroughly. Gently heat and bring to boiling. Adjust pH to 4.0 with 10*N* H_2SO_4. Distribute into tubes or flasks. Autoclave for 15 min at 15 psi pressure–121°C.

Use: For the cultivation of acidophilic *Bacillus* species such as *Bacillus acidocaldarius*.

Bacillus Broth, 1/4 Strength

Composition per liter:

Yeast extract	2.5g
Pancreatic digest of casein	1.0g

pH 7.2 ± 0.2 at 25°C

Preparation of Medium: Add components to distilled/deionized water and bring volume to 1.0L. Mix thoroughly. Distribute into tubes or flasks. Autoclave for 15 min at 15 psi pressure–121°C.

Use: For the cultivation of *Bacillus megaterium*.

Bacillus cereus Agar Base with Egg Yolk Emulsion and Polymyxin

Composition per liter:

Agar	15.0g
Sodium pyruvate	10.0g
Mannitol	10.0g
Na_2HPO_4	2.5g
NaCl	2.0g
Peptone	1.0g
KH_2PO_4	0.25g
Bromthymol Blue	0.12g
$MgSO_4 \cdot 7H_2O$	0.1g
Egg yolk emulsion	100.0mL
Selective supplement solution	10.0mL

pH 7.2 ± 0.2 at 25°C

Source: This medium, without egg yolk emulsion, is available as a premixed powder from HiMedia.

Selective Supplement Solution:
Composition per 10.0mL:

Polymyxin B	100,000 U

Preparation of Selective Supplement Solution: Add components to distilled/deionized water and bring volume to 10.0mL. Mix thoroughly. Filter sterilize.

Egg Yolk Emulsion:
Composition per liter:

Egg yolks	30.0mL
NaCl, 0.9% solution	70.0mL

Preparation of Egg Yolk Emulsion: Soak eggs with 1:100 dilution of saturated mercuric chloride solution for 1 min. Crack 11 eggs and separate yolks from whites. Mix egg yolks. Measure 30.0mL of egg yolk emulsion and add to 70.0mL of 0.9% sterile NaCl solution. Mix thoroughly. Warm to 45°–50°C.

Preparation of Medium: Add components, except egg yolk emulsion, and selective supplement solution, to distilled/deionized water and bring volume to 890.0mL. Mix thoroughly. Gently heat and bring to boiling. Autoclave for 15 min at 15 psi pressure–121°C. Cool to 45°–50°C. Aseptically add 100.0mL egg yolk emulsion and 10.0mL sterile selective supplement solution. Mix well. Pour into sterile Petri dishes or sterile tubes.

Use: For the isolation, detection, and enumeration of *Bacillus cereus.*

Bacillus cereus HiVeg Agar Base with Egg Yolk Emulsion

Composition per liter:

Agar	15.0g
Sodium pyruvate	10.0g
Mannitol	10.0g
Na_2HPO_4	2.5g
NaCl	2.0g
Plant peptone	1.0g
KH_2PO_4	0.25g
Bromthymol Blue	0.12g
$MgSO_4 \cdot 7H_2O$	0.1g
Egg yolk emulsion	100.0mL

pH 7.2 ± 0.2 at 25°C

Source: This medium, without egg yolk emulsion, is available as a premixed powder from HiMedia.

Egg Yolk Emulsion:
Composition per liter:

Egg yolks	30.0mL
NaCl, 0.9% solution	70.0mL

Preparation of Egg Yolk Emulsion: Soak eggs with 1:100 dilution of saturated mercuric chloride solution for 1 min. Crack 11 eggs and separate yolks from whites. Mix egg yolks. Measure 30.0mL of egg yolk emulsion and add to 70.0mL of 0.9% sterile NaCl solution. Mix thoroughly. Warm to 45°–50°C.

Preparation of Medium: Add components, except egg yolk emulsion, to distilled/deionized water and bring volume to 900.0mL. Mix thoroughly. Gently heat and bring to boiling. Autoclave for 15 min at 15 psi pressure–121°C. Cool to 45°–50°C. Aseptically add 100.0mL egg yolk emulsion. Mix well. Pour into sterile Petri dishes or sterile tubes.

Use: For the isolation, detection, and enumeration of *Bacillus cereus.*

Bacillus cereus Medium (BCM)

Composition per 110.0mL:

Agar	2.0g
D-Mannitol	1.0g
$(NH_4)_2PO_4$	0.1g
KCl	0.02g
$MgSO_4 \cdot 7H_2O$	0.02g
Yeast extract	0.02g
Bromcresol Purple	4.0mg
Egg yolk emulsion, 20%	10.0mL

pH 7.0 ± 0.2 at 25°C

Egg Yolk Emulsion, 20%:
Composition per 100.0mL:

Chicken egg yolks	11
Whole chicken egg	1
NaCl (0.9% solution)	80.0mL

Preparation of Egg Yolk Emulsion, 20%: Soak eggs with 1:100 dilution of saturated mercuric chloride solution for 1 min. Crack eggs and separate yolks from whites. Mix egg yolks with 1 chicken egg. Measure 20.0mL of egg yolk emulsion and add to 80.0mL of 0.9% NaCl solution. Mix thoroughly. Filter sterilize. Warm to 45°–50°C.

Preparation of Medium: Add components—except egg yolk emulsion, 20%—to distilled/deionized water and bring volume to 100.0mL. Mix thoroughly. Gently heat and bring to boiling. Autoclave for 15 min at 15 psi pressure–121°C. Cool to 45°–50°C. Aseptically add 10.0mL of sterile egg yolk emulsion, 20%. Mix thoroughly. Pour into sterile Petri dishes or distribute into sterile tubes.

Use: For the cultivation of *Bacillus cereus.*

Bacillus cereus Motility Medium

See: **BC Motility Medium**

Bacillus cereus Selective Agar Base

Composition per liter:

Agar	15.0g
Sodium pyruvate	10.0g
Mannitol	10.0g
Na_2HPO_4	2.5g
NaCl	2.0g
Peptone	1.0g
KH_2PO_4	0.25g
Bromthymol Blue	0.12g
$MgSO_4 \cdot 7H_2O$	0.1g
Egg yolk emulsion	25.0mL
Polymyxin B solution	10.0mL

pH 7.2 ± 0.2 at 25°C

Source: This medium is available as a premixed powder from Oxoid Unipath.

Egg Yolk Emulsion:
Composition:

Chicken egg yolks	11
Whole chicken egg	1

Preparation of Egg Yolk Emulsion: Soak eggs with 1:100 dilution of saturated mercuric chloride solution for 1 min. Crack eggs and separate yolks from whites. Mix egg yolks with 1 chicken egg.

Polymyxin B Solution:
Composition per 10.0mL:

Polymyxin B	100,000U

Preparation of Polymyxin B Solution: Add polymyxin B to distilled/deionized water and bring volume to 10.0mL. Mix thoroughly. Filter sterilize.

Preparation of Medium: Add components, except egg yolk emulsion and polymyxin B solution, to distilled/deionized water and bring volume to 965.0mL. Gently heat and bring to boiling. Distribute into tubes or flasks. Autoclave for 15 min at 15 psi pressure–121°C. Cool to 50°C. Aseptically add sterile polymyxin B and 25.0mL of sterile egg yolk emulsion. Mix thoroughly. Pour into sterile Petri dishes or leave in tubes.

Use: For the selection and presumptive identification of *Bacillus cereus*. Also for the isolation and enumeration of these bacteria. *Bacillus cereus* grows as moderate-sized (5mm) crenated colonies, which are turquoise, surrounded by a precipitate of egg yolk, which is also turquoise.

Bacillus coagulans Medium

Composition per liter:

Agar	20.0g
Glucose	5.0g
Proteose peptone	5.0g
Yeast extract	5.0g
K_2HPO_4	4.0g
$MnSO_4 \cdot 4H_2O$ solution	10.0mL
$CaCl_2$ solution	10.0mL

pH 5.0 ± 0.2 at 25°C

$MnSO_4 \cdot 4H_2O$ Solution:

Composition per 10.0mL:

$MnSO_4 \cdot 4H_2O$	0.05mg

Preparation of $MnSO_4 \cdot 4H_2O$ Solution: Add $MnSO_4 \cdot 4H_2O$ to distilled/deionized water and bring volume to 10.0mL. Mix thoroughly. Filter sterilize.

$CaCl_2$ Solution:

Composition per 10.0mL:

$CaCl_2$	0.045mg

Preparation of $CaCl_2$ Solution: Add $CaCl_2$ to distilled/deionized water and bring volume to 10.0mL. Mix thoroughly. Filter sterilize.

Preparation of Medium: Add components, except $MnSO_4 \cdot 4H_2O$ solution and $CaCl_2$ solution, to distilled/deionized water and bring volume to 980.0mL. Mix thoroughly. Gently heat and bring to boiling. Autoclave for 15 min at 15 psi pressure–121°C. Avoid overheating. Cool to 45°–50°C. Aseptically add sterile $MnSO_4 \cdot 4H_2O$ solution and $CaCl_2$ solution. Mix thoroughly. Pour into sterile Petri dishes or distribute into sterile tubes.

Use: For the cultivation of *Bacillus coagulans*.

Bacillus cycloheptanicus Agar

Composition per 1001.0mL:

Solution A	500.0mL
Solution C	500.0mL
Solution B (Trace elements solution SL-6)	1.0mL

pH 4.0 ± 0.2 at 25°C

Solution A:

Composition per liter:

Yeast extract	5.0g
Glucose	5.0g
KH_2PO_4	3.0g
$MgSO_4 \cdot 7H_2O$	0.5g
$CaCl_2 \cdot 2H_2O$	0.25g
$(NH_4)_2SO_4$	0.2g

Preparation of Solution A: Add components to distilled/deionized water and bring volume to 1.0L. Mix thoroughly. Adjust pH to 4.0. Autoclave for 15 min at 15 psi pressure–121°C.

Solution C:

Composition per 500.0mL:

Agar	15.0g

Preparation of Solution C: Add agar to distilled/deionized water and bring volume to 500.0mL. Gently heat and bring to boiling. Autoclave for 15 min at 15 psi pressure–121°C. Cool to 50°–55°C.

Solution B (Trace Elements Solution SL-6):

Composition per liter:

$MnCl_2 \cdot 4H_2O$	0.5g
H_3BO_3	0.3g
$CoCl_2 \cdot 6H_2O$	0.2g
$ZnSO_4 \cdot 7H_2O$	0.1g
$Na_2MoO_4 \cdot 2H_2O$	0.03g
$NiCl_2 \cdot 6H_2O$	0.02g
$CuCl_2 \cdot 2H_2O$	0.01g

Preparation of Solution B (Trace Elements Solution SL-6): Add components to distilled/deionized water and bring volume to 1.0L. Mix thoroughly. Autoclave for 15 min at 15 psi pressure–121°C.

Preparation of Medium: Aseptically combine 500.0mL of sterile solution A, 500.0mL of sterile solution C, and 1.0mL of sterile solution B. Mix thoroughly. Pour into sterile Petri dishes or distribute into sterile tubes.

Use: For the cultivation and maintenance of *Bacillus cycloheptanicus, Alicyclobacillus acidoterrestris,* and *Alicyclobacillus cycloheptanicus.*

Bacillus cycloheptanicus Broth

Composition per 1001.0mL:

Solution A	1.0L
Solution B (Trace elements solution SL-6)	1.0mL

pH 4.0 ± 0.2 at 25°C

Solution A:

Composition per liter:

Yeast extract	5.0g
Glucose	5.0g
KH_2PO_4	3.0g
$MgSO_4 \cdot 7H_2O$	0.5g
$CaCl_2 \cdot 2H_2O$	0.25g
$(NH_4)_2SO_4$	0.2g

Preparation of Solution A: Add components to distilled/deionized water and bring volume to 1.0L. Mix thoroughly. Adjust pH to 4.0. Autoclave for 15 min at 15 psi pressure–121°C.

Solution B (Trace Elements Solution SL-6):

Composition per liter:

$MnCl_2 \cdot 4H_2O$	0.5g
H_3BO_3	0.3g
$CoCl_2 \cdot 6H_2O$	0.2g
$ZnSO_4 \cdot 7H_2O$	0.1g
$Na_2MoO_4 \cdot 2H_2O$	0.03g
$NiCl_2 \cdot 6H_2O$	0.02g
$CuCl_2 \cdot 2H_2O$	0.01g

Preparation of Solution B (Trace Elements Solution SL-6): Add components to distilled/deionized water and bring volume to 1.0L. Mix thoroughly. Autoclave for 15 min at 15 psi pressure–121°C.

Preparation of Medium: Aseptically combine 1.0L of sterile solution A and 1.0mL of sterile solution B. Mix thoroughly. Aseptically distribute into sterile tubes or flasks.

Use: For the cultivation of *Bacillus cycloheptanicus, Alicyclobacillus acidoterrestris,* and *Alicyclobacillus cycloheptanicus.*

Bacillus fastidiosus Agar

Composition per liter:

Allantoin	20.0g
Agar	15.0g
K_2HPO_4	0.8g
$MgSO_4 \cdot 7H_2O$	0.5g
KH_2PO_4	0.2g
$CaCl_2 \cdot 2H_2O$	50.0mg
$FeSO_4 \cdot 7H_2O$	10.0mg
$MnSO_4 \cdot 4H_2O$	1.0mg

Preparation of Medium: Add components to distilled/deionized water and bring volume to 1.0L. Mix thoroughly. Gently heat and bring to boiling. Distribute into tubes or flasks. Autoclave for 15 min at 15 psi pressure–121°C. Pour into sterile Petri dishes or leave in tubes.

Use: For the cultivation of *Bacillus fastidiosus*.

Bacillus fastidiosus Medium

Composition per liter:

Agar	15.0g
$Na_2HPO_4 \cdot 12H_2O$	6.0g
Yeast extract	2.5g
Uric acid	1.0g
Mineral solution	100.0mL

pH 7.0 ± 0.2 at 25°C

Mineral Solution:

Composition per 100.0mL:

KH_2PO_4	0.1g
$MgSO_4 \cdot 7H_2O$	0.03g
$CaCl_2$	0.01g
NaCl	0.01g
$FeCl_3 \cdot 6H_2O$	1.0mg

Preparation of Mineral Solution: Add components to distilled/deionized water and bring volume to 100.0mL. Mix thoroughly.

Preparation of Medium: Add components to distilled/deionized water and bring volume to 1.0L. Mix thoroughly. Gently heat and bring to boiling. Distribute into tubes or flasks. Autoclave for 15 min at 15 psi pressure–121°C. Pour into sterile Petri dishes or leave in tubes.

Use: For the cultivation of *Bacillus fastidiosus*.

Bacillus filiformis Medium (DSMZ Medium 992)

Composition per liter:

Yeast extract	10.0g
Sodium citrate	3.0g
KCl	2.0g
$MgSO_4 \cdot 7H_2O$	1.0g
Sodium chloride solution	100.0mL
Sodium carbonate solution	10.0mL
Iron sulfate solution	1.0mL
Manganese chloride solution	1.0mL

pH 9.0 ± 0.2 at 25°C

Iron Sulfate Solution:

Composition per liter:

$FeSO_4 \cdot 7H_2O$	50.0g

Preparation of Iron Sulfate Solution: Add $FeSO_4 \cdot 7H_2O$ to distilled/deionized water and bring volume to 1.0L. Mix thoroughly.

Sodium Chloride Solution:

Composition per 100.0mL:

NaCl	100.0g

Preparation of Sodium Chloride Solution: Add NaCl to distilled/deionized water and bring volume to 100.0mL. Mix thoroughly. Autoclave for 15 min at 15 psi pressure–121°C.

Manganese Chloride Solution:

Composition per liter:

$MnCl_2 \cdot 4H_2O$	0.36g

Preparation of Manganese Chloride Solution: Add $MnCl_2 \cdot 4H_2O$ to distilled/deionized water and bring volume to 1.0L. Mix thoroughly.

Sodium Carbonate Solution:

Composition per 10.0mL:

Na_2CO_3	3.0g

Preparation of Sodium Carbonate Solution: Add Na_2CO_3 to distilled/deionized water and bring volume to 10.0mL. Mix thoroughly. Autoclave for 15 min at 15 psi pressure–121°C.

Preparation of Medium: Add components, except sodium chloride and sodium carbonate solutions, to distilled/deionized water and bring volume to 890.0mL. Mix thoroughly. Gently heat and bring to boiling. Autoclave for 15 min at 15 psi pressure–121°C. Cool to 50°C. Aseptically add 100.0mL sterile sodium chloride solution and 10.0mL sterile sodium carbonate solution. Mix thoroughly. Aseptically distribute into sterile tubes or flasks.

Use: For the cultivation of a *Bacillus filiformis*.

Bacillus halodenitrificans Agar (LMG Medium 142)

Composition per liter:

NaCl	100.0g
Agar	15.0g
Sodium acetate·$3H_2O$	10.0g
Na_2HPO_4	3.8g
KH_2PO_4	1.3g
$(NH_4)_2SO_4$	1.0g
$Mg(NO_3)_2 \cdot 6H_2O$	1.0g
Yeast extract	1.0g
Magnesium nitrate solution	100.0mL

pH 7.2 ± 0.2 at 25°C

Magnesium Nitrate Solution:

Composition per 100.0mL:

$Mg(NO_3)_2 \cdot 6H_2O$	1.0g

Preparation of Magnesium Nitrate Solution: Add $Mg(NO_3)_2 \cdot 6H_2O$ to distilled/deionized water and bring volume to 100.0mL. Mix thoroughly. Filter sterilize.

Preparation of Medium: Add components, except magnesium nitrate solution, to distilled/deionized water and bring volume to 900.0mL. Mix thoroughly. Adjust pH to 7.2 with KOH. Autoclave for 15 min at 15 psi pressure–121°C. Cool to 45°–50°C. Aseptically add 100.0mL sterile magnesium nitrate solution. Mix thoroughly. Aseptically pour into sterile Petri dishes or distribute into sterile tubes.

Use: For the cultivation of *Bacillus halodenitrificans*.

Bacillus mascerans Medium (TSBY Salt Medium) (LMG 199)

Composition per liter:

NaCl	18.0g
Pancreatic digest of casein	17.0g
$MgCl_2 \cdot H_2O$	4.0g
$MgSO_4 \cdot 7H_2O$	3.45g
Yeast extract	3.0g
Papaic digest of soybean meal	3.0g
K_2HPO_4	2.5g
Glucose	2.5g
KCl	0.34g
NH_4Cl	0.25g
$CaCl_2 \cdot 2H_2O$	0.14g

pH 7.2 ± 0.2 at 25°C

Preparation of Medium: Add components to distilled/deionized water and bring volume to 1.0L. Mix thoroughly. Adjust pH to 7.2. Distribute into tubes or flasks. Autoclave for 15 min at 15 psi pressure–121°C.

Use: For the cultivation of *Bacillus mascerans, Carnobacterium alterfunditum, and Carnobacterium funditum.*

Bacillus Medium

Composition per liter:

Agar	25.0g
Peptone	6.0g
Pancreatic digest of casein	3.0g
Yeast extract	3.0g
Beef extract	1.5g
$MnSO_4 \cdot 4H_2O$	1.0μg

pH 7.0 ± 0.2 at 25°C

Preparation of Medium: Add components to distilled/deionized water and bring volume to 1.0L. Mix thoroughly. Gently heat and bring to boiling. Distribute into tubes or flasks. Autoclave for 15 min at 15 psi pressure–121°C. Pour into sterile Petri dishes or leave in tubes.

Use: For the cultivation of *Bacillus* species.

Bacillus Medium

Composition per liter:

$(NH_4)_2HPO_4$	1.0g
$MgSO_4 \cdot 7H_2O$	0.2g
KCl	0.2g
Yeast extract	0.2g
Glucose solution	50.0mL
Bromcresol Purple solution	15.0mL

pH 7.0 ± 0.2 at 25°C

Glucose Solution:

Composition per 100.0mL:

Glucose	10.0g

Preparation of Glucose Solution: Add glucose to distilled/deionized water and bring volume to 100.0mL. Mix thoroughly. Filter sterilize.

Bromcresol Purple Solution:

Composition per 20.0mL:

Bromcresol Purple	0.32g
Ethanol (95% solution)	20.0mL

Preparation of Bromcresol Purple Solution: Add Bromcresol Purple to 20.0mL of ethanol. Mix thoroughly.

Preparation of Medium: Add components, except glucose solution, to distilled/deionized water and bring volume to 1.0L. Mix thoroughly. Gently heat and bring to boiling. Distribute 9.5mL volumes into test tubes that contain an inverted Durham tube. Autoclave for 20 min at 15 psi pressure–121°C. Cool to 25°C. Aseptically add 0.5mL of sterile glucose to each tube. Mix thoroughly.

Use: For cultivation and differentiation of *Bacillus* species based on acid and gas production from glucose.

Bacillus Medium (ATCC Medium 21)

Composition per liter:

Glycerol	20.0g
L-Glutamic acid	4.0g
Citric acid	2.0g
K_2HPO_4	0.5g
Ferric ammonium citrate	0.5g
$MgSO_4$	0.5g

pH 7.4 ± 0.2 at 25°C

Preparation of Medium: Add components to tap water and bring volume to 1.0L. Mix thoroughly. Gently heat and bring to boiling. Distribute into tubes or flasks. Autoclave for 15 min at 15 psi pressure–121°C.

Use: For the cultivation of *Bacillus licheniformis*.

Bacillus Medium (ATCC Medium 455)

Composition per liter:

Soluble starch	30.0g
Agar	20.0g
Polypeptone™	5.0g
Yeast extract	5.0g

Preparation of Medium: Add components to distilled/deionized water and bring volume to 1.0L. Mix thoroughly. Gently heat and bring to boiling. Distribute into tubes or flasks. Autoclave for 15 min at 15 psi pressure–121°C. Swirl medium to resuspend starch. Pour into sterile Petri dishes or leave in tubes.

Use: For the cultivation and maintenance of *Bacillus subtilis*. Also used to detect amylase-producing microorganisms.

Bacillus Medium (ATCC Medium 552)

Composition per liter:

Peptone	10.0g
Lactose	5.0g
NaCl	5.0g
Beef extract	3.0g
K_2HPO_4	2.0g

pH 7.2 ± 0.2 at 25°C

Preparation of Medium: Add components to distilled/deionized water and bring volume to 1.0L. Mix thoroughly. Gently heat and bring to boiling. Distribute into tubes or flasks. Autoclave for 15 min at 15 psi pressure–121°C.

Use: For the cultivation and maintenance of *Bacillus* species.

Bacillus pasteurii Agar

Composition per liter:

Agar	15.0g
Peptone	5.0g
NaCl	5.0g
Yeast extract	4.0g
Beef extract	1.0g
Urea solution	50.0mL

pH 8.0 ± 0.2 at 25°C

Urea Solution:

Composition per 100.0mL:

Urea	20.0g

Preparation of Urea Solution: Add urea to distilled/deionized water and bring volume to 100.0mL. Mix thoroughly. Filter sterilize. Warm to 50°–55°C.

Preparation of Medium: Add components, except urea solution, to distilled/deionized water and bring volume to 950.0mL. Mix thoroughly. Gently heat and bring to boiling. Autoclave for 15 min at 15 psi pressure–121°C. Cool to 50°–55°C. Aseptically add 50.0mL of sterile urea solution. Mix thoroughly. Pour into sterile Petri dishes or distribute into sterile tubes.

Use: For the cultivation and maintenance of *Bacillus pasteurii.*

Bacillus pasteurii Agar

Composition per liter:

Urea	20.0g
Agar	15.0g
Peptone	5.0g
Meat extract	3.0g

pH 7.0 ± 0.2 at 25°C

Preparation of Medium: Add components to distilled/deionized water and bring volume to 1.0L. Gently heat and bring to boiling. Adjust pH to 7.0. Autoclave for 15 min at 15 psi pressure–121°C. Pour into sterile Petri dishes or distribute into sterile tubes.

Use: For the cultivation and maintenance of *Bacillus pasteurii* and *Sporosarcina ureae.*

Bacillus pasteurii Medium

Composition per liter:

Urea	20.0g
Agar	15.0g
Peptone	5.0g
NaCl	5.0g
Yeast extract	2.0g
Beef extract	1.0g

pH 7.4 ± 0.2 at 25°C

Preparation of Medium: Add components to distilled/deionized water and bring volume to 1.0L. Mix thoroughly. Gently heat and bring to boiling. Distribute into tubes or flasks. Autoclave for 15 min at 15 psi pressure–121°C. Pour into sterile Petri dishes or leave in tubes.

Use: For the cultivation and maintenance of *Bacillus pasteurii.*

Bacillus pasteurii NH_4 YE Medium (Ammonium Yeast Extract Medium)

Composition per liter:

Yeast extract	20.0g
Agar	20.0g
$(NH_4)_2SO_4$	10.0g

pH 9.0 ± 0.2 at 25°C

Preparation of Medium: Add each component to a separate flask and bring volume of each to 333.0mL with 0.13m Tris buffer, pH 9.0. Autoclave ingredients separately for 15 min at 15 psi pressure–121°C. No growth occurs if components are sterilized together. Cool to 50°–55°C and aseptically combine solutions. Pour into sterile Petri dishes.

Use: For the cultivation and maintenance of *Bacillus pasteurii.*

Bacillus pasteurii Sporulation Agar

Composition per liter:

Urea	20.0g
Agar	15.0g
Peptone	5.0g
Meat extract	3.0g
$MnSO_4 \cdot H_2O$	10.0mg

pH 7.0 ± 0.2 at 25°C

Preparation of Medium: Add components to distilled/deionized water and bring volume to 1.0L. Gently heat and bring to boiling. Adjust pH to 7.0. Autoclave for 15 min at 15 psi pressure–121°C. Pour into sterile Petri dishes or distribute into sterile tubes.

Use: For the induction of sporulation in various species, including *Bacillus pasteurii* and *Sporosarcina ureae.*

Bacillus polymyxa Agar

Composition per liter:

Agar	20.0g
Starch, soluble	10.0g
Peptone	5.0g
Yeast extract	5.0g

pH 7.2 ± 0.2 at 25°C

Preparation of Medium: Add components to distilled/deionized water and bring volume to 1.0L. Mix thoroughly. Gently heat and bring to boiling. Distribute into tubes or flasks. Autoclave for 15 min at 15 psi pressure–121°C. Pour into sterile Petri dishes or leave in tubes.

Use: For the cultivation and maintenance of *Bacillus macerans, Bacillus polymyxa,* and *Bacillus thermoglucosidasius.*

Bacillus popilliae Maintenance Medium

Composition per liter:

Agar	20.0g
Yeast extract	15.0g
Pancreatic digest of casein	5.0g
K_2HPO_4	3.0g
Glucose solution	10.0mL

pH 7.2 ± 0.2 at 25°C

Glucose Solution:

Composition per 10.0mL:

Glucose	2.0g

Preparation of Glucose Solution: Add glucose to distilled/deionized water and bring volume to 10.0mL. Mix thoroughly. Filter sterilize.

Preparation of Medium: Add components, except glucose solution, to distilled/deionized water and bring volume to 990.0mL. Mix thoroughly. Gently heat and bring to boiling. Autoclave for 15 min at 15 psi pressure–121°C. Cool to 45°–50°C. Aseptically add sterile glucose solution. Mix thoroughly. Pour into sterile Petri dishes or distribute into sterile tubes.

Use: For the cultivation and maintenance of *Bacillus popilliae.*

Bacillus popilliae Medium

Composition per liter:

Yeast extract	10.0g
Acid hydrolysate of casein	7.95g
K_2HPO_4	3.0g
Beef extract	1.36g
Trehalose	1.0g
Starch	0.68g

pH 7.3 ± 0.1 at 25°C

Preparation of Medium: Add components to distilled/deionized water and bring volume to 1.0L. Mix thoroughly. Gently heat until dissolved. Do not overheat. Filter sterilize. Aseptically distribute into sterile tubes or flasks.

Use: For the cultivation of *Bacillus popilliae.*

Bacillus popilliae Medium

Composition per liter:

Yeast extract	15.0g
K_2HPO_4	3.0g

pH 7.2 ± 0.2 at 25°C

Preparation of Medium: Add components to distilled/deionized water and bring volume to 1.0L. Mix thoroughly. Distribute into tubes or flasks. Autoclave for 15 min at 15 psi pressure–121°C.

Use: For the cultivation of *Bacillus popilliae.*

Bacillus Pullulan Salts

Composition per liter:

Pullulan	2.5g
NaCl	1.0g
NH_4Cl	1.0g
KH_2PO_4	0.5g
$MgSO_4 \cdot 7H_2O$	0.5g
Yeast extract	0.1g
$CaCl_2 \cdot 2H_2O$	0.05g
Trace mineral solution	10.0mL
Vitamin solution	10.0mL

pH 6.0 ± 0.2 at 25°C

Trace Mineral Solution:

Composition per liter:

$CoCl_2 \cdot 6H_2O$	0.2g
$FeSO_4 \cdot 7H_2O$	0.13g
$ZnCl_2 \cdot 2H_2O$	0.1g
$MnCl_2 \cdot 4H_2O$	0.1g
$CaCl_2 \cdot 2H_2O$	20.0mg
Na_2SeO_3	20.0mg
$Na_2WO_4 \cdot 2H_2O$	20.0mg
$NaMoO_4 \cdot 2H_2O$	1.0mg
H_3BO_3	0.5mg
$CuSO_4 \cdot 5H_2O$	0.4mg
KI	0.1mg

Preparation of Trace Mineral Solution: Add components to distilled/deionized water and bring volume to 1.0L. Mix thoroughly.

Vitamin Solution:

Composition per liter:

Pyridoxine·HCl	10.0mg
Thiamine·HCl	5.0mg
Riboflavin	5.0mg
Nicotinic acid	5.0mg
Calcium pantothenate	5.0mg
p-Aminobenzoic acid	5.0mg
Thioctic acid	5.0mg
Biotin	2.0mg
Folic acid	2.0mg
Cyanocobalamin	0.1mg

Preparation of Vitamin Solution: Add components to distilled/deionized water and bring volume to 1.0L. Mix thoroughly. Filter sterilize.

Preparation of Medium: Add components, except vitamin solution, to distilled/deionized water and bring volume to 990.0mL. Mix thoroughly. Gently heat and bring to boiling. Adjust pH to 6.0. Autoclave for 15 min at 15 psi pressure–121°C. Cool to 25°C. Aseptically add sterile vitamin solution. Mix thoroughly. Aseptically distribute into sterile tubes or flasks.

Use: For the cultivation and maintenance of *Bacillus* species that can degrade pullulan.

Bacillus racemilacticus Agar

Composition per liter:

Agar	15.0g
$CaCO_3$	5.0g
Glucose	5.0g
Peptone	5.0g
Yeast extract	5.0g

pH 6.8 ± 0.2 at 25°C

Preparation of Medium: Add components to distilled/deionized water and bring volume to 1.0L. Mix thoroughly. Gently heat and bring to boiling. Distribute into tubes or flasks. Autoclave for 15 min at 15 psi pressure–121°C. Pour into sterile Petri dishes or leave in tubes.

Use: For the cultivation and maintenance of *Bacillus kaustophilus* and *Bacillus racemilacticus.*

Bacillus racemilacticus Agar

Composition per liter:

Agar	15.0g
$CaCO_3$	5.0g
Glucose	5.0g
Peptone	5.0g
Yeast extract	5.0g

pH 6.8 ± 0.2 at 25°C

Preparation of Medium: Add components to distilled/deionized water and bring volume to 1.0L. Gently heat and bring to boiling. Adjust pH to 6.8. Autoclave for 15 min at 15 psi pressure–121°C. Pour into sterile Petri dishes or distribute into sterile tubes.

Use: For the cultivation and maintenance of *Bacillus racemilacticus, Bacillus coagulans, Bacillus laevolacticus,* and other *Bacillus* species.

Bacillus racemilacticus Agar (YEPG with 0.5% $CaCO_3$)

Composition per liter:

Agar	15.0g
$CaCO_3$	5.0g
Glucose	5.0g
Peptone	5.0g
Yeast extract	5.0g

Preparation of Medium: Add components to distilled/deionized water and bring volume to 1.0L. Mix thoroughly. Gently heat and bring to boiling. Distribute into tubes or flasks. Autoclave for 15 min at 15 psi pressure–121°C. Pour into sterile Petri dishes or leave in tubes.

Use: For the cultivation and maintenance of *Bacillus racemilacticus* and other *Bacillus* species.

Bacillus schlegelii Agar (LMG Medium 85)

Composition per liter:

Agar	30.0g
$Na_2HPO_4 \cdot 12\ H_2O$	9.0g
KH_2PO_4	1.5g
Sodium pyruvate	1.5g
NH_4Cl	1.0g
$MgSO_4 \cdot 7H_2O$	0.2g
$MnSO_4 \cdot H_2O$	10.0mg
$CaCl_2 \cdot 2H_2O$	10.0mg
Ferric ammonium citrate	5.0mg
Trace elements solution	3.0mL

pH 7.1 ± 0.2 at 25°C

Trace Elements Solution:
Composition per liter:

H_3BO_3	0.3g
$CoCl_2 \cdot 6H_2O$	0.2g
$ZnSO_4 \cdot 7H_2O$	0.1g
$Na_2MoO_4 \cdot 2H_2O$	30.0mg
$MnCl_2 \cdot 4H_2O$	30.0mg
$NiCl_2 \cdot 6H_2O$	20.0mg
$CuCl_2 \cdot 2H_2O$	10.0mg

Preparation of Trace Elements Solution: Add components to distilled/deionized water and bring volume to 1.0L. Mix thoroughly.

Preparation of Medium: Add components to distilled/deionized water and bring volume to 1.0L. Mix thoroughly. Distribute into tubes or flasks. Gently heat and bring to boiling. Autoclave for 15 min at 15 psi pressure–121°C. Pour into sterile Petri dishes or leave in tubes.

Use: For the cultivation of *Bacillus schlegelii.*

Bacillus schlegelii Agar

Composition per liter:

Noble agar	30.0g
$Na_2HPO_4 \cdot 2H_2O$	4.5g
KH_2PO_4	1.5g
NH_4Cl	1.0g
$MgSO_4 \cdot 7H_2O$	0.2g
$MnSO_4 \cdot H_2O$	0.01g
$CaCl_2 \cdot 2H_2O$	0.01g
Ferric ammonium citrate	5.0mg
Agar solution	200.0mL
Pyruvate solution	100.0mL
Vrace elements solution SL-6	3.0mL

pH 7.1 ± 0.2 at 25°C

Agar Solution:
Composition per 200.0mL:

Noble agar	30.0g

Preparation of Agar Solution: Add agar to distilled/deionized water and bring volume to 200.0mL. Mix thoroughly. Gently heat and bring to boiling. Autoclave for 15 min at 15 psi pressure–121°C. Cool to 45°–50°C.

Pyruvate Solution:
Composition per 100.0mL:

Sodium pyruvate	1.5g

Preparation of Pyruvate Solution: Add sodium pyruvate to distilled/deionized water and bring volume to 100.0mL. Mix thoroughly. Filter sterilize. Warm to 45°–50°C.

Trace Elements Solution SL-6 :
Composition per liter:

H_3BO_3	0.3g
$CoCl_2 \cdot 6H_2O$	0.2g
$ZnSO_4 \cdot 7H_2O$	0.1g
$MnCl_2 \cdot 4H_2O$	0.03g
$Na_2MoO_4 \cdot H_2O$	0.03g
$NiCl_2 \cdot 6H_2O$	0.02g
$CuCl_2 \cdot 2H_2O$	0.01g

Preparation of Trace Elements Solution SL-6 : Add components to distilled/deionized water and bring volume to 1.0L. Adjust pH to 3.4.

Preparation of Medium: Add components, except sodium pyruvate solution and agar solution, to distilled/deionized water and bring volume to 700.0mL. Mix thoroughly. Gently heat and bring to boiling. Adjust pH to 7.1. Autoclave for 15 min at 15 psi pressure–121°C. Cool to 50°C. Add sodium pyruvate solution and agar solution. Mix thoroughly. Pour into sterile Petri dishes or distribute into sterile tubes.

Use: For the cultivation and maintenance of *Bacillus schlegelii.*

Bacillus schlegelii Broth

Composition per liter:

$Na_2HPO_4 \cdot 2H_2O$	4.5g
KH_2PO_4	1.5g
NH_4Cl	1.0g
$MgSO_4 \cdot 7H_2O$	0.2g
$MnSO_4 \cdot H_2O$	0.01g
$CaCl_2 \cdot 2H_2O$	0.01g
Ferric ammonium citrate	5.0mg
Pyruvate solution	100.0mL
SL-6 trace elements	3.0mL

pH 7.1 ± 0.2 at 25°C

Pyruvate Solution:
Composition per 100.0mL:

Sodium pyruvate	1.5g

Preparation of Pyruvate Solution: Add sodium pyruvate to distilled/deionized water and bring volume to 100.0mL. Mix thoroughly. Filter sterilize.

Trace Elements Solution SL-6 :
Composition per liter:

H_3BO_3	0.3g
$CoCl_2 \cdot 6H_2O$	0.2g
$ZnSO_4 \cdot 7H_2O$	0.1g
$MnCl_2 \cdot 4H_2O$	0.03g
$Na_2MoO_4 \cdot H_2O$	0.03g
$NiCl_2 \cdot 6H_2O$	0.02g
$CuCl_2 \cdot 2H_2O$	0.01g

Preparation of Trace Elements Solution SL-6 : Add components to distilled/deionized water and bring volume to 1.0L. Mix thoroughly. Adjust pH to 3.4.

Preparation of Medium: Add components, except sodium pyruvate solution, to distilled/deionized water and bring volume to 900.0mL. Mix thoroughly. Gently heat and bring to boiling. Adjust pH to 7.1. Autoclave for 15 min at 15 psi pressure–121°C. Cool to 50°C. Aseptically add sodium pyruvate solution. Aseptically distribute into sterile tubes or flasks.

Use: For the cultivation and maintenance of *Bacillus schlegelii.*

Bacillus schlegelii Chemolithotrophic Growth Medium (DSMZ Medium 261)

Composition per liter:

$Na_2HPO_4 \cdot 2H_2O$	4.5g
KH_2PO_4	1.5g
NH_4Cl	1.0g
$MgSO_4 \cdot 7H_2O$	0.2g
$MnSO_4 \cdot 2H_2O$	0.01g
$CaCl_2 \cdot 2H_2O$	0.01g
Ferric ammonium citrate	5.0mg
Trace elements solution SL-6	3.0mL

pH 7.1 ± 0.2 at 25°C

Trace Elements Solution SL-6:

Composition per liter:

$MnCl_2 \cdot 4H_2O$	0.5g
H_3BO_3	0.3g
$CoCl_2 \cdot 6H_2O$	0.2g
$ZnSO_4 \cdot 7H_2O$	0.1g
$Na_2MoO_4 \cdot 2H_2O$	0.03g
$NiCl_2 \cdot 6H_2O$	0.02g
$CuCl_2 \cdot 2H_2O$	0.01g

Preparation of Trace Elements Solution SL-6: Add components to distilled/deionized water and bring volume to 1.0L. Mix thoroughly.

Preparation of Medium: Add components to distilled/deionized water and bring volume to 1.0L. Mix thoroughly. Adjust pH to 7.1. Distribute into tubes or flasks. Autoclave for 15 min at 15 psi pressure–121°C.

Use: For the chemolithotrophic cultivation of *Bacillus schlegelii.* Incubation is at 65°C under an atmosphere of 5% O_2, 10% CO_2, 45% H_2.

Bacillus schlegelii Heterotrophic Growth Medium (DSMZ Medium 260)

Composition per liter:

$Na_2HPO_4 \cdot 2H_2O$	4.5g
Na-pyruvate	1.5g
KH_2PO_4	1.5g
NH_4Cl	1.0g
$MgSO_4 \cdot 7H_2O$	0.2g
$MnSO_4 \cdot 2H_2O$	0.01g
$CaCl_2 \cdot 2H_2O$	0.01g
Ferric ammonium citrate	5.0mg
Trace elements solution SL-6	3.0mL

pH 7.1 ± 0.2 at 25°C

Trace Elements Solution SL-6:

Composition per liter:

$MnCl_2 \cdot 4H_2O$	0.5g
H_3BO_3	0.3g
$CoCl_2 \cdot 6H_2O$	0.2g
$ZnSO_4 \cdot 7H_2O$	0.1g
$Na_2MoO_4 \cdot 2H_2O$	0.03g
$NiCl_2 \cdot 6H_2O$	0.02g
$CuCl_2 \cdot 2H_2O$	0.01g

Preparation of Trace Elements Solution SL-6: Add components to distilled/deionized water and bring volume to 1.0L. Mix thoroughly.

Preparation of Medium: Add components to distilled/deionized water and bring volume to 1.0L. Mix thoroughly. Adjust pH to 7.1. Distribute into tubes or flasks. Autoclave for 15 min at 15 psi pressure–121°C.

Use: For the heterotrophic cultivation of *Bacillus schlegelii.* Incubation is at 65°C.

Bacillus schlegelii Medium (LMG Medium 85)

Composition per liter:

$Na_2HPO_4 \cdot 12\ H_2O$	9.0g
KH_2PO_4	1.5g
Sodium pyruvate	1.5g
NH_4Cl	1.0g
$MgSO_4 \cdot 7H_2O$	0.2g
$MnSO_4 \cdot H_2O$	10.0mg
$CaCl_2 \cdot 2H_2O$	10.0mg
Ferric ammonium citrate	5.0mg
Trace elements solution	3.0mL

pH 7.1 ± 0.2 at 25°C

Trace Elements Solution:

Composition per liter:

H_3BO_3	0.3g
$CoCl_2 \cdot 6H_2O$	0.2g
$ZnSO_4 \cdot 7H_2O$	0.1g
$Na_2MoO_4 \cdot 2H_2O$	30.0mg
$MnCl_2 \cdot 4H_2O$	30.0mg
$NiCl_2 \cdot 6H_2O$	20.0mg
$CuCl_2 \cdot 2H_2O$	10.0mg

Preparation of Trace Elements Solution: Add components to distilled/deionized water and bring volume to 1.0L. Mix thoroughly.

Preparation of Medium: Add components to distilled/deionized water and bring volume to 1.0L. Mix thoroughly. Distribute in 30–50 mL amounts in Erlenmeyer flasks. Autoclave for 15 min at 15 psi pressure–121°C. Incubate without agitation.

Use: For the cultivation of *Bacillus schlegelii.*

Bacillus selenitireducens Medium (DSMZ Medium 968)

Composition per liter:

NaCl	90.0g
Na_2CO_3	10.6g
$NaHCO_3$	4.2g
Na-lactate	1.70g
$NaNO_3$	1.25g
Yeast extract	0.2g
K_2HPO_4	0.15g
$(NH_4)SO_4$	0.1g
KH_2PO_4	0.08g
$MgSO_4$	25.0mg
Resazurin	0.5mg

Cysteine solution	10.0mL
$Na_2S·9H_2O$ solution	10.0mL
Trace elements solution SL-10	1.0mL
Selenite tungstate solution	1.0mL

pH 9.8 ± 0.2 at 25°C

Cysteine Solution:

Composition per 10.0mL:

L-Cysteine·HCl·H_2O	0.25g

Preparation of Cysteine Solution: Add L-cysteine·HCl·H_2O to distilled/deionized water and bring volume to 10.0mL. Mix thoroughly. Sparge with 100% N_2. Autoclave for 15 min at 15 psi pressure–121°C.

$Na_2S·9H_2O$ Solution:

Composition per 10.0mL:

$Na_2S·9H_2O$	0.25g

Preparation of $Na_2S·9H_2O$ Solution: Add $Na_2S·9H_2O$ to distilled/deionized water and bring volume to 10.0mL. Sparge with N_2. Autoclave for 15 min at 15 psi pressure–121°C. Cool to 25°C.

Trace Elements Solution SL-10:

Composition per liter:

$FeCl_2·4H_2O$	1.5g
$CoCl_2·6H_2O$	190.0mg
$MnCl_2·4H_2O$	100.0mg
$ZnCl_2$	70.0mg
$Na_2MoO_4·2H_2O$	36.0mg
$NiCl_2·6H_2O$	24.0mg
H_3BO_3	6.0mg
$CuCl_2·2H_2O$	2.0mg
HCl (25% solution)	10.0mL

Preparation of Trace Elements Solution SL-10: Add $FeCl_2·4H_2O$ to 10.0mL of HCl solution. Mix thoroughly. Add distilled/deionized water and bring volume to 1.0L. Add remaining components. Mix thoroughly. Sparge with 80% N_2 + 20% CO_2. Autoclave for 15 min at 15 psi pressure–121°C.

Selenite Tungstate Solution

Composition per liter:

NaOH	0.5g
$Na_2WO_4·2H_2O$	4.0mg
$Na_2SeO_3·5H_2O$	3.0mg

Preparation of Selenite Tungstate Solution: Add components to distilled/deionized water and bring volume to 1.0L. Mix thoroughly. Sparge with 100% N_2. Filter sterilize.

Preparation of Medium: Add components, except $NaHCO_3$, Na_2CO_3, cysteine solution, and $Na_2S·9H_2O$ solution, to distilled/deionized water and bring volume to 980.0mL. Mix thoroughly. Gently heat and bring to boiling. Boil for 3 min. Cool to room temperature while sparging with 80% N_2 gas. Add solid $NaHCO_3$ and Na_2CO_3. Adjust pH to 9.8. Distribute to anaerobe tubes or bottles. Autoclave for 15 min at 15 psi pressure–121°C. Cool to room temperature. Aseptically and anaerobically add per liter 10.0mL sterile cysteine solution and 10.0mL sterile $Na_2S·9H_2O$ solution. Mix thoroughly.

Use: For the cultivation of *Bacillus selenitireducens* and *Bacillus arseniciselenatis.*

Bacillus stearothermophilus Broth

Composition per liter:

Pancreatic digest of casein	10.0g
Yeast extract	5.0g
K_2HPO_4	2.0g

pH 7.2 ± 0.2 at 25°C

Preparation of Medium: Add components to distilled/deionized water and bring volume to 1.0L. Mix thoroughly. Distribute into tubes or flasks. Autoclave for 15 min at 15 psi pressure–121°C.

Use: For the cultivation of *Bacillus stearothermophilus.*

Bacillus stearothermophilus Defined Broth

Composition per 100.0mL:

Mineral salts solution	10.0mL
Potassium phosphate buffer	5.0mL
L-Glutamate·HCl (1% solution)	4.0mL
L-Leucine (1% solution)	1.64mL
L-Lysine·HCl (1% solution)	1.4mL
L-Serine (1% solution)	1.4mL
L-Aspartate (1% solution)	1.3mL
L-Valine (1% solution)	1.26mL
Biotin (0.01% solution)	1.0mL
Glucose (20% solution)	1.0mL
L-Isoleucine (1% solution)	1.0mL
L-Proline (1% solution)	1.0mL
Nicotinic acid (0.01% solution)	1.0mL
Thiamine·HCl (0.01% solution)	1.0mL
L-Phenylalanine (1% solution)	0.86mL
L-Alanine (1% solution)	0.84mL
L-Threonine (1% solution)	0.84mL
L-Arginine·HCl (1% solution)	0.64mL
L-Tyrosine (1% solution)	0.56mL
L-Methionine (1% solution)	0.52mL
Glycine (1% solution)	0.5mL
L-Asparagine·H_2O (1% solution)	0.5mL
L-Cystine (1% solution)	0.5mL
L-Glutamine (1% solution)	0.5mL
L-Histidine·HCl·H_2O (1% solution)	0.42mL
L-Tryptophan (1% solution)	0.3mL
$CaCl_2$ (5% solution)	0.01mL
$FeCl_3·6H_2O$ (0.05% solution)	0.01mL
$MnCl_2$ (10mm solution)	0.01mL
$ZnSO_4·7H_2O$ (5% solution)	0.01mL

pH 7.3 ± 0.2 at 25°C

Mineral Salts Solution:

Composition per liter:

NaCl	10.0g
NH_4Cl	10.0g
$MgSO_4$	4.0g

Preparation of Mineral Salts Solution: Add components to distilled/deionized water and bring volume to 1.0L. Mix thoroughly.

Potassium Phosphate Buffer:

Composition per 500.0mL:

K_2HPO_4	125.0g
KH_2PO_4	30.0g

Preparation of Potassium Phosphate Buffer: Add components to distilled/deionized water and bring volume to 500.0mL. Mix thoroughly.

Preparation of Medium: Add components to distilled/deionized water and bring volume to 100.0mL. Mix thoroughly. Filter sterilize.

Use: For the cultivation of *Bacillus stearothermophilus* in a chemically defined medium.

Bacillus stearothermophilus Sporulation Broth

Composition per liter:

Agar	20.0g
Pancreatic digest of gelatin	5.0g
Yeast extract	4.0g
Beef extract	3.0g
$MnCl_2 \cdot 4H_2O$	10.0μg

pH 7.2 ± 0.2 at 25°C

Preparation of Medium: Add components to distilled/deionized water and bring volume to 1.0L. Mix thoroughly. Gently heat and bring to boiling. Distribute into tubes or flasks. Autoclave for 15 min at 15 psi pressure–121°C. Pour into sterile Petri dishes or leave in tubes.

Use: For the cultivation and sporulation of *Bacillus stearothermophilus*.

Bacillus thermoalcalophilus Medium

Composition per liter:

Yeast extract	10.0g
Sodium acetate	3.0g
KCl	1.8g
Na_2SO_4	0.4g
K_2HPO_4	0.3g
KH_2PO_4	0.3g
$MgSO_4$	0.2g
$FeSO_4$	0.01g

pH 8.2 ± 0.2 at 25°C

Preparation of Medium: Add components to distilled/deionized water and bring volume to 1.0L. Mix thoroughly. Adjust pH to 8.2. Distribute into tubes or flasks. Autoclave for 15 min at 15 psi pressure–121°C.

Use: For the cultivation of *Bacillus thermoalcalophilus*.

Bacillus thermoantarcticus Medium

Composition per liter:

Yeast extract	6.0g
NaCl	3.0g
Soil extract	500.0mL

pH 5.6–5.8 at 25°C

Soil Extract:

Composition per liter:

Garden soil, air dried	400.0g

Preparation of Soil Extract: Pass 400.0g of air-dried garden soil through a coarse sieve. Add soil to 960.0mL of tap water. Mix thoroughly. Autoclave for 60 min at 15 psi pressure–121°C. Cool to room temperature. Allow residue to settle. Decant supernatant solution. Filter through Whatman filter paper. Distribute into bottles in 200.0mL volumes. Autoclave for 15 min at 15 psi pressure–121°C. Store at room temperature until clear.

Preparation of Medium: Add components to distilled/deionized water and bring volume to 1.0L. Mix thoroughly. Adjust pH to 5.6–5.8. Distribute into tubes or flasks. Autoclave for 15 min at 15 psi pressure–121°C.

Use: For the cultivation of *Bacillus thermoantarcticus*.

Bacillus thermoglucosidasius Agar

Composition per liter:

Agar	30.0g
Starch	10.0g
Peptone	5.0g
Beef extract	3.0g
K_2HPO_4	3.0g
Yeast extract	3.0g

pH 7.0 ± 0.2 at 25°C

Preparation of Medium: Add components to distilled/deionized water and bring volume to 1.0L. Mix thoroughly. Adjust pH to 7.0. Gently heat and bring to boiling. Distribute into tubes or flasks. Autoclave for 20 min at 15 psi pressure–121°C. Pour into sterile Petri dishes or leave in tubes.

Use: For the cultivation and maintenance of *Bacillus thermoglucosidasius*.

Bacillus thermoglucosidasius Agar

Composition per liter:

Agar	30.0g
Soluble starch	10.0g
Peptone	5.0g
Beef extract	3.0g
KH_2PO_4	3.0g
Yeast extract	3.0g

pH 7.0 ± 0.2 at 25°C

Preparation of Medium: Add components to distilled/deionized water and bring volume to 1.0L. Gently heat and bring to boiling. Adjust pH to 7.0. Autoclave for 15 min at 15 psi pressure–121°C. Pour into sterile Petri dishes or distribute into sterile tubes.

Use: For the cultivation and maintenance of *Bacillus thermoglucosidasius*.

Bacillus thermoleovorans Medium

Composition per liter:

n-Heptadecane	1.0g
$(NH_4)_2HPO_4$	1.0g
Yeast extract	1.0g
KCl	0.2g
$MgSO_4 \cdot 7H_2O$	0.2g

Preparation of Medium: Add components to distilled/deionized water and bring volume to 1.0L. Mix thoroughly. Distribute into tubes or flasks. Autoclave for 15 min at 15 psi pressure–121°C.

Use: For the cultivation of *Bacillus thermoleovorans*.

Bacillus thuringiensis Medium

Composition per liter:

Glucose	3.0g
$(NH_4)_2SO_4$	2.0g
Yeast extract	2.0g
$K_2HPO_4 \cdot 3H_2O$	0.5g
$MgSO_4 \cdot 7H_2O$	0.2g
$CaCl_2 \cdot 2H_2O$	0.08g
$MnSO_4 \cdot 4H_2O$	0.05g

pH 7.3 ± 0.2 at 25°C

Preparation of Medium: Add components to distilled/deionized water and bring volume to 1.0L. Mix thoroughly. Adjust pH to 7.3.

Distribute into tubes or flasks. Autoclave for 15 min at 15 psi pressure–121°C.

Use: For the cultivation of *Bacillus thuringiensis*.

Bacillus tusciae Medium

Composition per liter:

$Na_2HPO_4 \cdot 2H_2O$	2.9g
KH_2PO_4	2.3g
NH_4Cl	1.0g
$MgSO_4 \cdot 7H_2O$	0.5g
$NaHCO_3$	0.5g
$Fe(NH_4)$ citrate	0.05g
$CaCl_2 \cdot 2H_2O$	0.01g
$MnSO_4 \cdot H_2O$	0.01g
Ferric ammonium citrate solution	20.0mL
Trace elements solution SL-6	5.0mL

pH 4.0 ± 0.2 at 25°C

Ferric Ammonium Citrate Solution:

Composition per 20.0mL:

Ferric ammonium citrate	0.05g

Preparation of Ferric Ammonium Citrate Solution: Add ferric ammonium citrate to distilled/deionized water and bring volume to 20.0mL. Mix thoroughly. Autoclave for 15 min at 15 psi pressure–121°C.

Trace Elements Solution SL-6:

Composition per liter:

$MnCl_2 \cdot 4H_2O$	0.5g
H_3BO_3	0.3g
$CoCl_2 \cdot 6H_2O$	0.2g
$ZnSO_4 \cdot 7H_2O$	0.1g
$Na_2MoO_4 \cdot 2H_2O$	0.03g
$NiCl_2 \cdot 6H_2O$	0.02g
$CuCl_2 \cdot 2H_2O$	0.01g

Preparation of Trace Elements Solution SL-6: Add components to distilled/deionized water and bring volume to 1.0L. Mix thoroughly. Autoclave for 15 min at 15 psi pressure–121°C.

Preparation of Medium: Add components, except ferric ammonium citrate solution and trace elements solution SL-6, to distilled/deionized water and bring volume to 975.0mL. Mix thoroughly. Gently heat and bring to boiling. Adjust pH to 4.0. Autoclave for 15 min at 15 psi pressure–121°C. Cool to 50°–55°C. Aseptically add 20.0mL of sterile ferric ammonium citrate solution and 5.0mL of sterile trace elements solution SL-6. Mix thoroughly. Aseptically distribute into sterile tubes or flasks. For chemolithotropic growth, incubate the culture under 2% O_2 + 10% CO_2 + 60% H_2 + 28% N_2.

Use: For the chemolithotrophic growth of *Bacillus tusciae*.

Bacillus tusciae Medium

Composition per liter:

Agar	15.0g
$Na_2HPO_4 \cdot 2H_2O$	2.9g
KH_2PO_4	2.3g
NH_4Cl	1.0g
$MgSO_4 \cdot 7H_2O$	0.5g
$NaHCO_3$	0.5g
$Fe(NH_4)$ citrate	0.05g
$CaCl_2 \cdot 2H_2O$	0.01g
$MnSO_4 \cdot H_2O$	0.01g
Ferric ammonium citrate solution	20.0mL
Carbon source	10.0mL
Trace elements solution SL-6	5.0mL

pH 4.0 ± 0.2 at 25°C

Ferric Ammonium Citrate Solution:

Composition per 20.0mL:

Ferric ammonium citrate	0.05g

Preparation of Ferric Ammonium Citrate Solution: Add ferric ammonium citrate to distilled/deionized water and bring volume to 20.0mL. Mix thoroughly. Autoclave for 15 min at 15 psi pressure–121°C.

Carbon Source:

Composition per 10.0mL:

Carbohydrate	2.0g
Organic acid (alternate)	1.0g

Preparation of Carbon Source: Add either carbohydrate or organic acid to distilled/deionized water and bring volume to 10.0mL. Mix thoroughly. Filter sterilize.

Trace Elements Solution SL-6:

Composition per liter:

$MnCl_2 \cdot 4H_2O$	0.5g
H_3BO_3	0.3g
$CoCl_2 \cdot 6H_2O$	0.2g
$ZnSO_4 \cdot 7H_2O$	0.1g
$Na_2MoO_4 \cdot 2H_2O$	0.03g
$NiCl_2 \cdot 6H_2O$	0.02g
$CuCl_2 \cdot 2H_2O$	0.01g

Preparation of Trace Elements Solution SL-6: Add components to distilled/deionized water and bring volume to 1.0L. Mix thoroughly. Autoclave for 15 min at 15 psi pressure–121°C.

Preparation of Medium: Add components, except ferric ammonium citrate solution, trace elements solution SL-6, and carbon source, to distilled/deionized water and bring volume to 965.0mL. Mix thoroughly. Gently heat and bring to boiling. Adjust pH to 4.0. Autoclave for 15 min at 15 psi pressure–121°C. Cool to 50°–55°C. Aseptically add 20.0mL of sterile ferric ammonium citrate solution, 10.0mL of sterile carbon source, and 5.0mL of sterile trace elements solution SL-6. Mix thoroughly. Aseptically distribute into sterile tubes or flasks.

Use: For the heterotrophic growth of *Bacillus tusciae*.

Bacillus Xylose Salts

Composition per liter:

Yeast extract	5.0g
Xylose	5.0g
NaCl	1.0g
NH_4Cl	1.0g
KH_2PO_4	0.5g
$MgSO_4 \cdot 7H_2O$	0.5g
$CaCl_2 \cdot 2H_2O$	0.05g
Trace mineral solution	10.0mL
Vitamin solution	10.0mL

pH 4.0 ± 0.2 at 25°C

Trace Mineral Solution:

Composition per liter:

$CoCl_2 \cdot 6H_2O$	0.2g
$FeSO_4 \cdot 7H_2O$	0.13g
$ZnCl_2 \cdot 2H_2O$	0.1g
$MnCl_2 \cdot 4H_2O$	0.1g

$CaCl_2 \cdot 2H_2O$	20.0mg
Na_2SeO_3	20.0mg
$Na_2WO_4 \cdot 2H_2O$	20.0mg
$NaMoO_4 \cdot 2H_2O$	1.0mg
H_3BO_3	0.5mg
$CuSO_4 \cdot 5H_2O$	0.4mg
KI	0.1mg

Preparation of Trace Mineral Solution: Add components to distilled/deionized water and bring volume to 1.0L. Mix thoroughly.

Vitamin Solution:

Composition per liter:

Pyridoxine·HCl	10.0mg
Thiamine·HCl	5.0mg
Riboflavin	5.0mg
Nicotinic acid	5.0mg
Calcium pantothenate	5.0mg
p-Aminobenzoic acid	5.0mg
Thioctic acid	5.0mg
Biotin	2.0mg
Folic acid	2.0mg
Cyanocobalamin	0.1mg

Preparation of Vitamin Solution: Add components to distilled/deionized water and bring volume to 1.0L. Mix thoroughly.

Preparation of Medium: Add components to distilled/deionized water and bring volume to 1.0L. Mix thoroughly. Gently heat and bring to boiling. Adjust pH of medium to 4.0. Distribute into tubes or flasks. Autoclave for 15 min at 15 psi pressure–121°C.

Use: For the cultivation and maintenance of *Bacillus* species that can utilize xylose as a carbon source.

Bacterial Cell Agar (BCA)

Composition per liter:

Tryptose	17.36g
Agar	15.0g
NaCl	8.68g
Beef extract	5,2g
Yeast extract	1.7g

pH 7.3 ± 0.2 at 25°C

Preparation of Medium: Add components, except agar, to distilled/deionized water and bring volume to 1.0L. Mix thoroughly. Autoclave for 15 min at 15 psi pressure–121°C. Cool to 30°C. Inoculate with a culture of *Aeromonas hydrophila.* Incubate with shaking at 30°C for 72 hr. Centrifuge culture in 40.0mL volumes at 10,000 × g for 10 min. Wash the cells four times in sterile 0.85% saline. Resuspend the cell pellet in 25.0mL of distilled/deionized water. Autoclave for 15 min at 15 psi pressure–121°C. Cool to 45°–50°C. In a separate flask, add 15.0g of agar to 1.0L of distilled/deionized water. Mix thoroughly. Gently heat and bring to boiling. Autoclave for 15 min at 15 psi pressure–121°C. Cool to 45°–50°C. Aseptically combine 25.0mL of washed cells and 250.0mL of cooled, sterile agar solution. Mix thoroughly. Pour into sterile Petri dishes.

Use: For the cultivation of freshwater *Myxobacterium* species.

Bacterium Medium

Composition per liter:

Agar	20.0g
Peptone	6.0g
Yeast extract	3.0g
Beef extract	1.5g
Glucose	1.0g

Preparation of Medium: Add components to distilled/deionized water and bring volume to 1.0L. Mix thoroughly. Gently heat and bring to boiling. Distribute into tubes or flasks. Autoclave for 15 min at 15 psi pressure–121°C. Pour into sterile Petri dishes or leave in tubes.

Use: An archaic medium used for the cultivation and growth of bacteria originally classified in the genus *Bacterium* but now classified in the genera *Brevibacterium* and *Kurthia.*

Bacteroides Bile Esculin Agar (BBE Agar)

Composition per liter:

Oxgall	20.0g
Pancreatic digest of casein	15.0g
Agar	15.0g
Papaic digest of soybean meal	5.0g
NaCl	5.0g
Esculin	1.0g
Ferric ammonium citrate	0.5g
Gentamicin solution	2.5mL
Hemin solution	2.5mL
Vitamin K_1 solution	1.0mL

pH 7.0 ± 0.2 at 25°C

Source: This medium is available as a premixed powder from BD Diagnostic Systems.

Gentamicin Solution:

Composition per 10.0mL:

Gentamicin	0.4mg

Preparation of Gentamicin Solution: Add gentamicin to 10.0mL of distilled/deionized water. Mix thoroughly. Filter sterilize.

Hemin Solution:

Composition per 100.0mL:

Hemin	0.5g
NaOH (1*N* solution)	10.0mL

Preparation of Hemin Solution: Add components to 100.0mL of distilled/deionized water. Mix thoroughly. Autoclave for 15 min at 15 psi pressure–121°C. Cool to 45°–50°C.

Vitamin K_1 Solution:

Composition per 100.0mL:

Vitamin K_1	1.0g
Ethanol	99.0mL

Preparation of Vitamin K_1 Solution: Add vitamin K_1 to 99.0mL of absolute ethanol. Mix thoroughly. Filter sterilize.

Preparation of Medium: Add components, except hemin solution, gentamicin solution, and vitamin K_1 solution, to distilled/deionized water and bring volume to 994.0mL. Mix thoroughly. Gently heat and bring to boiling. Autoclave for 15 min at 15 psi pressure–121°C. Cool to 45°–50°C. Aseptically add 2.5mL of sterile hemin solution, 2.5mL of sterile gentamicin solution, and 1.0mL of sterile vitamin K_1 solution.

Use: For the selection and presumptive identification of the *Bacteroides fragilis* group. For the differentiation of *Bacteroides* species based on the hydrolysis of esculin and presence of catalase. After incubation for 48 hr, bacteria of the *Bacteroides fragilis* group appear as gray, circular, raised colonies larger than 1.0mm. Esculin hydrolysis is indicated by the presence of a blackened zone around the colonies.

Bacteroides cellulosolvens Medium

Composition per liter:

Cellobiose or cellulose	5.0g
$NaHCO_3$	2.0g
NH_4Cl	0.68g
K_2HPO_4	0.3g
L-Cysteine·HCl·H_2O	0.25g
$Na_2S·9H_2O$	0.25g
KH_2PO_4	0.18g
$(NH_4)_2SO_4$	0.15g
$MgSO_4·7H_2O$	0.12g
$CaCl_2·2H_2O$	0.06g
$FeSO_4·7H_2O$	0.02g
Resazurin	1.0mg
Trace elements solution	10.0mL
Vitamin solution	10.0mL

pH 7.0 ± 0.2 at 25°C

Trace Elements Solution:

Composition per liter:

$MgSO_4·7H_2O$	3.0g
Nitrilotriacetic acid	1.5 g
$CaCl_2·2H_2O$	1.0g
NaCl	1.0g
$MnSO_4·2H_2O$	0.5 g
$CoSO_4·7H_2O$	0.18 g
$ZnSO_4·7H_2O$	0.18 g
$FeSO_4·7H_2O$	0.1g
$NiCl_2·6H_2O$	0.025 g
$KAl(SO_4)_2·12H_2O$	0.02g
$CuSO_4·5H_2O$	0.01g
H_3BO_3	0.01g
$Na_2MoO_4·2H_2O$	0.01g
$Na_2SeO_3·5H_2O$	0.3 mg

Preparation of Trace Elements Solution: Add nitrilotriacetic acid to approximately 500.0mL distilled/deionized water. Dissolve by adding KOH and adjust pH to 6.5. Add remaining components. Bring volume to 1.0L with additional distilled/deionized water. Adjust pH to 7.0 with KOH.

Vitamin Solution:

Composition per liter:

Pyridoxine·HCl	10.0mg
Calcium DL-pantothenate	5.0mg
Lipoic acid	5.0mg
Nicotinic acid	5.0mg
p-Aminobenzoic acid	5.0mg
Riboflavin	5.0mg
Thiamine·HCl	5.0mg
Biotin	2.0mg
Folic acid	2.0mg
Vitamin B_{12}	0.1mg

Preparation of Vitamin Solution: Add components to distilled/deionized water and bring volume to 1.0L. Mix thoroughly.

Cellobiose Solution:

Composition per 50.0mL:

D-Cellobiose (or cellulose)	5.0g

Preparation of Cellobiose Solution: Add cellobiose (or cellulose) to distilled/deionized water and bring volume to 50.0mL. Mix thoroughly. Filter sterilize.

Preparation of Medium: Add components, except cellobiose solution, to distilled/deionized water and bring volume to 950.0mL. Mix thoroughly. Autoclave for 15 min at 15 psi pressure–121°C. Aseptically add 50.0mL of sterile cellobiose (or cellulose) solution. Mix thoroughly. Aseptically distribute into sterile tubes or flasks.

Use: For the cultivation and maintenance of *Bacteroides cellosolvens*.

Bacteroides HiVeg Agar Base with Selective Supplement (BBE)

Composition per liter:

Plant hydrolysate	25.0g
Agar	15.0g
Papaic digest of soybean meal	10.0g
NaCl	5.0g
Synthetic detergent No. II	2.0g
Esculin	1.0g
Ferric ammonium citrate	0.5g
$Fe_4(P_2O_7)_3·H_2O$	0.01g
Vitamin K_1	0.01g
Selective supplement solution	10.0mL

pH 7.2 ± 0.2 at 25°C

Source: This medium, without selective supplement, is available as a premixed powder from HiMedia.

Selective Supplement Solution:

Composition per 10.0mL:

Gentamicin	0.1mg

Preparation of Selective Supplement Solution: Add gentamicin to distilled/deionized water and bring volume to 10.0mL. Mix thoroughly. Filter sterilize.

Preparation of Medium: Add components, except selective supplement, to distilled/deionized water and bring volume to 990.0mL. Mix thoroughly. Heat with frequent agitation and boil for 1 min to completely dissolve. Autoclave for 15 min at 15 psi pressure–121°C. Cool to 50°–55°C. Add 10.0mL of sterile selective supplement. Mix thoroughly. Pour into sterile Petri dishes or leave in tubes.

Use: For the selection and presumptive identification of the *Bacteroides fragilis* group. For the differentiation of *Bacteroides* species based on the hydrolysis of esculin and presence of catalase. After incubation for 48 hr, bacteria of the *Bacteroides fragilis* group appear as gray, circular, raised colonies larger than 1.0mm. Esculin hydrolysis is indicated by the presence of a blackened zone around the colonies.

Bacteroides Medium

Composition per liter:

Pancreatic digest of casein	27.0g
Yeast extract	3.0g
K_2HPO_4	2.5g
K_2CO_3	2.0g
NaCl	2.0g
Hemin solution	10.0mL
Vitamin K_1 solution	0.2mL

Hemin Solution:

Composition per 100.0mL:

Hemin	1.0g
NaOH (1*N* solution)	20.0mL

Preparation of Hemin Solution: Add hemin to 20.0mL of 1*N* NaOH solution. Mix thoroughly. Bring volume to 100.0mL with distilled/deionized water.

Vitamin K_1 Solution:

Composition per 100.0mL:

Vitamin K_1	1.0g
Ethanol	99.0mL

Preparation of Vitamin K_1 Solution: Add vitamin K_1 to 99.0mL of absolute ethanol. Mix thoroughly.

Preparation of Medium: Add components to distilled/deionized water and bring volume to 1.0L. Mix thoroughly. Distribute into tubes or flasks. Autoclave for 15 min at 15 psi pressure–121°C.

Use: For the cultivation of *Bacteroides asaccharolyticus* and *Bacteroides melaninogenicus*.

Bacteroides nodosus Agar

Composition per liter:

Agar	14.0g
Liver hydrolysate	10.0g
Proteose peptone No. 3	10.0g
Trypsin 1:250	10.0g
NaCl	5.0g
Yeast extract	2.0g
L-Cysteine·HCl solution	2.5mL

pH 7.4 ± 0.2 at 25°C

L-Cysteine·HCl Solution:

Composition per 10.0mL:

L-Cysteine HCl	1.0g

Preparation of L-Cysteine·HCl Solution: Dissolve 1.0g of L-cysteine·HCl in distilled/deionized water and bring volume to 10.0mL. Mix thoroughly. Filter sterilize. Warm to 50°C.

Preparation of Medium: Add components, except agar and L-cysteine·HCl solution, to distilled/deionized water and bring volume to 997.5mL. Mix thoroughly. Adjust pH to 8.5. Gently heat and bring to boiling. Boil for 5 min. Filter and allow to cool to 25°C. Adjust pH to 7.4. Add 14.0g of agar. Gently heat and bring to boiling. Autoclave for 15 min at 15 psi pressure–121°C. Cool to 50°–55°C. Aseptically add 2.5mL of sterile L-cysteine·HCl solution. Mix thoroughly. Pour into sterile Petri dishes or distribute into sterile tubes.

Use: For the cultivation and maintenance of *Bacteroides nodosus*.

Bacteroides vulgatus Medium (LMG Medium 204)

Composition per liter:

Special peptone	23.0g
Agar	15.0g
Glucose	5.0g
NaCl	5.0g
Soluble starch	1.0g
Cysteine hydrochloride	0.3g
Horse blood, sterile defibrinated	50.0mL

pH 7.1 ± 0.2 at 25°C

Preparation of Medium: Add components, except horse blood, to 950.0mL distilled/deionized water and bring volume to 1.0L. Mix thoroughly. Gently heat and bring to boiling. Autoclave for 15 min at 15 psi pressure–121°C. Cool to 45°–50°C. Aseptically add 50.0mL sterile horse blood. Mix thoroughly. Pour into sterile Petri dishes or distribute into sterile tubes.

Use: For the cultivation and maintenance of *Bacteroides vulgatus*.

BAF Agar

Composition per liter:

Glucose	30.0g
Agar	15.0g
Peptone	2.0g
KH_2PO_4	0.5g
$MgSO_4 \cdot 7H_2O$	0.5g
Yeast extract	0.2g
$CaCl_2 \cdot 2H_2O$	100.0mg
$FeCl_3 \cdot 6H_2O$	10.0mg
$MnSO_4$	5.0mg
$ZnSO_4 \cdot 7H_2O$	1.0mg
Folic acid	100.0μg
Inositol	50.0μg
Thiamine·HCl	50.0μg
Biotin	1.0μg

pH 5.8 ± 0.2 at 25°C

Preparation of Medium: Add components to distilled/deionized water and bring volume to 1.0L. Gently heat and bring to boiling. Adjust pH to 5.8. Distribute into tubes or flasks. Autoclave for 15 min at 15 psi pressure–121°C. Pour into sterile Petri dishes or leave in tubes.

Use: For the cultivation of a wide variety of bacteria.

BAGG Broth (Buffered Azide Glucose Glycerol Broth)

Composition per liter:

Pancreatic digest of casein	10.0g
Peptic digest of animal tissue	10.0g
Glucose	5.0g
NaCl	5.0g
K_2HPO_4	4.0g
KH_2PO_4	1.5g
NaN_3	0.5g
Bromcresol Purple	0.015g
Glycerol	5.0mL

pH 6.9 ± 0.2 at 25°C

Source: This medium is available as a premixed powder from BD Diagnostic Systems.

Caution: Sodium azide is toxic. Azides also react with metals and disposal must be highly diluted.

Preparation of Medium: Add 5.0mL of glycerol to 900.0mL of distilled/deionized water. Add remaining components and bring volume to 1.0L. Mix thoroughly. Gently heat and bring to boiling. Distribute into tubes in 10.0mL volumes. Autoclave for 15 min at 10 psi pressure–116°C.

Use: For the cultivation of fecal streptococci from a variety of clinical and nonclinical specimens. It is recommended for qualitative presumptive and confirmatory tests for fecal streptococci.

BAGG Broth Base with Glycerol (Buffered Azide Glucose Glycerol Broth Base)

Composition per liter:

Tryptose	20.0g
Glucose	5.0g

NaCl	5.0g
K_2HPO_4	4.0g
KH_2PO_4	1.5g
NaN_3	0.5g
Bromcresol Purple	0.015g
Glycerol	5.0mL

pH 6.9 ± 0.2 at 25°C

Source: This medium without glycerol is available as a premixed powder from HiMedia.

Caution: Sodium azide is toxic. Azides also react with metals and disposal must be highly diluted.

Preparation of Medium: Add 5.0mL of glycerol to 900.0mL of distilled/deionized water. Add remaining components and bring volume to 1.0L. Mix thoroughly. Gently heat and bring to boiling. Distribute into tubes in 10.0mL volumes. Autoclave for 15 min at 10 psi pressure–115°C.

Use: For the cultivation of fecal streptococci from a variety of clinical and nonclinical specimens. It is recommended for qualitative presumptive and confirmatory tests for fecal streptococci.

BAGG HiVeg Broth Base with Glycerol (Buffered Azide Glucose Glycerol HiVeg Broth Base)

Composition per liter:

Plant hydrolysate No. 1	20.0g
Glucose	5.0g
NaCl	5.0g
K_2HPO_4	4.0g
KH_2PO_4	1.5g
NaN_3	0.5g
Bromcresol Purple	0.015g
Glycerol	5.0mL

pH 6.9 ± 0.2 at 25°C

Source: This medium without glycerol is available as a premixed powder from HiMedia.

Caution: Sodium azide is toxic. Azides also react with metals and disposal must be highly diluted.

Preparation of Medium: Add 5.0mL of glycerol to 900.0mL of distilled/deionized water. Add remaining components and bring volume to 1.0L. Mix thoroughly. Gently heat and bring to boiling. Distribute into tubes in 10.0mL volumes. Autoclave for 15 min at 10 psi pressure–115°C.

Use: For the cultivation of fecal streptococci from a variety of clinical and nonclinical specimens. It is recommended for qualitative presumptive and confirmatory tests for fecal streptococci.

Baird-Parker Agar

Composition per liter:

Agar	17.0g
Glycine	12.0g
Sodium pyruvate	10.0g
Pancreatic digest of casein	10.0g
Beef extract	5.0g
LiCl	5.0g
Yeast extract	1.0g

pH 7.0 ± 0.2 at 25°C

Source: This medium is available as a premixed powder from Oxoid Unipath and BD Diagnostic Systems.

Preparation of Medium: Add components to distilled/deionized water and bring volume to 1.0L. Mix thoroughly. Gently heat and bring to boiling. Autoclave for 15 min at 15 psi pressure–121°C. Cool to 45°–50°C. Pour into sterile Petri dishes.

Use: Used as a base for the preparation of egg-tellurite-glycine-pyruvate agar for the selective isolation and enumeration of coagulase-positive staphylococci from food, skin, soil, air, and other materials.

Baird-Parker Agar

Composition per liter:

Agar	17.0g
Glycine	12.0g
Sodium pyruvate	10.0g
Pancreatic digest of casein	10.0g
Beef extract	5.0g
LiCl	5.0g
Yeast extract	1.0g
Sulfamethazine solution	10.0mL

pH 7.0 ± 0.2 at 25°C

Sulfamethazine Solution:

Composition per 10.0mL:

Sulfamethazine	0.05g

Preparation of Sulfamethazine Solution: Add sulfamethazine to distilled/deionized water and bring volume to 10.0mL. Mix thoroughly. Filter sterilize.

Preparation of Medium: Add components, except sulfamethazine solution, to distilled/deionized water and bring volume to 990.0mL. Mix thoroughly. Gently heat and bring to boiling. Autoclave for 15 min at 15 psi pressure–121°C. Cool to 45°–50°C. Aseptically add sterile sulfamethazine solution. Mix thoroughly. Pour into sterile Petri dishes or distribute into sterile tubes.

Use: Used as a base for the preparation of egg-tellurite-glycine-pyruvate agar for the selective isolation and enumeration of coagulase-positive staphylococci from food, skin, soil, air, and other materials.

Baird-Parker Agar Base with Egg Yolk Tellurite Enrichment

Composition per liter:

Agar	20.0g
Glycine	12.0g
Casein enzymatic hydrolysate	10.0g
Sodium pyruvate	10.0g
Plant extract	5.0g
LiCl	5.0g
Yeast extract	1.0g
Egg yolk tellurite enrichment	50.0mL

pH 7.0 ± 0.2 at 25°C

Source: This medium is available as a premixed powder from HiMedia.

Caution: Lithium chloride is harmful. Avoid bodily contact and inhalation of vapors. On contact with skin wash with plenty of water immediately.

Egg Yolk Tellurite Enrichment:

Composition per 100.0mL:

Chicken egg yolks	10
K_2TeO_3	0.15g
NaCl (0.9% solution)	50.0mL

Preparation of Egg Yolk Tellurite Enrichment: Soak eggs with 1:100 dilution of saturated mercuric chloride solution for 1 min. Crack 11 eggs and separate yolks from whites. Mix egg yolks. Measure 30.0mL of egg yolk emulsion and add to 70.0mL of 0.9% NaCl solution. Mix thoroughly. Add 0.15g K_2TeO_3. Filter sterilize. Warm to 45°–50°C.

Caution: Potassium tellurite is toxic.

Preparation of Medium: Add components, except egg yolk tellurite enrichment, to distilled/deionized water and bring volume to 950.0mL. Mix thoroughly. Gently heat and bring to boiling. Autoclave for 15 min at 15 psi pressure–121°C. Cool to 45°–50°C. Aseptically add 50 mL of egg yolk tellurite enrichment. Mix well. Pour into sterile Petri dishes or sterile tubes.

Use: For the selective isolation and enumeration of coagulase-positive staphylococci.

Baird-Parker Agar Base, HiVeg with Egg Yolk Tellurite Enrichment

Composition per liter:

Agar	20.0g
Glycine	12.0g
Plant hydrolysate	10.0g
Sodium pyruvate	10.0g
Plant extract	5.0g
LiCl	5.0g
Yeast extract	1.0g
Egg yolk tellurite enrichment	50.0mL

pH 7.0 ± 0.2 at 25°C

Source: This medium is available as a premixed powder from HiMedia.

Caution: Lithium chloride is harmful. Avoid bodily contact and inhalation of vapors. On contact with skin wash with plenty of water immediately.

Egg Yolk Tellurite Enrichment:

Composition per 100.0mL:

Chicken egg yolks	10
K_2TeO_3	0.15g
NaCl (0.9% solution)	50.0mL

Preparation of Egg Yolk Tellurite Enrichment: Soak eggs with 1:100 dilution of saturated mercuric chloride solution for 1 min. Crack 11 eggs and separate yolks from whites. Mix egg yolks. Measure 30.0mL of egg yolk emulsion and add to 70.0mL of 0.9% NaCl solution. Mix thoroughly. Add 0.15g K_2TeO_3. Filter sterilize. Warm to 45°–50°C.

Caution: Potassium tellurite is toxic.

Preparation of Medium: Add components, except egg yolk tellurite enrichment, to distilled/deionized water and bring volume to 950.0mL. Mix thoroughly. Gently heat and bring to boiling. Autoclave for 15 min at 15 psi pressure–121°C. Cool to 45°–50°C. Aseptically add 50 mL of egg yolk tellurite enrichment. Mix well. Pour into sterile Petri dishes or sterile tubes.

Use: For the selective isolation and enumeration of coagulase-positive staphylococci.

Baird-Parker Agar, Supplemented

Composition per liter:

Agar	17.0g
Glycine	12.0g
Sodium pyruvate	10.0g
Pancreatic digest of casein	10.0g
Beef extract	5.0g
LiCl	5.0g
Yeast extract	1.0g
RPF supplement	100.0mL

pH 7.0 ± 0.2 at 25°C

RPF Supplement:

Composition per 100.0mL:

Bovine fibrinogen	3.75g
Trypsin inhibitor	25.0mg
K_2TeO_3	25.0mg
Rabbit plasma	25.0mL

Caution: Potassium tellurite is toxic.

Preparation of RPF Supplement: Add components to distilled/deionized water and bring volume to 100.0mL. Mix thoroughly. Filter sterilize.

Preparation of Medium: Add components, except RPF supplement, to distilled/deionized water and bring volume to 900.0mL. Mix thoroughly. Gently heat and bring to boiling. Autoclave for 15 min at 15 psi pressure–121°C. Cool to 45°–50°C. Aseptically add 100.0mL of filter-sterilized RPF supplement. Mix thoroughly but gently. Pour into sterile Petri dishes.

Use: For the selective isolation and enumeration of coagulase-positive staphylococci from food, skin, soil, air, and other materials. For the differentiation and identification of staphylococci on the basis of their ability to coagulate plasma. Colonies surrounded by an opaque zone of coagulated plasma are diagnostic for *Staphylococcus aureus*.

Baird-Parker Egg Yolk Agar (ISO)

Composition per 1050.0mL:

Agar	20.0g
L-Glycine	12.0g
Pancreatic digest of casein	10.0g
Sodium pyruvate	10.0g
Meat extract	5.0g
LiCl	5.0g
Yeast extract	1.0g
Egg yolk tellurite enrichment	50.0mL

pH 7.2 ± 0.2 at 25°C

Source: This medium is available as a premixed powder from Oxoid Unipath.

Caution: Lithium chloride is harmful. Avoid bodily contact and inhalation of vapors. On contact with skin wash with plenty of water immediately.

Egg Yolk Tellurite Enrichment:

Composition per 100.0mL:

Chicken egg yolks	10
K_2TeO_3	0.15g
NaCl (0.9% solution)	50.0mL

Preparation of Egg Yolk Tellurite Enrichment: Soak eggs with 1:100 dilution of saturated mercuric chloride solution for 1 min. Crack 11 eggs and separate yolks from whites. Mix egg yolks. Measure 30.0mL of egg yolk emulsion and add to 70.0mL of 0.9% NaCl solution. Mix thoroughly. Add 0.15g K_2TeO_3. Filter sterilize. Warm to 45°–50°C.

Caution: Potassium tellurite is toxic.

Preparation of Medium: Add components, except egg yolk tellurite enrichment, to distilled/deionized water and bring volume to 950.0mL. Mix thoroughly. Gently heat and bring to boiling. Autoclave for 15 min at 15 psi pressure–121°C. Cool to 45°–50°C. Aseptically add 50 mL of egg yolk tellurite enrichment. Mix well. Pour into sterile Petri dishes or sterile tubes.

Use: For the selective isolation and enumeration of coagulase-positive staphylococci. A selective medium for the isolation and enumeration of coagulase-positive staphylococci from food, with formulation conforming to that recommended in ISO 6888-1:1999.

Baird-Parker Medium (BAM M17)

Composition per liter:

Agar 20.0g
Glycine 12.0g
Sodium pyruvate 10.0g
Pancreatic digest of casein 10.0g
Beef extract 5.0g
LiCl·6H_2O 5.0g
Yeast extract 1.0g
Egg yolk tellurite enrichment 50.0mL

pH 7.0 ± 0.2 at 25°C

Egg Yolk Tellurite Enrichment:

Composition per 100.0mL:

Chicken egg yolks 10
K_2TeO_3 .... 0.15g
NaCl (0.9% solution) 50.0mL

Preparation of Egg Yolk Tellurite Enrichment: Soak eggs with 1:100 dilution of saturated mercuric chloride solution for 1 min. Crack 11 eggs and separate yolks from whites. Mix egg yolks. Measure 30.0mL of egg yolk emulsion and add to 70.0mL of 0.9% NaCl solution. Mix thoroughly. Add 0.15g K_2TeO_3. Filter sterilize. Warm to 45°–50°C.

Caution: Potassium tellurite is toxic.

Source: This medium is available as a premixed powder from BD Diagnostic Systems.

Preparation of Medium: Add components, except EY tellurite enrichment, to distilled/deionized water and bring volume to 950.0mL. Mix thoroughly. Gently heat and bring to boiling. Autoclave for 15 min at 15 psi pressure–121°C. Cool to 48°–50°C. Aseptically add 50.0mL of sterile EY tellurite enrichment. Mix thoroughly. Pour into sterile Petri dishes. The medium must be densely opaque. Dry plates before use. Plates can be stored for up to 5 days at 20–25°C before use.

Use: For the selective isolation and enumeration of coagulase-positive staphylococci from foods.

Baird-Parker Medium (BAM M17)

Composition per liter:

Agar 20.0g
Glycine 12.0g
Sodium pyruvate 10.0g
Pancreatic digest of casein 10.0g
Beef extract 5.0g
LiCl·6H_2O 5.0g
Yeast extract 1.0g
Egg yolk tellurite enrichment 50.0mL

pH 7.0 ± 0.2 at 25°C

Egg Yolk Tellurite Enrichment:

Composition per 100.0mL:

Chicken egg yolks 10
K_2TeO_3 .... 0.15g
NaCl (0.9% solution) 50.0mL

Preparation of Egg Yolk Tellurite Enrichment: Soak eggs with 1:100 dilution of saturated mercuric chloride solution for 1 min. Crack 11 eggs and separate yolks from whites. Mix egg yolks. Measure 30.0mL of egg yolk emulsion and add to 70.0mL of 0.9% NaCl solution. Mix thoroughly. Add 0.15g K_2TeO_3. Filter sterilize. Warm to 45°–50°C.

Caution: Potassium tellurite is toxic.

Source: This medium is available as a premixed powder from BD Diagnostic Systems.

Preparation of Medium: Add components, except egg yolk tellurite enrichment, to distilled/deionized water and bring volume to 950.0mL. Mix thoroughly. Gently heat and bring to boiling. Autoclave for 15 min at 15 psi pressure–121°C. Cool to 48°–50°C. Aseptically add 50.0mL of sterile egg yolk tellurite enrichment. Mix thoroughly. Pour into sterile Petri dishes. The medium must be densely opqaue. Dry plates before use. Plates can be stored for up to 5 days at 20–25°C before use.

Use: For the selective isolation and enumeration of coagulase-positive staphylococci from foods.

Balamuth Medium

Composition per 200.0mL:

Dehydrated egg yolk 36.0g
Dried liver concentrate 1.0g
Rice starch 0.2g
Potassium phosphate buffer, pH 7.5 .... 125.0mL
NaCl solution 125.0mL

pH 7.3 ± 0.2 at 25°C

NaCl Solution

Composition per 200.0mL:

NaCl 1.6g

Preparation of NaCl Solution: Add NaCl to distilled/deionized water and bring volume to 200.0mL. Mix thoroughly.

Potassium Phosphate Buffer, 0.067*M*

Composition per 200.0mL:

K_2HPO_4 (1*M* solution) 8.6mL
KH_2PO_4 (1*M* solution) 4.66mL

Preparation of Potassium Phosphate Buffer: Combine the K_2HPO_4 and KH_2PO_4 solutions. Bring volume to 200.0mL with distilled/deionized water. Adjust pH to 7.5.

Preparation of Medium: Add dehydrated egg yolk to 36.0mL of distilled/deionized water. Add 125.0mL of 0.8% NaCl. Mix thoroughly in a blender. Heat in a covered, double boiler until infusion reaches 80°C and maintain at this temperature for 20 min. Add 20.0mL of distilled/deionized H_2O. Filter through a layer of cheesecloth. To 90–100.0mL of filtrate add 0.8% NaCl solution to bring volume to 125.0mL. Autoclave for 20 min at 15 psi pressure–121°C. Cool to 4°C.

Filter. To filtrate, add an equal volume of 0.067*M* potassium phosphate buffer, pH 7.5. Add 1.0g of dried liver concentrate. Mix thoroughly. Distribute into tubes or flasks in 10.0mL volumes. Autoclave for 20 min at 15 psi pressure–121°C. Prior to inoculation, add 0.01g of rice starch to each tube.

Use: For the cultivation and maintenance of *Entamoeba histolytica* and other intestinal protozoa.

BAM Agar (ATCC Medium 1655)

Composition per liter:

Agar	30.0g
Glucose	5.0g
KH_2PO_4	3.0g
Yeast extract	1.0g
$MgSO_4 \cdot 7H_2O$	0.5g
$CaCl_2 \cdot 2H_2O$	0.25g
$(NH_4)_2SO_4$	0.2g
Trace elements	1.0mL

pH 4.0 ± 0.2 at 25°C

Trace Elements:

Composition per liter:

$CaCl_2 \cdot 2H_2O$	0.66g
$Na_2MoO_4 \cdot 2H_2O$	0.3g
$ZnSO_4 \cdot 7H_2O$	0.18g
$CoCl_2 \cdot 6H_2O$	0.18g
$CuSO_4 \cdot 5H_2O$	0.16g
$MnSO_4 \cdot 4H_2O$	0.15g
H_3BO_3	0.1g

Preparation of Trace Elements: Add components to 1.0L of distilled/deionized water. Mix thoroughly.

Preparation of Medium: Add components, except agar, to distilled/deionized water and bring volume to 800.0mL. Mix thoroughly. Gently heat and bring to boiling. Adjust medium to pH 4.0 with H_2SO_4. Add agar to 200.0mL of distilled/deionized water. Autoclave agar separately to avoid acid hydrolysis. Autoclave for 15 min at 15 psi pressure–121°C. Mix the two solutions together. Pour into sterile Petri dishes or distribute into sterile tubes.

Use: For the cultivation and maintenance of *Bacillus acidoterrestris*.

BAM Broth

Composition per liter:

Glucose	5.0g
KH_2PO_4	3.0g
Yeast extract	1.0g
$MgSO_4 \cdot 7H_2O$	0.5g
$CaCl_2 \cdot 2H_2O$	0.25g
$(NH_4)_2SO_4$	0.2g
Trace elements	1.0mL

pH 4.0 ± 0.2 at 25°C

Trace Elements:

Composition per liter:

$CaCl_2 \cdot 2H_2O$	0.66g
$Na_2MoO_4 \cdot 2H_2O$	0.3g
$ZnSO_4 \cdot 7H_2O$	0.18g
$CoCl_2 \cdot 6H_2O$	0.18g
$CuSO_4 \cdot 5H_2O$	0.16g
$MnSO_4 \cdot 4H_2O$	0.15g
H_3BO_3	0.1g

Preparation of Trace Elements: Add components to 1.0L of distilled/deionized water. Mix thoroughly.

Preparation of Medium: Add components to distilled/deionized water and bring volume to 1.0L. Mix thoroughly. Gently heat and bring to boiling. Adjust medium to pH 4.0 with H_2SO_4. Distribute into tubes or flasks. Autoclave for 15 min at 15 psi pressure–121°C.

Use: For the cultivation and maintenance of *Bacillus acidoterrestris*.

BAM SM Agar (ATCC Medium 1656)

Composition per liter:

Agar	20.0g
Yeast extract	6.0g
Glucose	5.0g
KH_2PO_4	3.0g
$MgSO_4 \cdot 7H_2O$	0.5g
$CaCl_2 \cdot 2H_2O$	0.25g
$(NH_4)_2SO_4$	0.2g
Trace elements	1.0mL

pH 4.0 ± 0.2 at 25°C

Trace Elements:

Composition per liter:

$CaCl_2 \cdot 2H_2O$	0.66g
$Na_2MoO_4 \cdot 2H_2O$	0.3g
$ZnSO_4 \cdot 7H_2O$	0.18g
$CoCl_2 \cdot 6H_2O$	0.18g
$CuSO_4 \cdot 5H_2O$	0.16g
$MnSO_4 \cdot 4H_2O$	0.15g
H_3BO_3	0.1g

Preparation of Trace Elements: Add components to 1.0L of distilled/deionized water. Mix thoroughly.

Preparation of Medium: Add components, except agar, to distilled/deionized water and bring volume to 800.0mL. Mix thoroughly. Gently heat and bring to boiling. Adjust medium to pH 4.0 with H_2SO_4. Add agar to 200.0mL of distilled/deionized water. Autoclave agar separately to avoid acid hydrolysis. Autoclave for 15 min at 15 psi pressure–121°C. Mix the two solutions together. Pour into sterile Petri dishes or distribute into sterile tubes.

Use: For the cultivation and maintenance of *Bacillus cycloheptanicus*.

BAM SM Agar, Modified

Composition per liter:

Agar	30.0g
Glucose	5.0g
KH_2PO_4	3.0g
Yeast extract	1.0g
$MgSO_4 \cdot 7H_2O$	0.5g
$CaCl_2 \cdot 2H_2O$	0.25g
$(NH_4)_2SO_4$	0.2g
Trace elements	1.0mL

pH 4.0 ± 0.2 at 25°C

Trace Elements:

Composition per liter:

$CaCl_2 \cdot 2H_2O$	0.66g
$Na_2MoO_4 \cdot 2H_2O$	0.30g
$ZnSO_4 \cdot 7H_2O$	0.18g

$CoCl_2·6H_2O$	0.18g
$CuSO_4·5H_2O$	0.16g
$MnSO_4·4H_2O$	0.15g
H_3BO_3	0.10g

Preparation of Trace Elements: Add components to 1.0L of distilled/deionized water. Mix thoroughly.

Preparation of Medium: Add components, except agar, to distilled/deionized water and bring volume to 800.0mL. Mix thoroughly. Gently heat and bring to boiling. Adjust medium to pH 4.0 with H_2SO_4. Add agar to 200.0mL of distilled/deionized water. Autoclave agar separately to avoid acid hydrolysis. Autoclave for 15 min at 15 psi pressure–121°C. Mix the two solutions together. Pour into sterile Petri dishes or distribute into sterile tubes.

Use: For the cultivation and maintenance of *Bacillus cycloheptanicus*.

BAM SM Broth

Composition per liter:

Yeast extract	6.0g
Glucose	5.0g
KH_2PO_4	3.0g
$MgSO_4·7H_2O$	0.5g
$CaCl_2·2H_2O$	0.25g
$(NH_4)_2SO_4$	0.2g
Trace elements	1.0mL

pH 4.0 ± 0.2 at 25°C

Trace Elements:

Composition per liter:

$CaCl_2·2H_2O$	0.66g
$Na_2MoO_4·2H_2O$	0.3g
$ZnSO_4·7H_2O$	0.18g
$CoCl_2·6H_2O$	0.18g
$CuSO_4·5H_2O$	0.16g
$MnSO_4·4H_2O$	0.15g
H_3BO_3	0.1g

Preparation of Trace Elements: Add components to 1.0L of distilled/deionized water. Mix thoroughly.

Preparation of Medium: Add components to distilled/deionized water and bring volume to 1.0L. Mix thoroughly. Gently heat and bring to boiling. Adjust medium to pH 4.0 with H_2SO_4. Distribute into tubes or flasks. Autoclave for 15 min at 15 psi pressure–121°C.

Use: For the cultivation and maintenance of *Bacillus cycloheptanicus*.

Bandoni's MYP Medium

Composition per liter:

Agar	15.0g
Malt extract	7.0g
Papaic digest of soybean meal	1.0g
Yeast extract	0.5g

Preparation of Medium: Add components to distilled/deionized water and bring volume to 1.0L. Mix thoroughly. Gently heat and bring to boiling. Distribute into tubes or flasks. Autoclave for 15 min at 15 psi pressure–121°C. Pour into sterile Petri dishes or leave in tubes.

Use: For the cultivation and maintenance of *Coleosporium tussilaginis* and *Cystofilobasidium capitatum*.

Basal Medium (DSMZ Medium 1001)

Composition per liter:

$NaHCO_3$	2.5g
NH_4Cl	0.25g
NaH_2PO_4	0.6g
KCl	1.0g
Iron nitrilotriacetic acid solution	20.0mL
Vitamin mix	10.0mL
Mineral mix	10.0mL
Sodium acetate solution	10.0mL

pH 6.9 ± 0.2 at 25°C

Mineral Mix:

Composition per liter:

$MgSO_4·7H_2O$	3.0g
Nitrilotriacetic acid	1.5g
NaCl	1.0g
$MnSO_4·2H_2O$	0.5g
$ZnCl_2$	0.13g
$CoCl_2·6H_2O$	0.1g
$CaCl_2·2H_2O$	0.1g
$FeSO_4·7H_2O$	0.1g
$Na_2MoO_4·4H_2O$	0.025g
$NaWO_4·2H_2O$	0.025g
$NiCl_2·6H_2O$	0.024g
$CuSO_4·5H_2O$	0.01g
$KAl(SO_4)_2·12H_2O$	0.01g
H_3BO_3	0.01g

Preparation of Mineral Mix: Add nitrilotriacetic acid to 500.0mL of distilled/deionized water. Dissolve by adjusting pH to 6.5 with KOH. Add remaining components. Add distilled/deionized water to 1.0L. Mix thoroughly.

Vitamin Mix:

Composition per liter:

Pyridoxine-HCl	10.0mg
Thiamine-HCl·2H₂O	5.0mg
Riboflavin	5.0mg
Nicotinic acid	5.0mg
D-Ca-pantothenate	5.0mg
p-Aminobenzoic acid	5.0mg
Thioctic acid	5.0mg
Biotin	2.0mg
Folic acid	2.0mg
Vitamin B_{12}	0.1mg

Preparation of Vitamin Mix: Add components to distilled/deionized water and bring volume to 1.0L. Mix thoroughly. Sparge with 80% H_2 + 20% CO_2. Filter sterilize.

Sodium Acetate Solution:

Composition per 100.0mL:

Sodium acetate	13.6g

Preparation of Sodium Acetate Solution: Add sodium acetate to distilled/deionized water and bring volume to 80.0mL with distilled/deionized water. Mix thoroughly. Bring volume to 100.0mL with distilled/deionized water. Sparge with 100% N_2 for 45 min. Seal in bottle. Autoclave for 15 min at 15 psi pressure–121°C.

Iron Nitriloacetic Acid Solution:

Composition per 100.0mL:

$FeCl_3·6H_2O$	13.5g
Sodium nitrilotriacetic acid (NTA)	12.8g
$NaHCO_3$	8.2g

Preparation of Iron Nitriloacetic Acid: Add $NaHCO_3$ to distilled/deionized water and bring volume to 70.0mL with distilled/deionized water. Mix thoroughly. Add NTA. Mix thoroughly. Add $FeCl_3·6H_2O$. Adjust pH to 6.5 using 10*N* NaOH. Bring volume to 100.0mL with distilled/deionized water. Stir for about 15 minutes to allow components to go into solution. Sparge with 100% N_2 for 45 min. Filter sterilize. Aseptically and anoxically dispense into sterile serum bottles.

Preparation of Medium: Add components, except iron nitriloacetic acid solution and sodium acetate solution, to distilled/deionized water and bring volume to 1.0L. Mix thoroughly. Bubble the medium with 80:20 N_2:CO_2 (final pH should be 6.8 to 7.0). Approximately 10.0mL of media (anaerobic culture tube) should be gassed for 5 min in the aqueous phase (bubbled) and the headspace gassed for 1 min e prior to sealing the container. Autoclave for 15 min at 15 psi pressure–121°C. Add electron donor (acetate-final conc. of 10mM-) and electron acceptor (Fe(III)NTA (final conc. of 10*mM*), from sterile, anaerobic stock solutions using a sterile syringe and needle flushed with anaerobic gas. This medium should not be exposed to direct sunlight!

Use: For the cultivation of a *Rhodoferax ferrireducens*.

Basal Mineral Medium

Composition per liter:

NH_4Cl	0.8g
K_2HPO_4	0.7g
$MgSO_4·7H_2O$	0.01g
Disodium EDTA	9.2mg
$FeSO_4·7H_2O$	7.0mg
CaSO4·$2H_2O$	2.0mg
H_3BO_3	0.1mg
$ZnSO_4·7H_2O$	0.1mg
$MnSO_4·4H_2O$	0.02mg
$Co(NO_3)_2$	0.01mg
$NaMoO_4·2H_2O$	0.01mg
$CuSO_4·5H_2O$	0.5μg

Preparation of Medium: Add components to distilled/deionized water and bring volume to 1.0L. Mix thoroughly. Filter sterilize.

Use: For the cultivation of *Beggiatoa* species.

Basal Synthetic Medium

Composition per liter:

L-Glutamic acid	20.0g
$(NH_4)_2SO_4$	4.0g
K_2HPO_4	1.88g
KH_2PO_4	0.57g
$MgSO_4·7H_2O$	0.2g
Salt solution	10.0mL

Salt Solution:

Composition per liter:

$FeCl_3·6H_2O$	0.6g
$MnCl_2·4H_2O$	0.6g
$ZnCl_2$	0.6g
$CuSO_4·5H_2O$	0.6g
$CaCl_2·2H_2O$	0.6g
NaCl	0.6g

Preparation of Salt Solution: Add components to 1.0L of distilled/deionized water. Mix thoroughly.

Preparation of Medium: Add components to distilled/deionized water and bring volume to 1.0L. Mix thoroughly. Gently heat and bring to boiling. Distribute into tubes or flasks. Autoclave for 15 min at 15 psi pressure–121°C.

Use: For the cultivation and maintenance of *Acinetobacter lwoffii*.

Basal Thermophile Medium

Composition per liter:

Solution 1	850.0mL
Solution 2	100.0mL
Solution 3	50.0mL

Solution 1:

Composition per 850.0mL:

Pancreatic digest of casein	10.0g
K_2HPO_4	1.5g
NH_4Cl	0.9g
KH_2PO_4	0.75g
$MgCl_2·6H_2O$	0.2g
Trace elements solution	9.0mL
Wolfe's vitamin solution	5.0mL
Resazurin (0.2% solution)	1.0mL
$FeSO_4·7H_2O$ (10% solution)	0.03mL

Preparation of Solution 1: Add components to distilled/deionized water and bring volume to 850.0mL. Mix thoroughly. Autoclave for 45 min at 15 psi pressure–121°C. Cool to 45°–50°C.

Solution 2:

Composition per 100.0mL:

Yeast extract	3.0g

Preparation of Solution 2: Add yeast extract to distilled/deionized water and bring volume to 100.0mL. Mix thoroughly. Autoclave for 45 min at 15 psi pressure–121°C. Cool to 45°–50°C.

Solution 3:

Composition per 50.0mL:

Glucose	5.0g

Preparation of Solution 3: Add glucose to distilled/deionized water and bring volume to 50.0mL. Mix thoroughly. Autoclave for 45 min at 15 psi pressure–121°C. Cool to 45°–50°C.

Trace Elements Solution:

Composition per liter:

Nitrilotriacetic acid	12.5g
NaCl	1.0g
$FeCl_3·4H_2O$	0.2g
$MnCl_2·4H_2O$	0.1g
$CaCl_2·2H_2O$	0.1g
$ZnCl_2$	0.1g
$CuCl_2$	0.02g
Na_2SeO_3	0.02g
$CoCl_2·6H_2O$	0.017g
H_3BO_3	0.01g
$Na_2MoO_4·2H_2O$	0.01g

Preparation of Trace Elements Solution: Add nitrilotriacetic acid to 100.0mL of distilled/deionized water. Adjust pH to 6.5 with

KOH. Add remaining components and bring volume to 1.0L. Mix thoroughly.

Wolfe's Vitamin Solution:

Composition per liter:

Pyridoxine·HCl	10.0mg
Thiamine·HCl	5.0mg
Riboflavin	5.0mg
Nicotinic acid	5.0mg
Calcium pantothenate	5.0mg
p-Aminobenzoic acid	5.0mg
Thioctic acid	5.0mg
Biotin	2.0mg
Folic acid	2.0mg
Cyanocobalamin	0.1mg

Preparation of Wolfe's Vitamin Solution: Add components to distilled/deionized water and bring volume to 1.0L. Mix thoroughly.

$Na_2S·9H_2O$ Solution:

Composition per 100.0mL:

$Na_2S·9H_2O$	10.0g

Preparation of $Na_2S·9H_2O$ Solution: Add $Na_2S·9H_2O$ to distilled/deionized water and bring volume to 100.0mL. Mix thoroughly. Autoclave for 15 min at 15 psi pressure–121°C.

Preparation of Medium: Aseptically combine solution 1, solution 2, and solution 3 under 100% N_2. Distribute into tubes in 10.0mL volumes under 100% N_2. Immediately prior to inoculation, aseptically add 0.1mL of sterile $Na_2S·9H_2O$ solution to each tube.

Use: For the cultivation and maintenance of *Clostridium* species, *Fervidobacterium nodosum*, and *Thermoanaerobium brockii*.

Base Agar

See: **Antibiotic Medium 2**

Base Agar with Low pH

See: **Antibiotic Medium 8**

Base Cholesterol Medium

Composition per liter:

Casitone	10.0g
Yeast extract	10.0g
Cholesterol, ash free	2.0g
$CaCl_2$	1.0g
Lecithin, type IV	1.0g
Sodium thioglycolate	0.5g
Resazurin	1.0mg

Preparation of Medium: Prepare and dispense medium under 100% N_2. Add cholesterol and lecithin to distilled/deionized water and bring volume to 200.0mL of water. Mix thoroughly. Sparge with 100% N_2 for 10 min. Add other components to distilled/deionized water and bring volume to 800.0mL of water. Mix thoroughly. Combine the two solutions. Adjust pH to 7.5 with KOH. Gently heat and bring to boiling. Continue boiling while sparging with 100% N_2 until the resazurin turns colorless. Cool under 100% N_2. Distribute into tubes or flasks. Autoclave for 15 min at 15 psi pressure–121°C. Mix well after autoclaving.

Use: For the cultivation of *Eubacterium coprostanoligenes*.

Base Layer Agar with Nutrient Overlay Agar

Composition per 2.5L:

Fat substrate	50.0g
Nutrient agar	1.5L
Basal medium	1.0L

Fat Substrate:

Composition:

Fat	50.0g

Preparation of Fat Substrate: Tributyrin, corn oil, soybean oil, any cooking oil, lard, tallow, or triglycerides that do not contain antioxidants or other inhibitory substances may be used. Remove free fatty acids in the fat substrate by dissolving 50.0g of fat substrate in 500.0mL of petroleum ether. Pass the solution through an activated alumina column. Remove the petroleum ether by evaporation on a steam table under 100% N_2. Autoclave for 30 min at 15 psi pressure–121°C. Cool to 50°C.

Nutrient Agar:

Composition per liter:

Agar	15.0g
Pancreatic digest of gelatin	5.0g
Beef extract	3.0g

Preparation of Nutrient Agar: Add components to distilled/deionized water and bring volume to 1.0L. Mix thoroughly. Gently heat while stirring and bring to boiling. Distribute into tubes or flasks. Autoclave for 15 min at 15 psi pressure–121°C. Cool to 45°–50°C.

Source: The medium is available as a premixed powder from BD Diagnostic Systems.

Basal Medium:

Composition per liter:

Agar	15.0g
Victoria Blue B solution	200.0mL

Preparation of Basal Medium: Add agar to 800.0mL of distilled/deionized water. If tributyrin is used as the fat substrate, add agar to 1.0L of distilled/deionized water. Autoclave for 15 min at 15 psi pressure–121°C. Cool to 50°C. If tributyrin is not used as the fat substrate, aseptically add 200.0mL of Victoria Blue B solution. Mix thoroughly.

Victoria Blue B Solution:

Composition per 200.0mL:

Victoria Blue B	0.12g

Preparation of Victoria Blue B Solution: Add the Victoria Blue B to 200.0mL of distilled/deionized water. Mix thoroughly. Filter sterilize. Warm to 50°C.

Preparation of Medium: Aseptically combine 1.0L of sterile basal medium with 50.0g of sterile fat substrate in a warm, sterile blender container. Blend for 1 min until homogenized. Rapidly pour into sterile Petri dishes in 7.0mL volumes. Dry the surface of the plates by partially opening the lids in a laminar flow hood for 15 min. Add dilution of food samples to be tested. When the inoculum is dry, pour nutrient agar as an overlay onto each plate. Use 10–12mL of nutrient agar per plate.

Use: For the isolation, cultivation, and identification of lipolytic microorganisms from food.

Basic Cultivation Medium

Composition per liter:

Yeast extract	10.0g
Glucose	5.0g
$(NH_4)_2PO_4$	1.5g

K_2HPO_4	1.0g
$MgSO_4·7H_2O$	0.2g
$Fe_2(SO_4)_3·5H_2O$	0.01g
$ZnSO_4·7H_2O$	0.002g

pH 7.0 ± 0.2 at 25°C

Preparation of Medium: Add components to distilled/deionized water and bring volume to 1.0L. Mix thoroughly. Distribute into tubes or flasks. Autoclave for 15 min at 15 psi pressure–121°C.

Use: For the cultivation of a wide variety of microorganisms.

Basic Mineral Medium

Composition per liter:

NH_4NO_3	2.5g
$Na_2HPO_4·2H_2O$	1.0g
$MgSO_4·7H_2O$	0.5g
$Fe(SO_4)_3·5H_2O$	0.01g
$Co(NO_3)_2·6H_2O$	0.005g
$CaCl_2·2H_2O$	1.0mg
KH_2PO_4	0.5mg
$MnSO_4·2H_2O$	0.1mg
$(NH_4)_6Mo_7O_{24}·4H_2O$	0.1mg

Preparation of Medium: Add components to distilled/deionized water and bring volume to 1.0L. Mix thoroughly. Distribute into tubes or flasks. Autoclave for 15 min at 15 psi pressure–121°C.

Use: To supply the mineral nutrients necessary for the cultivation of a wide variety of microorganisms. Various carbon sources can be added as sterilized solutions for testing carbon utilization capabilities.

BBE Agar

See: ***Bacteroides*** **Bile Esculin Agar**

BBGS Agar

See: **Bile Salts Brilliant Green Starch Agar**

BC Medium (Medium for *Acetivibrio cellulolyticus*)

Composition per liter:

Cellulose powder	3.0g
$NaHCO_3$	2.0g
Mineral solution 1	75.0mL
Mineral solution 2	75.0mL
Cysteine-sulfide reducing solution	12.8mL
$FeSO_4·7H_2O$ solution	10.0mL
Vitamin mixture	10.0mL
Wolfe's mineral solution	10.0mL
Resazurin (0.1% solution)	1.0mL

pH 7.6 ± 0.2 at 25°C

Caution: This medium contains sodium sulfide and may produce toxic H_2S gas. Prepare in a chemical fume hood.

Mineral Solution 1:

Composition per liter:

K_2HPO_4	3.9g

Preparation of Mineral Solution 1: Add K_2HPO_4 to distilled/deionized water and bring volume to 1.0L. Mix thoroughly.

Mineral Solution 2:

Composition per liter:

NH_4Cl	12.0g
Na_2SO_4	2.5g
KH_2PO_4	2.4g
$MgSO_4·7H_2O$	1.2g
$CaCl_2·2H_2O$	0.8g

Preparation of Mineral Solution 2: Add components to distilled/deionized water and bring volume to 1.0L. Mix thoroughly.

$FeSO_4·7H_2O$ Solution:

Composition per 100.0mL:

$FeSO_4·7H_2O$	0.2g

Preparation of $FeSO_4·7H_2O$ Solution: Dissolve $FeSO_4·7H_2O$ in 100.0mL of distilled/deionized water. Add three drops of concentrated HCl. Mix thoroughly.

Vitamin Mixture:

Composition per liter:

Pyridoxine·HCl	10.0mg
Thiamine·HCl	5.0mg
Cyanocobalamin	5.0mg
Lipoic acid (thioctic acid)	5.0mg
Biotin	2.0mg
p-Aminobenzoic acid	0.5mg

Preparation of Vitamin Mixture: Add components to distilled/deionized water and bring volume to 1.0L. Store below –20°C.

Wolfe's Mineral Solution:

Composition per liter

$MgSO_4·7H_2O$	3.0g
Nitriloacetic acid	1.5g
$MnSO_4·H_2O$	0.5g
NaCl	1.0g
$FeSO_4·7H_2O$	0.1g
$CoCl_2·6H_2O$	0.1g
$CaCl_2$	0.1g
$ZnSO_4·7H_2O$	0.1g
$CuSO_4·5H_2O$	0.01g
$AlK(SO_4)_2·12H_2O$	0.01g
H_3BO_3	0.01g
$Na_2MoO_4·2H_2O$	0.01g

Preparation of Wolfe's Mineral Solution: Add nitrilotriacetic acid to 500.0mL of distilled/deionized water and adjust to pH 6.5 with KOH to dissolve. Bring volume to 1.0L with distilled/deionized water. Add remaining components one at a time.

Cysteine-Sulfide Reducing Solution:

Composition per 200.0mL:

L-Cysteine·HCl·H_2O	2.5g
$Na_2S·9H_2O$	2.5g

Preparation of Cysteine-Sulfide Reducing Solution: Add L-cysteine·HCl·H_2O to 50.0mL of distilled/deionized water. Quickly adjust pH to 10 with fresh 3*N* NaOH and flush under 100% N_2. Add $Na_2S·9H_2O$. Bring volume to 200.0mL with distilled/deionized water. Boil under 100% N_2. Transfer anaerobically to tubes or flasks and stopper. Autoclave for 15 min at 15 psi pressure–121°C.

Preparation of Medium: Add cellulose and $NaHCO_3$ to distilled/deionized water and bring volume to 800.0mL. Add all other components except cysteine-sulfide reducing solution. Heat and boil under 90% N_2 + 10% CO_2. Cool and continue flushing under 90% N_2 + 10% CO_2. The pH should be 7.6 at room temperature; do not adjust. Add 8.0mL of cysteine-sulfide reducing solution. Add 4.8mL more of cysteine-sulfide reducing solution. Distribute anaerobically into tubes in 7.0mL volumes and cap.

Use: For the cultivation and maintenance of *Acetivibrio cellulolyticus*, *Acetivibrio cellulosolvens, Bacteroides cellulosolvens,* and other cellulose-degrading microorganisms.

BC Motility Medium
(*Bacillus cereus* Motility Medium)

Composition per liter:

Pancreatic digest of casein	10.0g
Glucose	5.0g
Agar	3.0g
Na_2HPO_4	2.5g
Yeast extract	2.5g

pH 7.4 ± 0.2 at 25°C

Preparation of Medium: Add components to distilled/deionized water and bring volume to 1.0L. Mix thoroughly. Gently heat and bring to boiling. Distribute into tubes in 2.0mL volumes. Autoclave for 15 min at 15 psi pressure–121°C.

Use: For the cultivation and observation of motility of *Bacillus cereus*.

BC Motility Test HiVeg Medium

Composition per liter:

Plant hydrolysate	10.0g
Glucose	5.0g
Agar	3.0g
Na_2HPO_4	2.5g
Yeast extract	2.5g

pH 7.4 ± 0.2 at 25°C

Source: This medium is available as a premixed powder from HiMedia.

Preparation of Medium: Add components to distilled/deionized water and bring volume to 1.0L. Mix thoroughly. Gently heat and bring to boiling. Distribute into tubes in 2.0mL volumes. Autoclave for 15 min at 15 psi pressure–121°C.

Use: For the cultivation and observation of motility of *Bacillus cereus*.

BCA

See: **Bacterial Cell Agar**

BCG Glucose Agar
(Snyder Test Agar)

Composition per liter:

Agar	20.0g
Glucose	20.0g
Peptic digest of animal tissue	20.0g
NaCl	5.0g
Bromcresol Green	0.02

pH 4.8 ± 0.2 at 25°C

Source: This medium is available as a premixed powder from HiMedia.

Preparation of Medium: Add components to distilled/deionized water and bring volume to 1.0L. Mix thoroughly. Gently heat and bring to boiling. Distribute into tubes in 10.0mL volumes. Autoclave for 15 min at 15 psi pressure–121°C. Do not overheat. Pour into sterile Petri dishes or leave in tubes.

Use: For the cultivation and enumeration of lactobacilli in saliva and indication of dental caries activity.

BCG Glucose HiVeg Agar
(Snyder Test HiVeg Agar)

Composition per liter:

Agar	20.0g
Glucose	20.0g
Plant peptone	20.0g
NaCl	5.0g
Bromcresol Green	0.02

pH 4.8 ± 0.2 at 25°C

Source: This medium is available as a premixed powder from HiMedia.

Preparation of Medium: Add components to distilled/deionized water and bring volume to 1.0L. Mix thoroughly. Gently heat and bring to boiling. Distribute into tubes in 10.0mL volumes. Autoclave for 15 min at 15 psi pressure–121°C. Do not overheat. Pour into sterile Petri dishes or leave in tubes.

Use: For the cultivation and enumeration of lactobacilli in saliva and indication of dental caries activity.

BCM

See: ***Bacillus cereus*** **Medium**

BCM *Bacillus cereus* Group Plating Medium

Composition per liter:

Proprietary

Source: This medium is available from Biosynth International, Inc.

Use: For detection of *Bacillus cereus* in food. The medium contains 5-bromo-4-chloro-3-indoxyl-myoinositol-1-phosphate as a chromogenic substrate, which changes from colorless to turquoise upon enzymatic cleavage. *B. cereus*, *B. mycoides*, *B. thuringiensis*, and *B. weihenstephanensis* secrete phosphatidylinositol phospholipase C and so grow as turquoise colonies with species-specific morphologies.

BCM O157:H7(+) Plating Medium

Composition per liter:

Proprietary

Source: This medium is available from Biosynth International, Inc.

Use: For detection of this highly pathogenic EHEC serovar BCM *O157:H7(+)*.

BCP Azide Broth
(Bromcresol Purple Azide Broth)

Composition per liter:

Casein peptone	10.0g
Yeast extract	10.0g
D-Glucose	5.0g
NaCl	5.0g
K_2HPO_4	2.7g
KH_2PO_4	2.7g
NaN_3	0.5g
Bromcresol Purple	0.032g

pH 6.9 ± 0.2 at 25°C

Caution: Sodium azide is toxic. Azides also react with metals and disposal must be highly diluted.

Preparation of Medium: Add components to distilled/deionized water to 1.0L. Mix thoroughly. Gently heat to boiling. Distribute into tubes or flasks. Autoclave for 15 min at 15 psi pressure–121°C.

Use: For use in the confirmation test for the presence of fecal streptococci in water and wastewater.

BCP Broth

See: **Bromcresol Purple Dextrose Broth**

BCP D Agar
(Bromcresol Purple Deoxycholate Agar)

Composition per liter:

Agar	25.0g
Lactose	10.0g
Sucrose	10.0g
Pancreatic digest of casein	7.5g
Thiopeptone	7.5g
NaCl	5.0g
Yeast extract	2.0g
Sodium citrate	2.0g
Sodium deoxycholate	1.0g
Bromcresol Purple	0.02g

pH 7.2 ± 0.2 at 25°C

Preparation of Medium: Add components to distilled/deionized water and bring volume to 1.0L. Mix thoroughly. Gently heat and bring to boiling. Pour into sterile Petri dishes without sterilization. Do not autoclave. Use the same day.

Use: For the isolation, cultivation, and differentiation of Gram-negative enteric bacilli from clinical and nonclinical specimens. For the isolation, cultivation, and identification of microorganisms from fecal specimens. For the isolation and cultivation of *Salmonella, Shigella,* and other nonlactose- and nonsucrose-fermenting microorganisms. Nonlactose/nonsucrose fermenting microorganisms appear as colorless or blue colonies. Lactose/sucrose-fermenting microorganisms, such as coliform bacteria, appear as yellow-opaque white colonies surrounded by a zone of precipitated deoxycholate.

BCP DCLS Agar
(Bromcresol Purple Deoxycholate Citrate Lactose Sucrose Agar)

Composition per liter:

Agar	14.0g
Sodium citrate	10.0g
Lactose	7.5g
Sucrose	7.5g
Pancreatic digest of casein	7.5g
Peptone	7.5g
NaCl	5.0g
$Na_2S_2O_3 \cdot 5H_2O$	5.0g
Yeast extract	3.0g
Meat extract	3.0g
Sodium deoxycholate	2.5g
Bromcresol Purple	0.02g

pH 7.2 ± 0.2 at 25°C

Preparation of Medium: Add components to distilled/deionized water and bring volume to 1.0L. Mix thoroughly. Gently heat and bring to boiling. Pour into sterile Petri dishes without sterilization. Do not autoclave. Use the same day.

Use: For the differential isolation of Gram-negative enteric bacilli from clinical and nonclinical specimens. For the isolation and identification of microorganisms from fecal specimens. For the isolation of *Salmonella, Shigella,* and other nonlactose- and nonsucrose-fermenting microorganisms. Nonlactose/nonsucrose-fermenting microorganisms appear as colorless or blue colonies. Lactose/sucrose-fermenting microorganisms, such as coliform bacteria, appear as yellow-opaque white colonies surrounded by a zone of precipitated deoxycholate.

BCP MS G Agar

See: **Bromocresol Purple Milk Solids Glucose Agar**

BCYE Agar with Cysteine
(BCYE Alpha Base)
(Buffered Charcoal Yeast Extract Agar)

Composition per liter:

Agar	15.0g
Yeast extract	10.0g
ACES buffer (2-[(2-amino-2-oxoethyl)-amino]-ethane sulfonic acid)	10.0g
Charcoal, activated	2.0g
α-Ketoglutarate	1.0g
L-Cysteine·HCl·H_2O	0.4g
$Fe_4(P_2O_7)_3 \cdot 9H_2O$	0.25g
L-Cysteine solution	4.0mL

pH 6.9 ± 0.2 at 25°C

Source: This medium is available as a premixed powder from BD Diagnostic Systems.

L-Cysteine Solution:

Composition per 10.0mL:

L-cysteine·HCl·H_2O	0.4g

Preparation of L-Cysteine Solution: Add L-cysteine·HCl·H_2O to distilled/deionized water and bring volume to 10.0mL. Mix thoroughly. Filter sterilize.

Preparation of Medium: Add components, except L-cysteine solution, to distilled/deionized water and bring volume to 1.0L. Mix thoroughly. Adjust medium to pH 6.9 with 1*N* KOH. Heat gently and bring to boil for 1 min. Autoclave for 15 min at 15 psi pressure–121°C. Cool to 50°–55°C. Aseptically add 4.0mL of L-cysteine solution. Mix thoroughly. Pour into sterile Petri dishes with constant agitation to keep charcoal in suspension.

Use: For the isolation, cultivation, and maintenance of *Legionella pneumophila* and other *Legionella* species from environmental and clinical specimens.

BCYE Differential Agar
(Buffered Charcoal Yeast Extract Differential Agar)

Composition per liter:

Agar	15.0g
Yeast extract	10.0g
ACES buffer (2-[(2-amino-2-oxoethyl)-amino]-ethane sulfonic acid)	10.0g
Charcoal, activated	2.0g
α-Ketoglutarate	1.0g
L-Cysteine·HCl·H_2O	0.4g
$Fe_4(P_2O_7)_3 \cdot 9H_2O$	0.25g

Bromcresol Purple 0.01g
Bromthymol Blue 0.01g

pH 6.9 ± 0.2 at 25°C

Source: This medium is available as a premixed powder from BD Diagnostic Systems.

Preparation of Medium: Add components, except L-cysteine·HCl·H_2O, to distilled/deionized water and bring volume to 1.0L. Mix thoroughly. Adjust medium to pH 6.9 with 1*N* KOH. Heat gently and bring to boiling for 1 min. Autoclave for 15 min at 15 psi pressure–121°C. Cool to 50°–55°C. Add 4.0mL of a 10% solution of L-cysteine·HCl·H_2O that has been filter sterilized. Mix thoroughly. Pour into sterile Petri dishes with constant agitation to keep charcoal in suspension.

Use: For the isolation, cultivation, and maintenance of *Legionella pneumophila* and other *Legionella* species from environmental and clinical specimens. For the presumptive differential identification of *Legionella* species based on colony color and morphology. *Legionella pneumophila* appears as light blue/green colonies. *Legionella micdadei* appears as blue/gray or dark blue colonies.

BCYE Medium, Diphasic Blood Culture (Buffered Charcoal Yeast Extract Medium, Diphasic Blood Culture)

Composition per liter:

Agar phase 1.0L
Broth phase 1.0L

pH 6.9 ± 0.2 at 25°C

Agar Phase:

Composition per liter:

Agar 20.0g
ACES buffer (2-[(2-amino-2-oxoethyl)-amino]-ethane sulfonic acid) 10.0g
Yeast extract 10.0g
Charcoal, activated, acid washed 4.0g
KOH 2.8g
α-Ketoglutarate 1.0g
L-Cysteine·HCl·H_2O solution 10.0mL
$Fe_4(P_2O_7)_3$·$9H_2O$ solution 10.0mL

L-Cysteine·HCl·H_2O Solution:

Composition per 10.0mL:

L-Cysteine·HCl·H_2O 0.4g

Preparation of L-Cysteine·HCl·H_2O Solution: Add L-cysteine·HCl·H_2O to distilled/deionized water and bring volume to 10.0mL. Mix thoroughly. Filter sterilize.

$Fe_4(P_2O_7)_3$·$9H_2O$ Solution:

Composition per 10.0mL:

$Fe_4(P_2O_7)_3$·$9H_2O$ 0.25g

Preparation of $Fe_4(P_2O_7)_3$·$9H_2O$ Solution: Add $Fe_4(P_2O_7)_3$·$9H_2O$ to distilled/deionized water and bring volume to 10.0mL. Mix thoroughly. Filter sterilize.

Preparation of Agar Phase: Add components, except L-cysteine·HCl·H_2O solution and $Fe_4(P_2O_7)_3$ solution, to distilled/deionized water and bring volume to 980.0mL. Mix thoroughly. Adjust medium to pH 6.9 with 1*N* KOH. Heat gently and bring to boiling for 1 min. Autoclave for 15 min at 15 psi pressure–121°C. Cool to 50°–55°C. Aseptically add the L-cysteine·HCl·H_2O solution and $Fe_4(P_2O_7)_3$·$9H_2O$ solution. Mix thoroughly.

Broth Phase:

Composition per liter:

ACES buffer (2-[(2-amino-2-oxoethyl)-amino]-ethane sulfonic acid) 10.0g
Yeast extract 10.0g
Charcoal, activated, acid washed 4.0g
KOH 2.4g
α-Ketoglutarate 1.0g
Sodium polyaneolsulfonate 0.3g
L-Cysteine·HCl·H_2O solution 10.0mL
$Fe_4(P_2O_7)_3$·$9H_2O$ solution 10.0mL

L-Cysteine·HCl·H_2O Solution:

Composition per 10.0mL:

L-Cysteine·HCl·H_2O 0.4g

Preparation of L-Cysteine·HCl·H_2O Solution: Add L-cysteine·HCl·H_2O to distilled/deionized water and bring volume to 10.0mL. Mix thoroughly. Filter sterilize.

$Fe_4(P_2O_7)_3$·$9H_2O$ Solution:

Composition per 10.0mL:

$Fe_4(P_2O_7)_3$·$9H_2O$ 0.25g

Preparation of $Fe_4(P_2O_7)_3$·$9H_2O$ Solution: Add $Fe_4(P_2O_7)_3$·$9H_2O$ to distilled/deionized water and bring volume to 10.0mL. Mix thoroughly. Filter sterilize.

Preparation of Broth Phase: Add components, except L-cysteine·HCl·H_2O solution and $Fe_4(P_2O_7)_3$ solution, to distilled/deionized water and bring volume to 980.0mL. Mix thoroughly. Adjust medium to pH 6.9 with 1*N* KOH. Heat gently and bring to boiling for 1 min. Autoclave for 15 min at 15 psi pressure–121°C. Cool to 50–55°C. Aseptically add the cysteine·HCl·H_2O solution and $Fe_4(P_2O_7)_3$·$9H_2O$ solution. Mix thoroughly.

Preparation of Medium: Aseptically distribute cooled sterile agar phase into sterile blood culture bottles in 100.0mL volumes. Allow bottles to cool in a slanted position. Aseptically add 50.0mL of sterile broth phase to each blood culture bottle.

Use: For the isolation and cultivation of *Legionella pneumophila* and other *Legionella* species from blood samples.

BCYE Selective Agar with CCVC (Buffered Charcoal Yeast Extract Selective Agar with Cephalothin, Colistin, Vancomycin, and Cycloheximide)

Composition per 1014.0mL:

Agar 15.0g
Yeast extract 10.0g
ACES buffer (2-[(2-amino-2-oxoethyl)-amino]-ethane sulfonic acid) 10.0g
Charcoal, activated 2.0g
α-Ketoglutarate 1.0g
$Fe_4(P_2O_7)_3$·$9H_2O$ 0.25g
Antibiotic solution 10.0mL
Cysteine·HCl·H_2O solution 4.0mL

pH 6.9 ± 0.2 at 25°C

Source: This medium is available as a premixed powder from BD Diagnostic Systems.

L-Cysteine·HCl·H_2O Solution:

Composition per 10.0mL:

L-Cysteine·HCl·H_2O 1.0g

Preparation of L-Cysteine·HCl·H_2O Solution: Add L-cysteine·HCl·H_2O to distilled/deionized water and bring volume to 10.0mL. Mix thoroughly. Filter sterilize.

Antibiotic Solution:
Composition per 10.0mL:

Cycloheximide	80.0mg
Colistin	16.0mg
Cephalothin	4.0mg
Vancomycin	0.5mg

Preparation of Antibiotic Solution: Add components to distilled/deionized water and bring volume to 10.0mL. Mix thoroughly. Filter sterilize.

Caution: Cycloheximide is toxic. Avoid skin contact or aerosol formation and inhalation.

Preparation of Medium: Add components, except L-cysteine and antibiotic solutions, to distilled/deionized water and bring volume to 1.0L. Mix thoroughly. Adjust medium to pH 6.9 with 1*N* KOH. Heat gently and bring to boil for 1 min. Autoclave for 15 min at 15 psi pressure–121°C. Cool to 50°–55°C. Add 4.0mL of L-cysteine·HCl·H_2O solution and 10.0mL of sterile antibiotic solution. Mix thoroughly. Pour into sterile Petri dishes with constant agitation to keep charcoal in suspension.

Use: For the isolation, cultivation, and maintenance of *Legionella pneumophila* and other *Legionella* species from environmental and clinical specimens. For the selective recovery of *Legionella pneumophila* while reducing contaminating microorganisms from environmental water samples.

BCYE Selective Agar with GPVA (Buffered Charcoal Yeast Extract Selective Agar with Glycine, Polymyxin B, Vancomycin, and Anisomycin)

Composition per 1014.0mL:

Agar	15.0g
Yeast extract	10.0g
ACES buffer (2-[(2-amino-2-oxoethyl)-amino]-ethane sulfonic acid)	10.0g
Charcoal, activated	2.0g
α-Ketoglutarate	1.0g
$Fe_4(P_2O_7)_3·9H_2O$	0.25g
Antibiotic solution	10.0mL
L-Cysteine·HCl·H_2O solution	4.0mL

pH 6.9 ± 0.2 at 25°C

L-Cysteine·HCl·H_2O Solution:
Composition per 10.0mL:

L-Cysteine·HCl·H_2O	1.0g

Preparation of L-Cysteine·HCl·H_2O Solution: Add L-cysteine·HCl·H_2O to distilled/deionized water and bring volume to 10.0mL. Mix thoroughly. Filter sterilize.

Antibiotic Solution:
Composition per 10.0mL:

Glycine	3.0g
Anisomycin	0.08g
Vancomycin	5.0mg
Polymyxin B	100,000U

Preparation of Antibiotic Solution: Add components to distilled/deionized water and bring volume to 10.0mL. Mix thoroughly. Filter sterilize.

Preparation of Medium: Add components, except L-cysteine·HCl·H_2O solution and antibiotic solution, to distilled/deionized water and bring volume to 1.0L. Mix thoroughly. Adjust medium to pH 6.9 with 1*N* KOH. Heat gently and bring to boil for 1 min. Autoclave for 15 min at 15 psi pressure–121°C. Cool to 50°–55°C. Add 4.0mL of L-cysteine·HCl·H_2O solution and 10.0mL of sterile antibiotic solution. Mix thoroughly. Pour into sterile Petri dishes with constant agitation to keep charcoal in suspension.

Use: For the isolation, cultivation, and maintenance of *Legionella pneumophila* and other *Legionella* species from environmental and clinical specimens. For the selective recovery of *Legionella pneumophila* while reducing contaminating microorganisms from potable water samples.

BCYE Selective Agar with GVPC (Buffered Charcoal Yeast Extract Selective Agar with Glycine, Vancomycin, Polymyxin B, and Cycloheximide)

Composition per 1014.0mL:

Agar	15.0g
Yeast extract	10.0g
ACES buffer (2-[(2-amino-2-oxoethyl)-amino]-ethane sulfonic acid)	10.0g
Charcoal, activated	2.0g
α-Ketoglutarate	1.0g
$Fe_4(P_2O_7)_3·9H_2O$	0.25g
Antibiotic solution	10.0mL
L-Cysteine·HCl·H_2O solution	4.0mL

pH 6.9 ± 0.2 at 25°C

Source: This medium is available as a premixed powder from Oxoid Unipath.

L-Cysteine·HCl·H_2O Solution:
Composition per 10.0mL:

L-Cysteine·HCl·H_2O	1.0g

Preparation of L-Cysteine·HCl·H_2O Solution: Add L-cysteine·HCl·H_2O to distilled/deionized water and bring volume to 10.0mL. Mix thoroughly. Filter sterilize.

Antibiotic Solution:
Composition per 10.0mL:

Glycine	3.0g
Cycloheximide	0.08g
Vancomycin	1.0mg
Polymyxin B	79,200U

Preparation of Antibiotic Solution: Add components to distilled/deionized water and bring volume to 10.0mL. Mix thoroughly. Filter sterilize.

Caution: Cycloheximide is toxic. Avoid skin contact or aerosol formation and inhalation.

Preparation of Medium: Add components, except L-cysteine·HCl·H_2O solution and antibiotic solution, to distilled/deionized water and bring volume to 1.0L. Mix thoroughly. Adjust medium to pH 6.9 with 1*N* KOH. Heat gently and bring to boil for 1 min. Autoclave for 15 min at 15 psi pressure–121°C. Cool to 50°–55°C. Add 4.0mL of L-cysteine·HCl·H_2O solution and 10.0mL of sterile antibiotic solution. Mix thoroughly. Pour into sterile Petri dishes with constant agitation to keep charcoal in suspension.

Use: For the isolation, cultivation, and maintenance of *Legionella pneumophila* and other *Legionella* species from environmental and clinical specimens. For the selective recovery of *Legionella pneumo-*

phila while reducing contaminating microorganisms from potable water samples.

BCYE Selective Agar with PAC (Buffered Charcoal Yeast Extract Selective Agar with Polymyxin B, Anisomycin, and Cefamandole)

Composition per 1014.0mL:

Agar	15.0g
Yeast extract	10.0g
ACES buffer (2-[(2-amino-2-oxoethyl)-amino]-ethane sulfonic acid)	10.0g
Charcoal, activated	2.0g
α-Ketoglutarate	1.0g
$Fe_4(P_2O_7)_3 \cdot 9H_2O$	0.25g
Antibiotic solution	10.0mL
L-Cysteine·HCl·H_2O solution	4.0mL

pH 6.9 ± 0.2 at 25°C

Source: This medium is available as a premixed powder from BD Diagnostic Systems.

L-Cysteine·HCl·H_2O Solution:

Composition per 10.0mL:

L-Cysteine·HCl·H_2O	1.0g

Preparation of L-Cysteine·HCl·H_2O Solution: Add L-cysteine·HCl·H_2O to distilled/deionized water and bring volume to 10.0mL. Mix thoroughly. Filter sterilize.

Antibiotic Solution:

Composition per 10.0mL:

Polymyxin B	80,000 U
Anisomycin	80.0mg
Cefamandole	2.0mg

Preparation of Antibiotic Solution: Add components to distilled/deionized water and bring volume to 10.0mL. Mix thoroughly. Filter sterilize.

Preparation of Medium: Add components, except L-cysteine·HCl·H_2O solution and antibiotic solution, to distilled/deionized water and bring volume to 1.0L. Mix thoroughly. Adjust medium to pH 6.9 with 1*N* KOH. Heat gently and bring to boiling for 1 min. Autoclave for 15 min at 15 psi pressure–121°C. Cool to 50°–55°C. Add 4.0mL of L-cysteine·HCl·H_2O solution and 10.0mL of sterile antibiotic solution. Mix thoroughly. Pour into sterile Petri dishes with constant agitation to keep charcoal in suspension.

Use: For the isolation, cultivation, and maintenance of *Legionella pneumophila* and other *Legionella* species from environmental and clinical specimens. For the selective recovery of *Legionella pneumophila* while reducing contaminating microorganisms from potable water samples.

BCYE Selective Agar with PAV (Buffered Charcoal Yeast Extract Selective Agar with Polymyxin B, Anisomicin, and Vancomycin) (Wadowsky–Yee Medium)

Composition per 1014.0mL:

Agar	15.0g
Yeast extract	10.0g
ACES buffer (2-[(2-amino-2-oxoethyl)-amino]-ethane sulfonic acid)	10.0g
Charcoal, activated	2.0g
α-Ketoglutarate	1.0g
$Fe_4(P_2O_7)_3 \cdot 9H_2O$	0.25g
Antibiotic solution	10.0mL
L-Cysteine·HCl·H_2O solution	4.0mL

pH 6.9 ± 0.2 at 25°C

Source: This medium is available as a premixed powder from BD Diagnostic Systems.

L-Cysteine·HCl·H_2O Solution:

Composition per 10.0mL:

L-Cysteine·HCl·H_2O	1.0g

Preparation of L-Cysteine·HCl·H_2O Solution: Add L-cysteine·HCl·H_2O to distilled/deionized water and bring volume to 10.0mL. Mix thoroughly. Filter sterilize.

Antibiotic Solution:

Composition per 10.0mL:

Anisomycin	80.0mg
Vancomycin	0.5mg
Polymyxin B	40,000 U

Preparation of Antibiotic Solution: Add components to distilled/deionized water and bring volume to 10.0mL. Mix thoroughly. Filter sterilize.

Preparation of Medium: Add components, except L-cysteine and antibiotic solution, to distilled/deionized water and bring volume to 1.0L. Mix thoroughly. Adjust medium to pH 6.9 with 1*N* KOH. Heat gently and bring to boil for 1 min. Autoclave for 15 min at 15 psi pressure–121°C. Cool to 50°–55°C. Add 4.0mL of L-cysteine·HCl·H_2O solution and 10.0mL of sterile antibiotic solution. Mix thoroughly. Pour into sterile Petri dishes with constant agitation to keep charcoal in suspension.

Use: For the isolation, cultivation, and maintenance of *Legionella pneumophila* and other *Legionella* species from environmental and clinical specimens. For the selective recovery of *Legionella pneumophila* while reducing contaminating microorganisms from potable water samples.

BCYEα Agar, Modified

See: ***Legionella*** **Agar Base**

BCYEα with Alb (Buffered Charcoal Yeast Extract Agar with Albumin)

Composition per liter:

Agar	15.0g
Yeast extract	10.0g
ACES buffer (2-[(2-amino-2-oxoethyl)-amino]-ethane sulfonic acid)	10.0g
Charcoal, activated	2.0g
α-Ketoglutarate	1.0g
Bovine serum albumin solution	10.0mL
L-Cysteine·HCl·H_2O solution	10.0mL
$Fe_4(P_2O_7)_3 \cdot 9H_2O$ solution	10.0mL

pH 6.9 ± 0.2 at 25°C

Bovine Serum Albumin Solution:

Composition per 10.0mL:

Bovine serum albumin	0.1g

Preparation of Bovine Serum Albumin Solution: Add bovine serum albumin to distilled/deionized water and bring volume to 10.0mL. Mix thoroughly. Filter sterilize.

L-Cysteine·HCl·H_2O Solution:

Composition per 10.0mL:

L-Cysteine·HCl·H_2O 0.4g

Preparation of L-Cysteine·HCl·H_2O Solution: Add L-cysteine·HCl·H_2O to distilled/deionized water and bring volume to 10.0mL. Mix thoroughly. Filter sterilize.

$Fe_4(P_2O_7)_3$·9H_2O Solution:

Composition per 10.0mL:

$Fe_4(P_2O_7)_3$·9H_2O 0.25g

Preparation of $Fe_4(P_2O_7)_3$·9H_2O Solution: Add $Fe_4(P_2O_7)_3$·9H_2O to distilled/deionized water and bring volume to 10.0mL. Mix thoroughly. Filter sterilize.

Preparation of Medium: Add components—except $Fe_4(P_2O_7)_3$·9H_2O solution, L-cysteine·HCl·H_2O solution, and bovine serum albumin solution—to distilled/deionized water and bring volume to 970.0mL. Mix thoroughly. Adjust medium to pH 6.9 with 1*N* KOH. Heat gently and bring to boiling for 1 min. Autoclave for 15 min at 15 psi pressure–121°C. Cool to 50°–55°C. Aseptically add 10.0mL of sterile bovine serum albumin solution, the $Fe_4(P_2O_7)_3$·9H_2O solution, and the L-cysteine·HCl·H_2O solution. Mix thoroughly. Pour into sterile Petri dishes with constant agitation to keep charcoal in suspension.

Use: For the isolation, cultivation, and maintenance of *Legionella pneumophila* and other *Legionella* species from environmental and clinical specimens.

BCYEα without L-Cysteine (Buffered Charcoal Yeast Extract Agar without L-Cysteine)

Composition per liter:

Agar 15.0g
Yeast extract 10.0g
ACES buffer (2-[(2-amino-2-oxoethyl)-amino]-ethane sulfonic acid) 10.0g
Charcoal, activated 2.0g
α-Ketoglutarate 1.0g
$Fe_4(P_2O_7)_3$·9H_2O solution 10.0mL

pH 6.9 ± 0.2 at 25°C

$Fe_4(P_2O_7)_3$·9H_2O Solution:

Composition per 10.0mL:

$Fe_4(P_2O_7)_3$·9H_2O 0.25g

Preparation of $Fe_4(P_2O_7)_3$·9H_2O Solution: Add $Fe_4(P_2O_7)_3$·9H_2O to distilled/deionized water and bring volume to 10.0mL. Mix thoroughly. Filter sterilize.

Preparation of Medium: Add components, except $Fe_4(P_2O_7)_3$·9H_2O solution, to distilled/deionized water and bring volume to 990.0mL. Mix thoroughly. Adjust medium to pH 6.9 with 1*N* KOH. Heat gently and bring to boiling for 1 min. Autoclave for 15 min at 15 psi pressure–121°C. Cool to 50°–55°C. Aseptically add 10.0mL of sterile $Fe_4(P_2O_7)_3$·9H_2O solution. Mix thoroughly. Pour into sterile Petri dishes with constant agitation to keep charcoal in suspension.

Use: For the isolation, cultivation, and maintenance of *Legionella pneumophila* and other *Legionella* species from environmental and clinical specimens.

BCYT

See: Methanosarcina **Medium**

Bdellovibrio Medium

Composition per Petri dish:

Base layer agar 10.0mL
Semisolid agar 10.0mL
Host medium 1.0mL

Host Medium:

Composition per liter:

Yeast extract 3.0g
Peptone 0.6g

Preparation of Host Medium: Add components to distilled/deionized water and bring volume to 1.0L. Mix thoroughly. Adjust pH to 7.2. Distribute into tubes in 10.0mL volumes. Autoclave for 15 min at 15 psi pressure–121°C.

Base Layer Agar:

Composition per liter:

Agar 19.0g
Yeast extract 3.0g
Peptone 0.6g

Preparation of Base Layer Agar: Add components to distilled/deionized water and bring volume to 1.0L. Mix thoroughly. Gently heat and bring to boiling. Adjust pH to 7.2. Distribute into tubes in 10.0mL volumes. Autoclave for 15 min at 15 psi pressure–121°C.

Semisolid Agar:

Composition per liter:

Agar 6.0g
Yeast extract 3.0g
Peptone 0.6g

Preparation of Semisolid Agar: Add components to distilled/deionized water and bring volume to 1.0L. Mix thoroughly. Gently heat and bring to boiling. Adjust pH to 7.2. Distribute into tubes in 10.0mL volumes. Autoclave for 15 min at 15 psi pressure–121°C.

Preparation of Medium: Inoculate appropriate bacterial host into 10.0mL of host medium. Hosts include *Erwinia amylovora, Escherichia coli, Serratia marcescens*, or *Pseudomonas putida*. Incubate host culture for 24–48 hr at 30°C. Melt the base layer agar and semisolid agar. Pour the base layer agar into a sterile Petri dish. Allow base layer agar to solidify. Cool the semisolid agar to 40°–45°C. Add 1.0mL of the previously grown host culture. Mix thoroughly. Pour over the solidified base layer agar.

Use: For the cultivation of *Bdellovibrio bacteriovorus* and *Bdellovibrio starrii.*

B.D.G. Broth, Hajna

Composition per liter:

Tryptose 20.0g
Glucose 5.0g
NaCl 5.0g
K_2HPO_4 .... 4.0g
KH_2PO_4 .... 1.5g
Sodium deoxycholate 0.1g

pH 7.0 ± 0.2 at 25°C

Source: This medium is available from HiMedia.

Preparation of Medium: Add components to distilled/deionized water and bring volume to 1.0L. Mix thoroughly. Distribute into tubes with inverted Durham tubes. Autoclave for 15 min at 15 psi pressure–121°C.

Use: For the selective enrichment and cultivation of enteric bacilli from food and in treated drinking water.

Bean Agar

Composition per liter:

Dry white beans	250.0g
Agar	20.0g

Preparation of Medium: Soak beans in 500.0mL of distilled/deionized water for 12 hr. Autoclave for 20 min at 15 psi pressure–121°C. Filter broth through cotton. Bring volume of filtrate to 1.0L with distilled/deionized water. Add 20.0g of agar to the filtrate. Gently heat and bring to boiling. Distribute into tubes or flasks. Autoclave for 15 min at 15 psi pressure–121°C. Pour into sterile Petri dishes or leave in tubes.

Use: For the cultivation and maintenance of *Arthroderma melis* and *Rhynchosporium secalis.*

Beef Extract Agar

Composition per liter:

Agar	15.0g
Peptone	5.0g
Beef extract	3.0g

pH 7.4 ± 0.2 at 25°C

Preparation of Medium: Add components to distilled/deionized water and bring volume to 1.0L. Mix thoroughly. Heat gently and bring to boiling. Distribute into tubes or flasks. Autoclave for 15 min at 15 psi pressure–121°C. Pour into Petri dishes or leave in tubes.

Use: For the cultivation and maintenance of a wide variety of microorganisms. Recommended for the culture of microorganisms from milk and water.

Beef Extract Agar (ATCC Medium 225)

Composition per liter:

Agar	25.0g
Beef extract	10.0g
Peptone	10.0g
NaCl	5.0g

pH 7.2 ± 0.2 at 25°C

Preparation of Medium: Add components to distilled/deionized water and bring volume to 1.0L. Mix thoroughly. Heat gently and bring to boiling. Distribute into tubes or flasks. Autoclave for 15 min at 15 psi pressure–121°C. Pour into Petri dishes or leave in tubes.

Use: For the cultivation and maintenance of a wide variety of microorganisms, including *Alcaligenes* species, *Pseudomonas aeruginosa,* and *Bacillus sphaericus*.

Beef Extract Agar, HiVeg

Composition per liter:

Agar	15.0g
Plant peptone	10.0g
NaCl	5.0g
Plant extract	3.0g

pH 7.6 ± 0.2 at 25°C

Source: This medium is available as a premixed powder from HiMedia.

Preparation of Medium: Add components to distilled/deionized water and bring volume to 1.0L. Mix thoroughly. Heat gently and bring to boiling. Distribute into tubes or flasks. Autoclave for 15 min at 15 psi pressure–121°C. Pour into Petri dishes or leave in tubes.

Use: For the cultivation and maintenance of a wide variety of microorganisms, including *Alcaligenes* species, *Pseudomonas aeruginosa,* and *Bacillus sphaericus*.

Beef Extract Broth

Composition per liter:

Peptone	5.0g
Beef extract	3.0g

pH 7.4 ± 0.2 at 25°C

Preparation of Medium: Add components to distilled/deionized water and bring volume to 1.0L. Mix thoroughly. Heat gently and bring to boiling. Distribute into tubes or flasks. Autoclave for 15 min at 15 psi pressure–121°C.

Use: For the cultivation and maintenance of a wide variety of microorganisms. Recommended for the culture of microorganisms from milk and water.

Beef Extract Broth (ATCC Medium 225)

Composition per liter:

Beef extract	10.0g
Peptone	10.0g
NaCl	5.0g

pH 7.2 ± 0.2 at 25°C

Preparation of Medium: Add components to distilled/deionized water and bring volume to 1.0L. Mix thoroughly. Heat gently and bring to boiling. Distribute into tubes or flasks. Autoclave for 15 min at 15 psi pressure–121°C.

Use: For the cultivation of a wide variety of microorganisms, including *Alcaligenes* species, *Pseudomonas aeruginosa,* and *Bacillus sphaericus*.

Beef Extract Broth, HiVeg

Composition per liter:

Plant peptone	10.0g
NaCl	5.0g
Plant extract	3.0g

pH 7.2 ± 0.2 at 25°C

Source: This medium is available as a premixed powder from HiMedia.

Preparation of Medium: Add components to distilled/deionized water and bring volume to 1.0L. Mix thoroughly. Heat gently and bring to boiling. Distribute into tubes or flasks. Autoclave for 15 min at 15 psi pressure–121°C.

Use: For the cultivation and maintenance of a wide variety of microorganisms. Recommended for the culture of microorganisms from milk and water.

Beef Extract Peptone Serum Medium

Composition per liter:

Agar	25.0g
Beef extract	10.0g
Peptone	10.0g
NaCl	1.0g
Bovine serum	50.0mL

pH 8.5 ± 0.2 at 25°C

Preparation of Medium: Add components, except bovine serum, to distilled/deionized water and bring volume to 950.0mL. Mix thoroughly. Adjust pH to 8.5. Heat gently and bring to boiling. Autoclave for 15 min at 15 psi pressure–121°C. Cool to 50°–55°C. Aseptically add 50.0mL of sterile bovine serum. Pour into sterile Petri dishes or leave in tubes.

Use: For the cultivation and maintenance of *Serratia marcescens.*

Beef Extract V

Composition per liter:

Beef extract	24.0g

pH 9.0 at 25°C

Preparation of Medium: Add component to distilled/deionized water and bring volume to 1.0L. Mix thoroughly. Adjust pH to 9.0 with NaOH. Autoclave for 15 min at 15 psi pressure—118°–121°C.

Use: For use in the elution of viruses that have been adsorbed onto filters during the filtration of water and wastewater samples.

Beef Extract with Sodium Chloride

Composition per liter:

Beef extract	10.0g
NaCl	5.0g

Preparation of Medium: Add components to distilled/deionized water and bring volume to 1.0L. Mix thoroughly. Distribute into tubes or flasks. Autoclave for 15 min at 15 psi pressure–121°C.

Use: For the cultivation of *Bacillus megaterium.*

Beef Infusion Agar

Composition per liter:

Ground defatted beef	453.6g
Agar	20.0g
Peptone	10.0g
NaCl	5.0g

pH 7.6 ± 0.2 at 25°C

Preparation of Medium: Add ground beef to 1.0L of distilled/deionized water. Let stand overnight at 4°C. Gently heat and bring to 80°–90°C for 60 min. Let stand for 2 hr. Filter through muslin. To filtrate, add peptone and salt. Mix thoroughly. Adjust pH to 7.6 with 4% NaOH. Filter through Whatman #1 filter paper. Bring volume of filtrate to 1.0L. Add agar. Gently heat and bring to boiling. Distribute into tubes or flasks. Autoclave for 15 min at 15 psi pressure–121°C. Pour into sterile Petri dishes or leave in tubes.

Use: For the cultivation of a variety of microorganisms.

Beef Infusion Broth

Composition per liter:

Ground beef, defatted	453.6g
Peptone	10.0g
NaCl	5.0g

pH 7.6 ± 0.2 at 25°C

Preparation of Medium: Add ground beef to 1.0L of distilled/deionized water. Let stand overnight at 4°C. Gently heat and bring to 80°–90°C for 60 min. Let stand for 2 hr. Filter through muslin. To filtrate add peptone and salt. Mix thoroughly. Adjust pH to 7.6 with 4% NaOH. Filter through Whatman #1 filter paper. Bring volume of filtrate to 1.0L. Add agar. Gently heat and bring to boiling. Distribute into tubes or flasks. Autoclave for 15 min at 15 psi pressure–121°C.

Use: For the cultivation of a variety of microorganisms.

Beef Liver Medium for Anaerobes

Composition per liter:

Beef liver, minced	500.0g
Peptone	10.0g
K_2HPO_4	1.0g

pH 8.0 ± 0.2 at 25°C

Preparation of Medium: Add beef liver to 1.0L of tap water. Soak for 12–24 hr at 4°C. Skim fat off top. Autoclave for 10 min at 15 psi pressure–121°C. Filter through cheesecloth. Save meat. To filtrate, add peptone and K_2HPO_4. Adjust pH to 8.0. Filter through paper. Add tap water and bring volume to 1.0L. Add a small amount of $CaCO_3$ to a flask or test tube. Add 0.5 inch of reserved liver. Cover meat with 2 inches of broth. Cap tubes and autoclave for 15 min at 15 psi pressure–121°C.

Use: For the cultivation and maintenance of a variety of *Clostridium* species.

Beggiatoa Agar

Composition per 1010.0mL:

Agar	10.0g
Sodium acetate	0.5g
Pancreatic digest of gelatin	0.31g
Beef extract	0.19g
NH_4Cl	0.45mg
$MgSO_4 \cdot 7H_2O$	0.2mg
K_2HPO_4	0.1mg
$CaSO_4$ (saturated solution)	20.0mL
Catalase solution	10.0mL
Trace elements solution	5.0mL

pH 7.4 ± 0.2 at 25°C

Catalase Solution:

Composition per 10.0mL:

Catalase	15,000–35,000U

Preparation of Catalase Solution: Add catalase to distilled/deionized water and bring volume to 10.0mL. Mix thoroughly. Filter sterilize.

Trace Elements Solution:

Composition per liter:

$FeSO_4 \cdot 7H_2O$	0.7g
EDTA	0.2g
$ZnSO_4 \cdot 7H_2O$	0.01g
$MnSO_4 \cdot 4H_2O$	0.002g
H_3BO_3	10.0mg
$CO(NO_3)_2$	1.0mg
$Na_2MoO_4 \cdot 2H_2O$	1.0mg
$CuSO_4 \cdot 5H_2O$	5.0μg

Preparation of Trace Elements Solution: Add $FeSO_4 \cdot 7H_2O$ to 10.0mL of HCl solution. Mix thoroughly. Add distilled/deionized water and bring volume to 1.0L. Add remaining components. Mix thoroughly.

Preparation of Medium: Add components, except catalase solution, to distilled/deionized water and bring volume to 1.0L. Mix thoroughly. Gently heat and bring to boiling. Adjust pH to 7.4. Autoclave for 15 min at 15 psi pressure–121°C. Aseptically add 10.0mL of sterile catalase solution (freshly prepared). Mix thoroughly. Pour into sterile Petri dishes or distribute into sterile tubes.

Use: For the cultivation and maintenance of *Beggiatoa alba.*

Beggiatoa and *Thiothrix* Medium

Composition per liter:

$CaSO_4 \cdot 2H_2O$ (saturated solution)	20.0mL
NH_4Cl (4% solution)	5.0mL
Trace elements	5.0mL
K_2HPO_4 (1% solution)	1.0mL
$MgSO_4 \cdot 7H_2O$ (1% solution)	1.0mL

Trace Elements:

Composition per liter:

EDTA solution	20.0mL
$Co(NO_3)_2$ (0.01% solution)	10.0mL
$CuSO_4 \cdot 5H_2O$ (0.00005% solution)	10.0mL
H_3BO_3 (0.1% solution)	10.0mL
$MnSO_4 \cdot 4H_2O$ (0.02% solution)	10.0mL
$Na_2MoO_4 \cdot 2H_2O$ (0.01% solution)	10.0mL
$ZnSO_4 \cdot 7H_2O$ (0.1% solution)	10.0mL

Preparation of Trace Elements: Add components to distilled/deionized water and bring volume to 1.0L. Mix thoroughly.

EDTA Solution:

Composition per 100.0mL:

$FeSO_4$	7.0g
EDTA	2.0g
HCl, concentrated	1.0mL

Preparation of EDTA Solution: Add EDTA and $FeSO_4$ to concentrated HCl. Mix thoroughly. Carefully add to distilled/deionized water and bring volume to 100.0mL.

Preparation of Medium: Add components to distilled/deionized water and bring volume to 1.0L. Mix thoroughly. Distribute into tubes or flasks. Autoclave for 15 min at 15 psi pressure–121°C.

Use: For the cultivation of *Beggiatoa* species and myxotrophic *Thiothrix* species.

Beggiatoa Broth

Composition per 1010.0mL:

Sodium acetate	0.5g
Pancreatic digest of gelatin	0.31g
Beef extract	0.19g
NH_4Cl	0.45mg
$MgSO_4 \cdot 7H_2O$	0.2mg
K_2HPO_4	0.1mg
$CaSO_4$ (saturated solution)	20.0mL
Catalase solution	10.0mL
Trace elements solution	5.0mL

pH 7.4 ± 0.2 at 25°C

Catalase Solution:

Composition per 10.0mL:

Catalase	15,000–35,000U

Preparation of Catalase Solution: Add catalase to distilled/deionized water and bring volume to 10.0mL. Mix thoroughly. Filter sterilize.

Trace Elements Solution:

Composition per liter:

$FeSO_4 \cdot 7H_2O$	0.7g
EDTA	0.2g
$ZnSO_4 \cdot 7H_2O$	0.01g
$MnSO_4 \cdot 4H_2O$	0.002g
H_3BO_3	10.0mg
$CO(NO_3)_2$	1.0mg
$Na_2MoO_4 \cdot 2H_2O$	1.0mg
$CuSO_4 \cdot 5H_2O$	5.0μg

Preparation of Trace Elements Solution: Add $FeSO_4 \cdot 7H_2O$ to 10.0mL of HCl solution. Mix thoroughly. Add distilled/deionized water and bring volume to 1.0L. Add remaining components. Mix thoroughly.

Preparation of Medium: Add components, except catalase solution, to distilled/deionized water and bring volume to 1.0L. Mix thoroughly. Adjust pH to 7.4. Autoclave for 15 min at 15 psi pressure–121°C. Aseptically add 10.0mL of sterile catalase solution (freshly prepared). Mix thoroughly. Aseptically distribute into sterile tubes or flasks.

Use: For the cultivation of *Beggiatoa alba.*

Beggiatoa Medium (ATCC Medium 138)

Composition per liter:

Yeast extract	2.0g
Agar	2.0g
Sodium acetate	0.5g
$CaCl_2$	0.1g
Catalase	10,000U

pH 7.2 ± 0.2 at 25°C

Preparation of Medium: Add components, except catalase, to tap water and bring volume to 1.0L. Mix thoroughly. Autoclave for 15 min at 15 psi pressure–121°C. Cool to 45°–50°C. Aseptically add 10,000 units of sterile catalase.

Use: For the cultivation and maintenance of *Beggiatoa alba* and *Vitreoscilla* species.

Beggiatoa Medium (ATCC Medium 1193)

Composition per liter:

Sodium sulfide	0.5g
Sodium acetate	0.01g
Yeast extract	0.01g
Nutrient broth	0.01g

pH 7.5 ± 0.2 at 25°C

Preparation of Medium: Add components to distilled/deionized water and bring volume to 1.0L. Mix thoroughly. Autoclave for 15 min at 15 psi pressure–121°C. Distribute into tubes or flasks.

Use: For the cultivation of *Beggiatoa alba.*

Beijerinckia Agar

Composition per liter:

Agar	15.0g
K_2HPO_4	0.8g
KH_2PO_4	0.2g
$MgSO_4 \cdot 7H_2O$	0.1g
$FeSO_4 \cdot 7H_2O$	20.0mg
$Na_2MoO_4 \cdot 2H_2O$	5.0mg
$ZnSO_4 \cdot 6H_2O$	5.0mg
$CuSO_4 \cdot 6H_2O$	4.0mg
$MnSO_4 \cdot 6H_2O$	2.0mg
Glucose solution	50.0mL

pH 6.5 ± 0.2 at 25°C

Glucose Solution:
Composition per 50.0mL:

D-Glucose	10.0g

Preparation of Glucose Solution: Add glucose to distilled/deionized water and bring volume to 50.0mL. Mix thoroughly. Autoclave for 15 min at 15 psi pressure–121°C.

Preparation of Medium: Add components, except glucose solution, to distilled/deionized water and bring volume to 950.0mL. Mix thoroughly. Gently heat and bring to boiling. Autoclave for 15 min at 15 psi pressure–121°C. Aseptically add 10.0mL of sterile glucose solution. Mix thoroughly. Pour into sterile Petri dishes or distribute into sterile tubes

Use: For the cultivation and maintenance of *Beijerinckia derxii, Beijerinckia fluminensis, Beijerinckia indica, Beijerinckia mobilis, Beijerinckia* species, and *Clostridium barkeri.*

Beijerinckia Medium

Composition per liter:

Glucose	20.0g
KH_2PO_4	1.0g
$MgSO_4·7H_2O$	0.5g

pH 5.0 ± 0.2 at 25°C

Preparation of Medium: Add components to distilled/deionized water and bring volume to 1.0L. Mix thoroughly. Distribute into tubes or flasks. Autoclave for 15 min at 15 psi pressure–121°C.

Use: For the cultivation of *Beijerinckia* species.

Beijerinckia Medium

Composition per liter:

Glucose	20.0g
K_2HPO_4	0.8g
$MgSO_4·7H_2O$	0.5g
KH_2PO_4	0.2g
$CaCl_2$	0.05g
$FeCl_3·6H_2O$	0.025g
$Na_2MoO_4·2H_2O$	5.0mg

pH 6.9 ± 0.2 at 25°C

Preparation of Medium: Add components to distilled/deionized water and bring volume to 1.0L. Mix thoroughly. Distribute into tubes or flasks. Autoclave for 15 min at 15 psi pressure–121°C.

Use: For the isolation and cultivation of *Beijerinckia* species.

Beijerinckia Medium

Composition per liter:

Sucrose	20.0g
Agar	15.0g
KH_2PO_4	0.8g
$MgSO_4·7H_2O$	0.5g
K_2HPO_4	0.2g
$FeCl_3·6H_2O$	0.1g
$Na_2MoO_4·2H_2O$	5.0mg

pH 6.5 ± 0.2 at 25°C

Source: This medium is available from HiMedia.

Preparation of Medium: Add components to distilled/deionized water and bring volume to 1.0L. Mix thoroughly. Distribute into tubes or flasks. Autoclave for 15 min at 15 psi pressure–121°C.

Use: For the isolation and cultivation of *Beijerinckia* species.

Beijerinckia Medium, Modified

Composition per liter:

Agar	15.0g
Glucose	10.0g
K_2HPO_4	0.8g
KH_2PO_4	0.2g
$MgSO_4·7H_2O$	0.1g
$FeSO_4·7H_2O$	20.0mg
$MnSO_4·H_2O$	1.3mg
$ZnSO_4·7H_2O$	5.0mg
$CuSO_4·5H_2O$	4.0mg
$Na_2MoO_4·2H_2O$	5.0mg

pH 6.5 ± 0.2 at 25°C

Preparation of Medium: Add components to distilled/deionized water and bring volume to 1.0L. Mix thoroughly. Gently heat and bring to boiling. Distribute into tubes or flasks. Autoclave for 15 min at 15 psi pressure–121°C. Pour into sterile Petri dishes or leave in tubes.

Use: For the isolation and cultivation of *Beijerinckia derxii, Beijerinckia fluminensis, Beijerinckia indica,* and *Beijerinckia mobilis.*

Beijerinck's *Thiobacillus* Medium

Composition per liter:

Noble agar	20.0g
Na_2HPO_4	0.2g
$MgCl_2$	0.1g
NH_4Cl	0.1g
$Na_2S_2O_3$ solution	100.0mL
$NaHCO_3$ solution	10.0mL

pH 7.0–7.2 at 25°C

$Na_2S_2O_3$ Solution:
Composition per 100.0mL:

$Na_2S_2O_3$	5.0g

Preparation of $Na_2S_2O_3$ Solution: Add $Na_2S_2O_3$ to distilled/deionized water and bring volume to 100.0mL. Mix thoroughly. Filter sterilize.

$NaHCO_3$ Solution:
Composition per 10.0mL:

$NaHCO_3$	1.0g

Preparation of $NaHCO_3$ Solution: Add $NaHCO_3$ to distilled/deionized water and bring volume to 10.0mL. Mix thoroughly. Filter sterilize.

Preparation of Medium: Add components, except $Na_2S_2O_3$ solution and $NaHCO_3$ solution, to distilled/deionized water and bring volume to 890.0mL. Mix thoroughly. Autoclave for 15 min at 15 psi pressure–121°C. Aseptically add 100.0mL of sterile $Na_2S_2O_3$ solution and 10.0mL of sterile $NaHCO_3$ solution. Mix thoroughly. Pour into sterile Petri dishes or leave in tubes.

Use: For the cultivation and maintenance of *Thiobacillus thermophilica.*

Bennett's Agar

Composition per liter:

Agar	15.0g
Glucose	10.0g
N-Z amine, type A	2.0g
Beef extract	1.0g
Yeast extract	1.0g

pH 7.3 ± 0.2 at 25°C

Preparation of Medium: Add components to distilled/deionized water and bring volume to 1.0L. Mix thoroughly. Adjust pH to 7.3. Gently heat and bring to boiling. Distribute into tubes or flasks. Autoclave for 15 min at 15 psi pressure–121°C. Pour into sterile Petri dishes or leave in tubes.

Use: For the cultivation and maintenance of *Actinomadura umbrina, Micromonospora purpurea, Microtetraspora helvata, Nocardia salmonicolor,* and *Streptomyces* species.

Bennett's Agar with Maltose

Composition per liter:

Agar	15.0g
Maltose, technical	10.0g
N-Z amine, type A	2.0g
Beef extract	1.0g
Yeast extract	1.0g

pH 7.3 ± 0.2 at 25°C

Preparation of Medium: Add components to distilled/deionized water and bring volume to 1.0L. Mix thoroughly. Adjust pH to 7.3. Gently heat and bring to boiling. Distribute into tubes or flasks. Autoclave for 15 min at 15 psi pressure–121°C. Pour into sterile Petri dishes or leave in tubes.

Use: For the cultivation and maintenance of *Streptomyces* species.

Bennett's Agar with Sucrose

Composition per liter:

Agar	15.0g
Sucrose	10.0g
N-Z amine, type A	2.0g
Beef extract	1.0g
Yeast extract	1.0g

pH 7.3 ± 0.2 at 25°C

Preparation of Medium: Add components to distilled/deionized water and bring volume to 1.0L. Mix thoroughly. Adjust pH to 7.3. Gently heat and bring to boiling. Distribute into tubes or flasks. Autoclave for 15 min at 15 psi pressure–121°C. Pour into sterile Petri dishes or leave in tubes.

Use: For the cultivation and maintenance of *Actinomadura madurae, Excellospora viridilutea, Geodermatophilus obscurus, Intrasporangium calvum, Kibdelosporangium aridum, Microbispora thermodiastatica, Micromonospora coerulea, Micromonospora echinospora, Micromonospora purpureochromogenes, Micromonospora rosaria, Microtetraspora flexuosa, Promicromonospora enterophila, Saccharomonospora glauca, Streptomyces cacaoi, Thermoactinomyces dichotomicus, Thermoactinomyces glaucus,* and *Thermomonospora chromogena.*

Bennet's HiVeg Agar

Composition per liter:

Agar	15.0g
Glucose	10.0g
Plant hydrolysate	2.0g
Plant extract	1.0g
Yeast extract	1.0g

pH 7.3 ± 0.2 at 25°C

Source: This medium is available as a premixed powder from HiMedia.

Preparation of Medium: Add components to distilled/deionized water and bring volume to 1.0L. Mix thoroughly. Adjust pH to 7.3. Gently heat and bring to boiling. Distribute into tubes or flasks. Autoclave for 10 min at 15 psi pressure–121°C. Pour into sterile Petri dishes or leave in tubes.

Use: For the cultivation and maintenance of *Actinomadura umbrina, Micromonospora purpurea, Microtetraspora helvata, Nocardia salmonicolor,* and *Streptomyces* species.

Bennett's Medium

Composition per liter:

Agar	15.0g
Glucose	10.0g
Pancreatic digest of casein	2.0g
Yeast extract	1.0g
Beef extract	1.0g

pH 7.0 ± 0.2 at 25°C

Preparation of Medium: Add components to distilled/deionized water and bring volume to 1.0L. Mix thoroughly. Heat gently to boiling. Distribute into tubes or flasks. Autoclave for 15 min at 15 psi pressure–121°C. Pour into sterile Petri dishes or leave in tubes.

Use: For the cultivation and maintenance of a variety of soil microorganisms, such as *Streptomyces* species, *Nocardia* species, *Flexibacter* species, *Micromonospora* species, and others.

Bennett's Modified Agar Medium

Composition per liter:

Meer agar (washed agar)	20.0g
Dextrin	10.0g
Pancreatic digest of casein	2.0g
Yeast extract	1.0g
Beef extract	1.0g
$CoCl_2 \cdot 6H_2O$	0.01g

pH 7.0 ± 0.2 at 25°C

Preparation of Medium: Add components to distilled/deionized water and bring volume to 1.0L. Mix thoroughly. Heat gently to boiling. Distribute into tubes or flasks. Autoclave for 15 min at 15 psi pressure–121°C. Pour into sterile Petri dishes or leave in tubes.

Use: For the cultivation and maintenance of *Streptomyces* species.

Benzene Sulfonate Medium

Composition per liter:

Agar	15.0g
Sodium benzene sulfonate	1.0g
$(NH_4)_2SO_4$	1.0g
K_2HPO_4	0.7g
KH_2PO_4	0.3g
$MgSO_4 \cdot 7H_2O$	0.2g
$CaCl_2$	10.0mg
$FeSO_4 \cdot 7H_2O$	5.0mg
$ZnSO_4 \cdot 7H_2O$	70.0μg
$CuSO_4$	50.0μg
H_3BO_3	10.0μg
$MoO_3 \cdot 2H_2O$	10.0μg
$MnSO_4 \cdot 5H_2O$	2.0μg

Preparation of Medium: Add components to distilled/deionized water and bring volume to 1.0L. Mix thoroughly. Heat gently to boiling. Distribute into tubes or flasks. Autoclave for 15 min at 15 psi pressure–121°C. Pour into sterile Petri dishes or leave in tubes.

Use: For the cultivation and maintenance of *Comamonas testosteroni*.

Benzoate Medium

Composition per liter:

Noble agar	20.0g
NaCl	5.0g
$(NH_4)_2HPO_4$	3.0g
Sodium benzoate	3.0g
KH_2PO_4	1.2g
Yeast extract	0.5g
$MgSO_4 \cdot 7H_2O$	0.2g
Benzoate solution	25.0mL

Benzoate Solution:

Composition per 25.0mL:

Sodium benzoate	3.0g

Preparation of Benzoate Solution: Add sodium benzoate to distilled/deionized water and bring volume to 25.0mL. Mix thoroughly. Filter sterilize.

Preparation of Medium: Add components except benzoate solution to distilled/deionized water and bring volume to 975.0mL. Mix thoroughly. Heat gently to boiling. Autoclave for 15 min at 15 psi pressure–121°C. Cool to 45°–50°C. Aseptically add 25.0mL sterile benzoate solution. Mix thoroughly and pour into sterile Petri dishes or leave in tubes.

Use: For the cultivation of *Pseudomonas putida* and other microorganisms which can utilize benzoate as a carbon source.

Benzoate Medium II

Composition per 1.5L:

Noble agar	30.0g
$(NH_4)_2HPO_4$	3.0g
NaCl	1.67g
KH_2PO_4	1.2g
Yeast extract	0.5g
$MgSO_4 \cdot 7H_2O$	0.2g
$FeSO_4 \cdot 7H_2O$	0.1g
Benzoate solution	25.0mL

Benzoate Solution:

Composition per 25.0mL:

Sodium benzoate	1.0g

Preparation of Benzoate Solution: Add sodium benzoate to distilled/deionized water and bring volume to 25.0mL. Mix thoroughly. Filter sterilize.

Preparation of Medium: Add components, except agar and sodium benzoate, to distilled/deionized water and bring volume to 600.0mL. Mix thoroughly. Autoclave for 15 min at 15 psi pressure–121°C. Cool to 45°–50°C. In a separate flask, add agar to distilled/deionized water and bring volume to 375.0mL. Mix thoroughly. Gently heat and bring to boiling. Autoclave for 15 min at 15 psi pressure–121°C. Cool to 45°–50°C. Aseptically combine the two autoclave-sterilized solutions. Mix thoroughly. Aseptically add the sterile benzoate solution. Mix thoroughly. Pour into sterile Petri dishes or leave in tubes.

Use: For the cultivation of *Pseudomonas putida* and other microorganisms that can utilize benzoate as a carbon source.

Benzoate Minimal Salts Medium

Composition per liter:

K_2HPO_4	10.0g
$NaNH_4HPO_4 \cdot 4H_2O$	3.5g
$MgSO_4 \cdot 7H_2O$	0.2g
Citric acid, anhydrous	0.2g
Benzoate solution	25.0mL

pH 7.0 ± 0.2 at 25°C

Benzoate Solution:

Composition per 25.0mL:

Sodium benzoate	2.5g

Preparation of Benzoate Solution: Add sodium benzoate to distilled/deionized water and bring volume to 25.0mL. Mix thoroughly. Filter sterilize.

Preparation of Medium: Add components to distilled/deionized water and bring volume to 975.0mL. Mix thoroughly. Adjust pH to 7.0. Autoclave for 15 min at 15 psi pressure–121°C. Cool to 45°C. Aseptically add 25.0mL of sterile benzoate solution. Mix thoroughly. Aseptically distribute into sterile tubes or flasks.

Use: For the cultivation of microorganisms that can utilize benzoate as a carbon source.

Benzoate Nitrate Salts Medium (BNS)

Composition per liter:

Solution A	700.0mL
Solution B	300.0mL

pH 8.2 ± 0.2 at 25°C

Solution A:

Composition per 700.0mL:

KNO_3	2.0g
Sodium benzoate	1.0g
NH_4Cl	0.3g
Phosphate buffer solution	200.0mL

Phosphate Buffer Solution:

Composition per 200.0mL:

K_2HPO_4	5.12g
KH_2PO_4	1.5g

Preparation of Phosphate Buffer: Add components to distilled/deionized water and bring volume to 200.0mL. Mix thoroughly. Adjust pH to 9.0 with KOH.

Preparation of Solution A: Add components to distilled/deionized water and bring volume to 700.0mL. Mix thoroughly. Autoclave for 15 min at 15 psi pressure–121°C. Cool to room temperature.

Solution B:

Composition per 300.0mL:

$MgSO_4 \cdot 7H_2O$	0.2g
$CaCl_2$	10.0mg
Trace metals solution	1.0mL

Preparation of Solution B: Add components to distilled/deionized water and bring volume to 300.0mL. Mix thoroughly. Autoclave for 15 min at 15 psi pressure–121°C. Cool to room temperature.

Trace Metals Solution:

Composition per 300.0mL:

$MnSO_4 \cdot H_2O$	50.0mg
$ZnSO_4 \cdot 7H_2O$	50.0mg
$Co(NO_3)_2 \cdot 6H_2O$	10.0mg

$CuSO_4$	10.0mg
$Na_2B_4O_7 \cdot 10H_2O$	10.0mg
$Na_2MoO_4 \cdot 2H_2O$	0.2mg
Ferric EDTA solution	10.0mL

Preparation of Trace Metals Solution: Add components to distilled/deionized water and bring volume to 1.0L. Mix thoroughly.

Ferric EDTA Solution:
Composition per 550.0mL:

EDTA	17.9g
$FeSO_4 \cdot 7H_2O$	13.7g
KOH	3.23g

Preparation of Ferric EDTA Solution: Add EDTA and KOH to distilled/deionized water and bring volume to 186.0mL. Mix thoroughly. In a separate flask, add $FeSO_4 \cdot 7H_2O$ to distilled/deionized water and bring volume to 364.0mL. Mix thoroughly. Combine the two solutions. Sparge with air overnight to oxidize the Fe^{2+} to Fe^{3+}. Store in the dark.

Preparation of Medium: Aseptically combine 700.0mL of sterile solution A with 300.0mL of sterile solution B. Adjust pH to 8.2. Aseptically distribute into sterile screw-capped tubes. Fill tubes completely.

Use: For the cultivation of *Alcaligenes xylosoxydans.*

Betabacterium Medium

Composition per liter:

Pancreatic digest of casein	10.0g
Agar	10.0g
Yeast extract	5.0g
Glucose	5.0g
K_2HPO_4	2.0g
Liver extract	100.0mL

pH 7.2 ± 0.2 at 25°C

Preparation of Medium: Add 1 pound of finely ground beef liver to 2.0L of distilled/deionized water. Autoclave for 2.5–3 hr at 15 psi pressure–121°C under flowing steam. The liquid should become fluorescent yellow. Filter through sterile cheesecloth. Save solids and dry at 50°C. Add a few pieces of the dried liver to sterile test tubes or flasks. Prepare basal medium by adding components to distilled/deionized water and bring volume to 1.0L. Autoclave for 15 min at 15 psi pressure–121°C. Aseptically add sterile basal medium to each test tube or flask containing liver. Commercial liver extract may be used at a concentration of 0.1%.

Use: For the growth and maintenance of *Lactobacillus* species. *Betabacterium* is an archaic name that was used to describe several bacteria as a subgenus of the *Lactobacillus* group.

BG Sulfa Agar
(Brilliant Green Sulfapyridine Agar)

Composition per liter:

Agar	20.0g
Proteose peptone No. 3	10.0g
Lactose	10.0g
Sucrose	10.0g
NaCl	5.0g
Yeast extract	3.0g
Sodium sulfapyridine	1.0g
Brilliant Green	0.125g

pH 6.9 ± 0.2 at 25°C

Source: This medium is available as a premixed powder from BD Diagnostic Systems.

Preparation of Medium: Add components to distilled/deionized water and bring volume to 1.0L. Mix thoroughly. Heat gently to boiling. Distribute into tubes or flasks. Autoclave for no longer than 15 min at 15 psi pressure–121°C. Pour into sterile Petri dishes if desired.

Use: For the selective isolation of *Salmonella* species other than *Salmonella typhi* from food, dairy products, eggs and egg products, and feed. *Salmonella* appear as red, pink, or white colonies surrounded by zones of bright red.

BG Sulfa HiVeg Agar
(Brilliant Green Sulfa HiVeg Agar)

Composition per liter:

Agar	20.0g
Plant peptone No. 3	10.0g
Lactose	10.0g
Sucrose	10.0g
NaCl	5.0g
Yeast extract	3.0g
Sodium sulphapyridine	1.0g
Phenol Red	0.08g
Brilliant Green	12.5mg

pH 6.9 ± 0.2 at 25°C

Source: This medium is available as a premixed powder from HiMedia.

Preparation of Medium: Add components to distilled/deionized water and bring volume to 1.0L. Mix thoroughly. Heat gently to boiling. Distribute into tubes or flasks. Autoclave for no longer than 15 min at 15 psi pressure–121°C. Pour into sterile Petri dishes if desired.

Use: For the selective isolation of *Salmonella* species other than *Salmonella typhi* from food, dairy products, eggs and egg products, and feed. *Salmonella* appear as red, pink, or white colonies surrounded by zones of bright red.

BG 11 Agar
(Medium BG 11 for Cyanobacteria)

Composition per liter:

Agar	10.0g
$NaNO_3$	1.5g
$MgSO_4 \cdot 7H_2O$	0.075g
K_2HPO_4	0.04g
$CaCl_2 \cdot 2H_2O$	0.036g
Na_2CO_3	0.02g
Citric acid	6.0mg
Ferric ammonium citrate	6.0mg
Disodium EDTA	1.0mg
Trace metal mix A5	1.0mL

pH 7.1 ± 0.2 at 25°C

Trace Metal Mix A5:
Composition per liter:

H_3BO_3	2.86g
$MnCl_2 \cdot 4H_2O$	1.81g
$Na_2MoO_4 \cdot 2H_2O$	0.39g
$ZnSO_4 \cdot 7H_2O$	0.222g
$CuSO_4 \cdot 5H_2O$	0.079g
$Co(NO_3)_2 \cdot 6H_2O$	0.049g

Preparation of Trace Metal Mix A5: Add components to distilled/deionized water and bring volume to 1.0L. Mix thoroughly.

Preparation of Medium: Add components to distilled/deionized water and bring volume to 1.0L. Mix thoroughly. Heat gently to boiling. Distribute into tubes or flasks. Autoclave for 15 min at 15 psi pressure–121°C. For solid medium, pour into sterile Petri dishes or leave in tubes.

Use: For the cultivation and maintenance of a variety of cyanobacteria, including *Anabaena* species, *Calothrix* species, *Chaemisiphon* species, *Chorogloeopsis* species, *Chroococcidiopsis* species, *Cylindrospermum* species, *Dermocarpa* species, *Fischerella* species, *Gloebacter* species, *Gloeocapsa* species, *Gloeothece* species, *Nostoc* species, *Oscillatoria* species, *Phormidium* species, *Pleurocapsa* species, *Pseudanabaena* species, *Scytonema* species, *Spirulina* species, *Synechococcus* species, and *Synechocystis* species.

BG 11 Marine Agar
(Medium BG 11 for Marine Cyanobacteria)

Composition per liter:

Agar	10.0g
NaCl	10.0g
$NaNO_3$	1.5g
$MgSO_4 \cdot 7H_2O$	0.075g
K_2HPO_4	0.04g
$CaCl_2 \cdot 2H_2O$	0.036g
Na_2CO_3	0.02g
Citric acid	6.0mg
Ferric ammonium citrate	6.0mg
EDTA disodium salt	1.0mg
Vitamin B_{12} solution	100.0mL
Trace metal mix A5	1.0mL

pH 7.1 ± 0.2 at 25°C

Trace Metal Mix A5:

Composition per liter:

H_3BO_3	2.86g
$MnCl_2 \cdot 4H_2O$	1.81g
$Na_2MoO_4 \cdot 2H_2O$	0.39g
$ZnSO_4 \cdot 7H_2O$	0.222g
$CuSO_4 \cdot 5H_2O$	0.079g
$Co(NO_3)_2 \cdot 6H_2O$	0.049g

Preparation of Trace Metal Mix A5: Add components to distilled/deionized water and bring volume to 1.0L. Mix thoroughly.

Vitamin B_{12} Solution:

Composition per 100.0mL:

Vitamin B_{12}	1.0μg

Preparation of Vitamin B_{12} Solution: Add vitamin B_{12} to distilled/deionized water and bring volume to 100.0mL. Mix thoroughly. Filter sterilize.

Preparation of Medium: Add components, except vitamin B_{12} solution, to distilled/deionized water and bring volume to 900.0mL. Mix thoroughly. Heat gently to boiling. Autoclave for 15 min at 15 psi pressure–121°C. Aseptically add 100.0mL of sterile vitamin B_{12} solution. Mix thoroughly. Pour into sterile Petri dishes or leave in tubes.

Use: For the cultivation and maintenance of *Synechococcus* species. For the isolation of cyanobacteria from freshwater habitats.

BG 11 Marine Broth
(Medium BG 11 for Marine Cyanobacteria)

Composition per liter:

NaCl	10.0g
$NaNO_3$	1.5g
$MgSO_4 \cdot 7H_2O$	0.075g
K_2HPO_4	0.04g
$CaCl_2 \cdot 2H_2O$	0.036g
Na_2CO_3	0.02g
Citric acid	6.0mg
Ferric ammonium citrate	6.0mg
EDTA disodium salt	1.0mg
Vitamin B_{12} solution	100.0mL
Trace metal mix A5	1.0mL

pH 7.1 ± 0.2 at 25°C

Trace Metal Mix A5:

Composition per liter:

H_3BO_3	2.86g
$MnCl_2 \cdot 4H_2O$	1.81g
$Na_2MoO_4 \cdot 2H_2O$	0.39g
$ZnSO_4 \cdot 7H_2O$	0.222g
$CuSO_4 \cdot 5H_2O$	0.079g
$Co(NO_3)_2 \cdot 6H_2O$	0.049g

Preparation of Trace Metal Mix A5: Add components to distilled/deionized water and bring volume to 1.0L. Mix thoroughly.

Vitamin B_{12} Solution:

Composition per 100.0mL:

Vitamin B_{12}	1.0μg

Preparation of Vitamin B_{12} Solution: Add vitamin B_{12} to distilled/deionized water and bring volume to 100.0mL. Mix thoroughly. Filter sterilize.

Preparation of Medium: Add components, except vitamin B_{12} solution, to distilled/deionized water and bring volume to 900.0mL. Mix thoroughly. Heat gently to boiling. Autoclave for 15 min at 15 psi pressure–121°C. Aseptically add 100.0mL of sterile vitamin B_{12} solution. Mix thoroughly. Distribute into sterile tubes or flasks.

Use: For the cultivation and maintenance of *Synechococcus* species. For the isolation of cyanobacteria from freshwater habitats.

BG 11 Medium
(Medium BG 11 for Cyanobacteria)

Composition per liter:

Agar	10.0g
$NaNO_3$	1.5g
$MgSO_4 \cdot 7H_2O$	0.075g
K_2HPO_4	0.04g
$CaCl_2 \cdot 2H_2O$	0.036g
Na_2CO_3	0.02g
Citric acid	6.0mg
Ferric ammonium citrate	6.0mg
EDTA disodium salt	1.0mg
Trace metal mix A5	1.0mL

pH 7.1 ± 0.2 at 25°C

Trace Metal Mix A5:

Composition per liter:

H_3BO_3	2.86g
$MnCl_2 \cdot 4H_2O$	1.81g
$Na_2MoO_4 \cdot 2H_2O$	0.39g

$ZnSO_4 \cdot 7H_2O$	0.222g
$CuSO_4 \cdot 5H_2O$	0.079g
$Co(NO_3)_2 \cdot 6H_2O$	0.049g

Preparation of Trace Metal Mix A5: Add components to distilled/deionized water and bring volume to 1.0L. Mix thoroughly.

Preparation of Medium: Add components to distilled/deionized water and bring volume to 1.0L. Mix thoroughly. Gently heat and bring to boiling. Distribute into tubes or flasks. Autoclave for 15 min at 15 psi pressure–121°C. Pour into sterile Petri dishes or leave in tubes.

Use: For the cultivation and maintenance of *Anabaena* species, *Calothrix* species, *Chaemisiphon* species, *Chorogloeopsis* species, *Chroococcidiopsis* species, *Crinalium epipsammum*, *Cylindrospermum* species, *Dermocarpa* species, *Fischerella* species, *Gloebacter violaceus*, *Gloeocapsa* species, *Gloeothece* species, *Hapalosiphon fontinalis*, *Nostoc* species, *Oscillatoria* species, *Phormidium* species, *Pleurocapsa* species, *Pseudanabaena* species, *Scytonema* species, *Spirulina* species, *Synechococcus* species, *Synechocystis* species, and *Tolypothrix tenuis*.

BG 11 Uracil Agar

Composition per liter:

Agar	10.0g
Uracil	2.8g
$NaNO_3$	1.5g
$MgSO_4 \cdot 7H_2O$	0.075g
K_2HPO_4	0.04g
$CaCl_2 \cdot 2H_2O$	0.036g
Na_2CO_3	0.02g
Citric acid	6.0mg
Ferric ammonium citrate	6.0mg
EDTA disodium salt	1.0mg
Trace metal mix A5	1.0mL

pH 7.1 ± 0.2 at 25°C

Trace Metal Mix A5:
Composition per liter:

H_3BO_3	2.86g
$MnCl_2 \cdot 4H_2O$	1.81g
$Na_2MoO_4 \cdot 2H_2O$	0.39g
$ZnSO_4 \cdot 7H_2O$	0.222g
$CuSO_4 \cdot 5H_2O$	0.079g
$Co(NO_3)_2 \cdot 6H_2O$	0.049g

Preparation of Trace Metal Mix A5: Add components to distilled/deionized water and bring volume to 1.0L. Mix thoroughly.

Preparation of Medium: Add components to distilled/deionized water and bring volume to 1.0L. Mix thoroughly. Heat gently to boiling. Distribute into tubes or flasks. Autoclave for 15 min at 15 psi pressure–121°C. Pour into sterile Petri dishes.

Use: For the cultivation and maintenance of *Anabaena variabilis*.

BG 11 Uracil Broth

Composition per liter:

Uracil	2.8g
$NaNO_3$	1.5g
$MgSO_4 \cdot 7H_2O$	0.075g
K_2HPO_4	0.04g
$CaCl_2 \cdot 2H_2O$	0.036g
Na_2CO_3	0.02g
Citric acid	6.0mg
Ferric ammonium citrate	6.0mg
EDTA disodium salt	1.0mg
Trace metal mix A5	1.0mL

pH 7.1 ± 0.2 at 25°C

Trace Metal Mix A5:
Composition per liter:

H_3BO_3	2.86g
$MnCl_2 \cdot 4H_2O$	1.81g
$Na_2MoO_4 \cdot 2H_2O$	0.39g
$ZnSO_4 \cdot 7H_2O$	0.222g
$CuSO_4 \cdot 5H_2O$	0.079g
$Co(NO_3)_2 \cdot 6H_2O$	0.049g

Preparation of Trace Metal Mix A5: Add components to distilled/deionized water and bring volume to 1.0L. Mix thoroughly.

Preparation of Medium: Add components to distilled/deionized water and bring volume to 1.0L. Mix thoroughly. Heat gently to boiling. Distribute into tubes or flasks. Autoclave for 15 min at 15 psi pressure–121°C.

Use: For the cultivation and maintenance of *Anabena variabilis*.

BHI

See: **Brain Heart Infusion**

BHI Agar

See: **Brain Heart Infusion Agar**

BHI Broth

See: **Brain Heart Infusion Broth**

BHI Glucose Medium

Composition per liter:

Agar	12.0g
Pancreatic digest of gelatin	7.25g
Glucose	6.5g
Brain heart, solids from infusion	3.0g
Peptic digest of animal tissue	3.0g
NaCl	2.5g
Na_2HPO_4	1.25g

pH 7.4 ± 0.2 at 25°C

Preparation of Medium: Add components to distilled/deionized water and bring volume to 1.0L. Mix thoroughly. Gently heat and bring to boiling. Distribute into tubes or flasks. Autoclave for 15 min at 15 psi–121°C. Pour into sterile Petri dishes or leave in tubes.

Use: For the cultivation and maintenance of *Actinomadura pelletieri, Actinoplanes missouriensis, Actinoplanes philippinensis, Agromyces ramosus, Corynebacterium minutissimum, Dermatophilus congolensis, Intrasporangium calvum, Mycobacterium diernhoferi, Mycobacterium* species, *Nocardia asteroides, Nocardia brevicatena, Nocardia calcarea, Nocardia otitidiscaviarum, Pseudonocardia thermophila, Saccharopolyspora rectivirgula, Streptococcus iniae,* and *Streptococcus pyogenes.*

BHI with Glucose (DSMZ Medium 215b)

Composition per liter:

Pancreatic digest of gelatin	14.5g
Glucose	8.0g
Brain heart, solids from infusion	6.0g

Peptic digest of animal tissue..6.0g
NaCl..5.0g
Na_2HPO_4..2.5g

pH 7.4 ± 0.2 at 25°C

Preparation of Medium: Add components to distilled/deionized water and bring volume to 1.0L. Mix thoroughly. Distribute into tubes or flasks. Autoclave for 15 min at 15 psi pressure–121°C.

Use: For the cultivation of *Corynebacterium* spp., *Streptomyces flocculus, Mycobacterium spp., Nocardia* spp., *Rhodococcus* spp., *Dermatophilus congolensis,* and *Gordonia amicalis.*

BHI with Glycerol and Reducing Agents (DSMZ Medium 215c)

Composition per liter:

Pancreatic digest of gelatin..14.5g
Brain heart, solids from infusion 6.0g
Peptic digest of animal tissue..6.0g
NaCl..5.0g
Glucose 3.0g
Na_2HPO_4..2.5g
Glycerol solution...10.0mL
L-Cysteine·HCl–Na_2S solution...10.0mL

pH 7.4 ± 0.2 at 25°C

Glycerol Solution:

Composition per 100.0mL:

Glycerol 87.0g

Preparation of Glycerol Solution: Add glycerol to distilled/deionized water and bring volume to 100.0mL. Mix thoroughly.

L-Cysteine·HCl–Na_2S Solution:

Composition per 100.0mL:

L-Cysteine·HCl..2.5g
$Na_2S \cdot 9H_2O$..2.5g

Preparation of L-Cysteine·HCl–Na_2S Solution: Add L-cysteine·HCl to distilled/deionized water and bring volume to 80.0mL. Mix thoroughly. Adjust pH to 11 with NaOH. Add $Na_2S \cdot 9H_2O$. Mix thoroughly. Bring volume to 100.0mL with distilled/deionized water. Gently heat and bring to boiling under 100% N_2. Cool to 25°C under 100% N_2. Autoclave for 15 min at 15 psi pressure–121°C.

Preparation of Medium: Add components, except L-cysteine·HCl–Na_2S solution, to distilled/deionized water and bring volume to 990.0mL. Mix thoroughly. Gently heat and bring to boiling under 100% N_2. Cool to 25°C under 100% N_2. Autoclave for 15 min at 15 psi pressure–121°C. Aseptically and anaerobically under 100% N_2 add 10.0mL of L-cysteine·HCl–Na_2S solution. Mix thoroughly. Aseptically under 100% N_2 distribute to tubes. Alternately distribute 10.0mL amounts of the medium without L-cysteine·HCl–Na_2S solution to tubes prior to autoclaving. Autoclave for 15 min at 15 psi pressure–121°C. Aseptically and anaerobically add 1.0mL of L-cysteine·HCl–Na_2S solution to each tube.

Use: For the cultivation of *Clostridium* sp.

BHI Medium (DSMZ Medium 215)

Composition per liter:

Pancreatic digest of gelatin..14.5g
Brain heart, solids from infusion 6.0g
Peptic digest of animal tissue..6.0g
NaCl..5.0g
Na_2HPO_4..2.5g

pH 7.4 ± 0.2 at 25°C

Preparation of Medium: Add components to distilled/deionized water and bring volume to 1.0L. Mix thoroughly. Distribute into tubes or flasks. Autoclave for 15 min at 15 psi pressure–121°C.

Use: For the cultivation of *Enterococcus hirae, Lodobacter fluviatilis, Yersinia* spp., *Tatumella ptyseos, Mycobacterium vanbaalenii, Oligella urethralis=Moraxella urethralis, Moraxella (Branhamella) catarrhalis, Campylobacter sputorum, Helcococcus kunzii, Bacillus sporothermodurans, Haemophilus actinomycetemcomitans, Escherichia coli, Pelczaria aurantia, Bacillus* spp., *Comamonas nitrativorans, Salmonella bongori (Salmonella choleraesuis* subsp. *bongori), Sphingomonas sanguinis (Sphingomonas sanguis), Arsenophonus nasoniae, Streptococcus orisratti, Listeria* spp., *Jonesia denitrificans=Listeria denitrificans, Propionibacterium propionicus=Arachnia propionica, Corynebacterium* spp., and *Nocardiopsis tropica.*

BHI/1 Medium

Composition per liter:

Pancreatic digest of gelatin..14.5g
Casein hydrolysate...10.0g
Glucose 8.0g
Brain heart, solids from infusion 6.0g
Peptic digest of animal tissue 6.0g
NaCl..5.0g
Na_2HPO_4..2.5g

pH 7.4 ± 0.2 at 25°C

Preparation of Medium: Add components to distilled/deionized water and bring volume to 1.0L. Mix thoroughly. Distribute into tubes or flasks. Autoclave for 15 min at 15 psi–121°C.

Use: For the cultivation and maintenance of *Actinomyces israelii* and *Propionibacterium propionicus.*

BHI/2 Medium

Composition per liter:

Pancreatic digest of gelatin..14.5g
Casein hydrolysate...10.0g
Glucose 8.0g
Brain heart, solids from infusion 6.0g
Peptic digest of animal tissue 6.0g
NaCl..5.0g
Yeast extract..5.0g
Na_2HPO_4..2.5g

pH 7.4 ± 0.2 at 25°C

Preparation of Medium: Add components to distilled/deionized water and bring volume to 1.0L. Mix thoroughly. Distribute into tubes or flasks. Autoclave for 15 min at 15 psi pressure–121°C.

Use: For the cultivation and maintenance of *Actinomyces georgiae, Actinomyces gerencseriae, Actinomyces naeslundii, Actinomyces odontolyticus,* and *Actinomyces viscosus.*

BHI/3 Medium

Composition per liter:

Pancreatic digest of gelatin..14.5g
Casein hydrolysate...10.0g
Brain heart, solids from infusion 6.0g
Peptic digest of animal tissue 6.0g

Starch	5.0g
NaCl	5.0g
Glucose	3.0g
Na_2HPO_4	2.5g

pH 7.4 ± 0.2 at 25°C

Preparation of Medium: Add components to distilled/deionized water and bring volume to 1.0L. Mix thoroughly. Distribute into tubes or flasks. Autoclave for 15 min at 15 psi–121°C.

Use: For the cultivation and maintenance of *Actinomyces bovis, Actinomyces gerencseriae, Actinomyces naeslundii, Actinomyces odontolyticus, Actinomyces viscosus,* and *Streptomyces* species.

BHI with Serum and Glucose

See: **Brain Heart Infusion with Serum and Glucose**

BHIS

See: **Brain Heart Infusion, Supplemented**

BHIV Agar, 1/10

See: **Brain Heart Infusion Agar, 1/10 with Vitamins**

BHIY Media

Composition per liter:

Beef heart, infusion from	250.0g
Calf brains, infusion from	200.0g
Yeast extract	20.0g
Agar	15.0g
Proteose peptone	10.0g
Sodium phosphate	2.5g
Dextrose	0.2g

pH 7.0 ± 0.2 at 25°C

Preparation of Medium: Add components to distilled/deionized water and bring volume to 1.0L. Mix thoroughly. Gently heat and bring to boiling. Distribute into tubes or flasks. Autoclave for 15 min at 15 psi pressure–121°C. Pour into sterile Petri dishes or leave in tubes.

Use: For the cultivation and maintenance of *Streptococcus bovis* and *Streptococcus equinus.*

Bicarbonate Agar

Composition per 100.0mL:

Soybean-casein digest agar	90.0mL
Sodium bicarbonate solution	10.0mL

pH 7.3 ± 0.2 at 25°C

Soybean-Casein Digest Agar :

Composition per liter:

Agar	15.0g
Pancreatic digest of casein	15.0g
Papaic digest of soybean meal	5.0g
NaCl	5.0g

Preparation of Soybean-Casein Digest Agar: Add components to distilled/deionized water and bring volume to 1.0L. Mix thoroughly. Gently heat and bring to boiling. Autoclave for 15 min at 15 psi pressure–121°C.

Sodium Bicarbonate Solution:

Composition per 10.0mL:

$NaHCO_3$	0.7g

Preparation of Sodium Bicarbonate Solution: Add $NaHCO_3$ to distilled/deionized water and bring volume to 10.0mL. Mix thoroughly. Filter sterilize. Use freshly prepared solution.

Preparation of Medium: To 90.0mL of cooled, sterile soybean-casein digest agar, aseptically add 10.0mL of sterile sodium bicarbonate solution. Mix thoroughly. Pour into sterile Petri dishes.

Use: For the cultivation of *Vibrio* species from foods.

Bifidobacterium Medium

Composition per liter:

Glucose	20.0g
Pancreatic digest of casein	20.0g
Yeast extract	10.0g
Peptone	10.0g
Tomato juice	333.0mL
Tween™ 80	2.0mL

pH 6.8 ± 0.2 at 25°C

Preparation of Medium: Combine 333.0mL of tomato juice with 666.0mL of distilled/deionized water. Bring to boiling. Filter through paper. Add remaining components to filtrate. Mix thoroughly. Bring volume to 1.0L with distilled/deionized water. Distribute into tubes or flasks. Autoclave for 30 min at 15 psi pressure–110°C.

Use: For the cultivation of *Bifidobacterium infantis.*

Bifidobacterium Agar

Composition per liter:

Peptone, special	23.0g
Agar	15.0g
Glucose	5.0g
NaCl	5.0g
Starch, soluble	1.0g
L-Cysteine hydrochloride	0.3g

pH 6.8 ± 0.2 at 25°C

Source: This medium is available from HiMedia.

Preparation of Medium: Add components to distilled/deionized water and bring volume to 1.0L. Distribute into tubes or flasks. Autoclave for 15 min at 15 psi pressure–121°C. Pour into Petri dishes or leave in tubes.

Use: For the cultivation of *Bifidobacterium* spp.

Bifidobacterium Broth

Composition per liter:

Glucose	20.0g
Casein enzymatic hydrolysate	20.0g
Tomato juice, solids	16.65g
Peptic digest of animal tissue	10.0g
Yeast extract	10.0g
Tween™ 80	2.0mL

pH 6.8 ± 0.2 at 25°C

Source: This medium is available from HiMedia.

Preparation of Medium: Add components to distilled/deionized water and bring volume to 1.0L. Distribute into tubes or flasks. Autoclave for 15 min at 15 psi pressure–121°C.

Use: For the cultivation of *Bifidobacterium infantis.*

Bifidobacterium Medium

Composition per liter:

Tryptic digest of casein	10.0g
Glucose	10.0g
Beef extract	5.0g
Yeast extract	5.0g
K_2HPO_4	3.0g
Tween™ 80	1.0mL
Sodium ascorbate solution	25.0mL
L-Cysteine·HCl solution	25.0mL

pH 6.8 ± 0.2 at 25°C

Sodium Ascorbate Solution:

Composition per 25.0mL:

Sodium ascorbate	10.0g

Preparation of Sodium Ascorbate Solution: Add sodium ascorbate to distilled/deionized water and bring volume to 25.0mL. Mix thoroughly. Filter sterilize.

L-Cysteine·HCl Solution:

Composition per 25.0mL:

L-Cysteine·HCl	0.5g

Preparation of L-Cysteine·HCl Solution: Add L-cysteine·HCl to distilled/deionized water and bring volume to 25.0mL. Mix thoroughly. Filter sterilize.

Preparation of Medium: Add components, except sodium ascorbate solution and L-cysteine·HCl solution, to distilled/deionized water and bring volume to 1.0L. Mix thoroughly. Autoclave for 15 min at 15 psi pressure–121°C. Aseptically add 25.0mL of sterile sodium ascorbate solution and 25.0mL of sterile L-cysteine·HCl solution. Mix thoroughly. Aseptically distribute into sterile tubes or flasks. Medium that has not been freshly prepared should be heated in a steamer for 10 min prior to the addition of ascorbate and L-cysteine.

Use: For the cultivation and maintenance of *Bifidobacterium adolescentis, Bifidobacterium angulatum, Bifidobacterium animalis, Bifidobacterium asteroides, Bifidobacterium bifidum, Bifidobacterium boum, Bifidobacterium breve, Bifidobacterium catenulatum, Bifidobacterium choerinum, Bifidobacterium coryneforme, Bifidobacterium cuniculi, Bifidobacterium dentium, Bifidobacterium gallicum, Bifidobacterium indicum, Bifidobacterium infantis, Bifidobacterium longum, Bifidobacterium magnum, Bifidobacterium merycicum, Bifidobacterium minimum, Bifidobacterium pseudocatenulatum, Bifidobacterium pseudolongum, Bifidobacterium pullorum, Bifidobacterium ruminantium, Bifidobacterium saeculare, Bifidobacterium subtile, Bifidobacterium suis,* and *Bifidobacterium thermophilum.*

Bifidobacterium Medium

Composition per liter:

Special peptone	23.0g
Agar	15.0g
NaCl	5.0g
Glucose	5.0g
Starch, soluble	1.0g
L-Cysteine·HCl	0.3g

Preparation of Medium: Add components to distilled/deionized water and bring volume to 1.0L. Mix thoroughly. Gently heat and bring to boiling. Distribute into tubes or flasks. Autoclave for 15 min at 15 psi pressure–121°C. Pour into sterile Petri dishes or leave in tubes.

Use: For the cultivation and maintenance of numerous *Bifidobacterium* species.

BiGGY Agar (Bismuth Sulfite Glucose Glycerin Yeast Extract Agar) (Nickerson Medium)

Composition per liter:

Agar	16.0g
Glucose	10.0g
Glycine	10.0g
Bismuth ammonium citrate	5.0g
Na_2SO_3	3.0g
Yeast extract	1.0g

pH 6.8 ± 0.2 at 25°C

Source: This medium is available as a premixed powder from Oxoid Unipath and BD Diagnostic Systems.

Preparation of Medium: Add components to distilled/deionized water and bring volume to 1.0L. Mix thoroughly and heat with frequent agitation until boiling. Distribute into tubes or flasks. Do not autoclave. Cool to approximately 45°–50°C. If desired, add 2mg/L of neomycin sulfate. Swirl to disperse the insoluble material and pour into sterile Petri dishes.

Use: For the detection, isolation, and presumptive identification of *Candida* species. Addition of neomycin helps inhibit bacterial species. *Candida albicans* appears as brown to black colonies with no pigment diffusion and no sheen. *Candida tropicalis* appears as dark brown colonies with black centers, black pigment diffusion, and a sheen. *Candida krusei* appears as shiny, wrinkled, brown to black colonies with yellow pigment diffusion. *Candida pseudotropicalis* appears as flat, shiny red to brown colonies with no pigment diffusion. *Candida parakrusei* appears as flat, shiny, wrinkled, dark reddish-brown colonies with light reddish-brown peripheries and a yellow fringe. *Candida stellatoidea* appears as flat dark brown colonies with a light fringe.

Bile Broth Base, HiVeg with Streptokinase

Composition per liter:

Plant peptone	20.0g
NaCl	5.0g
Synthetic detergent No. V	5.0g
Streptokinase solution	1.0mL

pH 7.1 ± 0.2 at 25°C

Source: This medium, without streptokinase solution, is available as a premixed powder from HiMedia.

Streptokinase Solution:

Composition per 1.0mL:

Streptokinase	100,000 units

Streptokinase Solution: Add streptokinase to distilled/deionized water and bring volume to 1.0mL. Mix thoroughly. Filter sterilize.

Preparation of Medium: Add components to distilled/deionized water and bring volume to 1.0L. Mix thoroughly. Heat gently and bring to boiling. Distribute into tubes or flasks. Autoclave for 15 min at 15 psi pressure–121°C. Cool to 40°C. Aseptically add 1.0mL of streptokinase solution. Mix thoroughly. If desired carbohydrate may also be added to this medium prior to sterilization.

Use: For the culture of blood clots from patients with suspected enteric fever.

Bile Broth Base with Streptokinase

Composition per liter:

Peptone....20.0g
NaCl....5.0g
Synthetic detergent No. V....5.0g
Streptokinase solution....1.0mL

pH 7.1 ± 0.2 at 25°C

Source: This medium, without streptokinase solution, is available as a premixed powder from HiMedia.

Streptokinase Solution:

Composition per 1.0mL:

Streptokinase....100,000 units

Streptokinase Solution: Add streptokinase to distilled/deionized water and bring volume to 1.0mL. Mix thoroughly. Filter sterilize.

Preparation of Medium: Add components to distilled/deionized water and bring volume to 1.0L. Mix thoroughly. Heat gently and bring to boiling. Distribute into tubes or flasks. Autoclave for 15 min at 15 psi pressure–121°C. Cool to 40°C. Aseptically add 1.0mL of streptokinase solution. Mix thoroughly. If desired carbohydrate may also be added to this medium prior to sterilization.

Use: For the culture of blood clots from patients with suspected enteric fever.

Bile Esculin Agar

Composition per liter:

Oxgall....20.0g
Agar....15.0g
Pancreatic digest of gelatin....5.0g
Beef extract....3.0g
Esculin....1.0g
Ferric citrate....0.5g
Horse serum....50.0mL

pH 6.8 ± 0.2 at 25°C

Source: This medium is available as a premixed powder from Oxoid Unipath and BD Diagnostic Systems.

Preparation of Medium: Add components, except horse serum, to distilled/deionized water and bring volume to 950.0L. Mix thoroughly and heat with frequent agitation until boiling. Autoclave for 15 min at 15 psi pressure–121°C. Cool to 45°–50°C. Aseptically add 50.0mL of filter sterilized horse serum. Distribute into sterile Petri dishes or test tubes. Cool tubes in a slanted position.

Use: For differentiation between group D streptococci and nongroup D streptococci. To differentiate members of the Enterobacteriaceae, particularly *Klebsiella, Enterobacter*, and *Serratia,* from other enteric bacteria. To differentiate *Listeria monocytogenes*. Bile tolerance and esculin hydrolysis (seen as a dark brown to black complex) are presumptive for enterococci (group D streptococci).

Bile Esculin Agar

Composition per liter:

Esculin....1.0g
Bile esculin agar base....1.0L

pH 6.6 ± 0.2 at 25°C

Bile Esculin Agar Base:

Composition per liter:

Oxgall....40.0g
Agar....15.0g
Peptone....5.0g
Beef extract....3.0g
Ferric citrate....0.5g

Source: This medium is available as a premixed powder from BD Diagnostic Systems.

Preparation of Bile Esculin Agar Base: Add components to distilled/deionized water and bring volume to 1.0L. Mix thoroughly.

Preparation of Medium: Add desired amount of esculin—typically 1.0g—to bile esculin agar base. Mix thoroughly and heat with frequent agitation until boiling. Autoclave for 15 min at 15 psi pressure–121°C. Cool to 45°–50°C. Distribute into sterile Petri dishes or test tubes. Cool tubes in a slanted position.

Use: For the isolation and presumptive identification of group D streptococci.

Bile Esculin Agar (BAM M18)

Composition per liter:

Oxgall....40.0g
Agar....15.0g
Pancreatic digest of gelatin....5.0g
Beef extract....3.0g
Esculin....1.0g
Ferric citrate....0.5g

pH 6.6 ± 0.2 at 25°C

Preparation of Medium: Add components to distilled/deionized water and bring volume to 1.0L. Mix thoroughly and heat with frequent agitation until boiling. Autoclave for 15 min at 15 psi pressure–121°C. Distribute into sterile Petri dishes or test tubes. Cool tubes in a slanted position.

Use: For differentiation between group D streptococci and nongroup D streptococci. To differentiate members of the Enterobacteriaceae, particularly *Klebsiella, Enterobacter*, and *Serratia,* from other enteric bacteria. To differentiate *Listeria monocytogenes*. Bile tolerance and esculin hydrolysis (seen as a dark brown to black complex) are presumptive for enterococci (group D streptococci).

Bile Esculin Agar

Composition per liter:

Bile salts....40.0g
Agar....15.0g
Pancreatic digest of animal tissue....5.0g
Beef extract....3.0g
Esculin....1.0g
Ferric citrate....0.5g

pH 6.6 ± 0.2 at 25°C

Preparation of Medium: Add components to distilled/deionized water and bring volume to 1.0L. Mix thoroughly and heat with frequent agitation until boiling. Autoclave for 15 min at 15 psi pressure–121°C. Distribute into sterile Petri dishes or test tubes.

Use: For the isolation and identification of *Yersinia enterocolitica.*

Bile Esculin Agar, HiVeg

Composition per liter:

Plant peptone....25.0g
Agar....15.0g
Plant hydrolysate....15.0g
Plant extract....6.0g
Synthetic detergent No. II....2.0g

Esculin 1.0g
Ferric citrate 0.5g

pH 6.6 ± 0.2 at 25°C

Source: This medium is available as a premixed powder from HiMedia.

Preparation of Medium: Add components to distilled/deionized water and bring volume to 1.0L. Mix thoroughly and heat with frequent agitation until boiling. Autoclave for 15 min at 15 psi pressure–121°C. Distribute into sterile Petri dishes or test tubes. Cool tubes in a slanted position.

Use: For the isolation and presumptive identification of group D streptococci.

Bile Esculin Agar with Kanamycin

Composition per liter:

Oxgall 20.0g
Agar 15.0g
Beef extract 3.0g
Esculin 1.0g
Ferric citrate 0.5g
Hemin 10.0mg
Vitamin K_1 .... 10.0mg
Horse serum 50.0mL
Kanamycin solution 10.0mL

pH 7.1 ± 0.2 at 25°C

Source: This medium is available as a premixed powder from BD Diagnostic Systems.

Kanamycin Solution:

Composition per 10.0mL:

Kanamycin 1.0g

Preparation of Kanamycin Solution: Add kanamycin to distilled/deionized water and bring volume to 10.0mL. Mix thoroughly. Filter sterilize.

Preparation of Medium: Add components to distilled/deionized water and bring volume to 1.0L. Mix thoroughly and heat with frequent agitation until boiling. Autoclave for 15 min at 15 psi pressure–121°C. Cool to 45°–50°C. Aseptically add 50.0mL of 5% filter-sterilized horse serum and 10.0mL of sterile kanamycin solution. Distribute into test tubes or flasks. Cool tubes in a slanted position.

Use: For the selective isolation and/or presumptive identification of bacteria of the *Bacteroides fragilis* group from specimens containing mixed flora. Examine colonies with a long-wavelength UV light. Pigmented colonies of the *Bacteroides* group will fluoresce red-orange. Growth on this medium with blackening of the medium is presumptive for *Bacteroides fragilis*.

Bile Esculin Azide Agar

Composition per liter:

Pancreatic digest of casein 17.0g
Agar 15.0g
Oxgall 10.0g
NaCl 5.0g
Yeast extract 5.0g
Proteose peptone No. 3 .... 3.0g
Esculin 1.0g
Ferric ammonium citrate 0.5g
NaN_3 .... 0.15g

pH 7.1 ± 0.2 at 25°C

Source: This medium is available as a premixed powder from BD Diagnostic Systems.

Caution: Sodium azide is toxic. Azides also react with metals and disposal must be highly diluted.

Preparation: Add components to distilled/deionized water and bring volume to 1.0L. Mix thoroughly and heat with frequent agitation until boiling. Distribute into tubes or flasks. Autoclave for 15 min at 15 psi pressure–121°C. Cool to 45°–50°C. Pour into sterile Petri dishes or leave in tubes. Cool tubes in a slanted position.

Use: For the isolation and presumptive identification of group D streptococci.

Bile Esculin Azide HiVeg Agar

Composition per liter:

Plant hydrolysate 20.0g
Agar 15.0g
Plant extract 5.0g
Plant peptone No. 3 .... 5.0g
NaCl 5.0g
Synthetic detergent No. II 5.0g
Esculin 1.0g
Ferric ammonium citrate 0.5g
NaN_3 .... 0.15g

pH 7.1 ± 0.2 at 25°C

Source: This medium is available as a premixed powder from HiMedia.

Caution: Sodium azide is toxic. Azides also react with metals and disposal must be highly diluted.

Preparation: Add components to distilled/deionized water and bring volume to 1.0L. Mix thoroughly and heat with frequent agitation until boiling. Distribute into tubes or flasks. Autoclave for 15 min at 15 psi pressure–121°C. Cool to 45°–50°C. Pour into sterile Petri dishes or leave in tubes. Cool tubes in a slanted position.

Use: For the isolation and presumptive identification of group D streptococci.

Bile Esculin HiVeg Agar Base with Esculin

Composition per liter:

Plant peptone 22.0g
Agar 15.0g
Plant hydrolysate 15.0g
Plant extract 6.0g
Synthetic detergent No. II 5.0g
Ferric citrate 0.5g
Esculin solution 4.0mL

pH 6.8 ± 0.2 at 25°C

Source: This medium, without esculin, is available as a premixed powder from HiMedia.

Esculin Solution:

Composition per 4.0mL:

Esculin 1.0g

Esculin Solution: Add esculin to distilled/deionized water and bring volume to 4.0mL. Mix thoroughly. Filter sterilize.

Preparation of Medium: Add components, except esculin solution, to distilled/deionized water and bring volume to 1.0L. Mix thoroughly. Gently heat and bring to boiling. Autoclave for 15 min at 15 psi pressure–121°C. Cool to 50°C. Aseptically add 4.0mL of sterile esculin solution. Mix thoroughly. Pour into sterile Petri dishes or distribute into sterile test tubes.

Use: For the isolation and presumptive identification of group D streptococci.

Bile Esculin HiVeg Agar with Kanamycin

Composition per liter:

Plant peptone no. 2	17.0g
Agar	15.0g
Plant extract	6.0g
Synthetic detergent	5.0g
Esculin	1.0g
Ferric citrate	0.5g
Kanamycin	0.1g
$Fe_4(P_2O_7)_3 \cdot H_2O$	0.01g
Vitamin K_1	0.01g

pH 7.1 ± 0.2 at 25°C

Source: This medium is available as a premixed powder from HiMedia.

Preparation: Add components to distilled/deionized water and bring volume to 1.0L. Mix thoroughly and heat with frequent agitation until boiling. Distribute into tubes or flasks. Autoclave for 15 min at 15 psi pressure–121°C. Cool to 45°–50°C. Pour into sterile Petri dishes or leave in tubes. Cool tubes in a slanted position.

Use: For the selective isolation and/or presumptive identification of bacteria of the *Bacteroides fragilis* group from specimens containing mixed flora. Examine colonies with a long-wavelength UV light. Pigmented colonies of the *Bacteroides* group will fluoresce red-orange. Growth on this medium with blackening of the medium is presumptive for *Bacteroides fragilis*.

Bile Oxalate Sorbose Broth (BOS Broth)

Composition per liter:

Na_2HPO_4	9.14g
Sodium oxalate	5.0g
Bile salts	2.0g
NaCl	1.0g
$CaCl_2 \cdot 2H_2O$	0.01g
$MgSO_4 \cdot 7H_2O$	0.01g
Asparagine solution	100.0mL
Methionine solution	100.0mL
Sorbose solution	100.0mL
Yeast extract solution	10.0mL
Sodium pyruvate solution	10.0mL
Metanil Yellow solution	10.0mL
Sodium nitrofurantoin solution	10.0mL
Irgasan® solution	1.0mL

pH 7.6 ± 0.2 at 25°C

Asparagine Solution:

Composition per 100.0mL:

Asparagine	1.0g

Preparation of Asparagine Solution: Add asparagine to distilled/deionized water and bring volume to 100.0mL. Mix thoroughly. Filter sterilize.

Methionine Solution:

Composition per 100.0mL:

Methionine	1.0g

Preparation of Methionine Solution: Add methionine to distilled/deionized water and bring volume to 100.0mL. Mix thoroughly. Filter sterilize.

Sorbose Solution:

Composition per 100.0mL:

Sorbose	10.0g

Preparation of Sorbose Solution: Add sorbose to distilled/deionized water and bring volume to 100.0mL. Mix thoroughly. Filter sterilize.

Yeast Extract Solution:

Composition per 10.0mL:

Yeast extract	0.025g

Preparation of Yeast Extract Solution: Add yeast extract to distilled/deionized water and bring volume to 10.0mL. Mix thoroughly. Filter sterilize.

Sodium Pyruvate Solution:

Composition per 10.0mL:

Sodium pyruvate	0.05g

Preparation of Sodium Pyruvate Solution: Add sodium pyruvate to distilled/deionized water and bring volume to 10.0mL. Mix thoroughly. Filter sterilize.

Metanil Yellow Solution:

Composition per 10.0mL:

Metanil Yellow	0.025g

Preparation of Metanil Yellow Solution: Add Metanil Yellow to distilled/deionized water and bring volume to 10.0mL. Mix thoroughly. Filter sterilize.

Sodium Nitrofurantoin Solution:

Composition per 10.0mL:

Sodium nitrofurantoin	0.01g

Preparation of Sodium Nitrofurantoin Solution: Add sodium nitrofurantoin to distilled/deionized water and bring volume to 10.0mL. Mix thoroughly. Filter sterilize.

Irgasan® Solution:

Composition per 10.0mL:

Irgasan	0.04g
Ethanol (95% solution)	10.0mL

Preparation of Irgasan Solution: Add Irgasan to 10.0mL of ethanol. Mix thoroughly. Filter sterilize.

Preparation of Medium: Add components, except asparagine solution, methionine solution, sorbose solution, yeast extract solution, sodium pyruvate solution, Metanil Yellow solution, sodium nitrofurantoin solution, and Irgasan solution, to distilled/deionized water and bring volume to 659.0mL. Mix thoroughly. Gently heat and bring to boiling. Autoclave for 15 min at 15 psi pressure–121°C. Cool to 45°–50°C. Aseptically add 100.0mL of sterile asparagine solution, 100.0mL of sterile methionine solution, 100.0mL of sterile sorbose solution, 10.0mL of sterile yeast extract solution, 10.0mL of sterile sodium pyruvate solution, 10.0mL of sterile Metanil Yellow solution, 10.0mL of sterile sodium nitrofurantoin solution, and 1.0mL of sterile Irgasan solution. Mix thoroughly. Pour into sterile Petri dishes or distribute into sterile tubes.

Use: For the isolation and cultivation of *Yersinia enterocolitica* from foods.

Bile Peptone Transport Medium

Composition per liter:

Casein enzymatic hydrolysate	10.0g
NaCl	10.0g
Sodium taurocholate	5.0g

pH 8.5 ± 0.2 at 25°C

Source: This medium is available from HiMedia.

Preparation of Medium: Add components to distilled/deionized water and bring volume to 1.0L. Mix thoroughly. Distribute into tubes. Autoclave for 15 min at 15 psi pressure–121°C.

Use: For the transport and preservation of *Vibrio cholerae.*

Bile Salt Agar with Streptokinase

Composition per liter:

Agar	18.0g
Peptone	10.0g
Meat extract	10.0g
NaCl	5.0g
Sodium taurocholate	5.0g
Streptokinase solution	1.0mL

pH 8.2 ± 0.2 at 25°C

Source: This medium, without streptokinase solution, is available as a premixed powder from HiMedia.

Streptokinase Solution:

Composition per 1.0mL:

Streptokinase	100,000 U

Streptokinase Solution: Add streptokinase to distilled/deionized water and bring volume to 1.0mL. Mix thoroughly. Filter sterilize.

Preparation of Medium: Add components to distilled/deionized water and bring volume to 1.0L. Mix thoroughly. Heat gently and bring to boiling. Distribute into tubes or flasks. Autoclave for 15 min at 15 psi pressure–121°C. Cool to 40°C. Aseptically add 1.0mL of streptokinase solution. Mix thoroughly. If desired carbohydrate may also be added to this medium prior to sterilization.

Use: For the isolation and cultivation of bile tolerant enteric bacilli.

Bile Salts Brilliant Green Starch Agar (BBGS Agar)

Composition per liter:

Agar	15.0g
Soluble starch	10.0g
Proteose peptone	10.0g
Beef extract	5.0g
Bile salts	5.0g
Brilliant Green (0.05% solution)	1.0mL

pH 7.2 ± 0.2 at 25°C

Preparation of Medium: Add components to distilled/deionized water and bring volume to 1.0L. Mix thoroughly. Gently heat while stirring and bring to boiling. Distribute into tubes or flasks. Autoclave for 15 min at 15 psi pressure–121°C. Pour into sterile Petri dishes or leave in tubes.

Use: For the isolation and cultivation of *Aeromonas hydrophila* from foods.

Bile Salts Gelatin Agar

Composition per 100.0mL:

Gelatin	3.0g
Agar	1.5g
Pancreatic digest of casein	1.0g
NaCl	1.0g
Sodium taurocholate	0.5g
Na_2CO_3	0.1g
Water	100.0mL

pH 8.5 ± 0.2 at 25°C

Preparation of Medium: Add components to distilled/deionized water and bring volume to 1.0L. Mix thoroughly. Gently heat and bring to boiling. Distribute into tubes or flasks. Autoclave for 15 min at 15 psi pressure–121°C. Pour into sterile Petri dishes or leave in tubes.

Use: For the cultivation of *Vibrio cholerae.*

BIN Medium

Composition per liter:

Beef heart, infusion from	250.0g
Calf brains, infusion from	200.0g
Agar	15.0g
Proteose peptone	10.0g
NaCl	5.0g
Na_2HPO_4	2.5g
Glucose	2.0g
Irgasan solution	4.0mL
Crystal Violet solution	1.0mL
Sodium cholate solution	1.0mL
Sodium deoxycholate solution	1.0mL
Nystatin solution	1.0mL

pH 7.4 ± 0.2 at 25°C

Sodium Cholate Solution:

Composition per 100.0mL:

Sodium cholate	5.0g

Preparation of Sodium Cholate Solution: Add sodium cholate to distilled/deionized water and bring volume to 100.0mL. Mix thoroughly. Gently heat while stirring and bring to boiling. Autoclave for 15 min at 15 psi pressure–121°C. Cool to 25°C.

Sodium Deoxycholate Solution:

Composition per 100.0mL:

Sodium deoxycholate	5.0g

Preparation of Sodium Deoxycholate Solution: Add sodium deoxycholate to distilled/deionized water and bring volume to 100.0mL. Mix thoroughly. Gently heat while stirring and bring to boiling. Autoclave for 15 min at 15 psi pressure–121°C. Cool 25°C.

Irgasan Solution:

Composition per 50.0mL:

Irgasan DP300	10.0mg
Ethanol, 90%	50.0mL

Preparation of Irgasan Solution: Add irgasan to 90% ethanol and bring volume to 50.0mL. Mix thoroughly.

Crystal Violet Solution:

Composition per 10.0mL:

Crystal Violet	10.0mg

Preparation of Crystal Violet Solution: Add Crystal Violet to distilled/deionized water and bring volume to 10.0mL. Mix thoroughly. Gently heat while stirring and bring to boiling. Autoclave for 15 min at 15 psi pressure–121°C. Cool to 25°C.

Nystatin Solution:

Composition per 10.0mL:

Nystatin 2.5g

Preparation of Nystatin Solution: Add nystatin to distilled/deionized water and bring volume to 10.0mL. Mix thoroughly. Filter sterilize.

Preparation of Medium: Add components, except irgasan solution, Crystal Violet solution, sodium cholate solution, sodium deoxycholate solution, and nystatin solution, to distilled/deionized water and bring volume to 992.0mL. Mix thoroughly. Gently heat while stirring and bring to boiling. Autoclave for 15 min at 15 psi pressure–121°C. Cool to 85°C. Aseptially add 4.0mL irgasam solution. Mix thoroughly to volatilize the ethanol. Cool to 50°C. Aseptially add 1.0mL each of Crystal Violet solution, sodium cholate solution, sodium deoxycholate solution, and nystatin solution. Mix thoroughly. Pour into sterile Petri dishes.

Use: For the efficient detection of *Yersinia pestis* from clinical and other specimens. The formulation of this medium is based on brain heart infusion agar, to which the selective agents irgasan, cholate salts, Crystal Violet, and nystatin are introduced to enhance efficiency of recovery of *Y. pestis.*

Biosynth Chromogenic Medium for *Listeria monocytogenes* (BCM for *Listeria monocytogenes)* (BAM M17a)

Composition per liter:

Proprietary

Source: This medium is available from Biosynth International, Inc.

Use: To differentiate *Listeria monocytogenes* and *L. ivanovii* from other *Listeria* spp. Supplements render the medium selective. Differential activity for all *Listeria* species is based upon a chromogenic substrate included in the medium. This is a complete test system with a fluorogenic selective enrichment broth and a chromogenic plating medium both detecting the virulence factor phosphatidylinositol specific phospholipase C (PI-PLC). The medium contains a substrate for phosphatidylinositol-specific phospholipase C (PlcA) enzymes. The selective enrichment broth is fluorogenic. The plating medium for rapid detection and enumeration of pathogenic *Listeria* combines cleavage of the chromogenic PI-PLC substrate with the additional production of a white precipitate surrounding the target colonies.

Biotin Assay Medium

Composition per liter:

Glucose 40.0g
Sodium acetate 20.0g
Vitamin assay casamino acids 12.0g
K_2HPO_4 .... 1.0g
KH_2PO_4 .... 1.0g
$MgSO_4 \cdot 7H_2O$ 0.4g
DL-Tryptophane 0.2g
L-Cystine 0.2g
Adenine sulfate 0.02g
$FeSO_4$ 0.02g
Guanine·HCl 0.02g
$MgSO_4 \cdot 7H_2O$ 0.02g
NaCl 0.02g
Uracil 0.02g
Calcium pantothenate 2.0mg
Niacin 2.0mg
Pyridoxine·HCl 2.0mg
Riboflavin 2.0mg
Thiamine·HCl 2.0mg
p-Aminobenzoic acid 0.2mg

pH 6.7 ± 0.2 at 25°C

Source: This medium is available as a premixed powder from BD Diagnostic Systems.

Preparation of Medium: Add components to distilled/deionized water and bring volume to 1.0L. Mix thoroughly. Gently heat and bring to boiling. Continue boiling for 2–3 min. Distribute into tubes in 5.0mL volumes. Add standard solution or test solutions to each tube. Adjust the volume of each tube to 10.0mL with distilled/deionized water. Autoclave for 15 min at 15 psi pressure–121°C.

Use: For use in the microbiological assay of biotin using *Lactobacillus plantarum* as the test microorganism.

Biphasic Medium for *Neisseria*

Composition per liter:

Glucose starch agar 1.0L
Glucose starch broth 1.0L

pH 7.3 ± 0.2 at 25°C

Glucose Starch Agar:

Composition per liter:

Agar 20.0g
Gelatin 20.0g
Proteose peptone No. 3 .... 15.0g
Soluble starch 10.0g
NaCl 5.0g
Glucose 2.0g
Na_2HPO_4 .... 3.0g

Preparation of Glucose Starch Agar: Add components to distilled/deionized water and bring volume to 1.0L. Mix thoroughly. Gently heat and bring to boiling. Autoclave for 15 min at 15 psi pressure–121°C. Cool to 50°C.

Glucose Starch Broth:

Composition per liter:

Gelatin 20.0g
Proteose peptone No. 3 .... 15.0g
Soluble starch 10.0g
NaCl 5.0g
Glucose 2.0g
Na_2HPO_4 .... 3.0g

Preparation of Glucose Starch Broth: Add components to distilled/deionized water and bring volume to 1.0L. Mix thoroughly. Gently heat and bring to boiling. Autoclave for 15 min at 15 psi pressure–121°C. Cool to 25°C.

Preparation of Medium: Aseptically distibute glucose starch agar into flasks in 100–125mL volumes. Allow agar to solidify. Overlay agar with 25.0mL of sterile glucose starch broth.

Use: For selective isolation and cultivation of *Neisseria* species.

Biphenyl Agar (DSMZ Medium 457d)

Composition per liter:

Agar 15.0g
Na_2HPO_4 .... 2.44g

KH_2PO_4	1.52g
$(NH_4)_2SO_4$	0.5g
Biphenyl	0.25g
$MgSO_4 \cdot 7H_2O$	0.2g
$CaCl_2 \cdot 2H_2O$	0.05g
Trace elements solution SL-4	10.0mL

pH 6.9 ± 0.2 at 25°C

Trace Elements Solution SL-4:
Composition per liter:

EDTA	0.5g
$FeSO_4 \cdot 7H_2O$	0.2g
Trace elements solution SL-6	100.0mL

Trace Elements Solution SL-6:
Composition per liter:

H_3BO_3	0.3g
$CoCl_2 \cdot 6H_2O$	0.2g
$ZnSO_4 \cdot 7H_2O$	0.1g
$MnCl_2 \cdot 4H_2O$	0.03g
$Na_2MoO_4 \cdot H_2O$	0.03g
$NiCl_2 \cdot 6H_2O$	0.02g
$CuCl_2 \cdot 2H_2O$	0.01g

Preparation of Trace Elements Solution SL-6: Add components to distilled/deionized water and bring volume to 1.0L. Mix thoroughly. Adjust pH to 3.4.

Preparation of Trace Elements Solution SL-4: Add components to distilled/deionized water and bring volume to 1.0L. Mix thoroughly.

Biphenyl Solution:
Composition per liter:

Biphenyl	10.0g

Preparation of Biphenyl Solution: Add biphenyl to 1.0L ethanol. Mix thoroughly. Filter sterilize using a cellulose filter membrane.

Preparation of Medium: Add components, except biphenyl solution, to 1.0L distilled/deionized water. Adjust pH to 6.9. Heat and gently bring to boiling. Autoclave for 15 min at 15 psi pressure–121°C. Cool to 50°C. Add an aliquot of the biphenyl solution to the lid of a sterile Petri dish so that the final concentration will be approximately 0.25g/L biphenyl, and let the ethanol evaporate so that the crystals of biphenyl coat the lid of the Petri dish. Aseptically add sterile agar medium to the crystal-layered Petri dish.

Use: For the cultivation of biphenyl-utilizing bacteria.

Bird Seed Agar (*Guizotia abyssinica* Creatinine Agar) (Niger Seed Agar)/(Staib Agar)

Composition per liter:

Agar	15.0g
Glucose	15.0g
Creatinine	5.0g
KH_2PO_4	3.0g
Biphenyl	1.0g
Chloramphenicol	0.5g
Guizotia abyssinica seed (niger seed) extract	1000.0mL

pH 6.7 ± 0.2 at 25°C

Preparation of Medium: Prepare seed extract by grinding 50.0g of *Guizotia abyssinica* seed in 1.0L of distilled/deionized water. Boil for 30 min. Filter through cheesecloth and filter paper. Add remaining components to seed filtrate. Mix thoroughly and heat with frequent agitation until boiling. Distribute into flasks or tubes. Autoclave for 25 min at 15 psi pressure–110°C.

Use: For the selective isolation and differentiation of *Cryptococcus neoformans* from other *Cryptococcus* species and other yeasts.

Bismuth Sulfite Agar

Composition per liter:

Agar	20.0g
$Bi_2(SO_3)_3$	8.0g
Pancreatic digest of casein	5.0g
Peptic digest of animal tissue	5.0g
Beef extract	5.0g
Glucose	5.0g
Na_2HPO_4	4.0g
$FeSO_4 \cdot 7H_2O$	0.3g

pH 7.5 ± 0.2 at 25°C

Source: This medium is available as a premixed powder from Oxoid Unipath and BD Diagnostic Systems.

Preparation of Medium: Add components to distilled/deionized water and bring volume to 1.0L. Mix thoroughly and heat with frequent agitation until boiling. Boil for 1 min. Do not autoclave. Cool to 45°–50°C. Pour into sterile Petri dishes while gently shaking flask to disperse precipitate. Use plates the same day as prepared.

Use: For the selective isolation and identification of *Salmonella typhi* and other enteric bacilli. *Salmonella typhi* appears as flat, black, "rabbit-eye" colonies surrounded by a zone of black with a metallic sheen.

Bismuth Sulfite Agar

Composition per liter:

Agar	20.0g
Peptic digest of animal tissue	10.0g
Bismuth sulfite indicator	8.0g
Glucose	5.0g
Beef extract	5.0g
Na_2HPO_4	4.0g
$FeSO_4$	0.3g
Brilliant Green	0.025g

pH 7.7 ± 0.2 at 25°C

Source: This medium is available from HiMedia.

Preparation of Medium: Add components to distilled/deionized water and bring volume to 1.0L. Mix thoroughly and heat with frequent agitation until boiling. Boil for 1 min. Do not autoclave. Cool to 45°–50°C. Mix thoroughly. The sensitivity of the medium depends largely upon uniform dispersion of precipitated bismuth sulfite in the final gel which should be dispersed before pouring the plates. Pour into sterile Petri dishes while gently shaking flask to disperse precipitate. Use plates the same day as prepared.

Use: For the selective isolation and identification of *Salmonella typhi* and other enteric bacilli. *Salmonella typhi* appears as flat, black, "rabbit-eye" colonies surrounded by a zone of black with a metallic sheen.

Bismuth Sulfite Agar, HiVeg

Composition per liter:

Agar	20.0g
Plant peptone	10.0g
Bismuth sulfite indicator	8.0g
Glucose	5.0g

Plant extract 5.0g
Na_2HPO_4 .... 4.0g
$FeSO_4$ 0.3g
Brilliant Green 0.025g

pH 7.7 ± 0.2 at 25°C

Preparation of Medium: Add components to distilled/deionized water and bring volume to 1.0L. Mix thoroughly and heat with frequent agitation until boiling. Boil for 1 min. Do not autoclave. Cool to 45°–50°C. Mix thoroughly. The sensitivity of the medium depends largely upon uniform dispersion of precipitated bismuth sulfite in the final gel which should be dispersed before pouring the plates. Pour into sterile Petri dishes while gently shaking flask to disperse precipitate. Use plates the same day as prepared.

Use: For the selective isolation and identification of *Salmonella typhi* and other enteric bacilli. *Salmonella typhi* appears as flat, black, "rabbit-eye" colonies surrounded by a zone of black with a metallic sheen.

Bismuth Sulfite Agar, Modified

Composition per liter:

Agar 12.7g
Bismuth sulfite indicator 8.0g
Glucose 5.0g
Beef extract 5.0g
Peptic digest of animal tissue 5.0g
Na_2HPO_4 .... 4.0g
$FeSO_4$ 0.3g
Brilliant Green 0.016

pH 7.5 ± 0.2 at 25°C

Source: This medium is available as a premixed powder from HiMedia.

Preparation of Medium: Add components to distilled/deionized water and bring volume to 1.0L. Mix thoroughly and heat with frequent agitation until boiling. Boil for 1 min. Do not autoclave. Cool to 45°–50°C. Mix thoroughly. The sensitivity of the medium depends largely upon uniform dispersion of precipitated bismuth sulfite in the final gel which should be dispersed before pouring the plates. Pour into sterile Petri dishes while gently shaking flask to disperse precipitate. Use plates the same day as prepared.

Use: For the selective isolation and identification of *Salmonella typhi* and other enteric bacilli. *Salmonella typhi* appears as flat, black, "rabbit-eye" colonies surrounded by a zone of black with a metallic sheen.

Bismuth Sulfite Agar, Modified, HiVeg

Composition per liter:

Agar 12.7g
Bismuth sulfite indicator 8.0g
Glucose 5.0g
Plant extract 5.0g
Plant peptone 5.0g
Na_2HPO_4 .... 4.0g
$FeSO_4$ 0.3g
Brilliant Green 0.016

pH 7.5 ± 0.2 at 25°C

Source: This medium is available as a premixed powder from HiMedia.

Preparation of Medium: Add components to distilled/deionized water and bring volume to 1.0L. Mix thoroughly and heat with frequent agitation until boiling. Boil for 1 min. Do not autoclave. Cool to 45°–50°C. Mix thoroughly. The sensitivity of the medium depends largely upon uniform dispersion of precipitated bismuth sulfite in the final gel which should be dispersed before pouring the plates. Pour into sterile Petri dishes while gently shaking flask to disperse precipitate. Use plates the same day as prepared.

Use: For the selective isolation and identification of *Salmonella typhi* and other enteric bacilli. *Salmonella typhi* appears as flat, black, "rabbit-eye" colonies surrounded by a zone of black with a metallic sheen.

Bismuth Sulfite Agar Wilson and Blair (BAM 19)

Composition per liter:

Agar 20.0g
Pancreatic digest of casein 10.0g
$Bi_2(SO_3)_3$ 8.0g
Beef extract 5.0g
Glucose 5.0g
Na_2HPO_4 .... 4.0g
$FeSO_4 \cdot 7H_2O$ 0.3g
Brilliant Green 0.025g

pH 7.7 ± 0.2 at 25°C

Preparation of Medium: Add components to distilled/deionized water and bring volume to 1.0L. Mix thoroughly and heat with frequent agitation until boiling. Boil for 1 min. Do not autoclave. Cool to 45°–50°C. Pour into sterile Petri dishes while gently shaking flask to disperse precipitate. Let plates dry for about 2h with lids partially removed. Use plates the within one day of preparation; medium loses selectivity after 48h.

Use: For the selective isolation and identification of *Salmonella typhi* and other enteric bacilli. *Salmonella typhi* appears as flat, black, "rabbit-eye" colonies surrounded by a zone of black with a metallic sheen.

Bismuth Sulfite Broth (m-Bismuth Sulfite Broth)

Composition per liter:

$Bi_2(SO_3)_3$ 16.0g
Pancreatic digest of casein 10.0g
Peptic digest of animal tissue 10.0g
Beef extract 10.0g
Glucose 10.0g
Na_2HPO_4 .... 8.0g
$FeSO_4 \cdot 7H_2O$ 0.6g

pH 7.7 ± 0.2 at 25°C

Preparation of Medium: Add components to distilled/deionized water and bring volume to 1.0L. Mix thoroughly and heat with frequent agitation until boiling. Boil for 1 min. Do not autoclave. Cool to 45°–50°C. Mix to disperse the precipitate and aseptically distribute into sterile tubes or flasks. Use 2.0–2.2mL of medium for each membrane filter.

Use: For the selective isolation of *Salmonella typhi* and other enteric bacilli and for the detection of *Salmonella* by the membrane filter method.

Bismuth Sulfite Glucose Glycerin Yeast Extract Agar

See: **BiGGY Agar**

BL Agar
(Glucose Blood Liver Agar)

Composition per liter:

Agar	15.0g
Glucose	10.0g
Proteose peptone No. 3	10.0g
Pancreatic digest of casein	5.0g
Yeast extract	5.0g
Meat extract	3.0g
Phytone™	3.0g
Tween™ 80	1.0g
Soluble starch	0.5g
Liver extract	150.0mL
Horse blood	50.0mL
L-Cysteine·HCl solution	10.0mL
Solution A	10.0mL
Solution B	5.0mL

pH 7.2 ± 0.2 at 25°C

Liver Extract:

Composition per 170.0mL:

Liver powder	10.0g

Preparation of Liver Extract: Add 10.0g of liver powder to 170mL of distilled/deionized water. Gently heat to 60°C. Maintain at 50°–60°C for 1 hr. Gently bring to boiling. Boil for 5 min. Adjust pH to 7.2. Filter through Whatman #2 filter paper.

L-Cysteine·HCl Solution:

Composition per 10.0mL:

L-Cysteine·HCl	0.5g

Preparation of L-Cysteine·HCl Solution: Dissolve 0.5g of L-cysteine·HCl in distilled/deionized water and bring volume to 10.0mL. Mix thoroughly. Filter sterilize. Warm to 50°C.

Solution A:

Composition per 100.0mL:

K_2HPO_4	10.0g
KH_2PO_4	10.0g

Preparation of Solution A: Add components to distilled/deionized water and bring volume to 100.0mL. Mix thoroughly. Autoclave for 15 min at 15 psi pressure–121°C. Cool to 50°–55°C.

Solution B:

Composition per 100.0mL:

$MgSO_4 \cdot 7H_2O$	4.0g
NaCl	0.2g
$FeSO_4 \cdot 7H_2O$	0.2g
$MnSO_4 \cdot H_2O$	0.2g

Preparation of Solution B: Add components to distilled/deionized water and bring volume to 100.0mL. Mix thoroughly. Autoclave for 15 min at 15 psi pressure–121°C. Cool to 50°–55°C.

Preparation of Medium: Add components, except liver extract, horse blood, L-cysteine·HCl solution, solution A, and solution B, to distilled/deionized water and bring volume to 775.0mL. Mix thoroughly. Gently heat and bring to boiling. Autoclave for 15 min at 15 psi pressure–121°C. Cool to 50°–55°C. Aseptically add 150.0mL of sterile liver extract, 50.0mL of sterile horse blood, 10.0mL of sterile L-cysteine·HCl solution, 10.0 mL of sterile solution A, and 5.0mL of sterile solution B. Mix thoroughly. Pour into sterile Petri dishes or distribute into sterile tubes.

Use: For the cultivation and maintenance of *Atopobium minutum, Bacteroides distasonis, Bacteroides ovatus, Bacteroides thetaiotaomicron, Bacteroides uniformis, Bacteroides vulgatus,* numerous *Bifidobacterium* species, *Campylobacter divergens, Carnobacterium piscicola,* numerous *Clostridium* species, numerous *Lactobacillus* species, *Lactococcus lactis, Leuconostoc lactis, Leuconostoc mesenteroides,* and *Propionibacterium thoenii.*

Blaser's Agar
See: ***Campylobacter*** **Selective Medium, Blaser-Wang**

Blaser's *Campylobacter* Agar
See: ***Campylobacter*** **Agar, Blaser's**

Blaser-Wang *Campylobacter* Medium
See: **Blaser-Wang**

Blaser-Wang *Campylobacter* Medium
See: ***Campylobacter*** **Selective Medium, Blaser-Wang**

Blastobacter denitrificans Agar
(LMG Medium 157)

Composition per liter:

Agar	15.0g
Tryptone	2.0g
Lab Lemco beef extract	0.5g
Yeast extract	0.5g
Sodium acetate	0.2g
Glucose solution	10.0mL

pH 7.3 ± 0.2 at 25°C

Glucose Solution:

Composition per 10.0mL:

Glucose	2.5g

Preparation of Glucose Solution: Add glucose to 10.0mL of distilled/deionized water. Mix thoroughly. Filter sterilize.

Preparation of Medium: Add components, except glucose solution, to distilled/deionized water and bring volume to 990.0mL. Mix thoroughly. Autoclave for 15 min at 15 psi pressure–121°C. Cool to 45°–50°C. Aseptically add 10.0mL glucose solution. Mix thoroughly. Aseptically pour into sterile Petri dishes or distribute into sterile tubes.

Use: For the cultivation of *Blastobacter denitrificans.*

Blastobacter denitrificans Agar

Composition per liter:

Agar	15.0g
Pancreatic digest of casein	2.0g
Beef extract	0.5g
Yeast extract	0.5g
Sodium acetate	0.2g
Glucose solution	10.0mL

pH 7.2 ± 0.2 at 25°C

Glucose Solution:

Composition per 10.0mL:

Glucose	2.5g

Preparation of Glucose Solution: Add glucose to distilled/deionized water and bring volume to 10.0L. Mix thoroughly. Filter sterilize.

Preparation of Medium: Add components, except glucose solution, to distilled/deionized water and bring volume to 990.0mL. Mix thoroughly. Gently heat and bring to boiling. Autoclave for 15 min at 15 psi pressure–121°C. Cool to 50°–55°C. Aseptically add 10.0mL of sterile glucose solution. Mix thoroughly. Pour into sterile Petri dishes or distribute into sterile tubes.

Use: For the cultivation of *Blastobacter denitrificans*.

Blastobacter Enrichment Medium

Composition per liter:

Agar	18.0g
Peptone	0.5g
$MgSO_4 \cdot 7H_2O$	0.13g
$KH_2PO_4 \cdot 3H_2O$	0.13g

pH 7.2 ± 0.2 at 25°C

Preparation of Medium: Add components to tap water and bring volume to 1.0L. Mix thoroughly. Gently heat and bring to boiling. Distribute into tubes or flasks. Autoclave for 15 min at 15 psi pressure–121°C. Pour into sterile Petri dishes or leave in tubes.

Use: For the enrichment and cultivation of *Blastobacter* species.

Blastobacter Medium

Composition per liter:

Agar	15.0g
Peptone	10.0g
Yeast extract	10.0g
NaCl	5.0g

pH 7.2 ± 0.2 at 25°C

Preparation of Medium: Add components to distilled/deionized water and bring volume to 1.0L. Mix thoroughly and heat with frequent agitation until boiling. Autoclave for 15 min at 15 psi pressure–121°C.

Use: For the cultivation and maintenance of *Blastobacter natatorius* and other *Blastobacter* species.

Blastococcus aggregatus Medium

Composition per 1001.0mL:

Tryptone	2.0g
Yeast extract	2.0g
Tris(hydroxymethyl)amino methane·HCl buffer	1.0g
KNO_3	0.5g
Sodium glycerophosphate	0.1g
Artificial seawater	1.0L
Trace elements solution	1.0mL

pH 7.0 ± 0.2 at 25°C

Artificial Seawater:

Composition per liter:

Commercially available marine aquarium salts mixture	variable

Preparation of Artificial Seawater: Add commercially available marine aquarium salts mixture. Prepare according to manufacturer's recommendations. Mix thoroughly.

Trace Elements Solution:

Composition per liter:

H_3BO_3	2.85g
$MnCl_2 \cdot 4H_2O$	1.8g
Sodium tartrate	1.77g
$FeSO_4$	1.36g
$CoCl_2 \cdot 6H_2O$	40.4mg
$CuCl_2 \cdot 2H_2O$	26.9mg
$Na_2MoO_4 \cdot 2H_2O$	25.2mg
$ZnCl_2$	20.8mg

Preparation of Trace Elements Solution: Add components to distilled/deionized water and bring volume to 1.0L. Mix thoroughly.

Preparation of Medium: Combine components. Mix thoroughly. Adjust pH to 7.0. Distribute into tubes or flasks. Autoclave for 15 min at 15 psi pressure–121°C.

Use: For the cultivation of *Blastococcus aggregatus*.

Blastocystis Egg Medium

Composition per 1300.0mL:

Homogenized whole egg	783.0mL
Stone's modification of Locke's solution	217.0mL
Horse serum, heat inactivated	300.0mL

Homogenized Whole Egg:

Composition per liter:

Whole eggs	18–24

Preparation of Homogenized Whole Egg: Use fresh fertile eggs, less than 1 week old. Scrub the shells with soap. Let stand in a soap solution for 30 min. Rinse in running water. Soak eggs in 70% ethanol for 15 min. Break the eggs into a sterile container. Homogenize by shaking. Filter through four layers of sterile cheesecloth into a sterile graduated cylinder. Measure out 1.0L.

Stone's Modification of Locke's Solution:

Composition per liter:

NaCl	8.0g
Na_2HPO_4	2.0g
$NaHCO_3$	0.4g
KH_2PO_4	0.3g
$CaCl_2$	0.2g
KCl	0.2g
$MgCl_2 \cdot 6H_2O$	0.01g

Preparation of Stone's Modification of Locke's Solution: Add components to distilled/deionized water and bring volume to 1.0L. Mix thoroughly.

Preparation of Medium: Distribute homogenized whole egg in 4.0mL volumes into 16 × 125mm screw-capped test tubes. Place tubes in a slanted position. Inspissate at 80°C (moist heat) for 10 min. Allow to cool. Add 4.5mL of Stone's modification of Locke's solution to the surface of the solidified egg in each tube. Close tubes with a rubber stopper. Place tubes in a press. Autoclave for 15 min at 15 psi pressure–121°C. Cool to room temperature. Aseptically replace rubber stoppers with sterile screw caps. Prior to use, aseptically add 1.5mL of heat-inactivated sterile horse serum to each tube.

Use: For the cultivation of *Anophryoides* species, *Blastocystis hominis*, other *Blastocystis* species, *Endolimax nana*, and *Metanophrys* species.

BLE HiVeg Broth Base with *Listeria* Selective Supplement (Buffered *Listeria* Enrichment HiVeg Broth Base with *Listeria* Selective Supplement)

Composition per liter:

Plant hydrolysate	17.0g
Na_2HPO_4, anhydrous	9.6g

Yeast extract	6.0g
NaCl	5.0g
Papaic digest of soybean meal	3.0g
KH_2PO_4	2.5g
Glucose	2.5g
Sodium pyruvate	1.0g
Listeria selective supplement	5.0mL

pH 7.3 ± 0.2 at 25°C

***Listeria* Selective Supplement:**

Composition per 5.0mL:

Cycloheximide	50.0mg
Nalidixic acid	40.0mg
Acriflavin hydrochloride	15.0mg

Preparation of *Listeria* Selective Supplement: Add components to distilled/deionized water and bring volume to 5.0mL. Mix thoroughly. Filter sterilize.

Caution: Cycloheximide is toxic. Avoid skin contact or aerosol formation and inhalation.

Source: This medium, without *Listeria* selective supplement, is available as a premixed powder from HiMedia.

Preparation of Medium: Add components, except *Listeria* selective supplement, to distilled/deionized water and bring volume to 1.0L. Mix thoroughly. Gently heat and bring to boiling. Autoclave for 15 min at 15 psi pressure–121°C. Cool to 45°–50°C. Aseptically add 50.0mL of sterile *Listeria* selective supplement. Mix thoroughly.

Use: For the enrichment and isolation of *Listeria monocytogenes*.

Blood Agar

Composition per liter:

Agar	15.0g
Pancreatic digest of casein	15.0g
Papaic digest of soybean meal	5.0g
NaCl	5.0g
Sheep blood, defibrinated	50.0mL

pH 7.6 ± 0.2 at 25°C

Preparation of Medium: Add components, except sheep blood, to distilled/deionized water and bring volume to 950.0mL. Mix thoroughly. Gently heat and bring to boiling. Autoclave for 15 min at 15 psi pressure–121°C. Cool to 45°–50°C. Aseptically add 50.0mL of sterile sheep blood. Mix thoroughly. Pour into sterile Petri dishes in 20.0mL volumes.

Use: For the cultivation of fastidious microorganisms.

Blood Agar Base

Composition per liter:

Agar	15.0g
Beef extract	10.0g
Peptone	10.0g
NaCl	5.0g
Sheep blood, defibrinated	50.0mL

pH 7.3 ± 0.2 at 25°C

Source: This medium is available as a premixed powder from Oxoid Unipath.

Preparation of Medium: Add components, except sheep blood, to distilled/deionized water and bring volume to 950.0mL. Mix thoroughly. Heat with frequent agitation and boil for 1 min to completely dissolve. Autoclave for 15 min at 15 psi pressure–121°C. Cool to 45°–50°C. Aseptically add 50.0mL of sterile, defibrinated sheep blood. Mix thoroughly and pour into sterile Petri dishes.

Use: For the isolation, cultivation, and detection of hemolytic activity of streptococci and other fastidious microorganisms.

Blood Agar Base (ATCC Medium 368)

Composition per liter:

Beef heart, infusion from	500.0g
Agar	15.0g
Tryptose	10.0g
NaCl	5.0g

pH 6.8 ± 0.2 at 25°C

Source: This medium is available as a premixed powder from BD Diagnostic Systems.

Preparation of Medium: Add components to distilled/deionized water and bring volume to 1.0L. Mix thoroughly. Heat with frequent agitation and boil for 1 min to completely dissolve. Autoclave for 15 min at 15 psi pressure–121°C. Cool the basal medium to 45°–50°C. Aseptically add sterile, defibrinated blood to a final concentration of 5%. Mix thoroughly and pour into sterile Petri dishes.

Use: For the isolation, cultivation, and detection of hemolytic activity of staphylococci, streptococci, and other fastidious microorganisms.

Blood Agar Base (BAM M20a)

Composition per liter:

Beef heart, infusion from	500.0g
Agar	15.0g
Tryptose	10.0g
NaCl	5.0g
Sheep blood, defibrinated	50.0mL

pH 6.8 ± 0.2 at 25°C

Preparation of Medium: Add components, except sheep blood, to distilled/deionized water and bring volume to 950.0mL. Mix thoroughly. Heat with frequent agitation and boil for 1 min to completely dissolve. Autoclave for 15 min at 15 psi pressure–121°C. Cool to 45°–50°C. Aseptically add 50.0mL of sterile, defibrinated sheep blood. Mix thoroughly and pour into sterile Petri dishes.

Use: For the isolation, cultivation, and detection of hemolytic activity of staphylococci, streptococci, and other fastidious microorganisms.

Blood Agar Base (Infusion Agar)

Composition per liter:

Agar	15.0g
Pancreatic digest of casein	13.0g
NaCl	5.0g
Yeast extract	5.0g
Heart muscle, solids from infusion	2.0g
Sheep blood, defibrinated	50.0mL

pH 7.3 ± 0.2 at 25°C

Source: This medium is available as a premixed powder from BD Diagnostic Systems.

Preparation of Medium: Add components, except sheep blood, to distilled/deionized water and bring volume to 950.0mL. Mix thoroughly. Heat with frequent agitation and boil for 1 min to completely dissolve. Autoclave for 15 min at 15 psi pressure–121°C. Cool to 45°–50°C. Aseptically add 50.0mL of sterile, defibrinated sheep blood. Mix thoroughly and pour into sterile Petri dishes.

Use: For the isolation, cultivation, and detection of hemolytic activity of streptococci and other fastidious microorganisms.

Blood Agar Base (Infusion Agar) (FDA Medium M21)

Composition per liter:

Heart muscle, infusion from	375.0g
Agar	15.0g
Thiotone	10.0g
NaCl	5.0g

pH 7.3 ± 0.2 at 25°C

Preparation of Medium: Add components to distilled/deionized water and bring volume to 1.0L. Mix thoroughly. Gently heat and bring to boiling. Distribute into tubes or flasks. Autoclave for 20 min at 15 psi pressure–121°C. Pour into sterile Petri dishes or leave in tubes.

Use: For the cultivation of a variety of microorganisms. For the preparation of blood agar by the addition of sterile blood.

Blood Agar Base with Blood

Composition per liter:

Agar	15.0g
Beef extract	10.0g
Tryptose	10.0g
NaCl	5.0g
Sheep blood, defibrinated	50.0mL

pH 7.3 ± 0.2 at 25°C

Source: This medium without blood is available as a premixed powder from HiMedia.

Preparation of Medium: Add components, except sheep blood, to distilled/deionized water and bring volume to 950.0mL. Mix thoroughly. Heat with frequent agitation and boil for 1 min to completely dissolve. Autoclave for 15 min at 15 psi pressure–121°C. Cool to 45°–50°C. Aseptically add 50.0mL of sterile, defibrinated sheep blood. Mix thoroughly and pour into sterile Petri dishes.

Use: For the isolation, cultivation, and detection of hemolytic activity of streptococci and other fastidious microorganisms.

Blood Agar Base, HiVeg with Blood

Composition per liter:

Agar	15.0g
Plant hydrolysate No. 1	10.0g
Plant infusion	10.0g
NaCl	5.0g
Sheep blood, defibrinated	50.0mL

pH 7.3 ± 0.2 at 25°C

Source: This medium without blood is available as a premixed powder from HiMedia.

Preparation of Medium: Add components, except sheep blood, to distilled/deionized water and bring volume to 950.0mL. Mix thoroughly. Heat with frequent agitation and boil for 1 min to completely dissolve. Autoclave for 15 min at 15 psi pressure–121°C. Cool to 45°–50°C. Aseptically add 50.0mL of sterile, defibrinated sheep blood. Mix thoroughly and pour into sterile Petri dishes.

Use: For the isolation, cultivation, and detection of hemolytic activity of streptococci and other fastidious microorganisms.

Blood Agar Base, Sheep

Composition per liter:

Pancreatic digest of casein	14.0g
Agar	12.5g
NaCl	5.0g
Peptone	4.5g
Yeast extract	4.5g
Sheep blood, defibrinated	70.0mL

ph 7.3 ± 0.2 at 25°C

Source: This medium is available as a premixed powder from Oxoid Unipath.

Preparation: Add components to distilled/deionized water and bring volume to 1.0L. Mix thoroughly. Autoclave for 15 min at 15 psi pressure–121°C. Cool the basal medium to 45°–50°C. Aseptically add 70.0mL of sterile, defibrinated sheep blood. Pour into sterile Petri dishes.

Use: For giving improved hemolytic reactions with sheep blood.

Blood Agar Base with Low pH, HiVeg with Blood

Composition per liter:

Agar	15.0g
Plant hydrolysate No. 1	10.0g
Plant infusion	10.0g
NaCl	5.0g
Sheep blood, defibrinated	50.0mL

pH 6. 8 ± 0.2 at 25°C

Source: This medium without blood is available as a premixed powder from HiMedia.

Preparation of Medium: Add components, except sheep blood, to distilled/deionized water and bring volume to 950.0mL. Mix thoroughly. Heat with frequent agitation and boil for 1 min to completely dissolve. Autoclave for 15 min at 15 psi pressure–121°C. Cool to 45°–50°C. Aseptically add 50.0mL of sterile, defibrinated sheep blood. Mix thoroughly and pour into sterile Petri dishes.

Use: For the isolation and growth of a wide variety of microorganisms. For the detection of the hemolytic reactions of streptococci and other fastidious microorganisms. The slightly acid pH of this medium enhances distinct hemolytic reactions.

Blood Agar Base with Peptone

Composition per liter:

Agar	15.0g
Beef extract	10.0g
Peptone	10.0g
NaCl	5.0g

pH 7.3 ± 0.2 at 25°C

Preparation of Medium: Add components to distilled/deionized water and bring volume to 1.0L. Mix thoroughly. Gently heat and bring to boiling. Distribute into tubes or flasks. Autoclave for 15 min at 15 psi pressure–121°C. Pour into sterile Petri dishes or leave in tubes.

Use: For use as a base to which blood can be added; for the isolation, cultivation, and detection of hemolytic activity of streptococci and other fastidious microorganisms.

Blood Agar Base with 2.5% Sodium Chloride

Composition per liter:

Beef heart, infusion from	500.0g
NaCl	30.0g
Agar	15.0g
Tryptose	10.0g

pH 6.8 ± 0.2 at 25°C

Preparation of Medium: Add components to distilled/deionized water and bring volume to 1.0L. Mix thoroughly. Heat with frequent agitation and boil for 1 min to completely dissolve. Autoclave for 15 min at 15 psi pressure–121°C. Cool the basal medium to 45°–50°C. Aseptically add sterile, defibrinated blood to a final concentration of 5%. Mix thoroughly and pour into sterile Petri dishes.

Use: For the cultivation of *Paracoccus halodenitrificans*.

Blood Agar Base with 3.5% Sodium Chloride

Composition per liter:

Beef heart, infusion from	500.0g
NaCl	40.0g
Agar	15.0g
Tryptose	10.0g

pH 6.8 ± 0.2 at 25°C

Preparation of Medium: Add components to distilled/deionized water and bring volume to 1.0L. Mix thoroughly. Heat with frequent agitation and boil for 1 min to completely dissolve. Autoclave for 15 min at 15 psi pressure–121°C. Cool the basal medium to 45°–50°C. Aseptically add sterile, defibrinated blood to a final concentration of 5%. Mix thoroughly and pour into sterile Petri dishes.

Use: For the cultivation of *Vibrio costicola.*

Blood Agar Base with Special Peptone

Composition per liter:

Agar	15.0g
Beef extract	10.0g
Special peptone	10.0g
NaCl	5.0g
Sheep blood, defibrinated	50.0mL

pH 7.3 ± 0.2 at 25°C

Source: Special peptone (L72) is available from Oxoid Unipath.

Preparation of Medium: Add components, except sheep blood, to distilled/deionized water and bring volume to 950.0mL. Mix thoroughly. Heat with frequent agitation and boil for 1 min to completely dissolve. Autoclave for 15 min at 15 psi pressure–121°C. Cool to 45°–50°C. Aseptically add 50.0mL of sterile, defibrinated sheep blood. Mix thoroughly and pour into sterile Petri dishes.

Use: For the isolation, cultivation, and detection of hemolytic activity of streptococci and other fastidious microorganisms.

Blood Agar Base No. 2 (BAM M22)

Composition per 1004.0mL:

Agar	12.0g
Proteose peptone	15.0g
NaCl	5.0g
Yeast extract	5.0g
Liver digest	2.5g
Horse blood, defibrinated	50.0mL
FBP solution	4.0mL

pH 7.4 ± 0.2 at 25°C

FBP Solution:

Composition per 30.0mL:

$FeSO_4$	0.25g
$NaHSO_3$	0.25g
Sodium pyruvate	0.25g

Preparation of FBP Solution: Add components to distilled/deionized water and bring volume to 30.0mL. Mix thoroughly. Filter sterilize.

Preparation of Medium: Add components, except horse blood and FBP solution, to distilled/deionized water and bring volume to 950.0mL. Mix thoroughly. Gently heat and bring to boiling. Autoclave for 15 min at 15 psi pressure–121°C. Cool to 48°C. Aseptically add 50.0mL of sterile horse blood. Mix thoroughly. Aseptically add 4.0mL sterile FBP solution. Mix thoroughly. Pour into sterile Petri dishes in 20.0mL volumes.

Use: For the cultivation of *Brucella* spp. and other fastidious bacteria.

Blood Agar Base No. 2, HiVeg with Blood

Composition per liter:

Agar	15.0g
Plant peptone No. 3	15.0g
NaCl	5.0g
Yeast extract	5.0g
Plant extract No. 2	2.5g
Horse blood, defibrinated	70.0mL
Selective supplement	2 vials

pH 7.4 ± 0.2 at 25°C

Source: This medium without blood or supplement is available as a premixed powder from HiMedia.

Preparation of Medium: Add components, except horse blood and selective supplement, to distilled/deionized water and bring volume to 930.0mL. Mix thoroughly. Gently heat and bring to boiling. Autoclave for 15 min at 15 psi pressure–121°C. Cool to 45°–50°C. Aseptically add 70.0mL of sterile horse blood. Mix thoroughly. Aseptically add 2 vials of rehydrated selective supplement. For *Brucella* spp. use *Brucella* selective supplement. For *Campylobacter* spp. use *Campylobacter* supplement-I (Blaser-Wang), or *Campylobacter* Supplement II (Butzler), or *Campylobacter* Supplement III (Skirrow), or *Campylobacter* Growth Supplement. For streptococci use Strepto supplement. Mix thoroughly. Pour into sterile Petri dishes in 20.0mL volumes.

Use: For the cultivation of *Brucella* spp., *Campylobacter* spp., *Streptococcus* spp., and other fastidious bacteria.

Blood Agar Base No. 2 with 1.2% Agar, HiVeg™

Composition per liter:

Plant peptone No. 3	15.0g
Agar	12.0g
NaCl	5.0g
Yeast extract	5.0g
Plant extract No. 2	2.5g
Horse blood, defibrinated	70.0mL

pH 7.4 ± 0.2 at 25°C

Source: This medium without blood is available as a premixed powder from HiMedia.

Preparation of Medium: Add components, except horse blood and selective supplement, to distilled/deionized water and bring volume to 930.0mL. Mix thoroughly. Gently heat and bring to boiling. Autoclave for 15 min at 15 psi pressure–121°C. Cool to 45°–50°C. Aseptically add 70.0mL of sterile horse blood. Mix thoroughly. Aseptically add 2 vials of rehydrated selective supplement. For *Brucella* spp. use *Brucella* selective supplement. For *Campylobacter* spp. use *Campylobacter* supplement-I (Blaser-Wang), or *Campylobacter* Supplement II (Butzler), or *Campylobacter* Supplement III (Skirrow), or *Campylobacter* Growth Supplement. For streptococci use Strepto supplement. Mix thoroughly. Pour into sterile Petri dishes in 20.0mL volumes.

Use: For the cultivation of *Brucella* spp., *Campylobacter* spp., *Streptococcus* spp., and other fastidious bacteria.

Blood Agar, Diphasic

Composition per 800.0mL:

Lean beef, desiccated	25.0g
Agar	10.0g
Neopeptone	10.0g
NaCl	2.5g
Locke solution	200.0mL
Rabbit blood, defibrinated	100.0mL

pH 7.2–7.4 at 25°C

Locke Solution:

Composition per liter:

NaCl	8.0g
Glucose	2.5g
KH_2PO_4	0.3g
KCl	0.2g
$CaCl_2 \cdot 2H_2O$	0.2g

Preparation of Locke Solution: Add components to distilled/deionized water and bring volume to 1.0L. Mix thoroughly. Filter sterilize.

Preparation of Medium: Add beef to 500.0mL of distilled/deionized water. Let stand for 60 min. Gently heat and bring to 80°C for 5 min. Filter through Whatman #1 filter paper. To filtrate, add remaining components, except Locke solution and rabbit blood. Mix thoroughly. Adjust pH to 7.2–7.4 with NaOH. Autoclave for 20 min at 15 psi pressure–121°C. Cool to 45°–50°C. Aseptically add sterile rabbit blood. Mix thoroughly. Aseptically distribute into sterile tubes in 5.0mL volumes. Allow tubes to cool in a slanted position. Immediately prior to inoculation, overlay agar in each tube with 2.0mL of sterile Locke solution.

Use: For the cultivation of *Trypanosoma* species and *Leishmania* species.

Blood Agar, Diphasic Base Medium

Composition per 750.0mL:

Beef	25.0g
Agar	10.0g
Neopeptone	10.0g
NaCl	2.5g

pH 7.2–7.4 at 25°C

Preparation of Medium: Trim beef to remove fat. Add 25.0g of lean beef to 250.0mL of distilled/deionized water. Gently heat and bring to boiling. Boil for 2–3 min. Filter through Whatman #2 filter paper. Add agar, neopeptone, and NaCl to filtrate. Bring volume to 750.0mL with distilled/deionized water. Mix thoroughly. Adjust pH to 7.2–7.4. Gently heat and bring to boiling. Autoclave for 15 min at 15 psi pressure–121°C. Pour into sterile Petri dishes or distribute into sterile tubes.

Use: For the cultivation of *Trypanosoma* species.

Blood Agar with Low pH

Composition per liter:

Beef heart, solids from infusion	500.0g
Agar	15.0g
Tryptose	10.0g
NaCl	5.0g
Sheep blood, defibrinated	50.0mL

pH 6. 8 ± 0.2 at 25°C

Source: This medium is available as a premixed powder from BD Diagnostic Systems.

Preparation of Medium: Add components, except sheep blood, to distilled/deionized water and bring volume to 950.0mL. Mix thoroughly. Heat with frequent agitation and boil for 1 min to completely dissolve. Autoclave for 15 min at 15 psi pressure–121°C. Cool to 45°–50°C. Aseptically add 50.0mL of sterile, defibrinated sheep blood. Mix thoroughly and pour into sterile Petri dishes.

Use: For the isolation and growth of a wide variety of microorganisms. For the detection of the hemolytic reactions of streptococci and other fastidious microorganisms. The slightly acid pH of this medium enhances distinct hemolytic reactions.

Blood Agar No. 2

Composition per liter:

Proteose peptone	15.0g
Agar	12.0g
NaCl	5.0g
Yeast extract	5.0g
Liver digest	2.5g

pH 7.4 ± 0.2 at 25°C

Source: This medium is available as a premixed powder from BD Diagnostic Systems and Oxoid Unipath.

Preparation of Medium: Add components to distilled/deionized water and bring volume to 1.0L. Mix thoroughly. Heat with frequent agitation and boil for 1 min to completely dissolve. Autoclave for 15 min at 15 psi pressure–121°C. Cool the basal medium to 45°–50°C. Aseptically add sterile, defibrinated blood to a final concentration of 7%. Pour into sterile Petri dishes.

Use: For the isolation, cultivation, and detection of hemolytic activity of streptococci, pneumococci, and other particularly fastidious microorganisms.

Blood Base Agar (LMG Medium 45)

Composition per liter:

Agar	15.0g
Lab-Lemco beef extract	10.0g
Special peptones	10.0g
NaCl	5.0g

pH 7.1 ± 0.2 at 25°C

Source: Special peptones is available as a premixed powder from Oxoid Unipath.

Preparation of Medium: Add components to distilled/deionized water and bring volume to 1.0L. Mix thoroughly. Gently heat and bring

to boiling. Distribute into tubes or flasks. Autoclave for 15 min at 15 psi pressure–121°C. Pour into sterile Petri dishes or leave in tubes.

Use: For the cultivation and maintenance of heterotrophic bacteria.

Blood Base Agar with Charcoal (LMG Medium 46)

Composition per liter:

Agar	15.0g
Lab-Lemco beef extract	10.0g
Special peptones	10.0g
NaCl	5.0g
Charcoal	2.0g

pH 7.1 ± 0.2 at 25°C

Source: Special peptones is available as a premixed powder from Oxoid Unipath.

Preparation of Medium: Add components to distilled/deionized water and bring volume to 1.0L. Mix thoroughly. Gently heat and bring to boiling. Distribute into tubes or flasks. Autoclave for 15 min at 15 psi pressure–121°C. Pour into sterile Petri dishes or leave in tubes.

Use: For the cultivation and maintenance of various bacteria.

Blood Base Agar with Horse Blood (LMG Medium 47)

Composition per liter:

Agar	15.0g
Lab-Lemco beef extract	10.0g
Special peptones	10.0g
NaCl	5.0g
Horse blood, sterile defibrinated	50.0mL

pH 7.1 ± 0.2 at 25°C

Source: Special peptones is available as a premixed powder from Oxoid Unipath.

Preparation of Medium: Add components, except horse blood, to 950.0mL distilled/deionized water and bring volume to 1.0L. Mix thoroughly. Gently heat and bring to boiling. Autoclave for 15 min at 15 psi pressure–121°C. Cool to 45°–50°C. Aseptically add 50.0mL sterile horse blood. Mix thoroughly. Pour into sterile Petri dishes or distribute into sterile tubes.

Use: For the cultivation and maintenance of fastidious bacteria.

Blood Base Agar with Horse Blood, Fumarate, and Formate (LMG Medium 48)

Composition per liter:

Agar	15.0g
Lab-Lemco beef extract	10.0g
Special peptones, Oxoid	10.0g
NaCl	5.0g
Sodium fumarate	3.0g
Sodium formate	2.0g
Horse blood, sterile defibrinated	50.0mL

pH 7.2 ± 0.2 at 25°C

Preparation of Medium: Add components, except horse blood, to 950.0mL distilled/deionized water and bring volume to 1.0L. Mix thoroughly. Gently heat and bring to boiling. Autoclave for 15 min at 15 psi pressure–121°C. Cool to 45°–50°C. Aseptically add 50.0mL sterile horse blood. Mix thoroughly. Pour into sterile Petri dishes or distribute into sterile tubes.

Use: For the cultivation and maintenance of fastidious bacteria.

Blood Free *Campylobacter* Selectivity HiVeg Agar Base

Composition per liter:

Agar	12.0g
Plant extract	10.0g
Plant peptone	10.0g
NaCl	5.0g
Charcoal, bacteriological	4.0g
Plant hydrolysate	3.0g
Synthetic detergent No. III	1.0g
$FeSO_4$	0.25g
Sodium pyruvate	0.25g
Sodium deoxycholate solution	10.0mL
Cefazolin solution	1.0mL

pH 7.4 ± 0.2 at 25°C

Source: This medium, without deoxycholate and cefazolin solutions, is available as a premixed powder from HiMedia.

Sodium Deoxycholate Solution:

Composition per 100.0mL:

Sodium deoxycholate	10.0g

Preparation of Sodium Deoxycholate Solution: Add sodium deoxycholate to distilled/deionized water and bring volume to 100.0mL. Mix thoroughly. Gently heat while stirring and bring to boiling. Autoclave for 15 min at 15 psi pressure–121°C. Cool to 25°C.

Cefazolin Solution:

Composition per 10.0mL:

Cefazolin	0.1g

Preparation of Cefazolin Solution: Add cefazolin to distilled/deionized water and bring volume to 10.0mL. Mix thoroughly. Filter sterilize.

Preparation of Medium: Add components, except cefazolin solution and sodium deoxycholate solution, to distilled/deionized water and bring volume to 990.0mL. Mix thoroughly. Heat with frequent agitation and boil for 1 min to completely dissolve. Autoclave for 15 min at 15 psi pressure–121°C. Cool to 50°–55°C. Add 10.0mL of sterile sodium deoxycholate solution and 1.0mL of sterile cefazolin solution. Mix thoroughly. Pour into sterile Petri dishes.

Use: For the selective isolation of *Campylobacter* species, especially *Campylobacter jejuni* from human feces.

Blood Glucose Cystine Agar

Composition per 100.0mL:

Nutrient agar	85.0mL
Glucose cystine solution	10.0mL
Human blood, fresh	5.0mL

pH 6.8 ± 0.2 at 25°C

Nutrient Agar:

Composition per liter:

Agar	15.0g
Pancreatic digest of gelatin	5.0g
Beef extract	3.0g

Source: Nutrient agar is available as a premixed powder from BD Diagnostic Systems.

Preparation of Nutrient Agar: Add components to distilled/deionized water and bring volume to 1.0L. Mix thoroughly. Gently heat while stirring and bring to boiling. Distribute into tubes or flasks. Autoclave for 15 min at 15 psi pressure–121°C. Cool to 45°–50°C.

Glucose Cystine Solution:
Composition per 50.0mL:

Glucose	12.5g
L-Cystine·HCl	0.5g

Preparation of Glucose Cystine Solution: Add components to distilled/deionized water and bring volume to 50.0mL. Mix thoroughly. Filter sterilize.

Preparation of Medium: To 85.0mL of cooled, sterile agar solution, aseptically add 10.0mL of sterile glucose cystine solution and 5.0mL of human blood. Mix thoroughly. Pour into sterile Petri dishes or distribute into sterile tubes.

Use: For the cultivation of *Francisella tularensis*.

BM Medium (DSMZ Medium 1192)

Composition per liter:

NaCl	19.45g
$MgCl_2$	8.8g
Peptone	5.0g
Na_2SO_3	3.24g
$CaCl_2$	1.8g
Yeast extract	1.0g
KCl	0.55g
$NaHCO_3$	0.16g
Ferric citrate	0.1g
KBr	0.08g
$SrCl_2$	0.03g
H_3BO_3	0.02g
Na_2HPO_4	8.0mg
Na_2SiO_3	4.0mg
NaF	2.4mg
NH_4NO_3	1.6mg
Biotin	0.02mg
Vitamin B_{12}	0.001mg
Methanol	4.0mL

pH 7.6 ± 0.2 at 25°C

Preparation of Medium: Add components to distilled/deionized water and bring volume to 1.0L. Mix thoroughly. Gently heat while stirring and bring to boiling. Distribute into tubes or flasks. Autoclave for 15 min at 15 psi pressure–121°C.

Use: For the cultivation of *Bacillus methanolicus*.

Bosea Medium (DSMZ Medium 1052)

Composition per liter:

Agar	15.0g
Yeast extract	10.0g
ACES	10.0g
Activated charcoal	2.0g

pH 6.9 ± 0.2 at 25°C

Preparation of Medium: Add components, except agar, to distilled/deionized water and bring volume to 1.0L. Mix thoroughly. Adjust pH to 6.9. Add agar. Gently heat while stirring and bring to boiling. Distribute into tubes or flasks. Autoclave for 15 min at 15 psi pressure–121°C. Pour into sterile Petri dishes or leave in tubes.

Use: For the cultivation of *Bosea* spp.

Brain Heart Infusion Agar (BAM M24 Medium 2)

Composition per liter:

Agar	15.0g
Pancreatic digest of gelatin	14.5g
Brain heart, solids from infusion	6.0g
Peptic digest of animal tissue	6.0g
NaCl	5.0g
Glucose	3.0g
Na_2HPO_4	2.5g

pH 7.4 ± 0.2 at 25°C

Source: This medium is available as a premixed powder from BD Diagnostic Systems.

Preparation of Medium: Add components to distilled/deionized water and bring volume to 1.0L. Mix thoroughly. Distribute into tubes or flasks while shaking to distribute precipitate. Autoclave for 15 min at 15 psi pressure–121°C. Mix thoroughly. Pour into sterile Petri dishes.

Use: For the cultivation of a wide variety of fastidious microorganisms, including bacteria, yeasts, and molds.

Brain Heart Infusion Agar 0.7% (BHI Agar 0.7%) (BAM M23)

Composition per liter:

Pancreatic digest of gelatin	14.5g
Agar	7.0g
Brain heart, solids from infusion	6.0g
Peptic digest of animal tissue	6.0g
NaCl	5.0g
Glucose	3.0g
Na_2HPO_4	2.5g

pH 5.3 ± 0.2 at 25°C

Source: This medium without agar is available as a premixed powder from BD Diagnostic Systems.

Preparation of Medium: Add components, except agar, to distilled/deionized water and bring volume to 1.0L. Mix thoroughly. Adjust pH to 5.3 with 1*N* HCl. Mix thoroughly. Add agar. Gently heat and bring to boiling. Distribute into tubes. Autoclave for 10 min at 15 psi pressure–121°C.

Use: For the detection of staphylococcal enterotoxin.

Brain Heart Infusion Broth (BHI Broth) (BAM M24 Medium 2)

Composition per liter:

Pancreatic digest of gelatin	14.5g
Brain heart, solids from infusion	6.0g
Peptic digest of animal tissue	6.0g
NaCl	5.0g
Glucose	3.0g
Na_2HPO_4	2.5g

pH 7.4 ± 0.2 at 25°C

Source: This medium is available as a premixed powder from BD Diagnostic Systems.

Preparation of Medium: Add components to distilled/deionized water and bring volume to 1.0L. Mix thoroughly. Distribute into tubes or flasks while shaking to distribute precipitate. Autoclave for 15 min at 15 psi pressure–121°C.

Use: For the cultivation of a wide variety of microorganisms, including bacteria, yeasts, and molds, especially fastidious species.

Blue-Green Agar

Composition per liter:

Agar	10.0g
$NaNO_3$	1.5g
$MgSO_4 \cdot 7H_2O$	0.075g
K_2HPO_4	0.04g
$CaCl_2 \cdot 2H_2O$	0.036g
Na_2CO_3	0.02g
Citric acid	6.0mg
Ferric ammonium citrate	6.0mg
EDTA disodium salt	1.0mg
Vitamin B_{12} solution	50.0mL
Trace metal mix A5	1.0mL

pH 7.1 ± 0.2 at 25°C

Trace Metal Mix A5:

Composition per liter:

H_3BO_3	2.86g
$MnCl_2 \cdot 4H_2O$	1.81g
$Na_2MoO_4 \cdot 2H_2O$	0.39g
$ZnSO_4 \cdot 7H_2O$	0.222g
$CuSO_4 \cdot 5H_2O$	0.079g
$Co(NO_3)_2 \cdot 6H_2O$	0.049g

Preparation of Trace Metal Mix A5: Add components to distilled/deionized water and bring volume to 1.0L. Mix thoroughly.

Vitamin B_{12} Solution:

Composition per 50.0mL:

Vitamin B_{12}	0.01g

Preparation of Vitamin B_{12} Solution: Add vitamin B_{12} to distilled/deionized water and bring volume to 50.0mL. Mix thoroughly. Filter sterilize.

Preparation of Medium: Add components, except vitamin B_{12} solution, to glass-distilled water and bring volume to 950.0mL. Mix thoroughly. Heat gently and bring to boiling. Autoclave for 15 min at 15 psi pressure–121°C. Cool the basal medium to 45°–50°C. Add vitamin B_{12} solution. Mix thoroughly. Pour into sterile Petri dishes or distribute into sterile tubes.

Use: For the cultivation and maintenance of *Synechococcus* species.

Blue-Green Broth

Composition per liter:

$NaNO_3$	1.5g
$MgSO_4 \cdot 7H_2O$	0.075g
K_2HPO_4	0.04g
$CaCl_2 \cdot 2H_2O$	0.036g
Na_2CO_3	0.02g
Citric acid	6.0mg
Ferric ammonium citrate	6.0mg
EDTA disodium salt	1.0mg
Vitamin B_{12} solution	50.0mL
Trace metal mix A5	1.0mL

pH 7.1 ± 0.2 at 25°C

Trace Metal Mix A5:

Composition per liter:

H_3BO_3	2.86g
$MnCl_2 \cdot 4H_2O$	1.81g
$Na_2MoO_4 \cdot 2H_2O$	0.39g
$ZnSO_4 \cdot 7H_2O$	0.222g
$CuSO_4 \cdot 5H_2O$	0.079g
$Co(NO_3)_2 \cdot 6H_2O$	0.049g

Preparation of Trace Metal Mix A5: Add components to distilled/deionized water and bring volume to 1.0L. Mix thoroughly.

Vitamin B_{12} Solution:

Composition per 50.0mL:

Vitamin B_{12}	0.01g

Preparation of Vitamin B_{12} Solution: Add vitamin B_{12} solution to distilled/deionized water and bring volume to 50.0mL. Mix thoroughly. Filter sterilize.

Preparation of Medium: Add components, except vitamin B_{12}, to glass distilled water and bring volume to 950.0mL. Mix thoroughly. Heat gently and bring to boiling. Autoclave for 15 min at 15 psi pressure–121°C. Cool the basal medium to 45°–50°C. Add vitamin B_{12} solution. Mix thoroughly. Distribute into sterile tubes or flasks.

Use: For the cultivation and maintenance of *Synechococcus* species.

Blue-Green Nitrogen-Fixing Agar

Composition per liter:

Noble agar	10.0g
$MgSO_4 \cdot 7H_2O$	0.075g
K_2HPO_4	0.04g
$CaCl_2 \cdot 2H_2O$	0.036g
Na_2CO_3	0.02g
Citric acid	6.0mg
Ferric ammonium citrate	6.0mg
EDTA disodium salt	1.0mg
Trace metal mix A5	1.0mL

pH 7.1 ± 0.2 at 25°C

Trace Metal Mix A5:

Composition per liter:

H_3BO_3	2.86g
$MnCl_2 \cdot 4H_2O$	1.81g
$Na_2MoO_4 \cdot 2H_2O$	0.39g
$ZnSO_4 \cdot 7H_2O$	0.222g
$CuSO_4 \cdot 5H_2O$	0.079g
$Co(NO_3)_2 \cdot 6H_2O$	0.049g

Preparation of Trace Metal Mix A5: Add components to distilled/deionized water and bring volume to 1.0L. Mix thoroughly.

Preparation of Medium: Add components to glass-distilled water and bring volume to 1.0L. Mix thoroughly. Heat gently and bring to boiling. Autoclave for 15 min at 15 psi pressure–121°C. Check pH after autoclaving and readjust if necessary. Pour into sterile Petri dishes or distribute into sterile tubes.

Use: For the cultivation and maintenance of *Calothrix*, *Fischerella*, and *Nostoc* species.

Blue-Green Nitrogen-Fixing Broth

Composition per liter:

$MgSO_4 \cdot 7H_2O$	0.075g
K_2HPO_4	0.04g
$CaCl_2 \cdot 2H_2O$	0.036g
Na_2CO_3	0.02g
Citric acid	6.0mg
Ferric ammonium citrate	6.0mg
EDTA disodium salt	1.0mg
Trace metal mix A5	1.0mL

pH 7.1 ± 0.2 at 25°C

Trace Metal Mix A5:

Composition per liter:

H_3BO_3	2.86g
$MnCl_2 \cdot 4H_2O$	1.81g
$Na_2MoO_4 \cdot 2H_2O$	0.39g
$ZnSO_4 \cdot 7H_2O$	0.222g
$CuSO_4 \cdot 5H_2O$	0.079g
$Co(NO_3)_2 \cdot 6H_2O$	0.049g

Preparation of Trace Metal Mix A5: Add components to distilled/deionized water and bring volume to 1.0L. Mix thoroughly.

Preparation of Medium: Add components to glass-distilled water and bring volume to 1.0L. Mix thoroughly. Heat gently and bring to boiling. Autoclave for 15 min at 15 psi pressure–121°C. Check pH after autoclaving and readjust if necessary. Aseptically distribute into sterile tubes or flasks.

Use: For the cultivation and maintenance of *Calothrix*, *Fischerella*, and *Nostoc* species.

BMM Agar

Composition per liter:

Agar	15.0g
$FeSO_4 \cdot 7H_2O$	10.0g
K_2HPO_4	7.0g
$MnSO_4 \cdot H_2O$	6.2g
$(NH_4)_2SO_4$	3.0g
KH_2PO_4	2.0g
NaCl	2.0g
$MgSO_4 \cdot 7H_2O$	0.5g
Yeast extract	0.1g
Thiamine·HCl	100.0μg
Biotin	10.0μg
Methanol	10.0mL

pH 7.2 ± 0.2 at 25°C

Preparation of Medium: Add components to distilled/deionized water and bring volume to 1.0L. Mix thoroughly. Gently heat and bring to boiling. Adjust pH to 7.2. Distribute into tubes or flasks. Autoclave for 15 min at 15 psi pressure–121°C. Pour into sterile Petri dishes or leave in tubes.

Use: For the cultivation and maintenance of *Butyribacterium methylotrophicum.*

BMM Broth

Composition per liter:

$FeSO_4 \cdot 7H_2O$	10.0g
K_2HPO_4	7.0g
$MnSO_4 \cdot H_2O$	6.2g
$(NH_4)_2SO_4$	3.0g
KH_2PO_4	2.0g
NaCl	2.0g
$MgSO_4 \cdot 7H_2O$	0.5g
Yeast extract	0.1g
Thiamine·HCl	100.0μg
Biotin	10.0μg
Methanol	10.0mL

pH 7.2 ± 0.2 at 25°C

Preparation of Medium: Add components to distilled/deionized water and bring volume to 1.0L. Mix thoroughly. Adjust pH to 7.2. Distribute into tubes or flasks. Autoclave for 15 min at 15 psi pressure–121°C.

Use: For the cultivation and maintenance of *Butyribacterium methylotrophicum.*

BMPA-α Medium (Edelstein BMPA-α Medium)

Composition per liter:

Agar	13.0g
Yeast extract	10.0g
ACES buffer (2-[(2-amino-2-oxoethyl)-amino]-ethane sulfonic acid)	2.0g
Charcoal, activated	2.0g
α-Ketoglutarate	0.2g
$Fe_4(P_2O_7)_3 \cdot 9H_2O$	0.05g
Antibiotic inhibitor	10.0mL
L-Cysteine·HCl·H_2O solution	10.0mL

pH 6.9 ± 0.2 at 25°C

Source: This medium is available as premixed vials from Oxoid Unipath.

Antibiotic Inhibitor:

Composition per 10.0mL:

Anisomycin	0.08g
Cefamandole	4.0mg
Polymyxin B	80,000U

Preparation of Antibiotic Inhibitor: Add components to distilled/deionized water and bring volume to 10.0mL. Mix thoroughly. Filter sterilize.

L-Cysteine·HCl·H_2O Solution:

Composition per 10.0mL:

L-Cysteine·HCl·H_2O	0.08g

Preparation of L-Cysteine·HCl·H_2O Solution: Add L-cysteine·HCl·H_2O to distilled/deionized water and bring volume to 10.0mL. Mix thoroughly. Filter sterilize.

Preparation of Medium: Add components, except antibiotic inhibitor and L-cysteine·HCl·H_2O solution , to distilled/deionized water and bring volume to 980.0mL. Mix thoroughly. Adjust medium to pH 6.9 with 1*N* KOH. Heat gently and bring to boiling for 1 min. Autoclave for 15 min at 15 psi pressure–121°C. Cool to 50°–55°C. Add 10.0mL of the sterile L-cysteine·HCl·H_2O solution and 10.0mL of the sterile antibiotic solution. Mix thoroughly. Pour into sterile Petri dishes with constant agitation to keep charcoal in suspension.

Use: For the selective isolation and cultivation of *Legionella pneumophila* and other *Legionella* species.

BMPA-α Medium (Semiselective Medium for *Legionella pneumophila*)

Composition per liter:

Agar	15.0g
Yeast extract	10.0g

ACES buffer (2-[(2-amino-2-oxoethyl)-amino]-ethane sulfonic acid) 10.0g
Charcoal, activated 2.0g
α-Ketoglutarate 1.0g
$Fe_4(P_2O_7)_3{\cdot}9H_2O$ 0.25g
Antibiotic inhibitor 10.0mL
L-Cysteine·HCl·H_2O solution 10.0mL
pH 6.9 ± 0.2 at 25°C

Antibiotic Inhibitor:
Composition per 10.0mL:
Anisomycin 0.08g
Cefamandole 4.0mg
Polymyxin B 80,000U

Preparation of Antibiotic Inhibitor: Add components to distilled/deionized water and bring volume to 10.0mL. Mix thoroughly. Filter sterilize.

L-Cysteine·HCl·H_2O Solution:
Composition per 10.0mL:
L-Cysteine·HCl·H_2O 0.4g

Preparation of L-Cysteine·HCl·H_2O Solution: Add L-cysteine·HCl·H_2O to distilled/deionized water and bring volume to 10.0mL. Mix thoroughly. Filter sterilize.

Preparation of Medium: Add components, except antibiotic inhibitor and L-cysteine·HCl·H_2O solution , to distilled/deionized water and bring volume to 980.0mL. Mix thoroughly. Adjust medium to pH 6.9 with 1*N* KOH. Heat gently and bring to boiling for 1 min. Autoclave for 15 min at 15 psi pressure–121°C. Cool to 50°–55°C. Add 10.0mL of the sterile L-cysteine·HCl·H_2O solution and 10.0mL of the sterile antibiotic solution. Mix thoroughly. Pour into sterile Petri dishes with constant agitation to keep charcoal in suspension.

Use: For the selective isolation and cultivation of *Legionella pneumophila* and other *Legionella* species.

BMS Agar

Composition per liter:
Agar 20.0g
L-Malic acid 2.5g
Sucrose 2.5g
KOH 2.0g
Potato extract solution 950.0mL
Bromthymol Blue 1.0mL
Vitamin solution 1.0mL
pH 7.0 ± 0.2 at 25°C

Potato Extract Solution:
Composition per liter:
Potatoes, washed, peeled, and sliced 200.0g

Preparation of Potato Extract Solution: Wash, peel, and slice several large potatoes. Place the poltato slices in a gauze bag. Place the bag with the potatoes in 1.0L of distilled/deionized water. Boil for 30 min. Filter through cotton.

Bromthymol Blue Solution:
Composition per 100.0mL:
Bromthymol Blue 0.5g
Ethanol, 95% 100.0mL

Preparation of Bromthymol Blue Solution: Add Bromthymol Blue to 100.0mL of 95% ethanol. Mix thorougly.

Vitamin Solution:
Composition per 100.0mL:
Biotin 10.0g
Pyridoxine 20.0mg

Preparation of Vitamin Solution: Add components to distilled/deionized water and bring volume to 100.0mL. Mix thoroughly. Filter sterilize.

Preparation of Medium: Dissolve 2.5g of L-malic acid in 50.0mL of distilled/deionized water. Add 1.0mL of Bromthymol Blue solution. Adjust pH to 7.0 by adding KOH so that the solution is green. Add 950.0mL of potato extract solution. Mix thoroughly. Add 20.0g of agar and 2.5g of sucrose. Mix thoroughly. Gently heat and bring to boiling. Autoclave for 15 min at 15 psi pressure–121°C. Cool to 50°–55°C. Aseptically add 1.0mL of sterile vitamin solution. Mix thoroughly. Pour into sterile Petri dishes or distribute into sterile tubes.

Use: For the cultivation and maintenance of *Azospirillum lipoferum.*

BNS

See: **Benzoate Nitrate Salts Medium**

Bogoriella Medium (DSMZ Medium 785)

Composition per liter:
NaCl 40.0g
Na_2CO_3 .... 10.0g
Glucose 10.0g
Peptone 5.0g
Yeast extract 5.0g
KH_2PO_4 .... 1.0g
$MgSO_4{\cdot}7H_2O$ 0.2g
pH 9.6 ± 0.2 at 25°C

Preparation of Medium: Add components to distilled/deionized water and bring volume to 1.0L. Mix thoroughly. Distribute into tubes or flasks. Autoclave for 15 min at 15 psi pressure–121°C.

Use: For the cultivation of *Bogoriella caseilytica.*

Bolton Broth

Composition per 505mL:
Bolton selective enrichment broth base 500.0mL
Horse blood, lysed 25.0mL
Bolton selective supplement soution 5.0mL
pH 7.4 ± 0.2 at 25°C

Bolton Selective Enrichment Broth Base:
Composition per liter:
Peptone 10.0g
Lactalbumin hydrolysate 5.0g
Yeast extract 5.0g
NaCl 5.0g
α-Ketoglutarate 1.0g
Na-pyruvate 0.5g
Na-metabisulfite 0.5g
Na_2CO_3 .... 0.6g
Hemin 0.01g

Preparation of Bolton Selective Enrichment Broth Base: Add components to distilled/deionized water and bring volume to 1.0L. Mix thoroughly. Autoclave for 15 min at 15 psi pressure–121°C. Cool to 50°C.

Bolton Broth Supplement Solution:
Composition per 5.0mL

Vancomycin	10.0mg
Cefoperazone	10.0mg
Trimethoprim	10.0mg
Cycloheximide	10.0mg
Ethanol	2.5mL

Preparation of Bolton Supplement Solution: Add antibiotics to 2.5mL ethanol. Mix thoroughly. Bring volume to 5.0mL with distilled/deionized water. Mix thoroughly. Filter sterilize.

Caution: Cycloheximide is toxic. Avoid skin contact or aerosol formation and inhalation.

Preparation of Medium: Aseptically combine 500.0mL warm Bolton selective enrichment broth base, 25.0mL lysed horse blood, and 5.0mL Bolton slective supplement soution.

Use: For the enrichment of *Campylobacter* spp. from foods.

Bonner-Addicott Medium

Composition per liter:

Agar	25.0g
Glucose	20.0g
$Ca(NO_3)_2 \cdot 4H_2O$	0.236g
KNO_3	0.081g
KCl	0.065g
$MgSO_4 \cdot 7H_2O$	0.036g
KH_2PO_4	0.012g
Ferric tartrate	1.0mg

Preparation of Medium: Add components to distilled/deionized water and bring volume to 1.0L. Mix thoroughly. Gently heat and bring to boiling. Distribute into tubes or flasks. Autoclave for 15 min at 15 psi pressure–121°C. Pour into sterile Petri dishes or leave in tubes.

Use: For the cultivation of a variety of fungi.

Bordetella pertussis Selective Medium with Bordet-Gengou Agar Base

Composition per 1210.0mL:

Bordet-Gengou agar base	1.0L
Horse blood, defibrinated	200.0mL
Cephalexin solution	10.0mL

pH 6.7± 0.2 at 25°C

Source: This medium is available as a premixed powder from Oxoid Unipath.

Bordet-Gengou Agar Base:
Composition per liter:

Agar	20.0g
NaCl	5.5g
Pancreatic digest of casein	5.0g
Peptic digest of animal tissue	5.0g

Preparation of Bordet-Gengou Agar Base: Add components to 1.0L of 1% glycerol solution. Autoclave for 15 min at 15 psi pressure–121°C. Cool to 50°C.

Cephalexin Solution:
Composition per 10.0mL:

Cephalexin	0.04g

Preparation of Cephalexin Solution: Add cephalexin to distilled/deionized water and bring volume to 10.0mL. Mix thoroughly. Filter sterilize.

Preparation of Medium: Aseptically add 10.0mL of sterile cephalexin solution and 200.0mL of defibrinated horse blood to 1.0L Bordet-Gengou agar base. Mix thoroughly and pour into sterile Petri dishes.

Use: For the selective isolation and presumptive identification of *Bordetella pertussis* and *Bordetella parapertussis. Bordetella pertussis* appears as small, nearly transparent, "bisected pearl-like" colonies.

Bordetella pertussis Selective Medium with Charcoal Agar Base

Composition per 1110.0mL:

Charcoal agar base	1.0L
Horse blood, defibrinated	100.0mL
Cephalexin solution	10.0mL

pH 6.7± 0.2 at 25°C

Source: This medium is available as a premixed powder from Oxoid Unipath.

Charcoal Agar Base:
Composition per liter:

Agar	12.0g
Beef extract	10.0g
Starch	10.0g
NaCl	5.0g
Pancreatic digest of casein	5.0g
Peptic digest of animal tissue	5.0g
Charcoal	4.0g
Nicotinic acid	1.0mg

Preparation of Charcoal Agar Base: Add components of charcoal agar base to distilled/deionized water and bring volume to 1.0L. Autoclave for 15 min at 15 psi pressure–121°C. Cool to 50°C.

Cephalexin Solution:
Composition per 10.0mL:

Cephalexin	0.04g

Preparation of Cephalexin Solution: Add cephalexin to distilled/deionized water and bring volume to 10.0mL. Mix thoroughly. Filter sterilize.

Preparation of Medium: Aseptically add 10.0mL of sterile cephalexin solution and 100.0mL of defibrinated horse blood to charcoal agar base. Mix thoroughly and pour into sterile Petri dishes.

Use: For the selective isolation and presumptive identification of *Bordetella pertussis* and *Bordetella parapertussis. Bordetella pertussis* appears as small, pale, shiny colonies.

Bordet-Gengou Agar

Composition per liter:

Agar	20.0g
Glycerol	10.0g
NaCl	5.5g
Pancreatic digest of casein	5.0g
Peptic digest of animal tissue	5.0g
Potato, solids from infusion	4.5g
Rabbit blood	200.0mL

pH 6.7± 0.2 at 25°C

Source: This medium is available as a premixed powder from Oxoid Unipath and BD Diagnostic Systems.

Preparation of Medium: Add 10.0g of glycerol to 980.0mL of distilled/deionized water. Add other components, except rabbit blood, to the glycerol solution. Mix thoroughly. Heat with occasional agitation of the medium. Boil for 1 min. Autoclave for 15 min at 15 psi pressure–121°C. Cool medium to 50°C. Aseptically add 200.0mL of rabbit blood (prewarmed to 35°C) to a concentration of 15–30%. 150.0–200.0mL of sterile, defibrinated horse blood may be used in place of rabbit blood. Mix thoroughly and pour plates or prepare slants.

Use: For the detection and isolation of *Bordetella pertussis* and *Bordetella parapertussis* from clinical specimens. The medium is rendered selective by the addition of methicillin. *Bordetella pertussis* appears as small (<1mm), smooth, pearl-like colonies surrounded by a narrow zone of hemolysis. *Bordetella parapertussis* appears as brown, nonshiny colonies with a green-black coloration on the reverse side. *Bordetella bronchiseptica* appears as brown, nonshiny, moderately sized colonies with a roughly pitted surface.

Bordet-Gengou Agar Base with Rabbit Blood and Glycerol

Composition per liter:

Potatoes, infusion from	125.0g
Agar	20.0g
Peptic digest of animal tissue	10.0g
NaCl	5.5g
Rabbit blood	200.0mL
Glycerol	10.0mL

pH 6.7± 0.2 at 25°C

Source: This medium without glycerol and blood is available as a premixed powder from HiMedia.

Preparation of Medium: Add 10.0g of glycerol to 790.0mL of distilled/deionized water. Add other components, except rabbit blood, to the glycerol solution. Mix thoroughly. Heat with occasional agitation of the medium. Boil for 1 min. Autoclave for 15 min at 15 psi pressure–121°C. Cool medium to 50°C. Aseptically add 200.0mL of rabbit blood (prewarmed to 35°C). Note: 150.0–200.0mL of sterile, defibrinated horse blood may be used in place of rabbit blood. Mix thoroughly and pour plates or prepare slants.

Use: For the detection and isolation of *Bordetella pertussis* and *Bordetella parapertussis* from clinical specimens. *Bordetella pertussis* appears as small (<1mm), smooth, pearl-like colonies surrounded by a narrow zone of hemolysis. *Bordetella parapertussis* appears as brown, nonshiny colonies with a green-black coloration on the reverse side. *Bordetella bronchiseptica* appears as brown, nonshiny, moderately sized colonies with a roughly pitted surface.

Bordet-Gengou Agar Base with 1.6% Agar

Composition per liter:

Potatoes, infusion from	125.0g
Agar	16.0g
Peptic digest of animal tissue	10.0g
NaCl	5.5g
Rabbit blood	200.0mL
Glycerol	10.0mL

pH 6.7± 0.2 at 25°C

Source: This medium without glycerol and blood is available as a premixed powder from HiMedia.

Preparation of Medium: Add 10.0g of glycerol to 790.0mL of distilled/deionized water. Add other components, except rabbit blood, to the glycerol solution. Mix thoroughly. Heat with occasional agitation of the medium. Boil for 1 min. Autoclave for 15 min at 15 psi pressure–121°C. Cool medium to 50°C. Aseptically add 200.0mL of rabbit blood (prewarmed to 35°C). 150.0–200.0mL of sterile, defibrinated horse blood may be used in place of rabbit blood. Mix thoroughly and pour plates or prepare slants.

Use: For the detection and isolation of *Bordetella pertussis* and *Bordetella parapertussis* from clinical specimens. *Bordetella pertussis* appears as small (<1mm), smooth, pearl-like colonies surrounded by a narrow zone of hemolysis. *Bordetella parapertussis* appears as brown, nonshiny colonies with a green-black coloration on the reverse side. *Bordetella bronchiseptica* appears as brown, nonshiny, moderately sized colonies with a roughly pitted surface.

Bordet-Gengou HiVeg Agar Base with 1.6% Agar

Composition per liter:

Potatoes, infusion from	125.0g
Agar	16.0g
Plant peptone	10.0g
NaCl	5.5g
Rabbit blood	200.0mL
Glycerol	10.0mL

pH 6.7± 0.2 at 25°C

Source: This medium without glycerol and blood is available as a premixed powder from HiMedia.

Preparation of Medium: Add 10.0g of glycerol to 790.0mL of distilled/deionized water. Add other components, except rabbit blood, to the glycerol solution. Mix thoroughly. Heat with occasional agitation of the medium. Boil for 1 min. Autoclave for 15 min at 15 psi pressure–121°C. Cool medium to 50°C. Aseptically add 200.0mL of rabbit blood (prewarmed to 35°C). 150.0–200.0mL of sterile, defibrinated horse blood may be used in place of rabbit blood. Mix thoroughly and pour plates or prepare slants.

Use: For the detection and isolation of *Bordetella pertussis* and *Bordetella parapertussis* from clinical specimens. *Bordetella pertussis* appears as small (<1mm), smooth, pearl-like colonies surrounded by a narrow zone of hemolysis. *Bordetella parapertussis* appears as brown, nonshiny colonies with a green-black coloration on the reverse side. *Bordetella bronchiseptica* appears as brown, nonshiny, moderately sized colonies with a roughly pitted surface.

Bordet-Gengou HiVeg Agar Base with Rabbit Blood and Glycerol

Composition per liter:

Potatoes, infusion from	125.0g
Agar	20.0g
Plant peptone	10.0g
NaCl	5.5g
Rabbit blood	200.0mL
Glycerol	10.0mL

pH 6.7± 0.2 at 25°C

Source: This medium without glycerol and blood is available as a premixed powder from HiMedia.

Preparation of Medium: Add 10.0g of glycerol to 790.0mL of distilled/deionized water. Add other components, except rabbit blood, to the glycerol solution. Mix thoroughly. Heat with occasional agitation of the medium. Boil for 1 min. Autoclave for 15 min at 15 psi pressure–

121°C. Cool medium to 50°C. Aseptically add 200.0mL of rabbit blood (prewarmed to 35°C). Note: 150.0–200.0mL of sterile, defibrinated horse blood may be used in place of rabbit blood. Mix thoroughly and pour plates or prepare slants.

Use: For the detection and isolation of *Bordetella pertussis* and *Bordetella parapertussis* from clinical specimens. *Bordetella pertussis* appears as small (<1mm), smooth, pearl-like colonies surrounded by a narrow zone of hemolysis. *Bordetella parapertussis* appears as brown, nonshiny colonies with a green-black coloration on the reverse side. *Bordetella bronchiseptica* appears as brown, nonshiny, moderately sized colonies with a roughly pitted surface.

Bordet-Gengou Medium (ATCC Medium 35)

Composition per liter:

Agar	20.0g
Glycerol	10.0g
Proteose peptone	10.0g
NaCl	5.5g
Pancreatic digest of casein	5.0g
Peptic digest of animal tissue	5.0g
Potato, solids from infusion	4.5g
Rabbit blood	150.0mL

pH 6.7± 0.2 at 25°C

Source: This medium is available as a premixed powder from Oxoid Unipath and BD Diagnostic Systems.

Preparation of Medium: Add 10.0g of glycerol to 980.0mL of distilled/deionized water. Add other components, except rabbit blood, to the glycerol solution. Mix thoroughly. Heat with occasional agitation of the medium. Boil for 1 min. Autoclave for 15 min at 15 psi pressure–121°C. Cool medium to 50°C. Aseptically add 150.0mL of rabbit blood (prewarmed to 35°C). Mix thoroughly. Pour into sterile Petri dishes or distribute into sterile tubes. Allow tubes to cool in a slanted position.

Use: For the detection and isolation of *Bordetella pertussis* and *Bordetella parapertussis* from clinical specimens. The medium is rendered selective by the addition of methicillin. *Bordetella pertussis* appears as small (<1mm), smooth, pearl-like colonies surrounded by a narrow zone of hemolysis. *Bordetella parapertussis* appears as brown, nonshiny colonies with a green-black coloration on the reverse side. *Bordetella bronchiseptica* appears as brown, nonshiny, moderately sized colonies with a roughly pitted surface.

Bordet-Gengou Medium (LMG Medium 23)

Composition per liter:

Agar	20.0g
Proteose peptone	10.0g
NaCl	5.5g
Pancreatic digest of casein	5.0g
Peptic digest of animal tissue	5.0g
Potato, solids from 125g infusion	4.5g
Rabbit blood, sterile defibrinated	150.0mL
Glycerol	10.0mL

pH 6.7± 0.2 at 25°C

Preparation of Medium: Add 10.0mL of glycerol to 850.0mL of distilled/deionized water. Add the other components, except rabbit blood, to the glycerol solution. Mix thoroughly. Heat with occasional agitation of the medium. Boil for 1 min. Autoclave for 15 min at 15 psi pressure–121°C. Cool medium to 45–50°C. Aseptically add 150.0mL of rabbit blood (prewarmed to 35°C). Mix thoroughly. Pour into sterile Petri dishes or distribute into sterile tubes.

Use: For the detection and isolation of *Pseudomonas pertucinogena.*

Borrelia Medium

Composition per 370.0mL:

Solution 4	240.0mL
Solution 1	80.0mL
Solution 2	34.0mL
Rabbit serum, sterile	10.0mL
Solution 3	4.0mL
Solution 5	0.7mL

Solution 1:

Composition per liter:

$Na_2HPO_4 \cdot 7H_2O$	26.52g
Glucose	12.75g
Proteose peptone No.2	5.95g
Pancreatic digest of casein	2.55g
NaCl	1.2g
Sodium pyruvate	1.06g
$NaH_2PO_4 \cdot H_2O$	1.03g
KCl	0.85g
$MgCl_2 \cdot 6H_2O$	0.68g
N-acetylglucosamine	0.53g
Sodium citrate·$2H_2O$	0.47g

Preparation of Solution 1: Add components to distilled/deionized water and bring volume to 1.0L. Mix thoroughly. Store at –20° C.

Solution 2:

Composition per 100.0mL:

Bovine albumin fraction V	10.0g

Preparation of Solution 2: Add bovine albumin to distilled/deionized water and bring volume to 100.0mL. Mix thoroughly. Adjust pH to 7.8 with NaOH. Store at –20°C.

Solution 3:

Composition per 100.0mL:

$NaHCO_3$	4.5g

Preparation of Solution 3: Add $NaHCO_3$ to distilled/deionized water and bring volume to 100.0mL. Mix thoroughly. Prepare solution freshly.

Solution 4:

Composition per 100.0mL:

Gelatin	7.0g

Preparation of Solution 4: Add gelatin to distilled/deionized water and bring volume to 100.0mL. Mix thoroughly. Autoclave for 15 min at 10 psi pressure–115°C. Store at 4°C.

Solution 5:

Composition per 100.0mL:

Phenol Red	0.5g

Preparation of Solution 5: Add Phenol Red to distilled/deionized water and bring volume to 100.0mL. Mix thoroughly. Store at 4°C.

Preparation of Medium: Combine 80.0mL of solution 1, 34.0mL of solution 2, 4.0mL of solution 3, 0.7mL of solution 5, and 1.3mL of distilled/deionized water. Mix thoroughly. Filter sterilize under pressure. Aseptically distribute into sterile borosilicate screw-capped tubes in 6.0mL volumes. Melt solution 4 by immersing tube in warm water. Add

2.0mL of solution 4 to each screw-capped tube. Add 0.5mL of sterile rabbit serum to each screw-capped tube.

Use: For the cultivation of *Borrelia hermsii*, *Borrelia turicatae*, and *Borrelia parkeri*.

BOS Broth
See: **Bile Oxalate Sorbose Broth**

Bosea thiooxidans Medium (DSMZ Medium 763)

Composition per liter:

$Na_2S_2O_3 \cdot 5H_2O$	5.0g
Na-succinate	5.0g
Na_2HPO_4	4.0g
KH_2PO_4	1.5g
$MgCl_2$	0.1g
Na-glutamate	0.5g
Yeast extract	0.1g

pH 7.5–8.5 at 25°C

Preparation of Medium: Add components to distilled/deionized water and bring volume to 1.0L. Mix thoroughly. Distribute into tubes or flasks. Autoclave for 15 min at 15 psi pressure–121°C.

Use: For the cultivation of *Bosea thiooxidans*.

Botrytis Separation Agar

Composition per liter:

Agar	20.0g
Pancreatic digest of casein	5.0g
Glycerol	5.0g
$NaNO_3$	3.0g
Yeast extract	3.0g
Sorbose	2.5g
KCl	1.0g
$MgSO_4$	0.5g
KH_2PO_4	0.15g

Preparation of Medium: Add components to distilled/deionized water and bring volume to 1.0L. Mix thoroughly. Gently heat and bring to boiling. Distribute into tubes or flasks. Autoclave for 15 min at 15 psi pressure–121°C. Pour into sterile Petri dishes or leave in tubes.

Use: For the cultivation and differentiation of *Botrytis* species. *Botrytis cinerea* will grow equally well with and without sorbose. *Botrytis alli* is inhibited by sorbose.

Bouillon Medium

Composition per liter:

Peptone	15.0g
Meat extract	5.0g
NaCl	5.0g
K_2HPO_4	5.0g

pH 7.0 ± 0.2 at 25°C

Preparation of Medium: Add components to distilled/deionized water and bring volume to 1.0L. Mix thoroughly. Heat with frequent agitation and bring to boiling. Distribute into tubes or flasks. Autoclave for 15 min at 15 psi pressure–121°C.

Use: For the general cultivation of heterotrophic microorganisms.

Bovine Serum Albumin Tween™ 80 Agar (BSA Tween™ 80 Agar)

Composition per liter:

Basal medium	900.0mL
Albumin supplement	100.0mL

Basal Medium:

Composition per liter:

Agar	11.0g
Na_2HPO_4	1.0g
NaCl	1.0g
KH_2PO_4	0.3g
Glycerol (10% solution)	1.0mL
NH_4Cl (25% solution)	1.0mL
Sodium pyruvate (10% solution)	1.0mL
Thiamine (0.5% solution	1.0mL

Preparation of Basal Medium: Add components to distilled/deionized water and bring volume to 1.0L. Mix thoroughly. Adjust pH to 7.4. Autoclave for 15 min at 15 psi pressure–121°C. Cool to 25°C.

Albumin Supplement:

Composition per 100.0mL:

Bovine albumin	10.0g
Tween™ 80 (10% solution)	12.5mL
$FeSO_4$ (0.5% solution)	10.0mL
$MgCl_2$–$CaCl_2$ solution	1.0mL
Cyanocobalamin (0.02% solution)	1.0mL
$ZnSO_4$ (0.4% solution)	1.0mL

Preparation of Albumin Supplement: Add components to distilled/deionized water and bring volume to 100.0mL. Mix thoroughly. Adjust pH to 7.4. Filter sterilize.

$MgCl_2$–$CaCl_2$ Solution:

Composition per 100.0mL:

$CaCl_2 \cdot 2H_2O$	1.5g
$MgCl_2 \cdot 6H_2O$	1.5g

Preparation of $MgCl_2$–$CaCl_2$ Solution: Add components to distilled/deionized water and bring volume to 100.0mL. Mix thoroughly.

Preparation of Medium: To 900.0mL of cooled, sterile basal medium, aseptically add 100.0mL of sterile albumin supplement. Mix thoroughly. Aseptically distribute into sterile tubes or flasks.

Use: For the cultivation and maintenance of *Leptospira* species.

Bovine Serum Albumin Tween™ 80 Broth (BSA Tween™ 80 Broth)

Composition per liter:

Basal medium	900.0mL
Albumin supplement	100.0mL

pH 7.4 ± 0.2 at 25°C

Basal Medium:

Composition per liter:

Na_2HPO_4	1.0g
NaCl	1.0g
KH_2PO_4	0.3g
Glycerol (10% solution)	1.0mL
NH_4Cl (25% solution)	1.0mL
Sodium pyruvate (10% solution)	1.0mL
Thiamine (0.5% solution	1.0mL

Preparation of Basal Medium: Add components to distilled/deionized water and bring volume to 1.0L. Mix thoroughly. Adjust pH to 7.4. Autoclave for 15 min at 15 psi pressure–121°C. Cool to 25°C.

Albumin Supplement:
Composition per 100.0mL:

Bovine albumin	10.0g
Tween™ 80 (10% solution)	12.5mL
$FeSO_4$ (0.5% solution)	10.0mL
$MgCl_2$–$CaCl_2$ solution	1.0mL
Cyanocobalamin (0.02% solution)	1.0mL
$ZnSO_4$ (0.4% solution)	1.0mL

Preparation of Albumin Supplement: Add components to distilled/deionized water and bring volume to 100.0mL. Mix thoroughly. Adjust pH to 7.4. Filter sterilize.

$MgCl_2$–$CaCl_2$ Solution:
Composition per 100.0mL:

$CaCl_2 \cdot 2H_2O$	1.5g
$MgCl_2 \cdot 6H_2O$	1.5g

Preparation of $MgCl_2$–$CaCl_2$ Solution: Add components to distilled/deionized water and bring volume to 100.0mL. Mix thoroughly.

Preparation of Medium: To 900.0mL of cooled, sterile basal medium, aseptically add 100.0mL of sterile albumin supplement. Mix thoroughly. Aseptically distribute into sterile tubes or flasks.

Use: For the isolation and cultivation of *Leptospira* species.

Bovine Albumin Tween™ 80 Medium, Ellinghausen and McCullough, Modified (Albumin Fatty Acid Broth, *Leptospira* Medium)

Composition per liter:

Basal medium	900.0mL
Albumin fatty acid supplement	100.0mL

Basal Medium:
Composition per liter:

Na_2HPO_4, anhydrous	1.0g
NaCl	1.0g
KH_2PO_4, anhydrous	0.3g
NH_4Cl (25% solution)	1.0mL
Glycerol (10% solution)	1.0mL
Sodium pyruvate (10% solution)	1.0mL
Thiamine·HCl (0.5% solution)	1.0mL

pH 7.4 ± 0.2 at 25°C

Preparation of Basal Medium: Add components to distilled/deionized water and bring volume to 1.0L. Mix thoroughly. Adjust pH to 7.4. Gently heat and bring to boiling. Autoclave for 15 min at 15 psi pressure–121°C. Cool to 25°C.

Albumin Fatty Acid Supplement:
Composition per 200.0mL:

Bovine albumin fraction V	20.0g
Polysorbate (Tween™) 80 (10% solution)	25.0mL
$FeSO_4 \cdot 7H_2O$ (0.5% solution)	20.0mL
$CaCl_2 \cdot 2H_2O$ (1.5% solution)	2.0mL
$MgCl_2 \cdot 2H_2O$ (1.5% solution)	2.0mL
Vitamin B_{12} (0.2% solution)	2.0mL
$ZnSO_4 \cdot 7H_2O$ (0.4% solution)	2.0mL
$CuSO_4 \cdot 5H_2O$ (0.3% solution)	0.2mL

Preparation of Albumin Fatty Acid Supplement: Add bovine albumin to 100.0mL of distilled/deionized water. Mix thoroughly. Add remaining components while stirring. Adjust pH to 7.4. Bring volume to 200.0mL with distilled/deionized water. Filter sterilize. Store this supplement at –20°C.

Preparation of Medium: Aseptically combine 100.0mL of sterile albumin fatty acid supplement and 900.0mL of sterile basal medium. Mix thoroughly. Aseptically distribute into sterile tubes or flasks.

Use: For the cultivation of *Leptospira* species.

Bovine Albumin Tween™ 80 Semisolid Medium, Ellinghausen and McCullough, Modified (Albumin Fatty Acid Semisolid Medium, Modified)

Composition per liter:

Basal medium	900.0mL
Albumin fatty acid supplement	100.0mL

Basal Medium:
Composition per liter:

Agar	2.2g
Na_2HPO_4, anhydrous	1.0g
NaCl	1.0g
KH_2PO_4, anhydrous	0.3g
NH_4Cl (25% solution)	1.0mL
Glycerol (10% solution)	1.0mL
Sodium pyruvate (10% solution)	1.0mL
Thiamine·HCl (0.5% solution)	1.0mL

pH 7.4 ± 0.2 at 25°C

Preparation of Basal Medium: Add components to distilled/deionized water and bring volume to 1.0L. Mix thoroughly. Adjust pH to 7.4. Gently heat and bring to boiling. Autoclave for 15 min at 15 psi pressure–121°C. Cool to 25°C.

Albumin Fatty Acid Supplement:
Composition per 200.0mL:

Bovine albumin fraction V	20.0g
Polysorbate (Tween™) 80 (10% solution)	25.0mL
$FeSO_4 \cdot 7H_2O$ (0.5% solution)	20.0mL
$CaCl_2 \cdot 2H_2O$ (1.5% solution)	2.0mL
$MgCl_2 \cdot 2H_2O$ (1.5% solution)	2.0mL
Vitamin B_{12} (0.2% solution)	2.0mL
$ZnSO_4 \cdot 7H_2O$ (0.4% solution)	2.0mL
$CuSO_4 \cdot 5H_2O$ (0.3% solution)	0.2mL

Preparation of Albumin Fatty Acid Supplement: Add bovine albumin to 100.0mL of distilled/deionized water. Mix thoroughly. Add remaining components while stirring. Adjust pH to 7.4. Bring volume to 200.0mL with distilled/deionized water. Filter sterilize. Store this supplement at –20°C.

Preparation of Medium: Aseptically combine 100.0mL of sterile albumin fatty acid supplement and 900.0mL of sterile basal medium. Mix thoroughly. Aseptically distribute into sterile tubes or flasks.

Use: For the cultivation of *Leptospira* species.

Bovine Serum Albumin Tween™ 80 Soft Agar (BSA Tween™ 80 Soft Agar) (Semisolid BSA Tween™ 80 Medium)

Composition per liter:

Basal medium	900.0mL
Albumin supplement	100.0mL

Basal Medium:

Composition per liter:

Agar	2.0g
Na_2HPO_4	1.0g
NaCl	1.0g
KH_2PO_4	0.3g
Glycerol (10% solution)	1.0mL
NH_4Cl (25% solution)	1.0mL
Sodium pyruvate (10% solution)	1.0mL
Thiamine (0.5% solution)	1.0mL

Preparation of Basal Medium: Add components to distilled/deionized water and bring volume to 1.0L. Mix thoroughly. Adjust pH to 7.4. Autoclave for 15 min at 15 psi pressure–121°C. Cool to 25°C.

Albumin Supplement:

Composition per 100.0mL:

Bovine albumin	10.0g
Tween™ 80 (10% solution)	12.5mL
$FeSO_4$ (0.5% solution)	10.0mL
$CaCl_2$–$MgCl_2$ solution	1.0mL
Cyanocobalamin (0.02% solution)	1.0mL
$ZnSO_4$ (0.4% solution)	1.0mL

Preparation of Albumin Supplement: Add components to distilled/deionized water and bring volume to 100.0mL. Mix thoroughly. Adjust pH to 7.4. Filter sterilize.

$CaCl_2$–$MgCl_2$ Solution:

Composition per 100.0mL:

$CaCl_2 \cdot 2H_2O$	1.5g
$MgCl_2 \cdot 6H_2O$	1.5g

Preparation of $CaCl_2$–$MgCl_2$ Solution: Add components to distilled/deionized water and bring volume to 100.0mL. Mix thoroughly.

Preparation of Medium: To 900.0mL of cooled, sterile basal medium, aseptically add 100.0mL of sterile albumin supplement. Mix thoroughly. Aseptically distribute into sterile tubes or flasks.

Use: For the cultivation of *Leptospira* species.

BPHD Medium (DSMZ Medium 738)

Composition per liter:

NH_4Cl	1.0g
$Na_2HPO_4 \cdot 2H_2O$	0.42g
$Na_2H_2PO_4 \cdot H_2O$	0.18g
$MgCl_2 \cdot 6H_2O$	0.1g
$CaCl_2 \cdot 2H_2O$	0.1g
KCl	0.1g
$FeSO_4 \cdot 7H_2O$	0.04g
Phenol solution	20.0mL
Vitamin solution	10.0mL
Trace elements solution	10.0mL

pH 6.5 ± 0.2 at 25°C

Trace Elements Solution:

Composition per liter:

$MgSO_4 \cdot 7H_2O$	3.0g
Nitrilotriacetic acid	1.5g
NaCl	1.0g
$MnSO_4 \cdot 2H_2O$	0.5g
$CoSO_4 \cdot 7H_2O$	0.18g
$ZnSO_4 \cdot 7H_2O$	0.18g
$CaCl_2 \cdot 2H_2O$	0.1g
$FeSO_4 \cdot 7H_2O$	0.1g
$NiCl_2 \cdot 6H_2O$	0.025g
$KAl(SO_4)_2 \cdot 12H_2O$	0.02g
H_3BO_3	0.01g
$Na_2MoO_4 \cdot 4H_2O$	0.01g
$CuSO_4 \cdot 5H_2O$	0.01g
$Na_2SeO_3 \cdot 5H_2O$	0.3mg

Preparation of Trace Elements Solution: Add nitrilotriacetic acid to 500.0mL of distilled/deionized water. Dissolve by adjusting pH to 6.5 with KOH. Add remaining components. Add distilled/deionized water to 1.0L. Mix thoroughly.

Vitamin Solution:

Composition per liter:

Pyridoxine-HCl	10.0mg
Thiamine-HCl·$2H_2O$	5.0mg
Riboflavin	5.0mg
Nicotinic acid	5.0mg
D-Ca-pantothenate	5.0mg
p-Aminobenzoic acid	5.0mg
Lipoic acid	5.0mg
Biotin	2.0mg
Folic acid	2.0mg
Vitamin B_{12}	0.1mg

Preparation of Vitamin Solution: Add components to distilled/deionized water and bring volume to 1.0L. Mix thoroughly. Filter sterilize.

Phenol Solution:

Composition per liter:

Phenol	4.7g

Preparation of Phenol Solution: Add phenol to distilled/deionized water and bring volume to 1.0L. Mix thoroughly. Filter sterilize.

Preparation of Medium: Add components, except phenol solution and vitamin solution, to distilled/deionized water and bring volume to 970.0mL. Mix thoroughly. Adjust pH to 6.5. Autoclave for 15 min at 15 psi pressure–121°C. Aseptically add 20.0mL phenol solution and 10.0mL vitamin solution. Mix thoroughly. Aseptically distribute into sterile tubes or bottles.

Use: For the cultivation of *Bacillus* sp.

BPL Agar (Brilliant Green Phenol Red Agar)

Composition per liter:

Lactose	15.0g
Agar	13.0g
NaCl	5.0g
Meat peptone	7.0g
Phenol Red	0.04g
Brilliant Green	5.0mg

pH 6.5 ± 0.2 at 25°C

Source: This medium is available as a premixed powder from HiMedia.

Preparation of Medium: Add components to distilled/deionized water and bring volume to 1.0L. Mix thoroughly. Gently heat and bring to boiling. Distribute into tubes or flasks. Autoclave for 15 min at 15 psi pressure–121°C. Pour into sterile Petri dishes or leave in tubes.

Use: For the cultivation of *Salmonella* species, with the exception of *S. typhi*, from feces, urine, meat, milk, and other materials.

BPL HiVeg Agar (Brilliant Green Phenol Red HiVeg Agar)

Composition per liter:

Lactose	15.0g
Agar	13.0g
Plant peptone No. 1	7.0g
NaCl	5.0g
Phenol Red	0.04g
Brilliant Green	5.0mg

pH 6.5 ± 0.2 at 25°C

Source: This medium is available as a premixed powder from HiMedia.

Preparation of Medium: Add components to distilled/deionized water and bring volume to 1.0L. Mix thoroughly. Gently heat and bring to boiling. Distribute into tubes or flasks. Autoclave for 15 min at 15 psi pressure–121°C. Pour into sterile Petri dishes or leave in tubes.

Use: For the cultivation of *Salmonella* species, with the exception of *S. typhi*, from feces, urine, meat, milk and other materials.

B.Q.Vaccine Medium HiVeg with Glucose (Thioglycollate Broth with Plant Extract No. 2)

Composition per liter:

Plant peptone	10.0g
Plant extract No. 2	5.0g
Plant infusion	5.0g
NaCl	5.0g
K_2HPO_4	4.0g
Na-thioglycollate	1.0g
Glucose solution	50.0mL

pH 8.2 ± 0.2 at 25°C

Source: This medium without glucose solution is available as a premixed powder from HiMedia.

Glucose Solution:

Composition per 100.0mL:

Glucose	20.0g

Preparation of Glucose Solution: Add glucose to distilled/deionized water and bring volume to 100.0mL. Mix thoroughly. Filter sterilize.

Preparation of Medium: Add components, except glucose solution, to distilled/deionized water and bring volume to 950.0mL. Mix thoroughly. Autoclave for 15 min at 15 psi pressure–121°C. Aseptically add 50.0mL of sterile glucose solution. Mix thoroughly. Aseptically distribute into sterile tubes or flasks.

Use: For the cultivation of anaerobic organisms on a large scale.

B.Q.Vaccine Medium with Glucose (Thioglycollate Broth with Liver Extract)

Composition per liter:

Liver tissues	250.0g
Muscle tissues	250.0g
Peptic digest of animal tissue	10.0g
NaCl	5.0g
K_2HPO_4	4.0g
Na-thioglycollate	1.0g
Glucose solution	50.0mL

pH 8.2 ± 0.2 at 25°C

Source: This medium without glucose solution is available as a premixed powder from HiMedia.

Glucose Solution:

Composition per 100.0mL:

Glucose	20.0g

Preparation of Glucose Solution: Add glucose to distilled/deionized water and bring volume to 100.0mL. Mix thoroughly. Filter sterilize.

Preparation of Medium: Add components, except glucose solution, to distilled/deionized water and bring volume to 950.0mL. Mix thoroughly. Autoclave for 15 min at 15 psi pressure–121°C. Aseptically add 50.0mL of sterile glucose solution. Mix thoroughly. Aseptically distribute into sterile tubes or flasks.

Use: For the cultivation of anaerobic organisms on a large scale.

Brackish Acetate

Composition per liter:

Sodium acetate	1.0g
KNO_3	1.0g
$NaH_2PO_4 \cdot 2H_2O$	0.05g
Artificial seawater	250.0mL
Modified Hutner's basal salts	20.0mL
Vitamin solution	10.0mL

pH 7.2 ± 0.2 at 25°C

Artificial Seawater:

Composition per liter:

NaCl	23.5g
$MgCl_2$	5.0g
Na_2SO_4	3.9g
$CaCl_2$	1.1g
KCl	0.66g
$NaHCO_3$	0.19g
KBr	0.1g
H_3BO_3	0.026g
$SrCl_2$	0.024g
NaF	3.0mg

Preparation of Artificial Seawater: Add components to distilled/deionized water and bring volume to 100.0mL. Mix thoroughly.

Modified Hutner's Basal Salts:

Composition per liter:

$MgSO_4.7H_2O$	29.7g
Nitrilotriacetic acid	10.0g
$CaCl_2 \cdot 2H_2O$	3.34g
$FeSO_4 \cdot 7H_2O$	0.1g
$(NH_4)_2MoO_4$	9.25mg
Metals "44"	50.0mL

Preparation of Modified Hutner's Basal Salts: Dissolve the nitrilotriacetic acid first and neutralize the solution with KOH. Add the other components and adjust the pH to 7.2 with KOH or H_2SO_4. There may be a slight precipitate. Store at 5°C.

Metals "44":

Composition per 100.0mL:

$ZnSO_4 \cdot 7H_2O$	1.1g
$FeSO_4 \cdot 7H_2O$	0.5g
EDTA	0.25g
$MnSO_4 \cdot 7H_2O$	0.154g
$CuSO_4 \cdot 5H_2O$	0.04g

$Co(NO_3)_2 \cdot 6H_2O$	0.025g
$Na_2B_4O_7 \cdot 10H_2O$	0.018g

Preparation of Metals "44": Add components to distilled/deionized water and bring volume to 100.0mL. Mix thoroughly. Autoclave for 15 min at 15 psi pressure–121°C. Add aseptically to sterile basal medium.

Vitamin Solution:
Composition per liter:

Thiamine·HCl	5.0mg
D-Calcium pantothenate	5.0mg
Riboflavin	5.0mg
Biotin	2.0mg
Folic acid	2.0mg
Vitamin B_{12}	0.1mg

Preparation of Vitamin Solution: Add components to distilled/deionized water and bring volume to 1.0L. Mix thoroughly. Filter sterilize and add aseptically to sterile basal medium.

Preparation of Medium: Add a few drops of H_2SO_4 to the distilled water to retard precipitation of the metal salts. Add components, except for Metals "44" and vitamin solutions to 250.0mL of artificial seawater and 720.0mL of distilled/deionized water. Adjust pH to 7.2. Distribute into tubes or flasks. Sterilize by autoclaving for 15 min at 15 psi pressure–121°C. Aseptically add Metals "44" and vitamin solution. Mix thoroughly. Aseptically distribute into sterile tubes or flasks.

Use: For the cultivation of *Filomicrobium fusiforme*.

Brackish *Prosthecomicrobium* Medium

Composition per liter:

Agar	15.0g
Peptone	0.25g
Yeast extract	0.25g
Glucose	0.25g
Artificial seawater	250.0mL
Modified Hutner's basal salts	20.0mL
Vitamin solution	10.0mL

pH 7.2 ± 0.2 at 25°C

Artificial Seawater:
Composition per liter:

NaCl	23.477g
$MgCl_2$	4.981g
Na_2SO_4	3.917g
$CaCl_2$	1.102g
KCl	0.664g
$NaHCO_3$	0.192g
KBr	0.096g
H_3BO_3	0.026g
$SrCl_2$	0.024g
NaF	3.0mg

Preparation of Artificial Seawater: Add components to distilled/deionized water and bring volume to 100.0mL. Mix thoroughly.

Modified Hutner's Basal Salts:
Composition per liter:

$MgSO_4.7H_2O$	29.7g
Nitrilotriacetic acid	10.0g
$CaCl_2 \cdot 2H_2O$	3.34g
$FeSO_4 \cdot 7H_2O$	0.1g
$(NH_4)_2MoO_4$	9.25mg
Metals "44"	50.0mL

Preparation of Modified Hutner's Basal Salts: Dissolve the nitrilotriacetic acid first and neutralize the solution with KOH. Add the other ingredients and readjust the pH with KOH and/or H_2SO_4 to 7.2. There may be a slight precipitate. Store at 5°C.

Metals "44":
Composition per 100.0mL:

$ZnSO_4 \cdot 7H_2O$	1.1g
$FeSO_4 \cdot 7H_2O$	0.5g
EDTA	0.25g
$MnSO_4 \cdot 7H_2O$	0.154g
$CuSO_4 \cdot 5H_2O$	0.04g
$Co(NO_3)_2 \cdot 6H_2O$	0.025g
$Na_2B_4O_7 \cdot 10H_2O$	0.018g

Preparation of Metals "44": Add components to distilled/deionized water and bring volume to 100.0mL. Mix thoroughly. Autoclave for 15 min at 15 psi pressure–121°C. Add aseptically to sterile basal medium.

Vitamin Solution:
Composition per liter:

Thiamine·HCl	5.0mg
D-Calcium pantothenate	5.0mg
Riboflavin	5.0mg
Biotin	2.0mg
Folic acid	2.0mg
Vitamin B_{12}	0.1mg

Preparation of Vitamin Solution: Add components to distilled/deionized water and bring volume to 1.0L. Mix thoroughly. Filter sterilize and add aseptically to sterile basal medium.

Preparation of Medium: Add a few drops of H_2SO_4 to the distilled water to retard precipitation of the metal salts. Add components, except Metals "44" and vitamin solution, to 250.0mL of artificial seawater and 720.0mL of distilled/deionized water. Adjust pH to 7.2. Distribute into tubes or flasks. Sterilize by autoclaving for 15 min at 15 psi pressure–121°C. Aseptically add Metals "44" and vitamin solution. Mix thoroughly. Aseptically distribute into sterile tubes or flasks.

Use: For the cultivation of *Prosthecomicrobium litoralum*.

Brackish Water *Ameba* Medium

Composition per liter:

Agar	10.0g
Malt extract	0.1g
Yeast extract	0.1g
Artificial seawater	167.0mL

Artificial Seawater:
Composition per liter:

NaCl	27.5g
$MgCl_2 \cdot 6H_2O$	5.38g
$MgSO_4 \cdot 7H_2O$	6.78g
KCl	0.72g
$NaHCO_3$	0.2g
$CaCL_2 \cdot 2H_2O$	1.4g

Preparation of Artificial Seawater: Add components to distilled/deionized water and bring volume to 1.0L. Mix thoroughly.

Preparation of Medium: Add components to distilled/deionized water and bring volume to 1.0L. Mix thoroughly. Gently heat and bring to boiling. Distribute into tubes or flasks. Autoclave for 15 min at 15 psi pressure–121°C. Pour into sterile Petri dishes or leave in tubes.

Use: For the cultivation of *Flabellula hoguae*.

Brain Heart CC Agar
(Brain Heart Cycloheximide Chloramphenicol Agar)

Composition per liter:

Pancreatic digest of casein	16.0g
Agar	13.5g
Brain heart, solids from infusion	8.0g
Peptic digest of animal tissue	5.0g
NaCl	5.0g
Na_2HPO_4	2.5g
Glucose	2.0g
Cycloheximide	0.5g
Chloramphenicol	0.05g

pH 7.4 ± 0.2 at 25°C

Source: This medium is available as a premixed powder from BD Diagnostic Systems.

Caution: Cycloheximide is toxic. Avoid skin contact or aerosol formation and inhalation.

Preparation of Medium: Add components to distilled/deionized water and bring volume to 1.0L. Mix thoroughly. Distribute into tubes or flasks while shaking to distribute precipitate. Autoclave for 15 min at 15 psi pressure–118°C.

Use: For the selective isolation of fastidious pathogenic fungi such as *Histoplasma capsulatum* and *Blastomyces dermatiditis* from specimens heavily contaminated with bacteria and other fungi. It may also be used as a base supplemented with sheep blood and gentamicin for enrichment and additional selectivity.

Brain Heart CC Agar, HiVeg
(Brain Heart Cycloheximide Chloramphenicol Agar, HiVeg)

Composition per liter:

Agar	15.0g
Plant infusion	10.0g
Plant peptone No. 3	10.0g
Plant special infusion	7.5g
NaCl	5.0g
Na_2HPO_4	2.5g
Glucose	2.0g
Cycloheximide	0.5g
Chloramphenicol	0.05g

pH 7.4 ± 0.2 at 25°C

Source: This medium is available as a premixed powder from HiMedia.

Caution: Cycloheximide is toxic. Avoid skin contact or aerosol formation and inhalation.

Preparation of Medium: Add components to distilled/deionized water and bring volume to 1.0L. Mix thoroughly. Distribute into tubes or flasks while shaking to distribute precipitate. Autoclave for 15 min at 15 psi pressure–118°C.

Use: For the selective isolation of fastidious pathogenic fungi such as *Histoplasma capsulatum* and *Blastomyces dermatiditis* from specimens heavily contaminated with bacteria and other fungi. This medium may also be used as a base supplemented with sheep blood and gentamicin for enrichment and additional selectivity.

Brain Heart Infusion
See: **BHI**

Brain Heart Infusion
(BHI)

Composition per liter:

Pancreatic digest of gelatin	14.5g
Brain heart, solids from infusion	6.0g
Peptic digest of animal tissue	6.0g
NaCl	5.0g
Glucose	3.0g
Na_2HPO_4	2.5g

pH 7.4 ± 0.2 at 25°C

Source: This medium is available as a premixed powder from Oxoid Unipath and BD Diagnostic Systems.

Preparation of Medium: Add components to distilled/deionized water and bring volume to 1.0L. Mix thoroughly. Distribute into tubes or flasks. Autoclave for 15 min at 15 psi pressure–121°C.

Use: For the cultivation of fastidious and nonfastidious microorganisms, including aerobic and anaerobic bacteria, from a variety of clinical and nonclinical specimens. It is particularly useful for culturing streptococci, pneumococci, and meningococci. It is also used for the preparation of inocula for use in antimicrobial susceptibility tests and as a base for blood culture.

Brain Heart Infusion Agar
(BHI Agar)

Composition per liter:

Beef heart infusion	250.0g
Calf brain infusion	200.0g
Agar	13.5g
Proteose peptone	10.0g
NaCl	5.0g
$Na_2HPO_4 \cdot 12H_2O$	2.5g
Glucose	2.0g

pH 7.4 ± 0.2 at 25°C

Preparation of Medium: Add components to distilled/deionized water and bring volume to 1.0L. Mix thoroughly. Gently heat and bring to boiling. Distribute into tubes or flasks. Autoclave for 15 min at 15 psi pressure–121°C. Pour into sterile Petri dishes or leave in tubes.

Use: For the cultivation of a variety of fastidious and nonfastidious aerobic and anaerobic microorganisms.

Brain Heart Infusion Agar

Composition per liter:

Pancreatic digest of casein	16.0g
Agar	13.5g
Brain heart, solids from infusion	8.0g
Peptic digest of animal tissue	5.0g
NaCl	5.0g
Glucose	2.0g
Na_2HPO_4	2.5g

pH 7.4 ± 0.2 at 25°C

Source: This medium is available as a premixed powder from Oxoid Unipath and BD Diagnostic Systems.

Preparation of Medium: Add components to distilled/deionized water and bring volume to 1.0L. Mix thoroughly. Distribute into tubes or flasks while shaking to distribute precipitate. Autoclave for 15 min at 15 psi pressure–121°C.

Use: For the cultivation of a wide variety of fastidious microorganisms, including bacteria, yeasts, and molds. With the addition of 10% sheep blood, it is used for the isolation and cultivation of many fungal species, including systemic fungi, from clinical and nonclinical specimens. The addition of gentamicin and chloramphenicol with 10% sheep blood produces a selective medium used for the isolation of pathogenic fungi from specimens heavily contaminated with bacteria and saprophytic fungi. It is recommended for the isolation of *Histoplasma capsulatum* and other pathogenic fungi, including *Coccidioides immitis*.

Brain Heart Infusion Agar (BAM M24 Medium 2)

Composition per liter:

Agar	15.0g
Pancreatic digest of gelatin	14.5g
Brain heart, solids from infusion	6.0g
Peptic digest of animal tissue	6.0g
NaCl	5.0g
Glucose	3.0g
Na_2HPO_4	2.5g

pH 7.4 ± 0.2 at 25°C

Source: This medium is available as a premixed powder from BD Diagnostic Systems.

Preparation of Medium: Add components to distilled/deionized water and bring volume to 1.0L. Mix thoroughly. Distribute into tubes or flasks while shaking to distribute precipitate. Autoclave for 15 min at 15 psi pressure–121°C. Mix thoroughly. Pour into sterile Petri dishes.

Use: For the cultivation of a wide variety of fastidious microorganisms, including bacteria, yeasts, and molds.

Brain Heart Infusion Agar 0.7% (BHI Agar 0.7%) (BAM M23)

Composition per liter:

Beef heart infusion	250.0g
Calf brain infusion	200.0g
Proteose peptone	10.0g
Agar	7.0g
NaCl	5.0g
$Na_2HPO_4 \cdot 12H_2O$	2.5g
Glucose	2.0g

pH 5.3 ± 0.2 at 25°C

Source: This medium without agar is available as a premixed powder from BD Diagnostic Systems.

Preparation of Medium: Add components, except agar, to distilled/deionized water and bring volume to 1.0L. Mix thoroughly. Adjust pH to 5.3 with 1*N* HCl. Mix thoroughly. Add agar. Gently heat and bring to boiling. Distribute into tubes. Autoclave for 10 min at 15 psi pressure–121°C.

Use: For the detection of staphylococcal enterotoxin.

Brain Heart Infusion Agar 0.7% (BHI Agar 0.7%) (BAM M23)

Composition per liter:

Pancreatic digest of gelatin	14.5g
Agar	7.0g
Brain heart, solids from infusion	6.0g
Peptic digest of animal tissue	6.0g
NaCl	5.0g
Glucose	3.0g
Na_2HPO_4	2.5g

pH 5.3 ± 0.2 at 25°C

Source: This medium without agar is available as a premixed powder from BD Diagnostic Systems.

Preparation of Medium: Add components, except agar, to distilled/deionized water and bring volume to 1.0L. Mix thoroughly. Adjust pH to 5.3 with 1*N* HCl. Mix thoroughly. Add agar. Gently heat and bring to boiling. Distribute into tubes. Autoclave for 10 min at 15 psi pressure–121°C.

Use: For the detection of staphylococcal enterotoxin.

Brain Heart Infusion with 0.7% Agar

Composition per liter:

Beef heart infusion	250.0g
Calf brain infusion	200.0g
Proteose peptone	10.0g
Agar	7.0g
NaCl	5.0g
$Na_2HPO_4 \cdot 12H_2O$	2.5g
Glucose	2.0g

pH 5.3 ± 0.2 at 25°C

Preparation of Medium: Add components to distilled/deionized water and bring volume to 1.0L. Mix thoroughly. Gently heat and bring to boiling. Adjust pH to 5.3 with 1*N* HCl. Distribute into tubes in 25.0mL volumes. Autoclave for 10 min at 15 psi pressure–121°C.

Use: For the cultivation of *Staphylococcal* species for the production of enterotoxin.

Brain Heart Infusion Agar with 1% Agar, HiVeg

Composition per liter:

Agar	15.0g
Plant infusion	10.0g
Plant peptone No. 3	10.0g
Plant special infusion	7.5g
NaCl	5.0g
Na_2HPO_4	2.5g
Glucose	2.0g

pH 7.4 ± 0.2 at 25°C

Source: This medium is available as a premixed powder from HiMedia.

Preparation of Medium: Add components to distilled/deionized water and bring volume to 1.0L. Mix thoroughly. Gently heat and bring to boiling. Distribute into tubes or flasks. Autoclave for 15 min at 15 psi pressure–121°C. Pour into sterile Petri dishes or leave in tubes.

Use: For the cultivation of a variety of fastidious pathogenic bacteria, yeasts, and molds.

Brain Heart Infusion Agar with 1% Agar, HiVeg

Composition per liter:

Agar	10.0g
Plant infusion	10.0g
Plant peptone No. 3	10.0g
Plant special infusion	7.5g
NaCl	5.0g

Na_2HPO_4 .... 2.5g
Glucose 2.0g

pH 7.4 ± 0.2 at 25°C

Source: This medium is available as a premixed powder from HiMedia.

Preparation of Medium: Add components to distilled/deionized water and bring volume to 1.0L. Mix thoroughly. Gently heat and bring to boiling. Distribute into tubes or flasks. Autoclave for 15 min at 15 psi pressure–121°C. Pour into sterile Petri dishes or leave in tubes.

Use: For the cultivation of a variety of fastidious pathogenic bacteria, yeasts, and molds.

Brain Heart Infusion Agar with 1% Agar, HiVeg with Penicillin

Composition per liter:

Agar 15.0g
Plant infusion 10.0g
Plant peptone No. 3 .... 10.0g
Plant special infusion 7.5g
NaCl 5.0g
Na_2HPO_4 .... 2.5g
Glucose 2.0g
Penicillin solution 2.0mL

pH 7.4 ± 0.2 at 25°C

Source: This medium is available as a premixed powder from HiMedia.

Penicillin Solution:

Composition per 2.0mL:

Penicillin 0.1g

Preparation of Penicillin Solution: Add penicillin to ethanol and bring volume to 2.0mL. Mix thoroughly. Filter sterilize.

Preparation of Medium: Add components to distilled/deionized water and bring volume to 1.0L. Mix thoroughly. Gently heat and bring to boiling. Autoclave for 15 min at 15 psi pressure–121°C. Cool to 45–50°C. Aseptically add 2.0mL penicillin solution. Mix thoroughly. Pour into sterile Petri dishes or leave in tubes.

Use: For the cultivation of a variety of fastidious pathogenic bacteria, yeasts, and molds.

Brain Heart Infusion Agar with Chloramphenicol

Composition per liter:

Pancreatic digest of casein 16.0g
Agar 13.5g
Brain heart, solids from infusion 8.0g
Peptic digest of animal tissue 5.0g
NaCl 5.0g
Glucose 2.0g
Na_2HPO_4 .... 2.5g
Sheep blood, defibrinated 50.0mL
Chloramphenicol solution 10.0mL

pH 7.4 ± 0.2 at 25°C

Chloramphenicol Solution:

Composition per 10.0mL:

Chloramphenicol 0.05g

Preparation of Chloramphenicol Solution: Add chloramphenicol to distilled/deionized water and bring volume to 10.0mL. Mix thoroughly. Filter sterilize.

Preparation of Medium: Add components, except chloramphenicol solution and sheep blood, to distilled/deionized water and bring volume to 940.0mL. Mix thoroughly. Gently heat and bring to boiling. Autoclave for 15 min at 15 psi pressure–121°C. Cool to 45°–50°C. Aseptically add sterile chloramphenicol solution and sheep blood. Mix thoroughly. Pour into sterile Petri dishes or distribute into sterile tubes.

Use: For the isolation and cultivation of a wide variety of fungal species, especially systemic fungi, from clinical and nonclinical specimens. For the selective isolation of pathogenic fungi from specimens heavily contaminated with bacteria and saprophytic fungi. For the maintenance of fungal species on slant cultures.

Brain Heart Infusion Agar with Cysteine (LMG Medium 282)

Composition per 1010.1mL:

Brain heart infusion agar 1.0L
L-Cysteine solution 10.0mL

pH 7.4 ± 0.2 at 25°C

Brain Heart Infusion Agar:

Composition per liter:

Beef heart infusion 250.0g
Calf brain infusion 200.0g
Agar 13.5g
Proteose peptone 10.0g
NaCl 5.0g
$Na_2HPO_4 \cdot 12H_2O$ 2.5g
Glucose 2.0g

Preparation of Brain Heart Infusion Agar: Add components to distilled/deionized water and bring volume to 1.0L. Mix thoroughly. Gently heat and bring to boiling. Distribute into tubes or flasks. Autoclave for 15 min at 15 psi pressure–121°C. Cool to 45°–50°C.

L-Cysteine Solution:

Composition per 10.0mL:

L-Cysteine·HCl·H_2O 0.5g

Preparation of L-Cysteine Solution: Add L-cysteine·HCl·H_2O to distilled/deionized water and bring volume to 10.0mL. Mix thoroughly. Sparge with 100% N_2. Autoclave for 15 min at 15 psi pressure–121°C.

Preparation of Medium: Aseptically add 10.0mL sterile L-cysteine solution to 1.0L sterile brain heart infusion agar. Mix thoroughly. Pour into sterile Petri dishes or distribute into sterile tubes.

Use: For the cultivation of *Clostridium lactatifermentans.*

Brain Heart Infusion Agar with Kanamycin

Composition per liter:

Pancreatic digest of casein 16.0g
Agar 13.5g
Brain/heart, solids from infusion 8.0g
Peptic digest of animal tissue 5.0g
NaCl 5.0g
Glucose 2.0g
Na_2HPO_4 .... 2.5g
Sheep blood, defibrinated 50.0mL
Kanamycin solution 10.0mL

pH 7.4 ± 0.2 at 25°C

Kanamycin Solution:

Composition per 10.0mL:

Kanamycin 25.0mg

Preparation of Kanamycin Solution: Add kanamycin to distilled/deionized water and bring volume to 10.0mL. Mix thoroughly. Filter sterilize.

Preparation of Medium: Add components, except kanamycin solution and sheep blood, to distilled/deionized water and bring volume to 940.0mL. Mix thoroughly. Gently heat and bring to boiling. Autoclave for 15 min at 15 psi pressure–121°C. Cool to 45°–50°C. Aseptically add sterile kanamycin solution and sheep blood. Mix thoroughly. Pour into sterile Petri dishes or distribute into sterile tubes.

Use: For the cultivation of fastidious fungi.

Brain Heart Infusion Agar with Penicillin and Streptomycin

Composition per liter:

Pancreatic digest of casein	16.0g
Agar	13.5g
Brain heart, solids from infusion	8.0g
Peptic digest of animal tissue	5.0g
NaCl	5.0g
Glucose	2.0g
Na_2HPO_4	2.5g
Streptomycin	40.0mg
Penicillin	20,000U
Sheep blood, defibrinated	50.0mL

pH 7.4 ± 0.2 at 25°C

Preparation of Medium: Add components, except sheep blood, to distilled/deionized water and bring volume to 950.0mL. Mix thoroughly and while stirring bring to a boil for 1 min to completely dissolve. Autoclave for 15 min at 15 psi pressure–121°C. Cool to 50°C. Aseptically add 50.0mL of defibrinated sheep blood. Mix thoroughly. Pour into sterile Petri dishes while agitating gently to distribute the precipitate through the medium.

Use: For the isolation and cultivation of a wide variety of fungal species, especially systemic fungi, from clinical and nonclinical specimens. For the selective isolation of pathogenic fungi from specimens heavily contaminated with bacteria and saprophytic fungi. For the maintenance of fungal species on slant cultures.

Brain Heart Infusion Agar with 10% Sheep Blood, Gentamicin, and Chloramphenicol

Composition per liter:

Pancreatic digest of casein	16.0g
Agar	13.5g
Brain heart, solids from infusion	8.0g
Peptic digest of animal tissue	5.0g
NaCl	5.0g
Glucose	2.0g
Na_2HPO_4	2.5g
Sheep blood, defibrinated	100.0mL
Antibiotic solution	10.0mL

pH 7.4 ± 0.2 at 25°C

Antibiotic Solution:

Composition per 10.0mL:

Chloramphenicol	0.05g
Gentamicin	0.05g

Preparation of Antibiotic Solution: Add components to distilled/deionized water and bring volume to 10.0mL. Mix thoroughly. Filter sterilize.

Preparation of Medium: Add components, except antibiotic solution and sheep blood, to distilled/deionized water and bring volume to 890.0mL. Mix thoroughly. Gently heat and bring to boiling. Autoclave for 15 min at 15 psi pressure–121°C. Cool to 45°–50°C. Aseptically add sterile antibiotic solution and sheep blood. Mix thoroughly. Pour into sterile Petri dishes or distribute into sterile tubes.

Use: For the isolation and cultivation of a wide variety of fungal species, especially systemic fungi, from clinical and nonclinical specimens. For the selective isolation of pathogenic fungi from specimens heavily contaminated with bacteria and saprophytic fungi. For the maintenance of fungal species on slant cultures.

Brain Heart Infusion Agar with Tween™ 80 (ATCC Medium 1941)

Composition per liter:

Pancreatic digest of casein	16.0g
Agar	13.5g
Brain heart, solids from infusion	8.0g
Peptic digest of animal tissue	5.0g
NaCl	5.0g
Glucose	2.0g
Na_2HPO_4	2.5g
Tween™ 80	1.0g

pH 7.4 ± 0.2 at 25°C

Preparation of Medium: Add components to distilled/deionized water and bring volume to 1.0L. Mix thoroughly. Distribute into tubes or flasks while shaking to distribute precipitate. Autoclave for 15 min at 15 psi pressure–121°C. Pour into sterile Petri dishes or leave in tubes.

Use: For the cultivation of *Helicococcus kunzii.*

Brain Heart Infusion Agar with Vitamin B_{12} (LMG Medium 207)

Composition per 1100.1mL:

Brain heart infusion agar	1.0L
Vitamin B_{12} solution	100.0mL

pH 7.4 ± 0.2 at 25°C

Brain Heart Infusion Agar:

Composition per liter:

Beef heart infusion	250.0g
Calf brain infusion	200.0g
Agar	13.5g
Proteose peptone	10.0g
NaCl	5.0g
$Na_2HPO_4 \cdot 12H_2O$	2.5g
Glucose	2.0g

Preparation of Brain Heart Infusion Agar: Add components to distilled/deionized water and bring volume to 1.0L. Mix thoroughly. Gently heat and bring to boiling. Distribute into tubes or flasks. Autoclave for 15 min at 15 psi pressure–121°C. Cool to 45°–50°C.

Vitamin B_{12} Solution:

Composition per 100.0mL:

Vitamin B_{12}	0.1mg

Preparation of Vitamin B_{12} Solution: Add Vitamin B_{12} to 100.0mL of distilled/deionized water. Mix thoroughly. Filter sterilize.

Preparation of Medium: Aseptically add 100.01mL vitamin B_{12} solution to 1.0L sterile brain heart infusion agar. Mix thoroughly. Pour into sterile Petri dishes or distribute into sterile tubes.

Use: For the cultivation of *Bacillus sporothermodurans.*

Brain Heart Infusion Agar, 1/10 with Vitamins (1/10 BHIV Agar)

Composition per liter:

Agar	15.0g
Pancreatic digest of gelatin	1.45g
Brain heart, solids from infusion	0.6g
Peptic digest of animal tissue	0.6g
NaCl	0.5g
Glucose	0.3g
Na_2HPO_4	0.25g
Vitamin solution	10.0mL

pH 7.4 ± 0.2 at 25°C

Vitamin Solution:

Composition per 10.0mL:

Thiamine	100.0μg
Biotin	50.0μg

Preparation of Vitamin Solution: Add components to distilled/deionized water and bring volume to 10.0mL. Mix thoroughly. Filter sterilize.

Preparation of Medium: Add components, except vitamin solution, to distilled/deionized water and bring volume to 990.0mL. Mix thoroughly. Gently heat and bring to boiling. Autoclave for 15 min at 15 psi pressure–121°C. Cool to 50°C. Aseptically add 10.0mL of sterile vitamin solution. Mix thoroughly. Pour into sterile Petri dishes or distribute into sterile tubes.

Use: For the cultivation of *Smittium mucronatum.*

Brain Heart Infusion with Agar, Yeast Extract, Sodium Chloride, Inactivated Horse Serum, and Penicillin

Composition per liter:

NaCl	20.0g
Pancreatic digest of gelatin	14.5g
Agar	12.0g
Brain heart, solids from infusion	6.0g
Peptic digest of animal tissue	6.0g
Yeast extract	5.0g
NaCl	5.0g
Glucose	3.0g
Na_2HPO_4	2.5g
Horse serum, inactivated	100.0mL
Penicillin solution	10.0mL

pH 7.4 ± 0.2 at 25°C

Penicillin Solution:

Composition per 10.0mL:

Penicillin	1,000,000U

Preparation of Penicillin Solution: Add penicillin to distilled/deionized water and bring volume to 10.0mL. Mix thoroughly. Filter sterilize.

Preparation of Medium: Add components, except penicillin solution and inactivated horse serum, to distilled/deionized water and bring volume to 890.0mL. Mix thoroughly. Gently heat and bring to boiling. Autoclave for 15 min at 15 psi pressure–121°C. Cool to 45°–50°C. Aseptically add sterile penicillin solution and horse serum. Mix thoroughly. Pour into sterile Petri dishes or distribute into sterile tubes.

Use: For the cultivation of fastidious fungi.

Brain Heart Infusion with Agar, Yeast Extract, Sucrose, Horse Serum, and Penicillin

Composition per liter:

Sucrose	100.0g
Pancreatic digest of gelatin	14.5g
Agar	12.0g
Brain heart, solids from infusion	6.0g
Peptic digest of animal tissue	6.0g
Yeast extract	5.0g
NaCl	5.0g
Glucose	3.0g
Na_2HPO_4	2.5g
Horse serum	100.0mL
Penicillin solution	10.0mL

pH 7.4 ± 0.2 at 25°C

Penicillin Solution:

Composition per 10.0mL:

Penicillin	100,000U

Preparation of Penicillin Solution: Add penicillin to distilled/deionized water and bring volume to 10.0mL. Mix thoroughly. Filter sterilize.

Preparation of Medium: Add components, except penicillin solution and horse serum, to distilled/deionized water and bring volume to 890.0mL. Mix thoroughly. Gently heat and bring to boiling. Autoclave for 15 min at 15 psi pressure–121°C. Cool to 45°–50°C. Aseptically add sterile penicillin solution and horse serum. Mix thoroughly. Pour into sterile Petri dishes or distribute into sterile tubes.

Use: For the cultivation of fastidious fungi.

Brain Heart Infusion with Agar, Yeast Extract, Sucrose, Inactivated Horse Serum, and Penicillin

Composition per liter:

Sucrose	100.0g
Pancreatic digest of gelatin	14.5g
Agar	12.0g
Brain heart, solids from infusion	6.0g
Peptic digest of animal tissue	6.0g
Yeast extract	5.0g
NaCl	5.0g
Glucose	3.0g
Na_2HPO_4	2.5g
Horse serum, inactivated	100.0mL
Penicillin solution	10.0mL

pH 7.4 ± 0.2 at 25°C

Penicillin Solution:

Composition per 10.0mL:

Penicillin	1,000,000U

Preparation of Penicillin Solution: Add penicillin to distilled/deionized water and bring volume to 10.0mL. Mix thoroughly. Filter sterilize.

Preparation of Medium: Add components, except penicillin solution and inactivated horse serum, to distilled/deionized water and bring volume to 890.0mL. Mix thoroughly. Gently heat and bring to boiling. Autoclave for 15 min at 15 psi pressure–121°C. Cool to 45°–50°C. Aseptically add sterile penicillin solution and horse serum. Mix thoroughly. Pour into sterile Petri dishes or distribute into sterile tubes.

Use: For the cultivation of fastidious fungi.

Brain Heart Infusion Agar with Yeast Extract

Composition per liter:

Yeast extract	20.0g
Pancreatic digest of casein	16.0g
Agar	13.5g
Brain heart, solids from infusion	8.0g
Peptic digest of animal tissue	5.0g
NaCl	5.0g
Glucose	2.0g
Na_2HPO_4	2.5g

pH 7.4 ± 0.2 at 25°C

Source: This medium is available as a premixed powder from Oxoid Unipath and BD Diagnostic Systems.

Preparation of Medium: Add components to distilled/deionized water and bring volume to 1.0L. Mix thoroughly. Distribute into tubes or flasks while shaking to distribute precipitate. Autoclave for 15 min at 15 psi pressure–121°C. Pour into sterile Petri dishes or leave in tubes.

Use: For the cultivation of *Mycoplasma equirhinis*.

Brain Heart Infusion Blood Agar

Composition per liter:

Blood agar	1.1L
Overlay solution	1.0L

pH 7.4 ± 0.2 at 25°C

Blood Agar:

Composition per liter:

Agar	18.0g
Pancreatic digest of gelatin	14.5g
Brain heart, solids from infusion	6.0g
Peptic digest of animal tissue	6.0g
NaCl	5.0g
Glucose	3.0g
Na_2HPO_4	2.5g
Rabbit blood, defibrinated	100.0mL

Preparation of Blood Agar: Add components, except rabbit blood, to distilled/deionized water and bring volume to 900.0mL. Mix thoroughly. Gently heat and bring to boiling. Adjust pH to 7.4. Distribute 5.0mL of agar solution into tubes. Autoclave for 25 min at 15 psi pressure–121°C. Cool to 50°C. Aseptically add 0.5mL of sterile rabbit blood to each tube. Mix thoroughly. Allow tubes to cool in a slanted position.

Overlay Solution:

Composition per liter:

Pancreatic digest of gelatin	14.5g
Brain heart, solids from infusion	6.0g
Peptic digest of animal tissue	6.0g
NaCl	5.0g
Glucose	3.0g
Na_2HPO_4	2.5g

Preparation of Overlay Solution: Add components to distilled/deionized water and bring volume to 1.0L. Mix thoroughly. Distribute into tubes or flasks. Autoclave for 15 min at 15 psi pressure–121°C.

Preparation of Medium: Prior to inoculation, add 0.5mL of overlay solution to the surface of solidified blood agar slants.

Use: For the cultivation of *Blastocrithidia leptocoridis*, *Leishmania aethiopica*, *Leishmania amazonensis*, *Leishmania aristedesi*, *Leishmania braziliensis*, *Leishmania chagasi*, *Leishmania donovani*, *Leishmania enriettii*, *Leishmania garnhami*, *Leishmania gerbilli*, *Leishmania guyanensis*, *Leishmania hertigi*, *Leishmania infantum*, *Leishmania major*, *Leishmania mexicana*, *Leishmania panamensis*, *Leishmania pifanoi*, *Leishmania tropica*, and *Trypanosoma conorrhini*.

Brain Heart Infusion Broth (BHI Broth)

Composition per liter:

Beef heart infusion	250.0g
Calf brain infusion	200.0g
Proteose peptone	10.0g
NaCl	5.0g
$Na_2HPO_4 \cdot 12H_2O$	2.5g
Glucose	2.0g

pH 7.4 ± 0.2 at 25°C

Preparation of Medium: Add components to distilled/deionized water and bring volume to 1.0L. Mix thoroughly. Distribute into tubes or flasks. Autoclave for 15 min at 15 psi pressure–121°C.

Use: For the cultivation of a variety of fastidious and nonfastidious aerobic and anaerobic microorganisms.

Brain Heart Infusion Broth

Composition per liter:

Pancreatic digest of gelatin	14.5g
Beef heart, infusion from (solids)	6.0g
Peptic digest of animal tissue	6.0g
NaCl	6.0g
Glucose	3.0g
Na_2HPO_4	2.5g

pH 7.4 ± 0.2 at 25°C

Preparation of Medium: Add components to distilled/deionized water and bring volume to 1.0L. Mix thoroughly. Distribute into tubes or flasks while shaking to distribute precipitate. Autoclave for 15 min at 15 psi pressure–121°C.

Use: For the cultivation of a wide variety of microorganisms, including bacteria, yeasts, and molds, especially fastidious species.

Brain Heart Infusion Broth, HiVeg

Composition per liter:

Plant special infusion	12.5g
Plant peptone	10.0g
NaCl	5.0g
Plant infusion	5.0g
Na_2HPO_4	2.5g
Glucose	2.0g

pH 7.4 ± 0.2 at 25°C

Source: This medium is available as a premixed powder from HiMedia.

Preparation of Medium: Add components to distilled/deionized water and bring volume to 1.0L. Mix thoroughly. Distribute into tubes or flasks. Autoclave for 15 min at 15 psi pressure–121°C.

Use: For the cultivation of a microorganisms, including streptococci.

Brain Heart Infusion Broth with 6.5% NaCl

Composition per liter:

NaCl	65.0g
Pancreatic digest of gelatin	14.5g
Beef heart, infusion from (solids)	6.0g
Peptic digest of animal tissue	6.0g
Glucose	3.0g
Na_2HPO_4	2.5g

pH 7.4 ± 0.2 at 25°C

Preparation of Medium: Add components to distilled/deionized water and bring volume to 1.0L. Mix thoroughly. Distribute into tubes or flasks while shaking to distribute precipitate. Autoclave for 15 min at 15 psi pressure–121°C.

Use: For the differentiation of enterococci from nonenterococcal species.

Brain Heart Infusion Broth with 6.5% NaCl

Composition per liter:

Beef heart infusion	250.0g
Calf brain infusion	200.0g
NaCl	65.0g
Proteose peptone	10.0g
Na_2HPO_4	2.5g
Glucose	2.0g

pH 7.4 ± 0.2 at 25°C

Preparation of Medium: Add components to distilled/deionized water and bring volume to 1.0L. Mix thoroughly. Distribute into tubes or flasks while shaking to distribute precipitate. Autoclave for 15 min at 15 psi pressure–121°C.

Use: For the differentiation of enterococci from nonenterococcal species.

Brain Heart Infusion Broth with 6.5% Sodium Chloride, HiVeg

Composition per liter:

NaCl	65.0g
Plant infusion	10.0g
Plant peptone No. 3	10.0g
Plant special infusion	7.5g
Glucose	2.0g
Na_2HPO_4	2.0g

pH 7.4 ± 0.2 at 25°C

Source: This medium is available from HiMedia.

Preparation of Medium: Add components to distilled/deionized water and bring volume to 1.0L. Mix thoroughly. Distribute into tubes or flasks. Autoclave for 15 min at 15 psi pressure–121°C.

Use: For the cultivation of pathogenic *Vibrio* spp. For the differentiation of enterococci from nonenterococcal species.

Brain Heart Infusion with 0.1% Agar, HiVeg

Composition per liter:

Plant infusion	10.0g
Plant peptone No. 3	10.0g
Plant special infusion	7.5g
NaCl	5.0g
Na_2HPO_4	2.5g
Glucose	2.0g
Agar	1.0g

pH 7.4 ± 0.2 at 25°C

Source: This medium is available der from HiMedia.

Preparation of Medium: Add components to distilled/deionized water and bring volume to 1.0L. Mix thoroughly. Distribute into tubes or flasks. Autoclave for 15 min at 15 psi pressure–121°C.

Use: For the cultivation of a variety of fastidious and nonfastidious aerobic and anaerobic microorganisms, including streptococci.

Brain Heart Infusion with Casein

Composition per liter:

Pancreatic digest of gelatin	14.5g
Brain heart, solids from infusion	6.0g
Peptic digest of animal tissue	6.0g
NaCl	5.0g
Casein	5.0g
Glucose	3.0g
Na_2HPO_4	2.5g

pH 7.4 ± 0.2 at 25°C

Preparation of Medium: Add components to distilled/deionized water and bring volume to 1.0L. Mix thoroughly. Distribute into tubes or flasks. Autoclave for 15 min at 15 psi pressure–121°C.

Use: For the cultivation of *Serratia marcescens*.

Brain Heart Infusion Casein Starch

Composition per liter:

Pancreatic digest of gelatin	14.5g
Casein hydrolysate	10.0g
Brain heart, solids from infusion	6.0g
Peptic digest of animal tissue	6.0g
NaCl	5.0g
Glucose	3.0g
Na_2HPO_4	2.5g
Soluble starch	1.0g

pH 7.4 ± 0.2 at 25°C

Preparation of Medium: Add components to distilled/deionized water and bring volume to 1.0L. Mix thoroughly. Distribute into tubes or flasks. Autoclave for 15 min at 15 psi pressure–121°C.

Use: For the cultivation of *Actinomyces odontolyticus*.

Brain Heart Infusion with Chicken Serum

Composition per liter:

Yeast extract	20.0g
Pancreatic digest of casein	16.0g
Brain heart, solids from infusion	8.0g
Peptic digest of animal tissue	5.0g
NaCl	5.0g
Na_2HPO_4	2.5g
Glucose	2.0g
Chicken serum, heat inactivated	50.0mL
Nicotinamide adenine dinucleotide solution	10.0mL

pH 7.2 ± 0.2 at 25°C

Nicotinamide Adenine Dinucleotide Solution:

Composition per 10.0mL:

Nicotinamide adenine dinucleotide	0.1g

Preparation of Nicotinamide Adenine Dinucleotide Solution: Add 0.1g of nicotinamide adenine dinucleotide to distilled/deinonized water and bring volume to 10.0mL Filter sterilize.

Preparation of Medium: Add components, except chicken serum and nicotinamide adenine dinucleotide solution, to distilled/deionized water and bring volume to 940.0mL. Autoclave for 15 min at 15 psi pressure–121°C. Cool to 50°–55°C. Aseptically add 10.0mL. of nicotinamide adenine dinucleotide solution and 50.0mL of heat-inactivated chicken serum. Mix thoroughly. Aseptically distribute into sterile tubes or flasks.

Use: For the cultivation of *Haemophilus paragallinarum* and *Pasteurella avium.*

Brain Heart Infusion with Cystine

Composition per liter:

Pancreatic digest of gelatin	14.5g
Brain heart, solids from infusion	6.0g
Peptic digest of animal tissue	6.0g
NaCl	5.0g
Glucose	3.0g
Na_2HPO_4	2.5g
Cystine	1.0g

pH 7.4 ± 0.2 at 25°C

Preparation of Medium: Add components to distilled/deionized water and bring volume to 1.0L. Mix thoroughly. Distribute into tubes or flasks. Autoclave for 15 min at 15 psi pressure–121°C.

Use: For the cultivation of *Blastomyces dermatitidis, Histoplasma capsulatum, Mucor hiemalis, Paracoccidioides brasiliensis,* and *Sporothrix schenckii.*

Brain Heart Infusion with Erythromycin (BHI with Erythromycin) (DSMZ Medium 215a)

Composition per liter:

Pancreatic digest of gelatin	14.5g
Brain heart, solids from infusion	6.0g
Peptic digest of animal tissue	6.0g
NaCl	5.0g
Glucose	3.0g
Na_2HPO_4	2.5g
Erythromycin solution	10.0mL

pH 7.4 ± 0.2 at 25°C

Erythromycin Solution:

Composition per 10.0mL:

Erythromycin	10.0mg

Preparation of Erythromycin Solution: Add erythromycin to distilled/deionized water and bring volume to 10.0mL. Mix thoroughly. Filter sterilize.

Preparation of Medium: Add components, except erythromycin solution, to distilled/deionized water and bring volume to 990.0mL. Mix thoroughly. Gently heat and bring to boiling. Autoclave for 15 min at 15 psi pressure–121°C. Cool to 45°–50°C. Aseptically add sterile erythromycin solution. Mix thoroughly. Pour into sterile Petri dishes or distribute into sterile tubes.

Use: For the isolation and cultivation of *Nocardia crassostreae.*

Brain Heart Infusion with Glucose and Horse Serum (LMG Medium 183)

Composition per 1300.0mL:

Brain heart infusion agar	1.0L
Horse serum	200.0mL
Glucose solution	100.0mL

pH 7.4 ± 0.2 at 25°C

Brain Heart Infusion Agar:

Composition per liter:

Beef heart infusion	250.0g
Calf brain infusion	200.0g
Agar	13.5g
Proteose peptone	10.0g
NaCl	5.0g
$Na_2HPO_4 \cdot 12H_2O$	2.5g
Glucose	2.0g

Preparation of Brain Heart Infusion Agar: Add components to distilled/deionized water and bring volume to 1.0L. Mix thoroughly. Gently heat and bring to boiling. Distribute into tubes or flasks. Autoclave for 15 min at 15 psi pressure–121°C. Cool to 45°–50°C.

Glucose Solution:

Composition per 100.0mL:

Glucose	10.0g

Preparation of Glucose Solution: Add glucose to distilled/deionized water and bring volume to 100.0mL. Mix thoroughly. Filter sterilize.

Preparation of Medium: Aseptically add 200.0mL sterile horse serum and 100.0mL sterile glucose solution to 1.0L sterile brain heart infusion agar. Mix thoroughly. Pour into sterile Petri dishes or distribute into sterile tubes.

Use: For the cultivation of *Streptococcus pyogenes.*

Brain Heart Infusion with Horse Blood (LMG Medium 231)

Composition per liter:

Beef heart infusion	250.0g
Calf brain infusion	200.0g
Agar	13.5g
Proteose peptone	10.0g
NaCl	5.0g
$Na_2HPO_4 \cdot 12H_2O$	2.5g
Glucose	2.0g
Horse blood	50.0mL

pH 7.4 ± 0.2 at 25°C

Preparation of Medium: Add components, except horse blood, to distilled/deionized water and bring volume to 950.0L. Mix thoroughly. Gently heat and bring to boiling. Autoclave for 15 min at 15 psi pressure–121°C. Cool to 45°–50°C. Aseptically add 50.0mL sterile horse blood. Mix thoroughly. Pour into sterile Petri dishes or distribute into sterile tubes.

Use: For the cultivation of *Helcococcus kunzii* and *Globicatella sulfidifaciens.*

Brain Heart Infusion with PABA (Brain Heart Infusion with *p*-Aminobenzoic Acid)

Composition per liter:

Pancreatic digest of gelatin	14.5g
Brain heart, solids from infusion	6.0g
Peptic digest of animal tissue	6.0g
NaCl	5.0g
Glucose	3.0g
Na_2HPO_4	2.5g
p-Aminobenzoic acid	0.05g

pH 7.4 ± 0.2 at 25°C

Source: This medium is available as a premixed powder from BD Diagnostic Systems.

Preparation of Medium: Add components to distilled/deionized water and bring volume to 1.0L. Mix thoroughly. The addition of 1.0g agar to the medium enhances the growth of anaerobic and microaerophilic microorganisms. Heat with frequent agitation and boil for 1 min to dissolve. Distribute into tubes or flasks. Autoclave for 15 min at 15 psi pressure–121°C.

Use: For the detection of microorganisms in the blood of patients who have received sulfonamide therapy.

Brain Heart Infusion with PABA and Agar (Brain Heart Infusion with *p*-Aminobenzoic Acid and Agar)

Composition per liter:

Pancreatic digest of gelatin	14.5g
Brain heart, solids from infusion	6.0g
Peptic digest of animal tissue	6.0g
NaCl	5.0g
Glucose	3.0g
Na_2HPO_4	2.5g
Agar	1.0g
p-Aminobenzoic acid	0.05g

pH 7.4 ± 0.2 at 25°C

Source: This medium is available as a premixed powder from BD Diagnostic Systems.

Preparation of Medium: Add components to distilled/deionized water and bring volume to 1.0L. Mix thoroughly. The addition of 1.0g of agar to the medium enhances the growth of anaerobic and microaerophilic microorganisms. Heat with frequent agitation and boil for 1 min to dissolve. Distribute into tubes or flasks. Autoclave for 15 min at 15 psi pressure–121°C.

Use: For the detection of microorganisms in the blood of patients who have received sulfonamide therapy.

Brain Heart Infusion with PABA and Agar, HiVeg

Composition per liter:

Plant infusion	10.0g
Plant peptone	10.0g
Plant special infusion	7.5g
Na_2HPO_4	2.5g
NaCl	5.0g
Glucose	2.0g
Agar	1.0g
p-Aminobenzoic acid (PABA)	0.05g

pH 7.4 ± 0.2 at 25°C

Source: This medium is available as a premixed powder from HiMedia.

Preparation of Medium: Add components to distilled/deionized water and bring volume to 1.0L. Mix thoroughly. The addition of 1.0g agar to the medium enhances the growth of anaerobic and microaerophilic microorganisms. Heat with frequent agitation and boil for 1 min to dissolve. Distribute into tubes or flasks. Autoclave for 15 min at 15 psi pressure–121°C.

Use: For the detection of microorganisms in the blood of patients who have received sulfonamide therapy.

Brain Heart Infusion with Paraamino Benzoic Acid, HiVeg

Composition per liter:

Plant infusion	10.0g
Plant peptone No. 3	10.0g
Plant special infusion	7.5g
NaCl	5.0g
Na_2HPO_4	2.5g
Glucose	2.0g
p-Aminobenzoic acid (PABA)	0.05g

pH 7.4 ± 0.2 at 25°C

Source: This medium is available as a premixed powder from HiMedia.

Preparation of Medium: Add components to distilled/deionized water and bring volume to 1.0L. Mix thoroughly. The addition of 1.0g agar to the medium enhances the growth of anaerobic and microaerophilic microorganisms. Heat with frequent agitation and boil for 1 min to dissolve. Distribute into tubes or flasks. Autoclave for 15 min at 15 psi pressure–121°C.

Use: For the detection of microorganisms in the blood of patients who have received sulfonamide therapy.

Brain Heart Infusion with Rabbit Serum

Composition per liter:

Yeast extract	20.0g
Pancreatic digest of casein	16.0g
Brain heart, solids from infusion	8.0g
Peptic digest of animal tissue	5.0g
NaCl	5.0g
Na_2HPO_4	2.5g
Glucose	2.0g
Rabbit serum, heat inactivated	50.0mL
Nicotinamide adenine dinucleotide solution	10.0mL

pH 7.2 ± 0.2 at 25°C

Nicotinamide Adenine Dinucleotide Solution:

Composition per 10.0mL:

Nicotinamide adenine dinucleotide	0.1g

Preparation of Nicotinamide Adenine Dinucleotide Solution: Add 0.1g of nicotinamide adenine dinucleotide to distilled/deinonized water and bring volume to 10.0mL. Filter sterilize.

Preparation of Medium: Add components, except nicotinamide adenine dinucleotide solution and rabbit serum, to distilled/deionized water and bring volume to 940.0mL. Autoclave for 15 min at 15 psi pressure–121°C. Cool to 50°–55°C. Aseptically add 10.0mL of a 0.1% filter-sterilized solution of nicotinamide adenine dinucleotide. Aseptically add 50.0mL of heat-inactivated rabbit serum. Distribute aseptically into tubes or flasks.

Use: For the cultivation and maintenance of *Actinobacillus* species.

Brain Heart Infusion with Rabbit Serum and Yeast Extract

Composition per liter:

Pancreatic digest of casein	16.0g
Brain heart, solids from infusion	8.0g
Peptic digest of animal tissue	5.0g
NaCl	5.0g
Glucose	2.0g
Na_2HPO_4	2.5g
Rabbit serum, heat inactivated	200.0mL
Fresh yeast extract (25% solution)	100.0mL

pH 7.2 ± 0.2 at 25°C

Fresh Yeast Extract Solution:

Composition per 100.0mL:

Baker's yeast, live, pressed, starch-free	25.0g

Preparation of Fresh Yeast Extract Solution: Add the live Baker's yeast to 100.0mL of distilled/deionized water. Autoclave for 90 min at 15 psi pressure–121°C. Allow to stand. Remove supernatant solution. Adjust pH to 6.6–6.8.

Preparation of Medium: Add components, except rabbit serum, to distilled/deionized water and bring volume to 800.0mL. Autoclave for 15 min at 15 psi pressure–121°C. Cool to 50°–55°C. Aseptically add 200.0mL of sterile, heat-inactivated rabbit serum. Distribute aseptically into tubes or flasks.

Use: For the cultivation and maintenance of *Mycoplasma equirhinis.*

Brain Heart Infusion with Serum and Glucose (BHI with Serum and Glucose)

Composition per liter:

Pancreatic digest of gelatin	14.5g
Brain heart, solids from infusion	6.0g
Peptic digest of animal tissue	6.0g
NaCl	5.0g
Glucose	3.0g
Na_2HPO_4	2.5g
Glucose	2.0g
Bovine serum, heat inactivated	100.0mL

pH 7.4 ± 0.2 at 25°C

Preparation of Medium: Prepare and dispense medium anaerobically under 80% N_2 + 10% CO_2 + 10% H_2. Add components, except heat-inactivated fetal bovine serum, to distilled/deionized water and bring volume to 900.0mL. Mix thoroughly. Sparge with 80% N_2 + 10% CO_2 + 10% H_2 gas mixture. Gently heat and bring to boiling. Autoclave for 15 min at 15 psi pressure–121°C. Cool to 45°–50°C. Aseptically add sterile fetal bovine serum. Mix thoroughly. Pour into sterile Petri dishes or distribute into sterile tubes under an atmosphere of 80% N_2 + 10% CO_2 + 10% H_2.

Use: For the cultivation of *Actinomyces hordeovulneris, Actinomyces hyovaginalis, Serpulina hyodysenteriae, Serpulina pilosicoli,* and *Treponema phagedenis.*

Brain Heart Infusion with 3% Sodium Chloride

Composition per liter:

NaCl	30.0g
Pancreatic digest of gelatin	14.5g
Brain heart, solids from infusion	6.0g
Peptic digest of animal tissue	6.0g
Glucose	3.0g
Na_2HPO_4	2.5g

pH 7.4 ± 0.2 at 25°C

Preparation of Medium: Add components to distilled/deionized water and bring volume to 1.0L. Mix thoroughly. Distribute into tubes or flasks. Autoclave for 15 min at 15 psi pressure–121°C.

Use: For the cultivation of *Vibrio parahaemolyticus.*

Brain Heart Infusion with 5% Sodium Chloride

Composition per liter:

NaCl	50.0g
Pancreatic digest of gelatin	14.5g
Brain heart, solids from infusion	6.0g
Peptic digest of animal tissue	6.0g
Glucose	3.0g
Na_2HPO_4	2.5g

pH 7.4 ± 0.2 at 25°C

Preparation of Medium: Add components to distilled/deionized water and bring volume to 1.0L. Mix thoroughly. Distribute into tubes or flasks. Autoclave for 15 min at 15 psi pressure–121°C.

Use: For the cultivation of *Pediococcus halophilus.*

Brain Heart Infusion with Sucrose and Horse Serum

Composition per liter:

Sucrose	171.0g
Pancreatic digest of gelatin	14.5g
Brain heart, solids from infusion	6.0g
Peptic digest of animal tissue	6.0g
NaCl	5.0g
Glucose	3.0g
Na_2HPO_4	2.5g
Horse serum	100.0mL

pH 7.4 ± 0.2 at 25°C

Preparation of Medium: Add components, except horse serum, to distilled/deionized water and bring volume to 900.0mL. Mix thoroughly. Autoclave for 15 min at 15 psi pressure–121°C. Cool to 25°C. Aseptically add 100.0mL sterile horse serum. Mix thoroughly. Distribute into sterile tubes or flasks.

Use: For the cultivation of fastidious fungi.

Brain Heart Infusion, Supplemented (BHIS)

Composition per liter:

Brain heart infusion broth, dehydrated	37.0g
Yeast extract	5.0g
Supplement solution	10.2mL
Resazurin (0.25% solution)	4.0mL

pH 7.4 ± 0.2 at 25°C

Supplement Solution:

Composition per 10.2mL:

L-Cysteine·HCl·H_2O	0.5g
Hemin solution	10.0mL
Vitamin K_1 solution	0.2mL

Preparation of Supplement Solution: Add L-cysteine·HCl·H_2O and vitamin K_1 to 10.0mL of hemin solution. Mix thoroughly.

Hemin Solution:

Composition per 100.0mL:

Hemin	50.0mg
1*N* NaOH	1.0mL

Preparation of Hemin Solution: Add components to 100.0mL of distilled/deionized water and bring volume to 1.0L. Mix thoroughly.

Vitamin K_1 Solution:

Composition per 30.0mL:

Ethanol (95% solution)	30.0mL
Vitamin K_1	0.15mL

Preparation of Vitamin K_1 Solution: Add vitamin K_1 to 30.0mL of 95% ethanol. Mix thoroughly.

Preparation of Medium: Add components (brain heart infusion broth powder, yeast extract, and resazurin) to distilled/deionized water and bring volume to 990.0mL. Mix thoroughly and while stirring bring to boiling for 1 min. Cool to 45°–50°C. Add supplement solution containing hemin solution, vitamin K_1 solution, and L-cysteine. Autoclave for 15 min at 15 psi pressure–121°C. Pour into sterile Petri dishes while agitating gently to distribute the precipitate through the medium.

Use: For the cultivation of *Centipeda periodontii.*

Brain Heart Infusion with Thiamine (LMG Medium 97)

Composition per liter:

NaCl	30.0g
Pancreatic digest of gelatin	14.5g
Brain heart, solids from infusion	6.0g
Peptic digest of animal tissue	6.0g
Glucose	3.0g
Na_2HPO_4	2.5g
Thiamine·HCl	1.0mg

pH 7.4 ± 0.2 at 25°C

Preparation of Medium: Add components to distilled/deionized water and bring volume to 1.0L. Mix thoroughly. Distribute into tubes or flasks. Autoclave for 15 min at 15 psi pressure–121°C.

Use: For the cultivation of *Bacillus larvae.*

Brain Heart Infusion Soil Extract Medium

Composition per liter:

Yeast extract	20.0g
Pancreatic digest of casein	16.0g
Brain heart, solids from infusion	8.0g
Peptic digest of animal tissue	5.0g
NaCl	5.0g
Na_2HPO_4	2.5g
Glucose	2.0g
Soil extract	250.0mL
Vitamin B_{12} solution	1.0mL

pH 7.2 ± 0.2 at 25°C

Soil Extract:

Composition per 400.0mL:

African Violet soil	1.0g
Na_2CO_3	1.0g

Preparation of Soil Extract: Autoclave for 60 min at 15 psi pressure–121°C. Filter through paper before using in medium.

Vitamin B_{12} Solution:

Composition per 1.0mL:

Vitamin B_{12}	2.0μg

Preparation of Vitamin B_{12} Solution: Add vitamin B_{12} to distilled/deionized water and bring volume to 1.0mL. Mix thoroughly. Filter sterilize.

Preparation of Medium: Add components, except glucose, yeast extract, and vitamin B_{12} solution, to tap water and bring volume to 799.0mL. Mix thoroughly. Autoclave for 15 min at 15 psi pressure–121°C. Add yeast extract and glucose to 200.0mL of tap water. Filter sterilize and add aseptically to cooled, sterile basal medium. Aseptically add 1.0mL of vitamin B_{12} solution. Mix thoroughly. Aseptically distribute into sterile tubes or flasks.

Use: For the cultivation of a wide variety of microorganisms, including bacteria, yeasts, and molds, especially fastidious species from soil. For the isolation of *Histoplasma capsulatum* and other pathogenic fungi, including *Coccidioides immitis*.

Brain Liver Heart Semisolid Medium

Composition per liter:

Beef heart, infusion from	250.0g
Calf brains, infusion from	200.0g
Liver, infusion from	50.0g
Proteose peptone	10.0g
NaCl	5.0g
Neopeptone	3.25g
Pancreatic digest of casein	3.25g
Na_2HPO_4	2.5g
Glucose	2.0g
Agar	1.75g

pH 7.3 ± 0.2 at 25°C

Preparation of Medium: Add components to distilled/deionized water and bring volume to 1.0L. Mix thoroughly. Gently heat and bring to boiling. Distribute into tubes or flasks. Autoclave for 15 min at 15 psi pressure–121°C. Leave in tubes.

Use: For the cultivation of fastidious microorganisms. For the cultivation of *Actinomyces bovis*, *Actinomyces israelii*, *Actinomyces naeslundii*, *Clostridium symbiosum*, and *Eubacterium limosum*.

Brettanomyces Agar Base with Selective Supplement

Composition per liter:

Agar	20.0g
Glucose	10.0g
Peptic digest of animal tissue	5.0g
Yeast extract	3.0g
Malt extract	3.0g
Chloramphenical	0.1g
Cycloheximide	0.1g
Thiamine hydrochloride	0.01g
Selective supplement solution	10.0mL

Source: This medium is available from HiMedia.

Caution: Cycloheximide is toxic. Avoid skin contact or aerosol formation and inhalation.

Selective Supplement Solution:

Composition per 10.0mL:

Chlortetracycline	0.1g
Gentamicin	0.05g

Preparation of Selective Supplement Solution: Add components to distilled/deionized water and bring volume to 10.0mL. Mix thoroughly. Filter sterilize.

Preparation of Medium: Add components, except selective supplement solution, to distilled/deionized water and bring volume to 990.0mL. Mix thoroughly. Autoclave for 15 min at 15 psi pressure–121°C. Cool to 50°C. Aseptically add selective supplement solution. . Mix thoroughly. Pour into Petri dishes or aseptically distribute into sterile tubes.

Use: For the cultivation of *Brettanomyces* spp. and the detection of this yeast in soft drinks.

Brevibacillus levickii Medium (DSMZ Medium 1064)

Composition per liter:

Agar	18.0g
$CaCl_2 \cdot 2H_2O$	12.5g
$MnSO_4 \cdot H_2O$	2.5g
Yeast extract	2.0g
KH_2PO_4	1.5g
$(NH_4)_2SO_4$	1.25g
$MgSO_4 \cdot 7H_2O$	0.1g

pH 5.5 ± 0.2 at 25°C

Preparation of Medium: Add components to distilled/deionized water and bring volume to 1.0L. Mix thoroughly. Adjust pH to 5.5. Gently heat while stirring and bring to boiling. Distribute into tubes or flasks. Autoclave for 15 min at 15 psi pressure–121°C. Pour into sterile Petri dishes or leave in tubes.

Use: For the cultivation of *Brevibacillus levickii.*

Brevibacterium Medium (ATCC Medium 159)

Composition per liter:

Agar	25.0g
Glucose	20.0g
$CaCO_3$	20.0g
Yeast extract	10.0g

Preparation of Medium: Add components to distilled/deionized water and bring volume to 1.0L. Mix thoroughly. Gently heat to boiling. Distribute into tubes or flasks. Autoclave for 15 min at 15 psi pressure–121°C. Pour into sterile Petri dishes or leave in tubes.

Use: For the cultivation and maintenance of *Brevibacterium* species.

Brevibacterium Medium (ATCC Medium 677)

Composition per liter:

Agar	30.0g
KH_2PO_4	2.0g
Na_2HPO_4	2.0g
$(NH_4)_2SO_4$	2.0g
Yeast extract	2.0g
Tween™ 60	2.0g
$MgSO_4 \cdot 7H_2O$	0.2g
$FeSO_4 \cdot 7H_2O$	0.1g
$MnSO_4$	0.01g
n-Hexadecane	50.0mL

pH 7.0 ± 0.2 at 25°C

Preparation of Medium: Add components to distilled/deionized water and bring volume to 1.0L. Mix thoroughly. Blend for 30 min in a blender to disperse the *n*-hexadecane. Distribute into tubes or flasks. Autoclave for 20 min at 15 psi pressure–121°C.

Use: For the cultivation and maintenance of *Brevibacterium alkanophilum* and other microorganisms that can utilize hexadecane as a carbon source.

Brevibacterium Medium (ATCC Medium 681)

Composition per liter:

Glucose	10.0g
Peptone	5.0g
Yeast extract	5.0g

pH 5.0–6.0 at 25°C

Preparation of Medium: Add components to distilled/deionized water and bring volume to 1.0L. Mix thoroughly. Distribute into tubes or flasks. Autoclave for 15 min at 15 psi pressure–121°C.

Use: For the cultivation and maintenance of *Brevibacterium* spp. and *Enterobacter cloacae.*

Brevundimonas Agar (LMG Medium 221)

Composition per liter:

Agar	15.0g
Yeast extract	10.0g
Peptone	2.0g
$MgSO_4 \cdot 7H_2O$	0.2g
$CaCl_2$	0.1g
Riboflavin solution	5.0mL

Riboflavin Solution:

Composition per 10.0mL:

Riboflavin	2.0mg

Preparation of Riboflavin Solution: Add riboflavin to 10.0mL of distilled/deionized water. Mix thoroughly. Filter sterilize.

Preparation of Medium: Add components, except riboflavin solution , to distilled/deionized water and bring volume to 995.0mL. Mix thoroughly. Gently heat and bring to boiling. Autoclave for 15 min at 15 psi pressure–121°C. Cool to 45°–50°C. Aseptically add 5.0mL sterile riboflavin solution. Mix thoroughly. Pour into sterile Petri dishes or distribute into sterile tubes.

Use: For the cultivation of *Brevundimonas* spp. and *Caulobacter henricii.*

Brewer Anaerobic Agar

Composition per liter:

Agar	20.0g
Proteose peptone No. 3	10.0g
Glucose	10.0g
Pancreatic digest of casein	5.0g
Yeast extract	5.0g
NaCl	5.0g
Sodium thioglycollate	2.0g
Sodium formaldehyde sulfoxylate	1.0g
Resazurin	2.0mg

pH 7.2 ± 0.2 at 25°C

Source: This medium is available as a premixed powder from BD Diagnostic Systems.

Preparation of Medium: Add components to distilled/deionized water and bring volume to 1.0L. Mix thoroughly. Distribute into tubes or flasks. Autoclave for 15 min at 15 psi pressure–121°C.

Use: For the cultivation and maintenance of anaerobic and microaerophilic microorganisms.

Brewer Thioglycollate HiVeg Medium

Composition per liter:

Plant infusion	17.5g
Plant peptone No. 3	10.0g
Glucose	5.0g
NaCl	5.0g
K_2HPO_4	2.0g
Na-thioglycollate	1.0g
Agar	0.5g
Methylene Blue	2.0mg

pH 7.2 ± 0.2 at 25°C

Source: This medium is available as a premixed powder from HiMedia.

Preparation of Medium: Add components to distilled/deionized water and bring volume to 1.0L. Mix thoroughly. Gently heat to boiling. Distribute into tubes or flasks. Autoclave for 15 min at 15 psi pressure–121°C.

Use: For the cultivation and maintenance of anaerobic and microaerophilic microorganisms. For testing the sterility of biological products and materials.

Brewer Thioglycollate HiVeg Medium, Modified

Composition per liter:

Plant hydrolysate	17.5g
Glucose	10.0g
NaCl	5.0g
Papaic digest of soybean meal	2.5g
K_2HPO_4	2.0g
Na-thioglycollate	1.0g
Agar	0.5g
Methylene Blue	2.0mg

pH 7.2 ± 0.2 at 25°C

Source: This medium is available as a premixed powder from HiMedia.

Preparation of Medium: Add components to distilled/deionized water and bring volume to 1.0L. Mix thoroughly. Gently heat to boiling. Distribute into tubes or flasks. Autoclave for 15 min at 15 psi pressure–121°C.

Use: For the cultivation and maintenance of anaerobic and microaerophilic microorganisms. For testing the sterility of biological products and materials.

Brewer Thioglycollate Medium

Composition per liter:

Beef, infusion from	500.0g
Proteose peptone	10.0g
NaCl	5.0g
Glucose	5.0g
K_2HPO_4	2.0g
Sodium thioglycolate	0.5g
Agar	0.5g
Methylene Blue	2.0mg

pH 7.2 ± 0.2 at 25°C

Source: This medium is available as a premixed powder from BD Diagnostic Systems.

Preparation of Medium: Add components to distilled/deionized water and bring volume to 1.0L. Mix thoroughly. Gently heat to boiling. Distribute into tubes or flasks. Autoclave for 15 min at 15 psi pressure–121°C.

Use: For the cultivation and maintenance of anaerobic and microaerophilic microorganisms. For testing the sterility of biological products and materials.

Brewer Thioglycollate Medium, Modified

Composition per liter:

Tryptic digest of casein	17.0g
Glucose	10.0g
NaCl	5.0g
Enzymatic hydrolysate of soybean meal	3.0g
K_2HPO_4	2.0g
Sodium thioglycolate	1.0g
Agar	0.5g
Methylene Blue	2.0mg

pH 7.2 ± 0.2 at 25°C

Source: This medium is available as a premixed powder from BD Diagnostic Systems.

Preparation of Medium: Add components to distilled/deionized water and bring volume to 1.0L. Mix thoroughly. Gently heat to boiling. Distribute into tubes or flasks. Autoclave for 15 min at 15 psi pressure–121°C.

Use: For the cultivation and maintenance of anaerobic and microaerophilic microorganisms. For testing the sterility of biological products and materials.

Brewer Thioglycollate Medium, Modified

Composition per liter:

Casein enzymatic hydrolysate	17.5g
Glucose	10.0g
NaCl	5.0g
Papaic digest of soybean meal	2.5g
K_2HPO_4	2.0g
Na-thioglycollate	1.0g
Agar	0.5g
Methylene Blue	2.0mg

pH 7.2 ± 0.2 at 25°C

Source: This medium is available as a premixed powder from HiMedia.

Preparation of Medium: Add components to distilled/deionized water and bring volume to 1.0L. Mix thoroughly. Gently heat to boiling. Distribute into tubes or flasks. Autoclave for 15 min at 15 psi pressure–121°C.

Use: For the cultivation and maintenance of anaerobic and microaerophilic microorganisms. For testing the sterility of biological products and materials.

Brigg's Liver Broth pH 7.0

Composition per liter:

Glucose	20.0g
Neopeptone	15.0g
Yeast extract	6.0g
NaCl	5.0g
Tween™ 80	1.0g
Soluble starch	0.5g
L-Cysteine·HCl·H_2O	0.2g
Tomato juice solution	400.0mL
Liver extract	75.0mL

pH 7.0 ± 0.2 at 25°C

Tomato Juice Solution:

Composition per 500.0mL:

Tomato juice	250.0mL

Preparation of Tomato Juice Solution: Add 250.0mL of tomato juice to distilled/deionized water and bring volume to 500.0mL. Mix thoroughly. Adjust pH to 7.0 with 10% NaOH. Filter through Whatman #2 filter paper.

Liver Extract:

Composition per 170.0mL:

Liver powder	10.0g

Preparation of Liver Extract: Add 10.0g of liver powder to 170.0mL of distilled/deionized water. Gently heat to 60°C. Maintain at 50°–60°C for 1 hr. Gently bring to boiling. Boil for 5 min. Adjust pH to 7.2. Filter through Whatman #2 filter paper.

Preparation of Medium: Add components to distilled/deionized water and bring volume to 1.0L. Mix thoroughly. Distribute into tubes or flasks. Autoclave for 15 min at 15 psi pressure–121°C.

Use: For the cultivation of *Lactobacillus acetotolerans.*

Brigg's Liver Tomato Broth

Composition per liter:

Glucose	20.0g
Neopeptone	15.0g
Yeast extract	6.0g
NaCl	5.0g
Tween™ 80	1.0g
Soluble starch	0.5g
L-Cysteine·HCl	0.2g
Tomato juice	400.0mL
Liver extract	75.0mL

pH 5.0 ± 0.2 at 25°C

Preparation of Medium: Add components to distilled/deionized water and bring volume to 1.0L. (Note: 3.0g of proteolysed liver may be used instead of the liver extract.) Mix thoroughly. Distribute into tubes or flasks. Autoclave for 15 min at 15 psi pressure–121°C.

Use: For the cultivation of *Lactobacillus acetotolerans.*

BRILA MUG Broth (Brillant Green 2%-Bile MUG Broth)

Composition per liter:

Ox-bile (dried)	20.0g
Peptone	10.0g
Lactose	10.0g
L-Tryptophan	1.0g
Brillant Green	0.133g
4-Methylumbelliferyl-ß-D-glucuronide	0.1g

pH 7.2 ± 0.2 at 37°C

Source: This medium is available from Fluka, Sigma-Aldrich.

Preparation of Medium: Add components to distilled/deionized water and bring volume to 1.0L. Mix thoroughly. Distribute into test tubes that contain an inverted Durham tube in 10.0mL volumes. Autoclave for 15 min at 15 psi pressure–121°C.

Use: For the detection of *E. coli* and coliforms. Bile and Brilliant Green extensively inhibit the growth of accompanying flora, in particular Gram-positive microorganisms. The presence of *E. coli* results in fluorescence in the UV. A positive indole test and possibly gas formation from lactose fermentation provide confirmation. β-D-glucoronidase, which is produced by *E. coli*, cleaves 4-methylumbelliferyl-β-D-glucuronide to 4-methylumbelliferone and glucuronide. The fluorogen 4-methylumbelliferone can be detected under a long wavelength UV lamp.The broth can be used in conjunction with the MPN method for *E. coli* and coliform enumeration in the water of bathing areas.

Brilliance™ *Bacillus cereus* Agar (Chromogenic *Bacillus cereus* Agar)

Composition per liter:

Agar	13.0g
Peptone	10.0g
Sodium pyruvate	10.0g
Yeast extract	4.0g
Na_2HPO_4	2.52g
Chromogenic mix	1.2g
KH_2PO_4	0.28g
Bacillus cereus selective supplement	10.0mL

pH 7.2 ± 0.2 at 25°C

Source: This medium is available as a premixed powder from Oxoid Unipath.

***Bacillus cereus* Selective Supplement:**

Composition per 10.0mL:

Trimethoprim	10.0mg
Polymyxin B	106,000 U

Preparation of *Bacillus cereus* Selective Supplement Solution: Add components to distilled/deionized water and bring volume to 10.0mL. Mix thoroughly. Filter sterilize.

Preparation of Medium: Add components, except *Bacillus cereus* selective supplement. to distilled/deionized water and bring volume to 1.0L. Mix thoroughly. Gently heat while stirring and bring to boiling. Autoclave for 15 min at 15 psi pressure–121°C. Cool to 50°C. Aseptically add 5.0mL *Bacillus cereus* selective supplement. Mix thouroughly. Pour into sterile Petri dishes.

Use: For the isolation and differentiation of *Bacillus cereus* from food samples.

Brilliance™ *Candida* Agar

Composition per liter:

Chromogenic mix	13.6g
Agar	13.6g
Peptone	4.0g
Selective supplement solution	10.0mL

pH 6.0 ± 0.2 at 25°C

Source: This medium is available as a premixed powder from Oxoid Unipath.

Selective Supplement Solution:

Composition per 10.0mL:

Chloramphenicol	0.5g

Preparation of Selective Supplement Solution: Add chloramphenicol to distilled/deionized water and bring volume to 10.0mL. Mix thoroughly. Filter sterilize.

Preparation of Medium: Add components, except selective supplement solution, to distilled/deionized water and bring volume to 1.0L. Mix thoroughly. Add 10.0mL selective supplement solution. Gently heat while stirring and bring to boiling. Do not autoclave. Cool to 45°C. Mix thoroughly. Pour into sterile Petri dishes.

Use: For the rapid isolation and identification of clinically important *Candida* species. The green color of *Candida albicans* and *Candida dubliniensis* is caused by the same chromogenic reaction as the dark blue color of *Candida tropicalis. Candida glabrata, Candida kefyr, Candida parapsilosis,* and *Candida lusitaniae* appear as a variety of beige/brown/yellow colors, due to the mixture of natural pigmentation and some alkaline phosphatase activity.

Brilliance™ *E. coli*/Coliform Agar (Chromogenic *E. coli/*Coliform Agar)

Composition per liter:

Chromogenic mix	20.3g
Agar	15.0g
Peptone	5.0g
NaCl	5.0g
Na_2HPO_4	3.5g
Yeast extract	3.0g
Lactose	2.5g
KH_2PO_4	1.5g
Neutral Red	0.03g

pH 7.0 ± 0.2 at 25°C

Source: This medium is available as a premixed powder from Oxoid Unipath.

Preparation of Medium: Add components to distilled/deionized water and bring volume to 1.0L. Mix thoroughly. Gently heat and bring to boiling. Distribute into tubes or flasks. Autoclave for 15 min at 15 psi pressure–121°C. Pour into sterile Petri dishes or leave in tubes.

Use: For the presumptive identification of *Escherichia coli* and coliforms from food and environmental samples. *E. coli* forms pink colonies.

Brilliance™ *E. coli*/Coliform Selective Agar

Composition per liter:

Agar	10.6g
Peptone	8.0g
NaCl	5.0g
Na_2HPO_4	2.2g
KH_2PO_4	1.8g
Chromogenic mix	0.35g
Sodium lauryl sulfate	0.1g

pH 6.7 ± 0.2 at 25°C

Source: This medium is available as a premixed powder from Oxoid Unipath.

Preparation of Medium: Add components to distilled/deionized water and bring volume to 1.0L. Mix thoroughly. Gently heat and bring to boiling. Do not autoclave. Cool to 45°C. Pour into sterile Petri dishes or leave in tubes.

Use: For the presumptive identification of *Escherichia coli* and coliforms from food and environmental samples. *E. coli* forms pink-purple colonies.

Brilliance™ *Enterobacter sakazakii* Agar (DFI Agar) (Druggan, Forsythe and Iverson Agar)

Composition per liter:

Agar	15.0g
Tryptone	15.0g
Soya peptone	5.0g
NaCl	5.0g
Ferric ammonium citrate	1.0g
Sodium desoxycholate	1.0g
Sodium thiosulphate	1.0g
Chromogen	0.1g

pH 7.3 ± 0.2 at 25°C

Source: This medium is available as a premixed powder from Oxoid Unipath.

Preparation of Medium: Add components to distilled/deionized water and bring volume to 1.0L. Mix thoroughly. Gently heat and bring to boiling. Distribute into tubes or flasks. Autoclave for 15 min at 15 psi pressure–121°C. Pour into sterile Petri dishes or leave in tubes.

Use: For the differentiation and enumeration of *Enterobacter sakazakii* from infant formula and other food samples. A chromogenic medium for the isolation and differentiation of *Enterobacter sakazakii* (now *Cronobacter sakazakii*) from food and dairy samples, according to the formulation by Druggan, Forsythe, and Iverson

Brilliance™ ESBL Agar

Composition per liter:

Agar	15.0g
Peptones	12.0g
NaCl	5.0g
Phosphate buffers	4.0g
Chromogenic mix	4.0g
Antibiotic mix	0.28g

pH 7.2 ± 0.2 at 25°C

Source: This medium is available as a premixed powder from Oxoid Unipath.

Preparation of Medium: Add components to distilled/deionized water and bring volume to 1.0L. Mix thoroughly. Gently heat while stirring and bring to boiling. Do not autoclave. Cool to 45°C. Mix thoroughly. Pour into sterile Petri dishes.

Use: For the detection of Extended Spectrum β-Lactamase-producing organisms. The medium provides presumptive identification of ESBL-producing *E. coli* and the *Klebsiella, Enterobacter, Serratia,* and *Citrobacter* group (KESC), direct from clinical samples, in 24 hours.

Brilliance™ *Listeria* Agar (Oxoid Chromogenic *Listeria* Agar) (OCLA)

Composition per liter:

Peptone	18.5g
LiCl	15.0g

Agar	14.0g
NaCl	9.5g
Yeast extract	4.0g
Maltose	4.0g
Sodium pyruvate	2.0g
X-glucoside chromogenic mix	0.2g
Differential lecithin solution	40.0mL
Selective supplement solution	20.0mL

pH 7.2 ± 0.2 at 25°C

Source: This medium is available as a premixed powder from Oxoid Unipath.

Differential Lecithin Solution:
Composition per 40.0mL:

Lecithin	Proprietary

Preparation of Differential Lecithin Solution: Available as premixed solution.

Selective Supplement Solution:
Composition per 20.0mL:

Nalidixic acid	26.0mg
Polymyxin B	10.0mg
Ceftazidime	6.0mg
Amphotericin	10.0mg

Preparation of Selective Supplement Solution: Add components to distilled/deionized water and bring volume to 20.0mL. Mix thoroughly. Filter sterilize.

Preparation of Medium: Add components, except differential lecithin solution and selective supplement solution, to distilled/deionized water and bring volume to 940.0mL. Mix thoroughly. Gently heat while stirring and bring to boiling. Autoclave for 15 min at 15 psi pressure–121°C. Cool to 46°C. Aseptially add differential lecithin solution and selective supplement solution. Mix thoroughly. Pour into sterile Petri dishes.

Use: For the isolation, enumeration, and presumptive identification of *Listeria* species and *Listeria monocytogenes* from food samples. A chromogenic agar for the selective growth and differentiation of *Listeria monocytogenes* and *Listeria* spp.

Brilliance™ MRSA Agar

Composition per liter:

Peptone mix	25.0g
Salt mix	25.0g
Agar	15.0g
Kaolin	13.0g
Chromogenic mix	2.0g
Antibiotic cocktail	4.0mL

pH 7.2 ± 0.2 at 25°C

Source: This medium is available as a premixed powder from Oxoid Unipath.

Preparation of Medium: Add components, except antibiotic cocktail, to distilled/deionized water and bring volume to 1.0L. Mix thoroughly. Gently heat while stirring and bring to boiling. Autoclave for 15 min at 15 psi pressure–121°C. Cool to 50°C. Aseptially add antibiotic cocktail. Mix thoroughly. Pour into sterile Petri dishes.

Use: For universal MRSA screening. Oxoid Brilliance MRSA Agar incorporates a novel chromogen that yields a blue color as a result of phosphatase activity, indicative of many staphylococci including *Staphylococcus aureus*. To allow the medium to differentiate MRSA accurately, it contains a combination of antibacterial compounds designed to inhibit the growth of a wide variety of competitor organisms and MSSA. Also included are compounds to suppress the expression of phosphatase activity in other staphylococci, thus ensuring a high level of sensitivity and specificity.

Brilliance™ *Salmonella* Agar

Composition per liter:

Chromogenic mix	25.0g
Agar	15.0g
Inhibigen™ mix	14.0g
Salmonella selective supplement solution	10.0mL

pH 7.3 ± 0.2 at 25°C

Source: This medium is available as a premixed powder from Oxoid Unipath.

***Salmonella* Selective Supplement Solution:**
Composition per 10.0mL:

Novobiocin	10.0mg
Cefsulodin	24.0mg

Preparation of Selective Supplement Solution: Add components to distilled/deionized water and bring volume to 10.0mL. Mix thoroughly. Filter sterilize.

Preparation of Medium: Add components, except *Salmonella* selective supplement solution, to distilled/deionized water and bring volume to 1.0L. Mix thoroughly. Add 10.0mL *Salmonella* selective supplement. It is critical that the selective supplement is added prior to heating. Gently heat while stirring and bring to boiling. Do not autoclave. Cool to 50°C. Mix thoroughly. Pour into sterile Petri dishes.

Use: For the presumptice detection and identification of *Salmonella* spp. from foods and clinical specimens. The Inhibigen contained in this medium specifically targets *E.coli*, a particular benefit when testing fecal samples. Additional compounds are added to suppress growth of other competing flora. Differentiation of *Salmonella* from the other organisms that grow on Brilliance *Salmonella* Agar is achieved through the inclusion of two chromogens that target specific enzymes: caprylate esterase and β-glucosidase. The action of the enzymes on the chromogens results in a build-up of color within the colony. The color produced depends on which enzymes the organisms possess. The action of caprylate esterase present in all salmonellae results in a purple colony. Some Enterobacteriaceae species also produce caprylate esterase, but these are differentiated from *Salmonella* by a β-glucosidase substrate. This results in blue colonies, which are easy to distinguish from the purple *Salmonella* colonies.

Brilliance™ UTI Agar

Composition per liter:

Chromogenic mix	26.3g
Agar	15.0g
Peptone	15.0g

pH 7.0 ± 0.2 at 25°C

Source: This medium is available as a premixed powder from Oxoid Unipath.

Preparation of Medium: Add components to distilled/deionized water and bring volume to 1.0L. Mix thoroughly. Gently heat and bring to boiling. Distribute into tubes or flasks. Autoclave for 15 min at 15 psi pressure–121°C. Cool to 50°C. Mix thoroughly. Pour into sterile Petri dishes or leave in tubes.

Use: For the presumptive identification and differentiation of all the main microorganisms that cause urinary tract infections (UTIs). Brilliance UTI Agar contains two specific chromogenic substrates which are cleaved by enzymes produced by *Enterococcus* spp., *Escherichia coli,* and coliforms. In addition, it contains phenylalanine and tryptophan, which provide an indication of tryptophan deaminase activity, indicating the presence of *Proteus* spp., *Morganella* spp., and *Providencia* spp.

Brilliance™ UTI Clarity Agar

Composition per liter:

Chromogenic mix	17.0g
Agar	10.0g
Peptone	9.0g
Tryptophan	1.0g

pH 7.0 ± 0.2 at 25°C

Source: This medium is available as a premixed powder from Oxoid Unipath.

Preparation of Medium: Add components to distilled/deionized water and bring volume to 1.0L. Mix thoroughly. Gently heat and bring to boiling. Distribute into tubes or flasks. Autoclave for 15 min at 15 psi pressure–121°C. Pour into sterile Petri dishes or leave in tubes.

Use: For the detection and differentiation of coliform bacteria. For the presumptive identification of the main pathogens which cause infection of the urinary tract. Brilliance UTI Clarity Agar contains two chromogenic substrates which are cleaved by enzymes produced by *E. coli, Enterococcus* spp., and coliforms. Of the two chromogens included in the medium, one is metabolized by β-galactosidase, an enzyme produced by *E. coli*, which grow as pink colonies. The other is cleaved by β-glucosidase enzyme activity, allowing the specific detection of enterococci which form blue or turquoise colonies. Cleavage of both the chromogens gives dark blue or purple colonies, and indicates the organism is a coliform. The tryptophan in the medium is an indicator of tryptophan deaminase activity, resulting in colonies of *Proteus, Morganella,* and *Providencia* spp. with brown halos.

Brilliant Green Agar

Composition per liter:

Agar	20.0g
Lactose	10.0g
Sucrose	10.0g
Peptic digest of animal tissue	5.0g
Pancreatic digest of casein	5.0g
NaCl	5.0g
Phenol Red	0.08g
Brilliant Green	0.0125g

pH 6.9 ± 0.2 at 25°C

Source: This medium is available as a premixed powder from Oxoid Unipath and BD Diagnostic Systems.

Preparation of Medium: Add components to distilled/deionized water and bring volume to1.0L. Mix thoroughly. Gently heat and bring to boiling. Distribute into tubes or flasks. Autoclave for 15 min at 15 psi pressure–121°C. Pour into sterile Petri dishes.

Use: For the selective isolation of *Salmonella* other than *Salmonella typhi* from feces and other specimens, and food and dairy products. *Salmonella* other than *Salmonella typhi* appear as red/pink/white colonies surrounded by a zone of red in the agar, indicating nonlactose/sucrose fermentation. *Proteus* or *Pseudomonas* species may appear as small red colonies. Lactose- or sucrose-fermenting bacteria appear as yellow-green colonies surrounded by a zone of yellow-green in the agar.

Brilliant Green Agar Base with Phosphates and Sulfa Supplement

Composition per liter:

Agar	12.0g
Sucrose	10.0g
Plant peptone	10.0g
Lactose	10.0g
Plant extract	5.0g
Yeast extract	3.0g
Na_2HPO_4	1.0g
NaH_2PO_4	0.6g
Phenol Red	0.09g
Brilliant Green	4.7mg

pH 6.9 ± 0.2 at 25°C

Source: This medium, without sulfa supplement, is available as a premixed powder from HiMedia.

Sulfa Supplement Solution:

Composition per 10.0mL:

Sodium sulfacetamide	1.0g
Sodium mandelate	0.25g

Preparation of Sulfa Supplement Solution: Add components to distilled/deionized water and bring volume to 10.0mL. Mix thoroughly. Filter sterilize.

Preparation of Medium: Add components, except sulfa supplement solution, to distilled/deionized water and bring volume to 990.0mL. Mix thoroughly. Autoclave for 15 min at 15 psi pressure–121°C. Aseptically add 10.0mL of sterile sulfa supplement solution. Mix thoroughly. Pour into sterile Petri dishes.

Use: For the selective isolation of *Salmonella* other than *Salmonella typhi* from feces and other specimens while inhibiting *Escherichia coli, Proteus,* and *Pseudomonas* species.

Brilliant Green Agar, Modified

Composition per liter:

Agar	12.0g
Lactose	10.0g
Sucrose	10.0g
Beef extract	5.0g
Peptone	5.0g
NaCl	5.0g
Yeast extract	3.0g
Na_2HPO_4	1.0g
NaH_2PO_4	0.6g
Phenol Red	0.09g
Brilliant Green	4.7mg

pH 6.9 ± 0.2 at 25°C

Source: This medium is available as a premixed powder from Oxoid Unipath.

Preparation of Medium: Add components to distilled/deionized water and bring volume to1.0L. Mix thoroughly. Gently heat and bring to boiling. Do not autoclave. Cool to 45°–50°C. Addition of 1.0g of sodium sulfacetamide and 250.0mg of sodium mandelate enhances inhibition of contaminating microorganisms. Pour into sterile Petri dishes.

Use: For the selective isolation of *Salmonella* other than *Salmonella typhi* from feces and other specimens, and food and dairy products. *Salmonella* other than *Salmonella typhi* appear as red/pink/white colonies surrounded by a zone of red in the agar, indicating nonlactose/sucrose fermentation. *Proteus* or *Pseudomonas* species may appear as small red colonies. Lactose- or sucrose-fermenting bacteria appear as yellow-green colonies surrounded by a zone of yellow-green in the agar.

Brilliant Green Agar with Sulfadiazine

Composition per liter:

Agar	20.0g
Lactose	10.0g
Sucrose	10.0g
Pancreatic digest of casein	5.0g
Peptic digest of animal tissue	5.0g
NaCl	5.0g
Yeast extract	3.0g
Phenol Red	0.08g
Sulfadiazine	0.08g
Brilliant Green	0.0125g

pH 6.9 ± 0.2 at 25°C

Source: This medium is available as a premixed powder from BD Diagnostic Systems.

Preparation of Medium: Add components to distilled/deionized water and bring volume to1.0L. Mix thoroughly. Gently heat and bring to boiling. Distribute into tubes or flasks. Autoclave for 15 min at 15 psi pressure–121°C. Pour into sterile Petri dishes.

Use: For the selective detection of *Salmonella* in foods, especially from egg products. *Salmonella* other than *Salmonella typhi* appear as red/pink colonies surrounded by a zone of red in the agar indicating nonlactose/sucrose fermentation. *Proteus* or *Pseudomonas* species may appear as small red colonies. Lactose- or sucrose-fermenting bacteria appear as yellow-green colonies surrounded by a zone of yellow-green in the agar.

Brilliant Green Bile Agar

Composition per liter:

Noble agar	10.15g
Pancreatic digest of gelatin	8.25g
Lactose	1.9g
Na_2SO_3	0.205g
Basic Fuchsin	0.078g
Erioglaucine	0.065g
$FeCl_3$	0.0295g
KH_2PO_4	0.015g
Oxgall, dehydrated	2.95mg
Brilliant Green	0.03mg

pH 6.9 ± 0.2 at 25°C

Source: This medium is available as a premixed powder from BD Diagnostic Systems.

Caution: Basic Fuchsin is a potential carcinogen and care must be taken to avoid inhalation of the powdered dye and contamination of the skin.

Preparation of Medium: Add components to distilled/deionized water and bring volume to1.0L. For plating 10.0mL samples, prepare the medium double strength. Mix thoroughly. Gently heat and bring to boiling. Distribute into tubes or flasks. Autoclave for 15 min at 15 psi pressure–121°C. Pour into sterile Petri dishes. Care should be taken to avoid exposure of the prepared medium to light.

Use: For the detection and enumeration of coliform bacteria in materials of sanitary importance such as water, sewage, and foods. *Escherichia coli* appears as dark red colonies with a pink halo. *Enterobacter* species appear as pink colonies.

Brilliant Green Bile Broth (Brilliant Green Lactose Bile Broth)

Composition per liter:

Oxgall, dehydrated	20.0g
Lactose	10.0g
Pancreatic digest of gelatin	10.0g
Brilliant Green	0.013g

pH 7.2 ± 0.2 at 25°C

Source: This medium is available as a premixed powder from BD Diagnostic Systems and Oxoid Unipath.

Preparation of Medium: Add components to distilled/deionized water and bring volume to1.0L. Mix thoroughly. Distribute into tubes containing inverted Durham tubes, in 10.0mL amounts for testing 1.0mL or less of sample. Autoclave for 12 min (not longer than 15 min) at 15 psi pressure–121°C. After sterilization, cool the broth rapidly. Medium is sensitive to light.

Use: For the detection of coliform microorganisms in foods, dairy products, water, and wastewater, as well as in other materials of sanitary importance. Turbidity in the broth and gas in the Durham tube are positive indications of *Escherichia coli*.

Brilliant Green Bile Broth with MUG

Composition per liter:

Oxgall, dehydrated	20.0g
Lactose	10.0g
Pancreatic digest of gelatin	10.0g
MUG (4-Methyl umbelliferyl-β-D-glucuronide)	0.05g
Brilliant Green	0.013g

pH 7.2 ± 0.2 at 25°C

Source: This medium is available as a premixed powder from BD Diagnostic Systems.

Preparation of Medium: Add components to distilled/deionized water and bring volume to1.0L. Mix thoroughly. Distribute into tubes containing inverted Durham tubes, in 10.0mL amounts for testing 1.0mL or less of sample. Autoclave for 12 min (not longer than 15 min) at 15 psi pressure–121°C. After sterilization, cool the broth rapidly.

Use: For the detection of coliform microorganisms in foods, dairy products, water, and wastewater, as well as in other materials of sanitary importance. The presence of *Escherichia coli* and other coliforms is determined by the presence of fluorescence in the tube.

Brilliant Green 2%-Bile Broth, Fluorocult® (Fluorocult Brilliant Green 2%-Bile Broth) (BRILA)

Composition per liter:

Ox bile, dried	20.0g
Peptone	10.0g
Lactose	10.0g
L-Tryptophan	1.0g
4-Methylumbelliferyl-β-D-glucuronide	0.1g
Brilliant Green	0.0133g

pH 7.2 ± 0.2 at 25°C

Source: This medium is available from Merck.

Preparation of Medium: Add components to distilled/deionized water and bring volume to 1.0L. Mix thoroughly. Gently heat and bring to boiling. Cool. Distribute into test tubes containing inverted Durham tubes.

Autoclave for 15 min at 15 psi pressure–121°C. Do not autoclave longer. The prepared broth is clear and green.

Use: For the cultivation of *Escherichia coli.* Bile and Brilliant Green almost completely inhibit the growth of undesired microbial flora, in particular Gram-positive microorganisms. *E. coli* shows a positive fluorescence under UV light (366 nm). A positive indole reaction and, if necessary, gas formation due to fermenting lactose, confirm the findings.

Brilliant Green Broth (m-Brilliant Green Broth)

Composition per liter:

Proteose peptone No. 3	20.0g
Lactose	20.0g
Sucrose	20.0g
NaCl	10.0g
Yeast extract	6.0g
Phenol Red	0.16g
Brilliant Green	0.025g

pH 6.9 ± 0.2 at 25°C

Source: This medium is available as a premixed powder from BD Diagnostic Systems.

Preparation of Medium: Add components to distilled/deionized water and bring volume to 1.0L. Mix thoroughly. Gently heat with frequent mixing. Boil for 1 min. Cool to 25°C. Add 2.0mL to each sterile absorbent filter used.

Use: For the selective isolation and differentiation of *Salmonella* from polluted water by the membrane filter method.

Brilliant Green HiVeg Agar

Composition per liter:

Agar	10.15g
Plant peptone	8.25g
Lactose	1.9g
Basic Fuchsin	776.0mg
Erioglaucine	649.0mg
$FeCl_3$	295.0mg
Na_2SO_3	205.0mg
KH_2PO_4	15.3mg
Synthetic detergent No. II	2.95mg
Brilliant Green	29.5μg

pH 6.9 ± 0.2 at 25°C

Source: This medium is available as a premixed powder from HiMedia.

Preparation of Medium: Add components to distilled/deionized water and bring volume to1.0L. Mix thoroughly. Gently heat and bring to boiling. Distribute into tubes or flasks. Autoclave for 15 min at 15 psi pressure–121°C. Pour into sterile Petri dishes.

Use: For the selective isolation of *Salmonella* other than *Salmonella typhi* from feces and other specimens, and food and dairy products. *Salmonella* other than *Salmonella typhi* appear as red/pink/white colonies surrounded by a zone of red in the agar, indicating nonlactose/sucrose fermentation. *Proteus* or *Pseudomonas* species may appear as small red colonies. Lactose- or sucrose-fermenting bacteria appear as yellow-green colonies surrounded by a zone of yellow-green in the agar.

Brilliant Green HiVeg Agar Base with Sulfa Supplement

Composition per liter:

Agar	20.0g
Plant peptone No. 3	10.0g
Lactose	10.0g
Sucrose	10.0g
NaCl	5.0g
Yeast extract	3.0g
Phenol Red	0.08g
Brilliant Green	125.0mg
Sulfa supplement solution	10.0mL

pH 6.9 ± 0.2 at 25°C

Source: This medium, without sulfa supplement, is available as a premixed powder from HiMedia.

Sulfa Supplement Solution:

Composition per 10.0mL:

Sodium sulfacetamide	1.0g
Sodium mandelate	0.25g

Preparation of Sulfa Supplement Solution: Add components to distilled/deionized water and bring volume to 10.0mL. Mix thoroughly. Filter sterilize.

Preparation of Medium: Add components, except sulfa supplement solution, to distilled/deionized water and bring volume to 990.0mL. Mix thoroughly. Autoclave for 15 min at 15 psi pressure–121°C. Aseptically add 10.0mL of sterile sulfa supplement solution. Mix thoroughly. Aseptically distribute into sterile tubes or flasks.

Use: For the selective isolation of *Salmonella* other than *Salmonella typhi* from feces and other specimens.

Brilliant Green HiVeg Agar Base Modified with Sulfa Supplement

Composition per liter:

Agar	12.0g
Plant peptone No. 3	10.0g
Lactose	10.0g
Sucrose	10.0g
NaCl	5.0g
Yeast extract	3.0g
Phenol Red	0.08g
Brilliant Green	125.0mg
Sulfa supplement solution	10.0mL

pH 6.9 ± 0.2 at 25°C

Source: This medium, without sulfa supplement, is available as a premixed powder from HiMedia.

Sulfa Supplement Solution:

Composition per 10.0mL:

Sodium sulfacetamide	1.0g
Sodium mandelate	0.25g

Preparation of Sulfa Supplement Solution: Add components to distilled/deionized water and bring volume to 10.0mL. Mix thoroughly. Filter sterilize.

Preparation of Medium: Add components, except sulfa supplement solution, to distilled/deionized water and bring volume to 990.0mL. Mix thoroughly. Autoclave for 15 min at 15 psi pressure–121°C. Aseptically add 10.0mL of sterile sulfa supplement solution. Mix thoroughly. Pour into sterile Petri dishes.

Use: For the selective isolation of *Salmonella* other than *Salmonella typhi* from feces and other specimens.

Brilliant Green HiVeg Broth 2%

Composition per liter:

Plant peptone	25.0g
Lactose	10.0g
Synthetic detergent No. II	5.0g
Brilliant Green	13.3mg

pH 7.2 ± 0.2 at 25°C

Source: This medium is available as a premixed powder from HiMedia.

Preparation of Medium: Add components to distilled/deionized water and bring volume to 1.0L. Mix thoroughly. Gently heat with frequent mixing. Autoclave for 15 min at 15 psi pressure–121°C.

Use: For the selective isolation and differentiation of *Salmonella* from polluted water by the membrane filter method.

Brilliant Green Lactose Bile Broth (BAM M25)

Composition per liter:

Oxgall, dehydrated	20.0g
Lactose	10.0g
Pancreatic digest of gelatin	10.0g
Brilliant Green	0.0133g

pH 7.2 ± 0.1 at 25°C

Source: This medium is available as a premixed powder from BD Diagnostic Systems and Oxoid Unipath.

Preparation of Medium: Add lactose and pancreatic digest of gelatin to distilled/deionized water and bring volume to 500.0mL. Mix thoroughly. Add 20.0g oxgall dissolved in 200.0mL distilled/deionized water. The pH of this solution should be 7.0–7.5. Mix thoroughly. Bring volume to 975.0mL with distilled/deionized water. Adjust pH to 7.4. Add 13.3mL of 0.1% aqueous Brilliant Green in distilled/deionized water. Adjust volume to 1.0L with distilled/deionized water. Distribute into tubes containing inverted Durham tubes, in 10.0mL amounts for testing 1.0mL or less of sample. Make sure that the fluid level covers the inverted vials. Autoclave for 15 min at 15 psi pressure–121°C. After sterilization, cool the broth rapidly. Medium is sensitive to light.The final pH should be 7.2 ± 0.1 at 25°C.

Use: For the detection of coliform microorganisms in foods, dairy products, water, and wastewater as well as in other materials of sanitary importance. Turbidity in the broth and gas in the Durham tube are positive indications of *Escherichia coli*.

Brilliant Green Lactose Bile Broth

See: **Brilliant Green Bile Broth**

Brilliant Green Phenol Red Agar

Composition per liter:

Agar	15.0g
Lactose	15.0g
Peptone	10.0g
Meat extract	5.0g
NaCl	5.0g
Phenol Red	0.08g
Brilliant Green	0.0125g

pH 6.9 ± 0.2 at 25°C

Preparation of Medium: Add components to distilled/deionized water and bring volume to 1.0L. Mix thoroughly. Gently heat and bring to boiling. Distribute into tubes or flasks. Autoclave for 15 min at 15 psi pressure–121°C. Pour into sterile Petri dishes or leave in tubes.

Use: For the cultivation of *Salmonella* species.

Brilliant Green Sulfa Agar

See: **BG Sulfa Agar**

Brilliant Green Sulfapyridine Agar

See: **BG Sulfa Agar**

Brochothrix thermosphacta Medium

Composition per liter:

Peptone	20.0g
Glycerol	15.0g
Agar	13.0g
Yeast extract	2.0g
K_2HPO_4	1.0g
$MgSO_4 \cdot 7H_2O$	1.0g

pH 7.0 ± 0.2 at 25°C

Preparation of Medium: Add components to distilled/deionized water and bring volume to 1.0L. Mix thoroughly. Gently heat and bring to boiling. Distribute into tubes or flasks. Autoclave for 15 min at 15 psi pressure–121°C. Pour into sterile Petri dishes or leave in tubes.

Use: For the isolation and cultivation of *Brochothrix thermosphacta* from meats and meat products.

Brodie Medium

Composition per liter:

Agar	20.0g
Maltose	5.0g
Glucose	2.0g
Yeast extract	2.0g
Glycerol	1.0g
$MgSO_4$	0.5g
$Ca(NO_3)_2$	0.5g
KH_2PO_4	0.5g
Peptone	0.2g
DL-Asparagine	0.2g
$FeSO_4$	Trace

Preparation of Medium: Add components to distilled/deionized water and bring volume to 1.0L. Mix thoroughly. Gently heat and bring to boiling. Distribute into tubes or flasks. Autoclave for 15 min at 15 psi pressure–121°C. Pour into sterile Petri dishes or leave in tubes.

Use: For the cultivation and maintenance of *Cyathus* species and *Nidula niveo tormentosa*.

Brolacin Agar

See: **CLED Agar**

BROLACIN MUG Agar (Bromothymol Blue Lactose Cystine MUG Agar) (C.L.E.D. MUG Agar)

Composition per liter:

Agar	12.0g
Lactose	10.0g
Universal peptone	4.0g

Casein peptone 4.0g
Meat extract 3.0g
L-cystine 0.128g
4-Methylumbelliferyl-β-D-glucuronide 0.1g
Bromthymol Blue 0.02g

pH 7.3 ± 0.2 at 37°C

Source: This medium is available from Fluka, Sigma-Aldrich.

Preparation of Medium: Add components to distilled/deionized water and bring volume to 1.0L. Mix thoroughly. Gently heat while stirring and bring to boiling. Autoclave for 15 min at 15 psi pressure–121°C. Cool to 50°C. Pour into sterile Petri dishes.

Use: For the enumeration, isolation, and identification of microorganisms in urine. Growth of all urinary microorganisms is favored. Lactose catabolism produces a color change of Bromothymol Blue to yellow. Alkalization gives a color change to deep-blue. β-D-Glucoronidase, which is produced by *E. coli*, cleaves 4-methylumbelliferyl-β-D-glucuronide to 4-methylumbelliferone and glucuronide. The fluorogen 4-methylumbelliferone can be detected under a long wavelength UV lamp, permitting differentiation of *E. coli* colonies.

Bromcresol Purple Azide Broth

See: **BCP Azide Broth**

Bromcresol Purple Broth

Composition per liter:

Peptone 10.0g
NaCl 5.0g
Beef extract 3.0g
Bromcresol Purple 0.04g
Carbohydrate solution 10.0mL

pH 7.0 ± 0.2 at 25°C

Carbohydrate Solution:

Composition per 10.0mL:

Carbohydrate 5.0g

Preparation of Carbohydrate Solution: Add carbohydrate to distilled/deionized water and bring volume to 10.0mL. Mix thoroughly. Filter sterilize.

Preparation of Medium: Add components to distilled/deionized water and bring volume to 1.0L. Mix thoroughly. Gently heat and bring to boiling. Distribute into test tubes that contain an inverted Durham tube. Autoclave for 10 min at 15 psi pressure–121°C.

Use: For the differentiation of a variety of microorganisms based on their fermentation of specific carbohydrates. Bacteria that ferment the specific carbohydrate turn the medium yellow. When bacteria produce gas, the gas is trapped in the Durham tube.

Bromcresol Purple Broth (BAM M26)

Composition per liter:

Peptone 10.0g
NaCl 5.0g
Beef extract 3.0g
Bromcresol Purple 0.04g
Carbohydrate solution 50.0mL

pH 7.0 ± 0.2 at 25°C

Carbohydrate Solution:

Composition per 50.0mL:

Carbohydrate 25.0g

Preparation of Carbohydrate Solution: Add carbohydrate to distilled/deionized water and bring volume to 50.0mL. Mix thoroughly. Filter sterilize.

Preparation of Medium: Add components, except carbohydrate solution, to distilled/deionized water and bring volume to 950.0mL. Mix thoroughly. Gently heat and bring to boiling. Autoclave for 10 min at 15 psi pressure–121°C. Cool to 25°C. Aseptically add 50.0mL of carbohydrate solution. Aseptically distribute into tubes or flasks. Alternately distribute the medium without the carbohydrate solution into test tubes that contain an inverted Durham tube prior to autoclaving. Then autoclave for 10 min at 15 psi pressure–121°. Cool to 25°C, and aseptically add the carbohydrate solution to each tube to yield a final carbohydrate concentration of 5%.

Use: For the differentiation of a variety of microorganisms based on their fermentation of specific carbohydrates. Bacteria that ferment the specific carbohydrate turn the medium yellow. When bacteria produce gas, the gas is trapped in the Durham tube.

Bromcresol Purple Deoxycholate Agar

See: **BCP D Agar**

Bromcresol Purple Broth with Sodium Chloride (BAM M26)

Composition per liter:

NaCl 25.0g
Peptone 10.0g
Beef extract 3.0g
Bromcresol Purple 0.04g
Carbohydrate solution 50.0mL

pH 7.0 ± 0.2 at 25°C

Carbohydrate Solution:

Composition per 50.0mL:

Carbohydrate 25.0g

Preparation of Carbohydrate Solution: Add carbohydrate to distilled/deionized water and bring volume to 50.0mL. Mix thoroughly. Filter sterilize.

Preparation of Medium: Add components, except carbohydrate solution, to distilled/deionized water and bring volume to 950.0mL. Mix thoroughly. Gently heat and bring to boiling. Autoclave for 10 min at 15 psi pressure–121°C. Cool to 25°C. Aseptically add 50.0mL of carbohydrate solution. Aseptically distribute into tubes or flasks. Alternately distribute the medium without the carbohydrate solution into test tubes that contain an inverted Durham tube prior to autoclaving. Then autoclave for 10 min at 15 psi pressure–121°C. Cool to 25°C, and aseptically add the carbohydrate solution to each tube to yield a final carbohydrate concentration of 5%.

Use: For the differentiation of a halophilic *Vibrio* spp. Bacteria that ferment the specific carbohydrate turn the medium yellow. When bacteria produce gas, the gas is trapped in the Durham tube.

Bromcresol Purple Deoxycholate Citrate Lactose Sucrose Agar

See: **BCP DCLS Agar**

Bromcresol Purple Dextrose Broth (BCP Broth)

Composition per liter:

Glucose 10.0g
Peptone 5.0g

Beef extract	3.0g
Bromcresol Purple solution	2.0mL

pH 7.0 ± 0.2 at 25°C

Bromcresol Purple Solution:
Composition per 10.0mL:

Bromcresol Purple	0.16g
Ethanol (95% solution)	10.0mL

Preparation of Bromcresol Purple Solution: Add Bromcresol Purple to 10.0mL of ethanol. Mix thoroughly.

Preparation of Medium: Add components to distilled/deionized water and bring volume to 1.0L. Mix thoroughly. Distribute into tubes in 12–15mL volumes. Autoclave for 15 min at 15 psi pressure–121°C.

Use: For the cultivation and differentiation of bacteria based on their ability to ferment glucose. Bacteria that ferment glucose turn the medium yellow.

Bromcresol Purple Milk Solids Glucose Agar (BCP MS G Agar)

Composition per 2.0L:

Skim milk powder	80.0g
Glucose	40.0g
Agar	30.0g
Bromcresol Purple solution	2.0mL

pH 6.6 ± 0.2 at 25°C

Bromcresol Purple Solution:
Composition per 10.0mL:

Bromcresol Purple	0.16g
Ethanol (95% solution)	10.0mL

Preparation of Bromcresol Purple Solution: Add Bromcresol Purple to 10.0mL of ethanol. Mix thoroughly.

Preparation of Medium: Add skim milk powder and Bromcresol Purple solution to distilled/deionized water and bring volume to 1.0L. Mix thoroughly. Autoclave for 8 min at 11 psi pressure–116°C. Cool to 45°–50°C. In a separate flask, add glucose to distilled/deionized water and bring volume to 200.0mL. Mix thoroughly. Autoclave for 8 min at 11 psi pressure–116°C. Cool to 45°–50°C. In a third flask, add agar to distilled/deionized water and bring volume to 800.0mL. Mix thoroughly. Gently heat and bring to boiling. Autoclave for 15 min at 15 psi pressure–121°C. Cool to 45°–50°C. Aseptically combine the three sterile solutions. Mix thoroughly. Aseptically adjust the pH to 6.6 with sterile 1*N* HCl. Aseptically distribute into sterile tubes. Allow tubes to cool in a slanted position.

Use: For the cultivation and differentiation of *Trychophyton mentagrophytes, Trychophyton rubrum,* and *Microsporum persicolor.*

Bromcresol Purple Milk Yeast Extract with CCG

Composition per liter:

Milk solution	1.0L
Agar solution	900.0mL
Yeast extract solution	40.0mL
Chloramphenicol solution	10.0mL
Cycloheximide solution	10.0mL
Gentamicin solution	0.8mL

Milk Solution:
Composition per liter:

Skim milk powder	80.0g
Bromcresol Purple solution	2.0mL

Preparation of Milk Solution: Add components to distilled/deionized water and bring volume to 1.0L. Mix thoroughly. Autoclave for 8 min at 11 psi pressure–116°C. Cool to 45°–50°C.

Bromcresol Purple Solution:
Composition per 10.0mL:

Bromcresol Purple	0.16g
Ethanol (95% solution)	10.0mL

Preparation of Bromcresol Purple Solution: Add Bromcresol Purple to 10.0mL of ethanol. Mix thoroughly.

Agar Solution:
Composition per 900.0mL:

Agar	30.0g

Preparation of Agar Solution: Add agar to distilled/deionized water and bring volume to 900.0mL. Mix thoroughly. Gently heat and bring to boiling. Autoclave for 15 min at 15 psi pressure–121°C. Cool to 45°–50°C.

Yeast Extract Solution:
Composition per 100.0mL:

Yeast extract	40.0g

Preparation of Yeast Extract Solution: Add yeast extract to distilled/deionized water and bring volume to 100.0mL. Mix thoroughly. Filter sterilize.

Chloramphenicol Solution:
Composition per 10.0mL:

Chloramphenicol	0.1g

Preparation of Chloramphenicol Solution: Add chloramphenicol to distilled/deionized water and bring volume to 10.0mL. Mix thoroughly. Filter sterilize.

Cycloheximide Solution:
Composition per 10.0mL:

Cycloheximide	0.2g

Preparation of Cycloheximide Solution: Add cycloheximide to distilled/deionized water and bring volume to 10.0mL. Mix thoroughly. Filter sterilize.

Caution: Cycloheximide is toxic. Avoid skin contact or aerosol formation and inhalation.

Gentamicin Solution:
Composition per 10.0mL:

Gentamicin	0.5g

Preparation of Gentamicin Solution: Add gentamicin to distilled/deionized water and bring volume to 10.0mL. Mix thoroughly. Filter sterilize.

Preparation of Medium: Aseptically combine the sterile milk solution and sterile agar solution. Aseptically add 40.0mL of sterile yeast extract solution, 10.0mL of sterile chloramphenicol solution, 10.0mL of sterile cycloheximide solution, and 0.8mL of sterile gentamicin solution. Mix thoroughly. Aseptically distribute into sterile tubes. Allow tubes to cool in a slanted position.

Use: For the isolation, cultivation, and differentiation of *Trichophyton verrucosum* and *Trichophyton schoenleinii.*

Bromo Cresol Purple Azide HiVeg Broth

Composition per liter:

Plant hydrolysate	10.0g
Yeast extract	10.0g

NaCl	5.0g
D-Glucose	5.0g
K_2HPO_4	2.7g
KH_2PO_4	2.7g
NaN_3	0.5g
Bromo Cresol Purple	32.0mg

pH 7.0 ± 0.2 at 25°C

Source: This medium is available as a premixed powder from HiMedia.

Caution: Sodium azide is toxic. Azides also react with metals and disposal must be highly diluted.

Preparation of Medium: Add components to distilled/deionized water to 1.0L. Mix thoroughly. Gently heat to boiling. Distribute into tubes or flasks. Autoclave for 15 min at 10 psi pressure–115°C.

Use: For use in the confirmation test for the presence of fecal streptococci in water and wastewater.

Bromo Cresol Purple HiVeg Broth Base

Composition per liter:

Plant peptone	10.0g
Carbohydrate (test compound)	10.0g
NaCl	5.0g
Plant extract	3.0g
Bromo Cresol Purple	0.04g

pH 7.0 ± 0.2 at 25°C

Source: This medium without carbohydrate is available as a premixed powder from HiMedia.

Preparation of Medium: Add components to distilled/deionized water and bring volume to 1.0L. Mix thoroughly. Distribute into tubes in 12–15mL volumes. Autoclave for 10 min at 15 psi pressure–121°C. Carbohydrate solutions are added to test bacterial fermentative abilities.

Use: For the cultivation and differentiation of bacteria based on their ability to ferment various carbohydrates. Bacteria that ferment the carbohydrate turn the medium yellow.

Bromthymol Blue Agar

Composition per liter:

Agar	11.0g
Peptone	10.0g
NaCl	5.0g
Yeast extract	5.0g
Lactose (33% solution)	27.0mL
Bromthymol Blue (1% solution)	10.0mL
Sodium thiosulfate (50% solution)	2.0mL
Glucose (33% solution)	1.2mL
Maranil solution (5% solution)	1.0mL

pH 7.7–7.8 at 25°C

Preparation of Medium: Add agar, peptone, NaCl, and yeast extract to distilled/deionized water and bring volume to 1.0L. Mix thoroughly. Adjust pH to 8.0. Autoclave for 20 min at 15 psi pressure–121°C. Cool to 45°–50°C. Filter sterilize separately the lactose solution, Bromthymol Blue solution, sodium thiosulfate solution, glucose solution, and maranil solution. To the cooled, sterile agar solution aseptically add 27.0mL of sterile lactose solution, 10.0mL of sterile Bromthymol Blue solution, 2.0mL of sterile sodium thiosulfate solution, 1.2mL of sterile glucose solution, and 1.0mL of sterile maranil solution. Mix thoroughly. Adjust pH to 7.7–7.8. Pour into sterile Petri dishes or distribute into sterile tubes.

Use: For the selective isolation and cultivation of members of the Enterobacteriaceae.

Bromthymol Blue Broth

Composition per 101.45mL:

Pancreatic digest of casein	0.7g
NaCl	0.5g
Beef extract	0.3g
Yeast extract	0.3g
Beef heart, solids from infusion	0.2g
Horse serum	10.0mL
Bromthymol Blue solution	1.0mL
Ampicillin solution	1.0mL
Urea solution	0.25mL
Nystatin solution	0.1mL
Tripeptide solution	0.1mL

pH 6.0 ± 0.2 at 25°C

Bromthymol Blue Solution:

Composition per 50.0mL:

Bromthymol Blue	0.2g
NaOH (0.01*N* solution)	32.0mL

Preparation of Bromthymol Blue Solution: Add Bromthymol Blue to NaOH solution. Mix thoroughly. Bring volume to 50.0mL with distilled/deionized water. Autoclave for 15 min at 15 psi pressure–121°C. Store at 25°C.

Ampicillin Solution:

Composition per 10.0mL:

Ampicillin	1.0g

Preparation of Ampicillin Solution: Add ampicillin to distilled/deionized water and bring volume to 10.0mL. Mix thoroughly. Filter sterilize.

Urea Solution:

Composition per 100.0mL:

Urea	10.0g

Preparation of Urea Solution: Add urea to distilled/deionized water and bring volume to 100.0mL. Filter sterilize. Store at –20°C.

Nystatin Solution:

Composition per 1.0mL:

Nystatin	50,000U

Preparation of Nystatin Solution: Add nystatin to distilled/deionized water and bring volume to 1.0mL. Filter sterilize.

Tripeptide Solution:

Composition per 10.0mL:

Glycyl-L-histidyl-L-lysine acetate	0.2mg

Preparation of Tripeptide Solution: Add glycyl-L-histidyl-L-lysine acetate to distilled/deionized water and bring volume to 10.0mL. Mix thoroughly. Filter sterilize. Store at –20°C.

Preparation of Medium: Add components—except horse serum, ampicillin solution, urea solution, nystatin solution, and tripeptide solution—to distilled/deionized water and bring volume to 90.0mL. Mix thoroughly. Gently heat and bring to boiling. Autoclave for 15 min at 15 psi pressure–121°C. Cool to 45°–50°C. Aseptically add 10.0mL of sterile horse serum, 1.0mL of sterile ampicillin solution, 0.25mL of sterile urea solution, 0.1mL of sterile nystatin solution, and 0.1mL of sterile tripeptide solution. Mix thoroughly. Pour into sterile Petri dishes.

Use: For the cultivation of *Ureaplasma* species from clinical specimens.

Bromthymol Blue Lactose Agar
See: **BTB Lactose Agar**

Brooks Agar

Composition per liter:

Agar	20.0g
Cornmeal	5.0g
Yeast extract	2.0g
KH_2PO_4	1.5g
Malt extract	1.0g
Pancreatic digest of casein	1.0g
$MgSO_4 \cdot 7H_2O$	0.5g

Preparation of Medium: Add cornmeal to distilled/deionized water and bring volume to 500.0mL. Gently heat and bring to boiling. Boil for 30 min. Filter through cotton. Bring volume of filtrate to 1.0L with distilled/deionized water. Add remaining components. Mix thoroughly. Gently heat while stirring and bring to boiling. Autoclave for 15 min at 15 psi pressure–121°C. Pour into sterile Petri dishes or distribute into sterile tubes.

Use: For the cultivation of *Aspergillus japonicus, Aspergillus sojae,* and *Phytophthora citrophthora.*

Brucella Agar

Composition per liter:

Agar	15.0g
Pancreatic digest of casein	10.0g
Peptic digest of animal tissue	10.0g
NaCl	5.0g
Yeast extract	2.0g
Glucose	1.0g
$NaHSO_3$	0.1g
Horse blood, defibrinated	100.0mL

pH 7.0 ± 0.2 at 25°C

Source: This medium is available as a premixed powder from BD Diagnostic Systems and Oxoid Unipath.

Preparation of Medium: Add components to distilled/deionized water and bring volume to 900.0mL. Mix thoroughly. Heat gently with frequent mixing. Boil for 1 min. Autoclave for 15 min at 15 psi pressure–121°C. Cool to 45°–50°C. Add 100.0mL of sterile defibrinated horse blood. Mix gently and pour into sterile Petri dishes.

Use: For the cultivation and maintenance of *Brucella* species. For the isolation and cultivation of nonfastidious and fastidious microorganisms from a variety of clinical and nonclinical specimens.

Brucella Agar Base *Campylobacter* Medium

Composition per 1100.0mL:

Cycloheximide (actidione)	0.05g
Sodium cephazolin	0.015g
Novobiocin	5.0mg
Bacitracin	25,000U
Colistin sulfate	10,000U
Brucella agar base	1.0L
Horse blood, defibrinated	100.0mL

***Brucella* Agar Base**
Composition per liter:

Agar	15.0g
Pancreatic digest of casein	10.0g
Peptic digest of animal tissue	10.0g
NaCl	5.0g
Yeast extract	2.0g
Glucose	1.0g
$NaHSO_3$	0.1g

Preparation of *Brucella* Agar Base: Add components to distilled/deionized water and bring volume to 1.0L. Mix thoroughly.

Optional Supplement:
Composition per 10.0mL:

Sodium pyruvate	0.25g
$NaHSO_3$	0.25g
$FeSO_4 \cdot 7H_2O$	0.25g

Preparation of Optional Supplement: Add components to distilled/deionized water and bring volume to 10.0mL. Filter sterilize.

Caution: Cycloheximide is toxic. Avoid skin contact or aerosol formation and inhalation.

Preparation of Medium: Add components to 1.0L of prepared *Brucella* agar base. Mix thoroughly. Autoclave for 15 min at 15 psi pressure–121°C. Cool to 45°–50°C. Add 100.0mL of sterile, defibrinated horse blood. Addition of 10.0mL of optional supplement will improve growth. Mix thoroughly. Pour into sterile Petri dishes.

Use: For the selective isolation and cultivation of *Campylobacter jejuni* from fecal specimens or rectal swabs.

Brucella Agar

Composition per liter:

Agar	15.0g
Pancreatic digest of casein	10.0g
Peptic digest of animal tissue	10.0g
NaCl	5.0g
Yeast extract	2.0g
Glucose	1.0g
$NaHSO_3$	0.1g
Horse blood, defibrinated	100.0mL

pH 7.0 ± 0.2 at 25°C

Source: This medium is available as a premixed powder from BD Diagnostic Systems and Oxoid Unipath.

Preparation of Medium: Add components to distilled/deionized water and bring volume to 900.0mL. Mix thoroughly. Heat gently with frequent mixing. Boil for 1 min. Autoclave for 15 min at 15 psi pressure–121°C. Cool to 45°–50°C. Add 100.0mL of sterile defibrinated horse blood. Mix gently and pour into sterile Petri dishes.

Use: For the cultivation and maintenance of *Brucella* species. For the isolation and cultivation of nonfastidious and fastidious microorganisms from a variety of clinical and nonclinical specimens.

Brucella Agar with 1.0% Glucose

Composition per liter:

Agar	15.0g
Peptic digest of animal tissue	10.0g
Glucose	10.0g
NaCl	5.0g
Meat extract	5.0g
Sheep blood, defibrinated	100.0mL

Horse serum	50.0mL
Vitamin K_1 solution	1.0mL

pH 7.5 ± 0.2 at 25°C

Source: This medium is available from HiMedia.

Vitamin K_1 Solution:

Composition per 20.0mL:

Vitamin K_1	0.2g
Ethanol, absolute	20.0mL

Preparation of Vitamin K_1 Solution: Add vitamin K_1 to 20.0mL of ethanol. Mix thoroughly. Filter sterilize.

Preparation of Medium: Add components, except sheep blood, horse serum, and vitamin K_1 solution, to distilled/deionized water and bring volume to 849.0mL. Mix thoroughly. Gently heat and bring to boiling. Autoclave for 15 min at 15 psi pressure–121°C. Cool to 45°–50°C. Aseptically add 100.0mL of sterile sheep blood, 50.0mL of horse serum, and 1.0mL of sterile vitamin K_1 solution. Mix thoroughly. Pour into sterile Petri dishes or distribute into sterile tubes.

Use: For the cultivation of *Brucella* species.

Brucella Albimi Broth

See: ***Brucella*** **Broth**

Brucella Albimi Broth with 0.16% Agar

Composition per liter:

Pancreatic digest of casein	10.0g
Peptic digest of animal tissue	10.0g
NaCl	5.0g
Yeast extract	2.0g
Agar	1.6g
Glucose	1.0g
$NaHSO_3$	0.1g
Horse blood, defibrinated	100.0mL

pH 7.0 ± 0.2 at 25°C

Preparation of Medium: Add components, except horse blood, to distilled/deionized water and bring volume to 900.0mL. Mix thoroughly. Heat gently with frequent mixing. Boil for 1 min. Autoclave for 15 min at 15 psi pressure–121°C. Cool to 45°–50°C. Aseptically add 100.0mL of sterile defibrinated horse blood. Mix thoroughly. Aseptically distribute into sterile tubes or flasks.

Use: For the cultivation and maintenance of *Campylobacter* species.

Brucella Albimi Broth with 0.16% Agar and 1% Glycine (ATCC Medium 2161)

Composition per liter:

Pancreatic digest of casein	10.0g
Peptic digest of animal tissue	10.0g
Glycine	10.0g
NaCl	5.0g
Yeast extract	2.0g
Agar	1.6g
Glucose	1.0g
$NaHSO_3$	0.1g
Horse blood, defibrinated	100.0mL

pH 7.0 ± 0.2 at 25°C

Preparation of Medium: Add components, except horse blood, to distilled/deionized water and bring volume to 900.0mL. Mix thoroughly. Heat gently with frequent mixing. Boil for 1 min. Autoclave for 15 min at 15 psi pressure–121°C. Cool to 45°–50°C. Aseptically add 100.0mL of sterile defibrinated horse blood. Mix thoroughly. Aseptically distribute into sterile tubes or flasks.

Use: For the cultivation and maintenance of *Campylobacter* species.

Brucella Albimi Broth with Agar and 1.5% Sodium Chloride (ATCC Medium 2160)

Composition per liter:

NaCl	15.0g
Pancreatic digest of casein	10.0g
Peptic digest of animal tissue	10.0g
Yeast extract	2.0g
Agar	1.6g
Glucose	1.0g
$NaHSO_3$	0.1g
Horse blood, defibrinated	100.0mL

pH 7.0 ± 0.2 at 25°C

Preparation of Medium: Add components, except horse blood, to distilled/deionized water and bring volume to 900.0mL. Mix thoroughly. Heat gently with frequent mixing. Boil for 1 min. Autoclave for 15 min at 15 psi pressure–121°C. Cool to 45°–50°C. Aseptically add 100.0mL of sterile defibrinated horse blood. Mix thoroughly. Aseptically distribute into sterile tubes or flasks.

Use: For the cultivation and maintenance of *Campylobacter nitrofigilis*.

Brucella Albimi Broth with Formate and Fumarate

Composition per 1050.0mL:

Pancreatic digest of casein	10.0g
Peptic digest of animal tissue	10.0g
NaCl	5.0g
Yeast extract	2.0g
Glucose	1.0g
$NaHSO_3$	0.1g
Horse blood, defibrinated	100.0mL
Formate-fumarate solution	50.0mL

pH 7.0 ± 0.2 at 25°C

Formate-Fumarate Solution:

Composition per 100.0mL:

Sodium formate	6.0g
Fumaric acid	6.0g

Preparation of Formate-Fumarate Solution: Add components to distilled/deionized water and bring volume to 100.0mL. Mix thoroughly. Adjust pH to 7.0. Filter sterilize.

Preparation of Medium: Add components, except formate-fumarate solution and horse blood, to distilled/deionized water and bring volume to 900.0mL. Mix thoroughly. Heat gently with frequent mixing. Boil for 1 min. Autoclave for 15 min at 15 psi pressure–121°C. Cool to 45°–50°C. Add 100.0mL of sterile defibrinated horse blood. Mix gently and aseptically distribute into sterile tubes in 5.0mL volumes. Aseptically add 0.25mL of formate-fumarate solution to each tube containing 5.0mL of medium immediately prior to inoculation.

Use: For the cultivation and maintenance of *Campylobacter mucosalis*.

Brucella Albimi Broth with Sheep Blood

Composition per liter:

Pancreatic digest of casein	10.0g
Peptic digest of animal tissue	10.0g

NaCl	5.0g
Yeast extract	2.0g
Glucose	1.0g
$NaHSO_3$	0.1g
Sheep blood, defibrinated	100.0mL

pH 7.0 ± 0.2 at 25°C

Preparation of Medium: Add components, except sheep blood, to distilled/deionized water and bring volume to 900.0mL. Mix thoroughly. Heat gently with frequent mixing. Boil for 1 min. Autoclave for 15 min at 15 psi pressure–121°C. Cool to 45°–50°C. Aseptically add 100.0mL of sterile defibrinated sheep blood. Mix thoroughly. Aseptically distribute into sterile tubes or flasks.

Use: For the cultivation and maintenance of *Helicobacter nemestrinae* and *Helicobacter pylori.*

Brucella Albimi Medium, Semisolid

Composition per liter:

Pancreatic digest of casein	10.0g
Peptic digest of animal tissue	10.0g
Glycine	10.0g
NaCl	8.5g
Yeast extract	2.0g
Agar	1.6g
Glucose	1.0g
L-Cysteine·HCl·H_2O	0.2g
$NaHSO_3$	0.1g

pH 7.0 ± 0.2 at 25°C

Preparation of Medium: Add components to distilled/deionized water and bring volume to 1.0L. Mix thoroughly. Adjust pH to 7.0. Gently heat and bring to boiling. Distribute into tubes in 10.0mL volumes. Autoclave for 15 min at 15 psi pressure–121°C. Allow tubes to cool in an upright position.

Use: For the cultivation and identification of *Campylobacter* species.

Brucella Anaerobic Blood Agar

Composition per liter:

Vitamin K_1	0.01g
Anaerobic agar base	1000.0mL
Sheep blood, sterile, defibrinated	50.0mL

Anaerobic Agar Base

Composition per liter:

Pancreatic digest of casein	17.5g
Agar	15.0g
Glucose	10.0g
Papaic digest of soybean meal	2.5g
NaCl	2.5g
Sodium thioglycolate	2.0g
Sodium formaldehyde sulfoxylate	1.0g
L-Cystine·HCl·H_2O	0.4g
Methylene Blue	0.002g

pH 7.0 ± 0.2 at 25°C

Preparation of Anaerobic Agar Base: Add components to distilled/deionized water and bring volume to 1.0L. Mix thoroughly. Autoclave for 15 min at 15 psi pressure–121°C. Cool to 45°–50°C.

Preparation of Medium: To 950.0mL of cooled, sterile anaerobic agar base, aseptically add 10.0mg of vitamin K_1 and 50.0mL of sterile, defibrinated sheep blood.

Use: For the isolation of anaerobes.

Brucella Blood Agar with Hemin and Vitamin K_1

Composition per liter:

Agar	15.0g
Pancreatic digest of casein	10.0g
Peptic digest of animal tissue	10.0g
NaCl	5.0g
Yeast extract	2.0g
Glucose	1.0g
$NaHSO_3$	0.1g
Vitamin K_1	1.0mL
Hemin	1.0mL
Sheep blood, defibrinated	50.0mL

Source: This medium is available as a prepared medium from BD Diagnostic Systems.

Vitamin K_1 Solution:

Composition per 100.0mL:

Vitamin K_1	1.0g
Ethanol, absolute	99.0mL

Preparation of Vitamin K_1 Solution: Add vitamin K_1 to 99.0mL of absolute ethanol. Mix thoroughly. Filter sterilize.

Hemin Solution:

Composition per 100.0mL:

Hemin	1.0g
NaOH (1*N* solution)	20.0mL

Preparation of Hemin Solution: Add hemin to 20.0mL of 1*N* NaOH solution. Mix thoroughly. Bring volume to 100.0mL with distilled/deionized water.

Preparation of Medium: Add components, except vitamin K_1 solution and sheep blood, to distilled/deionized water and bring volume to 949.0mL. Mix thoroughly. Gently heat and bring to boiling. Autoclave for 15 min at 15 psi pressure–121°C. Cool to 45°–50°C. Aseptically add 1.0mL of sterile vitamin K_1 solution and 50.0mL of sterile defibrinated sheep blood. Mix gently and pour into sterile Petri dishes.

Use: For the isolation and cultivation of anaerobic microorganisms from clinical and nonclinical specimens. After growth on agar plates, colonies should be examined under a dissecting microscope under long-wave UV light. Members of the pigmented *Bacteroides* group appear as red/orange fluorescent colonies.

Brucella Blood Culture Broth

Composition per liter:

Sucrose	100.0g
Hemin	0.5g
Sodium polyanetholsulfonate (SPS)	0.25g
Brucella broth base	1000.0mL
Vitamin K_1 solution	1.0mL

pH 7.0 ± 0.2 at 25°C

***Brucella* Broth Base:**

Composition per liter:

Pancreatic digest of casein	10.0g
Peptic digest of animal tissue	10.0g
NaCl	5.0g
Yeast extract	2.0g
Glucose	1.0g
$NaHSO_3$	0.1g

Preparation of *Brucella* Broth Base: Add components to distilled/deionized water and bring volume to 1.0L. Mix thoroughly.

Vitamin K_1 Solution:

Composition per 100.0mL:

Vitamin K_1	1.09g
Ethanol, absolute	99.0mL

Preparation of Vitamin K_1 Solution: Add vitamin K_1 to 99.0mL of absolute ethanol. Store in the dark at 4°C.

Preparation of Medium: Add components, except vitamin K_1 solution, to prepared *Brucella* broth base. Autoclave for 15 min at 15 psi pressure–121°C. Cool to 45°–50°C. Aseptically add 1.0mL of vitamin K_1 solution. Distribute into sterile tubes or flasks.

Use: For the isolation and cultivation of microorganisms from blood. Especially useful for the cultivation of anaerobes.

Brucella Broth (*Brucella* Albimi Broth)

Composition per liter:

Pancreatic digest of casein	10.0g
Peptic digest of animal tissue	10.0g
NaCl	5.0g
Yeast extract	2.0g
Glucose	1.0g
$NaHSO_3$	0.1g
Horse blood, defibrinated	50.0mL

pH 7.0 ± 0.2 at 25°C

Source: This medium is available as a premixed powder from BD Diagnostic Systems.

Preparation of Medium: Add components, except horse blood, to distilled/deionized water and bring volume to 950.0mL. Mix thoroughly. Heat gently with frequent mixing. Boil for 1 min. Autoclave for 15 min at 15 psi pressure–121°C. Cool to 45°–50°C. Aseptically add 50.0mL of sterile horse blood. Mix thoroughly. Aseptically distribute into sterile tubes or flasks.

Use: For the cultivation and maintenance of *Campylobacter coli, Campylobacter fecalis*, and *Brucella* species. Also used for the isolation and cultivation of a wide variety of fastidious and nonfastidious microorganisms.

Brucella Broth with Additives (ATCC Medium 489)

Composition per liter:

Pancreatic digest of casein	10.0g
Peptic digest of animal tissue	10.0g
NaCl	3.5g
Yeast extract	2.0g
Glucose	1.0g
$NaHSO_3$	0.1g
Horse serum, inactivated	100.0mL
Fresh yeast extract solution	50.0mL

pH 7.0 ± 0.2 at 25°C

Fresh Yeast Extract Solution:

Composition per 100.0mL:

Baker's yeast, live, pressed, starch-free	25.0g

Preparation of Fresh Yeast Extract Solution: Add the live Baker's yeast to 100.0mL of distilled/deionized water. Autoclave for 90 min at 15 psi pressure–121°C. Allow to stand. Remove supernatant solution. Adjust pH to 6.6–6.8.

Preparation of Medium: Add components, except horse serum and fresh yeast extract solution, to distilled/deionized water and bring volume to 850.0mL. Mix thoroughly. Gently heat and bring to boiling. Autoclave for 15 min at 15 psi pressure–121°C. Cool to 45°–50°C. Aseptically add 100.0mL of sterile horse serum and 50.0mL of sterile fresh yeast extract solution. Mix thoroughly. Aseptically distribute into sterile tubes or flasks.

Use: For the cultivation of *Corynebacterium* species.

Brucella Broth with Additives (ATCC Medium 490)

Composition per liter:

NaCl	30.0g
Pancreatic digest of casein	10.0g
Peptic digest of animal tissue	10.0g
Yeast extract	2.0g
Glucose	1.0g
$NaHSO_3$	0.1g
Horse serum, inactivated	100.0mL
Fresh yeast extract solution	50.0mL

pH 7.0 ± 0.2 at 25°C

Fresh Yeast Extract Solution:

Composition per 100.0mL:

Baker's yeast, live, pressed, starch-free	25.0g

Preparation of Fresh Yeast Extract Solution: Add the live Baker's yeast to 100.0mL of distilled/deionized water. Autoclave for 90 min at 15 psi pressure–121°C. Allow to stand. Remove supernatant solution. Adjust pH to 6.6–6.8.

Preparation of Medium: Add components, except horse serum and fresh yeast extract solution, to distilled/deionized water and bring volume to 850.0mL. Mix thoroughly. Gently heat and bring to boiling. Autoclave for 15 min at 15 psi pressure–121°C. Cool to 45°–50°C. Aseptically add 100.0mL of horse serum and 50.0mL of sterile fresh yeast extract solution. Mix thoroughly. Aseptically distribute into sterile tubes or flasks.

Use: For the cultivation of salt-tolerant *Corynebacterium* species.

Brucella Broth with 0.16% Agar (ATCC Medium 1116)

Composition per liter:

Pancreatic digest of casein	10.0g
Peptic digest of animal tissue	10.0g
NaCl	5.0g
Yeast extract	2.0g
Agar	1.6g
Glucose	1.0g
$NaHSO_3$	0.1g
Horse blood, defibrinated	50.0mL

pH 7.0 ± 0.2 at 25°C

Source: This medium without agar is available as a premixed powder from BD Diagnostic Systems.

Preparation of Medium: Add components, except horse blood, to distilled/deionized water and bring volume to 950.0mL. Mix thoroughly. Heat gently with frequent mixing. Boil for 1 min. Autoclave for 15 min at 15 psi pressure–121°C. Cool to 45°–50°C. Aseptically add 50.0mL of sterile horse blood. Mix thoroughly. Aseptically distribute into sterile tubes or flasks.

Use: For the cultivation and maintenance of *Campylobacter fetus* subsp. *fetus* and *Campylobacter jejuni* subsp. *jejuni*.

Brucella Broth Base *Campylobacter* Medium

Composition per liter:

Cycloheximide (Actidione®)	50.0mg
Sodium cephazolin	15.0mg
Novobiocin	5.0mg
Bacitracin	25,000U
Colistin sulfate	10,000U
Brucella broth base	900.0mL
Horse blood, defibrinated	100.0mL

pH 7.0 ± 0.2 at 25°C

Caution: Cycloheximide is toxic. Avoid skin contact or aerosol formation and inhalation.

***Brucella* Broth Base:**

Composition per liter:

Pancreatic digest of casein	10.0g
Peptic digest of animal tissue	10.0g
NaCl	5.0g
Yeast extract	2.0g
Glucose	1.0g
$NaHSO_3$	0.1g

Preparation of *Brucella* Broth Base: Add components to distilled/deionized water and bring volume to 1.0L. Mix thoroughly.

Optional Supplement:

Composition per 10.0mL:

Sodium pyruvate	0.25g
$NaHSO_3$	0.25g
$FeSO_4 \cdot 7H_2O$	0.25g

Preparation of Optional Supplement: Add components to distilled/deionized water and bring volume to 10.0mL. Filter sterilize.

Preparation of Medium: Add components, except horse blood, to 900.0mL of prepared *Brucella* broth base. Mix thoroughly. Autoclave for 15 min at 15 psi pressure–121°C. Cool to 45°–50°C. Aseptically add 100.0mL of sterile, defibrinated horse blood. Addition of 10.0mL of optional supplement will improve growth. Mix thoroughly. Pour into sterile Petri dishes.

Use: For the selective isolation and cultivation of *Campylobacter jejuni* from fecal specimens or rectal swabs. Addition of the optional supplement improves growth.

Brucella Broth, Modified

Composition per liter:

Pancreatic digest of casein	10.0g
Peptic digest of animal tissue	10.0g
NaCl	5.0g
$MgSO_4 \cdot 7H_2O$	2.46g
Yeast extract	2.0g
$CaCl_2$	1.1g
Glucose	1.0g
$NaHSO_3$	0.1g
Horse blood, defibrinated	100.0mL

pH 7.0 ± 0.2 at 25°C

Preparation of Medium: Add components, except horse blood, to distilled/deionized water and bring volume to 900.0mL. Mix thoroughly. Gently heat and bring to boiling. Autoclave for 15 min at 15 psi pressure–121°C. Cool to 45°–50°C. Aseptically add sterile horse blood. Mix thoroughly. Aseptically distribute into sterile tubes or flasks.

Use: For the cultivation and maintenance of *Campylobacter coli* and *Campylobacter fecalis*.

Brucella Broth with Formate and Fumarate

See: ***Brucella*** **Albimi Broth with Formate and Fumarate**

Brucella Broth with Sheep Blood

See: ***Brucella*** **Albimi Broth with Sheep Blood**

Brucella FBP Agar

Composition per liter:

Agar	15.0g
Pancreatic digest of casein	10.0g
Peptic digest of animal tissue	10.0g
NaCl	5.0g
Yeast extract	2.0g
Glucose	1.0g
$NaHSO_3$	0.1g
FBP solution	30.0mL

pH 7.0 ± 0.2 at 25°C

FBP Solution:

Composition per 30.0mL:

$FeSO_4$	0.25g
$NaHSO_3$	0.25g
Sodium pyruvate	0.25g

Preparation of FBP Solution: Add components to distilled/deionized water and bring volume to 30.0mL. Mix thoroughly. Filter sterilize.

Preparation of Medium: Add components, except FBP solution, to distilled/deionized water and bring volume to 970.0mL. Mix thoroughly. Gently heat and bring to boiling. Autoclave for 15 min at 15 psi pressure–121°C. Cool to 45°–50°C. Aseptically add 30.0mL of sterile FBP solution. Mix thoroughly. Pour into sterile Petri dishes or distribute into sterile tubes.

Use: For the cultivation of *Brucella* species.

Brucella FBP Broth

Composition per liter:

Pancreatic digest of casein	10.0g
Peptic digest of animal tissue	10.0g
NaCl	5.0g
Yeast extract	2.0g
Glucose	1.0g
$NaHSO_3$	0.1g
FBP solution	30.0mL

pH 7.0 ± 0.2 at 25°C

FBP Solution:

Composition per 30.0mL:

$FeSO_4$	0.25g
Sodium metabisulfite, anhydrous	0.25g
Sodium pyruvate, anhydrous	0.25g

Preparation of FBP Solution: Add components to distilled/deionized water and bring volume to 30.0mL. Mix thoroughly. Filter sterilize.

Preparation of Medium: Add components, except FBP solution, to distilled/deionized water and bring volume to 970.0mL. Mix thoroughly. Gently heat and bring to boiling. Autoclave for 15 min at 15 psi pressure–121°C. Cool to 45°–50°C. Aseptically add 30.0mL of sterile FBP solution. Mix thoroughly. Aseptically distribute into sterile tubes.

Use: For the cultivation of *Brucella* species.

Brucella HiVeg Agar Base with Blood and Selective Supplement

Composition per liter:

Agar	15.0g
Plant hydrolysate	10.0g
Plant peptone	10.0g
NaCl	5.0g
Yeast extract	2.0g
Glucose	1.0g
$NaHSO_3$	0.1g
Horse blood, defibrinated	100.0mL
Selective supplement	10.0mL

pH 7.0 ± 0.2 at 25°C

Source: This medium, without horse blood or selective supplement, is available as a premixed powder from HiMedia.

Selective Supplement:
Composition per 10.0mL:

Cycloheximide	0.1g
Vancomycin	20.0mg
Nalidixic acid	5.0mg
Nystatin	1,000,000 U
Bacitracin	250,000 U
Polymyxin B sulfate	50,000 U

Preparation of Selective Supplement: Add components to distilled/deionized water and bring volume to 10.0mL. Mix thoroughly. Filter sterilize.

Caution: Cycloheximide is toxic. Avoid skin contact or aerosol formation and inhalation.

Preparation of Medium: Add components, except blood and selective supplement, to distilled/deionized water and bring volume to 900.0mL. Mix thoroughly. Heat gently with frequent mixing. Boil for 1 min. Autoclave for 15 min at 15 psi pressure–121°C. Cool to 45°–50°C. Add 100.0mL of sterile defibrinated horse blood and 10.0mL sterile selective supplement. Mix thoroughly. Pour into sterile Petri dishes.

Use: For the cultivation and maintenance of *Brucella* species. For the isolation and cultivation of nonfastidious and fastidious microorganisms from a variety of clinical and nonclinical specimens.

Brucella HiVeg Agar Base, Modified with Blood and Selective Supplement

Composition per liter:

Agar	15.0g
Plant hydrolysate	15.0g
Plant peptone	5.0g
NaCl	5.0g
Yeast extract	2.0g
Glucose	1.0g
Sodium citrate	1.0g
$NaHSO_3$	0.1g
Horse blood, defibrinated	100.0mL
Selective supplement	10.0mL

pH 7.0 ± 0.2 at 25°C

Source: This medium, without horse blood or selective supplement, is available as a premixed powder from HiMedia.

Selective Supplement:
Composition per 10.0mL:

Cycloheximide	0.1g
Vancomycin	20.0mg
Nalidixic acid	5.0mg
Nystatin	1,000,000 U
Bacitracin	250,000 U
Polymyxin B sulfate	50,000 U

Preparation of Selective Supplement: Add components to distilled/deionized water and bring volume to 10.0mL. Mix thoroughly. Filter sterilize.

Caution: Cycloheximide is toxic. Avoid skin contact or aerosol formation and inhalation.

Preparation of Medium: Add components, except blood and selective supplement, to distilled/deionized water and bring volume to 900.0mL. Mix thoroughly. Heat gently with frequent mixing. Boil for 1 min. Autoclave for 15 min at 15 psi pressure–121°C. Cool to 45°–50°C. Add 100.0mL of sterile defibrinated horse blood and 10.0mL sterile selective supplement. Pour into sterile Petri dishes.

Use: For the cultivation and maintenance of *Brucella* species. For the isolation and cultivation of nonfastidious and fastidious microorganisms from a variety of clinical and nonclinical specimens.

Brucella HiVeg Broth Base with Blood and Selective Supplement

Composition per liter:

Plant hydrolysate	10.0g
Plant peptone	10.0g
NaCl	5.0g
Yeast extract	2.0g
Glucose	1.0g
$NaHSO_3$	0.1g
Horse blood, defibrinated	100.0mL
Selective supplement	10.0mL

pH 7.0 ± 0.2 at 25°C

Source: This medium, without horse blood or selective supplement, is available as a premixed powder from HiMedia.

Selective Supplement:
Composition per 10.0mL:

Cycloheximide	0.1g
Vancomycin	20.0mg
Nalidixic acid	5.0mg
Nystatin	1,000,000 U
Bacitracin	250,000 U
Polymyxin B sulfate	50,000 U

Preparation of Selective Supplement: Add components to distilled/deionized water and bring volume to 10.0mL. Mix thoroughly. Filter sterilize.

Caution: Cycloheximide is toxic. Avoid skin contact or aerosol formation and inhalation.

Preparation of Medium: Add components, except blood and selective supplement, to distilled/deionized water and bring volume to

900.0mL. Mix thoroughly. Heat gently with frequent mixing. Boil for 1 min. Autoclave for 15 min at 15 psi pressure–121°C. Cool to 45°–50°C. Add 100.0mL of sterile defibrinated horse blood and 10.0mL sterile selective supplement.

Use: For the cultivation and maintenance of *Brucella* species.

Brucella Medium Base

Composition per liter:

Agar	15.0g
Glucose	10.0g
Peptone	10.0g
Beef extract	5.0g
NaCl	5.0g

pH 7.5 ± 0.2 at 25°C

Preparation: Add components to distilled/deionized water and bring volume to 1.0L. Mix thoroughly. Heat gently and bring to boiling. Distribute into tubes or flasks. Autoclave for 15 min at 15 psi pressure–121°C. Pour into sterile Petri dishes or leave in tubes.

Use: For the isolation of *Campylobacter* species.

Brucella Selective Medium

Composition per liter:

Beef heart, infusion from	500.0g
Agar	15.0g
Tryptose	10.0g
NaCl	5.0g
Glucose	2.5g
Gelatin	1.0g
Sheep blood	100.0mL
Antibiotic solution	10.0mL

pH 7.4 ± 0.2 at 25°C

Antibiotic Solution:

Composition per 10.0mL:

Cycloheximide	1.0g
Bacitracin	250,000U
Circulin	250,000U
Polymyxin B	100,000U

Preparation of Antibiotic Solution: Add components to distilled/deionized water and bring volume to 10.0mL. Mix thoroughly. Filter sterilize.

Caution: Cycloheximide is toxic. Avoid skin contact or aerosol formation and inhalation.

Preparation of Medium: Add components, except sheep blood and antibiotic solution, to distilled/deionized water and bring volume to 890.0mL. Mix thoroughly. Gently heat and bring to boiling. Autoclave for 15 min at 15 psi pressure–121°C. Cool to 45°–50°C. Aseptically add 100.0mL of sterile sheep blood and 10.0mL of sterile antibiotic solution. Mix thoroughly. Pour into sterile Petri dishes or distribute into sterile tubes.

Use: For the selective isolation and cultivation of *Brucella* species.

Brucella Selective Medium with Blood and Serum

Composition per liter:

Brain heart, infusion from	500.0g
Agar	15.0g
Tryptose	10.0g
Glucose	2.5g
NaCl	5.0g
Gelatin	1.0g
Sheep blood, defibrinated	100.0mL
Horse serum	50.0mL
Vitamin K_1 solution	1.0mL

pH 7.4 ± 0.2 at 25°C

Source: This medium is available from HiMedia.

Vitamin K_1 Solution:

Composition per 20.0mL:

Vitamin K_1	0.2g
Ethanol, absolute	20.0mL

Preparation of Vitamin K_1 Solution: Add vitamin K_1 to 20.0mL of ethanol. Mix thoroughly. Filter sterilize.

Preparation of Medium: Add components, except sheep blood, horse serum, and vitamin K_1 solution, to distilled/deionized water and bring volume to 849.0mL. Mix thoroughly. Gently heat and bring to boiling. Autoclave for 15 min at 15 psi pressure–121°C. Cool to 45°–50°C. Aseptically add 100.0mL of sterile sheep blood, 50.0mL of horse serum, and 1.0mL of sterile vitamin K_1 solution. Mix thoroughly. Pour into sterile Petri dishes or distribute into sterile tubes.

Use: For the cultivation of *Brucella* species.

Brucella Semisolid Medium with Cysteine

Composition per liter:

Peptamin	10.0g
Pancreatic digest of casein	10.0g
Glycine	10.0g
NaCl	5.0g
Yeast extract	2.0g
Agar	1.8g
Glucose	1.0g
L-Cysteine·HCl·H_2O	0.2g
$NaHSO_3$	0.1g
Sodium citrate	0.1g
Neutral Red solution	10.0mL

pH 7.0 ± 0.2 at 25°C

Neutral Red Solution:

Composition per 100.0mL:

Neutral Red	0.2g
Ethanol	10.0mL

Preparation of Neutral Red Solution: Add Neutral Red to 10.0mL of ethanol. Bring volume to 100.0mL.

Preparation of Medium: Add components to distilled/deionized water and bring volume to 1.0L. Mix thoroughly. Gently heat and bring to boiling. Distribute into tubes in 10.0mL volumes. Autoclave for 15 min at 15 psi pressure–121°C.

Use: For the cultivation and differentiation of *Campylobacter* species based on H_2S production from cysteine.

Brucella Semisolid Medium with Glycine

Composition per liter:

Peptamine	10.0g
Pancreatic digest of casein	10.0g
Glycine	10.0g
NaCl	5.0g
Yeast extract	2.0g
Agar	1.8g
Glucose	1.0g
$NaHSO_3$	0.1g

Sodium citrate 0.1g
Neutral Red solution 10.0mL

pH 7.0 ± 0.2 at 25°C

Neutral Red Solution:

Composition per 100.0mL:

Neutral Red 0.2g
Ethanol 10.0mL

Preparation of Neutral Red Solution: Add Neutral Red to 10.0mL of ethanol. Bring volume to 100.0mL.

Preparation of Medium: Add components to distilled/deionized water and bring volume to 1.0L. Mix thoroughly. Gently heat and bring to boiling. Distribute into tubes in 10.0mL volumes. Autoclave for 15 min at 15 psi pressure–121°C.

Use: For the cultivation and differentiation of *Campylobacter* species based on glycine utilization.

Brucella Semisolid Medium with Nitrate

Composition per liter:

Peptamin 10.0g
Pancreatic digest of casein 10.0g
Glycine 10.0g
KNO_3 .... 10.0g
NaCl 5.0g
Yeast extract 2.0g
Agar 1.8g
Glucose 1.0g
$NaHSO_3$ 0.1g
Sodium citrate 0.1g

pH 7.0 ± 0.2 at 25°C

Preparation of Medium: Add components to distilled/deionized water and bring volume to 1.0L. Mix thoroughly. Gently heat and bring to boiling. Distribute into tubes in 10.0mL volumes. Autoclave for 15 min at 15 psi pressure–121°C.

Use: For the cultivation and differentiation of *Campylobacter* species based on nitrate reduction.

Brucella Semisolid Medium with Sodium Chloride

Composition per liter:

NaCl 35.0g
Peptamin 10.0g
Pancreatic digest of casein 10.0g
Yeast extract 2.0g
Agar 1.8g
Glucose 1.0g
$NaHSO_3$ 0.1g
Sodium citrate 0.1g
Neutral Red solution 10.0mL

pH 7.0 ± 0.2 at 25°C

Neutral Red Solution:

Composition per 100.0mL:

Neutral Red 0.2g
Ethanol 10.0mL

Preparation of Neutral Red Solution: Add Neutral Red to 10.0mL of ethanol. Bring volume to 100.0mL.

Preparation of Medium: Add components to distilled/deionized water and bring volume to 1.0L. Mix thoroughly. Gently heat and bring to boiling. Distribute into tubes in 10.0mL volumes. Autoclave for 15 min at 15 psi pressure–121°C.

Use: For the cultivation and differentiation of *Campylobacter* species based on glycine utilization.

Bryant and Burkey Agar

Composition per liter:

Casein enzymatic hydrolysate 15.0g
Beef extract 7.5g
Sodium lactate 5.0g
Yeast extract 5.0g
Agar 0.75g
Sodium acetate 5.0g
L-Cysteine hydrochloride 0.6g
Sodium thioglycolate 0.2g
Resazurin 2.5mg

pH 5.9 ± 0.2 at 25°C

Source: This medium is available from HiMedia.

Preparation of Medium: Add components to distilled/deionized water and bring volume to 1.0L. Mix thoroughly. Distribute into tubes. Autoclave for 15 min at 15 psi pressure–121°C.

Use: For the detection of lactate fermenting *Clostridium* spp. in milk and dairy products related to cheese alteration.

Bryant and Burkey Medium

Composition per liter:

Casein enzymatic hydrolysate 15.0g
Beef extract 7.5g
Sodium lactate 3.0g
Yeast extract 5.0g
Sodium acetate 5.0g
L-Cysteine hydrochloride 0.5g
Resazurin 2.5mg

pH 5.9 ± 0.2 at 25°C

Source: This medium is available from HiMedia.

Preparation of Medium: Add components to distilled/deionized water and bring volume to 1.0L. Mix thoroughly. Distribute into tubes. Autoclave for 15 min at 15 psi pressure–121°C.

Use: For the detection of lactate fermenting *Clostridium* spp. in milk and dairy products related to cheese alteration.

Bryant-Robinson Medium

Composition per 1010.0mL:

Glucose, cellobiose, or maltose 5.0g
L-Methionine 0.08g
Mineral solution 50.0mL
Na_2CO_3 solution 50.0mL
Hemin solution 10.0mL
L-Cysteine·HCl–Na_2S solution 10.0mL
Vitamin solution 5.0mL
VFA solution 4.5mL
Resazurin 1.0mL

pH 6.5 ± 0.2 at 25°C

Mineral Solution:

Composition per liter:

KH_2PO_4 .... 18.0g
NaCl 18.0g
$(NH_4)_2SO_4$ 8.0g
$CaCl_2·6H_2O$ 0.53g
$MgCl_2·6H_2O$ 0.4g

$CoCl_2 \cdot 6H_2O$ 0.2g
$MnCl_2 \cdot 4H_2O$ 0.2g
$FeSO_4 \cdot 7H_2O$ 0.08g

Preparation of Mineral Solution: Add components to distilled/deionized water and bring volume to 1.0L. Mix thoroughly.

Na_2CO_3 Solution:
Composition per 100.0mL:
Na_2CO_3 .... 8.0g

Preparation of Na_2CO_3 Solution: Add Na_2CO_3 to O_2-free distilled/deionized water. Mix thoroughly. Gas with 100% CO_2 for 15 min. Autoclave for 15 min at 15 psi pressure–121°C.

Hemin Solution:
Composition per 100.0mL:
Hemin 0.01g
NaOH (0.002% solution) 100.0mL

Preparation of Hemin Solution: Add hemin to 100.0mL of NaOH solution. Mix thoroughly.

L-Cysteine·HCl–Na_2S Solution:
Composition per 100.0mL:
L-Cysteine·HCl 2.5g
$Na_2S \cdot 9H_2O$ 2.5g

Preparation of L-Cysteine·HCl–Na_2S Solution: Add L-cysteine·HCl to distilled/deionized water and bring volume to 80.0mL. Mix thoroughly. Adjust pH to 11 with NaOH. Add $Na_2S \cdot 9H_2O$. Mix thoroughly. Bring volume to 100.0mL with distilled/deionized water. Gently heat and bring to boiling under 100% N_2. Cool to 25°C under 100% N_2. Autoclave for 15 min at 15 psi pressure–121°C.

Vitamin Solution:
Composition per 100.0mL:
Calcium pantothenate 0.02g
Nicotinamide 0.02g
Pyridoxine·HCl 0.02g
Riboflavin 0.02g
Thiamine·HCl 0.02g
p-Aminobenzoic acid 1.0mg
Biotin 0.25mg
Folic acid 0.25mg
Vitamin B_{12} 0.1mg

Preparation of Vitamin Solution: Add components to distilled/deionized water and bring volume to 100.0mL. Mix thoroughly. Filter sterilize.

VFA (Volatile Fatty Acid) Solution:
Composition per liter:
Acetic acid 36.0mL
DL-α-Methylbutyric acid 2.0mL
Isovaleric acid 2.0mL
n-Valeric acid 2.0mL
Isobutyric acid 1.8mL

Preparation of VFA Solution: Add components to distilled/deionized water and bring volume to 1.0L. Mix thoroughly.

Preparation of Medium: Add components, except L-cysteine·HCl–Na_2S solution, to distilled/deionized water and bring volume to 1.0L. Mix thoroughly. Gently heat and bring to boiling. Continue boiling until resazurin turns colorless, indicating reduction. Anaerobically distribute into tubes in 10.0mL volumes. Cap with butyl rubber stoppers. Place tubes in a press. Autoclave for 15 min at 15 psi pressure–121°C. Immediately prior to inoculation, aseptically and anaerobically add 0.1mL of L-cysteine·HCl–Na_2S solution per tube.

Use: For the cultivation of *Bacteroides* species from rumens.

BS Medium

Composition per 1135.0mL:
$NaHCO_3$ 2.2g
NH_4Cl 0.25g
KH_2PO_4 .... 0.07g
Resazurin 0.5mg
$(NH_4)_2(Fe(SO_4)_2 \cdot 6H_2O$ 0.2mg
Na_2SeO_4 .... 0.1mg
$Na_2WO_4 \cdot 2H_2O$ 0.1mg
Marine medium/synthetic seawater mix 125.0mL
Wolfe's mineral solution 10.0mL
Yeast extract solution 10.0mL
KNO_3 solution 10.0mL

pH 7.0 ± 0.2 at 25°C

Marine Medium/Synthetic Seawater Mix:
Composition per liter:
NaCl 47.15g
$MgCl_2 \cdot 6H_2O$ 18.1g
$MgSO_4 \cdot 7H_2O$ 7.0g
Na_2SO_4 .... 3.24g
$CaCl_2 \cdot 2H_2O$ 3.13g
KCl 1.2g
Na_2CO_3 .... 0.1g
NaBr 0.1g
KBr 80.0mg
$SrCl_2 \cdot 6H_2O$ 72.0mg
H_3BO_3 .... 52.0mg
Na_2HPO_4 .... 8.1mg
NaF 2.4mg
Sodium silicate 0.4mg
KI 50.0μg

Preparation of Marine Medium/Synthetic Seawater Mix: Add components to distilled/deionized water and bring volume to 1.0L. Mix thoroughly.

Wolfe's Mineral Solution:
Composition per liter:
$MgSO_4 \cdot 7H_2O$ 3.0g
Nitrilotriacetic acid 1.5g
NaCl 1.0g
$MnSO_4 \cdot 2H_2O$ 0.5g
$CoCl_2 \cdot 6H_2O$ 0.1g
$ZnSO_4 \cdot 7H_2O$ 0.1g
$CaCl_2 \cdot 2H_2O$ 0.1g
$FeSO_4 \cdot 7H_2O$ 0.1g
$NiCl_2 \cdot 6H_2O$ 0.025g
$KAl(SO_4)_2 \cdot 12H_2O$ 0.02g
$CuSO_4 \cdot 5H_2O$ 0.01g
H_3BO_3 .... 0.01g
$Na_2MoO_4 \cdot 2H_2O$ 0.01g
$Na_2SeO_3 \cdot 5H_2O$ 0.3mg

Preparation of Wolfe's Mineral Solution: Add nitrilotriacetic acid to 500.0mL of distilled/deionized water. Adjust pH to 6.5 with KOH. Add remaining components. Add distilled/deionized water to 1.0L. Adjust pH to 6.8.

Yeast Extract Solution:

Composition per 10.0mL:

Yeast extract	0.5g

Preparation of Yeast Extract Solution: Add yeast extract to distilled/deionized water and bring volume to 10.0mL. Mix thoroughly. Sparge with 80% N_2 + 20% CO_2. Autoclave for 15 min at 15 psi pressure–121°C.

KNO_3 Solution:

Composition per 10.0mL:

KNO_3	1.0g

Preparation of KNO_3 Solution: Add KNO_3 to distilled/deionized water and bring volume to 10.0mL. Mix thoroughly. Sparge with 80% N_2 + 20% CO_2. Autoclave for 15 min at 15 psi pressure–121°C.

Preparation of Medium: Add components, except yeast extract solution and KNO_3 solution, to distilled/deionized water and bring volume to 980.0mL. Mix thoroughly. Adjust pH to 7.0 with H_2SO_4. Distribute 20.0mL volumes into 100.0mL bottles. Sparge with 80% N_2 + 20% CO_2. Autoclave for 15 min at 15 psi pressure–121°C. Aseptically and anaerobically add 0.2mL of sterile yeast extract solution and 0.2mL of sterile KNO_3 solution to each bottle. After inoculation, pressurize bottles to 2 bar with 80% N_2 + 20% CO_2.

Use: For the cultivation of *Pyrobaculum aerophilum.*

BSA Tween™ 80 Agar
See: **Bovine Serum Albumin Tween™ 80 Agar**

BSA Tween™ 80 Broth
See: **Bovine Serum Albumin Tween™ 80 Broth**

BSA Tween™ 80 Soft Agar
See: **Bovine Serum Albumin Tween™ 80 Soft Agar**

BSK Medium (Barbour-Stoenner-Kelly Medium)

Composition per 1260.0mL:

Bovine albumin fraction V	50.0g
HEPES (*N*-[2-hydroxyethyl]piperazine-*N′*-2-ethanesulfonic acid) buffer	6.0g
Neopeptone	5.0g
Glucose	5.0g
$NaHCO_3$	2.2g
Sodium pyruvate	0.8g
Sodium citrate	0.7g
N-Acetylglucosamine	0.4g
Gelatin solution	200.0mL
CMRL 1066, without glutamine, without bicarbonate, 10X	100.0mL
Rabbit serum	72.0mL

pH 7.6–7.65 at 25°C

Gelatin Solution:

Composition per 200.0mL:

Gelatin	14.0g

Preparation of Gelatin Solution: Add gelatin to distilled/deionized water and bring volume to 200.0mL. Heat gently to boiling. Mix thoroughly. Filter sterilize.

CMRL 1066 Medium without Glutamine, without Bicarbonate, 10X:

Composition per liter:

NaCl	6.8g
D-Glucose	1.0g
KCl	0.4g
L-Cysteine·HCl·H_2O	0.26g
$CaCl_2$, anhydrous	0.2g
$MgSO_4 \cdot 7H_2O$	0.2g
$NaH_2PO_4 \cdot H_2O$	0.14g
Sodium acetate·$3H_2O$	0.083g
L-Glutamic acid	0.075g
L-Arginine·HCl	0.07g
L-Lysine·HCl	0.07g
L-Leucine	0.06g
Glycine	0.05g
Ascorbic acid	0.05g
L-Proline	0.04g
L-Tyrosine	0.04g
L-Aspartic acid	0.03g
L-Threonine	0.03g
L-Alanine	0.025g
L-Phenylalanine	0.025g
L-Serine	0.025g
L-Valine	0.025g
L-Cystine	0.02g
L-Histidine·HCl·H_2O	0.02g
L-Isoleucine	0.02g
Phenol red	0.02g
L-Methionine	0.015g
Deoxyadenosine	0.01g
Deoxycytidine	0.01g
Deoxyguanosine	0.01g
Glutathione, reduced	0.01g
Thymidine	0.01g
Hydroxy-L-proline	0.01g
L-Tryptophan	0.01g
Nicotinamide adenine dinucleotide	7.0mg
Tween™ 80	5.0mg
Sodium glucoronate·H_2O	4.2mg
Coenzyme A	2.5mg
Cocarboxylase	1.0mg
Flavin adenine dinucleotide	1.0mg
Nicotinamide adenine dinucleotide phosphate	1.0mg
Uridine triphosphate	1.0mg
Choline chloride	0.5mg
Cholesterol	0.2mg
5-Methyldeoxycytidine	0.1mg
Inositol	0.05mg
p-Aminobenzoic acid	0.05mg
Niacin	0.025mg
Niacinamide	0.025mg
Pyridoxine	0.025mg
Pyridoxal·HCl	0.025mg
Biotin	0.01mg
D-Calcium pantothenate	0.01mg
Folic acid	0.01mg

Riboflavin 0.01mg
Thiamine·HCl 0.01mg

pH 7.2 ± 0.2 at 25°C

Preparation of CMRL 1066 Medium without Glutamine, without Bicarbonate, 10X: Add components to distilled/deionized water and bring volume to 1.0L. Mix thoroughly. Adjust pH to 7.2. Filter sterilize.

Preparation of Medium: Add components, except gelatin solution and rabbit serum, to 628.0mL of glass-distilled water. Mix thoroughly. Adjust pH to 7.6–7.65. Add 200.0mL of 7% aqueous gelatin solution. Filter sterilize entire medium. Aseptically add 72.0mL of sterile rabbit serum.

Use: For the cultivation of a wide variety of microorganisms in a chemically defined medium. For the cultivation of *Borrelia* and *Spirochaeta* species.

BSK Medium, Modified

Composition per 1264.0mL:

Bovine serum albumin, fraction V 50.0g
HEPES (*N*-[2-hydroxymethyl]piperazine-*N′* [ethane sulfonate]) buffer 6.0g
Neopeptone 5.0g
Glucose 5.0g
Yeastolate 2.54g
$NaHCO_3$ 2.2g
Sodium pyruvate 0.8g
Sodium citrate 0.7g
$MgSO_4 \cdot 7H_2O$ 0.6g
N-Acetylglucosamine 0.4g
$CaCl_2 \cdot 2H_2O$ 0.07g
CMRL 1066, 10X without glutamine or $NaHCO_3$ 100.0mL
Rabbit serum, heat inactivated 64.0mL

pH 7.5 ± 0.2 at 25°C

CMRL 1066, 10X without Glutamine or $NaHCO_3$: Composition per liter:

NaCl 6.8g
D-Glucose 1.0g
KCl 0.4g
L-Cysteine·HCl·H_2O 0.26g
$CaCl_2$, anhydrous 0.2g
$MgSO_4 \cdot 7H_2O$ 0.2g
$NaH_2PO_4 \cdot H_2O$ 0.14g
Sodium acetate·$3H_2O$ 0.083g
L-Glutamic acid 0.075g
L-Arginine·HCl 0.070g
L-Lysine·HCl 0.070g
L-Leucine 0.060g
Glycine 0.050g
Ascorbic acid 0.050g
L-Proline 0.040g
L-Tyrosine 0.040g
L-Aspartic acid 0.030g
L-Threonine 0.030g
L-Alanine 0.025g
L-Phenylalanine 0.025g
L-Serine 0.025g
L-Valine 0.025g
L-Cystine 0.020g
L-Histidine·HCl·H_2O 0.020g
L-Isoleucine 0.020g
Phenol Red 0.020g
L-Methionine 0.015g
Deoxyadenosine 0.010g
Deoxycytidine 0.010g
Deoxyguanosine 0.010g
Glutathione, reduced 0.010g
Thymidine 0.010g
Hydroxy-L-proline 0.010g
L-Tryptophan 0.010g
Nicotinamide adenine dinucleotide 7.0mg
Tween™ 80 .... 5.0mg
Sodium glucoronate·H_2O 4.2mg
Coenzyme A 2.5mg
Cocarboxylase 1.0mg
Flavin adenine dinucleotide 1.0mg
Nicotinamide adenine dinucleotide phosphate 1.0mg
Uridine triphosphate 1.0mg
Choline chloride 0.50mg
Cholesterol 0.20mg
5-Methyldeoxycytidine 0.10mg
Inositol 0.05mg
p-Aminobenzoic acid 0.05mg
Niacin 0.025mg
Niacinamide 0.025mg
Pyridoxine 0.025mg
Pyridoxal·HCl 0.025mg
Biotin 0.01mg
D-Calcium pantothenate 0.01mg
Folic acid 0.01mg
Riboflavin 0.01mg
Thiamine·HCl 0.01mg

Preparation of CMRL 1066, 10X Without Glutamine or $NaHCO_3$: Add components to distilled/deionized water and bring volume to 1.0L. Mix thoroughly. Adjust pH to 7.2. Filter sterilize.

Preparation of Medium: Add components, except CMRL 1066, 10X without glutamine or $NaHCO_3$ and rabbit serum, to distilled/deionized water and bring volume to 1100.0mL. Mix thoroughly. Adjust pH to 7.5 with NaOH. Filter sterilize. Aseptically add 100.0mL of sterile CMRL 1066, 10X without glutamine or $NaHCO_3$ and 64.0mL of sterile rabbit serum. Mix thoroughly. Aseptically distribute 10.0mL volumes into sterile 16 × 125.0mm test tubes.

Use: For the cultivation of *Borrelia afzelii, Borrelia burgdorferi,* and *Borrelia gorinii.*

BSK Medium, Revised

Composition per 1164.0mL:

Bovine serum albumin fraction V 50.0g
HEPES (*N*-[2-hydroxyethyl]piperazine-*N′*-2-ethanesulfonic acid) buffer 6.0g
Neopeptone 5.0g
Glucose 5.0g
TC-Yeastolate 2.54g
$NaHCO_3$ 2.2g
Sodium pyruvate 0.8g
Sodium citrate 0.7g
N-Acetylglucosamine 0.4g

CMRL 1066, without glutamine, without bicarbonate, 10X 100.0mL
Rabbit serum 64.0mL

pH 7.6–7.65 at 25°C

CMRL 1066 Medium without Glutamine, without Bicarbonate, 10X:

Composition per liter:

NaCl	6.8g
D-Glucose	1.0g
KCl	0.4g
L-Cysteine·HCl·H_2O	0.26g
$CaCl_2$, anhydrous	0.2g
$MgSO_4 \cdot 7H_2O$	0.2g
$NaH_2PO_4 \cdot H_2O$	0.14g
Sodium acetate·$3H_2O$	0.083g
L-Glutamic acid	0.075g
L-Arginine·HCl	0.070g
L-Lysine·HCl	0.070g
L-Leucine	0.060g
Glycine	0.050g
Ascorbic acid	0.050g
L-Proline	0.040g
L-Tyrosine	0.040g
L-Aspartic acid	0.030g
L-Threonine	0.030g
L-Alanine	0.025g
L-Phenylalanine	0.025g
L-Serine	0.025g
L-Valine	0.025g
L-Cystine	0.020g
L-Histidine·HCl·H_2O	0.020g
L-Isoleucine	0.020g
Phenol red	0.020g
L-Methionine	0.015g
Deoxyadenosine	0.010g
Deoxycytidine	0.010g
Deoxyguanosine	0.010g
Glutathione, reduced	0.010g
Thymidine	0.010g
Hydroxy-L-proline	0.010g
L-Tryptophan	0.010g
Nicotinamide adenine dinucleotide	7.0mg
Tween™ 80	5.0mg
Sodium glucoronate·H_2O	4.2mg
Coenzyme A	2.5mg
Cocarboxylase	1.0mg
Flavin adenine dinucleotide	1.0mg
Nicotinamide adenine dinucleotide phosphate	1.0mg
Uridine triphosphate	1.0mg
Choline chloride	0.50mg
Cholesterol	0.20mg
5-Methyldeoxycytidine	0.10mg
Inositol	0.05mg
p-Aminobenzoic acid	0.05mg
Niacin	0.025mg
Niacinamide	0.025mg
Pyridoxine	0.025mg
Pyridoxal·HCl	0.025mg
Biotin	0.01mg
Calcium DL-pantothenate	0.01mg
Folic acid	0.01mg
Riboflavin	0.01mg
Thiamine·HCl	0.01mg

pH 7.2 ± 0.2 at 25°C

Preparation of CMRL 1066 Medium without Glutamine, without Bicarbonate, 10X: Add components to distilled/deionized water and bring volume to 1.0L. Mix thoroughly. Adjust pH to 7.2. Filter sterilize.

Preparation of Medium: Add components, except rabbit serum and CMRL 1066, to 1.0L of glass-distilled/deionized water. Mix thoroughly. Adjust pH to 7.5 with NaOH. Filter sterilize. Aseptically add 100.0mL of sterile CMRL 1066 and 64.0mL of sterile rabbit serum. Adjust final pH to 7.5–7.6. Aseptically distribute into sterile tubes or flasks.

Use: For the cultivation of *Borrelia burgdorferi, Borrelia afzelii, Borrelia garinii, Borrelia anserina,* and *Borrelia japonica.*

BSL for *Corynebacterium* (Buffered Soy Lactose for *Corynebacterium)*

Composition per liter:

Agar	15.0g
Papaic digest of soybean meal	10.0g
Na_2HPO_4	6.0g
KH_2PO_4	3.0g
NH_4Cl	1.0g
$MgSO_4 \cdot 7H_2O$	0.2g
Lactose solution	100.0mL

pH 6.8–7.2 at 25°C

Lactose Solution:

Composition per 100.0mL:

Lactose	10.0g

Preparation of Lactose Solution: Add lactose to distilled/deionized water and bring volume to 100.0mL. Mix thoroughly. Autoclave for 15 min at 15 psi pressure–121°C. Cool to 45°–50°C.

Preparation of Medium: Add components, except lactose solution, to distilled/deionized water and bring volume to 900.0mL. Mix thoroughly. Heat gently with frequent mixing. Adjust pH to 6.8–7.2. Autoclave for 15 min at 15 psi pressure–121°C. Cool to 45°–50°C. Aseptically add sterile lactose solution. Mix thoroughly. Pour into sterile Petri dishes or distribute into sterile tubes.

Use: For the cultivation and maintenance of *Curtobacterium flaccumfaciens.*

BSR Medium

Composition per liter:

Beef heart, solids from infusion	500.0g
Sorbitol	70.0g
Sucrose	10.0g
Tryptose	10.0g
NaCl	5.0g
Fructose	1.0g
Glucose	1.0g
Phenol Red	0.02g
Horse serum	100.0mL

pH 7.6 ± 0.2 at 25°C

Preparation of Medium: Add components, except horse serum, to distilled/deionized water and bring volume to 900.0mL. Mix thoroughly. Autoclave for 15 min at 15 psi pressure–121°C. Cool to 45°–50°C.

Aseptically add 100.0mL of horse serum. Mix thoroughly. Aseptically distribute into sterile tubes or flasks.

Use: For the cultivation of *Spiroplasma citri*.

BSTSY Agar

Composition per liter:

Pancreatic digest of casein	17.0g
Agar	15.0g
NaCl	5.0g
Yeast extract	4.0g
Papaic digest of soybean meal	3.0g
K_2HPO_4	2.5g
Glucose	2.5g
Bovine serum	100.0mL

pH 7.3 ± 0.2 at 25°C

Preparation of Medium: Add components, except bovine serum, to distilled/deionized water and bring volume to 900.0mL. Mix thoroughly. Gently heat and bring to boiling. Autoclave for 15 min at 15 psi pressure–121°C. Cool to 45°–50°C. Aseptically add sterile bovine serum. Mix thoroughly. Pour into sterile Petri dishes or distribute into sterile tubes.

Use: For the isolation and cultivation of *Simonsiella* species and *Alysiella* species.

BT Medium (DSMZ Medium 816)

Composition per 1090mL:

NaCl	1.0g
KCl	0.5g
$MgCl_2·6H_2O$	0.4g
NH_4Cl	0.3g
KH_2PO_4	0.2g
$CaCl_2·2H_2O$	0.15g
Resazurin	0.5mg
$NaHCO_3$ solution	50.0mL
$Na_2S·9H_2O$ solution	13.0mL
Hydroxybenzoate solution	10.0mL
Yeast extract solution	5.0mL
Trypticase™ solution	5.0mL
Trace elements solution SL-10	1.0mL
Selenite-tungstate solution	1.0mL
Na_2CO_3 solution	variable

pH 7.6 ± 0.2 at 25°C

$Na_2S·9H_2O$ Solution:
Composition per 20.0mL:

$Na_2S·9H_2O$	0.6g

Preparation of $Na_2S·9H_2O$ Solution: Add $Na_2S·9H_2O$ to distilled/deionized water and bring volume to 20.0mL. Mix thoroughly. Autoclave under 100% N_2 for 15 min at 15 psi pressure–121°C. Cool to room temperature.

Selenite-Tungstate Solution:
Composition per liter:

NaOH	0.5g
$Na_2WO_4·2H_2O$	4.0mg
$Na_2SeO_3·5H_2O$	3.0mg

Preparation of Selenite-Tungstate Solution: Add components to distilled/deionized water and bring volume to 1.0L. Mix thoroughly. Sparge with 100% N_2. Filter sterilize.

Na_2CO_3 Solution:
Composition per 100.0mL:

Na_2CO_3	5.0g

Preparation of Na_2CO_3 Solution: Add Na_2CO_3 to distilled/deionized water and bring volume to 100.0mL. Mix thoroughly. Sparge with 80% N_2 + 20% CO_2. Filter sterilize.

$NaHCO_3$ Solution:
Composition per 100.0mL:

$NaHCO_3$	10.0g

Preparation of $NaHCO_3$ Solution: Add $NaHCO_3$ to distilled/deionized water and bring volume to 100.0mL. Mix thoroughly. Sparge with 80% N_2 + 20% CO_2. Filter sterilize.

Hydroxybenzoate Solution:
Composition per 10.0mL:

3-Hydroxybenzoic acid	2.8g

Preparation of Hydroxybenzoate Solution: Add 3-hydroxybenzoic acid to distilled/deionized water and bring volume to 10.0mL. Mix thoroughly. Neutralize with NaOH. Filter sterilize.

Trypticase™ Solution:
Composition per 10.0mL:

Trypticase™	1.0g

Preparation of Trypticase™ Solution: Add Trypticase™ to distilled/deionized water and bring volume to 10.0mL. Mix thoroughly. Autoclave under 100% N_2 for 15 min at 15 psi pressure–121°C. Cool to room temperature.

Yeast Extract Solution:
Composition per 10.0mL:

Yeast extract	1.0g

Preparation of Yeast Extract Solution: Add yeast extract to distilled/deionized water and bring volume to 10.0mL. Mix thoroughly. Autoclave under 100% N_2 for 15 min at 15 psi pressure–121°C. Cool to room temperature.

Trace Elements Solution SL-10:
Composition per liter:

$FeCl_2·4H_2O$	1.5g
$CoCl_2·6H_2O$	190.0mg
$MnCl_2·4H_2O$	100.0mg
$ZnCl_2$	70.0mg
$Na_2MoO_4·2H_2O$	36.0mg
$NiCl_2·6H_2O$	24.0mg
H_3BO_3	6.0mg
$CuCl_2·2H_2O$	2.0mg
HCl (25% solution)	10.0mL

Preparation of Trace Elements Solution SL-10: Add $FeCl_2·4H_2O$ to 10.0mL of HCl solution. Mix thoroughly. Add distilled/deionized water and bring volume to 1.0L. Add remaining components. Mix thoroughly. Sparge with 100% N_2. Autoclave for 15 min at 15 psi pressure–121°C.

Preparation of Medium: Prepare and dispense medium under 80% N_2 + 20% CO_2 gas atmosphere. Add components, except $NaHCO_3$ solution, $Na_2S·9H_2O$ solution, yeast extract solution, hydroxybenzoate solution, Trypticase™ solution, selenite-tungstate solution, and Na_2CO_3 solution, to distilled/deionized water and bring volume to 1.0mL. Mix thoroughly. Adjust pH to 7.2–7.6. Sparge with 80% N_2 + 20% CO_2. Autoclave for 15 min at 15 psi pressure–121°C. Aseptically and anaerobically add 50.0mL $NaHCO_3$ solution, 13.0mL $Na_2S·9H_2O$ solution, 10.0mL hydroxybenzoate solution, 5.0mL yeast extract solution,

5.0mL Trypticase™ solution, and 1.0mL selenite-tungstate solution. Mix thoroughly. Adjust pH to 7.6 Na_2CO_3 solution. Aseptically and anaerobically distribute into sterile tubes or bottles.

Use: For the cultivation of *Sporotomaculum hydroxybenzoicum.*

BTB Lactose Agar (Bromthymol Blue Lactose Agar)

Composition per liter:

Agar	15.0g
Lactose	10.0g
Proteose peptone	5.0g
Beef extract	3.0g
Bromthymol Blue	0.17g

pH 8.7–7.2 at 25°C

Preparation of Medium: Add components to distilled/deionized water and bring volume to 1.0L. Mix thoroughly. Heat gently with frequent mixing. Bring to boiling. Distribute into tubes or flasks. Autoclave for 15 min at 15 psi pressure–121°C. Pour into sterile Petri dishes if desired.

Use: For the isolation and cultivation of pathogenic staphylococci.

BTB Lactose HiVeg Agar (Bromthymol Blue Lactose HiVeg Agar)

Composition per liter:

Agar	15.0g
Lactose	10.0g
Plant peptone No. 3	5.0g
Plant extract	3.0g
Bromthymol Blue	0.17g

pH 8.6 ± 0.2 at 25°C

Source: This medium is available as a premixed powder from HiMedia.

Preparation of Medium: Add components to distilled/deionized water and bring volume to 1.0L. Mix thoroughly. Heat gently with frequent mixing. Bring to boiling. Distribute into tubes or flasks. Autoclave for 15 min at 15 psi pressure–121°C. Pour into sterile Petri dishes if desired.

Use: For the isolation and cultivation of pathogenic staphylococci.

B.T.B. Lactose Agar, Modified (Lactose Blue HiVeg Agar)

Composition per liter:

Lactose	15.5g
Agar	13.0g
NaCl	5.0g
Casein enzymatic hydrolysate	3.5g
Peptone	3.5g
Bromthymol Blue	0.04g

pH 7.0 ± 0.2 at 25°C

Source: This medium is available as a premixed powder from HiMedia.

Preparation of Medium: Add components to distilled/deionized water and bring volume to 1.0L. Mix thoroughly. Heat gently with frequent mixing. Bring to boiling. Distribute into tubes or flasks. Autoclave for 15 min at 15 psi pressure–121°C. Pour into sterile Petri dishes if desired.

Use: For the isolation and cultivation of pathogenic staphylococci.

B.T.B. Lactose HiVeg Agar, Modified (Lactose Blue HiVeg Agar)

Composition per liter:

Lactose	15.5g
Agar	13.0g
NaCl	5.0g
Plant hydrolysate	3.5g
Plant peptone	3.5g
Bromthymol Blue	0.04g

pH 7.0 ± 0.2 at 25°C

Source: This medium is available as a premixed powder from HiMedia.

Preparation of Medium: Add components to distilled/deionized water and bring volume to 1.0L. Mix thoroughly. Heat gently with frequent mixing. Bring to boiling. Distribute into tubes or flasks. Autoclave for 15 min at 15 psi pressure–121°C. Pour into sterile Petri dishes if desired.

Use: For the isolation and cultivation of pathogenic staphylococci.

BTB Teepol® Agar

Composition per liter:

NaCl	20.0g
Agar	15.0g
Peptone	10.0g
Sucrose	10.0g
Beef extract	5.0g
Bromthymol Blue	0.08g
Teepol	2.0mL

pH 7.8 ± 0.2 at 25°C

Preparation of Medium: Add components to distilled/deionized water and bring volume to 1.0L. Teepol may be substituted by 0.1mL of Tergitol™ 7. Mix thoroughly. Gently heat and bring to boiling. Adjust pH to 7.8. Autoclave for 15 min at 15 psi pressure–121°C. Pour into sterile Petri dishes.

Use: For the isolation and cultivation of *Vibrio anguillarum.*

BTU Medium

Composition per liter:

Ground meat, fat free	500.0g
Pancreatic digest of casein	30.0g
K_2HPO_4	5.0g
Yeast extract	5.0g
L-Cysteine·HCl	0.5g
Resazurin	1.0mg
NaOH (1*N* solution)	25.0mL
Formate-fumarate solution	4.26mL

pH 7.0 ± 0.2 at 25°C

Formate-Fumarate Solution:

Composition per 100.0mL:

Sodium formate	6.0g
Sodium fumarate	6.0g

Preparation of Formate-Fumarate Solution: Add components to distilled/deionized water and bring volume to 100.0mL. Mix thoroughly. Filter sterilize.

Preparation of Medium: Use lean beef or horse meat. Remove fat and connective tissue. Grind finely. Add ground meat and 25.0mL of NaOH solution to distilled/deionized water and bring volume to 1025.0mL. Gently heat and bring to boiling. Continue boiling for 15 min. without stirring. Cool to room temperature. Remove fat from surface. Filter and retain both meat particles and filtrate. Adjust volume of filtrate to 1.0L with distilled/deionized water. Add pancreatic digest of casein, K_2HPO_4, yeast extract, and resazurin. Gently heat and bring to boiling. Boil for 1–2 min. Add L-cysteine·HCl. Mix thoroughly. Distribute 7.0mL into tubes that contain meat particles (1 part meat particles to 5 parts fluid). Autoclave for 30 min at 15 psi pressure–121°C. Prior to inoculation, add 30.0µL of formate-fumarate solution for each milliliter of medium in the tubes.

Use: For the cultivation of *Bacteroides ureolyticus.*

Buffered Azide Glucose Glycerol Broth
See: **BAGG Broth**

Buffered Charcoal Yeast Extract Agar
See: **BCYE Agar**

Buffered Charcoal Yeast Extract Agar with Albumin
See: **BCYEα with Alb**

Buffered Charcoal Yeast Extract Agar without L-Cysteine
See: **BCYEα without L-Cysteine**

Buffered Charcoal Yeast Extract Differential Agar (DIFF/BCYE)

Composition per 1014.0mL:

Agar	17.0g
ACES (2-[(2-amino-2-oxoethyl)-amino]-ethane sulfonic acid) buffer	10.0g
Yeast extract	10.0g
Charcoal, activated	1.5g
$Fe_4(P_2O_7)_3 \cdot 9H_2O$	0.25g
Bromcresol Purple	0.01g
Bromthymol Blue	0.01g
Antibiotic solution	10.0mL
L-Cysteine·HCl·H_2O solution	4.0mL

pH 6.9 ± 0.2 at 25°C

Antibiotic Solution:
Composition per 10.0mL:

Vancomycin	1.0mg
Polymyxin B	50,000U

Preparation of Antibiotic Solution: Add components to distilled/deionized water and bring volume to 10.0mL. Mix thoroughly. Filter sterilize.

L-Cysteine·HCl·H_2O Solution:
Composition per 10.0mL:

L-Cysteine·HCl·H_2O	1.0g

Preparation of L-Cysteine·HCl·H_2O Solution: Add 1.0g of L-cysteine·HCl·H_2O to distilled/deionized water and bring volume to 10.0mL. Mix thoroughly. Filter sterilize.

Preparation of Medium: Add components, except L-cysteine·HCl·H_2O solution and antibiotic solution, to distilled/deionized water and bring volume to 1.0L. Mix thoroughly. Adjust medium to pH 6.9 with 1*N* KOH. Heat gently and bring to boil for 1 min. Autoclave for 15 min at 15 psi pressure–121°C. Cool to 50°–55°C. Add 4.0mL of sterile L-cysteine·HCl·H_2O solution and 10.0mL of sterile antibiotic solution. Mix thoroughly. Pour into sterile Petri dishes with constant agitation to keep charcoal in suspension.

Use: For the isolation, cultivation, and maintenance of *Legionella pneumophila* and other *Legionella* species from environmental and clinical specimens. For the selective recovery of *Legionella pneumophila* while reducing contaminating microorganisms from environmental water samples.

Buffered Charcoal Yeast Extract Medium, Diphasic Blood Culture
See: **BCYE Medium, Diphasic Blood Culture**

Buffered Charcoal Yeast Extract Selective Agar with Cephalothin, Colistin, Vancomycin, and Cycloheximide
See: **BCYE Selective Agar with CCVC**

Buffered Charcoal Yeast Extract Selective Agar with Glycine, Polymyxin B, Vancomycin, and Anisomycin
See: **BCYE Selective Agar with GPVA**

Buffered Charcoal Yeast Extract Selective Agar with Glycine, Vancomycin, Polymyxin B, and Cycloheximide
See: **BCYE Selective Agar with GVPC**

Buffered Charcoal Yeast Extract Selective Agar with Polymyxin B, Anisomycin, and Cefamandole
See: **BCYE Selective Agar with PAC**

Buffered Charcoal Yeast Extract Selective Agar with Polymyxin B, Anisomicin, and Vancomycin
See: **BCYE Selective Agar with PAV**

Buffered Clostridial Medium with Cellobiose

Composition per liter:

Meat extract	10.0g
Peptone	10.0g
Cellobiose	5.0g
Glucose	5.0g
NaCl	5.0g
Sodium acetate	3.0g
Yeast extract	3.0g
$NaHCO_3$	2.75g
Soluble starch	1.0g
L-Cysteine·HCl·H_2O	0.5g
Resazurin	1.0mg
Hemin solution	10.0mL
Vitamin K_1 solution	0.2mL

pH 7.0 ± 0.2 at 25°C

Hemin Solution:
Composition per 100.0mL:

Hemin	50.0mg
NaOH (1*N* solution)	1.0mL

Preparation of Hemin Solution: Dissolve hemin in 1.0mL of 1*N* NaOH solution. Bring volume to 100.0mL with distilled/deionized water. Mix thoroughly. Autoclave for 15 min at 15 psi pressure–121°C.

Vitamin K_1 Solution:
Composition per 30.15mL:

Ethanol (95% solution)	30.0mL
Vitamin K_1	0.15mL

Preparation of Vitamin K_1 Solution: Combine components. Mix thoroughly. Store at 4°C in the dark. Discard solution after 1 month.

Preparation of Medium: Prepare and dispense medium under 80% N_2 + 10% CO_2 + 10% H_2. Add components, except cellobiose, $NaHCO_3$, and L-cysteine·HCl·H_2O, to distilled/deionized water and bring volume to 1.0L. Mix thoroughly. Gently heat and bring to boiling. Continue boiling for 3 min. Cool to room temperature while sparging with 80% N_2 + 10% CO_2 + 10% H_2. Add cellobiose, $NaHCO_3$, and L-cysteine·HCl·H_2O, in that order. Mix thoroughly. Adjust pH to 7.0. Anaerobically distribute into tubes. Autoclave for 15 min at 15 psi pressure–121°C.

Use: For the cultivation of *Eubacterium xylanophilum* and *Clostridium termitidis.*

Buffered Enrichment Broth (BAM M52)

Composition per liter:

Na_2HPO_4	9.6g
KH_2PO_4	1.35g
Pyruvate solution	11.1mL
Nalidixic acid solution	8.0mL
Cycloheximide solution	5.0mL
Acriflavin solution	2.0mL

pH 7.3 ± 0.1 at 25°C

Nalidixic Acid Solution:
Composition per 10.0mL:

Nalidixic acid, sodium salt	0.05g

Preparation of Nalidixic Acid Solution: Add nalidixic acid to distilled/deionized water and bring volume to 10.0mL. Mix thoroughly. Filter sterilize.

Acriflavin Solution:
Composition per 10.0mL:

Acriflavin·HCl	0.05g

Preparation of Acriflavin Solution: Add acriflavin·HCl to distilled/deionized water and bring volume to 10.0mL. Mix thoroughly. Filter sterilize.

Cycloheximide Solution:
Composition per 10.0mL:

Cycloheximide	0.1g
Ethanol, 40%	10.0mL

Preparation of Cycloheximide Solution: Add cycloheximide to 40% ethanol and bring volume to 10.0mL. Mix thoroughly. Filter sterilize.

Caution: Cycloheximide is toxic. Avoid skin contact or aerosol formation and inhalation.

Pyruvate Solution:
Composition per 20.0mL:

Na-pyruvate	2.0g

Preparation of Pyruvate Solution: Add Na-pyruvate to distilled/deionized water and bring volume to 20.0mL. Mix thoroughly. Filter sterilize.

Preparation of Medium: Add components, except pyruvate solution, nalidixic acid solution, acriflavin solution, and cycloheximide solution, to distilled/deionized water and bring volume to 973.9.0L. Mix thoroughly. Gently heat and bring to boiling. Autoclave for 15 min at 15 psi pressure–121°C. Cool to 25°C. Aseptically add 11.1mL sterile pyruvate solution. Mix thoroughly. Aseptically add 8.0mL sterile nalidixic acid solution, 5.0mL sterile cycloheximide solution, and 2.0mL sterile acriflavin solution. Mix thoroughly. Aseptically distribute into sterile tubes or flasks.

Use: For the cultivation of of *Listeria* spp.

Buffered Glucose HiVeg Broth

Composition per liter:

Buffered plant peptone	7.0g
Glucose	5.0g
K_2HPO_4	5.0g

pH 7.2 ± 0.2 at 25°C

Source: This medium is available as a premixed powder from HiMedia.

Preparation of Medium: Add components to distilled/deionized water and bring volume to 1.0L. Mix thoroughly. Distribute into tubes or flasks. Autoclave for 15 min at 15 psi pressure–121°C.

Use: Used for the growth of bacteria for the Methyl Red and Voges Proskauer tests.

Buffered HiVeg Peptone Water

Composition per liter:

Plant peptone No. 3	10.0g
NaCl	5.0g
Na_2HPO_4	3.5g
KH_2PO_4	1.5g

pH 7.2 ± 0.2 at 25°C

Source: This medium is available as a premixed powder from HiMedia.

Preparation of Medium: Add components to distilled/deionized water and bring volume to 1.0L. Mix thoroughly. Distribute into tubes or flasks. Autoclave for 15 min at 15 psi pressure–121°C.

Use: Used as a preenrichment medium for the isolation of *Salmonella*, especially injured microorganisms, from various food sources.

Buffered HiVeg Peptone Water with Sodium Chloride

Composition per liter:

Na_2HPO_4	7.23g
NaCl	4.3g
KH_2PO_4	3.56g
Plant peptone	1.0g

pH 7.0 ± 0.2 at 25°C

Source: This medium is available as a premixed powder from HiMedia.

Preparation of Medium: Add components to distilled/deionized water and bring volume to 1.0L. Mix thoroughly. Distribute into tubes or flasks. Autoclave for 15 min at 15 psi pressure–121°C.

Use: Used as a preenrichment medium for the isolation of bacteria from various food sources.

Buffered *Listeria* Enrichment Broth Base with *Listeria* Selective Supplement

Composition per liter:

Casein enzymic hydrolysate	17.0g
Na_2HPO_4	9.6g

Yeast extract	6.0g
NaCl	5.0g
Papaic digest of soybean meal	3.0g
K_2HPO_4	2.5g
Glucose	2.5g
KH_2PO_4	1.35g
Sodium pyruvate	1.0g
Selective supplement solution	10.0mL

pH 7.2 ± 0.2 at 25°C

Source: This medium is available from HiMedia.

Selective Supplement Solution:

Composition per 10.0mL:

Cycloheximide	25.0g
Acriflavin hydrochloride	5.0mg
Nalidixic acid	5.0mg

Preparation of Selective Supplement Solution: Add components to distilled/deionized water and bring volume to 10.0mL. Mix thoroughly. Filter sterilize.

Caution: Cycloheximide is toxic. Avoid skin contact or aerosol formation and inhalation.

Preparation of Medium: Add components, except selective supplement solution, to distilled/deionized water and bring volume to 990.0mL. Mix thoroughly. Autoclave for 15 min at 15 psi pressure–121°C. Cool to 50°C. Aseptically add selective supplement solution. Mix thoroughly. Pour into Petri dishes or aseptically distribute into sterile tubes.

Use: For the enrichment and cultivation *Listeria monocytogenes.*

Buffered Marine Yeast Medium

Composition per liter:

NaCl	24.0g
Agar	20.0g
Yeast extract	5.0g
1*M* Phosphate buffer, pH 6.8	20.0mL
Hutner's mineral base	20.0mL
KOH (1*N*)	7.0mL

pH 6.8 ± 0.2 at 25°C

1*M* Phosphate Buffer, pH 6.8:

Composition per liter:

$K_2H_2PO_4$	85.4g
$NaH_2PO_4 \cdot H_2O$	70.4g

Preparation of 1*M* Phosphate Buffer, pH 6.8: Add components to distilled/deionized water and bring volume to 1.0L. Mix thoroughly. Adjust pH to 6.8.

Hutner's Mineral Base:

Composition per liter:

$MgSO_4.7H_2O$	29.7g
Nitrilotriacetic acid	10.0g
$CaCl_2 \cdot 2H_2O$	3.34g
$FeSO_4 \cdot 7H_2O$	0.01g
$(NH_4)_2MoO_4$	9.25mg
Metals "44"	50.0mL

Preparation of Hutner's Mineral Base: Initially add a few drops of H_2SO_4 to the distilled water to retard precipitation. Dissolve the nitrilotriacetic acid first and neutralize the solution with KOH. Add the other ingredients and adjust the pH to 7.2 with KOH and/or H_2SO_4. There may be a slight precipitate. Store at 5°C.

Metals "44":

Composition per 100.0mL:

$ZnSO_4 \cdot 7H_2O$	1.1g
$FeSO_4 \cdot 7H_2O$	0.5g
EDTA	0.25g
$MnSO_4 \cdot 7H_2O$	0.154g
$CuSO_4 \cdot 5H_2O$	0.04g
$Co(NO_3)_2 \cdot 6H_2O$	0.025g
$Na_2B_4O_7 \cdot 10H_2O$	0.018g

Preparation of Metals "44": Add components to distilled/deionized water and bring volume to 100.0mL. Mix thoroughly. Autoclave for 15 min at 15 psi pressure–121°C. Add aseptically to sterile basal medium.

Preparation of Medium: Add components to distilled/deionized water and bring volume to 1.0L. Mix thoroughly. Distribute into tubes or flasks. Autoclave for 15 min at 15 psi pressure–121°C.

Use: For the cultivation and maintenance of *Pseudomonas* species.

Buffered Peptone Water

Composition per liter:

Pancreatic digest of gelatin	10.0g
NaCl	5.0g
Na_2HPO_4	3.5g
KH_2PO_4	1.5g

pH 7.2 ± 0.2 at 25°C

Source: This medium is available as a premixed powder from BD Diagnostic Systems and Oxoid Unipath.

Preparation of Medium: Add components to distilled/deionized water and bring volume to 1.0L. Mix thoroughly. Distribute into tubes or flasks. Autoclave for 15 min at 15 psi pressure–121°C.

Use: Used as a preenrichment medium for the isolation of *Salmonella*, especially injured microorganisms, from various food sources.

Buffered S & H Agar

Composition per liter:

Agar	15.0g
Peptone	5.0g
Yeast extract	5.0g
Na_2HPO_4	2.7g
Citric acid·H_2O	1.15g
Glucose solution	40.0mL

pH 5.0 ± 0.2 at 25°C

Glucose Solution:

Composition per 50.0mL:

Glucose	25.0g

Preparation of Glucose Solution: Add 25.0g of glucose to 50.0mL of distilled/deionized water. Mix thoroughly and gently heat to dissolve. Filter sterilize.

Preparation of Medium: Add components, except glucose solution, to distilled/deionized water and bring volume to 960.0mL. Mix thoroughly. Gently heat to boiling. Adjust pH to 5.0 with HCl. Autoclave for 15 min at 15 psi pressure–121°C. Cool to 50°C. Aseptically add 40.0mL of sterile glucose solution to sterile basal medium. Pour into sterile Petri dishes or distribute into sterile tubes.

Use: For the cultivation and maintenance of *Acetobacter xylinum.*

Buffered S & H Broth

Composition per liter:

Peptone	5.0g
Yeast extract	5.0g
Na_2HPO_4	2.7g
Citric acid·H_2O	1.15g
Glucose solution	40.0mL

pH 5.0 ± 0.2 at 25°C

Glucose Solution:

Composition per 50.0mL:

Glucose	25.0g

Preparation of Glucose Solution: Add glucose to 50.0mL of distilled/deionized water. Mix thoroughly and gently heat to dissolve. Filter sterilize.

Preparation of Medium: Add components, except glucose solution, to distilled/deionized water and bring volume to 960.0mL. Mix thoroughly. Gently heat to boiling. Adjust pH to 5.0 with HCl. Autoclave for 15 min at 15 psi pressure–121°C. Cool to 50°C. Aseptically add 40.0mL of sterile glucose solution to sterile basal medium. Aseptically distribute into sterile tubes or flasks.

Use: For the cultivation of *Acetobacter xylinum.*

Buffered Soy Lactose for *Corynebacterium*

See: BSL for *Corynebacterium*

Buffered Tryptone Glucose Yeast Extract Broth

Composition per liter:

Yeast extract	20.0g
Casein enzymatic hydrolysate	50.0g
Peptic digest of animal tissue	5.0g
Na_2HPO_4	5.0g
Glucose	4.0g
Sodium thioglycolate	1.0g

pH 7.3 ± 0.2 at 25°C

Source: This medium is available from HiMedia.

Preparation of Medium: Add components to distilled/deionized water and bring volume to 1.0L. Mix thoroughly. Distribute into tubes or bottles. Autoclave for 15 min at 15 psi pressure–121°C.

Use: For the isolation of *Clostridium perfringens* from foods.

Buffered Yeast Agar

Composition per liter:

Glucose	20.0g
Agar	15.0g
Yeast extract	5.0g
$(NH_4)_2SO_4$	0.72g
$NH_4H_2PO_4$	0.26g

pH 5.5 ± 0.2 at 25°C

Source: This medium is available from HiMedia.

Preparation of Medium: Add components to distilled/deionized water and bring volume to 1.0L. Mix thoroughly. Distribute into tubes or bottles. Autoclave for 15 min at 15 psi pressure–121°C. Pour into Petri dishes or leave in tubes.

Use: For the cultivation of yeasts and molds from bottle washing operations.

Buffered Yeast Extract Broth

See: **BYEB**

Burke's Modified Nitrogen-Free Medium

Composition per liter:

$MgSO_4·7H_2O$	0.2g
Na_2HPO_4	0.19g
$NaHCO_3$	0.05g
$CaSO_4·2H_2O$	0.02g
KH_2PO_4	0.011g
$SrCl_2·6H_2O$	0.01g
NaCl	0.01g
Adenine	0.01g
$FeSO_4.7H_2O$	6.0mg
Na_2MoO_3	0.5mg

pH 7.8 ± 0.2 at 25°C

Preparation of Medium: Add components to distilled/deionized water and bring volume to 1.0L. Mix thoroughly. Distribute into tubes or flasks. Autoclave for 15 min at 15 psi pressure–121°C.

Use: For the cultivation of *Azotobacter vinelandii.*

Burke's Modified Nitrogen-Free Medium

Composition per liter:

Noble agar	15.0g
Glucose	10.0g
Cellulose	10.0g
K_2HPO_4	1.0g
$CaCl_2·2H_2O$	0.1g
$MgSO_4·7H_2O$	0.02g
$FeSO_4·7H_2O$	50.0mg
$Na_2MoO_4·2H_2O$	25.0mg
Vitamin B_{12}	0.1mg
Vitamin solution	1.0mL

pH 7.2–7.3 ± 0.2 at 25°C

Vitamin Solution:

Composition per 50.0mL:

Thiamine·HCl	843.3mg
Pantothenic acid	595.8mg
Nicotinic acid	307.8mg
p-Aminobenzoic acid	68.6mg
Pyridoxamine·2HCl	60.3mg
Biotin	50.0mg
Folic acid	11.0mg
Vitamin B_{12}	3.5mg

Preparation of Vitamin Solution: Add components to distilled/deionized water and bring volume to 50.0mL. Mix thoroughly. Filter sterilize.

Preparation of Medium: Add components, except glucose, cellulose, K_2HPO_4, and vitamin solution, to distilled/deionized water and bring volume to 850.0mL. Mix thoroughly. Adjust pH to 7.2–7.3. In three separate flasks, add glucose, cellulose, and K_2HPO_4 to 50.0mL of distilled/deionized water. Filter sterilize the vitamin solution. Autoclave the other solutions separately for 15 min at 15 psi pressure–121°C. Cool to 25°C. Aseptically combine all the solutions and mix thoroughly. Distribute into sterile tubes or flasks or pour into sterile Petri dishes.

Use: For the cultivation and maintenance of *Streptomyces* species.

Burke's Modified Nitrogen-Free Medium with Benzoate

Composition per liter:

Sodium benzoate	0.72g
$MgSO_4 \cdot 7H_2O$	0.2g
Na_2HPO_4	0.189g
$NaHCO_3$	0.05g
$CaSO_4 \cdot 2H_2O$	0.02g
KH_2PO_4	0.011g
$SrCl_2 \cdot 6H_2O$	0.01g
NaCl	0.01g
Adenine	0.01g
$FeSO_4 . 7H_2O$	6.0mg
Na_2MoO_3	0.5mg

pH 7.8 ± 0.2 at 25°C

Preparation of Medium: Add components to distilled/deionized water and bring volume to 1.0L. Mix thoroughly. Distribute into tubes or flasks. Autoclave for 15 min at 15 psi pressure–121°C.

Use: For the cultivation of *Pseudomonas* species and other microorganisms which can utilize benzoate as sole carbon source.

Burkholderia cepacia Agar

Composition per liter:

Agar	12.0g
Sodium pyruvate	7.0g
Peptone	5.0g
KH_2PO_4	4.4g
Yeast extract	4.0g
Bile salts	1.5g
Na_2HPO_4	1.4g
$(NH_4)_2SO_4$	1.0g
$MgSO_4$	0.2
Phenol Red	0.02g
$Fe(NH_4)_2(SO_4)_2 \cdot 6H_2O$	0.01g
Crystal Violet	0.001g
Selective supplement solution	10.0mL

pH 6.2 ± 0.2 at 25°C

Source: This medium is available as a premixed powder from Oxoid Unipath.

Selective Supplement Solution:

Composition per 10.0mL:

Polymyxin B	150,000IU
Ticarcillin	100.0mg
Gentamicin	5.0mg

Preparation of Selective Supplement Solution: Add components to distilled/deionized water and bring volume to 10.0mL. Mix thoroughly. Filter sterilize.

Preparation of Medium: Add components, except selective supplement solution, to distilled/deionized water and bring volume to 990.0mL. Mix thoroughly. Gently heat while stirring and bring to boiling. Autoclave for 15 min at 15 psi pressure–121°C. Cool to 50°C. Aseptially add 10.0mL selective supplement solution. Mix thoroughly. Pour into sterile Petri dishes.

Use: For the selective isolation of *Burkholderia cepacia* from the respiratory secretions of patients with cystic fibrosis and for routine testing of non-sterile inorganic salt solutions containing preservative. Slow growing *B. cepacia* can be missed on conventional media such as blood or MacConkey agar due to overgrowth caused by other faster growing organisms found in the respiratory tract of CF patients such as mucoid *Klebsiella* species, *Pseudomonas aeruginosa,* and *Staphylococcus* species. This may lead to the infection being missed or wrongly diagnosed.

Burkholderia pseudomallei Selective Agar (BPSA)

Composition per liter:

Agar	15.0g
Pancreatic Digest of Casein	5.0g
Maltose	4.0g
Yeast Extract	2.5g
Glucose	1.0g
Neutral Red	0.1g
Gentamicin solution	10.0mL
Glycerol	10.0mL
Nile Blue solution	1.0mL

pH 7.0 ± 0.2 at 25°C

Gentamicin Solution:

Composition per 10.0mL:

Gentamicin	20.0mg

Preparation of Gentamicin Solution: Add gentamicin to distilled/deionized water and bring volume to 10.0mL. Mix thoroughly. Filter sterilize.

Nile Blue Solution:

Composition per 10.0mL:

Nile Blue	0.2g

Preparation of Nile Blue Solution: Add Nile blue to 10.0mL of a 1% solution of dimethyl sulfoxide. Mix thoroughly. Filter sterilize.

Preparation of Medium: Add components, except gentamcin solution, Nile Blue solution, and glycerol, to distilled/deionized water and bring volume to 979.0mL. Mix thoroughly. Gently heat while stirring and bring to boiling. Autoclave for 15 min at 15 psi pressure–121°C. Cool to 45°C. Aseptially add 10.0mL sterile gentamicin solution, 1.0mL sterile Nile blue solution, and 10.0mL filter-sterilized glycerol. Mix thoroughly for 5 min on a heated magnetic stirrer at 40°C. Pour into sterile Petri dishes.

Use: For the cultivation of *Burkholderia pseudomallei* from clinical specimens collected from non-sterile sites with improved recovery of the more easily inhibited strains of *B. pseudomallei.*

Burk's Medium

Composition per liter:

Sucrose	20.0g
$MgSO_4 \cdot 7H_2O$	0.2g
K_2HPO_4	0.8g
KH_2PO_4	0.25g
$CaSO_4$	0.13g
$FeCl_3$	1.45mg
Na_2MoO_3	0.253mg

Source: This medium is available from HiMedia.

Preparation of Medium: Add components to distilled/deionized water and bring volume to 1.0L. Mix thoroughly. Distribute into tubes with inverted Durham tubes. Autoclave for 15 min at 15 psi pressure–121°C.

Use: For thecultivation of nitrogen fixing bacteria, such as *Azotrobacter* spp., from soil.

Bushnell-Haas Agar

Composition per liter:

Agar	15.0g
KH_2PO_4	1.0g
K_2HPO_4	1.0g
NH_4NO_3	1.0g
$MgSO_4 \cdot 7H_2O$	0.2g
$FeCl_3$	0.05g
$CaCl_2 \cdot 2H_2O$	0.02g

pH 7.0 ± 0.2 at 25°C

Preparation of Medium: Add components to distilled/deionized water and bring volume to 1.0L. Mix thoroughly. Gently heat and bring to boiling. Distribute into tubes or flasks. Autoclave for 15 min at 15 psi pressure–121°C. Pour into sterile Petri dishes or leave in tubes. For use in cultivating hydrocarbon-utilizing bacteria, layer 0.1–1.0% hydrocarbon on agar surface or aseptically add sterile hydrocarbon to cooled agar prior to pouring plates.

Use: For examining fuels for microbial contamination and for studying hydrocarbon utilization by microorganisms. Also for the cultivation of *Nocardia* species.

Bushnell-Haas Broth

Composition per liter:

KH_2PO_4	1.0g
K_2HPO_4	1.0g
NH_4NO_3	1.0g
$MgSO_4 \cdot 7H_2O$	0.2g
$FeCl_3$	0.05g
$CaCl_2 \cdot 2H_2O$	0.02g

pH 7.0 ± 0.2 at 25°C

Source: This medium is available as a premixed powder from BD Diagnostic Systems.

Preparation of Medium: Add components to distilled/deionized water and bring volume to 1.0L. Mix thoroughly. Distribute into tubes or flasks. Autoclave for 15 min at 15 psi pressure–121°C. For use in cultivating hydrocarbon-utilizing bacteria, layer 0.1–1.0% hydrocarbon on broth surface or add directly to broth.

Use: For examining fuels for microbial contamination and for studying the hydrocarbon utilization by microorganisms. Also for the cultivation of *Nocardia* species.

Bushnell-Haas Medium

Composition per liter:

KH_2PO_4	1.0g
K_2HPO_4	1.0g
NH_4NO_3	1.0g
Cholesterol	0.3g
$MgSO_4 \cdot 7H_2O$	0.2g
$FeCl_3$	0.05g
$CaCl_2 \cdot 2H_2O$	0.02g

pH 7.0 ± 0.2 at 25°C

Source: This medium is available as a premixed powder from BD Diagnostic Systems.

Preparation of Medium: Add components to distilled/deionized water and bring volume to 1.0L. Mix thoroughly. Distribute into tubes or flasks. Autoclave for 15 min at 15 psi pressure–121°C. For use in cultivating hydrocarbon-utilizing bacteria, layer 0.1–1.0% hydrocarbon on broth surface or add directly to broth.

Use: For the cultivation of *Nocardia* species.

Butanediol Medium

Composition per liter:

$NaH_2PO_4 \cdot H_2O$	2.1g
1,4-Butanediol	1.0g
NaCl	1.0g
NH_4Cl	1.0g
$CaCl_2 \cdot 2H_2O$	0.5g
$MgSO_4 \cdot 7H_2O$	0.5g
K_2HPO_4	0.3g
Yeast extract	0.2g
Modified Wolfe's mineral solution	10.0mL

pH 7.0 ± 0.2 at 25°C

Modified Wolfe's Mineral Solution:

Composition per liter:

$MgSO_4 \cdot 7H_2O$	3.0g
Nitrilotriacetic acid	1.5g
NaCl	1.0g
$MnSO_4 \cdot H_2O$	0.5g
$CaCl_2$	0.1g
$CoCl_2 \cdot 6H_2O$	0.1g
$FeSO_4 \cdot 7H_2O$	0.1g
$ZnSO_4 \cdot 7H_2O$	0.1g
$AlK(SO_4)_2 \cdot 12H_2O$	0.01g
$CuSO_4 \cdot 5H_2O$	0.01g
H_3BO_3	0.01g
$Na_2MoO_4 \cdot 2H_2O$	0.01g
Na_2SeO_3	0.01g
$NaWO_4 \cdot 2H_2O$	0.01g
$NiC1_2 \cdot 6H_2O$	0.01g

Preparation of Modified Wolfe's Mineral Solution: Add nitrilotriacetic acid to 500.0mL of distilled/deionized water. Adjust pH to 6.5 with KOH. Add remaining components one at a time. Add distilled/deionized water to 1.0L. Adjust pH to 6.8.

Preparation of Medium: Add components to distilled/deionized water and bring volume to 1.0L. Mix thoroughly. Adjust pH to 7.0. Distribute into tubes or flasks. Autoclave for 15 min at 15 psi pressure–121°C.

Use: For the cultivation of *Pseudomonas putida.*

Butyrivibrio Species Medium

Composition per 1001.0mL:

Na_2CO_3	4.0g
Pancreatic digest of casein	2.0g
Yeast extract	2.0g
K_2HPO_4	.0.3g
Hemin	1.0mg
Resazurin	1.0mg
Rumen fluid, clarified	150.0mL
Minerals solution	75.0mL
Carbohydrate solution	20.0mL
L-Cysteine·HCl·H_2O solution	10.0mL
$Na_2S \cdot 9H_2O$ solution	10.0mL
Volatile fatty acid mixture	3.1mL

pH 6.7 ± 0.2 at 25°C

Minerals Solution:

Composition per liter:

NaCl	12.0g
KH_2PO_4	6.0g
$(NH_4)_2SO_4$	6.0g

$MgSO_4 \cdot 7H_2O$	2.5g
$CaCl_2 \cdot 2H_2O$	1.6g

Preparation of Minerals Solution: Add components to distilled/deionized water and bring volume to 1.0L. Mix thoroughly.

L-Cysteine·HCl·H_2O Solution:
Composition per 10.0mL:

L-Cysteine·HCl·H_2O	0.25g

Preparation of L-Cysteine·HCl·H_2O Solution: Add L-cysteine·HCl·H_2O to distilled/deionized water and bring volume to 10.0mL. Mix thoroughly. Sparge with 100% CO_2. Autoclave for 15 min at 15 psi pressure–121°C.

$Na_2S \cdot 9H_2O$ Solution:
Composition per 10.0mL:

$Na_2S \cdot 9H_2O$	0.25g

Preparation of $Na_2S \cdot 9H_2O$ Solution: Add $Na_2S \cdot 9H_2O$ to distilled/deionized water and bring volume to 10.0mL. Mix thoroughly. Sparge with 100% CO_2. Autoclave for 15 min at 15 psi pressure–121°C.

Carbohydrate Solution:
Composition per 20.0mL:

Glucose	1.0g
Cellobiose	1.0g
Glycerol	1.0g
Maltose	1.0g
Starch, soluble	1.0g

Preparation of Carbohydrate Solution: Add components to distilled/deionized water and bring volume to 20.0mL. Mix thoroughly. Sparge under 100% CO_2. Autoclave for 15 min at 15 psi pressure–121°C.

Volatile Fatty Acid Mixture:
Composition per 7.75mL:

Acetic acid	4.25mL
Propionic acid	1.50mL
Butyric acid	1.0mL
DL-2-Methyl butyric acid	0.25mL
iso-Butyric acid	0.25mL
iso-Valeric acid	0.25mL
n-Valeric acid	0.25mL

Preparation of Volatile Fatty Acid Mixture: Combine components. Mix thoroughly.

Preparation of Medium: Prepare and dispense medium under 100% CO_2. Add components, except carbohydrate solution, Na_2CO_3, L-cysteine·HCl·H_2O solution, and $Na_2S \cdot 9H_2O$ solution, to distilled/deionized water and bring volume to 960.0mL Mix thoroughly. Gently heat and bring to boiling. Continue boiling for 5 min. Cool to room temperature while sparging with 100% CO_2. Add Na_2CO_3. Continue sparging with 100% CO_2 until pH reaches 6.8. Distribute into rubber-stoppered tubes under 100% CO_2. Autoclave for 15 min at 15 psi pressure–121°C. Aseptically and anaerobically add 20.0mL of sterile carbohydrate solution, 10.0mL of sterile L-cysteine·HCl·H_2O solution, and 10.0mL of sterile $Na_2S \cdot 9H_2O$ solution or, using a syringe, inject the appropriate amount of sterile carbohydrate solution, sterile $Na_2S \cdot 9H_2O$ solution, and sterile L-cysteine·HCl·H_2O solution into individual tubes containing medium.

Use: For the cultivation of *Butyrivibrio* species.

Butzler Medium

See: ***Campylobacter*** **Selective Medium, Butzler's**

Butzler's *Campylobacter* Medium

See: ***Campylobacter*** **Selective Medium, Butzler's**

BY Agar Medium (ATCC Medium 2038)

Composition per liter:

Agar	15.0g
Yeast extract	5.0g
Pancreatic digest of casein	5.0g
Beef extract	5.0g
NaCl	2.5g
K_2HPO_4	0.1g
$MgSO_4 \cdot 7H_2O$	0.05g

pH 7.2 ± 0.2 at 25°C

Preparation of Medium: Add components to distilled/deionized water and bring volume to 1.0L. Mix thoroughly. Gently heat and bring to boiling. Distribute into tubes or flasks. Autoclave for 15 min at 15 psi pressure–121°C. Pour into sterile Petri dishes or leave in tubes.

Use: For the cultivation and maintenance of *Paracoccus thiocyanatus.*

BY+ Medium

Composition per liter:

Glucose	5.0g
Peptone	1.0g
Yeast extract	1.0g
Seawater	1.0L

Preparation of Medium: Combine components. Mix thoroughly. Gently heat and bring to boiling. Distribute into flasks or tubes. Autoclave for 15 min at 15 psi pressure–121°C.

Use: For the cultivation of *Arenariomyces triseptatus, Haliphthoros milfordensis, Haliphthoros philippinensis, Halosphaeria salina, Japonochytrium* species, *Lignincola laevis, Lindra thalassiae, Schizochytrium aggregatum, Thraustochytrium* species, and *Torpedospora radiata.*

BYE Agar

Composition per liter:

Pancreatic digest of casein	16.0g
Agar	13.5g
Brain heart, solids from infusion	8.0g
Peptic digest of animal tissue	5.0g
NaCl	5.0g
Glucose	2.0g
Na_2HPO_4	2.5g
Yeast extract	2.0g
Blood, human or animal, sterile	150.0mL

pH 7.8–8.0 ± 0.2 at 25°C

Preparation of Medium: Add components, except blood, to distilled/deionized water and bring volume to 850.0mL. Mix thoroughly. Autoclave for 15 min at 15 psi pressure–121°C. Cool to 45°–50°C. Aseptically add 150.0mL of sterile blood. Outdated, citrated, or heparinized blood (blood from a blood bank is acceptable). Pour into sterile Petri dishes.

Use: For the isolation and cultivation of *Mycoplasma* species and L-forms of bacteria. For the detection of *Mycoplasma* species in tissue culture and cell lines.

BYE HiVeg Agar with Blood

Composition per liter:

Agar	13.0g
Plant infusion	10.0g
Plant peptone No. 3	10.0g
Plant special infusion	7.5g
NaCl	5.0g
Na_2HPO_4	2.5g
Glucose	2.0g
Yeast extract	2.0g
Horse or human blood, sterile	150.0mL

pH 7.9 ± 0.2 at 25°C

Source: This medium, without blood, is available as a premixed powder from HiMedia.

Preparation of Medium: Add components, except blood, to distilled/deionized water and bring volume to 850.0mL. Mix thoroughly. Autoclave for 15 min at 15 psi pressure–121°C. Cool to 45°–50°C. Aseptically add 150.0mL of sterile blood. Outdated, citrated, or heparinized blood (blood from a blood bank is acceptable). Pour into sterile Petri dishes.

Use: For the isolation and cultivation of *Mycoplasma* species and L-forms of bacteria. For the detection of *Mycoplasma* species in tissue culture and cell lines.

BYE HiVeg Broth with Blood

Composition per liter:

Plant infusion	10.0g
Plant peptone No. 3	10.0g
Plant special infusion	7.5g
NaCl	5.0g
Na_2HPO_4	2.5g
Glucose	2.0g
Yeast extract	2.0g
Horse or human blood, sterile	150.0mL

pH 7.9 ± 0.2 at 25°C

Source: This medium, without blood, is available as a premixed powder from HiMedia.

Preparation of Medium: Add components, except blood, to distilled/deionized water and bring volume to 850.0mL. Mix thoroughly. Autoclave for 15 min at 15 psi pressure–121°C. Cool to 45°–50°C. Aseptically add 150.0mL of sterile blood. Outdated, citrated, or heparinized blood (blood from a blood bank is acceptable).

Use: For the cultivation of *Mycoplasma* species and L-forms of bacteria.

BYEB
(Buffered Yeast Extract Broth)

Composition per liter:

ACES buffer (2-[(2-amino-2-oxoethyl)-amino]-ethane sulfonic acid)	10.0g
Yeast extract	10.0g
α-Ketoglutarate	1.0g
L-Cysteine·HCl·H_2O	0.4g
$Fe_4(P_2O_7)_3 \cdot 9H_2O$	0.25g

pH 6.9 ± 0.2 at 25°C

Preparation of Medium: Add components to distilled/deionized water and bring volume to 1.0L. Mix thoroughly. Adjust pH to 6.9. Filter sterilize. Aseptically distribute into sterile tubes or flasks.

Use: For the cultivation of *Legionella pneumophila*.

C/10 Agar

Composition per liter:

Agar	15.0g
Pancreatic digest of casein	3.0g
$CaCl_2 \cdot 2H_2O$	1.36g

pH 7.2 ± 0.2 at 25°C

Preparation of Medium: Add components to distilled/deionized water and bring volume to 1.0L. Mix thoroughly. Gently heat and bring to boiling. Adjust pH t o 7.2. Distribute into tubes or flasks. Autoclave for 15 min at 15 psi pressure–121°C. Pour into sterile Petri dishes or leave in tubes.

Use: For the cultivation of *Cytophaga flevensis, Flexibacter filiformis, Myxococcus*

C/10 Medium Reichenbach

Composition per liter:

Agar	15.0g
Pancreatic digest of casein	3.0g
$CaCl_2$	1.0g

pH 7.2 ± 0.2 at 25°C

Preparation of Medium: Add components, except agar, to distilled/deionized water and bring volume to 1.0L. Adjust pH to 7.2. Add agar. Mix thoroughly. Gently heat to boiling. Distribute into tubes or flasks. Autoclave for 15 min at 15 psi pressure–121°C. Pour into sterile Petri dishes or leave in tubes.

Use: For the cultivation and maintenance of *Flexibacter filiformis. fulvus,* and *Myxococcus xanthus.*

C 3G *Spiroplasma* Medium

Composition per liter:

Sucrose	100.0g
Phenol Red	10.0mg
PPLO broth without Crystal Violet	500.0mL
Horse serum	150.0mL
Fresh yeast extract solution	50.0mL
CMRL-1066 medium	5.0mL

pH 7.5 ± 0.2 at 25°C

Source: PPLO broth without Crystal Violet is available as a premixed powder from BD Diagnostic Systems.

PPLO Broth without Crystal Violet:
Composition per 500.0mL:

Beef heart, infusion from	11.52g
Peptone	2.32g
NaCl	1.15g

Preparation of PPLO Broth without Crystal Violet: Add components to distilled/deionized water and bring volume to 500.0mL. Mix thoroughly.

Fresh Yeast Extract Solution:
Composition per 100.0mL:

Baker's yeast, live, pressed, starch-free	25.0g

Preparation of Fresh Yeast Extract Solution: Add the live Baker's yeast to 100.0mL of distilled/deionized water. Autoclave for 90 min at 15 psi pressure–121°C. Allow to stand. Remove supernatant solution. Adjust pH to 6.6–6.8. Filter sterilize.

CMRL-1066 Medium:
Composition per liter:

NaCl	6.8g
$NaHCO_3$	2.2g
D-Glucose	1.0g
KCl	0.4g
L-Cysteine·HCl·H_2O	0.26g
$CaCl_2$, anhydrous	0.2g
$MgSO_4 \cdot 7H_2O$	0.2g
$NaH_2PO_4 \cdot H_2O$	0.14g
L-Glutamine	0.1g
Sodium acetate·$3H_2O$	0.083g
L-Glutamic acid	0.075g
L-Arginine·HCl	0.07g
L-Lysine·HCl	0.07g
L-Leucine	0.06g
Glycine	0.05g
Ascorbic acid	0.05g
L-Proline	0.04g
L-Tyrosine	0.04g
L-Aspartic acid	0.03g
L-Threonine	0.03g
L-Alanine	0.025g
L-Phenylalanine	0.025g
L-Serine	0.025g
L-Valine	0.025g
L-Cystine	0.02g
L-Histidine·HCl·H_2O	0.02g
L-Isoleucine	0.02g
Phenol Red	0.02g
L-Methionine	0.015g
Deoxyadenosine	0.01g
Deoxycytidine	0.01g
Deoxyguanosine	0.01g
Glutathione, reduced	0.01g
Thymidine	0.01g
Hydroxy-L-proline	0.01g
L-Tryptophan	0.01g
Nicotinamide adenine dinucleotide	7.0mg
Tween™ 80	5.0mg
Sodium glucoronate·H_2O	4.2mg
Coenzyme A	2.5mg
Cocarboxylase	1.0mg
Flavin adenine dinucleotide	1.0mg
Nicotinamide adenine dinucleotide phosphate	1.0mg
Uridine triphosphate	1.0mg
Choline chloride	0.5mg
Cholesterol	0.2mg
5-Methyldeoxycytidine	0.1mg
Inositol	0.05mg
p-Aminobenzoic acid	0.05mg
Niacin	0.025mg
Niacinamide	0.025mg
Pyridoxine	0.025mg
Pyridoxal·HCl	0.025mg
Biotin	0.01mg
D-Calcium pantothenate	0.01mg
Folic acid	0.01mg
Riboflavin	0.01mg
Thiamine·HCl	0.01mg

Source: CMRL-1066 medium is available as a premixed powder from BD Diagnostics.

Preparation of CMRL-1066 Medium: Add components to distilled/deionized water and bring volume to 1.0L. Mix thoroughly. Adjust pH to 7.2. Filter sterilize.

Preparation of Medium: Add components—except horse serum, fresh yeast extract, and CMRL medium—to distilled/deionized water and bring volume to 795.0mL. Adjust pH to 7.5. Autoclave for 15 min at 15 psi pressure–121°C. Aseptically add 150.0mL of sterile horse serum, 50.0mL of sterile fresh yeast extract solution, and 5.0mL of sterile CMRL medium. Distribute into sterile tubes or flasks.

Use: For the cultivation and maintenance of *Spiroplasma* species.

C 3N *Spiroplasma* Medium

Composition per 100.0mL:

Sucrose	12.0g
Phenol Red	10.0mg
PPLO broth without Crystal Violet	50.0mL
Horse serum	20.0mL
Fresh yeast extract solution	5.0mL
CMRL-1066 medium	0.5mL

pH 7.5 ± 0.2 at 25°C

PPLO Broth without Crystal Violet:
Composition per 500.0mL:

Beef heart, infusion from	11.52g
Peptone	2.32g
NaCl	1.15g

Source: PPLO broth without Crystal Violet is available as a premixed powder from BD Diagnostic Systems.

Preparation of PPLO Broth without Crystal Violet: Add components to distilled/deionized water and bring volume to 500.0mL. Mix thoroughly.

Fresh Yeast Extract Solution:
Composition per 100.0mL:

Baker's yeast, live, pressed, starch-free	25.0g

Preparation of Fresh Yeast Extract Solution: Add the live Baker's yeast to 100.0mL of distilled/deionized water. Autoclave for 90 min at 15 psi pressure–121°C. Allow to stand. Remove supernatant solution. Adjust pH to 6.6–6.8. Filter sterilize.

CMRL-1066 Medium:
Composition per liter:

NaCl	6.8g
$NaHCO_3$	2.2g
D-Glucose	1.0g
KCl	0.4g
L-Cysteine·HCl·H_2O	0.26g
$CaCl_2$, anhydrous	0.2g
$MgSO_4 \cdot 7H_2O$	0.2g
$NaH_2PO_4 \cdot H_2O$	0.14g
L-Glutamine	0.1g
Sodium acetate·$3H_2O$	0.083g
L-Glutamic acid	0.075g
L-Arginine·HCl	0.07g
L-Lysine·HCl	0.07g

L-Leucine	0.06g
Glycine	0.05g
Ascorbic acid	0.05g
L-Proline	0.04g
L-Tyrosine	0.04g
L-Aspartic acid	0.03g
L-Threonine	0.03g
L-Alanine	0.025g
L-Phenylalanine	0.025g
L-Serine	0.025g
L-Valine	0.025g
L-Cystine	0.02g
L-Histidine·HCl·H_2O	0.02g
L-Isoleucine	0.02g
Phenol Red	0.02g
L-Methionine	0.015g
Deoxyadenosine	0.01g
Deoxycytidine	0.01g
Deoxyguanosine	0.01g
Glutathione, reduced	0.01g
Thymidine	0.01g
Hydroxy-L-proline	0.01g
L-Tryptophan	0.01g
Nicotinamide adenine dinucleotide	7.0mg
Tween™ 80	5.0mg
Sodium glucoronate·H_2O	4.2mg
Coenzyme A	2.5mg
Cocarboxylase	1.0mg
Flavin adenine dinucleotide	1.0mg
Nicotinamide adenine dinucleotide phosphate	1.0mg
Uridine triphosphate	1.0mg
Choline chloride	0.5mg
Cholesterol	0.2mg
5-Methyldeoxycytidine	0.1mg
Inositol	0.05mg
p-Aminobenzoic acid	0.05mg
Niacin	0.025mg
Niacinamide	0.025mg
Pyridoxine	0.025mg
Pyridoxal·HCl	0.025mg
Biotin	0.01mg
D-Calcium pantothenate	0.01mg
Folic acid	0.01mg
Riboflavin	0.01mg
Thiamine·HCl	0.01mg

Source: CMRL-1066 medium is available as a premixed powder from BD Diagnostics.

Preparation of CMRL-1066 Medium: Add components to distilled/deionized water and bring volume to 1.0L. Mix thoroughly. Adjust pH to 7.2. Filter sterilize.

Preparation of Medium: Add components—except horse serum, fresh yeast extract, and CMRL medium—to distilled/deionized water and bring volume to 75.0mL. Adjust pH to 7.5. Autoclave for 15 min at 15 psi pressure–121°C. Aseptically add 20.0mL of sterile horse serum, 5.0mL of sterile yeast extract, and 0.5mL of sterile CMRL medium. Distribute into sterile tubes or flasks.

Use: For the cultivation and maintenance of *Spiroplasma kunkelii.*

CA
See: **Carrot Decoction Agar**

CA YE Broth
See: **Casamino Acids Yeast Extract Salts Broth, Gorbach**

Cadmium Fluoride Acriflavin Tellurite Medium
See: **CFAT Medium**

CAE Agar Base with Triphenyltetrazolium Chloride (Citrate Azide *Enterococcus* HiVeg Agar Base)

Composition per liter:

Agar	15.0g
Casein enzymatic hydrolysate	15.0g
Sodium citrate	15.0g
KH_2PO_4	5.0g
Yeast extract	5.0g
Na_2CO_3	2.0g
Polysorbate 80	1.0g
NaN_3	0.4g
2,3,5-Triphenyltetrazolium chloride solution	10.0mL

pH 7.0 ± 0.2 at 25°C

Source: This medium, without triphenyltetrazolium chloride solution, is available as a premixed powder from HiMedia.

Caution: Sodium azide is toxic. Azides also react with metals and disposal must be highly diluted.

2,3,5-Triphenyltetrazolium Chloride Solution:
Composition per 10.0mL:

2,3,5-Triphenyltetrazolium chloride	0.1g

Preparation of 2,3,5-Triphenyltetrazolium Chloride Solution: Add 2,3,5-triphenyltetrazolium chloride to distilled/deionized water and bring volume to 10.0mL. Mix thoroughly. Filter sterilize.

Preparation of Medium: Add components, except 2,3,5-triphenyltetrazolium chloride solution, to distilled/deionized water and bring volume to 990.0mL. Mix thoroughly. Gently heat and bring to boiling. Autoclave for 10 min at 15 psi pressure–121°C. Cool to 45°–50°C. Aseptically add 10.0mL 2,3,5-triphenyltetrazolium chloride solution. Mix thoroughly. Pour into sterile Petri dishes or distribute into sterile tubes.

Use: For the isolation, cultivation, and enumeration of entercocci in water, sewage, and feces by the membrane filter method. For the direct plating of specimens for the detection and enumeration of fecal streptococci.

CAE HiVeg Agar Base with Triphenyltetrazolium Chloride (Citrate Azide *Enterococcus* HiVeg Agar Base)

Composition per liter:

Agar	15.0g
Plant hydrolysate	15.0g
Sodium citrate	15.0g
KH_2PO_4	5.0g
Yeast extract	5.0g
Na_2CO_3	2.0g
Polysorbate 80	1.0g
NaN_3	0.4g
2,3,5-Triphenyltetrazolium chloride solution	10.0mL

pH 7.0 ± 0.2 at 25°C

Source: This medium, without triphenyltetrazolium chloride solution, is available as a premixed powder from HiMedia.

Caution: Sodium azide is toxic. Azides also react with metals and disposal must be highly diluted.

2,3,5-Triphenyltetrazolium Chloride Solution:
Composition per 10.0mL:

2,3,5-Triphenyltetrazolium chloride	0.1g

Preparation of 2,3,5-Triphenyltetrazolium Chloride Solution: Add 2,3,5-triphenyltetrazolium chloride to distilled/deionized water and bring volume to 10.0mL. Mix thoroughly. Filter sterilize.

Preparation of Medium: Add components, except 2,3,5-triphenyltetrazolium chloride solution, to distilled/deionized water and bring volume to 990.0mL. Mix thoroughly. Gently heat and bring to boiling. Autoclave for 10 min at 15 psi pressure–121°C. Cool to 45°–50°C. Aseptically add 10.0mL 2,3,5-triphenyltetrazolium chloride solution. Mix thoroughly. Pour into sterile Petri dishes or distribute into sterile tubes.

Use: For the isolation, cultivation, and enumeration of entercocci in water, sewage, and feces by the membrane filter method. For the direct plating of specimens for the detection and enumeration of fecal streptococci.

Caffeic Acid Ferric Citrate Test Medium (CAFC Test Medium) (Caffeic Acid Agar)

Composition per liter:

Agar	20.0g
$(NH_4)_2SO_4$	5.0g
Glucose	5.0g
Yeast extract	2.0g
K_2HPO_4	0.8g
$MgSO_4 \cdot 3H_2O$	0.7g
Caffeic acid·1/2H_2O	0.18g
Chloramphenicol	0.05g
Ferric citrate solution	4.0mL

pH 6.5 ± 0.2 at 25°C

Ferric Citrate Solution:
Composition per 20.0mL:

Ferric citrate	100.0mg

Preparation of Ferric Citrate Solution: Add ferric citrate to 20.0mL of distilled/deionized water. Mix thoroughly.

Preparation of Medium: Add components, except chloramphenicol, to distilled/deionized water and bring volume to 1.0L. Mix thoroughly. Heat to boiling. Autoclave for 15 min at 15 psi pressure–121°C. Cool to 45°–50°C. Aseptically add 0.05g of chloramphenicol. Mix thoroughly. Pour into sterile Petri dishes.

Use: For the isolation and presumptive identification of *Cryptococcus neoformans*. *Cryptococcus neoformans* appears as dark brown colonies. All other *Cryptococcus* species appear as light brown or nonpigmented colonies.

Caffeine Medium

Composition per liter:

Agar	15.0g
Solution A	400.0mL
Solution B	400.0mL
Solution C	200.0mL

pH 5.0 ± 0.2 at 25°C

Solution A:
Composition per 400.0mL:

Na_2HPO_4	7.8g
KH_2PO_4	3.0g
Caffeine	1.0g
NaCl	0.58g

Preparation of Solution A: Add components to distilled/deionized water and bring volume to 400.0mL. Mix thoroughly. Adjust pH to 5.0.

Solution B:
Composition per 400.0mL:

$MgSO_4 \cdot 7H_2O$	0.12g
$CaCl_2 \cdot 2H_2O$	11.0mg

Preparation of Solution B: Add components to distilled/deionized water and bring volume to 400.0mL. Mix thoroughly.

Solution C:
Composition per 200.0mL:

$FeCl_3$	16.0mg

Preparation of Solution C: Add $FeCl_3$ to distilled/deionized water and bring volume to 200.0mL. Mix thoroughly.

Preparation of Medium: To 400.0mL of solution A, add 400.0mL of solution B and 200.0mL of solution C. Adjust pH to 5.0. Add agar. Mix thoroughly. Gently heat and bring to boiling. Autoclave for 15 min at 15 psi pressure–121°C. Pour into sterile Petri dishes or distribute into sterile tubes.

Use: For the cultivation of *Pseudomonas* species.

CAGV Medium

See: **Casamino Acid Glucose Medium**

CAL Agar (Cellobiose Arginine Lysine Agar) (*Yersinia* Isolation Agar)

Composition per liter:

Agar	20.0g
L-Arginine·HCl	6.5g
L-Lysine·HCl	6.5g
NaCl	5.0g
Cellobiose	3.5g
Yeast extract	3.0g
Sodium deoxycholate	1.5g
Neutral Red	0.03g

pH 7.3 ± 0.2 at 25°C

Preparation of Medium: Add components to distilled/deionized water and bring volume to 1.0L. Mix thoroughly. Heat to boiling. Do not autoclave. Pour into sterile Petri dishes.

Use: For the isolation and characterization of *Yersinia enterocolitica* from fecal specimens and enumeration of *Yersinia enterocolitica* from water and other liquid specimens.

CAL Broth (Cellobiose Arginine Lysine Broth)

Composition per liter:

L-Arginine·HCl	6.5g
L-Lysine·HCl	6.5g
NaCl	5.0g
Cellobiose	3.5g
Yeast extract	3.0g

Sodium deoxycholate 1.5g
Neutral Red 0.03g

pH 7.3 ± 0.2 at 25°C

Preparation of Medium: Add components to distilled/deionized water and bring volume to 1.0L. Mix thoroughly. Heat to boiling. Do not autoclave. Distribute into sterile tubes in 6.0–8.0mL volumes.

Use: For the isolation and characterization of *Yersinia enterocolitica* from fecal specimens and enumeration of *Yersinia enterocolitica* from water and other liquid specimens.

CAL HiVeg Agar (Cellobiose Arginine Lysine HiVeg Agar)

Composition per liter:

Agar 20.0g
L-Arginine 6.5g
L-Lysine hydrochloride 6.5g
NaCl 5.0g
Cellobiose 3.5g
Yeast extract 3.0g
Synthetic detergent No. III 1.5g
Neutral Red 0.03g

pH 7.1 ± 0.2 at 25°C

Source: This medium is available as a premixed powder from HiMedia.

Preparation of Medium: Add components to distilled/deionized water and bring volume to 1.0L. Mix thoroughly. Heat to boiling. Do not autoclave. Pour into sterile Petri dishes.

Use: For the isolation and characterization of *Yersinia enterocolitica* from fecal specimens and enumeration of *Y. enterocolitica* from water.

CAL HiVeg Broth (Cellobiose Arginine Lysine HiVeg Broth)

Composition per liter:

L-Arginine 6.5g
L-Lysine hydrochloride 6.5g
NaCl 5.0g
Cellobiose 3.5g
Yeast extract 3.0g
Synthetic detergent No. III 1.5g
Neutral Red 0.03g

pH 7.1 ± 0.2 at 25°C

Source: This medium is available as a premixed powder from HiMedia.

Preparation of Medium: Add components to distilled/deionized water and bring volume to 1.0L. Mix thoroughly. Heat to boiling. Do not autoclave. Distribute into sterile tubes.

Use: For the isolation and characterization of *Yersinia enterocolitica* from fecal specimens and enumeration of *Yersinia enterocolitica* from water and other liquid specimens.

Calcium Caseinate Agar

Composition per liter:

Agar 13.0g
Peptic digest of animal tissue 4.0g
Calcium caseinate 3.5g
Meat extract 2.0g
Casein enzymic hydrolysate 2.0g
$CaCl_2 \cdot 2H_2O$ 0.2g
Tri-potassium citrate 0.35g
Na_2HPO_4 .... 0.105g
KH_2PO_4 .... 0.035g
NaCl 5.0g

pH 7.0 ± 0.2 at 25°C

Source: This medium is available from HiMedia.

Preparation of Medium: Add components to distilled/deionized water and bring volume to 1.0L. Mix thoroughly. Distribute into tubes or flasks. Gently heat and bring to boiling. Boil for 10 min. Autoclave for 15 min at 15 psi pressure–121°C. Mix thoroughly while pouring into Petri dishes.

Use: For the detection and enumeration of proteolytic microorganisms in foodstuffs and other materials.

Calcium Caseinate Agar with Skim Milk

Composition per liter:

Agar 13.0g
Skim Milk 10.0g
Peptic digest of animal tissue 4.0g
Calcium caseinate 3.5g
Meat extract 2.0g
Casein enzymic hydrolysate 2.0g
$CaCl_2 \cdot 2H_2O$ 0.2g
Tri-potassium citrate 0.35g
Na_2HPO_4 .... 0.105g
KH_2PO_4 .... 0.035g
NaCl 5.0g

pH 7.0 ± 0.2 at 25°C

Source: This medium is available from HiMedia.

Preparation of Medium: Add components to distilled/deionized water and bring volume to 1.0L. Mix thoroughly. Distribute into tubes or flasks. Gently heat and bring to boiling. Boil for 10 min. Autoclave for 15 min at 15 psi pressure–121°C. Mix thoroughly while pouring into Petri dishes.

Use: For the detection and enumeration of proteolytic microorganisms in foodstuffs and other materials.

Caldicellulosiruptor Medium

Composition per liter:

Pancreatic digest of casein 2.0g
K_2HPO_4 .... 1.5g
Cellobiose 1.0g
Yeast extract 1.0g
NaCl 0.9g
NH_4Cl 0.9g
KH_2PO_4 .... 0.75g
L-Cysteine·HCl 0.75g
$MgCl_2 \cdot 6H_2O$ 0.4g
$FeCl_3 \cdot 6H_2O$ 2.5mg
Resazurin 0.5mg
Trace elements solution SL-10 .... 1.0mL

pH 7.2 ± 0.2 at 25°C

Trace Elements Solution SL-10:

Composition per liter:

$FeCl_2 \cdot 4H_2O$ 1.5g
$CoCl_2 \cdot 6H_2O$ 190.0mg
$MnCl_2 \cdot 4H_2O$ 100.0mg

$ZnCl_2$	70.0mg
$Na_2MoO_4 \cdot 2H_2O$	36.0mg
$NiCl_2 \cdot 6H_2O$	24.0mg
H_3BO_3	6.0mg
$CuCl_2 \cdot 2H_2O$	2.0mg
HCl (25% solution)	10.0mL

Preparation of Trace Elements Solution SL-10: Add $FeCl_2 \cdot 4H_2O$ to 10.0mL of HCl solution. Mix thoroughly. Add distilled/deionized water and bring volume to 1.0L. Add remaining components. Mix thoroughly. Sparge with 100% N_2. Autoclave for 15 min at 15 psi pressure–121°C.

Preparation of Medium: Prepare and dispense medium under 100% N_2. Add components to distilled/deionized water and bring volume to 1.0L. Mix thoroughly. Sparge with 100% N_2. Anaerobically distribute into tubes or flasks. Autoclave for 15 min at 15 psi pressure–121°C.

Use: For the cultivation of *Caldicellulosiruptor saccharolyticus.*

Caldicellulosiruptor Medium

Composition per 1001.0mL:

Pancreatic digest of casein	2.0g
K_2HPO_4	1.5g
Cellulose	1.0g
Cellobiose	1.0g
Yeast extract	1.0g
NaCl	0.9g
NH_4Cl	0.9g
KH_2PO_4	0.75g
$MgCl_2 \cdot 6H_2O$	0.4g
$FeCl_3 \cdot 6H_2O$	2.5mg
L-Cysteine·HCl	0.75g
Resazurin	0.5mg

pH 7.2 ± 0.2 at 25°C

Trace Elements Solution SL-10:
Composition per liter:

$FeCl_2 \cdot 4H_2O$	1.5g
$CoCl_2 \cdot 6H_2O$	190.0mg
$MnCl_2 \cdot 4H_2O$	100.0mg
$ZnCl_2$	70.0mg
$Na_2MoO_4 \cdot 2H_2O$	36.0mg
$NiCl_2 \cdot 6H_2O$	24.0mg
H_3BO_3	6.0mg
$CuCl_2 \cdot 2H_2O$	2.0mg
HCl (25% solution)	10.0mL

Preparation of Trace Elements Solution SL-10: Add $FeCl_2 \cdot 4H_2O$ to 10.0mL of HCl solution. Mix thoroughly. Add distilled/deionized water and bring volume to 1.0L. Add remaining components. Mix thoroughly. Sparge with 100% N_2. Autoclave for 15 min at 15 psi pressure–121°C.

Preparation of Medium: Prepare and dispense medium under 100% N_2. Add components to distilled/deionized water and bring volume to 1.0L. Mix thoroughly. Sparge with 100% N_2. Anaerobically distribute into tubes or flasks. Autoclave for 15 min at 15 psi pressure–121°C.

Use: For the cultivation of *Caldicellulosiruptor saccharolyticus.*

Caldisphaera Medium (DSMZ Medium 991)

Composition per liter:

Sulfur, powder	10.0g
$MnCl_2 \cdot 4H_2O$	1.8g
$(NH_4)_2SO_4$	1.3g
KH_2PO_4	0.28g
$MgSO_4 \cdot 7H_2O$	0.25g
$CaCl_2 \cdot 2H_2O$	0.07g
$FeCl_3 \cdot 6H_2O$	0.02g
$Na_2B_4O_7 \cdot 10H_2O$	4.5mg
Resazurin	1.0mg
$ZnSO_4 \cdot 7H_2O$	0.22mg
$CuCl_2 \cdot 2H_2O$	0.05mg
$Na_2MoO_4 \cdot 4H_2O$	0.03mg
$VOSO_4 \cdot 2H_2O$	0.03mg
Na_3-citrate·$2H_2O$	0.03mg
$CoSO_4$	0.01mg
Vitamin solution	10.0mL
$Na_2S \cdot 9H_2O$ solution	10.0mL
Yeast extract solution	5.0mL

pH 4.3 ± 0.2 at 25°C

Yeast Extract Solution:
Composition per 5.0mL:

Yeast extract	0.5g

Preparation of Yeast Extract Solution: Add yeast extract to distilled/deionized water and bring volume to 5.0mL. Mix thoroughly. Sparge under 100% N_2 gas for 3 min. Autoclave for 15 min at 15 psi pressure–121°C. Store under N_2 gas.

$Na_2S \cdot 9H_2O$ Solution:
Composition per 10.0mL:

$Na_2S \cdot 9H_2O$	0.5g

Preparation of $Na_2S \cdot 9H_2O$ Solution: Add $Na_2S \cdot 9H_2O$ to distilled/deionized water and bring volume to 10.0mL. Mix thoroughly. Autoclave under 100% N_2 for 15 min at 15 psi pressure–121°C. Cool to room temperature.

Vitamin Solution:
Composition per liter:

Pyridoxine-HCl	10.0mg
Thiamine-HCl·$2H_2O$	5.0mg
Riboflavin	5.0mg
Nicotinic acid	5.0mg
D-Ca-pantothenate	5.0mg
p-Aminobenzoic acid	5.0mg
Lipoic acid	5.0mg
Biotin	2.0mg
Folic acid	2.0mg
Vitamin B_{12}	0.1mg

Preparation of Vitamin Solution: Add components to distilled/deionized water and bring volume to 1.0L. Mix thoroughly. Sparge with 80% H_2 + 20% CO_2. Filter sterilize.

Preparation of Medium: Add components, except vitamin solution, yeast extract solution, sulfur, and $Na_2S \cdot 9H_2O$ solution, to distilled/deionized water and bring volume to 975.0mL. Mix thoroughly. Gently heat and bring to boiling. Boil for 3 min. Cool to room temperature under 80% N_2 + 20% CO_2. Adjust pH to 3.5 with 10*N* H_2SO_4. Dispense under same gas atmosphere in suitable culture vessels (e.g., 20.0mL of the medium in 120 mL serum bottles). Autoclave for 15 min at 15 psi pressure–121°C. Steam sulfur for 3 hr on each of 3 successive days. Aseptically mix the sterilized sulfur with the medium and add vitamins and yeast extract from sterile, anaerobic stock solutions. Prior to inoculation change atmosphere to 80% H_2 + 20% CO_2. Aseptically and anoxically add $Na_2S·9H_2O$. Adjust pH to 4.0–4.5 if necessary. After inoculation pressurize vials to 1 bar overpressure with 80% H_2 + 20% CO_2 gas mixture.

Use: For the cultivation of *Caldisphaera* spp.

Calditerrivibrio Medium (DSMZ Medium 1112)

Composition per liter:

$NaNO_3$	0.85g
Na-acetate	0.82g
NH_4Cl	0.54g
$MgCl_2·6H_2O$	0.2g
$CaCl_2·2H_2O$	0.15g
KH_2PO_4	0.14g
Resazurin	0.5mg
$NaHCO_3$ solution	10.0mL
Vitamin solution	10.0mL
$Na_2S·9H_2O$ solution	10.0mL
Trace element solution SL-10	1.0mL
Selenite/tungstate solution	1.0mL

pH 7.0 ± 0.2 at 25°C

$NaHCO_3$ Solution:

Composition per 10.0mL:

$NaHCO_3$	2.5g

Preparation of $NaHCO_3$ Solution: Add components to distilled/deionized water and bring volume to 10.0mL. Mix thoroughly. Sparge with 20% CO_2 + 80% H_2. Autoclave for 15 min at 15 psi pressure–121°C. Cool to room temperature.

$Na_2S·9H_2O$ Solution:

Composition per 10.0mL:

$Na_2S·9H_2O$	0.5g

Preparation of $Na_2S·9H_2O$ Solution: Add $Na_2S·9H_2O$ to distilled/deionized water and bring volume to 10.0mL. Mix thoroughly. Autoclave under 100% N_2 for 15 min at 15 psi pressure–121°C. Cool to room temperature.

Selenite/Tungstate Solution:

Composition per liter:

NaOH	0.5g
$Na_2WO_4·2H_2O$	4.0mg
$Na_2SeO_3·5H_2O$	3.0mg

Preparation of Selenite/Tungstate Solution: Add components to distilled/deionized water and bring volume to 1.0L. Mix thoroughly. Sparge with 100% N_2. Autoclave for 15 min at 15 psi pressure–121°C.

Vitamin Solution:

Composition per liter:

Pyridoxine-HCl	10.0mg
Thiamine-HCl·2H₂O	5.0mg
Riboflavin	5.0mg
Nicotinic acid	5.0mg
D-Ca-pantothenate	5.0mg
p-Aminobenzoic acid	5.0mg
Lipoic acid	5.0mg
Biotin	2.0mg
Folic acid	2.0mg
Vitamin B_{12}	0.1mg

Preparation of Vitamin Solution: Add components to distilled/deionized water and bring volume to 1.0L. Mix thoroughly. Sparge with 80% H_2 + 20% CO_2. Filter sterilize.

Trace Elements Solution SL-10:

Composition per liter:

$FeCl_2·4H_2O$	1.5g
$CoCl_2·6H_2O$	190.0mg
$MnCl_2·4H_2O$	100.0mg
$ZnCl_2$	70.0mg
$Na_2MoO_4·2H_2O$	36.0mg
$NiCl_2·6H_2O$	24.0mg
H_3BO_3	6.0mg
$CuCl_2·2H_2O$	2.0mg
HCl (25% solution)	10.0mL

Preparation of Trace Elements Solution SL-10: Add $FeCl_2·4H_2O$ to 10.0mL of HCl solution. Mix thoroughly. Add distilled/deionized water and bring volume to 1.0L. Add remaining components. Mix thoroughly. Sparge with 100% N_2. Autoclave for 15 min at 15 psi pressure–121°C.

Preparation of Medium: Add components, except bicarbonate solution, sulfide solution, and vitamin solution, to distilled/deionized water and bring volume to 970.0mL. Mix thoroughly. Gently heat and bring to boiling. Boils for several minutes. Cool to room temperature while sparging with 80% N_2 + 20% CO_2. Distribute into screw-capped tubes or bottles under an atmosphere of 80% N_2 + 20% CO_2. Autoclave for 15 min at 15 psi pressure–121°C. Add the bicarbonate solution, sulfide solution, and vitamin solution. Adjust the final pH to 7.0.

Use: For the cultivation of *Calditerrivibrio* spp.

Caldivirga Medium (DSMZ Medium 883)

Composition per liter:

$(NH_4)_2SO_4$	1.3g
Sulfur, powdered	10.0g
$Na_2S·9H_2O$	0.5 g
KH_2PO_4	0.28g
$MgSO_4·7H_2O$	0.25g
$CaCl_2·2H_2O$	0.07g
$FeCl_3·6H_2O$	0.02g
$Na_2B_4O_7·10H_2O$	4.5mg
$MnCl_2·4H_2O$	1.8mg
Resazurin	0.5mg
$ZnSO_4·7H_2O$	0.22mg
$CuCl_2·2H_2O$	0.05mg
$Na_2MoO_4·2H_2O$	0.03mg
$VOSO_4·2H_2O$	0.03mg
$CoSO_4$	0.01mg
$Na_2S·9H_2O$ solution	7.5mL
Yeast extract solution	5.0mL
Na_3-citrate·$2H_2O$	3.0mL
Vitamin solution	1.0mL

pH 4.0± 0.2 at 25°C

$Na_2S·9H_2O$ Solution:

Composition per 10.0mL:

$Na_2S·9H_2O$	0.3g

Preparation of $Na_2S·9H_2O$ Solution: Add $Na_2S·9H_2O$ to distilled/deionized water and bring volume to 10.0mL. Mix thoroughly. Autoclave under 100% N_2 for 15 min at 15 psi pressure–121°C. Cool to room temperature.

Yeast Extract Solution:

Composition per 10.0mL:

Yeast extract 1.0g

Preparation of Yeast Extract Solution: Add yeast extract to distilled/deionized water and bring volume to 10.0mL. Mix thoroughly. Sparge with 100% N_2. Autoclave under 100% N_2 for 15 min at 15 psi pressure–121°C. Cool to room temperature.

Vitamin Solution:

Composition per liter:

Pyridoxine-HCl 10.0mg
Thiamine-HCl·$2H_2O$ 5.0mg
Riboflavin 5.0mg
Nicotinic acid 5.0mg
D-Ca-pantothenate 5.0mg
p-Aminobenzoic acid 5.0mg
Lipoic acid 5.0mg
Biotin 2.0mg
Folic acid 2.0mg
Vitamin B_{12} 0.1mg

Preparation of Vitamin Solution: Add components to distilled/deionized water and bring volume to 1.0L. Mix thoroughly. Sparge with 80% H_2 + 20% CO_2. Filter sterilize.

Preparation of Medium: Add components, except sulfur, vitamin solution, yeast extract solution, and $Na_2S·9H_2O$ solution, to distilled/deionized water and bring volume to 1.0L. Mix thoroughly. Adjust pH to 3.5 with 4*N* H_2SO_4. Sparge with 100% N_2 for 30 min. Distribute into anaerobe tubes or bottles containing sulfur, 50.0mg sulfur per 5.0mL medium. Autoclave for 20 min at 105°C. Cool to 25°C. Before use inject for every 10.0mL medium, 0.05mL sterile vitamin solution, 0.05mL sterile yeast extract, and 0.075mL sterile $Na_2S·9H_2O$ solution. Mix thoroughly. Final pH of the completed medium is around 4.0. After inoculation, pressurize the tubes to 100 kPa overpressure with 80% H_2 + 20% CO_2 gas mixture.

Use: For the cultivation of *Caldivirga maquilingensis*.

Caloramator proteoclasticus Medium (DSMZ Medium 788)

Composition per liter:

NH_4Cl 2.40g
$Na_2HPO_4·2H_2O$ 0.522g
KH_2PO_4 .... 0.48g
$MgCl_2·6H_2O$ 0.40g
Yeast extract 0.20g
Resazurin 0.5mg
$NaHCO_3$ solution 40.0mL
$Na_2S·9H_2O$ solution 10.0mL
Calcium chloride solution 10.0mL
Glucose solution 10.0mL
Seven vitamin solution 1.0mL
Trace elements solution SL-10 .... 1.0mL
Selenite-tungstate solution 1.0mL

pH 7.3 ± 0.2 at 25°C

Glucose Solution:

Composition per 10.0mL:

Glucose 5.0g

Preparation of Glucose Solution: Add glucose to distilled/deionized water and bring volume to 10.0mL. Mix thoroughly. Sparge with 100% N_2. Filter sterilize.

Calcium Chloride Solution:

Composition per 10.0mL:

$CaCl_2·2H_2O$ 0.2g

Preparation of Calcium Chloride Solution: Add $CaCl_2·2H_2O$ to distilled/deionized water and bring volume to 10.0mL. Mix thoroughly. Sparge with 100% N_2. Autoclave for 15 min at 15 psi pressure–121°C. Cool to 25°C.

Selenite-Tungstate Solution:

Composition per liter:

NaOH 0.5g
$Na_2WO_4·2H_2O$ 4.0mg
$Na_2SeO_3·5H_2O$ 3.0mg

Preparation of Selenite-Tungstate Solution: Add components to distilled/deionized water and bring volume to 1.0L. Mix thoroughly. Sparge with 100% N_2. Filter sterilize.

Seven Vitamin Solution:

Composition per liter:

Pyridoxine hydrochloride 300.0mg
Thiamine-HCl·$2H_2O$ 200.0mg
Nicotinic acid 200.0mg
Vitamin B_{12} 100.0mg
Calcium pantothenate 100.0mg
p-Aminobenzoic acid 80.0mg
D(+)-Biotin 20.0mg

Preparation of Seven Vitamin Solution: Add components to distilled/deionized water and bring volume to 1.0L. Sparge with 100% N_2. Mix thoroughly. Filter sterilize.

$NaHCO_3$ Solution:

Composition per 100.0mL:

$NaHCO_3$ 10.0g

Preparation of $NaHCO_3$ Solution: Add $NaHCO_3$ to distilled/deionized water and bring volume to 100.0mL. Mix thoroughly. Sparge with 80% N_2 + 20% CO_2. Filter sterilize.

$Na_2S·9H_2O$ Solution:

Composition per 10.0mL:

$Na_2S·9H_2O$ 0.3g

Preparation of $Na_2S·9H_2O$ Solution: Add $Na_2S·9H_2O$ to distilled/deionized water and bring volume to 10.0mL. Mix thoroughly. Autoclave under 100% N_2 for 15 min at 15 psi pressure–121°C. Cool to room temperature.

Trace Elements Solution SL-10:

Composition per liter:

$FeCl_2·4H_2O$ 1.5g
$CoCl_2·6H_2O$ 190.0mg
$MnCl_2·4H_2O$ 100.0mg
$ZnCl_2$ 70.0mg
$Na_2MoO_4·2H_2O$ 36.0mg
$NiCl_2·6H_2O$ 24.0mg
H_3BO_3 .... 6.0mg
$CuCl_2·2H_2O$ 2.0mg
HCl (25% solution) 10.0mL

Preparation of Trace Elements Solution SL-10: Add $FeCl_2·4H_2O$ to 10.0mL of HCl solution. Mix thoroughly. Add distilled/deionized water and bring volume to 1.0L. Add remaining components. Mix thoroughly. Sparge with 80% N_2 + 20% CO_2.

Preparation of Medium: Prepare and dispense medium under 80% N_2 + 20% CO_2. Add components, except seven vitamin solution,

$NaHCO_3$ solution, calcium chloride solution, glucose solution, and $Na_2S{\cdot}9H_2O$ solution, to distilled/deionized water and bring volume to 929.0mL. Mix thoroughly. Sparge with N_2 + 20% CO_2. Autoclave for 15 min at 15 psi pressure–121°C. Cool to 25°C. Aseptically and anaerobically add 1.0mL sterile seven vitamin solution, 40.0mL of sterile $NaHCO_3$ solution, 10.0mL sterile glucose solution, 10.0mL sterile calcium chloride solution, and 10.0mL of sterile $Na_2S{\cdot}9H_2O$ solution. Mix thoroughly. Adjust pH to 7.2–7.4. Aseptically and anaerobically distribute into sterile tubes or flasks.

Use: For the cultivation of *Caloramator proteoclasticus.*

Caloramator viterbensis Medium (DSMZ Medium 947)

Composition per liter:

Glycerol	3.0g
KH_2PO_4	2.0g
$NaHCO_3$	1.0g
$(NH_4)_2SO_4$	0.5g
NH_4Cl	0.5g
Yeast extract	0.3g
$MgCl_2{\cdot}6H_2O$	0.04g
$CaCl_2{\cdot}2H_2O$	0.04g
Resazurin	0.5mg
Cysteine solution	10.0mL
$Na_2S{\cdot}9H_2O$ solution	10.0mL
Trace elements solution SL-10	1.0mL
Selenite-tungstate solution	1.0mL
Vitamin solution	0.2mL

pH 6.3 ± 0.2 at 25°C

Vitamin Solution:

Composition per liter:

Pyridoxine-HCl	10.0mg
Thiamine-HCl·$2H_2O$	5.0mg
Riboflavin	5.0mg
Nicotinic acid	5.0mg
D-Ca-pantothenate	5.0mg
p-Aminobenzoic acid	5.0mg
Lipoic acid	5.0mg
Biotin	2.0mg
Folic acid	2.0mg
Vitamin B_{12}	0.1mg

Preparation of Vitamin Solution: Add components to distilled/deionized water and bring volume to 1.0L. Mix thoroughly. Sparge with 80% H_2 + 20% CO_2. Filter sterilize.

Selenite-Tungstate Solution:

Composition per liter:

NaOH	0.5g
$Na_2WO_4{\cdot}2H_2O$	4.0mg
$Na_2SeO_3{\cdot}5H_2O$	3.0mg

Preparation of Selenite-Tungstate Solution: Add components to distilled/deionized water and bring volume to 1.0L. Mix thoroughly. Sparge with 100% N_2. Filter sterilize.

$Na_2S{\cdot}9H_2O$ Solution:

Composition per 100.0mL:

$Na_2S{\cdot}9H_2O$	1.3g

Preparation of $Na_2S{\cdot}9H_2O$ Solution: Add $Na_2S{\cdot}9H_2O$ to distilled/deionized water and bring volume to 100.0mL. Mix thoroughly. Sparge with 100% N_2. Autoclave for 15 min at 15 psi pressure–121°C.

Trace Elements Solution SL-10:

Composition per liter:

$FeCl_2{\cdot}4H_2O$	1.5g
$CoCl_2{\cdot}6H_2O$	190.0mg
$MnCl_2{\cdot}4H_2O$	100.0mg
$ZnCl_2$	70.0mg
$Na_2MoO_4{\cdot}2H_2O$	36.0mg
$NiCl_2{\cdot}6H_2O$	24.0mg
H_3BO_3	6.0mg
$CuCl_2{\cdot}2H_2O$	2.0mg
HCl (25% solution)	10.0mL

Preparation of Trace Elements Solution SL-10: Add $FeCl_2{\cdot}4H_2O$ to 10.0mL of HCl solution. Mix thoroughly. Add distilled/deionized water and bring volume to 1.0L. Add remaining components. Mix thoroughly. Sparge with 100% N_2. Autoclave for 15 min at 15 psi pressure–121°C.

Cysteine Solution:

Composition per 100.0mL:

L-Cysteine·HCl·H_2O	1.3g

Preparation of Cysteine Solution: Add L-cysteine·HCl·H_2O to distilled/deionized water and bring volume to 100.0mL. Mix thoroughly. Sparge with 100% N_2. Autoclave for 15 min at 15 psi pressure–121°C.

Preparation of Medium: Prepare and dispense medium under 80% N_2 + 20% CO_2 gas atmosphere. Add components, except $NaHCO_3$, $Na_2S{\cdot}9H_2O$ solution, and cysteine solution, to distilled/deionized water and bring volume to 980.0mL. Mix thoroughly. Gently heat and bring to boiling. Cool to room temperature while sparging with 80% N_2 + 20% CO_2. Add solid $NaHCO_3$. Adjust pH to 6.3 by equilibrating with the same gas mixture. Distribute to tubes or bottles. Autoclave for 15 min at 15 psi pressure–121°C. Aseptically and anaerobically add 10.0mL sterile cysteine solution and 10.0mL sterile $Na_2S{\cdot}9H_2O$ solution. Mix thoroughly. Adjust pH to 6.8. Aseptically and anaerobically distribute into sterile tubes or bottles.

Use: For the cultivation of *Caloramator viterbiensis.*

Calymmatobacterium granulomatis Semidefined Medium

Composition per liter:

Papaic digest of soybean meal	20.0g
NaCl	2.5g
K_2HPO_4	1.5g
Sodium thioglycolate	0.6g
L-Cystine	0.4g

pH 7.2 ± 0.2 at 25°C

Preparation of Medium: Add components to distilled/deionized water and bring volume to 1.0L. Mix thoroughly. Adjust pH to 7.2. Distribute into screw-capped tubes in 20–22mL volumes. Autoclave for 15 min at 15 psi pressure–121°C. Tighten screw caps.

Use: For the cultivation of *Calymmatobacterium granulomatis.*

CAMG Broth

Composition per liter:

K_2HPO_4	5.0g
Pancreatic digest of casein	5.0g
Yeast extract	5.0g
Glucose	2.0g
Tween™ 80	0.5mL

pH 7.0 ± 0.1 at 25°C

Preparation of Medium: Prepare and dispense anaerobically under an atmosphere of 80% N_2 + 10% CO_2 + 10% H_2. Add components to distilled/deionized water and bring volume to 1.0L. Mix thoroughly. Sparge with a gas mixture of 80% N_2 + 10% CO_2 + 10% H_2. Adjust pH to 7.5. Distribute into tubes or flasks. Autoclave for 15 min at 15 psi pressure–121°C.

Use: For the cultivation of *Actinomyces naeslundii.*

Caminicella Medium (DSMZ Medium 964)

Composition per liter:

Sea salt	30.0g
Sulfur, powdered	12.0g
PIPES buffer	6.05g
D(+)-glucose	5.0g
Peptone	1.0g
Yeast extract	0.5g
Resazurin	0.5mg
$Na_2S \cdot 9H_2O$ solution	10.0mL

pH 7.5 ± 0.2 at 25°C

$Na_2S \cdot 9H_2O$ Solution:

Composition per 10.0mL:

$Na_2S \cdot 9H_2O$	0.5g

Preparation of $Na_2S \cdot 9H_2O$ Solution: Add $Na_2S \cdot 9H_2O$ to distilled/deionized water and bring volume to 10.0mL. Mix thoroughly. Sparge with 100% N_2. Autoclave for 15 min at 15 psi pressure–121°C.

Preparation of Medium: Add components, except $Na_2S \cdot 9H_2O$ solution, to distilled/deionized water and bring volume to 990.0mL. Mix thoroughly. Gently heat and bring to boiling. Boil for 3 min. Cool to 25°C while sparging with 80% N_2 + 20% CO_2. Adjust pH to 7.2–7.4. Distribute to anaerobe tubes or bottles under 80% N_2 + 20% CO_2. Autoclave for 15 min at 15 psi pressure–121°C. Aseptically and anaerobically add per liter of medium, 10.0mL sterile $Na_2S \cdot 9H_2O$ solution. Mix thoroughly. The final pH should be 7.5.

Use: For the cultivation of *Caminicella sporogenes.*

Camphor Minimal Medium

Composition per liter:

Agar	20.0g
K_2HPO_4	4.4g
NH_4Cl	2.1g
KH_2PO_4	1.7g
100× salt solution	10.0mL

100X Salt Solution:

Composition per liter:

$MgSO_4$	19.5g
$FeSO_4 \cdot 7H_2O$	5.0g
$MnSO_4 \cdot H_2O$	5.0g
Ascorbic acid	1.0g
$CaCl_2 \cdot 2H_2O$	0.3g

Preparation of 100X Salt Solution: Add components to distilled/deionized water and bring volume to 1.0L. Mix thoroughly.

Preparation of Medium: Add components to distilled/deionized water and bring volume to 1.0L. Gently heat and bring to boiling. Autoclave for 15 min at 15 psi pressure–121°C. Pour into sterile Petri dishes. Allow to cool to room temperature. Invert Petri dishes. Spread 0.2mL of 2*M* D-(+) camphor solution in methylene chloride (CH_2Cl_2) on the inside cover of each plate.

Use: For the cultivation and maintenance of *Pseudomonas putida.*

Campy BAP Medium

See: ***Campylobacter*** **Selective Medium, Blaser-Wang**

Campy Cefex Agar

See: ***Campylobacter*** **Isolation Agar B**

Campy THIO Medium

Composition per liter:

Pancreatic digest of casein	20.0g
Agar	15.0g
NaCl	2.5g
K_2HPO_4	1.5g
Sodium thioglycolate	0.6g
L-Cystine	0.4g
Na_2SO_3	0.2g
Antibiotic supplement	10.0mL

Antibiotic Supplement:

Composition per 10.0mL:

Cephalothin	15.0mg
Vancomycin	10.0mg
Trimethoprim	5.0mg
Amphotericin B	2.0mg
Polymyxin B	2500U

Preparation of Antibiotic Supplement: Add components to 10.0mL of distilled/deionized water. Filter sterilize.

Preparation of Medium: Add components, except antibiotic solution, to distilled/deionized water and bring volume to 990.0mL. Mix thoroughly. Gently heat and bring to boiling. Autoclave for 15 min at 15 psi pressure–121°C. Cool to 45°–50°C. Aseptically add 10.0mL of sterile antibiotic solution. Mix thoroughly. Aseptically distribute into sterile screw-capped tubes in 3.0mL volumes for 1.5cm swabs or 5.0mL volumes for 3.0cm swabs.

Use: For the maintenance—as a holding or transport medium—of *Campylobacter* species isolated from clinical specimens on swabs.

Campylo Thioglycollate HiVeg Medium Base with Selective Supplement

Composition per liter:

Plant hydrolysate	20.0g
NaCl	2.5g
Agar	1.6g
K_2HPO_4	1.5g
Na-thioglycollate	0.6g
L-Cystine	0.4g
Na_2SO_3	0.2g
Selective supplement	10.0mL

pH 7.3 ± 0.2 at 25°C

Source: This medium, without selective supplement, is available as a premixed powder from HiMedia.

Selective Supplement:

Composition per 10.0mL:

Cephalothin	15.0mg
Vancomycin	10.0mg
Trimethoprim	5.0mg
Amphotericin B	2.0mg
Polymyxin B sulfate	2500 U

Preparation of Selective Supplement: Add components to distilled/deionized water and bring volume to 10.0mL. Mix thoroughly. Filter sterilize.

Preparation of Medium: Add components, except selective supplement and sodium deoxycholate solution, to distilled/deionized water and bring volume to 990.0mL. Mix thoroughly. Heat with frequent agitation and boil for 1 min to completely dissolve. Autoclave for 15 min at 15 psi pressure–121°C. Cool to 50°–55°C. Add 10.0mL of sterile selective supplement. Mix thoroughly.

Use: For the cultivation of *Campylobacter* spp.

Campylobacter Agar

Composition per liter:

Agar	15.0g
Polypeptone™	10.0g
Meat extract	5.0g
NaCl	5.0g
Yeast extract	5.0g
Sodium L-glutamate	2.0g
Sodium succinate·$6H_2O$	2.0g
$MgCl_2 \cdot 6H_2O$	1.0g
Sheep blood, defibrinated	50.0mL

pH 7.0 ± 0.2 at 25°C

Preparation of Medium: Add components, except sheep blood, to distilled/deionized water and bring volume to 950.0mL. Mix thoroughly. Gently heat and bring to boiling. Adjust pH to 7.0. Autoclave for 15 min at 15 psi pressure–121°C. Cool to 50°–55°C. Aseptically add 50.0mL of sterile defibrinated sheep blood. Mix thoroughly. Pour into sterile Petri dishes or distribute into sterile tubes.

Use: For the cultivation of *Moraxella lincolnii.*

Campylobacter Agar

Composition per liter:

NaCl	150.0g
Agar	20.0g
Yeast extract	10.0g
Glucose	4.0g
KNO_3	2.0g

pH 7.0 ± 0.2 at 25°C

Preparation of Medium: Add components to distilled/deionized water and bring volume to 1.0L. Mix thoroughly. Gently heat and bring to boiling. Distribute into tubes or flasks. Autoclave for 15 min at 15 psi pressure–121°C. Pour into sterile Petri dishes or leave in tubes.

Use: For the cultivation of *Arhodomonas aquaeolei.*

Campylobacter Agar with 5 Antimicrobics and 10% Sheep Blood

Composition per liter:

Agar	15.0g
Pancreatic digest of casein	10.0g
Peptic digest of animal tissue	10.0g
NaCl	5.0g
Yeast extract	2.0g
Glucose	1.0g
$NaHSO_3$	0.1g
Sheep blood, defibrinated	100.0mL
Antibiotic supplement	10.0mL

pH 7.2 ± 0.2 at 25°C

Source: This medium is available as a prepared medium from BD Diagnostic Systems.

Antibiotic Supplement:

Composition per 10.0mL:

Cephalothin	0.015g
Vancomycin	0.01g
Trimethoprim	5.0mg
Amphotericin B	2.0mg
Polymyxin B	2500U

Preparation of Antibiotic Supplement: Add components to 10.0mL of distilled/deionized water. Filter sterilize.

Preparation of Medium: Add components, except sheep blood and antibiotic solution, to distilled/deionized water and bring volume to 890.0mL. Mix thoroughly. Gently heat and bring to boiling. Autoclave for 15 min at 15 psi pressure–121°C. Cool to 45°–50°C. Aseptically add 100.0mL of sterile sheep blood and 10.0mL of sterile antibiotic solution. Mix thoroughly. Pour into sterile Petri dishes or distribute into sterile tubes.

Use: For the primary selective isolation and cultivation of *Campylobacter jejuni* from human fecal specimens.

Campylobacter Agar, Blaser's (Blaser's *Campylobacter* Agar)

Composition per liter:

Campylobacter agar base	990.0mL
Supplement B	10.0mL

pH 7.4 ± 0.2 at 25°C

Campylobacter Agar Base:

Composition per liter:

Proteose peptone	15.0g
Agar	12.0g
NaCl	5.0g
Yeast extract	5.0g
Liver digest	2.5g

Source: *Campylobacter* agar base and *Campylobacter* antimicrobic supplement B are available as a premixed powder from BD Diagnostic Systems.

Preparation of *Campylobacter* Agar Base: Add components to distilled/deionized water and bring volume to 990.0mL. Mix thoroughly. Gently heat and bring to boiling. Autoclave for 15 min at 15 psi pressure–121°C. Cool to 45°–50°C.

Supplement B:

Composition per 10.0mL:

Cephalothin	15.0mg
Vancomycin	10.0mg
Trimethoprim	5.0mg
Amphotericin B	2.0mg
Polymyxin B	2500U

Preparation of Supplement B: Add components to 10.0mL of distilled/deionized water. Filter sterilize.

Preparation of Medium: Prepare 990.0mL of *Campylobacter* agar base. Autoclave and cool to 45°–50°C. Aseptically add 10.0mL of sterile supplement B. Mix thoroughly. Pour into sterile Petri dishes.

Use: For the selective isolation of *Campylobacter jejuni* from fecal specimens, food, and environmental specimens.

Campylobacter Agar, Skirrow's (Skirrow's *Campylobacter* Agar)

Composition per liter:

Campylobacter agar base	990.0mL
Supplement S	10.0mL

pH 7.4 ± 0.2 at 25°C

***Campylobacter* Agar Base:**

Composition per liter:

Proteose peptone	15.0g
Agar	12.0g
NaCl	5.0g
Yeast extract	5.0g
Liver digest	2.5g

Source: *Campylobacter* agar base and *Campylobacter* antimicrobic supplement S are available as a premixed powder from BD Diagnostic Systems.

Preparation of *Campylobacter* Agar Base: Add components to distilled/deionized water and bring volume to 990.0mL. Mix thoroughly. Gently heat and bring to boiling. Autoclave for 15 min at 15 psi pressure–121°C. Cool to 45°–50°C.

Supplement S:

Composition per 10.0mL:

Vancomycin	10.0mg
Trimethoprim	5.0mg
Polymyxin B	2500U

Preparation of Supplement S: Add components to 10.0mL of distilled/deionized water. Filter sterilize.

Preparation of Medium: Prepare 990.0mL of *Campylobacter* agar base. Autoclave and cool to 45°–50°C. Aseptically add 10.0mL of sterile supplement S. Mix thoroughly. Pour into sterile Petri dishes.

Use: For the selective isolation of *Campylobacter jejuni* from fecal specimens, food, and environmental specimens.

Campylobacter Blood-Free Agar Base, Modified (CCDA, Modified)

Composition per liter:

Peptone	20.0g
Agar	12.0g
NaCl	5.0g
Activated charcoal	4.0g
Casein hydrolysate	3.0g
Na-desoxycholate	1.0g
Na-pyruvate	0.25g
$FeSO_4$	0.25g
Cefoperazone-amphotericin B solution	10.0mL

pH 7.4 ± 0.2 at 25°C

Cefoperazone-Amphotericin B Solution:

Composition per 10.0mL:

Cefoperazone	0.016g
Amphotericn B	0.005g

Preparation of Cefoperazone-Amphotericin B Solution: Add cefoperazone and amphotericin B to distilled/deionized water and bring volume to 10.0mL. Mix thoroughly. Filter sterilize.

Preparation of Medium: Add components except cefoperazone-amphotericin B solution to distilled/deionized water and bring volume to 990.0mL. Mix thoroughly. Autoclave for 15 min at 15 psi pressure–121°C. Cool to 45°–50°C. Aseptically add 10.0mL of cefoperazone-amphotericin B solution. Mix thoroughly. Pour into sterile Petri dishes or aseptically distribute into tubes or flasks.

Use: For the isolation of *Campylobacter* spp. from foods.The use of *Campylobacter* Blood-Free Selective Agar is specified by the UK Ministry of Agriculture, Fisheries and Food (MAFF) in a validated method for isolation of *Campylobacter* from foods. Amphotericin largely reduces the growth of yeasts and molds. Cefoperazone especially inhibits Enterobacteriaceae.

Campylobacter Blood-Free Selective Agar

Composition per liter:

Agar	12.0g
Beef extract	10.0g
Peptone	10.0g
Charcoal	4.0g
Casein hydrolysate	3.0g
Sodium deoxycholate	1.0g
$Fe_2SO_4 \cdot H_2O$	0.25g
Sodium pyruvate	0.25g
Cefoperazone solution	10.0mL

pH 7.4 ± 0.2 at 25°C

Cefoperazone Solution:

Composition per 10.0mL:

Sodium cefoperazone	0.032g

Preparation of Cefoperazone Solution: Add sodium cefoperazone to distilled/deionized water and bring volume to 10.0mL. Mix thoroughly. Filter sterilize.

Preparation of Medium: Add components, except cefoperazone solution, to distilled/deionized water and bring volume to 990.0mL. Mix thoroughly. Heat with frequent agitation and boil for 1 min to completely dissolve. Autoclave for 15 min at 15 psi pressure–121°C. Cool to 50°–55°C. Add 10.0mL of sterile cefoperazone solution. Addition of 10.0mg/L of amphotericin B improves the selectivity of the medium. Mix thoroughly. Pour into sterile Petri dishes.

Use: For the selective isolation of *Campylobacter* species, especially *Campylobacter jejuni, Campylobacter coli*, and *Campylobacter laridis*.

Campylobacter Charcoal Differential Agar (CCDA) (Preston Blood-Free Medium)

Composition per liter:

Agar	12.0g
Beef extract	10.0g
Peptone	10.0g
NaCl	5.0g
Charcoal	4.0g
Casein hydrolysate	3.0g
Sodium deoxycholate	1.0g
$FeSO_4$	0.25g
Sodium pyruvate	0.25g
Cefoperazone solution	10.0mL

pH 7.5 ± 0.2 at 25°C

Cefoperazone Solution:

Composition per 10.0mL:

Sodium cefoperazone	0.032g

Preparation of Cefoperazone Solution: Add sodium cefoperazone to distilled/deionized water and bring volume to 10.0mL. Mix thoroughly. Filter sterilize.

Preparation of Medium: Add components, except cefoperazone solution, to distilled/deionized water and bring volume to 990.0mL. Mix thoroughly. Gently heat and bring to boiling. Autoclave for 15 min at 15 psi pressure–121°C. Cool to 45°–50°C. Aseptically add 10.0mL of sterile cefoperazone solution. Mix thoroughly. Pour into sterile Petri dishes or distribute into sterile tubes.

Use: For the cultivation of *Campylobacter* species.

Campylobacter Enrichment Broth

Composition per liter:

Beef extract	10.0g
Peptone	10.0g
Yeast extract	6.0g
NaCl	5.0g
Horse blood, laked	50.0mL
FBP solution	4.0mL
Antibiotic solution	4.0mL

pH 7.5 ± 0.2 at 25°C

Horse Blood, Laked:

Composition per 50.0mL:

Horse blood, fresh	50.0mL

Preparation of Horse Blood, Laked: Add blood to a sterile polypropylene bottle. Freeze overnight at –20°C. Thaw at 8°C. Refreeze at –20°C. Thaw again at 8°C.

FBP Solution:

Composition per 100.0mL:

$FeSO_4$	6.25g
$NaHSO_3$	6.25g
Sodium pyruvate	6.25g

Preparation of FBP Solution: Add components to distilled/deionized water and bring volume to 100.0mL. Mix thoroughly. Filter sterilize.

Antibiotic Solution:

Composition per 10.0mL:

Cycloheximide	0.1g
Sodium cefoperazone	0.03g
Trimethoprim lactate	0.0125g
Rifampicin	0.01g

Preparation of Antibiotic Solution: Add components to distilled/deionized water and bring volume to 10.0mL. Mix thoroughly. Filter sterilize.

Caution: Cycloheximide is toxic. Avoid skin contact or aerosol formation and inhalation.

Preparation of Medium: Add components—except laked horse blood, FBP solution, and antibiotic solution—to distilled/deionized water and bring volume to 942.0mL. Mix thoroughly. Gently heat and bring to boiling. Autoclave for 15 min at 15 psi pressure–121°C. Cool to 45°–50°C. Aseptically add 50.0mL of sterile laked horse blood, 4.0mL of FBP solution, and 4.0mL of antibiotic solution. Mix thoroughly. Pour into sterile Petri dishes or distribute into sterile tubes.

Use: For the isolation and cultivation of *Campylobacter* species from dairy products.

Campylobacter Enrichment Broth (FDA Medium M29)

Composition per 1024.0mL:

Basal medium	950.0mL
Horse blood, lysed	50.0mL
Cefoperazone solution	8.0mL
FBP solution	4.0mL
Trimethoprim lactate solution	4.0mL
Vancomycin solution	4.0mL
Cycloheximide solution	4.0mL

pH 7.5 ± 0.2 at 25°C

Basal Medium:

Composition per 950.0mL:

Beef extract	10.0g
Peptone	10.0g
Yeast extract	6.0g
NaCl	5.0g

Preparation of Basal Medium: Add components to distilled/deionized water and bring volume to 950.0mL. Mix thoroughly. Autoclave for 15 min at 15 psi pressure–121°C. Cool to 45°–50°C.

FBP Solution:

Composition per 100.0mL:

$FeSO_4 \cdot 7H_2O$	6.25g
$Na_2S_2O_5$	6.25g
Sodium pyruvate	6.25g

Preparation of FBP: Add components to distilled/deionized water and bring volume to 100.0mL. Mix thoroughly. Filter sterilize.

Cefoperazone Solution:

Composition per 10.0mL:

Cefoperazone	0.037g

Preparation of Cefoperazone Solution: Add cefoperazone to distilled/deionized water and bring volume to 10.0mL. Mix thoroughly. Filter sterilize.

Trimethoprim Lactate Solution:

Composition per 10.0mL:

Trimethoprim lactate	0.031g

Preparation of Trimethoprim Lactate Solution: Add trimethoprim lactate to distilled/deionized water and bring volume to 10.0mL. Mix thoroughly. Filter sterilize.

Vancomycin Solution:

Composition per 10.0mL:

Vancomycin	0.025g

Preparation of Vancomycin Solution: Add vancomycin to distilled/deionized water and bring volume to 10.0mL. Mix thoroughly. Filter sterilize.

Cycloheximide Solution:

Composition per 10.0mL:

Cycloheximide	0.025g

Preparation of Cycloheximide Solution: Add cycloheximide to distilled/deionized water and bring volume to 10.0mL. Mix thoroughly. Filter sterilize.

Caution: Cycloheximide is toxic. Avoid skin contact or aerosol formation and inhalation.

Preparation of Medium: To 950.0mL of cooled sterile basal medium, aseptically add 50.0mL of lysed (fresh, frozen, and thawed) horse blood, 4.0mL of sterile FBP solution, 8.0mL of sterile cefoperazone solution, 4.0mL of sterile trimethoprim lactate solution, 4.0mL of sterile vancomycin solution, and 4.0mL of sterile cycloheximide solution. Mix thoroughly. Aseptically distribute into sterile screw-capped tubes or bottles. Close caps tightly to reduce O_2 absorption. Use within 2 weeks.

Use: For the selective isolation and cultivation of *Campylobacter* species.

Campylobacter Enrichment Broth (FDA Medium M29)

Composition per 1020.0mL:

Basal medium	950.0mL
Horse blood, lysed	50.0mL
FBP solution	4.0mL
Cefoperazone solution	4.0mL
Trimethoprim lactate solution	4.0mL
Vancomycin solution	4.0mL
Cycloheximide solution	4.0mL

pH 7.5 ± 0.2 at 25°C

Basal Medium:

Composition per 950.0mL:

Beef extract	10.0g
Peptone	10.0g
Yeast extract	6.0g
NaCl	5.0g

Preparation of Basal Medium: Add components to distilled/deionized water and bring volume to 950.0mL. Mix thoroughly. Autoclave for 15 min at 15 psi pressure–121°C. Cool to 45°–50°C.

FBP Solution:

Composition per 100.0mL:

$FeSO_4 \cdot 7H_2O$	6.25g
$Na_2S_2O_5$	6.25g
Sodium pyruvate	6.25g

Preparation of FBP: Add components to distilled/deionized water and bring volume to 100.0mL. Mix thoroughly. Filter sterilize.

Cefoperazone Solution:

Composition per 10.0mL:

Cefoperazone	0.037g

Preparation of Cefoperazone Solution: Add cefoperazone to distilled/deionized water and bring volume to 10.0mL. Mix thoroughly. Filter sterilize.

Trimethoprim Lactate Solution:

Composition per 10.0mL:

Trimethoprim lactate	0.031g

Preparation of Trimethoprim Lactate Solution: Add trimethoprim lactate to distilled/deionized water and bring volume to 10.0mL. Mix thoroughly. Filter sterilize.

Vancomycin Solution:

Composition per 10.0mL:

Vancomycin	0.025g

Preparation of Vancomycin Solution: Add vancomycin to distilled/deionized water and bring volume to 10.0mL. Mix thoroughly. Filter sterilize.

Cycloheximide Solution:

Composition per 10.0mL:

Cycloheximide	0.025g

Preparation of Cycloheximide Solution: Add cycloheximide to distilled/deionized water and bring volume to 10.0mL. Mix thoroughly. Filter sterilize.

Caution: Cycloheximide is toxic. Avoid skin contact or aerosol formation and inhalation.

Preparation of Medium: To 950.0mL of cooled, sterile basal medium, aseptically add 50.0mL of lysed (fresh, frozen, and thawed) horse blood, 4.0mL of sterile FBP solution, 4.0mL of sterile cefoperazone solution, 4.0mL of sterile trimethoprim lactate solution, 4.0mL of sterile vancomycin solution, and 4.0mL of sterile cycloheximide solution. Mix thoroughly. Aseptically distribute into sterile screw-capped tubes or bottles. Close caps tightly to reduce O_2 absorption. Use within 2 weeks.

Use: For the selective isolation and cultivation of *Campylobacter* species.

Campylobacter Enrichment Broth (FDA Medium M29)

Composition per 1020.0mL:

Basal medium	950.0mL
Horse blood, lysed	50.0mL
FBP solution	4.0mL
Cefoperazone solution	4.0mL
Trimethoprim lactate solution	4.0mL
Rifampicin solution	4.0mL
Cycloheximide solution	4.0mL

pH 7.5 ± 0.2 at 25°C

Basal Medium:

Composition per 950.0mL:

Beef extract	10.0g
Peptone	10.0g
Yeast extract	6.0g
NaCl	5.0g

Preparation of Basal Medium: Add components to distilled/deionized water and bring volume to 950.0mL. Mix thoroughly. Autoclave for 15 min at 15 psi pressure–121°C. Cool to 45°–50°C.

FBP Solution:

Composition per 100.0mL:

$FeSO_4 \cdot 7H_2O$	6.25g
$Na_2S_2O_5$	6.25g
Sodium pyruvate	6.25g

Preparation of FBP Solution: Add components to distilled/deionized water and bring volume to 100.0mL. Mix thoroughly. Filter sterilize.

Cefoperazone Solution:

Composition per 10.0mL:

Cefoperazone	0.037g

Preparation of Cefoperazone Solution: Add cefoperazone to distilled/deionized water and bring volume to 10.0mL. Mix thoroughly. Filter sterilize.

Trimethoprim Lactate Solution:

Composition per 10.0mL:

Trimethoprim lactate	0.031g

Preparation of Trimethoprim Lactate Solution: Add trimethoprim lactate to distilled/deionized water and bring volume to 10.0mL. Mix thoroughly. Filter sterilize.

Rifampicin Solution:

Composition per 100.0mL:

Rifampicin	0.25g
Ethanol, absolute	50.0mL

Preparation of Rifampicin Solution: Add rifampicin to 50.0mL of ethanol. Mix thoroughly. Bring volume to 100.0mL with distilled/deionized water. Filter sterilize.

Cycloheximide Solution:

Composition per 10.0mL:

Cycloheximide	0.025g

Preparation of Cycloheximide Solution: Add cycloheximide to distilled/deionized water and bring volume to 10.0mL. Mix thoroughly. Filter sterilize.

Caution: Cycloheximide is toxic. Avoid skin contact or aerosol formation and inhalation.

Preparation of Medium: To 950.0mL of cooled sterile basal medium, aseptically add 50.0mL of lysed (fresh, frozen, and thawed) horse blood, 4.0mL of sterile FBP solution, 4.0mL of sterile cefoperazone solution, 4.0mL of sterile trimethoprim lactate solution, 4.0mL of sterile rifampicin solution, and 4.0mL of sterile cycloheximide solution. Mix thoroughly. Aseptically distribute into sterile screw-capped tubes or bottles. Close caps tightly to reduce O_2 absorption. Use within 2 weeks.

Use: For the selective isolation and cultivation of *Campylobacter* species from dairy products.

Campylobacter Enrichment Broth

Composition per liter:

Beef extract	10.0g
Peptone	10.0g
Yeast extract	6.0g
NaCl	5.0g
Horse blood, laked	50.0mL
FBP solution	4.0mL
Antibiotic solution	4.0mL

pH 7.5 ± 0.2 at 25°C

Horse Blood, Laked:

Composition per 50.0mL:

Horse blood, fresh	50.0mL

Preparation of Horse Blood, Laked: Add blood to a sterile polypropylene bottle. Freeze overnight at –20°C. Thaw at 8°C. Refreeze at –20°C. Thaw again at 8°C.

FBP Solution:

Composition per 100.0mL:

$FeSO_4$	6.25g
$NaHSO_3$	6.25g
Sodium pyruvate	6.25g

Preparation of FBP Solution: Add components to distilled/deionized water and bring volume to 100.0mL. Mix thoroughly. Filter sterilize.

Antibiotic Solution:

Composition per 10.0mL:

Cycloheximide	0.1g
Sodium cefoperazone	0.03g
Trimethoprim lactate	0.0125g
Vancomycin	0.01g

Preparation of Antibiotic Solution: Add components to distilled/deionized water and bring volume to 10.0mL. Mix thoroughly. Filter sterilize.

Caution: Cycloheximide is toxic. Avoid skin contact or aerosol formation and inhalation.

Preparation of Medium: Add components—except horse blood, FBP solution, and antibiotic solution—to distilled/deionized water and bring volume to 942.0mL. Mix thoroughly. Gently heat and bring to boiling. Autoclave for 15 min at 15 psi pressure–121°C. Cool to 45°–50°C. Aseptically add 50.0mL of sterile horse blood, 4.0mL of FBP solution, and 4.0mL of antibiotic solution. Mix thoroughly. Pour into sterile Petri dishes or distribute into sterile tubes.

Use: For the isolation and cultivation of *Campylobacter* species from foods.

Campylobacter Enrichment Broth (BAM M28a)

Composition per 1016.0mL:

Peptone	10.0g
Yeast extract	5.0g
Lactalbumin hydrolysate	5.0g
NaCl	5.0g
α-Ketoglutamic acid	1.0g
Sodium pyruvate	0.5g
Na_2CO_3	0.6g
$Na_2S_2O_5$	0.5g
Hemin	0.01g
Cefoperazone solution	4.0mL
Trimethoprim lactate solution	4.0mL
Vancomycin solution	4.0mL
Cycloheximide or amphotericin B solution	4.0mL

pH 7.4 ± 0.2 at 25°C

Cefoperazone Solution:

Composition per 10.0mL:

Cefoperazone	0.05g

Preparation of Cefoperazone Solution: Add cefoperazone to distilled/deionized water and bring volume to 10.0mL. Mix thoroughly. Filter sterilize. Can be stored for 5 days at 4°C, 14 days at –20°C, and 5 months at –70°C.

Trimethoprim Solution:

Composition per 10.0mL:

Trimethoprim lactate	0.066g

Preparation of Trimethoprim Solution: Add trimethoprim lactate to distilled/deionized water and bring volume to 10.0mL. Mix thoroughly. Filter sterilize. Alternately add 0.05g trimethoprim hydrochloride to 3.0mL 0.05*N* HCl. Heat to 50°C. Stirr until dissolved. Add distilled/deionized water and bring volume to 10.0mL. Mix thoroughly. Filter sterilize. Can be stored for 1 year at 4°C.

Vancomycin Solution:

Composition per 10.0mL:

Vancomycin	0.05g

Preparation of Vancomycin Solution: Add vancomycin to distilled/deionized water and bring volume to 10.0mL. Mix thoroughly. Filter sterilize. Can be stored for 2 months at 4°C.

Cycloheximide Solution:

Composition per 10.0mL:

Cycloheximide	0.025g
Ethanol	2.0mL

Preparation of Cycloheximide Solution: Add cycloheximide to 2.0mL ethanol to dissolve. Mix thoroughly. Add distilled/deionized water and bring volume to 10.0mL. Mix thoroughly. Filter sterilize. Can be stored for 1 year at 4°C.

Caution: Cycloheximide is toxic. Avoid skin contact or aerosol formation and inhalation.

Amphotericin B Solution:

Composition per 10.0mL:

Amphotericin B	0.005g

Preparation of Amphotericin B Solution: Add Amphotericin B to distilled/deionized water and bring volume to 10.0mL. Mix thoroughly. Filter sterilize. Can be stored for 1 year at –20°C.

Preparation of Medium: Add components, except antimicrobic solutions, to distilled/deionized water and bring volume to 1.0L. Mix thoroughly. Autoclave for 15 min at 15 psi pressure–121°C. Cool to 25°C. Aseptically add 4.0mL of sterile cefoperazone solution, 4.0mL of sterile trimethoprim lactate solution, 4.0mL of sterile vancomycin solution, and either 4.0mL of sterile cycloheximide solution or 4.0mL of sterile amphotericin B solution. Mix thoroughly. Aseptically distribute into sterile screw-capped tubes or bottles. Close caps tightly to reduce O_2 absorption. Use within 2 weeks.

Use: For the selective isolation and cultivation of Campylobacter species. For the pre-enrichment of Campylobacter spp. in food samples. This medium aids resuscitation of sublethally injured cells and overcomes the damaging effects of food processing.

Campylobacter Enrichment HiVeg Broth Base with Blood and Antibiotics (Preston Enrichment HiVeg Broth Base)

Composition per liter:

Plant extract	10.0g
Plant peptone	10.0g
NaCl	5.0g
Horse blood, lysed	50.0mL
Antibiotic supplement	10.0mL

pH 7.5 ± 0.2 at 25°C

Source: This medium, without horse blood and antibiotic supplement, is available as a premixed powder from HiMedia.

Antibiotic Supplement:

Composition per 10.0mL:

Cycloheximide	0.1g
Rifampicin	0.01g
Trimethoprim lactate	0.01g
Polmyxin B	5000U

Preparation of Antibiotic Supplement: Add components to 10.0mL of 50:50 acetone:distilled/deionized water. Filter sterilize.

Caution: Cycloheximide is toxic. Avoid skin contact or aerosol formation and inhalation.

Preparation of Medium: Add components, except horse blood and antibiotic supplement, to distilled/deionized water and bring volume to 940.0mL. Mix thoroughly. Gently heat and bring to boiling. Autoclave for 15 min at 15 psi pressure–121°C. Cool to 45°–50°C. Aseptically add 50.0mL of lysed horse blood and 10.0mL of sterile antibiotic supplement. Mix thoroughly. Pour into sterile Petri dishes.

Use: For the selective isolation of *Campylobacter* species.

Campylobacter fecalis Medium

Composition per 1133.6mL:

Starch, soluble	2.4g
Yeast extract	2.4g
Blood agar base	1.0L
Bovine blood	120.0mL
Antibiotic solution	10.0mL
Sodium lactate (60% syrup)	3.6mL

pH 6.8 ± 0.2 at 25°C

Blood Agar Base:

Composition per liter:

Beef heart, solids from infusion	500.0g
Agar	15.0g
Tryptose	10.0g
NaCl	5.0g

Preparation of Blood Agar Base: Add components to distilled/deionized water and bring volume to 1.0L. Mix thoroughly. Gently heat while stirring and bring to boiling.

Antibiotic Solution:

Composition per 10.0mL:

Cycloheximide	0.12g
Albamycin	6.0mg
Bacitracin	6000U

Preparation of Antibiotic Solution: Add components to distilled/deionized water and bring volume to 10.0mL. Mix thoroughly. Filter sterilize.

Caution: Cycloheximide is toxic. Avoid skin contact or aerosol formation and inhalation.

Preparation of Medium: To 1.0L of blood agar base, add soluble starch, yeast extract, and sodium lactate. Mix thoroughly. Autoclave for 15 min at 15 psi pressure–121°C. Cool to 45°–50°C. Aseptically add sterile bovine blood and antibiotic solution. Mix thoroughly. Pour into sterile Petri dishes or distribute into sterile tubes.

Use: For the cultivation and isolation of *Campylobacter fecalis*.

Campylobacter fetus Medium

Composition per liter:

Proteose peptone	10.0g
NaCl	5.0g
Beef extract	3.0g
Bovine blood	50.0mL
Antibiotic solution	10.0mL

pH 7.2–7.4 at 25°C

Antibiotic Solution:

Composition per 10.0mL:

Novobiocin	2.0mg
Bacitracin	2000U

Preparation of Antibiotic Solution: Add components to distilled/deionized water and bring volume to 10.0mL. Mix thoroughly. Filter sterilize.

Preparation of Medium: Add components, except bovine blood and antibiotic solution, to distilled/deionized water and bring volume to 940.0mL. Mix thoroughly. Gently heat and bring to boiling. Autoclave for 15 min at 15 psi pressure–121°C. Cool to 45°–50°C. Aseptically add sterile bovine blood and antibiotic solution. Mix thoroughly. Aseptically distribute into sterile tubes or flasks.

Use: For the isolation and cultivation of *Campylobacter fetus*.

Campylobacter fetus Medium

Composition per 1160.0mL:

Fluid thioglycolate agar	1.0L
Sheep blood, defibrinated	150.0mL
Antibiotic solution	10.0mL

pH 7.1 ± 0.2 at 25°C

Fluid Thioglycolate Agar:

Composition per liter:

Agar	15.0g
Pancreatic digest of casein	15.0g
Glucose	5.5g
Yeast extract	5.0g
NaCl	2.5g
Agar	0.75g
L-Cystine	0.5g
Sodium thioglycolate	0.5g
Resazurin	1.0mg

Preparation of Fluid Thioglycolate Agar: Add components to distilled/deionized water and bring volume to 1.0L. Mix thoroughly. Gently heat and bring to boiling. Autoclave for 15 min at 15 psi pressure–121°C. Cool to 25°C.

Antibiotic Solution:

Composition per 10.0mL:

Cycloheximide	0.05g
Novobiocin	5.0mg
Bacitracin	25,000U
Polymyxin B sulfate	10,000U

Preparation of Antibiotic Solution: Add components to distilled/deionized water and bring volume to 10.0mL. Mix thoroughly. Filter sterilize.

Caution: Cycloheximide is toxic. Avoid skin contact or aerosol formation and inhalation.

Preparation of Medium: To 1.0L of cooled, sterile, fluid thioglycolate agar, aseptically add 150.0mL of sterile sheep blood and 10.0mL of sterile antibiotic solution.

Use: For the isolation and cultivation of *Campylobacter fetus* from human specimens.

Campylobacter fetus Selective Medium

Composition per liter:

Fluid thioglycolate agar	1.0L
Sheep blood, defibrinated	150.0mL
Antibiotic solution	10.0mL

Fluid Thioglycolate Agar:

Composition per liter:

Agar	15.0g
Pancreatic digest of casein	15.0g
Glucose	5.5g
Yeast extract	5.0g
NaCl	2.5g
Agar	0.75g
L-Cystine	0.5g
Sodium thioglycolate	0.5g
Resazurin	1.0mg

Preparation of Fluid Thioglycolate Agar: Add components to distilled/deionized water and bring volume to 1.0L. Mix thoroughly. Gently heat and bring to boiling. Autoclave for 15 min at 15 psi pressure–121°C. Cool to 25°C.

Antibiotic Solution:

Composition per 10.0mL:

Cycloheximide	0.05g
Cephalothin	0.02g
Novobiocin	5.0mg
Bacitracin	25,000U
Colistin	10,000U

Preparation of Antibiotic Solution: Add components to distilled/deionized water and bring volume to 10.0mL. Mix thoroughly. Filter sterilize.

Caution: Cycloheximide is toxic. Avoid skin contact or aerosol formation and inhalation.

Preparation of Medium: To 1.0L of cooled, sterile, fluid thioglycolate agar, aseptically add 150.0mL of sterile sheep blood and 10.0mL of sterile antibiotic solution.

Use: For the isolation and cultivation of *Campylobacter fetus*.

Campylobacter HiVeg Agar Base with Blood and Antibiotic Supplement

Composition per liter:

Plant peptone No. 3	15.0g
Agar	12.0g
NaCl	5.0g
Yeast extract	5.0g
Plant extract No. 2	2.5g
Horse blood, lysed	50.0mL
Antibiotic supplement	10.0mL

pH 7.4 ± 0.2 at 25°C

Source: This medium, without horse blood and antibiotic supplement, is available as a premixed powder from HiMedia.

Antibiotic Supplement:

Composition per 10.0mL:

Cycloheximide	0.1g
Rifampicin	0.01g
Trimethoprim lactate	0.01g
Polmyxin B	5000U

Preparation of Antibiotic Supplement: Add components to 10.0mL of 50:50 acetone:distilled/deionized water. Filter sterilize.

Caution: Cycloheximide is toxic. Avoid skin contact or aerosol formation and inhalation.

Preparation of Medium: Add components, except horse blood and antibiotic supplement, to distilled/deionized water and bring volume to 940.0mL. Mix thoroughly. Gently heat and bring to boiling. Autoclave for 15 min at 15 psi pressure–121°C. Cool to 45°–50°C. Aseptically add 50.0mL of lysed horse blood and 10.0mL of sterile antibiotic supplement. Mix thoroughly. Pour into sterile Petri dishes.

Use: For the cultivation of *Campylobacter* species. Normally used with a selective supplement to suppress the growth of other bacterial species.

Campylobacter HiVeg Agar Base with Blood and Selective Supplement

Composition per liter:

Plant peptone No. 3	15.0g
Agar	12.0g
NaCl	5.0g
Yeast extract	5.0g

Plant extract No. 2	2.5g
Horse blood, lysed	50.0mL
Selectrive supplement solution	10.0mL

pH 7.4 ± 0.2 at 25°C

Source: This medium, without horse blood and antibiotic supplement, is available as a premixed powder from HiMedia.

Selective Supplement Solution:

Composition per 10.0mL:

Cephalothin	15.0mg
Vancomycin	10.0mg
Trimethoprim	5.0mg
Amphotericin B	2.0mg
Polymyxin B sulfate	2500 U

Preparation of Selective Supplement Solution: Add components to distilled/deionized water and bring volume to 10.0mL. Mix thoroughly. Filter sterilize.

Preparation of Medium: Add components, except horse blood and selective supplement, to distilled/deionized water and bring volume to 940.0mL. Mix thoroughly. Gently heat and bring to boiling. Autoclave for 15 min at 15 psi pressure–121°C. Cool to 45°–50°C. Aseptically add 50.0mL of lysed horse blood and 10.0mL of sterile selective supplement. Mix thoroughly. Pour into sterile Petri dishes.

Use: For the cultivation of *Campylobacter* species. Normally used with a selective supplement to suppress the growth of other bacterial species.

Campylobacter Isolation Agar A

Composition per liter:

Agar	12.0g
Beef extract	10.0g
Peptone	10.0g
NaCl	5.0g
Charcoal	4.0g
Casein hydrolysate	3.0g
Yeast extract	2.0g
Sodium deoxycholate	1.0g
$FeSO_4$	0.25g
Sodium pyruvate	0.25g
Antibiotic solution	10.0mL

pH 7.4 ± 0.2 at 25°C

Antibiotic Solution:

Composition per 10.0mL:

Cycloheximide	0.1g
Sodium cefoperazone	0.03g

Preparation of Antibiotic Solution: Add components to distilled/deionized water and bring volume to 10.0mL. Mix thoroughly. Filter sterilize.

Caution: Cycloheximide is toxic. Avoid skin contact or aerosol formation and inhalation.

Preparation of Medium: Add components, except antibiotic solution, to distilled/deionized water and bring volume to 990.0mL. Mix thoroughly. Gently heat and bring to boiling. Autoclave for 15 min at 15 psi pressure–121°C. Cool to 45°–50°C. Aseptically add sterile antibiotic solution. Mix thoroughly. Pour into sterile Petri dishes. Swirl flask while pouring to distribute charcoal.

Use: For the isolation and cultivation of *Campylobacter* species.

Campylobacter Isolation Agar B (*Campy* Cefex Agar)

Composition per liter:

Agar	15.0g
Pancreatic digest of casein	10.0g
Peptic digest of animal tissue	10.0g
NaCl	5.0g
Yeast extract	2.0g
Glucose	1.0g
$FeSO_4$	0.5g
Sodium pyruvate	0.5g
$NaHSO_3$	0.35g
Horse blood, laked	50.0mL
Antibiotic solution	10.0mL

pH 7.0 ± 0.2 at 25°C

Horse Blood, Laked:

Composition per 50.0mL:

Horse blood, fresh	50.0mL

Preparation of Horse Blood, Laked: Add blood to a sterile polypropylene bottle. Freeze overnight at –20°C. Thaw at 8°C. Refreeze at –20°C. Thaw again at 8°C.

Antibiotic Solution:

Composition per 10.0mL:

Cycloheximide	0.1g
Sodium cefoperazone	0.033g

Preparation of Antibiotic Solution: Add components to distilled/deionized water and bring volume to 10.0mL. Mix thoroughly. Filter sterilize.

Caution: Cycloheximide is toxic. Avoid skin contact or aerosol formation and inhalation.

Preparation of Medium: Add components, except horse blood and antibiotic solution, to distilled/deionized water and bring volume to 940.0mL. Mix thoroughly. Gently heat and bring to boiling. Autoclave for 15 min at 15 psi pressure–121°C. Cool to 45°–50°C. Aseptically add sterile horse blood and antibiotic solution. Mix thoroughly. Pour into sterile Petri dishes or distribute into sterile tubes.

Use: For the isolation and cultivation of *Campylobacter* species.

Campylobacter Medium

Composition per liter:

Sodium aspartate	10.0g
$MgSO_4{\cdot}7H_2O$	1.0g
K_2HPO_4	0.75g
Yeast extract	0.2g
$CaCl_2{\cdot}2H_2O$	28.0mg
Resazurin	1.0mg
Phosphate-cysteine solution	100.0mL
Trace elements solution SL-10	1.0mL

pH 7.0 ± 0.2 at 25°C

Phosphate-Cysteine Solution:

Composition per 100.0mL:

NaH_2PO_4	0.25g
L-Cysteine·HCl	0.25g

Preparation of Phosphate-Cysteine Solution: Add components to distilled/deionized water and bring volume to 100.0mL. Mix thoroughly. Autoclave for 15 min at 15 psi pressure–121°C.

Trace Elements Solution SL-10:

Composition per liter:

$FeCl_2 \cdot 4H_2O$	1.5g
$CoCl_2 \cdot 6H_2O$	190.0mg
$MnCl_2 \cdot 4H_2O$	100.0mg
$ZnCl_2$	70.0mg
$Na_2MoO_4 \cdot 2H_2O$	36.0mg
$NiCl_2 \cdot 6H_2O$	24.0mg
H_3BO_3	6.0mg
$CuCl_2 \cdot 2H_2O$	2.0mg
HCl (25% solution)	10.0mL

Preparation of Trace Elements Solution SL-10: Add $FeCl_2 \cdot 4H_2O$ to 10.0mL of HCl solution. Mix thoroughly. Add distilled/deionized water and bring volume to 1.0L. Add remaining components. Mix thoroughly.

Preparation of Medium: Add components, except phosphate-cysteine solution, to distilled/deionized water and bring volume to 900.0mL. Mix thoroughly. Autoclave for 15 min at 15 psi pressure–121°C. Aseptically add 100.0mL of sterile phosphate-cysteine solution. Mix thoroughly. Aseptically distribute into sterile tubes or flasks.

Use: For the cultivation and maintenance of *Campylobacter* species, *Actinobacillus ureae, Erysipelothrix rhusiopathiae, Helicobacter pylori, Moraxella bovis, Moraxella nonliquefaciens, Moraxella osloensis, Pasteurella haemolytica,* and *Pasteurella multocida.*

Campylobacter mucosalis Medium

Composition per liter:

Agar	15.0g
Beef extract	10.0g
Special peptone	10.0g
NaCl	5.0g
Sodium fumarate	3.0g
Sodium formate	2.0g
Horse blood, defibrinated	50.0mL

pH 7.2 ± 0.2 at 25°C

Source: Special peptone (L72) is available from Oxoid Unipath.

Preparation of Medium: Add components, except horse blood, to distilled/deionized water and bring volume to 950.0mL. Mix thoroughly. Heat with frequent agitation and boil for 1 min to completely dissolve. Autoclave for 15 min at 15 psi pressure–121°C. Cool to 45°–50°C. Aseptically add 50.0mL of sterile, defibrinated horse blood. Mix thoroughly and pour into sterile Petri dishes.

Use: For the cultivation and maintenance of *Campylobacter mucosalis.*

Campylobacter Nitrate Broth

Composition per liter:

Beef heart, infusion from	500.0g
Tryptose	10.0g
NaCl	5.0g
KNO_3	2.0g

pH 7.0 ± 0.2 at 25°C

Source: This medium is available as a premixed powder from HiMedia.

Preparation of Medium: Add components to distilled/deionized water and bring volume to 1.0L. Mix thoroughly. Gently heat and bring to boiling. Distribute into tubes or flasks. Autoclave for 15 min at 15 psi pressure–121°C.

Use: For the cultivation of *Campylobacter* species.

Campylobacter Nitrate HiVeg Broth

Composition per liter:

Plant hydrolysate No. 1	10.0g
Plant infusion	10.0g
NaCl	5.0g
KNO_3	2.0g

pH 7.0 ± 0.2 at 25°C

Source: This medium is available as a premixed powder from HiMedia.

Preparation of Medium: Add components to distilled/deionized water and bring volume to 1.0L. Mix thoroughly. Gently heat and bring to boiling. Distribute into tubes or flasks. Autoclave for 15 min at 15 psi pressure–121°C.

Use: For the cultivation of *Campylobacter* species.

Campylobacter rectus Medium

Composition per liter:

Yeast extract	11.0g
Pancreatic digest of casein	9.0g
Beef extract	3.0g
NaCl	2.0g
Na_2HPO_4	.0.4g
Na_2CO_3	0.25g
Resazurin	1.0mg
Formate-fumarate solution	100.0mL
Hemin solution	10.0mL

pH 7.5 ± 0.2 at 25°C

Formate-Fumarate Solution:

Composition per 100.0mL:

Sodium fumarate	3.0g
Sodium formate	2.0g

Preparation of Formate-Fumarate Solution: Add components to distilled/deionized water and bring volume to 100.0mL. Mix thoroughly. Filter sterilize. Sparge with 100% N_2 for 3–4 min.

Hemin Solution:

Composition per 10.0mL:

Hemin	5.0mg
NaOH (1*N* solution)	0.1mL

Preparation of Hemin Solution: Add hemin to NaOH solution to dissolve. Add distilled/deionized water and bring volume to 10.0mL. Mix thoroughly. Sparge with 100% N_2 for 3–4 min. Autoclave under 100% N_2 for 15 min at 15 psi pressure–121°C.

Preparation of Medium: Add components, except formate-fumarate solution and hemin solution, to distilled/deionized water and bring volume to 890.0mL. Sparge with 100% N_2 for 10 min. Autoclave under 100% N_2 for 15 min at 15 psi pressure–121°C. Aseptically and anaerobically add 100.0mL of sterile formate-fumarate solution and 10.0mL of sterile hemin solution under 100% N_2. Mix thoroughly. Aseptically and anaerobically distribute into sterile anaerobic tubes.

Use: For the cultivation and maintenance of *Campylobacter curvus* and *Campylobacter rectus.*

Campylobacter Selective Medium, Blaser-Wang (Blaser–Wang *Campylobacter* Medium) (Blaser's Agar) (Campy BAP Medium)

Composition per liter:

Brucella agar base	890.0mL
Sheep blood	100.0mL
Antibiotic supplement	10.0mL

***Brucella* Agar Base:**
Composition per 890.0mL:

Agar	15.0g
Glucose	10.0g
Pancreatic digest of casein	10.0g
NaCl	5.0g
Peptic digest of animal tissue	5.0g

pH 7.5 ± 0.2 at 25°C

Preparation of *Brucella* Agar Base: Add components to distilled/deionized water and bring volume to 890.0mL. Mix thoroughly. Gently heat and bring to boiling. Autoclave for 15 min at 15 psi pressure–121°C. Cool to 45°–50°C.

Antibiotic Supplement:
Composition per 10.0mL:

Cephalothin	15.0mg
Vancomycin	10.0mg
Trimethoprim	5.0mg
Amphotericin B	2.0mg
Polymyxin B	2500U

Preparation of Antibiotic Supplement: Add components to 10.0mL of distilled/deionized water. Filter sterilize.

Preparation of Medium: Prepare 890.0mL of *Brucella* agar base. Sterilize as directed. Cool to 50°–55°C and add 100.0mL of sheep blood or 50.0–70.0mL of laked horse blood. Laked blood is prepared by freezing whole blood overnight and thawing to room temperature. Aseptically add 10.0mL of sterile antibiotic supplement. Mix thoroughly. Pour into sterile Petri dishes.

Use: For the selective isolation of *Campylobacter* species.

Campylobacter Selective Medium, Blaser-Wang (Blaser–Wang *Campylobacter* Medium)

Composition per liter:

Columbia agar base	890.0mL
Sheep blood	100.0mL
Antibiotic supplement	10.0mL

pH 7.3 ± 0.2 at 25°C

Antibiotic Supplement:
Composition per 10.0mL:

Cephalothin	15.0mg
Vancomycin	10.0mg
Trimethoprim	5.0mg
Amphotericin B	2.0mg
Polymyxin B	2,500U

Preparation of Antibiotic Supplement: Add components to 10.0mL of distilled/deionized water. Filter sterilize.

Columbia Agar Base:
Composition per liter:

Special peptone	25.0g
Agar	10.0g
NaCl	5.0g
Starch	1.0g

Preparation of Columbia Agar Base: Add components to distilled/deionized water and bring volume to 890.0mL. Mix thoroughly. Gently heat and bring to boiling. Autoclave for 15 min at 15 psi pressure–121°C. Cool to 45°–50°C.

Preparation of Medium: To 890.0mL of cooled, sterile Columbia agar base, aseptically add 100.0mL of sheep blood or 50.0–70.0mL of laked horse blood. Laked blood is prepared by freezing whole blood overnight and thawing to room temperature. Aseptically add 10.0mL of sterile antibiotic supplement. Mix thoroughly. Pour into sterile Petri dishes.

Use: For the selective isolation of *Campylobacter* species.

Campylobacter Selective Medium, Butzler's (Butzler's *Campylobacter* Medium)

Composition per liter:

Brucella agar base	940.0mL
Sheep or horse blood, defibrinated	50.0mL
Antibiotic supplement	10.0mL

pH 7.5 ± 0.2 at 25°C

***Brucella* Agar Base:**
Composition per liter:

Agar	15.0g
Glucose	10.0g
Pancreatic digest of casein	10.0g
NaCl	5.0g
Peptic digest of animal tissue	5.0g

Preparation of *Brucella* Agar Base: Add components to distilled/deionized water and bring volume to 940.0mL. Mix thoroughly. Gently heat and bring to boiling. Autoclave for 15 min at 15 psi pressure–121°C. Cool to 45°–50°C.

Antibiotic Supplement:
Composition per 10.0mL:

Cycloheximide	50.0mg
Cephazolin	15.0mg
Novobiocin	5.0mg
Bacitracin	25,000U
Colistin sulfate	10,000U

Preparation of Antibiotic Supplement: Add components to 10.0mL of distilled/deionized water. Filter sterilize.

Caution: Cycloheximide is toxic. Avoid skin contact or aerosol formation and inhalation.

Preparation of Medium: To 940.0mL of cooled, sterile *Brucella* agar base, aseptically add 50.0mL of defibrinated sheep or horse blood and 10.0mL of sterile antibiotic supplement. Mix thoroughly. For enhanced growth, medium may also be supplemented with 0.25g of $Fe_2SO_4{\cdot}H_2O$, 0.25g of sodium metabisulfite, and 0.25g of sodium pyruvate. Pour into sterile Petri dishes.

Use: For the selective isolation of *Campylobacter* species.

Campylobacter Selective Medium, Butzler's (Butzler's *Campylobacter* Medium)

Composition per liter:

Columbia agar base	940.0mL
Blood, horse or sheep	50.0mL
Antibiotic supplement	10.0mL

pH 7.3 ± 0.2 at 25°C

Columbia Agar Base:
Composition per liter:

Peptone	25.0g
Agar	10.0g
NaCl	5.0g
Starch	1.0g

Preparation of Columbia Agar Base: Add components to distilled/deionized water and bring volume to 940.0mL. Mix thoroughly. Gently heat and bring to boiling. Autoclave for 15 min at 15 psi pressure–121°C. Cool to 45°–50°C.

Antibiotic Supplement:
Composition per 10.0mL:

Cycloheximide	50.0mg
Cephazolin	15.0mg
Novobiocin	5.0mg
Bacitracin	25,000U
Colistin sulfate	10,000U

Preparation of Antibiotic Supplement: Add components to 10.0mL of distilled/deionized water. Filter sterilize.

Caution: Cycloheximide is toxic. Avoid skin contact or aerosol formation and inhalation.

Preparation of Medium: To 940.0mL of cooled, sterile Columbia agar base, aseptically add 50.0mL of defibrinated sheep or horse blood and 10.0mL of sterile antibiotic supplement. Mix thoroughly. The medium may also be supplemented with 0.25g of $Fe_2SO_4{\cdot}H_2O$, 0.25g of sodium metabisulfite, and 0.25g of sodium pyruvate. Pour into sterile Petri dishes.

Use: For the selective isolation of *Campylobacter* species.

Campylobacter Selective Medium, Karmali's (Karmali's *Campylobacter* Medium)

Composition per liter:

Activated charcoal	4.0g
Columbia agar base	990.0mL
Antibiotic supplement	10.0mL

pH 7.4 ± 0.2 at 25°C

Source: This medium is available as a premixed powder from Oxoid Unipath.

Columbia Agar Base:
Composition per 990.0mL:

Peptone	25.0g
Agar	10.0g
NaCl	5.0g
Starch	1.0g

Preparation of Columbia Agar Base: Add components to distilled/deionized water and bring volume to 990.0mL. Mix thoroughly. Gently heat and bring to boiling. Autoclave for 15 min at 15 psi pressure–121°C. Cool to 45°–50°C.

Antibiotic Supplement:
Composition per 10.0mL:

Sodium pyruvate	0.05g
Cycloheximide	0.05g
Cefoperazone	0.016g
Hemin	0.016g
Vancomycin	0.01g

Preparation of Antibiotic Supplement: Add components to 10.0mL of distilled/deionized water. Filter sterilize.

Caution: Cycloheximide is toxic. Avoid skin contact or aerosol formation and inhalation.

Preparation of Medium: Prepare 990.0mL of Columbia agar base. Sterilize as directed. Cool to 50°–55°C. Add defibrinated sheep or horse blood to a final concentration of 5–7%. Add 10.0mL of sterile antibiotic supplement. Mix thoroughly. For enhanced growth, medium may also be supplemented with 0.25g of $Fe_2SO_4{\cdot}H_2O$, 0.25g of sodium metabisulfite, and 0.25g of sodium pyruvate. Pour into sterile Petri dishes. Swirl while pouring to keep charcoal in suspension.

Use: For the selective isolation of *Campylobacter* species.

Campylobacter Selective Medium, Preston's (Preston's *Campylobacter* Medium)

Composition per liter:

Campylobacter agar base	940.0mL
Horse blood, lysed	50.0mL
Antibiotic supplement	10.0mL

pH 7.5 ± 0.2 at 25°C

***Campylobacter* Agar Base:**
Composition per liter:

Agar	12.0g
Beef extract	10.0g
Peptone	10.0g
NaCl	5.0g

Preparation of *Campylobacter* Agar Base: Add components to distilled/deionized water and bring volume to 940.0mL. Mix thoroughly. Gently heat and bring to boiling. Autoclave for 15 min at 15 psi pressure–121°C. Cool to 45°–50°C.

Antibiotic Supplement:
Composition per 10.0mL:

Cycloheximide	0.1g
Rifampicin	0.01g
Trimethoprim lactate	0.01g
Polmyxin B	5000U

Preparation of Antibiotic Supplement: Add components to 10.0mL of 50:50 acetone:distilled/deionized water. Filter sterilize.

Caution: Cycloheximide is toxic. Avoid skin contact or aerosol formation and inhalation.

Preparation of Medium: To 940.0mL of cooled, sterile *Campylobacter* agar base, aseptically add 50.0mL of lysed horse blood and 10.0mL of sterile antibiotic supplement. Mix thoroughly. Pour into sterile Petri dishes.

Use: For the selective isolation of *Campylobacter* species.

Campylobacter sputorum subspecies *bubulus* Medium

Composition per liter:

Agar	1.5g
Brilliant Green	0.01g
Ethyl Violet	1.25mg
Brucella broth base	1.0L
Antibiotic solution	10.0mL

pH 7.0 ± 0.2 at 25°C

***Brucella* Broth Base:**
Composition per liter:

Pancreatic digest of casein	10.0g
Peptic digest of animal tissue	10.0g
NaCl	5.0g

Yeast extract	2.0g
Glucose	1.0g
$NaHSO_3$	0.1g

Preparation of *Brucella* Broth Base: Add components to distilled/deionized water and bring volume to 1.0L. Mix thoroughly.

Antibiotic Solution:

Composition per 10.0mL:

Cycloheximide	0.1g
Bacitracin	20,000U

Preparation of Antibiotic Solution: Add components to distilled/deionized water and bring volume to 10.0mL. Mix thoroughly. Filter sterilize.

Caution: Cycloheximide is toxic. Avoid skin contact or aerosol formation and inhalation.

Preparation of Medium: To 1.0L of *Brucella* broth base, add agar, Brilliant Green, and Ethyl Violet. Mix thoroughly. Autoclave for 15 min at 15 psi pressure–121°C. Cool to 45°–50°C. Aseptically add sterile antibiotic solution. Mix thoroughly. Aseptically distribute into sterile tubes or flasks.

Use: For the cultivation and isolation of *Campylobacter sputorum* subspecies *bubulus*.

Campylobacter sputorum subspecies *mucosalis* Medium

Composition per liter:

Yeast extract	2.8g
KNO_3	1.0g
Fluid thioglycolate broth without glucose	1.0L

Fluid Thioglycolate Broth without Glucose:

Composition per liter:

Pancreatic digest of casein	15.0g
Yeast extract	5.0g
NaCl	2.5g
Agar	0.75g
L-Cystine	0.5g
Sodium thioglycolate	0.5g
Resazurin	1.0mg

Preparation of Fluid Thioglycolate Broth without Glucose: Add components to distilled/deionized water and bring volume to 1.0L. Mix thoroughly.

Preparation of Medium: Combine components. Mix thoroughly. Gently heat and bring to boiling. Autoclave for 15 min at 15 psi pressure–121°C. Cool to 25°C. Aseptically distribute into sterile tubes or flasks.

Use: For the cultivation and isolation of *Campylobacter sputorum* subspecies *mucosalis*.

Campylobacter Thioglycolate Medium with 5 Antimicrobics

Composition per liter:

Pancreatic digest of casein	17.0g
Glucose	6.0g
Papaic digest of soybean meal	3.0g
NaCl	2.5g
Agar	1.6g
Sodium thioglycolate	0.5g
Na_2SO_3	0.1g
Antibiotic supplement solution	10.0mL

pH 7.0 ± 0.2 at 25°C

Antibiotic Supplement Solution:

Composition per 10.0mL:

Cephalothin	0.015g
Vancomycin	0.01g
Trimethoprim	5.0mg
Amphotericin B	2.0mg
Polymyxin B	2500U

Preparation of Antibiotic Supplement Solution: Add components to 10.0mL of distilled/deionized water. Mix thoroughly. Filter sterilize.

Preparation of Medium: Add components, except cephalothin, vancomycin, trimethoprim, amphotericin, and polymyxin B, to distilled deionized water and bring volume to 990.0mL. Mix thoroughly. Gently heat and bring to boiling. Autoclave for 15 min at 15 psi pressure–121°C. Cool to 45°–50°C. Add 10.0mL of sterile antibiotic supplement. Mix thoroughly. Pour into sterile Petri dishes.

Use: For the maintenence—as a holding medium or transport medium—of fecal specimens or swabs suspected of containing *Campylobacter jejuni* or other *Campylobacter* species when immediate inoculation of *Campylobacter* growth medium is unavailable.

Candida Agar

Composition per liter:

Agar	20.0g
Glucose	10.0g
Peptic digest of animal tissue	5.0g
Yeast extract	3.0g
Malt extract	3.0g
Aniline Blue	0.1g

pH 6.2 ± 0.2 at 25°C

Source: This medium is available from HiMedia.

Preparation of Medium: Add components to distilled/deionized water and bring volume to 1.0L. Mix thoroughly. Distribute into tubes or flasks. Autoclave for 15 min at 15 psi pressure–121°C. Pour into Petri dishes or leave in tubes.

Use: For the isolation and differentiation of *Candida albicans*.

Candida BCG Agar Base (*Candida* Bromcresol Green Agar Base)

Composition per liter:

Glucose	40.0g
Agar	15.0g
Peptone	10.0g
Yeast extract	1.0g
Bromcresol Green	0.02g
Neomycin solution	10.0mL

pH 6.1 ± 0.1 at 25°C

Source: This medium is available as a premixed powder from BD Diagnostic Systems.

Neomycin Solution:

Composition per 10.0mL:

Neomycin	0.5g

Preparation of Neomycin Solution: Add neomycin to distilled/deionized water and bring volume to 10.0mL. Mix thoroughly. Filter sterilize.

Preparation of Medium: Add components, except neomycin solution, to distilled/deionized water and bring volume to 1.0L. Mix thoroughly and heat gently until boiling. Autoclave for 15 min at 15 psi pressure–121°C. Cool to 50°–55°C. Aseptically add 10.0mL of sterile neomycin solution. Mix thoroughly. Pour into sterile Petri dishes or leave in tubes.

Use: For the selective isolation and identification of *Candida* species. It is a highly differential medium that is used for demonstrating morphological and biochemical reactions characterizing different *Candida* species. *Candida albicans* appears as blunt conical colonies with smooth edges and yellow to blue-green color. *Candida stellatoidea* appears as convex colonies with smooth edges and yellow to green color. *Candida tropicalis* appears as convex colonies with wavy edges and yellow-green to green color with a dark blue-green base. *Candida pseudotropicalis* appears as convex, shiny colonies with smooth edges and green color with a light green edge. *Candida krusei* appears as low conical colonies with spreading edges and blue-green color. *Candida stellatoidea* appears as convex colonies with smooth edges and yellow to green color.

Candida BCG HiVeg Agar Base with Neomycin

Composition per liter:

Glucose	40.0g
Agar	15.0g
Plant peptone	10.0g
Yeast extract	1.0g
Bromcresol Green	0.02g
Neomycin solution	10.0mL

pH 6.1 ± 0.1 at 25°C

Source: This medium, without neomycin solution, is available as a premixed powder from HiMedia.

Neomycin Solution:

Composition per 10.0mL:

Neomycin	0.5g

Preparation of Neomycin Solution: Add neomycin to distilled/deionized water and bring volume to 10.0mL. Mix thoroughly. Filter sterilize.

Preparation of Medium: Add components, except neomycin solution, to distilled/deionized water and bring volume to 990.0mL. Mix thoroughly and heat gently until boiling. Autoclave for 15 min at 15 psi pressure–121°C. Cool to 50°C. Aseptically add 10.0mL of sterile neomycin solution. Mix thoroughly. Pour into sterile Petri dishes or leave in tubes.

Use: For the selective isolation and identification of *Candida* species. It is a highly differential medium that is used for demonstrating morphological and biochemical reactions characterizing different *Candida* species. *Candida albicans* appears as blunt conical colonies with smooth edges and yellow to blue-green color. *Candida stellatoidea* appears as convex colonies with smooth edges and yellow to green color. *Candida tropicalis* appears as convex colonies with wavy edges and yellow-green to green color with a dark blue-green base. *Candida pseudotropicalis* appears as convex, shiny colonies with smooth edges and green color with a light green edge. *Candida krusei* appears as low conical colonies with spreading edges and blue-green color. *Candida stellatoidea* appears as convex colonies with smooth edges and yellow to green color.

Candida Bromcresol Green Agar Base

See: ***Candida* BCG Agar Base**

Candia Diagnostic Agar

Composition per liter:

Glucose	40.0g
Agar	15.0g
Peptone, mycological	10.0g
Ammonium 4-{2-[4-(2-acetamido-2-deoxy-β-D-glucopyranosyloxy)-3-methoxyphenyl]-vinyl}-1-(propan-3-yl-oate)-quinolium bromide	0.32g

pH 6.9 ± 0.2 at 25°C

Preparation of Medium: Add components to distilled/deionized water and bring volume to 1.0L. Mix thoroughly and heat with frequent agitation until boiling. Boil until components are fully dissolved. Do not autoclave. Cool to 45°–50°C. Pour into sterile Petri dishes.

Use: For the rapid isolation and identification of *Candida* species. *Candida albicans* and *Candida dubliniensis* produce white colonies with deep-red spots on a yellow transparent background. Colonies of *Candida tropicalis* and *Candida kefyr* are uniformly pink, and colonies of other *Candida* spp., including *Candida glabrata* and *Candida parapsilosis,* appear white.

Candida HiVeg Medium with Antibiotics

Composition per liter:

Agar	15.0g
Glucose	5.0g
K_2HPO_4	5.0g
Na_2SO_3	5.0g
Bismuth sulfite indicator	3.0g
Plant peptone No. 4	2.5g
Antibiotic solution	10.0mL

Source: This medium, without antibiotics, is available as a premixed powder from HiMedia.

Antibiotic Solution:

Composition per 10.0mL:

Streptomycin	25.0mg
Penicillin	300 U

Preparation of Antibiotic Solution: Add components to 10.0mL distilled/deionized water. Filter sterilize.

Preparation of Medium: Add components, except antibiotic solution, to distilled/deionized water and bring volume to 990.0mL. Mix thoroughly. Gently heat and bring to boiling. Do not autoclave. Cool to 45°–50°C. Aseptically add 10.0mL of sterile antibiotic solution. Mix thoroughly. Pour into sterile Petri dishes.

Use: For the cultivation of *Candida* species.

Candida Isolation Agar

Composition per liter:

Agar	20.0g
Glucose	10.0g
Peptone	5.0g
Yeast extract	3.0g
Malt extract	3.0g
Aniline Blue	0.1g

pH 5.9 ± 0.5 at 25°C

Source: This medium is available as a premixed powder from BD Diagnostic Systems.

Preparation of Medium: Add components to distilled/deionized water and bring volume to 1.0L. Mix thoroughly. Gently heat and bring to boiling. Distribute into tubes or flasks. Autoclave for 15 min at 15 psi pressure–121°C. Pour into sterile Petri dishes or leave in tubes.

Use: For the isolation and differentiation of *Candida albicans*. *Candida albicans* turns the medium blue.

Candida Medium with Antibiotics

Composition per liter:

Agar	15.0g
Glucose	5.0g
K_2HPO_4	5.0g
Na_2SO_3	5.0g
Bismuth sulfite indicator	3.0g
Mycological peptone	2.5g
Antibiotic solution	10.0mL

Source: This medium, without antibiotics, is available as a premixed powder from HiMedia.

Antibiotic Solution:

Composition per 10.0mL:

Streptomycin	25.0mg
Penicillin	300 U

Preparation of Antibiotic Solution: Add components to 10.0mL distilled/deionized water. Filter sterilize.

Preparation of Medium: Add components, except antibiotic solution, to distilled/deionized water and bring volume to 990.0mL. Mix thoroughly. Gently heat and bring to boiling. Do not autoclave. Cool to 45°–50°C. Aseptically add 10.0mL of sterile antibiotic solution. Mix thoroughly. Pour into sterile Petri dishes.

Use: For the cultivation of *Candida* species.

CandiSelect 4™

Composition per liter:

Proprietary

Source: Available from BioRad.

Preparation of Medium: Preprepared plates.

Use: For the direct identification of *Candida albicans* and for the presumptive identification of *Candida tropicalis, Candida glabrata*, and *Candida krusei.*

Cantharellus Agar

Composition per liter:

Agar	12.0g
Glucose	4.0g
Sodium succinate	1.35g
KI	1.21g
Na_2MoO_4	1.21g
$MnSO_4 \cdot H_2O$	0.845g
NH_4Cl	0.58g
KH_2PO_4	0.2g
$CuSO_4 \cdot 5H_2O$	0.125g
$CoCl_2 \cdot 6H_2O$	0.12g
$MgSO_4 \cdot H_2O$	0.1g
$CaCl_2 \cdot H_2O$	26.5mg
NaCl	20.0mg
m-Inositol	10.0mg
EDTA	9.3mg
$FeSO_4 \cdot 7H_2O$	6.95mg
$ZnSO_4 \cdot 7H_2O$	1.44mg
H_3BO_4	0.31mg
Calcium DL-pantothenate	0.1mg
Nicotinic acid	0.1mg
p-Aminobenzoic acid	0.1mg
Pyrdoxine HCl	0.1mg
Riboflavin	0.1mg
Thiamine HCl	0.1mg
Biotin	0.025mg
Tomato roots	variable

Preparation of Medium: Add components, except tomato root, to distilled/deionized water and bring volume to 1.0L. Mix thoroughly. Gently heat and bring to boiling. Distribute into tubes or flasks. Autoclave for 15 min at 15 psi pressure–121°C. Pour into sterile Petri dishes or leave in tubes. Add onto the agar a piece of fresh tomato root near the inoculum.

Use: For the cultivation and maintenance of *Cantharellus cibarius*.

Capnocytophaga Medium

Composition per liter:

Pancreatic digest of casein	17.0g
KNO_3	3.0g
NaCl	3.0g
Yeast extract	3.0g
Hemin	3.0mg
Glucose solution	20.0mL

pH 7.0 ± 0.2 at 25°C

Glucose Solution:

Composition per 20.0mL:

D-Glucose	3.0g

Preparation of Glucose Solution: Add glucose to distilled/deionized water and bring volume to 20.0mL. Mix thoroughly. Autoclave for 15 min at 15 psi pressure–121°C.

Preparation of Medium: Add components, except glucose solution, to distilled/deionized water and bring volume to 990.0mL. Mix thoroughly. Autoclave for 15 min at 15 psi pressure–121°C. Aseptically add 10.0mL of sterile glucose solution. Mix thoroughly. Aseptically distribute into sterile tubes or flasks.

Use: For the cultivation and maintenance of *Capnocytophaga* species.

Capnocytophaga II Medium (DSMZ Medium 779)

Composition per liter:

Proteose peptone no.3	10.0g
Yeast extract	5.0g
Na_2HPO_4	4.0g
Lab-Lemco meat extract	2.4g
Glucose	1.5g
Starch, soluble	0.5g
$Cysteine\text{-}HCl \cdot H_2O$	0.5g
Horse blood	50.0mL

pH 6.8 ± 0.2 at 25°C

Preparation of Medium: Add components, except horse blood, to distilled/deionized water and bring volume to 950.0L. Mix thoroughly. Adjust pH to 7.6–7.8. Autoclave for 15 min at 15 psi pressure–121°C. Cool to room temperature. Aseptically add 50.0mL horse blood. Mix thoroughly. Aseptically distribute into tubes or flasks. Incubate under 95% air + 5% CO_2 or anaerobically under 95% N_2 + 5% CO_2.

Use: For the cultivation of *Capnocytophaga haemolytica* and *Capnocytophaga granulosa.*

Caprylate Thallous Agar
See: **CT Agar**

Carbohydrate Consumption Broth Base with Carbohydrate

Composition per liter:

Proteose peptone	10.0g
NaCl	5.0g
Beef extract	1.0g
Bromcresol Purple	0.1g
Carbohydrate solution	50.0mL

pH 6.8 ± 0.2 at 25°C

Source: This medium, without carbohydrate solution, is available as a premixed powder from HiMedia.

Carbohydrate Solution:
Composition per 100.0mL:

Carbohydrate	10.0g

Preparation of Carbohydrate Solution: Add carbohydrate to distilled/deionized water and bring volume to 100.0mL. Adonitol, arabinose, cellobiose, glucose, dulcitol, fructose, galactose, inositol, lactose, maltose, mannitol, raffinose, rhamnose, salicin, sorbitol, sucrose, trehalose, xylose, or other carbohydrates may be used. Mix thoroughly. Filter sterilize.

Preparation of Medium: Add components, except carbohydrate solution, to distilled/deionized water and bring volume to 1.0L. Mix thoroughly. Gently heat and bring to boiling. Distribute in 10.0mL volumes into test tubes containing inverted Durham tubes. Autoclave for 15 min at 15 psi pressure–121°C. Cool to 25°C. Add 0.5mL of sterile carbohydrate solution to each tube.

Use: For the determination of carbohydrate fermentation reactions of microorganisms, particularly members of the Enterobacteriaceae.

Carbohydrate Fermentation Broth

Composition per liter:

Peptone	10.0g
NaCl	5.0g
Meat extract	3.0g
Carbohydrate solution	50.0mL
Andrade's indicator	10.0mL

pH 7.1 ± 0.2 at 25°C

Andrade's Indicator:
Composition per 100.0mL:

Acid Fuchsin	0.1 g
NaOH (1*N* solution)	16.0mL

Preparation of Andrade's Indicator: Add components to distilled/deionized water and bring volume to 100.0mL. Mix thoroughly.

Carbohydrate Solution:
Composition per 100.0mL:

Carbohydrate	10.0g

Preparation of Carbohydrate Solution: Add carbohydrate to distilled/deionized water and bring volume to 100.0mL. Adonitol, arabinose, cellobiose, glucose, dulcitol, fructose, galactose, inositol, lactose, maltose, mannitol, raffinose, rhamnose, salicin, sorbitol, sucrose, trehalose, xylose, or other carbohydrates may be used. Mix thoroughly. Filter sterilize.

Caution: Acid Fuchsin is a potential carcinogen and care must be taken to avoid inhalation of the powdered dye and contact with the skin.

Preparation of Medium: Add components, except carbohydrate solution, to distilled/deionized water and bring volume to 1.0L. Mix thoroughly. Gently heat and bring to boiling. Distribute in 10.0mL volumes into test tubes containing inverted Durham tubes. Autoclave for 15 min at 15 psi pressure–121°C. Cool to 25°C. Add 0.5mL of sterile carbohydrate solution to each tube.

Use: For the determination of carbohydrate fermentation reactions of microorganisms, particularly members of the Enterobacteriaceae. A Durham tube is used to collect gas produced during the fermentation reaction. Acid production is indicated by a pink reaction.

Carbohydrate Consumption HiVeg Broth Base with Carbohydrate

Composition per liter:

Plant peptone No. 3	10.0g
NaCl	5.0g
Plant extract	1.0g
Bromcresol Purple	0.1g
Carbohydrate solution	50.0mL

pH 6.8 ± 0.2 at 25°C

Source: This medium, without carbohydrate solution, is available as a premixed powder from HiMedia.

Carbohydrate Solution:
Composition per 100.0mL:

Carbohydrate	10.0g

Preparation of Carbohydrate Solution: Add carbohydrate to distilled/deionized water and bring volume to 100.0mL. Adonitol, arabinose, cellobiose, glucose, dulcitol, fructose, galactose, inositol, lactose, maltose, mannitol, raffinose, rhamnose, salicin, sorbitol, sucrose, trehalose, xylose, or other carbohydrates may be used. Mix thoroughly. Filter sterilize.

Preparation of Medium: Add components, except carbohydrate solution, to distilled/deionized water and bring volume to 1.0L. Mix thoroughly. Gently heat and bring to boiling. Distribute in 10.0mL volumes into test tubes containing inverted Durham tubes. Autoclave for 15 min at 15 psi pressure–121°C. Cool to 25°C. Add 0.5mL of sterile carbohydrate solution to each tube.

Use: For the determination of carbohydrate fermentation reactions of microorganisms, particularly members of the Enterobacteriaceae.

Carbohydrate Fermentation Broth

Composition per liter:

Peptone	10.0g
NaCl	5.0g
Meat extract	3.0g
Carbohydrate solution	50.0mL
Andrade's indicator	10.0mL

pH 7.1 ± 0.2 at 25°C

Andrade's Indicator:
Composition per 100.0mL:

Acid Fuchsin	0.1 g
NaOH (1*N* solution)	16.0mL

Preparation of Andrade's Indicator: Add components to distilled/deionized water and bring volume to 100.0mL. Mix thoroughly.

Carbohydrate Solution:

Composition per 100.0mL:

Carbohydrate	10.0g

Preparation of Carbohydrate Solution: Add carbohydrate to distilled/deionized water and bring volume to 100.0mL. Adonitol, arabinose, cellobiose, glucose, dulcitol, fructose, galactose, inositol, lactose, maltose, mannitol, raffinose, rhamnose, salicin, sorbitol, sucrose, trehalose, xylose, or other carbohydrates may be used. Mix thoroughly. Filter sterilize.

Caution: Acid Fuchsin is a potential carcinogen and care must be taken to avoid inhalation of the powdered dye and contact with the skin.

Preparation of Medium: Add components, except carbohydrate solution, to distilled/deionized water and bring volume to 1.0L. Mix thoroughly. Gently heat and bring to boiling. Distribute in 10.0mL volumes into test tubes containing inverted Durham tubes. Autoclave for 15 min at 15 psi pressure–121°C. Cool to 25°C. Add 0.5mL of sterile carbohydrate solution to each tube.

Use: For the determination of carbohydrate fermentation reactions of microorganisms, particularly members of the Enterobacteriaceae. A Durham tube is used to collect gas produced during the fermentation reaction. Acid production is indicated by a pink reaction.

Carbohydrate Medium Base

See: **CHO Medium Base**

Carbon Assimilation Medium

Composition per liter:

Agar solution	500.0mL
Mineral base medium	500.0mL

pH 6.5 ± 0.1 at 25°C

Agar Solution:

Composition per liter:

Agar	32.0g

Preparation of Agar Solution: Add agar to distilled/deionized water and bring volume to 1.0L. Mix thoroughly. Gently heat and bring to boiling. Autoclave for 15 min at 15 psi pressure–121°C. Cool to 45°–50°C.

Mineral Base Medium:

Composition per 500.0mL:

Carbohydrate	10.0g
NaCl	5.0g
NH_4HPO_4	1.0g
K_2HPO_4	1.0g
$MgSO_4 \cdot 7H_2O$, anhydrous	0.1g

Preparation of Mineral Base Medium: Add components to distilled/deionized water and bring volume to 500.0mL. Mix thoroughly. Gently heat until dissolved. Filter sterilize. Warm to 45°–50°C.

Preparation of Medium: Combine 500.0mL of cooled, sterile agar solution and 500.0mL of sterile mineral base medium. Mix thoroughly. Aseptically distribute into sterile tubes. Allow tubes to cool in a slanted position.

Use: For the cultivation and differentiation of microorganisms based on their ability to utilize a particular carbon source.

Carbon Assimilation Medium, Auxanographic Method for Yeast Identification

Composition per liter:

Noble agar	20.0g
$(NH_4)_2SO_4$	0.5g
KH_2PO_4	0.1g
$MgSO_4 \cdot 7H_2O$	0.05g
NaCl	0.01g
$CaCl_2 \cdot 2H_2O$	0.01g
DL-Methionine	2.0mg
DL-Tryptophan	2.0mg
L-Histidine·HCl	1.0mg
Inositol	0.2mg
KI	0.01mg
H_3BO_3	0.05mg
$ZnSO_4 \cdot 7H_2O$	0.04mg
$MnSO_4 \cdot 4H_2O$	0.04mg
Thiamine·HCl	0.04mg
Pyroxidine·HCl	0.04mg
Niacin	0.04mg
Calcium pantothenate	0.04mg
p-Aminobenzoic acid	0.02mg
Riboflavin	0.02mg
$FeCl_3$	0.02mg
$Na_2MoO_4 \cdot 4H_2O$	0.02mg
$CuSO_4 \cdot 5H_2O$	4.0μg
Folic acid	0.2μg
Biotin	0.2μg

pH 4.5 ± 0.2 at 25°C

Preparation of Medium: Add components to distilled/deionized water and bring volume to 1.0L. Mix thoroughly. Gently heat and bring to boiling. Distribute into screw-capped tubes in 20.0mL volumes. Autoclave for 15 min at 15 psi pressure–121°C.

Use: For carbohydrate assimilation tests by the auxanographic method for the identification of yeasts.

Carbon Monoxide Oxidizers Agar, Modified

Composition per 1001.0mL:

Agar	12.0g
$Na_2HPO_4 \cdot 12H_2O$	4.5g
Sodium acetate	3.0g
NH_4Cl	1.5g
KH_2PO_4	0.75g
$MgSO_4 \cdot 7H_2O$	0.2g
$CaCl_2 \cdot 2H_2O$	30.0mg
Ferric ammonium citrate	18.0mg
Trace elements solution	1.0mL
$NaHCO_3$ solution	10.0mL
Thiamine·HCl solution	10.0mL

Trace Elements Solution:

Composition per liter:

$Na_2MoO_4 \cdot 2H_2O$	0.9g
H_3BO_3	0.3g
$CoCl_2 \cdot 6H_2O$	0.2g
$ZnSO_4 \cdot 7H_2O$	0.1g
$MnCl_2 \cdot 4H_2O$	30.0mg
Na_2SeO_3	20.0mg
$NiCl_2 \cdot 6H_2O$	20.0mg
$CuCl_2 \cdot 2H_2O$	10.0mg

Preparation of Trace Elements Solution: Add components to distilled/deinonized water and bring volume to 1.0L. Mix thoroughly.

$NaHCO_3$ Solution:

Composition per 10.0mL:

$NaHCO_3$ 1.0g

Preparation of $NaHCO_3$ Solution: Add $NaHCO_3$ to distilled/deionized water and bring volume to 10.0mL. Filter sterilize.

Thiamine·HCl Solution:

Composition per 10.0mL:

Thiamine·HCl 20.0μg

Preparation of Thiamine·HCl Solution: Add thiamine·HCl to distilled/deionized water and bring volume to 10.0mL. Filter sterilize.

Preparation of Medium: Add components, except $NaHCO_3$ solution and thiamine·HCl solution, to distilled/deionized water and bring volume to 980.0mL. Mix thoroughly. Gently heat and bring to boiling. Autoclave for 15 min at 15 psi pressure–121°C. Cool to 50°–55°C. Aseptically add 10.0mL of sterile $NaHCO_3$ solution and 10.0mL of sterile thiamine·HCl solution. Mix thoroughly. Pour into sterile Petri dishes or distribute into sterile tubes.

Use: For the cultivation of *Carbophilus carboxidus* and *Zavarzinia compransoris.*

Carbon Utilization Test

Composition per liter:

Ionagar 10.0g
NH_4Cl 1.0g
$MgSO_4·7H_2O$ 0.5g
Ferric ammonium citrate 0.05g
$CaCl_2$ 0.5mg
Sodium potassium phosphate
buffer (0.33*M* solution, pH 6.8) 1.0L
Carbon source 10.0mL

pH 6.8 ± 0.2 at 25°C

Carbon Source:

Composition per 10.0mL:

Carbon source 1.0g

Preparation of Carbon Source: Add carbon source to distilled/deionized water and bring volume to 10.0mL. Mix thoroughly. Filter sterilize.

Preparation of Medium: Add components, except carbon source, to distilled/deionized water and bring volume to 990.0mL. Mix thoroughly. Gently heat and bring to boiling. Autoclave for 15 min at 15 psi pressure 121°C. Cool to 45° 50°C. Aseptically add sterile carbon source. Mix thoroughly. Pour into sterile Petri dishes or distribute into sterile tubes.

Use: For the cultivation and differentiation of *Pseudomonas* species based on their ability to utilize a specific carbon source.

Carbonate-Buffered Medium CMB4 with Glucose

Composition per 1002.0mL:

NaCl 4.0g
$NaHCO_3$ 2.5g
Glucose 1.0g
$MgCl_2·6H_2O$ 0.8g
KCl 0.5g
NH_4Cl 0.3g
KH_2PO_4 .... 0.2g
Resazurin 1.0mg
Modified Wolfe's mineral solution 10.0mL
Wolfe's vitamin solution 10.0mL
Sulfide-calcium solution 2.0mL

pH 6.9 ± 0.2 at 25°C

Modified Wolfe's Mineral Solution:

Composition per liter:

$MgSO_4·7H_2O$ 3.0g
Nitrilotriacetic acid 1.5g
NaCl 1.0g
$MnSO_4·H_2O$ 0.5g
$CaCl_2$ 0.1g
$CoCl_2·6H_2O$ 0.1g
$FeSO_4·7H_2O$ 0.1g
$ZnSO_4·7H_2O$ 0.1g
$AlK(SO_4)_2·12H_2O$ 0.01g
$CuSO_4·5H_2O$ 0.01g
H_3BO_3 .... 0.01g
$Na_2MoO_4·2H_2O$ 0.01g
Na_2SeO_3 .... 0.01g
$NaWO_4·2H_2O$ 0.01g
$NiCl_2·6H_2O$ 0.01g

Preparation of Modified Wolfe's Mineral Solution: Add nitrilotriacetic acid to 500.0mL of distilled/deionized water. Adjust pH to 6.5 with KOH. Add remaining components one at a time. Add distilled/deionized water to 1.0L. Adjust pH to 6.8.

Wolfe's Vitamin Solution:

Composition per liter:

Pyridoxine·HCl 10.0mg
p-Aminobenzoic acid 5.0mg
Lipoic acid 5.0mg
Nicotinic acid 5.0mg
Riboflavin 5.0mg
Thiamine·HCl 5.0mg
Calcium DL-pantothenate 5.0mg
Biotin 2.0mg
Folic acid 2.0mg
Vitamin B_{12} 0.1mg

Preparation of Wolfe's Vitamin Solution: Add components to distilled/deionized water and bring volume to 1.0L. Mix thoroughly.

Sulfide-Calcium Solution:

Composition per liter:

$Na_2S·9H_2O$ 36.0g
$CaCl_2·2H_2O$ 15.0g

Preparation of Sulfide-Calcium Solution: Add components to distilled/deionized water and bring volume to 1.0L. Mix thoroughly. Adjust pH to 7.2. Sparge with 100% N_2. Anaerobically distribute into tubes. Autoclave at 121°C for 15 min.

Preparation of Medium: Prepare and dispense medium under 80% N_2 + 20% CO_2. Add components, except $NaHCO_3$ and sulfide-calcium solution, to distilled/deionized water and bring volume to 1.0L. Mix thoroughly. Gently heat and bring to boiling. Continue boiling for 3 min. Cool to room temperature while sparging with 80% N_2 + 20% CO_2. Add $NaHCO_3$. Mix thoroughly. Anaerobically distribute 10.0mL volumes into anaerobic tubes. Autoclave for 15 min at 15 psi pressure–121°C. Aseptically and anaerobically add 0.02mL of sterile sulfide-calcium solution to each tube. Mix thoroughly. Adjust pH to 6.9.

Use: For the cultivation of *Spirochaeta thermophila.*

Carboxydobacterium Medium

Composition per liter:

$Na_2HPO_4 \cdot 12H_2O$	9.0g
KH_2PO_4	1.5g
NH_4Cl	1.5g
$MgSO_4 \cdot 7H_2O$	0.2g
$CaCl_2 \cdot 2H_2O$	20.0mg
Ferric ammonium citrate	1.2mg
TS2 trace elements solution	1.0mL

TS2 Trace Elements Solution:

Composition per liter:

$Na_2MoO_4 \cdot 2H_2O$	0.9g
H_3BO_3	0.3g
$CoCl_2 \cdot 6H_2O$	0.2g
$ZnSO_4 \cdot 7H_2O$	0.1g
$MnCl_2 \cdot 4H_2O$	30.0mg
Na_2SeO_3	20.0mg
$NiCl_2 \cdot 6H_2O$	20.0mg
$CuCl_2 \cdot 2H_2O$	10.0mg

Preparation of TS2 Trace Elements Solution: Add components to distilled/deionized water and bring volume to 1.0L. Mix thoroughly.

Caution: Carbon monoxide (CO) is a toxic gas.

Preparation of Medium: Add components to distilled/deionized water and bring volume to 1.0L. Mix thoroughly. Distribute into tubes or flasks. Autoclave for 15 min at 15 psi pressure–121°C. After inoculation, incubate in an atmosphere of 50% CO + 50% air.

Use: For the autotrophic cultivation of *Oligotropha carboxidovorans.*

Carboxydobacterium Medium

Composition per liter:

$Na_2HPO_4 \cdot 12H_2O$	9.0g
Sodium acetate	3.0g
KH_2PO_4	1.5g
NH_4Cl	1.5g
$MgSO_4 \cdot 7H_2O$	0.2g
$CaCl_2 \cdot 2H_2O$	20.0mg
Ferric ammonium citrate	1.2mg
TS2 trace elements solution	1.0mL

TS2 Trace Elements Solution:

Composition per liter:

$Na_2MoO_4 \cdot 2H_2O$	0.9g
H_3BO_3	0.3g
$CoCl_2 \cdot 6H_2O$	0.2g
$ZnSO_4 \cdot 7H_2O$	0.1g
$MnCl_2 \cdot 4H_2O$	30.0mg
Na_2SeO_3	20.0mg
$NiCl_2 \cdot 6H_2O$	20.0mg
$CuCl_2 \cdot 2H_2O$	10.0mg

Preparation of TS2 Trace Elements Solution: Add components to distilled/deionized water and bring volume to 1.0L. Mix thoroughly.

Preparation of Medium: Add components to distilled/deionized water and bring volume to 1.0L. Mix thoroughly. Distribute into tubes or flasks. Autoclave for 15 min at 15 psi pressure–121°C. After inoculation, incubate in air.

Use: For the organotrophic cultivation of *Oligotropha carboxidovorans.*

Carboxydobrachium pacificum Medium (DSMZ Medium 902)

Composition per 1050mL:

NaCl	20.0g
$MgSO_4 \cdot 7H_2O$	3.9g
KCl	0.7g
$CaCl_2 \cdot 2H_2O$	0.4g
NH_4Cl	0.3g
Na_2HPO_4	0.15g
Yeast extract	0.05g
Na_2SiO_3	0.03g
Resazurin	0.5mg
Vitamin solution	10.0mL
$NaHCO_3$ solution	10.0mL
$Na_2S \cdot 9H_2O$ solution	10.0mL
L-Cysteine solution	10.0mL
Pyruvate solution	10.0mL
Trace elements solution SL-10	1.0mL

pH 7.0 ± 0.2 at 25°C

Pyruvate Solution:

Composition per 10.0mL:

Na-pyruvate	2.5g

Preparation of Pyruvate Solution: Add pyruvate to distilled/deionized water and bring volume to 10.0mL. Mix thoroughly. Sparge with 100% N_2. Autoclave for 15 min at 15 psi pressure–121°C.

L-Cysteine Solution:

Composition per 10.0mL:

L-Cysteine·HCl·H_2O	0.45g

Preparation of L-Cysteine Solution: Add L-cysteine·HCl·H_2O to distilled/deionized water and bring volume to 10.0mL. Mix thoroughly. Sparge with 100% N_2. Autoclave for 15 min at 15 psi pressure–121°C.

$Na_2S \cdot 9H_2O$ Solution:

Composition per 10.0mL:

$Na_2S \cdot 9H_2O$	0.45g

Preparation of $Na_2S \cdot 9H_2O$ Solution: Add $Na_2S \cdot 9H_2O$ to distilled/deionized water and bring volume to 10.0mL. Mix thoroughly. Sparge with 100% N_2. Autoclave for 15 min at 15 psi pressure–121°C.

$NaHCO_3$ Solution:

Composition per 10.0mL:

$NaHCO_3$	0.5g

Preparation of $NaHCO_3$ Solution: Add $NaHCO_3$ to distilled/deionized water and bring volume to 10.0mL. Mix thoroughly. Autoclave for 15 min at 15 psi pressure–121°C. Cool to 25°C. Must be prepared freshly.

Trace Elements Solution SL-10:

Composition per liter:

$FeCl_2 \cdot 4H_2O$	1.5g
$CoCl_2 \cdot 6H_2O$	190.0mg
$MnCl_2 \cdot 4H_2O$	100.0mg
$ZnCl_2$	70.0mg
$Na_2MoO_4 \cdot 2H_2O$	36.0mg
$NiCl_2 \cdot 6H_2O$	24.0mg
H_3BO_3	6.0mg
$CuCl_2 \cdot 2H_2O$	2.0mg
HCl (25% solution)	10.0mL

Preparation of Trace Elements Solution SL-10: Add $FeCl_2·4H_2O$ to 10.0mL of HCl solution. Mix thoroughly. Add distilled/deionized water and bring volume to 1.0L. Add remaining components. Mix thoroughly. Sparge with 100% N_2. Autoclave for 15 min at 15 psi pressure–121°C.

Vitamin Solution:
Composition per liter:

Pyridoxine-HCl	10.0mg
Thiamine-HCl·$2H_2O$	5.0mg
Riboflavin	5.0mg
Nicotinic acid	5.0mg
D-Ca-pantothenate	5.0mg
p-Aminobenzoic acid	5.0mg
Lipoic acid	5.0mg
Biotin	2.0mg
Folic acid	2.0mg
Vitamin B_{12}	0.1mg

Preparation of Vitamin Solution: Add components to distilled/deionized water and bring volume to 1.0L. Mix thoroughly. Sparge with 80% H_2 + 20% CO_2. Filter sterilize.

Preparation of Medium: Prepare and dispense medium under 100% N_2. Add components, except vitamin solution, $NaHCO_3$ solution, pyruvate solution, L-cysteine-HCl·H_2O solution, and $Na_2S·9H_2O$ solution, to distilled/deionized water and bring volume to 1.0L. Mix thoroughly. Adjust pH to 7.0. Sparge with 100% N_2 for at least 30 min. Distribute into anaerobe tubes or bottles. Autoclave for 15 min at 15 psi pressure–121°C. Aseptically and anaerobically add per liter, 10.0mL vitamin solution, 10.0mL $NaHCO_3$ solution, 10.0mL pyruvate solution, 10.0mL L-cysteine-HCl·H_2O solution, and 10.0mL $Na_2S·9H_2O$. Mix thoroughly. The final pH should be 7.0.

Use: For the cultivation of *Carboxydibrachium pacificum (Carboxydobrachium pacificum).*

Carboxydothermus Medium

Composition per 1030.0mL:

$MgCl_2·6H_2O$	0.52g
KCl	0.33g
KH_2PO_4	0.33g
NH_4Cl	0.33g
$CaCl_2·2H_2O$	0.29g
Resazurin	0.5mg
Trace elements solution SL-4	10.0mL
Vitamin solution	10.0mL
Yeast extract solution	10.0mL
$Na_2S·9H_2O$ solution	10.0mL

pH 6.9 ± 0.2 at 25°C

Trace Elements Solution SL-4:
Composition per liter:

EDTA	0.5g
$FeSO_4·7H_2O$	0.2g
Trace elements solution SL-6	100.0mL

Preparation of Trace Elements Solution SL-4: Add components to distilled/deionized water and bring volume to 1.0L. Mix thoroughly.

Trace Elements Solution SL-6:
Composition per liter:

$MnCl_2·4H_2O$	0.5g
H_3BO_3	0.3g
$CoCl_2·6H_2O$	0.2g
$ZnSO_4·7H_2O$	0.1g
$Na_2MoO_4·2H_2O$	0.03g
$NiCl_2·6H_2O$	0.02g
$CuCl_2·2H_2O$	0.01g

Preparation of Trace Elements Solution SL-6: Add components to distilled/deionized water and bring volume to 1.0L. Mix thoroughly.

Vitamin Solution:
Composition per liter:

Pyridoxine·HCl	10.0mg
Calcium DL-pantothenate	5.0mg
Lipoic acid	5.0mg
Nicotinic acid	5.0mg
p-Aminobenzoic acid	5.0mg
Riboflavin	5.0mg
Thiamine·HCl	5.0mg
Biotin	2.0mg
Folic acid	2.0mg
Vitamin B_{12}	0.1mg

Preparation of Vitamin Solution: Add components to distilled/deionized water and bring volume to 1.0L. Mix thoroughly. Filter sterilize. Sparge with 100% N_2.

Yeast Extract Solution:
Composition per 10.0mL:

Yeast extract	10.0mg

Preparation of Yeast Extract Solution: Add yeast extract to distilled/deionized water and bring volume to 10.0mL. Mix thoroughly. Autoclave under 100% N_2 for 15 min at 15 psi pressure–121°C.

$Na_2S·9H_2O$ Solution:
Composition per 10.0mL:

$Na_2S·9H_2O$	0.7g

Preparation of $Na_2S·9H_2O$ Solution: Prepare and dispense solution anaerobically under 100% N_2. Add $Na_2S·9H_2O$ to distilled/deionized water and bring volume to 10.0mL. Mix thoroughly. Adjust pH to 7.0. Autoclave for 15 min at 15 psi–121°C.

Preparation of Medium: Add components, except vitamin solution, yeast extract solution, and $Na_2S·9H_2O$ solution, to distilled/deionized water and bring volume to 1.0L. Mix thoroughly. Sparge under 100% N_2 for 10 min. Autoclave under 100% N_2 for 15 min at 15 psi pressure–121°C. Aseptically and anaerobically combine with 10.0mL of sterile vitamin solution, 10.0mL of sterile yeast extract solution, and 10.0mL of sterile $Na_2S·9H_2O$ solution. Mix thoroughly. Pressurize inoculation flask with CO (carbon monoxide) gas at 2 bar pressure.

Caution: Carbon monoxide is a toxic gas.

Use: For the cultivationof *Carboxydothermus hydrogenoformans.*

Carboxymethyl Cellulose Medium (DSMZ Medium 1111)

Composition per liter:

Carboxymethyl cellulose	15.0g
Agar	6.0g
Casitone	2.0g
$(NH_4)_2SO_4$	1.0g
KH_2PO_4	1.0g
$MgSO_4·7H_2O$	1.0g
$CaCl_2·2H_2O$	1.0g
$FeCl_3·6H_2O$	0.2g
KH_2PO_4 solution	10.0mL

pH 7.0 ± 0.2 at 25°C

KH_2PO_4 Solution:

Composition per 10.0mL:

KH_2PO_4	1.0g

Preparation of KH_2PO_4 Solution: Add components to distilled/deionized water and bring volume to 10.0mL. Mix thoroughly. Autoclave for 15 min at 15 psi pressure–121°C.

Preparation of Medium: Add components, except KH_2PO_4 solution, to distilled/deionized water and bring volume to 990.0L. Mix thoroughly. Gently heat while stirring and bring to boiling. Distribute into tubes or flasks. Autoclave for 15 min at 15 psi pressure–121°C. Aeptically add 10.0mL KH_2PO_4 solution. Mix thoroughly. Pour into Petri dishes or leave in tubes.

Use: For the cultivation of *Cellvibrio japonicus*.

Cardiobacterium hominis Medium

Composition per liter:

Glucose	5.0g
Leucine	0.43g
Threonine	0.28g
Glutamic acid	0.2g
Valine	0.19g
Glycine	0.18g
Arginine	0.16g
Histidine	0.13g
Proline	0.1g
Tyrosine	0.04g
Buffered salts solution	100.0mL
Vitamin solution	10.0mL

pH 7.0 ± 0.2 at 25°C

Buffered Salts Solution:

Composition per liter:

Na_2PHO_4	284.0.g
KH_2PO_4	272.0.g
NaCl	5.0g
$FeSO_4 \cdot 7H_2O$	4.0g
$MgSO_4 \cdot 7H_2O$	4.0g
$ZnSO_4 \cdot 7H_2O$	0.4g
$MnSO_4 \cdot H_2O$	0.3g
$CuSO_4 \cdot 5H_2O$	0.05g

Preparation of Buffered Salts Solution: Add components to distilled/deionized water and bring volume to 1.0L. Mix thoroughly.

Vitamin Solution:

Composition per liter:

Pyridoxine·HCl	2.0mg
Calcium pantothenate	1.0mg
Nicotinamide	1.0mg
Thiamine·HCl	1.0mg
Biotin	0.1mg

Preparation of Vitamin Solution: Add components to distilled/deionized water and bring volume to 1.0L. Mix thoroughly.

Preparation of Medium: Add components to distilled/deionized water and bring volume to 1.0L. Mix thoroughly. Adjust pH to 7.0. Filter sterilize.

Use: For the isolation and cultivation of *Cardiobacterium hominis*.

Cardiobacterium hominis Medium

Composition per liter:

K_2HPO_4	7.0g
Yeast extract	5.0g
KH_2PO_4	3.0g
$(NH_4)_2SO_4$	0.1g
$MgSO_4 \cdot 7H_2O$	0.01g

pH 7.0 ± 0.2 at 25°C

Preparation of Medium: Add components to distilled/deionized water and bring volume to 1.0L. Mix thoroughly. Distribute into tubes or flasks. Autoclave for 15 min at 15 psi pressure–121°C.

Use: For the cultivation of *Cardiobacterium hominis*.

Carnation Leaf Agar (ATCC Medium 2041)

Composition per liter:

Carnation leaves, dried	Variable
Agar	20.0g

Preparation of Carnation Leaves: Harvest young carnation leaves, *Dianthus caryophyllus*, from actively growing disbudded plants that are free from pesticide residues. Cut the leaves into pieces approximately 5mm square and dry them in an oven at 40–55°C for 2 hr. (When dry, the leaves should be green and crisp. Loss of pigmentation indicates that the drying temperature was too high.) Place the leaf pieces in aluminum canisters 5cm deep and 9cm in diameter and sterlize with 2.5 megarads of gamma irradiation from a cesium-135 source for 4 days.

Preparation of Medium: Add agar to distilled/deionized water and bring volume to 1.0L. Mix thoroughly. Gently heat and bring to boiling. Distribute into tubes or flasks. Autoclave for 15 min at 15 psi pressure–121°C. Pour into sterile Petri dishes or leave in tubes. Float several sterile leaf pieces on each agar surface. Leave medium at room temperature for 3 to 4 days before use to check for growth of possible contaminants on the leaf pieces.

Use: For the cultivation and maintenance of *Cryptosporiopsis abietina* and *Phomopsis occulta*.

Carnitine Chloride Medium

Composition per liter:

Noble agar	15.0g
DL-Carnitine chloride	10.0g
Na_2HPO_4	10.0g
KH_2PO_4	5.5g
$(NH_4)_2HPO_4$	2.0g
$NH_4H_2PO_4$	1.5g
$MgSO_4 \cdot 7H_2O$	0.2g
Yeast extract	0.05g
$CaCl_2$	0.015g
$Fe_2(SO_4)_3$	0.6mg
$CuSO_4 \cdot 5H_2O$	0.2mg
$MnSO_4 \cdot H_2O$	0.2mg
$ZnSO_4 \cdot 7H_2O$	0.2mg

pH 7.0 ± 0.1 at 25°C

Preparation of Medium: Add components to distilled/deionized water and bring volume to 1.0L. Adjust pH to 7.0 with NaOH. Mix thoroughly. Heat gently until boiling. Distribute into tubes or flasks. Autoclave for 15 min at 15 psi pressure–121°C. Pour into sterile Petri dishes or leave in tubes.

Use: For the cultivation and maintenance of bacteria that can use carnitine as a carbon source.

Carnitine Medium for *Torulopsis*

Composition per liter:

Glucose	20.0g
Agar	15.0g
L-Asparagine·H_2O	1.0g
KH_2PO_4	0.5g
$MgSO_4$·$7H_2O$	0.5g
NaCl	0.1g
L-Phenylalanine	80.0mg
DL-Tryptophan	50.0mg
DL-Methionine	20.0mg
Adenine	10.0mg
Cytosine	10.0mg
Inositol	10.0mg
Calcium D-(+)-pantothenate	2.0mg
Thiamine·HCl	2.0mg
Pyridoxine·HCl	2.0mg
Nicotinic acid	2.0mg
DL-Carnitine·HCl	1.0mg
Choline	1.0mg
Biotin	20.0μg

pH 5.0 ± 0.2 at 25°C

Preparation of Medium: Add components to distilled/deionized water and bring volume to 1.0L. Gently heat and bring to boiling. Adjust pH to 5.0. Distribute into tubes or flasks. Autoclave for 15 min at 15 psi pressure–121°C. Pour into sterile Petri dishes or leave in tubes.

Use: For the cultivation and maintenance of *Candida pintolopesii.*

Carnobacterium Medium

Composition per liter:

Agar	15.0g
Beef extract	10.0g
Peptone	10.0g
NaCl	5.0g
Glucose	5.0g
Yeast extract	3.0g

pH 6.8 ± 0.2 at 25°C

Preparation of Medium: Add components to distilled/deionized water and bring volume to 1.0L. Mix thoroughly. Distribute into tubes or flasks. Autoclave for 15 min at 15 psi pressure–121°C.

Use: For the cultivation of *Carnobacterium piscicola.*

Carr's Ethanol Medium (LMG Medium 228)

Composition per liter:

Yeast extract	30.0g
Agar	20.0g
Bromcresol Blue	22.0mg
Ethanol	20.0mL

Preparation of Medium: Add components, except ethanol, to 980.0mL distilled/deionized water. Mix thoroughly. Gently heat and bring to boiling. Autoclave for 15 min at 15 psi pressure–121°C. Cool to 45°C. Aseptically add 20.0mL sterile ethanol. Mix thoroughly. Pour into sterile Petri dishes or distribute into sterile tubes.

Use: For the cultivation of *Acetobacter pasteurianus.*

Carrot Decoction Agar

Composition per liter:

Carrots	100.0g
Agar	15.0g

Preparation of Medium: Peel and slice carrots. Add to 1.0L of distilled/deionized water. Autoclave for 30 min at 15 psi pressure–121°C. Filter solids through cheesecloth. Add agar to filtrate. Mix thoroughly. Bring volume to 1.0L with distilled/deionized water. Gently heat and bring to boiling. Distribute into tubes or flasks. Autoclave for 15 min at 15 psi pressure–121°C. Pour into sterile Petri dishes or leave in tubes.

Use: For the cultivation and maintenance of *Tuberculina maxima* and *Tuberculina persicina.*

Carrot Potato Dextrose Agar (ATCC Medium 1829)

Composition per liter:

Agar	25.0g
Glucose	20.0g
Pancreatic digest of casein	2.5g
Yeast extract	0.5g
$MgSO_4$·$7H_2O$	0.3g
$CaCO_3$	0.2g
Potatoes, infusion from	500.0mL
Carrot juice (any commercial brand)	15.0 mL

pH 5.6 ± 0.2 at 25°C

Source: Potato dextrose agar, without carrot juice, is available as a premixed powder from BD Diagnostic Systems.

Potato Infusion:

Composition per 500.0mL:

Potatoes	300.0g

Preparation of Potato Infusion: Peel and dice potatoes. Add 500.0mL of distilled/deionized water. Gently heat and bring to boiling. Continue boiling for 30 min. Filter through cheesecloth. Reserve filtrate.

Preparation of Medium: Add components to distilled/deionized water and bring volume to 1.0L. Mix thoroughly. Gently heat and bring to boiling. Distribute into tubes or flasks. Autoclave for 15 min at 15 psi pressure–121°C. Pour into sterile Petri dishes or leave in tubes.

Use: For the cultivation of yeasts and molds and to induce sporulationi.

Cary and Blair Transport Medium

Composition per liter:

Agar	5.0g
NaCl	5.0g
Sodium thioglycolate	1.5g
Na_2HPO_4	1.1g
$CaCl_2$ solution	9.0mL

pH 8.0 ± 0.5 at 25°C

Source: This medium is available as a premixed powder from BD Diagnostic Systems and Oxoid Unipath.

$CaCl_2$ Solution:

Composition per 10.0mL:

$CaCl_2$	0.1g

Preparation of $CaCl_2$ Solution: Add $CaCl_2$ to distilled/deionized water and bring volume to 10.0mL. Mix thoroughly. Filter sterilize.

Preparation of Medium: Add components to distilled/deionized water and bring volume to 1.0L. Mix thoroughly and heat gently until boiling. Cool to 50°C. Add 9.0mL of a 1% $CaCl_2$ solution. Adjust the pH to 8.4. Distribute into screw-capped tubes in 7.0mL volumes. Sterilize under flowing steam for 15 min. After sterilization, tighten the screwcaps.

Use: For the maintenance—as a holding medium or transport medium—of clinical specimens during collection or shipment.

Cary and Blair Transport Medium, Modified

Composition per liter:

Agar	5.0g
NaCl	5.0g
Sodium thioglycolate	1.5g
L-Cysteine·HCl·H_2O	0.5g
$CaCl_2$·$2H_2O$	0.1g
Na_2HPO_4	0.1g
$NaHSO_3$	0.1g
Resazurin solution	4.0mL

pH 8.4 ± 0.2 at 25°C

Resazurin Solution:

Composition per 380.0mL:

Resazurin	0.05g
Ethanol (95% solution)	200.0mL

Preparation of Resazurin Solution: Add resazurin to 200.0mL of ethanol. Mix thoroughly. Bring volume to 380.0mL with distilled/deionized water.

Preparation of Medium: Add components, except L-cysteine·HCl·H_2O, to distilled/deionized water and bring volume to 1.0L. Mix thoroughly. Gas the solution with 100% CO_2 for 10–15 min. Add the L-cysteine·HCl·H_2O. Mix thoroughly. Adjust pH to 8.4. Anaerobically distribute into tubes under 100% N_2. Cap tubes with butyl rubber stoppers. Autoclave for 15 min at 0 psi pressure–100°C on 3 consecutive days.

Use: For the maintenance—as a holding medium—of clinical specimens during collection or shipment.

Caryophanon latum Medium

Composition per liter:

Papaic digest of soybean meal	2.0g
Pancreatic digest of casein	2.0g
Yeast extract	2.0g
K_2HPO_4	1.0g
Sodium acetate	1.0g
$MgSO_4$·$7H_2O$	0.27g
Sodium glutamate	0.1g
Thiamine·HCl	0.2mg
Biotin	0.05mg
Tris/HCl-buffer 10m*M*, pH 7.8	1000.0mL

pH 7.2 ± 0.2 at 25°C

Preparation of Medium: Combine components. Mix thoroughly. Distribute into tubes or flasks. Autoclave for 15 min at 15 psi pressure–121°C.

Use: For the cultivation and maintenance of *Caryophanon latum* and *Vitreoscilla stercoraria.*

Caryophanon Medium

Composition per liter:

Agar	15.0g
Yeast extract	2.0g
Sodium acetate	1.0g
Pancreatic digest of casein	1.0g

pH 7.5 ± 0.2 at 25°C

Preparation of Medium: Add components to distilled/deionized water and bring volume to 1.0L. Mix thoroughly. Gently heat and bring to boiling. Distribute into tubes or flasks. Autoclave for 15 min at 15 psi pressure–121°C. Pour into sterile Petri dishes or leave in tubes.

Use: For the cultivation and maintenance of *Caryophanon tenue* and other *Caryophanon* species.

CAS Medium

Composition per liter:

Pancreatic digest of casein	10.0g
$MgSO_4$·$7H_2O$	1.0g
K_2HPO_4	0.25g

pH 6.8 ± 0.2 at 25°C

Preparation of Medium: Add components to distilled/deionized water and bring volume to 1.0L. Mix thoroughly. Distribute into tubes or flasks. Autoclave for 15 min at 15 psi pressure–121°C.

Use: For the cultivation of myxobacteria.

Casamino Acids Glucose Medium (CAGV Medium)

Composition per liter:

Agar	20.0g
Glucose	1.0g
Vitamin-free casamino acids	1.0g
Solution A (mineral salts)	20.0mL
Vitamin solution	10.0mL

pH 7.2 ± 0.2 at 25°C

Solution A (Mineral Salts):

Composition per liter:

$MgSO_4$·$7H_2O$	29.7g
$NaMoO_4$·$2H_2O$	12.67g
Nitrilotriacetic acid	10.0g
$CaCl_2$·$2H_20$	3.34g
$FeSO_4$·$7H_2O$	0.1g
Solution B (metallic salts)	50.0mL

Preparation of Solution A (Mineral Salts): Add nitrilotriacetic acid to 500.0mL of distilled/deionized water. Dissolve by adjusting pH to 6.5 with KOH. Add remaining components. Readjust pH to 7.2 with H_2SO_4 or KOH. Add distilled/deionized water to 1.0L.

Solution B (Metallic Salts):

Composition per 100.0mL:

$ZnSO_4$·$7H_2O$	1.1g
$FeSO_4$·$7H_20$	0.5g
Ethylenediaminetetraacetic acid	0.3g
$MnSO_4$·H_2O	0.3g
$CuSO_4$·$5H_2O$	0.04g
$CoCL_2$·$6H_20$	0.02g
$Na_2B_4O_7$·$10H_20$	0.02g

Preparation of Solution B (Metallic Salts): Add a few drops of H_2SO_4 to distilled/deionized water to inhibit precipitate formation.

Add components to acidified distilled/deionized water and bring volume to 100.0mL. Mix thoroughly.

Vitamin Solution:

Composition per liter:

Pyridoxine·HCL	0.01g
Calcium pantothenate	5.0mg
Nicotinamide	5.0mg
Riboflavin	5.0mg
Thiamine·HCl	5.0mg
Biotin	2.0mg
Folic acid	2.0mg
Vitamin B_{12}	0.1mg

Preparation of Vitamin Solution: Add components to distilled/deionized water and bring volume to 1.0L. Mix thoroughly. Filter sterilize.

Preparation of Medium: Add components to distilled/deionized water and bring volume to 1.0L. Mix thoroughly. Gently heat and bring to boiling. Distribute into tubes or flasks. Autoclave for 15 min at 15 psi pressure–121°C. Pour into sterile Petri dishes or leave in tubes.

Use: For the cultivation and maintenance of *Microcyclus aquaticus*.

Casamino Acids Medium

Composition per liter:

Casamino acids	1.0g
Glucose	1.0g
Biotin	0.02mg
Modified Hutner's basal salts	20.0mL

Modified Hutner's Basal Salts:

Composition per liter:

$MgSO_4·7H_2O$	29.7g
Nitrilotriacetic acid	10.0g
$CaCl_2·2H_2O$	3.34g
$FeSO_4·7H_2O$	0.1g
$(NH_4)_2MoO_4$	9.25mg
Metals "44"	50.0mL

Preparation of Modified Hutner's Basal Salts: Add nitrilotriacetic acid to 500.0mL of distilled/deionized water. Dissolve by adjusting pH to 6.5 with KOH. Add remaining components. Add distilled/deionized water to 1.0L.

Metals "44":

Composition per 100.0mL:

$ZnSO_4·7H_2O$	1.1g
$FeSO_4·7H_2O$	0.5g
EDTA	0.25g
$MnSO_4·7H_2O$	0.154g
$CuSO_4·5H_2O$	0.04g
$Co(NO_3)_2·6H_2O$	0.025g
$Na_2B_4O_7·10H_2O$	0.018g

Preparation of Metals "44": Acidify distilled/deionized water with a drop of H_2SO_4 to retard precipitation of salts. Add components to distilled/deionized water and bring volume to 100.0mL.

Preparation of Medium: Add components to distilled/deionized water and bring volume to 1.0L. Mix thoroughly. Distribute into tubes or flasks. Autoclave for 15 min at 15 psi pressure–121°C.

Use: For cultivation of *Ancylobacter aquaticus* and *Enhydrobacter aerosaccus*.

Casamino Acids Peptone Czapek's Agar

Composition per liter:

Sucrose	30.0g
Agar	15.0g
Peptone	2.0g
Casamino acids	1.0g
K_2HPO_4	1.0g
KCl	0.5g
$MgSO_4·7H_2O$	0.5g
$FeSO_4·7H_2O$	0.01g

Preparation of Medium: Add components to distilled/deionized water and bring volume to 1.0L. Mix thoroughly. Gently heat and bring to boiling. Distribute into tubes or flasks. Autoclave for 15 min at 15 psi pressure–121°C. Pour into sterile Petri dishes or leave in tubes.

Use: For the isolation and cultivation of *Actinomadura* species, *Actinopolyspora* species, *Excellospora* species, and *Microspora* species.

Casamino Acids and Yeast Extract Agar

Composition per liter:

NaCl	200.0g
$MgSO_4·7H_2O$	20.0g
Agar	15.0g
Yeast extract	10.0g
Casamino acids	7.5g
Trisodium citrate	3.0g
KCl	2.0g
$FeSO_4$ solution	1.0mL

pH 7.4 ± 0.2 at 25°C

$FeSO_4$ Solution:

Composition per 10.0mL:

$FeSO_4·7H_2O$	2.5g

Preparation of $FeSO_4$ Solution: Add $FeSO_4·7H_2O$ to 0.001*M* HCl and bring volume to 50.0mL. Mix thoroughly. Filter sterilize.

Preparation of Medium: Add components to distilled/deionized water and bring volume to 1.0L. Mix thoroughly. Adjust pH to 7.4. Distribute into tubes or flasks. Autoclave for 15 min at 15 psi pressure–121°C.

Use: For the cultivation of *Actinopolyspora halophila*.

Casamino Acids Yeast Extract Broth (CYE Broth)

Composition per liter:

Casamino acids	30.0g
Yeast extract	4.0g
K_2HPO_4	0.5g

pH 7.4 ± 0.2 at 25°C

Preparation of Medium: Add components to distilled/deionized water and bring volume to 1.0L. Mix thoroughly. Distribute into tubes or flasks. Autoclave for 15 min at 15 psi pressure–121°C.

Use: For the cultivation of *Vibrio* species from foods.

Casamino Acids Yeast Extract Lincomycin Medium

Composition per liter:

Casamino acids	20.0g
K_2HPO_4	8.71g
Yeast extract	6.0g
NaCl	2.5g

Lincomycin solution 5.0mL
Trace salts solution 1.0mL

pH 8.5 ± 0.2 at 25°C

Lincomycin Solution:
Composition per 5.0mL:
Lincomycin 45.0mg

Preparation of Lincomycin Solution: Add lincomycin to distilled/deionized water and bring volume to 5.0mL. Mix thoroughly. Filter sterilize.

Trace Salts Solution:
Composition per liter:
$MgSO_4 \cdot 7H_2O$ 50.0g
$MnCl_2 \cdot 4H_2O$ 5.0g
$FeCl_2$ 5.0g

Preparation of Trace Salts Solution: Add components to distilled/deionized water and bring volume to 1.0L. Add sufficient 0.1*N* H_2SO_4 to dissolve components. Mix thoroughly. Filter sterilize.

Preparation of Medium: Add components, except trace salts solution and lincomycin solution, to distilled/deionized water and bring volume to 994.0mL. Mix thoroughly. Adjust pH to 8.5. Autoclave for 15 min at 15 psi pressure–121°C. Cool to 25°C. Aseptically add 1.0mL of sterile trace salts solution and 5.0mL of sterile lincomycin solution. Mix thoroughly. Aseptically distribute into sterile tubes or flasks.

Use: For the cultivation of heat-labile, toxin-producing enterotoxigenic *Escherichia coli.*

Casamino Acids Yeast Extract Salts Broth, Gorbach (CA YE Broth)

Composition per liter:
Casamino acids 20.0g
K_2HPO_4 .... 8.71g
Yeast extract 6.0g
NaCl 2.5g
Trace salts solution 1.0mL

pH 8.5 ± 0.2 at 25°C

Trace Salts Solution:
Composition per liter:
$MgSO_4 \cdot 7H_2O$ 50.0g
$MnCl_2 \cdot 4H_2O$ 5.0g
$FeCl_2$ 5.0g

Preparation of Trace Salts Solution: Add components to distilled/deionized water and bring volume to 1.0L. Add sufficient 0.1*N* H_2SO_4 to dissolve components. Mix thoroughly. Filter sterilize.

Preparation of Medium: Add components, except trace salts solution, to distilled/deionized water and bring volume to 999.0mL. Mix thoroughly. Adjust pH to 8.5. Autoclave for 15 min at 15 psi pressure–121°C. Cool to 25°C. Aseptically add 1.0mL of sterile trace salts solution. Mix thoroughly. Aseptically distribute into sterile tubes or flasks.

Use: For the cultivation of enterotoxigenic *Escherichia coli.*

Casamino Peptone Czapek Medium

Composition per liter:
Sucrose 30.0g
Agar 15.0g
Peptone 2.0g
Casamino acids 1.0g
K_2HPO_4 .... 1.0g
KCl 0.5g
$MgSO_4 \cdot 7H_2O$ 0.5g
$FeSO_4 \cdot 7H_2O$ 0.01g

Preparation of Medium: Add components to distilled/deionized water and bring volume to 1.0L. Mix thoroughly. Gently heat to boiling. Distribute into tubes or flasks. Autoclave for 15 min at 15 psi pressure–121°C. Pour into sterile Petri dishes or leave in tubes.

Use: For the cultivation and maintenance of *Actinoplanes* species, *Pseudonocardia compacta*, and *Streptomyces* species.

Casein Agar

Composition per liter:
Agar 10.0g
Skim milk 50.0mL

Preparation of Medium: Add components to distilled/deionized water and bring volume to 1.0L. Mix thoroughly. Gently heat and bring to boiling. Distribute into tubes or flasks. Autoclave for 15 min at 15 psi pressure–121°C. Pour into sterile Petri dishes or leave in tubes.

Use: For the cultivation and differentiation of aerobic actinomycetes based on casein utilization. Bacteria that utilize casein, such as *Streptomyces* and *Actinomadura* species, appear as colonies surrounded by a clear zone. *Nocardia asteroides, Nocardia caviae,* and *Mycobacterium fortuitum* do not utilize casein.

Casein Hydrolysate Yeast Extract HiVeg Broth (CAYE HiVeg Broth)

Composition per liter:
Plant acid hydrolysate 30.0g
Yeast extract 4.0g
Glucose 2.0g
K_2HPO_4 .... 0.5g

pH 7.0 ± 0.2 at 25°C

Source: This medium is available as a premixed powder from HiMedia.

Preparation of Medium: Add components to distilled/deionized water and bring volume to 1.0L. Mix thoroughly. Gently heat and bring to boiling. Distribute into tubes or flasks. Autoclave for 15 min at 15 psi pressure–121°C.

Use: For cultivation of *Vibrio cholerae* while testing enterotoxigenicity.

Casein Hydrolysate Yeast Extract Salts HiVeg Broth Base with Tracer Salts (CAYES)

Composition per liter:
Plant acid hydrolysate 20.0g
K_2HPO_4 .... 8.71g
Yeast extract 6.0g
NaCl 2.5g
Tracer salts solution 1.0mL

pH 7.0 ± 0.2 at 25°C

Source: This medium, without tracer salts solution, is available as a premixed powder from HiMedia.

Tracer Salts Solution:
Composition per 10.0mL:
$MgSO_4$ 0.5g
$MnCl_2$ 0.05g

$FeCl_3$ 0.05g
Sulfuric acid, 1*N* 10.0mL

Preparation of Tracer Salts Solution: Add components to 0.1*N* sulfuric acid and bring volume to 10.0mL. Mix thoroughly. Filter sterilize.

Preparation of Medium: Add components to distilled/deionized water and bring volume to 1.0L. Mix thoroughly. Gently heat and bring to boiling. Distribute into tubes or flasks. Autoclave for 15 min at 15 psi pressure–121°C.

Use: For agar dilution susceptibility tests with imidazole antifungal agents.

Casein Medium

Composition per liter:

NaCl	250.0g
Agar	20.0g
$MgCl_2 \cdot 6H_2O$	20.0g
Casein hydrolysate	5.0g
Yeast extract	5.0g
KCl	2.0g
$CaCl_2 \cdot 2H_2O$	0.2g

pH 7.4 ± 0.2 at 25°C

Preparation of Medium: Add components to distilled/deionized water and bring volume to 950.0mL. Mix thoroughly. Gently heat to boiling. Adjust pH to 7.4. Bring volume to 1.0L with distilled/deionized water. Distribute into tubes or flasks. Autoclave for 15 min at 15 psi pressure–121°C. Pour into sterile Petri dishes or leave in tubes.

Use: For the cultivation and maintenance of *Halobacterium* species and other halophilic bacteria.

Casein Soya Agar, Modified

Composition per liter:

Pancreatic digest of casein	14.5g
Agar	14.0g
Papaic digest of soybean meal	5.0g
NaCl	5.0g
Growth factors	1.5g

pH 7.3 ± 0.2 at 25°C

Source: This medium is available from HiMedia.

Preparation of Medium: Add components to distilled/deionized water and bring volume to 1.0L. Mix thoroughly. Gently heat and bring to boiling. Distribute into tubes or flasks. Autoclave for 15 min at 15 psi pressure–121°C. Do not overheat. Pour into sterile Petri dishes or leave in tubes.

Use: For use as a general-purpose medium for cultivation of various microorganisms.

Casein Soya Agar, Modified with Blood

Composition per liter:

Pancreatic digest of casein	14.5g
Agar	14.0g
Papaic digest of soybean meal	5.0g
NaCl	5.0g
Growth factors	1.5g
Sheep blood, sterile defibrinated	50.0mL

pH 7.3 ± 0.2 at 25°C

Source: This medium is available from HiMedia.

Preparation of Medium: Add components to distilled/deionized water and bring volume to 1.0L. Mix thoroughly. Gently heat and bring to boiling. Distribute into tubes or flasks. Autoclave for 15 min at 15 psi pressure–121°C. Do not overheat. Cool to 50°C. Aseptically add blood. Mix thoroughly. Pour into sterile Petri dishes or leave in tubes.

Use: For use as a general-purpose medium for cultivation of various fastidious microorganisms.

Casein Soya Peptone Medium, HiVeg (Tryptone Soya Agar, HiVeg)

Composition per liter:

Agar	15.0g
Plant peptone	15.0g
NaCl	5.0g
Plant hydrolysate	5.0g

pH 7.3 ± 0.2 at 25°C

Source: This medium is available as a premixed powder from HiMedia.

Preparation of Medium: Add components to distilled/deionized water and bring volume to 1.0L. Mix thoroughly. Gently heat and bring to boiling. Distribute into tubes or flasks. Autoclave for 15 min at 15 psi pressure–121°C. Pour into sterile Petri dishes or leave in tubes.

Use: For the cultivation and maintenance of a wide variety of microorganisms.

Casein Yeast Extract Glucose Agar (CYG Agar)

Composition per liter:

Agar	20.0g
Glucose	5.0g
Casein hydrolysate	5.0g
Yeast extract	5.0g

pH 7.0 ± 0.2 at 25°C

Preparation of Medium: Add components to distilled/deionized water and bring volume to 1.0L. Mix thoroughly. Gently heat and bring to boiling. Distribute into tubes or flasks. Autoclave for 15 min at 15 psi pressure–121°C.

Use: For agar dilution susceptibility tests with imidazole antifungal agents.

Casein Yeast Extract Glucose Broth (CYG Broth)

Composition per liter:

Casein hydrolysate	5.0g
Glucose	5.0g
Yeast extract	5.0g

pH 7.0 ± 0.2 at 25°C

Preparation of Medium: Add components to distilled/deionized water and bring volume to 1.0L. Mix thoroughly. Gently heat and bring to boiling. Distribute into tubes or flasks. Autoclave for 15 min at 15 psi pressure–121°C.

Use: For agar dilution susceptibility tests with imidazole antifungal agents.

Casein Yeast Magnesium Agar

Composition per liter:

Agar	15.0g
Casein enzymatic hydrolysate	10.0g

NaCl 5.0g
Yeast extract 5.0g
$MgSO_4$ 0.98g

pH 7.0 ± 0.2 at 25°C

Source: This medium is available as a premixed powder from HiMedia.

Preparation of Medium: Add components to tap water and bring volume to 1.0L. Mix thoroughly. Gently heat and bring to boiling. Distribute into tubes or flasks. Autoclave for 15 min at 15 psi pressure–121°C. Pour into sterile Petri dishes or leave in tubes.

Use: For the cultivation of recombinant strains of *Escherichia coli.*

Casein Yeast Magnesium HiVeg Agar

Composition per liter:

Agar 15.0g
Plant hydrolysate 10.0g
NaCl 5.0g
Yeast extract 5.0g
$MgSO_4$ 0.98g

pH 7.0 ± 0.2 at 25°C

Source: This medium is available as a premixed powder from HiMedia.

Preparation of Medium: Add components to tap water and bring volume to 1.0L. Mix thoroughly. Gently heat and bring to boiling. Distribute into tubes or flasks. Autoclave for 15 min at 15 psi pressure–121°C. Pour into sterile Petri dishes or leave in tubes.

Use: For the cultivation of recombinant strains of *Escherichia coli.*

Casein Yeast Magnesium HiVeg Broth

Composition per liter:

Plant hydrolysate 10.0g
NaCl 5.0g
Yeast extract 5.0g
$MgSO_4$ 0.98g

pH 7.0 ± 0.2 at 25°C

Source: This medium is available as a premixed powder from HiMedia.

Preparation of Medium: Add components to tap water and bring volume to 1.0L. Mix thoroughly. Gently heat and bring to boiling. Distribute into tubes or flasks. Autoclave for 15 min at 15 psi pressure–121°C.

Use: For the cultivation of recombinant strains of *Escherichia coli.*

Casitone Agar

Composition per liter:

Pancreatic digest of casein 20.0g
Agar 15.0g
$MgSO_4 \cdot 7H_2O$ 1.0g
Potassium phosphate buffer
(0.01*M* solution, pH 7.2) 1.0L

pH 7.2 ± 0.2 at 25°C

Preparation of Medium: Combine components. Mix thoroughly. Gently heat to boiling. Distribute into tubes or flasks. Adjust pH to 7.2. Autoclave for 15 min at 15 psi pressure–121°C. Pour into sterile Petri dishes or leave in tubes.

Use: For the cultivation and maintenance of *Myxococcus* species.

Casitone Agar

Composition per liter:

Agar 12.0g
Beef extract 1.0g
Pancreatic digest of casein (Casitone) 1.0g
Glucose 1.0g

Preparation of Medium: Add components to distilled/deionized water and bring volume to 1.0L. Mix thoroughly. Gently heat and bring to boiling. Distribute into tubes or flasks. Autoclave for 15 min at 15 psi pressure–121°C. Pour into sterile Petri dishes or leave in tubes.

Use: For the cultivation and maintenance of *Myxococcus xanthus.*

Casitone Agar

Composition per liter:

Agar 15.0g
Casitone 3.0g
$CaCl_2$ 0.1g

pH 7.2 ± 0.2 at 25°C

Preparation of Medium: Add components to distilled/deionized water and bring volume to 1.0L. Mix thoroughly. Bring pH to 7.2. Gently heat and bring to boiling. Distribute into tubes or flasks. Autoclave for 15 min at 15 psi pressure–121°C. Pour into sterile Petri dishes or leave in tubes.

Use: For the cultivation of *Myxococcus species.*

Casitone Glycerol Yeast Autolysate Broth (CGY Autolysate Broth)

Composition per liter:

Pancreatic digest of casein 5.0g
Yeast autolysate 1.0g
Glycerol 10.0mL

Preparation of Medium: Add components to distilled/deionized water and bring volume to 1.0L. Mix thoroughly. Distribute into tubes or flasks. Autoclave for 15 min at 15 psi pressure–121°C.

Use: For the isolation, cultivation, and enumeration of iron and sulfur bacteria from the *Sphaerotilus* group.

Casitone Glycerol Yeast Autolysate HiVeg Broth Base with Glycerol (CGY)

Composition per liter:

Plant hydrolysate 5.0g
Yeast autolysate 1.0g
Glycerol 10.0mL

pH 7.2 ± 0.2 at 25°C

Source: This medium is available as a premixed powder from HiMedia.

Preparation of Medium: Add components to distilled/deionized water and bring volume to 1.0L. Mix thoroughly. Bring pH to 7.2. Gently heat and bring to boiling. Distribute into tubes or flasks. Autoclave for 15 min at 15 psi pressure–121°C.

Use: For the maintenance of iron bacteria especially those belonging to the *Sphaerotilus-Leptothrix* group.

Casitone Yeast Extract Agar

Composition per liter:

Agar	15.0g
Pancreatic digest of casein	5.0g
Yeast extract	3.0g
$MgSO_4 \cdot 7H_2O$	1.0g

Preparation of Medium: Add components to distilled/deionized water and bring volume to 1.0L. Mix thoroughly. Gently heat to boiling. Distribute into tubes or flasks. Autoclave for 15 min at 15 psi pressure–121°C. Pour into sterile Petri dishes or leave in tubes.

Use: For the cultivation and maintenance of *Chitinophaga pinensis*.

Casitone Yeast Extract Glucose Agar

Composition per liter:

Agar	15.0g
Casitone	10.0g
Glucose	5.0g
Yeast extract	5.0g
NaCl	5.0g

pH 7.3 ± 0.2 at 25°C

Preparation of Medium: Add components to distilled/deionized water and bring volume to 1.0L. Mix thoroughly. Gently heat and bring to boiling. Distribute into tubes or flasks. Autoclave for 15 min at 15 psi pressure–121°C. Pour into sterile Petri dishes or leave in tubes.

Use: For the cultivation and maintenance of *Arthrobacter ilicis*.

Casman Agar Base

Composition per liter:

Noble agar	14.0g
Proteose peptone No. 3	10.0g
Tryptose	10.0g
NaCl	5.0g
Beef extract	3.0g
Cornstarch	1.0g
Glucose	0.5g
p-Aminobenzoic acid	0.05g
Nicotinamide	0.05g
Blood	50.0mL
Water-lysed blood solution	1.5mL

pH 7.3 ± 0.2 at 25°C

Water-Lysed Blood Solution:

Composition per 8.0mL:

Blood	2.0mL

Preparation of Water-Lysed Blood Solution: Add blood to distilled/deionized water and bring volume to 8.0mL. Mix thoroughly. Filter sterilize.

Preparation of Medium: Add components, except blood and water-lysed blood solution, to distilled/deionized water and bring volume to 948.5mL. Mix thoroughly. Gently heat to boiling. Autoclave for 15 min at 15 psi pressure–121°C. Cool to 50°C. Aseptically add 50.0mL of sterile blood and 1.5mL of sterile water-lysed blood solution (one part blood to three parts water). Water-lysed blood may be omitted if sterile blood is partially lysed due to storage. Mix thoroughly. Pour into sterile Petri dishes or distribute into sterile tubes.

Use: For the isolation of fastidious bacteria from clinical specimens. For the cultivation under reduced oxygen tension of fastidious microorganisms such as *Haemophilus influenzae*, *Neisseria meningitidis*, and *Neisseria gonorrhoeae*.

Casman Agar Base with Rabbit Blood (Casman-Medium) (DSMZ Medium 439)

Composition per liter:

Noble agar	14.0g
Proteose peptone No. 3	10.0g
Tryptose	10.0g
NaCl	5.0g
Beef extract	3.0g
Cornstarch	1.0g
Glucose	0.5g
p-Aminobenzoic acid	0.05g
Nicotinamide	0.05g
Rabbit blood	50.0mL
Water-lysed blood solution	1.5mL

pH 7.3 ± 0.2 at 25°C

Source: Casman agar base is available as a premixed powder from BD Diagnostic Systems.

Water-Lysed Blood Solution:

Composition per 8.0mL:

Rabbit blood	2.0mL

Preparation of Water-Lysed Blood Solution: Add blood to distilled/deionized water and bring volume to 8.0mL. Mix thoroughly. Filter sterilize.

Preparation of Medium: Add components, except rabbit blood and water-lysed blood solution, to distilled/deionized water and bring volume to 950.0L. Mix thoroughly. Gently heat to boiling. Autoclave for 15 min at 15 psi pressure–121°C. Cool to 50°C. Aseptically add 50.0mL of sterile rabbit blood and 1.5mL of sterile water-lysed blood solution. Water-lysed blood may be omitted if sterile blood is partially lysed due to storage. Mix thoroughly. Pour into sterile Petri dishes or distribute into sterile tubes.

Use: For the cultivation and maintenance of *Gardnerella vaginalis*.

Casman HiVeg Agar Base with Blood

Composition per liter:

Agar	14.0g
Plant hydrolysate No. 1	10.0g
Plant peptone No. 3	10.0g
NaCl	5.0g
Plant extract	3.0g
Corn starch	1.0g
Glucose	0.5g
Nicotinamide	0.05g
p-Amino benzoic acid (PABA)	0.05g
Blood	50.0mL
Water-lysed blood solution	1.5mL

pH 7.3 ± 0.2 at 25°C

Source: This medium, without blood and water-lysed blood solution, is available as a premixed powder from HiMedia.

Water-Lysed Blood Solution:

Composition per 8.0mL:

Blood	2.0mL

Preparation of Water-Lysed Blood Solution: Add blood to distilled/deionized water and bring volume to 8.0mL. Mix thoroughly. Filter sterilize.

Preparation of Medium: Add components, except blood and water-lysed blood solution, to distilled/deionized water and bring volume to 948.5mL. Mix thoroughly. Gently heat to boiling. Autoclave for 15 min at 15 psi pressure–121°C. Cool to 50°C. Aseptically add 50.0mL of sterile blood and 1.5mL of sterile water-lysed blood solution (one part blood to three parts water). Water-lysed blood may be omitted if sterile blood is partially lysed due to storage. Mix thoroughly. Pour into sterile Petri dishes or distribute into sterile tubes.

Use: For the isolation of fastidious bacteria from clinical specimens. For the cultivation under reduced oxygen tension of fastidious microorganisms such as *Haemophilus influenzae*, *Neisseria meningitidis*, and *Neisseria gonorrhoeae*.

Casman HiVeg Broth Base with Blood

Composition per liter:

Plant hydrolysate No. 1	10.0g
Plant peptone No. 3	10.0g
NaCl	5.0g
Plant extract	3.0g
Corn starch	1.0g
Glucose	0.5g
Nicotinamide	0.05g
p-Amino benzoic acid (PABA)	0.05g
Blood	50.0mL
Water-lysed blood solution	1.5mL

pH 7.3 ± 0.2 at 25°C

Source: This medium, without blood and water-lysed blood solution, is available as a premixed powder from HiMedia.

Water-Lysed Blood Solution:
Composition per 8.0mL:

Blood	2.0mL

Preparation of Water-Lysed Blood Solution: Add blood to distilled/deionized water and bring volume to 8.0mL. Mix thoroughly. Filter sterilize.

Preparation of Medium: Add components, except blood and water-lysed blood solution, to distilled/deionized water and bring volume to 948.5mL. Mix thoroughly. Gently heat to boiling. Autoclave for 15 min at 15 psi pressure–121°C. Cool to 50°C. Aseptically add 50.0mL of sterile blood and 1.5mL of sterile water-lysed blood solution (one part blood to three parts water). Water-lysed blood may be omitted if sterile blood is partially lysed due to storage. Mix thoroughly.

Use: For the cultivation of fastidious bacteria from clinical specimens. For the cultivation under reduced oxygen tension of fastidious microorganisms such as *Haemophilus influenzae*, *Neisseria meningitidis*, and *Neisseria gonorrhoeae*.

CASO Agar
See: **Tryptone Soya Agar**

CASO Bouillon
See: **Tryptone Soya Agar**

CASO MUG Agar

Composition per liter:

Casein peptone	16.0g
Agar	13.0g
NaCl	6.0g
Soy peptone	5.0g
Tryptophan	1.0g
4-Methylumbelliferyl-β-D-glucuronide	0.07g

pH 7.3 ± 0.2 at 25°C

Source: This medium is available from Fluka, Sigma-Aldrich.

Preparation of Medium: Add components to distilled/deionized water and bring volume to 1.0L. Mix thoroughly. Gently heat while stirring and bring to boiling. Autoclave for 15 min at 15 psi pressure–121°C. Cool to 50°C. Pour into sterile Petri dishes.

Use: This universal medium without indicator or inhibitor is intended for a broad range of application, including enumeration and cultivation of a wide variety of microorganisms. It is also suitable for the cultivation of more fastidious microorganisms. β-D-glucoronidase, which is produced by *E. coli*, cleaves 4-methylumbelliferyl-β-D-glucuronide to 4-methylumbelliferone and glucuronide. The fluorogen 4-methylumbelliferone can be detected under a long wavelength UV lamp. A positive indole reaction provides confirmation.

Castenholz Agar, Modified (DSMZ Medium 86a)

Composition per liter:

Agar	25.0g
$NaNO_3$	0.69g
Na_2HPO_4	0.14g
KNO_3	0.103g
$MgSO_4 \cdot 7H_2O$	0.1g
Nitrilotriacetic acid	0.1g
$CaSO_4 \cdot 2H_2O$	0.06g
NaCl	8.0mg
$MnSO_4 \cdot H_2O$	2.2mg
$ZnSO_4 \cdot 7H_2O$	0.5mg
H_3BO_3	0.5mg
$FeCl_3$	0.47mg
$CoCl_2 \cdot 6H_2O$	46.0μg
$CuSO_4 \cdot 5H_2O$	25.0μg
$Na_2MoO_4 \cdot 2H_2O$	25.0μg

pH 7.8 ± 0.2 at 25°C

Preparation of Medium: Add nitrilotriacetic acid to 500.0mL of distilled/deionized water. Dissolve by adjusting pH to 6.5 with KOH. Add remaining components. Mix thoroughly. Readjust pH to 7.8. Bring volume to 1.0L with distilled/deionized water. Mix thoroughly. Gently heat and bring to boil. Autoclave for 15 min at 15 psi pressure–121°C. Pour into plates or aseptically distribute to sterile tubes or flasks.

Use: For the cultivation of *Meiothermus taiwanensis.*

Castenholz D Medium (Medium D)

Composition per liter:

$NaNO_3$	0.7g
Na_2HPO_4	0.11g
KNO_3	0.1g
$MgSO_4 \cdot 7H_2O$	0.1g
Nitrilotriacetic acid	0.1g
$CaSO_4 \cdot 2H_2O$	0.06g
NaCl	8.0mg
$FeCl_3$ solution	1.0mL
Micronutrient solution	0.5mL

pH 7.5 ± 0.2 at 25°C

$FeCl_3$ Solution:

Composition per liter:

$FeCl_3 \cdot 6H_2O$	2.28g

Preparation of $FeCl_3$ Solution: Add $FeCl_3 \cdot 6H_2O$ to distilled/deionized water and bring volume to 1.0L. Mix thoroughly.

Micronutrient Solution:

Composition per liter:

$MnSO_4 \cdot H_2O$	2.28g
H_3BO_3	0.5g
$ZnSO_4 \cdot 7H_2O$	0.5g
$CoCl_2 \cdot 6H_2O$	0.025g
$CuSO_4 \cdot 5H_2O$	0.025g
$Na_2MoO_4 \cdot 2H_2O$	0.025g
H_2SO_4	0.5mL

Preparation of Micronutrient Solution: Add components to distilled/deionized water and bring volume to 1.0L. Mix thoroughly.

Preparation of Medium: Add nitrilotriacetic acid to 500.0mL of distilled/deionized water. Dissolve by adjusting pH to 6.5 with KOH. Add remaining components. Mix thoroughly. Readjust pH to 7.5. Bring volume to 1.0L with distilled/deionized water. Mix thoroughly. Distribute into tubes or flasks. Autoclave for 15 min at 15 psi pressure–121°C.

Use: For the isolation of cyanobacteria, including thermophilic species. For the cultivation of *Chloroflexus* species and *Fischerella* species.

Castenholz D Medium, Modified (Medium D, Modified)

Composition per liter:

NaCl	160.0g
$NaNO_3$	0.69g
Na_2HPO_4	0.111g
KNO_3	0.103g
$MgSO_4 \cdot 7H_2O$	0.1g
Nitrilotriacetic acid	0.1g
$CaSO_4 \cdot 2H_2O$	0.06g
$FeCl_3$	0.3mg
Trace metal solution, Castenholz	1.0mL

pH 7.5 ± 0.2 at 25°C

Trace Metal Solution, Castenholz:

Composition per liter:

$MnSO_4 \cdot H_2O$	2.28g
H_3BO_3	0.5g
$ZnSO_4 \cdot 7H_2O$	0.5g
$Co(NO_3)_2 \cdot 6H_2O$	0.025g
$CuSO_4 \cdot 5H_2O$	0.025g
$Na_2MoO_4 \cdot 2H_2O$	0.025g
H_2SO_4	0.5mL

Preparation of Trace Metal Solution, Castenholz: Add components to distilled/deionized water and bring volume to 1.0L. Mix thoroughly.

Preparation of Medium: Add nitrilotriacetic acid to 500.0mL of distilled/deionized water. Dissolve by adjusting pH to 6.5 with KOH. Add remaining components. Mix thoroughly. Readjust pH to 7.5. Bring volume to 1.0L with distilled/deionized water. Mix thoroughly. Distribute into screw-capped tubes or flasks. Autoclave for 15 min at 15 psi pressure–121°C.

Use: For the isolation of halophilic cyanobacteria.

Castenholz DG Medium (Medium DG)

Composition per liter:

Glycyl-glycine buffer	0.8g
$NaNO_3$	0.7g
Na_2HPO_4	0.11g
KNO_3	0.1g
$MgSO_4 \cdot 7H_2O$	0.1g
Nitrilotriacetic acid	0.1g
$CaSO_4 \cdot 2H_2O$	0.06g
NaCl	8.0mg
$FeCl_3$ solution	1.0mL
Micronutrient solution	0.5mL

pH 7.5 ± 0.2 at 25°C

$FeCl_3$ Solution:

Composition per liter:

$FeCl_3 \cdot 6H_2O$	2.28g

Preparation of $FeCl_3$ Solution: Add $FeCl_3 \cdot 6H_2O$ to distilled/deionized water and bring volume to 1.0L. Mix thoroughly.

Micronutrient Solution:

Composition per liter:

$MnSO_4 \cdot H_2O$	2.28g
H_3BO_3	0.5g
$ZnSO_4 \cdot 7H_2O$	0.5g
$CoCl_2 \cdot 6H_2O$	0.025g
$CuSO_4 \cdot 5H_2O$	0.025g
$Na_2MoO_4 \cdot 2H_2O$	0.025g
H_2SO_4	0.5mL

Preparation of Micronutrient Solution: Add components to distilled/deionized water and bring volume to 1.0L. Mix thoroughly.

Preparation of Medium: Add nitrilotriacetic acid to 500.0mL of distilled/deionized water. Dissolve by adjusting pH to 6.5 with KOH. Add remaining components. Mix thoroughly. Readjust pH to 8.1. Bring volume to 1.0L with distilled/deionized water. Mix thoroughly. Distribute into tubes or flasks. Autoclave for 15 min at 15 psi pressure–121°C.

Use: For the isolation of cyanobacteria, including thermophilic species.

Castenholz DGN Medium (Medium DGN)

Composition per liter:

Glycyl-glycine buffer	0.8g
$NaNO_3$	0.7g
NH_4Cl	0.2g
Na_2HPO_4	0.11g
KNO_3	0.1g
$MgSO_4 \cdot 7H_2O$	0.1g
Nitrilotriacetic acid	0.1g
$CaSO_4 \cdot 2H_2O$	0.06g
NaCl	8.0mg
$FeCl_3$ solution	1.0mL
Micronutrient solution	0.5mL

pH 7.5 ± 0.2 at 25°C

$FeCl_3$ Solution:

Composition per liter:

$FeCl_3 \cdot 6H_2O$	2.28g

Preparation of $FeCl_3$ Solution: Add $FeCl_3 \cdot 6H_2O$ to distilled/deionized water and bring volume to 1.0L. Mix thoroughly.

Micronutrient Solution:

Composition per liter:

$MnSO_4 \cdot H_2O$	2.28g
H_3BO_3	0.5g
$ZnSO_4 \cdot 7H_2O$	0.5g
$CoCl_2 \cdot 6H_2O$	0.025g
$CuSO_4 \cdot 5H_2O$	0.025g
$Na_2MoO_4 \cdot 2H_2O$	0.025g
H_2SO_4	0.5mL

Preparation of Micronutrient Solution: Add components to distilled/deionized water and bring volume to 1.0L. Mix thoroughly.

Preparation of Medium: Add nitrilotriacetic acid to 500.0mL of distilled/deionized water. Dissolve by adjusting pH to 6.5 with KOH. Add remaining components. Mix thoroughly. Readjust pH to 8.2. Bring volume to 1.0L with distilled/deionized water. Mix thoroughly. Distribute into tubes or flasks. Autoclave for 15 min at 15 psi pressure–121°C.

Use: For the isolation of cyanobacteria, including thermophilic species.

Castenholz Medium

Composition per liter:

Tryptone	1.0g
Yeast extract	1.0g
$NaNO_3$	689.0mg
$Na_2HPO_4 \cdot 2H_2O$	140.0mg
KNO_3	103.0mg
$MgSO_4 \cdot 7H_2O$	100.0mg
Nitrilotriacetic acid	100.0mg
$CaSO_4 \cdot 2H_2O$	60.0mg
NaCl	8.0mg
$MnSO_4 \cdot H_2O$	2.2mg
H_3BO_3	0.5mg
$ZnSO_4 \cdot 7H_2O$	0.5mg
$FeCl_3 \cdot 6H_2O$	0.47mg
$CoCl_2 \cdot 6H_2O$	46.0μg
$CuSO_4 \cdot 5H_2O$	25.04μg
$Na_2MoO_4 \cdot 2H_2O$	25.0μg

pH 8.2 ± 0.2 at 25°C

Preparation of Medium: Combine components. Mix thoroughly. Adjust pH to 8.2 with NaOH. Distribute into tubes or flasks. Autoclave for 15 min at 15 psi pressure–121°C.

Use: For the cultivation of *Thermus aquaticus*.

Castenholz Medium, Modified (DSMZ Medium 86a)

Composition per liter:

$NaNO_3$	0.69g
Na_2HPO_4	0.14g
KNO_3	0.103g
$MgSO_4 \cdot 7H_2O$	0.1g
Nitrilotriacetic acid	0.1g
$CaSO_4 \cdot 2H_2O$	0.06g
NaCl	8.0mg
$MnSO_4 \cdot H_2O$	2.2mg
$ZnSO_4 \cdot 7H_2O$	0.5mg
H_3BO_3	0.5mg
$FeCl_3$	0.47mg
$CoCl_2 \cdot 6H_2O$	46.0μg
$CuSO_4 \cdot 5H_2O$	25.0μg
$Na_2MoO_4 \cdot 2H_2O$	25.0μg

pH 7.8 ± 0.2 at 25°C

Preparation of Medium: Add nitrilotriacetic acid to 500.0mL of distilled/deionized water. Dissolve by adjusting pH to 6.5 with KOH. Add remaining components. Mix thoroughly. Readjust pH to 7.8. Bring volume to 1.0L with distilled/deionized water. Mix thoroughly. Distribute into screw-capped tubes or flasks. Autoclave for 15 min at 15 psi pressure–121°C.

Use: For the cultivation of *Meiothermus taiwanensis*.

Castenholz ND Medium (Medium ND)

Composition per liter:

Na_2HPO_4	0.11g
$MgSO_4 \cdot 7H_2O$	0.1g
Nitrilotriacetic acid	0.1g
$CaSO_4 \cdot 2H_2O$	0.06g
NaCl	8.0mg
$FeCl_3$ solution	1.0mL
Micronutrient solution	0.5mL

pH 7.5 ± 0.2 at 25°C

$FeCl_3$ Solution:

Composition per liter:

$FeCl_3 \cdot 6H_2O$	2.28g

Preparation of $FeCl_3$ Solution: Add $FeCl_3 \cdot 6H_2O$ to distilled/deionized water and bring volume to 1.0L. Mix thoroughly.

Micronutrient Solution:

Composition per liter:

$MnSO_4 \cdot H_2O$	2.28g
H_3BO_3	0.5g
$ZnSO_4 \cdot 7H_2O$	0.5g
$CoCl_2 \cdot 6H_2O$	0.025g
$CuSO_4 \cdot 5H_2O$	0.025g
$Na_2MoO_4 \cdot 2H_2O$	0.025g
H_2SO_4	0.5mL

Preparation of Micronutrient Solution: Add components to distilled/deionized water and bring volume to 1.0L. Mix thoroughly.

Preparation of Medium: Add nitrilotriacetic acid to 500.0mL of distilled/deionized water. Dissolve by adjusting pH to 6.5 with KOH. Add remaining components. Mix thoroughly. Readjust pH to 8.2. Bring volume to 1.0L with distilled/deionized water. Mix thoroughly. Distribute into tubes or flasks. Autoclave for 15 min at 15 psi pressure–121°C.

Use: For the isolation of cyanobacteria, including thermophilic species, that require reduced nitrogen concentrations.

Castenholz TYE Medium (Castenholz Trypticase™ Yeast Extract Medium)

Composition per liter:

Castenholz salts, 2X	500.0mL
1% TYE	100.0mL

pH 7.6 ± 0.2 at 25°C

Castenholz Salts, 2X:
Composition per liter:

Agar	30.0g
$NaNO_3$	1.4g
Na_2HPO_4	0.22g
KNO_3	0.21g
Nitrilotriacetic acid	0.2g
$MgSO_4{\cdot}7H_2O$	0.2g
$CaSO_4{\cdot}2H_2O$	0.12g
NaCl	0.016g
$FeCl_3$ (0.03% solution)	2.0mL
Nitsch's trace elements	2.0mL

Preparation of Castenholz Salts, 2X: Add components to distilled/deionized water and bring volume to 1.0L. Mix thoroughly. Gently heat and bring to boiling. Adjust pH to 8.2. Autoclave for 15 min at 15 psi pressure–121°C.

Nitsch's Trace Elements:
Composition per liter:

$MnSO_4$	2.2g
H_3BO_3	0.5g
$ZnSO_4$	0.5g
$CoCl_2{\cdot}6H_2O$	0.046g
Na_2MoO_4	0.025g
$CuSO_4$	0.016g
H_2SO_4	0.5mL

Preparation of Nitsch's Trace Elements: Add components to distilled/deionized water and bring volume to 1.0L. Mix thoroughly.

1% TYE
Composition per liter:

Pancreatic digest of casein	10.0g
Yeast extract	10.0g

Preparation of 1% TYE: Add components to distilled/deionized water and bring volume to 1.0L. Mix thoroughly. Autoclave for 15 min at 15 psi pressure–121°C.

Preparation of Medium: Aseptically combine 500.0mL of sterile Castenholz salts, 2X, 100.0mL of sterile 1% TYE, and 400.0mL of sterile distilled/deionized water. Adjust pH to 7.6.

Use: For the cultivation and maintenance of *Thermonema lapsum* and *Thermus* species.

Castenholz TYE Medium with 2% Trypticase™ Yeast Extract

Composition per liter:

Castenholz salts, 2X	500.0mL
2% TYE	100.0mL

pH 7.6 ± 0.2 at 25°C

Castenholz Salts, 2X:
Composition per liter:

Agar	30.0g
$NaNO_3$	1.4g
Na_2HPO_4	0.22g
KNO_3	0.21g
$MgSO_4{\cdot}7H_2O$	0.2g
Nitrilotriacetic acid	0.2g
$CaSO_4{\cdot}2H_2O$	0.12g
NaCl	0.016g
$FeCl_3$ solution (0.03% solution)	2.0mL
Nitsch's trace elements	2.0mL

Preparation of Castenholz Salts, 2X: Add components to distilled/deionized water and bring volume to 1.0L. Mix thoroughly. Gently heat and bring to boiling. Adjust pH to 8.2. Autoclave for 15 min at 15 psi pressure–121°C.

Nitsch's Trace Elements:
Composition per liter:

$MnSO_4$	2.2g
H_3BO_3	0.5g
$ZnSO_4$	0.5g
$CoCl_2{\cdot}6H_2O$	0.046g
Na_2MoO_4	0.025g
$CuSO_4$	0.016g
H_2SO_4	0.5mL

Preparation of Nitsch's Trace Elements: Add components to distilled/deionized water and bring volume to 1.0L. Mix thoroughly.

2% TYE
Composition per liter:

Pancreatic digest of casein	20.0g
Yeast extract	20.0g

Preparation of 2% TYE: Add components to distilled/deionized water and bring volume to 1.0L. Mix thoroughly. Autoclave for 15 min at 15 psi pressure–121°C.

Preparation of Medium: Aseptically combine 500.0mL of sterile Castenholz salts, 2X, 100.0mL of sterile 2% TYE, and 400.0mL of sterile distilled/deionized water. Adjust pH to 7.6.

Use: For the cultivation and maintenance of *Thermus* species.

CAT Medium (*Campylobacter* Blood Free Preson Agar with Cefoperazone, Amphotericin, and Teicoplanin)

Composition per liter:

Agar	12.0g
Beef extract	10.0g
Peptone	10.0g
NaCl	5.0g
Charcoal	4.0g
Casein hydrolysate	3.0g
Sodium deoxycholate	1.0g
$FeSO_4$	0.25g
Sodium pyruvate	0.25g
Selective supplement solution	10.0mL

pH 7.5 ± 0.2 at 25°C

Source: This medium is available as a premixed powder from Oxoid Unipath.

Selective Supplement Solution:
Composition per 10.0mL:

Amphotericin	10.0mg
Sodium cefoperazone	8.0mg
Teicoplanin	4.0mg

Preparation of Selective Supplement Solution: Add sodium cefoperazone to distilled/deionized water and bring volume to 10.0mL. Mix thoroughly. Filter sterilize.

Preparation of Medium: Add components, except selective supplement solution, to distilled/deionized water and bring volume to 990.0mL. Mix thoroughly. Gently heat and bring to boiling. Autoclave for 15 min at 15 psi pressure–121°C. Cool to 45°–50°C. Aseptically add 10.0mL of sterile selective supplement solution. Mix thoroughly. Pour into sterile Petri dishes or distribute into sterile tubes.

Use: For the cultivation of *Campylobacter* species. For the isolation of Campylobacter spp., especially *Campylobacter upsaliensis*.

Catenococcus Agar

Composition per 1001.0mL:

Agar, noble	20.0g
NaCl	20.0g
K_2HPO_4	5.54g
NH_4Cl	0.5g
$MgSO_4 \cdot 7H_2O$	0.3g
KH_2PO_4	1.84g
Sodium acetate	0.82g
$CaCl_2 \cdot 2H_2O$	0.1g
Yeast extract	0.05g
Trace elements solution SL-6	1.0mL

pH 6.8 ± 0.2 at 25°C

Trace Elements Solution SL-6:
Composition per liter:

$MnCl_2 \cdot 4H_2O$	0.5g
H_3BO_3	0.3g
$CoCl_2 \cdot 6H_2O$	0.2g
$ZnSO_4 \cdot 7H_2O$	0.1g
$Na_2MoO_4 \cdot 2H_2O$	0.03g
$NiCl_2 \cdot 6H_2O$	0.02g
$CuCl_2 \cdot 2H_2O$	0.01g

Preparation of Trace Elements Solution SL-6: Add components to distilled/deionized water and bring volume to 1.0L. Mix thoroughly.

Preparation of Medium: Add components to distilled/deionized water and bring volume to 1.0L. Mix thoroughly. Gently heat and bring to boiling. Adjust pH to 6.8. Distribute into tubes or flasks. Autoclave for 15 min at 15 psi pressure–121°C. Pour into sterile Petri dishes or leave in tubes.

Use: For the cultivation of *Catenococcus thiocyclus*.

Catenococcus Medium

Composition per 1003.0mL:

Agar	15.0g
NaCl	15.0g
K_2HPO_4	1.0g
NH_4Cl	0.5g
KH_2PO_4	0.15g
$CaCl_2 \cdot 2H_2O$ solution	10.0mL
$MgSO_4 \cdot 7H_2O$ solution	10.0mL
Sodium acetate solution	10.0mL
Yeast extract solution	2.0mL
Trace elements solution SL-4	1.0mL

pH 7.0 ± 0.2 at 25°C

$CaCl_2 \cdot 2H_2O$ Solution:
Composition per 10.0mL:

$CaCl_2 \cdot 2H_2O$	0.1g

Preparation of $CaCl_2 \cdot 2H_2O$ Solution: Add $CaCl_2 \cdot 2H_2O$ to distilled/deionized water and bring volume to 10.0mL. Mix thoroughly. Autoclave for 15 min at 15 psi pressure–121°C.

$MgSO_4 \cdot 7H_2O$ Solution:
Composition per 10.0mL:

$MgSO_4 \cdot 7H_2O$	1.0g

Preparation of $MgSO_4 \cdot 7H_2O$ Solution: Add $MgSO_4 \cdot 7H_2O$ to distilled/deionized water and bring volume to 10.0mL. Mix thoroughly. Autoclave for 15 min at 15 psi pressure–121°C.

Sodium Acetate Solution:
Composition per 10.0mL:

Sodium acetate	0.82g

Preparation of Sodium Acetate Solution: Add sodium acetate to distilled/deionized water and bring volume to 10.0mL. Mix thoroughly. Autoclave for 15 min at 15 psi pressure–121°C.

Yeast Extract Solution:
Composition per 10.0mL:

Yeast extract	1.0g

Preparation of Yeast Extract Solution: Add yeast extract to distilled/deionized water and bring volume to 10.0mL. Mix thoroughly. Autoclave for 15 min at 15 psi pressure–121°C.

Trace Elements Solution SL-4:
Composition per liter:

EDTA	0.5g
$FeSO_4 \cdot 7H_2O$	0.2g
Trace elements solution SL-6	100.0mL

Trace Elements Solution SL-6:
Composition per liter:

$MnCl_2 \cdot 4H_2O$	0.5g
H_3BO_3	0.3g
$CoCl_2 \cdot 6H_2O$	0.2g
$ZnSO_4 \cdot 7H_2O$	0.1g
$Na_2MoO_4 \cdot 2H_2O$	0.03g
$NiCl_2 \cdot 6H_2O$	0.02g
$CuCl_2 \cdot 2H_2O$	0.01g

Preparation of Trace Elements Solution SL-6: Add components to distilled/deionized water and bring volume to 1.0L. Mix thoroughly.

Preparation of Trace Elements Solution SL-4: Add components to distilled/deionized water and bring volume to 1.0L. Mix thoroughly. Filter sterilize.

Preparation of Medium: Add components, except $CaCl_2 \cdot 2H_2O$ solution, $MgSO_4 \cdot 7H_2O$ solution, sodium acetate solution, yeast extract solution, and trace elements solution SL-4, to distilled/deionized water and bring volume to 970.0mL. Mix thoroughly. Gently heat and bring to boiling. Autoclave for 15 min at 15 psi pressure–121°C. Cool to 50°–55°C. Aseptically add 10.0mL of sterile $CaCl_2 \cdot 2H_2O$ solution, 10.0mL of sterile $MgSO_4 \cdot 7H_2O$ solution, 10.0mL of sterile sodium acetate solution, 2.0mL of sterile yeast extract solution, and 1.0mL of sterile trace elements solution SL-4. Mix thoroughly. Pour into sterile Petri dishes or distribute into sterile tubes.

Use: For the cultivation of *Catenococcus thiocyclus*.

Caulobacter Medium

Composition per liter:

Agar	10.0g
Peptone	2.0g
Yeast extract	1.0g
$MgSO_4 \cdot 7H_2O$	0.2g
Riboflavin	1.0mg

pH 7.0 ± 0.2 at 25°C

Preparation of Medium: Add components to tap water and bring volume to 1.0L. Mix thoroughly. Gently heat and bring to boiling. Distribute into tubes or flasks. Autoclave for 15 min at 15 psi pressure–121°C. Pour into sterile Petri dishes or leave in tubes.

Use: For the cultivation of *Caulobacter* species from fresh water.

Caulobacter Medium

Composition per liter:

Agar	10.0g
Peptone	2.0g
Yeast extract	1.0g
$MgSO_4 \cdot 7H_2O$	0.2g

pH 7.0 ± 0.2 at 25°C

Source: This medium is available from HiMedia.

Preparation of Medium: Add components to tap water and bring volume to 1.0L. Mix thoroughly. Gently heat and bring to boiling. Distribute into tubes or flasks. Autoclave for 15 min at 15 psi pressure–121°C. Pour into sterile Petri dishes or leave in tubes.

Use: For the cultivation of *Caulobacter* species from fresh water.

Caulobacter Medium

Composition per liter:

Agar	10.0g
Peptone	0.5g
Seawater, filtered	1.0L

pH 7.0 ± 0.2 at 25°C

Preparation of Medium: Combine components. Mix thoroughly. Gently heat and bring to boiling. Distribute into tubes or flasks. Autoclave for 15 min at 15 psi pressure–121°C. Pour into sterile Petri dishes or leave in tubes.

Use: For the cultivation of *Caulobacter* species from marine isolates.

Caulobacter Medium

Composition per liter:

Glucose	1.0g
Peptone	1.0g
Yeast extract	1.0g
Salt solution	100.0mL

Salt Solution:

Composition per 100.0mL:

EDTA	0.1g
KNO_3	0.1g
K_2HPO_4	0.066g
$MgSO_4$	0.033g
$FeSO_4 \cdot 7H_2O$	9.3mg
$NaBO_3 \cdot 4H_2O$	2.63mg
$MgCl_2 \cdot 4H_2O$	1.81mg
$CaCl_2$	1.2mg
$(NH_4)_6Mo_7O_{24} \cdot 7H_2O$	1.0mg
$ZnSO_4 \cdot 7H_2O$	0.22mg
$CuSO_4 \cdot 5H_2O$	0.079mg
$Co(NO_3)_2 \cdot H_2O$	0.02mg

Preparation of Salt Solution: Add components to distilled/deionized water and bring volume to 100.0mL. Mix thoroughly.

Preparation of Medium: Add components to distilled/deionized water and bring volume to 1.0L. Mix thoroughly. Distribute into tubes or flasks. Autoclave for 15 min at 15 psi pressure–121°C.

Use: For the enrichment of *Stella* species from polluted waters.

Caulobacter Medium

Composition per liter:

Agar	10.0g
Peptone	2.0g
Yeast extract	1.0g
$MgSO_4 \cdot 7H_2O$	0.2g

Preparation of Medium: Add components to tap water and bring volume to 1.0L. Mix thoroughly. Gently heat to boiling. Distribute into tubes or flasks. Autoclave for 15 min at 15 psi pressure–121°C. Pour into sterile Petri dishes or leave in tubes.

Use: For the cultivation and maintenance of *Asticcacaulis excentricus, Caulobacter* species, *Labrys monachus, Pedomicrobium* species, *Pirellula staleyi, Pseudomonas carboxydohydrogena,* and *Stella* species.

Caulobacter Medium II

Composition per liter:

Peptone	10.0g
Yeast extract	3.0g
Seawater	1.0L

pH 7.2–7.4 at 25°C

Preparation of Medium: Add components to filtered aged seawater and bring volume to 1.0L. Mix thoroughly. Adjust pH to 7.2–7.4. Distribute into tubes or flasks. Autoclave for 15 min at 15 psi pressure–121°C.

Use: For the cultivation of *Caulobacter halobacteroides* and *Caulobacter maris.*

Caulobacter Medium with Riboflavin

Composition per liter:

Peptone	10.0g
Yeast extract	3.0g
Riboflavin	1.0mg
Seawater	1.0L

pH 7.2–7.4 at 25°C

Preparation of Medium: Add components to filtered aged seawater and bring volume to 1.0L. Mix thoroughly. Adjust pH to 7.3. Distribute into tubes or flasks. Autoclave for 15 min at 15 psi pressure–121°C.

Use: For the cultivation of *Caulobacter vibrioides.*

CBI Agar

See: ***Clostridium botulinum*** **Isolation Agar**

CC Medium

Composition per liter:

Agar	20.0g
KH_2PO_4	4.0g
Potato starch	0.5g
Solution 3	100.0mL
Solution 1	10.0mL

pH 7.3 ± 0.2 at 25°C

Solution 1:

Composition per liter:

$MgSO_4 \cdot 7H_2O$	20.0g
$CaCl_2 \cdot 2H_2O$	2.0g
$FeSO_4 \cdot 7H_2O$	0.4g
H_3BO_3	0.02g
$MnSO_4 \cdot 2H_2O$	0.015g
$NaMoO_4 \cdot 2H_2O$	0.015g

KI	0.01g
$ZnSO_4$	4.0mg
$CoCl_2 \cdot 4H_2O$	0.4mg
$CuSO_4 \cdot 5H_2O$	0.4mg

Preparation of Solution 1: Add components to distilled/deionized water and bring volume to 1.0L. Mix thoroughly. Adjust pH with 10.0mL of 10% HCl solution.

Solution 3:

Composition per 100.0mL:

Pancreatic digest of casein	12.0g
Yeast extract	12.0g
L-Cysteine·HCl	0.5g
L-Asparagine	0.03g
DL-Tryptophan	0.02g
Solution 2	12.0mL

Preparation of Solution 3: Add components to distilled/deionized water and bring volume to 100.0mL. Mix thoroughly. Filter sterilize.

Solution 2:

Composition per 100.0mL:

p-Aminobenzoic acid	0.02g
Calcium pantothenate	0.02g
m-Inositol	0.02g
Pyridoxine·HCl	0.02g
Thiamine·HCl	0.02g
Nicotinamide	0.01g
Nicotinic acid	0.01g
Folic acid	5.0mg
Biotin	1.0mg
Vitamin B_{12}	1.0mL

Preparation of Solution 2: Add components to distilled/deionized water and bring volume to 100.0mL. Mix thoroughly.

Preparation of Medium: Add KH_2PO_4 to distilled/deionized water and bring volume to 250.0mL. Mix thoroughly. Adjust pH to 7.6 with NaOH. Add 10.0mL of solution 1. In a separate flask, add potato starch to 70.0mL of boiling distilled/deionized water. Add potato starch solution to other solution. Add agar. Bring volume to 900.0mL of distilled/deionized water. Autoclave for 15 min at 15 psi pressure–121°C. Cool to 45°–50°C. Aseptically add 100.0mL of sterile solution 3. Mix thoroughly. Pour into sterile Petri dishes or distribute into sterile tubes.

Use: For the isolation and cultivation of *Actinomycetes* species.

CCD Agar with Pyruvate and Cefazolin (Blood-free Selective Medium)

Composition per liter:

Agar	12.0g
Beef extract	10.0g
Peptone	10.0g
NaCl	5.0g
Charcoal, bacteriological	4.0g
Casein hydrolysate	3.0g
Sodium deoxycholate solution	10.0mL
$FeSO_4$ solution	5.0mL
Sodium pyruvate solution	5.0mL
Cefazolin solution	1.0mL

pH 7.4 ± 0.2 at 25°C

Source: This medium, without deoxycholate and cefazolin solutions, is available as a premixed powder from HiMedia.

$FeSO_4$ Solution:

Composition per 10.0mL:

$FeSO_4$	0.5g

Preparation of $FeSO_4$ Solution: Add $FeSO_4$ to distilled/deionized water and bring volume to 10.0mL. Mix thoroughly. Gently heat while stirring and bring to boiling. Autoclave for 15 min at 15 psi pressure–121°C. Cool 25°C.

Sodium Pyruvate Solution:

Composition per 10.0mL:

Sodium pyruvate	0.5g

Preparation of Sodium Pyruvate Solution: Add sodium pyruvate to distilled/deionized water and bring volume to 10.0mL. Mix thoroughly. Gently heat while stirring and bring to boiling. Autoclave for 15 min at 15 psi pressure–121°C. Cool 25°C.

Sodium Deoxycholate Solution:

Composition per 100.0mL:

Sodium deoxycholate	10.0g

Preparation of Sodium Deoxycholate Solution: Add sodium deoxycholate to distilled/deionized water and bring volume to 100.0mL. Mix thoroughly. Gently heat while stirring and bring to boiling. Autoclave for 15 min at 15 psi pressure–121°C. Cool 25°C.

Cefazolin Solution:

Composition per 10.0mL:

Cefazolin	0.1g

Preparation of Cefazolin Solution: Add cefazolin to distilled/deionized water and bring volume to 10.0mL. Mix thoroughly. Filter sterilize.

Preparation of Medium: Add components, except cefazolin solution and sodium deoxycholate solution, to distilled/deionized water and bring volume to 990.0mL. Mix thoroughly. Heat with frequent agitation and boil for 1 min to completely dissolve. Autoclave for 15 min at 15 psi pressure–121°C. Cool to 50°–55°C. Add 10.0mL of sterile sodium deoxycholate solution and 1.0mL of sterile cefazolin solution. Mix thoroughly. Pour into sterile Petri dishes.

Use: For the selective isolation of *Campylobacter* species, especially *Campylobacter jejuni* from human feces.

CCDA

See: ***Campylobacter*** **Charcoal Differential Agar**

CCFA

See: ***Clostridium difficile*** **Agar**

CCVC Medium (Cephalothin Cycloheximide Vancomycin Colistin Medium

Composition per liter:

BCYE-alpha base	990.0mL
Antibiotic supplement solution	10.0mL

pH 6.9 ± 0.2 at 25°C

Source: This medium is available as a premixed powder from BD Diagnostic Systems.

BCYE-Alpha Base:

Composition per liter:

Agar	15.0g
Yeast extract	10.0g

ACES buffer (2-[(2-amino-2-oxoethyl)-amino]-ethane sulfonic acid) 10.0g
Charcoal, activated 2.0g
α-Ketoglutarate 1.0g
$Fe_4(P_2O_7)_3 \cdot 9H_2O$ 0.25g
L-Cysteine·HCl·H_2O solution 10.0mL

L-Cysteine·HCl·H_2O Solution:
Composition per 10.0mL:
L-Cysteine·HCl·H_2O 0.4g

Preparation of L-Cysteine·HCl·H_2O Solution: Add L-cysteine·HCl·H_2O to distilled/deionized water and bring volume to 10.0mL. Mix thoroughly. Filter sterilize.

Preparation of BCYE-Alpha Base: Add components, except L-cysteine·HCl·H_2O solution, to distilled/deionized water and bring volume to 990.0mL. Mix thoroughly. Adjust medium to pH 6.9 with 1*N* KOH. Heat gently and bring to boiling for 1 min. Autoclave for 15 min at 15 psi pressure–121°C. Cool to 50°–55°C. Add 4.0mL of L-cysteine·HCl·H_2O solution. Mix thoroughly.

Antibiotic Supplement Solution:
Composition per 10.0mL:
Cycloheximide 80.0mg
Colistin 16.0mg
Cephalothin 4.0mg
Vancomycin 0.5mg

Preparation of Antibiotic Supplement Solution: Add components to 10.0mL of distilled/deionized water. Filter sterilize.

Caution: Cycloheximide is toxic. Avoid skin contact or aerosol formation and inhalation.

Preparation of Medium: To cooled BCYE-alpha base, add 10.0mL sterile antibiotic supplement. Mix thoroughly. Adjust pH to 6.9 with sterile 1*N* KOH. Pour into sterile Petri dishes with constant agitation to keep charcoal in suspension.

Use: For the selective isolation and cultivation of *Legionella* species from environmental samples.

CCY Modified Medium

Composition per liter:
Yeast extract 30.0g
Casamino acids 20.0g
Na_2HPO4 .... 2.48g
KH_2PO_4 .... 0.41g
$MgSO_4 \cdot 7H_2O$ 20.0mg
$MnSO_4 \cdot H_2O$ 7.5mg
Citric acid 6.4mg
$FeSO_4 \cdot 7H_2O$ 6.4mg
Sodium pyruvate solution 100.00mL

pH 7.3 ± 0.2 at 25°C

Sodium Pyruvate Solution:
Composition per 100.0mL:
Sodium pyruvate 23.2g

Preparation of Sodium Pyruvate Solution: Add sodium pyruvate to distilled/deionized water and bring volume to 100.0mL. Mix thoroughly. Filter sterilize.

Preparation of Medium: Add components, except sodium pyruvate solution, to distilled/deionized water and bring volume to 900.0mL. Mix thoroughly. Adjust pH to 7.3. Autoclave for 15 min at 15 psi pressure–121°C. Aseptically add 100.0mL of sterile sodium pyruvate solution. Mix thoroughly. Aseptically distribute into sterile tubes or flasks.

Use: For the cultivation of *Staphylococcus aureus.*

CDC Anaerobe Blood Agar

Composition per liter:
Agar 20.0g
Pancreatic digest of casein 15.0g
Papaic digest of soybean meal 5.0g
NaCl 5.0g
Yeast extract 5.0g
L-Cystine 0.4g
Sheep blood, defibrinated 50.0mL
Vitamin K_1 solution 1.0mL
Hemin solution 0.5mL

pH 7.5 ± 0.2 at 25°C

Source: This medium is available as a prepared medium from BD Diagnostic Systems.

Vitamin K_1 Solution:
Composition per 100.0mL:
Vitamin K_1 .... 1.0g
Ethanol 99.0mL

Preparation of Vitamin K_1 Solution: Add vitamin K_1 to 99.0mL of absolute ethanol. Mix thoroughly. Filter sterilize.

Hemin Solution:
Composition per 100.0mL:
Hemin 1.0g
NaOH (1*N* solution) 20.0mL

Preparation of Hemin Solution: Add hemin to 20.0mL of 1*N* NaOH solution. Mix thoroughly. Bring volume to 100.0mL with distilled/deionized water.

Preparation of Medium: Add components, except vitamin K_1 and sheep blood, to distilled/deionized water and bring volume to 949.0mL. Mix thoroughly. Heat gently and bring to boiling for 1 min. Autoclave for 15 min at 15 psi pressure–121°C. Cool to 50°–55°C. Aseptically add 1.0mL of vitamin K_1 solution and 50.0mL of sterile, defibrinated sheep blood. Mix thoroughly. Pour into sterile Petri dishes.

Use: For the isolation and cultivation of fastidious and slow-growing, obligate anaerobic bacteria from a variety of clinical and nonclinical specimens. For the isolation and cultivation of *Actinomyces israelii, Bacteroides melaninogenicus, Bacteroides thetaiotaomicron, Clostridium haemolyticum,* and *Fusobacterium necrophorum.*

CDC Anaerobe Blood Agar with Kanamycin and Vancomycin

Composition per liter:
Agar 20.0g
Pancreatic digest of casein 15.0g
NaCl 5.0g
Papaic digest of soybean meal 5.0g
Yeast extract 5.0g
L-Cystine 0.4g
Sheep blood, defibrinated 50.0mL
Antibiotic solution 10.0mL
Vitamin K_1 solution 1.0mL
Hemin solution 0.5mL

pH 7.5 ± 0.2 at 25°C

Source: This medium is available as a prepared medium from BD Diagnostic Systems.

Antibiotic Solution:

Composition per 10.0mL:

Kanamycin	0.1g
Vancomycin	7.5mg

Preparation of Antibiotic Solution: Add components to distilled/deionized water and bring volume to 10.0mL. Mix thoroughly. Filter sterilize.

Vitamin K_1 Solution:

Composition per 100.0mL:

Vitamin K_1	1.0g
Ethanol	99.0mL

Preparation of Vitamin K_1 Solution: Add vitamin K_1 to 99.0mL of absolute ethanol. Mix thoroughly. Filter sterilize.

Hemin Solution:

Composition per 100.0mL:

Hemin	1.0g
NaOH (1*N* solution)	20.0mL

Preparation of Hemin Solution: Add hemin to 20.0mL of 1*N* NaOH solution. Mix thoroughly. Bring volume to 100.0mL with distilled/deionized water.

Preparation of Medium: Add components, except vitamin K_1 solution and sheep blood, to distilled/deionized water and bring volume to 949.0mL. Mix thoroughly. Heat gently and bring to boiling for 1 min. Autoclave for 15 min at 15 psi pressure–121°C. Cool to 50°–55°C. Aseptically add 1.0mL of sterile vitamin K_1 solution and 50.0mL of sterile, defibrinated sheep blood. Mix thoroughly. Pour into sterile Petri dishes.

Use: For the selective isolation of fastidious and slow-growing, obligate anaerobic Gram-negative bacteria, especially *Bacteroides* species, from a variety of clinical and nonclinical specimens.

CDC Anaerobe Blood Agar with Phenylethyl Alcohol (CDC Anaerobe Blood Agar with PEA)

Composition per liter:

Agar	20.0g
Pancreatic digest of casein	15.0g
NaCl	5.0g
Papaic digest of soybean meal	5.0g
Yeast extract	5.0g
L-Cystine	0.4g
Sheep blood, defibrinated	50.0mL
Vitamin K_1 solution	10.0mL
Hemin solution	0.5mL

pH 7.5 ± 0.2 at 25°C

Source: This medium is available as a prepared medium from BD Diagnostic Systems.

Vitamin K_1 Solution:

Composition per 100.0mL:

Vitamin K_1	0.1g
Phenylethyl alcohol	25.0g
Ethanol	74.0mL

Preparation of Vitamin K_1 Solution: Add components to 74.0mL of absolute ethanol. Mix thoroughly. Filter sterilize.

Hemin Solution:

Composition per 100.0mL:

Hemin	1.0g
NaOH (1*N* solution)	20.0mL

Preparation of Hemin Solution: Add hemin to 20.0mL of 1*N* NaOH solution. Mix thoroughly. Bring volume to 100.0mL with distilled/deionized water.

Preparation of Medium: Add components, except vitamin K_1 solution and sheep blood, to distilled/deionized water and bring volume to 940.0mL. Mix thoroughly. Heat gently and bring to boiling for 1 min. Autoclave for 15 min at 15 psi pressure–121°C. Cool to 50°–55°C. Aseptically add 1.0mL of vitamin K_1 solution and 50.0mL of sterile, defibrinated sheep blood. Mix thoroughly. Pour into sterile Petri dishes.

Use: For the selective isolation of fastidious and slow-growing, obligate anaerobic bacteria from a variety of clinical and nonclinical specimens.

CDC Anaerobe Laked Blood Agar with Kanamycin and Vancomycin (CDC Anaerobe Laked Blood Agar with KV)

Composition per liter:

Agar	20.0g
Pancreatic digest of casein	15.0g
Papaic digest of soybean meal	5.0g
Yeast extract	5.0g
L-Cystine	0.4g
Sheep blood, defibrinated, laked	50.0mL
Antibiotic solution	10.0mL
Vitamin K_1 solution	1.0mL
Hemin solution	0.5mL

pH 7.5 ± 0.2 at 25°C

Source: This medium is available as a prepared medium from BD Diagnostic Systems.

Antibiotic Solution:

Composition per 10.0mL:

Kanamycin	0.1g
Vancomycin	7.5mg

Preparation of Antibiotic Solution: Add components to distilled/deionized water and bring volume to 10.0mL. Mix thoroughly. Filter sterilize.

Vitamin K_1 Solution:

Composition per 100.0mL:

Vitamin K_1	1.0g
Ethanol	99.0mL

Preparation of Vitamin K_1 Solution: Add vitamin K_1 to 99.0mL of absolute ethanol. Mix thoroughly. Filter sterilize.

Hemin Solution:

Composition per 100.0mL:

Hemin	1.0g
NaOH (1*N* solution)	20.0mL

Preparation of Hemin Solution: Add hemin to 20.0mL of 1*N* NaOH solution. Mix thoroughly. Bring volume to 100.0mL with distilled/deionized water.

Preparation of Medium: Add components, except antibiotic solution, vitamin K_1, and laked sheep blood, to distilled/deionized water and bring volume to 939.0mL. Mix thoroughly. Heat gently and bring to boiling for 1 min. Autoclave for 15 min at 15 psi pressure–121°C. Cool to 50°–55°C. Aseptically add 1.0mL of sterile vitamin K_1 solution and 10.0mL of sterile antibiotic solution. Mix thoroughly. Aseptically add 50.0mL of sterile, defibrinated, laked sheep blood. Laked blood is prepared by freezing whole blood overnight and thawing to room temperature. Mix thoroughly. Pour into sterile Petri dishes.

Use: For the selective isolation of fastidious and slow-growing, obligate anaerobic bacteria from a variety of clinical and nonclinical specimens.

CDC Modified McClung-Toabe Egg Yolk Agar

See: **McClung-Toabe Egg Yolk Agar, CDC Modified**

Cefiximine Rhamnose Sorbitol MacConkey Agar (CR-SMAC Agar Base)

Composition per liter:

Peptone	20.0g
Agar	15.0g
Sorbitol	10.0g
NaCl	5.0g
Rhamnose	5.0g
Bile Salts No. 3	1.5g
Neutral Red	0.03g
Crystal Violet	0.001g
Selective supplement solution	10.0mL

pH 7.1 ± 0.2 at 25°C

Source: This medium is available as a premixed powder from Oxoid Unipath.

Selective Supplement Solution:

Composition per 10.0mL:

Cefiximine	0.05mg

Preparation of Selective Supplement Solution: Add cefiximine to distilled/deionized water and bring volume to 10.0mL. Mix thoroughly. Filter sterilize.

Preparation of Medium: Add components, except selective supplement solution, to distilled/deionized water and bring volume to 990.0mL. Mix thoroughly. Gently heat while stirring and bring to boiling. Autoclave for 15 min at 15 psi pressure–121°C. Cool to 50°C. Aseptially add selective supplement solution. Mix thoroughly. Pour into sterile Petri dishes.

Use: For the detection of *Escherichia coli* O157:H7. This is a elective, differential medium based on Sorbitol MacConkey Agar with added rhamnose and cefixime. This medium provides a selective base with improved differentiation of *E. coli* O157. The addition of rhamnose aids in the differentiation of *Escherichia coli* O157 from background flora. Cefixime reduces the level of competing flora, particularly *Proteus* spp., that often account for large numbers of non-sorbitol fermenting colonies. *E. coli* O157 do not usually ferment sorbitol or rhamnose, so will appear as straw colored colonies. However, rhamnose is fermented by most sorbitol negative *E. coli* of other serogroups. These colonies will be pink/red and will not be counted as presumptive *E. coli* O157 colonies.

Cefoperazone Vancomycin Amphotericin Medium

See: **CVA Medium**

Cefsulodin Irgasan® Novobiocin Agar (CIN Agar) (*Yersinia* Selective Agar) (BAM M35)

Composition per 1008.0mL:

Basal medium	757.0mL
Desoxycholate solution	200.0mL
Cefsulodin solution	10.0mL
Novobiocin solution	10.0mL
Crystal Violet solution	10.0mL
Strontium chloride solution	10.0mL
Neutral Red solution	10.0mL
NaOH, 5*N*	1.0mL
Irgasan solution	1.0mL

pH 7.4 ± 0.2 at 25°C

Basal Medium:

Composition per 757.0mL:

Mannitol	20.0g
Special peptone	20.0g
Agar	12.0g
Sodium pyruvate	2.0g
Yeast extract	2.0g
NaCl	1.0g
Magnesium sulfate solution	1.0mL

Preparation of Basal Medium: Add components to distilled/deionized water and bring volume to 757.0mL. Mix thoroughly. Gently heat and bring to boiling with stirring. Cool to about 80°C by placing in a 50°C water bath for about 10 min.

Magnesium Sulfate Solution:

Composition per 10mL:

$MgSO_4 \cdot 7H_2O$	0.1g

Preparation of Magnesium Sulfate Solution: Add $MgSO_4 \cdot 7H_2O$ to distilled/deionized water and bring volume to 10.0mL. Mix thoroughly.

Irgasan Solution:

Composition per 10mL:

Irgasan (triclosan)	0.04g

Preparation of Irgasan Solution: Add irgasan to 95% ethanol and bring volume to 10.0mL. Mix thoroughly. Can be stored for 4 weeks at –20°C.

Desoxycholate Solution:

Composition per 200.0mL:

Na-desoxycholate	0.5g

Preparation of Desoxycholate Solution: Add desoxycholate to distilled/deionized water and bring volume to 200.0mL. Mix thoroughly. Gently heat and bring to boiling with stirring. Cool to 50–55°C.

Neutral Red Solution:

Composition per 10.0mL:

Neutral Red	30.0mg

Preparation of Neutral Red Solution: Add Neutral Red to 10.0mL of distilled/deionized water. Mix thoroughly. Autoclave for 15 min at 15 psi pressure–121°C. Cool to 25°C.

Crystal Violet Solution:

Composition per 10.0mL:

Crystal Violet	1.0mg

Preparation of Crystal Violet Solution: Add Crystal Violet to 10.0mL of distilled/deionized water. Mix thoroughly. Autoclave for 15 min at 15 psi pressure–121°C. Cool to 25°C.

Cefsulodin Solution:
Composition per 10.0mL:

Cefsulodin	15.0mg

Preparation of Cefsulodin Solution: Add cefsulodin to 10.0mL of distilled/deionized water. Mix thoroughly. Filter sterilize.

Novobiocin Solution:
Composition per 10.0mL:

Novobiocin	2.5mg

Preparation of Novobiocin Solution: Add novobiocin to 10.0mL of distilled/deionized water. Mix thoroughly. Filter sterilize.

Strontium Chloride Solution:
Composition per 10.0mL:

$SrCl_2 \cdot 6H_2O$	1.0g

Preparation of Strontium Chloride: Add strontium chloride to 10.0mL of distilled/deionized water. Mix thoroughly. Filter sterilize.

Preparation of Medium: Add 1.0mL irgasan solution to 757.0mL basal medium. Mix thoroughly. Cool to 50–55°C. Add 200.0mL desoxychlolate solution. Mix thoroughly. Solution should remain clear. Aseptically add 1.0mL 5*N* NaOH, 10.0mL Neutral Red solution, 10.0mL Crystal Violet solution, 10.0mL cefsulodin solution, and 10.0mL novobiocin solution. Mix thoroughly. Slowly add 10.0mL strontium chloride solution while continuously stirring. Adjust pH to 7.4 with 5*N* NaOH. Pour into sterile Petri dishes or distribute into sterile tubes.

Use: For the selective isolation and differentiation of *Yersinia enterocolitica* based on mannitol fermentation. *Yersinia enterocolitica* appears as "bull's eye" colonies with deep red centers surrounded by a transparent periphery.

Cefsulodin Irgasan® Novobiocin Agar

See: **CIN Agar**

Celery Decoction Agar

Composition per liter:

Potatoes, infusion from	50.0g
Agar	15.0g
Glucose	5.0g
Celery decoction	320.0mL

pH 5.6 ± 0.2 at 25°C

Celery Decoction:
Composition per 500.0mL:

Celery leaves and petioles	50.0g

Preparation of Celery Decoction: Coarsely cut celery into pieces. Add celery pieces to 500.0mL of distilled/deionized water. Puree in a blender at high speed for 2 min. Strain through a wire sieve. Gently heat and bring to boiling. Maintain at 100°C for 5 min. Strain through a very fine sieve or muslin.

Preparation of Medium: Add components to distilled/deionized water and bring volume to 1.0L. Mix thoroughly. Adjust pH to 5.6 with 20% H_2SO_4. Gently heat and bring to boiling. Distribute into tubes or flasks. Autoclave for 15 min at 15 psi pressure–121°C. Pour into sterile Petri dishes or leave in tubes.

Use: For the cultivation and maintenance of *Septoria apiicola.*

Celery Decoction Agar

Composition per liter:

Potatoes, infusion from	200.0g
Glucose	20.0g
Agar	15.0g
Potato dextrose broth	4.0g
Celery decoction	320.0mL

pH 5.6 ± 0.2 at 25°C

Potatoes, Infusion From:
Composition per 500.0mL:

Potatoes	300.0g

Preparation of Potatoes, Infusion From: Peel and dice potatoes. Add 500.0mL of distilled/deionized water. Gently heat and bring to boiling. Continue boiling for 30 min. Filter through cheesecloth.

Celery Decoction:
Composition per 500.0mL:

Celery leaves and petioles	50.0g

Preparation of Celery Decoction: Coarsely cut celery into pieces. Add 500.0mL of distilled/deionized water. Puree in a blender at high speed for 2 min. Strain through a kitchen-type sieve. Gently heat and bring to boiling. Continue boiling for 5 min. Strain through a very fine sieve or muslin.

Preparation of Medium: Add components to distilled/deionized water and bring volume to 1.0L. Mix thoroughly. Adjust pH to 5.6 with 20% H_2SO_4. Gently heat and bring to boiling. Distribute into tubes or flasks. Autoclave for 15 min at 15 psi pressure–121°C. Pour into sterile Petri dishes or leave in tubes.

Use: For the cultivation of *Septoria apiicola.*

Cell Growth Medium

Composition per 2250.0mL:

Eagle's minimal essential medium with Hanks' salts (MEMH)	1.0L
L 15 medium, modified Leibovitz	1.0L
Fetal calf serum	200.0mL
$NaHCO_3$ solution	50.0mL

pH 7.5 at 25°C

MEMH:
Composition per liter:

NaCl	8.0g
Glucose	1.0g
KCl	0.4g
$CaCl_2 \cdot 2H_2O$	0.14g
$MgSO_4 \cdot 7H_2O$	0.1g
KH_2PO_4	0.06g
Na_2HPO_4	0.05g
L-Isoleucine	0.026g
L-Leucine	0.026g
L-Lysine	0.026g
L-Threonine	0.024g
L-Valine	0.0235g
L-Tyrosine	0.018g
L-Arginine	0.0174g
L-Phenylalanine	0.0165g
L-Cystine	0.012g
L-Histidine	8.0mg
L-Methionine	7.5mg
Phenol Red	5.0mg

L-Tryptophan 4.0mg
Inositol 1.8mg
Biotin 1.0mg
Folic acid 1.0mg
Calcium pantothenate 1.0mg
Choline chloride 1.0mg
Nicotinamide 1.0mg
Pyridoxal·HCl 1.0mg
Thiamine·HCl 1.0mg
Riboflavin 0.1mg

Preparation of MEMH: Add components to distilled/deionized water and bring volume to 1.0L. Mix thoroughly. Filter sterilize.

L 15 Medium, Modified Leibovitz:
Composition per liter:

NaCl 8.0g
DL-Threonine 0.6g
Sodium pyruvate 0.6g
DL-Alanine 0.5g
L-Arginine, free base 0.5g
KCl 0.4g
L-Asparagine·H_2O 0.3g
L-Histidine, free base 0.3g
L-Glutamine 0.3g
L-Isoleucine 0.3g
L-Phenylalanine 0.3g
L-Tyrosine 0.3g
DL-Methionine 0.2g
DL-Valine 0.2g
Glycine 0.2g
L-Serine 0.2g
Na_2HPO_4, anhydrous 0.2g
$CaCl_2$, anhydrous 0.1g
L-Cysteine, free base 0.1g
L-Leucine·HCl 0.1g
Streptomycin 0.1g
$MgCl_2$, anhydrous 0.094g
D-Galactose 0.09g
KH_2PO_4 .... 0.06g
Gentamicin 0.05g
L-Tryptophan 0.02g
Phenol Red 0.01g
i-Inositol 2.0mg
Choline chloride 1.0mg
D-Calcium pantothenate 1.0mg
Folic acid 1.0mg
Nicotinamide 1.0mg
Pyridoxine·HCl 1.0mg
Thiamine monophosphate·$2H_2O$ 1.0mg
Riboflavin-5-phosphate 0.1mg
Penicillin G 100,000U

pH 7.5 ± 0.2 at 25°C

Preparation of L 15 Medium, Modified Leibovitz: Add components to distilled/deionized water and bring volume to 1.0L. Mix thoroughly. Filter sterilize. Store at 5°C.

$NaHCO_3$ Solution:
Composition per 100.0mL:

$NaHCO_3$ 7.5g

Preparation of $NaHCO_3$ Solution: Add $NaHCO_3$ to distilled/deionized water and bring volume to 100.0mL. Mix thoroughly. Filter sterilize.

Preparation of Medium: Aseptically combine 1.0L of sterile Eagle's minimal essential medium with Hanks' salts (MEMH) and 1.0L of sterile L 15 medium, modified Leibovitz. Mix thoroughly. Immediately prior to use, aseptically add 200.0mL of fetal calf serum and 50.0mL of $NaHCO_3$ solution. Mix thoroughly. Aseptically distribute into sterile containers.

Use: For the cultivation of mammalian HeLa or Vero tissue culture cells to test the cytopathic effects of *Escherichia coli.*

Cellobiose Arginine Lysine Agar

See: **CAL Agar/Broth**

Cellobiose Polymyxin B Colistin Agar, Modified

Composition per liter:

Solution 1 .... 900.0mL
Solution 2 .... 100.0mL

pH 7.6 ± 0.2 at 25°C

Solution 1:
Composition per 900.0mL:

NaCl 20.0g
Agar 15.0g
Peptone 10.0g
Beef extract 5.0g
1000× dye stock solution 1.0mL

Preparation of Solution 1: Add components to distilled/deionized water and bring volume to 900.0mL. Mix thoroughly. Adjust pH to 7.6. Gently heat and bring to boiling. Do not autoclave. Cool to 48°–55°C.

1000X Dye Stock Solution:
Composition per 100.0mL:

Bromthymol Blue 4.0g
Cresol Red 4.0g
Ethanol (95% solution) 100.0mL

Preparation of 1000X Dye Stock Solution: Add Bromthymol Blue and Cresol Red to 100.0mL of ethanol. Mix thoroughly.

Solution 2:
Composition per 100.0mL:

Cellobiose 10.0g
Colistin 400,000U
Polymyxin B 100,000U

Preparation of Solution 2: Add cellobiose to distilled/deionized water and bring volume to 100.0mL. Mix thoroughly. Gently heat until dissolved. Cool to 25°C. Add colistin and polymyxin B. Mix thoroughly.

Preparation of Medium: Combine cooled solution 1 and solution 2. Mix thoroughly. Do not autoclave. Pour into sterile Petri dishes.

Use: For the cultivation of *Vibrio* species from foods.

Cellobiose Polymyxin Colistin Agar (CPC Agar)

Composition per liter:

Solution A 900.0mL
Solution B 100.0mL

pH 7.6 ± 0.2 at 25°C

Solution A:
Composition per 900.0mL:

NaCl 20.0g
Agar 15.0g

Peptone	10.0g
Beef extract	5.0g
Bromthymol Blue	0.04g
Cresol Red	0.04g

Preparation of Solution A: Add components to distilled/deionized water and bring volume to 900.0mL. Mix thoroughly. Adjust pH to 7.6. Gently heat and bring to boiling. Autoclave for 15 min at 15 psi pressure–121°C. Cool to 50°–55°C.

Solution B:

Composition per 100.0mL:

Cellobiose	15.0g
Colistin	1,360,000U
Polymyxin B	100,000U

Preparation of Solution B: Add components to distilled/deionized water and bring volume to 100.0mL. Mix thoroughly. Filter sterilize.

Preparation of Medium: Aseptically combine 900.0mL of cooled, sterile solution A and 100.0mL of sterile solution B. Mix thoroughly. Pour into sterile Petri dishes. Use within 7 days.

Use: For the cultivation and identification of *Vibrio* species from foods.

Cellulase Solution (BAM M187)

Composition per 100.0mL:

Cellulase	1.0g

Preparation of Medium: Add cellulase to distilled/deionized water and bring volume to 100.0mL. Mix thoroughly. Filter sterilize.

Use: For the pre-treatment of guar gum in the analysis of *Salmonella* spp.The cellulase reduces the viscosity of the guar gum so that it can be pipetted into broth or agar media.

Cellulolytic Agar with Sea Salts

Composition per liter:

Agar	20.0g
NH_4Cl	2.0g
K_2HPO_4	1.65g
Yeast extract	1.2g
L-Cysteine·HCl·H_2O	0.5g
Cellulose suspension	200.0mL
Filtered seawater	200.0mL
Mineral solution	150.0mL
Resazurin (0.1% solution)	1.0mL

pH 7.2 ± 0.2 at 25°C

Cellulose Suspension:

Composition per 200.0mL:

Cellulose powder, Whatman CF11	8.0g

Preparation of Cellulose Suspension: Add cellulose powder to 200.0mL of distilled/deionized water and mix thoroughly.

Mineral Solution:

Composition per liter:

NaCl	6.0g
$(NH_4)_2SO_4$	6.0g
$CaCl_2$	0.6g
$MgSO_4$	0.6g

Preparation of Mineral Solution: Add components to distilled/deionized water and bring volume to 1.0L. Mix thoroughly.

Preparation of Medium: Prepare and dispense medium anaerobically under 100% N_2. Add components to distilled/deionized water and bring volume to 1.0L. Mix thoroughly. Adjust pH to 7.2 with 5*M* NaOH. Distribute into tubes or flasks. Autoclave for 15 min at 15 psi pressure–121°C.

Use: For the cultivation and maintenance of *Clostridium papyrosolvens* and other marine bacteria that can utilize cellulose as a carbon source.

Cellulolytic Agar for Thermophiles

Composition per liter:

Agar	30.0g
K_2HPO_4	1.65g
NH_4SO_4	1.6g
Yeast extract	1.0g
NaCl	0.96g
L-Cysteine·HCl·H_2O	0.5g
$CaCl_2$	0.096g
$MgSO_4$	0.096g
Cellulose suspension	200.0mL
Resazurin (0.1% solution)	1.0mL

pH 7.2 ± 0.2 at 25°C

Cellulose Suspension:

Composition per 200.0mL:

Cellulose powder, Whatman CF11	8.0g

Preparation of Cellulose Suspension: Add cellulose powder to 200.0mL of distilled/deionized water and mix thoroughly.

Preparation of Medium: Prepare and dispense medium anaerobically in 100% N_2. Add components to distilled/deionized water and bring volume to 1.0L. Mix thoroughly. Adjust pH to 7.2 with 5*M* NaOH. Distribute into tubes or flasks. Autoclave for 15 min at 15 psi pressure–121°C.

Use: For cultivation of *Clostridium stercorarium* and other bacteria that can utilize cellulose as a carbon source.

Cellulolytic Broth for Thermophiles

Composition per liter:

K_2HPO_4	1.65g
NH_4SO_4	1.6g
Yeast extract	1.0g
NaCl	0.96g
L-Cysteine·HCl·H_2O	0.5g
$CaCl_2$	0.096g
$MgSO_4$	0.096g
Resazurin (0.1% solution)	1.0mL

pH 7.2 ± 0.2 at 25°C

Preparation of Medium: Prepare and dispense medium anaerobically in 100% N_2. Add components to distilled/deionized water and bring volume to 1.0L. Mix thoroughly. Adjust pH to 7.2 with 5*M* NaOH. Distribute into tubes or flasks that contain cellulose as a strip (4.5cm × 1.0cm) of Whatman #1 filter paper. Autoclave for 15 min at 15 psi pressure–121°C.

Use: For the cultivation of *Clostridium stercorarium* and other bacteria that can utilize cellulose as a carbon source.

Cellulolytic Broth with Sea Salts

Composition per liter:

K_2HPO_4	1.65g
NH_4Cl	1.0g

Yeast extract	0.6g
L-Cysteine·HCl·H_2O	0.5g
Filtered seawater	200.0mL
Mineral solution	150.0mL
Resazurin (0.1% solution)	1.0mL

pH 7.2 ± 0.2 at 25°C

Mineral Solution:

Composition per liter:

NaCl	6.0g
$(NH_4)_2SO_4$	6.0g
$CaCl_2$	0.6g
$MgSO_4$	0.6g

Preparation of Mineral Solution: Add components to distilled/deionized water and bring volume to 1.0L. Mix thoroughly.

Preparation of Medium: Prepare and dispense medium anaerobically in 100% N_2 atmosphere. Add components to distilled/deionized water and bring volume to 1.0L. Adjust pH to 7.2 with 5*M* NaOH. Distribute into tubes or flasks that contain cellulose as a strip (4.5cm × 1.0cm) of Whatman No. 1 filter paper. Autoclave for 15 min at 15 psi pressure–121°C.

Use: For the cultivation and maintenance of *Clostridium papyrosolvens* and other marine bacteria that can utilize cellulose as a carbon source.

Cellulolytic Clostridia Medium

Composition per liter:

Cellulose	20.0g
$CaCO_3$	2.0g
K_2HPO_4	1.0g
$(NH_4)_2SO_4$	1.0g
$MgSO_4·7H_2O$	0.5g
NaCl	0.5g
Resazurin	1.0mg

pH 7.1 ± 0.2 at 25°C

Preparation of Medium: Add components to distilled/deionized water and bring volume to 1.0L. Mix thoroughly. Distribute into tubes or flasks. Autoclave for 15 min at 15 psi pressure–121°C.

Use: For the isolation, cultivation, and enrichment of cellulolytic *Clostridium* species.

Cellulolytic Medium

Composition per liter:

$NaHCO_3$	2.06g
Cellulose	2.0g
NH_4Cl	0.68g
K_2HPO_4	0.296g
KH_2PO_4	0.18g
$(NH_4)_2SO_4$	0.15g
$MgSO_4·7H_2O$	0.12g
$CaCl_2·2H_2O$	61.0mg
$FeSO_4·7H_2O$	21.0mg
Nitrilotriacetic acid	15.0mg
NaCl	10.0mg
MnSO4·H_2O	5.0mg
$CoCl_2·H_2O$	1.0mg
Resazurin	1.0mg
$ZnSO_4·7H_2O$	1.0mg
$CuSO_4·5H_2O$	0.1mg
H_3BO_3	0.1mg
$KAl(SO_4)_2·12H_2O$	0.1mg
$Na_2MoO_4·2H_2O$	0.1mg
Wolfe's vitamin solution	10.0mL
L-Cysteine·HCl·H_2O solution	5.0mL

pH 7.5 ± 0.2 at 25°C

Wolfe's Vitamin Solution:

Composition per liter:

Pyridoxine·HCl	10.0mg
Thiamine·HCl	5.0mg
Riboflavin	5.0mg
Nicotinic acid	5.0mg
Calcium DL-pantothenate	5.0mg
p-Aminobenzoic acid	5.0mg
Thioctic acid	5.0mg
Biotin	2.0mg
Folic acid	2.0mg
Cyanocobalamin	100.0μg

Preparation of Wolfe's Vitamin Solution: Add components to distilled/deionized water and bring volume to 1.0L. Mix thoroughly. Filter sterilize.

L-Cysteine·HCl·H_2O Solution:

Composition per 20.0mL:

L-Cysteine·HCl·H_2O	1.0g

Preparation of L-Cysteine·HCl·H_2O Solution: Add L-cysteine·HCl·H_2O to distilled/deionized water and bring volume to 20.0mL. Mix thoroughly. Filter sterilize.

Preparation of Medium: Add components, except Wolfe's vitamin solution and L-cysteine·HCl·H_2O solution, to distilled/deionized water and bring volume to 985.0mL. Mix thoroughly. Autoclave for 15 min at 15 psi pressure–121°C. Aseptically add 10.0mL of sterile Wolfe's vitamin solution and 5.0mL of sterile L-cysteine·HCl·H_2O solution. Mix thoroughly. Aseptically distribute into sterile tubes or flasks.

Use: For the cultivation of cellulolytic bacteria.

Cellulolytic Medium with Rumen Fluid

Composition per liter:

Basal medium	975.0mL
Alkaline solution	25.0mL

pH 6.8 ± 0.2 at 25°C

Basal Medium:

Composition per 975.0mL:

Agar	15.0g
$NaHCO_3$	6.37g
Pancreatic digest of casein	5.0g
Cellobiose	5.0g
NaCl	0.9g
$(NH_4)_2SO_4$	0.9g
K_2HPO_4	0.45g
KH_2PO_4	0.45g
$MgSO_4·7H_2O$	0.18g
$CaCl_2$	0.09g
Resazurin	1.0mg
Rumen fluid, clarified	400.0mL

Preparation of Basal Medium: Add components to distilled/deionized water and bring volume to 975.0mL. Mix thoroughly. Gently heat and bring to boiling under a gas phase of 98% CO_2 + 2% H_2. Cool slightly.

Alkaline Solution:
Composition per 25.0mL:
L-Cysteine·HCl·H_2O 0.25g
$Na_2S·9H_2O$ 0.25g

Preparation of Alkaline Solution: Add components to 25.0mL of distilled/deionized water. Mix thoroughly. Prepare freshly.

Preparation of Medium: Prepare 975.0mL of basal medium. Heat to boiling and cool to 25°C. Add 25.0mL of freshly prepared alkaline solution. Distribute into tubes using anaerobic techniques under a gas phase of 98% CO_2 + 2% H_2. Autoclave for 15 min at 15 psi pressure–121°C. Adjust pH to 6.8.

Use: For the cultivation and maintenance of *Clostridium polysaccharolyticum.*

Cellulolytic Medium with Rumen Fluid and Soluble Starch

Composition per liter:
Basal medium 975.0mL
Alkaline solution 25.0mL
pH 6.8 ± 0.2 at 25°C

Basal Medium:
Composition per 975.0mL:
Agar 15.0g
$NaHCO_3$ 6.37g
Pancreatic digest of casein 5.0g
Cellobiose 5.0g
Soluble starch 5.0g
NaCl 0.9g
$(NH_4)_2SO_4$ 0.9g
K_2HPO_4 .... 0.45g
KH_2PO_4 .... 0.45g
$MgSO_4·7H_2O$ 0.18g
$CaCl_2$ 0.09g
Resazurin 1.0mg

Preparation of Basal Medium: Add components to distilled/deionized water and bring volume to 975.0mL. Mix thoroughly. Gently heat and bring to boiling under a gas phase of 98% CO_2 + 2% H_2. Cool slightly.

Alkaline Solution:
Composition per 25.0mL:
L-Cysteine·HCl·H_2O 0.25g
$Na_2S·9H_2O$ 0.25g

Preparation of Alkaline Solution: Add components to 25.0mL of distilled/deionized water. Mix thoroughly. Prepare freshly.

Preparation of Medium: Prepare 975.0mL of basal medium. Heat to boiling and cool to 25°C. Add 25.0mL of freshly prepared alkaline solution. Distribute into tubes using anaerobic techniques under a gas phase of 98% CO_2 + 2% H_2. Autoclave for 15 min at 15 psi pressure–121°C. Adjust pH to 6.8.

Use: For the cultivation of *Selenomonas ruminantium* and *Succinimonas amylolytica.*

Cellulomonas fermentans Medium

Composition per liter:
Yeast extract 5.0g
K_2HPO_4 .... 2.21g
KH_2PO_4 .... 1.5g
$(NH_4)_2SO_4$ 1.3g
$MgCl_2·6H_2O$ 0.1g
$CaCl_2·2H_2O$ 0.02g
$FeSO_4·7H_2O$ 1.25mg
Resazurin 1.0mg
Cellobiose solution 50.0mL
$NaHCO_3$ solution 10.0mL
L-Cysteine·HCl solution 10.0mL
pH 7.4 ± 0.2 at 25°C

Cellobiose Solution:
Composition per 50.0mL:
D-Cellobiose (or cellulose) 5.0g

Preparation of Cellobiose Solution: Add cellobiose (or cellulose) to distilled/deionized water and bring volume to 50.0mL. Mix thoroughly. Autoclave under 100% N_2 for 15 min at 15 psi pressure–121°C.

$NaHCO_3$ Solution:
Composition per 10.0mL:
$NaHCO_3$ 0.8g

Preparation of $NaHCO_3$ Solution: Add $NaHCO_3$ to distilled/deionized water and bring volume to 10.0mL. Mix thoroughly. Autoclave under 100% N_2 for 15 min at 15 psi pressure–121°C.

L-Cysteine·HCl Solution:
Composition per 10.0mL:
L-Cysteine·HCl 0.5g

Preparation of L-Cysteine·HCl Solution: Add L-cysteine·HCl to distilled/deionized water and bring volume to 10.0mL. Mix thoroughly. Autoclave under 100% N_2 for 15 min at 15 psi pressure–121°C.

Preparation of Medium: Add components, except cellobiose solution, $NaHCO_3$ solution, and L-cysteine·HCl solution, to distilled/deionized water and bring volume to 930.0mL. Mix thoroughly. Sparge under 100% N_2 for 10 min. Autoclave under 100% N_2 for 15 min at 15 psi pressure–121°C. Aseptically and anaerobically add 50.0mL of sterile cellobiose solution, 10.0mL of sterile $NaHCO_3$ solution, and 10.0mL of sterile L-cysteine·HCl solution. Mix thoroughly.

Use: For the cultivation and maintenance of *Cellulomonas fermentans*.

Cellulomonas PTYG Medium (*Cellulomonas* Peptone Tryptone Yeast Extract Glucose Medium)

Composition per liter:
Agar 15.0g
Glucose 5.0g
Peptone 5.0g
Pancreatic digest of casein 5.0g
Yeast extract 5.0g

Preparation of Medium: Add components to distilled/deionized water and bring volume to 1.0L. Mix thoroughly. Gently heat and bring to boiling. Distribute into tubes or flasks. Autoclave for 15 min at 15 psi pressure–121°C. Pour into sterile Petri dishes or leave in tubes.

Use: For the cultivation and maintenance of *Cellulomonas* species.

Cellulose Agar

Composition per liter:
Agar 20.0g
Cellulose, ball milled 10.0g
KH_2PO_4 .... 1.0g

$(NH_4)_2SO_4$	0.5g
L-Asparagine	0.5g
KCl	0.5g
Yeast extract	0.5g
$MgSO_4$	0.2g
$CaCl_2$	0.1g

pH 6.2 ± 0.2 at 25°C

Preparation of Medium: Add components to distilled/deionized water and bring volume to 1.0L. Gently heat and bring to boiling. Adjust pH to 6.2. Distribute into tubes or flasks. Autoclave for 15 min at 15 psi pressure–121°C. Pour into sterile Petri dishes or leave in tubes.

Use: For the cultivation and maintenance of *Achaetomium globosum, Chaetomium anahelicinum, Chaetomium apiculatum, Chaetomium atrobrunneum, Chaetomium carinthiacum, Chaetomium globosum, Chaetomium medusarum, Chaetomium quadrangulatum, Chaetomium reflexum, Chaetomium thielavioideum, Chaetomium undulatum, Chaetomium variosporum,* and *Chrysosporium pannorum.*

Cellulose Anaerobe Medium (LMG Medium 94)

Composition per liter:

$NaHCO_3$	2.1g
Cellulose	2.0g
NH_4Cl	0.68g
KH_2PO_4	0.18g
$(NH_4)_2SO_4$	0.15g
$MgSO_4 \cdot 7H_2O$	0.12g
K_2HPO_4	296.0mg
$CaCl_2 \cdot 2H_2O$	61.0mg
$FeSO_4 \cdot 7H_2O$	21.0mg
Nitrilotriacetic acid	15.0mg
NaCl	10.0mg
$MnSO_4 \cdot H_2O$	5.0mg
$CoCl_2 \cdot 6H_2O$	1.0mg
$ZnSO_4 \cdot 7H_2O$	1.0mg
Resazurin	1.0mg
$CuSO_4 \cdot 5H_2O$	0.1mg
$KAl(SO_4)_2 \cdot 12H_2O$	0.1mg
H_3BO_3	0.1mg
$Na_2MoO_4 \cdot 2H_2O$	0.1mg
Vitamin solution	5.0mL
Cysteine hydrochloride solution	5.0mL

pH 7.1 ± 0.2 at 25°C

Vitamin Solution:

Composition per liter:

Pyridoxine·HCl	10.0mg
Calcium DL-pantothenate	5.0mg
Lipoic acid	5.0mg
Nicotinic acid	5.0mg
p-Aminobenzoic acid	5.0mg
Riboflavin	5.0mg
Thiamine·HCl	5.0mg
Biotin	2.0mg
Folic acid	2.0mg
Vitamin B_{12}	0.1mg

Preparation of Vitamin Solution: Add components to distilled/deionized water and bring volume to 1.0L. Mix thoroughly. Filter sterilize.

L-Cysteine Solution:

Composition per 10.0mL:

L-Cysteine·HCl·H_2O	0.5g

Preparation of L-Cysteine Solution: Add L-cysteine·HCl·H_2O to distilled/deionized water and bring volume to 10.0mL. Mix thoroughly. Sparge with 100% N_2. Autoclave for 15 min at 15 psi pressure–121°C.

Preparation of Medium: Add nitrilotriacetic acid to 500.0mL of distilled/deionized water. Adjust pH to 6.5 with KOH. Add remaining components except vitamin solution and cysteine hydrochloride solution. Add 485.0mL distilled/deionized water. Adjust pH to 7.1. Autoclave for 15 min at 15 psi pressure–121°C. Cool to 50°C. Aseptically add 10.0mL of sterile vitamin solution and 5.0mL sterile cysteine hydrochloride solution. Mix thoroughly. Aseptically distribute to tubes or flasks.

Use: For the cultivation of cellulose-utilizing anerobic bacteria.

Cellulose Broth

Composition per liter:

Cellulose, powdered	1.0g
K_2HPO_4	1.0g
$(NH_4)_2SO_4$	1.0g
$MgSO_4 \cdot 7H_2O$	0.2g
$CaCl_2 \cdot 2H_2O$	0.1g
$FeCl_3$	0.02g

pH 7.0–7.5 at 25°C

Preparation of Medium: Add cellulose to 100.0mL of distilled/deionized water. Mix thoroughly. In a separate flask, add remaining components to distilled/deionized water and bring volume to 900.0mL. Mix thoroughly. Autoclave both solutions separately for 15 min at 15 psi pressure–121°C. Cool to 45°–50°C. Aseptically combine the two sterile solutions. Mix thoroughly. Aseptically distribute into sterile tubes or flasks.

Use: For the isolation and cultivation of *Cytophaga* species, *Herpetosiphon* species, *Saprospira* species, and *Flexithrix* species.

Cellulose Overlay Agar

Composition per plate:

Stan 5 agar	15.0mL
Cellulose overlay agar	5.0mL

Stan 5 Agar:

Composition per liter:

Solution B	650.0mL
Solution A	350.0mL

Solution A:

Composition per 350.0mL:

$CaCl_2 \cdot 2H_2O$	1.0g
$(NH_4)_2SO_4$	1.0g
$MgSO_4 \cdot 7H_2O$	1.0g
Trace elements solution	1.0mL

Preparation of Solution A: Add components to distilled/deionized water and bring volume to 350.0mL. Mix thoroughly. Gently heat and bring to boiling. Autoclave for 15 min at 15 psi pressure–121°C. Cool to 45°–50°C.

Trace Elements Solution:

Composition per liter:

EDTA	8.0g
$MnCl_2 \cdot 4H_2O$	0.1g

$CoCl_2$	0.02g
KBr	0.02g
$ZnCl_2$	0.02g
$CuSO_4$	0.01g
H_3BO_3	0.01g
$NaMoO_4 \cdot 2H_2O$	0.01g
$BaCl_2$	5.0mg
LiCl	5.0mg
$SnCl_2 \cdot 2H_2O$	5.0mg

Preparation of Trace Elements Solution: Add components to distilled/deionized water and bring volume to 1.0L. Mix thoroughly.

Solution B:

Composition per 650.0mL:

Agar	10.0g
K_2HPO_4	1.0g

Preparation of Solution B: Add components to distilled/deionized water and bring volume to 650.0mL. Mix thoroughly. Gently heat and bring to boiling. Autoclave for 15 min at 15 psi pressure–121°C. Cool to 45°–50°C.

Preparation of Stan 5 Agar: Aseptically combine 350.0mL of cooled, sterile solution A and 650.0mL of cooled, sterile solution B. Mix thoroughly.

Cellulose Overlay Agar:

Composition per liter:

Solution A	350.0mL
Solution B	650.0mL

Solution A:

Composition per 350.0mL:

$CaCl_2 \cdot 2H_2O$	1.0g
$(NH_4)_2SO_4$	1.0g
$MgSO_4 \cdot 7H_2O$	1.0g
Trace elements solution	1.0mL

Preparation of Solution A: Add components to distilled/deionized water and bring volume to 350.0mL. Mix thoroughly. Gently heat and bring to boiling. Autoclave for 15 min at 15 psi pressure–121°C. Cool to 45°–50°C.

Trace Elements Solution:

Composition per liter:

EDTA	8.0g
$MnCl_2 \cdot 4H_2O$	0.1g
$CoCl_2$	0.02g
KBr	0.02g
$ZnCl_2$	0.02g
$CuSO_4$	0.01g
H_3BO_3	0.01g
$NaMoO_4 \cdot 2H_2O$	0.01g
$BaCl_2$	5.0mg
LiCl	5.0mg
$SnCl_2 \cdot 2H_2O$	5.0mg

Preparation of Trace Elements Solution: Add components to distilled/deionized water and bring volume to 1.0L. Mix thoroughly.

Solution B:

Composition per 650.0mL:

Agar	10.0g
K_2HPO_4	1.0g

Preparation of Solution B: Add components to distilled/deionized water and bring volume to 650.0mL. Mix thoroughly. Gently heat and bring to boiling. Autoclave for 15 min at 15 psi pressure–121°C. Cool to 45°–50°C.

Preparation of Cellulose Overlay Agar: Aseptically combine 350.0mL of cooled, sterile solution A and 650.0mL of cooled, sterile solution B. Mix thoroughly.

Preparation of Medium: Pour cooled, sterile Stan 5 agar into sterile Petri dishes in 15.0mL volumes. Allow agar to solidify. Overlay each plate with 5.0mL of cellulose overlay agar.

Use: For the cultivation of myxobacteria.

Cellvibrio Medium

Composition per liter:

Agar	15.0g
$NaNO_3$	5.0g
K_2HPO_4	1.0g
$MgSO_4 \cdot 7H_2O$	0.5g
KCl	0.5g
$FeSO_4 \cdot 7H_2O$	10.0mg

pH 7.2 ± 0.2 at 25°C

Preparation of Medium: Add components to distilled/deionized water and bring volume to 1.0L. Mix thoroughly. Gently heat and bring to boiling. Distribute into tubes or flasks. Autoclave for 15 min at 15 psi pressure–121°C. Pour into sterile Petri dishes or leave in tubes.

Use: For the cultivation and maintenance of *Cellvibrio mixtus, Cytophaga aurantiaca, Cytophaga hutchinsonii,* and *Sporocytophaga myxococcoides.*

CENS Medium (DSMZ Medium 748)

Composition per liter:

Soytone	4.0g
NH_4Cl	1.0g
K_2HPO_4	0.9g
KH_2PO_4	0.6g
$Na_2S_2O_3 \cdot 5H_2O$	0.5g
$MgSO_4 \cdot 7H_2O$	0.2g
$CaCl_2 \cdot 2H_2O$	75.0mg
EDTA	5.0mg
Vitamin B_{12}	20.0μg
Biotin	15.0μg
Na-pyruvate solution	10.0mL
Chelated iron solution	2.0mL
True Blue trace elements	1.0mL

pH 7.0 ± 0.2 at 25°C

Na-pyruvate Solution:

Composition per 10.0mL:

Na-pyruvate	2.2g

Preparation of Na-pyruvate Solution: Add Na-pyruvate to distilled/deionized water and bring volume to 10.0mL. Mix thoroughly. Sparge with 100% N_2. Filter sterilize.

True Blue Trace Elements:

Composition per 250mL:

EDTA	2.5g
$MnCl_2 \cdot 4H_2O$	0.2g
H_3BO_3	0.1g
$Na_2MoO_4 \cdot 4H_2O$	0.1g
$ZnCl_2$	50.0mg
$NiCl_2 \cdot 6H_2O$	50.0mg

$CoCl_2 \cdot 2H_2O$ 20.0mg
$CuCl_2 \cdot 2H_2O$ 10.0mg
Na_2SeO_3 .... 5.0mg
$NaVO_3 \cdot H_2O$ 5.0mg

Preparation of True Blue Trace Elements: Add components one at a time to approximately 200.0mL distilled/deionized water. Bring final volume to 250.0mL with additional distilled/deionized water.

Chelated Iron Solution:
Composition per 900mL:
EDTA 2.0g
$FeCl_2 \cdot 4H_2O$ 1.0g
HCl 3.0mL

Preparation of Chelated Iron Solution: Add components to distilled/deionized water and bring volume to 900.0mL. Mix thoroughly.

Preparation of Medium: Add components, except Na-pyruvate solution, to distilled/deionized water and bring volume to 990.0L. Mix thoroughly. Autoclave for 15 min at 15 psi pressure–121°C. Cool to 25°C. Aseptically add 10.0mL sterile Na-pyruvate solution. Aseptically distribute into sterile tubes or flasks.

Use: For the cultivation of *Rhodocista centenaria (Rhodospirillum centenum).*

CENS Medium

Composition per liter:
Papaic digest of soybean meal 4.0g
NH_4Cl 1.0g
K_2HPO_4 .... 0.9g
KH_2PO_4 .... 0.6g
$Na_2S_2O_3 \cdot 5H_2O$ 0.5g
$MgSO_4 \cdot 7H_2O$ 0.2g
$CaCl_2 \cdot 2H_2O$ 75.0mg
EDTA 5.0mg
Vitamin B_{12} 20.0μg
D-Biotin 15.0μg
Sodium pyruvate solution 10.0mL
Chelated iron solution 2.0mL
True Blue trace elements 1.0mL
pH 6.8 ± 0.2 at 25°C

Sodium Pyruvate Solution:
Composition per 10.0mL:
Sodium pyruvate 2.2g

Preparation of Sodium Pyruvate Solution: Add sodium pyruvate to distilled/deionized water and bring volume to 10.0mL. Mix thoroughly. Filter sterilize.

Chelated Iron Solution:
Composition per 900.0mL:
EDTA 2.0g
$FeCl_2 \cdot 4H_2O$ 1.0g
HCl, concentrated 3.0mL

Preparation of Chelated Iron Solution: Add components to distilled/deionized water and bring volume to 900.0mL. Mix thoroughly.

True Blue Trace Elements:
Composition per 250.0mL:
EDTA 2.5g
$MnCl_2 \cdot 4H_2O$ 0.2g
H_3BO_3 .... 0.1g
$Na_2MoO_4 \cdot 2H_2O$ 0.1g
ZnCl2 .... 50.0mg
$NiCl_2 \cdot 6H_2O$ 50.0mg
$CoCl_2 \cdot 6H_2O$ 20.0mg
$CuCl_2 \cdot 2H_2O$ 10.0mg
Na_2SeO_3 .... 5.0mg
$NaVO_3$ 5.0mg

Preparation of True Blue Trace Elements: Add components, one at a time, to 200.0mL of distilled/deionized water. Mix thoroughly. Bring volume to 250.0mL with distilled/deionized water.

Preparation of Medium: Add components to distilled/deionized water and bring volume to 1.0L. Mix thoroughly. Adjust pH to 6.8. Distribute into tubes or flasks. Autoclave for 15 min at 15 psi pressure–121°C.

Use: For the cultivation of *Rhodospirillum centenum.*

Centenum Medium

Composition per liter of tap water:
Agar 20.0g
Yeast extract 10.0g
Sodium pyruvate 2.2g
K_2HPO_4 .... 1.0g
$MgSO_4$ 0.5g
Vitamin B_{12} 0.02mg
pH 7.0–7.2 at 25°C

Preparation of Medium: Add components to distilled/deionized water and bring volume to 1.0L. Mix thoroughly. Gently heat and bring to boiling. Distribute into tubes or flasks. Autoclave for 15 min at 15 psi pressure–121°C. Pour into sterile Petri dishes or leave in tubes.

Use: For the cultivation and maintenance of *Rhodospirillum* species.

Cephalothin Cycloheximide Vancomycin Colistin Medium
See: **CCVC Medium**

Ceratocystis Medium

Composition per liter:
Glucose 5.0g
Ammonium tartrate 5.0g
Asparagine 1.0g
KH_2PO_4 .... 1.0g
$MgSO_4 \cdot 7H_2O$ 0.5g
$CaCl_2$ 0.1g
NaCl 0.1g
Inositol 10.0mg
$ZnSO_4 \cdot 7H_2O$ 4.43mg
$MNSO_4 \cdot 4H_2O$ 4.05mg
$FeCl_3 \cdot 6H_2O$ 4.0mg
Pyridoxine 0.1mg
Thiamine 0.1mg

Preparation of Medium: Add components to distilled/deionized water and bring volume to 1.0L. Mix thoroughly. Gently heat and bring to boiling. Distribute into tubes or flasks. Autoclave for 15 min at 15 psi pressure–121°C.

Use: For the cultivation of *Ceratocystis multiannulata.*

Cereal Agar

Composition per liter:
Cereal, precooked mixed 100.0g
Agar 15.0g

Preparation of Medium: Add components to distilled/deionized water and bring volume to 1.0L. Mix thoroughly. Gently heat and bring to boiling. Autoclave for 15 min at 15 psi pressure–121°C. Pour into sterile Petri dishes or distribute into sterile tubes. Allow tubes to cool in a slanted position.

Use: For the cultivation and sporulation of fungi.

Cetrimide Agar, Non-USP

Composition per liter:

Beef heart, solids from infusion	500.0g
Agar	15.0g
Tryptose	10.0g
NaCl	5.0g
Cetrimide	0.9g

pH 7.2 ± 0.2 at 25°C

Preparation of Medium: Add components to distilled/deionized water and bring volume to 1.0L. Mix thoroughly. Gently heat and bring to boiling. Distribute into tubes or flasks. Autoclave for 15 min at 13 psi pressure–118°C. Pour into sterile Petri dishes or leave in tubes.

Use: For the selective isolation, cultivation, and identification of *Pseudomonas aeruginosa* and other Gram-negative, nonfermentative bacteria.

Cetrimide Agar, USP (Pseudosel® Agar)

Composition per liter:

Pancreatic digest of gelatin	20.0g
Agar	13.6g
K_2SO_4	10.0g
$MgCl_2$	1.4g
Cetrimide	0.3g
Glycerol	10.0mL

pH 7.2 ± 0.2 at 25°C

Source: This medium is available as a premixed powder from BD Diagnostic Systems.

Preparation of Medium: Add components to distilled/deionized water and bring volume to 1.0L. Mix thoroughly. Gently heat and bring to boiling. Distribute into tubes or flasks. Autoclave for 15 min at 13 psi pressure–118°C. Pour into sterile Petri dishes or leave in tubes.

Use: For the selective isolation, cultivation, and identification of *Pseudomonas aeruginosa* and other Gram-negative, nonfermentative bacteria.

Cetrimide Agar Base with Glycerol

Composition per liter:

Pancreatic digest of gelatin	20.0g
K_2SO_4	10.0g
$MgCl_2$	1.4g
Cetrimide	0.3g
Glycerol	10.0mL

pH 7.2 ± 0.2 at 25°C

Source: This medium, without glycerol, is available as a premixed powder from HiMedia.

Preparation of Medium: Add components to distilled/deionized water and bring volume to 1.0L. Mix thoroughly. Gently heat and bring to boiling. Distribute into tubes or flasks. Autoclave for 15 min at 13 psi pressure–118°C. Pour into sterile Petri dishes or leave in tubes.

Use: For the selective isolation, cultivation, and identification of *Pseudomonas aeruginosa* and other Gram-negative, nonfermentative bacteria.

Cetrimide Agar Base with Glycerol and Nalidixic Selective Supplement

Composition per liter:

Pancreatic digest of gelatin	20.0g
K_2SO_4	10.0g
$MgCl_2$	1.4g
Cetrimide	0.3g
Glycerol	10.0mL
Nalidixic selective supplement	5.0mL

pH 7.2 ± 0.2 at 25°C

Source: This medium, without glycerol and nalidixic acid supplement, is available as a premixed powder from HiMedia.

Nalidixic Selective Supplement:

Composition per 5.0mL:

Nalidixic acid	15.0mg

Preparation of Nalidixic Selective Supplement: Add nalidixic acid to distilled/deionized water and bring volume to 50.0mL. Mix thoroughly. Filter sterilize.

Preparation of Medium: Add components, except nalidixic selective supplement, to distilled/deionized water and bring volume to 1.0L. Mix thoroughly. Gently heat and bring to boiling. Distribute into tubes or flasks. Autoclave for 15 min at 15 psi pressure–121°C. Cool to 50°C. Aseptically add 5.0mL sterile nalidixic selective supplement. Mix thoroughly. Pour into sterile Petri dishes or leave in tubes.

Use: For the selective isolation, cultivation, and identification of *Pseudomonas aeruginosa* and other Gram-negative, nonfermentative bacteria.

Cetrimide HiVeg Agar Base with Glycerol

Composition per liter:

Plant peptone No. 2	20.0g
K_2SO_4	10.0g
$MgCl_2$	1.4g
Cetrimide	0.3g
Glycerol	10.0mL

pH 7.2 ± 0.2 at 25°C

Source: This medium, without glycerol, is available as a premixed powder from HiMedia.

Preparation of Medium: Add components to distilled/deionized water and bring volume to 1.0L. Mix thoroughly. Gently heat and bring to boiling. Distribute into tubes or flasks. Autoclave for 15 min at 13 psi pressure–118°C. Pour into sterile Petri dishes or leave in tubes.

Use: For the selective isolation, cultivation, and identification of *Pseudomonas aeruginosa* and other Gram-negative, nonfermentative bacteria.

Cetrimide HiVeg Agar Base with Glycerol and Nalidixic Selective Supplement

Composition per liter:

Plant peptone No. 2	20.0g
K_2SO_4	10.0g
$MgCl_2$	1.4g
Cetrimide	0.3g

Glycerol 10.0mL
Nalidixic selective supplement 5.0mL
pH 7.2 ± 0.2 at 25°C

Source: This medium, without glycerol and nalidixic acid supplement, is available as a premixed powder from HiMedia.

Nalidixic Selective Supplement:
Composition per 5.0mL:
Nalidixic acid 15.0mg

Preparation of Nalidixic Selective Supplement: Add nalidixic to distilled/deionized water and bring volume to 50.0mL. Mix thoroughly. Filter sterilize.

Preparation of Medium: Add components, except nalidixic selective supplement, to distilled/deionized water and bring volume to 1.0L. Mix thoroughly. Gently heat and bring to boiling. Distribute into tubes or flasks. Autoclave for 15 min at 15 psi pressure–121°C. Cool to 50°C. Aseptically add 5.0mL sterile nalidixic selective supplement. Mix thoroughly. Pour into sterile Petri dishes or leave in tubes.

Use: For the selective isolation, cultivation, and identification of *Pseudomonas aeruginosa* and other Gram-negative, nonfermentative bacteria.

CF Assay Medium (Citrovorum Factor Assay Medium)

Composition per liter:
Glucose 50.0g
Sodium acetate 40.0g
Vitamin assay casamino acids 10.0g
NH_4Cl 6.0g
K_2HPO_4 ..... 1.2g
KH_2PO_4 ..... 1.2g
$MgSO_4 \cdot 7H_2O$ 0.4g
DL-Alanine 0.2g
DL-Tryptophan 0.2g
L-Cystine 0.2g
L-Cysteine·HCl 0.2g
$MgSO_4 \cdot 7H_2O$ 0.04g
Adenine sulfate 0.02g
$FeSO_4$ 0.02g
Glycine 0.02g
Guanine·HCl 0.02g
NaCl 0.02g
Uracil 0.02g
Xanthine 0.02g
Pyridoxamine·HCl 6.0mg
Nicotinic acid 2.0mg
Pyridoxine·HCl 2.0mg
Calcium pantothenate 1.0mg
Riboflavin 1.0mg
Thiamine·HCl 1.0mg
Pyridoxal·HCl 600.0μg
p-Aminobenzoic acid 200.0μg
Folic acid 20.0μg
Biotin 2.0μg
pH 6.7 ± 0.2 at 25°C

Preparation of Medium: Add components to distilled/deionized water and bring volume to 1.0L. Mix thoroughly. Gently heat and bring to boiling. Continue boiling for 2–3 min. Allow precipitate to settle out. Distribute supernatant into tubes in 5.0mL volumes. Add standard solution or test solutions to each tube. Adjust the volume of each tube to 10.0mL with distilled/deionized water. Autoclave for 10 min at 15 psi pressure–121°C.

Use: For the microbiological assay of citrovorum factor using *Pediococcus acidilactici.*

CFAT Medium (Cadmium Fluoride Acriflavin Tellurite Medium)

Composition per liter:
Pancreatic digest of casein 17.0g
Agar 15.0g
Glucose 7.5g
NaCl 5.0g
Papaic digest of soybean meal 3.0g
K_2HPO_4 ..... 2.5g
NaF 0.8g
$CdSO_4$ 0.013g
K_2TeO_3 ..... 2.5mg
Neutral acriflavin 1.2mg
Basic Fuchsin 0.25mg
Sheep blood, defibrinated 50.0mL

Caution: Potassium tellurite is toxic.

Preparation of Medium: Add components, except sheep blood, to distilled/deionized water and bring volume to 950.0mL. Mix thoroughly. Gently heat and bring to boiling. Autoclave for 15 min at 15 psi pressure–121°C. Cool to 45°–50°C. Add 50.0mL of sterile, defibrinated sheep blood. Mix thoroughly. Pour into sterile Petri dishes or leave in tubes.

Use: For the isolation, cultivation, and enumeration of *Actinomyces viscosus* and *Actinomyces naeslundii* from clinical specimens, especially dental plaque.

CGY Agar

Composition per liter:
Agar 20.0g
Pancreatic digest of casein 5.0g
Yeast extract 1.0g
Glycerol 10.0mL
pH 7.2 ± 0.2 at 25°C

Preparation of Medium: Add components to distilled/deionized water and bring volume to 1.0L. Mix thoroughly. Gently heat and bring to boiling. Distribute into tubes or flasks. Autoclave for 15 min at 15 psi pressure–121°C. Pour into sterile Petri dishes or leave in tubes.

Use: For the cultivation and maintenance of *Bacillus pseudogordonae.*

CGY Autolysate Broth
See: **Casitone Glycerol Yeast Autolysate Broth**

CH 1 Medium

Composition per liter:
NaCl 250.0g
Tris(hydroxymethyl)amino methane buffer 12.0g
Glycerol 10.0g
Hy-Case SF 5.0g
Yeast extract 5.0g
Solution 1 ..... 50.0mL

Solution 1:

Composition per liter:

$MgCl_2 \cdot 6H_2O$	40.0g
KCl	4.0g
$CaCl_2 \cdot 2H_2O$	0.4g

pH 7.4 ± 0.2 at 25°C

Preparation of Solution 1: Add components to distilled/deionized water and bring volume to 1.0L. Mix thoroughly.

Preparation of Medium: Add components to distilled/deionized water and bring volume to 1.0L. Mix thoroughly. Adjust pH to 7.4. Distribute into tubes or flasks. Autoclave for 15 min at 15 psi pressure–121°C.

Use: For the cultivation of *Haloarcula vallismortis*.

Chaetomium Medium

Composition per liter:

Agar	15.0g
$NaNO_3$	2.0g
K_2HPO_4	1.2g
$MgSO_4 \cdot 7H_2O$	0.5g
KCl	0.5g
KH_2PO_4	0.14g
Yeast extract	0.02g
$Fe_2(SO_4)_3 \cdot H_2O$	0.01g

pH 7.2 ± 0.2 at 25°C

Preparation of Medium: Add components to distilled/deionized water and bring volume to 1.0L. Mix thoroughly. Gently heat and bring to boiling. Distribute into tubes or flasks. Autoclave for 15 min at 15 psi pressure–121°C. Pour into sterile Petri dishes or leave in tubes. Allow agar to cool. Place a strip of sterile filter paper on top of agar and inoculate filter strip.

Use: For the growth and maintenance of *Chaetomium* species.

Chalquist's Antigen Medium, Modified

Composition per liter:

Soluble starch	0.5g
Pancreatic digest of casein	0.05g
L-Cysteine·HCl·H_2O	0.01g
NAD (nicotinamide adeninedinucleotide)	0.01g
PPLO broth without Crystal Violet	90.0mL
Swine serum, inactivated	10.0mL
Phenol Red (1% solution)	0.25mL

pH 7.6 ± 0.2 at 25°C

PPLO Broth without Crystal Violet:

Composition per 500.0mL:

Beef heart, infusion from	11.52g
Peptone	2.32g
NaCl	1.15g

Source: PPLO broth without Crystal Violet is available as a premixed powder from BD Diagnostic Systems.

Preparation of PPLO Broth without Crystal Violet: Add components to distilled/deionized water and bring volume to 500.0mL. Mix thoroughly.

Preparation of Medium: Add components to distilled/deionized water and bring volume to 1.0L. Mix thoroughly. Distribute into tubes or flasks. Autoclave for 15 min at 15 psi pressure–121°C.

Use: For the cultivation of *Mycoplasma synoviae.*

Chapman Stone Agar

Composition per liter:

$(NH_4)_2SO_4$	75.0g
NaCl	55.0g
Gelatin	30.0g
Agar	15.0g
D-Mannitol	10.0g
Pancreatic digest of casein	10.0g
K_2HPO_4	5.0g
Yeast extract	2.0g

pH 7.0 ± 0.2 at 25°C

Source: This medium is available as a premixed powder from BD Diagnostic Systems.

Preparation of Medium: Add components to distilled/deionized water and bring volume to 1.0L. Mix thoroughly. Autoclave for 10 min at 15 psi pressure–121°C. Pour into sterile Petri dishes while the medium is still hot. Add 25.0mL of medium per Petri dish.

Use: For the isolation of staphylococci from a variety of specimens.

Charcoal Agar

Composition per liter:

Beef heart, solids from infusion	500.0g
Agar	18.0g
Peptone	10.0g
Soluble starch	10.0g
NaCl	5.0g
Charcoal, activated, acid washed	4.0g
Yeast extract	3.5g

pH 7.3 ± 0.2 at 25°C

Source: This medium is available as a premixed powder from BD Diagnostic Systems.

Preparation of Medium: Add components to distilled/deionized water and bring volume to 1.0L. Mix thoroughly. Gently heat and bring to boiling with frequent stirring. Autoclave for 15 min at 15 psi pressure–121°C. Cool to 45°–50°C. Pour into sterile Petri dishes or leave in tubes. Shake flask while dispensing to keep charcoal in suspension. Allow tubes to cool in a slanted position.

Use: For the cultivation and maintenance of fastidious microorganisms, especially *Bordetella pertussis,* for vaccine production.

Charcoal Agar

Composition per liter:

Agar	12.0g
Beef extract	10.0g
Peptone	10.0g
Starch	10.0g
NaCl	5.0g
Charcoal	4.0g
Nicotinic acid	0.001g

pH 7.4 ± 0.2 at 25°C

Source: This medium is available as a premixed powder from Oxoid Unipath.

Preparation of Medium: Add components to distilled/deionized water and bring volume to 1.0L. Mix thoroughly. Gently heat and bring to boiling with frequent stirring. Autoclave for 15 min at 15 psi pressure–121°C. Cool to 45°–50°C. This medium may be enriched by the addition of blood. Pour into sterile Petri dishes or distribute into tubes. Shake flask while dispensing to keep charcoal in suspension.

Use: For the cultivation and isolation of various bacteria; with the addition of blood, for the cultivation of fastidious bacteria.

Charcoal Agar Base, HiVeg with Blood and Selective Supplement

Composition per liter:

Agar	18.0g
Plant infusion	12.0g
Plant peptone	10.0g
Starch	10.0g
NaCl	5.0g
Charcoal	4.0g
Yeast extract	3.5g
Sheep blood, defibrinated	50.0mL
Selective supplement solution	10.0mL

pH 7.3 ± 0.2 at 25°C

Source: This medium, without blood or selective supplement, is available as a premixed powder from HiMedia.

Selective Supplement Solution:

Composition per 10.0mL:

Cefazolin	40.0mg

Preparation of Selective Supplement Solution: Add cefazolin to distilled/deionized water and bring volume to 10.0mL. Mix thoroughly. Filter sterilize.

Preparation of Medium: Add components, except blood and selective supplement solution, to distilled/deionized water and bring volume to 940.0mL. Mix thoroughly. Gently heat and bring to boiling with frequent stirring. Autoclave for 15 min at 15 psi pressure–121°C. Cool to 45°–50°C. Aseptically add 50.0 mL sterile sheep blood and 10.0mL sterile selective supplement solution. Mix thoroughly. Pour into sterile Petri plates or distribute into sterile tubes.

Use: For the cultivation and isolation of various bacteria; with the addition of blood, for the cultivation of fastidious bacteria. For the cultivation of *Bordetella pertussis* for vaccine production and the maintenance of stock cultures.

Charcoal Agar with Horse Blood

Composition per liter:

Agar	12.0g
Beef extract	10.0g
Peptone	10.0g
Starch	10.0g
NaCl	5.0g
Charcoal, bacteriological	4.0g
Nicotinic acid	1.0mg
Horse blood, defibrinated	100.0mL

pH 7.4 ± 0.2 at 25°C

Preparation of Medium: Add components to distilled/deionized water and bring volume to 900.0L. Mix thoroughly. Gently heat and bring to boiling with frequent stirring. Autoclave for 15 min at 15 psi pressure–121°C. Cool to 80°C. Aseptically add 100.0mL of sterile, defibrinated horse blood. Maintain at 80°C for 10 min to form chocolate agar. Pour into sterile Petri dishes or distribute into tubes. Shake flask while dispensing to keep charcoal in suspension.

Use: For the cultivation and isolation of *Haemophilus influenzae.*

Charcoal Agar with Horse Blood and Cepahalexin

Composition per liter:

Agar	12.0g
Beef extract	10.0g
Peptone	10.0g
Starch	10.0g
NaCl	5.0g
Charcoal	4.0g
Nicotinic acid	1.0mg
Horse blood, defibrinated	100.0mL
Cephalexin solution	10.0mL

pH 7.4 ± 0.2 at 25°C

Cephalexin Solution:

Composition per 10.0mL:

Cephalexin	0.04g

Preparation of Cephalexin Solution: Add cephalexin to distilled/deionized water and bring volume to 10.0mL. Mix thoroughly. Filter sterilize.

Preparation of Medium: Add components, except cephalexin solution and horse blood, to distilled/deionized water and bring volume to 890.0L. Mix thoroughly. Gently heat and bring to boiling with frequent stirring. Autoclave for 15 min at 15 psi pressure–121°C. Cool to 45°–50°C. Aseptically add 100.0mL of sterile, defibrinated horse blood and 10.0mL of sterile cephalexin solution. Pour into sterile Petri dishes or distribute into tubes. Shake flask while dispensing to keep charcoal in suspension.

Use: For the cultivation and isolation of *Bordetella pertussis*.

Charcoal Agar Slants

See: **Diphasic Medium for Amoeba**

Charcoal Blood Agar Base, HiVeg with Blood Penicillin and Damidodiphenyl Hydrochloride

Composition per liter:

Agar	12.0g
Plant extract	10.0g
Plant peptone	10.0g
Starch	10.0g
NaCl	5.0g
Charcoal	4.0g
Yeast extract	3.5g
Sheep blood, defibrinated	50.0mL
Penicillin solution	3.0mL
Diamidodiphenylamine hydrochloride solution	3.0mL

pH 7.5 ± 0.2 at 25°C

Source: This medium, without blood or selective supplement, is available as a premixed powder from HiMedia.

Penicllin Solution:

Composition per 10.0mL:

Penicillin	1000 units

Preparation of Penicillin Solution: Add penicillin to distilled/deionized water and bring volume to 10.0mL. Mix thoroughly. Filter sterilize.

Diamidodiphenylamine Hydrochloride Solution:

Composition per 10.0mL:

Diamido hydrochloride	0.01g
Diphenylamine hydrochloride	0.01g

Preparation of Diamidodiphenylamine Hydrochloride Solution: Add components to distilled/deionized water and bring volume to 10.0mL. Mix thoroughly. Filter sterilize.

Preparation of Medium: Add components, except blood, penicillin solution, and diamidodiphenylamine hydrochloride solution, to distilled/deionized water and bring volume to 950.0mL. Mix thoroughly. Gently heat and bring to boiling with frequent stirring. Autoclave for 15 min at 15 psi pressure–121°C. Cool to 45°–50°C. Aseptically add 50.0 mL sterile sheep blood and 3.0 mL sterile penicillin solution and 3.0 mL sterile diamidodiphenylamine hydrochloride solution. Mix thoroughly. Pour into sterile Petri plates or distribute into sterile tubes.

Use: For the cultivation and isolation of various bacteria; with the addition of blood, for the cultivation of fastidious bacteria. For the cultivation of *Bordetella pertussis* for vaccine production and the maintenance of stock cultures.

Charcoal Blood Medium

Composition per liter:

Beef heart, solids from infusion	500.0g
Agar	18.0g
Peptone	10.0g
Soluble starch	10.0g
NaCl	5.0g
Charcoal, activated, acid washed	4.0g
Yeast extract	3.5g
Horse or sheep blood, defibrinated	100.0mL
Cephalexin solution	10.0mL

pH 7.4 ± 0.2 at 25°C

Cephalexin Solution:

Composition per 10.0mL:

Cephalexin	0.04g

Preparation of Cephalexin Solution: Add cephalexin to distilled/deionized water and bring volume to 10.0mL. Mix thoroughly. Filter sterilize.

Preparation of Medium: Add components, except blood and cephalexin solution, to distilled/deionized water and bring volume to 890.0mL. Mix thoroughly. Gently heat and bring to boiling. Autoclave for 15 min at 15 psi pressure–121°C. Cool to 45°–50°C. Aseptically add sterile blood and cephalexin solution. Mix thoroughly. Pour into sterile Petri dishes or distribute into sterile tubes.

Use: For the cultivation of *Haemophilus influenzae.*

Charcoal HiVeg Agar Base with Niacin, Blood, and Selective Supplement

Composition per liter:

Agar	12.0g
Plant extract	10.0g
Plant peptone No. 2	10.0g
Starch	10.0g
NaCl	5.0g
Charcoal	4.0g
Niacin	1.0mg
Sheep blood, defibrinated	50.0mL
Selective supplement solution	10.0mL

pH 7.4 ± 0.2 at 25°C

Source: This medium, without blood or selective supplement, is available as a premixed powder from HiMedia.

Selective Supplement Solution:

Composition per 10.0mL:

Cefazolin	40.0mg

Preparation of Selective Supplement Solution: Add cefazolin to distilled/deionized water and bring volume to 10.0mL. Mix thoroughly. Filter sterilize.

Preparation of Medium: Add components, except blood and selective supplement solution, to distilled/deionized water and bring volume to 940.0mL. Mix thoroughly. Gently heat and bring to boiling with frequent stirring. Autoclave for 15 min at 15 psi pressure–121°C. Cool to 45°–50°C. Aseptically add 50.0 mL sterile sheep blood and 10.0mL sterile selective supplement solution. Mix thoroughly. Pour into sterile Petri plates or distribute into sterile tubes.

Use: For the cultivation and isolation of *Bordetella pertussis* and *Haemophilus influenzae.* For the cultivation of *Bordetella pertussis* for vaccine production and the maintenance of stock cultures.

Charcoal Yeast Extract Agar

See: **CYE Agar**

Charcoal Yeast Extract Agar, Buffered

See: **CYE Agar, Buffered**

Charcoal Yeast Extract Diphasic Blood Culture Medium

See: ***Legionella pneumophila*** **Medium**

Chase's Medium SP

Composition per liter:

Agar	10.0g
Proteose peptone	10.0g
K_2HPO_4	2.0g
$(NH_4)_2SO_4$	1.0g
Sucrose solution	100.0mL

pH 6.5 ± 0.2 at 25°C

Sucrose Solution:

Composition per 100.0mL:

Sucrose	10.0g

Preparation of Sucrose Solution: Add sucrose to distilled/deionized water and bring volume to 100.0mL. Mix thoroughly. Filter sterilize.

Preparation of Medium: Add components, except sucrose solution, to tap water and bring volume to 900.0mL. Mix thoroughly. Gently heat and bring to boiling. Autoclave for 15 min at 15 psi pressure–121°C. Cool to 45°–50°C. Aseptically add 100.0mL of sterile sucrose solution. Mix thoroughly. Pour into sterile Petri dishes or distribute into sterile tubes.

Use: For the cultivation and maintenance of ATCC strain 13949.

CHCA Salts Medium (Cyclohexane Carboxylic Acid Salts Medium)

Composition per liter:

K_2HPO_4	3.5g
KH_2PO_4	1.5g
Cyclohexane carboxylic acid	1.0g
NH_4NO_3	1.0g
$MgSO_4 \cdot 7H_2O$	0.5g
$FeSO_4 \cdot 7H_2O$	0.1g
Yeast extract	0.1g

$CaCl_2 \cdot 2H_2O$ 0.01g
$Na_2MoO_2 \cdot 2H_2O$ 0.01g
$ZnSO_4 \cdot 7H_2O$ 0.01g

pH 7.0 ± 0.2 at 25°C

Preparation of Medium: Add components to distilled/deionized water and bring volume to 1.0L. Mix thoroughly. Adjust pH to 7.0. Gently heat and bring to boiling. Distribute into tubes or flasks. Autoclave for 15 min at 15 psi pressure–121°C. Pour into sterile Petri dishes or leave in tubes.

Use: For the cultivation and maintenance of bacteria that can utilize cyclohexane carboxylic acid as a carbon source. For the cultivation and maintenance of *Arthrobacter globiformis*.

Cheese Agar

Composition per liter:

Cheese, ripened 100.0g
NaCl 50.0g
Agar 15.0g
Peptone 10.0g
Potassium citrate 10.0g
Sodium oxalate 2.0g

pH 7.4 ± 0.2 at 25°C

Preparation of Medium: Add the cheese and potassium citrate to distilled/deionized water and bring volume to 300.0mL. Gently heat and bring to 50°C to separate the fat. Discard the fat. In a separate flask, add the remaining components to distilled/deionized water and bring volume to 700.0mL. Gently heat and bring to boiling. Add the 300.0mL of aqueous suspension of cheese solids. Adjust pH to 7.4. Autoclave for 25 min at 15 psi pressure–121°C. Pour into sterile Petri dishes or distribute into sterile tubes.

Use: For the isolation and cultivation of *Brevibacterium linens*.

Cherry Agar, CBS Formula

Composition per liter:

Pulp of sour stone cherries 200.0g
Agar 20.0g

pH 3.8 ± 0.2 at 25°C

Preparation of Medium: Add 1.0L of distilled/deionized water to pulp of sour stone cherries. Gently heat and bring to boiling. Simmer gently for 2 hr. Strain through cloth. Autoclave filtrate for 30 min at 6 psi pressure–110°C. Add agar to distilled/deionized water and bring volume to 800.0mL. Gently heat and bring to boiling. Autoclave for 15 min at 15 psi pressure–121°C. To sterile agar solution, add 200.0mL of sterile cherry filtrate. Distribute into sterile tubes. Autoclave for 5 min at 15 psi pressure–121°C.

Use: For the cultivation of various fungi.

Cherry Decoction Agar

Composition per liter:

Agar 20.0g
Sucrose 10.0g
Cherry extract 90.0mL

pH 3.8–4.6 at 25°C

Preparation of Medium: Add agar to distilled/deionized water and bring volume to 500.0mL. Gently heat and bring to boiling. In a separate flask, add sucrose and cherry extract to distilled/deionized water and bring volume to 500.0mL. Mix thoroughly. Combine the two solutions. Adjust pH to 3.8–4.6. Distribute into tubes or flasks. Sterilize by tyndallization at 100°C for 30 min on 3 consecutive days. Pour into sterile Petri dishes or leave in tubes.

Use: For the cultivation and maintenance of *Acremonium sclerotigenum, Amylostereum laevigatum, Calcarisporium arbuscula, Helicoma morganii, Helicoon richonis, Inonotus radiatus, Phellinus igniarius, Phoma leveillei, Taphrina californica,* and *Taphrina populina.*

CHI 1776 Medium

Composition per liter:

Pancreatic digest of casein 25.0g
Yeast extract 7.5g
Glucose solution 25.0mL
Tris·HCl buffer, 1*M*, pH 7.5 .... 20.0mL
Diaminopimelic acid solution 10.0mL
Thymidine solution 10.0mL
$MgCl_2$ solution 5.0mL

Glucose Solution:
Composition per 100.0mL:

D-Glucose 20.0g

Preparation of Glucose Solution: Add glucose to distilled/deionized water and bring volume to 100.0mL. Mix thoroughly. Filter sterilize.

Diaminopimelic Acid Solution:
Composition per 100.0mL:

meso-Diaminopimelic acid 1.0g

Preparation of Diaminopimelic Acid Solution: Add *meso*-diaminopimelic acid to distilled/deionized water and bring volume to 100.0mL. Mix thoroughly. Filter sterilize.

Thymidine Solution:
Composition per 100.0mL:

Thymidine 0.4g

Preparation of Thymidine Solution: Add thymidine to distilled/deionized water and bring volume to 100.0mL. Mix thoroughly. Filter sterilize.

$MgCl_2$ Solution:
Composition per 100.0mL:

$MgCl_2$ 9.52g

Preparation of $MgCl_2$ Solution: Add $MgCl_2$ to distilled/deionized water and bring volume to 100.0mL. Mix thoroughly. Autoclave for 15 min at 15 psi pressure–121°C.

Preparation of Medium: Add components, except glucose solution, diaminopimelic acid solution, thymidine solution, and $MgCl_2$ solution, to distilled/deionized water and bring volume to 950.0mL. Mix thoroughly. Autoclave for 15 min at 15 psi pressure–121°C. Aseptically add 25.0mL of sterile glucose solution, 10.0mL of sterile diaminopimelic acid solution, 10.0mL of sterile thymidine solution, and 5.0mL of sterile $MgCl_2$ solution, Mix thoroughly. Distribute into sterile tubes or flasks.

Use: For the cultivation of *Escherichia coli*.

Chicken Soup Broth

Composition per 5.0L:

Chicken 2.5kg
Peppercorns 6
Cloves 3
Bay leaf 2
Celery, stalks including leaves 2

Onion, large.. 1
Carrot 1
Dill, fresh .... 1/4 cup
NaCl..0.1g

Preparation of Medium: Add a nice, whole chicken to a large pot. Add enough tap water to cover the chicken by about 1 in. Stud the whole, peeled onion with the three cloves. Add the onion and remaining ingredients to the pot. Rapidly heat and bring to boiling. Lower heat to a simmer and cook for 1 to 1.5 hr. Remove the chicken and vegetables from the broth. Remove skin and bones from the chicken. Cut up the meat into 1-inch pieces. Return the meat to the broth. If desired, slice the carrot and celery and return them to the broth.

Use: For the growth and nutrition of microbiologists.

China Blue Lactose Agar

Composition per liter:

Agar 12.0g
Lactose 10.0g
Peptone.. 5.0g
NaCl 5.0g
Beef extract 3.0g
China Blue 0.375g

pH 7.0 ± 0.2 at 25°C

Source: This medium is available as a premixed powder from Oxoid Unipath.

Preparation of Medium: Add components to distilled/deionized water and bring volume to 1.0L. Mix thoroughly. Gently heat and bring to boiling. Distribute into tubes or flasks. Autoclave for 15 min at 15 psi pressure–121°C. Pour into sterile Petri dishes or leave in tubes.

Use: For the cultivation, differentiation, and enumeration of bacteria from dairy products. Lactose-fermenting bacteria appear as blue colonies. Nonlactose-fermenting bacteria appear as colorless colonies.

China Blue Lactose Agar

Composition per liter:

Agar 15.0g
Lactose 10.0g
Peptic digest of animal tissue.. 5.0g
NaCl 5.0g
Plant extract 3.0g
China Blue 0.3g

pH 7.0 ± 0.2 at 25°C

Source: This medium is available as a premixed powder from HiMedia.

Preparation of Medium: Add components to distilled/deionized water and bring volume to 1.0L. Mix thoroughly. Gently heat and bring to boiling. Distribute into tubes or flasks. Autoclave for 15 min at 15 psi pressure–121°C. Pour into sterile Petri dishes or leave in tubes.

Use: For the cultivation, differentiation, and enumeration of bacteria from dairy products. Lactose-fermenting bacteria appear as blue colonies. Nonlactose-fermenting bacteria appear as colorless colonies.

China Blue Lactose HiVeg Agar

Composition per liter:

Agar 15.0g
Lactose 10.0g
Plant peptone.. 5.0g
NaCl 5.0g
Plant extract 3.0g
China Blue 0.3g

pH 7.0 ± 0.2 at 25°C

Source: This medium is available as a premixed powder from HiMedia.

Preparation of Medium: Add components to distilled/deionized water and bring volume to 1.0L. Mix thoroughly. Gently heat and bring to boiling. Distribute into tubes or flasks. Autoclave for 15 min at 15 psi pressure–121°C. Pour into sterile Petri dishes or leave in tubes.

Use: For the cultivation, differentiation, and enumeration of bacteria from dairy products. Lactose-fermenting bacteria appear as blue colonies. Nonlactose-fermenting bacteria appear as colorless colonies.

Chitin Agar

Composition per liter:

Agar 15.0g
Chitin, precipitated 3.0g
$(NH_4)_2SO_4$ 2.0g
Na_2HPO_4 .... 1.1g
KH_2PO_4 .... 0.7g
$MgSO_4 \cdot 7H_2O$ 0.2g
$FeSO_4$ 1.0mg
$MnSO_4$ 1.0mg

Chitin, Precipitated:

Composition per 2.5L:

Chitin 40.0g
HCl, concentrated 400.0mL

Preparation of Chitin, Precipitated: Add chitin to 400.0mL of cold concentrated HCl. Add this solution to 2.0L of distilled/deionized water at 5°C. Filter the solution through Whatman #1 filter paper. Dialyze the precipitated chitin against tap water for 12 hr. Adjust the pH to 7.0 with KOH.

Preparation of Medium: Add components to distilled/deionized water and bring volume to 1.0L. Mix thoroughly. Gently heat and bring to boiling. Distribute into tubes or flasks. Autoclave for 15 min at 15 psi pressure–121°C. Pour into sterile Petri dishes or leave in tubes.

Use: For the isolation and cultivation of *Cytophaga* species, *Herpetosiphon* species, *Saprospira* species, and *Flexithrix* species.

Chitin Agar

Composition per liter:

Agar 20.0g
Chitin 4.0g
K_2HPO_4 .... 0.7g
$MgSO_4 \cdot 7H_2O$ 0.5g
KH_2PO_4 .... 0.3g
$FeSO_4 \cdot 7H_2O$ 0.01g
$MnCl_2 \cdot 4H_2O$ 0.001g
$ZnSO_4 \cdot 7H_2O$ 0.001g

pH 8.0 ± 0.2 at 25°C

Preparation of Medium: Add components to distilled/deionized water and bring volume to 1.0L. Mix thoroughly. Gently heat and bring to boiling. Distribute into tubes or flasks. Autoclave for 15 min at 15 psi pressure–121°C. Pour into sterile Petri dishes or leave in tubes.

Use: For the selective isolation and cultivation of streptomycetes.

Chlamydia Growth Medium

Composition per 500.0mL:

Eagle minimum essential medium with Earle salts, 10X	50.0mL
Fetal calf serum	50.0mL
L-Glutamine solution	5.0mL

pH 7.4 ± 0.2 at 25°C

Eagle Minimum Essential Medium with Earle Salts, 10X:

Composition per liter:

NaCl	6.8g
Glucose	1.0g
KCl	0.4g
$CaCl_2 \cdot 2H_2O$	0.2g
$MgCl_2 \cdot 6H_2O$	0.2g
NaH_2PO_4	0.15g
L-Arginine	0.1g
L-Lysine	0.06g
L-Isoleucine	0.05g
L-Leucine	0.05g
L-Threonine	0.05g
L-Valine	0.05g
L-Tyrosine	0.04g
L-Phenylalanine	0.03g
L-Histidine	0.03g
L-Cystine	0.02g
L-Methionine	0.02g
L-Tryptophan	0.01g
i-Inositol	2.0mg
Calcium pantothenate	1.0mg
Choline chloride	1.0mg
Folic acid	1.0mg
Nicotinamide	1.0mg
Pyridoxal	1.0mg
Thiamine·HCl	1.0mg
Riboflavin	0.1mg

Preparation of Eagle Minimum Essential Medium with Earle Salts, 10X: Add components to distilled/deionized water and bring volume to 1.0L. Mix thoroughly. Adjust pH to 7.4 with 7.5% Na_2CO_3 solution. Filter sterilize.

Glutamine Solution:

Composition per 100.0mL:

L-Glutamine	2.92g
NaCl (0.85% solution)	100.0mL

Preparation of Glutamine Solution: Add the glutamine to the 0.85% NaCl solution. Mix thoroughly. Filter sterilize.

Preparation of Medium: Aseptically combine 50.0mL of sterile Eagle minimum essential medium with Earle salts, 10X, 50.0mL of fetal calf serum, and 5.0mL of sterile glutamine solution. Bring volume to 500.0mL with sterile distilled/deionized water. Mix thoroughly. Aseptically distribute into sterile tubes or flasks.

Use: For the cultivation of *Chlamydia* species.

Chlamydia Isolation Medium

Composition per 500.0mL:

Eagle minimum essential medium with Earle salts, 10X	50.0mL
Fetal calf serum	50.0mL
Selective supplement	10.0mL
L-Glutamine solution	5.0mL

pH 7.4 ± 0.2 at 25°C

Eagle Minimum Essential Medium with Earle Salts, 10X:

Composition per liter:

NaCl	6.8g
Glucose	1.0g
KCl	0.4g
$CaCl_2 \cdot 2H_2O$	0.2g
$MgCl_2 \cdot 6H_2O$	0.2g
NaH_2PO_4	0.15g
L-Arginine	0.1g
L-Lysine	0.06g
L-Isoleucine	0.05g
L-Leucine	0.05g
L-Threonine	0.05g
L-Valine	0.05g
L-Tyrosine	0.04g
L-Phenylalanine	0.03g
L-Histidine	0.03g
L-Cystine	0.02g
L-Methionine	0.02g
L-Tryptophan	0.01g
i-Inositol	2.0mg
Calcium pantothenate	1.0mg
Choline chloride	1.0mg
Folic acid	1.0mg
Nicotinamide	1.0mg
Pyridoxal	1.0mg
Thiamine·HCl	1.0mg
Riboflavin	0.1mg

Preparation of Eagle Minimum Essential Medium with Earle Salts, 10X: Add components to distilled/deionized water and bring volume to 1.0L. Mix thoroughly. Adjust pH to 7.4 with 7.5% Na_2CO_3 solution. Filter sterilize.

Selective Supplement:

Composition per 10.0mL:

Glucose	0.594g
Vancomycin	0.05g
Gentamicin	0.01g
Amphotericin B	2.0mg
Cycloheximide	2.0mg

Preparation of Selective Supplement: Add components to distilled/deionized water and bring volume to 10.0mL. Mix thoroughly. Filter sterilize.

Caution: Cycloheximide is toxic. Avoid skin contact or aerosol formation and inhalation.

Glutamine Solution:

Composition per 100.0mL:

L-Glutamine	2.92g
NaCl (0.85% solution)	100.0mL

Preparation of Glutamine Solution: Add the glutamine to the 0.85% NaCl solution. Mix thoroughly. Filter sterilize.

Preparation of Medium: Aseptically combine 50.0mL of sterile Eagle minimum essential medium with Earle salts, 10X, 50.0mL of fetal calf serum, 10.0mL of selective supplement, and 5.0mL of sterile glutamine solution. Bring volume to 500.0mL with sterile distilled/deionized water. Mix thoroughly. Aseptically distribute into sterile tubes or flasks.

Use: For the isolation and cultivation of *Chlamydia* species.

Chlamydomonas Enriched Medium

Composition per liter:

Agar	15.0g
Sodium acetate, anhydrous	2.0g
Tryptose	2.0g
Yeast extract	2.0g

Preparation of Medium: Add components to distilled/deionized water and bring volume to 1.0L. Mix thoroughly. Gently heat and bring to boiling. Distribute into tubes or flasks. Autoclave for 20 min at 15 psi pressure–121°C. Pour into sterile Petri dishes or leave in tubes.

Use: For the cultivation of *Chlamydomonas reinhardtii.*

Chlamydomonas Mutant Agar

Composition per liter:

Agar	18.0g
K_2HPO_4	1.162g
Sodium acetate	1.0g
NaH_2PO_4	0.92g
Sodium citrate	0.5g
$MgSO_4 \cdot 7H_2O$	0.3g
NH_4NO_3	0.3g
$CaCl_2$	0.04g
$FeCl_3 \cdot 6H_2O$	0.01g
Trace salts solution	10.0mL

pH 6.8 ± 0.2 at 25°C

Trace Salts Solution:

Composition per liter:

H_3BO_3	0.15g
$LiSO_4 \cdot 4H_2O$	0.1g
$MnSO_4 \cdot 4H_2O$	0.04g
$CoCl_2$	0.02g
$(NH_4)_6Mo_7O_{24}$	0.015g
$CuSO_4 \cdot 5H_2O$	0.004g

Preparation of Trace Salts Solution: Add components to distilled/deionized water and bring volume to 1.0L. Mix thoroughly.

Preparation of Medium: Add components to distilled/deionized water and bring volume to 1.0L. Mix thoroughly. Adjust pH to 6.8. Gently heat and bring to boiling. Distribute into tubes or flasks. Autoclave for 15 min at 15 psi pressure–121°C. Pour into sterile Petri dishes or leave in tubes.

Use: For the cultivation of *Chlamydomonas reinhardtii.*

Chlamydomonas Mutant Broth

Composition per liter:

K_2HPO_4	1.162g
Sodium acetate	1.0g
NaH_2PO_4	0.92g
Sodium citrate	0.5g
NH_4NO_3	0.3g
$MgSO_4 \cdot 7H_2O$	0.3g
$CaCl_2$	0.04g
$FeCl_3 \cdot 6H_2O$	0.01g
Trace salts solution	10.0mL

pH 6.8 ± 0.2 at 25°C

Trace Salts Solution:

Composition per liter:

H_3BO_3	0.15g
$LiSO_4 \cdot 4H_2O$	0.1g
$MnSO_4 \cdot 4H_2O$	0.04g
$CoCl_2$	0.02g
$(NH_4)_6Mo_7O_{24}$	0.015g
$CuSO_4 \cdot 5H_2O$	0.004g

Preparation of Trace Salts Solution: Add components to distilled/deionized water and bring volume to 1.0L. Mix thoroughly.

Preparation of Medium: Add components to distilled/deionized water and bring volume to 1.0L. Mix thoroughly. Adjust pH to 6.8. Distribute into tubes or flasks. Autoclave for 15 min at 15 psi pressure–121°C.

Use: For the cultivation of *Chlamydomonas reinhardtii.*

Chlamydospore Agar

Composition per liter:

Purified polysaccharide	20.0g
Agar	15.0g
KH_2PO_4	1.0g
$(NH_4)_2SO_4$	1.0g
Trypan Blue	0.1g
Biotin	5.0μg

pH 5.1 ± 0.2 at 25°C

Source: This medium is available as a premixed powder from BD Diagnostic Systems.

Preparation of Medium: Add components to distilled/deionized water and bring volume to 1.0L. Mix thoroughly. Gently heat and bring to boiling. Distribute into tubes or flasks. Autoclave for 15 min at 15 psi pressure–121°C. Pour into sterile Petri dishes or leave in tubes.

Use: For differentiating *Candida albicans* from other *Candida* species on the basis of chlamydospore formation.

Chloramphenicol Ampicillin LB Medium

Composition per liter:

NaCl	10.0g
Pancreatic digest of casein	10.0g
Yeast extract	5.0g
Ampicillin solution	10.0mL
Chloramphenicol solution	10.0mL

pH 7.0 ± 0.2 at 25°C

Ampicillin Solution:

Composition per 10.0mL:

Ampicillin	40.0mg

Preparation of Ampicillin Solution: Add ampicillin to distilled/deionized water and bring volume to 10.0mL. Mix thoroughly. Filter sterilize.

Chloramphenicol Solution:

Composition per 10.0mL:

Chloramphenicol	5.0mg

Preparation of Chloramphenicol Solution: Add chloramphenicol to distilled/deionized water and bring volume to 10.0mL. Mix thoroughly. Filter sterilize.

Preparation of Medium: Add components, except ampicillin solution and chloramphenicol solution, to distilled/deionized water and bring volume to 980.0mL. Mix thoroughly. Adjust pH to 7.0. Autoclave for 15 min at 15 psi pressure–121°C. Aseptically add 10.0mL of sterile ampicillin solution and 10.0mL of sterile chloramphenicol solution. Mix thoroughly. Aseptically distribute into sterile tubes or flasks.

Use: For the cultivation and maintenance of *Escherichia coli.*

Chloramphenicol Erythromycin LB Medium

Composition per liter:

NaCl	10.0g
Pancreatic digest of casein	10.0g
Yeast extract	5.0g
Erythromycin	10.0mL
Chloramphenicol	10.0mL

pH 7.0 ± 0.2 at 25°C

Erythromycin Solution:

Composition per 10.0mL:

Erythromycin	10mg

Preparation of Erythromycin Solution: Add erythromycin to distilled/deionized water and bring volume to 10.0mL. Mix thoroughly. Filter sterilize.

Chloramphenicol Solution:

Composition per 10.0mL:

Chloramphenicol	5.0mg

Preparation of Chloramphenicol Solution: Add chloramphenicol to distilled/deionized water and bring volume to 10.0mL. Mix thoroughly. Filter sterilize.

Preparation of Medium: Add components, except erythromycin solution and chloramphenicol solution, to distilled/deionized water and bring volume to 980.0mL. Mix thoroughly. Adjust pH to 7.0. Autoclave for 15 min at 15 psi pressure–121°C. Aseptically add 10.0mL of sterile ampicillin solution and 10.0mL of sterile chloramphenicol solution. Mix thoroughly. Aseptically distribute into sterile tubes or flasks.

Use: For the cultivation and maintenance of *Escherichia coli.*

Chloramphenicol L Broth Medium No. 1

Composition per liter:

Pancreatic digest of casein	10.0g
NaCl	5.0g
Yeast extract	5.0g
Glucose	1.0g
Chloramphenicol solution	10.0mL

pH 7.0 ± 0.2 at 25°C

Chloramphenicol Solution:

Composition per 10.0mL:

Chloramphenicol	5.0mg

Preparation of Chloramphenicol Solution: Add chloramphenicol to distilled/deionized water and bring volume to 10.0mL. Mix thoroughly. Filter sterilize.

Preparation of Medium: Add components, except chloramphenicol solution, to distilled/deionized water and bring volume to 990.0mL. Mix thoroughly. Adjust pH to 7.0. Autoclave for 15 min at 15 psi pressure–121°C. Aseptically add 10.0mL of sterile chloramphenicol solution and 10.0mL of sterile chloramphenicol solution. Mix thoroughly. Aseptically distribute into sterile tubes or flasks.

Use: For the cultivation and maintenance of *Escherichia coli.*

Chloramphenicol L Broth Medium No. 2

Composition per liter:

Pancreatic digest of casein	10.0g
NaCl	5.0g
Yeast extract	5.0g
Glucose	1.0g
Chloramphenicol solution	10.0mL

pH 7.0 ± 0.2 at 25°C

Chloramphenicol Solution:

Composition per 10.0mL:

Chloramphenicol	12.5mg

Preparation of Chloramphenicol Solution: Add chloramphenicol to distilled/deionized water and bring volume to 10.0mL. Mix thoroughly. Filter sterilize.

Preparation of Medium: Add components, except chloramphenicol solution, to distilled/deionized water and bring volume to 990.0mL. Mix thoroughly. Adjust pH to 7.0. Autoclave for 15 min at 15 psi pressure–121°C. Aseptically add 10.0mL of sterile chloramphenicol solution and 10.0mL of sterile chloramphenicol solution. Mix thoroughly. Aseptically distribute into sterile tubes or flasks.

Use: For the cultivation and maintenance of *Escherichia coli.*

Chloramphenicol L Broth Medium No. 3

Composition per liter:

Pancreatic digest of casein	10.0g
NaCl	5.0g
Yeast extract	5.0g
Glucose	1.0g
Chloramphenicol solution	10.0mL

pH 7.0 ± 0.2 at 25°C

Chloramphenicol Solution:

Composition per 10.0mL:

Chloramphenicol	50.0mg

Preparation of Chloramphenicol Solution: Add chloramphenicol to distilled/deionized water and bring volume to 10.0mL. Mix thoroughly. Filter sterilize.

Preparation of Medium: Add components, except chloramphenicol solution, to distilled/deionized water and bring volume to 990.0mL. Mix thoroughly. Adjust pH to 7.0. Autoclave for 15 min at 15 psi pressure–121°C. Aseptically add 10.0mL of sterile chloramphen icol solution and 10.0mL of sterile chloramphenicol solution. Mix thoroughly. Aseptically distribute into sterile tubes or flasks.

Use: For the cultivation and maintenance of *Escherichia coli.*

Chloramphenicol LB Medium No.1

Composition per liter:

NaCl	10.0g
Pancreatic digest of casein	10.0g
Yeast extract	5.0g
Chloramphenicol	50.0μg

Preparation of Medium: Adjust pH to 7.0. Autoclave at 15 psi pressure–121°C for 15 min. For solid medium, add 15.0g of agar.

Use: For the cultivation and maintenance of *Escherichia coli.*

Chloramphenicol LB Medium No. 2

Composition per liter:

NaCl	10.0g
Pancreatic digest of casein	10.0g

Yeast extract	5.0g
Chloramphenicol solution	10.0mL

pH 7.0 ± 0.2 at 25°C

Chloramphenicol Solution:

Composition per 10.0mL:

Chloramphenicol	5.0mg

Preparation of Chloramphenicol Solution: Add chloramphenicol to distilled/deionized water and bring volume to 10.0mL. Mix thoroughly. Filter sterilize.

Preparation of Medium: Add components, except chloramphenicol solution, to distilled/deionized water and bring volume to 990.0mL. Mix thoroughly. Adjust pH to 7.0. Autoclave for 15 min at 15 psi pressure–121°C. Aseptically add 10.0mL of sterile ampicillin solution and 10.0mL of sterile chloramphenicol solution. Mix thoroughly. Aseptically distribute into sterile tubes or flasks.

Use: For the cultivation and maintenance of *Escherichia coli*.

Chloramphenicol Yeast Glucose Agar

Composition per liter:

Glucose	20.0g
Agar	14.9g
Yeast extract	5.0g
Chloramphenicol	0.1g

pH 6.6 ± 0.2 at 25°C

Source: This medium is available from HiMedia.

Preparation of Medium: Add components to distilled/deionized water and bring volume to 1.0L. Mix thoroughly. Gently heat and bring to boiling. Distribute into tubes or flasks. Autoclave for 15 min at 15 psi pressure–121°C. Pour into sterile Petri dishes or leave in tubes.

Use: For the selective enumeration of yeasts and molds in milk and milk products. It is also recommended by ISO Committee under the specifications ISO 7954:1987.

Chlorella Agar

Composition per liter:

Agar	17.0g
KNO_3	2.5g
KH_2PO_4	2.45g
$MgSO_4 \cdot 7H_2O$	2.4g
K_2SO_4	0.217g
$FeSO_4 \cdot 7H_2O$	1.5mg
$MnSO_4 \cdot H_2O$	1.4mg
H_3BO_3	0.28mg
$ZnSO_4 \cdot 7H_2O$	0.22mg
$Na_2MoO_4 \cdot 2H_2O$	0.05mg
$CuSO_4 \cdot 5H_2O$	0.0078mg
Supplement solution	100.0mL

pH 4.5 ± 0.2 at 25°C

Source: This medium is available from HiMedia.

Supplement Solution:

Composition per 100.0mL:

Glucose	10.0g
Potassium citrate	32.0mg

Preparation of Supplement Solution: Add components to distilled/deionized water and bring volume to 100.0mL. Mix thoroughly. Filter sterilize.

Preparation of Medium: Add components, except supplement solution, to distilled/deionized water and bring volume to 900.0mL. Mix thoroughly. Autoclave for 15 min at 15 psi pressure–121°C. Cool to 50°C. Aseptically add supplement solution. Mix thoroughly. Pour into Petri dishes or aseptically distribute into sterile tubes.

Use: For the cultivation of *Chlorella* spp.

Chlorella Broth

Composition per liter:

KNO_3	2.5g
KH_2PO_4	2.45g
$MgSO_4 \cdot 7H_2O$	2.4g
K_2SO_4	0.217g
$FeSO_4 \cdot 7H_2O$	1.5mg
$MnSO_4 \cdot H_2O$	1.4mg
H_3BO_3	0.28mg
$ZnSO_4 \cdot 7H_2O$	0.22mg
$Na_2MoO_4 \cdot 2H_2O$	0.05mg
$CuSO_4 \cdot 5H_2O$	0.0078mg
Supplement solution	100.0mL

pH 4.5 ± 0.2 at 25°C

Source: This medium is available from HiMedia.

Supplement Solution:

Composition per 100.0mL:

Glucose	10.0g
Potassium citrate	32.0mg

Preparation of Supplement Solution: Add components to distilled/deionized water and bring volume to 100.0mL. Mix thoroughly. Filter sterilize.

Preparation of Medium: Add components, except supplement solution, to distilled/deionized water and bring volume to 900.0mL. Mix thoroughly. Autoclave for 15 min at 15 psi pressure–121°C. Cool to 50°C. Aseptically add supplement solution. Mix thoroughly. Aseptically distribute into sterile tubes.

Use: For the cultivation of *Chlorella* spp.

Chlorella Broth without Glucose and Citrate

Composition per liter:

KNO_3	2.5g
KH_2PO_4	2.45g
$MgSO_4 \cdot 7H_2O$	2.4g
K_2SO_4	0.217g
$FeSO_4 \cdot 7H_2O$	1.5mg
$MnSO_4 \cdot H_2O$	1.4mg
H_3BO_3	0.28mg
$ZnSO_4 \cdot 7H_2O$	0.22mg
$Na_2MoO_4 \cdot 2H_2O$	0.05mg
$CuSO_4 \cdot 5H_2O$	0.0078mg
Supplement solution	100.0mL

pH 4.5 ± 0.2 at 25°C

Source: This medium is available from HiMedia.

Preparation of Medium: Add components to distilled/deionized water and bring volume to 1.0L. Mix thoroughly. Distribute into tubes or flasks. Autoclave for 15 min at 15 psi pressure–121°C.

Use: For the cultivation of *Chlorella* spp.

Chlorinated Fatty Acid Medium

Composition per 1001.0mL:

$(NH_4)_2SO_4$	0.5g
$MgSO_4 \cdot 7H_2O$	0.1g
$Ca(NO_3)_2$	0.075g
2-Chloropropionate	0.54g
Phosphate buffer (20*mM* solution, pH 7.2)	1.0L
Trace elements solution	1.0mL

pH 7.2 ± 0.2 at 25°C

Trace Elements Solution:

Composition per liter:

$FeSO_4 \cdot 7H_2O$	1.0g
$MnSO_4 \cdot H_2O$	1.0g
$Co(NO_3)_2 \cdot 6H_2O$	0.25g
$CuCl_2 \cdot 2H_2O$	0.25g
$Na_2MoO_4 \cdot 2H_2O$	0.25g
$ZnCl_2$	0.25g
H_3BO_3	0.1g
NH_4VO_3	0.1g
$NiSO_4 \cdot 6H_2O$	0.1g

Preparation of Trace Elements Solution: Add components to distilled/deionized water and bring volume to 1.0L. Mix thoroughly.

Preparation of Medium: Combine components. Mix thoroughly. Distribute into tubes or flasks. Autoclave for 15 min at 15 psi pressure–121°C.

Use: For the cultivation of *Alcaligenes* species.

Chlorobiaceae Medium 1

Composition per 4990.0mL:

Solution 1	4.0L
O_2-free water	860.0mL
$NaHCO_3$ solution	100.0mL
$Na_2S \cdot 9H_2O$ solution	20.0mL
Trace elements solution	5.0mL
Vitamin B_{12} solution	5.0mL

pH 6.8 ± 0.2 at 25°C

Solution 1:

Composition per 4.0L:

$MgSO_4 \cdot 7H_2O$	2.5g
KCl	1.7g
KH_2PO_4	1.7g
NH_4Cl	1.7g
$CaCl_2 \cdot 2H_2O$	1.25g

Preparation of Solution 1: Add components to distilled/deionized water and bring volume to 4.0L. Mix thoroughly. Autoclave for 45 min at 15 psi pressure–121°C. Cool to 25°C under 100% N_2. Saturate with CO_2 by stirring under 100% CO_2 for 30 min.

O_2-Free Water:

Composition per 860.0mL:

H_2O	860.0mL

Preparation of O_2-Free Water: Autoclave H_2O for 15 min at 15 psi pressure–121°C. Cool to 25°C under 100% N_2.

$NaHCO_3$ Solution:

Composition per 100.0mL:

$NaHCO_3$	7.5g

Preparation of $NaHCO_3$ Solution: Add the $NaHCO_3$ to distilled/deionized water and bring volume to 100.0mL. Mix thoroughly. Gas with 100% CO_2 for 20 min. Filter sterilize with positive CO_2 pressure.

$Na_2S \cdot 9H_2O$ Solution:

Composition per 100.0mL:

$Na_2S \cdot 9H_2O$	10.0g

Preparation of $Na_2S \cdot 9H_2O$ Solution: Add $Na_2S \cdot 9H_2O$ to distilled/deionized water. Mix thoroughly. Gas with 100% N_2 for 15 min in a screw-capped bottle. Tightly close cap. Autoclave for 15 min at 15 psi pressure–121°C. Cool to 25°C.

Trace Elements Solution:

Composition per liter:

$FeCl_2 \cdot 4H_2O$	1.5g
$CoCl_2 \cdot 6H_2O$	0.19g
$MnCl_2 \cdot 4H_2O$	0.1g
$ZnCl_2$	0.07g
H_3BO_3	0.06g
$NaMoO_4 \cdot 2H_2O$	0.04g
$CuCl_2 \cdot 2H_2O$	0.02g
$NiCl_2 \cdot 6H_2O$	0.02g
HCl (25% solution)	6.5mL

Preparation of Trace Elements Solution: Add components to distilled/deionized water and bring volume to 1.0L. Mix thoroughly. Autoclave for 15 min at 15 psi pressure–121°C. Cool to 25°C.

Vitamin B_{12} Solution:

Composition per 100.0mL:

Vitamin B_{12}	2.0mg

Preparation of Vitamin B_{12} Solution: Add vitamin B_{12} to distilled/deionized water and bring volume to 100.0mL. Mix thoroughly. Filter sterilize.

Preparation of Medium: To 4.0L of sterile, CO_2-saturated solution 1, aseptically add the remaining components. Mix thoroughly. Adjust pH to 6.8. Aseptically distribute into sterile 100.0mL bottles using positive pressure of 95% N_2 + 5% CO_2. Completely fill bottles with medium except for a pea-sized air bubble.

Use: For the isolation and cultivation of members of the Chlorobiaceae.

Chlorobiaceae Medium 2

Composition per 1051.0mL:

Solution 1	950.0mL
$Na_2S \cdot 9H_2O$ solution	60.0mL
$NaHCO_3$ solution	40.0mL
Vitamin B_{12} solution	1.0mL

pH 6.8 ± 0.2 at 25°C

Solution 1:

Composition per 950.0mL:

KH_2PO_4	1.0g
NH_4Cl	0.5g
$MgSO_4 \cdot 7H_2O$	0.4g
$CaCl_2 \cdot 2H_2O$	0.05g
Trace elements solution SL-8	1.0mL

Preparation of Solution 1: Add components to distilled/deionized water and bring volume to 950.0mL. Mix thoroughly. Autoclave for 15 min at 15 psi pressure–121°C. Cool to 45°–50°C.

Trace Elements Solution SL-8:

Composition per liter:

Disodium EDTA	5.2g
$FeCl_2 \cdot 4H_2O$	1.5g
$CoCl_2 \cdot 6H_2O$	0.19g
$MnCl_2 \cdot 4H_2O$	0.1g
$ZnCl_2$	0.07g
H_3BO_3	0.06g
$NaMoO_4 \cdot 2H_2O$	0.04g
$CuCl_2 \cdot 2H_2O$	0.02g
$NiCl_2 \cdot 6H_20$	0.02g

Preparation of Trace Elements Solution SL-8: Add components to distilled/deionized water and bring volume to 1.0L. Mix thoroughly.

$Na_2S \cdot 9H_2O$ Solution:

Composition per 100.0mL:

$Na_2S \cdot 9H_2O$	5.0g

Preparation of $Na_2S \cdot 9H_2O$ Solution: Add $Na_2S \cdot 9H_2O$ to distilled/deionized water and bring volume to 100.0mL. Autoclave for 15 min at 15 psi pressure–121°C. Cool to 45°–50°C.

$NaHCO_3$ Solution:

Composition per 100.0mL:

$NaHCO_3$	5.0g

Preparation of $NaHCO_3$ Solution: Add $NaHCO_3$ to distilled/deionized water and bring volume to 100.0mL. Mix thoroughly. Filter sterilize.

Vitamin B_{12} Solution:

Composition per 100.0mL:

Vitamin B_{12}	2.0mg

Preparation of Vitamin B_{12} Solution: Add vitamin B_{12} to distilled/deionized water and bring volume to 100.0mL. Mix thoroughly. Filter sterilize.

Preparation of Medium: To 950.0mL of cooled, sterile solution 1, aseptically add 60.0mL of sterile $Na_2S \cdot 9H_2O$ solution, 40.0mL of sterile $NaHCO_3$ solution, and 1.0mL of sterile vitamin B_{12} solution. Mix thoroughly. Adjust pH to 6.8 with sterile H_2SO_4 or Na_2CO_3. Aseptically distribute into sterile 50.0mL or 100.0mL bottles with metal screw-caps and rubber seals. Completely fill bottles with medium except for a pea-sized air bubble.

Use: For the isolation and cultivation of freshwater and soil members of the Chlorobiaceae.

Chlorobium thiosulfatophilum Medium

Composition per 1050.0mL:

KH_2PO_4	1.0g
NH_4Cl	1.0g
$MgCl_2 \cdot 6H_2O$	0.5g
Solution A	20.0mL
Solution B	20.0mL
Solution C	10.0mL
Trace elements solution	1.0mL

pH 7.0 ± 0.2 at 25°C

Solution A:

Composition per 100.0mL:

$NaHCO_3$	10.0g

Preparation of Solution A: Add $NaHCO_3$ to distilled/deionized water and bring volume to 100.0mL. Mix thoroughly. Autoclave for 15 min at 15 psi pressure–121°C.

Solution B:

Composition per 100.0mL:

$Na_2S \cdot 9H_2O$	10.0g

Preparation of Solution B: Add $Na_2S \cdot 9H_2O$ to distilled/deionized water and bring volume to 100.0mL. Mix thoroughly. Autoclave for 15 min at 15 psi pressure–121°C.

Solution C:

Composition per 100.0mL:

$Na_2S_2O_3 \cdot 9H_2O$	10.0g

Preparation of Solution C: Add $Na_2S_2O_3 \cdot 9H_2O$ to distilled/deionized water and bring volume to 100.0mL. Mix thoroughly. Autoclave for 15 min at 15 psi pressure–121°C.

Trace Elements Solution:

Composition per liter:

$FeCl_3 \cdot 6H_2O$	2.7g
H_3BO_3	0.1g
$ZnSO_4 \cdot 7H_2O$	0.1g
$Co(NO_3)_2 \cdot 6H_2O$	50.0mg
$CuSO_4 \cdot 5H_2O$	5.0mg
$MnCl_2 \cdot 6H_2O$	5.0mg

Preparation of Trace Elements Solution: Add components to distilled/deionized water and bring volume to 1.0L. Mix thoroughly.

Preparation of Medium: Add components, except solution A, solution B, and solution C, to distilled/deionized water and bring volume to 1.0L. Mix thoroughly. Bring pH to 7.0–7.2 with H_3PO_4. Distribute into tubes or flasks. Autoclave for 15 min at 15 psi pressure–121°C. Aseptically add 0.2mL of sterile solution A, 0.2mL of sterile solution B, and 0.1mL of sterile solution C for each 10.0mL of medium. Mix thoroughly. Use immediately.

Use: For the cultivation and maintenance of *Chlorobium limnicola*.

Chlorobutane Medium

Composition per 1002.0mL:

NH_4NO_3	4.0g
KH_2PO_4	1.5g
$Na_2HPO_4 \cdot 12H_2O$	1.5g
$CaSO_4 \cdot 2H_2O$	10.0mg
$MgSO_4 \cdot 7H_2O$	10.0mg
$FeSO_4 \cdot 7H_2O$	5.0mg
Yeast extract	5.0mg
1-Chlorobutane	2.0mL

pH 7.0 ± 0.2 at 25°C

Preparation of 1-Chlorobutane: Filter sterilize.

Preparation of Medium: Add components, except 1-chlorobutane, to distilled/deionized water and bring volume to 1.0L. Mix thoroughly. Adjust pH to 7.0. Autoclave for 15 min at 15 psi pressure–121°C. Aseptically add 2.0mL of sterile 1-chlorobutane. Mix thoroughly. Aseptically distribute into sterile tubes or flasks.

Use: For the cultivation of *Corynebacterium* species.

Chloroflexus Agar

Composition per liter:

Agar	15.0g
Glycyl-glycine	0.5g

Yeast extract	0.5g
Na_2S	0.5g
NH_4Cl	0.2g
$MgSO_4·7H_2O$	0.1g
Nitrilotriacetic acid	0.1g
$NaNO_3$	0.689g
Na_2HPO_4	0.111g
KNO_3	0.103g
$CaSO_4·2H_2O$	0.06g
NaCl	8.0mg
$FeCl_3$ solution	1.0mL
Micronutrient solution	1.0mL

pH 8.2–8.4 at 25°C

$FeCl_3$ Solution:
Composition per liter:

$FeCl_3$	0.29g

Preparation of $FeCl_3$ Solution: Add $FeCl_3$ to distilled/deionized water and bring volume to 1.0L. Mix thoroughly.

Micronutrient Solution:
Composition per liter:

$MnSO_4·7H_2O$	2.28g
H_3BO_3	0.5g
$ZnSO_4·7H_2O$	0.5g
$CoCl_2·6H_2O$	0.045g
$CuSO_4·2H_2O$	0.025g
$Na_2MoO_4·2H_2O$	0.025g
H_2SO_4, concentrated	0.5mL

Preparation of Micronutrient Solution: Add components to distilled/deionized water and bring volume to 1.0L. Mix thoroughly.

Preparation of Medium: Add components, except Na_2S, to distilled/deionized water and bring volume to 1.0L. Mix thoroughly. Adjust pH to 8.2–8.4. Add Na_2S. Readjust pH to 8.2–8.4. Gently heat and bring to boiling. Distribute into tubes or flasks. Autoclave for 15 min at 15 psi pressure–121°C. Pour into sterile Petri dishes or leave in tubes.

Use: For the cultivation of *Chloroflexus aurantiacus*.

Chloroflexus aggregans Medium (DSMZ Medium 87a)

Composition per 1061.0mL:

Yeast extract	1.0g
Glycyl-glycine	1.0g
$NaNO_3$	0.5g
$Na_2HPO_4·7H_2O$	0.1g
$MgSO_4·7H_2O$	0.1g
KNO_3	0.1g
NaCl	0.1g
$CaCl_2·2H_2O$	0.05g
Neutralized sulfide solution	11.0mL
Ferric citrate solution	5.0mL
Trace elements solution SL-6	1.0mL

pH 8.2 ± 0.2 at 25°C

Ferric Citrate Solution:
Composition per 100.0mL:

Ferric citrate	0.1mg

Preparation of Ferric Citrate Solution: Add ferric citrate to distilled/deionized water and bring volume to 100.0mL. Mix thoroughly. Sparge under 100% N_2 gas for 3 min.

Trace Elements Solution SL-6:
Composition per liter:

$MnCl_2·4H_2O$	0.5g
H_3BO_3	0.3g
$CoCl_2·6H_2O$	0.2g
$ZnSO_4·7H_2O$	0.1g
$Na_2MoO_4·2H_2O$	0.03g
$NiCl_2·6H_2O$	0.02g
$CuCl_2·2H_2O$	0.01g

Preparation of Trace Elements Solution SL-6: Add components to distilled/deionized water and bring volume to 1.0L. Mix thoroughly.

Neutralized Sulfide Solution:
Composition per 100.0mL:

$Na_2S·9H_2O$	1.5g

Preparation of Neutralized Sulfide Solution: Add $Na_2S·9H_2O$ to distilled/deionized water in a 250mL screw-capped bottle fitted with a butyl rubber septum and bring volume to 100.0mL. Add a magnetic stir bar. Mix thoroughly. Sparge under 100% N_2 gas for 3 min. Autoclave for 15 min at 15 psi pressure–121°C. Cool to room temperature. Adjust pH to about 7.3 with sterile 2*M* H_2SO_4. Do not open the bottle to add H_2SO_4; use a sterile syringe. Stir the solution continuously to avoid precipitation of elemental sulfur. The final solution should be clear and yellow in color.

Preparation of Medium: Add components, except neutralized sulfide solution, to distilled/deionized water and bring volume to 1050.0mL. Mix thoroughly. Adjust pH to 8.2. Gently heat and bring to boiling. Continue boiling for 3–4 min under 100% N_2. Distribute 90.0mL of medium into 100mL screw-capped bottles with rubber septa under 100% N_2. Autoclave for 15 min at 15 psi pressure–121°C. Cool to room temperature. Using a sterile syringe, inject 1.0mL of neutralized sulfide solution into each bottle. Incubate the culture at 50°C at a light intensity of 300–500 lux. For heavy cell suspension supplement periodically with sterile yeast extract solution to yield a final concentration of 0.1%.

Use: For the growth and maintenance of *Chloroflexus aggregans* and *Roseiflexus castenholzii.*

Chloroflexus Broth

Composition per liter:

$NaNO_3$	0.689g
Glycyl-glycine	0.5g
Yeast extract	0.5g
Na_2S	0.5g
NH_4Cl	0.2g
Na_2HPO_4	0.111g
KNO_3	0.103g
$MgSO_4·7H_2O$	0.1g
Nitrilotriacetic acid	0.1g
$CaSO_4·2H_2O$	0.06g
NaCl	8.0mg
$FeCl_3$ solution	1.0mL
Micronutrient solution	1.0mL

pH 8.2–8.4 at 25°C

$FeCl_3$ Solution:
Composition per liter:

$FeCl_3$	0.29g

Preparation of $FeCl_3$ Solution: Add $FeCl_3$ to distilled/deionized water and bring volume to 1.0L. Mix thoroughly.

Micronutrient Solution:
Composition per liter:

$MnSO_4 \cdot 7H_2O$	2.28g
H_3BO_3	0.5g
$ZnSO_4 \cdot 7H_2O$	0.5g
$CoCl_2 \cdot 6H_2O$	0.045g
$CuSO_4 \cdot 2H_2O$	0.025g
$Na_2MoO_4 \cdot 2H_2O$	0.025g
H_2SO_4, concentrated	0.5mL

Preparation of Micronutrient Solution: Add components to distilled/deionized water and bring volume to 1.0L. Mix thoroughly.

Preparation of Medium: Add components, except Na_2S, to distilled/deionized water and bring volume to 1.0L. Mix thoroughly. Adjust pH to 8.2–8.4. Add Na_2S. Readjust pH to 8.2–8.4. Filter sterilize. Distribute into sterile tubes or flasks.

Use: For the cultivation of *Chloroflexus aurantiacus.*

Chloroflexus Medium, Modified

Composition per 1001.0 mL:

Glycyl-glycine	1.0g
Yeast extract	1.0g
$NaNO_3$	0.5g
KNO_3	0.1g
$MgSO_4 \cdot 7H_2O$	0.1g
$Na_2HPO_4 \cdot 2H_2O$	0.1g
NaCl	0.1g
$CaCl_2 \cdot 2H_2O$	0.05g
Neutralized sulfide solution	11.0mL
Ferric citrate solution	1.0mL
Trace elements solution SL-6	1.0mL

pH 8.2 ± 0.2 at 25°C

Neutralized Sulfide Solution:
Composition per 100.0mL:

$Na_2S \cdot 9H_2O$	1.5g

Preparation of Neutralized Sulfide Solution: Add $Na_2S \cdot 9H_2O$ to distilled/deionized water in a 250mL screw-capped bottle fitted with a butyl rubber septum and bring volume to 100.0mL. Add a magnetic stir bar. Mix thoroughly. Sparge under 100% N_2 gas for 3 min. Autoclave for 15 min at 15 psi pressure–121°C. Cool to room temperature. Adjust pH to about 7.3 with sterile 2*M* H_2SO_4. Do not open the bottle to add H_2SO_4; use a sterile syringe. Stir the solution continuously to avoid precipitation of elemental sulfur. The final solution should be clear and yellow in color.

Ferric Citrate Solution:
Composition per 100.0mL:

Ferric citrate	0.1g

Preparation of Ferric Citrate Solution: Add ferric citrate to distilled/deionized water and bring volume to 100.0mL. Mix thoroughly.

Trace Elements Solution SL-6:
Composition per liter:

$MnCl_2 \cdot 4H_2O$	0.5g
H_3BO_3	0.3g
$CoCl_2 \cdot 6H_2O$	0.2g
$ZnSO_4 \cdot 7H_2O$	0.1g
$Na_2MoO_4 \cdot 2H_2O$	0.03g
$NiCl_2 \cdot 6H_2O$	0.02g
$CuCl_2 \cdot 2H_2O$	0.01g

Preparation of Trace Elements Solution SL-6: Add components to distilled/deionized water and bring volume to 1.0L. Mix thoroughly.

Preparation of Medium: Add components, except neutralized sulfide solution, to distilled/deionized water and bring volume to 990.0mL. Mix thoroughly. Gently heat and bring to boiling. Continue boiling for 3–4 min under 100% N_2. Distribute 90.0mL of medium into 100mL screw-capped bottles under 100% N_2. Autoclave for 15 min at 15 psi pressure–121°C. Cool to room temperature. Using a sterile syringe, inject 1.0mL of neutralized sulfide solution into each bottle.

Use: For the growth and maintenance of *Chloroflexus aurantiacus.*

Chlorohydroxybenzoic Acid Medium

Composition per liter:

$K_2HPO_4 \cdot 3H_2O$	4.25g
NH_4Cl	2.0g
$NaH_2PO_4 \cdot H_2O$	1.0g
5-Chloro-2-hydroxybenzoic acid	0.5g
$MgSO_4 \cdot 7H_2O$	0.2g
Nitrilotriacetic acid	0.1g
$FeSO_4 \cdot 7H_2O$	0.012g
$MnSO_4 \cdot H_2O$	3.0mg
$ZnSO_4 \cdot 7H_2O$	3.0mg
$CoSO_4$	1.0mg

pH 7.0-7.4 at 25°C

Preparation of Medium: Add 5-chloro-2-hydroxybenzoic acid to 800.0mL of distilled/deionized water. Adjust pH to 7.0 with NaOH. Add remaining components and bring volume to 1.0L. Distribute into tubes or flasks. Autoclave for 15 min at 15 psi pressure–121°C.

Use: For the cultivation of bacteria that can utilize 5-chloro-hydroxybenzoic acid. For the cultivation of ATCC strain 35944.

CHO HiVeg Medium Base with Carbohydrate Solution

Composition per liter:

Plant hydrolysate	15.0g
Yeast extract	7.0g
NaCl	2.5g
Agar	0.75g
Na-thioglycollate	0.5g
L-Cystine	0.25
Ascorbic acid	0.1g
Bromthymol Blue	0.01g
Carbohydrate soltuion	6.25mL

pH 7.0 ± 0.2 at 25°C

Source: This medium is available as a premixed powder from HiMedia.

Carbohydrate Solution:
Composition per 100.0mL:

Carbohydrate	10.0g

Preparation of Carbohydrate Solution: Add carbohydrate to distilled/deionized water and bring volume to 100.0mL. Adonitol, arabinose, cellobiose, glucose, dulcitol, fructose, galactose, inositol, lactose, maltose, mannitol, raffinose, rhamnose, salicin, sorbitol, sucrose, trehalose, xylose, or other carbohydrates may be used. Mix thoroughly. Filter sterilize.

Preparation of Medium: Add components, except carbohydrate solution, to distilled/deionized water and bring volume to 1.0L. Mix thoroughly. Gently heat and bring to boiling. Distribute into tubes or flasks. Autoclave for 15 min at 15 psi pressure–121°C. Cool to 50°C. Aseptically add 6.25mL of sterile carbohydrate solution. Aseptically distribute into sterile tubes or leave in flasks.

Use: Used as a basal medium to which carbohydrates are added for fermentation studies of anaerobic bacteria.

CHO Medium

See: **Fermentation Broth**

CHO Medium Base (Carbohydrate Medium Base)

Composition per liter:

Pancreatic digest of casein	15.0g
Yeast extract	7.0g
NaCl	2.5g
Agar	0.75g
Sodium thioglycolate	0.5g
L-Cystine	0.25g
Ascorbic acid	0.1g
Bromthymol Blue	0.01g

pH 7.0 ± 0.2 at 25°C

Preparation of Medium: Add components to distilled/deionized water and bring volume to 1.0L. Mix thoroughly. Gently heat and bring to boiling. Distribute into tubes or flasks. Autoclave for 15 min at 15 psi pressure–121°C. Cool to 45°–50°C.

Use: Used as a basal medium to which carbohydrates are added for fermentation studies of anaerobic bacteria. Generally, 6.25mL of a 10% filter-sterilized solution of carbohydrate is added to the sterile basal medium.

Chocolate Agar

Composition per liter:

Agar	15.0g
Pantone	10.0g
Bitone	10.0g
NaCl	5.0g
Tryptic digest of beef heart	3.0g
Cornstarch	1.0g
Sheep blood, defibrinated	100.0mL
Supplement B	10.0mL

pH 7.3 ± 0.2 at 25°C

Supplement B:

Composition per 10.0mL:

Cephalothin	15.0mg
Vancomycin	10.0mg
Trimethoprim	5.0mg
Amphotericin B	2.0mg
Polymyxin B	2500U

Preparation of Supplement B: Add components to 10.0mL of distilled/deionized water. Mix thoroughly. Filter sterilize.

Source: Supplement B is available from BD Diagnostic Systems.

Preparation of Medium: Add components, except supplement B solution and sheep blood, to distilled/deionized water and bring volume to 890.0mL. Mix thoroughly. Gently heat until boiling. Autoclave for 15 min at 15 psi pressure–121°C. Cool to 45°–50°C. Aseptically add 100.0mL of sterile, defibrinated sheep blood. Gently heat while stirring and bring to 85°C for 5–10 min. Cool to 50°C. Aseptically add 10.0mL of sterile supplement B. Mix thoroughly. Pour into sterile Petri dishes or distribute into sterile tubes.

Use: For the isolation and cultivation of a variety of fastidious microorganisms.

Chocolate Agar

Composition per liter:

Proteose peptone No. 3	15.0g
Agar	10.0g
NaCl	5.0g
K_2HPO_4	4.0g
Cornstarch	1.0g
KH_2PO_4	1.0g
Hemoglobin solution	100.0mL
Supplement B	10.0mL

pH 7.0 ± 0.2 at 25°C

Source: This medium is available from BD Diagnostic Systems.

Supplement B:

Composition per 10.0mL:

Cephalothin	15.0mg
Vancomycin	10.0mg
Trimethoprim	5.0mg
Amphotericin B	2.0mg
Polymyxin B	2500U

Preparation of Supplement B: Add components to distilled/deionized water and bring volume to 10.0mL. Mix thoroughly. Filter sterilize.

Hemoglobin Solution:

Composition per 100.0mL:

Hemoglobin	10.0g

Preparation of Hemoglobin Solution: Add hemoglobin to distilled/deionized water and bring volume to 100.0mL. Mix thoroughly. Filter sterilize.

Preparation of Medium: Add components, except hemoglobin solution and supplement B, to distilled/deionized water and bring volume to 990.0mL. Mix thoroughly. Gently heat and bring to boiling. Autoclave for 15 min at 15 psi pressure–121°C. Cool to 45°–50°C. Aseptically add 100.0mL of sterile hemoglobin solution. Gently heat while stirring and bring to 85°C for 5–10 min. Cool to 50°C. Aseptically add 10.0mL of sterile supplement B. Mix thoroughly. Pour into sterile Petri dishes or distribute into sterile tubes.

Use: For the isolation and cultivation of fastidious microorganisms.

Chocolate Agar, Enriched

Composition per liter:

GC medium base	740.0mL
Hemoglobin solution	250.0mL
Supplement B	10.0mL

pH 7.3 ± 0.2 at 25°C

Source: This medium is available from BD Diagnostic Systems.

GC Medium Base:

Composition per 740.0mL:

Agar	20.0g
Proteose peptone No. 3	15.0g
NaCl	5.0g
K_2HPO_4	4.0g
Glucose	1.5g

Cornstarch	1.0g
KH_2PO_4	1.0g

pH 7.2 ± 0.2 at 25°C

Preparation of GC Medium Base: Add components to distilled/deionized water and bring volume to 740.0mL. Mix thoroughly. Gently heat until boiling. Autoclave for 15 min at 15 psi pressure–121°C. Cool to 45°–50°C.

Hemoglobin Solution:

Composition per 250.0mL:

Hemoglobin	10.0g

Preparation of Hemoglobin Solution: Add hemoglobin to distilled/deionized water and bring volume to 250.0mL. Mix thoroughly. Autoclave for 15 min at 15 psi pressure–121°C. Cool to 45°–50°C.

Supplement B:

Composition per 10.0mL:

Cephalothin	15.0mg
Vancomycin	10.0mg
Trimethoprim	5.0mg
Amphotericin B	2.0mg
Polymyxin B	2500U

Preparation of Supplement B: Add components to distilled/deionized water and bring volume to 10.0mL. Mix thoroughly. Filter sterilize.

Preparation of Medium: To 740.0mL of cooled sterile GC medium base, aseptically add 250.0mL of sterile hemoglobin solution and 10.0mL of sterile supplement B. Mix thoroughly. Pour into sterile Petri dishes or distribute into sterile tubes.

Use: For the cultivation of fastidious microorganisms, especially *Neisseria* species.

Chocolate Agar, Enriched

Composition per liter:

GC medium base	740.0mL
Hemoglobin solution	250.0mL
Supplement VX	10.0mL

pH 7.3 ± 0.2 at 25°C

Source: This medium is available from BD Diagnostic Systems.

GC Medium Base:

Composition per 740.0mL:

Proteose peptone No. 3	15.0g
Agar	20.0g
NaCl	5.0g
K_2HPO_4	4.0g
Glucose	1.5g
Cornstarch	1.0g
KH_2PO_4	1.0g

pH 7.2 ± 0.2 at 25°C

Preparation of GC Medium Base: Add components to distilled/deionized water and bring volume to 740.0mL. Mix thoroughly. Gently heat until boiling. Autoclave for 15 min at 15 psi pressure–121°C. Cool to 45°–50°C.

Hemoglobin Solution:

Composition per 250.0mL:

Hemoglobin	10.0g

Preparation of Hemoglobin Solution: Add hemoglobin to distilled/deionized water and bring volume to 250.0mL. Mix thoroughly. Autoclave for 15 min at 15 psi pressure–121°C. Cool to 45°–50°C.

Supplement VX:

Composition per 10.0mL:

Supplement VX contains essential growth factors.

Preparation of Supplement VX: Add components to distilled/deionized water and bring volume to 10.0mL. Mix thoroughly. Filter sterilize.

Preparation of Medium: To 740.0mL of cooled sterile GC medium base, aseptically add 250.0mL of sterile hemoglobin solution and 10.0mL of sterile supplement VX. Mix thoroughly. Pour into sterile Petri dishes or distribute into sterile tubes.

Use: For the cultivation of fastidious microorganisms, especially *Neisseria* species.

Chocolate Agar-*Bartonella* C-29 (ATCC Medium 2119)

Composition per 1010.0mL:

GC agar base solution	500.0 ml
Hemoglobin solution	500.0 ml
IsoVitaleX® enrichment	10.0mL

IsoVitaleX® Enrichment:

Composition per liter:

Glucose	100.0g
L-Cysteine·HCl	25.9g
L-Glutamine	10.0g
L-Cystine	1.1g
Adenine	1.0g
Nicotinamide adenine dinucleotide	0.25g
Vitamin B_{12}	0.1g
Thiamine pyrophosphate	0.1g
Guanine·HCl	0.03g
$Fe(NO_3)_3 \cdot 6H_2O$	0.02g
p-Aminobenzoic acid	0.013g
Thiamine·HCl	3.0mg

Preparation of IsoVitaleX®: Add components to distilled/deionized water and bring volume to 1.0L. Mix thoroughly. Filter sterilize.

GC Agar Base Solution:

Composition per 500.0mL:

Agar	10.0g
Pancreatic digest of casein	7.5g
Peptic digest of animal tissue	7.5g
NaCl	5.0g
K_2HPO_4	4.0g
Cornstarch	1.0g
KH_2PO_4	1.0g

Preparation of GC Agar Base: Add components to distilled/deionized water and bring volume to 500.0mL. Mix thoroughly. Gently heat until boiling. Autoclave for 15 min at 15 psi pressure–121°C. Cool to 45°–50°C.

Hemoglobin Solution:

Composition per 500.0mL:

Hemoglobin	10.0g

Preparation of Hemoglobin Solution: Add hemoglobin to distilled/deionized water and bring volume to 500.0mL. Mix thoroughly. Gently heat until boiling. Autoclave for 15 min at 15 psi pressure–121°C. Cool to 45°–50°C.

Preparation of Medium: Aseptically combine 500.0mL sterile, cooled GC agar base solution and 500.0mL cooled sterile hemoglobin solution. Aseptically add 10.0mL of sterile IsoVitaleX® enrichment. Mix thoroughly. Pour into sterile Petri dishes or distribute into sterile tubes.

Use: For the isolation and cultivation of fastidious microorganisms, especially *Neisseria* and *Haemophilus* species, from a variety of clinical specimens.

Chocolate Agar Base with Hemoglobin and Yeast Autolysate

Composition per liter:

Proteose peptone	20.0g
Agar	15.0g
Na_2HPO_4	5.0g
NaCl	5.0g
Glucose	0.5g
Hemoglobin solution	500.0mL
Yeast autolysate solution	20.0mL

pH 7.3 ± 0.2 at 25°C

Source: This medium is available from HiMedia.

Hemoglobin Solution:

Composition per 500.0mL:

Bovine hemoglobin	10.0g

Preparation of Hemoglobin Solution: Add bovine hemoglobin to distilled/deionized water and bring volume to 500.0mL. Mix thoroughly. Autoclave for 15 min at 15 psi pressure–121°C. Cool to 45°–50°C.

Yeast Autolysate Solution:

Composition per 20.0mL:

Yeast autolysate	10.0g
Glucose	1.0g
$NaHCO_3$	0.15g

Preparation of Yeast Autolysate Solution: Add components to distilled/deionized water and bring volume to 20.0mL. Mix thoroughly. Filter sterilize.

Preparation of Medium: Add components, except hemoglobin and yeast autolysate solutions, to distilled/deionized water and bring volume to 480.0mL. Mix thoroughly. Gently heat until boiling. Autoclave for 15 min at 15 psi pressure–121°C. Cool to 45°–50°C. Add 500.0mL sterile hemoglobin solution and 20.0mL sterile yeast autolysate solution. Mix thoroughly. Pour into sterile Petri dishes or distribute into sterile tubes.

Use: For the isolation of *Neisseria gonorrhoeae* from chronic and acute cases of gonococcal infections.

Chocolate Agar Base with Hemoglobin and Vitamino Growth Supplement

Composition per liter:

Proteose peptone	20.0g
Agar	15.0g
Na_2HPO_4	5.0g
NaCl	5.0g
Glucose	0.5g
Hemoglobin solution	500.0mL
Vitamino growth supplement solution	10.0mL

pH 7.3 ± 0.2 at 25°C

Source: This medium is available from HiMedia.

Hemoglobin Solution:

Composition per 500.0mL:

Bovine hemoglobin	10.0g

Preparation of Hemoglobin Solution: Add bovine hemoglobin to distilled/deionized water and bring volume to 500.0mL. Mix thoroughly. Autoclave for 15 min at 15 psi pressure–121°C. Cool to 45°–50°C.

Vitamino Growth Supplement Solution:

Composition per 10.0mL:

L-Glutamine	0.2g
Adenine sulfate	20.0mg
Guanine hydrochlroide	0.6mg
p-Aminobenzoic acid (PABA)	0.26mg
Vitamin B_{12}	0.2mg

Preparation of Vitamino Growth Supplement Solution: Add components to distilled/deionized water and bring volume to 10.0mL. Mix thoroughly. Filter sterilize.

Preparation of Medium: Add components, except hemoglobin and Vitamino growth supplement solutions, to distilled/deionized water and bring volume to 480.0mL. Mix thoroughly. Gently heat until boiling. Autoclave for 15 min at 15 psi pressure–121°C. Cool to 45°–50°C. Add 500.0mL sterile hemoglobin solution and 10.0mL sterile Vitamino growth supplement solution. Mix thoroughly. Pour into sterile Petri dishes or distribute into sterile tubes.

Use: For the isolation of *Neisseria gonorrhoeae* from chronic and acute cases of gonococcal infections.

Chocolate II Agar

Composition per liter:

Agar	12.0g
Casein enzymic hydrolysate	7.5g
Meat extract	7.5g
NaCl	5.0g
K_2HPO_4	4.0g
Corn starch	1.0g
KH_2PO_4	1.0g
Vitamin B_{12}	0.2mg
Hemoglobin solution	500.0mL

pH 7.3 ± 0.2 at 25°C

Source: This medium is available from HiMedia.

Hemoglobin Solution:

Composition per 500.0mL:

Bovine hemoglobin	10.0g

Preparation of Hemoglobin Solution: Add bovine hemoglobin to distilled/deionized water and bring volume to 500.0mL. Mix thoroughly. Autoclave for 15 min at 15 psi pressure–121°C. Cool to 45°–50°C.

Preparation of Medium: Add components, except hemoglobin solution, to distilled/deionized water and bring volume to 500.0mL. Mix thoroughly. Gently heat until boiling. Autoclave for 15 min at 15 psi pressure–121°C. Cool to 45°–50°C. Add 500.0mL sterile hemoglobin solution. Mix thoroughly. Pour into sterile Petri dishes or distribute into sterile tubes.

Use: For the isolation of *Neisseria* and *Haemophilus* species from a variety of clinical specimens.

Chocolate No. 2 Agar Base with Supplements

Composition per liter:

Agar	12.0g
Casein enzymic hydrolysate	7.5g

Meat extract	7.5g
NaCl	5.0g
K_2HPO_4	4.0g
Corn starch	1.0g
KH_2PO_4	1.0g
Hemoglobin solution	480.0mL
Supplement solution	40.0mL

pH 7.3 ± 0.2 at 25°C

Source: This medium is available from HiMedia.

Hemoglobin Solution:
Composition per 500.0mL:

Hemoglobin	10.0g

Preparation of Hemoglobin Solution: Add hemoglobin to distilled/deionized water and bring volume to 500.0mL. Mix thoroughly. Filter sterilize.

Supplement Solution:
Composition per 40.0mL:

p-Aminobenzoic acid	259.0mg
L-Glutamine	100.0mg
Adenine sulfate	10.0mg
NAD	2.5mg
Vitamin B_{12}	1.0mg
Cocarboxylase	1.0mg
Guanine·HCl	0.3mg
$Fe(NO_3)_3$	0.2mg
L-Cysteine·HCl	0.13mg
Thiamine·HCl	0.03mg

Preparation of Supplement Solution: Add components to distilled/deionized water and bring volume to 40.0mL. Mix thoroughly. Filter sterilize.

Preparation of Medium: Add components, except hemoglobin solution and supplement solution, to distilled/deionized water and bring volume to 480.0mL. Mix thoroughly. Autoclave for 15 min at 15 psi pressure–121°C. Cool to 50°C. Aseptically add hemoglobin and supplement solutions. Mix thoroughly. Pour into Petri dishes or aseptically distribute into sterile tubes.

Use: For the isolation of *Neisseria* spp. and *Haemophilus* spp. from a variety of clinical specimens.

Chocolate No. 2 Agar Base with Hemoglobin

Composition per liter:

Agar	12.0g
Hemoglobin	10.0g
Pancreatic digest of casein	7.5g
Selected meat peptone	7.5g
NaCl	5.0g
K_2HPO_4	4.0g
Cornstarch	1.0g
KH_2PO_4	1.0g

Preparation of Medium: Add components to distilled/deionized water and bring volume to 1.0L. Mix thoroughly. Gently heat to boiling. Autoclave for 15 min at 15 psi pressure–121°C. Pour into sterile Petri dishes or leave in tubes.

Use: For the isolation and cultivation of fastidious microorganisms.

Chocolate II Agar with Hemoglobin and IsoVitaleX® (GCII Agar with Hemoglobin and IsoVitaleX®)

Composition per liter:

GCII agar base	990.0mL
IsoVitaleX® enrichment	10.0mL

pH 7.3 ± 0.2 at 25°C

Source: This medium is available as a prepared medium from BD Diagnostic Systems.

GCII Agar Base:
Composition per liter:

Agar	12.0g
Hemoglobin	10.0g
Pancreatic digest of casein	7.5g
Selected meat peptone	7.5g
NaCl	5.0g
K_2HPO_4	4.0g
Cornstarch	1.0g
KH_2PO_4	1.0g

Preparation of GCII Agar Base: Add components to distilled/deionized water and bring volume to 1.0L. Mix thoroughly. Gently heat to boiling. Autoclave for 15 min at 15 psi pressure–121°C. Cool to 45°–50°C.

IsoVitaleX® Enrichment:
Composition per liter:

Glucose	100.0g
L-Cysteine·HCl	25.9g
L-Glutamine	10.0g
L-Cystine	1.1g
Adenine	1.0g
Nicotinamide adenine dinucleotide	0.25g
Vitamin B_{12}	0.1g
Thiamine pyrophosphate	0.1g
Guanine·HCl	0.03g
$Fe(NO_3)_3 \cdot 6H_2O$	0.02g
p-Aminobenzoic acid	0.013g
Thiamine·HCl	3.0mg

Preparation of IsoVitaleX®: Add components to distilled/deionized water and bring volume to 1.0L. Mix thoroughly. Filter sterilize.

Preparation of Medium: Aseptically add 10.0mL of sterile IsoVitaleX® enrichment to 990.0L of sterile, cooled GCII agar base. Mix thoroughly. Pour into sterile Petri dishes or distribute into sterile tubes.

Use: For the isolation and cultivation of fastidious microorganisms, especially *Neisseria* and *Haemophilus* species, from a variety of clinical specimens.

Chocolate HiVeg Agar Base with Hemoglobin and Yeast Autolysate

Composition per liter:

Plant peptone No. 3	20.0g
Agar	15.0g
Na_2HPO_4	5.0g
NaCl	5.0g
Glucose	0.5g
Hemoglobin solution	500.0mL
Yeast autolysate solution	20.0mL

pH 7.3 ± 0.2 at 25°C

Source: This medium, wihout hemoglobin or yeast autolysate, is available as a premixed powder from HiMedia.

Hemoglobin Solution:
Composition per 500.0mL:

Bovine hemoglobin	10.0g

Preparation of Hemoglobin Solution: Add bovine hemoglobin to distilled/deionized water and bring volume to 500.0mL. Mix thoroughly. Autoclave for 15 min at 15 psi pressure–121°C. Cool to 45°–50°C.

Yeast Autolysate Solution:
Composition per 20.0mL:

Yeast autolysate	10.0g
Glucose	1.0g
$NaHCO_3$	0.15g

Preparation of Yeast Autolysate Solution: Add components to distilled/deionized water and bring volume to 20.0mL. Mix thoroughly. Filter sterilize.

Preparation of Medium: Add components, except hemoglobin and yeast autolysate solutions, to distilled/deionized water and bring volume to 480.0mL. Mix thoroughly. Gently heat until boiling. Autoclave for 15 min at 15 psi pressure–121°C. Cool to 45°–50°C. Add 500.0mL sterile hemoglobin solution and 20.0mL sterile yeast autolysate solution. Mix thoroughly. Pour into sterile Petri dishes or distribute into sterile tubes.

Use: For the isolation of *Neisseria gonorrhoeae* from chronic and acute cases of gonococcal infections.

Chocolate HiVeg Agar Base with Hemoglobin and Vitamino Growth Supplement

Composition per liter:

Plant peptone No. 3	20.0g
Agar	15.0g
Na_2HPO_4	5.0g
NaCl	5.0g
Glucose	0.5g
Hemoglobin solution	500.0mL
Vitamino growth supplement solution	10.0mL

pH 7.3 ± 0.2 at 25°C

Source: This medium, wihout hemoglobin or vitamino growth supplement, is available as a premixed powder from HiMedia.

Hemoglobin Solution:
Composition per 500.0mL:

Bovine hemoglobin	10.0g

Preparation of Hemoglobin Solution: Add bovine hemoglobin to distilled/deionized water and bring volume to 500.0mL. Mix thoroughly. Autoclave for 15 min at 15 psi pressure–121°C. Cool to 45°–50°C.

Vitamino Growth Supplement Solution:
Composition per 10.0mL:

L-Glutamine	0.2g
Adenine sulfate	20.0mg
Guanine hydrochlroide	0.6mg
p-Aminobenzoic acid (PABA)	0.26mg
Vitamin B_{12}	0.2mg

Preparation of Vitamino Growth Supplement Solution: Add components to distilled/deionized water and bring volume to 10.0mL. Mix thoroughly. Filter sterilize.

Preparation of Medium: Add components, except hemoglobin and Vitamino growth supplement solutions, to distilled/deionized water and bring volume to 480.0mL. Mix thoroughly. Gently heat until boiling. Autoclave for 15 min at 15 psi pressure–121°C. Cool to 45°–50°C. Add 500.0mL sterile hemoglobin solution and 10.0mL sterile Vitamino growth supplement solution. Mix thoroughly. Pour into sterile Petri dishes or distribute into sterile tubes.

Use: For the isolation of *Neisseria gonorrhoeae* from chronic and acute cases of gonococcal infections.

Chocolate No. 2 HiVeg Agar Base with Hemoglobin

Composition per liter:

Agar	12.0g
Plant extract No.I	7.5g
Plant hydrolysate	7.5g
NaCl	5.0g
K_2HPO_4	4.0g
Corn starch	1.0g
KH_2PO_4	1.0g
Vitamin B_{12}	0.2mg
Hemoglobin solution	500.0mL

pH 7.3 ± 0.2 at 25°C

Source: This medium, wihout hemoglobin, is available as a premixed powder from HiMedia.

Hemoglobin Solution:
Composition per 500.0mL:

Bovine hemoglobin	10.0g

Preparation of Hemoglobin Solution: Add bovine hemoglobin to distilled/deionized water and bring volume to 500.0mL. Mix thoroughly. Autoclave for 15 min at 15 psi pressure–121°C. Cool to 45°–50°C.

Preparation of Medium: Add components, except hemoglobin solution, to distilled/deionized water and bring volume to 500.0mL. Mix thoroughly. Gently heat until boiling. Autoclave for 15 min at 15 psi pressure–121°C. Cool to 45°–50°C. Add 500.0mL sterile hemoglobin solution. Mix thoroughly. Pour into sterile Petri dishes or distribute into sterile tubes.

Use: For the isolation of *Neisseria* and *Haemophilus* species from a variety of clinical specimens.

Chocolate Tellurite Agar (Tellurite Blood Agar)

Composition per liter:

Agar	10.0g
Casein/meat (50/50) peptone	10.0g
Hemoglobin	10.0g
NaCl	5.0g
K_2HPO_4	4.0g
Cornstarch	1.0g
KH_2PO_4	1.0g
K_2TeO_3	0.1g
Bio-X enrichment	10.0mL

Bio-X Enrichment:
Composition per liter:

Glucose	100.0g
L-Cysteine·HCl	25.9g
L-Glutamate	10.0g
L-Cystine	1.1g
Adenine	1.0g
Cocarboxylase	0.1g
Guanine·HCl	0.03g

$FeNO_3$	0.02g
p-Aminobenzoic acid	0.013g
Vitamin B_{12}	0.01g
NAD (nicotinamide adenine dinucleotide)	250.0mg
Thiamine·HCl	3.0mg

pH 7.2 ± 0.2 at 25°C

Preparation of Bio-X Enrichment: Add components to distilled/deionized water and bring volume to 1.0L. Mix thoroughly. Filter sterilize.

Caution: Potassium tellurite is toxic.

Preparation of Medium: Add components, except Bio-X enrichment, to distilled/deionized water and bring volume to 990.0mL. Mix thoroughly. Gently heat and bring to boiling. Autoclave for 15 min at 15 psi pressure–121°C. Cool to 45°–50°C. Aseptically add filter-sterilized Bio-X enrichment. Mix thoroughly. Pour into sterile Petri dishes or distribute into sterile tubes.

Use: For the selective isolation and cultivation of *Corynebacterium* species. *Corynebacterium diphtheriae* appears as gray-black colonies.

Cholera HiVeg Medium Base with Tellurite and Blood

Composition per liter:

NaCl	20.0g
Agar	10.0g
Plant extract	10.0g
Plant peptone	10.0g
Sucrose	10.0g
Na_2CO_3	5.0g
Sodium lauryl sulfate	0.1g
Sheep blood, defibrinated	50.0mL
Tellurite solution	2.0mL

pH 8.5 ± 0.2 at 25°C

Source: This medium, without tellurite or blood, is available as a premixed powder from HiMedia.

Tellurite Solution:

Composition per 10.0mL:

K_2TeO_3	0.1g

Preparation of Tellurite Solution: Add K_2TeO_3 to distilled/deionized water and bring volume to 100.0mL. Mix thoroughly. Filter sterilize.

Caution: Potassium tellurite is toxic.

Preparation of Medium: Add components to distilled/deionized water and bring volume to 1.0L. Mix thoroughly. Gently heat and bring to boiling. Do not autoclave. Cool to 70°C. Aseptically add 2.0mL of sterile tellurite solution and 50.0mL of sterile defibrinated blood. Maintain at 70°C for several minutes. Cool to 50°C. Pour into sterile Petri dishes or leave in tubes.

Use: For the isolation of pathogenic vibrios, especially *Vibrio cholerae.* For the selective isolation of *Vibrio* species from specimens grossly contaminated with Enterobacteriaceae.

Cholera Medium Base with Tellurite and Blood

Composition per liter:

NaCl	20.0g
Agar	10.0g
Peptic digest of animal tissue	10.0g
Beef extract	10.0g
Sucrose	10.0g
Na_2CO_3	5.0g
Sodium lauryl sulfate	0.1g
Sheep blood, defibrinated	50.0mL
Tellurite solution	2.0mL

pH 8.5 ± 0.2 at 25°C

Source: This medium, without tellurite or blood, is available as a premixed powder from HiMedia.

Tellurite Solution:

Composition per 10.0mL:

K_2TeO_3	0.1g

Preparation of Tellurite Solution: Add K_2TeO_3 to distilled/deionized water and bring volume to 100.0mL. Mix thoroughly. Filter sterilize.

Caution: Potassium tellurite is toxic.

Preparation of Medium: Add components to distilled/deionized water and bring volume to 1.0L. Mix thoroughly. Gently heat and bring to boiling. Do not autoclave. Cool to 70°C. Aseptically add 2.0mL of sterile tellurite soltuion and 50.0mL of sterile defibrinated blood. Maintain at 70°C for several minutes. Cool to 50°C. Pour into sterile Petri dishes or leave in tubes.

Use: For the isolation of pathogenic vibrios, especially *Vibrio cholerae.* For the selective isolation of *Vibrio* species from specimens grossly contaminated with Enterobacteriaceae.

Cholera Medium TCBS

Composition per liter:

Sucrose	20.0g
Agar	14.0g
Peptone	10.0g
$Na_2S_2O_3$	10.0g
Sodium citrate	10.0g
NaCl	10.0g
Ox bile	8.0g
Yeast extract	5.0g
Ferric citrate	1.0g
Bromthymol Blue	0.04g
Thymol Blue	0.04g

pH 8.6 ± 0.2 at 25°C

Source: This medium is available as a premixed powder from Oxoid Unipath.

Preparation of Medium: Add components to distilled/deionized water and bring volume to 1.0mL. Mix thoroughly. Gently heat while stirring and bring to boiling. Do not autoclave. Cool to 45°C. Pour into sterile Petri dishes.

Use: For the isolation of pathogenic vibrios, especially *Vibrio cholerae.* This medium is suitable for the growth of *Vibrio cholerae*, *Vibrio parahaemolyticus*, and most other *Vibrios*. Most of the Enterobacteriaceae encountered in feces are totally suppressed for at least 24 hours. Slight growth of *Proteus* species and *Enterococcus faecalis* may occur but the colonies are easily distinguished from vibrio colonies. While inhibiting non-vibrios, it promotes rapid growth of pathogenic vibrios after overnight incubation at 35°C. *Vibrio cholerae* El Tor biotype forms yellow colonies, *Vibrio parahaemolyticus* forms blue-green colonies, *Vibrio alginolyticus* forms yellow colonies, *Vibrio metschnikovii* forms yellow colonies, *Vibrio fluvialis* forms yellow colonies, *Vibrio vulnificus* forms blue-green colonies, *Vibrio mimicus* forms blue-green colonies, *Enterococcus* species form yellow colonies, *Proteus* species form yellow-green colonies, *Pseudomonas* species form blue-green colonies and some strains of *Aeromonas hydrophila* produce yellow colonies, but *Plesimonas shigelloides* does not usually grow well on this medium.

Cholesterol Medium

Composition per 1030.0mL:

Solution A	500.0mL
Solution B	500.0mL
Amino acid solution	20.0mL
Vitamin solution	10.0mL

pH 6.8 ± 0.2 at 25°C

Solution A:

Composition per liter:

$(NH_4)_2SO_4$	5.0g
KH_2PO_4	1.0g
$MgSO_4 \cdot 7H_2O$	0.5g
$CaCl_2 \cdot 2H_2O$	0.1g
NaCl	0.1g
Wolfe's mineral solution	10.0mL

Preparation of Solution A: Add components to distilled/deionized water and bring volume to 1.0L. Mix thoroughly. Autoclave for 15 min at 15 psi pressure–121°C. Cool to 45°–50°C.

Wolfe's Mineral Solution:

Composition per liter:

$MgSO_4 \cdot 7H_2O$	3.0g
Nitrilotriacetic acid	1.5g
NaCl	1.0g
$MnSO_4 \cdot H_2O$	0.5g
$CaCl_2$	0.1g
$CoCl_2 \cdot 6H_2O$	0.1g
$FeSO_4 \cdot 7H_2O$	0.1g
$ZnSO_4 \cdot 7H_2O$	0.1g
$AlK(SO_4)_2 \cdot 12H_2O$	0.01g
$CuSO_4 \cdot 5H_2O$	0.01g
H_3BO_3	0.01g
$Na_2MoO_4 \cdot 2H_2O$	0.01g

Preparation of Wolfe's Mineral Solution: Add nitrilotriacetic acid to 500.0mL of distilled/deionized water. Dissolve by adjusting pH to 6.5 with KOH. Add distilled/deionized water to 1.0L. Add remaining components.

Solution B:

Composition per liter:

Noble agar	15.0g
Cholesterol	2.0g
Tween™ 80	1.0g
Yeast extract	0.5g

Preparation of Solution B: Add components to distilled/deionized water and bring volume to 1.0L. Mix thoroughly. Gently heat to boiling. Autoclave for 15 min at 15 psi pressure–121°C. Cool to 45°–50°C.

Amino Acid Solution:

Composition per 100.0mL:

L-Histidine	0.5g
DL-Methionine	0.1g
DL-Tryptophan	0.1g

Preparation of Amino Acid Solution: Add components to distilled/deionized water and bring volume to 100.0mL. Filter sterilize.

Vitamin Solution:

Composition per liter:

myo-Inositol	200.0mg
Calcium pantothenate	40.0mg
Niacin	40.0mg
Pyridoxine·HCl	40.0mg
Thiamine	40.0mg
p-Aminobenzoic acid	20.0mg
Riboflavin	20.0mg
Biotin	200.0μg
Folic acid	200.0μg

Preparation of Vitamin Solution: Add components to distilled/deionized water and bring volume to 1.0L. Filter sterilize.

Preparation of Medium: Combine cooled, sterile solution A and cooled, sterile solution B. Aseptically add filter-sterilized amino acid solution and vitamin solution. Adjust pH to 6.8. Pour into sterile Petri dishes or distribute into sterile tubes.

Use: For the cultivation of ATCC strain 31384.

Cholic Acid Medium

Composition per liter:

Noble agar	15.0g
K_2HPO_4	3.5g
Cholic acid	2.0g
$(NH_4)_2SO_4$	2.0g
KH_2PO_4	1.5g
$MgSO_4 \cdot 7H_2O$	0.1g
$CaCl_2 \cdot 2H_2O$	0.01g
$FeSO_4 \cdot 7H_2O$	0.5mg

pH 7.0 ± 0.2 at 25°C

Preparation of Medium: Add components to distilled/deionized water and bring volume to 1.0L. Mix thoroughly. Adjust pH to 7.0. Gently heat and bring to boiling. Autoclave for 15 min at 15 psi pressure–121°C. Pour into sterile Petri dishes or distribute into sterile tubes.

Use: For the cultivation and maintenance of *Nocardia* species and other bacteria that can utilize cholic acid as a carbon source.

Choline Assay Medium

Composition per liter:

Sucrose	40.0g
Potassium sodium tartrate	11.4g
$(NH_4)_2NO_3$	2.0g
KH_2PO_4	2.0g
$MgSO_4 \cdot 7H_2O$	1.0g
$CaCl_2 \cdot 2H_2O$	0.2g
NaCl	0.2g
$ZnSO_4 \cdot 7H_2O$	0.02g
$FeSO_4$	1.1mg
Na_3BO_3	0.7mg
$(NH_4)_2MoO_3$	0.5mg
CuCl	0.3mg
$MgSO_4 \cdot 7H_2O$	0.11mg
Biotin	0.01mg

pH5.5 ± 0.2 at 25°C

Source: This medium is available as a premixed powder from BD Diagnostic Systems.

Preparation of Medium: Add components to distilled/deionized water and bring volume to 1.0L. Mix thoroughly. Gently heat and bring to boiling. Continue boiling for 2–3 min. Allow precipitate to settle out. Distribute supernatant into 125.0mL flasks in 10.0mL volumes. Add standard solution or test solutions to each flask. Adjust the volume of each flask to 20.0mL with distilled/deionized water. Autoclave for 10 min at 15 psi pressure–121°C.

Use: For the microbiological assay of choline using *Neurospora crassa* as the test microorganism.

Choline Medium

Composition per liter:

NaCl	30.0g
Choline chloride	5.0g
K_2HPO_4	1.0g
$MgSO_4$	0.5g
$FeSO_4$	0.01g

pH 7.4 ± 0.2 at 25°C

Preparation of Medium: Add components to distilled/deionized water and bring volume to 1.0L. Mix thoroughly. Adjust pH to 7.4. Distribute into tubes or flasks. Autoclave for 15 min at 15 psi pressure–121°C.

Use: For the cultivation of *Achromobacter holinophagum.*

Chondromyces VYZ Agar

Composition per liter:

Agar	15.0g
Fresh baker's yeast cake	5.0g
$CaCl_2$	1.0g

pH 7.2 ± 0.2 at 25°C

Preparation of Medium: Add components to distilled/deionized water and bring volume to 1.0L. Mix thoroughly. Adjust pH to 7.2. Distribute into tubes or flasks. Autoclave for 15 min at 15 psi pressure–121°C.

Use: For the cultivation of *Chondromyces* species.

Chopped Liver Broth

Composition per liter:

Fresh beef liver	500.0g
Peptone	10.0g
K_2HPO_4	1.0g
Soluble starch	1.0g

pH 7.0 ± 0.2 at 25°C

Source: This medium is available as a premixed powder from Hi-Media.

Preparation of Medium: Grind fresh beef liver. Add to 1.0L of distilled/deionized water. Gently heat and bring to boiling. Continue boiling for 60 min. Cool to 25°C. Adjust pH to 7.0. Gently heat and bring to boiling. Continue boiling for 10 min. Filter through cheesecloth. Save chopped liver particles. To filtrate, add remaining components. Bring volume to 1.0L with distilled/deionized water. Adjust pH to 7.0. Filter through Whatman #1 filter paper. Add chopped liver particles to test tubes to a depth of 1.2–2.5 cm. Add 10.0mL of broth to each tube. Autoclave for 15 min at 15 psi pressure–121°C.

Use: For the isolation and cultivation of *Clostridium botulinum, Clostridium perfringens*, and other anaerobic bacteria from foods.

Chopped Liver HiVeg Broth

Composition per liter:

Plant infusion	100.0g
Plant peptone	10.0g
K_2HPO_4	1.0g
Starch, soluble	1.0g

pH 7.0 ± 0.2 at 25°C

Source: This medium is available as a premixed powder from Hi-Media.

Preparation of Medium: Add components to distilled/deionized water and bring volume to 1.0L. Mix thoroughly. Gently heat and bring to boiling. Distribute into tubes or flasks. Autoclave for 20 min at 15 psi pressure–121°C.

Use: For the isolation and cultivation of *Clostridium botulinum, Clostridium perfringens*, and other anaerobic bacteria from foods.

Chopped Meat Agar

Composition per liter:

Ground meat, fat free	500.0g
Pancreatic digest of casein	30.0g
Agar	15.0g
K_2HPO_4	5.0g
Yeast extract	5.0g
L-Cysteine·HCl	0.5g
Resazurin	1.0mg
NaOH (1*N* solution)	25.0mL

pH 7.0 ± 0.2 at 25°C

Preparation of Medium: Use lean beef or horse meat. Remove fat and connective tissue. Grind finely. Add ground meat and 25.0mL of NaOH solution to distilled/deionized water and bring volume to 1025.0mL. Gently heat and bring to boiling. Continue boiling for 15 min without stirring. Cool to room temperature. Remove fat from surface. Filter and retain both meat particles and filtrate. Adjust volume of filtrate to 1.0L with distilled/deionized water. Add pancreatic digest of casein, agar, K_2HPO_4, yeast extract, and resazurin. Gently heat and bring to boiling. Boil for 1–2 min. Add L-cysteine·HCl. Mix thoroughly. Distribute 7.0mL into tubes that contain meat particles (1 part meat particles to 5 parts fluid). Autoclave for 30 min at 15 psi pressure–121°C.

Use: For the cultivation of various anaerobes.

Chopped Meat Broth

Composition per liter:

Ground meat, fat free	500.0g
Pancreatic digest of casein	30.0g
K_2HPO_4	5.0g
Yeast extract	5.0g
L-Cysteine·HCl	0.5g
Resazurin	1.0mg
NaOH (1*N* solution)	25.0mL

pH 7.0 ± 0.2 at 25°C

Preparation of Medium: Use lean beef or horse meat. Remove fat and connective tissue. Grind finely. Add ground meat and 25.0mL of NaOH solution to distilled/deionized water and bring volume to 1025.0mL. Gently heat and bring to boiling. Continue boiling for 15 min without stirring. Cool to room temperature. Remove fat from surface. Filter and retain both meat particles and filtrate. Adjust volume of filtrate to 1.0L with distilled/deionized water. Add pancreatic digest of casein, K_2HPO_4, yeast extract, and resazurin. Gently heat and bring to boiling. Boil for 1–2 min. Add L-cysteine·HCl. Mix thoroughly. Distribute 7.0mL into tubes that contain meat particles (1 part meat particles to 5 parts fluid). Autoclave for 30 min at 15 psi pressure–121°C.

Use: For the cultivation of various anaerobes.

Chopped Meat Broth with Carbohydrates (DSMZ Medium 110)

Composition per liter:

Ground meat, fat free	500.0g
Pancreatic digest of casein	30.0g
K_2HPO_4	5.0g
Yeast extract	5.0g
Glucose	4.0g
Cellobiose	1.0g
Maltose	1.0g
Starch, soluble	1.0g
L-Cysteine·HCl	0.5g
Resazurin	1.0mg
NaOH (1*N* solution)	25.0mL

pH 7.0 ± 0.2 at 25°C

Preparation of Medium: Use lean beef or horse meat. Remove fat and connective tissue. Grind finely. Add ground meat and 25.0mL of NaOH solution to distilled/deionized water and bring volume to 1025.0mL. Gently heat and bring to boiling. Continue boiling for 15 min without stirring. Cool to room temperature. Remove fat from surface. Filter and retain both meat particles and filtrate. Adjust volume of filtrate to 1.0L with distilled/deionized water. Add pancreatic digest of casein, K_2HPO_4, yeast extract, and resazurin. Gently heat and bring to boiling. Boil for 1–2 min. Add glucose, cellobiose, maltose, and soluble starch. Add L-cysteine·HCl. Mix thoroughly. Distribute 7.0mL into tubes that contain meat particles (1 part meat particles to 5 parts fluid). Autoclave for 30 min at 15 psi pressure–121°C.

Use: For the cultivation of numerous anaerobes, including *Actinomyces suis, Anaerobiospirillum succiniciproducens, Bacteroides distasonis, Bacteroides eggerthii, Bacteroides fragilis, Bacteroides helcogenes, Bacteroides macacae, Bacteroides pyogenes, Bacteroides splanchnicus, Bacteroides suis, Centipeda periodontii,* numerous *Clostridium* species, *Eubacterium brachy, Eubacterium eligens, Eubacterium hallii, Eubacterium limosum, Eubacterium plautii, Eubacterium ruminantium, Eubacterium saburreum, Eubacterium siraeum, Eubacterium* species, *Eubacterium tarantellus, Eubacterium tenue, Eubacterium ventriosum, Megamonas hypermegas, Peptococcus niger, Peptostreptococcus productus, Prevotella buccae, Prevotella buccalis, Prevotella denticola, Prevotella intermedia, Prevotella oralis,* and *Selenomonas sputigena.*

Chopped Meat Broth with Formate and Fumarate (LMG Medium 69)

Composition per liter:

Ground meat, fat free	500.0g
Pancreatic digest of casein	30.0g
K_2HPO_4	5.0g
Yeast extract	5.0g
L-Cysteine·HCl	0.5g
Resazurin	1.0mg
NaOH (1*N* solution)	25.0mL
Formate-fumarate solution	7.7mL

pH 7.0 ± 0.2 at 25°C

Formate-Fumarate Solution:

Composition per 100.0mL:

Sodium formate	6.0g
Fumaric acid	6.0g

Preparation of Formate-Fumarate Solution: Add components to distilled/deionized water and bring volume to 100.0mL. Adjust pH to 7.0. Filter sterilize.

Preparation of Medium: Use lean beef or horse meat. Remove fat and connective tissue. Grind finely. Add ground meat and 25.0mL of NaOH solution to distilled/deionized water and bring volume to 1025.0mL. Gently heat and bring to boiling. Continue boiling for 15 min without stirring. Cool to room temperature. Remove fat from surface. Filter and retain both meat particles and filtrate. Adjust volume of filtrate to 1.0L with distilled/deionized water. Add pancreatic digest of casein, agar, K_2HPO_4, yeast extract, and resazurin. Gently heat and bring to boiling. Boil for 1–2 min. Add L-cysteine·HCl. Mix thoroughly. Distribute 6.5mL into tubes that contain meat particles (1 part meat particles to 5 parts fluid). Autoclave for 30 min at 15 psi pressure–121°C. Cool to 25°C. Aseptically add 50 μL of sterile formate/fumarate solution to each tube prior to inoculation.

Use: For the cultivation of various *Clostridium* spp.

Chopped Meat Broth with Vitamin K_1 (LMG Medium 70)

Composition per liter:

Ground meat, fat free	500.0g
Pancreatic digest of casein	30.0g
K_2HPO_4	5.0g
Yeast extract	5.0g
L-Cysteine·HCl	0.5g
Resazurin	1.0mg
NaOH (1*N* solution)	25.0mL
Vitamin K_1 solution	7.7mL

pH 7.0 ± 0.2 at 25°C

Vitamin K_1 Solution:

Composition per 50.0mL:

Ethanol (20% solution)	50.0mL
Vitamin K_1	0.7gL

Preparation of Vitamin K_1 Solution: Mix components. Filter sterilize. Store solution protected from light at 5°C. Discard after one month.

Preparation of Medium: Use lean beef or horse meat. Remove fat and connective tissue. Grind finely. Add ground meat and 25.0mL of NaOH solution to distilled/deionized water and bring volume to 1025.0mL. Gently heat and bring to boiling. Continue boiling for 15 min without stirring. Cool to room temperature. Remove fat from surface. Filter and retain both meat particles and filtrate. Adjust volume of filtrate to 1.0L with distilled/deionized water. Add pancreatic digest of casein, agar, K_2HPO_4, yeast extract, and resazurin. Gently heat and bring to boiling. Boil for 1–2 min. Add L-cysteine·HCl. Mix thoroughly. Distribute 6.5mL into tubes that contain meat particles (1 part meat particles to 5 parts fluid). Autoclave for 30 min at 15 psi pressure–121°C. Cool to 25°C. Aseptically add 50 μL of sterile vitamin K_1 solution to each tube prior to inoculation.

Use: For the cultivation of various *Clostridium* spp.

Chopped Meat Carbohydrate Medium

Composition per 1240.0mL:

Peptone	30.0g
K_2HPO_4	5.0g
Yeast extract	5.0g
Cellobiose	1.0g
Maltose	1.0g
Starch	1.0g
L-Cysteine·HCl·H_2O	0.5g
Chopped meat extract filtrate	1.0L
Chopped meat extract solids	200.0mL
Resazurin (0.025% solution)	4.0mL

pH 7.0 ± 0.2 at 25°C

Chopped Meat Extract:
Composition per liter:

Beef or horse meat	500.0g
NaOH (1*N* solution)	25.0mL

Preparation of Chopped Meat Extract: Use lean beef or horse meat. Remove fat and connective tissue. Grind. Add meat and NaOH to distilled/deionized water and bring volume to 1.0L. Gently heat and bring to boiling while stirring. Cool to 25°C. Remove fat from surface. Filter. Reserve ground meat particles and filtrate. Add distilled/deionized water to filtrate and bring volume to 1.0L.

Preparation of Medium: To 1.0L of chopped meat extract filtrate, add the remaining components, except the L-cysteine·HCl·H_2O and chopped meat solids. Mix thoroughly. Gently heat to boiling. Cool to room temperature. Add the L-cysteine·HCl·H_2O. Adjust pH to 7.0. Distribute 1 part chopped meat solids (by volume) and 5 parts of liquid (by volume) into tubes under O_2-free 97% N_2 + 3% H_2. Cap with rubber stoppers and place tubes in a press. Autoclave for 15 min at 15 psi pressure–121°C with fast exhaust.

Use: For the cultivation of anaerobic bacteria, including *Clostridium* species, *Eubacterium* species, and *Gemmiger formicilis*.

Chopped Meat Carbohydrate Medium with Rumen Fluid (ATCC Medium 1016)

Composition per 1390.0mL:

Peptone	30.0g
K_2HPO_4	5.0g
Yeast extract	5.0g
Cellobiose	1.0g
Maltose	1.0g
Starch	1.0g
L-Cysteine·HCl·H_2O	0.5g
Chopped meat extract filtrate	1.0L
Chopped meat extract solids	200.0mL
Rumen fluid	150.0mL
Resazurin (0.025% solution)	4.0mL

pH 7.0 ± 0.2 at 25°C

Chopped Meat Extract:
Composition per liter:

Beef or horse meat	500.0g
NaOH (1*N* solution)	25.0mL

Preparation of Chopped Meat Extract: Use lean beef or horse meat. Remove fat and connective tissue. Grind. Add meat and NaOH to distilled/deionized water and bring volume to 1.0L. Gently heat and bring to boiling while stirring. Cool to 25°C. Remove fat from surface. Filter. Reserve ground meat particles and filtrate. Add distilled/deionized water to filtrate and bring volume to 1.0L.

Preparation of Medium: To 1.0L of chopped meat extract filtrate, add the remaining components, except L-cysteine·HCl·H_2O and chopped meat solids. Mix thoroughly. Gently heat to boiling. Cool to room temperature. Add the L-cysteine·HCl·H_2O. Adjust pH to 7.0. Distribute 1 part chopped meat solids (by volume) and 5 parts of liquid (by volume) into tubes under O_2-free 97% N_2 + 3% H_2. Cap with rubber stoppers and place tubes in a press. Autoclave for 15 min at 15 psi pressure–121°C with fast exhaust.

Use: For the cultivation of anaerobic bacteria, including *Butyrivibrio crossotus, Eubacterium* species, and *Ruminococcus* species.

Chopped Meat Carbohydrate Medium with Rumen Fluid

Composition per 1390.0mL:

Peptone	30.0g
K_2HPO_4	5.0g
Yeast extract	5.0g
Glucose	4.0g
Cellobiose	1.0g
Maltose	1.0g
Starch	1.0g
L-Cysteine·HCl·H_2O	0.5g
Chopped meat extract filtrate	1.0L
Chopped meat extract solids	200.0mL
Rumen fluid	150.0mL
Resazurin (0.025% solution)	4.0mL

pH 7.0 ± 0.2 at 25°C

Chopped Meat Extract:
Composition per liter:

Beef or horse meat	500.0g
NaOH (1*N* solution)	25.0mL

Preparation of Chopped Meat Extract: Use lean beef or horse meat. Remove fat and connective tissue. Grind. Add meat and NaOH to distilled/deionized water and bring volume to 1.0L. Gently heat and bring to boiling while stirring. Cool to 25°C. Remove fat from surface. Filter. Reserve ground meat particles and filtrate. Add distilled/deionized water to filtrate and bring volume to 1.0L.

Preparation of Medium: To 1.0L of chopped meat extract filtrate, add the remaining components, except L-cysteine·HCl·H_2O and chopped meat solids. Mix thoroughly. Gently heat to boiling. Cool to room temperature. Add the L-cysteine·HCl·H_2O. Adjust pH to 7.0. Distribute 1 part chopped meat solids (by volume) and 5 parts of liquid (by volume) into tubes under O_2-free 97% N_2 + 3% H_2. Cap with rubber stoppers and place tubes in a press. Autoclave for 15 min at 15 psi pressure–121°C with fast exhaust.

Use: For the cultivation of anaerobic bacteria, including *Fusobacterium prausnitzii, Eubacterium* species, and *Prevotella ruminicola.*

Chopped Meat Carbohydrate Medium with Tween™ 80

Composition per 1240.0mL:

Peptone	30.0g
K_2HPO_4	5.0g
Yeast extract	5.0g
Cellobiose	1.0g
Maltose	1.0g
Starch	1.0g

Tween™ 80 .... 1.0g
L-Cysteine·HCl·H_2O 0.5g
Chopped meat extract filtrate 1.0L
Chopped meat extract solids 200.0mL
Resazurin (0.025% solution) 4.0mL
pH 7.0 ± 0.2 at 25°C

Chopped Meat Extract:
Composition per liter:
Beef or horse meat 500.0g
NaOH (1*N* solution) 25.0mL

Preparation of Chopped Meat Extract: Use lean beef or horse meat. Remove fat and connective tissue. Grind. Add meat and NaOH to distilled/deionized water and bring volume to 1.0L. Gently heat and bring to boiling while stirring. Cool to 25°C. Remove fat from surface. Filter. Reserve ground meat particles and filtrate. Add distilled/deionized water to filtrate and bring volume to 1.0L.

Preparation of Medium: To 1.0L of chopped meat extract filtrate, add the remaining components, except L-cysteine·HCl·H_2O and chopped meat solids. Mix thoroughly. Gently heat to boiling. Cool to room temperature. Add the L-cysteine·HCl·H_2O. Adjust pH to 7.0. Distribute 1 part chopped meat solids (by volume) and 5 parts of liquid (by volume) into tubes under O_2-free 97% N_2 + 3% H_2. Cap with rubber stoppers and place tubes in a press. Autoclave for 15 min at 15 psi pressure–121°C with fast exhaust.

Use: For the cultivation of *Coprococcus* species and *Peptostreptococcus micros*.

Chopped Meat Glucose Agar (LMG Medium 68)

Composition per liter:
Ground meat, fat free 500.0g
Pancreatic digest of casein 30.0g
Agar 15.0g
Glucose 10.0g
K_2HPO_4 .... 5.0g
Yeast extract 5.0g
L-Cysteine·HCl 0.5g
Resazurin 1.0mg
NaOH (1*N* solution) 25.0mL
pH 7.0 ± 0.2 at 25°C

Preparation of Medium: Use lean beef or horse meat. Remove fat and connective tissue. Grind finely. Add ground meat and 25.0mL of NaOH solution to distilled/deionized water and bring volume to 1025.0mL. Gently heat and bring to boiling. Continue boiling for 15 min without stirring. Cool to room temperature. Remove fat from surface. Filter and retain both meat particles and filtrate. Adjust volume of filtrate to 1.0L with distilled/deionized water. Add pancreatic digest of casein, agar, K_2HPO_4, yeast extract, and resazurin. Gently heat and bring to boiling. Boil for 1–2 min. Add L-cysteine·HCl. Mix thoroughly. Distribute 7.0mL into tubes that contain meat particles (1 part meat particles to 5 parts fluid). Autoclave for 30 min at 15 psi pressure–121°C.

Use: For the cultivation of various *Clostridium* spp.

Chopped Meat Glucose Broth (LMG Medium 68)

Composition per liter:
Ground meat, fat free 500.0g
Pancreatic digest of casein 30.0g
Glucose 10.0g
K_2HPO_4 .... 5.0g
Yeast extract 5.0g
L-Cysteine·HCl 0.5g
Resazurin 1.0mg
NaOH (1*N* solution) 25.0mL
pH 7.0 ± 0.2 at 25°C

Preparation of Medium: Use lean beef or horse meat. Remove fat and connective tissue. Grind finely. Add ground meat and 25.0mL of NaOH solution to distilled/deionized water and bring volume to 1025mL. Gently heat and bring to boiling. Continue boiling for 15 min without stirring. Cool to room temperature. Remove fat from surface. Filter and retain both meat particles and filtrate. Adjust volume of filtrate to 1.0L with distilled/deionized water. Add pancreatic digest of casein, agar, K_2HPO_4, yeast extract, and resazurin. Gently heat and bring to boiling. Boil for 1–2 min. Add L-cysteine·HCl. Mix thoroughly. Distribute 7.0mL into tubes that contain meat particles (1 part meat particles to 5 parts fluid). Autoclave for 30 min at 15 psi pressure–121°C.

Use: For the cultivation of various *Clostridium* spp.

Chopped Meat Glucose Medium

Composition per 1240.0mL:
Peptone 30.0g
K_2HPO_4 .... 5.0g
Yeast extract 5.0g
Glucose 5.0g
L-Cysteine·HCl·H_2O 0.5g
Chopped meat extract filtrate 1.0L
Chopped meat extract solids 200.0mL
Resazurin (0.025% solution) 4.0mL
pH 7.0 ± 0.2 at 25°C

Chopped Meat Extract:
Composition per liter:
Beef or horse meat 500.0g
NaOH (1*N* solution) 25.0mL

Preparation of Chopped Meat Extract: Use lean beef or horse meat. Remove fat and connective tissue. Grind. Add meat and NaOH to distilled/deionized water and bring volume to 1.0L. Gently heat and bring to boiling while stirring. Cool to 25°C. Remove fat from surface. Filter. Reserve ground meat particles and filtrate. Add distilled/deionized water to filtrate and bring volume to 1.0L.

Preparation of Medium: To 1.0L of chopped meat extract filtrate, add the remaining components, except L-cysteine·HCl·H_2O and chopped meat solids. Mix thoroughly. Gently heat to boiling. Cool to room temperature. Add the L-cysteine·HCl·H_2O. Adjust pH to 7.0. Distribute 1 part chopped meat solids (by volume) and 5 parts of liquid (by volume) into tubes under O_2-free 97% N_2 + 3% H_2. Cap with rubber stoppers and place tubes in a press. Autoclave for 15 min at 15 psi pressure–121°C with fast exhaust.

Use: For the cultivation of *Clostridium* species and *Selenomonas noxia*.

Chopped Meat Glucose Medium with NaCl

Composition per 1205.0mL:
NaCl 30.0g
Peptone 30.0g
K_2HPO_4 .... 5.0g
Yeast extract 5.0g

Glucose 5.0g
L-Cysteine·HCl·H_2O 0.5g
Chopped meat extract filtrate 1.0L
Chopped meat extract solids 200.0mL
Resazurin (0.025% solution) 4.0mL
pH 7.0 ± 0.2 at 25°C

Chopped Meat Extract:
Composition per liter:
Beef or horse meat 500.0g
NaOH (1*N* solution) 25.0mL

Preparation of Chopped Meat Extract: Use lean beef or horse meat. Remove fat and connective tissue. Grind. Add meat and NaOH to distilled/deionized water and bring volume to 1.0L. Gently heat and bring to boiling while stirring. Cool to 25°C. Remove fat from surface. Filter. Reserve ground meat particles and filtrate. Add distilled/deionized water to filtrate and bring volume to 1.0L.

Preparation of Medium: To 1.0L of chopped meat extract filtrate, add the remaining components, except the L-cysteine·HCl·H_2O and chopped meat solids. Mix thoroughly. Gently heat to boiling. Cool to room temperature. Add the L-cysteine·HCl·H_2O. Adjust pH to 7.0. Distribute 1 part chopped meat solids (by volume) and 5 parts of liquid (by volume) into tubes under O_2-free 97% N_2 + 3% H_2. Cap with rubber stoppers and place tubes in a press. Autoclave for 15 min at 15 psi pressure–121°C with fast exhaust.

Use: For the cultivation and maintenance of anaerobic halophilic bacteria.

Chopped Meat Medium

Composition per 1205.0mL:
Peptone 30.0g
K_2HPO_4 .... 5.0g
Yeast extract 5.0g
L-Cysteine·HCl·H_2O 0.5g
Chopped meat extract filtrate 1.0L
Chopped meat extract solids 200.0mL
Resazurin (0.025% solution) 4.0mL
pH 7.0 ± 0.2 at 25°C

Chopped Meat Extract:
Composition per liter:
Beef or horse meat 500.0g
NaOH (1*N* solution) 25.0mL

Preparation of Chopped Meat Extract: Use lean beef or horse meat. Remove fat and connective tissue. Grind. Add meat and NaOH to distilled/deionized water and bring volume to 1.0L. Gently heat and bring to boiling while stirring. Cool to 25°C. Remove fat from surface. Filter. Reserve ground meat particles and filtrate. Add distilled/deionized water to filtrate and bring volume to 1.0L.

Preparation of Medium: To 1.0L of chopped meat extract filtrate, add the remaining components, except the L-cysteine·HCl·H_2O and chopped meat solids. Mix thoroughly. Gently heat to boiling. Cool to room temperature. Add the L-cysteine·HCl·H_2O. Adjust pH to 7.0. Distribute 1 part chopped meat solids (by volume) and 5 parts of liquid (by volume) into tubes under O_2-free 97% N_2 + 3% H_2. Cap with rubber stoppers and place tubes in a press. Autoclave for 15 min at 15 psi pressure–121°C with fast exhaust.

Use: For the cultivation and maintenance of a variety of anaerobic bacteria, including *Bacteroides* species, *Bifidobacterium* species, *Capnocytophaga* species, *Clostridium* species, *Eubacterium* species, *Fusobacterium* species, *Peptostreptococcus* species, *Prevotella* species, *Propionibacterium* species, *Ruminococcus* species, and others.

Chopped Meat Medium with 10% Fetal Calf Serum

Composition per 1230.0mL:
Peptone 30.0g
K_2HPO_4 .... 5.0g
Yeast extract 5.0g
L-Cysteine·HCl·H_2O 0.5g
Chopped meat extract filtrate 1.0L
Chopped meat extract solids 200.0mL
Fetal calf serum 100.0mL
Resazurin (0.025% solution) 4.0mL
pH 7.0 ± 0.2 at 25°C

Chopped Meat Extract:
Composition per liter:
Beef or horse meat 500.0g
NaOH (1*N* solution) 25.0mL

Preparation of Chopped Meat Extract: Use lean beef or horse meat. Remove fat and connective tissue. Grind. Add meat and NaOH to distilled/deionized water and bring volume to 1.0L. Gently heat and bring to boiling while stirring. Cool to 25°C. Remove fat from surface. Filter. Reserve ground meat particles and filtrate. Add distilled/deionized water to filtrate and bring volume to 1.0L.

Preparation of Medium: To 1.0L of chopped meat extract filtrate, add the remaining components, except the L-cysteine·HCl·H_2O and chopped meat solids. Mix thoroughly. Gently heat to boiling. Cool to room temperature. Add the L-cysteine·HCl·H_2O. Adjust pH to 7.0. Distribute 1 part chopped meat solids (by volume) and 5 parts of liquid (by volume) into tubes under O_2-free 97% N_2 + 3% H_2. Cap with rubber stoppers and place tubes in a press. Autoclave for 15 min at 15 psi pressure–121°C with fast exhaust.

Use: For the cultivation and maintenance of *Actinomyces hordeovulneris*.

Chopped Meat Medium with Formate and Fumarate

Composition per 1230.0mL:
Peptone 30.0g
K_2HPO_4 .... 5.0g
Yeast extract 5.0g
L-Cysteine·HCl·H_2O 0.5g
Chopped meat extract filtrate 1.0L
Chopped meat extract solids 200.0mL
Resazurin (0.025% solution) 4.0mL
Formate-fumarate solution 0.05mL
pH 7.0 ± 0.2 at 25°C

Chopped Meat Extract:
Composition per liter:
Beef or horse meat 500.0g
NaOH (1*N* solution) 25.0mL

Preparation of Chopped Meat Extract: Use lean beef or horse meat. Remove fat and connective tissue. Grind. Add meat and NaOH to distilled/deionized water and bring volume to 1.0L. Gently heat and bring to boiling while stirring. Cool to 25°C. Remove fat from surface. Filter. Reserve ground meat particles and filtrate. Add distilled/deionized water to filtrate and bring volume to 1.0L.

Formate-Fumarate Solution:
Composition per 100.0mL:

Sodium formate	6.0g
Fumaric acid	6.0g

Preparation of Formate-Fumarate Solution: Add components to distilled/deionized water and bring volume to 100.0mL. Adjust pH to 7.0. Filter sterilize.

Preparation of Medium: To 1.0L of chopped meat extract filtrate, add the remaining components, except the L-cysteine·HCl·H_2O, formate-fumarate solution, and chopped meat solids. Mix thoroughly. Gently heat to boiling. Cool to room temperature. Add the L-cysteine·HCl·H_2O. Adjust pH to 7.0. Distribute 1 part chopped meat solids (by volume) and 5 parts of liquid (by volume) into tubes under O_2-free 97% N_2 + 3% H_2. Cap with rubber stoppers and place tubes in a press. Autoclave for 15 min at 15 psi pressure–121°C with fast exhaust. Prior to inoculation, add 0.05mL of formate-fumarate solution to each tube containing approximately 6.5mL of chopped meat medium.

Use: For the cultivation and maintenance of *Bacteroides ureolyticus* and *Wolinella* species.

Chopped Meat Medium with 1% Glucose

Composition per 1230.0mL:

Peptone	30.0g
Glucose	10.0g
K_2HPO_4	5.0g
Yeast extract	5.0g
L-Cysteine·HCl·H_2O	0.5g
Chopped meat extract filtrate	1.0L
Chopped meat extract solids	200.0mL
Resazurin (0.025% solution)	4.0mL

pH 7.0 ± 0.2 at 25°C

Chopped Meat Extract:
Composition per liter:

Beef or horse meat	500.0g
NaOH (1*N* solution)	25.0mL

Preparation of Chopped Meat Extract: Use lean beef or horse meat. Remove fat and connective tissue. Grind. Add meat and NaOH to distilled/deionized water and bring volume to 1.0L. Gently heat and bring to boiling while stirring. Cool to 25°C. Remove fat from surface. Filter. Reserve ground meat particles and filtrate. Add distilled/deionized water to filtrate and bring volume to 1.0L.

Preparation of Medium: To 1.0L of chopped meat extract filtrate, add the remaining components, except the L-cysteine·HCl·H_2O and chopped meat solids. Mix thoroughly. Gently heat to boiling. Cool to room temperature. Add the L-cysteine. Adjust pH to 7.0. Distribute 1 part chopped meat solids (by volume) and 5 parts of liquid (by volume) into tubes under O_2-free 97% N_2 + 3% H_2. Cap with rubber stoppers and place tubes in a press. Autoclave for 15 min at 15 psi pressure–121°C with fast exhaust.

Use: For the cultivation and maintenance of anaerobic bacteria, including *Bacteroides disiens, Coprococcus eutastus, Eubacterium formicigenerans, Prevotella disiens, Ruminococcus torques,* and *Streptococcus hansenii.*

Chopped Meat Medium with Menadione

Composition per 1230.0mL:

Peptone	30.0g
K_2HPO_4	5.0g
Yeast extract	5.0g
L-Cysteine·HCl·H_2O	0.5g
Chopped meat extract filtrate	1.0L
Chopped meat extract solids	200.0mL
Resazurin (0.025% solution)	4.0mL
Menadione solution	0.25mL

pH 7.0 ± 0.2 at 25°C

Chopped Meat Extract:
Composition per liter:

Beef or horse meat	500.0g
NaOH (1*N* solution)	25.0mL

Preparation of Chopped Meat Extract: Use lean beef or horse meat. Remove fat and connective tissue. Grind. Add meat and NaOH to distilled/deionized water and bring volume to 1.0L. Gently heat and bring to boiling while stirring. Cool to 25°C. Remove fat from surface. Filter. Reserve ground meat particles and filtrate. Add distilled/deionized water to filtrate and bring volume to 1.0L.

Menadione Solution:
Composition per liter:

Menadione (vitamin K_3)	50.0μg
Ethanol (20% solution)	25.0mL

Preparation of Menadione Solution: Dissolve menadione in ethanol. Filter sterilize.

Preparation of Medium: To 1.0L of chopped meat extract filtrate, add the remaining components, except the L-cysteine·HCl·H_2O, menadione solution, and chopped meat solids. Mix thoroughly. Gently heat to boiling. Cool to room temperature. Add the L-cysteine·HCl·H_2O. Adjust pH to 7.0. Distribute 1 part chopped meat solids (by volume) and 5 parts of liquid (by volume) into tubes under O_2-free 97% N_2 + 3% H_2. Cap with rubber stoppers and place tubes in a press. Autoclave for 15 min at 15 psi pressure–121°C with fast exhaust. Prior to inoculation, add 0.25mL of menadione solution to each tube containing approximately 5.0mL of chopped meat medium.

Use: For the cultivation and maintenance of *Bacteroides gingivalis, Bacteroides macacae,* and *Porphyromonas gingivalis.*

Chopped Meat Medium, Modified (DSMZ Medium 797)

Composition per 1230.0mL:

Chopped meat extract solids	200.0g
Trypticase™	30.0g
Agar	20.0g
K_2HPO_4	5.0g
Yeast extract	5.0g
L-Cysteine·HCl·H_2O	0.5g
Chopped meat extract filtrate	1.0L
Hemin solution	10.0mL
Resazurin (0.025% solution)	4.0mL
Vitamin K_1 solution	0.2mL

pH 7.0 ± 0.2 at 25°C

Chopped Meat Extract:
Composition per liter:

Beef or horse meat	500.0g
NaOH (1*N* solution)	25.0mL

Preparation of Chopped Meat Extract: Use lean beef or horse meat. Remove fat and connective tissue. Grind. Add meat and NaOH to distilled/deionized water and bring volume to 1.0L. Gently heat and bring to boiling while stirring. Cool to 25°C. Remove fat from surface.

Filter. Reserve both ground meat particles and filtrate. Add distilled/deionized water to filtrate and bring volume to 1.0L.

Hemin Solution:

Composition per 100.0mL:

Hemin	0.05g
NaOH (1*N* solution)	1.0mL

Preparation of Hemin Solution: Add components to distilled/deionized water and bring volume to 100.0mL. Mix thoroughly.

Vitamin K_1 Solution:

Composition per 30.0mL:

Ethanol (95% solution)	30.0mL
Vitamin K_1	0.15mL

Preparation of Vitamin K_1 Solution: Mix components. Store solution protected from light at 5°C. Discard after 1 month.

Preparation of Medium: To 1.0L of chopped meat extract filtrate, add the remaining components, except the L-cysteine·HCl·H_2O, hemin solution, vitamin K_1 solution, and chopped meat solids. Mix thoroughly. Gently heat to boiling. Boil for 5 min. Cool to 25°C while sparging with 80% N_2 + 10% H_2 + 10% CO_2. Add the L-cysteine·HCl·H_2O, hemin solution, and vitamin K_1 solution. Adjust pH to 7.0. Distribute 1 part chopped meat solids (by volume) and 5 parts of liquid (by volume) into tubes under 80% N_2 + 10% H_2 + 10% CO_2. Autoclave for 15 min at 15 psi pressure–121°C.

Use: For the cultivation of *Dialister pneumosintes*.

Chopped Meat Medium, Modified

Composition per 1230.0mL:

Pancreatic digest of casein	30.0g
Peptone	30.0g
Agar	20.0g
K_2HPO_4	5.0g
Yeast extract	5.0g
L-Cysteine·HCl·H_2O	0.5g
Chopped meat extract filtrate	1.0L
Chopped meat extract solids	200.0mL
Hemin solution	10.0mL
Resazurin (0.025% solution)	4.0mL
Vitamin K_1 solution	0.2mL

pH 7.0 ± 0.2 at 25°C

Chopped Meat Extract:

Composition per liter:

Beef or horse meat	500.0g
NaOH (1*N* solution)	25.0mL

Preparation of Chopped Meat Extract: Use lean beef or horse meat. Remove fat and connective tissue. Grind. Add meat and NaOH to distilled/deionized water and bring volume to 1.0L. Gently heat and bring to boiling while stirring. Cool to 25°C. Remove fat from surface. Filter. Reserve ground meat particles and filtrate. Add distilled/deionized water to filtrate and bring volume to 1.0L.

Hemin Solution:

Composition per 100.0mL:

Hemin	0.05g
NaOH (1*N* solution)	1.0mL

Preparation of Hemin Solution: Add components to distilled/deionized water and bring volume to 100.0mL. Mix thoroughly.

Vitamin K_1 Solution:

Composition per 30.0mL:

Ethanol (95% solution)	30.0mL
Vitamin K_1	0.15mL

Preparation of Vitamin K_1 Solution: Mix components. Store solution protected from light at 5°C. Discard after 1 month.

Preparation of Medium: To 1.0L of chopped meat extract filtrate, add the remaining components, except the L-cysteine·HCl·H_2O, hemin solution, vitamin K_1 solution, and chopped meat solids. Mix thoroughly. Gently heat to boiling. Cool to room temperature. Add the L-cysteine·HCl·H_2O, hemin solution, and vitamin K_1 solution. Adjust pH to 7.0. Distribute 1 part chopped meat solids (by volume) and 5 parts of liquid (by volume) into tubes under O_2-free 97% N_2 + 3% H_2. Cap with rubber stoppers and place tubes in a press. Autoclave for 15 min at 15 psi pressure–121°C with fast exhaust.

Use: For the cultivation and maintenance of a variety of anaerobic bacteria, including *Actinomyces* species, *Bacteroides* species, *Clostridium* species, *Eubacterium* species, *Fusobacterium* species, *Peptostreptococcus* species, *Porphyromonas* species, *Prevotella* species, *Propionibacterium* species, *Selenomonas* species, and others.

Chopped Meat Medium, Modified with Arginine

Composition per 1230.0mL:

Pancreatic digest of casein	30.0g
Peptone	30.0g
Agar	20.0g
Arginine	5.0g
K_2HPO_4	5.0g
Yeast extract	5.0g
L-Cysteine·HCl·H_2O	0.5g
Chopped meat extract filtrate	1.0L
Chopped meat extract solids	200.0mL
Hemin solution	10.0mL
Resazurin (0.025% solution)	4.0mL
Vitamin K_1 solution	0.2mL

pH 7.0 ± 0.2 at 25°C

Chopped Meat Extract:

Composition per liter:

Beef or horse meat	500.0g
NaOH (1*N* solution)	25.0mL
Tween™ 80	25.0mL

Preparation of Chopped Meat Extract: Use lean beef or horse meat. Remove fat and connective tissue. Grind. Add meat and NaOH to distilled/deionized water and bring volume to 1.0L. Gently heat and bring to boiling while stirring. Cool to 25°C. Remove fat from surface. Filter. Reserve ground meat particles and filtrate. Add distilled/deionized water to filtrate and bring volume to 1.0L.

Hemin Solution:

Composition per 100.0mL:

Hemin	0.05g
NaOH (1*N* solution)	1.0mL

Preparation of Hemin Solution: Add components to distilled/deionized water and bring volume to 100.0mL. Mix thoroughly.

Vitamin K_1 Solution:

Composition per 30.0mL:

Ethanol (95% solution)	30.0mL
Vitamin K_1	0.15mL

Preparation of Vitamin K_1 Solution: Mix components. Store solution protected from light at 5°C. Discard after 1 month.

Preparation of Medium: To 1.0L of chopped meat extract filtrate, add the remaining components, except the L-cysteine·HCl·H_2O, hemin solution, vitamin K_1 solution, and chopped meat solids. Mix thoroughly. Gently heat to boiling. Cool to room temperature. Add the L-cysteine·HCl·H_2O, hemin solution, and vitamin K_1 solution. Adjust pH to 7.0. Distribute 1 part chopped meat solids (by volume) and 5 parts of liquid (by volume) into tubes under O_2-free 97% N_2 + 3% H_2. Cap with rubber stoppers and place tubes in a press. Autoclave for 15 min at 15 psi pressure–121°C with fast exhaust.

Use: For the cultivation and maintenance of *Eubacterium lentum*.

Chopped Meat Medium, Modified with Formate and Fumarate

Composition per 1230.0mL:

Pancreatic digest of casein	30.0g
Peptone	30.0g
Agar	20.0g
K_2HPO_4	5.0g
Yeast extract	5.0g
L-Cysteine·HCl·H_2O	0.5g
Chopped meat extract filtrate	1.0L
Chopped meat extract solids	200.0mL
Hemin solution	10.0mL
Resazurin (0.025% solution)	4.0mL
Formate-fumarate solution	0.25mL
Vitamin K_1 solution	0.2mL

pH 7.0 ± 0.2 at 25°C

Chopped Meat Extract:

Composition per liter:

Beef or horse meat	500.0g
NaOH (1*N* solution)	25.0mL

Preparation of Chopped Meat Extract: Use lean beef or horse meat. Remove fat and connective tissue. Grind. Add meat and NaOH to distilled/deionized water and bring volume to 1.0L. Gently heat and bring to boiling while stirring. Cool to 25°C. Remove fat from surface. Filter. Reserve ground meat particles and filtrate. Add distilled/deionized water to filtrate and bring volume to 1.0L.

Hemin Solution:

Composition per 100.0mL:

Hemin	0.05g
NaOH (1*N* solution)	1.0mL

Preparation of Hemin Solution: Add components to distilled/deionized water and bring volume to 100.0mL. Mix thoroughly.

Formate-Fumarate Solution:

Composition per 100.0mL:

Sodium formate	6.0g
Fumaric acid	6.0g

Preparation of Formate-Fumarate Solution: Add components to distilled/deionized water and bring volume to 100.0mL. Adjust pH to 7.0. Filter sterilize.

Vitamin K_1 Solution:

Composition per 30.0mL:

Ethanol (95% solution)	30.0mL
Vitamin K_1	0.15mL

Preparation of Vitamin K_1 Solution: Mix components. Store solution protected from light at 5°C. Discard after one month.

Preparation of Medium: To 1.0L of chopped meat extract filtrate, add the remaining components, except the L-cysteine·HCl·H_2O, hemin solution, vitamin K_1 solution, formate-fumarate solution, and chopped meat solids. Mix thoroughly. Gently heat to boiling. Cool to room temperature. Add the L-cysteine·HCl·H_2O, hemin solution, and vitamin K_1 solution. Adjust pH to 7.0. Distribute 1 part chopped meat solids (by volume) and 5 parts of liquid (by volume) into tubes under O_2-free 97% N_2 + 3% H_2. Cap with rubber stoppers and place tubes in a press. Autoclave for 15 min at 15 psi pressure–121°C with fast exhaust. Prior to inoculation, add 0.25mL of formate-fumarate solution to each tube containing approximately 5.0mL of chopped meat medium, modified.

Use: For the cultivation and maintenance of *Bacteroides gracilis, Bacteroides ureolyticus, Campylobacter mucosalis,* and *Wolinella succinogenes.*

Chopped Meat Medium, Modified with Tween™ 80

Composition per 1230.0mL:

Pancreatic digest of casein	30.0g
Peptone	30.0g
Agar	20.0g
K_2HPO_4	5.0g
Yeast extract	5.0g
L-Cysteine·HCl·H_2O	0.5g
Chopped meat extract filtrate	1.0L
Chopped meat extract solids	200.0mL
Hemin solution	10.0mL
Resazurin (0.025% solution)	4.0mL
Vitamin K_1 solution	0.2mL

pH 7.0 ± 0.2 at 25°C

Chopped Meat Extract:

Composition per liter:

Beef or horse meat	500.0g
NaOH (1*N* solution)	25.0mL
Tween™ 80	25.0mL

Preparation of Chopped Meat Extract: Use lean beef or horse meat. Remove fat and connective tissue. Grind. Add meat and NaOH to distilled/deionized water and bring volume to 1.0L. Gently heat and bring to boiling while stirring. Cool to 25°C. Remove fat from surface. Filter. Reserve ground meat particles and filtrate. Add distilled/deionized water to filtrate and bring volume to 1.0L.

Hemin Solution:

Composition per 100.0mL:

Hemin	0.05g
NaOH (1*N* solution)	1.0mL

Preparation of Hemin Solution: Add components to distilled/deionized water and bring volume to 100.0mL. Mix thoroughly.

Vitamin K_1 Solution:

Composition per 30.0mL:

Ethanol (95% solution)	30.0mL
Vitamin K_1	0.15mL

Preparation of Vitamin K_1 Solution: Mix components. Store solution protected from light at 5°C. Discard after 1 month.

Preparation of Medium: To 1.0L of chopped meat extract filtrate, add the remaining components, except the L-cysteine·HCl·H_2O, hemin solution, vitamin K_1 solution, and chopped meat solids. Mix thoroughly. Gently heat to boiling. Cool to room temperature. Add the L-cysteine·HCl·H_2O, hemin solution, and vitamin K_1 solution. Adjust pH to 7.0. Distribute 1 part chopped meat solids (by volume) and 5 parts of liquid (by volume) into tubes under O_2-free 97% N_2 + 3% H_2. Cap with rubber stoppers and place tubes in a press. Autoclave for 15 min at 15 psi pressure–121°C with fast exhaust.

Use: For the cultivation and maintenance of *Lactobacillus* species and *Eubacterium biforme*.

Chopped Meat Medium with 10% Reduced Filtered Rumen Fluid

Composition per 1330.0mL:

Peptone	30.0g
K_2HPO_4	5.0g
Yeast extract	5.0g
L-Cysteine·HCl·H_2O	0.5g
Chopped meat extract filtrate	1.0L
Chopped meat extract solids	200.0mL
Rumen fluid, reduced and filtered	100.0mL
Resazurin (0.025% solution)	4.0mL

pH 7.0 ± 0.2 at 25°C

Chopped Meat Extract:
Composition per liter:

Beef or horse meat	500.0g
NaOH (1*N* solution)	25.0mL

Preparation of Chopped Meat Extract: Use lean beef or horse meat. Remove fat and connective tissue. Grind. Add meat and NaOH to distilled/deionized water and bring volume to 1.0L. Gently heat and bring to boiling while stirring. Cool to 25°C. Remove fat from surface. Filter. Reserve ground meat particles and filtrate. Add distilled/deionized water to filtrate and bring volume to 1.0L.

Preparation of Medium: To 1.0L of chopped meat extract filtrate, add the remaining components, except the L-cysteine·HCl·H_2O and chopped meat solids. Mix thoroughly. Gently heat to boiling. Cool to room temperature. Add the L-cysteine·HCl·H_2O. Adjust pH to 7.0. Distribute 1 part chopped meat solids (by volume) and 5 parts of liquid (by volume) into tubes under O_2-free 97% N_2 + 3% H_2. Cap with rubber stoppers and place tubes in a press. Autoclave for 15 min at 15 psi pressure–121°C

Use: For the cultivation and maintenance of *Eubacterium hallii*.

Chopped Meat Medium for *Treponema* spp. (DSMZ Medium 78a)

Composition per 1055.0mL:

Ground meat, fat free	500.0g
Pancreatic digest of casein	30.0g
K_2HPO_4	5.0g
Yeast extract	5.0g
L-Cysteine·HCl	0.5g
Resazurin	1.0mg
Amino acid solution	50.0mL
NaOH (1*N* solution)	25.0mL
Vitamin solution	5.0mL

pH 7.0 ± 0.2 at 25°C

Amino Acid Solution:
Composition per liter:

L-Glutamine	0.7g
L-Histidine	0.6g
L-Serine	0.5g

Preparation of Amino Acid Solution: Add components to distilled/deionized water and bring volume to 1.0L. Mix thoroughly.

Vitamin Solution:
Composition per liter:

Vitamin B_{12}	50.0mg
Pantothenic acid	50.0mg
Riboflavin	50.0mg
α-Lipoic acid	50.0mg
p-Aminobenzoic acid	50.0mg
Thiamine-HCl·$2H_2O$	50.0mg
Nicotinic acid	25.0mg
Nicotine amide	25.0mg
Biotin	20.0mg
Folic acid	20.0mg
Pyridoxamine-HCl	10.0mg

Preparation of Vitamin Solution: Add components to distilled/deionized water and bring volume to 1.0L. Mix thoroughly. Filter sterilize.

Preparation of Medium: Use lean beef or horse meat. Remove fat and connective tissue. Grind finely. Add ground meat and 25.0mL of NaOH solution to distilled/deionized water and bring volume to 1025.0mL. Gently heat and bring to boiling. Continue boiling for 15 min without stirring. Cool to room temperature. Remove fat from surface. Filter and retain both meat particles and filtrate. Adjust volume of filtrate to 1.0L with distilled/deionized water. Add pancreatic digest of casein, K_2HPO_4, yeast extract, and resazurin. Gently heat and bring to boiling. Boil for 1–2 min. Add L-cysteine·HCl. Mix thoroughly. Autoclave for 30 min at 15 psi pressure–121°C. Aseptically add amino acid and vitamin solutions. Mix thoroughly. Aseptically distribute 7.0mL amounts into tubes that contain meat particles (1 part meat particles to 5 parts fluid).

Use: For the cultivation of *Treponema brennaborense*.

Chopped Meat Medium with 0.025% Tween™ 80 (ATCC Medium 1228)

Composition per 1230.0mL:

Peptone	30.0g
K_2HPO_4	5.0g
Yeast extract	5.0g
L-Cysteine·HCl·H_2O	0.5g
Tween™ 80	0.25g
Chopped meat extract filtrate	1.0L
Chopped meat extract solids	200.0mL
Resazurin (0.025% solution)	4.0mL

pH 7.0 ± 0.2 at 25°C

Chopped Meat Extract:
Composition per liter:

Beef or horse meat	500.0g
NaOH (1*N* solution)	25.0mL

Preparation of Chopped Meat Extract: Use lean beef or horse meat. Remove fat and connective tissue. Grind. Add meat and NaOH to distilled/deionized water and bring volume to 1.0L. Gently heat and bring to boiling while stirring. Cool to 25°C. Remove fat from surface. Filter. Reserve ground meat particles and filtrate. Add distilled/deionized water to filtrate and bring volume to 1.0L.

Preparation of Medium: To 1.0L of chopped meat extract filtrate, add the remaining components, except the L-cysteine·HCl·H_2O and chopped meat solids. Mix thoroughly. Gently heat to boiling. Cool to room temperature. Add the L-cysteine·HCl·H_2O. Adjust pH to 7.0. Distribute 1 part chopped meat solids (by volume) and 5 parts of liquid (by volume) into tubes under O_2-free 97% N_2 + 3% H_2. Cap with rubber stoppers and place tubes in a press. Autoclave for 15 min at 15 psi pressure–121°C with fast exhaust.

Use: For the cultivation and maintenance of *Eubacterium* and *Lactobacillus* species.

Chopped Meat Medium with 1% Tween™ 80 (ATCC Medium 737)

Composition per 1230.0mL:

Peptone	30.0g
Tween™ 80	10.0g
K_2HPO_4	5.0g
Yeast extract	5.0g
L-Cysteine·HCl·H_2O	0.5g
Chopped meat extract filtrate	1.0L
Chopped meat extract solids	200.0mL
Resazurin (0.025% solution)	4.0mL

pH 7.0 ± 0.2 at 25°C

Chopped Meat Extract:

Composition per liter:

Beef or horse meat	500.0g
NaOH (1*N* solution)	25.0mL

Preparation of Chopped Meat Extract: Use lean beef or horse meat. Remove fat and connective tissue. Grind. Add meat and NaOH to distilled/deionized water and bring volume to 1.0L. Gently heat and bring to boiling while stirring. Cool to 25°C. Remove fat from surface. Filter. Reserve ground meat particles and filtrate. Add distilled/deionized water to filtrate and bring volume to 1.0L.

Preparation of Medium: To 1.0L of chopped meat extract filtrate, add the remaining components, except the L-cysteine·HCl·H_2O and chopped meat solids. Mix thoroughly. Gently heat to boiling. Cool to room temperature. Add the L-cysteine·HCl·H_2O. Adjust pH to 7.0. Distribute 1 part chopped meat solids (by volume) and 5 parts of liquid (by volume) into tubes under O_2-free 97% N_2 + 3% H_2. Cap with rubber stoppers and place tubes in a press. Autoclave for 15 min at 15 psi pressure–121°C with fast exhaust.

Use: For the cultivation and maintenance of *Eubacterium biforme* and *Lactobacillus* species.

Christensen Agar

Composition per liter:

Agar	15.0g
NaCl	5.0g
Sodium citrate	3.0g
KH_2PO_4	1.0g
L-Cysteine·HCl·H_2O	0.1g
Phenol Red	12.0mg

pH 6.9 ± 0.2 at 25°C

Preparation of Medium: Add components to distilled/deionized water and bring volume to 1.0L. Mix thoroughly. Gently heat and bring to boiling. Dispense into tubes or flasks. Autoclave for 15 min at 15 psi pressure–121°C. Pour into sterile Petri dishes or leave in tubes. Allow tubes to cool in a slanted position.

Use: For the differentiation of enteric pathogens, especially members of the Enterobacteriaceae, and coliforms based on their ability to utilize citrate as a carbon source. Bacteria that can utilize citrate turn the medium pink-red.

Christensen Agar

Composition per liter:

Agar	15.0g
NaCl	5.0g
Sodium citrate	3.0g
KH_2PO_4	1.0g
Yeast extract	0.5g
Glucose	0.2g
L-Cysteine·HCl·H_2O	0.1g
Phenol Red	12.0mg

pH 6.9 ± 0.2 at 25°C

Source: This medium is available as a premixed powder from BD Diagnostic Systems.

Preparation of Medium: Add components to distilled/deionized water and bring volume to 1.0L. Mix thoroughly. Gently heat and bring to boiling. Dispense into tubes or flasks. Autoclave for 15 min at 15 psi pressure–121°C. Pour into sterile Petri dishes or leave in tubes. Allow tubes to cool in a slanted position.

Use: For the differentiation of enteric pathogens, especially members of the Enterobacteriaceae, and coliforms based on their ability to utilize citrate as a carbon source. Bacteria that can utilize citrate turn the medium pink-red.

Christensen Citrate Agar (BAM M39)

Composition per liter:

Agar	15.0g
NaCl	5.0g
Sodium citrate	3.0g
KH_2PO_4	1.0g
Yeast extract	0.5g
Ferric ammonium citrate	0.4g
L-Cysteine·HCl·H_2O	0.1g
$Na_2S_2O_5$	0.08g
Phenol Red	12.0mg

pH 6.9 ± 0.2 at 25°C

Preparation of Medium: Add components to distilled/deionized water and bring volume to 1.0L. Mix thoroughly. Gently heat and bring to boiling. Dispense into tubes or flasks. Autoclave for 15 min at 15 psi pressure–121°C. Pour into sterile Petri dishes or leave in tubes. Allow tubes to cool in a slanted position.

Use: For the differentiation of enteric pathogens, especially members of the Enterobacteriaceae, and coliforms based on their ability to utilize citrate as a carbon source. Bacteria that can utilize citrate turn the medium pink-red.

Christensen Citrate Agar, Modified (Citrate Agar)

Composition per liter:

Agar	12.0g
NaCl	5.0g
Sodium citrate	3.8g
KH_2PO_4	1.0g
Yeast extract	0.5g
Glucose	0.2g
L-Cysteine·HCl·H_2O	0.1g
Phenol Red	0.02g

pH 6.7 ± 0.2 at 25°C

Preparation of Medium: Add components to distilled/deionized water and bring volume to 1.0L. Mix thoroughly. Gently heat and bring to boiling. Dispense into tubes or flasks. Autoclave for 15 min at 15 psi pressure–121°C. Pour into sterile Petri dishes or leave in tubes. Allow tubes to cool in a slanted position.

Use: For the differentiation of enteric pathogens, especially members of the Enterobacteriaceae, and coliforms based on their ability to utilize citrate as a carbon source. Bacteria that can utilize citrate turn the medium pink-red.

Christensen Citrate Sulfide Medium

Composition per liter:

Agar	15.0g
NaCl	5.0g
Sodium citrate·$2H_2O$	3.0g
KH_2HPO_4	1.0g
Yeast extract	0.5g
Ferric citrate	0.2g
Ammonium citrate	0.2g
Glucose	0.2g
L-Cysteine·HCl·H_2O	0.1g
$Na_2S_2O_3 \cdot 5H_2O$	0.08g
Phenol Red	0.012g

pH 6.7± 0.2 at 25°C

Preparation of Medium: Add components to distilled/deionized water and bring volume to 1.0L. Mix thoroughly. Gently heat and bring to boiling. Dispense into tubes or flasks. Autoclave for 15 min at 15 psi pressure–121°C. Pour into sterile Petri dishes or leave in tubes. Allow tubes to cool in a slanted position.

Use: For the differentiation of enteric pathogens, especially members of the Enterobacteriaceae, and coliforms based on their ability to utilize citrate as a carbon source and production of H_2S. Bacteria that can utilize citrate turn the medium pink-red. H_2S production appears as a blackening of the butt of the tube.

Christensen Citrate Sulfite Agar

Composition per liter:

Agar	14.0g
NaCl	5.0g
Sodium citrate·$2H_2O$	3.0g
KH_2HPO_4	1.0g
Yeast extract	0.5g
Ferric ammonium citrate	0.4g
Ammonium citrate	0.2g
Glucose	0.2g
L-Cysteine·HCl·H_2O	0.1g
$Na_2S_2O_3 \cdot 5H_2O$	0.08g
Phenol Red	0.012g

pH 6.7± 0.2 at 25°C

Preparation of Medium: Add components to distilled/deionized water and bring volume to 1.0L. Mix thoroughly. Gently heat and bring to boiling. Dispense into tubes or flasks. Autoclave for 15 min at 15 psi pressure–121°C. Pour into sterile Petri dishes or leave in tubes. Allow tubes to cool in a slanted position.

Use: For the differentiation of enteric pathogens, especially members of the Enterobacteriaceae, and coliforms based on their ability to utilize citrate as a carbon source and production of H_2S. Bacteria that can utilize citrate turn the medium pink-red. H_2S production appears as a blackening of the butt of the tube.

Christensen's Urea Agar

Composition per liter:

Agar	15.0g
NaCl	5.0g
KH_2PO_4	2.0g
Peptone	1.0g
Glucose	1.0g
Phenol Red	0.012g
Urea solution	100.0mL

pH 6.8 ± 0.1 at 25°C

Urea Solution:

Composition per 100.0mL:

Urea	20.0g

Preparation of Urea: Add urea to 100.0mL of distilled/deionized water. Mix thoroughly. Filter sterilize.

Preparation of Medium: Add components, except urea solution, to distilled/deionized water and bring volume to 900.0mL. Mix thoroughly. Gently heat and bring to boiling. Autoclave for 15 min at 15 psi pressure–121°C. Cool to 50–55°C. Aseptically add 100.0mL of sterile urea solution. Mix thoroughly. Pour into Petri dishes or distribute into sterile tubes. Allow tubes to solidify in a slanted position.

Use: For the differentiation of a variety of microorganisms, especially members of the Enterobacteriaceae, aerobic actinomycetes, streptococci, and nonfermenting Gram-negative bacteria, on the basis of urease production.

Christensen's Urea Agar with Sodium Chloride (BAM M40)

Composition per liter:

NaCl	20.0g
Agar	15.0g
KH_2PO_4	2.0g
Peptone	1.0g
Glucose	1.0g
Phenol Red	0.012g
Urea solution	100.0mL

pH 6.8 ± 0.1 at 25°C

Urea Solution:

Composition per 100.0mL:

Urea	20.0g

Preparation of Urea: Add urea to 100.0mL of distilled/deionized water. Mix thoroughly. Filter sterilize.

Preparation of Medium: Add components, except urea solution, to distilled/deionized water and bring volume to 900.0mL. Mix thoroughly. Gently heat and bring to boiling. Autoclave for 15 min at 15 psi pressure–121°C. Cool to 50–55°C. Aseptically add 100.0mL of sterile urea solution. Mix thoroughly. Pour into Petri dishes or distribute into sterile tubes. Allow tubes to solidify in a slanted position.

Use: For the differentiation of halophilic *Vibrio* spp. on the basis of urease production.

Christensen Urea Agar Base
See: **Urea Agar**

Christensen Urea Broth
See: **Urea Broth**

Christopher's Semisolid *Brucella* Medium Base

Composition per liter:

Casein enzymic hydrolysate	10.0g
Peptic digest of animal tissue	10.0g
NaCl	5.0g
Yeast extract	2.0g
Agar	1.5g
Glucose	1.0g
Sodium pyruvate	0.5g
$NaHSO_3$	0.1g

pH 7.0 ± 0.2 at 25°C

Preparation of Medium: Add components to distilled/deionized water and bring volume to 1.0L. Mix thoroughly. Gently heat and bring to boiling. Distribute into tubes or flasks. Autoclave for 15 min at 15 psi pressure–121°C.

Use: For the selective enrichment of *Campylobacter* species from food.

CHROMagar™ *Candida*

Composition per liter:

Glucose	20.0g
Agar	15.0g
Peptone	10.0g
Chromogenic mix	2.0g
Chloramphenicol	0.5g

Source: CHROMagar *Candida* is available from CHROMagar Microbiology. Prepared medium is also available from BD Diagnostic Systems.

Preparation of Medium: Add components to distilled/deionized water and bring volume to 1.0L. Mix thoroughly. Gently heat in a boiling water bath or steam bath. Shake periodically during heating to dissolve components. Heat long enough with shaking every 5 min to ensure complete dissolution. Do not overheat. Cool to 45–50°C. Pour into sterile Petri dishes.

Use: For the differentiation of *Candida* spp. Specific *Candida* spp. give characteristic color reactions, e.g., *Candida albicans* produce distinctive green colonies and *Candida tropicalis* produce distinctive dark blue-gray colonies.

CHROMagar™ *E. coli*

Composition per liter:
Proprietary

Source: CHROMagar *E. coli* is available from CHROMagar Microbiology.

Preparation of Medium: Add components to distilled/deionized water and bring volume to 1.0L. Mix thoroughly. Gently heat in a boiling water bath or steam bath. Shake periodically during heating to dissolve components. Heat long enough with shaking every 5 min to ensure complete dissolution. Do not overheat. Adding tellurite can increase specificity. Cool to 45–50°C. Pour into sterile Petri dishes.

Use: For the differentiation and presumptive identification of *Escherichia coli* which forms blue colonies.

CHROMagar™ ECC

Composition per liter:
Proprietary

Source: CHROMagar EEC is available from CHROMagar Microbiology.

Preparation of Medium: Add components to distilled/deionized water and bring volume to 1.0L. Mix thoroughly. Gently heat in a boiling water bath or steam bath. Shake periodically during heating to dissolve components. Heat long enough with shaking every 5 min to ensure complete dissolution. Do not overheat. Adding tellurite can increase specificity. Cool to 45–50°C. Pour into sterile Petri dishes.

Use: For the differentiation and presumptive identification of *Escherichia coli* and other coliform bacteria which form red colonies.

CHROMagar™ *Listeria*

Composition per liter:
Proprietary

Source: CHROMagar *Listeria* is available from CHROMagar Microbiology.

Preparation of Medium: Add components to distilled/deionized water and bring volume to 1.0L. Mix thoroughly. Gently heat in a boiling water bath or steam bath. Shake periodically during heating to dissolve components. Heat long enough with shaking every 5 min to ensure complete dissolution. Do not overheat. Adding tellurite can increase specificity. Cool to 45–50°C. Pour into sterile Petri dishes.

Use: For the differentiation and presumptive identification of *Listeria monocytogenes* which form blue colonies surrounded by white halos.

CHROMagar™ *Malassezia*

Composition per liter:
Proprietary

Source: CHROMagar *Malassezia* is available from CHROMagar Microbiology.

Preparation of Medium: Add components to distilled/deionized water and bring volume to 1.0L. Mix thoroughly. Gently heat in a boiling water bath or steam bath. Shake periodically during heating to dissolve components. Heat long enough with shaking every 5 min to ensure complete dissolution. Do not overheat. Cool to 45–50°C. Aseptically add glycerol (1.0mL per liter) and Tween 60 (0.5mL per liter). Mix thoroughly. Pour into sterile Petri dishes.

Use: For the differentiation and presumptive identification of *Malassezia* species.

CHROMagar™ MRSA

Composition per liter:

Chromopeptone	40.0g
NaCl	25.0g
Agar	14.0g
Chromogenic mix	0.5g
Inhibitory agents	0.07g
Cefoxitin	6.0mg

Source: CHROMagar MRSA is available from CHROMagar Microbiology. Prepared medium is also available from BD Diagnostic Systems.

Preparation of Medium: Add components to distilled/deionized water and bring volume to 1.0L. Mix thoroughly. Gently heat in a boiling water bath or steam bath. Shake periodically during heating to dissolve components. Heat long enough with shaking every 5 min to ensure complete dissolution. Do not overheat. Adding tellurite can increase specificity. Cool to 45–50°C. Pour into sterile Petri dishes.

Use: For the qualitative direct detection of nasal colonization by methicillin resistant *Staphylococcus aureus* (MRSA) to aid in the prevention and control of MRSA infections in healthcare settings.

CHROMagar™ O157

Composition per liter:

Proprietary

Source: CHROMagar O157 is available from CHROMagar Microbiology. Prepared medium is also available from BD Diagnostic Systems.

Preparation of Medium: Add components to distilled/deionized water and bring volume to 1.0L. Mix thoroughly. Gently heat in a boiling water bath or steam bath. Shake periodically during heating to dissolve components. Heat long enough with shaking every 5 min to ensure complete dissolution. Do not overheat. Adding tellurite can increase specificity. Cool to 45–50°C. Pour into sterile Petri dishes.

Use: For the differentiation and presumptive identification of *Escherichia coli* O157.

CHROMagar™ Orientation

Composition per liter:

Peptone	16.0g
Meat extract	16.0g
Yeast extract	16.0g
Agar	15.0g
Chromogenic mix	2.0g

pH 7.0 ± 0.2 at 25°C

Source: CHROMagar Orientation is available from CHROMagar Microbiology. Prepared medium is also available from BD Diagnostic Systems.

Preparation of Medium: Add components to distilled/deionized water and bring volume to 1.0L. Mix thoroughly. Gently heat in a boiling water bath or steam bath. Shake periodically during heating to dissolve components. Heat long enough with shaking every 5 min to ensure complete dissolution. Do not overheat. Cool to 45–50°C. Pour into sterile Petri dishes.

Use: For the differentiation and presumptive identification of Gram-negative bacteria and *Enterococcus* spp. For use in identifying urinary tract pathogens. Isolates produce characteristic diagnostic colors, e.g., *Escherichia coli* produces pinto red colonies.

CHROMagar™ *Pseudomonas*

Composition per liter:

Proprietary

Source: CHROMagar *Pseudomonas* is available from CHROMagar Microbiology.

Preparation of Medium: Add components to distilled/deionized water and bring volume to 1.0L. Mix thoroughly. Gently heat in a boiling water bath or steam bath. Shake periodically during heating to dissolve components. Heat long enough with shaking every 5 min to ensure complete dissolution. Do not overheat. Cool to 45–50°C. Pour into sterile Petri dishes.

Use: For the detection of *Pseudomonas*. For the simultaneous detection and enumeration of *Pseudomonas aeruginosa* with markedly different coloring (blue colonies).

CHROMagar™ *Salmonella*

Composition per liter:

Proprietary

Source: CHROMagar *Salmonella* is available from CHROMagar Microbiology. Prepared medium is also available from BD Diagnostic Systems.

Preparation of Medium: Add components to distilled/deionized water and bring volume to 1.0L. Mix thoroughly. Gently heat in a boiling water bath or steam bath. Shake periodically during heating to dissolve components. Heat long enough with shaking every 5 min to ensure complete dissolution. Do not overheat. Cool to 45–50°C. Pour into sterile Petri dishes.

Use: For the differentiation and presumptive identification of *Salmonella* spp.

CHROMagar™ *Salmonella* Plus

Composition per liter:

Proprietary

Source: CHROMagar *Salmonella* Plus is available from CHROMagar Microbiology.

Preparation of Medium: Add components to distilled/deionized water and bring volume to 1.0L. Mix thoroughly. Gently heat in a boiling water bath or steam bath. Shake periodically during heating to dissolve components. Heat long enough with shaking every 5 min to ensure complete dissolution. Do not overheat. Cool to 45–50°C. Pour into sterile Petri dishes.

Use: For the isolation of specimens, allowing direct detection of *Salmonella* including *S.Typhi, S.Paratyphi* and lactose positive *Salmonella* by colony color according to ISO 6579:2003 norm.

CHROMagar™ *Staph. aureus*

Composition per liter:

Proprietary

Source: CHROMagar *Staph. aureus* is available from CHROMagar Microbiology. Prepared medium is also available from BD Diagnostic Systems.

Preparation of Medium: Add components to distilled/deionized water and bring volume to 1.0L. Mix thoroughly. Gently heat in a boiling water bath or steam bath. Shake periodically during heating to dissolve components. Heat long enough with shaking every 5 min to ensure complete dissolution. Do not overheat. Adding an antibiotic such as tobramycin or methicillin can be used to identify resistant strains. Cool to 45–50°C. Pour into sterile Petri dishes.

Use: For the differentiation and presumptive identification of *Staphylococcus aureus*.

CHROMagar™ *Strep*B

Composition per liter:

Proprietary

Source: CHROMagar *Strep*B is available from CHROMagar Microbiology.

Preparation of Medium: Add components to distilled/deionized water and bring volume to 1.0L. Mix thoroughly. Gently heat in a boiling water bath or steam bath. Shake periodically during heating to dissolve components. Heat long enough with shaking every 5 min to ensure complete dissolution. Do not overheat. Cool to 45–50°C. Pour into sterile Petri dishes.

Use: For the differentiation and presumptive identification of *Streptococcus* B (*Streptococcus agalactiae*) based upon color formation. *Streptococcus* B forms mauve to pink colonies.

CHROMagar™ *Vibrio*

Composition per liter:

Proprietary

Source: CHROMagar *Vibrio* is available from CHROMagar Microbiology.

Preparation of Medium: Add components to distilled/deionized water and bring volume to 1.0L. Mix thoroughly. Gently heat in a boiling water bath or steam bath. Shake periodically during heating to dissolve components. Heat long enough with shaking every 5 min to ensure complete dissolution. Do not overheat. Adding tellurite can increase specificity. Cool to 45–50°C. Pour into sterile Petri dishes.

Use: For the differentiation and presumptive identification of *Vibrio parahaemolyiticus* which form mauve colonies; *Vibrio cholerae* form turquoise blue colonies and *Vibrio alginolyitcus* colonies are colorless.

CHROMagar™ VRE

Composition per liter:

Proprietary

Source: CHROMagar VRE is available from CHROMagar Microbiology.

Preparation of Medium: Add components to distilled/deionized water and bring volume to 1.0L. Mix thoroughly. Gently heat in a boiling water bath or steam bath. Shake periodically during heating to dissolve components. Heat long enough with shaking every 5 min to ensure complete dissolution. Do not overheat. Cool to 45–50°C. Pour into sterile Petri dishes.

Use: For the differentiation and presumptive identification of vancomycin resistant *Enterococcus* (*Enterococcus faecalis*/*E. facecium*). Vancomycin resistant *Enterococcus* strains form rose to mauve colonies.

Chromatiaceae Medium 1

Composition per 4990.0mL:

Solution 1	4.0L
O_2-free water	860.0mL
$NaHCO_3$ solution	100.0mL
$Na_2S·9H_2O$ solution	20.0mL
Trace elements solution	5.0mL
Vitamin B_{12} solution	5.0mL

pH 7.3 ± 0.2 at 25°C

Solution 1:

Composition per 4.0L:

$MgSO_4·7H_2O$	2.5g
KCl	1.7g
KH_2PO_4	1.7g
NH_4Cl	1.7g
$CaCl_2·2H_2O$	1.25g

Preparation of Solution 1: Add components to distilled/deionized water and bring volume to 4.0L. Mix thoroughly. Autoclave for 45 min at 15 psi pressure–121°C. Cool to 25°C under 100% N_2. Saturate with CO_2 by stirring under 100% CO_2 for 30 min.

O_2-Free Water:

Composition per 860.0mL:

H_2O	860.0mL

Preparation of O_2-Free Water: Autoclave H_2O for 15 min at 15 psi pressure–121°C. Cool to 25°C under 100% N_2.

$NaHCO_3$ Solution:

Composition per 100.0mL:

$NaHCO_3$	7.5g

Preparation of $NaHCO_3$ Solution: Add the $NaHCO_3$ to distilled/deionized water and bring volume to 100.0mL. Mix thoroughly. Gas with 100% CO_2 for 20 min. Filter sterilize with positive CO_2 pressure.

$Na_2S·9H_2O$ Solution:

Composition per 100.0mL:

$Na_2S·9H_2O$	10.0g

Preparation of $Na_2S·9H_2O$ Solution: Add $Na_2S·9H_2O$ to distilled/deionized water. Mix thoroughly. Gas with 100% N_2 for 15 min in a screw-capped bottle. Tightly close cap. Autoclave for 15 min at 15 psi pressure–121°C. Cool to 25°C.

Trace Elements Solution:

Composition per liter:

$FeCl_2·4H_2O$	1.5g
$CoCl_2·6H_2O$	0.19g
$MnCl_2·4H_2O$	0.1g
$ZnCl_2$	0.07g
H_3BO_3	0.06g
$NaMoO_4·2H_2O$	0.04g
$CuCl_2·2H_2O$	0.02g
$NiCl_2·6H_20$	0.02g
HCl (25% solution)	6.5mL

Preparation of Trace Elements Solution: Add components to distilled/deionized water and bring volume to 1.0L. Mix thoroughly. Autoclave for 15 min at 15 psi pressure–121°C. Cool to 25°C.

Vitamin B_{12} Solution:

Composition per 100.0mL:

Vitamin B_{12}	2.0mg

Preparation of Vitamin B_{12} Solution: Add components to distilled/deionized water and bring volume to 100.0mL. Mix thoroughly. Filter sterilize.

Preparation of Medium: To 4.0L of sterile, CO_2-saturated solution 1, aseptically add the remaining components. Mix thoroughly. Adjust pH to 7.3. Aseptically distribute into sterile 100.0mL bottles using positive pressure of 95% N_2 + 5% CO_2. Completely fill bottles with medium except for a pea-sized air bubble.

Use: For the isolation and cultivation of members of the Chlorobiaceae.

Chromatiaceae Medium 2

Composition per 1051.0mL:

Solution 1	950.0mL
$Na_2S·9H_2O$ solution	60.0mL
$NaHCO_3$ solution	40.0mL
Vitamin B_{12} solution	1.0mL

pH 7.3 ± 0.2 at 25°C

Solution 1:

Composition per 950.0mL:

KH_2PO_4	1.0g
NH_4Cl	0.5g

$MgSO_4 \cdot 7H_2O$ 0.4g
$CaCl_2 \cdot 2H_2O$ 0.05g
Trace elements solution SL-8 .... 1.0mL

Preparation of Solution 1: Add components to distilled/deionized water and bring volume to 950.0mL. Mix thoroughly. Autoclave for 15 min at 15 psi pressure–121°C. Cool to 45°–50°C.

Trace Elements Solution SL-8:
Composition per liter:
Disodium EDTA 5.2g
$FeCl_2 \cdot 4H_2O$ 1.5g
$CoCl_2 \cdot 6H_2O$ 0.19g
$MnCl_2 \cdot 4H_2O$ 0.1g
$ZnCl_2$ 0.07g
H_3BO_3 .... 0.06g
$NaMoO_4 \cdot 2H_2O$ 0.04g
$CuCl_2 \cdot 2H_2O$ 0.02g
$NiCl_2 \cdot 6H_20$ 0.02g

Preparation of Trace Elements Solution SL-8: Add components to distilled/deionized water and bring volume to 1.0L. Mix thoroughly.

$Na_2S \cdot 9H_2O$ Solution:
Composition per 100.0mL:
$Na_2S \cdot 9H_2O$ 5.0g

Preparation of $Na_2S \cdot 9H_2O$ Solution: Add $Na_2S \cdot 9H_2O$ to distilled/deionized water and bring volume to 100.0mL. Autoclave for 15 min at 15 psi pressure–121°C. Cool to 45°–50°C.

$NaHCO_3$ Solution:
Composition per 100.0mL:
$NaHCO_3$ 5.0g

Preparation of $NaHCO_3$ Solution: Add $NaHCO_3$ to distilled/deionized water and bring volume to 100.0mL. Mix thoroughly. Filter sterilize.

Vitamin B_{12} Solution:
Composition per 100.0mL:
Vitamin B_{12} 2.0mg

Preparation of Vitamin B_{12} Solution: Add vitamin B_{12} to distilled/deionized water and bring volume to 100.0mL. Mix thoroughly. Filter sterilize.

Preparation of Medium: To 950.0mL of cooled, sterile solution 1, aseptically add 60.0mL of sterile $Na_2S \cdot 9H_2O$ solution, 40.0mL of sterile $NaHCO_3$ solution, and 1.0mL of sterile vitamin B_{12} solution. Mix thoroughly. Adjust pH to 7.3 with sterile H_2SO_4 or Na_2CO_3. Aseptically distribute into sterile 50.0mL or 100.0mL bottles with metal screwcaps and rubber seals. Completely fill bottles with medium except for a pea-sized air bubble.

Use: For the isolation and cultivation of freshwater and soil members of the Chromatiaceae.

Chromatium Medium (ATCC Medium 37)

Composition per 127.0mL:
Solution 1 .... 76.2mL
Solution 2 + Solution 3 .... 44.8mL
Solution 4 .... 6.0mL

Solution 1:
Composition per 2.5 L:
$CaCl_2$ 2.0g

Preparation of Solution 1: Add $CaCl_2$ to distilled/deionized water and bring volume to 2.5L. Distribute in 80.0mL volumes into 127.0mL screw-capped bottles. Autoclave for 15 min at 15 psi pressure–121°C.

Solution 2:
Composition per 100.0mL:
Sodium ascorbate 2.4g
KC1 .... 1.0g
KH_2PO_4 .... 1.0g
$MgCl_2 \cdot 6H_2O$ 0.8g
NH_4Cl 0.8g
Heavy metal solution 50.0mL
Vitamin solution 15.0mL
Vitamin B_{12} solution 3.0mL

Preparation of Solution 2: Add components to distilled/deionized water and bring volume to 100.0mL. Mix thoroughly.

Heavy Metal Solution:
Composition per liter:
Ethylenediamine tetraacetate (EDTA) 1.5g
$FeSO_4 \cdot 7H_2O$ 0.2g
$ZnSO_4 \cdot 7H_2O$ 0.1g
$MnCl_2 \cdot 4H_2O$ 0.02g
Modified Hoagland trace elements solution 6.0mL

Preparation of Heavy Metal Solution: Dissolve EDTA in approximately 800.0mL of distilled/deionized water. Add remaining components and bring volume to 1.0L. Mix thoroughly.

Modified Hoagland Trace Elements Solution:
Composition per 3.6L:
H_3BO_3 .... 11.0g
$MnCl_2 \cdot 4H_2O$ 7.0g
$AlCl_3$ 1.0g
$CoCl_2$ 1.0g
$CuCl_2$ 1.0g
KI 1.0g
$NiCl_2$ 1.0g
$ZnCl_2$ 1.0g
$BaCl_2$ 0.5g
KBr 0.5g
LiCl 0.5g
Na_2MoO_4 .... 0.5g
$SeCl_4$ 0.5g
$SnCl_2 \cdot 2H_2O$ 0.5g
$NaVO_3 \cdot H_2O$ 0.1g

Preparation of Modified Hoagland Trace Elements Solution: Prepare each component as a separate solution. Dissolve each salt in approximately 100.0mL of distilled/deionized water. Adjust the pH of each solution to below 7.0. Combine all the salt solutions and bring the volume to 3.6L with distilled/deionized water. Adjust the pH to 3–4. A yellow precipitate may form after mixing. After a few days, it will turn into a fine white precipitate. Mix the solution thoroughly before using.

Vitamin B_{12} Solution:
Composition per 100.0mL:
Vitamin B_{12} (cyanocobalamin) 2.0mg

Preparation of Vitamin B_{12} Solution: Add vitamin B_{12} to distilled/deionized water and bring volume to 100.0mL. Mix thoroughly.

Vitamin Solution:
Composition per 100.0mL:
Pyridoxamine·2HCl 5.0mg

Nicotinic acid	2.0mg
Thiamine	1.0mg
Pantothenic acid	0.5mg
Biotin	0.2mg
p-Aminobenzoic acid	0.1mg

Preparation of Vitamin Solution: Add components to distilled/deionized water and bring volume to 100.0mL. Mix thoroughly.

Solution 3:

Composition per 900.0mL:

$NaHCO_3$	4.5g

Preparation of Solution 3: Add $NaHCO_3$ to distilled/deionized water and bring volume to 900.0mL. Mix thoroughly. Bubble 100% CO_2 through the solution for 30 min. After CO_2 saturation of solution 3, add solution 2 and immediately filter the mixture through a Seitz filter (or a Millipore) using positive CO_2 pressure to push the liquid through.

Solution 4:

Composition per 200.0mL:

$Na_2S·9H_2O$	3.0g

Preparation of Solution 4: Add $Na_2S·9H_2O$ to distilled/deionized water and bring volume to 200.0mL. Add a magnetic stir bar to the flask. Autoclave for 15 min at 15 psi pressure–121°C. On a magnetic stirrer, slowly add 2.0mL of sterile 2*M* H_2SO_4. This partially neutralizes the solution. The solution should turn yellow. H_2S gas will be liberated. Neutralization and distribution of the solution should be done as rapidly as possible under adequate ventilation.

Preparation of Medium: To the 80.0mL of sterile solution 1 in screw-capped bottles, add combined solutions 2 and 3 immediately after filtration and fill bottles to capacity. Mix thoroughly. Aseptically remove 6.0mL of the medium from the bottles and replace it with 6.0mL of neutralized solution 4. Let stand for 24 hr. The medium should form a fine white precipitate before using. To inoculate, remove 6.0mL of the completed medium from the bottles and replace it with 6.0mL of inoculum.

Use: For the cultivation and maintenance of *Chromatium tepidum.*

Chromatium Medium (ATCC Medium 1449)

Composition per liter:

KH_2PO_4	0.5g
NH_4Cl	0.4g
$MgSO_4·7H_2O$	0.2g
$CaCl_2·2H_2O$	0.05g
Disodium EDTA	0.01g
Trace elements	1.0mL
$NaHCO_3$ solution	50.0mL
$Na_2S·9H_2O$ solution	50.0mL
Sodium pyruvate solution	50.0mL

pH 7.0 ± 0.2 at 25°C

Trace Elements:

Composition per liter:

Disodium EDTA	5.2g
$FeCl_2·4H_2O$	1.5g
$CoCl_2·6H_2O$	0.19g
$Na_2MoO_4·2H_2O$	0.188g
$MnCl_2·4H_2O$	0.1g
$ZnCl_2$	0.07g
$VOSO_4·2H_2O$	0.03g
$NiCl_2·6H_2O$	0.025g
H_3BO_3	6.0mg
$CuCl_2·2H_2O$	2.0mg
Na_2SeO_3	2.0mg

Preparation of Trace Elements: Add components to distilled/deionized water and bring volume to 1.0L. Mix thoroughly.

$NaHCO_3$ Solution:

Composition per 50.0mL:

$NaHCO_3$	2.0g

Preparation of $NaHCO_3$ Solution: Add $NaHCO_3$ to distilled/deionized water and bring volume to 50.0mL. Filter sterilize. Use freshly prepared solution.

$Na_2S·9H_2O$ Solution:

Composition per 50.0mL:

$Na_2S·9H_2O$	1.0g

Preparation of $Na_2S·9H_2O$ Solution: Add $Na_2S·9H_2O$ to distilled/deionized water and bring volume to 50.0mL. Autoclave for 15 min at 15 psi pressure–121°C. Use freshly prepared solution.

Sodium Pyruvate Solution:

Composition per 50.0mL:

Sodium pyruvate	0.5g

Preparation of Sodium Pyruvate Solution: Add $NaHCO_3$ to distilled/deionized water and bring volume to 50.0mL. Filter sterilize. Use freshly prepared solution. Sodium acetate may be substituted for the sodium pyruvate.

Preparation of Medium: Add components, except $NaHCO_3$ solution, $Na_2S·9H_2O$ solution, and sodium pyruvate solution, to distilled deionized water and bring volume to 850.0mL. Autoclave for 15 min at 15 psi pressure–121°C. Cool to room temperature. Add the sterile $NaHCO_3$ solution, the sterile $Na_2S·9H_2O$ solution, and the sterile sodium pyruvate solution, in that order. Adjust the pH to 7.0. Distribute into screw-capped tubes or flasks. Fill to capacity.

Use: For the cultivation and maintenance of *Chromatium* species.

Chromatium Medium (ATCC Medium 2010)

Composition per 127.0mL:

Solution 1	76.2mL
Solution 2 + Solution 3	44.8mL
Solution 4	6.0mL

Solution 1:

Composition per 2.5L:

NaCl	125.0g
$CaCl_2$	2.0g

Preparation of Solution 1: Add NaCl and $CaCl_2$ to distilled/deionized water and bring volume to 2.5L. Distribute in 76.2mL volumes into 127.0mL screw-capped bottles. Autoclave for 15 min at 15 psi pressure–121°C.

Solution 2:

Composition per 100.0mL:

Sodium ascorbate	2.4g
KCl	1.0g
KH_2PO_4	1.0g
$MgCl_2·6H_2O$	0.8g
NH_4Cl	0.8g
Heavy metal solution	50.0mL

Vitamin solution	15.0mL
Vitamin B_{12} solution	3.0mL

Preparation of Solution 2: Add components to distilled/deionized water and bring volume to 100.0mL. Mix thoroughly.

Solution 3:

Composition per 900.0mL:

$NaHCO_3$	4.5g

Preparation of Solution 3: Add $NaHCO_3$ to distilled/deionized water and bring volume to 900.0mL. Mix thoroughly. Bubble 100% CO_2 through the solution for 30 min. After CO_2 saturation of solution 3, add solution 2 and immediately filter the mixture through a Seitz filter (or a Millipore) using positive CO_2 pressure to push the liquid through.

Solution 4:

Composition per 200.0mL:

$Na_2S{\cdot}9H_2O$	3.0g

Preparation of Solution 4: Add $Na_2S{\cdot}9H_2O$ to distilled/deionized water and bring volume to 200.0mL. Add a magnetic stir bar to the flask. Autoclave for 15 min at 15 psi pressure–121°C. On a magnetic stirrer, slowly add 2.0mL of sterile 2*M* H_2SO_4. This partially neutralizes the solution. The solution should turn yellow. H_2S gas will be liberated. Neutralization and distribution of the solution should be done as rapidly as possible under adequate ventilation.

Heavy Metal Solution:

Composition per liter:

Ethylenediamine tetraacetate (EDTA)	1.5g
$FeSO_4{\cdot}7H_2O$	0.2g
$ZnSO_4{\cdot}7H_2O$	0.1g
$MnCl_2{\cdot}4H_2O$	0.02g
Modified Hoagland trace elements solution	6.0mL

Preparation of Heavy Metal Solution: Dissolve EDTA in approximately 800.0mL of distilled/deionized water. Add remaining components. Bring volume to 1.0L with distilled/deionized water. Mix thoroughly.

Vitamin B_{12} Solution:

Composition per 100.0mL:

Vitamin B_{12} (cyanocobalamin)	2.0mg

Preparation of Vitamin B_{12} Solution: Add vitamin B_{12} to distilled/deionized water and bring volume to 100.0mL. Mix thoroughly.

Vitamin Solution:

Composition per 100.0mL:

Pyridoxamine·2HCl	5.0mg
Nicotinic acid	2.0mg
Thiamine	1.0mg
Pantothenic acid	0.5mg
Biotin	0.2mg
p-Aminobenzoic acid	0.1mg

Preparation of Vitamin Solution: Add components to distilled/deionized water and bring volume to 100.0mL. Mix thoroughly.

Modified Hoagland Trace Elements Solution:

Composition per 3.6L:

H_3BO_3	11.0g
$MnCl_2{\cdot}4H_2O$	7.0g
$AlCl_3$	1.0g
$CoCl_2$	1.0g
$CuCl_2$	1.0g
KI	1.0g
$NiCl_2$	1.0g
$ZnCl_2$	1.0g
$BaCl_2$	0.5g
KBr	0.5g
LiCl	0.5g
Na_2MoO_4	0.5g
$SeCl_4$	0.5g
$SnCl_2{\cdot}2H_2O$	0.5g
$NaVO_3{\cdot}H_2O$	0.1g

Preparation of Modified Hoagland Trace Elements Solution: Prepare each component as a separate solution. Dissolve each salt in approximately 100.0mL of distilled/deionized water. Adjust the pH of each solution to below 7.0. Combine all the salt solutions and bring the volume to 3.6L with distilled/deionized water. Adjust the pH to 3–4. A yellow precipitate may form after mixing. After a few days, it will turn into a fine white precipitate. Mix the solution thoroughly before using.

Preparation of Medium: To the 80.0mL of sterile solution 1 in screw-capped bottles, add combined solutions 2 and 3 immediately after filtration and fill bottles to capacity. Mix thoroughly. Aseptically remove 6.0mL of the medium from the bottles and replace it with 6.0mL of neutralized solution 4. Let stand for 24 hr. The medium should form a fine white precipitate before using. To inoculate, remove 6.0mL of the completed medium from the bottles and replace it with 6.0mL of inoculum.

Use: For the cultivation of *Ectothiorhodospira mobilis* and *Ectothiorhodospira marismortui.*

Chromatium Medium with Sodium Chloride

Composition per 1051.0mL:

Solution 1	950.0mL
$Na_2S{\cdot}9H_2O$ solution	60.0mL
$NaHCO_3$ solution	40.0mL
Vitamin B_{12} solution	1.0mL

pH 3–4 at 25°C

Solution 1:

Composition per 950.0mL:

NaCl	30.0g
KH_2PO_4	1.0g
NH_4Cl	0.5g
$MgSO_4{\cdot}7H_2O$	0.4g
$CaCl_2{\cdot}2H_2O$	0.05g
Trace elements solution SL-8	1.0mL

Preparation of Solution 1: Add components to distilled/deionized water and bring volume to 950.0mL. Mix thoroughly. Autoclave for 15 min at 15 psi pressure–121°C. Cool to 45°–50°C.

Trace Elements Solution SL-8:

Composition per liter:

Disodium EDTA	5.2g
$FeCl_2{\cdot}4H_2O$	1.5g
$CoCl_2{\cdot}6H_2O$	0.19g
$MnCl_2{\cdot}4H_2O$	0.10g
$ZnCl_2$	0.07g
H_3BO_3	0.06g
$NaMoO_4{\cdot}2H_2O$	0.04g
$CuCl_2{\cdot}2H_2O$	0.02g
$NiCl_2{\cdot}6H_20$	0.02g

Preparation of Trace Elements Solution SL-8: Add components to distilled/deionized water and bring volume to 1.0L. Mix thoroughly.

$Na_2S·9H_2O$ Solution:

Composition per 100.0mL:

$Na_2S·9H_2O$	5.0g

Preparation of $Na_2S·9H_2O$ Solution: Add $Na_2S·9H_2O$ to distilled/deionized water and bring volume to 100.0mL. Autoclave for 15 min at 15 psi pressure–121°C. Cool to 45°–50°C.

$NaHCO_3$ Solution:

Composition per 100.0mL:

$NaHCO_3$	5.0g

Preparation of $NaHCO_3$ Solution: Add $NaHCO_3$ to distilled/deionized water and bring volume to 100.0mL. Mix thoroughly. Filter sterilize.

Vitamin B_{12} Solution:

Composition per 100.0mL:

Vitamin B_{12}	2.0mg

Preparation of Vitamin B_{12} Solution: Add vitamin B_{12} to distilled/deionized water and bring volume to 100.0mL. Mix thoroughly. Filter sterilize.

Preparation of Medium: To 950.0mL of cooled, sterile solution 1, aseptically add 60.0mL of sterile $Na_2S·9H_2O$ solution, 40.0mL of sterile $NaHCO_3$ solution, and 1.0mL of sterile vitamin B_{12} solution. Mix thoroughly. Adjust pH to 7.3 with sterile H_2SO_4 or Na_2CO_3. Aseptically distribute into sterile 50.0mL or 100.0mL bottles with metal screw-caps and rubber seals. Completely fill bottles with medium except for a pea-sized air bubble.

Use: For the cultivation of *Ectothiorhodospira marismortui* and *Ectothiorhodospira mobilis.*

Chromatium salexigens Medium

Composition per 4990.0mL:

Solution A	4000.0mL
Solution B	860.0mL
Solution C (Vitamin B_{12} solution)	5.0mL
Solution D (Trace elements solution SL-12B)	5.0mL
Solution E	100.0mL
Solution F	20.0mL

pH 7.3 ± 0.2 at 25°C

Solution A:

Composition per 4000.0mL:

NaCl	100.0g
$MgCl_2·6H_2O$	3.0g
$MgSO_4$	2.5g
KH_2PO_4	1.7g
NH_4Cl	1.7g
KCl	1.7g
$CaCl_2·2H_2O$	1.25g
Sodium acetate	0.5g
$Na_2S_2O_3$	0.5g

Preparation of Solution A: Add components to distilled/deionized water and bring volume to 4.0L. Mix thoroughly. Adjust pH to 6.0. Dispense into a 5.0L flask with four openings at the top (two openings are in a central silicon rubber stopper and two openings are gas-tight screw caps). Add a teflon-coated magnetic stir bar to the flask. Autoclave for 45 min at 15 psi pressure–121°C. Cool to room temperature under 100% N_2 at 0.05–0.1 atm pressure (use a manometer to measure low pressure).

Solution B:

Composition per 860.0mL:

Distilled/deionized water	860.0mL

Preparation of Solution B: Add 860.0mL of distilled/deionized water to a cotton-stoppered flask. Autoclave for 20 min at 15 psi–121°C. Cool to room temperature under 100% N_2 in an anaerobic jar.

Solution C (Vitamin B_{12} Solution):

Composition per 5.0mL:

Vitamin B_{12}	1.0mg

Preparation of Solution C (Vitamin B_{12} Solution): Add vitamin B_{12} to distilled/deionized water and bring volume to 5.0mL. Mix thoroughly. Filter sterilize.

Solution D (Trace Elements Solution SL-12B):

Composition per liter:

Disodium ethylendiamine-tetraacetate (Disodium EDTA)	3.0g
$FeSO_4·7H_2O$	1.1g
H_3BO_3	0.3g
$CoCl_2·6H_2O$	0.19g
$MnCl_2·2H_2O$	50.0mg
$ZnCl_2$	42.0mg
$NiCl_2·6H_2O$	24.0mg
$Na_2MoO_4·2H_2O$	18.0mg
$CuCl_2·2H_2O$	2.0mg

Preparation of Solution D (Trace Elements Solution SL-12B): Add components to distilled/deionized water and bring volume to 1.0L. Mix thoroughly. Autoclave for 15 min at 15 psi–121°C.

Solution E:

Composition per 100.0mL:

$NaHCO_3$	7.5g

Preparation of Solution E: Add $NaHCO_3$ to distilled/deionized water and bring volume to 100.0mL. Mix thoroughly. Sparge with 100% CO_2 until saturated. Filter sterilize under 100% CO_2 in a sterile, gas-tight 100.0mL screw-capped bottle.

Solution F:

Composition per 100.0mL:

$Na_2S·9H_2O$	10.0g

Preparation of Solution F: Add $Na_2S·9H_2O$ to distilled/deionized water and bring volume to 100.0mL. Mix thoroughly. Dispense into a screw-capped bottle. Sparge with 100% N_2 for 3–4 min. Autoclave for 15 min at 15 psi pressure–121°C.

Preparation of Medium: Saturate cooled solution A under 100% CO_2 at 0.05–0.1 atm pressure for 30 min with magnetic stirring. Add 860.0mL of solution B, 5.0mL of solution C, 5.0mL of solution D, 100.0mL of solution E, and 20.0mL of solution F through one of the screw-capped openings under 95% N_2 and 5% CO_2 with magnetic stirring. Adjust pH to 7.3 with sterile 2*M* HCl or sterile 2*M* Na_2CO_3 solution. Aseptically and anaerobically distribute the medium through the medium outlet tube into sterile 100.0mL bottles under 95% N_2 and 5% CO_2 at 0.05–0.1 atm pressure. Leave a small gas bubble in each bottle to accommodate pressure changes. After 24 hr, the iron in the medium will precipitate out of solution as black flocs.

Use: For the growth and maintenance of *Chromatium salexigens.*

Chromatium tepidum Medium

Composition per 1001.0mL:

NH_4Cl	400.0mg
NaCl	400.0mg
$MgSO_4 \cdot 7H_2O$	200.0mg
$CaCl_2 \cdot 2H_2O$	50.0mg
Disodium ethylendiamine-tetraacetate (Disodium EDTA)	10.0mg
Ammonium acetate (or sodium pyruvate)	0.5mg
KH_2PO_4	0.5mg
$NaHCO_3$ solution	20.0mL
$Na_2S \cdot 9H_2O$ solution	20.0mL
Trace elements solution	1.0mL

pH 7.0 ± 0.2 at 25°C

$NaHCO_3$ Solution:

Composition per 20.0mL:

$NaHCO_3$	2.0g

Preparation of $NaHCO_3$ Solution: Add $NaHCO_3$ to distilled/deionized water and bring volume to 20.0mL. Mix thoroughly. Autoclave for 15 min at 15 psi pressure–121°C.

$Na_2S \cdot 9H_2O$ Solution:

Composition per 20.0mL:

$Na_2S \cdot 9H_2O$	1.0g

Preparation of $Na_2S \cdot 9H_2O$ Solution: Add $Na_2S \cdot 9H_2O$ to distilled/deionized water and bring volume to 20.0mL. Mix thoroughly. Autoclave for 15 min at 15 psi pressure–121°C.

Trace Elements Solutions:

Composition per liter:

Disodium ethylendiamine-tetraacetate (Disodium EDTA)	5.2g
$FeCl_2 \cdot 4H_2O$	1.5g
$CoCl_2 \cdot 6H_2O$	190.0mg
$Na_2MoO_4 \cdot 2H_2O$	188.0mg
$MnCl_2 \cdot 4H_2O$	100.0mg
$ZnCl_2$	70.0mg
$VOSO_4 \cdot 2H_2O$	30.0mg
$NiCl_2 \cdot 6H_2O$	25.0mg
H_3BO_3	6.0mg
$CuCl_2 \cdot 2H_2O$	2.0mg
$Na_2WO_4 \cdot 2H_2O$	2.0mg

Preparation of Trace Elements Solutions: Add components to distilled/deionized water and bring volume to 1.0L. Mix thoroughly.

Preparation of Medium: Add components, except $NaHCO_3$ solution and $Na_2S \cdot 9H_2O$ solution, to distilled/deionized water and bring volume to 960.0mL. Mix thoroughly. Autoclave for 15 min at 15 psi pressure–121°C. Aseptically add 20.0mL of sterile $NaHCO_3$ solution and 20.0mL of sterile $Na_2S \cdot 9H_2O$ solution. Mix thoroughly. Adjust pH to 7.0. Distribute into sterile screw-capped tubes or bottles so that medium completely fills container.

Use: For the growth and maintenance of *Chromatium tepidum.*

Chromatium/Thiocapsa Medium

Composition per 1060.0mL:

KH_2PO_4	1.0g
NH_4Cl	1.0g
$MgCl_2 \cdot 6H_2O$	0.5g
Solution A	20.0mL
Solution B	20.0mL
Solution C	10.0mL
Solution D	10.0mL
Trace elements solution	1.0mL

pH 7.0 ± 0.2 at 25°C

Solution A:

Composition per 100.0mL:

$NaHCO_3$	10.0g

Preparation of Solution A: Add $NaHCO_3$ to distilled/deionized water and bring volume to 100.0mL. Mix thoroughly. Autoclave for 15 min at 15 psi pressure–121°C.

Solution B:

Composition per 100.0mL:

$Na_2S \cdot 9H_2O$	10.0g

Preparation of Solution B: Add $Na_2S \cdot 9H_2O$ to distilled/deionized water and bring volume to 100.0mL. Mix thoroughly. Autoclave for 15 min at 15 psi pressure–121°C.

Solution C:

Composition per 100.0mL:

$Na_2S_2O_3 \cdot 9H_2O$	10.0g

Preparation of Solution C: Add $Na_2S_2O_3 \cdot 9H_2O$ to distilled/deionized water and bring volume to 100.0mL. Mix thoroughly. Autoclave for 15 min at 15 psi pressure–121°C.

Solution D:

Composition per 100.0mL:

Sodium malate	10.0g

Preparation of Solution D: Add sodium malate to distilled/deionized water and bring volume to 100.0mL. Mix thoroughly. Autoclave for 15 min at 15 psi pressure–121°C.

Trace Elements Solution:

Composition per liter:

$FeCl_3 \cdot 6H_2O$	2.7g
H_3BO_3	0.1g
$ZnSO_4 \cdot 7H_2O$	0.1g
$Co(NO_3)_2 \cdot 6H_2O$	50.0mg
$CuSO_4 \cdot 5H_2O$	5.0mg
$MnCl_2 \cdot 6H_2O$	5.0mg

Preparation of Trace Elements Solution: Add components to distilled/deionized water and bring volume to 1.0L. Mix thoroughly.

Preparation of Medium: Add components, except solution A, solution B, solution C, and solution D, to distilled/deionized water and bring volume to 1.0L. Mix thoroughly. Bring pH to 7.0–7.2 with H_3PO_4. Distribute into tubes or flasks. Autoclave for 15 min at 15 psi pressure–121°C. Aseptically add 0.2mL of sterile solution A, 0.2mL of sterile solution B, 0.1mL of sterile solution C, and 0.1mL of sterile solution D for each 10.0mL of medium. Mix thoroughly. Use immediately.

Use: For the cultivation of *Chlorobium limnicola* and *Chromatium* species.

Chromobacterium Medium

Composition per liter:

NaCl	30.0g
$MgCl_2$	10.8g
$MgSO_4$	5.4g
Peptone	5.0g

$CaCl_2$	1.0g
KCl	0.7g

pH 7.0 ± 0.2 at 25°C

Preparation of Medium: Add components to distilled/deionized water and bring volume to 1.0L. Mix thoroughly. Distribute into tubes or flasks. Autoclave for 15 min at 15 psi pressure–121°C.

Use: For the cultivation and maintenance of *Chromobacterium* species and *Alteromonas luteoviolacea*.

Chromogenic *E. coli/*Coliform Medium

Composition per liter:

Chromogenic mix	20.3g
Agar	15.0g
Peptone	5.0g
NaCl	5.0g
Na_2HPO_4	3.5g
Yeast extract	3.0g
Lactose	2.5g
NaH_2PO_4	1.5g
Neutral Red	0.03g

pH 6.8 ± 0.2 at 25°C

Preparation of Medium: Add components to distilled/deionized water and bring volume to 1.0L. Mix thoroughly. Gently heat and bring to boiling. Dispense into tubes or flasks. Autoclave for 15 min at 15 psi pressure–121°C. Pour into sterile Petri dishes or leave in tubes.

Use: For the differentiation between *Escherichia coli* and other coliforms in cultures produced from food samples. Agar base uses two enzyme substrates to differentiate between *E. coli* and other coliforms. One chromogenic substrate is cleaved by the enzyme glucuronidase which is specific for *E. coli* and produced by approximately 97% of strains. The second chromogenic substrate is cleaved by galactosidase, an enzyme produced by the majority of coliforms. This results in purple *E. coli* colonies, as they are able to cleave both chromogenic substrates and pink coliform colonies as they are only able to cleave the galactosidase chromogen.

Chromogenic *Enterobacter sakazakii* Agar, DFI Formulation

Composition per liter:

Agar	15.0g
Tryptone	15.0g
Soya peptone	5.0g
NaCl	5.0g
Ferric ammonium citrate	1.0g
Sodium deoxycholate	1.0g
$Na_2S_2O_3$	1.0g
Chromogen	0.1g

pH 7.3 ± 0.2 at 25°C

Preparation of Medium: Add components to distilled/deionized water and bring volume to 1.0L. Mix thoroughly. Gently heat and bring to boiling. Dispense into tubes or flasks. Autoclave for 15 min at 15 psi pressure–121°C. Pour into sterile Petri dishes or leave in tubes. Allow tubes to cool in a slanted position.

Use: For the differentiation and enumeration of *Enterobacter sakazakii* from infant formula and other food samples. The enzyme α-glucosidase, present in *E. sakazakii*, hydrolyzes the substrate 5-bromo-4-chloro-3-indolyl-α,D-glucopyranoside, thus producing blue-green colonies on this pale yellow medium. *Proteus vulgaris* is also weakly α-glucosidase positive and could grow to give colonies of a similar color to *E. sakazakii*. However, on this medium, *Proteus* spp. grow as grey colonies: they produce hydrogen sulphide in the presence of ferric ions forming ferrous sulphide. Deoxycholate inhibits the growth of most Gram-positive organisms.

Chromogenic *Candida* Agar

Composition per liter:

Chromogenic mix	13.6g
Agar	13.6g
Peptone	4.0g
Selective supplement solution	10.0mL

pH 6.0 ± 0.2 at 25°C

Source: This medium is available from Oxoid Unipath.

Selective Supplement Solution:

Composition per 10.0mL:

Chloramphenicol	500.0mg

Preparation of Selective Supplement Solution: Add chloramphenicol to distilled/deionized water and bring volume to 10.0mL. Mix thoroughly. Filter sterilize.

Preparation of Medium: Add components to distilled/deionized water and bring volume to 1.0mL. Mix thoroughly. Gently heat while stirring and bring to boiling. Do not autoclave. Cool to 45°C. Pour into sterile Petri dishes.

Use: For the rapid isolation and identification of clinically important *Candida* species. The medium incorporates two chromogens that indicate the presence of the target enzymes: X-NAG (5-bromo-4-chloro-3-indolyl N acetyl ß-D-glucosaminide) detects the activity of hexosaminidase. BCIP (5-bromo-6-chloro-3-indolyl phosphate p-toluidine salt) detects alkaline phosphatase activity. An opaque agent has been incorporated into the formulation to improve the color definition on the agar. The broad-spectrum antibacterial agent chloramphenicol is added to the agar to inhibit bacterial growth on the plates.

Chromogenic *Listeria* Agar

Composition per liter:

Peptone	18.5g
LiCl	15.0g
Agar	14.0g
NaCl	9.5g
Yeast extract	4.0g
Maltose	4.0g
Sodium pyruvate	2.0g
X-glucoside chromogenic mix	0.2g
Differential lecithin solution	40.0mL
Selective supplement solution	20.0mL

pH 7.2 ± 0.2 at 25°C

Source: This medium is available from Oxoid Unipath.

Differential Lecithin Solution:

Composition per 40.0mL:

Lecithin	Proprietary

Preparation of Differential Lecithin Solution: Available as premixed solution.

Selective Supplement Solution:

Composition per 20.0mL:

Nalidixic acid	26.0mg
Polymyxin B	10.0mg
Amphotericin	10.0mg
Ceftazidime	6.0mg

Preparation of Selective Supplement Solution: Add components to distilled/deionized water and bring volume to 20.0mL. Mix thoroughly. Filter sterilize.

Preparation of Medium: Add components, except differential lecithin solution and selective supplement solution, to distilled/deionized water and bring volume to 940.0mL. Mix thoroughly. Gently heat while stirring and bring to boiling. Autoclave for 15 min at 15 psi pressure–121°C. Cool to 46°C. Aseptially add differential lecithin solution and selective supplement solution. Mix thoroughly. Pour into sterile Petri dishes.

Use: For the isolation, enumeration, and presumptive identification of *Listeria* spp. and *Listeria monocytogenes*. This selective medium contains the substrate lecithin, which permits differentiation of *L. monocytogenes* and *L. Ivanovii* from other *Listeria* species. Differential activity for all *Listeria* species is due to the addition of a chromogenic substrate.

Chromogenic *Listeria* Agar (ISO)

Composition per liter:

Enzymatic digest of animal tissue	18.0g
Agar	12.0g
LiCl	10.0g
Yeast extract	10.0g
Enzymatic digest of casein	6.0g
NaCl	5.0g
Na_2HPO_4, anhydrous	2.5g
Glucose	2.0g
Sodium pyruvate	2.0g
Magnesium glycerophosphage	1.0g
$MgSO_4$, anhydrous	0.5g
X-glucoside chromogenic mix	0.05g
L-α-phosphotidylinositol	40.0mL
Selective supplement solution	20.0mL

pH 7.2 ± 0.2 at 25°C

Source: This medium is available from Oxoid Unipath.

Selective Supplement Solution:

Composition per 20.0mL:

Nalidixic acid	20.0mg
Ceftazidime	20.0mg
Amphotericin	10.0mg
Polymyxin B	76,700 U

Preparation of Selective Supplement Solution: Add components to distilled/deionized water and bring volume to 20.0mL. Mix thoroughly. Filter sterilize.

Preparation of Medium: Add components, except L-α-phosphotidylinositol and selective supplement solution, to distilled/deionized water and bring volume to 940.0mL. Mix thoroughly. Gently heat while stirring and bring to boiling. Autoclave for 15 min at 15 psi pressure–121°C. Cool to 46°C. Aseptially add L-α-phosphotidylinositol and selective supplement solution. Mix thoroughly. Pour into sterile Petri dishes.

Use: For the isolation, enumeration, and presumptive identification of *Listeria* spp. and *Listeria monocytogenes*. This selective medium contains the substrate lecithin, which permits differentiation of *L. monocytogenes* and *L. Ivanovii* from other *Listeria* species.

Chromogenic *Listeria* Agar (ISO) Modified

Composition per liter:

Enzymatic digest of animal tissue	18.0g
Agar	12.0g
LiCl	10.0g
Yeast extract	10.0g
Enzymatic digest of casein	6.0g
NaCl	5.0g
Na_2HPO_4, anhydrous	2.5g
Glucose	2.0g
Sodium pyruvate	2.0g
Magnesium glycerophosphage	1.0g
$MgSO_4$, anhydrous	0.5g
X-glucoside chromogenic mix	0.05g
Differential lecithin solution	40.0mL
Selective supplement solution	20.0mL

pH 7.2 ± 0.2 at 25°C

Source: This medium is available from Oxoid Unipath.

Differential Lecithin Solution:

Composition per 40.0mL:

Lecithin	Proprietary

Preparation of Differential Lecithin Solution: Available as premixed solution.

Selective Supplement Solution:

Composition per 20.0mL:

Nalidixic acid	20.0mg
Ceftazidime	20.0mg
Amphotericin	10.0mg
Polymyxin B	76,700 U

Preparation of Selective Supplement Solution: Add components to distilled/deionized water and bring volume to 20.0mL. Mix thoroughly. Filter sterilize.

Preparation of Medium: Add components, except differential lecithin solution and selective supplement solution, to distilled/deionized water and bring volume to 940.0mL. Mix thoroughly. Gently heat while stirring and bring to boiling. Autoclave for 15 min at 15 psi pressure–121°C. Cool to 46°C. Aseptially add differential lecithin solution and selective supplement solution. Mix thoroughly. Pour into sterile Petri dishes.

Use: For the isolation, enumeration, and presumptive identification of *Listeria* spp. and *Listeria monocytogenes*. This selective medium contains the substrate lecithin, which permits differentiation of *L. monocytogenes* and *L. Ivanovii* from other *Listeria* species.

Chromogenic *Salmonella* Esterase Agar (CSE Agar)

Composition per liter:

Agar	12.0g
Lactose	14.65
Peptone	4.0g
Tryptone	4.0g
Tween™ 20	3.0g
Lab Lemco	3.0g
Na_3-citrate dihydrate	0.5g
L-cysteine	0.128g
Tris	0.06g
SLA-octonoate solution	50.0mL

Novobiocin solution 10.0mL
Ethyl 4-dimethylaminobenzoate solution 10.0mL

pH 7.0 ± 0.2 at 25°C

Novobiocin Solution:

Composition per 10.0mL:

Novobiocin 70.0mg

Preparation of Novobiocin Solution: Add novobiocin to distilled/deionized water and bring volume to 10.0mL. Mix thoroughly. Filter sterilize.

Ethyl 4-dimethylaminobenzoate Solution:

Composition per 10.0mL:

Ethyl 4-dimethylaminobenzoate 0.35g
Methanol 8.0mL

Preparation of Ethyl 4-dimethylaminobenzoate Solution: Add ethyl 4-dimethylaminobenzoate to 8.0mL methanol. Mix thoroughly. Bring volume to 10.0mL with distilled/deionized water. Mix thoroughly. Filter sterilize.

SLA-Octonoate Solution:

Composition per 50.0mL:

4-[2-(4-octanoyloxy-3,5-dimethoxyphenyl)-vinyl]-quinolinium-1-(propan-3-yl carboxylic acid) bromide (SLPA-octanoate; bromide form) 0.3223g

Preparation of SLA-Octonoate Solution: Add SLA-octonoate to distilled/deionized water and bring volume to 50.0mL. Mix thoroughly. Filter sterilize.

Preparation of Medium: Add components, except novobiocin solution, SLA-octonoate solution, and ethyl 4-dimethylaminobenzoate solution, to distilled/deionized water and bring volume to 920.0mL. Mix thoroughly. Gently heat and bring to boiling. Autoclave for 15 min at 15 psi pressure–121°C. Cool to 50°C. Aseptically add 10.0mL novobiocin solution, 50.0mL SLA-octonoate solution, and 10.0mL ethyl 4-dimethylaminobenzoate solution. Mix thoroughly. Pour into sterile Petri dishes.

Use: For the detection of *Salmonella* spp. in clinical specimens. For the differentiation of *Salmonella* spp.

Chromogenic Substrate Broth

Composition per liter:

NaCl 10.0g
HEPES (*N*-[2-hydroxyethyl] piperazine-*N*′-[2-ethane-sulfonic acid]) buffer 6.9g
$(NH_4)_2SO_4$ 5.0g
o-Nitrophenyl-β-D-galactopyranoside 0.5g
Solanium 0.5g
$MgSO_4$ 0.1g
4-Methylumbelliferyl-β-D-glucuronide 0.075g
$CaCl_2$ 0.05g
Na_2SO_3 .... 0.04g
Amphotericin B 1.0mg
$MnSO_4$ 0.5mg
$ZnSO_4$ 0.5mg

Preparation of Medium: Add components to distilled/deionized water and bring volume to 1.0L. Mix thoroughly. Distribute into tubes or flasks. Autoclave for 15 min at 15 psi pressure–121°C.

Use: For the detection of coliform bacteria based on their hydrolysis of chromogenic substrates by production of β-D-galactopyranosidase. Bacteria that produce β-D-galactopyranosidase turn the medium yellow.

Chromogenic Urinary Tract Infection (UTI) Medium

Composition per liter:

Chromogenic mix 26.3g
Peptone 15.0g
Agar 15.0g

pH 6.8 ± 0.2 at 25°C

Source: This medium is available as a premixed powder from Oxoid Unipath.

Preparation of Medium: Add components to distilled/deionized water and bring volume to 1.0L. Mix thoroughly. Gently heat while stirring and bring to boiling. Autoclave for 15 min at 15 psi pressure–121°C. Cool to 50°C. Pour into sterile Petri dishes.

Use: For the presumptive identification and differentiation of all the main microorganisms that cause urinary tract infections (UTIs). The medium contains two specific chromogenic substrates which are cleaved by enzymes produced by *Enterococcus* spp., *Escherichia coli,* and coliforms. In addition, it contains phenylalanine and tryptophan which provide an indication of tryptophan deaminase activity, indicating the presence of *Proteus* spp., *Morganella* spp., and *Providencia* spp. It is based on electrolyte deficient CLED Medium which provides a valuable non-inhibitory diagnostic agar for plate culture of other urinary organisms, while preventing the swarming of *Proteus* spp. One chromogen, X-Gluc, is targeted towards β-glucosidase, and allows the specific detection of enterococci through the formation of blue colonies. The other chromogen, Red-Gal, is cleaved by the enzyme β-galactosidase which is produced by *Escherichia coli*, resulting in pink colonies. Cleavage of both chromogens occurs in the presence of coliforms, resulting in purple colonies. The medium also contains tryptophan which acts as an indicator of tryptophan deaminase activity, resulting in colonies of *Proteus, Morganella,* and *Providencia* spp. appearing brown.

Chromogenic UTI Medium, Clear

Composition per liter:

Peptone 15.0g
Agar 15.0g
Chromogenic mix 13.0g

pH 7.0 ± 0.2 at 25°C

Source: This medium is available as a premixed powder from Oxoid Unipath.

Preparation of Medium: Add components to distilled/deionized water and bring volume to 1.0L. Mix thoroughly. Gently heat while stirring and bring to boiling. Autoclave for 15 min at 15 psi pressure–121°C. Cool to 50°C. Pour into sterile Petri dishes.

Use: For the presumptive identification and differentiation of all the main microorganisms that cause urinary tract infections (UTIs). This medium uses the same chromogenic substrates as the existing opaque Chromogenc UTI Medium but has a clear background to make multiple sample testing easier. The medium contains two specific chromogenic substrates which are cleaved by enzymes produced by *Enterococcus* spp., *E. coli,* and coliforms. In addition, it contains tryptophan which indicates tryptophan deaminase activity (TDA), indicating the presence of *Proteus* spp. It is based on Cystine Lactose Electrolyte Deficient (CLED) Medium which provides a valuable non-inhibitory diagnostic agar for plate culture of other urinary organisms, while preventing the swarming of *Proteus* spp. The chromogen, X-glucoside, is targeted towards ß-glucosidase enzyme activity, and allows the specific detection of enterococci through the formation of blue colonies. The other chromogen, Red-Galactoside, is cleaved by the enzyme ß-galactosidase which is produced by *E. coli,* resulting in pink colonies. Cleavage of both the chromogens by members of the coliform group results in purple colonies. The medium also contains tryptophan which acts as an indicator of tryptophan deaminase activity (TDA), resulting in halos around the colonies of *Proteus, Morganella,* and *Providencia* spp.

Chrysiogenes Medium (DSMZ Medium 818)

Composition per liter:

NaCl	1.2g
$NaHCO_3$	0.6g
$MgCl_2·6H_2O$	0.4g
KCl	0.3g
NH_4Cl	0.3g
Na_2SO_4	0.3g
KH_2PO_4	0.2g
$CaCl_2·2H_2O$	0.15g
Resazurin	0.5mg
Vitamin solution	10.0mL
Na-acetate solution	10.0mL
KNO_3 solution	10.0mL
Seven vitamin solution	1.0mL
Trace elements solution SL-12	1.0mL

pH 7.4–7.8 at 25°C

KNO_3 Solution:
Composition per 10.0mL:

KNO_3	0.5g

Preparation of KNO_3 Solution: Add KNO_3 to distilled/deionized water and bring volume to 10.0mL. Mix thoroughly. Sparge with 100% N_2. Filter sterilize.

Na-Acetate Solution:
Composition per 10.0mL:

Na-acetate	0.5g

Preparation of Na-Acetate Solution: Add Na-acetate to distilled/deionized water and bring volume to 10.0mL. Mix thoroughly. Sparge with 100% N_2. Filter sterilize.

Seven Vitamin Solution:
Composition per liter:

Pyridoxine hydrochloride	300.0mg
Thiamine-HCl·$2H_2O$	200.0mg
Nicotinic acid	200.0mg
Vitamin B_{12}	100.0mg
Calcium pantothenate	100.0mg
p-Aminobenzoic acid	80.0mg
D(+)-Biotin	20.0mg

Preparation of Seven Vitamin Solution: Add components to distilled/deionized water and bring volume to 1.0L. Sparge with 100% N_2. Mix thoroughly. Filter sterilize.

Vitamin Solution:
Composition per liter:

Pyridoxine-HCl	10.0mg
Thiamine-HCl·$2H_2O$	5.0mg
Riboflavin	5.0mg
Nicotinic acid	5.0mg
D-Ca-pantothenate	5.0mg
p-Aminobenzoic acid	5.0mg
Lipoic acid	5.0mg
Biotin	2.0mg
Folic acid	2.0mg
Vitamin B_{12}	0.1mg

Preparation of Vitamin Solution: Add components to distilled/deionized water and bring volume to 1.0L. Mix thoroughly. Sparge with 80% H_2 + 20% CO_2. Filter sterilize.

Trace Elements Solution SL-12:
Composition per liter:

Na_2-EDTA	5.2g
$FeCl_2·4H_2O$	1.5g
$CoCl_2·6H_2O$	190.0mg
$MnCl_2·4H_2O$	100.0mg
$ZnCl_2$	70.0mg
$Na_2MoO_4·2H_2O$	36.0mg
$NiCl_2·6H_2O$	24.0mg
H_3BO_3	6.0mg
$CuCl_2·2H_2O$	2.0mg
HCl (25% solution)	10.0mL

Preparation of Trace Elements Solution SL-12: Add $FeCl_2·4H_2O$ to 10.0mL of HCl solution. Mix thoroughly. Add distilled/deionized water and bring volume to 1.0L. Add remaining components. Mix thoroughly. Adjust pH to 6.5. Sparge with 80% N_2 + 20% CO_2.

Preparation of Medium: Prepare and dispense medium under 100% N_2. Add components, except vitamin solution, seven vitamin solution, KNO_3 solution, and Na-acetate solution, to distilled/deionized water and bring volume to 969.0mL. Mix thoroughly. Sparge with 100% N_2. Adjust pH to 7.6. Autoclave for 15 min at 15 psi pressure–121°C. Aseptically and anaerobically add 10.0mL sterile KNO_3 solution, 10.0mL sterile vitamin solution, 1.0mL sterile seven vitamin solution, and 10.0mL sterile Na-acetate solution. Mix thoroughly. Aseptically and anaerobically distribute into sterile tubes or bottles.

Use: For the cultivation of *Chrysiogenes arsenatis.*

Chu's Medium No. 10

Composition per liter:

$Ca(NO_3)_2·4H_2O$	0.04g
$Na_2SiO_3·5H_2O$	0.025g
$MgSO_4·7H_2O$	0.025g
Na_2CO_3	0.02g
$FeCl_3$	0.008g
K_2HPO_4	0.005g

Source: This medium is available from HiMedia.

Preparation of Medium: Add components to distilled/deionized water and bring volume to 1.0L. Mix thoroughly. Gently heat to boiling. Distribute into tubes or flasks. Autoclave for 15 min at 15 psi pressure–121°C.

Use: For the cultivation and maintenance of cyanobacteria.

Chu's No. 10 Medium

Composition per liter:

Agar	15.0g
$Ca(NO_3)_2·4H_2O$	0.232g
$Na_2SiO_3·5H_2O$	0.044g
$MgSO_4·7H_2O$	0.025g
Na_2CO_3	0.02g
K_2HPO_4	0.01g

Citric acid	3.5mg
Ferric citrate	3.5mg

Preparation of Medium: Add components to distilled/deionized water and bring volume to 1.0L. Mix thoroughly. Gently heat to boiling. Distribute into tubes or flasks. Autoclave for 15 min at 15 psi pressure–121°C. Pour into sterile Petri dishes or leave in tubes.

Use: For the cultivation and maintenance of *Anabaena* species and *Plectomena boryanum.*

Chu's No. 10 Medium, Modified

Composition per liter:

Agar	15.0g
$Ca(NO_3)_2 \cdot 4H_2O$	0.232g
$Na_2SiO_3 \cdot 5H_2O$	0.044g
$MgSO_4 \cdot 7H_2O$	0.025g
Na_2CO_3	0.02g
K_2HPO_4	0.01g
Citric acid	3.5mg
Ferric citrate	3.5mg
Metal solution	1.0mL

Metal Solution:

Composition per liter:

H_3BO_3	2.4g
$MnCl_2 \cdot 4H_2O$	1.4g
$ZnCl_2$	0.4g
$CoCl_2 \cdot 6H_2O$	0.02g
$CuCl_2 \cdot 2H_2O$	0.1mg

Preparation of Metal Solution: Add components to distilled/deionized water and bring volume to 1.0L. Mix thoroughly.

Preparation of Medium: Add components to distilled/deionized water and bring volume to 1.0L. Mix thoroughly. Gently heat to boiling. Distribute into tubes or flasks. Autoclave for 15 min at 15 psi pressure–121°C. Pour into sterile Petri dishes or leave in tubes.

Use: For the cultivation and maintenance of *Anabaena* species and *Plectomena boryanum.*

Chu's No. 11 Medium, Modified

Composition per liter:

$NaNO_3$	1.5g
$MgSO_4 \cdot 7H_2O$	0.08g
$Na_2SiO_3 \cdot 9H_2O$	0.06g
$CaCl_2 \cdot 2H_2O$	0.04g
$K_2HPO_4 \cdot 3H_2O$	0.04g
Na_2CO_3	0.02g
Citric acid	6.0mg
Ferric ammonium citrate	6.0mg
EDTA	1.0mg
Seawater	999.0mL
Trace metal solution A5 with cobalt	1.0mL

pH 7.5 ± 0.2 at 25°C

Trace Metal Solution A5 with Cobalt:

Composition per liter:

H_3BO_3	2.86g
$MnCl_2 \cdot 4H_2O$	1.81g
$Na_2MoO_4 \cdot 2H_2O$	0.39g
$ZnSO_4 \cdot 7H_2O$	0.222g
$CuSO_4 \cdot H_2O$	0.079g
$Co(NO_3)_2 \cdot 6H_2O$	0.049g

Preparation of Trace Metal Solution A5 with Cobalt: Add components to distilled/deionized water and bring volume to 1.0L. Mix thoroughly.

Preparation of Medium: Add components to seawater and bring volume to 1.0L. Mix thoroughly. Gently heat and bring to boiling. Distribute into tubes or flasks. Autoclave for 15 min at 15 psi pressure–121°C.

Use: For the isolation and cultivation of cyanobacteria from marine habitats.

CIN Agar
(*Yersinia* Selective Agar)
(Cefsulodin Irgasan® Novobiocin Agar)

Composition per liter:

Mannitol	20.0g
Agar	12.0g
Pancreatic digest of gelatin	10.0g
Beef extract	5.0g
Peptic digest of animal tissue	5.0g
Sodium pyruvate	2.0g
Yeast extract	2.0g
NaCl	1.0g
Sodium deoxycholate	0.5g
Neutral Red	0.03g
Cefsulodin	0.015g
Irgasan®(triclosan)	4.0mg
Novobiocin	2.5mg
Crystal Violet	1.0mg

pH 7.4 ± 0.2 at 25°C

Source: This medium is available as a premixed powder from BD Diagnostic Systems.

Preparation of Medium: Add components, except cefsulodin and novobiocin, to distilled/deionized water and bring volume to 1.0L. Heat, mixing continuously, until boiling. Do not autoclave. Cool to 45°–50°C. Aseptically add cefsulodin and novobiocin. Mix thoroughly. Pour into sterile Petri dishes or distribute into sterile tubes.

Use: For the selective isolation and differentiation of *Yersinia enterocolitica* from a variety of clinical and nonclinical specimens based on mannitol fermentation. *Yersinia enterocolitica* appears as "bull's eye" colonies with deep red centers surrounded by a transparent periphery.

Cinnamate Medium

Composition per liter:

$NaHCO_3$	2.5g
$MgCl_2 \cdot 6H_2O$	2.03g
Cinnamic acid	1.48g
KH_2PO_4	1.36g
NH_4Cl	0.53g
$Na_2S \cdot 9H_2O$	0.24g
$CaCl_2 \cdot 2H_2O$	0.15g
Yeast extract	0.05g
Modified Wolfe's metals	10.0mL
Wolfe's vitamin solution	10.0mL

pH 7.5–7.7 at 25°C

Modified Wolfe's Metals:

Composition per liter:

Na_2SeO_3	10.0mg
$NaWO_4 \cdot 2H_2O$	10.0mg

$NiCl_2 \cdot 6H_2O$	10.0mg
Wolfe's metals solution	1.0L

Preparation of Modified Wolfe's Metals: Combine the components. Mix thoroughly.

Wolfe's Metals Solution:

Composition per liter:

$MgSO_4 \cdot 7H_2O$	3.0g
Nitrilotriacetic acid	1.5g
NaCl	1.0g
$MnSO_4 \cdot H_2O$	0.5g
$CaCl_2$	0.1g
$CoCl_2 \cdot 6H_2O$	0.1g
$FeSO_4 \cdot 7H_2O$	0.1g
$ZnSO_4 \cdot 7H_2O$	0.1g
$AlK(SO_4)_2 \cdot 12H_2O$	0.01g
$CuSO_4 \cdot 5H_2O$	0.01g
H_3BO_3	0.01g
$Na_2MoO_4 \cdot 2H_2O$	0.01g

Preparation of Wolfe's Metals Solution: Add nitrilotriacetic acid to 500.0mL of distilled/deionized water. Adjust pH to 6.5 with KOH. Add distilled/deionized water to 1.0L. Add remaining components. Mix thoroughly.

Wolfe's Vitamin Solution:

Composition per liter:

Pyridoxine·HCl	0.01g
p-Aminobenzoic acid	5.0mg
Calcium pantothenate	5.0mg
Nicotinic acid	5.0mg
Riboflavin	5.0mg
Thiamine·HCl	5.0mg
Thioctic acid	5.0mg
Biotin	2.0mg
Folic acid	2.0mg
Cyanocobalamin	100.0μg

Preparation of Wolfe's Vitamin Solution: Add components to distilled/deionized water and bring volume to 1.0L. Mix thoroughly.

Preparation of Medium: Add all components, except $NaHCO_3$ and $Na_2S \cdot 9H_2O$, to distilled/deionized water and bring volume to 1.0L. Gently heat and bring to boiling under 90% N_2 + 10% CO_2. Cool medium to room temperature while continuing to gas with 90% N_2 + 10% CO_2. Add $NaHCO_3$ and $Na_2S \cdot 9H_2O$. Adjust pH to 7.5–7.7. Distribute into tubes under 90% N_2 + 10% CO_2 using anaerobic techniques. Autoclave for 15 min at 15 psi pressure–121°C.

Use: For the cultivation of anaerobic bacteria that can utilize cinnamic acid as a carbon source. As a basal medium for the cultivation of *Formivibrio citricus*.

Citrate Agar

See: **Simmons Citrate Agar**

See: **Christensen Citrate Agar, Modified**

Citrate Azide Tween Carbonate Base

Composition per liter:

Agar	15.0g
Casein enzymic hydrolysate	15.0g
Sodium citrate	15.0g
Yeast extract	5.0g
KH_2PO_4	5.0g
Tween 80	1.0g
Selective supplement solution	10.0mL

pH 7.0 ± 0.2 at 25°C

Source: This medium is available from HiMedia.

Selective Supplement Solution:

Composition per 10.0mL:

Na_2CO_3	1.0g
NaN_3	0.2g
2,3,5, Triphenyltetrazolium chloride	0.05g

Preparation of Selective Supplement Solution: Add components to distilled/deionized water and bring volume to 10.0mL. Mix thoroughly. Filter sterilize.

Caution: Sodium azide is toxic. Azides also react with metals and disposal must be highly diluted.

Preparation of Medium: Add components, except selective supplement solution, to distilled/deionized water and bring volume to 990.0mL. Mix thoroughly. Autoclave for 15 min at 15 psi pressure–121°C. Cool to 50°C. Aseptically add selective supplement solution. Mix thoroughly. Pour into Petri dishes or aseptically distribute into sterile tubes.

Use: For the identification of enterococci in meat, meat products, dairy products, and other foodstuffs.

Citrate Medium, Koser's Modified

Composition per liter:

NaCl	5.0g
Citric acid	2.0g
$(NH_4)H_2PO_4$	1.0g
K_2HPO_4	1.0g
$MgSO_4 \cdot 7H_2O$	0.2g

pH 6.8 ± 0.2 at 25°C

Preparation of Medium: Add components to distilled/deionized water and bring volume to 1.0L. Mix thoroughly. Adjust pH to 6.8. Distribute into tubes in 5.0mL volumes. Autoclave for 15 min at 15 psi pressure–121°C.

Use: For the cultivation and differentiation of bacteria based on their ability to utilize citrate as a carbon source.

Citrate Phosphate Buffered Glucose Medium

Composition per liter:

Solution A	750.0mL
Solution C	200.0mL
Solution B	50.0mL

pH 3.5 ± 0.2 at 25°C

Solution A:

Composition per 750.0mL:

$(NH_4)_2SO_4$	3.0g
Citric acid, anhydrous	1.92g
Na_2HPO_4	1.23g
$MgSO_4 \cdot 7H_2O$	0.5g
KCl	0.1g
$Ca(NO_3)_2 \cdot 4H_2O$	0.02g
$FeSO_4 \cdot 7H_2O$	0.01g

Preparation of Solution A: Add components to distilled/deionized water and bring volume to 750.0mL. Mix thoroughly.

Solution B:

Composition per 50.0mL:

Glucose	10.0g
Yeast extract	1.0g

Preparation of Solution B: Add components to distilled/deionized water and bring volume to 50.0mL. Mix thoroughly.

Solution C:

Composition per 200.0mL:

Agarose (electrophoresis grade)	6.0g

Preparation of Solution C: Add agarose to distilled/deionized water and bring volume to 200.0mL. Mix thoroughly.

Preparation of Medium: Prepare solutions A, B, and C. Autoclave solutions separately for 15 min at 15 psi pressure–121°C. Cool to 50°–55°C. Combine solutions A, B, and C. Mix thoroughly. Immediately distribute into sterile tubes or flasks.

Use: For the cultivation and maintenance of *Acidiphilium organovorum.*

Citrobacter diversus Medium

Composition per liter:

Component 1	500.0mL
Component 2	250.0mL
Component 3	250.0mL

Component 1:

Composition per 500.0mL:

Agar	7.5g
Pancreatic digest of casein	7.5g
Papaic digest of soybean meal	2.5g

Preparation of Component 1: Add components to distilled/deionized water and bring volume to 500.0mL. Mix thoroughly. Gently heat and bring to boiling. Autoclave for 15 min at 15 psi pressure–121°C. Cool to 45°–50°C.

Component 2:

Composition per 250.0mL:

Agar	7.5g
Pancreatic digest of casein	7.5g
Papaic digest of soybean meal	2.5g
L-Tyrosine	1.0g

Preparation of Component 2: Add components to distilled/deionized water and bring volume to 250.0mL. Mix thoroughly. Gently heat and bring to boiling. Autoclave for 15 min at 15 psi pressure–121°C. Cool to 45°–50°C.

Component 3:

Composition per 250.0mL:

L-Tyrosine	1.0g

Preparation of Component 3: Add components to distilled/deionized water and bring volume to 250.0mL. Mix thoroughly. Gently heat and bring to boiling. Autoclave for 15 min at 15 psi pressure–121°C. Rapidly cool in ice water (0°C) for 20 min. A fine, white crystalline precipitate should form.

Preparation of Medium: Pour sterile component 1 into sterile Petri dishes in 18.0mL volumes. Allow agar to solidify. Warm component 3 to 50°C for 20 min. Aseptically combine component 2 and component 3. Mix thoroughly. Pour 7.0mL of mixture over the solidified component 1 agar.

Use: For the isolation and cultivation of *Citrobacter diversus.*

Citrovorum Factor Assay Medium
See: **CF Assay Medium**

CK Agar

Composition per liter:

Agar	15.0g
Glucose	5.0g
KNO_3	2.0g
$CaCl_2$	1.5g
$MgSO_4 \cdot 7H_2O$	1.5g
K_2HPO_4	0.25g
Ferric citrate	0.02g

Preparation of Medium: Add components to distilled/deionized water and bring volume to 1.0L. Mix thoroughly. Gently heat and bring to boiling. Distribute into tubes or flasks. Autoclave for 15 min at 15 psi pressure–121°C. Pour into sterile Petri dishes or leave in tubes.

Use: For the cultivation of myxobacteria.

CK1 Medium

Composition per liter:

$MgSO_4 \cdot 7H_2O$	3.0g
KNO_3	2.0g
$CaCl_2$	1.4g
Ferric citrate	0.02g
Glucose solution	100.0mL
K_2HPO_4 solution	10.0mL

Glucose Solution:

Composition per 100.0mL:

D-Glucose	10.0g

Preparation of Glucose Solution: Add D-glucose to distilled/deionized water and bring volume to 100.0mL. Mix thoroughly. Autoclave for 15 min at 15 psi pressure–121°C. Cool to 25°C.

K_2HPO_4 Solution:

Composition per 10.0mL:

K_2HPO_4	2.5mg

Preparation of K_2HPO_4 Solution: Add K_2HPO_4 to distilled/deionized water and bring volume to 10.0mL. Mix thoroughly. Autoclave for 15 min at 15 psi pressure–121°C. Cool to 25°C.

Preparation of Medium: Add components, except glucose solution and K_2HPO_4 solution, to distilled/deionized water and bring volume to 890.0mL. Mix thoroughly. Autoclave for 15 min at 15 psi pressure–121°C. Cool to 25°C. Aseptically add sterile glucose solution and K_2HPO_4 solution. Mix thoroughly. Aseptically distribute into sterile tubes or flasks.

Use: For the cultivation of myxobacteria.

Clausen HiVeg Medium

Composition per liter:

Plant hydrolysate	15.0g
Glucose	6.0g
Yeast extract	6.0g
Papaic digest of soybean meal	3.0g
NaCl	2.5g
K_2HPO_4	2.0g
L-Asparagine	1.25g
Sodium citrate	1.0g
Agar	0.75g

Na-thioglycollate	0.5g
L-Cystine	0.5g
$MgSO_4$	0.4g
Sodium dithionate	0.4g
Lecithin	0.3g
$CaCl_2$	4.0mg
$MnCl_2$	2.0mg
$CoSO_4$	1.0mg
$CuSO_4$	1.0mg
$FeSO_4$	1.0mg
Resazurin	1.0mg
$ZnSO_4$	1.0mg
Glycerol	5.0mL
Polysorbate 80	3.0mL

pH 7.1 ± 0.2 at 25°C

Source: This medium, without glycerol or polysorbate 80, is available as a premixed powder from HiMedia.

Preparation of Medium: Add components to distilled/deionized water and bring volume to 1.0L. Mix thoroughly. Gently heat to boiling. Autoclave for 15 min at 15 psi pressure–121°C. Pour into sterile Petri dishes or leave in tubes. The medium must not be resterilized.

Use: For sterility testing by the membrane filter method or the tube dilution method to determine the presence of microbial contamination in a variety of specimens.

Clausen Medium

Composition per liter:

Casein enzymatic hydrolysate	15.0g
Glucose	6.0g
Yeast extract	6.0g
Papaic digest of soybean meal	3.0g
NaCl	2.5g
K_2HPO_4	2.0g
L-Asparagine	1.25g
Sodium citrate	1.0g
Agar	0.75g
Na-thioglycollate	0.5g
L-Cystine	0.5g
$MgSO_4$	0.4g
Sodium dithionate	0.4g
Lecithin	0.3g
$CaCl_2$	4.0mg
$MnCl_2$	2.0mg
$CoSO_4$	1.0mg
$CuSO_4$	1.0mg
$FeSO_4$	1.0mg
Resazurin	1.0mg
$ZnSO_4$	1.0mg
Glycerol	5.0mL
Polysorbate 80	3.0mL

pH 7.1 ± 0.2 at 25°C

Source: This medium, without glycerol or polysorbate 80, is available as a premixed powder from HiMedia.

Preparation of Medium: Add components to distilled/deionized water and bring volume to 1.0L. Mix thoroughly. Gently heat to boiling. Autoclave for 15 min at 15 psi pressure–121°C. Pour into sterile Petri dishes or leave in tubes. The medium must not be resterilized.

Use: For sterility testing by the membrane filter method or the tube dilution method to determine the presence of microbial contamination in a variety of specimens.

Clausen Medium (Dithionite Thioglycolate, HS T, Broth)

Composition per liter:

Pancreatic digest of casein	15.0g
Glucose	6.0g
Yeast extract	6.0g
Glycerol	5.0g
Papaic digest of soybean meal	3.0g
Tween™ 80	3.0g
NaCl	2.5g
K_2HPO_4	2.0g
L-Asparagine	1.25g
Sodium citrate	1.0g
Agar	0.75g
L-Cysteine	0.5g
Sodium thioglycolate	0.5g
$MgSO_4 \cdot 7H_2O$	0.4g
Sodium dithionite	0.4g
Lecithin	0.3g
$CaCl_2 \cdot 2H_2O$	4.0mg
$MnCl_2 \cdot 4H_2O$	2.0mg
$CoSO_4 \cdot 7H_2O$	1.0mg
$CuSO_4 \cdot 5H_2O$	1.0mg
$FeSO_4 \cdot 7H_2O$	1.0mg
$ZnSO_4 \cdot 7H_2O$	1.0mg
Resazurin	1.0mg

pH 7.1 ± 0.2 at 25°C

Source: This medium is available as a premixed powder from Oxoid Unipath.

Preparation of Medium: Add Tween™ 80 and glycerol to distilled/deionized water and bring volume to 1.0L. Add remaining components. Gently heat and bring to boiling. Distribute into tubes or flasks. Autoclave for 15 min at 15 psi pressure–121°C. The medium must not be resterilized.

Use: For sterility testing by the membrane filter method or the tube dilution method to determine the presence of microbial contamination in a variety of specimens.

Clavibacter Medium

Composition per liter:

Agar	15.0g
Glucose	6.0g
Peptone	5.0g
NaCl	5.0g
Yeast extract	4.0g
Beef extract	1.0g

pH 7.2 ± 0.2 at 25°C

Preparation of Medium: Add components to distilled/deionized water and bring volume to 1.0L. Mix thoroughly. Gently heat and bring to boiling. Distribute into tubes or flasks. Autoclave for 15 min at 15 psi pressure–121°C. Pour into sterile Petri dishes or leave in tubes.

Use: For the cultivation and maintenance of *Clavibacter michiganensis.*

Clavibacter michiganensis Medium (LMG Medium 39)

Composition per liter:

Agar	15.0g
Glucose	10.0g
Yeast extract	5.0g
Peptone	5.0g
Casein hydrolysate	0.1g

pH 7.0 ± 0.2 at 25°C

Preparation of Medium: Add components to 1.0L soil extract. Mix thoroughly. Gently heat and bring to boiling. Distribute into tubes or flasks. Autoclave for 15 min at 15 psi pressure–121°C. Pour into sterile Petri dishes or leave in tubes.

Use: For the cultivation and maintenance of *Clavibacter michiganensis* subsp. *sepedonicus.*

Clavicorona Medium

Composition per liter:

Agar	15.0g
Malt extract	7.5g
Peptone	1.0g
Yeast extract	0.5g

Preparation of Medium: Add components to distilled/deionized water and bring volume to 1.0L. Mix thoroughly. Gently heat and bring to boiling. Distribute into tubes or flasks. Autoclave for 15 min at 15 psi pressure–121°C. Pour into sterile Petri dishes or leave in tubes.

Use: For the cultivation and maintenance of *Clavicorona divaricata, Clavicorona pyxidata, Favolus arcularius, Pistillaria micans, Pistillaria setipes,* and *Tremella mesenterica.*

CLED Agar (Cystine Lactose Electrolyte Deficient Agar) (Brolacin Agar)

Composition per liter:

Agar	15.0g
Lactose	10.0g
Pancreatic digest of casein	4.0g
Pancreatic digest of gelatin	4.0g
Beef extract	3.0g
L-Cystine	0.128g
Bromthymol Blue	0.02g

pH 7.3 ± 0.2 at 25°C

Preparation of Medium: Add components to distilled/deionized water and bring volume to 1.0L. Mix thoroughly. Gently heat while stirring and bring to boiling. Autoclave for 15 min at 15 psi pressure–121°C. Cool to 50°–55°C. Pour into sterile Petri dishes or distribute into sterile tubes.

Use: For the isolation, enumeration, and presumptive identification of microorganisms from urine.

CLED Agar with Andrade's Indicator (Cystine Lactose Electrolyte Deficient Agar with Andrade's Indicator)

Composition per liter:

Agar	15.0g
Pancreatic digest of casein	10.0g
Peptone	4.0g
Beef extract	3.0g
L-Cystine	0.128g
Bromthymol Blue	0.02g
Andrade's indicator	10.0mL

pH 7.5 ± 0.2 at 25°C

Source: This medium is available as a premixed powder from Oxoid Unipath.

Caution: Acid Fuchsin is a potential carcinogen and care must be taken to avoid inhalation of the powdered dye and contamination of the skin.

Andrade's Indicator:

Composition per 100.0mL:

NaOH (1*N* solution)	16.0mL
Acid Fuchsin	0.1g

Preparation of Andrade's Indicator: Add Acid Fuchsin to NaOH solution and bring volume to 100.0mL with distilled/deionized water.

Preparation of Medium: Add components to distilled/deionized water and bring volume to 1.0L. Mix thoroughly. Gently heat while stirring and bring to boiling. Autoclave for 15 min at 15 psi pressure–121°C. Cool to 50°–55°C. Pour into sterile Petri dishes or distribute into sterile tubes.

Use: For the differentiation of microorganisms based on colony characteristics.

CLED HiVeg Agar with Andrade's Indicator

Composition per liter:

Agar	15.0g
Lactose	10.0g
Plant hydrolysate	4.0g
Plant peptone	4.0g
Plant extract	3.0g
L-Cystine	0.128g
Andrade's indicator	0.1g
Bromo Thymol Blue	0.02g

pH 7.5 ± 0.2 at 25°C

Source: This medium is available as a premixed powder from HiMedia.

Caution: Acid Fuchsin in Andrade indicator is a potential carcinogen and care must be taken to avoid inhalation of the powdered dye and contamination of the skin.

Preparation of Medium: Add components to distilled/deionized water and bring volume to 1.0L. Mix thoroughly. Gently heat while stirring and bring to boiling. Autoclave for 15 min at 15 psi pressure–121°C. Cool to 50°–55°C. Pour into sterile Petri dishes or distribute into sterile tubes.

Use: For the isolation and differentiation of urinary pathogens on the basis of lactose fermentation.

CLED HiVeg Agar with Bromthymol Blue

Composition per liter:

Agar	15.0g
Lactose	10.0g
Plant hydrolysate	4.0g
Plant peptone	4.0g
Plant extract	3.0g

L-Cystine 0.128g
Bromthymol Blue 0.02g

pH 7.3 ± 0.2 at 25°C

Source: This medium is available as a premixed powder from HiMedia.

Preparation of Medium: Add components to distilled/deionized water and bring volume to 1.0L. Mix thoroughly. Gently heat while stirring and bring to boiling. Autoclave for 15 min at 15 psi pressure–121°C. Cool to 50°–55°C. Pour into sterile Petri dishes or distribute into sterile tubes.

Use: For the isolation and differentiation of urinary pathogens on the basis of lactose fermentation.

C.L.E.D. HiVeg Agar Base without Indicator

Composition per liter:

Agar 15.0g
Lactose 10.0g
Plant hydrolysate 4.0g
Plant peptone 4.0g
Plant extract 3.0g
L-Cystine 0.128g

pH 7.3 ± 0.2 at 25°C

Source: This medium is available as a premixed powder from HiMedia.

Preparation of Medium: Add components to distilled/deionized water and bring volume to 1.0L. Mix thoroughly. Gently heat while stirring and bring to boiling. Autoclave for 15 min at 15 psi pressure–121°C. Cool to 50°–55°C. Pour into sterile Petri dishes or distribute into sterile tubes.

Use: For the isolation, enumeration and presumptive identification of bacterial flora in the urinary tract.

Clostridia Medium

Composition per liter:

Sodium L-lactate 10.0g
Sodium acetate 8.0g
K_2HPO_4 .... 0.5g
$(NH_4)_2 \cdot 7H_2O$ 0.5g
Sodium thioglycolate 0.5g
Yeast extract 0.5g
$MgSO_4 \cdot 7H_2O$ 0.1g
$FeSO_4 \cdot 7H_2O$ 0.02g
p-Aminobenzoate 100.0μg
Biotin 0.1μg

pH 6.0–7.0 at 25°C

Preparation of Medium: Add components to distilled/deionized water and bring volume to 1.0L. Mix thoroughly. Adjust pH to 6.0–7.0. Distribute into tubes or flasks. Autoclave for 20 min at 15 psi pressure–121°C.

Use: For the isolation and cultivation of *Clostridium* species that ferment lactate and acetate.

Clostridial Agar

Composition per liter:

Casein enzymatic hydrolysate 17.0g
Agar 14.5g
Glucose 6.0g
Papaic digest of soybean meal 3.0g
NaCl 2.5g
Na-thioglycolate 1.8g
Sodium formaldehyde sulphoxylate 1.0g
L-Cystine 0.25g
NaN_3 .... 0.2g
Neomycin sulfate 0.15g

pH 7.0 ± 0.2 at 25°C

Source: This medium is available as a premixed powder from HiMedia.

Preparation of Medium: Add components to distilled/deionized water and bring volume to 1.0L. Mix thoroughly. Gently heat while stirring and bring to boiling. Autoclave for 15 min at 15 psi pressure–121°C. Cool to 50°–55°C. Pour into sterile Petri dishes or distribute into sterile tubes.

Use: For the selective isolation of pathogenic Clostridia from mixed flora.

Clostridial HiVeg Agar

Composition per liter:

Plant hydrolysate 17.0g
Agar 14.5g
Glucose 6.0g
Papaic digest of soybean meal 3.0g
NaCl 2.5g
Na-thioglycolate 1.8g
Sodium formaldehyde sulphoxylate 1.0g
L-Cystine 0.25g
NaN_3 .... 0.2g
Neomycin sulfate 0.15g

pH 7.0 ± 0.2 at 25°C

Source: This medium is available as a premixed powder from HiMedia.

Preparation of Medium: Add components to distilled/deionized water and bring volume to 1.0L. Mix thoroughly. Gently heat while stirring and bring to boiling. Autoclave for 15 min at 15 psi pressure–121°C. Cool to 50°–55°C. Pour into sterile Petri dishes or distribute into sterile tubes.

Use: For the selective isolation of pathogenic Clostridia from mixed flora.

Clostridium acetobutylicum Medium

Composition per liter:

Potato flakes, dried 40.0g
Glucose 6.0g
$CaCO_3$ 2.0g
L-Cysteine·HCl·H_2O 0.5g
Resazurin 1.0mg

Preparation of Medium: Add potato flakes and $CaCO_3$ to distilled/deionized water and bring volume to 1.0L. Autoclave for 30 min at 15 psi pressure–121°C. Filter through cheesecloth. To filtrate, add glucose, L-cysteine·HCl·H_2O, and resazurin. Autoclave for 20 min at 15 psi pressure–121°C. Distribute into sterile tubes or flasks.

Use: For the growth and maintenance of *Clostridium acetobutylicum, Clostridium beijerinckii, Clostridium butyricum, Clostridium roseum,* and other *Clostridium* species.

Clostridium acidurici Medium

Composition per liter:

Uric acid 2.0g
Yeast extract 1.0g
K_2HPO_4 .... 0.91g
KOH 0.67g

Sodium thioglycolate	0.5g
$MgSO_4 \cdot 7H_2O$	0.25g
$CaCl_2 \cdot 2H_2O$	0.015g
$FeSO_4 \cdot 7H_2O$	6.0mg
$NaHCO_3$ solution	25.0mL
Sodium thioglycolate solution	25.0mL

pH 7.0–7.5 at 25°C

$NaHCO_3$ Solution:

Composition per 25.0mL:

$NaHCO_3$	5.0g

Preparation of $NaHCO_3$ Solution: Add $NaHCO_3$ to distilled/deionized water and bring volume to 25.0mL. Mix thoroughly. Filter sterilize.

Sodium Thioglycolate Solution:

Composition per 25.0mL:

Sodium thioglycolate	0.5g

Preparation of Sodium Thioglycolate Solution: Add sodium thioglycolate to distilled/deionized water and bring volume to 25.0mL. Mix thoroughly. Autoclave solution separately for 15 min at 15 psi pressure–121°C.

Preparation of Medium: Add K_2HPO_4 and KOH to distilled/deionized water and bring volume to 1.0L. Add uric acid. Gently heat until boiling. Add the remaining components, except $NaHCO_3$ and sodium thioglycolate. Mix thoroughly. Adjust pH to 7.0–7.5. Distribute into tubes or flasks. Autoclave for 15 min at 15 psi pressure–121°C. Add 0.25mL of sterile $NaHCO_3$ solution and 0.25mL of sterile sodium thioglycolate solution for each 10.0mL of sterile basal medium.

Use: For the cultivation and maintenance of *Clostridium acidurici, Clostridium purinolyticum,* and other bacteria that can utilize uric acid as a carbon source.

Clostridium aerotolerans Medium

Composition per liter:

Agar	15.0g
Xylan	5.0g
Yeast extract	5.0g
Na_2CO_3	4.0g
NaCl	0.45g
$(NH_4)_2SO_4$	0.45g
K_2HPO_4	0.225g
KH_2PO_4	0.225g
L-Cysteine·HCl·H_2O	0.125g
$Na_2S \cdot 9H_2O$	0.125g
$MgSO_4 \cdot 7H_2O$	0.09g
$CaCl_2$	0.045g

pH 7.0 ± 0.2 at 25°C

Preparation of Medium: Prepare medium anaerobically under 100% CO_2. Add components to distilled/deionized water and bring volume to 1.0L. Mix thoroughly. Gently heat while stirring and bring to boiling. Autoclave for 15 min at 15 psi pressure–121°C. Cool to 50°–55°C. Adjust pH to 7.0. Pour into sterile Petri dishes or distribute into sterile tubes.

Use: For the cultivation and maintenance of *Clostridium aerotolerans.*

Clostridium aldrichii Agar

Composition per liter:

Agar	15.0g
Cellobiose	10.0g
NH_4Cl	2.0g
K_2HPO_4	1.65g
NaCl	0.9g
$(NH_4)_2SO_4$	0.9g
L-Cysteine·HCl·H_2O	0.5g
KCl	0.5g
$MgSO_4 \cdot 7H_2O$	0.5g
Trypticase™	0.5g
Yeast extract	0.5g
$Na_2S \cdot 9H_2O$	0.2g
$CaCl_2$	0.09g
Resazurin	0.5mg
Wolfe's mineral solution	10.0mL
Wolfe's vitamin solution	10.0mL
$NaHCO_3$	variable

pH 7.0 ± 0.2 at 25°C

Wolfe's Mineral Solution:

Composition per liter:

$MgSO_4 \cdot 7H_2O$	3.0g
Nitrilotriacetic acid	1.5g
NaCl	1.0g
$MnSO_4 \cdot H_2O$	0.5g
$CaCl_2$	0.1g
$CoCl_2 \cdot 6H_2O$	0.1g
$FeSO_4 \cdot 7H_2O$	0.1g
$ZnSO_4 \cdot 7H_2O$	0.1g
$AlK(SO_4)_2 \cdot 12H_2O$	0.01g
$CuSO_4 \cdot 5H_2O$	0.01g
H_3BO_3	0.01g
$Na_2MoO_4 \cdot 2H_2O$	0.01g

Preparation of Wolfe's Mineral Solution: Add nitrilotriacetic acid to approximately 500.0mL of water and adjust to pH 6.5 with KOH to dissolve the compound. Bring volume to 1.0L with remaining water and add remaining compounds one at a time.

Wolfe's Vitamin Solution:

Composition per liter:

Pyridoxine·HCl	10.0mg
p-Aminobenzoic acid	5.0mg
Lipoic acid	5.0mg
Nicotinic acid	5.0mg
Riboflavin	5.0mg
Thiamine·HCl	5.0mg
Calcium DL-pantothenate	5.0mg
Biotin	2.0mg
Folic acid	2.0mg
Vitamin B_{12}	0.1mg

Preparation of Wolfe's Vitamin Solution: Add components to distilled/deionized water and bring volume to 1.0L. Mix thoroughly.

Preparation of Medium: Prepare and dispense medium under 80% N_2 + 20% CO_2. Add components, except $NaHCO_3$, to distilled/deionized water and bring volume to 1.0L. Mix thoroughly. Gently heat and bring to boiling. Continue boiling for 3 min. Cool to room temperature while sparging with 80% N_2 + 20% CO_2. Add sufficient $NaHCO_3$ to bring the pH to 7.0. Distribute into tubes or flasks. Autoclave for 15 min at 15 psi pressure–121°C. Pour into sterile Petri dishes or leave in tubes.

Use: For the cultivation of *Clostridium aldrichii.*

Clostridium aldrichii Broth

Composition per liter:

Cellobiose	10.0g
NH_4Cl	2.0g
K_2HPO_4	1.65g
NaCl	0.9g
$(NH_4)_2SO_4$	0.9g
L-Cysteine·HCl·H_2O	0.5g
KCl	0.5g
$MgSO_4·7H_2O$	0.5g
Trypticase™	0.5g
Yeast extract	0.5g
$Na_2S·9H_2O$	0.2g
$CaCl_2$	0.09g
Resazurin	0.5mg
Wolfe's mineral solution	10.0mL
Wolfe's vitamin solution	10.0mL
$NaHCO_3$	variable

pH 7.0 ± 0.2 at 25°C

Wolfe's Mineral Solution:

Composition per liter:

$MgSO_4·7H_2O$	3.0g
Nitrilotriacetic acid	1.5g
NaCl	1.0g
$MnSO_4·H_2O$	0.5g
$CaCl_2$	0.1g
$CoCl_2·6H_2O$	0.1g
$FeSO_4·7H_2O$	0.1g
$ZnSO_4·7H_2O$	0.1g
$AlK(SO_4)_2·12H_2O$	0.01g
$CuSO_4·5H_2O$	0.01g
H_3BO_3	0.01g
$Na_2MoO_4·2H_2O$	0.01g

Preparation of Wolfe's Mineral Solution: Add nitrilotriacetic acid to approximately 500.0mL of water and adjust to pH 6.5 with KOH to dissolve the compound. Bring volume to 1.0L with remaining water and add remaining compounds one at a time.

Wolfe's Vitamin Solution:

Composition per liter:

Pyridoxine·HCl	10.0mg
p-Aminobenzoic acid	5.0mg
Lipoic acid	5.0mg
Nicotinic acid	5.0mg
Riboflavin	5.0mg
Thiamine·HCl	5.0mg
Calcium DL-pantothenate	5.0mg
Biotin	2.0mg
Folic acid	2.0mg
Vitamin B_{12}	0.1mg

Preparation of Wolfe's Vitamin Solution: Add components to distilled/deionized water and bring volume to 1.0L. Mix thoroughly.

Preparation of Medium: Prepare and dispense medium under 80% N_2 + 20% CO_2. Add components, except $NaHCO_3$, to distilled/deionized water and bring volume to 1.0L. Mix thoroughly. Gently heat and bring to boiling. Continue boiling for 3 min. Cool to room temperature while sparging with 80% N_2 + 20% CO_2. Add sufficient $NaHCO_3$ to bring the pH to 7.0. Anaerobically distribute into tubes. Autoclave for 15 min at 15 psi pressure–121°C.

Use: For the cultivation of *Clostridium aldrichii.*

Clostridium Alginate Medium

Composition per liter of seawater:

Agar	15.0g
Sodium alginate	10.0g
K_2HPO_4	2.0g
Peptone	1.0g
Yeast extract	1.0g
Seawater	1.0L

pH 7.0–7.5 at 25°C

Preparation of Medium: Add K_2HPO_4 to 1.0L of seawater. Mix thoroughly. Gently heat while stirring to dissolve. Filter solution twice. Add remaining components. Mix thoroughly. Adjust pH to 7.0–7.5. Autoclave for 15 min at 15 psi pressure–121°C. Pour into sterile Petri dishes or distribute into sterile tubes.

Use: For the cultivation and maintenance of *Clostridium alginolyticum* and other bacteria that can utilize alginate as a carbon source.

Clostridium alkalicellum Medium (DSMZ Medium 1036)

Composition per liter:

NaCl	10.0g
$NaHCO_3$	7.6g
Na_2CO_3	1.0g
NH_4Cl	0.5g
KH_2PO_4	0.2g
KCl	0.2g
Yeast extract	0.2
$MgSO_4·7H_2O$	0.1g
Cellosbiose solution	100.0mL
$Na_2S·9H_2O$ solution	10.0mL
Trace element solution SL-10	1.0mL
Selenite/tungstate solution	1.0mL

pH 8.9 ± 0.2 at 25°C

Cellobiose Solution:

Composition per 100.0mL:

Cellobiose	3.0g

Preparation of Cellobiose Solution: Add cellobiose to distilled/deionized water and bring volume to 100.0mL. Mix thoroughly. Sparge with 100% N_2. Filter sterilize.

$Na_2S·9H_2O$ Solution:

Composition per 10.0mL:

$Na_2S·9H_2O$	0.5g

Preparation of $Na_2S·9H_2O$ Solution: Add $Na_2S·9H_2O$ to distilled/deionized water and bring volume to 10.0mL. Mix thoroughly. Autoclave under 100% N_2 for 15 min at 15 psi pressure–121°C. Cool to room temperature.

Selenite/Tungstate Solution:

Composition per liter:

NaOH	0.5g
$Na_2WO_4·2H_2O$	4.0mg
$Na_2SeO_3·5H_2O$	3.0mg

Preparation of Selenite/Tungstate Solution: Add components to distilled/deionized water and bring volume to 1.0L. Mix thoroughly. Sparge with 100% N_2. Autoclave for 15 min at 15 psi pressure–121°C.

Vitamin Solution:

Composition per liter:

Pyridoxine-HCl	10.0mg
Thiamine-HCl·$2H_2O$	5.0mg

Riboflavin	5.0mg
Nicotinic acid	5.0mg
D-Ca-pantothenate	5.0mg
p-Aminobenzoic acid	5.0mg
Lipoic acid	5.0mg
Biotin	2.0mg
Folic acid	2.0mg
Vitamin B_{12}	0.1mg

Preparation of Vitamin Solution: Add components to distilled/deionized water and bring volume to 1.0L. Mix thoroughly. Sparge with 80% H_2 + 20% CO_2. Filter sterilize.

Trace Elements Solution SL-10:
Composition per liter:

$FeCl_2 \cdot 4H_2O$	1.5g
$CoCl_2 \cdot 6H_2O$	190.0mg
$MnCl_2 \cdot 4H_2O$	100.0mg
$ZnCl_2$	70.0mg
$Na_2MoO_4 \cdot 2H_2O$	36.0mg
$NiCl_2 \cdot 6H_2O$	24.0mg
H_3BO_3	6.0mg
$CuCl_2 \cdot 2H_2O$	2.0mg
HCl (25% solution)	10.0mL

Preparation of Trace Elements Solution SL-10: Add $FeCl_2 \cdot 4H_2O$ to 10.0mL of HCl solution. Mix thoroughly. Add distilled/deionized water and bring volume to 1.0L. Add remaining components. Mix thoroughly. Sparge with 100% N_2. Autoclave for 15 min at 15 psi pressure–121°C.

Preparation of Medium: Add components, except cellobiose and sulfide solutions, to distilled/deionized water and bring volume to 890.0mL. Mix thoroughly. Sparge with 100% N_2 for 30–60 min. Dispense under 100% N_2 gas atmosphere. Autoclave under 100% N_2 for 15 min at 15 psi pressure–121°C. Aseptically and anoxically add cellobiose and sulfide solutions. Adjust final pH of the medium to pH 8.8–9.0.

Use: For the cultivation of *Clostridium alkalicellum.*

Clostridium aminobutyricum Medium

Composition per liter:

K_2HPO_4	7.05g
Yeast extract	3.0g
KH_2PO_4	1.29g
$MgCl_2 \cdot 6H_2O$	0.2g
$CaCl_2 \cdot 2H_2O$	0.01g
$FeCl_3 \cdot 6H_2O$	0.01g
Methylene Blue	2.0mg
$MnSO_4 \cdot H_2O$	1.0mg
$Na_2MoO_4 \cdot 2H_2O$	1.0mg
γ-Aminobutyrate solution	100.0mL
Na_2CO_3 solution	100.0mL
$Na_2S \cdot 9H_2O$ solution	50.0mL

pH 7.4–7.7 at 25°C

γ-Aminobutyrate Solution:
Composition per 100.0mL:

γ-Aminobutyrate	5.0g

Preparation of γ-Aminobutyrate Solution: Add γ-aminobutyrate to distilled/deionized water and bring volume to 100.0mL. Mix thoroughly. Filter sterilize.

Na_2CO_3 Solution:
Composition per 100.0mL:

Na_2CO_3	2.0g

Preparation of Na_2CO_3 Solution: Add Na_2CO_3 to distilled/deionized water and bring volume to 100.0mL. Mix thoroughly. Autoclave for 15 min at 15 psi pressure–121°C.

$Na_2S \cdot 9H_2O$ Solution:
Composition per 50.0mL:

$Na_2S \cdot 9H_2O$	0.3g

Preparation of $Na_2S \cdot 9H_2O$ Solution: Add $Na_2S \cdot 9H_2O$ to distilled/deionized water and bring volume to 50.0mL. Mix thoroughly. Autoclave for 15 min at 15 psi pressure–121°C.

Preparation of Medium: Add components, except γ-aminobutyrate solution, Na_2CO_3 solution, and $Na_2S \cdot 9H_2O$ solution, to distilled/deionized water and bring volume to 750.0mL. Autoclave for 15 min at 15 psi pressure–121°C. Cool under 80% N_2 + 10% CO_2 + 10% H_2. Aseptically add the sterile γ-aminobutyrate solution, Na_2CO_3 solution, and $Na_2S \cdot 9H_2O$ solution. Adjust pH to 7.4–7.7. Distribute using anaerobic technique into tubes or flasks.

Use: For the cultivation and maintenance of *Clostridium aminobutyricum* and other bacteria which can utilize aminobutyric acid as a carbon source.

Clostridium aminobutyricum Medium

Composition per liter:

K_2HPO_4	7.0g
γ-Aminobutyric acid	5.0g
Yeast extract	3.0g
Agar	1.5g
KH_2PO_4	1.3g
$MgCl_2 \cdot 6H_2O$	0.2g
$CaCl_2 \cdot 2H_2O$	0.01g
$FeCl_3 \cdot 6H_2O$	0.01g
$Na_2MoO_4 \cdot 2H_2O$	1.0mg
Resazurin	1.0mg
$NaHCO_3$ solution	20.0mL
$Na_2S \cdot 9H_2O$ solution	20.0mL

pH 7.2 ± 0.2 at 25°C

$NaHCO_3$ Solution:
Composition per 20.0mL:

$NaHCO_3$	1.0g

Preparation of $NaHCO_3$ Solution: Add $NaHCO_3$ to distilled/deionized water and bring volume to 20.0mL. Mix thoroughly. Autoclave for 15 min at 15 psi pressure–121°C.

$Na_2S \cdot 9H_2O$ Solution:
Composition per 20.0mL:

$Na_2S \cdot 9H_2O$	0.3g

Preparation of $Na_2S \cdot 9H_2O$ Solution: Add $Na_2S \cdot 9H_2O$ to distilled/deionized water and bring volume to 20.0mL. Mix thoroughly. Autoclave for 15 min at 15 psi pressure–121°C.

Preparation of Medium: Add components, except $NaHCO_3$ solution and $Na_2S \cdot 9H_2O$ solution, to distilled/deionized water and bring volume to 960.0mL. Mix thoroughly. Sparge under 80% N_2 + 20% CO_2. Autoclave for 15 min at 15 psi pressure–121°C. Aseptically and anaerobically add 20.0mL of sterile $NaHCO_3$ solution and 20.0mL of sterile $Na_2S \cdot 9H_2O$ solution. Mix thoroughly.

Use: For the growth and maintenance of *Clostridium aminobutyricum* and other *Clostridium* species.

Clostridium aminovalericum Medium

Composition per liter:

K_2HPO_4	9.77g
5-Aminovaleric acid	6.0g
Yeast extract	5.0g
Mannitol	1.0g
KH_2PO_4	0.54g
Sodium thioglycolate	0.5g
$MgSO_4$	0.06g
$CaSO_4 \cdot 2H_2O$	0.034g
Trace elements solution SL-6	1.0mL

pH 7.9 ± 0.2 at 25°C

Trace Elements Solution SL-6:

Composition per liter:

H_3BO_3	0.3g
$CoCl_2 \cdot 6H_2O$	0.2g
$ZnSO_4 \cdot 7H_2O$	0.1g
$MnCl_2 \cdot 4H_2O$	0.03g
$Na_2MoO_4 \cdot H_2O$	0.03g
$NiCl_2 \cdot 6H_2O$	0.02g
$CuCl_2 \cdot 2H_2O$	0.01g

Preparation of Trace Elements Solution SL-6: Add components to distilled deionized water and bring volume to 1.0L. Mix thoroughly. Adjust pH to 3.4.

Preparation of Medium: Prepare medium under 100% N_2. Add components to distilled/deionized water and bring volume to 1.0L. Autoclave for 15 min at 15 psi pressure–121°C. Adjust pH to 7.9. Distribute into tubes or flasks under 100% N_2.

Use: For the cultivation and maintenance of *Clostridium aminovalericum* and other bacteria that can utilize aminovaleric acid as a carbon source.

Clostridium beijerinckii Agar

Composition per liter:

Agar	20.0g
K_2HPO_4	5.0g
Proteose peptone No. 3	5.0g
Sodium thioglycolate	5.0g
Yeast extract	5.0g
Polygalacturonic acid solution	50.0mL

pH 7.5 ± 0.2 at 25°C

Polygalacturonic Acid Solution:

Composition per liter:

Polygalacturonic acid (or pectin)	4.6g

Preparation of Polygalacturonic Acid Solution: Dissolve polygalacturonic acid or pectin in small amounts of distilled/deionized water neutralized with 10% NaOH to pH 7.2. Mix intensively. Bring volume to 50.0mL with distilled/deionized water.

Preparation of Medium: Add components to distilled/deionized water and bring volume to 1.0L. Mix thoroughly. Adjust pH to 7.5. Gently heat and bring to boiling. Distribute into tubes or flasks. Autoclave for 15 min at 15 psi pressure–121°C. Pour into sterile Petri dishes or leave in tubes.

Use: For the cultivation and maintenance of *Clostridium beijerinckii*.

Clostridium beijerinckii Broth

Composition per liter:

K_2HPO_4	5.0g
Proteose peptone No. 3	5.0g
Sodium thioglycolate	5.0g
Yeast extract	5.0g
Polygalacturonic acid solution	50.0mL

pH 7.5 ± 0.2 at 25°C

Polygalacturonic Acid Solution:

Composition per liter:

Polygalacturonic acid (or pectin)	4.6g

Preparation of Polygalacturonic Acid Solution: Dissolve polygalacturonic acid or pectin in small amounts of distilled/deionized water neutralized with 10% NaOH to pH 7.2. Mix intensively. Bring volume to 50.0mL with distilled/deionized water.

Preparation of Medium: Add components to distilled/deionized water and bring volume to 1.0L. Mix thoroughly. Adjust pH to 7.5. Distribute into tubes or flasks. Autoclave for 15 min at 15 psi pressure–121°C.

Use: For the cultivation of *Clostridium beijerinckii*.

Clostridium botulinum Isolation Agar (CBI Agar)

Composition per 1033.0mL:

Egg yolk agar base	900.0mL
Egg yolk emulsion, 50%	100.0mL
Cycloserine solution	25.0mL
Sulfamethoxazole solution	4.0mL
Trimethoprim solution	4.0mL

pH7.4 ± 0.2 at 25°C

Egg Yolk Agar Base:

Composition per 900.0mL:

Pancreatic digest of casein	40.0g
Agar	20.0g
Na_2HPO_4	5.0g
Yeast extract	5.0g
Glucose	2.0g
NaCl	2.0g
$MgSO_4 \cdot 7H_2O$ solution	0.2mL

Preparation of Egg Yolk Agar Base: Add components to distilled/deionized water and bring volume to 900.0mL. Mix thoroughly. Gently heat to boiling. Autoclave for 15 min at 15 psi pressure–121°C. Cool to 45°–50°C.

$MgSO_4 \cdot 7H_2O$ Solution:

Composition per 100.0mL:

$MgSO_4 \cdot 7H_2O$	5.0g

Preparation of $MgSO_4 \cdot 7H_2O$ Solution: Add $MgSO_4 \cdot 7H_2O$ to distilled/deionized water and bring volume to 100.0mL. Mix thoroughly.

Cycloserine Solution:

Composition per 100.0mL:

Cycloserine	1.0g

Preparation of Cycloserine Solution: Add cycloserine to distilled/deionized water and bring volume to 100.0mL. Mix thoroughly. Filter sterilize.

Sulfamethoxazole Solution:
Composition per 100.0mL:
Sulfamethoxazole 1.9g

Preparation of Sulfamethoxazole Solution: Add sulfamethoxazole to distilled/deionized water and bring volume to 50.0mL. Add sufficient 10% NaOH to dissolve. Bring volume to 100.0mL with distilled/deionized water. Mix thoroughly. Filter sterilize.

Trimethoprim Solution:
Composition per 100.0mL:
Trimethoprim 0.1g

Preparation of Trimethoprim Solution: Add trimethoprim to distilled/deionized water and bring volume to 50.0mL. Gently heat to 55°C. Add sufficient 0.05*N* HCl to dissolve. Bring volume to 100.0mL with distilled/deionized water. Mix thoroughly. Filter sterilize.

Egg Yolk Emulsion, 50%:
Composition per 100.0mL:
Chicken egg yolks 11
Whole chicken egg 1
NaCl (0.9% solution) 50.0mL

Preparation of Egg Yolk Emulsion, 50%: Soak eggs with 1:100 dilution of saturated mercuric chloride solution for 1 min. Crack eggs and separate yolks from whites. Mix egg yolks with 1 chicken egg. Beat to form emulsion. Measure 50.0mL of egg yolk emulsion and add to 50.0mL of 0.9% NaCl solution. Mix thoroughly. Filter sterilize. Warm to 45°–50°C.

Preparation of Medium: Aseptically add warmed, sterile egg yolk emulsion, 50%, and sterile cycloserine solution, sterile sulfamethoxazole solution, and sterile trimethoprim solution to cooled, sterile egg yolk agar base. Mix thoroughly. Pour into sterile Petri dishes.

Use: For isolation, cultivation, and differentiation based on lipase activity of *Clostridium botulinum* types A, B, and F from fecal specimens associated with foodborne and infant botulism. *Clostridium botulinum* types A, B, and F appear as raised colonies surrounded by an opaque zone. Other *Clostridium* species and *Clostridium botulinum* type G appear as pinpoint colonies with no opaque zone.

Clostridium botulinum Isolation HiVeg Agar

Composition per liter:
Plant hydrolysate 40.0g
Agar 20.0g
Na_2HPO_4 .... 5.0g
Yeast extract 5.0g
Glucose 2.0g
NaCl 2.0g
$MgSO_4$ 0.01g
Egg yolk emulsion, 50% 100.0mL
Cycloserine solution 25.0mL
Sulfamethoxazole solution 4.0mL
Trimethoprim solution 4.0mL
pH 7.4 ± 0.2 at 25°C

Source: This medium, without egg yolk emulsion, cycloserine solution, sulfmethoxazole solution, and trimethoprim solution, is available as a premixed powder (*C. botulinum* Isolation HiVeg Agar Base) from HiMedia.

Cycloserine Solution:
Composition per 100.0mL:
Cycloserine 1.0g

Preparation of Cycloserine Solution: Add cycloserine to distilled/deionized water and bring volume to 100.0mL. Mix thoroughly. Filter sterilize.

Sulfamethoxazole Solution:
Composition per 100.0mL:
Sulfamethoxazole 1.9g

Preparation of Sulfamethoxazole Solution: Add sulfamethoxazole to distilled/deionized water and bring volume to 50.0mL. Add sufficient 10% NaOH to dissolve. Bring volume to 100.0mL with distilled/deionized water. Mix thoroughly. Filter sterilize.

Trimethoprim Solution:
Composition per 100.0mL:
Trimethoprim 0.1g

Preparation of Trimethoprim Solution: Add trimethoprim to distilled/deionized water and bring volume to 50.0mL. Gently heat to 55°C. Add sufficient 0.05*N* HCl to dissolve. Bring volume to 100.0mL with distilled/deionized water. Mix thoroughly. Filter sterilize.

Egg Yolk Emulsion, 50%:
Composition per 100.0mL:
Chicken egg yolks 11
Whole chicken egg 1
NaCl (0.9% solution) 50.0mL

Preparation of Egg Yolk Emulsion, 50%: Soak eggs with 1:100 dilution of saturated mercuric chloride solution for 1 min. Crack eggs and separate yolks from whites. Mix egg yolks with 1 chicken egg. Beat to form emulsion. Measure 50.0mL of egg yolk emulsion and add to 50.0mL of 0.9% NaCl solution. Mix thoroughly. Filter sterilize. Warm to 45°–50°C.

Preparation of Medium: Add components, except egg yolk emulsion, cycloserine solution, sulfmethoxazole solution, and trimethoprim solution, to distilled/deionized water and bring volume to 900.0mL. Mix thoroughly. Gently heat to boiling. Autoclave for 15 min at 15 psi pressure–121°C. Cool to 45°–50°C. Aseptically add warmed, sterile egg yolk emulsion, 50%, and sterile cycloserine solution, sterile sulfamethoxazole solution, and sterile trimethoprim solution. Mix thoroughly. Pour into sterile Petri dishes.

Use: For isolation, cultivation, and differentiation based on lipase activity of *Clostridium botulinum* types A, B, and F from fecal specimens associated with foodborne and infant botulism. *Clostridium botulinum* types A, B, and F appear as raised colonies surrounded by an opaque zone. Other *Clostridium* species and *Clostridium botulinum* type G appear as pinpoint colonies with no opaque zone.

Clostridium Broth Base

Composition per liter:
Casein peptone 15.0g
Meat extract 10.0g
Yeast extract 5.0g
Sodium acetate 5.0g
L-Cysteine 0.5g
Lactate solution 10.0mL
pH 6.0 ± 0.2 at 25°C

Source: This medium is available from HiMedia.

Lactate Solution:
Composition per 10.0mL:
Sodium lactate 5.0g

Preparation of Lactate Solution: Add components to distilled/deionized water and bring volume to 10.0mL. Mix thoroughly. Filter sterilize.

Preparation of Medium: Add components, except lactate solution, to distilled/deionized water and bring volume to 990.0mL. Mix thoroughly. Autoclave for 15 min at 15 psi pressure–121°C. Cool to 50°C. Aseptically add sterile lactate solution, Mix thoroughly. Aseptically distribute into sterile tubes.

Use: For the identification of spores of *Clostridium tyrobutyricum*, which is usually responsible for "late blowing" in cheese.

Clostridium bryantii Medium

Composition per liter:

NaCl	21.0g
$MgCl_2 \cdot 6H_2O$	3.1g
Na_2SO_4	3.0g
KCl	0.5g
NH_4Cl	0.3g
KH_2PO_4	0.2g
$CaCl_2 \cdot 2H_2O$	0.15g
Resazurin	1.0mg
$NaHCO_3$ solution	20.0mL
$Na_2S \cdot 9H_2O$ solution	20.0mL
Sodium caproate solution	20.0mL
Vitamin solution	20.0mL
Trace elements solution SL-10	1.0mL

pH 7.2 ± 0.2 at 25°C

$NaHCO_3$ Solution:

Composition per 20.0mL:

$NaHCO_3$	5.0g

Preparation of $NaHCO_3$ Solution: Add $NaHCO_3$ to distilled/deionized water and bring volume to 20.0mL. Mix thoroughly. Sparge with 80% N_2 + 20% CO_2. Autoclave for 15 min at 15 psi pressure–121°C.

$Na_2S \cdot 9H_2O$ Solution:

Composition per 20.0mL:

$Na_2S \cdot 9H_2O$	0.4g

Preparation of $Na_2S \cdot 9H_2O$ Solution: Add $Na_2S \cdot 9H_2O$ to distilled/deionized water and bring volume to 20.0mL. Mix thoroughly. Sparge with 80% N_2 + 20% CO_2. Autoclave for 15 min at 15 psi pressure–121°C.

Sodium Caproate Solution:

Composition per 20.0mL:

Sodium caproate	1.4g

Preparation of Sodium Caproate Solution: Add sodium caproate to distilled/deionized water and bring volume to 20.0mL. Mix thoroughly. Sparge with 80% N_2 + 20% CO_2. Autoclave for 15 min at 15 psi pressure–121°C.

Vitamin Solution:

Composition per 20.0mL:

Thiamine·HCl	100.0μg
p-Aminobenzoic acid	40.0μg
D(+)-Biotin	10.0μg

Preparation of Vitamin Solution: Add components to distilled/deionized water and bring volume to 20.0mL. Mix thoroughly. Filter sterilize. Sparge with 80% N_2 + 20% CO_2.

Trace Elements Solution SL-10:

Composition per liter:

$FeCl_2 \cdot 4H_2O$	1.5g
$CoCl_2 \cdot 6H_2O$	190.0mg
$MnCl_2 \cdot 4H_2O$	100.0mg
$ZnCl_2$	70.0mg
$Na_2MoO_4 \cdot 2H_2O$	36.0mg
$NiCl_2 \cdot 6H_2O$	24.0mg
H_3BO_3	6.0mg
$CuCl_2 \cdot 2H_2O$	2.0mg
HCl (25% solution)	10.0mL

Preparation of Trace Elements Solution SL-10: Add $FeCl_2 \cdot 4H_2O$ to 10.0mL of HCl solution. Mix thoroughly. Add distilled/deionized water and bring volume to 1.0L. Add remaining components. Mix thoroughly. Sparge with 80% N_2 + 20% CO_2. Autoclave for 15 min at 15 psi pressure–121°C.

Preparation of Medium: Add components, except $NaHCO_3$ solution, $Na_2S \cdot 9H_2O$ solution, sodium caproate solution, and trace elements solution SL-10, to distilled/deionized water and bring volume to 919.0mL. Mix thoroughly. Sparge with 80% N_2 + 20% CO_2. Autoclave for 15 min at 15 psi pressure–121°C. Aseptically and anaerobically add 20.0mL of sterile $NaHCO_3$ solution, 20.0mL of sterile $Na_2S \cdot 9H_2O$ solution, 20.0mL of sterile sodium caproate solution, and 1.0mL of sterile trace elements solution SL-10. Mix thoroughly. Aseptically and anaerobically distribute into sterile screw-capped bottles.

Use: For the cultivation and maintenance of *Syntrophospora bryantii*.

Clostridium bryantii Medium

Composition per liter:

NaCl	21.0g
$MgCl_2 \cdot 6H_2O$	3.1g
KCl	0.5g
NH_4Cl	0.3g
KH_2PO_4	0.2g
$CaCl_2 \cdot 2H_2O$	0.15g
Resazurin	1.0mg
$NaHCO_3$ solution	20.0mL
$Na_2S \cdot 9H_2O$ solution	20.0mL
Sodium caproate solution	20.0mL
Vitamin solution	20.0mL
Trace elements solution SL-10	1.0mL

pH 7.2 ± 0.2 at 25°C

$NaHCO_3$ Solution:

Composition per 20.0mL:

$NaHCO_3$	5.0g

Preparation of $NaHCO_3$ Solution: Add $NaHCO_3$ to distilled/deionized water and bring volume to 20.0mL. Mix thoroughly. Sparge with 80% N_2 + 20% CO_2. Autoclave for 15 min at 15 psi pressure–121°C.

$Na_2S \cdot 9H_2O$ Solution:

Composition per 20.0mL:

$Na_2S \cdot 9H_2O$	0.4g

Preparation of $Na_2S \cdot 9H_2O$ Solution: Add $Na_2S \cdot 9H_2O$ to distilled/deionized water and bring volume to 20.0mL. Mix thoroughly. Sparge with 80% N_2 + 20% CO_2. Autoclave for 15 min at 15 psi pressure–121°C.

Sodium Caproate Solution:

Composition per 20.0mL:

Sodium caproate	1.4g

Preparation of Sodium Caproate Solution: Add sodium caproate to distilled/deionized water and bring volume to 20.0mL. Mix thoroughly. Sparge with 80% N_2 + 20% CO_2. Autoclave for 15 min at 15 psi pressure–121°C.

Vitamin Solution:
Composition per 20.0mL:

Thiamine·HCl	100.0μg
p-Aminobenzoic acid	40.0μg
D(+)-Biotin	10.0μg

Preparation of Vitamin Solution: Add components to distilled/deionized water and bring volume to 20.0mL. Mix thoroughly. Filter sterilize. Sparge with 80% N_2 + 20% CO_2.

Trace Elements Solution SL-10:
Composition per liter:

$FeCl_2 \cdot 4H_2O$	1.5g
$CoCl_2 \cdot 6H_2O$	190.0mg
$MnCl_2 \cdot 4H_2O$	100.0mg
$ZnCl_2$	70.0mg
$Na_2MoO_4 \cdot 2H_2O$	36.0mg
$NiCl_2 \cdot 6H_2O$	24.0mg
H_3BO_3	6.0mg
$CuCl_2 \cdot 2H_2O$	2.0mg
HCl (25% solution)	10.0mL

Preparation of Trace Elements Solution SL-10: Add $FeCl_2 \cdot 4H_2O$ to 10.0mL of HCl solution. Mix thoroughly. Add distilled/deionized water and bring volume to 1.0L. Add remaining components. Mix thoroughly. Sparge with 80% N_2 + 20% CO_2. Autoclave for 15 min at 15 psi pressure–121°C.

Preparation of Medium: Add components, except $NaHCO_3$ solution, $Na_2S \cdot 9H_2O$ solution, sodium caproate solution, and trace elements solution SL-10, to distilled/deionized water and bring volume to 919.0mL. Mix thoroughly. Sparge with 80% N_2 + 20% CO_2. Autoclave for 15 min at 15 psi pressure–121°C. Aseptically and anaerobically add 20.0mL of sterile $NaHCO_3$ solution, 20.0mL of sterile $Na_2S \cdot 9H_2O$ solution, 20.0mL of sterile sodium caproate solution, and 1.0mL of sterile trace elements solution SL-10. Mix thoroughly. Aseptically and anaerobically distribute into sterile screw-capped bottles.

Use: For the cultivation and maintenance of *Syntrophospora bryantii*.

Clostridium caminithermalis Medium (DSMZ Medium 986)

Composition per liter:

Sea salt (Sigma)	30.0g
Glucose	4.0g
NH_4Cl	1.0g
Peptone	0.5g
Yeast extract	0.5g
KH_2PO_4	0.3g
K_2HPO_4	0.3g
Resazurin	1.0mg
Glucose solution	100.0mL
Vitamin solution	10.0mL
L-Cysteine·HCl solution	10.0mL
$NaHCO_3$ solution	10.0mL
$Na_2S \cdot 9H_2O$ solution	10.0mL

pH 7.0 ± 0.2 at 25°C

Glucose Solution:
Composition per 100.0mL:

Glucose	4.0g

Preparation of Glucose Solution: Add glucose to distilled/deionized water and bring volume to 100.0mL. Mix thoroughly. Sparge with 100% N_2. Filter sterilize.

$Na_2S \cdot 9H_2O$ Solution:
Composition per 10.0mL:

$Na_2S \cdot 9H_2O$	0.25g

Preparation of $Na_2S \cdot 9H_2O$ Solution: Add $Na_2S \cdot 9H_2O$ to distilled/deionized water and bring volume to 10.0mL. Mix thoroughly. Autoclave under 100% N_2 for 15 min at 15 psi pressure–121°C. Cool to room temperature.

L-Cysteine·HCl Solution:
Composition per 10.0mL:

L-Cysteine·HCl	0.25g

Preparation of L-Cysteine Solution: Add L-cysteine·HCl to distilled/deionized water and bring volume to 10.0mL. Mix thoroughly. Sparge with 100% N_2. Autoclave under 100% N_2 for 15 min at 15 psi pressure–121°C.

Vitamin Solution:
Composition per liter:

Pyridoxine-HCl	10.0mg
Thiamine-HCl·$2H_2O$	5.0mg
Riboflavin	5.0mg
Nicotinic acid	5.0mg
D-Ca-pantothenate	5.0mg
p-Aminobenzoic acid	5.0mg
Lipoic acid	5.0mg
Biotin	2.0mg
Folic acid	2.0mg
Vitamin B_{12}	0.1mg

Preparation of Vitamin Solution: Add components to distilled/deionized water and bring volume to 1.0L. Mix thoroughly. Sparge with 80% H_2 + 20% CO_2. Filter sterilize.

$NaHCO_3$ Solution:
Composition per 10.0mL:

$NaHCO_3$	2.0g

Preparation of $NaHCO_3$ Solution: Add components to distilled/deionized water and bring volume to 10.0mL. Mix thoroughly. Sparge with 20% CO_2 + 80% H_2. Autoclave for 15 min at 15 psi pressure–121°C. Cool to room temperature.

Preparation of Medium: Add components, except vitamin, glucose, bicarbonate, cysteine and sulfide solutions, to distilled/deionized water and bring volume to 860.0mL. Mix thoroughly. Gently heat and bring to boiling. Boil for 3 min. Cool while sparging with 100% N_2. Dispense under 100% N_2 gas atmosphere. Autoclave under 100% N_2 for 15 min at 15 psi pressure–121°C. Aseptically and anoxically add vitamin, glucose, bicarbonate, cysteine, and sulfide solutions. Adjust final pH of the medium to pH 7.0.

Use: For the cultivation of *Clostridium caminithermalis*.

Clostridium cellobioparum Agar

Composition per 1025.0mL:

Ground meat, fat free	500.0g
Pancreatic digest of casein	30.0g
Agar	15.0g
K_2HPO_4	5.0g
Yeast extract	5.0g
Glucose	4.0g

Cellobiose	1.0g
Maltose	1.0g
Soluble starch	1.0g
L-Cysteine·HCl·H_2O	0.5g
Resazurin	1.0mg
NaOH (1*N* solution)	25.0mL

pH 7.0 ± 0.2 at 25°C

Preparation of Medium: Remove fat and connective tissue from lean beef or horse meat. Grind meat finely. Add ground meat to 25.0mL of 1*N* NaOH. Add 1.0L of distilled/deionized water. Gently heat and bring to boiling. Continue boiling for 15 min while stirring. Cool to room temperature. Skim fat off surface. Filter suspension and retain the filtrate and the meat particles. Bring volume of filtrate to 1.0L with distilled/deionized water. Add remaining components, except L-cysteine·HCl·H_2O. Mix thoroughly. Gently heat and bring to boiling. Cool to 50°–55°C. Add L-cysteine·HCl·H_2O. Adjust pH to 7.0. Distribute 7.0mL of agar into tubes containing meat particles (use 1 part meat particles to 4–5 parts fluid). Autoclave for 15 min at 15 psi pressure–121°C.

Use: For the cultivation of *Clostridium cellobioparum.*

Clostridium cellobioparum Broth

Composition per 1025.0mL:

Ground meat, fat free	500.0g
Pancreatic digest of casein	30.0g
K_2HPO_4	5.0g
Yeast extract	5.0g
Glucose	4.0g
Cellobiose	1.0g
Maltose	1.0g
Soluble starch	1.0g
L-Cysteine·HCl·H_2O	0.5g
Resazurin	1.0mg
NaOH (1*N* solution)	25.0mL

pH 7.0 ± 0.2 at 25°C

Preparation of Medium: Remove fat and connective tissue from lean beef or horse meat. Grind meat finely. Add ground meat to 25.0mL of 1*N* NaOH. Add 1.0L of distilled/deionized water. Gently heat and bring to boiling. Continue boiling for 15 min while stirring. Cool to room temperature. Skim fat off surface. Filter suspension and retain the filtrate and the meat particles. Bring volume of filtrate to 1.0L with distilled/deionized water. Add remaining components, except L-cysteine·HCl·H_2O. Mix thoroughly. Gently heat and bring to boiling. Cool to room temperature. Add L-cysteine·HCl·H_2O. Adjust pH to 7.0. Distribute 7.0mL of broth into tubes containing meat particles (use 1 part meat particles to 4–5 parts fluid). Autoclave for 15 min at 15 psi pressure–121°C.

Use: For the cultivation of *Clostridium cellobioparum.*

Clostridium cellobioparum Medium

Composition per 1010.0mL:

NaCl	1.0g
K_2HPO_4	0.5g
KH_2PO_4	0.5g
$(NH_4)_2SO_4$	0.5g
$CaCl_2$·$2H_2O$	0.1g
$MgSO_4$·$7H_2O$	0.1g
Resazurin	1.0mg
Rumen fluid, clarified	300.0mL
Cellobiose solution	50.0mL
$NaHCO_3$ solution	30.0mL
Na_2S·$9H_2O$ solution	20.0mL
L-Cysteine·HCl solution	10.0mL

pH 6.8 ± 0.2 at 25°C

Cellobiose Solution:
Composition per 50.0mL:

D-Cellobiose	5.0g

Preparation of Cellobiose Solution: Add cellobiose to distilled/deionized water and bring volume to 50.0mL. Mix thoroughly. Sparge under 100% N_2 gas for 3 min. Filter sterilize. Store under N_2 gas.

$NaHCO_3$ Solution:
Composition per 30.0mL:

$NaHCO_3$	10.0g

Preparation of $NaHCO_3$ Solution: Add $NaHCO_3$ to distilled/deionized water and bring volume to 30.0mL. Mix thoroughly. Sparge with 100% N_2. Autoclave for 15 min at 15 psi pressure–121°C.

Na_2S·$9H_2O$ Solution:
Composition per 20.0mL:

Na_2S·$9H_2O$	0.25g

Preparation of Na_2S·$9H_2O$ Solution: Add Na_2S·$9H_2O$ to distilled/deionized water and bring volume to 20.0mL. Mix thoroughly. Sparge with 100% N_2. Autoclave for 15 min at 15 psi pressure–121°C.

L-Cysteine·HCl Solution:
Composition per 10.0mL:

L-Cysteine·HCl	0.25g

Preparation of L-Cysteine·HCl Solution: Add L-cysteine·HCl to distilled/deionized water and bring volume to 10.0mL. Mix thoroughly. Autoclave under 100% N_2 for 15 min at 15 psi pressure–121°C.

Preparation of Medium: Add components, except cellobiose solution, $NaHCO_3$ solution, Na_2S·$9H_2O$ solution, and L-cysteine·HCl solution, to distilled/deionized water and bring volume to 900.0mL. Mix thoroughly. Sparge with 100% CO_2. Autoclave for 15 min at 15 psi pressure–121°C. Aseptically and anaerobically add 50.0mL of sterile cellobiose solution, 30.0mL of sterile $NaHCO_3$ solution, 20.0mL of sterile Na_2S·$9H_2O$ solution, and 10.0mL of sterile L-cysteine·HCl solution. Mix thoroughly. Aseptically and anaerobically distribute into sterile screw-capped bottles.

Use: For the cultivation and maintenance of *Clostridium cellobioparum* and *Clostridium polysaccharolyticum.*

Clostridium Cellulolytic Medium

Composition per liter:

Agar	20.0g
Cellulose	7.5g
K_2HPO_4·$3H_2O$	2.9g
Yeast extract	2.0g
KH_2PO_4	1.5g
$(NH_4)_2SO_4$	1.3g
$FeSO_4$	1.25g
L-Cysteine·HCl·H_2O	1.0g
$MgCl_2$·$6H_2O$	1.0g
$CaCl_2$·$2H_2O$	0.15g
Resazurin	2.0mg

pH 7.5 ± 0.2 at 25°C

Preparation of Medium: Add components, except L-cysteine·HCl·H_2O, to distilled/deionized water and bring volume to 1.0L. Mix thoroughly. Heat to boiling. Adjust pH to 7.5. Prereduce under

100% N_2. Add L-cysteine·HCl·H_2O. Distribute into tubes under 100% N_2. Cap tubes with rubber stoppers. Autoclave for 15 min at 15 psi pressure–121°C.

Use: For the cultivation and maintenance of *Clostridium cellulolyticum* and other bacteria that can degrade cellulose.

Clostridium Cellulose Medium

Composition per liter:

Agar	20.0g
Filter paper (or 5.0g Avicel)	10.0g
$CaCO_3$	5.0g
Polypeptone™	5.0g
$Na_2CO_3·10H_2O$	4.0g
K_2HPO_4	2.2g
Yeast extract	2.0g
KH_2PO_4	1.5g
$(NH_4)_2SO_4$	1.3g
$MgCl_2·6H_2O$	1.0g
L-Cysteine·HCl·H_2O	0.5g
$CaCl_2$	0.15g
$FeSO_4·7H_2O$	6.0mg

pH 7.0 ± 0.2 at 25°C

Preparation of Medium: Add components to distilled/deionized water and bring volume to 1.0L. Mix thoroughly. Heat to boiling. Autoclave for 15 min at 15 psi pressure–121°C. Pour into sterile Petri dishes or distribute into sterile tubes.

Use: For the cultivation and maintenance of *Clostridium cellulolyticum* and other bacteria that can degrade cellulose.

Clostridium cellulovorans Medium

Composition per liter:

$K_2HPO_4·3H_2O$	1.0g
NH_4Cl	1.0g
KCl	0.5g
$MgSO_4·7H_2O$	0.5g
Pancreatic digest of casein	0.5g
Yeast extract	0.5g
L-Cysteine·HCl·H_2O	0.15g
Resazurin	1.0mg
Cellulose, MN 300 or cellobiose solution	50.0mL
Na_2CO_3 solution	30.0mL
Rumen fluid, clarified	20.0mL
$Na_2S·9H_2O$ solution	20.0mL
Trace elements solution SL-10	1.0mL

pH 7.0 ± 0.2 at 25°C

Cellobiose Solution:

Composition per 50.0mL:

Cellulose, MN 300 or D-cellobiose	5.0g

Preparation of Cellobiose Solution: Add cellulose or cellobiose to distilled/deionized water and bring volume to 50.0mL. Mix thoroughly. Sparge under 100% N_2 gas for 3 min. Filter sterilize. Store under N_2 gas.

Na_2CO_3 Solution:

Composition per 30.0mL:

Na_2CO_3	1.0g

Preparation of Na_2CO_3 Solution: Add Na_2CO_3 to distilled/deionized water and bring volume to 30.0mL. Mix thoroughly. Sparge with 100% N_2. Autoclave for 15 min at 15 psi pressure–121°C.

$Na_2S·9H_2O$ Solution:

Composition per 20.0mL:

$Na_2S·9H_2O$	0.15g

Preparation of $Na_2S·9H_2O$ Solution: Add $Na_2S·9H_2O$ to distilled/deionized water and bring volume to 20.0mL. Mix thoroughly. Sparge with 100% N_2. Autoclave for 15 min at 15 psi pressure–121°C.

Trace Elements Solution SL-10:

Composition per liter:

$FeCl_2·4H_2O$	1.5g
$CoCl_2·6H_2O$	190.0mg
$MnCl_2·4H_2O$	100.0mg
$ZnCl_2$	70.0mg
$Na_2MoO_4·2H_2O$	36.0mg
$NiCl_2·6H_2O$	24.0mg
H_3BO_3	6.0mg
$CuCl_2·2H_2O$	2.0mg
HCl (25% solution)	10.0mL

Preparation of Trace Elements Solution SL-10: Add $FeCl_2·4H_2O$ to 10.0mL of HCl solution. Mix thoroughly. Add distilled/deionized water and bring volume to 1.0L. Add remaining components. Mix thoroughly.

Preparation of Medium: Add components, except cellobiose solution, Na_2CO_3 solution, and $Na_2S·9H_2O$ solution, to distilled/deionized water and bring volume to 900.0mL. Mix thoroughly. Sparge with 100% N_2. Autoclave for 15 min at 15 psi pressure–121°C. Aseptically and anaerobically add 50.0mL of sterile cellobiose solution, 30.0mL of sterile Na_2CO_3 solution, and 20.0mL of sterile $Na_2S·9H_2O$ solution. Mix thoroughly. Aseptically and anaerobically distribute into sterile screw-capped bottles.

Use: For the cultivation and maintenance of *Clostridium cellulovorans*.

Clostridium chartatabidum Medium

Composition per 1001.0mL:

Pancreatic digest of casein	2.0g
Glucose	0.5g
Glycerol	0.5g
Maltose	0.5g
Starch, soluble	0.5g
Yeast extract	0.5g
K_2HPO_4	0.3g
Hemin	1.0mg
Resazurin	1.0mg
Rumen fluid	200.0mL
Cellobiose solution	50.0mL
Mineral solution	38.0mL
Na_2CO_3 solution	30.0mL
L-Cysteine·HCl·H_2O solution	10.0mL
$Na_2S·9H_2O$ solution	10.0mL

pH 6.7 ± 0.2 at 25°C

Cellobiose Solution:

Composition per 50.0mL:

Cellulose, MN 300 or D-Cellobiose	0.5g

Preparation of Cellobiose Solution: Add cellulose or cellobiose to distilled/deionized water and bring volume to 50.0mL. Mix thoroughly. Sparge under 100% CO_2 gas for 3 min. Filter sterilize.

Mineral Solution:
Composition per liter:

NaCl	12.0g
KH_2PO_4	6.0g
$(NH_4)_2SO_4$	6.0g
$MgSO_4 \cdot 7H_2O$	2.5g
$CaCl_2 \cdot 2H_2O$	0.6g

Preparation of Mineral Solution: Add components to distilled/deionized water and bring volume to 1.0L. Mix thoroughly.

Na_2CO_3 Solution:
Composition per 30.0mL:

Na_2CO_3	4.0g

Preparation of Na_2CO_3 Solution: Add Na_2CO_3 to distilled/deionized water and bring volume to 30.0mL. Mix thoroughly. Sparge with 100% CO_2. Autoclave for 15 min at 15 psi pressure–121°C.

$Na_2S \cdot 9H_2O$ Solution:
Composition per 10.0mL:

$Na_2S \cdot 9H_2O$	0.25g

Preparation of $Na_2S \cdot 9H_2O$ Solution: Add $Na_2S \cdot 9H_2O$ to distilled/deionized water and bring volume to 10.0mL. Mix thoroughly. Sparge with 100% CO_2. Autoclave for 15 min at 15 psi pressure–121°C.

L-Cysteine·HCl Solution:
Composition per 10.0mL:

L-Cysteine·HCl	0.25g

Preparation of L-Cysteine·HCl Solution: Add L-cysteine·HCl to distilled/deionized water and bring volume to 10.0mL. Mix thoroughly. Autoclave under 100% CO_2 for 15 min at 15 psi pressure–121°C.

Preparation of Medium: Add components, except cellobiose solution, Na_2CO_3 solution, $Na_2S \cdot 9H_2O$ solution, and L-cysteine·HCl·H_2O solution, to distilled/deionized water and bring volume to 900.0mL. Mix thoroughly. Sparge with 100% CO_2. Autoclave for 15 min at 15 psi pressure–121°C. Aseptically and anaerobically add 50.0mL of sterile cellobiose solution, 30.0mL of sterile Na_2CO_3 solution, 10.0mL of sterile $Na_2S \cdot 9H_2O$ solution, and 10.0mL of sterile L-cysteine·HCl·H_2O solution. Mix thoroughly. Aseptically and anaerobically distribute into sterile screw-capped bottles under 100% CO_2.

Use: For the cultivation and maintenance of *Clostridium chartatabidum.*

Clostridium chauvoei Blood Agar

Composition per 100.0mL:

Liver extract	3.0g
Agar	1.6g
Glucose	1.0g
VL broth base	94.0mL
Sheep blood, defibrinated	5.0mL

pH 7.2–7.4 at 25°C

VL Broth Base:
Composition per liter:

Pancreatic digest of casein	10.0g
NaCl	5.0g
Yeast extract	5.0g
Meat extract	2.0g
Agar	0.6g
L-Cysteine·HCl·H_2O	0.4g

Preparation of VL Broth Base: Add components to distilled/deionized water and bring volume to 1.0L. Mix thoroughly. Gently heat until dissolved. Adjust pH to 7.2–7.4.

Preparation of Medium: Add liver extract, glucose, and agar to 94.0mL of VL broth base. Mix thoroughly. Gently heat and bring to boiling. Autoclave for 15 min at 15 psi pressure–121°C. Cool to 45°–50°C. Aseptically add sterile sheep blood. Mix thoroughly. Pour into sterile Petri dishes or distribute into sterile tubes.

Use: For the isolation and cultivation of *Clostridium chauvoei.*

Clostridium CK Medium (DSMZ Medium 869)

Composition per liter:

Pancreatic digest of casein	3.2g
NaCl	0.9g
Papaic digest of soybean meal	0.6g
K_2HPO_4	0.5g
Glucose	0.5g
Resazurin	0.5mg
Glucose solution	10.0mL

pH 5.5 ± 0.2 at 25°C

Glucose Solution:
Composition per 10.0mL:

Glucose	2.0g

Preparation of Glucose Solution: Add glucose to distilled/deionized water and bring volume to 10.0mL. Mix thoroughly. Sparge with 100% N_2. Filter sterilize.

Preparation of Medium: Add components, except glucose solution, to distilled/deionized water and bring volume to 990.0mL. Mix thoroughly. Adjust pH to 5.5. Sparge with 100% N_2. Autoclave for 15 min at 15 psi pressure–121°C. Cool to 25°C. Aseptically and anaerobically add 10.0mL sterile glucose solution. Aseptically and anaerobically distribute into tubes or flasks.

Use: For the cultivation of *Clostridium akagii, Clostridium acidisoli,* and *Clostridium uliginosum.*

Clostridium difficile Agar

Composition per liter:

Clostridum difficile agar base	920.0mL
Horse blood, defibrinated	70.0mL
Clostridium difficile selective supplement	10.0mL

pH 7.4 ± 0.2 at 25°C

Source: This medium is available as a premixed powder from Oxoid Unipath.

***Clostridum difficile* Agar Base:**
Composition per 920.0mL:

Proteose peptone	40.0g
Agar	15.0g
Fructose	6.0g
Na_2HPO_4	5.0g
NaCl	2.0g
KH_2PO_4	1.0g
$MgSO_4 \cdot 7H_2O$	0.1g

Preparation of *Clostridium difficile* Agar Base: Add components to distilled/deionized water and bring volume to 920.0mL. Mix thoroughly. Gently heat to boiling. Autoclave for 15 min at 15 psi pressure–121°C. Cool to 45°–50°C.

Clostridium difficile **Selective Supplement:**
Composition per 10.0mL:

D-Cycloserine	500.0mg
Cefoxitin	16.0mg

Preparation of *Clostridium difficile* Selective Supplement: Add components to distilled/deionized water and bring volume to 10.0mL. Mix thoroughly. Filter sterilize.

Preparation of Medium: Add 10.0mL of sterile *Clostridium difficile* selective supplement and 70.0mL of sterile, defibrinated horse blood to 920.0mL of cooled, sterile *Clostridium difficile* agar base. Mix thoroughly. Pour into sterile Petri dishes or distribute into sterile tubes.

Use: For the selective isolation and cultivation of *Clostridium difficile* from clinical and nonclinical specimens.

Clostridium difficile Agar (Cycloserine Cefoxitin Fructose Agar) (CCFA)

Composition per liter:

Peptic digest of animal tissue	32.0g
Agar	20.0g
Fructose	6.0g
Na_2HPO_4	5.0g
NaCl	2.0g
KH_2PO_4	1.0g
Cycloserine	0.25g
$MgSO_4$	0.1g
Neutral Red	0.03g
Cefoxitin solution	10.0mL

pH 7.2 ± 0.2 at 25°C

Source: This medium is available as a premixed powder from BD Diagnostic Systems.

Cefoxitin Solution:
Composition per 10.0mL:

Cefoxitin	16.0mg

Preparation of Cefoxitin Solution: Add cefoxitin to distilled/deionized water and bring volume to 10.0mL. Mix thoroughly. Filter sterilize.

Preparation of Medium: Add components to distilled/deionized water and bring volume to 990.0mL. Mix thoroughly. Gently heat to boiling. Autoclave for 15 min at 15 psi pressure–121°C. Cool to 45°–50°C. Aseptically add 10.0mL of sterile cefoxitin solution. Mix thoroughly. Pour into sterile Petri dishes or distribute into sterile tubes.

Use: For the selective isolation and cultivation of *Clostridium difficile* from clinical and nonclinical specimens.

Clostridium difficile HiVeg Agar Base

Composition per liter:

Plant peptone No. 3	40.0g
Agar	15.0g
Fructose	6.0g
Na_2HPO_4	5.0g
NaCl	2.0g
KH_2PO_4	1.0g
$MgSO_4$	0.1g
Horse blood, defibrinated	70.0mL
Clostridium difficile selective supplement	10.0mL

pH 7.4 ± 0.2 at 25°C

Source: This medium, without blood or selective supplement, is available as a premixed powder from HiMedia.

Clostridium difficile **Selective Supplement:**
Composition per 10.0mL:

D-Cycloserine	500.0mg
Cefoxitin	16.0mg

Preparation of *Clostridium difficile* Selective Supplement: Add components to distilled/deionized water and bring volume to 10.0mL. Mix thoroughly. Filter sterilize.

Preparation of Medium: Add components, except blood and selective supplement, to distilled/deionized water and bring volume to 920.0mL. Mix thoroughly. Gently heat to boiling. Autoclave for 15 min at 15 psi pressure–121°C. Cool to 45°–50°C. Add 10.0mL of sterile *Clostridium difficile* selective supplement and 70.0mL of sterile, defibrinated horse blood. Mix thoroughly. Pour into sterile Petri dishes or distribute into sterile tubes.

Use: For the selective isolation and cultivation of *Clostridium difficile* from fecal specimens.

Clostridium estertheticum Medium

Tryptose	10.0g
Beef extract	10.0g
Glucose	5.0g
NaCl	5.0g
Yeast extract	3.0g
Sodium acetate	3.0g
Soluble starch	1.0g
L-Cysteine·HCl·H_2O	0.5g
Agar	0.5g
Glucose solution	90.0mL
$NaHCO_3$ solution	10.0mL

pH 6.8 ± 0.2 at 25°C

Glucose Solution:
Composition per 90.0mL:

D-Glucose	10.0g

Preparation of Glucose Solution: Add glucose to distilled/deionized water and bring volume to 90.0mL. Mix thoroughly. Sparge with 100% N_2. Autoclave for 15 min at 15 psi pressure–121°C.

$NaHCO_3$ Solution:
Composition per 10.0mL:

$NaHCO_3$	2.0g

Preparation of $NaHCO_3$ Solution: Add $NaHCO_3$ to distilled/deionized water and bring volume to 10.0mL. Mix thoroughly. Sparge with 80% N_2 + 20% CO_2. Autoclave for 15 min at 15 psi pressure–121°C.

Preparation of Medium: Prepare and dispense medium under 80% N_2 + 20% CO_2 gas. Add components, except glucose solution and $NaHCO_3$ solution, to distilled/deionized water and bring volume to 900.0mL. Mix thoroughly. Adjust pH to 6.8. Sparge with 80% N_2 + 20% CO_2 gas. Autoclave for 15 min at 15 psi pressure–121°C. Aseptically and anaerobically add 90.0mL of sterile glucose solution and 10.0mL of sterile $NaHCO_3$ solution. Mix thoroughly. Aseptically and anaerobically distribute into sterile tubes or bottles.

Use: For the cultivation of *Clostridium estertheticum.*

Clostridium formicoaceticum Agar

Composition per liter:

Agar	15.0g
K_2HPO_4	10.0g
Yeast extract	5.0g
Sodium thioglycolate	0.75g
Pyridoxine·HCl	1.0mg
Resazurin	1.0mg
Fructose solution	50.0mL
$NaHCO_3$ solution	30.0mL
Trace elements solution SL-4	10.0mL

pH 8.0 ± 0.2 at 25°C

Trace Elements Solution SL-4:

Composition per liter:

EDTA	0.5g
$FeSO_4 \cdot 7H_2O$	0.2g
Trace elements solution SL-6	100.0mL

Trace Elements Solutions SL-6:

Composition per liter:

$MnCl_2 \cdot 4H_2O$	0.5g
H_3BO_3	0.3g
$CoCl_2 \cdot 6H_2O$	0.2g
$ZnSO_4 \cdot 7H_2O$	0.1g
$Na_2MoO_4 \cdot 2H_2O$	0.03g
$NiCl_2 \cdot 6H_2O$	0.02g
$CuCl_2 \cdot 2H_2O$	0.01g

Preparation of Trace Elements Solution SL-6: Add components to distilled/deionized water and bring volume to 1.0L. Mix thoroughly.

Preparation of Trace Elements Solution SL-4: Add components to distilled/deionized water and bring volume to 1.0L. Mix thoroughly.

Fructose Solution:

Composition per 50.0mL:

Fructose	5.0g

Preparation of Fructose Solution: Add fructose to distilled/deionized water and bring volume to 50.0mL. Mix thoroughly. Sparge under 100% N_2 gas for 3 min. Filter sterilize. Store under N_2 gas.

$NaHCO_3$ Solution:

Composition per 30.0mL:

$NaHCO_3$	10.0g

Preparation of $NaHCO_3$ Solution: Add $NaHCO_3$ to distilled/deionized water and bring volume to 30.0mL. Mix thoroughly. Sparge with 100% CO_2. Autoclave for 15 min at 15 psi pressure–121°C.

Preparation of Medium: Add components, except fructose solution and $NaHCO_3$ solution, to distilled/deionized water and bring volume to 920.0mL. Mix thoroughly. Gently heat and bring to boiling. Cool to 50°C while sparging with 100% CO_2. Autoclave for 15 min at 15 psi pressure–121°C. Aseptically and anaerobically add 50.0mL of sterile fructose solution and 30.0mL of sterile $NaHCO_3$ solution. Mix thoroughly. Adjust pH to 8.0. Aseptically and anaerobically pour into sterile Petri dishes or distribute into sterile screw-capped bottles under 100% CO_2.

Use: For the cultivation and maintenance of *Clostridium formicoaceticum.*

Clostridium formicoaceticum Broth

Composition per liter:

K_2HPO_4	10.0g
Yeast extract	5.0g
Sodium thioglycolate	0.75g
Pyridoxine·HCl	1.0mg
Resazurin	1.0mg
Fructose solution	50.0mL
$NaHCO_3$ solution	30.0mL
Trace elements solution SL-4	10.0mL

pH 8.0 ± 0.2 at 25°C

Trace Elements Solution SL-4:

Composition per liter:

EDTA	0.5g
$FeSO_4 \cdot 7H_2O$	0.2g
Trace elements solution SL-6	100.0mL

Trace Elements Solution SL-6:

Composition per liter:

$MnCl_2 \cdot 4H_2O$	0.5g
H_3BO_3	0.3g
$CoCl_2 \cdot 6H_2O$	0.2g
$ZnSO_4 \cdot 7H_2O$	0.1g
$Na_2MoO_4 \cdot 2H_2O$	0.03g
$NiCl_2 \cdot 6H_2O$	0.02g
$CuCl_2 \cdot 2H_2O$	0.01g

Preparation of Trace Elements Solution SL-6: Add components to distilled/deionized water and bring volume to 1.0L. Mix thoroughly.

Preparation of Trace Elements Solution SL-4: Add components to distilled/deionized water and bring volume to 1.0L. Mix thoroughly.

Fructose Solution:

Composition per 50.0mL:

Fructose	5.0g

Preparation of Fructose Solution: Add fructose to distilled/deionized water and bring volume to 50.0mL. Mix thoroughly. Sparge under 100% N_2 gas for 3 min. Filter sterilize. Store under N_2 gas.

$NaHCO_3$ Solution:

Composition per 30.0mL:

$NaHCO_3$	10.0g

Preparation of $NaHCO_3$ Solution: Add $NaHCO_3$ to distilled/deionized water and bring volume to 30.0mL. Mix thoroughly. Sparge with 100% CO_2. Autoclave for 15 min at 15 psi pressure–121°C.

Preparation of Medium: Add components, except fructose solution and $NaHCO_3$ solution, to distilled/deionized water and bring volume to 920.0mL. Mix thoroughly. Sparge with 100% CO_2. Autoclave for 15 min at 15 psi pressure–121°C. Aseptically and anaerobically add 50.0mL of sterile fructose solution and 30.0mL of sterile $NaHCO_3$ solution. Mix thoroughly. Adjust pH to 8.0. Aseptically and anaerobically distribute into sterile screw-capped bottles under 100% CO_2.

Use: For the cultivation of *Clostridium formicoaceticum.*

Clostridium grantii Medium

Composition per liter:

NaCl	19.45g
$MgCl_2$	8.8g
Peptone	5.0g
Na_2SO_3	3.24g
$NaHCO_3$	2.0g
$CaCl_2$	1.8g
Yeast extract	1.0g
KCl	0.55g
Ferric citrate	0.1g

KBr	0.08g
$SrCl_2$	0.03g
H_3BO_3	0.02g
Na_2HPO_4	8.0mg
Na_2SiO_3	4.0mg
NaF	2.4mg
NH_4NO_3	1.6mg
Resazurin	0.5mg
Glucose solution	20.0mL
L-Cysteine·HCl·H_2O solution	10.0mL
$Na_2S·9H_2O$ solution	10.0mL
$NaHCO_3$ solution	10.0mL

pH 6.6 ± 0.2 at 25°C

Glucose Solution:

Composition per 20.0mL:

D-Glucose	5.0g

Preparation of Glucose Solution: Add glucose to distilled/deionized water and bring volume to 20.0mL. Mix thoroughly. Sparge with 100% N_2. Autoclave for 15 min at 15 psi pressure–121°C.

L-Cysteine·HCl·H_2O Solution:

Composition per 10.0mL:

L-Cysteine·HCl·H_2O	0.3g

Preparation of L-Cysteine·HCl·H_2O Solution: Add L-cysteine·HCl·H_2O to distilled/deionized water and bring volume to 10.0mL. Mix thoroughly. Sparge with 100% N_2. Autoclave for 15 min at 15 psi pressure–121°C.

$Na_2S·9H_2O$ Solution:

Composition per 10.0mL:

$Na_2S·9H_2O$	0.3g

Preparation of $Na_2S·9H_2O$ Solution: Add $Na_2S·9H_2O$ to distilled/deionized water and bring volume to 10.0mL. Mix thoroughly. Sparge with 100% N_2. Autoclave for 15 min at 15 psi pressure–121°C.

$NaHCO_3$ Solution:

Composition per 10.0mL:

$NaHCO_3$	0.16g

Preparation of $NaHCO_3$ Solution: Add $NaHCO_3$ to distilled/deionized water and bring volume to 10.0mL. Mix thoroughly. Sparge with 80% N_2 + 20% CO_2. Autoclave for 15 min at 15 psi pressure–121°C.

Preparation of Medium: Add components, except glucose solution, L-cysteine·HCl·H_2O solution, $Na_2S·9H_2O$ solution, and $NaHCO_3$ solution, to distilled/deionized water and bring volume to 950.0mL. Mix thoroughly. Gently heat and bring to boiling. Cool to room temperature while sparging with 80% N_2 + 20% CO_2. Anaerobically distribute 9.5mL volumes into anaerobic tubes. Autoclave for 15 min at 15 psi pressure–121°C. Aseptically add 0.2mL of sterile glucose solution, 0.1mL of sterile L-Cysteine·HCl·H_2O solution, 0.1mL of sterile $Na_2S·9H_2O$ solution, and 0.1mL of sterile $NaHCO_3$ solution to each tube. Mix thoroughly.

Use: For the cultivation of *Clostridium grantii.*

Clostridium halophilum Medium

Composition per liter:

Solution A	900.0mL
Solution B	80.0mL
Solution C	10.0mL
Solution D	10.0mL

pH 8.3 ± 0.2 at 25°C

Solution A:

Composition per 900.0mL:

NaCl	60.0g
Betaine	5.86g
$MgSO_4·7H_2O$	5.0g
L-Alanine	2.2g
NH_4Cl	1.0g
Yeast extract	1.0g
$CaCl_2·2H_2O$	25.0mg
Resazurin	1.0mg
$Na_2SeO_3·5H_2O$	15.0μg
Wolfe's vitamin solution	10.0mL
Trace elements solution SL-10	1.0mL

Wolfe's Vitamin Solution:

Composition per liter:

Pyridoxine·HCl	10.0mg
p-Aminobenzoic acid	5.0mg
Lipoic acid	5.0mg
Nicotinic acid	5.0mg
Riboflavin	5.0mg
Thiamine·HCl	5.0mg
Calcium DL-pantothenate	5.0mg
Biotin	2.0mg
Folic acid	2.0mg
Vitamin B_{12}	0.1mg

Preparation of Wolfe's Vitamin Solution: Add components to distilled/deionized water and bring volume to 1.0L. Mix thoroughly.

Trace Elements Solution SL-10:

Composition per liter:

$FeCl_2·4H_2O$	1.5g
$CoCl_2·6H_2O$	190.0mg
$MnCl_2·4H_2O$	100.0mg
$ZnCl_2$	70.0mg
$Na_2MoO_4·2H_2O$	36.0mg
$NiCl_2·6H_2O$	24.0mg
H_3BO_3	6.0mg
$CuCl_2·2H_2O$	2.0mg
HCl (25% solution)	10.0mL

Preparation of Trace Elements Solution SL-10: Add $FeCl_2·4H_2O$ to 10.0mL of HCl solution. Mix thoroughly. Add distilled/deionized water and bring volume to 1.0L. Add remaining components. Mix thoroughly.

Preparation of Solution A: Prepare and dispense medium under 80% N_2 + 20% CO_2. Add components to distilled/deionized water and bring volume to 900.0mL. Mix thoroughly. Gently heat and bring to boiling. Continue boiling for 3 min. Cool to room temperature while sparging with 80% N_2 + 20% CO_2.

Solution B:

Composition per 80.0mL:

$NaHCO_3$	5.0g

Preparation of Solution B: Add $NaHCO_3$ to distilled/deionized water and bring volume to 80.0mL. Mix thoroughly. Sparge with 100% N_2 for 20 min.

Solution C:

Composition per 10.0mL:

K_2HPO_4	0.358g
KH_2PO_4	0.223g

Preparation of Solution C: Add components to distilled/deionized water and bring volume to 10.0mL. Mix thoroughly. Sparge with 100% N_2. Autoclave for 15 min at 15 psi pressure–121°C.

Solution D:

Composition per 10.0mL:

$Na_2S·9H_2O$	0.3g

Preparation of Solution D: Add components to distilled/deionized water and bring volume to 10.0mL. Mix thoroughly. Sparge with 100% N_2. Autoclave for 15 min at 15 psi pressure–121°C.

Preparation of Medium: Prepare and dispense medium under 80% N_2 + 20% CO_2. Anaerobically combine 900.0mL of cooled solution A with 80.0mL of sparged solution B. Mix thoroughly. Adjust pH to 8.3. Anaerobically distribute 9.8mL volumes into anaerobe tubes. Autoclave for 15 min at 15 psi pressure–121°C. Aseptically and anaerobically add 0.1mL of sterile solution C and 0.1mL of sterile solution D to each tube. Mix thoroughly.

Use: For the cultivation of *Clostridium halophilum.*

Clostridium histolyticum Medium (DSMZ Medium 139)

Composition per liter:

Proteose peptone	50.0g
Na_2HPO_4	9.0g
Pancreatic digest of casein	8.5g
NaCl	2.5g
KH_2PO_4	1.92g
Papaic digest of soybean meal	1.5g
K_2HPO_4	1.25g
Glucose	1.25g
$MgSO_4·7H_2O$	0.08g

pH 7.2 ± 0.2 at 25°C

Preparation of Medium: Add components to distilled/deionized water and bring volume to 1.0L. Mix thoroughly. Distribute into tubes or flasks. Autoclave for 15 min at 15 psi pressure–121°C.

Use: For the cultivation of *Clostridium histolyticum.*

Clostridium HiVeg Broth Base with Lactate

Composition per liter:

Plant hydrolysate	15.0g
Plant extract No. 1	10.0g
Sodium acetate	5.0g
Yeast extract	5.0g
L-Cysteine	0.5g
Lactate solution	10.0mL

pH 6.0 ± 0.2 at 25°C

Source: This medium, without lactate, is available as a premixed powder from HiMedia.

Lactate Solution:

Composition per10.0mL:

Sodium lactate	5.0g

Preparation of Lactate Solution: Add sodium lactate to distilled/deionized water and bring volume to 10.0mL. Mix thoroughly. Heat if necessary to completely dissolve.

Preparation of Medium: Add components to distilled/deionized water and bring volume to 1.0L. Mix thoroughly. Gently heat while stirring and bring to boiling. Distribute into tubes or flasks. Autoclave for 15 min at 15 psi pressure–121°C.

Use: For the identification of spores of *Clostridium tyrobutyricum* which is usually responsible for "late blowing" in cheese.

Clostridium hydroxybenzoicum Medium

Composition per 1055.0mL:

NaCl	10.0g
Yeast extract	10.0g
L-Arginine·HCl	2.1g
L-Lysine·HCl	1.8g
Glycine	0.75g
NH_4Cl	0.5g
L-Cysteine·HCl	0.4g
$Na_2S·9H_2O$	0.4g
$MgCl_2·7H_2O$	0.1g
$CaCl_2·2H_2O$	0.025g
Resazurin	1.0mg
$Na_2WO_4·2H_2O$	0.05mg
Sodium/potassium phosphate buffer (0.02*M* solution, pH 7.0)	1.0L
Wolfe's vitamin solution	50.0mL
Wolfe's mineral solution	5.0mL

pH 7.0 ± 0.2 at 25°C

Wolfe's Vitamin Solution:

Composition per liter:

Pyridoxine·HCl	10.0mg
p-Aminobenzoic acid	5.0mg
Lipoic acid	5.0mg
Nicotinic acid	5.0mg
Riboflavin	5.0mg
Thiamine·HCl	5.0mg
Calcium DL-pantothenate	5.0mg
Biotin	2.0mg
Folic acid	2.0mg
Vitamin B_{12}	0.1mg

Preparation of Wolfe's Vitamin Solution: Add components to distilled/deionized water and bring volume to 1.0L. Mix thoroughly.

Wolfe's Mineral Solution:

Composition per liter:

$MgSO_4·7H_2O$	3.0g
Nitrilotriacetic acid	1.5g
NaCl	1.0g
$MnSO_4·2H_2O$	0.5g
$CoCl_2·6H_2O$	0.1g
$ZnSO_4·7H_2O$	0.1g
$CaCl_2·2H_2O$	0.1g
$FeSO_4·7H_2O$	0.1g
$NiCl_2·6H_2O$	0.025g
$KAl(SO_4)_2·12H_2O$	0.02g
$CuSO_4·5H_2O$	0.01g
H_3BO_3	0.01g
$Na_2MoO_4·2H_2O$	0.01g
$Na_2SeO_3·5H_2O$	0.3mg

Preparation of Wolfe's Mineral Solution: Add nitrilotriacetic acid to 500.0mL of distilled/deionized water. Adjust pH to 6.5 with

KOH. Add remaining components. Add distilled/deionized water to 1.0L. Adjust pH to 6.8.

Preparation of Medium: Prepare and dispense medium under 100% N_2. Combine components. Mix thoroughly. Sparge with 100% N_2. Anaerobically distribute into tubes or flasks. Autoclave for 15 min at 15 psi pressure–121°C.

Use: For the cultivation of *Clostridium hydroxybenzoicum.*

Clostridium kluyveri Agar

Composition per 100.0mL:

Potassium acetate	1.0g
Sodium thioglycolate	50.0mg
K_2HPO_4	31.0mg
$NH_4 \cdot Cl$	25.0mg
KH_2PO_4	23.0mg
$MgSO_4 \cdot 7H_2O$	20.0mg
$FeSO_4 \cdot 7H_2O$	2.0mg
$MnSO_4 \cdot H_2O$	2.0mg
$CaCl_2 \cdot 2H_2O$	1.0mg
$Na_2MoO_4 \cdot 2H_2O$	0.2mg
Agar	50.0μg
Resazurin	50.0μg
p-Aminobenzoic acid	20.0μg
Biotin	10.0μg
$CaCO_3$	variable
Ethanol	2.0mL

pH 7.0 ± 0.2 at 25°C

Preparation of Medium: Add components, except sodium thioglycolate, ethanol, and $CaCO_3$, to distilled/deionized water and bring volume to 100.0mL. Mix thoroughly. Gently heat and bring to boiling. Continue boiling for 5 min. Cool rapidly to 50°C. Add sodium thioglycolate and ethanol. Distribute into tubes containing a small amount of $CaCO_3$. Autoclave for 15 min at 15 psi pressure–121°C. Store anaerobically.

Use: For the cultivation and maintenance of *Clostridium kluyveri.*

Clostridium kluyveri Medium

Composition per liter:

Potassium acetate	5.0g
Sodium thioglycolate	0.5g
K_2HPO_4	0.3g
NH_4Cl	0.25g
KH_2PO_4	0.2g
$MgSO_4 \cdot 7H_2O$	0.2g
$CaCl_2 \cdot 2H_2O$	0.01g
$FeSO_4 \cdot 7H_2O$	5.0mg
$MnSO_4 \cdot 4H_2O$	2.0mg
$Na_2MoO_4 \cdot 2H_2O$	2.0mg
p-Aminobenzoate acid	0.2mg
Biotin	0.01mg
Ethanol	20.0mL
Acetic acid, glacial	2.5mL

pH 7.0 ± 0.2 at 25°C

Preparation of Medium: Add components, except sodium thioglycolate, to distilled/deionized water and bring volume to 1.0L. Gently heat and bring to boiling. Mix thoroughly. Add sodium thioglycolate immediately prior to sterilization. Mix thoroughly. Autoclave for 15 min at 15 psi pressure–121°C. Adjust pH to 7.0 with sterile 60% K_2CO_3 solution.

Use: For the isolation and cultivation of *Clostridium kluyveri.*

Clostridium kluyveri Medium

Composition per liter:

Part A	965.0mL
Part B	35.0mL

pH 7.0 ± 0.2 at 25°C

Part A:

Composition per 965.0mL:

Sodium acetate·$3H_2O$	7.5g
$(NH_4)_2SO_4$	2.65g
Agar	2.0g
Yeast extract	2.0g
Sodium thioglycolate	0.5g
p-Aminobenzoic acid	0.1mg
Biotin	5.0μg
Potassium phosphate buffer (2*M*, pH 7.0)	10.0mL
Salt solution	10.0mL

Preparation of Part A: Add components to distilled/deionized water and bring volume to 965.0mL. Mix thoroughly. Gently heat and bring to boiling. Autoclave for 15 min at 15 psi pressure–121°C. Cool to room temperature.

Salt Solution:

Composition per 100.0mL:

$MgSO_4 \cdot H_2O$	2.5g
$CaCl_2$	0.15g
$FeSO_4 \cdot 7H_2O$	0.15g
$MnSO_4 \cdot 2H_2O$	0.02g
$Na_2MoO_4 \cdot 2H_2O$	0.02g

Preparation of Salt Solution: Add components to distilled/deionized water and bring volume to 100.0mL. Mix thoroughly.

Part B:

Composition per liter:

K_2CO_3 (1*M* solution)	20.0mL
Ethanol (95% solution)	15.0mL

Preparation of Part B: Prepare a 1*M* solution of K_2CO_3 and filter sterilize. Filter sterilize 25.0mL of 95% ethanol solution. Aseptically combine 20.0mL of sterile K_2CO_3 solution and 15.0mL of sterile ethanol.

Preparation of Medium: Add 35.0mL of sterile Part B to 965.0mL of sterile, cooled Part A. Adjust pH to 7.0 with HCl. Aseptically distribute into tubes under 97% N_2 + 3% H_2. Cap with rubber stoppers.

Use: For the cultivation and maintenance of *Clostridium kluyveri.*

Clostridium lentocellum Agar

Composition per 1201.0mL:

Agar	30.0g
K_2HPO_4	1.65g
NH_4SO_4	1.6g
Yeast extract	1.0g
NaCl	0.96g
L-Cysteine·HCl	0.5g
$CaCl_2$	96.0mg
$MgSO_4$	96.0mg

Cellulose suspension 200.0mL
Resazurin (0.1% solution) 1.0mL

pH 7.2 ± 0.2 at 25°C

Cellulose Suspension:

Composition per 100.0mL:

Whatman CF cellulose powder 4.0g

Preparation of Cellulose Suspension: Add cellulose powder to distilled/deionized water and bring volume to 100.0mL. Mix thoroughly.

Preparation of Medium: Add components to distilled/deionized water and bring volume to 1.0L. Mix thoroughly. Adjust pH to 7.2. Gently heat and bring to boiling. Distribute into tubes or flasks. Autoclave for 15 min at 15 psi pressure–121°C. Pour into sterile Petri dishes or leave in tubes.

Use: For the cultivation of *Clostridium lentocellum* and other *Clostridium* species.

Clostridium litorale Medium

Composition per liter:

Solution A 900.0mL
Solution B 80.0mL
Solution C 10.0mL
Solution D 10.0mL

pH 8.3 ± 0.2 at 25°C

Solution A:

Composition per 900.0mL:

NaCl 10.0g
Betaine 5.86g
L-Alanine 2.2g
NH_4Cl 1.0g
Yeast extract 1.0g
$MgSO_4 \cdot 7H_2O$ 0.5g
$CaCl_2 \cdot 2H_2O$ 25.0mg
Resazurin 1.0mg
$Na_2SeO_3 \cdot 5H_2O$ 15.0μg
Wolfe's vitamin solution 10.0mL
Trace elements solution SL-10 .... 1.0mL

Wolfe's Vitamin Solution:

Composition per liter:

Pyridoxine·HCl 10.0mg
p-Aminobenzoic acid 5.0mg
Lipoic acid 5.0mg
Nicotinic acid 5.0mg
Riboflavin 5.0mg
Thiamine·HCl 5.0mg
Calcium DL-pantothenate 5.0mg
Biotin 2.0mg
Folic acid 2.0mg
Vitamin B_{12} 0.1mg

Preparation of Wolfe's Vitamin Solution: Add components to distilled/deionized water and bring volume to 1.0L. Mix thoroughly.

Trace Elements Solution SL-10:

Composition per liter:

$FeCl_2 \cdot 4H_2O$ 1.5g
$CoCl_2 \cdot 6H_2O$ 190.0mg
$MnCl_2 \cdot 4H_2O$ 100.0mg
$ZnCl_2$ 70.0mg
$Na_2MoO_4 \cdot 2H_2O$ 36.0mg
$NiCl_2 \cdot 6H_2O$ 24.0mg
H_3BO_3 .... 6.0mg
$CuCl_2 \cdot 2H_2O$ 2.0mg
HCl (25% solution) 10.0mL

Preparation of Trace Elements Solution SL-10: Add $FeCl_2 \cdot 4H_2O$ to 10.0mL of HCl solution. Mix thoroughly. Add distilled/deionized water and bring volume to 1.0L. Add remaining components. Mix thoroughly.

Preparation of Solution A: Prepare and dispense medium under 80% N_2 + 20% CO_2. Add components to distilled/deionized water and bring volume to 900.0mL. Mix thoroughly. Gently heat and bring to boiling. Continue boiling for 3 min. Cool to room temperature while sparging with 80% N_2 + 20% CO_2.

Solution B:

Composition per 80.0mL:

$NaHCO_3$ 5.0g

Preparation of Solution B: Add $NaHCO_3$ to distilled/deionized water and bring volume to 80.0mL. Mix thoroughly. Sparge with 100% N_2 for 20 min.

Solution C:

Composition per 10.0mL:

K_2HPO_4 .... 0.358g
KH_2PO_4 .... 0.223g

Preparation of Solution C: Add components to distilled/deionized water and bring volume to 10.0mL. Mix thoroughly. Sparge with 100% N_2. Autoclave for 15 min at 15 psi pressure–121°C.

Solution D:

Composition per 10.0mL:

$Na_2S \cdot 9H_2O$ 0.3g

Preparation of Solution D: Add $Na_2S \cdot 9H_2O$ to distilled/deionized water and bring volume to 10.0mL. Mix thoroughly. Sparge with 100% N_2. Autoclave for 15 min at 15 psi pressure–121°C.

Preparation of Medium: Prepare and dispense medium under 80% N_2 + 20% CO_2. Anaerobically combine 900.0mL of cooled solution A with 80.0mL of sparged solution B. Mix thoroughly. Adjust pH to 8.3. Anaerobically distribute 9.8mL volumes into anaerobe tubes. Autoclave for 15 min at 15 psi pressure–121°C. Aseptically and anaerobically add 0.1mL of sterile solution C and 0.1mL of sterile solution D to each tube. Mix thoroughly.

Use: For the cultivation of *Clostridium litorale.*

Clostridium ljungdahlii Medium (DSMZ Medium 879)

Composition per liter:

NH_4Cl 1.0g
Yeast extract 1.0g
NaCl 0.8g
$MgSO_4 \cdot 7H_2O$ 0.2g
KCl 0.1g
KH_2PO_4 .... 0.1g
$CaCl_2 \cdot 2H_2O$ 0.02g
$Na_2WO_4 \cdot 2H_2O$ 0.20mg
Fructose solution 50.0mL
Trace elements solution 10.0mL
Vitamin solution 10.0mL
$NaHCO_3$ solution 10.0mL

L-Cysteine solution	10.0mL
$Na_2S·9H_2O$ solution	10.0mL

pH 5.9 ± 0.2 at 25°C

$NaHCO_3$ Solution:

Composition per 10.0mL:

$NaHCO_3$	1.0g

Preparation of $NaHCO_3$ Solution: Add $NaHCO_3$ to distilled/deionized water and bring volume to 10.0mL. Mix thoroughly. Sparge with 80% N_2 + 20% CO_2. Filter sterilize.

Fructose Solution:

Composition per 50.0mL:

Fructose	5.0g

Preparation of Fructose Solution: Add fructose to distilled/deionized water and bring volume to 50.0mL. Mix thoroughly. Sparge with 100% N_2. Filter sterilize.

$Na_2S·9H_2O$ Solution:

Composition per 10.0mL:

$Na_2S·9H_2O$	0.3g

Preparation of $Na_2S·9H_2O$ Solution: Add $Na_2S·9H_2O$ to distilled/deionized water and bring volume to 10.0mL. Sparge with N_2. Autoclave for 15 min at 15 psi pressure–121°C. Cool to 25°C. Store anaerobically.

L-Cysteine Solution:

Composition per 10.0mL:

L-Cysteine·HCl·H_2O	0.3g

Preparation of L-Cysteine Solution: Add L-cysteine·HCl·H_2O to distilled/deionized water and bring volume to 10.0mL. Mix thoroughly. Sparge with 100% N_2. Autoclave for 15 min at 15 psi pressure–121°C.

Trace Elements Solution:

Composition per liter:

$MgSO_4·7H_2O$	3.0g
Nitrilotriacetic acid	1.5g
NaCl	1.0g
$MnSO_4·2H_2O$	0.5g
$CoSO_4·7H_2O$	0.18g
$ZnSO_4·7H_2O$	0.18g
$CaCl_2·2H_2O$	0.1g
$FeSO_4·7H_2O$	0.1g
$NiCl_2·6H_2O$	0.025g
$KAl(SO_4)_2·12H_2O$	0.02g
H_3BO_3	0.01g
$Na_2MoO_4·4H_2O$	0.01g
$CuSO_4·5H_2O$	0.01g
$Na_2SeO_3·5H_2O$	0.3mg

Preparation of Trace Elements Solution: Add nitrilotriacetic acid to 500.0mL of distilled/deionized water. Dissolve by adjusting pH to 6.5 with KOH. Add remaining components. Add distilled/deionized water to 1.0L. Mix thoroughly.

Vitamin Solution:

Composition per liter:

Pyridoxine-HCl	10.0mg
Thiamine-HCl·$2H_2O$	5.0mg
Riboflavin	5.0mg
Nicotinic acid	5.0mg
D-Ca-pantothenate	5.0mg
p-Aminobenzoic acid	5.0mg
Lipoic acid	5.0mg
Biotin	2.0mg
Folic acid	2.0mg
Vitamin B_{12}	0.1mg

Preparation of Vitamin Solution: Add components to distilled/deionized water and bring volume to 1.0L. Mix thoroughly. Sparge with 80% H_2 + 20% CO_2. Filter sterilize.

Preparation of Medium: Prepare and dispense medium under 80% N_2 + 20% CO_2 gas atmosphere. Add components, except $NaHCO_3$ solution, fructose solution, L-cysteine solution, $Na_2S·9H_2O$ solution, vitamin solution, and trace elements solution SL-10, to distilled/deionized water and bring volume to 900.0mL. Mix thoroughly. Gently heat and bring to boiling. Boil for 10 min. Cool to room temperature while sparging with 80% N_2 + 20% CO_2. Autoclave for 15 min at 15 psi pressure–121°C. Aseptically and anaerobically add 50.0mL fructose solution, 10.0mL $NaHCO_3$ solution, 10.0mL L-cysteine solution, 10.0mL $Na_2S·9H_2O$ solution, 10.0mL vitamin solution, and 10.0mL trace elements solution. Mix thoroughly. Final pH is 5.9. Aseptically and anaerobically distribute into sterile tubes or bottles.

Use: For the cultivation of *Clostridium ljungdahlii.*

Clostridium longisporum Medium (DSMZ Medium 741)

Composition per liter:

Agar	13.5g
Beef extract	10.0g
Pancreatic digest of casein	10.0g
NaCl	5.0g
Glucose	5.0g
Yeast extract	3.0g
Sodium acetate	3.0g
Soluble starch	1.0g
L-Cysteine·HCl·H_2O	0.5g

pH 6.8 ± 0.2 at 25°C

Preparation of Medium: Prepare and dispense medium under 100% CO_2. Add components to distilled/deionized water and bring volume to 1.0L. Mix thoroughly. Gently heat and bring to boiling. Distribute into tubes or flasks. Autoclave for 15 min at 10 psi pressure–115°C. Pour into sterile Petri dishes or leave in tubes.

Use: For the cultivation of *Clostridium longisporum.*

Clostridium Medium

Composition per liter:

Sodium L-glutamate	10.0g
Sodium thioglycolate	0.5g
Yeast extract	0.5g
K_2HPO_4	0.2g
$MgSO_4·7H_2O$	0.1g

pH 7.6 ± 0.2 at 25°C

Preparation of Medium: Add components to distilled/deionized water and bring volume to 1.0L. Mix thoroughly. Distribute into tubes or flasks. Autoclave for 15 min at 15 psi pressure–121°C.

Use: For the enrichment and isolation of glutamate-fermenting *Clostridium* species.

Clostridium Medium

Composition per liter:

Crotonic acid	6.0g
Pancreatic digest of casein	1.0g

Yeast extract	1.0g
Sodium thioglycolate	0.5g
NaOH	0.30g
$(NH_4)_2HPO_4$	0.15g
K_2HPO_4	0.10g
NH_4Cl	0.05g
$CaCl_2 \cdot 2H_2O$	0.04g
$MgCl_2 \cdot 6H_2O$	0.033g
$(NH_4)_6Mo_7O_{24} \cdot 4H_2O$	10.0mg
p-Aminobenzoic acid	0.80mg
$FeSO_4 \cdot 7H_2O$	0.6mg
$MgSO_4 \cdot 7H_2O$	0.6mg
$MnSO_4 \cdot 2H_2O$	0.4mg
Resazurin	1.0mg
Biotin	0.04mg
K_2CO_3 solution	80.0mL
$KHCO_3$ solution	3.0mL

pH 6.8 ± 0.2 at 25°C

K_2CO_3 Solution:
Composition per 80.0mL:

K_2CO_3	40.0g

Preparation of K_2CO_3 Solution: Add K_2CO_3 to distilled/deionized water and bring volume to 80.0mL. Mix thoroughly. Sparge with 100% N_2. Autoclave for 15 min at 15 psi pressure–121°C.

$KHCO_3$ Solution:
Composition per 10.0mL:

$KHCO_3$	2.0g

Preparation of $KHCO_3$ Solution: Add $KHCO_3$ to distilled/deionized water and bring volume to 10.0mL. Mix thoroughly. Sparge with 100% N_2. Autoclave for 15 min at 15 psi pressure–121°C.

Preparation of Medium: Add components, except K_2CO_3 solution and $KHCO_3$ solution, to distilled/deionized water and bring volume to 917.0mL. Mix thoroughly. Sparge with 100% N_2. Autoclave for 15 min at 15 psi pressure–121°C. Aseptically and anaerobically add 80.0mL of sterile K_2CO_3 solution and 3.0mL of sterile $KHCO_3$ solution. Mix thoroughly. Aseptically and anaerobically distribute into sterile screw-capped bottles under 100% N_2.

Use: For the cultivation of *Clostridium halophilum* and *Clostridium litorale.*

Clostridium Medium

Composition per liter:

Proteose peptone	50.0g
Na_2HPO_4	9.0g
Pancreatic digest of casein	8.5g
NaCl	2.5g
KH_2PO_4	1.92g
Papaic digest of soybean meal	1.5g
K_2HPO_4	1.25g
Glucose	1.25g
$MgSO_4 \cdot 7H_2O$	0.08g

pH 7.2 ± 0.2 at 25°C

Preparation of Medium: Add components to distilled/deionized water and bring volume to 1.0L. Mix thoroughly. Distribute into tubes or flasks. Sparge with 80% N_2 + 20% CO_2. Autoclave for 15 min at 15 psi pressure–121°C.

Use: For the cultivation of *Clostridium halophilum* and *Clostridium litorale.*

Clostridium Medium

Composition per liter:

Uric acid	2.0g
Yeast extract	1.2g
$MgSO_4 \cdot 7H_2O$	0.05g
$CaCl_2 \cdot 2H_2O$	5.0mg
$FeSO_4 \cdot 7H_2O$	2.0mg
Resazurin	1.0mg
KOH (10*N* solution)	3.0mL
$K_2HPO_4 \cdot 3H_2O$ (70% solution)	1.5mL
Mercaptoacetic acid	1.5mL

pH 7.2 ± 0.2 at 25°C

Preparation of Medium: Add KOH solution and $K_2HPO_4 \cdot 3H_2O$ solution to distilled/deionized water and bring volume to 500.0mL. Gently heat and bring to boiling. Mix thoroughly. Add uric acid slowly. Cool to 45°–50°C. Add remaining components. Add mercaptoacetic acid immediately prior to sterilization. Bring volume to 1.0L with distilled/deionized water. Mix thoroughly. Autoclave for 15 min at 15 psi pressure–121°C. Adjust pH to 7.2 with sterile 60% K_2CO_3 solution.

Use: For the isolation and cultivation of purine-fermenting *Clostridium* species.

Clostridium Medium (ATCC Medium 39)

Composition per liter:

K_2HPO_4	7.0g
γ-Aminobutyric acid	5.0g
Yeast extract	3.0g
Agar	1.5g
KH_2PO_4	1.3g
$MgCl_2 \cdot 6H_2O$	0.2g
$CaCl_2 \cdot 2H_2O$	0.01g
$FeCl_3 \cdot 6H_2O$	0.01g
Methylene Blue	2.0mg
$MnSO_4$	1.0mg
Na_2MoO_4	1.0mg
$Na_2S \cdot 9H_2O$ solution	10.0mL

$Na_2S \cdot 9H_2O$ Solution:
Composition per 20.0mL:

$Na_2S \cdot 9H_2O$	0.6g

Preparation of $Na_2S \cdot 9H_2O$ Solution: Add $Na_2S \cdot 9H_2O$ to distilled/deionized water and bring volume to 20.0mL. Autoclave for 15 min at 15 psi pressure–121°C. Use freshly prepared solution.

Preparation of Medium: Add components, except $Na_2S \cdot 9H_2O$ solution, to distilled/deionized water and bring volume to 1.0L. Mix thoroughly. Gently heat to boiling. Autoclave for 15 min at 15 psi pressure–121°C. Cool to 45°–50°C. Distribute anaerobically into sterile tubes. Aseptically add 0.1mL of sterile 1.5% $Na_2S \cdot 9H_2O$ solution to each 5.0mL of the medium. Cap with rubber stoppers.

Use: For the cultivation and maintenance of a variety of *Clostridium* species.

Clostridium Medium (ATCC Medium 40)

Composition per liter:

K_2HPO_4	7.0g
δ-Aminovaleric acid·HCl (neutralized)	5.0g
Agar	1.5g

KH_2PO_4	1.3g
Yeast extract	1.0g
$MgCl_2 \cdot 6H_2O$	0.2g
$CaCl_2 \cdot 2H_2O$	0.01g
$FeCl_3 \cdot 6H_2O$	0.01g
Methylene Blue	2.0mg
$MnSO_4$	1.0mg
Na_2MoO_4	1.0mg
$Na_2S \cdot 9H_2O$ solution	20.0mL

$Na_2S \cdot 9H_2O$ Solution:
Composition per 100.0mL:

$Na_2S \cdot 9H_2O$	1.5g

Preparation of $Na_2S \cdot 9H_2O$ Solution: Add $Na_2S \cdot 9H_2O$ to distilled/deionized water and bring volume to 100.0mL. Autoclave for 15 min at 15 psi pressure–121°C. Use freshly prepared solution.

Preparation of Medium: Add components, except $Na_2S \cdot 9H_2O$ solution, to distilled/deionized water and bring volume to 1.0L. Mix thoroughly. Gently heat to boiling. Autoclave for 15 min at 15 psi pressure–121°C. Cool to 45°–50°C. Distribute anaerobically into sterile tubes. Aseptically add 0.1mL of sterile $Na_2S \cdot 9H_2O$ solution to each 5.0mL of the medium. Cap with rubber stoppers.

Use: For the cultivation and maintenance of a variety of *Clostridium* species.

Clostridium Medium (ATCC Medium 43)

Composition per liter:

Agar	15.0g
Yeast extract	5.0g
L-Arginine·HCl	2.0g
L-Lysine·HCl	2.0g
NH_4Cl	2.0g
Sodium formate	2.0g
K_2HPO_4	1.75g
$MgSO_4 \cdot 7H_2O$	0.2g
$CaCl_2 \cdot 2H_2O$	0.01g
$FeSO_4 \cdot 7H_2O$	0.01g
Methylene Blue	2.0mg
$Na_2S \cdot 9H_2O$ solution	30.0mL

$Na_2S \cdot 9H_2O$ Solution:
Composition per 100.0mL:

$Na_2S \cdot 9H_2O$	1.0g

Preparation of $Na_2S \cdot 9H_2O$ Solution: Add $Na_2S \cdot 9H_2O$ to distilled/deionized water and bring volume to 100.0mL. Autoclave for 15 min at 15 psi pressure–121°C. Use freshly prepared solution.

Preparation of Medium: Add components, except $Na_2S \cdot 9H_2O$ solution, to tap water and bring volume to 1.0L. Mix thoroughly. Gently heat to boiling. Autoclave for 15 min at 15 psi pressure–121°C. Cool to 45°–50°C. Distribute anaerobically into sterile tubes. Aseptically add 0.15mL of sterile $Na_2S \cdot 9H_2O$ solution to each 5.0mL of the medium. Cap with rubber stoppers.

Use: For the cultivation and maintenance of a variety of *Clostridium* species.

Clostridium Medium (ATCC Medium 163)

Composition per liter:

Agar	20.0g
Sodium glutamate	17.0g
Yeast extract	6.0g
Sodium thioglycolate	0.5g
Phosphate buffer (1.0*M*, pH 7.4)	40.0mL
$MgSO_4$ (2.0*M* solution)	0.5mL
$FeSO_4$ (0.2*M* solution)	0.2mL
$CaCl_2$ (1.0*M* solution)	0.1mL
$CoCl_2$ (0.1*M* solution)	0.1mL
$MnCl_2$ (0.1*M* solution)	0.1mL
Na_2MoO_4 (0.1*M* solution)	0.1mL

Preparation of Medium: Add components to distilled/deionized water and bring volume to 1.0L. Mix thoroughly. Gently heat to boiling. Autoclave for 15 min at 15 psi pressure–121°C. Pour into sterile Petri dishes or distribute into sterile tubes.

Use: For the cultivation and maintenance of a variety of *Clostridium* species.

Clostridium Medium (ATCC Medium 511)

Composition per liter:

Yeast extract	4.0g
Alanine	3.0g
Peptone	3.0g
L-Cysteine	0.2g
$MgSO_4$	0.05g
$FeSO_4$	0.01g
Potassium phosphate buffer (1.0*M*, pH 7.1)	5.0mL
$CaSO_4$ (saturated solution)	2.5mL

Preparation of Medium: Add components to distilled/deionized water and bring volume to 1.0L. Mix thoroughly. Distribute into tubes or flasks. Autoclave for 15 min at 15 psi pressure–121°C.

Use: For the cultivation of a variety of *Clostridium* species.

Clostridium Medium (ATCC Medium 568)

Composition per liter:

Na_2CO_3	10.0g
Fructose	3.0g
K_2HPO_4	2.0g
Yeast extract	2.0g
$(NH_4)_2SO_4$	1.0g
$MgSO_4 \cdot 7H_2O$	0.5g
Sodium thioglycolate	0.05g
$CaSO_4$	0.015g
$FeSO_4 \cdot 7H_2O$	2.5mg
$MnSO_4 \cdot H_2O$	0.5mg
$Na_2MoO_4 \cdot 2H_2O$	0.5mg

pH 7.8 ± 0.2 at 25°C

Preparation of Medium: Add components to distilled/deionized water and bring volume to 1.0L. Mix thoroughly. Distribute into tubes or flasks. Autoclave for 15 min at 15 psi pressure–121°C.

Use: For the cultivation of a variety of *Clostridium* species.

Clostridium Medium (ATCC Medium 591)

Composition per liter:

Solution 1	600.0mL
Solution 2	400.0mL

pH 8.0 ± 0.2 at 25°C

Solution 1:
Composition per 600.0mL:
Peptone 5.0g

Preparation of Solution 1: Add peptone to distilled/deionized water and bring volume to 600.0mL. Mix thoroughly. Autoclave for 15 min at 15 psi pressure–121°C.

Solution 2:
Composition per 400.0mL:
$NaHCO_3$ 20.0g
Fructose 10.0g
K_2HPO_4 .... 10.0g
Sodium thioglycolate 0.75g
Vitamin solution 14.0mL
Trace elements solution 10.0mL

Preparation of Solution 2: Add components, except sodium thioglycolate, to distilled/deionized water and bring volume to 400.0mL. Mix thoroughly. Gas with 100% CO_2. Add sodium thioglycolate. Adjust pH to 8.0. Filter sterilize.

Vitamin Solution:
Composition per 100.0mL:
Thiamine 0.1g
Nicotinic acid 0.05g
Pyridoxine 0.05g
Pantothenic acid 0.025g
p-Aminobenzoic acid 5.0mg
Vitamin B_{12} 2.0mg
Biotin 1.0mg

Preparation of Vitamin Solution: Add components to distilled/deionized water and bring volume to 100.0mL. Mix thoroughly.

Trace Elements Solution:
Composition per liter:
EDTA 0.5g
$FeSO_4 \cdot 7H_2O$ 0.2g
H_3BO_3 .... 0.03g
$CoCl_2 \cdot 6H_2O$ 0.02g
$ZnSO_4 \cdot 7H_2O$ 0.01g
$MnCl_2 \cdot 4H_2O$ 3.0mg
$Na_2MoO_4 \cdot 2H_2O$ 3.0mg
$NiCl_2 \cdot 6H_2O$ 2.0mg
$CuCl_2 \cdot 2H_2O$ 1.0mg

Preparation of Trace Elements Solution: Add components to distilled/deionized water and bring volume to 1.0L. Mix thoroughly.

Preparation of Medium: Aseptically combine 600.0mL of sterile solution 1 and 400.0mL of sterile solution 2. Distribute into sterile tubes or flasks.

Use: For the cultivation and maintenance of a variety of *Clostridium* species.

Clostridium M1 Medium

Composition per liter:
NaCl 60.0g
Betaine·H_2O 6.0g
$MgSO_4 \cdot 7H_2O$ 5.0g
$NaHCO_3$ 5.0g
L-Alanine 2.2g
NH_4Cl 1.0g
Yeast extract 1.0g
$CaCl_2 \cdot 2H_2O$ 25.0mg
Resazurin 1.0mg
$Na_2SeO_3 \cdot 5H_2O$ 15.0μg
Phosphate solution 100.0mL
Vitamin solution 10.0mL
$Na_2S \cdot 9H_2O$ solution 10.0mL
Trace elements solution SL-10 .... 1.0mL
pH 7.3 ± 0.2 at 25°C

Phosphate Solution:
Composition per 100.0mL:
K_2HPO_4 .... 0.358g
KH_2PO_4 .... 0.223g

Preparation of Phosphate Solution: Add components to distilled/deionized water and bring volume to 100.0mL. Mix thoroughly. Sparge with 100% N_2. Autoclave for 15 min at 15 psi pressure–121°C.

$Na_2S \cdot 9H_2O$ Solution:
Composition per 10.0mL:
$Na_2S \cdot 9H_2O$ 0.3g

Preparation of $Na_2S \cdot 9H_2O$ Solution: Add $Na_2S \cdot 9H_2O$ to distilled/deionized water and bring volume to 10.0mL. Mix thoroughly. Sparge with 100% N_2. Autoclave for 15 min at 15 psi pressure–121°C.

Vitamin Solution:
Composition per liter:
Pyridoxine·HCl 10.0mg
Calcium DL-pantothenate 5.0mg
Lipoic acid 5.0mg
Nicotinic acid 5.0mg
p-Aminobenzoic acid 5.0mg
Riboflavin 5.0mg
Thiamine·HCl 5.0mg
Biotin 2.0mg
Folic acid 2.0mg
Vitamin B_{12} 0.1mg

Preparation of Vitamin Solution: Add components to distilled/deionized water and bring volume to 1.0L. Mix thoroughly. Filter sterilize. Sparge with 100% N_2.

Trace Elements Solution SL-10:
Composition per liter:
$FeCl_2 \cdot 4H_2O$ 1.5g
$CoCl_2 \cdot 6H_2O$ 190.0mg
$MnCl_2 \cdot 4H_2O$ 100.0mg
$ZnCl_2$ 70.0mg
$Na_2MoO_4 \cdot 2H_2O$ 36.0mg
$NiCl_2 \cdot 6H_2O$ 24.0mg
H_3BO_3 .... 6.0mg
$CuCl_2 \cdot 2H_2O$ 2.0mg
HCl (25% solution) 10.0mL

Preparation of Trace Elements Solution SL-10: Add $FeCl_2 \cdot 4H_2O$ to 10.0mL of HCl solution. Mix thoroughly. Add distilled/deionized water and bring volume to 1.0L. Add remaining components. Mix thoroughly.

Preparation of Medium: Add components, except phosphate solution and $Na_2S \cdot 9H_2O$ solution, to distilled/deionized water and bring volume to 890.0mL. Mix thoroughly. Sparge with 80% N_2 + 20% CO_2. Autoclave for 15 min at 15 psi pressure–121°C. Aseptically and anaerobically add 100.0mL of sterile phosphate solution and 10.0mL of sterile $Na_2S \cdot 9H_2O$ solution. Mix thoroughly. Aseptically and anaerobically distribute into sterile screw-capped bottles under 80% N_2 + 20% CO_2.

Use: For the cultivation of *Clostridium halophilum, Clostridium litorale,* and *Clostridium* species.

Clostridium M1 Medium

Composition per liter:

NaCl	10.0g
Betaine·H_2O	6.0g
$NaHCO_3$	5.0g
L-Alanine	2.2g
NH_4Cl	1.0g
Yeast extract	1.0g
$MgSO_4 \cdot 7H_2O$	0.5g
$CaCl_2 \cdot 2H_2O$	25.0mg
Resazurin	1.0mg
$Na_2SeO_3 \cdot 5H_2O$	15.0μg
Phosphate solution	100.0mL
Vitamin solution	10.0mL
$Na_2S \cdot 9H_2O$ solution	10.0mL
Trace elements solution SL-10	1.0mL

pH 7.3 ± 0.2 at 25°C

Phosphate Solution:

Composition per 100.0mL:

K_2HPO_4	0.358g
KH_2PO_4	0.223g

Preparation of Phosphate Solution: Add components to distilled/deionized water and bring volume to 100.0mL. Mix thoroughly. Sparge with 100% N_2. Autoclave for 15 min at 15 psi pressure–121°C.

$Na_2S \cdot 9H_2O$ Solution:

Composition per 10.0mL:

$Na_2S \cdot 9H_2O$	0.3g

Preparation of $Na_2S \cdot 9H_2O$ Solution: Add $Na_2S \cdot 9H_2O$ to distilled/deionized water and bring volume to 10.0mL. Mix thoroughly. Sparge with 100% N_2. Autoclave for 15 min at 15 psi pressure–121°C.

Vitamin Solution:

Composition per liter:

Pyridoxine·HCl	10.0mg
Calcium DL-pantothenate	5.0mg
Lipoic acid	5.0mg
Nicotinic acid	5.0mg
p-Aminobenzoic acid	5.0mg
Riboflavin	5.0mg
Thiamine·HCl	5.0mg
Biotin	2.0mg
Folic acid	2.0mg
Vitamin B_{12}	0.1mg

Preparation of Vitamin Solution: Add components to distilled/deionized water and bring volume to 1.0L. Mix thoroughly. Filter sterilize. Sparge with 100% N_2.

Trace Elements Solution SL-10:

Composition per liter:

$FeCl_2 \cdot 4H_2O$	1.5g
$CoCl_2 \cdot 6H_2O$	190.0mg
$MnCl_2 \cdot 4H_2O$	100.0mg
$ZnCl_2$	70.0mg
$Na_2MoO_4 \cdot 2H_2O$	36.0mg
$NiCl_2 \cdot 6H_2O$	24.0mg
H_3BO_3	6.0mg
$CuCl_2 \cdot 2H_2O$	2.0mg
HCl (25% solution)	10.0mL

Preparation of Trace Elements Solution SL-10: Add $FeCl_2 \cdot 4H_2O$ to 10.0mL of HCl solution. Mix thoroughly. Add distilled/deionized water and bring volume to 1.0L. Add remaining components. Mix thoroughly.

Preparation of Medium: Add components, except phosphate solution and $Na_2S \cdot 9H_2O$ solution, to distilled/deionized water and bring volume to 890.0mL. Mix thoroughly. Sparge with 80% N_2 + 20% CO_2. Autoclave for 15 min at 15 psi pressure–121°C. Aseptically and anaerobically add 100.0mL of sterile phosphate solution and 10.0mL of sterile $Na_2S \cdot 9H_2O$ solution. Mix thoroughly. Aseptically and anaerobically distribute into sterile screw-capped bottles under 80% N_2 + 20% CO_2.

Use: For the cultivation and maintenance of *Clostridium halophilum* and *Clostridium litorale.*

Clostridium methylpentosum Medium

Composition per liter:

L-Cysteine·HCl	1.0g
Resazurin	1.0mg
Mineral solution A	100.0mL
Mineral solution B	100.0mL
Mineral solution C	100.0mL
$NaHCO_3$ solution	20.0mL
Rhamnose solution	20.0mL
Vitamin solution	10.0mL

pH 6.8 ± 0.2 at 25°C

Mineral Solution A:

Composition per 100.0mL:

KH_2PO_4	0.9g
NaCl	0.9g
NH_4Cl	0.9g

Preparation of Mineral Solution A: Add components to distilled/deionized water and bring volume to 100.0mL. Mix thoroughly.

Mineral Solution B:

Composition per 100.0mL:

$CaCl_2 \cdot 2H_2O$	20.0mg
$MnCl_2 \cdot 4H_2O$	20.0mg
$CoCl_2 \cdot 6H_2O$	5.0mg

Preparation of Mineral Solution B: Add components to distilled/deionized water and bring volume to 100.0mL. Mix thoroughly.

Mineral solution C:

Composition per 100.0mL

$MgSO_4 \cdot 7H_2O$	20.0mg
$FeSO_4 \cdot 7H_2O$	5.0mg
$CuSO_4 \cdot H_2O$	2.0mg
$ZnSO_4 \cdot 7H_2O$	2.0mg

Preparation of Mineral Solution C: Add components to distilled/deionized water and bring volume to 100.0mL. Mix thoroughly.

$NaHCO_3$ Solution:

Composition per 20.0mL:

$NaHCO_3$	1.0g

Preparation of $NaHCO_3$ Solution: Add $NaHCO_3$ to distilled/deionized water and bring volume to 20.0mL. Mix thoroughly. Sparge with 100% N_2. Autoclave for 15 min at 15 psi pressure–121°C.

Rhamnose Solution:

Composition per 20.0mL:

Rhamnose	2.0g

Preparation of Rhamnose Solution: Add rhamnose to distilled/deionized water and bring volume to 20.0mL. Mix thoroughly. Sparge under 100% N_2 gas for 3 min. Filter sterilize. Store under N_2 gas.

Vitamin Solution:

Composition per liter:

Pyridoxine·HCl	10.0mg
Calcium DL-pantothenate	5.0mg
Lipoic acid	5.0mg
Nicotinic acid	5.0mg
p-Aminobenzoic acid	5.0mg
Riboflavin	5.0mg
Thiamine·HCl	5.0mg
Biotin	2.0mg
Folic acid	2.0mg
Vitamin B_{12}	0.1mg

Preparation of Vitamin Solution: Add components to distilled/deionized water and bring volume to 1.0L. Mix thoroughly.

Preparation of Medium: Add components, except $NaHCO_3$ solution and rhamnose solution, to distilled/deionized water and bring volume to 960.0mL. Mix thoroughly. Adjust pH to 6.5 with KOH. Sparge under 100% N_2 for 3–4 min. Autoclave for 15 min at 15 psi pressure–121°C. Aseptically and anaerobically add 20.0mL of sterile $NaHCO_3$ solution and 20.0mL of sterile rhamnose solution. Mix thoroughly. Aseptically and anaerobically distribute into sterile screw-capped bottles under 100% N_2. Final pH of the medium should be 6.8.

Use: For the cultivation of *Clostridium methylpentosum.*

Clostridium neopropionicum Medium

Composition per liter

$KHCO_3$	4.0g
Ethanol	1.0g
NH_4Cl	1.0g
NaCl	0.6g
Pancreatic digest of casein	0.5g
Yeast extract	0.5g
KH_2PO_4	0.3g
$MgCl_2·6H_2O$	0.1g
$CaCl_2·2H_2O$	0.08g
Resazurin	1.0mg
Trace elements solution	10.0mL
Vitamin solution	10.0mL
L-Cysteine·HCl·H_2O solution	10.0mL
$Na_2S·9H_2O$ solution	10.0mL

pH 6.8 ± 0.2 at 25°C

Trace Elements Solution:

Composition per liter:

Nitrilotriacetic acid	12.8g
$FeCl_3·6H_2O$	1.35g
NaCl	1.0g
$NiCl_2·6H_2O$	0.12g
$MnCl_2·4H_2O$	0.1g
$CaCl_2·2H_2O$	0.1g
$ZnCl_2$	0.1g
$Na_2SeO_3·5H_2O$	0.026g
$CuCl_2·2H_2O$	0.025g
$CoCl_2·6H_2O$	0.024g
$Na_2MoO_4·2H_2O$	0.024g
H_3BO_3	0.01g

Preparation of Trace Elements Solution: Add nitrilotriacetic acid to approximately 500.0mL of distilled/deionized water. Dissolve by adding KOH and adjust pH to 6.5. Add remaining components. Bring volume to 1.0L with additional distilled/deionized water. Adjust pH to 7.0 with KOH.

Vitamin Solution:

Composition per liter:

Pyridoxine·HCl	10.0mg
Calcium DL-pantothenate	5.0mg
Lipoic acid	5.0mg
Nicotinic acid	5.0mg
p-Aminobenzoic acid	5.0mg
Riboflavin	5.0mg
Thiamine·HCl	5.0mg
Biotin	2.0mg
Folic acid	2.0mg
Vitamin B_{12}	0.1mg

Preparation of Vitamin Solution: Add components to distilled/deionized water and bring volume to 1.0L. Mix thoroughly. Filter sterilize. Sparge with 80% N_2 + 20% CO_2.

L-Cysteine·HCl Solution:

Composition per 10.0mL:

L-Cysteine·HCl	0.25g

Preparation of L-Cysteine·HCl Solution: Add L-cysteine·HCl to distilled/deionized water and bring volume to 10.0mL. Mix thoroughly. Autoclave under 80% N_2 + 20% CO_2 for 15 min at 15 psi pressure–121°C.

$Na_2S·9H_2O$ Solution:

Composition per 10.0mL:

$Na_2S·9H_2O$	0.3g

Preparation of $Na_2S·9H_2O$ Solution: Add $Na_2S·9H_2O$ to distilled/deionized water and bring volume to 10.0mL. Mix thoroughly. Sparge with 80% N_2 + 20% CO_2. Autoclave for 15 min at 15 psi pressure–121°C.

Preparation of Medium: Add components, except vitamin solution, L-cysteine·HCl·H_2O solution, and $Na_2S·9H_2O$ solution, to distilled/deionized water and bring volume to 960.0mL. Mix thoroughly. Sparge under 80% N_2 + 20% CO_2 for 3–4 min. Autoclave for 15 min at 15 psi pressure–121°C. Aseptically and anaerobically add 20.0mL of sterile vitamin solution, 10.0mL of sterile L-cysteine·HCl·H_2O solution, and 10.0mL of sterile $Na_2S·9H_2O$ solution. Mix thoroughly. Aseptically and anaerobically distribute into sterile screw-capped bottles under 80% N_2 + 20% CO_2.

Use: For the cultivation and maintenance of *Clostridium neopropionicum.*

Clostridium noterae Medium

Composition per liter:

Yeast extract	2.0g
NH_4Cl	1.0g
NaCl	0.45g
$K_2HPO_4·3H_2O$	0.4g
L-Cysteine·HCl·H_2O	0.15g
Na_2CO_3 solution	30.0mL
Trace metals solution	10.0mL
$Na_2S·9H_2O$ solution	10.0mL

pH 7.9 ± 0.1 at 25°C

Na_2CO_3 Solution:

Composition per 50.0mL:

Na_2CO_3	5.0g

Preparation of Na_2CO_3 Solution: Add Na_2CO_3 to distilled/deionized water and bring volume to 50.0mL. Autoclave for 15 min at 15 psi pressure–121°C. Use freshly prepared solution.

Trace Metals Solution:
Composition per liter:

$Na_2EDTA{\cdot}2H_2O$	0.5g
$CoCl_2{\cdot}6H_2O$	0.15g
$FeSO_4{\cdot}7H_2O$	0.1g
$MnCl_2{\cdot}4H_2O$	0.1g
$ZnCl_2$	0.1g
$AlCl_3{\cdot}6H_2O$	0.04g
$CuCl_2{\cdot}2H_2O$	0.02g
$NiSO_4{\cdot}6H_2O$	0.02g
H_2SeO_3	0.01g
H_3BO_3	0.01g
$Na_2MoO_4{\cdot}2H_2O$	0.01g

Preparation of Trace Metals Solution: Add components to distilled/deionized water and bring volume to 1.0L. Mix thoroughly.

$Na_2S{\cdot}9H_2O$ Solution:
Composition per 10.0mL:

$Na_2S{\cdot}9H_2O$	0.15g

Preparation of $Na_2S{\cdot}9H_2O$ Solution: Add $Na_2S{\cdot}9H_2O$ to distilled/deionized water and bring volume to 10.0mL. Autoclave for 15 min at 15 psi pressure–121°C. Use freshly prepared solution.

Preparation of Medium: Add components, except Na_2CO_3 solution and $Na_2S{\cdot}9H_2O$ solution, to distilled/deionized water and bring volume to 1.0L. Mix thoroughly. Adjust pH to 7.0 with 10*M* NaOH. Gently heat to boiling. Distribute under O_2-free 100% N_2 gas into tubes in 5.0mL volumes. Cap with rubber stoppers. Autoclave for 15 min at 15 psi pressure–121°C. Prior to inoculation, add to each tube 0.15mL of Na_2CO_3 solution and 0.05mL of $Na_2S{\cdot}9H_2O$ solution. Incubate under 80% H_2 + 20% CO_2 to provide conditions for H_2 fixation.

Use: For the cultivation and maintenance of *Clostridium noterae.*

Clostridium novyi Blood Agar

Composition per 100.0mL:

Agar	2.0g
Glucose	1.0g
Neopeptone	1.0g
Proteolyzed liver	0.5g
Yeast extract	0.5g
Horse blood, defibrinated	10.0mL
Reducing solution	0.75mL
Salts solution	0.5mL

pH 7.6–7.8 at 25°C

Salts Solution:
Composition per 100.0mL:

$MgSO_4{\cdot}7H_2O$	4.0g
$MnSO_4{\cdot}4H_2O$	0.2g
HCl	0.05g
$FeCl_3$	0.04g

Preparation of Salts Solution: Add components to distilled/deionized water and bring volume to 100.0mL. Mix thoroughly.

Reducing Solution:
Composition per 10.0mL:

L-Cysteine·HCl·H_2O	0.12g
Dithiothreitol	0.12g
Glutamine	0.06g

Preparation of Reducing Solution: Add components to distilled/deionized water and bring volume to 10.0mL. Mix thoroughly. Adjust pH to 7.6–7.8. Filter sterilize.

Preparation of Medium: Add agar to distilled/deionized water and bring volume to 50.0mL. Mix thoroughly. Gently heat and bring to boiling. In another flask, add neopeptone, yeast extract, liver extract, and salts solution to distilled/deionized water and bring volume to 50.0mL. Mix thoroughly. Gently heat until dissolved. Combine the two solutions. Distribute into screw-capped bottles in 18.0mL volumes. Autoclave for 10 min at 10 psi pressure–115°C. Cool to 45°–50°C. Medium may be stored at 4°C at this point. Immediately prior to inoculation, aseptically add 2.0mL of horse blood and 0.15mL of sterile reducing solution to each tube of melted agar at 50°C. Mix thoroughly. Pour the contents of each tube into a sterile Petri dish.

Use: For the cultivation of *Clostridium novyi.*

Clostridium oroticum Medium

Composition per liter:

K_2HPO_4	6.95g
Pancreatic digest of casein	5.0g
Sodium orotate	2.5g
KH_2PO_4	1.36g
Yeast extract	0.5g
Riboflavin	15.0mg

pH 7.5 ± 0.2 at 25°C

Preparation of Medium: Add components to distilled/deionized water and bring volume to 1.0L. Mix thoroughly. Adjust pH to 7.5. Distribute into tubes or flasks. Autoclave for 15 min at 15 psi pressure–121°C.

Use: For the cultivation of *Clostridium oroticum.*

Clostridium papyrosolvens Medium

Composition per liter:

K_2HPO_4	1.65g
NH_4Cl	1.0g
Yeast extract	0.6g
L-Cysteine·HCl	0.5g
Resazurin	1.0mg
Seawater, filtered	200.0mL
Mineral salt solution	150.0mL
Cellobiose solution	50.0mL

pH 7.2 ± 0.2 at 25°C

Mineral Salt Solution:
Composition per liter:

$(NH_4)_2SO_4$	6.0g
NaCl	6.0g
$MgSO_4{\cdot}7H_2O$	1.2g
$CaCl_2{\cdot}2H_2O$	0.8g

Preparation of Mineral Salt Solution: Add components to distilled/deionized water and bring volume to 1.0L. Mix thoroughly.

Cellobiose Solution:
Composition per 50.0mL:

D-Cellobiose	5.0g

Preparation of Cellobiose Solution: Add cellobiose to distilled/deionized water and bring volume to 50.0mL. Mix thoroughly. Sparge under 100% N_2 gas for 3 min. Filter sterilize.

Preparation of Medium: Add components, except cellobiose solution, to distilled/deionized water and bring volume to 950.0mL. Mix thoroughly. Adjust pH to 7.2 with 5*N* NaOH. Sparge with 100% N_2. Autoclave for 15 min at 15 psi pressure–121°C. Aseptically and anaerobically add 50.0mL of sterile cellobiose solution. Mix thoroughly. Aseptically and anaerobically distribute into sterile screw-capped bottles under 100% N_2.

Use: For the cultivation and maintenance of *Clostridium papyrosolvens.*

Clostridium papyrosolvens Medium

Composition per liter:

Paper strips, sterile	3.0g
K_2HPO_4	1.65g
NH_4Cl	1.0g
Yeast extract	0.6g
L-Cysteine·HCl	0.5g
Resazurin	1.0mg
Seawater, filtered	200.0mL
Mineral salt solution	150.0mL

pH 7.2 ± 0.2 at 25°C

Mineral Salt Solution:

Composition per liter:

$(NH_4)_2SO_4$	6.0g
NaCl	6.0g
$MgSO_4·7H_2O$	1.2g
$CaCl_2·2H_2O$	0.8g

Preparation of Mineral Salt Solution: Add components to distilled/deionized water and bring volume to 1.0L. Mix thoroughly.

Preparation of Medium: Add components, except paper strips, to distilled/deionized water and bring volume to 1.0L. Mix thoroughly. Adjust pH to 7.2 with 5*N* NaOH. Sparge with 100% N_2. Autoclave for 15 min at 15 psi pressure–121°C. Aseptically and anaerobically add 3.0g of sterile paper strips (filter paper, Kleenex, or lens tissue). Mix thoroughly. Aseptically and anaerobically distribute into sterile screw-capped bottles under 100% N_2.

Use: For the cultivation and maintenance of *Clostridium papyrosolvens.*

Clostridium perfringens Agar, OPSP (Perfringens Agar, OPSP)

Composition per liter:

Pancreatic digest of casein	15.0g
Agar	10.0g
Liver extract	7.0g
Papaic digest of soybean meal	5.0g
Yeast extract	5.0g
Tris(hydroxymethyl)aminomethane buffer	1.5g
Ferric ammonium citrate	1.0g
$Na_2S_2O_5$	1.0g
Antibiotic inhibitor	10.0mL

pH 7.3 ± 0.2 at 25°C

Source: This medium is available as a premixed powder from Oxoid Unipath.

Antibiotic Inhibitor:

Composition per 10.0mL:

Sodium sulfadiazine	0.1g
Oleandomycin phosphate	0.5mg
Polymyxin B	10,000U

Preparation of Antibiotic Inhibitor: Add components to distilled/deionized water and bring volume to 10.0mL. Mix thoroughly. Filter sterilize.

Preparation of Medium: Add components, except antibiotic inhibitor, to distilled/deionized water and bring volume to 990.0mL. Mix thoroughly. Gently heat and bring to boiling. Autoclave for 15 min at 15 psi pressure–121°C. Cool to 45°–50°C. Aseptically add sterile antibiotic inhibitor. Mix thoroughly. Pour into sterile Petri dishes or distribute into sterile tubes.

Use: For the presumptive identification and enumeration of *Clostridium perfringens* in foods.

Clostridium perfringens Sporulation Broth

Composition per liter:

Tryptose	15.0g
Na_2HPO_4	11.0g
Starch, soluble	3.0g
Yeast extract	3.0g
Na-thioglycollate	1.0g
$MgSO_4$	0.1g

pH 7.8 ± 0.2 at 25°C

Source: This medium is available as a premixed powder from HiMedia.

Preparation of Medium: Add components to distilled/deionized water and bring volume to 1.0L. Mix thoroughly. Gently heat and bring to boiling. Autoclave for 15 min at 15 psi pressure–121°C. Mix thoroughly. Pour into sterile Petri dishes or distribute into sterile tubes.

Use: For the production of *Clostridium perfringens* spores.

Clostridium perfringens Sporulation HiVeg Broth

Composition per liter:

Plant hydrolysate No. 1	15.0g
Na_2HPO_4	11.0g
Starch, soluble	3.0g
Yeast extract	3.0g
Na-thioglycollate	1.0g
$MgSO_4$	0.1g

pH 7.8 ± 0.2 at 25°C

Source: This medium is available as a premixed powder from HiMedia.

Preparation of Medium: Add components to distilled/deionized water and bring volume to 1.0L. Mix thoroughly. Gently heat and bring to boiling. Autoclave for 15 min at 15 psi pressure–121°C. Mix thoroughly. Pour into sterile Petri dishes or distribute into sterile tubes.

Use: For the production of *Clostridium perfringens* spores.

Clostridium pfennigii Medium

Composition per 1001.0mL:

Solution A	890.0mL
Solution B	100.0mL
Solution C	10.0mL
Solution D	1.0mL

pH 7.0–7.2 at 25°C

Solution A:

Composition per 890.0mL:

Sodium vanillate	2.0g
Yeast extract	2.0g

Resazurin	1.0mg
Rumen fluid, clarified	267.0mL
Mineral solution	50.0mL
Vitamin solution	5.0mL
Trace elements solution SL-10	1.0mL

Preparation of Solution A: Add components to distilled/deionized water and bring volume to 890.0mL. Mix thoroughly. Adjust pH to 6.9. Sparge with 80% N_2 + 20% CO_2 for 20 min. Distribute 8.9mL into anaerobic tubes under 80% N_2 + 20% CO_2. Autoclave under 80% N_2 + 20% CO_2 for 15 min at 15 psi pressure–121°C.

Mineral Solution:
Composition per liter:

KH_2PO_4	10.0g
NaCl	8.0g
NH_4Cl	8.0g
$MgCl_2 \cdot 6H_2O$	6.6g
$CaCl_2 \cdot 2H_2O$	1.0g

Preparation of Mineral Solution: Add components to distilled/deionized water and bring volume to 1.0L. Mix thoroughly.

Vitamin Solution:
Composition per liter:

Pyridoxine·HCl	6.2mg
Nicotinic acid	2.5mg
p-Aminobenzoic acid	1.25mg
Thiamine·HCl	1.25mg
Pantothenic acid	0.62mg
Biotin	0.25mg

Preparation of Vitamin Solution: Add components to distilled/deionized water and bring volume to 1.0L. Adjust pH to 7.0. Mix thoroughly.

Trace Elements Solution SL-10:
Composition per liter:

$FeCl_2 \cdot 4H_2O$	1.5g
$CoCl_2 \cdot 6H_2O$	190.0mg
$MnCl_2 \cdot 4H_2O$	100.0mg
$ZnCl_2$	70.0mg
$Na_2MoO_4 \cdot 2H_2O$	36.0mg
$NiCl_2 \cdot 6H_2O$	24.0mg
H_3BO_3	6.0mg
$CuCl_2 \cdot 2H_2O$	2.0mg
HCl (25% solution)	10.0mL

Preparation of Trace Elements Solution SL-10: Add $FeCl_2 \cdot 4H_2O$ to 10.0mL of HCl solution. Mix thoroughly. Add distilled/deionized water and bring volume to 1.0L. Add remaining components. Mix thoroughly.

Solution B:
Composition per 100.0mL:

$NaHCO_3$	5.0g

Preparation of Solution B: Add $NaHCO_3$ to distilled/deionized water and bring volume to 100.0mL. Mix thoroughly. Filter sterilize. Sparge with 80% N_2 + 20% CO_2 for 20 min.

Solution C:
Composition per 10.0mL:

L-Cysteine	0.24g

Preparation of Solution C: Add L-cysteine to distilled/deionized water and bring volume to 10.0mL. Mix thoroughly. Autoclave under 80% N_2 + 20% CO_2 for 15 min at 15 psi pressure–121°C.

Solution D:
Composition per 1.0mL:

$Na_2S \cdot 9H_2O$	78.0mg

Preparation of Solution D: Add $Na_2S \cdot 9H_2O$ to distilled/deionized water and bring volume to 1.0mL. Mix thoroughly. Autoclave under 80% N_2 + 20% CO_2 for 15 min at 15 psi pressure–121°C.

Preparation of Medium: To each tube containing 8.9mL of sterile solution A, add (using a syringe) 1.0mL of sterile solution B, 0.1mL of sterile solution C, and 0.01mL of sterile solution D.

Use: For the cultivation and maintenance of *Clostridium pfennigii.*

Clostridium propionicum Medium

Composition per 1007.5mL:

Yeast extract	4.0g
L-Alanine	3.0g
Peptone	3.0g
L-Cysteine·HCl	0.3g
$MgSO_4 \cdot 7H_2O$	0.1g
$FeSO_4 \cdot 7H_2O$	0.018g
Resazurin	1.0mg
Potassium phosphate buffer solution, 1*M*, pH 7.1	5.0mL
$CaSO_4$, saturated solution	2.5mL

Preparation of Medium: Add components to distilled/deionized water and bring volume to 1.0L. Mix thoroughly. Bring pH to 7.1. Sparge with 100% N_2 for 20 min. Distribute into tubes or bottles under 100% N_2. Autoclave under 100% N_2 for 15 min at 15 psi pressure–121°C.

Use: For the cultivation and maintenance of *Clostridium propionicum.*

Clostridium Selective Agar (Clostrisel Agar)

Composition per liter:

Pancreatic digest of casein	17.0g
Agar	14.0g
Glucose	6.0g
Papaic digest of soybean meal	3.0g
NaCl	2.5g
Sodium thioglycolate	1.8g
Sodium formaldehyde sulfoxylate	1.0g
L-Cystine	0.25g
NaN_3	0.15g
Neomycin sulfate	0.15g

pH 7.0 ± 0.2 at 25°C

Source: This medium is available as a premixed powder from BD Diagnostic Systems.

Preparation of Medium: Add components to distilled/deionized water and bring volume to 1.0L. Mix thoroughly. Gently heat while stirring and bring to boiling. Distribute into tubes or flasks. Autoclave for 15 min at 15 psi pressure–118°C. Pour into sterile Petri dishes or leave in tubes.

Caution: Sodium azide is toxic. Azides also react with metals and disposal must be highly diluted.

Use: For the selective isolation of pathogenic *Clostridium* species from specimens containing mixed flora, e.g., from wounds, fecal specimens, soil, and other specimens.

Clostridium sphenoides Medium

Composition per liter:

Agar	15.0g
Trisodium citrate·$2H_2O$	14.7g
Yeast extract	4.0g
KH_2PO_4	3.4g
K_2HPO_4	2.0g
Peptone	2.0g
NaCl	0.6g
L-Cysteine·HCl	0.3g
$(NH_4)_2SO_4$	0.3g
$MgSO_4 \cdot 7H_2O$	0.2g
$CaCl_2 \cdot 2H_2O$	0.06g
Resazurin	1.0mg

pH 6.7–7.0 ± 0.2 at 25°C

Preparation of Medium: Add components, except L-cysteine·HCl, to distilled/deionized water and bring volume to 1.0L. Mix thoroughly. Gently heat and bring to boiling. Add L-cysteine·HCl. Distribute anaerobically into tubes in 5.0mL volumes. Autoclave for 20 min at 15 psi pressure–121°C. Cool to 45°–50°C. Inoculate with serial dilution of mud specimens before agar solidifies.

Use: For the isolation of *Clostridium sphenoides* from mud.

Clostridium sticklandii Medium

Composition per liter:

Yeast extract	5.0g
L-Arginine·HCl	2.0g
L-Lysine·HCl	2.0g
NH_4Cl	2.0g
Sodium formate	2.0g
K_2HPO_4	1.75g
$MgSO_4 \cdot 7H_2O$	0.2g
$CaCl_2 \cdot 2H_2O$	10.0mg
$FeSO_4 \cdot 7H_2O$	10.0mg
$Na_2S \cdot H_2O$ solution	10.0mL

pH 7.0 ± 0.2 at 25°C

$Na_2S \cdot 9H_2O$ Solution:

Composition per 10.0mL:

$Na_2S \cdot 9H_2O$	0.3g

Preparation of $Na_2S \cdot 9H_2O$ Solution: Add $Na_2S \cdot 9H_2O$ to distilled/deionized water and bring volume to 10.0mL. Mix thoroughly. Autoclave for 15 min at 15 psi pressure–121°C.

Preparation of Medium: Add components, except $Na_2S \cdot H_2O$ solution, to distilled/deionized water and bring volume to 1.0L. Mix thoroughly. Adjust pH to 7.0. Distribute into tubes or flasks. Autoclave for 15 min at 15 psi pressure–121°C. Aseptically add 0.1mL of sterile $Na_2S \cdot H_2O$ solution to each 10.0mL of medium.

Use: For the cultivation of *Clostridium sticklandii.*

Clostridium sticklandii Medium

Composition per liter:

Tryptone	20.0g
Yeast extract	10.0g
K_2HPO_4	1.04g
KH_2PO_4	0.68g
$Na_2S \cdot 9H_2O$ solution	0.15g

pH 7.0 ± 0.2 at 25°C

$Na_2S \cdot 9H_2O$ Solution:

Composition per 10.0mL:

$Na_2S \cdot 9H_2O$	0.3g

Preparation of $Na_2S \cdot 9H_2O$ Solution: Add $Na_2S \cdot 9H_2O$ to distilled/deionized water and bring volume to 10.0mL. Mix thoroughly. Autoclave for 15 min at 15 psi pressure–121°C.

Preparation of Medium: Add components, except $Na_2S \cdot 9H_2O$ solution, to distilled/deionized water and bring volume to 990.0mL. Mix thoroughly. Adjust pH to 7.0. Autoclave for 15 min at 15 psi pressure–121°C. Aseptically add 10.0mL of sterile $Na_2S \cdot 9H_2O$ solution. Mix thoroughly. Aseptically distribute into sterile tubes or flasks.

Use: For the cultivation of *Clostridium sticklandii.*

Clostridium termitidis Medium

Composition per liter:

NaCl	1.0g
KCl	0.5g
Yeast extract	0.5g
$MgCl_2 \cdot 6H_2O$	0.4g
NH_4Cl	0.3g
KH_2PO_4	0.2g
$CaCl_2 \cdot 2H_2O$	0.15g
Resazurin	1.0mg
Trace elements solution SL-10	1.0mL
Cellobiose solution	50.0mL
$NaHCO_3$ solution	20.0mL
$Na_2S \cdot 9H_2O$ solution	10.0mL

pH 7.0 ± 0.2 at 25°C

Trace Elements Solution SL-10:

Composition per liter:

$FeCl_2 \cdot 4H_2O$	1.5g
$CoCl_2 \cdot 6H_2O$	190.0mg
$MnCl_2 \cdot 4H_2O$	100.0mg
$ZnCl_2$	70.0mg
$Na_2MoO_4 \cdot 2H_2O$	36.0mg
$NiCl_2 \cdot 6H_2O$	24.0mg
H_3BO_3	6.0mg
$CuCl_2 \cdot 2H_2O$	2.0mg
HCl (25% solution)	10.0mL

Preparation of Trace Elements Solution SL-10: Add $FeCl_2 \cdot 4H_2O$ to 10.0mL of HCl solution. Mix thoroughly. Add distilled/deionized water and bring volume to 1.0L. Add remaining components. Mix thoroughly.

Cellobiose Solution:

Composition per 50.0mL:

D-Cellobiose	5.0g

Preparation of Cellobiose Solution: Add cellobiose to distilled/deionized water and bring volume to 50.0mL. Mix thoroughly. Sparge under 100% N_2 gas for 3 min. Filter sterilize.

$NaHCO_3$ Solution:

Composition per 20.0mL:

$NaHCO_3$	4.5g

Preparation of $NaHCO_3$ Solution: Add $NaHCO_3$ to distilled/deionized water and bring volume to 20.0mL. Mix thoroughly. Sparge with 100% N_2. Autoclave for 15 min at 15 psi pressure–121°C.

$Na_2S·9H_2O$ Solution:
Composition per 10.0mL:

$Na_2S·9H_2O$	0.3g

Preparation of $Na_2S·9H_2O$ Solution: Add $Na_2S·9H_2O$ to distilled/deionized water and bring volume to 10.0mL. Mix thoroughly. Sparge with 100% N_2. Autoclave for 15 min at 15 psi pressure–121°C.

Preparation of Medium: Add components, except cellobiose solution, $NaHCO_3$ solution, and $Na_2S·9H_2O$ solution, and bring volume to 920.0mL. Mix thoroughly. Sparge with 80% N_2 + 20% CO_2 until pH reaches below 6.0. Autoclave for 15 min at 15 psi pressure–121°C. Aseptically and anaerobically add 50.0mL of sterile cellobiose solution, 20.0mL of sterile $NaHCO_3$ solution, and 10.0mL of sterile $Na_2S·9H_2O$ solution. Mix thoroughly. Aseptically and anaerobically distribute into sterile screw-capped bottles under 80% N_2 + 20% CO_2.

Use: For the cultivation and maintenance of *Clostridium termitidis.*

Clostridium thermoaceticum Medium (TYE-CO) (DSMZ Medium 316)

Composition per liter:

Trypticase™	10.0g
Yeast extract	3.0g
$Na_2HPO_4·12H_2O$	2.8g
NH_4Cl	1.0g
KH_2PO_4	0.3g
$MgCl_2·6H_2O$	0.2g
$FeSO_4·7H_2O$	1.0mg
Resazurin	1.0mg
Trace elements solution	10.0mL
$Na_2S·9H_2O$ solution	10.0mL
Vitamin solution	5.0mL

pH 7.0 ± 0.2 at 25°C

Vitamin Solution:
Composition per liter:

Pyridoxine-HCl	10.0mg
Thiamine-HCl·2H$_2$O	5.0mg
Riboflavin	5.0mg
Nicotinic acid	5.0mg
D-Ca-pantothenate	5.0mg
p-Aminobenzoic acid	5.0mg
Lipoic acid	5.0mg
Biotin	2.0mg
Folic acid	2.0mg
Vitamin B_{12}	0.1mg

Preparation of Vitamin Solution: Add components to distilled/deionized water and bring volume to 1.0L. Mix thoroughly. Sparge with 80% H_2 + 20% CO_2. Filter sterilize.

Trace Elements Solution:
Composition per liter:

$MgSO_4·7H_2O$	3.0g
Nitrilotriacetic acid	1.5g
NaCl	1.0g
$MnSO_4·2H_2O$	0.5g
$CoSO_4·7H_2O$	0.18g
$ZnSO_4·7H_2O$	0.18g
$CaCl_2·2H_2O$	0.1g
$FeSO_4·7H_2O$	0.1g
$NiCl_2·6H_2O$	0.025g
$KAl(SO_4)_2·12H_2O$	0.02g
H_3BO_3	0.01g
$Na_2MoO_4·4H_2O$	0.01g
$CuSO_4·5H_2O$	0.01g
$Na_2SeO_3·5H_2O$	0.3mg

Preparation of Trace Elements Solution: Add nitrilotriacetic acid to 500.0mL of distilled/deionized water. Dissolve by adjusting pH to 6.5 with KOH. Add remaining components. Add distilled/deionized water to 1.0L. Mix thoroughly.

$Na_2S·9H_2O$ Solution:
Composition per 10.0mL:

$Na_2S·9H_2O$	0.6g

Preparation of $Na_2S·9H_2O$ Solution: Add $Na_2S·9H_2O$ to distilled/deionized water and bring volume to 10.0mL. Mix thoroughly. Autoclave under 100% N_2 for 15 min at 15 psi pressure–121°C. Cool to room temperature.

Preparation of Medium: Add components, except vitamin solution, and $Na_2S·9H_2O$ solution, to distilled/deionized water and bring volume to 985.0mL. Mix thoroughly. Sparge with 100% CO_2. Autoclave for 15 min at 15 psi pressure–121°C. Cool to 25°C while sparging with 100% CO_2. Aseptically and anaerobically add 10.0mL vitamin solution, and 10.0mL of sterile $Na_2S·9H_2O$ solution. Mix thoroughly. Aseptically and anaerobically distribute into sterile tubes or flasks.

Use: For the cultivation of *Moorella thermoacetica=Clostridium thermoaceticum.*

Clostridium thermoaceticum Medium (TYE-CO)

Composition per liter:

Pancreatic digest of casein	10.0g
Yeast extract	3.0g
$Na_2HPO_4·12H_2O$	2.8g
$FeSO_4·7H_2O$	1.0g
NH_4Cl	1.0g
KH_2PO_4	0.3g
$MgCl_2·6H_2O$	0.2g
Resazurin	1.0mg
Trace elements solution	10.0mL
$Na_2S·9H_2O$ solution	10.0mL
Vitamin solution	5.0mL

pH 7.0 ± 0.2 at 25°C

Trace Elements Solution:
Composition per liter:

$MgSO_4·7H_2O$	3.0g
Nitrilotriacetic acid	1.5 g
$CaCl_2·2H_2O$	1.0g
NaCl	1.0g
$MnSO_4·2H_2O$	0.5 g
$CoSO_4·7H_2O$	0.18 g
$ZnSO_4·7H_2O$	0.18 g
$FeSO_4·7H_2O$	0.1g
$NiCl_2·6H_2O$	0.025 g
$KAl(SO_4)_2·12H_2O$	0.02g
$CuSO_4·5H_2O$	0.01g
H_3BO_3	0.01g
$Na_2MoO_4·2H_2O$	0.01g
$Na_2SeO_3·5H_2O$	0.3 mg

Preparation of Trace Elements Solution: Add nitrilotriacetic acid to approximately 500.0mL distilled/deionized water. Dissolve by adding KOH and adjust pH to 6.5. Add remaining components. Bring volume to 1.0L with additional distilled/deionized water. Adjust pH to 7.0 with KOH.

Vitamin Solution:

Composition per liter:

Pyridoxine·HCl	10.0mg
Calcium DL-pantothenate	5.0mg
Lipoic acid	5.0mg
Nicotinic acid	5.0mg
p-Aminobenzoic acid	5.0mg
Riboflavin	5.0mg
Thiamine·HCl	5.0mg
Biotin	2.0mg
Folic acid	2.0mg
Vitamin B_{12}	0.1mg

Preparation of Vitamin Solution: Add components to distilled/deionized water and bring volume to 1.0L. Adjust pH to 7.0. Mix thoroughly.

$Na_2S·9H_2O$ Solution:

Composition per 10.0mL:

$Na_2S·9H_2O$	0.6g

Preparation of $Na_2S·9H_2O$ Solution: Add $Na_2S·9H_2O$ to distilled/deionized water and bring volume to 10.0mL. Mix thoroughly. Sparge with 100% N_2. Autoclave for 15 min at 15 psi pressure–121°C.

Preparation of Medium: Add components, except $Na_2S·9H_2O$ solution, to distilled/deionized water and bring volume to 990.0mL. Mix thoroughly. Sparge with 100% N_2 for 15–20 min. Before autoclaving, sparge with 100% CO (carbon monoxide). Autoclave for 15 min at 15 psi pressure–121°C. Aseptically and anaerobically add 10.0mL of sterile $Na_2S·9H_2O$ solution.

Caution: CO is toxic.

Use: For the cultivation and maintenance of *Clostridium thermoaceticum.*

Clostridium thermoaceticum Medium

Composition per 1010.0mL:

Solution A	100.0mL
Solution B	600.0mL
Solution C	300.0mL
Solution D	10.0mL

pH 6.9 ± 0.2 at 25°C

Solution A:

Composition per 100.0mL:

Glucose	18.0g

Preparation of Solution A: Add glucose to distilled/deionized water and bring volume to 100.0mL. Mix thoroughly. Sparge with 100% N_2 for 5–10 min. Autoclave for 15 min at 15 psi pressure–121°C.

Solution B:

Composition per 600.0mL:

Pancreatic digest of casein	5.0g
Yeast extract	5.0g
Pyruvic acid	1.8g
$(NH_4)_2SO_4$	1.0g
$MgSO_4·7H_2O$	0.25g
$Fe(NH_4)_2(SO_4)_2·6H_2O$	0.04g
$Co(NO_3)_2·6H_2O$	0.03g
$Na_2WO_4·2H_2O$	3.3mg
$Na_2MoO_4·2H_2O$	2.4mg
$ZnCl_2$	1.4mg
Resazurin	1.0mg
$Na_2SeO_3·5H_2O$	0.3mg
$NiCl_2·6H_2O$	0.2mg

Preparation of Solution B: Add components to distilled/deionized water and bring volume to 600.0mL. Mix thoroughly. Sparge with 100% CO_2 for 5–10 min. Autoclave for 15 min at 15 psi pressure–121°C.

Solution C:

Composition per 300.0mL:

$NaHCO_3$	16.8g
K_2HPO_4	7.0g
KH_2PO_4	5.5g

Preparation of Solution C: Add components to distilled/deionized water and bring volume to 300.0mL. Mix thoroughly. Sparge with 100% CO_2 for 5–10 min. Autoclave for 15 min at 15 psi pressure–121°C.

Solution D:

Composition per 10.0mL:

L-Cysteine·HCl solution (5%)	5.0mL
$Na_2S·9H_2O$ solution (5%)	5.0mL

Preparation of Solution D: Combine 5.0mL of L-cysteine·HCl solution and 5.0mL of $Na_2S·9H_2O$ solution. Mix thoroughly. Sparge with 100% N_2 for 5–10 min. Autoclave for 15 min at 15 psi pressure–121°C.

L-Cysteine·HCl Solution:

Composition per 10.0mL:

L-Cysteine·HCl	0.5g

Preparation of L-Cysteine·HCl Solution: Add L-cysteine·HCl to distilled/deionized water and bring volume to 10.0mL. Mix thoroughly. Autoclave under 80% N_2 + 20% CO_2 for 15 min at 15 psi pressure–121°C.

$Na_2S·9H_2O$ Solution:

Composition per 10.0mL:

$Na_2S·9H_2O$	0.5g

Preparation of $Na_2S·9H_2O$ Solution: Add $Na_2S·9H_2O$ to distilled/deionized water and bring volume to 10.0mL. Mix thoroughly.

Preparation of Medium: Aseptically and anaerobically combine 100.0mL of sterile solution A, 600.0mL of sterile solution B, 300.0mL of sterile solution C, and 10.0mL of sterile solution D. Mix thoroughly. Aseptically and anaerobically distribute into tubes or bottles.

Use: For the cultivation and maintenance of *Clostridium thermoaceticum.*

Clostridium thermoaceticum Medium

Composition per liter:

Pancreatic digest of casein	5.0g
Yeast extract	5.0g
$(NH_4)_2SO_4$	0.5g
$MgSO_4·7H_2O$	0.1g
$Fe(NH_4)_2(SO_4)_2$	0.04g
$Na_2MoO_4·2H_2O$	2.4mg
Resazurin	1.0mg
Phosphate solution	100.0mL
Glucose solution	100.0mL
$NaHCO_3$ solution	100.0mL

L-Cysteine·HCl solution....................10.0mL
$Na_2S \cdot 9H_2O$ solution....................10.0mL

pH 6.9 ± 0.2 at 25°C

Phosphate Solution:

Composition per 100.0mL:

K_2HPO_4....................7.0g
KH_2PO_4....................4.5g

Preparation of Phosphate Solution: Add components to distilled/deionized water and bring volume to 100.0mL. Mix thoroughly. Sparge with 100% N_2. Autoclave for 15 min at 15 psi pressure–121°C.

Glucose Solution:

Composition per 100.0mL:

D-Glucose....................18.0g

Preparation of Glucose Solution: Add glucose to distilled/deionized water and bring volume to 100.0mL. Mix thoroughly. Sparge with 100% N_2. Autoclave for 15 min at 15 psi pressure–121°C.

$NaHCO_3$ Solution:

Composition per 100.0mL:

$NaHCO_3$....................10.0g

Preparation of $NaHCO_3$ Solution: Add $NaHCO_3$ to distilled/deionized water and bring volume to 100.0mL. Mix thoroughly. Sparge with 100% CO_2. Autoclave for 15 min at 15 psi pressure–121°C.

L-Cysteine·HCl Solution:

Composition per 10.0mL:

L-Cysteine·HCl....................0.3g

Preparation of L-Cysteine·HCl Solution: Add L-cysteine·HCl to distilled/deionized water and bring volume to 10.0mL. Mix thoroughly. Sparge with 100% N_2. Autoclave for 15 min at 15 psi pressure–121°C.

$Na_2S \cdot 9H_2O$ Solution:

Composition per 10.0mL:

$Na_2S \cdot 9H_2O$....................0.3g

Preparation of $Na_2S \cdot 9H_2O$ Solution: Add $Na_2S \cdot 9H_2O$ to distilled/deionized water and bring volume to 10.0mL. Mix thoroughly. Sparge with 100% N_2. Autoclave for 15 min at 15 psi pressure–121°C.

Preparation of Medium: Add components, except phosphate solution, glucose solution, $NaHCO_3$ solution, L-cysteine·HCl solution, and $Na_2S \cdot 9H_2O$ solution, to distilled/deionized water and bring volume to 680.0mL. Mix thoroughly. Sparge with 100% CO_2. Autoclave for 15 min at 15 psi pressure–121°C. Aseptically and anaerobically add 100.0mL of sterile phosphate solution, 100.0mL of sterile glucose solution, 100.0mL of sterile $NaHCO_3$ solution, 10.0mL of sterile L-cysteine·HCl solution, and 10.0mL of sterile $Na_2S \cdot 9H_2O$ solution. Mix thoroughly. Check that final pH is 6.9.

Use: For the cultivation and maintenance of *Clostridium thermoaceticum.*

Clostridium thermoaceticum II Medium (DSMZ Medium 527)

Composition per 1010mL:

Solution B....................600.0mL
Solution C....................300.0mL
Solution A....................100.0mL
Solution D....................10.0mL

pH 6.9 ± 0.2 at 25°C

Solution A:

Composition per 100.0mL:

Glucose....................18.0g

Preparation of Solution A: Add glucose to distilled/deionized water and bring volume to 100.0mL. Mix thoroughly. Sparge with 100% N_2 gas mixture. Autoclave for 15 min at 15 psi pressure–121°C. Cool to 25°C.

Solution B:

Composition per 600.0mL:

Yeast extract....................5.0g
Tryptone....................5.0g
Pyruvic acid....................1.8g
$(NH_4)_2SO_4$....................1.0g
$MgSO_4 \cdot 7H_2O$....................0.25g
$Fe(NH_4)_2(SO_4)_2 \cdot 6H_2O$....................0.04g
$Co(NO_3)_2 \cdot 6H_2O$....................0.03g
$Na_2WO_4 \cdot 2H_2O$....................3.3mg
$Na_2MoO_4 \cdot 4H_2O$....................2.4mg
$ZnCl_2$....................1.4mg
Resazurin....................1.0mg
$Na_2SeO_3 \cdot 5H_2O$....................0.3mg
$NiCl_2 \cdot 6H_2O$....................0.2mg

Preparation of Solution B: Add components to distilled/deionized water and bring volume to 600.0mL. Mix thoroughly. Sparge with 100% CO_2 gas mixture. Autoclave for 15 min at 15 psi pressure–121°C. Cool to 25°C.

Solution C:

Composition per 300.0mL:

$NaHCO_3$....................16.8g
K_2HPO_4....................7.0g
KH_2PO_4....................5.5g

Preparation of Solution C: Add components to distilled/deionized water and bring volume to 300.0mL. Mix thoroughly. Sparge with 100% CO_2 gas mixture. Autoclave for 15 min at 15 psi pressure–121°C. Cool to 25°C.

Solution D:

Composition per 10.0mL:

Cysteine solution....................5.0mL
$Na_2S \cdot 9H_2O$ solution....................5.0mL

Preparation of Solution D: Combine 5.0mL cysteine soluiton and 5.0mL $Na_2S \cdot 9H_2O$ solution. Mix thoroughly. Sparge with 100% N_2 gas mixture. Autoclave for 15 min at 15 psi pressure–121°C. Cool to 25°C.

$Na_2S \cdot 9H_2O$ Solution:

Composition per 10.0mL:

$Na_2S \cdot 9H_2O$....................0.5g

Preparation of $Na_2S \cdot 9H_2O$ Solution: Add $Na_2S \cdot 9H_2O$ to distilled/deionized water and bring volume to 10.0mL. Sparge with N_2.

Cysteine Solution:

Composition per 100.0mL:

L-Cysteine·HCl·H_2O....................5.0g

Preparation of Cysteine Solution: Add L-cysteine·HCl·H_2O to distilled/deionized water and bring volume to 100.0mL. Mix thoroughly. Sparge with 100% N_2.

Preparation of Medium: Aseptically and anaerobically combine 100.0mL solution A, 600.0mL solution B, 300.0mL solution C, and 10.0mL solution D. Aseptically and anaerobically distribute into sterile tubes or flasks.

Use: For the cultivation of *Moorella thermoacetica=Clostridium thermoaceticum.*

Clostridium thermocellum Medium (LMG Medium 42)

Composition per liter:

Agar	30.0g
Cellulose	10.0g
Sodium-beta-glycerophosphate	6.0g
K_2HPO_4	5.5g
Yeast extract	4.5g
$MgCl_2 \cdot 6H_2O$	2.6g
KH_2PO_4	1.43g
$(NH_4)_2SO_4$	1.3g
$CaCl_2 \cdot 2H_2O$	0.13g
Glutathione	0.25g
$FeSO_4 \cdot 7H_2O$	1.1mg
Resazurin	1.0mg

pH 7.1 ± 0.2 at 25°C

Preparation of Medium: Add components to distilled/deionized water and bring volume to 1.0L under 95% N_2+ 5% CO_2 gas atmosphere. Mix thoroughly and sparge with 95% N_2+ 5% CO_2 gas. Gently heat and bring to boiling. Distribute into tubes or flasks. Autoclave for 15 min at 15 psi pressure–121°C. Pour into sterile Petri dishes or leave in tubes.

Use: For the cultivation and maintenance of *Clostridium thermocellum.*

Clostridium thermocellum Medium (LMG Medium 42)

Composition per liter:

Agar	30.0g
Sodium-beta-glycerophosphate	6.0g
K_2HPO_4	5.5g
Yeast extract	4.5g
$MgCl_2 \cdot 6H_2O$	2.6g
KH_2PO_4	1.43g
$(NH_4)_2SO_4$	1.3g
$CaCl_2 \cdot 2H_2O$	0.13g
Glutathione	0.25g
$FeSO_4 \cdot 7H_2O$	1.1mg
Resazurin	1.0mg
Cellobiose solution	50.0mL

pH 7.1 ± 0.2 at 25°C

Cellobiose Solution:

Composition per 100.0mL:

Cellobiose	10.0g

Preparation of Cellobiose Solution: Add cellobiose to 100.0mL of distilled/deionized water. Mix thoroughly. Sparge with 95% N_2+ 5% CO_2 gas. Filter sterilize.

Preparation of Medium: Add components, except cellobiose solution, to 950.0mL distilled/deionized water under 95% N_2+ 5% CO_2 gas atmosphere. Mix thoroughly and sparge with 95% N_2+ 5% CO_2 gas. Gently heat and bring to boiling. Autoclave for 15 min at 15 psi pressure–121°C. Aseptically add 50.0mL sterile cellobiose solution. Pour into sterile Petri dishes or leave in tubes.

Use: For the cultivation and maintenance of *Clostridium thermocellum.*

Clostridium thermocellum Medium

Composition per liter:

Filter paper	18.75g
$Na_2HPO_4 \cdot 12H_2O$	4.2g
Yeast extract	2.0g
KH_2PO_4	1.5g
NH_4Cl	0.5g
$MgCl_2 \cdot 6H_2O$	0.18g
Reducing solution	40.0mL
Wolfe's modified mineral elixir	5.0mL
Resazurin (0.1% solution)	1.0mL
Vitamin solution	0.5mL

Caution: This medium contains Na_2S, and H_2S production will occur, especially upon prolonged boiling. H_2S is hazardous and preparation of this medium should be done in a chemical fume hood.

Reducing Solution:

Composition per 200.0mL:

L-Cysteine·HCl·H_2O	2.5g
$Na_2S \cdot 9H_2O$	2.5g
NaOH (0.2*N* solution)	200.0mL

Preparation of Reducing Solution: Gently heat the NaOH solution and bring to boiling. Gas with 95% N_2 + 5% H_2. Cool to room temperature. Add the L-cysteine·HCl·H_2O and $Na_2S \cdot 9H_2O$. Anaerobically distribute into tubes. Cap with rubber stoppers. Autoclave for 15 min at 15 psi pressure–121°C.

Vitamin Solution:

Composition per 500.0mL:

Pyridoxine HCl	0.1g
p-Aminobenzoic acid	0.05g
Calcium pantothenate	0.05g
Nicotinic acid	0.05g
Thioctic acid	0.05g
Biotin	0.02g
Folic acid	0.02g
Riboflavin	5.0mg
Thiamine·HCl	5.0mg
Vitamin B_{12}	1.0mg

Preparation of Vitamin Solution: Add components to distilled/deionized water and bring volume to 500.0mL. Mix thoroughly. Store solution in the dark at –10°C.

Wolfe's Modified Mineral Elixir:

Composition per liter:

$MgSO_4 \cdot 7H_2O$	3.0g
Nitrilotriacetic acid	1.5g
NaCl	1.0g
$MnSO_4 \cdot H_2O$	0.5g
$CaCl_2$, anhydrous	0.1g
$Co(NO_3)_2 \cdot 6H_2O$	0.1g
$FeSO_4 \cdot 7H_2O$	0.1g
$ZnSO_4 \cdot 7H_2O$	0.1g
$AlK(SO_4)_2$, anhydrous	0.01g
$CuSO_4 \cdot 5H_2O$	0.01g
H_3BO_3	0.01g
$Na_2MoO_4 \cdot 2H_2O$	0.01g
Na_2SeO_3, anhydrous	1.0mg

Preparation of Wolfe's Modified Mineral Elixir: Add nitrilotriacetic acid to 500.0mL of distilled/deionized water. Dissolve by adjusting pH to 6.5 with KOH. Add remaining components. Add distilled/deionized water to 1.0L.

Preparation of Medium: Add components, except reducing solution, to distilled/deionized water and bring volume to 1.0L. If medium is to be distributed into tubes, omit bulk filter paper and substitute one Whatman #1 filter paper strip (8mm × 70mm) per tube of broth. Gently heat and bring to boiling under 95% N_2 + 5% H_2. Continue boiling until color changes from blue to pink. Add the reducing solution. The pink color will disappear, indicating that the solution has been reduced. Distribute into tubes or flasks under 95% N_2 + 5% H_2 using anaerobic techniques. If tubes are used, remember to add Whatman #1 filter paper strips prior to the addition of broth. Cap tubes with rubber stoppers. Autoclave for 15 min at 15 psi pressure–121°C.

Use: For the cultivation and maintenance of *Clostridium thermocellum*.

Clostridium thermocellum Medium

Composition per liter:

Cellulose	10.0g
$H_2HPO_4·3H_2O$	7.2g
Sodium-β-glycerophosphate	6.0g
Yeast extract	4.5g
$MgCl_2·6H_2O$	2.6g
KH_2PO_4	1.43g
$(NH_4)_2SO_4$	1.3g
Glutathione	0.25g
$CaCl_2·2H_2O$	0.13g
$FeSO_4·7H_2O$	1.1mg
Resazurin	1.0mg

pH 7.0–7.2 at 25°C

Preparation of Medium: Add components to distilled/deionized water and bring volume to 1.0L. Mix thoroughly. Adjust pH to 7.0–7.2. Distribute into tubes or flasks. Autoclave for 15 min at 15 psi pressure–121°C.

Use: For the cultivation of *Clostridium thermocellum.*

Clostridium thermocellum Medium

Composition per liter:

$H_2HPO_4·3H_2O$	7.2g
Sodium-β-glycerophosphate	6.0g
Cellobiose	5.0g
Yeast extract	4.5g
$MgCl_2·6H_2O$	2.6g
KH_2PO_4	1.43g
$(NH_4)_2SO_4$	1.3g
Glutathione	0.25g
$CaCl_2·2H_2O$	0.13g
$FeSO_4·7H_2O$	1.1mg
Resazurin	1.0mg

pH 7.0–7.2 at 25°C

Preparation of Medium: Add components to distilled/deionized water and bring volume to 1.0L. Mix thoroughly. Adjust pH to 7.0–7.2. Distribute into tubes or flasks. Autoclave for 15 min at 15 psi pressure–121°C.

Use: For the cultivation of *Clostridium thermocellum.*

Clostridium thermocellum Medium

Composition per liter:

Agar	30.0g
Morpholinopropane sulfonic acid	10.0g
Yeast extract	6.0g
Urea	2.0g
KH_2PO_4	1.0g
K_2HPO_4	1.0g
$MgCl_2·6H_2O$	0.5g
$CaCl_2·2H_2O$	0.05g
$FeSO_4·7H_2O$	1.25mg
Resazurin	1.0mg
Glucose solution	50,0mL
L-Cysteine·HCl solution	20.0mL

pH 7.2 ± 0.2 at 25°C

Glucose Solution:

Composition per 50.0mL:

D-Glucose	5.0g

Preparation of Glucose Solution: Add glucose to distilled/deionized water and bring volume to 50.0mL. Mix thoroughly. Filter sterilize. Sparge with 100% N_2 gas. Warm to 50°–55°C.

L-Cysteine·HCl Solution:

Composition per 10.0mL:

L-Cysteine·HCl	0.3g

Preparation of L-Cysteine·HCl Solution: Add L-cysteine·HCl to distilled/deionized water and bring volume to 10.0mL. Mix thoroughly. Filter sterilize. Sparge with 100% N_2 gas. Warm to 50°–55°C.

Preparation of Medium: Prepare and dispense medium under 100% N_2. Add components, except glucose solution and L-cysteine·HCl solution, to distilled/deionized water and bring volume to 930.0mL. Mix thoroughly. Gently heat and bring to boiling. Autoclave for 15 min at 15 psi pressure–121°C. Cool to 50°–55°C. Aseptically and anaerobically add 50.0mL of sterile glucose solution and 20.0mL of sterile L-cysteine·HCl solution. Mix thoroughly. Pour into sterile Petri dishes or distribute into sterile tubes.

Use: For the cultivation and maintenance of *Clostridium thermocellum*.

Clostridium thermocellum Medium

Composition per liter:

Cellulose	10.0g
$K_2HPO_4·3H_2O$	2.9g
Cellobiose	2.0g
Yeast extract	2.0g
KH_2PO_4	1.5g
$(NH_4)_2·SO_4$	1.3g
$MgCl_2·6H_2O$	1.0g
$CaCl_2$	0.15g
$FeSO_4·7H_2O$ (5%)	25.0μg
Reductant solution	50.0mL
Resazurin (0.2%)	1.0mL

pH 7.8 ± 0.2 at 25°C

Reductant Solution:

Composition per 50.0mL:

$NaHCO_3$	5.0g
L-Cysteine·HCl	0.5g

Preparation of Reductant Solution: Add components to distilled/deionized water and bring volume to 50.0mL. Mix thoroughly. Filter sterilize.

Preparation of Medium: Add components, except reductant solution, to distilled/deionized water and bring volume to 950.0mL. Mix thoroughly. Adjust pH to 7.8. Autoclave for 15 min at 15 psi pressure–

121°C. Aseptically add 50.0mL of sterile reductant solution. Mix thoroughly. Aseptically distribute into sterile tubes or flasks.

Use: For the cultivation of *Clostridium celerecrescens, Clostridium papyrosolvens, Clostridium stercorarium,* and *Clostridium thermocellum.*

Clostridium thermohydrosulfuricum Medium

Composition per liter:

Sucrose	10.0g
Pancreatic digest of casein	10.0g
Yeast extract	2.0g
$FeSO_4 \cdot 7H_2O$	0.2g
Na_2SO_3	0.2g
$Na_2S_2O_3 \cdot 5H_2O$	0.08g
Resazurin	1.0mg

pH 6.8 ± 0.2 at 25°C

Preparation of Medium: Add components to distilled/deionized water and bring volume to 1.0L. Mix thoroughly. Sparge with 100% N_2 for 15 min. Autoclave for 30 min at 15 psi pressure–121°C.

Use: For the cultivation and maintenance of *Clostridium thermohydrosulfuricum, Clostridium thermosaccharolyticum, Thermoanaerobacter ethanolicus,* and *Thermoanaerobacter thermohydrosulfuricus.*

Clostridium thermolacticum Medium

Composition per 1020.0mL:

$KHCO_3$	4.5g
NaCl	2.25g
Sucrose	2.0g
Yeast extract	2.0g
$MgSO_4 \cdot 7H_2O$	0.5g
NH_4Cl	0.5g
K_2HPO_4	0.348g
$CaCl_2 \cdot 2H_2O$	0.25g
KH_2PO_4	0.227g
$FeSO_4 \cdot 7H_2O$	2.0mg
Resazurin	1.0mg
L-Cysteine·HCl·H_2O solution	10.0mL
$Na_2S \cdot 9H_2O$ solution	10.0mL
Wolfe's vitamin solution	10.0mL
Trace elements solution SL-6	3.0mL

pH 7.0–7.2 at 25°C

Wolfe's Vitamin Solution:

Composition per liter:

Pyridoxine·HCl	10.0mg
p-Aminobenzoic acid	5.0mg
Calcium pantothenate	5.0mg
Nicotinic acid	5.0mg
Riboflavin	5.0mg
Thiamine·HCl	5.0mg
Thioctic acid	5.0mg
Biotin	2.0mg
Folic acid	2.0mg
Cyanocobalamin	100.0μg

Preparation of Medium: Add components to distilled/deionized water and bring volume to 1.0L. Mix thoroughly.

Trace Elements Solution SL-6:

Composition per liter:

H_3BO_3	0.3g
$CoCl_2 \cdot 6H_2O$	0.2g
$ZnSO_4 \cdot 7H_2O$	0.1g
$MnCl_2 \cdot 4H_2O$	0.03g
$Na_2MoO_4 \cdot H_2O$	0.03g
$NiCl_2 \cdot 6H_2O$	0.02g
$CuCl_2 \cdot 2H_2O$	0.01g

Preparation of Trace Elements Solution SL-6: Add components to distilled/deionized water and bring volume to 1.0L. Mix thoroughly. Adjust pH to 3.4.

L-Cysteine·HCl·H_2O Solution:

Composition per 10.0mL:

L-Cysteine·HCl·H_2O	0.3g

Preparation of L-Cysteine·HCl·H_2O Solution: Add L-cysteine·HCl·H_2O to distilled/deionized water and bring volume to 10.0mL. Gas tubes under 100% N_2 and tightly seal. Autoclave for 15 min at 15 psi pressure–121°C. Use freshly prepared solution.

$Na_2S \cdot 9H_2O$ Solution:

Composition per 10.0mL:

$Na_2S \cdot 9H_2O$	0.3g

Preparation of $Na_2S \cdot 9H_2O$ Solution: Add $Na_2S \cdot 9H_2O$ to distilled/deionized water and bring volume to 10.0mL. Gas tube under 100% N_2 and tightly seal. Autoclave for 15 min at 15 psi pressure–121°C. Use freshly prepared solution.

Preparation of Medium: Prepare anaerobically under 80% N_2 + 20% CO_2. Add components, except L-cysteine·HCl·H_2O solution and $Na_2S \cdot 9H_2O$ solution, to distilled/deionized water and bring volume to 1.0L. Mix thoroughly. Distribute into tubes using anaerobic techniques. Autoclave for 15 min at 15 psi pressure–121°C. Prior to inoculation of cultures, inject 0.1mL of sterile L-cysteine·HCl·H_2O solution and 0.1mL of sterile $Na_2S \cdot 9H_2O$ solution per 10.0mL of medium.

Use: For the cultivation and maintenance of *Clostridium thermolacticum.*

Clostridium thermosuccinogenes Medium

Composition per 1011.0mL:

Inulin	5.0g
NaCl	1.2g
$MgCl_2 \cdot 6H_2O$	0.4g
KCl	0.3g
NH_4Cl	0.27g
KH_2PO_4	0.21g
$CaCl_2 \cdot 2H_2O$	0.15g
Na_2SO_4	0.1g
Resazurin	1.0mg
Na_2HPO_4 solution	20.0mL
Vitamin solution	10.0mL
Yeast extract solution	10.0mL
Casamino acids solution	10.0mL
$NaHCO_3$ solution	10.0mL
$Na_2S \cdot 9H_2O$ solution	10.0mL
Trace elements solution SL-10	1.0mL

pH 7.0 ± 0.2 at 25°C

Na_2HPO_4 Solution:

Composition per 20.0mL:

Na_2HPO_4	2.66g

Preparation of Na_2HPO_4 Solution: Add Na_2HPO_4 to distilled/deionized water and bring volume to 20.0mL. Mix thoroughly. Sparge with 100% N_2. Autoclave for 15 min at 15 psi pressure–121°C.

Vitamin Solution:
Composition per liter:

Pyridoxine·HCl	10.0mg
Calcium DL-pantothenate	5.0mg
Lipoic acid	5.0mg
Nicotinic acid	5.0mg
p-Aminobenzoic acid	5.0mg
Riboflavin	5.0mg
Thiamine·HCl	5.0mg
Biotin	2.0mg
Folic acid	2.0mg
Vitamin B_{12}	0.1mg

Preparation of Vitamin Solution: Add components to distilled/deionized water and bring volume to 1.0L. Adjust pH to 7.0. Mix thoroughly. Sparge with 80% N_2 + 20% CO_2. Autoclave for 15 min at 15 psi pressure–121°C.

Yeast Extract Solution:
Composition per 10.0mL:

Yeast extract	0.03 g

Preparation of Yeast Extract Solution: Add yeast extract to distilled/deionized water and bring volume to 10.0mL. Mix thoroughly. Autoclave under 100% N_2 for 15 min at 15 psi pressure–121°C.

Casamino Acids Solution:
Composition per 10.0mL:

Casamino acids	0.03g

Preparation of Casamino Acids Solution: Add casamino acids to distilled/deionized water and bring volume to 10.0mL. Mix thoroughly. Autoclave under 100% N_2 for 15 min at 15 psi pressure–121°C.

$NaHCO_3$ Solution:
Composition per 10.0mL:

$NaHCO_3$	1.0mg

Preparation of $NaHCO_3$ Solution: Add $NaHCO_3$ to distilled/deionized water and bring volume to 10.0mL. Mix thoroughly. Sparge with 80% N_2 + 20% CO_2. Autoclave for 15 min at 15 psi pressure–121°C.

$Na_2S{\cdot}9H_2O$ Solution:
Composition per 10.0mL:

$Na_2S{\cdot}9H_2O$	0.15mg

Preparation of $Na_2S{\cdot}9H_2O$ Solution: Add $Na_2S{\cdot}9H_2O$ to distilled/deionized water and bring volume to 10.0mL. Mix thoroughly. Sparge with 100% N_2. Autoclave for 15 min at 15 psi pressure–121°C.

Trace Elements Solution SL-10:
Composition per liter:

$FeCl_2{\cdot}4H_2O$	1.5g
$CoCl_2{\cdot}6H_2O$	190.0mg
$MnCl_2{\cdot}4H_2O$	100.0mg
$ZnCl_2$	70.0mg
$Na_2MoO_4{\cdot}2H_2O$	36.0mg
$NiCl_2{\cdot}6H_2O$	24.0mg
H_3BO_3	6.0mg
$CuCl_2{\cdot}2H_2O$	2.0mg
HCl (25% solution)	10.0mL

Preparation of Trace Elements Solution SL-10: Add $FeCl_2{\cdot}4H_2O$ to 10.0mL of HCl solution. Mix thoroughly. Add distilled/deionized water and bring volume to 1.0L. Add remaining components. Mix thoroughly. Sparge with 100% N_2. Autoclave for 15 min at 15 psi pressure–121°C.

Preparation of Medium: Add components, except Na_2HPO_4 solution, vitamin solution, yeast extract solution, casamino acids solution, $NaHCO_3$ solution, $Na_2S{\cdot}9H_2O$ solution, and trace elements solution SL-10, to distilled/deionized water and bring volume to 930.0mL. Mix thoroughly. Sparge with 80% N_2 + 100% CO_2. Autoclave for 15 min at 15 psi pressure–121°C. Aseptically and anaerobically add 20.0mL of sterile Na_2HPO_4 solution, 10.0mL of sterile vitamin solution, 10.0mL of sterile yeast extract solution, 10.0mL of sterile casamino acids solution, 10.0mL of sterile $NaHCO_3$ solution, 10.0mL of sterile $Na_2S{\cdot}9H_2O$ solution, and 1.0mL of sterile trace elements solution SL-10. Aseptically and anaerobically distribute into tubes or bottles.

Use: For the cultivation and maintenance of *Clostridium thermosuccinogenes.*

Clostridium thermosulfurogenes Medium

Composition per 1015.0mL:

$Na_2HPO_4{\cdot}12H_2O$	5.3g
NH_4Cl	1.0g
Yeast extract	1.0g
KH_2PO_4	0.3g
$MgCl_2{\cdot}6H_2O$	0.2g
$FeSO_4{\cdot}7H_2O$	1.5mg
Resazurin	1.0mg
Glucose solution	50.0mL
Trace elements solution	10.0mL
$Na_2S{\cdot}9H_2O$ solution	10.0mL
Vitamin solution	5.0mL

pH 6.0 ± 0.2 at 25°C

Glucose Solution:
Composition per 50.0mL:

D-Glucose	5.0g

Preparation of Glucose Solution: Add glucose to distilled/deionized water and bring volume to 50.0mL. Mix thoroughly. Sparge with 100% N_2. Autoclave for 15 min at 15 psi pressure–121°C.

Trace Elements Solution:
Composition per liter:

Nitrilotriacetic acid	12.5g
NaCl	1.0g
$FeCl_3{\cdot}4H_2O$	0.2g
$MnCl_2{\cdot}4H_2O$	0.1g
$CaCl_2{\cdot}2H_2O$	0.1g
$ZnCl_2$	0.02g
$CuCl_2$	0.02g
Na_2SeO_3	0.02g
$CoCl_2{\cdot}6H_2O$	0.017g
H_3BO_3	0.01g
$Na_2MoO_4{\cdot}2H_2O$	0.01g

Preparation of Trace Elements Solution: Add nitrilotriacetic acid to 500.0mL of distilled/deionized water. AdjustpH to 6.5 with KOH. Add remaining components. Add distilled/deionized water to 1.0L.

Vitamin Solution:
Composition per liter:

Pyridoxine·HCl	10.0mg
Calcium DL-pantothenate	5.0mg
Lipoic acid	5.0mg
Nicotinic acid	5.0mg
p-Aminobenzoic acid	5.0mg
Riboflavin	5.0mg

Thiamine·HCl 5.0mg
Biotin 2.0mg
Folic acid 2.0mg
Vitamin B_{12} 0.1mg

Preparation of Vitamin Solution: Add components to distilled/deionized water and bring volume to 1.0L. Adjust pH to 7.0. Mix thoroughly. Sparge with 100% N_2.

$Na_2S·9H_2O$ Solution:
Composition per 10.0mL:
$Na_2S·9H_2O$ 0.5g

Preparation of $Na_2S·9H_2O$ Solution: Add $Na_2S·9H_2O$ to distilled/deionized water and bring volume to 10.0mL. Mix thoroughly. Sparge with 100% N_2. Autoclave for 15 min at 15 psi pressure–121°C. Adjust pH to 7.0 with 1*N* HCl before use.

Preparation of Medium: Add components, except glucose solution and $Na_2S·9H_2O$ solution, and bring volume to 940.0mL. Mix thoroughly. Sparge with 80% N_2 + 100% CO_2. Autoclave for 15 min at 15 psi pressure–121°C. Aseptically and anaerobically add 50.0mL of sterile glucose solution and 10.0mL of sterile $Na_2S·9H_2O$ solution. Mix thoroughly. Aseptically and anaerobically distribute into tubes or bottles.

Use: For the cultivation and maintenance of *Thermoanaerobacterium thermosulfurigenes*.

Clostridium ultunense Medium (DSMZ Medium 727)

Composition per liter:
Yeast extract 10.0g
Peptone 10.0g
Lab Lemco powder 5.0g
Na_2HPO_4 .... 0.43g
Starch, soluble 0.4g
KH_2PO_4 .... 0.4g
Na-acetate 0.4g
NH_4Cl 0.3g
NaCl 0.3g
$CaCl_2·2H_2O$ 0.1g
$MgCl_2·6H_2O$ 0.1g
Resazurin 0.5mg
$NaHCO_3$ solution 50.0mL
Vitamin solution 10.0mL
$Na_2S·9H_2O$ solution 10.0mL
Trace elements solution 1.0mL
Selenite-tungstate solution 1.0mL
pH 7.0 ± 0.2 at 25°C

Selenite–Tungstate Solution
Composition per liter:
NaOH 0.5g
$Na_2WO_4·2H_2O$ 4.0mg
$Na_2SeO_3·5H_2O$ 3.0mg

Preparation of Selenite–Tungstate Solution: Add components to distilled/deionized water and bring volume to 1.0L. Mix thoroughly. Sparge with 100% N_2. Filter sterilize.

$Na_2S·9H_2O$ Solution:
Composition per 10.0mL:
$Na_2S·9H_2O$ 0.3g

Preparation of $Na_2S·9H_2O$ Solution: Add $Na_2S·9H_2O$ to distilled/deionized water and bring volume to 10.0mL. Mix thoroughly. Autoclave under 100% N_2 for 15 min at 15 psi pressure–121°C. Cool to room temperature.

$NaHCO_3$ Solution:
Composition per 100.0mL:
$NaHCO_3$ 10.0g

Preparation of $NaHCO_3$ Solution: Add $NaHCO_3$ to distilled/deionized water and bring volume to 100.0mL. Mix thoroughly. Sparge with 80% N_2 + 20% CO_2. Filter sterilize.

Vitamin Solution:
Composition per liter:
Pyridoxine-HCl 10.0mg
Thiamine-HCl·$2H_2O$ 5.0mg
Riboflavin 5.0mg
Nicotinic acid 5.0mg
D-Ca-pantothenate 5.0mg
p-Aminobenzoic acid 5.0mg
Lipoic acid 5.0mg
Biotin 2.0mg
Folic acid 2.0mg
Vitamin B_{12} 0.1mg

Preparation of Vitamin Solution: Add components to distilled/deionized water and bring volume to 1.0L. Mix thoroughly. Sparge with 80% H_2 + 20% CO_2. Filter sterilize.

Trace Elements Solution:
Composition per liter:
$FeCl_2·4H_2O$ 1.5g
Na_2-EDTA 0.5g
$CoCl_2·6H_2O$ 190.0mg
$MnCl_2·4H_2O$ 100.0mg
$ZnCl_2$ 70.0mg
$Na_2MoO_4·2H_2O$ 36.0mg
$NiCl_2·6H_2O$ 24.0mg
H_3BO_3 .... 6.0mg
$CuCl_2·2H_2O$ 2.0mg
HCl (25% solution) 10.0mL

Preparation of Trace Elements Solution: Add $FeCl_2·4H_2O$ to 10.0mL of HCl solution. Mix thoroughly. Add distilled/deionized water and bring volume to 1.0L. Add remaining components. Mix thoroughly. Sparge with 80% N_2 + 20% CO_2. Autoclave for 15 min at 15 psi pressure–121°C.

Preparation of Medium: Prepare and dispense medium under 80% N_2 + 20% CO_2 gas atmosphere. Add components, except $NaHCO_3$ solution, $Na_2S·9H_2O$ solution, vitamin solution, selenite-tungstate solution, and trace elements solution, to distilled/deionized water and bring volume to 928.0mL. Mix thoroughly. Adjust pH to 7.0. Sparge with 80% N_2 + 20% CO_2. Autoclave for 15 min at 15 psi pressure–121°C. Aseptically and anaerobically add 50.0mL $NaHCO_3$ solution, 10.0mL $Na_2S·9H_2O$ solution, 10.0mL vitamin solution, 1.0mL selenite–tungstate solution, and 1.0mL trace elements solution. Mix thoroughly. Aseptically and anaerobically distribute into sterile tubes or bottles.

Use: For the cultivation of *Clostridium ultunense.*

Clostridium vincentii Medium (DSMZ Medium 769)

Composition per liter:
Yeast extract 1.0g
Trypticase™ 0.4g
NH_4NO_3 .... 0.1g

Resazurin	0.5mg
Sea water, natural	300.0mL
$NaHCO_3$ solution	20.0mL
Phosphate solution	20.0mL
Lactose solution	20.0mL
Vitamin solution	10.0mL
$Na_2S·9H_2O$ solution	10.0mL
Cysteine solution	10.0mL

pH 6.5 ± 0.2 at 25°C

Vitamin Solution:

Composition per liter:

Pyridoxine-HCl	10.0mg
Thiamine-HCl·$2H_2O$	5.0mg
Riboflavin	5.0mg
Nicotinic acid	5.0mg
D-Ca-pantothenate	5.0mg
p-Aminobenzoic acid	5.0mg
Lipoic acid	5.0mg
Biotin	2.0mg
Folic acid	2.0mg
Vitamin B_{12}	0.1mg

Preparation of Vitamin Solution: Add components to distilled/deionized water and bring volume to 1.0L. Mix thoroughly. Sparge with 80% H_2 + 20% CO_2. Filter sterilize.

Cysteine Solution:

Composition per 10.0mL:

L-Cysteine·HCl·H_2O	0.3g

Preparation of Cysteine Solution: Add L-cysteine·HCl·H_2O to distilled/deionized water and bring volume to 10.0mL. Mix thoroughly. Sparge with 100% N_2. Autoclave for 15 min at 15 psi pressure–121°C. Cool to room temperature.

Lactose Solution:

Composition per 100.0mL:

Lactose	10.0g

Preparation of Lactose Solution: Add lactose to distilled/deionized water and bring volume to 100.0mL. Mix thoroughly. Sparge with 100% N_2. Autoclave for 15 min at 15 psi pressure–121°C. Cool to room temperature.

Phosphate Solution:

Composition liter:

$Na_2HPO_4·12H_2O$	43.0g
NaH_2PO_4	5.44g

Preparation of Phosphate Solution: Add components to distilled/deionized water and bring volume to 1.0L. Mix thoroughly. Sparge with 100% N_2. Adjust pH to 6.5. Autoclave for 15 min at 15 psi pressure–121°C. Cool to room temperature.

$Na_2S·9H_2O$ Solution:

Composition per 10.0mL:

$Na_2S·9H_2O$	0.3g

Preparation of $Na_2S·9H_2O$ Solution: Add $Na_2S·9H_2O$ to distilled/deionized water and bring volume to 10.0mL. Mix thoroughly. Autoclave under 100% N_2 for 15 min at 15 psi pressure–121°C. Cool to room temperature.

$NaHCO_3$ Solution:

Composition per 100.0mL:

$NaHCO_3$	5.0g

Preparation of $NaHCO_3$ Solution: Add $NaHCO_3$ to distilled/deionized water and bring volume to 100.0mL. Mix thoroughly. Sparge with 80% N_2 + 20% CO_2. Filter sterilize.

Preparation of Medium: Prepare and dispense medium under 80% N_2 + 20% CO_2 gas atmosphere. Add components, except $NaHCO_3$ solution, $Na_2S·9H_2O$ solution, lactose solution, vitamin solution, phosphate solution, and cysteine solution, to distilled/deionized water and bring volume to 910.0mL. Mix thoroughly. Adjust pH to 6.8. Sparge with 80% N_2 + 20% CO_2. Autoclave for 15 min at 15 psi pressure–121°C. Aseptically and anaerobically add 20.0mL $NaHCO_3$ solution, 10.0mL $Na_2S·9H_2O$ solution, 20.0mL phosphate solution, 10.0mL vitamin solution, 10.0mL cysteine solution, and 20.0mL lactose solution. Mix thoroughly. Adjust pH to 6.5. Sparge with 80% N_2 + 20% CO_2. Aseptically and anaerobically distribute into sterile tubes or bottles.

Use: For the cultivation of *Clostridium vincentii.*

Clostrisel Agar

See: ***Clostridium*** **Selective Agar**

CM

See: **Coliform Medium**

CM Agar

Composition per liter:

Agar	20.0g
Polypeptone™	10.0g
Yeast extract	10.0g
NaCl	5.0g

pH 7.0 ± 0.2 at 25°C

Preparation of Medium: Add components to distilled/deionized water and bring volume to 1.0L. Mix thoroughly. Gently heat until boiling. Distribute into tubes or flasks. Autoclave for 15 min at 15 psi pressure–121°C. Pour into sterile Petri dishes or leave in tubes.

Use: For the cultivation and maintenance of *Bacillus subtilis*.

CM-DYA

See: **Cornmeal Agar with Dextrose and Yeast Extract**

CM 3 Agar

Composition per liter:

Agar	15.0g
Cellobiose	6.0g
Sodium citrate	3.0g
K_2HPO_4	2.9g
Yeast extract	2.0g
KH_2PO_4	1.5g
$(NH_4)_2SO_4$	1.3g
$MgCl_2·6H_2O$	1.0g
$CaCl_2$	0.15g
L-Cysteine·HCl solution	44.0mL
Resazurin solution	2.0mL
$FeSO_4$solution	25.0μL

pH 7.2 ± 0.2 at 25°C

L-Cysteine·HCl Solution:

Composition per 100.0mL:

L-Cysteine·HCl	2.5g

Preparation of L-Cysteine·HCl Solution: Add L-cysteine·HCl to distilled/deionized water and bring volume to 100.0mL. Mix thoroughly. Autoclave for 15 min at 15 psi pressure–121°C.

Resazurin Solution:

Composition per 10.0mL:

Resazurin	0.01g

Preparation of Resazurin Solution: Add resazurin to distilled/deionized water and bring volume to 10.0mL. Mix thoroughly.

$FeSO_4$ Solution:

Composition per 10.0mL:

$FeSO_4$	0.5g

Preparation of $FeSO_4$ Solution: Add $FeSO_4$ to distilled/deionized water and bring volume to 10.0mL. Mix thoroughly.

Preparation of Medium: Add components, except L-cysteine·HCl solution, to distilled/deionized water and bring volume to 956.0mL. Mix thoroughly. Adjust pH to 7.2. Gently heat and bring to boiling. Autoclave for 15 min at 15 psi pressure–121°C. Cool to 50°–55°C. Aseptically add 44.0mL of sterile L-cysteine·HCl solution. Mix thoroughly. Pour into sterile Petri dishes or distribute into sterile tubes.

Note: Cellobiose may be replaced by 5.0g of Avicel (microcrystalline cellulose) or 5.0g of cellulose powder Whatman CF-11.

Use: For the cultivation and maintenance of *Clostridium celerecrescens, Clostridium papyrosolvens,* and *Clostridium thermocellum.*

CM 3 Broth

Composition per liter:

Cellobiose	6.0g
Sodium citrate	3.0g
K_2HPO_4	2.9g
Yeast extract	2.0g
KH_2PO_4	1.5g
$(NH_4)_2SO_4$	1.3g
$MgCl_2·6H_2O$	1.0g
$CaCl_2$	0.15g
L-Cysteine·HCl solution	22.0mL
Resazurin solution	2.0mL
$FeSO_4$ solution	25.0μL

pH 7.2 ± 0.2 at 25°C

L-Cysteine·HCl Solution:

Composition per 100.0mL:

L-Cysteine·HCl	2.5g

Preparation of L-Cysteine·HCl Solution: Add L-cysteine·HCl to distilled/deionized water and bring volume to 100.0mL. Mix thoroughly. Autoclave for 15 min at 15 psi pressure–121°C.

Resazurin Solution:

Composition per 10.0mL:

Resazurin	0.01g

Preparation of Resazurin Solution: Add resazurin to distilled/deionized water and bring volume to 10.0mL. Mix thoroughly.

$FeSO_4$ Solution:

Composition per 10.0mL:

$FeSO_4$	0.5g

Preparation of $FeSO_4$ Solution: Add $FeSO_4$ to distilled/deionized water and bring volume to 10.0mL. Mix thoroughly.

Preparation of Medium: Add components, except L-cysteine·HCl solution, to distilled/deionized water and bring volume to 978.0mL. Mix thoroughly. Adjust pH to 7.2. Gently heat and bring to boiling. Autoclave for 15 min at 15 psi pressure–121°C. Cool to 50°–55°C. Aseptically add 22.0mL of sterile L-cysteine·HCl solution. Mix thoroughly. Pour into sterile Petri dishes or distribute into sterile tubes.

Note: Cellobiose may be replaced by sterile strips of filter paper.

Use: For the cultivation of *Clostridium celerecrescens, Clostridium papyrosolvens,* and *Clostridium thermocellum.*

CM3 Medium

Composition per liter:

3-(*N*-Morpholino)propanesulfonic acid (MOPS) buffer	20.0g
Cellobiose (or Cellulose MN 300)	10.0g
K_2HPO_4	4.4g
Urea	1.5g
L-Cysteine·HCl·H_2O	1.0g
$MgSO_4·7H_2O$	0.5g
$(NH_4)_2SO_4$	0.4g
$CaCl_2·2H_2O$	0.05g
$FeSO_4·7H_2O$	1.0mg

pH 7.1 ± 0.2 at 25°C

Preparation of Medium: Add components to distilled/deionized water and bring volume to 1.0L. Mix thoroughly. Distribute into tubes or flasks. Autoclave for 15 min at 15 psi pressure–121°C.

Use: For the cultivation of *Clostridium* species.

CM3 Medium

Composition per 1040.0mL:

$K_2HPO_4·3H_2O$	2.9g
Yeast extract	2.0g
KH_2PO_4	1.5g
$(NH_4)_2SO_4$	1.3g
$FeSO_4·7H_2O$	1.25g
$CaCl_2·2H_2O$	0.75g
L-Cysteine·HCl	0.5g
$MgCl_2·6H_2O$	0.2g
Resazurin	1.0mg
Cellobiose solution	50.0mL
Na_2CO_3 solution	40.0mL
Trace elements solution SL-10	1.0mL

pH 7.2 ± 0.2 at 25°C

Cellobiose Solution:

Composition per 50.0mL:

D-Cellobiose	6.0g

Preparation of Cellobiose Solution: Add cellobiose to distilled/deionized water and bring volume to 50.0mL. Mix thoroughly. Sparge under 100% N_2 gas for 3 min. Filter sterilize.

Na_2CO_3 Solution:

Composition per 40.0mL:

Na_2CO_3	2.0g

Preparation of Na_2CO_3 Solution: Add Na_2CO_3 to distilled/deionized water and bring volume to 40.0mL. Mix thoroughly. Sparge with 100% CO_2. Autoclave for 15 min at 15 psi pressure–121°C.

Trace Elements Solution SL-10:

Composition per liter:

$FeCl_2·4H_2O$	1.5g
$CoCl_2·6H_2O$	190.0mg

$MnCl_2 \cdot 4H_2O$ 100.0mg
$ZnCl_2$ 70.0mg
$Na_2MoO_4 \cdot 2H_2O$ 36.0mg
$NiCl_2 \cdot 6H_2O$ 24.0mg
H_3BO_3 .... 6.0mg
$CuCl_2 \cdot 2H_2O$ 2.0mg
HCl (25% solution) 10.0mL

Preparation of Trace Elements Solution SL-10: Add $FeCl_2 \cdot 4H_2O$ to 10.0mL of HCl solution. Mix thoroughly. Add distilled/deionized water and bring volume to 1.0L. Add remaining components. Mix thoroughly.

Preparation of Medium: Add components, except cellobiose solution and Na_2CO_3 solution, to distilled/deionized water and bring volume to 1.0L. Mix thoroughly. Adjust pH to 6.0 with 6*N* HCl. Sparge with 100% N_2. Anaerobically distribute 10.0mL volumes into screw-capped tubes. Aseptically and anaerobically add 0.4mL of sterile Na_2CO_3 solution to each tube containing 10.0mL of medium and 0.5mL of sterile cellobiose solution to each tube. pH of the medium after the addition of Na_2CO_3 solution should be 7.2.

Use: For the cultivation of *Clostridium aldrichii, Clostridium celerecrescens, Clostridium cellulolyticum, Clostridium lentocellum, Clostridium populeti,* and *Clostridium thermopalmarium.*

CM3 Medium

Composition per 1040.0mL:
$K_2HPO_4 \cdot 3H_2O$ 2.9g
Yeast extract 2.0g
KH_2PO_4 .... 1.5g
$(NH_4)_2SO_4$ 1.3g
$FeSO_4 \cdot 7H_2O$ 1.25g
$CaCl_2 \cdot 2H_2O$ 0.75g
L-Cysteine·HCl 0.5g
$MgCl_2 \cdot 6H_2O$ 0.2g
Resazurin 1.0mg
Glucose solution 50.0mL
Na_2CO_3 solution 40.0mL
$Na_2S \cdot 9H_2O$ solution 10.0mL
Trace elements solution SL-10 .... 1.0mL
pH 6.0–7.0 at 25°C

Glucose Solution:
Composition per 50.0mL:
D-Glucose 6.0g

Preparation of Glucose Solution: Add glucose to distilled/deionized water and bring volume to 50.0mL. Mix thoroughly. Sparge under 100% N_2 gas for 3 min. Filter sterilize.

Na_2CO_3 Solution:
Composition per 40.0mL:
Na_2CO_3 .... 2.0g

Preparation of Na_2CO_3 Solution: Add Na_2CO_3 to distilled/deionized water and bring volume to 40.0mL. Mix thoroughly. Sparge with 100% CO_2. Autoclave for 15 min at 15 psi pressure–121°C.

$Na_2S \cdot 9H_2O$ Solution:
Composition per 10.0mL:
$Na_2S \cdot 9H_2O$ 0.5g

Preparation of $Na_2S \cdot 9H_2O$ Solution: Add $Na_2S \cdot 9H_2O$ to distilled/deionized water and bring volume to 10.0mL. Mix thoroughly. Sparge with 100% N_2. Autoclave for 15 min at 15 psi pressure–121°C.

Trace Elements Solution SL-10:
Composition per liter:
$FeCl_2 \cdot 4H_2O$ 1.5g
$CoCl_2 \cdot 6H_2O$ 190.0mg
$MnCl_2 \cdot 4H_2O$ 100.0mg
$ZnCl_2$ 70.0mg
$Na_2MoO_4 \cdot 2H_2O$ 36.0mg
$NiCl_2 \cdot 6H_2O$ 24.0mg
H_3BO_3 .... 6.0mg
$CuCl_2 \cdot 2H_2O$ 2.0mg
HCl (25% solution) 10.0mL

Preparation of Trace Elements Solution SL-10: Add $FeCl_2 \cdot 4H_2O$ to 10.0mL of HCl solution. Mix thoroughly. Add distilled/deionized water and bring volume to 1.0L. Add remaining components. Mix thoroughly.

Preparation of Medium: Add components, except glucose solution and Na_2CO_3 solution, to distilled/deionized water and bring volume to 1.0L. Mix thoroughly. Adjust pH to 6.0 with 6*N* HCl. Sparge with 100% N_2. Anaerobically distribute 10.0mL volumes into screw-capped tubes. Aseptically and anaerobically add to each tube containing 10.0mL of medium, 0.4mL of sterile Na_2CO_3 solution, 0.5mL of sterile glucose solution, and 0.5mL of sterile $Na_2S \cdot 9H_2O$ solution. After addition of the Na_2CO_3 solution, the pH of the medium should be 6.0–7.0.

Use: For the cultivation and maintenance of *Clostridium aldrichii, Clostridium celerecrescens, Clostridium cellulolyticum, Clostridium lentocellum, Clostridium populeti,* and *Clostridium thermopalmarium.*

CM4 Medium

Composition per liter:
Cellobiose 6.0g
Yeast extract 5.0g
K_2HPO_4 .... 2.9g
KH_2PO_4 .... 1.5g
$(NH_4)_2SO_4$ 1.3g
NaCl 1.0g
$MgCl_2$ 0.75g
Sodium thioglycolate 0.5g
$CaCl_2$ 0.0132g
Resazurin (1.0% solution) 0.2mL
$FeSO_4$ (1.25% solution) 0.1mL

Preparation of Medium: Add components to distilled/deionized water and bring volume to 1.0L. Mix thoroughly. Gently heat until boiling. Boil until color changes from red to colorless, indicating a reduced state. Cool. Distribute into tubes or flasks under 97% N_2 + 3% H_2. Cap with rubber stoppers. Autoclave for 15 min at 15 psi pressure–121°C. Pour into sterile Petri dishes or leave in tubes.

Use: For the cultivation and maintenance of *Clostridium* species and other bacteria that can utilize cellobiose as a carbon source.

CM plus YE Agar, Modified

Composition per 1001.0mL:
NaCl 150.0g
$MgSO_4 \cdot 7H_2O$ 20.0g
Agar 15.0g
Yeast extract 10.0g
Vitamin assay casamino acids 7.5g
Trisodium citrate·$2H_2O$ 3.0g

KCl	2.0g
Fe^{2+} solution	1.0mL

pH 7.4 ± 0.2 at 25°C

Fe^{2+} Solution:

Composition per 100.0mL:

$FeSO_4 \cdot 7H_2O$	4.98g

Preparation of Fe^{2+} Solution: Add $FeSO_4 \cdot 7H_2O$ to distilled/deionized water and bring volume to 100.0mL. Mix thoroughly. Filter sterilize.

Preparation of Medium: Add components, except Fe^{2+} solution, to distilled/deionized water and bring volume to 1.0L. Mix thoroughly. Autoclave for 15 min at 15 psi pressure–121°C. Aseptically add 1.0mL of sterile Fe^{2+} solution. Mix thoroughly. Aseptically distribute into sterile tubes or flasks.

Use: For the cultivation of *Actinopolyspora mortivallis.*

CM plus YE Medium

Composition per liter:

NaCl	200.0g
$MgSO_4 \cdot 7H_2O$	20.0g
Yeast extract	10.0g
Casamino acids, vitamin free	7.5g
Sodium citrate	3.0g
KCl	2.0g
$FeSO_4 \cdot 7H_2O$ (4.98% solution)	1.0mL

pH 7.4 ± 0.2 at 25°C

Preparation of Medium: Add components to distilled/deionized water and bring volume to 1.0L. Mix thoroughly. Gently heat until boiling. Adjust pH to 7.4 with NaOH. Distribute into tubes or flasks. Autoclave for 10 min at 15 psi pressure–121°C.

Use: For the cultivation and maintenance of *Actinopolyspora halophila* and *Haloarcula japonica.*

CM + YE Medium B, Modified (DSMZ Medium 910)

Composition per liter:

NaCl	100.0g
Agar	15.0g
Yeast extract	10.0g
$MgSO_4 \cdot 7H_2O$	10.0g
Vitamin assay casamino acids	7.5g
$Na_3Citrate \cdot 2H_2O$	3.0g
KCl	2.0g
Fe^{2+} solution	1.0mL

pH 7.4 ± 0.2 at 25°C

Fe^{2+} Solution:

Composition per 100.0mL:

$FeSO_4 \cdot 7H_2O$	4.98g

Preparation of Fe^{2+} Solution: Add $FeSO_4 \cdot 7H_2O$ to distilled/deionized water and bring volume to 100.0mL. Mix thoroughly.

Preparation of Medium: Add components to distilled/deionized water and bring volume to 1.0L. Mix thoroughly. Adjust pH to 7.4. Gently heat and bring to boiling. Distribute into tubes or flasks. Autoclave for 15 min at 15 psi pressure–121°C. Pour into sterile Petri dishes or leave in tubes.

Use: For the cultivation and maintenance of *Nocardiopsis kunsanensis.*

CMA

See: **Cornmeal Agar**

CMA with Lupine

Composition per liter:

Agar	20.0g
Cornmeal polenta	15.0g
Lupine stems	variable

pH 7.0 ± 0.2 at 25°C

Preparation of Medium: Add cornmeal polenta to distilled/deionized water and bring volume to 1.0L. Mix thoroughly. Gently heat and bring to boiling. Continue boiling for 30 min. Filter through Whatman #1 filter paper. Add agar to filtrate. Gently heat and bring to boiling. Distribute 6.0mL volumes into tubes. Cut lupine stems into 8.0cm-long pieces. Add 2–3 lupine stems per tube. Autoclave for 15 min at 15 psi pressure–121°C. Allow tubes to cool in a slanted position.

Use: For the cultivation of *Glomerella cingulata.*

CMA with Sterile Carrot

Composition per liter:

Agar	20.0g
Cornmeal polenta	15.0g
Carrot	variable

pH 7.0 ± 0.2 at 25°C

Preparation of Medium: Add cornmeal polenta to distilled/deionized water and bring volume to 1.0L. Mix thoroughly. Gently heat and bring to boiling. Continue boiling for 30 min. Filter through Whatman #1 filter paper. Add agar to filtrate. Gently heat and bring to boiling. Distribute 6.0mL volumes into tubes. Cut carrots into 8.0cm-long pieces. Add 2–3 carrot slices per tube. Autoclave for 15 min at 15 psi pressure–121°C. Allow tubes to cool in a slanted position.

Use: For the cultivation of *Pyrenochaeta fallax.*

CML Medium (Cooked Meat Liver Medium)

Composition per liter:

Cooked meat	57.0g
Glucose	10.0g
Tryptose	10.0g
Liver infusion broth	10.0mL

pH 6.9 ± 0.2 at 25°C

Liver Infusion Broth:

Composition per liter:

Beef liver, infusion from	500.0g
Proteose peptone	10.0g
NaCl	5.0g

Preparation of Liver Infusion Broth: Add components to distilled/deionized water and bring volume to 1.0L. Mix thoroughly.

Preparation of Medium: Add cooked meat to distilled/deionized water and bring volume to 1.0L. Chill to 4°C until liquid is clear. Filter through cheesecloth. Add remaining components to filtrate. Distribute into tubes or flasks. Autoclave for 10 min at 15 psi pressure–121°C.

Use: For the cultivation and maintenance of *Fusobacterium varium.*

CMRL-1066 Medium with Glutamine, 10X (Connaught Medical Research Laboratories Medium with Glutamine, 10X)

Composition per liter:

NaCl	6.8g
$NaHCO_3$	2.2g
D-Glucose	1.0g
KCl	0.4g
L-Cysteine·HCl·H_2O	0.26g
$CaCl_2$, anhydrous	0.2g
$MgSO_4·7H_2O$	0.2g
$NaH_2PO_4·H_2O$	0.14g
L-Glutamine	0.1g
Sodium acetate·$3H_2O$	0.083g
L-Glutamic acid	0.075g
L-Arginine·HCl	0.07g
L-Lysine·HCl	0.07g
L-Leucine	0.06g
Glycine	0.05g
Ascorbic acid	0.05g
L-Proline	0.04g
L-Tyrosine	0.04g
L-Aspartic acid	0.03g
L-Threonine	0.03g
L-Alanine	0.025g
L-Phenylalanine	0.025g
L-Serine	0.025g
L-Valine	0.025g
L-Cystine	0.02g
L-Histidine·HCl·H_2O	0.02g
L-Isoleucine	0.02g
Phenol Red	0.02g
L-Methionine	0.015g
Deoxyadenosine	0.01g
Deoxycytidine	0.01g
Deoxyguanosine	0.01g
Glutathione, reduced	0.01g
Thymidine	0.01g
Hydroxy-L-proline	0.01g
L-Tryptophan	0.01g
Nicotinamide adenine dinucleotide	7.0mg
Tween™ 80	5.0mg
Sodium glucoronate·H_2O	4.2mg
Coenzyme A	2.5mg
Cocarboxylase	1.0mg
Flavin adenine dinucleotide	1.0mg
Nicotinamide adenine dinucleotide phosphate	1.0mg
Uridine triphosphate	1.0mg
Choline chloride	0.5mg
Cholesterol	0.2mg
5-Methyldeoxycytidine	0.1mg
Inositol	0.05mg
p-Aminobenzoic acid	0.05mg
Niacin	0.025mg
Niacinamide	0.025mg
Pyridoxine	0.025mg
Pyridoxal·HCl	0.025mg
Biotin	0.01mg
D-Calcium pantothenate	0.01mg
Folic acid	0.01mg
Riboflavin	0.01mg
Thiamine·HCl	0.01mg

pH 7.2 ± 0.2 at 25°C

Source: This medium is available as a premixed powder from BD Diagnostics.

Preparation of Medium: Add components to distilled/deionized water and bring volume to 1.0L. Mix thoroughly. Adjust pH to 7.2. Filter sterilize.

Use: For the cultivation of a wide variety of microorganisms in a chemically defined basal medium.

CMYG

See: **Cornmeal Yeast Glucose Agar**

CN Screen Medium (*Cryptococcus neoformans* Screen Medium)

Composition per liter:

Agar	15.0g
K_2HPO_4	4.0g
$MgSO_4·7H_2O$	2.5g
Glucose	1.25g
Asparagine	1.0g
Glutamine	1.0g
Glycine	1.0g
Thiamine·HCl	1.0g
Tryptophan	1.0g
EDTA	0.6g
Biotin	0.51g
Dihydroxyphenylalanine (Dopa)	0.2g
Phenol Red	0.2g

pH 5.5–5.6 ± 0.2 at 25°C

Preparation of Medium: Add components to distilled/deionized water and bring volume to 1.0L. Mix thoroughly. Gently heat until boiling. Distribute into tubes or flasks. Autoclave for 15 min at 15 psi pressure–121°C.

Use: For the screening of yeast isolates for the presumptive identification of *Cryptococcus neoformans*. *Cryptococcus neoformans* forms black colonies.

CNS Agar

Composition per liter:

Agar	15.0g
LiCl	10.0g
Pancreatic digest of gelatin	5.0g
Beef extract	3.0g
K_2HPO_4	2.0g
Yeast extract	2.0g
KH_2PO_4	0.5g
Glucose solution	50.0mL
$MgSO_4·7H_2O$ solution	10.0mL
Antibiotic solution	10.0mL
Bravo 500	0.082mL

pH 6.9 ± 0.2 at 25°C

Glucose Solution:

Composition per 50.0mL:

Glucose	5.0g

Preparation of Glucose Solution: Add glucose to distilled/deionized water and bring volume to 50.0mL. Mix thoroughly. Filter sterilize.

$MgSO_4 \cdot 7H_2O$ Solution:
Composition per 10.0mL:

$MgSO_4 \cdot 7H_2O$	0.25g

Preparation of $MgSO_4 \cdot 7H_2O$ Solution: Add $MgSO_4 \cdot 7H_2O$ to distilled/deionized water and bring volume to 10.0mL. Mix thoroughly. Filter sterilize.

Antibiotic Solution:
Composition per 10.0mL:

Cycloheximide	0.04g
Polymyxin B sulfate	0.032g
Nalidixic acid	0.025g

Preparation of Antibiotic Solution: Add components to distilled/deionized water and bring volume to 10.0mL. Mix thoroughly. Filter sterilize.

Caution: Cycloheximide is toxic. Avoid skin contact or aerosol formation and inhalation.

Preparation of Medium: Add components—except glucose solution, $MgSO_4 \cdot 7H_2O$ solution, and antibiotic solution—to distilled/deionized water and bring volume to 930.0mL. Mix thoroughly. Gently heat and bring to boiling. Autoclave for 15 min at 15 psi pressure–121°C. Cool to 45°–50°C. Aseptically add 50.0mL of sterile glucose solution, 10.0mL of sterile $MgSO_4 \cdot 7H_2O$ solution, and 10.0mL of sterile antibiotic solution. Mix thoroughly. Pour into sterile Petri dishes or distribute into sterile tubes.

Use: For the isolation and cultivation of *Corynbacterium nebraskense.*

Coagulase Agar Base

Composition per liter:

Agar	25.0g
Brain heart infusion	10.5g
Pancreatic digest of casein	10.5g
D-Mannitol	10.0g
Brain heart infusion	5.0g
NaCl	3.5g
Papaic digest of soybean meal	3.5g
Bromcresol Purple	0.02g
Rabbit plasma	100.0mL

pH 7.4 ± 0.2 at 25°C

Preparation of Medium: Add components, except rabbit plasma, to distilled/deionized water and bring volume to 1.0L. Mix thoroughly. Gently heat, while stirring, until boiling. Distribute into tubes or flasks. Autoclave for 15 min at 15 psi pressure–121°C. Cool to 45°–50°C. Add rabbit plasma to a final concentration of 7–15%. Mix thoroughly. Pour into sterile Petri dishes in 18.0mL volume per plate.

Use: For the cultivation and differentiation of *Staphylococcus aureus* from other *Staphylococcus* species based on coagulase production.

Coagulase Mannitol Agar

Composition per liter:

Agar	14.5g
Pancreatic digest of casein	10.5g
D-Mannitol	10.0g
Brain heart infusion	5.0g
NaCl	3.5g
Papaic digest of soybean meal	3.5g
Bromcresol Purple	0.02g
Rabbit plasma with 0.15% EDTA	100.0mL

pH 7.3 ± 0.2 at 25°C

Source: This medium is available as a premixed powder from BD Diagnostic Systems.

Preparation of Medium: Add components, except rabbit plasma, to distilled/deionized water and bring volume to 1.0L. Mix thoroughly. Gently heat while stirring until boiling. Distribute into tubes or flasks. Autoclave for 15 min at 15 psi pressure–121°C. Cool to 45°–50°C. For detection of coagulase activity add rabbit plasma with 0.15% EDTA to a final concentration of 7–15%. Mix thoroughly. Pour into sterile Petri dishes.

Use: For the cultivation and differentiation of *Staphylococcus aureus* from other *Staphylococcus* species based on coagulase production and mannitol fermentation.

Coagulase Mannitol HiVeg Agar Base with Plasma

Composition per liter:

Agar	14.5g
Plant hydrolysate	10.5g
Mannitol	10.0g
Plant special infusion	5.0g
Papaic digest of soybean meal	3.5g
NaCl	3.5g
Bromcresol Purple	0.02
Rabbit plasma	100.0–150.0mL

pH 7.3 ± 0.2 at 25°C

Source: This medium is available as a premixed powder from HiMedia.

Preparation of Medium: Add components, except rabbit plasma, to distilled/deionized water and bring volume to 1.0L. Mix thoroughly. Gently heat while stirring until boiling. Distribute into tubes or flasks. Autoclave for 15 min at 15 psi pressure–121°C. Cool to 45°–50°C. For detection of coagulase activity add rabbit plasma with 0.15% EDTA to a final concentration of 7–15%. Mix thoroughly. Pour into sterile Petri dishes.

Use: For the cultivation and differentiation of *Staphylococcus aureus* from other *Staphylococcus* species based on coagulase production and mannitol fermentation. For the primary isolation and identification of pathogenic Staphylococci from clinical specimens or for classifying pure cultures.

Coagulase Mannitol Broth Base with Plasma

Composition per liter:

Heart muscle, infusion from	375.0g
D-Mannitol	10.0g
Peptic digest of animal tissue	10.0g
NaCl	5.0g
Phenol Red	0.025g
Rabbit plasma, sterile, pretested normal	120.0–150.0mL

pH 7.3 ± 0.2 at 25°C

Source: This medium is available as a premixed powder from HiMedia.

Preparation of Medium: Add components, except rabbit plasma, to distilled/deionized water and bring volume to 1.0L. Mix thoroughly. Gently heat while stirring until boiling. Distribute into tubes or flasks. Autoclave for 15 min at 15 psi pressure–121°C. Cool to 45°–50°C. For detection of coagulase activity add rabbit plasma with 0.15% EDTA to a final concentration of 12–15%. Mix thoroughly.

Use: For the cultivation and differentiation of *Staphylococcus aureus* from other *Staphylococcus* species based on coagulase production and mannitol fermentation. For the simultaneous detection of coagulase production and mannitol fermentation in the differentiation of Staphylococci.

Coagulase Mannitol HiVeg Broth Base with Plasma

Composition per liter:

D-Mannitol	10.0g
Plant infusion	10.0g
Plant peptone	10.0g
NaCl	5.0g
Phenol Red	0.025g
Rabbit plasma, strerile, pretested normal	120.0-150.0mL

pH 7.3 ± 0.2 at 25°C

Source: This medium is available as a premixed powder from HiMedia.

Preparation of Medium: Add components, except rabbit plasma, to distilled/deionized water and bring volume to 1.0L. Mix thoroughly. Gently heat while stirring until boiling. Distribute into tubes or flasks. Autoclave for 15 min at 15 psi pressure–121°C. Cool to 45°–50°C. For detection of coagulase activity add rabbit plasma with 0.15% EDTA to a final concentration of 12–15%. Mix thoroughly.

Use: For the cultivation and differentiation of *Staphylococcus aureus* from other *Staphylococcus* species based on coagulase production and mannitol fermentation. For the simultaneous detection of coagulase production and mannitol fermentation in the differentiation of Staphylococci.

Coal Medium

Composition per 1011.0mL:

Coal, Pittsburgh seam	10.0g
$NaHCO_3$	3.5g
Yeast extract	2.0g
NaCl	0.4g
NH_4Cl	0.4g
$MgCl_2 \cdot 6H_2O$	0.33g
$Na_2S \cdot 9H_2O$	0.3g
$CaCl_2 \cdot 2H_2O$	0.05g
$Na_2SeO_3 \cdot 5H_2O$	3.0μg
KH_2PO_4	1.0mg
Resazurin	1.0mg
Wolfe's vitamin solution	10.0mL
Trace elements solution SL-10	1.0mL

pH 7.3 ± 0.1 at 25°C

Wolfe's Vitamin Solution:

Composition per liter:

Pyridoxine·HCl	10.0mg
p-Aminobenzoic acid	5.0mg
Lipoic acid	5.0mg
Nicotinic acid	5.0mg
Riboflavin	5.0mg
Thiamine·HCl	5.0mg
Calcium DL-pantothenate	5.0mg
Biotin	2.0mg
Folic acid	2.0mg
Vitamin B_{12}	0.1mg

Preparation of Wolfe's Vitamin Solution: Add components to distilled/deionized water and bring volume to 1.0L. Mix thoroughly.

Trace Elements Solution SL-10:

Composition per liter:

$FeCl_2 \cdot 4H_2O$	1.5g
$CoCl_2 \cdot 6H_2O$	190.0mg
$MnCl_2 \cdot 4H_2O$	100.0mg
$ZnCl_2$	70.0mg
$Na_2MoO_4 \cdot 2H_2O$	36.0mg
$NiCl_2 \cdot 6H_2O$	24.0mg
H_3BO_3	6.0mg
$CuCl_2 \cdot 2H_2O$	2.0mg
HCl (25% solution)	10.0mL

Preparation of Trace Elements Solution SL-10: Add $FeCl_2 \cdot 4H_2O$ to 10.0mL of HCl solution. Mix thoroughly. Add distilled/deionized water and bring volume to 1.0L. Add remaining components. Mix thoroughly.

Preparation of Medium: Prepare and dispense medium under 80% N_2 + 20% CO_2. Add components, except $NaHCO_3$, to distilled/deionized water and bring volume to 1.0L. Mix thoroughly. Gently heat and bring to boiling. Continue boiling for 3 min. Cool to room temperature while sparging with 80% N_2 + 20% CO_2. Add $NaHCO_3$. Mix thoroughly. Adjust pH to 7.3. Anaerobically distribute 10.0mL volumes into anaerobic tubes. Autoclave for 15 min at 15 psi pressure–121°C. Adjust pH to 7.3.

Use: For the cultivation of unidentified bacterium ATCC 55237.

COBA
(Colistin Oxolinic Acid Blood Agar)

Composition per liter:

Columbia agar base	930.0mL
Horse blood, defibrinated, sterile	50.0mL
Colistin sulfate solution	10.0mL
Oxolinic acid solution	10.0mL

pH 7.3 ± 0.2 at 25°C

Columbia Agar Base:

Composition per 930.0mL:

Agar	13.5g
Pancreatic digest of casein	10.0g
Peptic digest of animal tissue	10.0g
NaCl	5.0g
Beef extract	3.0g
Yeast extract	3.0g
Cornstarch	1.0g

Preparation of Columbia Agar Base: Add components to distilled/deionized water and bring volume to 930.0mL. Mix thoroughly. Gently heat until boiling. Autoclave for 15 min at 15 psi pressure–121°C. Cool to 45°–50°C.

Colistin Sulfate Solution:

Composition per 10.0mL:

Colistin sulfate	10.0mg

Preparation of Colistin Sulfate Solution: Add colistin sulfate to distilled/deionized water and bring volume to 10.0mL. Mix thoroughly. Filter sterilize.

Oxolinic Acid Solution:

Composition per 10.0mL:

Oxolinic acid	5.0–10.0mg

Preparation of Oxolinic Acid Solution: Add oxolinic acid to distilled/deionized water and bring volume to 10.0mL. Mix thoroughly. Filter sterilize.

Preparation of Medium: To 930.0mL of sterile, cooled Columbia agar base, add sterile colistin sulfate, sterile oxolinic acid, and sterile, defibrinated horse blood. Mix thoroughly. Pour into sterile Petri dishes.

Use: For the isolation and cultivation of streptococci in pure culture from mixed flora in clinical specimens.

Colby and Zatman Agar

Composition per liter:

Agar, noble	20.0g
K_2HPO_4	1.2g
KH_2PO_4	0.62g
$(NH_4)_2SO_4$	0.5g
$MgSO_4 \cdot 7H_2O$	0.2g
NaCl	0.1g
$CaCl_2 \cdot 6H_2O$	0.05g
$ZnSO_4 \cdot 7H_2O$	70.0μg
H_3BO_3	10.0μg
$MnSO_4 \cdot 5H_2O$	10.0μg
$Na_2MoO_4 \cdot 2H_2O$	10.0μg
$CoCl_2 \cdot 6H_2O$	5.0μg
$CuSO_4 \cdot 5H_2O$	5.0μg
$FeCl_3 \cdot 6H_2O$	1.0mg
Trimethylamine solution	10.0mL

pH 7.0 ± 0.2 at 25°C

Trimethylamine Solution

Composition per 10.0mL:

Trimethylamine	1.0g

Preparation of Trimethylamine Solution: Add trimethylamine to distilled/deionized water and bring volume to 10.0mL. Mix thoroughly. Filter sterilize.

Preparation of Medium: Add components, except trimethylamine solution, to distilled/deionized water and bring volume to 990.0mL. Mix thoroughly. Adjust pH to 7.0. Gently heat and bring to boiling. Autoclave for 15 min at 15 psi pressure–121°C. Cool to 50°C. Aseptically add 10.0mL of sterile trimethylamine solution. Mix thoroughly. Pour into sterile Petri dishes or distribute into sterile tubes.

Use: For the cultivation of *Aminobacter aminovorans, Bacillus* species, *Hyphomicrobium aestuarii, Hyphomicrobium facilis, Hyphomicrobium* species, *Hyphomicrobium variabile, Hyphomicrobium zavarzinii, Methylobacterium extorquens, Methylobacterium* species, and *Methylophilus methylotrophus*.

Colby and Zatman Medium

Composition per liter:

Agar	15.0g
K_2HPO_4	1.2g
KH_2PO_4	0.62g
$(NH_4)_2SO_4$	0.5g
$MgSO_4 \cdot 7H_2O$	0.2g
NaCl	0.1g
$CaCl_2 \cdot 2H_2O$	34.0mg
$FeCl_3 \cdot H_2O$	1.0mg
Trace elements solution	1.0mL
Trimethylamine, 10% solution, filter sterilized	10.0mL

pH 7.0 ± 0.2 at 25°C

Trace Elements Solution:

Composition per liter:

$ZnSO_4 \cdot 7H_2O$	70.0mg
H_3BO_3	10.0mg
$Na_2MoO_4 \cdot 2H_2O$	10.0mg
$MnSO_4 \cdot H_2O$	7.0mg
$CoCl_2 \cdot H_2O$	5.0mg
$CuSO_4 \cdot 5H_2O$	5.0mg

Preparation of Medium: Add components, except trimethylamine solution, to distilled/deionized water and bring volume to 990.0mL. Mix thoroughly. Gently heat and bring to boiling. Autoclave for 15 min at 15 psi pressure–121°C. Cool to 50°–55°C. Aseptically add 10.0mL of sterile trimethylamine solution. Mix thoroughly. Pour into sterile Petri dishes or distribute into sterile tubes.

Use: For the cultivation and maintenance of *Aminomonas aminovorus*.

Colby and Zatman Thiamine Medium

Composition per liter:

Agar, noble	20.0g
K_2HPO_4	1.2g
KH_2PO_4	0.62g
$(NH_4)_2 \cdot SO_4$	0.5g
$MgSO_4 \cdot 7H_2O$	0.2g
NaCl	0.1g
$CaCl_2 \cdot 6H_2O$	0.05g
$FeCl_3 \cdot 6H_2O$	1.0mg
$ZnSO_4 \cdot 7H_2O$	70.0μg
H_3BO_3	10.0μg
$MnSO_4 \cdot 5H_2O$	10.0μg
$Na_2MoO_4 \cdot 2H_2O$	10.0μg
$CoCl_2 \cdot 6H_2O$	5.0μg
$CuSO_4 \cdot 5H_2O$	5.0μg
Thiamine	0.5mg
Trimethylamine solution	10.0mL

pH 7.0 ± 0.2 at 25°C

Trimethylamine Solution

Composition per 10.0mL:

Trimethylamine	1.0g

Preparation of Trimethylamine Solution: Add trimethylamine to distilled/deionized water and bring volume to 10.0mL. Mix thoroughly. Filter sterilize.

Preparation of Medium: Add components, except trimethylamine solution, to distilled/deionized water and bring volume to 990.0mL. Mix thoroughly. Adjust pH to 7.0. Gently heat and bring to boiling. Autoclave for 15 min at 15 psi pressure–121°C. Cool to 50°C. Aseptically add 10.0mL of sterile trimethylamine solution. Mix thoroughly. Pour into sterile Petri dishes or distribute into sterile tubes.

Use: For the cultivation of *Aminobacter aminovorans, Bacillus* species, *Hyphomicrobium aestuarii, Hyphomicrobium facilis, Hyphomicrobium* species, *Hyphomicrobium variabile, Hyphomicrobium zavarzinii, Methylobacterium extorquens, Methylobacterium* species, and *Methylophilus methylotrophus*.

Cold Filterable Tryptone Soya Broth (Cold Filterable TSB) (Irradiated Tryptone Soya Broth)

Composition per liter:

Pancreatic digest of casein	17.0g
Papaic digest of soybean meal	3.0g

NaCl	5.0g
K_2HPO_4	2.5g
Glucose	2.5g

pH 7.3± 0.2 at 25°C

Source: This medium is available from Oxoid Unipath.

Preparation of Medium: Ready to use as a sterile gamma irradiated solution.

Use: For the cultivation of a wide variety of microorganisms. For microbiological Media Fill Trials (MFT) for the pharmaceutical industry.

Cold Filterable Vegetable Peptone Broth (cfVPB)

Composition per liter:

Vegetable peptone	18.0g
NaCl	5.0g
Yeast extract	3.0g
Glucose	2.5g
K_2HPO_4	2.5g

pH 7.0 ± 0.2 at 25°C

Source: Available in polyethene bags from Oxoid Unipath.

Preparation of Medium: Ready to use as a sterile gamma irradiated solution.

Use: For use in process simulations in the pharmaceutical industry, either as a liquid placebo, or as a growth medium for a solid placebo, added downstream of processing.

Coletsos Medium

Composition per 1625mL:

Potato starch	10.0g
Gelatin	4.0g
Asparagine	2.25g
KH_2PO_4	1.5g
Na-glutamate	1.0g
Na-pyruvate	1.0g
Mg-citrate	0.375g
Litmus	0.25g
Malachite green	0.25g
$MgSO_4$	0.15g
Activated carbon	0.1g
Oligonucleotide mixture	3.0mg
Egg mixture	625.0mL
Glycerol	7.5mL

Egg Mixture:

Composition per liter:

Whole eggs	18–24

Preparation of Egg Mixture: Use fresh eggs, less than 1 week old. Scrub the shells with soap. Let stand in a soap solution for 30 min. Rinse in running water. Soak eggs in 70% ethanol for 15 min. Break the eggs into a sterile container. Separate egg whites from egg yolks. Combine 8 parts egg white with 2 parts egg yolk. Homogenize by shaking. Filter through four layers of sterile cheesecloth into a sterile graduated cylinder. Bring volume to 1.0L distilled/deionized water.

Preparation of Medium: Add glycerol to 600.0mL of distilled/deionized water. Mix thoroughly. Add remaining components, except egg mixture. Bring volume to 1.0L. Mix thoroughly. Gently heat while stirring and bring to boiling. Autoclave for 15 min at 15 psi pressure–121°C. Cool to 50°C. Aseptically add 625.0mL of egg mixture. Mix thoroughly. Distribute into sterile screw-capped tubes. Place tubes in a slanted position. Inspissate at 85°C (moist heat) for 45 min.

Use: For the cultivation of *Mycobacterium tuberculosis.*

Coletsos Selective Medium

Composition per 1625mL:

Potato starch	10.0g
Gelatin	4.0g
Asparagine	2.25g
KH_2PO_4	1.5g
Na-glutamate	1.0g
Na-pyruvate	1.0g
Mg-citrate	0.375g
Litmus	0.25g
Malachite green	0.25g
$MgSO_4$	0.15g
Activated carbon	0.1g
Oligonucleotide mixture	3.0mg
Egg mixture solution	625.0mL
Glycerol	7.5mL
Nalidixic acid solution	1.0mL
Lincomycin solution	1.0mL
Cycloheximide solution	1.0mL

Nalidixic Acid Solution:

Composition per 100.0mL:

Nalidixic acid	0.5g

Preparation of Nalidixic Acid Solution: Add nalidixic acid to distilled/deionized water and bring volume to 100.0mL. Mix thoroughly. Filter sterilize.

Cycloheximide Solution:

Composition per 100.0mL:

Cycloheximide	1.5g
Ethanol	40.0mL

Preparation of Cycloheximide Solution: Add cycloheximide to 40.0mL of ethanol. Mix thoroughly. Bring volume to 100.0mL with distilled/deionized water. Filter sterilize.

Caution: Cycloheximide is toxic. Avoid skin contact or aerosol formation and inhalation.

Lincomycin Solution:

Composition per 100.0mL:

Lincomycin	0.5g

Preparation of Lioncomycin Solution: Add lincomycin to distilled/deionized water and bring volume to 100.0mL. Mix thoroughly. Filter sterilize.

Egg Mixture Solution:

Composition per liter:

Whole eggs	18–24

Preparation of Egg Mixture Solution: Use fresh eggs, less than 1 week old. Scrub the shells with soap. Let stand in a soap solution for 30 min. Rinse in running water. Soak eggs in 70% ethanol for 15 min. Break the eggs into a sterile container. Separate egg whites from egg yolks. Combine 8 parts egg white with 2 parts egg yolk. Homogenize by shaking. Filter through four layers of sterile cheesecloth into a sterile graduated cylinder. Bring volume to 1.0L with distilled/deionized water.

Preparation of Medium: Add glycerol to 600.0mL of distilled/deionized water. Mix thoroughly. Add remaining components, except egg mixture, lincomycin solution, cycloheximide solution, and nalidixic acid solution. Mix thoroughly. Bring volume to 1.0L. Gently heat while stirring and bring to boiling. Autoclave for 15 min at 15 psi pressure–121°C. Cool to 50°C. Aseptically add 625.0mL of egg mixture. Mix thoroughly. Aseptically add 1.0mL cycloheximide solution, 1.0mL lincomycin solution, and 1.0mL nalidixic acid solution. Distribute into sterile screw-capped tubes. Place tubes in a slanted position. Inspissate at 85°C (moist heat) for 45 min.

Use: For the isolation and cultivation of *Mycobacterium tuberculosis.*

Coli ID

Composition per liter:

Proprietary

Source: This medium is available from bioMérieux.

Use: A selective chromogenic medium for the detection and enumeration of *E. coli* at 44°C, and simultaneous enumeration of *E. coli* and other coliforms at 37°C, from food products.

Coliform Agar, Chromocult®
(Chromocult Coliform Agar)

Composition per liter:

Agar	10.0g
NaCl	5.0g
Peptone	3.0g
Na_2HPO_4	2.7g
NaH_2PO_4	2.2g
Tryptophan	1.0g
Na-pyruvate	1.0g
Chromogenic mixture	0.4g
Tergitol 7	0.15g

pH 7.0 ± 0.2 at 25°C

Source: This medium is available from Merck.

Preparation of Medium: Add components to distilled/deionized water and bring volume to 1.0L. Mix well and warm gently until dissolved. Autoclave for 15 min at 15 psi pressure–121°C. Pour into sterile Petri dishes. Some turbidity may occur, but this does not effect the performance.

Use: For the detection of *E. coli* and coliform bacteria in foods. The interaction of selected peptones, pyruvate, sorbitol, and phosphate buffer guarantees rapid colony growth, even for sublethally injured coliforms. The growth of Gram-positive bacteria as well as some Gram-negative bacteria is largely inhibited by the content of Tergitol 7 which has no negative effect on the growth of the coliform bacteria. A combination of two chromogenic substrates allows for the simultaneous detection of total coliforms and *E. coli.*The characteristic enzyme for coliforms, β-D-galactosidase, cleaves the Salmon-GAL substrate and causes a salmon to red color of the coliform colonies. The substrate X-glucuronide is used for the identification of β-D-glucuronidase, which is characteristic for *E. coli. E. coli* cleaves both Salmon-GAL and X-glucuronide, so that positive colonies take on a dark-blue to violet color. These are easily distinguished from other coliform colonies which have a salmon to red color. As part of an additional confirmation of *E. coli*, the inclusion of tryptophan improves the indole reaction, thereby increasing detection reliability when it is used in combination with the Salmon-GAL and X-glucuronide reaction.

Coliform Agar ES, Chromocult®
(Chromocult Coliform Agar ES)
(Chromocult Enhanced Selectivity Agar)

Composition per liter:

Agar	10.0g
MOPS	10.0g
KCl	7.5g
Peptone	5.0g
Bile salts	1.15g
Na-propionate	0.5g
6-Chloro-3-indoxyl-β-D-galactopyranoside	0.15g
5-Bromo-4-chloro-3-indoxyl-β-D-glucuronic acid	0.1g
Isopropyl-β-D-thiogalactopyranoside	0.1g

pH 7.0 ± 0.2 at 25°C

Source: This medium is available from Merck.

Preparation of Medium: Add components to distilled/deionized water and bring volume to 1.0L. Mix thoroughly and heat with frequent agitation until components are completely dissolved (approximately 45 min). Do not autoclave. Cool to 45°–50°C. Pour into sterile Petri dishes. The plates should be clear and colorless.

Use: For the detection of *E.coli* and total coliforms. The combination of suitable peptones and the buffering using MOPS allows rapid growth of coliforms and an optimal transformation of the chromogenic substrates. The amount of bile salts and propionate largely inhibit growth of Gram-positive and Gram-negative accompanying flora. The simultaneous detection of total coliforms and *E.coli* is achieved using the combination of two chromogrenic substrates. The substrate Salmon™--β-D-GAL is split by β-D-galactosidase, characteristic for coliforms, resulting in a salmon to red coloration of coliform colonies. The detection of the β-D-glucuronidase, characteristic for *E. coli,* is cleaved via the substrate X-β-D-glucuronide, causing a blue coloration of positive colonies. As *E. coli* splits Salmon™-β-D-GAL as well as X-β-D-glucuronide, the colonies turn to a dark violet color and can be easily differentiated from the other coliforms being salmon-red.

Coliform HiVeg Broth

Composition per liter:

Lactose	20.0g
Synthetic detergent	20.0g
Plant peptone No. 3	10.0g
Yeast extract	6.0g
Sodium lauryl sulfate	1.0g
Bromcresol Purple	0.35g
Synthetic detergent No. III	0.1g

pH 7.0 ± 0.2 at 25°C

Preparation of Medium: Add components to distilled/deionized water and bring volume to 1.0L. Mix thoroughly. Distribute into flasks or tubes. Autoclave for 25 min at 15 psi pressure–121°C.

Use: For the isolation and cultivation of coliform bacteria from cream, yogurt, and raw milk.

Coliform Medium
(CM)

Composition per liter:

Bile salts No. 3	20.0g
Lactose	20.0g
Proteose peptone No. 3	10.0g
Yeast extract	6.0g
Sodium lauryl sulfate	1.0g

Sodium deoxycholate.......................................0.1g
Bromcresol Purple solution.......................................10.0mL

pH 7.0 ± 0.2 at 25°C

Bromcresol Purple Solution:

Composition per 100.0mL:

Bromcresol Purple.......................................0.35g
NaOH (0.1*N* solution).......................................2.0mL

Preparation of Bromcresol Purple Solution: Combine Bromcresol Purple and NaOH solution. Mix thoroughly. Bring volume to 100.0mL with distilled/deionized water. Filter sterilize.

Preparation of Medium: Add components, except Bromcresol Purple solution, to distilled/deionized water and bring volume to 990.0mL. Mix thoroughly. Add 10.0mL of Bromcresol Purple solution. Mix thoroughly. Adjust pH to 7.0 with 1*N* HCl. Distribute into flasks in 95.0mL volumes. Autoclave for 15 min at 15 psi pressure–121°C.

Use: For the isolation and cultivation of coliform microorganisms from cream.

Coliform Medium (CM)

Composition per liter:

Bile salts No. 3.......................................20.0g
Lactose.......................................20.0g
Proteose peptone No. 3.......................................10.0g
Yeast extract.......................................6.0g
Sodium lauryl sulfate.......................................1.0g
Sodium deoxycholate.......................................0.1g
Bromcresol Purple solution.......................................10.0mL

pH 6.8 ± 0.2 at 25°C

Bromcresol Purple Solution:

Composition per 100.0mL:

Bromcresol Purple.......................................0.35g
NaOH (0.1*N* solution).......................................2.0mL

Preparation of Bromcresol Purple Solution: Combine Bromcresol Purple and NaOH solution. Mix thoroughly. Bring volume to 100.0mL with distilled/deionized water. Filter sterilize.

Preparation of Medium: Add components, except Bromcresol Purple solution, to distilled/deionized water and bring volume to 990.0mL. Mix thoroughly. Add 10.0mL of Bromcresol Purple solution. Mix thoroughly. Adjust pH to 6.8 with 1*N* NH_4OH. Distribute into flasks in 95.0mL volumes. Autoclave for 25 min at 15 psi pressure–121°C.

Use: For the isolation and cultivation of coliform microorganisms from yogurt and raw milk.

Coliform Medium, Modified (MCM)

Composition per liter:

Lactose.......................................20.0g
Tris(hydroxymethyl)aminomethane buffer.......................................12.1g
Proteose peptone No. 3.......................................10.0g
Yeast extract.......................................6.0g
Bile salts No. 3.......................................1.0g
Sodium lauryl sulfate.......................................1.0g
Sodium deoxycholate.......................................0.1g
Bromcresol Purple solution.......................................10.0mL

pH 7.0 ± 0.2 at 25°C

Bromcresol Purple Solution:

Composition per 100.0mL:

Bromcresol Purple.......................................0.35g
NaOH (0.1*N* solution).......................................2.0mL

Preparation of Bromcresol Purple Solution: Combine Bromcresol Purple and NaOH solution. Mix thoroughly. Bring volume to 100.0mL with distilled/deionized water. Filter sterilize.

Preparation of Medium: Add components, except Bromcresol Purple solution, to distilled/deionized water and bring volume to 990.0mL. Mix thoroughly. Add 10.0mL of Bromcresol Purple solution. Mix thoroughly. Adjust pH to 7.0 with 1*N* HCl. Distribute into flasks in 95.0mL volumes. Autoclave for 15 min at 15 psi pressure–121°C.

Use: For the isolation and cultivation of coliform microorganisms from cream.

Coliform Medium, Modified (MCM)

Composition per liter:

Lactose.......................................20.0g
Tris(hydroxymethyl)aminomethane buffer.......................................12.1g
Proteose peptone No. 3.......................................10.0g
Yeast extract.......................................6.0g
Bile salts No. 3.......................................1.0g
Sodium lauryl sulfate.......................................1.0g
Sodium deoxycholate.......................................0.1g
Bromcresol Purple solution.......................................10.0mL

pH 6.8 ± 0.2 at 25°C

Bromcresol Purple Solution:

Composition per 100.0mL:

Bromcresol Purple.......................................0.35g
NaOH (0.1*N* solution).......................................2.0mL

Preparation of Bromcresol Purple Solution: Combine Bromcresol Purple and NaOH solution. Mix thoroughly. Bring volume to 100.0mL with distilled/deionized water. Filter sterilize.

Preparation of Medium: Add components, except Bromcresol Purple solution, to distilled/deionized water and bring volume to 990.0mL. Mix thoroughly. Add 10.0mL of Bromcresol Purple solution. Mix thoroughly. Adjust pH to 6.8 with 1*N* NH_4OH. Distribute into flasks in 95.0mL volumes. Autoclave for 25 min at 15 psi pressure–121°C.

Use: For the isolation and cultivation of coliform microorganisms from yogurt.

Coliform PA Broth

Composition per liter:

Casein enzymic hydrolysate.......................................10.0g
Lactose.......................................7.5g
Pancreatic digest of gelatin.......................................5.0g
Beef extract.......................................3.0g
K_2HPO_4.......................................1.375g
KH_2PO_4.......................................1.375g
NaCl.......................................2.5g
Sodium lauryl sulphate.......................................0.05g
Bromocresol Purple.......................................8.5mg

pH 6.8 ± 0.2 at 25°C

Source: This medium is available from HiMedia.

Preparation of Medium: Add components to distilled/deionized water and bring volume to 1.0L. Mix thoroughly. Gently heat and bring to boiling. Distribute into tubes or flasks. Autoclave for 12 min at 15 psi pressure–121°C.

Use: For the determination of presence or absence of coliforms during detection of pollution in treated water from treatment plants or distribution systems.

Collimonas Medium (DSMZ Medium 1035)

Composition per liter:

NaCl	5.6g
Pancreatic digest of casein	1.8g
KH_2PO_4	1.0g
Papaic digest of soybean meal	0.6g

pH 6.5 ± 0.2 at 25°C

Preparation of Medium: Add components to distilled/deionized water and bring volume to 1.0L. Mix thoroughly. Adjust pH to 6.5. Distribute into tubes or flasks. Autoclave for 15 min at 15 psi pressure–121°C.

Use: For the cultivation of *Collimonas* spp.

Colloidal Chitin Agar

Composition per liter:

Agar	20.0g
Chitin, colloidal	4.0g
K_2HPO_4	0.7g
$MgSO_4 \cdot 5H_2O$	0.5g
KH_2PO_4	0.3g
$FeSO_4 \cdot 7H_2O$	0.01g
$MnCl_2$	1.0mg
$ZnSO_4$	1.0mg

pH 7.0 ± 0.2 at 25°C

Preparation of Medium: Add components to distilled/deionized water and bring volume to 1.0L. Mix thoroughly. Gently heat and bring to boiling. Distribute into tubes or flasks. Autoclave for 15 min at 15 psi pressure–121°C. Pour into sterile Petri dishes or leave in tubes.

Use: For the isolation and cultivation of *Micromonospora* species from water, soil, or sediment. For the germination of spores of *Micromonospora* species.

Colonization Medium

Composition per 700.0mL:

Mannose	1.0g
Pancreatic digest of gelatin	0.2g
Brain heart, solids from infusion	0.08g
Peptic digest of animal tissue	0.08g
NaCl	0.07g
Glucose	0.04g
Na_2HPO_4	0.03g
Bile salts No. 3	0.1g
Dulbecco's phosphate-buffered saline	700.0mL

pH 7.4 ± 0.2 at 25°C

Dulbecco's Phosphate-Buffered Saline:

Composition per liter:

NaCl	8.0g
$Na_2HPO_4 \cdot 7H_2O$	2.16g
KCl	0.2g
KH_2PO_4	0.2g
$CaCl_2$	0.1g
$MnCl_2 \cdot 6H_2O$	0.1g

Preparation of Dulbecco's Phosphate-Buffered Saline: Add components to distilled/deionized water and bring volume to 1.0L. Mix thoroughly.

Preparation of Medium: Combine components. Mix thoroughly. Filter sterilize. Aseptically distribute into sterile tubes or flasks.

Use: For the differentiation of enterotoxigenic *Escherichia coli* from foods based on the HeLa cell test for colonization.

Colonization Medium with Dulbecco's Phosphate Buffer

Composition per liter:

NaCl	8.0g
Mannose	1.4g
K_2HPO_4	1.15g
Brain heart infusion powder	0.7g
KCl	0.2g
KH_2PO_4	0.2g
Bile salts mixture	0.14g

pH 7.5 ± 0.2 at 25°C

Source: This medium is available from HiMedia.

Preparation of Medium: Add components to distilled/deionized water and bring volume to 1.0L. Mix thoroughly. Filter sterilize. Do not autoclave or heat. Aseptically distribute into tubes.

Use: For preparation of the solution of enterotoxigenic *Escherichia coli* used for colonization test in HeLa cell lines.

Columbia Agar

Composition per liter:

Columbia agar base	950.0mL
Sheep blood	50.0mL

pH 7.3 ± 0.2 at 25°C

Columbia Agar Base:

Composition per liter:

Agar	13.5g
Pancreatic digest of casein	12.0g
NaCl	5.0g
Peptic digest of animal tissue	5.0g
Beef extract	3.0g
Yeast extract	3.0g
Cornstarch	1.0g

Preparation of Columbia Agar Base: Add components to distilled/deionized water and bring volume to 1.0L. Mix thoroughly. Gently heat until boiling. Autoclave for 15 min at 15 psi pressure–121°C. Cool to 45°–50°C.

Preparation of Medium: To 950.0mL of cooled, sterile Columbia agar base, aseptically add 50.0mL of sterile, defibrinated sheep blood. Mix thoroughly. Pour into sterile Petri dishes or distribute into sterile tubes.

Use: For the isolation and cultivation of nonfastidious and fastidious microorganisms from a variety of clinical and nonclinical specimens.

Columbia Blood Agar

Composition per liter:

Columbia blood agar base	950.0mL
Sheep blood	50.0mL

pH 7.3 ± 0.2 at 25°C

Columbia Blood Agar Base:
Composition per liter:

Agar	15.0g
Pantone	10.0g
Bitone	10.0g
NaCl	5.0g
Tryptic digest of beef heart	3.0g
Cornstarch	1.0g

Source: Columbia blood agar base is available as a premixed powder from BD Diagnostic Systems.

Preparation of Columbia Blood Agar Base: Add components to distilled/deionized water and bring volume to 1.0L. Mix thoroughly. Gently heat until boiling. Autoclave for 15 min at 15 psi pressure–121°C. Cool to 45°–50°C.

Preparation of Medium: To 950.0mL of cooled, sterile Columbia blood agar base, aseptically add 50.0mL of sterile, defibrinated sheep blood. Mix thoroughly. Pour into sterile Petri dishes or distribute into sterile tubes.

Use: With the addition of blood or other enrichments, used for the isolation and cultivation of fastidious microorganisms.

Columbia Blood Agar (DSMZ Medium 693)

Composition per liter:

Columbia blood agar base	950.0mL
Sheep blood	50.0mL

pH 7.3 ± 0.2 at 25°C

Source: This medium is available as a premixed powder from Oxoid Unipath.

Columbia Blood Agar Base:
Composition per liter:

Special peptone	23.0g
Agar	10.0g
NaCl	5.0g
Starch	1.0g

Preparation of Columbia Blood Agar Base: Add components to distilled/deionized water and bring volume to 1.0L. Mix thoroughly. Gently heat until boiling. Autoclave for 15 min at 15 psi pressure–121°C. Cool to 45°–50°C.

Preparation of Medium: To 950.0mL of cooled, sterile Columbia blood agar base, aseptically add 50.0mL of sterile, defibrinated sheep blood. Mix thoroughly. Pour into sterile Petri dishes or distribute into sterile tubes.

Use: For the cultivation of *Corynebacterium* spp., *Actinomyces* spp., *Arcanobacterium* spp., *Streptococcus pneumoniae, Lactobacillus iners, Isobaculum melis, Nocardia paucivorans,* and a variety of fastidious microorganisms.

Columbia Blood Agar Base

Composition per liter:

Agar	15.0g
Pantone	10.0g
Bitone	10.0g
NaCl	5.0g
Tryptic digest of beef heart	3.0g
Cornstarch	1.0g

Source: Columbia blood agar base is available as a premixed powder from BD Diagnostic Systems.

Preparation of Medium: Add components to distilled/deionized water and bring volume to 1.0L. Mix thoroughly. Gently heat until boiling. Autoclave for 15 min at 15 psi pressure–121°C. Pour into sterile Petri dishes or distribute into sterile tubes.

Use: For the cultivation of *Balneatrix alpica.*

Columbia Blood Agar Base with 1% Agar, HiVeg with Blood

Composition per liter:

Plant special peptone	23.3g
Agar	10.0g
NaCl	5.0g
Corn starch	1.0g
Sheep blood, defibrinated	50.0mL

pH 7.3 ± 0.2 at 25°C

Source: This medium is available as a premixed powder from Hi-Media.

Preparation of Medium: Add components to distilled/deionized water and bring volume to 1.0L. Mix thoroughly. Gently heat until boiling. Autoclave for 15 min at 15 psi pressure–121°C. Cool to 45°–50°C. For Columbia Blood Agar: Add 5% sterile defibrinated sheep blood to sterile cool base. For Chocolate Agar: Add 10% sterile defibrinated sheep blood to sterile cool base. Heat to 80°C for 10 min with constant agitation. For Selective Medium: Add desired quantity of antimicrobial agent to sterile base. Mix thoroughly. Pour into sterile Petri dishes or distribute into sterile tubes.

Use: For the isolation and cultivation of fastidious bacteria from a variety of clinical and nonclinical specimens.

Columbia Blood Agar Base, HiVeg with Blood

Composition per liter:

Plant special peptone	23.0g
Agar	15.0g
NaCl	5.0g
Corn starch	1.0g
Sheep blood, defibrinated	50.0mL

pH 7.3 ± 0.2 at 25°C

Source: This medium is available as a premixed powder from Hi-Media.

Preparation of Medium: Add components to distilled/deionized water and bring volume to 950.0mL. Mix thoroughly. Gently heat until boiling. Autoclave for 15 min at 15 psi pressure–121°C. Cool to 45°–50°C. For Columbia Blood Agar: Add 5% sterile defibrinated sheep blood to sterile cool base. For Chocolate Agar: Add 10% sterile defibrinated sheep blood to sterile cool base. Heat to 80°C for 10 min with constant agitation. For Selective Medium: Add desired quantity of antimicrobial agent to sterile base. Mix thoroughly. Pour into sterile Petri dishes or distribute into sterile tubes.

Use: For the isolation and cultivation of fastidious microorganisms from a variety of clinical and nonclinical specimens.

Columbia Blood Agar Base with Horse Blood (LMG Medium 151)

Composition per liter:

Columbia blood agar base	950.0mL
Horse blood	50.0mL

pH 7.3 ± 0.2 at 25°C

Columbia Blood Agar Base:
Composition per liter:

Special peptone	23.0g
Agar	10.0g
NaCl	5.0g
Starch	1.0g

Preparation of Columbia Blood Agar Base: Add components to distilled/deionized water and bring volume to 1.0L. Mix thoroughly. Gently heat until boiling. Autoclave for 15 min at 15 psi pressure–121°C. Cool to 45°–50°C.

Preparation of Medium: To 950.0mL of cooled, sterile Columbia blood agar base, aseptically add 50.0mL of sterile, defibrinated horse blood. Mix thoroughly. Pour into sterile Petri dishes or distribute into sterile tubes.

Use: For the cultivation of *Arcanobacterium* spp., *Paenibacillus* spp., *Corynebacterium* spp., *Lactobacillus iners, Globicatella sanguinis, Gemella morbillorum, Enterococcus cecorum, Enterococcus columbae, Enterococcus parauberis, Enterococcus pseudoavium, Enterococcus raffinosus, Enterococcus saccharolyticus*, and other bacteria.

Columbia Blood Agar Base with Horse Blood (LMG Medium 210)

Composition per liter:

Columbia blood agar base	950.0mL
Horse blood	50.0mL

pH 7.3 ± 0.2 at 25°C

Columbia Blood Agar Base:
Composition per liter:

Agar	15.0g
Pancreatic digest of casein	10.0g
Proteose peptone No. 3	5.0g
Yeast extract	5.0g
NaCl	5.0g
Beef heart digest	3.0g
Corn starch	1.0g

Preparation of Columbia Blood Agar Base: Add components to distilled/deionized water and bring volume to 1.0L. Mix thoroughly. Gently heat until boiling. Autoclave for 15 min at 15 psi pressure–121°C. Cool to 45°–50°C.

Preparation of Medium: To 950.0mL of cooled, sterile Columbia blood agar base, aseptically add 50.0mL of sterile, defibrinated horse blood. Mix thoroughly. Pour into sterile Petri dishes or distribute into sterile tubes.

Use: For the cultivation of *Actinomyces* spp., *Streptococcus dysgalactiae,* and *Actinobaculum* spp.

Columbia Blood Agar Base with Horse Blood and Charcoal (DSMZ Medium 429a)

Composition per liter:

Columbia blood agar base with charcoal	960.0mL
Horse blood	40.0mL

pH 7.3 ± 0.2 at 25°C

Columbia Blood Agar Base with Charcoal:
Composition per liter:

Agar	15.0g
Pancreatic digest of casein	10.0g
Proteose peptone No. 3	5.0g
Yeast extract	5.0g
NaCl	5.0g
Beef heart digest	3.0g
Charcoal	2.0g
Cornstarch	1.0g

Preparation of Columbia Blood Agar Base with Charcoal: Add components to distilled/deionized water and bring volume to 1.0L. Mix thoroughly. Gently heat until boiling. Autoclave for 15 min at 15 psi pressure–121°C. Cool to 45°–50°C.

Preparation of Medium: To 960.0mL of cooled, sterile Columbia blood agar base, aseptically add 40.0mL of sterile, defibrinated horse blood. Mix thoroughly. Pour into sterile Petri dishes or distribute into sterile tubes.

Use: For the cultivation of *Neisseria gonorrhoeae.*

Columbia Blood Agar Base with Rabbit Serum

Composition per 1020.0mL:

Starch, soluble	9.0g
Resazurin	10.0μg
Columbia blood agar base	1.0L
Rabbit serum	20.0mL

pH 7.3 ± 0.2 at 25°C

Columbia Blood Agar Base:
Composition per liter:

Special peptone	23.0g
Agar	10.0g
NaCl	5.0g
Starch	1.0g

Source: Columbia Blood Agar Base is available as a premixed powder from Oxoid Unipath.

Preparation of Columbia Blood Agar Base: Add components to distilled/deionized water and bring volume to 1.0L. Mix thoroughly. Gently heat and bring to boiling. Autoclave for 15 min at 15 psi pressure–121°C. Cool to 45°–50°C.

Preparation of Medium: Combine 1.0L of Columbia blood agar base with 9.0g of soluble starch and 10.0μg of resazurin. Mix thoroughly. Gently heat and bring to boiling. Autoclave for 15 min at 15 psi pressure–121°C. Cool to 45°–50°C. Aseptically add 20.0mL of sterile rabbit serum. Mix thoroughly. Pour into sterile Petri dishes or distribute into sterile tubes.

Use: For the cultivation of *Mobiluncus curtisii* and *Mobiluncus mulieris*.

Columbia Broth

Composition per liter:

Bitone	10.0g
Pancreatic digest of casein	5.0g
Peptic digest of animal tissue	5.0g
NaCl	5.0g
Tryptic digest of beef heart	3.0g
Tris(hydroxymethyl)aminomethane·HCl	2.86g
Glucose	2.5g
Tris(hydroxymethyl)aminomethane	0.83g
Na_2CO_3	0.6g
L-Cysteine·HCl	0.1g

$MgSO_4$, anhydrous 0.1g
$FeSO_4$ 0.02g

pH 7.5 ± 0.2 at 25°C

Source: This medium is available as a premixed powder from BD Diagnostic Systems.

Preparation of Medium: Add components to distilled/deionized water and bring volume to 1.0L. Mix thoroughly. Gently heat until boiling. Distribute into tubes or flasks. Autoclave for 15 min at 15 psi pressure–121°C.

Use: For the cultivation and isolation of fastidious bacteria from clinical specimens or as a general purpose broth.

Columbia Broth

Composition per liter:

Pancreatic digest of casein	10.0g
Peptic digest of animal tissue	8.0g
NaCl	5.0g
Yeast extract	5.0g
Tris(hydroxymethyl) aminomethane·HCl buffer	2.86g
Glucose	2.5g
Tris(hydroxymethyl) aminomethane buffer	0.83g
L-Cysteine·HCl·H_2O	0.1g
$MgSO_4 \cdot 7H_2O$	0.05g
$FeSO_4$	0.012g

pH 7.4 ± 0.2 at 25°C

Source: This medium is available as a premixed powder from BD Diagnostic Systems.

Preparation of Medium: Add components to distilled/deionized water and bring volume to 1.0L. Mix thoroughly. Gently heat until boiling. Distribute into tubes or flasks. Autoclave for 15 min at 15 psi pressure–121°C.

Use: For the cultivation of a wide variety of microorganisms. Used as a general purpose medium.

Columbia Broth Base, HiVeg with Blood

Composition per liter:

Plant peptone No. 5	10.0g
Plant special peptone	10.0g
NaCl	5.0g
Plant infusion	3.0g
Tris(hydroxymethyl)aminomethane	2.86
Glucose	2.5g
Na_2CO_3	0.6g
L-Cystine hydrochloride	0.1g
$MgSO_4$	0.1g
$FeSO_4$	0.02g
Sheep blood, defibrinated	50.0mL

pH 7.5 ± 0.2 at 25°C

Source: This medium, without blood, is available as a premixed powder from HiMedia.

Preparation of Medium: Add components to distilled/deionized water and bring volume to 1.0L. Mix thoroughly. Gently heat until boiling. Distribute into tubes or flasks. Autoclave for 15 min at 15 psi pressure–121°C. Cool to 45°–50°C. Add blood and/or selective antimicrobics. Mix well.

Use: For the cultivation and isolation of fastidious bacteria from clinical specimens.

Columbia Broth Base, HiVeg with SPS

Composition per liter:

Plant peptone No. 5	10.0g
Plant special peptone	10.0g
NaCl	5.0g
Plant infusion	3.0g
Tris (hydroxymethyl) aminomethane	2.86
Glucose	2.5g
Na_2CO_3	0.6g
L-Cystine hydrochloride	0.1g
$MgSO_4$	0.1g
$FeSO_4$	0.02g
SPS (sodium polystyrene sulfonate)	0.1mL

pH 7.5 ± 0.2 at 25°C

Source: This medium, without SPS, is available as a premixed powder from HiMedia.

Preparation of Medium: Add components to distilled/deionized water and bring volume to 1.0L. Mix thoroughly. Gently heat until boiling. Distribute into tubes or flasks. Autoclave for 15 min at 15 psi pressure–121°C. Cool to 45°–50°C. Mix thoroughly.

Use: For the cultivation and isolation of fastidious bacteria from clinical specimens. For blood cultures, the SPS inhibits lysozyme activity and interferes with phagocytosis and destroys the aminoglycosides.

Columbia CNA Agar (Columbia Colistin Nalidixic Acid Agar)

Composition per liter:

Columbia blood agar base	950.0L
Sheep blood	50.0mL

pH 7.3 ± 0.2 at 25°C

Source: This medium is available as a premixed powder from BD Diagnostic Systems.

Columbia Blood Agar Base:
Composition per liter:

Agar	13.5g
Pancreatic digest of casein	12.0g
NaCl	5.0g
Peptic digest of animal tissue	5.0g
Beef extract	3.0g
Yeast extract	3.0g
Cornstarch	1.0g
Nalidixic acid	15.0mg
Colistin	10.0mg

Preparation of Columbia Blood Agar Base: Add components to distilled/deionized water and bring volume to 1.0L. Mix thoroughly. Gently heat until boiling. Autoclave for 15 min at 15 psi pressure–121°C. Cool to 45°–50°C.

Preparation of Medium: To 950.0mL of cooled, sterile Columbia blood agar base, aseptically add 50.0mL of sterile, defibrinated sheep blood. Mix thoroughly. Pour into sterile Petri dishes or distribute into sterile tubes.

Use: For the selective isolation, cultivation, and differentiation of Gram-positive cocci from clinical and nonclinical specimens.

Columbia C.N.A. Agar Base with Blood

Composition per liter:

Peptone, special	23.0g
Agar	15.0g
NaCl	5.0g
Corn starch	1.0g
Sheep blood, defibrinated	50.0mL

pH 7.3 ± 0.2 at 25°C

Source: This medium, without blood, is available as a premixed powder from HiMedia.

Preparation of Medium: Add components to distilled/deionized water and bring volume to 1.0L. Mix thoroughly. Gently heat until boiling. Autoclave for 15 min at 15 psi pressure–121°C. Cool to 45°–50°C. Aseptically add 50.0mL of sterile, defibrinated sheep blood to 950.0mL of cooled, sterile agar base. Mix thoroughly. Pour into sterile Petri dishes or distribute into sterile tubes.

Use: For the selective isolation, cultivation, and differentiation of Gram-positive cocci from clinical and nonclinical specimens.

Columbia C.N.A. Agar Base with Blood

Composition per liter:

Biopeptone	20.0g
Agar	15.0g
NaCl	5.0g
Tryptic digest of beef heart	3.0g
Cornstarch	1.0g
Nalidixic acid	0.015g
Colistin sulfate	0.01g
Sheep blood, defibrinated	50.0mL

pH 7.3 ± 0.2 at 25°C

Source: This medium, without blood, is available as a premixed powder from HiMedia.

Preparation of Medium: Add components to distilled/deionized water and bring volume to 1.0L. Mix thoroughly. Gently heat until boiling. Autoclave for 15 min at 15 psi pressure–121°C. Cool to 45°–50°C. Aseptically add 50.0mL of sterile, defibrinated sheep blood to 950.0mL of cooled, sterile agar base. Mix thoroughly. Pour into sterile Petri dishes or distribute into sterile tubes.

Use: For the selective isolation, cultivation, and differentiation of Gram-positive cocci from clinical and nonclinical specimens.

Columbia C.N.A. Agar Base with 1% Agar and Blood

Composition per liter:

Biopeptone	20.0g
Agar	10.0g
NaCl	5.0g
Tryptic digest of beef heart	3.0g
Corn starch	1.0g
Nalidixic acid	0.015g
Colistin sulfate	0.01g
Sheep blood, defibrinated	50.0mL

pH 7.3 ± 0.2 at 25°C

Source: This medium, without blood, is available as a premixed powder from HiMedia.

Preparation of Medium: Add components to distilled/deionized water and bring volume to 1.0L. Mix thoroughly. Gently heat until boiling. Autoclave for 15 min at 15 psi pressure–121°C. Cool to 45°–50°C. Aseptically add 50.0mL of sterile, defibrinated sheep blood to 950.0mL of cooled, sterile agar base. Mix thoroughly. Pour into sterile Petri dishes or distribute into sterile tubes.

Use: For the selective isolation, cultivation, and differentiation of Gram-positive cocci from clinical and nonclinical specimens.

Columbia CNA Agar, Modified with Sheep Blood

Composition per liter:

Columbia blood agar base	950.0mL
Sheep blood, defribinated	50.0mL

pH 7.3 ± 0.2 at 25°C

Source: This medium is available as a premixed powder from BD Diagnostic Systems.

Columbia Blood Agar Base:

Composition per liter:

Agar	13.5g
Pancreatic digest of casein	12.0g
NaCl	5.0g
Peptic digest of animal tissue	5.0g
Beef extract	3.0g
Yeast extract	3.0g
Cornstarch	1.0g
Nalidixic acid	5.0mg
Colistin	10.0mg

Preparation of Columbia Blood Agar Base: Add components to distilled/deionized water and bring volume to 1.0L. Mix thoroughly. Gently heat until boiling. Autoclave for 15 min at 15 psi pressure–121°C. Cool to 45°–50°C.

Preparation of Medium: To 950.0L of cooled, sterile Columbia blood agar base, aseptically add 50.0mL of sterile, defibrinated sheep blood. Mix thoroughly. Pour into sterile Petri dishes or distribute into sterile tubes.

Use: For the selective isolation, cultivation, and differentiation of Gram-positive cocci from clinical and nonclinical materials.

Columbia C.N.A. HiVeg Agar Base with 1% Agar

Composition per liter:

Plant peptone No. 5	20.0g
Agar	10.0g
NaCl	5.0g
Plant infusion	3.0g
Corn starch	1.0g
Nalidixic acid	0.015g
Colistin sulfate	0.01g
Sheep blood, defibrinated	50.0mL

pH 7.3 ± 0.2 at 25°C

Source: This medium, without blood, is available as a premixed powder from HiMedia.

Preparation of Medium: Add components to distilled/deionized water and bring volume to 1.0L. Mix thoroughly. Gently heat until boiling. Autoclave for 15 min at 15 psi pressure–121°C. Cool to 45°–50°C. Aseptically add 50.0mL of sterile, defibrinated sheep blood to 950.0mL of cooled, sterile agar base. Mix thoroughly. Pour into sterile Petri dishes or distribute into sterile tubes.

Use: For the selective isolation, cultivation, and differentiation of Gram-positive cocci from clinical and nonclinical specimens.

Columbia C.N.A. HiVeg Agar Base with Blood

Composition per liter:

Plant peptone No. 5	20.0g
Agar	15.0g
NaCl	5.0g
Plant infusion	3.0g
Cornstarch	1.0g
Nalidixic acid	0.015g
Colistin sulfate	0.01g
Sheep blood, defibrinated	50.0mL

pH 7.3 ± 0.2 at 25°C

Source: This medium, without blood, is available as a premixed powder from HiMedia.

Preparation of Medium: Add components to distilled/deionized water and bring volume to 1.0L. Mix thoroughly. Gently heat until boiling. Autoclave for 15 min at 15 psi pressure–121°C. Cool to 45°–50°C. Aseptically add 50.0mL of sterile, defibrinated sheep blood to 950.0mL of cooled, sterile agar base. Mix thoroughly. Pour into sterile Petri dishes or distribute into sterile tubes.

Use: For the selective isolation, cultivation, and differentiation of Gram-positive cocci from clinical and nonclinical specimens.

Columbia Colistin Nalidixic Acid Agar

See: **Columbia CNA Agar**

Colwella psychroerythrus Medium

Composition per liter:

NaCl	29.0g
$MgCl_2 \cdot 6H_2O$	8.0g
Pancreatic digest of casein	8.0g
KH_2PO_4	5.4g
$CaCl_2 \cdot 6H_2O$	33.0mg
$FeCl_2 \cdot 4H_2O$	2.0mg

pH 7.0 ± 0.2 at 25°C

Preparation of Medium: Add components to distilled/deionized water and bring volume to 1.0L. Mix thoroughly. Distribute into tubes or flasks. Autoclave for 15 min at 15 psi pressure–121°C.

Use: For the cultivation of *Aminobacter aminovorans, Bacillus* species, *Hyphomicrobium aestuarii, Hyphomicrobium facilis, Hyphomicrobium* species, *Hyphomicrobium variabile, Hyphomicrobium zavarzinii, Methylobacterium extorquens, Methylobacterium* species, and *Methylophilus methylotrophus.*

Complex Medium

Composition per liter:

NaCl	250.0g
$MgSO_4 \cdot 7H_2O$	20.0g
Yeast extract	10.0g
Casamino acids	7.5g
Trisodium citrate	3.0g
KCl	2.0g

pH 7.5–7.8 at 25°C

Preparation of Medium: Add components to distilled/deionized water and bring volume to 1.0L. Mix thoroughly. Autoclave for 5 min at 15 psi pressure–121°C. Filter through Whatman #1 filter paper. Adjust pH of filtrate to 7.4. Distribute into tubes or flasks. Autoclave for 15 min at 15 psi pressure–121°C.

Use: For the isolation and cultivation of *Actinomadura* species, *Actinopolyspora* species, *Excellospora* species, and *Microspora* species.

Congo Red Acid Morpholinepropanesulfonic Acid Pigmentation Agar (CRAMP Agar)

Composition per liter:

Agarose	14.0g
Morpholinepropanesulfonic acid	8.4g
NaCl	2.9g
Casamino acids	2.0g
Galactose	2.0g
Tricine (*n*-Tris-hydroxymethyl-methylglycine) buffer	1.8g
$Na_2S_2O_3 \cdot 5H_2O$	0.6g
NH_4Cl	0.5g
K_2HPO_4	0.24g
$MgSO_4 \cdot 7H_2O$	0.1g
Congo Red	5.0mg

pH 5.3 ± 0.2 at 25°C

Preparation of Medium: Add components to distilled/deionized water and bring volume to 1.0L. Mix thoroughly. Gently heat and bring to boiling. Adjust pH to 5.3. Distribute into tubes or flasks. Autoclave for 15 min at 15 psi pressure–121°C. Pour into sterile Petri dishes or leave in tubes.

Use: For the cultivation of *Yersinia* species with plasmids.

Congo Red Agar (CR Agar)

Composition per liter:

GC agar base	890.0mL
Hemoglobin solution	100.0mL
Supplement solution	10.0mL
Congo Red (0.01% solution)	0.1mL

pH 7.2 ± 0.2 at 25°C

GC Agar Base:

Composition per 890.0mL:

Agar	10.0g
Pancreatic digest of casein	7.5g
Peptic digest of animal tissue	7.5g
NaCl	5.0g
K_2HPO_4	4.0g
Cornstarch	1.0g
KH_2PO_4	1.0g

Preparation of GC Agar Base: Add components to distilled/deionized water and bring volume to 890.0mL. Mix thoroughly. Gently heat until boiling. Autoclave for 15 min at 15 psi pressure–121°C. Cool to 45°–50°C.

Hemoglobin Solution:

Composition per 100.0mL:

Hemoglobin	2.0g

Preparation of Hemoglobin Solution: Add hemoglobin to distilled/deionized water and bring volume to 100.0mL. Mix thoroughly. Autoclave for 15 min at 15 psi pressure–121°C. Cool to 50°C.

Congo Red Solution:

Composition per 100.0mL:

Congo Red	0.01g

Preparation of Congo Red Solution: Add Congo Red to 100.0mL of distilled/deionized water. Mix thoroughly. Autoclave for 15 min at 15 psi pressure–121°C.

Supplement Solution:
Composition per liter:

Glucose	100.0g
L-Cysteine·HCl	25.9g
L-Glutamine	10.0g
L-Cystine	1.1g
Adenine	1.0g
Nicotinamide adenine dinucleotide	0.25g
Vitamin B_{12}	0.1g
Thiamine pyrophosphate	0.1g
Guanine·HCl	0.03g
$Fe(NO_3)_3·6H_2O$	0.02g
p-Aminobenzoic acid	0.013g
Thiamine·HCl	3.0mg

Source: The supplement solution IsoVitaleX® enrichment is available from BD Diagnostic Systems. This enrichment may be replaced by supplement VX from BD Diagnostic Systems.

Preparation of Supplement Solution: Add components to distilled/deionized water and bring volume to 1.0L. Mix thoroughly. Filter sterilize.

Preparation of Medium: To 890.0mL of sterile, cooled GC agar base aseptically add 100.0mL of sterile, cooled hemoglobin solution, 10.0mL of sterile supplement solution, and 0.1mL of sterile Congo Red solution. Mix thoroughly. Pour into sterile Petri dishes.

Use: For the isolation and differentiation of virulent and avirulent strains of *Shigella, Vibrio cholerae, Escherichia coli,* and *Neisseria meningitidis.* Used for the detection and differentiation of "iron-responsive" avirulent mutants. Used in the preparation of live vaccines. Used for the differentiation of sensitive *Neisseria gonorrhoeae* (no growth) from other *Neisseria* species (growth) that are resistant to Congo Red.

Congo Red Agar (CR Agar)

Composition per liter:

Soybean-casein digest agar	890.0mL
Hemoglobin solution	100.0mL
Supplement solution	10.0mL
Congo Red (0.01% solution)	0.1mL

pH 7.3 ± 0.2 at 25°C

Soybean-Casein Digest Agar:
Composition per 890.0mL:

Pancreatic digest of casein	17.0g
Agar	15.0g
NaCl	5.0g
Papaic digest of soybean meal	3.0g
Glucose	2.5g
K_2HPO_4	2.5g

Preparation of Soybean-Casein Digest Agar: Add components to distilled/deionized water and bring volume to 890.0mL. Mix thoroughly. Gently heat until boiling. Autoclave for 15 min at 15 psi pressure–121°C. Cool to 45°–50°C.

Hemoglobin Solution:
Composition per 100.0mL:

Hemoglobin	2.0g

Preparation of Hemoglobin Solution: Add hemoglobin to distilled/deionized water and bring volume to 100.0mL. Mix thoroughly. Autoclave for 15 min at 15 psi pressure–121°C. Cool to 50°C.

Congo Red Solution:
Composition per 100.0mL:

Congo Red	0.01g

Preparation of Congo Red Solution: Add Congo Red to 100.0mL of distilled/deionized water. Mix thoroughly. Autoclave for 15 min at 15 psi pressure–121°C.

Supplement Solution:
Composition per liter:

Glucose	100.0g
L-Cysteine·HCl	25.9g
L-Glutamine	10.0g
L-Cystine	1.1g
Adenine	1.0g
Nicotinamide adenine dinucleotide	0.25g
Vitamin B_{12}	0.1g
Thiamine pyrophosphate	0.1g
Guanine·HCl	0.03g
$Fe(NO_3)_3·6H_2O$	0.02g
p-Aminobenzoic acid	0.013g
Thiamine·HCl	3.0mg

Preparation of Supplement Solution: Add components to distilled/deionized water and bring volume to 1.0L. Mix thoroughly. Filter sterilize.

Source: The supplement solution IsoVitaleX® enrichment is available from BD Diagnostic Systems. This enrichment may be replaced by supplement VX from BD Diagnostic Systems.

Preparation of Medium: To 890.0mL of sterile, cooled soybean-casein digest agar, aseptically add 100.0mL of sterile, cooled hemoglobin solution, 10.0mL of sterile supplement solution, and 0.1mL of sterile Congo Red solution. Mix thoroughly. Pour into sterile Petri dishes.

Use: For the isolation and differentiation of virulent and avirulent strains of *Shigella, Vibrio cholerae, Escherichia coli,* and *Neisseria meningitidis.* Used for the detection and differentiation of "iron-responsive" avirulent mutants. Used in the preparation of live vaccines. Used for the differentiation of sensitive *Neisseria gonorrhoeae* (no growth) from other *Neisseria* species (growth) that are resistant to Congo Red.

Congo Red BHI Agarose Medium

Composition per liter:

Agarose	15.0g
Pancreatic digest of gelatin	14.5g
Brain heart, solids from infusion	6.0g
Peptic digest of animal tissue	6.0g
NaCl	5.0g
Glucose	3.0g
Na_2HPO_4	2.5g
Congo Red	0.075g

pH 7.4 ± 0.2 at 25°C

Preparation of Medium: Add components to distilled/deionized water and bring volume to 1.0L. Mix thoroughly. Gently heat and bring to boiling. Distribute into tubes or flasks. Autoclave for 15 min at 15 psi pressure–121°C. Pour into sterile Petri dishes in 20.0mL volumes.

Use: For the isolation, cultivation, and detection of virulent strains of *Yersinia enterocolitica.*

Congo Red BHI Agarose Medium (CRBHO Medium) (BAM M41)

Composition per liter:

Pancreatic digest of gelatin	14.5g
Agarose	12.0g
Brain heart, solids from infusion	6.0g
Peptic digest of animal tissue	6.0g
NaCl	5.0g
Glucose	3.0g
Na_2HPO_4	2.5g
$MgCl_2$	1.0g
Congo Red solution	20.0mL

pH 7.4 ± 0.2 at 25°C

Preparation of Medium: Add components to distilled/deionized water and bring volume to 1.0L. Mix thoroughly. Gently heat and bring to boiling. Distribute into tubes or flasks. Autoclave for 15 min at 15 psi pressure–121°C. Pour into sterile Petri dishes in 20.0mL volumes.

Congo Red Solution:
Composition per 100.0mL:

Congo Red	375.0mg

Preparation of Congo Red Solution: Add Congo Red to 100.0mL of distilled/deionized water. Mix thoroughly. Autoclave for 15 min at 15 psi pressure–121°C. Cool to 25°C.

Use: For the isolation, cultivation, and detection of virulent strains of *Yersinia enterocolitica.*

Congo Red Magnesium Oxalate Agar (CRMOX Agar)

Composition per liter:

Solution 1	825.0mL
Solution 2	80.0mL
Solution 3	80.0mL
Solution 4	10.0mL
Solution 5	5.0mL

pH 7.3 ± 0.2 at 25°C

Solution 1:
Composition per 825.0mL:

Pancreatic digest of casein	15.0g
Agar	15.0g
Papaic digest of soybean meal	5.0g
NaCl	5.0g

pH 7.3 ± 0.2 at 25°C

Preparation of Solution 1: Add components to distilled/deionized water and bring volume to 825.0mL. Mix thoroughly. Gently heat and bring to boiling. Autoclave for 15 min at 15 psi pressure–121°C. Do not overheat.

Solution 2:
Composition per liter:

$MgCl_2 \cdot 6H_2O$	50.8g

Preparation of Solution 2: Add $MgCl_2 \cdot 6H_2O$ to distilled/deionized water and bring volume to 1.0L. Mix thoroughly. Autoclave for 15 min at 15 psi pressure–121°C.

Solution 3:
Composition per liter:

Sodium oxalate	33.2g

Preparation of Solution 3: Add sodium oxalate to distilled/deionized water and bring volume to 1.0L. Mix thoroughly. Autoclave for 15 min at 15 psi pressure–121°C.

Solution 4:
Composition per 100.0mL:

D-Galactose	20.0g

Preparation of Solution 4: Add D-galactose to distilled/deionized water and bring volume to 100.0mL. Mix thoroughly. Filter sterilize.

Solution 5:
Composition per 10.0mL:

Congo Red	0.1g

Preparation of Solution 5: Add Congo Red to distilled/deionized water and bring volume to 10.0mL. Mix thoroughly. Autoclave for 15 min at 15 psi pressure–121°C.

Preparation of Medium: Aseptically combine 80.0mL of sterile solution 2, 80.0mL of sterile solution 3, 10.0mL of sterile solution 4, and 5.0mL of sterile solution 5. Mix thoroughly. Warm to 50°C. Add this mixture to 825.0mL of cooled, sterile solution 1. Mix thoroughly. Pour into sterile Petri dishes.

Use: For the cultivation and identification of pathogenic serotypes of *Yersinia enterocolitica.* For the determination of whether *Yersinia* strains contain the *Yersinia* virulence plasmid.

Connaught Medical Research Laboratories Medium with Glutamine, 10X

See: **CMRL-1066 Medium with Glutamine, 10X**

Conradi Drigalski Agar

Composition per liter:

Agar	15.0g
Casein	10.0g
Lactose	10.0g
Peptone	10.0g
NaCl	5.0g
Bromcresol Purple	0.03g
Crystal Violet	4.0mg

pH 6.8 ± 0.2 at 25°C

Preparation of Medium: Add components to distilled/deionized water and bring volume to 1.0L. Mix thoroughly. Gently heat until boiling. Distribute into tubes or flasks. Autoclave for 15 min at 15 psi pressure–121°C. Pour into sterile Petri dishes or leave in tubes.

Use: For the isolation and cultivation of Gram-negative enteric bacilli.

Converse Liquid Medium, Levine Modification

Composition per liter:

Ionagar No. 2 or Noble agar	10.0g
Glucose	4.0g
Ammonium acetate	1.23g
K_2HPO_4	0.52g
Tamol	0.5g
$MgSO_4 \cdot 7H_2O$	0.4g
KH_2PO_4	0.4g
NaCl	0.014g
Na_2CO_3	0.012g
$CaCl_2 \cdot 2H_2O$	0.002g
$ZnSO_4 \cdot 7H_2O$	0.002g

Preparation of Medium: Add components to distilled/deionized water and bring volume to 1.0L. Mix thoroughly. Gently heat and bring to boiling. Autoclave for 15 min at 15 psi pressure–121°C. Pour into sterile Petri dishes in 15.0mL volumes.

Use: For the cultivation and induction of spherules of *Coccidioides immitis*.

Cooke Rose Bengal Agar

Composition per liter:

Agar	20.0g
Glucose	10.0g
Enzymatic hydrolysate of soybean meal	5.0g
KH_2PO_4	1.0g
$MgSO_4 \cdot 7H_2O$	0.5g
Rose Bengal	35.0mg

pH 6.0 ± 0.2 at 25°C

Source: This medium is available as a premixed powder from BD Diagnostic Systems.

Preparation of Medium: Add components to distilled/deionized water and bring volume to 1.0L. Mix thoroughly. Gently heat until boiling. Distribute into tubes or flasks. Autoclave for 15 min at 15 psi pressure–121°C. Pour into sterile Petri dishes or leave in tubes.

Use: For the isolation of fungi.

Cooked Meat Liver Medium

See: **CML Medium**

Cooked Meat Medium (LMG Medium 140)

Composition per liter:

Heart muscle	454.0g
Peptone	40.0g
Beef extract	10.0g
NaCl	5.0g
Yeast extract	5.0g
K_2HPO_4	5.0g
Glucose	2.0g
Resazurin solution	4.0mL

pH 7.0 ± 0.2 at 25°C

Resazurin Solution:

Composition per 100.0mL:

Resazurin	0.025g

Preparation of Resazurin Solution: Add resazurin to distilled/deionized water and bring volume to 100.0mL. Mix thoroughly.

Preparation of Medium: Finely chop beef heart. Add approximately 1.5g of heart particles to test tubes. Add remaining components to distilled/deionized water and bring volume to 1.0L. Mix thoroughly. Distribute into tubes in 10.0mL volumes. Autoclave for 15 min at 15 psi pressure–121°C. Slowly cool tubes to prevent expulsion of meat particles.

Use: For the cultivation and maintenance of *Peptostreptococcus magnus*.

Cooked Meat Medium

Composition per liter:

Beef heart	454.0g
Proteose peptone	20.0g
NaCl	5.0g
Glucose	2.0g

pH 7.2 ± 0.2 at 25°C

Source: This medium is available as a premixed powder from BD Diagnostic Systems.

Preparation of Medium: Finely chop beef heart. Add approximately 1.5g of heart particles to test tubes. Add remaining components to distilled/deionized water and bring volume to 1.0L. Mix thoroughly. Distribute into tubes in 10.0mL volumes. Autoclave for 15 min at 15 psi pressure–121°C. Slowly cool tubes to prevent expulsion of meat particles.

Use: For the cultivation and maintenance of anaerobic microorganisms.

Cooked Meat Medium

Composition per liter:

Heart muscle	454.0g
Beef extract	10.0g
Peptone	10.0g
NaCl	5.0g
Glucose	2.0g

pH 7.2 ± 0.2 at 25°C

Source: This medium is available as a premixed powder from Oxoid Unipath.

Preparation of Medium: Finely chop beef heart. Add approximately 1.5g of heart particles to test tubes. Add remaining components to distilled/deionized water and bring volume to 1.0L. Mix thoroughly. Distribute into tubes in 10.0mL volumes. Autoclave for 15 min at 15 psi pressure–121°C. Slowly cool tubes to prevent expulsion of meat particles.

Use: For the cultivation and maintenance of aerobic and anaerobic microorganisms. For the cultivation of anaerobes, especially pathogenic clostridia.

Cooked Meat Medium

Composition per liter:

Heart tissue granules	98.0g
Peptic digest of animal tissue	20.0g
NaCl	5.0g
Glucose	2.0g

pH 7.2 ± 0.2 at 25°C

Source: This medium is available as a premixed powder from BD Diagnostic Systems.

Preparation of Medium: Add approximately 1.0g of heart tissue granules to test tubes. Add remaining components to distilled/deionized water and bring volume to 1.0L. Mix thoroughly. Distribute into tubes in 10.0mL volumes. Autoclave for 15 min at 15 psi pressure–121°C. Slowly cool tubes to prevent expulsion of meat particles.

Use: For the cultivation of anaerobes, especially pathogenic clostridia.

Cooked Meat Medium with Glucose, Hemin, and Vitamin K

Composition per liter:

Heart tissue granules	98.0g
Peptic digest of animal tissue	20.0g
NaCl	5.0g

Glucose	5.0g
Yeast extract	5.0g
Hemin	5.0mg
Vitamin K	1.0mg

pH 7.2 ± 0.2 at 25°C

Source: This medium is available as a premixed powder from BD Diagnostic Systems.

Preparation of Medium: Add approximately 1.0g of heart tissue granules to test tubes. Add remaining components to distilled/deionized water and bring volume to 1.0L. Mix thoroughly. Distribute into tubes in 10.0mL volumes. Autoclave for 15 min at 15 psi pressure–121°C. Slowly cool tubes to prevent expulsion of meat particles.

Use: For the cultivation of anaerobes, especially pathogenic *Clostridia.*

Cooked Meat Medium with Glucose, Yeast Extract, and Cysteine

Composition per liter:

Heart muscle	454.0g
Glucose	12.0g
Beef extract	10.0g
Peptone	10.0g
NaCl	5.0g
K_2HPO_4	5.0g
L-Cysteine·HCl	0.5g
Resazurin	1.0mg

pH 7.2 ± 0.2 at 25°C

Source: Cooked meat medium is available as a premixed powder from Oxoid Unipath.

Preparation of Medium: Finely chop beef heart. Add approximately 1.5g of heart particles to test tubes. Add remaining components to distilled/deionized water and bring volume to 1.0L. Mix thoroughly. Distribute into tubes in 10.0mL volumes. Autoclave for 15 min at 15 psi pressure–121°C. Slowly cool tubes to prevent expulsion of meat particles.

Use: For the cultivation and maintenance of *Clostridium sphenoides.*

Cooked Meat Medium with Peptone and Yeast Extract

Composition per liter:

Heart muscle	454.0g
Peptone	40.0g
Beef extract	10.0g
NaCl	5.0g
Yeast extract	5.0g
Glucose	2.0g

pH 7.2 ± 0.2 at 25°C

Preparation of Medium: Finely chop beef heart. Add approximately 1.5g of heart particles to test tubes. Add remaining components to distilled/deionized water and bring volume to 1.0L. Mix thoroughly. Distribute into tubes in 10.0mL volumes. Autoclave for 15 min at 15 psi pressure–121°C. Slowly cool tubes to prevent expulsion of meat particles.

Use: For the cultivation and maintenance of *Peptostreptococcus magnus.*

Cooked Meat Medium, Modified

Composition per liter:

Cooked meat medium	66.0g
Solution A	1.0L

pH 6.8 ± 0.2 at 25°C

Cooked Meat Medium:
Composition per 481g:

Beef heart	454.0g
Proteose peptone	20.0g
NaCl	5.0g
Glucose	2.0g

Source: Cooked meat medium is available in dehydrated form from BD Diagnostic Systems.

Solution A:
Composition per liter:

Pancreatic digest of casein	10.0g
Glucose	2.0g
Soluble starch	1.0g
Sodium thioglycolate	1.0g
Neutral Red (1% aqueous)	5.0mL

Preparation of Solution A: Add components to distilled/deionized water and bring volume to 1.0L. Mix thoroughly. Gently heat until dissolved.

Preparation of Medium: Add 1.0g of dehydrated cooked meat medium to each of 66 test tubes. Add 15.0mL of solution A to each test tube. Allow meat particles to rehydrate. Autoclave for 15 min at 15 psi pressure–121°C.

Use: For the cultivation of a variety of anaerobic microorganisms.

Cooked Meat Medium, Modified (BAM M43)

Composition per tube:

Cooked meat medium	1.0g
Diluent	1.0L

pH 6.8 ± 0.2 at 25°C

Cooked Meat Medium:
Composition per 481g:

Beef heart	454.0g
Proteose peptone	20.0g
NaCl	5.0g
Glucose	2.0g

Source: Cooked meat medium is available in dehydrated form from BD Diagnostic Systems.

Diluent:
Composition per liter:

Pancreatic digest of casein	10.0g
Glucose	2.0g
Soluble starch	1.0g
Sodium thioglycolate	1.0g
Neutral Red (1% aqueous)	5.0mL

Preparation of Diluent: Add components to distilled/deionized water and bring volume to 1.0L. Mix thoroughly. Gently heat until dissolved.

Preparation of Medium: Add 1.0g of dehydrated cooked meat medium and 15.0mL diluent to 20 × 150mm test tubes. Let meat particles rehydrate. Gently heat and bring to boiling. Autoclave for 15 min at 15 psi pressure–121°C.

Cook's *Cytophaga* Agar

Composition per liter:

Agar 10.0g
Pancreatic digest of casein 2.0g

pH 7.3 ± 0.2 at 25°C

Preparation of Medium: Add components to distilled/deionized water and bring volume to 1.0L. Mix thoroughly. Gently heat and bring to boiling. Distribute into tubes or flasks. Autoclave for 15 min at 15 psi pressure–121°C. Pour into sterile Petri dishes or leave in tubes.

Use: For the cultivation of *Lysobacter antibioticus, Lysobacter brunescens, Lysobacter enzymogenes, Lysobacter gummosus*, and other *Lysobacter* species.

Cook's *Cytophaga* Agar for *Lysobacter*

Composition per liter:

Agar 12.0g
Pancreatic digest of casein 2.0g

pH 7.0 ± 0.2 at 25°C

Preparation of Medium: Add components to distilled/deionized water and bring volume to 1.0L. Mix thoroughly. Gently heat and bring to boiling. Distribute into tubes or flasks. Autoclave for 15 min at 15 psi pressure–121°C. Pour into sterile Petri dishes or leave in tubes.

Use: For the cultivation and maintenance of *Lysobacter gummosus.*

Coprinus Medium

Composition per 1026.0mL:

Agar 20.0g
Glucose 20.0g
Asparagine 2.0g
Pancreatic digest of casein 0.75g
Yeast extract 0.75g
Malt extract 0.60g
Salt solution 25.0mL
Thiamine solution 1.0mL

pH 6.8 ± 0.2 at 25°C

Salt Solution:

Composition per 500.0mL:

Na_2HPO_4 .... 45.0g
KH_2PO_4 .... 20.0g
Ammonium tartrate 10.0g
$Na_2SO_4 \cdot 10H_2O$ 5.6g

Preparation of Salt Solution: Add components to distilled/deionized water and bring volume to 500.0mL. Mix thoroughly. Filter sterilize.

Thiamine Solution:

Composition per 100.0mL:

Thiamine 10.0mg

Preparation of Thiamine Solution: Add thiamine to distilled/deionized water and bring volume to 100.0mL. Mix thoroughly. Filter sterilize.

Preparation of Medium: Add components, except salt solution and thiamine solution, to distilled/deionized water and bring volume to 1.0L. Mix thoroughly. Gently heat and bring to boiling. Autoclave for 15 min at 15 psi pressure–121°C. Cool to 45°–50°C. Aseptically add 25.0mL of sterile salt solution and 1.0mL of sterile thiamine solution. Mix thoroughly. Pour into sterile Petri dishes or distribute into sterile tubes.

Use: For the cultivation and maintenance of *Coprinus cinereus, Dendrophoma obscurans*, and *Trichophyton violaceum.*

Coprothermobacter proteolyticus Medium

Composition per 1168.1mL:

Yeast extract 2.0g
Trypticase™ 2.0g
NaOH solution 1.0L
Gelatin solution 113.0mL
Na_2S solution 22.6mL
Solution A 10.0mL
Mineral salts solution 10.0mL
Solution B 2.0mL
Resazurin solution 0.5mL

NaOH Solution:

Composition per liter:

NaOH 4.0g

Preparation of NaOH Solution: Add NaOH to distilled/deionized water and bring volume to 1.0L. Mix thoroughly.

Gelatin Solution:

Composition per 100.0mL:

Gelatin 3.0g

Preparation of Gelatin Solution: Gently heat 100.0mL of distilled/deionized water to 80°C. Sparge with 100% N_2 for 15 min. Add the gelatin. Mix thoroughly. Sparge with 100% N_2 for 10 min. Autoclave for 15 min at 15 psi pressure–121°C.

Na_2S Solution:

Na_2S 2.5g
Distilled water .... 100 ml

Preparation of Na_2S Solution: Gently heat 100.0mL of distilled/deionized water to 100°C. Boil for 5 min. Sparge with 100% N_2 for 15 min. Add the Na_2S. Mix thoroughly. Sparge with 100% N_2 for 10 min. Autoclave for 15 min at 15 psi pressure–121°C.

Solution A:

Composition per liter:

NH_4Cl 100.0g
$MgCl_2 \cdot H_2O$ 100.0g
$CaCl_2 \cdot 2H_2O$ 40.0g

Preparation of Solution A: Add components to distilled/deionized water and bring volume to 1.0L. Mix thoroughly. Adjust pH to 4 with HCl.

Mineral Salts Solution:

Composition per liter:

$EDTA \cdot 2H_2O$ 0.5g
$CoCl_2 \cdot H_2O$ 0.15g
$MnCl_2 \cdot 4H_2O$ 0.1g
$FeSO_4 \cdot 7H_2O$ 0.1g
$ZnCl_2$ 0.1g
$AlCl_3 \cdot H_2O$ 40mg
$Na_2WO_4 \cdot 2H_2O$ 30mg
$CuCl_2 \cdot 2H_2O$ 20mg
$NiSO_4 \cdot H_2O$ 20mg
H_2SeO_3 .... 10mg
H_3BO_4 .... 10mg
$NaMoO_4 \cdot 2H_2O$ 10mg

Preparation of Mineral Salts Solution: Add components to distilled/deionized water and bring volume to 1.0L. Mix thoroughly. Adjust pH to 3 with HCl.

Solution B:
Composition per liter:
$K_2HPO_4 \cdot 3H_2O$ 200.0g

Preparation of Solution B: Add $K_2HPO_4 \cdot 3H_2O$ to distilled/deionized water and bring volume to 1.0L. Mix thoroughly.

Resazurin Solution:
Composition per 100.0mL:
Resazurin 0.2g

Preparation of Resazurin Solution: Add resazurin to distilled/deionized water and bring volume to 100.0mL. Mix thoroughly.

Preparation of Medium: Sparge 1.0L of NaOH solution with 100% CO_2 for 30 min. Add 2.0g of yeast extract and 2.0g of Trypticase™. Mix thoroughly. Add 10.0mL of solution A, 2.0mL of solution B, 0.5mL of resazurin solution, and 10.0mL of mineral salts solution with pipets which have been flushed a few times with 100% N_2. Mix thoroughly. Anaerobically distribute 9.0mL volumes into anaerobic tubes fitted with butyl rubber stoppers. Autoclave for 15 min at 15 psi pressure–121°C. One hour prior to inoculation, add 1.0mL of sterile gelatin solution and 0.2mL of sterile Na_2S solution to each 9.0mL of medium.

Use: For the cultivation of *Coprothermobacter proteolyticus.*

Corn Meal Agar

Composition per liter:
Corn meal, infusion from 50.0g
Agar 15.0g
pH 6.0 ± 0.2 at 25°C

Source: This medium is available from HiMedia.

Preparation of Medium: Add components to distilled/deionized water and bring volume to 1.0L. Mix thoroughly. Gently heat and bring to boiling. Distribute into tubes or flasks. Autoclave for 15 min at 15 psi pressure–121°C. Pour into sterile Petri dishes or leave in tubes.

Use: For chlamydospore production by *Candida albicans* and the maintenance of fungal stock cultures.

Corn Meal Agar with Glucose

Composition per liter:
Agar 15.0g
Corn meal, infusion from 50.0g
Glucose 2.0g
pH 6.0 ± 0.2 at 25°C

Source: This medium is available from HiMedia.

Preparation of Medium: Add components to distilled/deionized water and bring volume to 1.0L. Mix thoroughly. Gently heat and bring to boiling. Distribute into tubes or flasks. Autoclave for 15 min at 15 psi pressure–121°C. Pour into sterile Petri dishes or leave in tubes.

Use: For the cultivation of phytopathological and other fungi.

Corn Meal HiVeg Peptone Yeast Agar

Composition per liter:
Agar 20.0g
Cellulose 20.0g
Glucose 10.0g
Plant peptone 10.0g
Yeast extract 4.0g
pH 6.5 ± 0.2 at 25°C

Source: This medium is available as a premixed powder from HiMedia.

Preparation of Medium: Add components to distilled/deionized water and bring volume to 1.0L. Mix thoroughly. Gently heat until boiling. Distribute into tubes or flasks. Autoclave for 15 min at 15 psi pressure–121°C. Pour into sterile Petri dishes or leave in tubes.

Use: For the cultivation and maintenance of fungi.

Corn Milk Medium

Composition per liter:
Skim milk 20.0g
Agar 15.0g
Yeast extract 12.5g
Peptone 10.0g
Beef extract 5.0g
K_2HPO_4 .... 5.0g
NaCl 5.0g
$MgSO_4 \cdot 7H_2O$ 1.0g
Corn steep liquor 7.0mL
pH 7.0 ± 0.2 at 25°C

Preparation of Medium: Add components to distilled/deionized water and bring volume to 1.0L. Mix thoroughly. Gently heat until boiling. Distribute into tubes or flasks. Autoclave for 15 min at 15 psi pressure–121°C. Pour into sterile Petri dishes or leave in tubes.

Use: For the cultivation and maintenance of *Bacillus subtilis.*

Corn Oil Medium

Composition per liter:
Agar 20.0g
Glucose 20.0g
Pancreatic digest of casein 5.0g
Peptic digest of animal tissue 5.0g
pH 6.8–7.0 at 25°C

Preparation of Medium: Add components, except corn oil, to distilled/deionized water and bring volume to 1.0L. Mix thoroughly. Gently heat and bring to boiling. Distribute into tubes. Autoclave for 15 min at 15 psi pressure–121°C. Allow to cool in a slanted position. Add a few drops of sterile corn oil to the surface of the slants.

Use: For the cultivation and maintenance of *Pityrosporum ovale.*

Corn Steep Liquor Medium

Composition per liter:
Glucose 60.0g
Corn steep liquor 40.0g
Urea 8.0g
KH_2PO_4 .... 5.0g
Fumaric acid 1.0g
$MgSO_4 \cdot 7H_2O$ 0.5g
Hutner's mineral base 20.0mL
pH 7.0 ± 0.2 at 25°C

Hutner's Mineral Base:
Composition per liter:
$MgSO_4 \cdot 7H_2O$ 29.7g
Nitrilotriacetic acid 10.0g
$CaCl_2 \cdot 2H_2O$ 3.34g

$FeSO_4 \cdot 7H_2O$ 99.0mg
$(NH_4)_2MoO_4$ 9.25mg
Metals "44" 50.0mL

Preparation of Hutner's Mineral Base: Add nitrilotriacetic acid to 500.0mL of distilled/deionized water. Dissolve by adjusting pH to 6.5 with KOH. Add remaining components. Add distilled/deionized water to 1.0L.

Metals "44":
Composition per 100.0mL:
$ZnSO_4 \cdot 7H_2O$ 1.1g
$FeSO_4 \cdot 7H_2O$ 0.5g
EDTA 0.25g
$MnSO_4 \cdot 7H_2O$ 0.154g
$CuSO_4 \cdot 5H_2O$ 0.04g
$Co(NO_3)_2 \cdot 6H_2O$ 0.025g
$Na_2B_4O_7 \cdot 10H_2O$ 0.018g

Preparation of Metals "44": Add components to distilled/deionized water and bring volume to 100.0mL. Mix thoroughly.

Preparation of Medium: Add components to distilled/deionized water and bring volume to 1.0L. Mix thoroughly. Distribute into tubes or flasks. Autoclave for 15 min at 15 psi pressure–121°C.

Use: For the cultivation of *Pseudomonas* species.

Corn Steep Starch Nutrient Agar

Composition per liter:
Soluble starch 10.0g
Agar 7.5g
Pancreatic digest of gelatin 2.5g
Beef extract 1.5g
Corn steep liquor 1.0mL

pH 6.8 ± 0.2 at 25°C

Preparation of Medium: Add components to distilled/deionized water and bring volume to 1.0L. Mix thoroughly. Gently heat and bring to boiling. Distribute into tubes or flasks. Autoclave for 15 min at 15 psi pressure–121°C. Pour into sterile Petri dishes or leave in tubes.

Use: For the cultivation of *Clostridium thermoamylolyticum.*

Cornmeal Agar (ATCC Medium 307)

Composition per liter:
Cornmeal 50.0g
Agar 7.5g

Preparation of Medium: Add cornmeal to distilled/deionized water and bring volume to 800.0mL. Leave overnight in refrigerator. Heat to 60°C for 1 hr. Bring volume to 1.0L with distilled/deionized water. Add agar. Gently heat and bring to boiling. Autoclave for 15 min at 15 psi pressure–121°C. Pour into sterile Petri dishes or distribute into sterile tubes.

Use: For the cultivation and maintenance of numerous fungi.

Cornmeal Agar (CMA)

Composition per liter:
Agar 20.0g
Cornmeal polenta 15.0g

pH 7.0 ± 0.2 at 25°C

Preparation of Medium: Add cornmeal polenta to distilled/deionized water and bring volume to 1.0L. Mix thoroughly. Gently heat and bring to boiling. Continue boiling for 30 min. Filter through Whatman #1 filter paper. Add agar to filtrate. Gently heat and bring to boiling. Distribute into tubes or flasks. Autoclave for 10 min at 15 psi pressure–121°C. Pour into sterile Petri dishes or leave in tubes.

Use: For the cultivation and maintenance of many filamentous fungi.

Cornmeal Agar

Composition per liter:
Agar 15.0g
Cornmeal, solids from infusion 2.0g

pH 5.6–6.0 at 25°C

Source: This medium is available as a premixed powder from BD Diagnostic Systems and Oxoid Unipath.

Preparation of Medium: Add components to distilled/deionized water and bring volume to 1.0L. Mix thoroughly. Gently heat until boiling. Distribute into tubes or flasks. Autoclave for 15 min at 15 psi pressure–121°C. Pour into sterile Petri dishes or leave in tubes.

Use: For the cultivation and maintenance of fungi.

Cornmeal Agar with Dextrose

Composition per liter:
Agar 15.0g
Cornmeal, solids from infusion 2.0g
Glucose 2.0g
Tween™ 80 .... 1.0g

pH 5.6–6.0 at 25°C

Source: This medium is available as a premixed powder from BD Diagnostic Systems.

Preparation of Medium: Add components to distilled/deionized water and bring volume to 1.0L. Mix thoroughly. Gently heat until boiling. Distribute into tubes or flasks. Autoclave for 15 min at 15 psi pressure–121°C. Pour into sterile Petri dishes or leave in tubes.

Use: For the cultivation of phytopathological and other fungi.

Cornmeal Agar with Polysorbate 80

Composition per liter:
Agar 15.0g
Cornmeal, solids from infusion 2.0g
Tween™ 80 .... 1.0g

pH 5.6–6.0 at 25°C

Source: This medium is available as a premixed powder from BD Diagnostic Systems and Oxoid Unipath.

Preparation of Medium: Add components to distilled/deionized water and bring volume to 1.0L. Mix thoroughly. Gently heat until boiling. Distribute into tubes or flasks. Autoclave for 15 min at 15 psi pressure–121°C. Pour into sterile Petri dishes or leave in tubes.

Use: For the cultivation and maintenance of fungi. For the production of chlamydospores by *Candida albicans* and the cultivation of phytopathological fungi.

Cornmeal Agar with Polysorbate 80
See: **Cornmeal Agar**

Cornmeal Agar, Quarter-strength (ATCC Medium 2221)

Composition per liter:

Agar	15.0g
Cornmeal infusion	250.0mL

pH 5.6–6.0 at 25°C

Cornmeal Infusion:

Composition per liter:

Yellow cornmeal	50.0g

Preparation of Cornmeal Infusion: Add cornmeal to distilled/deionized water and bring volume to 1.0L. Gently heat and bring to boiling. Simmer for 10 minutes. Filter through cheesecloth. Return volume to 1.0 liter.

Preparation of Medium: Add agar to 250.0mL cornmeal infusion and bring volume to 1.0L with distilled/deionized water. Gently heat and bring to boiling. Autoclave for 15 min at 15 psi pressure–121°C. Pour into sterile Petri dishes or distribute into sterile tubes.

Use: For the cultivation and maintenance of fungi.

Cornmeal Agar with Soil Extract

Composition per liter:

Cornmeal	50.0g
Agar	7.5g
Soil extract	50.0mL

Soil Extract:

Composition per 200.0mL:

African Violet soil	77.0g
Na_2CO_3	0.2g

Preparation of Soil Extract: Add components to 200.0mL of distilled/deionized water. Mix thoroughly. Autoclave for 60 min at 15 psi pressure–121°C. Filter through paper and reserve filtrate.

Preparation of Medium: Add cornmeal to distilled/deionized water and bring volume to 800.0mL. Leave overnight in refrigerator. Heat to 60°C for 1 hr. Add 50.0mL of soil extract. Bring volume to 1.0L with distilled/deionized water. Add agar. Gently heat and bring to boiling. Autoclave for 15 min at 15 psi pressure–121°C. Pour into sterile Petri dishes or distribute into sterile tubes.

Use: For the cultivation and maintenance of *Helicodendron tubulosum, Microsporum distortum, Mortierella humilis, Mortierella hygrophila, Mortierella minutissima,* and *Nigrospora sphaerica.*

Cornmeal Agar with Strep100 and Tet100 (ATCC Medium 2285)

Composition per liter:

Agar	15.0g
Cornmeal, solids from infusion	2.0g
Antibiotic solution	10.0mL

pH 5.6–6.0 at 25°C

Source: This medium without antibiotics is available as a premixed powder from BD Diagnostic Systems and Oxoid Unipath.

Preparation of Medium: Add components except antibiotic solution to 990.0mL distilled/deionized water. Mix thoroughly. Gently heat until boiling. Distribute into tubes or flasks. Autoclave for 15 min at 15 psi pressure–121°C. Cool to 45–50°C. Aseptically add 10.0mL sterile antibiotic solution. Mix thoroughly. Pour into sterile Petri dishes or leave in tubes.

Antibiotic Solution:

Composition per 10.0mL:

Tetracycline	0.1g
Streptomycin sulfate	0.1g

Preparation of Antibiotic Solution: Add components to distilled/deionized water and bring volume to 10.0mL. Mix thoroughly. Filter sterilize.

Use: For the cultivation and maintenance of fungi.

Cornmeal *Phytophthora* Isolation Medium No. 1

Composition per liter:

Agar	15.0g
Cornmeal, solids from infusion	2.0g
Vancomycin	0.2g
Pentachloronitrobenzene (PCNB)	0.1g
Pimaricin	0.01g

pH 5.6–6.0 at 25°C

Preparation of Medium: Add components, except pimaricin and vancomycin, to distilled/deionized water and bring volume to 1.0L. Mix thoroughly. Gently heat until boiling. Autoclave for 15 min at 15 psi pressure–121°C. Aseptically add pimaricin and vancomycin. Mix thoroughly. Pour into sterile Petri dishes.

Use: For the cultivation of *Phytophthora* species.

Cornmeal *Phytophthora* Isolation Medium No. 2

Composition per liter:

Agar	15.0g
Cornmeal, solids from infusion	2.0g
Vancomycin	0.3g
Pentachloronitrobenzene (PCNB)	0.025g
Pimaricin	5.0mg

pH 5.6–6.0 at 25°C

Preparation of Medium: Add components, except pimaricin and vancomycin, to distilled/deionized water and bring volume to 1.0L. Mix thoroughly. Gently heat until boiling. Autoclave for 15 min at 15 psi pressure–121°C. Aseptically add pimaricin and vancomycin. Mix thoroughly. Pour into sterile Petri dishes.

Use: For the cultivation of *Phytophthora* species.

Cornmeal and V8 Juice Agar (ATCC Medium 309)

Composition per liter:

Agar	7.5g
$CaCO_3$	3.0g
Cornmeal extract	800.0mL
V8 juice	200.0mL

pH 5.6–6.0 at 25°C

Cornmeal Extract:

Composition per 800.0mL:

Yellow cornmeal	50.0g

Preparation of Cornmeal Extract: Add 50.0g of yellow cornmeal to 800 ml of water. Leave in1 hone hour. Filter out cornmeal through cheesecloth. Bring volume back to 800.0mL.

Preparation of Medium: Combine components. Mix thoroughly. Gently heat and bring to boiling. Autoclave for 15 min at 15 psi pressure–121°C. Pour into sterile Petri dishes or distribute into sterile tubes.

Use: For the cultivation and maintenance of fungi.

Cornmeal Yeast Extract Seawater Agar (ATCC Medium 2422)

Composition per liter:

Instant ocean	17.5g
Agar	15.0g
Yeast extract	1.0g
Cornmeal infusion	400.0mL

pH 7.2–7.5 at 25°C

Cornmeal Infusion:

Composition per liter:

Yellow cornmeal	50.0g

Preparation of Cornmeal Infusion: Add cornmeal to distilled/deionized water and bring volume to 1.0L. Gently heat and bring to boiling. Simmer for 10 minutes. Filter through cheesecloth. Return volume to 1.0 liter.

Preparation of Medium: Add instant ocean, agar, and yeast extract to 400.0mL cornmeal infusion and bring volume to 1.0L with distilled/deionized water. Gently heat and bring to boiling. Autoclave for 15 min at 15 psi pressure–121°C. Pour into sterile Petri dishes or distribute into sterile tubes.

Use: For the cultivation and maintenance of fungi.

Cornmeal Yeast Glucose Agar (CMYG)

Composition per liter:

Agar	15.0g
Cornmeal, solids from infusion	2.0g
Glucose	2.0g
Yeast extract	1.0g

pH 5.6–6.0 at 25°C

Preparation of Medium: Add components to distilled/deionized water and bring volume to 1.0L. Mix thoroughly. Gently heat until boiling. Distribute into tubes or flasks. Autoclave for 15 min at 15 psi pressure–121°C. Pour into sterile Petri dishes or leave in tubes.

Use: For the cultivation of numerous filamentous fungi.

Cornstarch Soluble Medium (CSSM)

Composition per liter:

Cornstarch	42.0g
n-Butanol	18.0g
Yeast extract	10.0g
Asparagine·H_2O	2.0g
$(NH_4)_2SO_4$	2.0g
NaCl	1.0g
KH_2PO_4	0.75g
K_2HPO_4	0.75g
L-Cysteine·HCl·H_2O	0.5g
$MgSO_4$	0.02g
$FeSO_4 \cdot 7H_2O$	0.01g
$MnSO_4 \cdot H_2O$	0.01g

Preparation of Medium: Add components to distilled/deionized water and bring volume to 1.0L. Mix thoroughly. Gently heat until boiling. Boil and cool under 80% N_2 + 10% H_2 + 10% CO_2. Distribute anaerobically into tubes under the same gas mixture. Cap with butyl rubber stoppers. Autoclave for 15 min at 15 psi pressure–121°C.

Use: For the cultivation and maintenance of *Clostridium thermoamylolyticum.*

Cornstarch Soluble Medium (CSSM)/(ATCC Medium 1500)

Composition per liter:

Cornstarch	42.0g
Yeast extract	10.0g
Asparagine·H_2O	2.0g
$(NH_4)_2SO_4$	2.0g
NaCl	1.0g
KH_2PO_4	0.75g
K_2HPO_4	0.75g
L-Cysteine·HCl·H_2O	0.5g
$MgSO_4$	0.02g
$FeSO_4 \cdot 7H_2O$	0.01g
$MnSO_4 \cdot H_2O$	0.01g

Preparation of Medium: Add components to distilled/deionized water and bring volume to 1.0L. Mix thoroughly. Gently heat until boiling. Boil and cool under 80% N_2 + 10% H_2 + 10% CO_2. Distribute anaerobically into tubes under the same gas mixture. Cap with butyl rubber stoppers. Autoclave for 15 min at 15 psi pressure–121°C.

Use: For the cultivation and maintenance of *Clostridium thermoamylolyticum.*

Corynebacterium Agar

Composition per liter:

Agar	15.0g
Beef extract	10.0g
Peptone	10.0g
NaCl	5.0g

pH 7.2 ± 0.2 at 25°C

Preparation of Medium: Add components to distilled/deionized water and bring volume to 1.0L. Mix thoroughly. Adjust pH to 7.2. Gently heat and bring to boiling. Distribute into tubes or flasks. Autoclave for 15 min at 15 psi pressure–121°C. Pour into sterile Petri dishes or leave in tubes.

Use: For the cultivation and maintenance of *Brevibacterium helvolum, Brevibacterium linens, Brochothrix thermosphacta, Cellulomonas cellasea, Corynebacterium ammoniagenes, Corynebacterium callunae, Corynebacterium glutamicum,* other *Corynebacterium* species, *Curtobacterium flaccumfaciens, Deinococcus radiodurans, Microbacterium laevaniformans, Mycobacterium vaccae, Rhodococcus equi, Rhodococcus fascians, Sporolactobacillus inulinus,* and *Streptococcus mutans.*

Corynebacterium Agar

Composition per liter:

Agar	15.0g
Beef extract	10.0g
Peptone	10.0g
NaCl	5.0g
$MnSO_4$	10.0mg

pH 7.2 ± 0.2 at 25°C

Preparation of Medium: Add components to distilled/deionized water and bring volume to 1.0L. Mix thoroughly. Adjust pH to 7.2. Gently heat and bring to boiling. Distribute into tubes or flasks. Autoclave for 15 min at 15 psi pressure–121°C. Pour into sterile Petri dishes or leave in tubes.

Use: For the cultivation, maintenance, and sporulation of *Bacillus* species.

Corynebacterium Agar

Composition per liter:

Agar	15.0g
Tryptic digest of casein	10.0g
Glucose	5.0g
NaCl	5.0g
Yeast extract	5.0g

pH 7.2–7.4 at 25°C

Preparation of Medium: Add components to distilled/deionized water and bring volume to 1.0L. Mix thoroughly. Adjust pH to 7.2–7.4. Gently heat and bring to boiling. Distribute into tubes or flasks. Autoclave for 15 min at 15 psi pressure–121°C. Pour into sterile Petri dishes or leave in tubes.

Use: For the cultivation of a wide variety of bacteria including *Arthrobacter atrocyaneus, Arthrobacter aurescens, Arthrobacter citreus, Arthrobacter crystallopoietes, Arthrobacter globiformis, Arthrobacter histidinolovorans, Arthrobacter ilicis, Arthrobacter nicontinovorans, Arthrobacter nicotianae, Arthrobacter oxydans, Arthrobacter pascens, Arthrobacter polychromogenes, Arthrobacter protophormiae, Arthrobacter ramosus, Arthrobacter* species, *Arthrobacter sulfureus, Arthrobacter uratoxydans, Arthrobacter ureafaciens, Arthrobacter viscosus, Aureobacterium barkeri, Aureobacterium liquefaciens, Aureobacterium saperdae, Aureobacterium* species, *Aureobacterium testaceum, Brevibacterium acetylicum, Brevibacterium casei, Brevibacterium epidermidis, Brevibacterium iodinum, Brevibacterium linens, Brevibacterium liquefaciens, Brevibacterium oxydans, Brevibacterium* species, *Brevibacterium stationis, Brochothrix thermosphacta, Cellulomonas biazotea, Cellulomonas cellasea, Cellulomonas cellulans, Cellulomonas fimi, Cellulomonas flavigena, Cellulomonas gelida, Cellulomonas turbata, Cellulomonas uda, Clavibacter michiganensis, Clavibacter xyli, Corynebacterium ammoniagenes, Corynebacterium bovis, Corynebacterium callunae, Corynebacterium flavescens, Corynebacterium glutamicum, Corynebacterium hoagii, Corynebacterium mycetoides, Corynebacterium renale, Corynebacterium* species, *Corynebacterium variabilis, Corynebacterium vitarumen, Curtobacterium albidum, Curtobacterium citreum, Curtobacterium flaccumfaciens, Curtobacterium luteum, Curtobacterium pusillum, Deinococcus proteolyticus, Deinococcus radiodurans, Enterococcus casseliflavus, Enterococcus faecalis, Enterococcus faecium, Enterococcus hirae, Kurthia gibsonii, Kurthia zopfii, Lactococcus lactis, Microbacterium imperiale, Microbacterium lacticum, Microbacterium laevaniformans, Micrococcus agilis, Micrococcus kristinae, Micrococcus lylae, Micrococcus nishinomiyaensis, Micrococcus roseus, Micrococcus sedentarius, Micrococcus* species, *Micrococcus varians, Nocardia corynebacteroides, Nocardia* species, *Nocardioides jensenii, Nocardioides simplex, Planococcus kocurii, Rathayibacter rathayi, Rhodococcus equi, Rhodococcus fascians, Staphylococcus arlettae, Staphylococcus aureus, Staphylococcus auricularis, Staphylococcus capitis, Staphylococcus caprae, Staphylococcus carnosus, Staphylococcus caseolyticus, Staphylococcus chromogenes, Staphylococcus cohnii, Staphylococcus epidermidis, Staphylococcus equorum, Staphylococcus gallinarum, Staphylococcus haemolyticus, Staphylococcus hominis, Staphylococcus hyicus, Staphylococcus intermedius, Staphylococcus kloosii, Staphylococcus lentus, Staphylococcus saprophyticus, Staphylococcus sciuri, Staphylococcus simulans, Staphylococcus* species, *Staphylococcus warneri, Staphylococcus xylosus, Stomatococcus mucilaginosus, Streptococcus bovis, Streptococcus canis, Streptococcus equinus, Streptococcus oralis, Streptococcus salivarius, Streptococcus sanguis, Terrabacter tumescens,* and *Tsukamurella paurometabolum.*

Corynebacterium Agar with Blood

Composition per liter:

Agar	15.0g
Tryptic digest of casein	10.0g
Glucose	5.0g
NaCl	5.0g
Yeast extract	5.0g
Blood, defibrinated	50.0mL

Preparation of Medium: Add components, except defibrinated blood, to distilled/deionized water and bring volume to 950.0mL. Mix thoroughly. Gently heat and bring to boiling. Autoclave for 15 min at 15 psi pressure–121°C. Cool to 50°–55°C. Aseptically add 50.0mL of defibrinated blood. Mix thoroughly. Pour into sterile Petri dishes or distribute into sterile tubes.

Use: For the cultivation of *Brevibacterium incertum, Corynebacterium bovis, Corynebacterium kutscheri, Moraxella bovis, Streptococcus acidominimus, Streptococcus intestinalis, Streptococcus oralis,* and various other *Streptococcus* species.

Corynebacterium Agar with Salt

Composition per liter:

NaCl	65.0g
Agar	15.0g
Tryptic digest of casein	10.0g
Glucose	5.0g
Yeast extract	5.0g

Preparation of Medium: Add components to distilled/deionized water and bring volume to 1.0L. Mix thoroughly. Gently heat and bring to boiling. Distribute into tubes or flasks. Autoclave for 15 min at 15 psi pressure–121°C. Pour into sterile Petri dishes or leave in tubes.

Use: For the cultivation of *Micrococcus halobius.*

Corynebacterium Broth

Composition per liter:

Tryptic digest of casein	10.0g
Glucose	5.0g
NaCl	5.0g
Yeast extract	5.0g

pH 7.2–7.4 at 25°C

Preparation of Medium: Add components to distilled/deionized water and bring volume to 1.0L. Mix thoroughly. Adjust pH to 7.2–7.4. Distribute into tubes or flasks. Autoclave for 15 min at 15 psi pressure–121°C.

Use: For the cultivation and maintenance of *Cellulomonas fimi, Clavibacter michiganensis, Corynebacterium* species, *Enterococcus faecalis, Enterococcus hirae, Lactococcus lactis, Micrococcus kristinae, Micrococcus* species, *Micrococcus varians, Staphylococcus warneri,* and *Streptococcus salivarius.*

Corynebacterium diphtheriae Virulence Test Medium

See: **K-L Virulence Agar**

Corynebacterium Liquid Enrichment Medium

Composition per 2000.0mL:

Fosfomycin	0.15g
Glucose 6-phosphate	0.03g
Solution A	985.0mL

Bovine serum 100.0mL
Nystatin solution 1.15mL
L-Cystine (1% solution) 1.0mL
Egg yolk emulsion .... 10 eggs

pH 7.4 ± 0.2 at 25°C

Solution A:

Composition per liter:

Meat extract 9.0g
Proteose peptone No. 3 .... 9.0g
NaCl 2.7g
Glucose 1.8g
$Na_2HPO_4·12H_2O$ 1.8g
K_2TeO_3 (2% solution) 75.0mL
L-Cystine (1% solution) 10.0mL

Caution: Potassium tellurite is toxic.

Preparation of Solution A: Add components to distilled/deionized water and bring volume to 985.0mL. Mix thoroughly. Filter sterilize.

Egg Yolk Emulsion:

Composition:

Chicken egg yolks 9
Whole chicken egg 1

Preparation of Egg Yolk Emulsion: Soak eggs with 1:100 dilution of saturated mercuric chloride solution for 1 min. Crack eggs and separate yolks from whites. Mix egg yolks with 1 chicken egg. Filter sterilize.

Nystatin Solution:

Composition per 10.0mL:

Nystatin 10,000U

Preparation of Nystatin Solution: Add nystatin to distilled/deionized water and bring volume to 10.0mL. Mix thoroughly. Filter sterilize.

L-Cystine Solution:

Composition per 10.0mL:

L-Cystine 0.1g

Preparation of L-Cystine Solution: Add L-cystine to distilled/deionized water and bring volume to 10.0mL. Mix thoroughly. Filter sterilize.

Preparation of Medium: To 985.0mL of sterile solution A, aseptically add the remaining components. Mix thoroughly. Aseptically distribute into sterile tubes in 2.0–3.0mL volumes.

Use: For the isolation and cultivation of *Corynebacterium diphtheriae*.

Corynebacterium Medium with Blood (DSMZ Medium 240)

Composition per liter:

Agar 15.0g
Casein peptone, tryptic digest 10.0g
Yeast extract 5.0g
Glucose 5.0g
NaCl 5.0g
Distilled water 1000.0mL
Sheep or horse blood, defibrinated 50.0mL

pH 7.3 ± 0.2 at 25°C

Preparation of Medium: Add components, except sheep or horse blood, to distilled/deionized water and bring volume to 950.0mL. Mix thoroughly. Gently heat and bring to boiling. Autoclave for 15 min at 15 psi pressure–121°C. Cool to 45°–50°C. Aseptically add sterile sheep or horse blood. Mix thoroughly. Pour into sterile Petri dishes or distribute into sterile tubes.

Use: For the cultivation of *Streptococcus alactolyticus, Corynebacterium* spp., *Desemzia incerta=Brevibacterium incertum,* and *Moraxella bovis.*

Corynebacterium Medium CII

Composition per liter:

$CaCO_3$ 20.0g
Agar 15.0g
Sucrose 10.0g
Yeast extract 4.0g

Preparation of Medium: Add components to distilled/deionized water and bring volume to 1.0L. Mix thoroughly. Gently heat and bring to boiling. Distribute into tubes or flasks. Autoclave for 15 min at 15 psi pressure–121°C. Pour into sterile Petri dishes or leave in tubes.

Use: For the cultivation of *Clavibacter michiganensis.*

Corynebacterium Medium with Salt (DSMZ Medium 229)

Composition per liter:

NaCl 65.0g
Agar 15.0g
Casein peptone, tryptic digest 10.0g
Yeast extract 5.0g
Glucose 5.0g

pH 7.3 ± 0.2 at 25°C

Preparation of Medium: Add components to distilled/deionized water and bring volume to 1.0L. Mix thoroughly. Gently heat and bring to boiling. Distribute into tubes or flasks. Autoclave for 15 min at 15 psi pressure–121°C. Pour into sterile Petri dishes or leave in tubes.

Use: For the cultivation and maintenance of *Nesterenkonia halobia=Micrococcus halobius.*

Costein's LDS Test Medium

Composition per liter:

Meat peptone 4.5g
Papaic digest of soybean meal 2.0g
Yeast extract 3.0g
NaCl 5.0g
D-Glucose 1.0g
L-Lysine monohydrochloride 10.0g
$Na_2S_2O_3$ 0.2g
$Fe(NH_4)_2(SO_4)_2·6H_2O$ 0.2g
Bromocresol Purple 0.032
Agar 6.0g

pH 5.6 ± 0.2 at 25°C

Source: This medium is available from HiMedia.

Preparation of Medium: Add components to distilled/deionized water and bring volume to 1.0L. Mix thoroughly. Distribute into tubes. Overlay with viscous parrafin. Autoclave for 15 min at 15 psi pressure–121°C. Allow tubes to solidify in a vertical position.

Use: For the identification of members of Enterobacteriaceae on the basis of lysine decarboxylase and hydrogen sulfide production.

Cow Manure Agar

Composition per liter:

Cow manure	50.0g
Agar	15.0g

Preparation of Medium: Add cow manure to tap water and bring volume to 1.0L. Gently heat and bring to boiling. Boil for 1 hr. Filter through cheesecloth. Filter through Whatman filter paper. Bring volume to 1.0L with tap water. Add agar. Mix thoroughly. Gently heat and bring to boiling. Distribute into tubes or flasks. Autoclave for 15 min at 15 psi pressure–121°C. Pour into sterile Petri dishes or leave in tubes.

Use: For the cultivation of *Streptomyces* species.

CP Medium

Composition per liter:

Peptone	2.5g
Starch	2.0g
$NaNO_3$	0.38g
Tris(hydroxymethyl)aminomethane buffer	0.25g
K_2HPO_4	0.038g
$MgSO_4 \cdot 7H_2O$	0.038g
$CaCl_2 \cdot 2H_2O$	0.017g
NaCl	0.013g
TC vitamins minimal eagle, 100X	5.0mL
Solution 1	1.0mL
Solution 2	1.0mL
Solution 3	1.0mL
Solution 4	1.0mL
Vitamin B_{12} solution	0.2mL

pH 8.7 ± 0.2 at 25°C

TC Vitamins Minimal Eagle 100X:

Composition per liter:

Inositol	2.0mg
Choline chloride	1.0mg
Folic acid	1.0mg
Nicotinamide	1.0mg
Calcium pantothenate	1.0mg
Pyridoxal	1.0mg
Thiamine·HCl	1.0mg
Riboflavin	0.1mg

Preparation of TC Vitamins Minimal Eagle, 100X: Add components to distilled/deionized water and bring volume to 1.0L. Mix thoroughly. Filter sterilize.

Solution 1:

Composition per 100.0mL:

EDTA	5.0g
KOH	3.1g

Preparation of Solution 1: Add components to distilled/deionized water and bring volume to 100.0mL. Mix thoroughly.

Solution 2:

Composition per liter:

$FeSO_4 \cdot 7H_2O$	4.98g

Preparation of Solution 2: Add $FeSO_4 \cdot 7H_2O$ to distilled/deionized water acidified with 1.0mL of H_2SO_4. Bring volume to 1.0L. Mix thoroughly.

Solution 3:

Composition per 100.0mL:

H_3BO_3	1.14g

Preparation of Solution 3: Add H_3BO_3 to distilled/deionized water and bring volume to 100.0mL. Mix thoroughly.

Solution 4:

Composition per 100.0mL:

$ZnSO_4 \cdot 7H_2O$	0.88g
$MnCl_2 \cdot 4H_2O$	0.144g
MoO_3	0.071g
$CoNO_3 \cdot 6H_2O$	0.049g
$CuSO_4 \cdot 5H_2O$	0.016g

Preparation of Solution 4: Add components to distilled/deionized water and bring volume to 100.0mL. Mix thoroughly.

Vitamin B_{12} Solution

Composition per 10.0mL:

Vitamin B_{12}	10.0mg

Preparation of Vitamin B_{12} Solution: Add vitamin B_{12} to distilled/deionized water and bring volume to 10.0mL. Mix thoroughly. Filter sterilize.

Preparation of Medium: Add components, except for vitamin solutions, to distilled/deionized water and bring volume to 995.0mL. Mix thoroughly. Distribute into tubes or flasks. Autoclave for 15 min at 15 psi pressure–121°C. Aseptically add vitamin solutions. Mix thoroughly.

Use: For the cultivation of *Lysobacter* species.

CP Medium for *Coprothermobacter proteolyticus*

Composition per 1010.0mL:

$NaHCO_3$	8.4g
Pancreatic digest of casein	2.0g
Yeast extract	2.0g
$MgCl_2 \cdot 6H_2O$	1.0g
NH_4Cl	1.0g
$CaCl_2 \cdot 2H_2O$	0.4g
$K_2HPO_4 \cdot 3H_2O$	0.4g
Resazurin	0.5mg
Gelatin solution	100.0mL
$Na_2S \cdot 9H_2O$ solution	10.0mL
Wolfe's mineral solution	10.0mL

pH 7.0 ± 0.2 at 25°C

Gelatin Solution:

Composition per 100.0mL:

Gelatin	3.0g

Preparation of Gelatin Solution: Add gelatin to distilled/deionized water and bring volume to 100.0mL. Mix thoroughly. Sparge with 100% N_2. Autoclave for 15 min at 15 psi pressure–121°C.

$Na_2S \cdot 9H_2O$ Solution:

Composition per 10.0mL:

$Na_2S \cdot 9H_2O$	0.5g

Preparation of $Na_2S \cdot 9H_2O$ Solution: Add $Na_2S \cdot 9H_2O$ to distilled/deionized water and bring volume to 10.0mL. Mix thoroughly. Sparge with 100% N_2. Autoclave for 15 min at 15 psi pressure–121°C. Before use, neutralize to pH 7.0 with sterile HCl.

Wolfe's Mineral Solution:

Composition per liter:

$MgSO_4 \cdot 7H_2O$	3.0g
Nitrilotriacetic acid	1.5g
NaCl	1.0g
$MnSO_4 \cdot 2H_2O$	0.5g

$CoCl_2 \cdot 6H_2O$	0.1g
$ZnSO_4 \cdot 7H_2O$	0.1g
$CaCl_2 \cdot 2H_2O$	0.1g
$FeSO_4 \cdot 7H_2O$	0.1g
$NiCl_2 \cdot 6H_2O$	0.025g
$KAl(SO_4)_2 \cdot 12H_2O$	0.02g
$CuSO_4 \cdot 5H_2O$	0.01g
H_3BO_3	0.01g
$Na_2MoO_4 \cdot 2H_2O$	0.01g
$Na_2SeO_3 \cdot 5H_2O$	0.3mg

Preparation of Wolfe's Mineral Solution: Add nitrilotriacetic acid to 500.0mL of distilled/deionized water. Adjust pH to 6.5 with KOH. Add remaining components. Add distilled/deionized water to 1.0L. Adjust pH to 6.8.

Preparation of Medium: Prepare medium anaerobically under 100% CO_2. Add components, except gelatin solution, $Na_2S \cdot 9H_2O$ solution, and Wolfe's mineral solution, to distilled/deionized water and bring volume to 880.0mL. Mix thoroughly. Gently heat and bring to boiling. Cool to room temperature while sparging with 100% CO_2. Sparge with 100% CO_2 for 20 min. Adjust pH to 7.0. Autoclave for 15 min at 15 psi pressure–121°C. Aseptically and anaerobically add 100.0mL of sterile gelatin solution, 10.0mL of sterile $Na_2S \cdot 9H_2O$ solution, and 10.0mL of sterile Wolfe's mineral solution to each tube. Mix thoroughly.

Use: For the cultivation of *Coprothermobacter proteolyticus.*

CP Medium for *Thermobacteroides leptospartum*

Composition per 1010.0mL:

$NaHCO_3$	8.4g
Pancreatic digest of casein	2.0g
Yeast extract	2.0g
$MgCl_2 \cdot 6H_2O$	1.0g
NH_4Cl	1.0g
$CaCl_2 \cdot 2H_2O$	0.4g
$K_2HPO_4 \cdot 3H_2O$	0.4g
Resazurin	0.5mg
Glucose solution	100.0mL
$Na_2S \cdot 9H_2O$ solution	10.0mL
Wolfe's mineral solution	10.0mL

pH 7.0 ± 0.2 at 25°C

Glucose Solution:

Composition per 100.0mL:

D-Glucose	5.0g

Preparation of Glucose Solution: Add glucose to distilled/deionized water and bring volume to 100.0mL. Mix thoroughly. Sparge with 100% N_2. Autoclave for 15 min at 15 psi pressure–121°C.

$Na_2S \cdot 9H_2O$ Solution:

Composition per 10.0mL:

$Na_2S \cdot 9H_2O$	0.5g

Preparation of $Na_2S \cdot 9H_2O$ Solution: Add $Na_2S \cdot 9H_2O$ to distilled/deionized water and bring volume to 10.0mL. Mix thoroughly. Sparge with 100% N_2. Autoclave for 15 min at 15 psi pressure–121°C. Before use, neutralize to pH 7.0 with sterile HCl.

Wolfe's Mineral Solution:

Composition per liter:

$MgSO_4 \cdot 7H_2O$	3.0g
Nitrilotriacetic acid	1.5g
NaCl	1.0g
$MnSO_4 \cdot 2H_2O$	0.5g
$CoCl_2 \cdot 6H_2O$	0.1g
$ZnSO_4 \cdot 7H_2O$	0.1g
$CaCl_2 \cdot 2H_2O$	0.1g
$FeSO_4 \cdot 7H_2O$	0.1g
$NiCl_2 \cdot 6H_2O$	0.025g
$KAl(SO_4)_2 \cdot 12H_2O$	0.02g
$CuSO_4 \cdot 5H_2O$	0.01g
H_3BO_3	0.01g
$Na_2MoO_4 \cdot 2H_2O$	0.01g
$Na_2SeO_3 \cdot 5H_2O$	0.3mg

Preparation of Wolfe's Mineral Solution: Add nitrilotriacetic acid to 500.0mL of distilled/deionized water. Adjust pH to 6.5 with KOH. Add remaining components. Add distilled/deionized water to 1.0L. Adjust pH to 6.8.

Preparation of Medium: Prepare medium anaerobically under 100% CO_2. Add components, except glucose solution, $Na_2S \cdot 9H_2O$ solution, and Wolfe's mineral solution, to distilled/deionized water and bring volume to 880.0mL. Mix thoroughly. Gently heat and bring to boiling. Cool to room temperature while sparging with 100% CO_2. Sparge with 100% CO_2 for 20 min. Adjust pH to 7.0. Autoclave for 15 min at 15 psi pressure–121°C. Aseptically and anaerobically add 100.0mL of sterile glucose solution, 10.0mL of sterile $Na_2S \cdot 9H_2O$ solution, and 10.0mL of sterile Wolfe's mineral solution to each tube. Mix thoroughly.

Use: For the cultivation of *Thermobacteroides leptospartum.*

CPC Agar

See: **Cellobiose Polymyxin Colistin Agar**

CPC Agar Base with Cellobiose, Colistin, and Polymyxin B

Composition per liter:

NaCl	20.0g
Agar	15.0g
Cellobiose	15.0g
Peptic digest of animal tissue	10.0g
Beef extract	5.0g
Bromthymol Blue	0.04g
Cresol Red	0.04g
Cellobiose colistin polymyxin B solution	100.0mL

pH 7.6 ± 0.2 at 25°C

Source: This medium, without cellobiose colistin polymyxin B solution, is available as a premixed powder from HiMedia.

Cellobiose Colistin Polymyxin B Solution:

Composition per 100.0mL:

Cellobiose	15.0g
Colistin	1,360,000U
Polymyxin B	100,000U

Preparation of Cellobiose Colistin Polymyxin B Solution: Add components to distilled/deionized water and bring volume to 100.0mL. Mix thoroughly. Filter sterilize.

Preparation of Medium: Add components, except cellobiose colistin polymyxin B solution, to tap water and bring volume to 1.0L. Mix thoroughly. Distribute into tubes or flasks. Autoclave for 15 min at 15 psi pressure–121°C. Aseptically add 100.0mL of sterile cellobiose colistin polymyxin B solution to 900.0 mL of the cooled agar base. Mix thoroughly. Pour into sterile Petri dishes. Use within 7 days.

Use: For the cultivation and identification of *Vibrio* species from foods.

CPC HiVeg Agar Base with Cellobiose, Colistin, and Polymyxin B

Composition per liter:

NaCl	20.0g
Agar	15.0g
Cellobiose	15.0g
Plant peptone	10.0g
Plant extract	5.0g
Bromthymol Blue	0.04g
Cresol Red	0.04g
Cellobiose colistin polymyxin B solution	100.0mL

pH 7.6 ± 0.2 at 25°C

Source: This medium, without cellobiose colistin polymyxin B solution, is available as a premixed powder from HiMedia.

Cellobiose Colistin Polymyxin B Solution:

Composition per 100.0mL:

Cellobiose	15.0g
Colistin	1,360,000U
Polymyxin B	100,000U

Preparation of Cellobiose Colistin Polymyxin B Solution: Add components to distilled/deionized water and bring volume to 100.0mL. Mix thoroughly. Filter sterilize.

Preparation of Medium: Add components, except cellobiose colistin polymyxin B solution, to tap water and bring volume to 1.0L. Mix thoroughly. Distribute into tubes or flasks. Autoclave for 15 min at 15 psi pressure–121°C. Aseptically add 100.0mL of sterile cellobiose colistin polymyxin B solution to 900.0 mL of the cooled agar base. Mix thoroughly. Pour into sterile Petri dishes. Use within 7 days.

Use: For the cultivation and identification of *Vibrio* species from foods.

CPC Medium

Composition per liter:

Sucrose	30.0g
Peptone	2.0g
Casein hydrolysate	1.0g
$K_2HPO_4 \cdot 3H_2O$	1.0g
KCl	0.5g
$MgSO_4 \cdot 7H_2O$	0.5g
$FeSO_4 \cdot 7H_2O$	0.1g

pH 7.2 ± 0.2 at 25°C

Preparation of Medium: Add components to tap water and bring volume to 1.0L. Mix thoroughly. Distribute into tubes or flasks. Autoclave for 15 min at 15 psi pressure–121°C.

Use: For the cultivation of *Actinoplanes* species.

CR Agar

See: **Congo Red Agar**

Craig's Medium

Composition per liter:

Casein acid hydrolysate	30.0g
Yeast extract	4.0g
K_2HPO_4	0.5g
Glucose solution	20.0mL

pH 7.2 ± 0.2 at 25°C

Source: This medium is available from HiMedia.

Glucose Solution:

Composition per 100.0mL:

Glucose	20.0g

Preparation of Glucose Solution: Add glucose to distilled/deionized water and bring volume to 100.0mL. Mix thoroughly. Filter sterilize.

Preparation of Medium: Add components, except glucose solution, to distilled/deionized water and bring volume to 980.0mL. Mix thoroughly. Autoclave for 15 min at 15 psi pressure–121°C. Cool to 50°C. Aseptically add glucose solution. Mix thoroughly. Aseptically distribute into sterile tubes.

Use: For the enrichment and cultivation of *Vibrio cholerae* during testing of enterotoxigenicity.

CRAMP Agar

See: **Congo Red Acid Morpholinepropanesulfonic Acid Pigmentation Agar**

CRAMP HiVeg Agar Base

Composition per liter:

Agarose	14.0g
Morpholine propane sulfonic acid	8.4g
Tricine	1.8g
NaCl	2.9g
Galactose	2.0g
Plant acid hydrolysate	2.0g
$Na_2S_2O_3$	0.6g
NH_4Cl	0.5g
K_2HPO_4	0.24g
$MgSO_4$	0.0986g
Congo Red	0.005g

pH 5.3 ± 0.2 at 25°C

Source: This medium is available as a premixed powder from HiMedia.

Preparation of Medium: Add components to distilled/deionized water and bring volume to 1.0L. Mix thoroughly. Gently heat and bring to boiling. Adjust pH to 5.3. Distribute into tubes or flasks. Autoclave for 15 min at 15 psi pressure–121°C. Pour into sterile Petri dishes or leave in tubes.

Use: For the cultivation of *Yersinia* species with plasmids.

CREA (Creatine Agar)

Composition per liter:

Sucrose	30.0g
Agar	15.0g
Creatine·H_2O	3.0g
$K_3PO_4 \cdot 7H_2O$	1.6g
Bromcresol Purple	50.0mg
Minerals solution	10.0mL
Trace minerals solution	1.0mL

Minerals Solution:

Composition per 100.0mL:

KCl	5.0g
$MgSO_4 \cdot 7H_2O$	5.0g
$FeSO_4 \cdot 7H_2O$	0.1g

Preparation of Minerals Solution: Add components to distilled/deionized water and bring volume to 100.0mL. Mix thoroughly.

Trace Minerals Solution:

Composition per 100.0mL:

$ZnS0_4 \cdot 7H_2O$	1.0g
$CuSO_4 \cdot 5H_2O$	0.5g

Preparation of Trace Minerals Solution: Add components to distilled/deionized water and bring volume to 100.0mL. Mix thoroughly.

Preparation of Medium: Add components to distilled/deionized water and bring volume to 1.0L. Mix thoroughly. Gently heat and bring to boiling. Distribute into tubes or flasks. Autoclave for 15 min at 15 psi pressure–121°C. Pour into sterile Petri dishes or leave in tubes.

Use: For the cultivation and maintenance of *Penicillium* species.

Creatinine Agar

Composition per liter:

Agar	15.0g
$Na_2HPO_4 \cdot 12H_2O$	9.0g
NaCl	5.0g
KH_2PO_4	1.5g
Creatinine	1.0g
Meat extract	1.0g
Yeast extract	1.0g
$MgSO_4 \cdot 7H_2O$	0.2g
$MnCl_2 \cdot 4H_2O$	20.0mg
$CaCl_2$	1.2mg
Glucose solution	100.0mL

Glucose Solution:

Composition per 100.0mL:

Glucose	5.0g

Preparation of Glucose Solution: Add glucose to distilled/deionized water and bring volume to 100.0mL. Mix thoroughly. Filter sterilize. Warm to 50°C.

Preparation of Medium: Add components, except glucose solution, to distilled/deionized water and bring volume to 900.0mL. Mix thoroughly. Gently heat and bring to boiling. Autoclave for 15 min at 15 psi pressure–121°C. Cool to 50°–55°C. Aseptically add 100.0mL of sterile glucose solution. Mix thoroughly. Pour into sterile Petri dishes or distribute into sterile tubes.

Use: For the cultivation and maintenance of *Pseudomonas* species and other bacteria that can utilize creatinine.

Creatinine Medium

Composition per liter:

Creatinine	5.0g
Agar	2.0g
Fumaric acid	2.0g
K_2HPO_4	2.0g
Yeast extract	1.0g
Salt solution	10.0mL

pH 6.8 ± 0.2 at 25°C

Salt Solution:

Composition per liter:

$MgSO_4$	12.2g
$FeSO_4 \cdot 7H_2O$	2.8g
$MnSO_4 \cdot H_2O$	1.7g
$CaCl_2 \cdot 2H_2O$	0.76g
NaCl	0.6g
$Na_2MoO_4 \cdot 2H_2O$	0.1g
$ZnSO_4 \cdot 7H_2O$	0.06g
HCl (0.1*N* solution)	1.0L

Preparation of Salt Solution: Dissolve salts in 1.0L of 0.1*N* HCl solution. Mix thoroughly.

Preparation of Medium: Add components to distilled/deionized water and bring volume to 1.0L. Mix thoroughly. Adjust pH to 6.8 with NaOH or KOH. Gently heat until boiling. Distribute into tubes or flasks. Autoclave for 15 min at 15 psi pressure–121°C.

Use: For the cultivation and maintenance of *Pseudomonas* species.

Creatinine Medium (LMG 107)

Composition per liter:

Creatinine	5.0g
Fumaric acid	2.0g
K_2HPO_4	2.0g
Yeast extract	1.0g
Salt solution	10.0mL

pH 6.8 ± 0.2 at 25°C

Salt Solution:

Composition per liter:

$MgSO_4 \cdot 7H_2O$	25.0g
$FeSO_4 \cdot 7H_2O$	2.8g
$MnSO_4 \cdot H_2O$	1.7g
$CaCl_2 \cdot 2H_2O$	0.76g
NaCl	0.6g
$Na_2MoO_4 \cdot 2H_2O$	0.1g
$ZnSO_4 \cdot 7H_2O$	60.0mg
HCl (0.1*M* solution)	1.0L

Preparation of Salt Solution: Add components to distilled/deionized water and bring volume to 1.0L. Mix thoroughly.

Preparation of Medium: Add components to distilled/deionized water and bring volume to 1.0L. Mix thoroughly. Distribute into tubes or flasks. Autoclave for 15 min at 15 psi pressure–121°C.

Use: For the cultivation and maintenance of *Flavobacterium filamentosum.*

Creatinine/NMH Medium

Composition per 1100.0mL:

Yeast extract	5.0g
NaCl	1.0g
$NaHCO_3$	1.0g
$MgSO_4 \cdot 7H_2O$	0.5g
$Na_2S \cdot 9H_2O$	0.5g
$MnCl_2 \cdot 4H_2O$	0.06g
$CaSO_4 \cdot 2H_2O$	0.05g
$FeSO_4 \cdot 7H_2O$	0.01g
$Na_2SeO_3 \cdot 5H_2O$	26.0μg
Vitamin B_{12}	20.0μg
Resazurin	1.0mg

Phosphate solution 100.0mL
Creatinine solution 100.0mL
Trace elements solution SL-4 ....10.0mL
Vitamin solution 10.0mL
L-Cysteine·HCl solution 10.0mL
$Na_2S \cdot 9H_2O$ solution 10.0mL

pH 7.0 ± 0.2 at 25°C

Phosphate Solution:
Composition per 100.0mL:
K_2HPO_4 .... 5.33g
KH_2PO_4 .... 2.64g

Preparation of Phosphate Solution: Add components to distilled/deionized water and bring volume to 100.0mL. Mix thoroughly. Sparge with 100% N_2. Autoclave for 15 min at 15 psi pressure–121°C.

Creatinine Solution:
Composition per 100.0mL:
Creatinine 5.5g

Preparation of Creatinine Solution: Add creatinine to distilled/deionized water and bring volume to 100.0mL. Mix thoroughly. Sparge with 100% N_2. Filter sterilize.

Trace Elements Solution SL-4:
Composition per liter:
EDTA 0.5g
$FeSO_4 \cdot 7H_2O$ 0.2g
Trace elements solution SL-6 ....100.0mL

Trace Elements Solution SL-6:
Composition per liter:
$MnCl_2 \cdot 4H_2O$ 0.5g
H_3BO_3 .... 0.3g
$CoCl_2 \cdot 6H_2O$ 0.2g
$ZnSO_4 \cdot 7H_2O$ 0.1g
$Na_2MoO_4 \cdot 2H_2O$ 0.03g
$NiCl_2 \cdot 6H_2O$ 0.02g
$CuCl_2 \cdot 2H_2O$ 0.01g

Preparation of Trace Elements Solution SL-6: Add components to distilled/deionized water and bring volume to 1.0L. Mix thoroughly.

Preparation of Trace Elements Solution SL-4: Add components to distilled/deionized water and bring volume to 1.0L. Mix thoroughly.

Vitamin Solution:
Composition per liter:
Pyridoxine·HCl 10.0mg
Calcium DL-pantothenate 5.0mg
Lipoic acid 5.0mg
Nicotinic acid 5.0mg
p-Aminobenzoic acid 5.0mg
Riboflavin 5.0mg
Thiamine·HCl 5.0mg
Biotin 2.0mg
Folic acid 2.0mg
Vitamin B_{12} 0.1mg

Preparation of Vitamin Solution: Add components to distilled/deionized water and bring volume to 1.0L. Mix thoroughly.

L-Cysteine·HCl Solution:
Composition per 10.0mL:
L-Cysteine·HCl 0.5g

Preparation of L-Cysteine·HCl Solution: Add L-cysteine·HCl to distilled/deionized water and bring volume to 10.0mL. Mix thoroughly. Sparge with 100% N_2. Autoclave for 15 min at 15 psi pressure–121°C.

$Na_2S \cdot 9H_2O$ Solution:
Composition per 10.0mL:
$Na_2S \cdot 9H_2O$ 0.5g

Preparation of $Na_2S \cdot 9H_2O$ Solution: Add $Na_2S \cdot 9H_2O$ to distilled/deionized water and bring volume to 10.0mL. Mix thoroughly. Sparge with 100% N_2. Autoclave for 15 min at 15 psi pressure–121°C.

Preparation of Medium: Add components, except creatinine solution, phosphate solution, L-cysteine·HCl solution, and $Na_2S \cdot 9H_2O$ solution, to distilled/deionized water and bring volume to 880.0mL. Mix thoroughly. Adjust pH to 7.0. Autoclave for 15 min at 15 psi pressure–121°C. Aseptically and anaerobically add 100.0mL of sterile phosphate solution, 10.0mL of sterile L-cysteine·HCl solution, and 10.0mL of sterile $Na_2S \cdot 9H_2O$ solution. Immediately prior to use, aseptically and anaerobically add 100.0mL of sterile creatinine solution. Mix thoroughly. Aseptically and anaerobically distribute into tubes or bottles.

Use: For the cultivation and maintenance of *Clostridium* species.

CreDm1 Medium

Composition per 1002.0mL:
Solution A 980.0mL
Solution D (Vitamin solution) 10.0mL
Solution E 10.0mL
Solution B (Trace elements solution SL-10) 1.0mL
Solution C (Selentite-tungstate solution) 1.0mL

pH 6.7–6.9 at 25°C

Solution A:
Composition per 980.0mL:
KH_2PO_4 .... 1.4g
NH_4Cl 0.5g
$MgCl_2 \cdot 6H_2O$ 0.2g
$CaCl_2 \cdot 2H_2O$ 0.15g
Yeast extract 50.0mg

Preparation of Solution A: Add components to distilled/deionized water and bring volume to 980.0mL. Mix thoroughly. Autoclave for 15 min at 15 psi pressure–121°C.

Solution B (Trace Elements Solution SL-10):
Composition per liter:
$FeCl_2 \cdot 4H_2O$ 1.5g
$CoCl_2 \cdot 6H_2O$ 190.0mg
$MnCl_2 \cdot 4H_2O$ 100.0mg
$ZnCl_2$ 70.0mg
$Na_2MoO_4 \cdot 2H_2O$ 36.0mg
$NiCl_2 \cdot 6H_2O$ 24.0mg
H_3BO_3 .... 6.0mg
$CuCl_2 \cdot 2H_2O$ 2.0mg
HCl (25% solution) 10.0mL

Preparation of Solution B (Trace Elements Solution SL-10): Add $FeCl_2 \cdot 4H_2O$ to 10.0mL of HCl solution. Mix thoroughly. Add distilled/deionized water and bring volume to 1.0L. Add remaining components. Mix thoroughly. Autoclave for 15 min at 15 psi pressure–121°C.

Solution C (Selenite-Tungstate Solution):
Composition per liter:
NaOH 0.5g
$Na_2WO_4 \cdot 2H_2O$ 4.0mg
$Na_2SeO_3 \cdot 5H_2O$ 3.0mg

Preparation of Solution C (Selenite-Tungstate Solution): Add components to distilled/deionized water and bring volume to 1.0L. Mix thoroughly. Autoclave for 15 min at 15 psi pressure–121°C.

Solution D (Vitamin Solution):

Composition per liter:

Pyridoxine·HCl	10.0mg
Calcium DL-pantothenate	5.0mg
Lipoic acid	5.0mg
Nicotinic acid	5.0mg
p-Aminobenzoic acid	5.0mg
Riboflavin	5.0mg
Thiamine·HCl	5.0mg
Biotin	2.0mg
Folic acid	2.0mg
Vitamin B_{12}	0.1mg

Preparation of Solution D (Vitamin Solution): Add components to distilled/deionized water and bring volume to 1.0L. Mix thoroughly. Filter sterilize.

Solution E:

Composition per 10.0mL:

Disodium-DL-malate	1.6g

Preparation of Solution E: Add disodium-DL-malate to distilled/deionized water and bring volume to 10.0mL. Mix thoroughly. Autoclave for 15 min at 15 psi pressure–121°C.

Preparation of Medium: Aseptically combine 980.0mL of sterile solution A with 1.0mL of sterile solution B, 1.0mL of sterile solution C, 10.0mL of sterile solution D, and 10.0mL of sterile solution E, in that order. Mix thoroughly. Adjust pH to 6.7–6.9. Aseptically distribute into sterile tubes or flasks.

Use: For the cultivation of *Campylobacter* species.

Crithidia Medium

Composition per liter:

Sucrose	15.0g
Pancreatic digest of casein	6.0g
Yeast extract	1.0g
Liver concentrate	0.1g
Hemin solution	5.0mL

pH 7.8 ± 0.2 at 25°C

Hemin Solution:

Composition per 2.5mL:

Hemin	25.0mg
Triethanolamine (TEA)	2.5mL

Preparation of Hemin Solution: Add hemin to 2.5mL triethanolamine. Mix thoroughly.

Preparation of Medium: Add components to distilled/deionized water and bring volume to 1.0L. Mix thoroughly. Adjust pH to 7.8. Distribute into tubes or flasks. Autoclave for 15 min at 15 psi pressure–121°C. Slight hemin precipitate may occur.

Use: For the cultivation of *Crithidia acanthocephali*, *Crithidia deanei*, *Crithidia fasciculata*, *Crithidia harmosa*, *Crithidia hutneri*, *Crithidia luciliae*, *Crithidia mellificae*, *Crithidia oncopelti*, *Crithidia* species, *Herpetomonas samuelpessoai*, *Leptomonas pyrrhocoris*, and *Phytomonas davidi*.

CRMOX Agar

See: **Congo Red-Magnesium Oxalate Agar**

Crossley Milk Medium

Composition per liter:

Skim milk powder	100.0g
Peptone	10.0g
Bromcresol Purple	0.1g

pH 5.8 ± 0.2 at 25°C

Source: This medium is available as a premixed powder from Oxoid Unipath.

Preparation of Medium: Add components to a very small volume of distilled/deionized water and mix to a paste. Gradually add more distilled/deionized water and bring volume to 1.0L. Distribute in 10.0mL volumes into tubes. Autoclave for 5 min at 15 psi pressure–121°C.

Use: For the routine examination of canned food samples for anaerobic bacteria.

Cryptoanaerobacter Medium (DSMZ Medium 1022)

Composition per liter:

Solution A	650.0mL
Clostridium sporogenes supernatant	350.0mL

pH 7.5–8.0 at 25°C

Solution A:

Composition per 650.0mL:

Yeast extract	5.0g
$NaHCO_3$	4.0g
Casamino acids	1.0g
4-Hydroxybenzoic acid	0.45g
KH_2PO_4	0.4g
NH_4Cl	0.4g
Resazurin	0.5mg
Vitamin solution	10.0mL
Magnesium chloride solution	10.0mL
Calcium chloride solution	10.0mL
Trace element solution SL-10	1.0mL
Selenite/tungstate solution	1.0mL

Selenite/Tungstate Solution:

Composition per liter:

NaOH	0.5g
$Na_2WO_4 \cdot 2H_2O$	4.0mg
$Na_2SeO_3 \cdot 5H_2O$	3.0mg

Preparation of Selenite/Tungstate Solution: Add components to distilled/deionized water and bring volume to 1.0L. Mix thoroughly. Sparge with 100% N_2. Autoclave for 15 min at 15 psi pressure–121°C.

Trace Elements Solution SL-10:

Composition per liter:

$FeCl_2 \cdot 4H_2O$	1.5g
$CoCl_2 \cdot 6H_2O$	190.0mg
$MnCl_2 \cdot 4H_2O$	100.0mg
$ZnCl_2$	70.0mg
$Na_2MoO_4 \cdot 2H_2O$	36.0mg
$NiCl_2 \cdot 6H_2O$	24.0mg
H_3BO_3	6.0mg
$CuCl_2 \cdot 2H_2O$	2.0mg
HCl (25% solution)	10.0mL

Preparation of Trace Elements Solution SL-10: Add $FeCl_2 \cdot 4H_2O$ to 10.0mL of HCl solution. Mix thoroughly. Add distilled/deionized water and bring volume to 1.0L. Add remaining components. Mix thor-

oughly. Sparge with 100% N_2. Autoclave for 15 min at 15 psi pressure–121°C.

Vitamin Solution:

Composition per liter:

Pyridoxine-HCl	10.0mg
Thiamine-HCl·$2H_2O$	5.0mg
Riboflavin	5.0mg
Nicotinic acid	5.0mg
D-Ca-pantothenate	5.0mg
p-Aminobenzoic acid	5.0mg
Lipoic acid	5.0mg
Biotin	2.0mg
Folic acid	2.0mg
Vitamin B_{12}	0.1mg

Preparation of Vitamin Solution: Add components to distilled/deionized water and bring volume to 1.0L. Mix thoroughly. Sparge with 80% H_2 + 20% CO_2. Filter sterilize.

Magnesium Chloride Solution:

Composition per 10.0mL:

$MgCl_2 \cdot 6H_2O$	0.08g

Preparation of Magnesium Chloride Solution: Add $MgCl_2 \cdot 6H_2O$ to distilled/deionized water and bring volume to 1.0L. Mix thoroughly. Sparge with 100% N_2. Autoclave for 15 min at 15 psi pressure–121°C.

Calcium Chloride Solution:

Composition per 10.0mL:

$CaCl_2 \cdot 2H_2O$	0.06g

Preparation of Calcium Chloride Solution: Add $CaCl_2 \cdot 2H_2O$ to distilled/deionized water and bring volume to 1.0L. Mix thoroughly. Sparge with 100% N_2. Autoclave for 15 min at 15 psi pressure–121°C.

Preparation of Solution A: Add components, except bicarbonate, magnesium chloride and calcium chloride solution, to distilled/deionized water and bring volume to 630.0mL. Mix thoroughly. Adjust pH to 7.0–7.5. Gently heat and bring to boiling. Boil for 3 min. Cool while sparging with 80% N_2 + 20% CO_2. Add the solid bicarbonate. Adjust the pH to 7.8. Dispense under 80% N_2 + 20% CO_2 gas atmosphere into anaerobic vials. Autoclave for 15 min at 15 psi pressure–121°C. Aseptically and anoxically add magnesium chloride and calcium chloride solutions. Adjust final pH of the medium to pH 7.7. Note: It may be necessary to add 10–20 mg sodium dithionite per liter (e.g., from 5% (w/v) solution, freshly prepared under N_2 and filter sterilized), if the solution is not completely reduced after inoculation.

***Clostridium sporogenes* Supernatant:**

Composition per liter:

Yeast extract	5.0g
$NaHCO_3$	4.0g
Casamino acids	1.0g
KH_2PO_4	0.4g
NH_4Cl	0.4g
Resazurin	0.5mg
Vitamin solution	10.0mL
Magnesium chloride solution	10.0mL
Calcium chloride solution	10.0mL
$Na_2S \cdot 9H_2O$ solution	10.0mL
Trace element solution SL-10	1.0mL
Selenite-tungstate solution	1.0mL
Clostridium sporogenes	Variable

$Na_2S \cdot 9H_2O$ Solution:

Composition per 10.0mL:

$Na_2S \cdot 9H_2O$	0.3g

Preparation of $Na_2S \cdot 9H_2O$ Solution: Add $Na_2S \cdot 9H_2O$ to distilled/deionized water and bring volume to 10.0mL. Mix thoroughly. Autoclave under 100% N_2 for 15 min at 15 psi pressure–121°C. Cool to room temperature.

Preparation of *Clostridium sporogenes* Supernatant: Add components, except bicarbonate, $Na_2S \cdot 9H_2O$ solution, magnesium chloride and calcium chloride solution, to distilled/deionized water and bring volume to 980.0mL. Mix thoroughly. Adjust pH to 7.0–7.5. Gently heat and bring to boiling. Boil for 3 min. Cool while sparging with 80% N_2 + 20% CO_2. Add the solid bicarbonate. Adjust the pH to 7.8. Dispense under 80% N_2 + 20% CO_2 gas atmosphere into anaerobic bottles. Autoclave for 15 min at 15 psi pressure–121°C. Aseptically and anoxically add $Na_2S \cdot 9H_2O$ solution, magnesium chloride and calcium chloride solutions. Adjust final pH to pH 7.0. Inoculate with *Clostridium* sp. DSM 754. Incubate for 5 to 8 days at 37°C. Disrupt cells of the grown culture by autoclaving for 20 min at 15 psi pressure–121°C. Centrifuge autoclaved culture at 18,000g for 20 min. Discard cell pellet. Store the supernatant in screw-capped bottles at –20°C. Before use sterilize the supernatant by autoclaving under 100% N_2 for 15 min at 15 psi pressure–121°C. Cool to room temperature under 100% N_2.

Preparation of Medium: Aseptically and anoxically combine 650.0mL of solution A with 350.0mL of *Clostridum sporogenes* supernatant.

Use: For the cultivation of *Cryptoanaerobacter* spp.

Cryptococcus neoformans Screen Medium

See: **CN Screen Medium**

CRYS Medium

Composition per liter:

Stock extract	500.0mL
2× PP medium	500.0mL

Stock Extract:

Composition per 500.0mL:

Cerophyll	5.0g
Brown rice	5.0g
Yeast extract	5.0g
Dried seaweed	5.0g

Preparation of Stock Extract: Add components to distilled/deionized water and bring volume to 500.0mL. Mix thoroughly. Gently heat and bring to boiling. Continue boiling for 5 min. Filter three times through Whatman #1 filter paper while still hot. Cool to room temperature. Adjust pH to 7.2. Bring volume to 500.0mL with distilled/deionized water. Autoclave for 15 min at 15 psi pressure–121°C.

2× PP Medium:

Composition per 500.0mL:

Proteose peptone	10.0g
Pancreatic digest of peptone	10.0g
Ribonucleic acid from *Torula* yeast	1.0g
Asolectin	0.2g
Artificial seawater	167.0mL
Vitamin solution	2.0mL

Artificial Seawater:

Composition per 167.0mL:

Aqua-Marin sea salts	6.95g

Source: Aqua-Marin sea salts are available from Aquatrol, Inc., Anaheim, CA.

Preparation of Artificial Seawater: Add Aqua-Marin sea salts to distilled/deionized water and bring volume to 167.0mL Mix thoroughly. Filter sterilize.

Vitamin Solution:

Composition per 100.0mL:

Thiamine·HCl	150.0mg
Calcium D-(+)-pantothenate	100.0mg
Folic acid	50.0mg
Nicotinamide	50.0mg
Pyridoxal·HCl	50.0mg
Riboflavin	50.0mg
DL-6-Thioctic acid	1.0mg
Biotin solution	10.0mL

Biotin Solution:

Composition per 10.0mL:

Biotin	0.01mg

Preparation of Biotin Solution: Add biotin to 10.0mL of absolute ethanol. Mix thoroughly.

Preparation of Vitamin Solution: Add components to distilled/deionized water and bring volume to 100.0mL. Mix thoroughly. Filter sterilize. For long-term storage, preserve under nitrogen at –20°C.

Preparation of 2× PP Medium: Add asolectin to 200.0mL of distilled/deionized water. Gently heat to 80°C. Mix thoroughly. Add other components, except artificial seawater and vitamin solution, to distilled/deionized water and bring volume to 331.0mL. Mix thoroughly. Adjust pH to 7.2. Autoclave for 15 min at 15 psi pressure–121°C. Aseptically add 167.0mL of sterile artificial seawater and 2.0mL of sterile vitamin solution. Mix thoroughly.

Preparation of Medium: Aseptically combine 500.0mL of sterile stock extract with 500.0mL of sterile 2× PP medium. Mix thoroughly. Aseptically distribute into sterile tubes or flasks.

Use: For the cultivation of *Pseudocohnilembus marinus*.

Crystal Violet Agar

Composition per liter:

Agar	15.0g
Lactose	10.0g
Proteose peptone	5.0g
Beef extract	3.0g
Crystal Violet	3.3mg

pH 6.8 ± 0.1 at 25°C

Preparation of Medium: Add components to distilled/deionized water and bring volume to 1.0L. Mix thoroughly. Gently heat until boiling. Distribute into tubes or flasks. Autoclave for 15 min at 15 psi pressure–121°C. Pour into sterile Petri dishes or leave in tubes.

Use: For the differentiation of pathogenic staphylococci from nonpathogenic staphylococci. Hemolytic and coagulating strains of *Staphylococcus aureus* appear as purple or yellow colonies. Nonhemolytic and noncoagulating strains of *Staphylococcus* species appear as white colonies.

Crystal Violet Azide Esculin Agar

Composition per liter:

Agar	15.0g
Glucose	5.0g
NaCl	5.0g
Proteose peptone	5.0g
Pancreatic digest of casein	5.0g
Meat extract	3.0g
Esculin	1.0g
NaN_3	1.0g
Crystal Violet	0.1g
Bovine blood, citrated	100.0mL

pH 7.5 ± 0.2 at 25°C

Caution: Sodium azide is toxic. Azides also react with metals and disposal must be highly diluted.

Preparation of Medium: Add components, except citrated bovine blood, to distilled/deionized water and bring volume to 900.0mL. Mix thoroughly. Gently heat and bring to boiling. Autoclave for 15 min at 15 psi pressure–121°C. Cool to 45°–50°C. Aseptically add sterile, citrated bovine blood. Mix thoroughly. Pour into sterile Petri dishes or distribute into sterile tubes.

Use: For the cultivation of *Erysipelothrix rhusiopathiae.*

Crystal Violet Esculin Agar

Composition per liter:

Agar	15.0g
Glucose	5.0g
NaCl	5.0g
Proteose peptone	5.0g
Pancreatic digest of casein	5.0g
Meat extract	3.0g
Esculin	1.0g
Crystal Violet	2.0mg
Blood, citrated	100.0mL

pH 7.5 ± 0.2 at 25°C

Preparation of Medium: Add components, except citrated blood, to distilled/deionized water and bring volume to 900.0mL. Mix thoroughly. Gently heat and bring to boiling. Autoclave for 15 min at 15 psi pressure–121°C. Cool to 45°–50°C. Aseptically add citrated blood. Mix thoroughly. Pour into sterile Petri dishes or distribute into sterile tubes.

Use: For the cultivation of *Erysipelothrix rhusiopathiae.*

Crystal Violet Lactose Agar

Composition per liter:

Agar	15.0g
Lactose	10.0g
Proteose peptone	5.0g
Beef extract	3.0g
Crystal Violet	3.3mg

pH 6.8 ± 0.2 at 25°C

Source: This medium is available as a premixed powder from HiMedia.

Preparation of Medium: Add components to distilled/deionized water and bring volume to 1.0L. Mix thoroughly. Gently heat and bring to boiling. Distribute into tubes or flasks. Autoclave for 15 min at 15 psi pressure–121°C. Pour into sterile Petri dishes or leave in tubes.

Use: For the differentiation of pure cultures of pathogenic and nonpathogenic staphylococci.

Crystal Violet Lactose Broth

Composition per liter:

Lactose	5.0g
Peptic digest of animal tissue	5.0g
K_2HPO_4	5.0g
KH_2PO_4	1.0g
Crystal Violet	1.43mg

pH 7.4 ± 0.2 at 25°C

Source: This medium is available as a premixed powder from HiMedia.

Preparation of Medium: Add components to distilled/deionized water and bring volume to 1.0L. Mix thoroughly. Gently heat and bring to boiling. Distribute into tubes or flasks. Autoclave for 15 min at 15 psi pressure–121°C. Pour into sterile Petri dishes or leave in tubes.

Use: For the differentiation of pure cultures of pathogenic and nonpathogenic staphylococci.

Crystal Violet Lactose HiVeg Agar

Composition per liter:

Agar	15.0g
Lactose	10.0g
Plant peptone No. 3	5.0g
Plant extract	3.0g
Crystal Violet	3.3mg

pH 6.8 ± 0.2 at 25°C

Source: This medium is available as a premixed powder from HiMedia.

Preparation of Medium: Add components to distilled/deionized water and bring volume to 1.0L. Mix thoroughly. Gently heat and bring to boiling. Distribute into tubes or flasks. Autoclave for 15 min at 15 psi pressure–121°C. Pour into sterile Petri dishes or leave in tubes.

Use: For the differentiation of pure cultures of pathogenic and nonpathogenic staphylococci.

Crystal Violet Pectate Medium (CVP Medium)

Composition per liter:

Sodium polypectate	9.0g
Agar	2.0g
$NaNO_3$	1.0g
NaOH (1*N* solution)	4.5mL
$CaCl_2 \cdot H_2O$ (10% solution)	3.0mL
Crystal Violet (0.075% solution)	1.0mL
Sodium lauryl sulfate (10% solution)	0.5mL

pH 7.2 ± 0.2 at 25°C

Preparation of Medium: In a preheated blender, add 500.0mL of boiling distilled/deionized water and the components, except sodium polypectate and sodium lauryl sulfate solution. Blend at high speed for 15 sec. Continue blending at low speed and slowly add 9.0g of sodium polypectate. Pour the incomplete medium into a 2L flask and add 0.5mL of sodium lauryl sulfate solution. Mix thoroughly. Cap flask with an aluminum foil seal rather than cotton. Autoclave for 25 min at 15 psi pressure–121°C. Pour medium quickly into sterile Petri dishes. Allow plates to dry at 25°C for 48 hr before use.

Use: For the cultivation of pectinolytic microorganisms, such as *Erwinia* species, from foods.

Crystal Violet Pectate Medium (CVP Medium)

Composition per liter:

Sodium polypectate	18.0g
Agar	4.0g
$NaNO_3$	2.0g
$CaCl_2 \cdot 2H_2O$	0.6g
NaOH	0.36g
Sodium lauryl sulfate	0.1g
Crystal Violet	1.5mg

pH 7.2 ± 0.2 at 25°C

Source: This medium is available as a premixed powder from HiMedia.

Preparation of Medium: Add components to distilled/deionized water and bring volume to 1.0L. Mix thoroughly. Gently heat and bring to boiling. Distribute into tubes or flasks. Autoclave for 15 min at 15 psi pressure–121°C.

Use: For the cultivation of pectinolytic microorganisms, such as *Erwinia* species, from foods.

Crystal Violet Tetrazolium Agar Base

Composition per liter:

Agar	15.0g
Casein enzymatic hydrolysate	5.0g
Yeast extract	2.5g
Glucose	1.0g
Crystal Violet	1.0mg
2,3,5-Triphenyltetrazolium chloride solution	10.0mL

pH 7.0 ± 0.2 at 25°C

Source: This medium, without triphenyltetrazolium chloride solution, is available as a premixed powder from HiMedia.

Caution: Sodium azide is toxic. Azides also react with metals and disposal must be highly diluted.

2,3,5-Triphenyltetrazolium Chloride Solution:
Composition per 10.0mL:

2,3,5-Triphenyltetrazolium chloride	0.1g

Preparation of 2,3,5-Triphenyltetrazolium Chloride Solution: Add 2,3,5-triphenyltetrazolium chloride to distilled/deionized water and bring volume to 10.0mL. Mix thoroughly. Filter sterilize.

Preparation of Medium: Add components, except 2,3,5-triphenyltetrazolium chloride solution, to distilled/deionized water and bring volume to 990.0mL. Mix thoroughly. Gently heat and bring to boiling. Autoclave for 10 min at 15 psi pressure–121°C. Cool to 45°–50°C. Aseptically add 10.0mL 2,3,5-triphenyltetrazolium chloride solution. Mix thoroughly. Pour into sterile Petri dishes or distribute into sterile tubes.

Use: For the detection of Gram-negative psychrotrophic bacteria causing food spoilage.

Crystal Violet Tetrazolium HiVeg Agar Base

Composition per liter:

Agar	15.0g
Plant hydrolysate	5.0g
Yeast extract	2.5g
Glucose	1.0g

Crystal Violet 1.0mg
2,3,5-Triphenyltetrazolium chloride solution 10.0mL
pH 7.0 ± 0.2 at 25°C

Source: This medium, without triphenyltetrazolium chloride solution, is available as a premixed powder from HiMedia.

Caution: Sodium azide is toxic. Azides also react with metals and disposal must be highly diluted.

2,3,5-Triphenyltetrazolium Chloride Solution:

Composition per 10.0mL:

2,3,5-Triphenyltetrazolium chloride 0.1g

Preparation of 2,3,5-Triphenyltetrazolium Chloride Solution: Add 2,3,5-triphenyltetrazolium chloride to distilled/deionized water and bring volume to 10.0mL. Mix thoroughly. Filter sterilize.

Preparation of Medium: Add components, except 2,3,5-triphenyltetrazolium chloride solution, to distilled/deionized water and bring volume to 990.0mL. Mix thoroughly. Gently heat and bring to boiling. Autoclave for 10 min at 15 psi pressure–121°C. Cool to 45°–50°C. Aseptically add 10.0mL 2,3,5-triphenyltetrazolium chloride solution. Mix thoroughly. Pour into sterile Petri dishes or distribute into sterile tubes.

Use: For the detection of Gram-negative psychrotrophic bacteria causing food spoilage.

CS Vitamin B_{12} Agar

Composition per liter:

Glucose 20.0g
K_2SO_4 .... 20.0g
Agar 15.0g
Sodium acetate 12.0g
Vitamin assay casamino acids 10.0g
Papaic digest of soybean meal 5.0g
Sodium thioglycolate 1.7g
K_2HPO_4 .... 1.0g
KH_2PO_4 .... 1.0g
Ribonucleic acid 1.0g
Sorbitan monooleate complex 1.0g
$MgSO_4 \cdot 7H_2O$ 0.4g
DL-Tryptophan 0.2g
L-Cystine 0.2g
$FeSO_4$ 0.02g
$MgSO_4 \cdot 7H_2O$ 0.02g
NaCl 0.02g
Adenine sulfate 0.018g
Guanine·HCl 0.012g
Uracil 0.01g
Xanthine 0.01g
Pyridoxal 4.0mg
Pyridoxine 4.0mg
Calcium pentothenate 2.0mg
Niacin 2.0mg
Riboflavin 2.0mg
Thiamine·HCl 2.0mg
Folic acid 1.0mg
Biotin 1.0μg
Lactobacillus leichmannii suspension 1.0mL
pH 6.2 ± 0.1 at 25°C

Preparation of Medium: Add components, except *Lactobacillus leichmannii* suspension, to distilled/deionized water and bring volume to 1.0L. Mix thoroughly. Gently heat and bring to boiling. Autoclave for 15 min at 15 psi pressure–121°C. Cool to 45°–50°C. Inoculate medium with 1.0mL of *Lactobacillus leichmannii* suspension. Mix thoroughly. Pour into sterile 150mm Petri dishes in 50.0mL volumes. Allow agar surface to dry before using.

Use: For the microbiological assay of vitamin B_{12} by the cup plate or disk method using *Lactobacillus leichmannii* as the test microorganism.

CSSM

See: **Cornstarch Soluble Medium**

CT Agar (Caprylate Thallous Agar)

Composition per liter:

Solution A 500.0mL
Solution B 500.0mL
pH 7.2 ± 0.2 at 25°C

Solution A:

Composition per 500.0mL:

K_2HPO_4 .... 2.61g
KH_2PO_4 .... 0.68g
Thallous sulfate 0.25g
$MgSO_4 \cdot 7H_2O$ 0.12g
$CaCl_2 \cdot 2H_2O$ 0.016g
Trace elements solution 10.0mL
Yeast extract 2.0mL
Caprylic acid 1.1mL

Preparation of Solution A: Add components to distilled/deionized water and bring volume to 500.0mL. Mix thoroughly. Adjust pH to 7.2 with NaOH. Autoclave for 20 min at 10 psi pressure–115°C.

Trace Elements Solution:

Composition per liter:

H_3PO_4 .... 1.96g
$FeSO_4 \cdot 7H_2O$ 0.056g
$ZnSO_4 \cdot 4H_2O$ 0.029g
$CuSO_4 \cdot 5H_2O$ 0.025g
$MnSO_4 \cdot 4H_2O$ 0.022g
H_3BO_3 .... 6.2mg
$Co(NO_3)_2 \cdot 6H_2O$ 3.0mg

Preparation of Trace Elements Solution: Add components to distilled/deionized water and bring volume to 1.0L. Mix thoroughly. Store at 4°C.

Solution B:

Composition per liter:

Agar 15.0g
NaCl 7.0g
$(NH_4)_2SO_4$ 1.0g

Preparation of Solution B: Add components to distilled/deionized water and bring volume to 500.0mL. Mix thoroughly. Gently heat and bring to boiling. Adjust pH to 7.2. Autoclave for 20 min at 10 psi pressure–115°C.

Preparation of Medium: Aseptically combine 500.0mL of sterile solution A and 500.0mL of sterile solution B. Mix thoroughly. Pour into sterile Petri dishes in 25.0–30.0mL volumes.

Use: For the isolation and cultivation of the *Serratia* species.

CT Agar

Composition per liter:

Agar	20.0g
Pancreatic digest of casein	20.0g
$MgSO_4 \cdot 7H_2O$	2.0g
Potassium phosphate buffer (0.02*M* solution, pH 7.6)	500.0mL

pH 7.6 ± 0.2 at 25°C

Preparation of Medium: Add agar, pancreatic digest of casein, and $MgSO_4 \cdot 7H_2O$ to distilled/deionized water and bring volume to 500.0mL. Mix thoroughly. Gently heat and bring to boiling. Autoclave agar–pancreatic digest of casein-$MgSO_4 \cdot 7H_2O$ solution and potassium phosphate buffer solution separately for 15 min at 15 psi pressure–121°C. Cool to 25°C. Aseptically combine the two solutions. Aseptically add sterile components. Mix thoroughly. Pour into sterile Petri dishes or distribute into sterile tubes.

Use: For the cultivation of myxobacteria.

CT Broth

Composition per liter:

Pancreatic digest of casein	20.0g
$MgSO_4 \cdot 7H_2O$	2.0g
Potassium phosphate buffer (0.02*M* solution, pH 7.6)	500.0mL

pH 7.6 ± 0.2 at 25°C

Preparation of Medium: Add pancreatic digest of casein and $MgSO_4 \cdot 7H_2O$ to distilled/deionized water and bring volume to 500.0mL. Mix thoroughly. Autoclave pancreatic digest of casein-$MgSO_4 \cdot 7H_2O$ solution and potassium phosphate buffer solution separately for 15 min at 15 psi pressure–121°C. Cool to 25°C. Aseptically combine the two solutions. Aseptically distribute into sterile tubes or flasks.

Use: For the cultivation of myxobacteria.

CT Medium

Composition per liter:

Agar	15.0g
Pancreatic digest of casein	10.0g
Yeast extract	3.5g
$MgSO_4$	0.96g

Preparation of Medium: Add components to distilled/deionized water and bring volume to 1.0L. Mix thoroughly. Gently heat until boiling. Distribute into tubes or flasks. Autoclave for 15 min at 15 psi pressure–121°C. Pour into sterile Petri dishes or leave in tubes.

Use: For the cultivation and maintenance of *Stigmatella aurantiaca.*

CTA Agar (Cystine Trypticase™ Agar)

Composition per liter:

Pancreatic digest of casein	20.0g
Agar	14.0g
NaCl	5.0g
L-Cystine	0.5g
Na_2SO_3	0.5g
Phenol Red	0.017g

pH 7.3 ± 0.2 at 25°C

Preparation of Medium: Add components to distilled/deionized water and bring volume to 1.0L. Mix thoroughly. Gently heat until boiling. Distribute into tubes or flasks. Autoclave for 15 min at 15 psi pressure–118°C. Pour into sterile Petri dishes or leave in tubes. Two drops of sterile rabbit serum added per tube prior to solidification enhances the recovery of *Corynebacterium diphtheriae.*

Use: For the cultivation and maintenance of a variety of fastidious microorganisms, including *Corynebacterium diphtheriae.* For carbohydrate fermentation tests in the differentiation of *Neisseria* species.

CTA Medium (Cystine Trypticase™ Agar Medium) (Cystine Tryptic Agar)

Composition per liter:

Pancreatic digest of casein	20.0g
NaCl	5.0g
Carbohydrate	5.0g
Agar	2.5g
L-Cystine	0.5g
Na_2SO_3	0.5g
Phenol Red	0.017g

pH 7.3 ± 0.2 at 25°C

Source: The medium is available as a premixed powder from BD Diagnostic Systems.

Preparation of Medium: Add components to distilled/deionized water and bring volume to 1.0L. Mix thoroughly. Adjust pH to 7.3. Gently heat until boiling. Distribute into tubes or flasks. Autoclave for 15 min at 15 psi pressure–118°C. Cool tubes in an upright position. Store at room temperature.

Use: For the cultivation and maintenance of a variety of fastidious microorganisms. For the detection of bacterial motility. Used, with added specific carbohydrate, for fermentation reactions of fastidious microorganisms, especially *Neisseria* species, pneumococci, streptococci, and nonspore-forming anaerobes.

CTA Medium with Yeast Extract and Rabbit Serum (Cystine Trypticase™ Agar Medium with Yeast Extract and Rabbit Serum)

Composition per liter:

Yeast extract	50.0g
Pancreatic digest of casein	20.0g
NaCl	5.0g
Carbohydrate	5.0g
Agar	2.5g
L-Cystine	0.5g
Na_2SO_3	0.5g
Phenol Red	0.017g
Rabbit serum	250.0mL

pH 7.3 ± 0.2 at 25°C

Preparation of Medium: Add components, except rabbit serum, to distilled/deionized water and bring volume to 750.0mL. Mix thoroughly. Adjust pH to 7.3. Gently heat until boiling. Autoclave for 15 min at 15 psi pressure–118°C. Cool to 50°C. Aseptically add sterile rabbit serum. Mix thoroughly. Distribute into sterile tubes. Store at room temperature. Do not refrigerate.

Use: For the cultivation and maintenance of fastidious microorganisms, especially mycoplasmas and related microorganisms.

CTLM Medium

Composition per 1100.0mL:

Beef liver, infusion from	125.0g
Tryptose	25.0g
Proteose peptone	2.5g
L-Cysteine·HCl	1.75g
Maltose	1.25g
NaCl	1.25g
Agar	1.15g
L-Ascorbic acid	0.25g
$NaHCO_3$	0.075g
Horse serum, heat inactivated	100.0mL
Ringer's salt solution, 10×	75.0mL

pH 6.0 ± 0.2 at 25°C

Ringer's Salt Solution, 10×:

Composition per 100.0mL:

NaCl	9.0g
KCl	0.42g
$CaCl_2$	0.24g

Preparation of Ringer's Salt Solution, 10×: Add components to distilled/deionized water and bring volume to 100.0mL. Mix thoroughly.

Preparation of Medium: Add components, except horse serum, to distilled/deionized water and bring volume to 1.0L. Mix thoroughly. Adjust pH to 6.0. Gently heat and bring to boiling. Autoclave for 25 min at 15 psi pressure–121°C. Cool to 25°C. Aseptically add 100.0mL of sterile, heat-inactivated horse serum. Mix thoroughly. Aseptically distribute into sterile, screw-capped tubes or flasks.

Use: For the cultivation of *Trichomonas vaginalis*.

CTLM Medium

Composition per 1100.0mL:

Beef liver, infusion from	125.0g
Tryptose	25.0g
Proteose peptone	2.5g
L-Cysteine·HCl	1.75g
Maltose	1.25g
NaCl	1.25g
Agar	1.15g
L-Ascorbic acid	0.25g
$NaHCO_3$	0.075g
Horse serum, heat inactivated	100.0mL
Ringer's salt solution, 10×	75.0mL

pH 7.0 ± 0.2 at 25°C

Ringer's Salt Solution, 10×:

Composition per 100.0mL:

NaCl	9.0g
KCl	0.42g
$CaCl_2$	0.24g

Preparation of Ringer's Salt Solution, 10×: Add components to distilled/deionized water and bring volume to 100.0mL. Mix thoroughly.

Preparation of Medium: Add components, except horse serum, to distilled/deionized water and bring volume to 1.0L. Mix thoroughly. Adjust pH to 7.0. Gently heat and bring to boiling. Autoclave for 25 min at 15 psi pressure–121°C. Cool to 25°C. Aseptically add 100.0mL of sterile, heat-inactivated horse serum. Mix thoroughly. Aseptically distribute into sterile, screw-capped tubes or flasks.

Use: For the cultivation of *Monocercomonas colubrorum, Tetratrichomonas gallinarum, Tritrichomonas foetus,* and *T. mobilensis.*

CTLM Medium

Composition per 1100.0mL:

Beef liver, infusion from	125.0g
Tryptose	25.0g
Proteose peptone	2.5g
L-Cysteine·HCl	1.75g
Maltose	1.25g
NaCl	1.25g
Agar	1.15g
L-Ascorbic acid	0.25g
$NaHCO_3$	0.075g
Horse serum, heat inactivated	100.0mL
Ringer's salt solution, 10×	75.0mL

pH 7.3 ± 0.2 at 25°C

Ringer's Salt Solution, 10×:

Composition per 100.0mL:

NaCl	9.0g
KCl	0.42g
$CaCl_2$	0.24g

Preparation of Ringer's Salt Solution, 10×: Add components to distilled/deionized water and bring volume to 100.0mL. Mix thoroughly.

Preparation of Medium: Add components, except horse serum, to distilled/deionized water and bring volume to 1.0L. Mix thoroughly. Adjust pH to 7.3. Gently heat and bring to boiling. Autoclave for 25 min at 15 psi pressure–121°C. Cool to 25°C. Aseptically add 100.0mL of sterile, heat-inactivated horse serum. Mix thoroughly. Aseptically distribute into sterile, screw-capped tubes or flasks.

Use: For the cultivation of *Trichomonas gallinae*.

CTT Medium

Composition per liter:

Agar	15.0g
Pancreatic digest of casein	10.0g
Tris(hydroxymethyl)aminomethane buffer	1.21g
Potassium phosphate buffer (1 m*M*, pH 7.6)	1.0L
Magnesium sulfate solution	10.0mL

pH 7.6 ± 0.2 at 25°C

Magnesium Sulfate Solution:

Composition per 10.0mL:

$MgSO_4 \cdot 7H_2O$	2.0g

Preparation of Magnesium Sulfate Solution: Add $MgSO_4 \cdot 7H_2O$ to 10.0mL of distilled/deionized water. Mix thoroughly.

Preparation of Medium: Combine components. Mix thoroughly. Adjust pH to 7.6. Gently heat and bring to boiling. Distribute into tubes or flasks. Autoclave for 15 min at 15 psi pressure–121°C. Pour into sterile Petri dishes or leave in tubes.

Use: For the cultivation of myxobacteria.

Cultivation Medium for Chlamydiae (DSMZ Medium 1193)

Composition per 101.0mL:

IM medium	90.0mL
Fetal bovine serum	10.0mL
Amino acids, 100x	1.0mL

pH 7.4 ± 0.2 at 25°C

IM medium:

Composition per 100.0mL:

Pancreatic digest of gelatin	0.05g
Bile salts No. 3	0.05g
Brain heart, solids from infusion	0.02g
Peptic digest of animal tissue	0.02g
NaCl	0.017g
Glucose	0.01g
Na_2HPO_4	8.0mg
Earle's balanced salts solution	80.0mL
Fetal bovine serum, heat inactivated (2 hr at 55°C)	20.0mL

pH 7.4 ± 0.2 at 25°C

Earle's Balanced Salts Solution:

Composition per liter:

NaCl	6.8g
$NaHCO_3$	2.2g
Glucose	1.0g
KCl	0.4g
$CaCl_2·2H_2O$	0.265g
$MgSO_4·7H_2O$	0.2g
$NaH_2PO_4·H_2O$	0.14g

Preparation of Earle's Balanced Salts Solution: Add components to distilled/deionized water and bring volume to 1.0L. Mix thoroughly. Filter sterilize.

Preparation of IM: Combine components. Mix thoroughly. Filter sterilize. Store at 4°–10°C.

Preparation of Medium: Combine components. Mix thoroughly. Filter sterilize. Store for no longer than 4 weeks at room temperature to facilitate detection of contamination. Prepare a 25 cm^2 flask and seed with either cells L929 (ACC 2) or HeLa (ACC 57) cells. Incubate at 37°C plus 5% CO_2. When a confluent layer has formed, infection can be carried out. Exchange medium with 6.0mL of IM with the addition of 0.001mg/mL cycloheximide (final concentration)) and add 0.5–1.0mL of EB stock solution (thawed quickly to 37°C). Centrifuge for 1 h onto the cell layer at 1600 rpm at 20°C. Incubate at 37°C + 5% CO_2. Control cells daily and look for inclusions. Not all chlamydiae form well-visible inclusions; ultimately, immunofluorescence or *in situ* hybridization techniques are necessary to visualize inclusions.

Use: For the screening for *Chlamydia* using cell line cultures to test for infectivity.

CVA Medium (Cefoperazone Vancomycin Amphotericin Medium)

Composition per liter:

Agar	15.0g
Casein peptone	10.0g
Meat peptone	10.0g
NaCl	5.0g
Yeast autolysate	2.0g
Glucose	1.0g
$NaHSO_3$	0.1g
Sheep blood, defibrinated	50.0mL
CVA antibiotic solution	10.0mL

pH 7.0 ± 0.2 at 25°C

CVA Antibiotic Solution:

Composition per 10.0mL:

Cefoperazone	20.0mg
Vancomycin	10.0mg
Amphotericin B	2.0mg

Preparation of CVA Antibiotic Solution: Add components to distilled/deionized water and bring volume to 10.0mL. Mix thoroughly. Filter sterilize.

Preparation of Medium: Add components, except CVA antibiotic solution and sheep blood, to distilled/deionized water and bring volume to 940.0mL. Mix thoroughly. Gently heat until boiling. Autoclave for 15 min at 15 psi pressure–121°C. Cool to 45°–50°C. Aseptically add sterile CVA antibiotic solution and sterile, defibrinated sheep blood. Mix thoroughly. Pour into sterile Petri dishes.

Use: For the isolation and cultivation of *Campylobacter* species from clinical specimens.

CVP Medium

See: **Crystal Violet Pectate Medium**

CY Agar

Composition per liter:

Agar	15.0g
Pancreatic digest of casein	3.0g
$CaCl_2·2H_2O$	1.0g
Yeast extract	1.0g
Cyanocobalamin	0.5mg

pH 7.2 ± 0.2 at 25°C

Preparation of Medium: Add components to distilled/deionized water and bring volume to 1.0L. Mix thoroughly. Gently heat and bring to boiling. Distribute into tubes or flasks. Autoclave for 15 min at 15 psi pressure–121°C. Pour into sterile Petri dishes or leave in tubes.

Use: For the cultivation of myxobacteria.

CYA Agar

See: **Czapek Yeast Autolysate Agar**

CYA Agar with Arginine and *p*-Aminobenzoic Acid (ATCC Medium 2033)

Composition per liter:

Agar	15.0g
Yeast extract	5.0g
$NaNO_3$	3.0g
K_2HPO_4	1.0g
KCl	0.5g
$MgSO_4·7H_2O$	0.5g
Arginine	0.2g
$FeSO_4·7H_2O$	0.01g
p-Aminobenzoic acid	1mg
Sucrose solution	100.0mL

pH 7.3 ± 0.2 at 25°C

Sucrose Solution:

Composition per 100.0mL:

Sucrose	30.0g

Preparation of Sucrose Solution: Add sucrose to distilled/deionized water and bring volume to 100.0mL. Mix thoroughly. Autoclave for 15 min at 15 psi pressure–121°C. Cool to 50°C.

Preparation of Medium: Add components, except sucrose solution, to distilled/deionized water and bring volume to 900.0mL. Mix thoroughly. Autoclave for 15 min at 15 psi pressure–121°C. Cool to 50°C. Aseptically add sterile sucrose solution. Mix thoroughly. Pour into sterile Petri dishes or leave in tubes.

Use: For the isolation and cultivation of heat-resistant filamentous fungi (molds) from foods.

CYC Agar

Composition per liter:

Sucrose	30.0g
Agar	16.0g
Vitamin assay casamino acids	6.0g
$NaNO_3$	2.0g
Yeast extract	2.0g
Magnesium glycerophosphate	0.5g
KCl	0.5g
K_2SO_4	0.35g
$FeSO_4$	0.01g

pH 6.8 ± 0.2 at 25°C

Preparation of Medium: Add components to distilled/deionized water and bring volume to 1.0L. Mix thoroughly. Gently heat and bring to boiling. Distribute into tubes or flasks. Autoclave for 15 min at 15 psi pressure–121°C. Pour into sterile Petri dishes or leave in tubes.

Use: For the cultivation and maintenance of *Pseudonocardia thermophila, Saccharomonospora caesia, Saccharomonospora viridis, Saccharopolyspora hirsuta, Saccharopolyspora rectivirgula, Streptomyces thermogriseoviolaceus, Streptomyces thermohygroscopicus, Streptomyces thermovulgaris, Thermoactinomyces candidus, Thermoactinomyces putidus, Thermoactinomyces sacchari, Thermoactinomyces thalpophilus, Thermoactinomyces vulgaris,* and *Thermomonospora fusca.*

CYC Medium

Composition per liter:

Sucrose	30.0g
Casamino acids, vitamin free	6.0g
$NaNO_3$	3.0g
Yeast extract	2.0g
K_2HPO_4	1.0g
$MgSO_4 \cdot 7H_2O$	0.5g
KCl	0.5g
$FeSO_4 \cdot 7H_2O$	0.01g
Antibiotic solution	10.0mL

pH 7.2 ± 0.2 at 25°C

Antibiotic Solution:

Composition per 10.0mL:

Cycloheximide	0.05g
Novobiocin	0.025g

Preparation of Antibiotic Solution: Add components to distilled/deionized water and bring volume to 10.0mL. Mix thoroughly. Filter sterilize.

Caution: Cycloheximide is toxic. Avoid skin contact or aerosol formation and inhalation.

Preparation of Medium: Add components, except antibiotic solution, to distilled/deionized water and bring volume to 990.0mL. Mix thoroughly. Gently heat and bring to boiling. Autoclave for 15 min at 15 psi pressure–121°C. Cool to 45°–50°C. Aseptically add sterile antibiotic solution. Mix thoroughly. Aseptically distribute into sterile tubes.

Use: For the isolation and cultivation of *Thermoactinomyces* species.

CYC Medium, Cross and Attwell Modification (DSMZ Medium 550)

Composition per liter:

Sucrose	30.0g
Agar	15.0g
Casamino acids	6.1g
$NaNO_3$	3.0g
Yeast extract	2.0g
K_2HPO_4	1.0g
$MgSO_4 \cdot 7H_2O$	0.5g
KCl	0.5g
Tryptophan	0.02g
$FeSO_4 \cdot 7H_2O$	0.01g

pH 7.2 ± 0.2 at 25°C

Preparation of Medium: Add components to distilled/deionized water and bring volume to 1.0L. Mix thoroughly. Distribute into tubes or flasks. Autoclave for 15 min at 15 psi pressure–121°C. Pour into sterile Petri dishes or leave in tubes.

Use: For the cultivation and maintenance of *Thermomonospora curvata, Thermobifida fusca, Thermoactinomyces vulgaris, Saccharopolyspora thermophila,* and *Thermobifida alba.*

Cyclohexanecarboxylic Acid Agar

Composition per liter:

Agar, noble	15.0g
Cyclohexanecarboxylic acid	5.0g
$(NH_4)_2SO_4$	1.0g
KH_2PO_4	1.0g
K_2HPO_4	1.0g
$MgSO_4 \cdot 7H_2O$	0.2g
Yeast extract	0.1g
$FeSO_4 \cdot 7H_2O$	10.0mg
$CaCl_2 \cdot 2H_2O$	2.0mg
$MnSO_4 \cdot 4H_2O$	2.0mg
$ZnSO_4 \cdot 7H_2O$	2.0mg

pH7.2 ± 0.2 at 25°C

Preparation of Medium: Add components to distilled/deionized water and bring volume to 1.0L. Mix thoroughly. Adjust pH to 7.2. Gently heat and bring to boiling. Distribute into tubes or flasks. Autoclave for 15 min at 15 psi pressure–121°C. Pour into sterile Petri dishes or leave in tubes.

Use: For the cultivation and maintenance of *Corynebacterium cyclohexanicum* and *Saccharomyces cerevisiae.*

Cyclohexanecarboxylic Acid Broth

Composition per liter:

Cyclohexanecarboxylic acid	5.0g
$(NH_4)_2SO_4$	1.0g
KH_2PO_4	1.0g
K_2HPO_4	1.0g
$MgSO_4 \cdot 7H_2O$	0.2g
Yeast extract	0.1g
$FeSO_4 \cdot 7H_2O$	10.0mg
$CaCl_2 \cdot 2H_2O$	2.0mg
$MnSO_4 \cdot 4H_2O$	2.0mg
$ZnSO_4 \cdot 7H_2O$	2.0mg

pH 7.2 ± 0.2 at 25°C

Preparation of Medium: Add components to distilled/deionized water and bring volume to 1.0L. Mix thoroughly. Adjust pH to 7.2. Distribute into tubes or flasks. Autoclave for 15 min at 15 psi pressure–121°C.

Use: For the cultivation of *Corynebacterium cyclohexanicum* and *Saccharomyces cerevisiae.*

Cyclohexanecarboxylic Acid Medium

Composition per liter:

K_2HPO_4	3.5g
Cyclohexanecarboxylic acid	2.0g
KH_2PO_4	1.5g
NH_4NO_3	1.0g
$MgSO_4 \cdot 7H_2O$	0.5g
Yeast extract	0.1g
$CaCl_2 \cdot 2H_2O$	0.01g
$FeCl_3 \cdot 6H_2O$	0.01g
$NaMoO_4 \cdot 7H_2O$	0.01g
$ZnSO_4 \cdot 7H_2O$	0.01g

pH 7.0 ± 0.2 at 25°C

Preparation of Medium: Add components to distilled/deionized water and bring volume to 1.0L. Mix thoroughly. Distribute into tubes or flasks. Autoclave for 15 min at 15 psi pressure–121°C.

Use: For the cultivation and maintenance of *Alcaligenes faecalis* and other bacteria that can utilize cyclohexanecarboxylic acid as a carbon source.

Cyclohexanecarboxylic Acid Salts Medium

See: **CHCA Salts Medium**

Cyclohexanone Medium

Composition per liter:

NH_4NO_3	3.0g
K_2HPO_4	0.25g
$MgSO_4 \cdot 7H_2O$	0.2g
$CaCl_2 \cdot 2H_2O$	0.01g
$FeCl_3 \cdot 6H_2O$	1.0mg
Cyclohexanonc	1.0mL

Preparation of Medium: Add components, except cyclohexanone, to distilled/deionized water and bring volume to 999.0mL. Mix thoroughly. Distribute into tubes or flasks. Autoclave for 20 min at 15 psi pressure–121°C. Filter sterilize cyclohexanone. Aseptically add 1.0mL of cyclohexanone. Mix thoroughly.

Use: For the cultivation and maintenance of *Nocardia* species and other bacteria that can utilize cyclohexanone as a carbon source.

Cycloheximide Agar

See: **Actidione® Agar**

Cycloheximide Chloramphenicol Agar

See: **Mycosel™ Agar**
See: **Mycobiotic Agar**

Cycloserine Cefoxitin Egg Yolk Fructose Agar

Composition per liter:

Proteose peptone No. 2	40.0g
Agar	25.0g
Fructose	6.0g
Na_2HPO_4	5.0g
NaCl	2.0g
KH_2PO_4	1.0g
$MgSO_4 \cdot 7H_2O$	0.1g
Egg yolk emulsion	100.0mL
Antibiotic solution	10.0mL
Neutral Red solution	3.0mL
Hemin solution	1.0mL

Egg Yolk Emulsion:

Composition:

Chicken egg yolks	11
Whole chicken egg	1

Preparation of Egg Yolk Emulsion: Soak eggs with 1:100 dilution of saturated mercuric chloride solution for 1 min. Crack eggs. Separate yolks from whites for 11 eggs. Mix egg yolks with 1 chicken egg.

Antibiotic Solution:

Composition per 10.0mL:

Cycloserine	0.5g
Cefoxitin	0.016g

Preparation of Antibiotic Solution: Add components to distilled/deionized water and bring volume to 10.0mL. Mix thoroughly. Filter sterilize.

Neutral Red Solution:

Composition per 10.0mL:

Neutral Red	0.1g
Ethanol	10.0mL

Preparation of Neutral Red Solution: Add Neutral Red to 10.0mL of ethanol. Mix thoroughly.

Hemin Solution:

Composition per 100.0mL:

Hemin	0.5g
NaOH (1*N* solution)	10.0mL

Preparation of Hemin Solution: Add hemin to 10.0mL of 1*N* NaOH solution. Mix thoroughly. Bring volume to 100.0mL with distilled/deionized water.

Preparation of Medium: Add components, except egg yolk emulsion and antibiotic solution, to distilled/deionized water and bring volume to 890.0mL. Mix thoroughly. Gently heat and bring to boiling. Autoclave for 15 min at 15 psi pressure–121°C. Cool to 45°–50°C. Aseptically add sterile egg yolk emulsion and antibiotic solution. Mix thoroughly. Pour into sterile Petri dishes.

Use: For the selective isolation and cultivation of *Clostridium difficile* from feces.

Cycloserine Cefoxitin Fructose Agar

See: ***Clostridium difficile* Agar**

CYE-ACES Agar

See: **CYE Agar, Buffered**

CYE-ACES Agar (DSMZ Medium 585)

Composition per liter:

Solution A	490.0mL
Solution B	490.0mL
Soluiton C	10.0mL
Solution D	10.0mL

pH 6.9 ± 0.1 at 25°C

Solution A:

Composition per 490mL:

Yeast extract	10.0g
ACES (*N*-2-acetamido-2-aminoethane-sulfonic acid)	10.0g
Activated charcoal	2.0g

Preparation of Solution A: Add components to distilled/deionized water and bring volume to 490.0mL. Mix thoroughly. Gently heat and bring to boiling. Autoclave for 15 min at 15 psi pressure–121°C. Cool to 50°C.

Solution B:

Composition per 490mL:

Agar	15.0g

Preparation of Solution B: Add agar to distilled/deionized water and bring volume to 490.0mL. Mix thoroughly. Gently heat and bring to boiling. Autoclave for 15 min at 15 psi pressure–121°C. Cool to 50°C.

Solution C:

Composition per 10.0mL:

L-Cysteine·HCl·H_2O	0.4g

Preparation of Solution C: Add L-cysteine·HCl·H_2O to distilled/deionized water and bring volume to 10.0mL. Mix thoroughly. Filter sterilize.

Solution D:

Composition per 10.0mL:

$Fe_4(PO_4)_2$	0.25g

Preparation of Solution D: Add $Fe_4(PO_4)_2$ to distilled/deionized water and bring volume to 10.0mL. Heat to 50–55°C to dissolve $Fe_4(PO_4)_2$. Mix thoroughly. Filter sterilize. Store in the dark. Do not use if chemical loses its green color and becomes brown or yellow.

Preparation of Medium: Add 10.0mL solution C and then 10.0mL solution D to 490.0mL solution A. Adjust the pH 6.9 ± 0.05 at 50°C by adding 4.0 to 4.5 mL of sterile 1.0*N* KOH. The pH of the medium is critical. Finally, add 490.0mL solution B. Mix thoroughly. Swirl medium in flask during dispensing to Petri dishes or tubes to keep charcoal suspended. Pour into Petri dishes or aseptically distribute into sterile tubes.

Use: For the cultivation of *Afipia clevelandensis, Afipia broomeae, Afipia felis, Legionella pneumophila, Legionella longbeachae,* and *Xylella fastidiosa.*

CYE Agar (Charcoal Yeast Extract Agar)

Composition per liter:

Agar	17.0g
Yeast extract	10.0g
Charcoal, activated, acid-washed	2.0g
L-Cysteine·HCl·H_2O solution	10.0mL
$Fe_4(P_2O_7)_3$ solution	10.0mL

pH 6.9 ± .05 at 50°C

L-Cysteine·HCl·H_2O Solution:

Composition per 10.0mL:

L-Cysteine·HCl·H_2O	0.4g

Preparation of L-Cysteine·HCl·H_2O solution: Add L-cysteine·HCl·H_2O to distilled/deionized water and bring volume to 10.0mL. Mix thoroughly. Filter sterilize.

$Fe_4(P_2O_7)_3$ Solution:

Composition per liter:

$Fe_4(P_2O_7)_3$	0.25g

Preparation of $Fe_4(P_2O_7)_3$ Solution: Add soluble $Fe_4(P_2O_7)_3$ to distilled/deionized water and bring volume to 10.0mL. Mix thoroughly. Filter sterilize. The soluble $Fe_4(P_2O_7)_3$ must be kept dry and in the dark. Do not use if brown or yellow. Prepare solutions freshly. Do not heat over 60°C to dissolve. The mixture dissolves readily in a 50°C water bath.

Preparation of Medium: Add components, except L-cysteine·HCl·H_2O solution and $Fe_4(P_2O_7)_3$ solution, to distilled/deionized water and bring volume to 980.0mL. Mix thoroughly. Gently heat to boiling. Autoclave for 15 min at 15 psi pressure–121°C. Cool to 50°C. Add 10.0mL of sterile L-cysteine·HCl·H_2O solution and 10.0mL of sterile $Fe_4(P_2O_7)_3$ solution. Adjust pH to 6.9 at 50°C by adding 4.0–4.5mL of 1.0*N* KOH. This is a critical step. Mix thoroughly. Pour in 20.0mL volumes into sterile Petri dishes. Swirl medium while pouring to keep charcoal in suspension.

Use: For the cultivation and maintenance of *Legionella* species and *Tatlockia micdadei.*

CYE Agar, Buffered (Charcoal Yeast Extract Agar, Buffered)

Composition per liter:

Agar	17.0g
ACES buffer (*N*-2-acetamido-2-aminoethane sulfonic acid)	10.0g
Yeast extract	10.0g
Charcoal, activated, acid-washed	2.0g
L-Cysteine·HCl·H_2O solution	10.0mL
$Fe_4(P_2O_7)_3$ solution	10.0mL

pH 6.9 ± .05 at 50°C

L-Cysteine·HCl·H_2O Solution:

Composition per 10.0mL:

L-Cysteine·HCl·H_2O	0.4g

Preparation of L-Cysteine·HCl·H_2O Solution: Add L-cysteine·HCl·H_2O to distilled/deionized water and bring volume to 10.0mL. Mix thoroughly. Filter sterilize.

$Fe_4(P_2O_7)_3$ Solution:

Composition per liter:

$Fe_4(P_2O_7)_3$	0.25g

Preparation of $Fe_4(P_2O_7)_3$ Solution: Add soluble $Fe_4(P_2O_7)_3$ to distilled/deionized water and bring volume to 10.0mL. Mix thoroughly. Filter sterilize. The soluble $Fe_4(P_2O_7)_3$ must be kept dry and in the dark. Do not use if brown or yellow. Prepare solutions freshly. Do not heat over 60°C to dissolve. The mixture dissolves readily in a 50°C water bath.

Preparation of Medium: Add components, except L-cysteine·HCl·H_2O solution and $Fe_4(P_2O_7)_3$ solution, to distilled/deionized water and bring volume to 980.0mL. Mix thoroughly. Gently heat to boiling. Autoclave for 15 min at 15 psi pressure–121°C. Cool to 50°C. Add 10.0mL of sterile L-cysteine·HCl·H_2O solution and 10.0mL of sterile $Fe_4(P_2O_7)_3$ solution. Adjust pH to 6.9 at 50°C by adding 4.0–4.5mL of 1.0 *N* KOH. This is a critical step. Mix thoroughly. Pour in 20.0mL volumes into sterile Petri dishes. Swirl medium while pouring to keep charcoal in suspension.

Use: For the cultivation and maintenance of *Legionella* species and *Xylella fastidiosa.*

CYE Broth
See: **Casamino Acids Yeast Extract Broth**

CYE DBCM
See: ***Legionella pneumophila* Medium**
Charcoal Yeast Extract Diphasic Blood Culture Medium

CYG Agar
See: **Casein Yeast Extract Glucose Agar**

CYG Agar

Composition per liter:

Agar	15.0g
Pancreatic digest of casein	3.0g
$CaCl_2 \cdot 2H_2O$	1.0g
Yeast extract	1.0g
Cyanocobalamin	0.5mg
Glucose solution	100.0mL

pH 7.2 ± 0.2 at 25°C

Glucose Solution:

Composition per 100.0mL:

D-Glucose	5.0g

Preparation of Glucose Solution: Add D-glucose to distilled/deionized water and bring volume to 100.0mL. Mix thoroughly. Autoclave for 15 min at 15 psi pressure–121°C. Cool to 25°C.

Preparation of Medium: Add components, except glucose solution, to distilled/deionized water and bring volume to 900.0mL. Mix thoroughly. Gently heat and bring to boiling. Autoclave for 15 min at 15 psi pressure–121°C. Cool to 45°–50°C. Aseptically add sterile glucose solution. Mix thoroughly. Pour into sterile Petri dishes or distribute into sterile tubes.

Use: For the isolation and cultivation of *Cytophaga* species, *Herpetosiphon* species, *Saprospira* species, and *Flexithrix* species.

Cylindrocladium Isolation Medium

Composition per liter:

Agar	20.0g
Glucose	15.0g
KH_2PO_4	1.0g
KNO_3	0.5g
$MgSO_4 \cdot 7H_2O$	0.5g
Yeast extract	0.5g
Chloramphenicol solution	10.0mL
Chlortetracycline solution	10.0mL
Thiabendazole solution	10.0mL
Tergitol NPX® (Union Carbide)	1.0mL

Chloramphenicol Solution:

Composition per 10.0mL:

Chloramphenicol	0.1g
Ethanol (95% solution)	10.0mL

Preparation of Chloramphenicol Solution: Add chloramphenicol to 10.0mL of ethanol. Mix thoroughly. Filter sterilize.

Chlortetracycline Solution:

Composition per 10.0mL:

Chlortetracycline	0.04g
Ethanol, absolute	5.0mL

Preparation of Chlortetracycline Solution: Add chlortetracycline to 5.0mL of ethanol. Mix thoroughly. Bring volume to 10.0mL with distilled/deionized water. Filter sterilize.

Thiabendazole Solution:

Composition per 10.0mL:

Thiabendazole	1.0mg

Preparation of Thiabendazole Solution: Add thiabendazole to distilled/deionized water and bring volume to 10.0mL. Mix thoroughly. Filter sterilize.

Preparation of Medium: Filter sterilize tergitol NPX. Add components—except tergitol NPX, thiabendazole solution, chloramphenicol solution, and chlortetracycline solution—to distilled/deionized water and bring volume to 969.0mL. Mix thoroughly. Gently heat and bring to boiling. Autoclave for 15 min at 15 psi pressure–121°C. Cool to 45°–50°C. Aseptically add sterile tergitol NPX, thiabendazole solution, chloramphenicol solution, and chlortetracycline solution. Mix thoroughly. Pour into sterile Petri dishes or distribute into sterile tubes.

Use: For the isolation and cultivation of *Cylindrocladium* species.

CYM Agar

Composition per liter:

Agar	20.0g
Peptone	2.0g
Glucose	2.0g
Yeast extract	2.0g
K_2HPO_4	1.0g
$MgSO_4$	0.5g
KH_2PO_4	0.46g

Preparation of Medium: Add components to distilled/deionized water and bring volume to 1.0L. Mix thoroughly. Gently heat and bring to boiling. Distribute into tubes or flasks. Autoclave for 15 min at 15 psi pressure–121°C. Pour into sterile Petri dishes or leave in tubes.

Use: For the cultivation of *Sordaria brevicollis.*

CYM Medium

Composition per liter:

Agar	20.0g
Glucose	20.0g
Peptone	2.0g
Yeast extract	2.0g
K_2HPO_4	1.0g
$MgSO_4$	0.5g
KH_2PO_4	0.46g

Preparation of Medium: Add components to distilled/deionized water and bring volume to 1.0L. Mix thoroughly. Gently heat and bring to boiling. Distribute into tubes or flasks. Autoclave for 15 min at 15 psi pressure–121°C. Pour into sterile Petri dishes or leave in tubes.

Use: For the cultivation and maintenance of most *Agaricus* species, *Kretzchmaria clavus, Phellinus igniarius, Phellinus nigricans, Phlebia chrysocrea, Phlebia livida, Tricholoma bakamatsutake, Tricholoma fulvocastaneum, Tricholoma matsutake,* and *Tricholoma ponderosum.*

CYS Medium (DSMZ Medium 1108)

Composition per liter:

Gelrite	8.0g
NZ Case (Wako pure chemicals, Japan)	3.0g
NaCl	3.0g
Yeast extract	2.0g

Soluble starch	1.0g
$MgCl_2$	0.125g
$CaCl_2$	0.025g
$FeSO_4 \cdot 7H_2O$	0.01g
Solution A	0.1mL
Solution B	0.1mL
Solution C	0.1mL
Solution D	0.1mL
Solution E	0.1mL
Solution F	0.1mL
Solution G	0.1mL

pH 7.5 ± 0.2 at 25°C

Solution A:

Composition per 100.0mL:

$Na_2MoO_4 \cdot 4H_2O$	1.2g

Preparation of Solution A: Add $Na_2MoO_4 \cdot 4H_2O$ to 100.0mL of distilled/deionized water. Mix thoroughly.

Solution B:

Composition per 100.0mL:

$VOSO_4 \cdot 2H_2O$	0.1mg

Preparation of Solution B: Add $VOSO_4 \cdot 2H_2O$ to 100.0mL of distilled/deionized water. Mix thoroughly.

Solution C:

Composition per 100.0mL:

$MnCl_2 \cdot 4H_2O$	0.5g

Preparation of Solution C: Add $MnCl_2 \cdot 4H_2O$ to 100.0mL of distilled/deionized water. Mix thoroughly.

Solution D:

Composition per 100.0mL:

$ZnSO_4 \cdot 7H_2O$	0.06g

Preparation of Solution D: Add $ZnSO_4 \cdot 7H_2O$ to 100.0mL of distilled/deionized water. Mix thoroughly.

Solution E:

Composition per 100.0mL:

$CuSO_4 \cdot 5H_2O$	0.015g

Preparation of Solution E: Add $CuSO_4 \cdot 5H_2O$ to 100.0mL of distilled/deionized water. Mix thoroughly.

Solution F:

Composition per 100.0mL:

$CoCl_2 \cdot 6H_2O$	0.8g

Preparation of Solution F: Add $CoCl_2 \cdot 6H_2O$ to 100.0mL of distilled/deionized water. Mix thoroughly.

Solution G:

Composition per 100.0mL:

$NiCl_2 \cdot 6H_2O$	0.02g

Preparation of Solution G: Add $NiCl_2 \cdot 6H_2O$ to 100.0mL of distilled/deionized water. Mix thoroughly.

Preparation of Medium: Add components, except solutions A–G, to distilled/deionized water and bring volume to 900.0mL. Mix thoroughly. Adjust pH to 7.5. Individually and in order add solutions A–G. After the addition of each solution mix thoroughly. Bring final volume to 1.0L with distilled/deionized water. Autoclave for 15 min at 15 psi pressure–121°C.

Use: For the cultivation of *Saccharomyces cerevisiae.*

Cystine Heart Agar

Composition per liter:

Beef heart, solids from infusion	500.0g
Agar	15.0g
Glucose	10.0g
Proteose peptone	10.0g
NaCl	5.0g
L-Cystine	1.0g
Hemoglobin solution	100.0mL

pH 6.8 ± 0.2 at 25°C

Source: This medium is available as a premixed powder from BD Diagnostic Systems.

Hemoglobin Solution:

Composition per 100.0mL:

Hemoglobin	2.0g

Preparation of Hemoglobin Solution: Add hemoglobin to cold distilled/deionized water and bring volume to 100.0mL. Mix thoroughly by shaking for 10–15 min. Autoclave for 15 min at 15 psi pressure–121°C. Cool to 50°–60°C.

Preparation of Medium: Add components, except hemoglobin solution, to distilled/deionized water and bring volume to 900.0mL. Mix thoroughly. Gently heat until boiling. Autoclave for 15 min at 15 psi pressure–121°C. Cool to 50–60°C. Aseptically add 100.0mL of sterile cooled hemoglobin solution. Mix thoroughly. Pour into sterile Petri dishes or distribute into sterile tubes.

Use: For the cultivation and maintenance of *Francisella tularensis* and *Francisella philomiragia.* Without the hemoglobin enrichment, it supports excellent growth of Gram-negative cocci and other pathogenic microorganisms.

Cystine Heart Agar with Rabbit Blood

Composition per liter:

Beef heart, solids from infusion	500.0g
Agar	15.0g
Glucose	10.0g
Proteose peptone	10.0g
NaCl	5.0g
L-Cystine	1.0g
Rabbit blood, defibrinated	50.0mL

pH 6.8 ± 0.2 at 25°C

Source: This medium is available as a premixed powder from BD Diagnostic Systems.

Preparation of Medium: Add components, except rabbit blood, to distilled/deionized water and bring volume to 950.0mL. Mix thoroughly. Gently heat until boiling. Autoclave for 15 min at 15 psi pressure–121°C. Cool to 50°–60°C. Aseptically add 50.0mL of sterile, defibrinated rabbit blood. Mix thoroughly. Pour into sterile Petri dishes or distribute into sterile tubes.

Use: For the cultivation and maintenance of *Francisella tularensis* and *Francisella philomiragia.* Without the hemoglobin enrichment, it supports excellent growth of Gram-negative cocci and other pathogenic microorganisms.

Cystine HiVeg Agar Base with Hemoglobin

Composition per liter:

Agar	15.0g
Plant infusion	10.0g
Plant peptone No. 3	10.0g

Glucose 10.0g
NaCl 5.0g
L-Cystine 1.0g
Hemoglobin solution 100.0mL

pH 6.8 ± 0.2 at 25°C

Source: This medium, wihout hemoglobin, is available as a premixed powder from HiMedia.

Hemoglobin Solution:
Composition per 100.0mL:
Bovine hemoglobin 2.0g

Preparation of Hemoglobin Solution: Add bovine hemoglobin to distilled/deionized water and bring volume to 100.0mL. Mix thoroughly. Autoclave for 15 min at 15 psi pressure–121°C. Cool to 45°–50°C.

Preparation of Medium: Add components, except hemoglobin solution, to distilled/deionized water and bring volume to 900.0mL. Mix thoroughly. Gently heat until boiling. Autoclave for 15 min at 15 psi pressure–121°C. Cool to 45°–50°C. Add 100.0mL sterile hemoglobin solution. Mix thoroughly. Pour into sterile Petri dishes or distribute into sterile tubes.

Use: For the cultivation of Gram-negative cocci and other fastidious pathogens. For the cultivation of *Francicella tularensis*.

Cystine Lactose Electrolyte Deficient Agar
See: **CLED Agar**

Cystine Lactose Electrolyte Deficient Agar with Andrade Indicator
See: **CLED Agar with Andrade Indicator**

Cystine Tellurite Blood Agar

Composition per liter:
Heart infusion agar 900.0mL
K_2TeO_3 solution 75.0mL
Rabbit blood 25.0mL
L-Cystine 22.0mg

pH 7.4 ± 0.2 at 25°C

Heart Infusion Agar:
Composition per 900.0mL:
Beef heart, solids from infusion 500.0g
Agar 20.0g
Tryptose 10.0g
Yeast extract 5.0g
NaCl 5.0g

Preparation of Heart Infusion Agar: Add components to distilled/deionized water and bring volume to 900.0mL. Mix thoroughly. Autoclave for 15 min at 15 psi pressure–121°C. Cool to 45°–50°C.

K_2TeO_3 Solution:
Composition per 100.0mL:
K_2TeO_3 .... 0.3g

Preparation of K_2TeO_3 Solution: Add K_2TeO_3 to distilled/deionized water and bring volume to 100.0mL. Mix thoroughly. Autoclave for 15 min at 15 psi pressure–121°C.

Caution: Potassium tellurite is toxic.

Preparation of Medium: Add sterile K_2TeO_3 solution, sterile rabbit blood, and sterile, solid L-cystine to sterile, cooled heart infusion agar. Mix thoroughly. Pour into sterile Petri dishes or distribute into sterile tubes.

Use: For the isolation, differentiation, and cultivation of *Corynebacterium diphtheriae*. *Corynebacterium diphtheriae* appears as dark gray to black colonies.

Cystine Tryptic Agar
See: **CTA Agar**

Cystine Trypticase™ Agar
See: **CTA Agar**

Cystine Trypticase™ Agar Medium
See: **CTA Medium**

Cystine Trypticase™ Agar Medium with Yeast Extract and Rabbit Serum
See: **CTA Medium with Yeast Extract and Rabbit Serum**

Cystine Tryptone Agar

Composition per liter:
Casein enzymatic hydrolysate 20.0g
NaCl 5.0g
Agar 2.5g
L-Cystine 0.5g
Na_2SO_3 .... 0.5g
Phenol Red 17.0mg

pH 7.0 ± 0.2 at 25°C

Source: This medium is available as a premixed powder from HiMedia.

Preparation of Medium: Add components to distilled/deionized water and bring volume to 1.0L. Mix thoroughly. Gently heat and bring to boiling. Distribute into tubes or flasks. Autoclave for 15 min at 15 psi pressure–121°C. Pour into sterile Petri dishes or leave in tubes.

Use: For the maintenance, subculturing, and detection of motility of various bacteria.

Cystine Tryptone Agar, HiVeg

Composition per liter:
Plant hydrolysate 20.0g
NaCl 5.0g
Agar 2.5g
L-Cystine 0.5g
Na_2SO_3 .... 0.5g
Phenol Red 17.0mg

pH 7.0 ± 0.2 at 25°C

Source: This medium is available as a premixed powder from HiMedia.

Preparation of Medium: Add components to distilled/deionized water and bring volume to 1.0L. Mix thoroughly. Gently heat and bring to boiling. Distribute into tubes or flasks. Autoclave for 15 min at 15 psi pressure–121°C. Pour into sterile Petri dishes or leave in tubes.

Use: For the maintenance, subculturing, and detection of motility of various bacteria.

Cystine Tryptone Agar, HiVeg with Carbohydrate

Composition per liter:
Plant hydrolysate 20.0g
NaCl 5.0g

Agar ... 2.5g
L-Cystine ... 0.5g
Na_2SO_3 ... 0.5g
Phenol Red ... 17.0mg
Carbohydrate solution ... 50.0mL
pH 7.3 ± 0.2 at 25°C

Source: This medium, without carbohydrate solution, is available as a premixed powder from HiMedia.

Carbohydrate Solution:
Composition per 100.0mL:
Carbohydrate ... 10.0g

Preparation of Carbohydrate Solution: Add carbohydrate to distilled/deionized water and bring volume to 100.0mL. Adonitol, arabinose, cellobiose, glucose, dulcitol, fructose, galactose, inositol, lactose, maltose, mannitol, raffinose, rhamnose, salicin, sorbitol, sucrose, trehalose, xylose, or other carbohydrates may be used. Mix thoroughly. Filter sterilize.

Preparation of Medium: Add components, except carbohydrate solution, to distilled/deionized water and bring volume to 1.0L. Mix thoroughly. Gently heat and bring to boiling. Distribute into tubes or flasks. Autoclave for 15 min at 15 psi pressure–121°C. Cool to 50°C. Aseptically add 50.0mL of sterile carbohydrate solution. Mix thoroughly. Pour into sterile Petri dishes or leave in tubes.

Use: For fermentation studies of various bacteria.

CYT Agar

Composition per liter:
Agar ... 15.0g
Pancreatic digest of casein ... 1.0g
$CaCl_2 \cdot 2H_2O$ ... 0.5g
$MgSO_4 \cdot 7H_2O$ ... 0.5g
Yeast extract ... 0.5g
pH 7.2 ± 0.2 at 25°C

Preparation of Medium: Add components to distilled/deionized water and bring volume to 1.0L. Mix thoroughly. Gently heat and bring to boiling. Distribute into tubes or flasks. Autoclave for 15 min at 15 psi pressure–121°C. Pour into sterile Petri dishes or leave in tubes.

Use: For the isolation and cultivation of *Cytophaga* species, *Herpetosiphon* species, *Saprospira* species, and *Flexithrix* species.

Cytophaga Agarase Agar (ATCC Medium 793)

Composition per liter:
Agar ... 15.0g
KH_2PO_4 ... 1.0g
$MgSO_4 \cdot 7H_2O$ ... 0.5g
NH_4Cl ... 0.5g
$CaCl_2 \cdot H_2O$ ... 0.02g
Vishniac and Santer trace element mixture ... 0.2mL
pH 7.2 ± 0.2 at 25°C

Vishniac and Santer Trace Element Mixture:
Composition per liter:
Ethylenediamine tetraacetic acid (EDTA) ... 50.0g
$ZnSO_4 \cdot 7H_2O$ ... 22.0g
$CaCl_2$ ... 5.54g
$MnCl_2 \cdot 4H_2O$ ... 5.06g
$FeSO_4 \cdot 7H_2O$ ... 4.99g
$CoCl_2 \cdot 6H_2O$ ... 1.61g
$CuSO_4 \cdot 5H_2O$ ... 1.57g
$(NH_4)_6Mo_7O_{24} \cdot 4H_2O$ ... 1.1g

Preparation of Vishniac and Santer Trace Element Mixture: Add components to distilled/deionized water and bring volume to 1.0L. Adjust pH to 6.0 with KOH. Mix thoroughly.

Preparation of Medium: Add components to distilled/deionized water and bring volume to 1.0L. Adjust pH to 7.2. Mix thoroughly. Distribute into tubes or flasks. Autoclave for 15 min at 15 psi pressure–121°C. Pour into sterile Petri dishes or leave in tubes.

Use: For the cultivation and maintenance of *Cytophaga flevensis*.

Cytophaga Agarase Broth

Composition per liter:
Agar ... 1.0g
KH_2PO_4 ... 1.0g
$MgSO_4 \cdot 7H_2O$ ... 0.5g
NH_4Cl ... 0.5g
$CaCl_2 \cdot H2O$ ... 0.02g
Vishniac and Santer trace element mixture ... 0.2mL
pH 7.2 ± 0.2 at 25°C

Vishniac and Santer Trace Element Mixture:
Composition per liter:
Ethylenediamine tetraacetic acid (EDTA) ... 50.0g
$ZnSO_4 \cdot 7H_2O$ ... 22.0g
$CaCl_2$ ... 5.54g
$MnCl_2 \cdot 4H_2O$ ... 5.06g
$FeSO_4 \cdot 7H_2O$ ... 4.99g
$CoCl_2 \cdot 6H_2O$ ... 1.61g
$CuSO_4 \cdot 5H_2O$ ... 1.57g
$(NH_4)_6Mo_7O_{24} \cdot 4H_2O$ ... 1.1g

Preparation of Vishniac and Santer Trace Element Mixture: Add components to distilled/deionized water and bring volume to 1.0L. Adjust pH to 6.0 with KOH. Mix thoroughly.

Preparation of Medium: Add components to distilled/deionized water and bring volume to 1.0L. Adjust pH to 7.2. Mix thoroughly. Distribute into tubes or flasks. Autoclave for 15 min at 15 psi pressure–121°C.

Use: For the cultivation of *Cytophaga flevensis*.

Cytophaga agarovorans Agar (LMG Medium 99)

Composition per 1001.0mL:
NaCl ... 30.0g
Agar ... 15.0g
KH_2PO_4 ... 1.0g
$MgSO_4 \cdot 7H_2O$ ... 1.0g
NH_4Cl ... 1.0g
Yeast extract ... 1.0g
$CaCl_2 \cdot 2H_2O$ ... 50.0mg
$FeCl_3 \cdot H_2O$ ... 1.25mg
Glucose solution ... 10.0mL
$NaHCO_3$ solution ... 10.0mL
$Na_2S \cdot 9H_2O$ solution ... 1.0mL

Glucose Solution:
Composition per 10.0mL:
D-Glucose ... 1.0g

Preparation of Glucose Solution: Add glucose to distilled/deionized water and bring volume to 10.0mL. Mix thoroughly. Filter sterilize.

$NaHCO_3$ Solution:
Composition per 100.0mL:

$NaHCO_3$	5.0g

Preparation of $NaHCO_3$ Solution: Add $NaHCO_3$ to distilled/deionized water and bring volume to 100.0mL. Mix thoroughly. Filter sterilize.

$Na_2S·9H_2O$ Solution:
Composition per 10.0mL:

$Na_2S·9H_2O$	1.0g

Preparation of $Na_2S·9H_2O$ Solution: Add $Na_2S·9H_2O$ to distilled/deionized water and bring volume to 10.0mL. Filter sterilize.

Preparation of Medium: Add components, except glucose solution, $NaHCO_3$ solution, and $Na_2S·9H_2O$ solution, to distilled/deionized water and bring volume to 980.0mL. Mix thoroughly. Gently heat and bring to boiling. Autoclave for 15 min at 15 psi pressure–121°C. Cool to 50°–55°C. Aseptically add 10.0mL of sterile glucose solution, 10.0mL of sterile $NaHCO_3$ solution, and 1.0mL of sterile $Na_2S·9H_2O$ solution. Mix thoroughly. Pour into sterile Petri dishes or distribute into sterile tubes.

Use: For the cultivation of *Cytophaga agarovorans.*

Cytophaga agarovorans Broth (LMG Medium 99)

Composition per 1001.0mL:

NaCl	30.0g
KH_2PO_4	1.0g
$MgSO_4·7H_2O$	1.0g
NH_4Cl	1.0g
Yeast extract	1.0g
Agar	1.0g
$CaCl_2·2H_2O$	50.0mg
$FeCl_3·H_2O$	1.25mg
Glucose solution	10.0mL
$NaHCO_3$ solution	10.0mL
$Na_2S·9H_2O$ solution	1.0mL

Glucose Solution:
Composition per 10.0mL:

D-Glucose	1.0g

Preparation of Glucose Solution: Add glucose to distilled/deionized water and bring volume to 10.0mL. Mix thoroughly. Filter sterilize.

$NaHCO_3$ Solution:
Composition per 100.0mL:

$NaHCO_3$	5.0g

Preparation of $NaHCO_3$ Solution: Add $NaHCO_3$ to distilled/deionized water and bring volume to 100.0mL. Mix thoroughly. Filter sterilize.

$Na_2S·9H_2O$ Solution:
Composition per 10.0mL:

$Na_2S·9H_2O$	1.0g

Preparation of $Na_2S·9H_2O$ Solution: Add $Na_2S·9H_2O$ to distilled/deionized water and bring volume to 10.0mL. Filter sterilize.

Preparation of Medium: Add components, except glucose solution, $NaHCO_3$ solution, and $Na_2S·9H_2O$ solution, to distilled/deionized water and bring volume to 980.0mL. Add 1.0g of agar as a detoxifying agent. Mix thoroughly. Autoclave for 15 min at 15 psi pressure–121°C. Aseptically add 10.0mL of sterile glucose solution, 10.0mL of sterile $NaHCO_3$ solution, and 1.0mL of sterile $Na_2S·9H_2O$ solution. Mix thoroughly. Aseptically distribute into sterile tubes or flasks.

Use: For the cultivation and maintenance of *Cytophaga agarovorans.*

Cytophaga fermentans Medium

Composition per liter:

NaCl	30.0g
Agar	5.0g
$NaHCO_3$	5.0g
KH_2PO_4	1.0g
NH_4Cl	1.0g
$MgCl_2·6H_2O$	0.5g
Yeast extract	0.3g
$Na_2S·9H_2O$	0.1g
$CaCl_2$	0.04g
Ferric citrate (4m*M* solution)	5.0mL
Trace elements solution	2.0mL

pH 7.0 ± 0.2 at 25°C

Trace Elements Solution:
Composition per 100.0mL:

H_3BO_3	0.28g
$MnSO_4·6H_2O$	0.21g
$Na_2MoO_4·2H_2O$	0.075g
$Zn(NO_3)_2·6H_2O$	0.025g
$CoCl_2·6H_2O$	0.02g
$Cu(NO_3)_2·3H_2O$	0.02g

Preparation of Trace Elements Solution: Add components to distilled/deionized water and bring volume to 100.0mL. Mix thoroughly.

Preparation of Medium: Add components to distilled/deionized water and bring volume to 1.0L. Mix thoroughly. Gently heat and bring to boiling. Distribute into tubes or flasks. Autoclave for 15 min at 15 psi pressure–121°C.

Use: For the cultivation of agar-digesting *Cytophaga fermentans.*

Cytophaga hutchinsonii Agar

Composition per liter:

Agar	15.0g
Pancreatic digest of casein	3.0g
$CaCl_2·2H_2O$	1.36g
Yeast extract	1.0g
Cellobiose solution	50.0mL

pH 7.2 ± 0.2 at 25°C

Cellobiose Solution:
Composition per 50.0mL:

D-Cellobiose	6.0g

Preparation of Cellobiose Solution: Add cellobiose to distilled/deionized water and bring volume to 50.0mL. Mix thoroughly. Filter sterilize.

Preparation of Medium: Add components, except cellobiose solution, to distilled/deionized water and bring volume to 950.0mL. Mix thoroughly. Gently heat and bring to boiling. Autoclave for 15 min at 15 psi pressure–121°C. Cool to 50°–55°C. Aseptically add 50.0mL of sterile cellobiose solution. Mix thoroughly. Pour into sterile Petri dishes or distribute into sterile tubes.

Use: For the cultivation and maintenance of *Cytophaga aurantiaca* and *Cytophaga hutchinsonii.*

Cytophaga hutchinsonii Broth

Composition per liter:

Pancreatic digest of casein	3.0g
$CaCl_2 \cdot 2H_2O$	1.36g
Yeast extract	1.0g
Filter paper strips	variable

pH 7.2 ± 0.2 at 25°C

Preparation of Medium: Add components, except filter paper strips, to distilled/deionized water and bring volume to 1.0L. Mix thoroughly. Gently heat and bring to boiling. Distribute in 5.0mL volumes into tubes. Add a strip of filter paper about 7.0cm in length to each tube. Autoclave for 15 min at 15 psi pressure–121°C.

Use: For the cultivation of *Cytophaga aurantiaca* and *Cytophaga hutchinsonii.*

Cytophaga Marine Medium (DSMZ Medium 172)

Composition per liter:

NaCl	24.7g
Agar	15.0g
$MgSO_4 \cdot 7H_2O$	6.3g
$MgCl_2 \cdot 6H_2O$	4.6g
$CaCl_2 \cdot 2H_2O$	1.2g
Yeast extract	1.0g
Tryptone	1.0g
KCl	0.7g
Sodium bicarbonate solution	10.0mL
Calcium chloride solution	10.0mL

pH 7.0 ± 0.2 at 25°C

Sodium Bicarbonate Solution:

Composition per 10.0mL:

$NaHCO_3$	0.2g

Preparation of Sodium Bicarbonate Solution: Add $NaHCO_3$ to distilled/deionized water and bring volume to 10.0mL. Mix thoroughly. Autoclave for 15 min at 15 psi pressure–121°C.

Calcium Chloride Solution:

Composition per 10.0mL:

$CaCl_2 \cdot 2H_2O$	1.2g

Preparation of Calcium Chloride Solution: Add $CaCl_2 \cdot 2H_2O$ to distilled/deionized water and bring volume to 10.0mL. Mix thoroughly. Autoclave for 15 min at 15 psi pressure–121°C.

Preparation of Medium: Add components, except sodium bicarbonate and calcium chloride solution, to distilled/deionized water and bring volume to 900.0mL. Mix thoroughly. Gently heat and bring to boiling. Autoclave for 15 min at 15 psi pressure–121°C. Cool to 45–50°C. Aseptically add 10.0mL of sterile bicarbonate solution and 10.0mL sterile calcium chloride solution. Adjust pH to 7.2. Mix thoroughly. Pour into sterile Petri dishes or distribute into sterile tubes.

Use: For the cultivation and maintenance of *Cellulophaga lytica, Cytophaga latercula, Marinilabilia salmonicolor, Saprospira grandis, Cytophaga marinoflava, Persicobacter diffluens, Flammeovirga aprica, Flexibacter tractuosus, Microscilla spp., Marinilabilia salmonicolor, Flexibacter litoralis, Flexithrix dorotheae,* and *Cellulophaga lytica.*

Cytophaga Medium

Composition per liter:

NaCl	30.0g
Agar	15.0g
KH_2PO_4	1.0g
NH_4Cl	1.0g
Yeast extract	1.0g
$MgSO_4$	0.5g
$CaCl_2$	0.04g
$FeCl_3 \cdot 6H_2O$	1.25mg
$NaHCO_3$ solution	100.0mL
Glucose solution	10.0mL
$Na_2S \cdot 9H_2O$ solution	1.0mL

Glucose Solution:

Composition per 100.0mL:

Glucose	10.0g

Preparation of Glucose Solution: Add glucose to distilled/deionized water and bring volume to 100.0mL. Mix thoroughly. Autoclave for 15 min at 15 psi pressure–121°C.

$NaHCO_3$ Solution:

Composition per 100.0mL:

$NaHCO_3$	5.0g

Preparation of $NaHCO_3$ Solution: Add $NaHCO_3$ to distilled/deionized water and bring volume to 100.0mL. Mix thoroughly. Filter sterilize.

$Na_2S \cdot 9H_2O$ Solution:

Composition per 100.0mL:

$Na_2S \cdot 9H_2O$	10.0g

Preparation of $Na_2S \cdot 9H_2O$ Solution: Add $Na_2S \cdot 9H_2O$ to distilled/deionized water and bring volume to 100.0mL. Mix thoroughly. Autoclave for 15 min at 15 psi pressure–121°C.

Preparation of Medium: Add components, except $NaHCO_3$ solution, glucose solution, and $Na_2S \cdot 9H_2O$ solution, to distilled/deionized water and bring volume to 889.0mL. Autoclave for 15 min at 15 psi pressure–121°C. Cool to 50°C. Aseptically add sterile $NaHCO_3$ solution, sterile glucose solution, and sterile $Na_2S \cdot 9H_2O$ solution. Mix thoroughly. Pour into sterile Petri dishes or distribute into sterile tubes.

Use: For the cultivation of *Cytophaga agarovorans.*

Cytophaga Medium (ATCC Medium 420)

Composition per liter:

NaCl	20.0g
Yeast extract	10.0g
Agar	3.0g
$MgSO_4 \cdot 7H_2O$	1.0g
NH_4Cl	1.0g
K_2HPO_4	0.2g
$FeCl_3$	1.0μg

pH 7.5 ± 0.2 at 25°C

Preparation of Medium: Add components to tap water and bring volume to 1.0L. Adjust pH to 7.5. Mix thoroughly. Distribute into tubes or flasks. Autoclave for 15 min at 15 psi pressure–121°C.

Use: For the cultivation of *Cytophaga fermentans.*

Cytophaga Medium (ATCC Medium 1299)

Composition per liter:

Agar	11.0g
Pancreatic digest of casein	0.5g
Yeast extract	0.5g
Beef extract	0.2g
Sodium acetate	0.2g

pH 7.2 ± 0.2 at 25°C

Preparation of Medium: Add components to distilled/deionized water and bring volume to 1.0L. Mix thoroughly. Distribute into tubes or flasks. Autoclave for 15 min at 15 psi pressure–121°C. Pour into sterile Petri dishes or leave in tubes.

Use: For the cultivation and maintenance of *Cytophaga* species and *Flavobacterium branchiophilum.*

Cytophaga Medium, Modified

Composition per liter:

NaCl	20.0g
Agar	15.0g
Yeast extract	10.0g
$MgSO_4 \cdot 7H_2O$	1.0g
NH_4Cl	1.0g
K_2HPO_4	0.2g
$FeCl_3$	1.0μg

pH 7.5 ± 0.2 at 25°C

Preparation of Medium: Add components to tap water and bring volume to 1.0L. Adjust pH to 7.5. Mix thoroughly. Distribute into tubes or flasks. Autoclave for 15 min at 15 psi pressure–121°C.

Use: For the cultivation of *Cytophaga fermentans.*

Cytophaga Spirochete Medium

See: ***Cytophaga*** **Medium**
See: **Spirochete Medium**

Cytosine Nutrient Agar

Composition per liter:

Agar	15.0g
Pancreatic digest of gelatin	5.0g
Beef extract	3.0g
Cytosine	20.0mg

pH 6.8 ± 0.2 at 25°C

Preparation of Medium: Add components to distilled/deionized water and bring volume to 1.0L. Mix thoroughly. Gently heat and bring to boiling. Distribute into tubes or flasks. Autoclave for 15 min at 15 psi pressure–121°C. Pour into sterile Petri dishes or leave in tubes.

Use: For the cultivation of *Escherichia coli.*

Cystine Tellurite Blood Agar

Composition per 120.0mL:

Heart infusion agar	100.0mL
K_2TeO_3 solution	15.0mL
Sheep blood	5.0mL
L-Cystine	5.0mg

pH 7.4 ± 0.2 at 25°C

Heart Infusion Agar:

Composition per liter:

Beef heart, infusion from	500.0g
Agar	20.0g
Tryptose	10.0g
Yeast extract	5.0g
NaCl	5.0g

Preparation of Heart Infusion Agar: Add components to distilled/deionized water and bring volume to 1.0L. Mix thoroughly. Autoclave for 15 min at 15 psi pressure–121°C. Cool to 45°–50°C.

K_2TeO_3 Solution:

Composition per 100.0mL:

K_2TeO_3	0.3g

Preparation of K_2TeO_3 Solution: Add K_2TeO_3 to distilled/deionized water and bring volume to 100.0mL. Mix thoroughly. Autoclave for 15 min at 15 psi pressure–121°C.

Caution: Potassium tellurite is toxic.

Preparation of Medium: Add sterile K_2TeO_3 solution, sterile, defibrinated sheep blood, and sterile, solid L-cystine to sterile, cooled heart infusion agar. Mix thoroughly. Pour into sterile Petri dishes or distribute into sterile tubes.

Use: For the isolation, differentiation, and cultivation of *Corynebacterium diphtheriae. Corynebacterium diphtheriae* appears as dark gray to black colonies.

CYU 2%

Composition per liter:

Basal solution	960.0mL
Yeast extract solution	30.0mL
Trace metal solution	10.0mL

Basal Solution:

Composition per 960.0mL:

Glucose	20.0g
K_2HPO_4	9.2g
Urea	5.0g
$(NH_4)_2SO_4$	2.2g
KH_2PO_4	1.3g
Trisodium citrate·$2H_2O$	1.0g
$MgSO_4 \cdot 7H_2O$ solution	1.66mL

$MgSO_4 \cdot 7H_2O$ Solution:

Composition per 10.0mL:

$MgSO_4 \cdot 7H_2O$	2.5g

Preparation of $MgSO_4 \cdot 7H_2O$ Solution: Add $MgSO_4 \cdot 7H_2O$ to distilled/deionized water and bring volume to 10.0mL. Mix thoroughly.

Preparation of Basal Solution: Add components to distilled/deionized water and bring volume to 960.0mL. Mix thoroughly. Filter sterilize.

Yeast Extract Solution:

Composition per 100.0mL:

Yeast extract	10.0g

Preparation of Yeast Extract Solution: Add yeast extract to distilled/deionized water and bring volume to 100.0mL. Mix thoroughly. Autoclave for 15 min at 15 psi pressure–121°C. Cool to 25°C.

Trace Metal Solution:

Composition per liter:

$CaCl_2 \cdot 2H_2O$	3972.0mg
$FeSO_4 \cdot 7H_2O$	1250.0mg
H_3BO_3	1140.0mg
$ZnSO_4 \cdot 7H_2O$	882.0mg
$CuCl_2 \cdot 5H_2O$	157.0mg
$MnCl_2 \cdot 4H_2O$	140.0mg
$NaMoO_4 \cdot 2H_2O$	119.0mg
Vanadyl sulfate·$2H_2O$	100.0mg
$CsCl_2 \cdot 6H_2O$	49.0mg
$FeCl_3 \cdot 6H_2O$	29.0mg
$CdSO_4 \cdot 8H_2O$	10.0mg
$Ni(NO_3)_2 \cdot 6H_2O$	10.0mg
HCl, concentrated	10.0mL

Preparation of Trace Metal Solution: Add 10.0mL of concentrated HCl to 900.0mL of distilled/deionized water. Mix thoroughly. Add remaining components in the order shown. Mix thoroughly after adding each component. Bring volume to 990.0mL with distilled/deionized water. Mix thoroughly. Filter sterilize.

Preparation of Medium: Aseptically combine 960.0mL of sterile basal solution, 30.0mL of sterile yeast extract solution, and 30.0mL of sterile trace metal solution. Aseptically distribute into sterile tubes or flasks.

Use: For the cultivation of *Neospongicoccum excentricum.*

CZA

See: **Czapek Agar**

CZA200

See: **Czapek Agar with 20% Sucrose**

Czapek Agar (ATCC Medium 312)

Composition per liter:

Sucrose	30.0g
Agar	15.0g
$NaNO_3$	3.0g
K_2HPO_4	1.0g
KCl	0.5g
$MgSO_4 \cdot 7H_2O$	0.5g
$FeSO_4 \cdot 7H_2O$	0.01g

pH 7.3 ± 0.2 at 25°C

Preparation of Medium: Add components, except sucrose, to distilled/deionized water and bring volume to 900.0mL. Mix thoroughly. Distribute into tubes or flasks. In a separate flask, add sucrose to distilled/deionized water and bring volume to 100.0mL. Mix thoroughly. Autoclave both solutions separately for 15 min at 15 psi pressure–121°C. Cool to 50°C. Combine the sterile solutions. Mix thoroughly. Pour into sterile Petri dishes or distribute into sterile tubes.

Use: For the cultivation and maintenance of *Streptomyces* species. For the cultivation of Actinoplanaceae.

Czapek Agar

See: **Czapek Yeast Autolysate Agar**

Czapek Agar with Peptone (ATCC Medium 522)

Composition per liter:

Sucrose	30.0g
Agar	15.0g
Peptone	5.0g
$NaNO_3$	3.0g
K_2HPO_4	1.0g
KCl	0.5g
$MgSO_4 \cdot 7H_2O$	0.5g
$FeSO_4 \cdot 7H_2O$	0.01g

pH 7.3 ± 0.2 at 25°C

Preparation of Medium: Add components, except sucrose, to distilled/deionized water and bring volume to 900.0mL. Mix thoroughly. Distribute into tubes or flasks. In a separate flask, add sucrose to distilled/deionized water and bring volume to 100.0mL. Mix thoroughly. Autoclave both solutions separately for 15 min at 15 psi pressure–121°C. Cool to 50°C. Combine the sterile solutions. Mix thoroughly. Pour into sterile Petri dishes or distribute into sterile tubes.

Use: For the cultivation and maintenance of *Streptomyces* species. For the cultivation of Actinoplanaceae.

Czapek Agar with Sucrose

Composition per liter:

Sucrose	170.0g
Agar	15.0g
$NaNO_3$	3.0g
K_2HPO_4	1.0g
KCl	0.5g
$MgSO_4 \cdot 7H_2O$	0.5g
$FeSO_4 \cdot 7H_2O$	0.01g

pH 7.3 ± 0.2 at 25°C

Preparation of Medium: Add components, except sucrose, to distilled/deionized water and bring volume to 700.0mL. Mix thoroughly. In a separate flask, add sucrose to distilled/deionized water and bring volume to 300.0mL. Mix thoroughly. Autoclave both solutions separately for 15 min at 15 psi pressure–121°C. Cool to 45–50°C. Combine the sterile solutions. Mix thoroughly. Pour into sterile Petri dishes or distribute into sterile tubes.

Use: For the cultivation of *Aspergillus echinulatus, Aspergillus flavipes, Aspergillus penicilloides,* numerous *Eurotium* species, *Penicillium citreonigrum,* and *Penicillium thomii.*

Czapek Agar with 20% Sucrose (CZA200)

Composition per liter:

Sucrose	200.0g
Agar	15.0g
$NaNO_3$	3.0g
K_2HPO_4	1.0g
KCl	0.5g
$MgSO_4 \cdot 7H_20$	0.5g
$FeSO_4 \cdot 7H_2O$	0.01g

pH 7.3 ± 0.2 at 25°C

Preparation of Medium: Add components, except sucrose, to distilled/deionized water and bring volume to 700.0mL. Mix thoroughly. In a separate flask, add sucrose to distilled/deionized water and bring volume to 300.0mL. Mix thoroughly. Autoclave both solutions separately for 15 min at 15 psi pressure–121°C. Cool to 45–50°C. Combine the sterile solutions. Mix thoroughly. Pour into sterile Petri dishes or distribute into sterile tubes.

Use: For the cultivation of *Aspergillus repens (Eurotium repens).*

Czapek Dox Agar

Composition per liter:

Sucrose	30.0g
Agar	15.0g
$NaNO_3$	3.0g
K_2HPO_4	1.0g
$MgSO_4 \cdot 7H_2O$	0.5g
KCl	0.5g
$FeSO_4 \cdot 7H_2O$	0.01g

pH 7.3 ± 0.2 at 25°C

Preparation of Medium: Add components to distilled/deionized water and bring volume to 1.0L. Mix thoroughly. Distribute into tubes or flasks. Autoclave for 15 min at 15 psi pressure–121°C. Pour into sterile Petri dishes or leave in tubes.

Use: For the cultivation and maintenance of *Actinoplanes* species, *Amorphosporangium auranticolor, Ampullariella* species, *Spirillospora albida,* and *Streptomyces armeniacus.*

Czapek Dox Agar

Composition per liter:

Sucrose	30.0g
Agar	15.0g
$NaNO_3$	2.0g
K_2HPO_4	1.0g
$MgSO_4 \cdot 7H_2O$	0.5g
KCl	0.5g
$FeSO_4 \cdot 7H_2O$	0.01g

pH 7.3 ± 0.2 at 25°C

Source: This medium is available from HiMedia.

Preparation of Medium: Add components to distilled/deionized water and bring volume to 1.0L. Mix thoroughly. Distribute into tubes or flasks. Autoclave for 15 min at 15 psi pressure–121°C. Pour into sterile Petri dishes or leave in tubes.

Use: For the cultivation and maintenance of *Actinoplanes* species, *Amorphosporangium auranticolor, Ampullariella* species, *Spirillospora albida,* and *Streptomyces armeniacus.*

Czapek Dox Agar with 3% Glucose

Composition per liter:

Glucose	30.0g
Agar	15.0g
$NaNO_3$	3.0g
K_2HPO_4	1.0g
KCl	0.5g
$MgSO_4 \cdot 7H_2O$	0.5g
$FeSO_4 \cdot 7H_2O$	0.01g

pH 7.3 ± 0.2 at 25°C

Preparation of Medium: Add components to distilled/deionized water and bring volume to 1.0L. Mix thoroughly. Gently heat and bring to boiling. Distribute into tubes or flasks. Autoclave for 15 min at 15 psi pressure–121°C. Pour into sterile Petri dishes or leave in tubes.

Use: For the cultivation and maintenance of *Microbispora rosea* and *Streptomyces* species.

Czapek Dox Agar with 20% Sucrose

Composition per liter:

Sucrose	200.0g
Agar	20.0g
$NaNO_3$	2.0g
K_2HPO_4	1.0g
KCl	0.5g
$MgSO_4 \cdot 7H_2O$	0.5g
$FeSO_4$	0.01g
$ZnSO_4$	0.01g
$CuSO_4$	0.005g

Preparation of Medium: Add components to distilled/deionized water and bring volume to 1.0L. Mix thoroughly. Gently heat and bring to boiling. Distribute into tubes or flasks. Autoclave for 20 min at 15 psi pressure–121°C. Pour into sterile Petri dishes or leave in tubes.

Use: For the cultivation and maintenance of *Aspergillus brunneus, Aspergillus equitis, Aspergillus hollandicus, Aspergillus nidulellus, Aspergillus reptans*, and *Aspergillus rubrobrunneus.*

Czapek Dox Agar, Modified

Composition per liter:

Sucrose	30.0g
Agar	12.0g
$NaNO_3$	2.0g
Magnesium glycerophosphate	0.5g
KCl	0.5g
K_2SO_4	0.35g
$FeSO_4$	0.01g

pH 6.8 ± 0.2 at 25°C

Source: This medium is available as a premixed powder from Oxoid Unipath and HiMedia.

Preparation of Medium: Add components to distilled/deionized water and bring volume to 1.0L. Mix thoroughly. Distribute into tubes or flasks. Autoclave for 15 min at 15 psi pressure–121°C. Pour into sterile Petri dishes or leave in tubes.

Use: For the cultivation and maintenance of numerous fungal species. For chlamydospore production by *Candida albicans.*

Czapek Dox Broth

Composition per liter:

Sucrose	30.0g
$NaNO_3$	3.0g
K_2HPO_4	1.0g
$MgSO_4 \cdot 7H_2O$	0.5g
KCl	0.5g
$FeSO_4 \cdot 7H_2O$	0.01g

pH 7.3 ± 0.2 at 25°C

Source: This medium is available as a premixed powder from BD Diagnostic Systems.

Preparation of Medium: Add components to distilled/deionized water and bring volume to 1.0L. Mix thoroughly. Distribute into tubes or flasks. Autoclave for 15 min at 15 psi pressure–121°C.

Use: For the cultivation and maintenance of a variety of fungal and bacterial species that can use nitrate as sole nitrogen source.

Czapek Dox Liquid Medium, Modified

Composition per liter:

Sucrose	30.0g
$NaNO_3$	2.0g
Magnesium glycerophosphate	0.5g
KCl	0.5g
K_2SO_4	0.35g
$FeSO_4$	0.01g

pH 6.8 ± 0.2 at 25°C

Source: This medium is available as a premixed powder from Oxoid Unipath.

Preparation of Medium: Add components to distilled/deionized water and bring volume to 1.0L. Mix thoroughly. Distribute into tubes or flasks. Autoclave for 15 min at 15 psi pressure–121°C.

Use: For the cultivation of fungi and bacteria capable of utilizing sodium nitrate as the sole source of nitrogen.

Czapek Malt Agar

Composition per liter:

Malt extract	40.0g
Sucrose	30.0g
Agar	15.0g
KNO_3	2.0g
K_2HPO_4	1.0g
KCl	0.5g
$MgSO_4 \cdot 7H_2O$	0.5g
$FeSO_4 \cdot 7H2O$	0.1g

pH 6.8 ± 0.2 at 25°C

Source: This medium is available from HiMedia.

Preparation of Medium: Add components to distilled/deionized water and bring volume to 1.0L. Mix thoroughly. Gently heat and bring to boiling. Distribute into tubes or flasks. Autoclave for 15 min at 15 psi pressure–121°C. Pour into sterile Petri dishes or leave in tubes.

Use: For the isolation and cultivation of saprophytic fungi.

Czapek Peptone Agar

Composition per liter:

Sucrose	30.0g
Agar	15.0g
Peptone	5.0g
KNO_3	2.0g
Yeast extract	2.0g
K_2HPO_4	1.0g
KCl	0.5g
$MgSO_4 \cdot 7H_2O$	0.5g
$FeSO_4 \cdot 7H_2O$	0.01g

pH 7.3 ± 0.2 at 25°C

Preparation of Medium: Add components to distilled/deionized water and bring volume to 1.0L. Mix thoroughly. Gently heat and bring to boiling. Distribute into tubes or flasks. Autoclave for 15 min at 15 psi pressure–121°C. Pour into sterile Petri dishes or leave in tubes.

Use: For the cultivation of *Actinoplanes campanulatus, Actinoplanes digitatis, Actinoplanes italicus, Actinoplanes lobatus, Actinoplanes regularis, Actinoplanes utahensis, Ampullariella campanulata, Micromonospora brunnea, Micromonospora purpurea, Micromonospora purpureochromogenes, Micromonospora* species, *Microtetraspora glauca, Micromonospora chalcea, Nocardia brevicatena, Pilimelia anulata, Pilimelia terevasa, Planomonospora parontospora, Promicromonospora citrea, Saccharomonospora caesia, Saccharomonospora viridis, Spirillospora albida, Sporichthya polymorpha, Streptomyces yerevanensis, Thermoactinomyces thalpophilus, Thermoactinomyces vulgaris,* and *Thermomonospora chromogena.*

Czapek Peptone Yeast Agar

Composition per liter:

Sucrose	30.0g
Agar	15.0g
Peptone	5.0g
$NaNO_3$	3.0g
Yeast extract	2.0g
K_2HPO_4	1.0g
KCl	0.5g
$MgSO_4 \cdot 7H_20$	0.5g
$FeSO_4 \cdot 7H_2O$	0.01g

pH 7.3 ± 0.2 at 25°C

Preparation of Medium: Add components, except sucrose, to distilled/deionized water and bring volume to 900.0mL. Mix thoroughly. Distribute into tubes or flasks. In a separate flask, add sucrose to distilled/deionized water and bring volume to 100.0mL. Mix thoroughly. Autoclave both solutions separately for 15 min at 15 psi pressure–121°C. Cool to 50°C. Combine the sterile solutions. Mix thoroughly. Pour into sterile Petri dishes or distribute into sterile tubes.

Use: For the cultivation of various actinomycetes.

Czapek Solution Agar

Composition per liter:

Sucrose	30.0g
Agar	15.0g
$NaNO_3$	2.0g
K_2HPO_4	1.0g
KCl	0.5g
$MgSO_4 \cdot 7H_20$	0.5g
$FeSO_4 \cdot 7H_2O$	0.01g

pH 7.3 ± 0.2 at 25°C

Source: This medium is available as a premixed powder from BD Diagnostic Systems.

Preparation of Medium: Add components to distilled/deionized water and bring volume to 1.0L. Mix thoroughly. Distribute into tubes or flasks. Autoclave for 15 min at 15 psi pressure–121°C. Pour into sterile Petri dishes or leave in tubes.

Use: For the cultivation of *Aspergillus, Penicillium,* and other fungi. For the cultivation and maintenance of microorganisms that can utilize nitrate as sole nitrogen source.

Czapek Solution Agar with Sucrose

Composition per liter:

Sucrose	200.0g
Agar	20.0g
$NaNO_3$	3.0g
K_2HPO_4	1.0g
KCl	0.5g
$MgSO_4 \cdot 7H_2O$	0.5g
$FeSO_4 \cdot 7H_2O$	10.0mg

Preparation of Medium: Add components to distilled/deionized water and bring volume to 1.0L. Mix thoroughly. Gently heat and bring to boiling. Distribute into tubes or flasks. Autoclave for 15 min at 15 psi pressure–121°C. Pour into sterile Petri dishes or leave in tubes.

Use: For the cultivation and maintenance of osmophilic fungi.

Czapek Yeast Autolysate Agar (CYA Agar) (Czapek Agar)

Composition per liter:

Agar	15.0g
Yeast extract	5.0g
$NaNO_3$	3.0g
K_2HPO_4	1.0g
KCl	0.5g
$MgSO_4 \cdot 7H_2O$	0.5g
$FeSO_4 \cdot 7H_2O$	0.01g
Sucrose solution	100.0mL

pH 7.3 ± 0.2 at 25°C

Sucrose Solution:

Composition per 100.0mL:

Sucrose	30.0g

Preparation of Sucrose Solution: Add sucrose to distilled/deionized water and bring volume to 100.0mL. Mix thoroughly. Autoclave for 15 min at 15 psi pressure–121°C. Cool to 50°C.

Preparation of Medium: Add components, except sucrose solution, to distilled/deionized water and bring volume to 900.0mL. Mix thoroughly. Autoclave for 15 min at 15 psi pressure–121°C. Cool to 50°C. Aseptically add sterile sucrose solution. Mix thoroughly. Pour into sterile Petri dishes or leave in tubes.

Use: For the isolation and cultivation of heat-resistant filamentous fungi (molds) from foods.

Czapek Yeast Extract Agar

Composition per liter:

Sucrose	30.0g
Agar	15.0g
Yeast extract	5.0g
K_2HPO_4	1.0g
Czapek concentrate	10.0mL

Czapek Concentrate:

Composition per liter:

$NaNO_3$	30.0 g
KCl	5.0 g
$MgSO_4 \cdot 7H_2O$	5.0 g
$FeSO_4 \cdot 7H2O$	0.1 g
$ZnSO_4 \cdot 7H_2O$	0.1 g
$CuSO_4 \cdot 5H_2O$	0.05 g

Preparation of Czapek Concentrate: Add components to distilled/deionized water and bring volume to 1.0L. Mix thoroughly.

Preparation of Medium: Add components to distilled/deionized water and bring volume to 1.0L. Mix thoroughly. Gently heat and bring to boiling. Distribute into tubes or flasks. Autoclave for 15 min at 15 psi pressure–121°C. Pour into sterile Petri dishes or leave in tubes.

Use: For the cultivation and maintenance of *Aspergillus niger.*

Czapek Yeast Extract Agar

Composition per liter:

Sucrose	30.0g
Agar	15.0g
Yeast extract	5.0g
$NaNO_3$	3.0g
K_2HPO_4	1.0g
$MgSO_4 \cdot 7H_2O$	0.5g
KCl	0.5g
$FeSO_4 \cdot 7H_2O$	0.01g
Trace metal solution	1.0mL

pH 7.3 ± 0.2 at 25°C

Trace Metal Solution:

Composition per 100.0mL:

$ZnSO_4 \cdot 7H_2O$	1.0g
$CuSO_4 \cdot 5H_2O$	0.5g

Preparation of Trace Metal Solution: Add components to 100.0mL distilled/deionized water. Mix thoroughly.

Preparation of Medium: Add components to distilled/deionized water and bring volume to 1.0L. Mix thoroughly. Adjust pH to 6.2. Gently heat and bring to boiling. Distribute into tubes or flasks. Autoclave for 15 min at 15 psi pressure–121°C. Pour into sterile Petri dishes or leave in tubes.

Use: For the cultivation and maintenance of a variety of fungal and bacterial species that can use nitrate as sole nitrogen source.

Czapek Yeast Extract Agar

Composition per liter:

Sucrose	30.0g
Agar	15.0g
Yeast extract	5.0g
K_2HPO_4	1.0g
KNO_3	0.3g
KCl	0.05g
$MgSO_4 \cdot 7H_2O$	0.05g
$FeSO_4 \cdot 7H2O$	1.0mg
$ZnSO_4 \cdot 7H_2O$	1.0mg
$CuSO_4 \cdot 5H_2O$	0.5mg

pH 6.8 ± 0.2 at 25°C

Source: This medium is available from HiMedia.

Preparation of Medium: Add components to distilled/deionized water and bring volume to 1.0L. Mix thoroughly. Gently heat and bring to boiling. Distribute into tubes or flasks. Autoclave for 15 min at 15 psi pressure–121°C. Pour into sterile Petri dishes or leave in tubes.

Use: For the isolation and cultivation of *Aspergillus niger.*

CZYA

See: **Czapek Yeast Autolysate Agar**

DA Medium

Composition per liter:

NaCl	116.9g
Agar	15.0g
Tris-HCl	6.024g
$NaHCO_3$	1.68g
$MgSO_4 \cdot 7H_2O$	1.232g
KNO_3	0.505g
$CaCl_2$	0.033g
KH_2PO_4	0.014g
H_3BO_3	6.0mg
$MnCl_2 \cdot 4H_2O$	99.0μg
$ZnCl_2$	14.0μg
$CoCl_2 \cdot 6H_2O$	4.76μg

$CuCl_2 \cdot 2H_2O$ 34.0ng
$FeCl_3$ solution 50.0mL

pH 7.5 ± 0.2 at 25°C

$FeCl_3$ Solution:
Composition per 50.0mL:

EDTA 5.84mg
$FeCl_3$ 0.32mg

Preparation of $FeCl_3$ Solution: Add components to distilled/deionized water and bring volume to 50.0mL. Mix thoroughly.

Preparation of Medium: Add components to distilled/deionized water and bring volume to 1.0L. Mix thoroughly. Gently heat and bring to boiling. Distribute into screw-capped tubes. Autoclave for 15 min at 15 psi pressure–121°C. Allow to cool in a slanted position.

Use: For the cultivation of *Dunaliella bardawil.*

Dap Nutrient Agar

Composition per liter:

Urea 20.0g
Agar 15.0g
Peptone 5.0g
Meat extract 3.0g
DL-α,ε-Diaminopimelic acid 0.1g

pH 7.0 ± 0.2 at 25°C

Preparation of Medium: Add components to distilled/deionized water and bring volume to 1.0L. Mix thoroughly. Gently heat and bring to boiling. Adjust pH to 7.0. Autoclave for 15 min at 15 psi pressure–121°C. Pour into sterile Petri dishes or distribute into sterile tubes.

Use: For the cultivation and maintenance of *Bacillus megaterium.*

Dap Nutrient Agar

Composition per liter:

Urea 20.0g
Agar 15.0g
Peptone 5.0g
Meat extract 3.0g
DL-α,ε-Diaminopimelic acid 0.1g
$MnSO_4 \cdot H_2O$ 10.0mg

pH 7.0 ± 0.2 at 25°C

Preparation of Medium: Add components to distilled/deionized water and bring volume to 1.0L. Mix thoroughly. Gently heat and bring to boiling. Adjust pH to 7.0. Autoclave for 15 min at 15 psi pressure–121°C. Pour into sterile Petri dishes or distribute into sterile tubes.

Use: For the sporulation of *Bacillus megaterium.*

Davis and Mingioli Glucose Minimal Medium

Composition per liter:

Agar 15.0g
K_2HPO_4 .... 7.0g
KH_2PO_4 .... 3.0g
$(NH_4)_2SO_4$ 1.0g
Sodium citrate·$3H_2O$ 0.5g
$MgSO_4 \cdot 7H_2O$ 0.1g
L-Arginine 0.02g
L-Tryptophan 0.02g
Glucose solution 10.0mL

pH 7.0 ± 0.2 at 25°C

Glucose Solution:
Composition per 10.0mL:

Glucose 2.0g

Preparation of Glucose Solution: Add glucose to distilled/deionized water and bring volume to 10.0mL. Mix thoroughly. Filter sterilize.

Preparation of Medium: Add components, except glucose solution, to distilled/deionized water and bring volume to 990.0mL. Gently heat and bring to boiling. Autoclave for 15 min at 15 psi pressure–121°C. Cool to 45°–50°C. Aseptically add sterile glucose solution. Mix thoroughly. Pour into sterile Petri dishes or distribute into sterile tubes.

Use: For the cultivation and maintenance of *Escherichia coli.*

Davis and Mingioli Medium A

Composition per liter:

K_2HPO_4 .... 7.0g
KH_2PO_4 .... 3.0g
$(NH_4)_2SO_4$ 1.0g
Sodium citrate·$3H_2O$ 0.5g
$MgSO_4 \cdot 7H_2O$ 0.1g
Glucose solution 10.0mL
Amino acid solution 10.0mL

pH 7.0 ± 0.2 at 25°C

Glucose Solution:
Composition per 10.0mL:

Glucose 2.5g

Preparation of Glucose Solution: Add glucose to distilled/deionized water and bring volume to 10.0mL. Mix thoroughly. Filter sterilize.

Amino Acid Solution:
Composition per 10.0mL:

L-Leucine 40.0mg
L-Histidine 20.0mg
L-Methionine 20.0mg

Preparation of Amino Acid Solution: Add components to distilled/deionized water and bring volume to 10.0mL. Mix thoroughly. Filter sterilize.

Preparation of Medium: Add components, except glucose solution and amino acid solution, to distilled/deionized water and bring volume to 980.0mL. Gently heat and bring to boiling. Autoclave for 15 min at 15 psi pressure–121°C. Cool to 45°–50°C. Aseptically add 10.0mL of sterile glucose solution and 10.0mL of sterile amino acid solution. Mix thoroughly. Pour into sterile Petri dishes or distribute into sterile tubes.

Use: For the cultivation of *Escherichia coli.*

Davis and Mingioli Medium, Modified

Composition per liter:

K_2HPO_4 .... 7.0g
KH_2PO_4 .... 2.0g
$(NH_4)_2SO_4$ 1.0g
Na citrate·$2H_2O$ 0.5g
$MgSO_4 \cdot 7H_2O$ 0.1g
Glucose solution 10.0mL

Streptomycin solution 10.0mL
Additives solution 10.0mL

pH 7.2 ± 0.2 at 25°C

Glucose Solution:
Composition per 10.0mL:
Glucose 4.0g

Preparation of Glucose Solution: Add glucose to distilled/deionized water and bring volume to 10.0mL. Mix thoroughly. Filter sterilize.

Streptomycin Solution:
Composition per 10.0mL:
Streptomycin 4.0g

Preparation of Streptomycin Solution: Add streptomycin to distilled/deionized water and bring volume to 10.0mL. Mix well. Filter sterilize.

Additives Solution:
Composition per 10.0mL:
DL-Threonine 0.1g
DL- or L-Leucine 0.1g
Thiamine·HCl 0.5mg

Preparation of Additives Solution: Add components to distilled/deionized water and bring volume to 10.0mL. Mix thoroughly. Filter sterilize.

Preparation of Medium: Add components—except glucose solution, streptomycin solution, and additives solution—to distilled/deionized water and bring volume to 970.0mL. Mix thoroughly. Adjust pH to 7.2. Gently heat and bring to boiling. Autoclave for 25 min at 15 psi pressure–121°C. Cool to 45°–50°C. Aseptically add sterile glucose solution, streptomycin solution, and additives solution. Mix thoroughly. Aseptically distribute into sterile tubes or flasks.

Use: For the cultivation and maintenance of *Escherichia coli.*

Davis and Mingioli Medium, Modified

Composition per liter:
Agar 15.0g
K_2HPO_4 .... 7.0g
Lactose 2.0g
$(NH_4)_2SO_4$ 1.0g
KH_2PO_4 .... 0.91g
$MgSO_4 \cdot 7H_2O$ 0.1g
Tap water 10.0mL

pH 7.0 ± 0.2 at 25°C

Preparation of Medium: Add components to distilled/deionized water and bring volume to 1.0L. Mix thoroughly. Adjust pH to 7.0. Distribute into tubes or flasks. Autoclave for 20 min at 15 psi pressure–121°C. Pour into sterile Petri dishes or leave in tubes.

Use: For the cultivation and maintenance of *Escherichia coli.*

Davis and Mingioli Medium with Proline

Composition per liter:
K_2HPO_4 .... 7.0g
KH_2PO_4 .... 3.0g
Glucose 2.0g
$(NH_4)_2SO_4$ 1.0g
L-Proline 0.5g
Sodium citrate·$3H_2O$ 0.5g
$MgSO_4 \cdot 7H_2O$ 0.1g
Glucose solution 1.0mL

pH 7.0 ± 0.2 at 25°C

Glucose Solution:
Composition per 100.0mL:
Glucose 20.0g

Preparation of Glucose Solution: Add glucose to distilled/deionized water and bring volume to 100.0mL. Mix thoroughly. Filter sterilize.

Preparation of Medium: Add components, except glucose solution, to distilled/deionized water and bring volume to 999.0mL. Gently heat and bring to boiling. Autoclave for 15 min at 15 psi pressure–121°C. Cool to 25°C. Aseptically add 1.0mL of sterile glucose solution. Mix thoroughly. Aseptically distribute into sterile tubes or flasks.

Use: For the cultivation and maintenance of *Escherichia coli.*

Davis and Mingioli Medium with Vitamin B_1 and Asparagine

Composition per liter:
K_2HPO_4 .... 7.0g
KH_2PO_4 .... 3.0g
$(NH_4)_2SO_4$ 1.0g
Sodium citrate·$3H_2O$ 0.5g
L-Asparagine 0.4g
$MgSO_4 \cdot 7H_2O$ 0.1g
Vitamin B_1 .... 0.1mg
Glucose solution 1.0mL

pH 7.0 ± 0.2 at 25°C

Glucose Solution:
Composition per 100.0mL:
Glucose 20.0g

Preparation of Glucose Solution: Add glucose to distilled/deionized water and bring volume to 100.0mL. Mix thoroughly. Filter sterilize.

Preparation of Medium: Add components, except glucose solution, to distilled/deionized water and bring volume to 999.0mL. Gently heat and bring to boiling. Autoclave for 15 min at 15 psi pressure–121°C. Cool to 25°C. Aseptically add 1.0mL of sterile glucose solution. Mix thoroughly. Aseptically distribute into sterile tubes or flasks.

Use: For the cultivation and maintenance of *Escherichia coli.*

Davis Supplemented Minimal Medium

Composition per liter:
Agar 15.0g
K_2HPO_4 .... 7.0g
KH_2PO_4 .... 3.0g
Casein hydrolysate 2.0g
Yeast extract 2.0g
$(NH_4)_2SO_4$ 1.0g
Sodium citrate·$3H_2O$ 0.5g
$MgSO_4 \cdot 7H_2O$ 0.1g
Glucose solution 20.0mL

pH 7.0 ± 0.2 at 25°C

Glucose Solution
Composition per 100.0mL:
Glucose 10.0g

Preparation of Glucose Solution: Add glucose to distilled/deionized water and bring volume to 100.0mL. Mix thoroughly. Filter sterilize.

Preparation of Medium: Add components, except glucose solution, to distilled/deionized water and bring volume to 980.0mL. Gently heat and bring to boiling. Autoclave for 15 min at 15 psi pressure–121°C. Cool to 45°–50°C. Aseptically add 20.0mL of sterile glucose solution. Mix thoroughly. Pour into sterile Petri dishes or distribute into sterile tubes.

Use: For the cultivation and maintenance of *Escherichia coli*.

DCLS Agar (Deoxycholate Citrate Lactose Sucrose Agar)

Composition per liter:

Agar	12.0g
Sodium citrate·$3H_2O$	10.5g
Lactose	5.0g
$Na_2S_2O_3$	5.0g
Sucrose	5.0g
Pancreatic digest of casein	3.5g
Peptic digest of animal tissue	3.5g
Beef extract	3.0g
Sodium deoxycholate	2.5g
Neutral Red	0.03g

pH 7.2 ± 0.1 at 25°C

Source: This medium is available as a premixed powder from BD Diagnostic Systems and Oxoid.

Preparation of Medium: Add components to distilled/deionized water and bring volume to 1.0L. Mix thoroughly. Gently heat while stirring and bring to boiling. Do not overheat. Do not autoclave. Pour into sterile Petri dishes in 20.0mL volumes.

Use: For the selective isolation of *Salmonella* species, *Shigella* species, and *Vibrio* species from fecal specimens.

DCLS Agar

Composition per liter:

Agar	12.0g
Sodium citrate	10.0g
Proteose peptone	7.0g
Lactose	5.0g
$Na_2S_2O_3$	5.0g
Sucrose	5.0g
Beef extract	3.0g
Sodium deoxycholate	2.5g
Neutral Red	0.03

pH 7.2 ± 0.1 at 25°C

Source: This medium is available as a premixed powder from HiMedia.

Preparation of Medium: Add components to distilled/deionized water and bring volume to 1.0L. Mix thoroughly. Gently heat while stirring and bring to boiling. Do not overheat. Do not autoclave. Pour into sterile Petri dishes or distribute into sterile tubes.

Use: For the selective isolation of *Salmonella* species, *Shigella* species, and *Vibrio* species from fecal specimens.

DCLS Agar, Hajna

Composition per liter:

Agar	20.0g
Sodium citrate	10.0g
Lactose	7.5g
Sucrose	7.5g
Peptic digest of animal tissue	5.0g
Casein enzymatic hydrolysate	5.0g
NaCl	5.0g
$Na_2S_2O_3$	5.0g
Plant extract	3.0g
Beef extract	3.0g
Sodium deoxycholate	2.5g
Bromcresol Purple	0.02g

pH 7.2 ± 0.1 at 25°C

Source: This medium is available as a premixed powder from HiMedia.

Preparation of Medium: Add components to distilled/deionized water and bring volume to 1.0L. Mix thoroughly. Gently heat while stirring and bring to boiling. Do not overheat. Do not autoclave. Pour into sterile Petri dishes or distribute into sterile tubes.

Use: For the selective isolation of *Salmonella* species, *Shigella* species, and *Vibrio* species from fecal specimens.

DCLS HiVeg Agar

Composition per liter:

Agar	12.0g
Sodium citrate	10.0g
Plant peptone No. 3	8.0g
Lactose	5.0g
$Na_2S_2O_3$	5.0g
Sucrose	5.0g
Plant extract	3.0g
Synthetic detergent No. III	1.5g
Neutral Red	0.03

pH 7.2 ± 0.1 at 25°C

Source: This medium is available as a premixed powder from HiMedia.

Preparation of Medium: Add components to distilled/deionized water and bring volume to 1.0L. Mix thoroughly. Gently heat while stirring and bring to boiling. Do not overheat. Do not autoclave. Pour into sterile Petri dishes or distribute into sterile tubes.

Use: For the selective isolation of *Salmonella* species, *Shigella* species, and *Vibrio* species from fecal specimens.

DCLS HiVeg Agar, Hajna

Composition per liter:

Agar	20.0g
Sodium citrate	10.0g
Lactose	7.5g
Sucrose	7.5g
Plant peptone	6.0g
Plant hydrolysate	5.0g
NaCl	5.0g
$Na_2S_2O_3$	5.0g
Plant extract	3.0g
Yeast extract	3.0g

Synthetic detergent No. III.......... 1.5g
Bromresol Purple 0.02g

pH 7.2 ± 0.1 at 25°C

Source: This medium is available as a premixed powder from HiMedia.

Preparation of Medium: Add components to distilled/deionized water and bring volume to 1.0L. Mix thoroughly. Gently heat while stirring and bring to boiling. Do not overheat. Do not autoclave. Pour into sterile Petri dishes or distribute into sterile tubes.

Use: For the selective isolation of *Salmonella* species, *Shigella* species, and *Vibrio* species from fecal specimens.

DCMYBA

Composition per liter:

Agar 20.0g
Cornmeal.......... 15.0g
Supplement solution 100.0mL

Supplement Solution:

Composition per 100.0mL:

Glucose 20.0g
Yeast extract.......... 1.0g
Biotin 100.0μg

Preparation of Supplement Solution: Add components to distilled/deionized water and bring volume to 100.0mL. Mix thoroughly. Filter sterilize.

Preparation of Medium: Add cornmeal to distilled/deionized water and bring volume to 900.0mL. Mix thoroughly. Gently heat and bring to boiling. Maintain at 100°C for 30 min. Filter through Whatman filter paper. Add agar to filtrate and bring volume to 1.0L with distilled/deionized water. Mix thoroughly. Gently heat and bring to boiling. Autoclave for 15 min at 15 psi pressure–121°C. Cool to 50–55°C. Aseptically add 100.0mL of sterile supplement solution. Mix thoroughly. Pour into sterile Petri dishes or distribute into sterile tubes.

Use: For the cultivation of *Sordaria brevicollis*.

D/E-Neutralizing Agar

Composition per liter:

Agar 15.0g
Glucose 10.0g
Soybean lecithin.......... 7.0g
$Na_2S_2O_3 \cdot 5H_2O$ 6.0g
Polysorbate 80.......... 5.0g
Pancreatic digest of casein.......... 5.0g
$NaHSO_3$ 2.5g
Yeast extract.......... 2.5g
Sodium thioglycolate.......... 1.0g
Bromcresol Purple 0.02g

pH 7.6 ± 0.2 at 25°C

Source: This medium is available as a premixed powder from BD Diagnostic Systems.

Preparation of Medium: Add components to distilled/deionized water and bring volume to 1.0L. Mix thoroughly. Gently heat and bring to boiling. Distribute into flasks in 9.0mL volumes. Autoclave for 15 min at 15 psi pressure–121°C.

Use: For the neutralization and testing of antiseptics and disinfectants.

D/E-Neutralizing Broth (Dey/Engley-Neutralizing Broth)

Composition per liter:

Glucose 10.0g
Soybean lecithin.......... 7.0g
$Na_2S_2O_3 \cdot 5H_2O$ 6.0g
Tween™ 80.......... 5.0g
Pancreatic digest of casein.......... 5.0g
$NaHSO_3$ 2.5g
Yeast extract.......... 2.5g
Sodium thioglycolate.......... 1.0g
Bromcresol Purple 0.02g

pH 7.6± 0.2 at 25°C

Source: This medium is available as a premixed powder from BD Diagnostic Systems.

Preparation of Medium: Add components to distilled/deionized water and bring volume to 1.0L. Mix thoroughly. Distribute into tubes in 9.0mL volumes. Autoclave for 15 min at 15 psi pressure–121°C.

Use: For the neutralization and testing of antiseptics and disinfectants.

D/E-Neutralizing Broth Base (Dey/Engley-Neutralizing Broth Base)

Composition per liter:

Glucose 10.0g
Pancreatic digest of casein.......... 5.0g
Yeast extract.......... 2.5g
Bromcresol Purple 0.02g

pH 7.6± 0.2 at 25°C

Preparation of Medium: Add components to distilled/deionized water and bring volume to 1.0L. Mix thoroughly. Distribute into tubes in 9.0mL volumes. Autoclave for 15 min at 15 psi pressure–121°C.

Use: For the neutralization and testing of antiseptics and disinfectants.

Decarboxylase Basal Medium (BAM M44)

Composition per liter:

Peptone or gelysate.......... 5.0g
Yeast extract.......... 3.0g
Glucose 1.0g
Bromcresol Purple 0.02g

pH 6.5 ± 0.2 at 25°C

Preparation of Medium: Add components to distilled/deionized water and bring volume to 1.0L. Mix thoroughly. Adjust pH so that it will be 6.5 ± 0.2 after sterilization. Distribute into 16 × 150mm screw-capped tubes in 5.0mL volumes. Autoclave medium with loosely capped tubes for 10 min at 15 psi pressure–121°C. Screw the caps on tightly for storage and after inoculation.

Use: For the cultivation and differentiation of bacteria based on their ability to decarboxylate the amino acid. As the basal medium for arginine broth, lysine broth, and ornithine broth. Bacteria that decarboxylate arginine, lysine, or ornithine turn the medium turbid purple. The unsupplemented decarboylase basal medium is used as a control.

Decarboxylase Basal Medium with Sodium Chloride (BAM M44)

Composition per liter:

Peptone or gelysate.......... 5.0g
Yeast extract.......... 3.0g

Glucose 1.0g
Bromcresol Purple 0.02g

pH 6.5 ± 0.2 at 25°C

Preparation of Medium: Add components to distilled/deionized water and bring volume to 1.0L. Mix thoroughly. Adjust pH so that it will be 6.5 ± 0.2 after sterilization. Distribute into 16 × 150mm screw-capped tubes in 5.0mL volumes. Autoclave medium with loosely capped tubes for 10 min at 15 psi pressure–121°C. Screw the caps on tightly for storage and after inoculation.

Use: For the cultivation and differentiation of *Vibrio* spp. based on their ability to decarboxylate the amino acid. As the basal medium for arginine broth, lysine broth, and ornithine broth. Bacteria that decarboxylate arginine, lysine, or ornithine turn the medium turbid purple. The unsupplemented decarboylase basal medium is used as a control.

Decarboxylase Base, Møller

Composition per liter:

Amino acid 10.0g
Beef extract 5.0g
Peptone 5.0g
Glucose 0.5g
Bromcresol Purple 0.01g
Cresol Red 5.0mg
Pyridoxal 5.0mg
Mineral oil 200.0mL

pH 6.0 ± 0.2 at 25°C

Source: This medium is available as a premixed powder from BD Diagnostic Systems.

Preparation of Medium: Add components, except mineral oil, to distilled/deionized water and bring volume to 1.0L. For amino acid, use L-arginine, L-lysine, or L-ornithine. Mix thoroughly. Distribute into screw-capped tubes in 5.0mL volumes. Autoclave medium and mineral oil separately for 15 min at 15 psi pressure–121°C. After inoculation, overlay medium with 1.0mL of sterile mineral oil per tube.

Use: For the cultivation and differentiation of bacteria based on their ability to decarboxylate the amino acid. Bacteria that decarboxylate arginine, lysine, or ornithine turn the medium turbid purple.

Decarboxylase HiVeg Agar Base

Composition per liter:

Agar 15.0g
Plant peptone 5.0g
Yeast extract 3.0g
Glucose 1.0g
Bromcresol Purple 0.02
Amino acid solution 100.0mL

pH 6.5 ± 0.2 at 25°C

Source: This medium wihout amino acid is available as a premixed powder from HiMedia.

Amino Acid Solution:

Composition per 100.0mL:

L-arginine, L-lysine, or L-ornithine 10.0g

Preparation of Amino Acid Solution: Add amino acid to distilled/deionized water and bring volume to 100.0mL. Mix thoroughly.

Preparation of Medium: Add components and bring volume to 900.0L. For amino acid, use L-arginine, L-lysine, or L-ornithine and add 100.0ml of a 10% solution. Mix thoroughly. Distribute into screw-capped tubes in 5.0mL volumes. Autoclave for 15 min at 15 psi pressure–121°C.

Use: As a basal medium for the cultivation and differentiation of bacteria based on their ability to decarboxylate amino acids. The medium is supplemented with specific L-amino acids for testing decarboxylase activity on that amino acid. Amino acids are added to a final concentration of 0.5 percent.

Decarboxylase HiVeg Broth Base, Moeller

Composition per liter:

Plant extract 5.0g
Plant peptone 5.0g
Glucose 0.5g
Bromcresol Purple 0.01g
Cresol Red 5.0mg
Pyridoxal 5.0mg
Amino acid solution 100.0mL

pH 6.0 ± 0.2 at 25°C

Source: This medium wihout amino acid is available as a premixed powder from HiMedia.

Amino Acid Solution:

Composition per 100.0mL:

L-arginine, L-lysine, or L-ornithine 10.0g

Preparation of Amino Acid Solution: Add amino acid to distilled/deionized water and bring volume to 100.0mL. Mix thoroughly.

Preparation of Medium: Add components and bring volume to 900.0L. For amino acid, use L-arginine, L-lysine, or L-ornithine and add 100.0ml of a 10% solution. Mix thoroughly. Distribute into screw-capped tubes in 5.0mL volumes. Autoclave for 15 min at 15 psi pressure–121°C.

Use: As a basal medium for the cultivation and differentiation of bacteria based on their ability to decarboxylate amino acids. The medium is supplemented with specific L-amino acids for testing decarboxylase activity on that amino acid. Bacteria that decarboxylate arginine, lysine, or ornithine turn the medium turbid purple.

Decarboxylase Medium Base, Falkow

Composition per liter:

Amino acid (arginine, lysine, or ornithine) 5.0g
Peptone 5.0g
Yeast extract 3.0g
Glucose 1.0g
Bromcresol Purple 0.02g
Mineral oil 200.0mL

pH 6.8 ± 0.2 at 25°C

Source: This medium is available as a premixed powder from BD Diagnostic Systems.

Preparation of Medium: Add components, except mineral oil, to distilled/deionized water and bring volume to 1.0L. For amino acid, use L-arginine, L-lysine, or L-ornithine. Mix thoroughly. Distribute into screw-capped tubes in 5.0mL volumes. Autoclave medium and mineral oil separately for 15 min at 15 psi pressure–121°C. After inoculation, overlay medium with 1.0mL of sterile mineral oil per tube.

Use: For the cultivation and differentiation of bacteria based on their ability to decarboxylate a specific amino acid. Bacteria that decarboxylate arginine, lysine, or ornithine turn the medium turbid purple.

Decarboxylase Medium, Ornithine Modified

Composition per liter:

L-Ornithine	10.0g
Meat peptone	5.0g
Yeast extract	3.0g
Bromcresol Purple solution	5.0mL

pH 5.5 ± 0.2 at 25°C

Bromcresol Purple Solution:

Composition per 100.0mL:

Bromcresol Purple	0.2g
Ethanol	50.0mL

Preparation of Bromcresol Purple Solution: Add Bromcresol Purple to ethanol. Mix thoroughly. Bring volume to 100.0mL with distilled/deionized water. Mix thoroughly. Filter sterilize.

Preparation of Medium: Add components to distilled/deionized water and bring volume to 1.0L. Mix thoroughly. Gently heat until dissolved. Adjust pH to 5.5 with HCl or NaOH. Distribute into screw-capped tubes. Autoclave for 15 min at 15 psi pressure–121°C.

Use: For the cultivation and differentiation of bacteria based on their ability to decarboxylate ornithine. Bacteria that decarboxylate ornithine turn the medium turbid purple.

Decarboxylase Test HiVeg Medium Base (Falkow)

Composition per liter:

Plant peptone	5.0g
Yeast extract	3.0g
Glucose	1.0g
Bromcresol Purple	0.02
Amino acid solution	100.0mL

pH 6.8 ± 0.2 at 25°C

Source: This medium wihout amino acid is available as a premixed powder from HiMedia.

Amino Acid Solution:

Composition per 100.0mL:

L-arginine, L-lysine, or L-ornithine	10.0g

Preparation of Amino Acid Solution: Add amino acid to distilled/deionized water and bring volume to 100.0mL. Mix thoroughly.

Preparation of Medium: Add components and bring volume to 900.0L. For amino acid, use L-arginine, L-lysine, or L-ornithine and add 100.0ml of a 10% solution. Mix thoroughly. Distribute into screw-capped tubes in 5.0mL volumes. Autoclave for 15 min at 15 psi pressure–121°C.

Use: As a basal medium for the cultivation and differentiation of bacteria based on their ability to decarboxylate amino acids. The medium is supplemented with specific L-amino acids for testing decarboxylase activity on that amino acid. Amino acids are added to a final concentration of 0.5 percent. Bacteria that decarboxylate arginine, lysine, or ornithine turn the medium turbid purple.

Deep Liver Broth

Composition per liter:

Pancreatic digest of casein	10.0g
Glucose	5.0g
Yeast extract	5.0g
K_2HPO_4	2.0g
Liver infusion	1.0g

pH 7.4 ± 0.2 at 25°C

Preparation of Medium: Add components to distilled/deionized water and bring volume to 1.0L. Mix thoroughly. Distribute into tubes or flasks. Autoclave for 15 min at 15 psi pressure–121°C.

Use: For the cultivation of *Bacillus polymyxa* and *Leuconostoc mesenteroides*.

Deferribacter Medium (DSMZ Medium 935)

Composition per liter:

NaCl	25.0g
Sulfur, powdered	10.0g
Na-acetate	2.0g
KNO_3	0.5g
NH_4Cl	0.33g
KCl	0.33g
$CaCl_2 \cdot 2H_2O$	0.33g
$MgCl_2 \cdot 6H_2O$	0.33g
KH_2PO_4	0.33g
Yeast extract	0.15g
$NaHCO_3$ solution	10.0mL
Trace elements solution	10.0mL
Vitamin solution	10.0mL

pH 7.0 ± 0.2 at 25°C

$NaHCO_3$ Solution:

Composition per 10.0mL:

$NaHCO_3$	0.3g

Preparation of $NaHCO_3$ Solution: Add $NaHCO_3$ to distilled/deionized water and bring volume to 10.0mL. Mix thoroughly. Autoclave for 15 min at 15 psi pressure–121°C. Cool to 25°C. Must be prepared freshly.

Trace Elements Solution:

Composition per liter:

$MgSO_4 \cdot 7H_2O$	3.0g
Nitrilotriacetic acid	1.5g
NaCl	1.0g
$MnSO_4 \cdot 2H_2O$	0.5g
$CoSO_4 \cdot 7H_2O$	0.18g
$ZnSO_4 \cdot 7H_2O$	0.18g
$CaCl_2 \cdot 2H_2O$	0.1g
$FeSO_4 \cdot 7H_2O$	0.1g
$NiCl_2 \cdot 6H_2O$	0.025g
$KAl(SO_4)_2 \cdot 12H_2O$	0.02g
H_3BO_3	0.01g
$Na_2MoO_4 \cdot 4H_2O$	0.01g
$CuSO_4 \cdot 5H_2O$	0.01g
$Na_2SeO_3 \cdot 5H_2O$	0.3mg

Preparation of Trace Elements Solution: Add nitrilotriacetic acid to 500.0mL of distilled/deionized water. Dissolve by adjusting pH to 6.5 with KOH. Add remaining components. Add distilled/deionized water to 1.0L. Mix thoroughly.

Vitamin Solution:

Composition per liter:

Pyridoxine-HCl	10.0mg
Thiamine-HCl·$2H_2O$	5.0mg
Riboflavin	5.0mg
Nicotinic acid	5.0mg
D-Ca-pantothenate	5.0mg
p-Aminobenzoic acid	5.0mg
Lipoic acid	5.0mg

Biotin 2.0mg
Folic acid 2.0mg
Vitamin B_{12} 0.1mg

Preparation of Vitamin Solution: Add components to distilled/deionized water and bring volume to 1.0L. Mix thoroughly. Sparge with 80% H_2 + 20% CO_2. Filter sterilize.

Preparation of Medium: Add components, except vitamin solution, $NaHCO_3$ solution, and sulfur, to 980.0mL distilled/deionized water. Gently heat and bring to boiling. Boil for 3 min. Cool to room temperature while sparging with 100% N_2. Adjust pH to 7.0. Anaerobically under 100% N_2 distribute into tubes of bottles containing the sulfur (0.1g sulfur per 10mL medium). Autoclave for 20 min at 110°C. Aseptically and anaerobically add 10.0mL sterile vitamin solution and 10.0mL sterile $NaHCO_3$ solution. Mix thoroughly. The final pH should be 7.0.

Use: For the cultivation of *Deferribacter desulfuricans* and *Deferribacter thermophilus.*

Defined Glucose Medium EMSY-1

Composition per liter:
Na_2HPO_4 .... 1.79g
KH_2PO_4 .... 1.7g
Citric acid 0.5g
NH_4Cl 0.43g
$MgSO_4 \cdot 7H_2O$ 0.41g
$CaCl_2 \cdot 2H_2O$ 0.04g
NaCl 0.03g
$FeCl_3 \cdot 6H_2O$ 4.84mg
Glucose solution 100.0mL
Yeast extract solution 10.0mL
TK6-3 solution 1.0mL
pH 7.2 ± 0.2 at 25°C

Glucose Solution:
Composition per 100.0mL:
Glucose 10.0g

Preparation of Glucose Solution: Add glucose to distilled/deionized water and bring volume to 100.0mL. Mix thoroughly. Filter sterilize.

Yeast Extract Solution:
Composition per 10.0mL:
Yeast extract 0.4g

Preparation of Yeast Extract Solution: Add yeast extract to distilled/deionized water and bring volume to 10.0mL. Mix thoroughly. Filter sterilize.

TK6-3 Solution:
Composition per liter:
$ZnSO_4 \cdot 7H_2O$ 1.45g
$CuSO_4 \cdot 5H_2O$ 0.76g
$MnSO_4 \cdot H_2O$ 0.31g
H_3BO_3 .... 0.19g
$Na_2MoO_4 \cdot 2H_2O$ 0.17g
KI 0.04g
H_2SO_4 (1*N* solution) 1.0mL

Preparation of TK6-3 Solution: Add components to distilled/deionized water and bring volume to 1.0L. Mix thoroughly.

Preparation of Medium: Add components, except glucose solution and yeast extract solution, to distilled/deionized water and bring volume to 890.0mL. Mix thoroughly. Gently heat and bring to boiling. Autoclave for 15 min at 15 psi pressure–121°C. Cool rapidly to 25°C. Aseptically add 100.0mL of sterile glucose solution and 10.0mL of sterile yeast extract solution. Mix thoroughly. Aseptically distribute into sterile tubes or flasks.

Use: For the cultivation and maintenance of *Xanthomonas campestris*.

Defined Medium with Povidone Iodine

Composition per 1025.0mL:
Basal solution 1.0L
Solution B 10.0mL
Solution C 10.0mL
Solution A 5.0mL

Basal Solution:
Composition per liter:
Agar 20.0g
Na_2HPO_4 .... 4.8g
KH_2PO_4 .... 4.4g
NH_4Cl 1.0g
$MgSO_4 \cdot 7H_2O$ 0.5g

Preparation of Basal Solution: Add components to distilled/deionized water and bring volume to 1.0L. Mix thoroughly. Gently heat and bring to boiling. Autoclave for 15 min at 15 psi pressure–121°C. Cool to 45°–50°C.

Solution A:
Composition per 100.0mL:
Ferric ammonium citrate 1.0g
$CaCl_2 \cdot 2H_2O$ 0.1g

Preparation of Solution A: Add components to distilled/deionized water and bring volume to 100.0mL. Mix thoroughly. Filter sterilize.

Solution B:
Composition per 100.0mL:
D-Glucose 10.0g

Preparation of Solution B: Add glucose to distilled/deionized water and bring volume to 100.0mL. Mix thoroughly. Filter sterilize.

Solution C:
Composition per 100.0mL:
Povidone-iodine 0.1g

Preparation of Solution C: Add povidone-iodine to distilled/deionized water and bring volume to 100.0mL. Mix thoroughly. Filter sterilize.

Preparation of Medium: To 1.0L of cooled, sterile basal solution, aseptically add 5.0mL of sterile solution A, 10.0mL of sterile solution B, and 10.0mL of sterile solution C. Mix thoroughly. Pour into sterile Petri dishes or distribute into sterile tubes.

Use: For the cultivation and maintenance of *Pseudomonas aeruginosa* and *Pseudomonas cepacia.*

Defined Medium for *Rhodopseudomonas*

Composition per liter:
Malic acid 4.0g
$(NH_4)_2SO_4$ 1.0g
K_2HPO_4 .... 0.9g
KH_2PO_4 .... 0.6g
$MgSO_4 \cdot 7H_2O$ 0.2g
$CaCl_2 \cdot 2H_2O$ 0.075g
EDTA 0.02g
$FeSO_4 \cdot 7H_2O$ 0.012g
Thiamine 1.0mg

Biotin 0.015mg
Trace elements 1.0mL

pH 6.8 ± 0.2 at 25°C

Trace Elements:

Composition per 250.0mL:

H_3BO_3 .... 0.7g
$MnSO_4 \cdot H_2O$ 0.4g
$Na_2MoO_4 \cdot 2H_2O$ 0.19g
$ZnSO_4 \cdot 7H_2O$ 0.06g
$CoCl_2 \cdot 6H_2O$ 0.05g
$Cu(NO_3)_2 \cdot 3H_2O$ 0.01g

Preparation of Trace Elements: Add components to distilled/deionized water and bring volume to 250.0mL. Mix thoroughly.

Preparation of Medium: Add components to distilled/deionized water and bring volume to 1.0L. Mix thoroughly. Adjust pH to 6.8. Distribute into tubes or flasks. Autoclave for 15 min at 15 psi pressure–121°C.

Use: For the cultivation and maintenance of *Rhodobacter capsulatus*.

Dehalobacter restrictus Medium (DSMZ Medium 732)

Composition per 1046mL:

Solution A 900.0mL
Solution B 100.0mL
Solution G 15.0mL
Solution C 10.0mL
Solution E 10.0mL
Solution F 10.0mL
Solution D 1.0mL

pH 7.2 ± 0.2 at 25°C

Solution A:

Composition per liter:

K_2HPO_4 .... 0.653g
Na-acetate 0.460g
$NaH_2PO_4 \cdot H_2O$ 0.173g
Peptone 0.1g
Resazurin 0.5mg

Preparation of Solution A: Add components to distilled/deionized water and bring volume to 1.0 L. Mix thoroughly. Gently heat and bring to boiling. Cool to room temperature under 80% H_2 + 20% CO_2 gas. Distribute 9ml volumes into 50mL serum bottles under 80% H_2 + 20% CO_2 gas. Pressurize closed bottles with H_2 + CO_2 gas to 0.5 bar overpressure. Autoclave for 15 min at 15 psi pressure–121°C.

Solution B:

Composition per 100.0mL:

$NaHCO_3$ 3.730g
NH_4HCO_3 .... 0.443g

Preparation of Solution B: Add components to distilled/deionized water and bring volume to 100.0mL in bottles. Mix thoroughly. Flush solution with 80% N_2 + 20% CO_2 gas. Close bottles. Autoclave for 15 min at 15 psi pressure–121°C.

Solution C:

Composition per 10.0mL:

$MgCl_2 \cdot 6H_2O$ 0.12g
$CaCl_2 \cdot 2H_2O$ 0.11g

Preparation of Solution C: Add components to distilled/deionized water and bring volume to 10.0mL in bottles. Mix thoroughly. Flush solution with 100% N_2 for 20 min. Close bottles. Autoclave for 15 min at 15 psi pressure–121°C.

Solution D:

Composition per 10.0mL:

Na_2-EDTA 5.0mg
$FeCl_2 \cdot 4H_2O$ 5.0mg
$AlCl_3$ 0.1mg
Trace elements solution SL-10 .... 10.0mL

Trace Elements Solution SL-10:

Composition per liter:

$FeCl_2 \cdot 4H_2O$ 1.5g
$CoCl_2 \cdot 6H_2O$ 190.0mg
$MnCl_2 \cdot 4H_2O$ 100.0mg
$ZnCl_2$ 70.0mg
$Na_2MoO_4 \cdot 2H_2O$ 36.0mg
$NiCl_2 \cdot 6H_2O$ 24.0mg
H_3BO_3 .... 6.0mg
$CuCl_2 \cdot 2H_2O$ 2.0mg
HCl (25% solution) 10.0mL

Preparation of Trace Elements Solution SL-10: Add $FeCl_2 \cdot 4H_2O$ to 10.0mL of HCl solution. Mix thoroughly. Add distilled/deionized water and bring volume to 1.0L. Add remaining components. Mix thoroughly. Sparge with 100% N_2. Autoclave for 15 min at 15 psi pressure–121°C.

Preparation of Solution D: Add Na_2-EDTA, $FeCl_2 \cdot 4H_2O$, and $AlCl_3$ to 10.0mL trace solution SL-10 in a Hungate bottle. Mix thoroughly. Flush solution with 100% N_2 for 20 min. Close bottles. Autoclave for 15 min at 15 psi pressure–121°C.

Solution E:

Composition per liter:

Vitamin solution 900.0mL
Seven vitamin solution 100.0mL

Vitamin Solution:

Composition per liter:

Pyridoxine-HCl 10.0mg
Thiamine-HCl·$2H_2O$ 5.0mg
Riboflavin 5.0mg
Nicotinic acid 5.0mg
D-Ca-pantothenate 5.0mg
p-Aminobenzoic acid 5.0mg
Lipoic acid 5.0mg
Biotin 2.0mg
Folic acid 2.0mg
Vitamin B_{12} 0.1mg

Preparation of Vitamin Solution: Add components to distilled/deionized water and bring volume to 1.0L. Mix thoroughly.

Seven Vitamin Solution:

Composition per liter:

Pyridoxine hydrochloride 300.0mg
Thiamine-HCl·$2H_2O$ 200.0mg
Nicotinic acid 200.0mg
Vitamin B_{12} 100.0mg
Calcium pantothenate 100.0mg
p-Aminobenzoic acid 80.0mg
D(+)-Biotin 20.0mg

Preparation of Seven Vitamin Solution: Add components to distilled/deionized water and bring volume to 1.0L. Sparge with 100% N_2. Mix thoroughly.

Preparation of Solution E: Combine 900.0mL vitamin solution and 100.0mL seven vitamin solution. Mix thoroughly. Sparge with 100% N_2. Filter sterilize.

Solution F:

Composition per 10.0mL:

$Na_2S·9H_2O$	0.3g

Preparation of Solution F: Add $Na_2S·9H_2O$ to distilled/deionized water and bring volume to 10.0mL. Mix thoroughly. Sparge with 100% N_2. Autoclave for 15 min at 15 psi pressure–121°C.

Solution G:

Hexadecane	45.0mL
Tetrachloroethene	5.0mL

Preparation of Solution G: Using a syringe, aseptically inject 5.0mL sterile tetrachloroethene to the 45.0mL sterile hexadecane in the 100mL serum bottle. Sparge with 100% N_2. Autoclave for 15 min at 15 psi pressure–121°C.

Hexadecane:

Hexadecane	45.0mL

Preparation of Hexadecane: Add hexadecane to a 100mL serum bottle. Sparge with 100% N_2. Autoclave for 15 min at 15 psi pressure–121°C.

Tetrachloroethene:

Tetrachloroethene	10.0mL

Preparation of Tetrachloroethene: Add tetrachloroethene to a 10mL serum bottle. Sparge with 100% N_2. Autoclave for 15 min at 15 psi pressure–121°C.

Preparation of Medium: Add 9mL sterile solution A to a 50 mL sterile bottle. Then add by injection 1.0mL sterile solution B, 0.1mL sterile solution C, 0.01mL sterile solution D, 0.1mL sterile solution E, and 0.1mL sterile solution F. Inoculate the culture into the medium. Then add by injection 0.15mL sterile solution G.

Use: For the cultivation of *Dehalobacter restrictus* and *Sulfurospirillum halorespiran.*

Dehalobacter restrictus Medium (DSMZ Medium 732)

Composition per 1056mL:

Solution A	900.0mL
Solution B	100.0mL
Solution G	15.0mL
Solution C	10.0mL
Solution E	10.0mL
Solution F	10.0mL
Solution H	10.0mL
Solution D	1.0mL

pH 7.2 ± 0.2 at 25°C

Solution A:

Composition per liter:

K_2HPO_4	0.653g
Na-acetate	0.460g
$NaH_2PO_4·H_2O$	0.173g
Peptone	0.1g
Resazurin	0.5mg

Preparation of Solution A: Add components to distilled/deionized water and bring volume to 1.0 L. Mix thoroughly. Gently heat and bring to boiling. Cool to room temperature under 80% N_2 + 20% CO_2 gas. Distribute 9ml volumes into 50mL serum bottles under 80% N_2 + 20% CO_2 gas. Pressurize closed bottles with N_2 + CO_2 gas to 0.5 bar overpressure. Autoclave for 15 min at 15 psi pressure–121°C.

Solution B:

Composition per 100.0mL:

$NaHCO_3$	3.730g
NH_4HCO_3	0.443g

Preparation of Solution B: Add components to distilled/deionized water and bring volume to 100.0mL in bottles. Mix thoroughly. Flush solution with 80% N_2 + 20% CO_2 gas. Close bottles. Autoclave for 15 min at 15 psi pressure–121°C.

Solution C:

Composition per 10.0mL:

$MgCl_2·6H_2O$	0.12g
$CaCl_2·2H_2O$	0.11g

Preparation of Solution C: Add components to distilled/deionized water and bring volume to 10.0mL in bottles. Mix thoroughly. Flush solution with 100% N_2 for 20 min. Close bottles. Autoclave for 15 min at 15 psi pressure–121°C.

Solution D:

Composition per 10.0mL:

Na_2-EDTA	5.0mg
$FeCl_2·4H_2O$	5.0mg
$AlCl_3$	0.1mg
Trace elements solution SL-10	10.0mL

Trace Elements Solution SL-10:

Composition per liter:

$FeCl_2·4H_2O$	1.5g
$CoCl_2·6H_2O$	190.0mg
$MnCl_2·4H_2O$	100.0mg
$ZnCl_2$	70.0mg
$Na_2MoO_4·2H_2O$	36.0mg
$NiCl_2·6H_2O$	24.0mg
H_3BO_3	6.0mg
$CuCl_2·2H_2O$	2.0mg
HCl (25% solution)	10.0mL

Preparation of Trace Elements Solution SL-10: Add $FeCl_2·4H_2O$ to 10.0mL of HCl solution. Mix thoroughly. Add distilled/deionized water and bring volume to 1.0L. Add remaining components. Mix thoroughly. Sparge with 100% N_2. Autoclave for 15 min at 15 psi pressure–121°C.

Preparation of Solution D: Add Na_2-EDTA, $FeCl_2·4H_2O$, and $AlCl_3$ to 10.0mL trace solution SL-10 in a Hungate bottle. Mix thoroughly. Flush solution with 100% N_2 for 20 min. Close bottles. Autoclave for 15 min at 15 psi pressure–121°C.

Solution E:

Composition per liter:

Vitamin solution	900.0mL
Seven vitamin solution	100.0mL

Vitamin Solution:

Composition per liter:

Pyridoxine-HCl	10.0mg
Thiamine-HCl·2H$_2$O	5.0mg
Riboflavin	5.0mg
Nicotinic acid	5.0mg
D-Ca-pantothenate	5.0mg
p-Aminobenzoic acid	5.0mg
Lipoic acid	5.0mg

Biotin	2.0mg
Folic acid	2.0mg
Vitamin B_{12}	0.1mg

Preparation of Vitamin Solution: Add components to distilled/deionized water and bring volume to 1.0L. Mix thoroughly.

Seven Vitamin Solution:
Composition per liter:

Pyridoxine hydrochloride	300.0mg
Thiamine-HCl·$2H_2O$	200.0mg
Nicotinic acid	200.0mg
Vitamin B_{12}	100.0mg
Calcium pantothenate	100.0mg
p-Aminobenzoic acid	80.0mg
D(+)-Biotin	20.0mg

Preparation of Seven Vitamin Solution: Add components to distilled/deionized water and bring volume to 1.0L. Sparge with 100% N_2. Mix thoroughly.

Preparation of Solution E: Combine 900.0mL vitamin solution and 100.0mL seven vitamin solution. Mix thoroughly. Sparge with 100% N_2. Filter sterilize.

Solution F:
Composition per 10.0mL:

$Na_2S·9H_2O$	0.3g

Preparation of Solution F: Add $Na_2S·9H_2O$ to distilled/deionized water and bring volume to 10.0mL. Mix thoroughly. Sparge with 100% N_2. Autoclave for 15 min at 15 psi pressure–121°C.

Solution G:

Hexadecane	45.0mL
Tetrachloroethene	5.0mL

Hexadecane:

Hexadecane	45.0mL

Preparation of Hexadecane: Add hexadecane to a 100mL serum bottle. Sparge with 100% N_2. Autoclave for 15 min at 15 psi pressure–121°C.

Tetrachloroethene:

Tetrachloroethene	10.0mL

Preparation of Tetrachloroethene: Add tetrachloroethene to a 10mL serum bottle. Sparge with 100% N_2. Autoclave for 15 min at 15 psi pressure–121°C.

Preparation of Solution G: Using a syringe, aseptically inject 5.0mL sterile tetrachloroethene to the 45.0mL sterile hexadecane in the 100mL serum bottle. Sparge with 100% N_2. Autoclave for 15 min at 15 psi pressure–121°C.

Solution H:
Composition per 10.0mL:

Na-lactate	2.5g

Preparation of Solution H: Add Na-lactate to distilled/deionized water and bring volume to 10.0mL. Mix thoroughly. Sparge with 100% N_2. Autoclave for 15 min at 15 psi pressure–121°C.

Preparation of Medium: Add 9mL sterile solution A to a 50mL sterile bottle. Then add by injection 1.0mL sterile solution B, 0.1mL sterile solution C, 0.01mL sterile solution D, 0.1mL sterile solution E, 0.1mL sterile solution F, and 0.1mL solution H. Inoculate the culture into the medium. Then add by injection 0.15mL sterile solution G.

Use: For the cultivation of *Sulfurospirillum halorespirans* DSM 13726.

Deleya halophila Medium

Composition per liter:

NaCl	81.0g
$MgSO_4$	19.6g
Yeast extract	10.0g
Proteose peptone No.3	5.0g
$MnCl_2$	4.0g
KCl	2.0g
Glucose	1.0g
$CaCl_2$	0.47g
NaBr	0.026g
$NaHCO_3$ solution	10.0mL

pH 7.5 ± 0.2 at 25°C

$NaHCO_3$ Solution
Composition per 10.0mL:

$NaHCO_3$	0.06g

Preparation of $NaHCO_3$ Solution: Add $NaHCO_3$ to distilled/deionized water and bring volume to 10.0mL. Mix thoroughly. Filter sterilize.

Preparation of Medium: Add components, except $NaHCO_3$ solution, to distilled/deionized water and bring volume to 990.0mL. Mix thoroughly. Adjust pH to 7.5. Autoclave for 15 min at 15 psi pressure–121°C. Cool to 50°C. Aseptically add 10.0mL of sterile $NaHCO_3$ solution. Mix thoroughly. Aseptically distribute into sterile tubes or flasks.

Use: For the cultivation of *Deleya halophila.*

DeMan, Rogosa, Sharpe Agar

See: **MRS Agar**

DeMan, Rogosa, Sharpe Broth

See: **MRS Broth**

Demi-Fraser Broth

Composition per liter:

NaCl	20.0g
Tryptose	10.0g
Na_2HPO_4	9.6g
Beef extract	5.0g
Yeast extract	5.0g
LiCl	3.0g
KH_2PO_4	1.35g
Esculin	1.0g
Acriflavin·HCl	12.5mg
Nalidixic acid	10.0mg
Ferric ammonium citrate supplement	10.0mL

pH 7.2 ± 0.2 at 25°C

Source: This medium is available as a premixed powder and supplement from BD Diagnostic Systems.

Ferric Ammonium Citrate Supplement:
Composition per 10.0mL:

Ferric ammonium citrate	0.5g

Preparation of Ferric Ammonium Citrate Supplement: Add ferric ammonium citrate to distilled/deionized water and bring volume to 10.0mL. Mix thoroughly. Filter sterilize.

Preparation of Medium: Add components, except ferric ammonium citrate supplement, to distilled/deionized water and bring volume to 990.0mL. Mix thoroughly. Autoclave for 15 min at 15 psi pressure–121°C. Aseptically add 10.0mL of sterile ferric ammonium citrate supplement. Mix thoroughly. Aseptically distribute into sterile tubes or flasks.

Use: For the cultivation of *Listeria* species from food and environmental samples.

Denitrovibrio Medium (DSMZ Medium 881)

Composition per 1032.0mL:

NaCl	20.0g
$MgCl_2 \cdot 6H_2O$	3.0g
KH_2PO_4	1.0g
$NaNO_3$	0.7g
KCl	0.5g
NH_4Cl	0.25g
$CaCl_2 \cdot 2H_2O$	0.15g
Na_2SO_4	0.02g
Resazurin	0.5mg
Na-acetate solution	10.0mL
$NaHCO_3$ solution	10.0mL
$Na_2S \cdot 9H_2O$ solution	10.0mL
Seven vitamin solution	1.0mL
Trace elements solution SL-10	1.0mL

pH 6.8–7.2 at 25°C

Na-Acetate Solution:

Composition per 10.0mL:

Na-acetate	1.64g

Preparation of Na-Acetate Solution: Add Na-acetate to distilled/deionized water and bring volume to 10.0mL. Mix thoroughly. Sparge with 100% N_2. Filter sterilize.

$Na_2S \cdot 9H_2O$ Solution:

Composition per 10.0mL:

$Na_2S \cdot 9H_2O$	0.5g

Preparation of $Na_2S \cdot 9H_2O$ Solution: Add $Na_2S \cdot 9H_2O$ to distilled/deionized water and bring volume to 10.0mL. Mix thoroughly. Autoclave under 100% N_2 for 15 min at 15 psi pressure–121°C. Cool to room temperature.

$NaHCO_3$ Solution:

Composition per 10.0mL:

$NaHCO_3$	2.5g

Preparation of $NaHCO_3$ Solution: Add $NaHCO_3$ to distilled/deionized water and bring volume to 10.0mL. Mix thoroughly. Sparge with 80% N_2 + 20% CO_2. Filter sterilize.

Seven Vitamin Solution:

Composition per liter:

Pyridoxine hydrochloride	300.0mg
Thiamine-HCl·$2H_2O$	200.0mg
Nicotinic acid	200.0mg
Vitamin B_{12}	100.0mg
Calcium pantothenate	100.0mg
p-Aminobenzoic acid	80.0mg
D(+)-Biotin	20.0mg

Preparation of Seven Vitamin Solution: Add components to distilled/deionized water and bring volume to 1.0L. Sparge with 100% N_2. Mix thoroughly. Filter sterilize.

Trace Elements Solution SL-10:

Composition per liter:

$FeCl_2 \cdot 4H_2O$	1.5g
$CoCl_2 \cdot 6H_2O$	190.0mg
$MnCl_2 \cdot 4H_2O$	100.0mg
$ZnCl_2$	70.0mg
$Na_2MoO_4 \cdot 2H_2O$	36.0mg
$NiCl_2 \cdot 6H_2O$	24.0mg
H_3BO_3	6.0mg
$CuCl_2 \cdot 2H_2O$	2.0mg
HCl (25% solution)	10.0mL

Preparation of Trace Elements Solution SL-10: Add $FeCl_2 \cdot 4H_2O$ to 10.0mL of HCl solution. Mix thoroughly. Add distilled/deionized water and bring volume to 1.0L. Add remaining components. Mix thoroughly. Sparge with 80% N_2 + 20% CO_2. Autoclave for 15 min at 15 psi pressure–121°C.

Preparation of Medium: Prepare and dispense medium under 100% N_2 gas atmosphere. Add components, except $NaHCO_3$ solution, Na-acetate solution, $Na_2S \cdot 9H_2O$ solution, seven vitamin solution, and trace elements solution SL-10, to distilled/deionized water and bring volume to 949.0mL. Mix thoroughly. Adjust pH to 6.8–7.2. Sparge with 80% N_2 + 20% CO_2. Autoclave for 15 min at 15 psi pressure–121°C. Aseptically and anaerobically add 10.0mL $NaHCO_3$ solution, 10.0mL Na-acetate solution, 10.0mL $Na_2S \cdot 9H_2O$ solution, 1.0mL seven vitamin solution, and 1.0mL trace elements solution SL-10. Mix thoroughly. Aseptically and anaerobically distribute into sterile tubes or bottles.

Use: For the cultivation of *Denitrovibrio acetiphilus.*

Deoxycholate Agar

Composition per liter:

Agar	16.0g
Lactose	10.0g
NaCl	5.0g
Pancreatic digest of casein	5.0g
Peptic digest of animal tissue	5.0g
K_2HPO_4	2.0g
Ferric citrate	1.0g
Sodium citrate	1.0g
Sodium deoxycholate	1.0g
Neutral Red	0.033g

pH 7.3 ± 0.2 at 25°C

Source: This medium is available as a premixed powder from BD Diagnostic Systems.

Preparation of Medium: Add components to distilled/deionized water and bring volume to 1.0L. Mix thoroughly. Gently heat and bring to boiling. Do not autoclave. Cool to 45°–50°C. Pour into sterile Petri dishes.

Use: For the selective isolation, cultivation, enumeration, and differentiation of Gram-negative enteric microorganisms from a variety of clinical and nonclinical specimens. *Escherichia coli* appears as large, flat, rose-red colonies. *Enterobacter* and *Klebsiella* species appear as large, mucoid, pale colonies with a pink center. *Proteus* and *Salmonella* species appear as large, colorless to tan colonies. *Shigella* species appear as colorless to pink colonies. *Pseudomonas* species appear as irregular colorless to brown colonies.

Deoxycholate Agar (Desoxycholate Agar)

Composition per liter:

Agar	15.0g
Lactose	10.0g
Peptone	10.0g

NaCl	5.0g
K_2HPO_4	2.0g
Ferric citrate	1.0g
Sodium citrate	1.0g
Sodium deoxycholate	1.0g
Neutral Red	0.03g

pH 7.3 ± 0.2 at 25°C

Source: This medium is available as a premixed powder from Oxoid Unipath and BD Diagnostic Systems.

Preparation of Medium: Add components to distilled/deionized water and bring volume to 1.0L. Mix thoroughly. Gently heat and bring to boiling. Do not autoclave. Cool to 50°C. Pour into sterile Petri dishes.

Use: For the selective isolation, cultivation, enumeration, and differentiation of Gram-negative enteric microorganisms from a variety of clinical and nonclinical specimens. *Escherichia coli* appears as large, flat, rose-red colonies. *Enterobacter* and *Klebsiella* species appear as large, mucoid, pale colonies with a pink center. *Proteus* and *Salmonella* species appear as large, colorless to tan colonies. *Shigella* species appear as colorless to pink colonies. *Pseudomonas* species appear as irregular colorless to brown colonies.

Deoxycholate Agar

Composition per liter:

Agar	15.0g
Peptic digest of animal tissue	10.0g
Lactose	10.0g
NaCl	5.0g
K_2HPO_4	2.0g
Ferric citrate	1.0g
Sodium citrate	1.0g
Sodium deoxycholate	1.0g
Neutral Red	0.03g

pH 7.5 ± 0.2 at 25°C

Source: This medium is available as a premixed powder from HiMedia.

Preparation of Medium: Add components to distilled/deionized water and bring volume to 1.0L. Mix thoroughly. Gently heat and bring to boiling. Do not autoclave. Cool to 45°–50°C. Pour into sterile Petri dishes.

Use: For the selective isolation, cultivation, enumeration, and differentiation of Gram-negative enteric microorganisms from a variety of clinical and nonclinical specimens. *Escherichia coli* appears as large, flat, rose-red colonies. *Enterobacter* and *Klebsiella* species appear as large, mucoid, pale colonies with a pink center. *Proteus* and *Salmonella* species appear as large, colorless to tan colonies. *Shigella* species appear as colorless to pink colonies. *Pseudomonas* species appear as irregular colorless to brown colonies.

Deoxycholate Agar, HiVeg

Composition per liter:

Agar	15.0g
Plant peptone	10.0g
Lactose	10.0g
NaCl	5.0g
K_2HPO_4	2.0g
Ferric citrate	1.0g
Sodium citrate	1.0g
Synthetic detergent No. III	1.0g
Neutral Red	0.03g

pH 7.5 ± 0.2 at 25°C

Source: This medium is available as a premixed powder from HiMedia.

Preparation of Medium: Add components to distilled/deionized water and bring volume to 1.0L. Mix thoroughly. Gently heat and bring to boiling. Do not autoclave. Cool to 45°–50°C. Pour into sterile Petri dishes.

Use: For the selective isolation, cultivation, enumeration, and differentiation of Gram-negative enteric microorganisms from a variety of clinical and nonclinical specimens. *Escherichia coli* appears as large, flat, rose-red colonies. *Enterobacter* and *Klebsiella* species appear as large, mucoid, pale colonies with a pink center. *Proteus* and *Salmonella* species appear as large, colorless to tan colonies. *Shigella* species appear as colorless to pink colonies. *Pseudomonas* species appear as irregular colorless to brown colonies.

Deoxycholate Citrate Agar

Composition per liter:

Sodium citrate	50.0g
Agar	15.0g
Lactose	10.0g
Beef extract	5.0g
Peptone	5.0g
$Na_2S_2O_3 \cdot 5H_2O$	5.0g
Sodium deoxycholate	2.5g
Ferric citrate	1.0g
Neutral Red	0.025g

pH 7.3 ± 0.2 at 25°C

Source: This medium is available as a premixed powder from Oxoid Unipath.

Preparation of Medium: Add components to distilled/deionized water and bring volume to 1.0L. Mix thoroughly. Gently heat and bring to boiling. Do not autoclave. Cool to 45°–50°C. Pour into sterile Petri dishes. Dry the agar surface before use.

Use: For the selective isolation and cultivation of enteric pathogens, especially *Salmonella* and *Shigella* species.

Deoxycholate Citrate Agar

Composition per liter:

Sodium citrate	20.0g
Agar	17.0g
Lactose	10.0g
Meat, solids from infusion	10.0g
Peptic digest of animal tissue	10.0g
Sodium deoxycholate	5.0g
Ferric citrate	1.0g
Neutral Red	0.02g

pH 7.3 ± 0.2 at 25°C

Source: This medium is available as a premixed powder from BD Diagnostic Systems.

Preparation of Medium: Add components to distilled/deionized water and bring volume to 1.0L. Mix thoroughly. Gently heat and bring to boiling. Do not autoclave. Cool to 45°–50°C. Pour into sterile Petri dishes. Dry the agar surface before use.

Use: For the selective isolation and cultivation of enteric pathogens, especially *Salmonella* and *Shigella* species.

Deoxycholate Citrate Agar (Desoxycholate Citrate Agar)

Composition per liter:

Pork infusion	330.0g
Sodium citrate	20.0g
Agar	13.5g
Lactose	10.0g
Proteose peptone No. 3	10.0g
Sodium deoxycholate	5.0g
Ferric ammonium citrate	2.0g
Neutral Red	0.02g

pH 7.5 ± 0.2 at 25°C

Source: This medium is available as a premixed powder from BD Diagnostic Systems.

Preparation of Medium: Add components to distilled/deionized water and bring volume to 1.0L. Mix thoroughly. Gently heat and bring to boiling. Do not autoclave. Cool to 45°–50°C. Pour into sterile Petri dishes. Dry the agar surface before use.

Use: For the selective isolation and cultivation of enteric pathogens, especially *Salmonella* and *Shigella* species.

Deoxycholate Citrate Agar

Composition per liter:

Sodium citrate	20.0g
Agar	13.0g
Proteose peptone	10.0g
Heart infusion solids	10.0g
Lactose	10.0g
Sodium deoxycholate	5.0g
Ferric ammonium citrate	2.0g
Neutral Red	0.02g

pH 7.5 ± 0.2 at 25°C

Source: This medium is available as a premixed powder from HiMedia.

Preparation of Medium: Add components to distilled/deionized water and bring volume to 1.0L. Mix thoroughly. Gently heat and bring to boiling. Do not autoclave. Cool to 45°–50°C. Pour into sterile Petri dishes. Dry the agar surface before use. Avoid excessive heating as it is detrimental to the medium.

Use: For the selective isolation and cultivation of enteric pathogens, especially *Salmonella* and *Shigella* species.

Deoxycholate Citrate Agar, HiVeg

Composition per liter:

Sodium citrate	20.0g
Agar	13.5g
Plant peptone No. 3	13.0g
Plant infusion	10.0g
Lactose	10.0g
Synthetic detergent No. III	2.0g
Ferric ammonium citrate	2.0g
Neutral Red	0.02g

pH 7.5 ± 0.2 at 25°C

Source: This medium is available as a premixed powder from HiMedia.

Preparation of Medium: Add components to distilled/deionized water and bring volume to 1.0L. Mix thoroughly. Gently heat and bring to boiling. Do not autoclave. Cool to 45°–50°C. Pour into sterile Petri dishes. Dry the agar surface before use. Avoid excessive heating as it is detrimental to the medium.

Use: For the selective isolation and cultivation of enteric pathogens, especially *Salmonella* and *Shigella* species.

Deoxycholate Citrate Agar, Hynes

Composition per liter:

Agar	12.0g
Lactose	10.0g
Sodium citrate	8.5g
$Na_2S_2O_3 \cdot 5H_2O$	5.4g
Beef extract powder	5.0g
Peptone	5.0g
Sodium deoxycholate	5.0g
Ferric citrate	1.0g
Neutral Red	0.02g

pH 7.3 ± 0.2 at 25°C.

Source: This medium is available as a premixed powder from Oxoid Unipath.

Preparation of Medium: Add components to distilled/deionized water and bring volume to 1.0L. Mix thoroughly. Gently heat and bring to boiling. Do not autoclave. Cool to 45°–50°C. Pour into sterile Petri dishes. Dry the agar surface before use.

Use: For the selective isolation, cultivation, and differentiation of enteric pathogens, especially *Salmonella* and *Shigella* species. Lactose-fermenting bacteria appear as pink colonies that may or may not be surrounded by a zone of precipitated deoxycholate. Nonlactose-fermenting bacteria appear as colorless colonies that are surrounded by a clear orange-yellow zone.

Deoxycholate Citrate Lactose Sucrose Agar
See: **DCLS Agar**

Deoxycholate Lactose Agar

Composition per liter:

Agar	15.0g
Lactose	10.0g
NaCl	5.0g
Pancreatic digest of casein	5.0g
Peptic digest of animal tissue	5.0g
Sodium citrate	2.0g
Sodium deoxycholate	0.5g
Neutral Red	0.033g

pH 7.1 ± 0.2 at 25°C

Source: This medium is available as a premixed powder from BD Diagnostic Systems.

Preparation of Medium: Add components to distilled/deionized water and bring volume to 1.0L. Mix thoroughly. Gently heat and bring to boiling. Do not autoclave. Cool to 45°–50°C. Pour into sterile Petri dishes. Dry the agar surface before use.

Use: For the selective isolation, cultivation, and differentiation of enteric pathogens, especially *Salmonella* and *Shigella* species. Lactose-fermenting bacteria appear as pink colonies that may or may not be surrounded by a zone of precipitated deoxycholate. Nonlactose-fermenting bacteria appear as colorless colonies that are surrounded by a clear orange-yellow zone. Also used for the enumeration of coliform bacteria from water, milk, and dairy products.

Deoxycholate Lactose HiVeg Agar

Composition per liter:

Agar	15.0g
Plant special peptone	10.0g
Lactose	10.0g
NaCl	5.0g
Sodium citrate	2.0g
Synthetic detergent No. III	0.5g
Neutral Red	0.03g

pH 7.1 ± 0.2 at 25°C

Source: This medium is available as a premixed powder from HiMedia.

Preparation of Medium: Add components to distilled/deionized water and bring volume to 1.0L. Mix thoroughly. Gently heat and bring to boiling. Do not autoclave. Cool to 45°–50°C. Pour into sterile Petri dishes. Dry the agar surface before use.

Use: For the selective isolation, cultivation, and differentiation of enteric pathogens, especially *Salmonella* and *Shigella* species. Lactose-fermenting bacteria appear as pink colonies that may or may not be surrounded by a zone of precipitated deoxycholate. Nonlactose-fermenting bacteria appear as colorless colonies that are surrounded by a clear orange-yellow zone. Also used for the enumeration of coliform bacteria from water, milk, and dairy products.

Deoxycholate Lactose Sucrose Sorbitol Agar

Composition per liter:

Sodium citrate	20.0g
Agar	15.0g
D-Sorbitol	10.0g
Lactose	10.0g
Sucrose	5.0g
Pancreatic digest of casein	5.0g
Yeast extract	5.0g
Sodium deoxycholate	2.5g
Ferric citrate	1.0g
Neutral Red	0.02g

pH 7.4 ± 0.2 at 25°C

Preparation of Medium: Add components to distilled/deionized water and bring volume to 1.0L. Mix thoroughly. Gently heat and bring to boiling. Do not overheat. Adjust pH to 7.4. Do not autoclave. Pour into sterile Petri dishes or distribute into sterile tubes.

Use: For the isolation and cultivation of *Hafnia* species.

Dermabacter Medium

Composition per liter:

Pancreatic digest of casein	10.0g
Glucose	5.0g
NaCl	5.0g
Yeast extract	5.0g

pH 7.4 ± 0.2 at 25°C

Preparation of Medium: Add components to distilled/deionized water and bring volume to 1.0L. Mix thoroughly. Distribute into tubes or flasks. Autoclave for 15 min at 15 psi pressure–121°C.

Use: For the cultivation and maintenance of *Dermabacter hominus*.

Dermasel Agar Base

Composition per liter:

Glucose	20.0g
Agar	14.5g
Papaic digest of soybean meal	10.0g
Antibiotic inhibitor	10.0mL

pH 6.8–7.0 at 25°C

Source: This medium is available as a premixed powder from Oxoid Unipath.

Antibiotic Inhibitor:
Composition per 10.0mL:

Cycloheximide	0.4g
Chloramphenicol	0.05g
Acetone	10.0mL

Preparation of Antibiotic Inhibitor: Add cycloheximide and chloramphenicol to 10.0mL of acetone. Mix thoroughly.

Caution: Cycloheximide is toxic. Avoid skin contact or aerosol formation and inhalation.

Preparation of Medium: Add components to distilled/deionized water and bring volume to 990.0mL. Mix thoroughly. Gently heat and bring to boiling. Do not overheat. Add antibiotic inhibitor. Mix thoroughly. Autoclave for 10 min at 15 psi pressure–121°C. Pour into sterile Petri dishes.

Use: For the isolation and cultivation of dermatophytic fungi isolated from hair, nails, or skin scrapings.

Dermatophyte Test Medium Agar

See: **DTM Agar**

Dermatophyte Test Medium

Composition per liter:

Agar	20.0g
Enzymatic digest of soybean meal	10.0g
Glucose	10.0g
Cycloheximide	0.5g
Phenol Red	0.2g
Selective supplement solution	10.0mL

pH 5.5 ± 0.2 at 25°C

Source: This medium is available from Acumedia, Neogen Corp.

Caution: Cycloheximide is toxic. Avoid skin contact or aerosol formation and inhalation.

Selective Supplement Solution:
Composition per 10.0mL:

Gentamicin	0.1g
Chlortetracycline	0.1g

Preparation of Selective Supplement Solution: Add components to distilled/deionized water and bring volume to 10.0mL. Mix thoroughly. Filter sterilize.

Preparation of Medium: Add components, except selective supplement solution, to distilled/deionized water and bring volume to 990.0mL. Mix thoroughly. Distribute into tubes or flasks. Gently heat and bring to boiling. Autoclave for 15 min at 15 psi pressure–121°C. Cool to 50°C. Aseptically add 10.0mL selective supplement solution. Mix thoroughly. Pour into sterile Petri dishes or leave in tubes.

Use: For the selective isolation of dermatophytic fungi.

Dermatophyte Test Medium Base

Composition per liter:

Agar	20.0g
Glucose	10.0g

Papaic digest of soybean meal	10.0g
Cycloheximide	0.5g
Phenol Red	0.2g
Gentamycin sulfate	0.1g
Chlortetracycline	0.1g

pH 5.5 ± 0.2 at 25°C

Source: This medium is available as a premixed powder from BD Diagnostic Systems.

Caution: Cycloheximide is toxic. Avoid skin contact or aerosol formation and inhalation.

Preparation of Medium: Add components, except gentamycin sulfate and chlortetracycline, to distilled/deionized water and bring volume to 1.0L. Mix thoroughly. Gently heat while stirring and bring to boiling. Autoclave for 15 min at 15 psi pressure–121°C. Cool to 45°–50°C. Aseptically add gentamycin sulfate and chlortetracycline. Mix thoroughly. Pour into sterile Petri dishes.

Use: For the selective isolation and cultivation of pathogenic fungi from cutaneous sources.

Dermocystidium Medium

Composition per liter:

NaCl	48.0g
Agar	36.0g
$MgSO_4 \cdot 7H_2O$	16.0g
Glucose	8.0 g
Casein hydrolysate or sodium glutamate	4.0g
Tris(hydroxymethyl)aminomethane buffer	4.0g
KCl	1.4g
$CaCl_2$	0.94g
K_2HPO_4	0.86g
Thiamine·HCl	400.0μg
Cyanocobalamine	6.0μg
Trace metal mix stock	20.0mL

Trace Metal Mix Stock:
Composition per 100.0mL:

H_3BO_3	114.0mg
EDTA	100.0mg
$FeCl_3 \cdot 6H_2O$	96.8mg
$MnCl_2 \cdot 4H_2O$	36.0mg
$Na_2MoO_4 \cdot 2H_2O$	23.0mg
$ZnCl_2$	13.4mg
$CuCl_2 \cdot 2H_2O$	536.0μg
$CoCl_2 \cdot 6H_2O$	400.0μg

pH 7.4 ± 0.3 at 25°C

Preparation of Trace Metal Mix Stock: Add components to distilled/deionized water and bring volume to 100.0mL. Mix thoroughly.

Preparation of Medium: Add components to distilled/deionized water and bring volume to 1.0L. Mix thoroughly. Gently heat and bring to boiling. Adjust pH to 7.4 with concentrated HCl. Bring volume to 2.0L with distilled/deionized water. Distribute into tubes or flasks. Autoclave for 15 min at 15 psi pressure–121°C. Pour into sterile Petri dishes or leave in tubes.

Use: For the cultivation and maintenance of *Dermocystidium* species.

Derxia gummosa Medium

Composition per liter:

Agar	20.0g
Starch	20.0g
$MgSO_4 \cdot 7H_2O$	0.2g
KH_2PO_4	0.15g
$NaHCO_3$	0.1g
K_2HPO_4	0.05g
$CaCl_2$	0.02g
$Na_2MoO_4 \cdot 2H_2O$	2.0mg
Bromthymol Blue solution	5.0mL
$FeCl_3 \cdot 6H_2O$ (10% solution)	0.1mL

pH 6.9 ± 0.2 at 25°C

Bromthymol Blue Solution:
Composition per 10.0mL:

Bromthymol Blue	0.5g
Ethanol	10.0mL

Preparation of Bromthymol Blue Solution: Add Bromthymol Blue to 10.0mL of ethanol. Mix thoroughly.

Preparation of Medium: Add components to distilled/deionized water and bring volume to 1.0L. Mix thoroughly. Distribute into tubes or flasks. Autoclave for 15 min at 15 psi pressure–121°C.

Use: For the cultivation of *Derxia gummosa.*

Derxia Medium

Composition per liter:

Agar	20.0g
Glucose	20.0g
NH_4Cl	2.0g
K_2HPO_4	1.0g
$MgSO_4 \cdot 7H_2O$	0.2g
$CaSO_4$	5.0mg
$FeSO_4 \cdot 7H_2O$	5.0mg
$Na_2MoO_4 \cdot 2H_2O$	0.5mg

pH 6.7 ± 0.2 at 25°C

Preparation of Medium: Add components to distilled/deionized water and bring volume to 1.0L. Mix thoroughly. Gently heat and bring to boiling. Adjust pH to 6.7. Distribute into tubes or flasks. Autoclave for 15 min at 15 psi pressure–121°C. Pour into sterile Petri dishes or leave in tubes.

Use: For the cultivation and maintenance of *Derxia gummosa*.

Desoxycholate Agar

See: **Deoxycholate Agar**

Desoxycholate Citrate Agar

See: **Deoxycholate Citrate Agar**

Desulfacinum hydrothermale Medium (DSMZ Medium 875)

Composition per 1004mL:

Solution A	920.0mL
Soluiton C ($NaHCO_3$ solution)	50.0mL
Solution F	13.0mL
Solution D (Seven vitamin solution)	10.0mL
Solution E	10.0mL
Soluiton B (Trace elements solution SL-10)	1.0mL

pH 7.0–7.3 at 25°C

Solution A:
Composition per 920mL:

NaCl	10.4g
$MgSO_4 \cdot 7H_2O$	2.72g

$MgCl_2 \cdot 6H_2O$	2.24g
$CaCl_2 \cdot 2H_2O$	0.56g
KCl	0.29g
NH_4Cl	0.1g
KH_2PO_4	0.08g
Resazurin	0.5mg

Preparation of Solution A: Add components to distilled/deionized water and bring volume to 920.0mL. Mix thoroughly. Sparge with 100% N_2. Autoclave for 15 min at 15 psi pressure–121°C.

Solution B (Trace Elements Solution SL-10):
Composition per liter:

$FeCl_2 \cdot 4H_2O$	1.5g
$CoCl_2 \cdot 6H_2O$	190.0mg
$MnCl_2 \cdot 4H_2O$	100.0mg
$ZnCl_2$	70.0mg
$Na_2MoO_4 \cdot 2H_2O$	36.0mg
$NiCl_2 \cdot 6H_2O$	24.0mg
H_3BO_3	6.0mg
$CuCl_2 \cdot 2H_2O$	2.0mg
HCl (25% solution)	10.0mL

Preparation of Solution B (Trace Elements Solution SL-10): Add $FeCl_2 \cdot 4H_2O$ to 10.0mL of HCl solution. Mix thoroughly. Add distilled/deionized water and bring volume to 1.0L. Add remaining components. Mix thoroughly. Sparge with 100% N_2. Autoclave for 15 min at 15 psi pressure–121°C.

Solution C ($NaHCO_3$ Solution:)
Composition per 100.0mL:

$NaHCO_3$	5.0g

Preparation of Solution C ($NaHCO_3$ Solution): Add $NaHCO_3$ to distilled/deionized water and bring volume to 100.0mL. Mix thoroughly. Sparge with 80% N_2 + 20% CO_2. Autoclave for 15 min at 15 psi pressure–121°C.

Solution D (Seven Vitamin Solution):
Composition per liter:

Pyridoxine hydrochloride	300.0mg
Thiamine-HCl·$2H_2O$	200.0mg
Nicotinic acid	200.0mg
Vitamin B_{12}	100.0mg
Calcium pantothenate	100.0mg
p-Aminobenzoic acid	80.0mg
D(+)-Biotin	20.0mg

Preparation of Solution D (Seven Vitamin Solution): Add components to distilled/deionized water and bring volume to 1.0L. Sparge with 100% N_2. Mix thoroughly. Filter sterilize.

Solution E:
Composition per 10.0mL:

Na-lactate	2.5g

Preparation of Solution E: Add Na-lactate to distilled/deionized water and bring volume to 10.0mL. Mix thoroughly. Sparge with 100% N_2. Autoclave for 15 min at 15 psi pressure–121°C.

Solution F:
Composition per 100.0mL:

$Na_2S \cdot 9H_2O$	3.0g

Preparation of Solution F: Add $Na_2S \cdot 9H_2O$ to distilled/deionized water and bring volume to 100.0mL. Mix thoroughly. Sparge with 100% N_2. Autoclave for 15 min at 15 psi pressure–121°C.

Preparation of Medium: Prepare and dispense medium under 80% N_2 + 20% CO_2 gas atmosphere. Add 50.0mL sterile solution C, 13.0mL sterile solution F, 10.0mL sterile solution D, 10.0mL sterile solution E, and 1.0mL sterile solution B to 920.0mL sterile solution A. Mix thoroughly. The pH of the completed medium should be 7.0–7.3. Aseptically and anaerobically distribute into sterile tubes or bottles.

Use: For the cultivation of *Desulfacinum hydrothermale.*

Desulfacinum Medium (DSMZ Medium 1100)

Composition per liter:

NaCl	7.0g
$MgCl_2 \cdot 6H_2O$	3.1g
Na_2SO_3	3.0g
KCl	0.5g
Yeast extract	0.5g
NH_4Cl	0.3g
KH_2PO_4	0.2g
$CaCl_2 \cdot 2H_2O$	0.1g
Resazurin	0.5mg
Na-lactate solution	10.0mL
$NaHCO_3$ solution	10.0mL
$Na_2S \cdot 9H_2O$ solution	10.0mL
Trace elements solution SL-10 with EDTA	1.0mL
Selenite/tungstate solution	1.0mL

pH 7.2 ± 0.2 at 25°C

Na-lactate Solution:
Composition per 10.0mL:

Na-lactate	2.0g

Preparation of Na-lactate Solution: Add components to distilled/deionized water and bring volume to 10.0mL. Mix thoroughly. Sparge with 100% N_2. Autoclave for 15 min at 15 psi pressure–121°C. Cool to room temperature.

$NaHCO_3$ Solution:
Composition per 10.0mL:

$NaHCO_3$	1.5g

Preparation of $NaHCO_3$ Solution: Add components to distilled/deionized water and bring volume to 10.0mL. Mix thoroughly. Sparge with 20% CO_2 + 80% H_2. Autoclave for 15 min at 15 psi pressure–121°C. Cool to room temperature.

$Na_2S \cdot 9H_2O$ Solution:
Composition per 10.0mL:

$Na_2S \cdot 9H_2O$	0.5g

Preparation of $Na_2S \cdot 9H_2O$ Solution: Add $Na_2S \cdot 9H_2O$ to distilled/deionized water and bring volume to 10.0mL. Mix thoroughly. Adjust to pH 7.0. Autoclave under 100% N_2 for 15 min at 15 psi pressure–121°C. Cool to room temperature.

Selenite/Tungstate Solution:
Composition per liter:

NaOH	0.5g
$Na_2WO_4 \cdot 2H_2O$	4.0mg
$Na_2SeO_3 \cdot 5H_2O$	3.0mg

Preparation of Selenite/Tungstate Solution: Add components to distilled/deionized water and bring volume to 1.0L. Mix thoroughly. Sparge with 100% N_2. Autoclave for 15 min at 15 psi pressure–121°C.

Trace Elements Solution SL-10 with EDTA:
Composition per liter:
$FeCl_2 \cdot 4H_2O$ 1.5g
Na_2-EDTA 0.5g
$CoCl_2 \cdot 6H_2O$ 190.0mg
$MnCl_2 \cdot 4H_2O$ 100.0mg
$ZnCl_2$ 70.0mg
$Na_2MoO_4 \cdot 2H_2O$ 36.0mg
$NiCl_2 \cdot 6H_2O$ 24.0mg
H_3BO_3 .... 6.0mg
$CuCl_2 \cdot 2H_2O$ 2.0mg
HCl (25% solution) 10.0mL

Preparation of Trace Elements Solution SL-10 with EDTA: Add $FeCl_2 \cdot 4H_2O$ to 10.0mL of HCl solution. Mix thoroughly. Add distilled/deionized water and bring volume to 1.0L. Add remaining components. Mix thoroughly. Sparge with 100% N_2. Autoclave for 15 min at 15 psi pressure–121°C. Adjust pH to 7.0.

Preparation of Medium: Add components, except bicarbonate, lactate, and sulfite solution, to distilled/deionized water and bring volume to 970.0mL. Mix thoroughly. Gently heat and bring to boiling. Boil for 1 min. Cool while sparging with 100% N_2. Add the solid bicarbonate. Dispense under 100% N_2 into culture vessels. Autoclave for 15 min at 15 psi pressure–121°C. Aseptically and anoxically add bicarbonate, lactate, and sulfide solutions. Adjust final pH of the medium to pH 7.2.

Use: For the cultivation of *Desulfacinum* spp.

Desulfatirhabdium Medium (DSMZ Medium 1086)

Composition per liter:
Na_2SO_4 .... 2.8g
Na_2HPO_4 .... 0.53g
KH_2PO_4 .... 0.41g
NH_4Cl 0.3g
NaCl 0.3g
$CaCl_2 \cdot 2H_2O$ 0.11g
$MgCl_2 \cdot 6H_2O$ 0.1g
Yeast extract 0.02g
Crotonate solution 10.0mL
Benzoate solution 10.0mL
Vitamin solution 10.0mL
$Na_2S \cdot 9H_2O$ solution 10.0mL
$NaHCO_3$ solution 10.0mL
Trace elements solution SL-10 .... 1.0mL
Selenite/tungstate solution 1.0mL
pH 7.1 ± 0.2 at 25°C

$NaHCO_3$ Solution:
Composition per 10.0mL:
$NaHCO_3$ 4.0g

Preparation of $NaHCO_3$ Solution: Add $NaHCO_3$ to distilled/deionized water and bring volume to 10.0mL. Mix thoroughly. Sparge with 20% CO_2 + 80% H_2. Autoclave for 15 min at 15 psi pressure–121°C. Cool to room temperature.

Vitamin Solution:
Composition per liter:
Pyridoxine-HCl 10.0mg
Thiamine-HCl·$2H_2O$ 5.0mg
Riboflavin 5.0mg
Nicotinic acid 5.0mg
D-Ca-pantothenate 5.0mg
p-Aminobenzoic acid 5.0mg
Lipoic acid 5.0mg
Biotin 2.0mg
Folic acid 2.0mg
Vitamin B_{12} 0.1mg

Preparation of Vitamin Solution: Add components to distilled/deionized water and bring volume to 1.0L. Mix thoroughly. Sparge with 80% H_2 + 20% CO_2. Filter sterilize.

Crotonate Solution:
Composition per 10.0mL:
Na-crotonate 1.7g

Preparation of Crotonate Solution: Add components to distilled/deionized water and bring volume to 10.0mL. Mix thoroughly. Sparge with 100% N_2. Autoclave for 15 min at 15 psi pressure–121°C.

Benzoate Solution:
Composition per 10.0mL:
Na-benzoate 0.43g

Preparation of Benzoate Solution: Add components to distilled/deionized water and bring volume to 10.0mL. Mix thoroughly. Sparge with 100% N_2. Autoclave for 15 min at 15 psi pressure–121°C.

$Na_2S \cdot 9H_2O$ Solution:
Composition per 10.0mL:
$Na_2S \cdot 9H_2O$ 0.5g

Preparation of $Na_2S \cdot 9H_2O$ Solution: Add $Na_2S \cdot 9H_2O$ to distilled/deionized water and bring volume to 10.0mL. Mix thoroughly. Autoclave under 100% N_2 for 15 min at 15 psi pressure–121°C. Cool to room temperature.

Selenite/Tungstate Solution:
Composition per liter:
NaOH 0.5g
$Na_2WO_4 \cdot 2H_2O$ 4.0mg
$Na_2SeO_3 \cdot 5H_2O$ 3.0mg

Preparation of Selenite/Tungstate Solution: Add components to distilled/deionized water and bring volume to 1.0L. Mix thoroughly. Sparge with 100% N_2. Autoclave for 15 min at 15 psi pressure–121°C.

Trace Elements Solution SL-10:
Composition per liter:
$FeCl_2 \cdot 4H_2O$ 1.5g
$CoCl_2 \cdot 6H_2O$ 190.0mg
$MnCl_2 \cdot 4H_2O$ 100.0mg
$ZnCl_2$ 70.0mg
$Na_2MoO_4 \cdot 2H_2O$ 36.0mg
$NiCl_2 \cdot 6H_2O$ 24.0mg
H_3BO_3 .... 6.0mg
$CuCl_2 \cdot 2H_2O$ 2.0mg
HCl (25% solution) 10.0mL

Preparation of Trace Elements Solution SL-10: Add $FeCl_2 \cdot 4H_2O$ to 10.0mL of HCl solution. Mix thoroughly. Add distilled/deionized water and bring volume to 1.0L. Add remaining components. Mix thoroughly. Sparge with 100% N_2. Autoclave for 15 min at 15 psi pressure–121°C.

Preparation of Medium: Add components, except bicarbonate, vitamins, crotonate, benzoate, and sulfide solutions, to distilled/deionized water and bring volume to 950.0mL. Mix thoroughly. Gently heat and bring to boiling. Boil for 1 min. Cool while sparging with 80% N_2 + 20% CO_2. Dispense under 80% N_2 + 20% CO_2 into culture vessels.

Autoclave for 15 min at 15 psi pressure–121°C. Aseptically and anoxically add vitamins, crotonate, benzoate, and sulfide. Adjust the final pH of the medium to 7.0–7.2. After inoculation pressurize the vessels with 80% N_2 + 20% CO_2 to 0.7 bar overpressure.

Use: For the cultivation of *Desulfatirhabdium* spp.

Desulfitobacterium dehalogenans Medium

Composition per liter:

Solution A	955.0mL
Solution B	25.0mL
Solution C	20.0mL

Solution A:

Composition per 955.0mL:

Na_2HPO_4	2.2g
Yeast extract	2.0g
3-Chloro-4-hydroxyphenylacetic acid	1.5g
L-Cysteine·HCl·H_2O	0.7g
NH_4Cl	0.5g
KH_2PO_4	0.44g
$MgCl_2 \cdot 6H_2O$	0.2g
$CaCl_2$	25.0mg
Wolfe's mineral solution	10.0mL

Wolfe's Mineral Solution:

$MgSO_4 \cdot 7H_2O$	3.0g
Nitrilotriacetic acid	1.5g
NaCl	1.0g
$MnSO_4 \cdot H_2O$	0.5g
$FeSO_4 \cdot 7H_2O$	0.1g
$CoCl_2 \cdot 6H_2O$	0.1g
$CaCl_2$	0.1g
$ZnSO_4 \cdot 7H_2O$	0.1g
$CuSO_4 \cdot 5H_2O$	0.01g
$AlK(SO_4)_2 \cdot 12H_2O$	0.01g
H_3BO_3	0.01g
$Na_2MoO_4 \cdot 2H_2O$	0.01g

Preparation of Wolfe's Mineral Solution: Add nitrilotriacetic acid to 500.0mL of distilled/deionized water. Adjust pH to 6.5 with KOH. Add remaining components one at a time. Add distilled/deionized water to 1.0L.

Preparation of Solution A: Add components, except L-cysteine·HCl·H_2O, to distilled/deionized water and bring volume to 955.0mL. Mix thoroughly. Gently heat and bring to boiling. Cool to room temperature while sparging with 90% N_2 + 10% CO_2. Adjust pH to 7.3. Add L-cysteine·HCl·H_2O. Autoclave for 15 min at 15 psi pressure–121°C.

Solution B:

Composition per 25.0mL:

Sodium pyruvate	2.2g

Preparation of Solution B: Add sodium pyruvate to distilled/deionized water and bring volume to 25.0mL. Mix thoroughly. Filter sterilize. Sparge with 100% N_2.

Solution C:

Composition per 20.0mL:

$NaHCO_3$	1.0g

Preparation of Solution C: Add $NaHCO_3$ to distilled/deionized water and bring volume to 20.0mL. Mix thoroughly. Filter sterilize. Sparge with 100% CO_2.

Preparation of Medium: Aseptically and anaerobically combine 955.0mL of sterile solution A with 25.0mL of sterile solution B and 20.0mL of sterile solution C. Mix thoroughly. Aseptically and anaerobically distribute into sterile tubes or flasks.

Use: For the cultivation of *Desulfitobacterium dehalogenans.*

Desulfitobacterium dehalogenans Medium

Composition per liter:

KH_2PO_4	5.4g
Sodium pyruvate	2.2g
3-Chloro-4-hydroxyphenylacetic acid	1.9g
Yeast extract	1.0g
NH_4Cl	0.5g
$MgCl_2 \cdot 6H_2O$	90.0mg
$CaCl_2$	25.0mg
Reducing solution	20.0mL
Wolfe's vitamin solution	10.0mL
Modified Wolfe's mineral solution	5.0mL

pH 7.5 ± 0.2 at 25°C

Reducing Solution:

Composition per liter:

L-Cysteine·HCl·H_2O	12.5g
$Na_2S \cdot 9H_2O$	12.5g
NaOH	8.0g

Preparation of Reducing Solution: Add components to distilled/deionized water and bring volume to 1.0L. Mix thoroughly. Sparge with 100% N_2. Anaerobically distribute into anaerobic tubes. Autoclave for 15 min at 15 psi pressure–121°C.

Wolfe's Vitamin Solution:

Composition per liter:

Pyridoxine·HCl	10.0mg
p-Aminobenzoic acid	5.0mg
Lipoic acid	5.0mg
Nicotinic acid	5.0mg
Riboflavin	5.0mg
Thiamine·HCl	5.0mg
Calcium DL-pantothenate	5.0mg
Biotin	2.0mg
Folic acid	2.0mg
Vitamin B_{12}	0.1mg

Preparation of Wolfe's Vitamin Solution: Add components to distilled/deionized water and bring volume to 1.0L. Mix thoroughly.

Modified Wolfe's Mineral Solution:

Composition per liter:

$MgSO_4 \cdot 7H_2O$	3.0g
Nitrilotriacetic acid	1.5g
NaCl	1.0g
$MnSO_4 \cdot H_2O$	0.5g
$CaCl_2$	0.1g
$CoCl_2 \cdot 6H_2O$	0.1g
$FeSO_4 \cdot 7H_2O$	0.1g
$ZnSO_4 \cdot 7H_2O$	0.1g
$AlK(SO_4)_2 \cdot 12H_2O$	0.01g
$CuSO_4 \cdot 5H_2O$	0.01g
H_3BO_3	0.01g
$Na_2MoO_4 \cdot 2H_2O$	0.01g
Na_2SeO_3	0.01g
$NaWO_4 \cdot 2H_2O$	0.01g
$NiCl_2 \cdot 6H_2O$	0.01g

Preparation of Modified Wolfe's Mineral Solution: Add nitrilotriacetic acid to 500.0mL of distilled/deionized water. Adjust pH to 6.5 with KOH. Add remaining components one at a time. Add distilled/deionized water to 1.0L. Adjust pH to 6.8.

Preparation of Medium: Prepare and dispense medium under 100% N_2. Add components, except reducing solution, to distilled/deionized water and bring volume to 980.0mL. Mix thoroughly. Gently heat and bring to boiling. Continue boiling for 3 min. Cool to room temperature while sparging with 100% N_2. Add reducing solution. Mix thoroughly. Adjust pH to 7.5. Anaerobically distribute into anaerobic tubes. Autoclave for 15 min at 15 psi pressure–121°C.

Use: For the cultivation of *Desulfitobacterium dehalogenans.*

Desulfitobacterium hafniense Medium

Composition per 1005.0mL:

$NaHCO_3$	2.6g
NH_4Cl	1.0g
Yeast extract	1.0g
$K_2HPO_4 \cdot 3H_2O$	0.4g
$MgCl_2 \cdot 6H_2O$	0.1g
NaCl	0.1g
$CaCl_2 \cdot 2H_2O$	0.05g
Resazurin	0.5mg
$Na_2S \cdot 9H_2O$ solution	10.0mL
Sodium pyruvate solution	10.0mL
Wolfe's vitamin solution	10.0mL
$Na_2S_2O_3$ solution	5.0mL
Selenite-tungstate solution	1.0mL
Trace elements solution SL-10 with EDTA	1.0mL

pH 7.5 ± 0.2 at 25°C

$Na_2S \cdot 9H_2O$ Solution:

Composition per 10.0mL:

$Na_2S \cdot 9H_2O$	0.3g

Preparation of $Na_2S \cdot 9H_2O$ Solution: Add $Na_2S \cdot 9H_2O$ to distilled/deionized water and bring volume to 10.0mL. Mix thoroughly. Sparge with 100% N_2. Autoclave for 15 min at 15 psi pressure–121°C.

Sodium Pyruvate Solution:

Composition per 10.0mL:

Sodium pyruvate	2.5g

Preparation of Sodium Pyruvate Solution: Add sodium pyruvate to distilled/deionized water and bring volume to 10.0mL. Mix thoroughly. Sparge with 100% N_2. Autoclave for 15 min at 15 psi pressure–121°C.

Wolfe's Vitamin Solution:

Composition per liter:

Pyridoxine·HCl	10.0mg
p-Aminobenzoic acid	5.0mg
Lipoic acid	5.0mg
Nicotinic acid	5.0mg
Riboflavin	5.0mg
Thiamine·HCl	5.0mg
Calcium DL-pantothenate	5.0mg
Biotin	2.0mg
Folic acid	2.0mg
Vitamin B_{12}	0.1mg

Preparation of Wolfe's Vitamin Solution: Add components to distilled/deionized water and bring volume to 1.0L. Mix thoroughly. Filter sterilize. Sparge with 100% N_2.

$Na_2S_2O_3$ Solution:

Composition per 10.0mL:

$Na_2S_2O_3 \cdot 5H_2O$	2.5g

Preparation of $Na_2S_2O_3$ Solution: Add $Na_2S_2O_3 \cdot 5H_2O$ to distilled/deionized water and bring volume to 10.0mL. Mix thoroughly. Sparge with 100% N_2. Autoclave for 15 min at 15 psi pressure–121°C.

Selenite-Tungstate Solution:

Composition per liter:

NaOH	0.5g
$Na_2WO_4 \cdot 2H_2O$	4.0mg
$Na_2SeO_3 \cdot 5H_2O$	3.0mg

Preparation of Selenite-Tungstate Solution: Add components to distilled/deionized water and bring volume to 1.0L. Mix thoroughly.

Trace Elements Solution SL-10 with EDTA:

Composition per liter:

$FeCl_2 \cdot 4H_2O$	1.5g
Disodium EDTA	0.5g
$CoCl_2 \cdot 6H_2O$	190.0mg
$MnCl_2 \cdot 4H_2O$	100.0mg
$ZnCl_2$	70.0mg
$Na_2MoO_4 \cdot 2H_2O$	36.0mg
$NiCl_2 \cdot 6H_2O$	24.0mg
H_3BO_3	6.0mg
$CuCl_2 \cdot 2H_2O$	2.0mg
HCl (25% solution)	10.0mL

Preparation of Trace Elements Solution SL-10 with EDTA: Add $FeCl_2 \cdot 4H_2O$ to 10.0mL of HCl solution. Mix thoroughly. Add distilled/deionized water and bring volume to 1.0L. Add remaining components. Mix thoroughly.

Preparation of Medium: Prepare and dispense medium under 80% N_2 + 20% CO_2. Add components, except $NaHCO_3$, $Na_2S \cdot 9H_2O$ solution, sodium pyruvate solution, vitamin solution, and $Na_2S_2O_3 \cdot 5H_2O$ solution, to distilled/deionized water and bring volume to 970.0mL. Mix thoroughly. Gently heat and bring to boiling. Continue boiling for 3 min. Cool to room temperature while sparging with 80% N_2 + 20% CO_2. Add $NaHCO_3$. Mix thoroughly. Adjust pH to 7.0. Anaerobically distribute 9.7mL volumes into anaerobic tubes. Autoclave for 15 min at 15 psi pressure–121°C. Aseptically and anaerobically add 0.1mL of sterile $Na_2S \cdot 9H_2O$ solution, 0.1mL of sterile sodium pyruvate solution, 0.1mL of sterile vitamin solution, and 0.05mL of sterile $Na_2S_2O_3 \cdot 5H_2O$ solution to each tube. Mix thoroughly.

Use: For the cultivation of *Desulfitobacterium hafniense.*

Desulfitobacterium Medium (DSMZ Medium 663)

Composition per liter:

KH_2PO_4	5.44g
Yeast extract	1.0g
NH_4Cl	0.5g
$MgCl_2 \cdot 2H_2O$	0.18g
$CaCl_2 \cdot 2H_2O$	0.032g
Resazurin	0.5mg
Vitamin solution	20.0mL
Na-pyruvate solution	10.0mL
Na-thiosulfate solution	10.0mL
Cysteine solution	10.0mL

$Na_2S \cdot 9H_2O$ solution	10.0mL
Trace elements solution	5.0mL

pH 7.5 ± 0.2 at 25°C

Na-pyruvate Solution:

Composition per 10.0mL:

Na-pyruvate	2.5g

Preparation of Na-pyruvate Solution: Add Na-pyruvate to distilled/deionized water and bring volume to 10.0mL. Mix thoroughly. Sparge with 100% N_2. Filter sterilize.

Na-thiosulfate Solution:

Composition per 10.0mL:

$Na_2S_2O_3 \cdot 5H_2O$	2.5g

Preparation of Na-thiosulfate Solution: Add $Na_2S_2O_3 \cdot 5H_2O$ to distilled/deionized water and bring volume to 10.0mL. Mix thoroughly. Sparge with 100% N_2. Filter sterilize.

$Na_2S \cdot 9H_2O$ Solution:

Composition per 10.0mL:

$Na_2S \cdot 9H_2O$	0.4g

Preparation of $Na_2S \cdot 9H_2O$ Solution: Add $Na_2S \cdot 9H_2O$ to distilled/deionized water and bring volume to 10.0mL. Mix thoroughly. Autoclave under 100% N_2 for 15 min at 15 psi pressure–121°C. Cool to room temperature.

Cysteine Solution:

Composition per 10.0mL:

L-Cysteine·HCl·H_2O	0.4g

Preparation of Cysteine Solution: Add L-cysteine·HCl·H_2O to distilled/deionized water and bring volume to 10.0mL. Mix thoroughly. Sparge with 100% N_2. Autoclave for 15 min at 15 psi pressure–121°C.

Trace Elements Solution:

Composition per liter:

Nitrilotriacetic acid	12.8g
$FeCl_3 \cdot 6H_2O$	1.35g
NaCl	1.0g
$CoCl_2 \cdot 4H_2O$	0.24g
$NiCl_2 \cdot 6H_2O$	0.12g
$MnCl_2 \cdot 4H_2O$	0.1g
$CaCl_2 \cdot 2H_2O$	0.1g
$ZnCl_2$	0.1g
$Na_2SeO_3 \cdot 5H_2O$	0.026g
$CuCl_2 \cdot 2H_2O$	0.025g
$Na_2MoO_4 \cdot 4H_2O$	0.024g
H_3BO_3	0.01g

Preparation of Trace Elements Solution: Add nitrilotriacetic acid to 500.0mL of distilled/deionized water. Dissolve by adjusting pH to 6.5 with KOH. Add remaining components. Add distilled/deionized water to 1.0L. Mix thoroughly. Adjust pH to 6.8.

Vitamin Solution:

Composition per liter:

Pyridoxine-HCl	10.0mg
Thiamine-HCl·$2H_2O$	5.0mg
Riboflavin	5.0mg
Nicotinic acid	5.0mg
D-Ca-pantothenate	5.0mg
p-Aminobenzoic acid	5.0mg
Lipoic acid	5.0mg
Biotin	2.0mg
Folic acid	2.0mg
Vitamin B_{12}	0.1mg

Preparation of Vitamin Solution: Add components to distilled/deionized water and bring volume to 1.0L. Mix thoroughly. Sparge with 80% H_2 + 20% CO_2. Filter sterilize.

Preparation of Medium: Prepare and dispense medium under an oxygen-free atmosphere of 100% N_2. Add components, except vitamin solution, cysteine solution, Na-pyruvate solution, Na-thiosulfate solution, and $Na_2S \cdot 9H_2O$ solution, to distilled/deionized water and bring volume to 940.0mL. Mix thoroughly. Sparge with 100% N_2. Autoclave for 15 min at 15 psi pressure–121°C. Cool to 25°C. Aseptically and anaerobically add 20.0mL sterile vitamin solution, 10.0mL of sterile cysteine solution, 10.0mL sterile Na-pyruvate solution, 10.0mL sterile Na-thiosulfate solution, and 10.0mL of sterile $Na_2S \cdot 9H_2O$ solution. Mix thoroughly. Adjust pH to 7.5. Aseptically and anaerobically distribute into sterile tubes or flasks.

Use: For the cultivation of *Desulfitobacterium dehalogenans.*

Desulfitobacterium PCE Medium (DSMZ Medium 717)

Composition per liter:

$(NH_4)H_2PO_4$	2.88g
$MgSO_4 \cdot 7H_2O$	0.1g
Yeast extract	0.1g
$Ca(NO_3)_2 \cdot 4H_2O$	0.05g
Resazurin	0.1mg
$NaHCO_3$ solution	50.0mL
KOH solution	20.0mL
Na-lactate solution	20.0mL
Na-fumarate solution	20.0mL
Vitamin solution	10.0mL
$Na_2S \cdot 9H_2O$ solution	3.3mL
Seven vitamin solution	1.0mL
Trace elements solution SL-10	1.0mL
Selenite-tungstate solution	1.0mL

pH 7.1 ± 0.2 at 25°C

Selenite-Tungstate Solution:

Composition per liter:

NaOH	0.5g
$Na_2WO_4 \cdot 2H_2O$	4.0mg
$Na_2SeO_3 \cdot 5H_2O$	3.0mg

Preparation of Selenite-Tungstate Solution: Add components to distilled/deionized water and bring volume to 1.0L. Mix thoroughly. Sparge with 100% N_2. Filter sterilize.

$Na_2S \cdot 9H_2O$ Solution:

Composition per 10.0mL:

$Na_2S \cdot 9H_2O$	0.5g

Preparation of $Na_2S \cdot 9H_2O$ Solution: Add $Na_2S \cdot 9H_2O$ to distilled/deionized water and bring volume to 10.0mL. Mix thoroughly. Autoclave under 100% N_2 for 15 min at 15 psi pressure–121°C. Cool to room temperature.

$NaHCO_3$ Solution:

Composition per 100.0mL:

$NaHCO_3$	5.0g

Preparation of $NaHCO_3$ Solution: Add $NaHCO_3$ to distilled/deionized water and bring volume to 100.0mL. Mix thoroughly. Sparge with 80% N_2 + 20% CO_2. Filter sterilize.

KOH Solution:

Composition per 100.0mL:

KOH 10.0g

Preparation of KOH Solution: Add KOH to distilled/deionized water and bring volume to 100.0mL. Mix thoroughly. Sparge with 100% N_2. Filter sterilize.

Trace Elements Solution SL-10:

Composition per liter:

$FeCl_2 \cdot 4H_2O$ 1.5g
$CoCl_2 \cdot 6H_2O$ 190.0mg
$MnCl_2 \cdot 4H_2O$ 100.0mg
$ZnCl_2$ 70.0mg
$Na_2MoO_4 \cdot 2H_2O$ 36.0mg
$NiCl_2 \cdot 6H_2O$ 24.0mg
H_3BO_3 .... 6.0mg
$CuCl_2 \cdot 2H_2O$ 2.0mg
HCl (25% solution) 10.0mL

Preparation of Trace Elements Solution SL-10: Add $FeCl_2 \cdot 4H_2O$ to 10.0mL of HCl solution. Mix thoroughly. Add distilled/deionized water and bring volume to 1.0L. Add remaining components. Mix thoroughly. Sparge with 80% N_2 + 20% CO_2. Autoclave for 15 min at 15 psi pressure–121°C.

Vitamin Solution:

Composition per liter:

Pyridoxine-HCl 10.0mg
Thiamine-HCl·$2H_2O$ 5.0mg
Riboflavin 5.0mg
Nicotinic acid 5.0mg
D-Ca-pantothenate 5.0mg
p-Aminobenzoic acid 5.0mg
Lipoic acid 5.0mg
Biotin 2.0mg
Folic acid 2.0mg
Vitamin B_{12} 0.1mg

Preparation of Vitamin Solution: Add components to distilled/deionized water and bring volume to 1.0L. Mix thoroughly. Sparge with 80% H_2 + 20% CO_2. Filter sterilize.

Seven Vitamin Solution:

Composition per liter:

Pyridoxine hydrochloride 300.0mg
Thiamine-HCl·$2H_2O$ 200.0mg
Nicotinic acid 200.0mg
Vitamin B_{12} 100.0mg
Calcium pantothenate 100.0mg
p-Aminobenzoic acid 80.0mg
D(+)-Biotin 20.0mg

Preparation of Seven Vitamin Solution: Add components to distilled/deionized water and bring volume to 1.0L. Sparge with 100% N_2. Mix thoroughly. Filter sterilize.

Na-lactate Solution:

Composition per 100.0mL:

Na-lactate 25.0g

Preparation of Na-lactate Solution: Add Na-lactate to distilled/deionized water and bring volume to 100.0mL. Mix thoroughly. Sparge with 100% N_2. Autoclave for 15 min at 15 psi pressure–121°C. Cool to room temperature.

Na-fumarate Solution:

Composition per 100.0mL:

Na-fumarate 16.0g

Preparation of Na-fumarate Solution: Add Na-fumarate to distilled/deionized water and bring volume to 100.0mL. Mix thoroughly. Sparge with 100% N_2. Autoclave for 15 min at 15 psi pressure–121°C. Cool to room temperature.

Preparation of Medium: Prepare and dispense medium under 80% N_2 + 20% CO_2 gas atmosphere. Add components, except $NaHCO_3$ solution, $Na_2S \cdot 9H_2O$ solution, KOH solution, Na-lactate solution, Na-fumarate solution, vitamin solution, seven vitamin solution, selenite-tungstate solution, and trace elements solution SL-10, to distilled/deionized water and bring volume to 873.7mL. Mix thoroughly. Adjust pH to 7.0–7.2. Sparge with 80% N_2 + 20% CO_2. Autoclave for 15 min at 15 psi pressure–121°C. Aseptically and anaerobically add 50.0mL $NaHCO_3$ solution, 3.3mL $Na_2S \cdot 9H_2O$ solution, 20.0mL KOH solution, 20.0mL Na-lactate solution, 20.0mL Na-fumarate solution, 10.0mL vitamin solution, 1.0mL seven vitamin solution, 1.0mL selenite-tungstate solution, and 1.0mL trace elements solution SL-10. Mix thoroughly. Aseptically and anaerobically distribute into sterile tubes or bottles.

Use: For the cultivation of *Desulfitobacterium* spp. and *Desulfitobacterium hafniense.*

Desulfitobacterium PCE Medium

Composition per 1001.0mL:

$(NH_4)H_2PO_4$ 2.88g
$MgSO_4 \cdot 7H_2O$ 0.1g
Yeast extract 0.1g
$Ca(NO_3)_2 \cdot 4H_2O$ 0.05g
Resazurin 0.1mg
$NaHCO_3$ solution 50.0mL
KOH solution 20.0mL
Sodium fumarate solution 20.0mL
Sodium-L-lactate solution 20.0mL
Wolfe's vitamin solution 10.0mL
Seven vitamin solution 1.0mL
Selenite-tungstate solution 1.0mL
Wolfe's mineral solution 1.0mL

pH 7.0–7.2 at 25°C

$NaHCO_3$ Solution:

Composition per 50.0mL:

$NaHCO_3$ 2.5g

Preparation of $NaHCO_3$ Solution: Add $NaHCO_3$ to distilled/deionized water and bring volume to 50.0mL. Mix thoroughly. Sparge with 80% N_2 + 20% CO_2. Autoclave for 15 min at 15 psi pressure–121°C.

KOH Solution:

Composition per 20.0mL:

KOH 2.0g

Preparation of KOH Solution: Add KOH to distilled/deionized water and bring volume to 20.0mL. Mix thoroughly. Sparge with 100% N_2. Autoclave for 15 min at 15 psi pressure–121°C.

Sodium Fumarate Solution:

Composition per 20.0mL:

Sodium fumarate 3.2g

Preparation of Sodium Fumarate Solution: Add sodium fumarate to distilled/deionized water and bring volume to 20.0mL. Mix thoroughly. Sparge with 100% N_2. Autoclave for 15 min at 15 psi pressure–121°C.

Sodium L-Lactate Solution:
Composition per 20.0mL:

Sodium L-lactate 3.2g

Preparation of Sodium L-Lactate Solution: Add sodium L-lactate to distilled/deionized water and bring volume to 20.0mL. Mix thoroughly. Sparge with 100% N_2. Autoclave for 15 min at 15 psi pressure–121°C.

Wolfe's Vitamin Solution:
Composition per liter:

Pyridoxine·HCl	10.0mg
p-Aminobenzoic acid	5.0mg
Lipoic acid	5.0mg
Nicotinic acid	5.0mg
Riboflavin	5.0mg
Thiamine·HCl	5.0mg
Calcium DL-pantothenate	5.0mg
Biotin	2.0mg
Folic acid	2.0mg
Vitamin B_{12}	0.1mg

Preparation of Wolfe's Vitamin Solution: Add components to distilled/deionized water and bring volume to 1.0L. Mix thoroughly. Filter sterilize.

Seven Vitamin Solution:
Composition per liter:

Pyridoxine·HCl	0.3g
Thiamine·HCl	0.2g
Nicotinic acid	0.2g
Calcium DL-pantothenate	0.1g
Vitamin B_{12}	0.1g
p-Aminobenzoic acid	80.0mg
Biotin	20.0mg

Preparation of Seven Vitamin Solution: Add components to distilled/deionized water and bring volume to 1.0L. Mix thoroughly.

Selenite-Tungstate Solution:
Composition per liter:

NaOH	0.5g
$Na_2WO_4·2H_2O$	4.0mg
$Na_2SeO_3·5H_2O$	3.0mg

Preparation of Selenite-Tungstate Solution: Add components to distilled/deionized water and bring volume to 1.0L. Mix thoroughly. Sparge with 100% N_2. Autoclave for 15 min at 15 psi pressure–121°C.

Wolfe's Mineral Solution:
Composition per liter:

$MgSO_4·7H_2O$	3.0g
Nitrilotriacetic acid	1.5g
NaCl	1.0g
$MnSO_4·2H_2O$	0.5g
$CoCl_2·6H_2O$	0.1g
$ZnSO_4·7H_2O$	0.1g
$CaCl_2·2H_2O$	0.1g
$FeSO_4·7H_2O$	0.1g
$NiCl_2·6H_2O$	0.025g
$KAl(SO_4)_2·12H_2O$	0.02g
$CuSO_4·5H_2O$	0.01g
H_3BO_3	0.01g
$Na_2MoO_4·2H_2O$	0.01g
$Na_2SeO_3·5H_2O$	0.3mg

Preparation of Wolfe's Mineral Solution: Add nitrilotriacetic acid to 500.0mL of distilled/deionized water. Adjust pH to 6.5 with KOH. Add remaining components. Add distilled/deionized water to 1.0L. Adjust pH to 6.8.

Preparation of Medium: Prepare and dispense medium under 80% N_2 + 20% CO_2 gas mixture. Add components, except $NaHCO_3$ solution, KOH solution, sodium fumarate solution, sodium-L-lactate solution, Wolfe's vitamin solution, and seven vitamins solution, to distilled/deionized water and bring volume to 880.0mL. Mix thoroughly. Adjust pH to 7.0–7.2. Sparge with 100% N_2. Autoclave for 15 min at 15 psi pressure–121°C. Aseptically and anaerobically add 50.0mL of sterile $NaHCO_3$ solution, 20.0mL of sterile KOH solution, 20.0mL of sterile sodium fumarate solution, 20.0mL of sterile sodium-L-lactate solution, 10.0mL of sterile Wolfe's vitamin solution, and 1.0mL of sterile seven vitamins solution. Mix thoroughly. Aseptically and anaerobically distribute into sterile tubes or bottles.

Use: For the cultivation of *Desulfitobacterium* species.

Desulfitobacterium PCE II Medium (DSMZ Medium 1062)

Composition per liter:

NaCl	1.0g
KCl	0.5g
$MgCl_2·6H_2O$	0.4g
NH_4Cl	0.25g
KH_2PO_4	0.2g
$CaCl_2·2H_2O$	0.15g
Resazurin	0.5mg
Trace elements solution	10.0mL
Pyruvate solution	10.0mL
Fumarate solution	10.0mL
Yeast extract solution	10.0mL
Ferrous sulfate solution	10.0mL
$NaHCO_3$ solution	10.0mL
Selenite/tungstate solution	1.0mL
Vitamin solution	1.0mL

pH 7.5 ± 0.2 at 25°C

Ferrous Sulfate Solution:
Composition per 10.0mL:

$FeSO_4·7H_2O$ 22.0mg

Preparation of Ferrous Sulfate Solution: Add $FeSO_4·7H_2O$ to distilled/deionized water and bring volume to 10.0mL. Mix thoroughly. Sparge with 100% N_2. Autoclave for 15 min at 15 psi pressure–121°C.

Pyruvate Solution:
Composition per 10.0mL:

Na-pyruvate 4.5g

Preparation of Pyruvate Solution: Add Na-pyruvate to distilled/deionized water and bring volume to 10.0mL. Mix thoroughly. Sparge with 100% N_2. Autoclave for 15 min at 15 psi pressure–121°C.

Fumarate Solution:
Composition per 10.0mL:

Na_2-fumarate 6.5g

Preparation of Fumarate Solution: Add Na_2-fumarate to distilled/deionized water and bring volume to 10.0mL. Mix thoroughly. Sparge with 100% N_2. Autoclave for 15 min at 15 psi pressure–121°C.

Yeast Extract Solution:

Composition per 10.0mL:

Yeast extract	2.0g

Preparation of Yeast Extract Solution: Add yeast extract to distilled/deionized water and bring volume to 10.0mL. Mix thoroughly. Sparge with 100% N_2. Autoclave for 15 min at 15 psi pressure–121°C.

$NaHCO_3$ Solution:

Composition per 10.0mL:

$NaHCO_3$	3.0g

Preparation of $NaHCO_3$ Solution: Add $NaHCO_3$ to distilled/deionized water and bring volume to 10.0mL. Mix thoroughly. Sparge with 20% CO_2 + 80% H_2. Autoclave for 15 min at 15 psi pressure–121°C. Cool to room temperature.

Vitamin Solution:

Nicotinamide	9.0mg
Calcium DL-pantothenate	5.0mg
Riboflavin	5.0mg
Thiamine·HCl	5.0mg
Biotin	2.0mg
Folic acid	2.0mg
Cyanocobalamin	0.1mg

Preparation of Vitamin Solution: Add components to distilled/deionized water and bring volume to 1.0L. Mix thoroughly. Filter sterilize.

Trace Elements Solution:

Composition per liter:

$MgSO_4·7H_2O$	3.0g
Nitrilotriacetic acid	1.5g
NaCl	1.0g
$MnSO_4·2H_2O$	0.5g
$CoSO_4·7H_2O$	0.18g
$ZnSO_4·7H_2O$	0.18g
$CaCl_2·2H_2O$	0.1g
$FeSO_4·7H_2O$	0.1g
$NiCl_2·6H_2O$	0.025g
$KAl(SO_4)_2·12H_2O$	0.02g
H_3BO_3	0.01g
$Na_2MoO_4·4H_2O$	0.01g
$CuSO_4·5H_2O$	0.01g
$Na_2SeO_3·5H_2O$	0.3mg

Preparation of Trace Elements Solution: Add nitrilotriacetic acid to 500.0mL of distilled/deionized water. Dissolve by adjusting pH to 6.5 with KOH. Add remaining components. Add distilled/deionized water to 1.0L. Mix thoroughly.

Selenite/Tungstate Solution:

Composition per liter:

NaOH	0.5g
$Na_2WO_4·2H_2O$	4.0mg
$Na_2SeO_3·5H_2O$	3.0mg

Preparation of Selenite/Tungstate Solution: Add components to distilled/deionized water and bring volume to 1.0L. Mix thoroughly. Sparge with 100% N_2. Autoclave for 15 min at 15 psi pressure–121°C.

Preparation of Medium: Add components, except bicarbonate, iron sulfate, vitamins, and organic substrates, to distilled/deionized water and bring volume to 940.0mL. Gently heat and bring to boiling. Boil for 1 min. Mix thoroughly. Cool to room temperature while sparging with 80% N_2 + 20% CO_2. Dispense into culture vessels under an atmosphere of 80% N_2 + 20% CO_2. Autoclave for 15 min at 15 psi pressure–121°C. Add ferrous iron sulfate, vitamins, pyruvate, fumarate, and yeast extract from sterile anoxic solutions. Adjust the final pH of the medium to 7.5. Addition of 10–20 mg sodium dithionite per liter (e.g., from 5% (w/v) solution, freshly prepared under N_2 and filter sterilized) may stimulate growth at the beginning.

Use: For the cultivation of *Desulfitobacterium* spp.

Desulfobacca Medium (DSMZ Medium 728)

Composition per liter:

$NaHCO_3$	4.0g
Na_2SO_4	3.0g
Na-acetate	1.64g
$Na_2HPO_4·2H_2O$	0.53g
KH_2PO_4	0.41g
NH_4Cl	0.3g
NaCl	0.3g
$CaCl_2·2H_2O$	0.11g
$MgCl_2·6H_2O$	0.1g
Resazurin	0.5mg
$Na_2S·9H_2O$ solution	10.0mL
$CaCl_2$ solution	10.0mL
Trace elements solution SL-10	1.0mL
Selenite-tungstate solution	1.0mL
Seven vitamin solution	0.2mL

pH 7.1 ± 0.2 at 25°C

Seven Vitamin Solution:

Composition per liter:

Pyridoxine hydrochloride	300.0mg
Thiamine-HCl·$2H_2O$	200.0mg
Nicotinic acid	200.0mg
Vitamin B_{12}	100.0mg
Calcium pantothenate	100.0mg
p-Aminobenzoic acid	80.0mg
D(+)-Biotin	20.0mg

Preparation of Seven Vitamin Solution: Add components to distilled/deionized water and bring volume to 1.0L. Sparge with 100% N_2. Mix thoroughly. Filter sterilize.

Trace Elements Solution SL-10:

Composition per liter:

$FeCl_2·4H_2O$	1.5g
$CoCl_2·6H_2O$	190.0mg
$MnCl_2·4H_2O$	100.0mg
$ZnCl_2$	70.0mg
$Na_2MoO_4·2H_2O$	36.0mg
$NiCl_2·6H_2O$	24.0mg
H_3BO_3	6.0mg
$CuCl_2·2H_2O$	2.0mg
HCl (25% solution)	10.0mL

Preparation of Trace Elements Solution SL-10: Add $FeCl_2·4H_2O$ to 10.0mL of HCl solution. Mix thoroughly. Add distilled/deionized water and bring volume to 1.0L. Add remaining components. Mix thoroughly. Sparge with 80% N_2 + 20% CO_2. Autoclave for 15 min at 15 psi pressure–121°C.

$Na_2S·9H_2O$ Solution:

Composition per 10.0mL:

$Na_2S·9H_2O$	0.5g

Preparation of $Na_2S{\cdot}9H_2O$ Solution: Add $Na_2S{\cdot}9H_2O$ to distilled/deionized water and bring volume to 10.0mL. Mix thoroughly. Autoclave under 100% N_2 for 15 min at 15 psi pressure–121°C. Cool to room temperature.

Selenite-Tungstate Solution:
Composition per liter:

NaOH	0.5g
$Na_2WO_4{\cdot}2H_2O$	4.0mg
$Na_2SeO_3{\cdot}5H_2O$	3.0mg

Preparation of Selenite-Tungstate Solution: Add components to distilled/deionized water and bring volume to 1.0L. Mix thoroughly. Sparge with 100% N_2. Filter sterilize.

$CaCl_2$ Solution:
Composition per 10.0mL:

$CaCl_2{\cdot}2H_2O$	0.11g

Preparation of $CaCl_2$ Solution: Add $CaCl_2{\cdot}2H_2O$ to distilled/deionized water and bring volume to 10.0mL. Mix thoroughly. Sparge with 80% N_2 + 20% CO_2. Filter sterilize.

Preparation of Medium: Prepare and dispense medium under 80% N_2 + 20% CO_2 gas atmosphere. Add components, except $Na_2S{\cdot}9H_2O$ solution, $CaCl_2$ solution, seven vitamin solution, selenite-tungstate solution, and trace elements solution SL-10, to distilled/deionized water and bring volume to 977.8mL. Mix thoroughly. Adjust pH to 7.0–7.2. Sparge with 80% N_2 + 20% CO_2. Autoclave for 15 min at 15 psi pressure–121°C. Aseptically and anaerobically add 10.0mL $Na_2S{\cdot}9H_2O$ solution, 10.0mL $CaCl_2$ solution, 0.2mL seven vitamin solution, 1.0mL selenite-tungstate solution, and 1.0mL trace elements solution SL-10. Mix thoroughly. Aseptically and anaerobically distribute into sterile tubes or bottles.

Use: For the cultivation of *Desulfobacca acetoxidans.*

Desulfobacter Medium

Composition per 1001.0mL:

Solution A	870.0mL
Solution C	100.0mL
Solution D	10.0mL
Solution E (Vitamin solution)	10.0mL
Solution F	10.0mL
Solution B (Trace elements solution SL-10)	1.0mL

pH 7.1–7.4 at 25°C

Solution A:
Composition per 870.0mL:

NaCl	21.0g
$MgCl_2{\cdot}6H_2O$	3.1g
Na_2SO_4	3.0g
KCl	0.5g
NH_4Cl	0.3g
KH_2PO_4	0.2g
$CaCl_2{\cdot}2H_2O$	0.15g
Resazurin	1.0mg

Preparation of Solution A: Add components to distilled/deionized water and bring volume to 870.0mL. Mix thoroughly. Gently heat and bring to boiling. Continue boiling for 3–4 min. Allow to cool to room temperature while gassing under 80% N_2 + 20% CO_2. Continue gassing until pH reaches below 6.0. Seal the flask under 80% N_2 + 20% CO_2. Autoclave for 15 min at 15 psi pressure–121°C.

Solution B (Trace Elements Solution SL-10):
Composition per liter:

$FeCl_2{\cdot}4H_2O$	1.5g
$CoCl_2{\cdot}6H_2O$	190.0mg
$MnCl_2{\cdot}4H_2O$	100.0mg
$ZnCl_2$	70.0mg
$Na_2MoO_4{\cdot}2H_2O$	36.0mg
$NiCl_2{\cdot}6H_2O$	24.0mg
H_3BO_3	6.0mg
$CuCl_2{\cdot}2H_2O$	2.0mg
HCl (25% solution)	10.0mL

Preparation of Solution B (Trace Elements Solution SL-10): Add $FeCl_2{\cdot}4H_2O$ to 10.0mL of HCl solution. Mix thoroughly. Add distilled/deionized water and bring volume to 1.0L. Add remaining components. Mix thoroughly. Gas under 100% N_2. Autoclave for 15 min at 15 psi pressure–121°C.

Solution C:
Composition per 100.0mL:

$NaHCO_3$	5.0g

Preparation of Solution C: Add $NaHCO_3$ to distilled/deionized water and bring volume to 100.0mL. Mix thoroughly. Filter sterilize. Gas under 80% N_2 + 20% CO_2.

Solution D:
Composition per 10.0mL:

Sodium acetate·$3H_20$	2.5g

Preparation of Solution D: Add sodium acetate·$3H_20$· to distilled/deionized water and bring volume to 10.0mL. Mix thoroughly. Gas under 100% N_2. Autoclave for 15 min at 15 psi pressure–121°C.

Solution E (Vitamin Solution):
Composition per liter:

Pyridoxine·HCl	10.0mg
Calcium DL-pantothenate	5.0mg
Lipoic acid	5.0mg
Nicotinic acid	5.0mg
p-Aminobenzoic acid	5.0mg
Riboflavin	5.0mg
Thiamine·HCl	5.0mg
Biotin	2.0mg
Folic acid	2.0mg
Vitamin B_{12}	0.1mg

Preparation of Solution E (Vitamin Solution): Add components to distilled/deionized water and bring volume to 1.0L. Mix thoroughly. Gas under 100% N_2. Autoclave for 15 min at 15 psi pressure–121°C.

Solution F:
Composition per 10.0mL:

$Na_2S{\cdot}9H_2O$	0.4g

Preparation of Solution F: Add $Na_2S{\cdot}9H_2O$ to distilled/deionized water and bring volume to 10.0mL. Mix thoroughly. Gas under 100% N_2. Autoclave for 15 min at 15 psi pressure–121°C.

Preparation of Medium: Aseptically and anaerobically combine solution A with solution B, solution C, solution D, solution E, and solution F, in that order. Mix thoroughly. Anaerobically distribute into sterile tubes or flasks under 80% N_2 + 20% CO_2.

Use: For the cultivation and maintenance of *Desulfobacter* species and *Malonomonas rubra.*

Desulfobacter postgatei Medium (DSMZ Medium 193)

Composition per 1001.0mL:

Solution A	870.0mL
Solution C	100.0mL
Solution D	10.0mL
Solution E (Vitamin solution)	10.0mL
Solution F	10.0mL
Solution B (Trace elements solution SL-10)	1.0mL

pH 7.1–7.4 at 25°C

Solution A:
Composition per 870.0mL:

NaCl	7.0g
Na_2SO_4	3.0g
$MgCl_2 \cdot 6H_2O$	1.3g
KCl	0.5g
NH_4Cl	0.3g
KH_2PO_4	0.2g
$CaCl_2 \cdot 2H_2O$	0.15g
Resazurin	1.0mg

Preparation of Solution A: Add components to distilled/deionized water and bring volume to 870.0mL. Mix thoroughly.

Solution B (Trace Elements Solution SL-10):
Composition per liter:

$FeCl_2 \cdot 4H_2O$	1.5g
$CoCl_2 \cdot 6H_2O$	190.0mg
$MnCl_2 \cdot 4H_2O$	100.0mg
$ZnCl_2$	70.0mg
$Na_2MoO_4 \cdot 2H_2O$	36.0mg
$NiCl_2 \cdot 6H_2O$	24.0mg
H_3BO_3	6.0mg
$CuCl_2 \cdot 2H_2O$	2.0mg
HCl (25% solution)	10.0mL

Preparation of Solution B (Trace Elements Solution SL-10): Add $FeCl_2 \cdot 4H_2O$ to 10.0mL of HCl solution. Mix thoroughly. Add distilled/deionized water and bring volume to 1.0L. Add remaining components. Mix thoroughly. Sparge with 100% N_2. Autoclave for 15 min at 15 psi pressure–121°C.

Solution C:
Composition per 100.0mL:

$NaHCO_3$	5.0g

Preparation of Solution C: Add $NaHCO_3$ to distilled/deionized water and bring volume to 100.0mL. Mix thoroughly. Filter sterilize. Flush with 80% N_2 + 20% CO_2 to remove dissolved oxygen.

Solution D:
Composition per 10.0mL:

Na-acetate·$3H_2O$	2.5g

Preparation of Solution D: Add Na-acetate·$3H_2O$ to distilled/deionized water and bring volume to 10.0mL. Mix thoroughly. Sparge with 100% N_2. Autoclave for 15 min at 15 psi pressure–121°C.

Solution E (Vitamin Solution):
Composition per liter:

Pyridoxine-HCl	10.0mg
Thiamine-HCl·$2H_2O$	5.0mg
Riboflavin	5.0mg
Nicotinic acid	5.0mg
D-Ca-pantothenate	5.0mg
p-Aminobenzoic acid	5.0mg
Lipoic acid	5.0mg
Biotin	2.0mg
Folic acid	2.0mg
Vitamin B_{12}	0.10mg

Solution E (Vitamin Solution): Add components to distilled/deionized water and bring volume to 1.0L. Mix thoroughly. Sparge with 100% N_2. Autoclave for 15 min at 15 psi pressure–121°C.

Solution F:
Composition per 10.0mL:

$Na_2S \cdot 9H_2O$	0.4g

Preparation of Solution F: Add $Na_2S \cdot 9H_2O$ to distilled/deionized water and bring volume to 10.0mL. Mix thoroughly. Sparge with 100% N_2. Autoclave for 15 min at 15 psi pressure–121°C.

Preparation of Medium: Gently heat solution A and bring to boiling. Boil solution A for a few minutes. Cool to room temperature. Gas with 80% N_2 + 20% CO_2 gas mixture to reach a pH below 6. Autoclave for 15 min at 15 psi pressure–121°C. Cool to room temperature. Sequentially add 1.0mL solution B, 100.0mL solution C, 10.0mL solution D, 10.0mL solution E, and 10.0mL solution F. Distribute anaerobically under 80% N_2 + 20% CO_2 into appropriate vessels. Addition of 10–20mg sodium dithionite per liter from a 5% (w/v) solution, freshly prepared under N_2 and filter-sterilized, may stimulate growth.

Use: For the cultivation of *Desulfobacter postgatei, Paracoccus solventivorans,* and *Desulfotomaculum* spp.

Desulfobacter postgatei Medium

Composition per 1001.0mL:

Solution A	870.0mL
Solution C	100.0mL
Solution D	10.0mL
Solution E (Vitamin solution)	10.0mL
Solution F	10.0mL
Solution B (Trace elements solution SL-10)	1.0mL

pH 7.1–7.4 at 25°C

Solution A:
Composition per 870.0mL:

NaCl	7.0g
Na_2SO_4	3.0g
$MgCl_2 \cdot 6H_2O$	1.3g
KCl	0.5g
NH_4Cl	0.3g
KH_2PO_4	0.2g
$CaCl_2 \cdot 2H_2O$	0.15g
Resazurin	1.0mg

Preparation of Solution A: Add components to distilled/deionized water and bring volume to 870.0mL. Mix thoroughly. Gently heat and bring to boiling. Continue boiling for 3–4 min. Allow to cool to room temperature while gassing under 80% N_2 + 20% CO_2. Continue gassing until pH reaches below 6.0. Seal the flask under 80% N_2 + 20% CO_2. Autoclave for 15 min at 15 psi pressure–121°C.

Solution B (Trace Elements Solution SL-10):
Composition per liter:

$FeCl_2 \cdot 4H_2O$	1.5g
$CoCl_2 \cdot 6H_2O$	190.0mg
$MnCl_2 \cdot 4H_2O$	100.0mg
$ZnCl_2$	70.0mg
$Na_2MoO_4 \cdot 2H_2O$	36.0mg

$NiCl_2 \cdot 6H_2O$	24.0mg
H_3BO_3	6.0mg
$CuCl_2 \cdot 2H_2O$	2.0mg
HCl (25% solution)	10.0mL

Preparation of Solution B (Trace Elements Solution SL-10): Add $FeCl_2 \cdot 4H_2O$ to 10.0mL of HCl solution. Mix thoroughly. Add distilled/deionized water and bring volume to 1.0L. Add remaining components. Mix thoroughly. Gas under 100% N_2. Autoclave for 15 min at 15 psi pressure–121°C.

Solution C:
Composition per 100.0mL:

$NaHCO_3$	5.0g

Preparation of Solution C: Add $NaHCO_3$ to distilled/deionized water and bring volume to 100.0mL. Mix thoroughly. Filter sterilize. Gas under 80% N_2 + 20% CO_2.

Solution D:
Composition per 10.0mL:

Sodium acetate·$3H_20$	2.5g

Preparation of Solution D: Add sodium acetate·$3H_20$· to distilled/deionized water and bring volume to 10.0mL. Mix thoroughly. Gas under 100% N_2. Autoclave for 15 min at 15 psi pressure–121°C.

Solution E (Vitamin Solution):
Composition per liter:

Pyridoxine·HCl	10.0mg
Calcium DL-pantothenate	5.0mg
Lipoic acid	5.0mg
Nicotinic acid	5.0mg
p-Aminobenzoic acid	5.0mg
Riboflavin	5.0mg
Thiamine·HCl	5.0mg
Biotin	2.0mg
Folic acid	2.0mg
Vitamin B_{12}	0.1mg

Preparation of Solution E (Vitamin Solution): Add components to distilled/deionized water and bring volume to 1.0L. Mix thoroughly. Gas under 100% N_2. Autoclave for 15 min at 15 psi pressure–121°C.

Solution F:
Composition per 10.0mL:

$Na_2S \cdot 9H_2O$	0.4g

Preparation of Solution F: Add $Na_2S \cdot 9H_2O$ to distilled/deionized water and bring volume to 10.0mL. Mix thoroughly. Gas under 100% N_2. Autoclave for 15 min at 15 psi pressure–121°C.

Preparation of Medium: Aseptically and anaerobically combine solution A with solution B, solution C, solution D, solution E, and solution F, in that order. Mix thoroughly. Anaerobically distribute into sterile tubes or flasks under 80% N_2 + 20% CO_2.

Use: For the cultivation and maintenance of *Desulfobacter postgatei*.

Desulfobacter spp. Medium (DSMZ Medium 195)

Composition per 991.0mL:

Solution A	870.0mL
Solution C	100.0mL
Solution D	10.0mL
Solution E (Vitamin solution)	10.0mL
Solution F	10.0mL
Solution B (Trace elements solution SL-10)	1.0mL

pH 7.1–7.4 at 25°C

Solution A:

NaCl	21.0g
$MgCl_2 \cdot 6H_2O$	3.1g
Na_2SO_4	3.0g
KCl	0.5g
NH_4Cl	0.3g
KH_2PO_4	0.2g
$CaCl_2 \cdot 2H_2O$	0.15g
Resazurin	1.0mg

Preparation of Solution A: Add components to distilled/deionized water and bring volume to 870.0mL. Mix thoroughly.

Solution B (Trace Elements Solution SL-10):
Composition per liter:

$FeCl_2 \cdot 4H_2O$	1.5g
$CoCl_2 \cdot 6H_2O$	190.0mg
$MnCl_2 \cdot 4H_2O$	100.0mg
$ZnCl_2$	70.0mg
$Na_2MoO_4 \cdot 2H_2O$	36.0mg
$NiCl_2 \cdot 6H_2O$	24.0mg
H_3BO_3	6.0mg
$CuCl_2 \cdot 2H_2O$	2.0mg
HCl (25% solution)	10.0mL

Preparation of Solution B (Trace Elements Solution SL-10): Add $FeCl_2 \cdot 4H_2O$ to 10.0mL of HCl solution. Mix thoroughly. Add distilled/deionized water and bring volume to 1.0L. Add remaining components. Mix thoroughly. Sparge with 100% N_2. Autoclave for 15 min at 15 psi pressure–121°C.

Solution C:
Composition per 100.0mL:

$NaHCO_3$	5.0g

Preparation of Solution C: Add $NaHCO_3$ to distilled/deionized water and bring volume to 100.0mL. Mix thoroughly. Filter sterilize. Flush with 80% N_2 + 20% CO_2 to remove dissolved oxygen.

Solution D:
Composition per 10.0mL:

Na-acetate·$3H_2O$	2.5g

Preparation of Solution D: Add Na-acetate·$3H_2O$ to distilled/deionized water and bring volume to 10.0mL. Mix thoroughly. Sparge with 100% N_2. Autoclave for 15 min at 15 psi pressure–121°C.

Solution E (Vitamin Solution):
Composition per liter:

Pyridoxine-HCl	10.0mg
Thiamine-HCl·$2H_2O$	5.0mg
Riboflavin	5.0mg
Nicotinic acid	5.0mg
D-Ca-pantothenate	5.0mg
p-Aminobenzoic acid	5.0mg
Lipoic acid	5.0mg
Biotin	2.0mg
Folic acid	2.0mg
Vitamin B_{12}	0.10mg

Preparation of Solution E (Vitamin Solution): Add components to distilled/deionized water and bring volume to 1.0L. Mix thoroughly. Sparge with 100% N_2. Autoclave for 15 min at 15 psi pressure–121°C.

Solution F:

Composition per 10.0mL:

$Na_2S \cdot 9H_2O$ 0.4g

Preparation of Solution F: Add $Na_2S \cdot 9H_2O$ to distilled/deionized water and bring volume to 10.0mL. Mix thoroughly. Sparge with 100% N_2. Autoclave for 15 min at 15 psi pressure–121°C.

Preparation of Medium: Gently heat solution A and bring to boiling. Boil solution A for a few minutes. Cool to room temperature. Gas with 80% N_2 + 20% CO_2 gas mixture to reach a pH below 6. Autoclave for 15 min at 15 psi pressure–121°C. Cool to room temperature. Sequentially add 1.0mL solution B, 100.0mL solution C, 10.0mL solution D, 10.0mL solution E, and 10.0mL solution F. Distribute anaerobically under 80% N_2 + 20% CO_2 into appropriate vessels. Addition of 10–20mg sodium dithionite per liter from a 5% (w/v) solution, freshly prepared under N_2 and filter sterilized, may stimulate growth.

Use: For the cultivation of *Desulfobacter* sp., *Malonomonas rubra, Desulfuromusa kysingii, Desulfuromusa bakii, Desulfuromusa succinoxidans, Desulfobacter vibrioformis, Desulfacinum infernum, Desulfocapsa sulfexigens, Desulforhopalus singaporensis, Desulforhopalus* sp., *Desulfobacterium* sp., and *Desulfobulbus* sp.

Desulfobacter sp. Medium (DSMZ Medium 195)

Composition per 991.0mL:

Solution A 870.0mL
Solution C 100.0mL
Solution D 10.0mL
Solution E (Vitamin solution) 10.0mL
Solution F 10.0mL
Solution B (Trace elements solution SL-10) 1.0mL

pH 7.1–7.4 at 25°C

Solution A:

NaCl 21.0g
$MgCl_2 \cdot 6H_2O$ 3.1g
Na_2SO_4 .... 3.0g
KCl 0.5g
NH_4Cl 0.3g
KH_2PO_4 .... 0.2g
$CaCl_2 \cdot 2H_2O$ 0.15g
Resazurin 1.0mg

Preparation of Solution A: Add components to distilled/deionized water and bring volume to 870.0mL. Mix thoroughly.

Solution B (Trace Elements Solution SL-10):

Composition per liter:

$FeCl_2 \cdot 4H_2O$ 1.5g
$CoCl_2 \cdot 6H_2O$ 190.0mg
$MnCl_2 \cdot 4H_2O$ 100.0mg
$ZnCl_2$ 70.0mg
$Na_2MoO_4 \cdot 2H_2O$ 36.0mg
$NiCl_2 \cdot 6H_2O$ 24.0mg
H_3BO_3 .... 6.0mg
$CuCl_2 \cdot 2H_2O$ 2.0mg
HCl (25% solution) 10.0mL

Preparation of Solution B (Trace Elements Solution SL-10): Add $FeCl_2 \cdot 4H_2O$ to 10.0mL of HCl solution. Mix thoroughly. Add distilled/deionized water and bring volume to 1.0L. Add remaining components. Mix thoroughly. Sparge with 100% N_2. Autoclave for 15 min at 15 psi pressure–121°C.

Solution C:

Composition per 100.0mL:

$NaHCO_3$ 5.0g

Preparation of Solution C: Add $NaHCO_3$ to distilled/deionized water and bring volume to 100.0mL. Mix thoroughly. Filter sterilize. Flush with 80% N_2 + 20% CO_2 to remove dissolved oxygen.

Solution D:

Composition per 100.0mL:

Resorcinol 1.1g

Preparation of Solution D: Add resorcinol to distilled/deionized water and bring volume to 100.0mL. Mix thoroughly. Sparge with 100% N_2. Filter sterilize. Use freshly prepared.

Solution E (Vitamin Solution):

Composition per liter:

Pyridoxine-HCl 10.0mg
Thiamine-HCl·$2H_2O$ 5.0mg
Riboflavin 5.0mg
Nicotinic acid 5.0mg
D-Ca-pantothenate 5.0mg
p-Aminobenzoic acid 5.0mg
Lipoic acid 5.0mg
Biotin 2.0mg
Folic acid 2.0mg
Vitamin B_{12} 0.10mg

Preparation of Solution E (Vitamin Solution): Add components to distilled/deionized water and bring volume to 1.0L. Mix thoroughly. Sparge with 100% N_2. Autoclave for 15 min at 15 psi pressure–121°C.

Solution F:

Composition per 10.0mL:

$Na_2S \cdot 9H_2O$ 0.4g

Preparation of Solution F: Add $Na_2S \cdot 9H_2O$ to distilled/deionized water and bring volume to 10.0mL. Mix thoroughly. Sparge with 100% N_2. Autoclave for 15 min at 15 psi pressure–121°C.

Preparation of Medium: Gently heat solution A and bring to boiling. Boil solution A for a few minutes. Cool to room temperature. Gas with 80% N_2 + 20% CO_2 gas mixture to reach a pH below 6. Autoclave for 15 min at 15 psi pressure–121°C. Cool to room temperature. Sequentially add 1.0mL solution B, 100.0mL solution C, 10.0mL solution D, 10.0mL solution E, and 10.0mL solution F. Distribute anaerobically under 80% N_2 + 20% CO_2 into appropriate vessels. During growth the culture is fed once with the same amount of resorcinol solution.

Use: For the cultivation of unclassified bacterium DSM 4661.

Desulfobacterium anilini Medium (DSMZ Medium 476)

Composition per 1002.0mL:

Solution A 870.0mL
Solution C 100.0mL
Solution D 10.0mL
Solution E (Seven vitamin solution) 10.0mL
Solution F 10.0mL
Solution B (Trace elements solution SL-10) 1.0mL
Solution G 1.0mL

pH 7.1–7.4 at 25°C

Solution A:

Composition per 870.0mL:

NaCl	7.0g
Na_2SO_4	3.0g
$MgCl_2·6H_2O$	1.3g
KH_2PO_4	0.2g
NH_4Cl	0.3g
KCl	0.5g
$CaCl_2·2H_2O$	0.15g
Resazurin	1.0mg

Preparation of Solution A: Add components to distilled/deionized water and bring volume to 870.0mL. Mix thoroughly.

Solution B (Trace Elements Solution SL-10):

Composition per liter:

$FeCl_2·4H_2O$	1.5g
$CoCl_2·6H_2O$	190.0mg
$MnCl_2·4H_2O$	100.0mg
$ZnCl_2$	70.0mg
$Na_2MoO_4·2H_2O$	36.0mg
$NiCl_2·6H_2O$	24.0mg
H_3BO_3	6.0mg
$CuCl_2·2H_2O$	2.0mg
HCl (25% solution)	10.0mL

Preparation of Solution B (Trace Elements Solution SL-10): Add $FeCl_2·4H_2O$ to 10.0mL of HCl solution. Mix thoroughly. Add distilled/deionized water and bring volume to 1.0L. Add remaining components. Mix thoroughly. Sparge with 100% N_2. Autoclave for 15 min at 15 psi pressure–121°C.

Solution C:

Composition per 100.0mL:

$NaHCO_3$	5.0g

Preparation of Solution C: Add $NaHCO_3$ to distilled/deionized water and bring volume to 100.0mL. Mix thoroughly. Filter sterilize. Flush with 80% N_2 + 20% CO_2 to remove dissolved oxygen.

Solution D:

Composition per 10.0mL:

Phenol	0.1g

Preparation of Solution D: Prepare freshly. Add phenol to distilled/deionized water and bring volume to 10.0mL. Mix thoroughly. Sparge with 100% N_2. Filter sterilize.

Solution E (Seven Vitamin Solution):

Composition per liter:

Pyridoxine hydrochloride	300.0mg
Thiamine-HCl·$2H_2O$	200.0mg
Nicotinic acid	200.0mg
Vitamin B_{12}	100.0mg
Calcium pantothenate	100.0mg
p-Aminobenzoic acid	80.0mg
D(+)-Biotin	20.0mg

Solution E (Seven Vitamin Solution): Add components to distilled/deionized water and bring volume to 1.0L. Mix thoroughly. Sparge with 100% N_2. Autoclave for 15 min at 15 psi pressure–121°C.

Solution F:

Composition per 10.0mL:

$Na_2S·9H_2O$	0.4g

Preparation of Solution F: Add $Na_2S·9H_2O$ to distilled/deionized water and bring volume to 10.0mL. Mix thoroughly. Sparge with 100% N_2. Autoclave for 15 min at 15 psi pressure–121°C.

Solution G (Selenite-Tungstate Solution):

Composition per liter:

NaOH	0.5g
$Na_2WO_4·2H_2O$	4.0mg
$Na_2SeO_3·5H_2O$	3.0mg

Preparation of Solution G (Selenite-Tungstate Solution): Add components to distilled/deionized water and bring volume to 1.0L. Mix thoroughly. Sparge with 100% N_2. Filter sterilize.

Preparation of Medium: Gently heat solution A and bring to boiling. Boil solution A for a few minutes. Cool to room temperature. Gas with 80% N_2 + 20% CO_2 gas mixture to reach a pH below 6. Autoclave for 15 min at 15 psi pressure–121°C. Cool to room temperature. Sequentially add 1.0mL solution B, 100.0mL solution C, 10.0mL solution D, 10.0mL solution E, 10.0mL solution F, and 1.0mL solution G. Distribute anaerobically under 80% N_2 + 20% CO_2 into appropriate vessels.

Use: For the cultivation of *Desulfobacterium anilini.*

Desulfobacterium anilini Medium

Composition per 1011.0mL:

Solution A	870.0mL
Solution C	100.0mL
Solution D	10.0mL
Solution E (Vitamin solution)	10.0mL
Solution F	10.0mL
Solution G	10.0mL
Solution B (Trace elements solution SL-10)	1.0mL

pH 7.1–7.4 at 25°C

Solution A:

Composition per 870.0mL:

NaCl	7.0g
Na_2SO_4	3.0g
$MgCl_2·6H_2O$	1.3g
KCl	0.5g
NH_4Cl	0.3g
KH_2PO_4	0.2g
$CaCl_2·2H_2O$	0.15g
Resazurin	1.0mg

Preparation of Solution A: Add components to distilled/deionized water and bring volume to 870.0mL. Mix thoroughly. Gently heat and bring to boiling. Continue boiling for 3–4 min. Allow to cool to room temperature while gassing under 80% N_2 + 20% CO_2. Continue gassing until pH reaches below 6.0. Seal the flask under 80% N_2 + 20% CO_2. Autoclave for 15 min at 15 psi pressure–121°C.

Solution B (Trace Elements Solution SL-10):

Composition per liter:

$FeCl_2·4H_2O$	1.5g
$CoCl_2·6H_2O$	190.0mg
$MnCl_2·4H_2O$	100.0mg
$ZnCl_2$	70.0mg
$Na_2MoO_4·2H_2O$	36.0mg
$NiCl_2·6H_2O$	24.0mg
H_3BO_3	6.0mg
$CuCl_2·2H_2O$	2.0mg
HCl (25% solution)	10.0mL

Preparation of Solution B (Trace Elements Solution SL-10): Add $FeCl_2 \cdot 4H_2O$ to 10.0mL of HCl solution. Mix thoroughly. Add distilled/deionized water and bring volume to 1.0L. Add remaining components. Mix thoroughly. Gas under 100% N_2. Autoclave for 15 min at 15 psi pressure–121°C.

Solution C:

Composition per 100.0mL:

$NaHCO_3$	5.0g

Preparation of Solution C: Add $NaHCO_3$ to distilled/deionized water and bring volume to 100.0mL. Mix thoroughly. Filter sterilize. Gas under 80% N_2 + 20% CO_2.

Solution D:

Composition per 10.0mL:

Sodium acetate·$3H_2O$	2.5g

Preparation of Solution D: Add sodium acetate to distilled/deionized water and bring volume to 10.0mL. Mix thoroughly. Gas under 100% N_2. Autoclave for 15 min at 15 psi pressure–121°C.

Solution E (Vitamin Solution):

Composition per liter:

Pyridoxine·HCl	10.0mg
Calcium DL-pantothenate	5.0mg
Lipoic acid	5.0mg
Nicotinic acid	5.0mg
p-Aminobenzoic acid	5.0mg
Riboflavin	5.0mg
Thiamine·HCl	5.0mg
Biotin	2.0mg
Folic acid	2.0mg
Vitamin B_{12}	0.1mg

Preparation of Solution E (Vitamin Solution): Add components to distilled/deionized water and bring volume to 1.0L. Mix thoroughly. Gas under 100% N_2. Autoclave for 15 min at 15 psi pressure–121°C.

Solution F:

Composition per 10.0mL:

$Na_2S \cdot 9H_2O$	0.4g

Preparation of Solution F: Add $Na_2S \cdot 9H_2O$ to distilled/deionized water and bring volume to 10.0mL. Mix thoroughly. Gas under 100% N_2. Autoclave for 15 min at 15 psi pressure–121°C.

Solution G:

Composition per 10.0mL:

Phenol	94.0mg

Preparation of Solution G: Add phenol to distilled/deionized water and bring volume to 10.0mL. Mix thoroughly. Filter sterilize. Gas under 80% N_2 + 20% CO_2. Prepare solution freshly.

Preparation of Medium: Aseptically and anaerobically combine solution A with solution B, solution C, solution D, solution E, solution F, and solution G, in that order. Mix thoroughly. Anaerobically distribute into sterile tubes or flasks under 80% N_2 + 20% CO_2.

Use: For the cultivation of *Desulfobacterium anilini.*

Desulfobacterium anilini Medium

Composition per 1002.0mL:

Solution A	870.0mL
Solution C	100.0mL
Solution D	10.0mL
Solution E (Vitamin solution)	10.0mL
Solution F	10.0mL
Solution B (Trace elements solution SL-10)	1.0mL
Solution G	1.0mL

pH 7.1–7.4 at 25°C

Solution A:

Composition per 870.0mL:

NaCl	7.0g
Na_2SO_4	3.0g
$MgCl_2 \cdot 6H_2O$	1.3g
KCl	0.5g
NH_4Cl	0.3g
KH_2PO_4	0.2g
$CaCl_2 \cdot 2H_2O$	0.15g
Resazurin	1.0mg

Preparation of Solution A: Add components to distilled/deionized water and bring volume to 870.0mL. Mix thoroughly. Gently heat and bring to boiling. Continue boiling for 3–4 min. Allow to cool to room temperature while gassing under 80% N_2 + 20% CO_2. Continue gassing until pH reaches below 6.0. Seal the flask under 80% N_2 + 20% CO_2. Autoclave for 15 min at 15 psi pressure–121°C.

Solution B (Trace Elements Solution SL-10):

Composition per liter:

$FeCl_2 \cdot 4H_2O$	1.5g
$CoCl_2 \cdot 6H_2O$	190.0mg
$MnCl_2 \cdot 4H_2O$	100.0mg
$ZnCl_2$	70.0mg
$Na_2MoO_4 \cdot 2H_2O$	36.0mg
$NiCl_2 \cdot 6H_2O$	24.0mg
H_3BO_3	6.0mg
$CuCl_2 \cdot 2H_2O$	2.0mg
HCl (25% solution)	10.0mL

Preparation of Solution B (Trace Elements Solution SL-10): Add $FeCl_2 \cdot 4H_2O$ to 10.0mL of HCl solution. Mix thoroughly. Add distilled/deionized water and bring volume to 1.0L. Add remaining components. Mix thoroughly. Gas under 100% N_2. Autoclave for 15 min at 15 psi pressure–121°C.

Solution C:

Composition per 100.0mL:

$NaHCO_3$	5.0g

Preparation of Solution C: Add $NaHCO_3$ to distilled/deionized water and bring volume to 100.0mL. Mix thoroughly. Filter sterilize. Gas under 80% N_2 + 20% CO_2.

Solution D:

Composition per 10.0mL:

Phenol	0.094g

Preparation of Solution D: Add phenol to distilled/deionized water and bring volume to 10.0mL. Mix thoroughly. Gas under 100% N_2. Filter sterilize.

Solution E (Vitamin Solution):

Composition per 100.0mL

Nicotinic acid amide	35.0mg
Thiamine dichloride	30.0mg
p-Aminobenzoic acid	20.0mg
Biotin	10.0mg
Calcuim pantothenate	10.0mg
Pyridoxal·HCl	10.0mg
Vitamin B_{12}	5.0mg

Preparation of Solution E (Vitamin Solution): Add components to distilled/deionized water and bring volume to 100.0mL. Mix thoroughly. Gas under 100% N_2. Autoclave for 15 min at 15 psi pressure–121°C.

Solution F:
Composition per 10.0mL:
$Na_2S{\cdot}9H_2O$ 0.4g

Preparation of Solution F: Add $Na_2S{\cdot}9H_2O$ to distilled/deionized water and bring volume to 10.0mL. Mix thoroughly. Gas under 100% N_2. Autoclave for 15 min at 15 psi pressure–121°C.

Solution G:
Composition per liter:
NaOH 0.5g
$Na_2WO_4{\cdot}2H_2O$ 4.0mg
$Na_2SeO_3{\cdot}5H_2O$ 3.0mg

Preparation of Solution G: Add components to distilled/deionized water and bring volume to 1.0L. Mix thoroughly. Gas under 100% N_2. Autoclave for 15 min at 15 psi pressure–121°C.

Preparation of Medium: Aseptically and anaerobically combine solution A with solution B, solution C, solution D, solution E, solution F, and solution G, in that order. Mix thoroughly. Anaerobically distribute into sterile tubes or flasks under 80% N_2 + 20% CO_2

Use: For the cultivation and maintenance of *Desulfobacterium anilini.*

Desulfobacterium catecholicum Medium

Composition per 1001.0mL:
Solution A 870.0mL
Solution C 100.0mL
Solution D 10.0mL
Solution E (Vitamin solution) 10.0mL
Solution F 10.0mL
Solution B (Trace elements solution SL-10) 1.0mL
pH 7.1–7.4 at 25°C

Solution A:
Composition per 870.0mL:
Na_2SO_4 .... 3.0g
NaCl 1.0g
KCl 0.5g
$MgCl_2{\cdot}6H_2O$ 0.4g
NH_4Cl 0.3g
KH_2PO_4 .... 0.2g
$CaCl_2{\cdot}2H_2O$ 0.15g
Resazurin 1.0mg

Preparation of Solution A: Add components to distilled/deionized water and bring volume to 870.0mL. Mix thoroughly. Gently heat and bring to boiling. Continue boiling for 3–4 min. Allow to cool to room temperature while gassing under 80% N_2 + 20% CO_2. Continue gassing until pH reaches below 6.0. Seal the flask under 80% N_2 + 20% CO_2. Autoclave for 15 min at 15 psi pressure–121°C.

Solution B (Trace Elements Solution SL-10):
Composition per liter:
$FeCl_2{\cdot}4H_2O$ 1.5g
$CoCl_2{\cdot}6H_2O$ 190.0mg
$MnCl_2{\cdot}4H_2O$ 100.0mg
$ZnCl_2$ 70.0mg
$Na_2MoO_4{\cdot}2H_2O$ 36.0mg
$NiCl_2{\cdot}6H_2O$ 24.0mg
H_3BO_3 .... 6.0mg
$CuCl_2{\cdot}2H_2O$ 2.0mg
HCl (25% solution) 10.0mL

Preparation of Solution B (Trace Elements Solution SL-10): Add $FeCl_2{\cdot}4H_2O$ to 10.0mL of HCl solution. Mix thoroughly. Add distilled/deionized water and bring volume to 1.0L. Add remaining components. Mix thoroughly. Gas under 100% N_2. Autoclave for 15 min at 15 psi pressure–121°C.

Solution C:
Composition per 100.0mL:
$NaHCO_3$ 2.5g

Preparation of Solution C: Add $NaHCO_3$ to distilled/deionized water and bring volume to 100.0mL. Mix thoroughly. Filter sterilize. Gas under 80% N_2 + 20% CO_2.

Solution D:
Composition per 10.0mL:
Sodium benzoate 0.37g
Catechol 0.055g

Preparation of Solution D: Add sodium benzoate and catechol to distilled/deionized water and bring volume to 10.0mL. Mix thoroughly. Gas under 100% N_2. Filter sterilize.

Solution E (Vitamin Solution):
Composition per liter:
Pyridoxine·HCl 10.0mg
Calcium DL-pantothenate 5.0mg
Lipoic acid 5.0mg
Nicotinic acid 5.0mg
p-Aminobenzoic acid 5.0mg
Riboflavin 5.0mg
Thiamine·HCl 5.0mg
Biotin 2.0mg
Folic acid 2.0mg
Vitamin B_{12} 0.1mg

Preparation of Solution E (Vitamin Solution): Add components to distilled/deionized water and bring volume to 1.0L. Mix thoroughly. Gas under 100% N_2. Autoclave for 15 min at 15 psi pressure–121°C.

Solution F:
Composition per 10.0mL:
$Na_2S{\cdot}9H_2O$ 0.4g

Preparation of Solution F: Add $Na_2S{\cdot}9H_2O$ to distilled/deionized water and bring volume to 10.0mL. Mix thoroughly. Gas under 100% N_2. Autoclave for 15 min at 15 psi pressure–121°C.

Preparation of Medium: Aseptically and anaerobically combine solution A with solution B, solution C, solution D, solution E, and solution F, in that order. Mix thoroughly. Anaerobically distribute into sterile tubes or flasks under 80% N_2 + 20% CO_2.

Use: For the cultivation and maintenance of *Desulfobacterium catecholicum.*

Desulfobacterium cetonicum Medium

Composition per 1011.0mL:
Solution A 950.0mL
Solution C 10.0mL
Solution D 40.0mL
Solution E 10.0mL
Solution B (Trace elements solution SL-10) 1.0mL
pH 7.2–7.4 at 25°C

Solution A:
Composition per 950.0mL:
NaCl 10.0g
Na_2SO_4 .... 2.8g

KH_2PO_4	0.7g
$MgCl_2 \cdot 6H_2O$	0.3g
NH_4Cl	0.3g
$CaCl_2 \cdot 2H_2O$	0.05g

Preparation of Solution A: Add components to distilled/deionized water and bring volume to 950.0mL. Mix thoroughly. Sparge with 80% N_2 + 20% CO_2. Autoclave for 15 min at 15 psi pressure–121°C.

Solution B (Trace Elements Solution SL-10):
Composition per liter:

$FeCl_2 \cdot 4H_2O$	1.5g
$CoCl_2 \cdot 6H_2O$	190.0mg
$MnCl_2 \cdot 4H_2O$	100.0mg
$ZnCl_2$	70.0mg
$Na_2MoO_4 \cdot 2H_2O$	36.0mg
$NiCl_2 \cdot 6H_2O$	24.0mg
H_3BO_3	6.0mg
$CuCl_2 \cdot 2H_2O$	2.0mg
HCl (25% solution)	10.0mL

Preparation of Solution B (Trace Elements Solution SL-10): Add $FeCl_2 \cdot 4H_2O$ to 10.0mL of HCl solution. Mix thoroughly. Add distilled/deionized water and bring volume to 1.0L. Add remaining components. Mix thoroughly. Sparge with 100% N_2. Autoclave for 15 min at 15 psi pressure–121°C.

Solution C:
Composition per 10.0mL:

Sodium butyrate	1.2g

Preparation of Solution C: Add sodium butyrate to distilled/deionized water and bring volume to 10.0mL. Mix thoroughly. Sparge with 100% N_2. Autoclave for 15 min at 15 psi pressure–121°C.

Solution D:
Composition per 40.0mL:

$NaHCO_3$	2.0g

Preparation of Solution D: Add $NaHCO_3$ to distilled/deionized water and bring volume to 40.0mL. Mix thoroughly. Sparge with 100% N_2. Autoclave for 15 min at 15 psi pressure–121°C.

Solution E:
Composition per 10.0mL:

$Na_2S \cdot 9H_2O$	0.3g

Preparation of Solution E: Add $Na_2S \cdot 9H_2O$ to distilled/deionized water and bring volume to 10.0mL. Mix thoroughly. Sparge with 100% N_2. Autoclave for 15 min at 15 psi pressure–121°C.

Preparation of Medium: Prepare and dispense medium under 80% N_2 + 20% CO_2. Aseptically and anaerobically combine 950.0mL of sterile solution A with 1.0mL of sterile solution B, 10.0mL of sterile solution C, 40.0mL of sterile solution D, and 10.0mL of sterile solution E. Mix thoroughly. Check that final pH is 7.2–7.4.

Use: For the cultivation of *Desulfobacterium cetonicum.*

Desulfobacterium indolicum Medium

Composition per 1002.4mL:

Solution A	900.0mL
Solution C	50.0mL
Solution D	30.0mL
Solution E (Wolfe's vitamin solution)	10.0mL
Solution G	10.0mL
Solution B (Trace elements solution SL-10)	1.0mL
Solution F	1.0mL
Solution H	0.4mL

pH 7.6 ± 0.2 at 25°C

Solution A:
Composition per 900.0mL:

NaCl	21.0g
$MgCl_2 \cdot 6H_2O$	3.0g
Na_2SO_4	3.0g
KCl	0.5g
NH_4Cl	0.3g
KH_2PO_4	0.2g
$CaCl_2 \cdot 2H_2O$	0.15g
Resazurin	1.0mg

Preparation of Solution A: Prepare and dispense solution anaerobically under 80% N_2 + 20% CO_2. Add components to distilled/deionized water and bring volume to 900.0mL. Mix thoroughly. Gently heat and bring to boiling. Continue boiling until resazurin turns colorless, indicating reduction. Cap with rubber stoppers. Autoclave for 15 min at 15 psi pressure–121°C. Cool to 45°–50°C.

Solution B (Trace Elements Solution SL-10):
Composition per liter:

$FeCl_2 \cdot 4H_2O$	1.5g
$CoCl_2 \cdot 6H_2O$	0.19g
$MnCl_2 \cdot 4H_2O$	0.1g
$ZnCl_2$	0.07g
$Na_2MoO_4 \cdot 2H_2O$	0.036g
$NiCl_2 \cdot 6H_2O$	0.024g
H_3BO_3	6.0mg
$CuCl_2 \cdot 2H_2O$	2.0mg
HCl (25% solution)	10.0mL

Preparation of Solution B (Trace Elements Solution SL-10): Add the $FeCl_2 \cdot 4H_2O$ to 10.0mL of HCl solution. Mix thoroughly. Bring volume to approximately 900.0mL with distilled/deionized water. Mix thoroughly. Adjust pH to 6.0 with NaOH. Bring volume to 1.0L with distilled/deionized water. Filter sterilize. Aseptically gas under 100% N_2 for 20 min.

Solution C:
Composition per 50.0mL:

$NaHCO_3$	2.5g

Preparation of Solution C: Add $NaHCO_3$ to distilled/deionized water and bring volume to 50.0mL. Mix thoroughly. Filter sterilize. Aseptically gas under 80% N_2 + 20% CO_2 for 20 min.

Solution D:
Composition per 107.7mL:

Indole	0.3g
NaCl (30% solution)	7.0mL
$MgCl_2 \cdot 6H_2O$ (40% solution)	0.7mL

Preparation of Solution D: Prepare and dispense all solutions anaerobically under 100% N_2. Add indole to distilled/deionized water and bring volume to 100.0mL. Mix thoroughly. Gently heat while stirring until dissolved. Prepare the NaCl solution and the $MgCl_2 \cdot 6H_2O$ solution separately. Autoclave the three solutions separately for 15 min at 15 psi pressure–121°C. Cool to 45°–50°C. To 100.0mL of sterile indole solution, aseptically and anaerobically add 7.0mL of sterile NaCl solution and 0.7mL of sterile $MgCl_2 \cdot 6H_2O$ solution. Mix thoroughly.

Solution E (Wolfe's Vitamin Solution):
Composition per liter:

Pyridoxine·HCl	0.01g

Thiamine·HCl	5.0mg
Riboflavin	5.0mg
Nicotinic acid	5.0mg
Calcium pantothenate	5.0mg
p-Aminobenzoic acid	5.0mg
Thioctic acid	5.0mg
Biotin	2.0mg
Folic acid	2.0mg
Cyanocobalamin	0.1mg

Preparation of Solution E (Wolfe's Vitamin Solution): Add components to distilled/deionized water and bring volume to 1.0L. Mix thoroughly. Filter sterilize. Aseptically gas under 100% N_2 for 20 min.

Solution F:

Composition per liter:

$Na_2SeO_3 \cdot 5H_2O$	3.0mg
NaOH (0.01*M* solution)	1.0L

Preparation of Solution F: Add $Na_2SeO_3 \cdot 5H_2O$ to 1.0L of NaOH solution. Mix thoroughly. Filter sterilize. Aseptically gas under 100% N_2 for 20 min.

Solution G:

Composition per 10.0mL:

$Na_2S \cdot 9H_2O$	0.4g

Preparation of Solution G: Add $Na_2S \cdot 9H_2O$ to distilled/deionized water and bring volume to 10.0mL. Gas under 100% N_2 for 20 min. Cap with a rubber stopper. Autoclave for 15 min at 15 psi pressure–121°C. Cool to 25°C.

Solution H:

Composition per 10.0mL:

$Na_2S_2O_4$	0.5g

Preparation of Solution H: Add $Na_2S_2O_4$ to distilled/deionized water and bring volume to 10.0mL. Mix thoroughly. Filter sterilize. Aseptically gas under 100% N_2 for 20 min. Prepare solution freshly.

Preparation of Medium: To 900.0mL of cooled, sterile solution A, aseptically and anaerobically add in the following order: 1.0mL of sterile solution B, 50.0mL of sterile solution C, 10.0mL of sterile solution E, 1.0mL of sterile solution F, and 10.0mL of sterile solution G. Mix thoroughly. Immediately prior to inoculation, aseptically and anaerobically add 30.0mL of sterile solution D and 0.4mL of sterile solution H. Mix thoroughly. Aseptically and anaerobically distribute into sterile tubes or flasks.

Use: For the cultivation and maintenance of *Desulfobacterium indolicum*.

Desulfobacterium Medium

Composition per 1002.4mL:

Solution A	930.0mL
Solution C	50.0mL
Solution D (Wolfe's vitamin solution)	10.0mL
Solution F	10.0mL
Solution B (Trace elements solution SL-10)	1.0mL
Solution E	1.0mL
Solution G	0.4mL

pH 7.0 ± 0.2 at 25°C

Solution A:

Composition per 930.0mL:

NaCl	21.0g
$MgCl_2 \cdot 6H_2O$	3.0g
Na_2SO_4	3.0g
KCl	0.5g
NH_4Cl	0.3g
KH_2PO_4	0.2g
$CaCl_2 \cdot 2H_2O$	0.15g
Resazurin	1.0mg

Preparation of Solution A: Prepare and dispense solution anaerobically under 80% N_2 + 20% CO_2. Add components to distilled/deionized water and bring volume to 930.0mL. Mix thoroughly. Gently heat and bring to boiling. Continue boiling until resazurin turns colorless, indicating reduction. Cap with rubber stoppers. Autoclave for 15 min at 15 psi pressure–121°C. Cool to 45°–50°C.

Solution B (Trace Elements Solution SL-10):

Composition per liter:

$FeCl_2 \cdot 4H_2O$	1.5g
$CoCl_2 \cdot 6H_2O$	0.19g
$MnCl_2 \cdot 4H_2O$	0.1g
$ZnCl_2$	0.07g
$Na_2MoO_4 \cdot 2H_2O$	0.036g
$NiCl_2 \cdot 6H_2O$	0.024g
H_3BO_3	6.0mg
$CuCl_2 \cdot 2H_2O$	2.0mg
HCl (25% solution)	10.0mL

Preparation of Solution B (Trace Elements Solution SL-10): Add the $FeCl_2 \cdot 4H_2O$ to 10.0mL of HCl solution. Mix thoroughly. Bring volume to approximately 900.0mL with distilled/deionized water. Mix thoroughly. Adjust pH to 6.0 with NaOH. Bring volume to 1.0L with distilled/deionized water. Filter sterilize. Aseptically gas under 100% N_2 for 20 min.

Solution C:

Composition per 50.0mL:

$NaHCO_3$	2.5g

Preparation of Solution C: Add $NaHCO_3$ to distilled/deionized water and bring volume to 50.0mL. Mix thoroughly. Filter sterilize. Aseptically gas under 80% N_2 + 20% CO_2 for 20 min.

Solution D (Wolfe's Vitamin Solution):

Composition per liter:

Pyridoxine·HCl	0.01g
Thiamine·HCl	5.0mg
Riboflavin	5.0mg
Nicotinic acid	5.0mg
Calcium pantothenate	5.0mg
p-Aminobenzoic acid	5.0mg
Thioctic acid	5.0mg
Biotin	2.0mg
Folic acid	2.0mg
Cyanocobalamin	0.1mg

Preparation of Solution D (Wolfe's Vitamin Solution): Add components to distilled/deionized water and bring volume to 1.0L. Mix thoroughly. Filter sterilize. Aseptically gas under 100% N_2 for 20 min.

Solution E:

Composition per liter:

$Na_2SeO_3 \cdot 5H_2O$	3.0mg
NaOH (0.01*M* solution)	1.0L

Preparation of Solution E: Add $Na_2SeO_3 \cdot 5H_2O$ to 1.0L of NaOH solution. Mix thoroughly. Filter sterilize. Aseptically gas under 100% N_2 for 20 min.

Solution F:

Composition per 10.0mL:

$Na_2S \cdot 9H_2O$ 0.4g

Preparation of Solution F: Add $Na_2S \cdot 9H_2O$ to distilled/deionized water and bring volume to 10.0mL. Gas under 100% N_2 for 20 min. Cap with a rubber stopper. Autoclave for 15 min at 15 psi pressure–121°C. Cool to 25°C.

Solution G:

Composition per 10.0mL:

$Na_2S_2O_4$ 0.5g

Preparation of Solution G: Add $Na_2S_2O_4$ to distilled/deionized water and bring volume to 10.0mL. Mix thoroughly. Filter sterilize. Aseptically gas under 100% N_2 for 20 min. Prepare solution freshly.

Preparation of Medium: To 900.0mL of cooled, sterile solution A, aseptically and anaerobically add in the following order: 1.0mL of sterile solution B, 50.0mL of sterile solution C, 10.0mL of sterile solution D, 1.0mL of sterile solution E, and 10.0mL of sterile solution F. Mix thoroughly. Immediately prior to inoculation, aseptically and anaerobically add 0.4mL of sterile solution G. Mix thoroughly. Aseptically and anaerobically distribute into sterile tubes or flasks.

Use: For the cultivation and maintenance of *Desulfobacterium autotrophicum*.

Desulfobacterium Medium with Lactate

Composition per 1002.4mL:

Solution A 930.0mL
Solution C 50.0mL
Solution D (Wolfe's vitamin solution) 10.0mL
Solution F 10.0mL
Solution B (Trace elements solution SL-10) 1.0mL
Solution E 1.0mL
Solution G 0.4mL

pH 7.0 ± 0.2 at 25°C

Solution A:

Composition per 930.0mL:

NaCl 21.0g
$MgCl_2 \cdot 6H_2O$ 3.0g
Na_2SO_4 .... 3.0g
Lactic acid, sodium salt 1.1g
KCl 0.5g
NH_4Cl 0.3g
KH_2PO_4 .... 0.2g
$CaCl_2 \cdot 2H_2O$ 0.15g
Resazurin 1.0mg

Preparation of Solution A: Prepare and dispense solution anaerobically under 80% N_2 + 20% CO_2. Add components to distilled/deionized water and bring volume to 930.0mL. Mix thoroughly. Gently heat and bring to boiling. Continue boiling until resazurin turns colorless, indicating reduction. Cap with rubber stoppers. Autoclave for 15 min at 15 psi pressure–121°C. Cool to 45°–50°C.

Solution B (Trace Elements Solution SL-10):

Composition per liter:

$FeCl_2 \cdot 4H_2O$ 1.5g
$CoCl_2 \cdot 6H_2O$ 0.19g
$MnCl_2 \cdot 4H_2O$ 0.10g
$ZnCl_2$ 0.070g
$Na_2MoO_4 \cdot 2H_2O$ 0.036g
$NiCl_2 \cdot 6H_2O$ 0.024g
H_3BO_3 .... 6.0mg
$CuCl_2 \cdot 2H_2O$ 2.0mg
HCl (25% solution) 10.0mL

Preparation of Solution B (Trace Elements Solution SL-10): Add the $FeCl_2 \cdot 4H_2O$ to 10.0mL of HCl solution. Mix thoroughly. Bring volume to approximately 900.0mL with distilled/deionized water. Mix thoroughly. Adjust pH to 6.0 with NaOH. Bring volume to 1.0L with distilled/deionized water. Filter sterilize. Aseptically gas under 100% N_2 for 20 min.

Solution C:

Composition per 50.0mL:

$NaHCO_3$ 2.5g

Preparation of Solution C: Add $NaHCO_3$ to distilled/deionized water and bring volume to 50.0mL. Mix thoroughly. Filter sterilize. Aseptically gas under 80% N_2 + 20% CO_2 for 20 min.

Solution D (Wolfe's Vitamin Solution):

Composition per liter:

Pyridoxine·HCl 0.01g
Thiamine·HCl 5.0mg
Riboflavin 5.0mg
Nicotinic acid 5.0mg
Calcium pantothenate 5.0mg
p-Aminobenzoic acid 5.0mg
Thioctic acid 5.0mg
Biotin 2.0mg
Folic acid 2.0mg
Cyanocobalamin 0.1mg

Preparation of Solution D (Wolfe's Vitamin Solution): Add components to distilled/deionized water and bring volume to 1.0L. Mix thoroughly. Filter sterilize. Aseptically gas under 100% N_2 for 20 min.

Solution E:

Composition per liter:

$Na_2SeO_3 \cdot 5H_2O$ 3.0mg
NaOH (0.01*M* solution) 1.0L

Preparation of Solution E: Add $Na_2SeO_3 \cdot 5H_2O$ to 1.0L of NaOH solution. Mix thoroughly. Filter sterilize. Aseptically gas under 100% N_2 for 20 min.

Solution F:

Composition per 10.0mL:

$Na_2S \cdot 9H_2O$ 0.4g

Preparation of Solution F: Add $Na_2S \cdot 9H_2O$ to distilled/deionized water and bring volume to 10.0mL. Gas under 100% N_2 for 20 min. Cap with a rubber stopper. Autoclave for 15 min at 15 psi pressure–121°C. Cool to 25°C.

Solution G:

Composition per 10.0mL:

$Na_2S_2O_4$ 0.5g

Preparation of Solution G: Add $Na_2S_2O_4$ to distilled/deionized water and bring volume to 10.0mL. Mix thoroughly. Filter sterilize. Aseptically gas under 100% N_2 for 20 min. Prepare solution freshly.

Preparation of Medium: To 900.0mL of cooled, sterile solution A, aseptically and anaerobically add, in the following order, 1.0mL of sterile solution B, 50.0mL of sterile solution C, 10.0mL of sterile solution D, 1.0mL of sterile solution E, and 10.0mL of sterile solution F. Mix thoroughly. Immediately prior to inoculation, aseptically and anaerobically add 0.4mL of sterile solution G. Mix thoroughly. Aseptically and anaerobically distribute into sterile tubes or flasks.

Use: For the cultivation of *Desulfobacterium autotrophicum.*

Desulfobacterium Medium, Modified

Composition per 1002.4mL:

Solution A	920.0mL
Solution C	50.0mL
Solution D	10.0mL
Solution E (Wolfe's vitamin solution)	10.0mL
Solution G	10.0mL
Solution B (Trace elements solution SL-10)	1.0mL
Solution F	1.0mL
Solution H	0.4mL

pH 7.0 ± 0.2 at 25°C

Solution A:

Composition per 920.0mL:

NaCl	21.0g
$MgCl_2 \cdot 6H_2O$	3.0g
Na_2SO_4	3.0g
KCl	0.5g
NH_4Cl	0.3g
KH_2PO_4	0.2g
$CaCl_2 \cdot 2H_2O$	0.15g
Resazurin	1.0mg

Preparation of Solution A: Prepare and dispense solution anaerobically under 80% N_2 + 20% CO_2. Add components to distilled/deionized water and bring volume to 920.0mL. Mix thoroughly. Gently heat and bring to boiling. Continue boiling until resazurin turns colorless, indicating reduction. Cap with rubber stoppers. Autoclave for 15 min at 15 psi pressure–121°C. Cool to 45°–50°C.

Solution B (Trace Elements Solution SL-10):

Composition per liter:

$FeCl_2 \cdot 4H_2O$	1.5g
$CoCl_2 \cdot 6H_2O$	0.19g
$MnCl_2 \cdot 4H_2O$	0.1g
$ZnCl_2$	0.07g
$Na_2MoO_4 \cdot 2H_2O$	0.036g
$NiCl_2 \cdot 6H_2O$	0.024g
H_3BO_3	6.0mg
$CuCl_2 \cdot 2H_2O$	2.0mg
HCl (25% solution)	10.0mL

Preparation of Solution B (Trace Elements Solution SL-10): Add the $FeCl_2 \cdot 4H_2O$ to 10.0mL of HCl solution. Mix thoroughly. Bring volume to approximately 900.0mL with distilled/deionized water. Mix thoroughly. Adjust pH to 6.0 with NaOH. Bring volume to 1.0L with distilled/deionized water. Filter sterilize. Aseptically gas under 100% N_2 for 20 min.

Solution C:

Composition per 50.0mL:

$NaHCO_3$	2.5g

Preparation of Solution C: Add $NaHCO_3$ to distilled/deionized water and bring volume to 50.0mL. Mix thoroughly. Filter sterilize. Aseptically gas under 80% N_2 + 20% CO_2 for 20 min.

Solution D:

Composition per 10.0mL:

Sodium acetate·$3H_2O$	2.5g

Preparation of Solution D: Prepare and dispense solution anaerobically under 80% N_2 + 20% CO_2. Add sodium acetate to distilled/deionized water and bring volume to 10.0mL. Mix thoroughly. Cap with rubber stopper. Autoclave for 15 min at 15 psi pressure–121°C. Cool to 45°–50°C.

Solution E (Wolfe's Vitamin Solution):

Composition per liter:

Pyridoxine·HCl	0.01g
Thiamine·HCl	5.0mg
Riboflavin	5.0mg
Nicotinic acid	5.0mg
Calcium pantothenate	5.0mg
p-Aminobenzoic acid	5.0mg
Thioctic acid	5.0mg
Biotin	2.0mg
Folic acid	2.0mg
Cyanocobalamin	0.1mg

Preparation of Solution E (Wolfe's Vitamin Solution): Add components to distilled/deionized water and bring volume to 1.0L. Mix thoroughly. Filter sterilize. Aseptically gas under 100% N_2 for 20 min.

Solution F:

Composition per liter:

$Na_2SeO_3 \cdot 5H_2O$	3.0mg
NaOH (0.01*M* solution)	1.0L

Preparation of Solution F: Add $Na_2SeO_3 \cdot 5H_2O$ to 1.0L of NaOH solution. Mix thoroughly. Filter sterilize. Aseptically gas under 100% N_2 for 20 min.

Solution G:

Composition per 10.0mL:

$Na_2S \cdot 9H_2O$	0.4g

Preparation of Solution G: Add $Na_2S \cdot 9H_2O$ to distilled/deionized water and bring volume to 10.0mL. Gas under 100% N_2 for 20 min. Cap with a rubber stopper. Autoclave for 15 min at 15 psi pressure–121°C. Cool to 25°C.

Solution H:

Composition per 10.0mL:

$Na_2S_2O_4$	0.5g

Preparation of Solution H: Add $Na_2S_2O_4$ to distilled/deionized water and bring volume to 10.0mL. Mix thoroughly. Filter sterilize. Aseptically gas under 100% N_2 for 20 min. Prepare solution freshly.

Preparation of Medium: To 920.0mL of cooled, sterile solution A, aseptically and anaerobically add in the following order: 1.0mL of sterile solution B, 50.0mL of sterile solution C, 10.0mL of sterile solution D, 10.0mL of sterile solution E, 1.0mL of sterile solution F, and 10.0mL of sterile solution G. Mix thoroughly. Immediately prior to inoculation, aseptically and anaerobically add 0.4mL of sterile solution H. Mix thoroughly. Aseptically and anaerobically distribute into sterile tubes or flasks.

Use: For the cultivation and maintenance of *Desulfobacter curvatus* and *Desulfobacter latus.*

Desulfobacterium oleovorans Medium

Composition per 1154.0mL:

Solution A	1.0L
Solution H	67.0mL
Solution D	50.0mL
Soultion I	13.0mL
Solution E	10.0mL
Solution G	10.0mL
Solution C (Selenite-tungstate solution)	2.0mL

Solution B (Trace elements solution SL-10) 1.0mL
Solution F 1.0mL

pH 7.2 ± 0.2 at 25°C

Solution A:

Composition per liter:

Na_2SO_4 .... 4.0g
NaCl 1.0g
$MgCl_2 \cdot 6H_2O$ 0.4g
NH_4Cl 0.25g
KH_2PO_4 .... 0.2g
$CaCl_2 \cdot 2H_2O$ 0.1g

Preparation of Solution A: Add components to distilled/deionized water and bring volume to 1.0L. Mix thoroughly. Gently heat and bring to boiling. Continue boiling for 3–4 min. Allow to cool to room temperature while gassing under 80% N_2 + 20% CO_2. Continue gassing until pH reaches below 6.0. Seal the flask under 80% N_2 + 20% CO_2. Autoclave for 15 min at 15 psi pressure–121°C.

Solution B (Trace Elements Solution SL-10):

Composition per liter:

$FeCl_2 \cdot 4H_2O$ 1.5g
$CoCl_2 \cdot 6H_2O$ 190.0mg
$MnCl_2 \cdot 4H_2O$ 100.0mg
$ZnCl_2$ 70.0mg
$Na_2MoO_4 \cdot 2H_2O$ 36.0mg
$NiCl_2 \cdot 6H_2O$ 24.0mg
H_3BO_3 .... 6.0mg
$CuCl_2 \cdot 2H_2O$ 2.0mg
HCl (25% solution) 10.0mL

Preparation of Solution B (Trace Elements Solution SL-10): Add $FeCl_2 \cdot 4H_2O$ to 10.0mL of HCl solution. Mix thoroughly. Add distilled/deionized water and bring volume to 1.0L. Add remaining components. Mix thoroughly. Gas under 100% N_2. Autoclave for 15 min at 15 psi pressure–121°C.

Solution C (Selenite-Tungstate Solution):

Composition per liter:

NaOH 0.5g
$Na_2WO_4 \cdot 2H_2O$ 4.0mg
$Na_2SeO_3 \cdot 5H_2O$ 3.0mg

Preparation of Solution C (Selenite-Tungstate Solution): Add components to distilled/deionized water and bring volume to 1.0L. Mix thoroughly. Gas under 100% N_2. Autoclave for 15 min at 15 psi pressure–121°C.

Solution D:

Composition per 50.0mL:

$NaHCO_3$ 2.5g

Preparation of Solution D: Add $NaHCO_3$ to distilled/deionized water and bring volume to 50.0mL. Mix thoroughly. Gas under 80% N_2 + 20% CO_2. Autoclave for 15 min at 15 psi pressure–121°C.

Solution E:

Composition per liter:

Pyridoxine·HCl 10.0mg
Calcium DL-pantothenate 5.0mg
Lipoic acid 5.0mg
Nicotinic acid 5.0mg
p-Aminobenzoic acid 5.0mg
Riboflavin 5.0mg
Thiamine·HCl 5.0mg
Biotin 2.0mg
Folic acid 2.0mg
Vitamin B_{12} 0.1mg

Preparation of Solution E: Add components to distilled/deionized water and bring volume to 1.0L. Mix thoroughly. Gas under 100% N_2. Filter sterilize.

Solution F:

Composition per 10.0mL:

Vitamin B_{12} 0.5mg

Preparation of Solution F: Add vitamin B_{12} to distilled/deionized water and bring volume to 10.0mL. Mix thoroughly. Gas under 100% N_2. Filter sterilize.

Solution G:

Composition per 80.0mL:

Stearic acid 2.85g
NaOH (4.0*M* solution) 2.5mL

Preparation of Solution G: Add components to distilled/deionized water and bring volume to 80.0mL. Mix thoroughly. Gas under 100% N_2. In a closed bottle, heat in a boiling water bath. Shake until stearic acid dissolves. Autoclave for 15 min at 15 psi pressure–121°C. On storage, solution will solidify and should be remelted before use.

Solution H:

Composition per liter:

NaCl 286.4g
$MgCl_2 \cdot 6H_2O$ 44.7g
$CaCl_2 \cdot 2H_2O$ 2.2g

Preparation of Solution H: Add components to distilled/deionized water and bring volume to 80.0mL. Mix thoroughly. Gas under 100% N_2. Autoclave for 15 min at 15 psi pressure–121°C.

Solution I:

Composition per 20.0mL:

$Na_2S \cdot 9H_2O$ 0.6g

Preparation of Solution I: Add $Na_2S \cdot 9H_2O$ to distilled/deionized water and bring volume to 10.0mL. Mix thoroughly. Gas under 100% N_2. Autoclave for 15 min at 15 psi pressure–121°C.

Preparation of Medium: To 1.0L of sterile solution A, add in order: 1.0mL of sterile solution B, 2.0mL of sterile solution C, 50.0mL of sterile solution D, 10.0mL of sterile solution E, 1.0mL of sterile solution F, 10.0mL of sterile solution G, 67.0mL of sterile solution H, and 13.0mL of sterile solution I. Mix thoroughly. Final pH of the medium should be 7.2. Prior to inoculation, add 10.0–20.0mg of sodium dithionate to 1.0L of medium.

Use: For the cultivation and maintenance of *Desulfobacterium oleovorans.*

Desulfobacterium phenolicum Medium

Composition per 1002.4mL:

Solution A 930.0mL
Solution C 50.0mL
Solution E (Wolfe's vitamin solution) 10.0mL
Solution G 10.0mL
Solution D 4.0mL
Solution B (Trace elements solution SL-10) 1.0mL
Solution F 1.0mL
Solution H 0.4mL

pH 7.0 ± 0.2 at 25°C

Solution A:

Composition per 920.0mL:

NaCl	21.0g
$MgCl_2 \cdot 6H_2O$	3.0g
Na_2SO_4	3.0g
KCl	0.5g
NH_4Cl	0.3g
KH_2PO_4	0.2g
$CaCl_2 \cdot 2H_2O$	0.15g
Resazurin	1.0mg

Preparation of Solution A: Prepare and dispense solution anaerobically under 80% N_2 + 20% CO_2. Add components to distilled/deionized water and bring volume to 920.0mL. Mix thoroughly. Gently heat and bring to boiling. Continue boiling until resazurin turns colorless, indicating reduction. Cap with rubber stoppers. Autoclave for 15 min at 15 psi pressure–121°C. Cool to 45°–50°C.

Solution B (Trace Elements Solution SL-10):

Composition per liter:

$FeCl_2 \cdot 4H_2O$	1.5g
$CoCl_2 \cdot 6H_2O$	0.19g
$MnCl_2 \cdot 4H_2O$	0.1g
$ZnCl_2$	0.07g
$Na_2MoO_4 \cdot 2H_2O$	0.036g
$NiCl_2 \cdot 6H_2O$	0.024g
H_3BO_3	6.0mg
$CuCl_2 \cdot 2H_2O$	2.0mg
HCl (25% solution)	10.0mL

Preparation of Solution B (Trace Elements Solution SL-10): Add $FeCl_2 \cdot 4H_2O$ to 10.0mL of HCl solution. Mix thoroughly. Bring volume to 900.0mL with distilled/deionized water. Mix thoroughly. Adjust pH to 6.0 with NaOH. Bring volume to 1.0L with distilled/deionized water. Filter sterilize. Aseptically gas under 100% N_2 for 20 min.

Solution C:

Composition per 50.0mL:

$NaHCO_3$	2.5g

Preparation of Solution C: Add $NaHCO_3$ to distilled/deionized water and bring volume to 50.0mL. Mix thoroughly. Filter sterilize. Aseptically gas under 80% N_2 + 20% CO_2 for 20 min.

Solution D:

Composition per 10.0mL:

Sodium benzoate	1.0g
Phenol	0.1g

Preparation of Solution D: Add components to distilled/deionized water and bring volume to 10.0mL. Mix thoroughly. Filter sterilize. Aseptically gas under 100% N_2 for 20 min.

Solution E (Wolfe's Vitamin Solution):

Composition per liter:

Pyridoxine·HCl	0.01g
Thiamine·HCl	5.0mg
Riboflavin	5.0mg
Nicotinic acid	5.0mg
Calcium pantothenate	5.0mg
p-Aminobenzoic acid	5.0mg
Thioctic acid	5.0mg
Biotin	2.0mg
Folic acid	2.0mg
Cyanocobalamin	0.1mg

Preparation of Solution E (Wolfe's Vitamin Solution): Add components to distilled/deionized water and bring volume to 1.0L. Mix thoroughly. Filter sterilize. Aseptically gas under 100% N_2 for 20 min.

Solution F:

Composition per liter:

$Na_2SeO_3 \cdot 5H_2O$	3.0mg
NaOH (0.01*M* solution)	1.0L

Preparation of Solution F: Add $Na_2SeO_3 \cdot 5H_2O$ to 1.0L of NaOH solution. Mix thoroughly. Filter sterilize. Aseptically gas under 100% N_2 for 20 min.

Solution G:

Composition per 10.0mL:

$Na_2S \cdot 9H_2O$	0.4g

Preparation of Solution G: Add $Na_2S \cdot 9H_2O$ to distilled/deionized water and bring volume to 10.0mL. Gas under 100% N_2 for 20 min. Cap with a rubber stopper. Autoclave for 15 min at 15 psi pressure–121°C. Cool to 25°C.

Solution H:

Composition per 10.0mL:

$Na_2S_2O_4$	0.5g

Preparation of Solution H: Add $Na_2S_2O_4$ to distilled/deionized water and bring volume to 10.0mL. Mix thoroughly. Filter sterilize. Aseptically gas under 100% N_2 for 20 min. Prepare solution freshly.

Preparation of Medium: To 920.0mL of cooled, sterile solution A, aseptically and anaerobically add in the following order: 1.0mL of sterile solution B, 50.0mL of sterile solution C, 10.0mL of sterile solution D, 10.0mL of sterile solution E, 1.0mL of sterile solution F, and 10.0mL of sterile solution G. Mix thoroughly. Immediately prior to inoculation aseptically and anaerobically add 0.4mL of sterile solution H. Mix thoroughly. Aseptically and anaerobically distribute into sterile tubes or flasks.

Use: For the cultivation and maintenance of *Desulfobacterium phenolicum*.

Desulfobacula toluolica Medium (DSMZ Medium 383b)

Composition per 1013.5mL:

Soultion A	930.0mL
Solution C	50.0mL
Solution D	10.0mL
Solution E	10.0mL
Solution G	10.0mL
Solution B	1.0mL
Solution F	1.0mL
Sodium-tungstate solution	1.0mL
Vitamin B_{12} solution	0.5mL

pH 7.0 at 25°C

Solution A:

Composition per 930.0mL:

NaCl	21.0g
Na_2SO_4	3.0g
$MgCl_2 \cdot 6H_2O$	3.0g
KCl	0.5g
NH_4Cl	0.3g
KH_2PO_4	0.2g
$CaCl_2 \cdot 2H_2O$	0.15g
Resazurin	1.0mg

Preparation of Solution A: Add components to distilled/deionized water and bring volume to 930.0mL. Mix thoroughly. Sparge with 80% N_2 + 20% CO_2 gas until saturated. Autoclave for 15 min at 15 psi pressure–121°C. Cool to 25°C.

Solution B:

Composition per liter:

Na-EDTA	5.2g
$FeCl_2·4H_2O$	1.5g
H_3BO_3	300.0mg
$CoCl_2·6H_2O$	190.0mg
$MnCl_2·4H_2O$	100.0mg
$ZnCl_2$	70.0mg
$Na_2MoO_4·2H_2O$	36.0mg
$NiCl_2·6H_2O$	24.0mg
$CuCl_2·2H_2O$	2.0mg
HCl (25% solution)	7.7mL

Preparation of Solution B: Add $FeCl_2·4H_2O$ to 7.7mL of HCl solution. Mix thoroughly. Add distilled/deionized water and bring volume to 1.0L. Add remaining components. Mix thoroughly. Adjust pH to 6.0. Sparge with 100% N_2. Autoclave for 15 min at 15 psi pressure–121°C.

Solution C:

Composition per 100.0mL:

$NaHCO_3$	5.0g

Preparation of Solution C: Add $NaHCO_3$ to distilled/deionized water and bring volume to 100.0mL. Mix thoroughly. Sparge with 100% CO_2 until saturated, approximately 20 minutes. Filter sterilize under 100% CO_2 into a sterile, gas-tight 100.0mL screw-capped bottle.

Solution D:

Composition per 10.0mL:

Na-benzoate	0.4g

Preparation of Solution D: Add Na-benzoate to distilled/deionized water and bring volume to 10.0mL. Sparge with N_2. Filter sterilize. Store anaerobically.

Solution E:

Composition per liter:

Pyridoxine-HCl	10.0mg
Thiamine-HCl·$2H_2O$	5.0mg
Riboflavin	5.0mg
Nicotinic acid	5.0mg
D-Ca-pantothenate	5.0mg
p-Aminobenzoic acid	5.0mg
Lipoic acid	5.0mg
Biotin	2.0mg
Folic acid	2.0mg
Vitamin B_{12}	0.1mg

Preparation of Solution E: Add components to distilled/deionized water and bring volume to 1.0L. Mix thoroughly. Sparge with 100% N_2. Filter sterilize.

Solution F:

Composition per liter:

NaOH	0.5g
$Na_2SeO_3·5H_2O$	3.0mg

Preparation of Solution F: Add components to distilled/deionized water and bring volume to 1.0L. Mix thoroughly. Sparge with 100% N_2. Filter sterilize.

Solution G:

Composition per 10.0mL:

$Na_2S·9H_2O$	0.4g

Preparation of Solution G: Add $Na_2S·9H_2O$ to distilled/deionized water and bring volume to 10.0mL. Sparge with N_2. Autoclave for 15 min at 15 psi pressure–121°C. Cool to 25°C. Store anaerobically.

Vitamin B_{12} Solution:

Composition per 100.0mL:

Vitamin B_{12}	10.0mg

Vitamin B_{12} Solution: Add vitamin B_{12} to distilled/deionized water and bring volume to 100.0mL. Mix thoroughly. Sparge under 100% N_2 gas for 3 min. Filter sterilize.

Selenite-Tungstate Solution:

Composition per liter:

NaOH	0.5g
$Na_2WO_4·2H_2O$	4.0mg
$Na_2SeO_3·5H_2O$	3.0mg

Preparation of Selenite-Tungstate Solution: Add components to distilled/deionized water and bring volume to 1.0L. Mix thoroughly. Sparge with 100% N_2. Filter sterilize.

Preparation of Medium: Add solution B, solution C, solution D, solution E, Vitamin B_{12} solution, solution F, selenite-tungstate solution, and solution G to solution A in that order under N_2 gas. Adjust the pH to 7.0.

Use: For the cultivation of *Desulfobacula toluolica.*

Desulfobulbus Medium

Composition per liter:

Sodium propionate	1.85g
Na_2SO_4	1.5g
$(NH_4)_2SO_4$	1.24g
NaCl	0.6g
L-Cysteine·HCl·H_2O	0.5g
KH_2PO_4	0.3g
$MgSO_4·7H_2O$	0.12g
$CaCl_2·2H_2O$	0.08g
Trace minerals	10.0mL
Vitamin solution	10.0mL

pH 7.0 ± 0.2 at 25°C

Trace Minerals:

Composition per liter:

Nitrilotriacetic acid	12.8g
$FeSO_4·7H_2O$	0.3g
$CoCl_2·2H_2O$	0.1g
$MnCl_2·4H_2O$	0.1g
$ZnCl_2$	0.1g
$CuCl_2$	0.02g
H_3BO_3	0.01g
Na_2MoO_4	0.01g
$NiSO_4·6H_2O$	2.6mg
Na_2SeO_3	1.7mg

Preparation of Trace Minerals: Add nitrilotriacetic acid to 500.0mL of distilled/deionized water. Dissolve by adjusting pH to 6.5 with KOH. Add remaining components. Add distilled/deionized water to 1.0L.

Vitamin Solution:

Composition per liter:

Pyridoxine·HCl	0.01g
Thiamine·HCl	5.0mg
Riboflavin	5.0mg
Nicotinic acid	5.0mg
Calcium pantothenate	5.0mg
p-Aminobenzoic acid	5.0mg
Thioctic acid	5.0mg
Biotin	2.0mg
Folic acid	2.0mg
Cyanocobalamin	0.1mg

Preparation of Vitamin Solution: Add components to distilled/deionized water and bring volume to 1.0L. Mix thoroughly.

Preparation of Medium: Add components, except L-cysteine·HCl·H_2O, to distilled/deionized water and bring volume to 1.0L. Mix thoroughly. Gently heat and bring to boiling. Cool to 25°C under 85% N_2 + 15% CO_2. Add L-cysteine·HCl·H_2O. Mix thoroughly. Adjust pH to 7.2 with $KHCO_3$. Anaerobically distribute into tubes or flasks under 85% N_2 + 15% CO_2. Autoclave for 15 min at 15 psi pressure–121°C. Adjust pH to 7.0, if necessary.

Use: For the cultivation and maintenance of *Desulfobulbus elongatus*.

Desulfobulbus Medium

Composition per 1001.0mL:

Solution A	870.0mL
Solution C	100.0mL
Solution D	10.0mL
Solution E (Vitamin solution)	10.0mL
Solution F	10.0mL
Solution B (Trace elements solution SL-10)	1.0mL

pH 7.1–7.4 at 25°C

Solution A:

Composition per 870.0mL:

Na_2SO_4	3.0g
NaCl	1.0g
KCl	0.5g
$MgCl_2·6H_2O$	0.4g
NH_4Cl	0.3g
KH_2PO_4	0.2g
$CaCl_2·2H_2O$	0.15g
Resazurin	1.0mg

Preparation of Solution A: Add components to distilled/deionized water and bring volume to 870.0mL. Mix thoroughly. Gently heat and bring to boiling. Continue boiling for 3–4 min. Allow to cool to room temperature while gassing under 80% N_2 + 20% CO_2. Continue gassing until pH reaches below 6.0. Seal the flask under 80% N_2 + 20% CO_2. Autoclave for 15 min at 15 psi pressure–121°C.

Solution B (Trace Elements Solution SL-10):

Composition per liter:

$FeCl_2·4H_2O$	1.5g
$CoCl_2·6H_2O$	190.0mg
$MnCl_2·4H_2O$	100.0mg
$ZnCl_2$	70.0mg
$Na_2MoO_4·2H_2O$	36.0mg
$NiCl_2·6H_2O$	24.0mg
H_3BO_3	6.0mg
$CuCl_2·2H_2O$	2.0mg
HCl (25% solution)	10.0mL

Preparation of Solution B (Trace Elements Solution SL-10): Add $FeCl_2·4H_2O$ to 10.0mL of HCl solution. Mix thoroughly. Add distilled/deionized water and bring volume to 1.0L. Add remaining components. Mix thoroughly. Gas under 100% N_2. Autoclave for 15 min at 15 psi pressure–121°C.

Solution C:

Composition per 100.0mL:

$NaHCO_3$	5.0g

Preparation of Solution C: Add $NaHCO_3$ to distilled/deionized water and bring volume to 100.0mL. Mix thoroughly. Filter sterilize. Gas under 80% N_2 + 20% CO_2.

Solution D:

Composition per 10.0mL:

Sodium propionate	2.5g

Preparation of Solution D: Add sodium propionate to distilled/deionized water and bring volume to 10.0mL. Mix thoroughly. Gas under 100% N_2. Autoclave for 15 min at 15 psi pressure–121°C.

Solution E (Vitamin Solution):

Composition per liter:

Pyridoxine·HCl	10.0mg
Calcium DL-pantothenate	5.0mg
Lipoic acid	5.0mg
Nicotinic acid	5.0mg
p-Aminobenzoic acid	5.0mg
Riboflavin	5.0mg
Thiamine·HCl	5.0mg
Biotin	2.0mg
Folic acid	2.0mg
Vitamin B_{12}	0.1mg

Preparation of Solution E (Vitamin Solution): Add components to distilled/deionized water and bring volume to 1.0L. Mix thoroughly. Gas under 100% N_2. Autoclave for 15 min at 15 psi pressure–121°C.

Solution F:

Composition per 10.0mL:

$Na_2S·9H_2O$	0.4g

Preparation of Solution F: Add $Na_2S·9H_2O$ to distilled/deionized water and bring volume to 10.0mL. Mix thoroughly. Gas under 100% N_2. Autoclave for 15 min at 15 psi pressure–121°C.

Preparation of Medium: Aseptically and anaerobically combine 870.0mL of sterile solution A with 1.0mL of sterile solution B, 100.0mL of sterile solution C, 10.0mL of sterile solution D, 10.0mL of sterile solution E, and 10.0mL of sterile solution F, in that order. Mix thoroughly. Anaerobically distribute into sterile tubes or flasks under 100% N_2.

Use: For the cultivation and maintenance of *Desulfobulbus* species.

Desulfobulbus Medium

Composition per 1001.0mL:

Solution A	870.0mL
Solution C	100.0mL
Solution D	10.0mL
Solution E (Vitamin solution)	10.0mL
Solution F	10.0mL
Solution B (Trace elements solution SL-10)	1.0mL

pH 7.1–7.4 at 25°C

Solution A:

Composition per 870.0mL:

NaCl	7.0g
Na_2SO_4	3.0g
$MgCl_2 \cdot 6H_2O$	1.3g
KCl	0.5g
NH_4Cl	0.3g
KH_2PO_4	0.2g
$CaCl_2 \cdot 2H_2O$	0.15g
Resazurin	1.0mg

Preparation of Solution A: Add components to distilled/deionized water and bring volume to 870.0mL. Mix thoroughly. Gently heat and bring to boiling. Continue boiling for 3–4 min. Allow to cool to room temperature while gassing under 80% N_2 + 20% CO_2. Continue gassing until pH reaches below 6.0. Seal the flask under 80% N_2 + 20% CO_2. Autoclave for 15 min at 15 psi pressure–121°C.

Solution B (Trace Elements Solution SL-10):

Composition per liter:

$FeCl_2 \cdot 4H_2O$	1.5g
$CoCl_2 \cdot 6H_2O$	190.0mg
$MnCl_2 \cdot 4H_2O$	100.0mg
$ZnCl_2$	70.0mg
$Na_2MoO_4 \cdot 2H_2O$	36.0mg
$NiCl_2 \cdot 6H_2O$	24.0mg
H_3BO_3	6.0mg
$CuCl_2 \cdot 2H_2O$	2.0mg
HCl (25% solution)	10.0mL

Preparation of Solution B (Trace Elements Solution SL-10): Add $FeCl_2 \cdot 4H_2O$ to 10.0mL of HCl solution. Mix thoroughly. Add distilled/deionized water and bring volume to 1.0L. Add remaining components. Mix thoroughly. Gas under 100% N_2. Autoclave for 15 min at 15 psi pressure–121°C.

Solution C:

Composition per 100.0mL:

$NaHCO_3$	5.0g

Preparation of Solution C: Add $NaHCO_3$ to distilled/deionized water and bring volume to 100.0mL. Mix thoroughly. Filter sterilize. Gas under 80% N_2 + 20% CO_2.

Solution D:

Composition per 10.0mL:

Sodium propionate	1.5g

Preparation of Solution D: Add sodium propionate to distilled/deionized water and bring volume to 10.0mL. Mix thoroughly. Gas under 100% N_2. Autoclave for 15 min at 15 psi pressure–121°C.

Solution E (Vitamin Solution):

Composition per liter:

Pyridoxine·HCl	10.0mg
Calcium DL-pantothenate	5.0mg
Lipoic acid	5.0mg
Nicotinic acid	5.0mg
p-Aminobenzoic acid	5.0mg
Riboflavin	5.0mg
Thiamine·HCl	5.0mg
Biotin	2.0mg
Folic acid	2.0mg
Vitamin B_{12}	0.1mg

Preparation of Solution E (Vitamin Solution): Add components to distilled/deionized water and bring volume to 1.0L. Mix thoroughly. Gas under 100% N_2. Autoclave for 15 min at 15 psi pressure–121°C.

Solution F:

Composition per 10.0mL:

$Na_2S \cdot 9H_2O$	0.4g

Preparation of Solution F: Add $Na_2S \cdot 9H_2O$ to distilled/deionized water and bring volume to 10.0mL. Mix thoroughly. Gas under 100% N_2. Autoclave for 15 min at 15 psi pressure–121°C.

Preparation of Medium: Aseptically and anaerobically combine solution A with solution B, solution C, solution D, solution E, and solution F, in that order. Mix thoroughly. Anaerobically distribute into sterile tubes or flasks under 80% N_2 + 20% CO_2.

Use: For the cultivation and maintenance of *Desulfobulbus* species and *Streptomyces* species.

Desulfobulbus spp. Medium (DSMZ Medium 196)

Composition per 1001.0mL:

Solution A	870.0mL
Solution C	100.0mL
Solution D	10.0mL
Solution E (Vitamin solution)	10.0mL
Solution F	10.0mL
Solution B (Trace elements solution SL-10)	1.0mL

pH 7.1–7.4 at 25°C

Solution A:

Composition per 870.0mL:

NaCl	21.0g
$MgCl_2 \cdot 6H_2O$	3.1g
Na_2SO_4	3.0g
KH_2PO_4	0.2g
NH_4Cl	0.3g
KCl	0.5g
$CaCl_2 \cdot 2H_2O$	0.15g
Resazurin	1.0mg

Preparation of Solution A: Add components to distilled/deionized water and bring volume to 870.0mL Mix thoroughly.

Solution B (Trace Elements Solution SL-10):

Composition per liter:

$FeCl_2 \cdot 4H_2O$	1.5g
$CoCl_2 \cdot 6H_2O$	190.0mg
$MnCl_2 \cdot 4H_2O$	100.0mg
$ZnCl_2$	70.0mg
$Na_2MoO_4 \cdot 2H_2O$	36.0mg
$NiCl_2 \cdot 6H_2O$	24.0mg
H_3BO_3	6.0mg
$CuCl_2 \cdot 2H_2O$	2.0mg
HCl (25% solution)	10.0mL

Preparation of Solution B (Trace Elements Solution SL-10): Add $FeCl_2 \cdot 4H_2O$ to 10.0mL of HCl solution. Mix thoroughly. Add distilled/deionized water and bring volume to 1.0L. Add remaining components. Mix thoroughly. Sparge with 100% N_2. Autoclave for 15 min at 15 psi pressure–121°C.

Solution C:

Composition per 100.0mL:

$NaHCO_3$	5.0g

Preparation of Solution C: Add $NaHCO_3$ to distilled/deionized water and bring volume to 100.0mL. Mix thoroughly. Filter sterilize. Flush with 80% N_2 + 20% CO_2 to remove dissolved oxygen.

Solution D:

Composition per 10.0mL:

Na-propionate	1.5g

Preparation of Solution D: Add Na-propionate to distilled/deionized water and bring volume to 10.0mL. Mix thoroughly. Sparge with 100% N_2. Autoclave for 15 min at 15 psi pressure–121°C.

Solution E (Vitamin Solution):

Composition per liter:

Pyridoxine-HCl	10.0mg
Thiamine-HCl·$2H_2O$	5.0mg
Riboflavin	5.0mg
Nicotinic acid	5.0mg
D-Ca-pantothenate	5.0mg
p-Aminobenzoic acid	5.0mg
Lipoic acid	5.0mg
Biotin	2.0mg
Folic acid	2.0mg
Vitamin B_{12}	0.10mg

Preparation of Solution E (Vitamin Solution): Add components to distilled/deionized water and bring volume to 1.0L. Mix thoroughly. Sparge with 100% N_2. Autoclave for 15 min at 15 psi pressure–121°C.

Solution F:

Composition per 10.0mL:

$Na_2S·9H_2O$	0.4g

Preparation of Solution F: Add $Na_2S·9H_2O$ to distilled/deionized water and bring volume to 10.0mL. Mix thoroughly. Sparge with 100% N_2. Autoclave for 15 min at 15 psi pressure–121°C.

Preparation of Medium: Gently heat solution A and bring to boiling. Boil solution A for a few minutes. Cool to room temperature. Gas with 80% N_2 + 20% CO_2 gas mixture to reach a pH below 6. Autoclave for 15 min at 15 psi pressure–121°C. Cool to room temperature. Sequentially add 1.0mL solution B, 100.0mL solution C, 10.0mL solution D, 10.0mL solution E, and 10.0mL solution F. Distribute anaerobically under 80% N_2 + 20% CO_2 into appropriate vessels. Addition of 10-20mg sodium dithionite per liter from a 5% (w/v) solution, freshly prepared under N_2 and filter sterilized, may stimulate growth.

Use: For the cultivation of *Desulfosarcina variabilis.*

Desulfocapsa sulfoexigens Medium (DSMZ Medium 195b)

Composition per 1001.0mL:

Solution A	890.0mL
Solution C	100.0mL
Solution D	10.0mL
Solution B (Trace elements solution SL-10)	1.0mL

pH 7.2 ± 0.2 at 25°C

Solution A:

NaCl	21.0g
$MgCl_2·6H_2O$	3.1g
Na_2SO_4	3.0g
$FeCl_3·6H_2O$	2.7g
KCl	0.5g
NH_4Cl	0.3g
KH_2PO_4	0.2g
$CaCl_2·2H_2O$	0.15g
Resazurin	1.0mg

Preparation of Solution A: Dissolve 2.7g $FeCl_3·6H_2O$ in 890.0ml distilled/deionized water. Adjust pH to 7 with 1*N* NaOH. Add remaining components. Mix thoroughly. Sparge with 80% N_2 + 20% CO_2 gas mixture.

Solution B (Trace Elements Solution SL-10):

Composition per liter:

$FeCl_2·4H_2O$	1.5g
$CoCl_2·6H_2O$	190.0mg
$MnCl_2·4H_2O$	100.0mg
$ZnCl_2$	70.0mg
$Na_2MoO_4·2H_2O$	36.0mg
$NiCl_2·6H_2O$	24.0mg
H_3BO_3	6.0mg
$CuCl_2·2H_2O$	2.0mg
HCl (25% solution)	10.0mL

Preparation of Solution B (Trace Elements Solution SL-10): Add $FeCl_2·4H_2O$ to 10.0mL of HCl solution. Mix thoroughly. Add distilled/deionized water and bring volume to 1.0L. Add remaining components. Mix thoroughly. Sparge with 100% N_2. Autoclave for 15 min at 15 psi pressure–121°C.

Solution C:

Composition per 100.0mL:

$NaHCO_3$	5.0g

Preparation of Solution C: Add $NaHCO_3$ to distilled/deionized water and bring volume to 100.0mL. Mix thoroughly. Filter sterilize. Flush with 80% N_2 + 20% CO_2 to remove dissolved oxygen.

Solution D:

Composition per 10.0mL:

Na-thiosulfate	5.0g

Preparation of Solution D: Add Na-thiosulfate to distilled/deionized water and bring volume to 10.0mL. Mix thoroughly. Filter sterilize. Flush with 80% N_2 + 20% CO_2 to remove dissolved oxygen.

Preparation of Medium: Gently heat solution A and bring to boiling. Boil solution A for a few minutes. Cool to room temperature. Gas with 80% N_2 + 20% CO_2 gas mixture to reach a pH below 6. Autoclave for 15 min at 15 psi pressure–121°C. Cool to room temperature. Sequentially add 1.0mL solution B, 100.0mL solution C, and 10.0mL solution D. Adjust pH to 7.2 with sodium bicarbonate or sodium carbonate. Distribute anaerobically under 80% N_2 + 20% CO_2 into appropriate vessels. Alternately distribute solution A to tubes prior to autoclaving. Dispense 8.9 mL amounts under 80% N_2 + 20% CO_2 into Hungate tubes. Seal and autoclave for 15 min at 15 psi pressure–121°C. Before use aseptically add appropriate amounts of remaining solutions to each tube from sterile anaerobic solutions.

Use: For the cultivation of *Desulfocapsa sulfoexigens.*

Desulfococcus amylolyticus Medium

Composition per 1011.0mL:

Sulfur, powdered	10.0g
Starch	5.0g
$NaHCO_3$	0.8g
$MgCl_2·6H_2O$	0.7g
$Na_2S·9H_2O$	0.5g
NaCl	0.5g
$CaCl_2·2H_2O$	0.44g
KCl	0.33g

KH_2PO_4	0.33g
NH_4Cl	0.33g
Yeast extract	0.2g
Resazurin	1.0mg
Vitamin solution	10.0mL
Trace elements solution SL-10	1.0mL

pH 6.2–6.4 at 25°C

Vitamin Solution:

Composition per liter:

Pyridoxine-HCl	10.0mg
Calcium DL-pantothenate	5.0mg
Lipoic acid	5.0mg
Nicotinic acid	5.0mg
p-Aminobenzoic acid	5.0mg
Riboflavin	5.0mg
Thiamine-HCl	5.0mg
Biotin	2.0mg
Folic acid	2.0mg
Vitamin B_{12}	0.1mg

Preparation of Vitamin Solution: Add components to distilled/deionized water and bring volume to 1.0L. Mix thoroughly. Gas under 100% N_2. Autoclave for 15 min at 15 psi pressure–121°C.

Trace Elements Solution SL-10:

Composition per liter:

$FeCl_2 \cdot 4H_2O$	1.5g
$CoCl_2 \cdot 6H_2O$	190.0mg
$MnCl_2 \cdot 4H_2O$	100.0mg
$ZnCl_2$	70.0mg
$Na_2MoO_4 \cdot 2H_2O$	36.0mg
$NiCl_2 \cdot 6H_2O$	24.0mg
H_3BO_3	6.0mg
$CuCl_2 \cdot 2H_2O$	2.0mg
HCl (25% solution)	10.0mL

Preparation of Trace Elements Solution SL-10: Add $FeCl_2 \cdot 4H_2O$ to 10.0mL of HCl. Mix thoroughly. Add distilled/deionized water and bring volume to 1.0L. Add remaining components. Mix thoroughly. Gas under 100% N_2. Autoclave for 15 min at 15 psi pressure–121°C.

Preparation of Medium: Add components to distilled/deionized water and bring volume to 1.0L. Mix thoroughly. Gas under 80% N_2 + 20% CO_2. Autoclave for 15 min at 15 psi pressure–121°C.

Use: For the cultivation and maintenance of *Desulfurococcus amylolyticus*.

Desulfococcus Medium

Composition per 1001.0mL:

Solution A	870.0mL
Solution C	100.0mL
Solution D	10.0mL
Solution E (Vitamin solution)	10.0mL
Solution F	10.0mL
Solution B (Trace Elements Solution SL-10)	1.0mL

pH 7.1–7.4 at 25°C

Solution A:

Composition per 870.0mL:

NaCl	7.0g
Na_2SO_4	3.0g
$MgCl_2 \cdot 6H_2O$	1.3g
KCl	0.5g
NH_4Cl	0.3g
KH_2PO_4	0.2g
$CaCl_2 \cdot 2H_2O$	0.15g
Resazurin	1.0mg
$Na_2SeO_3 \cdot 5H_2O$	3.0μg

Preparation of Solution A: Add components to distilled/deionized water and bring volume to 870.0mL. Mix thoroughly. Gently heat and bring to boiling. Continue boiling for 3–4 min. Allow to cool to room temperature while gassing under 80% N_2 + 20% CO_2. Continue gassing until pH reaches below 6.0. Seal the flask under 80% N_2 + 20% CO_2. Autoclave for 15 min at 15 psi pressure–121°C.

Solution B (Trace Elements Solution SL-10):

Composition per liter:

$FeCl_2 \cdot 4H_2O$	1.5g
$CoCl_2 \cdot 6H_2O$	190.0mg
$MnCl_2 \cdot 4H_2O$	100.0mg
$ZnCl_2$	70.0mg
$Na_2MoO_4 \cdot 2H_2O$	36.0mg
$NiCl_2 \cdot 6H_2O$	24.0mg
H_3BO_3	6.0mg
$CuCl_2 \cdot 2H_2O$	2.0mg
HCl (25% solution)	10.0mL

Preparation of Solution B (Trace Elements Solution SL-10): Add $FeCl_2 \cdot 4H_2O$ to 10.0mL of HCl solution. Mix thoroughly. Add distilled/deionized water and bring volume to 1.0L. Add remaining components. Mix thoroughly. Gas under 100% N_2. Autoclave for 15 min at 15 psi pressure–121°C.

Solution C:

Composition per 100.0mL:

$NaHCO_3$	5.0g

Preparation of Solution C: Add $NaHCO_3$ to distilled/deionized water and bring volume to 100.0mL. Mix thoroughly. Filter sterilize. Gas under 80% N_2 + 20% CO_2.

Solution D:

Composition per 10.0mL:

Sodium benzoate	0.6g

Preparation of Solution D: Add sodium benzoate to distilled/deionized water and bring volume to 10.0mL. Mix thoroughly. Gas under 100% N_2. Autoclave for 15 min at 15 psi pressure–121°C.

Solution E (Vitamin Solution):

Composition per liter:

Pyridoxine·HCl	10.0mg
Calcium DL-pantothenate	5.0mg
Lipoic acid	5.0mg
Nicotinic acid	5.0mg
p-Aminobenzoic acid	5.0mg
Riboflavin	5.0mg
Thiamine·HCl	5.0mg
Biotin	2.0mg
Folic acid	2.0mg
Vitamin B_{12}	0.1mg

Preparation of Solution E (Vitamin Solution): Add components to distilled/deionized water and bring volume to 1.0L. Mix thoroughly. Gas under 100% N_2. Autoclave for 15 min at 15 psi pressure–121°C.

Solution F:

Composition per 10.0mL:

$Na_2S \cdot 9H_2O$	0.4g

Preparation of Solution F: Add $Na_2S{\cdot}9H_2O$ to distilled/deionized water and bring volume to 10.0mL. Mix thoroughly. Gas under 100% N_2. Autoclave for 15 min at 15 psi pressure–121°C.

Preparation of Medium: Aseptically and anaerobically combine solution A with solution B, solution C, solution D, solution E, and solution F, in that order. Mix thoroughly. Anaerobically distribute into sterile tubes or flasks under 80% N_2 + 20% CO_2.

Use: For the cultivation and maintenance of *Desulfococcus multivorans.*

Desulfococcus multivorans Medium

Composition per liter:

NaCl	10.0g
$MgSO_4{\cdot}7H_2O$	2.0g
$CaSO_4$	1.0g
NH_4Cl	1.0g
Yeast extract	1.0g
K_2HPO_4	0.5g
Sodium lactate (70% solution)	3.5mL

Preparation of Medium: Add components, except $FeSO_4{\cdot}7H_2O$, ascorbic acid, and thioglycollic acid, to distilled/deionized water and bring volume to 1.0L. Mix thoroughly. Sparge with 80% N_2 + 20% CO_2 for 10–15 min. Add $FeSO_4{\cdot}7H_2O$, ascorbic acid, and thioglycollic acid. Mix thoroughly. Continue to sparge with 80% N_2 + 20% CO_2 and adjust pH to 7.4. Anaerobically distribute into tubes or flasks. Autoclave for 10 min at 10 psi pressure–115°C.

Use: For the cultivation of *Desulfococcus multivorans.*

Desulfococcus niacini Medium

Composition per 1001.0mL:

Solution A	870.0mL
Solution C	100.0mL
Solution D	10.0mL
Solution E (Vitamin solution)	10.0mL
Solution F	10.0mL
Solution B (Trace elements solution SL-10)	1.0mL

pH 7.4 ± 0.2 at 25°C

Solution A:

Composition per 870.0mL:

NaCl	7.0g
Na_2SO_4	3.0g
$MgCl_2{\cdot}6H_2O$	1.3g
KCl	0.5g
NH_4Cl	0.3g
KH_2PO_4	0.2g
$CaCl_2{\cdot}2H_2O$	0.15g
Resazurin	1.0mg
$Na_2SeO_3{\cdot}5H_2O$	3.0μg

Preparation of Solution A: Add components to distilled/deionized water and bring volume to 870.0mL. Mix thoroughly. Gently heat and bring to boiling. Continue boiling for 3–4 min. Allow to cool to room temperature while gassing under 80% N_2 + 20% CO_2. Continue gassing until pH reaches below 6.0. Seal the flask under 80% N_2 + 20% CO_2. Autoclave for 15 min at 15 psi pressure–121°C.

Solution B (Trace Elements Solution SL-10):

Composition per liter:

$FeCl_2{\cdot}4H_2O$	1.5g
$CoCl_2{\cdot}6H_2O$	190.0mg
$MnCl_2{\cdot}4H_2O$	100.0mg
$ZnCl_2$	70.0mg
$Na_2MoO_4{\cdot}2H_2O$	36.0mg
$NiCl_2{\cdot}6H_2O$	24.0mg
H_3BO_3	6.0mg
$CuCl_2{\cdot}2H_2O$	2.0mg
HCl (25% solution)	10.0mL

Preparation of Solution B (Trace Elements Solution SL-10): Add $FeCl_2{\cdot}4H_2O$ to 10.0mL of HCl solution. Mix thoroughly. Add distilled/deionized water and bring volume to 1.0L. Add remaining components. Mix thoroughly. Gas under 100% N_2. Autoclave for 15 min at 15 psi pressure–121°C.

Solution C:

Composition per 100.0mL:

$NaHCO_3$	5.0g

Preparation of Solution C: Add $NaHCO_3$ to distilled/deionized water and bring volume to 100.0mL. Mix thoroughly. Filter sterilize. Gas under 80% N_2 + 20% CO_2.

Solution D:

Composition per 10.0mL:

Sodium nicotinate	0.82g

Preparation of Solution D: Add sodium nicotinate to distilled/deionized water and bring volume to 10.0mL. Mix thoroughly. Gas under 100% N_2. Autoclave for 15 min at 15 psi pressure–121°C.

Solution E (Vitamin Solution):

Composition per liter:

Pyridoxine·HCl	10.0mg
Calcium DL-pantothenate	5.0mg
Lipoic acid	5.0mg
Nicotinic acid	5.0mg
p-Aminobenzoic acid	5.0mg
Riboflavin	5.0mg
Thiamine·HCl	5.0mg
Biotin	2.0mg
Folic acid	2.0mg
Vitamin B_{12}	0.1mg

Preparation of Solution E (Vitamin Solution): Add components to distilled/deionized water and bring volume to 1.0L. Mix thoroughly. Gas under 100% N_2. Autoclave for 15 min at 15 psi pressure–121°C.

Solution F:

Composition per 10.0mL:

$Na_2S{\cdot}9H_2O$	0.4g

Preparation of Solution F: Add $Na_2S{\cdot}9H_2O$ to distilled/deionized water and bring volume to 10.0mL. Mix thoroughly. Gas under 100% N_2. Autoclave for 15 min at 15 psi pressure–121°C.

Preparation of Medium: Aseptically and anaerobically combine solution A with solution B, solution C, solution D, solution E, and solution F, in that order. Mix thoroughly. Adjust pH to 7.4. Anaerobically distribute into sterile tubes or flasks under 80% N_2 + 20% CO_2.

Use: For the cultivation and maintenance of *Desulfococcus niacini.*

Desulfoglaeba Medium (DSMZ Medium 1074)

Composition per liter:

NaCl	20.0g
Na_2SO_4	3.5g
$MgCl_2{\cdot}6H_2O$	3.0g

KCl	0.5g
NH_4Cl	0.25g
KH_2PO_4	0.2g
$CaCl_2 \cdot 2H_2O$	0.15g
Yeast extract	0.05g
Resazurin	0.5mg
Wolfe's vitamin solution	10.0mL
$NaHCO_3$ solution	10.0mL
$Na_2S \cdot 9H_2O$ solution	10.0mL
L-Cysteine·HCl solution	10.0mL
Wolfe's Mineral Elixir	1.0mL
n-Decane	1.0mL

pH 7.1 ± 0.2 at 25°C

L-Cysteine·HCl Solution:
Composition per 10.0mL:

L-Cysteine·HCl	0.1g

Preparation of L-Cysteine Solution: Add L-cysteine·HCl to distilled/deionized water and bring volume to 10.0mL. Mix thoroughly. Sparge with 100% N_2. Autoclave under 100% N_2 for 15 min at 15 psi pressure–121°C.

$Na_2S \cdot 9H_2O$ Solution:
Composition per 10.0mL:

$Na_2S \cdot 9H_2O$	0.1g

Preparation of $Na_2S \cdot 9H_2O$ Solution: Add $Na_2S \cdot 9H_2O$ to distilled/deionized water and bring volume to 10.0mL. Mix thoroughly. Autoclave under 100% N_2 for 15 min at 15 psi pressure–121°C. Cool to room temperature.

$NaHCO_3$ Solution:
Composition per 10.0mL:

$NaHCO_3$	3.0g

Preparation of $NaHCO_3$ Solution: Add components to distilled/deionized water and bring volume to 10.0mL. Mix thoroughly. Sparge with 20% CO_2 + 80% H_2. Autoclave for 15 min at 15 psi pressure–121°C. Cool to room temperature.

Wolfe's Mineral Elixir:
Composition per liter:

$MgSO_4 \cdot 7H_2O$	30.0g
NaCl	10.0g
$MnSO_4 \cdot 2H_2O$	5.0g
$(NH_4)_2NiSO_4 \cdot 6H_2O$	2.8g
$CoCl_2 \cdot 6H_2O$	1.8g
$ZnSO_4 \cdot 7H_2O$	1.8g
$FeSO_4 \cdot 7H_2O$	1.0g
$CaCl_2 \cdot 2H_2O$	1.0g
$KAl(SO_4)_2 \cdot 12H_2O$	0.18g
$CuSO_4 \cdot 5H_2O$	0.1g
H_3BO_3	0.1g
$Na_2MoO_4 \cdot 2H_2O$	0.1g
Na_2SeO_4	0.1g
$Na_2WO_4 \cdot 2H_2O$	0.1g

Preparation of Wolfe's Mineral Elixir: Adjust pH of 1.0L of distilled/deionized water to 1.0 with dilute H_2SO_4. Add remaining components one at a time. Mix thoroughly to dissolve.

Wolfe's Vitamin Solution:
Composition per liter:

Pyridoxine·HCl	10.0mg
Thiamine·HCl	5.0mg
Riboflavin	5.0mg
Nicotinic acid	5.0mg
Calcium pantothenate	5.0mg
p-Aminobenzoic acid	5.0mg
Thioctic acid	5.0mg
Biotin	2.0mg
Folic acid	2.0mg
Cyanocobalamin	100.0μg

Preparation of Wolfe's Vitamin Solution: Add components to distilled/deionized water and bring volume to 1.0L. Mix thoroughly. Filter sterilize.

Preparation of Medium: Add components, except bicarbonate, vitamins, and reducing agents, to distilled/deionized water and bring volume to 960.0mL. Adjust pH to 7.2. Gently heat and bring to boiling. Boil for 1 min. Mix thoroughly. Cool to room temperature while sparging with 80% N_2 + 20% CO_2. Dispense into culture vessels under an atmosphere of 80% N_2 + 20% CO_2 (e.g., Balch tubes or serum vials). Add n-decane to each vessel. Close the vessels using butyl rubber stoppers fixed with aluminum crimps. Autoclave for 15 min at 15 psi pressure–121°C. Add vitamins and reducing agents from sterile, anoxic stock solutions. Adjust pH to 7.1. Use 20% as inoculum and incubate vials in an inverted position.

Use: For the cultivation of *Desulfoglaeba* spp.

Desulfohalobium Medium

Composition per 1010.0mL:

NaCl	100.0g
$MgCl_2 \cdot 6H_2O$	20.0g
KCl	4.0g
Na_2SO_4	3.0g
$CaCl_2 \cdot 2H_2O$	2.7g
NH_4Cl	1.0g
Sodium acetate	1.0g
Trypticase™	1.0g
Yeast extract	1.0g
K_2HPO_4	0.3g
KH_2PO_4	0.3g
Sodium (L)-lactate	2.5g
Resazurin	1.0mg
$Na_2SeO_3 \cdot 5H_2O$	3.0μg
$Na_2S \cdot 9H_2O$ solution	10.0mL
Trace elements solution SL-10	1.0mL

pH 7.0 ± 0.2 at 25°C

Trace Elements Solution SL-10:
Composition per liter:

$FeCl_2 \cdot 4H_2O$	1.5g
$CoCl_2 \cdot 6H_2O$	190.0mg
$MnCl_2 \cdot 4H_2O$	100.0mg
$ZnCl_2$	70.0mg
$Na_2MoO_4 \cdot 2H_2O$	36.0mg
$NiCl_2 \cdot 6H_2O$	24.0mg
H_3BO_3	6.0mg
$CuCl_2 \cdot 2H_2O$	2.0mg
HCl (25% solution)	10.0mL

Preparation of Trace Elements Solution SL-10: Add $FeCl_2 \cdot 4H_2O$ to 10.0mL of HCl solution. Mix thoroughly. Add distilled/deionized water and bring volume to 1.0L. Add remaining components. Mix thoroughly. Gas under 100% N_2. Autoclave for 15 min at 15 psi pressure–121°C.

$Na_2S \cdot 9H_2O$ Solution:

Composition per 10.0mL:

$Na_2S \cdot 9H_2O$ 1.0mg

Preparation of $Na_2S \cdot 9H_2O$ Solution: Add $Na_2S \cdot 9H_2O$ to distilled/deionized water and bring volume to 10.0mL. Mix thoroughly. Gas under 100% N_2. Autoclave for 15 min at 15 psi pressure–121°C.

Preparation of Medium: Add components, except $Na_2S \cdot 9H_2O$ solution, to distilled/deionized water and bring volume to 1.0L. Mix thoroughly. Gas under 100% N_2. Autoclave for 15 min at 15 psi pressure–121°C. Aseptically and anaerobically add 10.0mL of sterile $Na_2S \cdot 9H_2O$ solution. Mix thoroughly. Aseptically and anaerobically distribute into sterile tubes or flasks.

Use: For the cultivation and maintenance of *Methanohalophilus oregonense.*

Desulfohalobium utahense Medium (DSMZ Medium 1055)

Composition per liter:

NaCl 10.0g
$MgSO_4 \cdot 7H_2O$ 10.0g
KCl 6.0g
NH_4Cl 1.0g
KH_2PO_4 .... 0.1g
$CaCl_2 \cdot 2H_2O$ 0.4g
Yeast extract 0.5g
Resazurin 0.5mg
$NaHCO_3$ solution 10.0mL
Lactate solution 10.0mL
$Na_2S \cdot 9H_2O$ solution 10.0mL
Trace elements solution SL-10 .... 1.0mL
Selenite-tungstate solution 1.0mL

pH 7.1 ± 0.2 at 25°C

Lactate Solution:

Composition per 10.0mL:

Na-(L)-lactate 2.5g

Preparation of Lactate Solution: Add Na-(L)-lactate to distilled/deionized water and bring volume to 10.0mL. Mix thoroughly. Autoclave under 100% N_2 for 15 min at 15 psi pressure–121°C. Cool to room temperature.

$Na_2S \cdot 9H_2O$ Solution:

Composition per 10.0mL:

$Na_2S \cdot 9H_2O$ 0.1g

Preparation of $Na_2S \cdot 9H_2O$ Solution: Add $Na_2S \cdot 9H_2O$ to distilled/deionized water and bring volume to 10.0mL. Mix thoroughly. Autoclave under 100% N_2 for 15 min at 15 psi pressure–121°C. Cool to room temperature.

$NaHCO_3$ Solution:

Composition per 10.0mL:

$NaHCO_3$ 4.0g

Preparation of $NaHCO_3$ Solution: Add $NaHCO_3$ to distilled/deionized water and bring volume to 10.0mL. Mix thoroughly. Sparge with 20% CO_2 + 80% H_2. Autoclave for 15 min at 15 psi pressure–121°C. Cool to room temperature.

Selenite/Tungstate Solution:

Composition per liter:

NaOH 0.5g
$Na_2WO_4 \cdot 2H_2O$ 4.0mg
$Na_2SeO_3 \cdot 5H_2O$ 3.0mg

Preparation of Selenite/Tungstate Solution: Add components to distilled/deionized water and bring volume to 1.0L. Mix thoroughly. Sparge with 100% N_2. Autoclave for 15 min at 15 psi pressure–121°C.

Trace Elements Solution SL-10:

Composition per liter:

$FeCl_2 \cdot 4H_2O$ 1.5g
$CoCl_2 \cdot 6H_2O$ 190.0mg
$MnCl_2 \cdot 4H_2O$ 100.0mg
$ZnCl_2$ 70.0mg
$Na_2MoO_4 \cdot 2H_2O$ 36.0mg
$NiCl_2 \cdot 6H_2O$ 24.0mg
H_3BO_3 .... 6.0mg
$CuCl_2 \cdot 2H_2O$ 2.0mg
HCl (25% solution) 10.0mL

Preparation of Trace Elements Solution SL-10: Add $FeCl_2 \cdot 4H_2O$ to 10.0mL of HCl solution. Mix thoroughly. Add distilled/deionized water and bring volume to 1.0L. Add remaining components. Mix thoroughly. Sparge with 100% N_2. Autoclave for 15 min at 15 psi pressure–121°C.

Preparation of Medium: Add components, except bicarbonate, vitamins, and reducing agents, to distilled/deionized water and bring volume to 960.0mL. Adjust pH to 7.2. Gently heat and bring to boiling. Boil for 1 min. Mix thoroughly. Cool to room temperature while sparging with 80% N_2 + 20% CO_2. Dispense into culture vessels under an atmosphere of 80% N_2 + 20% CO_2. Autoclave for 15 min at 15 psi pressure–121°C. Aseptically add bicarbonate, sodium lactate, and sulfide. Adjust the final pH of the medium to 7.0–7.2.

Use: For the cultivation of *Desulfohalobium utahense.*

Desulfohalobium utahense Medium with Malate (DSMZ Medium 1055)

Composition per liter:

NaCl 10.0g
$MgSO_4 \cdot 7H_2O$ 10.0g
KCl 6.0g
NH_4Cl 1.0g
KH_2PO_4 .... 0.1g
$CaCl_2 \cdot 2H_2O$ 0.4g
Yeast extract 0.5g
Resazurin 0.5mg
$NaHCO_3$ solution 10.0mL
Malate solution 10.0mL
$Na_2S \cdot 9H_2O$ solution 10.0mL
Trace elements solution SL-10 .... 1.0mL
Selenite/tungstate solution 1.0mL

pH 7.1 ± 0.2 at 25°C

Malate Solution:

Composition per 10.0mL:

Na_2(DL)-malate 1.0g

Preparation of Malate Solution: Add Na_2(DL)-malate to distilled/deionized water and bring volume to 10.0mL. Mix thoroughly. Autoclave under 100% N_2 for 15 min at 15 psi pressure–121°C. Cool to room temperature.

$Na_2S \cdot 9H_2O$ Solution:

Composition per 10.0mL:

$Na_2S \cdot 9H_2O$ 0.1g

Preparation of $Na_2S \cdot 9H_2O$ Solution: Add $Na_2S \cdot 9H_2O$ to distilled/deionized water and bring volume to 10.0mL. Mix thoroughly.

Autoclave under 100% N_2 for 15 min at 15 psi pressure–121°C. Cool to room temperature.

$NaHCO_3$ Solution:

Composition per 10.0mL:

$NaHCO_3$	4.0g

Preparation of $NaHCO_3$ Solution: Add $NaHCO_3$ to distilled/deionized water and bring volume to 10.0mL. Mix thoroughly. Sparge with 20% CO_2 + 80% H_2. Autoclave for 15 min at 15 psi pressure–121°C. Cool to room temperature.

Selenite/Tungstate Solution:

Composition per liter:

NaOH	0.5g
$Na_2WO_4 \cdot 2H_2O$	4.0mg
$Na_2SeO_3 \cdot 5H_2O$	3.0mg

Preparation of Selenite/Tungstate Solution: Add components to distilled/deionized water and bring volume to 1.0L. Mix thoroughly. Sparge with 100% N_2. Autoclave for 15 min at 15 psi pressure–121°C.

Trace Elements Solution SL-10:

Composition per liter:

$FeCl_2 \cdot 4H_2O$	1.5g
$CoCl_2 \cdot 6H_2O$	190.0mg
$MnCl_2 \cdot 4H_2O$	100.0mg
$ZnCl_2$	70.0mg
$Na_2MoO_4 \cdot 2H_2O$	36.0mg
$NiCl_2 \cdot 6H_2O$	24.0mg
H_3BO_3	6.0mg
$CuCl_2 \cdot 2H_2O$	2.0mg
HCl (25% solution)	10.0mL

Preparation of Trace Elements Solution SL-10: Add $FeCl_2 \cdot 4H_2O$ to 10.0mL of HCl solution. Mix thoroughly. Add distilled/deionized water and bring volume to 1.0L. Add remaining components. Mix thoroughly. Sparge with 100% N_2. Autoclave for 15 min at 15 psi pressure–121°C.

Preparation of Medium: Add components, except bicarbonate, vitamins, and reducing agents, to distilled/deionized water and bring volume to 960.0mL. Adjust pH to 7.2. Gently heat and bring to boiling. Boil for 1 min. Mix thoroughly. Cool to room temperature while sparging with 80% N_2 + 20% CO_2. Dispense into culture vessels under an atmosphere of 80% N_2 + 20% CO_2. Autoclave for 15 min at 15 psi pressure–121°C. Aseptically add bicarbonate, sodium lactate, and sulfide. Adjust the final pH of the medium to 7.0–7.2.

Use: For the cultivation of *Desulfovermiculus halophilus.*

Desulfoluna Medium (DSMZ Medium 1099)

Composition per liter:

NaCl	20.0g
$MgSO_4 \cdot 7H_2O$	2.0g
NH_4Cl	1.0g
Na_2SO_4	1.0g
KH_2PO_4	0.5g
Yeast extract	0.5g
$CaCl_2 \cdot 2H_2O$	0.1g
Resazurin	1.0mg
L-Cysteine·HCl solution	10.0mL
Na-lactate, 60% solution	2.9mL
Trace elements solution SL-10	1.0mL

pH 6.5 ± 0.2 at 25°C

Trace Elements Solution SL-10:

Composition per liter:

$FeCl_2 \cdot 4H_2O$	1.5g
$CoCl_2 \cdot 6H_2O$	190.0mg
$MnCl_2 \cdot 4H_2O$	100.0mg
$ZnCl_2$	70.0mg
$Na_2MoO_4 \cdot 2H_2O$	36.0mg
$NiCl_2 \cdot 6H_2O$	24.0mg
H_3BO_3	6.0mg
$CuCl_2 \cdot 2H_2O$	2.0mg
HCl (25% solution)	10.0mL

Preparation of Trace Elements Solution SL-10: Add $FeCl_2 \cdot 4H_2O$ to 10.0mL of HCl solution. Mix thoroughly. Add distilled/deionized water and bring volume to 1.0L. Add remaining components. Mix thoroughly. Sparge with 100% N_2. Autoclave for 15 min at 15 psi pressure–121°C.

L-Cysteine·HCl Solution:

Composition per 10.0mL:

L-Cysteine·HCl	0.5g

Preparation of L-Cysteine Solution: Add L-cysteine·HCl to distilled/deionized water and bring volume to 10.0mL. Mix thoroughly. Sparge with 100% N_2. Autoclave under 100% N_2 for 15 min at 15 psi pressure–121°C.

Preparation of Medium: Add components, except cyseine solution, to distilled/deionized water and bring volume to 990.0mL. Gently heat and bring to boiling. Boil for 1 min. Mix thoroughly. Cool to room temperature while sparging with 100% N_2. Dispense into culture vessels under an atmosphere of 100% N_2. Autoclave for 15 min at 15 psi pressure–121°C. Aseptically add 10.0mL cysteine solution. Adjust the final pH of the medium to 6.5.

Use: For the cultivation of *Desulfoluna* spp.

Desulfomicrobium WHB Medium

Composition per 1003.0mL:

Solution A	870.0mL
Solution C	100.0mL
Solution D	10.0mL
Solution E (Vitamin solution)	10.0mL
Solution F	10.0mL
Solution B (Trace elements solution SL-10)	1.0mL
Solution G	1.0mL
Solution H	1.0mL

pH 7.1–7.4 at 25°C

Solution A:

Composition per 870.0mL:

NaCl	21.0g
$MgCl_2 \cdot 6H_2O$	3.1g
Na_2SO_4	3.0g
KCl	0.5g
NH_4Cl	0.3g
KH_2PO_4	0.2g
$CaCl_2 \cdot 2H_2O$	0.15g
Resazurin	1.0mg

Preparation of Solution A: Add components to distilled/deionized water and bring volume to 870.0mL. Mix thoroughly. Gently heat and bring to boiling. Continue boiling for 3–4 min. Allow to cool to room temperature while gassing under 80% N_2 + 20% CO_2. Continue gassing until pH reaches below 6.0. Seal the flask under 80% N_2 + 20% CO_2. Autoclave for 15 min at 15 psi pressure–121°C.

Solution B (Trace Elements Solution SL-10):

Composition per liter:

$FeCl_2 \cdot 4H_2O$	1.5g
$CoCl_2 \cdot 6H_2O$	190.0mg
$MnCl_2 \cdot 4H_2O$	100.0mg
$ZnCl_2$	70.0mg
$Na_2MoO_4 \cdot 2H_2O$	36.0mg
$NiCl_2 \cdot 6H_2O$	24.0mg
H_3BO_3	6.0mg
$CuCl_2 \cdot 2H_2O$	2.0mg
HCl (25% solution)	10.0mL

Preparation of Solution B (Trace Elements Solution SL-10): Add $FeCl_2 \cdot 4H_2O$ to 10.0mL of HCl solution. Mix thoroughly. Add distilled/deionized water and bring volume to 1.0L. Add remaining components. Mix thoroughly. Sparge with 100% N_2. Autoclave for 15 min at 15 psi pressure–121°C.

Solution C:

Composition per 100.0mL:

$NaHCO_3$	5.0g

Preparation of Solution C: Add $NaHCO_3$ to distilled/deionized water and bring volume to 100.0mL. Mix thoroughly. Filter sterilize. Sparge with 80% N_2 + 20% CO_2.

Solution D:

Composition per 10.0mL:

Sodium lactate	4.0g

Preparation of Solution D: Add sodium lactate to distilled/deionized water and bring volume to 10.0mL. Mix thoroughly. Sparge with 100% N_2. Autoclave for 15 min at 15 psi pressure–121°C.

Solution E (Vitamin Solution):

Composition per liter:

Pyridoxine·HCl	10.0mg
Calcium DL-pantothenate	5.0mg
Lipoic acid	5.0mg
Nicotinic acid	5.0mg
p-Aminobenzoic acid	5.0mg
Riboflavin	5.0mg
Thiamine·HCl	5.0mg
Biotin	2.0mg
Folic acid	2.0mg
Vitamin B_{12}	0.1mg

Preparation of Solution E (Vitamin Solution): Add components to distilled/deionized water and bring volume to 1.0L. Mix thoroughly. Filter sterilize. Sparge with 100% N_2.

Solution F:

Composition per 10.0mL:

$Na_2S \cdot 9H_2O$	0.4g

Preparation of Solution F: Add $Na_2S \cdot 9H_2O$ to distilled/deionized water and bring volume to 10.0mL. Mix thoroughly. Sparge with 100% N_2. Autoclave for 15 min at 15 psi pressure–121°C.

Solution G (Selenite-Tungstate Solution):

Composition per liter:

NaOH	0.5g
$Na_2WO_4 \cdot 2H_2O$	4.0mg
$Na_2SeO_3 \cdot 5H_2O$	3.0mg

Preparation of Solution G (Selenite-Tungstate Solution): Add components to distilled/deionized water and bring volume to 1.0L. Mix thoroughly. Sparge with 100% N_2. Autoclave for 15 min at 15 psi pressure–121°C.

Solution H (Seven Vitamins):

Composition per liter:

Pyridoxine·HCl	0.3g
Thiamine·HCl	0.2g
Nicotinic acid	0.2g
Calcium DL-pantothenate	0.1g
Vitamin B_{12}	0.1g
p-Aminobenzoic acid	80.0mg
Biotin	20.0mg

Preparation of Solution H (Seven Vitamins): Add components to distilled/deionized water and bring volume to 1.0L. Mix thoroughly. Filter sterilize. Sparge with 100% N_2.

Preparation of Medium: Aseptically and anaerobically combine solution A with solution B, solution C, solution D, solution E, solution F, solution G, and solution H, in that order. Mix thoroughly. Anaerobically distribute into sterile tubes or flasks under 80% N_2 + 20% CO_2.

Use: For the cultivation of *Desulfomicrobium* species.

Desulfomonile Medium

Composition per 1002.0mL:

Solution A	870.0mL
Solution C	100.0mL
Solution D	10.0mL
Solution E (Vitamin solution)	10.0mL
Solution F	10.0mL
Solution B (Trace elements solution SL-10)	1.0mL
Solution G	1.0mL

pH 6.8–7.0 at 25°C

Solution A:

Composition per 870.0mL:

Na_2SO_4	3.0g
NaCl	1.0g
KCl	0.5g
$MgCl_2 \cdot 6H_2O$	0.4g
NH_4Cl	0.3g
KH_2PO_4	0.2g
$CaCl_2 \cdot 2H_2O$	0.15g
Resazurin	1.0mg

Preparation of Solution A: Add components to distilled/deionized water and bring volume to 870.0mL. Mix thoroughly. Gently heat and bring to boiling. Continue boiling for 3–4 min. Allow to cool to room temperature while gassing under 80% N_2 + 20% CO_2. Continue gassing until pH reaches below 6.0. Seal the flask under 80% N_2 + 20% CO_2. Autoclave for 15 min at 15 psi pressure–121°C.

Solution B (Trace Elements Solution SL-10):

Composition per liter:

$FeCl_2 \cdot 4H_2O$	1.5g
$CoCl_2 \cdot 6H_2O$	190.0mg
$MnCl_2 \cdot 4H_2O$	100.0mg
$ZnCl_2$	70.0mg
$Na_2MoO_4 \cdot 2H_2O$	36.0mg
$NiCl_2 \cdot 6H_2O$	24.0mg
H_3BO_3	6.0mg
$CuCl_2 \cdot 2H_2O$	2.0mg
HCl (25% solution)	10.0mL

Preparation of Solution B (Trace Elements Solution SL-10): Add $FeCl_2 \cdot 4H_2O$ to 10.0mL of HCl solution. Mix thoroughly. Add distilled/deionized water and bring volume to 1.0L. Add remaining components. Mix thoroughly. Gas under 100% N_2. Autoclave for 15 min at 15 psi pressure–121°C.

Solution C:

Composition per 100.0mL:

$NaHCO_3$	2.5g

Preparation of Solution C: Add $NaHCO_3$ to distilled/deionized water and bring volume to 100.0mL. Mix thoroughly. Filter sterilize. Gas under 80% N_2 + 20% CO_2.

Solution D:

Composition per 10.0mL:

Sodium pyruvate	4.0g

Preparation of Solution D: Add sodium pyruvate to distilled/deionized water and bring volume to 10.0mL. Mix thoroughly. Gas under 100% N_2. Autoclave for 15 min at 15 psi pressure–121°C.

Solution E (Vitamin Solution):

Composition per liter:

Nicotinamide	50.0mg
1,4-Naphthoquinone	20.0mg
Pyridoxine·HCl	5.0mg
Calcium DL-pantothenate	5.0mg
Thioctic acid	5.0mg
p-Aminobenzoic acid	5.0mg
Riboflavin	5.0mg
Thiamine·HCl	5.0mg
Biotin	5.0mg
Folic acid	5.0mg
Vitamin B_{12}	5.0mg
Hemin	5.0mg

Preparation of Solution E (Vitamin Solution): Add 1,4-naphthoquinone and hemin to 10.0mL of 0.1*N* NaOH. Mix thoroughly. Add remaining components and bring volume to 1.0L with distilled/deionized water. Mix thoroughly. Gas under 100% N_2. Autoclave for 15 min at 15 psi pressure–121°C.

Solution F:

Composition per 10.0mL:

$Na_2S \cdot 9H_2O$	0.4g

Preparation of Solution F: Add $Na_2S \cdot 9H_2O$ to distilled/deionized water and bring volume to 10.0mL. Mix thoroughly. Gas under 100% N_2. Autoclave for 15 min at 15 psi pressure–121°C.

Solution G:

Composition per 10.0mL:

NaOH	0.5g
$Na_2WO_4 \cdot 2H_2O$	4.0mg
$Na_2SeO_3 \cdot 5H_2O$	3.0mg

Preparation of Solution G: Add components to distilled/deionized water and bring volume to 10.0mL. Mix thoroughly. Gas under 100% N_2. Autoclave for 15 min at 15 psi pressure–121°C.

Preparation of Medium: Aseptically and anaerobically combine 870.0mL of sterile solution A with 1.0mL of sterile solution B, 100.0mL of sterile solution C, 10.0mL of sterile solution D, 10.0mL of sterile solution E, and 10.0mL of sterile solution F, in that order. Mix thoroughly. Anaerobically distribute into sterile tubes or flasks under 100% N_2. Add 50.0mg/L of sodium dithionite prior to inoculation.

Use: For the cultivation and maintenance of *Desulfomonile tiedjei.*

Desulfomonile tiedjei Medium

Composition per liter:

$NaHCO_3$	3.0g
PIPES (piperazine-*N,N′*-bis-2-ethanesulfonic acid) buffer	1.5g
Na_2SO_4	1.42g
Yeast extract	1.0g
Mineral solution	20.0mL
Trace metal solution	10.0mL
$Na_2S_2O_4$ solution	10.0mL
Vitamin solution	10.0mL
Sodium pyruvate solution	10.0mL
Resazurin (0.1% solution)	1.0mL

pH 7.3 ± 0.2 at 25°C

Mineral Solution:

Composition per liter:

NH_4Cl	50.0g
NaCl	40.0g
$MgCl_2 \cdot 6H_2O$	8.3g
KCl	5.0g
KH_2PO_4	5.0g
$CaCl_2 \cdot 2H_2O$	1.0g

Preparation of Mineral Solution: Add components to distilled/deionized water and bring volume to 1.0L. Mix thoroughly.

Trace Metal Solution:

Composition per liter:

Nitrilotriacetic acid	2.0g
$MnSO_4.H_2O$	1.0g
$Fe(NH_4)_2(SO_4)_2 \cdot 6H_2O$	0.8g
$CoCl_2 \cdot 6H_2O$	0.2g
$ZnSO_4 \cdot 7H_2O$	0.2g
$CuCl_2 \cdot 2H_2O$	0.02g
$Na_2MoO_4 \cdot H_2O$	0.02g
Na_2SeO_4	0.02g
Na_2WO_4	0.02g
$NiCl_2 \cdot 6H_2O$	0.02g

Preparation of Trace Metal Solution: Add nitrilotriacetic acid to 500.0mL of distilled/deionized water. Dissolve by adjusting pH to 6.5 with KOH. Add remaining components. Add distilled/deionized water to 1.0L.

Vitamin Solution:

Composition per liter:

Nicotinamide	0.05g
1,4-Naphthoquinone	0.02g
p-Aminobenzoic acid	5.0mg
Biotin	5.0mg
Calcium pantothenate	5.0mg
Cyanocobalamin	5.0mg
Folic acid	5.0mg
Hemin	5.0mg
Pyridoxine·HCl	5.0mg
Riboflavin	5.0mg
Thioctic acid	5.0mg
NaOH (0.1*N* solution)	5.0mL

Preparation of Vitamin Solution: Add thioctic acid, 1,4-naphthoquinone, and hemin to 5.0mL of 0.1*N* NaOH solution. Mix thoroughly. Bring volume to 1.0L with distilled/deionized water. Add remaining components. Mix thoroughly.

$Na_2S_2O_4$ Solution:
Composition per 10.0mL:

$Na_2S_2O_4$	0.087g

Preparation of $Na_2S_2O_4$ Solution: Add $Na_2S_2O_4$ to distilled/deionized water and bring volume to 10.0mL. Filter sterilize. Prepare freshly.

Sodium Pyruvate Solution:
Composition per 10.0mL:

Sodium pyruvate	4.4g

Preparation of Sodium Pyruvate Solution: Add sodium pyruvate to distilled/deionized water and bring volume to 10.0mL. Filter sterilize.

Preparation of Medium: Add PIPES buffer, Na_2SO_4, yeast extract, mineral solution, and trace metal solution to distilled/deionized water and bring volume to 970.0mL. Mix thoroughly. Adjust pH to 7.3 with HCl. Add $NaHCO_3$ and resazurin. Gently heat and bring to boiling under 80% N_2 + 20% CO_2. Replace headspace with 2 atm pressure of the same gas phase. Autoclave for 15 min at 15 psi pressure–121°C. Cool to 25°C. Anaerobically and aseptically add sterile vitamin solution, sodium pyruvate solution, and $Na_2S_2O_4$ solution. Mix thoroughly.

Use: For the cultivation and maintenance of *Desulfomonile tiedjei.*

Desulfomusa hansenii Medium (DSMZ Medium 916)

Solution A	870.0mL
Solution D	50.0mL
Solution F	50.0mL
Solution E (Vitamin solution)	10.0mL
Solution G	10.0mL
Solution H	10.0mL
Solution B (Trace elements solution SL-10)	1.0mL
Solution C (Selenite-tungstate solution)	1.0mL

pH 7.2 ± 0.2 at 25°C

Solution A:
Composition per 870.0mL:

NaCl	20.0g
Na_2SO_4	1.42g
KCl	0.67g
NH_4Cl	0.1g
KH_2PO_4	0.01g
$MgSO_4 \cdot 7H_2O$	0.02g
Resazurin	0.5mg

Preparation of Solution A: Add components to 870.0mL of distilled/deionized water. Mix thoroughly. Gently heat and bring to boiling. Boil for 3 min. Cool to room temperature while sparging with 80% N_2 + 20% CO_2. Autoclave for 15 min at 15 psi pressure–121°C. Cool to room temperature.

Solution B (Trace Elements Solution SL-10):
Composition per liter:

$FeCl_2 \cdot 4H_2O$	1.5g
$CoCl_2 \cdot 6H_2O$	190.0mg
$MnCl_2 \cdot 4H_2O$	100.0mg
$ZnCl_2$	70.0mg
$Na_2MoO_4 \cdot 2H_2O$	36.0mg
$NiCl_2 \cdot 6H_2O$	24.0mg
H_3BO_3	6.0mg
$CuCl_2 \cdot 2H_2O$	2.0mg
HCl (25% solution)	10.0mL

Preparation of Solution B (Trace Elements Solution SL-10): Add $FeCl_2 \cdot 4H_2O$ to 10.0mL of HCl solution. Mix thoroughly. Add distilled/deionized water and bring volume to 1.0L. Add remaining components. Mix thoroughly. Sparge with 100% N_2. Autoclave for 15 min at 15 psi pressure–121°C.

Solution C (Selenite-Tungstate Solution):
Composition per liter:

NaOH	0.5g
$Na_2WO_4 \cdot 2H_2O$	4.0mg
$Na_2SeO_3 \cdot 5H_2O$	3.0mg

Preparation of Solution C (Selenite-Tungstate Solution): Add components to distilled/deionized water and bring volume to 1.0L. Sparge with 100% N_2. Autoclave for 15 min at 15 psi pressure–121°C.

Solution D:
Composition per 50.0mL:

$NaHCO_3$	5.0g

Preparation of Solution D: Add $NaHCO_3$ to distilled/deionized water and bring volume to 50.0mL. Mix thoroughly. Sparge with 100% N_2. Autoclave for 15 min at 15 psi pressure–121°C.

Solution E (Vitamin Solution):
Composition per liter:

Pyridoxine-HCl	10.0mg
Thiamine-HCl·$2H_2O$	5.0mg
Riboflavin	5.0mg
Nicotinic acid	5.0mg
D-Ca-pantothenate	5.0mg
p-Aminobenzoic acid	5.0mg
Lipoic acid	5.0mg
Biotin	2.0mg
Folic acid	2.0mg
Vitamin B_{12}	0.1mg

Preparation of Solution E (Vitamin Solution): Add components to distilled/deionized water and bring volume to 1.0L. Mix thoroughly. Sparge with 100% N_2. Filter sterilize.

Solution F:
Composition per 50.0mL:

$MgCl_2 \cdot 6H_2O$	10.6g
$CaCl_2 \cdot 2H_2O$	1.52g

Preparation of Solution F: Add components to distilled/deionized water and bring volume to 50.0mL. Mix thoroughly. Sparge with 100% N_2. Autoclave for 15 min at 15 psi pressure–121°C.

Solution G:
Composition per 10.0mL:

$FeCl_2 \cdot 4H_2O$	0.52g

Preparation of Solution G: Add $FeCl_2 \cdot 4H_2O$ to distilled/deionized water and bring volume to 10.0mL. Mix thoroughly. Adjust pH to 2.0. Sparge with 100% N_2. Autoclave for 15 min at 15 psi pressure–121°C.

Solution H:
Composition per 10.0mL:

Na-propionate	0.96g

Preparation of Solution H: Add Na-propionate to distilled/deionized water and bring volume to 10.0mL. Mix thoroughly. Sparge with 100% N_2. Autoclave for 15 min at 15 psi pressure–121°C.

Preparation of Medium: Aseptically and anaerobically under 80% N_2 + 20% CO_2 sequentially add to 860.0mL sterile solution A, 1.0mL solution B, 1.0mL solution C, 50.0mL solution D, 10.0mL solution E,

50.0mL solution F, 10.0mL solution G, and 10.0mL solution H. Aseptically and anaerobically distribute under 80% N_2 + 20% CO_2 into appropriate vessels. The final pH should be 7.2. Addition of 10–20mg sodium dithionite per liter from a 5% (w/v) solution, freshly prepared under N_2 and filter sterilized, may stimulate growth.

Use: For the cultivation of *Desulfomusa hansenii.*

Desulfonatronospira Medium (DSMZ Medium 1101)

Composition per liter:

Na_2CO_3	95.0g
$NaHCO_3$	15.0g
NaCl	15.0g
K_2HPO_4	0.5g
Ammonium chloride solution	10.0mL
Magnesium chloride solution	10.0mL
Yeast extract solution	10.0mL
Pyruvate solution	10.0mL
Thiosulfate solution	10.0mL
$Na_2S{\cdot}9H_2O$ solution	10.0mL
Vitamin solution	10.0mL
Trace elements solution SL-10 with -EDTA	1.0mL
Selenite/tungstate solution	1.0mL

pH 10.0 ± 0.2 at 25°C

Trace Elements Solution SL-10 with EDTA:

Composition per liter:

$FeCl_2{\cdot}4H_2O$	1.5g
Na_2-EDTA	0.5g
$CoCl_2{\cdot}6H_2O$	190.0mg
$MnCl_2{\cdot}4H_2O$	100.0mg
$ZnCl_2$	70.0mg
$Na_2MoO_4{\cdot}2H_2O$	36.0mg
$NiCl_2{\cdot}6H_2O$	24.0mg
H_3BO_3	6.0mg
$CuCl_2{\cdot}2H_2O$	2.0mg
HCl (25% solution)	10.0mL

Preparation of Trace Elements Solution SL-10 with EDTA: Add $FeCl_2{\cdot}4H_2O$ to 10.0mL of HCl solution. Mix thoroughly. Add distilled/deionized water and bring volume to 1.0L. Add remaining components. Mix thoroughly. Sparge with 100% N_2. Autoclave for 15 min at 15 psi pressure–121°C. Adjsut pH to 7.0.

Selenite/Tungstate Solution:

Composition per liter:

NaOH	0.5g
$Na_2WO_4{\cdot}2H_2O$	4.0mg
$Na_2SeO_3{\cdot}5H_2O$	3.0mg

Preparation of Selenite/Tungstate Solution: Add components to distilled/deionized water and bring volume to 1.0L. Mix thoroughly. Sparge with 100% N_2. Autoclave for 15 min at 15 psi pressure–121°C.

Vitamin Solution:

Composition per liter:

Pyridoxine-HCl	10.0mg
Thiamine-HCl·$2H_2O$	5.0mg
Riboflavin	5.0mg
Nicotinic acid	5.0mg
D-Ca-pantothenate	5.0mg
p-Aminobenzoic acid	5.0mg
Lipoic acid	5.0mg
Biotin	2.0mg
Folic acid	2.0mg
Vitamin B_{12}	0.1mg

Preparation of Vitamin Solution: Add components to distilled/deionized water and bring volume to 1.0L. Mix thoroughly. Sparge with 80% H_2 + 20% CO_2. Filter sterilize.

$Na_2S{\cdot}9H_2O$ Solution:

Composition per 10.0mL:

$Na_2S{\cdot}9H_2O$	0.24g

Preparation of $Na_2S{\cdot}9H_2O$ Solution: Add $Na_2S{\cdot}9H_2O$ to distilled/deionized water and bring volume to 10.0mL. Mix thoroughly. Autoclave under 100% N_2 for 15 min at 15 psi pressure–121°C. Cool to room temperature.

Ammonium Chloride Solution:

Composition per 10.0mL:

NH_4Cl	0.2g

Preparation of Ammonium Chloride Solution: Add NH_4Cl to distilled/deionized water and bring volume to 10.0mL. Mix thoroughly. Sparge with 100% N_2. Autoclave for 15 min at 15 psi pressure–121°C.

Magnesium Chloride Solution:

Composition per 10.0mL:

$MgCl_2{\cdot}6H_2O$	0.2g

Preparation of Magnesium Chloride Solution: Add $MgCl_2{\cdot}6H_2O$ to distilled/deionized water and bring volume to 10.0mL. Mix thoroughly. Sparge with 100% N_2. Autoclave for 15 min at 15 psi pressure–121°C.

Yeast Extract Solution:

Composition per 10.0mL:

Yeast extract	0.5g

Preparation of Yeast Extract Solution: Add yeast extract to distilled/deionized water and bring volume to 10.0mL. Mix thoroughly. Sparge with 100% N_2. Autoclave for 15 min at 15 psi pressure–121°C.

Lactate Solution:

Composition per 10.0mL:

Na-lactate	2.2g

Preparation of Lactate Solution: Add Na-lactate to distilled/deionized water and bring volume to 10.0mL. Mix thoroughly. Sparge with 100% N_2. Autoclave for 15 min at 15 psi pressure–121°C.

Sulfite Solution:

Composition per 10.0mL:

Na_2SO_3	1.3g

Preparation of Sulfite: Add Na_2SO_3 to distilled/deionized water and bring volume to 10.0mL. Mix thoroughly. Sparge with 100% N_2. Filter sterilize.

Preparation of Medium: Add components, except ammonium chloride, magnesium chloride, yeast extract, trace elements, vitamins, lactate, sulfite, and sulfide solutions, to distilled/deionized water and bring volume to 930.0mL. Gently heat and bring to boiling. Boil for 1 min. Mix thoroughly. Cool to room temperature while sparging with 100% N_2. Dispense into culture vessels under an atmosphere of 100% N_2. Autoclave for 15 min at 15 psi pressure–121°C. Aseptically add the ammonium chloride, magnesium chloride, yeast extract, trace elements, vitamins, lactate, sulfite, and sulfide solutions. Adjust the pH to 10.

Use: For the cultivation of *Desulfonatronospira* spp.

Desulfonatronospira Medium (DSMZ Medium 1101)

Composition per liter:

Na_2CO_3	95.0g
$NaHCO_3$	15.0g
NaCl	15.0g
K_2HPO_4	0.5g
Ammonium chloride solution	10.0mL
Magnesium chloride solution	10.0mL
Yeast extract solution	10.0mL
Lactate solution	10.0mL
Sulfite solution	10.0mL
$Na_2S \cdot 9H_2O$ solution	10.0mL
Vitamin solution	10.0mL
Trace elements solution SL-10 with EDTA	1.0mL
Selenite/tungstate solution	1.0mL

pH 10.0 ± 0.2 at 25°C

Trace Elements Solution SL-10 with EDTA:

Composition per liter:

$FeCl_2 \cdot 4H_2O$	1.5g
Na_2-EDTA	0.5g
$CoCl_2 \cdot 6H_2O$	190.0mg
$MnCl_2 \cdot 4H_2O$	100.0mg
$ZnCl_2$	70.0mg
$Na_2MoO_4 \cdot 2H_2O$	36.0mg
$NiCl_2 \cdot 6H_2O$	24.0mg
H_3BO_3	6.0mg
$CuCl_2 \cdot 2H_2O$	2.0mg
HCl (25% solution)	10.0mL

Preparation of Trace Elements Solution SL-10 with EDTA: Add $FeCl_2 \cdot 4H_2O$ to 10.0mL of HCl solution. Mix thoroughly. Add distilled/deionized water and bring volume to 1.0L. Add remaining components. Mix thoroughly. Sparge with 100% N_2. Autoclave for 15 min at 15 psi pressure–121°C. Adjsut pH to 7.0.

Selenite/Tungstate Solution:

Composition per liter:

NaOH	0.5g
$Na_2WO_4 \cdot 2H_2O$	4.0mg
$Na_2SeO_3 \cdot 5H_2O$	3.0mg

Preparation of Selenite/Tungstate Solution: Add components to distilled/deionized water and bring volume to 1.0L. Mix thoroughly. Sparge with 100% N_2. Autoclave for 15 min at 15 psi pressure–121°C.

Vitamin Solution:

Composition per liter:

Pyridoxine-HCl	10.0mg
Thiamine-HCl·$2H_2O$	5.0mg
Riboflavin	5.0mg
Nicotinic acid	5.0mg
D-Ca-pantothenate	5.0mg
p-Aminobenzoic acid	5.0mg
Lipoic acid	5.0mg
Biotin	2.0mg
Folic acid	2.0mg
Vitamin B_{12}	0.1mg

Preparation of Vitamin Solution: Add components to distilled/deionized water and bring volume to 1.0L. Mix thoroughly. Sparge with 80% H_2 + 20% CO_2. Filter sterilize.

Lactate Solution:

Composition per 10.0mL:

Na-lactate	2.2g

Preparation of Lactate Solution: Add Na-lactate to distilled/deionized water and bring volume to 10.0mL. Mix thoroughly. Sparge with 100% N_2. Autoclave for 15 min at 15 psi pressure–121°C.

Sulfite Solution:

Composition per 10.0mL:

Na_2SO_3	1.3g

Preparation of Sulfite: Add Na_2SO_3 to distilled/deionized water and bring volume to 10.0mL. Mix thoroughly. Sparge with 100% N_2. Filter sterilize.

$Na_2S \cdot 9H_2O$ Solution:

Composition per 10.0mL:

$Na_2S \cdot 9H_2O$	0.24g

Preparation of $Na_2S \cdot 9H_2O$ Solution: Add $Na_2S \cdot 9H_2O$ to distilled/deionized water and bring volume to 10.0mL. Mix thoroughly. Autoclave under 100% N_2 for 15 min at 15 psi pressure–121°C. Cool to room temperature.

Ammonium Chloride Solution:

Composition per 10.0mL:

NH_4Cl	0.2g

Preparation of Ammonium Chloride Solution: Add NH_4Cl to distilled/deionized water and bring volume to 10.0mL. Mix thoroughly. Sparge with 100% N_2. Autoclave for 15 min at 15 psi pressure–121°C.

Magnesium Chloride Solution:

Composition per 10.0mL:

$MgCl_2 \cdot 6H_2O$	0.2g

Preparation of Magnesium Chloride Solution: Add $MgCl_2 \cdot 6H_2O$ to distilled/deionized water and bring volume to 10.0mL. Mix thoroughly. Sparge with 100% N_2. Autoclave for 15 min at 15 psi pressure–121°C.

Yeast Extract Solution:

Composition per 10.0mL:

Yeast extract	0.5g

Preparation of Yeast Extract Solution: Add yeast extract to distilled/deionized water and bring volume to 10.0mL. Mix thoroughly. Sparge with 100% N_2. Autoclave for 15 min at 15 psi pressure–121°C.

Pyruvate Solution:

Composition per 10.0mL:

Na-pyruvate	2.2g

Preparation of Pyruvate Solution: Add Na-pyruvate to distilled/deionized water and bring volume to 10.0mL. Mix thoroughly. Sparge with 100% N_2. Autoclave for 15 min at 15 psi pressure–121°C.

Thiosulfate Solution:

Composition per 10.0mL:

$Na_2S_2O_3$	5.0g

Preparation of Thiosulfate Solution: Add $Na_2S_2O_3$ to distilled/deionized water and bring volume to 10.0mL. Mix thoroughly. Sparge with 100% N_2. Filter sterilize.

Preparation of Medium: Add components, except ammonium chloride, magnesium chloride, yeast extract, trace elements, vitamin, pyruvate, thiosulfate, and sulfide solutions, to distilled/deionized water and bring volume to 930.0mL. Gently heat and bring to boiling. Boil for 1 min. Mix thoroughly. Cool to room temperature while sparging with 100% N_2. Dispense into culture vessels under an atmosphere of 100% N_2. Autoclave for 15 min at 15 psi pressure–121°C. Aseptically add the ammonium chloride, magnesium chloride, yeast extract, trace elements, vitamin, pyruvate, thiosulfate, and sulfide solutions. Adjust the pH to 10.

Use: For the cultivation of *Desulfonatronospira delicata*.

Desulfonatronovibrio Medium (DSMZ Medium 742)

Composition per liter:

$NaHCO_3$	15.0g
Na_2CO_3	10.0g
NaCl	10.0g
Na_2SO_4	3.0g
NH_4Cl	1.0g
Na_2HPO_4	0.2g
KCl	0.2g
Resazurin	0.5mg
Yeast extract solution	10.0mL
Na-formate solution	10.0mL
Trace elements solution	10.0mL
Vitamin solution	10.0mL
$Na_2S·9H_2O$ solution	10.0mL

pH 9.6 ± 0.2 at 25°C

Na-Formate Solution:

Composition per 10.0mL:

Na-formate	5.0g

Preparation of Na-Formate Solution: Add Na-formate to distilled/deionized water and bring volume to 10.0mL. Mix thoroughly. Sparge with 100% N_2. Autoclave for 15 min at 15 psi pressure–121°C.

Yeast Extract Solution:

Composition per 10.0mL:

Yeast extract	1.5g

Preparation of Yeast Extract Solution: Add yeast extract to distilled/deionized water and bring volume to 10.0mL. Mix thoroughly. Sparge with 100% N_2. Autoclave under 100% N_2 for 15 min at 15 psi pressure–121°C. Cool to room temperature.

$Na_2S·9H_2O$ Solution:

Composition per 10.0mL:

$Na_2S·9H_2O$	1.0g

Preparation of $Na_2S·9H_2O$ Solution: Add $Na_2S·9H_2O$ to distilled/deionized water and bring volume to 10.0mL. Sparge with N_2. Autoclave for 15 min at 15 psi pressure–121°C. Cool to 25°C.

Trace Elements Solution:

Composition per liter:

$MgSO_4·7H_2O$	3.0g
Nitrilotriacetic acid	1.5g
NaCl	1.0g
$MnSO_4·2H_2O$	0.5g
$CoSO_4·7H_2O$	0.18g
$ZnSO_4·7H_2O$	0.18g
$CaCl_2·2H_2O$	0.1g
$FeSO_4·7H_2O$	0.1g
$NiCl_2·6H_2O$	0.025g
$KAl(SO_4)_2·12H_2O$	0.02g
H_3BO_3	0.01g
$Na_2MoO_4·4H_2O$	0.01g
$CuSO_4·5H_2O$	0.01g
$Na_2SeO_3·5H_2O$	0.3mg

Preparation of Trace Elements Solution: Add nitrilotriacetic acid to 500.0mL of distilled/deionized water. Dissolve by adjusting pH to 6.5 with KOH. Add remaining components. Add distilled/deionized water to 1.0L. Mix thoroughly. Filter sterilize.

Vitamin Solution:

Composition per liter:

Pyridoxine-HCl	10.0mg
Thiamine-HCl·$2H_2O$	5.0mg
Riboflavin	5.0mg
Nicotinic acid	5.0mg
D-Ca-pantothenate	5.0mg
p-Aminobenzoic acid	5.0mg
Lipoic acid	5.0mg
Biotin	2.0mg
Folic acid	2.0mg
Vitamin B_{12}	0.1mg

Preparation of Vitamin Solution: Add components to distilled/deionized water and bring volume to 1.0L. Mix thoroughly. Sparge with 100% N_2. Filter sterilize.

Preparation of Medium: Prepare and dispense medium under an oxygen-free atmosphere of 100% N_2. Add components, except vitamin solution, yeast extract solution, Na-formate solution, trace elements solution, and $Na_2S·9H_2O$ solution, to distilled/deionized water and bring volume to 950.0mL. Mix thoroughly. Sparge with 100% N_2. Autoclave for 15 min at 15 psi pressure–121°C. Cool to 25°C. Aseptically and anaerobically add 10.0mL sterile vitamin solution, 10.0mL of sterile yeast extract solution, 10.0mL sterile Na-formate solution, 10.0mL sterile trace elements solution, and 10.0mL of sterile $Na_2S·9H_2O$ solution. Mix thoroughly. Adjust pH to 9.6. Aseptically and anaerobically distribute into sterile tubes or flasks.

Use: For the cultivation of *Desulfonatronovibrio hydrogenovorans*.

Desulfonatronum Medium (DSMZ Medium 813)

Composition per 1010.0mL:

NaCl	10.0g
Na_2SO_4	5.0g
Na_2CO_3	3.5g
NH_4Cl	1.0g
Yeast extract	1.0g
$Na_2S·9H_2O$	0.5g
KH_2PO_4	0.2g
KCl	0.2g
$MgCl_2·6H_2O$	0.1g
Resazurin	0.5mg
Na-formate solution	10.0mL
Vitamin solution	10.0mL
Trace elements solution SL-10	1.0mL

pH 8.9 ± 0.2 at 25°C

Na-Formate Solution:

Composition per 10.0mL:

Na-formate	4.0g

Preparation of Na-Formate Solution: Add Na-formate to distilled/deionized water and bring volume to 10.0mL. Mix thoroughly. Sparge with 100% N_2. Autoclave for 15 min at 15 psi pressure–121°C.

Vitamin Solution:

Composition per liter:

Pyridoxine-HCl	10.0mg
Thiamine-HCl·$2H_2O$	5.0mg
Riboflavin	5.0mg
Nicotinic acid	5.0mg
D-Ca-pantothenate	5.0mg
p-Aminobenzoic acid	5.0mg
Lipoic acid	5.0mg
Biotin	2.0mg
Folic acid	2.0mg
Vitamin B_{12}	0.1mg

Preparation of Vitamin Solution: Add components to distilled/deionized water and bring volume to 1.0L. Mix thoroughly. Sparge with 80% H_2 + 20% CO_2. Filter sterilize.

Trace Elements Solution SL-10:

Composition per liter:

$FeCl_2·4H_2O$	1.5g
$CoCl_2·6H_2O$	190.0mg
$MnCl_2·4H_2O$	100.0mg
$ZnCl_2$	70.0mg
$Na_2MoO_4·2H_2O$	36.0mg
$NiCl_2·6H_2O$	24.0mg
H_3BO_3	6.0mg
$CuCl_2·2H_2O$	2.0mg
HCl (25% solution)	10.0mL

Preparation of Trace Elements Solution SL-10: Add $FeCl_2·4H_2O$ to 10.0mL of HCl solution. Mix thoroughly. Add distilled/deionized water and bring volume to 1.0L. Add remaining components. Mix thoroughly. Sparge with 100% N_2. Autoclave for 15 min at 15 psi pressure–121°C.

Preparation of Medium: Prepare and dispense medium under an oxygen-free atmosphere of 100% N_2. Add components, except Na-formate solution and $Na_2S·9H_2O$, to distilled/deionized water and bring volume to 1.0L. Mix thoroughly. Sparge with 100% N_2 for 30 min. Add the $Na_2S·9H_2O$. Mix thoroughly. Adjust pH to 8.8–9.0. Dispense into Hungate tubes under 100% N_2. Autoclave for 15 min at 15 psi pressure–121°C. Cool to 25°C. Aseptically and anaerobically add Na-formate solution, 0.1mL per 10.0mL medium.

Use: For the cultivation of *Desulfonatronum lacustre.*

Desulfonema ishimotoi Medium (DSMZ Medium 739)

Composition per 1007.6mL:

Solution A	850.0mL
Agar solution	50.0mL
Solution C	50.0mL
Solution H	20.0mL
Solution D	10.0mL
Solution F (Vitamin solution)	10.0mL
Solution I	10.0mL
Solution G (Artificial sediment)	6.6mL
Solution E	1.0mL
Solution B (Trace elements solution SL-10)	1.0mL

pH 6.9 ± 0.2 at 25°C

Solution A:

NaCl	21.0g
$MgCl_2·6H_2O$	5.5g
Na_2SO_4	3.0g
$CaCl_2·2H_2O$	1.35g
KCl	0.5g
NH_4Cl	0.3g
KH_2PO_4	0.2g
Resazurin	0.5mg
$Na_2SeO_3·5H_2O$	3μg

Preparation of Solution A: Add components to distilled/deionized water and bring volume to 800.0mL. Mix thoroughly.

Solution B (Trace Elements Solution SL-10):

Composition per liter:

$FeCl_2·4H_2O$	1.5g
$CoCl_2·6H_2O$	190.0mg
$MnCl_2·4H_2O$	100.0mg
$ZnCl_2$	70.0mg
$Na_2MoO_4·2H_2O$	36.0mg
$NiCl_2·6H_2O$	24.0mg
H_3BO_3	6.0mg
$CuCl_2·2H_2O$	2.0mg
HCl (25% solution)	10.0mL

Preparation of Solution B (Trace Elements Solution SL-10): Add $FeCl_2·4H_2O$ to 10.0mL of HCl solution. Mix thoroughly. Add distilled/deionized water and bring volume to 1.0L. Add remaining components. Mix thoroughly. Sparge with 100% N_2. Autoclave for 15 min at 15 psi pressure–121°C.

Solution C:

Composition per 100.0mL:

$NaHCO_3$	5.0g

Preparation of Solution C: Add $NaHCO_3$ to distilled/deionized water and bring volume to 100.0mL. Mix thoroughly. Filter sterilize. Flush with 80% N_2 + 20% CO_2 to remove dissolved oxygen.

Solution D:

Composition per 10.0mL:

Na-acetate·$3H_2O$	2.5g
Isobutyric acid	0.18g
Na_2-succinate	0.1g

Preparation of Solution D: Add components to distilled/deionized water and bring volume to 10.0mL. Mix thoroughly. Sparge with 100% N_2. Filter sterilize.

Solution E:

Composition per 10.0mL:

Na_2-succinate	1.0g

Preparation of Solution E: Add Na_2-succinate to distilled/deionized water and bring volume to 10.0mL. Mix thoroughly. Sparge with 100% N_2. Autoclave for 15 min at 15 psi pressure–121°C.

Solution F (Vitamin Solution):

Composition per liter:

Pyridoxine-HCl	10.0mg
Vitamin B_{12}	5.1mg
Thiamine-HCl·$2H_2O$	5.0mg
Riboflavin	5.0mg
Nicotinic acid	5.0mg
D-Ca-pantothenate	5.0mg
p-Aminobenzoic acid	5.0mg
Lipoic acid	5.0mg
Biotin	2.0mg
Folic acid	2.0mg

Preparation of Solution F (Vitamin Solution): Add components to distilled/deionized water and bring volume to 1.0L. Mix thoroughly. Filter sterilize. Sparge with 100% N_2.

Solution G (Artificial Sediment):

Composition per 6.6mL:

$AlCl_3·6H_2O$, 4.9% (w/v)	5.0mL
Na_2CO_3, 10.6% (w/v)	1.6mL

Preparation of Solution G (Artificial Sediment): Combine components. Mix thoroughly. Sparge with 100% N_2. Autoclave for 15 min at 15 psi pressure–121°C.

Solution H:

Composition per 20.0mL:

Rumen fluid, clarified	20.0mL

Preparation of Solution H: Sparge with 100% N_2. Autoclave for 15 min at 15 psi pressure–121°C.

Solution I:

Composition per 10.0mL:

$Na_2S·9H_2O$	0.4g

Preparation of Solution I: Add $Na_2S·9H_2O$ to distilled/deionized water and bring volume to 10.0mL. Mix thoroughly. Sparge with 100% N_2. Autoclave for 15 min at 15 psi pressure–121°C.

Preparation of Medium: Gently heat solution A and bring to boiling. Boil solution A for a few minutes. Cool to room temperature. Gas with 80% N_2 + 20% CO_2 gas mixture to reach a pH below 6. Autoclave for 15 min at 15 psi pressure–121°C. Add 50.0mL hot agar solution. Mix thoroughly. Cool to 50°C. Sequentially add 1.0mL solution B, 100.0mL solution C, 10.0mL solution D, 1.0mL solution E, 10.0mL solution F, 6.6mL solution G, 20.0mL solution H, and 10.0mL solution I. Distribute anaerobically under 80% N_2 + 20% CO_2 into appropriate vessels. The pH should be 6.9. Addition of 10–20mg sodium dithionite per liter from a 5% (w/v) solution, freshly prepared under N_2 and filter sterilized, may stimulate growth.

Use: For the cultivation of *Desulfonema ishimotoi.*

Desulfonema limicola Medium (DSMZ Medium 201)

Composition per 1007.6mL:

Solution A	850.0mL
Solution C	100.0mL
Solution H	20.0mL
Solution D	10.0mL
Solution F (Vitamin solution)	10.0mL
Solution I	10.0mL
Solution G (Artificial sediment)	6.6mL
Solution E	1.0mL
Solution B (Trace elements solution SL-10)	1.0mL

pH 7.6 ± 0.2 at 25°C

Solution A:

NaCl	13.0g
Na_2SO_4	3.0g
$MgCl_2·6H_2O$	2.2g
KCl	0.5g
KH_2PO_4	0.2g
NH_4Cl	0.3g
$CaCl_2·2H_2O$	0.15g
Resazurin	0.5mg

Preparation of Solution A: Add components to distilled/deionized water and bring volume to 870.0mL. Mix thoroughly.

Solution B (Trace Elements Solution SL-10):

Composition per liter:

$FeCl_2·4H_2O$	1.5g
$CoCl_2·6H_2O$	190.0mg
$MnCl_2·4H_2O$	100.0mg
$ZnCl_2$	70.0mg
$Na_2MoO_4·2H_2O$	36.0mg
$NiCl_2·6H_2O$	24.0mg
H_3BO_3	6.0mg
$CuCl_2·2H_2O$	2.0mg
HCl (25% solution)	10.0mL

Preparation of Solution B (Trace Elements Solution SL-10): Add $FeCl_2·4H_2O$ to 10.0mL of HCl solution. Mix thoroughly. Add distilled/deionized water and bring volume to 1.0L. Add remaining components. Mix thoroughly. Sparge with 100% N_2. Autoclave for 15 min at 15 psi pressure–121°C.

Solution C:

Composition per 100.0mL:

$NaHCO_3$	5.0g

Preparation of Solution C: Add $NaHCO_3$ to distilled/deionized water and bring volume to 100.0mL. Mix thoroughly. Filter sterilize. Flush with 80% N_2 + 20% CO_2 to remove dissolved oxygen.

Solution D:

Composition per 10.0mL:

Na-acetate·$3H_2O$	2.5g

Preparation of Solution D: Add Na-acetate·$3H_2O$ to distilled/deionized water and bring volume to 10.0mL. Mix thoroughly. Sparge with 100% N_2. Autoclave for 15 min at 15 psi pressure–121°C.

Solution E:

Composition per 10.0mL:

Na_2-succinate	1.0g

Preparation of Solution E: Add Na_2-succinate to distilled/deionized water and bring volume to 10.0mL. Mix thoroughly. Sparge with 100% N_2. Autoclave for 15 min at 15 psi pressure–121°C.

Solution F (Vitamin Solution):

Composition per liter:

Pyridoxine-HCl	10.0mg
Thiamine-HCl·$2H_2O$	5.0mg
Riboflavin	5.0mg
Nicotinic acid	5.0mg
D-Ca-pantothenate	5.0mg
p-Aminobenzoic acid	5.0mg
Lipoic acid	5.0mg
Biotin	2.0mg
Folic acid	2.0mg
Vitamin B_{12}	0.10mg

Preparation of Solution F (Vitamin Solution): Add components to distilled/deionized water and bring volume to 1.0L. Mix thoroughly. Filter sterilize. Sparge with 100% N_2.

Solution G (Artificial Sediment):

Composition per 6.6mL:

$AlCl_3·6H_2O$, 4.9% (w/v)	5.0mL
Na_2CO_3, 10.6% (w/v)	1.6mL

Preparation of Solution G (Artificial Sediment): Combine components. Mix thoroughly. Sparge with 100% N_2. Autoclave for 15 min at 15 psi pressure–121°C.

Solution H:

Composition per 20.0mL:

Rumen fluid, clarified 20.0mL

Preparation of Solution H: Sparge with 100% N_2. Autoclave for 15 min at 15 psi pressure–121°C.

Solution I:

Composition per 10.0mL:

$Na_2S{\cdot}9H_2O$ 0.4g

Preparation of Solution I: Add $Na_2S{\cdot}9H_2O$ to distilled/deionized water and bring volume to 10.0mL. Mix thoroughly. Sparge with 100% N_2. Autoclave for 15 min at 15 psi pressure–121°C.

Preparation of Medium: Gently heat solution A and bring to boiling. Boil solution A for a few minutes. Cool to room temperature. Gas with 80% N_2 + 20% CO_2 gas mixture to reach a pH below 6. Autoclave for 15 min at 15 psi pressure–121°C. Cool to room temperature. Sequentially add 1.0mL solution B, 100.0mL solution C, 10.0mL solution D, 1.0mL solution E, 10.0mL solution F, 6.6mL solution G, 20.0mL solution H, and 10.0mL solution I. Distribute anaerobically under 80% N_2 + 20% CO_2 into appropriate vessels. The pH should be 7.6. Addition of 10–20mg sodium dithionite per liter from a 5% (w/v) solution, freshly prepared under N_2 and filter-sterilized, may stimulate growth.

Use: For the cultivation of *Desulfonema limicola.*

Desulfonema limicola Medium (DSMZ Medium 201)

Composition per 1007.6mL:

Solution A 800.0mL
Solution C 100.0mL
Agar solution 50.0mL
Solution H 20.0mL
Solution D 10.0mL
Solution F (Vitamin solution) 10.0mL
Solution I 10.0mL
Solution G (Artificial sediment) 6.6mL
Solution E 1.0mL
Solution B (Trace elements solution SL-10) 1.0mL

pH 7.6 ± 0.2 at 25°C

Solution A:

NaCl 13.0g
Na_2SO_4 .... 3.0g
$MgCl_2{\cdot}6H_2O$ 2.2g
KCl 0.5g
KH_2PO_4 .... 0.2g
NH_4Cl 0.3g
$CaCl_2{\cdot}2H_2O$ 0.15g
Resazurin 0.5mg

Preparation of Solution A: Add components to distilled/deionized water and bring volume to 870.0mL. Mix thoroughly.

Solution B (Trace Elements Solution SL-10):

Composition per liter:

$FeCl_2{\cdot}4H_2O$ 1.5g
$CoCl_2{\cdot}6H_2O$ 190.0mg
$MnCl_2{\cdot}4H_2O$ 100.0mg
$ZnCl_2$ 70.0mg
$Na_2MoO_4{\cdot}2H_2O$ 36.0mg
$NiCl_2{\cdot}6H_2O$ 24.0mg
H_3BO_3 .... 6.0mg
$CuCl_2{\cdot}2H_2O$ 2.0mg
HCl (25% solution) 10.0mL

Preparation of Solution B (Trace Elements Solution SL-10): Add $FeCl_2{\cdot}4H_2O$ to 10.0mL of HCl solution. Mix thoroughly. Add distilled/deionized water and bring volume to 1.0L. Add remaining components. Mix thoroughly. Sparge with 100% N_2. Autoclave for 15 min at 15 psi pressure–121°C.

Solution C:

Composition per 100.0mL:

$NaHCO_3$ 5.0g

Preparation of Solution C: Add $NaHCO_3$ to distilled/deionized water and bring volume to 100.0mL. Mix thoroughly. Filter sterilize. Flush with 80% N_2 + 20% CO_2 to remove dissolved oxygen.

Solution D:

Composition per 10.0mL:

Na-acetate·$3H_2O$ 2.5g

Preparation of Solution D: Add Na-acetate·$3H_2O$ to distilled/deionized water and bring volume to 10.0mL. Mix thoroughly. Sparge with 100% N_2. Autoclave for 15 min at 15 psi pressure–121°C.

Solution E:

Composition per 10.0mL:

Na_2-succinate 1.0g

Preparation of Solution D: Add Na_2-succinate to distilled/deionized water and bring volume to 10.0mL. Mix thoroughly. Sparge with 100% N_2. Autoclave for 15 min at 15 psi pressure–121°C.

Solution F (Vitamin Solution):

Composition per liter:

Pyridoxine-HCl 10.0mg
Thiamine-HCl·$2H_2O$ 5.0mg
Riboflavin 5.0mg
Nicotinic acid 5.0mg
D-Ca-pantothenate 5.0mg
p-Aminobenzoic acid 5.0mg
Lipoic acid 5.0mg
Biotin 2.0mg
Folic acid 2.0mg
Vitamin B_{12} 0.10mg

Preparation of Solution F (Vitamin Solution): Add components to distilled/deionized water and bring volume to 1.0L. Mix thoroughly. Filter sterilize. Sparge with 100% N_2.

Solution G (Artificial Sediment):

Composition per 6.6mL:

$AlCl_3{\cdot}6H_2O$, 4.9% (w/v) 5.0mL
Na_2CO_3, 10.6% (w/v) 1.6mL

Preparation of Solution G (Artificial Sediment): Combine components. Mix thoroughly. Sparge with 100% N_2. Autoclave for 15 min at 15 psi pressure–121°C.

Solution H:

Composition per 20.0mL:

Rumen fluid, clarified 20.0mL

Preparation of Solution H: Sparge with 100% N_2. Autoclave for 15 min at 15 psi pressure–121°C.

Solution I:

Composition per 10.0mL:

$Na_2S{\cdot}9H_2O$ 0.4g

Preparation of Solution I: Add $Na_2S{\cdot}9H_2O$ to distilled/deionized water and bring volume to 10.0mL. Mix thoroughly. Sparge with 100% N_2. Autoclave for 15 min at 15 psi pressure–121°C.

Agar Solution:

Composition per liter:

Agar 20.0g

Preparation of Agar Solution: Wash agar three times for 1 hr with distilled water at room temperature. Add 20.0g washed agar to distilled/deionized water and bring volume to 1.0L. Mix thoroughly. Autoclave for 15 min at 15 psi pressure–121°C. Cool to 50°C.

Preparation of Medium: Gently heat solution A and bring to boiling. Boil solution A for a few minutes. Add 50.0mL agar solution. Cool to room temperature. Gas with 80% N_2 + 20% CO_2 gas mixture to reach a pH below 6. Autoclave for 15 min at 15 psi pressure–121°C. Cool to room temperature. Sequentially add 1.0mL solution B, 100.0mL solution C, 10.0mL solution D, 1.0mL solution E, 10.0mL solution F, 6.6mL solution G, 20.0mL solution H, and 10.0mL solution I. Distribute anaerobically under 80% N_2 + 20% CO_2 into appropriate vessels. The pH should be 7.6. Addition of 10–20mg sodium dithionite per liter from a 5% (w/v) solution, freshly prepared under N_2 and filter-sterilized, may stimulate growth.

Use: For the cultivation of *Desulfonema limicola.*

Desulfonema limicola Medium

Composition per 1009.0mL:

Solution A 850.0mL
Solution C 100.0mL
Solution H 20.0mL
Solution D 10.0mL
Solution F (Wolfe's vitamin solution) 10.0mL
Solution I 10.0mL
Solution G 6.6mL
Solution B (Trace elements solution SL-10) 1.0mL
Solution E 1.0mL
Solution J 0.4mL

pH 7.6 ± 0.2 at 25°C

Solution A:

Composition per 920.0mL:

NaCl 13.0g
$MgCl_2{\cdot}6H_2O$ 2.2g
Na_2SO_4 .... 3.0g
KCl 0.5g
NH_4Cl 0.3g
KH_2PO_4 .... 0.2g
$CaCl_2{\cdot}2H_2O$ 0.15g
Resazurin 0.5mg

Preparation of Solution A: Prepare and dispense solution anaerobically under 80% N_2 + 20% CO_2. Add components to distilled/deionized water and bring volume to 920.0mL. Mix thoroughly. Gently heat and bring to boiling. Continue boiling until resazurin turns colorless, indicating reduction, and a pH of 6.0 is reached. Cap with rubber stoppers. Autoclave for 15 min at 15 psi pressure–121°C. Cool to 25°C.

Solution B (Trace Elements Solution SL-10):

Composition per liter:

$FeCl_2{\cdot}4H_2O$ 1.5g
$CoCl_2{\cdot}6H_2O$ 0.19g
$MnCl_2{\cdot}4H_2O$ 0.1g
$ZnCl_2$ 0.07g
$Na_2MoO_4{\cdot}2H_2O$ 0.036g
$NiCl_2{\cdot}6H_2O$ 0.024g
H_3BO_3 .... 6.0mg
$CuCl_2{\cdot}2H_2O$ 2.0mg
HCl (25% solution) 10.0mL

Preparation of Solution B (Trace Elements Solution SL-10): Add the $FeCl_2{\cdot}4H_2O$ to 10.0mL of HCl solution. Mix thoroughly. Bring volume to approximately 900.0mL with distilled/deionized water. Mix thoroughly. Adjust pH to 6.0 with NaOH. Bring volume to 1.0L with distilled/deionized water. Filter sterilize. Aseptically gas under 100% N_2 for 20 min.

Solution C:

Composition per 100.0mL:

$NaHCO_3$ 5.0g

Preparation of Solution C: Add $NaHCO_3$ to distilled/deionized water and bring volume to 100.0mL. Mix thoroughly. Filter sterilize. Aseptically gas under 80% N_2 + 20% CO_2 for 20 min.

Solution D:

Composition per 10.0mL:

Sodium acetate·$3H_2O$ 2.5g

Preparation of Solution D: Prepare and dispense solution anaerobically under 80% N_2 + 20% CO_2. Add sodium acetate to distilled/deionized water and bring volume to 10.0mL. Mix thoroughly. Cap with a rubber stopper. Autoclave for 15 min at 15 psi pressure–121°C. Cool to 25°C.

Solution E:

Composition per 1.0mL:

Disodium succinate 0.1g

Preparation of Solution E: Add disodium succinate to distilled/deionized water and bring volume to 1.0mL. Mix thoroughly. Gas with 80% N_2 + 20% CO_2. Cap with rubber stopper. Autoclave for 15 min at 15 psi pressure–121°C. Cool to 25°C.

Solution F (Wolfe's Vitamin Solution):

Composition per liter:

Pyridoxine·HCl 0.01g
Thiamine·HCl 5.0mg
Riboflavin 5.0mg
Nicotinic acid 5.0mg
Calcium pantothenate 5.0mg
p-Aminobenzoic acid 5.0mg
Thioctic acid 5.0mg
Biotin 2.0mg
Folic acid 2.0mg
Cyanocobalamin 0.1mg

Preparation of Solution F (Wolfe's Vitamin Solution): Add components to distilled/deionized water and bring volume to 1.0L. Mix thoroughly. Filter sterilize. Aseptically gas under 100% N_2 for 20 min.

Solution G:

Composition per 6.6mL:

$AlCl_3{\cdot}6H_2O$ (4.9% solution) 5.0mL
Na_2CO_3 (10.6% solution) 1.6mL

Preparation of Solution G: Combine both solutions. Mix thoroughly. Gas with 100% N_2. Cap with a rubber stopper. Autoclave for 15 min at 15 psi pressure–121°C. Cool to 25°C.

Solution H:

Composition per 10.0mL:

Rumen fluid, clarified 20.0mL

Preparation of Solution H: Gas rumen fluid under 100% N_2 for 20 min. Cap with a rubber stopper. Autoclave for 15 min at 15 psi pressure–121°C. Cool to 25°C.

Solution I:

Composition per 10.0mL:

$Na_2S\cdot9H_2O$ 0.4g

Preparation of Solution I: Add $Na_2S\cdot9H_2O$ to distilled/deionized water and bring volume to 10.0mL. Gas under 100% N_2 for 20 min. Cap with a rubber stopper. Autoclave for 15 min at 15 psi pressure–121°C. Cool to 25°C.

Solution J:

Composition per 10.0mL:

$Na_2S_2O_4$ 0.5g

Preparation of Solution J: Add $Na_2S_2O_4$ to distilled/deionized water and bring volume to 10.0mL. Mix thoroughly. Filter sterilize. Aseptically gas under 100% N_2 for 20 min. Prepare solution freshly.

Preparation of Medium: To 850.0mL of cooled, sterile solution A, aseptically and anaerobically add in the following order: 1.0mL of sterile solution B, 100.0mL of sterile solution C, 10.0mL of sterile solution D, 1.0mL of sterile solution E, 10.0mL of sterile solution F, 6.6mL of sterile solution G, 20.0mL of sterile solution H, and 10.0mL of sterile solution I. Mix thoroughly. Immediately prior to inoculation, aseptically and anaerobically add 0.4mL of sterile solution J. Mix thoroughly. Aseptically and anaerobically distribute into sterile tubes or flasks.

Use: For the cultivation and maintenance of *Desulfonema limicola.*

Desulfonema magnum Medium (DSMZ Medium 202)

Composition per 957.6mL:

Solution A 850.0mL
Solution C 50.0mL
Solution H 20.0mL
Solution D 10.0mL
Solution F (Vitamin solution) 10.0mL
Solution I 10.0mL
Solution G (Artificial sediment) 6.6mL
Solution E 1.0mL
Solution B (Trace elements solution SL-10) 1.0mL
pH 6.9 ± 0.2 at 25°C

Solution A:

NaCl 21.0g
$MgCl_2\cdot6H_2O$ 5.5g
Na_2SO_4 .... 3.0g
$CaCl_2\cdot2H_2O$ 1.35g
KCl 0.5g
KH_2PO_4 .... 0.2g
NH_4Cl 0.3g
Resazurin 0.5mg
$Na_2SeO_3\cdot5H_2O$ 3μg

Preparation of Solution A: Add components to distilled/deionized water and bring volume to 870.0mL Mix thoroughly.

Solution B (Trace Elements Solution SL-10):

Composition per liter:

$FeCl_2\cdot4H_2O$ 1.5g
$CoCl_2\cdot6H_2O$ 190.0mg
$MnCl_2\cdot4H_2O$ 100.0mg
$ZnCl_2$ 70.0mg
$Na_2MoO_4\cdot2H_2O$ 36.0mg
$NiCl_2\cdot6H_2O$ 24.0mg
H_3BO_3 .... 6.0mg
$CuCl_2\cdot2H_2O$ 2.0mg
HCl (25% solution) 10.0mL

Preparation of Solution B (Trace Elements Solution SL-10): Add $FeCl_2\cdot4H_2O$ to 10.0mL of HCl solution. Mix thoroughly. Add distilled/deionized water and bring volume to 1.0L. Add remaining components. Mix thoroughly. Sparge with 100% N_2. Autoclave for 15 min at 15 psi pressure–121°C.

Solution C:

Composition per 100.0mL:

$NaHCO_3$ 5.0g

Preparation of Solution C: Add $NaHCO_3$ to distilled/deionized water and bring volume to 100.0mL. Mix thoroughly. Filter sterilize. Flush with 80% N_2 + 20% CO_2 to remove dissolved oxygen.

Solution D:

Composition per 10.0mL:

Na-acetate 0.6g

Preparation of Solution D: Add Na-acetate·$3H_2O$ to distilled/deionized water and bring volume to 10.0mL. Mix thoroughly. Sparge with 100% N_2. Autoclave for 15 min at 15 psi pressure–121°C.

Solution E:

Composition per 10.0mL:

Na_2-succinate 1.0g

Preparation of Solution D: Add Na_2-succinate to distilled/deionized water and bring volume to 10.0mL. Mix thoroughly. Sparge with 100% N_2. Autoclave for 15 min at 15 psi pressure–121°C.

Solution F (Vitamin Solution):

Composition per liter:

Pyridoxine-HCl 10.0mg
Vitamin B_{12} 5.1mg
Thiamine-HCl·$2H_2O$ 5.0mg
Riboflavin 5.0mg
Nicotinic acid 5.0mg
D-Ca-pantothenate 5.0mg
p-Aminobenzoic acid 5.0mg
Lipoic acid 5.0mg
Biotin 2.0mg
Folic acid 2.0mg

Preparation of Solution F (Vitamin Solution): Add components to distilled/deionized water and bring volume to 1.0L. Mix thoroughly. Filter sterilize. Sparge with 100% N_2.

Solution G (Artificial Sediment):

Composition per 6.6mL:

$AlCl_3\cdot6H_2O$, 4.9% (w/v) 5.0mL
Na_2CO_3, 10.6% (w/v) 1.6mL

Preparation of Solution G (Artificial Sediment): Combine components. Mix thoroughly. Sparge with 100% N_2. Autoclave for 15 min at 15 psi pressure–121°C.

Solution H:

Composition per 20.0mL:

Rumen fluid, clarified 20.0mL

Preparation of Solution H: Sparge with 100% N_2. Autoclave for 15 min at 15 psi pressure–121°C.

Solution I:

Composition per 10.0mL:

$Na_2S \cdot 9H_2O$ 0.4g

Preparation of Solution I: Add $Na_2S \cdot 9H_2O$ to distilled/deionized water and bring volume to 10.0mL. Mix thoroughly. Sparge with 100% N_2. Autoclave for 15 min at 15 psi pressure–121°C.

Preparation of Medium: Gently heat solution A and bring to boiling. Boil solution A for a few minutes. Cool to room temperature. Gas with 80% N_2 + 20% CO_2 gas mixture to reach a pH below 6. Autoclave for 15 min at 15 psi pressure–121°C. Cool to room temperature. Sequentially add 1.0mL solution B, 100.0mL solution C, 10.0mL solution D, 1.0mL solution E, 10.0mL solution F, 6.6mL solution G, 20.0mL solution H, and 10.0mL solution I. Distribute anaerobically under 80% N_2 + 20% CO_2 into appropriate vessels. The pH should be 6.9. Addition of 10–20mg sodium dithionite per liter from a 5% (w/v) solution, freshly prepared under N_2 and filter sterilized, may stimulate growth.

Use: For the cultivation of *Desulfonema magnum.*

Desulfonema magnum Medium

Composition per 1001.0mL:

Solution A 890.0mL
Solution C 50.0mL
Solution J 20.0mL
Solution D 10.0mL
Solution G (Wolfe's vitamin solution) 10.0mL
Solution K 10.0mL
Solution I 6.6mL
Solution B (Trace elements solution SL-10) 1.0mL
Solution E 1.0mL
Solution F 1.0mL
Solution H 1.0mL
Solution L 0.4mL

pH 6.9 ± 0.2 at 25°C

Solution A:

Composition per 890.0mL:

NaCl 21.0g
$MgCl_2 \cdot 6H_2O$ 5.5g
Na_2SO_4 .... 3.0g
$CaCl_2 \cdot 2H_2O$ 1.35g
KCl 0.5g
NH_4Cl 0.3g
KH_2PO_4 .... 0.2g
Resazurin 0.5mg

Preparation of Solution A: Prepare and dispense solution anaerobically under 80% N_2 + 20% CO_2. Add components to distilled/deionized water and bring volume to 890.0mL. Mix thoroughly. Gently heat and bring to boiling. Continue boiling until resazurin turns colorless, indicating a reduction, and a pH of 6.0 is reached. Cap with a rubber stopper. Autoclave for 15 min at 15 psi pressure–121°C. Cool to 25°C.

Solution B (Trace Elements Solution SL-10):

Composition per liter:

$FeCl_2 \cdot 4H_2O$ 1.5g
$CoCl_2 \cdot 6H_2O$ 0.19g
$MnCl_2 \cdot 4H_2O$ 0.1g
$ZnCl_2$ 0.07g
$Na_2MoO_4 \cdot 2H_2O$ 0.036g
$NiCl_2 \cdot 6H_2O$ 0.024g
H_3BO_3 .... 6.0mg
$CuCl_2 \cdot 2H_2O$ 2.0mg
HCl (25% solution) 10.0mL

Preparation of Solution B (Trace Elements Solution SL-10): Add the $FeCl_2 \cdot 4H_2O$ to 10.0mL of HCl solution. Mix thoroughly. Bring volume to approximately 900.0mL with distilled/deionized water. Mix thoroughly. Adjust pH to 6.0 with NaOH. Bring volume to 1.0L with distilled/deionized water. Filter sterilize. Aseptically gas under 100% N_2 for 20 min.

Solution C:

Composition per 50.0mL:

$NaHCO_3$ 2.5g

Preparation of Solution C: Add $NaHCO_3$ to distilled/deionized water and bring volume to 50.0mL. Mix thoroughly. Filter sterilize. Aseptically gas under 80% N_2 + 20% CO_2 for 20 min.

Solution D:

Composition per 10.0mL:

Sodium benzoate 0.6g

Preparation of Solution D: Add sodium benzoate to distilled/deionized water and bring volume to 10.0mL. Mix thoroughly. Gas with 100% N_2 for 10 min. Cap with a rubber stopper. Autoclave for 15 min at 15 psi pressure–121°C. Cool to 25°C.

Solution E:

Composition per 1.0mL:

$Na_2SeO_3 \cdot 5H_2O$ 3.5μg

Preparation of Solution E: Add $Na_2SeO_3 \cdot 5H_2O$ to distilled/deionized water and bring volume to 1.0mL. Mix thoroughly. Gas with 100% N_2 for 10 min. Cap with a rubber stopper. Autoclave for 15 min at 15 psi pressure–121°C. Cool to 25°C.

Solution F:

Composition per 1.0mL:

Disodium succinate 0.1g

Preparation of Solution F: Add disodium succinate to distilled/deionized water and bring volume to 1.0mL. Mix thoroughly. Gas with 100% N_2. Cap with a rubber stopper. Autoclave for 15 min at 15 psi pressure–121°C. Cool to 25°C.

Solution G (Wolfe's Vitamin Solution):

Composition per liter:

Pyridoxine·HCl 0.01g
Thiamine·HCl 5.0mg
Riboflavin 5.0mg
Nicotinic acid 5.0mg
Calcium pantothenate 5.0mg
p-Aminobenzoic acid 5.0mg
Thioctic acid 5.0mg
Biotin 2.0mg
Folic acid 2.0mg
Cyanocobalamin 0.1mg

Preparation of Solution G (Wolfe's Vitamin Solution): Add components to distilled/deionized water and bring volume to 1.0L. Mix thoroughly. Filter sterilize. Aseptically gas under 100% N_2 for 20 min.

Solution H:

Composition per 1.0mL:

Vitamin B_{12} 0.05mg

Preparation of Solution H: Add Vitamin B_{12} to distilled/deionized water and bring volume to 1.0mL. Mix thoroughly. Gas with 100% N_2. Cap with a rubber stopper. Autoclave for 15 min at 15 psi pressure–121°C. Cool to 25°C.

Solution I:

Composition per 6.6mL:

$AlCl_3 \cdot 6H_2O$ (4.9% solution)	5.0mL
Na_2CO_3 (10.6% solution)	1.6mL

Preparation of Solution I: Combine both solutions. Mix thoroughly. Gas with 100% N_2. Cap with a rubber stopper. Autoclave for 15 min at 15 psi pressure–121°C. Cool to 25°C.

Solution J:

Composition per 10.0mL:

Rumen fluid, clarified	20.0mL

Preparation of Solution J: Gas rumen fluid under 100% N_2 for 20 min. Cap with a rubber stopper. Autoclave for 15 min at 15 psi pressure–121°C. Cool to 25°C.

Solution K:

Composition per 10.0mL:

$Na_2S \cdot 9H_2O$	0.4g

Preparation of Solution K: Add $Na_2S \cdot 9H_2O$ to distilled/deionized water and bring volume to 10.0mL. Gas under 100% N_2 for 20 min. Cap with a rubber stopper. Autoclave for 15 min at 15 psi pressure–121°C. Cool to 25°C.

Solution L:

Composition per 10.0mL:

$Na_2S_2O_4$	0.5g

Preparation of Solution L: Add $Na_2S_2O_4$ to distilled/deionized water and bring volume to 10.0mL. Mix thoroughly. Filter sterilize. Aseptically gas under 100% N_2 for 20 min. Use freshly prepared solution.

Preparation of Medium: To 890.0mL of cooled, sterile solution A, aseptically and anaerobically add in the following order: 1.0mL of sterile solution B, 50.0mL of sterile solution C, 10.0mL of sterile solution D, 1.0mL of sterile solution E, 1.0mL of sterile solution F, 10.0mL of sterile solution G, 1.0mL of sterile solution H, 6.6mL of sterile solution I, 20.0mL of sterile solution J, and 10.0mL of sterile solution K. Mix thoroughly. Immediately prior to inoculation, aseptically and anaerobically add 0.4mL of sterile solution L. Mix thoroughly. Aseptically and anaerobically distribute into sterile tubes or flasks.

Use: For the cultivation and maintenance of *Desulfonema magnum*.

Desulforhabdus amnigenus Medium (DSMZ Medium 708)

Composition per liter:

Na_2SO_4	2.8g
$Na_2HPO_4 \cdot 2H_2O$	0.53g
KH_2PO_4	0.41g
NH_4Cl	0.3g
NaCl	0.3g
$CaCl_2 \cdot 2H_2O$	0.11g
$MgCl_2 \cdot 6H_2O$	0.1g
Resazurin	0.5mg
$Na_2S \cdot 9H_2O$ solution	10.0mL
Vitamin solution	10.0mL
$NaHCO_3$ solution	10.0mL
Na-propionate solution	10.0mL
Selenite-tungstate solution	1.0mL
Seven vitamin solution	1.0mL
Trace elements solution SL-10	1.0mL

pH 7.2–7.6 at 25°C

Na-propionate Solution:

Composition per 10.0mL:

Na-propionate	2.0g

Preparation of Na-propionate Solution: Add Na-propionate to distilled/deionized water and bring volume to 10.0mL. Mix thoroughly. Sparge with 100% N_2. Autoclave for 15 min at 15 psi pressure–121°C.

Selenite-Tungstate Solution:

Composition per liter:

NaOH	0.5g
$Na_2WO_4 \cdot 2H_2O$	4.0mg
$Na_2SeO_3 \cdot 5H_2O$	3.0mg

Preparation of Selenite-Tungstate Solution: Add components to distilled/deionized water and bring volume to 1.0L. Mix thoroughly. Sparge with 100% N_2. Filter sterilize.

$Na_2S \cdot 9H_2O$ Solution:

Composition per 10.0mL:

$Na_2S \cdot 9H_2O$	0.5g

Preparation of $Na_2S \cdot 9H_2O$ Solution: Add $Na_2S \cdot 9H_2O$ to distilled/deionized water and bring volume to 10.0mL. Mix thoroughly. Autoclave under 100% N_2 for 15 min at 15 psi pressure–121°C. Cool to room temperature.

$NaHCO_3$ Solution:

Composition per 10.0mL:

$NaHCO_3$	4.0g

Preparation of $NaHCO_3$ Solution: Add $NaHCO_3$ to distilled/deionized water and bring volume to 10.0mL. Mix thoroughly. Sparge with 80% N_2 + 20% CO_2. Filter sterilize.

Trace Elements Solution SL-10:

Composition per liter:

$FeCl_2 \cdot 4H_2O$	1.5g
$CoCl_2 \cdot 6H_2O$	190.0mg
$MnCl_2 \cdot 4H_2O$	100.0mg
$ZnCl_2$	70.0mg
$Na_2MoO_4 \cdot 2H_2O$	36.0mg
$NiCl_2 \cdot 6H_2O$	24.0mg
H_3BO_3	6.0mg
$CuCl_2 \cdot 2H_2O$	2.0mg
HCl (25% solution)	10.0mL

Preparation of Trace Elements Solution SL-10: Add $FeCl_2 \cdot 4H_2O$ to 10.0mL of HCl solution. Mix thoroughly. Add distilled/deionized water and bring volume to 1.0L. Add remaining components. Mix thoroughly. Sparge with 80% N_2 + 20% CO_2. Autoclave for 15 min at 15 psi pressure–121°C.

Vitamin Solution:

Composition per liter:

Pyridoxine-HCl	10.0mg
Thiamine-HCl·$2H_2O$	5.0mg
Riboflavin	5.0mg
Nicotinic acid	5.0mg
D-Ca-pantothenate	5.0mg
p-Aminobenzoic acid	5.0mg
Lipoic acid	5.0mg

Biotin	2.0mg
Folic acid	2.0mg
Vitamin B_{12}	0.1mg

Preparation of Vitamin Solution: Add components to distilled/deionized water and bring volume to 1.0L. Mix thoroughly. Sparge with 80% H_2 + 20% CO_2. Filter sterilize.

Seven Vitamin Solution:

Composition per liter:

Pyridoxine hydrochloride	300.0mg
Thiamine-HCl·$2H_2O$	200.0mg
Nicotinic acid	200.0mg
Vitamin B_{12}	100.0mg
Calcium pantothenate	100.0mg
p-Aminobenzoic acid	80.0mg
D(+)-Biotin	20.0mg

Preparation of Seven Vitamin Solution: Add components to distilled/deionized water and bring volume to 1.0L. Sparge with 100% N_2. Mix thoroughly. Filter sterilize.

Preparation of Medium: Prepare and dispense medium under 80% N_2 + 20% CO_2 gas atmosphere. Add components, except $NaHCO_3$ solution, $Na_2S·9H_2O$ solution, Na-propionate solution, vitamin solution, seven vitamin solution, selenite-tungstate solution, and trace elements solution SL-10, to distilled/deionized water and bring volume to 947.0mL. Mix thoroughly. Adjust pH to 7.2–7.6. Sparge with 80% N_2 + 20% CO_2. Autoclave for 15 min at 15 psi pressure–121°C. Aseptically and anaerobically add 10.0mL $NaHCO_3$ solution, 10.0mL $Na_2S·9H_2O$ solution, 10.0mL Na-propionate solution, 10.0mL vitamin solution, 1.0mL seven vitamin solution, 1.0mL selenite-tungstate solution, and 1.0mL trace elements solution SL-10. Mix thoroughly. Aseptically and anaerobically distribute into sterile tubes or bottles.

Use: For the cultivation of *Desulforhabdus amnigena (Desulforhabdus amnigenus).*

Desulforhabdus amnigenus Medium

Composition per 1001.0mL:

Na_2SO_4	2.8g
$Na_2HPO_4·2H_2O$	0.53g
KH_2PO_4	0.41g
NaCl	0.3g
NH_4Cl	0.3g
$CaCl_2·2H_2O$	0.11g
$MgCl_2·6H_2O$	0.1g
Resazurin	0.5mg
Sodium propionate solution	20.0mL
$NaHCO_3$ solution	10.0mL
Wolfe's vitamin solution	10.0mL
$Na_2S·9H_2O$ solution	10.0mL
Selenite-tungstate solution	1.0mL
Trace elements solution SL-10	1.0mL
Seven vitamin solution	1.0mL

pH 7.2–7.6 at 25°C

Sodium Propionate Solution:

Composition per 20.0mL:

Sodium propionate	2.0g

Preparation of Sodium Propionate Solution: Add sodium propionate to distilled/deionized water and bring volume to 20.0mL. Mix thoroughly. Sparge with 100% N_2. Autoclave for 15 min at 15 psi pressure–121°C.

$NaHCO_3$ Solution:

Composition per 10.0mL:

$NaHCO_3$	4.0g

Preparation of $NaHCO_3$ Solution: Add $NaHCO_3$ to distilled/deionized water and bring volume to 10.0mL. Mix thoroughly. Sparge with 80% N_2 + 20% CO_2. Autoclave for 15 min at 15 psi pressure–121°C.

$Na_2S·9H_2O$ Solution:

Composition per 10.0mL:

$Na_2S·9H_2O$	0.5g

Preparation of $Na_2S·9H_2O$ Solution: Add $Na_2S·9H_2O$ to distilled/deionized water and bring volume to 10.0mL. Mix thoroughly. Sparge with 100% N_2. Autoclave for 15 min at 15 psi pressure–121°C. Before use, neutralize to pH 7.0 with sterile HCl.

Wolfe's Vitamin Solution:

Composition per liter:

Pyridoxine·HCl	10.0mg
p-Aminobenzoic acid	5.0mg
Lipoic acid	5.0mg
Nicotinic acid	5.0mg
Riboflavin	5.0mg
Thiamine·HCl	5.0mg
Calcium DL-pantothenate	5.0mg
Biotin	2.0mg
Folic acid	2.0mg
Vitamin B_{12}	0.1mg

Preparation of Wolfe's Vitamin Solution: Add components to distilled/deionized water and bring volume to 1.0L. Mix thoroughly.

Selenite-Tungstate Solution:

Composition per liter:

NaOH	0.5g
$Na_2WO_4·2H_2O$	4.0mg
$Na_2SeO_3·5H_2O$	3.0mg

Preparation of Selenite-Tungstate Solution: Add components to distilled/deionized water and bring volume to 1.0L. Mix thoroughly. Sparge with 100% N_2. Autoclave for 15 min at 15 psi pressure–121°C.

Trace Elements Solution SL-10:

Composition per liter:

$FeCl_2·4H_2O$	1.5g
$CoCl_2·6H_2O$	190.0mg
$MnCl_2·4H_2O$	100.0mg
$ZnCl_2$	70.0mg
$Na_2MoO_4·2H_2O$	36.0mg
$NiCl_2·6H_2O$	24.0mg
H_3BO_3	6.0mg
$CuCl_2·2H_2O$	2.0mg
HCl (25% solution)	10.0mL

Preparation of Trace Elements Solution SL-10: Add $FeCl_2·4H_2O$ to 10.0mL of HCl solution. Mix thoroughly. Add distilled/deionized water and bring volume to 1.0L. Add remaining components. Mix thoroughly.

Seven Vitamin Solution:

Composition per liter:

Pyridoxine·HCl	0.3g
Thiamine·HCl	0.2g
Nicotinic acid	0.2g
Calcium DL-pantothenate	0.1g
Vitamin B_{12}	0.1g

p-Aminobenzoic acid	80.0mg
Biotin	20.0mg

Preparation of Seven Vitamin Solution: Add components to distilled/deionized water and bring volume to 1.0L. Mix thoroughly.

Preparation of Medium: Prepare and dispense medium under 80% N_2 + 20% CO_2 gas mixture. Add components, except sodium propionate solution, $NaHCO_3$ solution, selenite-tungstate solution, and $Na_2S{\cdot}9H_2O$ solution, to distilled/deionized water and bring volume to 960.0mL. Mix thoroughly. Adjust pH to 7.2–7.6. Sparge with 80% N_2 + 20% CO_2 gas mixture. Autoclave for 15 min at 15 psi pressure–121°C. Aseptically and anaerobically add 20.0mL of sterile sodium propionate solution, 10.0mL of sterile $NaHCO_3$ solution, 1.0mL of sterile selenite-tungstate solution, and 10.0mL of sterile $Na_2S{\cdot}9H_2O$ solution. Mix thoroughly. Aseptically and anaerobically distribute into sterile tubes or bottles.

Use: For the cultivation of *Desulforhabdus amnigenus.*

Desulfosarcina Medium

Composition per 1001.0mL:

Solution A	870.0mL
Solution C	100.0mL
Solution D	10.0mL
Solution E (Vitamin solution)	10.0mL
Solution F	10.0mL
Solution B (Trace elements solution SL-10)	1.0mL

pH 7.1–7.4 at 25°C

Solution A:
Composition per 870.0mL:

NaCl	7.0g
Na_2SO_4	3.0g
$MgCl_2{\cdot}6H_2O$	1.3g
KCl	0.5g
NH_4Cl	0.3g
KH_2PO_4	0.2g
$CaCl_2{\cdot}2H_2O$	0.15g
Resazurin	1.0mg
$Na_2SeO_3{\cdot}5H_2O$	3.0μg

Preparation of Solution A: Add components to distilled/deionized water and bring volume to 870.0mL. Mix thoroughly. Gently heat and bring to boiling. Continue boiling for 3–4 min. Allow to cool to room temperature while gassing under 80% N_2 + 20% CO_2. Continue gassing until pH reaches below 6.0. Seal the flask under 80% N_2 + 20% CO_2. Autoclave for 15 min at 15 psi pressure–121°C.

Solution B (Trace Elements Solution SL-10):
Composition per liter:

$FeCl_2{\cdot}4H_2O$	1.5g
$CoCl_2{\cdot}6H_2O$	190.0mg
$MnCl_2{\cdot}4H_2O$	100.0mg
$ZnCl_2$	70.0mg
$Na_2MoO_4{\cdot}2H_2O$	36.0mg
$NiCl_2{\cdot}6H_2O$	24.0mg
H_3BO_3	6.0mg
$CuCl_2{\cdot}2H_2O$	2.0mg
HCl (25% solution)	10.0mL

Preparation of Solution B (Trace Elements Solution SL-10): Add $FeCl_2{\cdot}4H_2O$ to 10.0mL of HCl solution. Mix thoroughly. Add distilled/deionized water and bring volume to 1.0L. Add remaining components. Mix thoroughly. Gas under 100% N_2. Autoclave for 15 min at 15 psi pressure–121°C.

Solution C:
Composition per 100.0mL:

$NaHCO_3$	5.0g

Preparation of Solution C: Add $NaHCO_3$ to distilled/deionized water and bring volume to 100.0mL. Mix thoroughly. Filter sterilize. Gas under 80% N_2 + 20% CO_2.

Solution D:
Composition per 10.0mL:

Sodium benzoate	0.6g

Preparation of Solution D: Add sodium benzoate to distilled/deionized water and bring volume to 10.0mL. Mix thoroughly. Gas under 100% N_2. Autoclave for 15 min at 15 psi pressure–121°C.

Solution E (Vitamin Solution):
Composition per liter:

Pyridoxine·HCl	10.0mg
Calcium DL-pantothenate	5.0mg
Lipoic acid	5.0mg
Nicotinic acid	5.0mg
p-Aminobenzoic acid	5.0mg
Riboflavin	5.0mg
Thiamine·HCl	5.0mg
Biotin	2.0mg
Folic acid	2.0mg
Vitamin B_{12}	0.1mg

Preparation of Solution E (Vitamin Solution): Add components to distilled/deionized water and bring volume to 1.0L. Mix thoroughly. Gas under 100% N_2. Autoclave for 15 min at 15 psi pressure–121°C.

Solution F:
Composition per 10.0mL:

$Na_2S{\cdot}9H_2O$	0.4g

Preparation of Solution F: Add $Na_2S{\cdot}9H_2O$ to distilled/deionized water and bring volume to 10.0mL. Mix thoroughly. Gas under 100% N_2. Autoclave for 15 min at 15 psi pressure–121°C.

Preparation of Medium: Aseptically and anaerobically combine solution A with solution B, solution C, solution D, solution E, and solution F, in that order. Mix thoroughly. Anaerobically distribute into sterile tubes or flasks under 80% N_2 + 20% CO_2.

Use: For the cultivation and maintenance of *Desulfosarcina variabilis*.

Desulfosarcina variabilis Medium

Composition per 1009.0mL:

Solution A	850.0mL
Solution C	100.0mL
Solution G	20.0mL
Solution D	10.0mL
Solution E (Wolfe's vitamin solution)	10.0mL
Solution H	10.0mL
Solution F	6.6mL
Solution B (Trace elements solution SL-10)	1.0mL
Solution I	0.4mL

pH 7.6 ± 0.2 at 25°C

Solution A:
Composition per 920.0mL:

NaCl	13.5g
Na_2SO_4	3.0g
$MgCl_2{\cdot}6H_2O$	2.2g
KCl	0.5g

NH_4Cl	0.3g
KH_2PO_4	0.2g
$CaCl_2 \cdot 2H_2O$	0.15g
$Na_2SeO_3 \cdot 5H_2O$	3.0μg
Resazurin	0.5mg

Preparation of Solution A: Prepare and dispense solution anaerobically under 80% N_2 + 20% CO_2. Add components to distilled/deionized water and bring volume to 920.0mL. Mix thoroughly. Gently heat and bring to boiling. Continue boiling until resazurin turns colorless, indicating reduction, and a pH of 6.0 is reached. Cap with rubber stoppers. Autoclave for 15 min at 15 psi pressure–121°C. Cool to 25°C.

Solution B (Trace Elements Solution SL-10):

Composition per liter:

$FeCl_2 \cdot 4H_2O$	1.5g
$CoCl_2 \cdot 6H_2O$	0.19g
$MnCl_2 \cdot 4H_2O$	0.10g
$ZnCl_2$	0.070g
$Na_2MoO_4 \cdot 2H_2O$	0.036g
$NiCl_2 \cdot 6H_2O$	0.024g
H_3BO_3	6.0mg
$CuCl_2 \cdot 2H_2O$	2.0mg
HCl (25% solution)	10.0mL

Preparation of Solution B (Trace Elements Solution SL-10): Add the $FeCl_2 \cdot 4H_2O$ to 10.0mL of HCl solution. Mix thoroughly. Bring volume to approximately 900.0mL with distilled/deionized water. Mix thoroughly. Adjust pH to 6.0 with NaOH. Bring volume to 1.0L with distilled/deionized water. Filter sterilize. Aseptically gas under 100% N_2 for 20 min.

Solution C:

Composition per 100.0mL:

$NaHCO_3$	5.0g

Preparation of Solution C: Add $NaHCO_3$ to distilled/deionized water and bring volume to 100.0mL. Mix thoroughly. Filter sterilize. Aseptically gas under 80% N_2 + 20% CO_2 for 20 min.

Solution D:

Composition per 10.0mL:

Sodium benzoate	0.6g

Preparation of Solution D: Prepare and dispense solution anaerobically under 80% N_2 + 20% CO_2. Add sodium benzoate to distilled/deionized water and bring volume to 10.0mL. Mix thoroughly. Cap with a rubber stopper. Autoclave for 15 min at 15 psi pressure–121°C. Cool to 25°C.

Solution E (Wolfe's Vitamin Solution):

Composition per liter:

Pyridoxine·HCl	0.01g
Thiamine·HCl	5.0mg
Riboflavin	5.0mg
Nicotinic acid	5.0mg
Calcium pantothenate	5.0mg
p-Aminobenzoic acid	5.0mg
Thioctic acid	5.0mg
Biotin	2.0mg
Folic acid	2.0mg
Cyanocobalamin	0.1mg

Preparation of Solution E (Wolfe's Vitamin Solution): Add components to distilled/deionized water and bring volume to 1.0L. Mix thoroughly. Filter sterilize. Aseptically gas under 100% N_2 for 20 min.

Solution F:

Composition per 6.6mL:

$AlCl_3 \cdot 6H_2O$ (4.9% solution)	5.0mL
Na_2CO_3 (10.6% solution)	1.6mL

Preparation of Solution F: Combine both solutions. Mix thoroughly. Gas with 100% N_2. Cap with a rubber stopper. Autoclave for 15 min at 15 psi pressure–121°C. Cool to 25°C.

Solution G:

Composition per 10.0mL:

Rumen fluid, clarified	20.0mL

Preparation of Solution G: Gas rumen fluid under 100% N_2 for 20 min. Cap with a rubber stopper. Autoclave for 15 min at 15 psi pressure–121°C. Cool to 25°C.

Solution H:

Composition per 10.0mL:

$Na_2S \cdot 9H_2O$	0.4g

Preparation of Solution H: Add $Na_2S \cdot 9H_2O$ to distilled/deionized water and bring volume to 10.0mL. Gas under 100% N_2 for 20 min. Cap with a rubber stopper. Autoclave for 15 min at 15 psi pressure–121°C. Cool to 25°C.

Solution I:

Composition per 10.0mL:

$Na_2S_2O_4$	0.5g

Preparation of Solution I: Add $Na_2S_2O_4$ to distilled/deionized water and bring volume to 10.0mL. Mix thoroughly. Filter sterilize. Aseptically gas under 100% N_2 for 20 min. Prepare solution freshly.

Preparation of Medium: To 850.0mL of cooled, sterile solution A, aseptically and anaerobically add in the following order: 1.0mL of sterile solution B, 100.0mL of sterile solution C, 10.0mL of sterile solution D, 10.0mL of sterile solution E, 6.6mL of sterile solution F, 20.0mL of sterile solution G, and 10.0mL of sterile solution H. Mix thoroughly. Immediately prior to inoculation, aseptically and anaerobically add 0.4mL of sterile solution I. Mix thoroughly. Aseptically and anaerobically distribute into sterile tubes or flasks.

Use: For the cultivation and maintenance of *Desulfosarcina variabilis.*

Desulfotomaculum acetoxidans Medium

Composition per 1011.0mL:

Solution A	1.0L
Solution B	10.0mL
Vitamin solution	1.0mL

pH 7.0 ± 0.2 at 25°C

Solution A:

Composition per liter:

$NaHCO_3$	4.5g
Na_2SO_4	2.84g
Sodium acetate	1.4g
Sodium butyrate	1.4g
NaCl	1.17g
Yeast extract	1.0g
$MgCl_2 \cdot 6H_2O$	0.4g
KCl	0.3g
NH_4Cl	0.27g
KH_2PO_4	0.2g
$CaCl_2 \cdot 2H_2O$	0.15g

Resazurin 0.5mg
Trace elements solution 1.0mL

Preparation of Solution A: Add components, except $NaHCO_3$ and vitamin solution, and bring volume to 1.0L. Mix thoroughly. Gently heat and bring to boiling. Continue boiling for 3–4 min. Allow to cool to room temperature while gassing under O_2-free 80% N_2 + 20% CO_2. Add $NaHCO_3$ and continue gassing with O_2-free 80% N_2 + 20% CO_2 until pH reaches 6.9–7.1. Seal the flask under 80% N_2 + 20% CO_2. Autoclave for 15 min at 15 psi pressure–121°C.

Trace Elements Solution:
Composition per liter:

$FeCl_2·4H_2O$ 1.5g
$CoCl_2·6H_2O$ 120.0mg
$MnCl_2·4H_2O$ 100.0mg
$ZnCl_2$ 68.0mg
H_3BO_3 .... 62.0mg
$Na_2MoO_4·2H_2O$ 24.0mg
$NiCl_2·6H_2O$ 24.0mg
$CuCl_2·2H_2O$ 17.0mg
HCl (25% solution) 10.0mL

Preparation of Trace Elements Solution: Add $FeCl_2·4H_2O$ to 10.0mL of HCl solution. Mix thoroughly. Add distilled/deionized water and bring volume to 1.0L. Add remaining components. Mix thoroughly. Gas under 100% N_2.

Vitamin Solution:
Composition per 100.0mL:

Thiamine·HCl 10.0mg
p-Aminobenzoic acid 4.0mg
D(+)-Biotin 1.0mg

Preparation of Vitamin Solution: Add components to distilled/deionized water and bring volume to 100.0mL. Mix thoroughly. Filter sterilize. Gas under 100% N_2.

Solution B:
Composition per 10.0mL:

$Na_2S·9H_2O$ 0.36g

Preparation of Solution B: Add $Na_2S·9H_2O$ to distilled/deionized water and bring volume to 10.0mL. Mix thoroughly. Gas under 100% N_2. Autoclave for 15 min at 15 psi pressure–121°C.

Preparation of Medium: To 1.0L of sterile solution A, add 10.0mL of sterile solution B and 1.0mL of sterile vitamin solution. Mix thoroughly.

Use: For the cultivation and maintenance of *Desulfotomaculum acetoxidans.*

Desulfotomaculum acetoxidans Medium

Composition per 1020.0mL:

Solution A 1.0L
$Na_2S·9H_2O$ solution 10.0mL
Wolfe's vitamin solution 10.0mL

Solution A:
Composition per liter:

$NaHCO_3$ 4.5g
Na_2SO_4 .... 2.84g
Sodium acetate 1.4g
Sodium butyrate 1.4g
NaCl 1.17g
Yeast extract 1.0g
$MgCl_2·6H_2O$ 0.4g
KCl 0.3g
NH_4Cl 0.27g
KH_2PO_4 .... 0.2g
$CaCl_2·2H_2O$ 0.15g
Resazurin 0.5mg
Trace elements solution SL-7 .... 1.0mL

Trace Elements Solution SL-7:
Composition per 1010.0mL:

$FeCl_2·4H_2O$ 1.5g
$CoCl_2·6H_2O$ 190.0mg
$MnCl_2·4H_2O$ 100.0mg
$ZnCl_2$ 70.0mg
H_3BO_3 .... 62.0mg
$Na_2MoO_4·2H_2O$ 36.0mg
$NiCl_2·6H_2O$ 24.0mg
$CuCl_2·2H_2O$ 17.0mg
Hydrochloric acid, 25% 10.0mL

Preparation of Trace Elements Solution SL-7: Add $FeCl_2·4H_2O$ to 10.0mL of HCl solution. Mix thoroughly. Add distilled/deionized water and bring volume to 1.0L. Add remaining components. Mix thoroughly.

Preparation of Solution A: Prepare and dispense medium under 80% N_2 + 20% CO_2. Add components, except $NaHCO_3$ and bring volume to 1.0L. Mix thoroughly. Gently heat and bring to boiling. Cool to room temperature while sparging with 80% N_2 + 20% CO_2. Add $NaHCO_3$. Mix thoroughly. Continue sparging with 80% N_2 + 20% CO_2 until pH stabilizes at 6.9–7.1. Anaerobically distribute into tubes in 10.0mL amounts under an atmosphere of 80% N_2 + 20% CO_2. Autoclave for 15 min at 15 psi pressure–121°C. Cool to room temperature.

$Na_2S·9H_2O$ Solution:
Composition per 10.0mL:

$Na_2S·9H_2O$ 0.36g

Preparation of $Na_2S·9H_2O$ Solution: Add $Na_2S·9H_2O$ to distilled/deionized water and bring volume to 10.0mL. Mix thoroughly. Sparge with 100% N_2. Autoclave for 15 min at 15 psi pressure–121°C. Before use, neutralize to pH 7.0 with sterile HCl.

Wolfe's Vitamin Solution:
Composition per liter:

Pyridoxine·HCl 10.0mg
p-Aminobenzoic acid 5.0mg
Lipoic acid 5.0mg
Nicotinic acid 5.0mg
Riboflavin 5.0mg
Thiamine·HCl 5.0mg
Calcium DL-pantothenate 5.0mg
Biotin 2.0mg
Folic acid 2.0mg
Vitamin B_{12} 0.1mg

Preparation of Wolfe's Vitamin Solution: Add components to distilled/deionized water and bring volume to 1.0L. Mix thoroughly. Filter sterilize. Sparge with 100% N_2.

Preparation of Medium: Aseptically and anaerobically add 0.1mL of sterile $Na_2S·9H_2O$ solution and 0.1mL of sterile Wolfe's vitamin solution to each tube containing 10.0mL of solution A. Prepare immediately prior to use.

Use: For the cultivation of *Desulfotomaculum acetoxidans.*

Desulfotomaculum alkaliphilum Medium (DSMZ Medium 866)

Composition per liter:

$NaHCO_3$	8.0g
Na_2SO_4	5.0g
Na-formate	5.0g
NaCl	5.0g
Yeast extract	1.5g
NH_4Cl	1.0g
Na_2CO_3	0.5g
$Na_2S·9H_2O$	0.5g
KH_2PO_4	0.2g
KCl	0.2g
$MgCl_2·6H_2O$	0.1g
Vitamin solution	10.0mL
Trace elements solution SL-10	1.0mL

pH 8.7–9.0 at 25°C

Vitamin Solution:

Composition per liter:

Pyridoxine-HCl	10.0mg
Thiamine-HCl·$2H_2O$	5.0mg
Riboflavin	5.0mg
Nicotinic acid	5.0mg
D-Ca-pantothenate	5.0mg
p-Aminobenzoic acid	5.0mg
Lipoic acid	5.0mg
Biotin	2.0mg
Folic acid	2.0mg
Vitamin B_{12}	0.1mg

Preparation of Vitamin Solution: Add components to distilled/deionized water and bring volume to 1.0L. Mix thoroughly. Sparge with 80% H_2 + 20% CO_2. Filter sterilize.

Trace Elements Solution SL-10:

Composition per liter:

$FeCl_2·4H_2O$	1.5g
$CoCl_2·6H_2O$	190.0mg
$MnCl_2·4H_2O$	100.0mg
$ZnCl_2$	70.0mg
$Na_2MoO_4·2H_2O$	36.0mg
$NiCl_2·6H_2O$	24.0mg
H_3BO_3	6.0mg
$CuCl_2·2H_2O$	2.0mg
HCl (25% solution)	10.0mL

Preparation of Trace Elements Solution SL-10: Add $FeCl_2·4H_2O$ to 10.0mL of HCl solution. Mix thoroughly. Add distilled/deionized water and bring volume to 1.0L. Add remaining components. Mix thoroughly. Sparge with 80% N_2 + 20% CO_2. Autoclave for 15 min at 15 psi pressure–121°C.

Preparation of Medium: Prepare and dispense medium anaerobically under N_2. Add components, except $NaHCO_3$ and $Na_2S·9H_2O$, to distilled/deionized water and bring volume to 1.0L. Mix thoroughly. Gently heat and bring to boiling. Boil for 5 min. Cool to room temperature while sparging with 100% N_2. Add $NaHCO_3$ and $Na_2S·9H_2O$. Mix thoroughly. Distribute into anaerobe tubes or bottles. Autoclave for 15 min at 15 psi pressure–121°C.

Use: For the cultivation of *Desulfotomaculum alkaliphilum.*

Desulfotomaculum geothermicum Medium

Composition per 1001.0mL:

Solution A	870.0mL
Solution C	100.0mL
Solution D	10.0mL
Solution E (Vitamin solution)	10.0mL
Solution F	10.0mL
Solution B (Trace elements solution SL-10)	1.0mL

pH 7.1–7.4 at 25°C

Solution A:

Composition per 870.0mL:

NaCl	21.0g
$MgCl_2·6H_2O$	3.1g
Na_2SO_4	3.0g
KCl	0.5g
NH_4Cl	0.3g
KH_2PO_4	0.2g
$CaCl_2·2H_2O$	0.15g
Resazurin	1.0mg

Preparation of Solution A: Add components to distilled/deionized water and bring volume to 870.0mL. Mix thoroughly. Gently heat and bring to boiling. Continue boiling for 3–4 min. Allow to cool to room temperature while gassing under 80% N_2 + 20% CO_2. Continue gassing until pH reaches below 6.0. Seal the flask under 80% N_2 + 20% CO_2. Autoclave for 15 min at 15 psi pressure–121°C.

Solution B (Trace Elements Solution SL-10):

Composition per liter:

$FeCl_2·4H_2O$	1.5g
$CoCl_2·6H_2O$	190.0mg
$MnCl_2·4H_2O$	100.0mg
$ZnCl_2$	70.0mg
$Na_2MoO_4·2H_2O$	36.0mg
$NiCl_2·6H_2O$	24.0mg
H_3BO_3	6.0mg
$CuCl_2·2H_2O$	2.0mg
HCl (25% solution)	10.0mL

Preparation of Solution B (Trace Elements Solution SL-10): Add $FeCl_2·4H_2O$ to 10.0mL of HCl solution. Mix thoroughly. Add distilled/deionized water and bring volume to 1.0L. Add remaining components. Mix thoroughly. Gas under 100% N_2. Autoclave for 15 min at 15 psi pressure–121°C.

Solution C:

Composition per 100.0mL:

$NaHCO_3$	5.0g

Preparation of Solution C: Add $NaHCO_3$ to distilled/deionized water and bring volume to 100.0mL. Mix thoroughly. Filter sterilize. Gas under 80% N_2 + 20% CO_2.

Solution D:

Composition per 10.0mL:

Sodium lactate	2.5g

Preparation of Solution D: Add sodium lactate to distilled/deionized water and bring volume to 10.0mL. Mix thoroughly. Gas under 100% N_2. Autoclave for 15 min at 15 psi pressure–121°C.

Solution E (Vitamin Solution):

Composition per liter:

Pyridoxine·HCl	10.0mg
Calcium DL-pantothenate	5.0mg
Lipoic acid	5.0mg

Nicotinic acid	5.0mg
p-Aminobenzoic acid	5.0mg
Riboflavin	5.0mg
Thiamine·HCl	5.0mg
Biotin	2.0mg
Folic acid	2.0mg
Vitamin B_{12}	0.1mg

Preparation of Solution E (Vitamin Solution): Add components to distilled/deionized water and bring volume to 1.0L. Mix thoroughly. Gas under 100% N_2. Autoclave for 15 min at 15 psi pressure–121°C.

Solution F:
Composition per 10.0mL:

$Na_2S·9H_2O$	0.05g

Preparation of Solution F: Add $Na_2S·9H_2O$ to distilled/deionized water and bring volume to 10.0mL. Mix thoroughly. Gas under 100% N_2. Autoclave for 15 min at 15 psi pressure–121°C.

Preparation of Medium: Aseptically and anaerobically combine solution A with solution B, solution C, solution D, solution E, and solution F, in that order. Mix thoroughly. Anaerobically distribute into sterile tubes or flasks under 80% N_2 + 20% CO_2.

Use: For the cultivation and maintenance of *Desulfotomaculum geothermicum.*

Desulfotomaculum Groll Medium (DSMZ Medium 124a)

Composition per liter:

$NaHCO_3$	4.5g
Na_2SO_4	2.84g
Na-acetate	1.4g
Na-butyrate	1.4g
NaCl	1.17g
Yeast extract	1.0g
$MgCl_2·6H_2O$	0.4g
KCl	0.3g
NH_4Cl	0.27g
KH_2PO_4	0.2g
$CaCl_2·2H_2O$	0.15g
Resazurin	0.5mg
$Na_2S·9H_2O$ solution	10.0mL
Substrate solution	10.0mL
Selenite solution	10.0mL
Vitamin solution	1.0mL
Trace elements solution	1.0mL

pH 7.0 ± 0.2 at 25°C

$Na_2S·9H_2O$ Solution:
Composition per 100.0mL:

$Na_2S·9H_2O$	3.6g

Preparation of $Na_2S·9H_2O$ Solution: Add $Na_2S·9H_2O$ to distilled/deionized water and bring volume to 100.0mL. Sparge with N_2. Autoclave for 15 min at 15 psi pressure–121°C. Cool to 25°C. Store anaerobically.

Trace Elements Solution:
Composition per liter:

$FeCl_2·4H_2O$	1.5g
$CoCl_2·6H_2O$	120.0mg
$MnCl_2·4H_2O$	100.0mg
$ZnCl_2$	68.0mg
$Na_2MoO_4·2H_2O$	24.0mg
$NiCl_2·6H_2O$	24.0mg
H_3BO_3	62.0mg
$CuCl_2·2H_2O$	17.0mg
HCl (0.5*M*)	1.0L

Preparation of Trace Elements Solution: Add $FeCl_2·4H_2O$ to 1.0L of 0.5*M* HCl. Mix thoroughly. Add remaining components. Mix thoroughly. Sparge with 100% N_2. Autoclave for 15 min at 15 psi pressure–121°C.

Vitamin Solution:

Thiamine-HCl·$2H_2O$	10.0mg
p-Aminobenzoic acid	4.0mg
D(+)-Biotin	1.0mg

Preparation of Vitamin Solution: Add components to distilled/deionized water and bring volume to 100.0mL. Mix thoroughly. Sparge with 80% N_2 + 20% CO_2. Filter sterilize.

Selenite Solution:
Composition 10.0mL:

Sodium selenite	3.0μg

Preparation of Selenite Solution: Add sodium selenite to distilled/deionized water and bring volume to 10.0mL. Mix thoroughly. Sparge with 80% N_2 + 20% CO_2. Autoclave for 15 min at 15 psi pressure–121°C.

Substrate Solution:
Composition per 10.0mL:

Sodium benzoate	0.6g

Preparation of Substrate Solution: Add sodium benzoate to distilled/deionized water and bring volume to 10.0mL. Mix thoroughly. Sparge with 80% N_2 + 20% CO_2. Autoclave for 15 min at 15 psi pressure–121°C.

Preparation of Medium: Prepare and dispense medium under an oxygen-free 80% N_2 + 20% CO_2 gas mixture. Add components, except $NaHCO_3$, substrate solution, selenite solution, vitamin solution, and $Na_2S·9H_2O$ solution, to distilled/deionized water and bring volume to 1.0L. Mix thoroughly. Sparge with 80% N_2 + 20% CO_2. Heat gently and bring to boiling. Cool to room temperature. Add $NaHCO_3$. Continue sparging with 80% N_2 + 20% CO_2 until an equilibrium pH of 6.9-7.1 is reached. Autoclave for 15 min at 15 psi pressure–121°C. Cool to 25°C while sparging with 80% N_2 + 20% CO_2. Aseptically and anaerobically add 10.0mL of sterile substrate solution, 10.0mL of sterile selenite solution, 1.0mL of sterile vitamin solution, and 10.0mL of sterile $Na_2S·9H_2O$ solution. Mix thoroughly. Aseptically and anaerobically distribute into sterile tubes or flasks. Alternately the medium can be distributed to tubes under anaerobic conditions and autoclaved in tubes prior to addition of substrate solution, selenite solution, vitamin solution, and $Na_2S·9H_2O$ solution. Appropriate amounts of these solutions can then be added to each tube to yield the desired concentrations. Additions are performed aseptically and anaerobically under an oxygen-free 80% N_2 + 20% CO_2 gas mixture.

Use: For the cultivation of *Desulfotomaculum gibsoniae.*

Desulfotomaculum halophilum Medium (DSMZ Medium 815)

Composition per liter:

Iron, powder	150.0g
NaCl	40.0g
$MgCl_2·6H_2O$	8.0g
$CaCl_2·2H_2O$	6.0g
Na-lactate	3.6g

Na_2SO_4	3.0g
MOPS	3.0g
KCl	2.0g
Yeast extract	1.0g
NH_4Cl	0.3g
KH_2PO_4	0.2g
$SrCl_2·6H_2O$	0.1g
Resazurin	0.5mg
$NaHCO_3$ solution	50.0mL
Trace elements solution SL-12	1.0mL

pH 7.2 ± 0.2 at 25°C

Trace Elements Solution SL-12:

Composition per liter:

Na_2-EDTA	5.2g
$FeCl_2·4H_2O$	1.5g
$CoCl_2·6H_2O$	190.0mg
$MnCl_2·4H_2O$	100.0mg
$ZnCl_2$	70.0mg
$Na_2MoO_4·2H_2O$	36.0mg
$NiCl_2·6H_2O$	24.0mg
H_3BO_3	6.0mg
$CuCl_2·2H_2O$	2.0mg
HCl (25% solution)	10.0mL

Preparation of Trace Elements Solution SL-12: Add $FeCl_2·4H_2O$ to 10.0mL of HCl solution. Mix thoroughly. Add distilled/deionized water and bring volume to 1.0L. Add remaining components. Mix thoroughly. Adjust pH to 6.0. Sparge with 80% N_2 + 20% CO_2. Autoclave for 15 min at 15 psi pressure–121°C.

$NaHCO_3$ Solution:

Composition per 100.0mL:

$NaHCO_3$	10.0g

Preparation of $NaHCO_3$ Solution: Add $NaHCO_3$ to distilled/deionized water and bring volume to 100.0mL. Mix thoroughly. Sparge with 80% N_2 + 20% CO_2. Filter sterilize.

Preparation of Medium: Prepare and dispense medium under 100% N_2. Add components, except iron and $NaHCO_3$ solution, to distilled/deionized water and bring volume to 950.0L. Mix thoroughly. Sparge with 100% N_2. Adjust pH to 6.0. Dispense under 100% N_2 into tubes or bottles containing 1.5g iron per 10.0mL medium. Autoclave for 30 min at 105°C. Cool to 25°C. Aseptically and anaerobically add sterile $NaHCO_3$ solution, 0.5mL per 10.0mL medium. Final pH is 7.2.

Use: For the cultivation of *Desulfotomaculum halophilum* and *Desulfocella halophila.*

Desulfotomaculum halophilum Medium (DSMZ Medium 815)

Composition per liter:

NaCl	40.0g
$MgCl_2·6H_2O$	8.0g
$CaCl_2·2H_2O$	6.0g
Na-pyruvate	3.6g
Na_2SO_4	3.0g
MOPS	3.0g
KCl	2.0g
Yeast extract	1.0g
NH_4Cl	0.3g
KH_2PO_4	0.2g
$SrCl_2·6H_2O$	0.1g
Resazurin	0.5mg
$NaHCO_3$ solution	50.0mL
Trace elements solution SL-12	1.0mL

pH 7.2 ± 0.2 at 25°C

Trace Elements Solution SL-12:

Composition per liter:

Na_2-EDTA	5.2g
$FeCl_2·4H_2O$	1.5g
$CoCl_2·6H_2O$	190.0mg
$MnCl_2·4H_2O$	100.0mg
$ZnCl_2$	70.0mg
$Na_2MoO_4·2H_2O$	36.0mg
$NiCl_2·6H_2O$	24.0mg
H_3BO_3	6.0mg
$CuCl_2·2H_2O$	2.0mg
HCl (25% solution)	10.0mL

Preparation of Trace Elements Solution SL-12: Add $FeCl_2·4H_2O$ to 10.0mL of HCl solution. Mix thoroughly. Add distilled/deionized water and bring volume to 1.0L. Add remaining components. Mix thoroughly. Adjust pH to 6.0. Sparge with 80% N_2 + 20% CO_2. Autoclave for 15 min at 15 psi pressure–121°C.

$NaHCO_3$ Solution:

Composition per 100.0mL:

$NaHCO_3$	10.0g

Preparation of $NaHCO_3$ Solution: Add $NaHCO_3$ to distilled/deionized water and bring volume to 100.0mL. Mix thoroughly. Sparge with 80% N_2 + 20% CO_2. Filter sterilize.

Preparation of Medium: Prepare and dispense medium under 100% N_2. Add components, except $NaHCO_3$ solution, to distilled/deionized water and bring volume to 950.0mL. Mix thoroughly. Sparge with 100% N_2. Adjust pH to 6.0. Dispense under 100% N_2 into tubes or bottles. Autoclave for 30 min at 105°C. Cool to 25°C. Aseptically and anaerobically add sterile $NaHCO_3$ solution, 0.5mL per 10.0mL medium. Final pH is 7.2.

Use: For the cultivation of *Desulfocella halophila* DSM 11763.

Desulfotomaculum sapomandens Medium

Composition per 1019.7mL:

Solution A	966.0mL
Solution B	20.0mL
Solution C	10.0mL
Solution D	10.0mL
Solution F	10.0mL
Solution H	1.7mL
Solution G	1.0mL
Solution E	1.0mL

pH 7.2–7.5 at 25°C

Solution A:

Composition per 966.0mL:

Na_2SO_4	3.0g
NaCl	1.0g
$MgCl_2·6H_2O$	0.4g
NH_4Cl	0.3g
KH_2PO_4	0.2g
$CaCl_2·2H_2O$	0.15g
Resazurin	1.0mg
Trace elements solution SL-10	1.0mL

Trace Elements Solution SL-10:

Composition per liter:

$FeCl_2 \cdot 4H_2O$	1.5g
$CoCl_2 \cdot 6H_2O$	190.0mg
$MnCl_2 \cdot 4H_2O$	100.0mg
$ZnCl_2$	70.0mg
$Na_2MoO_4 \cdot 2H_2O$	36.0mg
$NiCl_2 \cdot 6H_2O$	24.0mg
H_3BO_3	6.0mg
$CuCl_2 \cdot 2H_2O$	2.0mg
HCl (25% solution)	10.0mL

Preparation of Trace Elements Solution SL-10: Add $FeCl_2 \cdot 4H_2O$ to 10.0mL of HCl solution. Mix thoroughly. Add distilled/deionized water and bring volume to 1.0L. Add remaining components. Mix thoroughly. Gas under 100% N_2. Autoclave for 15 min at 15 psi pressure–121°C.

Solution B:

Composition per 20.0mL:

$NaHCO_3$	1.0g

Preparation of Solution B: Add $NaHCO_3$ to distilled/deionized water and bring volume to 20.0mL. Mix thoroughly. Gas under 100% N_2. Autoclave for 15 min at 15 psi pressure–121°C.

Solution C:

Composition per 10.0mL:

Ethanol	1.0mL

Preparation of Solution C: Add ethanol to distilled/deionized water and bring volume to 10.0mL. Mix thoroughly. Filter sterilize.

Solution D:

Composition per 10.0mL:

Sodium benzoate	0.7g

Preparation of Solution D: Add sodium benzoate to distilled/deionized water and bring volume to 10.0mL. Mix thoroughly. Filter sterilize.

Solution E:

Composition per 1.0mL:

Rumen fluid, clarified	1.0mL

Preparation of Solution E: Gas under 100% N_2. Autoclave for 15 min at 15 psi pressure–121°C.

Solution F:

Composition per liter:

Pyridoxine·HCl	10.0mg
Calcium DL-pantothenate	5.0mg
Lipoic acid	5.0mg
Nicotinic acid	5.0mg
p-Aminobenzoic acid	5.0mg
Riboflavin	5.0mg
Thiamine·HCl	5.0mg
Biotin	2.0mg
Folic acid	2.0mg
Vitamin B_{12}	0.1mg

Preparation of Solution F: Add components to distilled/deionized water and bring volume to 1.0L. Mix thoroughly. Gas under 100% N_2. Filter sterilize.

Solution G:

Composition per 1.0mL:

Sodium dithionite	0.025g

Preparation of Solution G: Add sodium dithionite to distilled/deionized water and bring volume to 1.0mL. Mix thoroughly. Gas under 100% N_2. Filter sterilize.

Solution H:

Composition per 10.0mL:

$Na_2S \cdot 9H_2O$	0.3g

Preparation of Solution H: Add $Na_2S \cdot 9H_2O$ to distilled/deionized water and bring volume to 10.0mL. Mix thoroughly. Gas under 100% N_2. Autoclave for 15 min at 15 psi pressure–121°C.

Preparation of Medium: Aseptically and anaerobically combine in the following order: 966.0mL of sterile solution A with 20.0.0mL of sterile solution B, 10.0mL of sterile solution C, 10.0mL of sterile solution D, 1.0mL of sterile solution E, 10.0mL of sterile solution F, 1.0mL of sterile solution G, and 1.7mL of sterile solution H. Mix thoroughly.

Use: For the cultivation and maintenance of *Desulfotomaculum sapomandens.*

Desulfotomaculum spp. Medium I (DSMZ Medium 63a)

Composition per liter:

Soultion A	980.0mL
Solution B	10.0mL
Solution C	10.0mL

pH 6.5–7.0 at 25°C

Solution A:

Composition per 980.0mL:

Na-pyruvate	5.0g
Na-acetate	2.0g
$MgSO_4 \cdot 7H_2O$	2.0g
Yeast extract	1.0g
NH_4Cl	1.0g
Na_2SO_4	1.0g
K_2HPO_4	0.5g
$CaCl_2 \cdot 2H_2O$	0.1g
Resazurin	1.0mg

Preparation of Solution A: Add components to distilled/deionized water and bring volume to 980.0mL. Mix thoroughly.

Solution B:

Composition per 10.0mL:

$FeSO_4 \cdot 7H_2O$	0.5g

Preparation of Solution B: Add $FeSO_4 \cdot 7H_2O$ to distilled/deionized water and bring volume to 10.0mL. Mix thoroughly.

Solution C:

Composition per 10.0mL:

Na-thioglycolate	0.1g
Ascorbic acid	0.1g

Preparation of Solution C: Add components to distilled/deionized water and bring volume to 10.0mL. Mix thoroughly.

Preparation of Medium: Bring solution A to a boil for a few minutes. Cool to room temperature while gassing with oxygen-free N_2 gas. Add solutions B and C. Adjust pH to 6.5–7.0. Immediately distribute under N_2 into anaerobic tubes. During distribution continuously swirl the medium to keep the grey precipitate suspended. Autoclave for 15 min at 15 psi pressure–121°C.

Use: For the cultivation of *Desulfotomaculum* spp.

Desulfotomaculum species Medium II

Composition per 1001.0mL:

Solution A	870.0mL
Solution C	100.0mL
Solution D	10.0mL
Solution E (Seven vitamin solution)	10.0mL
Solution F	10.0mL
Solution B (Trace elements solution SL-10)	1.0mL

pH 7.1–7.4 at 25°C

Solution A:

Composition per 870.0mL:

NaCl	7.0g
$MgCl_2 \cdot 6H_2O$	1.3g
Na_2SO_4	0.7g
KCl	0.5g
NH_4Cl	0.3g
KH_2PO_4	0.2g
$CaCl_2 \cdot 2H_2O$	0.15g
Resazurin	1.0mg

Preparation of Solution A: Add components to distilled/deionized water and bring volume to 870.0mL. Mix thoroughly. Gently heat and bring to boiling. Continue boiling for 3–4 min. Allow to cool to room temperature while gassing under 80% N_2 + 20% CO_2. Continue gassing until pH reaches below 6.0. Seal the flask under 80% N_2 + 20% CO_2. Autoclave for 15 min at 15 psi pressure–121°C.

Solution B (Trace Elements Solution SL-10):

Composition per liter:

$FeCl_2 \cdot 4H_2O$	1.5g
$CoCl_2 \cdot 6H_2O$	190.0mg
$MnCl_2 \cdot 4H_2O$	100.0mg
$ZnCl_2$	70.0mg
$Na_2MoO_4 \cdot 2H_2O$	36.0mg
$NiCl_2 \cdot 6H_2O$	24.0mg
H_3BO_3	6.0mg
$CuCl_2 \cdot 2H_2O$	2.0mg
HCl (25% solution)	10.0mL

Preparation of Solution B (Trace Elements Solution SL-10): Add $FeCl_2 \cdot 4H_2O$ to 10.0mL of HCl solution. Mix thoroughly. Add distilled/deionized water and bring volume to 1.0L. Add remaining components. Mix thoroughly. Sparge with 100% N_2. Autoclave for 15 min at 15 psi pressure–121°C.

Solution C:

Composition per 100.0mL:

$NaHCO_3$	5.0g

Preparation of Solution C: Add $NaHCO_3$ to distilled/deionized water and bring volume to 100.0mL. Mix thoroughly. Filter sterilize. Sparge with 80% N_2 + 20% CO_2.

Solution D:

Composition per 10.0mL:

3,4,5-Trimethoxybenzoate	0.42g

Preparation of Solution D: Add 3,4,5-trimethoxybenzoic acid to distilled/deionized water and bring volume to 10.0mL. Mix thoroughly. Sparge with 100% N_2. Autoclave for 15 min at 15 psi pressure–121°C.

Solution E (Seven Vitamin Solution):

Composition per liter:

Pyridoxine·HCl	0.3g
Thiamine·HCl	0.2g
Nicotinic acid	0.2g
Calcium DL-pantothenate	0.1g
Vitamin B_{12}	0.1g
p-Aminobenzoic acid	80.0mg
Biotin	20.0mg

Preparation of Solution E (Seven Vitamin Solution): Add components to distilled/deionized water and bring volume to 1.0L. Mix thoroughly. Sparge with 100% N_2. Autoclave for 15 min at 15 psi pressure–121°C.

Solution F:

Composition per 10.0mL:

$Na_2S \cdot 9H_2O$	0.4g

Preparation of Solution F: Add $Na_2S \cdot 9H_2O$ to distilled/deionized water and bring volume to 10.0mL. Mix thoroughly. Sparge with 100% N_2. Autoclave for 15 min at 15 psi pressure–121°C.

Preparation of Medium: Aseptically and anaerobically combine solution A with solution B, solution C, solution D, solution E, and solution F, in that order. Mix thoroughly. Anaerobically distribute into sterile tubes or flasks under 80% N_2 + 20% CO_2.

Use: For the cultivation of *Desulfotomaculum* species.

Desulfotomaculum thermosapovorans Medium

Composition per 1015.0mL:

Solution A	900.0mL
Solution B	100.0mL
Solution C	10.0mL
Solution D	5.0mL
Solution E	0.5mL

pH 7.0–7.2 at 25°C

Solution A:

Composition per 900.0mL:

NaCl	8.0g
Na_2SO_4	3.0g
Sodium butyrate	2.2g
NH_4Cl	1.0g
KCl	0.5g
Sodium acetate	0.5g
$MgCl_2 \cdot 6H_2O$	0.4g
Yeast extract	0.4g
KH_2PO_4	0.2g
$CaCl_2 \cdot 2H_2O$	0.15g
$Na_2SeO_3 \cdot 5H_2O$	0.003mg
Trace elements solution SL-10	1.5mL

Trace Elements Solution SL-10:

Composition per liter:

$FeCl_2 \cdot 4H_2O$	1.5g
$CoCl_2 \cdot 6H_2O$	190.0mg
$MnCl_2 \cdot 4H_2O$	100.0mg
$ZnCl_2$	70.0mg
$Na_2MoO_4 \cdot 2H_2O$	36.0mg
$NiCl_2 \cdot 6H_2O$	24.0mg
H_3BO_3	6.0mg
$CuCl_2 \cdot 2H_2O$	2.0mg
HCl (25% solution)	10.0mL

Preparation of Trace Elements Solution SL-10: Add $FeCl_2 \cdot 4H_2O$ to 10.0mL of HCl solution. Mix thoroughly. Add distilled/deionized water and bring volume to 1.0L. Add remaining components. Mix thoroughly. Sparge with 100% N_2. Autoclave for 15 min at 15 psi pressure–121°C.

Preparation of Solution A: Add components to distilled/deionized water and bring volume to 900.0mL. Mix thoroughly. Sparge with 80% N_2 + 20% CO_2.

Solution B:

Composition per 100.0mL:

$NaHCO_3$	5.0g

Preparation of Solution B: Add $NaHCO_3$ to distilled/deionized water and bring volume to 10.0mL. Mix thoroughly. Sparge with 100% N_2. Autoclave for 15 min at 15 psi pressure–121°C.

Solution C:

Composition per liter:

Pyridoxine·HCl	10.0mg
p-Aminobenzoic acid	5.0mg
Lipoic acid	5.0mg
Nicotinic acid	5.0mg
Riboflavin	5.0mg
Thiamine·HCl	5.0mg
Calcium DL-pantothenate	5.0mg
Biotin	2.0mg
Folic acid	2.0mg
Vitamin B_{12}	0.1mg

Preparation of Solution C: Add components to distilled/deionized water and bring volume to 1.0L. Mix thoroughly. Filter sterilize.

Solution D:

Composition per 5.0mL:

$Na_2S·9H_2O$	0.2g

Preparation of Solution D: Add $Na_2S·9H_2O$ to distilled/deionized water and bring volume to 5.0mL. Mix thoroughly. Sparge with 100% N_2. Autoclave for 15 min at 15 psi pressure–121°C. Before use, neutralize to pH 7.0 with sterile HCl.

Solution E:

Composition per 1.0mL:

$Na_2S_2O_4$ (sodium dithionite)	20.0mg

Preparation of Solution E: Add $Na_2S_2O_4$ to distilled/deionized water and bring volume to 1.0mL. Mix thoroughly. Sparge with 100% N_2. Autoclave for 15 min at 15 psi pressure–121°C.

Preparation of Medium: Prepare medium anaerobically under 80% N_2 + 20% CO_2. Aseptically and anaerobically combine 900.0mL of sterile solution A, 100.0mL of sterile solution B, 10.0mL of sterile solution C, 5.0mL of sterile solution D, and 0.5 mL of sterile solution E. Mix thoroughly. Adjust pH to 7.0–7.2.

Use: For the cultivation of *Desulfotomaculum thermosapovorans.*

Desulfovibrio aespoeensis Medium (DSMZ Medium 721)

Composition per 1004.0mL:

Solution A	870.0mL
Solution C	100.0mL
Solution D	10.0mL
Solution E (Vitamin solution)	10.0mL
Solution F	10.0mL
Selenite-tungstate solution	2.0mL
Solution B (Trace elements solution SL-10)	1.0mL
Seven vitamin solution	1.0mL

pH 7.3–7.5 at 25°C

Solution A:

Composition per 870.0mL:

NaCl	7.0g
Na_2SO_4	3.0g
$MgCl_2·6H_2O$	1.3g
KCl	0.5g
NH_4Cl	0.3g
KH_2PO_4	0.2g
$CaCl_2·2H_2O$	0.15g
Resazurin	1.0mg

Preparation of Solution A: Add components to distilled/deionized water and bring volume to 870.0mL. Mix thoroughly.

Solution B (Trace Elements Solution SL-10):

Composition per liter:

$FeCl_2·4H_2O$	1.5g
$CoCl_2·6H_2O$	190.0mg
$MnCl_2·4H_2O$	100.0mg
$ZnCl_2$	70.0mg
$Na_2MoO_4·2H_2O$	36.0mg
$NiCl_2·6H_2O$	24.0mg
H_3BO_3	6.0mg
$CuCl_2·2H_2O$	2.0mg
HCl (25% solution)	10.0mL

Preparation of Solution B (Trace Elements Solution SL-10): Add $FeCl_2·4H_2O$ to 10.0mL of HCl solution. Mix thoroughly. Add distilled/deionized water and bring volume to 1.0L. Add remaining components. Mix thoroughly. Sparge with 100% N_2. Autoclave for 15 min at 15 psi pressure–121°C.

Solution C:

Composition per 100.0mL:

$NaHCO_3$	5.0g

Preparation of Solution C: Add $NaHCO_3$ to distilled/deionized water and bring volume to 100.0mL. Mix thoroughly. Filter sterilize. Flush with 80% N_2 + 20% CO_2 to remove dissolved oxygen.

Solution D:

Composition per 10.0mL:

Na-lactate	2.5g

Preparation of Solution D: Add Na-lactate to distilled/deionized water and bring volume to 10.0mL. Mix thoroughly. Sparge with 100% N_2. Autoclave for 15 min at 15 psi pressure–121°C.

Solution E (Vitamin Solution):

Composition per liter:

Pyridoxine-HCl	10.0mg
Thiamine-HCl·$2H_2O$	5.0mg
Riboflavin	5.0mg
Nicotinic acid	5.0mg
D-Ca-pantothenate	5.0mg
p-Aminobenzoic acid	5.0mg
Lipoic acid	5.0mg
Biotin	2.0mg
Folic acid	2.0mg
Vitamin B_{12}	0.10mg

Preparation of Solution E (Vitamin Solution): Add components to distilled/deionized water and bring volume to 1.0L. Mix thoroughly. Sparge with 100% N_2. Autoclave for 15 min at 15 psi pressure–121°C.

Solution F:

Composition per 10.0mL:

$Na_2S{\cdot}9H_2O$ 0.4g

Preparation of Solution F: Add $Na_2S{\cdot}9H_2O$ to distilled/deionized water and bring volume to 10.0mL. Mix thoroughly. Sparge with 100% N_2. Autoclave for 15 min at 15 psi pressure–121°C.

Seven Vitamin Solution:

Composition per liter:

Pyridoxine hydrochloride 300.0mg
Thiamine-HCl·$2H_2O$ 200.0mg
Nicotinic acid 200.0mg
Vitamin B_{12} 100.0mg
Calcium pantothenate 100.0mg
p-Aminobenzoic acid 80.0mg
D(+)-Biotin 20.0mg

Preparation of Seven Vitamin Solution: Add components to distilled/deionized water and bring volume to 1.0L. Sparge with 100% N_2. Mix thoroughly. Filter sterilize.

Selenite Tungstate Solution

Composition per liter:

NaOH 0.5g
$Na_2WO_4{\cdot}2H_2O$ 4.0mg
$Na_2SeO_3{\cdot}5H_2O$ 3.0mg

Preparation of Selenite Tungstate Solution: Add components to distilled/deionized water and bring volume to 1.0L. Mix thoroughly. Sparge with 100% N_2. Filter sterilize.

Preparation of Medium: Gently heat solution A and bring to boiling. Boil solution A for a few minutes. Cool to room temperature. Gas with 80% N_2 + 20% CO_2 gas mixture to reach a pH below 6. Autoclave for 15 min at 15 psi pressure–121°C. Cool to room temperature. Sequentially add 1.0mL solution B, 100.0mL solution C, 10.0mL solution D, 10.0mL solution E, 10.0mL solution F, 2.0mL selenite tungstate solution, and 1.0mL seven vitamin solution. Distribute anaerobically under 80% N_2 + 20% CO_2 into appropriate vessels.

Use: For the cultivation of *Desulfovibrio aespoeensis.*

Desulfovibrio alcoholovorans Medium

Composition per liter:

Solution A 980.0mL
Solution B 10.0mL
Solution C 10.0mL

pH 7.8 ± 0.2 at 25°C

Solution A:

Composition per 980.0mL:

$MgSO_4{\cdot}7H_2O$ 2.0g
1,2-Propanediol 1.5g
Na_2SO_4 .... 1.0g
NH_4Cl 1.0g
Yeast extract 1.0g
K_2HPO_4 .... 0.5g
$CaCl_2{\cdot}2H_2O$ 0.1g
$Na_2SeO_3{\cdot}5H_2O$ 3.0mg
Resazurin 1.0mg

Preparation of Solution A: Add components to distilled/deionized water and bring volume to 980.0mL. Mix thoroughly. Gently heat and bring to boiling. Continue boiling for 3–4 min. Allow to cool to room temperature while gassing under 100% N_2.

Solution B:

Composition per 10.0mL:

$FeSO_4{\cdot}7H_2O$ 0.5g

Preparation of Solution B: Add $FeSO_4{\cdot}7H_2O$ to distilled/deionized water and bring volume to 10.0mL. Mix thoroughly. Gas under 100% N_2. Autoclave for 15 min at 15 psi pressure–121°C.

Solution C:

Composition per 10.0mL:

Ascorbic acid 0.1g
Sodium thioglycolate 0.1g

Preparation of Solution C: Add components to distilled/deionized water and bring volume to 10.0mL. Mix thoroughly. Gas under 100% N_2. Autoclave for 15 min at 15 psi pressure–121°C.

Preparation of Medium: To 980.0mL of cooled solution A, anaerobically add 10.0mL of solution B and 10.0mL of solution C. Mix thoroughly. Adjust pH to 7.8 with NaOH. Distribute into tubes or flasks. During distribution, swirl the medium to keep the precipitate in suspension. Autoclave for 15 min at 15 psi pressure–121°C.

Use: For the cultivation and maintenance of *Desulfovibrio alcoholovorans*.

Desulfovibrio asponium Medium

Composition per 1004.0mL:

Solution A 870.0mL
Solution C 100.0mL
Solution D 10.0mL
Solution E (Vitamin solution) 10.0mL
Solution F 10.0mL
Solution G (Selenite-tungstate solution) 2.0mL
Solution B (Trace elements solution SL-10) 1.0mL
Solution H (Seven vitamin solution) 1.0mL

pH 7.3–7.5 at 25°C

Solution A:

Composition per 870.0mL:

NaCl 7.0g
Na_2SO_4 .... 3.0g
$MgCl_2{\cdot}6H_2O$ 1.3g
KCl 0.5g
NH_4Cl 0.3g
KH_2PO_4 .... 0.2g
$CaCl_2{\cdot}2H_2O$ 0.15g
Resazurin 1.0mg

Preparation of Solution A: Add components to distilled/deionized water and bring volume to 870.0mL. Mix thoroughly. Gently heat and bring to boiling. Continue boiling for 3-4 min. Allow to cool to room temperature while gassing under 80% N_2 + 20% CO_2. Continue gassing until pH reaches below 6.0. Seal the flask under 80% N_2 + 20% CO_2. Autoclave for 15 min at 15 psi pressure–121°C.

Solution B (Trace Elements Solution SL-10):

Composition per liter:

$FeCl_2{\cdot}4H_2O$ 1.5g
$CoCl_2{\cdot}6H_2O$ 190.0mg
$MnCl_2{\cdot}4H_2O$ 100.0mg
$ZnCl_2$ 70.0mg
$Na_2MoO_4{\cdot}2H_2O$ 36.0mg
$NiCl_2{\cdot}6H_2O$ 24.0mg
H_3BO_3 .... 6.0mg

$CuCl_2 \cdot 2H_2O$ 2.0mg
HCl (25% solution) 10.0mL

Preparation of Solution B (Trace Elements Solution SL-10): Add $FeCl_2 \cdot 4H_2O$ to 10.0mL of HCl solution. Mix thoroughly. Add distilled/deionized water and bring volume to 1.0L. Add remaining components. Mix thoroughly. Sparge with 100% N_2. Autoclave for 15 min at 15 psi pressure–121°C.

Solution C:

Composition per 100.0mL:

$NaHCO_3$ 5.0g

Preparation of Solution C: Add $NaHCO_3$ to distilled/deionized water and bring volume to 100.0mL. Mix thoroughly. Filter sterilize. Sparge with 80% N_2 + 20% CO_2.

Solution D:

Composition per 10.0mL:

Sodium lactate 2.5g

Preparation of Solution D: Add sodium lactate to distilled/deionized water and bring volume to 10.0mL. Mix thoroughly. Sparge with 100% N_2. Autoclave for 15 min at 15 psi pressure–121°C.

Solution E (Vitamin Solution):

Composition per liter:

Pyridoxine·HCl 10.0mg
Calcium DL-pantothenate 5.0mg
Lipoic acid 5.0mg
Nicotinic acid 5.0mg
p-Aminobenzoic acid 5.0mg
Riboflavin 5.0mg
Thiamine·HCl 5.0mg
Biotin 2.0mg
Folic acid 2.0mg
Vitamin B_{12} 0.1mg

Preparation of Solution E (Vitamin Solution): Add components to distilled/deionized water and bring volume to 1.0L. Mix thoroughly. Sparge with 100% N_2. Autoclave for 15 min at 15 psi pressure–121°C.

Solution F:

Composition per 10.0mL:

$Na_2S \cdot 9H_2O$ 0.4g

Preparation of Solution F: Add $Na_2S \cdot 9H_2O$ to distilled/deionized water and bring volume to 10.0mL. Mix thoroughly. Sparge with 100% N_2. Autoclave for 15 min at 15 psi pressure–121°C.

Solution G (Selenite-Tungstate Solution):

Composition per liter:

NaOH 0.5g
$Na_2WO_4 \cdot 2H_2O$ 4.0mg
$Na_2SeO_3 \cdot 5H_2O$ 3.0mg

Preparation of Solution G (Selenite-Tungstate Solution): Add components to distilled/deionized water and bring volume to 1.0L. Mix thoroughly. Sparge with 100% N_2. Autoclave for 15 min at 15 psi pressure–121°C.

Solution H (Seven Vitamin Solution):

Composition per liter:

Pyridoxine·HCl 0.3g
Thiamine·HCl 0.2g
Nicotinic acid 0.2g
Calcium DL-pantothenate 0.1g
Vitamin B_{12} 0.1g
p-Aminobenzoic acid 80.0mg
Biotin 20.0mg

Preparation of Solution H (Seven Vitamin Solution): Add components to distilled/deionized water and bring volume to 1.0L. Mix thoroughly. Filter sterilize. Sparge with 100% N_2.

Preparation of Medium: Aseptically and anaerobically combine solution A with solution B, solution C, solution D, solution E, and solution F, in that order. Mix thoroughly. Anaerobically distribute into sterile tubes or flasks under 80% N_2 + 20% CO_2.

Use: For the cultivation of *Desulfovibrio asponium.*

Desulfovibrio baarsii Medium

Composition per 1009.0mL:

Solution A 850.0mL
Solution C 100.0mL
Solution G 20.0mL
Solution D 10.0mL
Solution E (Wolfe's vitamin solution) 10.0mL
Solution H 10.0mL
Solution F 6.6mL
Solution B (Trace elements solution SL-10) 1.0mL
Solution I 0.4mL

pH 7.6 ± 0.2 at 25°C

Solution A:

Composition per 920.0mL:

Na_2SO_4 .... 3.0g
NaCl 1.0g
KCl 0.5g
$MgCl_2 \cdot 6H_2O$ 0.4g
NH_4Cl 0.3g
KH_2PO_4 .... 0.2g
$CaCl_2 \cdot 2H_2O$ 0.15g
Resazurin 0.5mg

Preparation of Solution A: Prepare and dispense solution anaerobically under 80% N_2 + 20% CO_2. Add components to distilled/deionized water and bring volume to 920.0mL. Mix thoroughly. Gently heat and bring to boiling. Continue boiling until resazurin turns colorless, indicating reduction, and a pH of 6.0 is reached. Cap with rubber stoppers. Autoclave for 15 min at 15 psi pressure–121°C. Cool to 25°C.

Solution B (Trace Elements Solution SL-10):

Composition per liter:

$FeCl_2 \cdot 4H_2O$ 1.5g
$CoCl_2 \cdot 6H_2O$ 0.19g
$MnCl_2 \cdot 4H_2O$ 0.10g
$ZnCl_2$ 0.070g
$Na_2MoO_4 \cdot 2H_2O$ 0.036g
$NiCl_2 \cdot 6H_2O$ 0.024g
H_3BO_3 .... 6.0mg
$CuCl_2 \cdot 2H_2O$ 2.0mg
HCl (25% solution) 10.0mL

Preparation of Solution B (Trace Elements Solution SL-10): Add the $FeCl_2 \cdot 4H_2O$ to 10.0mL of HCl solution. Mix thoroughly. Bring volume to approximately 900.0mL with distilled/deionized water. Mix thoroughly. Adjust pH to 6.0 with NaOH. Bring volume to 1.0L with distilled/deionized water. Filter sterilize. Aseptically gas under 100% N_2 for 20 min.

Solution C:

Composition per 100.0mL:

$NaHCO_3$	5.0g

Preparation of Solution C: Add $NaHCO_3$ to distilled/deionized water and bring volume to 100.0mL. Mix thoroughly. Filter sterilize. Aseptically gas under 80% N_2 + 20% CO_2 for 20 min.

Solution D:

Composition per 10.0mL:

Sodium butyrate	0.7g
Sodium caproate	0.3g
Sodium octanoate	0.15g

Preparation of Solution D: Prepare and dispense solution anaerobically under 80% N_2 + 20% CO_2. Add components to distilled/deionized water and bring volume to 10.0mL. Mix thoroughly. Cap with a rubber stopper. Autoclave for 15 min at 15 psi pressure–121°C. Cool to 25°C.

Solution E (Wolfe's Vitamin Solution):

Composition per liter:

Pyridoxine·HCl	0.01g
Thiamine·HCl	5.0mg
Riboflavin	5.0mg
Nicotinic acid	5.0mg
Calcium pantothenate	5.0mg
p-Aminobenzoic acid	5.0mg
Thioctic acid	5.0mg
Biotin	2.0mg
Folic acid	2.0mg
Cyanocobalamin	0.1mg

Preparation of Solution E (Wolfe's Vitamin Solution): Add components to distilled/deionized water and bring volume to 1.0L. Mix thoroughly. Filter sterilize. Aseptically gas under 100% N_2 for 20 min.

Solution F:

Composition per 6.6mL:

$AlCl_3 \cdot 6H_2O$ (4.9% solution)	5.0mL
Na_2CO_3 (10.6% solution)	1.6mL

Preparation of Solution F: Combine both solutions. Mix thoroughly. Gas with 100% N_2. Cap with a rubber stopper. Autoclave for 15 min at 15 psi pressure–121°C. Cool to 25°C.

Solution G:

Composition per 10.0mL:

Rumen fluid, clarified	20.0mL

Preparation of Solution G: Gas rumen fluid under 100% N_2 for 20 min. Cap with a rubber stopper. Autoclave for 15 min at 15 psi pressure–121°C. Cool to 25°C.

Solution H:

Composition per 10.0mL:

$Na_2S \cdot 9H_2O$	0.4g

Preparation of Solution H: Add $Na_2S \cdot 9H_2O$ to distilled/deionized water and bring volume to 10.0mL. Gas under 100% N_2 for 20 min. Cap with a rubber stopper. Autoclave for 15 min at 15 psi pressure–121°C. Cool to 25°C.

Solution I:

Composition per 10.0mL:

$Na_2S_2O_4$	0.5g

Preparation of Solution I: Add $Na_2S_2O_4$ to distilled/deionized water and bring volume to 10.0mL. Mix thoroughly. Filter sterilize. Aseptically gas under 100% N_2 for 20 min. Prepare solution freshly.

Preparation of Medium: To 850.0mL of cooled, sterile solution A, aseptically and anaerobically add in the following order: 1.0mL of sterile solution B, 100.0mL of sterile solution C, 10.0mL of sterile solution D, 10.0mL of sterile solution E, 6.6mL of sterile solution F, 20.0mL of sterile solution G, and 10.0mL of sterile solution H. Mix thoroughly. Immediately prior to inoculation, aseptically and anaerobically add 0.4mL of sterile solution I. Mix thoroughly. Aseptically and anaerobically distribute into sterile tubes or flasks.

Use: For the cultivation and maintenance of *Desulfovibrio baarsii.*

Desulfovibrio Brackish Medium (DSMZ Medium 410)

Composition per liter:

Soultion A	980.0mL
Solution B	10.0mL
Solution C	10.0mL

Solution A:

Composition per 980.0mL:

NaCl	10.0g
$MgSO_4 \cdot 7H_2O$	2.0g
DL-Na-lactate	2.0g
Yeast extract	1.0g
NH_4Cl	1.0g
Na_2SO_4	1.0g
K_2HPO_4	0.5g
$CaCl_2 \cdot 2H_2O$	0.1g
Resazurin	1.0mg

pH 7.8 ± 0.2 at 25°C

Preparation of Solution A: Add components to distilled/deionized water and bring volume to 980.0mL. Mix thoroughly.

Solution B:

Composition per 10.0mL:

$FeSO_4 \cdot 7H_2O$	0.5g

Preparation of Solution B: Add $FeSO_4 \cdot 7H_2O$ to distilled/deionized water and bring volume to 10.0mL. Mix thoroughly.

Solution C:

Composition per 10.0mL:

Na-thioglycolate	0.1g
Ascorbic acid	0.1g

Preparation of Solution C: Add components to distilled/deionized water and bring volume to 10.0mL. Mix thoroughly.

Preparation of Medium: Bring solution A to a boil for a few minutes. Cool to room temperature while gassing with oxygen-free N_2 gas. Add solutions B and C. Adjust pH to 7.8 with NaOH. Immediately distribute under N_2 into anaerobic tubes. During distribution continuously swirl the medium to keep the grey precipitate suspended. Autoclave for 15 min at 15 psi pressure–121°C.

Use: For the cultivation of *Desulfovibrio giganteus (Desulfobacter giganteus).*

Desulfovibrio carbinolicus Medium

Composition per 1001.0mL:

Solution A	870.0mL
Solution C	100.0mL

Solution D	10.0mL
Solution E (Vitamin solution)	10.0mL
Solution F	10.0mL
Solution B (Trace elements solution SL-10)	1.0mL

pH 7.1–7.4 at 25°C

Solution A:

Composition per 870.0mL:

Na_2SO_4	3.0g
NaCl	1.0g
KCl	0.5g
$MgCl_2·6H_2O$	0.4g
NH_4Cl	0.3g
KH_2PO_4	0.2g
$CaCl_2·2H_2O$	0.15g
Yeast extract	0.1g
Casamino acids	0.1g
Resazurin	1.0mg

Preparation of Solution A: Add components to distilled/deionized water and bring volume to 870.0mL. Mix thoroughly. Gently heat and bring to boiling. Continue boiling for 3–4 min. Allow to cool to room temperature while gassing under 80% N_2 + 20% CO_2. Continue gassing until pH reaches below 6.0. Seal the flask under 80% N_2 + 20% CO_2. Autoclave for 15 min at 15 psi pressure–121°C.

Solution B (Trace Elements Solution SL-10):

Composition per liter:

$FeCl_2·4H_2O$	1.5g
$CoCl_2·6H_2O$	190.0mg
$MnCl_2·4H_2O$	100.0mg
$ZnCl_2$	70.0mg
$Na_2MoO_4·2H_2O$	36.0mg
$NiCl_2·6H_2O$	24.0mg
H_3BO_3	6.0mg
$CuCl_2·2H_2O$	2.0mg
HCl (25% solution)	10.0mL

Preparation of Solution B (Trace Elements Solution SL-10): Add $FeCl_2·4H_2O$ to 10.0mL of HCl solution. Mix thoroughly. Add distilled/deionized water and bring volume to 1.0L. Add remaining components. Mix thoroughly. Gas under 100% N_2. Autoclave for 15 min at 15 psi pressure–121°C.

Solution C:

Composition per 100.0mL:

$NaHCO_3$	5.0g

Preparation of Solution C: Add $NaHCO_3$ to distilled/deionized water and bring volume to 100.0mL. Mix thoroughly. Filter sterilize. Gas under 80% N_2 + 20% CO_2.

Solution D:

Composition per 10.0mL:

Sodium propionate	0.7g

Preparation of Solution D: Add sodium propionate to distilled/deionized water and bring volume to 10.0mL. Mix thoroughly. Gas under 100% N_2. Autoclave for 15 min at 15 psi pressure–121°C.

Solution E (Vitamin Solution):

Composition per liter:

Pyridoxine·HCl	10.0mg
Calcium DL-pantothenate	5.0mg
Lipoic acid	5.0mg
Nicotinic acid	5.0mg
p-Aminobenzoic acid	5.0mg
Riboflavin	5.0mg
Thiamine·HCl	5.0mg
Biotin	2.0mg
Folic acid	2.0mg
Vitamin B_{12}	0.1mg

Preparation of Solution E (Vitamin Solution): Add components to distilled/deionized water and bring volume to 1.0L. Mix thoroughly. Gas under 100% N_2. Autoclave for 15 min at 15 psi pressure–121°C.

Solution F:

Composition per 10.0mL:

$Na_2S·9H_2O$	0.4g

Preparation of Solution F: Add $Na_2S·9H_2O$ to distilled/deionized water and bring volume to 10.0mL. Mix thoroughly. Gas under 100% N_2. Autoclave for 15 min at 15 psi pressure–121°C.

Preparation of Medium: Aseptically and anaerobically combine 870.0mL of sterile solution A with 1.0mL of sterile solution B, 100.0mL of sterile solution C, 10.0mL of sterile solution D, 10.0mL of sterile solution E, and 10.0mL of sterile solution F, in that order. Mix thoroughly. Anaerobically distribute into sterile tubes or flasks under 100% N_2.

Use: For the cultivation and maintenance of *Desulfovibrio carbinolicus*.

Desulfovibrio Choline Medium (DSMZ Medium 272)

Composition per liter:

Soultion A	980.0mL
Solution B	10.0mL
Solution C	10.0mL

Solution A:

Composition per 980.0mL:

Choline hydrochloride	5.0g
$MgSO_4·7H_2O$	2.0g
Yeast extract	1.0g
NH_4Cl	1.0g
Na_2SO_4	1.0g
K_2HPO_4	0.5g
$CaCl_2·2H_2O$	0.1g
Resazurin	1.0mg

pH 7.8 ± 0.2 at 25°C

Preparation of Solution A: Add components to distilled/deionized water and bring volume to 980.0mL. Mix thoroughly.

Solution B:

Composition per 10.0mL:

$FeSO_4·7H_2O$	0.5g

Preparation of Solution B: Add $FeSO_4·7H_2O$ to distilled/deionized water and bring volume to 10.0mL. Mix thoroughly.

Solution C:

Composition per 10.0mL:

Na-thioglycolate	0.1g
Ascorbic acid	0.1g

Preparation of Solution C: Add components to distilled/deionized water and bring volume to 10.0mL. Mix thoroughly.

Preparation of Medium: Bring solution A to a boil for a few minutes. Cool to room temperature while gassing with oxygen-free N_2 gas. Add solutions B and C. Adjust pH to 7.8 with NaOH. Immediately distribute under N_2 into anaerobic tubes. During distribution continuously swirl the medium to keep the grey precipitate suspended. Autoclave for 15 min at 15 psi pressure–121°C.

Use: For the cultivation of *Desulfovibrio* spp.

Desulfovibrio gabonensis Medium

Composition per 1002.0mL:

NaCl	50.0g
$MgCl_2·6H_2O$	3.3g
Na_2SO_4	3.0g
$MgSO_4·7H_2O$	1.6g
KCl	0.3g
NH_4Cl	0.3g
KH_2PO_4	0.2g
$CaCl_2·2H_2O$	0.1g
Yeast extract	0.1g
Resazurin	0.5mg
Sodium lactate solution	10.0mL
$NaHCO_3$ solution	10.0mL
$Na_2S·9H_2O$ solution	10.0mL
Trace elements solution SL-10 with EDTA	1.0mL
Seven vitamin solution	1.0mL

pH 7.0–7.2 at 25°C

Sodium Lactate Solution:

Composition per 10.0mL:

Sodium lactate	2.5g

Preparation of Sodium Lactate Solution: Add sodium lactate to distilled/deionized water and bring volume to 10.0mL. Mix thoroughly. Sparge with 100% N_2. Autoclave for 15 min at 15 psi pressure–121°C.

$NaHCO_3$ Solution:

Composition per 10.0mL:

$NaHCO_3$	2.5g

Preparation of $NaHCO_3$ Solution: Add $NaHCO_3$ to distilled/deionized water and bring volume to 10.0mL. Mix thoroughly. Sparge with 80% N_2 + 20% CO_2. Autoclave for 15 min at 15 psi pressure–121°C.

$Na_2S·9H_2O$ Solution:

Composition per 10.0mL:

$Na_2S·9H_2O$	0.2g

Preparation of $Na_2S·9H_2O$ Solution: Add $Na_2S·9H_2O$ to distilled/deionized water and bring volume to 10.0mL. Mix thoroughly. Sparge with 100% N_2. Autoclave for 15 min at 15 psi pressure–121°C.

Trace Elements Solution SL-10 with EDTA:

Composition per liter:

Disodium EDTA	3.0g
$FeCl_2·4H_2O$	1.5g
$CoCl_2·6H_2O$	190.0mg
$MnCl_2·4H_2O$	100.0mg
$ZnCl_2$	70.0mg
$Na_2MoO_4·2H_2O$	36.0mg
$NiCl_2·6H_2O$	24.0mg
H_3BO_3	6.0mg
$CuCl_2·2H_2O$	2.0mg

Preparation of Trace Elements Solution SL-10 with EDTA: Add components to distilled/deionized water and bring volume to 1.0L. Mix thoroughly. Adjust pH to 6.0.

Seven Vitamin Solution:

Composition per liter:

Pyridoxine·HCl	0.3g
Thiamine·HCl	0.2g
Nicotinic acid	0.2g
Calcium DL-pantothenate	0.1g
Vitamin B_{12}	0.1g
p-Aminobenzoic acid	80.0mg
Biotin	20.0mg

Preparation of Seven Vitamin Solution: Add components to distilled/deionized water and bring volume to 1.0L. Mix thoroughly. Filter sterilize. Sparge with 100% N_2.

Preparation of Medium: Prepare and dispense medium under 80% N_2% + 20% CO_2. Add components, except sodium lactate solution, $NaHCO_3$ solution, $Na_2S·9H_2O$ solution, trace elements solution SL-10 with EDTA, and seven vitamin solution, to distilled/deionized water and bring volume to 970.0mL. Mix thoroughly. Gently heat and bring to boiling. Continue boiling for 3 min. Cool to room temperature while sparging with 80% N_2 + 20% CO_2. Anaerobically distribute 9.7mL volumes into anaerobic tubes. Autoclave for 15 min at 15 psi pressure–121°C. Aseptically add 0.1mL of sterile sodium lactate solution, 0.1mL of sterile $NaHCO_3$ solution, 0.1mL of sterile $Na_2S·9H_2O$ solution, 0.01mL of sterile trace elements solution SL-10 with EDTA, and 0.01mL of sterile seven vitamin solution to each tube. Mix thoroughly.

Use: For the cultivation of *Desulfovibrio gabonensis.*

Desulfovibrio giganteus Medium

Composition per 1001.0mL:

Solution A	870.0mL
Solution C	100.0mL
Solution D	10.0mL
Solution E (Vitamin solution)	10.0mL
Solution F	10.0mL
Solution B (Trace elements solution SL-10)	1.0mL

pH 7.5 ± 0.2 at 25°C

Solution A:

Composition per 870.0mL:

NaCl	20.0g
Na_2SO_4	3.0g
KCl	0.5g
$MgCl_2·6H_2O$	0.4g
NH_4Cl	0.3g
KH_2PO_4	0.2g
$CaCl_2·2H_2O$	0.15g
Resazurin	1.0mg

Preparation of Solution A: Add components to distilled/deionized water and bring volume to 870.0mL. Mix thoroughly. Gently heat and bring to boiling. Continue boiling for 3–4 min. Allow to cool to room temperature while gassing under 80% N_2 + 20% CO_2. Continue gassing until pH reaches below 6.0. Seal the flask under 80% N_2 + 20% CO_2. Autoclave for 15 min at 15 psi pressure–121°C.

Solution B (Trace Elements Solution SL-10):

Composition per liter:

$FeCl_2·4H_2O$	1.5g
$CoCl_2·6H_2O$	190.0mg
$MnCl_2·4H_2O$	100.0mg
$ZnCl_2$	70.0mg
$Na_2MoO_4·2H_2O$	36.0mg
$NiCl_2·6H_2O$	24.0mg

H_3BO_3 .... 6.0mg
$CuCl_2 \cdot 2H_2O$ 2.0mg
HCl (25% solution) 10.0mL

Preparation of Solution B (Trace Elements Solution SL-10): Add $FeCl_2 \cdot 4H_2O$ to 10.0mL of HCl solution. Mix thoroughly. Add distilled/deionized water and bring volume to 1.0L. Add remaining components. Mix thoroughly. Gas under 100% N_2. Autoclave for 15 min at 15 psi pressure–121°C.

Solution C:
Composition per 100.0mL:
$NaHCO_3$ 5.0g

Preparation of Solution C: Add $NaHCO_3$ to distilled/deionized water and bring volume to 100.0mL. Mix thoroughly. Filter sterilize. Gas under 80% N_2 + 20% CO_2.

Solution D:
Composition per 10.0mL:
Sodium lactate 1.5g

Preparation of Solution D: Add sodium acetate to distilled/deionized water and bring volume to 10.0mL. Mix thoroughly. Gas under 100% N_2. Autoclave for 15 min at 15 psi pressure–121°C.

Solution E (Vitamin Solution):
Composition per liter:
Pyridoxine·HCl 10.0mg
Calcium DL-pantothenate 5.0mg
Lipoic acid 5.0mg
Nicotinic acid 5.0mg
p-Aminobenzoic acid 5.0mg
Riboflavin 5.0mg
Thiamine·HCl 5.0mg
Biotin 2.0mg
Folic acid 2.0mg
Vitamin B_{12} 0.1mg

Preparation of Solution E (Vitamin Solution): Add components to distilled/deionized water and bring volume to 1.0L. Mix thoroughly. Gas under 100% N_2. Autoclave for 15 min at 15 psi pressure–121°C.

Solution F:
Composition per 10.0mL:
$Na_2S \cdot 9H_2O$ 0.4g

Preparation of Solution F: Add $Na_2S \cdot 9H_2O$ to distilled/deionized water and bring volume to 10.0mL. Mix thoroughly. Gas under 100% N_2. Autoclave for 15 min at 15 psi pressure–121°C.

Preparation of Medium: Aseptically and anaerobically combine solution A with solution B, solution C, solution D, solution E, and solution F, in that order. Mix thoroughly. Anaerobically distribute into sterile tubes or flasks under 80% N_2 + 20% CO_2.

Use: For the cultivation and maintenance of *Desulfovibrio giganteus.*

Desulfovibrio gigas Medium

Composition per 1001.0mL:
Solution A 950.0mL
Solution B 40.0mL
Solution C 6.0mL
Solution D (Vitamin solution) 5.0mL
pH 7.2 ± 0.2 at 25°C

Solution A:
Composition per 950.0mL:
Na_2SO_4 .... 2.0g
Sodium (L)-lactate 2.0g
KH_2PO_4 .... 1.0g
NH_4Cl 0.5g
$MgSO_4 \cdot 7H_2O$ 0.4g
$CaCl_2 \cdot 2H_2O$ 0.1g
H_2SO_4 (1*M* solution) 1.0mL
Trace elements solution SL-6 .... 1.0mL

Trace Elements Solution SL-6:
Composition per liter:
$MnCl_2 \cdot 4H_2O$ 0.5g
H_3BO_3 .... 0.3g
$CoCl_2 \cdot 6H_2O$ 0.2g
$ZnSO_4 \cdot 7H_2O$ 0.1g
$Na_2MoO_4 \cdot 2H_2O$ 0.03g
$NiCl_2 \cdot 6H_2O$ 0.02g
$CuCl_2 \cdot 2H_2O$ 0.01g

Preparation of Trace Elements Solution SL-6: Add components to distilled/deionized water and bring volume to 1.0L. Mix thoroughly. Gas under 100% N_2. Autoclave for 15 min at 15 psi pressure–121°C.

Preparation of Solution A: Add components to distilled/deionized water and bring volume to 950.0mL. Mix thoroughly. Gently heat and bring to boiling. Continue boiling for 3–4 min. Allow to cool to room temperature while gassing under 80% N_2 + 20% CO_2. Seal the flask under 80% N_2 + 20% CO_2. Autoclave for 15 min at 15 psi pressure–121°C.

Solution B:
Composition per 40.0mL:
$NaHCO_3$ 2.0g

Preparation of Solution B: Add $NaHCO_3$ to distilled/deionized water and bring volume to 40.0mL. Mix thoroughly. Filter sterilize. Gas under 80% N_2 + 20% CO_2.

Solution C:
Composition per 10.0mL:
$Na_2S \cdot 9H_2O$ 0.5g

Preparation of Solution C: Add $Na_2S \cdot 9H_2O$ to distilled/deionized water and bring volume to 10.0mL. Mix thoroughly. Gas under 100% N_2. Autoclave for 15 min at 15 psi pressure–121°C.

Solution D (Vitamin Solution):
Composition per liter:
Pyridoxine·HCl 62.5g
Nicotinic acid 25.0mg
p-Aminobenzoic acid 12.5mg
Thiamine·HCl 12.5mg
Calcium DL-pantothenate 6.5mg
Biotin 2.5mg

Preparation of Solution D (Vitamin Solution): Add components to distilled/deionized water and bring volume to 1.0L. Mix thoroughly. Gas under 100% N_2. Autoclave for 15 min at 15 psi pressure–121°C.

Preparation of Medium: Aseptically and anaerobically combine 950.0mL of sterile solution A with 40.0mL of sterile solution B, 6.0mL of sterile solution C, and 5.0mL of sterile solution D. Adjust pH to 7.2. Mix thoroughly. Anaerobically distribute into sterile tubes or flasks under 80% N_2 + 20% CO_2

Use: For the cultivation and maintenance of *Desulfovibrio gigas*.

Desulfovibrio halophilus Medium

Composition per 1154.0mL:

Solution A	1.0L
Solution H	67.0mL
Solution D	50.0mL
Soultion I	13.0mL
Solution E	10.0mL
Solution G	10.0mL
Solution C (Selenite-tungstate solution)	2.0mL
Solution B (Trace elements solution SL-10)	1.0mL
Solution F	1.0mL

pH 6.8 ± 0.2 at 25°C

Solution A:

Composition per liter:

Na_2SO_4	4.0g
NH_4Cl	0.25g
KH_2PO_4	0.2g
$CaCl_2 \cdot 2H_2O$	0.1g

Preparation of Solution A: Add components to distilled/deionized water and bring volume to 1.0L. Mix thoroughly. Gently heat and bring to boiling. Continue boiling for 3–4 min. Allow to cool to room temperature while gassing under 80% N_2 + 20% CO_2. Continue gassing until pH reaches below 6.0. Seal the flask under 80% N_2 + 20% CO_2. Autoclave for 15 min at 15 psi pressure–121°C.

Solution B (Trace Elements Solution SL-10):

Composition per liter:

$FeCl_2 \cdot 4H_2O$	1.5g
$CoCl_2 \cdot 6H_2O$	190.0mg
$MnCl_2 \cdot 4H_2O$	100.0mg
$ZnCl_2$	70.0mg
$Na_2MoO_4 \cdot 2H_2O$	36.0mg
$NiCl_2 \cdot 6H_2O$	24.0mg
H_3BO_3	6.0mg
$CuCl_2 \cdot 2H_2O$	2.0mg
HCl (25% solution)	10.0mL

Preparation of Solution B (Trace Elements Solution SL-10): Add $FeCl_2 \cdot 4H_2O$ to 10.0mL of HCl solution. Mix thoroughly. Add distilled/deionized water and bring volume to 1.0L. Add remaining components. Mix thoroughly. Gas under 100% N_2. Autoclave for 15 min at 15 psi pressure–121°C.

Solution C (Selenite-Tungstate Solution):

Composition per liter:

NaOH	0.5g
$Na_2WO_4 \cdot 2H_2O$	4.0mg
$Na_2SeO_3 \cdot 5H_2O$	3.0mg

Preparation of Solution C (Selenite-Tungstate Solution): Add components to distilled/deionized water and bring volume to 1.0L. Mix thoroughly. Gas under 100% N_2. Autoclave for 15 min at 15 psi pressure–121°C.

Solution D:

Composition per 50.0mL:

$NaHCO_3$	2.5g

Preparation of Solution D: Add $NaHCO_3$ to distilled/deionized water and bring volume to 50.0mL. Mix thoroughly. Gas under 80% N_2 + 20% CO_2. Autoclave for 15 min at 15 psi pressure–121°C.

Solution E:

Composition per liter:

Pyridoxine·HCl	10.0mg
Calcium DL-pantothenate	5.0mg
Lipoic acid	5.0mg
Nicotinic acid	5.0mg
p-Aminobenzoic acid	5.0mg
Riboflavin	5.0mg
Thiamine·HCl	5.0mg
Biotin	2.0mg
Folic acid	2.0mg
Vitamin B_{12}	0.1mg

Preparation of Solution E: Add components to distilled/deionized water and bring volume to 1.0L. Mix thoroughly. Gas under 100% N_2. Filter sterilize.

Solution F:

Composition per 10.0mL:

Vitamin B_{12}	0.5mg

Preparation of Solution F: Add vitamin B_{12} to distilled/deionized water and bring volume to 10.0mL. Mix thoroughly. Gas under 100% N_2. Filter sterilize.

Solution G:

Composition per 80.0mL:

Sodium-(L)-lactate	2.25g

Preparation of Solution G: Add sodium-(L)-lactate to distilled/deionized water and bring volume to 80.0mL. Mix thoroughly. Gas under 100% N_2. In a closed bottle, heat in a boiling water bath. Shake until stearic acid dissolves. Autoclave for 15 min at 15 psi pressure–121°C. On storage, solution will solidify and should be remelted before use.

Solution H:

Composition per liter:

NaCl	70.4g
$MgCl_2 \cdot 6H_2O$	3.0g
$CaCl_2 \cdot 2H_2O$	2.2g

Preparation of Solution H: Add components to distilled/deionized water and bring volume to 80.0mL. Mix thoroughly. Gas under 100% N_2. Autoclave for 15 min at 15 psi pressure–121°C.

Solution I:

Composition per 20.0mL:

$Na_2S \cdot 9H_2O$	0.15g

Preparation of Solution I: Add $Na_2S \cdot 9H_2O$ to distilled/deionized water and bring volume to 10.0mL. Mix thoroughly. Gas under 100% N_2. Autoclave for 15 min at 15 psi pressure–121°C.

Preparation of Medium: To 1.0L of sterile solution A, add in the following order: 1.0mL of sterile solution B, 2.0mL of sterile solution C, 50.0mL of sterile solution D, 10.0mL of sterile solution E, 1.0mL of sterile solution F, 10.0mL of sterile solution G, 67.0mL of sterile solution H, and 13.0mL of sterile solution I. Mix thoroughly. Final pH of medium should be 7.2. Prior to inoculation, add 10.0–20.0mg of sodium dithionate to 1.0L of medium.

Use: For the cultivation and maintenance of *Desulfovibrio halophilus*.

Desulfovibrio halophilus Medium

Composition per liter:

NaCl	70.0g
$MgCl_2 \cdot 6H_2O$	3.0g
Na_2SO_4	3.0g

$NaHCO_3$	2.5g
KCl	0.3g
NH_4Cl	0.3g
KH_2PO_4	0.2g
$Na_2S{\cdot}9H_2O$	0.2g
$CaCl_2{\cdot}2H_2O$	0.15g
Wolfe's vitamin solution	10.0mL
Sodium lactate	3.7mL
Trace elements solution SL-10	1.0mL

pH 6.9–7.1 at 25°C

Trace Elements Solution SL-10:
Composition per liter:

$FeCl_2{\cdot}4H_2O$	1.5g
$CoCl_2{\cdot}6H_2O$	190.0mg
$MnCl_2{\cdot}4H_2O$	100.0mg
$ZnCl_2$	70.0mg
$Na_2MoO_4{\cdot}2H_2O$	36.0mg
$NiCl_2{\cdot}6H_2O$	24.0mg
H_3BO_3	6.0mg
$CuCl_2{\cdot}2H_2O$	2.0mg
HCl (25% solution)	10.0mL

Preparation of Trace Elements Solution SL-10: Add $FeCl_2{\cdot}4H_2O$ to 10.0mL of HCl solution. Mix thoroughly. Add distilled/deionized water and bring volume to 1.0L. Add remaining components. Mix thoroughly.

Wolfe's Vitamin Solution:
Composition per liter:

Pyridoxine·HCl	10.0mg
p-Aminobenzoic acid	5.0mg
Lipoic acid	5.0mg
Nicotinic acid	5.0mg
Riboflavin	5.0mg
Thiamine·HCl	5.0mg
Calcium DL-pantothenate	5.0mg
Biotin	2.0mg
Folic acid	2.0mg
Vitamin B_{12}	0.1mg

Preparation of Wolfe's Vitamin Solution: Add components to distilled/deionized water and bring volume to 1.0L. Mix thoroughly.

Preparation of Medium: Prepare and dispense medium under 90% N_2 + 10% CO_2. Add components, except $NaHCO_3$ and $Na_2S{\cdot}9H_2O$, to distilled/deionized water and bring volume to 1.0L. Mix thoroughly. Gently heat and bring to boiling. Continue boiling for 3 min. Cool to room temperature while sparging with 90% N_2 + 10% CO_2. Add $NaHCO_3$ and $Na_2S{\cdot}9H_2O$. Mix thoroughly. Anaerobically distribute into tubes. Autoclave for 15 min at 15 psi pressure–121°C.

Use: For the cultivation of *Desulfovibrio halophilus.*

Desulfovibrio inopinatus Medium (DSMZ Medium 799)

Composition per 1008.0mL:

Solution A	870.0mL
Solution C	100.0mL
Solution D	10.0mL
Solution E (Vitamin solution)	10.0mL
Solution F	10.0mL
Yeast extract solution	5.0mL
Solution B (Trace elements solution SL-10)	1.0mL
Seven vitamin solution	1.0mL
Selenite-tungstate solution	1.0mL

pH 7.1–7.4 at 25°C

Solution A:
Composition per 870.0mL:

NaCl	7.0g
Na_2SO_4	3.0g
$MgCl_2{\cdot}6H_2O$	1.3g
KCl	0.5g
NH_4Cl	0.3g
KH_2PO_4	0.2g
$CaCl_2{\cdot}2H_2O$	0.15g
Resazurin	1.0mg

Preparation of Solution A: Add components to distilled/deionized water and bring volume to 870.0mL. Mix thoroughly.

Solution B (Trace Elements Solution SL-10):
Composition per liter:

$FeCl_2{\cdot}4H_2O$	1.5g
$CoCl_2{\cdot}6H_2O$	190.0mg
$MnCl_2{\cdot}4H_2O$	100.0mg
$ZnCl_2$	70.0mg
$Na_2MoO_4{\cdot}2H_2O$	36.0mg
$NiCl_2{\cdot}6H_2O$	24.0mg
H_3BO_3	6.0mg
$CuCl_2{\cdot}2H_2O$	2.0mg
HCl (25% solution)	10.0mL

Preparation of Solution B (Trace Elements Solution SL-10): Add $FeCl_2{\cdot}4H_2O$ to 10.0mL of HCl solution. Mix thoroughly. Add distilled/deionized water and bring volume to 1.0L. Add remaining components. Mix thoroughly. Sparge with 100% N_2. Autoclave for 15 min at 15 psi pressure–121°C.

Solution C:
Composition per 100.0mL:

$NaHCO_3$	5.0g

Preparation of Solution C: Add $NaHCO_3$ to distilled/deionized water and bring volume to 100.0mL Mix thoroughly. Filter sterilize. Flush with 80% N_2 + 20% CO_2 to remove dissolved oxygen.

Solution D:
Composition per 10.0mL:

Na-pyruvate	2.5g

Preparation of Solution D: Add Na-pyruvate to distilled/deionized water and bring volume to 10.0mL. Mix thoroughly. Sparge with 100% N_2. Autoclave for 15 min at 15 psi pressure–121°C.

Solution E (Vitamin Solution):
Composition per liter:

Pyridoxine-HCl	10.0mg
Thiamine-HCl·$2H_2O$	5.0mg
Riboflavin	5.0mg
Nicotinic acid	5.0mg
D-Ca-pantothenate	5.0mg
p-Aminobenzoic acid	5.0mg
Lipoic acid	5.0mg
Biotin	2.0mg
Folic acid	2.0mg
Vitamin B_{12}	0.10mg

Preparation of Solution E (Vitamin Solution): Add components to distilled/deionized water and bring volume to 1.0L. Mix thor-

oughly. Sparge with 100% N_2. Autoclave for 15 min at 15 psi pressure–121°C.

Solution F:

Composition per 10.0mL:

$Na_2S \cdot 9H_2O$ 0.4g

Preparation of Solution F: Add $Na_2S \cdot 9H_2O$ to distilled/deionized water and bring volume to 10.0mL. Mix thoroughly. Sparge with 100% N_2. Autoclave for 15 min at 15 psi pressure–121°C.

Seven Vitamin Solution:

Composition per liter:

Pyridoxine hydrochloride 300.0mg
Thiamine-HCl·$2H_2O$ 200.0mg
Nicotinic acid 200.0mg
Vitamin B_{12} 100.0mg
Calcium pantothenate 100.0mg
p-Aminobenzoic acid 80.0mg
D(+)-Biotin 20.0mg

Preparation of Seven Vitamin Solution: Add components to distilled/deionized water and bring volume to 1.0L. Sparge with 100% N_2. Mix thoroughly. Filter sterilize.

Selenite-Tungstate Solution:

Composition per liter:

NaOH 0.5g
$Na_2WO_4 \cdot 2H_2O$ 4.0mg
$Na_2SeO_3 \cdot 5H_2O$ 3.0mg

Preparation of Selenite-Tungstate Solution: Add components to distilled/deionized water and bring volume to 1.0L. Mix thoroughly. Sparge with 100% N_2. Filter sterilize.

Yeast Extract Solution:

Composition per 10.0mL:

Yeast extract 1.0g

Preparation of Yeast Extract Solution: Add yeast extract to distilled/deionized water and bring volume to 10.0mL. Mix thoroughly. Sparge with 100% N_2. Autoclave under 100% N_2 for 15 min at 15 psi pressure–121°C. Cool to room temperature.

Preparation of Medium: Gently heat solution A and bring to boiling. Boil solution A for a few minutes. Cool to room temperature. Gas with 80% N_2 + 20% CO_2 gas mixture to reach a pH below 6. Autoclave for 15 min at 15 psi pressure–121°C. Cool to room temperature. Sequentially add 1.0mL solution B, 100.0mL solution C, 10.0mL solution D, 10.0mL solution E, 10.0mL solution F, 5.0mL yeast extract solution, 1.0mL selenite-tungstate solution, and 1.0ml seven vitamin solution. Distribute aseptically and anaerobically under 80% N_2 + 20% CO_2 into sterile tubes or bottles.

Use: For the cultivation of *Desulfovibrio inopinatus.*

Desulfovibrio magneticus Medium (DSMZ Medium 896)

Composition per liter:

Na-fumarate 0.58g
Na-pyruvate 0.44g
KH_2PO_4 .... 0.2g
NH_4Cl 0.06g
Cysteine-HCl·H_2O 0.05g
Vitamin solution 8.0mL
Trace elements solution 4.0mL
Fe(III)quinate solution 2.0mL

pH 7.0 ± 0.2 at 25°C

Trace Elements Solution:

Composition per liter:

$MgSO_4 \cdot 7H_2O$ 3.0g
Nitrilotriacetic acid 1.5g
NaCl 1.0g
$MnSO_4 \cdot 2H_2O$ 0.5g
$CoSO_4 \cdot 7H_2O$ 0.18g
$ZnSO_4 \cdot 7H_2O$ 0.18g
$CaCl_2 \cdot 2H_2O$ 0.1g
$FeSO_4 \cdot 7H_2O$ 0.1g
$NiCl_2 \cdot 6H_2O$ 0.025g
$KAl(SO_4)_2 \cdot 12H_2O$ 0.02g
H_3BO_3 .... 0.01g
$Na_2MoO_4 \cdot 4H_2O$ 0.01g
$CuSO_4 \cdot 5H_2O$ 0.01g
$Na_2SeO_3 \cdot 5H_2O$ 0.3mg

Preparation of Trace Elements Solution: Add nitrilotriacetic acid to 500.0mL of distilled/deionized water. Dissolve by adjusting pH to 6.5 with KOH. Add remaining components. Add distilled/deionized water to 1.0L. Mix thoroughly.

Vitamin Solution:

Composition per liter:

Pyridoxine-HCl 10.0mg
Thiamine-HCl·$2H_2O$ 5.0mg
Riboflavin 5.0mg
Nicotinic acid 5.0mg
D-Ca-pantothenate 5.0mg
p-Aminobenzoic acid 5.0mg
Lipoic acid 5.0mg
Biotin 2.0mg
Folic acid 2.0mg
Vitamin B_{12} 0.1mg

Preparation of Vitamin Solution: Add components to distilled/deionized water and bring volume to 1.0L. Mix thoroughly. Sparge with 80% H_2 + 20% CO_2. Filter sterilize.

Ferric Quinate Solution:

Composition per 100.0mL:

$FeCl_3 \cdot 6H_2O$ 0.45g
Quinic acid 0.19g

Preparation of Ferric Quinate Solution: Add components to distilled/deionized water and bring volume to 100.0mL. Sparge with N_2. Autoclave for 15 min at 15 psi pressure–121°C. Cool to 25°C.

Preparation of Medium: Add components, except vitamin solution and ferric quinate solution, to distilled/deionized water and bring volume to 990.0mL. Purge medium with N_2 gas for 10 min. Mix thoroughly. Autoclave for 15 min at 15 psi pressure–121°C. Cool to 25°C. Aseptically and anerobically add 8.0mL vitamin solution and 2.0mL ferric quinate solution. Mix thoroughly. Adjust pH to 7.0. Purge medium with N_2 gas for 10 min. Under the same atmosphere, aseptically distribute medium to sterile tubes or bottles.

Use: For the cultivation of *Desulfovibrio magneticus.*

Desulfovibrio Marine Medium (DSMZ Medium 163)

Composition per liter:

Solution A 980.0mL
Solution B 10.0mL
Solution C 10.0mL

pH 7.8 ± 0.2 at 25°C

Solution A:

Composition per 980.0mL:

NaCl 25.0g
DL-Na-lactate 2.0g
$MgSO_4 \cdot 7H_2O$ 2.0g
Yeast extract 1.0g
NH_4Cl 1.0g
Na_2SO_4 .... 1.0g
K_2HPO_4 .... 0.5g
$CaCl_2 \cdot 2H_2O$ 0.1g
Resazurin 1.0mg

Preparation of Solution A: Add components to 980.0mL distilled/deionized water. Mix thoroughly.

Solution B:

Composition per 10.0mL:

$FeSO_4 \cdot 7H_2O$ 0.5g

Preparation of Solution B: Add $FeSO_4 \cdot 7H_2O$ to 10.0mL distilled/deionized water. Mix thoroughly.

Solution C:

Composition per 10.0mL:

Na-thioglycolate 0.1g
Ascorbic acid 0.1g

Preparation of Solution C: Add components to 10.0mL distilled/deionized water. Mix thoroughly.

Preparation of Medium: Bring solution A to the boil for a few minutes. Cool to room temperature while gassing with oxygen-free N_2 gas. Add solutions B and C. Mix thoroughly. Adjust pH to 7.8 with NaOH. Distribute under N_2 into anaerobic tubes. During distribution continuously swirl the medium to keep the grey precipitate suspended. Autoclave for 15 min at 15 psi pressure–121°C.

Use: For the cultivation of *Desulfovibrio vulgaris, Desulfovibrio desulfuricans, Desulfovibrio senezii,* and *Desulfovibrio vietnamensis.*

Desulfovibrio Medium (DSMZ Medium 63)

Composition per liter:

Soultion A 980.0mL
Solution B 10.0mL
Solution C 10.0mL

Solution A:

Composition per 980.0mL:

$MgSO_4 \cdot 7H_2O$ 2.0g
DL-Na-lactate 2.0g
Yeast extract 1.0g
NH_4Cl 1.0g
Na_2SO_4 .... 1.0g
K_2HPO_4 .... 0.5g
$CaCl_2 \cdot 2H_2O$ 0.1g
Resazurin 1.0mg

pH 7.8 ± 0.2 at 25°C

Preparation of Solution A: Add components to distilled/deionized water and bring volume to 980.0mL. Mix thoroughly.

Solution B:

Composition per 10.0mL:

$FeSO_4 \cdot 7H_2O$ 0.5g

Preparation of Solution B: Add $FeSO_4 \cdot 7H_2O$ to distilled/deionized water and bring volume to 10.0mL. Mix thoroughly.

Solution C:

Composition per 10.0mL:

Na-thioglycolate 0.1g
Ascorbic acid 0.1g

Preparation of Solution C: Add components to distilled/deionized water and bring volume to 10.0mL. Mix thoroughly.

Preparation of Medium: Bring solution A to a boil for a few minutes. Cool to room temperature while gassing with oxygen-free N_2 gas. Add solutions B and C. Adjust pH to 7.8 with NaOH. Immediately distribute under N_2 into anaerobic tubes. During distribution continuously swirl the medium to keep the grey precipitate suspended. Autoclave for 15 min at 15 psi pressure–121°C.

Use: For the cultivation of *Desulfovibrio* spp.

Desulfovibrio Medium (DSMZ Medium 63b)

Composition per liter:

Soultion A 980.0mL
Solution B 10.0mL
Solution C 10.0mL

Solution A:

Composition per 980.0mL:

$MgSO_4 \cdot 7H_2O$ 2.0g
DL-Na-lactate 2.0g
Yeast extract 1.0g
NH_4Cl 1.0g
Na_2SO_4 .... 1.0g
K_2HPO_4 .... 0.5g
$CaCl_2 \cdot 2H_2O$ 0.1g
Resazurin 1.0mg
Seawater 980.0mL

pH 7.8 ± 0.2 at 25°C

Preparation of Solution A: Add components to filtered aged seawater and bring volume to 980.0mL. Mix thoroughly.

Solution B:

Composition per 10.0mL:

$FeSO_4 \cdot 7H_2O$ 0.5g

Preparation of Solution B: Add $FeSO_4 \cdot 7H_2O$ to distilled/deionized water and bring volume to 10.0mL. Mix thoroughly.

Solution C:

Composition per 10.0mL:

Na-thioglycolate 0.1g
Ascorbic acid 0.1g

Preparation of Solution C: Add components to distilled/deionized water and bring volume to 10.0mL. Mix thoroughly.

Preparation of Medium: Bring solution A to a boil for a few minutes. Cool to room temperature while gassing with oxygen-free N_2 gas. Add solutions B and C. Adjust pH to 7.8 with NaOH. Immediately distribute under N_2 into anaerobic tubes. During distribution continuously swirl the medium to keep the grey precipitate suspended. Autoclave for 15 min at 15 psi pressure–121°C.

Use: For the cultivation of *Desulfovibrio indonesiensis.*

Desulfovibrio Medium (DSMZ Medium 63b)

Composition per liter:

Soultion A	980.0mL
Solution B	10.0mL
Solution C	10.0mL

Solution A:

Composition per 980.0mL:

NaCl	15.0g
$MgSO_4 \cdot 7H_2O$	2.0g
DL-Na-lactate	2.0g
Yeast extract	1.0g
NH_4Cl	1.0g
Na_2SO_4	1.0g
K_2HPO_4	0.5g
$CaCl_2 \cdot 2H_2O$	0.1g
Resazurin	1.0mg

pH 7.8 ± 0.2 at 25°C

Preparation of Solution A: Add components to aged tapwater and bring volume to 980.0mL. Mix thoroughly.

Solution B:

Composition per 10.0mL:

$FeSO_4 \cdot 7H_2O$	0.5g

Preparation of Solution B: Add $FeSO_4 \cdot 7H_2O$ to distilled/deionized water and bring volume to 10.0mL. Mix thoroughly.

Solution C:

Composition per 10.0mL:

Na-thioglycolate	0.1g
Ascorbic acid	0.1g

Preparation of Solution C: Add components to distilled/deionized water and bring volume to 10.0mL. Mix thoroughly.

Preparation of Medium: Bring solution A to a boil for a few minutes. Cool to room temperature while gassing with oxygen-free N_2 gas. Add solutions B and C. Adjust pH to 7.8 with NaOH. Immediately distribute under N_2 into anaerobic tubes. During distribution continuously swirl the medium to keep the grey precipitate suspended. Autoclave for 15 min at 15 psi pressure–121°C.

Use: For the cultivation of *Desulfovibrio indonesiensis.*

Desulfovibrio Medium (LMG Medium 104)

Composition per liter:

Solution A	980.0mL
Solution B	10.0mL
Solution C	10.0mL

pH 7.8 ± 0.2 at 25°C

Solution A:

Composition per 980.0mL:

$MgSO_4 \cdot 7H_2O$	2.0g
DL-Sodium lactate	2.0g
Yeast extract	1.0g
NH_4Cl	1.0g
Na_2SO_4	1.0g
K_2HPO_4	0.5g
$CaCl_2 \cdot 2H_2O$	0.1g
Resazurin	1.0mg

Preparation of Solution A: Add components to 980.0mL distilled/deionized water. Mix thoroughly. Adjust pH to 7.4.

Solution B:

Composition per 10.0mL:

$FeSO_4 \cdot 7H_2O$	0.5g

Preparation of Solution B: Add $FeSO_4 \cdot 7H_2O$ to distilled/deionized water and bring volume to 10.0mL. Mix thoroughly.

Solution C:

Composition per 10.0mL:

Sodium thioglycolate	0.1g
Ascorbic acid	0.1g

Preparation of Solution C: Add components to distilled/deionized water and bring volume to 10.0mL. Mix thoroughly.

Preparation of Medium: Mix 980.0mL solution A, 10.0mL solution B, and 10.0mL solution C. Adjust pH to 7.8 Distribute into tubes or flasks. Autoclave for 15 min at 15 psi pressure–121°C. Pour into sterile Petri dishes or leave in tubes.

Use: For the cultivation of *Desulfovibrio desulfuricans* subsp. *Desulfuricans* and *Desulfovibrio vulgaris* subsp. *Vulgaris.*

Desulfovibrio Medium

Composition per 1056.5mL:

$(NH_4)_2SO_4$	5.3g
Sodium acetate	2.0g
NaCl	1.0g
KH_2PO_4	0.5g
$MgSO_4 \cdot 7H_2O$	0.2g
$CaCl_2 \cdot 2H_2O$	0.1g
Na_2CO_3 solution	50.0mL
Solution 1	10.0mL
Solution 2	1.0mL

pH 7.2 ± 0.2 at 25°C

Solution 1:

Composition per liter:

Nitrilotriacetic acid	12.8g
$FeCl_2 \cdot 4H_2O$	0.3g
$CoCl_2 \cdot 6H_2O$	0.17g
$MnCl_2 \cdot 4H_2O$	0.1g
$ZnCl_2$	0.1g
$CuCl_2$	0.02g
H_3BO_3	0.01g
$Na_2MoO_4 \cdot 2H_2O$	0.01g

Preparation of Solution 1: Add nitrilotriacetic acid to 500.0mL of distilled/deionized water. Dissolve by adjusting pH to 6.5 with NaOH. Add remaining components. Readjust pH to 7.2 with H_2SO_4 or NaOH. Add distilled/deionized water to 1.0L.

Solution 2:

Composition per 100.0mL:

Resazurin	0.2g

Preparation of Solution 2: Add resazurin to distilled/deionized water and bring volume to 100.0mL. Mix thoroughly.

Na_2CO_3 Solution:

Composition per 100.0mL:

Na_2CO_3	8.0g

Preparation of Na_2CO_3 Solution: Add Na_2CO_3 to distilled/deionized water and bring volume to 100.0mL. Mix thoroughly. Filter sterilize. Gas with 100% N_2 for 20 min.

Preparation of Medium: Add components—except Na_2CO_3 solution, HCl solution, and $Na_2S_2O_4$ solution—to distilled/deionized water and bring volume to 1.0L. Mix thoroughly. Gently heat and bring to boiling. Autoclave for 15 min at 15 psi pressure–121°C. Cool to 45°–50°C. Anaerobically and aseptically add 50.0mL of sterile Na_2CO_3 solution, 5.5mL of sterile HCl solution, and 1.0mL of sterile $Na_2S_2O_4$ solution. Mix thoroughly. Anaerobically and aseptically distribute into sterile tubes or flasks.

Use: For the isolation, cultivation, and enrichment of *Desulfovibrio* species.

Desulfovibrio Medium

Composition per liter of tap water:

Agar	15.0g
Glucose	5.0g
Peptone	5.0g
Beef extract	3.0g
$MgSO_4$	1.5g
Na_2SO_4	1.5g
Yeast extract	0.2g
$Fe(NH_4)_2(SO_4)_2$	0.1g

pH 7.0 ± 0.2 at 25°C

Preparation of Medium: Sterilize by autoclaving for 15 min at 15 psi–121°C.

Use: For the cultivation and maintenance of *Desulfomaculum nigrificans, Desulfovibrio desulfuricans,* and *Desulfovibrio gigas.*

Desulfovibrio Medium

Composition per liter:

Solution A	980.0mL
Solution B	10.0mL
Solution C	10.0mL

pH 7.8 ± 0.2 at 25°C

Solution A:

Composition per 980.0mL:

DL-Sodium lactate	2.0g
$MgSO_4 \cdot 7H_2O$	2.0g
Na_2SO_4	1.0g
NH_4Cl	1.0g
Yeast extract	1.0g
K_2HPO_4	0.5g
$CaCl_2 \cdot 2H_2O$	0.1g
Resazurin	1.0mg

Preparation of Solution A: Add components to distilled/deionized water and bring volume to 980.0mL. Mix thoroughly. Adjust pH to 7.4.

Solution B:

Composition per 10.0mL:

$FeSO_4 \cdot 7H_2O$	0.5g

Preparation of Solution B: Add $FeSO_4 \cdot 7H_2O$ to distilled/deionized water and bring volume to 10.0mL. Mix thoroughly.

Solution C:

Composition per 10.0mL:

Ascorbic acid	0.1g
Sodium thioglycolate	0.1g

Preparation of Solution C: Add components to distilled/deionized water and bring volume to 10.0mL. Mix thoroughly.

Preparation of Medium: Combine 980.0mL of solution A, 10.0mL of solution B, and 10.0mL of solution C. Mix thoroughly. Adjust pH to 7.8. Distribute into tubes or flasks. Autoclave for 15 min at 15 psi pressure–121°C.

Use: For the cultivation and maintenance of *Desulfovibrio desulfuricans, Desulfovibrio giganteus,* and *Desulfovibrio vulgaris.*

Desulfovibrio Medium

Composition per liter:

Solution A	980.0mL
Solution B	10.0mL
Solution C	10.0mL

pH 7.8 ± 0.2 at 25°C

Solution A:

Composition per 980.0mL:

Choline·HCl	5.0g
K_2HPO_4	4.0g
$MgSO_4 \cdot 7H_2O$	2.0g
Na_2SO_4	1.0g
NH_4Cl	1.0g
Yeast extract	1.0g
$CaCl_2 \cdot 2H_2O$	0.1g
Resazurin	1.0mg

Preparation of Solution A: Add components to distilled/deionized water and bring volume to 980.0mL. Mix thoroughly. Gently heat and bring to boiling. Continue boiling for 3–4 min. Allow to cool to room temperature while gassing under 100% N_2.

Solution B:

Composition per 10.0mL:

$FeSO_4 \cdot 7H_2O$	0.5g

Preparation of Solution B: Add components to distilled/deionized water and bring volume to 10.0mL. Mix thoroughly. Gas under 100% N_2. Autoclave for 15 min at 15 psi pressure–121°C.

Solution C:

Composition per 10.0mL:

Ascorbic acid	0.1g
Sodium thioglycolate	0.1g

Preparation of Solution C: Add components to distilled/deionized water and bring volume to 10.0mL. Mix thoroughly. Gas under 100% N_2. Autoclave for 15 min at 15 psi pressure–121°C.

Preparation of Medium: Dissolve the ingredients of each solution in the appropriate quantities of water. Bring solution A to a boil for a few minutes, then cool to room temperature while gassing with oxygen-free N_2 gas. Add solutions B and C, adjust pH to 7.8 with NaOH, and distribute under N_2 in anaerobic tubes. During distribution, continuously swirl the medium to keep the grey precipitate suspended. Autoclave for 15 min at 121°C.

Use: For the cultivation and maintenance of *Desulfobacterium macestii, Desulfomicrobium apsheronum, Desulfomonas pigra, Desulfotomaculum species, Desulfovibrio* species, and *Thermodesulfobacterium mobile.*

Desulfovibrio Medium with Lactate

Composition per liter:

Agar	15.0g
Lactate	10.0g

Glucose ... 5.0g
Peptone ... 5.0g
Beef extract ... 3.0g
$MgSO_4$ ... 1.5g
Na_2SO_4 ... 1.5g
Yeast extract ... 0.2g
$Fe(NH_4)_2(SO_4)_2$ ... 0.1g

pH 7.0 ± 0.2 at 25°C

Preparation of Medium: Add components to tap water and bring volume to 1.0L. Mix thoroughly. Gently heat and bring to boiling. Distribute into tubes or flasks. Autoclave for 15 min at 15 psi pressure–121°C. Pour into sterile Petri dishes or leave in tubes.

Use: For the cultivation and maintenance of *Desulfovibrio desulfuricans.*

Desulfovibrio Medium with Sodium Chloride

Composition per liter:

NaCl ... 30.0g
Agar ... 15.0g
Glucose ... 5.0g
Peptone ... 5.0g
Beef extract ... 3.0g
$MgSO_4$ ... 1.5g
Na_2SO_4 ... 1.5g
Yeast extract ... 0.2g
$Fe(NH_4)_2(SO_4)_2$ ... 0.1g

pH 7.0 ± 0.2 at 25°C

Preparation of Medium: Add components to tap water and bring volume to 1.0L. Mix thoroughly. Gently heat and bring to boiling. Distribute into tubes or flasks. Autoclave for 15 min at 15 psi pressure–121°C. Pour into sterile Petri dishes or leave in tubes.

Use: For the cultivation and maintenance of *Desulfovibrio desulfuricans* and *Desulfovibrio salexigens.*

Desulfovibrio MG-1 Medium (DSMZ Medium 615)

Composition per liter:

Na_2SO_4 ... 4.5g
Glycerol ... 2.0g
NH_4Cl ... 1.0g
Yeast extract ... 1.0g
Na_3-citrate·$2H_2O$ ... 0.6g
KH_2PO_4 ... 0.5g
Na-thioglycolate ... 0.1g
$MgSO_4·7H_2O$ ... 0.06g
$CaCl_2·2H_2O$ ... 0.04g
$FeSO_4·7H_2O$ ... 4.0mg
Resazurin ... 0.5mg

pH 6.9 ± 0.2 at 25°C

Preparation of Medium: Prepare and dispense medium under 100% N_2 gas atmosphere. Add components to distilled/deionized water and bring volume to 1.0L. Mix thoroughly. Adjust pH to 7.5. Distribute into tubes or flasks. Autoclave for 15 min at 15 psi pressure–121°C.

Use: For the cultivation of *Desulfovibrio* sp.

Desulfovibrio MG-1 Medium

Composition per liter:

Na_2SO_4 ... 4.5g
Glycerol ... 2.0g
NH_4Cl ... 1.0g
Yeast extract ... 1.0g
Trisodium citrate·$2H_2O$ ... 0.6g
KH_2PO_4 ... 0.5g
Sodium thioglycolate ... 0.1g
$MgSO_4·7H_2O$ ... 0.06g
$CaCl_2·2H_2O$ ... 0.04g
$FeSO_4·7H_2O$ ... 4.0mg
Resazurin ... 0.5mg

pH 7.5 ± 0.2 at 25°C

Preparation of Medium: Prepare and dispense medium under 100% N_2. Add components to distilled/deionized water and bring volume to 1.0L. Mix thoroughly. Gently heat and bring to boiling. Continue boiling for 5 min. Cool to room temperature while sparging with 100% N_2. Anaerobically distribute into tubes or flasks. Autoclave for 15 min at 15 psi pressure–121°C.

Use: For the cultivation of *Desulfovibrio* species.

Desulfovibrio sapovorans Medium

Composition per 1009.0mL:

Solution A ... 850.0mL
Solution C ... 100.0mL
Solution G ... 20.0mL
Solution D ... 10.0mL
Solution E (Wolfe's vitamin solution) ... 10.0mL
Solution H ... 10.0mL
Solution F ... 6.6mL
Solution B (Trace elements solution SL-10) ... 1.0mL
Solution I ... 0.4mL

pH 7.7 ± 0.2 at 25°C

Solution A:

Composition per 920.0mL:

Na_2SO_4 ... 3.0g
NaCl ... 1.0g
KCl ... 0.5g
$MgCl_2·6H_2O$ ... 0.4g
NH_4Cl ... 0.3g
KH_2PO_4 ... 0.2g
$CaCl_2·2H_2O$ ... 0.15g
Resazurin ... 0.5mg

Preparation of Solution A: Prepare and dispense solution anaerobically under 90% N_2 + 10% CO_2. Add components to distilled/deionized water and bring volume to 920.0mL. Mix thoroughly. Gently heat and bring to boiling. Continue boiling until resazurin turns colorless, indicating reduction, and a pH of 6.0 is reached. Cap with rubber stoppers. Autoclave for 15 min at 15 psi pressure–121°C. Cool to 25°C.

Solution B (Trace Elements Solution SL-10):

Composition per liter:

$FeCl_2·4H_2O$ ... 1.5g
$CoCl_2·6H_2O$ ... 0.19g
$MnCl_2·4H_2O$ ... 0.10g
$ZnCl_2$ ... 0.070g
$Na_2MoO_4·2H_2O$ ... 0.036g
$NiCl_2·6H_2O$ ... 0.024g
H_3BO_3 ... 6.0mg
$CuCl_2·2H_2O$ ... 2.0mg
HCl (25% solution) ... 10.0mL

Preparation of Solution B (Trace Elements Solution SL-10): Add the $FeCl_2·4H_2O$ to 10.0mL of HCl solution. Mix thoroughly.

Bring volume to approximately 900.0mL with distilled/deionized water. Mix thoroughly. Adjust pH to 6.0 with NaOH. Bring volume to 1.0L with distilled/deionized water. Filter sterilize. Aseptically gas under 100% N_2 for 20 min.

Solution C:

Composition per 100.0mL:

$NaHCO_3$	5.0g

Preparation of Solution C: Add $NaHCO_3$ to distilled/deionized water and bring volume to 100.0mL. Mix thoroughly. Filter sterilize. Aseptically gas under 90% N_2 + 10% CO_2 for 20 min.

Solution D:

Composition per 10.0mL:

Sodium butyrate	0.7g
Sodium caproate	0.3g
Sodium octanoate	0.15g

Preparation of Solution D: Prepare and dispense solution anaerobically under 90% N_2 + 10% CO_2. Add components to distilled/deionized water and bring volume to 10.0mL. Mix thoroughly. Cap with a rubber stopper. Autoclave for 15 min at 15 psi pressure–121°C. Cool to 25°C.

Solution E (Wolfe's Vitamin Solution):

Composition per liter:

Pyridoxine·HCl	0.01g
Thiamine·HCl	5.0mg
Riboflavin	5.0mg
Nicotinic acid	5.0mg
Calcium pantothenate	5.0mg
p-Aminobenzoic acid	5.0mg
Thioctic acid	5.0mg
Biotin	2.0mg
Folic acid	2.0mg
Cyanocobalamin	0.1mg

Preparation of Solution E (Wolfe's Vitamin Solution): Add components to distilled/deionized water and bring volume to 1.0L. Mix thoroughly. Filter sterilize. Aseptically gas under 100% N_2 for 20 min.

Solution F:

Composition per 6.6mL:

$AlCl_3 \cdot 6H_2O$ (4.9% solution)	5.0mL
Na_2CO_3 (10.6% solution)	1.6mL

Preparation of Solution F: Combine both solutions. Mix thoroughly. Gas with 100% N_2. Cap with a rubber stopper. Autoclave for 15 min at 15 psi pressure–121°C. Cool to 25°C.

Solution G:

Composition per 10.0mL:

Rumen fluid, clarified	20.0mL

Preparation of Solution G: Gas rumen fluid under 100% N_2 for 20 min. Cap with a rubber stopper. Autoclave for 15 min at 15 psi pressure–121°C. Cool to 25°C.

Solution H:

Composition per 10.0mL:

$Na_2S \cdot 9H_2O$	0.4g

Preparation of Solution H: Add $Na_2S \cdot 9H_2O$ to distilled/deionized water and bring volume to 10.0mL. Gas under 100% N_2 for 20 min. Cap with a rubber stopper. Autoclave for 15 min at 15 psi pressure–121°C. Cool to 25°C.

Solution I:

Composition per 10.0mL:

$Na_2S_2O_4$	0.5g

Preparation of Solution I: Add $Na_2S_2O_4$ to distilled/deionized water and bring volume to 10.0mL. Mix thoroughly. Filter sterilize. Aseptically gas under 100% N_2 for 20 min. Prepare solution freshly.

Preparation of Medium: Prepare and dispense medium under 90% N_2 + 10% CO_2. To 850.0mL of cooled, sterile solution A, aseptically and anaerobically add in the following order: 1.0mL of sterile solution B, 100.0mL of sterile solution C, 10.0mL of sterile solution D, 10.0mL of sterile solution E, 6.6mL of sterile solution F, 20.0mL of sterile solution G, and 10.0mL of sterile solution H. Mix thoroughly. Immediately prior to inoculation, aseptically and anaerobically add 0.4mL of sterile solution I. Mix thoroughly. Aseptically and anaerobically distribute into sterile tubes or flasks.

Use: For the cultivation and maintenance of *Desulfovibrio sapovorans*.

Desulfovibrio sax Medium (DSMZ Medium 383a)

Composition per 1022.6mL:

Soultion A	930.0mL
Solution C	50.0mL
Solution E	20.0mL
Solution D	10.0mL
Solution G	10.0mL
Solution B	1.0mL
Solution F	1.0mL
Vitamin B_{12} solution	0.5mL
Yeast extract solution	0.1mL

pH 7.3 at 25°C

Solution A:

Composition per 930.0mL:

NaCl	21.0g
Na_2SO_4	3.0g
$MgCl_2 \cdot 6H_2O$	3.0g
KCl	0.5g
NH_4Cl	0.3g
KH_2PO_4	0.2g
$CaCl_2 \cdot 2H_2O$	0.15g
Resazurin	1.0mg

Preparation of Solution A: Add components to distilled/deionized water and bring volume to 930.0mL. Mix thoroughly. Sparge with 80% N_2 + 20% CO_2 gas until saturated. Autoclave for 15 min at 15 psi pressure–121°C. Cool to 25°C.

Solution B:

Composition per liter:

$FeCl_2 \cdot 4H_2O$	1.5g
H_3BO_3	300.0mg
$CoCl_2 \cdot 6H_2O$	190.0mg
$MnCl_2 \cdot 4H_2O$	100.0mg
$ZnCl_2$	70.0mg
$Na_2MoO_4 \cdot 2H_2O$	36.0mg
$NiCl_2 \cdot 6H_2O$	24.0mg
$CuCl_2 \cdot 2H_2O$	2.0mg
HCl (25% solution)	7.7mL

Preparation of Solution B: Add $FeCl_2 \cdot 4H_2O$ to 10.0mL of HCl solution. Mix thoroughly. Add distilled/deionized water and bring volume to 1.0L. Add remaining components. Mix thoroughly. Sparge with 100% N_2. Autoclave for 15 min at 15 psi pressure–121°C.

Solution C:

Composition per 100.0mL:

$NaHCO_3$	5.0g

Preparation of Solution C: Add $NaHCO_3$ to distilled/deionized water and bring volume to 100.0mL. Mix thoroughly. Sparge with 100% CO_2 until saturated, approximately 20 min. Filter sterilize under 100% CO_2 into a sterile, gas-tight 100.0mL screw-capped bottle.

Solution D:

Composition per 10.0mL:

Na-benzoate	0.5g

Preparation of Solution D: Add Na-benzoate to distilled/deionized water and bring volume to 10.0mL. Sparge with N_2. Filter sterilize. Store anaerobically.

Solution E:

Composition per liter:

Pyridoxine-HCl	10.0mg
Thiamine-HCl·$2H_2O$	5.0mg
Riboflavin	5.0mg
Nicotinic acid	5.0mg
D-Ca-pantothenate	5.0mg
p-Aminobenzoic acid	5.0mg
Lipoic acid	5.0mg
Biotin	2.0mg
Folic acid	2.0mg
Vitamin B_{12}	0.1mg

Preparation of Solution E: Add components to distilled/deionized water and bring volume to 1.0L. Mix thoroughly. Sparge with 100% N_2. Filter sterilize.

Solution F:

Composition per liter:

NaOH	0.5g
$Na_2SeO_3·5H_2O$	3.0mg

Preparation of Solution F: Add components to distilled/deionized water and bring volume to 1.0L. Mix thoroughly. Sparge with 100% N_2. Filter sterilize.

Solution G:

Composition per 10.0mL:

$Na_2S·9H_2O$	0.4g

Preparation of Solution G: Add $Na_2S·9H_2O$ to distilled/deionized water and bring volume to 10.0mL. Sparge with N_2. Autoclave for 15 min at 15 psi pressure–121°C. Cool to 25°C. Store anaerobically.

Vitamin B_{12} Solution:

Composition per 100.0mL:

Vitamin B_{12}	10.0mg

Vitamin B_{12} Solution: Add vitamin B_{12} to distilled/deionized water and bring volume to 100.0mL. Mix thoroughly. Sparge under 100% N_2 gas for 3 min. Filter sterilize.

Yeast Extract Solution:

Composition per 10.0mL:

Yeast extract	1.0g

Preparation of Yeast Extract Solution: Add yeast extract to distilled/deionized water and bring volume to 10.0mL. Mix thoroughly. Sparge with 100% N_2. Autoclave under 100% N_2 for 15 min at 15 psi pressure–121°C. Cool to room temperature.

Preparation of Medium: Add solution B, solution C, solution D, solution E, Vitamin B_{12} solution, yeast extract solution, solution F, and solution G to solution A in that order under N_2 gas. Adjust the pH to 7.3.

Use: For the cultivation of *Desulfotignum balticum (Desulfoarculus* sp.*)*.

Desulfovibrio SHV Medium

Composition per 1003.0mL:

Solution A	870.0mL
Solution C	100.0mL
Solution D	10.0mL
Solution E (Vitamin solution)	10.0mL
Solution F	10.0mL
Solution B (Trace elements solution SL-10)	1.0mL
Solution G (Selenite-tungstate solution)	1.0mL
Solution H (Seven vitamin solution)	1.0mL

pH 7.1–7.4 at 25°C

Solution A:

Composition per 870.0mL:

NaCl	7.0g
Na_2SO_4	3.0g
$MgCl_2·6H_2O$	1.3g
KCl	0.5g
NH_4Cl	0.3g
KH_2PO_4	0.2g
$CaCl_2·2H_2O$	0.15g
Resazurin	1.0mg

Preparation of Solution A: Add components to distilled/deionized water and bring volume to 870.0mL. Mix thoroughly. Gently heat and bring to boiling. Continue boiling for 3–4 min. Allow to cool to room temperature while gassing under 80% N_2 + 20% CO_2. Continue gassing until pH reaches below 6.0. Seal the flask under 80% N_2 + 20% CO_2. Autoclave for 15 min at 15 psi pressure–121°C.

Solution B (Trace Elements Solution SL-10):

Composition per liter:

$FeCl_2·4H_2O$	1.5g
$CoCl_2·6H_2O$	190.0mg
$MnCl_2·4H_2O$	100.0mg
$ZnCl_2$	70.0mg
$Na_2MoO_4·2H_2O$	36.0mg
$NiCl_2·6H_2O$	24.0mg
H_3BO_3	6.0mg
$CuCl_2·2H_2O$	2.0mg
HCl (25% solution)	10.0mL

Preparation of Solution B (Trace Elements Solution SL-10): Add $FeCl_2·4H_2O$ to 10.0mL of HCl solution. Mix thoroughly. Add distilled/deionized water and bring volume to 1.0L. Add remaining components. Mix thoroughly. Sparge with 100% N_2. Autoclave for 15 min at 15 psi pressure–121°C.

Solution C:

Composition per 100.0mL:

$NaHCO_3$	5.0g

Preparation of Solution C: Add $NaHCO_3$ to distilled/deionized water and bring volume to 100.0mL. Mix thoroughly. Filter sterilize. Sparge with 80% N_2 + 20% CO_2.

Solution D:

Composition per 10.0mL:

Sodium lactate	4.0g

Preparation of Solution D: Add sodium lactate to distilled/deionized water and bring volume to 10.0mL. Mix thoroughly. Sparge with 100% N_2. Autoclave for 15 min at 15 psi pressure–121°C.

Solution E (Vitamin Solution):

Composition per liter:

Pyridoxine·HCl	10.0mg
Calcium DL-pantothenate	5.0mg
Lipoic acid	5.0mg
Nicotinic acid	5.0mg
p-Aminobenzoic acid	5.0mg
Riboflavin	5.0mg
Thiamine·HCl	5.0mg
Biotin	2.0mg
Folic acid	2.0mg
Vitamin B_{12}	0.1mg

Preparation of Solution E (Vitamin Solution): Add components to distilled/deionized water and bring volume to 1.0L. Mix thoroughly. Sparge with 100% N_2. Autoclave for 15 min at 15 psi pressure–121°C.

Solution F:

Composition per 10.0mL:

$Na_2S·9H_2O$	0.4g

Preparation of Solution F: Add $Na_2S·9H_2O$ to distilled/deionized water and bring volume to 10.0mL. Mix thoroughly. Sparge with 100% N_2. Autoclave for 15 min at 15 psi pressure–121°C.

Solution G (Selenite-Tungstate Solution):

Composition per liter:

NaOH	0.5g
$Na_2WO_4·2H_2O$	4.0mg
$Na_2SeO_3·5H_2O$	3.0mg

Preparation of Solution G (Selenite-Tungstate Solution): Add components to distilled/deionized water and bring volume to 1.0L. Mix thoroughly. Sparge with 100% N_2. Autoclave for 15 min at 15 psi pressure–121°C.

Solution H (Seven Vitamin Solution) :

Composition per liter:

Pyridoxine·HCl	0.3g
Thiamine·HCl	0.2g
Nicotinic acid	0.2g
Calcium DL-pantothenate	0.1g
Vitamin B_{12}	0.1g
p-Aminobenzoic acid	80.0mg
Biotin	20.0mg

Preparation of Solution H (Seven Vitamin Solution): Add components to distilled/deionized water and bring volume to 1.0L. Mix thoroughly.

Preparation of Medium: Aseptically and anaerobically combine solution A with solution B, solution C, solution D, solution E, solution F, solution G, and solution H, in that order. Mix thoroughly. Anaerobically distribute into sterile tubes or flasks under 80% N_2 + 20% CO_2.

Use: For the cultivation of *Desulfovibrio* species.

Desulfovibrio sp. Medium (DSMZ Medium 200)

Composition per 1001.0mL:

Solution A	870.0mL
Solution C	100.0mL
Solution D	10.0mL
Solution E (Vitamin solution)	10.0mL
Solution F	10.0mL
Solution B (Trace elements solution SL-10)	1.0mL

pH 6.8–7.0 at 25°C

Solution A:

Composition per 870.0mL:

NaCl	20.0g
$MgCl_2·6H_2O$	3.1g
Na_2SO_4	3.0g
KH_2PO_4	0.2g
NH_4Cl	0.3g
KCl	0.5g
$CaCl_2·2H_2O$	0.15g
Resazurin	1.0mg

Preparation of Solution A: Add components to distilled/deionized water and bring volume to 870.0mL. Mix thoroughly.

Solution B (Trace Elements Solution SL-10):

Composition per liter:

$FeCl_2·4H_2O$	1.5g
$CoCl_2·6H_2O$	190.0mg
$MnCl_2·4H_2O$	100.0mg
$ZnCl_2$	70.0mg
$Na_2MoO_4·2H_2O$	36.0mg
$NiCl_2·6H_2O$	24.0mg
H_3BO_3	6.0mg
$CuCl_2·2H_2O$	2.0mg
HCl (25% solution)	10.0mL

Preparation of Solution B (Trace Elements Solution SL-10): Add $FeCl_2·4H_2O$ to 10.0mL of HCl solution. Mix thoroughly. Add distilled/deionized water and bring volume to 1.0L. Add remaining components. Mix thoroughly. Sparge with 100% N_2. Autoclave for 15 min at 15 psi pressure–121°C.

Solution C:

Composition per 100.0mL:

$NaHCO_3$	5.0g

Preparation of Solution C: Add $NaHCO_3$ to distilled/deionized water and bring volume to 100.0mL. Mix thoroughly. Filter sterilize. Flush with 80% N_2 + 20% CO_2 to remove dissolved oxygen.

Solution D:

Composition per 10.0mL:

Na-butyrate	0.7g
Na-caproate	0.3g
Na-octanoate	0.15g

Preparation of Solution D: Add components to distilled/deionized water and bring volume to 10.0mL. Mix thoroughly. Sparge with 100% N_2. Autoclave for 15 min at 15 psi pressure–121°C.

Solution E (Vitamin Solution):

Composition per liter:

Pyridoxine-HCl	10.0mg
Thiamine-HCl·$2H_2O$	5.0mg
Riboflavin	5.0mg
Nicotinic acid	5.0mg
D-Ca-pantothenate	5.0mg
p-Aminobenzoic acid	5.0mg
Lipoic acid	5.0mg
Biotin	2.0mg
Folic acid	2.0mg
Vitamin B_{12}	0.10mg

Preparation of Solution E (Vitamin Solution): Add components to distilled/deionized water and bring volume to 1.0L. Mix thoroughly. Sparge with 100% N_2. Autoclave for 15 min at 15 psi pressure–121°C.

Solution F:

Composition per 10.0mL:

$Na_2S·9H_2O$ 0.4g

Preparation of Solution F: Add $Na_2S·9H_2O$ to distilled/deionized water and bring volume to 10.0mL. Mix thoroughly. Sparge with 100% N_2. Autoclave for 15 min at 15 psi pressure–121°C.

Preparation of Medium: Gently heat solution A and bring to boiling. Boil solution A for a few minutes. Cool to room temperature. Gas with 80% N_2 + 20% CO_2 gas mixture to reach a pH below 6. Autoclave for 15 min at 15 psi pressure–121°C. Cool to room temperature. Sequentially add 1.0mL solution B, 100.0mL solution C, 10.0mL solution D, 10.0mL solution E, and 10.0mL solution F. Distribute anaerobically under 80% N_2 + 20% CO_2 into appropriate vessels.

Use: For the cultivation of *Desulfovibrio* sp.

Desulfovibrio sulfodismutans Medium (DSMZ Medium 386)

Composition per 1002.0mL:

Solution A 920.0mL
Solution C 50.0mL
Solution D 10.0mL
Solution F 10.0mL
Solution G 10.0mL
Solution B 1.0mL
Solution E 1.0mL

pH 7.1–7.4 at 25°C

Solution A:

Composition per 920.0mL:

NaCl 1.0g
KCl 0.5g
$MgCl_2·6H_2O$ 0.4g
KH_2PO_4 .... 0.2g
NH_4Cl 0.3g
$CaCl_2·2H_2O$ 0.15g

Preparation of Solution A: Add components to distilled/deionized water and bring volume to 920.0mL. Mix thoroughly. Sparge with 80% N_2 + 20% CO_2 gas until saturated. Autoclave for 15 min at 15 psi pressure–121°C. Cool to 25°C.

Solution B:

Composition per liter:

$FeCl_2·4H_2O$ 1.5g
H_3BO_3 .... 300.0mg
$CoCl_2·6H_2O$ 190.0mg
$MnCl_2·4H_2O$ 100.0mg
$ZnCl_2$ 70.0mg
$Na_2MoO_4·2H_2O$ 36.0mg
$NiCl_2·6H_2O$ 24.0mg
$CuCl_2·2H_2O$ 2.0mg
HCl (25% solution) 7.7mL

Preparation of Solution B: Add $FeCl_2·4H_2O$ to 10.0mL of HCl solution. Mix thoroughly. Add distilled/deionized water and bring volume to 1.0L. Add remaining components. Mix thoroughly. Sparge with 100% N_2. Autoclave for 15 min at 15 psi pressure–121°C.

Solution C:

Composition per 100.0mL:

$NaHCO_3$ 5.0g

Preparation of Solution C: Add $NaHCO_3$ to distilled/deionized water and bring volume to 100.0mL. Mix thoroughly. Sparge with 80% N_2 + 20% CO_2 gas until saturated, approximately 20 min. Filter sterilize under 100% CO_2 into a sterile, gas-tight 100.0mL screw-capped bottle.

Solution D:

Composition per 10.0mL:

Na-acetate·$3H_2O$ 0.3g

Preparation of Solution D: Add Na-acetate·$3H_2O$ to distilled/deionized water and bring volume to 10.0mL. Sparge with N_2. Filter sterilize. Store anaerobically.

Solution E:

Composition per 1.0mL:

Ca-D-pantothenate 50.0μg
D(+)-Biotin 10.0μg

Preparation of Solution E: Add components to distilled/deionized water and bring volume to 1.0mL. Mix thoroughly. Sparge with 100% N_2. Filter sterilize.

Solution F:

Composition per 10.0mL:

$Na_2S·9H_2O$ 0.4g

Preparation of Solution F: Add $Na_2S·9H_2O$ to distilled/deionized water and bring volume to 10.0mL. Sparge with N_2. Autoclave for 15 min at 15 psi pressure–121°C. Cool to 25°C. Store anaerobically.

Solution G:

Composition per liter:

NaOH 0.5g
$Na_2SeO_3·5H_2O$ 3.0mg

Preparation of Solution G: Add components to distilled/deionized water and bring volume to 1.0L. Mix thoroughly. Sparge with 100% N_2. Filter sterilize.

Preparation of Medium: Add solution B, solution C, solution D, solution E, solution F, and solution G to solution A in that order under 80% N_2 + 20% CO_2 gas. Adjust the pH to 7.1–7.4. When growth has started feed culture again with same amount of solution G. After a further 2 days repeat feeding once more.

Use: For the cultivation of *Desulfovibrio sulfodismutans.*

Desulfovibrio sulfodismutans Medium

Composition per 1002.0mL:

Solution A 920.0mL
Solution C 50.0mL
Solution D 10.0mL
Solution F 10.0mL
Solution G 10.0mL
Solution B (Trace elements solution SL-10) 1.0mL
Solution E 1.0mL

pH 7.1–7.4 at 25°C

Solution A:

Composition per 920.0mL:

NaCl 1.0g
KCl 0.5g
$MgCl_2·6H_2O$ 0.4g

NH_4Cl	0.3g
KH_2PO_4	0.2g
$CaCl_2 \cdot 2H_2O$	0.15g

Preparation of Solution A: Prepare and dispense solution anaerobically under 80% N_2 + 20% CO_2. Add components to distilled/deionized water and bring volume to 920.0mL. Mix thoroughly. Gently heat and bring to boiling. Continue boiling until resazurin turns colorless, indicating reduction. Cap with rubber stoppers. Autoclave for 15 min at 15 psi pressure–121°C. Cool to 45°–50°C.

Solution B (Trace Elements Solution SL-10):
Composition per liter:

$FeCl_2 \cdot 4H_2O$	1.5g
$CoCl_2 \cdot 6H_2O$	0.19g
$MnCl_2 \cdot 4H_2O$	0.1g
$ZnCl_2$	0.07g
$Na_2MoO_4 \cdot 2H_2O$	0.036g
$NiCl_2 \cdot 6H_2O$	0.024g
H_3BO_3	6.0mg
$CuCl_2 \cdot 2H_2O$	2.0mg
HCl (25% solution)	10.0mL

Preparation of Solution B (Trace Elements Solution SL-10): Add the $FeCl_2 \cdot 4H_2O$ to 10.0mL of HCl solution. Mix thoroughly. Bring volume to approximately 900.0mL with distilled/deionized water. Mix thoroughly. Adjust pH to 6.0 with NaOH. Bring volume to 1.0L with distilled/deionized water. Filter sterilize. Aseptically gas under 100% N_2 for 20 min.

Solution C:
Composition per 50.0mL:

$NaHCO_3$	2.5g

Preparation of Solution C: Add $NaHCO_3$ to distilled/deionized water and bring volume to 50.0mL. Mix thoroughly. Filter sterilize. Aseptically gas under 80% N_2 + 20% CO_2 for 20 min.

Solution D:
Composition per 10.0mL:

Na-acetate·$3H_2O$	0.3g

Preparation of Solution D: Prepare and dispense solution anaerobically under 80% N_2 + 20% CO_2. Add Na-acetate·$3H_2O$ to distilled/deionized water and bring volume to 10.0mL. Mix thoroughly. Cap with rubber stopper. Autoclave for 15 min at 15 psi pressure–121°C. Cool to 45°–50°C.

Solution E:
Composition per liter:

Calcium pantothenate	0.05mg
Biotin	0.01mg

Preparation of Solution E: Add components to distilled/deionized water and bring volume to 1.0L. Mix thoroughly. Filter sterilize. Aseptically gas under 100% N_2 for 20 min.

Solution F:
Composition per 10.0mL:

$Na_2S \cdot 9H_2O$	0.4g

Preparation of Solution F: Add $Na_2S \cdot 9H_2O$ to distilled/deionized water and bring volume to 10.0mL. Gas under 100% N_2 for 20 min. Cap with a rubber stopper. Autoclave for 15 min at 15 psi pressure–121°C. Cool to 25°C.

Solution G:
Composition per 10.0mL:

$Na_2S_2O_5$	1.05g

Preparation of Solution G: Add $Na_2S_2O_5$ to distilled/deionized water and bring volume to 10.0mL. Mix thoroughly. Filter sterilize. Aseptically gas under 100% N_2 for 20 min. Prepare solution freshly.

Preparation of Medium: To 920.0mL of cooled, sterile solution A, aseptically and anaerobically add in the following order: 1.0mL of sterile solution B, 50.0mL of sterile solution C, 10.0mL of sterile solution D, 1.0mL of sterile solution E, 10.0mL of sterile solution F, and 10.0mL of sterile solution G. Mix thoroughly. Aseptically and anaerobically distribute into sterile tubes or flasks.

Use: For the cultivation and maintenance of *Desulfovibrio sulfodismutans*.

Desulfovibrio zosterae Medium (DSMZ Medium 383c)

Composition per 1023.5mL:

Soultion A	930.0mL
Solution C	50.0mL
Solution E	20.0mL
Solution D	10.0mL
Solution G	10.0mL
Solution B	1.0mL
Solution F	1.0mL
Selenite-tungstate solution	1.0mL
Vitamin B_{12} solution	0.5mL

pH 7.3 at 25°C

Solution A:
Composition per 930.0mL:

NaCl	21.0g
Na_2SO_4	3.0g
$MgCl_2 \cdot 6H_2O$	3.0g
KCl	0.5g
NH_4Cl	0.3g
KH_2PO_4	0.2g
$CaCl_2 \cdot 2H_2O$	0.15g
Resazurin	1.0mg

Preparation of Solution A: Add components to distilled/deionized water and bring volume to 930.0mL. Mix thoroughly. Sparge with 80% N_2 + 20% CO_2 gas until saturated. Autoclave for 15 min at 15 psi pressure–121°C. Cool to 25°C.

Solution B:
Composition per liter:

$FeCl_2 \cdot 4H_2O$	1.5g
H_3BO_3	300.0mg
$CoCl_2 \cdot 6H_2O$	190.0mg
$MnCl_2 \cdot 4H_2O$	100.0mg
$ZnCl_2$	70.0mg
$Na_2MoO_4 \cdot 2H_2O$	36.0mg
$NiCl_2 \cdot 6H_2O$	24.0mg
$CuCl_2 \cdot 2H_2O$	2.0mg
HCl (25% solution)	7.7mL

Preparation of Solution B: Add $FeCl_2 \cdot 4H_2O$ to 10.0mL of HCl solution. Mix thoroughly. Add distilled/deionized water and bring volume to 1.0L. Add remaining components. Mix thoroughly. Sparge with 100% N_2. Autoclave for 15 min at 15 psi pressure–121°C.

Solution C:
Composition per 100.0mL:

$NaHCO_3$	5.0g

Preparation of Solution C: Add $NaHCO_3$ to distilled/deionized water and bring volume to 100.0mL. Mix thoroughly. Sparge with 100% CO_2 until saturated, approximately 20 min. Filter sterilize under 100% CO_2 into a sterile, gas-tight 100.0mL screw-capped bottle.

Solution D:

Composition per 10.0mL:

Na-lactate	1.25g

Preparation of Solution D: Add Na-lactate to distilled/deionized water and bring volume to 10.0mL. Sparge with N_2. Filter sterilize. Store anaerobically.

Solution E:

Composition per liter:

Pyridoxine hydrochloride	300.0mg
Thiamine-HCl·$2H_2O$	200.0mg
Nicotinic acid	200.0mg
Vitamin B_{12}	100.0mg
Calcium pantothenate	100.0mg
p-Aminobenzoic acid	80.0mg
D(+)-Biotin	20.0mg

Preparation of Solution E: Add components to distilled/deionized water and bring volume to 1.0L. Mix thoroughly. Sparge with 100% N_2. Filter sterilize.

Solution F:

Composition per liter:

NaOH	0.5g
Na_2SeO_3·$5H_2O$	3.0mg

Preparation of Solution F: Add components to distilled/deionized water and bring volume to 1.0L. Mix thoroughly. Sparge with 100% N_2. Filter sterilize.

Solution G:

Composition per 10.0mL:

Na_2S·$9H_2O$	0.4g

Preparation of Solution G: Add Na_2S·$9H_2O$ to distilled/deionized water and bring volume to 10.0mL. Sparge with N_2. Autoclave for 15 min at 15 psi pressure–121°C. Cool to 25°C. Store anaerobically.

Vitamin B_{12} Solution:

Composition per 100.0mL:

Vitamin B_{12}	10.0mg

Vitamin B_{12} Solution: Add vitamin B_{12} to distilled/deionized water and bring volume to 100.0mL. Mix thoroughly. Sparge under 100% N_2 gas for 3 min. Filter sterilize.

Selenite-Tungstate Solution

Composition per liter:

NaOH	0.5g
Na_2WO_4·$2H_2O$	4.0mg
Na_2SeO_3·$5H_2O$	3.0mg

Preparation of Selenite-Tungstate Solution: Add components to distilled/deionized water and bring volume to 1.0L. Mix thoroughly. Sparge with 100% N_2. Filter sterilize.

Preparation of Medium: Add solution B, solution C, solution D, solution E, Vitamin B_{12} solution, selenite-tungstate solution, solution F, and solution G to solution A in that order under N_2 gas. Adjust the pH to 7.3.

Use: For the cultivation of *Desulfovibrio zosterae (Desulfovibrio* sp.*)*.

Desulfovigra adipica Medium (DSMZ Medium 868)

Composition per 2.0L:

Solution A	940.0mL
Solution E	50.0mL
Solution K	50.0mL
Solution G	20.0mL
Solution F	10.0mL
Solution M (Vitamin solution)	10.0mL
Solution H	7.0mL
Solution B	1.0mL
Solution C	1.0mL
Solution D	1.0mL
Solution I	1.0mL
Solution J (Trace elements solution SL-10)	1.0mL
Solution L (Selenite-tungstate solution)	1.0mL
Solution N	variable

pH 7.1 ± 0.2 at 25°C

Solution A:

Composition per 940.0mL:

NaCl	1.0g
KCl	0.5g
$MgCl_2$·$6H_2O$	0.4g
NH_4Cl	0.25g
KH_2PO_4	0.2g
$CaCl_2$·$2H_2O$	0.15g
Resazurin	0.5mg

Preparation of Solution A: Prepare under 80% N_2 + 20% CO_2 gas atmosphere. Add components to distilled/deionized water and bring volume to 940.0mL. Mix thoroughly. Adjust pH to 7.2. Sparge with 80% N_2 + 20% CO_2. Autoclave for 15 min at 15 psi pressure–121°C. Cool to 25°C.

Solution B:

Composition per liter:

$FeCl_2$·$4H_2O$	1.5g
$CoCl_2$·$6H_2O$	190.0mg
$MnCl_2$·$4H_2O$	100.0mg
$ZnCl_2$	70.0mg
Na_2MoO_4·$2H_2O$	36.0mg
$NiCl_2$·$6H_2O$	24.0mg
H_3BO_3	6.0mg
$CuCl_2$·$2H_2O$	2.0mg
HCl (25% solution)	10.0mL

Preparation of Solution B: Add $FeCl_2$·$4H_2O$ to 10.0mL of HCl solution. Mix thoroughly. Add distilled/deionized water and bring volume to 1.0L. Add remaining components. Mix thoroughly. Sparge with 80% N_2 + 20% CO_2. Autoclave for 15 min at 15 psi pressure–121°C. Cool to 25°C.

Solution C:

Composition per liter:

Pyridoxine hydrochloride	300.0mg
Thiamine-HCl·$2H_2O$	200.0mg
Nicotinic acid	200.0mg
Vitamin B_{12}	100.0mg
Calcium pantothenate	100.0mg
p-Aminobenzoic acid	80.0mg
D(+)-Biotin	20.0mg

Preparation of Solution C: Add components to distilled/deionized water and bring volume to 1.0L. Sparge with 100% N_2. Mix thoroughly. Filter sterilize.

Solution D:

Composition per liter:

NaOH	0.5g
$Na_2WO_4 \cdot 2H_2O$	4.0mg
$Na_2SeO_3 \cdot 5H_2O$	3.0mg

Preparation of Solution D: Add components to distilled/deionized water and bring volume to 1.0L. Mix thoroughly. Sparge with 100% N_2. Filter sterilize.

Solution E:

Composition per 100.0mL:

$NaHCO_3$	5.0g

Preparation of Solution E: Add $NaHCO_3$ to distilled/deionized water and bring volume to 100.0mL. Mix thoroughly. Sparge with 100% N_2 gas mixture. Autoclave for 15 min at 15 psi pressure–121°C. Cool to 25°C.

Solution F:

Composition per 10.0mL:

Yeast extract	1.0g

Preparation of Solution F: Add yeast extract to distilled/deionized water and bring volume to 10.0mL. Mix thoroughly. Sparge with 100% N_2 gas mixture. Autoclave for 15 min at 15 psi pressure–121°C. Cool to 25°C.

Solution G:

Composition per 20.0mL:

$Na_2S \cdot 9H_2O$	0.625g

Preparation of Solution G: Add $Na_2S \cdot 9H_2O$ to distilled/deionized water and bring volume to 20.0mL. Mix thoroughly. Autoclave under 100% N_2 for 15 min at 15 psi pressure–121°C. Cool to 25°C.

Solution H:

Composition per 10.0mL:

Na_2SO_4	1.0g

Preparation of Solution H: Add Na_2SO_4 to distilled/deionized water and bring volume to 10.0mL. Mix thoroughly. Sparge with 100% N_2 gas mixture. Autoclave for 15 min at 15 psi pressure–121°C. Cool to 25°C.

Solution I:

Composition per 10.0mL:

Propanol	1.0g

Preparation of Solution I: Add propanol to distilled/deionized water and bring volume to 10.0mL. Mix thoroughly. Sparge with 100% N_2 gas mixture. Filter sterilize.

Solution J (Trace Elements Solution SL-10):

Composition per liter:

$FeCl_2 \cdot 4H_2O$	1.5g
$CoCl_2 \cdot 6H_2O$	190.0mg
$MnCl_2 \cdot 4H_2O$	100.0mg
$ZnCl_2$	70.0mg
$Na_2MoO_4 \cdot 2H_2O$	36.0mg
$NiCl_2 \cdot 6H_2O$	24.0mg
H_3BO_3	6.0mg
$CuCl_2 \cdot 2H_2O$	2.0mg
HCl (25% solution)	10.0mL

Preparation of Solution J (Trace Elements Solution SL-10): Add $FeCl_2 \cdot 4H_2O$ to 10.0mL of HCl solution. Mix thoroughly. Add distilled/deionized water and bring volume to 1.0L. Add remaining components. Mix thoroughly. Sparge with 80% N_2 + 20% CO_2. Autoclave for 15 min at 15 psi pressure–121°C.

Solution K:

Composition per 100.0mL:

$NaHCO_3$	5.0g

Preparation of Solution K: Add $NaHCO_3$ to distilled/deionized water and bring volume to 100.0mL. Mix thoroughly. Sparge with 100% N_2 gas mixture. Autoclave for 15 min at 15 psi pressure–121°C. Cool to 25°C.

Solution L (Selenite–Tungstate Solution):

Composition per liter:

NaOH	0.5g
$Na_2WO_4 \cdot 2H_2O$	4.0mg
$Na_2SeO_3 \cdot 5H_2O$	3.0mg

Preparation of Solution L (Selenite–Tungstate Solution): Add components to distilled/deionized water and bring volume to 1.0L. Mix thoroughly. Sparge with 100% N_2. Filter sterilize.

Solution M (Vitamin Solution):

Composition per liter:

Pyridoxine-HCl	10.0mg
Thiamine-HCl·$2H_2O$	5.0mg
Riboflavin	5.0mg
Nicotinic acid	5.0mg
D-Ca-pantothenate	5.0mg
p-Aminobenzoic acid	5.0mg
Lipoic acid	5.0mg
Biotin	2.0mg
Folic acid	2.0mg
Vitamin B_{12}	0.1mg

Preparation of Solution M (Vitamin Solution): Add components to distilled/deionized water and bring volume to 1.0L. Mix thoroughly. Sparge with 80% H_2 + 20% CO_2. Filter sterilize.

Solution N:

Composition per 100.0mL:

Na_2CO_3	5.0g

Preparation of Solution N: Add Na_2CO_3 to distilled/deionized water and bring volume to 100.0mL. Mix thoroughly. Sparge with 100% N_2 gas mixture. Filter sterilize.

Preparation of Medium: Prepare and dispense medium under 80% N_2 + 20% CO_2 gas atmosphere. Sequentially add 1.0mL solution B, 1.0mL solution C, 1.0mL solution D, 50.0mL solution E, 10.0mL solution F, 20.0mL solution G, 7.0mL solution H, 1.0mL solution I, 1.0mL solution J, 50.0mL solution K, 1.0mL solution L, and 10.0mL solution M, to 940.0mL solution A. Mix thoroughly. Adjust pH to 7.1 with solution N. Distribute anaerobically under 80% N_2 + 20% CO_2 into appropriate vessels.

Use: For the cultivation of *Desulfovirga adipica (Desulfobacterium* sp.*)*.

Desulfurella Medium

Composition per liter:

Sulfur, powdered	10.0g
Sodium acetate	5.0g
$CaCl_2 \cdot 2H_2O$	0.33g

KCl	0.33g
KH_2PO_4	0.33g
$MgCl_2 \cdot 6H_2O$	0.33g
NH_4Cl	0.33g
Yeast extract	0.1g
Resazurin	1.0 mg
$NaHCO_3$ solution	40.0mL
$Na_2S \cdot 9H_2O$ solution	10.0mL
Wolfe's vitamin solution	10.0mL
Trace elements solution SL-10	1.0 mL

pH 6.8–7.0 at 25°C

$NaHCO_3$ Solution:
Composition per 40.0mL:

$NaHCO_3$	2.0g

Preparation of $NaHCO_3$ Solution: Add $NaHCO_3$ to distilled/deionized water and bring volume to 40.0mL. Mix thoroughly. Gas under 100% N_2. Autoclave for 15 min at 15 psi pressure–121°C.

$Na_2S \cdot 9H_2O$ Solution:
Composition per 10.0mL:

$Na_2S \cdot 9H_2O$	0.5g

Preparation of $Na_2S \cdot 9H_2O$ Solution: Add $Na_2S \cdot 9H_2O$ to distilled/deionized water and bring volume to 10.0mL. Mix thoroughly. Sparge with 100% N_2. Autoclave for 15 min at 15 psi pressure–121°C.

Wolfe's Vitamin Solution:
Composition per liter:

Pyridoxine·HCl	10.0mg
Calcium D-(+)-pantothenate	5.0mg
Nicotinic acid	5.0mg
p-Aminobenzoic acid	5.0mg
Riboflavin	5.0mg
Thiamine·HCl	5.0mg
Thioctic acid	5.0mg
Biotin	2.0mg
Folic acid	2.0mg
Cyanocobalamine	100.0μg

Preparation of Wolfe's Vitamin Solution: Add components to distilled/deionized water and bring volume to 1.0L. Mix thoroughly. Filter sterilize.

Trace Elements Solution SL-10:
Composition per liter:

$FeCl_2 \cdot 4H_2O$	1.5g
$CoCl_2 \cdot 6H_2O$	190.0mg
$MnCl_2 \cdot 4H_2O$	100.0mg
$ZnCl_2$	70.0mg
$Na_2MoO_4 \cdot 2H_2O$	36.0mg
$NiCl_2 \cdot 6H_2O$	24.0mg
H_3BO_3	6.0mg
$CuCl_2 \cdot 2H_2O$	2.0mg
HCl (25% solution)	10.0mL

Preparation of Trace Elements Solution SL-10: Add $FeCl_2 \cdot 4H_2O$ to 10.0mL of HCl solution. Mix thoroughly. Add distilled/deionized water and bring volume to 1.0L. Add remaining components. Mix thoroughly. Sparge with 100% N_2. Autoclave for 15 min at 15 psi pressure–121°C.

Preparation of Medium: Prepare and dispense medium under 80% N_2 + 20% CO_2. Add components, except $NaHCO_3$ solution, Wolfe's vitamin solution, and $Na_2S \cdot 9H_2O$ solution, to distilled/deionized water and bring volume to 940.0mL. Mix thoroughly. Adjust pH to 5.9. Do not autoclave. Sterilize medium by heating to 100°C for 1 hr on 3 consecutive days. Prior to inoculation, aseptically and anaerobically add 40.0mL of sterile $NaHCO_3$ solution, 10.0mL of sterile Wolfe's vitamin solution, and 10.0mL of sterile $Na_2S \cdot 9H_2O$ solution. Mix thoroughly. Final pH should be 6.8–7.0.

Use: For the cultivation of *Desulfurella* species, especially *Desulfurella acetivorans*.

Desulfurella II Medium (DSMZ Medium 480c)

Composition per liter:

MOPS [3-(*N*-morpholino) propane sulfonic acid]	3.0g
Sulfur, powder	1.0g
NH_4Cl	0.33g
$CaCl_2 \cdot 2H_2O$	0.33g
$MgCl_2 \cdot 6H_2O$	0.33g
KCl	0.33g
KH_2PO_4	0.33g
Yeast extract	0.1g
Resazurin	1.0mg
$NaHCO_3$ solution	40.0mL
$Na_2S \cdot 9H_2O$ solution	10.0mL
Vitamin solution	10.0mL
Substrate solution	10.0mL
Trace elements solution SL-10	1.0mL

pH 6.9 ± 0.2 at 25°C

$NaHCO_3$ Solution:
Composition per 10.0mL:

$NaHCO_3$	2.0g

Preparation of $NaHCO_3$ Solution: Add $NaHCO_3$ to distilled/deionized water and bring volume to 10.0mL. Mix thoroughly. Autoclave for 15 min at 15 psi pressure–121°C. Cool to 25°C. Must be prepared freshly.

Trace Elements Solution SL-10:
Composition per liter:

$FeCl_2 \cdot 4H_2O$	1.5g
$CoCl_2 \cdot 6H_2O$	190.0mg
$MnCl_2 \cdot 4H_2O$	100.0mg
$ZnCl_2$	70.0mg
$Na_2MoO_4 \cdot 2H_2O$	36.0mg
$NiCl_2 \cdot 6H_2O$	24.0mg
H_3BO_3	6.0mg
$CuCl_2 \cdot 2H_2O$	2.0mg
HCl (25% solution)	10.0mL

Preparation of Trace Elements Solution SL-10: Add $FeCl_2 \cdot 4H_2O$ to 10.0mL of HCl solution. Mix thoroughly. Add distilled/deionized water and bring volume to 1.0L. Add remaining components. Mix thoroughly. Sparge with 100% N_2. Autoclave for 15 min at 15 psi pressure–121°C.

Vitamin Solution:
Composition per liter:

Pyridoxine-HCl	10.0mg
Thiamine-HCl·$2H_2O$	5.0mg
Riboflavin	5.0mg
Nicotinic acid	5.0mg
D-Ca-pantothenate	5.0mg
p-Aminobenzoic acid	5.0mg
Lipoic acid	5.0mg

Biotin 2.0mg
Folic acid....... 2.0mg
Vitamin B_{12} 0.1mg

Preparation of Vitamin Solution: Add components to distilled/deionized water and bring volume to 1.0L. Mix thoroughly. Sparge with 80% H_2 + 20% CO_2. Filter sterilize.

$Na_2S·9H_2O$ Solution:

Composition per 10.0mL:

$Na_2S·9H_2O$ 0.5g

Preparation of $Na_2S·9H_2O$ Solution: Add $Na_2S·9H_2O$ to distilled/deionized water and bring volume to 10.0mL. Sparge with N_2. Autoclave for 15 min at 15 psi pressure–121°C. Cool to 25°C.

Substrate Solution:

Composition per 10.0mL:

Na-lactate 2.5g

Preparation of Substrate Solution: Add Na-lactate to distilled/deionized water and bring volume to 10.0mL. Sparge with N_2. Autoclave for 15 min at 15 psi pressure–121°C. Cool to 25°C. Store anaerobically.

Preparation of Medium: Prepare and dispense medium under an oxygen-free 80% N_2 + 20% CO_2 gas mixture. Add components, except sulfur, substrate solution, vitamin solution, $NaHCO_3$ solution, and $Na_2S·9H_2O$ solution, to 930.0mL distilled/deionized water. Mix thoroughly. Sparge for 30 min with 80% N_2 + 20% CO_2. Adjust pH to 5.9 with concentrated NaOH. Distribute under 80% N_2 + 20% CO_2 into anaerobic tubes or bottles containing sulfur powder (100mg S per 10.0mL medium). Autoclave 20 min at 110°C. Sparge with 80% N_2 + 20% CO_2. Aseptically and anaerobically add 10.0mL sterile $Na_2S·9H_2O$ solution, 10.0mL sterile vitamin solution, 10.0mL substrate solution, and 40.0mL sterile $NaHCO_3$ solution per liter of medium.

Use: For the cultivation of *Thermoproteus uzoniensis* and *Desulfurella kamchatkensis* DSM 10409.

Desulfurella II Medium (DSMZ Medium 480c)

Composition per liter:

MOPS [3-(*N*-morpholino) propane sulfonic acid] 3.0g
Sulfur, powder....... 1.0g
NH_4Cl 0.33g
$CaCl_2·2H_2O$....... 0.33g
$MgCl_2·6H_2O$....... 0.33g
KCl....... 0.33g
KH_2PO_4....... 0.33g
Yeast extract....... 0.1g
Resazurin 1.0mg
$NaHCO_3$ solution 40.0mL
$Na_2S·9H_2O$ solution 10.0mL
Vitamin solution....... 10.0mL
Substrate solution....... 10.0mL
Trace elements solution SL-10 .... 1.0mL

pH 6.9 ± 0.2 at 25°C

$NaHCO_3$ Solution:

Composition per 10.0mL:

$NaHCO_3$....... 2.0g

Preparation of $NaHCO_3$ Solution: Add $NaHCO_3$ to distilled/deionized water and bring volume to 10.0mL. Mix thoroughly. Autoclave for 15 min at 15 psi pressure–121°C. Cool to 25°C. Must be prepared freshly.

Trace Elements Solution SL-10:

Composition per liter:

$FeCl_2·4H_2O$....... 1.5g
$CoCl_2·6H_2O$ 190.0mg
$MnCl_2·4H_2O$ 100.0mg
$ZnCl_2$....... 70.0mg
$Na_2MoO_4·2H_2O$ 36.0mg
$NiCl_2·6H_2O$....... 24.0mg
H_3BO_3....... 6.0mg
$CuCl_2·2H_2O$ 2.0mg
HCl (25% solution)....... 10.0mL

Preparation of Trace Elements Solution SL-10: Add $FeCl_2·4H_2O$ to 10.0mL of HCl solution. Mix thoroughly. Add distilled/deionized water and bring volume to 1.0L. Add remaining components. Mix thoroughly. Sparge with 100% N_2. Autoclave for 15 min at 15 psi pressure–121°C.

Vitamin Solution:

Composition per liter:

Pyridoxine-HCl....... 10.0mg
Thiamine-HCl·$2H_2O$....... 5.0mg
Riboflavin 5.0mg
Nicotinic acid....... 5.0mg
D-Ca-pantothenate 5.0mg
p-Aminobenzoic acid....... 5.0mg
Lipoic acid 5.0mg
Biotin 2.0mg
Folic acid 2.0mg
Vitamin B_{12}....... 0.1mg

Preparation of Vitamin Solution: Add components to distilled/deionized water and bring volume to 1.0L. Mix thoroughly. Sparge with 80% H_2 + 20% CO_2. Filter sterilize.

$Na_2S·9H_2O$ Solution:

Composition per 10.0mL:

$Na_2S·9H_2O$....... 0.5g

Preparation of $Na_2S·9H_2O$ Solution: Add $Na_2S·9H_2O$ to distilled/deionized water and bring volume to 10.0mL. Sparge with N_2. Autoclave for 15 min at 15 psi pressure–121°C. Cool to 25°C. Store anaerobically.

Substrate Solution:

Composition per 10.0mL:

Na-pyruvate 2.5g

Preparation of Substrate Solution: Add Na-pyruvate to distilled/deionized water and bring volume to 10.0mL. Sparge with N_2. Autoclave for 15 min at 15 psi pressure–121°C. Cool to 25°C. Store anaerobically.

Preparation of Medium: Prepare and dispense medium under an oxygen-free 80% N_2 + 20% CO_2 gas mixture. Add components, except sulfur, substrate solution, vitamin solution, $NaHCO_3$ solution, and $Na_2S·9H_2O$ solution, to 930.0mL distilled/deionized water. Mix thoroughly. Sparge for 30 min with 80% N_2 + 20% CO_2. Adjust pH to 5.9 with concentrated NaOH. Distribute under 80% N_2 + 20% CO_2 into anaerobic tubes or bottles containing sulfur powder (100mg S per 10.0mL medium). Autoclave for 20 min at 110°C. Sparge with 80% N_2 + 20% CO_2. Aseptically and anaerobically add the appropriate amounts to achieve concentrations of 10.0mL sterile vitamin solution,

10.0mL substrate solution, 10.0mL $Na_2S{\cdot}9H_2O$ solution, and 40.0mL sterile $NaHCO_3$ solution per liter of medium.

Use: For the cultivation of *Desulfurella propionica* DSM 10410.

Desulfurella multipotens Medium (DSMZ Medium 480a)

Composition per liter:

Sulfur, powder	10.0g
Na-butyrate	5.0g
NH_4Cl	0.33g
$CaCl_2{\cdot}2H_2O$	0.33g
$MgCl_2{\cdot}6H_2O$	0.33g
KCl	0.33g
KH_2PO_4	0.33g
Yeast extract	0.1g
Resazurin	1.0mg
$NaHCO_3$ solution	40.0mL
$Na_2S{\cdot}9H_2O$ solution	10.0mL
Vitamin solution	10.0mL
Trace elements solution SL-10	1.0mL

pH 6.9 ± 0.2 at 25°C

$NaHCO_3$ Solution:

Composition per 10.0mL:

$NaHCO_3$	2.0g

Preparation of $NaHCO_3$ Solution: Add $NaHCO_3$ to distilled/deionized water and bring volume to 10.0mL. Mix thoroughly. Autoclave for 15 min at 15 psi pressure–121°C. Cool to 25°C. Must be prepared freshly.

Trace Elements Solution SL-10:

Composition per liter:

$FeCl_2{\cdot}4H_2O$	1.5g
$CoCl_2{\cdot}6H_2O$	190.0mg
$MnCl_2{\cdot}4H_2O$	100.0mg
$ZnCl_2$	70.0mg
$Na_2MoO_4{\cdot}2H_2O$	36.0mg
$NiCl_2{\cdot}6H_2O$	24.0mg
H_3BO_3	6.0mg
$CuCl_2{\cdot}2H_2O$	2.0mg
HCl (25% solution)	10.0mL

Preparation of Trace Elements Solution SL-10: Add $FeCl_2{\cdot}4H_2O$ to 10.0mL of HCl solution. Mix thoroughly. Add distilled/deionized water and bring volume to 1.0L. Add remaining components. Mix thoroughly. Sparge with 100% N_2. Autoclave for 15 min at 15 psi pressure–121°C.

Vitamin Solution:

Composition per liter:

Pyridoxine-HCl	10.0mg
Thiamine-HCl·$2H_2O$	5.0mg
Riboflavin	5.0mg
Nicotinic acid	5.0mg
D-Ca-pantothenate	5.0mg
p-Aminobenzoic acid	5.0mg
Lipoic acid	5.0mg
Biotin	2.0mg
Folic acid	2.0mg
Vitamin B_{12}	0.1mg

Preparation of Vitamin Solution: Add components to distilled/deionized water and bring volume to 1.0L. Mix thoroughly. Sparge with 80% H_2 + 20% CO_2. Filter sterilize.

$Na_2S{\cdot}9H_2O$ Solution:

Composition per 10.0mL:

$Na_2S{\cdot}9H_2O$	0.5g

Preparation of $Na_2S{\cdot}9H_2O$ Solution: Add $Na_2S{\cdot}9H_2O$ to distilled/deionized water and bring volume to 10.0mL. Sparge with N_2. Autoclave for 15 min at 15 psi pressure–121°C. Cool to 25°C. Store anaerobically.

Preparation of Medium: Prepare and dispense medium under an oxygen-free 80% N_2 + 20% CO_2 gas mixture. Add components, except vitamin solution, $NaHCO_3$ solution, and $Na_2S{\cdot}9H_2O$ solution, to 940.0mL distilled/deionized water. Mix thoroughly. Sparge with 80% N_2 + 20% CO_2. Adjust pH to 5.9 with concentrated NaOH. Sterilize medium by heating for 1 hr at 90fl100°C on 3 subsequent days. Sparge with 80% N_2 + 20% CO_2. Before use, aseptically and anaerobically add 10.0mL sterile vitamin solution, 10.0mL sterile $Na_2S{\cdot}9H_2O$ solution, and 40.0mL sterile $NaHCO_3$ solution. Mix thoroughly. Aseptically and anaerobically distribute into sterile tubes or flasks.

Use: For the cultivation of *Desulfurella multipotens*.

Desulfurococcus Medium

Composition per 1300.0mL:

Solution C	500.0mL
Solution B	450.0mL
Solution A	300.0mL
Solution D	50.0mL

Solution A:

Composition per 300.0mL:

$(NH_4)_2SO_4$	1.3g
KH_2PO_4	0.28g
$MgSO_4{\cdot}7H_2O$	0.25g
$CaCl_2{\cdot}2H_2O$	0.07g
$FeSO_4{\cdot}7H_2O$	0.028g
$Na_2B_4O_7{\cdot}10H_2O$	4.5mg
$MnCl_2{\cdot}4H_2O$	1.8mg
$ZnSO_4{\cdot}7H_2O$	0.22mg
$CuCl_2{\cdot}2H_2O$	0.05mg
$Na_2MoO_4{\cdot}2H_2O$	0.03mg
$VOSO_4{\cdot}2H_2O$	0.03mg
$CoSO_4{\cdot}7H_2O$	0.01mg

Preparation of Solution A: Add components to distilled/deionized water and bring volume to 300.0mL. Mix thoroughly. Gently heat and bring to boiling. Autoclave for 15 min at 15 psi pressure–121°C. Cool to 25°C. Gas under 100% N_2 for 20 min.

Solution B:

Composition per 450.0mL:

Sulfur	5.0g

Preparation of Solution B: Add sulfur to distilled/deionized water and bring volume to 450.0mL. Autoclave for 30 min at 0 psi pressure–100°C on 3 consecutive days. Gas under 100% N_2 for 20 min.

Solution C:

Composition per 500.0mL:

Pancreatic digest of casein	2.0g
Yeast extract	2.0g
Resazurin	1.0mg

Preparation of Solution C: Add components to distilled/deionized water and bring volume to 500.0mL. Mix thoroughly. Gently heat and bring to boiling. Autoclave for 15 min at 15 psi pressure–121°C. Cool to 25°C. Gas under 100% N_2 for 20 min.

Solution D:

Composition per 50.0mL:

$Na_2S·9H_2O$	0.5g

Preparation of Solution D: Add $Na_2S·9H_2O$ to distilled/deionized water and bring volume to 50.0mL. Mix thoroughly. Autoclave for 15 min at 15 psi pressure–121°C. Cool to 25°C. Gas under 100% N_2 for 20 min.

Preparation of Medium: Aseptically combine solutions A–D under nitrogen gas. Seal containers with butyl rubber stoppers.

Use: For the cultivation and maintenance of *Desulfurococcus mobilis* and *Desulfurococcus mucosus*.

Desulfurococcus Medium (DSMZ Medium 184)

Composition per liter:

Sulfur, powdered	5.0g
$(NH_4)_2SO_4$	1.3g
Yeast extract	1.0g
KH_2PO_4	0.28g
$MgSO_4·7H_2O$	0.25g
$CaCl_2·2H_2O$	0.07g
$FeCl_3·6H_2O$	0.02g
$Na_2B_4O_7·10H_2O$	4.5mg
$MnCl_2·4H_2O$	1.8mg
Resazurin	1.0mg
$ZnSO_4·7H_2O$	0.22mg
$CuCl_2·2H_2O$	0.05mg
$Na_2MoO_4·4H_2O$	0.03mg
$VOSO_4·2H_2O$	0.03mg
$CoSO_4$	0.01mg
$Na_2S·9H_2O$ solution	10.0mL

pH 5.8 ± 0.2 at 25°C

$Na_2S·9H_2O$ Solution:

Composition per 10.0mL:

$Na_2S·9H_2O$	0.5g

Preparation of $Na_2S·9H_2O$ Solution: Add $Na_2S·9H_2O$ to distilled/deionized water and bring volume to 10.0mL. Mix thoroughly. Sparge with 100% N_2. Autoclave for 15 min at 15 psi pressure–121°C.

Preparation of Sulfur: Add 10.0g of powdered sulfur to a flask and sterilize by steaming for 3 hr on 3 consecutive days.

Preparation of Medium: Prepare and dispense medium under 100% N_2. Add components, except sulfur, and $Na_2S·9H_2O$ solution, to distilled/deionized water and bring volume to 990.0mL. Mix thoroughly. Sparge with 100% N_2. Autoclave for 15 min at 15 psi pressure–121°C. Aseptically add 5.0g of sterile sulfur and 10.0mL of sterile $Na_2S·9H_2O$ solution. Adjust pH to 5.5 with 10*N* sterile H_2SO_4. Mix thoroughly. Aseptically and anaerobically distribute into sterile tubes or flasks.

Use: For the cultivation of *Desulfurococcus mobilis* and *Desulfurococcus mucosus*.

Desulfurococcus Medium

Composition per 1010.0mL:

Sulfur, powdered	5.0g
$(NH_4)_2SO_4$	1.3g
Yeast extract	1.0g
KH_2PO_4	0.28g
$MgSO_4·7H_2O$	0.25g
$CaCl_2·2H_2O$	0.07g
$FeCl_3·6H_2O$	0.02g
$Na_2B_4·10H_2O$	4.5mg
$MnCl_2·4H_2O$	1.8mg
Resazurin	1.0mg
$ZnSO_4·7H_2O$	0.22mg
$CuCl_2·2H_2O$	0.05mg
$Na_2MoO_4·2H_2O$	0.03mg
$VOSO_4·2H_2O$	0.03mg
$CoSO_4$	0.01mg
$Na_2S·9H_2O$ solution	10.0mL

pH 5.5 ± 0.2 at 25°C

Preparation of Sulfur: Sterilize 5.0g of powdered sulfur by steaming for 3 hr on 3 consecutive days.

$Na_2S·9H_2O$ Solution:

Composition per 10.0mL:

$Na_2S·9H_2O$	0.5g

Preparation of $Na_2S·9H_2O$ Solution: Add $Na_2S·9H_2O$ to distilled/deionized water and bring volume to 10.0mL. Mix thoroughly. Gas under 100% N_2. Autoclave for 15 min at 15 psi pressure–121°C.

Preparation of Medium: Add components, except $Na_2S·9H_2O$ solution and sulfur, to distilled/deionized water and bring volume to 1.0L. Mix thoroughly. Sparge with 100% N_2. Autoclave for 15 min at 15 psi pressure–121°C. Aseptically and anaerobically add 10.0mL of sterile $Na_2S·9H_2O$ solution and 5.0g of sterile powdered sulfur. Mix thoroughly. Final pH of medium should be 5.5.

Use: For the cultivation and maintenance of *Desulfurococcus mobilis* and *Desulfurococcus mucosus*.

Desulfuromonas acetexigenes Medium (DSMZ Medium 647)

Composition per liter:

Sulfur, powdered	5.0g
KH_2PO_4	1.0g
NH_4Cl	0.5g
$MgSO_4·7H_2O$	0.4g
$CaCl_2·2H_2O$	0.1g
$NaHCO_3$ solution	33.0mL
Na-pyruvate solution	10.0mL
$Na_2S·9H_2O$ solution	10.0mL
Seven vitamin solution	1.0mL
Trace elements solution SL-10	1.0mL

pH 7.2 ± 0.2 at 25°C

$Na_2S·9H_2O$ Solution:

Composition per 10.0mL:

$Na_2S·9H_2O$	0.5g

Preparation of $Na_2S·9H_2O$ Solution: Add $Na_2S·9H_2O$ to distilled/deionized water and bring volume to 10.0mL. Mix thoroughly. Autoclave under 100% N_2 for 15 min at 15 psi pressure–121°C. Cool to room temperature.

$NaHCO_3$ Solution:

Composition per 100.0mL:

$NaHCO_3$	5.0g

Preparation of $NaHCO_3$ Solution: Add $NaHCO_3$ to distilled/deionized water and bring volume to 100.0mL. Mix thoroughly. Sparge with 80% N_2 + 20% CO_2. Filter sterilize.

Na-pyruvate Solution:

Composition per 10.0mL:

Na-pyruvate	0.6g

Preparation of Na-pyruvate Solution: Add Na-pyruvate to distilled/deionized water and bring volume to 10.0mL. Mix thoroughly. Sparge with 100% N_2. Filter sterilize.

Seven Vitamin Solution:

Composition per liter:

Pyridoxine hydrochloride	300.0mg
Thiamine-HCl·$2H_2O$	200.0mg
Nicotinic acid	200.0mg
Vitamin B_{12}	100.0mg
Calcium pantothenate	100.0mg
p-Aminobenzoic acid	80.0mg
D(+)-Biotin	20.0mg

Preparation of Seven Vitamin Solution: Add components to distilled/deionized water and bring volume to 1.0L. Sparge with 100% N_2. Mix thoroughly. Filter sterilize.

Trace Elements Solution SL-10:

Composition per liter:

$FeCl_2·4H_2O$	1.5g
$CoCl_2·6H_2O$	190.0mg
$MnCl_2·4H_2O$	100.0mg
$ZnCl_2$	70.0mg
$Na_2MoO_4·2H_2O$	36.0mg
$NiCl_2·6H_2O$	24.0mg
H_3BO_3	6.0mg
$CuCl_2·2H_2O$	2.0mg
HCl (25% solution)	10.0mL

Preparation of Trace Elements Solution SL-10: Add $FeCl_2·4H_2O$ to 10.0mL of HCl solution. Mix thoroughly. Add distilled/deionized water and bring volume to 1.0L. Add remaining components. Mix thoroughly. Sparge with 80% N_2 + 20% CO_2. Autoclave for 15 min at 15 psi pressure–121°C.

Preparation of Medium: Sulfur is sterilized by steaming for 3 hr on each of 3 successive days. Prepare and dispense medium under 80% N_2 + 20% CO_2 gas atmosphere. Add components, except sulfur, $NaHCO_3$ solution, Na-pyruvate solution, $Na_2S·9H_2O$ solution, seven vitamin solution, and trace elements solution SL-10, to distilled/deionized water and bring volume to 945.0mL. Mix thoroughly. Adjust pH to 7.2. Sparge with 80% N_2 + 20% CO_2. Autoclave for 15 min at 15 psi pressure–121°C. Aseptically and anaerobically add 5.0g sterile sulfur, 33.0mL $NaHCO_3$ solution, 10.0mL Na-pyruvate solution, 10.0mL $Na_2S·9H_2O$ solution, 1.0mL seven vitamin solution, and 1.0mL trace elements solution SL-10. Mix thoroughly. Aseptically and anaerobically distribute into sterile tubes or bottles.

Use: For the cultivation of *Desulfuromonas thiophila* and *Desulfuromonas acetexigens.*

Desulfuromonas acetexigenes Medium (DSMZ Medium 647)

Composition per liter:

NaCl	20.0g
Sulfur, powdered	5.0g
$MgCl_2·6H_2O$	2.0g
KH_2PO_4	1.0g
$MgSO_4·7H_2O$	1.0g
NH_4Cl	0.5g
$CaCl_2·2H_2O$	0.1g
$NaHCO_3$ solution	50.0mL
Na-pyruvate solution	10.0mL
$Na_2S·9H_2O$ solution	10.0mL
Seven vitamin solution	1.0mL
Trace elements solution SL-10	1.0mL

pH 7.2 ± 0.2 at 25°C

$Na_2S·9H_2O$ Solution:

Composition per 10.0mL:

$Na_2S·9H_2O$	0.5g

Preparation of $Na_2S·9H_2O$ Solution: Add $Na_2S·9H_2O$ to distilled/deionized water and bring volume to 10.0mL. Mix thoroughly. Autoclave under 100% N_2 for 15 min at 15 psi pressure–121°C. Cool to room temperature.

$NaHCO_3$ Solution:

Composition per 100.0mL:

$NaHCO_3$	5.0g

Preparation of $NaHCO_3$ Solution: Add $NaHCO_3$ to distilled/deionized water and bring volume to 100.0mL. Mix thoroughly. Sparge with 80% N_2 + 20% CO_2. Filter sterilize.

Na-pyruvate Solution:

Composition per 10.0mL:

Na-pyruvate	0.6g

Preparation of Na-pyruvate Solution: Add Na-pyruvate to distilled/deionized water and bring volume to 10.0mL. Mix thoroughly. Sparge with 100% N_2. Filter sterilize.

Seven Vitamin Solution:

Composition per liter:

Pyridoxine hydrochloride	300.0mg
Thiamine-HCl·$2H_2O$	200.0mg
Nicotinic acid	200.0mg
Vitamin B_{12}	100.0mg
Calcium pantothenate	100.0mg
p-Aminobenzoic acid	80.0mg
D(+)-Biotin	20.0mg

Preparation of Seven Vitamin Solution: Add components to distilled/deionized water and bring volume to 1.0L. Sparge with 100% N_2. Mix thoroughly. Filter sterilize.

Trace Elements Solution SL-10:

Composition per liter:

$FeCl_2·4H_2O$	1.5g
$CoCl_2·6H_2O$	190.0mg
$MnCl_2·4H_2O$	100.0mg
$ZnCl_2$	70.0mg
$Na_2MoO_4·2H_2O$	36.0mg
$NiCl_2·6H_2O$	24.0mg
H_3BO_3	6.0mg
$CuCl_2·2H_2O$	2.0mg
HCl (25% solution)	10.0mL

Preparation of Trace Elements Solution SL-10: Add $FeCl_2·4H_2O$ to 10.0mL of HCl solution. Mix thoroughly. Add distilled/deionized water and bring volume to 1.0L. Add remaining components. Mix thoroughly. Sparge with 80% N_2 + 20% CO_2. Autoclave for 15 min at 15 psi pressure–121°C.

Preparation of Medium: Sulfur is sterilized by steaming for 3 hr on each of 3 successive days. Prepare and dispense medium under 80% N_2 + 20% CO_2 gas atmosphere. Add components, except sulfur, $NaHCO_3$ solution, Na-pyruvate solution, $Na_2S·9H_2O$ solution, seven vitamin solution, and trace elements solution SL-10, to distilled/deionized water and bring volume to 928.0mL. Mix thoroughly. Adjust pH to 7.2. Sparge with 80% N_2 + 20% CO_2. Autoclave for 15 min at 15 psi pressure–121°C. Aseptically and anaerobically add 5.0g sterile sulfur, 50.0mL $NaHCO_3$ solution, 10.0mL Na-pyruvate solution, 10.0mL $Na_2S·9H_2O$ solution, 1.0mL seven vitamin solution, and 1.0mL trace elements solution SL-10. Mix thoroughly. Aseptically and anaerobically distribute into sterile tubes or bottles.

Use: For the cultivation of *Desulfuromonas acetexigenes* DSM8988.

Desulfuromonas acetoxidans Medium

Composition per 1001.0mL:

Fumaric acid	1.5g
KH_2PO_4	1.0g
NH_4Cl	0.5g
Sodium acetate	0.5g
$MgSO_4·7H_2O$	0.4g
$CaCl_2·2H_2O$	0.1g
Resazurin	0.5mg
$NaHCO_3$ solution	40.0mL
Trace elements solution SL-4	10.0mL
$Na_2S·9H_2O$ solution	6.0mL
Vitamin solution	5.0mL

pH 7.5 ± 0.2 at 25°C

$NaHCO_3$ Solution:

Composition per 40.0mL:

$NaHCO_3$	2.0g

Preparation of $NaHCO_3$ Solution: Add $NaHCO_3$ to distilled/deionized water and bring volume to 40.0mL. Mix thoroughly. Gas under 80% N_2 + 20% CO_2. Autoclave for 15 min at 15 psi pressure–121°C.

Trace Elements Solution SL-4:

Composition per liter:

EDTA	0.5g
$FeSO_4·7H_2O$	0.2g
Trace elements solution SL-6	100.0mL

Trace Elements Solution SL-6:

Composition per liter:

$MnCl_2·4H_2O$	0.5g
H_3BO_3	0.3g
$CoCl_2·6H_2O$	0.2g
$ZnSO_4·7H_2O$	0.1g
$Na_2MoO_4·2H_2O$	0.03g
$NiCl_2·6H_2O$	0.02g
$CuCl_2·2H_2O$	0.01g

Preparation of Trace Elements Solution SL-6: Add components to distilled/deionized water and bring volume to 1.0L. Mix thoroughly. Gas under 100% N_2. Autoclave for 15 min at 15 psi pressure–121°C.

Preparation of Trace Elements Solution SL-4: Add components to distilled/deionized water and bring volume to 100.0mL. Mix thoroughly.

$Na_2S·9H_2O$ Solution:

Composition per 6.0mL:

$Na_2S·9H_2O$	0.3g

Preparation of $Na_2S·9H_2O$ Solution: Add $Na_2S·9H_2O$ to distilled/deionized water and bring volume to 10.0mL. Mix thoroughly. Gas under 100% N_2. Autoclave for 15 min at 15 psi pressure–121°C.

Vitamin Solution:

Composition per liter:

Pyridoxine·HCl	62.5g
Nicotinic acid	25.0mg
p-Aminobenzoic acid	12.5mg
Thiamine·HCl	12.5mg
Calcium DL-pantothenate	6.5mg
Biotin	2.5mg

Preparation of Vitamin Solution: Add components to distilled/deionized water and bring volume to 1.0L. Mix thoroughly. Adjust pH to 7.5. Gas under 100% N_2. Filter sterilize.

Preparation of Medium: Add components, except $NaHCO_3$ solution, $Na_2S·9H_2O$ solution, and vitamin solution, to distilled/deionized water and bring volume to 1.0L. Mix thoroughly. Adjust pH to 5.0. Sparge with 100% N_2. Autoclave for 15 min at 15 psi pressure–121°C. Aseptically and anaerobically add 40.0mL of sterile $NaHCO_3$ solution, 6.0mL of sterile $Na_2S·9H_2O$ solution, and 5.0mL of sterile vitamin solution. Mix thoroughly. Aseptically and anaerobically distribute into sterile screw-capped bottles or tubes. Fill completely, leaving only a small gas bubble.

Use: For the cultivation and maintenance of *Desulfuromonas acetoxidans*.

Desulfuromonas Medium

Composition per 1051.0mL:

Elemental sulfur slurry	10.0g
Solution 1	1.0L
Solution 3	40.0mL
Solution 4	6.0mL
Solution 5	5.0mL
Solution 2	1.0mL

pH 7.2 ± 0.2 at 25°C

Solution 1:

Composition per liter:

NaCl	20.0g
$MgCl_2·6H_2O$	3.0g
KH_2PO_4	1.0g
NH_4Cl	0.3g
$CaCl_2·2H_2O$	0.1g
HCl (2*N* solution)	4.0mL

Preparation of Solution 1: Add components to distilled/deionized water and bring volume to 1.0L. Mix thoroughly. Autoclave for 15 min at 15 psi pressure–121°C. Cool to 25°C.

Solution 2:

Composition per liter:

Disodium EDTA	5.2g
$CoCl_2·6H_2O$	1.9g
$FeCl_2·4H_2O$	1.5g
$MnCl_2·4H_2O$	1.0g
$ZnCl_2$	0.7g
H_3BO_3	0.62g
$Na_2MoO_4·2H_2O$	0.36g
$NiCl_2·6H_2O$	0.24g
$CuCl_2·2H_2O$	0.17g

pH 6.5 ± 0.2 at 25°C

Preparation of Solution 2: Add components to distilled/deionized water and bring volume to 1.0L. Mix thoroughly. Adjust pH to 6.5. Autoclave for 15 min at 15 psi pressure–121°C. Cool to 25°C.

Solution 3:

Composition per 100.0mL:

$NaHCO_3$	10.0g

Preparation of Solution 3: Add $NaHCO_3$ to distilled/deionized water and bring volume to 100.0mL. Mix thoroughly. Autoclave for 15 min at 15 psi pressure–121°C. Cool to 25°C.

Solution 4:

Composition per 100.0mL:

$Na_2S{\cdot}9H_2O$	5.0g

Preparation of Solution 4: Add $Na_2S{\cdot}9H_2O$ to distilled/deionized water and bring volume to 100.0mL. Mix thoroughly. Autoclave for 15 min at 15 psi pressure–121°C. Cool to 25°C.

Solution 5:

Composition per 200.0mL:

Pyridoxamine·HCl	0.01g
Nicotinic acid	4.0mg
p-Aminobenzoic acid	2.0mg
Thiamine	2.0mg
Cyanocobalamin	1.0mg
Pantothenic acid	1.0mg
Biotin	0.5mg

Preparation of Solution 5: Add components to distilled/deionized water and bring volume to 200.0mL. Mix thoroughly. Filter sterilize.

Elemental Sulfur Slurry:

Composition per 10.0g:

Sulfur flowers	10.0g

Preparation of Elemental Sulfur Slurry: Add highly purified sulfur flowers to a mortar and grind to a fine powder. Add sufficient distilled/deionized water to produce a slurry. Distribute into 100.0mL screw-capped bottles in 20.0mL volumes. Autoclave for 30 min at 10 psi pressure–115°C. Decant supernatant solution. Reserve sulfur slurry.

Preparation of Medium: To 1.0L of cooled, sterile solution 1, aseptically add 1.0mL of sterile solution 2, 40.0mL of sterile solution 3, 6.0mL of sterile solution 4, and 5.0mL of sterile solution 5. Mix thoroughly. Adjust pH to 7.2. Aseptically distribute into sterile 50.0mL screw-capped bottles. Fill bottles completely with medium except for a pea-sized air bubble. Aseptically add a pea-sized piece of sulfur slurry to each 50.0mL of medium.

Use: For the isolation and cultivation of marine *Desulfuromonas* species.

Desulfuromonas Medium

Composition per 1031.0mL:

Elemental sulfur slurry	10.0g
Solution 1	1.0L
Solution 3	20.0mL
Solution 4	6.0mL
Solution 5	5.0mL
Solution 2	1.0mL

pH 7.2 ± 0.2 at 25°C

Solution 1:

Composition per liter:

KH_2PO_4	1.0g
$MgCl_2{\cdot}6H_2O$	0.4g
NH_4Cl	0.3g
$CaCl_2{\cdot}2H_2O$	0.1g
HCl (2*N* solution)	4.0mL

Preparation of Solution 1: Add components to distilled/deionized water and bring volume to 1.0L. Mix thoroughly. Autoclave for 15 min at 15 psi pressure–121°C. Cool to 25°C.

Solution 2:

Composition per liter:

Disodium EDTA	5.2g
$CoCl_2{\cdot}6H_2O$	1.9g
$FeCl_2{\cdot}4H_2O$	1.5g
$MnCl_2{\cdot}4H_2O$	1.0g
$ZnCl_2$	0.7g
H_3BO_3	0.62g
$Na_2MoO_4{\cdot}2H_2O$	0.36g
$NiCl_2{\cdot}6H_2O$	0.24g
$CuCl_2{\cdot}2H_2O$	0.17g

Preparation of Solution 2: Add components to distilled/deionized water and bring volume to 1.0L. Mix thoroughly. Adjust pH to 6.5. Autoclave for 15 min at 15 psi pressure–121°C. Cool to 25°C.

Solution 3:

Composition per 100.0mL:

$NaHCO_3$	10.0g

Preparation of Solution 3: Add $NaHCO_3$ to distilled/deionized water and bring volume to 100.0mL. Mix thoroughly. Autoclave for 15 min at 15 psi pressure–121°C. Cool to 25°C.

Solution 4:

Composition per 100.0mL:

$Na_2S{\cdot}9H_2O$	5.0g

Preparation of Solution 4: Add $Na_2S{\cdot}9H_2O$ to distilled/deionized water and bring volume to 100.0mL. Mix thoroughly. Autoclave for 15 min at 15 psi pressure–121°C. Cool to 25°C.

Solution 5:

Composition per 200.0mL:

Pyridoxamine·HCl	0.01g
Nicotinic acid	4.0mg
p-Aminobenzoic acid	2.0mg
Thiamine	2.0mg
Cyanocobalamin	1.0mg
Pantothenic acid	1.0mg
Biotin	0.5mg

Preparation of Solution 5: Add components to distilled/deionized water and bring volume to 200.0mL. Mix thoroughly. Filter sterilize.

Elemental Sulfur Slurry:

Composition per 10.0g:

Sulfur flowers	10.0g

Preparation of Elemental Sulfur Slurry: Add highly purified sulfur flowers to a mortar and grind to a fine powder. Add sufficient distilled/deionized water to produce a slurry. Distribute into 100.0mL screw-capped bottles in 20.0mL volumes. Autoclave for 30 min at 10 psi pressure–115°C. Decant supernatant solution. Reserve sulfur slurry.

Preparation of Medium: To 1.0L of cooled, sterile solution 1, aseptically add 1.0mL of sterile solution 2, 40.0mL of sterile solution 3, 6.0mL of sterile solution 4, and 5.0mL of sterile solution 5. Mix thoroughly. Adjust pH to 7.2. Aseptically distribute into sterile 50.0mL screw-capped bottles. Fill bottles completely with medium except for a pea-sized air bubble. Aseptically add a pea-sized piece of sulfur slurry to each 50.0mL of medium.

Use: For the isolation and cultivation of freshwater *Desulfuromonas* species.

Desulfuromonas Medium

Composition per 1003.0mL:

Solution A	950.0mL
Solution C	40.0mL
Solution D	10.0mL
Solution B	3.0mL

pH 7.2 ± 0.2 at 25°C

Solution A:

Composition per 950.0mL:

NaCl	20.0g
Disodium DL-malate	2.66g
$MgCl_2 \cdot 6H_2O$	2.0g
Na_2SO_4	1.77g
$MgSO_4 \cdot 7H_2O$	1.0g
KH_2PO_4	0.5g
Yeast extract	0.5g
NH_4Cl	0.3g
$CaCl_2 \cdot 2H_2O$	0.1g
Resazurin	1.0mg
Trace elements solution SL-4	10.0mL

Trace Elements Solution SL-4:

Composition per liter:

EDTA	0.5g
$FeSO_4 \cdot 7H_2O$	0.2g
Trace elements solution SL-6	100.0mL

Trace Elements Solution SL-6:

Composition per liter:

$MnCl_2 \cdot 4H_2O$	0.5g
H_3BO_3	0.3g
$CoCl_2 \cdot 6H_2O$	0.2g
$ZnSO_4 \cdot 7H_2O$	0.1g
$Na_2MoO_4 \cdot 2H_2O$	0.03g
$NiCl_2 \cdot 6H_2O$	0.02g
$CuCl_2 \cdot 2H_2O$	0.01g

Preparation of Trace Elements Solution SL-6: Add components to distilled/deionized water and bring volume to 1.0L. Mix thoroughly. Gas under 100% N_2. Autoclave for 15 min at 15 psi pressure–121°C.

Preparation of Trace Elements Solution SL-4: Add components to distilled/deionized water and bring volume to 1.0L. Mix thoroughly.

Preparation of Solution A: Add components to distilled/deionized water and bring volume to 950.0mL. Mix thoroughly. Adjust pH to 6.0 with 2*N* NaOH. Gently heat and bring to boiling. Continue boiling for 3–4 min. Allow to cool to room temperature while gassing under 100% N_2. Autoclave for 15 min at 15 psi pressure–121°C.

Solution B:

Composition per 3.0mL:

Ethanol	0.3mL

Preparation of Solution B: Add ethanol to distilled/deionized water and bring volume to 3.0mL. Sparge under 100% N_2. Autoclave for 15 min at 15 psi pressure–121°C.

Solution C:

Composition per 40.0mL:

$NaHCO_3$	1.85g

Preparation of Solution C: Add $NaHCO_3$ to distilled/deionized water and bring volume to 40.0mL. Mix thoroughly. Gas under 80% N_2 + 20% CO_2. Filter sterilize.

Solution D:

Composition per 10.0mL:

$Na_2S \cdot 9H_2O$	0.3g

Preparation of Solution D: Add $Na_2S \cdot 9H_2O$ to distilled/deionized water and bring volume to 10.0mL. Mix thoroughly. Gas under 100% N_2. Autoclave for 15 min at 15 psi pressure–121°C.

Preparation of Medium: To 950.0mL of sterile solution A, aseptically and anaerobically add 3.0mL of sterile solution B, 40.0mL of sterile solution C, and 10.0mL of sterile solution D. Mix thoroughly. Aseptically and anaerobically distribute into tubes or flasks.

Use: For the cultivation and maintenance of *Desulfuromonas* species.

Desulfuromonas spp. Medium (DSMZ Medium 148)

Composition per 1001.0mL:

Fumaric acid	1.5g
KH_2PO_4	1.0g
NH_4Cl	0.5g
Na-acetate	0.5g
$MgSO_4 \cdot 7H_2O$	0.4g
$CaCl_2 \cdot 2H_2O$	0.1g
Resazurin	0.5mg
$NaHCO_3$ solution	40.0mL
Trace elements solution SL-4	10.0mL
$Na_2S \cdot 9H_2O$ solution	6.0mL
Vitamin solution	5.0mL

pH 7.5 ± 0.2 at 25°C

$NaHCO_3$ Solution:

Composition per 100.0mL:

$NaHCO_3$	5.0g

Preparation of $NaHCO_3$ Solution: Add $NaHCO_3$ to distilled/deionized water and bring volume to 100.0mL. Mix thoroughly. Autoclave for 15 min at 15 psi pressure–121°C. Cool to 25°C. Must be prepared freshly.

Trace Elements Solution SL-4:

Composition per liter:

EDTA	0.5g
$FeSO_4 \cdot 7H_2O$	0.2g
Trace elements solution SL-6	100.0mL

Trace Elements Solution SL-6:

Composition per liter:

H_3BO_3	0.3g
$CoCl_2 \cdot 6H_2O$	0.2g
$ZnSO_4 \cdot 7H_2O$	0.1g
$MnCl_2 \cdot 4H_2O$	0.03g
$Na_2MoO_4 \cdot H_2O$	0.03g
$NiCl_2 \cdot 6H_2O$	0.02g
$CuCl_2 \cdot 2H_2O$	0.01g

Preparation of Trace Elements Solution SL-6: Add components to distilled/deionized water and bring volume to 1.0L. Mix thoroughly. Adjust pH to 3.4.

Preparation of Trace Elements Solution SL-4: Add components to distilled/deionized water and bring volume to 1.0L. Mix thoroughly.

Vitamin Solution:

Composition per liter:

Pyridoxine-HCl	62.5mg
Nicotinic acid	25.0mg
Thiamine-HCl·2H_2O	12.5mg
p-Aminobenzoic acid	12.5mg
Ca-pantothenate	6.5mg
Biotin	2.5mg

Preparation of Vitamin Solution: Add components to distilled/deionized water and bring volume to 1.0L. Mix thoroughly. Filter sterilize.

$Na_2S·9H_2O$ Solution:

Composition per 100.0mL:

$Na_2S·9H_2O$	5.0g

Preparation of $Na_2S·9H_2O$ Solution: Add $Na_2S·9H_2O$ to distilled/deionized water and bring volume to 100.0mL. Mix thoroughly. Sparge with 100% N_2. Autoclave for 15 min at 15 psi pressure–121°C. Before use, neutralize to pH 7.0 with sterile HCl.

Preparation of Medium: Add components, except vitamin solution, $NaHCO_3$ solution, and $Na_2S·9H_2O$ solution, to 950.0mL distilled/deionized water. Adjust pH to 5.0. Autoclave for 15 min at 15 psi pressure–121°C. Aseptically and anaerobically add 5.0mL vitamin solution, 40.0mL $NaHCO_3$ solution, and 6.0mL $Na_2S·9H_2O$ solution. Mix thoroughly. Adjust pH to 7.5 with sterile 2*M* Na_2CO_3 and 2*M* H_2SO_4, respectively. Aseptically distribute into sterile screw-capped bottles or tubes. Fill completely, leaving only a small gas bubble.

Use: For the cultivation of *Desulfuromonas acetexigenes.*

Desulfuromonas succinoxidans Medium

Composition per liter:

NaCl	20.0g
$MgCl_2·6H_2O$	2.0g
KH_2PO_4	1.0g
$MgSO_4·7H_2O$	1.0g
NH_4Cl	0.3g
$CaCl_2·2H_2O$	0.1g
Resazurin	0.5mg
$NaHCO_3$ solution	50.0mL
Disodium fumarate solution	10.0mL
Sodium acetate solution	10.0mL
$Na_2S·9H_2O$ solution	10.0mL
Trace elements solution SL-10	1.0mL
Seven vitamin solution	1.0mL

pH 7.2 ± 0.2 at 25°C

$NaHCO_3$ Solution:

Composition per 50.0mL:

$NaHCO_3$	2.5g

Preparation of $NaHCO_3$ Solution: Add $NaHCO_3$ to distilled/deionized water and bring volume to 50.0mL. Mix thoroughly. Sparge with 80% N_2 + 20% CO_2. Autoclave for 15 min at 15 psi pressure–121°C.

Disodium Fumarate Solution:

Composition per 10.0mL:

Disodium fumarate	1.6g

Preparation of Disodium Fumarate Solution: Add disodium fumarate to distilled/deionized water and bring volume to 10.0mL. Mix thoroughly. Sparge with 100% N_2. Autoclave for 15 min at 15 psi pressure–121°C.

Sodium Acetate Solution:

Composition per 10.0mL:

Sodium acetate	0.8g

Preparation of Sodium Acetate Solution: Add sodium acetate to distilled/deionized water and bring volume to 10.0mL. Mix thoroughly. Sparge with 100% N_2. Autoclave for 15 min at 15 psi pressure–121°C.

$Na_2S·9H_2O$ Solution:

Composition per 10.0mL:

$Na_2S·9H_2O$	0.5g

Preparation of $Na_2S·9H_2O$ Solution: Add $Na_2S·9H_2O$ to distilled/deionized water and bring volume to 10.0mL. Mix thoroughly. Sparge with 100% N_2. Autoclave for 15 min at 15 psi pressure–121°C.

Trace Elements Solution SL-10:

Composition per liter:

$FeCl_2·4H_2O$	1.5g
$CoCl_2·6H_2O$	190.0mg
$MnCl_2·4H_2O$	100.0mg
$ZnCl_2$	70.0mg
$Na_2MoO_4·2H_2O$	36.0mg
$NiCl_2·6H_2O$	24.0mg
H_3BO_3	6.0mg
$CuCl_2·2H_2O$	2.0mg
HCl (25% solution)	10.0mL

Preparation of Trace Elements Solution SL-10: Add $FeCl_2·4H_2O$ to 10.0mL of HCl solution. Mix thoroughly. Add distilled/deionized water and bring volume to 1.0L. Add remaining components. Mix thoroughly.

Seven Vitamin Solution:

Composition per liter:

Pyridoxine·HCl	0.3g
Thiamine·HCl	0.2g
Nicotinic acid	0.2g
Calcium DL-pantothenate	0.1g
Vitamin B_{12}	0.1g
p-Aminobenzoic acid	80.0mg
Biotin	20.0mg

Preparation of Seven Vitamin Solution: Add components to distilled/deionized water and bring volume to 1.0L. Mix thoroughly. Filter sterilize. Sparge with 100% N_2.

Preparation of Medium: Prepare and dispense medium under 80% N_2 + 20% CO_2. Add components, except $NaHCO_3$ solution, disodium fumarate solution, sodium acetate solution, $Na_2S·9H_2O$ solution, and seven vitamin solution, to distilled/deionized water and bring volume to 920.0mL. Mix thoroughly. Sparge with 80% N_2 + 20% CO_2. Autoclave for 15 min at 15 psi pressure–121°C. Aseptically and anaerobically add 50.0mL of sterile $NaHCO_3$ solution, 10.0mL of sterile disodium fumarate solution, 10.0mL of sterile sodium acetate solution, 10.0mL of sterile $Na_2S·9H_2O$ solution, and 1.0mL of sterile seven vitamin solution. Mix thoroughly.

Use: For the cultivation of *Desulfuromonas succinoxidans.*

Desultobacterium Medium (DSMZ Medium 383)

Composition per 1012.0mL:

Soultion A	930.0mL
Solution C	50.0mL
Solution D	10.0mL
Solution E	10.0mL
Solution G	10.0mL
Solution B	1.0mL
Solution F	1.0mL

pH 7.0 at 25°C

Solution A:

Composition per 930.0mL:

NaCl	21.0g
Na_2SO_4	3.0g
$MgCl_2 \cdot 6H_2O$	3.0g
KCl	0.5g
NH_4Cl	0.3g
KH_2PO_4	0.2g
$CaCl_2 \cdot 2H_2O$	0.15g
Resazurin	1.0mg

Preparation of Solution A: Add components to distilled/deionized water and bring volume to 930.0mL. Mix thoroughly. Sparge with 80% N_2 + 20% CO_2 gas until saturated. Autoclave for 15 min at 15 psi pressure–121°C. Cool to 25°C.

Solution B:

Composition per liter:

$FeCl_2 \cdot 4H_2O$	1.5g
$CoCl_2 \cdot 6H_2O$	190.0mg
$MnCl_2 \cdot 4H_2O$	100.0mg
$ZnCl_2$	70.0mg
$Na_2MoO_4 \cdot 2H_2O$	36.0mg
$NiCl_2 \cdot 6H_2O$	24.0mg
H_3BO_3	300.0mg
$CuCl_2 \cdot 2H_2O$	2.0mg
HCl (25% solution)	7.7mL

Preparation of Solution B: Add $FeCl_2 \cdot 4H_2O$ to 10.0mL of HCl solution. Mix thoroughly. Add distilled/deionized water and bring volume to 1.0L. Add remaining components. Mix thoroughly. Sparge with 100% N_2. Autoclave for 15 min at 15 psi pressure–121°C.

Solution C:

Composition per 100.0mL:

$NaHCO_3$	5.0g

Preparation of Solution C: Add $NaHCO_3$ to distilled/deionized water and bring volume to 100.0mL. Mix thoroughly. Sparge with 100% CO_2 until saturated, approximately 20 min. Filter sterilize under 100% CO_2 into a sterile, gas-tight 100.0mL screw-capped bottle.

Solution D:

Composition per 10.0mL:

Na_2-benzoate	0.5g

Preparation of Solution D: Add Na_2-benzoate to distilled/deionized water and bring volume to 10.0mL. Sparge with N_2. Filter sterilize. Store anaerobically.

Solution E:

Composition per liter:

Pyridoxine-HCl	10.0mg
Thiamine-HCl·$2H_2O$	5.0mg
Riboflavin	5.0mg
Nicotinic acid	5.0mg
D-Ca-pantothenate	5.0mg
p-Aminobenzoic acid	5.0mg
Lipoic acid	5.0mg
Biotin	2.0mg
Folic acid	2.0mg
Vitamin B_{12}	0.1mg

Preparation of Solution E: Add components to distilled/deionized water and bring volume to 1.0L. Mix thoroughly. Sparge with 100% N_2. Filter sterilize.

Solution F:

Composition per liter:

NaOH	0.5g
$Na_2SeO_3 \cdot 5H_2O$	3.0mg

Preparation of Solution F: Add components to distilled/deionized water and bring volume to 1.0L. Mix thoroughly. Sparge with 100% N_2. Filter sterilize.

Solution G:

Composition per 10.0mL:

$Na_2S \cdot 9H_2O$	0.4g

Preparation of Solution G: Add $Na_2S \cdot 9H_2O$ to distilled/deionized water and bring volume to 10.0mL. Sparge with N_2. Autoclave for 15 min at 15 psi pressure–121°C. Cool to 25°C. Store anaerobically.

Preparation of Medium: Add solutions B, C, D, E, F, and G to solution A in that order under N_2 gas. Adjust the pH to 7.0.

Use: For the cultivation of *Desulfotobacterium* spp.

Dethiobacter Medium (DSMZ Medium 1104)

Composition per liter:

Na_2CO_3	22.0g
$NaHCO_3$	8.0g
NaCl	6.0g
Sulfur, powdered	1.5g
K_2HPO_4	0.5g
Thiosulfate solution	10.0mL
Vitamin solution	10.0mL
Ammonium chloride solution	10.0mL
Magnesium chloride solution	10.0mL
$Na_2S \cdot 9H_2O$ solution	10.0mL
Yeast extract solution	10.0mL
Acetate solution	10.0mL
Trace elements solution SL-10 with EDTA	1.0mL
Selenite/tungstate solution	1.0mL
Ethanol	0.6mL

pH 9.5 ± 0.2 at 25°C

Ammonium Chloride Solution:

Composition per 10.0mL:

NH_4Cl	0.2g

Preparation of Ammonium Chloride Solution: Add NH_4Cl to distilled/deionized water and bring volume to 10.0mL. Mix thoroughly. Sparge with 100% N_2. Autoclave for 15 min at 15 psi pressure–121°C.

Magnesium Chloride Solution:

Composition per 10.0mL:

$MgCl_2 \cdot 6H_2O$	0.2g

Preparation of Magnesium Chloride Solution: Add $MgCl_2·6H_2O$ to distilled/deionized water and bring volume to 10.0mL. Mix thoroughly. Sparge with 100% N_2. Autoclave for 15 min at 15 psi pressure–121°C.

$Na_2S·9H_2O$ Solution:

Composition per 10.0mL:

$Na_2S·9H_2O$	0.24g

Preparation of $Na_2S·9H_2O$ Solution: Add $Na_2S·9H_2O$ to distilled/deionized water and bring volume to 10.0mL. Mix thoroughly. Autoclave under 100% N_2 for 15 min at 15 psi pressure–121°C. Cool to room temperature.

Thiosulfate Solution:

Composition per 10.0mL:

$Na_2S_2O_3$	2.5g

Preparation of Thiosulfate Solution: Add $Na_2S_2O_3$ to distilled/deionized water and bring volume to 10.0mL. Mix thoroughly. Sparge with 100% N_2. Filter sterilize.

Trace Elements Solution SL-10 with EDTA:

Composition per liter:

$FeCl_2·4H_2O$	1.5g
Na_2-EDTA	0.5g
$CoCl_2·6H_2O$	190.0mg
$MnCl_2·4H_2O$	100.0mg
$ZnCl_2$	70.0mg
$Na_2MoO_4·2H_2O$	36.0mg
$NiCl_2·6H_2O$	24.0mg
H_3BO_3	6.0mg
$CuCl_2·2H_2O$	2.0mg
HCl (25% solution)	10.0mL

Preparation of Trace Elements Solution SL-10 with EDTA: Add $FeCl_2·4H_2O$ to 10.0mL of HCl solution. Mix thoroughly. Add distilled/deionized water and bring volume to 1.0L. Add remaining components. Mix thoroughly. Sparge with 100% N_2. Autoclave for 15 min at 15 psi pressure–121°C. Adjust pH to 7.0.

Selenite/Tungstate Solution:

Composition per liter:

NaOH	0.5g
$Na_2WO_4·2H_2O$	4.0mg
$Na_2SeO_3·5H_2O$	3.0mg

Preparation of Selenite/Tungstate Solution: Add components to distilled/deionized water and bring volume to 1.0L. Mix thoroughly. Sparge with 100% N_2. Autoclave for 15 min at 15 psi pressure–121°C.

Vitamin Solution:

Composition per liter:

Pyridoxine-HCl	10.0mg
Thiamine-HCl·$2H_2O$	5.0mg
Riboflavin	5.0mg
Nicotinic acid	5.0mg
D-Ca-pantothenate	5.0mg
p-Aminobenzoic acid	5.0mg
Lipoic acid	5.0mg
Biotin	2.0mg
Folic acid	2.0mg
Vitamin B_{12}	0.1mg

Preparation of Vitamin Solution: Add components to distilled/deionized water and bring volume to 1.0L. Mix thoroughly. Sparge with 80% H_2 + 20% CO_2. Filter sterilize.

Yeast Extract Solution:

Composition per 10.0mL:

Yeast extract	0.05g

Preparation of Yeast Extract Solution: Add yeast extract to distilled/deionized water and bring volume to 10.0mL. Mix thoroughly. Sparge with 100% N_2. Autoclave for 15 min at 15 psi pressure–121°C.

Acetate Solution:

Composition per 10.0mL:

Sodium acetate	0.4g

Preparation of Acetate Solution: Add sodium acetate to distilled/deionized water and bring volume to 10.0mL. Mix thoroughly. Sparge with 100% N_2. Autoclave for 15 min at 15 psi pressure–121°C.

Preparation of Medium: Add components, except ammonium chloride, magnesium chloride, yeast extract, trace elements, vitamin, ethanol, acetate, sulfur, thiosulfate, and sulfide solutions, to distilled/deionized water and bring volume to 930.0mL. Gently heat and bring to boiling. Boil for 1 min. Mix thoroughly. Cool to room temperature while sparging with 100% N_2. Dispense into culture vessels under an atmosphere of 100% N_2. Autoclave for 15 min at 15 psi pressure–121°C. Aseptically add the ammonium chloride, magnesium chloride, yeast extract, trace elements, vitamin, ethanol, acetate, sulfur, thiosulfate, and sulfide solutions. Sulfur is sterilized by steaming for 3 h on each of 3 successive days and added as powder. Adjust pH of final medium to 9.5. After inoculation add sterile H_2 gas to 0.5 bar overpressure.

Use: For the cultivation of *Dethiobacter* spp.

Dethiobacter Medium (DSMZ Medium 1104)

Composition per liter:

Na_2CO_3	22.0g
$NaHCO_3$	8.0g
NaCl	6.0g
Sulfur, powdered	1.5g
K_2HPO_4	0.5g
Thiosulfate solution	10.0mL
Vitamin solution	10.0mL
Ammonium chloride solution	10.0mL
Magnesium chloride solution	10.0mL
$Na_2S·9H_2O$ solution	10.0mL
Yeast extract solution	10.0mL
Acetate solution	10.0mL
Butyrate solution	10.0mL
Trace element ssolution SL-10 with EDTA	1.0mL
Selenite/tungstate solution	1.0mL

pH 9.5 ± 0.2 at 25°C

Ammonium Chloride Solution:

Composition per 10.0mL:

NH_4Cl	0.2g

Preparation of Ammonium Chloride Solution: Add NH_4Cl to distilled/deionized water and bring volume to 10.0mL. Mix thoroughly. Sparge with 100% N_2. Autoclave for 15 min at 15 psi pressure–121°C.

Magnesium Chloride Solution:

Composition per 10.0mL:

$MgCl_2·6H_2O$	0.2g

Preparation of Magnesium Chloride Solution: Add $MgCl_2·6H_2O$ to distilled/deionized water and bring volume to 10.0mL. Mix thor-

oughly. Sparge with 100% N_2. Autoclave for 15 min at 15 psi pressure–121°C.

$Na_2S·9H_2O$ Solution:

Composition per 10.0mL:

$Na_2S·9H_2O$ 0.24g

Preparation of $Na_2S·9H_2O$ Solution: Add $Na_2S·9H_2O$ to distilled/deionized water and bring volume to 10.0mL. Mix thoroughly. Autoclave under 100% N_2 for 15 min at 15 psi pressure–121°C. Cool to room temperature.

Thiosulfate Solution:

Composition per 10.0mL:

$Na_2S_2O_3$ 2.5g

Preparation of Thiosulfate Solution: Add $Na_2S_2O_3$ to distilled/deionized water and bring volume to 10.0mL. Mix thoroughly. Sparge with 100% N_2. Filter sterilize.

Trace Elements Solution SL-10 with EDTA:

Composition per liter:

$FeCl_2·4H_2O$ 1.5g
Na_2-EDTA 0.5g
$CoCl_2·6H_2O$ 190.0mg
$MnCl_2·4H_2O$ 100.0mg
$ZnCl_2$ 70.0mg
$Na_2MoO_4·2H_2O$ 36.0mg
$NiCl_2·6H_2O$ 24.0mg
H_3BO_3 .... 6.0mg
$CuCl_2·2H_2O$ 2.0mg
HCl (25% solution) 10.0mL

Preparation of Trace Elements Solution SL-10 with EDTA: Add $FeCl_2·4H_2O$ to 10.0mL of HCl solution. Mix thoroughly. Add distilled/deionized water and bring volume to 1.0L. Add remaining components. Mix thoroughly. Sparge with 100% N_2. Autoclave for 15 min at 15 psi pressure–121°C. Adjsut pH to 7.0.

Selenite/Tungstate Solution:

Composition per liter:

NaOH 0.5g
$Na_2WO_4·2H_2O$ 4.0mg
$Na_2SeO_3·5H_2O$ 3.0mg

Preparation of Selenite/Tungstate Solution: Add components to distilled/deionized water and bring volume to 1.0L. Mix thoroughly. Sparge with 100% N_2. Autoclave for 15 min at 15 psi pressure–121°C.

Vitamin Solution:

Composition per liter:

Pyridoxine-HCl 10.0mg
Thiamine-HCl·$2H_2O$ 5.0mg
Riboflavin 5.0mg
Nicotinic acid 5.0mg
D-Ca-pantothenate 5.0mg
p-Aminobenzoic acid 5.0mg
Lipoic acid 5.0mg
Biotin 2.0mg
Folic acid 2.0mg
Vitamin B_{12} 0.1mg

Preparation of Vitamin Solution: Add components to distilled/deionized water and bring volume to 1.0L. Mix thoroughly. Sparge with 80% H_2 + 20% CO_2. Filter sterilize.

Yeast Extract Solution:

Composition per 10.0mL:

Yeast extract 0.05g

Preparation of Yeast Extract Solution: Add yeast extract to distilled/deionized water and bring volume to 10.0mL. Mix thoroughly. Sparge with 100% N_2. Autoclave for 15 min at 15 psi pressure–121°C.

Acetate Solution:

Composition per 10.0mL:

Sodium acetate 0.4g

Preparation of Acetate Solution: Add sodium acetate to distilled/deionized water and bring volume to 10.0mL. Mix thoroughly. Sparge with 100% N_2. Autoclave for 15 min at 15 psi pressure–121°C.

Butyrate Solution:

Composition per 10.0mL:

Sodium butyrate 1.1g

Preparation of Butyrate Solution: Add sodium butyrate to distilled/deionized water and bring volume to 10.0mL. Mix thoroughly. Sparge with 100% N_2. Autoclave for 15 min at 15 psi pressure–121°C.

Preparation of Medium: Add components, except ammonium chloride, magnesium chloride, yeast extract, trace elements, vitamin, ethanol, acetate, sulfur, thiosulfate, and sulfide solutions, to distilled/deionized water and bring volume to 920.0mL. Gently heat and bring to boiling. Boil for 1 min. Mix thoroughly. Cool to room temperature while sparging with 100% N_2. Dispense into culture vessels under an atmosphere of 100% N_2. Autoclave for 15 min at 15 psi pressure–121°C. Aseptically add the ammonium chloride, magnesium chloride, yeast extract, trace elements, vitamin, ethanol, acetate, sulfur, thiosulfate, and sulfide solutions. Sulfur is sterilized by steaming for 3 h on each of 3 successive days and added as powder. Adjust pH of final medium to 9.5. After inoculation add sterile H_2 gas to 0.5 bar overpressure.

Use: For the cultivation of *Desulfurivibrio alkaliphilus.*

Dethiosulfovibrio II Medium (DSMZ Medium 906)

Composition per liter:

NaCl 20.0g
Yeast extract 5.0g
$MgCl_2·6H_2O$ 3.0g
Na_3-citrate·$5H_2O$ 3.0g
Peptone 2.0g
KH_2PO_4 .... 1.0g
Resazurin 0.5mg
Calcium chloride solution 10.0mL
$Na_2S·9H_2O$ solution 10.0mL
$Na_2S_2O_3$ solution 10.0mL
Trace elements solution SL-10 .... 1.0mL

pH 6.7–6.8 at 25°C

$Na_2S_2O_3$ Solution:

Composition per 10.0mL:

$Na_2S_2O_3·5H_2O$ 2.5g

Preparation of $Na_2S_2O_3$ Solution: Add $Na_2S_2O_3·5H_2O$ to distilled/deionized water and bring volume to 10.0mL. Mix thoroughly. Autoclave under 100% N_2 for 15 min at 15 psi pressure–121°C. Cool to room temperature.

$Na_2S·9H_2O$ Solution:

Composition per 10.0mL:

$Na_2S·9H_2O$ 0.5g

Preparation of $Na_2S \cdot 9H_2O$ Solution: Add $Na_2S \cdot 9H_2O$ to distilled/deionized water and bring volume to 10.0mL. Mix thoroughly. Sparge with 100% N_2. Autoclave for 15 min at 15 psi pressure–121°C.

Trace Elements Solution SL-10:

Composition per liter:

$FeCl_2 \cdot 4H_2O$	1.5g
$CoCl_2 \cdot 6H_2O$	190.0mg
$MnCl_2 \cdot 4H_2O$	100.0mg
$ZnCl_2$	70.0mg
$Na_2MoO_4 \cdot 2H_2O$	36.0mg
$NiCl_2 \cdot 6H_2O$	24.0mg
H_3BO_3	6.0mg
$CuCl_2 \cdot 2H_2O$	2.0mg
HCl (25% solution)	10.0mL

Preparation of Trace Elements Solution SL-10: Add $FeCl_2 \cdot 4H_2O$ to 10.0mL of HCl solution. Mix thoroughly. Add distilled/deionized water and bring volume to 1.0L. Add remaining components. Mix thoroughly. Sparge with 80% N_2 + 20% CO_2. Autoclave for 15 min at 15 psi pressure–121°C.

Calcium Chloride Solution:

Composition per 10.0mL:

$CaCl_2 \cdot 2H_2O$	0.2g

Preparation of Calcium Chloride Solution: Add $CaCl_2 \cdot 2H_2O$ to distilled/deionized water and bring volume to 10.0mL. Mix thoroughly. Sparge with 100% N_2. Autoclave for 15 min at 15 psi pressure–121°C.

Preparation of Medium: Add components, except calcium chloride solution, trace elements solution SL-10, $Na_2S \cdot 9H_2O$ solution, and $Na_2S_2O_3$ solution, to distilled/deionized water and bring volume to 969.0mL. Mix thoroughly. Sparge with 100% N_2 for 30 min. Distribute under 100% N_2 into anaerobe tubes or bottles. Autoclave for 15 min at 15 psi pressure–121°C. Aseptically and anaerobically per 1.0L of medium add 10.0mL calcium chloride solution, 1.0mL trace elements solution SL-10, 10.0mL $Na_2S \cdot 9H_2O$ solution, and 10.0mL $Na_2S_2O_3$ solution. Mix thoroughly. The final pH should be 6.7–6.8.

Use: For the cultivation of *Dethiosulfovibrio* spp.

Dethiosulfovibrio peptidovorans Medium (DSMZ Medium 786)

Composition per 1085.0mL:

NaCl	30.0g
Trypticase™	5.0g
$MgCl_2 \cdot 6H_2O$	3.0g
Yeast extract	1.0g
NH_4Cl	1.0g
L-Cysteine	0.5g
Na-acetate	0.5g
K_2HPO_4	0.3g
KH_2PO_4	0.3g
$CaCl_2 \cdot 2H_2O$	0.1g
KCl	0.1g
Resazurin	0.5mg
$NaHCO_3$ solution	50.0mL
Na-thiosulfate solution	20.0mL
$Na_2S \cdot 9H_2O$ solution	15.0mL
Trace elements solution	10.0mL

pH 7.3 ± 0.2 at 25°C

Trace Elements Solution:

Composition per liter:

$MgSO_4 \cdot 7H_2O$	3.0g
Nitrilotriacetic acid	1.5g
NaCl	1.0g
$MnSO_4 \cdot 2H_2O$	0.5g
$CoSO_4 \cdot 7H_2O$	0.18g
$ZnSO_4 \cdot 7H_2O$	0.18g
$CaCl_2 \cdot 2H_2O$	0.1g
$FeSO_4 \cdot 7H_2O$	0.1g
$NiCl_2 \cdot 6H_2O$	0.025g
$KAl(SO_4)_2 \cdot 12H_2O$	0.02g
H_3BO_3	0.01g
$Na_2MoO_4 \cdot 4H_2O$	0.01g
$CuSO_4 \cdot 5H_2O$	0.01g
$Na_2SeO_3 \cdot 5H_2O$	0.3mg

Preparation of Trace Elements Solution: Add nitrilotriacetic acid to 500.0mL of distilled/deionized water. Dissolve by adjusting pH to 6.5 with KOH. Add remaining components. Add distilled/deionized water to 1.0L. Mix thoroughly.

$Na_2S \cdot 9H_2O$ Solution:

Composition per 20.0mL:

$Na_2S \cdot 9H_2O$	0.6g

Preparation of $Na_2S \cdot 9H_2O$ Solution: Add $Na_2S \cdot 9H_2O$ to distilled/deionized water and bring volume to 20.0mL. Mix thoroughly. Autoclave under 100% N_2 for 15 min at 15 psi pressure–121°C. Cool to room temperature.

$NaHCO_3$ Solution:

Composition per 100.0mL:

$NaHCO_3$	10.0g

Preparation of $NaHCO_3$ Solution: Add $NaHCO_3$ to distilled/deionized water and bring volume to 100.0mL. Mix thoroughly. Sparge with 80% N_2 + 20% CO_2. Filter sterilize.

Na-thiosulfate Solution:

Composition per 20.0mL:

$Na_2S_2O_3 \cdot 5H_2O$	5.0g

Preparation of Na-thiosulfate Solution: Add $Na_2S_2O_3 \cdot 5H_2O$ to distilled/deionized water and bring volume to 20.0mL. Mix thoroughly. Sparge with 100% N_2. Filter sterilize.

Preparation of Medium: Prepare and dispense medium under 80% N_2 + 20% CO_2. Add components, except Na-thiosulfate solution, $NaHCO_3$ solution, and $Na_2S \cdot 9H_2O$ solution, to distilled/deionized water and bring volume to 1.0L. Mix thoroughly. Sparge with 80% N_2 + 20% CO_2. Autoclave for 15 min at 15 psi pressure–121°C. Cool to 25°C. Aseptically and anaerobically add 20.0mL Na-thiosulfate solution, 50.0mL $NaHCO_3$ solution, and 15.0mL $Na_2S \cdot 9H_2O$ solution. Mix thoroughly. Adjust pH to 7.3. Aseptically and anaerobically distribute into sterile tubes or flasks.

Use: For the cultivation of *Dethiosulfovibrio peptidovorans.*

DEV Lactose Peptone MUG Broth

Composition per liter:

Lactose	10.0g
Meat peptone	10.0g
NaCl	5.0g
Tryptophan	1.0g

4-Methylumbelliferyl-β-D-glucuronide 0.1g
Bromocresol Purple 0.01g
pH 7.2 ± 0.2 at 37°C

Source: This medium is available from Fluka, Sigma-Aldrich.

Preparation of Medium: Add components to distilled/deionized water and bring volume to 1.0L. Mix thoroughly. Distribute into test tubes that contain an inverted Durham tube in 10.0mL volumes. Autoclave for 15 min at 15 psi pressure–121°C.

Use: For the enrichment and titer determination of coliform bacteria in connection with the bacteriological examination of water. The presence of *E. coli* can be demonstrated by fluorescence in the UV and a positive indole test.

Dextran Agar

Composition per liter:
Minimal mineral base solution 700.0mL
Agar solution 200.0mL
Dextran-deoxyglucose solution 100.0mL
pH 4.0 ± 0.2 at 25°C

Minimal Mineral Base Solution:
Composition per 700.0mL:
$(NH_4)_2SO_4$ 5.0g
KH_2PO_4 .... 1.5g
$CaCl_2$ 0.1g
$MnSO_4$ 0.1g
NaCl 0.1g
Yeast extract 0.1g

Preparation of Minimal Mineral Base Solution: Add components to distilled/deionized water and bring volume to 700.0mL. Mix thoroughly. Gently heat and bring to boiling. pH is 4.0 before autoclaving. Autoclave for 15 min at 15 psi pressure–121°C. Cool to 45°–50°C.

Agar Solution:
Composition per 200.0mL:
Agar 10.0g

Preparation of Agar Solution: Add agar to distilled/deionized water and bring volume to 200.0mL. Mix thoroughly. Gently heat and bring to boiling. Autoclave for 15 min at 15 psi pressure–121°C. Cool to 45°–50°C.

Dextran-Deoxyglucose Solution:
Composition per 100.0mL:
Dextran 10.0g
2-Deoxy-D-glucose 0.5g

Preparation of Dextran-Deoxyglucose Solution: Add components to distilled/deionized water and bring volume to 100.0mL. Mix thoroughly. Filter sterilize.

Preparation of Medium: Aseptically combine 700.0mL of sterile minimal mineral base solution, 200.0mL of sterile agar solution, and 100.0mL of sterile dextran-deoxyglucose solution. Mix thoroughly. Pour into sterile Petri dishes or distribute into sterile tubes.

Use: For the cultivation and maintenance of *Lipomyces starkeyi.*

Dextran Broth

Composition per liter:
Minimal mineral base solution 900.0mL
Dextran-deoxyglucose solution 100.0mL
pH 4.0 ± 0.2 at 25°C

Minimal Mineral Base Solution:
Composition per 900.0mL:
$(NH_4)_2SO_4$ 5.0g
KH_2PO_4 .... 1.5g
$CaCl_2$ 0.1g
$MnSO_4$ 0.1g
NaCl 0.1g
Yeast extract 0.1g

Preparation of Minimal Mineral Base Solution: Add components to distilled/deionized water and bring volume to 900.0mL. Mix thoroughly. Gently heat and bring to boiling. pH is 4.0 before autoclaving. Autoclave for 15 min at 15 psi pressure–121°C. Cool to room temperature.

Dextran-Deoxyglucose Solution:
Composition per 100.0mL:
Dextran 10.0g
2-Deoxy-D-glucose 0.5g

Preparation of Dextran-Deoxyglucose Solution: Add components to distilled/deionized water and bring volume to 100.0mL. Mix thoroughly. Filter sterilize.

Preparation of Medium: Combine 900.0mL of sterile minimal mineral base solution with 100.0mL of sterile dextran-deoxyglucose solution. Mix thoroughly. Distribute into sterile flasks or tubes.

Use: For the cultivation and maintenance of *Lipomyces starkeyi.*

Dextran Medium (DSMZ Medium 1050)

Composition per liter:
Agar 15.0g
Dextran 10.0g
Polypepton 5.0g
NH_4NO_3 .... 1.0g
KH_2PO_4 .... 1.0g
$MgSO_4 \cdot 7H_2O$ 0.5g
$CaCl_2 \cdot 2H_2O$ 0.25g
pH 6.8 ± 0.2 at 25°C

Preparation of Medium: Add components to distilled/deionized water and bring volume to 1.0L. Mix thoroughly. Adjust pH to 6.8. Gently heat while stirring and bring to boiling. Distribute into tubes or flasks. Autoclave for 15 min at 15 psi pressure–121°C. Pour into sterile Petri dishes or leave in tubes.

Use: For the cultivation of *Arthrobacter dextranyliticum.*

Dextrin Fuchsin Sulfite Agar

See: ***Aeromonas*** **Differential Agar**

Dextrose Agar

Composition per liter:
Agar 15.0g
Glucose 10.0g
NaCl 5.0g
Pancreatic digest of casein 5.0g
Peptic digest of animal tissue 5.0g
Beef extract 3.0g
pH 6.9 ± 0.2 at 25°C

Source: This medium is available as a premixed powder from BD Diagnostic Systems.

Preparation of Medium: Add components to distilled/deionized water and bring volume to 1.0L. Mix thoroughly. Gently heat and bring to boiling. Distribute into tubes or flasks. Autoclave for 15 min at 15 psi pressure–121°C. Pour into sterile Petri dishes or leave in tubes.

Use: For the cultivation and enumeration of microorganisms from foods. For use as a base for the preparation of blood agar.

Dextrose Agar

Composition per liter:

Agar	15.0g
Glucose	10.0g
Tryptose	10.0g
NaCl	5.0g
Beef extract	3.0g

pH 7.3 ± 0.2 at 25°C

Source: This medium is available as a premixed powder from BD Diagnostic Systems.

Preparation of Medium: Add components to distilled/deionized water and bring volume to 1.0L. Mix thoroughly. Gently heat and bring to boiling. Distribute into tubes or flasks. Autoclave for 15 min at 15 psi pressure–121°C. Pour into sterile Petri dishes or leave in tubes.

Use: For the cultivation of a wide variety of microorganisms. For use as a base for the preparation of blood agar and for general laboratory procedures.

Dextrose Agar

Composition per liter:

Agar	15.0g
Glucose	10.0g
Tryptose	10.0g
NaCl	5.0g
Beef extract	3.0g

pH 7.3 ± 0.2 at 25°C

Source: This medium is available as a premixed powder from HiMedia.

Preparation of Medium: Add components to distilled/deionized water and bring volume to 1.0L. Mix thoroughly. Gently heat and bring to boiling. Distribute into tubes or flasks. Autoclave for 20 min at 15 psi pressure–121°C. Pour into sterile Petri dishes or leave in tubes.

Use: For the cultivation and maintenance of a wide variety of microorganisms.

Dextrose Ascitic Fluid Semisolid Agar

Composition per liter:

Pancreatic digest of casein	2.66g
NaCl	1.33g
Agar	0.5g
Phenol Red	4.8mg
Ascitic fluid	50.0mL
Glucose solution	15.0mL

pH 7.4 ± 0.2 at 25°C

Glucose Solution:

Composition per 15.0mL:

Glucose	3.0g

Preparation of Glucose Solution: Add glucose to distilled/deionized water and bring volume to 15.0mL. Mix thoroughly. Filter sterilize.

Preparation of Medium: Add components, except ascitic fluid and glucose solution, to distilled/deionized water and bring volume to 935.0mL. Mix thoroughly. Gently heat and bring to boiling. Autoclave for 15 min at 15 psi pressure–121°C. Cool to 45°–50°C. Aseptically add sterile ascitic fluid and glucose solution. Mix thoroughly. Aseptically distribute into sterile tubes.

Use: For the isolation and cultivation of microorganisms from spinal fluid.

Dextrose Broth

Composition per liter:

Tryptose	10.0g
Glucose	5.0g
NaCl	5.0g
Beef extract	3.0g

pH 7.2 ± 0.2 at 25°C

Source: This medium is available as a premixed powder from BD Diagnostic Systems and Oxoid Unipath.

Preparation of Medium: Add components to distilled/deionized water and bring volume to 1.0L. Mix thoroughly. Distribute into tubes or flasks. Autoclave for 15 min at 15 psi pressure–121°C.

Use: For the isolation and enrichment of fastidious or damaged microorganisms.

Dextrose Broth

Composition per liter:

Pancreatic digest of casein	10.0g
Glucose	5.0g
NaCl	5.0g

pH 7.3 ± 0.2 at 25°C

Source: This medium is available as a premixed powder from BD Diagnostic Systems.

Preparation of Medium: Add components to distilled/deionized water and bring volume to 1.0L. Mix thoroughly. Distribute into tubes or flasks. Autoclave for 15 min at 15 psi pressure–121°C.

Use: For the cultivation and differentiation of microorganisms based on their ability to ferment glucose. If desired, a Durham tube may be added to the test tubes to determine gas production.

Dextrose HiVeg Agar

Composition per liter:

Agar	15.0g
Glucose	10.0g
Plant hydrolysate No. 1	10.0g
NaCl	5.0g
Plant extract	3.0g

pH 7.3 ± 0.2 at 25°C

Source: This medium is available as a premixed powder from HiMedia.

Preparation of Medium: Add components to distilled/deionized water and bring volume to 1.0L. Mix thoroughly. Gently heat and bring to boiling. Distribute into tubes or flasks. Autoclave for 20 min at 15 psi pressure–121°C. Pour into sterile Petri dishes or leave in tubes.

Use: For the cultivation and maintenance of a wide variety of microorganisms.

Dextrose HiVeg Agar with Blood

Composition per liter:

Agar 15.0g
Glucose 10.0g
Plant hydrolysate No. 1 .... 10.0g
NaCl 5.0g
Plant extract 3.0g
Sheep blood, defibrinated 50.0mL

pH 7.3 ± 0.2 at 25°C

Source: This medium wtihout sheep blood is available as a premixed powder from HiMedia.

Preparation of Medium: Add components, except sheep blood, to distilled/deionized water and bring volume to 950.0mL. Mix thoroughly. Gently heat and bring to boiling. Autoclave for 15 min at 15 psi pressure–121°C. Cool to 50°C. Aseptically add 50.0mL sterile blood. Pour into sterile Petri dishes or leave in tubes.

Use: For the cultivation and maintenance of a wide variety of microorganisms.

Dextrose HiVeg Agar Base, Emmons (Sabouraud Glucose HiVeg Agar Base, Modified)

Composition per liter:

Glucose 20.0g
Agar 17.0g
Plant special peptone 10.0g

pH 6.9 ± 0.2 at 25°C

Source: This medium is available as a premixed powder from HiMedia.

Preparation of Medium: Add components to tap water and bring volume to 1.0L. Mix thoroughly. Gently heat and bring to boiling. Distribute into tubes or flasks. Autoclave for 15 min at 15 psi pressure–121°C. Pour into sterile Petri dishes or leave in tubes.

Use: For the cultivation of yeasts and molds. For the cultivation of dermatophytes and other pathogenic and nonpathogenic fungi from clinical and nonclinical specimens. For the cultivation of yeast and filamentous fungi.

Dextrose HiVeg Broth

Composition per liter:

Plant hydrolysate No. 1 .... 10.0g
Glucose 5.0g
NaCl 5.0g
Plant extract 3.0g

pH 7.0 ± 0.2 at 25°C

Source: This medium is available as a premixed powder from HiMedia.

Preparation of Medium: Add components to distilled/deionized water and bring volume to 1.0L. Mix thoroughly. Distribute into tubes or flasks. Autoclave for 15 min at 15 psi pressure–121°C.

Use: For antibiotic sensitivity testing using tube dilution method. For the cultivation and maintenance of a wide variety of microorganisms.

Dextrose HiVeg Broth with Blood

Composition per liter:

Plant hydrolysate No. 1 .... 10.0g
NaCl 5.0g
Glucose 5.0g
Plant extract 3.0g
Sheep blood, defibrinated 50.0mL

pH 7.3 ± 0.2 at 25°C

Source: This medium wtihout sheep blood is available as a premixed powder from HiMedia.

Preparation of Medium: Add components, except sheep blood, to distilled/deionized water and bring volume to 950.0mL. Mix thoroughly. Gently heat and bring to boiling. Autoclave for 15 min at 15 psi pressure–121°C. Cool to 50°C. Aseptically add 50.0mL sterile blood. Aseptically distribute into tubes or flasks.

Use: For the cultivation and maintenance of a wide variety of microorganisms.

Dextrose HiVeg Peptone Agar

Composition per liter:

Plant peptone 20.0g
Agar 15.0g
Glucose 10.0g
NaCl 5.0g

pH 7.2 ± 0.2 at 25°C

Source: This medium is available as a premixed powder from HiMedia.

Preparation of Medium: Add components to distilled/deionized water and bring volume to 1.0L. Mix thoroughly. Gently heat and bring to boiling. Distribute into tubes or flasks. Autoclave for 15 min at 15 psi pressure–121°C. Pour into sterile Petri dishes or leave in tubes.

Use: For the cultivation and maintenance of a wide variety of microorganisms.

Dextrose HiVeg Peptone Broth

Composition per liter:

Plant peptone 20.0g
Glucose 10.0g
NaCl 5.0g

pH 7.2 ± 0.2 at 25°C

Source: This medium is available as a premixed powder from HiMedia.

Preparation of Medium: Add components to distilled/deionized water and bring volume to 1.0L. Mix thoroughly. Distribute into tubes or flasks. Autoclave for 15 min at 15 psi pressure–121°C.

Use: For the cultivation and maintenance of a wide variety of microorganisms.

Dextrose Peptone Agar

Composition per liter:

Peptic digest of animal tissue 20.0g
Agar 15.0g
Glucose 10.0g
NaCl 5.0g

pH 7.2 ± 0.2 at 25°C

Source: This medium is available as a premixed powder from HiMedia.

Preparation of Medium: Add components to distilled/deionized water and bring volume to 1.0L. Mix thoroughly. Gently heat and bring to boiling. Distribute into tubes or flasks. Autoclave for 15 min at 15 psi pressure–121°C. Pour into sterile Petri dishes or leave in tubes.

Use: For the cultivation and maintenance of a wide variety of microorganisms.

Dextrose Proteose No. 3 Agar

Composition per liter:

Proteose peptone No. 3	20.0g
Agar	13.0g
NaCl	5.0g
Glucose	2.0g
Tellurite blood solution	50.0mL

pH 7.4 ± 0.2 at 25°C

Tellurite Blood Solution:

Composition per 60.0mL:

Sheep blood, defibrinated	50.0mL
Chapman tellurite solution	10.0mL

Preparation of Tellurite Blood Solution: Aseptically combine 10.0mL of Chapman tellurite solution with 50.0mL of sterile, defibrinated sheep blood. Mix thoroughly.

Chapman Tellurite Solution:

Composition per 100.0mL:

K_2TeO_3	1.0g

Preparation of Chapman Tellurite Solution: Add K_2TeO_3 to distilled/deionized water and bring volume to 100.0mL. Mix thoroughly. Filter sterilize.

Caution: Potassium tellurite is toxic.

Preparation of Medium: Add components, except tellurite blood solution, to distilled/deionized water and bring volume to 940.0mL. Mix thoroughly. Gently heat and bring to boiling. Autoclave for 15 min at 15 psi pressure–121°C. Cool to 75°–80°C. Aseptically add 50.0mL of sterile tellurite blood solution. Mix thoroughly. Maintain at 75°–80°C for 10–15 min or until the agar becomes chocolatized. Cool slowly to 50°C. Pour into sterile Petri dishes or distribute into sterile tubes.

Use: For propagating pure cultures of *Neisseria gonorrhoeae* and other fastidious microorganisms.

Dextrose Proteose Peptone HiVeg Agar Base with Tellurite and Blood

Composition per liter:

Plant peptone No. 3	20.0g
Agar	15.0g
NaCl	5.0g
Glucose	2.0g
Sheep blood, defibrinated	50.0mL
Tellurite solution	2.0mL

pH 7.4 ± 0.2 at 25°C

Source: This medium, without tellurite or blood, is available as a premixed powder from HiMedia.

Tellurite Solution:

Composition per 10.0mL:

K_2TeO_3	0.1g

Preparation of Tellurite Solution: Add K_2TeO_3 to distilled/deionized water and bring volume to 100.0mL. Mix thoroughly. Filter sterilize.

Caution: Potassium tellurite is toxic.

Preparation of Medium: Add components, except tellurite solution and blood, to distilled/deionized water and bring volume to 950.0mL. Mix thoroughly. Gently heat and bring to boiling. Autoclave for 15 min at 15 psi pressure–121°C. Cool to 50°C. Aseptically add 2.0mL of sterile tellurite soltuion and 50.0mL of sterile defibrinated blood. Pour into sterile Petri dishes or leave in tubes.

Use: For the cultivation and maintenance of a wide variety of microorganisms. For use as a base for the preparation of blood agar and for general laboratory procedures. For the isolation of *Corynebacterium diphtheriae.*

Dextrose Proteose Peptone Agar Base with Tellurite and Blood

Composition per liter:

Proteose peptone	20.0g
Agar	15.0g
NaCl	5.0g
Glucose	2.0g
Sheep blood, defibrinated	50.0mL
Tellurite solution	2.0mL

pH 7.4 ± 0.2 at 25°C

Source: This medium, without tellurite or blood, is available as a premixed powder from HiMedia.

Tellurite Solution:

Composition per 10.0mL:

K_2TeO_3	0.1g

Preparation of Tellurite Solution: Add K_2TeO_3 to distilled/deionized water and bring volume to 100.0mL. Mix thoroughly. Filter sterilize.

Caution: Potassium tellurite is toxic.

Preparation of Medium: Add components, except tellurite solution and blood, to distilled/deionized water and bring volume to 950.0mL. Mix thoroughly. Gently heat and bring to boiling. Autoclave for 15 min at 15 psi pressure–121°C. Cool to 50°C. Aseptically add 2.0mL of sterile tellurite soltuion and 50.0mL of sterile defibrinated blood. Pour into sterile Petri dishes or leave in tubes.

Use: For the cultivation and maintenance of a wide variety of microorganisms. For use as a base for the preparation of blood agar and for general laboratory procedures. For the isolation of *Corynebacterium diphtheriae.*

Dextrose Soil Agar (DSA)

Composition per liter:

Soil	150.0g
Agar	20.0g
Glucose	5.0g

Preparation of Medium: Add soil to distilled/deionized water and bring volume to 1.0L. Mix thoroughly. Autoclave for 60 min at 15 psi pressure–121°C. Filter through Whatman #1 filter paper. Bring volume of filtrate to 1.0L with distilled/deionized water. Mix thoroughly. Add agar and glucose. Gently heat and bring to boiling. Distribute into tubes or flasks. Autoclave for 15 min at 15 psi pressure–121°C. Pour into sterile Petri dishes or leave in tubes

Use: For the cultivation and maintenance of *Chaetomium globosum.*

Dextrose Starch Agar

Composition per liter:

Gelatin	20.0g
Proteose peptone	15.0g

Agar	10.0g
Starch	10.0g
Glucose	5.0g
NaCl	5.0g
Na_2HPO_4	3.0g

pH 7.3 ± 0.2 at 25°C

Source: This medium is available as a premixed powder from BD Diagnostic Systems.

Preparation of Medium: Add components to distilled/deionized water and bring volume to 1.0L. Mix thoroughly. Gently heat while stirring and bring to boiling. Distribute into tubes or flasks. Autoclave for 15 min at 15 psi pressure–121°C. Pour into sterile Petri dishes or leave in tubes.

Use: For the cultivation and maintenance of *Neisseria gonorrhoeae*, *Neisseria animalis*, and other fastidious microorganisms.

Dextrose Sucrose Cellulose Agar (DSA Cellulose)

Composition per liter:

Agar	20.0g
Cellulose, powdered	10.0g
KH_2PO_4	1.0g
KNO_3	1.0g
$MgSO_4 \cdot 7H_2O$	0.5g
KCL	0.5g
Amidon	0.2g
Glucose	0.2g
Sucrose	0.2g

Preparation of Medium: Add components to distilled/deionized water and bring volume to 1.0L. Mix thoroughly. Gently heat and bring to boiling. Distribute into tubes or flasks. Autoclave for 30 min at 15 psi pressure–121°C. Pour into sterile Petri dishes or leave in tubes.

Use: For the cultivation of *Bipolaris sorghicola* and *Codinaea simplex*.

Dextrose Tryptone Agar

Composition per liter:

Agar	15.0g
Pancreatic digest of casein	10.0g
Glucose	5.0g
Bromcresol Purple	0.04g

pH 6.9 ± 0.2 at 25°C

Source: This medium is available as a premixed powder from BD Diagnostic Systems.

Preparation of Medium: Add components to distilled/deionized water and bring volume to 1.0L. Mix thoroughly. Gently heat and bring to boiling. Distribute into tubes or flasks. Autoclave for 15 min at 15 psi pressure–121°C. Pour into sterile Petri dishes or leave in tubes.

Use: For the isolation and cultivation of mesophilic and thermophilic aerobic microorganisms in food.

Dextrose Tryptone Agar

Composition per liter:

Agar	12.0g
Pancreatic digest of casein	10.0g
Glucose	5.0g
Bromcresol Purple	0.04g

pH 6.9 ± 0.2 at 25°C

Source: This medium is available as a premixed powder from Oxoid Unipath.

Preparation of Medium: Add components to distilled/deionized water and bring volume to 1.0L. Mix thoroughly. Gently heat and bring to boiling. Distribute into tubes or flasks. Autoclave for 15 min at 15 psi pressure–121°C. Pour into sterile Petri dishes or leave in tubes.

Use: For the isolation, cultivation, and enumeration of "flat-sour" thermophiles and mesophiles in food. Acid-producing microorganisms such as "flat-sour" thermophiles appear as yellow colonies surrounded by a yellow zone.

Dextrose Tryptone Broth

Composition per liter:

Pancreatic digest of casein	10.0g
Glucose	5.0g
Bromcresol Purple	0.04g

pH 6.9 ± 0.2 at 25°C

Source: This medium is available as a premixed powder from Oxoid Unipath.

Preparation of Medium: Add components to distilled/deionized water and bring volume to 1.0L. Mix thoroughly. Distribute into tubes or flasks. Autoclave for 15 min at 15 psi pressure–121°C.

Use: For the isolation and cultivation of "flat-sour" thermophiles and mesophiles in food. Acid-producing microorganisms such as "flat-sour" thermophiles turn the medium yellow.

Dextrose Tryptone Broth (m-Dextrose Tryptone Broth)

Composition per liter:

Pancreatic digest of casein	20.0g
Glucose	10.0g
Bromcresol Purple	0.04g

pH 6.7 ± 0.2 at 25°C

Preparation of Medium: Add components to distilled/deionized water and bring volume to 1.0L. Mix thoroughly. Distribute into tubes or flasks. Autoclave for 15 min at 15 psi pressure–121°C.

Use: For the isolation, cultivation, and enumeration of "flat-sour" thermophiles and mesophiles in food by the membrane filter technique. Acid-producing microorganisms such as "flat-sour" thermophiles turn the medium yellow.

Dextrose Tryptone HiVeg Agar

Composition per liter:

Agar	15.0g
Plant hydrolysate	10.0g
Glucose	5.0g

pH 6.7 ± 0.2 at 25°C

Source: This medium is available as a premixed powder from HiMedia.

Preparation of Medium: Add components to distilled/deionized water and bring volume to 1.0L. Mix thoroughly. Gently heat and bring to boiling. Distribute into tubes or flasks. Autoclave for 15 min at 15 psi pressure–121°C. Pour into sterile Petri dishes or leave in tubes.

Use: For the isolation, cultivation, and enumeration of "flat-sour" thermophiles and mesophiles in food.

Dextrose Tryptone HiVeg Agar, Modified

Composition per liter:

Agar	15.0g
Plant hydrolysate	10.0g
Glucose	5.0g
K_2HPO_4	1.25
Yeast extract	1.0g
Bromcresol Purple	0.04g

pH 6.7 ± 0.2 at 25°C

Source: This medium is available as a premixed powder from HiMedia.

Preparation of Medium: Add components to distilled/deionized water and bring volume to 1.0L. Mix thoroughly. Gently heat and bring to boiling. Distribute into tubes or flasks. Autoclave for 15 min at 15 psi pressure–121°C. Pour into sterile Petri dishes or leave in tubes.

Use: For the isolation, cultivation, and enumeration of "flat-sour" thermophiles and mesophiles in food. Acid-producing microorganisms such as "flat-sour" thermophiles appear as yellow colonies surrounded by a yellow zone.

Dextrose Tryptone HiVeg Broth

Composition per liter:

Plant hydrolysate	10.0g
Glucose	5.0g
Bromcresol Purple	0.04g

pH 6.7 ± 0.2 at 25°C

Source: This medium is available as a premixed powder from HiMedia.

Preparation of Medium: Add components to distilled/deionized water and bring volume to 1.0L. Mix thoroughly. Distribute into tubes or flasks. Autoclave for 15 min at 15 psi pressure–121°C.

Use: For the isolation and cultivation of "flat-sour" thermophiles and mesophiles in food. Acid-producing microorganisms such as "flat-sour" thermophiles turn the medium yellow.

Dextrose Tryptone HiVeg Broth, Modified

Composition per liter:

Plant hydrolysate	10.0g
Glucose	5.0g
K_2HPO_4	1.25
Yeast extract	1.0g
Bromcresol Purple	0.04g

pH 6.7 ± 0.2 at 25°C

Source: This medium is available as a premixed powder from HiMedia.

Preparation of Medium: Add components to distilled/deionized water and bring volume to 1.0L. Mix thoroughly. Distribute into tubes or flasks. Autoclave for 15 min at 15 psi pressure–121°C.

Use: For the isolation, cultivation, and enumeration of "flat-sour" thermophiles and mesophiles in food. Acid-producing microorganisms such as "flat-sour" thermophiles turn the medium yellow.

Dextrose Yeast Asparagine Agar (DYAA)

Composition per liter:

Agar	20.0g
Glucose	10.0g
Yeast extract	1.0g
Asparagine	0.5g
$K_2HPO_4 \cdot 3H_2O$	0.5g
$MgSO_4 \cdot 7H_2O$	0.25g
$FeCl_3$ solution	0.5mL

$FeCl_3$ Solution:

Composition per 10.0mL:

$FeCl_3$	1.0g

Preparation of $FeCl_3$ Solution: Add $FeCl_3$ to distilled/deionized water and bring volume to 10.0mL. Mix thoroughly.

Preparation of Medium: Add components to distilled/deionized water and bring volume to 1.0L. Mix thoroughly. Gently heat and bring to boiling. Distribute into tubes or flasks. Autoclave for 15 min at 15 psi pressure–121°C. Pour into sterile Petri dishes or leave in tubes.

Use: For the cultivation of *Aciculoconidium aculeatum,* many *Acremonium* species, *Acrodontium simplex, Plectosphaerella cucumerina,* and many other filamentous fungi.

Dextrose Yeast Extract Peptone (DYPA)

Composition per liter:

Agar	20.0g
Glucose	20.0g
Peptone	10.0g
Yeast extract	5.0g

Preparation of Medium: Add components to distilled/deionized water and bring volume to 1.0L. Mix thoroughly. Gently heat and bring to boiling. Distribute into tubes or flasks. Autoclave for 15 min at 15 psi pressure–121°C. Pour into sterile Petri dishes or leave in tubes.

Use: For the cultivation of *Candida maltosa, Candida ethanolica, Candida boidinii, Candida tropicalis, Pichia membranaefaciens, Debaryomyces hansenii, Dekkera intermedia, Dekkera bruxellensis, Dekkera abstinens, Dekkera anomala, Dekkera custersiana, Dekkera lambica, Dekkera naardenensis, Saccharomyces servazzii, Williopsis californica, Zygosaccharomyces rouxii*, and other fungi.

Dey-Engley Neutralizing Agar
See: **D-E Neutralizing Agar**

Dey-Engley Neutralizing Agar (D-E HiVeg Agar Disinfectant Testing)

Composition per liter:

Agar	15.0g
Glucose	10.0g
Lecithin	7.0g
$Na_2S_2O_3$	6.0g
Casein enzymatic hydrolysate	5.0g
Polysorbate 80	5.0g
$NaHSO_3$	2.5g
Yeast extract	2.5g
Na-thioglycollate	1.0g
Bromcresol Purple	0.02g

pH 7.6 ± 0.2 at 25°C

Source: This medium is available as a premixed powder from HiMedia.

Preparation of Medium: Add components to distilled/deionized water and bring volume to 1.0L. Mix thoroughly. Gently heat and bring to boiling. Distribute into flasks in 9.0mL volumes. Autoclave for 15 min at 15 psi pressure–121°C.

Use: For the neutralization and testing of antiseptics and disinfectants.

Dey-Engley Neutralizing Broth

See: **D-E Neutralizing Broth**

Dey-Engley Neutralizing Broth Base

Composition per liter:

Glucose	10.0g
Casein enzymatic hydrolysate	5.0g
Yeast extract	2.5g

pH 7.6± 0.2 at 25°C

Source: This medium is available as a premixed powder from HiMedia.

Preparation of Medium: Add components to distilled/deionized water and bring volume to 1.0L. Mix thoroughly. Distribute into tubes in 9.0mL volumes. Autoclave for 15 min at 15 psi pressure–121°C.

Use: For the neutralization and testing of antiseptics and disinfectants.

Dey-Engley Neutralizing HiVeg Agar (D-E HiVeg Agar Disinfectant Testing)

Composition per liter:

Agar	15.0g
Glucose	10.0g
Lecithin	7.0g
$Na_2S_2O_3$	6.0g
Plant hydrolysate	5.0g
Polysorbate 80	5.0g
$NaHSO_3$	2.5g
Yeast extract	2.5g
Na-thioglycollate	1.0g
Bromcresol Purple	0.02g

pH 7.6 ± 0.2 at 25°C

Source: This medium is available as a premixed powder from HiMedia.

Preparation of Medium: Add components to distilled/deionized water and bring volume to 1.0L. Mix thoroughly. Gently heat and bring to boiling. Distribute into flasks in 9.0mL volumes. Autoclave for 15 min at 15 psi pressure–121°C.

Use: For the neutralization and testing of antiseptics and disinfectants.

Dey-Engley Neutralizing HiVeg Broth

Composition per liter:

Glucose	10.0g
Lecithin	7.0g
$Na_2S_2O_3$	6.0g
Plant hydrolysate	5.0g
Polysorbate 80	5.0g
$NaHSO_3$	2.5g
Yeast extract	2.5g
Na-thioglycollate	1.0g
Bromcresol Purple	0.02g

pH 7.6± 0.2 at 25°C

Source: This medium is available as a premixed powder from HiMedia.

Preparation of Medium: Add components to distilled/deionized water and bring volume to 1.0L. Mix thoroughly. Distribute into tubes in 9.0mL volumes. Autoclave for 15 min at 15 psi pressure–121°C.

Use: For the neutralization and testing of antiseptics and disinfectants.

Dey-Engley Neutralizing HiVeg Broth Base

Composition per liter:

Glucose	10.0g
Plant hydrolysate	5.0g
Yeast extract	2.5g

pH 7.6± 0.2 at 25°C

Source: This medium is available as a premixed powder from HiMedia.

Preparation of Medium: Add components to distilled/deionized water and bring volume to 1.0L. Mix thoroughly. Distribute into tubes in 9.0mL volumes. Autoclave for 15 min at 15 psi pressure–121°C.

Use: For the neutralization and testing of antiseptics and disinfectants.

DG18 Agar

See: **Dichloran Glycerol Agar**

Diagnostic Sensitivity Test Agar (DST Agar)

Composition per liter:

Agar	12.0g
Proteose peptone	10.0g
Veal infusion solids	10.0g
NaCl	3.0g
Na_2HPO_4	2.0g
Glucose	2.0g
Sodium acetate	1.0g
Adenine sulfate	0.01g
Guanine·HCl	0.01g
Uracil	0.01g
Xanthine	0.01g
Thiamine	0.02mg
Horse blood, defibrinated	70.0mL

pH 7.4 ± 0.2 at 25°C

Source: This medium is available as a premixed powder from Oxoid Unipath.

Preparation of Medium: Add components, except horse blood, to distilled/deionized water and bring volume to 930.0mL. Mix thoroughly. Gently heat and bring to boiling. Autoclave for 15 min at 15 psi pressure–121°C. Cool to 45°–50°C. Aseptically add sterile horse blood. Mix thoroughly. Pour into sterile Petri dishes or distribute into sterile tubes.

Use: For antimicrobial testing of various pathogenic microorganisms. DSTA is primarily used for susceptibility tests rather than the primary isolation of organisms from clinical samples. An essential requirement for satisfactory antimicrobial susceptibility media is that the reactive levels of thymidine and thymine must be sufficiently reduced to avoid antagonism of trimethoprim and sulphonamides. DSTA meets this requirement and in the presence of lysed horse blood (or defibrinated horse blood if the plates are stored long enough to allow some lysis of the erythrocytes) the level of thymidine will be further reduced. This is caused by the action of the enzyme thymidine phosphorylase which is released from lysed horse erythrocytes. Thymidine is an essential growth factor for thymidine-dependent organisms and they will not grow in its absence or they will grow poorly in media containing reduced levels.

Dialister Medium (DSMZ Medium 1107)

Composition per liter:

Trypticase	17.0g
Yeast extract	3.0g
NaCl	3.0g
KNO_3	3.0g
Lactate solution	10.0mL
Hemin solution	10.0mL
Glucose soltuion	10.0mL
L-Cysteine solution	5.0mL
DTT solution	5.0mL
Vitamin K_1 solution	5.0mL
Formiate solution	50.0mL
Fumarate solution	50.0mL
Horse blood, sterile	50.0mL

pH 7.0 ± 0.2 at 25°C

L-Cystine Solution:

Composition per 5.0mL:

L-Cystine	0.25g
NaOH (1*N* solution)	5.0mL

Preparation of L-Cystine Solution: Add L-cystine to 5.0mL of NaOH solution. Mix thoroughly. Sparge with 100% N_2. Filter sterilize.

Vitamin K_1 Solution:

Composition per 100.0mL:

Vitamin K_1	0.5g
Ethanol	99.0mL

Preparation of Vitamin K_1 Solution: Add vitamin K_1 to 99.0mL of absolute ethanol. Mix thoroughly. Sparge with 100% N_2. Filter sterilize. Store refrigerated.

Hemin Solution:

Composition per 100.0mL:

Hemin	50.0mg
NaOH (1*N* solution)	20.0mL

Preparation of Hemin Solution: Add hemin to 20.0mL of 1*N* NaOH solution. Mix thoroughly. Bring volume to 100.0mL with distilled/deionized water. Sparge with 100% N_2. Filter sterilize.

Lactate Solution:

Composition per 10.0mL:

Sodium lactate	5.0g

Preparation of Lactate Solution: Add sodium lactate to distilled/deionized water and bring volume to 10.0mL. Mix thoroughly. Sparge with 100% N_2. Filter sterilize.

Glucose Solution:

Composition per 10.0mL:

Glucose	4.0g

Preparation of Glucose Solution: Add glucose to distilled/deionized water and bring volume to 10.0mL. Mix thoroughly. Sparge with 100% N_2. Filter sterilize.

Formiate Solution:

Composition per 10.0mL:

Na-formiate	0.6g

Preparation of Formiate Solution: Add Na-formiate to distilled/deionized water and bring volume to 10.0mL. Mix thoroughly. Sparge with 100% N_2. Filter sterilize.

Fumarate Solution:

Composition per 10.0mL:

Na-fumarate	0.6g

Preparation of Fumarate Solution: Add Na-fumarate to distilled/deionized water and bring volume to 10.0mL. Mix thoroughly. Sparge with 100% N_2. Filter sterilize.

DTT Solution:

Composition per 10.0mL:

DTT	0.15g

Preparation of DTT Solution: Add DTT to distilled/deionized water and bring volume to 10.0mL. Mix thoroughly. Sparge with 100% N_2. Filter sterilize.

$NaHCO_3$ Solution:

Composition per 10.0mL:

$NaHCO_3$	1.0g

Preparation of $NaHCO_3$ Solution: Add $NaHCO_3$ to distilled/deionized water and bring volume to 10.0mL. Mix thoroughly. Sparge with 20% CO_2 + 80% H_2. Filter sterilize.

Preparation of Medium: Add components, except horse blood, lactate, hemin, glucose, cysteine, DTT, Vitamin K1, formiate, and fumarate solutions, to distilled/deionized water and bring volume to 810.0mL. Mix thoroughly. Adjust pH to 7.8. Gently heat while stirring and bring to boiling. Distribute into tubes or bottles. Autoclave for 15 min at 15 psi pressure–121°C. Aseptically and anaerobically add horse blood, lactate, hemin, glucose, cysteine, DTT, vitamin K_1, formiate, and fumarate solutions. Incubate in oxygen-free, 5–10% CO_2 containing atmosphere.

Use: For the cultivation of *Dialister fulvus, Dialister vulgaris,* and other *Dialister* spp.

Diamalt Agar

Composition per liter:

Diamalt	150.0g
Agar	20.0g

Preparation of Medium: Add components to distilled/deionized water and bring volume to 1.0L. Mix thoroughly. Gently heat and bring to boiling. Distribute into tubes or flasks. Autoclave for 15 min at 15 psi pressure–121°C. Pour into sterile Petri dishes or leave in tubes.

Use: For the isolation and cultivation of yeasts.

Diaminopimelic Acid Medium

Composition per liter:

Pancreatic digest of gelatin	5.0g
Beef extract	3.0g
Diaminopimelic acid	0.05g

pH 6.9 ± 0.2 at 25°C

Preparation of Medium: Add components to distilled/deionized water and bring volume to 1.0L. Mix thoroughly. Distribute into tubes or flasks. Autoclave for 15 min at 15 psi pressure–121°C.

Use: For the cultivation and maintenance of *Bacillus megaterium*.

Diamonds Medium, Modified

Composition per liter:

Pancreatic digest of casein	20.0g
Yeast extract	1.0g
L-Cysteine·HCl·H_2O	0.5g
Maltose	0.5g
L-Ascorbic acid	0.02g
Horse serum, inactivated	100.0mL
Antibiotic inhibitor	10.0mL

pH 6.5 ± 0.2 at 25°C

Antibiotic Inhibitor:

Composition per 10.0mL:

Streptomycin sulfate	0.15g
Amphotericin B	0.2mg
Penicillin G	100,000U

Preparation of Antibiotic Inhibitor: Add components to distilled/deionized water and bring volume to 10.0mL. Mix thoroughly. Filter sterilize.

Preparation of Medium: Add components, except antibiotic inhibitor and horse serum, to distilled/deionized water and bring volume to 890.0mL. Mix thoroughly. Gently heat and bring to boiling. Autoclave for 15 min at 15 psi pressure–121°C. Cool to 25°C. Aseptically add sterile antibiotic inhibitor and horse serum. Mix thoroughly. Aseptically distribute into sterile tubes in 5.0mL volumes.

Use: For the cultivation of *Trichomonas* species.

Diazotrophic Medium (RBA)

Composition per 1008.0mL:

Solution A	903.0mL
Solution B	50.0mL
Solution C	50.0mL
Solution D	5.0mL

pH 7.3 ± 0.2 at 25°C

Solution A:

Composition per 903.0mL:

Agar	15.0g
K_2HPO_4	0.9g
$CaCl_2·2H_2O$	0.1g
KH_2PO_4	0.1g
$MgSO_4·7H_2O$	0.1g
NaCl	0.1g
$FeSO_4·7H_2O$	0.01g
$MnSO_4·H_2O$	5.0mg
$NaVO_3·2H_2O$	5.0mg
$Na_2MoO_4·2H_2O$	0.5mg
Trace elements solution SL-6	3.0mL

Trace Elements Solution SL-6:

Composition per liter:

$MnCl_2·4H_2O$	0.5g
H_3BO_3	0.3g
$CoCl_2·6H_2O$	0.2g
$ZnSO_4·7H_2O$	0.1g
$Na_2MoO_4·2H_2O$	0.03g
$NiCl_2·6H_2O$	0.02g
$CuCl_2·2H_2O$	0.01g

Preparation of Trace Elements Solution SL-6: Add components to distilled/deionized water and bring volume to 1.0L. Mix thoroughly. Adjust pH to 7.3. Autoclave for 15 min at 15 psi pressure–121°C.

Preparation of Solution A: Add components to distilled/deionized water and bring volume to 903.0mL. Mix thoroughly. Gently heat and bring to boiling. Adjust pH to 7.3. Autoclave for 15 min at 15 psi pressure–121°C. Cool to 50°–55°C. Pour into sterile Petri dishes or distribute into sterile tubes.

Solution B:

Composition per 50.0mL:

DL-Malate	2.0g
Disodium succinate	1.0g
Yeast extract	0.05g

Preparation of Solution B: Add components to distilled/deionized water and bring volume to 50.0mL. Mix thoroughly. Adjust pH to 7.3. Autoclave for 15 min at 15 psi pressure–121°C.

Solution C:

Composition per 50.0mL:

D-Mannitol	2.0g
D-Glucose	2.0g
Sodium pyruvate	1.0g

Preparation of Solution C: Add components to distilled/deionized water and bring volume to 50.0mL. Mix thoroughly. Adjust pH to 7.3. Filter sterilize.

Solution D:

Composition per liter:

Pyridoxine·HCl	62.5g
Nicotinic acid	25.0mg
p-Aminobenzoic acid	12.5mg
Thiamine·HCl	12.5mg
Calcium DL-pantothenate	6.5mg
Biotin	2.5mg

Preparation of Solution D: Add components to distilled/deionized water and bring volume to 1.0L. Mix thoroughly. Adjust pH to 7.5. Gas under 100% N_2. Filter sterilize.

Preparation of Medium: To 903.0mL of sterile solution A, aseptically add 50.0mL of sterile solution B, 50.0mL of sterile solution C, and 5.0mL of sterile solution D. Mix thoroughly. Final pH should be 7.3. Pour into sterile Petri dishes or distribute into sterile tubes.

Use: For the cultivation and maintenance of *Arthrobacter* species, *Azomonas* species, *Azorhizophilus paspali,* and *Azotobacter* species.

Dibenzothiophene Mineral Medium

Composition per liter:

Beef extract	10.0g
Na_2HPO_4	3.0g
KH_2PO_4	2.0g
NH_4Cl	2.0g
Dibenzothiophene	0.5g
$MgCl_2·6H_2O$	0.2g
$FeCl_3·6H_2O$	0.028g

Preparation of Medium: Add components to distilled/deionized water and bring volume to 1.0L. Mix thoroughly. Distribute into tubes or flasks. Autoclave for 15 min at 15 psi pressure–121°C.

Use: For the cultivation of bacteria that can metabolize dibenzothiophene.

Dichloran Glycerol Agar (DG18 Agar)

Composition per liter:

Agar	15.0g
Glucose	10.0g
Peptone	5.0g
KH_2PO_4	1.0g
$MgSO_4 \cdot 7H_2O$	0.5g
Dichloran	2.0mg
Chloramphenicol solution	10.0mL

pH 5.6 ± 0.2 at 25°C

Source: This medium is available as a premixed powder from Oxoid Unipath.

Chloramphenicol Solution:
Composition per 10.0mL:

Chloramphenicol	0.1g

Preparation of Chloramphenicol Solution: Add chloramphenicol to distilled/deionized water and bring volume to 10.0mL. Mix thoroughly. Filter sterilize.

Preparation of Medium: Add components, except chloramphenicol solution, to distilled/deionized water and bring volume to 990.0mL. Mix thoroughly. Gently heat and bring to boiling. Autoclave for 15 min at 15 psi pressure–121°C. Cool to 45°–50°C. Aseptically add sterile chloramphenicol solution. Mix thoroughly. Pour into sterile Petri dishes or distribute into sterile tubes.

Use: For the enumeration and isolation of xerophilic molds from dried and semidried foods.

Dichloran 18% Glycerol Agar (DG18 Agar) (BAM M184)

Composition per liter:

Glycerol	220.0g
Agar	15.0g
Glucose	10.0g
Peptone	5.0g
KH_2PO_4	1.0g
$MgSO_4 \cdot 7H_2O$	0.5g
Dichloran	2.0mg
Chloramphenicol	0.1g

pH 5.6 ± 0.2 at 25°C

Preparation of Medium: Add components, except glycerol, to distilled/deionized water and bring volume to 950.0mL. Mix thoroughly. Gently heat and bring to boiling. Cool to 50°C. Add 220.0g of glycerol. Mix thoroughly. Autoclave for 15 min at 15 psi pressure–121°C. The final pH should be 5.6 and the final a_w should be 0.955. Pour into sterile Petri dishes or distribute into sterile tubes.

Use: Used as a general purpose medium for the enumeration of molds from foods. For the enumeration and isolation of xerophilic molds from dried and semidried foods.This medium is preferred when the a_w of the analyzed food is 0.95 or lower. The low water activity of this medium reduces interference by bacteria and fast-growing fungi.

Dichloran HiVeg Medium Base with Rose Bengal and Selective Supplement

Composition per liter:

Agar	15.0g
Glucose	10.0g
Plant peptone	5.0g
KH_2PO_4	1.0g
$MgSO_4$	0.5g
Rose Bengal	0.025g
Bromcresol Purple	0.02g
Dichloran	0.002
Selective supplement	10.0mL

pH 5.6 ± 0.2 at 25°C

Source: This medium, wthout selective supplement, is available as a premixed powder from HiMedia.

Selective Supplement:
Composition per 10.0mL:

Chloramphenicol	0.1mg

Preparation of Selective Supplement: Add chloramphenicol to distilled/deionized water and bring volume to 10.0mL. Mix thoroughly. Filter sterilize.

Preparation of Medium: Add components, except selective supplement, to distilled/deionized water and bring volume to 1.0L. Mix thoroughly. Gently heat and bring to boiling. Autoclave for 15 min at 15 psi pressure–121°C. The final pH should be 5.6. Temper in a water bath at 45°C. Aseptically add 10.0mL sterile selective supplement. Mix thoroughly. Pour into sterile Petri dishes or distribute into sterile tubes. This medium is light sensitive and should be stored in a dark, cool place until used. It is intended for spread plates only.

Use: For the analysis of food samples containing spreader molds, e.g., Mucor and Rhizopus. The dichloran and Rose Bengal slow down the growth of fast-growing fungi, thus allowing the detection of other fungi with slower growth rates.

Dichloran Medium Base with Rose Bengal and Selective Supplement

Composition per liter:

Agar	15.0g
Glucose	10.0g
Peptic digest of animal tissue	5.0g
KH_2PO_4	1.0g
$MgSO_4$	0.5g
Rose Bengal	0.025g
Bromcresol Purple	0.02g
Dichloran	0.002
Selective supplement	10.0mL

pH 5.6 ± 0.2 at 25°C

Source: This medium, wthout selective supplement, is available as a premixed powder from HiMedia.

Selective Supplement:
Composition per 10.0mL:

Chloramphenicol	0.1mg

Preparation of Selective Supplement: Add chloramphenicol to distilled/deionized water and bring volume to 10.0mL. Mix thoroughly. Filter sterilize.

Preparation of Medium: Add components, except selective supplement, to distilled/deionized water and bring volume to 1.0L. Mix thoroughly. Gently heat and bring to boiling. Autoclave for 15 min at 15 psi pressure–121°C. The final pH should be 5.6. Temper in a water bath at 45°C. Aseptically add 10.0mL sterile selective supplement. Mix thoroughly. Pour into sterile Petri dishes or distribute into sterile tubes. This medium is light sensitive and should be stored in a dark, cool place until used. It is intended for spread plates only.

Use: For the analysis of food samples containing spreader molds, e.g., Mucor and Rhizopus. The dichloran and Rose Bengal slow down the growth of fast-growing fungi, thus allowing the detection of other fungi with slower growth rates.

Dichloran Rose Bengal Chloramphenicol Agar (DRBC Agar) (BAM M183)

Composition per liter:

Agar	15.0g
Glucose	10.0g
Peptone	5.0g
KH_2PO_4	1.0g
$MgSO_4·7H_2O$	0.5g
Chloramphenicol	0.1g
Dichloran solution	1.0mL
Rose Bengal solution	0.5mL

pH 5.6 ± 0.2 at 25°C

Dichloran Solution:

Composition per 10.0mL:

Dichloran (2,6-dichloro-4-nitroaniline)	0.2g

Preparation of Dichloran Solution: Add dichloran to 10.0mL of distilled/deionized water. Mix thoroughly.

Rose Bengal Solution:

Composition per 100.0mL:

Rose Bengal	2.0g

Preparation of Rose Bengal Solution: Add Rose Bengal to 100.0mL of distilled/deionized water. Mix thoroughly.

Preparation of Medium: Add components to distilled/deionized water and bring volume to 1.0L. Mix thoroughly. Gently heat and bring to boiling. Mix thoroughly. Autoclave for 15 min at 15 psi pressure–121°C. The final pH should be 5.6. Temper in a water bath at 45°C. Pour into sterile Petri dishes or distribute into sterile tubes. This medium is light sensitive and should be stored in a dark, cool place until used. It is intended for spread plates only.

Use: For the analysis of food samples containing spreader molds, e.g. Mucor and Rhizopus. The dichloran and Rose Bengal slow down the growth of fast-growing fungi, thus allowing the detection of other fungi with slower growth rates.

Dichloran Rose Bengal Chloramphenicol Agar

See: **DRBC Agar**

Dichloroacetic Acid Medium No. 1

Composition per liter:

Yeast extract	10.0g
Glucose	5.0g
$(NH_4)_2PO_4$	1.5g
K_2HPO_4	1.0g
2,4-Dichloroacetic acid	0.75g
$MgSO_4·7H_2O$	0.2g
$Fe_2(SO_4)_3·5H_2O$	0.01g
$ZnSO_4·7H_2O$	2.0mg

pH 7.0 ± 0.2 at 25°C

Preparation of Medium: Add components to distilled/deionized water and bring volume to 1.0L. Mix thoroughly. Distribute into tubes or flasks. Autoclave for 15 min at 15 psi pressure–121°C.

Use: For the cultivation of *Rhodococcus* species.

Dichloroacetic Acid Medium No. 2

Composition per liter:

Yeast extract	10.0g
Glucose	5.0g
$(NH_4)_2PO_4$	1.5g
K_2HPO_4	1.0g
$MgSO_4·7H_2O$	0.2g
$Fe_2(SO_4)_3·5H_2O$	0.01g
$ZnSO_4·7H_2O$	0.002g
2,4-Dichloroacetic acid	10.0mg

pH 7.0 ± 0.2 at 25°C

Preparation of Medium: Add components to distilled/deionized water and bring volume to 1.0L. Mix thoroughly. Distribute into tubes or flasks. Autoclave for 15 min at 15 psi pressure–121°C.

Use: For the cultivation of *Pseudomonas* species.

Dichloromethane Medium for *Hyphomicrobium*

Composition per liter:

$K_2HPO_4·3H_2O$	4.1g
KH_2PO_4	1.4g
$MgSO_4·7H_2O$	0.2g
$(NH_4)_2SO_4$	0.2g
Dichloromethane (methylene chloride)	1.0mL
Trace elements solution	1.0mL

pH 7.2 ± 0.2 at 25°C

Trace Elements Solution:

Composition per liter:

$Ca(NO_3)_2$	25.0g
$FeSO_4·7H_2O$	1.0g
H_3BO_3	1.0g
$MnSO_4·H_2O$	1.0g
$Co(NO_3)_2·6H_2O$	0.25g
$CuCl_2·2H_2O$	0.25g
$(NH_4)_6Mo_7O_{24}·4H_2O$	0.25g
$ZnCl_2$	0.25g
NH_4VO_3	0.1g

Preparation of Trace Elements Solution: Add components to distilled/deionized water and bring volume to 1.0L. Mix thoroughly.

Preparation of Medium: Filter sterilize dichloromethane. Add components, except dichloromethane, to distilled/deionized water and bring volume to 999.0mL. Mix thoroughly. Gently heat and bring to boiling. Adjust pH to 7.2. Autoclave for 15 min at 15 psi pressure–121°C. Cool to 45°–50°C. Aseptically add sterile dichloromethane. Mix thoroughly. Aseptically distribute into sterile tubes or flasks.

Use: For the cultivation and maintenance of *Hyphomicrobium* species.

Dichotomicrobium thermohalophilum Agar

Composition per liter:

Agar	18.0g
Disodium DL-malate	1.0g

Yeast extract	1.0g
Artificial seawater, 3×	960.0mL
Hutner's basal salts solution	20.0mL
$NaHCO_3$ solution	20.0mL

pH 7.0–7.2 at 25°C

Artificial Seawater, 3×:
Composition per liter:

NaCl	70.43g
$MgCl_2·6H_2O$	31.86g
Na_2SO_4	11.75g
$CaCl_2·2H_2O$	4.35g
$NaHCO_3$	2.88g
KCl	1.99g
KBr	0.29g
H_3BO_3	0.08g

Preparation of Artificial Seawater, 3×: Add components to distilled/deionized water and bring volume to 1.0L. Mix thoroughly.

Hutner's Basal Salts Solution:
Composition per liter:

$MgSO_4·7H_2O$	29.7g
Nitrilotriacetic acid	10.0g
$CaCl_2·2H_2O$	3.335g
$FeSO_4·7H_2O$	99.0mg
$(NH_4)_6MoO_7O_{24}·4H_2O$	9.25mg
"Metals 44"	50.0mL

"Metals 44":
Composition per 100.0mL:

$ZnSO_4·7H_2O$	1.095g
$FeSO_4·7H_2O$	0.5g
Sodium EDTA	0.25g
$MnSO_4·H2O$	0.154g
$CuSO_4·5H_2O$	39.2mg
$Co(NO_3)_2·6H_2O$	24.8mg
$Na_2B_4O_7·10H_2O$	17.7mg

Preparation of "Metals 44": Add sodium EDTA to distilled/deionized water and bring volume to 90.0mL. Mix thoroughly. Add a few drops of concentrated H_2SO_4 to retard precipitation of heavy metal ions. Add remaining components. Mix thoroughly. Bring volume to 100.0mL with distilled/deionized water.

Preparation of Hutner's Basal Salts Solution: Add nitrilotriacetic acid to 500.0mL of distilled/deionized water. Adjust pH to 6.5 with KOH. Add remaining components. Add distilled/deionized water to 1.0L. Adjust pH to 6.8.

$NaHCO_3$ Solution:
Composition per 20.0mL:

$NaHCO_3$	3.0g

Preparation of $NaHCO_3$ Solution: Add $NaHCO_3$ to distilled/deionized water and bring volume to 20.0mL. Mix thoroughly. Filter sterilize.

Preparation of Medium: Add components, except $NaHCO_3$ solution, to distilled/deionized water and bring volume to 980.0mL. Mix thoroughly. Gently heat and bring to boiling. Autoclave for 15 min at 15 psi pressure–121°C. Cool to 50°–55°C. Aseptically add 20.0mL of sterile $NaHCO_3$ solution. Mix thoroughly. Pour into sterile Petri dishes or distribute into sterile tubes.

Use: For the cultivation and maintenance of *Dichotomicrobium thermohalophilum.*

Dichotomicrobium thermohalophilum Broth

Composition per liter:

Disodium DL-malate	1.0g
Yeast extract	1.0g
Artificial seawater, 3×	960.0mL
Hutner's basal salts solution	20.0mL
$NaHCO_3$ solution	20.0mL

pH 7.0–7.2 at 25°C

Artificial Seawater, 3×:
Composition per liter:

NaCl	70.43g
$MgCl_2·6H_2O$	31.86g
Na_2SO_4	11.75g
$CaCl_2·2H_2O$	4.35g
$NaHCO_3$	2.88g
KCl	1.99g
KBr	0.29g
H_3BO_3	0.08g

Preparation of Artificial Seawater, 3×: Add components to distilled/deionized water and bring volume to 1.0L. Mix thoroughly.

Hutner's Basal Salts Solution:
Composition per liter:

$MgSO_4·7H_2O$	29.7g
Nitrilotriacetic acid	10.0g
$CaCl_2·2H_2O$	3.335g
$FeSO_4·7H_2O$	99.0mg
$(NH_4)_6MoO_7O_{24}·4H_2O$	9.25mg
"Metals 44"	50.0mL

"Metals 44":
Composition per 100.0mL:

$ZnSO_4·7H_2O$	1.095g
$FeSO_4·7H_2O$	0.5g
Sodium EDTA	0.25g
$MnSO_4·H2O$	0.154g
$CuSO_4·5H_2O$	39.2mg
$Co(NO_3)_2·6H_2O$	24.8mg
$Na_2B_4O_7·10H_2O$	17.7mg

Preparation of "Metals 44": Add sodium EDTA to distilled/deionized water and bring volume to 90.0mL. Mix thoroughly. Add a few drops of concentrated H_2SO_4 to retard precipitation of heavy metal ions. Add remaining components. Mix thoroughly. Bring volume to 100.0mL with distilled/deionized water.

Preparation of Hutner's Basal Salts Solution: Add nitrilotriacetic acid to 500.0mL of distilled/deionized water. Adjust pH to 6.5 with KOH. Add remaining components. Add distilled/deionized water to 1.0L. Adjust pH to 6.8.

$NaHCO_3$ Solution:
Composition per 20.0mL:

$NaHCO_3$	3.0g

Preparation of $NaHCO_3$ Solution: Add $NaHCO_3$ to distilled/deionized water and bring volume to 20.0mL. Mix thoroughly. Filter sterilize.

Preparation of Medium: Add components, except $NaHCO_3$ solution, to distilled/deionized water and bring volume to 980.0mL. Mix thoroughly. Autoclave for 15 min at 15 psi pressure–121°C. Aseptically add 20.0mL of sterile $NaHCO_3$ solution. Mix thoroughly. Aseptically distribute into sterile tubes or flasks.

Use: For the cultivation of *Dichotomicrobium thermohalophilum.*

Dictyoglomus Medium

Composition per liter:

Soluble starch	5.0g
$Na_2HPO_4 \cdot 12H_2O$	4.2g
Polypeptone™	2.0g
Yeast extract	2.0g
KH_2PO_4	1.5g
L-Cysteine·HCl·H_2O	1.0g
Na_2CO_3	1.0g
NH_4Cl	0.5g
$MgCl_2 \cdot 6H_2O$	0.38g
$CaCl_2$	0.05g
$Fe(NH_4)_2(SO_4)_2 \cdot 6H_2O$	0.039g
Resazurin	2.0mg
Trace metals	10.0mL
Wolfe's vitamin solution	10.0mL

pH 7.2 ± 0.2 at 25°C

Trace Metals:

Composition per liter:

$CoCl_2 \cdot 6H_2O$	0.29g
$ZnSO_4 \cdot 7H_2O$	0.28g
$Na_2MoO_4 \cdot 2H_2O$	0.24g
$MnCl_2 \cdot 4H_2O$	0.2g
Na_2SeO_3	0.017g

Preparation of Trace Metals: Add components to distilled/deionized water and bring volume to 1.0L. Adjust pH to 6.0 with KOH. Mix thoroughly.

Wolfe's Vitamin Solution:

Composition per liter:

Pyridoxine·HCl	0.01g
Thiamine·HCl	5.0mg
Riboflavin	5.0mg
Nicotinic acid	5.0mg
Calcium pantothenate	5.0mg
p-Aminobenzoic acid	5.0mg
Thioctic acid	5.0mg
Biotin	2.0mg
Folic acid	2.0mg
Cyanocobalamin	0.1mg

Preparation of Wolfe's Vitamin Solution: Add components to distilled/deionized water and bring volume to 1.0L. Mix thoroughly. Filter sterilize.

Preparation of Medium: Prepare and dispense medium under 100% N_2. Add components, except Wolfe's vitamin solution, to distilled/deionized water and bring volume to 990.0mL. Mix thoroughly. Gently heat and bring to boiling. Continue boiling until resazurin turns colorless. Autoclave for 15 min at 15 psi pressure–121°C. Cool to 25°C under 100% N_2. Aseptically add sterile Wolfe's vitamin solution. Mix thoroughly. Adjust pH to 7.2 if necessary. Aseptically and anaerobically distribute into sterile tubes or flasks.

Use: For the cultivation and maintenance of *Dictyoglomus thermophilum.*

Dictyostelium Medium

Composition per liter:

Glucose	15.4g
Agar	15.0g
Peptone	14.3g
Yeast extract	7.15g
$Na_2HPO_4 \cdot 12H_2O$	1.28g
KH_2PO_4	0.49g

pH 6.7 ± 0.2 at 25°C

Preparation of Medium: Add components to distilled/deionized water and bring volume to 1.0L. Mix thoroughly. Gently heat and bring to boiling. Adjust pH to 6.7. Distribute into tubes or flasks. Autoclave for 15 min at 15 psi pressure–121°C. Pour into sterile Petri dishes or leave in tubes.

Use: For the cultivation of *Dictyostelium discoideum* and *Fusarium acuminatum.*

Diethyl Phosphonate Agar

Composition per liter:

Agar	12.0g
Tris(hydroxymethyl)methylamine	6.0g
p-Hydroxybenzoate, Na salt	0.75g
KCl	0.2g
$MgSO_4 \cdot 7H_2O$	0.2g
NH_4Cl	0.2g
Diethyl phosphonate solution	100.0mL

pH 7.4 ± 0.2 at 25°C

Diethyl Phosphonate Solution:

Composition per 100.0mL:

Diethyl phosphonate	0.015g

Source: Diethyl phosphonate is available from Eastman Organic Chemical Division, Rochester, NY.

Preparation of Diethyl Phosphonate Solution: Add diethyl phosphonate to distilled/deionized water and bring volume to 100.0mL. Mix thoroughly. Filter sterilize.

Preparation of Medium: Add components, except diethyl phosphonate solution, to distilled/deionized water and bring volume to 900.0mL. Mix thoroughly. Gently heat and bring to boiling. Autoclave for 15 min at 15 psi pressure–121°C. Cool to 25°C. Aseptically add 100.0mL diethyl phosphate solution. Mix. Pour into sterile Petri dishes or distribute into sterile tubes.

Use: For the cultivation and maintenance of *Comamonas acidovorans.*

DIFF/BCYE

See: **Buffered Charcoal Yeast Extract Differential Agar**

Differential Agar for Group D Streptococci

Composition per liter:

NaCl	65.0g
Agar	13.5g
Casein enzymic hydrolysate	16.0g
Glucose	10.0g
Brain heart infusion	8.0g
Peptic digest of animal tissue	5.0g
Na_2HPO_4	2.5g
Bromcresol Purple	0.02g

pH 7.4 ± 0.2 at 25°C

Source: This medium is available from HiMedia.

Preparation of Medium: Add components to distilled/deionized water and bring volume to 1.0L. Mix thoroughly. Gently heat and bring to boiling. Distribute into tubes or flasks. Autoclave for 15 min at 15 psi pressure–121°C. Pour into sterile Petri dishes or leave in tubes. For tubes, allow to solidify in slanted position.

Use: For the differentiation and identification of Group D streptococci.

Differential Agar Medium A8 for *Ureaplasma urealyticum*

Composition per 103.1mL:

Basal agar	80.0mL
Horse serum, unheated	20.0mL
Fresh yeast extract solution	1.0mL
Urea solution	1.0mL
CVA enrichment	0.5mL
L-Cysteine·HCl·H_2O solution	0.5mL
GHL tripeptide solution	0.1mL

pH 5.5 ± 0.2 at 25°C

Basal Agar:

Composition per 80.0mL:

Tryptic soy broth	2.4g
Noble agar	1.05g
Putrescine·2HCl	0.17g
$CaCl_2 \cdot 2H_2O$	0.015g

Preparation of Basal Agar: Add components to distilled/deionized water and bring volume to 80.0mL. Mix thoroughly. Adjust pH to 5.5 with 2*N* HCl. Gently heat and bring to boiling. Autoclave for 15 min at 15 psi pressure–121°C. Cool to 50°–55°C.

Fresh Yeast Extract Solution:

Composition per 100.0mL:

Baker's yeast, live, pressed, starch-free	25.0g

Preparation of Fresh Yeast Extract Solution: Add the live Baker's yeast to 100.0mL of distilled/deionized water. Autoclave for 90 min at 15 psi pressure–121°C. Allow to stand. Remove supernatant solution. Adjust pH to 6.6–6.8. Filter sterilize.

Urea Solution:

Composition per 30.0mL:

Urea	3.0g

Preparation of Urea Solution: Add urea to distilled/deionized water and bring volume to 30.0mL. Mix thoroughly. Filter sterilize.

CVA Enrichment:

Composition per liter:

Glucose	100.0g
L-Cysteine·HCl·H_2O	25.9g
L-Glutamine	10.0g
Adenine	1.0g
L-Cystine·2HCl	1.0g
Nicotinamide adenine dinucleotide	0.25g
Cocarboxylase	0.1g
Guanine·HCl	0.03g
$Fe(NO_3)_3$	0.02g
Vitamin B_{12}	0.01g
p-Aminobenzoic acid	0.013g
Thiamine·HCl	3.0mg

Preparation of CVA Enrichment: Add components to distilled/deionized water and bring volume to 1.0L. Mix thoroughly. Filter sterilize.

L-Cysteine·HCl·H_2O Solution:

Composition per 50.0mL:

L-Cysteine·HCl·H_2O	1.0g

Preparation of L-Cysteine·HCl·H_2O Solution: Add L-cysteine·HCl·H_2O to distilled/deionized water and bring volume to 50.0mL. Mix thoroughly. Filter sterilize.

GHL Tripeptide Solution:

Composition per 10.0mL:

GHL tripeptide	0.2mg

Preparation of GHL Tripeptide Solution: Add GHL tripeptide (glycyl-L-histidyl-L-lysine acetate) to distilled/deionized water and bring volume to 10.0mL. Mix thoroughly. Filter sterilize.

Preparation of Medium: To 80.0mL of cooled, sterile basal agar, aseptically add 20.0mL of sterile horse serum, 1.0mL of sterile fresh yeast extract solution, 1.0mL of sterile urea solution, 0.5mL of sterile CVA enrichment, 0.5mL of sterile L-cysteine·HCl·H_2O solution, and 0.1mL of sterile GHL tripeptide solution. Mix thoroughly. Pour into sterile Petri dishes in 20.0mL volumes.

Use: For the cultivation and maintenance of *Ureaplasma urealyticum*.

Differential Broth for Lactic Streptococci

Composition per liter:

Sodium citrate	20.0g
Arginine	5.0g
Pancreatic digest of casein	5.0g
Yeast extract	5.0g
K_2HPO_4	1.0g
Bromcresol Purple	0.02g
Skim milk (11% solution)	35.0mL

pH 6.2 ± 0.2 at 25°C

Preparation of Medium: Add components, except skim milk solution, to distilled/deionized water and bring volume to 800.0mL. Mix thoroughly. Add 35.0mL of skim milk solution. Bring volume to 1.0L with distilled/deionized water. Place medium in a steam bath for 15 min. Cool to 25°C. Adjust pH to 6.2. Distribute 7.0mL volumes into screw-capped tubes that contain an inverted Durham tube. Autoclave for 15 min at 15 psi pressure–121°C. Allow autoclave to cool below 70°C before opening door.

Use: For the cultivation and differentiation of *Lactobacillus lactis, Lactobacillus lactis* subspecies *cremoris,* and *Lactobacillus lactis* subspecies *diacetylactis*. Lactose-fermenting bacteria such as *Lactobacillus lactis* subspecies *cremoris* turn the medium yellow. Arginine-utilizing bacteria such as *Lactobacillus lactis* initially turn the medium yellow but then turn it back to violet. Citrate-utilizing bacteria such as *Lactobacillus lactis* subspecies *diacetylactis* turn the medium violet and produce CO_2 that is trapped as a bubble in the Durham tube.

Differential Buffered Charcoal Yeast Extract Agar Base with Selective Supplement

Composition per liter:

Agar	15.0g
ACES buffer	10.0g
Yeast extract	10.0g
Charcoal, activated	1.5g
L-Cysteine·HCl	0.4g
Ferric pyrophosphate, soluble	0.25g
α-Ketoglutarate	0.2g
Bromcresol Purple	0.01g
Bromthymol Blue	0.01g
Selective supplement	10.0mL

pH 6.9 ± 0.2 at 25°C

Source: This medium, wthout selective supplement, is available as a premixed powder from HiMedia.

Selective Supplement:

Composition per 10.0mL:

Vancomycin	0.1g
Polymyxin B	50,000 units

Preparation of Selective Supplement: Add components to distilled/deionized water and bring volume to 10.0mL. Mix thoroughly. Filter sterilize.

Preparation of Medium: Add components, except selective supplement, to distilled/deionized water and bring volume to 1.0L. Mix thoroughly. Gently heat and bring to boiling. Autoclave for 15 min at 15 psi pressure–121°C. Cool to 50°C. Aseptically add 10.0mL sterile selective supplement. Mix thoroughly. Pour into sterile Petri dishes or distribute into sterile tubes.

Use: For the selective isolation and differentiation of *Legionella* species.

Differential Reinforced Clostridial HiVeg Broth Base with Ferric Citrate and Sodium Sulfite

Composition per liter:

Plant extract	10.0g
Plant peptone	10.0g
Sodium acetate, hydrated	5.0g
Yeast extract	1.5g
Glucose	1.0g
Starch	1.0g
L-Cysteine·HCl	0.5g
Ferric citrate sodium sulfite solution	20.0mL

pH 7.2 ± 0.2 at 25°C

Source: This medium, wthout sodium sulfite and ferric citrate, is available as a premixed powder from HiMedia.

Ferric Citrate Sodium Sulfite Solution:

Composition per 100.0mL:

Ferric citrate solution	50.0mL
Sodium sulfite solution	50.0mL

Preparation of Ferric Citrate Sodium Sulfite Solution: Aseptically combine 50.0mL of sterile ferric citrate solution and 50.0mL of sterile sodium sulfite solution. Mix thoroughly. Filter sterilize.

Ferric Citrate Solution:

Composition per 100.0mL:

Ferric citrate	7.0g

Preparation of Ferric Citrate Solution: Add ferric citrate to distilled/deionized water and bring volume to 100.0mL. Mix thoroughly. Filter sterilize

Sodium Sulfite Solution:

Composition per 100.0mL:

Na_2SO_3	7.0g

Preparation of Sodium Sulfite Solution: Add Na_2SO_3 to distilled/deionized water and bring volume to 100.0mL. Mix thoroughly. Filter sterilize.

Preparation of Medium: Add components, except ferric citrate-sodium sulfite solution, to distilled/deionized water and bring volume to 980.0mL. Mix thoroughly. Gently heat and bring to boiling. Autoclave for 15 min at 15 psi pressure–121°C. Cool to 25°C. Just before using, aseptically add 20.0mL sterile ferric citrate sodium sulfite solution. Mix thoroughly. Aseptically distribute into tubes or flasks.

Use: For the cultivation and enumeration of *Clostridium* species from water.

Dihydrolase Broth Base with Arginine

Composition per liter:

NaCl	30.0g
Yeast extract	6.0g
Peptic digest of animal tissue	5.0g
Glucose	2.0g
Bromcresol Purple	0.032g
L-Arginine solution	50.0mL

pH 6.5 ± 0.2 at 25°C

Source: This medium is available as a premixed powder from HiMedia.

Arginine Solution:

Composition per 100.0mL:

L-Arginine	10.0g

Preparation of Arginine Solution: Add L-arginine to distilled/deionized water and bring volume to 100.0mL. Mix thoroughly. Filter sterilize.

Preparation of Medium: Add components, except arginine solution, to distilled/deionized water and bring volume to 1.0L. Mix thoroughly. Gently heat until dissolved. Distribute into screw-capped tubes in 5.0mL volumes. Autoclave for 15 min at 10 psi pressure–115°C. Aseptically add 50.0mL sterile arginine solution.

Use: For the differentiation of Gram-negative enteric bacteria based on the production of arginine dihydrolase.

Dihydrolase HiVeg Broth Base with Arginine

Composition per liter:

NaCl	30.0g
Yeast extract	6.0g
Plant peptone	5.0g
Glucose	2.0g
Bromcresol Purple	0.032g
L-Argnine solution	50.0mL

pH 6.5 ± 0.2 at 25°C

Source: This medium is available as a premixed powder from HiMedia.

Arginine Solution:

Composition per 100.0mL:

L-Arginine	10.0g

Preparation of Arginine Solution: Add L-arginine to distilled/deionized water and bring volume to 100.0mL. Mix thoroughly. Filter sterilize.

Preparation of Medium: Add components, except arginine solution, to distilled/deionized water and bring volume to 1.0L. Mix thoroughly. Gently heat until dissolved. Distribute into screw-capped tubes in 5.0mL volumes. Autoclave for 15 min at 10 psi pressure–115°C. Aseptically add 50.0mL sterile arginine solution.

Use: For the differentiation of Gram-negative enteric bacteria based on the production of arginine dihydrolase.

Dilute Peptone Water

Composition per liter:

NaCl	1.0g
Peptone	1.0g

pH 7.0 ± 0.2 at 25°C

Preparation of Medium: Add components to distilled/deionized water and bring volume to 1.0L. Mix thoroughly. Distribute into tubes or flasks. Autoclave for 15 min at 15 psi pressure–121°C.

Use: For the cultivation of various heterotrophic bacteria.

Dilute Potato Medium (DSMZ Medium 789)

Composition per liter:

Glucose	1.0g
Na_2HPO_4	0.12g
$Ca(NO_3)_2 \cdot 4H_2O$	0.05g
Peptone	0.05g
Potato decoction	100.0mL

pH 7.3 ± 0.2 at 25°C

Potato Decoction:

Diced potato	20.0g

Preparation of Potato Decoction: Add diced potatoes to distilled/deionized water and bring volume to 1.0L. Boil for 30 min. Filter to remove solid potatoes. Bring volume to 1.0L with distilled/deionized water.

Preparation of Medium: Add components to distilled/deionized water and bring volume to 1.0L. Mix thoroughly. Distribute into tubes or flasks. Autoclave for 15 min at 15 psi pressure–121°C.

Use: For the cultivation of various fungi.

Dilute Potato Medium

Composition per 1090.0mL:

Glucose	1.0g
Na_2HPO_4	0.12g
$Ca(NO_3)_2 \cdot 4H_2O)$	0.05g
Peptone	0.05g
Potato decoction	100.0mL

pH 6.8 ± 0.2 at 25°C

Potato Decoction:

Composition per liter:

Potato	20.0g

Preparation of Potato Decoction: Peel and dice potato. Add to 1.0L of distilled/deionized water. Gently heat and bring to boiling. Continue boiling for 30 min. Filter through Whatman #1 filter paper. Bring volume of filtrate to 1.0L with distilled/deionized water.

Preparation of Medium: Add components to distilled/deionized water and bring volume to 1090.0mL. Mix thoroughly. Adjust pH to 6.8. Distribute into tubes or flasks. Autoclave for 15 min at 15 psi pressure–121°C.

Use: For the cultivation and maintenance of *Rhizobacter daucus*.

Dinoflagellate Medium

Composition per 1020.0mL:

Seawater solution	1.0L
Basal solution	20.0mL

pH 7.8 ± 0.2 at 25°C

Seawater Solution:

Composition per 1100.0mL:

Seawater	1010.0mL

Preparation of Seawater Solution: Add seawater to distilled/deionized water and bring volume to 1100.0mL. Mix thoroughly. Adjust pH to 7.8. Autoclave for 15 min at 15 psi pressure–121°C. Cool to 25°C.

Basal Solution:

Composition per 100.0mL:

Buffer salts solution	25.0mL
Fe solution	25.0mL
Vitamin solution	25.0mL
Metal solution	25.0mL

Preparation of Basal Solution: Adjust final pH to 7.8.

Buffer Salts Solution:

Composition per 25.0mL:

Tris-HCl	500.0mg
$NaNO_3$	350.0mg
Sodium glycerophosphate·$6H_2O$	50.0mg

Preparation of Buffer Salts Solution: Add components to distilled/deionized water and bring volume to 25.0mL. Adjust ph to 7.8. Mix thoroughly.

Fe Solution:

Composition per 500.0mL:

$Fe(NH_4)_2(SO_4)_2 \cdot 6H_2O$	351.0mg
EDTA	330.0mg

Preparation of Fe Solution: Add components to distilled/deionized water and bring volume to 500.0mL. Mix thoroughly.

Vitamin Solution:

Composition per 25.0mL:

Vitamin B_{12}	10.0μg
Biotin	5.0μg
Thiamine	0.5mg

Preparation of Vitamin Solution: Add components to distilled/deionized water and bring volume to 25.0mL. Mix thoroughly.

Metal Solution:

Composition per 25.0mL:

H_3BO_3	114.0mg
EDTA	100.0mg
$MnSO_4 \cdot 4H_2O$	16.4mg
$FeCl_3 \cdot 6H_2O$	4.9mg
$ZnSO_4 \cdot 7H_2O$	2.2mg
$CoSO_4 \cdot 7H_2O$	0.48mg

Preparation of Metal Solution: Add components, in the order listed, to distilled/deionized water and bring volume to 25.0mL. Mix thoroughly. Adjust pH to 7.5.

Preparation of Basal Solution: Combine 25.0mL buffer salts solution, 25.0mL Fe solution, 25.0mL vitamin solution, and 25.0mL metal solution. Adjust pH to 7.8. Mix thoroughly. Autoclave for 15 min at 15 psi pressure–121°C. Cool to 25°C.

Preparation of Medium: Aseptically combine 20.0mL of sterile basal solution with 1.0L of sterile seawater solution. Mix thoroughly. Aseptically distribute into sterile, screw-capped tubes or flasks.

Use: For the cultivation of *Amphidinium carteri*.

Diphasic Blood Agar Base Medium (ATCC Medium 449)

Composition per 500.0mL:

Beef	25.0g
Agar	10.0g

Neopeptone 10.0g
NaCl 2.5g

Preparation of Medium: Trim beef to remove fat. Add 25.0g of lean beef to 250.0mL of distilled/deionized water. Gently heat and bring to boiling. Boil for 2–3 min. Filter through Whatman #2 filter paper. Add agar, neopeptone, and NaCl to filtrate. Bring volume to 500.0mL with distilled/deionized water. Mix thoroughly. Adjust pH to 7.2–7.4. Gently heat and bring to boiling. Autoclave for 20 min at 15 psi pressure–121°C. Cool to 50°–55°C. Add as required to other diphasic blood agars.

Use: As the base medium for diphasic blood agars.

Diphasic Blood Agar Medium with 10% Blood

Composition per 1120.0mL:

Blood agar, diphasic base medium 630.0mL
Locke's solution 420.0mL
Rabbit blood, defibrinated 70.0mL

pH 7.2–7.4 at 25°C

Blood Agar, Diphasic Base Medium:
Composition per 750.0mL:

Beef 25.0g
Agar 10.0g
Neopeptone 10.0g
NaCl 2.5g

Preparation of Blood Agar, Diphasic Base Medium: Trim beef to remove fat. Add 25.0g of lean beef to 250.0mL of distilled/deionized water. Gently heat and bring to boiling. Boil for 2–3 min. Filter through Whatman #2 filter paper. Add agar, neopeptone, and NaCl to filtrate. Bring volume to 750.0mL with distilled/deionized water. Mix thoroughly. Adjust pH to 7.2–7.4. Gently heat and bring to boiling. Autoclave for 15 min at 15 psi pressure–121°C. Cool to 50°–55°C.

Locke's Solution:
Composition per liter:

NaCl 8.0g
Glucose 2.5g
KH_2PO_4 .... 0.3g
$CaCl_2$ 0.2g
KCl 0.2g

Preparation of Locke's Solution: Add components to distilled/deionized water and bring volume to 1.0L. Mix thoroughly. Autoclave for 15 min at 15 psi pressure–121°C. Cool to 50°–55°C.

Preparation of Medium: Aseptically combine 630.0mL of sterile blood agar, diphasic base medium, with 70.0mL of sterile defibrinated rabbit blood warmed to 50°–55°C. Mix thoroughly. Aseptically distribute 5.0mL volumes into16 × 125mm screw-capped test tubes. Allow to cool in a slanted position. Overlay the agar in each tube with 3.0mL of sterile Locke's solution.

Use: For the cultivation of *Leishmania braziliensis, Leishmania enriettii, Leishmania tropica, Trypanosoma conorrhini, Trypanosoma cruzi,* and *Trypanosoma rangeli.*

Diphasic Blood Agar Medium with 30% Blood

Composition per 1450.0mL:

Blood agar, diphasic base medium 700.0mL
Locke's solution 450.0mL
Rabbit blood, defibrinated 300.0mL

pH 7.2–7.4 at 25°C

Blood Agar, Diphasic Base Medium:
Composition per 750.0mL:

Beef 25.0g
Agar 10.0g
Neopeptone 10.0g
NaCl 2.5g

Preparation of Blood Agar, Diphasic Base Medium: Trim beef to remove fat. Add 25.0g of lean beef to 250.0mL of distilled/deionized water. Gently heat and bring to boiling. Boil for 2–3 min. Filter through Whatman #2 filter paper. Add agar, neopeptone, and NaCl to filtrate. Bring volume to 750.0mL with distilled/deionized water. Mix thoroughly. Adjust pH to 7.2–7.4. Gently heat and bring to boiling. Autoclave for 15 min at 15 psi pressure–121°C. Cool to 50°–55°C.

Locke's Solution:
Composition per liter:

NaCl 8.0g
Glucose 2.5g
KH_2PO_4 .... 0.3g
$CaCl_2$ 0.2g
KCl 0.2g

Preparation of Locke's Solution: Add components to distilled/deionized water and bring volume to 1.0L. Mix thoroughly. Autoclave for 15 min at 15 psi pressure–121°C. Cool to 50°–55°C.

Preparation of Medium: Aseptically combine 700.0mL of sterile blood agar, diphasic base medium, with 300.0mL of sterile defibrinated rabbit blood warmed to 50°–55°C. Mix thoroughly. Aseptically distribute 5.0mL volumes into 16 × 125mm screw-capped test tubes. Allow to cool in a slanted position. Overlay the agar in each tube with 3.0mL of sterile Locke's solution.

Use: For the cultivation and maintenance of *Blastocrithidia culicis, Crithidia deanei, Crithidia flexonema, Crithidia luciliae, Crithidia mellificae, Endotrypanum* species, *Herpetomonas anglusteri, Herpetomonas mariadeanei, Herpetomonas megaseliae, Herpetomonas muscarum, Herpetomonas roitmani, Leishmania braziliensis, Leishmania donovani, Leishmania peruviana, Leishmania tarentolae, Leptomonas collosoma, Leptomonas costoris, Leptomonas lactosovorans, Leptomonas mirabilis, Leptomonas pulexsimulantis, Leptomonas samueli, Leptomonas seymouri, Trypanosoma avium, Trypanosoma bennetti, Trypanosoma cervi, Trypanosoma chattoni, Trypanosoma conorrhini, Trypanosoma cruzi, Trypanosoma cyclops, Trypanosoma fallisi, Trypanosoma lewisi, Trypanosoma lucknowi, Trypanosoma mega, Trypanosoma musculi, Trypanosoma neveulemairei, Trypanosoma ranarum, Trypanosoma rotatorium,* and *Trypanosoma tamiasi.*

Diphasic Blood Culture Buffered Charcoal Yeast Extract Medium

See: ***Legionella pneumophila*** **Medium Charcoal Yeast Extract Diphasic Blood Culture Medium**

Diphasic Medium for *Amoeba* (Charcoal Agar Slants)

Composition per liter:

Agar slants 1.0L
Buffered saline overlay 1.0L

pH 7.4 ± 0.2 at 25°C

Agar Slants:
Composition per liter:

Agar 10.0g
Charcoal, activated 10.0g

Pancreatic digest of casein	5.0g
KH_2PO_4	4.0g
Na_2HPO_4	3.0g
Asparagine	2.0g
Sodium citrate	1.0g
Ferric ammonium citrate	0.1g
$MgSO_4 \cdot 7H_2O$	0.1g
Cholesterol solution	25.0mL
Glycerol	10.0mL

Cholesterol Solution:

Composition per 25.0mL:

Cholesterol	0.25g
Acetone	25.0mL

Preparation of Cholesterol Solution: Add cholesterol to 25.0mL of acetone. Mix thoroughly.

Preparation of Agar Slants: Add components, except agar, charcoal, and cholesterol solution, to distilled/deionized water and bring volume to 1.0L. Mix thoroughly. Gently heat to dissolve. Do not boil. Add agar, charcoal, and cholesterol solution. Mix thoroughly. Gently heat and bring to boiling. Distribute into tubes in 3.0mL volumes. Autoclave for 15 min at 15 psi pressure–121°C. Resuspend charcoal. Allow tubes to cool in a slanted position with short butts or no butts.

Buffered Saline Overlay:

Composition per liter:

NaCl	5.0g
Solution B	810.0mL
Solution A	190.0mL

Solution A:

Composition per liter:

KH_2PO_4, anhydrous	9.07g

Preparation of Solution A: Add KH_2PO_4 to distilled/deionized water and bring volume to 1.0L. Mix thoroughly.

Solution B:

Composition per liter:

Na_2HPO_4, anhydrous	9.46g

Preparation of Solution B: Add Na_2HPO_4 to distilled/deionized water and bring volume to 1.0L. Mix thoroughly.

Preparation of Buffered Saline Overlay: Combine 810.0mL of solution A and 190.0mL of solution B. Add the NaCl. Mix thoroughly. Autoclave for 15 min at 15 psi pressure–121°C. Cool to 25°C. Store at 4°C.

Preparation of Medium: To each agar slant, aseptically add 3.0mL of sterile, buffered saline overlay.

Use: For the cultivation and maintenance of *Amoebae* species.

Diphosphothiamine Medium

Composition per liter:

Proteose peptone	20.0g
Glucose	10.0g
NaCl	5.0g
Tween™ 40	0.05g
Diphosphothiamine	1.0mg

pH 7.3 ± 0.2 at 25°C

Preparation of Medium: Add components, except diphosphothiamine, to distilled/deionized water and bring volume to 1.0L. Mix thoroughly. Gently heat until dissolved. Autoclave for 15 min at 15 psi pressure–121°C. Cool to 45°–50°C. Aseptically add 1.0mg of diphosphothiamine. Mix thoroughly. Aseptically distribute into sterile tubes or flasks.

Use: For the cultivation of *Haemophilus piscium.*

Diphtheria Virulence HiVeg Agar Base with Tellurite and Diphtheria Virulence Supplement

Composition per liter:

Plant peptone No. 3	20.0g
Agar	15.0g
NaCl	2.5g
Diptheria virulence supplement	200.0mL
Tellurite solution	50.0mL

pH 7.8± 0.2 at 25°C

Source: This medium, without tellurite or diphtheria virulence supplement, is available as a premixed powder from HiMedia.

Tellurite Solution:

Composition per 100.0mL:

K_2TeO_3	1.0g

Preparation of Tellurite Solution: Add K_2TeO_3 to distilled/deionized water and bring volume to 100.0mL. Mix thoroughly. Filter sterilize.

Caution: Potassium tellurite is toxic.

Diphtheria Virulence Supplement:

Composition per 260.0mL:

Horse serum	200.0mL
Potassium tellurite solution	60.0mL

Preparation of Diphtheria Virulence Supplement: Aseptically combine sterile horse serum and sterile tellurite solution. Mix thoroughly.

Preparation of Medium: Add components, except tellurite solution and diphtheria virulence supplement, to distilled/deionized water and bring volume to 1.0L. Mix thoroughly. Gently heat and bring to boiling. Distribute into tubes or flasks. Autoclave for 15 min at 15 psi pressure–121°C. Cool to 55–60°C. Aseptically add 2.0mL of sterile diphtheria virulence supplement and 0.5mL sterile tellurite solution to each Petri dish. Quickly add 10.0mL sterile Diphtheria Virulence HiVeg Base Agar to each Petri dish. Before the medium solidifies, place a filter paper strip saturated with potent diphtheria antitoxin across the diameter of the plate. Allow the strip to sink to the bottom of the Petri plate. Inoculate the plate with a heavy inoculum across the strip.

Use: For the detection of diphtheria toxin producing strains of *Corynebacterium diphtheriae.* For testing the toxigenicity of *Corynebacterium diphtheriae.* The reaction of antitoxin forms the actual basis for the detection of the diphtheria toxin.

Disinfectant Test Broth (*Staphylococcus aureus* Enrichment Broth)

Composition per liter:

Peptic digest of animal tissue	10.0g
Beef infusion	5.0g
NaCl	5.0g

pH 6.8 ± 0.2 at 25°C

Source: This medium is available as a premixed powder from HiMedia.

Preparation of Medium: Add components to distilled/deionized water and bring volume to 1.0L. Mix thoroughly. Distribute into tubes or flasks. Autoclave for 15 min at 15 psi pressure–121°C.

Use: For the determination of phenol coefficients of disinfectants.

Disinfectant Test Broth AOAC

Composition per liter:

Peptic digest of animal tissue	10.0g
Beef extract	5.0g
NaCl	5.0g

pH 6.8 ± 0.2 at 25°C

Source: This medium is available as a premixed powder from BD Diagnostic Systems.

Preparation of Medium: Add components to distilled/deionized water and bring volume to 1.0L. Mix thoroughly. Distribute into tubes or flasks. Autoclave for 15 min at 13 psi pressure–118°C.

Use: For the determination of phenol coefficients of disinfectants.

Disinfectant Test HiVeg Broth

Composition per liter:

Plant peptone	10.0g
Plant extract	5.0g
NaCl	5.0g

pH 6.8 ± 0.2 at 25°C

Source: This medium is available as a premixed powder from HiMedia.

Preparation of Medium: Add components to distilled/deionized water and bring volume to 1.0L. Mix thoroughly. Distribute into tubes or flasks. Autoclave for 15 min at 15 psi pressure–121°C.

Use: For the determination of phenol coefficients of disinfectants.

Disinfectant Test Medium

Composition per liter:

Peptic digest of animal tissue	5.0g
Proteose peptone	5.0g
NaCl	5.0g
Beef extract	5.0g
Yeast extract	5.0g

pH 6.8 ± 0.2 at 25°C

Source: This medium is available as a premixed powder from HiMedia.

Preparation of Medium: Add components to distilled/deionized water and bring volume to 1.0L. Mix thoroughly. Distribute into tubes or flasks. Autoclave for 15 min at 15 psi pressure–121°C.

Use: For the determination of phenol coefficients of disinfectants.

Dithionite Thioglycolate, HS T, Broth

See: **Clausen Medium**

Dixon Agar

Composition per liter:

Malt extract	30.0g
Oxbile	20.0g
Agar	15.0g
Mycological peptone	5.0g
Glycerol mono-oleate	2.50g
Tween™ 40	10.0mL

pH 5.4 ± 0.2 at 25°C

Preparation of Medium: Add components to distilled/deionized water and bring volume to 1.0L. Mix thoroughly. Gently heat while stirring until boiling. Distribute into tubes or flasks. Autoclave for 10 min at 15 psi pressure–115°C. Do not overheat or agar will not harden. If a lower pH (3.5) is desired, cool medium to 55°C and aseptically add 100.0mL of sterile lactic acid. Pour into sterile Petri dishes or distribute into sterile tubes.

Use: For the cultivation and maintenance of *Malassezia* species.

DM Medium

Composition per liter:

Starch, soluble	5.0g
$MgSO_4 \cdot 7H_2O$	0.5g
K_2HPO_4	0.25g

Preparation of Medium: Add components to distilled/deionized water and bring volume to 1.0L. Mix thoroughly. Distribute into tubes or flasks. Autoclave for 15 min at 15 psi pressure–121°C.

Use: For the cultivation of myxobacteria.

DMA Medium

Composition per liter:

$NaHCO_3$	3.0g
KH_2PO_4	0.85g
K_2HPO_4	0.8g
NH_4Cl	0.5g
$FeSO_4 \cdot 7H_2O$	1.0mg
Resazurin	0.5mg
Glucose solution	100.0mL
Wolfe's mineral solution	10.0mL
Wolfe's vitamin solution	10.0mL
$Na_2S \cdot 9H_2O$ solution	10.0mL
$MgSO_4 \cdot 7H_2O$ solution	10.0mL

Glucose Solution:

Composition per 100.0mL:

D-Glucose	2.0g

Preparation of Glucose Solution: Add glucose to distilled/deionized water and bring volume to 100.0mL. Mix thoroughly. Sparge with 100% N_2. Autoclave for 15 min at 15 psi pressure–121°C.

Wolfe's Mineral Solution:

Composition per liter:

$MgSO_4 \cdot 7H_2O$	3.0g
Nitrilotriacetic acid	1.5g
NaCl	1.0g
$MnSO_4 \cdot 2H_2O$	0.5g
$CoCl_2 \cdot 6H_2O$	0.1g
$ZnSO_4 \cdot 7H_2O$	0.1g
$CaCl_2 \cdot 2H_2O$	0.1g
$FeSO_4 \cdot 7H_2O$	0.1g
$NiCl_2 \cdot 6H_2O$	0.025g
$KAl(SO_4)_2 \cdot 12H_2O$	0.02g
$CuSO_4 \cdot 5H_2O$	0.01g
H_3BO_3	0.01g
$Na_2MoO_4 \cdot 2H_2O$	0.01g
$Na_2SeO_3 \cdot 5H_2O$	0.3mg

Preparation of Wolfe's Mineral Solution: Add nitrilotriacetic acid to 500.0mL of distilled/deionized water. Adjust pH to 6.5 with KOH. Add remaining components. Add distilled/deionized water to 1.0L. Sparge with 100% N_2. Adjust pH to 6.8. Filter sterilize.

Wolfe's Vitamin Solution:

Composition per liter:

Pyridoxine·HCl	10.0mg
p-Aminobenzoic acid	5.0mg
Lipoic acid	5.0mg
Nicotinic acid	5.0mg
Riboflavin	5.0mg
Thiamine·HCl	5.0mg
Calcium DL-pantothenate	5.0mg
Biotin	2.0mg
Folic acid	2.0mg
Vitamin B_{12}	0.1mg

Preparation of Wolfe's Vitamin Solution: Add components to distilled/deionized water and bring volume to 1.0L. Mix thoroughly. Filter sterilize.

$MgSO_4 \cdot 7H_2O$ Solution:

Composition per 10.0mL:

$MgSO_4 \cdot 7H_2O$	0.25g

Preparation of $MgSO_4 \cdot 7H_2O$ Solution: Add $MgSO_4 \cdot 7H_2O$ to distilled/deionized water and bring volume to 10.0mL. Mix thoroughly. Sparge with 100% N_2. Autoclave for 15 min at 15 psi pressure–121°C.

$Na_2S \cdot 9H_2O$ Solution:

Composition per 10.0mL:

$Na_2S \cdot 9H_2O$	0.3g

Preparation of $Na_2S \cdot 9H_2O$ Solution: Add $Na_2S \cdot 9H_2O$ to distilled/deionized water and bring volume to 10.0mL. Mix thoroughly. Sparge with 100% N_2. Autoclave for 15 min at 15 psi pressure–121°C. Before use, neutralize to pH 7.0 with sterile HCl.

Preparation of Medium: Prepare and dispense medium under 80% N_2 + 20% CO_2 gas mixture. Add components, except glucose solution, Wolfe's mineral solution, Wolfe's vitamin solution, $Na_2S \cdot 9H_2O$ solution, and $MgSO_4 \cdot 7H_2O$ solution, to distilled/deionized water and bring volume to 860.0mL. Mix thoroughly. Sparge with 80% N_2 + 20% CO_2 gas mixture for 30 min. Autoclave for 15 min at 15 psi pressure–121°C. Aseptically and anaerobically add 100.0mL of sterile glucose solution, 10.0mL of sterile Wolfe's mineral solution, 10.0mL of sterile Wolfe's vitamin solution, 10.0mL of sterile $Na_2S \cdot 9H_2O$ solution, and 10.0mL of sterile $MgSO_4 \cdot 7H_2O$ solution. Mix thoroughly. Aseptically and anaerobically distribute into sterile tubes or bottles.

Use: For the cultivation of unclassified bacterium DSMZ 8827.

DNase Agar

Composition per liter:

Tryptose	20.0g
Agar	12.0g
NaCl	5.0g
Deoxyribonucleic acid	2.0g

pH 7.3 ± 0.2 at 25°C

Source: This medium is available as a premixed powder from Oxoid Unipath.

Preparation of Medium: Add components to distilled/deionized water and bring volume to 1.0L. Mix thoroughly. Gently heat and bring to boiling. Distribute into tubes or flasks. Autoclave for 15 min at 15 psi pressure–121°C. Pour into sterile Petri dishes or leave in tubes.

Use: For the differentiation of microorganisms, especially *Staphylococcus* species and *Serratia marcescens*, based on their production of deoxyribo-nuclease.

DNase Medium

Composition per liter:

Agar	15.0g
Pancreatic digest of casein	10.0g
Peptic digest of animal tissue	10.0g
L-Arabinose	10.0g
NaCl	5.0g
Deoxyribonucleic acid	2.0g
Methyl Green	0.09g
Phenol Red	0.05g
Antibiotic solution	10.0mL

pH 7.3 ± 0.2 at 25°C

Antibiotic Solution:

Composition per 10.0mL:

Cephalothin	0.01g
Ampicillin	5.0mg
Colistimethate	5.0mg
Amphotericin B	2.5mg

Preparation of Antibiotic Solution: Add components to distilled/deionized water and bring volume to 10.0mL. Mix thoroughly. Filter sterilize.

Preparation of Medium: Add components, except antibiotic solution, to distilled/deionized water and bring volume to 990.0mL. Mix thoroughly. Gently heat and bring to boiling. Autoclave for 15 min at 15 psi pressure–121°C. Cool to 45°–50°C. Aseptically add sterile components. Mix thoroughly. Pour into sterile Petri dishes or distribute into sterile tubes.

Use: For the isolation and cultivation of *Serratia marcescens.*

DNase Test Agar

Composition per liter:

Agar	15.0g
Pancreatic digest of casein	15.0g
NaCl	5.0g
Papaic digest of soybean meal	5.0g
Deoxyribonucleic acid	2.0g

pH 7.3 ± 0.2 at 25°C

Source: This medium is available as a premixed powder from BD Diagnostic Systems.

Preparation of Medium: Add components to distilled/deionized water and bring volume to 1.0L. Mix thoroughly. Gently heat while stirring and bring to boiling. Distribute into tubes or flasks. Autoclave for 15 min at 13 psi pressure–118°C. Pour into sterile Petri dishes or leave in tubes.

Use: For the differentiation of microorganisms, especially *Staphylococcus* species and *Serratia marcescens*, based on their production of deoxyribonuclease.

DNase Test Agar with Methyl Green

Composition per liter:

Agar	15.0g
Pancreatic digest of casein	10.0g

Peptic digest of animal tissue	10.0g
NaCl	5.0g
Deoxyribonucleic acid	2.0g
Methyl Green	0.05g

pH 7.3 ± 0.2 at 25°C

Source: This medium is available as a premixed powder from BD Diagnostic Systems.

Preparation of Medium: Add components to distilled/deionized water and bring volume to 1.0L. Mix thoroughly. Gently heat while stirring and bring to boiling. Distribute into tubes or flasks. Autoclave for 15 min at 13 psi pressure–118°C. Pour into sterile Petri dishes or leave in tubes.

Use: For the differentiation of microorganisms, especially *Staphylococcus* species and *Serratia marcescens*, based on their production of deoxyribonuclease.

DNase Test Agar with Toluidine Blue

Composition per liter:

Agar	15.0g
Pancreatic digest of casein	10.0g
Peptic digest of animal tissue	10.0g
NaCl	5.0g
Deoxyribonucleic acid	2.0g
Toluidine Blue	0.1g

pH 7.3 ± 0.2 at 25°C

Preparation of Medium: Add components to distilled/deionized water and bring volume to 1.0L. Mix thoroughly. Gently heat while stirring and bring to boiling. Distribute into tubes or flasks. Autoclave for 15 min at 13 psi pressure–118°C. Pour into sterile Petri dishes or leave in tubes.

Use: For the differentiation of microorganisms, especially *Staphylococcus* species and *Serratia marcescens*, based on their production of deoxyribonuclease.

DNase Test HiVeg Agar Base

Composition per liter:

Agar	15.0g
Plant hydrolysate	15.0g
Papaic digest of soybean meal	5.0g
NaCl	5.0g
Deoxyribonucleic acid (DNA)	2.0g

pH 7.3 ± 0.2 at 25°C

Source: This medium is available as a premixed powder from HiMedia.

Preparation of Medium: Add components to distilled/deionized water and bring volume to 1.0L. Mix thoroughly. Gently heat while stirring and bring to boiling. Distribute into tubes or flasks. Autoclave for 15 min at 13 psi pressure–118°C. Pour into sterile Petri dishes or leave in tubes.

Use: For the differentiation of microorganisms, especially *Staphylococcus* species and *Serratia marcescens*, based on their production of deoxyribonuclease.

DNase Test HiVeg Agar Base without DNA

Composition per liter:

Agar	15.0g
Plant hydrolysate	15.0g
Papaic digest of soybean meal	5.0g
NaCl	5.0g

pH 6.8 ± 0.2 at 25°C

Source: This medium is available as a premixed powder from HiMedia.

Preparation of Medium: Add components to distilled/deionized water and bring volume to 1.0L. Mix thoroughly. Distribute into tubes or flasks. Autoclave for 15 min at 15 psi pressure–121°C. Pour into sterile Petri dishes or leave in tubes

Use: As a base medium for the differentiation of microorganisms, especially *Staphylococcus* species and *Serratia marcescens*, based on their production of deoxyribonuclease.

DNase Test HiVeg Agar with Toluidine Blue

Composition per liter:

Plant hydrolysate No. 1	20.0g
Agar	15.0g
NaCl	5.0g
Deoxyribonucleic acid (DNA)	2.0g
Toluidine Blue	0.1g

pH 7.3 ± 0.2 at 25°C

Source: This medium is available as a premixed powder from HiMedia.

Preparation of Medium: Add components to distilled/deionized water and bring volume to 1.0L. Mix thoroughly. Gently heat while stirring and bring to boiling. Distribute into tubes or flasks. Autoclave for 15 min at 13 psi pressure–118°C. Pour into sterile Petri dishes or leave in tubes.

Use: For the differentiation of microorganisms, especially *Staphylococcus* species and *Serratia marcescens*, based on their production of deoxyribonuclease.

DNB Medium

Composition per liter:

Nutrient broth	2.4g
Yeast extract	1.5g

Preparation of Medium: Add components to distilled/deionized water and bring volume to 1.0L. Mix thoroughly. Distribute into tubes or flasks. Autoclave for 15 min at 15 psi pressure–121°C.

Use: For the cultivation of *Bdellovibrio bacteriovorus* and ATCC strain 43826.

Doepel Medium

Composition per liter:

Pancreatic digest of casein	8.0g
Yeast extract	8.0g
Glucose	5.0g
K_2HPO_4	2.0g
$MgSO_4$	0.3g

pH 7.2 ± 0.2 at 25°C

Preparation of Medium: Add components to distilled/deionized water and bring volume to 1.0L. Mix thoroughly. Adjust pH to 7.2. Distribute into tubes or flasks. Autoclave for 15 min at 15 psi pressure–121°C.

Use: For the cultivation of *Clavibacter toxicus*.

Dorset Egg Medium

Composition per liter:

Homogenized whole egg	950.0mL
Glycerol	50.0mL

pH 6.8–7.4 at 25°C

Source: This medium is available as a prepared medium from BD Diagnostic Systems.

Homogenized Whole Egg:
Composition per liter:

Whole eggs	18–24

Preparation of Homogenized Whole Egg: Use fresh eggs, less than 1 week old. Scrub the shells with soap. Let stand in a soap solution for 30 min. Rinse in running water. Soak eggs in 70% ethanol for 15 min. Break the eggs into a sterile container. Homogenize by shaking. Filter through four layers of sterile cheesecloth into a sterile graduated cylinder. Measure out 1.0L.

Preparation of Medium: Filter sterilize glycerol. Combine glycerol and homogenized whole egg. Mix thoroughly. Distribute into sterile screw-capped tubes. Place tubes in a slanted position. Inspissate at 85°C (moist heat) for 45 min.

Use: For the maintenance of *Mycobacterium* species.

Double-Strength Crude *Lactobacillus* Medium

Composition per 1025.0mL:

Yeast extract (Basamine)	20.0g
Sucrose	20.0g
Casein hydrolysate	15.0g
Potassium acetate	3.0g
Histidine·HCl·H_2O	2.0g
Ascorbic acid	1.0g
Pyridoxamine·HCl	33.0μg
Salts solution A	20.0mL
Salts solution B	5.0mL

pH 5.4 ± 0.2 at 25°C

Salts Solution A:
Composition per liter:

$K_2HPO_4 \cdot 3H_2O$	16.5g
$KH_2PO_4 \cdot H_2O$	16.5g

Preparation of Salts Solution A: Add components to distilled/deionized water and bring volume to 1.0L. Mix thoroughly. Autoclave for 15 min at 15 psi pressure–121°C.

Salts Solution B:
Composition per liter:

$MgSO_4 \cdot 7H_2O$	8.0g
$FeSO_4 \cdot 7H_2O$	0.4g
NaCl	0.4g
HCl, concentrated	0.1mL
$MnSO_4 \cdot H_2O$	0.1mL

Preparation of Salts Solution B: Add components to distilled/deionized water and bring volume to 1.0L. Mix thoroughly. Autoclave for 15 min at 15 psi pressure–121°C.

Preparation of Medium: Add components, except salts solution A and salts solution B, to distilled/deionized water and bring volume to 975.0mL. Mix thoroughly. Autoclave for 15 min at 15 psi pressure–121°C. Aseptically add 20.0mL of sterile salts solution A and 5.0mL of sterile salts solution B. Mix thoroughly. Adjust pH to 5.4. Mix thoroughly. Aseptically distribute into sterile tubes or flasks.

Use: For the cultivation of *Lactobacillus* species.

Double-Strength Crude Medium for *Lactobacillus*

Composition per liter:

Sucrose	20.0g
Yeast extract	20.0g
Casein hydrolysate	15.0g
Potassium acetate	3.0g
Histidine·HCl	2.0g
Ascorbic acid	1.0g
Pyridoxamine·HCl	33.0μg
Salts solution A	20.0mL
Salts solution B	5.0mL

pH 5.4 ± 0.2 at 25°C

Salts Solution A:
Composition per 100.0mL:

$KH_2PO_4 \cdot H_2O$	16.5g
$K_2HPO_4 \cdot 3H_2O$	16.5g

Preparation of Salts Solution A: Add components to distilled/deionized water and bring volume to 100.0mL. Mix thoroughly.

Salts Solution B:
Composition per 100.0mL:

$MgSO_4 \cdot 7H_2O$	8.0g
$FeSO_4 \cdot 7H_2O$	0.4g
$MnSO_4 \cdot H2O$	0.4g
NaCl	0.4g
HCl, concentrated	0.1mL

Preparation of Salts Solution B: Add components to distilled/deionized water and bring volume to 100.0mL. Mix thoroughly.

Preparation of Medium: Add components to distilled/deionized water and bring volume to 1.0L. Mix thoroughly. Gently heat and bring to boiling. Adjust pH to 5.4 with acetic acid. Distribute into tubes or flasks. Autoclave for 15 min at 15 psi pressure–121°C.

Use: For the cultivation and maintenance of *Lactobacillus* species.

Double Sugar Agar, Russell (Russell Double Sugar Agar)

Composition per liter:

Agar	15.0g
Lactose	10.0g
Casein enzymic hydrolysate	7.5g
NaCl	5.0g
Beef extract	3.0g
Peptic digest of animal tissue	2.5g
Glucose	1.0g
Phenol Red	0.025g

pH 7.3 ± 0.2 at 25°C

Source: This medium is available from HiMedia.

Preparation of Medium: Add components to distilled/deionized water and bring volume to 1.0L. Mix thoroughly. Gently heat and bring to boiling. Distribute into tubes or flasks. Autoclave for 15 min at 15 psi pressure–121°C. Allow the tubes to solidify in slanting position to form a generous butt.

Use: For the differentiation of Gram-negative enteric bacilli on the basis of their ability to ferment glucose and lactose with or without gas formation.

Double Sugar HiVeg Agar

Composition per liter:

Agar	15.0g
Lactose	10.0g
Plant hydrolysate	7.5g
NaCl	5.0g
Plant extract	3.0g
Plant peptone	2.5g
Glucose	1.0g
Phenol Red	0.025g

pH 7.3 ± 0.2 at 25°C

Source: This medium is available as a premixed powder from HiMedia.

Preparation of Medium: Add components to distilled/deionized water and bring volume to 1.0L. Mix thoroughly. Gently heat and bring to boiling. Distribute into tubes. Autoclave for 15 min at 15 psi pressure–121°C. Allow tubes to cool in a slanted position.

Use: For the identification of Gram-negative enteric bacilli based on their fermentation of glucose and lactose. Bacteria that ferment both glucose and lactose produce a yellow slant and yellow butt. Bacteria that ferment glucose but do not ferment lactose produce a red slant and a yellow butt. Bacteria that ferment neither glucose nor lactose produce an unchanged pink-orange color.

Doyle and Roman Enrichment Medium

Composition per liter:

Pancreatic digest of casein	10.0g
Peptic digest of animal tissue	10.0g
NaCl	5.0g
Sodium succinate	3.0g
Yeast extract	2.0g
Glucose	1.0g
$NaHSO_3$	0.1g
L-Cysteine·HCl·H_2O	0.1g
Horse blood, lysed	70.0mL
Antibiotic solution	10.0mL

pH 7.0 ± 0.2 at 25°C

Antibiotic Solution:

Composition per 10.0mL:

Cycloheximide	0.05g
Vancomycin	0.015g
Trimethoprim lactate	5.0mg
Polymyxin B	200,000U

Preparation of Antibiotic Solution: Add components to distilled/deionized water and bring volume to 10.0mL. Mix thoroughly. Filter sterilize.

Caution: Cycloheximide is toxic. Avoid skin contact or aerosol formation and inhalation.

Preparation of Medium: Add components, except antibiotic solution and horse blood, to distilled/deionized water and bring volume to 920.0mL. Mix thoroughly. Gently heat and bring to boiling. Autoclave for 15 min at 15 psi pressure–121°C. Cool to 45°–50°C. Aseptically add sterile antibiotic solution and horse blood. Mix thoroughly. Aseptically distribute into sterile flasks in 90.0–100.0mL volumes.

Use: For the cultivation and enrichment of *Campylobacter* species from foods.

Doyle's Enrichment Broth Base with Antibiotic Solution

Composition per liter:

Casein enzymatic hydrolysate	10.0g
Peptic digest of animal tissue	10.0g
NaCl	5.0g
Sodium succinate	3.0g
Yeast extract	2.0g
Glucose	1.0g
L-Cysteine·HCl	0.1g
$NaHSO_3$	0.1g
Antibiotic solution	10.0mL

pH 7.0 ± 0.2 at 25°C

Source: This medium, without antibiotic solution, is available as a premixed powder from HiMedia.

Antibiotic Solution:

Composition per 10.0mL:

Cycloheximide	0.05g
Vancomycin	0.015g
Trimethoprim lactate	5.0mg
Polymyxin B	200,000U

Preparation of Antibiotic Solution: Add components to distilled/deionized water and bring volume to 10.0mL. Mix thoroughly. Filter sterilize.

Caution: Cycloheximide is toxic. Avoid skin contact or aerosol formation and inhalation.

Preparation of Medium: Add components, except antibiotic solution, to distilled/deionized water and bring volume to 990.0mL. Mix thoroughly. Gently heat and bring to boiling. Autoclave for 15 min at 15 psi pressure–121°C. Cool to 45°–50°C. Aseptically add a selective sterile antibiotic solution. Generally 50.0mL of defibrinated horse blood is also added as an enrichment. Mix thoroughly.

Use: For the cultivation and enrichment of *Campylobacter* species from foods.

Doyle's Enrichment HiVeg Broth Base with Antibiotic Solution

Composition per liter:

Plant hydrolysate	10.0g
Plant peptone	10.0g
NaCl	5.0g
Sodium succinate	3.0g
Yeast extract	2.0g
Glucose	1.0g
L-Cysteine·HCl	0.1g
$NaHSO_3$	0.1g
Antibiotic solution	10.0mL

pH 7.0 ± 0.2 at 25°C

Source: This medium, without antibiotic solution, is available as a premixed powder from HiMedia.

Antibiotic Solution:

Composition per 10.0mL:

Cycloheximide	0.05g
Vancomycin	0.015g
Trimethoprim lactate	5.0mg
Polymyxin B	200,000U

Preparation of Antibiotic Solution: Add components to distilled/deionized water and bring volume to 10.0mL. Mix thoroughly. Filter sterilize.

Caution: Cycloheximide is toxic. Avoid skin contact or aerosol formation and inhalation.

Preparation of Medium: Add components, except antibiotic solution, to distilled/deionized water and bring volume to 990.0mL. Mix thoroughly. Gently heat and bring to boiling. Autoclave for 15 min at 15 psi pressure–121°C. Cool to 45°–50°C. Aseptically add a selective sterile antibiotic solution. Generally 50.0mL of defibrinated horse blood is also added as an enrichment. Mix thoroughly.

Use: For the cultivation and enrichment of *Campylobacter* species from foods.

DP Agar

Composition per liter:

Agar	20.0g
Pancreatic digest of casein	10.0g
Yeast extract	10.0g
K_2HPO_4	7.0g
Glucose	5.0g
KH_2PO_4	3.0g
Trisodium citrate·$3H_2O$	0.5g
$MgSO_4·7H_2O$	0.1g

Preparation of Medium: Add components to distilled/deionized water and bring volume to 1.0L. Mix thoroughly. Gently heat and bring to boiling. Distribute into tubes or flasks. Autoclave for 15 min at 15 psi pressure–121°C. Pour into sterile Petri dishes or leave in tubes.

Use: For the cultivation of *Escherichia coli.*

DPA with Calcium Carbonate

Composition per liter:

Agar	20.0g
Glucose	20.0g
Peptone	10.0g
$CaCO_3$	5.0g

Preparation of Medium: Add components to distilled/deionized water and bring volume to 1.0L. Mix thoroughly. Gently heat and bring to boiling. Distribute into tubes or flasks. Autoclave for 30 min at 15 psi pressure–121°C. Pour into sterile Petri dishes or leave in tubes.

Use: For the cultivation and maintenance of *Dekkera* species.

DPM Medium (DSMZ Medium 737)

Composition per liter:

Na-propionate	1.20g
KH_2PO_4	1.0g
$MgSO_4·7H_2O$	0.1g
$CaCl_2·2H_2O$	0.01g
Chelated iron solution	1.8mL
Trace elements solution	1.0mL

pH 6.8 ± 0.2 at 25°C

Chelated Iron Solution:

Composition per liter:

Na_2-EDTA	7.56g
$FeSO_4·5H_2O$	5.54g

Preparation of Chelated Iron Solution: Add components to distilled/deionized water and bring volume to 1.0L. Mix thoroughly.

Trace Elements Solution:

Composition per liter:

H_3BO_3	2.860
$MnCl_2·4H_2O$	1.81g
$ZnSO_4·7H_2O$	0.22g
$CuSO_4·5H_2O$	0.08g
$Na_2MoO_4·4H_2O$	0.025g
$CoCl_2$	0.025g

Preparation of Trace Elements Solution: Add components to distilled/deionized water and bring volume to 1.0L. Mix thoroughly.

Preparation of Medium: Add components to distilled/deionized water and bring volume to 1.0L. Mix thoroughly. Distribute into tubes or flasks. Autoclave for 15 min at 15 psi pressure–121°C.

Use: For the cultivation of *Frankia* sp.

DRBC Agar (Dichloran Rose Bengal Chloramphenicol Agar)

Composition per liter:

Agar	15.0g
Glucose	10.0g
Peptone	5.0g
KH_2PO_4	1.0g
$MgSO_4·7H_2O$	0.5g
Rose Bengal	0.025g
Dichloran	0.002g
Chloramphenicol solution	10.0mL

pH 5.6 ± 0.2 at 25°C

Source: This medium is available as a premixed powder from Oxoid Unipath.

Chloramphenicol Solution:

Composition per 10.0mL:

Chloramphenicol	0.1g

Preparation of Chloramphenicol Solution: Add chloramphenicol to distilled/deionized water and bring volume to 10.0mL. Mix thoroughly. Filter sterilize.

Preparation of Medium: Add components, except chloramphenicol solution, to distilled/deionized water and bring volume to 990.0mL. Mix thoroughly. Gently heat and bring to boiling. Autoclave for 15 min at 15 psi pressure–121°C. Cool to 45°–50°C. Aseptically add sterile chloramphenicol solution. Mix thoroughly. Pour into sterile Petri dishes or distribute into sterile tubes.

Use: For the isolation, cultivation, and enumeration of yeasts and molds associated with food spoilage.

Drigalski Litmus Lactose Agar

Composition per liter:

Agar	10.0g
Lactose	10.0g
Meat peptone	5.0g
Beef extract	3.0g
Litmus	1.0g

pH 7.4 ± 0.2 at 25°C

Source: This medium is available as a premixed powder from Sigma Aldrich.

Preparation of Medium: Add components to distilled/deionized water and bring volume to 1.0L. Mix thoroughly. Gently heat and bring to boiling. Distribute into tubes or flasks. Autoclave for 15 min at 15 psi pressure–121°C. Pour into sterile Petri dishes or leave in tubes.

Use: For the maintenance of lactic acid bacteria and differentiation of bacteria on the basis of lactose fermentation. Bacteria that ferment lactose appear as red colonies and others as dark blue-purple colonies.

Drigalski Litmus Lactose Agar

Composition per liter:

Lactose	15.0g
Agar	13.0g
Peptic digest of animal tissue	7.0g
NaCl	5.0g
Litmus	1.2g

pH 7.4 ± 0.2 at 25°C

Source: This medium is available as a premixed powder from HiMedia.

Preparation of Medium: Add components to distilled/deionized water and bring volume to 1.0L. Mix thoroughly. Gently heat and bring to boiling. Distribute into tubes or flasks. Autoclave for 15 min at 15 psi pressure–121°C. Pour into sterile Petri dishes or leave in tubes.

Use: For the maintenance of lactic acid bacteria and differentiation of bacteria on the basis of lactose fermentation. Bacteria that ferment lactose appear as red colonies and others as dark blue-purple colonies.

Drigalski Litmus Lactose Agar, Modified

Composition per liter:

Agar	16.0g
Lactose	10.0g
Peptic digest of animal tissue	10.0g
Beef extract	4.0g
Bromthymol Blue	0.04g

pH 7.4 ± 0.2 at 25°C

Source: This medium is available as a premixed powder from HiMedia.

Preparation of Medium: Add components to distilled/deionized water and bring volume to 1.0L. Mix thoroughly. Gently heat and bring to boiling. Distribute into tubes or flasks. Autoclave for 15 min at 15 psi pressure–121°C. Pour into sterile Petri dishes or leave in tubes.

Use: For the maintenance of lactic acid bacteria and differentiation of bacteria on the basis of lactose fermentation. Bacteria that ferment lactose appear as red colonies and others as dark blue-purple colonies.

Drigalski Litmus Lactose HiVeg Agar

Composition per liter:

Lactose	15.0g
Agar	13.0g
Plant peptone	7.0g
NaCl	5.0g
Litmus	1.2g

pH 7.4 ± 0.2 at 25°C

Source: This medium is available as a premixed powder from HiMedia.

Preparation of Medium: Add components to distilled/deionized water and bring volume to 1.0L. Mix thoroughly. Gently heat and bring to boiling. Distribute into tubes or flasks. Autoclave for 15 min at 15 psi pressure–121°C. Pour into sterile Petri dishes or leave in tubes.

Use: For the maintenance of lactic acid bacteria and differentiation of bacteria on the basis of lactose fermentation. Bacteria that ferment lactose appear as red colonies and others as dark blue-purple colonies.

DS Sporulation Medium, Modified

See: **Duncan-Strong Sporulation Medium, Modified**

DSA

See: **Dextrose Soil Agar**

DSA Cellulose

See: **Dextrose Sucrose Cellulose Agar**

DSIC Medium, Modified (DSMZ Medium 747)

Composition per 994.0mL:

Solution A	910.0mL
Solution B	70.0mL
Solution C	14.0mL

pH 7.0–7.1 at 25°C

Solution A:

Composition per 960.0mL:

NaCl	125.0g
K_2SO_4	2.5g
Na-acetate	2.0g
Yeast extract	0.75g
KH_2PO_4	0.6g
NH_4Cl	0.5g
$Na_2S2O_3{\cdot}5H_2O$	0.1g
MOPS buffer	10.0mL
Vitamin B_{12} solution	1.0mL
Trace elements solution SL-10	1.0mL

Vitamin B_{12} Solution:

Composition per 100.0mL:

Vitamin B_{12}	10.0mg

Vitamin B_{12} Solution: Add vitamin B_{12} to distilled/deionized water and bring volume to 100.0mL. Mix thoroughly. Sparge under 100% N_2 gas for 3 min.

Trace Elements Solution SL-10:

Composition per liter:

$FeCl_2{\cdot}4H_2O$	1.5g
$CoCl_2{\cdot}6H_2O$	190.0mg
$MnCl_2{\cdot}4H_2O$	100.0mg
$ZnCl_2$	70.0mg
$Na_2MoO_4{\cdot}2H_2O$	36.0mg
$NiCl_2{\cdot}6H_2O$	24.0mg
H_3BO_3	6.0mg
$CuCl_2{\cdot}2H_2O$	2.0mg
HCl (25% solution)	10.0mL

Preparation of Trace Elements Solution SL-10: Add $FeCl_2{\cdot}4H_2O$ to 10.0mL of HCl solution. Mix thoroughly. Add distilled/deionized water and bring volume to 1.0L. Add remaining components. Mix thoroughly. Sparge with 80% N_2 + 20% CO_2. Autoclave for 15 min at 15 psi pressure–121°C.

MOPS Buffer:
Composition per 10.0mL:

MOPS [3-(*N*-morpholino) propane sulfonic acid]	2.1g
Na-acetate	0.3g
EDTA	0.1g

Preparation of MOPS Buffer: Add components to distilled/deionized water and bring volume to 10.0mL. Mix thoroughly. Sparge with 100% N_2. Adjust to pH 7.2. Filter sterilize.

Preparation of Solution A: Add components to distilled/deionized water and bring volume to 960.0mL. Mix thoroughly. Sparge with 100% N_2. Gently heat and bring to boiling while continuing to sparge with 100% N_2. Distribute about 13mL aliquots in 15mL Hungate tubes. Autoclave for 15 min at 15 psi pressure–121°C. Cool to 25°C.

Solution B:
Composition per 70.0mL:

$MgCl_2 \cdot 6H_2O$	10.0g
$CaCl_2 \cdot 2H_2O$	0.2g

Preparation of Solution B: Add components to distilled/deionized water and bring volume to 70.0mL. Mix thoroughly. Adjust pH to 7.0. Sparge with 100% N_2. Gently heat and bring to boiling while continuing to sparge with 100% N_2. Distribute into a screw-capped bottle. Autoclave for 15 min at 15 psi pressure–121°C. Cool to 25°C.

Solution C:
Composition per 14.0mL:

$NaHCO_3$	1.0g

Preparation of Solution C: Add $NaHCO_3$ to distilled/deionized water and bring volume to 14.0mL. Mix thoroughly. Sparge with 80% N_2 + 20% CO_2. Filter sterilize.

Preparation of Medium: Inject 1.0 ml of solution B and 0.2 ml of solution C in each tube of solution A.

Use: For the cultivation of *Rhodovibrio sodomensis=Rhodospirillum sodomense.*

DSM 134

See: ***Haliscomenobacter*** **Medium**

DST Agar

See: **Diagnostic Sensitivity Test Agar**

DTC Agar

Composition per liter:

Agar	20.0g
Pancreatic digest of casein	15.0g
NaCl	5.0g
Papaic digest of soybean meal	5.0g
Deoxyribonucleic acid	2.0g
Toluidine Blue O	0.1g
Cephalothin solution	10.0mL

Cephalothin Solution:
Composition per 10.0mL:

Cephalothin	1.0g

Preparation of Cephalothin Solution: Add cephalothin to distilled/deionized water and bring volume to 10.0mL. Mix thoroughly. Filter sterilize.

Preparation of Medium: Add components, except cephalothin solution, to distilled/deionized water and bring volume to 990.0mL. Mix thoroughly. Gently heat and bring to boiling. Autoclave for 15 min at 15 psi pressure–121°C. Cool to 45°–50°C. Aseptically add sterile cephalothin solution. Mix thoroughly. Pour into sterile Petri dishes.

Use: For the isolation and cultivation of *Serratia* species. *Serratia* appear as colonies with red halos.

DTM Agar (Dermatophyte Test Medium Agar)

Composition per liter:

Agar	20.0g
Enzymatic digest of soybean meal	10.0g
Glucose	10.0g
Cycloheximide	0.5g
Phenol Red	0.2g
Chlortetracycline	0.1g
Gentamicin	0.1g

pH 7.3 ± 0.2 at 25°C

Source: Available as a prepared medium from BD Diagnostic Systems.

Caution: Cycloheximide is toxic. Avoid skin contact or aerosol formation and inhalation.

Preparation of Medium: Add components to distilled/deionized water and bring volume to 1.0L. Mix thoroughly. Gently heat and bring to boiling. Distribute into tubes or flasks. Autoclave for 15 min at 15 psi pressure–121°C. Pour into sterile Petri dishes or leave in tubes.

Use: For the isolation and cultivation of dermatophytic fungi.

Dubos Agar with Filter Paper

Composition per liter:

Agar	15.0g
K_2HPO_4	1.0g
KCl	0.5g
$MgSO_4 \cdot 7H_2O$	0.5g
$NaNO_3$	0.5g
$FeSO_4 \cdot 7H_2O$	0.01g

pH 7.2 ± 0.2 at 25°C

Preparation of Medium: Add components to distilled/deionized water and bring volume to 1.0L. Mix thoroughly. Gently heat and bring to boiling. Adjust pH to 7.2. Autoclave for 15 min at 15 psi pressure–121°C. Pour into sterile Petri dishes. Lay sterile strips of Whatman #1 filter paper on the surface of the agar.

Use: For the cultivation and maintenance of *Cytophaga hutchinsonii.*

Dubos Broth

Composition per liter:

Na_2HPO_4	2.5g
L-Asparagine	2.0g
KH_2PO_4	1.0g
Pancreatic digest of casein	0.5g
Tween™ 80	0.2g
$CaCl_2 \cdot 2H_2O$	0.5mg
$CuSO_4$	0.1mg
$ZnSO_4 \cdot 7H_2O$	0.1mg
Ferric ammonium citrate	0.05g
$MgSO_4 \cdot 7H_2O$	0.01g
Bovine serum albumin or bovine serum	20.0mL

pH 6.5 ± 0.2 at 25°C

Source: This medium is available as a premixed powder from BD Diagnostic Systems.

Preparation of Medium: Add components, except bovine serum or bovine serum albumin, to distilled/deionized water and bring volume to 980.0mL. Mix thoroughly. Gently heat and bring to boiling. Autoclave for 15 min at 15 psi pressure–121°C. Cool to 45°–50°C. Aseptically add sterile bovine serum or bovine serum albumin. Mix thoroughly. Aseptically distribute into sterile tubes.

Use: For the cultivation of *Mycobacterium tuberculosis* and other *Mycobacterium* species.

Dubos Broth Base with Serum and Glycerol

Composition per liter:

Na_2HPO_4	2.5g
L-Asparagine	2.0g
KH_2PO_4	1.0g
Casein enzymatic hydrolysate	0.5g
Polysorbate 80	0.2g
Ferric ammonium citrate	0.05g
$MgSO_4$	0.01g
$CaCl_2$	0.5mg
$CuSO_4$	0.1mg
$ZnSO_4$	0.1mg
Glycerol	50.0mL
Bovine serum or bovine albumin V	20.0mL

pH 6.5 ± 0.2 at 25°C

Source: This medium, without bovine serum or glycerol, is available as a premixed powder from HiMedia.

Preparation of Medium: Add components, except bovine serum, to distilled/deionized water and bring volume to 980.0mL. Mix thoroughly. Gently heat and bring to boiling. Autoclave for 15 min at 15 psi pressure–121°C. Cool to 45°–50°C. Aseptically add 20.0mL sterile bovine serum or bovine serum albumin. Mix thoroughly. Aseptically distribute into sterile tubes.

Use: For the cultivation of *Mycobacterium tuberculosis* and other *Mycobacterium* species.

Dubos Broth with Horse Serum

Composition per liter:

Na_2HPO_4	2.5g
L-Asparagine	2.0g
KH_2PO_4	1.0g
Pancreatic digest of casein	0.5g
Tween™ 80	0.2g
$CaCl_2 \cdot 2H_2O$	0.5mg
$CuSO_4$	0.1mg
$ZnSO_4 \cdot 7H_2O$	0.1mg
Ferric ammonium citrate	0.05g
$MgSO_4 \cdot 7H_2O$	0.01g
Horse serum	50.0mL

pH 6.5 ± 0.2 at 25°C

Preparation of Medium: Add components, except horse serum, to distilled/deionized water and bring volume to 950.0mL. Mix thoroughly. Gently heat and bring to boiling. Autoclave for 15 min at 15 psi pressure–121°C. Cool to 45°–50°C. Aseptically add sterile horse serum. Mix thoroughly. Aseptically distribute into sterile tubes.

Use: For the cultivation and maintenance of *Corynebacterium* species.

Dubos HiVeg Broth Base with Serum and Glycerol

Composition per liter:

Na_2HPO_4	2.5g
L-Asparagine	2.0g
KH_2PO_4	1.0g
Plant hydrolysate	0.5g
Polysorbate 80	0.2g
Ferric ammonium citrate	0.05g
$MgSO_4$	0.01g
$CaCl_2$	0.5mg
$CuSO_4$	0.1mg
$ZnSO_4$	0.1mg
Glycerol	50.0mL
Bovine serum or bovine albumin V	20.0mL

pH 6.5 ± 0.2 at 25°C

Source: This medium, without bovine serum or glycerol, is available as a premixed powder from HiMedia.

Preparation of Medium: Add components, except bovine serum, to distilled/deionized water and bring volume to 980.0mL. Mix thoroughly. Gently heat and bring to boiling. Autoclave for 15 min at 15 psi pressure–121°C. Cool to 45°–50°C. Aseptically add 20.0mL sterile bovine serum or bovine serum albumin. Mix thoroughly. Aseptically distribute into sterile tubes.

Use: For the cultivation of *Mycobacterium tuberculosis* and other *Mycobacterium* species.

Dubos Mineral Medium

Composition per liter:

K_2HPO_4	1.0g
KCl	0.5g
$MgSO_4 \cdot 7H_2O$	0.5g
$NaNO_3$	0.5g
$FeSO_4 \cdot 7H_2O$	0.01g

pH 7.2 ± 0.2 at 25°C

Preparation of Medium: Add components to distilled/deionized water and bring volume to 1.0L. Mix thoroughly. Distribute into tubes or flasks. Autoclave for 15 min at 15 psi pressure–121°C.

Use: For the isolation and cultivation of *Cytophaga* species, *Herpetosiphon* species, *Saprospira* species, and *Flexithrix* species.

Dubos Oleic Agar

Composition per liter:

Agar	15.0g
Na_2HPO_4	2.5g
KH_2PO_4	1.0g
L-Asparagine	1.0g
Pancreatic digest of casein	0.5g
Ferric ammonium citrate	0.05g
$MgSO_4 \cdot 7H_2O$	0.01g
$CaCl_2 \cdot 2H_2O$	0.5mg
$CuSO_4$	0.1mg
$ZnSO_4 \cdot 7H_2O$	0.1mg
Dubos oleic albumin complex	20.0mL
Penicillin solution	10.0mL

pH 6.6 ± 0.2 at 25°C

Source: This medium is available as a premixed powder from BD Diagnostic Systems.

Dubos Oleic Albumin Complex:

Composition per 100.0mL:

Bovine serum albumin, fraction V	5.0g
Oleic acid, sodium salt	0.05g
NaCl (0.85% solution)	100.0mL

Preparation of Dubos Oleic Albumin Complex: Add bovine serum albumin and oleic acid to 100.0mL of NaCl solution. Mix thoroughly. Filter sterilize.

Penicillin Solution:

Composition per 10.0mL:

Penicillin	10,000U

Preparation of Penicillin Solution: Add penicillin to distilled/deionized water and bring volume to 10.0mL. Mix thoroughly. Filter sterilize.

Preparation of Medium: Add components, except Dubos oleic albumin complex and penicillin solution, to distilled/deionized water and bring volume to 970.0mL. Mix thoroughly. Gently heat and bring to boiling. Autoclave for 15 min at 15 psi pressure–121°C. Cool to 45°–50°C. Aseptically add sterile Dubos oleic albumin complex and penicillin solution. Mix thoroughly. Pour into sterile Petri dishes or distribute into sterile tubes. Allow tubes to cool in a slanted position.

Use: For the isolation of *Mycobacterium tuberculosis* and determining its sensitivity to chemotherapeutic agents.

Dubos Oleic HiVeg Agar Base with Serum and Glycerol

Composition per liter:

Agar	15.0g
Na_2HPO_4	2.5g
L-Asparagine	1.0g
KH_2PO_4	1.0g
Plant hydrolysate	0.5g
Ferric ammonium citrate	0.05g
$MgSO_4$	0.01g
$CaCl_2$	0.5mg
$CuSO_4$	0.1mg
$ZnSO_4$	0.1mg
Glycerol	50.0mL
Bovine serum or bovine albumin V	20.0mL

pH 6.5 ± 0.2 at 25°C

Source: This medium, without bovine serum or glycerol, is available as a premixed powder from HiMedia.

Preparation of Medium: Add components, except bovine serum, to distilled/deionized water and bring volume to 980.0mL. Mix thoroughly. Gently heat and bring to boiling. Autoclave for 15 min at 15 psi pressure–121°C. Cool to 45°–50°C. Aseptically add 20.0mL sterile bovine serum or bovine serum albumin. Mix thoroughly. Aseptically distribute into sterile tubes.

Use: For the cultivation of *Mycobacterium tuberculosis* and other *Mycobacterium* species.

Dubos Oleic HiVeg Broth Base with Antibiotic and Oleic Albumin Supplement

Composition per liter:

Na_2HPO_4	2.5g
Asparagine	1.0g
KH_2PO_4	1.0g
Plant hydrolysate	0.5g
Ferric ammonium citrate	0.05g
$MgSO_4$	0.01g
$CaCl_2$	0.5mg
$CuSO_4$	0.1mg
$ZnSO_4$	0.1mg
Penicllin solution	10.0mL

pH 6.5 ± 0.2 at 25°C

Source: This medium, without penicillin and oleic albumin supplement, is available as a premixed powder from HiMedia.

Penicillin Solution:

Composition per 10.0mL:

Penicillin	10,000U

Preparation of Penicllin Solution: Add penicillin to distilled/deionized water and bring volume to 10.0mL. Mix thoroughly. Filter sterilize.

Oleic Albumin Supplement:

Composition per 100.0mL:

Oleic acid	10.0g
Albumin fraction IV	4.0g
NaCl	0.68g
NaOH	8.0mg

Preparation of Oleic Albumin Supplement: Add components to distilled/deionized water and bring volume to 100.0mL. Mix thoroughly. Filter sterilize.

Preparation of Medium: Add components, except penicllin solution and oleic albumin supplement, to distilled/deionized water and bring volume to 880.0mL. Mix thoroughly. Gently heat and bring to boiling. Autoclave for 15 min at 15 psi pressure–121°C. Cool to 45°–50°C. Aseptically add 10.0mL sterile penicillin solution and 100.0mL sterile oleic albumin solution. Mix thoroughly. Aseptically distribute into sterile tubes.

Use: For the cultivation of *Mycobacterium tuberculosis* and other *Mycobacterium* species.

Dubos Salts Agar

Composition per liter:

Agar	15.0g
K_2HPO_4	1.0g
KCl	0.5g
$MgSO_4 \cdot 7H_2O$	0.5g
$NaNO_3$	0.5g
$FeSO_4 \cdot 7H_2O$	0.01g
Filter paper strips, sterile	variable

pH 7.2 ± 0.2 at 25°C

Preparation of Medium: Add components, except filter paper strips, to distilled/deionized water and bring volume to 1.0L. Mix thoroughly. Gently heat and bring to boiling. Autoclave for 15 min at 15 psi pressure–121°C. Pour into sterile Petri dishes. Aseptically place sterile filter paper strips onto the surface of the solidified medium.

Use: For the cultivation and maintenance of *Alteromonas* species, *Cellvibrio mixtus, Cellvibrio* species, *Cytophaga aurantiaca, Cytophaga hutchinsonii, Pseudomonas* species, *and Sporocytophaga myxococcoides.*

Dubos Salts Agar with 1% Sodium Chloride

Composition per liter:

Agar	15.0g
NaCl	10.0g

K_2HPO_4	1.0g
KCl	0.5g
$MgSO_4 \cdot 7H_2O$	0.5g
$NaNO_3$	0.5g
$FeSO_4 \cdot 7H_2O$	10.0mg
Filter paper strips	1 strip per tube

pH 7.2 ± 0.2 at 25°C

Preparation of Medium: Add components to distilled/deionized water and bring volume to 1.0L. Mix thoroughly. Gently heat and bring to boiling. Distribute into tubes. Autoclave for 15 min at 15 psi pressure–121°C. Allow tubes to cool in a slanted position. Aseptically add a strip of sterile filter paper to the surface of each slant.

Use: For the cultivation of *Cytophaga* species.

Dubos Salts Agar with Yeast Extract

Composition per liter:

Agar	15.0g
K_2HPO_4	1.0g
Yeast extract	0.5g
KCl	0.5g
$MgSO_4 \cdot 7H_2O$	0.5g
$NaNO_3$	0.5g
$FeSO_4 \cdot 7H_2O$	0.01g
Filter paper strips, sterile	variable

pH 7.2 ± 0.2 at 25°C

Preparation of Medium: Add components, except filter paper strips, to distilled/deionized water and bring volume to 1.0L. Mix thoroughly. Gently heat and bring to boiling. Autoclave for 15 min at 15 psi pressure–121°C. Pour into sterile Petri dishes. Aseptically place sterile filter paper strips onto the surface of the solidified medium.

Use: For the cultivation and maintenance of *Cellulomonas* species and *Cellvibrio* species.

Dubos Salts Broth

Composition per liter:

K_2HPO_4	1.0g
KCl	0.5g
$MgSO_4 \cdot 7H_2O$	0.5g
$NaNO_3$	0.5g
$FeSO_4 \cdot 7H_2O$	0.01g
Filter paper strips, sterile	variable

pH 7.2 ± 0.2 at 25°C

Preparation of Medium: Add components, except filter paper strips, to distilled/deionized water and bring volume to 1.0L. Mix thoroughly. Distribute into tubes containing a filter paper strip (filter paper strip should protrude above the surface of the broth). Autoclave for 15 min at 15 psi pressure–121°C.

Use: For the cultivation of *Cytophaga aurantiaca* and *Pseudomonas* species.

Dubos Salts Broth with Yeast Extract

Composition per liter:

K_2HPO_4	1.0g
Yeast extract	0.5g
KCl	0.5g
$MgSO_4 \cdot 7H_2O$	0.5g
$NaNO_3$	0.5g
$FeSO_4 \cdot 7H_2O$	0.01g

pH 7.2 ± 0.2 at 25°C

Preparation of Medium: Add components to distilled/deionized water and bring volume to 1.0L. Mix thoroughly. Distribute into tubes or flasks. Autoclave for 15 min at 15 psi pressure–121°C.

Use: For the cultivation of *Cytophaga aurantiaca.*

Dubos Salts Broth with Yeast Extract

Composition per liter:

K_2HPO_4	1.0g
Yeast extract	0.5g
KCl	0.5g
$MgSO_4 \cdot 7H_2O$	0.5g
$NaNO_3$	0.5g
$FeSO_4 \cdot 7H_2O$	0.01g
Filter paper strips, sterile	variable

pH 7.2 ± 0.2 at 25°C

Preparation of Medium: Add components, except filter paper strips, to distilled/deionized water and bring volume to 1.0L. Mix thoroughly. Distribute into tubes containing a filter paper strip (filter paper strip should protrude above the surface of the broth). Autoclave for 15 min at 15 psi pressure–121°C.

Use: For the cultivation of *Cellulomonas* species and *Cellvibrio* species.

Dubos Salts Medium (DSMZ Medium 1161)

Composition per liter:

Agar	15.0g
K_2HPO_4	1.0g
$MgSO_4 \cdot 7H_2O$	0.5g
KCl	0.5g
$NaNO_3$	0.5g
$FeSO_4 \cdot 7H_2O$	10.0mg
Filter paper	Variable

pH 7.0 ± 0.2 at 25°C

Preparation of Medium: Add components, except filter paper, to distilled/deionized water and bring volume to 1.0L. Mix thoroughly. Adjust pH to 7.0. Gently heat while stirring and bring to boiling. Distribute into tubes or flasks. Autoclave for 15 min at 15 psi pressure–121°C. Pour into sterile Petri dishes or leave in tubes. For tubes allow to solidify on a slant. When the agar has solidified, place a strip of sterile filter paper on to the surface of each slope. Inoculate on to the filter paper.

Use: For the cultivation of *Cellvibrio* spp.

Ducreyi Medium, Revised

See: ***Haemophilus ducreyi*** **Medium, Revised**

Dulaney Slants

Composition per liter:

Egg yolks	50.0mL
Locke solution	50.0mL

Locke Solution:

Composition per 100.0mL:

NaCl	0.9g
Glucose	0.25g
KCl	0.042g

$CaCl_2 \cdot 2H_2O$ 0.024g
Na_2CO_3 .... 0.02g

Preparation of Locke Solution: Add components to distilled/deionized water and bring volume to 100.0mL. Mix thoroughly. Filter sterilize.

Preparation of Medium: Aseptically remove the yolks from 5–8 day old hen egg embryos. Add an equal volume of sterile Locke solution containing sterile glass beads. Mix thoroughly to homogenize. Aseptically distribute into sterile tubes. Inspissate tubes in a slanted position at 80°C (moist heat) for 15 min.

Use: For the cultivation of *Calymmatobacter granulomatis* from clinical specimens.

Dulcitol Selenite Broth (Selenite-F Broth with Dulcitol)

Composition per liter:
NaH_2PO_4 .... 10.0g
Peptic digest of animal tissue 5.0g
Dulcitol 4.0g
$HNaO_3Se$ 4.0g

pH 7.0 ± 0.2 at 25°C

Source: This medium is available from HiMedia.

Caution: Sodium hydrogen selenite is a very toxic, corrosive agent and causes teratogenicity. Upon contact with skin, wash immediately with a lot of water.

Preparation of Medium: Add sodium hydrogen selenite to distilled/deionized water and bring volume to 1.0L. Mix thoroughly. Add remaining components. Mix thoroughly. Gently heat if needed to get all compoents to dissolve. Distribute into tubes or flasks. Sterilize in a boiling water bath or free flowing steam for 10 min. Excessive heating is detrimental. Do not autoclave.

Use: For the selective enrichment of *Salmonella* species.

Duncan-Strong Sporulation Medium, Modified (DS Sporulation Medium, Modified) (Sporulation Medium, Modified)

Composition per liter:
Proteose peptone 15.0g
$Na_2HPO_4 \cdot 7H_2O$ 10.0g
Raffinose 4.0g
Yeast extract 4.0g
Sodium thioglycolate 1.0g

pH 7.8 ± 0.2 at 25°C

Preparation of Medium: Add components to distilled/deionized water and bring volume to 1.0L. Mix thoroughly. Gently heat and bring to boiling. Distribute into tubes or flasks. Autoclave for 15 min at 15 psi pressure–121°C. Adjust pH to 7.8 with filter-sterilized 0.66*M* Na_2CO_3. Pour into sterile Petri dishes or leave in tubes.

Use: For the cultivation and induction of sporulation of *Clostridium perfringens*.

Dung Extract Agar (DSMZ Medium 781)

Composition per liter:
Agar 15.0g
Malt extract 5.0g
$(NO_3)_2 \cdot 4H_2O$ 0.72g
$MgSO_4 \cdot 7H_2O$ 0.5g
K_2HPO_4 .... 0.25g
Peptone 0.1g
Dung extract 100.0mL

pH 6.9 ± 0.2 at 25°C

Dung Extract:
Composition per 150.0mL:
Horse dung .... variable

Preparation of Dung Extract: Add an average sized piece of horse dung to150.0mL water. Gently heat and bring to boiling. Boil for 2 hr in a water bath. Filter. Use immediately.

Preparation of Medium: Add components to distilled/deionized water and bring volume to 1.0L. Mix thoroughly. Gently heat and bring to boiling. Distribute into tubes or flasks. Autoclave for 15 min at 15 psi pressure–121°C. Pour into sterile Petri dishes or leave in tubes.

Use: For the cultivation and maintenance of *Panaeolus cyanescens*.

Dunkelberg Agar
See: **Peptone Starch Dextrose Agar**

Dunkelberg Carbohydrate Medium, Modified

Composition per 100.0mL:
Proteose peptone No. 3 .... 1.5g
Carbohydrate 1.0g
$Na_2HPO_4 \cdot 2H_2O$ 0.207g
Phenol Red 0.055g
$NaH_2PO_4 \cdot H_2O$ 0.038g
Horse serum 5.0mL

pH 7.4 ± 0.2 at 25°C

Preparation of Medium: Add components, except horse serum, to distilled/deionized water and bring volume to 95.0mL. For carbohydrate, use glucose, maltose, or starch. Mix thoroughly. Filter sterilize. Aseptically add sterile horse serum. Mix thoroughly. Aseptically distribute into sterile tubes or flasks.

Use: For the cultivation and differentiation of *Gardnerella vaginalis* based on its ability to ferment glucose, maltose, or starch.

Dunkelberg Maintenance Medium

Composition per liter:
Proteose peptone No. 3 .... 20.0g
Soluble starch 10.0g
Agar 8.0g
Glucose 2.0g
Na_2HPO_4 .... 1.0g
NaH_2PO_4 .... 1.0g

pH 6.8 ± 0.2 at 25°C

Preparation of Medium: Add starch to approximately 100.0mL of cold distilled/deionized water. Mix thoroughly. Add starch solution to 400.0mL of boiling distilled/deionized water. Add remaining components. Mix thoroughly. Bring volume to 1.0L with distilled/deionized water. Distribute into screw-capped tubes. Autoclave for 12 min at 8 psi pressure–112°C.

Use: For the cultivation and maintenance of *Gardnerella vaginalis*.

Dunkelberg Semisolid Carbohydrate Fermentation Medium

Composition per liter:

Proteose peptone No. 3	20.0g
Carbohydrate	10.0g
Agar	5.0g
Bromcresol Purple solution	1.0mL

pH 7.4 ± 0.2 at 25°C

Bromcresol Purple Solution:

Composition per 10.0mL:

Bromcresol Purple	0.16g
Ethanol (95% solution)	10.0mL

Preparation of Bromcresol Purple Solution: Add Bromcresol Purple to 10.0mL of ethanol. Mix thoroughly. Filter sterilize.

Preparation of Medium: Add components to distilled/deionized water and bring volume to 1.0L. For carbohydrate, use glucose, maltose, or starch. Mix thoroughly. Gently heat and bring to boiling. Filter sterilize. Aseptically distribute into sterile tubes or flasks.

Use: For the cultivation and differentiation of *Gardnerella vaginalis* based on its ability to ferment glucose, maltose, or starch.

DV Medium

Composition per 100.1mL:

NaCl	1.8g
$MgSO_4 \cdot 7H_2O$	0.5g
Tris-HCl	100.0mg
KCl	60.0mg
$NaNO_3$	50.0mg
$Na_2SiO_3 \cdot 9H_2O$	20.0mg
$CaCl_2$	10.0mg
Nitrilotriacetic acid (NTA)	10.0mg
K_2HPO_4	3.0mg
$FeCl_2$	0.01mg
Vitamin B_{12}	0.3 µg
Metal solution	3.0mL
Vitamin solution	0.1mL
Seawater, charcoal filtered	100.0mL

Metal Solution:

Composition per 25.0mL:

EDTA	25.0mg
$MnCl_2$	1.0mg
H_3BO_3	0.5mg
$FeCl_2$	0.25mg
$ZnCl_2$	125.0µg
$CoCl_2$	25.0µg

Preparation of Metal Solution: Add components to distilled/deionized water and bring volume to 25.0mL. Mix thoroughly.

Vitamin Solution:

Composition per 100.0mL:

Inositol	100.0mg
Thymine	50.0mg
Orotic acid	26.0mg
Thiamine·HCl	20.0mg
Calcium D-(+)-pantothenate	10.0mg
Nicotinic acid	10.0mg
Putrescine·2HCl	4.0mg
Pyridoxine·2HCl	4.0mg
Pyridoxamine·2HCl	2.0mg
p-Aminobenzoic acid	1.0mg
Riboflavin	0.5mg
Folic acid	0.025mg
Biotin	50.0µg
Folinic acid	20.0µg
Vitamin B_{12}	5.0µg

Preparation of Vitamin Solution: Add components to distilled/deionized water and bring volume to 25.0mL. Mix thoroughly. Filter sterilize.

Preparation of Medium: Add components, except vitamin solution, to charcoal-filtered seawater and bring volume to 100.0mL. Mix thoroughly. Autoclave for 15 min at 15 psi pressure–121°C. Aseptically add 0.1mL of sterile vitamin solution. Mix thoroughly. Aseptically distribute into sterile, screw-capped tubes or flasks.

Use: For the cultivation of *Dunaliella tertiolecta* and *Trichosphaerium* species.

DYA with Calcium Carbonate

Composition per liter:

Glucose	50.0g
Agar	20.0g
$CaCO_3$	5.0g
Yeast extract	5.0g

Preparation of Medium: Add components to distilled/deionized water and bring volume to 1.0L. Mix thoroughly. Gently heat and bring to boiling. Distribute into tubes or flasks. Autoclave for 15 min at 15 psi pressure–121°C. Pour into sterile Petri dishes or leave in tubes.

Use: For the cultivation of *Dekkera intermedia, Dekkera bruxellensis, Dekkera abstinens, Dekkera anomala, Dekkera custersiana, Dekkera lambica,* and *Dekkera naardenensis.*

DYAA

See: **Dextrose Yeast Asparagine Agar**

DYAA Cellulose

Composition per liter:

Agar	20.0g
Cellulose	5.0g
Glucose	5.0g
K_2HPO_4	1.0g
L-Asparagine	0.5g
$MgSO_4 \cdot 7H_2O$	0.5g
$(NH_4)_2SO_4$	0.5g
Yeast extract	0.5g
$CaCl_2$	0.1g

Preparation of Medium: Add components to distilled/deionized water and bring volume to 1.0L. Mix thoroughly. Gently heat and bring to boiling. Distribute into tubes or flasks. Autoclave for 15 min at 15 psi pressure–121°C. Pour into sterile Petri dishes or leave in tubes.

Use: For the cultivation of *Bipolaris sacchari* and *Drechslera biseptata.*

DYAA Cellulose Malt

Composition per liter:

Agar	20.0g
Malt extract	10.0g
Cellulose	5.0g
Glucose	5.0g

K_2HPO_4	1.0g
L-Asparagine	0.5g
$MgSO_4 \cdot 7H_2O$	0.5g
$(NH_4)_2SO_4$	0.5g
Yeast extract	0.5g
$CaCl_2$	0.1g

Preparation of Medium: Add components to distilled/deionized water and bring volume to 1.0L. Mix thoroughly. Gently heat and bring to boiling. Distribute into tubes or flasks. Autoclave for 15 min at 15 psi pressure–121°C. Pour into sterile Petri dishes or leave in tubes.

Use: For the cultivation of *Bipolaris sorghicola* and *Codinaea simplex.*

DYAA LUP

Composition per liter:

Agar	20.0g
Glucose	10.0g
Yeast extract	1.0g
Asparagine	0.5g
$K_2HPO_4 \cdot 3H_2O$	0.5g
$MgSO_4 \cdot 7H_2O$	0.25g
$FeCl_3$ solution	0.5mL
Lupine stems	variable

$FeCl_3$ Solution:

Composition per 10.0mL:

$FeCl_3$	1.0g

Preparation of $FeCl_3$ Solution: Add $FeCl_3$ to distilled/deionized water and bring volume to 10.0mL. Mix thoroughly.

Preparation of Medium: Add components, except lupine stems, to distilled/deionized water and bring volume to 1.0L. Mix thoroughly. Gently heat and bring to boiling. Distribute 6.0mL volumes into tubes. Cut lupine stems into 8.0cm-long pieces. Add 2–3 lupine stems per tube. Autoclave for 15 min at 15 psi pressure–121°C. Allow tubes to cool in a slanted position.

Use: For the cultivation of *Ceratocystis coerulescens, Ceratocystis microspora, Coniella pulchella, Phoma lini,* and *Phoma glomerata.*

DYPA

See: **Dextrose Yeast Extract Peptone**

E Agar (m-E Agar)

Composition per liter:

Yeast extract	30.0g
Agar	15.0g
NaCl	15.0g
Pancreatic digest of gelatin	10.0g
Esculin	1.0g
Nalidixic acid	0.25g
NaN_3	0.15g
Cycloheximide	0.05g
TTC solution	15.0mL

pH 7.1 ± 0.2 at 25°C

Caution: Cycloheximide is toxic. Avoid skin contact or aerosol formation and inhalation.

TTC Solution:

Composition per 15.0mL:

2,3,5-Triphenyltetrazolium chloride	0.15g

Preparation of TTC Solution: Add triphenyltetrazolium chloride to distilled/deionized water and bring volume to 15.0mL. Mix thoroughly. Filter sterilize.

Preparation of Medium: Add components, except TTC solution, to distilled/deionized water and bring volume to 1.0L. Mix thoroughly. Gently heat and bring to boiling. Autoclave for 15 min at 15 psi pressure–121°C. Cool to 45°–50°C. Aseptically add sterile TTC solution. Mix thoroughly. Pour into sterile Petri dishes or distribute into sterile tubes.

Caution: Sodium azide is toxic. Azides also react with metals and disposal must be highly diluted.

Use: For the isolation, cultivation, and enumeration of enterococci in water by the membrane filter method. It is used in conjunction with esculin iron agar.

E. coli O157:H7 MUG Agar

Composition per liter:

Casein peptone	20.0g
Agar	13.0g
Sorbitol	10.0g
NaCl	5.0g
Meat extract	2.0g
$Na_2S_2O_3$	2.0g
Na-deoxycholate	1.12g
Yeast extract	1.0g
Ammonium ferric citrate	0.5g
4-Methylumbelliferyl-β-D-glucuronide	0.1g
Bromthymol Blue	0.025g

pH 7.4 ± 0.2 at 37°C

Source: This medium is available from Fluka, Sigma-Aldrich.

Preparation of Medium: Add components to distilled/deionized water and bring volume to 1.0L. Mix thoroughly. Gently heat while stirring and bring to boiling. Autoclave for 15 min at 15 psi pressure–121°C. Cool to 50°C. Pour into sterile Petri dishes.

Use: For the isolation and differentiation of enterohemorrhagic (EHEC) *E. coli* O157:H7 strains from food and clinical specimens.

E Medium for Anaerobes

Composition per 100.0mL:

Glucose	0.05g
L-Cysteine·HCl·H_2O	0.05g
Maltose	0.05g
$(NH_4)_2SO_4$	0.05g
Peptone	0.05g
Soluble starch	0.05g
Yeast extract	0.05g
Salts solution	50.0mL
Rumen fluid	30.0mL
Resazurin solution	0.4mL

pH 7.0 ± 0.2 at 25°C

Salts Solution:

Composition per liter:

$NaHCO_3$	10.0g
NaCl	2.0g
K_2HPO_4	1.0g
KH_2PO_4	1.0g
$CaCl_2$, anhydrous	0.2g
$MgSO_4$	0.2g

Preparation of Salts Solution: Add $CaCl_2$ and $MgSO_4$ to approximately 300.0mL of distilled/deionized water. Mix thoroughly. Bring volume to 800.0mL with distilled/deionized water. Add remaining components. Mix thoroughly. Bring volume to 1.0L with distilled/deionized water. Mix thoroughly. Store at 4°C.

Rumen Fluid:

Composition per 100.0mL:

Rumen fluid 100.0mL

Preparation of Rumen Fluid: Obtain the rumen contents from a cow that has been fed an alfalfa-hay ration. Filter rumen contents through two layers of cheesecloth. Store under 100% CO_2 in the refrigerator. The particulate material will settle out. Use only the supernatant liquid.

Resazurin Solution:

Composition per 44.0mL:

Resazurin 0.011g

Preparation of Resazurin Solution: Add resazurin to distilled/deionized water and bring volume to 44.0mL. Mix thoroughly.

Preparation of Medium: Add components, except L-cysteine·HCl·H_2O, to distilled/deionized water and bring volume to 100.0mL. Mix thoroughly. Gently heat and bring to boiling. Continue boiling until resazurin turns colorless, indicating reduction. Cool in an ice-water bath under 100% CO_2. Add the L-cysteine·HCl·H_2O. Adjust pH to 7.0 with 8*N* NaOH or 5*N* HCl. Anaerobically distribute into tubes under O_2-free 100% N_2. Cap tubes with butyl rubber stoppers. Place tubes in a press. Autoclave for 12 min at 15 psi pressure–121°C with fast exhaust.

Use: For the cultivation and maintenance of *Bacteroides ruminicola, Bacteroides succinogenes, Butyrivibrio fibrisolvens, Clostridium methylpentosum, Eubacterium ruminantium, Lachnospira multipara, Micromonospora ruminantium, Prevotella ruminicola, Propionibacterium acidipropionici, Selenomonas ruminantium, Selenomonas suis, Succinivibrio dextrinosolvens, Treponema bryantii,* and *Treponema succinifaciens.*

E Medium for Anaerobes with 0.1% Cellobiose

Composition per 100.0mL:

Cellobiose 0.1g
Glucose 0.05g
L-Cysteine·HCl·H_2O 0.05g
Maltose 0.05g
$(NH_4)_2SO_4$ 0.05g
Peptone 0.05g
Soluble starch 0.05g
Yeast extract 0.05g
Salts solution 50.0mL
Rumen fluid 30.0mL
Resazurin solution 0.4mL

pH 7.0 ± 0.2 at 25°C

Salts Solution:

Composition per liter:

$NaHCO_3$ 10.0g
NaCl 2.0g
K_2HPO_4 .... 1.0g
KH_2PO_4 .... 1.0g
$CaCl_2$, anhydrous 0.2g
$MgSO_4$ 0.2g

Preparation of Salts Solution: Add $CaCl_2$ and $MgSO_4$ to approximately 300.0mL of distilled/deionized water. Mix thoroughly. Bring volume to 800.0mL with distilled/deionized water. Add remaining components. Mix thoroughly. Bring volume to 1.0L with distilled/deionized water. Mix thoroughly. Store at 4°C.

Rumen Fluid:

Composition per 100.0mL:

Rumen fluid 100.0mL

Preparation of Rumen Fluid: Obtain the rumen contents from a cow that has been fed an alfalfa-hay ration. Filter rumen contents through two layers of cheesecloth. Store under 100% CO_2 in the refrigerator. The particulate material will settle out. Use only the supernatant liquid.

Resazurin Solution:

Composition per 44.0mL:

Resazurin 0.011g

Preparation of Resazurin Solution: Add resazurin to distilled/deionized water and bring volume to 44.0mL. Mix thoroughly.

Preparation of Medium: Add components, except L-cysteine·HCl·H_2O, to distilled/deionized water and bring volume to 100.0mL. Mix thoroughly. Gently heat and bring to boiling. Continue boiling until resazurin turns colorless, indicating reduction. Cool in an ice-water bath under 100% CO_2. Add the L-cysteine·HCl·H_2O. Adjust pH to 7.0 with 8*N* NaOH or 5*N* HCl. Anaerobically distribute into tubes under O_2-free 100% N_2. Cap tubes with butyl rubber stoppers. Place tubes in a press. Autoclave for 12 min at 15 psi pressure–121°C with fast exhaust.

Use: For the cultivation and maintenance of *Eubacterium cellulosolvens* and *Fibrobacter inyrdyinslid.*

E Medium for Anaerobes with Filtered Rumen Fluid and 0.1% Cellobiose

Composition per 100.0mL:

Cellobiose 0.1g
Glucose 0.05g
L-Cysteine·HCl·H_2O 0.05g
Maltose 0.05g
$(NH_4)_2SO_4$ 0.05g
Peptone 0.05g
Soluble starch 0.05g
Yeast extract 0.05g
Salts solution 50.0mL
Rumen fluid, filtered 30.0mL
Resazurin solution 0.4mL

pH 7.0 ± 0.2 at 25°C

Salts Solution:

Composition per liter:

$NaHCO_3$ 10.0g
NaCl 2.0g
K_2HPO_4 .... 1.0g
KH_2PO_4 .... 1.0g
$CaCl_2$ (anhydrous) 0.2g
$MgSO_4$ 0.2g

Preparation of Salts Solution: Add $CaCl_2$ and $MgSO_4$ to approximately 300.0mL of distilled/deionized water. Mix thoroughly. Bring volume to 800.0mL with distilled/deionized water. Add remaining components. Mix thoroughly. Bring volume to 1.0L with distilled/deionized water. Mix thoroughly. Store at 4°C.

Rumen Fluid:

Composition per 100.0mL:

Rumen fluid 100.0mL

Preparation of Rumen Fluid: Obtain the rumen contents from a cow that has been fed an alfalfa-hay ration. Filter rumen contents through two layers of cheesecloth. Store under 100% CO_2 in the refrigerator. The particulate material will settle out. Use only the supernatant liquid. Filter through a 0.20μm filter.

Resazurin Solution:

Composition per 44.0mL:

Resazurin 0.011g

Preparation of Resazurin Solution: Add resazurin to distilled/deionized water and bring volume to 44.0mL. Mix thoroughly.

Preparation of Medium: Add components, except L-cysteine·HCl·H_2O, to distilled/deionized water and bring volume to 100.0mL. Mix thoroughly. Gently heat and bring to boiling. Continue boiling until resazurin turns colorless, indicating reduction. Cool in an ice-water bath under 100% CO_2. Add the L-cysteine·HCl·H_2O. Adjust pH to 7.0 with 8*N* NaOH or 5*N* HCl. Anaerobically distribute into tubes under O_2-free 100% N_2. Cap tubes with butyl rubber stoppers. Place tubes in a press. Autoclave for 12 min at 15 psi pressure–121°C with fast exhaust.

Use: For the cultivation and maintenance of the *Fibrobacter* species.

E Medium for Anaerobes, Modified

Composition per 103.6mL:

L-Cysteine·HCl·H_2O 0.05g
$(NH_4)_2SO_4$ 0.05g
Peptone 0.05g
Yeast extract 0.05g
Salts solution 50.0mL
Rumen fluid 30.0mL
Potassium phosphate buffer (1*M*, pH 6.5) 2.8mL
Hemin solution 1.0mL
Glucose-maltose solution 1.4mL
Starch solution 1.4mL
Resazurin (0.025% solution) 0.4mL
Vitamin K_3 solution 0.2mL

pH 6.5 ± 0.2 at 25°C

Salts Solution:

Composition per liter:

$NaHCO_3$ 10.0g
NaCl 2.0g
K_2HPO_4 .... 1.0g
KH_2PO_4 .... 1.0g
$CaCl_2$, anhydrous 0.2g
$MgSO_4$ 0.2g

Preparation of Salts Solution: Add $CaCl_2$ and $MgSO_4$ to approximately 300.0mL of distilled/deionized water. Mix thoroughly. Bring volume to 800.0mL with distilled/deionized water. Add remaining components. Mix thoroughly. Bring volume to 1.0L with distilled/deionized water. Mix thoroughly. Store at 4°C.

Rumen Fluid:

Composition per 100.0mL:

Rumen fluid 100.0mL

Preparation of Rumen Fluid: Obtain the rumen contents from a cow that has been fed an alfalfa-hay ration. Filter rumen contents through two layers of cheesecloth. Store under 100% CO_2 in the refrigerator. The particulate material will settle out. Use only the supernatant liquid.

Hemin Solution:

Composition per 100.0mL:

Hemin 0.05g
NaOH (1*N* solution) 1.0mL

Preparation of Hemin Solution: Add hemin to 1.0mL of 1*N* NaOH solution. Mix thoroughly. Bring volume to 100.0mL with distilled/deionized water. Autoclave for 15 min at 15 psi pressure–121°C. Cool to 45°–50°C.

Glucose-Maltose Solution:

Composition per 10.0mL:

Glucose 0.5g
Maltose 0.5g

Preparation of Glucose-Maltose Solution: Add components to distilled/deionized water and bring volume to 10.0mL. Mix thoroughly. Filter sterilize.

Starch Solution:

Composition per 10.0mL:

Starch, soluble 0.5g

Preparation of Starch Solution: Add starch to distilled/deionized water and bring volume to 10.0mL. Mix thoroughly. Autoclave for 15 min at 15 psi pressure–121°C.

Resazurin Solution:

Composition per 44.0mL:

Resazurin 0.011g

Preparation of Resazurin Solution: Add resazurin to distilled/deionized water and bring volume to 44.0mL. Mix thoroughly.

Vitamin K_3 Solution:

Composition per 25.0mL:

Vitamin K_3 (menadione) 0.0125g
Ethanol, absolute 25.0mL

Preparation of Vitamin K_3 Solution: Add vitamin K_3 to 99.0mL of ethanol. Mix thoroughly.

Preparation of Medium: Filter sterilize potassium phosphate buffer. Add components—except L-cysteine·HCl·H_2O, vitamin K_3 solution, potassium phosphate buffer, glucose-maltose solution, and starch solution—to distilled/deionized water and bring volume to 98.0mL. Mix thoroughly. Gently heat and bring to boiling. Continue boiling until resazurin turns colorless, indicating reduction. Cool in an ice-water bath under O_2-free 97% N_2 + 3% H_2. Add the L-cysteine·HCl·H_2O and vitamin K_3 solution. Mix thoroughly. Adjust pH to 6.5 with 8*N* NaOH or 5*N* HCl. Anaerobically distribute into tubes under O_2-free 97% N_2 + 3% H_2 in 7.0mL volumes. Cap tubes with butyl rubber stoppers. Place tubes in a press. Autoclave for 12 min at 15 psi pressure–121°C with fast exhaust. Immediately prior to inoculation, aseptically add 0.2mL of filter-sterilized potassium phosphate buffer, 0.1mL of sterile glucose-maltose solution, and 0.1mL of sterile starch solution to each tube. Mix thoroughly.

Use: For the cultivation and maintenance of *Bacteroides* species, *Butyrivibrio fibrisolvens, Clostridium methylpentosum, Eubacterium ruminantium, Lachnospira multipara, Micromonospora ruminantium, Prevotella ruminicola, Propionibacterium acidipropionici, Selenomonas* species, *Succinivibrio dextrinosolvens,* and *Treponema* species.

E Medium for Anaerobes with 0.3% Phloroglucinol

Composition per 110.4mL:

$(NH_4)_2SO_4$ 0.5g
L-Cysteine·HCl·H_2O 0.05g

Soluble starch....0.05g
Salts solution....50.0mL
Rumen fluid....30.0mL
Phloroglucinol solution....30.0mL
Resazurin solution....0.4mL

pH 6.6 ± 0.2 at 25°C

Salts Solution:

Composition per liter:

$NaHCO_3$....10.0g
NaCl....2.0g
K_2HPO_4....1.0g
KH_2PO_4....1.0g
$CaCl_2$, anhydrous....0.2g
$MgSO_4$....0.2g

Preparation of Salts Solution: Add $CaCl_2$ and $MgSO_4$ to approximately 300.0mL of distilled/deionized water. Mix thoroughly. Bring volume to 800.0mL with distilled/deionized water. Add remaining components. Mix thoroughly. Bring volume to 1.0L with distilled/deionized water. Mix thoroughly. Store at 4°C.

Rumen Fluid:

Composition per 100.0mL:

Rumen fluid....100.0mL

Preparation of Rumen Fluid: Obtain the rumen contents from a cow that has been fed an alfalfa-hay ration. Filter rumen contents through two layers of cheesecloth. Store under 100% CO_2 in the refrigerator. The particulate material will settle out. Use only the supernatant liquid.

Phloroglucinol Solution:

Composition per 100.0mL:

Phloroglucinol....1.0g

Preparation of Phloroglucinol Solution: Add phloroglucinol to distilled/deionized water and bring volume to 100.0mL. Mix thoroughly. Filter sterilize. Keep away from light.

Resazurin Solution:

Composition per 44.0mL:

Resazurin....0.011g

Preparation of Resazurin Solution: Add resazurin to distilled/deionized water and bring volume to 44.0mL. Mix thoroughly.

Preparation of Medium: Add components, except L-cysteine·HCl·H_2O, to distilled/deionized water and bring volume to 100.0mL. Mix thoroughly. Gently heat and bring to boiling. Continue boiling until resazurin turns colorless, indicating reduction. Cool in an ice-water bath under 100% CO_2. Add the L-cysteine·HCl·H_2O. Adjust pH to 6.6 with 8*N* NaOH or 5*N* HCl. Anaerobically distribute into tubes under O_2-free 100% N_2. Cap tubes with butyl rubber stoppers. Place tubes in a press. Autoclave for 12 min at 15 psi pressure–121°C with fast exhaust.

Use: For the cultivation and maintenance of *Coprococcus* species.

E Medium for Anaerobes with 0.2% Rutin

Composition per 110.4mL:

$(NH_4)_2SO_4$....0.5g
L-Cysteine·HCl·H_2O....0.05g
Soluble starch....0.05g
Salts solution....50.0mL
Rumen fluid....30.0mL
Rutin solution....30.0mL
Resazurin solution....0.4mL

pH 6.6 ± 0.2 at 25°C

Salts Solution:

Composition per liter:

$NaHCO_3$....10.0g
NaCl....2.0g
K_2HPO_4....1.0g
KH_2PO_4....1.0g
$CaCl_2$, anhydrous....0.2g
$MgSO_4$....0.2g

Preparation of Salts Solution: Add $CaCl_2$ and $MgSO_4$ to approximately 300.0mL of distilled/deionized water. Mix thoroughly. Bring volume to 800.0mL with distilled/deionized water. Add remaining components. Mix thoroughly. Bring volume to 1.0L with distilled/deionized water. Mix thoroughly. Store at 4°C.

Rumen Fluid:

Composition per 100.0mL:

Rumen fluid....100.0mL

Preparation of Rumen Fluid: Obtain the rumen contents from a cow that has been fed an alfalfa-hay ration. Filter rumen contents through two layers of cheesecloth. Store under 100% CO_2 at 4°C. The particulate material will settle out. Use the liquid.

Rutin Solution:

Composition per 100.0mL:

Rutin....0.2g

Preparation of Rutin Solution: Add rutin to distilled/deionized water and bring volume to 100.0mL. Mix thoroughly. Filter sterilize.

Resazurin Solution:

Composition per 44.0mL:

Resazurin....0.011g

Preparation of Resazurin Solution: Add resazurin to distilled/deionized water and bring volume to 44.0mL. Mix thoroughly.

Preparation of Medium: Add components, except L-cysteine·HCl·H_2O, to distilled/deionized water and bring volume to 100.0mL. Mix thoroughly. Gently heat and bring to boiling. Continue boiling until resazurin turns colorless, indicating reduction. Cool in an ice-water bath under 100% CO_2. Add the L-cysteine·HCl·H_2O. Adjust pH to 6.6 with 8*N* NaOH or 5*N* HCl. Anaerobically distribute into tubes under O_2-free 100% N_2. Cap tubes with butyl rubber stoppers. Place tubes in a press. Autoclave for 12 min at 15 psi pressure–121°C with fast exhaust.

Use: For the cultivation of *Butyrivibrio* species.

Eagle Medium

Composition per 99.1mL:

Eagle MEM in Hanks BSS....87.0mL
Fetal bovine serum....10.0mL
$NaHCO_3$ (7.5% solution)....1.0mL
Penicillin-streptomycin solution....1.0mL
Amphotericin B solution....0.1mL

pH 7.2–7.4 at 25°C

Eagle MEM in Hanks BSS:

Composition per liter:

NaCl....8.0g
Glucose....1.0g
KCl....0.4g
$CaCl_2 \cdot 2H_2O$....0.14g

$MgSO_4 \cdot 7H_2O$	0.1g
KH_2PO_4	0.06g
Na_2HPO_4	0.05g
L-Isoleucine	0.026g
L-Leucine	0.026g
L-Lysine	0.026g
L-Threonine	0.024g
L-Valine	0.0235g
L-Tyrosine	0.018g
L-Arginine	0.0174g
L-Phenylalanine	0.0165g
L-Cystine	0.012g
L-Histidine	8.0mg
L-Methionine	7.5mg
Phenol Red	5.0mg
L-Tryptophan	4.0mg
Inositol	1.8mg
Biotin	1.0mg
Folic acid	1.0mg
Calcium pantothenate	1.0mg
Choline chloride	1.0mg
Nicotinamide	1.0mg
Pyridoxal·HCl	1.0mg
Thiamine·HCl	1.0mg
Riboflavin	0.1mg

Preparation of Eagle MEM in Hanks BSS: Add components to distilled/deionized water and bring volume to 1.0L. Mix thoroughly.

Penicillin-Streptomycin Solution:
Composition per 1.0mL:

Streptomycin	0.01g
Penicillin	10,000U

Preparation of Penicillin-Streptomycin Solution: Add components to distilled/deionized water and bring volume to 1.0mL. Mix thoroughly.

Amphotericin B Solution:
Composition per 1.0mL:

Amphotericin B	1.0mg

Preparation of Amphotericin B Solution: Add amphotericin B to distilled/deionized water and bring volume to 1.0mL. Mix thoroughly.

Preparation of Medium: Combine components. Mix thoroughly. Filter sterilize.

Use: For the cultivation of animal tissue culture cell lines.

Eagle Medium

Composition per liter:

Hanks balanced salt solution (10X)	100.0mL
Calf serum	50.0mL
$NaHCO_3$ (7.5% solution)	29.6mL
Tissue culture amino acids (50X)	20.0mL
Tissue culture vitamins (100X)	10.0mL
Glutamine solution	10.0mL
Phenol Red (0.5% solution)	4.0mL
Penicillin solution	1.0mL
Streptomycin solution	0.4mL

pH 7.0 ± 0.2 at 25°C

Hanks Balanced Salt Solution (10X):
Composition per 100.0mL:

NaCl	8.0g
Glucose	1.0g
KCl	0.4g
$NaHCO_3$	0.35g
$CaCl_2 \cdot 2H_2O$	0.14g
$MgCl_2 \cdot 6H_2O$	0.1g
$MgSO_4 \cdot 7H_2O$	0.1g
Na_2HPO_4	0.06g
KH_2PO_4	0.06g
Phenol Red	0.02g

Preparation of Hanks Balanced Salt Solution (10X): Add components to distilled/deionized water and bring volume to 100.0mL. Mix thoroughly.

Tissue Culture Amino Acids (50X):
Composition per liter:

L-Arginine	0.1g
L-Lysine	0.058g
L-Isoleucine	0.052g
L-Leucine	0.052g
L-Threonine	0.048g
L-Valine	0.046g
L-Tyrosine	0.036g
L-Phenylalanine	0.032g
L-Histidine	0.031g
L-Cystine	0.024g
L-Methionine	0.015g
L-Tryptophan	0.01g

Preparation of Tissue Culture Amino Acids (50X): Add components to distilled/deionized water and bring volume to 1.0L. Mix thoroughly.

Tissue Culture Vitamins (100X):
Composition per liter:

Inositol	2.0mg
Calcium pantothenate	1.0mg
Choline chloride	1.0mg
Folic acid	1.0mg
Nicotinamide	1.0mg
Pyridoxal	1.0mg
Thiamine·HCl	1.0mg
Riboflavin	0.1mg

Preparation of Tissue Culture Vitamins (100X): Add components to distilled/deionized water and bring volume to 1.0L. Mix thoroughly.

Glutamine Solution:
Composition per 100.0mL:

L-Glutamine	2.9g

Preparation of Glutamine Solution: Add glutamine to distilled/deionized water and bring volume to 100.0mL. Mix thoroughly.

Penicillin Solution:
Composition per 1.0mL:

Penicillin	200,000U

Preparation of Penicillin Solution: Add penicillin to distilled/deionized water and bring volume to 1.0mL. Mix thoroughly.

Streptomycin Solution:
Composition per 1.0mL:

Streptomycin	0.5g

Preparation of Streptomycin Solution: Add streptomycin to distilled/deionized water and bring volume to 1.0mL. Mix thoroughly.

Preparation of Medium: Combine components. Mix thoroughly. Adjust pH to 7.0 with 1*N* NaOH. Filter sterilize.

Use: For the cultivation of animal tissue culture cell lines, especially for use with rhinoviruses.

Eagle Medium

Composition per 100.1mL:

Eagle MEM in Earle BSS	94.0mL
$NaHCO_3$ (7.5% solution)	3.0mL
Fetal bovine serum, inactivated	2.0mL
Penicillin-streptomycin solution	1.0mL
Amphotericin B solution	0.1mL

pH 7.2–7.4 at 25°C

Eagle MEM in Earle BSS:

Composition per liter:

NaCl	6.8g
Glucose	1.0g
KCl	0.4g
$CaCl_2 \cdot 2H_2O$	0.2g
$MgCl_2 \cdot 6H_2O$	0.2g
NaH_2PO_4	0.15g
L-Arginine	0.1g
L-Lysine	0.06g
L-Isoleucine	0.05g
L-Leucine	0.05g
L-Threonine	0.05g
L-Valine	0.05g
L-Tyrosine	0.04g
L-Phenylalanine	0.03g
L-Histidine	0.03g
L-Cystine	0.02g
L-Methionine	0.02g
L-Tryptophan	0.01g
i-Inositol	2.0mg
Calcium pantothenate	1.0mg
Choline chloride	1.0mg
Folic acid	1.0mg
Nicotinamide	1.0mg
Pyridoxal	1.0mg
Thiamine·HCl	1.0mg
Riboflavin	0.1mg

Preparation of Eagle MEM in Earle BSS: Add components to distilled/deionized water and bring volume to 1.0L. Mix thoroughly.

Penicillin-Streptomycin Solution:

Composition per 1.0mL:

Streptomycin	0.01g
Penicillin	10,000U

Preparation of Penicillin-Streptomycin Solution: Add components to distilled/deionized water and bring volume to 1.0mL. Mix thoroughly.

Amphotericin B Solution:

Composition per 1.0mL:

Amphotericin B	1.0mg

Preparation of Amphotericin B Solution: Add amphotericin B to distilled/deionized water and bring volume to 1.0mL. Mix thoroughly.

Preparation of Medium: Combine components. Mix thoroughly. Filter sterilize.

Use: For the cultivation of animal tissue culture cell lines.

Eagle Medium, Modified

Composition per liter:

Eagle MEM (10X)	100.0mL
Fetal bovine serum	100.0mL
Glucose solution	20.0mL
HEPES (*N*-2-hydroxyethyl piperazine-*N′*-2-ethanesulfonic acid) buffer, 1*M*, pH 7.2	20.0mL
Glutamine solution	10.0mL
$NaHCO_3$ (7.5% solution)	7.5mL
Gentamicin sulfate solution	0.2mL

pH 7.2 ± 0.2 at 25°C

Eagle MEM (10X):

Composition per 100.0mL:

Sterile salt solution	97.0mL
TC amino acids, minimal Eagle 50X	2.0mL
TC vitamins, minimal Eagle 100X	1.0mL

Preparation of Eagle MEM (10X): Combine components. Mix thoroughly. Filter sterilize.

Sterile Salt Solution:

Composition per 100.0mL:

NaCl	6.8g
Glucose	1.0g
KCl	0.4g
$CaCl_2$	0.2g
$MgCl_2$	0.2g
NaH_2PO_4	0.15g

Preparation of Sterile Salt Solution: Add components to distilled/deionized water and bring volume to 100.0mL. Mix thoroughly. Filter sterilize.

TC Amino Acids (50X):

Composition per liter:

L-Arginine	0.1g
L-Lysine	0.06g
L-Isoleucine	0.05g
L-Leucine	0.05g
L-Threonine	0.05g
L-Valine	0.05g
L-Tyrosine	0.04g
L-Phenylalanine	0.03g
L-Histidine	0.03g
L-Cystine	0.02g
L-Methionine	0.02g
L-Tryptophan	0.01g

Preparation of TC Amino Acids (50X): Add components to distilled/deionized water and bring volume to 1.0L. Mix thoroughly. Adjust pH to 7.2–7.4. Filter sterilize.

TC Vitamins, Minimal Eagle 100X:

Composition per liter:

Inositol	2.0mg
Calcium pantothenate	1.0mg
Choline chloride	1.0mg
Folic acid	1.0mg
Nicotinamide	1.0mg

Pyridoxal	1.0mg
Thiamine·HCl	1.0mg
Riboflavin	0.1mg

Preparation of TC Vitamins, Minimal Eagle 100X: Add components to distilled/deionized water and bring volume to 1.0L. Mix thoroughly. Filter sterilize.

Glucose Solution:

Composition per 100.0mL:

Glucose	27.0g

Preparation of Glucose Solution: Add glucose to distilled/deionized water and bring volume to 100.0mL. Mix thoroughly. Filter sterilize.

Glutamine Solution:

Composition per 10.0mL:

L-Glutamine	5.0g

Preparation of Glutamine Solution: Add glutamine to distilled/deionized water and bring volume to 10.0mL. Mix thoroughly. Filter sterilize.

Gentamicin Solution:

Composition per 1.0mL:

Gentamicin sulfate	0.05g

Preparation of Gentamicin Solution: Add gentamicin sulfate to distilled/deionized water and bring volume to 1.0mL. Mix thoroughly. Filter sterilize.

Preparation of Medium: Combine components. Mix thoroughly. Filter sterilize.

Use: For the cultivation of animal tissue culture cell lines, especially for McCoy cells.

Eagle's Minimal Essential Medium with Earle's Salts and Nonessential Amino Acids (MEM with Earle's Salts and Nonessential Amino Acids) (BAM M46)

Composition per liter:

NaCl	6.8g
$NaHCO_3$	2.2g
Glucose	1.0g
KCl	0.4g
$CaCl_2 \cdot 2H_2O$	0.265g
$MgSO_4 \cdot 7H_2O$	0.2g
L-Arginine·H_2O	0.15g
$NaH_2PO_4 \cdot H_2O$	0.14g
L-Arginine·HCl	0.126g
L-Lysine·HCl	72.5mg
L-Tyrosine, disodium salt	52.1mg
L-Leucine	52.0mg
L-Threonine	48.0mg
L-Valine	46.0mg
L-Histidine·HCl·H_2O	42.0mg
D-Phenylalanine	32.0mg
L-Cysteine·2HCl	31.29mg
L-Methionine	15.0mg
L-Glutamic acid	14.7mg
L-Aspartic acid	13.3mg
L-Proline	11.5mg
L-Serine	10.5mg
L-Tryptophan	10.0mg
L-Alanine	8.9mg
Phenol Red	10.0mg
L-Glycine	7.5mg
i-Inositol	2.0mg
D-Calcium pantothenate	1.0mg
Choline chloride	1.0mg
Folic acid	1.0mg
Nicotinamide	1.0mg
Pyridoxal·HCl	1.0mg
Thiamine·HCl	1.0mg
Riboflavin	0.1mg

pH 7.2 ± 0.2 at 25°C

Preparation of Medium: Add components to 1.0L of distilled/deionized water. Mix thoroughly. Filter sterilize.

Use: For the cultivation of animal cells in tissue culture, for example, cells for viral detection and identification by characteristic cytopathic effects.

Earle's Balanced Salts, Phenol Red-Free

Composition per liter:

NaCl	6.8g
$NaHCO_3$	2.2g
Glucose	1.0g
KCl	0.4g
$CaCl_2 \cdot 2H_2O$	0.265g
$MgSO_4 \cdot 7H_2O$	0.2g
$NaH_2PO_4 \cdot H_2O$	0.14g

pH 7.2 ± 0.2 at 25°C

Preparation of Medium: Add components to distilled/deionized water and bring volume to 1.0L. Mix thoroughly. Filter sterilize.

Use: For the preparation of tissue culture media where Phenol Red is not desired.

EB Motility Medium

Composition per liter:

Peptone or gelysate	10.0g
NaCl	5.0g
Agar	4.0g
Beef extract	3.0g

pH 7.4 ± 0.2 at 25°C

Preparation of Medium: Add components to distilled/deionized water and bring volume to 1.0L. Mix thoroughly. Gently heat and bring to boiling. Distribute into tubes in 8.0mL volumes. Autoclave for 15 min at 15 psi pressure–121°C.

Use: For the cultivation and differentiation of bacteria based on motility.

EBA Medium

Composition per 994.0mL:

Glycine	2.0g
NaCl	1.0g
KCl	0.5g
$MgSO_4 \cdot 7H_2O$	0.5g
NH_4Cl	0.25g
KH_2PO_4	0.2g
$CaCl_2 \cdot 2H_2O$	0.15g
Resazurin	1.0mg

$NaHCO_3$ solution	30.0mL
$Na_2S·9H_2O$ solution	10.0mL
Biotin solution	2.0mL
$Na_2SeO_3·5H_2O$ solution	1.0mL
Trace elements solution SL-10	1.0mL

pH 7.2–7.4 at 25°C

$NaHCO_3$ Solution:

Composition per 100.0mL:

$NaHCO_3$	8.4g

Preparation of $NaHCO_3$ Solution: Add $NaHCO_3$ to distilled/deionized water and bring volume to 100.0mL. Mix thoroughly. Filter sterilize. Gas under 80% N_2 + 20% CO_2.

$Na_2S·9H_2O$ Solution:

Composition per 10.0mL:

$Na_2S·9H_2O$	0.6g

Preparation of $Na_2S·9H_2O$ Solution: Add $Na_2S·9H_2O$ to distilled/deionized water and bring volume to 10.0mL. Mix thoroughly. Gas under 100% N_2. Autoclave for 15 min at 15 psi pressure–121°C.

Biotin Solution:

Composition per 10.0mL:

Biotin	0.1mg

Preparation of Biotin Solution: Add biotin to distilled/deionized water and bring volume to 10.0mL. Mix thoroughly. Gas under 100% N_2. Autoclave for 10 min at 15 psi pressure–121°C.

$Na_2SeO_3·5H_2O$ Solution:

Composition per liter:

$Na_2SeO_3·5H_2O$	0.26g
NaOH (10m*M* solution)	1.0L

Preparation of $Na_2SeO_3·5H_2O$ Solution: Add $Na_2SeO_3·5H_2O$ to 1.0L of 10m*M* NaOH solution. Mix thoroughly.

Trace Elements Solution SL-10:

Composition per liter:

$FeCl_2·4H_2O$	1.5g
$CoCl_2·6H_2O$	190.0mg
$MnCl_2·4H_2O$	100.0mg
$ZnCl_2$	70.0mg
$Na_2MoO_4·2H_2O$	36.0mg
$NiCl_2·6H_2O$	24.0mg
H_3BO_3	6.0mg
$CuCl_2·2H_2O$	2.0mg
HCl (25% solution)	10.0mL

Preparation of Trace Elements Solution SL-10: Add $FeCl_2·4H_2O$ to 10.0mL of HCl solution. Mix thoroughly. Add distilled/deionized water and bring volume to 1.0L. Add remaining components. Mix thoroughly. Gas under 100% N_2. Autoclave for 15 min at 15 psi pressure–121°C.

Preparation of Medium: Add components, except $NaHCO_3$ solution, $Na_2S·9H_2O$ solution, and biotin solution, to distilled/deionized water and bring volume to 952.0mL. Mix thoroughly. Sparge with 100% N_2. Autoclave for 15 min at 15 psi pressure–121°C. Aseptically and anaerobically add 30.0mL of sterile $NaHCO_3$ solution, 10.0mL of sterile $Na_2S·9H_2O$ solution, and 2.0mL of sterile biotin solution. Aseptically and anaerobically distibute into sterile flasks or tubes.

Use: For the cultivation and maintenance of *Eubacterium acidaminophilum.*

EC Broth
(*Escherichia coli* Broth)
(EC Medium)

Composition per liter:

Pancreatic digest of casein	20.0g
Lactose	5.0g
NaCl	5.0g
K_2HPO_4	4.0g
Bile salts mixture	1.5g
KH_2PO_4	1.5g

pH 6.9 ± 0.2 at 25°C

Source: This medium is available as a premixed powder from BD Diagnostic Systems.

Preparation of Medium: Add components to distilled/deionized water and bring volume to 1.0L. Mix thoroughly. Distribute into test tubes that contain an inverted Durham tube. Autoclave for 12 min at 15 psi pressure–121°C. Cool broth as quickly as possible.

Use: For the cultivation and differentiation of coliform bacteria at 37°C and of *Escherichia coli* at 45.5°C.

EC Broth
(*Escherichia coli* Broth)
(EC Medium)
(BAM M49)

Composition per liter:

Pancreatic digest of casein	20.0g
Lactose	5.0g
NaCl	5.0g
K_2HPO_4	4.0g
KH_2PO_4	1.5g
Bile salts No.3	1.12g
Novobiocin solution	10.0mL

pH 6.9 ± 0.2 at 25°C

Source: This medium is available as a premixed powder from Oxoid.

Novobiocin Solution:

Composition per 50.0mL:

Novobiocin	0.1g

Preparation of Novobiocin Solution: Add novobiocin to distilled/deionized water and bring volume to 50.0mL. Mix thoroughly. Filter sterilize.

Preparation of Medium: Add components, except novobiocin solution, to distilled/deionized water and bring volume to 990.0mL. Mix thoroughly. Autoclave for 15 min at 15 psi pressure–121°C. Cool to 50°C. Aseptically add 10.0mL novobiocin solution. Mix thoroughly. Aseptically distribute to sterile tubes or flasks.

Use: A selective enrichment broth for the growth of *E. coli* O157 from food and environmental samples.

EC Broth with MUG

Composition per liter:

Pancreatic digest of casein	20.0g
Lactose	5.0g
NaCl	5.0g
K_2HPO_4	4.0g
Bile salts mixture	1.5g

KH_2PO_4	1.5g
4-Methylumbelliferyl-β-D-glucuronide (MUG)	0.05g

pH 6.9 ± 0.2 at 25°C

Source: This medium is available as a premixed powder from BD Diagnostic Systems.

Preparation of Medium: Add components to distilled/deionized water and bring volume to 1.0L. Mix thoroughly. Distribute into test tubes that contain an inverted Durham tube in 10.0mL volumes. Autoclave for 15 min at 15 psi pressure–121°C.

Use: For the detection of *Escherichia coli* in water and food samples by a fluorogenic procedure.

EC HiVeg Broth

Composition per liter:

Plant hydrolysate no. 1	20.0g
Lactose	5.0g
NaCl	5.0g
K_2HPO_4	4.0g
KH_2PO_4	1.5g
Synthetic detergent	1.5g

pH 6.9 ± 0.2 at 25°C

Source: This medium is available as a premixed powder from HiMedia.

Preparation of Medium: Add components to distilled/deionized water and bring volume to 1.0L. Mix thoroughly. Distribute into test tubes that contain an inverted Durham tube. Autoclave for 15 min at 15 psi pressure–121°C. Cool broth as quickly as possible.

Use: For the cultivation and differentiation of coliform bacteria at 37°C and of *Escherichia coli* at 45.5°C. Recommended for selective enumeration of presumptive *Escherichia coli* by MPN technique. For the selective enumeration of faecal and nonfaecal coliforms in water, wastewater, and shell fish.

EC Medium, Modified with Novobiocin

Composition per liter:

Tryptone	20.0g
NaCl	5.0g
Lactose	5.0g
K_2HPO_4	4.0g
KH_2PO_4	1.5g
Bile salts	1.12g
Novobiocin supplement	10.0mL

pH 6.9 ± 0.2 at 25°C

Source: This medium is available as a premixed powder and supplement from BD Diagnostic Systems.

Novobiocin Supplement:

Composition per 10.0mL:

Sodium novobiocin	20.0mg

Preparation of Novobiocin Supplement: Add sodium novobiocin to distilled/deionized water and bring volume to 10.0mL. Mix thoroughly. Filter sterilize.

Preparation of Medium: Add components, except novobiocin supplement, to distilled/deionized water and bring volume to 990.0mL. Mix thoroughly. Autoclave for 15 min at 15 psi pressure–121°C. Aseptically add 10.0mL of sterile novobiocin supplement. Mix thoroughly. Aseptically distribute into sterile tubes or flasks.

Use: For the cultivation of *Escherichia coli* O157:H7.

ECD Agar

Composition per liter:

Casein enzymic hydrolysate	20.0g
Agar	15.0g
Yeast extract	5.0g
NaCl	5.0g
Na_2HPO_4	5.0g
KH_2PO_4	1.5g
Bile salts	1.5g

pH 7.2 ± 0.2 at 25°C

Preparation of Medium: Add components to distilled/deionized water and bring volume to 1.0L. Mix thoroughly. Gently heat and bring to boiling. Distribute into tubes or flasks. Autoclave for 15 min at 15 psi pressure–121°C. Pour into sterile Petri dishes or leave in tubes.

Use: For the selective detection of coliforms, especially *Escherichia coli* in water and food.

ECD Agar, Fluorocult®
(Fluorocult *ECD Agar)*

Composition per liter:

Peptone from casein	20.0g
Agar	15.0g
NaCl	5.0g
Lactose	5.0g
K_2HPO_4	4.0g
Bile salt mixture	1.5g
KH_2PO_4	1.5g
Tryptophan	1.0g
4-Methylumbelliferyl-ß-D-glucuronide	0.07g

pH 7.0 ± 0.2 at 25°C

Source: This medium is available from Merck.

Preparation of Medium: Add components to distilled/deionized water and bring volume to 1.0L. Mix thoroughly. Autoclave for 15 min at 15 psi pressure–121°C. Cool to 45°–50°C. Pour into sterile Petri dishes. The prepared broth is clear and yellowish brown.

Use: For the detection of *E. coli* in meats. The medium complies with the German-DIN-Norm 10110 for the examination of meat, with the regulations according to § 35 LMBG (06.00/36) for the examination of food, and with ISO Standard 6391 (1996) for the enumeration of *E. coli* in meat and meat products. The bile salt mixture of this medium largely inhibits the accompanying flora not usually found in the intestines. Using fluorescence under UV light and a positive indole reaction, *E. coli* colonies can be identified among the grown colonies.

ECD MUG Agar

Composition per liter:

Casein peptone	20.0g
Agar	15.0g
NaCl	5.0g
Lactose	5.0g
K_2HPO_4	4.0g
Bile salt mixture	1.5g
KH_2PO_4	1.5g
Tryptophan	1.0g
4-Methylumbelliferyl-β-D-glucuronide	0.07g

pH 7.0 ± 0.2 at 37°C

Source: This medium is available from Fluka, Sigma-Aldrich.

Preparation of Medium: Add components to distilled/deionized water and bring volume to 1.0L. Mix thoroughly. Gently heat while stirring and bring to boiling. Autoclave for 15 min at 15 psi pressure–121°C. Cool to 50°C. Pour into sterile Petri dishes.

Use: For detection of *Escherichia coli* in a variety of specimens. The bile-salt mixture in this *E. coli* Direct Agar extensively inhibits the non-obligatory intestinal accompanying flora. Fluorescence in the UV and a positive indole test demonstrate the presence of *E. coli* in the colonies.

Echinamoeba Agar (ATCC Medium 2339)

Composition per liter:

Agar	18.0g
$FeCl_2·6H_2O$	0.552g
$MgSO_4·7H_2O$	0.5g
$CaSO_4·2H_2O$	0.49g
$NaHCO_3$	0.37g
$CaCl_2·2H_2O$	0.24g
Yeast extract	0.2g
KCl	8.6mg
KNO_3	0.16mg
Mineral solution	10.0mL
Glycerol	1.0mL

pH 7.0 ± 0.2 at 25°C

Mineral Solution:
Composition per liter:

$ZnSO_4·7H_2O$	3.5g
$Na_2MoO_4·2H_2O$	0.4g
$MnCl_2·4H_2O$	0.25g
$(NH_4)_2Ni(SO_4)_2·6H_2O$	0.2g
LiCl	0.3g
$SnCl_2·2H_2O$	0.1

Preparation of Mineral Solution: Add components to distilled/deionized water and bring volume to 1.0L. Mix thoroughly.

Preparation of Medium: Add components to distilled/deionized water and bring volume to 1.0L. Mix thoroughly. Adjust pH to 7.0 with NaOH. Distribute into tubes or flasks. Autoclave for 15 min at 15 psi pressure–121°C.

Use: For the cultivation and maintenance of *Echinamoeba* spp.

Echinamoeba Agar

Composition per liter:

Agar	18.0g
$FeCl_2·6H_2O$	0.552g
$MgSO_4·7H_2O$	0.5g
$CaSO_4·2H_2O$	0.49g
$NaHCO_3$	0.37g
$CaCl_2·2H_2O$	0.24g
Yeast extract	0.2g
KCl	8.6mg
KNO_3	0.16mg
Mineral solution	10.0mL
Glycerol	1.0mL

pH 7.0 ± 0.2 at 25°C

Mineral Solution:
Composition per liter:

$ZnSO_4·7H_2O$	3.5g
$Na_2MoO_4·2H_2O$	0.4g
$MnCl_2·4H_2O$	0.25g
$(NH_4)_2Ni(SO_4)_2·6H_2O$	0.2g
LiCl	0.3g
$SnCl_2·2H_2O$	0.1g

Preparation of Mineral Solution: Add components to distilled/deionized water and bring volume to 1.0L. Mix thoroughly.

Preparation of Medium: Add components to distilled/deionized water and bring volume to 1.0L. Mix thoroughly. Adjust pH to 7.0 with NaOH. Gently heat and bring to boiling. Distribute into tubes or flasks. Autoclave for 15 min at 15 psi pressure–121°C. Pour into sterile Petri dishes or leave in tubes.

Use: For the cultivation and maintenance of *Echinamoeba* spp.

Echinamoeba Broth

Composition per liter:

$FeCl_2·6H_2O$	0.552g
$MgSO_4·7H_2O$	0.5g
$CaSO_4·2H_2O$	0.49g
$NaHCO_3$	0.37g
$CaCl_2·2H_2O$	0.24g
Yeast extract	0.2g
KCl	8.6mg
KNO_3	0.16mg
Mineral solution	10.0mL
Glycerol	1.0mL

pH 7.0 ± 0.2 at 25°C

Mineral Solution:
Composition per liter:

$ZnSO_4·7H_2O$	3.5g
$Na_2MoO_4·2H_2O$	0.4g
$MnCl_2·4H_2O$	0.25g
$(NH_4)_2Ni(SO_4)_2·6H_2O$	0.2g
LiCl	0.3g
$SnCl_2·2H_2O$	0.1g

Preparation of Mineral Solution: Add components to distilled/deionized water and bring volume to 1.0L. Mix thoroughly.

Preparation of Medium: Add components to distilled/deionized water and bring volume to 1.0L. Mix thoroughly. Adjust pH to 7.0 with NaOH. Distribute into tubes or flasks. Autoclave for 15 min at 15 psi pressure–121°C.

Use: For the cultivation and maintenance of *Echinamoeba* spp.

ECM Agar

Composition per liter:

Agar	15.0g
NaCl	6.0g
Escherichia coli cells, washed	1.0g
$MgSO_4·7H_2O$	0.5g

Preparation of Medium: Add components to distilled/deionized water and bring volume to 1.0L. Mix thoroughly. Gently heat and bring to boiling. Distribute into tubes or flasks. Autoclave for 15 min at 15 psi pressure–121°C. Pour into sterile Petri dishes or leave in tubes.

Use: For the cultivation of myxobacteria.

Ectothiorhodospira abdelmalekii Medium

Composition per 1010.0mL:

NaCl	120.0g
Na_2SO_4	15.0g

$NaHCO_3$ 10.0g
Na_2CO_3 .... 5.0g
Sodium acetate 1.0g
KH_2PO_4 .... 0.8g
NH_4Cl 0.8g
$MgCl_2 \cdot 6H_2O$ 0.1g
$Na_2S \cdot 9H_2O$ solution 10.0mL
Trace elements solution SLA 1.0mL
Vitamin solution 1.0mL

pH 8.5 ± 0.2 at 25°C

$Na_2S \cdot 9H_2O$ Solution:

Composition per 10.0mL:

$Na_2S \cdot 9H_2O$ 0.5g

Preparation of $Na_2S \cdot 9H_2O$ Solution: Add $Na_2S \cdot 9H_2O$ to distilled/deionized water and bring volume to 10.0mL. Mix thoroughly. Gas under 100% N_2. Autoclave for 15 min at 15 psi pressure–121°C.

Trace Elements Solution SLA:

Composition per liter:

$CuCl_2 \cdot 2H_2O$ 10.0g
$FeCl_2 \cdot 4H_2O$ 1.8g
H_3BO_3 .... 0.5g
$CoCl_2 \cdot 6H_2O$ 0.25g
$ZnCl_2$ 0.1g
$MnCl_2 \cdot 4H_2O$ 70.0mg
$Na_2MoO_4 \cdot 2H_2O$ 30.0mg
$Na_2SeO_3 \cdot 5H_2O$ 10.0mg
$NiCl_2 \cdot 6H_2O$ 10.0mg

Preparation of Trace Elements Solution SLA: Add components to distilled/deionized water and bring volume to 1.0L. Mix thoroughly. Adjust pH to 2.0–3.0.

Vitamin Solution VA:

Composition per 100.0mL:

Nicotinic acid amide 35.0mg
Thiamine dichloride 30.0mg
p-Aminobenzoic acid 20.0mg
Biotin 10.0mg
Calcuim DL-pantothenate 10.0mg
Pyridoxal·HCl 10.0mg
Vitamin B_{12} 5.0mg

Preparation of Vitamin Solution VA: Add components to distilled/deionized water and bring volume to 100.0mL. Mix thoroughly. Filter sterilize.

Preparation of Medium: Add components, except $Na_2S \cdot 9H_2O$ solution, to distilled/deionized water and bring volume to 1.0L. Mix thoroughly. Adjust pH to 8.5. Filter sterilize. Aseptically add 10.0mL of sterile $Na_2S \cdot 9H_2O$ solution. Mix thoroughly. Distribute into sterile tubes or flasks.

Use: For the cultivation and maintenance of *Ectothiorhodospira abdelmalekii.*

Ectothiorhodospira halochloris Medium

Composition per liter:

NaCl 180.0g
Na_2SO_4 .... 20.0g
$NaHCO_3$ 14.0g
Na_2CO_3 .... 6.0g
$Na_2S \cdot 9H_2O$ 1.0g
Sodium succinate 1.0g
NH_4Cl 0.8g
KH_2PO_4 .... 0.5g
Yeast extract 0.5g
$MgCl_2 \cdot 6H_2O$ 0.1g
$CaCl_2 \cdot 2H_2O$ 0.05g
Vitamin solution VA 1.0mL
Trace elements solution SLA 1.0mL

pH 8.5–8.7 at 25°C

Vitamin Solution VA:

Composition per liter:

Nicotinamide 0.04g
Thiamine dichloride 0.03g
p-Aminobenzoic acid 0.02g
Biotin 0.01g
Calcium pantothenate 0.01g
Pyridoxal chloride 0.01g
Vitamin B_{12} 5.0mg

Preparation of Vitamin Solution VA: Add components to distilled/deionized water and bring volume to 1.0L. Mix thoroughly.

Trace Elements Solution SLA:

Composition per liter:

$FeCl_2 \cdot 4H_2O$ 1.8g
H_3BO_3 .... 0.5g
$CoCl_2 \cdot 6H_2O$ 0.25g
$ZnCl_2$ 0.1g
$MnCl_2 \cdot 4H_2O$ 0.07g
$NaMoO_4 \cdot 2H_2O$ 0.03g
$CuCl_2 \cdot 2H_2O$ 0.01g
Na_2SeO_3 .... 0.01g
$NiCl_2 \cdot 6H_20$ 0.01g

Preparation of Trace Elements Solution SLA: Add components to distilled/deionized water and bring volume to 1.0L. Mix thoroughly. Adjust pH to 3.0 with 2*N* HCl.

Preparation of Medium: Add components, except trace elements solution SLA, to distilled/deionized water and bring volume to 999.0mL. Mix thoroughly. Filter sterilize. Aseptically add 1.0mL of sterile trace elements solution SLA. Mix thoroughly. Aseptically distribute into flasks or bottles. Completely fill bottles with medium except for a pea-sized air bubble.

Use: For the enrichment and isolation of *Ectothiorhodospira halochloris.*

Ectothiorhodospira halophila Medium

Composition per liter:

NaCl 200.0g
NH_4Cl 0.4g
$(NH_4)_2SO_4$ 0.1g
Na_2CO_3 solution 100.0mL
Tris buffer (1*M* solution, pH 7.5) 30.0mL
Solution C 5.0mL
Potassium phosphate buffer (1*M* solution, pH 7.5) 3.0mL
Additional solution 2.5mL

pH 7.4–8.0 at 25°C

Na_2CO_3 Solution:

Composition per 100.0mL:

Na_2CO_3 .... 10.0g

Preparation of Na_2CO_3 Solution: Add Na_2CO_3 to distilled/deionized water and bring volume to 100.0mL. Mix thoroughly. Autoclave for 15 min at 15 psi pressure–121°C. Cool to 25°C.

Solution C:

Composition per liter:

$MgCl_2 \cdot 6H_2O$	24.0g
$CaCl_2 \cdot 2H_2O$	3.3g
$FeCl_3 \cdot 4H_2O$	1.1g
$(NH_4)_6Mo_7O_{24} \cdot 4H_2O$	0.1g
Nitrilotriacetic acid	10.0mg
Trace elements solution	50.0mL

Preparation of Solution C: Add nitrilotriacetic acid to 500.0mL of distilled/deionized water. Dissolve by adjusting pH to 6.5 with KOH. Add remaining components. Readjust pH to 7.2 with H_2SO_4 or KOH. Add distilled/deionized water to 1.0L.

Trace Elements Solution:

Composition per 100.0mL:

$ZnCl_2$	0.52g
EDTA	0.25g
$MnCl_2 \cdot 4H_2O$	0.08g
$FeCl_3 \cdot 4H_2O$	0.03g
$Co(NO_3)_2 \cdot 6H_2O$	0.02g
$CuCl_2 \cdot 2H_2O$	0.02g
H_3BO_3	0.01g

Preparation of Trace Elements Solution: Add components to distilled/deionized water and bring volume to 1.0L. Mix thoroughly. Adjust pH to 3.0 with 2*N* HCl.

Additional Solution:

Composition per 50.0mL:

$NaS_2O_3 \cdot 6H_2O$	6.0g
Sodium succinate	5.0g
Sodium ascorbate	1.0g

Preparation of Additional Solution: Add components to distilled/deionized water and bring volume to 50.0mL. Mix thoroughly. Filter sterilize.

Preparation of Medium: Add components, except Na_2CO_3 solution and additional solution, to distilled/deionized water and bring volume to 900.0mL. Autoclave for 15 min at 15 psi pressure–121°C. Cool to 45°–50°C. Aseptically adjust pH to 7.4–7.8 with filter-sterilized HCl. Aseptically distribute into 50.0mL screw-capped bottles. Fill each bottle almost to the top, leaving a space of 2.8mL in the neck. Aseptically add 2.5mL of sterile additional solution to each bottle. Mix thoroughly.

Use: For the isolation and cultivation of *Ectothiorhodospira halophila.*

Ectothiorhodospira halophila Medium

Composition per liter:

NaCl	220.0g
Potassium succinate	1.0g
$Na_2S \cdot 9H_2O$	0.1g
K_2HPO_4 solution	20.0mL
$NaHCO_3$ solution	20.0mL
Solution C	20.0mL
$(NH_4)_2SO_4$ solution	5.0mL
Vitamin solution	0.5mL

pH 7.4–8.0 at 25°C

K_2HPO_4 Solution:

Composition per liter:

K_2HPO_4	125.0g

Preparation of K_2HPO_4 Solution: Add K_2HPO_4 to distilled/deionized water and bring volume to 1.0L. Mix thoroughly.

$NaHCO_3$ Solution:

Composition per liter:

$NaHCO_3$	100.0g

Preparation of $NaHCO_3$ Solution: Add $NaHCO_3$ to distilled/deionized water and bring volume to 1.0L. Mix thoroughly.

Solution C:

Composition per liter:

$MgCl_2 \cdot 6H_2O$	24.0g
$CaCl_2 \cdot 2H_2O$	3.3g
$FeCl_3 \cdot 4H_2O$	1.1g
$(NH_4)_6Mo_7O_{24} \cdot 4H_2O$	0.1g
Nitrilotriacetic acid	10.0mg
Trace elements solution	50.0mL

Preparation of Solution C: Add components to distilled/deionized water and bring volume to 1.0L. Mix thoroughly.

Trace Elements Solution:

Composition per 100.0mL:

$ZnCl_2$	0.52g
EDTA	0.25g
$MnCl_2 \cdot 4H_2O$	0.08g
$FeCl_3 \cdot 4H_2O$	0.03g
$Co(NO_3)_2 \cdot 6H_2O$	0.02g
$CuCl_2 \cdot 2H_2O$	0.02g
H_3BO_3	0.01g

Preparation of Trace Elements Solution: Add components to distilled/deionized water and bring volume to 1.0L. Mix thoroughly. Adjust pH to 3.0 with 2*N* HCl.

$(NH_4)_2SO_4$ Solution:

Composition per liter:

$(NH_4)_2SO_4$	100.0g

Preparation of $(NH_4)_2SO_4$ Solution: Add components to distilled/deionized water and bring volume to 1.0L. Mix thoroughly.

Vitamin Solution:

Composition per liter:

Nicotinic acid	2.0mg
Thiamine	1.0mg
p-Aminobenzoic acid	0.2mg
Biotin	0.02mg
Vitamin B_{12}	1.0μg

Preparation of Vitamin Solution: Add components to distilled/deionized water and bring volume to 1.0L. Mix thoroughly.

Preparation of Medium: Add components to distilled/deionized water and bring volume to 1.0L. Mix thoroughly. Adjust pH to 7.4–8.0. Filter sterilize. Aseptically distribute into flasks or bottles. Completely fill bottles with medium except for a pea-sized air bubble.

Use: For the isolation and cultivation of *Ectothiorhodospira halophila.*

Ectothiorhodospira Medium

Composition per liter:

NaCl	180.0g
Na_2SO_4	20.0g
$NaHCO_3$	14.0g
Na_2CO_3	6.0g
Sodium succinate	1.0g
NH_4Cl	0.8g
KH_2PO_4	0.5g
$MgCl_2 \cdot 6H_2O$	0.1g
$CaCl_2 \cdot 2H_2O$	0.05g
Feeding solution	10.0mL
Trace elements solution SLA	1.0mL
Vitamin solution VA	1.0mL

pH 8.5–8.7 at 25°C

Feeding Solution:

Composition per 100.0mL:

NaCl	10.0g
$NaHCO_3$	10.0g
$Na_2S \cdot 9H_2O$	5.0g

Preparation of Feeding Solution: Add components to distilled/deionized water and bring volume to 100.0mL. Mix thoroughly. Filter sterilize.

Trace Elements Solution SLA:

Composition per liter:

$FeCl_2 \cdot 4H_2O$	1.8g
H_3BO_3	0.5g
$CoCl_2 \cdot 6H_2O$	0.25g
$ZnCl_2$	0.1g
$MnCl_2 \cdot 4H_2O$	0.07g
$NaMoO_4 \cdot 2H_2O$	0.03g
$CuCl_2 \cdot 2H_2O$	0.01g
Na_2SeO_3	0.01g
$NiCl_2 \cdot 6H_2O$	0.01g

Preparation of Trace Elements Solution SLA: Add components to distilled/deionized water and bring volume to 1.0L. Mix thoroughly. Adjust pH to 3.0 with 2*N* HCl.

Vitamin Solution VA:

Composition per liter:

Nicotinamide	0.04g
Thiamine dichloride	0.03g
p-Aminobenzoic acid	0.02g
Biotin	0.01g
Calcium pantothenate	0.01g
Pyridoxal chloride	0.01g
Vitamin B_{12}	5.0mg

Preparation of Vitamin Solution VA: Add components to distilled/deionized water and bring volume to 1.0L. Mix thoroughly.

Preparation of Medium: Add components, except trace elements solution SLA and feeding solution, to distilled/deionized water and bring volume to 999.0mL. Mix thoroughly. Filter sterilize. Aseptically add 1.0mL of sterile trace elements solution SLA. Mix thoroughly. Aseptically distribute into flasks or bottles. Completely fill bottles with medium except for a pea-sized air bubble. Prior to inoculation, aseptically remove a sufficient amount of medium to permit the addition of feeding medium. Add 1.0mL of feeding solution per each 100.0mL of medium.

Use: For the isolation and cultivation of *Ectothiorhodospira halochloris* and *Ectothiorhodospira halophila.*

Ectothiorhodospira Medium

Composition per liter:

NaCl	130.0g
Na_2SO_4	10.0g
Sodium acetate	2.0g
KH_2PO_4	0.8g
Sodium carbonate buffer, (1*M*, pH 9.0)	200.0mL
$MgCl_2 \cdot 6H_2O$ solution	10.0mL
$Na_2S \cdot 9H_2O$ solution	6.0mL
$CaCl_2 \cdot 2H_2O$ solution	5.0mL
NH_4Cl solution	4.0mL
SLA trace elements	1.0mL
VA vitamin solution	1.0mL

pH 9.0 ± 0.2 at 25°C

$MgCl_2 \cdot 6H_2O$ Solution:

Composition per 10.0mL:

$MgCl_2 \cdot 6H_2O$	0.1g

Preparation of $MgCl_2 \cdot 6H_2O$ Solution: Add $MgCl_2 \cdot 6H_2O$ to distilled/deionized water and bring volume to 10.0mL. Mix thoroughly. Filter sterilize.

$Na_2S \cdot 9H_2O$ Solution:

Composition per 10.0mL:

$Na_2S \cdot 9H_2O$	0.5g

Preparation of $Na_2S \cdot 9H_2O$ Solution: Add $Na_2S \cdot 9H_2O$ to distilled/deionized water and bring volume to 10.0mL. Mix thoroughly. Filter sterilize. Use freshly prepared solution.

$CaCl_2 \cdot 2H_2O$ Solution:

Composition per 10.0mL:

$CaCl_2 \cdot 2H_2O$	0.1g

Preparation of $CaCl_2 \cdot 2H_2O$ Solution: Add $CaCl_2 \cdot 2H_2O$ to distilled/deionized water and bring volume to 10.0mL. Mix thoroughly. Filter sterilize.

NH_4Cl Solution:

Composition per 10.0mL:

NH_4Cl	2.0g

Preparation of NH_4Cl Solution: Add NH_4Cl to distilled/deionized water and bring volume to 10.0mL. Mix thoroughly. Filter sterilize.

SLA Trace Elements:

Composition per liter:

$FeCl_2 \cdot 4H_2O$	1.8g
H_3BO_3	0.5g
$CoCl_2 \cdot 6H_2O$	0.25g
$ZnCl_2$	0.1g
$MnCl_2 \cdot 4H_2O$	0.07g
$Na_2MoO_4 \cdot 2H_2O$	0.03g
$CuCl_2 \cdot 2H_2O$	0.01g
$Na_2SeO_3 \cdot 5H_2O$	0.01g
$NiCl_2 \cdot 6H_2O$	0.01g

Preparation of SLA Trace Elements: Add components to distilled/deionized water and bring volume to 1.0L. Mix thoroughly. Adjust pH to 2–3.

VA Vitamin Solution:

Composition per 500.0mL:

Nicotinamide	0.175g
Thiamine·HCl	0.15g
p-Aminobenzoic acid	0.1g
Biotin	0.05g
Calcium pantothenate	0.05g
Pyridoxine·2HCl	0.05g
Cyanocobalamin	0.025g

Preparation of VA Vitamin Solution: Add components to distilled/deionized water and bring volume to 500.0mL. Mix thoroughly.

Preparation of Medium: Add components—except $MgCl_2·6H_2O$ solution, $Na_2S·9H_2O$ solution, $CaCl_2·2H_2O$ solution, and NH_4Cl solution—to distilled/deionized water and bring volume to 975.0mL. Mix thoroughly. Gently heat and bring to boiling. Autoclave for 15 min at 15 psi pressure–121°C. Cool to 25°C. Aseptically add 10.0mL of sterile $MgCl_2·6H_2O$ solution, 6.0mL of sterile $Na_2S·9H_2O$ solution, 5.0mL of sterile $CaCl_2·2H_2O$ solution, and 4.0mL of sterile NH_4Cl solution. Mix thoroughly. Aseptically distribute into culture bottles. Incubate for 2 days before inoculation.

Use: For the cultivation and maintenance of *Ectothiorhodospira abdelmalekii* and *Ectothiorhodospira halochloris*.

Ectothiorhodospira Medium, Modified

Composition per liter:

NaCl	30.0g
Na_2SO_4	10.0g
Sodium acetate	2.0g
KH_2PO_4	0.8g
Sodium carbonate buffer (1*M*, pH 9.0)	200.0mL
$MgCl_2·6H_2O$ solution	10.0mL
$Na_2S·9H_2O$ solution	6.0mL
$CaCl_2·2H_2O$ solution	5.0mL
NH_4Cl solution	4.0mL
SLA trace elements	1.0mL
VA vitamin solution	1.0mL

pH 9.0 ± 0.2 at 25°C

$MgCl_2·6H_2O$ Solution:

Composition per 10.0mL:

$MgCl_2·6H_2O$	0.1g

Preparation of $MgCl_2·6H_2O$ Solution: Add $MgCl_2·6H_2O$ to distilled/deionized water and bring volume to 10.0mL. Mix thoroughly. Filter sterilize.

$Na_2S·9H_2O$ Solution:

Composition per 10.0mL:

$Na_2S·9H_2O$	0.5g

Preparation of $Na_2S·9H_2O$ Solution: Add $Na_2S·9H_2O$ to distilled/deionized water and bring volume to 10.0mL. Mix thoroughly. Filter sterilize. Use freshly prepared solution.

$CaCl_2·2H_2O$ Solution:

Composition per 10.0mL:

$CaCl_2·2H_2O$	0.1g

Preparation of $CaCl_2·2H_2O$ Solution: Add $CaCl_2·2H_2O$ to distilled/deionized water and bring volume to 10.0mL. Mix thoroughly. Filter sterilize.

NH_4Cl Solution:

Composition per 10.0mL:

NH_4Cl	2.0g

Preparation of NH_4Cl Solution: Add NH_4Cl to distilled/deionized water and bring volume to 10.0mL. Mix thoroughly. Filter sterilize.

SLA Trace Elements:

Composition per liter:

$FeCl_2·4H_2O$	1.8g
H_3BO_3	0.5g
$CoCl_2·6H_2O$	0.25g
$ZnCl_2$	0.1g
$MnCl_2·4H_2O$	0.07g
$Na_2MoO_4·2H_2O$	0.03g
$CuCl_2·2H_2O$	0.01g
$Na_2SeO_3·5H_2O$	0.01g
$NiCl_2·6H_2O$	0.01g

Preparation of SLA Trace Elements: Add components to distilled/deionized water and bring volume to 1.0L. Mix thoroughly. Adjust pH to 2–3.

VA Vitamin Solution:

Composition per 500.0mL:

Nicotinamide	0.175g
Thiamine·HCl	0.15g
p-Aminobenzoic acid	0.1g
Biotin	0.05g
Calcium pantothenate	0.05g
Pyridoxine·2HCl	0.05g
Cyanocobalamin	0.025g

Preparation of VA Vitamin Solution: Add components to distilled/deionized water and bring volume to 500.0mL. Mix thoroughly.

Preparation of Medium: Add components—except $MgCl_2·6H_2O$ solution, $Na_2S·9H_2O$ solution, $CaCl_2·2H_2O$ solution, and NH_4Cl solution—to distilled/deionized water and bring volume to 975.0mL. Mix thoroughly. Gently heat and bring to boiling. Autoclave for 15 min at 15 psi pressure–121°C. Cool to 25°C. Aseptically add 10.0mL of sterile $MgCl_2·6H_2O$ solution, 6.0mL of sterile $Na_2S·9H_2O$ solution, 5.0mL of sterile $CaCl_2·2H_2O$ solution, and 4.0mL of sterile NH_4Cl solution. Mix thoroughly. Aseptically distribute into culture bottles. Incubate for 2 days before inoculation.

Use: For the cultivation and maintenance of *Ectothiorhodospira vacuolata.*

Ectothiorhodospira vacuolata Medium

Composition per 1002.0mL:

NaCl	30.0g
$NaHCO_3$	3.0g
KH_2PO_4	1.0g
Sodium malate	1.0g
$Na_2SO_3·5H_2O$	0.5g
NH_4Cl	0.5g
$MgCl_2·6H_2O$	0.2g
$CaCl_2·2H_2O$	0.1g
$Na_2S·9H_2O$ solution	10.0mL
Trace elements solution SLA	1.0mL
Vitamin solution VA	1.0mL

pH 8.7 ± 0.2 at 25°C

$Na_2S{\cdot}9H_2O$ Solution:
Composition per 10.0mL:
$Na_2S{\cdot}9H_2O$ 0.5g

Preparation of $Na_2S{\cdot}9H_2O$ Solution: Add $Na_2S{\cdot}9H_2O$ to distilled/deionized water and bring volume to 10.0mL. Mix thoroughly. Gas under 100% N_2. Autoclave for 15 min at 15 psi pressure–121°C.

Trace Elements Solution SLA:
Composition per liter:
$CuCl_2{\cdot}2H_2O$ 10.0g
$FeCl_2{\cdot}4H_2O$ 1.8g
H_3BO_3 .... 0.5g
$CoCl_2{\cdot}6H_2O$ 0.25g
$ZnCl_2$ 0.1g
$MnCl_2{\cdot}4H_2O$ 70.0mg
$Na_2MoO_4{\cdot}2H_2O$ 30.0mg
$Na_2SeO_3{\cdot}5H_2O$ 10.0mg
$NiCl_2{\cdot}6H_2O$ 10.0mg

Preparation of Trace Elements Solution SLA: Add components to distilled/deionized water and bring volume to 1.0L. Mix thoroughly. Adjust pH to 2.0–3.0.

Vitamin Solution VA:
Composition per 100.0mL:
Nicotinic acid amide 35.0mg
Thiamine dichloride 30.0mg
p-Aminobenzoic acid 20.0mg
Biotin 10.0mg
Calcium DL-pantothenate 10.0mg
Pyridoxal·HCl 10.0mg
Vitamin B_{12} 5.0mg

Preparation of Vitamin Solution VA: Add components to distilled/deionized water and bring volume to 100.0mL. Mix thoroughly. Filter sterilize.

Preparation of Medium: Add components, except $Na_2S{\cdot}9H_2O$ solution, to distilled/deionized water and bring volume to 1.0L. Mix thoroughly. Adjust pH to 8.7. Filter sterilize. Aseptically add 10.0mL of sterile $Na_2S{\cdot}9H_2O$ solution. Mix thoroughly. Distribute into sterile tubes or flasks.

Use: For the cultivation and maintenance of *Ectothiorhodospira vacuolata.*

Ectothiorhodosynus Medium (DSMZ Medium 1002)

Composition per liter:
NaCl 20.0g
Na-acetate 1.0g
K_2HPO_4 .... 0.5g
$Na_2S_2O_3$ 0.5g
$MgCl_2{\cdot}6H_2O$ 0.2g
Yeast extract 0.1g
Vitamin B_{12} 0.02mg
Ammonium chloride solution 10.0mL
Calcium chloride solution 10.0mL
Sodium carbonate solution 10.0mL
Sodium bicarbonate solution 10.0mL
$Na_2S{\cdot}9H_2O$ solution 10.0mL
Trace elements SL-6 solution 1.0mL
pH 9.3 ± 0.2 at 25°C

Trace Elements SL-6 Solution:
Composition per liter:
H_3BO_3 .... 0.3g
$CoCl_2{\cdot}6H_2O$ 0.2g
$ZnSO_4{\cdot}7H_2O$ 0.1g
$MnCl_2{\cdot}4H_2O$ 0.03g
$Na_2MoO_4{\cdot}H_2O$ 0.03g
$NiCl_2{\cdot}6H_2O$ 0.02g
$CuCl_2{\cdot}2H_2O$ 0.01g

Preparation of Trace Elements SL-6 Solution: Add components to distilled/deionized water and bring volume to 1.0L. Mix thoroughly. Adjust pH to 3.4.

$Na_2S{\cdot}9H_2O$ Solution:
Composition per 10.0mL:
$Na_2S{\cdot}9H_2O$ 0.24g

Preparation of $Na_2S{\cdot}9H_2O$ Solution: Add $Na_2S{\cdot}9H_2O$ to distilled/deionized water and bring volume to 10.0mL. Mix thoroughly. Autoclave under 100% N_2 for 15 min at 15 psi pressure–121°C. Cool to room temperature.

Sodium Bicarbonate Solution:
Composition per 10.0mL:
$NaHCO_3$ 5.0g

Preparation of Sodium Bicarbonate Solution: Add components to distilled/deionized water and bring volume to 10.0mL. Mix thoroughly. Filter sterilize.

Sodium Carbonate Solution:
Composition per 10.0mL:
Na_2CO_3 .... 5.0g

Preparation of Sodium Carbonate Solution: Add components to distilled/deionized water and bring volume to 10.0mL. Mix thoroughly. Autoclave for 15 min at 15 psi pressure–121°C.

Calcium Chloride Solution:
Composition per 10.0mL:
$CaCl_2{\cdot}2H_2O$ 0.1g

Preparation of Calcium Chloride Solution: Add components to distilled/deionized water and bring volume to 1.0L. Mix thoroughly. Autoclave for 15 min at 15 psi pressure–121°C.

Ammonium Chloride Solution:
Composition per 10.0mL:
NH_4Cl 0.5g

Preparation of Ammonium Chloride Solution: Add components to distilled/deionized water and bring volume to 10.0mL. Mix thoroughly. Autoclave for 15 min at 15 psi pressure–121°C.

Preparation of Medium: Add components, except ammonium chloride, calcium chloride, carbonate, bicarbonate, and sulfide solutions, to distilled/deionized water and bring volume to 950.0mL. Mix thoroughly. Adjust pH to 9.3. Gently heat while stirring and bring to boiling. Distribute into tubes or flasks. Autoclave for 15 min at 15 psi pressure–121°C. Aseptically add the ammonium chloride, calcium chloride, carbonate, bicarbonate, and sulfide solutions. Mix thoroughly. Aseptically distribute into sterile tubes or flasks.

Use: For the cultivation of *Ectothiorhodosynus* spp.

Edelstein BMPA-α Medium
See: **BMPA-α Medium**

Edwards Medium HiVeg Base, Modified

Composition per liter:

Agar	15.0g
Plant extract	10.0g
Plant peptone	10.0g
NaCl	5.0g
Esculin	1.0g
Thallous sulfate	0.33g
Crystal Violet	1.3mg
Sterile bovine or sheep blood	50.0mL

pH 7.4 ± 0.2 at 25°C

Source: This medium, without blood, is available as a premixed powder from HiMedia.

Preparation of Medium: Add components, except blood, to distilled/deionized water and bring volume to 950.0mL. Mix thoroughly. Gently heat and bring to boiling. Autoclave for 20 min at 10 psi pressure–115°C. Cool to 45°–50°C. Aseptically add 50.0mL of sterile bovine blood or sheep blood. Mix thoroughly. Pour into sterile Petri dishes.

Use: For the selective isolation and cultivation of *Streptococcus agalactiae* and other streptococci involved in bovine mastitis.

Edwards and Bruner Semisolid Medium

Composition per liter:

Gelatin	80.0g
Peptic digest of animal tissue	10.0g
NaCl	5.0g
Agar	4.0g
Beef extract	3.0g

pH 6.9 ± 0.2 at 25°C

Preparation of Medium: Add components to distilled/deionized water and bring volume to 1.0L. Mix thoroughly. Gently heat and bring to boiling. Distribute into tubes or flasks. Autoclave for 15 min at 15 psi pressure–121°C. Pour into sterile Petri dishes or leave in tubes.

Use: For the detection of motility and separation of H and O phases of enteric bacilli.

Edwards Medium, Modified

Composition per liter:

Agar	15.0g
Beef extract	10.0g
Peptone	10.0g
NaCl	5.0g
Esculin	1.0g
Tl_2SO_4	0.33g
Crystal Violet	1.3mg
Bovine or sheep blood	50.0mL

pH 7.4 ± 0.2 at 25°C

Source: This medium is available as a premixed powder from Oxoid.

Preparation of Medium: Add components, except blood, to distilled/deionized water and bring volume to 950.0mL. Mix thoroughly. Gently heat and bring to boiling. Autoclave for 20 min at 10 psi pressure–115°C. Cool to 45°–50°C. Aseptically add 50.0mL of sterile bovine blood or sheep blood. Mix thoroughly. Pour into sterile Petri dishes.

Use: For the selective isolation and cultivation of *Streptococcus agalactiae* and other streptococci involved in bovine mastitis.

Edwards Medium, Modified

Composition per 1060.0mL:

Agar	15.0g
Lab Lemco powder	10.0g
Peptone	10.0g
NaCl	5.0g
Esculin	1.0g
Tl_2SO_4	0.33g
Crystal Violet	0.0013g
Blood, bovine or sheep	60.0mL

pH 7.4 ± 0.2 at 25°C

Source: This medium is available as a premixed powder from Oxoid Unipath.

Caution: Thallous sulfate is toxic.

Preparation of Medium: Add components, except blood, to distilled/deionized water and bring volume to 1.0L. Mix thoroughly. Gently heat while stirring and bring to boiling. Autoclave for 20 min at 15 psi pressure–121°C. Cool to 50°C. Aeptically add 60.0mL sterile bovine or sheep blood. Mix thoroughly. Pour into sterile Petri dishes.

Use: For the rapid isolation of *Streptococcus agalactiae* and other streptococci involved in bovine mastitis. Esculin differentiates the negative *Streptococcus agalactiae*, which form blue colonies, from esculin-positive Group D streptococci, which form black colonies.

EE Broth (Enterobacteriaceae Enrichment Broth)

Composition per liter:

Ox bile	20.0g
Peptone	10.0g
Na_2HPO_4	6.45g
Glucose	5.0g
KH_2PO_4	2.0g
Brilliant Green	0.0135g

pH 7.2 ± 0.2 at 25°C

Source: This medium is available as a premixed powder from Oxoid.

Preparation of Medium: Add components to distilled/deionized water and bring volume to 1.0L. Mix thoroughly. Distribute into flasks in 100.0mL volumes. Gently heat at 100°C for 30 min. Do not autoclave. Cool rapidly to 25°C.

Use: For the cultivation and enrichment of members of the Enterobacteriaceae in the examination of foods and animal feed. Used in conjunction with tryptone soy broth. Bacteria belonging to the Enterobacteriaceae turn this medium turbid and yellow-green.

EE Broth, HiVeg

Composition per liter:

Plant peptone	25.0g
Na_2HPO_4	6.45g
Glucose	5.0g
Synthetic detergent No. II	5.0g
KH_2PO_4	2.0g
Brilliant Green	13.5mg

pH 7.2 ± 0.2 at 25°C

Source: This medium is available as a premixed powder from HiMedia.

Preparation of Medium: Add components to distilled/deionized water and bring volume to 1.0L. Mix thoroughly. Distribute into flasks in 100.0mL volumes. Gently heat at 100°C for 30 min. Do not autoclave. Cool rapidly to 25°C.

Use: For the cultivation and enrichment of members of the Enterobacteriaceae in the examination of foods and animal feed. Used in conjunction with tryptone soy broth. Bacteria belonging to the Enterobacteriaceae turn this medium turbid and yellow-green.

EE Broth, Mossel (Enterobacteriaceae Enrichment Broth, Mossel)

Composition per liter:

Enzymatic hydrolysate of protein	10.0g
Na_2HPO_4	8.0g
Glucose	5.0g
KH_2PO_4	2.0g
Oxgall	0.1g
Brilliant Green	0.0135g

pH 7.2 ± 0.2 at 25°C

Source: This medium is available as a premixed powder from BD Diagnostic Systems.

Preparation of Medium: Add components to distilled/deionized water and bring volume to 1.0L. Mix thoroughly. Distribute into flasks in 120.0mL volumes. Gently heat at 100°C for 30 min. Do not autoclave. Cool rapidly to 25°C.

Use: For the cultivation and enrichment of members of the Enterobacteriaceae in the examination of foods and animal feed. Used in conjunction with tryptone soy broth. Bacteria belonging to the Enterobacteriaceae turn this medium turbid and yellow-green.

EE HiVeg Broth, Mossel

Composition per liter:

Oxbile	20.0g
Peptic digest of animal tissue	10.0g
Na_2HPO_4, dihydrate	6.45g
Glucose	5.0g
KH_2PO_4	2.0g
Brilliant Green	15.0mg

pH 7.2 ± 0.2 at 25°C

Source: This medium is available as a premixed powder from HiMedia.

Preparation of Medium: Add components to distilled/deionized water and bring volume to 1.0L. Mix thoroughly. Distribute into flasks in 100.0mL volumes. Gently heat at 100°C for 30 min. Do not autoclave. Cool rapidly to 25°C.

Use: For the cultivation and enrichment of members of the Enterobacteriaceae in the examination of foods and animal feed.

EE HiVeg Broth, Modified

Composition per liter:

Plant peptone No. 2	25.0g
Na_2HPO_4, dihydrate	8.0g
Glucose monohydrate	5.0g
Synthetic detergent No. II	5.0g
KH_2PO_4	2.0g
Brilliant Green	15.0mg

pH 7.2 ± 0.2 at 25°C

Source: This medium is available as a premixed powder from HiMedia.

Preparation of Medium: Add components to distilled/deionized water and bring volume to 1.0L. Mix thoroughly. Distribute into flasks in 100.0mL volumes. Gently heat at 100°C for 30 min. Do not autoclave. Cool rapidly to 25°C.

Use: For the cultivation and enrichment of members of the Enterobacteriaceae in the examination of foods and animal feed. Used in conjunction with tryptone soy broth. Bacteria belonging to the Enterobacteriaceae turn this medium turbid and yellow-green.

EG Agar

Composition per 1055.0mL:

Agar	15.0g
Proteose peptone No. 3	10.0g
Yeast extract	5.0g
Na_2HPO_4	4.0g
Lab Lemco meat extract	2.4g
Glucose	1.5g
L-Cysteine·HCl·H_2O	0.5g
Soluble starch	0.5g
L-Cystine	0.2g
Horse blood	50.0mL
HCl (1*N* solution)	50.0mL
10% Silicone SH 5535	5.0mL

pH 7.6–7.8 at 25°C

Source: Lab Lemco meat extract is available from Oxoid Unipath. 10% Silicone SH 5535 is available from Toray. Proteose peptone No. 3 is available from BD Diagnostic Systems.

Preparation of Medium: Add L-cystine to 50.0mL of 1*N* HCl. Mix thoroughly. Add remaining components and bring volume to 950.0mL. Mix thoroughly. Autoclave for 15 min at 15 psi pressure–121°C. Cool to 50°–55°C. Aseptically add 50.0mL of horse blood. Mix thoroughly. Pour into sterile Petri dishes or distribute into sterile tubes.

Use: For the cultivation of *Actinomyces naeslundii, Actinomyces viscosus, Bacteroides* species, *Bifidobacterium* species, *Campylobacter* species, *Clostridium* species, *Eubacterium* species, *Fusobacterium necrophorum, Fusobacterium nucleatum, Fusobacterium pseudonecrophorum, Lactobacillus* species, *Megasphaera cerevisiae, Megasphaera elsdenii, Mitsuokella multiacida, Peptostreptococcus* species, *Prevotella* species, *Propionibacterium* species, *Rikenella microfusus, Selenomonas* species, and *Wolinella recta.*

EG Medium

Pancreatic digest of casein	2.0g
Yeast extract	2.0g
Beef extract	1.0g
Sodium acetate	1.0g
$CaCl_2$	0.01g

Preparation of Medium: Add components to distilled/deionized water and bring volume to 1.0L. Mix thoroughly. Distribute into tubes or flasks. Autoclave for 20 min at 15 psi pressure–121°C.

Use: For the cultivation of *Khawkinea quartana.*

EG Sodium Chloride Medium No. 7 (Ethylene Glycol NaCl Medium No. 7)

Composition per liter:

NaCl	70.0g
Agar	15.0g
K_2HPO_4	7.5g

KH_2PO_4	1.0g
$(NH_4)_2SO_4$	0.8g
$MgSO_4 \cdot 7H_2O$	0.1g
$FeSO_4 \cdot 7H_2O$	0.01g
Ethylene glycol	10.0mL
Salt solution	1.0mL

Salt Solution:

Composition per liter:

$CaCl_2 \cdot 2H_2O$	6.0g
$ZnSO_4 \cdot 7H_2O$	4.4g
$MnSO_4 \cdot H_2O$	3.0g
$(NH_4)_6Mo_7O_{24} \cdot 4H_2O$	1.82g
$CuCl_2 \cdot 2H_2O$	0.2g

Preparation of Salt Solution: Add components to distilled/deionized water and bring volume to 1.0L. Mix thoroughly.

Preparation of Medium: Add components to distilled/deionized water and bring volume to 1.0L. Mix thoroughly. Gently heat and bring to boiling. Distribute into tubes or flasks. Autoclave for 15 min at 15 psi pressure–121°C. Pour into sterile Petri dishes or leave in tubes.

Use: For the cultivation of ATCC strain 27042.

Egg Meat Medium

Composition per liter:

Beef muscles	454.0g
Egg white	6.0g
$CaCO_3$	5.0g

pH 7.2 ± 0.2 at 25°C

Source: This medium is available from HiMedia.

Preparation of Medium: Add components to distilled/deionized water and bring volume to 1.0L. Mix thoroughly. Gently heat and bring to boiling. Distribute into tubes or flasks. Autoclave for 15 min at 15 psi pressure–121°C.

Use: For the determination of proteolytic activity of anaerobic microorganisms. For the maintenance of anaerobic bacteria.

Egg Tellurite Glycine Pyruvate Agar

See: **ETGPA**

Egg Yolk Agar

Composition per liter:

Proteose peptone No. 2	40.0g
Agar	25.0g
Na_2HPO_4	5.0g
Glucose	2.0g
NaCl	2.0g
KH_2PO_4	1.0g
$MgSO_4 \cdot 7H_2O$	0.1g
Egg yolk emulsion	100.0mL
Hemin solution	1.0mL

pH 7.6 ± 0.2 at 25°C

Hemin Solution:

Composition per 100.0mL:

Hemin	0.5g
NaOH (1*N* solution)	20.0mL

Preparation of Hemin Solution: Add hemin to 20.0mL of 1*N* NaOH solution. Mix thoroughly. Bring volume to 100.0mL with distilled/deionized water.

Egg Yolk Emulsion:

Composition:

Chicken egg yolks	11
Whole chicken egg	1

Preparation of Egg Yolk Emulsion: Soak eggs with 1:100 dilution of saturated mercuric chloride solution for 1 min. Crack eggs and separate yolks from whites. Mix egg yolks with 1 chicken egg.

Preparation of Medium: Add components, except egg yolk emulsion, to distilled/deionized water and bring volume to 900.0mL. Mix thoroughly. Gently heat and bring to boiling. Autoclave for 15 min at 15 psi pressure–121°C. Cool to 45°–50°C. Aseptically add sterile egg yolk emulsion. Mix thoroughly. Pour into sterile Petri dishes.

Use: For the isolation, cultivation, and differentiation of *Clostridium* species and some other anaerobic bacteria.

Egg Yolk Agar Base, HiVeg with Egg Yolk Emulsion

Composition per liter:

Plant peptone No. 3	40.0g
Agar	25.0g
Na_2HPO_4	5.0g
Glucose	2.0g
NaCl	2.0g
KH_2PO_4	1.0g
$MgSO_4$	0.1g
$Fe_4(P_2O_7)_3 \cdot H_2O$	5.0mg
Egg yolk emulsion	100.0mL

pH 7.6 ± 0.2 at 25°C

Source: This medium, without egg yolk emulsion, is available as a premixed powder from HiMedia.

Egg Yolk Emulsion:

Composition per liter:

Egg yolks	30.0mL
NaCl, 0.9% solution	70.0mL

Preparation of Egg Yolk Emulsion: Soak eggs with 1:100 dilution of saturated mercuric chloride solution for 1 min. Crack 11 eggs and separate yolks from whites. Mix egg yolks. Measure 30.0mL of egg yolk emulsion and add to 70.0mL of 0.9% sterile NaCl solution. Mix thoroughly. Warm to 45°–50°C.

Preparation of Medium: Add components, except egg yolk emulsion, to distilled/deionized water and bring volume to 900.0mL. Mix thoroughly. Gently heat and bring to boiling. Autoclave for 15 min at 15 psi pressure–121°C. Cool to 45°–50°C. Aseptically add sterile egg yolk emulsion. Mix thoroughly. Pour into sterile Petri dishes.

Use: For the isolation, cultivation, and differentiation of *Clostridium* species and some other anaerobic bacteria.

Egg Yolk Agar, Lombard-Dowell

See: **Lombard-Dowell Egg Yolk Agar**

Egg Yolk Agar, Modified

Composition per liter:

Agar	20.0g
Pancreatic digest of casein	15.0g
Vitamin K_1	10.0g
NaCl	5.0g
Papaic digest of soybean meal	5.0g
Yeast extract	5.0g
L-Cystine	0.4g

Hemin 5.0mg
Egg yolk emulsion 100.0mL

Source: This medium is available as a prepared medium from BD Diagnostic Systems.

Egg Yolk Emulsion:
Composition:
Chicken egg yolks 11
Whole chicken egg 1

Preparation of Egg Yolk Emulsion: Soak eggs with 1:100 dilution of saturated mercuric chloride solution for 1 min. Crack eggs and separate yolks from whites. Mix egg yolks with 1 chicken egg.

Preparation of Medium: Add components, except egg yolk emulsion, to distilled/deionized water and bring volume to 900.0mL. Mix thoroughly. Gently heat and bring to boiling. Autoclave for 15 min at 15 psi pressure–121°C. Cool to 45°–50°C. Aseptically add sterile egg yolk emulsion. Mix thoroughly. Pour into sterile Petri dishes.

Use: For the isolation, cultivation, and differentiation of *Clostridium* species and some other anaerobic bacteria.

Egg Yolk Agar with Neomycin
See: **Lombard-Dowell Neomycin Agar**

Egg Yolk Emulsion

Composition per 100.0mL:
Sterile saline 70.0mL
Egg yolk 30.0mL

Source: Sterile egg yolk emulsion is available from Fluka, Sigma-Aldrich.

Preparation of Medium: Use fresh eggs, less than 1 week old. Scrub the shells with soap. Let stand in a soap solution for 30 min. Rinse in running water. Soak eggs in 70% ethanol for 15 min. or soak eggs with 1:100 dilution of saturated mercuric chloride solution for 1 min. Crack eggs and separate yolks from whites, placing egg yolks into a sterile container. Use enough eggs to produce at least 30.0mL egg yolk. Homogenize by shaking. Add 0.9g NaCl to distilled/deionized water and bring volume to 100.0mL. Sterilze the saline solution by filtration or by autoclaving for 15 min at 15 psi pressure–121°C. If autoclaving is used, cool to 25°C. Aseptically add 30.0mL homogenized egg yolks to 70.0mL of sterile saline solution. Mix thoroughly.

Use: Sterile stabilized emulsion of egg yolk is recommended for use in various culture media.

Egg Yolk Emulsion

Composition per 100.0mL:
NaCl 0.45g
Egg yolk 50.0mL

Preparation of Medium: Use fresh eggs, less than 1 week old. Scrub the shells with soap. Let stand in a soap solution for 30 min. Rinse in running water. Soak eggs in 70% ethanol for 15 min. or soak eggs with 1:100 dilution of saturated mercuric chloride solution for 1 min. Crack eggs and separate yolks from whites, placing egg yolks into a sterile container. Use enough eggs to produce at least 50.0mL egg yolk. Homogenize by shaking. Add 0.45g NaCl to distilled/deionized water and bring volume to 50.0mL. Sterilize the saline solution by filtration or by autoclaving for 15 min at 15 psi pressure–121°C. If autoclaving is used, cool to 25°C. Aseptically add 50.0mL homogenized egg yolks to 50.0mL of the sterile NaCl solution. Mix thoroughly.

Use: Sterile stabilized emulsion of egg yolk is recommended for use in various culture media.

Egg Yolk Emulsion, 50% (BAM M51)

Composition per 100.0mL:
Chicken egg yolks .... variable
NaCl (0.85% solution) 40.0mL

Preparation of Egg Yolk Emulsion: Wash fresh eggs with a stiff brush and drain. Soak eggs in 70% ethanol for 1 h. Crack eggs aseptically and separate yolks from whites. Remove egg yolks with a sterile syringe or a wide-mouth pipet. Place 50.0mL of egg yolks into a sterile container. Add 50.0mL sterile 0.85% saline.

Use: For use in media requiring egg yolk emulsion.

Egg Yolk Tellurite Emulsion 20%

Composition per 100.0mL:
NaCl 0.425g
K_2TeO_3 .... 0.21g
Egg yolk 20.0mL

Preparation of Medium: Use fresh eggs, less than 1 week old. Scrub the shells with soap. Let stand in a soap solution for 30 min. Rinse in running water. Soak eggs in 70% ethanol for 15 min or soak eggs with 1:100 dilution of saturated mercuric chloride solution for 1 min. Crack eggs and separate yolks from whites, placing egg yolks into a sterile container. Use enough eggs to produce at least 20.0mL egg yolk. Homogenize by shaking. Add 0.45g NaCl and 0.21g K_2TeO_3 to distilled/deionized water and bring volume to 80.0mL. Sterilize the saline-tellurite solution by filtration or by autoclaving for 15 min at 15 psi pressure–121°C. If autoclaving is used, cool to 25°C. Aseptically add 20.0mL homogenized egg yolks to 80.0mL of the sterile saline-tellurite solution. Mix thoroughly.

Use: For use in various culture media. It may be added directly to nutrient media for the identification of *Clostridium, Bacillus,* and *Staphylococcus* species by their lipase activity.

EGGC (DSMZ Medium 1191)

Composition per liter:
Agar 15.0g
Glucose 0.6g
Na-acetate 0.3g
Casamino acids 0.3g
Salt solution 10.0mL
Vitamin solution 5.0mL
Phosphate soltution 2.0mL

pH 7.2 ± 0.2 at 25°C

Salt Solution:
Composition per liter:
$(NH_4)_2SO_4$ 1.0g
KCl 0.5g
$MgSO_4 \cdot 7H_2O$ 0.5g
$CaCl_2 \cdot 2H_2O$ 0.2g
$CaCO_3$ 0.2g
$FeCl_3 \cdot 6H_2O$ 0.005g

Preparation of Salt Solution: Add components to distilled/deionized water and bring volume to 1.0L. Mix thoroughly.

Vitamin Solution:

Composition per100.0mL:

Thiamine-HCl·$2H_2O$	50.0mg
Nicotinic acid	50.0mg
Pyridoxine-HCl	50.0mg
D-Ca-pantothenate	50.0mg
Riboflavin	10.0mg
Vitamin B_{12}	1.0mg
Folic acid	0.2mg
Biotin	0.1mg

Preparation of Vitamin Solution: Add components to distilled/deionized water and bring volume to 100.0mL. Mix thoroughly. Sparge with 80% H_2 + 20% CO_2. Filter sterilize.

Phosphate Solution:

Composition per 20.0mL:

Na_2HPO_4	2.48g
NaH_2PO_4	0.308g

Preparation of Phosphate Solution: Add components to distilled/deionized water and bring volume to 20.0mL. Mix thoroughly. Filter sterilize.

Preparation of Medium: Add components, except phosphate and vitamin solutions, to distilled/deionized water and bring volume to 980.0mL. Mix thoroughly. Adjust pH to 7.2. Gently heat while stirring and bring to boiling. Distribute into tubes or flasks. Autoclave for 15 min at 15 psi pressure–121°C. Cool to 50°C. Aseptically add the vitamin and phosphate solutions. Mix thoroughly. Pour into Petri dishes or aseptically distribute into sterile tubes.

Use: For the cultivation of *Thiothrix* spp.

EIA Substrate

Composition per liter:

Agar	15.0g
Esculin	1.0g
Ferric citrate	0.5g

pH 7.1 ± 0.1 at 25°C

Source: This medium is available as a premixed powder from BD Diagnostic Systems.

Preparation of Medium: Add components to distilled/deionized water and bring volume to 1.0L. Mix thoroughly. Gently heat and bring to boiling. Adjust pH to 7.1. Distribute into tubes or flasks. Autoclave for 15 min at 15 psi pressure–121°C. Pour into sterile Petri dishes.

Use: For the cultivation and enumeration of marine enterococci by the membrane filter method.

Eijkman Lactose HiVeg Broth

Composition per liter:

Plant hydrolysate No. 1	15.0g
NaCl	5.0g
K_2HPO_4	4.0g
Lactose	3.0g
KH_2PO_4	1.5g

pH 6.8 ± 0.1 at 25°C

Source: This medium is available as a premixed powder from HiMedia.

Preparation of Medium: Add components to distilled/deionized water and bring volume to 1.0L. Mix thoroughly. Distribute into test tubes that contain an inverted Durham tube. Autoclave for 15 min at 15 psi pressure–121°C.

Use: For the cultivation and differentiation of *Escherichia coli* from other coliform organisms based on their ability to ferment lactose and produce gas.

Eijkman Lactose Medium

Composition per liter:

Pancreatic digest of casein	15.0g
K_2HPO_4	10.0g
KH_2PO_4	4.0g
Lactose	3.0g
NaCl	2.5g

pH 6.8 ± 0.1 at 25°C

Preparation of Medium: Add components to distilled/deionized water and bring volume to 1.0L. Mix thoroughly. Distribute into test tubes that contain an inverted Durham tube. Autoclave for 15 min at 15 psi pressure–121°C.

Use: For the cultivation and differentiation of *Escherichia coli* from other coliform organisms based on their ability to ferment lactose and produce gas.

Eijkman Lactose Medium

Composition per liter:

Tryptose	15.0g
NaCl	5.0g
K_2HPO_4	4.0g
Lactose	3.0g
KH_2PO_4	1.5g

pH 6.8 ± 0.1 at 25°C

Preparation of Medium: Add components to distilled/deionized water and bring volume to 1.0L. Mix thoroughly. Distribute into test tubes that contain an inverted Durham tube. Autoclave for 15 min at 15 psi pressure–121°C.

Use: For the cultivation and differentiation of *Escherichia coli* from other coliform organisms based on their ability to ferment lactose and produce gas.

Ekho Lake Strains Medium (DSMZ Medium 621a)

Composition per liter:

Agar	15.0g
Peptone	0.25g
Yeast extract	0.25g
Artificial sea water	250.0mL
Hutner's basal salts solution	20.0mL
Glucose solution	10.0mL
Vitamin solution	5.0mL

pH 7.3 ± 0.2 at 25°C

Hutner's Basal Salts Solution:

Composition per liter:

$MgSO_4 \cdot 7H_2O$	29.7g
Nitrilotriacetic acid	10.0g
$CaCl_2 \cdot 2H_2O$	3.335g
$FeSO_4 \cdot 7H_2O$	99.0mg
$(NH_4)_6MoO_7O_{24} \cdot 4H_2O$	9.25mg
"Metals 44"	50.0mL

"Metals 44":
Composition per 100.0mL:

$ZnSO_4 \cdot 7H_2O$	1.095g
$FeSO_4 \cdot 7H_2O$	0.5g
Sodium EDTA	0.25g
$MnSO_4 \cdot H2O$	0.154g
$CuSO_4 \cdot 5H_2O$	39.2mg
$Co(NO_3)_2 \cdot 6H_2O$	24.8mg
$Na_2B_4O_7 \cdot 10H_2O$	17.7mg

Preparation of "Metals 44": Add sodium EDTA to distilled/deionized water and bring volume to 90.0mL. Mix thoroughly. Add a few drops of concentrated H_2SO_4 to retard precipitation of heavy metal ions. Add remaining components. Mix thoroughly. Bring volume to 100.0mL with distilled/deionized water.

Preparation of Hutner's Basal Salts Solution: Add nitrilotriacetic acid to 500.0mL of distilled/deionized water. Adjust pH to 6.5 with KOH. Add remaining components. Add distilled/deionized water to 1.0L. Adjust pH to 6.8.

Artificial Sea Water:
Composition per liter:

NaCl	23.477g
$MgCl_2 \cdot 6H_2O$	4.981g
Na_2SO_4	3.917g
$CaCl_2$	1.12g
KCl	664.0mg
$NaHCO_3$	192.0mg
H_3BO_3	26.0mg
$SrCl_2$	24.0mg
KBr	6.0mg
NaF	3.0mg

Preparation of Artificial Sea Water: Add components to distilled/deionized water and bring volume to 1.0L. Mix thoroughly. Filter sterilize.

Vitamin Solution:
Composition per liter:

Pyridoxine-HCl	20.0mg
Riboflavin	10.0mg
Nicotinamide	10.0mg
Thiamine-HCl·$2H_2O$	10.0mg
Ca-pantothenate	10.0mg
p-Aminobenzoic acid	10.0mg
Biotin	4.0mg
Folic acid	4.0mg
Vitamin B_{12}	0.2mg

Preparation of Vitamin Solution: Add components to distilled/deionized water and bring volume to 1.0L. Mix thoroughly. Filter sterilize. Store in the dark at 5°C.

Glucose Solution:
Composition per 10.0mL:

Glucose	0.25g

Preparation of Glucose Solution: Add glucose to distilled/deionized water and bring volume to 10.0mL. Mix thoroughly. Filter sterilize.

Preparation of Medium: Add components, except artificial sea water, glucose solution, and vitamin solution, to distilled/deionized water and bring volume to 735.0mL. Mix thoroughly. Adjust pH to 7.5. Autoclave for 20 min at 15 psi pressure–121°C. Cool to 60°C. Warm artificial sea water to 55°C. Aseptically add 250.0mL warm artificial sea water. Mix thoroughly. Adjust pH to 7.3. Aseptically add 10.0mL glucose solution and 5.0 mL vitamin solution. Mix thoroughly. Aseptically distribute into sterile tubes or flasks.

Use: For the cultivation of *Nocardioides aquaticus, Antarctobacter heliothermus, Sulfitobacter brevis, Roseovarius tolerans, Staleya guttiformis, Roseovarius tolerans, Friedmanniella lacustris,* and *Nesterenkonia lacusekhoensis.*

Elek Agar

See: **K-L Virulence Agar**

Elliker Agar

Composition per liter:

Pancreatic digest of casein	20.0g
Agar	15.0g
Glucose	5.0g
Lactose	5.0g
Sucrose	5.0g
Yeast extract	5.0g
NaCl	4.0g
Gelatin	2.5g
Sodium acetate	1.5g
Ascorbic acid	0.5g

pH 6.8 ± 0.2 at 25°C

Preparation of Medium: Add components to distilled/deionized water and bring volume to 1.0L. Mix thoroughly. Gently heat and bring to boiling. Distribute into tubes or flasks. Autoclave for 15 min at 15 psi pressure–121°C. Pour into sterile Petri dishes or leave in tubes.

Use: For the isolation and cultivation of lactic streptococci.

Elliker Agar

Composition per liter:

Pancreatic digest of casein	20.0g
Agar	15.0g
Yeast extract	10.0g
Gelatin	4.0g
Glucose	3.0g
Ascorbic acid	2.5g
Lactose	2.5g
NaCl	2.5g
Sodium acetate	2.5g
Sucrose	2.5g

pH 6.8 ± 0.1 at 25°C

Preparation of Medium: Add components to distilled/deionized water and bring volume to 1.0L. Mix thoroughly. Gently heat and bring to boiling. Autoclave for 15 min at 15 psi pressure–121°C. Pour into sterile Petri dishes or distribute into sterile tubes.

Use: For the cultivation of streptococci and lactobacilli of importance in the dairy industry.

Elliker Broth

Composition per liter:

Pancreatic digest of casein	20.0g
Yeast extract	10.0g
Gelatin	4.0g
Glucose	3.0g
Ascorbic acid	2.5g
Lactose	2.5g
NaCl	2.5g

Sodium acetate 2.5g
Sucrose 2.5g
pH 6.8 ± 0.1 at 25°C

Preparation of Medium: Add components to distilled/deionized water and bring volume to 1.0L. Mix thoroughly. Distribute into test tubes that contain an inverted Durham tube. Autoclave for 15 min at 15 psi pressure–121°C.

Use: For the cultivation of streptococci and lactobacilli of importance in the dairy industry.

Elliker Broth

Composition per liter:

Pancreatic digest of casein 20.0g
Glucose 5.0g
Lactose 5.0g
Sucrose 5.0g
Yeast extract 5.0g
NaCl 4.0g
Gelatin 2.5g
Sodium acetate 1.5g
Ascorbic acid 0.5g
pH 6.8 ± 0.2 at 25°C

Source: This medium is available as a premixed powder from BD Diagnostic Systems.

Preparation of Medium: Add components to distilled/deionized water and bring volume to 1.0L. Mix thoroughly. Distribute into test tubes that contain an inverted Durham tube. Autoclave for 15 min at 15 psi pressure–121°C.

Use: For the cultivation of streptococci and lactobacilli of importance in the dairy industry.

Elliker Broth (Lactobacilli Broth)

Composition per liter:

Casein enzymatic hydrolysate 22.5g
Glucose 5.0g
Lactose 5.0g
Saccharose 5.0g
Yeast extract 5.0g
NaCl 4.0g
Sodium acetate 1.5g
Ascorbic acid 0.5g
pH 6.8 ± 0.2 at 25°C

Source: This medium is available as a premixed powder from HiMedia.

Preparation of Medium: Add components to distilled/deionized water and bring volume to 1.0L. Mix thoroughly. Distribute into test tubes that contain an inverted Durham tube. Autoclave for 15 min at 15 psi pressure–121°C.

Use: For the cultivation of streptococci and lactobacilli of importance in the dairy industry.

Elliker HiVeg Broth (Lactobacilli HiVeg Broth)

Composition per liter:

Plant hydrolysate 22.5g
Glucose 5.0g
Lactose 5.0g
Saccharose 5.0g
Yeast extract 5.0g
NaCl 4.0g
Sodium acetate 1.5g
Ascorbic acid 0.5g
pH 6.8 ± 0.2 at 25°C

Source: This medium is available as a premixed powder from HiMedia.

Preparation of Medium: Add components to distilled/deionized water and bring volume to 1.0L. Mix thoroughly. Distribute into test tubes that contain an inverted Durham tube. Autoclave for 15 min at 15 psi pressure–121°C.

Use: For the cultivation of streptococci and lactobacilli of importance in the dairy industry.

Elliker Lactose Broth (ATCC Medium 2333)

Composition per liter:

Pancreatic digest of casein 20.0g
Lactose 10.0g
Yeast extract 5.0g
NaCl 4.0g
Gelatin 2.5g
Ascorbic acid 2.5g
Sodium acetate 1.5g
pH 7.0 ± 0.1 at 25°C

Preparation of Medium: Add components to distilled/deionized water and bring volume to 1.0L. Mix thoroughly. Distribute into test tubes that contain an inverted Durham tube. Autoclave for 15 min at 15 psi pressure–121°C.

Use: For the cultivation of streptococci and lactobacilli of importance in the dairy industry.

EMB Agar (Eosin Methylene Blue Agar)

Composition per liter:

Agar 13.5g
Pancreatic digest of casein 10.0g
Lactose 5.0g
Sucrose 5.0g
K_2HPO_4 .... 2.0g
Eosin Y 0.4g
Methylene Blue 0.065g
pH 7.2 ± 0.2 at 25°C

Source: This medium is available as a premixed powder from BD Diagnostic Systems.

Preparation of Medium: Add components to distilled/deionized water and bring volume to 1.0L. Mix thoroughly. Gently heat and bring to boiling. Distribute into tubes or flasks. Autoclave for 15 min at 15 psi pressure–121°C. Pour into sterile Petri dishes.

Use: For the isolation, cultivation, and differentiation of Gram-negative enteric bacteria based on lactose fermentation. Bacteria that ferment lactose, especially the coliform bacterium *Escherichia coli*, appear as colonies with a green metallic sheen or blue-black to brown color. Bacteria that do not ferment lactose appear as colorless or transparent, light purple colonies.

EMB Agar Base

Composition per liter:

Agar	15.0g
Peptone	10.0g
K_2HPO_4	2.0g
Eosin Y	0.4g
Methylene Blue	0.065g

pH 7.3 ± 0.2 at 25°C

Preparation of Medium: Add components to distilled/deionized water and bring volume to 1.0L. Mix thoroughly. Gently heat and bring to boiling. Distribute into tubes or flasks. Autoclave for 15 min at 15 psi pressure–121°C. Pour into sterile Petri dishes.

Use: For the isolation, cultivation, and differentiation of Gram-negative enteric bacteria based on lactose fermentation. Bacteria that ferment lactose, especially the coliform bacterium *Escherichia coli*, appear as colonies with a green metallic sheen or blue-black to brown color. Bacteria that do not ferment lactose appear as colorless or transparent, light purple colonies.

EMB Agar, Modified (Eosin Methylene Blue Agar, Modified)

Composition per liter:

Agar	15.0g
Lactose	10.0g
Pancreatic digest of gelatin	10.0g
K_2HPO_4	2.0g
Eosin Y	0.4g
Methylene Blue	0.065g

pH 6.8 ± 0.2 at 25°C

Source: This medium is available as a premixed powder from Oxoid.

Preparation of Medium: Add components to distilled/deionized water and bring volume to 1.0L. Mix thoroughly. Gently heat and bring to boiling. Distribute into tubes or flasks. Autoclave for 15 min at 15 psi pressure–121°C. Cool to 60°C. Shake medium to oxidize methylene blue. Pour into sterile Petri dishes. Swirl flask while pouring plates to distribute precipitate.

Use: For the isolation, cultivation, and differentiation of Gram-negative enteric bacteria based on lactose fermentation. Bacteria that ferment lactose, especially the coliform bacterium *Escherichia coli*, appear as colonies with a green metallic sheen or blue-black to brown color. Bacteria that do not ferment lactose appear as colorless or transparent, light purple colonies.

EMB HiVeg Agar

Composition per liter:

Agar	13.5g
Plant peptone	10.0g
Lactose	5.0g
Sucrose	5.0g
K_2HPO_4	2.0g
Eosin Y	0.4g
Methylene Blue	0.065g

pH 7.2 ± 0.2 at 25°C

Source: This medium is available as a premixed powder from HiMedia.

Preparation of Medium: Add components to distilled/deionized water and bring volume to 1.0L. Mix thoroughly. Gently heat and bring to boiling. Distribute into tubes or flasks. Autoclave for 15 min at 15 psi pressure–121°C. Pour into sterile Petri dishes.

Use: For the differential isolation of Gram-negative enteric bacilli from clinical and nonclinical specimens. For the isolation, cultivation, and differentiation of Gram-negative enteric bacteria based on lactose fermentation. Bacteria that ferment lactose, especially the coliform bacterium *Escherichia coli*, appear as colonies with a green metallic sheen or blue-black to brown color. Bacteria that do not ferment lactose appear as colorless or transparent, light purple colonies.

EMB HiVeg Agar, Levine

Composition per liter:

Agar	15.0g
Plant peptone	10.0g
Lactose	10.0g
K_2HPO_4	2.0g
Eosin Y	0.4g
Methylene Blue	0.065g

pH 7.2 ± 0.2 at 25°C

Source: This medium is available as a premixed powder from HiMedia.

Preparation of Medium: Add components to distilled/deionized water and bring volume to 1.0L. Mix thoroughly. Gently heat and bring to boiling. Distribute into tubes or flasks. Autoclave for 15 min at 15 psi pressure–121°C. Pour into sterile Petri dishes.

Use: For the isolation, enumeration, and differentiation of members of Enterobacteriaceae. For the isolation, cultivation, and differentiation of Gram-negative enteric bacteria based on lactose fermentation. Bacteria that ferment lactose, especially the coliform bacterium *Escherichia coli*, appear as colonies with a green metallic sheen or blue-black to brown color. Bacteria that do not ferment lactose appear as colorless or transparent, light purple colonies.

EMB HiVeg Broth

Composition per liter:

Plant peptone	10.0g
Lactose	5.0g
Sucrose	5.0g
K_2HPO_4	2.0g
Eosin Y	0.4g
Methylene Blue	0.065g

pH 7.2 ± 0.2 at 25°C

Source: This medium is available as a premixed powder from HiMedia.

Preparation of Medium: Add components to distilled/deionized water and bring volume to 1.0L. Mix thoroughly. Distribute into test tubes that contain an inverted Durham tube. Autoclave for 15 min at 15 psi pressure–121°C.

Use: For the differential cultivation of Gram-negative enteric bacilli from clinical and nonclinical specimens.

Emerson Agar (ATCC Medium 199)

Composition per liter:

Agar	20.0g
Glucose	10.0g
Beef extract	4.0g
Pancreatic digest of gelatin	4.0g

NaCl 2.5g
Yeast extract 1.0g

pH 7.0 ± 0.2 at 25°C

Source: This medium is available as a premixed powder from BD Diagnostic Systems.

Preparation of Medium: Add components to distilled/deionized water and bring volume to 1.0L. Mix thoroughly. Gently heat and bring to boiling. Distribute into tubes or flasks. Autoclave for 15 min at 13 psi pressure–118°C. Pour into sterile Petri dishes or leave in tubes.

Use: For the isolation, cultivation, and maintenance of members of the Actinomycetaceae, Streptomycetaceae, and molds. For the cultivation and maintenance of *Arthrobacter* species, *Microbispora rosea, Micromonospora coerulea, Mycobacterium* species, *Nocardia asteroides, Nocardiopsis dassonvillei, Pseudonocardia thermophila, Staphylococcus epidermidis, Streptomyces flaveus, Streptomyces olivaceus, Streptomyces thermoviolaceus, Streptomyces thermovulgaris,* and *Streptomyces vendargensis.*

Emerson Agar, Half Strength

Composition per liter:

Agar 20.0g
Soluble starch 7.5g
Yeast extract 2.0g
K_2HPO_4 .... 0.5g
$MgSO_4 \cdot 7H_2O$ 0.25g

pH 7.0 ± 0.2 at 25°C

Preparation of Medium: Add components to distilled/deionized water and bring volume to 1.0L. Mix thoroughly. Gently heat and bring to boiling. Distribute into tubes or flasks. Autoclave for 15 min at 15 psi pressure–121°C. Pour into sterile Petri dishes or leave in tubes.

Use: For the cultivation and maintenance of *Streptosporangium longisporum.*

Emerson HiVeg Agar

Composition per liter:

Agar 20.0g
Glucose 10.0g
Plant extract 4.0g
Plant peptone 4.0g
NaCl 2.5g
Yeast extract 1.0g
Cycloheximide 0.05g

pH 7.0 ± 0.2 at 25°C

Source: This medium, without cyclohexamide, is available as a premixed powder from HiMedia.

Caution: Cycloheximide is toxic. Avoid skin contact or aerosol formation and inhalation.

Preparation of Medium: Add components, except cycloheximide, to distilled/deionized water and bring volume to 1.0L. Mix thoroughly. Gently heat and bring to boiling. Add cycloheximide. Mix thoroughly. Distribute into tubes or flasks. Autoclave for 15 min at 15 psi pressure–121°C. Pour into sterile Petri dishes or leave in tubes.

Use: For the isolation, cultivation, and maintenance of members of the Actinomycetaceae, Streptomycetaceae, and molds.

Emerson YpSs Agar with 0.25% Seawater

Composition per liter:

Agar 20.0g
Soluble starch 15.0g
Yeast extract 4.0g
K_2HPO_4 .... 1.0g
$MgSO_4 \cdot 7H_2O$ 0.5g
Seawater 2.5mL

pH 7.0 ± 0.2 at 25°C

Preparation of Medium: Add components to distilled/deionized water and bring volume to 1.0L. Mix thoroughly. Gently heat and bring to boiling. Distribute into tubes or flasks. Autoclave for 15 min at 13 psi pressure–118°C. Pour into sterile Petri dishes or leave in tubes.

Use: For the cultivation and maintenance of *Halosphaeria quadricornuta.*

Emerson YpSs Agar

Composition per liter:

Agar 20.0g
Soluble starch 15.0g
Yeast extract 4.0g
K_2HPO_4 .... 1.0g
$MgSO_4 \cdot 7H_2O$ 0.5g

pH 7.0 ± 0.2 at 25°C

Source: This medium is available as a premixed powder from BD Diagnostic Systems.

Preparation of Medium: Add components to distilled/deionized water and bring volume to 1.0L. Mix thoroughly. Gently heat and bring to boiling. Distribute into tubes or flasks. Autoclave for 15 min at 13 psi pressure–118°C. Pour into sterile Petri dishes or leave in tubes.

Use: For the cultivation and maintenance of *Allomyces* and other fungi.

Emerson YpSs Agar with 25% Seawater and a Birch Stick

Composition per liter:

Agar 20.0g
Soluble starch 15.0g
Yeast extract 4.0g
K_2HPO_4 .... 1.0g
$MgSO_4 \cdot 7H_2O$ 0.5g
Seawater 250.0mL
Birch stick .... 1 per test tube

pH 7.0 ± 0.2 at 25°C

Preparation of Medium: Add components, except birch stick, to distilled/deionized water and bring volume to 1.0L. Mix thoroughly. Gently heat and bring to boiling. Distribute into tubes. Add one birch stick to each tube. Autoclave for 15 min at 13 psi pressure–118°C.

Use: For the cultivation and maintenance of *Helicascus kanaloanus, Lulworthia* species, *Pseudohalonectria adversaria, Pseudohalonectria falcata, Pseudohalonectria phialidica,* and *Ophiobolus* species.

Emerson YpSs Broth, 1/2 strength (ATCC Medium 2370)

Composition per liter:

Soluble starch 7.5g
Yeast extract 2.0g

K_2HPO_4 .... 0.5g
$MgSO_4 \cdot 7H_2O$ 0.25g

pH 7.0 ± 0.2 at 25°C

Preparation of Medium: Add components to distilled/deionized water and bring volume to 1.0L. Mix thoroughly. Distribute into tubes or flasks. Autoclave for 15 min at 15 psi pressure–121°C.

Use: For the cultivation and maintenance of *Lulworthia medusa.*

Emerson YpSs Broth, 1/4 strength (ATCC Medium 2371)

Composition per liter:

Soluble starch 3.75g
Yeast extract 1.0g
K_2HPO_4 .... 0.25g
$MgSO_4 \cdot 7H_2O$ 0.125g

pH 7.0 ± 0.2 at 25°C

Preparation of Medium: Add components to distilled/deionized water and bring volume to 1.0L. Mix thoroughly. Distribute into tubes or flasks. Autoclave for 15 min at 15 psi pressure–121°C.

Use: For the cultivation and maintenance of *Lulworthia medusa.*

Emerson's Yeast Starch Agar (EYS Agar)

Composition per liter:

Starch, soluble 15.0g
Agar 12.0g
Yeast extract 4.0g
Na_2HPO_4 .... 1.0g
$MgSO_4$ 0.5g

pH 6.8 ± 0.2 at 25°C

Preparation of Medium: Add components to distilled/deionized water and bring volume to 1.0L. Mix thoroughly. Gently heat and bring to boiling. Adjust pH to 6.8. Distribute into tubes or flasks. Autoclave for 15 min at 15 psi pressure–121°C. Pour into sterile Petri dishes or leave in tubes.

Use: For the cultivation and maintenance of *Microtetraspora recticatena* and *Saccharothrix coeruleoviolacea.*

Emmon's Modification of Sabouraud's Agar

Composition per liter:

Agar 20.0g
Glucose 20.0g
Neopeptone 10.0g

pH 6.8–7.0 at 25°C

Preparation of Medium: Add components to distilled/deionized water and bring volume to 1.0L. Mix thoroughly. Adjust pH to 6.8–7.0. Gently heat and bring to boiling. Distribute into tubes or flasks. Autoclave for 15 min at 15 psi pressure–121°C. Pour into sterile Petri dishes or leave in tubes.

Use: For the cultivation of *Prototheca moriformis*, *Prototheca stagnora*, *Prototheca wickerhamii*, and *Prototheca zopfii.*

Emmon's Modification of Sabouraud's Agar (Sabouraud's Agar, Modified)

Composition per liter:

Agar 20.0g
Glucose 20.0g
Pancreatic digest of casein 5.0g
Peptic digest of animal tissue 5.0g

pH 6.8–7.0 at 25°C

Source: This medium is available as a premixed powder from BD Diagnostic Systems.

Preparation of Medium: Add components to distilled/deionized water and bring volume to 1.0L. Mix thoroughly. Gently heat and bring to boiling. Distribute into tubes or flasks. Autoclave for 15 min at 15 psi pressure–121°C. Pour into sterile Petri dishes or leave in tubes.

Use: For the cultivation and maintenance of *Bacillus subtilis, Nocardia* species, and *Streptomyces albus*.

Emmon's Modification of Sabouraud's Agar with 0.5% Yeast Extract in Olive Oil

Composition per liter:

Agar 20.0g
Glucose 20.0g
Yeast extract 5.0g
Pancreatic digest of casein 5.0g
Peptic digest of animal tissue 5.0g
Olive oil 10.0mL

pH 6.8–7.0 at 25°C

Preparation of Medium: Add components to distilled/deionized water and bring volume to 1.0L. Mix thoroughly. Gently heat and bring to boiling. Distribute into tubes or flasks. Autoclave for 15 min at 15 psi pressure–121°C. Pour into sterile Petri dishes or leave in tubes.

Use: For the cultivation and maintenance of *Pityrosporum ovale.*

ENB Agar

Composition per liter:

Agar 15.0g
Pancreatic digest of gelatin 5.0g
NaCl 5.0g
Peptone 5.0g
K_2HPO_4 .... 3.5g
Beef extract 3.0g
KH_2PO_4 .... 1.5g
Glucose solution 10.0mL

pH 6.8 ± 0.2 at 25°C

Glucose Solution:
Composition per liter:

Glucose 1.0g

Preparation of Glucose Solution: Add glucose to distilled/deionized water and bring volume to 10.0mL. Mix thoroughly. Autoclave for 15 min at 15 psi pressure–121°C.

Preparation of Medium: Add components, except glucose solution, to distilled/deionized water and bring volume to 990.0mL. Mix thoroughly. Gently heat and bring to boiling. Adjust pH to 6.8. Autoclave for 15 min at 15 psi pressure–121°C. Aseptically add 10.0mL of sterile glucose solution. Mix thoroughly. Pour into sterile Petri dishes or distribute into sterile tubes.

Use: For the cultivation of *Escherichia coli.*

Endamoeba Medium
See: ***Entamoeba*** **Medium**

Endo Agar

Composition per liter:

Agar	15.0g
Lactose	10.0g
Peptic digest of animal tissue	10.0g
K_2HPO_4	3.5g
Na_2SO_3	2.5g
Basic Fuchsin	0.5g

pH 7.4 ± 0.2 at 25°C

Source: This medium is available as a premixed powder from BD Diagnostic Systems.

Caution: Basic Fuchsin is a potential carcinogen and care must be taken to avoid inhalation of the powdered dye and contact with the skin.

Preparation of Medium: Add components to distilled/deionized water and bring volume to 1.0L. Mix thoroughly. Gently heat and bring to boiling. Autoclave for 15 min at 15 psi pressure–121°C. Cool to 45°–50°C. Pour into sterile Petri dishes. Swirl flask while pouring plates to keep precipitate in suspension. Protect from the light.

Use: For the selective isolation, cultivation, and differentiation of coliform and other enteric microorganisms based on their ability to ferment lactose. Lactose-fermenting bacteria appear as dark red colonies with a gold metallic sheen. Lactose-nonfermenting bacteria appear as colorless or translucent colonies.

Endo Agar

Composition per liter:

Agar	10.0g
Lactose	10.0g
Peptic digest of animal tissue	10.0g
K_2HPO_4	3.5g
Na_2SO_3	2.5g
Basic Fuchsin solution	4.0mL

pH 7.5 ± 0.2 at 25°C

Source: This medium is available as a premixed powder from Oxoid.

Basic Fuchsin Solution:

Composition per 10.0mL:

Basic Fuchsin	1.0g
Ethanol (95% solution)	10.0mL

Preparation of Basic Fuchsin Solution: Add Basic Fuchsin to 10.0mL of ethanol. Mix thoroughly.

Caution: Basic Fuchsin is a potential carcinogen and care must be taken to avoid inhalation of the powdered dye and contact with the skin.

Preparation of Medium: Add components to distilled/deionized water and bring volume to 1.0L. Mix thoroughly. Gently heat and bring to boiling. Autoclave for 15 min at 15 psi pressure–121°C. Cool to 45°–50°C. Pour into sterile Petri dishes. Swirl flask while pouring plates to keep precipitate in suspension. Protect from the light.

Use: For the selective isolation, cultivation, and differentiation of coliform and other enteric microorganisms based on their ability to ferment lactose. Lactose-fermenting bacteria appear as dark red colonies with a gold metallic sheen. Lactose-nonfermenting bacteria appear as colorless or translucent colonies.

Endo Agar

Composition per liter:

Agar	15.0g
Peptic digest of animal tissue	10.0g
Lactose	10.0g
K_2HPO_4	3.5g
Na_2SO_3	2.5g
Basic Fuchsin	0.5g

pH 7.5 ± 0.2 at 25°C

Source: This medium is available as a premixed powder from HiMedia.

Caution: Basic Fuchsin is a potential carcinogen and care must be taken to avoid inhalation of the powdered dye and contact with the skin.

Preparation of Medium: Add components to distilled/deionized water and bring volume to 1.0L. Mix thoroughly. Gently heat and bring to boiling. Autoclave for 15 min at 15 psi pressure–121°C. Cool to 45°–50°C. Pour into sterile Petri dishes. Swirl flask while pouring plates to keep precipitate in suspension. Protect from the light.

Use: For the selective isolation, cultivation, and differentiation of coliform and other enteric microorganisms based on their ability to ferment lactose. Lactose-fermenting bacteria appear as dark red colonies with a gold metallic sheen. Lactose-nonfermenting bacteria appear as colorless or translucent colonies.

Endo Agar Base

Composition per liter:

Agar	12.0g
Peptic digest of animal tissue	10.0g
Lactose	10.0g
K_2HPO_4	3.5g
Na_2SO_3	2.5g
Basic Fuchsin solution	4.0mL

pH 7.5 ± 0.2 at 25°C

Source: This medium, without Basic Fuchsin solution, is available as a premixed powder from HiMedia.

Basic Fuchsin Solution:

Composition per 10.0mL:

Basic Fuchsin	1.0g
Ethanol (95% solution)	10.0mL

Preparation of Basic Fuchsin Solution: Add Basic Fuchsin to 10.0mL of ethanol. Mix thoroughly.

Caution: Basic Fuchsin is a potential carcinogen and care must be taken to avoid inhalation of the powdered dye and contact with the skin.

Preparation of Medium: Add components to distilled/deionized water and bring volume to 1.0L. Mix thoroughly. Gently heat and bring to boiling. Autoclave for 15 min at 15 psi pressure–121°C. Cool to 45°–50°C. Pour into sterile Petri dishes. Swirl flask while pouring plates to keep precipitate in suspension. Protect from the light.

Use: For the selective isolation, cultivation, and differentiation of coliform and other enteric microorganisms based on their ability to ferment lactose. Lactose-fermenting bacteria appear as dark red colonies with a gold metallic sheen. Lactose-nonfermenting bacteria appear as colorless or translucent colonies.

Endo Agar, Modified

Composition per liter:

Agar	12.5g
Peptic digest of animal tissue	10.0g
Lactose	10.0g
Na_2SO_3	3.3g
K_2HPO_4	2.5g
Basic Fuchsin	0.3g

pH 7.4 ± 0.2 at 25°C

Source: This medium is available as a premixed powder from Hi-Media.

Caution: Basic Fuchsin is a potential carcinogen and care must be taken to avoid inhalation of the powdered dye and contact with the skin.

Preparation of Medium: Add components to distilled/deionized water and bring volume to 1.0L. Mix thoroughly. Gently heat and bring to boiling. Autoclave for 15 min at 15 psi pressure–121°C. Cool to 45°–50°C. Pour into sterile Petri dishes. Swirl flask while pouring plates to keep precipitate in suspension. Protect from the light.

Use: For the selective isolation, cultivation, and differentiation of coliform and other enteric microorganisms based on their ability to ferment lactose. Lactose-fermenting bacteria appear as dark red colonies with a gold metallic sheen. Lactose-nonfermenting bacteria appear as colorless or translucent colonies.

Endo Agar, LES (Endo Agar, Laurance Experimental Station) (m-Endo Agar, LES) (m-LES, Endo Agar)

Composition per liter:

Agar	14.0g
Lactose	9.4g
Peptones (pancreatic digest of casein 65% and yeast extract 35%)	7.5g
NaCl	3.7g
Pancreatic digest of casein	3.7g
Peptic digest of animal tissue	3.7g
K_2HPO_4	3.3g
Na_2SO_3	1.6g
Yeast extract	1.2g
KH_2PO_4	1.0g
Basic Fuchsin	0.8g
Sodium lauryl sulfate	0.05g
Ethanol	20.0mL

pH 7.2 ± 0.2 at 25°C

Source: This medium is available as a premixed powder from BD Diagnostic Systems.

Caution: Basic Fuchsin is a potential carcinogen and care must be taken to avoid inhalation of the powdered dye and contact with the skin.

Preparation of Medium: Add ethanol to approximately 900.0mL of distilled/deionized water. Add remaining components. Bring volume to 1.0L with distilled/deionized water. Mix thoroughly. Gently heat and bring to boiling. Autoclave for 15 min at 15 psi pressure–121°C. Pour into sterile 60mm Petri dishes in 4.0mL volumes. Protect from the light.

Use: For the cultivation and enumeration of coliform bacteria by the membrane filter method.

Endo Agar, LES (m-Endo Agar, LES)

Composition per liter:

Agar	10.0g
Lactose	9.4g
Tryptose	7.5g
NaCl	3.7g
Peptone	3.7g
Pancreatic digest of casein	3.7g
K_2HPO_4	3.3g
Na_2SO_3	1.6g
Yeast extract	1.2g
KH_2PO_4	1.0g
Sodium deoxycholate	0.1g
Sodium lauryl sulfate	0.05g
Basic Fuchsin solution	8.0mL

pH 7.2 ± 0.2 at 25°C

Basic Fuchsin Solution:

Composition per 10.0mL:

Basic Fuchsin	1.0g
Ethanol (95% solution)	10.0mL

Caution: Basic Fuchsin is a potential carcinogen and care must be taken to avoid inhalation of the powdered dye and contact with the skin.

Preparation of Basic Fuchsin Solution: Add Basic Fuchsin to 10.0mL of ethanol. Mix thoroughly.

Preparation of Medium: Add components to distilled/deionized water and bring volume to 1.0L. Mix thoroughly. Gently heat and bring to boiling. Autoclave for 15 min at 15 psi pressure–121°C. Cool to 45°–50°C. Pour into sterile Petri dishes. Swirl flask while pouring plates to keep precipitate in suspension. Protect from the light.

Use: For the cultivation and enumeration of coliform bacteria from water by the membrane filter method.

Endo Agar with Sodium Chloride

Composition per liter:

Agar	12.0g
Lactose	10.0g
Peptone, special	8.0g
NaCl	3.0g
Na_2SO_3	2.5g
K_2HPO_4	2.0g
Basic Fuchsin	0.2g

pH 7.4 ± 0.2 at 25°C

Source: This medium is available as a premixed powder from Hi-Media.

Caution: Basic Fuchsin is a potential carcinogen and care must be taken to avoid inhalation of the powdered dye and contact with the skin.

Preparation of Medium: Add components to distilled/deionized water and bring volume to 1.0L. Mix thoroughly. Gently heat and bring to boiling. Autoclave for 15 min at 15 psi pressure–121°C. Cool to 45°–50°C. Pour into sterile Petri dishes. Swirl flask while pouring plates to keep precipitate in suspension. Protect from the light.

Use: For the selective isolation, cultivation, and differentiation of coliform and other enteric microorganisms based on their ability to ferment lactose.

Endo Broth (m-Endo Broth)

Composition per liter:

Lactose	12.5g
Peptone	10.0g
NaCl	5.0g
Pancreatic digest of casein	5.0g
Peptic digest of animal tissue	5.0g
K_2HPO_4	4.375g
Na_2SO_3	2.1g
Yeast extract	1.5g
KH_2PO_4	1.375g
Basic Fuchsin	1.05g
Sodium deoxycholate	0.1g
Ethanol (95% solution)	20.0mL

pH 7.2 ± 0.1 at 25°C

Source: This medium is available as a premixed powder from BD Diagnostic Systems.

Caution: Basic Fuchsin is a potential carcinogen and care must be taken to avoid inhalation of the powdered dye and contact with the skin.

Preparation of Medium: Add ethanol to approximately 900.0mL of distilled/deionized water. Add remaining components. Bring volume to 1.0L with distilled/deionized water. Mix thoroughly. Gently heat and bring to boiling. Rapidly cool broth below 45°C. Do not autoclave. Use 1.8–2.0mL for each filter pad. Protect from the light. Prepare broth freshly.

Use: For the cultivation and enumeration of coliform bacteria from water by the membrane filter method.

Endo HiVeg Agar

Composition per liter:

Agar	15.0g
Plant peptone	10.0g
Lactose	10.0g
K_2HPO_4	3.5g
Na_2SO_3	2.5g
Basic Fuchsin	0.5g

pH 7.5 ± 0.2 at 25°C

Source: This medium is available as a premixed powder from HiMedia.

Caution: Basic Fuchsin is a potential carcinogen and care must be taken to avoid inhalation of the powdered dye and contact with the skin.

Preparation of Medium: Add components to distilled/deionized water and bring volume to 1.0L. Mix thoroughly. Gently heat and bring to boiling. Autoclave for 15 min at 15 psi pressure–121°C. Cool to 45°–50°C. Pour into sterile Petri dishes. Swirl flask while pouring plates to keep precipitate in suspension. Protect from the light.

Use: For the selective isolation, cultivation, and differentiation of coliform and other enteric microorganisms based on their ability to ferment lactose. Lactose-fermenting bacteria appear as dark red colonies with a gold metallic sheen. Lactose-nonfermenting bacteria appear as colorless or translucent colonies.

Endo HiVeg Agar Base

Composition per liter:

Agar	12.0g
Plant peptone	10.0g
Lactose	10.0g
K_2HPO_4	3.5g
Na_2SO_3	2.5g
Basic Fuchsin solution	4.0mL

pH 7.5 ± 0.2 at 25°C

Source: This medium, without Basic Fuchsin solution, is available as a premixed powder from HiMedia.

Basic Fuchsin Solution:

Composition per 10.0mL:

Basic Fuchsin	1.0g
Ethanol (95% solution)	10.0mL

Preparation of Basic Fuchsin Solution: Add Basic Fuchsin to 10.0mL of ethanol. Mix thoroughly.

Caution: Basic Fuchsin is a potential carcinogen and care must be taken to avoid inhalation of the powdered dye and contact with the skin.

Preparation of Medium: Add components to distilled/deionized water and bring volume to 1.0L. Mix thoroughly. Gently heat and bring to boiling. Autoclave for 15 min at 15 psi pressure–121°C. Cool to 45°–50°C. Pour into sterile Petri dishes. Swirl flask while pouring plates to keep precipitate in suspension. Protect from the light.

Use: For the selective isolation, cultivation, and differentiation of coliform and other enteric microorganisms based on their ability to ferment lactose. Lactose-fermenting bacteria appear as dark red colonies with a gold metallic sheen. Lactose-nonfermenting bacteria appear as colorless or translucent colonies.

Endo HiVeg Agar, Modified

Composition per liter:

Agar	12.5g
Plant peptone	10.0g
Lactose	10.0g
Na_2SO_3	3.3g
K_2HPO_4	2.5g
Basic Fuchsin	0.3g

pH 7.4 ± 0.2 at 25°C

Source: This medium is available as a premixed powder from HiMedia.

Caution: Basic Fuchsin is a potential carcinogen and care must be taken to avoid inhalation of the powdered dye and contact with the skin.

Preparation of Medium: Add components to distilled/deionized water and bring volume to 1.0L. Mix thoroughly. Gently heat and bring to boiling. Autoclave for 15 min at 15 psi pressure–121°C. Cool to 45°–50°C. Pour into sterile Petri dishes. Swirl flask while pouring plates to keep precipitate in suspension. Protect from the light.

Use: For the selective isolation, cultivation, and differentiation of coliform and other enteric microorganisms based on their ability to ferment lactose. Lactose-fermenting bacteria appear as dark red colonies with a gold metallic sheen. Lactose-nonfermenting bacteria appear as colorless or translucent colonies.

Endo HiVeg Agar with NaCl

Composition per liter:

Agar	12.0g
Lactose	10.0g
Plant special peptone	8.0g

NaCl	3.0g
Na_2SO_3	2.5g
K_2HPO_4	2.0g
Basic Fuchsin	0.2g

pH 7.4 ± 0.2 at 25°C

Source: This medium is available as a premixed powder from Hi-Media.

Caution: Basic Fuchsin is a potential carcinogen and care must be taken to avoid inhalation of the powdered dye and contact with the skin.

Preparation of Medium: Add components to distilled/deionized water and bring volume to 1.0L. Mix thoroughly. Gently heat and bring to boiling. Autoclave for 15 min at 15 psi pressure–121°C. Cool to 45°–50°C. Pour into sterile Petri dishes. Swirl flask while pouring plates to keep precipitate in suspension. Protect from the light.

Use: For the selective isolation, cultivation, and differentiation of coliform and other enteric microorganisms based on their ability to ferment lactose.

Endothia Complete Agar

Composition per liter:

Agar	20.0g
Glucose	10.0g
Malt extract	7.5g
Yeast extract	2.5g
NH_4NO_3	1.5g
KH_2PO_4	1.0g
KCl	0.5g
Na_2SO_4	0.25g
$MgSO_4$	0.125g
$CaCl_2$	0.063g
Thiamine	2.0mg
Trace elements solution	1.0mL

pH 4.0–5.0 at 25°C

Trace Elements Solution:
Composition per liter:

$CuSO_4$	0.2g
$ZnCl_2$	0.2g
$MnCl_2$	0.07g
$FeCL_3$	0.05g
H_3BO_3	0.03g
Na_2MoO_4	0.02g

Preparation of Trace Elements Solution: Add components to distilled/deionized water and bring volume to 1.0L. Mix thoroughly.

Preparation of Medium: Add components to distilled/deionized water and bring volume to 1.0L. Mix thoroughly. Gently heat and bring to boiling. Distribute into tubes or flasks. Autoclave for 15 min at 15 psi pressure–121°C. Pour into sterile Petri dishes or leave in tubes.

Use: For the cultivation and maintenance of *Cryphonectria parasitica.*

Endothia Complete Broth

Composition per liter:

Glucose	10.0g
Malt extract	7.5g
Yeast extract	2.5g
NH_4NO_3	1.5g
KH_2PO_4	1.0g
KCl	0.5g
Na_2SO_4	0.25g
$MgSO_4$	0.125g
$CaCl_2$	0.063g
Thiamine	2.0mg
Trace elements solution	1.0mL

pH 4.0–5.0 at 25°C

Trace Elements Solution:
Composition per liter:

$CuSO_4$	0.2g
$ZnCl_2$	0.2g
$MnCl_2$	0.07g
$FeCL_3$	0.05g
H_3BO_3	0.03g
Na_2MoO_4	0.02g

Preparation of Trace Elements Solution: Add components to distilled/deionized water and bring volume to 1.0L. Mix thoroughly.

Preparation of Medium: Add components to distilled/deionized water and bring volume to 1.0L. Mix thoroughly. Gently heat and bring to boiling. Distribute into tubes or flasks. Autoclave for 15 min at 15 psi pressure–121°C.

Use: For the cultivation of *Cryphonectria parasitica.*

Enriched *Cytophaga* Agar

Composition per liter:

Agar	15.0g
Pancreatic digest of casein	2.0g
Beef extract	0.5g
Yeast extract	0.5g
Sodium acetate	0.2g

pH 7.2 ± 0.2 at 25°C

Preparation of Medium: Add components to distilled/deionized water and bring volume to 1.0L. Mix thoroughly. Gently heat and bring to boiling. Distribute into tubes or flasks. Autoclave for 15 min at 15 psi pressure–121°C. Pour into sterile Petri dishes or leave in tubes.

Use: For the cultivation of *Cytophaga arvensicola, Cytophaga johnsonae, Cytophaga psychrophila, Cytophaga* species, *Cytophaga succinicans, Flavobacterium aquatile, Flavobacterium branchiophila, Flexibacter aurantiacus, Flexibacter canadensis, Flexibacter* species, *Pseudomonas echinoides, Psychrobacter immobilis, Xanthobacter autotrophicus,* and *Zoogloea ramigera.*

Enriched *Cytophaga* Agar Medium (DSMZ Medium 1133)

Composition per liter:

Agar	15.0g
Tryptone	2.0g
Beef extract	0.5g
Yeast extract	0.5g
Sodium acetate	0.2g

pH 7.3 ± 0.2 at 25°C

Preparation of Medium: Add components to distilled/deionized water and bring volume to 1.0L. Mix thoroughly. Adjust pH to 7.3. Gently heat while stirring and bring to boiling. Distribute into tubes or flasks. Autoclave for 15 min at 15 psi pressure–121°C. Pour into Petri dishes or aseptically distribute into sterile tubes. For soft agar for the maintenance of active cultures reduce agar content to 4.0g/lL. Dispense in 3–4mL amounts in 7mL screw-capped bottles (bijou bottles).

Use: For the cultivation of *Cytophaga* spp.

Enriched *Cytophaga* Medium

Composition per liter:

Pancreatic digest of casein	2.0g
Beef extract	0.5g
Yeast extract	0.5g
Sodium acetate	0.2g

pH 7.2 ± 0.2 at 25°C

Preparation of Medium: Add components to distilled/deionized water and bring volume to 1.0L. Mix thoroughly. Gently heat and bring to boiling. Distribute into tubes or flasks. Autoclave for 15 min at 15 psi pressure–121°C. Pour into sterile Petri dishes or leave in tubes.

Use: For the cultivation of *Cytophaga columnaris*.

Enriched *Isonema* Medium

Composition per 1111.0mL:

Natural seawater	1.0L
Horse serum, inactivated	100.0mL
Enrichment solution	10.0mL
Vitamin solution	1.0mL

Enrichment Solution:

Composition per liter:

$NaNO_3$	4.667g
$Na_2SiO_3 \cdot 9H_2O$	3.0g
Sodium glycerophosphate	0.667g
$EDTA \cdot 2H_2O$	0.553g
H_3BO_3	0.38g
$Fe(NH_4)_2(SO_4)_2 \cdot 6H_2O$	0.234g
$MnSO_4 \cdot 4H_2O$	0.054g
$FeCl_3 \cdot 6H_2O$	0.016g
$ZnSO_4 \cdot 7H_2O$	7.3mg
$CoSO_4 \cdot 7H_2O$	1.6mg

Preparation of Enrichment Solution: Add $Na_2SiO_3 \cdot 9H_2O$ to distilled/deionized water. Mix thoroughly. Neutralize with 1*N* HCl. Add 500.0mL of distilled/deionized water. Mix thoroughly. Add remaining components and bring volume to 1.0L with distilled/deionized water. Mix thoroughly. Filter sterilize.

Vitamin Solution:

Composition per liter:

Thiamine	0.1g
Vitamin B_{12}	2.0mg
Biotin	1.0mg

Preparation of Vitamin Solution: Add components to distilled/deionized water and bring volume to 1.0L. Mix thoroughly. Filter sterilize.

Preparation of Medium: Allow natural seawater to age for 2 months. Filter sterilize. Aseptically add 10.0mL of sterile enrichment solution and 1.0mL of sterile vitamin solution. Mix thoroughly. Prior to inoculation, aseptically add 100.0mL of heat-inactivated horse serum. Mix thoroughly. Aseptically distribute into sterile tubes or flasks.

Use: For the cultivation of *Diplonema* species, *Gymnophrydium marinum, Isonema* species, and *Rhynchopus* species.

Enriched Nutrient Agar

Composition per liter:

Agar	15.0g
Heart, solids from infusion	12.5g
Yeast extract	3.3g
Peptone	2.1g
NaCl	2.1g
Beef extract	0.42g

pH 7.0 ± 0.2 at 25°C

Preparation of Medium: Add components to distilled/deionized water and bring volume to 1.0L. Mix thoroughly. Gently heat and bring to boiling. Distribute into tubes or flasks. Autoclave for 15 min at 15 psi pressure–121°C. Pour into sterile Petri dishes or leave in tubes.

Use: For the cultivation of *Micrococcus luteus*.

Enriched Nutrient Broth

Composition per liter:

Beef heart, infusion from	250.0g
Tryptose	5.0g
Pancreatic digest of gelatin	3.38g
NaCl	2.5g
Yeast extract	2.5g
Beef extract	2.02g

pH 7.2 ± 0.1 at 25°C

Preparation of Medium: Add components to distilled/deionized water and bring volume to 1.0L. Mix thoroughly. Gently heat and bring to boiling. Distribute into tubes or flasks. Autoclave for 15 min at 15 psi pressure–121°C. Pour into sterile Petri dishes or leave in tubes.

Use: For the cultivation and maintenance of *Bacillus licheniformis, Bacillus polymyxa, Bacillus subtilis, Escherichia coli, Listonella anguillarum, Micrococcus luteus, Pseudomonas aeruginosa, Salmonella choleraesuis, Staphylococcus aureus, Staphylococcus epidermidis, Streptococcus* species, and *Vibrio cholerae*.

Enriched Nutrient Broth

Composition per liter:

Beef heart, infusion from	300.0g
Tryptose	7.5g
NaCl	3.0g
Yeast extract	3.0g
Peptone	2.5g
NaCl	2.5g
Beef extract	0.5g

Preparation of Medium: Add components to distilled/deionized water and bring volume to 1.0L. Mix thoroughly. Distribute into tubes or flasks. Autoclave for 15 min at 15 psi pressure–121°C.

Use: For the cultivation of fastidious bacteria.

Enriched Thioglycollate HiVeg Broth

Composition per liter:

Plant hydrolysate	17.0g
Glucose	6.0g
Papaic digest of soybean meal	3.0g
NaCl	2.5g
Agar	0.7g
Na-thioglycollate	0.5g
L-Cystine	0.25g
Na_2SO_3	0.1g
$Fe_4(P_2O_7)_3 \cdot H_2O$	0.005g
Vitamin K_1	0.001g

pH 7.1 ± 0.2 at 25°C

Source: This medium is available as a premixed powder from HiMedia.

Preparation of Medium: Add components to distilled/deionized water and bring volume to 1.0L. Mix thoroughly. Distribute into tubes or flasks. Autoclave for 15 min at 15 psi pressure–121°C. Prepare freshly or boil and cool the medium just before use.

Use: For the cultivation of both aerobic and anaerobic organisms. For the performance of sterility tests of turbid or viscous specimens.

Enrichment Broth for *Aeromonas hydrophila*

Composition per liter:

NaCl	5.0g
Maltose	3.5g
Yeast extract	3.0g
Bile salts No. 3	1.0g
L-Cysteine·HCl·H_2O	0.3g
Bromthymol Blue	0.03g
Novobiocin	5.0mg

pH 7.0 ± 0.2 at 25°C

Preparation of Medium: Add components to distilled/deionized water and bring volume to 1.0L. Mix thoroughly. Distribute into tubes or flasks. Autoclave for 15 min at 15 psi pressure–121°C.

Use: For the cultivation and enrichment of *Aeromonas hydrophila.*

Enrichment Broth, pH 7.3

Composition per liter:

Pancreatic digest of casein	17.0g
Yeast extract	6.0g
NaCl	5.0g
Papaic digest of soybean meal	3.0g
Glucose	2.5g
K_2HPO_4	2.5g
Nalidixic acid solution	8.0mL
Cycloheximide solution	5.1mL
Acriflavin·HCl solution	3.0mL

pH 7.3 ± 0.2 at 25°C

Cycloheximide Solution:

Composition per 10.0mL:

Cycloheximide	0.1g
Ethanol, absolute	4.0mL

Preparation of Cycloheximide Solution: Add components to distilled/deionized water and bring volume to 10.0mL. Mix thoroughly. Filter sterilize.

Caution: Cycloheximide is toxic. Avoid skin contact or aerosol formation and inhalation.

Nalidixic Acid Solution:

Composition per 10.0mL:

Nalidixic acid	0.05g

Preparation of Nalidixic Acid Solution: Add nalidixic acid to distilled/deionized water and bring volume to 10.0mL. Mix thoroughly. Filter sterilize.

Acriflavin·HCl Solution:

Composition per 10.0mL:

Acriflavin·HCl	0.05g

Preparation of Acriflavin·HCl Solution: Add acriflavin·HCl to distilled/deionized water and bring volume to 10.0mL. Mix thoroughly. Filter sterilize.

Preparation of Medium: Add components—except cycloheximide solution, nalidixic acid solution, and acriflavin·HCl solution—to distilled/deionized water and bring volume to 983.9mL. Mix thoroughly. Gently heat and bring to boiling. Autoclave for 15 min at 15 psi pressure–121°C. Cool to 45°–50°C. Aseptically add 5.1mL of sterile cycloheximide solution, 8.0mL of sterile nalidixic acid solution, and 3.0mL of sterile acriflavin·HCl solution. Mix thoroughly. Aseptically distribute into sterile tubes.

Use: For the isolation, cultivation, and enrichment of a variety of microorganisms from nondairy foods.

Enrichment Broth, pH 7.3

Composition per liter:

Pancreatic digest of casein	17.0g
Yeast extract	6.0g
NaCl	5.0g
Papaic digest of soybean meal	3.0g
Glucose	2.5g
K_2HPO_4	2.5g
Nalidixic acid solution	8.0mL
Cycloheximide solution	5.1mL
Acriflavin·HCl solution	2.0mL

pH 7.3 ± 0.2 at 25°C

Cycloheximide Solution:

Composition per 10.0mL:

Cycloheximide	0.1g
Ethanol, absolute	4.0mL

Preparation of Cycloheximide Solution: Add components to distilled/deionized water and bring volume to 10.0mL. Mix thoroughly. Filter sterilize.

Caution: Cycloheximide is toxic. Avoid skin contact or aerosol formation and inhalation.

Nalidixic Acid Solution:

Composition per 10.0mL:

Nalidixic acid	0.05g

Preparation of Nalidixic Acid Solution: Add nalidixic acid to distilled/deionized water and bring volume to 10.0mL. Mix thoroughly. Filter sterilize.

Acriflavin·HCl Solution:

Composition per 10.0mL:

Acriflavin·HCl	0.05g

Preparation of Acriflavin·HCl Solution: Add acriflavin·HCl to distilled/deionized water and bring volume to 10.0mL. Mix thoroughly. Filter sterilize.

Preparation of Medium: Add components—except cycloheximide solution, nalidixic acid solution, and acriflavin·HCl solution—to distilled/deionized water and bring volume to 984.9mL. Mix thoroughly. Gently heat and bring to boiling. Autoclave for 15 min at 15 psi pressure–121°C. Cool to 45°–50°C. Aseptically add 5.1mL of cycloheximide solution, 8.0mL of nalidixic acid solution, and 2.0mL of acriflavin·HCl solution. Mix thoroughly. Aseptically distribute into sterile tubes.

Use: For the isolation, cultivation, and enrichment of a variety of microorganisms from milk and dairy products.

Enrichment Broth, pH 7.3 with Pyruvate

Composition per liter:

Pancreatic digest of casein	17.0g
Yeast extract	6.0g

NaCl 5.0g
Papaic digest of soybean meal 3.0g
Glucose 2.5g
K_2HPO_4 .... 2.5g
Pyruvate solution 11.1mL
Nalidixic acid solution 8.0mL
Cycloheximide solution 5.1mL
Acriflavin·HCl solution 3.0mL
pH 7.3 ± 0.2 at 25°C

Pyruvate Solution:
Composition per 20.0mL:
Sodium pyruvate 2.0g

Preparation of Pyruvate Solution: Add sodium pyruvate to distilled/deionized water and bring volume to 20.0mL. Mix thoroughly. Filter sterilize.

Cycloheximide Solution:
Composition per 10.0mL:
Cycloheximide 0.1g
Ethanol, absolute 4.0mL

Preparation of Cycloheximide Solution: Add components to distilled/deionized water and bring volume to 10.0mL. Mix thoroughly. Filter sterilize.

Caution: Cycloheximide is toxic. Avoid skin contact or aerosol formation and inhalation.

Nalidixic Acid Solution:
Composition per 10.0mL:
Nalidixic acid 0.05g

Preparation of Nalidixic Acid Solution: Add nalidixic acid to distilled/deionized water and bring volume to 10.0mL. Mix thoroughly. Filter sterilize.

Acriflavin·HCl Solution:
Composition per 10.0mL:
Acriflavin·HCl 0.05g

Preparation of Acriflavin·HCl Solution: Add acriflavin·HCl to distilled/deionized water and bring volume to 10.0mL. Mix thoroughly. Filter sterilize.

Preparation of Medium: Add components—except pyruvate solution, cycloheximide solution, nalidixic acid solution, and acriflavin·HCl solution—to distilled/deionized water and bring volume to 972.8mL. Mix thoroughly. Gently heat and bring to boiling. Autoclave for 15 min at 15 psi pressure–121°C. Cool to 45°–50°C. Aseptically add 11.1mL of sterile pyruvate solution. Mix thoroughly. Inoculate medium and incubate at 30°C for 6 hr. Aseptically add 5.1mL of sterile cycloheximide solution, 8.0mL of sterile nalidixic acid solution, and 3.0mL of sterile acriflavin·HCl solution. Mix thoroughly.

Use: For the isolation, cultivation, and enrichment of *Listeria* species from nondairy foods.

Enrichment Broth, pH 7.3 with Pyruvate

Composition per liter:
Pancreatic digest of casein 17.0g
NaCl 5.0g
Papaic digest of soybean meal 3.0g
Glucose 2.5g
K_2HPO_4 .... 2.5g
Yeast extract 6.0g
Pyruvate solution 11.1mL
Nalidixic acid solution 8.0mL
Cycloheximide solution 5.1mL
Acriflavin·HCl solution 2.0mL
pH 7.3 ± 0.2 at 25°C

Pyruvate Solution:
Composition per 20.0mL:
Sodium pyruvate 2.0g

Preparation of Pyruvate Solution: Add sodium pyruvate to distilled/deionized water and bring volume to 20.0mL. Mix thoroughly. Filter sterilize.

Cycloheximide Solution:
Composition per 10.0mL:
Cycloheximide 0.1g
Ethanol, absolute 4.0mL

Preparation of Cycloheximide Solution: Add components to distilled/deionized water and bring volume to 10.0mL. Mix thoroughly. Filter sterilize.

Caution: Cycloheximide is toxic. Avoid skin contact or aerosol formation and inhalation.

Nalidixic Acid Solution:
Composition per 10.0mL:
Nalidixic acid 0.05g

Preparation of Nalidixic Acid Solution: Add nalidixic acid to distilled/deionized water and bring volume to 10.0mL. Mix thoroughly. Filter sterilize.

Acriflavin·HCl Solution:
Composition per 10.0mL:
Acriflavin·HCl 0.05g

Preparation of Acriflavin·HCl Solution: Add acriflavin·HCl to distilled/deionized water and bring volume to 10.0mL. Mix thoroughly. Filter sterilize.

Preparation of Medium: Add components—except pyruvate solution, cycloheximide solution, nalidixic acid solution, and acriflavin·HCl solution—to distilled/deionized water and bring volume to 973.8mL. Mix thoroughly. Gently heat and bring to boiling. Autoclave for 15 min at 15 psi pressure–121°C. Cool to 45°–50°C. Aseptically add 11.1mL of sterile pyruvate solution. Mix thoroughly. Inoculate medium and incubate at 30°C for 6 hr. Aseptically add 5.1mL of sterile cycloheximide solution, 8.0mL of sterile nalidixic acid solution, and 2.0mL of sterile acriflavin·HCl solution. Mix thoroughly.

Use: For the isolation, cultivation, and enrichment of *Listeria* species from milk and dairy products.

Entamoeba dispar Axenic Culture Medium

Composition per liter:
YI-S solution 990.0mL
Gastric mucin stock suspension 10.0mL

YI-S Solution:
Composition per liter:
YI broth 880.0mL
Bovine serum, heat inactivated 100.0mL
Vitamin mixture 18 .... 20.0mL

Source: Vitamin mixture 18 is available from Biofluids, Rockville, MD.

YI Broth:

Composition per liter:

YI base stock	780.0mL
10× Glucose buffer stock	100.0mL

YI Base Stock:

Composition per 780.0mL:

Yeast extract	30.0g
L-Cysteine·HCl	1.0g
NaCl	1.0g
Ascorbic acid	0.2g
Ferric ammonium citrate	228.0mg

10× Glucose Buffer Stock:

Composition per 100.0.0mL:

Glucose	10.0g
K_2HPO_4	1.0g
KH_2PO_4	0.6g

Preparation of 10× Glucose Buffer Stock: Add components to distilled/deionized water and bring volume to 100.0mL. Mix thoroughly. Filter sterilize.

Preparation of YI Base Stock: Add components to 600.0mL of distilled/deionized water. Mix thoroughly. Bring volume to 780.0mL with distilled/deionized water. Adjust pH to 6.8 with 1*N* NaOH. Distribute in 78.0mL aliquots to 100.0mL screw-capped bottles. Autoclave for 15 min at 15 psi pressure–121°C. Cool to room temperature.

Preparation of YI Broth: Aseptically add 10.0mL of 10× glucose buffer stock to 78.0mL of cooled YI base stock. Adjust osmolarity with NaCl to 380.0milliosmols/kg.

Preparation of YI-S Solution: Aseptically add 2.0mL of vitamin mixture 18 and 10.0mL of heat-inactivated bovine serum to 88.0mL of YI broth. Distribute in 12.0mL aliquots to 16 x 125mm screw-capped test tubes. Store at 4°C in the dark with the caps screwed on tightly. Use within 96 hr.

Gastric Mucin Stock Suspension:

Composition per 100.0mL:

Gastric mucin	1.0g

Preparation of Gastric Mucin Stock Suspension: Add gastric mucin to distilled/deionized water and bring volume to 100.0mL. Mix thoroughly. Autoclave for 15 min at 15 psi pressure–121°C.

Preparation of Medium: Aseptically add 0.12mL of gastric mucin stock suspension (thoroughly mix by swirling before use) to 12.0mL of the complete YI-S solution in a 16 x 125mm screw-capped test tube. Store at 4°C in the dark with the caps screwed on tightly. Use within 96 hrs.

Use: For the cultivation of *Entamoeba dispar*.

Entamoeba HiVeg Medium

Composition per liter:

Agar	11.0g
Plant infusion No. 1	10.8g
Plant peptone No. 3	5.5g
Sodium glycerophosphate	3.0g
NaCl	2.7g

pH 7.0 ± 0.2 at 25°C

Source: This medium is available as a premixed powder from HiMedia.

Preparation of Medium: Add components to distilled/deionized water and bring volume to 1.0L. Mix thoroughly. Gently heat and bring to boiling. Distribute into tubes or flasks. Autoclave for 15 min at 15 psi pressure–121°C. Allow the tubes to solidify in a slanted position. Cover about half of the slant with fresh sterile horse seum saline mixture (1:6) and add a 5mm loopful of rice powder that has been sterilized in an oven at 160°C. Scorching should be prevented.

Use: For the cultivation of *Entamoeba histolytica.*

Entamoeba Medium (Endamoeba Medium)

Composition per liter:

Liver infusion	272.0g
Rice powder	14.2g
Agar	11.0g
Proteose peptone	5.5g
Sodium glycerophosphate	3.0g
NaCl	2.7g
Horse serum	50.0mL

pH 7.0 ± 0.2 at 25°C

Source: This medium is available as a premixed powder from BD Diagnostic Systems.

Rice Powder:

Composition per 15.0g:

Rice powder	15.0g

Preparation of Rice Powder: Sterilize rice powder at 160°C for 60 min. Do not overheat or rice powder will scorch.

Preparation of Medium: Add components, except horse serum and rice powder, to distilled/deionized water and bring volume to 994.0mL. Mix thoroughly. Gently heat and bring to boiling. Distribute into tubes in 7.0mL volumes. Autoclave for 15 min at 15 psi pressure–121°C. Allow tubes to cool in a slanted position. Aseptically add enough sterile horse serum to each tube to cover about half the slant. Aseptically add 0.1g of sterile rice powder to each tube.

Use: For the cultivation of *Entamoeba histolytica.*

Enteric Fermentation Base (Fermentation Base for *Campylobacter*)

Composition per liter:

Peptic digest of animal tissue	10.0g
NaCl	5.0g
Beef extract	3.0g
Carbohydrate solution	100.0mL
Andrade's indicator	10.0mL

pH 7.2 ± 0.1 at 25°C

Source: This medium is available as a premixed powder from BD Diagnostic Systems.

Carbohydrate Solution:

Composition per 100.0mL:

Carbohydrate	10.0g

Preparation of Carbohydrate Solution: Add carbohydrate to distilled/deionized water and bring volume to 100.0mL. Mix thoroughly. Filter sterilize. Glucose, lactose, mannitol, sucrose, adonitol, arabinose, cellobiose, dulcitol, glycerol, inositol, salicin, xylose, or other carbohydrates may be used. For the preparation of expensive carbohydrate solutions (adonitol, arabinose, cellobiose, dulcitol, glycerol, inositol, salicin, or xylose), 5.0g of carbohydrate per 100.0mL of distilled/deionized water may be used.

Andrade's Indicator:

Composition per 100.0mL:

NaOH (1*N* solution) 16.0mL
Acid Fuchsin 0.1g

Caution: Acid Fuchsin is a potential carcinogen and care must be taken to avoid inhalation of the powdered dye and contact with the skin.

Preparation of Andrade's Indicator: Add components to distilled/deionized water and bring volume to 100.0mL. Mix thoroughly.

Preparation of Medium: Add components, except carbohydrate solution, to distilled/deionized water and bring volume to 900.0mL. Mix thoroughly. Gently heat and bring to boiling. Distribute into tubes that contain an inverted Durham tube in 9.0mL volumes. Autoclave for 15 min at 15 psi pressure–121°C. Cool to 25°C. Aseptically add 1.0mL of sterile carbohydrate solution per tube. Mix thoroughly.

Use: For the cultivation and differentiation of a variety of bacteria based on their ability to ferment different carbohydrates. Bacteria that produce acid from carbohydrate fermentation turn the medium dark pink to red. Bacteria that produce gas have a bubble trapped in the Durham tube.

Enterobacter Medium

Composition per 800.0mL:

Casein hydrolysate 2.0g
K_2HPO_4 .... 1.4g
K_2SO_4 .... 1.0g
Yeast extract 1.0g
KH_2PO_4 .... 0.6g
$MgSO_4$ 0.5g
Glycerol 20.0mL

Preparation of Medium: Add components to distilled/deionized water and bring volume to 800.0mL. Mix thoroughly. Distribute into tubes or flasks. Autoclave for 15 min at 15 psi pressure–121°C.

Use: For the cultivation and maintenance of *Enterobacter* species and *Klebsiella pneumoniae.*

Enterobacteria Enrichment Broth, Mossel

Composition per liter:

Pancreatic digest of gelatin 10.0g
Na_2HPO_4, anhydrous 6.4g
Glucose, anhydrous 4.5g
Bile salts (sodium desoxycholate, sodium lauryl sulphate, sodium citrate) 3.055g
KH_2PO_4 .... 2.0g
Brilliant Green 0.015g

pH 7.2 ± 0.2 at 25°C

Source: This medium is available as a premixed powder from Oxoid Unipath.

Preparation of Medium: Add components to distilled/deionized water and bring volume to 1.0L. Mix thoroughly. Gently heat while stirring and bring to boiling. Maintain at 100°C for 30 min, typically using flowing steam. Do not autoclave. Cool rapidly in cold running water. MixAseptically distribute into tubes or flasks.

Use: For the enrichment of bile-tolerant Gram-negative bacteria in the microbiological examination of pharmaceutical products.

Enterobacteriaceae Enrichment Broth

See: **EE Broth**

Enterobacteriaceae Enrichment Broth, Mossel

See: **EE Broth, Mossel**

Enterococci Broth, Chromocult (Chromocult Enterococci Broth)

Composition per liter:

Peptone 8.6g
NaCl 6.4g
Tween 80 .... 2.2g
NaN_3 .... 0.6g
5-Bromo-4-chloro-3-indolyl-β-D-glucopyranoside (X-GLU) 0.04

pH 7.5 ± 0.2 at 25°C

Source: This medium is available from Merck.

Preparation of Medium: Add components to distilled/deionized water and bring volume to 1.0L. Mix well. Distribute into tubes. Autoclave for 15 min at 15 psi pressure–121°C. The prepared broth is clear and yellowish.

Use: For the detection of enterococci in food and water. The sodium azide present in this medium largely inhibits the growth of the accompanying, and especially the Gram-negative microbial flora while sparing the enterococci. The substrate X-GLU (5-bromo-4-chloro-3-indolyl-β-D-glucopyranoside) is cleaved, stimulated by selected peptones, by the enzyme β-D-glucosidase which is characteristic for enterococci. This results in an intensive blue-green color of the broth. Azide, at the same time, prevents a false positive result by most other β-D-glucosidase positive bacteria. Therefore, the color change of the broth largely confirms the presence of enterococci and group D streptococci.

Enterococci Confirmatory Agar

Composition per liter:

Agar 15.0g
Glucose 5.0g
Pancreatic digest of casein 5.0g
Yeast extract 5.0g
NaN_3 .... 0.4g
Methylene Blue 10.0mg
Enterococci confirmatory broth .... variable

pH 8.0 ± 0.2 at 25°C

Source: This medium is available as a premixed powder from BD Diagnostic Systems.

Caution: Sodium azide is toxic. Azides also react with metals and disposal must be highly diluted.

Preparation of Medium: Add components to distilled/deionized water and bring volume to 1.0L. Mix thoroughly. Gently heat and bring to boiling. Distribute into tubes. Autoclave for 15 min at 15 psi pressure–121°C. Allow tubes to cool in a slanted position. Add sufficient amount of Enterococci confirmatory broth (see below) to cover half the slant.

Use: For the identification of enterococci from water by the confirmatory test.

Enterococci Confirmatory Broth

Composition per liter:

NaCl 65.0g
Glucose 5.0g
Pancreatic digest of casein 5.0g
Yeast extract 5.0g
NaN_3 .... 0.4g

Methylene Blue 10.0mg
Penicillin 650U

pH 8.0 ± 0.2 at 25°C

Source: This medium is available as a premixed powder from BD Diagnostic Systems.

Caution: Sodium azide is toxic. Azides also react with metals and disposal must be highly diluted.

Preparation of Medium: Add components, except penicillin, to distilled/deionized water and bring volume to 1.0L. Mix thoroughly. Gently heat and bring to boiling. Autoclave for 15 min at 15 psi pressure–121°C. Cool to 25°C. Aseptically add penicillin. Mix thoroughly.

Use: For the identification of enterococci from water by the confirmatory test.

Enterococci Presumptive Broth

Composition per liter:

Glucose 5.0g
Pancreatic digest of casein 5.0g
Yeast extract 5.0g
NaN_3 .... 0.4g
Bromthymol Blue 32.0mg

pH 8.4 ± 0.2 at 25°C

Source: This medium is available as a premixed powder from BD Diagnostic Systems.

Caution: Sodium azide is toxic. Azides also react with metals and disposal must be highly diluted.

Preparation of Medium: Add components to distilled/deionized water and bring volume to 1.0L. Mix thoroughly. Distribute into tubes or flasks. Autoclave for 15 min at 15 psi pressure–121°C.

Use: For the isolation and identification of enterococci from water by the presumptive test. Bacteria that produce acid and turn the medium yellow and turbid after incubation at 45°C are presumptive enterococci.

Enterococcosel™ Agar

Composition per liter:

Pancreatic digest of casein 17.0g
Agar 13.5g
Oxgall 10.0g
NaCl 5.0g
Yeast extract 5.0g
Peptic digest of animal tissue 3.0g
Esculin 1.0g
Sodium citrate 1.0g
Ferric ammonium citrate 0.5g
NaN_3 .... 0.25g

pH 7.1 ± 0.2 at 25°C

Source: This medium is available as a premixed powder from BD Diagnostic Systems.

Caution: Sodium azide is toxic. Azides also react with metals and disposal must be highly diluted.

Preparation of Medium: Add components to distilled/deionized water and bring volume to 1.0L. Mix thoroughly. Gently heat while stirring and bring to boiling. Distribute into tubes or flasks. Autoclave for 15 min at 15 psi pressure–121°C. Pour into sterile Petri dishes or leave in tubes.

Use: For the rapid, selective isolation, cultivation, and enumeration of fecal group D streptococci (enterococci). For the cultivation of staphylococci and *Listeria monocytogenes*.

Enterococcosel™ Broth

Composition per liter:

Pancreatic digest of casein 17.0g
Oxgall 10.0g
NaCl 5.0g
Yeast extract 5.0g
Peptic digest of animal tissue 3.0g
Esculin 1.0g
Sodium citrate 1.0g
Ferric ammonium citrate 0.5g
NaN_3 .... 0.25g

pH 7.1 ± 0.2 at 25°C

Source: This medium is available as a premixed powder from BD Diagnostic Systems.

Caution: Sodium azide is toxic. Azides also react with metals and disposal must be highly diluted.

Preparation of Medium: Add components to distilled/deionized water and bring volume to 1.0L. Mix thoroughly. Gently heat while stirring until dissolved. Distribute into tubes or flasks. Autoclave for 15 min at 15 psi pressure–121°C.

Use: For the cultivation and differentiation of group D streptococci (enterococci).

Enterococcus Agar (m-*Enterococcus* Agar) (Azide Agar)

Composition per liter:

Pancreatic digest of casein 15.0g
Agar 10.0g
Papaic digest of soybean meal 5.0g
Yeast extract 5.0g
KH_2PO_4 .... 4.0g
Glucose 2.0g
NaN_3 .... 0.4g
Triphenyltetrazolium chloride 0.1g

pH 7.2 ± 0.2 at 25°C

Source: This medium is available as a premixed powder from BD Diagnostic Systems.

Caution: Sodium azide is toxic. Azides also react with metals and disposal must be highly diluted.

Preparation of Medium: Add components to distilled/deionized water and bring volume to 1.0L. Mix thoroughly. Gently heat and bring to boiling. Cool to 45°–50°C. Do not autoclave. Pour into sterile Petri dishes.

Use: For the isolation, cultivation, and enumeration of entercocci in water, sewage, and feces by the membrane filter method. For the direct plating of specimens for the detection and enumeration of fecal streptococci.

Enterococcus Confirmatory HiVeg Agar with Penicillin

Composition per liter:

Agar 15.0g
Glucose 5.0g
Plant hydrolysate 5.0g

Yeast extract 5.0g
NaN_3 .... 0.4g
Methylene Blue 0.01g
Penicillin solution 10.0mL

pH 8.0± 0.2 at 25°C

Source: This medium, without penicillin, is available as a premixed powder from HiMedia.

Caution: Sodium azide is toxic. Azides also react with metals and disposal must be highly diluted.

Penicillin Solution:

Composition per 10.0mL:

Penicillin 650U

Preparation of Penicillin Solution: Add penicillin to distilled/deionized water and bring volume to 10.0mL. Mix thoroughly. Filter sterilize.

Preparation of Medium: Add components, except penicillin solution, to distilled/deionized water and bring volume to 1.0L. Mix thoroughly. Gently heat and bring to boiling. Autoclave for 15 min at 15 psi pressure–121°C. Cool to 45°–50°C. Aseptically add 10.0mL sterile penicillin solution. Mix thoroughly. Pour into sterile Petri dishes or dispense into sterile tubes.

Use: For the isolation, cultivation, and enumeration of entercocci in water, sewage, and feces by the membrane filter method. For the direct plating of specimens for the detection and enumeration of fecal streptococci.

Enterococcus Confirmatory HiVeg Broth with Penicillin

Composition per liter:

NaCl 65.0g
Glucose 5.0g
Plant hydrolysate 5.0g
Yeast extract 5.0g
NaN_3 .... 0.4g
Methylene Blue 0.01g
Penicillin solution 10.0mL

pH 8.0 ± 0.2 at 25°C

Source: This medium, without penicillin, is available as a premixed powder from HiMedia.

Caution: Sodium azide is toxic. Azides also react with metals and disposal must be highly diluted.

Penicillin Solution:

Composition per 10.0mL:

Penicillin 650U

Preparation of Penicillin Solution: Add penicillin to distilled/deionized water and bring volume to 10.0mL. Mix thoroughly. Filter sterilize.

Preparation of Medium: Add components, except penicillin solution, to distilled/deionized water and bring volume to 1.0L. Mix thoroughly. Gently heat and bring to boiling. Autoclave for 15 min at 15 psi pressure–121°C. Cool to 45°–50°C. Aseptically add 10.0mL sterile penicillin solution. Mix thoroughly.

Use: For the cultivation of entercocci in water, sewage, and feces.

Enterococcus faecium Medium

Composition per liter:

Sucrose 97.3g
Brain heart, solids from infusion 37.0g
Agar 13.3g
NaCl 9.3g
Yeast extract 5.0g
$MgSO_4$ 0.25g

pH 7.4 ± 0.2 at 25°C

Preparation of Medium: Add components to distilled/deionized water and bring volume to 1.0L. Mix thoroughly. Gently heat and bring to boiling. Distribute into tubes or flasks. Autoclave for 15 min at 15 psi pressure–121°C. Pour into sterile Petri dishes or leave in tubes.

Use: For the cultivation and maintenance of *Enterococcus faecium.*

Enterococcus Presumptive HiVeg Broth with Penicillin

Composition per liter:

Glucose 5.0g
Plant hydrolysate 5.0g
Yeast extract 5.0g
NaN_3 .... 0.4g
Bromthymol Blue 0.032g
Penicillin solution 10.0mL

pH 8.4 ± 0.2 at 25°C

Source: This medium, without penicillin, is available as a premixed powder from HiMedia.

Caution: Sodium azide is toxic. Azides also react with metals and disposal must be highly diluted.

Penicillin Solution:

Composition per 10.0mL:

Penicillin 650U

Preparation of Penicillin Solution: Add penicillin to distilled/deionized water and bring volume to 10.0mL. Mix thoroughly. Filter sterilize.

Preparation of Medium: Add components, except penicillin solution, to distilled/deionized water and bring volume to 1.0L. Mix thoroughly. Gently heat and bring to boiling. Autoclave for 15 min at 15 psi pressure–121°C. Cool to 45°–50°C. Aseptically add 10.0mL sterile penicillin solution. Mix thoroughly.

Use: For the cultivation and presumptive identification of enterococci.

Eosin Methylene Blue Agar

See: **EMB Agar**

Eosin Methylene Blue Agar, Levine

See: **Levine EMB Agar**

Eosin Methylene Blue Agar, Modified

See: **EMB Agar, Modified**

Epoxysuccinic Acid Medium

See: **ESA Medium**

Erwinia amylovora Selective Medium

Composition per liter:

Agar 20.0g
Mannitol 10.0g
L-Asparagine 3.0g
Sodium taurocholate 2.5g

K_2HPO_4	2.0g
Nicotinic acid	0.5g
$MgSO_4 \cdot 7H_2O$	0.2g
Nitrilotriacetic acid	10.0mL
Actidione (cycloheximide) solution	10.0mL
Bromthymol Blue	9.0mL
Neutral Red	2.5mL
$TlNO_3$ solution	1.75mL
Tergitol™ 7	0.1mL

pH 7.2–7.3 at 25°C

Cycloheximide Solution:
Composition per 10.0mL:

Cycloheximide	0.05g

Preparation of Cycloheximide Solution: Add cycloheximide to distilled/deionized water and bring volume to 10.0mL. Mix thoroughly. Filter sterilize.

Caution: Cycloheximide is toxic. Avoid skin contact or aerosol formation and inhalation.

$TlNO_3$ Solution:
Composition per 10.0mL:

$TlNO_3$	0.1g

Preparation of $TlNO_3$ Solution: Add $TlNO_3$ to distilled/deionized water and bring volume to 10.0mL. Mix thoroughly. Filter sterilize.

Preparation of Medium: Add components, except $TlNO_3$ solution and cycloheximide solution, to distilled/deionized water and bring volume to 988.25mL. Mix thoroughly. Adjust pH to 7.2–7.3. Gently heat and bring to boiling. Autoclave for 15 min at 15 psi pressure–121°C. Cool to 45°–50°C. Aseptically add 1.75mL of sterile $TlNO_3$ solution and 10.0mL of sterile cycloheximide solution. Mix thoroughly. Pour into sterile Petri dishes or distribute into sterile tubes.

Use: For the isolation and cultivation of *Erwinia amylovora.*

Erwinia Fermentation Medium

Composition per liter:

Lactose	45.0g
K_2HPO_4	3.6g
KH_2PO_4	1.8g
Yeast extract	1.8g
$(NH_4)_2SO_4$	1.46g
$MgSO_4 \cdot 7H_2O$	0.6g
$CaCl_2 \cdot 2H_2O$	0.04g
$FeSO_4 \cdot 7H_2O$	1.9mg
$CoCl_2$	1.0mg
$CuSO_4 \cdot 5H_2O$	1.0mg
$MnSO_4 \cdot H_2O$	1.0mg
$Na_2MoO_4 \cdot 2H_2O$	1.0mg
$ZnSO_4 \cdot 7H_2O$	1.0mg

pH 7.0 ± 0.2 at 25°C

Preparation of Medium: Add components to distilled/deionized water and bring volume to 1.0L. Mix thoroughly. Adjust pH to 7.0 with NaOH (approximately 0.18g). Distribute into tubes or flasks. Autoclave for 15 min at 15 psi pressure–121°C.

Use: For the cultivation of *Rahnella aquatilis.*

Erwinia Medium D3

Composition per liter:

Agar	15.0g
Arabinose	10.0g
Sucrose	10.0g
LiCl	7.0g
Casein hydrolysate	5.0g
NaCl	5.0g
$MgSO_4 \cdot 7H_2O$	0.3g
Acid Fuchsin	0.1g
Bromthymol Blue	0.06g
Sodium dodecyl sulfate	0.05g

pH 7.0 ± 0.2 at 25°C

Caution: Acid Fuchsin is a potential carcinogen and care must be taken to avoid inhalation of the powdered dye and contact with the skin.

Preparation of Medium: Add components to distilled/deionized water and bring volume to 1.0L. Mix thoroughly. Gently heat and bring to boiling. Adjust pH to 8.2. Autoclave for 15 min at 15 psi pressure–121°C. Cool to 45°–50°C. The pH after autoclaving should be 7.0. Pour into sterile Petri dishes or distribute into sterile tubes.

Use: For the isolation and cultivation of *Erwinia* species.

Erwinia Selective Medium

Composition per liter:

Agar	15.0g
$(NH_4)_2SO_4$	5.0g
K_2HPO_4	2.0g
Eosin Y	0.4g
Methylene Blue	0.065g
Glycerol	10.0mL
Antibiotic solution	10.0mL

Antibiotic Solution:
Composition per 10.0mL:

Cycloheximide	0.25g
Novobiocin	0.04g
Neomycin sulfate	0.04g

Preparation of Antibiotic Solution: Add components to distilled/deionized water and bring volume to 10.0mL. Mix thoroughly. Filter sterilize.

Caution: Cycloheximide is toxic. Avoid skin contact or aerosol formation and inhalation.

Preparation of Medium: Add components, except antibiotic solution, to distilled/deionized water and bring volume to 990.0mL. Mix thoroughly. Gently heat and bring to boiling. Autoclave for 15 min at 15 psi pressure–121°C. Cool to 45°–50°C. Aseptically add sterile antibiotic solution. Mix thoroughly. Pour into sterile Petri dishes or distribute into sterile tubes.

Use: For the selective isolation and cultivation of *Erwinia* species.

Erwinia tracheiphila Agar

Composition per liter:

Agar	15.0g
Glucose	10.0g
Peptone	10.0g
Beef extract	5.0g
Yeast extract	1.0g

pH 7.4 ± 0.2 at 25°C

Preparation of Medium: Add components to distilled/deionized water and bring volume to 1.0L. Mix thoroughly. Gently heat and bring to boiling. Distribute into tubes or flasks. Autoclave for 15 min at 15 psi pressure–121°C. Pour into sterile Petri dishes or leave in tubes.

Use: For the cultivation and maintenance of *Erwinia tracheiphila.*

Erysipelothrix Medium

Composition per liter:

$Na_2HPO_4 \cdot 12H_2O$	18.0g
Glucose	6.0g
Peptone S	5.0g
Yeast extract	5.0g
L-Arginine·HCl	0.5g
Tween™ 80	0.5mL

pH 7.8–8.0 at 25°C

Preparation of Medium: Add components to distilled/deionized water and bring volume to 1.0L. Mix thoroughly. Distribute into tubes or flasks. Autoclave for 15 min at 15 psi pressure–121°C.

Use: For the cultivation of *Erysipelothrix* species.

Erythritol Agar

Composition per liter:

Agar	15.0g
Erythritol	2.0g
K_2HPO_4	1.15g
NH_4NO_3	1.0g
KH_2PO_4	0.625g
$MgSO_4 \cdot 7H_2O$	0.02g

Preparation of Medium: Add components to distilled/deionized water and bring volume to 1.0L. Mix thoroughly. Gently heat and bring to boiling. Distribute into tubes or flasks. Autoclave for 15 min at 15 psi pressure–121°C. Pour into sterile Petri dishes or leave in tubes.

Use: For the cultivation and maintenance of *Klebsiella pneumoniae.*

Erythritol Broth

Composition per liter:

Erythritol	2.0g
K_2HPO_4	1.15g
NH_4NO_3	1.0g
Yeast extract	1.0g
KH_2PO_4	0.625g
$MgSO_4 \cdot 7H_2O$	0.02g

Preparation of Medium: Add components to distilled/deionized water and bring volume to 1.0L. Mix thoroughly. Distribute into tubes or flasks. Autoclave for 15 min at 15 psi pressure–121°C.

Use: For the cultivation of *Klebsiella pneumoniae.*

Erythrobacter longus Medium

Composition per liter:

Peptone	2.0g
Proteose peptone No. 3	1.0g
Papaic digest of soybean meal	1.0g
Yeast extract	1.0g
Artificial seawater	700.0mL
Ferric citrate solution	2.0mL

pH 7.5 ± 0.2 at 25°C

Ferric Citrate Solution:

Composition per 10.0mL:

Ferric citrate	0.5g

Preparation of Ferric Citrate Solution: Add ferric citrate to distilled/deionized water and bring volume to 10.0mL. Mix thoroughly.

Preparation of Medium: Add components to distilled/deionized water and bring volume to 1.0L. Mix thoroughly. Distribute into tubes or flasks. Autoclave for 15 min at 15 psi pressure–121°C.

Use: For the cultivation of *Erythrobacter longus.*

Erythromicrobium roseococcus Medium (DSMZ Medium 767)

Composition per 1001.0mL:

Na-acetate	1.0g
Yeast extract	1.0g
Peptone	1.0g
$MgSO_4 \cdot 7H_2O$	0.5g
NH_4Cl	0.3g
K_2HPO_4	0.3g
KCl	0.3g
$CaCl_2 \cdot 2H_2O$	0.05g
Trace elements solution SL-6	1.0mL
Vitamin B_{12} solution	1.0mL

pH 7.5–7.8 at 25°C

Vitamin B_{12} Solution:

Composition per 100.0mL:

Vitamin B_{12}	2.0mg

Preparation of Vitamin B_{12} Solution: Add vitamin B_{12} to distilled/deionized water and bring volume to 100.0mL. Mix thoroughly. Filter sterilize.

Trace Elements Solution SL-6:

Composition per liter:

$MnCl_2 \cdot 4H_2O$	0.5g
H_3BO_3	0.3g
$CoCl_2 \cdot 6H_2O$	0.2g
$ZnSO_4 \cdot 7H_2O$	0.1g
$Na_2MoO_4 \cdot 2H_2O$	0.03g
$NiCl_2 \cdot 6H_2O$	0.02g
$CuCl_2 \cdot 2H_2O$	0.01g

Preparation of Trace Elements Solution SL-6: Add components to distilled/deionized water and bring volume to 1.0L. Mix thoroughly. Autoclave for 15 min at 15 psi pressure–121°C.

Preparation of Medium: Add components, except vitamin solution, to distilled/deionized water and bring volume to 1.0L. Mix thoroughly. Adjust pH to 7.5–7.8. Autoclave for 15 min at 15 psi pressure–121°C. Cool to room temperature. Aseptically add 10mL vitamin solution. Mix thoroughly. Aseptically distribute into sterile tubes or bottles.

Use: For the cultivation of *Erythrobacter litoralis, Erythromicrobium ramosum,* and *Roseococcus thiosulfatophilus.*

Erythromycin L Broth Medium

Composition per liter:

Pancreatic digest of casein	10.0g
NaCl	5.0g
Yeast extract	5.0g
Glucose	1.0g
Erythromycin solution	10.0mL

pH 7.0 ± 0.2 at 25°C

Erythromycin Solution:

Composition per 10.0mL:

Erythromycin	10.0mg

Preparation of Erythromycin Solution: Add erythromycin to distilled/deionized water and bring volume to 10.0mL. Mix thoroughly. Filter sterilize.

Preparation of Medium: Add components, except erythromycin solution, to distilled/deionized water and bring volume to 990.0mL. Mix thoroughly. Bring pH to 7.0. Autoclave for 15 min at 15 psi pressure–121°C. Aseptically add 10.0mL of sterile erythromycin solution. Mix thoroughly. Aseptically distribute into sterile tubes or flasks.

Use: For the cultivation of *Escherichia coli*.

Erythromycin LB Medium

Composition per liter:

Pancreatic digest of casein	10.0g
NaCl	10.0g
Yeast extract	5.0g
Erythromycin solution	10.0mL

pH 7.0 ± 0.2 at 25°C

Erythromycin Solution:
Composition per 10.0mL:

Erythromycin	50.0mg

Preparation of Erythromycin Solution: Add erythromycin to distilled/deionized water and bring volume to 10.0mL. Mix thoroughly. Filter sterilize.

Preparation of Medium: Add components, except erythromycin solution, to distilled/deionized water and bring volume to 990.0mL. Mix thoroughly. Bring pH to 7.0. Autoclave for 15 min at 15 psi pressure–121°C. Aseptically add 10.0mL of sterile erythromycin solution. Mix thoroughly. Aseptically distribute into sterile tubes or flasks.

Use: For the cultivation of *Escherichia coli*.

ESA Medium (Epoxysuccinic Acid Medium)

Composition per liter:

Agar	20.0g
$(NH_4)_2SO_4$	3.0g
KH_2PO_4	1.5g
Na_2HPO_4	1.5g
$MgSO_4·7H_2O$	0.5g
Yeast extract	0.5g
$CaCl_2·2H_2O$	0.01g
$FeSO_4·7H_2O$	0.01g
$MnSO_4·4H_2O$	0.001g
Epoxysuccinate solution	100.0mL

pH 7.0 ± 0.2 at 25°C

Epoxysuccinate Solution:
Composition per 100.0mL:

Sodium *cis*-epoxysuccinate	10.0g

Preparation of Epoxysuccinate Solution: Add sodium *cis*-epoxysuccinate to distilled/deionized water and bring volume to 100.0mL. Mix thoroughly. Filter sterilize.

Preparation of Medium: Add components, except epoxysuccinate solution, to distilled/deionized water and bring volume to 900.0mL. Mix thoroughly. Gently heat and bring to boiling. Autoclave for 15 min at 15 psi pressure–121°C. Cool to 50°–60°C. Aseptically add sterile epoxysuccinate solution. Mix thoroughly. Pour into sterile Petri dishes or distribute into sterile tubes.

Use: For the cultivation and maintenance of *Nocardia tartaricans*.

Escherichia coli Broth
See: ***EC*** **Broth**

Escherichia Medium (ATCC Medium 52)

Composition per liter:

Glucose	40.0g
Agar	30.0g
Peptone	10.0g

Preparation of Medium: Add components to distilled/deionized water and bring volume to 1.0L. Mix thoroughly. Gently heat and bring to boiling. Distribute into tubes or flasks. Autoclave for 15 min at 15 psi pressure–121°C. Pour into sterile Petri dishes or leave in tubes.

Use: For the cultivation and maintenance of *Escherichia coli*.

Escherichia Medium (ATCC Medium 53)

Composition per liter:

Agar	15.0g
Casein hydrolysate	6.0g
K_2HPO_4	0.2g
$MgSO_4·7H_2O$	0.2g
Asparagine	0.15g
$FeSO_4·7H_2O$	1.0μg
Vitamin B_{12} solution	10.0mL
Glycerol	2.0mL

pH 7.0 ± 0.2 at 25°C

Vitamin B_{12} Solution:
Composition per 10.0mL:

Vitamin B_{12}	0.04g

Preparation of Vitamin B_{12} Solution: Add vitamin B_{12} to distilled/deionized water and bring volume to 10.0mL. Mix thoroughly.

Preparation of Medium: Add components—except glycerol, agar, and vitamin B_{12} solution—to distilled/deionized water and bring volume to 900.0mL. Mix thoroughly. Gently heat and bring to boiling. Adjust pH to 7.0. Filter through Whatman #1 filter paper. Add glycerol and agar. Mix thoroughly. Gently heat and bring to boiling. Add vitamin B_{12} solution. Distribute into tubes or flasks. Autoclave for 15 min at 15 psi pressure–121°C. Pour into sterile Petri dishes or leave in tubes.

Use: For the cultivation and maintenance of *Escherichia coli*.

Escherichia Medium (ATCC Medium 57)

Composition per liter:

Agar	15.0g
K_2HPO_4	7.0g
Glycerol	5.0g
KH_2PO_4	3.0g
$(NH_4)_2SO_4$	1.5g
L-Lysine	0.1g
$MgSO_4$	0.1g
$CaCl_2$	0.01g
$FeSO_4·7H_2O$	0.5mg

pH 7.1 ± 0.2 at 25°C

Preparation of Medium: Add components to distilled/deionized water and bring volume to 1.0L. Mix thoroughly. Gently heat and bring

to boiling. Distribute into tubes or flasks. Autoclave for 15 min at 15 psi pressure–121°C. Pour into sterile Petri dishes or leave in tubes.

Use: For the cultivation and maintenance of *Escherichia coli.*

Escherichia Medium (ATCC Medium 60)

Composition per liter:

Agar	15.0g
K_2HPO_4	7.0g
KH_2PO_4	3.0g
Casein hydrolysate	2.0g
Sodium citrate, anhydrous	0.4g
$MgSO_4 \cdot 7H_2O$	0.1g
$(NH_4)_2SO_4$	0.1g
Glycerol	2.0mL

Preparation of Medium: Add components to distilled/deionized water and bring volume to 1.0L. Mix thoroughly. Gently heat and bring to boiling. Distribute into tubes or flasks. Autoclave for 15 min at 15 psi pressure–121°C. Pour into sterile Petri dishes or leave in tubes.

Use: For the cultivation and maintenance of *Escherichia coli.*

Escherichia Medium (ATCC Medium 62)

Composition per 700.0mL:

Agar	12.5g
Casein hydrolysate	6.0g
Glycerol	2.0g
K_2HPO_4	0.2g
$MgSO_4 \cdot 7H_2O$	0.2g
L-Asparagine	0.15g
$FeSO_4 \cdot 7H_2O$	5.0mg
Vitamin B_{12}	0.4mg

pH 7.2 ± 0.2 at 25°C

Preparation of Medium: Add components, except glycerol, agar, and vitamin B_{12}, to distilled/deionized water and bring volume to 500.0mL. Adjust pH to 7.2. Mix thoroughly. Gently heat and bring to boiling. Filter through Whatman #1 filter paper. Cool to 25°C. Add agar and glycerol. Bring volume to 1.0L with distilled/deionized water. Mix thoroughly. Gently heat and bring to boiling. Readjust pH to 7.2. Add vitamin B_{12}. Mix thoroughly. Distribute into tubes in 10.0mL volumes. Autoclave for 15 min at 15 psi pressure–121°C. Allow tubes to cool in a slanted position.

Use: For the cultivation and maintenance of *Escherichia coli.*

Escherichia Medium (ATCC Medium 271)

Composition per liter:

Pancreatic digest of casein	10.0g
NaCl	8.0g
Yeast extract	1.0g

Preparation of Medium: Add components to distilled/deionized water and bring volume to 1.0L. Mix thoroughly. Distribute into tubes or flasks. Autoclave for 15 min at 15 psi pressure–121°C.

Use: For the cultivation of *Escherichia coli.*

Escherichia Medium (ATCC Medium 765)

Composition per liter:

K_2HPO_4	11.67g
Casamino acids	11.0g
KH_2PO_4	4.49g
$(NH_4)_2SO_4$	1.98g
$MgSO_4 \cdot 7H_2O$	0.25g
$FeSO_4 \cdot 7H_2O$	0.5mg
Glycerol	20.25mL

Preparation of Medium: Add components to distilled/deionized water and bring volume to 1.0L. Mix thoroughly. Distribute into tubes or flasks. Autoclave for 15 min at 15 psi pressure–121°C.

Use: For the cultivation and maintenance of *Escherichia coli.*

Esculin Agar

Composition per liter:

Agar	15.0g
Pancreatic digest of casein	13.0g
NaCl	5.0g
Yeast extract	5.0g
Heart muscle, solids from infusion	2.0g
Esculin	1.0g
Ferric citrate	0.5g

pH 7.3 ± 0.2 at 25°C

Preparation of Medium: Add components to distilled/deionized water and bring volume to 1.0L. Mix thoroughly. Gently heat and bring to boiling. Distribute into screw-capped tubes in 3.0mL volumes. Autoclave for 15 min at 15 psi pressure–121°C. Allow tubes to cool in a slanted position.

Use: For the cultivation and differentiation of bacteria based on their ability to hydrolyze esculin and produce H_2S. Bacteria that hydrolyze esculin appear as colonies surrounded by a reddish-brown to dark brown zone. Bacteria that produce H_2S appear as black colonies.

Esculin Agar, Lombard-Dowell
See: **Lombard-Dowell Esculin Agar**

Esculin Agar, Modified CDC (BAM M53)

Composition per liter:

Heart muscle, infusion from	375.0g
Agar	15.0g
Thiotone	10.0g
NaCl	5.0g
Esculin	1.0g
Ferric citrate	0.5g

pH 7.0 ± 0.2 at 25°C

Preparation of Medium: Add components to distilled/deionized water and bring volume to 1.0L. Mix thoroughly. Gently heat and bring to boiling. Cool to 55°C. Adjust pH to 7.0. Distribute into tubes or leave in flask. Autoclave for 20 min at 15 psi pressure–121°C. Pour into sterile Petri dishes or leave in tubes. Allow tubes to cool in inclined position to produce slants.

Use: For the differentiation of *Enterobacter* spp.

Esculin Azide Broth

Composition per liter:

Peptic digest of animal tissue	20.0g
Bile salts	10.0g
Yeast extract	5.0g
Esculin	1.0g
Sodium citrate	1.0g
Ferric ammonium citrate	0.5g
NaN_3	0.25g

pH 7.2 ± 0.2 at 25°C

Source: This medium is available as a premixed powder from HiMedia.

Caution: Sodium azide is toxic. Azides also react with metals and disposal must be highly diluted.

Preparation of Medium: Add components to distilled/deionized water and bring volume to 1.0L. Mix thoroughly. Gently heat and bring to boiling. Cool to 45°–50°C. Do not autoclave.

Use: For the cultivation of entercocci in water, sewage, and feces.

Esculin Azide HiVeg Broth

Composition per liter:

Plant peptone	25.0g
Synthetic detergent	5.0g
Yeast extract	5.0g
Esculin	1.0g
Sodium citrate	1.0g
Ferric ammonium citrate	0.5g
NaN_3	0.25g

pH 7.2 ± 0.2 at 25°C

Source: This medium is available as a premixed powder from HiMedia.

Caution: Sodium azide is toxic. Azides also react with metals and disposal must be highly diluted.

Preparation of Medium: Add components to distilled/deionized water and bring volume to 1.0L. Mix thoroughly. Gently heat and bring to boiling. Cool to 45°–50°C. Do not autoclave.

Use: For the cultivation of entercocci in water, sewage, and feces.

Esculin Broth

Composition per liter:

Beef heart, solids from infusion	500.0g
Tryptose	10.0g
NaCl	5.0g
Agar	1.0g
Esculin	1.0g

pH 7.0 ± 0.2 at 25°C

Preparation of Medium: Add components to distilled/deionized water and bring volume to 1.0L. Mix thoroughly. Gently heat and bring to boiling. Distribute into screw-capped tubes in 7.0mL volumes. Autoclave for 15 min at 15 psi pressure–121°C.

Use: For the cultivation and differentiation of bacteria based on their ability to hydrolyze esculin. Bacteria that hydrolyze esculin turn the medium brown-black to black.

Esculin Iron Agar

Composition per liter:

Agar	15.0g
Esculin	1.0g
Ferric ammonium citrate	0.5g

pH 7.1 ± 0.2 at 25°C

Source: This medium is available as a premixed powder from BD Diagnostic Systems.

Preparation of Medium: Add components to distilled/deionized water and bring volume to 1.0L. Mix thoroughly. Gently heat and bring to boiling. Distribute into tubes or flasks. Autoclave for 15 min at 15 psi pressure–121°C. Pour into sterile Petri dishes.

Use: For the cultivation and identification of enterococci based on their ability to hydrolyze esculin. Used in conjunction with E agar and the membrane filter method.

Esculin Mannitol Agar

Composition per liter:

Agar	13.5g
Polypeptone™	10.0g
D-Mannitol	10.0g
Pancreatic digest of casein	5.0g
Yeast extract	5.0g
NaCl	5.0g
Heart peptone	3.0g
Cornstarch	1.0g
Esculin	1.0g
Ferric ammonium citrate	0.5g
Phenol Red	0.025g
Nalidixic acid solution	10.0mL
Colistin solution	10.0mL

pH 7.3 ± 0.2 at 25°C

Nalidixic Acid Solution:

Composition per 10.0mL:

Nalidixic acid	0.015g

Preparation of Nalidixic Acid Solution: Add nalidixic acid to distilled/deionized water and bring volume to 10.0mL. Mix thoroughly. Filter sterilize.

Colistin Solution:

Composition per 10.0mL:

Colistin	0.01g

Preparation of Colistin Solution: Add colistin to distilled/deionized water and bring volume to 10.0mL. Mix thoroughly. Filter sterilize.

Preparation of Medium: Add components, except nalidixic acid solution and colistin solution, to distilled/deionized water and bring volume to 980.0mL. Mix thoroughly. Gently heat and bring to boiling. Autoclave for 15 min at 15 psi pressure–121°C. Cool to 45°–50°C. Aseptically add sterile nalidixic acid solution and colistin solution. Mix thoroughly. Pour into sterile Petri dishes or distribute into sterile tubes.

Use: For the selective isolation, cultivation, and differentiation of *Staphylococcus aureus* and group D streptococci based on mannitol fermentation and hydrolysis of esculin. Bacteria that ferment mannitol appear as yellow colonies surrounded by a yellow zone. Bacteria that hydrolyze esculin appear as dark brown to black colonies surrounded by a dark brown to black zone.

Esculin Thallium Medium

Composition per liter:

Agar	15.0g
Beef extract	10.0g

Peptone	10.0g
NaCl	5.0g
Esculin	1.0g
Thallous sulfate	0.33g
Crystal Violet	1.3mg
Blood, bovine or sheep	50.0mL

pH 7.4 ± 0.2 at 25°C

Preparation of Medium: Add components, except blood, to distilled/deionized water and bring volume to 950.0mL. Mix thoroughly. Gently heat and bring to boiling. Autoclave for 15 min at 15 psi pressure–121°C. Cool to 45°–50°C. Aseptically add sterile blood. Mix thoroughly. Pour into sterile Petri dishes or distribute into sterile tubes.

Caution: Thallium salts are toxic.

Use: For the cultivation of *Streptococcus* species that cause bovine mastitis.

E.T. Medium

Composition per liter:

Beef heart, infusion from	250.0g
Liver, infusion from	250.0g
Peptone, special	20.0g
NaCl	5.0g
K_2HPO_4	4.0g

pH 7.1 ± 0.2 at 25°C

Source: This medium is available as a premixed powder from HiMedia.

Preparation of Medium: Add components to distilled/deionized water and bring volume to 1.0L. Mix thoroughly. Gently heat and bring to boiling. Distribute into tubes or flasks. Autoclave for 15 min at 15 psi pressure–121°C. Pour into sterile Petri dishes.

Use: For the mass cultivation of clostridia for enterotoxin production.

Ethanoligenes Medium (DSMZ Medium 1057)

Composition per liter:

Tryptone	4.0g
NaCl	4.0g
Beef extract	2.0g
K_2HPO_4	1.5g
Yeast extract	1.0g
L-cysteine·HCl	0.5g
$MgCl_2 \cdot 6H_2O$	0.2g
$FeSO_4 \cdot 7H_2O$	0.1g
Vitamin solution	10.0mL
Trace elements solution	1.0mL

Vitamin Solution:

Composition per liter:

Pyridoxine-HCl	10.0mg
Thiamine-HCl·$2H_2O$	5.0mg
Riboflavin	5.0mg
Nicotinic acid	5.0mg
D-Ca-pantothenate	5.0mg
p-Aminobenzoic acid	5.0mg
Lipoic acid	5.0mg
Biotin	2.0mg
Folic acid	2.0mg
Vitamin B_{12}	0.1mg

Preparation of Vitamin Solution: Add components to distilled/deionized water and bring volume to 1.0L. Mix thoroughly. Sparge with 100% N_2. Filter sterilize.

Glucose Solution:

Composition per 10.0mL:

Glucose	10.0g

Preparation of Glucose Solution: Add components to distilled/deionized water and bring volume to 10.0mL. Mix thoroughly. Sparge with 100% N_2. Filter sterilize.

Trace Elements Solution:

Composition per liter:

$MgSO_4 \cdot 7H_2O$	3.0g
Nitrilotriacetic acid	1.5g
NaCl	1.0g
$MnSO_4 \cdot 2H_2O$	0.5g
$CoSO_4 \cdot 7H_2O$	0.18g
$ZnSO_4 \cdot 7H_2O$	0.18g
$CaCl_2 \cdot 2H_2O$	0.1g
$FeSO_4 \cdot 7H_2O$	0.1g
$NiCl_2 \cdot 6H_2O$	0.025g
$KAl(SO_4)_2 \cdot 12H_2O$	0.02g
H_3BO_3	0.01g
$Na_2MoO_4 \cdot 4H_2O$	0.01g
$CuSO_4 \cdot 5H_2O$	0.01g
$Na_2SeO_3 \cdot 5H_2O$	0.3mg

Preparation of Trace Elements Solution: Add nitrilotriacetic acid to 500.0mL of distilled/deionized water. Dissolve by adjusting pH to 6.5 with KOH. Add remaining components. Add distilled/deionized water to 1.0L. Mix thoroughly.

Preparation of Medium: Add components, except glucose and vitamin solutions, to distilled/deionized water and bring volume to 980.0mL. Mix thoroughly. Adjust pH to 6.0. Gently heat while stirring and bring to boiling. Boil for 1 min. Autoclave for 15 min at 15 psi pressure–121°C. Cool to room temperature while sparging with N_2. Dispense into tubes or bottles under an atmosphere of 100% N_2. Prior to inoculation, aseptically and anoxically add the vitamin and glucose solutions.

Use: For the cultivation of *Ethanoligenes* spp.

ETGPA (Egg Tellurite Glycine Pyruvate Agar)

Composition per liter:

Agar	17.0g
Glycine	12.0g
Sodium pyruvate	10.0g
Pancreatic digest of casein	10.0g
Beef extract	5.0g
LiCl	5.0g
Yeast extract	1.0g
Egg yolk emulsion	50.0mL
K_2TeO_3 solution	10.0mL

pH 7.0 ± 0.2 at 25°C

Source: This medium is available as a premixed powder from BD Diagnostic Systems.

Egg Yolk Emulsion:

Composition:

Chicken egg yolks	11
Whole chicken egg	1

Preparation of Egg Yolk Emulsion: Soak egg with 1:100 dilution of saturated mercuric chloride solution for 1 min. Crack eggs and separate yolks from whites. Mix egg yolks with 1 chicken egg.

K_2TeO_3 Solution:

Composition per 100.0mL:

K_2TeO_3	1.0g

Preparation of K_2TeO_3 Solution: Add K_2TeO_3 to distilled/deionized water and bring volume to 100.0mL. Mix thoroughly. Filter sterilize.

Caution: Potassium tellurite is toxic.

Preparation of Medium: Add components to distilled/deionized water and bring volume to 940.0mL. Mix thoroughly. Gently heat and bring to boiling. Autoclave for 15 min at 15 psi pressure–121°C. Cool to 45°–50°C. Add 10.0mL of sterile 1% tellurite solution and 50.0mL of sterile egg yolk emulsion. If desired, add sulfamethazine to a final concentration of 50.0mg/mL. Mix thoroughly but gently and pour into sterile Petri dishes.

Use: For the selective isolation and enumeration of coagulase-positive staphylococci from food, skin, soil, air, and other materials. For the differentiation and identification of staphylococci on the basis of their ability to clear egg yolk. Addition of sulfamethazine inhibits the growth of *Proteus*. Gray-black colonies surrounded by a clear zone are diagnostic for *Staphylococcus aureus*.

Ethyl Violet Azide Broth (EVA Broth)

Composition per liter:

Pancreatic digest of casein	13.5g
Yeast extract	6.5g
Glucose	5.0g
NaCl	5.0g
K_2HPO_4	2.7g
KH_2PO_4	2.7g
NaN_3	0.4g
Ethyl Violet	0.83mg

pH 7.0 ± 0.2 at 25°C

Source: This medium is available as a premixed powder from BD Diagnostic Systems.

Caution: Sodium azide is toxic. Azides also react with metals and disposal must be highly diluted.

Preparation of Medium: Add components to distilled/deionized water and bring volume to 1.0L. Mix thoroughly. Gently heat and bring to boiling. Distribute into tubes in 10.0mL volumes. Autoclave for 15 min at 15 psi pressure–121°C.

Use: For the isolation, cultivation, and enumeration of enterococci from water and other specimens. Fecal enterococci turn the medium turbid with a purple sediment on the bottom of the tube.

Ethyl Violet Azide Broth (EVA Broth)

Composition per liter:

Tryptose	20.0g
Glucose	5.0g
NaCl	5.0g
K_2HPO_4	2.7g
KH_2PO_4	2.7g
NaN_3	0.4g
Ethyl Violet	0.83mg

pH 7.0 ± 0.2 at 25°C

Source: This medium is available as a premixed powder from BD Diagnostic Systems.

Caution: Sodium azide is toxic. Azides also react with metals and disposal must be highly diluted.

Preparation of Medium: Add components to distilled/deionized water and bring volume to 1.0L. Mix thoroughly. Gently heat and bring to boiling. Distribute into tubes in 10.0mL volumes. Autoclave for 15 min at 15 psi pressure–121°C.

Use: For the isolation, cultivation, and enumeration of enterococci from water and other specimens. Fecal enterococci turn the medium turbid with a purple sediment on the bottom of the tube.

Ethyl Violet Azide Broth (EVA Broth)

Composition per liter:

Tryptose	20.0g
Glucose	5.0g
NaCl	5.0g
K_2HPO_4	2.7g
KH_2PO_4	2.7g
NaN_3	0.3g
Ethyl Violet	0.5mg

pH 6.8 ± 0.2 at 25°C

Source: This medium is available as a premixed powder from Oxoid Unipath.

Preparation of Medium: Add components to distilled/deionized water and bring volume to 1.0L. Mix thoroughly. Gently heat and bring to boiling. Distribute into tubes in 10.0mL volumes. Autoclave for 15 min at 15 psi pressure–121°C.

Caution: Sodium azide is toxic. Azides also react with metals and disposal must be highly diluted.

Use: For the isolation, cultivation, and enumeration of enterococci from water and other specimens. Fecal enterococci turn the medium turbid with a purple sediment on the bottom of the tube.

Ethyl Violet Azide HiVeg Broth (E.V.A. HiVeg Broth)

Composition per liter:

Plant hydrolysate	20.0g
Glucose	5.0g
NaCl	5.0g
K_2HPO_4	2.7g
KH_2PO_4	2.7g
NaN_3	0.4g
Ethyl Violet	8.3mg

pH 7.0 ± 0.2 at 25°C

Source: This medium is available as a premixed powder from HiMedia.

Caution: Sodium azide is toxic. Azides also react with metals and disposal must be highly diluted.

Preparation of Medium: Add components to distilled/deionized water and bring volume to 1.0L. Mix thoroughly. Gently heat and bring to boiling. Distribute into tubes in 10.0mL volumes. Autoclave for 15 min at 15 psi pressure–121°C.

Use: For the isolation, cultivation, and enumeration of enterococci from water and other specimens. Fecal enterococci turn the medium turbid with a purple sediment on the bottom of the tube.

Ethylene Glycol NaCl Medium No. 7

See: **EG NaCl Medium No. 7**

Ethylene Production Agar for *Mucor*

Composition per liter:

Solution 3	700.0mL
Solution 1	100.0mL
Solution 2	100.0mL
Solution 4	100.0mL

Solution 1:

Composition per 100.0mL:

Agar, noble	10.0g
D-Glucose	5.0g
DL-Methionine	5.0g

Preparation of Solution 1: Add components to distilled/deionized water and bring volume to 100.0mL. Mix thoroughly. Gently heat and bring to boiling. Autoclave for 15 min at 15 psi pressure–121°C. Cool to 45°–50°C.

Solution 2:

Composition per 100.0mL:

$NaNO_3$	3.5g

Preparation of Solution 2: Add $NaNO_3$ to distilled/deionized water and bring volume to 100.0mL. Mix thoroughly. Autoclave for 15 min at 15 psi pressure–121°C. Cool to 45°–50°C.

Solution 3:

Composition per 700.0mL:

Citric acid	0.756g
$MgSO_4 \cdot 7H_2O$	0.5g
NaOH	0.432g
$CaCl_2$	0.1g
Ferric citrate·$5H_2O$	0.1g
$MnCl_2 \cdot 4H_2O$	0.05g
$ZnSO_4 \cdot 7H_2O$	0.05g
$CuCl_2 \cdot 2H_2O$	5.0mg
$Na_2MoO_4 \cdot 2H_2O$	5.0mg
$Na_2B_4O_7 \cdot 10H_2O$	2.0mg
$CoCl_2 \cdot 6H_2O$	0.2mg

Preparation of Solution 3: Add components to distilled/deionized water and bring volume to 700.0mL. Mix thoroughly. Autoclave for 15 min at 15 psi pressure–121°C. Cool to 45°–50°C.

Solution 4:

Composition per 100.0mL:

KH_2PO_4	5.0g

Preparation of Solution 4: Add KH_2PO_4 to distilled/deionized water and bring volume to 100.0mL. Mix thoroughly. Autoclave for 15 min at 15 psi pressure–121°C. Cool to 45°–50°C.

Preparation of Medium: Aseptically combine 100.0mL of sterile solution 1,100.0mL of sterile solution 2,700.0mL of sterile solution 3, and 100.0mL of sterile solution 4. Mix thoroughly. Pour into sterile Petri dishes or distribute into sterile tubes.

Use: For the cultivation and maintenance of ethylene-producing *Mucor* species.

Ethylene Production Broth for *Mucor*

Composition per liter:

Solution 1	100.0mL
Solution 2	100.0mL
Solution 3	700.0mL
Solution 4	100.0mL

Solution 1:

Composition per 100.0mL:

D-Glucose	5.0g
DL-Methionine	5.0g

Preparation of Solution 1: Add components to distilled/deionized water and bring volume to 100.0mL. Mix thoroughly. Autoclave for 15 min at 15 psi pressure–121°C.

Solution 2:

Composition per 100.0mL:

$NaNO_3$	3.5g

Preparation of Solution 2: Add $NaNO_3$ to distilled/deionized water and bring volume to 100.0mL. Mix thoroughly. Autoclave for 15 min at 15 psi pressure–121°C.

Solution 3:

Composition per 700.0mL:

Citric acid	0.756g
$MgSO_4 \cdot 7H_2O$	0.5g
NaOH	0.432g
$CaCl_2$	0.1g
Ferric citrate·$5H_2O$	0.1g
$MnCl_2 \cdot 4H_2O$	0.05g
$ZnSO_4 \cdot 7H_2O$	0.05g
$CuCl_2 \cdot 2H_2O$	5.0mg
$Na_2MoO_4 \cdot 2H_2O$	5.0mg
$Na_2B_4O_7 \cdot 10H_2O$	2.0mg
$CoCl_2 \cdot 6H_2O$	0.2mg

Preparation of Solution 3: Add components to distilled/deionized water and bring volume to 700.0mL. Mix thoroughly. Autoclave for 15 min at 15 psi pressure–121°C.

Solution 4:

Composition per 100.0mL:

KH_2PO_4	5.0g

Preparation of Solution 4: Add KH_2PO_4 to distilled/deionized water and bring volume to 100.0mL. Mix thoroughly. Autoclave for 15 min at 15 psi pressure–121°C.

Preparation of Medium: Aseptically combine 100.0mL of sterile solution 1, 100.0mL of sterile solution 2, 700.0mL of sterile solution 3, and 100.0mL of sterile solution 4. Mix thoroughly. Aseptically distribute into sterile flasks or tubes.

Use: For the cultivation of ethylene-producing *Mucor* species.

ETSA Medium

Composition per 998.0mL:

Agar	19.0g
Pancreatic digest of casein	15.0g
Papaic digest of soybean meal	5.0g
NaCl	5.0g
Yeast extract	1.0g
KNO_3	0.5g
Sodium formate	0.5g
Sodium succinate	0.5g

Sheep blood, defibrinated	30.0mL
L-Cysteine·HCl·H_2O solution	10.0mL
Dithiothreitol solution	10.0mL
Glucose solution	10.0mL
Na_2CO_3 solution	10.0mL
Menadione solution	2.0mL
Sodium fumarate solution	2.0mL
Sodium lactate (60% syrup)	1.3mL
Hemin solution	1.0mL

pH 7.3 ± 0.2 at 25°C

Hemin Solution:
Composition per 200.0mL:

KOH	1.12g
Hemin	0.2g
Ethanol (95% solution)	100.0mL

Preparation of Hemin Solution: Add KOH to distilled/deionized water and bring volume to 100.0mL. Mix thoroughly. Add ethanol. Mix thoroughly. Add hemin. Mix thoroughly.

Menadione Solution:
Composition per 100.0mL:

Menadione (vitamin K_3)	1.0g
Ethanol (95% solution)	50.0mL

Preparation of Menadione Solution: Add menadione to 50.0mL of ethanol. Mix thoroughly. Bring volume to 100.0mL with distilled/deionized water. Filter sterilize.

L-Cysteine·HCl·H_2O Solution:
Composition per 10.0mL:

L-Cysteine·HCl·H_2O	0.4g

Preparation of L-Cysteine·HCl·H_2O Solution: Add L-cysteine·HCl·H_2O to distilled/deionized water and bring volume to 10.0mL. Mix thoroughly. Filter sterilize.

Dithiothreitol Solution:
Composition per 10.0mL:

Dithiothreitol	0.05g

Preparation of Dithiothreitol Solution: Add dithiothreitol to distilled/deionized water and bring volume to 10.0mL. Mix thoroughly. Filter sterilize.

Glucose Solution:
Composition per 10.0mL:

D-Glucose	1.0g

Preparation of Glucose Solution: Add glucose to distilled/deionized water and bring volume to 10.0mL. Mix thoroughly. Filter sterilize.

Sodium Fumarate Solution:
Composition per 10.0mL:

Sodium fumarate	0.1g

Preparation of Sodium Fumarate Solution: Add sodium fumarate to distilled/deionized water and bring volume to 10.0mL. Mix thoroughly. Filter sterilize.

Na_2CO_3 Solution:
Composition per 10.0mL:

Na_2CO_3	0.4g

Preparation of Na_2CO_3 Solution: Add Na_2CO_3 to distilled/deionized water and bring volume to 10.0mL. Mix thoroughly. Filter sterilize.

Preparation of Medium: Add components—except menadione solution, L-cysteine·HCl·H_2O solution, dithiothreitol solution, glucose solution, sodium fumarate solution, Na_2CO_3 solution, and sheep blood—to distilled/deionized water and bring volume to 926.0mL. Mix thoroughly. Gently heat and bring to boiling. Autoclave for 15 min at 15 psi pressure–121°C. Cool to 50°–60°C. Aseptically add in the following order: 2.0mL of sterile menadione solution, 10.0mL of sterile L-cysteine·HCl·H_2O solution, 10.0mL of sterile dithiothreitol solution, 10.0mL of sterile glucose solution, 2.0mL of sterile sodium fumarate solution, 10.0mL of sterile Na_2CO_3 solution, and 30.0mL of sterile sheep blood. Mix thoroughly. Aseptically and anaerobically distribute into tubes under 80% N_2 + 10% CO_2 + 10% H_2. Cap tubes with butyl rubber stoppers. Place tubes in a press. Autoclave for 15 min at 15 psi pressure–121°C with fast exhaust.

Use: For the cultivation and maintenance of *Bacteroides pneumosintes, Falcivibrio grandis, Falcivibrio vaginalis, Mobiluncus curtisii, Mobiluncus mulieris, Serpula hyodysenteriae,* and *Treponema* species.

Eubacterium acidaminophilum Medium

Composition per liter:

KH_2PO_4	1.0g
NH_4Cl	0.5g
$MgCl_2·6H_2O$	0.4g
$CaCl_2·2H_2O$	0.1g
$NaHCO_3$ solution	20.0mL
Sodium selenite solution	10.0mL
$Na_2S·9H_2O$ solution	3.0mL
Vitamin solution	1.0mL
Trace elements solution SL-7	1.0mL

pH 7.2–7.3 at 25°C

Sodium Selenite Solution:
Composition per liter:

NaOH	0.5g
$Na_2SeO_3·5H_2O$	0.03g

Preparation of Sodium Selenite Solution: Add components to distilled/deionized water and bring volume to 1.0L. Mix thoroughly. Filter sterilize. Gas with 100% N_2 for 20 min.

$Na_2S·9H_2O$ Solution:
Composition per 10.0mL:

$Na_2S·9H_2O$	1.0g

Preparation of $Na_2S·9H_2O$ Solution: Add $Na_2S·9H_2O$ to distilled/deionized water and bring volume to 10.0mL. Mix thoroughly. Gas with 100% N_2 for 20 min. Autoclave for 15 min at 15 psi pressure–121°C. Cool to 45°–50°C.

$NaHCO_3$ Solution:
Composition per 100.0mL:

$NaHCO_3$	10.0g

Preparation of $NaHCO_3$ Solution: Add $NaHCO_3$ to distilled/deionized water and bring volume to 100.0mL. Mix thoroughly. Sparge with 80% N_2 + 20% CO_2. Filter sterilize.

Vitamin Solution:
Composition per 100.0mL:

p-Aminobenzoic acid	4.0mg
Biotin	1.0mg

Preparation of Vitamin Solution: Add components to distilled/deionized water and bring volume to 100.0mL. Mix thoroughly. Filter sterilize.

Trace Elements Solution SL-7:

Composition per liter:

$FeCl_2 \cdot 4H_2O$	1.5g
$CoCl_2 \cdot 6H_2O$	0.19g
$MnCl_2 \cdot 4H_2O$	0.1g
$ZnCl_2$	0.07g
$Na_2MoO_4 \cdot 2H_2O$	0.036g
$NiCl_2 \cdot 6H_2O$	0.024g
H_3BO_3	6.0mg
$CuCl_2 \cdot 2H_2O$	2.0mg
HCl (25% solution)	10.0mL

Preparation of Trace Elements Solution SL-7: Add the $FeCl_2 \cdot 4H_2O$ to the HCl. Add distilled/deionized water and bring volume to 1.0L. Add remaining components. Mix thoroughly. Autoclave for 15 min at 15 psi pressure–121°C under 100% N_2. Cool to 25°C.

Preparation of Medium: Add components—except $NaHCO_3$ solution, $Na_2S \cdot 9H_2O$ solution, vitamin solution, and sodium selenite solution—to distilled/deionized water and bring volume to 966.0mL. Mix thoroughly. Gently heat and bring to boiling. Autoclave for 15 min at 15 psi pressure–121°C. Cool to 45°–50°C under 80% N_2 and 20% CO_2. Aseptically add 20.0mL of sterile $NaHCO_3$ solution, 3.0mL of sterile $Na_2S \cdot 9H_2O$ solution, 1.0mL of sterile vitamin solution, and 10.0mL of sterile sodium selenite solution. Mix thoroughly. Adjust pH to 7.2–7.3 with dilute sterile HCl or Na_2CO_3, if necessary. Aseptically and anaerobically distribute into sterile tubes under 80% N_2 and 20% CO_2. Cap tubes with rubber stoppers.

Use: For the cultivation and maintenance of *Eubacterium acidaminophilum*.

Eubacterium aggregans Medium (DSMZ Medium 711a)

Composition per liter:

Yeast extract	1.0g
NH_4Cl	1.0g
NaCl	0.6g
Cysteine-HCl·H_2O	0.5g
K_2HPO_4	0.3g
KH_2PO_4	0.3g
Yeast extract	0.2g
$MgCl_2 \cdot 6H_2O$	0.2g
$CaCl_2 \cdot 2H_2O$	0.1g
KCl	0.1g
Resazurin	0.5mg
$NaHCO_3$ solution	40.0mL
Fructose solution	10.0mL
$Na_2S \cdot 9H_2O$ solution	10.0mL
Trace elements solution SL-10	1.5mL

pH 7.1 ± 0.2 at 25°C

Trace Elements Solution SL-10:

Composition per liter:

$FeCl_2 \cdot 4H_2O$	1.5g
$CoCl_2 \cdot 6H_2O$	190.0mg
$MnCl_2 \cdot 4H_2O$	100.0mg
$ZnCl_2$	70.0mg
$Na_2MoO_4 \cdot 2H_2O$	36.0mg
$NiCl_2 \cdot 6H_2O$	24.0mg
H_3BO_3	6.0mg
$CuCl_2 \cdot 2H_2O$	2.0mg
HCl (25% solution)	10.0mL

Preparation of Trace Elements Solution SL-10: Add $FeCl_2 \cdot 4H_2O$ to 10.0mL of HCl solution. Mix thoroughly. Add distilled/deionized water and bring volume to 1.0L. Add remaining components. Mix thoroughly. Sparge with 80% N_2 + 20% CO_2. Autoclave for 15 min at 15 psi pressure–121°C.

$Na_2S \cdot 9H_2O$ Solution:

Composition per 10.0mL:

$Na_2S \cdot 9H_2O$	0.3g

Preparation of $Na_2S \cdot 9H_2O$ Solution: Add $Na_2S \cdot 9H_2O$ to distilled/deionized water and bring volume to 10.0mL. Mix thoroughly. Autoclave under 100% N_2 for 15 min at 15 psi pressure–121°C. Cool to room temperature.

$NaHCO_3$ Solution:

Composition per 100.0mL:

$NaHCO_3$	10.0g

Preparation of $NaHCO_3$ Solution: Add $NaHCO_3$ to distilled/deionized water and bring volume to 100.0mL. Mix thoroughly. Sparge with 80% N_2 + 20% CO_2. Filter sterilize.

Fructose Solution:

Composition per 10.0mL:

Fructose	5.0g

Preparation of Fructose Solution: Add fructose to distilled/deionized water and bring volume to 10.0mL. Mix thoroughly. Sparge with 100% N_2. Autoclave for 15 min at 15 psi pressure–121°C. Cool to room temperature.

Preparation of Medium: Prepare and dispense medium under 80% N_2 + 20% CO_2 gas atmosphere. Add components, except $NaHCO_3$ solution, $Na_2S \cdot 9H_2O$ solution, fructose solution, and trace elements solution SL-10, to distilled/deionized water and bring volume to 938.5mL. Mix thoroughly. Adjust pH to 7.1. Sparge with 80% N_2 + 20% CO_2. Autoclave for 15 min at 15 psi pressure–121°C. Aseptically and anaerobically add 40.0mL $NaHCO_3$ solution, 10.0mL $Na_2S \cdot 9H_2O$ solution, 10.0mL fructose solution, and 1.5mL trace elements solution SL-10. Mix thoroughly. Aseptically and anaerobically distribute into sterile tubes or bottles.

Use: For the cultivation of *Clostridium methoxybenzovorans* and *Eubacterium aggregans*.

Eubacterium angustum Medium

Composition per liter:

$NaHCO_3$	5.0g
Tris(hydroxymethyl)aminomethane buffer	3.0g
Uric acid	3.0g
Yeast extract	1.0g
L-Cysteine·HCl·H_2O	0.5g
NH_4Cl	0.5g
K_2HPO_4	0.1g
$MgSO_4 \cdot 7H_2O$	0.05g
$Na_2S \cdot 9H_2O$	0.05g
Resazurin	1.0mg
Wolfe's mineral solution	10.0mL
Selenium solution	1.0mL

pH 7.9 ± 0.2 at 25°C

Selenium Solution:

Composition per liter:

$Na_2SeO_3 \cdot 5H_2O$	3.0mg
NaOH (0.01*M* solution)	1.0L

Preparation of Selenium Solution: Add $Na_2SeO_3 \cdot 5H_2O$ to 1.0L of NaOH solution. Mix thoroughly. Filter sterilize. Aseptically gas under 100% N_2 for 20 min.

Wolfe's Mineral Solution:

Composition per liter

$MgSO_4 \cdot 7H_2O$	3.0g
Nitrilotriacetic acid	1.5g
NaCl	1.0g
$MnSO_4 \cdot H_2O$	0.5g
$FeSO_4 \cdot 7H_2O$	0.1g
$CoCl_2 \cdot 6H_2O$	0.1g
$CaCl_2$	0.1g
$ZnSO_4 \cdot 7H_2O$	0.1g
$CuSO_4 \cdot 5H_2O$	0.01g
$AlK(SO_4)_2 \cdot 12H_2O$	0.01g
H_3BO_3	0.01g
$Na_2MoO_4 \cdot 2H_2O$	0.01g

Preparation of Wolfe's Mineral Solution: Add nitrilotriacetic acid to 500.0mL of distilled/deionized water. Dissolve by adjusting pH to 6.5 with KOH. Add remaining components. Add distilled/deionized water to 1.0L.

Preparation of Medium: Add the uric acid and the tris(hydroxymethyl)aminomethane buffer to 900.0mL of distilled/deionized water. Mix thoroughly. Gently heat while stirring until dissolved. Add remaining components, except L-cysteine·HCl·H_2O, $NaHCO_3$, and $Na_2S \cdot 9H_2O$. Gently heat and bring to boiling. Cool to 25°C under 80% N_2 + 10% CO_2 + 10% H_2. Add L-cysteine·HCl·H_2O, $NaHCO_3$, and $Na_2S \cdot 9H_2O$. Mix thoroughly. Anaerobically distribute into tubes under 80% N_2 + 10% CO_2 + 10% H_2. Cap tubes with rubber stoppers. Place tubes in a press. Autoclave for 15 min at 15 psi pressure–121°C with fast exhaust.

Use: For the cultivation and maintenance of *Eubacterium angustum.*

Eubacterium callanderi Medium

Composition per liter:

$NaHCO_3$	3.5g
Glucose	1.8g
Resazurin	1.0mg
Clarified rumen fluid	50.0mL
Pfennig's mineral solution	50.0mL
L-Cysteine-sulfide reducing agent	20.0mL
Wolfe's vitamin solution	10.0mL
Trace elements solution SL-7	1.0mL

Pfennig's Mineral Solution:

Composition per 100.0mL:

KH_2PO_4	1.0g
NaCl	0.8g
NH_4Cl	0.8g
$MgCl_2 \cdot 6H_2O$	0.66g
$CaCl_2 \cdot 2H_2O$	0.1g

Preparation of Pfennig's Mineral Solution: Add components to distilled/deionized water and bring volume to 100.0mL. Mix thoroughly.

L-Cysteine-Sulfide Reducing Agent:

Composition per 200.0mL:

L-Cysteine·HCl·H_2O	2.5g
$Na_2S \cdot 9H_2O$	2.5g
NaOH (3*N* solution)	13.4mL

Preparation of L-Cysteine-Sulfide Reducing Agent: Add L-cysteine·HCl·H_2O to distilled/deionized water and bring volume to 50.0mL. Rapidly adjust pH to 10 with 3*N* NaOH solution. Add $Na_2S \cdot 9H_2O$. Mix thoroughly. Bring volume to 200.0mL with distilled/deionized water. Gently heat and bring to boiling. Cool to room temperature. Distribute into tubes under 100% N_2. Autoclave for 15 min at 15 psi pressure–121°C.

Wolfe's Vitamin Solution:

Composition per liter:

Pyridoxine·HCl	10.0mg
p-Aminobenzoic acid	5.0mg
Lipoic acid	5.0mg
Nicotinic acid	5.0mg
Riboflavin	5.0mg
Thiamine·HCl	5.0mg
Calcium DL-pantothenate	5.0mg
Biotin	2.0mg
Folic acid	2.0mg
Vitamin B_{12}	0.1mg

Preparation of Wolfe's Vitamin Solution: Add components to distilled/deionized water and bring volume to 1.0L. Mix thoroughly.

Trace Elements Solution SL-7:

Composition per 1010.0mL:

$FeCl_2 \cdot 4H_2O$	1.5g
$CoCl_2 \cdot 6H_2O$	190.0mg
$MnCl_2 \cdot 4H_2O$	100.0mg
$ZnCl_2$	70.0mg
H_3BO_3	62.0mg
$Na_2MoO_4 \cdot 2H_2O$	36.0mg
$NiCl_2 \cdot 6H_2O$	24.0mg
$CuCl_2 \cdot 2H_2O$	17.0mg
Hydrochloric acid, 25%	10.0mL

Preparation of Trace Elements Solution SL-7: Add $FeCl_2 \cdot 4H_2O$ to 10.0mL of HCl solution. Mix thoroughly. Add distilled/deionized water and bring volume to 1.0L. Add remaining components. Mix thoroughly.

Preparation of Medium: Prepare and dispense medium under 80% N_2 + 20% CO_2. Add components, except $NaHCO_3$ and cysteine-sulfide reducing agent, to distilled/deionized water and bring volume to 980.0mL. Mix thoroughly. Gently heat and bring to boiling. Continue boiling for 3 min. Cool to room temperature while sparging with 80% N_2 + 20% CO_2. Add $NaHCO_3$. Mix thoroughly. Anaerobically distribute 10.0mL volumes into anaerobic tubes. Autoclave for 15 min at 15 psi pressure–121°C. Aseptically and anaerobically add 0.2mL of sterile cysteine-sulfide reducing agent to each tube. Mix thoroughly.

Use: For the cultivation of *Eubacterium callanderi.*

Eubacterium lentum Medium

Composition per 1205.0mL:

Peptone	30.0g
Arginine	5.0g
K_2HPO_4	5.0g
Yeast extract	5.0g
L-Cysteine·HCl·H_2O	0.5g
Chopped meat extract filtrate	1.0L
Chopped meat extract solids	200.0mL
Resazurin (0.025% solution)	4.0mL

pH 7.0 ± 0.2 at 25°C

Chopped Meat Extract:

Composition per liter:

Beef or horse meat	500.0g
NaOH (1*N* solution)	25.0mL

Preparation of Chopped Meat Extract: Use lean beef or horse meat. Remove fat and connective tissue. Grind. Add meat and NaOH to distilled/deionized water and bring volume to 1.0L. Gently heat and bring to boiling while stirring. Cool to 25°C. Remove fat from surface. Filter. Reserve ground meat particles and filtrate. Add distilled/deionized water to filtrate and bring volume to 1.0L.

Preparation of Medium: To 1.0L of chopped meat extract filtrate, add the remaining components, except L-cysteine·HCl·H_2O and chopped meat solids. Mix thoroughly. Gently heat to boiling. Cool to room temperature. Add the L-cysteine·HCl·H_2O. Adjust pH to 7.0. Distribute 1 part chopped meat solids (by volume) and 5 parts of liquid (by volume) into tubes under O_2-free 97% N_2 + 3% H_2. Cap with rubber stoppers and place tubes in a press. Autoclave for 15 min at 15 psi pressure–121°C with fast exhaust.

Use: For the cultivation and maintenance of *Eubacterium lentum.*

Eubacterium Medium

Composition per liter:

Pancreatic digest of casein	20.0g
Agar	15.0g
Meat extract	15.0g
Glucose	5.0g
$Na_2HPO_4 \cdot 12H_2O$	4.0g
L-Cysteine·HCl	0.5g

pH 7.4 ± 0.2 at 25°C

Preparation of Medium: Add components to distilled/deionized water and bring volume to 1.0L. Mix thoroughly. Gently heat and bring to boiling. Distribute into tubes or flasks. Autoclave for 15 min at 15 psi pressure–121°C. Pour into sterile Petri dishes or leave in tubes. Use freshly prepared medium.

Use: For the cultivation of *Eubacterium* species.

Eubacterium Medium

Composition per liter:

Beef brain powder	33.33g
Pancreatic digest of casein	15.0g
Yeast extract	10.0g
Glucose	5.5g
Yeast extract	5.0g
NaCl	2.5g
Sodium thioglycolate	1.8g
L-Cystine	0.5g

pH 7.0 ± 0.2 at 25°C

Preparation of Medium: Add components, except beef brain powder, to distilled/deionized water and bring volume to 1.0L. Mix thoroughly. Gently heat and bring to boiling under 97% N_2 + 3% H_2. Continue boiling for 15–20 min. Adjust pH to 7.0. Cool to 25°C under 97% N_2 + 3% H_2. Anaerobically distribute into tubes in 9.0mL volumes. Add 0.3g of beef brain powder to each tube. Cap tubes with rubber stoppers. Place tubes in a press. Autoclave for 15 min at 15 psi pressure–121°C with fast exhaust.

Use: For the cultivation of *Eubacterium* species.

Eubacterium oxidoreducens Medium

Composition per 1001.0mL:

Solution A	890.0mL
Solution B	100.0mL
Solution C	10.0mL
Solution D	1.0mL

pH 7.0–7.2 at 25°C

Solution A:

Composition per 890.0mL:

Crotonic acid	5.0g
Yeast extract	2.0g
Resazurin	1.0mg
Mineral solution	50.0mL
Vitamin solution	5.0mL
Trace elements solution SL-10	1.0mL

Preparation of Solution A: Add components to distilled/deionized water and bring volume to 890.0mL. Mix thoroughly. Adjust pH to 6.9. Sparge with 80% N_2 + 20% CO_2 for 20 min. Distribute 8.9mL into anaerobic tubes under 80% N_2 + 20% CO_2. Autoclave under 80% N_2 + 20% CO_2 for 15 min at 15 psi pressure–121°C.

Mineral Solution:

Composition per liter:

KH_2PO_4	10.0g
$MgCl_2 \cdot 6H_2O$	6.6g
NaCl	8.0g
NH_4Cl	8.0g
$CaCl_2 \cdot 2H_2O$	1.0g

Preparation of Mineral Solution: Add components to distilled/deionized water and bring volume to 1.0L. Mix thoroughly.

Vitamin Solution:

Composition per liter:

Pyridoxine·HCl	6.2mg
Nicotinic acid	2.5mg
p-Aminobenzoic acid	1.25mg
Thiamine·HCl	1.25mg
Pantothenic acid	0.62mg
Biotin	0.25mg

Preparation of Vitamin Solution: Add components to distilled/deionized water and bring volume to 1.0L. Adjust pH to 7.0. Mix thoroughly.

Trace Elements Solution SL-10:

Composition per liter:

$FeCl_2 \cdot 4H_2O$	1.5g
$CoCl_2 \cdot 6H_2O$	190.0mg
$MnCl_2 \cdot 4H_2O$	100.0mg
$ZnCl_2$	70.0mg
$Na_2MoO_4 \cdot 2H_2O$	36.0mg
$NiCl_2 \cdot 6H_2O$	24.0mg
H_3BO_3	6.0mg
$CuCl_2 \cdot 2H_2O$	2.0mg
HCl (25% solution)	10.0mL

Preparation of Trace Elements Solution SL-10: Add $FeCl_2 \cdot 4H_2O$ to 10.0mL of HCl solution. Mix thoroughly. Add distilled/deionized water and bring volume to 1.0L. Add remaining components. Mix thoroughly.

Solution B:

Composition per 100.0mL:

$NaHCO_3$	5.0g

Preparation of Solution B: Add $NaHCO_3$ to distilled/deionized water and bring volume to 100.0mL. Mix thoroughly. Filter sterilize. Sparge with 80% N_2 + 20% CO_2 for 20 min.

Solution C:
Composition per 10.0mL:

L-Cysteine	0.24g

Preparation of Solution C: Add L-cysteine to distilled/deionized water and bring volume to 10.0mL. Mix thoroughly. Autoclave under 80% N_2 + 20% CO_2 for 15 min at 15 psi pressure–121°C.

Solution D:
Composition per 1.0mL:

$Na_2S{\cdot}9H_2O$	78.0mg

Preparation of Solution D: Add $Na_2S{\cdot}9H_2O$ to distilled/deionized water and bring volume to 1.0mL. Mix thoroughly. Autoclave under 80% N_2 + 20% CO_2 for 15 min at 15 psi pressure–121°C.

Preparation of Medium: To each tube containing 8.9mL of sterile solution A, add (using a syringe) 1.0mL of sterile solution B, 0.1mL of sterile solution C, and 0.01mL of sterile solution D.

Use: For the cultivation and maintenance of *Eubacterium oxidoreducens*.

Euglena B_{12} Medium

Composition per liter:

Sucrose	30.0g
L-Glutamic acid	6.0g
Glycine	5.0g
DL-Aspartic acid	4.0g
DL-Malic acid	2.0g
Succinic acid	1.04g
$MgSO_4{\cdot}7H_2O$	0.8g
$(NH_4)_2CO_3$	0.72g
KH_2PO_4	0.6g
$CaCO_3$	0.16g
$FeCl_3$	0.06g
$ZnSO_4{\cdot}7H_2O$	0.04g
Thiamine·HCl	0.012g
$MnSO_4{\cdot}H_2O$	6.0mg
$CoSO_4$	5.0mg
$(NH_4)_2MoO_3$	1.34mg
H_3BO_3	1.14mg
$CuSO_4{\cdot}5H_2O$	0.62mg

pH 3.5 ± 0.1 at 25°C

Preparation of Medium: Add components to distilled/deionized water and bring volume to 1.0L. Mix thoroughly. Gently heat until dissolved. Distribute into tubes in 2.0mL volumes. Add standard solution or test solution to each tube. Bring volume of each tube to 4.0mL with distilled/deionized water. Autoclave for 15 min at 0 psi pressure–100°C.

Use: For the assay of vitamin B_{12} using *Euglena gracilis* as the test organism.

Eugon Agar

Composition per liter:

Agar	15.0g
Pancreatic digest of casein	15.0g
Glucose	5.5g
Papaic digest of soybean meal	5.0g
NaCl	4.0g
L-Cystine	0.3g
Na_2SO_3	0.2g

pH 7.0 ± 2.0 at 25°C

Source: This medium is available as a premixed powder from BD Diagnostic Systems.

Preparation of Medium: Add components to distilled/deionized water and bring volume to 1.0L. Mix thoroughly. Gently heat and bring to boiling. Distribute into tubes or flasks. Autoclave for 15 min at 13 psi pressure–118°C. Pour into sterile Petri dishes.

Use: For the cultivation and maintenance of a variety of fastidious microorganisms. For the cultivation and maintenance of *Bifidobacterium* species.

Eugon Agar with Fildes Enrichment (LMG Medium 75)

Composition per liter:

Agar	15.0g
Pancreatic digest of casein	15.0g
Glucose	5.5g
Papaic digest of soybean meal	5.0g
NaCl	4.0g
L-Cystine	0.7g
Na_2SO_3	0.2g
Fildes digested sheep blood	50.0mL

pH 7.0 ± 0.2 at 25°C

Fildes Enrichment Solution:
Composition per 206.0mL:

Pepsin	1.0g
NaCl (0.85% solution)	150.0mL
Sheep blood, defibrinated	50.0mL
HCl	6.0mL

pH 7.0–7.2 at 25°C

Source: Fildes enrichment solution is available from BD Diagnostic Systems.

Preparation of Fildes Enrichment Solution: Combine components. Mix thoroughly. Incubate at 56°C for 4 hr. Bring pH to 7.0 with 20% NaOH. Adjust pH to 7.2 with HCl. Do not autoclave. Add 0.25 mL of chloroform and store at 4°C. Before use, heat to 56°C to remove chloroform.

Preparation of Medium: Add components, except Fildes digested sheep blood, to 950.0mL distilled/deionized water. Mix thoroughly. Gently heat and bring to boiling. Autoclave for 15 min at 15 psi pressure–121°C. Cool to 50°fl55°C. Aseptically add 50.0mL sterile Fildes digested sheep blood. Mix thoroughly. Pour into sterile Petri dishes or distribute into sterile tubes.

Use: For the cultivation of *Taylorella equigenitalis*.

Eugon Agar with Fildes Enrichment

Composition per liter:

Agar	15.0g
Pancreatic digest of casein	15.0g
Glucose	5.5g
Papaic digest of soybean meal	5.0g
NaCl	4.0g
L-Cystine	0.3g
Na_2SO_3	0.2g
Fildes enrichment solution	50.0mL

pH 7.0 ± 2.0 at 25°C

Fildes Enrichment Solution:

Composition per 206.0mL:

Pepsin	1.0g
NaCl (0.85% solution)	150.0mL
Sheep blood, defibrinated	50.0mL
HCl	6.0mL

pH 7.0–7.2 at 25°C

Source: Fildes enrichment solution is available from BD Diagnostic Systems.

Preparation of Fildes Enrichment Solution: Combine components. Mix thoroughly. Incubate at 56°C for 4 hr. Bring pH to 7.0 with 20% NaOH. Adjust pH to 7.2 with HCl. Do not autoclave. Add 0.25 mL of chloroform and store at 4°C. Before use, heat to 56°C to remove chloroform.

Preparation of Medium: Add components, except Fildes enrichment solution, to distilled/deionized water and bring volume to 1.0L. Mix thoroughly. Gently heat and bring to boiling. Autoclave for 15 min at 13 psi pressure–118°C. Cool to 50°–55°C. Aseptically add 50.0mL of Fildes enrichment solution. Mix thoroughly. Pour into sterile Petri dishes or distribute into sterile tubes.

Use: For the cultivation and maintenance of *Taylorella equigenitalis*.

Eugon Blood Agar (Eugonagar™)

Composition per liter:

Agar	15.0g
Pancreatic digest of casein	15.0g
Glucose	5.5g
Papaic digest of soybean meal	5.0g
NaCl	4.0g
L-Cystine	0.3g
Na_2SO_3	0.2g
Sheep blood, defibrinated	100.0mL

pH 7.0 ± 2.0 at 25°C

Source: This medium is available as a premixed powder from BD Diagnostic Systems.

Preparation of Medium: Add components, except sheep blood, to distilled/deionized water and bring volume to 900.0mL. Mix thoroughly. Gently heat and bring to boiling. Autoclave for 15 min at 13 psi pressure–118°C. Cool to 45°–50°C. Aseptically add sterile sheep blood. Mix thoroughly. Pour into sterile Petri dishes or distribute into sterile tubes. If desired, medium may be chocolatized by maintaining at 80°–85°C for 20 min after the addition of sheep blood.

Use: For the cultivation and maintenance of fastidious microorganisms. For the cultivation and maintenance of *Bifidobacterium* species.

Eugon Broth (Eugonbroth™)

Composition per liter:

Pancreatic digest of casein	15.0g
Glucose	5.5g
Papaic digest of soybean meal	5.0g
NaCl	4.0g
L-Cystine	0.3g
Na_2SO_3	0.2g

pH 7.0 ± 0.2 at 25°C

Source: This medium is available as a premixed powder from BD Diagnostic Systems.

Preparation of Medium: Add components to distilled/deionized water and bring volume to 1.0L. Mix thoroughly. Gently heat while stirring and bring to boiling. Distribute into tubes or flasks. Autoclave for 15 min at 13 psi pressure–118°C.

Use: For the cultivation and maintenance of a variety of fastidious microorganisms.

Eugonic Agar

Composition per liter:

Agar	15.0g
Casein enzymatic hydrolysate	15.0g
Glucose	5.0g
Papaic digest of soybean meal	5.0g
NaCl	4.0g
L-Cystine	0.2g
Na_2SO_3	0.2g
Sheep blood, defibrinated	50.0mL

pH 7.0 ± 2.0 at 25°C

Source: This medium, without blood, is available as a premixed powder from HiMedia.

Preparation of Medium: Add components, except blood, to distilled/deionized water and bring volume to 950.0mL. Mix thoroughly. Gently heat and bring to boiling. Autoclave for 15 min at 15 psi pressure–121°C. Cool to 50°C. Aseptically add 50.0mL of sterile defibrinated blood. Pour into sterile Petri dishes or leave in tubes.

Preparation of Medium: Add components to distilled/deionized water and bring volume to 1.0L. Mix thoroughly. Gently heat and bring to boiling. Distribute into tubes or flasks. Autoclave for 15 min at 15 psi pressure–121°C. Mix thoroughly. Pour into sterile Petri dishes.

Use: For the cultivation and maintenance of a variety of fastidious microorganisms, e.g., *Brucella, Haemophilus, Neisseria, Pasteurella,* and *Lactobacillus* species.

Eugonic HiVeg Agar

Composition per liter:

Agar	15.0g
Plant hydrolysate	15.0g
Glucose	5.0g
Papaic digest of soybean meal	5.0g
NaCl	4.0g
L-Cystine	0.2g
Na_2SO_3	0.2g
Sheep blood, defibrinated	50.0mL

pH 7.0 ± 2.0 at 25°C

Source: This medium, without blood, is available as a premixed powder from HiMedia.

Preparation of Medium: Add components, except blood, to distilled/deionized water and bring volume to 950.0mL. Mix thoroughly. Gently heat and bring to boiling. Autoclave for 15 min at 15 psi pressure–121°C. Cool to 50°C. Aseptically add 50.0mL of sterile defibrinated blood. Pour into sterile Petri dishes or leave in tubes.

Preparation of Medium: Add components to distilled/deionized water and bring volume to 1.0L. Mix thoroughly. Gently heat and bring to boiling. Distribute into tubes or flasks. Autoclave for 15 min at 15 psi pressure–121°C. Mix thoroughly. Pour into sterile Petri dishes.

Use: For the cultivation and maintenance of a variety of fastidious microorganisms, e.g., *Brucella, Haemophilus, Neisseria, Pasteurella,* and *Lactobacillus* species.

Eugonic HiVeg Broth

Composition per liter:

Plant hydrolysate	15.0g
Glucose	5.0g
Papaic digest of soybean meal	5.0g
NaCl	4.0g
L-Cystine	0.2g
Na_2SO_3	0.2g
Sheep blood, defibrinated	50.0mL

pH 7.0 ± 2.0 at 25°C

Source: This medium, without blood, is available as a premixed powder from HiMedia.

Preparation of Medium: Add components, except blood, to distilled/deionized water and bring volume to 950.0mL. Mix thoroughly. Gently heat and bring to boiling. Autoclave for 15 min at 15 psi pressure–121°C. Cool to 50°C. Aseptically add 50.0mL of sterile defibrinated blood. Mix thoroughly. Dispense into sterile tubes or flasks.

Use: For the cultivation and maintenance of a variety of fastidious microorganisms, e.g., *Brucella, Haemophilus, Neisseria, Pasteurella,* and *Lactobacillus* species.

EVA Broth

See: **Ethyl Violet Azide Broth**

Exiguobacterium Medium

Composition per liter:

Beef extract	10.0g
Peptone	10.0g
NaCl	5.0g
Glucose	5.0g
Yeast extract	3.0g

pH 8.0 ± 0.2 at 25°C

Preparation of Medium: Add components to distilled/deionized water and bring volume to 1.0L. Mix thoroughly. Adjust pH to 8.0. Distribute into tubes or flasks. Autoclave for 15 min at 15 psi pressure–121°C.

Use: For the cultivation and maintenance of *Exiguobacterium aurantiacum.*

Extracted Hay Medium

Composition: Hay or grass 50.0g

Preparation of Medium: Add hay or grass to 1.0L of distilled/deionized water. Gently heat and bring to boiling. Continue boiling for 30 min. Rinse with cold water twice. Add 1.0L of distilled/deionized water, boil 30 min, and rinse. Repeat this process at least five times. Dry the extracted hay or grass. Add 10–30 blades of extracted hay or grass to a large test tube. Autoclave for 15 min at 15 psi pressure–121°C.

Use: For the isolation and cultivation of *Beggiatoa* species and myxotrophic *Thiothrix* species.

EYGA Agar

Composition per liter:

Agar	12.0g
K_2HPO_4	1.1g
Glucose	1.0g
Yeast extract	1.0g
KH_2PO_4	0.86g
$(NH_4)_2SO_4$	0.5g
$MgSO_4 \cdot 7H_2O$	0.2g
NaCl	0.1g
$CaCl_2$	0.025g
Vitamin B_{12}	2.0μg
EDTA/trace elements mix	3.0mL

pH 6.8 ± 0.2 at 25°C

EDTA/Trace Elements Mix:

Composition per 600.0mL:

EDTA	5.0g
$ZnSO_4 \cdot 7H_2O$	2.2g
$MnSO_4 \cdot 4H_2O$	0.57g
$FeSO_4 \cdot 7H_2O$	0.50g
$CoCl_2 \cdot 6H_2O$	0.161g
$CuSO_4 \cdot 5H_2O$	0.157g
$Na_2MoO_4 \cdot 2H_2O$	0.151g

Preparation of EDTA/Trace Elements Mix: Add components to distilled/deionized water and bring volume to 600.0mL. Mix thoroughly.

Preparation of Medium: Add components to distilled/deionized water and bring volume to 1.0L. Mix thoroughly. Adjust pH to 6.8. Distribute into tubes or flasks. Autoclave for 15 min at 15 psi pressure–121°C.

Use: For the cultivation and maintenance of *Arthrobacter* species.

EYS Agar

See: **Emerson's Yeast Starch Agar**

FAA Alternative Selective

See: **Fastidious Anaerobe Agar, Alternative Selective**

FAA Alternative Selective with Neomycin, Vancomycin, and Josamycin

See: **Fastidious Anaerobe Agar, Alternative Selective with Neomycin, Vancomycin, and Josamycin**

FAA Selective with Neomycin and Vancomycin

See: **Fastidious Anaerobe Agar, Selective with Neomycin and Vancomycin**

Falcivibrio Medium

Composition per liter:

Pancreatic digest of casein	10.0g
Gelatin peptone	10.0g
NaCl	5.0g
Yeast extract	5.0g
Glucose	1.0g
L-Arginine	1.0g
Sodium pyruvate	1.0g
Cysteine	0.3g
Hemin	5.0mg
Resazurin	1.0mg
Menadione	0.5mg
Serum, equine, bovine, or ovine	50.0mL

pH 7.1 ± 0.2 at 25°C

Preparation of Medium: Prepare medium under 100% N_2. Add components, except serum, to distilled/deionized water and bring volume to 950.0mL. Mix thoroughly. Autoclave for 15 min at 15 psi pressure–121°C. Aseptically and anaerobically add 50.0mL of sterile serum. Mix thoroughly. Aseptically and anaerobically distribute into sterile tubes or flasks.

Use: For the cultivation and maintenance of *Falcivibrio grandis* and *Falcivibrio vaginalis.*

Fastidious Anaerobe Agar (FAA)

Composition per liter:

Peptone	23.0g
Agar	12.0g
NaCl	5.0g
Glucose	1.0g
L-Arginine	1.0g
Sodium pyruvate	1.0g
Soluble starch	1.0g
L-Cysteine·HCl·H_2O	0.5g
Sodium succinate	0.5g
$NaHCO_3$	0.4g
$Na_4P_2O_7 \cdot 10H_2O$	0.25g
Sheep blood, defibrinated	50.0mL
Hemin solution	1.0mL
Vitamin K_1 solution	0.1mL

pH 7.2 ± 0.2 at 25°C

Vitamin K_1 Solution:

Composition per 100.0mL:

Vitamin K_1	1.0g
Ethanol	99.0mL

Preparation of Vitamin K_1 Solution: Add vitamin K_1 to 99.0mL of absolute ethanol. Mix thoroughly.

Hemin Solution:

Composition per 100.0mL:

Hemin	1.0g
NaOH (1*N* solution)	20.0mL

Preparation of Hemin Solution: Add hemin to 20.0mL of 1*N* NaOH solution. Mix thoroughly. Bring volume to 100.0mL with distilled/deionized water.

Preparation of Medium: Add components, except defibrinated sheep blood, to distilled/deionized water and bring volume to 950.0mL. Mix thoroughly. Gently heat and bring to boiling. Autoclave for 15 min at 15 psi pressure–121°C. Cool to 45°–50°C. Aseptically add 50.0mL of sterile defibrinated sheep blood. Mix thoroughly. Pour into sterile Petri dishes or distribute into sterile tubes.

Use: For the cultivation of a variety of fastidious anaerobes from clinical and nonclinical specimens.

Fastidious Anaerobe Agar, Alternative Selective (FAA Alternative Selective)

Composition per liter:

Peptone	23.0g
Agar	12.0g
NaCl	5.0g
Glucose	1.0g
L-Arginine	1.0g
Sodium pyruvate	1.0g
Soluble starch	1.0g
L-Cysteine·HCl·H_2O	0.5g
Sodium succinate	0.5g
$NaHCO_3$	0.4g
$Na_4P_2O_7 \cdot 10H_2O$	0.25g
Sheep blood, defibrinated	50.0mL
Hemin solution	1.0mL
Vitamin K_1 solution	0.1mL

pH 7.2 ± 0.2 at 25°C

Vitamin K_1 Solution:

Composition per 100.0mL:

Vitamin K_1	1.0g
Ethanol	99.0mL

Preparation of Vitamin K_1 Solution: Add vitamin K_1 to 99.0mL of absolute ethanol. Mix thoroughly.

Hemin Solution:

Composition per 100.0mL:

Hemin	1.0g
NaOH (1*N* solution)	20.0mL

Preparation of Hemin Solution: Add hemin to 20.0mL of 1*N* NaOH solution. Mix thoroughly. Bring volume to 100.0mL with distilled/deionized water.

Preparation of Medium: Add components, except defibrinated sheep blood, to distilled/deionized water and bring volume to 950.0mL. Mix thoroughly. Gently heat and bring to boiling. Autoclave for 15 min at 15 psi pressure–121°C. Cool to 45°–50°C. Aseptically add 50.0mL of sterile defibrinated sheep blood. Mix thoroughly. Pour into sterile Petri dishes or distribute into sterile tubes.

Use: For the cultivation of a variety of fastidious anaerobes from clinical and nonclinical specimens.

Fastidious Anaerobe Agar, Alternative Selective with Neomycin, Vancomycin, and Josamycin (FAA Alternative Selective Medium with Neomycin, Vancomycin, and Josamycin)

Composition per liter:

Peptone	23.0g
Agar	12.0g
NaCl	5.0g
Glucose	1.0g
L-Arginine	1.0g
Sodium pyruvate	1.0g
Soluble starch	1.0g
L-Cysteine·HCl·H_2O	0.5g
Sodium succinate	0.5g
$NaHCO_3$	0.4g
$Na_4P_2O_7 \cdot 10H_2O$	0.25g
Neomycin	0.1g
Sheep blood, defibrinated	50.0mL
Vancomycin solution	10.0mL
Josamycin solution	10.0mL
Hemin solution	1.0mL
Vitamin K_1 solution	0.1mL

pH 7.2 ± 0.2 at 25°C

Vitamin K_1 Solution:

Composition per 100.0mL:

Vitamin K_1	1.0g
Ethanol	99.0mL

Preparation of Vitamin K_1 Solution: Add vitamin K_1 to 99.0mL of absolute ethanol. Mix thoroughly.

Hemin Solution:

Composition per 100.0mL:

Hemin	1.0g
NaOH (1*N* solution)	20.0mL

Preparation of Hemin Solution: Add hemin to 20.0mL of 1*N* NaOH solution. Mix thoroughly. Bring volume to 100.0mL with distilled/deionized water.

Vancomycin Solution:
Composition per 10.0mL:

Vancomycin	5.0mg

Preparation of Vancomycin Solution: Add vancomycin to distilled/deionized water and bring volume to 10.0mL. Mix thoroughly. Filter sterilize.

Josamycin Solution:
Composition per 10.0mL:

Josamycin	3.0mg

Preparation of Josamycin Solution: Add josamycin to distilled/deionized water and bring volume to 10.0mL. Mix thoroughly. Filter sterilize.

Preparation of Medium: Add components, except defibrinated sheep blood, vancomycin solution, and josamycin solution, to distilled/deionized water and bring volume to 930.0mL. Mix thoroughly. Gently heat and bring to boiling. Autoclave for 15 min at 15 psi pressure–121°C. Cool to 45°–50°C. Aseptically add 50.0mL of sterile defibrinated sheep blood, 10.0mL vancomycin solution, and 10.0mL of josamycin solution. Mix thoroughly. Pour into sterile Petri dishes or distribute into sterile tubes.

Use: For the selective cultivation of *Fusobacterium* species from clinical and nonclinical specimens.

Fastidious Anaerobe Agar, Selective (FAA Selective)

Composition per liter:

Peptone	23.0g
Agar	12.0g
NaCl	5.0g
Glucose	1.0g
L-Arginine	1.0g
Sodium pyruvate	1.0g
Soluble starch	1.0g
L-Cysteine·HCl·H_2O	0.5g
Sodium succinate	0.5g
$NaHCO_3$	0.4g
$Na_4P_2O_7 \cdot 10H_2O$	0.25g
Sheep blood, defibrinated	50.0mL
Hemin solution	1.0mL
Vitamin K_1 solution	0.1mL

pH 7.2 ± 0.2 at 25°C

Vitamin K_1 Solution:
Composition per 100.0mL:

Vitamin K_1	1.0g
Ethanol	99.0mL

Preparation of Vitamin K_1 Solution: Add vitamin K_1 to 99.0mL of absolute ethanol. Mix thoroughly.

Hemin Solution:
Composition per 100.0mL:

Hemin	1.0g
NaOH (1*N* solution)	20.0mL

Preparation of Hemin Solution: Add hemin to 20.0mL of 1*N* NaOH solution. Mix thoroughly. Bring volume to 100.0mL with distilled/deionized water.

Preparation of Medium: Add components, except defibrinated sheep blood, to distilled/deionized water and bring volume to 950.0mL. Mix thoroughly. Gently heat and bring to boiling. Autoclave for 15 min at 15 psi pressure–121°C. Cool to 45°–50°C. Aseptically add 50.0mL of sterile defibrinated sheep blood. Mix thoroughly. Pour into sterile Petri dishes or distribute into sterile tubes.

Use: For the cultivation of a variety of fastidious anaerobes from clinical and nonclinical specimens.

Fastidious Anaerobe Agar, Selective with Neomycin and Vancomycin (FAA Selective with Neomycin and Vancomycin)

Composition per liter:

Peptone	23.0g
Agar	12.0g
NaCl	5.0g
Glucose	1.0g
L-Arginine	1.0g
Sodium pyruvate	1.0g
Soluble starch	1.0g
L-Cysteine·HCl·H_2O	0.5g
Sodium succinate	0.5g
$NaHCO_3$	0.4g
$Na_4P_2O_7 \cdot 10H_2O$	0.25g
Neomycin	0.1g
Sheep blood, defibrinated	50.0mL
Vancomycin solution	10.0mL
Hemin solution	1.0mL
Vitamin K_1 solution	0.1mL

pH 7.2 ± 0.2 at 25°C

Vitamin K_1 Solution:
Composition per 100.0mL:

Vitamin K_1	1.0g
Ethanol	99.0mL

Preparation of Vitamin K_1 Solution: Add vitamin K_1 to 99.0mL of absolute ethanol. Mix thoroughly.

Hemin Solution:
Composition per 100.0mL:

Hemin	1.0g
NaOH (1*N* solution)	20.0mL

Preparation of Hemin Solution: Add hemin to 20.0mL of 1*N* NaOH solution. Mix thoroughly. Bring volume to 100.0mL with distilled/deionized water.

Vancomycin Solution:
Composition per 10.0mL:

Vancomycin	7.5mg

Preparation of Vancomycin Solution: Add vancomycin to distilled/deionized water and bring volume to 10.0mL. Mix thoroughly. Filter sterilize.

Preparation of Medium: Add components, except defibrinated sheep blood and vancomycin solution, to distilled/deionized water and bring volume to 940.0mL. Mix thoroughly. Gently heat and bring to boiling. Autoclave for 15 min at 15 psi pressure–121°C. Cool to 45°–50°C. Aseptically add 50.0mL of sterile defibrinated sheep blood and 10.0mL of vancomycin solution. Mix thoroughly. Pour into sterile Petri dishes or distribute into sterile tubes.

Use: For the selective cultivation of *Fusobacterium* species from clinical and nonclinical specimens.

Fay and Barry Medium

Composition per liter:

Amino acid	10.0g
Peptone	5.0g
Yeast extract	3.0g
Bromcresol Purple solution	5.0mL

pH 5.5 ± 0.2 at 25°C

Bromcresol Purple Solution:

Composition per 100.0mL:

Bromcresol Purple	0.2g
Ethanol	50.0mL

Preparation of Bromcresol Purple Solution: Add Bromcresol Purple to 50.0mL of absolute ethanol. Add distilled/deionized water and bring volume to 100.0mL. Mix thoroughly.

Preparation of Medium: Add components to distilled/deionized water and bring volume to 1.0L. The amino acid may be L-arginine, L-ornithine, or L-lysine, depending on which amino acid decarboxylase activity is being measured. Mix thoroughly. Distribute into tubes or flasks. Autoclave for 15 min at 15 psi pressure–121°C.

Use: For the determination of decarboxylase activities of *Aeromonas* species.

Faybitch's Sucrose Gelatin Agar

Composition per liter:

Sucrose	100.0g
Gelatin	15.0g
Agar	10.0g

Preparation of Medium: Add components to distilled/deionized water and bring volume to 1.0L. Mix thoroughly. Gently heat and bring to boiling. Distribute into tubes or flasks. Autoclave for 15 min at 15 psi pressure–121°C. Pour into sterile Petri dishes or leave in tubes.

Use: For the growth of microbial cultures that are to be lyophilized.

FB Medium (DSMZ Medium 980)

Composition per liter:

NH_4Cl	0.54g
$MgCl_2·6H_2O$	0.2g
$CaCl_2·2H_2O$	0.15g
KH_2PO_4	0.14g
Resazurin	0.5mg
$NaHCO_3$ solution	10.0mL
Yeast extract solution	10.0mL
Na-crotonate solution	10.0mL
Cysteine solution	10.0mL
$Na_2S·9H_2O$ solution	10.0mL
Vitamin solution	10.0mL
Trace elements solution SL-9	1.0mL
Selenite tungstate solution	1.0mL

pH 7.0 ± 0.2 at 25°C

Na-Crotonate Solution:

Composition per 10.0mL:

Na-crotonate	0.86g

Preparation of Na-Crotonate Solution: Add Na-crotonate to distilled/deionized water and bring volume to 10.0mL. Mix thoroughly. Sparge with 100% N_2. Autoclave for 15 min at 15 psi pressure–121°C.

Cysteine Solution:

Composition per 10.0mL:

L-Cysteine·HCl·H_2O	0.25g

Preparation of Cysteine Solution: Add L-cysteine·HCl·H_2O to distilled/deionized water and bring volume to 10.0mL. Mix thoroughly. Sparge with 100% N_2. Autoclave for 15 min at 15 psi pressure–121°C.

Trace Elements Solution SL-9:

Composition per liter:

$MgSO_4·7H_2O$	3.0g
Nitrilotriacetic acid	1.5g
NaCl	1.0g
$MnSO_4·2H_2O$	0.5g
$CoSO_4·7H_2O$	0.18g
$ZnSO_4·7H_2O$	0.18g
$CaCl_2·2H_2O$	0.1g
$FeSO_4·7H_2O$	0.1g
$NiCl_2·6H_2O$	0.025g
$KAl(SO_4)_2·12H_2O$	0.02g
H_3BO_3	0.01g
$Na_2MoO_4·4H_2O$	0.01g
$CuSO_4·5H_2O$	0.01g
$Na_2SeO_3·5H_2O$	0.3mg

Preparation of Trace Elements Solution SL-9: Add nitrilotriacetic acid to 500.0mL of distilled/deionized water. Dissolve by adjusting pH to 6.5 with KOH. Add remaining components. Add distilled/deionized water to 1.0L. Mix thoroughly.

Vitamin Solution:

Composition per liter:

Pyridoxine-HCl	10.0mg
Thiamine-HCl·$2H_2O$	5.0mg
Riboflavin	5.0mg
Nicotinic acid	5.0mg
D-Ca-pantothenate	5.0mg
p-Aminobenzoic acid	5.0mg
Lipoic acid	5.0mg
Biotin	2.0mg
Folic acid	2.0mg
Vitamin B_{12}	0.1mg

Preparation of Vitamin Solution: Add components to distilled/deionized water and bring volume to 1.0L. Mix thoroughly. Sparge with 80% H_2 + 20% CO_2. Filter sterilize.

$Na_2S·9H_2O$ Solution:

Composition per 10.0mL:

$Na_2S·9H_2O$	0.25g

Preparation of $Na_2S·9H_2O$ Solution: Add $Na_2S·9H_2O$ to distilled/deionized water and bring volume to 10.0mL. Sparge with N_2. Autoclave for 15 min at 15 psi pressure–121°C. Cool to 25°C. Store anaerobically.

$NaHCO_3$ Solution:

Composition per 10.0mL:

$NaHCO_3$	2.5g

Preparation of $NaHCO_3$ Solution: Add $NaHCO_3$ to distilled/deionized water and bring volume to 10.0mL. Mix thoroughly. Sparge with 80% N_2 + 20% CO_2. Filter sterilize.

Yeast Extract Solution:

Composition per 10.0mL:

Yeast extract	0.2g

Preparation of Yeast Extract Solution: Add yeast extract to distilled/deionized water and bring volume to 10.0mL. Mix thoroughly. Sparge with 100% N_2. Filter sterilize.

Preparation of Medium: Add components, except Na-crotonate solution, yeast extract solution, cysteine solution, vitamin solution, $NaHCO_3$ solution, and $Na_2S{\cdot}9H_2O$ solution, to distilled/deionized water and bring volume to 940.0mL. Mix thoroughly. Gently heat and bring to boiling. Boil for 3 min. Cool to 25°C while sparging with 80% N_2 + 20% CO_2. Distribute to anaerobe tubes or bottles under 80% N_2 + 20% CO_2. Autoclave for 15 min at 15 psi pressure–121°C. Aseptically and anaerobically add per liter of medium, 10.0mL sterile cysteine solution, 10.0mL sterile Na-crotonate solution, 10.0mL sterile vitamin solution, 10.0mL sterile yeast extract solution, 10.0mL sterile $NaHCO_3$ solution, and 10.0mL sterile $Na_2S{\cdot}9H_2O$ solution. Mix thoroughly. The final pH should be 7.0.

Use: For the cultivation of *Sporotomaculum syntrophicum.*

FC Agar (Fecal Coliform Agar) (m-FC Agar) (m-Fecal Coliform Agar)

Composition per liter:

Agar	15.0g
Lactose	12.5g
NaCl	5.0g
Proteose peptone No. 3	5.0g
Yeast extract	3.0g
Bile salts	1.5g
Aniline Blue	0.1g
Rosolic acid solution	10.0mL

pH 7.4 ± 0.2 at 25°C

Source: This medium is available as a premixed powder from BD Diagnostic Systems.

Rosolic Acid Solution:

Composition per 100.0mL:

Rosolic acid	1.0g

Preparation of Rosolic Acid Solution: Add rosolic acid to 0.2*N* NaOH and bring volume to 100.0L. Mix thoroughly.

Preparation of Medium: Add 10.0mL rosolic acid solution to 950.0mL distilled/deionized water. Mix thoroughly. Add other components and bring volume to 1.0L with distilled/deionized water. Mix thoroughly. Gently heat and bring to boiling with frequent mixing. Do not autoclave. Pour into sterile Petri dishes or leave in tubes.

Use: For the cultivation of fecal coliform bacteria from waters and the enumeration of coliform bacteria using the membrane filtration method.

FC Agar (Fecal Coliform Agar) (m-FC Agar) (m-Fecal Coliform Agar)

Composition per liter:

Agar	15.0g
Lactose	12.5g
Tryptose	10.0g
NaCl	5.0g
Proteose peptone No. 3	5.0g
Yeast extract	3.0g
Bile salts	1.5g
Aniline Blue	0.1g
Rosolic acid solution	10.0mL

pH 7.4 ± 0.2 at 25°C

Rosolic Acid Solution:

Composition per 100.0mL:

Rosolic acid	1.0g

Preparation of Rosolic Acid Solution: Add rosolic acid to 0.2*N* NaOH and bring volume to 100.0L. Mix thoroughly.

Preparation of Medium: Add 10.0mL rosolic acid solution to 950.0mL of distilled/deionized water. Mix thoroughly. Add other components and bring volume to 1.0L with distilled/deionized water. Mix thoroughly. Gently heat and bring to boiling with frequent mixing. Do not autoclave. Pour into sterile Petri dishes or leave in tubes.

Use: For the cultivation of fecal coliform bacteria from waters and the enumeration of coliform bacteria using the membrane filtration method.

FC Broth (Fecal Coliform Broth) (m-FC Broth) (m-Fecal Coliform Broth)

Composition per liter:

Lactose	12.5g
Tryptose	10.0g
NaCl	5.0g
Proteose peptone No. 3	5.0g
Yeast extract	3.0g
Bile salts	1.5g
Aniline Blue	0.1g
Rosolic acid solution	10.0mL

pH 7.4 ± 0.2 at 25°C

Rosolic Acid Solution:

Composition per 100.0mL:

Rosolic acid	1.0g

Preparation of Rosolic Acid Solution: Add rosolic acid to 0.2*N* NaOH and bring volume to 100.0L. Mix thoroughly.

Preparation of Medium: Add 10.0mL of rosolic acid solution to 950.0mL of distilled/deionized water. Mix thoroughly. Add other components and bring volume to 1.0L with distilled/deionized water. Mix thoroughly. Gently heat and bring to boiling with frequent mixing. Do not autoclave. Pour into sterile Petri dishes or leave in tubes.

Use: For the cultivation of fecal coliform bacteria from waters and the enumeration of coliform bacteria using the membrane filtration method.

FC Broth (Fecal Coliform Broth) (m-FC Broth) (m-Fecal Coliform Broth)

Composition per liter:

Lactose	12.5g
NaCl	5.0g
Proteose peptone No. 3	5.0g
Yeast extract	3.0g
Bile salts	1.5g

Aniline Blue	0.1g
Rosolic acid solution	10.0mL

pH 7.4 ± 0.2 at 25°C

Source: This medium is available as a premixed powder from BD Diagnostic Systems.

Rosolic Acid Solution:

Composition per 100.0mL:

Rosolic acid	1.0g

Preparation of Rosolic Acid Solution: Add rosolic acid to 0.2*N* NaOH and bring volume to 100.0L. Mix thoroughly.

Preparation of Medium: Add 10.0mL of rosolic acid solution to 950.0mL of distilled/deionized water. Mix thoroughly. Add other components and bring volume to 1.0L with distilled/deionized water. Mix thoroughly. Gently heat and bring to boiling with frequent mixing. Do not autoclave. Pour into sterile Petri dishes or leave in tubes.

Use: For the cultivation of fecal coliform bacteria from waters and the enumeration of coliform bacteria using the membrane filtration method.

FCIC

See: **Fecal Coliform Agar, Modified**

FDA Agar (ATCC Medium 182) (AATCC Bacteriostasis Agar) (American Association of Textile Chemists and Colorists Bacteriostasis Agar)

Composition per liter:

Agar	15.0g
Peptic digest of animal tissue	10.0g
Beef extract	5.0g
NaCl	5.0g

pH 6.8 ± 0.1 at 25°C

Source: This medium is available as a premixed powder from BD Diagnostic Systems.

Preparation of Medium: Add components to distilled/deionized water and bring volume to 1.0L. Mix thoroughly. Gently heat and bring to boiling. Distribute into tubes or flasks. Autoclave for 15 min at 15 psi pressure–121°C. Pour into sterile Petri dishes or leave in tubes.

Use: For testing the antibacterial activities of antiseptics and disinfectants.

FDA Broth (AATCC Bacteriostasis Broth) (American Association of Textile Chemists and Colorists Bacteriostasis Broth)

Composition per liter:

Peptic digest of animal tissue	10.0g
Beef extract	5.0g
NaCl	5.0g

pH 6.8 ± 0.1 at 25°C

Source: This medium is available as a premixed powder from BD Diagnostic Systems.

Preparation of Medium: Add components to distilled/deionized water and bring volume to 1.0L. Mix thoroughly. Distribute into tubes or flasks. Autoclave for 15 min at 15 psi pressure–121°C.

Use: For testing the antibacterial activities of antiseptics and disinfectants.

Fe(III) Lactate Nutrient Agar

Composition per liter:

Agar	15.0g
Peptone	5.0g
NaCl	5.0g
Yeast extract	2.0g
Beef extract	1.0g
Fe(III)-lactate solution	25.0mL

pH 7.2 ± 0.2 at 25°C

Fe(III)-Lactate Solution:

Composition per 30.0mL:

$FeCl_3 \cdot 6H_2O$ solution	20.0mL
Sodium lactate solution	10.0mL

Preparation of Fe(III)-Lactate Solution: Aseptically combine the component solutions. Mix thoroughly.

$FeCl_3 \cdot 6H_2O$ Solution:

Composition per 100.0mL:

$FeCl_3 \cdot 6H_2O$	5.0g

Preparation of $FeCl_3 \cdot 6H_2O$ Solution: Add $FeCl_3 \cdot 6H_2O$ to distilled/deionized water and bring volume to 100.0mL. Mix thoroughly. Filter sterilize.

Sodium Lactate Solution

Composition per 100.0mL:

Sodium lactate	5.0g

Preparation of Sodium Lactate Solution: Add sodium lactate to distilled/deionized water and bring volume to 100.0mL. Mix thoroughly. Filter sterilize.

Preparation of Medium: Add components, except Fe(III)-lactate solution, to distilled/deionized water and bring volume to 975.0L. Mix thoroughly. Gently heat and bring to boiling. Autoclave for 15 min at 15 psi pressure–121°C. Cool to 50°–55°C. Aseptically add 25.0mL of filter-sterilized Fe(III)-lactate solution. Mix thoroughly. Pour into sterile Petri dishes or distribute into sterile tubes.

Use: For the cultivation of *Shewanella putrefaciens*.

Fecal Coliform Agar

See: **FC Agar**

Fecal Coliform Agar, Modified (m-Fecal Coliform Agar, Modified) (FCIC)

Composition per liter:

Agar	15.0g
Inositol	10.0g
Tryptose	10.0g
Proteose peptone No. 3	5.0g
NaCl	5.0g
Yeast extract	3.0g
Bile salts No. 3	1.5g
Aniline Blue	0.1g

pH 7.4 ± 0.2 at 25°C

Preparation of Medium: Add components to distilled/deionized water and bring volume to 1.0L. Mix thoroughly. Gently heat and bring to boiling. Do not autoclave. Cool to 50°C. Adjust pH to 7.4. Pour into sterile Petri dishes in 20.0mL volumes. Allow surface of plates to dry before using.

Use: For the isolation, cultivation, and enumeration of *Klebsiella* species using the membrane filter method.

Fecal Coliform Agar, Modified

Composition per liter:

Agar	15.0g
Lactose	12.5g
Tryptose	10.0g
Proteose peptone No. 3	5.0g
NaCl	5.0g
Yeast extract	3.0g
Bile salts No. 3	1.5g
Aniline Blue	0.1g

pH 7.4 ± 0.2 at 25°C

Preparation of Medium: Add components and bring volume to 1.0L. Mix thoroughly. Gently heat and bring to boiling. Do not autoclave. Cool to 50°C. Adjust pH to 7.4. Pour into sterile Petri dishes in 20.0mL volumes. Allow surface of plates to dry before using.

Use: For the isolation, cultivation, and identification of stressed fecal coliform microorganisms based on their ability to ferment lactose. Lactose-fermenting bacteria turn the medium blue.

Fecal Coliform Broth
See: **FC Broth**

Feeley-Gorman Agar
See: **F-G Agar**

Feeley-Gorman Agar with Selenium
See: **F-G Agar with Selenium**

Feeley-Gorman Broth
See: **F-G Broth**

Feeley Gorman HiVeg Agar (F.G. HiVeg Agar)

Composition per liter:

Plant acid hydrolysate	17.5g
Agar	17.0g
Plant extract	3.0g
Starch	1.5g
L-Cysteine·HCl	0.4g
$Fe_4(P_2O_7)_3 \cdot H_2O$, soluble	0.25g

pH 6.9 ± 0.05 at 25°C

Source: This medium is available as a premixed powder from HiMedia.

Preparation of Medium: Add components to distilled/deionized water and bring volume to 1.0L. Mix thoroughly. Gently heat and bring to boiling. Autoclave for 15 min at 15 psi pressure–121°C. Cool to 45°–50°C. Mix thoroughly. Adjust pH to 6.9. Pour into sterile Petri dishes or distribute into sterile tubes.

Use: For the isolation and cultivation of *Legionella pneumophila*.

Feeley Gorman HiVeg Broth (F.G. HiVeg Broth)

Composition per liter:

Plant acid hydrolysate	17.5g
Plant extract	3.0g
Starch	1.5g
L-Cysteine·HCl	0.4g
$Fe_4(P_2O_7)_3 \cdot H_2O$, soluble	0.25g

pH 6.9 ± 0.05 at 25°C

Source: This medium is available as a premixed powder from HiMedia.

Preparation of Medium: Add components to distilled/deionized water and bring volume to 1.0L. Mix thoroughly. Gently heat and bring to boiling. Autoclave for 15 min at 15 psi pressure–121°C. Cool to 45°–50°C. Mix thoroughly. Adjust pH to 6.9.

Use: For the cultivation of *Legionella pneumophila*.

Feodorov Medium

Composition per liter:

Mannitol or glucose	20.0g
Marine salts mixture	18.0g
$CaCO_3$	0.5g
K_2HPO_4	0.3g
$MgSO_4$	0.3g
$CaHPO_4$	0.2g
K_2SO_4	0.2g
$FeCl_3$	0.1g
Trace elements solution	1.0mL

Trace Elements Solution:

Composition per 100.0mL:

H_3BO_3	0.5g
$(NH_4)_6Mo_7O_{24} \cdot 4H_2O$	0.5g
KI	0.05g
NaBr	0.05g
$Al_2(SO_4)_3 \cdot 18H_2O$	0.03g
$ZnSO_4$	0.02g

Preparation of Trace Elements Solution: Add components to distilled/deionized water and bring volume to 100.0mL. Mix thoroughly.

Preparation of Medium: Add components to distilled/deionized water and bring volume to 1.0L. Mix thoroughly. Distribute into tubes or flasks. Autoclave for 15 min at 15 psi pressure–121°C.

Use: For the cultivation and maintenance of *Azotobacter vinelandii*.

Fermentation Basal Medium

Composition per liter:

Agar	15.0g
$(NH_4)_2HPO_4$	1.0g
$MgSO_4 \cdot 7H_2O$	0.2g
KCl	0.02g
Carbohydrate solution	100.0mL
Bromcresol Purple solution	20.0mL

pH 7.0 ± 0.2 at 25°C

Carbohydrate Solution:

Composition per 100.0mL:

Carbohydrate	10.0g

Preparation of Carbohydrate Solution: Add carbohydrate to distilled/deionized water and bring volume to 100.0mL. Mix thoroughly. Filter sterilize.

Bromcresol Purple Solution:

Composition per 100.0mL:

Bromcresol Purple	0.04g
Ethanol	50.0mL

Preparation of Bromcresol Purple Solution: Add Bromcresol Purple to 50.0mL of absolute ethanol. Add distilled/deionized water and bring volume to 100.0mL. Mix thoroughly.

Preparation of Medium: Add components, except carbohydrate solution, to distilled/deionized water and bring volume to 900.0mL. Mix thoroughly. Gently heat and bring to boiling. Autoclave for 15 min at 15 psi pressure–121°C. Cool to 45°–50°C. Aseptically add 100.0mL of sterile carbohydrate solution. Various carbohydrates are used for different fermentation tests. Mix thoroughly. Pour into sterile Petri dishes or distribute into sterile tubes.

Use: For the differentiation of aerobic actinomycetes based upon carbohydrate fermentation. Actinomycetes that produce acid from carbohydrates turn the medium yellow.

Fermentation Base for *Campylobacter*

See: **Enteric Fermentation Base**

Fermentation Broth (CHO Medium)

Composition per liter:

Pancreatic digest of casein	15.0g
Yeast extract	7.0g
NaCl	2.5g
Agar	0.75g
Sodium thioglycolate	0.5g
L-Cystine	0.25g
Ascorbic acid	0.1g
Bromthymol Blue	0.01g
Carbohydrate or starch solution	100.0mL

pH 7.0 ± 0.1 at 25°C

Source: This medium is available as a premixed powder from BD Diagnostic Systems.

Carbohydrate Solution:

Composition per 100.0mL:

Carbohydrate	6.0g

Preparation of Carbohydrate Solution: Add carbohydrate to distilled/deionized water and bring volume to 10.0mL. Mix thoroughly. Filter sterilize.

Starch Solution:

Composition per 100.0mL:

Starch	2.5g

Preparation of Starch Solution: Add starch to distilled/deionized water and bring volume to 100.0mL. Mix thoroughly. Filter sterilize.

Preparation of Medium: Add components, except carbohydrate solution, to distilled/deionized water and bring volume to 900.0mL. Mix thoroughly. Distribute into tubes or flasks. Autoclave for 15 min at 15 psi pressure–121°C. Cool to 45°–50°C. Aseptically add 100.0mL of sterile carbohydrate solution. Mix thoroughly. Aseptically distribute into sterile tubes or flasks. Loosen caps on tubes. Place in an anaerobic chamber under an atmosphere of 85% N_2, 10% H_2, and 5% CO_2. Fasten the caps securely or maintain in an anaerobic chamber.

Use: For the differentiation of anaerobic bacteria based upon carbohydrate fermentation. Bacteria that ferment carbohydrates turn the medium yellow.

Fermentation HiVeg Medium Base for *C. perfringens* with Salicin and Raffinose

Composition per liter:

Plant hydrolysate	10.0g
Plant special peptone	10.0g
Agar	2.0g
Na-thioglycollate	0.25g
Salicin solution	10.0mL
Raffinose solution	10.0mL

pH 7.0 ± 2.0 at 25°C

Source: This medium, without salicin and raffinose, is available as a premixed powder from HiMedia.

Salicin Solution:

Composition per 10.0mL:

Salicin	0.1g

Preparation of Salicin Solution: Add salicin to distilled/deionized water and bring volume to 10.0mL. Mix thoroughly. Filter sterilize.

Raffinose Solution:

Composition per 10.0mL:

Raffinose	0.1g

Preparation of Raffinose Solution: Add raffinose to distilled/deionized water and bring volume to 10.0mL. Mix thoroughly. Filter sterilize.

Preparation of Medium: Add components, except salicin and raffinose solutions, to distilled/deionized water and bring volume to 980.0mL. Mix thoroughly. Gently heat and bring to boiling. Autoclave for 15 min at 15 psi pressure–121°C. Cool to 45°–50°C. Aseptically add 10.0mL sterile salicin solution and 10.0mL sterile raffinose solution. Mix thoroughly. Aseptically distribute into tubes or flasks.

Use: For the cultivation of *Clostridium perfringens*.

Fermentation HiVeg Medium for Neisseriae with Carbohydrate

Composition per liter:

Plant hydrolysate	20.0g
NaCl	5.0g
Agar	3.5g
Cystine	0.5g
Na_2SO_3	0.5g
Phenol Red	0.017g
Carbohydrate solution	100.0mL

pH 6.9 ± 0.2 at 25°C

Source: This medium is available as a premixed powder from HiMedia.

Carbohydrate Solution:

Composition per 100.0mL:

Carbohydrate	10.0g

Preparation of Carbohydrate Solution: Add carbohydrate to distilled/deionized water and bring volume to 100.0mL. Adonitol, arabinose, cellobiose, glucose, dulcitol, fructose, galactose, inositol, lactose, maltose, mannitol, raffinose, rhamnose, salicin, sorbitol, sucrose, trehalose, xylose, or other carbohydrates may be used. Mix thoroughly. Filter sterilize.

Preparation of Medium: Add components, except carbohydrate solution, to distilled/deionized water and bring volume to 900.0mL. Mix thoroughly. Gently heat and bring to boiling. Distribute into tubes or flasks. Autoclave for 15 min at 12 psi pressure–118°C. Cool to 50°C. Aseptically add 100.0mL of sterile carbohydrate solution. Mix thoroughly.

Use: For the cultivation and identification of *Neisseria* spp. For studying fermentation reactions of fastidious organisms such as *Neisseria* species.

Fermentation HiVeg Medium for *Staphylococcus* and *Micrococcus*

Composition per liter:

Glucose	10.0g
Plant hydrolysate	10.0g
Agar	2.2g
Yeast extract	1.0g
Bromcresol Purple	0.04g

pH 7.0 ± 0.2 at 25°C

Source: This medium is available as a premixed powder from HiMedia.

Preparation of Medium: Add components to distilled/deionized water and bring volume to 1.0L. Mix thoroughly. Distribute into tubes or flasks. Autoclave for 15 min at 15 psi pressure–121°C.

Use: For the cultivation and identification of *Staphylococcus* and *Micrococcus* spp.

Fermentation Medium

Composition per liter:

Glucose or mannitol	10.0g
Pancreatic digest of casein	10.0g
Agar	2.2g
Yeast extract	1.0g
Bromcresol Purple	0.04g

pH 7.0 ± 0.2 at 25°C

Preparation of Medium: Add components to distilled/deionized water and bring volume to 1.0L. Mix thoroughly. Gently heat and bring to boiling. Distribute into tubes or flasks. Autoclave for 10 min at 15 psi pressure–121°C. Pour into sterile Petri dishes or leave in tubes.

Use: For differentiating *Staphylococcus* and *Micrococcus* species based upon the fermentation of glucose and mannitol.

Fermentation Medium Base for *C. perfringens* with Salicin and Raffinose

Composition per liter:

Casein enzymatic hydrolysate	10.0g
Peptone, special	10.0g
Agar	2.0g
Na-thioglycollate	0.25g
Salicin solution	10.0mL
Raffinose solution	10.0mL

pH 7.0 ± 2.0 at 25°C

Source: This medium, without salicin and raffinose, is available as a premixed powder from HiMedia.

Salicin Solution:

Composition per 10.0mL:

Salicin	0.1g

Preparation of Salicin Solution: Add salicin to distilled/deionized water and bring volume to 10.0mL. Mix thoroughly. Filter sterilize.

Raffinose Solution:

Composition per 10.0mL:

Raffinose	0.1g

Preparation of Raffinose Solution: Add raffinose to distilled/deionized water and bring volume to 10.0mL. Mix thoroughly. Filter sterilize.

Preparation of Medium: Add components, except salicin and raffinose solutions, to distilled/deionized water and bring volume to 980.0mL. Mix thoroughly. Gently heat and bring to boiling. Autoclave for 15 min at 15 psi pressure–121°C. Cool to 45°–50°C. Aseptically add 10.0mL sterile salicin solution and 10.0mL sterile raffinose solution. Mix thoroughly. Aseptically distribute into tubes or flasks.

Use: For the cultivation of *Clostridium perfringens*.

Fermentation Medium for Neisseriae with Carbohydrate

Composition per liter:

Casein enzymatic hydrolysate	20.0g
NaCl	5.0g
Agar	3.5g
Cystine	0.5g
Na_2SO_3	0.5g
Phenol Red	0.017g
Carbohydrate solution	100.0mL

pH 6.9 ± 0.2 at 25°C

Source: This medium is available as a premixed powder from HiMedia.

Carbohydrate Solution:

Composition per 100.0mL:

Carbohydrate	10.0g

Preparation of Carbohydrate Solution: Add carbohydrate to distilled/deionized water and bring volume to 100.0mL. Adonitol, arabinose, cellobiose, glucose, dulcitol, fructose, galactose, inositol, lactose, maltose, mannitol, raffinose, rhamnose, salicin, sorbitol, sucrose, trehalose, xylose, or other carbohydrates may be used. Mix thoroughly. Filter sterilize.

Preparation of Medium: Add components, except carbohydrate solution, to distilled/deionized water and bring volume to 900.0mL. Mix thoroughly. Gently heat and bring to boiling. Distribute into tubes or flasks. Autoclave for 15 min at 12 psi pressure–118°C. Cool to 50°C. Aseptically add 100.0mL of sterile carbohydrate solution. Mix thoroughly.

Use: For the cultivation and identification of *Neisseria* spp. For studying fermentation reactions of fastidious organisms such as *Neisseria* species.

Fermentation Medium for *Staphylococcus* and *Micrococcus*

Composition per liter:

Glucose	10.0g
Casein enzymatic hydrolysate	10.0g
Agar	2.2g
Yeast extract	1.0g
Bromcresol Purple	0.04g

pH 7.0 ± 0.2 at 25°C

Source: This medium is available as a premixed powder from HiMedia.

Preparation of Medium: Add components to distilled/deionized water and bring volume to 1.0L. Mix thoroughly. Distribute into tubes or flasks. Autoclave for 15 min at 15 psi pressure–121°C.

Use: For the cultivation and identification of *Staphylococcus* and *Micrococcus* spp.

Ferric Citrate Medium

Composition per liter:

Ferric citrate	13.7g
Sodium lactate (60% solution)	5.6g
$NaHCO_3$	2.5g
NH_4Cl	1.5g
NaH_2PO_4	0.6g
KCl	0.1g
Wolfe's mineral solution	10.0mL
Wolfe's vitamin solution	10.0mL

pH 7.0 ± 0.2 at 25°C

Wolfe's Vitamin Solution:

Composition per liter:

Pyridoxine·HCl	10.0mg
Calcium D-(+)-pantothenate	5.0mg
Nicotinic acid	5.0mg
p-Aminobenzoic acid	5.0mg
Riboflavin	5.0mg
Thiamine·HCl	5.0mg
Thioctic acid	5.0mg
Biotin	2.0mg
Folic acid	2.0mg
Cyanocobalamin	0.1mg

Preparation of Wolfe's Vitamin Solution: Add components to distilled/deionized water and bring volume to 1.0L. Mix thoroughly. Filter sterilize.

Wolfe's Mineral Solution:

Composition per liter:

$MgSO_4·7H_2O$	3.0g
Nitrilotriacetic acid	1.5g
NaCl	1.0g
$MnSO_4·H_2O$	0.5g
$CaCl_2$	0.1g
$CoCl_2·6H_2O$	0.1g
$FeSO_4·7H_2O$	0.1g
$ZnSO_4·7H_2O$	0.1g
$AlK(SO_4)_2·12H_2O$	0.01g
$CuSO_4·5H_2O$	0.01g
H_3BO_3	0.01g
$Na_2MoO_4·2H_2O$	0.01g

Preparation of Wolfe's Mineral Solution: Add nitrilotriacetic acid to 500.0mL of distilled/deionized water. Adjust pH to 6.5 with KOH. Add remaining components sequentially. Add distilled/deionized water to 1.0L. Mix thoroughly.

Preparation of Medium: Add ferric citrate to distilled/deionized water and bring volume to 1.0L. Gently heat and bring to boiling. Continue boiling until ferric citrate is dissolved. Cool to room temperature. Adjust to pH 6.6 with 10*N* NaOH. Add remaining components. Mix thoroughly. Sparge with 80% N_2 + 20% CO_2. Anaerobically distribute into tubes or flasks. Autoclave for 15 min at 15 psi pressure–121°C. Final pH should be 7.0.

Use: For the cultivation of *Aeromonas encheleia* and *Shewanella alga.*

Ferroglobus placidus Medium (DSMZ Medium 730)

Composition per 1020.0mL:

NaCl	18.0g
$NaHCO_3$	10.0g
$MgCl_2·6H_2O$	4.3g
KNO_3	1.0g
KCl	0.34g
NH_4Cl	0.24g
$CaCl_2·2H_2O$	0.14g
$K_2HPO_4·3H_2O$	0.14g
Resazurin	0.5mg
Trace elements solution	10.0mL
Vitamin solution	10.0mL
$Na_2S·9H_2O$ solution	10.0mL
Na-pyruvate solution	10.0mL

pH 7.0 ± 0.2 at 25°C

Na-pyruvate Solution:

Composition per 10.0mL:

Na-pyruvate	1.0g

Preparation of Na-pyruvate Solution: Add Na-pyruvate to distilled/deionized water and bring volume to 10.0mL. Mix thoroughly. Sparge with 100% N_2. Filter sterilize.

Trace Elements Solution:

Composition per liter:

$MgSO_4·7H_2O$	3.0g
Nitrilotriacetic acid	1.5g
NaCl	1.0g
$MnSO_4·2H_2O$	0.5g
$CoSO_4·7H_2O$	0.18g
$ZnSO_4·7H_2O$	0.18g
$CaCl_2·2H_2O$	0.1g
$FeSO_4·7H_2O$	0.1g
$NiCl_2·6H_2O$	0.025g
$KAl(SO_4)_2·12H_2O$	0.02g
H_3BO_3	0.01g
$Na_2MoO_4·4H_2O$	0.01g
$CuSO_4·5H_2O$	0.01g
$Na_2SeO_3·5H_2O$	0.3mg

Preparation of Trace Elements Solution: Add nitrilotriacetic acid to 500.0mL of distilled/deionized water. Dissolve by adjusting pH to 6.5 with KOH. Add remaining components. Add distilled/deionized water to 1.0L. Mix thoroughly.

Vitamin Solution:

Composition per liter:

Pyridoxine-HCl	10.0mg
Thiamine-HCl·2H_2O	5.0mg
Riboflavin	5.0mg
Nicotinic acid	5.0mg
D-Ca-pantothenate	5.0mg
p-Aminobenzoic acid	5.0mg
Lipoic acid	5.0mg
Biotin	2.0mg
Folic acid	2.0mg
Vitamin B_{12}	0.1mg

Preparation of Vitamin Solution: Add components to distilled/deionized water and bring volume to 1.0L. Mix thoroughly. Sparge with 80% H_2 + 20% CO_2. Filter sterilize.

$Na_2S·9H_2O$ Solution:

Composition per 10.0mL:

$Na_2S·9H_2O$	0.5g

Preparation of $Na_2S·9H_2O$ Solution: Add $Na_2S·9H_2O$ to distilled/deionized water and bring volume to 10.0mL. Sparge with N_2. Autoclave for 15 min at 15 psi pressure–121°C. Cool to 25°C. Store anaerobically.

Preparation of Medium: Add components, except $Na_2S·9H_2O$ solution, vitamin solution, and Na-pyruvate solution, to distilled/deionized water and bring volume to 990.0mL. Mix thoroughly. Flush medium with N_2 for 20 min. Adjust medium pH to 7.0 with 4*N* H_2SO_4. Add 10.0mL of $Na_2S·9H_2O$ solution. Mix thoroughly. Readjust the medium pH to 7.0 with H_2SO_4, while flushing the gas phase only with N_2. Dispense 10mL volumes into 100mL serum bottles with rubber stoppers under N_2. Replace gas phase by 80% H_2 + 20% CO_2 gas mixture and finally pressurize the bottles to 2 bar gas overpressure. Autoclave for 15 min at 15 psi pressure–121°C. Aseptically inject via syringe 0.1mL vitamin solution and 0.1mL Na-pyruvate solution into each tube containing 10.0mL of the autoclaved medium.

Use: For the cultivation of *Ferroglobus placidus.*

Ferroplasma acidiphilum Medium (DSMZ Medium 874)

Composition per 1001.6mL:

Solution A	950.0mL
Solution B	10.0mL
Solution C	1.6mL

pH 1.7 ± 0.2 at 25°C

Solution A:

Composition per 950.0mL:

$MgSO_4·7H_2O$	0.4g
$(NH_4)_2SO_4$	0.2g
KCl	0.1g
K_2HPO_4	0.1g

Preparation of Solution A: Add components to distilled/deionized water and bring volume to 950.0mL. Mix thoroughly.

Solution B:

Composition per 50.0mL:

$FeSO_4·7H_2O$	25.0g
H_2SO_4, 1*N*	40.0mL

Preparation of Solution B: Add components to distilled/deionized water and bring volume to 50.0mL. Mix thoroughly.

Solution C:

Composition per 10.0mL:

Yeast extract	1.0g

Preparation of Solution C: Add yeast extract to distilled/deionized water and bring volume to 10.0mL. Mix thoroughly. Autoclave for 15 min at 15 psi pressure–121°C. Cool to 25°C.

Preparation of Medium: Add 50.0mL solution B to 950.0mL solution A. Mix thoroughly. Adjust pH to 1.6–1.8 with H_2SO_4. Filter sterilize. Aseptically add 1.6 mL of sterile solution C. Pour into sterile Petri dishes or leave in tubes. Mix thoroughly. Aseptically distribute to sterile tubes or flasks.

Use: For the cultivation of *Ferroplasma acidiphilum (Ferriplasma acidophilum).*

Ferrous Sulfate/Yeast Extract Medium (DSMZ Medium 1190)

Composition per liter:

Yeast extract	0.2g
Basal salts solution	20.0mL
Ferrous sulfate solution	20.0mL
Trace elements solution	1.0mL

Basal Salts Solution:

Composition per liter:

$MgSO_4·7H_2O$	25.0g
$(NH_4)_2SO_4$	22.5g
$Na_2SO_4·10H_2O$	7.5g
KCl	2.5g
KH_2PO_4	2.5g
$Ca(NO_3)·4H_2O$	0.7g

Preparation of Basal Salts Solution: Add components to distilled/deionized water and bring volume to 1.0L. Mix thoroughly. Autoclave for 15 min at 15 psi pressure–121°C.

Trace Elements Solution:

Composition per liter:

$ZnSO_4·7H_2O$	10.0g
$CuSO_4·5H_2O$	1.0g
$CoO_4·7H_2O$	1.0g
$MnSO_4·2H_2O$	1.0g
$NiSO_4·6H_2O$	1.0g
Na_2SeO_4	1.0g
$Na_2WO_4·2H_2O$	1.0g
$Cr_2(SO_4)_3·15H_2O$	0.5g
H_3BO_3	0.6g
$Na_2MoO_4·2H_2O$	0.5g
$NaVO_3$	0.1g

Preparation of Trace Elements Solution: Adjust pH of 800.0mL of distilled/deionized water to 2.0 with dilute H_2SO_4. Add the above salts in order one at a time, allowing each to dissolve before adding the next. Maintain the pH at 2.0 by adding H_2SO_4 as necessary. After addition of vanadate, bring volume to 1.0L with water. Adjust final pH to 2.0. Autoclave for 15 min at 15 psi pressure–121°C. The vanadate will require several days to dissolve.

Ferrous Sulfate Solution:

Composition per 20.0mL:

$FeSO_4·7H_2O$	2.78g

Preparation of Ferrous Sulfate Solution: Adjust pH of 20.0mL of distilled/deionized water to 1.8 with H_2SO_4. Add $FeSO_4·7H_2O$. Mix thoroughly. Filter sterilize.

Preparation of Medium: Add components, except ferrous sulfate solution, to distilled/deionized water and bring volume to 980.0mL. Mix thoroughly. Adjust pH to 2.0 with H_2SO_4. Gently heat while stirring and bring to boiling. Boil for 1 min. Autoclave for 20 min at 15 psi pressure–121°C. Cool to room temperature. Aseptically add the ferrous sulfate solution.

Use: For the cultivation of *Sulfobacillus* spp.

Ferrous Sulfide Agar

Composition per 1200.0mL:

Agar layer	1.0L
Liquid overlay	200.0mL

Agar Layer:

Composition per liter:

Agar	30.0g
FeS washed precipitate supension	500.0mL

Preparation of Agar Layer: Add agar to distilled/deionized water and bring volume to 500.0mL. Mix thoroughly. Gently heat and bring to boiling. Autoclave for 15 min at 15 psi pressure–121°C. Cool to 45°–50°C. Heat FeS washed precipitate suspension to 45°–50°C. Mix thoroughly. Aseptically add 500.0mL of sterile FeS washed precipitate suspension to 500.0mL of sterile agar at 45°–50°C. Mix thoroughly.

FeS Washed Precipitate Suspension:

Composition per 500.0mL:

$Fe(NH_4)_2(SO_4)_2 \cdot 6H_2O$	78.4g
$Na_2S \cdot 9H_2O$	15.6g

Preparation of FeS Washed Precipitate Suspension: Add $Na_2S \cdot 9H_2O$ and $Fe(NH_4)_2(SO_4)_2 \cdot 6H_2O$ to 500.0mL boiling distilled/deionized water. Let precipitate settle from the hot solution in a completely filled and stoppered bottle. Wash precipitate four times by decanting supernatant and replacing each time with 500.0mL of boiling distilled/deionized water. Store FeS washed precipitate suspension in a completely filled 500.0mL glass-stoppered bottle.

Liquid Overlay:

Composition per liter:

$(NH_4)_2Cl$	1.0g
K_2HPO_4	0.5g
$MgSO_4 \cdot 7H_2O$	0.2g
$CaCl_2$	0.1g

Preparation of Liquid Overlay: Add components to distilled/deionized water and bring volume to 1.0L. Mix thoroughly. Autoclave for 15 min at 15 psi pressure–121°C. Cool to 25°C. Aseptically bubble 100% CO_2 for 15 sec.

Preparation of Medium: Aseptically distribute agar layer into sterile tubes in 10.0mL volumes. Allow tubes to cool in a slanted poistion. Aseptically add 2.0mL of sterile liquid overlay to each tube.

Use: For the enumeration, enrichment, and isolation of iron and sulfur bacteria, including *Gallionella ferruginea*.

Ferulate Medium

Composition per 1016.0mL:

Solution A	916.0mL
Solution B	70.0mL
Solution C	10.0mL
Solution D	10.0mL
Solution E	10.0mL

pH 7.2 ± 0.2 at 25°C

Solution A:

Composition per 916.0mL:

Pancreatic digest of casein	1.0g
Resazurin	1.0mg
Mineral solution	50.0mL
Rumen fluid, clarified	50.0mL
Vitamin solution	5.0mL
Trace elements solution SL-10	1.0mL

Mineral Solution:

Composition per liter:

Nitrilotriacetic acid	12.5g
NaCl	1.0g
$FeCl_3 \cdot 4H_2O$	0.2g
$MnCl_2 \cdot 4H_2O$	0.1g
$CaCl_2 \cdot 2H_2O$	0.1g
$ZnCl_2$	0.1g
$CuCl_2$	0.02g
Na_2SeO_3	0.02g
$CoCl_2 \cdot 6H_2O$	0.017g
H_3BO_3	0.01g
$Na_2MoO_4 \cdot 2H_2O$	0.01g

Preparation of Mineral Solution: Add nitrilotriacetic acid to 500.0mL of distilled/deionized water. Adjust pH to 6.5 with KOH. Add remaining components. Add distilled/deionized water to 1.0L. Mix thoroughly.

Vitamin Solution:

Composition per liter:

Nicotinic acid	2.5mg
Thiamine·HCl	1.25mg
p-Aminobenzoic acid	1.25mg
Pantothenic acid	0.62mg
Pyridoxine·HCl	6.2mg
Biotin	0.25mg

Preparation of Vitamin Solution: Add components to distilled/deionized water and bring volume to 1.0L. Mix thoroughly.

Trace Elements Solution SL-10:

Composition per liter:

$FeCl_2 \cdot 4H_2O$	1.5g
$CoCl_2 \cdot 6H_2O$	190.0mg
$MnCl_2 \cdot 4H_2O$	100.0mg
$ZnCl_2$	70.0mg
$Na_2MoO_4 \cdot 2H_2O$	36.0mg
$NiCl_2 \cdot 6H_2O$	24.0mg
H_3BO_3	6.0mg
$CuCl_2 \cdot 2H_2O$	2.0mg
HCl (25% solution)	10.0mL

Preparation of Trace Elements Solution SL-10: Add $FeCl_2 \cdot 4H_2O$ to 10.0mL of HCl solution. Mix thoroughly. Add distilled/deionized water and bring volume to 1.0L. Add remaining components. Mix thoroughly.

Preparation of Solution A: Add components to distilled/deionized water and bring volume to 916.0mL. Adjust pH to 7.2. Gently heat and bring to boiling. Continue boiling for a few minutes. Allow to cool to room temperature under 80% N_2 + 20% CO_2. Distribute into bottles under 80% N_2 + 20% CO_2. Autoclave for 15 min at 15 psi pressure–121°C.

Solution B:

Composition per 70.0mL:

$NaHCO_3$	3.5g

Preparation of Solution B: Add $NaHCO_3$ to distilled/deionized water and bring volume to 70.0mL. Mix thoroughly. Filter sterilize. Sparge with 80% N_2 + 20% CO_2 for 15 min.

Solution C:

Composition per 10.0mL:

L-Cysteine·HCl 0.3g

Preparation of Solution C: Add L-cysteine·HCl to distilled/deionized water and bring volume to 10.0mL. Mix thoroughly. Sparge with 100% N_2 for 3–4 min. Autoclave under 100% N_2 for 15 min at 15 psi pressure–121°C.

Solution D:

Composition per 10.0mL:

$Na_2S·9H_2O$ 0.3g

Preparation of Solution D: Add $Na_2S·9H_2O$ to distilled/deionized water and bring volume to 10.0mL. Mix thoroughly. Sparge with 100% N_2 for 3–4 min. Autoclave under 100% N_2 for 15 min at 15 psi pressure–121°C.

Solution E:

Composition per 10.0mL:

Sodium ferulate 1.5g

Preparation of Solution E: Add sodium ferulate to distilled/deionized water and bring volume to 10.0mL. Mix thoroughly. Filter sterilize. Sparge with 100% N_2 for 3–4 min.

Preparation of Medium: To 916.0mL of sterile solution A, add 70.0mL of sterile solution B, 10.0mL of sterile solution C, 10.0mL of sterile solution D, and 10.0mL of sterile solution E. Mix thoroughly. Anaerobically and aseptically distribute into sterile tubes or flasks.

Use: For the cultivation and maintenance of *Eubacterium callanderi.*

Fervidobacterium islandicum Medium

Composition per liter:

$(NH_4)_2SO_4$ 1.3g
Yeast extract 1.0g
KH_2PO_4 .... 0.28g
$MgSO_4·7H_2O$ 0.25g
$CaCl_2·2H_2O$ 0.07g
$FeCl_3·6H_2O$ 0.02g
$Na_2B_4·10H_2O$ 4.5mg
$MnCl_2·4H_2O$ 1.8mg
Resazurin 1.0mg
$ZnSO_4·7H_2O$ 0.22mg
$CuCl_2·2H_2O$ 0.05mg
$Na_2MoO_4·2H_2O$ 0.03mg
$VOSO_4·2H_2O$ 0.03mg
$CoSO_4$ 0.01mg
Glucose solution 20.0mL
$Na_2S·9H_2O$ solution 10.0mL

pH 7.0 ± 0.2 at 25°C

Glucose Solution:

Composition per 20.0mL:

Glucose 2.0g

Preparation of Glucose Solution: Add glucose to distilled/deionized water and bring volume to 20.0mL. Mix thoroughly. Sparge with 100% N_2 for 3–4 min. Autoclave for 15 min at 15 psi pressure–121°C.

$Na_2S·9H_2O$ Solution:

Composition per 10.0mL:

$Na_2S·9H_2O$ 0.5g

Preparation of $Na_2S·9H_2O$ Solution: Add $Na_2S·9H_2O$ to distilled/deionized water and bring volume to 10.0mL. Mix thoroughly. Gas under 100% N_2. Autoclave for 15 min at 15 psi pressure–121°C.

Preparation of Medium: Add components, except glucose solution and $Na_2S·9H_2O$ solution, to distilled/deionized water and bring volume to 970.0mL. Mix thoroughly. Adjust pH to 7.0. Sparge with 100% N_2. Autoclave for 15 min at 15 psi pressure–121°C. Aseptically and anaerobically add 20.0mL of sterile glucose solution and 10.0mL of sterile $Na_2S·9H_2O$ solution. Mix thoroughly.

Use: For the cultivation of *Fervidobacterium islandicum.*

Fervidobacterium Medium

Composition per liter:

Pancreatic digest of casein 10.0g
Glucose 5.0g
Yeast extract 3.0g
K_2HPO_4 .... 1.5g
NH_4Cl 0.9g
KH_2PO_4 .... 0.75g
$MgCl_2·6H_2O$ 0.2g
$Na_2S·9H_2O$ solution 10.0mL
Trace elements solution 9.0mL
Wolfe's vitamin solution 5.0mL
Resazurin (0.2% solution) 1.0mL
$FeSO_4·7H_2O$ (10% solution) 0.03mL

pH 7.0 ± 0.1 at 25°C

Trace Elements Solution:

Composition per liter:

Nitrilotriacetic acid 12.5g
NaCl 1.0g
$FeCl_3·4H_2O$ 0.2g
$CaCl_2·2H_2O$ 0.1g
$MnCl_2·4H_2O$ 0.1g
$ZnCl_2$ 0.1g
$CuCl_2$ 0.02g
Na_2SeO_3 .... 0.02g
$CoCl_2·6H_2O$ 0.017g
H_3BO_3 .... 0.01g
$Na_2MoO_4·2H_2O$ 0.01g

Preparation of Trace Elements Solution: Add nitrilotriacetic acid to 500.0mL of distilled/deionized water. Dissolve by adjusting pH to 6.5 with KOH. Add remaining components. Add distilled/deionized water to 1.0L. Filter sterilize. Maintain under an atmosphere of 100% N_2.

Wolfe's Vitamin Solution:

Composition per liter:

Pyridoxine·HCl 0.01g
Thiamine·HCl 5.0mg
Riboflavin 5.0mg
Nicotinic acid 5.0mg
Calcium pantothenate 5.0mg
p-Aminobenzoic acid 5.0mg
Thioctic acid 5.0mg
Biotin 2.0mg
Folic acid 2.0mg
Cyanocobalamin 0.1mg

Preparation of Wolfe's Vitamin Solution: Add components to distilled/deionized water and bring volume to 1.0L. Mix thoroughly. Filter sterilize. Maintain under an atmosphere of 100% N_2.

$Na_2S \cdot 9H_2O$ Solution:
Composition per 10.0mL:
$Na_2S \cdot 9H_2O$ 0.5g

Preparation of $Na_2S \cdot 9H_2O$ Solution: Add $Na_2S \cdot 9H_2O$ to distilled/deionized water and bring volume to 10.0mL. Mix thoroughly. Filter sterilize. Maintain under an atmosphere of 100% N_2.

Preparation of Medium: Add components, except sodium sulfide solution, trace elements solution, and Wolfe's vitamin solution, to distilled/deionized water and bring volume to 976.0mL. Mix thoroughly. Autoclave for 15 min at 15 psi pressure–121°C. Cool under an atmosphere of 100% N_2. Aseptically add 9.0mL of trace elements solution and 5.0mL of Wolfe's vitamin solution under an atmosphere of 100% N_2. Mix thoroughly. Aseptically distribute into sterile tubes or flasks under an atmosphere of 100% N_2. Add $Na_2S \cdot 9H_2O$ solution just prior to use to a concentration of 0.1%.

Use: For the cultivation and maintenance of *Clostridium* species, *Fervidobacterium nodosum, Fervidobacterium islandicum,* and *Thermoanaerobium brockii.*

F-G Agar
(Feeley-Gorman Agar)

Composition per liter:
Casein, acid hydrolyzed 17.5g
Agar 17.0g
Beef extract 3.0g
Starch 1.5g
L-Cysteine solution 10.0mL
$Fe_4(P_2O_7)_3$ solution 10.0mL
pH 6.9 ± 0.05 at 25°C

L-Cysteine Solution:
Composition per 10.0mL:
L-Cysteine·HCl·H_2O 0.4g

Preparation of L-Cysteine Solution: Add L-cysteine·HCl·H_2O to distilled/deionized water and bring volume to 10.0mL. Mix thoroughly. Filter sterilize.

$Fe_4(P_2O_7)_3$ Solution:
Composition per 10.0mL:
$Fe_4(P_2O_7)_3$ 0.25g

Preparation of $Fe_4(P_2O_7)_3$ Solution: Add $Fe_4(P_2O_7)_3$ to distilled/deionized water and bring volume to 10.0mL. Mix thoroughly. Filter sterilize.

Preparation of Medium: Add components, except L-cysteine solution and $Fe_4(P_2O_7)_3$ solution, to distilled/deionized water and bring volume to 980.0mL. Mix thoroughly. Gently heat and bring to boiling. Autoclave for 15 min at 15 psi pressure–121°C. Cool to 45°–50°C. Aseptically add 10.0mL of L-cysteine solution. Mix thoroughly. Aseptically add 10.0mL of $Fe_4(P_2O_7)_3$ solution. Mix thoroughly. Adjust pH to 6.9. Pour into sterile Petri dishes or distribute into sterile tubes.

Use: For the isolation and cultivation of *Legionella pneumophila.*

F-G Agar with Selenium
(Feeley-Gorman Agar with Selenium)

Composition per liter:
Casein, acid hydrolyzed 17.5g
Agar 17.0g
Beef extract 3.0g
Starch 1.5g
L-Cysteine solution 10.0mL
$Fe_4(P_2O_7)_3$ solution 10.0mL
$Na_2SeO_3 \cdot 5H_2O$ solution 10.0mL
pH 6.9 ± 0.05 at 25°C

L-Cysteine Solution:
Composition per 10.0mL:
L-Cysteine·HCl·H_2O 0.4g

Preparation of L-Cysteine Solution: Add L-cysteine·HCl·H_2O to distilled/deionized water and bring volume to 10.0mL. Mix thoroughly. Filter sterilize.

$Fe_4(P_2O_7)_3$ Solution:
Composition per 10.0mL:
$Fe_4(P_2O_7)_3$ 0.25g

Preparation of $Fe_4(P_2O_7)_3$ Solution: Add $Fe_4(P_2O_7)_3$ to distilled/deionized water and bring volume to 10.0mL. Mix thoroughly. Filter sterilize.

$Na_2SeO_3 \cdot 5H_2O$ Solution:
Composition per 10.0mL:
$Na_2SeO_3 \cdot 5H_2O$ 0.01g

Preparation of $Na_2SeO_3 \cdot 5H_2O$ Solution: Add $Na_2SeO_3 \cdot 5H_2O$ to distilled/deionized water and bring volume to 10.0mL. Mix thoroughly. Filter sterilize.

Preparation of Medium: Add components—except L-cysteine solution, $Fe_4(P_2O_7)_3$ solution, and $Na_2SeO_3 \cdot 5H_2O$ solution—to distilled/deionized water and bring volume to 970.0mL. Mix thoroughly. Gently heat and bring to boiling. Autoclave for 15 min at 15 psi pressure–121°C. Cool to 45°–50°C. Aseptically add 10.0mL of sterile L-cysteine solution. Mix thoroughly. Aseptically add 10.0mL of sterile $Fe_4(P_2O_7)_3$ solution and 10.0mL of sterile $Na_2SeO_3 \cdot 5H_2O$ solution. Mix thoroughly. Adjust pH to 6.9. Pour into sterile Petri dishes or distribute into sterile tubes.

Use: For the isolation and cultivation of *Legionella pneumophila.*

F-G Broth
(Feeley-Gorman Broth)

Composition per liter:
Casein, acid hydrolyzed 17.5g
Beef extract 3.0g
Starch 1.5g
L-Cysteine solution 10.0mL
$Fe_4(P_2O_7)_3$ solution 10.0mL
pH 6.9 ± 0.05 at 25°C

L-Cysteine Solution:
Composition per 10.0mL:
L-Cysteine·HCl·H_2O 0.4g

Preparation of L-Cysteine Solution: Add L-cysteine·HCl·H_2O to distilled/deionized water and bring volume to 10.0mL. Mix thoroughly. Filter sterilize.

$Fe_4(P_2O_7)_3$ Solution:
Composition per 10.0mL:
$Fe_4(P_2O_7)_3$ 0.25g

Preparation of $Fe_4(P_2O_7)_3$ Solution: Add $Fe_4(P_2O_7)_3$ to distilled/deionized water and bring volume to 10.0mL. Mix thoroughly. Filter sterilize.

Preparation of Medium: Add components, except L-cysteine solution and $Fe_4(P_2O_7)_3$ solution, to distilled/deionized water and bring volume to 980.0mL. Mix thoroughly. Gently heat and bring to boiling. Autoclave for 15 min at 15 psi pressure–121°C. Cool to 45°–50°C. Aseptically add 10.0mL of L-cysteine solution. Mix thoroughly. Aseptically add 10.0mL of $Fe_4(P_2O_7)_3$ solution. Mix thoroughly. Adjust pH to 6.9. Aseptically distribute into sterile tubes or flasks.

Use: For the cultivation of *Legionella pneumophila.*

FGTC Agar

Composition per liter:

Pancreatic digest of casein	15.0g
Agar	15.0g
Papaic digest of soybean meal	5.0g
NaCl	5.0g
KH_2PO_4	5.0g
Amylose Azure	3.0g
Galactose	1.0g
Thallous acetate	0.5g
MUG (4-Methylumbelliferyl-α-D-galactoside	0.1g
$NaHCO_3$ solution	20.0mL
Gentamicin solution	2.5mL
Tween™ 80	0.75mL

pH 7.3 ± 0.2 at 25°C

Gentamicin Solution:

Composition per 10.0mL:

Gentamicin	0.01g

Preparation of Gentamicin Solution: Add gentamicin to distilled/deionized water and bring volume to 10.0mL. Mix thoroughly.

$NaHCO_3$ Solution:

Composition per 20.0mL:

$NaHCO_3$	2.0g

Preparation of $NaHCO_3$ Solution: Add the $NaHCO_3$ to distilled/deionized water and bring volume to 20.0mL. Mix thoroughly. Filter sterilize. Use freshly prepared solution.

Preparation of Medium: Add components, except $NaHCO_3$ solution, to distilled/deionized water and bring volume to 980.0mL. Mix thoroughly. Gently heat and bring to boiling. Autoclave for 15 min at 15 psi pressure–121°C. Cool to 50°C. Aseptically add sterile $NaHCO_3$ solution. Mix thoroughly. Pour into sterile Petri dishes.

Use: For the cultivation, differentiation, and enumeration of *Enterococcus* species based on starch hydrolysis and production of fluorescence. Bacteria that hydrolyze starch, such as *Streptococcus bovis*, appear as colonies surrounded by a clear zone. Bacteria that produce fluorescence, such as *Streptococcus bovis* and *Enterococcus faecium*, appear as colonies surrounded by a zone of bright bluish fluorescence when viewed under a long-wave UV lamp. Other bacteria, such as *Enterococcus faecalis*, *Enterococcus avium*, or *Streptococcus equinus*, do not hydrolyze starch or produce fluorescence.

FGTC Agar Base with Bicarbonate, Gentamicin and Amylose Azure

Composition per liter:

Agar	15.0g
Casein enzymatic hydrolysate	15.0g
KH_2PO_4	5.0g
Papaic digest of soybean meal	5.0g
NaCl	5.0g
Galactose	1.0g
Polysorbate 80	0.75g
Thallous acetate	0.5g
4-Methylumbellifery β-D-glucuronide (MUG)	0.1g
$NaHCO_3$ solution	20.0mL
Amylose azure solution	10.0mL
Gentamicin solution	2.5mL

pH 7.3 ± 0.2 at 25°C

Source: This medium, without gentamicin, amylose azure, or $NaHCO_3$ solutions, is available as a premixed powder from HiMedia.

Gentamicin Solution:

Composition per 10.0mL:

Gentamicin	0.01g

Preparation of Gentamicin Solution: Add gentamicin to distilled/deionized water and bring volume to 10.0mL. Mix thoroughly. Filter sterilize.

Amylose Azure Solution:

Composition per 10.0mL:

Amylose azure	3.0g

Preparation of Amylose Azure Solution: Add amylose azure to distilled/deionized water and bring volume to 10.0mL. Mix thoroughly. Filter sterilize.

$NaHCO_3$ Solution:

Composition per 20.0mL:

$NaHCO_3$	2.0g

Preparation of $NaHCO_3$ Solution: Add the $NaHCO_3$ to distilled/deionized water and bring volume to 20.0mL. Mix thoroughly. Filter sterilize. Use freshly prepared solution.

Preparation of Medium: Add components, except $NaHCO_3$ solution, amylose azure solution, and gentamicin solution, to distilled/deionized water and bring volume to 970.0mL. Mix thoroughly. Gently heat and bring to boiling. Autoclave for 15 min at 15 psi pressure–121°C. Cool to 50°C. Aseptically add 20.0mL sterile $NaHCO_3$ solution, 10.0mL sterile amylose azure solution, and 2.5mL sterile gentamicin solution. Mix thoroughly. Pour into sterile Petri dishes.

Use: For the cultivation, differentiation, and enumeration of *Enterococcus* species based on starch hydrolysis and production of fluorescence. Bacteria that hydrolyze starch, such as *Streptococcus bovis*, appear as colonies surrounded by a clear zone. Bacteria that produce fluorescence, such as *Streptococcus bovis* and *Enterococcus faecium*, appear as colonies surrounded by a zone of bright bluish fluorescence when viewed under a long-wave UV lamp. Other bacteria, such as *Enterococcus faecalis*, *Enterococcus avium*, or *Streptococcus equinus*, do not hydrolyze starch or produce fluorescence.

FGTC HiVeg Agar Base with Bicarbonate, Gentamicin, and Amylose Azure

Composition per liter:

Agar	15.0g
Plant hydrolysate	15.0g
KH_2PO_4	5.0g
Papaic digest of soybean meal	5.0g
NaCl	5.0g
Galactose	1.0g
Polysorbate 80	0.75g
Thallous acetate	0.5g
4-Methylumbellifery β-D-glucuronide (MUG)	0.1g
$NaHCO_3$ solution	20.0mL

Amylose azure solution 10.0mL
Gentamicin solution...2.5mL

pH 7.3 ± 0.2 at 25°C

Source: This medium, without gentamicin, amylose azure, or $NaHCO_3$ solutions, is available as a premixed powder from HiMedia.

Gentamicin Solution:

Composition per 10.0mL:

Gentamicin..0.01g

Preparation of Gentamicin Solution: Add gentamicin to distilled/deionized water and bring volume to 10.0mL. Mix thoroughly. Filter sterilize.

Amylose Azure Solution:

Composition per 10.0mL:

Amylose azure 3.0g

Preparation of Amylose Azure Solution: Add amylose azure to distilled/deionized water and bring volume to 10.0mL. Mix thoroughly. Filter sterilize.

$NaHCO_3$ Solution:

Composition per 20.0mL:

$NaHCO_3$..2.0g

Preparation of $NaHCO_3$ Solution: Add the $NaHCO_3$ to distilled/deionized water and bring volume to 20.0mL. Mix thoroughly. Filter sterilize. Use freshly prepared solution.

Preparation of Medium: Add components, except $NaHCO_3$ solution, amylose azure solution, and gentamicin solution, to distilled/deionized water and bring volume to 970.0mL. Mix thoroughly. Gently heat and bring to boiling. Autoclave for 15 min at 15 psi pressure–121°C. Cool to 50°C. Aseptically add 20.0mL sterile $NaHCO_3$ solution, 10.0mL sterile amylose azure solution, and 2.5mL sterile gentamicin solution. Mix thoroughly. Pour into sterile Petri dishes.

Use: For the cultivation, differentiation, and enumeration of *Enterococcus* species based on starch hydrolysis and production of fluorescence. Bacteria that hydrolyze starch, such as *Streptococcus bovis*, appear as colonies surrounded by a clear zone. Bacteria that produce fluorescence, such as *Streptococcus bovis* and *Enterococcus faecium*, appear as colonies surrounded by a zone of bright bluish fluorescence when viewed under a long-wave UV lamp. Other bacteria, such as *Enterococcus faecalis*, *Enterococcus avium*, or *Streptococcus equinus*, do not hydrolyze starch or produce fluorescence.

Fibrobacter Medium

Composition per 1020.0mL:

Cellobiose 4.0g
Na_2CO_3 ....4.0g
Pancreatic digest of casein..1.0g
NaCl...0.6g
Yeast extract..0.5g
K_2HPO_4...0.3g
KH_2PO_4...0.3g
$(NH_4)_2SO_4$..0.3g
$MgSO_4 \cdot 7H_2O$ 0.12g
$CaCl_2 \cdot 2H_2O$...0.08g
Resazurin 1.0mg
Vitamin solution..20.0mL
$Na_2S \cdot 9H_2O$ solution 10.0mL
L-Cysteine·HCl·H_2O solution 10.0mL
VFA solution..4.65mL
Trace elements solution 1.0mL

pH 6.6 ± 0.2 at 25°C

Vitamin Solution:

Composition per 100.0mL:

Calcium D-(+)-pantothenate 20.0mg
Lipoic acid...20.0mg
Nicotinamide..20.0mg
Pyridoxal·HCl 20.0mg
Pyridoxamine·2HCl..20.0mg
Riboflavin 20.0mg
Thiamine·HCl 20.0mg
p-Aminobenzoic acid...1.0mg
Biotin 1.0mg
Cyanocobalamin 1.0mg

Preparation of Vitamin Solution: Add components to distilled/deionized water and bring volume to 100.0mL. Mix thoroughly.

$Na_2S \cdot 9H_2O$ Solution:

Composition per 10.0mL:

$Na_2S \cdot 9H_2O$...0.25g

Preparation of $Na_2S \cdot 9H_2O$ Solution: Add $Na_2S \cdot 9H_2O$ to distilled/deionized water and bring volume to 10.0mL. Mix thoroughly. Sparge with 100% N_2. Autoclave for 15 min at 15 psi pressure–121°C.

L-Cysteine·HCl·H_2O Solution:

Composition per 10.0mL:

L-Cysteine·HCl·H_2O 0.25g

Preparation of L-Cysteine·HCl·H_2O Solution: Add L-cysteine·HCl·H_2O to distilled/deionized water and bring volume to 10.0mL. Mix thoroughly. Sparge with 100% N_2. Autoclave for 15 min at 15 psi pressure–121°C.

Trace Elements Solution:

Composition per liter:

$FeSO_4 \cdot 7H_2O$..2.0g
$CoCl_2 \cdot 6H_2O$ 0.2g
H_3BO_3...0.2g
$ZnSO_4 \cdot 7H_2O$ 0.1g
$MnCl_2 \cdot 4H_2O$ 0.03g
$Na_2MoO_4 \cdot 2H_2O$ 0.03g
$NiCl_2 \cdot 6H_2O$..0.02g
$CuCl_2 \cdot 2H_2O$ 0.01g

Preparation of Trace Elements Solution: Add components to distilled/deionized water and bring volume to 1.0L. Mix thoroughly.

VFA Solution:

Composition per 310.0mL:

Acetic acid..17.0mL
Propionic acid 6.0mL
n-Butyric acid 4.0mL
Isobutyric acid 1.0mL
Isovaleric acid..1.0mL
DL-α-Methylbutyric acid 1.0mL
n-Valeric acid..1.0mL

Preparation of VFA Solution: Add volatile fatty acids to approximately 200.0mL of distilled/deionized water. Mix thoroughly. Adjust pH to 7.0 with NaOH pellets. Bring volume to 310.0mL with distilled/deionized water.

Preparation of Medium: Prepare and dispense medium under 100% CO_2. Add components, except Na_2CO_3, to distilled/deionized water and bring volume to 1.0L. Mix thoroughly. Gently heat and bring to boiling. Continue boiling for 3 min. Cool to room temperature while sparging with 100% CO_2. Add Na_2CO_3. Mix thoroughly. Continue sparging with 100% CO_2 for 10 min. Anaerobically distribute into anaerobic tubes. Autoclave for 15 min at 15 psi pressure–121°C. Prior to inoculation, aseptically and anaerobically add 10.0mL of sterile $Na_2S{\cdot}9H_2O$ solution and 10.0mL of sterile L-cysteine·HCl·H_2O solution per liter of medium. Mix thoroughly. Adjust pH to 6.6.

Use: For the cultivation of *Fibrobacter intestinalis* and *Fibrobacter succinogenes.*

Fildes Enrichment Agar

Composition per liter:

Agar	15.0g
Peptone	5.0g
Beef extract	3.0g
Fildes enrichment solution	50.0mL

Fildes Enrichment Solution:
Composition per 206.0mL:

Pepsin	1.0g
NaCl (0.85% solution)	150.0mL
Sheep blood, defibrinated	50.0mL
HCl	6.0mL

pH 7.0–7.2 at 25°C

Source: Fildes enrichment solution is available from BD Diagnostic Systems.

Preparation of Fildes Enrichment Solution: Combine components. Mix thoroughly. Incubate at 56°C for 4 hr. Bring pH to 7.0 with 20% NaOH. Adjust pH to 7.2 with HCl. Do not autoclave. Add 0.25 mL of chloroform and store at 4°C. Before use, heat to 56°C to remove chloroform.

Preparation of Medium: Add components, except Fildes enrichment solution, to distilled/deionized water and bring volume to 950.0mL. Mix thoroughly. Gently heat and bring to boiling. Autoclave for 15 min at 15 psi pressure–121°C. Cool to 56°C. Aseptically add 50.0mL of sterile Fildes enrichment solution. Mix thoroughly. Pour into sterile Petri dishes or distribute into sterile tubes.

Use: For the isolation and cultivation of *Haemophilus influenzae.*

Filobacillus milosensis Medium (DSMZ Medium 607a)

Composition per liter:

NaCl	100.0g
Peptone	0.75g
Yeast extract	0.75g
Glucose	0.75g
Artificial sea water	250.0mL
Tris/HCl (0.1*M,* pH 7.5)	50.0mL
Hutner's basal salts solution	20.0mL
Vitamin solution	10.0mL

pH 7.2 ± 0.2 at 25°C

Hutner's Basal Salts Solution:
Composition per liter:

$MgSO_4{\cdot}7H_2O$	29.7g
Nitrilotriacetic acid	10.0g
$CaCl_2{\cdot}2H_2O$	3.335g
$FeSO_4{\cdot}7H_2O$	99.0mg
$(NH_4)_6MoO_7O_{24}{\cdot}4H_2O$	9.25mg
"Metals 44"	50.0mL

"Metals 44":
Composition per 100.0mL:

$ZnSO_4{\cdot}7H_2O$	1.095g
$FeSO_4{\cdot}7H_2O$	0.5g
Sodium EDTA	0.25g
$MnSO_4{\cdot}H2O$	0.154g
$CuSO_4{\cdot}5H_2O$	39.2mg
$Co(NO_3)_2{\cdot}6H_2O$	24.8mg
$Na_2B_4O_7{\cdot}10H_2O$	17.7mg

Preparation of "Metals 44": Add sodium EDTA to distilled/deionized water and bring volume to 90.0mL. Mix thoroughly. Add a few drops of concentrated H_2SO_4 to retard precipitation of heavy metal ions. Add remaining components. Mix thoroughly. Bring volume to 100.0mL with distilled/deionized water.

Preparation of Hutner's Basal Salts Solution: Add nitrilotriacetic acid to 500.0mL of distilled/deionized water. Adjust pH to 6.5 with KOH. Add remaining components. Add distilled/deionized water to 1.0L. Adjust pH to 6.8.

Artificial Sea Water:
Composition per liter:

NaCli	23.477g
$MgCl_2{\cdot}6H_2O$	4.981g
Na_2SO_4	3.917g
$CaCl_2$	1.12g
KCl	664.0mg
$NaHCO_3$	192.0mg
H_3BO_3	26.0mg
$SrCl_2$	24.0mg
KBr	6.0mg
NaF	3.0mg

Preparation of Artificial Sea Water: Add components to distilled/deionized water and bring volume to 1.0L. Mix thoroughly. Filter sterilize.

Vitamin Solution:
Composition per liter:

Riboflavin	5.0mg
Nicotinamide	5.0mg
Thiamine-HCl·$2H_2O$	5.0mg
Ca-pantothenate	5.0mg
Biotin	2.0mg
Folic acid	2.0mg
Vitamin B_{12}	0.1mg

Preparation of Vitamin Solution: Add components to distilled/deionized water and bring volume to 1.0L. Mix thoroughly. Filter sterilize.

Preparation of Medium: Add components, except artificial sea water and vitamin solution, to distilled/deionized water and bring volume to 740.0mL. Mix thoroughly. Adjust pH to 7.2. Autoclave for 15 min at 15 psi pressure–121°C. Cool to room temperature. Aseptically add 250.0mL artificial sea water and 10.0mL vitamin solution. Mix thoroughly. Aseptically and anaerobically distribute into sterile tubes or bottles.

Use: For the cultivation of *Filobacillus milosensis.*

Fish Peptone Agar

Composition per liter:

Agar	5.0g
Maltose	5.0g
NaCl	5.0g

Peptone	5.0g
Pancreatic digest of casein	5.0g
Yeast extract	5.0g
Trout tissue extract solution	50.0mL

pH 7.0 ± 0.2 at 25°C

Trout Tissue Extract Solution:

Composition per liter:

Fish (brook trout)	500.0g
Pepsin	1.0g
HCl, concentrated	15.0mL

Preparation of Trout Tissue Extract Solution: Add 1.0L of distilled/deionized water to brook trout and blend for 20–30 min. Add 1.0g of pepsin and 15.0mL of concentrated HCl to digest the trout proteins. Incubate for 12 hr at 45°C. Adjust pH to 7.0. Allow solids to settle. Filter sterilize. Do not autoclave. Store at 5°C.

Preparation of Medium: Add components, except trout tissue extract solution, to distilled/deionized water and bring volume to 950.0L. Mix thoroughly. Gently heat and bring to boiling. Autoclave for 15 min at 13 psi pressure–118°C. Cool to 45°–50°C. Aseptically add 50.0mL of sterile trout tissue extract solution. Mix thoroughly. Pour into sterile Petri dishes or distribute into sterile tubes.

Use: For the cultivation and maintenance of *Aeromonas salmonicida.*

Fish Peptone Broth

Composition per liter:

Maltose	5.0g
NaCl	5.0g
Peptone	5.0g
Pancreatic digest of casein	5.0g
Yeast extract	5.0g
Trout tissue extract solution	50.0mL

pH 7.0 ± 0.2 at 25°C

Trout Tissue Extract Solution:

Composition per liter:

Fish (brook trout)	500.0g
Pepsin	1.0g
HCl, concentrated	15.0mL

Preparation of Trout Tissue Extract Solution: Add 1.0L of distilled/deionized water to brook trout and blend for 20–30 min. Add 1.0g of pepsin and 15.0mL of concentrated HCl to digest the trout proteins. Incubate for 12 hr at 45°C. Adjust pH to 7.0. Allow solids to settle. Filter sterilize. Do not autoclave. Store at 5°C.

Preparation of Medium: Add components, except trout tissue extract solution, to distilled/deionized water and bring volume to 950.0L. Mix thoroughly. Gently heat and bring to boiling. Autoclave for 15 min at 10 psi pressure–118°C. Cool to 45°–50°C. Aseptically add 50.0mL of sterile trout tissue extract solution. Mix thoroughly. Aseptically distribute into sterile tubes or flasks.

Use: For the cultivation of *Aeromonas salmonicida.*

Five g Agar (5g Agar)

Composition per liter:

Glycerol	50.0g
Agar	15.0g
Yeast extract	5.0g
$CaCO_3$	1.0g

Preparation of Medium: Add components to distilled/deionized water and bring volume to 1.0L. Mix thoroughly. Gently heat and bring to boiling. Distribute into tubes or flasks. Autoclave for 15 min at 15 psi pressure–121°C. Pour into sterile Petri dishes or leave in tubes.

Use: For the cultivation and maintenance of *Dermatophilus congolensis* and *Geodermatophilus obscurus.*

Flagella Broth

Composition per liter:

Tryptose or biosate	10.0g
NaCl	2.5g
K_2HPO_4	1.0g

pH 7.0 ± 0.1 at 25°C

Preparation of Medium: Add components to distilled/deionized water and bring volume to 1.0L. Mix thoroughly. Gently heat and bring to boiling. Distribute into tubes or flasks. Autoclave for 15 min at 15 psi pressure–121°C. Pour into sterile Petri dishes or leave in tubes.

Use: For the cultivation of flagella-producing bacteria.

Flavobacterium aquatile Medium (DSMZ Medium 102)

Composition per liter:

Agar	15.0g
Na-caseinate	2.0g
Proteose peptone	1.0g
Yeast extract	0.5g
K_2HPO_4	0.5g

pH 7.4 ± 0.2 at 25°C

Preparation of Medium: Add components to distilled/deionized water and bring volume to 1.0L. Mix thoroughly. Adjust pH to 7.4. Gently heat and bring to boiling. Distribute into tubes or flasks. Autoclave for 15 min at 15 psi pressure–121°C.

Use: For the cultivation and maintenance of *Flavobacterium aquatile.*

Flavobacterium M1 Agar

Composition per liter:

Agar	15.0g
Proteose peptone	5.0g
NaCl	3.0g
Beef extract	2.0g
Yeast extract	1.0g

pH 7.0–7.2 at 25°C

Preparation of Medium: Add components to distilled/deionized water and bring volume to 1.0L. Mix thoroughly. Gently heat and bring to boiling. Distribute into tubes or flasks. Autoclave for 15 min at 15 psi pressure–121°C. Pour into sterile Petri dishes or leave in tubes.

Use: For the cultivation and maintenance of *Flavobacterium indolthelicum.*

Flavobacterium Medium

Composition per liter:

Na_2SO_4	1.0g
Pancreatic digest of casein	1.0g
Yeast extract	1.0g

pH 6.0 ± 0.2 at 25°C

Preparation of Medium: Add components to distilled/deionized water and bring volume to 1.0L. Mix thoroughly. Adjust pH to 6.0 with

H_2SO_4. Distribute into tubes or flasks. Autoclave for 15 min at 15 psi pressure–121°C.

Use: For the cultivation and maintenance of *Flavobacterium acidurans*.

Flavobacterium Medium (ATCC Medium 65)

Composition per liter:

Agar	12.0g
Sodium caseinate	2.0g
Peptone	1.0g
K_2HPO_4	0.5g
Yeast extract	0.5g

pH 7.4 ± 0.2 at 25°C

Preparation of Medium: Add components to distilled/deionized water and bring volume to 1.0L. Mix thoroughly. Gently heat and bring to boiling. Distribute into tubes or flasks. Autoclave for 15 min at 15 psi pressure–121°C. Pour into sterile Petri dishes or leave in tubes.

Use: For the cultivation and maintenance of *Flavobacterium aquatile*.

Flavobacterium Medium (ATCC Medium 647)

Composition per liter:

Agar	12.0g
Beef extract	10.0g
Peptone	10.0g
NaCl	5.0g

pH 7.2–7.3 at 25°C

Preparation of Medium: Add components to distilled/deionized water and bring volume to 1.0L. Mix thoroughly. Gently heat and bring to boiling. Distribute into tubes or flasks. Autoclave for 15 min at 15 psi pressure–121°C. Pour into sterile Petri dishes or leave in tubes.

Use: For the isolation and cultivation of *Flavobacterium* species from food and food-processing equipment.

Flavobacterium Medium (ATCC Medium 1687)

Composition per liter:

Sodium glutamate	4.0g
K_2HPO_4	0.65g
$NaNO_3$	0.5g
KH_2PO_4	0.19g
$MgSO_4 \cdot 7H_2O$	0.1g
$FeSO_4$ solution	2.0mL

pH 7.4 ± 0.2 at 25°C

$FeSO_4$ Solution:

Composition per 10.0mL:

$FeSO_4 \cdot 7H_2O$	0.03g

Preparation of $FeSO_4$ Solution: Add $FeSO_4$ to distilled/deionized water and bring volume to 10.0mL. Mix thoroughly. Filter sterilize.

Preparation of Medium: Add components, except $FeSO_4$ solution, to distilled/deionized water and bring volume to 998.0mL. Mix thoroughly. Autoclave for 15 min at 15 psi pressure–121°C. Cool to 25°C. Aseptically add 2.0mL of sterile $FeSO_4$ solution. Mix thoroughly. Adjust pH to 7.4. Aseptically distribute into sterile tubes or flasks.

Use: For the cultivation of *Flavobacterium* species.

Flavobacterium Medium M1

Composition per liter:

Agar	12.0g
Proteose peptone	5.0g
NaCl	3.0g
Beef extract	2.0g
Yeast extract	0.2g

pH 7.2–7.4 at 25°C

Preparation of Medium: Add components to distilled/deionized water and bring volume to 1.0L. Mix thoroughly. Gently heat and bring to boiling. Distribute into tubes or flasks. Autoclave for 15 min at 15 psi pressure–121°C. Pour into sterile Petri dishes or leave in tubes.

Use: For the isolation and cultivation of *Flavobacterium* species.

Flavobacterium Medium with Thiamine

Composition per liter:

Agar	12.0g
Sodium caseinate	2.0g
Peptone	1.0g
K_2HPO_4	0.5g
Yeast extract	0.5g
Thiamine·HCl	10.0mg

pH 7.4 ± 0.2 at 25°C

Preparation of Medium: Add components to distilled/deionized water and bring volume to 1.0L. Mix thoroughly. Gently heat and bring to boiling. Distribute into tubes or flasks. Autoclave for 15 min at 15 psi pressure–121°C. Pour into sterile Petri dishes or leave in tubes.

Use: For the cultivation and maintenance of *Flavobacterium aquatile* and *Flavobacterium lutescens*.

Flavobacterium resinovorum Agar (LMG Medium 216)

Composition per liter:

Agar	15.0g
Lab-Lemco beef extract	10.0g
Peptone	10.0g
NaCl	5.0g

pH 7.3 ± 0.2 at 25°C

Preparation of Medium: Add components to tap water and bring volume to 1.0L. Mix thoroughly. Gently heat and bring to boiling. Distribute into tubes or flasks. Autoclave for 15 min at 15 psi pressure–121°C. Pour into sterile Petri dishes or leave in tubes.

Use: For the cultivation and maintenance of *Flavobacterium resinovorum*.

Flavobacterium tirrenicum Medium

Composition per liter:

Agar	15.0g
NaCl	10.0g
Peptone	5.0g
Meat extract	3.0g
Ethanolamine	2.0g

pH 7.0 ± 0.2 at 25°C

Preparation of Medium: Add components to distilled/deionized water and bring volume to 1.0L. Mix thoroughly. Gently heat and bring to boiling. Adjust pH to 7.0. Distribute into tubes or flasks. Autoclave for 15 min at 15 psi pressure–121°C. Pour into sterile Petri dishes or leave in tubes.

Use: For the cultivation and maintenance of *Flavobacterium* species.

Flegler's Mutinus Medium

Composition per liter:

Agar	20.0g
Glucose	5.0g
Malt extract	5.0g
KH_2PO_4	0.5g
$MgSO_4$	0.5g
NH_4NO_3	0.5g
Ferric citrate	5.0mg
Thiamine·HCl	0.1mg

Preparation of Medium: Add components to distilled/deionized water and bring volume to 1.0L. Mix thoroughly. Gently heat and bring to boiling. Distribute into tubes or flasks. Autoclave for 15 min at 15 psi pressure–121°C. Pour into sterile Petri dishes or leave in tubes.

Use: For the cultivation and maintenance of *Dictyophora indusiata* and *Dictyophora phalloidea.*

Fletcher *Leptospira* HiVeg Medium Base (*Leptospira* HiVeg Medium Base, Fletcher)

Composition per liter:

Agar	1.5g
NaCl	0.5g
Plant peptone	0.3g
Plant extract	0.2g
Rabbit serum	50.0mL

pH 7.9 ± 0.1 at 25°C

Source: This medium, without rabbit serum, is available as a premixed powder from BD Diagnostic Systems.

Preparation of Medium: Add components, except rabbit serum, to distilled/deionized water and bring volume to 950.0mL. Mix thoroughly. Gently heat and bring to boiling. Autoclave for 15 min at 15 psi pressure–121°C. Cool to 50°–55°C. Aseptically add 50.0mL of sterile rabbit serum. Mix thoroughly. Aseptically distribute into sterile tubes or flasks.

Use: For the isolation, cultivation, and maintenance of cultures of *Leptospira* species.

Fletcher Medium

Composition per liter:

Agar	1.5g
NaCl	0.5g
Peptone	0.3g
Beef extract	0.2g
Rabbit serum	50.0mL

pH 7.9 ± 0.1 at 25°C

Source: This medium is available as a premixed powder from BD Diagnostic Systems.

Preparation of Medium: Add components, except rabbit serum, to distilled/deionized water and bring volume to 950.0mL. Mix thoroughly. Gently heat and bring to boiling. Autoclave for 15 min at 15 psi pressure–121°C. Cool to 50°–55°C. Aseptically add 50.0mL of sterile rabbit serum. Mix thoroughly. Aseptically distribute into sterile tubes or flasks.

Use: For the isolation, cultivation, and maintenance of cultures of *Leptospira* species.

Fletcher Medium with Fluorouracil (Fluorouracil *Leptospira* Medium)

Composition per liter:

Agar	1.5g
NaCl	0.5g
Peptone	0.3g
Beef extract	0.2g
Rabbit serum	50.0mL
Fluorouracil solution	20.0mL

pH 7.9 ± 0.1 at 25°C

Fluorouracil Solution:

Composition per 100.0mL:

Fluorouracil	10.0g

Preparation of Fluorouracil Solution: Add fluorouracil to 50.0mL of distilled/deionized water. Add 1.0mL of 2*N* NaOH and bring volume to 100.0mL. Gently heat to 56°C for 2 hr. Adjust pH to 7.4–7.6 with NaOH. Mix thoroughly. Filter sterilize.

Preparation of Medium: Add components, except rabbit serum and fluorouracil solution, to distilled/deionized water and bring volume to 930.0mL. Mix thoroughly. Gently heat and bring to boiling. Autoclave for 15 min at 15 psi pressure–121°C. Cool to 50°–55°C. Aseptically add 80.0mL of sterile rabbit serum. Mix thoroughly. Aseptically distribute into sterile tubes or flasks. Immediately prior to use, add 0.1mL of fluorouracil solution per 5.0mL of medium.

Use: For the isolation, cultivation, and maintenance of cultures of *Leptospira* species.

Fletcher's Semisolid Medium

Composition per 2120.0mL:

Agar	1.5g
NaCl	0.5g
Peptone	0.3g
Beef extract	0.2g
Rabbit serum	240.0mL

pH 7.9 ± 0.1 at 25°C

Preparation of Medium: Add components, except rabbit serum, to distilled/deionized water and bring volume to 1880.0mL. Mix thoroughly. Gently heat and bring to boiling. Autoclave for 15 min at 15 psi pressure–121°C. Cool to 50°–55°C. Aseptically add 240.0mL of sterile rabbit serum. Mix thoroughly. Aseptically distribute into sterile tubes or flasks.

Use: For the isolation, cultivation, and maintenance of cultures of *Leptospira* species.

Flexibacter Agar

Composition per liter:

Agar	15.0g
Monosodium glutamate	5.0g
Pancreatic digest of casein	1.0g
Vitamin-free casamino acids	1.0g
Sodium glycerophosphate	0.1g
Vitamin B_{12}	1.0μg
Seawater	1.0L
Trace elements solution HO-LE	1.0mL

Trace Elements Solution HO-LE:

Composition per liter:

H_3BO_3	2.85g
$MnCl_2 \cdot 4H_2O$	1.8g

Sodium tartrate	1.77g
$FeSO_4 \cdot 7H_2O$	1.36g
$CoCl_2 \cdot 6H_2O$	0.04g
$CuCl_2.2H_2O$	0.027g
$Na_2MoO_4 \cdot 2H_2O$	0.025g
$ZnCl_2$	0.02g

Preparation of Trace Elements Solution HO-LE: Add components to distilled/deionized water and bring volume to 1.0L. Mix thoroughly. Filter sterilize.

Preparation of Medium: Add components, except traqce elements solution HO-LE, to filtered seawater and bring volume to 999.0mL. Mix thoroughly. Gently heat and bring to boiling. Autoclave for 15 min at 15 psi pressure–121°C. Cool to 45°–50°C. Aseptically add 1.0mL of sterile trace elements solution HO-LE. Mix thoroughly. Pour into sterile Petri dishes or distribute into sterile tubes.

Use: For the cultivation and maintenance of *Flexibacter polymorphus*.

Flexibacter Agar (LMG Medium 60)

Composition per liter:

Agar	15.0g
Casamino acids	1.0g
Tris buffer	1.0g
$MgSO_4 \cdot 7H_2O$	0.1g
KNO_3	0.1g
$CaCl_2 \cdot 2H_2O$	0.1g
Sodium-β-glycerophosphate	0.1g
Thiamine	1.0mg
Vitamin B_{12}	1.0μg
Glucose solution	10.0mL
Vitamin solution	10.0mL
Trace elements solution	1.0mL

pH 7.5 ± 0.2 at 25°C

Glucose Solution:

Composition per 100.0mL:

Glucose	10.0g

Preparation of Glucose Solution: Add glucose to distilled/deionized water and bring volume to 100.0mL. Mix thoroughly. Filter sterilize.

Trace Elements Solution:

Composition per liter:

H_3BO_3	2.85g
$FeSO_4 \cdot 7H_2O$	2.49g
$MnCl_2 \cdot 4H_2O$	1.8g
Sodium tartrate	1.77g
$CaCl_2 \cdot 2H_2O$	40.4mg
$CuCl_2 \cdot 2H_2O$	26.9mg
$ZnCl_2$	20.8mg
$Na_2MoO_4 \cdot 2H_2O$	25.2mg

Preparation of Trace Elements Solution: Add components to distilled/deionized water and bring volume to 1.0L. Mix thoroughly. Filter sterilize.

Vitamin Solution:

Composition per 10.0mL:

Thiamine	10.0mg
Vitamin B_{12}	10.0μg

Preparation of Vitamin Solution: Add components to distilled/deionized water and bring volume to 10.0mL. Mix thoroughly. Filter sterilize.

Preparation of Medium: Add components, except vitamin solution, glucose solution, and trace elements solution, to 979.0mL distilled/deionized water. Mix thoroughly. Gently heat and bring to boiling. Autoclave for 15 min at 15 psi pressure–121°C. Cool to 45°–50°C. Aseptically add 10.0mL of sterile vitamin solution, 10.0mL of sterile glucose solution, and 1.0mL of sterile trace elments solution. Mix thoroughly. Pour into sterile Petri dishes or distribute into sterile tubes.

Use: For the cultivation of *Flexibacter* spp.

Flexibacter Agar

Composition per liter:

Agar	15.0g
Sodium glutamate	5.0g
$MgSO_4 \cdot 7H_2O$	1.0g
Yeast extract	1.0g
Glucose solution	20.0mL

pH 7.2 ± 0.2 at 25°C

Glucose Solution:

Composition per 20.0mL:

Glucose	2.0g

Preparation of Glucose Solution: Add glucose to distilled/deionized water and bring volume to 20.0mL. Mix thoroughly. Sparge with 100% N_2 for 3–4 min. Autoclave for 15 min at 15 psi pressure–121°C.

Preparation of Medium: Add components, except glucose solution, to distilled/deionized water and bring volume to 980.0mL. Mix thoroughly. Gently heat and bring to boiling. Autoclave for 15 min at 15 psi pressure–121°C. Aseptically add 20.0mL of sterile glucose solution. Mix thoroughly. Pour into sterile Petri dishes or distribute into sterile tubes.

Use: For the cultivation of *Flexibacter elegans.*

Flexibacter Broth

Composition per liter:

Monosodium glutamate	5.0g
Pancreatic digest of casein	1.0g
Vitamin-free casamino acids	1.0g
Sodium glycerophosphate	0.1g
Vitamin B_{12}	1.0μg
Trace elements solution HO-LE	1.0mL

Trace Elements Solution HO-LE:

Composition per liter:

H_3BO_3	2.85g
$MnCl_2 \cdot 4H_2O$	1.8g
Sodium tartrate	1.77g
$FeSO_4 \cdot 7H_2O$	1.36g
$CoCl_2 \cdot 6H_2O$	0.04g
$CuCl_2.2H_2O$	0.027g
$Na_2MoO_4 \cdot 2H_2O$	0.025g
$ZnCl_2$	0.02g
Seawater	999.0mL

Preparation of Trace Elements Solution HO-LE: Add components to distilled/deionized water and bring volume to 1.0L. Mix thoroughly. Filter sterilize.

Preparation of Medium: Add components, except traqce elements solution HO-LE, to filtered seawater and bring volume to 999.0mL.

Mix thoroughly. Autoclave for 15 min at 15 psi pressure–121°C. Cool to 45°–50°C. Aseptically add 1.0mL of sterile trace elements solution HO-LE. Mix thoroughly. Aseptically distribute into sterile tubes or flasks.

Use: For the cultivation of *Flexibacter polymorphus*.

Flexibacter canadensis Agar

Composition per 1001.0mL:

Agar	10.0g
Casamino acids	1.0g
Tris	1.0g
$CaCl_2 \cdot 2H_2O$	0.1g
KNO_3	0.1g
$MgSO_4 \cdot 7H_2O$	0.1g
Sodium glycerophosphate	0.1g
Thiamine HCl	1.0mg
Vitamin B_{12}	1.0μg
Glucose solution	10.0mL
Trace elements solution SL-10	1.0mL

pH 7.5 ± 0.2 at 25°C

Glucose Solution:

Composition per 10.0mL:

Glucose	1.0g

Preparation of Glucose Solution: Add glucose to distilled/deionized water and bring volume to 10.0mL. Mix thoroughly. Sparge with 100% N_2 for 3–4 min. Autoclave for 15 min at 15 psi pressure–121°C.

Trace Elements Solution SL-10:

Composition per liter:

$FeCl_2 \cdot 4H_2O$	1.5g
$CoCl_2 \cdot 6H_2O$	190.0mg
$MnCl_2 \cdot 4H_2O$	100.0mg
$ZnCl_2$	70.0mg
$Na_2MoO_4 \cdot 2H_2O$	36.0mg
$NiCl_2 \cdot 6H_2O$	24.0mg
H_3BO_3	6.0mg
$CuCl_2 \cdot 2H_2O$	2.0mg
HCl (25% solution)	10.0mL

Preparation of Trace Elements Solution SL-10: Add $FeCl_2 \cdot 4H_2O$ to 10.0mL of HCl solution. Mix thoroughly. Add distilled/deionized water and bring volume to 1.0L. Add remaining components. Mix thoroughly.

Preparation of Medium: Add components, except glucose solution, to distilled/deionized water and bring volume to 990.0mL. Mix thoroughly. Gently heat and bring to boiling. Autoclave for 15 min at 15 psi pressure–121°C. Aseptically add 10.0mL of sterile glucose solution. Mix thoroughly. Pour into sterile Petri dishes or distribute into sterile tubes.

Use: For the cultivation of *Flexibacter canadensis*.

Flexibacter canadensis Medium

Composition per liter:

Agar	15.0g
Tryptone	2.0g
Beef extract	0.5g
Yeast extract	0.5g
Sodium acetate	0.2g

pH 7.2–7.4 at 25°C

Preparation of Medium: Add components to distilled/deionized water and bring volume to 1.0L. Mix thoroughly. Gently heat and bring to boiling. Distribute into tubes or flasks. Autoclave for 15 min at 15 psi pressure–121°C. Pour into sterile Petri dishes or leave in tubes.

Use: For the cultivation and maintenance of *Flexibacter canadensis*.

Flexibacter Medium

Composition per liter:

Agar	15.0g
Tris(hydroxymethyl)aminomethane buffer	1.0g
Casamino acids	1.0g
$MgSO_4 \cdot 7H_2O$	0.1g
KNO_3	0.1g
$CaCl_2 \cdot 2H_2O$	0.1g
Sodium β-glycerophosphate	0.1g
Thiamine	1.0mg
Cobalamin	1.0μg
Glucose solution	10.0mL
Trace elements solution HO-LE	1.0mL

pH 7.5 ± 0.2 at 25°C

Glucose Solution:

Composition per 10.0mL:

D-Glucose	1.0g

Preparation of Glucose Solution: Add glucose to distilled/deionized water and bring volume to 10.0mL. Mix thoroughly. Filter sterilize.

Trace Elements Solution HO-LE:

Composition per liter:

H_3BO_3	2.85g
$MnCl_2 \cdot 4H_2O$	1.8g
Sodium tartrate	1.77g
$FeSO_4 \cdot 7H_2O$	1.36g
$CoCl_2 \cdot 6H_2O$	0.04g
$CuCl_2 \cdot 2H_2O$	0.027g
$Na_2MoO_4 \cdot 2H_2O$	0.025g
$ZnCl_2$	0.020g

Preparation of Trace Elements Solution HO-LE: Add components to distilled/deionized water and bring volume to 1.0L. Mix thoroughly. Filter sterilize.

Preparation of Medium: Add components, except glucose solution, to distilled/deionized water and bring volume to 990.0mL. Mix thoroughly. Gently heat and bring to boiling. Autoclave for 15 min at 15 psi pressure–121°C. Aseptically add 10.0mL of sterile glucose solution. Mix thoroughly. Pour into sterile Petri dishes or distribute into sterile tubes.

Use: For the cultivation and maintenance of *Flexibacter* species.

Flexibacter Medium

Composition per liter:

Agar	15.0g
Tryptone	5.0g
Yeast extract	5.0g
Tris(hydroxymethyl)aminomethane buffer	1.0g
KNO_3	0.5g
Sodium β-glycerophosphate	0.1g
Seawater	1.0L
Trace elements solution HO-LE	1.0mL

pH 7.0 ± 0.2 at 25°C

Trace Elements Solution HO-LE:

Composition per liter:

H_3BO_3	2.85g
$MnCl_2 \cdot 4H_2O$	1.8g

Sodium tartrate 1.77g
$FeSO_4 \cdot 7H_2O$ 1.36g
$CoCl_2 \cdot 6H_2O$ 0.04g
$CuCl_2.2H_2O$ 0.027g
$Na_2MoO_4 \cdot 2H_2O$ 0.025g
$ZnCl_2$ 0.020g

Preparation of Trace Elements Solution HO-LE: Add components to distilled/deionized water and bring volume to 1.0L. Mix thoroughly. Filter sterilize.

Preparation of Medium: Add components to filtered, aged seawater and bring volume to 1.0L. Mix thoroughly. Distribute into tubes or flasks. Autoclave for 15 min at 15 psi pressure–121°C.

Use: For the cultivation and maintenance of *Cytophaga latercula, Flexibacter aurantiacus, Flexibacter flexilis, Flexibacter roseolus, Flexibacter ruber, Flexithrix dorotheae,* and *Saprospira grandis.*

Flexibacter Medium (ATCC Medium 1559)

Composition per liter:
Solution B 700.0mL
Solution A 300.0mL

Solution A:
Composition per 300.0mL:
Pancreatic digest of casein 0.5g
Yeast extract 0.5g
Beef extract 0.2g
Sodium acetate 0.2g

Preparation of Solution A: Add components to distilled/deionized water and bring volume to 1.0L. Mix thoroughly. Autoclave for 15 min at 15 psi pressure–121°C. Cool to 45°–50°C.

Solution B:
Aged seawater .... 700.0 mL

Preparation of Solution B: Allow seawater to sit for 7 days. Autoclave for 15 min at 15 psi pressure–121°C. Cool to 45°–50°C.

Preparation of Medium: Aseptically add 300.0mL of sterile solution A to 700.0mL of sterile solution B at 45°–50°C. Mix thoroughly. Aseptically distribute into sterile tubes or flasks.

Use: For the cultivation of *Flexibacter maritimus.*

Flexibacter polymorphus Medium (LMG 108)

Composition per liter:
Agar 15.0g
Monosodium glutamate 5.0g
Tryptone 1.0g
Casamino acids, vitamin-free 1.0g
Sodium glycerophosphate 0.1g
Vitamin B_{12} 1.0μg
Trace elements solution HO-LE 1.0mL
Seawater, filtered, aged 1.0L

Trace Elements Solution HO-LE:
Composition per liter:
H_3BO_3 .... 2.85g
$MnCl_2 \cdot 4H_2O$ 1.8g
Sodium tartrate 1.77g
$FeSO_4 \cdot 7H_2O$ 1.36g
$CoCl_2 \cdot 6H_2O$ 0.04g
$CuCl_2.2H_2O$ 0.027g
$Na_2MoO_4 \cdot 2H_2O$ 0.025g
$ZnCl_2$ 0.02g

Preparation of Trace Elements Solution HO-LE: Add components to distilled/deionized water and bring volume to 1.0L. Mix thoroughly.

Preparation of Medium: Combine components. Mix thoroughly. Distribute into tubes or flasks. Autoclave for 15 min at 15 psi pressure–121°C.

Use: For the cultivation and maintenance of *Flexibacter polymorphus.*

Flexibacterium Medium

Composition per 1060.0mL:
Yeast extract 1.0g
$Ca(NO_3)_2 \cdot 4H_2O$ 0.1g
K_2HPO_4 .... 0.02g
Seawater, filtered 1.0L
Glucose solution 50.0mL
Trace elements 10.0mL

pH 7.0 ± 0.2 at 25°C

Glucose Solution:
Composition per 50.0mL:
Glucose 1.0g

Preparation of Glucose Solution: Add glucose to distilled/deionized water and bring volume to 50.0mL. Mix thoroughly. Autoclave for 15 min at 15 psi pressure–121°C. Cool to 25°C.

Trace Elements:
Composition per liter:
$FeSO_4 \cdot 7H_2O$ 0.5mg
$ZnSO_4 \cdot 7H_2O$ 0.3mg
H_3BO_3 .... 0.1mg
$CoCl_2 \cdot 6H_2O$ 0.1mg
$CuSO_4 \cdot 5H_2O$ 0.1mg
$MnSO_4 \cdot 4H_2O$ 0.1mg
$Na_2MoO_4 \cdot 2H_2O$ 0.1mg

Preparation of Trace Elements: Add components to distilled/deionized water and bring volume to 1.0L. Mix thoroughly.

Preparation of Medium: Combine components, except glucose solution. Mix thoroughly. Adjust pH to 7.2. Gently heat and bring to boiling. Autoclave for 15 min at 15 psi pressure–121°C. Cool to 45°–50°C. Aseptically add sterile glucose solution. Mix thoroughly. Aseptically distribute into sterile tubes or bottles.

Use: For the cultivation of *Flexibacter litoralis* and *Flexibacter marinus.*

Flexibacterium Medium

Composition per 1050.0mL:
Tris(hydroxymethyl)aminomethane buffer 1.0g
Yeast extract 1.0g
$CaCl_2 \cdot 2H_2O$ 0.1g
KCl 0.1g
$MgSO_4 \cdot 7H_2O$ 0.1g
Sodium glycerophosphate 0.1g
$NaNO_3$ 0.1g
Cobalamin 1.0μg
Glucose solution 50.0mL
Trace elements 10.0mL

pH 7.5 ± 0.2 at 25°C

Glucose Solution:

Composition per 50.0mL:

Glucose	1.0g

Preparation of Glucose Solution: Add glucose to distilled/deionized water and bring volume to 50.0mL. Mix thoroughly. Autoclave for 15 min at 15 psi pressure–121°C. Cool to 25°C.

Trace Elements:

Composition per liter:

$FeSO_4 \cdot 7H_2O$	0.5mg
$ZnSO_4 \cdot 7H_2O$	0.3mg
H_3BO_3	0.1mg
$CoCl_2 \cdot 6H_2O$	0.1mg
$CuSO_4 \cdot 5H_2O$	0.1mg
$MnSO_4 \cdot 4H_2O$	0.1mg
$Na_2MoO_4 \cdot 2H_2O$	0.1mg

Preparation of Trace Elements: Add components to distilled/deionized water and bring volume to 1.0L. Mix thoroughly.

Preparation of Medium: Add components, except glucose solution, to distilled/deionized water and bring volume to 1.0L. Mix thoroughly. Adjust pH to 7.5. Autoclave for 15 min at 15 psi pressure–121°C. Cool to 45°–50°C. Aseptically add sterile glucose solution. Mix thoroughly. Aseptically distribute into sterile tubes or flasks.

Use: For the cultivation of *Saprospira thermalis*, *Flexibacter elegans*, and *Flexibacter rubrum*.

Flexibacterium Medium

Composition per 1050.0mL:

Tris(hydroxymethyl)aminomethane buffer	1.0g
Yeast extract	1.0g
Glycerol	1.0g
$CaCl_2 \cdot 2H_2O$	0.1g
KCl	0.1g
$MgSO_4 \cdot 7H_2O$	0.1g
Sodium glycerophosphate	0.1g
$NaNO_3$	0.1g
Cobalamin	1.0μg
Trace elements	10.0mL

pH 7.5 ± 0.2 at 25°C

Trace Elements:

Composition per liter:

$FeSO_4 \cdot 7H_2O$	0.5mg
$ZnSO_4 \cdot 7H_2O$	0.3mg
H_3BO_3	0.1mg
$CoCl_2 \cdot 6H_2O$	0.1mg
$CuSO_4 \cdot 5H_2O$	0.1mg
$MnSO_4 \cdot 4H_2O$	0.1mg
$Na_2MoO_4 \cdot 2H_2O$	0.1mg

Preparation of Trace Elements: Add components to distilled/deionized water and bring volume to 1.0L. Mix thoroughly.

Preparation of Medium: Add components to distilled/deionized water and bring volume to 1.0L. Mix thoroughly. Adjust pH to 7.5. Distribute into sterile tubes or flasks. Autoclave for 15 min at 15 psi pressure–121°C.

Use: For the cultivation of *Saprospira albida*.

Flexiligladius Medium

Composition per liter:

Beef heart, solids from infusion	50.0g
Agar	15.0g
Tryptose	1.0g
NaCl	0.5g

pH 7.4 ± 0.2 at 25°C

Preparation of Medium: Add components to distilled/deionized water and bring volume to 1.0L. Mix thoroughly. Gently heat and bring to boiling. Distribute into tubes or flasks. Autoclave for 15 min at 15 psi pressure–121°C. Pour into sterile Petri dishes or leave in tubes.

Use: For the cultivation and maintenance of *Ensifer adhaerens*.

Flexistipes Medium

Composition per liter:

NaCl	58.35g
$MgCl_2 \cdot 6H_2O$	37.80g
Na_2SO_4	9.72g
$CaCl_2 \cdot 2H_2O$	7.14g
KCl	1.65g
Yeast extract	1.0g
H_3BO_3	0.66g
$Na_2S \cdot 9H_2O$	0.5g
KBr	0.24g
$SrCl_2 \cdot 6H_2O$	0.17g
$Na_2HPO_4 \cdot 2H_2O$	0.03g
Sodium meta silicate	0.012g
NaF	0.007g
KNO_3	0.005g
Resazurin	0.5mg

pH 6.5 ± 0.2 at 25°C

Preparation of Medium: Prepare and dispense medium under 100% N_2. Add components to distilled/deionized water and bring volume to 1.0L. Mix thoroughly. Adjust pH to 6.5 with H_2SO_4. Distribute into tubes or flasks. Autoclave for 15 min at 15 psi pressure–121°C.

Use: For the cultivation and maintenence of *Flexistipes sinusarabici*.

Flexithrix Marine Agar (LMG Medium 61)

Composition per liter:

Agar	15.0g
Tryptone	5.0g
Yeast extract	5.0g
Tris buffer	1.0g
KNO_3	0.5g
Sodium-β-glycerophosphate	0.1g
Seawater	1.0L
Trace elements solution	1.0mL

pH 7.4 ± 0.2 at 25°C

Trace Elements Solution:

Composition per liter:

H_3BO_3	2.85g
$FeSO_4 \cdot 7H_2O$	2.49g
$MnCl_2 \cdot 4H_2O$	1.8g
Sodium tartrate	1.77g
$CaCl_2 \cdot 2H_2O$	40.4mg
$CuCl_2 \cdot 2H_2O$	26.9mg
$ZnCl_2$	20.8mg
$Na_2MoO_4 \cdot 2H_2O$	25.2mg

Preparation of Trace Elements Solution: Add components to distilled/deionized water and bring volume to 1.0L. Mix thoroughly.

Preparation of Medium: Add components to filtered aged seawater and bring volume to 1.0L. Mix thoroughly. Gently heat and bring to boiling. Distribute into tubes or flasks. Autoclave for 15 min at 15 psi pressure–121°C. Pour into sterile Petri dishes or leave in tubes.

Use: For the cultivation of *Flexithrix dorotheae.*

FlGlyM Medium (DSMZ Medium 298b)

Composition per liter:

NaCl	1.0g
KCl	0.5g
$MgCl_2 \cdot 6H_2O$	0.4g
NH_4Cl	0.25g
KH_2PO_4	0.2g
$CaCl_2 \cdot 2H_2O$	0.15g
Resazurin	1.0mg
$NaHCO_3$ solution	50.0mL
$Na_2S \cdot 9H_2O$ solution	10.0mL
Na-glycolate solution	10.0mL
L-Cysteine solution	10.0mL
Na-acetate solution	7.0mL
Trace elements solution SL-10	1.0mL
Vitamin solution	0.5mL

pH 7.2 ± 0.2 at 25°C

$Na_2S \cdot 9H_2O$ Solution:
Composition per 10.0mL:

$Na_2S \cdot 9H_2O$	0.36g

Preparation of $Na_2S \cdot 9H_2O$ Solution: Add $Na_2S \cdot 9H_2O$ to distilled/deionized water and bring volume to 10.0mL. Mix thoroughly. Autoclave under 100% N_2 for 15 min at 15 psi pressure–121°C. Cool to room temperature.

$NaHCO_3$ Solution:
Composition per 100.0mL:

$NaHCO_3$	10.0g

Preparation of $NaHCO_3$ Solution: Add $NaHCO_3$ to distilled/deionized water and bring volume to 100.0mL. Mix thoroughly. Sparge with 80% N_2 + 20% CO_2. Filter sterilize.

Na-Acetate Solution:
Composition per 10.0mL:

Na-acetate	0.25g

Preparation of Na-Acetate Solution: Add Na-acetate to distilled/deionized water and bring volume to 10.0mL. Mix thoroughly. Sparge with 100% N_2. Filter sterilize.

Na-Glycolate Solution:
Composition per 10.0mL:

Na-glycolate	1.0g

Preparation of Na-Glycolate Solution: Add Na-glycolate to distilled/deionized water and bring volume to 10.0mL. Mix thoroughly. Sparge with 100% N_2. Filter sterilize.

L-Cysteine Solution:
Composition per 10.0mL:

L-Cysteine·HCl·H_2O	0.3g

Preparation of L-Cysteine Solution: Add L-cysteine·HCl·H_2O to distilled/deionized water and bring volume to 10.0mL. Mix thoroughly. Sparge with 100% N_2. Autoclave for 15 min at 15 psi pressure–121°C.

Trace Elements Solution SL-10:
Composition per liter:

$FeCl_2 \cdot 4H_2O$	1.5g
H_3BO_3	300.0mg
$CoCl_2 \cdot 6H_2O$	190.0mg
$MnCl_2 \cdot 4H_2O$	100.0mg
$ZnCl_2$	70.0mg
$Na_2MoO_4 \cdot 2H_2O$	36.0mg
$NiCl_2 \cdot 6H_2O$	24.0mg
$CuCl_2 \cdot 2H_2O$	2.0mg
HCl (25% solution)	10.0mL

Preparation of Trace Elements Solution SL-10: Add $FeCl_2 \cdot 4H_2O$ to 10.0mL of HCl solution. Mix thoroughly. Add distilled/deionized water and bring volume to 1.0L. Add remaining components. Mix thoroughly. Sparge with 80% N_2 + 20% CO_2. Autoclave for 15 min at 15 psi pressure–121°C.

Vitamin Solution:
Composition per liter:

Pyridoxine hydrochloride	300.0mg
Thiamine-HCl·$2H_2O$	200.0mg
Nicotinic acid	200.0mg
Vitamin B_{12}	100.0mg
Calcium pantothenate	100.0mg
p-Aminobenzoic acid	80.0mg
Folic Acid	30.0mg
D(+)-Biotin	20.0mg
α-Lipoic acid	10.0mg

Preparation of Vitamin Solution: Add components to distilled/deionized water and bring volume to 1.0L. Mix thoroughly. Filter sterilize.

Preparation of Medium: Prepare and dispense medium under 80% N_2 + 20% CO_2 gas atmosphere. Add components, except $NaHCO_3$ solution, $Na_2S \cdot 9H_2O$ solution, L-cysteine solution, Na-acetate solution, Na-glycolate solution, vitamin solution, and trace elements solution SL-10, to distilled/deionized water and bring volume to 979.0mL. Mix thoroughly. Adjust pH to 7.2. Sparge with 80% N_2 + 20% CO_2. Autoclave for 15 min at 15 psi pressure–121°C. Aseptically and anaerobically add 10.0mL $NaHCO_3$ solution, 10.0mL $Na_2S \cdot 9H_2O$ solution, 10.0mL L-cysteine solution, 7.0mL Na-acetate solution, 10.0mL Na-glycolate solution, 1.0mL trace elements solution SL-10, and 0.5mL vitamin solution. Mix thoroughly. Aseptically and anaerobically distribute into sterile tubes or bottles. After inoculation, flush and repressurize the gas head space of culture bottles with sterile 80% N_2 + 20% CO_2 to 1 bar overpressure.

Use: For the cultivation of *Syntrophobotulus* spp.

FlGlyM Medium (DSMZ Medium 298b)

Composition per liter:

NaCl	1.0g
KCl	0.5g
$MgCl_2 \cdot 6H_2O$	0.4g
NH_4Cl	0.25g
KH_2PO_4	0.2g
$CaCl_2 \cdot 2H_2O$	0.15g
Resazurin	1.0mg
$NaHCO_3$ solution	50.0mL

$Na_2S{\cdot}9H_2O$ solution	10.0mL
Na-glyoxalate solution	10.0mL
L-Cysteine solution	10.0mL
Na-acetate solution	7.0mL
Trace elements solution SL-10	1.0mL
Vitamin solution	0.5mL

pH 7.2 ± 0.2 at 25°C

$Na_2S{\cdot}9H_2O$ Solution:

Composition per 10.0mL:

$Na_2S{\cdot}9H_2O$	0.36g

Preparation of $Na_2S{\cdot}9H_2O$ Solution: Add $Na_2S{\cdot}9H_2O$ to distilled/deionized water and bring volume to 10.0mL. Mix thoroughly. Autoclave under 100% N_2 for 15 min at 15 psi pressure–121°C. Cool to room temperature.

$NaHCO_3$ Solution:

Composition per 100.0mL:

$NaHCO_3$	10.0g

Preparation of $NaHCO_3$ Solution: Add $NaHCO_3$ to distilled/deionized water and bring volume to 100.0mL. Mix thoroughly. Sparge with 80% N_2 + 20% CO_2. Filter sterilize.

Na-Acetate Solution:

Composition per 10.0mL:

Na-acetate	0.25g

Preparation of Na-Acetate Solution: Add Na-acetate to distilled/deionized water and bring volume to 10.0mL. Mix thoroughly. Sparge with 100% N_2. Filter sterilize.

Na-Glyoxalate Solution:

Composition per 10.0mL:

Na-glyoxalate	0.6g

Preparation of Na-Glyoxalate Solution: Add Na-glyoxalate to distilled/deionized water and bring volume to 10.0mL. Mix thoroughly. Sparge with 100% N_2. Filter sterilize.

L-Cysteine Solution:

Composition per 10.0mL:

L-Cysteine·HCl·H_2O	0.3g

Preparation of L-Cysteine Solution: Add L-cysteine·HCl·H_2O to distilled/deionized water and bring volume to 10.0mL. Mix thoroughly. Sparge with 100% N_2. Autoclave for 15 min at 15 psi pressure–121°C.

Trace Elements Solution SL-10:

Composition per liter:

$FeCl_2{\cdot}4H_2O$	1.5g
H_3BO_3	300.0mg
$CoCl_2{\cdot}6H_2O$	190.0mg
$MnCl_2{\cdot}4H_2O$	100.0mg
$ZnCl_2$	70.0mg
$Na_2MoO_4{\cdot}2H_2O$	36.0mg
$NiCl_2{\cdot}6H_2O$	24.0mg
$CuCl_2{\cdot}2H_2O$	2.0mg
HCl (25% solution)	10.0mL

Preparation of Trace Elements Solution SL-10: Add $FeCl_2{\cdot}4H_2O$ to 10.0mL of HCl solution. Mix thoroughly. Add distilled/deionized water and bring volume to 1.0L. Add remaining components. Mix thoroughly. Sparge with 80% N_2 + 20% CO_2. Autoclave for 15 min at 15 psi pressure–121°C.

Vitamin Solution:

Composition per liter:

Pyridoxine hydrochloride	300.0mg
Thiamine-HCl·$2H_2O$	200.0mg
Nicotinic acid	200.0mg
Vitamin B_{12}	100.0mg
Calcium pantothenate	100.0mg
p-Aminobenzoic acid	80.0mg
Folic Acid	30.0mg
D(+)-Biotin	20.0mg
α-Lipoic acid	10.0mg

Preparation of Vitamin Solution: Add components to distilled/deionized water and bring volume to 1.0L. Mix thoroughly. Filter sterilize.

Preparation of Medium: Prepare and dispense medium under 80% N_2 + 20% CO_2 gas atmosphere. Add components, except $NaHCO_3$ solution, $Na_2S{\cdot}9H_2O$ solution, L-cysteine solution, Na-acetate solution, Na-glyoxalate solution, vitamin solution, and trace elements solution SL-10, to distilled/deionized water and bring volume to 979.0mL. Mix thoroughly. Adjust pH to 7.2. Sparge with 80% N_2 + 20% CO_2. Autoclave for 15 min at 15 psi pressure–121°C. Aseptically and anaerobically add 10.0mL $NaHCO_3$ solution, 10.0mL $Na_2S{\cdot}9H_2O$ solution, 10.0mL L-cysteine solution, 7.0mL Na-acetate solution, 10.0mL Na-glyoxalate solution, 1.0mL trace elements solution SL-10, and 0.5mL vitamin solution. Mix thoroughly. Aseptically and anaerobically distribute into sterile tubes or bottles. After inoculation, flush and repressurize the gas head space of culture bottles with sterile 80% N_2 + 20% CO_2 to 1 bar overpressure.

Use: For the cultivation of *Syntrophobotulus glycolicus.*

FLN Medium (Fluorescence Lactose Nitrate Medium)

Composition per liter:

Lactose	20.0g
Agar	15.0g
Proteose peptone No. 3	10.0g
KNO_3	2.0g
K_2HPO_4	1.5g
$MgSO_4{\cdot}7H_2O$	1.5g
$NaNO_2$	0.5g
Phenol Red	0.02g

pH 7.2 ±0.2 at 25°C

Preparation of Medium: Add components to distilled/deionized water and bring volume to 1.0L. Mix thoroughly. Gently heat and bring to boiling. Distribute into tubes or flasks. Autoclave for 15 min at 15 psi pressure–121°C. Pour into sterile Petri dishes or leave in tubes.

Use: For the differentiation of pseudomonads from other nonfermentative bacilli. Lactose fermentation is indicated by the medium turning yellow. *Pseudomonas cepacia* often produces acid from lactose. Denitrification from nitrate or nitrite is indicated by the formation of gas bubbles in the solid medium. *Pseudomonas aeruginosa*, *Pseudomonas mendocina*, and *Pseudomonas denitrificans* are positive for denitrification. Fluorescein production is indicated by fluorescence under UV light. *Pseudomonas aeruginosa* is positive for fluorescein production; *Pseudomonas denitrificans* does not produce fluorescein.

Flo Agar

Composition per liter:

Agar	14.0g
Pancreatic digest of casein	10.0g

Peptic digest of animal tissue 10.0g
K_2HPO_4 .... 1.5g
$MgSO_4·7H_2O$ 1.5g

pH 7.2 ± 0.2 at 25°C

Source: This medium is available as a premixed powder from BD Diagnostic Systems.

Preparation of Medium: Add components to distilled/deionized water and bring volume to 1.0L. Mix thoroughly. Gently heat and bring to boiling. Distribute into tubes or flasks. Autoclave for 15 min at 15 psi pressure–121°C. Pour into sterile Petri dishes or leave in tubes.

Use: For cultivation of fluorescent *Pseudomonas* species.

Fluconazole Testing Medium

Composition per liter:

Glucose 19.98g
$(NH_4)_2SO_4$ 4.99g
KH_2PO_4 .... 1.99g
$MgSO_4·7H_2O$ anhydrous 0.99g
L-Glutamine 0.58g
NaCl 0.2g
$CaCl_2·2H_2O$ 0.2g
L-Lysine monohydrochloride 0.073g
L-Isoleucine 0.052g
L-Leucine 0.052g
Threonine 0.0476g
Valine 0.047g
L-Arginine monohydrochloride 0.042g
L-Histidine 0.023g
Tryptophan 0.02g
DL-Methionine 0.0189g
Inositol 0.00397g
$ZnSO_4·7H_2O$ 0.0014g
H_3BO_3 .... 0.00099g
Nicotinic acid 0.00079g
Pyridoxine hydrochloride 0.00079g
Calcium D-pantothenic acid 0.00079g
Aneurine hydrochloride 0.00079g
$MnSO_4·2H_2O$ 0.00079g
$Na_2MoO_4·2H_2O$ 0.00047g
p-Amino benzoic acid (PABA) 0.000395g
Riboflavin 0.000395g
$FeCl_3$ 0.000395g
Folic acid 0.000395g
KI 0.0002g
$CuSO_4·5H_2O$ 0.00012g
Biotin crystalline 0.000004g
Agar solution 100.0mL

Source: This medium is available from HiMedia.

Agar Solution:

Composition per 100.0mL:

Agar 10.0g

Preparation of Agar Solution: Add agar to distilled/deionized water and bring volume to 100.0mL. Mix thoroughly. Adjust pH to 7.5 with phosphate buffer. Autoclave for 10 min at 10 psi pressure–115 °C. Cool to 50°C.

Preparation of Medium: Add components, except agar solution, to distilled/deionized water and bring volume to 900.0mL. Mix thoroughly. Filter sterilize. Aseptically add agar solution. Mix thoroughly. Pour into Petri dishes or aseptically distribute into sterile tubes.

Use: For fluconazole susceptibility testing using *Candida* species.

Fluid Casein Digest Soya Lecithin HiVeg Medium

Composition per liter:

Plant hydrolysate 20.0g
Soya lecithin 5.0g
Polysorbate 20 .... 40.0 ml

pH 7.3 ± 0.2 at 25°C

Source: This medium, without polysorbate 20, is available as a premixed powder from HiMedia.

Preparation of Medium: Add plant hydrolysate and soya lecithin to distilled/deionized water and bring volume to 960.0mL. Mix thoroughly. Heat as necessary until components are completely dissolved. Add 40.0mL polysorbate 20. Mix thorougly. Distribute into tubes or flasks. Autoclave for 15 min at 15 psi pressure–121°C.

Use: This is a highly nutritional medium containing neutralizing agents for neutralizing quaternary ammonium compounds used for the sanitary examination of surfaces.

Fluid Casein Digest Soya Lecithin Medium

Composition per liter:

Casein enzymatic hydrolysate 20.0g
Soya lecithin 5.0g
Polysorbate 20 .... 40.0 ml

pH 7.3 ± 0.2 at 25°C

Source: This medium, without polysorbate 20, is available as a premixed powder from HiMedia.

Preparation of Medium: Add plant hydrolysate and soya lecithin to distilled/deionized water and bring volume to 960.0mL. Mix thoroughly. Heat is necessary until components are completely dissolved. Add 40.0mL polysorbate 20. Mix thorougly. Distribute into tubes or flasks. Autoclave for 15 min at 15 psi pressure–121°C.

Use: This is a highly nutritional medium containing neutralizing agents for neutralizing quaternary ammonium compounds used for the sanitary examination of surfaces.

Fluid Lactose HiVeg Medium

Composition per liter:

Plant peptone No. 2 .... 5.0g
Lactose 5.0g
Plant extract 3.0g

pH 6.9 ± 0.2 at 25°C

Source: This medium is available as a premixed powder from HiMedia.

Preparation of Medium: Add components to distilled/deionized water and bring volume to 1.0L. Mix thoroughly. Distribute into tubes or flasks. Autoclave for 15 min at 15 psi pressure–121°C.

Use: As a pre-enrichment medium for the detection of coliform bacteria in water, dairy products, and foods.

Fluid Lactose HiVeg Medium with Soya Lecithin and Polysorbate 20

Composition per liter:

Plant peptone No. 2 .... 5.0g
Lactose 5.0g
Soya lecithin 5.0g

Plant extract 3.0g
Polysorbate 20.. 40 mL

pH 6.9 ± 0.2 at 25°C

Source: This medium is available as a premixed powder from HiMedia.

Preparation of Medium: Add components, except polysorbate 20, to distilled/deionized water and bring volume to 960.0mL. Mix thoroughly. Heat as necessary until components are completely dissolved. Add 40.0mL polysorbate 20. Mix thorougly. Distribute into tubes or flasks. Autoclave for 15 min at 15 psi pressure–121°C.

Use: For the cultivation of coliform bacteria in water, dairy products, and foods.

Fluid Lactose Medium

Composition per liter:

Pancreatic digest of gelatin 5.0g
Lactose 5.0g
Plant extract 3.0g

pH 6.9 ± 0.2 at 25°C

Source: This medium is available as a premixed powder from HiMedia.

Preparation of Medium: Add components to distilled/deionized water and bring volume to 1.0L. Mix thoroughly. Distribute into tubes or flasks. Autoclave for 15 min at 15 psi pressure–121°C.

Use: As a pre-enrichment medium for the detection of coliform bacteria in water, dairy products, and foods.

Fluid Sabouraud HiVeg Medium

Composition per liter:

Glucose 20.0g
Plant hydrolysate.. 5.0g
Plant peptone.. 5.0g

pH5.7 ± 0.2 at 25°C

Source: This medium is available as a premixed powder from HiMedia.

Preparation of Medium: Add components to distilled/deionized water and bring volume to 1.0L. Mix thoroughly. Distribute into tubes or flasks. Autoclave for 15 min at 15 psi pressure–121°C.

Use: For cultivation of yeasts, molds, and aciduric microorganisms. For sterility test medium for molds and lower bacteria in pharmaceutical preparations.

Fluid Sabouraud Medium

See also: **Antibiotic Medium 13**

Fluid Sabouraud Medium

Composition per liter:

Glucose 20.0g
Pancreatic digest of casein.. 5.0g
Peptic digest of animal tissue.. 5.0g

pH5.7 ± 0.2 at 25°C

Source: This medium is available as a premixed powder from BD Diagnostic Systems.

Preparation of Medium: Add components to distilled/deionized water and bring volume to 1.0L. Mix thoroughly. Distribute into tubes or flasks. Autoclave for 15 min at 15 psi pressure–121°C.

Use: For cultivation of yeasts, molds, and aciduric microorganisms.

Fluid Selenite Cystine HiVeg Medium (Selenite Cystine HiVeg Broth)

Composition per liter:

Na_2HPO_4.. 10.0g
Plant hydrolysate 5.0g
Lactose.. 4.0g
L-Cystine.. 0.01g
$HNaO_3Se$ 4.0g

pH 7.0 ± 0.2 at 25°C

Source: This medium is available as a premixed powder from HiMedia.

Caution: Sodium selenite is toxic and a potential teratogen and care must be taken to avoid inhalation of the powdered dye, contact with the skin, or ingestion, especially in pregnant laboratory workers.

Preparation of Medium: Add sodium hydrogen selenite to distilled/deionized water and bring volume to 1.0L. Add remaining components. Mix thoroughly. Gently heat. Do not autoclave. Distribute into sterile tubes in 10.0mL volumes. Sterilize for 15 min at 0 psi pressure–100°C. Do not autoclave.

Use: For the isolation and cultivation of *Salmonella* species from feces, dairy products, and other specimens.

Fluid Selenite Cystine Medium (Selenite Cystine Broth)

Composition per liter:

Na_2HPO_4.. 10.0g
Casein enzymatic hydrolysate 5.0g
Lactose.. 4.0g
L-Cystine.. 0.01g
$HNaO_3Se$ 4.0g

pH 7.0 ± 0.2 at 25°C

Source: This medium is available as a premixed powder from HiMedia.

Caution: Sodium selenite is toxic and a potential teratogen and care must be taken to avoid inhalation of the powdered dye, contact with the skin, or ingestion, especially in pregnant laboratory workers.

Preparation of Medium: Add sodium hydrogen selenite to distilled/deionized water and bring volume to 1.0L. Add remaining components. Mix thoroughly. Gently heat. Do not autoclave. Distribute into sterile tubes in 10.0mL volumes. Sterilize for 15 min at 0 psi pressure–100°C. Do not autoclave.

Use: For the isolation and cultivation of *Salmonella* species from feces, dairy products, and other specimens.

Fluid Tetrathionate HiVeg Medium without Iodine and BG (Tetrathionate HiVeg Broth Base without Iodine & BG)

Composition per liter:

$Na_2S_2O_3$ 30.0g
$CaCO_3$.. 10.0g
Plant hydrolysate 2.5g
Plant peptone 2.5g
Synthetic detergent 1.0g

Iodine-potassium iodide solution	20.0mL
Brilliant Green solution	10.0mL

pH 8.4 ± 0.2 at 25°C

Source: This medium, without iodine-potassium iodine solution and Brilliant Green solution, is available as a premixed powder from HiMedia.

Preparation of Tetrathionate Broth Base: Add components to distilled/deionized water and bring volume to 1.0L. Mix thoroughly. Gently heat and bring to boiling. A slight precipitate will remain. Do not autoclave. Cool to 25°C. Store at 4°C.

Iodine-Potassium Iodide Solution:

Composition per 20.0mL:

Iodine, resublimed	6.0g
KI	5.0g

Preparation of Iodine-Potassium Iodide Solution: Add KI to 5.0mL of sterile distilled/deionized water. Mix thoroughly. Add iodine. Mix thoroughly. Bring volume to 20.0mL with sterile distilled/deionized water.

Brilliant Green Solution:

Composition per 100.0mL:

Brilliant Green	0.1g

Preparation of Brilliant Green Solution: Add Brilliant Green to sterile distilled/deionized water. Mix thoroughly.

Preparation of Medium: Add components, except iodine-potassium iodine soulution and Brilliant Green solution to distilled/deionized water and bring volume to 1.0L. Mix thoroughly. Gently heat and bring to boiling. A slight precipitate will remain. Do not autoclave. Cool to 25°C. Add 20.0mL of iodine-potassium iodide solution, and 10.0mL of Brilliant Green solution. Mix thoroughly. Aseptically distribute into tubes in 10.0mL volumes. Do not heat medium after it has been mixed.

Use: For the selective isolation and cultivation of *Salmonella* species from foods.

Fluid Tetrathionate Medium without Iodine and BG (Tetrathionate Broth Base without Iodine & BG)

Composition per liter:

$Na_2S_2O_3$	30.0g
$CaCO_3$	10.0g
Casein enzymatic hydrolysate	2.5g
Peptic digest of animal tissue	2.5g
Bile salts	1.0g
Iodine-potassium iodide solution	20.0mL
Brilliant Green solution	10.0mL

pH 8.4 ± 0.2 at 25°C

Source: This medium, without iodine-potassium iodine solution and Brilliant Green solution, is available as a premixed powder from HiMedia.

Preparation of Tetrathionate Broth Base: Add components to distilled/deionized water and bring volume to 1.0L. Mix thoroughly. Gently heat and bring to boiling. A slight precipitate will remain. Do not autoclave. Cool to 25°C. Store at 4°C.

Iodine-Potassium Iodide Solution:

Composition per 20.0mL:

Iodine, resublimed	6.0g
KI	5.0g

Preparation of Iodine-Potassium Iodide Solution: Add KI to 5.0mL of sterile distilled/deionized water. Mix thoroughly. Add iodine. Mix thoroughly. Bring volume to 20.0mL with sterile distilled/deionized water.

Brilliant Green Solution:

Composition per 100.0mL:

Brilliant Green	0.1g

Preparation of Brilliant Green Solution: Add Brilliant Green to sterile distilled/deionized water. Mix thoroughly.

Preparation of Medium: Add components, except iodine-potassium iodine soulution and Brilliant Green solution, to distilled/deionized water and bring volume to 1.0L. Mix thoroughly. Gently heat and bring to boiling. A slight precipitate will remain. Do not autoclave. Cool to 25°C. Add 20.0mL of iodine-potassium iodide solution, and 10.0mL of Brilliant Green solution. Mix thoroughly. Aseptically distribute into tubes in 10.0mL volumes. Do not heat medium after it has been mixed.

Use: For the selective isolation and cultivation of *Salmonella* species from foods.

Fluid Thioglycollate Agar

Composition per liter:

Pancreatic digest of casein	15.0g
Glucose	5.0g
Yeast extract	5.0g
NaCl	2.5g
Agar	0.75g
L-Cystine	0.5g
Sodium thioglycolate	0.5g
Resazurin	1.0mg

pH 7.1 ± 0.2 at 25°C

Preparation of Medium: Add components to distilled/deionized water and bring volume to 1.0L. Mix thoroughly. Gently heat and bring to boiling. Distribute into tubes or flasks. Autoclave for 15 min at 15 psi pressure–121°C. Pour into sterile Petri dishes or leave in tubes.

Use: For the cultivation of anaerobic bacteria. For the cultivation of *Campylobacter fetus, Campylobacter jejuni, Leptotrichia buccalis,* and *Streptococcus* species.

Fluid Thioglycolate Agar with Calcium Carbonate

Composition per liter:

Agar	.75.0g
Pancreatic digest of casein	15.0g
$CaCO_3$	10.0g
Glucose	5.0g
Yeast extract	5.0g
NaCl	2.5g
L-Cystine	0.5g
Sodium thioglycolate	0.5g
Resazurin	0.001g

pH 7.1 ± 0.2 at 25°C

Preparation of Medium: Add components, except $CaCO_3$, to distilled/deionized water and bring volume to 1.0L. Mix thoroughly. Gently heat and bring to boiling. Distribute into tubes or flasks. Autoclave for 15 min at 15 psi pressure–121°C. Sterilize $CaCO_3$ by autoclaving for 15 min at 15 psi pressure–121°C. Aseptically add sterile $CaCO_3$ to sterile tubes or plates—0.1g of sterile $CaCO_3$ per 10.0mL of medium to be added. Pour medium into sterile Petri dishes or distribute into sterile tubes.

Use: For the rapid cultivation of anaerobic bacteria. For the cultivation and maintenance of *Campylobacter fetus, Campylobacter jejuni, Leptotrichia buccalis,* and *Streptococcus* species.

Fluid Thioglycollate HiVeg Medium

Composition per liter:

Plant hydrolysate	15.0g
Glucose	5.5g
Yeast extract	5.0g
NaCl	2.5g
Agar	0.75g
L-Cystine	0.5g
Na-thioglycollate	0.5g
Resazurin	0.001g

pH 7.1 ± 0.2 at 25°C

Source: This medium is available as a premixed powder from HiMedia.

Preparation of Medium: Add components to distilled/deionized water and bring volume to 1.0L. Mix thoroughly. Distribute into tubes or flasks. Autoclave for 15 min at 15 psi pressure–121°C. Prepare freshly or boil and cool the medium just before use.

Use: For the cultivation of facultative anaerobes. For the cultivation of both aerobic and anaerobic organisms in the performance of sterility tests.

Fluid Thioglycollate HiVeg Medium with HiVeg Extract

Composition per liter:

Plant hydrolysate	15.0g
Glucose	5.5g
Plant extract	5.0g
Yeast extract	5.0g
NaCl	2.5g
Agar	0.75g
L-Cystine	0.5g
Na-thioglycollate	0.5g
Resazurin	0.001g

pH 7.1 ± 0.2 at 25°C

Source: This medium is available as a premixed powder from HiMedia.

Preparation of Medium: Add components to distilled/deionized water and bring volume to 1.0L. Mix thoroughly. Distribute into tubes or flasks. Autoclave for 15 min at 15 psi pressure–121°C. Prepare freshly or boil and cool the medium just before use.

Use: For the cultivation of facultative anaerobes. For the cultivation of both aerobic and anaerobic organisms in the performance of sterility tests.

Fluid Thioglycolate Medium

Composition per liter:

Pancreatic digest of casein	15.0g
Glucose	5.5g
Yeast extract	5.0g
NaCl	2.5g
Agar	0.75g
L-Cystine	0.5g
Sodium thioglycolate	0.5g
Resazurin	1.0mg

pH 7.1 ± 0.2 at 25°C

Source: This medium is available as a premixed powder from BD Diagnostic Systems.

Preparation of Medium: Add components to distilled/deionized water and bring volume to 1.0L. Mix thoroughly. Gently heat and bring to boiling. Distribute into tubes or flasks. Autoclave for 15 min at 15 psi pressure–121°C. If medium becomes oxidized before use (resazurin turns red), heat in a boiling water bath to expel absorbed O_2. Cool to 25°C.

Use: For the cultivation of anaerobic, microaerophilic, and aerobic microorganisms. For use in sterility testing of a variety of specimens.

Fluid Thioglycolate Medium with Beef Extract

Composition per liter:

Pancreatic digest of casein	15.0g
Glucose	5.5g
Yeast extract	5.0g
Beef extract	5.0g
NaCl	2.5g
Agar	0.75g
L-Cystine	0.5g
Sodium thioglycolate	0.5g
Resazurin	1.0mg

pH 7.2 ± 0.2 at 25°C

Source: This medium is available as a premixed powder from BD Diagnostic Systems.

Preparation of Medium: Add components to distilled/deionized water and bring volume to 1.0L. Mix thoroughly. Gently heat and bring to boiling. Distribute into tubes or flasks. Autoclave for 15 min at 15 psi pressure–121°C. If medium becomes oxidized before use (resazurin turns red), heat in a boiling water bath to expel absorbed O_2. Cool to 25°C.

Use: For the cultivation of anaerobic, microaerophilic, and aerobic microorganisms. For use in sterility testing of a variety of specimens.

Fluid Thioglycolate Medium with K Agar

Composition per liter:

Pancreatic digest of casein	15.0g
Glucose	5.0g
Yeast extract	5.0g
KCl	2.5g
L-Cystine	0.5g
K agar	0.45g
Resazurin	1.0mg
Thioglycolic acid	0.3mL

pH 7.2 ± 0.2 at 25°C

Source: This medium is available as a premixed powder from BD Diagnostic Systems.

Preparation of Medium: Add components to distilled/deionized water and bring volume to 1.0L. Mix thoroughly. Gently heat and bring to boiling. Distribute into tubes or flasks. Autoclave for 15 min at 15 psi pressure–121°C. If medium becomes oxidized before use (resazurin turns red), heat in a boiling water bath to expel absorbed O_2. Cool to 25°C.

Use: For the cultivation of anaerobic, microaerophilic, and aerobic microorganisms. For use in sterility testing of a variety of specimens.

Fluid Thioglycolate Medium with Rabbit Serum

Composition per liter:

Pancreatic digest of casein	15.0g
Glucose	5.5g
Yeast extract	5.0g
NaCl	2.5g
Agar	0.75g
L-Cystine	0.5g
Sodium thioglycolate	0.5g
Resazurin	1.0mg
Rabbit serum	100.0mL

pH 7.1 ± 0.2 at 25°C

Preparation of Medium: Add components, except rabbit serum, to distilled/deionized water and bring volume to 900.0mL. Mix thoroughly. Gently heat and bring to boiling. Distribute into tubes in 9.0mL volumes. Autoclave for 15 min at 15 psi pressure–121°C. If medium becomes oxidized before use (resazurin turns red), heat in a boiling water bath to expel absorbed O_2. Cool to 25°C. Immediately prior to inoculation, aseptically and anaerobically add 1.0mL of sterile rabbit serum to each tube. Mix thoroughly.

Use: For the cultivation of anaerobic, microaerophilic, and aerobic microorganisms. For use in the sterility testing of a variety of specimens.

Fluid Thioglycolate Medium without Glucose or E_h Indicator

Composition per liter:

Pancreatic digest of casein	20.0g
NaCl	2.5g
Agar	0.75g
L-Cystine	0.5g
Sodium thioglycolate	0.5g

pH 7.1 ± 0.2 at 25°C

Source: This medium is available as a premixed powder from BD Diagnostic Systems.

Preparation of Medium: Add components to distilled/deionized water and bring volume to 1.0L. Mix thoroughly. Gently heat and bring to boiling. Distribute into tubes or flasks. Autoclave for 15 min at 15 psi pressure–121°C. Aseptically distribute into sterile screw-capped tubes or flasks.

Use: For cultivation of anaerobic bacteria. For use as a basal medium in carbohydrate fermentation tests for differentiating anaerobic bacteria. For carbohydrate fermentation tests, 1.0 mL of a 10% filter-sterilized carbohydrate solution is aseptically added to 9.0mL of fluid thioglycolate medium.

Fluorescence Denitrification Medium

See: **FN Medium**

Fluorescence Lactose Nitrate Medium

See: **FLN Medium**

Fluorescent Pectolytic Agar (FPA Medium)

Composition per liter:

Proteose peptone No. 3	20.0g
Agar	15.0g
Pectin	5.0g
K_2HPO_4	1.5g
$MgSO_4 \cdot 7H_2O$	0.73g
Antibiotic solution	10.0mL

Antibiotic Solution:

Composition per 10.0mL:

Cycloheximide	0.075g
Novobiocin	0.045g
Penicillin G	75,000U
Ethanol	1.0mL

Preparation of Antibiotic Solution: Add components to 1.0mL of ethanol. Mix thoroughly. Let stand for 30 min. Bring volume to 10.0mL with distilled/deionized water. Mix thoroughly. Filter sterilize.

Caution: Cycloheximide is toxic. Avoid skin contact or aerosol formation and inhalation.

Preparation of Medium: Add components, except antibiotic solution, to distilled/deionized water and bring volume to 990.0mL. Mix thoroughly. Gently heat and bring to boiling. Autoclave for 15 min at 15 psi pressure–121°C. Cool to 45°–50°C. Aseptically add sterile antibiotic solution. Mix thoroughly. Pour into sterile Petri dishes.

Use: For the cultivation of fluorescent *Pseudomonas* species that are pectinolytic.

Fluorocult® *E. coli* O157:H7 Agar (*E. coli* O157:H7 Agar, Fluorocult)

Composition per liter:

Peptone from casein	20.0g
Agar	13.0g
Sorbitol	10.0g
NaCl	5.0g
Meat extract	2.0g
$Na_2S_2O_3$	2.0g
Sodium deoxycholate	1.12g
Yeast extract	1.0g
Ammonium ferric citrate	0.5g
4-Methylumbelliferyl-β-D-glucuronide	0.1g
Bromthymol Blue	0.025g

pH 7.4 ± 0.2 at 25°C

Source: This medium is available from Merck.

Preparation of Medium: Add components to distilled/deionized water and bring volume to 1.0L. Mix thoroughly. Autoclave for 15 min at 15 psi pressure–121°C. Cool to 45°–50°C. Pour into sterile Petri dishes.

Use: For the isolation and differentiation of enterohemorrhagic (EHEC) *Escherichia coli* O157:H7 strains from foods. In contrast to most other *E. coli* strains, *E. coli* O157:H7 shows the following characteristics: no sorbitol-cleavage capacity within 48 hr and no formation of glucuronidase (MUG-negative/no fluorescence). Sodium deoxycholate inhibits the growth of the Gram-positive accompanying flora for the greater part. Sorbitol serves, together with the pH indicator Bromothymol Blue, to determine the degradation of sorbitol which, in the case of sorbitol-positive microorganisms, results in the colonies turning yellow in color. Sorbitol-negative strains, on the other hand, do not lead to any change in the color of the culture medium and thus proliferate as greenish colonies. Sodium thiosulfate and ammonium iron(III) citrate result in black-brown discoloration of the agar for colonies, in the presence of hydrogen-sulfide-forming pathogens, precipitating iron sulfide. *Proteus mirabilis* in particular, which displays biochemical properties similar to those of *E. coli* O157:H7, can thus be very easily differentiated from *E. coli* O157:H7 on account of the brownish discoloration. 4-Methylumbelliferyl-β-D-glucuronide (MUG) is converted into 4-methylumbelliferone by β-D-glucuronidase-forming pathogens; 4-methylumbelliferone fluoresces under UV light. The activity of β-D-glucuronidase is a highly specific characteristic of *E. coli.* In contrast to most *E. coli* strains, *E. coli* O157:H7 is not capable of forming β-D-glucoronidase. When irradiated with long-wave UV light, no fluorescence is formed.

Fluorouracil *Leptospira* Medium

See: **Fletcher Medium with Fluorouracil**

FM Medium

Composition per liter:

Agar	15.0g
Pancreatic digest of casein	15.0g
Glucose	5.0g
NaCl	5.0g
Yeast extract	5.0g
L-Cysteine·HCl·H_2O	0.75g
Crystal Violet	0.01g
Horse serum	50.0mL
Streptomycin solution	10.0mL

pH 7.2 ± 0.2 at 25°C

Streptomycin Solution:

Composition per 10.0mL:

Streptomycin	0.01g

Preparation of Streptomycin Solution: Add streptomycin to distilled/deionized water and bring volume to 10.0mL. Mix thoroughly. Filter sterilize.

Preparation of Medium: Add components, except horse serum and streptomycin solution, to distilled/deionized water and bring volume to 940.0mL. Mix thoroughly. Gently heat and bring to boiling. Autoclave for 15 min at 15 psi pressure–121°C. Cool to 45°–50°C. Aseptically add sterile horse serum and streptomycin solution. Mix thoroughly. Pour into sterile Petri dishes or distribute into sterile tubes.

Use: For the selective isolation and cultivation of *Fusobacterium* species.

FN Medium
(Fluorescence Denitrification Medium)

Composition per liter:

Agar	15.0g
Proteose peptone No. 3	10.0g
KNO_3	2.0g
K_2HPO_4	1.5g
$MgSO_4·7H_2O$	1.5g
$NaNO_2$	0.5g

pH 7.2 ±0.2 at 25°C

Preparation of Medium: Add components to distilled/deionized water and bring volume to 1.0L. Mix thoroughly. Gently heat and bring to boiling. Distribute into tubes or flasks. Autoclave for 15 min at 15 psi pressure–121°C. Pour into sterile Petri dishes or leave in tubes.

Use: For the differentiation of pseudomonads from other nonfermentative bacilli. Denitrification from nitrate or nitrite is indicated by the formation of gas bubbles in the solid medium. *Pseudomonas aeruginosa, Pseudomonas mendocina,* and *Pseudomonas denitrificans* are positive for denitrification. Fluorescein production is indicated by fluorescence under UV light. *Pseudomonas aeruginosa* is positive for fluorescein production; *Pseudomonas denitrificans* does not produce fluorescein.

FNA Medium
(Fluorescein Denitrification Agar)

Composition per liter:

Agar	15.0g
Peptic digest of animal tissue	5.0g
Casein enzymatic hydrolysate	5.0g
KNO_3	2.0g
$MgSO_4·7H_2O$	1.5g
K_2HPO_4	1.5g

Source: This medium is available from HiMedia.

Preparation of Medium: Add components to distilled/deionized water and bring volume to 1.0L. Mix thoroughly. Gently heat and bring to boiling. Distribute into tubes or flasks. Autoclave for 15 min at 15 psi pressure–121°C. Pour into sterile Petri dishes or leave in tubes.

Use: For the differentiation of *Pseudomonas* spp. based upon their ability to reduce nitrates or nitrites to nitrogen gas and the detection of fluorescein pigment production.

Folic Acid Agar

Composition per liter:

Noble agar	15.0g
$K_2HPO_4·3H_2O$	1.2g
Folic acid	1.0g
KH_2PO_4	0.5g
Salts A	5.0mL
Salts B	1.5mL

Salts A:

Composition per 100.0mL:

$MgSO_4·7H_2O$	1.0g
$CaCl_2·2H_2O$	0.1g
$FeSO_4·7H_2O$	0.1g

Preparation of Salts A Solution: Add components to distilled/deionized water and bring volume to 100.0mL. Mix thoroughly. Maintain for 3 days at 25°C to dissolve. Filter sterilize the supernatant.

Salts B:

Composition per 100.0mL:

$MnSO_4$	0.1g
Na_2MoO_4	0.1g

Preparation of Salts B Solution: Add components to distilled/deionized water and bring volume to 100.0mL. Mix thoroughly. Filter sterilize.

Preparation of Medium: Add components, except salts A solution and salts B solution, to distilled/deionized water and bring volume to 994.5mL. Mix thoroughly. Gently heat and bring to boiling. Autoclave for 15 min at 15 psi pressure–121°C. Cool to 45°–50°C. Aseptically add 5.0mL of sterile salts A solution and 1.5mL of sterile salts B solution. Mix thoroughly. Pour into sterile Petri dishes or distribute into sterile tubes.

Use: For the cultivation and maintenance of *Pseudomonas* species.

Folic Acid Assay HiVeg Medium

Composition per liter:

Glucose	40.0g
Sodium citrate	20.0g

Plant acid hydrolysate, vitamin free	12.0g
K_2HPO_4	1.0g
KH_2PO_4	1.0g
$MgSO_4$	0.4g
DL-Tryptophan	0.2g
L-Cystine	0.2g
Adenine sulfate	0.02g
NaCl	0.02g
$FeSO_4$	0.02g
Guanine hydrochloride	0.02g
$MnSO_4$	0.02g
Uracil	0.02g
Pyridoxine hydrochloride	4.0mg
Niacin	2.0mg
Riboflavin	2.0mg
Thiamine hydrochloride	2.0mg
Calcium pantothenate	0.4mg
p-Aminobenzoic acid (PABA)	0.2mg
Biotin	0.8.0μg

pH 6.8 ± 0.2 at 25°C

Source: This medium is available as a premixed powder from HiMedia.

Preparation of Medium: Add components to distilled/deionized water and bring volume to 1.0L. Mix thoroughly. Gently heat and bring to boiling with frequent mixing. Cool to 45°–50°C. Mix thoroughly. Distribute into tubes or flasks. Autoclave for 10 min at 15 psi pressure–121°C.

Use: For the microbiological assaying of folic acid using *Enterococcus hirae* as the test organism.

Folic Acid Assay Medium

Composition per liter:

Glucose	40.0g
Sodium citrate	20.0g
Casein acid hydrolysate, vitamin free	12.0g
K_2HPO_4	1.0g
KH_2PO_4	1.0g
$MgSO_4$	0.4g
DL-Tryptophan	0.2g
L-Cystine	0.2g
Adenine sulfate	0.02g
NaCl	0.02g
$FeSO_4$	0.02g
Guanine hydrochloride	0.02g
$MnSO_4$	0.02g
Uracil	0.02g
Pyridoxine hydrochloride	4.0mg
Niacin	2.0mg
Riboflavin)	2.0mg
Thiamine hydrochloride	2.0mg
Calcium pantothenate	0.4mg
p-Aminobenzoic acid (PABA)	0.2mg
Biotin	0.8.0μg

pH 6.8 ± 0.2 at 25°C

Source: This medium is available as a premixed powder from HiMedia.

Preparation of Medium: Add components to distilled/deionized water and bring volume to 1.0L. Mix thoroughly. Gently heat and bring to boiling with frequent mixing. Cool to 45°–50°C. Mix thoroughly. Distribute into tubes or flasks. Autoclave for 10 min at 15 psi pressure–121°C.

Use: For the microbiological assaying of folic acid using *Enterococcus hirae* as the test organism.

Folic Acid Assay Medium

Composition per liter:

Glucose	40.0g
Sodium citrate	20.0g
Vitamin assay casamino acids	12.0g
K_2HPO_4	1.0g
KH_2PO_4	1.0g
$MgSO_4 \cdot 7H_2O$	0.4g
L-Cystine	0.2g
DL-Tryptophan	0.2g
Adenine sulfate	0.02g
$FeSO_4 \cdot 7H_2O$	0.02g
Guanine·HCl	0.02g
$MnSO_4 \cdot 7H_2O$	0.02g
NaCl	0.02g
Uracil	0.02g
Pyridoxine·HCl	4.0mg
Niacin	2.0mg
Riboflavin	2.0mg
Thiamine·HCl	2.0mg
Calcium pantothenate	0.4mg
p-Aminobenzoic acid	0.2mg
Biotin	0.8μg

pH 6.8 ± 0.2 at 25°C

Source: This medium is available as a premixed powder from BD Diagnostic Systems.

Preparation of Medium: Add components to distilled/deionized water and bring volume to 1.0L. Mix thoroughly. Gently heat and bring to boiling with frequent mixing. Cool to 45°–50°C. Mix thoroughly. Distribute into tubes or flasks. Autoclave for 10 min at 15 psi pressure–121°C.

Use: For the microbiological assaying of folic acid using *Enterococcus hirae* as the test organism.

Folic Acid Assay Medium

Composition per liter:

Glucose	20.3g
Lactose	20.0g
Sodium acetate	10.0g
Vitamin assay casamino acids	5.0g
K_2HPO_4	2.5g
Polysorbate 80 (Tween™ 80)	2.0g
L-Alanine	0.4g
L-Asparagine	0.2g
$MgSO_4 \cdot 7H_2O$	0.2g
Mercaptoacetic acid	0.125g
L-Tryptophan	0.1g
$MnSO_4 \cdot H_2O$	0.08g
Adenine	0.04g
Guanine	0.04g
Uracil	0.04g
$(NH_4)_2SO_4 \cdot FeSO_4 \cdot 6H_2O$	0.03g
Glutathione	5.0mg
Pyridoxine·HCl	4.0mg
p-Aminobenzoic acid	2.0mg

Riboflavin 1.0mg
Calcium pantothenate 0.8mg
Niacin 0.8mg
Thiamine·HCl 0.4mg
Biotin 0.02mg

Source: This medium is available as a premixed powder from BD Diagnostic Systems.

Preparation of Medium: Add components to distilled/deionized water and bring volume to 1.0L. Mix thoroughly. Gently heat and bring to boiling. Distribute into tubes or flasks. Autoclave for 15 min at 0 psi pressure–100°C.

Use: For the determination of the folic acid content of pharmaceutical products and other materials based upon the extent of growth of the folic acid-requiring *Lactobacillus* test species.

Folic Acid Casei HiVeg Medium

Composition per liter:

Glucose 40.0g
Sodium acetate 40.0g
Vitamin free casein acid hydrolysate 10.0g
K_2HPO_4 .... 1.0g
KH_2PO_4 .... 1.0g
L-Asparagine 0.6g
L-Cysteine·HCl 0.5g
$MgSO_4$ 0.4g
DL-Tryptophan 0.2g
Sorbitan monooleate complex 0.1g
$FeSO_4$ 0.02g
NaCl 0.02g
Xanthine 0.02g
$MnSO_4$ 0.015g
Adenine sulfate 0.01g
Uracil 0.01g
Guanine hydrochloride 0.01g
Glutathione, reduced 5.0mg
Pyridoxine hydrochloride 4.0mg
p-Aminobenzoic acid (PABA) 2.0mg
Riboflavin 1.0mg
Calcium pantothenate 0.8mg
Nicotinic acid 0.8mg
Thiamine hydrochloride 0.4mg
Biotin 0.02mg

pH 6.7 ± 0.1 at 25°C

Source: This medium is available as a premixed powder from HiMedia.

Preparation of Medium: Add components to distilled/deionized water and bring volume to 1.0L. Mix thoroughly. Gently heat and bring to boiling with frequent mixing. Cool to 45°–50°C. Mix thoroughly. Distribute into tubes or flasks. Autoclave for 10 min at 15 psi pressure–121°C.

Use: For assaying concentrations of folic acid, especially folic acid in blood serum, using *Lactobacillus casei* subspecies *rhamnosus* as the test organism. The extent of growth of *Lactobacillus casei* is proportional to the amount of folic acid in the test substance.

Folic Acid Casei Medium

Composition per liter:

Glucose 40.0g
Sodium acetate 40.0g
Pancreatic digest of casein, charcoal treated 10.0g
K_2HPO_4 .... 1.0g
KH_2PO_4 .... 1.0g
L-Asparagine 0.6g
L-Cysteine·HCl·H_2O 0.5g
$MgSO_4 \cdot 7H_2O$ 0.4g
L-Tryptophan 0.2g
Sorbitan monooleate complex 0.1g
$FeSO_4 \cdot 7H_2O$ 0.02g
NaCl 0.02g
Xanthine 0.02g
$MnSO_4$ 0.015g
Adenine sulfate 0.01g
Guanine·HCl 0.01g
Uracil 0.01g
Glutathione, reduced 5.0mg
Pyridoxine·HCl 4.0mg
p-Aminobenzoic acid 2.0mg
Riboflavin 1.0mg
Calcium pantothenate 0.8mg
Niacin 0.8mg
Thiamine·HCl 0.4mg
Biotin 0.02mg

pH 6.7 ± 0.1 at 25°C

Source: This medium is available as a premixed powder from BD Diagnostic Systems.

Preparation of Medium: Add components to distilled/deionized water and bring volume to 1.0L. Mix thoroughly. Gently heat and bring to boiling with frequent mixing. Cool to 45°–50°C. Mix thoroughly. Distribute into tubes or flasks. Autoclave for 10 min at 15 psi pressure–121°C.

Use: For assaying concentrations of folic acid, especially folic acid in blood serum, using *Lactobacillus casei* subspecies *rhamnosus* as the test organism. The extent of growth of *Lactobacillus casei* is proportional to the amount of folic acid in the test substance.

Folic Acid Casei Medium with Chloramphenicol

Composition per liter:

Glucose 40.0g
Sodium acetate 40.0g
Pancreatic digest of casein, charcoal treated 10.0g
K_2HPO_4 .... 1.0g
KH_2PO_4 .... 1.0g
L-Asparagine 0.6g
L-Cysteine·HCl·H_2O 0.5g
$MgSO_4 \cdot 7H_2O$ 0.4g
L-Tryptophan 0.2g
Sorbitan monooleate complex 0.1g
$FeSO_4 \cdot 7H_2O$ 0.02g
NaCl 0.02g
Xanthine 0.02g
$MnSO_4 \cdot 7H_2O$ 0.015g
Adenine sulfate 0.01g
Guanine·HCl 0.01g
Uracil 0.01g
Glutathione, reduced 5.0mg
Pyridoxine·HCl 4.0mg
p-Aminobenzoic acid 2.0mg
Riboflavin 1.0mg
Calcium pantothenate 0.8mg

Niacin	0.8mg
Thiamine·HCl	0.4mg
Biotin	0.02mg
Chloramphenicol solution	10.0mL
Folic acid solution	10.0mL

pH 6.7 ± 0.1 at 25°C

Chloramphenicol Solution:
Composition per 10.0mL:

Chloramphenicol	0.02g

Preparation of Chloramphenicol Solution: Add chloramphenicol to distilled/deionized water and bring volume to 10.0mL. Mix thoroughly. Filter sterilize.

Folic Acid Solution:
Composition per 10.0mL:

Folic acid	0.1μg

Preparation of Folic Acid Solution: Add folic acid to distilled/deionized water and bring volume to 10.0mL. Mix thoroughly. Filter sterilize.

Preparation of Medium: Add components, except chloramphenicol solution and folic acid solution, to distilled/deionized water and bring volume to 980.0mL. Mix thoroughly. Gently heat and bring to boiling. Autoclave for 15 min at 15 psi pressure–121°C. Cool to 45°–50°C. Aseptically add 10.0mL of sterile chloramphenicol solution and 10.0mL of sterile folic acid solution. Mix thoroughly. Aseptically distribute into sterile tubes or flasks.

Use: For the cultivation and maintenance of *Lactobacillus casei*.

Folic Acid Culture Agar

Composition per liter:

Peptonized milk	15.0g
Agar	10.0g
Glucose	10.0g
Yeast extract	5.0g
KH_2PO_4	2.0g
Tomato juice	100.0mL
Polysorbate 80	1.0mL

pH 6.8 ± 0.2 at 25°C

Source: This medium is available from HiMedia.

Preparation of Medium: Add components to distilled/deionized water and bring volume to 1.0L. Mix thoroughly. Gently heat and bring to boiling. Distribute into tubes or flasks. Autoclave for 15 min at 15 psi pressure–121°C. Pour into sterile Petri dishes or leave in tubes.

Use: For the cultivation of *Enterococcus hirae* for use in the folic acid assay.

Folic Acid Culture HiVeg Agar

Composition per liter:

Plant hydrolysate No. 4	15.0g
Agar	10.0g
Glucose	10.0g
Tomato juice	5.0g
Yeast extract	5.0g
KH_2PO_4	2.0g
Polysorbate 80	1.0g

pH 6.7 ± 0.1 at 25°C

Source: This medium is available as a premixed powder from HiMedia.

Preparation of Medium: Add components to distilled/deionized water and bring volume to 1.0L. Mix thoroughly. Gently heat and bring to boiling with frequent mixing. Cool to 45°–50°C. Mix thoroughly. Distribute into tubes or flasks. Autoclave for 5 min at 15 psi pressure–121°C. Pour into sterile Petri dishes.

Use: For the microbiological assaying of folic acid using *Enterococcus hirae* as the test organism.

Folic Acid HiVeg Medium

Composition per liter:

Sodium citrate	52.0g
Glucose	40.0g
Plant acid hydrolysate, vitamin free	10.0g
K_2HPO_4	6.4g
L-Cystine hydrochloride	0.76g
L-Asparagine	0.6g
$MgSO_4$	0.4g
L-Tryptophan	0.2g
Polysorbate 80	0.1g
$FeSO_4$	0.02g
NaCl	0.02g
Xanthine	0.02g
$MnSO_4$	0.02g
Adenine sulfate	0.01g
Guanine hydrochloride	0.01g
Uracil	0.01g
Glutathione	5.2mg
Pyridoxine hydrochloride	4.0mg
p-Aminobenzoic acid (PABA)	1.0mg
Riboflavin	1.0mg
Nicotinic acid	0.8mg
Calcium pantothenate	0.8mg
Thiamine hydrochloride	0.4mg
Biotin	0.02mg

pH 6.8 ± 0.2 at 25°C

Source: This medium is available as a premixed powder from HiMedia.

Preparation of Medium: Add components to distilled/deionized water and bring volume to 1.0L. Mix thoroughly. Gently heat and bring to boiling with frequent mixing. Cool to 45°–50°C. Mix thoroughly. Distribute into tubes or flasks. Autoclave for 10 min at 15 psi pressure–121°C.

Use: For the microbiological assaying of folic acid using *Enterococcus hirae* as the test organism.

Folic Acid Inoculum HiVeg Medium

Composition per liter:

Plant hydrolysate No. 4	15.0g
Glucose	10.0g
Yeast extract	5.0g
KH_2PO_4	2.0g
Polysorbate 80	1.0g
Tomato juice	100 mL

pH 6.8 ± 0.2 at 25°C

Source: This medium is available as a premixed powder from HiMedia.

Preparation of Medium: Add components to distilled/deionized water and bring volume to 1.0L. Mix thoroughly. Gently heat and bring to boiling with frequent mixing. Cool to 45°–50°C. Mix thoroughly.

Distribute into tubes or flasks. Autoclave for 10 min at 15 psi pressure–121°C.

Use: For cultivating *Enterococcus hirae* as the test organism for the microbiological assaying of folic acid.

Folic AOAC Medium

Composition per liter:

Sodium citrate	52.0g
Glucose	40.0g
Vitamin assay casamino acids	10.0g
K_2HPO_4	6.4g
L-Cysteine·HCl	0.76g
L-Asparagine	0.6g
$MgSO_4 \cdot 7H_2O$	0.4g
L-Tryptophan	0.2g
Sorbitan monooleate complex	0.1g
$FeSO_4 \cdot 7H_2O$	0.02g
$MnSO_4 \cdot 7H_2O$	0.02g
NaCl	0.02g
Xanthine	0.02g
Adenine sulfate	0.01g
Guanine·HCl	0.01g
Uracil	0.01g
Glutathione	5.2mg
Pyridoxine·HCl	4.0mg
p-Aminobenzoic acid	1.0mg
Riboflavin	1.0mg
Calcium pantothenate	0.8mg
Nicotinic acid	0.8mg
Thiamine·HCl	0.4mg
Biotin	0.02mg

pH 6.7 ± 0.1 at 25°C

Source: This medium is available as a premixed powder from BD Diagnostic Systems.

Preparation of Medium: Add components to distilled/deionized water and bring volume to 1.0L. Mix thoroughly. Gently heat and bring to boiling with frequent mixing. Cool to 45°–50°C. Mix thoroughly. Distribute into tubes or flasks. Autoclave for 5 min at 15 psi pressure–121°C.

Use: For the microbiological assaying of folic acid using *Enterococcus hirae* as the test organism.

Fomes annosus Isolation Medium No. 1

Composition per liter:

Agar	20.0g
Peptone	5.0g
KH_2PO_4	0.5g
$MgSO_4$	0.25g
Pentachloronitrobenzene (PCNB)	0.19mg
Streptomycin	0.1mg
Ethanol (95% solution)	20.0mL
Lactic acid (50% solution)	2.0mL

Preparation of Medium: Filter sterilize ethanol and lactic acid. Add components, except ethanol and lactic acid, to distilled/deionized water and bring volume to 978.0mL. Mix thoroughly. Gently heat and bring to boiling. Autoclave for 15 min at 15 psi pressure–121°C. Cool to 45°–50°C. Aseptically add sterile ethanol and lactic acid. Mix thoroughly. Pour into sterile Petri dishes or distribute into sterile tubes.

Use: For the cultivation of *Fomes annosus*.

Fomes annosus Isolation Medium No. 2

Composition per liter:

Malt extract	20.0g
Agar	17.0g
Streptomycin sulfate solution	10.0mL
Phenylphenol solution	2.5mL
Lactic acid (50% solution)	1.0mL

Streptomycin Sulfate Solution:
Composition per 10.0mL:

Streptomycin sulfate	0.1g

Preparation of Streptomycin Sulfate Solution: Add streptomycin sulfate to distilled/deionized water and bring volume to 10.0mL. Mix thoroughly. Filter sterilize.

Phenylphenol Solution:
Composition per 20.0mL:

o-Phenylphenol	0.48g
Ethanol (95% solution)	20.0mL

Preparation of Phenylphenol Solution: Add *o*-phenylphenol to 20.0mL of ethanol. Mix thoroughly. Filter sterilize.

Preparation of Medium: Filter sterilize lactic acid. Add components—except lactic acid, phenylphenol solution, and streptomycin sulfate solution—to distilled/deionized water and bring volume to 986.5mL. Mix thoroughly. Gently heat and bring to boiling. Autoclave for 15 min at 15 psi pressure–121°C. Cool to 45°–50°C. Aseptically add sterile lactic acid, phenylphenol solution, and streptomycin sulfate solution. Mix thoroughly. Pour into sterile Petri dishes or distribute into sterile tubes.

Use: For the cultivation of *Fomes annosus*.

Forget Fredette Agar

Composition per liter:

Casein enzymatic hydrolysate	17.0g
Agar	10.0g
NaCl	5.0g
Papaic digest of soybean meal	3.0g
Glucose	2.5g
K_2HPO_4	2.5g
NaN_3	0.5g

pH 7.4 ± 0.2 at 25°C

Source: This medium is available from HiMedia.

Caution: Sodium azide is toxic. Azides also react with metals and disposal must be highly diluted.

Preparation of Medium: Add components to distilled/deionized water and bring volume to 1.0L. Mix thoroughly. Gently heat and bring to boiling. Distribute into tubes or flasks. Autoclave for 15 min at 15 psi pressure–121°C. Pour into sterile Petri dishes or leave in tubes.

Use: For the selective isolation of anaerobic microorganisms.

Formate Fumarate Medium

Composition per 900.0mL:

Agar	20.0g
Sodium fumarate	16.6g
Sodium formate	6.8g
Yeast extract	4.0g
Sodium thioglycolate solution	100.0mL

pH 7.3 ± 0.2 at 25°C

Sodium Thioglycolate Solution:

Composition per 100.0mL:

Sodium thioglycolate	0.5g

Preparation of Sodium Thioglycolate Solution: Add sodium thioglycolate to distilled/deionized water and bring volume to 100.0mL. Mix thoroughly. Filter sterilize. Sparge with 100% N_2 gas. Warm to 50°–55°C.

Preparation of Medium: Add components, except sodium thioglycolate solution, to distilled/deionized water and bring volume to 900.0mL. Mix thoroughly. Gently heat and bring to boiling. Continue boiling for 5 min. Cool to 50°–55°C while sparging with 80% N_2 + 20% CO_2. Anaerobically distribute 4.5mL volumes into anaerobic tubes. Autoclave for 15 min at 15 psi pressure–121°C. Cool to 50°–55°C. Prior to inoculation, add 0.5mL of sterile sodium thioglycolate solution to each tube.

Use: For the cultivation and identification of *Bacteroides gracilis, Bacteroides ureolyticus, Campylobacter* species, and *Wolinella* species.

Formate Ricinoleate Broth

Composition per liter:

Pancreatic digest of gelatin	5.0g
Lactose	5.0g
Sodium formate	5.0g
Sodium ricinoleate	1.0g

pH 7.4 ± 0.2 at 25°C

Source: This medium is available from HiMedia.

Preparation of Medium: Add components to distilled/deionized water and bring volume to 1.0L. Mix thoroughly. Gently heat and bring to boiling. Distribute into tubes with inverted Durham tubes. Autoclave for 15 min at 15 psi pressure–121°C.

Use: For the detection of coliform bacteria in milk, water, and other material of sanitary importance.

Formivibrio citricus Medium

Composition per liter:

Trisodium citrate	2.94g
$NaHCO_3$	2.5g
$MgCl_2 \cdot 6H_2O$	2.03g
KH_2PO_4	1.36g
NH_4Cl	0.53g
$Na_2S \cdot 9H_2O$	0.24g
$CaCl_2 \cdot 2H_2O$	0.15g
Modified Wolfe's metals solution	10.0mL
Wolfe's vitamin solution	10.0mL

pH 7.5–7.7 at 25°C

Modified Wolfe's Metals Solution:

Composition per liter

$NaWO_4 \cdot 2H_2O$	10.01g
Na_2SeO_3	0.01g
$NiCl_2 \cdot 6H_2O$	0.01g
Wolfe's mineral solution	1.0L

Preparation of Modified Wolfe's Metals Solution: Add components to Wolfe's mineral solution and bring volume to 10.0mL. Mix thoroughly.

Wolfe's Mineral Solution:

Composition per liter

$MgSO_4 \cdot 7H_2O$	3.0g
Nitriloacetic acid	1.5g
NaCl	1.0g
$MnSO_4 \cdot H_2O$	0.5g
$FeSO_4 \cdot 7H_2O$	0.1g
$CoCl_2 \cdot 6H_2O$	0.1g
$CaCl_2$	0.1g
$ZnSO_4 \cdot 7H_2O$	0.1g
$CuSO_4 \cdot 5H_2O$	0.01g
$AlK(SO_4)_2 \cdot 12H_2O$	0.01g
H_3BO_3	0.01g
$Na_2MoO_4 \cdot 2H_2O$	0.01g

Preparation of Wolfe's Mineral Solution: Add nitrilotriacetic acid to 500.0mL of distilled/deionized water. Dissolve by adjusting pH to 6.5 with KOH. Add remaining components. Add distilled/deionized water to 1.0L.

Wolfe's Vitamin Solution:

Composition per liter:

Pyridoxine·HCl	0.01g
Thiamine·HCl	5.0mg
Riboflavin	5.0mg
Nicotinic acid	5.0mg
Calcium pantothenate	5.0mg
p-Aminobenzoic acid	5.0mg
Thioctic acid	5.0mg
Biotin	2.0mg
Folic acid	2.0mg
Cyanocobalamin	0.1mg

Preparation of Wolfe's Vitamin Solution: Add components to distilled/deionized water and bring volume to 1.0L. Mix thoroughly.

Preparation of Medium: Add components, except $NaHCO_3$ and sodium sulfide, to distilled/deionized water and bring volume to 1.0L. Mix thoroughly. Gently heat and bring to boiling. Cool under an atmosphere of 90% N_2 + 10% CO_2. Add bicarbonate and sodium sulfide. Mix thoroughly. Adjust pH to 7.5–7.7. Anaerobically distribute into tubes under 90% N_2 + 10% CO_2. Cap tubes with rubber stoppers. Place tubes in a press. Autoclave at 121°C for 15 min with fast exhaust.

Use: For the cultivation and maintenance of *Formivibrio citricus*.

Fowells Acetate Agar

Composition per liter:

Agar	20.0g
Sodium acetate trihydrate	5.0g

pH 6.5–7.0 at 25°C

Preparation of Medium: Add sodium acetate trihydrate to distilled/deionized water and bring volume to 1.0L. Adjust pH to 6.5–7.0. Add agar. Mix thoroughly. Gently heat and bring to boiling. Distribute into tubes or flasks. Autoclave for 15 min at 15 psi pressure–121°C. Pour into sterile Petri dishes or leave in tubes.

Use: For the cultivation of fungi with the production of ascospores.

FPA Medium

See: **Fluorescent Pectolytic Agar**

FPA Medium

Composition per liter:

Proteose peptone No. 3	20.0g
Agar	15.0g
Pectin, citrus	5.0g
K_2HPO_4	1.5g

$MgSO_4 \cdot 7H_2O$ 1.5g
Antibiotic solution 10.0mL

pH 7.0 ± 0.2 at 25°C

Antibiotic Solution:
Composition per 10.0mL:
Cycloheximide 0.075g
Novobiocin 0.045g
Penicillin G 75,000U

Preparation of Antibiotic Solution: Add components to distilled/deionized water and bring volume to 10.0mL. Mix thoroughly. Filter sterilize.

Caution: Cycloheximide is toxic. Avoid skin contact or aerosol formation and inhalation.

Preparation of Medium: Add components, except agar and antibiotic solution, to distilled/deionized water and bring volume to 990.0mL. Mix thoroughly. Adjust pH to 7.0. Add agar. Mix thoroughly. Gently heat and bring to boiling. Autoclave for 15 min at 15 psi pressure–121°C. Cool to 45°–50°C. Aseptically add sterile antibiotic solution. Mix thoroughly. Pour into sterile Petri dishes or distribute into sterile tubes.

Use: For the isolation and cultivation of *Pseudomonas* species that cause soft rot.

FRAG Agar (Fragilis Agar)

Composition per 1025.0mL:
L-Cysteine·HCl·H_2O 0.5g
Basal solution 995.0mL
Glucuronic acid solution 25.0mL
Gentamicin solution 1.0mL
Hemin-vitamin K_1 solution 1.0mL
Ferric sulfate solution 1.0mL
Mineral solution 1.0mL
Phenol Red (1% solution) 1.0mL
Vitamin B_{12} solution 0.05mL

pH 7.0 ± 0.1 at 25°C

Basal Solution:
Composition per 995.0mL:
Oxgall 20.0g
Agar 15.4g
K_2HPO_4 .... 2.26g
Yeast extract 2.0g
Pancreatic digest of casein 1.4g
$(NH_4)_2HPO_4$ 1.0g
K_2HPO_4 .... 0.9g
Papaic digest of soybean meal 0.12g
NaCl 0.12g

Preparation of Basal Solution: Add components to distilled/deionized water and bring volume to 995.0mL. Mix thoroughly. Gently heat and bring to boiling. Autoclave for 15 min at 15 psi pressure–121°C. Cool to 45°–50°C.

Glucuronic Acid Solution:
Composition per 100.0mL:
D-Glucuronic acid 40.0g

Preparation of Glucuronic Acid Solution: Add glucuronic acid to distilled/deionized water and bring volume to 100.0mL. Mix thoroughly. Filter sterilize.

Gentamicin Solution:
Composition per 10.0mL:
Gentamicin 0.1mg

Preparation of Gentamicin Solution: Add gentamicin to distilled/deionized water and bring volume to 10.0mL. Mix thoroughly. Filter sterilize.

Vitamin K_1-Hemin Solution:
Composition per liter:
Vitamin K_1 solution 10.0mL
Hemin solution 10.0mL

Preparation of Vitamin K_1-Hemin Solution: Add components to distilled/deionized water and bring volume to 1.0L. Mix thoroughly. Filter sterilize.

Vitamin K_1 Solution:
Composition per 100.0mL:
Vitamin K_1 .... 1.0g
Ethanol 99.0mL

Preparation of Vitamin K_1 Solution: Add vitamin K_1 to 99.0mL of absolute ethanol. Mix thoroughly.

Hemin Solution:
Composition per 10.0mL:
Hemin 0.5g
NaOH 0.4g

Preparation of Hemin Solution: Add hemin and NaOH to distilled/deionized water and bring volume to 10.0mL. Mix thoroughly.

Ferric Sulfate Solution:
Composition per 100.0mL:
$FeSO_4 \cdot 9H_2O$ 0.04g

Preparation of Ferric Sulfate Solution: Add $FeSO_4 \cdot 9H_2O$ to distilled/deionized water and bring volume to 100.0mL. Mix thoroughly. Filter sterilize.

Mineral Solution:
Composition per 100.0mL:
NaCl 9.0g
$CaCl_2 \cdot 2H_2O$ 0.27g
$MgCl_2 \cdot 6H_2O$ 0.2g
$CoCl_2 \cdot 6H_2O$ 0.1g
$MnCl_2 \cdot 4H_2O$ 0.1g

Preparation of Mineral Solution: Add components to distilled/deionized water and bring volume to 100.0mL. Mix thoroughly. Filter sterilize.

Vitamin B_{12} Solution:
Composition per 10.0mL:
Vitamin B_{12} 0.1mg

Preparation of Vitamin B_{12} Solution: Add vitamin B_{12} to distilled/deionized water and bring volume to 10.0mL. Mix thoroughly. Filter sterilize.

Preparation of Medium: To 995.0mL of cooled, sterile basal solution, aseptically add 0.5g of L-cysteine·HCl·H_2O, 25.0mL of sterile glucuronic acid solution, 1.0mL of sterile gentamicin solution, 1.0mL of sterile hemin-vitamin K_1 solution, 1.0mL of sterile ferric sulfate solution, 1.0mL of sterile mineral solution, 1.0mL of Phenol Red solution, and 0.05mL of sterile vitamin B_{12} solution. Mix thoroughly. Pour into sterile Petri dishes or distribute into sterile tubes.

Use: For the isolation, cultivation, and differentiation of the *Bacteroides fragilis* group (*Bacteroides fragilis, Bacteroides thetaiotamicron,*

Bacteroides vulgatus, Bacteroides distasonis, Bacteriodes ovatus, and *Bacteroides uniformis*) from clinical specimens.

Fragilis Agar
See: **FRAG Agar**

Francisella tularensis Isolation Medium

Composition per liter:

Agar	10.0g
Glucose	10.0g
Pancreatic digest of casein	10.0g
Peptic digest of animal tissue	10.0g
L-Cysteine·HCl·H_2O	5.0g
NaCl	5.0g
Sodium thioglycolate	2.0g
Glucose	1.0g
Thiamine·HCl	5.0mg

pH 7.2 ± 0.2 at 25°C

Preparation of Medium: Add components, except agar, to distilled/deionized water and bring volume to 1.0L. Mix thoroughly. Adjust pH to 7.2. Add agar. Gently heat and bring to boiling. Autoclave for 20 min at 15 psi pressure–121°C. Cool to 45°–50°C. Pour into sterile Petri dishes.

Use: For the isolation and cultivation of *Francisella tularensis.*

Frankia Agar

Composition per liter:

Starch, soluble	20.0g
Agar	15.0g
Glucose	10.0g
N-Z-Amine™	5.0g
Yeast extract	5.0g
$CaCO_3$	1.0g

pH 7.0 ± 0.2 at 25°C

Preparation of Medium: Add components to distilled/deionized water and bring volume to 1.0L. Mix thoroughly. Gently heat and bring to boiling. Adjust pH to 7.0 with NaOH. Distribute into tubes or flasks. Autoclave for 15 min at 15 psi pressure–121°C. Pour into sterile Petri dishes or leave in tubes.

Use: For the cultivation and maintenance of *Frankia alni.*

Frankia Isolation Medium

Composition per liter:

Sucrose	40.0g
$Ca(NO_3)_2·4H_2O$	0.242g
KNO_3	0.085g
KCl	0.061g
$MgSO_4·7H_2O$	0.042g
KH_2PO_4	0.02g
$MnSO_4·H_2O$	4.5mg
$FeCl_3·6H_2O$	2.5mg
H_3BO_3	1.5mg
$ZnSO_4·7H_2O$	1.5mg
Nicotinic acid	0.5mg
Pyridoxine·HCl	0.5mg
$Na_2MoO_4·2H_2O$	0.25mg
Thiamine·HCl	0.1mg
$CuSO_4·5H_2O$	0.04mg
Mannitol solution	10.0mL
Supplement solution	10.0mL

pH 5.5 ± 0.2 at 25°C

Mannitol Solution:
Composition per 100.0mL:

Mannitol	11.84g

Preparation of Mannitol Solution: Add mannitol to distilled/deionized water and bring volume to 100.0mL. Mix thoroughly. Filter sterilize.

Supplement Solution:
Composition per 10.0mL:

L-Glutamic acid	0.185g
L-Arginine	0.174g
L-Glutamine	0.146g
L-Aspartic acid	0.133g
L-Asparagine	0.132g
Glycine	0.075g
Urea	0.06g
Naphthaloneacetic acid	2.0mg
Zeatin	1.0μg

Preparation of Supplement Solution: Add components to distilled/deionized water and bring volume to 10.0mL. Mix thoroughly. Filter sterilize.

Preparation of Medium: Add components, except mannitol solution and supplement solution, to distilled/deionized water and bring volume to 980.0mL. Mix thoroughly. Adjust pH to 5.5. Gently heat and bring to boiling. Autoclave for 15 min at 15 psi pressure–121°C. Cool to 45°–50°C. Aseptically add 10.0mL of sterile mannitol solution and 10.0mL of sterile supplement solution. Mix thoroughly. Aseptically distribute into sterile tubes or flasks.

Use: For the isolation and cultivation of *Frankia* species from root nodules.

Frankia Medium

Composition per liter:

Sucrose	20.0g
Agar	10.0g
Edamin	1.0g
Mannitol	1.0g
$CaCO_3$	0.5g
K_2HPO_4	0.5g
Yeast extract	0.5g
$MgCl_2·7H_2O$	0.2g
NaCl	0.1g
H_3BO_3	0.1g
$MnSO_4·H_2O$	0.025g
$ZnSO_4·7H_2O$	0.01g
Nicotinic acid	0.5mg
$Na_2MoO_4·2H_2O$	0.25mg
Pyridoxine·HCl	0.1mg
Thiamine·HCl	0.1mg
$CuSO_4·5H_2O$	0.025mg

pH 7.0 ± 0.2 at 25°C

Preparation of Medium: Add components to distilled/deionized water and bring volume to 1.0L. Mix thoroughly. Gently heat and bring to boiling. Adjust pH to 7.0. Distribute into tubes or flasks. Autoclave for 15 min at 15 psi pressure–121°C. Pour into sterile Petri dishes or leave in tubes.

Use: For the cultivation of *Frankia* species from root nodules.

Frankia Medium

Composition per 1200.0mL:

Na_2HPO_4	1.0g
NaCl	1.0g
KH_2PO_4	0.3g
Glycerol	1.0mL
NH_4Cl	1.0mL
Thiamine	1.0mL
Albumin fatty acid supplement	200.0mL

pH 7.4 ± 0.2 at 25°C

Albumin Fatty Acid Supplement:

Composition per 200.0mL:

Bovine albumin fraction V	20.0g
Tween™ 80	25.0mL
$FeSO_4 \cdot 7H_2O$	20.0mL
$CaCl_2 \cdot 2H_2O$	2.0mL
$MgCl_2 \cdot 6H_2O$	2.0mL
Vitamin B_{12}	2.0mL
$ZnSO_4 \cdot 7H_2O$	2.0mL
$CuSO_4 \cdot 5H_2O$	0.2mL

Preparation of Albumin Fatty Acid Supplement: Add components to distilled/deionized water and bring volume to 200.0mL. Mix thoroughly. Filter sterilize.

Preparation of Medium: Add components, except albumin fatty acid supplement, to distilled/deionized water and bring volume to 800.0mL. Mix thoroughly. Gently heat and bring to boiling. Autoclave for 15 min at 15 psi pressure–121°C. Cool to 45°–50°C. Aseptically add 200.0mL of sterile albumin fatty acid supplement. Mix thoroughly. Aseptically distribute into sterile tubes or flasks.

Use: For the cultivation and maintenance of *Frankia* species.

Frankia Medium

Composition per liter:

Basal medium	900.0mL
Albumin fatty acid supplement	100.0mL

pH 7.4 ± 0.2 at 25°C

Basal Medium:

Composition per liter:

Na_2HPO_4	1.0g
NaCl	1.0g
KH_2PO_4	0.3g
Glycerol (10% solution)	1.0mL
NH_4Cl (25% solution)	1.0mL
Thiamine (0.5% solution)	1.0mL

Preparation of Medium: Add components to distilled/deionized water and bring volume to 1.0L. Mix thoroughly. Adjust pH to 7.4. Distribute into tubes or flasks. Autoclave for 15 min at 15 psi pressure–121°C.

Albumin Fatty Acid Supplement:

Composition per 200.0mL:

Bovine albumin fraction V	20.0g
Tween™ 80 (10% solution)	25.0mL
$FeSO_4 \cdot 7H_2O$ (0.5% solution)	20.0mL
$CaCl_2 \cdot 2H_2O$ (1% solution)	2.0mL
$MgCl_2 \cdot 6H_2O$ (1% solution)	2.0mL
Vitamin B_{12} (0.2% solution)	2.0mL
$ZnSO_4 \cdot 7H_2O$ (0.4% solution)	2.0mL
$CuSO_4 \cdot 5H_2O$ (0.3% solution)	0.2mL

Preparation of Albumin Fatty Acid Supplement: Add albumin to 100.0mL of distilled/deionized water. Mix thoroughly. Slowly add Tween™ 80 and then other components. Adjust pH to 7.4. Add distilled/deionized water and bring volume to 200.0mL. Mix thoroughly. Autoclave for 15 min at 15 psi pressure–121°C.

Preparation of Medium: Aseptically combine 900.0mL of sterile basal medium and 100.0mL of sterile albumin fatty acid supplement. Mix thoroughly. Aseptically distribute into sterile tubes or flasks.

Use: For the cultivation and maintenance of *Frankia* species.

Fraser Broth

Composition per liter:

NaCl	20.0g
Na_2HPO_4	12.0g
Beef extract	5.0g
Proteose peptone	5.0g
Pancreatic digest of casein	5.0g
Yeast extract	5.0g
LiCl	3.0g
KH_2PO_4	1.35g
Esculin	1.0g
Fraser supplement solution	10.0mL

pH 7.2 ± 0.2 at 25°C

Source: This medium is available as a premixed powder from BD Diagnostic Systems.

Fraser Supplement Solution:

Composition per 10.0mL:

Ferric ammonium citrate	0.5g
Acriflavine·HCl	0.25g
Nalidixic acid	0.1g
Ethanol	5.0mL

Preparation of Fraser Supplement Solution: Add components to distilled/deionized water and bring volume to 10.0mL. Mix thoroughly. Filter sterilize.

Preparation of Medium: Add components, except Fraser supplement solution, to distilled/deionized water and bring volume to 990.0mL. Mix thoroughly. Gently heat and bring to boiling. Autoclave for 15 min at 15 psi pressure–121°C. Cool to 45°–50°C. Aseptically add sterile Fraser supplement solution. Mix thoroughly. Aseptically distribute into sterile tubes or flasks.

Use: For the isolation of *Listeria* species from food and environmental species.

Fraser Broth

Composition per liter:

NaCl	20.0g
Na_2HPO_4	12.0g
Proteose peptone	5.0g
Tryptone	5.0g
Lab Lemco powder	5.0g
LiCl	3.0g
KH_2PO_4	1.35g
Esculin	1.0g
Fraser supplement solution	10.0mL

pH 7.2 ± 0.2 at 25°C

Source: This medium is available as a premixed powder from Oxoid Unipath.

Fraser Supplement Solution:
Composition per 10.0mL:

Ferric ammonium citrate	0.5g
Acriflavine·HCl	0.25g
Nalidixic acid	0.1g
Ethanol	5.0mL

Preparation of Fraser Supplement Solution: Add components to distilled/deionized water and bring volume to 10.0mL. Mix thoroughly. Filter sterilize.

Preparation of Medium: Add components, except Fraser supplement solution, to distilled/deionized water and bring volume to 990.0mL. Mix thoroughly. Gently heat and bring to boiling. Autoclave for 15 min at 15 psi pressure–121°C. Cool to 45°–50°C. Aseptically add sterile Fraser supplement solution. Mix thoroughly. Aseptically distribute into sterile tubes or flasks.

Use: For the isolation of *Listeria* species from food and environmental species.

Fraser Broth, Half
See: Half Fraser Broth

Fraser HiVeg Broth Base

Composition per liter:

NaCl	20.0g
$Na_2HPO_4 \cdot 2H_2O$	12.0g
Plant extract No. 1	5.0g
Plant hydrolysate	5.0g
Plant peptone	5.0g
Yeast extract	5.0g
LiCl	3.0g
KH_2PO_4	1.35g
Esculin	1.0g
Fraser supplement solution	10.0mL

pH 7.2 ± 0.2 at 25°C

Source: This medium, without Fraser supplement solution, is available as a premixed powder from HiMedia.

Caution: Lithium chloride is harmful. Avoid bodily contact and inhalation of vapors. On contact with skin wash with plenty of water immediately.

Fraser Supplement Solution:
Composition per 10.0mL:

Ferric ammonium citrate	0.5g
Acriflavine·HCl	0.25g
Nalidixic acid	0.1g
Ethanol	5.0mL

Preparation of Fraser Supplement Solution: Add components to distilled/deionized water and bring volume to 10.0mL. Mix thoroughly. Filter sterilize.

Preparation of Medium: Add components, except Fraser supplement solution, to distilled/deionized water and bring volume to 990.0mL. Mix thoroughly. Gently heat and bring to boiling. Autoclave for 15 min at 15 psi pressure–121°C. Cool to 45°–50°C. Aseptically add sterile Fraser supplement solution. Mix thoroughly. Aseptically distribute into sterile tubes or flasks.

Use: For the isolation of *Listeria* species from food and environmental species.

Fraser Secondary Enrichment Broth

Composition per liter:

NaCl	20.0g
Na_2HPO_4	12.0g
Beef extract	5.0g
Proteose peptone	5.0g
Pancreatic digest of casein	5.0g
Yeast extract	5.0g
LiCl	3.0g
KH_2PO_4	1.35g
Esculin	1.0g
Acriflavin solution	10.0mL
Ferric ammonium citrate solution	10.0mL
Nalidixic acid solution	1.0mL

Ferric Ammonium Citrate Solution:
Composition per 10.0mL:

Ferric ammonium citrate	0.5g

Preparation of Ferric Ammonium Citrate Solution: Add ferric ammonium citrate to distilled/deionized water and bring volume to 10.0mL. Mix thoroughly. Filter sterilize.

Acriflavin Solution:
Composition per 10.0mL:

Acriflavin	0.025g

Preparation of Acriflavin Solution: Add acriflavin to distilled/deionized water and bring volume to 10.0mL. Mix thoroughly. Filter sterilize.

Nalidixic Acid Solution:
Composition per 10.0mL:

Nalidixic acid	0.04g
NaOH (0.1*N* solution)	10.0mL

Preparation of Nalidixic Acid Solution: Add nalidixic acid to 10.0mL of NaOH solution. Mix thoroughly. Filter sterilize.

Preparation of Medium: Add components, except acriflavin solution and ferric ammonium citrate solution, to distilled/deionized water and bring volume to 980.0mL. Mix thoroughly. Gently heat and bring to boiling. Distribute into tubes in 10.0mL volumes. Autoclave for 12 min at 15 psi pressure–121°C. Cool rapidly to 25°C. Immediately prior to inoculation, aseptically add 0.1mL of sterile acriflavin solution and 0.1mL of ferric ammonium citrate solution to each tube. Mix thoroughly.

Use: For the isolation, cultivation, and enrichment of *Listeria monocytogenes* from foods and environmental specimens based on esculin hydrolysis. Bacteria that hydrolyze esculin appear as black colonies.

Fraser Secondary Enrichment HiVeg Broth Base

Composition per liter:

NaCl	20.0g
Na_2HPO_4	12.0g
Plant extract	5.0g
Plant hydrolysate	5.0g
Plant peptone No. 3	5.0g
Yeast extract	5.0g
LiCl	3.0g
KH_2PO_4	1.35g
Esculin	1.0g

Ferric ammonium citrate	0.5g
Acriflavin solution	10.0mL
Ferric ammonium citrate solution	10.0mL
Nalidixic acid solution	1.0mL

pH 7.2 ± 0.2 at 25°C

Source: This medium, without acriflavin, ferric ammonim citrate, nad nalidixic acid solutions, is available as a premixed powder from HiMedia.

Caution: Lithium chloride is harmful. Avoid bodily contact and inhalation of vapors. On contact with skin wash with plenty of water immediately.

Ferric Ammonium Citrate Solution:
Composition per 10.0mL:

Ferric ammonium citrate	0.5g

Preparation of Ferric Ammonium Citrate Solution: Add ferric ammonium citrate to distilled/deionized water and bring volume to 10.0mL. Mix thoroughly. Filter sterilize.

Acriflavin Solution:
Composition per 10.0mL:

Acriflavin	0.025g

Preparation of Acriflavin Solution: Add acriflavin to distilled/deionized water and bring volume to 10.0mL. Mix thoroughly. Filter sterilize.

Nalidixic Acid Solution:
Composition per 10.0mL:

Nalidixic acid	0.04g
NaOH (0.1*N* solution)	10.0mL

Preparation of Nalidixic Acid Solution: Add nalidixic acid to 10.0mL of NaOH solution. Mix thoroughly. Filter sterilize.

Preparation of Medium: Add components, except acriflavin solution and ferric ammonium citrate solution, to distilled/deionized water and bring volume to 980.0mL. Mix thoroughly. Gently heat and bring to boiling. Distribute into tubes in 10.0mL volumes. Autoclave for 12 min at 15 psi pressure–121°C. Cool rapidly to 25°C. Immediately prior to inoculation, aseptically add 0.1mL of sterile acriflavin solution and 0.1mL of ferric ammonium citrate solution to each tube. Mix thoroughly.

Use: For the isolation, cultivation, and enrichment of *Listeria monocytogenes* from foods and environmental specimens based on esculin hydrolysis. Bacteria that hydrolyze esculin appear as black colonies.

Freezing Agar

Composition per liter:

Potatoes, unpeeled	200.0g
Agar	20.0g
Glucose	8.0g
Yeast extract	1.0g
Activated carbon	0.5g

Preparation of Medium: Thinly slice potatoes. Add to 500.0mL of tap water. Autoclave for 15 min at 15 psi pressure–121°C. Mash the potatoes in their liquid. Filter solids through cheesecloth. Add agar, glucose, yeast extract, and activated carbon to filtrate. Mix thoroughly. Bring volume to 1.0L with distilled/deionized water. Gently heat and bring to boiling. Distribute into tubes or flasks. Autoclave for 15 min at 15 psi pressure–121°C. Pour into sterile Petri dishes or leave in tubes.

Use: For the cultivation and maintenance of *Amauroascus species, Aphanoascus cinnabarinus, Arachniotus flavoluteus, Arachniotus hebridensis, Arachnotheca albicans, Auxarthron thaxteri, Auxarthron zuffianum, Dichotomomyces cejpii, Disarticulatus devroeyi, Disarticulatus indicus, Eleutherascus lectardii, Epidermophyton floccosum, Geomyces pannorus, Gymnascella citrina, Gymnoascoideus petalosporus, Gymnoascus intermedius, Kuehniella racovitzae, Lasiobolidium orbiculoides, Macronodus bifurcatus, Microascus trigonosporus, Onygena corvina, Onygena equina, Pectinotrichum llanense, Plunkettomyces littoralis, Preussia typharum, Pseudeurotuim desertorum, Pseudeurotuim ovalis, Pseudoarachniotus roseus, Pseudoarachniotus ruber, Pseudoarachniotus trochleosporus, Rollandina capitata, Rollandina hyalinospora,* and *Shanorella spirotricha.*

Freezing Medium

Composition per 11.5mL:

Bolton broth base	9.5mL
Fetal bovine serum	1.0mL
Glycerol solution	1.0mL

Bolton Broth Base:
Composition per liter:

Peptone	10.0g
Lactalbumin hydrolysate	5.0g
Yeast extract	5.0g
NaCl	5.0g
α-Ketoglutarate	1.0g
Na-pyruvate	0.5g
Na-metabisulfite	0.5g
Na_2CO_3	0.6g
Hemin	0.01g

Preparation of Bolton Broth Base: Add components to distilled/deionized water and bring volume to 1.0L. Mix thoroughly. Autoclave for 15 min at 15 psi pressure–121°C. Cool to room temperature.

Glycerol Solution:
Composition per 100.0mL:

Glycerol	10.0g

Preparation of Glycerol Solution: Add glycerol to distilled/deionized water and bring volume to 100.0mL. Mix thoroughly. Autoclave for 15 min at 15 psi pressure–121°C. Cool to room temperature.

Preparation of Medium: Aseptically combine 9.5mL Bolton broth base, 1.0mL sterile glycerol solution, and 1.0mL filter sterilized fetal bovine serum. Mix thoroughly.

Use: For the preservation of *Campylobacter* spp.

Freshwater *Amoeba* Medium

Composition per liter:

Agar	10.0g
Malt extract	0.1g
Yeast extract	0.1g

Preparation of Medium: Add components to distilled/deionized water and bring volume to 1.0L. Mix thoroughly. Gently heat and bring to boiling. Distribute into tubes or flasks. Autoclave for 15 min at 15 psi pressure–121°C. Pour into sterile Petri dishes or leave in tubes.

Use: For the cultivation of *Acanthamoeba astronyxis, Acanthamoeba castellanii, Acanthamoeba griffini, Acanthamoeba pearcei, Acanthamoeba polyphaga, Acanthamoeba rhysodes, Acanthamoeba stevensoni, Acanthamoeba tubiashi, Capsellina* species, *Cochliopodium actinophora, Cochliopodium bilimbosum, Hartmannella limax, Hartmannella vermiformis, Naegleria fowleri, Naegleria gruberi, Naegleria lovaniensis,*

Naegleria minor, Naegleria thorntoni, Paraflabellula reniformis, Protacanthamoeba caledonica, Rosculus species, *Saccamoeba limax, Tetramitus rostratus, Vahlkampfia inornata*, and *Vannella miroides*.

Frey *Mycoplasma* Broth Base

Composition per liter:

Casein enzymic hydrolysate	7.5g
Yeast extract	5.0g
NaCl	5.0g
Papaic digest of soybean meal	2.5g
Na_2HPO_4	1.6g
KCl	0.4g
$MgSO_4 \cdot 7H_2O$	0.2g
KH_2PO_4	0.1g
Horse serum, sterile inactivated	100.0mL

pH 7.7 ± 0.2 at 25°C

Source: This medium is available from HiMedia.

Preparation of Medium: Add components, except horse serum, to distilled/deionized water and bring volume to 900.0mL. Mix thoroughly. Autoclave for 15 min at 15 psi pressure–121°C. Cool to room temperature. Aseptically add horse serum. Mix thoroughly. Aseptically distribute into sterile tubes.

Use: For the cultivation of avian *Mycoplasma* spp.

Friis Medium (DSMZ Medium 1078)

Composition per 162.0mL:

Beef heart, infusion from	0.7g
Pancreatic digest of gelatin	0.32g
Peptone	0.27g
Brain heart, solids from infusion	0.13g
NaCl	0.1g
Na_2HPO_4	0.08g
Glucose	0.07g
NaCl	0.07g
Hank's balance salt solution (BSS)	50.0mL
Porcine serum, heat inactivated	30.0mL
Yeast extract solution	3.32mL
Phenol Red solution	0.133mL

pH 7.4 ± 0.2 at 25°C

Hanks' Balanced Salt Solution:

Composition per liter:

Solution A	25.0mL
Solution B	25.0mL

Solution A:

Composition per 25.0mL:

Na_2Cl	0.8g
KCl	0.4g
$CaCl_2$	1.4g
$MgCl_2 \cdot 6H_2O$	0.185g
$MgSO_4 \cdot 7H_2O$	0.1g

Preparation of Solution A: Add components to 25.0mL distilled/deionized water. Autoclave for 15 min at 15 psi pressure–121°C.

Solution B:

Composition per 25.0mL:

$Na_2HPO_4 \cdot 7H_2O$	0.15g
KH_2PO_4	0.06g

Preparation of Solution B: Add components to 25.0mL distilled/deionized water. Autoclave for 15 min at 15 psi pressure–121°C.

Preparation of Hanks' Balanced Salt Solution: Add solutions A and B to sterile distilled/deionized water and bring volume to 1.0L. Mix thoroughly.

Phenol Red Solution:

Composition per 15.0mL:

Phenol Red	0.1g

Preparation of Phenol Red Solution: Add components to 15.0mL distilled/deionized water. Filter sterilize.

Yeast Extract Solution:

Composition per 15.0mL:

Yeast extract	2.5g

Preparation of Yeast Extract Solution: Add components to 15.0mL distilled/deionized water. Autoclave for 15 min at 15 psi pressure–121°C.

Preparation of Medium: Add components, except porcine serum, yeast extract solution, and Phenol Red solution, to distilled/deionized water and bring volume to 78.0mL. Mix thoroughly. Adjust pH to 7.4. Gently heat while stirring and bring to boiling. Boil for 1 min. Autoclave for 15 min at 15 psi pressure–121°C. Cool to room temperature. Aseptically add porcine serum, yeast extract solution, and Phenol Red solution. Bring volume to 162.0mL with distilled/deionized water.

Use: For the cultivation of *Mycoplasma* spp., e.g., *Mycoplasma hyorhinis*.

Fructose Mineral Medium

Composition per liter:

Solution A	500.0mL
Solution B	300.0mL
Solution C	200.0mL

Solution A:

Composition per 500.0mL:

$Na_2HPO_4 \cdot 12H_2O$	9.0g
K_2PO_4	1.5g
NH_4Cl	1.0g
$MgSO_4 \cdot 7H_2O$	0.2g
Trace elements solution SL-6	1.0mL

Trace Elements Solution SL-6:

Composition per liter:

$MnCl_2 \cdot 4H_2O$	0.5g
H_3BO_3	0.3g
$CoCl_2 \cdot 6H_2O$	0.2g
$ZnSO_4 \cdot 7H_2O$	0.1g
$Na_2MoO_4 \cdot 2H_2O$	0.03g
$NiCl_2 \cdot 6H_2O$	0.02g
$CuCl_2 \cdot 2H_2O$	0.01g

Preparation of Trace Elements Solution SL-6: Add components to distilled/deionized water and bring volume to 1.0L. Mix thoroughly.

Preparation of Solution A: Add components to distilled/deionized water and bring volume to 500.0mL. Mix thoroughly. Autoclave for 15 min at 15 psi pressure–121°C.

Solution B:

Composition per 300.0mL:

$CaCl_2 \cdot 2H_2O$	0.01g
Ferric ammonium citrate	0.005g

Preparation of Solution B: Add components to distilled/deionized water and bring volume to 500.0mL. Mix thoroughly. Autoclave for 15 min at 15 psi pressure–121°C.

Solution C:
Composition per 200.0mL:

Fructose	5.0g

Preparation of Solution C: Add fructose to distilled/deionized water and bring volume to 200.0mL. Mix thoroughly. Filter sterilize.

Preparation of Medium: To 500.0mL of sterile solution A, aseptically add 300.0mL of sterile solution B and 200.0mL of sterile solution C. Mix thoroughly. Aseptically dispense into sterile tubes or flasks.

Use: For the general cultivation of a wide variety of bacteria.

FSM Selective Medium

Composition per liter:

Agar	18.0g
Peptone	10.0g
Glucose	4.0g
$(NH_4)_2SO_4$	1.32g
K_2HPO_4	1.18g
Casamino acids	1.0g
Yeast extract	1.0g
KH_2PO_4	0.44g
$MgSO_4 \cdot 7H_2O$	0.2g
$FeC_6H_5O_7 \cdot 5H_2O$	3.0mg
Citric acid	1.9mg
$ZnSO_4 \cdot 7H_2O$	1.6mg
$MnSO_4 \cdot H_2O$	1.5mg
2,3,5-Triphenyltetrazolium·HCl solution	10.0mL
Benomyl solution	10.0mL
Polymyxin B solution	10.0mL
Chloroneb solution	10.0mL
Dichloran solution	10.0mL
Bacitracin solution	10.0mL
Cycloheximide solution	10.0mL
Pentachloronitrobenzene solution	10.0mL
Pimaricin solution	10.0mL
Tyrothricin solution	10.0mL
Vancomycin solution	10.0mL
Chloromycetin solution	10.0mL
Penicillin G solution	10.0mL

2,3,5-Triphenyltetrazolium·HCl Solution:
Composition per 10.0mL:

2,3,5-Triphenyltetrazolium·HCl	0.5mg

Preparation of 2,3,5-Triphenyltetrazolium·HCl Solution: Add 2,3,5-triphenyltetrazolium·HCl to distilled/deionized water and bring volume to 10.0mL. Mix thoroughly. Autoclave for 7 min at 15 psi pressure–121°C.

Benomyl Solution:
Composition per 10.0mL:

Benomyl	0.5mg

Preparation of Benomyl Solution: Add benomyl to distilled/deionized water and bring volume to 10.0mL. Mix thoroughly. Filter sterilize.

Polymyxin B Solution:
Composition per 10.0mL:

Polymyxin B	0.1mg

Preparation of Polymyxin B Solution: Add polymyxin B to distilled/deionized water and bring volume to 10.0mL. Mix thoroughly. Filter sterilize.

Chloroneb Solution:
Composition per 10.0mL:

Chloroneb	0.1mg

Preparation of Chloroneb Solution: Add chloroneb to distilled/deionized water and bring volume to 10.0mL. Mix thoroughly. Filter sterilize.

Dichloran Solution:
Composition per 10.0mL:

Dichloran	0.1mg

Preparation of Dichloran Solution: Add dichloran to distilled/deionized water and bring volume to 10.0mL. Mix thoroughly. Filter sterilize.

Bacitracin Solution:
Composition per 10.0mL:

Bacitracin	0.05mg

Preparation of Bacitracin Solution: Add bacitracin to distilled/deionized water and bring volume to 10.0mL. Mix thoroughly. Filter sterilize.

Cycloheximide Solution:
Composition per 10.0mL:

Cycloheximide	0.05mg

Preparation of Cycloheximide Solution: Add cycloheximide to distilled/deionized water and bring volume to 10.0mL. Mix thoroughly. Filter sterilize.

Caution: Cycloheximide is toxic. Avoid skin contact or aerosol formation and inhalation.

Pentachloronitrobenzene Solution:
Composition per 10.0mL:

Pentachloronitrobenzene	0.03mg

Preparation of Pentachloronitrobenzene Solution: Add pentachloronitrobenzene to distilled/deionized water and bring volume to 10.0mL. Mix thoroughly. Filter sterilize.

Pimaricin Solution:
Composition per 10.0mL:

Pimaricin	0.02mg

Preparation of Pimaricin Solution: Add pimaricin to distilled/deionized water and bring volume to 10.0mL. Mix thoroughly. Filter sterilize.

Tyrothricin Solution:
Composition per 10.0mL:

Tyrothricin	0.02mg

Preparation of Tyrothricin Solution: Add tyrothricin to distilled/deionized water and bring volume to 10.0mL. Mix thoroughly. Filter sterilize.

Vancomycin Solution:
Composition per 10.0mL:

Vancomycin	0.01mg

Preparation of Vancomycin Solution: Add vancomycin to distilled/deionized water and bring volume to 10.0mL. Mix thoroughly. Filter sterilize.

Chloromycetin Solution:
Composition per 10.0mL:
Chloromycetin 5.0μg

Preparation of Chloromycetin Solution: Add chloromycetin to distilled/deionized water and bring volume to 10.0mL. Mix thoroughly. Filter sterilize.

Penicillin G Solution:
Composition per 10.0mL:
Penicillin G 1.0μg

Preparation of Penicillin G Solution: Add penicillin G to distilled/deionized water and bring volume to 10.0mL. Mix thoroughly. Filter sterilize.

Preparation of Medium: Add components—except 2,3,5-triphenyltetrazolium chloride solution, benomyl solution, polymyxin B solution, chloroneb solution, dichloran solution, bacitracin solution, cycloheximide solution, pentachloronitrobenzene solution, pimaricin solution, tyrothricin solution, vancomycin solution, chloromycetin solution, and penicillin G solution—to distilled/deionized water and bring volume to 870.0mL. Mix thoroughly. Gently heat and bring to boiling. Autoclave for 15 min at 15 psi pressure–121°C. Cool to 45°–50°C. Aseptically add 10.0mL each of sterile 2,3,5-triphenyltetrazolium chloride solution, benomyl solution, polymyxin B solution, chloroneb solution, dichloran solution, bacitracin solution, cycloheximide solution, pentachloronitrobenzene solution, pimaricin solution, tyrothricin solution, vancomycin solution, chloromycetin solution, and penicillin G solution. Mix thoroughly. Pour into sterile Petri dishes. Dry plates for 24 hr at 30°C.

Use: For the isolation and cultivation of *Pseudomonas solanacearum* from soil.

FTX Broth

Composition per 1001.0mL:
Sodium glutamate 10.0g
Glucose 2.0g
Tris 2.0g
Sodium glycerophosphate 0.1g
Artificial seawater 1.0L
Trace elements solution 1.0mL
pH 8.0 ± 0.2 at 25°C

Artificial Seawater:
Composition per liter:
NaCl 24.7g
$MgSO_4 \cdot 7H_2O$ 6.3g
$MgCl_2 \cdot 6H_2O$ 4.6g
$CaCl_2$ 1.0g
KCl 0.7g
$NaHCO_3$ 0.2g

Preparation of Artificial Seawater: Add components to distilled/deionized water and bring volume to 1.0L. Mix thoroughly.

Trace Elements Solution:
Composition per liter:
Disodium EDTA 8.0g
$MnCl_2 \cdot 4H_2O$ 0.1g
$CoCl_2 \cdot 6H_2O$ 0.02g
KBr 0.02g
KI 0.02g
$ZnCl_2$ 0.02g
$CuSO_4$ 0.01g
H_3BO_3 .... 0.01g
$Na_2MoO_4 \cdot 2H_2O$ 0.01g
LiCl 5.0mg
$SnCl_2 \cdot 2H_2O$ 5.0mg

Preparation of Trace Elements Solution: Add components to distilled/deionized water and bring volume to 1.0L. Mix thoroughly.

Preparation of Medium: Add components to 1.0L of artificial seawater. Mix thoroughly. Gently heat and bring to boiling. Distribute into tubes or flasks. Autoclave for 15 min at 15 psi pressure–121°C.

Use: For the isolation and cultivation of *Cytophaga* species, *Herpetosiphon* species, *Saprospira* species, and *Flexithrix* species.

Fuchsin Lactose Broth

Composition per liter:
Peptone, special 5.0g
Lactose 5.0g
Meat extract 3.0g
Basic Fuchsin 0.013g
pH 6.8 ± 0.2 at 25°C

Source: This medium is available from HiMedia.

Caution: Basic Fuchsin is a potential carcinogen and care should be taken to avoid inhalation of the powdered dye and contamination of the skin.

Preparation of Medium: Add components to distilled/deionized water and bring volume to 1.0L. Mix thoroughly. Gently heat and bring to boiling. Distribute into tubes with inverted Durham tubes. Autoclave for 15 min at 15 psi pressure–121°C.

Use: For the determination of coliform levels in the bacteriological examination of water and other materials.

FUF Medium (DSMZ Medium 318a)

Composition per liter:
$KHCO_3$ 2.0g
NH_4Cl 1.0g
NaCl 0.6g
KH_2PO_4 .... 0.3g
$MgCl_2 \cdot 6H_2O$ 0.1g
$CaCl_2 \cdot 2H_2O$ 0.08g
Resazurin 1.0mg
HEPES solution 50.0mL
Furoic acid solution 50.0mL
Trace elements solution 10.0mL
Vitamin solution 10.0mL
Cysteine solution 10.0mL
$Na_2S \cdot 9H_2O$ solution 10.0mL
pH 6.8 ± 0.2 at 25°C

Vitamin Solution:
Composition per liter:
Pyridoxine-HCl 10.0mg
Thiamine-HCl·$2H_2O$ 5.0mg
Riboflavin 5.0mg
Nicotinic acid 5.0mg
D-Ca-pantothenate 5.0mg
p-Aminobenzoic acid 5.0mg
Lipoic acid 5.0mg
Biotin 2.0mg

Folic acid 2.0mg
Vitamin B_{12} 0.1mg

Preparation of Vitamin Solution: Add components to distilled/deionized water and bring volume to 1.0L. Mix thoroughly. Sparge with 80% H_2 + 20% CO_2. Filter sterilize.

Cysteine Solution:
Composition per 10.0mL:
L-Cysteine·HCl·H_2O 0.3g

Preparation of Cysteine Solution: Add L-cysteine·HCl·H_2O to distilled/deionized water and bring volume to 10.0mL. Mix thoroughly. Sparge with 100% N_2. Autoclave for 15 min at 15 psi pressure–121°C.

$Na_2S{\cdot}9H_2O$ Solution:
Composition per 10.0mL:
$Na_2S{\cdot}9H_2O$ 0.3g

Preparation of $Na_2S{\cdot}9H_2O$ Solution: Add $Na_2S{\cdot}9H_2O$ to distilled/deionized water and bring volume to 10.0mL. Mix thoroughly. Autoclave under 100% N_2 for 15 min at 15 psi pressure–121°C. Cool to room temperature.

Trace Elements Solution:
Composition per liter:
Nitrilotriacetic acid 12.8g
$FeCl_3{\cdot}6H_2O$ 1.35g
NaCl 1.0g
$NiCl_2{\cdot}6H_2O$ 0.12g
$MgCl_2{\cdot}4H_2O$ 0.1g
$CaCl_2{\cdot}2H_2O$ 0.1g
$ZnCl_2$ 0.1g
$Na_2SeO_3{\cdot}5H_2O$ 0.026g
$CuCl_2{\cdot}2H_2O$ 0.025g
$CoCl_2{\cdot}6H_2O$ 0.024g
$Na_2MoO_4{\cdot}4H_2O$ 0.024g
H_3BO_3 .... 0.01g

Preparation of Trace Elements Solution: Add nitrilotriacetic acid to 200.0mL of distilled/deionized water. Dissolve by adjusting pH to 6.5 with KOH. Add remaining components. Add distilled/deionized water to 1.0L. Mix thoroughly.

Furoic Acid Solution:
Composition per 100.0mL:
2-Furoic acid 4.4g

Preparation of Furoic Acid Solution: Add furoic acid to distilled/deionized water and bring volume to 100.0mL. Mix thoroughly. Adjust pH to 7.0 with NaOH. Sparge with 100% N_2. Filter sterilize.

HEPES Solution:
Composition per 100.0mL:
HEPES 6.2g

Preparation of HEPES Solution: Add HEPES to distilled/deionized water and bring volume to 100.0mL. Mix thoroughly. Adjust pH to 7.0. Sparge with 100% N_2. Filter sterilize.

Preparation of Medium: Prepare and dispense medium under 80% N_2 + 20% CO_2 gas atmosphere. Add components, except cysteine solution, furoic acid solution, $Na_2S{\cdot}9H_2O$ solution, and vitamin solution, to distilled/deionized water and bring volume to 920.0mL. Mix thoroughly. Adjust pH to 7.0. Sparge with 80% N_2 + 20% CO_2. Autoclave for 15 min at 15 psi pressure–121°C. Aseptically and anaerobically add 10.0mL cysteine solution, 50.0mL furoic acid solution, 10.0mL $Na_2S{\cdot}9H_2O$ solution, and 10.0mL vitamin solution. Mix thoroughly. Aseptically and anaerobically distribute into sterile tubes or bottles. After inoculation, flush and repressurize the gas head space of culture bottles with sterile 80% N_2 + 20% CO_2 to 1 bar overpressure.

Use: For the cultivation of *Methanosarcina thermophila (Methanosarcina* sp.*)* and *Methanosarcina mazei=Methanococcus mazei (Methanosarcina frisia).*

Fumarate Medium (DSMZ Medium 195a)

Composition per 991.0mL:
Solution A 870.0mL
Solution C 100.0mL
Solution D 10.0mL
Solution E (Vitamin solution) 10.0mL
Solution B (Trace elements solution Sl-10) 1.0mL
pH 7.4 ± 0.2 at 25°C

Solution A:
Composition per 870.0mL:
NaCl 21.0g
$MgCl_2{\cdot}6H_2O$ 3.1g
Na_2SO_4 .... 3.0g
Na_2-fumarate 2.5g
KCl 0.5g
NH_4Cl 0.3g
KH_2PO_4 .... 0.2g
$CaCl_2{\cdot}2H_2O$ 0.15g
Resazurin 1.0mg

Preparation of Solution A: Add components to distilled/deionized water and bring volume to 870.0mL Mix thoroughly.

Solution B (Trace Elements Solution SL-10):
Composition per liter:
$FeCl_2{\cdot}4H_2O$ 1.5g
$CoCl_2{\cdot}6H_2O$ 190.0mg
$MnCl_2{\cdot}4H_2O$ 100.0mg
$ZnCl_2$ 70.0mg
$Na_2MoO_4{\cdot}2H_2O$ 36.0mg
$NiCl_2{\cdot}6H_2O$ 24.0mg
H_3BO_3 .... 6.0mg
$CuCl_2{\cdot}2H_2O$ 2.0mg
HCl (25% solution) 10.0mL

Preparation of Solution B (Trace Elements Solution SL-10): Add $FeCl_2{\cdot}4H_2O$ to 10.0mL of HCl solution. Mix thoroughly. Add distilled/deionized water and bring volume to 1.0L. Add remaining components. Mix thoroughly. Sparge with 100% N_2. Autoclave for 15 min at 15 psi pressure–121°C.

Solution C:
Composition per 100.0mL:
$NaHCO_3$ 5.0g

Preparation of Solution C: Add $NaHCO_3$ to distilled/deionized water and bring volume to 100.0mL. Mix thoroughly. Filter sterilize. Flush with 80% N_2 + 20% CO_2 to remove dissolved oxygen.

Solution D:
Composition per 100.0mL:
Resorcinol 1.1g

Preparation of Solution D: Add resorcinol to distilled/deionized water and bring volume to 100.0mL. Mix thoroughly. Sparge with 100% N_2. Filter sterilize. Use freshly prepared solution.

Solution E (Vitamin Solution):

Composition per liter:

Pyridoxine-HCl	10.0mg
Thiamine-HCl·$2H_2O$	5.0mg
Riboflavin	5.0mg
Nicotinic acid	5.0mg
D-Ca-pantothenate	5.0mg
p-Aminobenzoic acid	5.0mg
Lipoic acid	5.0mg
Biotin	2.0mg
Folic acid	2.0mg
Vitamin B_{12}	0.10mg

Preparation of Solution E (Vitamin Solution): Add components to distilled/deionized water and bring volume to 1.0L. Mix thoroughly. Sparge with 100% N_2. Autoclave for 15 min at 15 psi pressure–121°C.

Solution F:

Composition per 10.0mL:

$Na_2S·9H_2O$	0.4g

Preparation of Solution F: Add $Na_2S·9H_2O$ to distilled/deionized water and bring volume to 10.0mL. Mix thoroughly. Sparge with 100% N_2. Autoclave for 15 min at 15 psi pressure–121°C.

Preparation of Medium: Gently heat solution A and bring to boiling. Boil solution A for a few minutes. Cool to room temperature. Gas with 80% N_2 + 20% CO_2 gas mixture to reach a pH below 6. Autoclave for 15 min at 15 psi pressure–121°C. Cool to room temperature. Sequentially add 1.0mL solution B, 100.0mL solution C, 10.0mL solution D, 10.0mL solution E, and 10.0mL solution F. Adjust pH to 7.4 with sodium bicarbonate or sodium carbonate. Distribute anaerobically under 80% N_2 + 20% CO_2 into appropriate vessels. Addition of 10–20mg sodium dithionite per liter from a 5% (w/v) solution, freshly prepared under N_2 and filter-sterilized, may stimulate growth. During growth the culture can be fed with the resorcinol solution.

Use: For the cultivation of *Desulfuromusa kysingii, Desulfuromusa bakii,* and *Desulfuromusa succinoxidans.*

Fundibacter jadensis Medium (DSMZ Medium 821)

Composition per liter:

Peptone	2.5g
Meat extract	1.5g
Na-acetate	1.0g
Artificial sea water, concentrated	190.0mL

pH 7.2 ± 0.2 at 25°C

Artificial Sea Water, Concentrated:

Composition per 1.0L:

NaCl	99.4g
$MgSO_4·7H_2O$	23.76g
$MgCl_2·6H_2O$	18.12g
$CaCl_2·2H_2O$	5.2g
KCl	2.56g

Preparation of Artificial Sea Water, Concentrated: Add components to distilled/deionized water and bring volume to 1.0L. Mix thoroughly. Adjust pH to 7.1–7.3. Autoclave for 15 min at 15 psi pressure–121°C. Cool to room temperature.

Preparation of Medium: Add components, except concentrated artificial sea water, to distilled/deionized water and bring volume to 810.0mL. Mix thoroughly. Adjust pH to 7.1–7.3. Autoclave for 15 min at 15 psi pressure–121°C. Cool to 60°C. Add 190.0mL concentrated artificial sea water. Mix thoroughly. Aseptically distribute into tubes or flasks.

Use: For the cultivation of *Alcanivorax jadensis.*

Fundibacter jadensis Medium (DSMZ Medium 821)

Composition per liter:

Peptone	2.5g
Meat extract	1.5g
Artificial sea water, concentrated	190.0mL
Hexadecane	2.0mL

pH 7.2 ± 0.2 at 25°C

Artificial Sea Water, Concentrated:

Composition per 1.0L:

NaCl	99.4g
$MgSO_4·7H_2O$	23.76g
$MgCl_2·6H_2O$	18.12g
$CaCl_2·2H_2O$	5.2g
KCl	2.56g

Preparation of Artificial Sea Water, Concentrated: Add components to distilled/deionized water and bring volume to 1.0L. Mix thoroughly. Adjust pH to 7.1–7.3. Autoclave for 15 min at 15 psi pressure–121°C. Cool to room temperature.

Preparation of Medium: Add components, except concentrated artificial sea water, to distilled/deionized water and bring volume to 810.0mL. Mix thoroughly. Adjust pH to 7.1–7.3. Autoclave for 15 min at 15 psi pressure–121°C. Cool to 60°C. Add 190.0mL concentrated artificial sea water. Mix thoroughly. Aseptically distribute into tubes or flasks.

Use: For the cultivation of *Alcanivorax jadensis.*

Fungal Agar (Mycological agar)

Composition per liter:

Agar	15.0g
Papaic digest of soybean meal	10.0g
Glucose	10.0g

pH 7.0 ± 0.2 at 25°C

Source: This medium is available from HiMedia.

Preparation of Medium: Add components to distilled/deionized water and bring volume to 1.0L. Mix thoroughly. Gently heat and bring to boiling. Distribute into tubes or flasks. Autoclave for 15 min at 15 psi pressure–121°C. Pour into sterile Petri dishes or leave in tubes.

Use: For the cultivation of fungi and for production of chlamydospores.

Fungal Agar with Low pH (Mycological Agar with Low pH)

Composition per liter:

Agar	15.0g
Papaic digest of soybean meal	10.0g
Glucose	10.0g

pH 4.8 ± 0.2 at 25°C

Source: This medium is available from HiMedia.

Preparation of Medium: Add components to distilled/deionized water and bring volume to 1.0L. Mix thoroughly. Gently heat and bring

to boiling. Distribute into tubes or flasks. Autoclave for 15 min at 15 psi pressure–121°C. Pour into sterile Petri dishes or leave in tubes.

Use: For the selective enumeration and cultivation of saprophytic fungi and aciduric bacteria.

Fungal Broth (Mycological Broth)

Composition per liter:

Glucose 40.0g
Papaic digest of soybean meal 10.0g

pH 7.0 ± 0.2 at 25°C

Source: This medium is available from HiMedia.

Preparation of Medium: Add components to distilled/deionized water and bring volume to 1.0L. Mix thoroughly. Gently heat and bring to boiling. Distribute into tubes or flasks. Autoclave for 15 min at 15 psi pressure–121°C.

Use: For the cultivation of fungi.

Fungi Kimmig Agar

Composition per liter:

Glucose 19.0g
Agar 15.0g
Peptone 15.0g
NaCl 1.0g
Glycerol 5.0mL

pH 6.5 ± 0.2 at 25°C

Preparation of Medium: Add components to distilled/deionized water and bring volume to 1.0L. Mix thoroughly. Gently heat and bring to boiling. Mix thoroughly. Distribute into tubes or flasks. Autoclave for 15 min at 15 psi pressure–121°C. Pour into sterile Petri dishes.

Use: For the cultivation, isolation, identification, and strain preservation of fungi.

Fungi Kimmig Agar Base

Composition per liter:

Agar 15.0g
NaCl 11.4g
Glucose 10.0g
Peptic digest of animal tissue 9.3g
Casein enzymatic hydrolysate 4.3g
Glycerol 5.0mL

pH 6.5 ± 0.2 at 25°C

Source: This medium, without glycerol, is available as a premixed powder from HiMedia.

Preparation of Medium: Add components to distilled/deionized water and bring volume to 1.0L. Mix thoroughly. Gently heat and bring to boiling. Mix thoroughly. Distribute into tubes or flasks. Autoclave for 15 min at 15 psi pressure–121°C. Pour into sterile Petri dishes.

Use: For the cultivation, isolation, identification, and strain preservation of fungi.

Fungi Kimmig HiVeg Agar Base

Composition per liter:

Agar 15.0g
NaCl 11.4g
Glucose 10.0g
Plant peptone 9.3g
Plant hydrolysate 4.3g
Glycerol 5.0mL

pH 6.5 ± 0.2 at 25°C

Source: This medium without glycerol, is available as a premixed powder from HiMedia.

Preparation of Medium: Add components to distilled/deionized water and bring volume to 1.0L. Mix thoroughly. Gently heat and bring to boiling. Mix thoroughly. Distribute into tubes or flasks. Autoclave for 15 min at 15 psi pressure–121°C. Pour into sterile Petri dishes.

Use: For the cultivation, isolation, identification, and strain preservation of fungi.

Fungi Kimmig Selective Agar

Composition per liter:

Glucose 19.0g
Agar 15.0g
Peptone 15.0g
NaCl 1.0g
Glycerol 5.0mL
Selective solution 10.0mL

pH 6.5 ± 0.2 at 25°C

Selective Solution:

Composition per 10.0mL:

Cycloheximide 0.4g
Streptomycin 0.04mg
Penicillin 40,000U

Preparation of Selective Solution: Add components to distilled/deionized water and bring volume to 10.0mL. Mix thoroughly. Filter sterilize.

Caution: Cycloheximide is toxic. Avoid skin contact or aerosol formation and inhalation.

Preparation of Medium: Add components, except selective solution, to distilled/deionized water and bring volume to 990.0mL. Mix thoroughly. Gently heat and bring to boiling. Mix thoroughly. Autoclave for 15 min at 15 psi pressure–121°C. Cool to 45°–50°C. Aseptically add 10.0mL selective solution. Mix thoroughly. Distribute into tubes or flasks. Autoclave for 5 min at 15 psi pressure–121°C. Pour into sterile Petri dishes.

Use: For the cultivation, isolation, identification, and strain preservation of pathogenic fungi.

Fungi Kimmig Selective Agar

Composition per liter:

Glucose 19.0g
Agar 15.0g
Peptone 15.0g
NaCl 1.0g
Glycerol 5.0mL
Selective solution 10.0mL

pH 6.5 ± 0.2 at 25°C

Selective Solution:

Composition per 10.0mL:

Novobiocin 0.1g
Colistin 0.08g

Preparation of Selective Solution: Add components to distilled/deionized water and bring volume to 10.0mL. Mix thoroughly. Filter sterilize.

Preparation of Medium: Add components, except selective solution, to distilled/deionized water and bring volume to 990.0mL. Mix thoroughly. Gently heat and bring to boiling. Mix thoroughly. Autoclave for 15 min at 15 psi pressure–121°C. Cool to 45°–50°C. Aseptically add 10.0mL selective solution. Mix thoroughly. Distribute into tubes or flasks.

Use: For the cultivation, isolation, identification, and strain preservation of pathogenic fungi.

Fungobiotic Agar (Mycobio Agar)

Composition per liter:

Agar	15.0g
Papaic digest of soybean meal	10.0g
Glucose	10.0g
Cycloheximide	0.5g
Chloramphenicol	0.5g

pH 6.5 ± 0.2 at 25°C

Source: This medium is available from HiMedia.

Caution: Cycloheximide is toxic. Avoid skin contact or aerosol formation and inhalation.

Preparation of Medium: Add components to distilled/deionized water and bring volume to 1.0L. Mix thoroughly. Gently heat and bring to boiling. Distribute into tubes or flasks. Autoclave for 15 min at 15 psi pressure–121°C. Pour into sterile Petri dishes or leave in tubes.

Use: For the isolation of dermatophytic fungi.

Furoate Agar

Composition per liter:

Agar	20.0g
2-Furoic acid	2.0g
K_2HPO_4	1.0g
NH_4Cl	1.0g
$MgSO_4 \cdot 7H_2O$	0.1g

pH 7.0 ± 0.2 at 25°C

Preparation of Medium: Add components to distilled/deionized water and bring volume to 1.0L. Mix thoroughly. Gently heat and bring to boiling. Distribute into tubes or flasks. Autoclave for 15 min at 15 psi pressure–121°C. Pour into sterile Petri dishes or leave in tubes.

Use: For the cultivation and maintenance of *Bacillus megaterium* and *Pseudomonas* species.

Furunculosis Agar

Agar	15.0g
Tryptone	10.0g
Yeast extract	5.0g
NaCl	2.5g
Tyrosine	1.0g

pH 7.3 ± 0.2 at 25°C

Preparation of Medium: Add components to distilled/deionized water and bring volume to 1.0L. Mix thoroughly. Gently heat and bring to boiling. Mix thoroughly. Distribute into tubes or flasks. Autoclave for 15 min at 15 psi pressure–121°C. Pour into sterile Petri dishes.

Use: For the cultivation amd identification based upon pigment production of *Aeromonas salmonicida.*

Furunculosis Agar

Composition per liter:

Agar	15.0g
Casein enzymatic hydrolysate	10.0g
Yeast extract	5.0g
NaCl	2.5g
Tyrosine	1.0g

pH 7.3 ± 0.2 at 25°C

Source: This medium is available as a premixed powder from HiMedia.

Preparation of Medium: Add components to distilled/deionized water and bring volume to 1.0L. Mix thoroughly. Gently heat and bring to boiling. Mix thoroughly. Distribute into tubes or flasks. Autoclave for 15 min at 15 psi pressure–121°C. Pour into sterile Petri dishes or leave in tubes. Allow tubes to cool in a slanted position.

Use: For the cultivation amd identification based upon pigment production of *Aeromonas salmonicida.*

Furunculosis HiVeg Agar

Composition per liter:

Agar	15.0g
Plant hydrolysate	10.0g
Yeast extract	5.0g
NaCl	2.5g
Tyrosine	1.0g

pH 7.3 ± 0.2 at 25°C

Source: This medium is available as a premixed powder from HiMedia.

Preparation of Medium: Add components to distilled/deionized water and bring volume to 1.0L. Mix thoroughly. Gently heat and bring to boiling. Mix thoroughly. Distribute into tubes or flasks. Autoclave for 15 min at 15 psi pressure–121°C. Pour into sterile Petri dishes or leave in tubes. Allow tubes to cool in a slanted position.

Use: For the cultivation amd identification based upon pigment production of *Aeromonas salmonicida.*

Fusibacter paucivorans Medium (DSMZ Medium 853)

Composition per 1068.0mL:

NaCl	30.0g
Na-thiosulfate·$5H_2O$	3.16g
$MgCl_2 \cdot 6H_2O$	3.0g
Yeast extract	1.0g
Trypticase™	1.0g
NH_4Cl	1.0g
KCl	1.0g
Na-acetate·$3H_2O$	0.5g
Cysteine-HCl·H_2O	0.5g
K_2HPO_4	0.3g
KH_2PO_4	0.3g
$CaCl_2 \cdot 2H_2O$	0.1g
Resazurin	0.5mg
$NaHCO_3$ solution	40.0mL
Glucose solution	18.0mL
$Na_2S \cdot 9H_2O$ solution	10.0mL
Trace elements solution	10.0mL

pH 7.3 ± 0.2 at 25°C

Trace Elements Solution:
Composition per liter:

$MgSO_4 \cdot 7H_2O$	3.0g
Nitrilotriacetic acid	1.5g
NaCl	1.0g
$MnSO_4 \cdot 2H_2O$	0.5g
$CoSO_4 \cdot 7H_2O$	0.18g
$ZnSO_4 \cdot 7H_2O$	0.18g
$CaCl_2 \cdot 2H_2O$	0.1g
$FeSO_4 \cdot 7H_2O$	0.1g
$NiCl_2 \cdot 6H_2O$	0.025g
$KAl(SO_4)_2 \cdot 12H_2O$	0.02g
H_3BO_3	0.01g
$Na_2MoO_4 \cdot 4H_2O$	0.01g
$CuSO_4 \cdot 5H_2O$	0.01g
$Na_2SeO_3 \cdot 5H_2O$	0.3mg

Preparation of Trace Elements Solution: Add nitrilotriacetic acid to 500.0mL of distilled/deionized water. Dissolve by adjusting pH to 7.0 with KOH. Add remaining components. Add distilled/deionized water to 1.0L. Mix thoroughly.

$NaHCO_3$ Solution:
Composition per 100.0mL:

$NaHCO_3$	10.0g

Preparation of $NaHCO_3$ Solution: Add $NaHCO_3$ to distilled/deionized water and bring volume to 100.0mL. Mix thoroughly. Autoclave for 15 min at 15 psi pressure–121°C. Cool to 25°C. Must be prepared freshly.

$Na_2S \cdot 9H_2O$ Solution:
Composition per 10.0mL:

$Na_2S \cdot 9H_2O$	0.3g

Preparation of $Na_2S \cdot 9H_2O$ Solution: Add $Na_2S \cdot 9H_2O$ to distilled/deionized water and bring volume to 10.0mL. Mix thoroughly. Autoclave under 100% N_2 for 15 min at 15 psi pressure–121°C. Cool to room temperature.

Glucose Solution:
Composition per 50.0mL:

Glucose	10.0g

Preparation of Glucose Solution: Add glucose to distilled/deionized water and bring volume to 50.0mL. Mix thoroughly. Filter sterilize.

Preparation of Medium: Prepare and dispense medium under 80% N_2 + 20% CO_2 gas mixture. Add components, except glucose solution, $NaHCO_3$ solution, and $Na_2S \cdot 9H_2O$ solution, to distilled/deionized water and bring volume to 1.0L. Mix thoroughly. Adjust pH to 7.0. Sparge with 80% N_2 + 20% CO_2. Autoclave for 15 min at 15 psi pressure–121°C. Cool to 25°C. Aseptically and anaerobically add 18.0mL sterile glucose solution, 10.0mL sterile $Na_2S \cdot 9H_2O$ solution, and 40.0mL sterile $NaHCO_3$ solution. Mix thoroughly. Adjust pH to 7.3. Aseptically and anaerobically distribute into tubes or bottles.

Use: For the cultivation of *Fusibacter paucivorans.*

Fusobacterium Medium

Composition per liter:

Agar	15.0g
Pancreatic digest of casein	15.0g
Glucose	5.0g
NaCl	5.0g
Yeast extract	5.0g
L-Cysteine	0.75g
Crystal Violet	0.01g
Bovine serum	50.0mL
Streptomycin solution	10.0mL

pH 7.2 ± 0.2 at 25°C

Streptomycin Solution:
Composition per 10.0mL:

Streptomycin	0.01g

Preparation of Streptomycin Solution: Add streptomycin to distilled/deionized water and bring volume to 10.0mL. Mix thoroughly. Filter sterilize.

Preparation of Medium: Add components, except bovine serum and streptomycin solution, to distilled/deionized water and bring volume to 940.0mL. Mix thoroughly. Gently heat and bring to boiling. Autoclave for 15 min at 15 psi pressure–121°C. Cool to 45°–50°C. Aseptically add 50.0mL of sterile bovine serum and 10.0mL of sterile streptomycin solution. Mix thoroughly. Pour into sterile Petri dishes or distribute into sterile tubes.

Use: For the cultivation of *Fusobacterium* species.

Fusobacterium necrophorum Medium

Composition per 500.0mL:

Pancreatic digest of casein	16.0g
Agar	8.3g
Biosate	4.0g
$MgSO_4$	2.5g
Na_2HPO_4	2.5g
Thiotone	2.0g
Glucose	0.5g
Egg yolk emulsion, 50%	45.0mL
Crystal Violet solution	25.0mL
Phenylethyl alcohol solution	1.35mL

pH 7.3 ± 0.2 at 25°C

Egg Yolk Emulsion, 50%:
Composition per 100.0mL:

Chicken egg yolks	11
Whole chicken egg	1
NaCl (0.9% solution)	50.0mL

Preparation of Egg Yolk Emulsion, 50%: Soak eggs with 1:100 dilution of saturated mercuric chloride solution for 1 min. Crack 11 eggs, separating yolks from whites. Mix egg yolks with 1 chicken egg. Beat to form emulsion. Combine 50.0mL of egg yolk emulsion and 50.0mL of 0.9% NaCl solution. Mix thoroughly.

Crystal Violet Solution:
Composition per 25.0mL:

Crystal Violet	0.0115g

Preparation of Crystal Violet Solution: Aseptically add Crystal Violet to sterile distilled/deionized water and bring volume to 25.0mL. Mix thoroughly.

Phenylethyl Alcohol Solution:
Composition per 100.0mL:

Phenylethyl alcohol	0.27g

Preparation of Phenylethyl Alcohol Solution: Aseptically add phenylethyl alcohol to sterile distilled/deionized water and bring volume to 100.0mL. Mix thoroughly.

Preparation of Medium: Add components—except egg yolk emulsion, 50%, Crystal Violet solution, and phenylethyl alcohol solution—to distilled/deionized water and bring volume to 428.65mL. Mix thoroughly. Gently heat and bring to boiling. Autoclave for 15 min at 15 psi pressure–121°C. Cool to 45°–50°C. Aseptically add 45.0mL of sterile egg yolk emulsion, 50%, 25.0mL of Crystal Violet solution, and 1.35mL of sterile phenylethyl alcohol solution. Mix thoroughly. Pour into sterile Petri dishes or distribute into sterile tubes.

Use: For the isolation and cultivation of *Fusobacterium necrophorum*.

FWM Medium

Composition per 1013.0mL:

Solution A	940.0mL
Solution E ($NaHCO_3$ solution)	50.0mL
Solution F (Substrate solution)	10.0mL
Solution G ($Na_2S \cdot 9H_2O$ solution)	10.0mL
Solution B (Trace elements solution SL-10)	1.0mL
Solution C (Seven vitamin solution)	1.0mL
Solution D (Selenite-tungstate solution)	1.0mL

pH 7.2–7.4 at 25°C

Solution A:

Composition per 940.0mL:

NaCl	1.0g
KCl	0.5g
$MgCl_2 \cdot 6H_2O$	0.4g
NH_4Cl	0.25g
KH_2PO_4	0.2g
$CaCl_2 \cdot 2H_2O$	0.15g
Resazurin	0.5mg

Preparation of Solution A: Prepare and dispense under 80% N_2 + 20% CO_2. Add components to distilled/deionized water and bring volume to 940.0mL. Mix thoroughly. Autoclave for 15 min at 15 psi pressure–121°C.

Solution B (Trace Elements Solution SL-10):

Composition per liter:

$FeCl_2 \cdot 4H_2O$	1.5g
$CoCl_2 \cdot 6H_2O$	190.0mg
$MnCl_2 \cdot 4H_2O$	100.0mg
$ZnCl_2$	70.0mg
$Na_2MoO_4 \cdot 2H_2O$	36.0mg
$NiCl_2 \cdot 6H_2O$	24.0mg
H_3BO_3	6.0mg
$CuCl_2 \cdot 2H_2O$	2.0mg
HCl (25% solution)	10.0mL

Preparation of Solution B (Trace Elements Solution SL-10): Prepare and dispense under 100% N_2. Add $FeCl_2 \cdot 4H_2O$ to 10.0mL of HCl solution. Mix thoroughly. Add distilled/deionized water and bring volume to 1.0L. Add remaining components. Mix thoroughly. Autoclave for 15 min at 15 psi pressure–121°C.

Solution C (Seven Vitamin Solution):

Composition per liter:

Pyridoxine·HCl	300.0mg
Nicotinic acid	200.0mg
Thiamine·HCl	200.0mg
Calcium DL-pantothenate	100.0mg
Cyanocobalamine	100.0mg
p-Aminobenzoic acid	80.0mg
D(+) Biotin	20.0mg

Preparation of Solution C (Seven Vitamin Solution): Add components to distilled/deionized water and bring volume to 1.0L. Mix thoroughly. Filter sterilize. Sparge with 100% N_2.

Solution D (Selenite-Tungstate Solution):

Composition per liter:

NaOH	0.5g
$Na_2WO_4 \cdot 2H_2O$	4.0mg
$Na_2SeO_3 \cdot 5H_2O$	3.0mg

Preparation of Solution D (Selenite-Tungstate Solution): Add components to distilled/deionized water and bring volume to 1.0L. Mix thoroughly. Filter sterilize. Sparge with 100% N_2.

Solution E ($NaHCO_3$ Solution):

Composition per 50.0mL:

$NaHCO_3$	2.5g

Preparation of Solution E ($NaHCO_3$ Solution): Add $NaHCO_3$ to distilled/deionized water and bring volume to 50.0mL. Mix thoroughly. Gas under 100% N_2. Autoclave for 15 min at 15 psi pressure–121°C.

Solution F (Substrate Solution):

Composition per 10.0mL:

Sodium-DL-3-hydroxybutyrate	1.5g

Preparation of Solution F (Substrate Solution): Add sodium-DL-3-hydroxybutyrate to distilled/deionized water and bring volume to 50.0mL. Mix thoroughly. Gas under 100% N_2. Autoclave for 15 min at 15 psi pressure–121°C.

Solution G ($Na_2S \cdot 9H_2O$ Solution):

Composition per 10.0mL:

$Na_2S \cdot 9H_2O$	0.3g

Preparation of Solution G ($Na_2S \cdot 9H_2O$ Solution): Add $Na_2S \cdot 9H_2O$ to distilled/deionized water and bring volume to 10.0mL. Mix thoroughly. Gas under 100% N_2. Autoclave for 15 min at 15 psi pressure–121°C.

Preparation of Medium: To 940.0mL of sterile solution A, aseptically and anaerobically add 1.0mL of sterile solution B, 1.0mL of sterile solution C, 1.0mL of sterile solution D, 50.0mL of sterile solution E, 10.0mL of sterile solution F, and 10.0mL of sterile solution G. Mix thoroughly. Aseptically and anaerobically distribute into sterile tubes or flasks.

Use: For the cultivation and maintenance of *Clostridium homopropionicum* and *Desulfococcus biacutus*.

FWM Medium

Composition per 1013.0mL:

Solution A	940.0mL
Solution E ($NaHCO_3$ solution)	50.0mL
Solution F (Substrate solution)	10.0mL
Solution G ($Na_2S \cdot 9H_2O$ solution)	10.0mL
Solution B (Trace elements solution SL-10)	1.0mL
Solution C (Seven vitamin solution)	1.0mL
Solution D (Selenite-tungstate solution)	1.0mL

pH 7.2–7.4 at 25°C

Solution A:

Composition per 940.0mL:

NaCl	1.0g
KCl	0.5g
$MgCl_2 \cdot 6H_2O$	0.4g
NH_4Cl	0.25g
KH_2PO_4	0.2g
$CaCl_2 \cdot 2H_2O$	0.15g
Resazurin	0.5mg

Preparation of Solution A: Prepare and dispense under 80% N_2 + 20% CO_2. Add components to distilled/deionized water and bring volume to 940.0mL. Mix thoroughly. Autoclave for 15 min at 15 psi pressure–121°C.

Solution B (Trace Elements Solution SL-10):

Composition per liter:

$FeCl_2 \cdot 4H_2O$	1.5g
$CoCl_2 \cdot 6H_2O$	190.0mg
$MnCl_2 \cdot 4H_2O$	100.0mg
$ZnCl_2$	70.0mg
$Na_2MoO_4 \cdot 2H_2O$	36.0mg
$NiCl_2 \cdot 6H_2O$	24.0mg
H_3BO_3	6.0mg
$CuCl_2 \cdot 2H_2O$	2.0mg
HCl (25% solution)	10.0mL

Preparation of Solution B (Trace Elements Solution SL-10): Prepare and dispense under 100% N_2. Add $FeCl_2 \cdot 4H_2O$ to 10.0mL of HCl solution. Mix thoroughly. Add distilled/deionized water and bring volume to 1.0L. Add remaining components. Mix thoroughly. Autoclave for 15 min at 15 psi pressure–121°C.

Solution C (Seven Vitamin Solution):

Composition per liter:

Pyridoxine·HCl	300.0mg
Nicotinic acid	200.0mg
Thiamine·HCl	200.0mg
Calcium DL-pantothenate	100.0mg
Cyanocobalamine	100.0mg
p-Aminobenzoic acid	80.0mg
D(+)-Biotin	20.0mg

Preparation of Solution C (Seven Vitamin Solution): Add components to distilled/deionized water and bring volume to 1.0L. Mix thoroughly. Filter sterilize. Sparge with 100% N_2.

Solution D (Selenite-Tungstate Solution):

Composition per liter:

NaOH	0.5g
$Na_2WO_4 \cdot 2H_2O$	4.0mg
$Na_2SeO_3 \cdot 5H_2O$	3.0mg

Preparation of Solution D (Selenite-Tungstate Solution): Add components to distilled/deionized water and bring volume to 1.0L. Mix thoroughly. Filter sterilize. Sparge with 100% N_2.

Solution E ($NaHCO_3$ Solution):

Composition per 50.0mL:

$NaHCO_3$	2.5g

Preparation of Solution E ($NaHCO_3$ Solution): Add $NaHCO_3$ to distilled/deionized water and bring volume to 50.0mL. Mix thoroughly. Gas under 100% N_2. Autoclave for 15 min at 15 psi pressure–121°C.

Solution F (Substrate Solution):

Composition per 10.0mL:

Xylan or xylose	2.0g

Preparation of Solution F (Substrate Solution): Add 2.0g of xylan or 2.0g of xylose to distilled/deionized water and bring volume to 10.0mL. Mix thoroughly. Gas under 100% N_2. Autoclave for 15 min at 15 psi pressure–121°C.

Solution G ($Na_2S \cdot 9H_2O$ Solution):

Composition per 10.0mL:

$Na_2S \cdot 9H_2O$	0.3g

Preparation of Solution G ($Na_2S \cdot 9H_2O$ Solution): Add $Na_2S \cdot 9H_2O$ to distilled/deionized water and bring volume to 10.0mL. Mix thoroughly. Gas under 100% N_2. Autoclave for 15 min at 15 psi pressure–121°C.

Preparation of Medium: To 940.0mL of sterile solution A, aseptically and anaerobically add 1.0mL of sterile solution B, 1.0mL of sterile solution C, 1.0mL of sterile solution D, 50.0mL of sterile solution E, 10.0mL of sterile solution F, and 10.0mL of sterile solution G. Mix thoroughly. Aseptically and anaerobically distribute into sterile tubes or flasks.

Use: For the cultivation and maintenance of *Desulfococcus biacutus*.

FWN Medium, Modified with Fructose

Composition per 1020.0mL:

Solution A	950.0mL
Solution C ($NaHCO_3$ solution)	50.0mL
Solution B (Wolfe's vitamin solution)	10.0mL
Solution D ($Na_2S \cdot 9H_2O$ solution)	10.0mL

Solution A:

Composition per 950.0mL:

Fructose	2.0g
NaCl	1.0g
KCl	0.5g
$MgCl_2 \cdot 6H_2O$	0.4g
NH_4Cl	0.25g
KH_2PO_4	0.2g
$CaCl_2 \cdot 2H_2O$	0.15g
Resazurin	0.5mg
Modified Wolfe's mineral solution	10.0mL

Modified Wolfe's Mineral Solution:

Composition per liter:

$MgSO_4 \cdot 7H_2O$	3.0g
Nitrilotriacetic acid	1.5g
NaCl	1.0g
$MnSO_4 \cdot H_2O$	0.5g
$CaCl_2$	0.1g
$CoCl_2 \cdot 6H_2O$	0.1g
$FeSO_4 \cdot 7H_2O$	0.1g
$ZnSO_4 \cdot 7H_2O$	0.1g
$AlK(SO_4)_2 \cdot 12H_2O$	0.01g
$CuSO_4 \cdot 5H_2O$	0.01g
H_3BO_3	0.01g
$Na_2MoO_4 \cdot 2H_2O$	0.01g
Na_2SeO_3	0.01g
$NaWO_4 \cdot 2H_2O$	0.01g
$NiCl_2 \cdot 6H_2O$	0.01g

Preparation of Modified Wolfe's Mineral Solution: Add nitrilotriacetic acid to 500.0mL of distilled/deionized water. Adjust pH to 6.5 with KOH. Add remaining components one at a time. Add distilled/deionized water to 1.0L. Adjust pH to 6.8.

Preparation of Solution A: Prepare and dispense under 80% N_2 + 20% CO_2. Add components to distilled/deionized water and bring volume to 950.0mL. Mix thoroughly. Autoclave for 15 min at 15 psi pressure–121°C.

Solution B (Wolfe's Vitamin Solution):

Composition per liter:

Pyridoxine·HCl	10.0mg
p-Aminobenzoic acid	5.0mg
Lipoic acid	5.0mg

Nicotinic acid	5.0mg
Riboflavin	5.0mg
Thiamine·HCl	5.0mg
Calcium DL-pantothenate	5.0mg
Biotin	2.0mg
Folic acid	2.0mg
Vitamin B_{12}	0.1mg

Preparation of Solution B (Wolfe's Vitamin Solution): Add components to distilled/deionized water and bring volume to 1.0L. Mix thoroughly. Sparge with 100% N_2. Filter sterilize.

Solution C ($NaHCO_3$ Solution):
Composition per 50.0mL:

$NaHCO_3$	2.5g

Preparation of Solution C ($NaHCO_3$ Solution): Add $NaHCO_3$ to distilled/deionized water and bring volume to 50.0mL. Mix thoroughly. Sparge with 100% N_2. Autoclave for 15 min at 15 psi pressure–121°C.

Solution D ($Na_2S·9H_2O$ Solution):
Composition per 10.0mL:

$Na_2S·9H_2O$	0.3g

Preparation of Solution D ($Na_2S·9H_2O$ Solution): Add $Na_2S·9H_2O$ to distilled/deionized water and bring volume to 10.0mL. Mix thoroughly. Sparge with 100% N_2. Autoclave for 15 min at 15 psi pressure–121°C. Before use, neutralize to pH 7.0 with sterile HCl.

Preparation of Medium: To 950.0mL of sterile solution A, aseptically and anaerobically add 10.0mL of sterile solution B, 50.0mL of sterile solution C, and 10.0mL of sterile solution D. Mix thoroughly. Aseptically and anaerobically distribute into sterile tubes or flasks under 100% N_2.

Use: For the cultivation of *Clostridium homopropionicum.*

FWN Medium, Modified with Xylan

Composition per 1020.0mL:

Solution A	950.0mL
Solution C ($NaHCO_3$ solution)	50.0mL
Solution D ($Na_2S·9H_2O$ solution)	10.0mL
Solution B (Wolfe's vitamin solution)	10.0mL

Solution A:
Composition per 950.0mL:

Xylan	2.0g
NaCl	1.0g
KCl	0.5g
$MgCl_2·6H_2O$	0.4g
NH_4Cl	0.25g
KH_2PO_4	0.2g
$CaCl_2·2H_2O$	0.15g
Resazurin	0.5mg
Modified Wolfe's mineral solution	10.0mL

Modified Wolfe's Mineral Solution:
Composition per liter:

$MgSO_4·7H_2O$	3.0g
Nitrilotriacetic acid	1.5g
NaCl	1.0g
$MnSO_4·H_2O$	0.5g
$CaCl_2$	0.1g
$CoCl_2·6H_2O$	0.1g
$FeSO_4·7H_2O$	0.1g
$ZnSO_4·7H_2O$	0.1g
$AlK(SO_4)_2·12H_2O$	0.01g
$CuSO_4·5H_2O$	0.01g
H_3BO_3	0.01g
$Na_2MoO_4·2H_2O$	0.01g
Na_2SeO_3	0.01g
$NaWO_4·2H_2O$	0.01g
$NiCl_2·6H_2O$	0.01g

Preparation of Modified Wolfe's Mineral Solution: Add nitrilotriacetic acid to 500.0mL of distilled/deionized water. Adjust pH to 6.5 with KOH. Add remaining components one at a time. Add distilled/deionized water to 1.0L. Adjust pH to 6.8.

Preparation of Solution A: Prepare and dispense under 80% N_2 + 20% CO_2. Add components to distilled/deionized water and bring volume to 950.0mL. Mix thoroughly. Autoclave for 15 min at 15 psi pressure–121°C.

Solution B (Wolfe's Vitamin Solution):
Composition per liter:

Pyridoxine·HCl	10.0mg
p-Aminobenzoic acid	5.0mg
Lipoic acid	5.0mg
Nicotinic acid	5.0mg
Riboflavin	5.0mg
Thiamine·HCl	5.0mg
Calcium DL-pantothenate	5.0mg
Biotin	2.0mg
Folic acid	2.0mg
Vitamin B_{12}	0.1mg

Preparation of Solution B (Wolfe's Vitamin Solution): Add components to distilled/deionized water and bring volume to 1.0L. Mix thoroughly. Sparge with 100% N_2. Filter sterilize.

Solution C ($NaHCO_3$ Solution):
Composition per 50.0mL:

$NaHCO_3$	2.5g

Preparation of Solution C ($NaHCO_3$ Solution): Add $NaHCO_3$ to distilled/deionized water and bring volume to 50.0mL. Mix thoroughly. Sparge with 100% N_2. Autoclave for 15 min at 15 psi pressure–121°C.

Solution D ($Na_2S·9H_2O$ Solution):
Composition per 10.0mL:

$Na_2S·9H_2O$	0.3g

Preparation of Solution D ($Na_2S·9H_2O$ Solution): Add $Na_2S·9H_2O$ to distilled/deionized water and bring volume to 10.0mL. Mix thoroughly. Sparge with 100% N_2. Autoclave for 15 min at 15 psi pressure–121°C. Before use, neutralize to pH 7.0 with sterile HCl.

Preparation of Medium: To 950.0mL of sterile solution A, aseptically and anaerobically add 10.0mL of sterile solution B, 50.0mL of sterile solution C, and 10.0mL of sterile solution D. Mix thoroughly. Aseptically and anaerobically distribute into sterile tubes or flasks under 100% N_2.

Use: For the cultivation of *Cytophaga xylanolytica.*

FX A Broth

Composition per liter:

Pancreatic digest of casein	10.0g
Yeast extract	2.0g
$MgSO_4·7H_2O$	1.0g

pH 7.0 ± 0.2 at 25°C

Preparation of Medium: Add components to distilled/deionized water and bring volume to 1.0L. Mix thoroughly. Distribute into tubes or flasks. Autoclave for 15 min at 15 psi pressure–121°C.

Use: For the isolation and cultivation of *Cytophaga* species, *Herpetosiphon* species, *Saprospira* species, and *Flexithrix* species.

FX AG Broth

Composition per liter:

Pancreatic digest of casein	10.0g
Yeast extract	2.0g
$MgSO_4 \cdot 7H_2O$	1.0g
Glucose solution	100.0mL

pH 7.0 ± 0.2 at 25°C

Glucose Solution:

Composition per 100.0mL:

D-Glucose	2.0g

Preparation of Glucose Solution: Add D-glucose to distilled/deionized water and bring volume to 100.0mL. Mix thoroughly. Autoclave for 15 min at 15 psi pressure–121°C. Cool to 25°C.

Preparation of Medium: Add components, except glucose solution, to distilled/deionized water and bring volume to 900.0mL. Mix thoroughly. Gently heat and bring to boiling. Autoclave for 15 min at 15 psi pressure–121°C. Cool to 45°–50°C. Aseptically add sterile glucose solution. Mix thoroughly. Pour into sterile Petri dishes or distribute into sterile tubes.

Use: For the isolation and cultivation of *Cytophaga* species, *Herpetosiphon* species, *Saprospira* species, and *Flexithrix* species.

G Medium

Composition per liter:

$(NH_4)_2SO_4$	2.0g
Yeast extract	2.0g
Glucose	1.0g
K_2HPO_4	0.6g
KH_2PO_4	0.4g
$MgSO_4 \cdot 7H_2O$	0.2g
$CaCl_2$	0.08g
$MnSO_4 \cdot H_2O$	0.05g
$CuSO_4 \cdot 5H_2O$	5.0mg
$ZnSO_4 \cdot 7H_2O$	5.0mg
$FeSO_4 \cdot 7H_2O$	0.5mg

pH 7.8 ± 0.2 at 25°C

Preparation of Medium: Add components to distilled/deionized water and bring volume to 1.0L. Mix thoroughly. Distribute into tubes or flasks. Autoclave for 15 min at 15 psi pressure–121°C.

Use: For the cultivation and maintenance of *Bacillus cereus*.

GA Medium
See: **Gelatin Agar**

Gaizotia abyssinica Creatinine Agar
See: **Birdseed Agar**

GAM Agar

Composition per liter:

GAM agar	74.0g

Source: GAM agar is available from Nissui.

Preparation of Medium: Add GAM agar to distilled/deionized water and bring volume to 1.0L. Mix thoroughly. Gently heat and bring to boiling. Distribute into tubes or flasks. Autoclave for 15 min at 15 psi pressure–121°C. Pour into sterile Petri dishes or leave in tubes.

Use: For the cultivation and maintenance of *Fusobacterium necrophorum, Fusobacterium pseudonecrophorum, Pediococcus* species, *Peptostreptococcus hydrogenalis,* and *Peptostreptococcus tetradius.*

GAM Semisolid

Composition per liter:

GAM broth	74.0g
Agar	2.0g

Source: GAM broth is available from Nissui.

Preparation of Medium: Add components to distilled/deionized water and bring volume to 1.0L. Mix thoroughly. Gently heat and bring to boiling. Distribute into tubes or flasks. Autoclave for 15 min at 15 psi pressure–121°C. Pour into sterile Petri dishes or leave in tubes.

Use: For the cultivation and maintenance of *Actinomyces naeslundii, Actinomyces viscosus, Atopobium minutum, Bacteroides* species, *Bifidobacterium* species, *Campylobacter divergens, Clostridium* species, *Eubacterium alactolyticum, Fusobacterium necrophorum, Fusobacterium nucleatum, Fusobacterium pseudonecrophorum, Lactobacillus* species, *Lactococcus lactis, Leuconostoc lactis, Leuconostoc mesenteroides, Leuconostoc oenos, Mitsuokella multiacida, Pediococcus* species, *Peptostreptococcus hydrogenalis, Peptostreptococcus prevotii, Peptostreptococcus productus, Peptostreptococcus tetradius, Prevotella intermedia, Prevotella melaninogenica, Propionibacterium acidipropionici, Propionibacterium* species, *Rikenella microfusus, Selenomonas ruminantium,* and *Selenomonas sputigena.*

Gardnerella vaginalis Selective Medium

Composition per liter:

Columbia blood agar base	940.0mL
Rabbit or horse serum	50.0mL
Antibiotic inhibitor solution	10.0mL

pH 7.2 ± 0.2 at 25°C

Source: This medium is available as a premixed powder from Oxoid Unipath.

Columbia Blood Agar Base:

Composition per liter:

Special peptone	23.0g
Agar	10.0g
NaCl	5.0g
Starch	1.0g

Source: Columbia blood agar base is available as a premixed powder from Oxoid Unipath.

Preparation of Columbia Blood Agar Base: Add components to distilled/deionized water and bring volume to 1.0L. Mix thoroughly. Gently heat until boiling. Autoclave for 15 min at 15 psi pressure–121°C. Cool to 45°–50°C.

Antibiotic Inhibitor Solution:

Composition per 10.0mL:

Nalidixic acid	0.035g
Gentamicin sulfate	4.0mg
Amphotericin B	2.0mg
Ethanol	4.0mL

Preparation of Antibiotic Inhibitor Solution: Add components to distilled/deionized water and bring volume to 10.0mL. Mix thoroughly. Filter sterilize.

Preparation of Medium: To 940.0mL of cooled, sterile Columbia blood agar base, aseptically add 50.0mL of rabbit or horse blood serum and 10.0mL of sterile antibiotic inhibitor solution. Pour into sterile Petri dishes or distribute into sterile tubes.

Use: For the selective isolation, cultivation, and differentiation of *Gardnerella vaginalis* from clinical specimens, such as the vaginal discharge of patients with vaginitis. *Gardnerella vaginalis* exhibits β-hemolysis on this medium.

Gassner Agar (Water-Blue Metachrome-Yellow Lactose Agar)

Composition per liter:

Lactose	43.0g
Peptone	14.0g
Agar	13.0g
NaCl	5.0g
Metachrome Yellow	1.25g
Water Blue	0.62g

pH 7.2 ± 0.2 at 25°C

Preparation of Medium: Add components to distilled/deionized water and bring volume to 1.0L. Mix thoroughly. Gently heat and bring to boiling. Mix thoroughly. Distribute into tubes or flasks. Autoclave for 15 min at 15 psi pressure–121°C. Pour into sterile Petri dishes.

Use: For the detection and isolation of pathogenic Enterobacteriaceae. For use in the execution of the German Meat Inspection Law (Deutsches Fleischbeschaugesetz). This culture medium contains Metachrome Yellow, which primarily inhibits the accompanying Gram-positive microbial flora. It also contains lactose, which, when degraded to acid, is shown by the indicator Water Blue, which is deep blue in the acidic range and colorless in the alkaline range. The prepared culture medium is green; in the acidic pH range it becomes blue-green to blue. At alkaline pHs, however, the yellow color of the Metachrome Yellow becomes increasingly apparent.

Gassner Lactose Agar

Composition per liter:

Lactose	50.0g
Agar	13.0g
Meat peptone	7.0g
NaCl	5.0g
Metachrome Yellow	1.25g
Water Blue	0.625g

pH 7.2 ± 0.2 at 25°C

Source: This medium is available as a premixed powder from HiMedia.

Preparation of Medium: Add components to distilled/deionized water and bring volume to 1.0L. Mix thoroughly. Gently heat and bring to boiling. Mix thoroughly. Distribute into tubes or flasks. Autoclave for 15 min at 15 psi pressure–121°C. Pour into sterile Petri dishes.

Use: For the detection and isolation of pathogenic Enterobacteriaceae.

Gassner Lactose HiVeg Agar

Composition per liter:

Lactose	50.0g
Agar	13.0g
Plant peptone No. 1	7.0g
NaCl	5.0g
Metachrome Yellow	1.25g
Water Blue	0.625g

pH 7.2 ± 0.2 at 25°C

Source: This medium is available as a premixed powder from HiMedia.

Preparation of Medium: Add components to distilled/deionized water and bring volume to 1.0L. Mix thoroughly. Gently heat and bring to boiling. Mix thoroughly. Distribute into tubes or flasks. Autoclave for 15 min at 15 psi pressure–121°C. Pour into sterile Petri dishes.

Use: For the detection and isolation of pathogenic Enterobacteriaceae.

Gauze's Medium No. 1

Composition per liter:

Agar	30.0g
Soluble starch	20.0g
KNO_3	1.0g
K_2HPO_4	0.5g
$MgSO_4$	0.5g
NaCl	0.5g
$FeSO_4$	0.01g

Preparation of Medium: Add components to tap water and bring volume to 1.0L. Mix thoroughly. Gently heat and bring to boiling. Distribute into tubes or flasks. Autoclave for 15 min at 15 psi pressure–121°C. Pour into sterile Petri dishes or leave in tubes.

Use: For the cultivation and maintenance of *Streptomyces phaeopurpureus*.

Gauze's Synthetic Medium No.1 (DSMZ Medium 1048)

Composition per liter:

Agar	15.0g
KNO_3	1.0g
NaCl	0.5g
$MgSO_4 \cdot 7H_2O$	0.5g
K_2HPO_4	0.5g
$FeSO_4 \cdot 7H_2O$	0.01g

pH 7.4 ± 0.2 at 25°C

Preparation of Medium: Add components to distilled/deionized water and bring volume to 1.0L. Mix thoroughly. Adjust pH to 7.4. Gently heat while stirring and bring to boiling. Distribute into tubes or flasks. Autoclave for 15 min at 15 psi pressure–121°C. Pour into Petri dishes or aseptically distribute into sterile tubes.

Use: For the cultivation of *Nocariopsis* spp.

GBNA Medium (Gum Base Nalidixic Acid Medium)

Composition per liter:

Gellan gum	8.0g
Pancreatic digest of casein	5.7g
NaCl	1.7g
Papaic digest of soybean meal	1.0g
Glucose	0.83g
K_2HPO_4	0.83g
$MgCl_2 \cdot 6H_2O$	0.33g
Nalidixic acid	0.05g

pH 7.2 ± 0.2 at 25°C

Source: This medium is available as a premixed powder from BD Diagnostic Systems.

Preparation of Medium: Add components to tap water and bring volume to 1.0L. Mix thoroughly. Gently heat and bring to boiling. Distribute into tubes or flasks. Autoclave for 15 min at 15 psi pressure–121°C. Pour into sterile Petri dishes or leave in tubes.

Use: For the isolation and cultivation of *Listeria monocytogenes* from clinical and nonclinical specimens.

GBS Agar Base, Islam (Group B Streptococci Agar) (Islam GBS Agar)

Composition per liter:

Proteose peptone	23.0g
Agar	10.0g
Na_2HPO_4	5.75g
Soluble starch	5.0g
NaH_2PO_4	1.5g
Horse serum, heat inactivated	50.0mL

pH 7.5 ± 0.1 at 25°C

Source: This medium is available as a premixed powder from Oxoid Unipath.

Preparation of Medium: Add components, except horse serum, to distilled/deionized water and bring volume to 950.0mL. Mix thoroughly. Gently heat and bring to boiling. Autoclave for 15 min at 15 psi pressure–121°C. Cool to 45°–50°C. Aseptically add 50.0mL of sterile inactivated horse serum. Mix thoroughly. Pour into sterile Petri dishes or distribute into sterile tubes.

Use: For the isolation and detection of group B streptococci (GBS) in clinical specimens. The medium is designed to exploit the ability of most group B streptococci (GBS) to produce orange/red pigmented colonies when incubated under anaerobic conditions. There is a pigment-enhancing effect around a sulphonamide disc, which does not grow; the enhanced pigment effect can be seen over a radius of 10–20mm. Non-group B organisms able to grow on this medium do not produce the orange/red pigment.

GBS Medium, Rapid (Group B Streptococci Medium) (GBS Medium Base)

Starch	80.0g
Proteose peptone	23.0g
Na_2HPO_4	5.75g
NaH_2PO_4	1.5g
Horse serum, inactivated	50.0mL
Antibiotic inhibitor solution	10.0 mL

pH 7.5 ± 0.2 at 25°C

Source: This medium is available as a premixed powder from Oxoid Unipath and HiMedia.

Antibiotic Inhibitor Solution:
Composition per 10.0mL:

Metronidazole	10.0mg
Gentamicin	2.0mg

Preparation of Antibiotic Inhibitor Solution: Add components to distilled/deionized water and bring volume to 10.0mL. Mix thoroughly. Filter sterilize.

Preparation of Medium: Add components, except horse serum and antibiotic inhibitor solution, to distilled/deionized water and bring volume to 940.0mL. Mix thoroughly. Gently heat and bring to boiling. Autoclave for 15 min at 15 psi pressure–121°C. Cool to 45°–50°C. Aseptically add 50.0mL of sterile heat-inactivated horse serum and 10.0mL of sterile antibiotic inhibitor solution. Mix thoroughly. Aseptically distribute into sterile tubes or flasks. Cool to 5°C and hold at that temperature for 12 hr prior to use.

Use: For the rapid isolation and cultivation of group B streptococci from clinical specimens.

GC Agar (LMG Medium 236)

Composition 1010.0mL:

Solution A	500.0mL
Solution B	500.0mL
Supplement solution	10.0mL

pH 7.2 ± 0.2 at 25°C

Solution A:

Agar	5.0g
GC agar base, 2X	500.0mL

GC Agar Base, 2X:
Composition per 500.0mL:

Agar	10.0g
Pancreatic digest of casein	7.5g
Peptic digest of animal tissue	7.5g
NaCl	5.0g
K_2HPO_4	4.0g
Cornstarch	1.0g
KH_2PO_4	1.0g

Source: GC agar base is available as a premixed powder from BD Diagnostic Systems. This base may be replaced by GC medium base available from BD Diagnostic Systems.

Preparation of GC Agar Base, 2X: Add components to distilled/deionized water and bring volume to 500.0mL. Mix thoroughly.

Preparation of Solution A: Add 5.0g agar to 500.0mL GC agar base, 2X. Mix thoroughly. Gently heat until boiling. Autoclave for 15 min at 15 psi pressure–121°C. Cool to 45°–50°C.

Solution B:
Composition per 500.0mL:

Bovine hemoglobin	10.0g

Preparation of Solution B: Add bovine hemoglobin to distilled/deionized water and bring volume to 500.0mL. Mix thoroughly. Autoclave for 15 min at 15 psi pressure–121°C. Cool to 45°–50°C.

Supplement Solution:
Composition per liter:

Glucose	100.0g
L-Cysteine·HCl	25.9g
L-Glutamine	10.0g
L-Cystine	1.1g
Adenine	1.0g
Nicotinamide adenine dinucleotide	0.25g
Vitamin B_{12}	0.1g
Thiamine pyrophosphate	0.1g
Guanine·HCl	0.03g
$Fe(NO_3)_3 \cdot 6H_2O$	0.02g
p-Aminobenzoic acid	0.013g
Thiamine·HCl	3.0mg

Source: The supplement solution (IsoVitaleX® enrichment) is available from BD Diagnostic Systems. This enrichment may be replaced by supplement VX from BD Diagnostic Systems.

Preparation of Supplement Solution: Add components to distilled/deionized water and bring volume to 1.0L. Mix thoroughly. Filter sterilize.

Preparation of Medium: To 500.0mL of sterile GC agar base, aseptically add 500.0mL of sterile hemoglobin solution at 45°–50°C. Mix thoroughly. Aseptically add 10.0mL of sterile supplement solution. Mix thoroughly. Pour into sterile Petri dishes or distribute into sterile tubes.

Use: For the cultivation and maintenance of *Haemophilus parainfluenzae* and *Neisseria* spp.

GC Agar (ATCC Medium 814)

Composition per 1010.0mL:

GC agar base, 2X	500.0mL
Hemoglobin solution	500.0mL
Supplement solution	10.0mL

pH 7.2 ± 0.2 at 25°C

GC Agar Base, 2X:

Composition per 500.0mL:

Agar	10.0g
Pancreatic digest of casein	7.5g
Peptic digest of animal tissue	7.5g
NaCl	5.0g
K_2HPO_4	4.0g
Cornstarch	1.0g
KH_2PO_4	1.0g

Source: GC agar base is available as a premixed powder from BD Diagnostic Systems. This base may be replaced by GC medium base available from BD Diagnostic Systems.

Preparation of GC Agar Base, 2X: Add components to distilled/deionized water and bring volume to 500.0mL. Mix thoroughly. Gently heat until boiling. Autoclave for 15 min at 15 psi pressure–121°C. Cool to 45°–50°C.

Hemoglobin Solution:

Composition 500.0mL:

Bovine hemoglobin	10.0g

Preparation of Hemoglobin Solution: Add bovine hemoglobin to distilled/deionized water and bring volume to 500.0mL. Mix thoroughly. Autoclave for 15 min at 15 psi pressure–121°C. Cool to 45°–50°C.

Supplement Solution:

Composition per liter:

Glucose	100.0g
L-Cysteine·HCl	25.9g
L-Glutamine	10.0g
L-Cystine	1.1g
Adenine	1.0g
Nicotinamide adenine dinucleotide	0.25g
Vitamin B_{12}	0.1g
Thiamine pyrophosphate	0.1g
Guanine·HCl	0.03g
$Fe(NO_3)_3 \cdot 6H_2O$	0.02g
p-Aminobenzoic acid	0.013g
Thiamine·HCl	3.0mg

Source: The supplement solution (IsoVitaleX® enrichment) is available from BD Diagnostic Systems. This enrichment may be replaced by supplement VX from BD Diagnostic Systems.

Preparation of Supplement Solution: Add components to distilled/deionized water and bring volume to 1.0L. Mix thoroughly. Filter sterilize.

Preparation of Medium: To 500.0mL of sterile GC agar base, aseptically add 500.0mL of sterile hemoglobin solution at 45°–50°C. Mix thoroughly. Aseptically add 10.0mL of sterile supplement solution. Mix thoroughly. Pour into sterile Petri dishes or distribute into sterile tubes.

Use: For the isolation and cultivation of fastidious bacteria, especially *Neisseria* and *Haemophilus* species. For the cultivation and maintenance of *Branhamella catarrhalis, Campylobacter pylori, Eikenella corrodens, Helicobacter pylori, Moraxella nonliquefaciens, Morococcus cerebrosis, Oligella ureolytica, Oligella urethralis, Pasteurella volantium, Proteus mirabilis,* and *Taylorella equigenitalis*.

GC Agar (GC Medium) (ATCC Medium 1351)

Composition per liter:

GC agar base	950.0mL
Blood, defibrinated	50.0mL

pH 7.2 ± 0.2 at 25°C

GC Agar Base:

Composition per liter:

Agar	10.0g
Pancreatic digest of casein	7.5g
Peptic digest of animal tissue	7.5g
NaCl	5.0g
K_2HPO_4	4.0g
Cornstarch	1.0g
KH_2PO_4	1.0g

Source: GC agar base is available as a premixed powder from BD Diagnostic Systems. This base may be replaced by GC medium base available from BD Diagnostic Systems.

Preparation of GC Agar Base: Add components to distilled/deionized water and bring volume to 1.0L. Mix thoroughly. Gently heat until boiling. Autoclave for 15 min at 15 psi pressure–121°C. Cool to 75°–80°C.

Preparation of Medium: To 950.0mL of sterile GC agar base aseptically add 50.0mL sterile defibrinated blood with thorough mixing and maintain at 75°–80°C for 15–20 min until the medium is chocolatized. Pour into sterile Petri dishes or distribute into sterile tubes.

Use: For the isolation and cultivation of fastidious bacteria, especially *Neisseria* and *Haemophilus* species. For the cultivation and maintenance of *Branhamella catarrhalis, Campylobacter pylori, Eikenella corrodens, Helicobacter pylori, Moraxella nonliquefaciens, Morococcus cerebrosis, Oligella ureolytica, Oligella urethralis, Pasteurella volantium, Proteus mirabilis,* and *Taylorella equigenitalis*.

GC Agar with Ampicillin

Composition per 1020.0mL:

GC agar base, 2X	500.0mL
Hemoglobin solution	500.0mL

Supplement solution	10.0mL
Ampicillin solution	10.0mL

pH 7.2 ± 0.2 at 25°C

GC Agar Base, 2X:

Composition per 500.0mL:

Agar	10.0g
Pancreatic digest of casein	7.5g
Peptic digest of animal tissue	7.5g
NaCl	5.0g
K_2HPO_4	4.0g
Cornstarch	1.0g
KH_2PO_4	1.0g

Source: GC agar base is available as a premixed powder from BD Diagnostic Systems. This base may be replaced by GC medium base available from BD Diagnostic Systems.

Preparation of GC Agar Base, 2X: Add components to distilled/deionized water and bring volume to 500.0mL. Mix thoroughly. Gently heat until boiling. Autoclave for 15 min at 15 psi pressure–121°C. Cool to 45°–50°C.

Hemoglobin Solution:

Composition per 500.0mL:

Bovine hemoglobin	10.0g

Preparation of Hemoglobin Solution: Add bovine hemoglobin to distilled/deionized water and bring volume to 500.0mL. Mix thoroughly. Autoclave for 15 min at 15 psi pressure–121°C. Cool to 45°–50°C.

Supplement Solution:

Composition per liter:

Glucose	100.0g
L-Cysteine·HCl	25.9g
L-Glutamine	10.0g
L-Cystine	1.1g
Adenine	1.0g
Nicotinamide adenine dinucleotide	0.25g
Vitamin B_{12}	0.1g
Thiamine pyrophosphate	0.1g
Guanine·HCl	0.03g
$Fe(NO_3)_3 \cdot 6H_2O$	0.02g
p-Aminobenzoic acid	0.013g
Thiamine·HCl	3.0mg

Source: The supplement solution (IsoVitaleX® enrichment) is available from BD Diagnostic Systems. This enrichment may be replaced by supplement VX from BD Diagnostic Systems.

Preparation of Supplement Solution: Add components to distilled/deionized water and bring volume to 1.0L. Mix thoroughly. Filter sterilize.

Ampicillin Solution:

Composition per 10.0mL:

Ampicillin	0.02g

Preparation of Ampicillin Solution: Add ampicillin to distilled/deionized water and bring volume to 10.0mL. Mix thoroughly. Filter sterilize.

Preparation of Medium: To 500.0mL of sterile GC agar base, aseptically add 500.0mL of sterile hemoglobin solution at 45°–50°C. Mix thoroughly. Aseptically add 10.0mL of sterile supplement solution and 10.0mL of sterile ampicillin solution. Mix thoroughly. Pour into sterile Petri dishes or distribute into sterile tubes.

Use: For the cultivation and maintenance of *Branhamella catarrhalis, Haemophilus influenzae,* and *Haemophilus parainfluenzae.*

GC Agar with Ampicillin and Gentamicin

Composition per 1030.0mL:

GC agar base, 2X	500.0mL
Hemoglobin solution	500.0mL
Supplement solution	10.0mL
Ampicillin solution	10.0mL
Gentamicin solution	10.0mL

pH 7.2 ± 0.2 at 25°C

GC Agar Base, 2X:

Composition per 500.0mL:

Agar	10.0g
Pancreatic digest of casein	7.5g
Peptic digest of animal tissue	7.5g
NaCl	5.0g
K_2HPO_4	4.0g
Cornstarch	1.0g
KH_2PO_4	1.0g

Source: GC agar base is available as a premixed powder from BD Diagnostic Systems. This base may be replaced by GC medium base available from BD Diagnostic Systems.

Preparation of GC Agar Base, 2X: Add components to distilled/deionized water and bring volume to 500.0mL. Mix thoroughly. Gently heat until boiling. Autoclave for 15 min at 15 psi pressure–121°C. Cool to 45°–50°C.

Hemoglobin Solution:

Composition per 500.0mL:

Bovine hemoglobin	10.0g

Preparation of Hemoglobin Solution: Add bovine hemoglobin to distilled/deionized water and bring volume to 500.0mL. Mix thoroughly. Autoclave for 15 min at 15 psi pressure–121°C. Cool to 45°–50°C.

Supplement Solution:

Composition per liter:

Glucose	100.0g
L-Cysteine·HCl	25.9g
L-Glutamine	10.0g
L-Cystine	1.1g
Adenine	1.0g
Nicotinamide adenine dinucleotide	0.25g
Vitamin B_{12}	0.1g
Thiamine pyrophosphate	0.1g
Guanine·HCl	0.03g
$Fe(NO_3)_3 \cdot 6H_2O$	0.02g
p-Aminobenzoic acid	0.013g
Thiamine·HCl	3.0mg

Source: The supplement solution (IsoVitaleX® enrichment) is available from BD Diagnostic Systems. This enrichment may be replaced by supplement VX from BD Diagnostic Systems.

Preparation of Supplement Solution: Add components to distilled/deionized water and bring volume to 1.0L. Mix thoroughly. Filter sterilize.

Ampicillin Solution:

Composition per 10.0mL:

Ampicillin	0.01g

Preparation of Ampicillin Solution: Add ampicillin to distilled/deionized water and bring volume to 10.0mL. Mix thoroughly. Filter sterilize.

Gentamicin Solution:
Composition per 10.0mL:

Gentamicin	2.0mg

Preparation of Gentamicin Solution: Add gentamicin to distilled/deionized water and bring volume to 10.0mL. Mix thoroughly. Filter sterilize.

Preparation of Medium: To 500.0mL of sterile GC agar base, aseptically add 500.0mL of sterile hemoglobin solution at 45°–50°C. Mix thoroughly. Aseptically add 10.0mL of sterile supplement solution, 10.0mL of sterile ampicillin solution, and 10.0mL of sterile gentamicin solution. Mix thoroughly. Pour into sterile Petri dishes or distribute into sterile tubes.

Use: For the cultivation and maintenance of *Haemophilus parainfluenzae*.

GC Agar with Ampicillin and Tetracycline

Composition per 1030.0mL:

GC agar base, 2X	500.0mL
Hemoglobin solution	500.0mL
Supplement solution	10.0mL
Ampicillin solution	10.0mL
Tetracycline solution	10.0mL

pH 7.2 ± 0.2 at 25°C

GC Agar Base, 2X:
Composition per 500.0mL:

Agar	10.0g
Pancreatic digest of casein	7.5g
Peptic digest of animal tissue	7.5g
NaCl	5.0g
K_2HPO_4	4.0g
Cornstarch	1.0g
KH_2PO_4	1.0g

Source: GC agar base is available as a premixed powder from BD Diagnostic Systems. This base may be replaced by GC medium base available from BD Diagnostic Systems.

Preparation of GC Agar Base, 2X: Add components to distilled/deionized water and bring volume to 500.0mL. Mix thoroughly. Gently heat until boiling. Autoclave for 15 min at 15 psi pressure–121°C. Cool to 45°–50°C.

Hemoglobin Solution:
Composition per 500.0mL:

Bovine hemoglobin	10.0g

Preparation of Hemoglobin Solution: Add bovine hemoglobin to distilled/deionized water and bring volume to 500.0mL. Mix thoroughly. Autoclave for 15 min at 15 psi pressure–121°C. Cool to 45°–50°C.

Supplement Solution:
Composition per liter:

Glucose	100.0g
L-Cysteine·HCl	25.9g
L-Glutamine	10.0g
L-Cystine	1.1g
Adenine	1.0g
Nicotinamide adenine dinucleotide	0.25g
Vitamin B_{12}	0.1g
Thiamine pyrophosphate	0.1g
Guanine·HCl	0.03g
$Fe(NO_3)_3·6H_2O$	0.02g
p-Aminobenzoic acid	0.013g
Thiamine·HCl	3.0mg

Source: The supplement solution (IsoVitaleX® enrichment) is available from BD Diagnostic Systems. This enrichment may be replaced by supplement VX from BD Diagnostic Systems.

Preparation of Supplement Solution: Add components to distilled/deionized water and bring volume to 1.0L. Mix thoroughly. Filter sterilize.

Ampicillin Solution:
Composition per 10.0mL:

Ampicillin	0.01g

Preparation of Ampicillin Solution: Add ampicillin to distilled/deionized water and bring volume to 10.0mL. Mix thoroughly. Filter sterilize.

Tetracycline Solution:
Composition per 10.0mL:

Tetracycline	5.0mg

Preparation of Tetracycline Solution: Add tetracycline to distilled/deionized water and bring volume to 10.0mL. Mix thoroughly. Filter sterilize.

Preparation of Medium: To 500.0mL of sterile GC agar base, aseptically add 500.0mL of sterile hemoglobin solution at 45°–50°C. Mix thoroughly. Aseptically add 10.0mL of sterile supplement solution, 10.0mL of sterile ampicillin solution, and 10.0mL of sterile tetracycline solution. Mix thoroughly. Pour into sterile Petri dishes or distribute into sterile tubes.

Use: For the cultivation and maintenance of *Haemophilus parainfluenzae*.

GC Agar Base with Blood

Composition per liter:

Peptone, special	15.0g
Agar	10.0g
NaCl	5.0g
K_2HPO_4	4.0g
KH_2PO_4	1.0g
Cornstarch	1.0g
Blood, defibrinated	50.0mL

pH 7.2 ± 0.2 at 25°C

Source: This medium,without blood, is available as a premixed powder from HiMedia.

Preparation of Medium: Add components, except blood, to distilled/deionized water and bring volume to 950.0L. Mix thoroughly. Gently heat until boiling. Autoclave for 15 min at 15 psi pressure–121°C. Cool to 75°–80°C. Add 50.0mL sterile defibrinated blood with thorough mixing and maintain at 75°–80°C for 15–20 min until the medium is chocolatized. Pour into sterile Petri dishes or distribute into sterile tubes.

Use: For the isolation and cultivation of fastidious bacteria, especially *Neisseria* and *Haemophilus* species.

GC Agar Base with Kellogg's Supplement (ATCC Medium 1674)

Composition per liter:

GC agar base	990.0mL
Kellogg's supplement solution	10.0mL

pH 7.2 ± 0.2 at 25°C

GC Agar Base:

Composition per 990.0mL:

Agar	10.0g
Pancreatic digest of casein	7.5g
Peptic digest of animal tissue	7.5g
NaCl	5.0g
K_2HPO_4	4.0g
Cornstarch	1.0g
KH_2PO_4	1.0g

Source: GC agar base is available as a premixed powder from BD Diagnostic Systems. This base may be replaced by GC medium base available from BD Diagnostic Systems.

Preparation of GC Agar Base: Add components to distilled/deionized water and bring volume to 990.0mL. Mix thoroughly. Gently heat until boiling. Autoclave for 15 min at 15 psi pressure–121°C. Cool to 45°–50°C.

Kellogg's Supplement Solution:

Composition per 100.0mL:

Glucose	40.0g
L-Glutamine	0.5g
$Fe(NO_3)_3 \cdot 6H_2O$	0.05g
Cocarboxylase	2.0mg

Preparation of Kellogg's Supplement Solution: Add components to distilled/deionized water and bring volume to 100.0mL. Mix thoroughly. Filter sterilize.

Preparation of Medium: To 990.0mL of sterile GC agar base, aseptically add 10.0mL of sterile Kellogg's supplement solution warmed to 45°–50°C. Mix thoroughly. Pour into sterile Petri dishes or distribute into sterile tubes.

Use: For the cultivation and maintenance of *Arsenophonus nasoniae*.

GC Agar with Chloramphenicol, Tetracycline, and Ampicillin

Composition per 1040.0mL:

GC agar base, 2X	500.0mL
Hemoglobin solution	500.0mL
Supplement solution	10.0mL
Ampicillin solution	10.0mL
Tetracycline solution	10.0mL
Chloramphenicol solution	10.0mL

pH 7.2 ± 0.2 at 25°C

GC Agar Base, 2X:

Composition per 500.0mL:

Agar	10.0g
Pancreatic digest of casein	7.5g
Peptic digest of animal tissue	7.5g
NaCl	5.0g
K_2HPO_4	4.0g
Cornstarch	1.0g
KH_2PO_4	1.0g

Source: GC agar base is available as a premixed powder from BD Diagnostic Systems. This base may be replaced by GC medium base available from BD Diagnostic Systems.

Preparation of GC Agar Base, 2X: Add components to distilled/deionized water and bring volume to 500.0mL. Mix thoroughly. Gently heat until boiling. Autoclave for 15 min at 15 psi pressure–121°C. Cool to 45°–50°C.

Hemoglobin Solution:

Composition per 500.0mL:

Bovine hemoglobin	10.0g

Preparation of Hemoglobin Solution: Add bovine hemoglobin to distilled/deionized water and bring volume to 500.0mL. Mix thoroughly. Autoclave for 15 min at 15 psi pressure–121°C. Cool to 45°–50°C.

Supplement Solution:

Composition per liter:

Glucose	100.0g
L-Cysteine·HCl	25.9g
L-Glutamine	10.0g
L-Cystine	1.1g
Adenine	1.0g
Nicotinamide adenine dinucleotide	0.25g
Vitamin B_{12}	0.1g
Thiamine pyrophosphate	0.1g
Guanine·HCl	0.03g
$Fe(NO_3)_3 \cdot 6H_2O$	0.02g
p-Aminobenzoic acid	0.013g
Thiamine·HCl	3.0mg

Source: The supplement solution (IsoVitaleX® enrichment) is available from BD Diagnostic Systems. This enrichment may be replaced by supplement VX from BD Diagnostic Systems.

Preparation of Supplement Solution: Add components to distilled/deionized water and bring volume to 1.0L. Mix thoroughly. Filter sterilize.

Ampicillin Solution:

Composition per 10.0mL:

Ampicillin	0.01g

Preparation of Ampicillin Solution: Add ampicillin to distilled/deionized water and bring volume to 10.0mL. Mix thoroughly. Filter sterilize.

Tetracycline Solution:

Composition per 10.0mL:

Tetracycline	5.0mg

Preparation of Tetracycline Solution: Add tetracycline to distilled/deionized water and bring volume to 10.0mL. Mix thoroughly. Filter sterilize.

Chloramphenicol Solution:

Composition per 10.0mL:

Chloramphenicol	5.0mg

Preparation of Chloramphenicol Solution: Add chloramphenicol to distilled/deionized water and bring volume to 10.0mL. Mix thoroughly. Filter sterilize.

Preparation of Medium: To 500.0mL of sterile GC agar base, aseptically add 500.0mL of sterile hemoglobin solution at 45°–50°C. Mix thoroughly. Aseptically add 10.0mL of sterile supplement solution, 10.0mL of sterile ampicillin solution, 10.0mL of sterile chloramphenicol solution, and 10.0mL of sterile tetracycline solution. Mix thoroughly. Pour into sterile Petri dishes or distribute into sterile tubes.

Use: For the cultivation and maintenance of *Haemophilus parainfluenzae*.

GC Agar with Defined Supplements

GC agar base	990.0mL
Defined supplements solution	10.0mL

pH 7.2 ± 0.2 at 25°C

GC Agar Base:
Composition per liter:

Agar	10.0g
Pancreatic digest of casein	7.5g
Peptic digest of animal tissue	7.5g
NaCl	5.0g
K_2HPO_4	4.0g
Cornstarch	1.0g
KH_2PO_4	1.0g

Source: GC agar base is available as a premixed powder from BD Diagnostic Systems. This base may be replaced by GC medium base available from BD Diagnostic Systems.

Preparation of GC Agar Base: Add components to distilled/deionized water and bring volume to 1.0L. Mix thoroughly. Gently heat until boiling. Autoclave for 15 min at 15 psi pressure–121°C. Cool to 45°–50°C.

Defined Supplements Solution:
Composition per 100.0mL:

Glucose	40.0g
Glutamine	1.0g
$Fe(NO_3)_3 \cdot 6H_2O$	0.05g
Cocarboxylase	2.0mg

Preparation of Defined Supplements Solution: Add components to distilled/deionized water and bring volume to 100.0mL. Mix thoroughly. Filter sterilize.

Preparation of Medium: To 990.0mL of sterile GC agar base, aseptically add 10.0mL of sterile defined supplements solution. Mix thoroughly. Pour into sterile Petri dishes or distribute into sterile tubes.

Use: For the cultivation and maintenance of *Neisseria gonorrhoeae*.

GC Agar with Penicillin G

Composition per 1020.0mL:

GC medium base, 2×	500.0mL
Hemoglobin solution	500.0mL
Supplement solution	10.0mL
Penicillin G solution	10.0mL

pH 7.2 ± 0.2 at 25°C

GC Medium Base, 2X:
Composition per 500.0mL:

Proteose peptone No. 3	15.0g
Agar	10.0g
NaCl	5.0g
K_2HPO_4	4.0g
Cornstarch	1.0g
KH_2PO_4	1.0g

Source: GC medium base is available as a premixed powder from BD Diagnostic Systems.

Preparation of GC Medium Base, 2X: Add components to distilled/deionized water and bring volume to 500.0mL. Mix thoroughly. Gently heat until boiling. Autoclave for 15 min at 15 psi pressure–121°C. Cool to 45°–50°C.

Hemoglobin Solution:
Composition per 500.0mL:

Bovine hemoglobin	10.0g

Preparation of Hemoglobin Solution: Add bovine hemoglobin to distilled/deionized water and bring volume to 500.0mL. Mix thoroughly. Autoclave for 15 min at 15 psi pressure–121°C. Cool to 45°–50°C.

Supplement Solution:
Composition per liter:

Glucose	100.0g
L-Cysteine·HCl	25.9g
L-Glutamine	10.0g
L-Cystine	1.1g
Adenine	1.0g
Nicotinamide adenine dinucleotide	0.25g
Vitamin B_{12}	0.1g
Thiamine pyrophosphate	0.1g
Guanine·HCl	0.03g
$Fe(NO_3)_3 \cdot 6H_2O$	0.02g
p-Aminobenzoic acid	0.013g
Thiamine·HCl	3.0mg

Source: The supplement solution (IsoVitaleX® enrichment) is available from BD Diagnostic Systems. This enrichment may be replaced by supplement VX from BD Diagnostic Systems.

Preparation of Supplement Solution: Add components to distilled/deionized water and bring volume to 1.0L. Mix thoroughly. Filter sterilize.

Penicillin G Solution:
Composition per 10.0mL:

Penicillin G	0.05g

Preparation of Penicillin G Solution: Add penicillin G to distilled/deionized water and bring volume to 10.0mL. Mix thoroughly. Filter sterilize.

Preparation of Medium: To 500.0mL of sterile GC medium base, aseptically add 500.0mL of sterile hemoglobin solution at 45°–50°C. Mix thoroughly. Aseptically add 10.0mL of sterile supplement solution and 10.0mL of sterile penicillin G solution. Mix thoroughly. Pour into sterile Petri dishes or distribute into sterile tubes.

Use: For the cultivation and maintenance of *Neisseria gonorrhoeae*.

GC Agar with Streptomycin and Chloramphenicol

Composition per 1030.0mL:

GC agar base, 2X	500.0mL
Hemoglobin solution	500.0mL
Supplement solution	10.0mL
Streptomcyin solution	10.0mL
Chloramphenicol solution	10.0mL

pH 7.2 ± 0.2 at 25°C

GC Agar Base, 2X:
Composition per 500.0mL:

Agar	10.0g
Pancreatic digest of casein	7.5g
Peptic digest of animal tissue	7.5g

NaCl	5.0g
K_2HPO_4	4.0g
Cornstarch	1.0g
KH_2PO_4	1.0g

Source: GC agar base is available as a premixed powder from BD Diagnostic Systems. This base may be replaced by GC medium base available from BD Diagnostic Systems.

Preparation of GC Agar Base, 2X: Add components to distilled/deionized water and bring volume to 500.0mL. Mix thoroughly. Gently heat until boiling. Autoclave for 15 min at 15 psi pressure–121°C. Cool to 45°–50°C.

Hemoglobin Solution:
Composition per 500.0mL:

Bovine hemoglobin	10.0g

Preparation of Hemoglobin Solution: Add bovine hemoglobin to distilled/deionized water and bring volume to 500.0mL. Mix thoroughly. Autoclave for 15 min at 15 psi pressure–121°C. Cool to 45°–50°C.

Supplement Solution:
Composition per liter:

Glucose	100.0g
L-Cysteine·HCl	25.9g
L-Glutamine	10.0g
L-Cystine	1.1g
Adenine	1.0g
Nicotinamide adenine dinucleotide	0.25g
Vitamin B_{12}	0.1g
Thiamine pyrophosphate	0.1g
Guanine·HCl	0.03g
$Fe(NO_3)_3 \cdot 6H_2O$	0.02g
p-Aminobenzoic acid	0.013g
Thiamine·HCl	3.0mg

Source: The supplement solution (IsoVitaleX® enrichment) is available from BD Diagnostic Systems. This enrichment may be replaced by supplement VX from BD Diagnostic Systems.

Preparation of Supplement Solution: Add components to distilled/deionized water and bring volume to 1.0L. Mix thoroughly. Filter sterilize.

Streptomycin Solution:
Composition per 10.0mL:

Streptomycin	0.25g

Preparation of Streptomycin Solution: Add streptomycin to distilled/deionized water and bring volume to 10.0mL. Mix thoroughly. Filter sterilize.

Chloramphenicol Solution:
Composition per 10.0mL:

Chloramphenicol	25.0mg

Preparation of Chloramphenicol Solution: Add chloramphenicol to distilled/deionized water and bring volume to 10.0mL. Mix thoroughly. Filter sterilize.

Preparation of Medium: To 500.0mL of sterile GC agar base, aseptically add 500.0mL of sterile hemoglobin solution at 45°–50°C. Mix thoroughly. Aseptically add 10.0mL of sterile supplement solution, 10.0mL of sterile streptomy6cin solution, and 10.0mL of sterile chloramphenicol solution. Mix thoroughly. Pour into sterile Petri dishes or distribute into sterile tubes.

Use: For the cultivation of *Azorhizophilus paspali.*

GC Agar with Supplement A

Composition per 1020.0mL:

GC medium base, 2X	500.0mL
Hemoglobin solution	500.0mL
Supplement solution	10.0mL
Supplement A	10.0mL

pH 7.2 ± 0.2 at 25°C

GC Medium Base, 2X:
Composition per 500.0mL:

Proteose peptone No. 3	15.0g
Agar	10.0g
NaCl	5.0g
K_2HPO_4	4.0g
Cornstarch	1.0g
KH_2PO_4	1.0g

Source: GC medium base and Supplement A are available as a premixed powder from BD Diagnostic Systems.

Preparation of GC Medium Base, 2X: Add components to distilled/deionized water and bring volume to 500.0mL. Mix thoroughly. Gently heat until boiling. Autoclave for 15 min at 15 psi pressure–121°C. Cool to 45°–50°C.

Hemoglobin Solution:
Composition per 500.0mL:

Bovine hemoglobin	10.0g

Preparation of Hemoglobin Solution: Add bovine hemoglobin to distilled/deionized water and bring volume to 500.0mL. Mix thoroughly. Autoclave for 15 min at 15 psi pressure–121°C. Cool to 45°–50°C.

Supplement Solution:
Composition per liter:

Glucose	100.0g
L-Cysteine·HCl	25.9g
L-Glutamine	10.0g
L-Cystine	1.1g
Adenine	1.0g
Nicotinamide adenine dinucleotide	0.25g
Vitamin B_{12}	0.1g
Thiamine pyrophosphate	0.1g
Guanine·HCl	0.03g
$Fe(NO_3)_3 \cdot 6H_2O$	0.02g
p-Aminobenzoic acid	0.013g
Thiamine·HCl	3.0mg

Source: The supplement solution (IsoVitaleX® enrichment) is available from BD Diagnostic Systems. This enrichment may be replaced by supplement VX from BD Diagnostic Systems.

Preparation of Supplement Solution: Add components to distilled/deionized water and bring volume to 1.0L. Mix thoroughly. Filter sterilize.

Supplement A:
Composition per 10.0mL:

Supplement A contains yeast concentrate with Crystal Violet.

Preparation of Supplement A: Add components to distilled/deionized water and bring volume to 10.0mL. Mix thoroughly. Filter sterilize.

Preparation of Medium: To 500.0mL of sterile GC medium base, aseptically add 500.0mL of sterile hemoglobin solution at 45°–50°C. Mix thoroughly. Aseptically add 10.0mL of sterile supplement solution and 10.0mL of sterile supplement A solution. Mix thoroughly. Pour into sterile Petri dishes or distribute into sterile tubes.

Use: For the cultivation of *Neisseria gonorrhoeae*, other *Neisseria* species, and *Haemophilus* species.

GC Agar with Supplement A and with VCN Inhibitor

Composition per 1030.0mL:

GC medium base, 2X	500.0mL
Hemoglobin solution	500.0mL
Supplement solution	10.0mL
Supplement A	10.0mL
VCN inhibitor	10.0mL

pH 7.2 ± 0.2 at 25°C

GC Medium Base, 2X:

Composition per 500.0mL:

Proteose peptone No. 3	15.0g
Agar	10.0g
NaCl	5.0g
K_2HPO_4	4.0g
Cornstarch	1.0g
KH_2PO_4	1.0g

Source: GC medium base and Supplement A are available as a premixed powder from BD Diagnostic Systems.

Preparation of GC Medium Base, 2X: Add components to distilled/deionized water and bring volume to 500.0mL. Mix thoroughly. Gently heat until boiling. Autoclave for 15 min at 15 psi pressure–121°C. Cool to 45°–50°C.

Hemoglobin Solution:

Composition per 500.0mL:

Bovine hemoglobin	10.0g

Preparation of Hemoglobin Solution: Add bovine hemoglobin to distilled/deionized water and bring volume to 500.0mL. Mix thoroughly. Autoclave for 15 min at 15 psi pressure–121°C. Cool to 45°–50°C.

Supplement Solution:

Composition per liter:

Glucose	100.0g
L-Cysteine·HCl	25.9g
L-Glutamine	10.0g
L-Cystine	1.1g
Adenine	1.0g
Nicotinamide adenine dinucleotide	0.25g
Vitamin B_{12}	0.1g
Thiamine pyrophosphate	0.1g
Guanine·HCl	0.03g
$Fe(NO_3)_3 \cdot 6H_2O$	0.02g
p-Aminobenzoic acid	0.013g
Thiamine·HCl	3.0mg

Source: The supplement solution (IsoVitaleX® enrichment) is available from BD Diagnostic Systems. This enrichment may be replaced by supplement VX from BD Diagnostic Systems.

Preparation of Supplement Solution: Add components to distilled/deionized water and bring volume to 1.0L. Mix thoroughly. Filter sterilize.

Supplement A:

Composition per 10.0mL:

Supplement A contains yeast concentrate with Crystal Violet.

Preparation of Supplement A: Add components to distilled/deionized water and bring volume to 10.0mL. Mix thoroughly. Filter sterilize.

VCN Inhibitor:

Composition per 10.0mL:

Colistin	7.5mg
Vancomycin	3.0mg
Nystatin	12,500U

Preparation of VCN Inhibitor: Add components to distilled/deionized water and bring volume to 10.0mL. Mix thoroughly. Filter sterilize.

Preparation of Medium: To 500.0mL of sterile GC medium base, aseptically add 500.0mL of sterile hemoglobin solution at 45°–50°C. Mix thoroughly. Aseptically add 10.0mL of sterile supplement solution, 10.0mL of sterile supplement A solution, and 10.0mL of sterile VCN inhibitor. Mix thoroughly. Pour into sterile Petri dishes or distribute into sterile tubes.

Use: For the cultivation of *Neisseria gonorrhoeae*, other *Neisseria* species, and *Haemophilus* species.

GC Agar with Supplement A and VCTN Inhibitor

Composition per 1030.0mL:

GC medium base, 2X	500.0mL
Hemoglobin solution	500.0mL
Supplement solution	10.0mL
Supplement A	10.0mL
VCTN inhibitor	10.0mL

pH 7.2 ± 0.2 at 25°C

GC Medium Base, 2X:

Composition per 500.0mL:

Proteose peptone No. 3	15.0g
Agar	10.0g
NaCl	5.0g
K_2HPO_4	4.0g
Cornstarch	1.0g
KH_2PO_4	1.0g

Source: GC medium base and Supplement A are available as a premixed powder from BD Diagnostic Systems.

Preparation of GC Medium Base, 2X: Add components to distilled/deionized water and bring volume to 500.0mL. Mix thoroughly. Gently heat until boiling. Autoclave for 15 min at 15 psi pressure–121°C. Cool to 45°–50°C.

Hemoglobin Solution:

Composition per 500.0mL:

Bovine hemoglobin	10.0g

Preparation of Hemoglobin Solution: Add bovine hemoglobin to distilled/deionized water and bring volume to 500.0mL. Mix thoroughly. Autoclave for 15 min at 15 psi pressure–121°C. Cool to 45°–50°C.

Supplement Solution:

Composition per liter:

Glucose	100.0g
L-Cysteine·HCl	25.9g
L-Glutamine	10.0g
L-Cystine	1.1g
Adenine	1.0g
Nicotinamide adenine dinucleotide	0.25g

Vitamin B_{12}	0.1g
Thiamine pyrophosphate	0.1g
Guanine·HCl	0.03g
$Fe(NO_3)_3 \cdot 6H_2O$	0.02g
p-Aminobenzoic acid	0.013g
Thiamine·HCl	3.0mg

Source: The supplement solution (IsoVitaleX® enrichment) is available from BD Diagnostic Systems. This enrichment may be replaced by supplement VX from BD Diagnostic Systems.

Preparation of Supplement Solution: Add components to distilled/deionized water and bring volume to 1.0L. Mix thoroughly. Filter sterilize.

Supplement A:
Composition per 10.0mL:
Supplement A contains yeast concentrate with Crystal Violet.

Preparation of Supplement A: Add components to distilled/deionized water and bring volume to 10.0mL. Mix thoroughly. Filter sterilize.

VCTN Inhibitor:
Composition per liter:

Colistin	7.5mg
Trimethoprim lactate	5.0mg
Vancomycin	4.0mg
Nystatin	12,500U

Preparation of VCTN Inhibitor: Add components to distilled/deionized water and bring volume to 10.0mL. Mix thoroughly. Filter sterilize.

Preparation of Medium: To 500.0mL of sterile GC medium base, aseptically add 500.0mL of sterile hemoglobin solution at 45°–50°C. Mix thoroughly. Aseptically add 10.0mL of sterile supplement solution, 10.0mL of sterile supplement A, and 10.0mL of sterile VCTN inhibitor. Mix thoroughly. Pour into sterile Petri dishes or distribute into sterile tubes.

Use: For the cultivation of *Neisseria gonorrhoeae*, other *Neisseria* species, and *Haemophilus* species.

GC Agar with Supplement B

GC agar base	990.0mL
Supplement B	10.0mL

pH 7.2 ± 0.2 at 25°C

GC Agar Base:
Composition per liter:

Agar	10.0g
Pancreatic digest of casein	7.5g
Peptic digest of animal tissue	7.5g
NaCl	5.0g
K_2HPO_4	4.0g
Cornstarch	1.0g
KH_2PO_4	1.0g

Source: GC agar base is available as a premixed powder from BD Diagnostic Systems. This base may be replaced by GC medium base available from BD Diagnostic Systems.

Preparation of GC Agar Base: Add components to distilled/deionized water and bring volume to 1.0L. Mix thoroughly. Gently heat until boiling. Autoclave for 15 min at 15 psi pressure–121°C. Cool to 45°–50°C.

Supplement B:
Composition per 10.0mL:
Supplement B contains yeast concentrate, glutamine, coenzyme, cocarboxylase, hematin, and growth factors.

Preparation of Supplement B: Add components to distilled/deionized water and bring volume to 10.0mL. Mix thoroughly. Filter sterilize.

Preparation of Medium: To 990.0mL of sterile GC agar base, aseptically add 10.0mL of sterile defined supplement B. Mix thoroughly. Pour into sterile Petri dishes or distribute into sterile tubes.

Use: For the cultivation and maintenance of *Neisseria gonorrhoeae*.

GC HiVeg Agar Base with Blood

Composition per liter:

Plant special peptone	15.0g
Agar	10.0g
NaCl	5.0g
K_2HPO_4	4.0g
KH_2PO_4	1.0g
Cornstarch	1.0g
Blood, defibrinated	50.0mL

pH 7.2 ± 0.2 at 25°C

Source: This medium,without blood, is available as a premixed powder from HiMedia.

Preparation of Medium: Add components, except blood, to distilled/deionized water and bring volume to 950.0L. Mix thoroughly. Gently heat until boiling. Autoclave for 15 min at 15 psi pressure–121°C. Cool to 75°–80°C. Add 50.0mL sterile defibrinated blood with thorough mixing and maintain at 75°–80°C for 15–20 min until the medium is chocolatized. Pour into sterile Petri dishes or distribute into sterile tubes.

Use: For the isolation and cultivation of fastidious bacteria, especially *Neisseria* and *Haemophilus* species.

GC HiVeg Agar Base with Hemoglobin

Composition per liter:

Plant special peptone	15.0g
Agar	10.0g
NaCl	5.0g
K_2HPO_4	4.0g
KH_2PO_4	1.0g
Cornstarch	1.0g
Hemoglobin solution	100.0mL

pH 7.2 ± 0.2 at 25°C

Source: This medium,without hemoglobin solution, is available as a premixed powder from HiMedia.

Hemoglobin Solution:
Composition 100.0mL:

Bovine hemoglobin	2.0g

Preparation of Hemoglobin Solution: Add bovine hemoglobin to distilled/deionized water and bring volume to 100.0mL. Mix thoroughly. Autoclave for 15 min at 15 psi pressure–121°C. Cool to 45°–50°C.

Preparation of Medium: Add components, except hemoglobin, to distilled/deionized water and bring volume to 500.0L. Mix thoroughly. Gently heat until boiling. Autoclave for 15 min at 15 psi pressure–121°C. Cool to 45°–50°C. Add 100.0mL sterile hemoglobin solution.

Mix thoroughly. Note: Antibiotics may be added to increase selectivity. Pour into sterile Petri dishes or distribute into sterile tubes.

Use: For the isolation and cultivation of fastidious bacteria, especially *Neisseria* and *Haemophilus* species.

GC Medium, New York City Formulation

Composition per liter:

GC agar base	850.0mL
Horse blood, lysed	100.0mL
Yeast autolysate supplement	30.0mL
LCAT antibiotic solution	20.0mL

pH 7.3 ± 0.2 at 25°C

GC Agar Base:

Composition per 850.0mL:

Special peptone	15.0g
Agar	10.0g
NaCl	5.0g
K_2HPO_4	4.0g
Cornstarch	1.0g
KH_2PO_4	1.0g

pH 7.2 ± 0.2 at 25°C

Preparation of GC Agar Base: Add components of GC medium base and the hemoglobin to distilled/deionized water and bring volume to 850.0mL. Mix thoroughly. Gently heat until boiling. Autoclave for 15 min at 15 psi pressure–121°C. Cool to 45°–50°C.

Horse Blood, Lysed:

Composition per 100.0mL:

Saponin	0.5g
Horse blood, defibrinated	100.0mL

Preparation of Horse Blood, Lysed: Add saponin to defibrinated horse blood. Mix thoroughly. Allow blood to lyse.

Yeast Autolysate Supplement:

Composition per 30.0mL:

Yeast autolysate	10.0g
Glucose	1.0g
$NaHCO_3$	0.15g

Preparation of Yeast Autolysate Supplement: Add components to distilled/deionized water and bring volume to 30.0mL. Mix thoroughly. Filter sterilize.

LCAT Antibiotic Solution:

Composition per 20.0mL:

Colistin	6.0mg
Trimethoprim lactate	5.0mg
Lincomycin	1.0mg
Amphotericin B	1.0mg

Preparation of LCAT Antibiotic Solution: Add components to distilled/deionized water and bring volume to 20.0mL. Mix thoroughly. Filter sterilize.

Preparation of Medium: To 850.0mL of cooled sterile GC agar base, aseptically add 100.0mL of sterile lysed horse blood, 30.0mL of sterile yeast autolysate supplement, and 20.0mL of LCAT antibiotic solution. Mix thoroughly. Pour into sterile Petri dishes or distribute into sterile tubes.

Use: For the selective isolation and cultivation of fastidious microorganisms, especially *Neisseria* species.

GC Medium
See: **GC Agar**

GCII Agar

Composition per liter:

GCII agar base, 2X	490.0mL
Hemoglobin solution	490.0mL
Supplement solution	10.0mL

pH 7.2 ± 0.2 at 25°C

GCII Agar Base, 2X:

Composition per liter:

Agar	10.0g
Pancreatic digest of casein	7.5g
Selected meat peptone	7.5g
NaCl	5.0g
K_2HPO_4	4.0g
Cornstarch	1.0g
KH_2PO_4	1.0g

Source: GCII agar base is available as a premixed powder from BD Diagnostic Systems.

Preparation of GCII Agar Base, 2X: Add components to distilled/deionized water and bring volume to 500.0mL. Mix thoroughly. Gently heat until boiling. Autoclave for 15 min at 15 psi pressure–121°C. Cool to 45°–50°C.

Hemoglobin Solution:

Composition per 500.0mL:

Hemoglobin	10.0g

Preparation of Hemoglobin Solution: Add hemoglobin to distilled/deionized water and bring volume to 500.0mL. Mix thoroughly. Autoclave for 15 min at 15 psi pressure–121°C. Cool to 45°–50°C.

Supplement Solution:

Composition per liter:

Glucose	100.0g
L-Cysteine·HCl	25.9g
L-Glutamine	10.0g
L-Cystine	1.1g
Adenine	1.0g
Nicotinamide adenine dinucleotide	0.25g
Vitamin B_{12}	0.1g
Thiamine pyrophosphate	0.1g
Guanine·HCl	0.03g
$Fe(NO_3)_3 \cdot 6H_2O$	0.02g
p-Aminobenzoic acid	0.013g
Thiamine·HCl	3.0mg

Preparation of Supplement Solution: Add components to distilled/deionized water and bring volume to 1.0L. Mix thoroughly. Filter sterilize.

Preparation of Medium: To 490.0mL of sterile GCII agar base, aseptically add 490.0mL of sterile hemoglobin solution at 45°–50°C. Mix thoroughly. Aseptically add 10.0mL of sterile supplement solution. Mix thoroughly. Pour into sterile Petri dishes or distribute into sterile tubes.

Use: For the isolation and cultivation of fastidious microorganisms, especially *Neisseria* and *Haemophilus* species, from clinical specimens.

GCII Agar

Composition per liter:

GCII agar base	950.0mL
Blood, defibrinated	50.0mL

pH 7.2 ± 0.2 at 25°C

GCII Agar Base with Extra Agar:
Composition per liter:

Agar	12.0g
Pancreatic digest of casein	7.5g
Selected meat peptone	7.5g
NaCl	5.0g
K_2HPO_4	4.0g
Cornstarch	1.0g
KH_2PO_4	1.0g

Source: GCII agar base is available as a premixed powder from BD Diagnostic Systems.

Preparation of GCII Agar Base with Extra Agar: Add components to distilled/deionized water and bring volume to 1.0L. Mix thoroughly. Gently heat until boiling. Autoclave for 15 min at 15 psi pressure–121°C. Cool to 45°–50°C.

Preparation of Medium: To 950.0mL of sterile GCII agar base, aseptically add 50.0mL of sterile defibrinated blood with thorough mixing and maintain at 75°–80°C for 15–20 min until the medium is chocolatized. Pour into sterile Petri dishes or distribute into sterile tubes.

Use: For the isolation and cultivation of fastidious microorganisms, especially *Neisseria* and *Haemophilus* species, from clinical specimens.

GCII Agar with Hemoglobin and IsoVitaleX®
See: **Chocolate II Agar with Hemoglobin and IsoVitaleX®**

GC-Lect™ Agar

Composition per liter:

GCII agar base, 2X	500.0mL
Hemoglobin solution	500.0mL
Supplement solution	10.0mL
Selective agent solution	10.0mL

pH 7.2 ± 0.2 at 25°C

Source: This medium is available as a prepared medium from BD Diagnostic Systems.

GCII Agar Base, 2X with Extra Agar:
Composition per liter:

Agar	12.0g
Pancreatic digest of casein	7.5g
Selected meat peptone	7.5g
NaCl	5.0g
K_2HPO_4	4.0g
Cornstarch	1.0g
KH_2PO_4	1.0g

Source: GCII agar base is available as a premixed powder from BD Diagnostic Systems.

Preparation of GCII Agar Base, 2X with Extra Agar: Add components to distilled/deionized water and bring volume to 500.0mL. Mix thoroughly. Gently heat until boiling. Autoclave for 15 min at 15 psi pressure–121°C. Cool to 45°–50°C.

Hemoglobin Solution:
Composition per 500.0mL:

Hemoglobin	10.0g

Preparation of Hemoglobin Solution: Add hemoglobin to distilled/deionized water and bring volume to 500.0mL. Mix thoroughly. Autoclave for 15 min at 15 psi pressure–121°C. Cool to 45°–50°C.

Supplement Solution:
Composition per liter:

Glucose	100.0g
L-Cysteine·HCl	25.9g
L-Glutamine	10.0g
L-Cystine	1.1g
Adenine	1.0g
Nicotinamide adenine dinucleotide	0.25g
Vitamin B_{12}	0.1g
Thiamine pyrophosphate	0.1g
Guanine·HCl	0.03g
$Fe(NO_3)_3 \cdot 6H_2O$	0.02g
p-Aminobenzoic acid	0.013g
Thiamine·HCl	3.0mg

Source: The supplement solution (IsoVitaleX® enrichment) is available from BD Diagnostic Systems. This enrichment may be replaced by supplement VX from BD Diagnostic Systems.

Preparation of Supplement Solution: Add components to distilled/deionized water and bring volume to 1.0L. Mix thoroughly. Filter sterilize.

Selective Agent Solution:
Composition per 10.0mL:

Selective agents	0.017g

Preparation of Selective Agent Solution: Add selective agents to distilled/deionized water and bring volume to 10.0mL. Mix thoroughly. Filter sterilize.

Preparation of Medium: To 500.0mL of sterile GCII agar base, aseptically add 500.0mL of sterile hemoglobin solution at 45°–50°C. Mix thoroughly. Aseptically add 10.0mL of sterile supplement solution and 10.0mL of selective agents solution. Mix thoroughly. Pour into sterile Petri dishes or distribute into sterile tubes.

Use: For the isolation and cultivation of *Neisseria gonorrhoeae* from clinical specimens.

GC Medium with Chloramphenicol

Composition per 1020.0mL:

GC agar base, 2X	500.0mL
Hemoglobin solution	500.0mL
Supplement solution	10.0mL
Chloramphenicol solution	10.0mL

pH 7.2 ± 0.2 at 25°C

GC Agar Base, 2X:
Composition per 500.0mL:

Agar	10.0g
Pancreatic digest of casein	7.5g
Peptic digest of animal tissue	7.5g
NaCl	5.0g
K_2HPO_4	4.0g
Cornstarch	1.0g
KH_2PO_4	1.0g

Source: GC agar base is available as a premixed powder from BD Diagnostic Systems. This base may be replaced by GC medium base available from BD Diagnostic Systems.

Preparation of GC Agar Base, 2X: Add components to distilled/deionized water and bring volume to 500.0mL. Mix thoroughly. Gently heat until boiling. Autoclave for 15 min at 15 psi pressure–121°C. Cool to 45°–50°C.

Hemoglobin Solution:

Composition per 500.0mL:

Bovine hemoglobin	10.0g

Preparation of Hemoglobin Solution: Add bovine hemoglobin to distilled/deionized water and bring volume to 500.0mL. Mix thoroughly. Autoclave for 15 min at 15 psi pressure–121°C. Cool to 45°–50°C.

Supplement Solution:

Composition per liter:

Glucose	100.0g
L-Cysteine·HCl	25.9g
L-Glutamine	10.0g
L-Cystine	1.1g
Adenine	1.0g
Nicotinamide adenine dinucleotide	0.25g
Vitamin B_{12}	0.1g
Thiamine pyrophosphate	0.1g
Guanine·HCl	0.03g
$Fe(NO_3)_3 \cdot 6H_2O$	0.02g
p-Aminobenzoic acid	0.013g
Thiamine·HCl	3.0mg

Source: The supplement solution (IsoVitaleX® enrichment) is available from BD Diagnostic Systems. This enrichment may be replaced by supplement VX from BD Diagnostic Systems.

Preparation of Supplement Solution: Add components to distilled/deionized water and bring volume to 1.0L. Mix thoroughly. Filter sterilize.

Chloramphenicol Solution:

Composition per 10.0mL:

Chloramphenicol	8.0mg

Preparation of Chloramphenicol Solution: Add chloramphenicol to distilled/deionized water and bring volume to 10.0mL. Mix thoroughly. Filter sterilize.

Preparation of Medium: To 500.0mL of sterile GC agar base, aseptically add 500.0mL of sterile hemoglobin solution at 45°–50°C. Mix thoroughly. Aseptically add 10.0mL of sterile supplement solution and 10.0mL of sterile chloramphenicol solution. Mix thoroughly. Pour into sterile Petri dishes or distribute into sterile tubes.

Use: For the cultivation of *Neisseria gonorrhoeae.*

GCA Agar with Thiamine

Composition per liter:

Glucose	25.0g
Agar	14.0g
Papaic digest of soybean meal	10.0g
NaCl	5.0g
Pancreatic digest of heart muscle	3.0g
Cysteine·HCl·H_2O	1.0g
Thiamine	0.05mg
Rabbit blood, defibrinated	50.0mL

pH 6.8 ± 0.2 at 25°C

Preparation of Medium: Add components, except rabbit blood, to distilled/deionized water and bring volume to 950.0mL. Mix thoroughly. Gently heat and bring to boiling. Autoclave for 15 min at 15 psi pressure–121°C. Cool to 45°–50°C. Aseptically add sterile rabbit blood. Mix thoroughly. Pour into sterile Petri dishes or distribute into sterile tubes.

Use: For the isolation and cultivation of *Francisella tularensis.*

Gelatin Agar

Composition per liter:

Gelatin	30.0g
Agar	15.0g
Pancreatic digest of casein	10.0g
NaCl	10.0g

pH 7.2 ± 0.2 at 25°C

Preparation of Medium: Add components to distilled/deionized water and bring volume to 1.0L. Mix thoroughly. Gently heat and bring to boiling. Distribute into tubes or flasks. Autoclave for 15 min at 15 psi pressure–121°C.

Use: For the cultivation of bacteria isolated from foods and their differentiation based on proteolytic activity.

Gelatin Agar

Composition per liter:

Agar	15.0g
Gelatin	15.0g
Peptone	4.0g
Yeast extract	1.0g

pH 7.2 ± 0.2 at 25°C

Preparation of Medium: Add components to distilled/deionized water and bring volume to 1.0L. Mix thoroughly. Gently heat and bring to boiling. Distribute into tubes or flasks. Autoclave for 15 min at 15 psi pressure–121°C. Pour into sterile Petri dishes or leave in tubes.

Use: For the cultivation of a variety of heterotrophic bacteria based upon their utilization of gelatin.

Gelatin Agar (GA Medium)

Composition per liter:

Solution 1	950.0mL
Solution 2	50.0mL

pH 7.2 ± 0.2 at 25°C

Solution 1:

Composition per 950.0mL:

Gelatin	30.0g
Agar	15.0g
Pancreatic digest of casein	10.0g
NaCl	2.0g
D-Mannitol	1.0g
Glucose	1.0g
KNO_3	1.0g
Sodium acetate	1.0g
Sodium formate	1.0g
Sodium succinate	1.0g
Yeast extract	1.0g
Sodium lactate (60% solution)	5.0mL

Preparation of Solution 1: Add components to distilled/deionized water and bring volume to 950.0mL. Mix thoroughly. Gently heat and bring to boiling. Autoclave for 15 min at 15 psi pressure–121°C.

Solution 2:

Composition per 50.0mL:

Na_2HPO_4	1.0g
L-Cysteine·HCl·H_2O	0.5g
$Na_2CO_3 \cdot H_2O$	0.5g
Sucrose	0.5g

Dithiothreitol ... 0.1g
Menadione solution ... 2.0mL

Preparation of Solution 2: Add components to distilled/deionized water and bring volume to 50.0mL. Mix thoroughly. Filter sterilize.

Menadione Solution:
Composition per 100.0mL:
Menadione (vitamin K_3) ... 0.05g
Ethanol ... 99.0mL

Preparation of Menadione Solution: Add menadione to 99.0mL of absolute ethanol. Mix thoroughly.

Preparation of Medium: Aseptically combine sterile solution 1 with sterile solution 2. Mix thoroughly. Pour into sterile Petri dishes.

Use: For the cultivation and differentiation of microorganisms from dental plaque based on their ability to produce gelatinase. For the differentiation of aerobic, anaerobic, and facultative microorganisms of clinical significance.

Gelatin DEV Agar

Composition per liter:
Agar ... 15.0g
Peptone from meat ... 10.0g
Meat extract ... 10.0g
Gelatin ... 10.0g
NaCl ... 5.0g
pH 7.3 ± 0.2 at 25°C

Source: This medium is available from HiMedia.

Preparation of Medium: Add components to distilled/deionized water and bring volume to 1.0L. Mix thoroughly. Gently heat and bring to boiling. Distribute into tubes or flasks. Autoclave for 15 min at 15 psi pressure–121°C. Pour into sterile Petri dishes or leave in tubes.

Use: For determining the total microbial count and detecting gelatin-liquefying microorganisms in water.

Gelatin Infusion Broth

Composition per liter:
Beef heart, solids from infusion ... 500.0g
Gelatin ... 40.0g
Tryptose ... 10.0g
NaCl ... 5.0g
pH 7.4 ± 0.2 at 25°C

Preparation of Medium: Add components to distilled/deionized water and bring volume to 1.0L. Mix thoroughly. Gently heat and bring to boiling. Distribute into tubes or flasks. Autoclave for 15 min at 15 psi pressure–121°C. Pour into sterile Petri dishes or leave in tubes.

Use: For the cultivation and differentiation of a variety of heterotrophic bacteria based upon their production of gelatinase. The gelatinase liquefies the medium.

Gelatin Iron Agar

Composition per liter:
Gelatin ... 120.0g
Peptic digest of animal tissue ... 25.0g
Meat extract ... 7.5g
NaCl ... 5.0g
AGar ... 1.0g
$FeCl_2$ ... 0.5g
pH 7.0 ± 0.2 at 25°C

Source: This medium is available from HiMedia.

Preparation of Medium: Add components to distilled/deionized water and bring volume to 1.0L. Mix thoroughly. Gently heat and bring to boiling. Distribute into tubes or flasks. Autoclave for 15 min at 15 psi pressure–121°C. Pour into sterile Petri dishes or leave in tubes.

Use: For the detection of gelatin liquefaction and hydrogen sulfide production.

Gelatin Medium

Composition per liter:
Gelatin ... 4.0g
pH 7.0 ± 0.2 at 25°C

Preparation of Medium: Add gelatin to distilled/deionized water and bring volume to 1.0L. Mix thoroughly. Gently heat and bring to boiling. Distribute into tubes. Autoclave for 15 min at 15 psi pressure–121°C.

Use: For the cultivation and differentiation of *Nocardia* and *Streptomyces* species based on utilization of gelatin. *Nocardia asteroides* usually exhibits no growth. *Nocardia brasiliensis* shows good growth and round, compact colonies. *Streptomyces* species show varying degrees of growth.

Gelatin Medium

Composition per liter:
Gelatin ... 120.0g
Pancreatic digest of casein ... 13.0g
Sodium chloride ... 5.0g
Yeast extract ... 5.0g
Heart muscle, solids from infusion ... 2.0g
Sodium thioglycolate ... 0.5g
pH 7.0 ± 0.2 at 25°C

Preparation of Medium: Add components to distilled/deionized water and bring volume to 1.0L. Mix thoroughly. Gently heat and bring to boiling. Distribute into tubes. Autoclave for 15 min at 15 psi pressure–121°C. Pour into sterile Petri dishes or leave in tubes.

Use: For the cultivation of gelatin-utilizing *Clostridium* species.

Gelatin Metronidazole Cadmium Medium (GMC Medium)

Composition per liter:
Solution 1 ... 950.0mL
Solution 2 ... 50.0mL
pH 7.2 ± 0.2 at 25°C

Solution 1:
Composition per 950.0mL:
Gelatin ... 30.0g
Agar ... 15.0g
Pancreatic digest of casein ... 10.0g
NaCl ... 2.0g
D-Mannitol ... 1.0g
Glucose ... 1.0g
KNO_3 ... 1.0g
Sodium acetate ... 1.0g
Sodium formate ... 1.0g
Sodium succinate ... 1.0g
Yeast extract ... 1.0g
$CdSO_4 \cdot 8H_2O$ ... 0.02g

Metronidazole 0.01g
Sodium lactate (60% solution) 5.0mL

Preparation of Solution 1: Add components to distilled/deionized water and bring volume to 950.0mL. Mix thoroughly. Gently heat and bring to boiling. Autoclave for 15 min at 15 psi pressure–121°C.

Solution 2:
Composition per 50.0mL:
Na_2HPO_4 .... 1.0g
L-Cysteine·HCl·H_2O 0.5g
$Na_2CO_3 \cdot H_2O$ 0.5g
Sucrose 0.5g
Dithiothreitol 0.1g
Menadione solution 2.0mL

Preparation of Solution 2: Add components to distilled/deionized water and bring volume to 50.0mL. Mix thoroughly. Filter sterilize.

Menadione Solution:
Composition per 100.0mL:
Menadione (vitamin K_3) 0.05g
Ethanol 99.0mL

Preparation of Menadione Solution: Add menadione to 99.0mL of absolute ethanol. Mix thoroughly.

Preparation of Medium: Aseptically combine sterile solution 1 with sterile solution 2. Mix thoroughly. Pour into sterile Petri dishes.

Use: For the cultivation and differentiation of microorganisms from dental plaque based on their ability to produce gelatinase. For the differentiation of aerobic, anaerobic, and facultative microorganisms of clinical significance.

Gelatin Phosphate Salt Agar (GPS Agar)

Composition per liter:
Agar 15.0g
Gelatin 10.0g
NaCl 10.0g
K_2HPO_4 .... 5.0g
pH 7.2 ± 0.2 at 25°C

Preparation of Medium: Add components to distilled/deionized water and bring volume to 1.0L. Mix thoroughly. Gently heat and bring to boiling. Distribute into tubes or flasks. Autoclave for 15 min at 15 psi pressure–121°C. Pour into sterile Petri dishes.

Use: For the cultivation and differentiation of *Vibrio* species from foods.

Gelatin Phosphate Salt Broth (GPS Broth)

Composition per liter:
Gelatin 10.0g
NaCl 10.0g
K_2HPO_4 .... 5.0g
pH 7.2 ± 0.2 at 25°C

Preparation of Medium: Add components to distilled/deionized water and bring volume to 1.0L. Mix thoroughly. Gently heat and bring to boiling. Distribute into tubes or flasks. Autoclave for 15 min at 15 psi pressure–121°C.

Use: For the cultivation of *Vibrio* species from foods.

Gelatin Salt Agar

Composition per liter:
NaCl 30.0g
Agar 15.0g
Gelatin 15.0g
Peptone 4.0g
Yeast extract 1.0g
pH 7.2 ± 0.2 at 25°C

Preparation of Medium: Add components to distilled/deionized water and bring volume to 1.0L. Mix thoroughly. Gently heat and bring to boiling. Distribute into tubes or flasks. Autoclave for 15 min at 15 psi pressure–121°C. Pour into sterile Petri dishes or leave in tubes.

Use: For the cultivation and differentiation of *Vibrio* species from foods.

Gelatin Salt Agar (GS Agar) (BAM M55)

Composition per liter:
NaCl 30.0g
Agar 25.0g
Gelatin 15.0g
Peptone 4.0g
Yeast extract 1.0g
pH 7.2 ± 0.2 at 25°C

Preparation of Medium: Add components to distilled/deionized water and bring volume to 1.0L. Mix thoroughly. Gently heat and bring to boiling. Distribute into tubes or flasks. Autoclave for 15 min at 15 psi pressure–121°C. Pour into sterile Petri dishes or leave in tubes.

Use: For the cultivation and differentiation of *Vibrio* species from foods. The high concentration of agar inhibits the spreading of *V. alginolyticus* and some other *Vibrio* spp.

Gelatinase Test Medium

Composition per liter:
Gelatin 3.0g
ACES buffer 1.0g
Yeast extract 1.0g
Charcoal, activated 0.15g
α-Ketoglutarate monopotassium salt 0.1g
L-Cysteine·HCl·H_2O (4% solution) 1.0mL
KOH (85% solution) 1.0mL
$Fe_4(P_2O_7)_3$ solution 1.0mL
pH 6.9 ± 0.2 at 25°C

L-Cysteine·HCl·H_2O Solution:
Composition per 10.0mL:
L-Cysteine·HCl·H_2O 0.4g

Preparation of L-Cysteine·HCl·H_2O Solution: Add L-cysteine·HCl·H_2O to distilled/deionized water and bring volume to 10.0mL. Mix thoroughly. Filter sterilize.

$Fe_4(P_2O_7)_3$ Solution:
Composition per 10.0mL:
$Fe_4(P_2O_7)_3$ 0.15g

Preparation of $Fe_4(P_2O_7)_3$ Solution: Add $Fe_4(P_2O_7)_3$ to distilled/deionized water and bring volume to 10.0mL. Mix thoroughly. Filter sterilize.

Preparation of Medium: Add ACES buffer to distilled/deionized water and bring volume to 899.0mL. Mix thoroughly. Gently heat to 50°C. Add 1.0mL of KOH solution. Mix thoroughly. Add charcoal, yeast extract, and α-ketoglutarate. Add 80.0mL of distilled/deionized water to wash sides of flask. Mix thoroughly. Autoclave for 15 min at 15 psi pressure–121°C. Cool to 50°C. Aseptically add 10.0mL of sterile cysteine solution and 10.0mL of sterile $Fe_4(P_2O_7)_3$ solution. Mix thoroughly. Adjust pH to 6.9. Aseptically distribute into sterile screw-capped tubes.

Use: For the cultivation and differentiation of gelatinase-producing bacteria.

General Salts Medium for Estuarine Methanogens

Composition per 410.8mL:

Agar	8.0g
NaCl	3.6g
$NaHCO_3$	2.0g
Complete salts solution	200.0mL
Cysteine-sulfide reducing agent	16.0mL
Wolfe's mineral solution	4.0mL
Vitamin solution	4.0mL
Yeast extract-Trypticase™ solution	4.0mL
Sodium acetate (25% solution)	2.0mL
$Fe(NH_4)_2SO_4$ (0.2% solution)	0.4mL
Resazurin (0.1% solution)	0.4mL

pH 7.0 ± 0.2 at 25°C

Complete Salts Solution:
Composition per liter:

$MgSO_4 \cdot 7H_2O$	6.9g
$MgCl_2 \cdot 6H_2O$	5.5g
KCl	0.67g
NH_4Cl	0.5g
$CaCl_2 \cdot 2H_2O$	0.28g
K_2HPO_4	0.28g

Preparation of Complete Salts Solution: Add components to distilled/deionized water and bring volume to 1.0L. Mix thoroughly.

Cysteine-Sulfide Reducing Agent:
Composition per 400.0mL:

L-Cysteine·HCl·H_2O	5.0g
Na_2S (12.5% solution)	40.0mL
NaOH (1*N* solution)	30.0mL

Preparation of Cysteine-Sulfide Reducing Agent: Add distilled/deionized water to a 500.0mL round-bottomed flask. Add freshly prepared NaOH solution. Gently heat and bring to boiling under 100% N_2. Remove gassing probe. Add L-cysteine·HCl·H_2O. Add freshly prepared Na_2S solution. Renew gassing for several minutes. Cap with rubber stoppers. Distribute into 8.0mL/18mm Hungate tubes.

Yeast Extract-Trypticase™ Solution:
Composition per 100.0mL:

Yeast extract	20.0g
Pancreatic digest of casein	20.0g

Preparation of Yeast Extract-Trypticase™ Solution: Add components to distilled/deionized water and bring volume to 100.0mL. Mix thoroughly.

Wolfe's Mineral Solution:
Composition per liter:

$MgSO_4 \cdot 7H_2O$	3.0g
Nitrilotriacetic acid	1.5g
NaCl	1.0g
$MnSO_4 \cdot H_2O$	0.5g
$FeSO_4 \cdot 7H_2O$	0.1g
$CoCl_2 \cdot 6H_2O$	0.1g
$CaCl_2$	0.1g
$ZnSO_4 \cdot 7H_2O$	0.1g
$CuSO_4 \cdot 5H_2O$	0.01g
$AlK(SO_4)_2 \cdot 12H_2O$	0.01g
H_3BO_3	0.01g
$Na_2MoO_4 \cdot 2H_2O$	0.01g

Preparation of Wolfe's Mineral Solution: Add nitrilotriacetic acid to 500.0mL of distilled/deionized water. Dissolve by adjusting pH to 6.5 with KOH. Add remaining components. Add distilled/deionized water to 1.0L.

Vitamin Solution:
Composition per liter:

Pyridoxine-HCl	10.0mg
Thiamine-HCl·$2H_2O$	5.0mg
Riboflavin	5.0mg
Nicotinic acid	5.0mg
D-Ca-pantothenate	5.0mg
p-Aminobenzoic acid	5.0mg
Lipoic acid	5.0mg
Biotin	2.0mg
Folic acid	2.0mg
Vitamin B_{12}	0.1mg

Preparation of Vitamin Solution: Add components to distilled/deionized water and bring volume to 1.0L. Mix thoroughly. Sparge with 80% H_2 + 20% CO_2. Filter sterilize.

Preparation of Medium: Add components, except cysteine-sulfide reducing agent, to distilled/deionized water and bring volume to 410.8mL. Mix thoroughly. Adjust pH to 7.0. Gently heat and bring to boiling under 80% N_2 + 20% CO_2. Add cysteine-sulfide reducing agent. Continue boiling until resazurin turns colorless, indicating reduction. Distribute anaerobically into culture tubes with aluminum crimp seals. Autoclave for 15 min at 15 psi pressure–121°C.

Use: For the cultivation and maintenance of *Methanococcus deltae, Methanococcus vannielii, Methanococcus voltae, Methanogenium cariaci, Methanogenium marisnigri,* and *Methanogenium olentangyi.*

Gentamicin Sulfate Solution (BAM M57)

Composition per 100.0mL:

Gentamicin sulfate	0.5g

Preparation of Gentamicin Sulfate Solution: Add gentamicin sulfate to distilled/deionized water and bring volume to 100.0mL. Mix thoroughly. Filter sterilize.

Use: As an inhibitor to render media selective.

Geo Medium

Composition per liter:

Agar	15.0g
$CaCO_3$	1.0g
Glucose	1.0g
Starch, soluble	1.0g
Yeast extract	1.0g

pH 7.2 ± 0.2 at 25°C

Preparation of Medium: Add components to distilled/deionized water and bring volume to 1.0L. Mix thoroughly. Gently heat and bring to boiling. Adjust pH to 7.2. Distribute into tubes or flasks. Autoclave for 15 min at 15 psi pressure–121°C. Pour into sterile Petri dishes or leave in tubes.

Use: For the cultivation of *Geodermatophilus obscurus.*

Geoalkalibacter Medium (DSMZ Medium 1059)

Composition per liter:

Sulfur, powdered	10.0g
$NaHCO_3$	10.0g
Na_2CO_3	3.0g
NaCl	1.0g
NH_4Cl	0.5g
KCl	0.2g
KH_2PO_4	0.2g
$MgCl_2 \cdot 6H_2O$	0.1g
Yeast extract	0.1g
Resazurin	0.5mg
Sodium acetate solution	10.0mL
Trace elements solution SL-10	1.0mL
Selenite/tungstate solution	1.0mL

pH 9.1 ± 0.2 at 25°C

Sodium Acetate Solution:

Composition per 10.0mL:

Sodium acetate	2.5g

Preparation of Sodium Acetate Solution: Add components to distilled/deionized water and bring volume to 10.0mL. Mix thoroughly. Sparge with 100% N_2. Autoclave for 15 min at 15 psi pressure–121°C.

Trace Elements Solution SL-10:

Composition per liter:

$FeCl_2 \cdot 4H_2O$	1.5g
$CoCl_2 \cdot 6H_2O$	190.0mg
$MnCl_2 \cdot 4H_2O$	100.0mg
$ZnCl_2$	70.0mg
$Na_2MoO_4 \cdot 2H_2O$	36.0mg
$NiCl_2 \cdot 6H_2O$	24.0mg
H_3BO_3	6.0mg
$CuCl_2 \cdot 2H_2O$	2.0mg
HCl (25% solution)	10.0mL

Preparation of Trace Elements Solution SL-10: Add $FeCl_2 \cdot 4H_2O$ to 10.0mL of HCl solution. Mix thoroughly. Add distilled/deionized water and bring volume to 1.0L. Add remaining components. Sparge with 100% N_2. Autoclave for 15 min at 15 psi pressure–121°C.

Selenite/Tungstate Solution:

Composition per liter:

NaOH	0.5g
$Na_2WO_4 \cdot 2H_2O$	4.0mg
$Na_2SeO_3 \cdot 5H_2O$	3.0mg

Preparation of Selenite/Tungstate Solution: Add components to distilled/deionized water and bring volume to 1.0L. Mix thoroughly. Sparge with 100% N_2. Autoclave for 15 min at 15 psi pressure–121°C.

Preparation of Medium: Add components, except sulfur, carbonate, sodium bicarbonate, and sodium acetate solution, to distilled/deionized water and bring volume to 990.0mL. Gently heat and bring to boiling. Boil for 1 min. Cool to room temperature under N_2 gas atmosphere. Add solid carbonate and bicarbonate. Ajust pH to 9.0–9.2. Dispense under N_2 gas atmosphere into culture vessels. Autoclave for 15 min at 15 psi pressure–121°C. Sterilize sulfur separately in screw-capped tubes by steaming in a water bath for 3 hr on each of 3 successive days. Aseptically add the sterilized sulfur to the medium. Mix thoroughly. Aseptically and anoxically add the sodium acetate solution. Mix thorougly.

Use: For the cultivation of *Geoalkalibacter* spp.

Geobacter Medium (DSMZ Medium 579)

Composition per liter:

Fe(III) citrate	13.7g
$NaHCO_3$	2.5g
Na-acetate	2.5g
NH_4Cl	1.5g
NaH_2PO_4	0.6g
KCl	0.1g
$Na_2WO_4 \cdot 2H_2O$	0.25mg
Vitamin solution	10.0mL
Trace elements solution	10.0mL

pH 6.7–7.0 at 25°C

Trace Elements Solution:

Composition per liter:

$MgSO_4 \cdot 7H_2O$	3.0g
Nitrilotriacetic acid	1.5g
NaCl	1.0g
$MnSO_4 \cdot 2H_2O$	0.5g
$CoSO_4 \cdot 7H_2O$	0.18g
$ZnSO_4 \cdot 7H_2O$	0.18g
$CaCl_2 \cdot 2H_2O$	0.1g
$FeSO_4 \cdot 7H_2O$	0.1g
$NiCl_2 \cdot 6H_2O$	0.025g
$KAl(SO_4)_2 \cdot 12H_2O$	0.02g
H_3BO_3	0.01g
$Na_2MoO_4 \cdot 4H_2O$	0.01g
$CuSO_4 \cdot 5H_2O$	0.01g
$Na_2SeO_3 \cdot 5H_2O$	0.3mg

Preparation of Trace Elements Solution: Add nitrilotriacetic acid to 500.0mL of distilled/deionized water. Dissolve by adjusting pH to 6.5 with KOH. Add remaining components. Add distilled/deionized water to 1.0L. Mix thoroughly.

Vitamin Solution:

Composition per liter:

Pyridoxine-HCl	10.0mg
Thiamine-HCl·2H₂O	5.0mg
Riboflavin	5.0mg
Nicotinic acid	5.0mg
D-Ca-pantothenate	5.0mg
p-Aminobenzoic acid	5.0mg
Lipoic acid	5.0mg
Biotin	2.0mg
Folic acid	2.0mg
Vitamin B_{12}	0.1mg

Preparation of Vitamin Solution: Add components to distilled/deionized water and bring volume to 1.0L. Mix thoroughly. Sparge with 80% H_2 + 20% CO_2. Filter sterilize.

Preparation of Medium: Prepare and dispense medium under an oxygen-free 80% N_2 + 20% CO_2 gas mixture. Dissolve ferric citrate in

900.0mL distilled/deionized water by heating and adjust to pH 6.0. Add other components, except vitamin solution, to distilled/deionized water and bring volume to 990.0mL. Mix thoroughly. Sparge with 80% N_2 + 20% CO_2. Autoclave for 15 min at 15 psi pressure–121°C. Cool to 25°C. Aseptically and anaerobically add 10.0mL vitamin solution. Mix thoroughly. Aseptically and anaerobically distribute into sterile tubes or flasks.

Use: For the cultivation of *Geobacter metallireducens* and *Geobacter grbiciae.*

Geobacter Medium (DSMZ Medium 826)

Composition per 1050.0mL:

$NaHCO_3$	2.5g
NH_4Cl	1.5g
Na-acetate	0.82g
Na_2HPO_4	0.6g
KCl	0.1g
Resazurin	0.5mg
Na-fumarate solution	50.0mL
Trace elements solution	10.0mL
Vitamin solution	10.0mL
Selenite-tungstate solution	1.0mL

pH 6.8 ± 0.2 at 25°C

Na-Fumarate Solution:

Composition per 100.0mL:

Na-fumarate	16.0g

Preparation of Na-Fumarate Solution: Add Na-fumarate to distilled/deionized water and bring volume to 100.0mL. Mix thoroughly. Sparge with 100% N_2. Autoclave for 15 min at 15 psi pressure–121°C.

Selenite-Tungstate Solution

Composition per liter:

NaOH	0.5g
$Na_2WO_4 \cdot 2H_2O$	4.0mg
$Na_2SeO_3 \cdot 5H_2O$	3.0mg

Preparation of Selenite-Tungstate Solution: Add components to distilled/deionized water and bring volume to 1.0L. Mix thoroughly. Sparge with 100% N_2. Filter sterilize.

Trace Elements Solution:

Composition per liter:

$MgSO_4 \cdot 7H_2O$	3.0g
Nitrilotriacetic acid	1.5g
NaCl	1.0g
$MnSO_4 \cdot 2H_2O$	0.5g
$CoSO_4 \cdot 7H_2O$	0.18g
$ZnSO_4 \cdot 7H_2O$	0.18g
$CaCl_2 \cdot 2H_2O$	0.1g
$FeSO_4 \cdot 7H_2O$	0.1g
$NiCl_2 \cdot 6H_2O$	0.025g
$KAl(SO_4)_2 \cdot 12H_2O$	0.02g
H_3BO_3	0.01g
$Na_2MoO_4 \cdot 4H_2O$	0.01g
$CuSO_4 \cdot 5H_2O$	0.01g
$Na_2SeO_3 \cdot 5H_2O$	0.3mg

Preparation of Trace Elements Solution: Add nitrilotriacetic acid to 500.0mL of distilled/deionized water. Dissolve by adjusting pH to 6.5 with KOH. Add remaining components. Add distilled/deionized water to 1.0L. Mix thoroughly.

Vitamin Solution:

Composition per liter:

Pyridoxine-HCl	10.0mg
Thiamine-HCl·$2H_2O$	5.0mg
Riboflavin	5.0mg
Nicotinic acid	5.0mg
D-Ca-pantothenate	5.0mg
p-Aminobenzoic acid	5.0mg
Lipoic acid	5.0mg
Biotin	2.0mg
Folic acid	2.0mg
Vitamin B_{12}	0.1mg

Preparation of Vitamin Solution: Add components to distilled/deionized water and bring volume to 1.0L. Mix thoroughly. Sparge with 80% H_2 + 20% CO_2. Filter sterilize.

Preparation of Medium: Prepare and dispense medium under 80% N_2 + 20% CO_2 gas mixture. Add components, except $NaHCO_3$ and fumarate solution, to distilled/deionized water and bring volume to 1.0L. Mix thoroughly. Gently heat and bring to boiling. Boil for 5 min. Cool to room temperature while sparging with 80% N_2 + 20% CO_2. Add the $NaHCO_3$. Equilibrate with the 80% N_2 + 20% CO_2 gas mixture to reach a pH of 6.8. Distribute into anaerobe tubes under 80% N_2 + 20% CO_2. Autoclave for 15 min at 15 psi pressure–121°C. Cool to room temperature. Aseptically and anaerobically add 0.5mL Na-fumarate solution per 10.0mL medium. The pH should be 6.8.

Use: For the cultivation of *Geobacter sulfurreducens, Geobacter bremensis (Geobacter* sp.*), Geobacter chapellei, Geobacter hydrogenophilus,* and *Geothrix fermentans.*

Geobacter Medium

Composition per liter:

Ferric citrate	13.7g
Sodium acetate	2.5g
$NaHCO_3$	2.5g
NH_4Cl	1.5g
NaH_2PO_4	0.6g
KCl	0.1g
$Na_2WO_4 \cdot 2H_2O$	0.25mg
Trace elements solution	10.0mL
Vitamin solution	10.0mL

pH 6.7–7.0 at 25°C

Trace Elements Solution:

Composition per liter:

$MgSO_4 \cdot 7\ H_2O$	3.0g
Nitrilotriacetic acid	1.5g
$CaCl_2 \cdot 2\ H_2O$	.1.0g
NaCl	1.0g
$MnSO_4 \cdot 2\ H_2O$	0.5g
$CoSO_4 \cdot 7\ H_2O$	0.18g
$ZnSO_4 \cdot 7\ H_2O$	0.18g
$FeSO_4 \cdot 7\ H_2O$	0.1g
$NiCl_2 \cdot 6\ H_2O$	0.025g
$KAl(SO_4)_2 \cdot 12\ H_2O$	0.02g
$CuSO_4 \cdot 5\ H_2O$	0.01g
H_3BO_3	0.01g
$Na_2MoO_4 \cdot 2\ H_2O$	0.01g
$Na_2SeO_3 \cdot 5\ H_2O$	0.3 mg

Preparation of Trace Elements Solution: Dissolve nitrilotriacetic acid and adjust pH to 6.5 with KOH. Add minerals. Adjust final pH to 7.0 with KOH.

Vitamin Solution:

Composition per liter:

Pyridoxine-HCl	10.0mg
Calcium DL-pantothenate	5.0mg
Lipoic acid	5.0mg
Nicotinic acid	5.0mg
p-Aminobenzoic acid	5.0mg
Riboflavin	5.0mg
Thiamine-HCl	5.0mg
Biotin	2.0mg
Folic acid	2.0mg
Vitamin B_{12}	0.1mg

Preparation of Vitamin Solution: Add components to distilled/deionized water and bring volume to 1.0L. Mix thoroughly.

Preparation of Medium: Prepare and dispense medium under 80% N_2 + 20% CO_2. Add ferric citrate to distilled/deionized water and bring volume to 500.0mL. Mix thoroughly. Gently heat until ferric citrate dissolves. Adjust pH to 6.0. Add remaining components. Add distilled/deionized water and bring volume to 1.0L. Mix thoroughly. Distribute into tubes or flasks. Autoclave for 15 min at 15 psi pressure–121°C.

Use: For the cultivation and maintenance of *Geobacter metallireducens.*

Geobacter Medium

Composition per liter:

$NaHCO_3$	2.5g
NH_4Cl	1.5g
Sodium acetate	0.82g
NaH_2PO_4	0.6g
KCl	0.1g
Sodium fumarate solution	50.0mL
Modified Wolfe's mineral solution	10.0mL
Wolfe's vitamin solution	10.0mL

pH 6.8 ± 0.2 at 25°C

Sodium Fumarate Solution:

Composition per 50.0mL:

Sodium fumarate	8.0g

Preparation of Sodium Fumarate Solution: Add sodium fumarate to distilled/deionized water and bring volume to 50.0mL. Mix thoroughly. Filter sterilize. Sparge with 100% N_2.

Wolfe's Vitamin Solution:

Composition per liter:

Pyridoxine·HCl	10.0mg
p-Aminobenzoic acid	5.0mg
Lipoic acid	5.0mg
Nicotinic acid	5.0mg
Riboflavin	5.0mg
Thiamine·HCl	5.0mg
Calcium DL-pantothenate	5.0mg
Biotin	2.0mg
Folic acid	2.0mg
Vitamin B_{12}	0.1mg

Preparation of Wolfe's Vitamin Solution: Add components to distilled/deionized water and bring volume to 1.0L. Mix thoroughly.

Modified Wolfe's Mineral Solution:

Composition per liter:

$MgSO_4 \cdot 7H_2O$	3.0g
Nitrilotriacetic acid	1.5g
NaCl	1.0g
$MnSO_4 \cdot H_2O$	0.5g
$CaCl_2$	0.1g
$CoCl_2 \cdot 6H_2O$	0.1g
$FeSO_4 \cdot 7H_2O$	0.1g
$ZnSO_4 \cdot 7H_2O$	0.1g
$AlK(SO_4)_2 \cdot 12H_2O$	0.01g
$CuSO_4 \cdot 5H_2O$	0.01g
H_3BO_3	0.01g
$Na_2MoO_4 \cdot 2H_2O$	0.01g
Na_2SeO_3	0.01g
$NaWO_4 \cdot 2H_2O$	0.01g
$NiCl_2 \cdot 6H_2O$	0.01g

Preparation of Modified Wolfe's Mineral Solution: Add nitrilotriacetic acid to 500.0mL of distilled/deionized water. Adjust pH to 6.5 with KOH. Add remaining components one at a time. Add distilled/deionized water to 1.0L. Adjust pH to 6.8.

Preparation of Medium: Prepare and dispense medium under 80% N_2 + 20% CO_2. Add components, except sodium fumarate solution, to distilled/deionized water and bring volume to 950.0mL. Mix thoroughly. Gently heat and bring to boiling. Continue boiling for 3 min. Cool to room temperature while sparging with 80% N_2 + 20% CO_2. Anaerobically distribute 9.5mL volumes into anaerobic tubes. Autoclave for 15 min at 15 psi pressure–121°C. Aseptically and anaerobically add 0.5mL of sterile sodium fumarate solution to each tube. Mix thoroughly. Adjust pH to 6.8.

Use: For the cultivation of *Geobacter sulfurreducens.*

Geococcus nitratireducens and *Petrobacter succinimandens* Medium (DSMZ Medium 979)

Composition per liter:

Glucose	3.6g
NaCl	1.0g
NH_4Cl	1.0g
Yeast extract	1.0g
KH_2PO_4	0.3g
K_2HPO_4	0.3g
$MgCl_2 \cdot 6H_2O$	0.2g
Sodium acetate	0.16g
$CaCl_2 \cdot 2H_2O$	0.1g
KCl	0.1g
Trace elements solution SL-10	10.0mL
Vitamin solution	5.0mL

pH 6.8 ± 0.2 at 25°C

Trace Elements Solution SL-10:

Composition per liter:

$MgSO_4 \cdot 7H_2O$	3.0g
Nitrilotriacetic acid	1.5g
NaCl	1.0g
$MnSO_4 \cdot 2H_2O$	0.5g
$CoSO_4 \cdot 7H_2O$	0.18g
$ZnSO_4 \cdot 7H_2O$	0.18g
$CaCl_2 \cdot 2H_2O$	0.1g
$FeSO_4 \cdot 7H_2O$	0.1g

$NiCl_2 \cdot 6H_2O$	0.025g
$KAl(SO_4)_2 \cdot 12H_2O$	0.02g
H_3BO_3	0.01g
$Na_2MoO_4 \cdot 4H_2O$	0.01g
$CuSO_4 \cdot 5H_2O$	0.01g
$Na_2SeO_3 \cdot 5H_2O$	0.3mg

Preparation of Trace Elements Solution SL-10: Add nitrilotriacetic acid to 500.0mL of distilled/deionized water. Dissolve by adjusting pH to 6.5 with KOH. Add remaining components. Add distilled/deionized water to 1.0L. Mix thoroughly.

Vitamin Solution:

Composition per liter:

Pyridoxine-HCl	10.0mg
Thiamine-HCl·$2H_2O$	5.0mg
Riboflavin	5.0mg
Nicotinic acid	5.0mg
D-Ca-pantothenate	5.0mg
p-Aminobenzoic acid	5.0mg
Lipoic acid	5.0mg
Biotin	2.0mg
Folic acid	2.0mg
Vitamin B_{12}	0.1mg

Preparation of Vitamin Solution: Add components to distilled/deionized water and bring volume to 1.0L. Mix thoroughly. Sparge with 80% H_2 + 20% CO_2. Filter sterilize.

Preparation of Medium: Add components, except vitamin solution, to distilled/deionized water and bring volume to 1.0L. Gently heat and bring to boiling. Mix thoroughly. Cool to room temperature while sparging with 80% N_2 + 20% CO_2. Dispense into culture vessels under an atmosphere of 80% N_2 + 20% CO_2. Autoclave for 45 min at 6 psi pressure–110°C. Aseptically and anoxically add vitamin solution.

Use: For the cultivation of *Geococcus nitratireducens.*

Geococcus nitratireducens and *Petrobacter succinimandens* Medium (DSMZ Medium 979)

Composition per liter:

Succinic acid	2.36g
NaCl	1.0g
NH_4Cl	1.0g
Yeast extract	1.0g
KH_2PO_4	0.3g
K_2HPO_4	0.3g
$MgCl_2 \cdot 6H_2O$	0.2g
$CaCl_2 \cdot 2H_2O$	0.1g
KCl	0.1g
Trace elements solution SL-10	10.0mL
Vitamin solution	5.0mL

pH 6.8 ± 0.2 at 25°C

Trace Elements Solution SL-10:

Composition per liter:

$MgSO_4 \cdot 7H_2O$	3.0g
Nitrilotriacetic acid	1.5g
NaCl	1.0g
$MnSO_4 \cdot 2H_2O$	0.5g
$CoSO_4 \cdot 7H_2O$	0.18g
$ZnSO_4 \cdot 7H_2O$	0.18g
$CaCl_2 \cdot 2H_2O$	0.1g
$FeSO_4 \cdot 7H_2O$	0.1g
$NiCl_2 \cdot 6H_2O$	0.025g
$KAl(SO_4)_2 \cdot 12H_2O$	0.02g
H_3BO_3	0.01g
$Na_2MoO_4 \cdot 4H_2O$	0.01g
$CuSO_4 \cdot 5H_2O$	0.01g
$Na_2SeO_3 \cdot 5H_2O$	0.3mg

Preparation of Trace Elements Solution SL-10: Add nitrilotriacetic acid to 500.0mL of distilled/deionized water. Dissolve by adjusting pH to 6.5 with KOH. Add remaining components. Add distilled/deionized water to 1.0L. Mix thoroughly.

Vitamin Solution:

Composition per liter:

Pyridoxine-HCl	10.0mg
Thiamine-HCl·$2H_2O$	5.0mg
Riboflavin	5.0mg
Nicotinic acid	5.0mg
D-Ca-pantothenate	5.0mg
p-Aminobenzoic acid	5.0mg
Lipoic acid	5.0mg
Biotin	2.0mg
Folic acid	2.0mg
Vitamin B_{12}	0.1mg

Preparation of Vitamin Solution: Add components to distilled/deionized water and bring volume to 1.0L. Mix thoroughly. Sparge with 80% H_2 + 20% CO_2. Filter sterilize.

Preparation of Medium: Add components, except vitamin solution, to distilled/deionized water and bring volume to 1.0L. Gently heat and bring to boiling. Mix thoroughly. Cool to room temperature while sparging with 80% N_2 + 20% CO_2. Dispense into culture vessels under an atmosphere of 80% N_2 + 20% CO_2. Autoclave for 45 min at 6 psi pressure–110°C. Aseptically and anoxically add vitamin solution.

Use: For the cultivation of *Petrobacter succinimandens.*

Geodermatophilus obscurus Medium

Composition per liter:

Agar	20.0g
Malt extract, purified solids	15.0g
Starch, soluble	10.0g
Sucrose	10.0g
Yeast extract	5.0g
$CaCO_3$	2.0g

Preparation of Medium: Add components to distilled/deionized water and bring volume to 1.0L. Mix thoroughly. Gently heat and bring to boiling. Distribute into tubes or flasks. Autoclave for 15 min at 15 psi pressure–121°C. Pour into sterile Petri dishes or leave in tubes.

Use: For the isolation and cultivation of *Geodermatophilus obscurus.*

George's Medium, Modified

Composition per liter:

Agar	15.0g
Peptone	1.0g
KNO_3	0.2g
K_2HPO_4	0.02g
$MgSO_4 \cdot 7H_2O$	0.02g
Ferric citrate	0.035mg

Preparation of Medium: Add components to tap water and bring volume to 1.0L. Mix thoroughly. Gently heat and bring to boiling. Distribute into tubes. Autoclave for 15 min at 15 psi pressure–121°C. Pour into sterile Petri dishes or leave in tubes.

Use: For the cultivation of a variety of algae.

Geothermobacter Medium (DSMZ Medium 981)

Composition per liter:

NaCl	19.0g
$MgCl_2 \cdot 6H_2O$	9.0g
$NaHCO_3$	1.7g
KNO_3	1.01g
KCl	0.5g
KH_2PO_4	0.42g
$CaCl_2 \cdot 2H_2O$	0.3g
$MgSO_4 \cdot 7H_2O$	0.15g
Yeast extract	0.1g
$(NH_4)_2SO_4$	0.1g
NaBr	0.05g
$SrCl_2 \cdot 6H_2O$	0.02g
Resazurin	0.5mg
Vitamin solution	10.0mL
$NaHCO_3$ solution	10.0mL
DL-Malic acid solution	10.0mL
L-Cysteine solution	5.0mL
Trace elements solution SL-10	1.0mL
Selenite/tungstate solution	1.0mL

pH 6.2 ± 0.2 at 25°C

DL-Malic Acid Solution:

Composition per 10.0mL:

DL-Malic acid	1.34g

Preparation of DL-Malic Acid Solution: Add DL-malic acid to 10.0mL distilled/deionized water. Mix thoroughly. Sparge with 100% N_2. Filter sterilize.

L-Cystine Solution:

Composition per 5.0mL:

L-Cystine-HCl·$2H_2O$	0.05g
NaOH (1*N* solution)	5.0mL

Preparation of L-Cystine Solution: Add L-cystine to 5.0mL of NaOH solution. Mix thoroughly. Sparge with 100% N_2. Filter sterilize.

$NaHCO_3$ Solution:

Composition per 10.0mL:

$NaHCO_3$	1.7g

Preparation of $NaHCO_3$ Solution: Add $NaHCO_3$ to distilled/deionized water and bring volume to 10.0mL. Mix thoroughly. Sparge with 20% CO_2 + 80% H_2. Autoclave for 15 min at 15 psi pressure–121°C. Cool to room temperature.

Trace Elements Solution SL-10:

Composition per liter:

$MgSO_4 \cdot 7H_2O$	3.0g
Nitrilotriacetic acid	1.5g
NaCl	1.0g
$MnSO_4 \cdot 2H_2O$	0.5g
$CoSO_4 \cdot 7H_2O$	0.18g
$ZnSO_4 \cdot 7H_2O$	0.18g
$CaCl_2 \cdot 2H_2O$	0.1g
$FeSO_4 \cdot 7H_2O$	0.1g
$NiCl_2 \cdot 6H_2O$	0.025g
$KAl(SO_4)_2 \cdot 12H_2O$	0.02g
H_3BO_3	0.01g
$Na_2MoO_4 \cdot 4H_2O$	0.01g
$CuSO_4 \cdot 5H_2O$	0.01g
$Na_2SeO_3 \cdot 5H_2O$	0.3mg

Preparation of Trace Elements Solution SL-10: Add nitrilotriacetic acid to 500.0mL of distilled/deionized water. Dissolve by adjusting pH to 6.5 with KOH. Add remaining components. Add distilled/deionized water to 1.0L. Mix thoroughly.

Selenite/Tungstate Solution:

Composition per liter:

NaOH	0.5g
$Na_2WO_4 \cdot 2H_2O$	4.0mg
$Na_2SeO_3 \cdot 5H_2O$	3.0mg

Preparation of Selenite/Tungstate Solution: Add components to distilled/deionized water and bring volume to 970.0mL. Mix thoroughly. Sparge with 100% N_2. Autoclave for 15 min at 15 psi pressure–121°C.

Vitamin Solution:

Composition per liter:

Pyridoxine-HCl	10.0mg
Thiamine-HCl·$2H_2O$	5.0mg
Riboflavin	5.0mg
Nicotinic acid	5.0mg
D-Ca-pantothenate	5.0mg
p-Aminobenzoic acid	5.0mg
Lipoic acid	5.0mg
Biotin	2.0mg
Folic acid	2.0mg
Vitamin B_{12}	0.1mg

Preparation of Vitamin Solution: Add components to distilled/deionized water and bring volume to 1.0L. Mix thoroughly. Sparge with 80% H_2 + 20% CO_2. Filter sterilize.

Preparation of Medium: Add components, except bicarbonate, cysteine, vitamins, and malic acid, to distilled/deionized water and bring volume to 1.0L. Mix thoroughly. Gently heat and bring to boiling. Boil for 3 min. Cool to room temperature while sparging with 80% N_2 + 20% CO_2. Dispense into culture vessels under an atmosphere of 80% N_2 + 20% CO_2. Autoclave for 15 min at 15 psi pressure–121°C. Aseptically and anoxically add vitamins, yeast extract, malic acid, and cysteine solution. Adjust pH to 6.2.

Use: For the cultivation of *Geothermobacter* spp.

GFY Agar

Composition per liter:

Agar	15.0g
Glucose	5.0g
Fructose	5.0g
Yeast extract	5.0g
$CaCO_3$	3.0g
Vitamin B_{12}	2.0mg

Preparation of Medium: Add components to distilled/deionized water and bring volume to 1.0L. Mix thoroughly. Gently heat and bring to boiling. Distribute into tubes or flasks. Autoclave for 15 min at 15 psi pressure–121°C. Pour into sterile Petri dishes or leave in tubes.

Use: For the cultivation and maintenance of *Flexibacter* species.

Gilardi Motility Test and Maintenance Medium

See: **Motility Test and Maintenance Medium, Gilardi**

Gillies Agar No. 2
(Sucrose Salicin Agar)

Composition per liter:

Peptic digest of animal tissue	10.0g
Casein enzymic hydrolysate	10.0g
NaCl	5.0g
$Na_2H_2PO_4$	0.25g
Sucrose	10.0g
Salicin	10.0g
Bromthymol Blue	0.01g
$Na_2S_2O_3$	0.025g
Agar	3.0g

pH 7.4 ± 0.2 at 25°C

Source: This medium is available from HiMedia.

Preparation of Medium: Add components to distilled/deionized water and bring volume to 1.0L. Mix thoroughly. Gently heat and bring to boiling. Distribute into tubes. Autoclave for 15 min at 15 psi pressure–121°C. Allow the tubes to cool in an upright position. Suspend Kovacs reagent strips and lead acetate papers from the cap or the cotton plug over the medium but not touching the surface of the medium.

Use: For the detection of motility, hydrogen sulphide, indole production, and fermentation of sucrose and salicin for identification of *Salmonella* spp. and *Shigella* spp.

Ginger Beer Plant Medium
(DSMZ Medium 267)

Ginger, powdered	1.5g
Citric acid	1.5g
Soft brown sugar	100.0mL

Preparation of Medium: Add components to tap water and bring volume to 1.0L. Mix thoroughly. Distribute into tubes or flasks. Autoclave for 15 min at 15 psi pressure–121°C.

Use: For the cultivation of ginger beer plant.

Giolitti-Cantoni Broth
(Giolitti-Cantoni Broth Base with Tellurite)

Composition per 1030.0mL:

Mannitol	20.0g
Pancreatic digest of casein	10.0g
Beef extract	5.0g
LiCl	5.0g
NaCl	5.0g
Yeast extract	5.0g
Sodium pyruvate	3.0g
Glycine	1.2g
Chapman tellurite solution	0.3mL

pH 6.9 ± 0.2 at 25°C

Source: This medium is available as a premixed powder from Oxoid Unipath and BD Diagnostic Systems.

Chapman Tellurite Solution:

Composition per 100.0mL:

K_2TeO_3	3.5g

Preparation of Chapman Tellurite Solution: Add K_2TeO_3 to distilled/deionized water and bring volume to 100.0mL. Mix thoroughly. Filter sterilize.

Caution: Potassium tellurite is toxic.

Preparation of Medium: Add components, except Chapman tellurite solution, to distilled/deionized water and bring volume to 1.0L. Gently heat and bring to boiling. Mix thoroughly. Distribute into tubes in 10.0mL volumes. Autoclave for 15 min at 15 psi pressure–121°C. Cool rapidly to 25°C. Immediately prior to use, incubate tubes at 100°C to expel oxygen. Aseptically add 0.3mL of sterile Chapman tellurite solution to each tube. Mix thoroughly. Cool to 25°C.

Use: For the cultivation and enrichment of *Staphylococcus aureus* from foods.

Giolitti-Cantoni Broth Base with Tellurite
See: **Giolitti-Cantoni Broth**

Gisa Agar

Composition per liter:

Agar	18.0g
$(NH_4)_2SO_4$	5.0g
Calcium glucoisosaccharinate	2.0g
KH_2PO_4	1.0g
$MgSO_4 \cdot 7H_2O$	0.5g
NaCl	0.1g
$CaCl_2 \cdot 2H_2O$	0.1g
Inositol	2.0mg
KI	1.0mg
H_3BO_3	0.5mg
$ZnSO_4 \cdot 7H_2O$	0.4mg
$MnSO_4 \cdot 4H_2O$	0.4mg
Thiamine·HCl	0.4mg
Pyroxidine·HCl	0.4mg
Niacin	0.4mg
Calcium pantothenate	0.4mg
p-Aminobenzoic acid	0.2mg
Riboflavin	0.2mg
$FeCl_3$	0.2mg
$Na_2MoO_4 \cdot 4H_2O$	0.2mg
$CuSO_4 \cdot 5H_2O$	0.04mg
Folic acid	2.0μg
Biotin	2.0μg

pH 5.5 ± 0.2 at 25°C

Preparation of Medium: Add components to distilled/deionized water and bring volume to 1.0L. Mix thoroughly. Gently heat and bring to boiling. Distribute into tubes or flasks. Autoclave for 15 min at 15 psi pressure–121°C. Pour into sterile Petri dishes or leave in tubes.

Use: For the cultivation and maintenance of *Ancylobacter aquaticus, Pseudomonas* species, *Xanthobacter autotrophicus,* and *Xanthobacter* species.

Gisa Broth

Composition per liter:

$(NH_4)_2SO_4$	5.0g
Calcium glucoisosaccharinate	2.0g
KH_2PO_4	1.0g
$MgSO_4 \cdot 7H_2O$	0.5g
NaCl	0.1g
$CaCl_2 \cdot 2H_2O$	0.1g
Inositol	2.0mg
KI	1.0mg
H_3BO_3	0.5mg
$ZnSO_4 \cdot 7H_2O$	0.4mg
$MnSO_4 \cdot 4H_2O$	0.4mg

Thiamine·HCl	0.4mg
Pyroxidine·HCl	0.4mg
Niacin	0.4mg
Calcium pantothenate	0.4mg
p-Aminobenzoic acid	0.2mg
Riboflavin	0.2mg
$FeCl_3$	0.2mg
$Na_2MoO_4 \cdot 4H_2O$	0.2mg
$CuSO_4 \cdot 5H_2O$	0.04mg
Folic acid	2.0μg
Biotin	2.0μg

pH 5.5 ± 0.2 at 25°C

Preparation of Medium: Add components to distilled/deionized water and bring volume to 1.0L. Mix thoroughly. Distribute into tubes or flasks. Autoclave for 15 min at 15 psi pressure–121°C.

Use: For the cultivation and maintenance of *Ancylobacter aquaticus, Pseudomonas* species, *Xanthobacter autotrophicus,* and *Xanthobacter* species.

Gliding Medium

Composition per liter:

Pancreatic digest of casein	0.5g
Yeast extract	0.5g

pH 7.0 ± 0.2 at 25°C

Preparation of Medium: Add components to distilled/deionized water and bring volume to 1.0L. Mix thoroughly. Distribute into tubes or flasks. Autoclave for 15 min at 15 psi pressure–121°C.

Use: For the cultivation and maintenance of the gliding bacteria *Cytophaga* species and *Flexibacter columnaris*.

Gluconacetobacter johannae and Gluconacetobacter azotocaptans Medium (DSMZ Medium 920)

Composition per liter:

K_2HPO_4	4.81g
MES buffer	4.4g
Yeast extract	2.7g
Glucose	2.7g
Mannitol	1.8g
KH_2PO_4	0.65g

pH 6.7 ± 0.2 at 25°C

Preparation of Medium: Add components to distilled/deionized water and bring volume to 1.0L. Mix thoroughly. Distribute into tubes or flasks. Autoclave for 15 min at 15 psi pressure–121°C.

Use: For the cultivation of *Gluconacetobacter azotocaptans* and *Gluconacetobacter johannae.*

Gluconacetobacter rhaeticus Medium (DSMZ Medium 1044)

Composition per liter:

Glucose	50.0g
Agar	15.0g
Yeast extract	5.0g

pH 6.5 ± 0.2 at 25°C

Preparation of Medium: Add components to distilled/deionized water and bring volume to 1.0L. Mix thoroughly. Adjust pH to 6.5. Gently heat while stirring and bring to boiling. Distribute into tubes or flasks. Autoclave for 15 min at 15 psi pressure–121°C. Pour into Petri dishes or leave in tubes.

Use: For the cultivation of *Gluconacetobacter rhaeticus.*

Gluconacetobacter xylinus Medium (LMG Medium 129)

Composition per liter:

D-Glucose	50.0g
Yeast extract	5.0g

pH 6.9 ± 0.2 at 25°C

Preparation of Medium: Add components to distilled/deionized water and bring volume to 1.0L. Mix thoroughly. Distribute into tubes or flasks. Autoclave for 15 min at 15 psi pressure–121°C.

Use: For the cultivation of *Gluconacetobacter xylinus* subsp. *Xylinus.*

Gluconate Peptone Broth

Composition per liter:

Potassium gluconate	40.0g
Casein peptone	1.5g
K_2HPO_4	1.0g
Yeast extract	1.0g

pH 7.0 ± 0.2 at 25°C

Preparation of Medium: Add components to distilled/deionized water and bring volume to 1.0L. Mix thoroughly. Distribute into tubes or flasks. Autoclave for 15 min at 15 psi pressure–121°C.

Use: For the cultivation and differentiation of Gram-negative bacteria based on their ability to oxidize gluconate to 2-ketogluconate. For the differentiation of fluorescent *Pseudomonas* species. After inoculation with bacteria and 48 hr of growth in this medium, Benedict's reagent is added. Bacteria that produce the reducing sugar 2-ketogluconate turn the reagent yellow-orange to orange-red.

Gluconate Test HiVeg Medium

Composition per liter:

Potassium gluconate	40.0g
Plant peptone	1.5g
K_2HPO_4	1.0g
Yeast extract	1.0g

pH 7.0 ± 0.2 at 25°C

Source: This medium is available as a premixed powder from HiMedia.

Preparation of Medium: Add components to distilled/deionized water and bring volume to 1.0L. Mix thoroughly. Gently heat and bring to boiling. Mix thoroughly. Distribute into tubes or flasks. Autoclave for 15 min at 15 psi pressure–121°C.

Use: For the cultivation and differentiation of Gram-negative bacteria based on their ability to oxidize gluconate to 2-ketogluconate. For the differentiation of fluorescent *Pseudomonas* species. After inoculation with bacteria and 48 hr of growth in this medium, Benedict's reagent is added. Bacteria that produce the reducing sugar 2-ketogluconate turn the reagent yellow-orange to orange-red.

Gluconobacter Agar

Composition per liter:

Glucose	100.0g
$CaCO_3$	20.0g

Agar 15.0g
Yeast extract 10.0g

pH 6.8 ± 0.2 at 25°C

Preparation of Medium: Add components to distilled/deionized water and bring volume to 1.0L. Mix thoroughly. Adjust pH to 6.8. Gently heat and bring to boiling. Distribute into tubes or flasks. Autoclave for 15 min at 15 psi pressure–121°C. Pour into sterile Petri dishes or leave in tubes.

Use: For the cultivation and maintenance of *Acetobacter pasteurianus, Acetobacter* species, *Acetobacter xylinum,* and *Gluconobacter oxydans*.

Gluconobacter Broth

Composition per liter:

Glucose 100.0g
$CaCO_3$ 20.0g
Yeast extract 10.0g

pH 6.8 ± 0.2 at 25°C

Preparation of Medium: Add components to distilled/deionized water and bring volume to 1.0L. Mix thoroughly. Adjust pH to 6.8. Distribute into tubes or flasks. Autoclave for 15 min at 15 psi pressure–121°C.

Use: For the cultivation and maintenance of *Acetobacter pasteurianus, Acetobacter* species, *Acetobacter xylinum,* and *Gluconobacter oxydans*.

Gluconobacter oxydans Agar

Composition per liter:

Glucose 100.0g
$CaCO_3$ 20.0g
Agar 15.0g
Yeast extract 10.0g

pH 6.8 ± 0.2 at 25°C

Preparation of Medium: Add components to distilled/deionized water and bring volume to 1.0L. Mix thoroughly. Gently heat and bring to boiling. Distribute into tubes or flasks. Autoclave for 15 min at 15 psi pressure–121°C. Pour into sterile Petri dishes or leave in tubes.

Use: For the cultivation and maintenance of *Acetobacter aceti, Acetobacter pasteurianus, Acetobacter xylinum,* and *Gluconobacter oxydans.*

Glucose Agar

Composition per liter:

Glucose 50.0g
Agar 20.0g
KNO_3 .... 2.0g
KH_2PO_4 .... 1.0g
$MgSO_4·7H_2O$ 0.5g

pH 5.4 ± 0.2 at 25°C

Preparation of Medium: Add components to distilled/deionized water and bring volume to 1.0L. Mix thoroughly. Adjust pH to 5.4. Gently heat and bring to boiling. Distribute into tubes or flasks. Autoclave for 20 min at 15 psi pressure–121°C. Pour into sterile Petri dishes or leave in tubes.

Use: For the cultivation and maintenance of *Cladosporium cladosporioides, Cladosporium cucumerinum, Cladosporium herbarum, Cladosporium macrocarpum, Cladosporium sphaerospermum, Cladosporium uredinicola,* and *Hormoconis resinae.*

Glucose Agar

Composition per liter:

Agar 15.0g
Tryptone 10.0g
Glucose 10.0g
NaCl 5.0g
Yeast extract 1.5g
Bromcresol Purple 0.015g

pH 7.0 ± 0.2 at 25°C

Source: This medium is available from HiMedia.

Preparation of Medium: Add components to distilled/deionized water and bring volume to 1.0L. Mix thoroughly. Gently heat and bring to boiling. Distribute into tubes or flasks. Autoclave for 15 min at 15 psi pressure–121°C. Pour into sterile Petri dishes or leave in tubes.

Use: For determining the fermentation reactions of presumptive Enterobacteriaceae.

Glucose Agar with 25% Glucose

Composition per liter:

Glucose 250.0g
Agar 25.0g
Polypeptone™ 5.0g
Malt extract 3.0g
Yeast extract 3.0g

Preparation of Medium: Add components to distilled/deionized water and bring volume to 1.0L. Mix thoroughly. Gently heat and bring to boiling. Distribute into tubes or flasks. Autoclave for 15 min at 15 psi pressure–121°C. Pour into sterile Petri dishes or leave in tubes.

Use: For the cultivation and maintenance of osmophilic yeasts and bacteria.

Glucose Agar, 9K

Composition per liter:

$(NH_4)_2SO_4$ 3.0g
KH_2PO_4 .... 0.5g
$MgSO_4·7H_2O$ 0.5g
KCl 0.1g
$Ca(NO_3)_2$ 0.0125g
$FeSO_4·7H_2O$ 0.01mg
Agar solution 500.0mL
Glucose solution 100.0mL

pH 4.5 ± 0.2 at 25°C

Agar Solution:

Composition per 500.0mL:

Agar 15.0g

Preparation of Agar Solution: Add agar to distilled/deionized water and bring volume to 500.0mL. Mix thoroughly. Autoclave for 15 min at 15 psi pressure–121°C. Cool to 55°C.

Glucose Solution:

Composition per 100.0mL:

Glucose 10.0g

Preparation of Glucose Solution: Add glucose to distilled/deionized water and bring volume to 100.0mL. Mix thoroughly. Filter sterilize.

Preparation of Medium: Add components, except agar solution and glucose solution, to distilled/deionized water and bring volume to 400.0mL. Mix thoroughly. Adjust pH to 4.5 with H_2SO_4. Autoclave for

15 min at 15 psi pressure–121°C. Cool to 55°C. Aseptically add 500.0mL of sterile agar solution and 100.0mL of sterile glucose solution. Mix thoroughly. Pour into sterile Petri dishes or distribute into sterile tubes.

Use: For the cultivation of *Thiobacillus acidophilus*.

Glucose Asparagine Agar

Composition per liter:

Agar	15.0g
Glucose	10.0g
Asparagine	0.5g
K_2HPO_4	0.5g

pH 7.0 ± 0.2 at 25°C

Preparation of Medium: Add components to distilled/deionized water and bring volume to 1.0L. Mix thoroughly. Gently heat and bring to boiling. Distribute into tubes or flasks. Autoclave for 15 min at 15 psi pressure–121°C. Pour into sterile Petri dishes or leave in tubes.

Use: For the cultivation and maintenance of *Actinoplanes violaceus, Ampullariella kummingensis, Streptomyces avermitilis, Streptomyces calvus, Streptomyces hygroscopicus, Streptomyces tosaensis,* and *Streptoverticillium morookaense*.

Glucose Asparagine Agar 2

Composition per liter:

Agar	15.0g
Glucose	10.0g
Yeast extract	2.0g
L-Asparagine	1.0g
K_2HPO_4	0.5g

pH 7.3 ± 0.2 at 25°C

Preparation of Medium: Add components to distilled/deionized water and bring volume to 1.0L. Mix thoroughly. Gently heat and bring to boiling. Distribute into tubes or flasks. Autoclave for 15 min at 15 psi pressure–121°C. Pour into sterile Petri dishes or leave in tubes.

Use: For the cultivation and maintenance of *Actinomadura echinospora, Amycolata autotrophica, Amycolatopsis methanolica, Amycolatopsis rugosa, Amycolatopsis sulphurea, Gordona bronchialis, Gordona rubropertinctus, Gordona terrae, Microbispora rosea, Micromonospora narashino, Nocardia amarae, Nocardia diaphanozonaria, Nocardia farcinica, Nocardia otitidiscaviarum, Nocardia salmonicida, Nocardia violaceofusca, Nocardioides fastidiosa, Rhocococcus* species, *Streptomyces griseus,* and *Streptomyces lactamdurans*.

Glucose Azide Broth

Composition per liter:

Peptic digest of animal tissue	10.0g
Glucose	5.0g
K_2HPO_4	5.0g
NaCl	5.0g
Yeast extract	3.0g
KH_2PO_4	2.0g
NaN_3	0.25g
Bromcresol Purple	0.03g

pH 6.7 ± 0.2 at 25°C

Source: This medium is available as a premixed powder from HiMedia.

Caution: Sodium azide is toxic. Azides also react with metals and disposal must be highly diluted.

Preparation of Medium: Add components to distilled/deionized water and bring volume to 1.0L. Mix thoroughly. Gently heat and bring to boiling. Mix thoroughly. Distribute into tubes or flasks. Autoclave for 15 min at 15 psi pressure–121°C.

Use: For the enumeration of fecal streptococci by the MPN technique from water and sewage.

Glucose Azide HiVeg Broth

Composition per liter:

Plant peptone	10.0g
Glucose	5.0g
K_2HPO_4	5.0g
NaCl	5.0g
Yeast extract	3.0g
KH_2PO_4	2.0g
NaN_3	0.25g
Bromcresol Purple	0.03g

pH 6.7 ± 0.2 at 25°C

Source: This medium is available as a premixed powder from HiMedia.

Caution: Sodium azide is toxic. Azides also react with metals and disposal must be highly diluted.

Preparation of Medium: Add components to distilled/deionized water and bring volume to 1.0L. Mix thoroughly. Gently heat and bring to boiling. Mix thoroughly. Distribute into tubes or flasks. Autoclave for 15 min at 15 psi pressure–121°C.

Use: For the enumeration of fecal streptococci by the MPN technique from water and sewage.

Glucose Blood Liver Agar
See: **BL Agar**

Glucose Broth

Composition per 800.0mL:

Agar	10.0g
Beef extract	10.0g
Peptone	10.0g
NaCl	5.0g
Glucose solution	200.0mL

pH 7.0 ± 0.2 at 25°C

Glucose Solution:

Composition per 200.0mL:

Glucose	20.0g

Preparation of Glucose Solution: Add glucose to distilled/deionized water and bring volume to 200.0mL. Mix thoroughly. Filter sterilize.

Preparation of Medium: Add components, except glucose solution, to distilled/deionized water and bring volume to 800.0mL. Mix thoroughly. Gently heat and bring to boiling. Adjust pH to 7.0. Autoclave for 15 min at 15 psi pressure–121°C. Cool to 45°C. Aseptically add 200.0mL of sterile glucose solution. Mix thoroughly. Aseptically distribute into sterile tubes or flasks.

Use: For the cultivation of *Pseudomonas* species.

Glucose Broth

Composition per liter:

Casein enzymatic hydrolysate	10.0g
Glucose	5.0g
NaCl	5.0g

Source: This medium is available from HiMedia.

Preparation of Medium: Add components to distilled/deionized water and bring volume to1.0L. Mix thoroughly. Gently heat and bring to boiling. Cool to 25°. Adjust pH to 7.0. Distribute into sterile tubes or flasks. Autoclave for 15 min at 12 psi pressure–118°C.

Use: For the study of glucose fermentation where pH indicator is not desired.

Glucose Broth, Buffered

Composition per liter:

Peptone	5.0g
NaCl	5.0g
K_2HPO_4	3.68g
Yeast extract	2.0g
KH_2PO_4	1.32g
Lab Lemco beef extract	1.0g
Glucose	1.0g

pH 7.2 ± 0.2 at 25°C

Source: CM1 broth powder, which contains the peptone, NaCl, yeast extract, and Lab Lemco beef extract, is available from Oxoid Unipath.

Preparation of Medium: Add components to distilled/deionized water and bring volume to 1.0L. Mix thoroughly. Distribute into tubes or flasks. Autoclave for 15 min at 15 psi pressure–121°C.

Use: For the cultivation of *Aerococcus viridans, Carnobacterium alterfunditum, Carnobacterium funditum, Enterococcus durans, Enterococcus faecalis, Enterococcus faecium, Enterococcus hirae, Lactococcus lactis, Lactococcus plantarum, Streptococcus agalactiae, Streptococcus equismilis, Streptococcus mutans, Streptococcus pyogenes, Streptococcus salivarius,* and other *Streptococcus* species.

Glucose Broth, 9K

Composition per liter:

Glucose	10.0g
$(NH_4)_2SO_4$	3.0g
KH_2PO_4	0.5g
$MgSO_4 \cdot 7H_2O$	0.5g
KCl	0.1g
$Ca(NO_3)_2$	0.0125g
$FeSO_4 \cdot 7H_2O$	0.01mg
Glucose solution	100.0mL

pH 3.5 ± 0.2 at 25°C

Glucose Solution:

Composition per 100.0mL:

Glucose	10.0g

Preparation of Glucose Solution: Add glucose to distilled/deionized water and bring volume to 100.0mL. Mix thoroughly. Filter sterilize.

Preparation of Medium: Add components, except glucose solution, to distilled/deionized water and bring volume to 900.0mL. Mix thoroughly. Adjust pH to 3.5 with H_2SO_4. Autoclave for 15 min at 15 psi pressure–121°C. Cool to 25°C. Aseptically add 100.0mL of sterile glucose solution. Mix thoroughly. Aseptically distribute into sterile tubes or flasks.

Use: For the cultivation of *Thiobacillus acidophilus.*

Glucose Cysteine Agar Base with Thiamine and Hemoglobin

Composition per liter:

Glucose	25.0g
Papaic digest of soyabean meal	10.0g
Agar	14.0g
NaCl	5.0g
Meat peptone	3.0g
Cysteine·HCl	1.0g
Thiamine	5.0mg
Hemoglobin solution	500.0mL

pH 6.9 ± 0.2 at 25°C

Source: This medium,without hemoglobin solution, is available as a premixed powder from HiMedia.

Hemoglobin Solution:

Composition 500.0mL:

Bovine hemoglobin	10.0g

Preparation of Hemoglobin Solution: Add bovine hemoglobin to distilled/deionized water and bring volume to 500.0mL. Mix thoroughly. Autoclave for 15 min at 15 psi pressure–121°C. Cool to 45°–50°C.

Preparation of Medium: Add components, except hemoglobin, to distilled/deionized water and bring volume to 500.0L. Mix thoroughly. Gently heat until boiling. Autoclave for 15 min at 15 psi pressure–121°C. Cool to 45°–50°C. Add 500.0mL sterile hemoglobin solution. Mix thoroughly. Pour into sterile Petri dishes or distribute into sterile tubes.

Use: For thecultivation and enumeration of *Pasteurella tularensis.*

Glucose Cysteine HiVeg Agar Base with Thiamine and Hemoglobin

Composition per liter:

Glucose	25.0g
Papaic digest of soybean meal	10.0g
Agar	14.0g
NaCl	5.0g
Plant peptone No. 1	3.0g
Cysteine·HCl	1.0g
Thiamine	5.0mg
Hemoglobin solution	500.0mL

pH 6.9 ± 0.2 at 25°C

Source: This medium,without hemoglobin solution, is available as a premixed powder from HiMedia.

Hemoglobin Solution:

Composition 500.0mL:

Bovine hemoglobin	10.0g

Preparation of Hemoglobin Solution: Add bovine hemoglobin solution to distilled/deionized water and bring volume to 500.0mL. Mix thoroughly. Autoclave for 15 min at 15 psi pressure–121°C. Cool to 45°–50°C.

Preparation of Medium: Add components, except hemoglobin solution, to distilled/deionized water and bring volume to 500.0L. Mix thoroughly. Gently heat until boiling. Autoclave for 15 min at 15 psi pressure–121°C. Cool to 45°–50°C. Add 500.0mL sterile hemoglobin solution. Mix thoroughly. Pour into sterile Petri dishes or distribute into sterile tubes.

Use: For the cultivation and enumeration of *Pasteurella tularensis.*

Glucose Cysteine HiVeg Agar Base with Thiamine and Blood

Composition per liter:

Glucose	25.0g
Papaic digest of soybean meal	10.0g
Agar	14.0g
NaCl	5.0g
Plant peptone No. 1	3.0g
Cysteine·HCl	1.0g
Thiamine	5.0mg
Blood, defibrinated	50.0mL

pH 7.2 ± 0.2 at 25°C

Source: This medium,without blood, is available as a premixed powder from HiMedia.

Preparation of Medium: Add components, except blood, to distilled/deionized water and bring volume to 950.0L. Mix thoroughly. Gently heat until boiling. Autoclave for 15 min at 15 psi pressure–121°C. Cool to 75°–80°C. Add 50.0mL sterile defibrinated blood with thorough mixing and maintain at 75°–80°C for 15–20 min until the medium is chocolatized. Pour into sterile Petri dishes or distribute into sterile tubes.

Use: For the cultivation and enumeration of *Pasteurella tularensis*.

Glucose HiVeg Broth

Composition per liter:

Plant hydrolysate	10.0g
Glucose	5.0g
NaCl	5.0g

Source: This medium is available from HiMedia.

Preparation of Medium: Add components to distilled/deionized water and bring volume to1.0L. Mix thoroughly. Gently heat and bring to boiling. Cool to 25°. Adjust pH to 7.0. Distribute into sterile tubes or flasks. Autoclave for 15 min at 12 psi pressure–118°C.

Use: For the study of glucose fermentation where pH indicator is not desired.

Glucose HiVeg Peptone Agar

Composition per liter:

Plant peptone	20.0g
Agar	15.0g
Glucose	10.0g
NaCl	5.0g

pH 7.2 ± 0.2 at 25°C

Source: This medium is available as a premixed powder from HiMedia.

Preparation of Medium: Add components to distilled/deionized water and bring volume to 1.0L. Mix thoroughly. Gently heat and bring to boiling. Distribute into tubes or flasks. Autoclave for 15 min at 15 psi pressure–121°C. Pour into sterile Petri dishes or leave in tubes.

Use: For the cultivation of *Agrobacterium* species.

Glucose Medium Nakayama (DSMZ Medium 452)

Composition per liter:

Yeast extract	15.0g
Glucose	10.0g
Peptone	10.0g
Agar	10.0g
Solution A	10.0mL
Solution B	10.0mL
Solution C	1.0mL

pH 6.8 ± 0.2 at 25°C

Solution A:

Composition per 100.0mL:

KH_2PO_4	0.5g
K_2HPO_4	0.5g

Preparation of Solution A: Add components to distilled/deionized water and bring volume to 100.0mL. Mix thoroughly. Autoclave for 15 min at 15 psi pressure–121°C. Cool to 50°C.

Solution B:

Composition per 100.0mL:

$MgSO_4·7H_2O$	3.0g
NaCl	0.1g
$MnSO_4·5H_2O$	0.1g
$CuSO_4·5H_2O$	0.01g

Preparation of Solution B: Add components to distilled/deionized water and bring volume to 100.0mL. Mix thoroughly. Autoclave for 15 min at 15 psi pressure–121°C. Cool to 50°C.

Solution C:

Composition per 100.0mL:

Na_3-citrate	2.0g
$FeSO_4·7H_2O$	0.1g

Preparation of Solution C: Add components to distilled/deionized water and bring volume to 100.0mL. Mix thoroughly. Autoclave for 15 min at 15 psi pressure–121°C. Cool to 50°C.

Preparation of Medium: Add components, except solutions A, B, and C, to distilled/deionized water and bring volume to 979.0mL. Mix thoroughly. Adjust pH to 6.8. Gently heat and bring to boiling. Autoclave for 15 min at 15 psi pressure–121°C. Cool to 50°C. Aseptically add 10.0mL sterile solution A, 10.0mL sterile solution B, and 1.0mL sterile solution C. Mix thoroughly. Pour into sterile Petri dishes or aseptically dispense into sterile tubes.

Use: For the growth and maintenance of *Bacillus laevolacticus.*

Glucose Nitrogen-Free Salt Agar

Composition per liter:

Agar	15.0g
$CaCO_3$	1.0g
K_2HPO_4	1.0g
$MgSO_4·7H_2O$	0.2g
NaCl	0.2g
$FeSO_4·7H_2O$	0.1g
$Na_2MoO_4·2H_2O$	5.0mg
Glucose solution	100.0mL

pH 7.0 ± 0.2 at 25°C

Glucose Solution:

Composition per 100.0mL:

Glucose	10.0g

Preparation of Glucose Solution: Add glucose to distilled/deionized water and bring volume to 100.0mL. Mix thoroughly. Filter sterilize.

Preparation of Medium: Add components, except glucose solution, to distilled/deionized water and bring volume to 900.0mL. Mix thoroughly. Gently heat and bring to boiling. Autoclave for 15 min at

15 psi pressure–121°C. Cool to 45°–50°C. Aseptically add 100.0mL of sterile glucose solution. Mix thoroughly. Pour into sterile Petri dishes or distribute into sterile tubes.

Use: For the cultivation of *Azotobacter* species.

Glucose Nitrogen-Free Salt Solution

Composition per liter:

$CaCO_3$	1.0g
K_2HPO_4	1.0g
$MgSO_4 \cdot 7H_2O$	0.2g
NaCl	0.2g
$FeSO_4 \cdot 7H_2O$	0.1g
$Na_2MoO_4 \cdot 2H_2O$	5.0mg
Glucose solution	100.0mL

pH 7.0 ± 0.2 at 25°C

Glucose Solution:

Composition per 100.0mL:

Glucose	10.0g

Preparation of Glucose Solution: Add glucose to distilled/deionized water and bring volume to 100.0mL. Mix thoroughly. Filter sterilize.

Preparation of Medium: Add components, except glucose solution, to distilled/deionized water and bring volume to 900.0mL. Mix thoroughly. Gently heat and bring to boiling. Autoclave for 15 min at 15 psi pressure–121°C. Cool to 45°–50°C. Aseptically add 100.0mL of sterile glucose solution. Mix thoroughly. Aseptically distribute into sterile tubes or flasks.

Use: For the cultivation of *Azotobacter* species.

Glucose Nutrient Agar

Composition per liter:

Agar	15.0g
Pancreatic digest of casein	10.0g
Glucose	5.0g
K_2HPO_4	5.0g
NaCl	5.0g
Yeast extract	5.0g

Preparation of Medium: Add components to distilled/deionized water and bring volume to 1.0L. Mix thoroughly. Gently heat and bring to boiling. Distribute into tubes or flasks. Autoclave for 15 min at 15 psi pressure–121°C. Pour into sterile Petri dishes or leave in tubes.

Use: For the isolation and cultivation of *Brochothrix thermosphacta*.

Glucose Peptone Agar

Composition per liter:

Peptone	20.0g
Agar	15.0g
Glucose	10.0g
NaCl	5.0g

pH 7.2 ± 0.2 at 25°C

Preparation of Medium: Add components to distilled/deionized water and bring volume to 1.0L. Mix thoroughly. Gently heat and bring to boiling. Distribute into tubes or flasks. Autoclave for 15 min at 15 psi pressure–121°C. Pour into sterile Petri dishes or leave in tubes.

Use: For the cultivation of *Agrobacterium* species.

Glucose Peptone Medium

Composition per liter:

Agar	15.0g
Glucose	0.3g
Peptone	0.17g

Preparation of Medium: Add components to distilled/deionized water and bring volume to 1.0L. Mix thoroughly. Gently heat and bring to boiling. Distribute into tubes or flasks. Autoclave for 15 min at 15 psi pressure–121°C. Pour into sterile Petri dishes or leave in tubes.

Use: For the cultivation and maintenance of *Acytostelium ellipticum, Phlyctochytrium africanum, Rhizophylctis harderi,* and *Rhizophylctis rosea.*

Glucose Peptone Medium

Composition per liter:

Peptone	20.0g
$CaCO_3$	10.0g
Glucose	10.0g
Yeast extract	10.0g

Preparation of Medium: Add components to 1.0L of tap water. Mix thoroughly. Distribute into tubes or flasks. Autoclave for 15 min at 15 psi pressure–121°C.

Use: For the cultivation and maintenance of *Mycobacterium chitae.*

Glucose Peptone Yeast Extract Salts Medium

See: **GPY Salts Medium**

Glucose Phosphate Broth

Composition per liter:

Peptone	10.0g
K_2HPO_4	5.0g
Glucose	5.0g

pH 7.5 ± 0.2 at 25°C

Preparation of Medium: Add components, except glucose, to distilled/deionized water and bring volume to 1.0L. Mix thoroughly. Gently heat and bring to boiling. Filter while hot through Whatman filter paper. Cool to 25°C. Adjust pH to 7.5. Add 5.0g of glucose. Mix thoroughly. Distribute into sterile tubes or flasks. Autoclave for 10 min at 10 psi pressure–115°C.

Use: For the cultivation of a variety of nonfastidious heterotrophic microorganisms.

Glucose Phosphate Broth

Composition per liter:

Peptone	10.0g
K_2HPO_4	5.0g
Glucose	5.0g

pH 7.5 ± 0.2 at 25°C

Preparation of Medium: Add components, except glucose, to distilled/deionized water and bring volume to 1.0L. Mix thoroughly. Gently heat and bring to boiling. Filter while hot through Whatman filter paper. Cool to 25°C. Adjust pH to 7.5. Add 5.0g of glucose. Mix thoroughly. Distribute into sterile tubes or flasks. Autoclave for 10 min at 10 psi pressure–115°C.

Use: For the cultivation of a variety of nonfastidious heterotrophic microorganisms.

Glucose Salt Teepol Broth (GSTB)

Composition per liter:

NaCl	30.0g
Peptone	10.0g
Glucose	5.0g
Beef extract	3.0g
Methyl Violet	2.0mg
Sodium lauryl sulfate (Teepol—0.1% solution)	4.0mL

pH 8.8 ± 0.2 at 25°C

Preparation of Medium: Add components to distilled/deionized water and bring volume to 1.0L. Mix thoroughly. Adjust pH to 8.8. Distribute into tubes or flasks. Autoclave for 15 min at 15 psi pressure–121°C.

Use: For the cultivation of *Vibrio* species from foods.

Glucose Salt Teepol HiVeg Broth

Composition per liter:

NaCl	30.0g
Plant peptone	10.0g
Glucose	5.0g
Teepol	4.0g
Plant extract	3.0g
Methyl Violet	2.0mg

pH 8.8 ± 0.2 at 25°C

Source: This medium is available as a premixed powder from HiMedia.

Preparation of Medium: Add components to distilled/deionized water and bring volume to 1.0L. Mix thoroughly. Adjust pH to 8.8. Distribute into tubes or flasks. Autoclave for 15 min at 15 psi pressure–121°C.

Use: For the cultivation of *Vibrio* species from foods.

Glucose Salts Medium

Composition per 1000.5mL:

Glucose	5.0g
$(NH_4)_2SO_4$	1.0g
$MgSO_4 \cdot 7H_2O$	0.5g
NaCl	0.5g
$NaH_2PO_4 \cdot 12H_2O$	0.7mg
$NaH_2PO_4 \cdot 2H_2O$	0.3mg
Trace elements solution	0.5mL

pH 6.9 ± 0.2 at 25°C

Trace Elements Solution:

Composition per liter:

H_3BO_3	0.3g
$CoCl_2 \cdot 6H_2O$	0.2g
$ZnSO_4 \cdot 7H_2O$	0.1g
$MnCl_2 \cdot 4H_2O$	0.03g
$Na_2MoO_4 \cdot 2H_2O$	0.03g
$NiCl_2 \cdot 6H_2O$	0.02g
$CuCl_2 \cdot 2H_2O$	0.01g

Preparation of Trace Elements Solution: Add components to distilled/deionized water and bring volume to 1.0L. Mix thoroughly. Filter sterilize.

Preparation of Medium: Add components, except trace elements solution, to distilled/deionized water and bring volume to 1.0L. Mix thoroughly. Autoclave for 15 min at 15 psi pressure–121°C. Aseptically add 0.5mL of sterile trace elements solution. Mix thoroughly. Aseptically distribute into sterile tubes or flasks.

Use: For the cultivation of a wide variety of bacteria and fungi.

Glucose Starch Agar

Composition per liter:

Gelatin	20.0g
Proteose peptone	15.0g
Agar	10.0g
Glucose	10.0g
Starch, soluble	5.0g
NaCl	5.0g
Na_2HPO_4	3.0g

pH 7.2 ± 0.2 at 25°C

Source: This medium is available from HiMedia.

Preparation of Medium: Add components to distilled/deionized water and bring volume to 1.0L. Mix thoroughly. Gently heat and bring to boiling. Distribute into tubes or flasks. Autoclave for 15 min at 15 psi pressure–121°C. Pour into sterile Petri dishes or leave in tubes.

Use: For use as a basal medium with the addition of salicin, raffinose, and Phenol Red for the detection of *Clostridium* spp.

Glucose Tetrazolium Medium

Composition per liter:

Agar	15.0g
Pancreatic digest of gelatin	6.0g
Yeast extract	3.0g
Beef extract	1.5g
1,3,5-Triphenyl tetrazolium chloride	0.05g
Glucose solution	100.0mL

pH 6.6 ± 0.1 at 25°C

Glucose Solution:

Composition per 100.0mL:

Glucose	10.0g

Preparation of Glucose Solution: Add glucose to distilled/deionized water and bring volume to 100.0mL. Mix thoroughly. Filter sterilize.

Preparation of Medium: Add components, except glucose solution, to distilled/deionized water and bring volume to 900.0mL. Mix thoroughly. Gently heat and bring to boiling. Autoclave for 15 min at 15 psi pressure–121°C. Cool to 45°–50°C. Aseptically add glucose solution. Mix thoroughly. Pour into sterile Petri dishes or distribute into sterile tubes.

Use: For the cultivation and maintenance of *Streptococcus mutans*.

Glucose Tryptone Yeast Extract Medium

See: **GTYE Medium**

Glucose Yeast Broth with Sodium Chloride

Composition per liter:

NaCl	50.0g
Agar	15.0g
Yeast extract	3.0g
Glucose	1.0g

pH 7.0 ± 0.2 at 25°C

Preparation of Medium: Add components to distilled/deionized water and bring volume to 1.0L. Mix thoroughly. Gently heat and bring to boiling. Distribute into tubes or flasks. Autoclave for 15 min at 15 psi pressure–121°C. Pour into sterile Petri dishes or leave in tubes.

Use: For the cultivation and maintenance of *Pediococcus halophilus*.

Glucose Yeast Chalk Agar

Composition per liter:

Chalk	40.0g
Agar	15.0g
Glucose	5.0g
Yeast extract	5.0g

Preparation of Medium: Add components to distilled/deionized water and bring volume to 1.0L. Mix thoroughly. Gently heat and bring to boiling. Distribute into tubes or flasks. Autoclave for 15 min at 15 psi pressure–121°C. Pour into sterile Petri dishes or leave in tubes.

Use: For the cultivation and maintenance of *Xanthomonas* species.

Glucose Yeast Extract Agar

Composition per liter:

Agar	15.0g
Glucose	5.0g
Meat extract	5.0g
Peptone	5.0g
Yeast extract	5.0g

pH 6.5 ± 0.2 at 25°C

Preparation of Medium: Add components to distilled/deionized water and bring volume to 1.0L. Mix thoroughly. Gently heat and bring to boiling. Distribute into tubes or flasks. Autoclave for 15 min at 15 psi pressure–121°C. Pour into sterile Petri dishes or leave in tubes.

Use: For the isolation and cultivation of *Leuconostoc* species and *Pediococcus* species.

Glucose Yeast Extract Agar

Composition per liter:

$CaCO_3$	20.0g
Glucose	20.0g
Agar	17.0g
Yeast extract	10.0g

Preparation of Medium: Add components to distilled/deionized water and bring volume to 1.0L. Mix thoroughly. Gently heat and bring to boiling. Distribute into tubes or flasks. Autoclave for 15 min at 15 psi pressure–121°C. Pour into sterile Petri dishes or leave in tubes.

Use: For the cultivation of *Agrobacterium* species, *Clostridium* species, *Erwinia* species, *Pseudomonas* species, and *Xanthomonas campestris*.

Glucose Yeast Extract Agar

Composition per liter:

Agar	15.0g
Peptic digest of animal tissue	5.0g
Yeast extract	5.0g
Glucose	2.0g
KH_2PO_4	0.5g
K_2HPO_4	0.5g
$MgSO_4·7H_2O$	0.3g
NaCl	0.01g
$MnSO_4·H_2O$	0.01g
$ZnSO_4·7H_2O$	1.6mg
$CuSO_4·5H_2O$	1.6mg
$CoSO_4·7H_2O$	1.6mg

pH 7.2 ± 0.2 at 25°C

Source: This medium is available from HiMedia.

Preparation of Medium: Add components to distilled/deionized water and bring volume to 1.0L. Mix thoroughly. Gently heat and bring to boiling. Distribute into tubes or flasks. Autoclave for 15 min at 15 psi pressure–121°C. Pour into sterile Petri dishes or leave in tubes.

Use: For the enumeration and cultivation of lactobacilli in pharmaceutical preparations.

Glucose Yeast Extract Iron Agar (LMG Medium 153)

Composition per liter:

Agar	15.0g
Glucose	10.0g
Yeast extract	5.0g
Iron solution	20.0mL

pH 7.1 ± 0.2 at 25°C

Iron Solution:

Composition per 100.0mL:

$FeCl_3$	0.03g

Preparation of Iron Solution: Add $FeCl_3$ to distilled/deionized water and bring volume to 100.0mL. Mix thoroughly. Filter sterilize.

Preparation of Medium: Add components, except iron solution, to distilled/deionized water and bring volume to 980.0mL. Mix thoroughly. Autoclave for 15 min at 15 psi pressure–121°C. Cool to 45°–50°C. Aseptically add 20.0mL sterile iron solution. Mix thoroughly. Aseptically pour into sterile Petri dishes or distribute into sterile tubes.

Use: For the cultivation of *Pseudomonas denitrificans*. Note: Buffered Nutrient Agar is more commonly used for the culture of *P. denitrificans*.

Glucose Yeast Extract Medium (ATCC Medium 846)

Composition per liter:

Agar, noble	13.0g
Yeast extract	5.0g
KH_2PO_4	1.0g
NaCl	1.0g
Peptone	1.0g

pH 6.8 ± 0.2 at 25°C

Preparation of Medium: Add components to distilled/deionized water and bring volume to 1.0L. Mix thoroughly. Gently heat and bring to boiling. Distribute into tubes or flasks. Autoclave for 15 min at 15 psi pressure–121°C. Pour into sterile Petri dishes or leave in tubes.

Use: For the cultivation and maintenance of *Pseudomonas glathei*.

Glucose Yeast Extract Medium (ATCC Medium 985)

Composition per liter:

Agar	15.0g
Yeast extract	3.0g
Glucose	1.0g

pH 7.0 ± 0.2 at 25°C

Preparation of Medium: Add components to distilled/deionized water and bring volume to 1.0L. Mix thoroughly. Gently heat and bring to boiling. Distribute into tubes or flasks. Autoclave for 15 min at 15 psi pressure–121°C. Pour into sterile Petri dishes or leave in tubes.

Use: For the cultivation and maintenance of *Acinetobacter tartarogenes, Agrobacterium viscosum,* and *Pseudomonas* species.

Glucose Yeast Extract Medium (ATCC Medium 1742)

Composition per liter:

Agar	15.0g
Glucose	5.0g
Yeast extract	3.5g

Preparation of Medium: Add components to distilled/deionized water and bring volume to 1.0L. Mix thoroughly. Gently heat and bring to boiling. Distribute into tubes or flasks. Autoclave for 15 min at 15 psi pressure–121°C. Pour into sterile Petri dishes or leave in tubes.

Use: For the cultivation and maintenance of *Xanthobacter* species.

Glucose Yeast Extract Peptone Agar

Composition per liter:

Glucose	20.0g
Agar	15.0g
Peptone	10.0g
Yeast extract	3.0g

pH 6.8 ± 0.2 at 25°C

Source: This medium is available from HiMedia.

Preparation of Medium: Add components to distilled/deionized water and bring volume to 1.0L. Mix thoroughly. Gently heat and bring to boiling. Distribute into tubes or flasks. Autoclave for 15 min at 15 psi pressure–121°C. Pour into sterile Petri dishes or leave in tubes.

Use: For the isolation and cultivation of yeasts from soils.

Glucose Yeast Extract Peptone Agar with 2% Glucose

Composition per liter:

Glucose	20.0g
Agar	15.0g
Peptone	5.0g
Yeast extract	1.0g
K_2HPO_4	0.5g
$MgSO_4 \cdot 7H_2O$	0.5g
$CaCO_3$	1.0g
$FeSO_4 \cdot 7H_2O$	50.0mg

pH 6.8 ± 0.2 at 25°C

Preparation of Medium: Add components to distilled/deionized water and bring volume to 1.0L. Mix thoroughly. Adjust pH to 6.8. Gently heat and bring to boiling. Distribute into tubes or flasks. Autoclave for 15 min at 15 psi pressure–121°C. Pour into sterile Petri dishes or leave in tubes.

Use: For the cultivation and maintenance of *Lactobacillus acidophilus* and *Leclercia adecarboxylata.*

Glucose Yeast Extract Peptone Medium (GYP Medium)

Composition per liter:

Glucose	20.0g
Agar	10.0g
Peptone	5.0g
Yeast extract	5.0g
$CaCO_3$	0.1g

Preparation of Medium: Add components to distilled/deionized water and bring volume to 1.0L. Mix thoroughly. Gently heat and bring to boiling. Distribute into tubes or flasks. Autoclave for 15 min at 15 psi pressure–121°C. Pour into sterile Petri dishes or leave in tubes.

Use: For the isolation and cultivation of *Sporolactobacillus* species.

Glucose Yeast Extract Peptone Thioglycolate Medium

See: **GYPT Medium**

Glucose Yeast Medium with Calcium Carbonate

Composition per liter:

Agar	15.0g
$CaCO_3$	7.5g
Peptone	5.0g
Yeast extract	5.0g
Glucose	3.0g

pH 7.0 ± 0.2 at 25°C

Preparation of Medium: Add components to distilled/deionized water and bring volume to 1.0L. Mix thoroughly. Gently heat and bring to boiling. Distribute into tubes or flasks. Adjust pH to 6.3. Autoclave for 15 min at 15 psi pressure–121°C. Pour into sterile Petri dishes or leave in tubes.

Use: For the cultivation and maintenance of *Erwinia herbicola* and *Bacillus* species.

Glucose Yeast Plant Peptone Agar

Composition per liter:

Glucose	20.0g
Agar	15.0g
Plant peptone	10.0g
Yeast extract	5.0g

pH 7.2 ± 0.2 at 25°C

Source: This medium is available as a premixed powder from HiMedia.

Preparation of Medium: Add components to distilled/deionized water and bring volume to 1.0L. Mix thoroughly. Gently heat and bring to boiling. Distribute into tubes or flasks. Autoclave for 15 min at 15 psi pressure–121°C. Pour into sterile Petri dishes or leave in tubes.

Use: For the isolation of yeasts from soil specimens.

Glucose Yeast Peptone Medium

Composition per liter:

Agar	20.0g
Glucose	5.0g
Peptone	5.0g
Yeast extract	3.0g

pH 6.8 ± 0.2 at 25°C

Preparation of Medium: Add components to distilled/deionized water and bring volume to 1.0L. Mix thoroughly. Gently heat and bring to boiling. Adjust pH to 6.8. Distribute into tubes or flasks. Autoclave for 15 min at 15 psi pressure–121°C. Pour into sterile Petri dishes or leave in tubes.

Use: For the cultivation and maintenance of a variety of yeasts, including *Candida* species, *Cryptococcus albidosimilis*, *Cryptococcus albidus*,

Cryptococcus laurentii, *Cryptococcus vishniacii*, *Eeniella nana*, *Filobasidiella neoformans*, *Histoplasma capsulatum*, *Kluyveromyces lactis*, *Lipomyces kononenkoae*, *Rhodotorula matritensis*, *Saccharomyces bayanus*, *Saccharomyces cerevisiae*, *Saccharomyces douglasii*, *Saccharomyces paradoxus*, *Schizosaccharomyces pombe*, and *Zygosaccharomyces rouxii*.

Glucose Yeast Peptone Medium

Composition per liter:

Glucose	10.0g
Peptone	5.0g
Yeast extract	5.0g

pH 5.0 ± 1.0 at 25°C

Preparation of Medium: Add components to distilled/deionized water and bring volume to 1.0L. Mix thoroughly. Adjust pH to 5.0. Distribute into tubes or flasks. Autoclave for 15 min at 15 psi pressure–121°C.

Use: For the cultivation of *Enterobacter cloacae*.

Glutamate Medium

Composition per liter:

Solution A	500.0mL
Solution B	250.0mL
Solution C	250.0mL

Solution A:

Composition per 500.0mL:

Mannitol	10.0g
K_2HPO_4	0.22g

Preparation of Solution A: Add components to distilled/deionized water and bring volume to 500.0mL. Mix thoroughly. Autoclave for 15 min at 15 psi pressure–121°C. Cool to 25°C.

Solution B:

Composition per 250.0mL:

$MgSO_4 \cdot 7H_2O$	0.1g
$CaCl_2 \cdot 6H_2O$	0.08g
$FeCl_3 \cdot 6H_2O$	0.05g

Preparation of Solution B: Add components to distilled/deionized water and bring volume to 250.0mL. Mix thoroughly. Autoclave for 15 min at 15 psi pressure–121°C. Cool to 25°C.

Solution C:

Composition per 250.0mL:

Sodium glutamate	1.1g
Calcium pantothenate	0.5mg
Thiamine·HCl	0.1mg
Biotin	0.5μg

Preparation of Solution C: Add components to distilled/deionized water and bring volume to 250.0mL. Mix thoroughly. Filter sterilize.

Preparation of Medium: Aseptically combine 500.0mL of cooled, sterile solution A, 250.0mL of cooled, sterile solution B, and 250.0mL of sterile solution C. Mix thoroughly. Aseptically distribute into sterile tubes or flasks.

Use: For the isolation of *Rhizobium* species.

Glutamate Medium (ATCC Medium 820)

Composition per liter:

Agar	15.0g
Sodium glutamate	5.0g
KH_2PO_4	1.0g
$MgSO_4 \cdot 7H_2O$	0.2g
KCl	0.1g
Glucose solution	100.0mL

pH 6.5 ± 0.2 at 25°C

Glucose Solution:

Composition per 100.0mL:

Glucose	10.0g

Preparation of Glucose Solution: Add glucose to distilled/deionized water and bring volume to 100.0mL. Mix thoroughly. Filter sterilize.

Preparation of Medium: Add components, except glucose solution, to distilled/deionized water and bring volume to 900.0mL. Mix thoroughly. Gently heat and bring to boiling. Autoclave for 15 min at 15 psi pressure–121°C. Cool to 45°–50°C. Aseptically add 100.0mL of sterile glucose solution. Mix thoroughly. Pour into sterile Petri dishes or distribute into sterile tubes.

Use: For the cultivation and maintenance of *Pseudomonas* species.

Glutamate Medium (ATCC Medium 1372)

Composition per liter:

K_2HPO_4	6.0g
Sodium glutamate	5.0g
Peptone	4.0g
Yeast extract	4.0g

Preparation of Medium: Add components to distilled/deionized water and bring volume to 1.0L. Mix thoroughly. Gently heat and bring to boiling. Distribute into tubes or flasks. Autoclave for 15 min at 15 psi pressure–121°C. Pour into sterile Petri dishes or leave in tubes.

Use: For the cultivation and maintenance of *Escherichia coli*.

Glutamate Starch Phenol Red Agar Base

Composition per liter:

Starch, soluble	20.0g
L-Glutamate, sodium	10.0g
Agar	12.0g
KH_2PO_4	2.0g
$MgSO_4 \cdot 7H_2O$	0.5g
Phenol Red	0.36g
Selective supplement solution	10.0mL

pH 7.2 ± 0.2 at 25°C

Source: This medium is available from HiMedia.

Selective Supplement Solution:

Composition per 10.0mL:

Penicillin G100,00 U**Preparation of Selective Supplement Solution:** Add components to distilled/deionized water and bring volume to 10.0mL. Mix thoroughly. Filter sterilize.

Preparation of Medium: Add components, except selective supplement solution, to distilled/deionized water and bring volume to 990.0mL. Mix thoroughly. Autoclave for 15 min at 15 psi pressure–121°C. Cool to 50°C. Aseptically add selective supplement solution. Mix thoroughly. Pour into Petri dishes or aseptically distribute into sterile tubes.

Use: For the detection of *Pseudomonas* spp. and *Aeromonas* spp. in foodstuffs, wastewater, and equipment, in the food industry.

Glutamine Medium

Composition per liter:

Agar	15.0g
Pancreatic digest of gelatin	5.0g
Beef extract	3.0g
Glutamine-tryptophan solution	10.0mL

pH 6.8 ± 0.2 at 25°C

Glutamine-Tryptophan Solution:

Composition per liter:

L-Glutamine	100.0mg
L-Tryptophan	100.0mg

Preparation of Medium: Add components, except glutamine-tryptophan solution, to distilled/deionized water and bring volume to 990.0mL. Mix thoroughly. Gently heat while stirring and bring to boiling. Autoclave for 15 min at 15 psi pressure–121°C. Aseptically add 10.0mL of sterile glutamine-tryptophan solution. Mix thoroughly. Pour into sterile Petri dishes or distribute into sterile tubes.

Use: For the cultivation of *Escherichia coli.*

Glutarate Medium

Composition per liter:

Sodium glutarate	2.6g
NaCl	1.0g
KCl	0.5g
$MgCl_2 \cdot 6H_2O$	0.4g
NH_4Cl	0.25g
KH_2PO_4	0.2g
$CaCl_2 \cdot 2H_2O$	0.15g
Resazurin	1.0mg
Rumen fluid	20.0mL
$NaHCO_3$ solution	20.0mL
$Na_2S \cdot 9H_2O$ solution	10.0mL
Trace elements solution SL-10	1.0mL

pH 7.2 ± 0.2 at 25°C

Trace Elements Solution SL-10:

Composition per liter:

$FeCl_2 \cdot 4H_2O$	1.5g
$CoCl_2 \cdot 6H_2O$	190.0mg
$MnCl_2 \cdot 4H_2O$	100.0mg
$ZnCl_2$	70.0mg
$Na_2MoO_4 \cdot 2H_2O$	36.0mg
$NiCl_2 \cdot 6H_2O$	24.0mg
H_3BO_3	6.0mg
$CuCl_2 \cdot 2H_2O$	2.0mg
HCl (25% solution)	10.0mL

Preparation of Trace Elements Solution SL-10: Add $FeCl_2 \cdot 4H_2O$ to 10.0mL of HCl solution. Mix thoroughly. Add distilled/deionized water and bring volume to 1.0L. Add remaining components. Mix thoroughly. Gas under 80% N_2 + 20% CO_2. Autoclave for 15 min at 15 psi pressure–121°C.

$NaHCO_3$ Solution:

Composition per 20.0mL:

$NaHCO_3$	2.5g

Preparation of $NaHCO_3$ Solution: Add $NaHCO_3$ to distilled/deionized water and bring volume to 20.0mL. Mix thoroughly. Filter sterilize. Gas under 80% N_2 + 20% CO_2.

$Na_2S \cdot 9H_2O$ Solution:

Composition per 10.0mL:

$Na_2S \cdot 9H_2O$	0.36g

Preparation of $Na_2S \cdot 9H_2O$ Solution: Add $Na_2S \cdot 9H_2O$ to distilled/deionized water and bring volume to 10.0mL. Mix thoroughly. Gas under 100% N_2. Autoclave for 15 min at 15 psi pressure–121°C.

Preparation of Medium: Prepare and dispense medium under 80% N_2 + 20% CO_2. Add components, except $NaHCO_3$ solution and $Na_2S \cdot 9H_2O$ solution, to distilled/deionized water and bring volume to 970.0mL. Mix thoroughly. Sparge with 80% N_2 + 20% CO_2. Autoclave for 15 min at 15 psi pressure–121°C. Aseptically and anaerobically add 20.0mL of sterile $NaHCO_3$ solution and 10.0mL of sterile $Na_2S \cdot 9H_2O$ solution. Mix thoroughly. Aseptically and anaerobically distribute into sterile tubes or flasks.

Use: For the cultivation of a wide variety of bacteria that can utilize glutarate as a carbon source.

Glycerol Agar

Composition per 1070.0mL:

Agar	15.0g
Peptone	5.0g
Beef extract	3.0g
Soil extract	1.0L
Glycerol	70.0mL

pH 7.0 ± 0.2 at 25°C

Soil Extract:

Composition per liter:

Soil, air dried	1.0Kg

Preparation of Soil Extract: Sift soil through a #9 mesh screen. Add to 2.4L of tap water. Mix thoroughly. Autoclave for 60 min at 15 psi pressure–121°C. Cool to 25°C. Filter through Whatman #1 filter paper. Bring volume to 1.0L with tap water.

Preparation of Medium: Combine components. Mix thoroughly. Gently heat and bring to boiling. Autoclave for 15 min at 15 psi pressure–121°C. Pour into sterile Petri dishes or distribute into sterile tubes.

Use: For the selective isolation and cultivation of *Nocardia* species and *Rhodococcus* species.

Glycerol Agar

Composition per liter:

Beef heart, solids from infusion	250.0g
Glycerol	60.0g
Agar	15.0g
Pancreatic digest of gelatin	5.0g
Tryptose	5.0g
Beef extract	3.0g
NaCl	2.5g

pH 7.3 ± 0.2 at 25°C

Preparation of Medium: Add components to distilled/deionized water and bring volume to 1.0L. Mix thoroughly. Gently heat and bring to boiling. Distribute into tubes or flasks. Autoclave for 15 min at 15 psi pressure–121°C. Pour into sterile Petri dishes or leave in tubes.

Use: For the cultivation and maintenance of *Bacillus subtilis*, *Enterococcus faecalis, Erwinia chrysanthemi, Gordona rubropertinctus, Mycobacterium* species, *Nocardia brevicatena, Rhodococcus equi,* and *Rhodococcus rhodochrous.*

Glycerol Agar

Composition per liter:

Beef heart, infusion from	300.0g
Glycerol	60.0g
Agar	17.5g
Tryptose	7.0g
NaCl	3.0g
Peptone	2.5g
NaCl	2.5g
Yeast extract	1.0g
Beef extract	0.5g

pH 7.3 ± 0.2 at 25°C

Preparation of Medium: Add components to distilled/deionized water and bring volume to 1.0L. Mix thoroughly. Adjust pH to 7.3. Gently heat and bring to boiling. Distribute into tubes or flasks. Autoclave for 15 min at 15 psi pressure–121°C. Pour into sterile Petri dishes or leave in tubes.

Use: For the cultivation of fastidious bacteria.

Glycerol Arginine Agar

Composition per liter:

Agar	15.0g
Glycerol	12.5g
Arginine	1.0g
K_2HPO_4	1.0g
NaCl	1.0g
$MgSO_4 \cdot 7H_2O$	0.5g
$Fe_2(SO_4)_3 \cdot 6H_2O$	0.01g
$CuSO_4 \cdot 5H_2O$	1.0mg
$MnSO_4 \cdot H_2O$	1.0mg
$ZnSO_4 \cdot 7H_2O$	1.0mg

Preparation of Medium: Add components to distilled/deionized water and bring volume to 1.0L. Mix thoroughly. Gently heat and bring to boiling. Distribute into tubes or flasks. Autoclave for 15 min at 15 psi pressure–121°C. Pour into sterile Petri dishes or leave in tubes.

Use: For the selective isolation and cultivation of streptomycetes.

Glycerol Asparagine Agar

See: **ISP 5**

Glycerol Asparagine Meat Agar

Composition per liter:

Agar	20.0g
Glycerol	10.0g
Beef extract	10.0g
L-Asparagine	1.0g
K_2HPO_4	1.0g
Trace salts solution	1.0mL

pH 7.4 ± 0.2 at 25°C

Trace Salts Solution:

Composition per 100.0mL:

$FeSO_4 \cdot 7H_2O$	0.1g
$MnCl_2 \cdot 4H_2O$	0.1g
$ZnSO_4 \cdot 7H_2O$	0.1g

Preparation of Trace Salts Solution: Add components to distilled/deionized water and bring volume to 100.0mL. Mix thoroughly. Filter sterilize.

Preparation of Medium: Add components, except trace salts solution, to distilled/deionized water and bring volume to 999.0mL. Mix thoroughly. Gently heat and bring to boiling. Autoclave for 15 min at 15 psi pressure–121°C. Cool to 45°–50°C. Aseptically add 1.0mL of sterile trace salts solution. Mix thoroughly. Pour into sterile Petri dishes or distribute into sterile tubes.

Use: For the cultivation of *Actinomadura* species and *Streptomyces* species.

Glycerol Asparagine Medium

Composition per liter:

Agar	20.0g
Glycerol	10.0g
K_2HPO_4	1.0g
L-Asparagine	1.0g
Trace salts solution	1.0mL

pH 7.2–7.4 at 25°C

Trace Salts Solution:

Composition per 100.0mL:

$FeSO_4 \cdot 7H_2O$	0.1g
$MnCl_2 \cdot 4H_2O$	0.1g
$ZnSO_4 \cdot 7H_2O$	0.1g

Preparation of Trace Salts Solution: Add components to distilled/deionized water and bring the volume to 100.0mL. Mix thoroughly. Filter sterilize.

Preparation of Medium: Add components to distilled/deionized water and bring volume to 1.0L. Mix thoroughly. Gently heat and bring to boiling. Adjust pH to 7.0–7.4. Distribute into tubes or flasks. Autoclave for 15 min at 15 psi pressure–121°C. Pour into sterile Petri dishes or leave in tubes.

Use: For the cultivation and maintenance of *Streptomyces* species.

Glycerol Beef Extract Medium

Composition per liter:

Agar	15.0g
Peptone	10.0g
NaCl	5.0g
Beef extract	3.0g
Glycerol	40.0mL

Preparation of Medium: Add components to distilled/deionized water and bring volume to 1.0L. Mix thoroughly. Gently heat and bring to boiling. Distribute into tubes or flasks. Autoclave for 15 min at 15 psi pressure–121°C. Pour into sterile Petri dishes or leave in tubes.

Use: For the cultivation and maintenance of *Corynebacterium alkanolyticum, Pseudomonas mallei, Pseudomonas pseudomallei,* and *Rhodococcus* species.

Glycerol Calcium Malate Agar

Composition per liter:

Agar	15.0g
Calcium malate	10.0g
Glycerol	10.0g
K_2HPO_4	0.5g
NH_4Cl	0.5g

Preparation of Medium: Add components to distilled/deionized water and bring volume to 1.0L. Mix thoroughly. Gently heat and bring to boiling. Distribute into tubes or flasks. Autoclave for 15 min at 15 psi pressure–121°C. Pour into sterile Petri dishes or leave in tubes.

Use: For the cultivation and maintenance of *Actinoplanes awajinensis, Dactylosporangium aurantiacum, Dactylosporangium salmoneum, Dactylosporangium thailandense,* and *Planobispora rosea.*

Glycerol Chalk Agar

Composition per liter:

NaCl	30.0g
Agar	15.0g
Glycerol	10.0g
$CaCO_3$	5.0g
Peptone	5.0g
Yeast extract	3.0g

Preparation of Medium: Add components to distilled/deionized water and bring volume to 1.0L. Mix thoroughly. Gently heat and bring to boiling. Autoclave for 15 min at 15 psi pressure–121°C. Pour into sterile Petri dishes. Swirl flask while dispensing medium to keep $CaCO_3$ in suspension.

Use: For the cultivation of *Photobacterium* species and *Lucibacterium* species.

Glycerol Corn Steep Agar

Composition per liter:

Agar	12.0g
Corn steep powder	5.0g
Glycerol	5.0g
NaCl	3.0g
Peptone from casein	3.0g
Yeast extract	3.0g
Beef extract	1.0g

pH 7.2 ± 0.2 at 25°C

Preparation of Medium: Add components to distilled/deionized water and bring volume to 1.0L. Mix thoroughly. Gently heat and bring to boiling. Distribute into tubes or flasks. Autoclave for 15 min at 15 psi pressure–121°C. Pour into sterile Petri dishes or leave in tubes.

Use: For the cultivation and maintenance of *Saccharomonospora glauca.*

Glycerol-Enriched Medium

Composition per liter:

Glycerol	30.0g
Peptone	20.0g
Yeast extract	10.0g

Preparation of Medium: Add components to distilled/deionized water and bring volume to 1.0L. Mix thoroughly. Gently heat and bring to boiling. Distribute into tubes or flasks. Autoclave for 15 min at 15 psi pressure–121°C.

Use: For the cultivation and maintenance of *Saccharomyces cerevisiae.*

Glycerol-Enriched Medium with 2% Ethanol

Composition per liter:

Glycerol	30.0g
Peptone	20.0g
Yeast extract	10.0g
Ethanol (absolute)	20.0mL

Preparation of Medium: Add components, except ethanol, to distilled/deionized water and bring volume to 980.0mL. Mix thoroughly. Gently heat and bring to boiling. Autoclave for 15 min at 15 psi pressure–121°C. Cool to room temperature. Filter sterilize the ethanol. Aseptically add 20.0mL of sterile ethanol to the basal medium. Mix thoroughly. Distribute into sterile flasks or tubes.

Use: For the cultivation and maintenance of *Kluyveromyces lactis.*

Glycerol-Free Medium

Composition per liter:

L-Asparagine	4.0g
L-Glutamine	4.0g
Monosodium glutamate	4.0g
Na_2HPO_4	2.5g
Pancreatic digest of casein	1.0g
KH_2PO_4	1.0g
Triton® WR 1339	0.25g
Ferric ammonium citrate	0.05g
$CaCl_2$	1.0mg
$CuSO_4$	0.5mg
$ZnSO_4$	0.5mg

Preparation of Medium: Add components to distilled/deionized water and bring volume to 1.0L. Mix thoroughly. Distribute into tubes or flasks. Autoclave for 15 min at 15 psi pressure–121°C.

Use: For the cultivation of *Mycobacterium tuberculosis* Bacille Calmette-Guèrin (BCG) for vaccine production.

Glycerol Glycine Agar

Composition per liter:

Agar	15.0g
Glycine	2.5g
K_2HPO_4	1.0g
NaCl	1.0g
$CaCO_3$	0.1g
$FeSO_4$	0.1g
$MgSO_4$	0.1g
Glycerol	20.0mL

pH 7.0 ± 0.2 at 25°C

Preparation of Medium: Add components, except glycerol, to distilled/deionized water and bring volume to 1.0L. Mix thoroughly. Gently heat and bring to boiling. Add glycerol. Mix thoroughly. Distribute into tubes or flasks. Autoclave for 15 min at 15 psi pressure–121°C. Pour into sterile Petri dishes or leave in tubes.

Use: For the cultivation of *Streptomyces* species.

Glycerol Nutrient Agar

Composition per liter:

Agar	15.0g
Peptone	5.0g
NaCl	5.0g
Yeast extract	2.0g
Beef extract	1.0g
Glycerol	10.0mL

Preparation of Medium: Add components to distilled/deionized water and bring volume to 1.0L. Mix thoroughly. Gently heat and bring to boiling. Distribute into tubes or flasks. Autoclave for 15 min at 15 psi pressure–121°C. Pour into sterile Petri dishes or leave in tubes.

Use: For the cultivation of *Mycobacterium smegmatis.*

Glycerol Peptone Agar

See: **GP Agar**

Glycerol Soil Agar

Composition per liter:

Glycerol	20.0g
Agar	15.0g
Peptone	5.0g
Beef extract	3.0g
Soil extract	150.0mL

pH 7.0 ± 0.2 at 25°C

Soil Extract:

Composition per liter:

Soil	400.0g

Preparation of Soil Extract: Air dry garden soil and pass through a sieve. Weigh out 400.0g and add to 960.0mL of tap water. Autoclave for 60 min at 15 psi pressure–121°C. Cool to room temperature and allow soil to settle out. Decant supernatant solution. Filter through paper. Distribute into flasks in 200.0mL volumes. Autoclave for 15 min at 15 psi pressure–121°C.

Preparation of Medium: Add components to tap water and bring volume to 1.0L. Mix thoroughly. Gently heat and bring to boiling. Distribute into tubes or flasks. Autoclave for 15 min at 15 psi pressure–121°C. Pour into sterile Petri dishes or leave in tubes.

Use: For the cultivation and maintenance of *Actinomadura madurae, Amycolata autotrophica, Amycolata saturnea, Gordona bronchialis, Gordona terrae, Mycobacterium* species, *Nocardia* species, and *Tsukamurella paurometabolum.*

Glycine Cycloheximide Phenol Red Agar

Composition per liter:

Solution B	800.0mL
Solution A	200.0mL

Solution A:

Composition per 200.0mL:

Glycine	10.0g
$(NH_4)_2SO_4$	5.0g
KH_2PO_4	1.0g
$MgSO_4 \cdot 7H_2O$	0.5g
NaCl	0.1g
$CaCl_2 \cdot 2H_2O$	0.1g
DL-Methionine	0.02g
DL-Tryptophan	0.02g
L-Histidine·HCl	0.01g
Inositol	2.0mg
H_3BO_3	0.5mg
$ZnSO_4 \cdot 7H_2O$	0.4mg
$MnSO_4 \cdot 4H_2O$	0.4mg
Thiamine·HCl	0.4mg
Pyroxidine·HCl	0.4mg
Niacin	0.4mg
Calcium pantothenate	0.4mg
p-Aminobenzoic acid	0.2mg
Riboflavin	0.2mg
$FeCl_3$	0.2mg
$Na_2MoO_4 \cdot 4H_2O$	0.2mg
KI	0.1mg
$CuSO_4 \cdot 5H_2O$	0.04mg
Folic acid	2.0μg
Biotin	2.0μg
Cycloheximide solution	1.6mL

Preparation of Solution A: Add components to distilled/deionized water and bring volume to 200.0mL. Mix thoroughly. Filter sterilize.

Cycloheximide Solution:

Composition per 100.0mL:

Cycloheximide	0.1g

Preparation of Cycloheximide Solution: Add cylcohexamide to distilled/deionized water and bring volume to 100.0mL. Mix thoroughly. Filter sterilize.

Caution: Cycloheximide is toxic. Avoid skin contact or aerosol formation and inhalation.

Solution B:

Composition per 800.0mL:

Agar	20.0g
Phenol Red solution	30.0mL

Preparation of Solution B: Add components to distilled/deionized water and bring volume to 800.0mL. Mix thoroughly. Gently heat and bring to boiling. Autoclave for 15 min at 15 psi pressure–121°C. Cool to 50°–55°C.

Phenol Red Solution:

Composition per 100.0mL:

Phenol Red	0.5g

Preparation of Phenol Red Solution: Add Phenol Red to distilled/deionized water and bring volume to 100.0mL. Mix thoroughly.

Preparation of Medium: To 800.0mL of sterile solution B, aseptically add 200.0mL of sterile solution A at 55°C. Mix thoroughly. Pour into sterile Petri dishes or distribute into sterile tubes.

Use: For the selective cultivation and differentiation of fungi from clinical and nonclinical specimens. *Cryptococcus neoformans* turns the medium bright red.

Glycocholate Mineral Medium

Composition per liter:

Agar	15.0g
K_2HPO_4	3.5g
$(NH_4)_2SO_4$	2.0g
Sodium glycocholate	2.0g
KH_2PO_4	1.5g
$MgSO_4 \cdot 7H_2O$	0.1g
Yeast extract	0.1g
$CaCl_2 \cdot 2H_2O$	0.01g
$FeSO_4 \cdot 7H_2O$	0.5mg

pH 7.0 ± 0.2 at 25°C

Preparation of Medium: Add components to distilled/deionized water and bring volume to 1.0L. Mix thoroughly. Gently heat and bring to boiling. Distribute into tubes or flasks. Autoclave for 15 min at 15 psi pressure–121°C. Pour into sterile Petri dishes or leave in tubes.

Use: For the cultivation and maintenance of *Pseudomonas putida* and *Pseudomonas* species.

GMC Medium

See: **Gelatin Metronidazole Cadmium Medium**

G25N
(25% Glycerol Nitrate Agar)

Composition per liter :

Glycerol, analytical grade	250.0g
Agar	12.0g
Yeast autolysate or extract	3.7g
KH_2PO_4	0.75g
Czapek concentrate	7.5mL

Preparation of Medium: Add components to distilled/deionized water and bring volume to 1.0L. Mix thoroughly. Distribute into tubes or flasks. Autoclave for 15 min at 15 psi pressure–121°C. Pour into sterile Petri dishes or leave in tubes.

Use: For the cultivation and maintenance of *Penicillium* species.

GN Broth, Hajna

Composition per liter:

Pancreatic digest of casein	10.0g
Peptic digest of animal tissue	10.0g
NaCl	5.0g
Sodium citrate	5.0g
K_2HPO_4	4.0g
D-Mannitol	2.0g
KH_2PO_4	1.5g
Glucose	1.0g
Sodium deoxycholate	0.5g

pH 7.0 ± 0.2 at 25°C

Source: This medium is available as a premixed powder from BD Diagnostic Systems.

Preparation of Medium: Add components to distilled/deionized water and bring volume to 1.0L. Mix thoroughly. Gently heat and bring to boiling. Distribute into tubes or flasks. Autoclave for 15 min at 13 psi pressure–118°C.

Use: For the selective cultivation of *Salmonella* and *Shigella* species.

GN Broth, Hajna

Composition per liter:

Tryptose	20.0g
NaCl	5.0g
Sodium citrate	5.0g
K_2HPO_4	4.0g
Mannitol	2.0g
KH_2PO_4	1.5g
Glucose	1.0g
Sodium deoxycholate	0.5g

pH 7.0 ± 0.2 at 25°C

Source: This medium is available as a premixed powder from HiMedia.

Preparation of Medium: Add components to distilled/deionized water and bring volume to 1.0L. Mix thoroughly. Gently heat and bring to boiling. Distribute into tubes or flasks. Autoclave for 15 min at 10 psi pressure–115°C.

Use: For the selective cultivation of *Salmonella* and *Shigella* species.

GN HiVeg Broth

Composition per liter:

Plant hydrolysate No. 1	20.0g
NaCl	5.0g
Sodium citrate	5.0g
K_2HPO_4	4.0g
Mannitol	2.0g
KH_2PO_4	1.5g
Glucose	1.0g
Synthetic detergent No. III	0.5g

pH 7.0 ± 0.2 at 25°C

Source: This medium is available as a premixed powder from HiMedia.

Preparation of Medium: Add components to distilled/deionized water and bring volume to 1.0L. Mix thoroughly. Gently heat and bring to boiling. Distribute into tubes or flasks. Autoclave for 15 min at 10 psi pressure–115°C.

Use: For the selective cultivation of *Salmonella* and *Shigella* species.

GNYS Agar
(DSMZ Medium 1119)

Composition per liter:

Agar	18.0g
Sodium gluconate	3.0g
Yeast extract	1.0g
$(NH_4)_2SO_4$	1.0g
KH_2PO_4	1.0g
NaCl	0.2g
$MgCl_2·6H_2O$	0.2g
$CaCl_2·2H_2O$	0.05g
Trace elements solution SL-8	1.0mL

pH 5.0 ± 0.2 at 25°C

Trace Elements Solution SL-8:

Composition per liter:

Disodium EDTA	5.2g
$FeCl_2·4H_2O$	1.5g
$CoCl_2·6H_2O$	0.19g
$MnCl_2·4H_2O$	0.1g
$ZnCl_2$	0.07g
H_3BO_3	0.06g
$NaMoO_4·2H_2O$	0.04g
$CuCl_2·2H_2O$	0.02g
$NiCl_2·6H_20$	0.02g

Preparation of Trace Elements Solution SL-8: Add components to distilled/deionized water and bring volume to 1.0L. Mix thoroughly.

Preparation of Medium: Add components to distilled/deionized water and bring volume to 1.0L. Mix thoroughly. Adjust pH to 5.0. Gently heat while stirring and bring to boiling. Distribute into tubes or flasks. Autoclave for 15 min at 15 psi pressure–121°C. Pour into Petri dishes or leave in tubes.

Use: For the cultivation of *Acidisphaera rubrifaciens.*

GNYS Medium
(DSMZ Medium 1119)

Composition per liter:

Sodium gluconate	3.0g
Yeast extract	1.0g
$(NH_4)_2SO_4$	1.0g
KH_2PO_4	1.0g
NaCl	0.2g
$MgCl_2·6H_2O$	0.2g

$CaCl_2 \cdot 2H_2O$ 0.05g
Trace elements solution SL-8 .... 1.0mL
pH 4.3 ± 0.2 at 25°C

Trace Elements Solution SL-8:
Composition per liter:
Disodium EDTA 5.2g
$FeCl_2 \cdot 4H_2O$ 1.5g
$CoCl_2 \cdot 6H_2O$ 0.19g
$MnCl_2 \cdot 4H_2O$ 0.1g
$ZnCl_2$ 0.07g
H_3BO_3 .... 0.06g
$NaMoO_4 \cdot 2H_2O$ 0.04g
$CuCl_2 \cdot 2H_2O$ 0.02g
$NiCl_2 \cdot 6H_2O$ 0.02g

Preparation of Trace Elements Solution SL-8: Add components to distilled/deionized water and bring volume to 1.0L. Mix thoroughly.

Preparation of Medium: Add components to distilled/deionized water and bring volume to 1.0L. Mix thoroughly. Adjust pH to 4.0-4.5. Gently heat while stirring and bring to boiling. Distribute into tubes or flasks. Autoclave for 15 min at 15 psi pressure–121°C.

Use: For the cultivation of *Acidisphaera rubrifaciens.*

GÖ1 Medium

Composition per liter:
Sucrose 17.1g
Sodium acetate 10.0g
NaCl 2.25g
Pancreatic digest of casein 2.0g
Yeast extract 2.0g
$NaHCO_3$ 0.85g
$MgSO_4 \cdot 7H_2O$ 0.5g
NH_4Cl 0.5g
K_2HPO_4 .... 0.348g
$CaCl_2 \cdot 2H_2O$ 0.25g
KH_2PO_4 .... 0.227g
$FeSO_4 \cdot 7H_2O$ 2.0mg
Resazurin 1.0mg
$NaHCO_3$ solution 40.0mL
Vitamin solution 10.0mL
L-Cysteine·HCl·H_2O solution 10.0mL
$Na_2S \cdot 9H_2O$ solution 10.0mL
Trace elements solution SL-10 .... 1.0mL
pH 6.5–6.8 at 25°C

$NaHCO_3$ Solution:
Composition per 50.0mL:
$NaHCO_3$ 5.0g

Preparation of $NaHCO_3$ Solution: Add $NaHCO_3$ to distilled/deionized water and bring volume to 50.0mL. Mix thoroughly. Sparge with 80% N_2 + 20% CO_2. Autoclave for 15 min at 15 psi pressure–121°C.

Vitamin Solution:
Composition per liter:
Pyridoxine·HCl 10.0mg
DL-Calcium pantothenate 5.0mg
Lipoic acid 5.0mg
Nicotinic acid 5.0mg
p-Aminobenzoic acid 5.0mg
Riboflavin 5.0mg
Thiamine·HCl 5.0mg
Biotin 2.0mg
Folic acid 2.0mg
Vitamin B_{12} 0.1mg

Preparation of Vitamin Solution: Add components to distilled/deionized water and bring volume to 1.0L. Mix thoroughly. Filter sterilize. Sparge with 80% N_2 + 20% CO_2.

L-Cysteine·HCl·H_2O Solution:
Composition per 10.0mL:
L-Cysteine·HCl·H_2O 0.3g

Preparation of L-Cysteine·HCl·H_2O Solution: Add L-cysteine·HCl·H_2O to distilled/deionized water and bring volume to 10.0mL. Mix thoroughly. Autoclave under 100% N_2 for 15 min at 15 psi pressure–121°C.

$Na_2S \cdot 9H_2O$ Solution:
Composition per 10.0mL:
$Na_2S \cdot 9H_2O$ 0.3g

Preparation of $Na_2S \cdot 9H_2O$ Solution: Add $Na_2S \cdot 9H_2O$ to distilled/deionized water and bring volume to 10.0mL. Mix thoroughly. Sparge with 100% N_2. Autoclave for 15 min at 15 psi pressure–121°C.

Trace Elements Solution SL-10:
Composition per liter:
$FeCl_2 \cdot 4H_2O$ 1.5g
$CoCl_2 \cdot 6H_2O$ 190.0mg
$MnCl_2 \cdot 4H_2O$ 100.0mg
$ZnCl_2$ 70.0mg
$Na_2MoO_4 \cdot 2H_2O$ 36.0mg
$NiCl_2 \cdot 6H_2O$ 24.0mg
H_3BO_3 .... 6.0mg
$CuCl_2 \cdot 2H_2O$ 2.0mg
HCl (25% solution) 10.0mL

Preparation of Trace Elements Solution SL-10: Add $FeCl_2 \cdot 4H_2O$ to 10.0mL of HCl solution. Mix thoroughly. Add distilled/deionized water and bring volume to 1.0L. Add remaining components. Mix thoroughly. Sparge with 100% N_2.

Preparation of Medium: Add components, except $NaHCO_3$ solution, vitamin solution, L-cysteine·HCl·H_2O solution, and $Na_2S \cdot 9H_2O$ solution, to distilled/deionized water and bring volume to 920.0mL. Mix thoroughly. Sparge under 80% N_2 + 20% CO_2 for 3–4 min. Autoclave for 15 min at 15 psi pressure–121°C. Aseptically and anaerobically add 40.0mL of sterile $NaHCO_3$ solution, 20.0mL of sterile vitamin solution, 10.0mL of sterile L-cysteine·HCl·H_2O solution, and 10.0mL of sterile $Na_2S \cdot 9H_2O$ solution. Mix thoroughly. Aseptically and anaerobically distribute into sterile screw-capped bottles under 80% N_2 + 20% CO_2.

Use: For the cultivation of *Methanosarcina mazei.*

GOL CHI1 Medium (DSMZ Medium 298c)

Composition per liter:
NaCl 1.0g
KCl 0.5g
$MgCl_2 \cdot 6H_2O$ 0.4g
NH_4Cl 0.25g
KH_2PO_4 .... 0.2g
$CaCl_2 \cdot 2H_2O$ 0.15g
Resazurin 1.0mg
$NaHCO_3$ solution 10.0mL

Butanediol solution	10.0mL
$Na_2S·9H_2O$ solution	10.0mL
Vitamin solution	10.0mL
Quinic acid solution	10.0mL
Trace elements solution SL-10	1.0mL

pH 7.2 ± 0.2 at 25°C

$Na_2S·9H_2O$ Solution:
Composition per 10.0mL:

$Na_2S·9H_2O$	0.36g

Preparation of $Na_2S·9H_2O$ Solution: Add $Na_2S·9H_2O$ to distilled/deionized water and bring volume to 10.0mL. Mix thoroughly. Autoclave under 100% N_2 for 15 min at 15 psi pressure–121°C. Cool to room temperature.

$NaHCO_3$ Solution:
Composition per 10.0mL:

$NaHCO_3$	2.5g

Preparation of $NaHCO_3$ Solution: Add $NaHCO_3$ to distilled/deionized water and bring volume to 10.0mL. Mix thoroughly. Sparge with 80% N_2 + 20% CO_2. Filter sterilize.

Butanediol Solution:
Composition per 10.0mL:

2,3 butanediol	0.9g

Preparation of Butanediol Solution: Add butanediol to distilled/deionized water and bring volume to 10.0mL. Mix thoroughly. Sparge with 100% N_2. Filter sterilize.

Trace Elements Solution SL-10:
Composition per liter:

$FeCl_2·4H_2O$	1.5g
H_3BO_3	300.0mg
$CoCl_2·6H_2O$	190.0mg
$MnCl_2·4H_2O$	100.0mg
$ZnCl_2$	70.0mg
$Na_2MoO_4·2H_2O$	36.0mg
$NiCl_2·6H_2O$	24.0mg
$CuCl_2·2H_2O$	2.0mg
HCl (25% solution)	10.0mL

Preparation of Trace Elements Solution SL-10: Add $FeCl_2·4H_2O$ to 10.0mL of HCl solution. Mix thoroughly. Add distilled/deionized water and bring volume to 1.0L. Add remaining components. Mix thoroughly. Sparge with 80% N_2 + 20% CO_2. Autoclave for 15 min at 15 psi pressure–121°C.

Vitamin Solution:
Composition per liter:

Pyridoxine-HCl	10.0mg
Thiamine-HCl·$2H_2O$	5.0mg
Riboflavin	5.0mg
Nicotinic acid	5.0mg
D-Ca-pantothenate	5.0mg
p-Aminobenzoic acid	5.0mg
Lipoic acid	5.0mg
Biotin	2.0mg
Folic acid	2.0mg
Vitamin B_{12}	0.1mg

Preparation of Vitamin Solution: Add components to distilled/deionized water and bring volume to 1.0L. Mix thoroughly. Sparge with 80% H_2 + 20% CO_2. Filter sterilize.

Quinic Acid Solution:
Composition per 10.0mL:

Quinic acid	1.0g

Preparation of Quinic Acid Solution: Add quinic acid to distilled/deionized water and bring volume to 10.0mL. Mix thoroughly. Adjust pH to 7.0. Sparge with 80% N_2 + 20% CO_2. Filter sterilize.

Preparation of Medium: Prepare and dispense medium under 80% N_2 + 20% CO_2 gas atmosphere. Add components, except $NaHCO_3$ solution, butanediol solution, $Na_2S·9H_2O$ solution, vitamin solution, quinic acid solution, and trace elements solution SL-10, to distilled/deionized water and bring volume to 949.0mL. Mix thoroughly. Adjust pH to 7.2. Sparge with 80% N_2 + 20% CO_2. Autoclave for 15 min at 15 psi pressure–121°C. Aseptically and anaerobically add 10.0mL $NaHCO_3$ solution, 10.0mL butanediol solution, 10.0mL $Na_2S·9H_2O$ solution, 10.0mL vitamin solution, 10.0mL quinic acid solution, and 1.0mL trace elements solution SL-10. Mix thoroughly. Aseptically and anaerobically distribute into sterile tubes or bottles. After inoculation, flush and repressurize the gas head space of culture bottles with sterile 80% N_2 + 20% CO_2 to 1 bar overpressure.

Use: For the cultivation of *Propionivibrio limicola.*

Gonococcus Medium

Composition per 623.0mL:

Part II	500.0mL
Part I	123.0mL

pH 7.4 ± 0.2 at 25°C

Part I:
Composition per 123.0mL:

K_2HPO_4	10.5g
Glucose	7.5g
NaCl	5.25g
KH_2PO_4	4.5g
Sodium acetate	1.5g
L-Cysteine·HCl·H_2O	1.2g
Sodium citrate	1.13g
$NaHCO_3$	1.0g
K_2SO_4	0.9g
Na_2SO_4	0.75g
$MgCl_2·6H_2O$	0.45g
KCl	0.3g
NH_4Cl	0.3g
L-Arginine·HCl	0.25g
L-Proline	0.25g
Oxaloacetate	0.25g
L-Glutamic acid	0.19g
L-Methionine	0.19g
L-Asparagine·H_2O	0.13g
L-Isoleucine	0.13g
L-Serine	0.13g
L-Cystine	0.05g
Calcium pantothenate	0.02g
Thiamine·HCl	0.02g
Thiamine pyrophosphate chloride	0.02g
Nicotinamide adenine dinucleotide	0.01g
$CaCl_2·2H_2O$	5.0mg
$Fe(NO_3)_3·9H_2O$	5.0mg
Uracil	5.0mg
Biotin	4.0mg
Hypoxanthine	2.5mg
Sodium thioglycolate	0.025mg

Preparation of Part I: Add components to distilled/deionized water and bring volume to 123.0mL. Mix thoroughly. Adjust pH to 7.2 with 6*N* NaOH. Warm to 50°C for 45 min. Filter sterilize.

Part II:

Composition per 500.0mL:

Agar 10.0g
Soluble starch 7.5g

Preparation of Part II: Add components to distilled/deionized water and bring volume to 500.0mL. Mix thoroughly. Gently heat and bring to boiling. Autoclave for 15 min at 15 psi pressure–121°C. Cool to 45°–50°C.

Preparation of Medium: Aseptically combine 123.0mL of sterile part I and 500.0mL of cooled, sterile part II. Mix thoroughly. Pour into sterile Petri dishes.

Use: For the cultivation of *Neisseria gonorrhoeae.*

Gorbenko Medium

Composition per liter:

Agar 16.2g
Peptone 0.9g
NaCl 0.9g
Yeast extract 0.36g
Beef extract 0.18g
Lake water 1000.0mL

pH 7.4 ± 0.2 at 25°C

Preparation of Medium: Combine components. Mix thoroughly. Gently heat and bring to boiling. Distribute into tubes or flasks. Autoclave for 15 min at 15 psi pressure–121°C. Pour into sterile Petri dishes or leave in tubes.

Use: For the cultivation of *Gordona rubropertinctus, Gordona terrae,* and heterotrophic marine bacteria.

Gorham's Medium for Algae

Composition per liter:

$NaNO_3$ 0.496g
$MgSO_4 \cdot 7H_2O$ 0.075g
$Na_2SiO_3 \cdot 9H_2O$ 0.058g
K_2HPO_4 .... 0.039g
$CaCl_2 \cdot 2H_2O$ 0.036g
Na_2CO_3 .... 0.02g
Citric acid 6.0mg
EDTA 6.0mg
Ferric citrate 6.0mg

pH 7.5 ± 0.5 at 25°C

Preparation of Medium: Add components to distilled/deionized water and bring volume to 1.0L. Mix thoroughly. Distribute into tubes or flasks. Autoclave for 15 min at 15 psi pressure–121°C.

Use: For the cultivation and maintenance of *Anabaena flosaquae, Selenastrum capricornutum,* and *Microcystis aeruginosa.*

GP Agar
(Glycerol Peptone Agar)

Composition per liter:

Agar 15.0g
Peptone 10.0g
Glycerol 10.0g

Preparation of Medium: Add components to distilled/deionized water and bring volume to 1.0L. Mix thoroughly. Gently heat and bring to boiling. Distribute into tubes or flasks. Autoclave for 15 min at 15 psi pressure–121°C. Pour into sterile Petri dishes or leave in tubes.

Use: For the cultivation of actinomycetes and *Mycobacterium spp.*

GPHF Agar

Composition per liter:

Agar 12.0g
Glucose 10.0g
Beef extract 5.0g
Pancreatic digest of casein 5.0g
Yeast extract 5.0g
$CaCl_2 \cdot 2H_2O$ 0.74g

pH 7.2 ± 0.2 at 25°C

Preparation of Medium: Add components to distilled/deionized water and bring volume to 1.0L. Mix thoroughly. Gently heat and bring to boiling. Distribute into tubes or flasks. Autoclave for 15 min at 15 psi pressure–121°C. Pour into sterile Petri dishes or leave in tubes.

Use: For the cultivation and maintenance of *Actinoplanes caeruleus, Micromonospora* species, *Microtetraspora fusca,* and *Streptomyces yerevanensis.*

GPS Agar

See: **Gelatin Phosphate Salt Agar**

GPS Broth

See: **Gelatin Phosphate Salt Broth**

GPVA Medium

Composition per liter:

Agar 15.0g
Yeast extract 10.0g
ACES buffer (2-[(2-amino-2-oxoethyl)-amino]-ethane sulfonic acid) 10.0g
Glycine 3.0g
Charcoal, activated 2.0g
α-Ketoglutarate 1.0g
$Fe_4(P_2O_7)_3 \cdot 9H_2O$ 0.25g
Antibiotic inhibitor solution 10.0mL

pH 6.9 ± 0.2 at 25°C

Antibiotic Inhibitor Solution:

Composition per 10.0mL:

Anisomycin 0.08g
Vancomycin 5.0mg
Polymyxin B 100,000U

Preparation of Antibiotic Inhibitor Solution: Add components to distilled/deionized water and bring volume to 10.0mL. Mix thoroughly. Filter sterilize.

Preparation of Medium: Add components, except antibiotic inhibitor solution, to distilled/deionized water and bring volume to 990.0mL. Mix thoroughly. Gently heat and bring to boiling. Autoclave for 15 min at 15 psi pressure–121°C. Cool to 45°–50°C. Adjust pH to 6.9. Aseptically add 10.0mL of sterile antibiotic inhibitor solution. Mix thoroughly. Pour into sterile Petri dishes or distribute into sterile tubes.

Use: For the isolation and cultivation of *Legionella* species from environmental waters.

GPY/10 Medium

Composition per liter:

KH_2PO_4	1.5g
Glucose	1.0g
Peptone	1.0g
Na_2HPO_4	0.915g
$MgSO_4·7H_2O$	0.1g
Yeast extract	0.1g

pH 6.6–6.8 at 25°C

Preparation of Medium: Add components to distilled/deionized water and bring volume to 1.0L. Mix thoroughly. Gently heat and bring to boiling. Adjust pH to 6.6–6.8. Distribute into tubes or flasks. Autoclave for 15 min at 15 psi pressure–121°C. Pour into sterile Petri dishes or leave in tubes.

Use: For the cultivation of *Echinostelium minutum.*

GPY Salts Medium (Glucose Peptone Yeast Extract Salts Medium)

Composition per liter:

Glucose	1.0g
Peptone	0.5g
Yeast extract	0.1g
Modified Hutner's mineral base	20.0mL

Hutner's Mineral Base:

Composition per liter:

$MgSO_4·7H_2O$	29.7g
Nitrilotriacetic acid	10.0g
$CaCl_2·2H_2O$	3.34g
$FeSO_4·7H_2O$	99.0mg
$(NH_4)_2MoO_4$	9.25mg
Metals "44"	50.0mL

Preparation of Hutner's Mineral Base: Add nitrilotriacetic acid to 500.0mL of distilled/deionized water. Dissolve by adjusting pH to 6.5 with KOH. Add remaining components. Readjust pH to 7.2 with H_2SO_4 or KOH. Add distilled/deionized water to 1.0L. Store at 5°C.

Metals "44":

Composition per 100.0mL:

$ZnSO_4·7H_2O$	1.1g
$FeSO_4·7H_2O$	0.5g
EDTA	0.25g
$MnSO_4·7H_2O$	0.154g
$CuSO_4·5H_2O$	0.04g
$Co(NO_3)_2·6H_2O$	0.025g
$Na_2B_4O_7·10H_2O$	0.018g

Preparation of Metals "44": Add a few drops of H_2SO_4 to distilled/deionized water to inhibit precipitate formation. Add components to acidified distilled/deionized water and bring volume to 100.0mL. Mix thoroughly.

Preparation of Medium: Add components to distilled/deionized water and bring volume to 1.0L. Mix thoroughly. Distribute into tubes or flasks. Autoclave for 15 min at 15 psi pressure–121°C.

Use: For the cultivation and maintenance of *Prosthecobacter fusiformis*.

Grace's Insect Medium

Composition per liter:

Sucrose	26.68g
KCl	4.1g
$MgCl_2·6H_2O$	2.28g
$MgSO_4·7H_2O$	2.78g
L-Histidine	2.5g
DL-Serine	1.1g
$NaH_2PO_4·H_2O$	1.013g
$CaCl_2$	0.75g
L-Leucine	0.75g
D-Glucose	0.7g
L-Arginine·HCl	0.7g
Malic acid	0.67g
Glycine	0.65g
L-Lysine·HCl	0.625g
L-Glutamic acid	0.6g
L-Glutamine	0.6g
Fructose	0.4g
α-Ketoglutaric acid	0.37g
$NaHCO_3$	0.35g
L-Asparagine	0.35g
L-Aspartic acid	0.35g
L-Proline	0.35g
L-Alanine	0.225g
β-Alanine	0.2g
Choline chloride	0.2mg
L-Threonine	0.175g
L-Phenylalanine	0.15g
L-Tryptophan	0.1g
L-Tyrosine	0.1g
L-Valine	0.1g
Succinic acid	0.06g
Fumaric acid	0.055g
L-Isoleucine	0.05g
L-Methionine	0.05g
L-Cysteine	0.022g
p-Aminobenzoic acid	0.02mg
Calcium DL-pantothenate	0.02mg
Folic acid	0.02mg
i-Inositol	0.02mg
Nicotinic acid	0.02mg
Pyridoxine·HCl	0.02mg
Riboflavin	0.02mg
Thiamine·HCl	0.02mg
Biotin	0.01mg

Source: Grace's Insect Medium is available from BD Diagnostics.

Preparation of Medium: Add components to distilled/deionized water and bring volume to 1.0L. Mix thoroughly. Filter sterilize. Aseptically distribute into sterile tubes or flasks.

Use: For the cultivation of *Blastocrithidia culicis*.

Gracilibacillus Medium (DSMZ Medium 802)

Composition per liter:

NaCl	100.0g
$MgSO_4·7H_2O$	20.0g
Tris-HCl	12.0g
KCl	2.0g
Yeast extract	2.0g
Trypticase™ peptone	2.0g
NaBr	0.1g
Trace metals solution	2.0mL
Phosphate solution	10.0mL

Glucose solution 10.0mL
Calcium chloride solution 5.0mL
Iron chloride manganese chloride solution 2.0mL

pH 7.8 ± 0.2 at 25°C

Trace Metals Solution:

Composition per liter:

$FeCl_2 \cdot 4H_2O$ 2.0g
$CoCl_2 \cdot 6H_2O$ 250.0mg
$MnCl_2 \cdot 4H_2O$ 100.0mg
$ZnCl_2$ 70.0mg
$NiCl_2 \cdot 6H_2O$ 70.0mg
$AlCl_3 \cdot 6H_2O$ 60.0mg
$Na_2MoO_4 \cdot 2H_2O$ 40.0mg
H_3BO_3 .... 6.0mg
$Na_2WO_4 \cdot 2H_2O$ 6.0mg
$CuCl_2 \cdot 2H_2O$ 2.0mg
HCl (32% solution) 10.0mL

Preparation of Trace Metals Solution: Add $FeCl_2 \cdot 4H_2O$ to 10.0mL of HCl solution. Mix thoroughly. Add distilled/deionized water and bring volume to 1.0L. Add remaining components. Mix thoroughly.

Phosphate Solution:

Composition per liter:

KH_2PO_4 .... 50.0g

Preparation of Phosphate Solution: Add KH_2PO_4 to distilled/deionized water and bring volume to 1.0L. Mix thoroughly. Autoclave for 15 min at 15 psi pressure–121°C. Cool to 25°C.

Calcium Chloride Solution:

Composition per 10.0mL:

$CaCl_2 \cdot 2H_2O$ 0.1g

Preparation of Calcium Chloride Solution: Add $CaCl_2 \cdot 2H_2O$ to distilled/deionized water and bring volume to 10.0mL. Mix thoroughly. Sparge with 100% N_2. Autoclave for 15 min at 15 psi pressure–121°C. Cool to 25°C.

Iron Chloride Manganese Chloride Solution:

Composition per 10.0mL:

$FeCl_2 \cdot 4H_2O$ 0.2g
$MnCl_2 \cdot 4H_2O$ 0.2g

Preparation of Iron Chloride Manganese Chloride Solution: Add components to distilled/deionized water and bring volume to 10.0mL. Mix thoroughly. Sparge with 100% N_2. Autoclave for 15 min at 15 psi pressure–121°C. Cool to 25°C.

Glucose Solution:

Composition per 10.0mL:

Glucose 2.0g

Preparation of Glucose Solution: Add glucose to distilled/deionized water and bring volume to 10.0mL. Mix thoroughly. Sparge with 100% N_2. Filter sterilize.

Preparation of Medium: Add components, except glucose solution, phosphate solution, calcium chloride solution, and iron chloride manganese chloride solution, to distilled/deionized water and bring volume to 1.0L. Mix thoroughly. Adjust pH to 7.8. Autoclave for 15 min at 15 psi pressure–121°C. Cool to room temperature. Aseptically add 10.0mL glucose solution, 10.0mL phosphate solution, 5.0mL calcium chloride solution, and 2.0mL iron chloride manganese chloride solution. Mix thoroughly. Aseptically and anaerobically distribute into sterile tubes or bottles.

Use: For the cultivation of *Gracilibacillus* spp.

Grahamella Medium

Composition per liter:

Agar 20.0g
Saline (0.9% solution) 800.0mL
Rabbit serum, filter sterilized 90.0mL
Rabbit hemoglobin (2% solution) 45.0mL

pH 7.2 ± 0.2 at 25°C

Preparation of Medium: Add components, except rabbit serum and rabbit hemoglobin, to distilled/deionized water and bring volume to 900.0mL. Mix thoroughly. Gently heat and bring to boiling. Distribute into small tubes in 2.0mL volumes. Autoclave for 15 min at 15 psi pressure–121°C. Cool to 60°C. Aseptically add 0.2mL of sterile rabbit serum and 0.1mL of rabbit hemoglobin solution to each tube. Mix thoroughly. Autoclave for 30 min with a mixture of steam and air at 0 psi pressure–56°C.

Use: For the cultivation of *Grahamella* species.

Granada Medium

Composition per liter:

Starch, soluble 150.0g
Proteose peptone No. 3 .... 38.0g
NaCl 3.0g
Trimethoprim lactate 0.015g
Sodium phosphate
  buffer (0.06*M*, pH 7.4) 900.0mL
Horse serum, coagulated 100.0mL

pH 7.4 ± 0.2 at 25°C

Preparation of Medium: Add proteose peptone No. 3 and NaCl to 200.0mL of sodium phosphate buffer and bring to boiling. Add 400.0mL of cold sodium phosphate buffer, starch, and trimethoprim lactate. Mix thoroughly. Bring volume to 900.0mL with sodium phosphate buffer. Gently heat while stirring in a boiling water bath for exactly 20 min. Do not autoclave. Cool to 90°–95°C. Add horse serum. Mix thoroughly. Cool to 60°–65°C while stirring. Pour into sterile Petri dishes. Medium will solidify in 2–3 hr.

Use: For the early selective isolation and cultivation of Group B streptococci from clinical specimens.

Granulibacter Medium (DSMZ Medium 1186)

Composition per liter:

Glucose 50.0g
$CaCO_3$ 12.5g
Yeast extract 5.0g

pH 7.0 ± 0.2 at 25°C

Preparation of Medium: Add components to distilled/deionized water and bring volume to 1.0L. Mix thoroughly. Gently heat while stirring and bring to boiling. Distribute into tubes or flasks. Autoclave for 15 min at 15 psi pressure–121°C.

Use: For the cultivation of *Granulibacter* spp.

Gray's Agar

Composition per liter:

Glucose 30.0g
Agar 15.0g

Yeast extract 7.0g
KH_2PO_4 .... 5.0g

Preparation of Medium: Add components to distilled/deionized water and bring volume to 1.0L. Mix thoroughly. Gently heat and bring to boiling. Distribute into tubes or flasks. Autoclave for 15 min at 15 psi pressure–121°C. Pour into sterile Petri dishes or leave in tubes.

Use: For the cultivation and maintenance of *Emericellopsis minima* and *Verticillium lateritium.*

Green Top Agar

Composition per liter:
Agar 15.0g
Yeast extract 2.0g
Sodium acetate 1.0g
Pancreatic digest of casein 1.0g
Soil extract 50.0mL
pH 7.4 ± 0.2 at 25°C

Soil Extract:
Composition per 200.0mL:
African Violet soil 0.5g
Na_2CO_3 .... 0.5g

Preparation of Soil Extract: Add components to tap water and bring volume to 200.0mL. Autoclave for 60 min at 15 psi pressure–121°C. Filter through Whatman filter paper.

Preparation of Medium: Add components to distilled/deionized water and bring volume to 1.0L. Mix thoroughly. Gently heat and bring to boiling. Distribute into tubes or flasks. Autoclave for 15 min at 15 psi pressure–121°C. Pour into sterile Petri dishes or leave in tubes.

Use: For the cultivation and maintenance of *Bacillus macroides, Caryophanon latum, Lampropedia lyalina,* and *Vittreoscilla stercoraria.*

Green Yeast and Mold Broth (m-Green Yeast and Mold Broth)

Composition per liter:
Glucose 50.0g
Yeast extract 9.0g
Pancreatic digest of casein 5.0g
Peptic digest of animal tissue 5.0g
$MgSO_4 \cdot 7H_2O$ 2.1g
K_2HPO_4 .... 2.0g
α-Amylase 0.05g
Thiamine 0.05g
Bromcresol Green 0.026g
pH 4.6 ± 0.2 at 25°C

Source: This medium is available as a premixed powder from BD Diagnostic Systems and Acumedia.

Preparation of Medium: Add components to distilled/deionized water and bring volume to 1.0L. Mix thoroughly. Distribute into tubes or flasks. Autoclave for 10 min at 15 psi pressure–121°C.

Use: For the cultivation of fungi from beverages.

Group A Selective Strep Agar with Sheep Blood

Composition per liter:
Pancreatic digest of casein 14.5g
Agar 14.0g
NaCl 5.0g
Papaic digest of soybean meal 5.0g
Sheep blood 50.0mL
Growth factor solution 10.0mL
Selective agents solution 10.0mL
pH 7.4 ± 0.2 at 25°C

Growth Factor Solution:
Composition per 10.0mL:
Growth factors, BBL 1.5g

Preparation of Growth Factor Solution: Add growth factors to distilled/deionized water and bring volume to 10.0mL. Mix thoroughly. Filter sterilize.

Selective Agents Solution:
Composition per 10.0mL:
Selective agents 0.042g

Preparation of Selective Agents Solution: Add selective agents to distilled/deionized water and bring volume to 10.0mL. Mix thoroughly. Filter sterilize.

Preparation of Medium: Add components, except sheep blood, growth factor solution, and selective agents solution, to distilled/deionized water and bring volume to 930.0mL. Mix thoroughly. Gently heat and bring to boiling. Autoclave for 15 min at 15 psi pressure–121°C. Cool to 45°–50°C. Aseptically add 50.0mL of sheep blood, 10.0mL of sterile growth factor solution, and 10.0mL of sterile selective agents solution. Mix thoroughly. Pour into sterile Petri dishes or distribute into sterile tubes.

Use: For the selective cultivation and primary isolation of group A streptococci, especially *Streptococcus pyogenes,* from clinical specimens.

Group B Streptococci Agar
See: **GBS Agar Base, Islam**

Group B Streptococci Medium
See: **GBS Medium, Rapid**

GS Medium

Composition per liter:
Morpholinopropane sulfonic acid 10.0g
Yeast extract 6.0g
Glucose 5.0g
Urea 2.0g
L-Cysteine·HCl 1.0g
$K_2HPO_4 \cdot 3H_2O$ 1.0g
KH_2PO_4 .... 0.5g
$MgCl_2 \cdot 6H_2O$ 0.5g
$CaCl_2 \cdot 2H_2O$ 0.05g
$FeSO_4 \cdot 7H_2O$ 1.25mg
Resazurin 1.0mg
pH 7.2 ± 0.2 at 25°C

Glucose Solution:
Composition per 20.0mL:
Glucose 2.0g

Preparation of Glucose Solution: Add glucose to distilled/deionized water and bring volume to 20.0mL. Mix thoroughly. Filter sterilize. Sparge with 100% N_2.

Preparation of Medium: Add components, except glucose solution, to distilled/deionized water and bring volume to 980.0mL. Mix thoroughly. Sparge with 100% N_2. Autoclave for 25 min at 15 psi pressure–121°C. Aseptically and anaerobically add 20.0mL of sterile glucose solution. Mix thoroughly.

Use: For the cultivation and maintenance of *Clostridium thermocellum*.

GSP Agar

Composition per liter:

Starch, soluble	20.0g
Agar	15.0g
Sodium glutamate	10.0g
K_2HPO_4	2.0g
$MgSO_4 \cdot 7H_2O$	0.5g
Phenol Red	0.36g

pH 7.2–7.4 at 25°C

Preparation of Medium: Add components to distilled/deionized water and bring volume to 1.0L. Mix thoroughly. Distribute into tubes or flasks. Autoclave for 15 min at 15 psi pressure–121°C.

Use: For the selective isolation and cultivation of *Pseudomonas* species.

GSTB

See: **Glucose Salt Teepol Broth**

GTC Agar Base with Bicarbonate

Composition per liter:

Agar	15.0g
Casein enzymatic hydrolysate	15.0g
Papaic digst of soybean meal	5.0g
NaCl	5.0g
KH_2PO_4	5.0g
Glucose	1.0g
Esculin	1.0g
Thallous acetate	0.5g
Ferric citrate	0.5g
Polysorbate 80	0.75g
Bicarbonate solution	10.0mL

pH 7.3 ± 0.2 at 25°C

Source: This medium is available from HiMedia.

Bicarbonate Solution:

Composition per 10.0mL:

$NaHCO_3$	1.0g

Preparation of Bicarbonate Solution: Add components to distilled/deionized water and bring volume to 10.0mL. Mix thoroughly. Filter sterilize.

Preparation of Medium: Add components, except bicarbonate solution, to distilled/deionized water and bring volume to 990.0mL. Mix thoroughly. Autoclave for 15 min at 15 psi pressure–121°C. Cool to 50°C. Aseptically add bicarbonate solution. Mix thoroughly. Pour into Petri dishes or aseptically distribute into sterile tubes.

Use: For the recovery of enterococci from food.

GTYE Medium (Glucose Tryptone Yeast Extract Medium)

Composition per liter:

NaCl	30.0g
Agar	15.0g
Pancreatic digest of casein	10.0g
Yeast extract	10.0g
Glucose	5.0g
$MgSO_4 \cdot 7H_2O$	5.0g

Preparation of Medium: Add components to tap water and bring volume to 1.0L. Mix thoroughly. Gently heat and bring to boiling. Distribute into tubes or flasks. Autoclave for 15 min at 15 psi pressure–121°C. Pour into sterile Petri dishes or leave in tubes.

Use: For the cultivation of ATCC strain 21081.

Guanosine Medium

Composition per liter:

Glucose	20.0g
Peptone	10.0g
Yeast extract	10.0g
Guanosine	0.02g

Preparation of Medium: Add components to distilled/deionized water and bring volume to 1.0L. Mix thoroughly. Distribute into tubes or flasks. Autoclave for 15 min at 15 psi pressure–121°C.

Use: For the cultivation and maintenance of *Salmonella choleraesuis*.

Guizotia abyssinica Creatinine Agar

See: **Birdseed Agar**

Gum Base Nalidixic Acid Medium

See: **GBNA Medium**

Gum *Listeria* Medium (Gum Base Nalidixic Acid Medium)

Composition per liter:

Gellan gum	8.0g
Casein enzymic hydrolysate	5.7g
NaCl	1.7g
Papaic digest of soybean meal	1.0g
Dextrose	0.83g
K_2HPO_4	0.83g
$MgCl_2 \cdot 6H_2O$	0.33g
Nalidixic acid	0.05g

pH 7.2 ± 0.2 at 25°C

Source: This medium is available from HiMedia.

Preparation of Medium: Add components to distilled/deionized water and bring volume to 1.0L. Mix thoroughly. Gently heat and bring to boiling. Distribute into tubes or flasks. Autoclave for 15 min at 15 psi pressure–121°C. Pour into sterile Petri dishes or leave in tubes.

Use: For the isolation of *Listeria monocytogenes* from clinical and nonclinical specimens.

Gum Tragacanth Gum Arabic Medium

Composition per 100.0mL:

Gum tragacanth	2.0g
Gum arabic	1.0g

Preparation of Medium: Add components to distilled/deionized water and bring volume to 100.0mL. Mix thoroughly. Gently heat and bring to boiling. Autoclave for 15 min at 15 psi pressure–121°C. Pour into sterile Petri dishes.

Use: For the cultivation of aciduric flat sour sporeformers from foods. For the isolation and cultivation of *Desulfotomaculum nigrificans* from foods.

Guttman's IIB Medium

Composition per liter:

Sucrose	10.0g
Triethanolamine	5.0g
L-Glutamic acid	1.0g
$MgSO_4 \cdot 7H_2O$	0.8g
NaCl	0.5g
L-Arginine	0.3g
L-Histidine	0.2g
K_3PO_4	0.2g
DL-Methionine	0.2g
Salts solution	1.0mL
Vitamin solution	1.0mL

Salts Solution:

Composition per 100.0mL:

$CaCl_2$	0.3g
$Fe(NH_4)_2(SO_4)_2 \cdot 6H_2O$	0.2g
$MnSO_4 \cdot H_2O$	0.1g
$ZnSO_4 \cdot 7H_2O$	0.1g
$CuSO_4 \cdot 5H_2O$	0.04g
$CoSO_4 \cdot 7H_2O$	0.01g
H_3BO_3	0.01g
Ammonium molybdate	0.005g

Preparation of Salts Solution: Add components to distilled/deionized water and bring volume to 100.0mL. Mix thoroughly.

Vitamin Solution:

Composition per 100.0mL:

Calcium D-(+)-pantothenate	0.1g
Nicotinic acid	0.1g
Pyridoxamine·HCl	0.1g
Thiamine·HCl	0.1g
p-Aminobenzoic acid	0.001g
Biotin	0.001g

Preparation of Vitamin Solution: Add components to distilled/deionized water and bring volume to 100.0mL. Mix thoroughly. Filter sterilize. Store at 4°C.

Preparation of Medium: Add components, except vitamin solution, to distilled/deionized water and bring volume to 999.0mL. Mix thoroughly. Autoclave for 18 min at 15 psi pressure–121°C. Aseptically add 1.0mL of sterile vitamin solution. Mix thoroughly. Aseptically distribute into sterile screw-capped tubes.

Use: For the cultivation of *Crithidia oncopelti.*

GV Medium

Composition per 1051.0mL:

$NaHCO_3$	3.0g
NaCl	2.25g
Yeast extract	1.0g
NH_4Cl	0.5g
K_2HPO_4	0.348g
KH_2PO_4	0.227g
$FeSO_4 \cdot 7H_2O$	2.0mg
Resazurin	0.5mg
Vitamin solution	10.0mL
Substrate solution	10.0mL
L-Cysteine·HCl solution	10.0mL
$Na_2S \cdot 9H_2O$ solution	10.0mL
$MgSO_4 \cdot 7H_2O$ solution	10.0mL
$CaCl_2 \cdot 2H_2O$ solution	10.0mL
Selenite-tungstate solution	1.0mL
Trace elements solution SL-10	1.0mL

pH 7.0 ± 0.2 at 25°C

Vitamin Solution:

Composition per liter:

Pyridoxine-HCl	10.0mg
Calcium DL-pantothenate	5.0mg
Lipoic acid	5.0mg
Nicotinic acid	5.0mg
p-Aminobenzoic acid	5.0mg
Riboflavin	5.0mg
Thiamine·HCl	5.0mg
Biotin	2.0mg
Folic acid	2.0mg
Vitamin B_{12}	0.1mg

Preparation of Vitamin Solution: Add components to distilled/deionized water and bring volume to 1.0L. Mix thoroughly. Sparge with 80% N_2 + 20% CO_2. Autoclave for 15 min at 15 psi pressure–121°C.

Substrate Solution:

Composition per 10.0mL:

Sodium crotonate	10.0mg

Preparation of Substrate Solution: Add sodium crotonate to distilled/deionized water and bring volume to 10.0mL. Mix thoroughly. Sparge with 80% N_2 + 20% CO_2. Autoclave for 15 min at 15 psi pressure–121°C.

L-Cysteine·HCl Solution:

Composition per 10.0mL:

L-Cysteine·HCl	0.3g

Preparation of L-Cysteine·HCl Solution: Add L-cysteine·HCl to distilled/deionized water and bring volume to 10.0mL. Mix thoroughly. Sparge with 80% N_2 + 20% CO_2. Autoclave for 15 min at 15 psi pressure–121°C.

$Na_2S \cdot 9H_2O$ Solution:

Composition per 10.0mL:

$Na_2S \cdot 9H_2O$	0.3g

Preparation of $Na_2S \cdot 9H_2O$ Solution: Add $Na_2S \cdot 9H_2O$ to distilled/deionized water and bring volume to 10.0mL. Mix thoroughly. Gas under 80% N_2 + 20% CO_2. Autoclave for 15 min at 15 psi pressure–121°C.

$MgSO_4 \cdot 7H_2O$ Solution:

Composition per 10.0mL:

$MgSO_4 \cdot 7H_2O$	0.5g

Preparation of $MgSO_4 \cdot 7H_2O$ Solution: Add $MgSO_4 \cdot 7H_2O$ to distilled/deionized water and bring volume to 10.0mL. Mix thoroughly. Sparge with 80% N_2 + 20% CO_2. Autoclave for 15 min at 15 psi pressure–121°C.

$CaCl_2 \cdot 2H_2O$ Solution:

Composition per 10.0mL:

$CaCl_2 \cdot 2H_2O$	0.25g

Preparation of $CaCl_2 \cdot 2H_2O$ Solution: Add $CaCl_2 \cdot 2H_2O$ to distilled/deionized water and bring volume to 10.0mL. Mix thoroughly. Sparge with 80% N_2 + 20% CO_2. Autoclave for 15 min at 15 psi pressure–121°C.

Selenite-Tungstate Solution:

Composition per liter:

NaOH	0.5g

$Na_2WO_4 \cdot 2H_2O$ 4mg
$Na_2SeO_3 \cdot 5H_2O$ 3mg

Preparation of Selenite-Tungstate Solution: Add components to distilled/deionized water and bring volume to 1.0L. Mix thoroughly. Sparge with 80% N_2 + 20% CO_2. Autoclave for 15 min at 15 psi pressure–121°C.

Trace Elements Solution SL-10:
Composition per liter:
$FeCl_2 \cdot 4H_2O$ 1.5g
$CoCl_2 \cdot 6H_2O$ 190.0mg
$MnCl_2 \cdot 4H_2O$ 100.0mg
$ZnCl_2$ 70.0mg
$Na_2MoO_4 \cdot 2H_2O$ 36.0mg
$NiCl_2 \cdot 6H_2O$ 24.0mg
H_3BO_3 .... 6.0mg
$CuCl_2 \cdot 2H_2O$ 2.0mg
HCl (25% solution) 10.0mL

Preparation of Trace Elements Solution SL-10: Add $FeCl_2 \cdot 4H_2O$ to 10.0mL of HCl solution. Mix thoroughly. Add distilled/deionized water and bring volume to 1.0L. Add remaining components. Mix thoroughly. Gas under 80% N_2 + 20% CO_2. Autoclave for 15 min at 15 psi pressure–121°C.

Preparation of Medium: Prepare and dispense medium under 80% N_2 + 20% CO_2. Add components, except substrate solution, $MgSO_4 \cdot 7H_2O$ solution, $CaCl_2 \cdot 2H_2O$ solution, selenite-tungstate solution, L-cysteine·HCl solution, and $Na_2S \cdot 9H_2O$ solution, to distilled/deionized water and bring volume to 1.0L. Mix thoroughly. Sparge with 80% N_2 + 20% CO_2. Autoclave for 15 min at 15 psi pressure–121°C. Aseptically and anaerobically add 10.0mL of sterile $MgSO_4 \cdot 7H_2O$ solution, 10.0mL of sterile substrate solution, 10.0mL of sterile $CaCl_2 \cdot 2H_2O$ solution, 1.0mL of sterile selenite-tungstate solution, 10.0mL of sterile L-cysteine·HCl solution, and 10.0mL of sterile $Na_2S \cdot 9H_2O$ solution. Mix thoroughly. Adjust pH to 7.0 to 7.2. Aseptically and anaerobically distribute into sterile tubes or flasks.

Use: For the cultivation and maintenance of *Ilyobacter delafieldii.*

GV Medium

Composition per 1051.0mL:
$NaHCO_3$ 3.0g
NaCl 2.25g
Yeast extract 1.0g
NH_4Cl 0.5g
K_2HPO_4 .... 0.348g
KH_2PO_4 .... 0.227g
$FeSO_4 \cdot 7H_2O$ 2.0mg
Resazurin 0.5mg
Vitamin solution 10.0mL
Substrate solution 10.0mL
L-Cysteine·HCl solution 10.0mL
$Na_2S \cdot 9H_2O$ solution 10.0mL
$MgSO_4 \cdot 7H_2O$ solution 10.0mL
$CaCl_2 \cdot 2H_2O$ solution 10.0mL
Selenite-tungstate solution 1.0mL
Trace elements solution SL-10 .... 1.0mL
pH 7.0 ± 0.2 at 25°C

Vitamin Solution:
Composition per liter:
Pyridoxine-HCl 10.0mg
Calcium DL-pantothenate 5.0mg
Lipoic acid 5.0mg
Nicotinic acid 5.0mg
p-Aminobenzoic acid 5.0mg
Riboflavin 5.0mg
Thiamine-HCl 5.0mg
Biotin 2.0mg
Folic acid 2.0mg
Vitamin B_{12} 0.1mg

Preparation of Vitamin Solution: Add components to distilled/deionized water and bring volume to 1.0L. Mix thoroughly. Sparge with 80% N_2 + 20% CO_2. Autoclave for 15 min at 15 psi pressure–121°C.

Substrate Solution:
Composition per 10.0mL:
D-Glucose 2.0g

Preparation of Substrate Solution: Add D-glucose to distilled/deionized water and bring volume to 10.0mL. Mix thoroughly. Sparge with 80% N_2 + 20% CO_2. Autoclave for 15 min at 15 psi pressure–121°C.

L-Cysteine·HCl Solution:
Composition per 10.0mL:
L-Cysteine·HCl 0.3g

Preparation of L-Cysteine·HCl Solution: Add L-cysteine·HCl to distilled/deionized water and bring volume to 10.0mL. Mix thoroughly. Sparge with 80% N_2 + 20% CO_2. Autoclave for 15 min at 15 psi pressure–121°C.

$Na_2S \cdot 9H_2O$ Solution:
Composition per 10.0mL:
$Na_2S \cdot 9H_2O$ 0.3g

Preparation of $Na_2S \cdot 9H_2O$ Solution: Add $Na_2S \cdot 9H_2O$ to distilled/deionized water and bring volume to 10.0mL. Mix thoroughly. Gas under 80% N_2 + 20% CO_2. Autoclave for 15 min at 15 psi pressure–121°C.

$MgSO_4 \cdot 7H_2O$ Solution:
Composition per 10.0mL:
$MgSO_4 \cdot 7H_2O$ 0.5g

Preparation of $MgSO_4 \cdot 7H_2O$ Solution: Add $MgSO_4 \cdot 7H_2O$ to distilled/deionized water and bring volume to 10.0mL. Mix thoroughly. Sparge with 80% N_2 + 20% CO_2. Autoclave for 15 min at 15 psi pressure–121°C.

$CaCl_2 \cdot 2H_2O$ Solution:
Composition per 10.0mL:
$CaCl_2 \cdot 2H_2O$ 0.25g

Preparation of $CaCl_2 \cdot 2H_2O$ Solution: Add $CaCl_2 \cdot 2H_2O$ to distilled/deionized water and bring volume to 10.0mL. Mix thoroughly. Sparge with 80% N_2 + 20% CO_2. Autoclave for 15 min at 15 psi pressure–121°C.

Selenite-Tungstate Solution:
Composition per liter:
NaOH 0.5g
$Na_2WO_4 \cdot 2H_2O$ 4mg
$Na_2SeO_3 \cdot 5H_2O$ 3mg

Preparation of Selenite-Tungstate Solution: Add components to distilled/deionized water and bring volume to 1.0L. Mix thoroughly. Sparge with 80% N_2 + 20% CO_2. Autoclave for 15 min at 15 psi pressure–121°C.

Trace Elements Solution SL-10:
Composition per liter:

$FeCl_2 \cdot 4H_2O$	1.5g
$CoCl_2 \cdot 6H_2O$	190.0mg
$MnCl_2 \cdot 4H_2O$	100.0mg
$ZnCl_2$	70.0mg
$Na_2MoO_4 \cdot 2H_2O$	36.0mg
$NiCl_2 \cdot 6H_2O$	24.0mg
H_3BO_3	6.0mg
$CuCl_2 \cdot 2H_2O$	2.0mg
HCl (25% solution)	10.0mL

Preparation of Trace Elements Solution SL-10: Add $FeCl_2 \cdot 4H_2O$ to 10.0mL of HCl solution. Mix thoroughly. Add distilled/deionized water and bring volume to 1.0L. Add remaining components. Mix thoroughly. Gas under 80% N_2 + 20% CO_2. Autoclave for 15 min at 15 psi pressure–121°C.

Preparation of Medium: Prepare and dispense medium under 80% N_2 + 20% CO_2. Add components, except substrate solution, $MgSO_4 \cdot 7H_2O$ solution, $CaCl_2 \cdot 2H_2O$ solution, selenite-tungstate solution, L-cysteine·HCl solution, and $Na_2S \cdot 9H_2O$ solution, to distilled/deionized water and bring volume to 1.0L. Mix thoroughly. Sparge with 80% N_2 + 20% CO_2. Autoclave for 15 min at 15 psi pressure–121°C. Aseptically and anaerobically add 10.0mL of sterile substrate solution, 10.0mL of sterile $MgSO_4 \cdot 7H_2O$ solution, 10.0mL of sterile $CaCl_2 \cdot 2H_2O$ solution, 1.0mL of sterile selenite-tungstate solution, 10.0mL of sterile L-cysteine·HCl solution, and 10.0mL of sterile $Na_2S \cdot 9H_2O$ solution. Mix thoroughly. Adjust pH to 7.0–7.2. Aseptically and anaerobically distribute into sterile tubes or flasks.

Use: For the cultivation and maintenance of *Clostridium magnum* and *Clostridium* species.

GV Medium

Composition per 1061.0mL:

$NaHCO_3$	3.0g
NaCl	2.25g
Yeast extract	1.0g
NH_4Cl	0.5g
K_2HPO_4	0.348g
KH_2PO_4	0.227g
$FeSO_4 \cdot 7H_2O$	2.0mg
Resazurin	0.5mg
Substrate solution	20.0mL
Vitamin solution	10.0mL
L-Cysteine·HCl solution	10.0mL
$Na_2S \cdot 9H_2O$ solution	10.0mL
$MgSO_4 \cdot 7H_2O$ solution	10.0mL
$CaCl_2 \cdot 2H_2O$ solution	10.0mL
Selenite-tungstate solution	1.0mL
Trace elements solution SL-10	1.0mL

pH 7.0 ± 0.2 at 25°C

Substrate Solution:
Composition per 20.0mL:

Sodium citrate	6.0g

Preparation of Substrate Solution: Add sodium citrate to distilled/deionized water and bring volume to 20.0mL. Mix thoroughly. Sparge with 80% N_2 + 20% CO_2. Autoclave for 15 min at 15 psi pressure–121°C.

Vitamin Solution:
Composition per liter:

Pyridoxine-HCl	10.0mg
Calcium DL-pantothenate	5.0mg
Lipoic acid	5.0mg
Nicotinic acid	5.0mg
p-Aminobenzoic acid	5.0mg
Riboflavin	5.0mg
Thiamine-HCl	5.0mg
Biotin	2.0mg
Folic acid	2.0mg
Vitamin B_{12}	0.1mg

Preparation of Vitamin Solution: Add components to distilled/deionized water and bring volume to 1.0L. Mix thoroughly. Sparge with 80% N_2 + 20% CO_2. Autoclave for 15 min at 15 psi pressure–121°C.

L-Cysteine·HCl Solution:
Composition per 10.0mL:

L-Cysteine·HCl	0.3g

Preparation of L-Cysteine·HCl Solution: Add L-cysteine·HCl to distilled/deionized water and bring volume to 10.0mL. Mix thoroughly. Sparge with 80% N_2 + 20% CO_2. Autoclave for 15 min at 15 psi pressure–121°C.

$Na_2S \cdot 9H_2O$ Solution:
Composition per 10.0mL:

$Na_2S \cdot 9H_2O$	0.3g

Preparation of $Na_2S \cdot 9H_2O$ Solution: Add $Na_2S \cdot 9H_2O$ to distilled/deionized water and bring volume to 10.0mL. Mix thoroughly. Gas under 80% N_2 + 20% CO_2. Autoclave for 15 min at 15 psi pressure–121°C.

$MgSO_4 \cdot 7H_2O$ Solution:
Composition per 10.0mL:

$MgSO_4 \cdot 7H_2O$	0.5g

Preparation of $MgSO_4 \cdot 7H_2O$ Solution: Add $MgSO_4 \cdot 7H_2O$ to distilled/deionized water and bring volume to 10.0mL. Mix thoroughly. Sparge with 80% N_2 + 20% CO_2. Autoclave for 15 min at 15 psi pressure–121°C.

$CaCl_2 \cdot 2H_2O$ Solution:
Composition per 10.0mL:

$CaCl_2 \cdot 2H_2O$	0.25g

Preparation of $CaCl_2 \cdot 2H_2O$ Solution: Add $CaCl_2 \cdot 2H_2O$ to distilled/deionized water and bring volume to 10.0mL. Mix thoroughly. Sparge with 80% N_2 + 20% CO_2. Autoclave for 15 min at 15 psi pressure–121°C.

Selenite-Tungstate Solution:
Composition per liter:

NaOH	0.5g
$Na_2WO_4 \cdot 2H_2O$	4mg
$Na_2SeO_3 \cdot 5H_2O$	3mg

Preparation of Selenite-Tungstate Solution: Add components to distilled/deionized water and bring volume to 1.0L. Mix thoroughly. Sparge with 80% N_2 + 20% CO_2. Autoclave for 15 min at 15 psi pressure–121°C.

Trace Elements Solution SL-10:
Composition per liter:

$FeCl_2 \cdot 4H_2O$	1.5g
$CoCl_2 \cdot 6H_2O$	190.0mg
$MnCl_2 \cdot 4H_2O$	100.0mg
$ZnCl_2$	70.0mg

$Na_2MoO_4 \cdot 2H_2O$ 36.0mg
$NiCl_2 \cdot 6H_2O$ 24.0mg
H_3BO_3 .... 6.0mg
$CuCl_2 \cdot 2H_2O$ 2.0mg
HCl (25% solution) 10.0mL

Preparation of Trace Elements Solution SL-10: Add $FeCl_2 \cdot 4H_2O$ to 10.0mL of HCl solution. Mix thoroughly. Add distilled/deionized water and bring volume to 1.0L. Add remaining components. Mix thoroughly. Gas under 80% N_2 + 20% CO_2. Autoclave for 15 min at 15 psi pressure–121°C.

Preparation of Medium: Prepare and dispense medium under 80% N_2 + 20% CO_2. Add components, except substrate solution, $MgSO_4 \cdot 7H_2O$ solution, $CaCl_2 \cdot 2H_2O$ solution, selenite-tungstate solution, L-cysteine·HCl solution, and $Na_2S \cdot 9H_2O$ solution, to distilled/deionized water and bring volume to 1.0L. Mix thoroughly. Sparge with 80% N_2 + 20% CO_2. Autoclave for 15 min at 15 psi pressure–121°C. Aseptically and anaerobically add 20.0mL of sterile substrate solution, 10.0mL of sterile $MgSO_4 \cdot 7H_2O$ solution, 10.0mL of sterile $CaCl_2 \cdot 2H_2O$ solution, 1.0mL of sterile selenite-tungstate solution, 10.0mL of sterile L-cysteine·HCl solution, and 10.0mL of sterile $Na_2S \cdot 9H_2O$ solution. Mix thoroughly. Adjust pH to 7.0–7.2. Aseptically and anaerobically distribute into sterile tubes or flasks.

Use: For the cultivation and maintenance of *Campylobacter* species.

GV Medium

Composition per 1061.0mL:

NaHCO$_3$ 3.0g
NaCl 2.25g
Yeast extract 1.0g
NH_4Cl 0.5g
K_2HPO_4 .... 0.348g
KH_2PO_4 .... 0.227g
$FeSO_4 \cdot 7H_2O$ 2.0mg
Resazurin 0.5mg
Vitamin solution 10.0mL
Substrate solution 10.0mL
L-Cysteine·HCl solution 10.0mL
$Na_2S \cdot 9H_2O$ solution 10.0mL
$MgSO_4 \cdot 7H_2O$ solution 10.0mL
$CaCl_2 \cdot 2H_2O$ solution 10.0mL
Selenite-tungstate solution 1.0mL
Trace elements solution SL-10 .... 1.0mL

pH 7.0 ± 0.2 at 25°C

Vitamin Solution:

Composition per liter:

Pyridoxine-HCl 10.0mg
Calcium DL-pantothenate 5.0mg
Lipoic acid 5.0mg
Nicotinic acid 5.0mg
p-Aminobenzoic acid 5.0mg
Riboflavin 5.0mg
Thiamine-HCl 5.0mg
Biotin 2.0mg
Folic acid 2.0mg
Vitamin B_{12} 0.1mg

Preparation of Vitamin Solution: Add components to distilled/deionized water and bring volume to 1.0L. Mix thoroughly. Sparge with 80% N_2 + 20% CO_2. Autoclave for 15 min at 15 psi pressure–121°C.

Substrate Solution:

Composition per 20.0mL:

D-Glucose 10.0g

Preparation of Substrate Solution: Add sodium crotonate to distilled/deionized water and bring volume to 20.0mL. Mix thoroughly. Sparge with 80% N_2 + 20% CO_2. Autoclave for 15 min at 15 psi pressure–121°C.

L-Cysteine·HCl Solution:

Composition per 10.0mL:

L-Cysteine·HCl 0.3g

Preparation of L-Cysteine·HCl Solution: Add L-cysteine·HCl to distilled/deionized water and bring volume to 10.0mL. Mix thoroughly. Sparge with 80% N_2 + 20% CO_2. Autoclave for 15 min at 15 psi pressure–121°C.

$Na_2S \cdot 9H_2O$ Solution:

Composition per 10.0mL:

$Na_2S \cdot 9H_2O$ 0.3g

Preparation of $Na_2S \cdot 9H_2O$ Solution: Add $Na_2S \cdot 9H_2O$ to distilled/deionized water and bring volume to 10.0mL. Mix thoroughly. Gas under 80% N_2 + 20% CO_2. Autoclave for 15 min at 15 psi pressure–121°C.

$MgSO_4 \cdot 7H_2O$ Solution:

Composition per 10.0mL:

$MgSO_4 \cdot 7H_2O$ 0.5g

Preparation of $MgSO_4 \cdot 7H_2O$ Solution: Add $MgSO_4 \cdot 7H_2O$ to distilled/deionized water and bring volume to 10.0mL. Mix thoroughly. Sparge with 80% N_2 + 20% CO_2. Autoclave for 15 min at 15 psi pressure–121°C.

$CaCl_2 \cdot 2H_2O$ Solution:

Composition per 10.0mL:

$CaCl_2 \cdot 2H_2O$ 0.25g

Preparation of $CaCl_2 \cdot 2H_2O$ Solution: Add $CaCl_2 \cdot 2H_2O$ to distilled/deionized water and bring volume to 10.0mL. Mix thoroughly. Sparge with 80% N_2 + 20% CO_2. Autoclave for 15 min at 15 psi pressure–121°C.

Selenite-Tungstate Solution:

Composition per liter:

NaOH 0.5g
$Na_2WO_4 \cdot 2H_2O$ 4mg
$Na_2SeO_3 \cdot 5H_2O$ 3mg

Preparation of Selenite-Tungstate Solution: Add components to distilled/deionized water and bring volume to 1.0L. Mix thoroughly. Sparge with 80% N_2 + 20% CO_2. Autoclave for 15 min at 15 psi pressure–121°C.

Trace Elements Solution SL-10:

Composition per liter:

$FeCl_2 \cdot 4H_2O$ 1.5g
$CoCl_2 \cdot 6H_2O$ 190.0mg
$MnCl_2 \cdot 4H_2O$ 100.0mg
$ZnCl_2$ 70.0mg
$Na_2MoO_4 \cdot 2H_2O$ 36.0mg
$NiCl_2 \cdot 6H_2O$ 24.0mg
H_3BO_3 .... 6.0mg
$CuCl_2 \cdot 2H_2O$ 2.0mg
HCl (25% solution) 10.0mL

Preparation of Trace Elements Solution SL-10: Add $FeCl_2 \cdot 4H_2O$ to 10.0mL of HCl solution. Mix thoroughly. Add distilled/deionized

water and bring volume to 1.0L. Add remaining components. Mix thoroughly. Gas under 80% N_2 + 20% CO_2. Autoclave for 15 min at 15 psi pressure–121°C.

Preparation of Medium: Prepare and dispense medium under 80% N_2 + 20% CO_2. Add components, except substrate solution, $MgSO_4·7H_2O$ solution, $CaCl_2·2H_2O$ solution, selenite-tungstate solution, L-cysteine·HCl solution, and $Na_2S·9H_2O$ solution, to distilled/deionized water and bring volume to 1.0L. Mix thoroughly. Sparge with 80% N_2 + 20% CO_2. Autoclave for 15 min at 15 psi pressure–121°C. Aseptically and anaerobically add 20.0mL of sterile substrate solution, 10.0mL of sterile $MgSO_4·7H_2O$ solution, 10.0mL of sterile $CaCl_2·2H_2O$ solution, 1.0mL of sterile selenite-tungstate solution, 10.0mL of sterile L-cysteine·HCl solution, and 10.0mL of sterile $Na_2S·9H_2O$ solution. Mix thoroughly. Adjust pH to 7.0–7.2. Aseptically and anaerobically distribute into sterile tubes or flasks.

Use: For the cultivation and maintenance of *Clostridium quinii.*

GY Agar (ATCC Medium 1994)

Composition per liter:

Agar	15.0 g
Glucose	10.0 g
Yeast extract	2.5 g

Preparation of Medium: Add components to distilled/deionized water and bring volume to 1.0L. Mix thoroughly. Gently heat and bring to boiling. Distribute into tubes or flasks. Autoclave for 15 min at 15 psi pressure–121°C. Pour into sterile Petri dishes or leave in tubes.

Use: For the cultivation and maintenance of *Aspergillus nidulans.*

GY Double-Strength Agar with Uracil and Uridine

Composition per liter:

Glucose	20.0 g
Agar	15.0 g
Yeast extract	5.0 g
Uracil	1.5g
Uridine	1.5g

Preparation of Medium: Add components to distilled/deionized water and bring volume to 1.0L. Mix thoroughly. Gently heat and bring to boiling. Distribute into tubes or flasks. Autoclave for 15 min at 15 psi pressure–121°C. Pour into sterile Petri dishes or leave in tubes.

Use: For the cultivation and maintenance of *Aspergillus nidulans.*

GYE Medium

Composition per liter:

Glucose	15.0g
Yeast extract	5.0g
$CaCl_2·2H_2O$	0.2g

pH 7.0 ± 0.2 at 25°C

Preparation of Medium: Add components to tap water and bring volume to 1.0L. Mix thoroughly. Distribute into tubes or flasks. Autoclave for 15 min at 15 psi pressure–121°C.

Use: For the cultivation of *Bacillus thermoruber.*

GYEP Medium

Composition per liter:

Glucose	20.0g
Peptone	10.0g
Yeast extract	5.0g

Preparation of Medium: Add components to distilled/deionized water and bring volume to 1.0L. Mix thoroughly. Gently heat and bring to boiling. Distribute into tubes or flasks. Autoclave for 15 min at 15 psi pressure–121°C.

Use: For the cultivation of *Aspergillus nidulans* and *Basidiobolus microsporus.*

GYM+S Agar

Composition per liter:

Starch	20.0g
Agar	12.0g
Malt extract	10.0g
$CaCO_3$	4.0g
Glucose	4.0g
Yeast extract	4.0g

pH 7.2 ± 0.2 at 25°C

Preparation of Medium: Add components, except agar, to distilled/deionized water and bring volume to 1.0L. Mix thoroughly. Using pH indicator paper, adjust pH to 7.2 with KOH. Add agar. Gently heat and bring to boiling. Distribute into tubes or flasks. Autoclave for 15 min at 15 psi pressure–121°C. Pour into sterile Petri dishes or leave in tubes.

Use: For the cultivation and maintenance of *Actinomadura* species, *Actinoplanes* species, *Amycolata autotrophica, Amycolatopsis orientalis, Amycolatopsis sulphurea, Cellulomonas cellulans, Cellulomonas turbata, Gordona rubropertinctus, Kineosporia aurantiaca, Mycobacterium* species, *Nocardia* species, *Nocardioides albus, Nocardiopsis albus, Oerskovia* species, *Promicromonospora enterophila, Pseudonocardia compacta, Saccharomonospora viridis, Saccharopolyspora hirsuta, Saccharothrix coeruleofusca, Saccharothrix longispora, Sporichthya polymorpha, Streptomyces* species, *Streptosporangium corrugatum,* and *Thermoactinomyces dichotomicus.*

GYM + Seawater (DSMZ Medium 871)

Composition per liter:

Sea salts	32.0g
Agar	15.0g
Malt extract	10.0g
Glucose	4.0g
Yeast extract	4.0g
$CaCO_3$	2.0g

pH 7.0–7.4 at 25°C

Preparation of Medium: Add sea salts to distilled/deionized water and bring volume to 1.0L. Mix thoroughly. Add remaining components. Mix thoroughly. Distribute into tubes or flasks. Autoclave for 15 min at 15 psi pressure–121°C.

Use: For the cultivation of *Nocardiopsis aegyptia (Nocardiopsis* sp.*)* and *Nocardiopsis halotolerans (Nocardiopsis* sp.*).*

GYM Starch Agar (DSMZ Medium 214)

Composition per liter:

Starch	20.0g
Agar	12.0g
Malt extract	10.0g
Glucose	4.0g

Yeast extract	4.0g
$CaCO_3$	2.0g

pH 7.2 ± 0.2 at 25°C

Preparation of Medium: Add components, except agar, to distilled/deionized water and bring volume to 1.0L. Mix thoroughly. Adjust pH to 7.2. Add agar. Gently heat and bring to boiling. Distribute into tubes or flasks. Autoclave for 15 min at 15 psi pressure–121°C. Pour into sterile Petri dishes or leave in tubes.

Use: For the cultivation and maintenance of *Amycolatopsis orientalis* subsp. *orientalis, Nocardia* spp., *Mycobacterium* spp., *Pseudonocardia* spp., *Saccharothrix* spp., *Kineosporia aurantiaca, Kitasatospora setae, Oerskovia turbata (Cellulomonas turbata), Cellulosimicrobium cellulans,* and *Thermoactinomyces dichotomicus.*

GYM Starch Medium (DSMZ Medium 214)

Composition per liter:

Starch	20.0g
Malt extract	10.0g
Glucose	4.0g
Yeast extract	4.0g

pH 7.2 ± 0.2 at 25°C

Preparation of Medium: Add components to distilled/deionized water and bring volume to 1.0L. Mix thoroughly. Adjust pH to 7.2. Distribute into tubes or flasks. Autoclave for 15 min at 15 psi pressure–121°C.

Use: For the cultivation of *Amycolatopsis orientalis* subsp. *orientalis, Nocardia* spp., *Mycobacterium* spp., *Pseudonocardia* spp., *Saccharothrix* spp., *Kineosporia aurantiaca, Kitasatospora setae, Oerskovia turbata (Cellulomonas turbata), Cellulosimicrobium cellulans,* and *Thermoactinomyces dichotomicus.*

GYM *Streptomyces* Agar (DSMZ Medium 65)

Composition per liter:

Agar	12.0g
Malt extract	10.0g
Glucose	4.0g
Yeast extract	4.0g
$CaCO_3$	2.0g

pH 7.2 ± 0.2 at 25°C

Preparation of Medium: Add components, except agar, to distilled/deionized water and bring volume to 1.0L. Mix thoroughly. Adjust pH to 7.2. Add agar. Gently heat and bring to boiling. Distribute into tubes or flasks. Autoclave for 15 min at 15 psi pressure–121°C. Pour into sterile Petri dishes or leave in tubes.

Use: For the cultivation and maintenance of *Streptomyces* spp.

GYM *Streptomyces* Medium (DSMZ Medium 65)

Composition per liter:

Malt extract	10.0g
Glucose	4.0g
Yeast extract	4.0g

pH 7.2 ± 0.2 at 25°C

Preparation of Medium: Add components to distilled/deionized water and bring volume to 1.0L. Mix thoroughly. Adjust pH to 7.2. Distribute into tubes or flasks. Autoclave for 15 min at 15 psi pressure–121°C.

Use: For the cultivation and maintenance of *Streptomyces* spp.

GYP Agar

Composition per liter:

Glucose	40.0g
Agar	20.0g
Peptone	20.0g
Sodium acetate	20.0g
Yeast extract	20.0g
Solution A	10.0mL

pH 6.8 ± 0.2 at 25°C

Solution A:

Composition per 100.0mL:

$MgSO_4 \cdot 7H_2O$	4.0g
$FeSO_4 \cdot 7H_2O$	0.2g
$MnSO_4 \cdot H_2O$	0.2g
NaCl	0.2g

Preparation of Solution A: Add components to distilled/deionized water and bring volume to 100.0mL. Mix thoroughly.

Preparation of Medium: Add components to distilled/deionized water and bring volume to 1.0L. Mix thoroughly. Gently heat and bring to boiling. Distribute into tubes or flasks. Autoclave for 15 min at 15 psi pressure–121°C. Pour into sterile Petri dishes or leave in tubes.

Use: For the cultivation and maintenance of *Bacillus laevolacticus, Bacillus racemilacticus,* and *Sporolactobacillus inulinus.*

GYP Medium

See: **Glucose Yeast Extract Peptone Medium**

GYP Sodium Acetate Mineral Salts Broth (Glucose Yeast Peptone Sodium Acetate Mineral Salts Broth)

Composition per liter:

$FeSO_4 \cdot 7H_2O$	10.0g
Glucose	10.0g
$MnSO_4 \cdot H_2O$	10.0g
Peptone	10.0g
Sodium acetate	10.0g
Yeast extract	10.0g
$MgSO_4 \cdot 7H_2O$	0.2g
NaCl	10.0mg

pH 6.8 ± 0.2 at 25°C

Preparation of Medium: Add components to distilled/deionized water and bring volume to 1.0L. Mix thoroughly. Adjust pH to 6.8. Distribute into tubes or flasks. Autoclave for 15 min at 15 psi pressure–121°C.

Use: For the cultivation of *Lactobacillus pentosus, Lactobacillus plantarum, Pediococcus acidilactici,* and *Pediococcus pentosaceus.*

GYP Sodium Acetate Mineral Salts Broth with Sodium Chloride

Composition per liter:

NaCl	50.0g
$FeSO_4 \cdot 7H_2O$	10.0g
Glucose	10.0g

$MnSO_4 \cdot H_2O$	10.0g
Peptone	10.0g
Sodium acetate	10.0g
Yeast extract	10.0g
$MgSO_4 \cdot 7H_2O$	0.2g
NaCl	10.0mg

pH 6.8 ± 0.2 at 25°C

Preparation of Medium: Add components to distilled/deionized water and bring volume to 1.0L. Mix thoroughly. Adjust pH to 6.8. Distribute into tubes or flasks. Autoclave for 15 min at 15 psi pressure–121°C.

Use: For the cultivation of *Pediococcus halophilus.*

GYP Sodium Acetate Mineral Salts Broth with 5% Sodium Chloride (LMG Medium 244)

Composition per liter:

NaCl	50.0g
Glucose	10.0g
Yeast extract	10.0g
Peptone	10.0g
Sodium acetate	10.0g
$MgSO_4 \cdot 7H_2O$	0.2g
$MnSO_4 \cdot 4H_2O$	10.0mg
$FeSO_4 \cdot 7H_2O$	10.0mg

pH 6.8± 0.2 at 25°C

Preparation of Medium: Add components to distilled/deionized water and bring volume to 1.0L. Mix thoroughly. Distribute into tubes or flasks. Autoclave for 15 min at 15 psi pressure–121°C.

Use: For the cultivation of *Tetragenococcus halophilus* and *Tetragenococcus muriaticus.*

GYPT Medium (Glucose Yeast Extract Peptone Thioglycolate Medium)

Composition per liter:

Agar	8.0g
Yeast extract	6.0g
Glucose	5.0g
Peptone	2.0g
Sodium thioglycolate	0.5g

pH 7.4 ± 0.2 at 25°C

Preparation of Medium: Add components to distilled/deionized water and bring volume to 1.0L. Mix thoroughly. Adjust pH to 7.4. Gently heat and bring to boiling. Anaerobically distribute into tubes under 97% N_2 + 3% H_2. Cap tubes with rubber stoppers. Place tubes in a press. Autoclave for 15 min at 15 psi pressure–121°C with fast exhaust.

Use: For the cultivation of *Spirochaeta stenostrepta.*

H Agar

Composition per liter:

Agar	15.0g
Pancreatic digest of casein	10.0g
NaCl	8.0g

Preparation of Medium: Add components to distilled/deionized water and bring volume to 1.0L. Mix thoroughly. Gently heat and bring to boiling. Autoclave for 15 min at 15 psi pressure–121°C. Pour into sterile Petri dishes.

Preparation of Medium: Add components to distilled/deionized water and bring volume to 1.0L. Mix thoroughly. Distribute into tubes or flasks. Autoclave for 15 min at 15 psi pressure–121°C.

Use: For the cultivation of *Escherichia coli* and a variety of other bacteria.

H Agar (Hominis Agar)

Composition per 98.0mL:

Base agar	65.0mL
Horse serum	20.0mL
Yeast dialysate	10.0mL
Penicillin solution	2.0mL
Thallium acetate solution	1.0mL

pH 7.3 ± 0.2 at 25°C

Base Agar:

Composition per liter:

Papaic digest of soybean meal	20.0g
Agarose	10.0g
NaCl	5.0g
Phenol Red (2% solution)	1.0mL

Preparation of Base Agar: Add components to distilled/deionized water and bring volume to 1.0L. Mix thoroughly. Gently heat and bring to boiling. Adjust pH to 7.3. Autoclave for 15 min at 15 psi pressure–121°C. Cool to 45°–50°C.

Yeast Dialysate:

Composition per 10.0mL:

Active, dried yeast	450.0g

Preparation of Yeast Dialysate: Add active, dried yeast to distilled/deionized water and bring volume to 1250.0mL. Gently heat and bring to 40°C. Autoclave for 15 min at 15 psi pressure–121°C. Put into dialysis tubing. Dialyze against 1.0L of distilled/deionized water for 2 days at 4°C. Discard tubing and its contents. Autoclave dialysate for 15 min at 15 psi pressure–121°C. Store at –20°C.

Penicillin Solution:

Composition per 10.0mL:

Penicillin	100,000U

Preparation of Penicillin Solution: Add penicillin to distilled/deionized water and bring volume to 10.0mL. Mix thoroughly. Filter sterilize.

Thallium Acetate Solution:

Composition per 10.0mL:

Thallium acetate	0.33g

Preparation of Thallium Acetate Solution: Add thallium acetate to distilled/deionized water and bring volume to 10.0mL. Mix thoroughly. Filter sterilize.

Caution: Thallium salts are toxic.

Preparation of Medium: To 65.0mL of cooled, sterile base agar, aseptically add 10.0mL of sterile yeast dialysate, 20.0mL of horse serum, 2.0mL of sterile penicillin solution, and 1.0mL of sterile thallium acetate solution. Mix thoroughly. Pour into 10mm × 35mm Petri dishes in 5.0mL volumes. Allow plates to stand overnight at 25°C to remove excess surface moisture.

Use: For the isolation of *Mycoplasma pneumoniae* and *Mycoplasma hominis*.

H Broth

Composition per liter:

NaCl	5.0g
Pancreatic digest of casein	5.0g
Peptone	5.0g
Beef extract	3.0g
K_2HPO_4	2.5g
Glucose	1.0g

pH 7.2 ± 0.2 at 25°C

Preparation of Medium: Add components to distilled/deionized water and bring volume to 1.0L. Mix thoroughly. Distribute into tubes in 4.0mL volumes. Autoclave for 15 min at 10 psi pressure–115°C.

Use: For the preparation of the H agglutination antigen used in the differentiation and identification of *Salmonella* species types and subtypes.

H Broth (Hominis Broth)

Composition per 99.0mL:

Base broth	65.0mL
Horse serum	20.0mL
Yeast dialysate	10.0mL
Penicillin solution	2.0mL
Glucose solution	1.0mL
Thallium acetate solution	1.0mL

pH 7.3 ± 0.2 at 25°C

Base Broth:

Composition per liter:

Papaic digest of soybean meal	20.0g
NaCl	5.0g
Phenol Red (2% solution)	1.0mL

Preparation of Base Broth: Add components to distilled/deionized water and bring volume to 1.0L. Mix thoroughly. Gently heat and bring to boiling. Adjust pH to 7.3. Autoclave for 15 min at 15 psi pressure–121°C. Cool to 25°C.

Yeast Dialysate:

Composition per 10.0mL:

Dried yeast, active	450.0g

Preparation of Yeast Dialysate: Add active, dried yeast to distilled/deionized water and bring volume to 1250.0mL. Gently heat and bring to 40°C. Autoclave for 15 min at 15 psi pressure–121°C. Put into dialysis tubing. Dialyze against 1.0L of distilled/deionized water for 2 days at 4°C. Discard tubing and its contents. Autoclave dialysate for 15 min at 15 psi pressure–121°C. Store at –20°C.

Penicillin Solution:

Composition per 10.0mL:

Penicillin	100,000U

Preparation of Penicillin Solution: Add penicillin to distilled/deionized water and bring volume to 10.0mL. Mix thoroughly. Filter sterilize.

Glucose Solution:

Composition per 10.0mL:

D-Glucose	1.8g

Preparation of Glucose Solution: Add D-glucose to distilled/deionized water and bring volume to 10.0mL. Mix thoroughly. Filter sterilize.

Thallium Acetate Solution:

Composition per 10.0mL:

Thallium acetate	0.33g

Preparation of Thallium Acetate Solution: Add thallium acetate to distilled/deionized water and bring volume to 10.0mL. Mix thoroughly. Filter sterilize.

Caution: Thallium salts are toxic.

Preparation of Medium: To 65.0mL of cooled, sterile base broth, aseptically add 10.0mL of sterile yeast dialysate, 20.0mL of horse serum, 2.0mL of sterile penicillin solution, 1.0mL of sterile glucose solution, and 1.0mL of sterile thallium acetate solution. Mix thoroughly. Aseptically distribute into sterile screw-capped tubes in 5.0mL volumes. Screw caps down tightly.

Use: For the isolation and cultivation of *Mycoplasma pneumoniae*.

H Broth (Hominis Broth)

Composition per 100.0mL:

Base broth	65.0mL
Horse serum	20.0mL
Yeast dialysate	10.0mL
Penicillin solution	2.0mL
Arginine solution	2.0mL
Thallium acetate solution	1.0mL

pH 7.3 ± 0.2 at 25°C

Base Broth:

Composition per liter:

Papaic digest of soybean meal	20.0g
NaCl	5.0g
Phenol Red (2% solution)	1.0mL

Preparation of Base Broth: Add components to distilled/deionized water and bring volume to 1.0L. Mix thoroughly. Gently heat and bring to boiling. Adjust pH to 7.3. Autoclave for 15 min at 15 psi pressure–121°C. Cool to 25°C.

Yeast Dialysate:

Composition per 10.0mL:

Dried yeast, active	450.0g

Preparation of Yeast Dialysate: Add active, dried yeast to distilled/deionized water and bring volume to 1250.0mL. Gently heat and bring to 40°C. Autoclave for 15 min at 15 psi pressure–121°C. Put into dialysis tubing. Dialyze against 1.0L of distilled/deionized water for 2 days at 4°C. Discard tubing and its contents. Autoclave dialysate for 15 min at 15 psi pressure–121°C. Store at –20°C.

Penicillin Solution:

Composition per 10.0mL:

Penicillin	100,000U

Preparation of Penicillin Solution: Add penicillin to distilled/deionized water and bring volume to 10.0mL. Mix thoroughly. Filter sterilize.

Arginine Solution:

Composition per 10.0mL:

L-Arginine	1.74g

Preparation of Arginine Solution: Add L-arginine to distilled/deionized water and bring volume to 10.0mL. Mix thoroughly. Filter sterilize.

Thallium Acetate Solution:

Composition per 10.0mL:

Thallium acetate	0.33g

Caution: Thallium salts are toxic.

Preparation of Thallium Acetate Solution: Add thallium acetate to distilled/deionized water and bring volume to 10.0mL. Mix thoroughly. Filter sterilize.

Preparation of Medium: To 65.0mL of cooled, sterile base broth, aseptically add 10.0mL of sterile yeast dialysate, 20.0mL of horse serum, 2.0mL of sterile penicillin solution, 1.0mL of sterile glucose solution, and 1.0mL of sterile thallium acetate solution. Mix thoroughly. Aseptically distribute into sterile screw-capped tubes in 5.0mL volumes. Screw caps down tightly.

Use: For the isolation and cultivation of *Mycoplasma hominis*.

H Diphasic Medium

Composition per 197.0mL:

Base agar	65.0mL
Base broth	65.0mL
Horse serum	40.0mL
Yeast dialysate	20.0mL
Penicillin solution	4.0mL
Glucose solution	1.0mL
Thallium acetate solution	2.0mL

pH 7.3 ± 0.2 at 25°C

Base Agar:

Composition per liter:

Papaic digest of soybean meal	20.0g
Agarose	10.0g
NaCl	5.0g
Phenol Red (2% solution)	1.0mL

Preparation of Base Agar: Add components to distilled/deionized water and bring volume to 1.0L. Mix thoroughly. Gently heat and bring to boiling. Adjust pH to 7.3. Autoclave for 15 min at 15 psi pressure–121°C. Cool to 45°–50°C.

Base Broth:

Composition per liter:

Papaic digest of soybean meal	20.0g
NaCl	5.0g
Phenol Red (2% solution)	1.0mL

Preparation of Base Broth: Add components to distilled/deionized water and bring volume to 1.0L. Mix thoroughly. Gently heat and bring to boiling. Adjust pH to 7.3. Autoclave for 15 min at 15 psi pressure–121°C. Cool to 25°C.

Yeast Dialysate:

Composition per 10.0mL:

Dried yeast, active	450.0g

Preparation of Yeast Dialysate: Add active, dried yeast to distilled/deionized water and bring volume to 1250.0mL. Gently heat and bring to 40°C. Autoclave for 15 min at 15 psi pressure–121°C. Put into dialysis tubing. Dialyze against 1.0L of distilled/deionized water for 2 days at 4°C. Discard tubing and its contents. Autoclave dialysate for 15 min at 15 psi pressure–121°C. Store at –20°C.

Penicillin Solution:

Composition per 10.0mL:

Penicillin	100,000U

Preparation of Penicillin Solution: Add penicillin to distilled/deionized water and bring volume to 10.0mL. Mix thoroughly. Filter sterilize.

Glucose Solution:

Composition per 10.0mL:

D-Glucose	1.8g

Preparation of Glucose Solution: Add D-glucose to distilled/deionized water and bring volume to 10.0mL. Mix thoroughly. Filter sterilize.

Thallium Acetate Solution:

Composition per 10.0mL:

Thallium acetate	0.33g

Caution: Thallium salts are toxic.

Preparation of Thallium Acetate Solution: Add thallium acetate to distilled/deionized water and bring volume to 10.0mL. Mix thoroughly. Filter sterilize.

Preparation of Medium: To 65.0mL of cooled, sterile base agar, aseptically add 10.0mL of sterile yeast dialysate, 20.0mL of horse serum, 2.0mL of sterile penicillin solution, and 1.0mL of sterile thallium acetate solution. Mix thoroughly. Aseptically distribute into screw-capped tubes in 3.0mL volumes. Allow agar to solidify. To 65.0mL of cooled, sterile base broth, aseptically add 10.0mL of sterile yeast dialysate, 20.0mL of horse serum, 2.0mL of sterile penicillin solution, 1.0mL of sterile glucose solution, and 1.0mL of sterile thallium acetate solution. Mix thoroughly. Aseptically distribute 3.0mL of broth solution on top of the 3.0mL of solidified base agar in each tube. Screw caps down tightly.

Use: For the isolation and cultivation of *Mycoplasma pneumoniae*.

H Medium

Composition per liter:

Pancreatic digest of casein	10.0g
NaCl	8.0g

Preparation of Medium: Add components to distilled/deionized water and bring volume to 1.0L. Mix thoroughly. Distribute into tubes or flasks. Autoclave for 15 min at 15 psi pressure–121°C.

Use: For the cultivation of *Escherichia coli* and a variety of other bacteria.

H Top Agar

Composition per liter:

Pancreatic digest of casein	10.0g
NaCl	8.0g
Agar	7.0g

Preparation of Medium: Add components to distilled/deionized water and bring volume to 1.0L. Mix thoroughly. Gently heat and bring to boiling. Autoclave for 15 min at 15 psi pressure–121°C. Pour into sterile Petri dishes that contain H agar.

Use: For the cultivation of *Escherichia coli* and a variety of other bacteria.

HA

See: **Halophilic Agar**

HAEB

See: **Horie Arabinose Ethyl Violet Broth**

Haemophilus Agar

Composition per 1010.0mL:

Beef heart, infusion from	250.0g
Calf brain, infusion from	200.0g
Agar	13.5g
Proteose peptone	10.0g
NaCl	5.0g
$Na_2HPO_4 \cdot 12H_2O$	2.5g
Glucose	2.0g
β-NADH	1.0g
L-Cysteine·HCl	0.5g
Chicken serum, inactivated	10.0mL

pH 7.4 ± 0.2 at 25°C

Preparation of Medium: Add components, except β-NADH, L-cysteine·HCl, and chicken serum, to distilled/deionized water and bring volume to 1.0L. Mix thoroughly. Adjust pH to 7.4. Gently heat and bring to boiling. Autoclave for 15 min at 15 psi pressure–121°C. Cool to 50°C. Aseptically add 1.0g of β-NADH, 0.5g of L-cysteine·HCl, and 10.0mL of chicken serum. Mix thoroughly. Pour into sterile Petri dishes or distribute into sterile tubes.

Use: For the cultivation and maintenance of *Haemophilus paragallinarum.*

Haemophilus ducreyi Medium

Composition per liter:

Columbia blood agar base	675.0mL
Rabbit blood	300.0mL
Fresh yeast extract solution	25.0mL

pH 6.5–7.0 at 25°C

Columbia Blood Agar Base:

Composition per 675.0mL:

Agar	15.0g
Pantone	10.0g
Bitone	10.0g
NaCl	5.0g
Tryptic digest of beef heart	3.0g
Cornstarch	1.0g

Preparation of Columbia Blood Agar Base: Add components to distilled/deionized water and bring volume to 675.0mL. Mix thoroughly. Gently heat until boiling. Autoclave for 15 min at 15 psi pressure–121°C. Cool to 45°–50°C.

Fresh Yeast Extract Solution:

Composition per 100.0mL:

Baker's yeast, live, pressed, starch-free	25.0g

Preparation of Fresh Yeast Extract Solution: Add the live Baker's yeast to 100.0mL of distilled/deionized water. Autoclave for 90 min at 15 psi pressure–121°C. Allow to stand. Remove supernatant solution. Adjust pH to 6.6–6.8. Filter sterilize.

Preparation of Medium: To 675.0mL of cooled, sterile Columbia blood agar base, aseptically add rabbit blood and sterile fresh yeast extract solution. Aseptically adjust pH to 6.5–7.0.

Use: For the cultivation and maintenance of *Haemophilus ducreyi.*

Haemophilus ducreyi Medium, Revised (Ducreyi Medium, Revised)

Composition per 1010.0mL:

Solution B	500.0mL
Solution A	400.0mL
Solution C	110.0mL

pH 7.4 ± 0.2 at 25°C

Solution A:

Composition per 400.0mL:

Beef heart, infusion from	500.0g
Agar	15.0g
Tryptose	10.0g
NaCl	5.0g

Preparation of Solution A: Add components to distilled/deionized water and bring volume to 400.0L. Mix thoroughly. Gently heat and bring to boiling. Autoclave for 15 min at 15 psi pressure–121°C. Cool to 45°–50°C.

Solution B:

Composition per 500.0mL:

Hemoglobin	10.0g

Preparation of Solution B: Add hemoglobin to distilled/deionized water and bring volume to 500.0L. Mix thoroughly. Gently heat and bring to boiling. Autoclave for 15 min at 15 psi pressure–121°C. Cool to 45°–50°C.

Solution C:

Composition per 110.0mL:

Fetal bovine serum	100.0mL
Supplement solution	10.0mL

Supplement Solution:

Composition per liter:

Glucose	100.0g
L-Cysteine·HCl	25.9g
L-Glutamine	10.0g
L-Cystine	1.1g
Adenine	1.0g
Nicotinamide adenine dinucleotide	0.25g
Vitamin B_{12}	0.1g
Thiamine pyrophosphate	0.1g
Guanine·HCl	0.03g
$Fe(NO_3)_3 \cdot 6H_2O$	0.02g
p-Aminobenzoic acid	0.013g
Thiamine·HCl	3.0mg

Source: The supplement solution (IsoVitaleX® enrichment) is available from BD Diagnostic Systems. This enrichment may be replaced by supplement VX from BD Diagnostic Systems.

Preparation of Supplement Solution: Add components to distilled/deionized water and bring volume to 1.0L. Mix thoroughly. Filter sterilize.

Preparation of Solution C: Combine components. Mix thoroughly. Filter sterilize. Warm to 45°–50°C.

Preparation of Medium: Aseptically combine solution A, solution B, and solution C. Mix thoroughly. Pour into sterile Petri dishes or distribute into sterile tubes.

Use: For the cultivation and maintenance of *Haemophilus ducreyi.*

Haemophilus influenzae Defined Medium MI

Composition per liter:

NaCl	5.8g
K_2HPO_4	3.5g
Glycerol	3.0g
KH_2PO_4	2.7g
Inosine	2.0g
L-Glutamic acid	1.3g
K_2SO_4	1.0g
Sodium lactate	0.8g
L-Aspartic acid	0.5g
Nitrilotriethanol	0.4g
L-Arginine	0.3g
L-Leucine	0.3g
L-Cystine	0.2g
$MgCl_2$	0.2g
L-Tyrosine	0.2g
L-Methionine	0.1g
L-Serine	0.1g
Uracil	0.1g
L-Lysine	0.05g
Glycine	0.03g
$CaCl_2$	0.022g
Hypoxanthine	0.02g
Polyvinyl alcohol	0.02g
Tween™ 80	0.02g
Hemin	0.01g
L-Histidine	0.01g
Calcium pantothenate	4.0mg
Ethylenediaminetetraacetate	4.0mg
Nicotinamide adenine dinucleotide	4.0mg
Thiamine	4.0mg

Preparation of Medium: Add components to distilled/deionized water and bring volume to 1.0L. Mix thoroughly. Filter sterilize.

Use: For the cultivation of *Haemophilus influenzae* in a chemically defined medium.

Haemophilus influenzae Defined Medium MI-Cit

Composition per liter:

NaCl	5.8g
K_2HPO_4	3.5g
Glycerol	3.0g
KH_2PO_4	2.7g
Inosine	2.0g
L-Glutamic acid	1.3g
K_2SO_4	1.0g
Sodium lactate	0.8g
L-Aspartic acid	0.5g
Nitrilotriethanol	0.4g
L-Leucine	0.3g
L-Cystine	0.2g
$MgCl_2$	0.2g
L-Tyrosine	0.2g
Citrulline	0.15g
L-Methionine	0.1g
L-Serine	0.1g
L-Lysine	0.05g
Glycine	0.03g
$CaCl_2$	0.022g
Hypoxanthine	0.02g
Polyvinyl alcohol	0.02g
Tween™ 80	0.02g
Hemin	0.01g
L-Histidine	0.01g
Calcium pantothenate	4.0mg
Ethylenediaminetetraacetate	4.0mg
Nicotinamide adenine dinucleotide	4.0mg
Thiamine	4.0mg

Preparation of Medium: Add components to distilled/deionized water and bring volume to 1.0L. Mix thoroughly. Filter sterilize.

Use: For the cultivation of *Haemophilus influenzae* in a chemically defined medium.

Haemophilus Medium (DSMZ Medium 804)

Composition per liter:

Acid hydrolysate of casein	31.5g
Beef extract	5.4g
Yeast extract	5.0g
Starch	2.7g
NAD solution	1.0mL
Hemin solution	1.0mL

pH 7.3 ± 0.1 at 25°C

NAD Solution:

Composition per 10.0mL:

NAD	150.0mg

Preparation of NAD Solution: Add NAD to distilled/deionized water and bring volume to 10.0mL. Mix thoroughly. Filter sterilize.

Hemin Solution:

Composition per 10.0mL:

Hemin	150.0mg
NaOH (1*N* solution)	2.0mL

Preparation of Hemin Solution: Add hemin to 2.0mL of 1*N* NaOH solution. Mix thoroughly. Bring volume to 10.0mL with distilled/deionized water. Filter sterilize.

Preparation of Medium: Add components, except hemin solution and NAD solution, to distilled/deionized water and bring to 1.0L. Mix thoroughly. Gently heat and bring to boiling. Autoclave for 10 min at 10 psi pressure–115°C. Cool to 25°C. Aseptically add 1.0mL sterile NAD solution and 1.0mL sterile hemin solution. Mix thoroughly. Aseptically distribute into sterile tubes or flasks.

Use: For the cultivation of *Haemophilus influenzae.*

Haemophilus Medium

Composition per 1050.0mL:

Beef heart, infusion from	25.0g
Agar	14.0g
Peptone	5.0g
NaCl	2.5g
Glucose	1.0g
Yeast extract solution	100.0mL
Horse serum	50.0mL

Yeast Extract Solution:

Composition per liter:

Baker's yeast	250.0g

Preparation of Yeast Extract Solution: Add Baker's yeast to distilled/deionized water and bring volume to 1.0L. Mix thoroughly. Gently heat and bring to boiling. Filter through a paper filter. Adjust pH to

8.0. Filter through a Seitz filter. Store at −20°C. Check sterility before using.

Preparation of Medium: Add components, except yeast extract solution and horse serum, to distilled/deionized water and bring volume to 850.0mL. Mix thoroughly. Gently heat and bring to boiling. Autoclave for 15 min at 15 psi pressure–121°C. Cool to 50°–55°C. Aseptically add 100.0mL of sterile yeast extract solution and 50.0mL of sterile horse serum. Mix thoroughly. Pour into sterile Petri dishes or distribute into sterile tubes.

Use: For the cultivation and maintenance of *Actinobacillus pleuropneumoniae, Haemophilus actinomycetemcomitans, Haemophilus haemoglobinophilus, Haemophilus paragallinarum, Haemophilus paraphrophilus, Haemophilus parasuis, Haemophilus segnis, Pasteurella avium, Pasteurella volantinum,* and *Taylorella equigenitalis.*

Haemophilus somnus Agar

Composition per liter:

Agar	20.0g
Tryptose	10.0g
NaCl	5.0g
Beef extract	3.0g
Sheep blood, defibrinated	100.0mL
IsoVitaleX® enrichment solution	10.0mL
L-Cysteine·HCl solution	5.0mL

L-Cysteine·HCl Solution:

Composition per 10.0mL:

L-Cysteine·HCl	1.0g

Preparation of L-Cysteine·HCl Solution: Dissolve 1.0g of L-cysteine·HCl in distilled/deionized water and bring volume to 10.0mL. Mix thoroughly. Filter sterilize. Warm to 50°C.

IsoVitaleX® Enrichment Solution:

Composition per liter:

Glucose	100.0g
L-Cysteine·HCl	25.9g
L-Glutamine	10.0g
Adenine	1.0g
Thiamine pyrophosphate	0.1g
Vitamin B_{12}	0.1g
Guanine·HCl	0.03g
$Fe(NO_3)_3 \cdot 9H_2O$	0.02g
p-Aminobenzoic acid	0.013g
Thiamine·HCl	0.003g

Preparation of IsoVitaleX® Enrichment: Add components to distilled/deionized water and bring volume to 1.0L. Mix thoroughly. Filter sterilize. Warm to 50°C.

Preparation of Medium: Add components, except sheep blood, IsoVitaleX® enrichment solution, and L-cysteine·HCl solution, to distilled/deionized water and bring volume to 895.0mL. Mix thoroughly. Gently heat and bring to boiling. Autoclave for 15 min at 15 psi pressure–121°C. Cool to 50°–55°C. Warm defibrinated sheep blood to 50°C. Aseptically add 100.0mL of sterile defibrinated sheep blood, 10.0mL of sterileIsoVitaleX® enrichment solution, and 5.0mL of sterile L-cysteine·HCl solution. Mix thoroughly. Pour into sterile Petri dishes or distribute into sterile tubes.

Use: For the cultivation and maintenance of *Haemophilus somnus*.

Haemophilus Test Medium (HTM)

Composition per liter:

Beef infusion	300.0g
Acid hydrolysate of casein	17.5g
Agar	17.0g
Yeast extract	5.0g
Starch	1.5g
HTM supplement	10.0mL

pH 7.4 ± 0.2 at 25°C

Source: This medium is available as a premixed powder from Oxoid Unipath.

HTM Supplement:

Composition per 10.0mL:

Nicotinamide adenine dinucleotide	0.03g
Hematin	0.03g

Preparation of HTM Supplement: Add components to distilled/deionized water and bring volume to 10.0mL. Mix thoroughly. Filter sterilize.

Preparation of Medium: Add components, except HTM supplement, to distilled/deionized water and bring volume to 990.0mL. Mix thoroughly. Gently heat and bring to boiling. Autoclave for 15 min at 15 psi pressure–121°C. Cool to 45°–50°C. Aseptically add 10.0mL of sterile HTM supplement. Mix thoroughly. Pour into sterile Petri dishes or distribute into sterile tubes.

Use: For the susceptibility testing of *Haemophilus influenzae*. The medium forms part of the recommended methods of the United States National Committee for Clinical Laboratory Standards (NCCLS). *Haemophilus influenzae* require complex media for growth. These complex media have aggravated the routine susceptibility testing of *Haemophilus influenzae* because of antagonism between some essential nutrients and certain antimicrobial agents. This medium overcomes those limitations. The transparency of the medium allows zones of inhibition to be read easily through the bottom of the Petri dish. HTM contains low levels of antimicrobial antagonists, which allows testing of trimethoprim/sulphamethoxazole to be carried out.

Hagedorn and Holt Selective Medium

Composition per liter:

NaCl	20.0g
Agar	15.0g
Yeast extract	2.0g
Pancreatic digest of casein	1.7g
Agar	1.5g
NaCl	0.5g
Papaic digest of soybean meal	0.3g
K_2HPO_4	0.25g
Glucose	0.25g
Cycloheximide	0.1g
Methyl Red	0.15mg

pH 7.3 ± 0.2 at 25°C

Caution: Cycloheximide is toxic. Avoid skin contact or aerosol formation and inhalation.

Preparation of Medium: Add components to distilled/deionized water and bring volume to 1.0L. Mix thoroughly. Gently heat and bring to boiling. Distribute into tubes or flasks. Autoclave for 15 min at 15 psi pressure–121°C. Pour into sterile Petri dishes or leave in tubes.

Use: For the selective isolation of *Arthrobacter* species in soil.

Hagem's Modess Medium

Composition per liter:

Agar, noble	15.0g
Glucose	10.0g
DL-Asparagine	1.0g
Yeast extract	1.0g
Peptone	1.0g
$MgSO_4 \cdot 7H_2O$	0.5g
KH_2PO_4	0.35g
K_2HPO_4	0.15g
Thiamine·HCl	40.0mg
$FeCl_3 \cdot 6H_2O$ or ferric citrate	1.0mg

Preparation of Medium: Add components to distilled/deionized water and bring volume to 1.0L. Mix thoroughly. Gently heat and bring to boiling. Distribute into tubes or flasks. Autoclave for 15 min at 15 psi pressure–121°C. Pour into sterile Petri dishes or leave in tubes.

Use: For the cultivation and maintenance of *Agaricus macrosporus, Boletus rubinellus, Flammulina velutipes, Inonotus hispidus, Lactarius turpis, Nodulisporium tuberum, Odontia bicolor, Phellinus pomaceus, Phellinus tremulus, Phellinus weirii, Phlebia gigantea, Poria medula-panis, Pteridiospora spinosispora, Schizophyllum commune, Thanatephorus cucumeris, Thelephora terrestris, Trametes versicolor, Tuber albidum,* and *Tuber rufum.*

Half Fraser Broth

Composition per liter:

NaCl	20.0g
Na_2HPO_4	12.0g
Proteose peptone	5.0g
Tryptone	5.0g
Lab Lemco powder	5.0g
LiCl	3.0g
KH_2PO_4	1.35g
Esculin	1.0g
Half Fraser supplement solution	10.0mL

pH 7.2 ± 0.2 at 25°C

Source: This medium is available as a premixed powder from Oxoid Unipath.

Half Fraser Supplement Solution:
Composition per 10.0mL:

Ferric ammonium citrate	0.5g
Acriflavine·HCl	0.125g
Nalidixic acid	0.05g
Ethanol	5.0mL

Preparation of Half Fraser Supplement Solution: Add components to distilled/deionized water and bring volume to 10.0mL. Mix thoroughly. Filter sterilize.

Preparation of Medium: Add components, except h alf Fraser supplement solution, to distilled/deionized water and bring volume to 990.0mL. Mix thoroughly. Gently heat and bring to boiling. Autoclave for 15 min at 15 psi pressure–121°C. Cool to 45°–50°C. Aseptically add sterile half Fraser supplement solution. Mix thoroughly. Aseptically distribute into sterile tubes or flasks.

Use: For the isolation of *Listeria* species from food and environmental species. A primary selective enrichment broth for *Listeria* spp.

Half Fraser Broth without Ferric Ammonium Citrate

Composition per liter:

NaCl	20.0g
Na_2HPO_4	12.0g
Proteose peptone	5.0g
Tryptone	5.0g
Lab Lemco powder	5.0g
LiCl	3.0g
KH_2PO_4	1.35g
Esculin	1.0g
Half Fraser supplement solution without ferric ammonium citrate	10.0mL

pH 7.2 ± 0.2 at 25°C

Source: This medium is available as a premixed powder from Oxoid Unipath.

Half Fraser Supplement Solution without Ferric Ammonium Citrate:
Composition per 10.0mL:

Acriflavine·HCl	0.125g
Nalidixic acid	0.05g
Ethanol	5.0mL

Preparation of Half Fraser Supplement Solution without Ferric Ammonium Citrate: Add components to distilled/deionized water and bring volume to 10.0mL. Mix thoroughly. Filter sterilize.

Preparation of Medium: Add components, except half Fraser supplement solution without ferric ammonium citrate, to distilled/deionized water and bring volume to 990.0mL. Mix thoroughly. Gently heat and bring to boiling. Autoclave for 15 min at 15 psi pressure–121°C. Cool to 45°–50°C. Aseptically add sterile Half Fraser supplement solution without ferric ammonium citrate. Mix thoroughly. Aseptically distribute into sterile tubes or flasks.

Use: For the isolation of *Listeria* species from food and environmental specimens. A primary selective enrichment broth for *Listeria* spp. A pre-supplemented primary selective enrichment broth for *Listeria* spp.

Haliscomenobacter hydrossis Medium (LMG Medium 154)

Composition per liter:

Glutamic acid	1.31g
$MgSO_4 \cdot 7H_2O$	75.0mg
$CaCl_2 \cdot 2H_2O$	50.0mg
K_2HPO_4	40.0mg
$Na_2HPO_4 \cdot 2H_2O$	40.0mg
KH_2PO_4	27.0mg
$FeCl_3 \cdot 6H_2O$	5.0mg
$MnSO_4 \cdot H_2O$	3.0mg
Pancreatic digest of casein	1.7mg
NaCl	0.5mg
Papaic digest of soybean meal	0.3mg
K_2HPO_4	0.25mg
Glucose	0.25mg
Glucose solution	10.0mL
Vitamin solution	1.0mL
Trace elements solution	1.0mL

pH 7.5 ± 0.2 at 25°C

Glucose Solution:
Composition per 10.0mL:

Glucose	2.0g

Preparation of Glucose Solution: Add glucose to 10.0mL of distilled/deionized water. Mix thoroughly. Autoclave for 15 min at 15 psi pressure–121°C.

Vitamin Solution:

Composition per 10.0mL:

Thiamine	4.0mg
Vitamin B_{12}	0.1mg

Preparation of Vitamin Solution: Add Vitamin B_{12} and thiamine to 10.0mL of distilled/deionized water. Mix thoroughly. Filter sterilize.

Trace Elements Solution:

Composition per liter:

H_3BO_3	0.3g
$CoCl_2 \cdot 6H_2O$	0.2g
$ZnSO_4 \cdot 7H_2O$	0.1g
$Na_2MoO_4 \cdot 2H_2O$	30.0mg
$MnCl_2 \cdot 4H_2O$	30.0mg
$NiCl_2 \cdot 6H_2O$	20.0mg
$CuCl_2 \cdot 2H_2O$	10.0mg

Preparation of Trace Elements Solution: Add components to distilled/deionized water and bring volume to 1.0L. Mix thoroughly.

Preparation of Medium: Add components, except vitamin solution and glucose solution, to 989.0mL distilled/deionized water. Mix thoroughly. Autoclave for 15 min at 15 psi pressure–121°C. Cool to 25°C. Aseptically add 10.0mL sterile glucose solution and 1.0mL sterile vitamin solution. Mix thoroughly. Aseptically distribute to sterile tubes or flasks.

Use: For the cultivation of *Haliscomenobacter hydrossis.*

Haliscomenobacter Medium

Composition per liter:

Agar	10.0g
$(NH_4)_2SO_4$	0.5g
Glucose	0.15g
$CaCO_3$	0.1g
KCl	0.05g
K_2HPO_4	0.05g
$MgSO_4 \cdot 7H_2O$	0.05g
$Ca(NO_3)_2$	0.01g
Vitamin solution	10.0mL

Vitamin Solution:

Composition per 10.0mL:

Thiamine	0.4mg
Vitamin B_{12}	0.05mg

Preparation of Vitamin Solution: Add components to distilled/deionized water and bring volume to 10.0mL. Mix thoroughly. Filter sterilize.

Preparation of Medium: Add components, except vitamin solution, to distilled/deionized water and bring volume to 990.0mL. Mix thoroughly. Gently heat and bring to boiling. Autoclave for 15 min at 15 psi pressure–121°C. Cool to 45°–50°C. Aseptically add sterile vitamin solution. Mix thoroughly. Pour into sterile Petri dishes or distribute into sterile tubes.

Use: For the isolation of *Haliscomenobacter* species from activated sludge.

Haliscomenobacter Medium (DSM 134)

Composition per liter:

Glutamic acid	1.31g
$MgSO_4 \cdot 7H_2O$	0.075g
$CaCl_2 \cdot 2H_2O$	0.05g
K_2HPO_4	0.04g
$Na_2HPO_4 \cdot 2H_2O$	0.04g
KH_2PO_4	0.027g
$FeCl_3 \cdot 6H_2O$	5.0mg
$MnSO_4 \cdot H_2O$	3.0mg
Pancreatic digest of casein	1.7mg
NaCl	0.5mg
Papaic digest of soybean meal	0.3mg
K_2HPO_4	0.25mg
Vitamin solution	10.0mL
Glucose solution	5.0mL
Trace elements solution SL-6	1.0mL

pH 7.5 ± 0.2 at 25°C

Vitamin Solution:

Composition per 10.0mL:

Thiamine	0.4mg
Vitamin B_{12}	0.01mg

Preparation of Vitamin Solution: Add components to distilled/deionized water and bring volume to 10.0mL. Mix thoroughly. Filter sterilize.

Glucose Solution:

Composition per 5.0mL:

D-Glucose	2.0g

Preparation of Glucose Solution: Add glucose to distilled/deionized water and bring volume to 5.0mL. Mix thoroughly. Autoclave for 15 min at 15 psi pressure–121°C.

Trace Elements Solution SL-6:

Composition per liter:

H_3BO_3	0.3g
$CoCl_2 \cdot 6H_2O$	0.2g
$ZnSO_4 \cdot 7H_2O$	0.1g
$MnCl_2 \cdot 4H_2O$	0.03g
$Na_2MoO_4 \cdot H_2O$	0.03g
$NiCl_2 \cdot 6H_2O$	0.02g
$CuCl_2 . 2H_2O$	0.01g

Preparation of Trace Elements Solution SL-6: Add components to distilled/deionized water and bring volume to 1.0L. Mix thoroughly. Adjust pH to 3.4.

Preparation of Medium: Add components, except vitamin solution and glucose solution, to distilled/deionized water and bring volume to 985.0mL. Mix thoroughly. Adjust pH to 7.5. Gently heat and bring to boiling. Autoclave for 15 min at 15 psi pressure–121°C. Cool to 25°C. Aseptically add sterile vitamin solution and glucose solution. Mix thoroughly. Aseptically distribute into sterile tubes or flasks.

Use: For the cultivation and maintenance of *Haliscomenobacter hydrossis*.

Haloalkaliphilic Agar

Composition per liter:

Solution A	900.0mL
Solution B	100.0mL

pH 8.5–9.5 at 25°C

Solution A:

Composition per 900.0mL:

NaCl	200.0g
Agar	25.0g
Yeast extract	10.0g
Casamino acids	7.5g
Sodium citrate	3.0g
KCl	2.0g
$MgSO_4·7H_2O$	1.0g
$FeSO_4·7H_2O$	0.05g
$MnSO_4·4H_2O$	0.25mg

Preparation of Solution A: Add components, except NaCl, to distilled/deionized water and bring volume to 900.0mL. Mix thoroughly. Gently heat and bring to boiling. Add 200.0g of NaCl. Mix thoroughly. Autoclave for 15 min at 15 psi pressure–121°C. Cool to 50°–55°C.

Solution B:

Composition per 100.0mL:

Na_2CO_3	5.0g

Preparation of Solution B: Dissolve 5.0g of Na_2CO_3 in distilled/deionized water and bring volume to 100.0mL. Mix thoroughly. Autoclave for 15 min at 15 psi pressure–121°C. Cool to 50°–55°C.

Preparation of Medium: Aseptically mix 900.0mL of solution A and 100.0mL of solution B. Mix thoroughly. Aseptically adjust pH to 8.5–9.5. Pour into sterile Petri dishes or distribute into sterile tubes.

Use: For the cultivation and maintenance of *Natronobacterium gregoryi, Natronobacterium magadii, Natronobacterium pharaonis,* and *Natronococcus occultus.*

Haloalkaliphilic Growth Medium (DSMZ Medium 1150)

Composition per liter:

Glucose	5.0g
$Na_2B_4O_7·10H_2O$	4.0g
NH_4Cl	1.0g
$NaNO_3$	0.5g
KH_2PO_4	0.5g

pH 7.0 ± 0.2 at 25°C

Preparation of Medium: Add components to distilled/deionized water and bring volume to 1.0L. Mix thoroughly. Adjust pH to 10.0 with concentrated NaOH. Gently heat while stirring and bring to boiling. Distribute into tubes or flasks. Autoclave for 15 min at 15 psi pressure–121°C.

Use: For the cultivation of *Halomonas campisalis.*

Haloanaerobacter chitinovorans Medium

Composition per 1001.0mL:

NaCl	10.0g
$MgSO_4·7H_2O$	9.6g
$MgCl_2·6H_2O$	7.0g
KCl	3.8g
Na_2CO_3	1.0g
NH_4Cl	1.0g
Yeast extract	1.0g
$CaCl_2·2H_2O$	0.5g
$K_2HPO_4·3H_2O$	0.4g
Resazurin	0.001g
Na_2CO_3 solution	20.0mL
Substrate solution	20.0mL
$Na_2S·9H_2O$ solution	10.0mL
L-Cysteine·HCl·H_2O solution	10.0mL
Trace elements solution SL-6	1.0mL

pH 7.2 ± 0.2 at 25°C

Na_2CO_3 Solution:

Composition per 20.0mL:

Na_2CO_3	3.0g

Preparation of Na_2CO_3 Solution: Add Na_2CO_3 to distilled/deionized water and bring volume to 20.0mL. Mix thoroughly. Filter sterilize. Sparge with 100% N_2.

Substrate Solution:

Composition per 20.0mL:

Glucose or *N*-acetylglucosamine	5.0g

Preparation of Substrate Solution: Add glucose or *N*-acetylglucosamine to distilled/deionized water and bring volume to 20.0mL. Mix thoroughly. Sparge with 100% N_2. Autoclave for 15 min at 15 psi pressure–121°C.

$Na_2S·9H_2O$ Solution:

Composition per 10.0mL:

$Na_2S·9H_2O$	0.5g

Preparation of $Na_2S·9H_2O$ Solution: Add $Na_2S·9H_2O$ to distilled/deionized water and bring volume to 10.0mL. Mix thoroughly. Sparge with 100% N_2. Autoclave for 15 min at 15 psi pressure–121°C. Before use, neutralize to pH 7.0 with sterile HCl.

L-Cysteine·HCl·H_2O Solution:

Composition per 10.0mL:

L-Cysteine·HCl·H_2O	0.5g

Preparation of L-Cysteine·HCl·H_2O Solution: Add L-cysteine·HCl·H_2O to distilled/deionized water and bring volume to 10.0mL. Mix thoroughly. Sparge with 100% N_2. Autoclave for 15 min at 15 psi pressure–121°C.

Trace Elements Solution SL-6:

Composition per liter:

$MnCl_2·4H_2O$	0.5g
H_3BO_3	0.3g
$CoCl_2·6H_2O$	0.2g
$ZnSO_4·7H_2O$	0.1g
$Na_2MoO_4·2H_2O$	0.03g
$NiCl_2·6H_2O$	0.02g
$CuCl_2·2H_2O$	0.01g

Preparation of Trace Elements Solution SL-6: Add components to distilled/deionized water and bring volume to 1.0L. Mix thoroughly.

Preparation of Medium: Prepare and dispense medium under 100% N_2. Add components, except Na_2CO_3 solution, $Na_2S·9H_2O$ solution, and L-cysteine·HCl·H_2O solution, to distilled/deionized water and bring volume to 960.0mL. Mix thoroughly. Adjust pH to 7.2. Gently heat and bring to boiling. Cool to room temperature while sparging with 100% N_2. Autoclave for 15 min at 15 psi pressure–121°C. Aseptically and anaerobically add 20.0mL of sterile $NaHCO_3$ solution, 10.0mL of sterile $Na_2S·9H_2O$ solution, and 10.0mL of sterile L-cysteine·HCl·H_2O solution. Mix thoroughly. Aseptically distribute into sterile tubes or flasks.

Use: For the cultivation of *Haloanaerobacter chitinovorans.*

Haloanaerobacter chitinovorans Medium

Composition per 1001.0mL:

NaCl	10.0g
$MgSO_4 \cdot 7H_2O$	9.6g
$MgCl_2 \cdot 6H_2O$	7.0g
Chitin	5.0g
KCl	3.8g
Na_2CO_3	1.0g
NH_4Cl	1.0g
Yeast extract	1.0g
$CaCl_2 \cdot 2H_2O$	0.5g
$K_2HPO_4 \cdot 3H_2O$	0.4g
Resazurin	0.001g
Na_2CO_3 solution	20.0mL
$Na_2S \cdot 9H_2O$ solution	10.0mL
L-Cysteine·HCl·H_2O solution	10.0mL
Trace elements solution SL-6	1.0mL

pH 7.2 ± 0.2 at 25°C

Na_2CO_3 Solution:

Composition per 20.0mL:

Na_2CO_3	3.0g

Preparation of Na_2CO_3 Solution: Add Na_2CO_3 to distilled/deionized water and bring volume to 20.0mL. Mix thoroughly. Filter sterilize. Sparge with 100% N_2.

$Na_2S \cdot 9H_2O$ Solution:

Composition per 10.0mL:

$Na_2S \cdot 9H_2O$	0.5g

Preparation of $Na_2S \cdot 9H_2O$ Solution: Add $Na_2S \cdot 9H_2O$ to distilled/deionized water and bring volume to 10.0mL. Mix thoroughly. Sparge with 100% N_2. Autoclave for 15 min at 15 psi pressure–121°C. Before use, neutralize to pH 7.0 with sterile HCl.

L-Cysteine·HCl·H_2O Solution:

Composition per 10.0mL:

L-Cysteine·HCl·H_2O	0.5g

Preparation of L-Cysteine·HCl·H_2O Solution: Add L-cysteine·HCl·H_2O to distilled/deionized water and bring volume to 10.0mL. Mix thoroughly. Sparge with 100% N_2. Autoclave for 15 min at 15 psi pressure–121°C.

Trace Elements Solution SL-6:

Composition per liter:

$MnCl_2 \cdot 4H_2O$	0.5g
H_3BO_3	0.3g
$CoCl_2 \cdot 6H_2O$	0.2g
$ZnSO_4 \cdot 7H_2O$	0.1g
$Na_2MoO_4 \cdot 2H_2O$	0.03g
$NiCl_2 \cdot 6H_2O$	0.02g
$CuCl_2 \cdot 2H_2O$	0.01g

Preparation of Trace Elements Solution SL-6: Add components to distilled/deionized water and bring volume to 1.0L. Mix thoroughly.

Preparation of Medium: Prepare and dispense medium under 100% N_2. Add components, except Na_2CO_3 solution, $Na_2S \cdot 9H_2O$ solution, and L-cysteine·HCl·H_2O solution, to distilled/deionized water and bring volume to 940.0mL. Mix thoroughly. Adjust pH to 7.2. Gently heat and bring to boiling. Cool to room temperature while sparging with 100% N_2. Autoclave for 15 min at 15 psi pressure–121°C. Aseptically and anaerobically add 20.0mL of sterile $NaHCO_3$ solution,10.0mL of sterile $Na_2S \cdot 9H_2O$ solution, and 10.0mL of sterile L-cysteine·HCl·H_2O solution. Mix thoroughly. Aseptically distribute into sterile tubes or flasks.

Use: For the cultivation of *Haloanaerobacter chitinovorans.*

Haloanaerobium alcaliphilum Medium (DSMZ Medium 807)

Composition per liter:

NaCl	100.0g
$MgSO_4 \cdot 7H_2O$	17.0g
Trypticase™	10.0g
$NaHCO_3$	4.1g
Cysteine-HCl·H_2O	0.5g
Resazurin	1.0mg
Solution A	50.0mL
Solution B	50.0mL
Yeast extract solution	50.0mL
Glucose solution	20.0mL
Trace elements solution	10.0mL
$Na_2S \cdot 9H_2O$ solution	10.0mL

pH 7.0 ± 0.1 at 25°C

$Na_2S \cdot 9H_2O$ Solution:

Composition per 10.0mL:

$Na_2S \cdot 9H_2O$	0.25g

Preparation of $Na_2S \cdot 9H_2O$ Solution: Add $Na_2S \cdot 9H_2O$ to distilled/deionized water and bring volume to 10.0mL. Mix thoroughly. Autoclave under 100% N_2 for 15 min at 15 psi pressure–121°C. Cool to room temperature.

Glucose Solution:

Composition per 20.0mL:

Glucose	5.0g

Preparation of Glucose Solution: Add glucose to distilled/deionized water and bring volume to 20.0mL. Mix thoroughly. Filter sterilize.

Yeast Extract Solution:

Composition per 50.0mL:

Yeast extract	10.0g

Preparation of Yeast Extract Solution: Add yeast extract to distilled/deionized water and bring volume to 50.0mL. Mix thoroughly. Autoclave under 100% N_2 for 15 min at 15 psi pressure–121°C. Cool to room temperature.

Solution A

Composition per liter:

K_2HPO_4	6.0g

Preparation of Solution A: Add K_2HPO_4 to distilled/deionized water and bring volume to 1.0L. Mix thoroughly.

Solution B

Composition per liter:

NaCl	12.0g
KH_2PO_4	6.0g
$(NH_4)SO_4$	6.0g
$MgSO_4 \cdot 7H_2O$	2.6g
NH_4Cl	2.5g

$CaCl_2·2H_2O$ 0.28g
K_2HPO_4 .... 0.28g

Preparation of Solution B: Add components to distilled/deionized water and bring volume to 1.0L. Mix thoroughly.

Trace Elements Solution:
Composition per liter:

$MgSO_4·7H_2O$	3.0g
Nitrilotriacetic acid	1.5g
NaCl	1.0g
$MnSO_4·2H_2O$	0.5g
$CoSO_4·7H_2O$	0.18g
$ZnSO_4·7H_2O$	0.18g
$CaCl_2·2H_2O$	0.1g
$FeSO_4·7H_2O$	0.1g
$NiCl_2·6H_2O$	0.025g
$KAl(SO_4)_2·12H_2O$	0.02g
H_3BO_3	0.01g
$Na_2MoO_4·4H_2O$	0.01g
$CuSO_4·5H_2O$	0.01g
$Na_2SeO_3·5H_2O$	0.3mg

Preparation of Trace Elements Solution: Add nitrilotriacetic acid to 500.0mL of distilled/deionized water. Dissolve by adjusting pH to 6.5 with KOH. Add remaining components. Add distilled/deionized water to 1.0L. Mix thoroughly.

Preparation of Medium: Prepare and dispense medium under 80% N_2 + 20% CO_2 gas atmosphere. Add components, except $NaHCO_3$, $Na_2S·9H_2O$ solution, cysteine-HCl·H_2O, yeast extract solution, and glucose solution, to distilled/deionized water and bring volume to 920.0mL. Mix thoroughly. Gently heat and bring to boiling. Boil for 5 min. Cool to room temperature under 80% N_2 + 20% CO_2 gas atmosphere. Add 4.1g $NaHCO_3$ and 0.5g cysteine-HCl·H_2O. Adjust pH to 7.0. Autoclave for 15 min at 15 psi pressure–121°C. Aseptically and anaerobically add 20.0mL sterile glucose solution, 50.0mL sterile yeast extract solution, and 10.0mL sterile $Na_2S·9H_2O$ solution. Mix thoroughly. Aseptically and anaerobically distribute into sterile tubes or bottles.

Use: For the cultivation of *Halanaerobium alcaliphilum.*

Haloanaerobium congolense Medium (DSMZ Medium 933)

Composition per 1080.0mL:

NaCl	100.0g
$MgCl_2·6H_2O$	10.0g
Trypticase™	1.0g
NH_4Cl	1.0g
KCl	1.0g
Na-acetate	0.5g
Cysteine	0.5g
K_2HPO_4	0.3g
KH_2PO_4	0.3g
$CaCl_2·2H_2O$	0.1g
Resazurin	0.01g
Glucose solution	20.0mL
Thiosulfate solution	20.0mL
$Na_2S·9H_2O$ solution	20.0mL
$NaHCO_3$ solution	20.0mL
Trace elements solution	1.0mL

pH 7.0 ± 0.2 at 25°C

Trace Elements Solution:
Composition per liter:

$MgSO_4·7H_2O$	3.0g
Nitrilotriacetic acid	1.5g
NaCl	1.0g
$MnSO_4·2H_2O$	0.5g
$CoSO_4·7H_2O$	0.18g
$ZnSO_4·7H_2O$	0.18g
$CaCl_2·2H_2O$	0.1g
$FeSO_4·7H_2O$	0.1g
$NiCl_2·6H_2O$	0.025g
$KAl(SO_4)_2·12H_2O$	0.02g
H_3BO_3	0.01g
$Na_2MoO_4·4H_2O$	0.01g
$CuSO_4·5H_2O$	0.01g
$Na_2SeO_3·5H_2O$	0.3mg

Preparation of Trace Elements Solution: Add nitrilotriacetic acid to 500.0mL of distilled/deionized water. Dissolve by adjusting pH to 6.5 with KOH. Add remaining components. Add distilled/deionized water to 1.0L. Mix thoroughly.

$Na_2S·9H_2O$ Solution:
Composition per 10.0mL:
$Na_2S·9H_2O$ 0.2g

Preparation of $Na_2S·9H_2O$ Solution: Add $Na_2S·9H_2O$ to distilled/deionized water and bring volume to 10.0mL. Sparge with 100% N_2. Autoclave for 15 min at 15 psi pressure–121°C. Cool to 25°C.

$NaHCO_3$ Solution:
Composition per 50.0mL:
$NaHCO_3$ 5.0g

Preparation of $NaHCO_3$ Solution: Add $NaHCO_3$ to distilled/deionized water and bring volume to 50.0mL. Mix thoroughly. Sparge with 80% N_2 + 20% CO_2. Autoclave for 15 min at 15 psi pressure–121°C. Cool to 25°C.

Thiosulfate Solution:
Composition per 20.0mL:
$Na_2S_2O_3·5H_2O$ 5.0g

Preparation of Thiosulfate Solution: Add $Na_2S_2O_3·5H_2O$ to distilled/deionized water and bring volume to 20.0mL. Mix thoroughly. Sparge with 100% N_2. Autoclave for 15 min at 15 psi pressure–121°C. Cool to room temperature.

Glucose Solution:
Composition per 20.0mL:
Glucose 3.5g

Preparation of Glucose Solution: Add glucose to distilled/deionized water and bring volume to 20.0mL. Mix thoroughly. Sparge with 100% N_2. Filter sterilize.

Preparation of Medium: Prepare and dispense medium under 80% N_2 + 20% CO_2 gas atmosphere. Add components, except $NaHCO_3$ solution, glucose solution, $Na_2S·9H_2O$ solution, and thiosulfate solution, to distilled/deionized water and bring volume to 1.0L. Mix thoroughly. Gently heat and bring to boiling. Boil for 5 min. Cool to room temperature while sparging with 80% N_2 + 20% CO_2. Adjust pH to 7.0. Distribute into anaerobe tubes or bottles. Autoclave for 15 min at 15 psi pressure–121°C. Cool to room temperature. Aseptically and anaerobically add per 10.0mL medium, 0.2mL $NaHCO_3$ solution, 0.2mL glucose solution, 0.2mL $Na_2S·9H_2O$ solution, and 0.2mL thiosulfate solution. Mix thoroughly. The final pH should be 7.0.

Use: For the cultivation of *Thermococcus waiotapuensis.*

Haloanaerobium lacusroseus Medium (DSMZ Medium 764)

Composition per liter:

NaCl	200.0g
KCl	4.0g
$MgCl_2 \cdot 6H_2O$	2.0g
Yeast extract	1.0g
NH_4Cl	1.0g
Na-acetate	1.0g
Trypticase™	0.5g
K_2HPO_4	0.3g
KH_2PO_4	0.3g
$CaCl_2 \cdot 2H_2O$	0.2g
Resazurin	0.001g
Glucose solution	100.0mL
$NaHCO_3$ solution	50.0mL
$Na_2S \cdot 9H_2O$ solution	10.0mL
Dithionite solution	5.0mL
Trace elements soslution SL-6	1.0mL

pH 7.0 ± 0.2 at 25°C

Glucose Solution:

Composition per 100.0mL:

Glucose	17.4g

Preparation of Glucose Solution: Add glucose to distilled/deionized water and bring volume to 100.0mL. Mix thoroughly. Filter sterilize.

Trace Elements Solution SL-6:

Composition per liter:

$MnCl_2 \cdot 4H_2O$	0.5g
H_3BO_3	0.3g
$CoCl_2 \cdot 6H_2O$	0.2g
$ZnSO_4 \cdot 7H_2O$	0.1g
$Na_2MoO_4 \cdot 2H_2O$	0.03g
$NiCl_2 \cdot 6H_2O$	0.02g
$CuCl_2 \cdot 2H_2O$	0.01g

Preparation of Trace Elements Solution SL-6: Add components to distilled/deionized water and bring volume to 1.0L. Mix thoroughly. Autoclave for 15 min at 15 psi pressure–121°C.

$Na_2S \cdot 9H_2O$ Solution:

Composition per 10.0mL:

$Na_2S \cdot 9H_2O$	0.2g

Preparation of $Na_2S \cdot 9H_2O$ Solution: Add $Na_2S \cdot 9H_2O$ to distilled/deionized water and bring volume to 10.0mL. Mix thoroughly. Sparge with 100% N_2. Autoclave for 15 min at 15 psi pressure–121°C. Neutralize to pH 7.0 with sterile HCl.

$NaHCO_3$ Solution:

Composition per 100.0mL:

$NaHCO_3$	10.0g

Preparation of $NaHCO_3$ Solution: Add $NaHCO_3$ to distilled/deionized water and bring volume to 100.0mL. Mix thoroughly. Sparge with 80% N_2 + 20% CO_2. Filter sterilize.

Dithionite Solution

Composition per 10.0mL:

Na-dithionite	2.0mg

Preparation of Dithionite Solution: Add Na-dithionite to distilled/deionized water and bring volume to 10.0mL. Mix thoroughly. Sparge with 100% N_2. Filter sterilize.

Preparation of Medium: Prepare and dispense medium under 80% N_2 + 20% CO_2 gas atmosphere. Add components, except glucose solution, $NaHCO_3$ solution, dithionite solution, and $Na_2S \cdot 9H_2O$ solution, to distilled/deionized water and bring volume to 835.0mL. Mix thoroughly. Adjust pH to 7.0. Gently heat and bring to boiling. Cool while sparging with 80% N_2 + 20% CO_2. Distribute into Hungate tubes under 80% N_2 + 20% CO_2. Autoclave for 15 min at 15 psi pressure–121°C. Cool to room temperature. Aseptically and anaerobically inject glucose solution (0.25mL per 10mL medium), dithionite solution (0.05mL per 10mL medium), $NaHCO_3$ solution (0.5mL per 10mL medium), and $Na_2S \cdot 9H_2O$ solution (0.1mL per 10mL medium). Aseptically and anaerobically distribute into sterile tubes or bottles.

Use: For the cultivation of *Halanaerobium lacusrosei (Haloanaerobium lacusroseus).*

Haloanaerobium Medium

Composition per 1066.0mL:

NaCl	130.0g
Pancreatic digest of casein	10.0g
Yeast extract	10.0g
$MgSO_4 \cdot H_2O$	5.0g
KCl	1.0g
Thioglycolate-ascorbate reducing agent	30.9mL
Glucose solution	25.75mL
NaOH solution	10.3mL
Wolfe's vitamin solution	10.0mL
Wolfe's mineral solution	10.0mL

pH 7.0 ± 0.2 at 25°C

Glucose Solution:

Composition per 30.0mL:

D-Glucose	3.0g

Preparation of Glucose Solution: Add D-glucose to distilled/deionized water and bring volume to 30.0mL. Mix thoroughly. Filter sterilize.

NaOH Solution:

Composition per 20.0mL:

NaOH	1.6g

Preparation of NaOH Solution: Add NaOH to distilled/deionized water and bring volume to 20.0mL. Mix thoroughly. Autoclave for 15 min at 15 psi pressure–121°C.

Wolfe's Vitamin Solution:

Composition per liter:

Pyridoxine·HCl	0.01g
Thiamine·HCl	5.0mg
Riboflavin	5.0mg
Nicotinic acid	5.0mg
Calcium pantothenate	5.0mg
p-Aminobenzoic acid	5.0mg
Thioctic acid	5.0mg
Biotin	2.0mg
Folic acid	2.0mg
Cyanocobalamin	0.1mg

Preparation of Wolfe's Vitamin Solution: Add components to distilled/deionized water and bring volume to 1.0L. Mix thoroughly.

Wolfe's Mineral Solution:

Composition per liter

$MgSO_4 \cdot 7H_2O$	3.0g
Nitrilotriacetic acid	1.5g
NaCl	1.0g
$MnSO_4 \cdot H_2O$	0.5g
$FeSO_4 \cdot 7H_2O$	0.1g
$CoCl_2 \cdot 6H_2O$	0.1g
$CaCl_2$	0.1g
$ZnSO_4 \cdot 7H_2O$	0.1g
$CuSO_4 \cdot 5H_2O$	0.01g
$AlK(SO_4)_2 \cdot 12H_2O$	0.01g
H_3BO_3	0.01g
$Na_2MoO_4 \cdot 2H_2O$	0.01g

Preparation of Wolfe's Mineral Solution: Add nitrilotriacetic acid to 500.0mL of distilled/deionized water. Dissolve by adjusting pH to 6.5 with KOH. Add remaining components. Add distilled/deionized water to 1.0L.

Thioglycolate-Ascorbate Reducing Agent:

Composition per 100.0mL:

Ascorbic acid	1.0g
Sodium thioglycolate	1.0g

Preparation of Thioglycolate-Ascorbate Reducing Agent: Add components to distilled/deionized water and bring volume to 100.0mL. Mix thoroughly. Adjust pH to 7.0. Filter sterilize.

Preparation of Medium: Add components, except thioglycolate-ascorbate reducing agent, glucose, and NaOH solutions, to distilled/deionized water and bring volume to 990.0mL. Mix thoroughly. Gently heat and bring to boiling. Anaerobically distribute into tubes under 97% N_2 + 3% H_2 in 9.7mL volumes. Cap tubes with rubber stoppers. Autoclave for 15 min at 15 psi pressure–121°C. Cool to 25°C. Immediately prior to inoculation, aseptically add 0.3mL of sterile thioglycolate-ascorbate reducing agent, 0.25mL of sterile glucose solution, and 0.1mL of sterile NaOH solution to each tube.

Use: For the cultivation and maintenance of *Haloanaerobium praevalens.*

Haloanaerobium praevalens Medium

Composition per liter:

NaCl	130.0g
Agar	20.0g
Yeast extract	2.0g
Pancreatic digest of casein	2.0g
NH_4Cl	0.5g
$MgSO_4 \cdot 7H_2O$	0.5g
K_2HPO_4	0.35g
$CaCl_2 \cdot 2H_2O$	0.25g
KH_2PO_4	0.23g
$FeSO_4 \cdot 7H_2O$	2.0mg
$NaHCO_3$ solution	20.0mL
L-Cysteine-sulfide reducing agent	20.0mL
Wolfe's vitamin solution	10.0mL
Methanol	10.0mL
Resazurin (0.025% solution)	4.0mL
Trace elements solution SL-6	3.0mL

pH 6.8 ± 0.2 at 25°C

$NaHCO_3$ Solution:

Composition per 20.0mL:

$NaHCO_3$	850.0mg

Preparation of $NaHCO_3$ Solution: Add $NaHCO_3$ to distilled/deionized water and bring volume to 20.0mL. Mix thoroughly. Filter sterilize. Gas with 100% CO_2 for 20 min.

L-Cysteine-Sulfide Reducing Agent:

Composition per 20.0mL:

L-Cysteine·HCl·H_2O	0.3g
$Na_2S \cdot 9H_2O$	0.3g

Preparation of L-Cysteine-Sulfide Reducing Agent: Add L-cysteine·HCl·H_2O to 10.0mL of distilled/deionized water. Mix thoroughly. In a separate tube, add $Na_2S \cdot 9H_2O$ to 10.0mL of distilled/deionized water. Mix thoroughly. Gas both solutions with 100% N_2 and cap tubes. Autoclave both solutions for 15 min at 15 psi pressure–121°C using fast exhaust. Cool to 50°C. Aseptically combine the two solutions under 100% N_2.

Wolfe's Vitamin Solution:

Composition per liter:

Pyridoxine·HCl	10.0mg
Thiamine·HCl	5.0mg
Riboflavin	5.0mg
Nicotinic acid	5.0mg
Calcium pantothenate	5.0mg
p-Aminobenzoic acid	5.0mg
Thioctic acid	5.0mg
Biotin	2.0mg
Folic acid	2.0mg
Cyanocobalamin	100.0μg

Preparation of Wolfe's Vitamin Solution: Add components to distilled/deionized water and bring volume to 1.0L. Mix thoroughly. Filter sterilize.

Trace Elements Solution SL-6:

Composition per liter:

H_3BO_3	0.3g
$CoCl_2 \cdot 6H_2O$	0.2g
$ZnSO_4 \cdot 7H_2O$	0.1g
$MnCl_2 \cdot 4H_2O$	0.03g
$Na_2MoO_4 \cdot H_2O$	0.03g
$NiCl_2 \cdot 6H_2O$	0.02g
$CuCl_2 .2H_2O$	0.01g

Preparation of Trace Elements Solution SL-6: Add components to distilled/deionized water and bring volume to 1.0L. Mix thoroughly. Adjust pH to 3.4.

Preparation of Medium: Add components, except $NaHCO_3$ solution, L-cysteine-sulfide reducing agent, Wolfe's vitamin solution, and methanol, to distilled/deionized water and bring volume to 940.0mL. Mix thoroughly. Autoclave for 15 min at 15 psi pressure–121°C. Cool under 80% N_2 + 20% CO_2. Aseptically and anaerobically add the sterile $NaHCO_3$ solution, the sterile L-cysteine-sulfide reducing agent, the sterile Wolfe's vitamin solution, and filter-sterilized methanol. Mix thoroughly. Adjust pH to 6.8. Aseptically and anaerobically distribute into sterile tubes or flasks.

Use: For the cultivation and maintenance of *Haloanaerobium praevalens.*

Haloanaerobium salsugo Medium

Composition per liter:

NaCl	90.0g
Purified agar (if necessary)	20.0g
Casamino acids	5.0g

Yeast extract 5.0g
Dipotassium PIPES (piperazine-*N,N'*-bis[2-ethanesulfonic acid]) buffer 1.5g
Resazurin 1.0mg
Glucose solution 50.0mL
L-Cysteine-sulfide reducing solution 20.0mL
Mineral solution 20.0mL
Wolfe's vitamin solution 10.0mL
Modified Wolfe's mineral solution 5.0mL
pH 6.0–7.0 at 25°C

Glucose Solution:
Composition per 50.0mL:
D-Glucose 10.0g

Preparation of Glucose Solution: Add glucose to distilled/deionized water and bring volume to 50.0mL. Mix thoroughly. Sparge with 100% N_2. Autoclave for 15 min at 15 psi pressure–121°C.

Mineral Solution:
Composition per liter:
NH_4Cl 50.0g
NaCl 40.0g
$MgSO_4 \cdot 7H_2O$ 10.0g
KCl 5.0g
KH_2PO_4 .... 5.0g
$CaCl_2 \cdot 2H_2O$ 2.0g

Preparation of Mineral Solution: Add components to distilled/deionized water and bring volume to 1.0L. Mix thoroughly.

Wolfe's Vitamin Solution:
Composition per liter:
Pyridoxine·HCl 10.0mg
p-Aminobenzoic acid 5.0mg
Lipoic acid 5.0mg
Nicotinic acid 5.0mg
Riboflavin 5.0mg
Thiamine·HCl 5.0mg
Calcium DL-pantothenate 5.0mg
Biotin 2.0mg
Folic acid 2.0mg
Vitamin B_{12} 0.1mg

Preparation of Wolfe's Vitamin Solution: Add components to distilled/deionized water and bring volume to 1.0L. Mix thoroughly.

Modified Wolfe's Mineral Solution:
Composition per liter:
$MgSO_4 \cdot 7H_2O$ 3.0g
Nitrilotriacetic acid 1.5g
NaCl 1.0g
$MnSO_4 \cdot H_2O$ 0.5g
$CaCl_2$ 0.1g
$CoCl_2 \cdot 6H_2O$ 0.1g
$FeSO_4 \cdot 7H_2O$ 0.1g
$ZnSO_4 \cdot 7H_2O$ 0.1g
$AlK(SO_4)_2 \cdot 12H_2O$ 0.01g
$CuSO_4 \cdot 5H_2O$ 0.01g
H_3BO_3 .... 0.01g
$Na_2MoO_4 \cdot 2H_2O$ 0.01g
Na_2SeO_3 .... 0.01g
$NaWO_4 \cdot 2H_2O$ 0.01g
$NiCl_2 \cdot 6H_2O$ 0.01g

Preparation of Modified Wolfe's Mineral Solution: Add nitrilotriacetic acid to 500.0mL of distilled/deionized water. Adjust pH to 6.5 with KOH. Add remaining components one at a time. Add distilled/deionized water to 1.0L. Adjust pH to 6.8.

L-Cysteine-Sulfide Reducing Solution:
Composition per 200.0mL:
L-Cysteine·HCl·H_2O 5.0g
$Na_2S \cdot 9H_2O$ 5.0g
NaOH 1.25g

Preparation of L-Cysteine-Sulfide Reducing Solution: Add NaOH to distilled/deionized water and bring volume to 200.0mL. Mix thoroughly. Gently heat and bring to boiling. Cool to room temperature while sparging with 100% N_2. Add L-cysteine·HCl·H_2O and $Na_2S \cdot 9H_2O$. Mix thoroughly. Anaerobically distribute into tubes. Autoclave for 15 min at 15 psi pressure–121°C.

Preparation of Medium: Prepare and dispense medium under 100% N_2. Add components, except glucose solution and L-cysteine-sulfide reducing solution, to distilled/deionized water and bring volume to 950.0mL. Mix thoroughly. Gently heat and bring to boiling. Continue boiling for 3 min. Cool to room temperature while sparging with 100% N_2. Adjust pH to 6.0–7.0. Add 20.0mL of L-cysteine-sulfide reducing solution. Mix thoroughly. Anaerobically distribute 9.5mL volumes into anaerobic tubes. Autoclave for 15 min at 15 psi pressure–121°C. Aseptically and anaerobically add 0.5mL of sterile glucose solution to each tube. Mix thoroughly.

Use: For the cultivation of *Haloanaerobium salsugo.*

Haloarcula japonica Medium

Composition per liter:
NaCl 200.0g
$MgSO_4 \cdot 7H_2O$ 20.0g
Yeast extract 10.0g
Casamino acids 7.5g
Trisodium citrate·$2H_2O$ 3.0g
KCl 2.0g
$FeSO_4 \cdot 7H_2O$ 50.0mg
$MnCl_2 \cdot 4H_2O$ 0.36mg

Preparation of Medium: Add components to distilled/deionized water and bring volume to 1.0L. Mix thoroughly. Distribute into tubes or flasks. Autoclave for 15 min at 15 psi pressure–121°C.

Use: For the cultivation of *Haloarcula japonica.*

Haloarcula marismortui Medium

Composition per liter:
NaCl 208.0g
$MgSO_4 \cdot 7H_2O$ 46.6g
Yeast extract 10.0g
$CaCl_2$ 0.5g
$MnCl_2$ 0.125g

Preparation of Medium: Add components to distilled/deionized water and bring volume to 1.0L. Mix thoroughly. Distribute into tubes or flasks. Autoclave for 15 min at 15 psi pressure–121°C.

Use: For the cultivation of *Haloarcula marismortui.*

Haloarcula Medium

Composition per 1001.0mL:
NaCl 250.0g
$MgSO_4 \cdot 7H_2O$ 20.0g
Agar 15.0g
Sodium citrate 3.0g

KCl	2.0g
$CaCl_2$	0.2g
Peptone solution	100.0mL
Trace elements solution	1.0mL

pH 7.4 ± 0.1 at 25°C

Peptone Solution:

Composition per 100.0mL:

Peptone	10.0g

Preparation of Peptone Solution: Add peptone to distilled/deionized water and bring volume to 100.0mL. Mix thoroughly. Filter sterilize.

Trace Elements Solution:

Composition per 100.0mL:

$FeCl_2 \cdot 4H_2O$	0.36g
$MnCl_2 \cdot 4H_2O$	0.022g

Preparation of Trace Elements Solution: Add components to distilled/deionized water and bring volume to 100.0mL. Mix thoroughly. Filter sterilize.

Preparation of Medium: Add components, except peptone solution and trace elements solution, to distilled/deionized water and bring volume to 900.0mL. Mix thoroughly. Gently heat and bring to boiling. Adjust pH to 7.4 with NaOH. Autoclave for 15 min at 15 psi pressure–121°C. Cool to 45°–50°C. Aseptically add 100.0mL of sterile peptone solution and 1.0mL of sterile trace elements solution. Mix thoroughly. Pour into sterile Petri dishes or distribute into sterile tubes.

Use: For the cultivation and maintenance of *Haloanaerobium praevalens*.

Haloarcula vallismortis Synthetic Medium

Composition per 1029.0mL:

Basal salts solution	1.0L
Glucose solution	20.0mL
NH_4Cl solution	5.0mL
$FeSO_4 \cdot 6H_2O$ solution	2.0mL
K_2HPO_4 solution	2.0mL

pH 7.5 ± 0.2 at 25°C

Basal Salts Solution:

Composition per liter:

NaCl	200.0g
$MgSO_4 \cdot 7H_2O$	36.0g
Tris[hydroxymethyl]aminomethane	6.0g
KCl	4.0g
$CaCl_2 \cdot 2H_2O$	1.0g

Preparation of Basal Salts Solution: Add components to distilled/deionized water and bring volume to 1.0L. Mix thoroughly. Adjust pH to 7.5 with HCl. Autoclave for 15 min at 15 psi pressure–121°C.

$FeSO_4 \cdot 6H_2O$ Solution:

Composition per 100.0mL:

$FeSO_4 \cdot 6H_2O$	0.2g
HCl (1.0m*M* solution)	100.0mL

Preparation of $FeSO_4 \cdot 6H_2O$ Solution: Combine components. Mix thoroughly. Filter sterilize.

K_2HPO_4 Solution:

Composition per 100.0mL:

K_2HPO_4	5.0g

Preparation of K_2HPO_4 Solution: Combine components. Mix thoroughly. Filter sterilize.

NH_4Cl Solution:

Composition per 100.0mL:

NH_4Cl	20.0g

Preparation of NH_4Cl Solution: Combine components. Mix thoroughly. Filter sterilize.

Glucose Solution:

Composition per 100.0mL:

D-Glucose	25.0g

Preparation of Glucose Solution: Add glucose to distilled/deionized water and bring volume to 100.0mL. Mix thoroughly. Filter sterilize.

Preparation of Medium: Aseptically combine 1.0L of sterile basal salts solution with 20.0mL of sterile glucose solution, 5.0mL of sterile NH_4Cl solution, 2.0mL of sterile K_2HPO_4 solution, and 2.0mL of sterile $FeSO_4 \cdot 6H_2O$ solution. Mix thoroughly. Aseptically distribute into sterile tubes or flasks.

Use: For the cultivation of *Haloarcula vallismortis.*

Halobacillus Medium (DSMZ Medium 755)

Composition per liter:

NaCl	100.0g
$MgSO_4 \cdot 7H_2O$	5.0g
Peptone, casein digest	5.0g
Yeast extract	3.0g

pH 7.5 ± 0.2 at 25°C

Preparation of Medium: Add components to distilled/deionized water and bring volume to 1.0L. Mix thoroughly. Adjust pH to 7.5. Distribute into tubes or flasks. Autoclave for 15 min at 15 psi pressure–121°C.

Use: For the cultivation of *Halobacillus trueperi* and *Halobacillus litoralis.*

Halobacteria Agar

Composition per liter:

NaCl	200.0g
Agar	20.0g
$MgSO_4 \cdot 7H_2O$	20.0g
Casamino acids	5.0g
Yeast extract	5.0g
Trisodium citrate	3.0g
KCl	2.0g
Sodium glutamate	1.0g
$FeCl_2 \cdot 4H_2O$	36.0mg
$MnCl_2 \cdot 4H_2O$	0.36mg

pH 7.0 ± 0.2 at 25°C

Preparation of Medium: Add components to distilled/deionized water and bring volume to 1.0L. Mix thoroughly. Gently heat and bring to boiling. Distribute into tubes or flasks. Autoclave for 15 min at 15 psi pressure–121°C. Pour into sterile Petri dishes or leave in tubes.

Use: For the cultivation and maintenance of *Haloarcula* species, *Halobacterium* species, *Halococcus morrhuae,* and *Haloferax* species.

Halobacteria Medium (DSMZ Medium 372)

Composition per liter:

NaCl	200.0g
$MgSO_4 \cdot 7H_2O$	20.0g

Agar	20.0g
Yeast extract	5.0g
Casamino acids	5.0g
Na_3-citrate	3.0g
KCl	2.0g
Na-glutamate	1.0g
$FeCl_2 \cdot 4H_2O$	36.0mg
$MnCl_2 \cdot 4H_2O$	0.36mg

pH 7.1 ± 0.2 at 25°C

Preparation of Medium: Add components to distilled/deionized water and bring volume to 1.0L. Mix thoroughly. Gently heat and bring to boiling. Distribute into tubes or flasks. Autoclave for 15 min at 15 psi pressure–121°C. Pour into sterile Petri dishes or leave in tubes.

Use: For the isolation and cultivation of *Halorubrum* spp., *Haloarcula* spp., *Haloferax* spp., *Halococcus* spp., *Haloterrigena* spp., *Halogeometricum borinquense, Natrialba* spp., and *Halomicrobium mukohataei.*

Halobacteria Medium

Composition per liter:

NaCl	220.0g
Agar	10.0g
$MgSO_4 \cdot 7H_2O$	10.0g
Casein hydrolysate	5.0g
KCl	5.0g
Disodium citrate	3.0g
KNO_3	1.0g
Yeast extract	1.0g
$CaCl_2 \cdot 6H_2O$	0.2g

pH 7.2–7.4 at 25°C

Preparation of Medium: Add components to distilled/deionized water and bring volume to 1.0L. Mix thoroughly. Gently heat until dissolved. Adjust pH to 7.2–7.4. Distribute into tubes or flasks. Autoclave for 15 min at 15 psi pressure–121°C.

Use: For the cultivation and enumeration of halobacteria.

Halobacteriaceae Medium 1

Composition per liter:

Salt, crude solar	250.0g
$MgSO_4 \cdot 7H_2O$	20.0g
KCl	5.0g
Pancreatic digest of casein	5.0g
Yeast extract	5.0g
$CaCl_2 \cdot 6H_2O$	0.2g

pH 7.0 ± 0.2 at 25°C

Preparation of Medium: Add components to distilled/deionized water and bring volume to 1.0L. Mix thoroughly. Gently heat until dissolved. Adjust pH to 7.0. Distribute into tubes or flasks. Autoclave for 15 min at 15 psi pressure–121°C.

Use: For the axenic cultivation of members of the Halobacteriaceae.

Halobacteriaceae Medium 2

Composition per liter:

NaCl	250.0g
$MgSO_4 \cdot 7H_2O$	20.0g
Yeast extract	10.0g
Casamino acids	7.5g
Trisodium citrate	3.0g
KCl	2.0g
$FeCl_2$	2.3mg

pH 7.5–7.8 at 25°C

Preparation of Medium: Add components to distilled/deionized water and bring volume to 1.0L. Mix thoroughly. Gently heat until dissolved. Adjust pH to 7.5–7.8. Distribute into tubes or flasks. Autoclave for 15 min at 15 psi pressure–121°C.

Use: For the axenic cultivation of halobacteria and halococci.

Halobacteriaceae Medium 3

Composition per liter:

NaCl	240.0g
L-Glutamine	15.0g
KCl	5.0g
K_2SO_4	5.0g
$MgCl_2 \cdot 6H_2O$	5.0g
$MgSO_4$, anhydrous	5.0g
NH_4Cl	5.0g
Pancreatic digest of casein	5.0g
Yeast extract	5.0g
K_2HPO_4	0.5g
L-Arginine	0.5g
L-Isoleucine	0.25g
L-Leucine	0.25g
L-Lysine	0.25g
L-Proline	0.25g
L-Valine	0.25g
Cytidylic acid	0.2g
$CaCl_2 \cdot 2H_2O$	0.1g
L-Methionine	0.1g
L-Tyrosine	0.1g
L-Phenylalanine	0.05g
$FeCl_2 \cdot 6H_2O$	5.0mg

pH 6.8 ± 0.2 at 25°C

Preparation of Medium: Add components to distilled/deionized water and bring volume to 1.0L. Mix thoroughly. Gently heat until dissolved. Adjust pH to 6.8. Distribute into tubes or flasks. Autoclave for 15 min at 15 psi pressure–121°C.

Use: For the cultivation of some halobacteria and halococci.

Halobacteriaceae Medium 4

Composition per liter:

NaCl	250.0g
$MgSO_4 \cdot 7H_2O$	20.0g
NH_4Cl	5.0g
L-Glutamic acid	1.3g
DL-Valine	1.0g
Glycerol	1.0g
L-Lysine	0.85g
L-Leucine	0.8g
DL-Serine	0.61g
DL-Threonine	0.5g
DL-Isoleucine	0.44g
DL-Alanine	0.43g
L-Arginine	0.4g
DL-Methionine	0.37g
DL-Phenylalanine	0.26g
L-Tyrosine	0.2g
Adenylic acid	0.1g
KNO_3	0.1g

Uridylic acid 0.1g
Glycine 0.06g
KH_2PO_4 .... 0.05g
K_2HPO_4 .... 0.05g
L-Cysteine 0.05g
L-Proline 0.05g
Sodium citrate 0.05g
$FeCl_2$ 2.3mg
$CaCl_2 \cdot 2H_2O$ 0.7mg
$ZnSO_4 \cdot 7H_2O$ 0.44mg
$MnSO_4 \cdot H_2O$ 0.3mg
$CuSO_4 \cdot 5H_2O$ 0.05mg

pH 6.2 ± 0.2 at 25°C

Preparation of Medium: Add components to distilled/deionized water and bring volume to 1.0L. Mix thoroughly. Gently heat until dissolved. Adjust pH to 6.2. Distribute into tubes or flasks. Autoclave for 15 min at 15 psi pressure–121°C.

Use: For the cultivation of members of the Halobacteriaceae.

Halobacterium Agar

Composition per liter:

NaCl 250.0g
Agar 20.0g
$MgSO_4 \cdot 7H_2O$ 20.0g
Yeast extract 10.0g
Casamino acids 7.5g
Trisodium citrate 3.0g
KCl 2.0g
$FeSO_4 \cdot 7H_2O$ 0.05g
$MnSO_4 \cdot H_2O$ 0.2mg

pH 7.4 ± 0.2 at 25°C

Preparation of Medium: Add components, except agar, to distilled/deionized water and bring volume to 1.0L. Mix thoroughly. Add agar. Gently heat and bring to boiling. Distribute into tubes or flasks. Autoclave for 15 min at 15 psi pressure–121°C. Pour into sterile Petri dishes or leave in tubes.

Use: For the cultivation and maintenance of *Haloarcula* species, *Halobacterium* species, *Halococcus morrhuae, Haloferax mediterranei,* and *Haloferax volcanii.*

Halobacterium denitrificans Medium

Composition per liter:

NaCl 176.0g
Agar 20.0g
$MgCl_2 \cdot 6H_2O$ 20.0g
HEPES (*N*-2-hydroxyethylpiperazine-*N*′-2-ethanesulfonic acid) buffer 11.9g
Yeast extract 5.0g
Hy-Case SF (Humko-Sheffield) 2.0g
KCl 2.0g
$CaCl_2 \cdot 2H_2O$ 0.1g

pH 6.7 ± 0.2 at 25°C

Preparation of Medium: Add components to distilled/deionized water and bring volume to 1.0L. Mix thoroughly. Gently heat and bring to boiling. Adjust pH to 6.7. Distribute into tubes or flasks. Autoclave for 15 min at 15 psi pressure–121°C. Pour into sterile Petri dishes or leave in tubes.

Use: For the aerobic cultivation and maintenance of *Haloferax (Halobacterium) denitrificans.*

Halobacterium denitrificans Medium

Composition per liter:

NaCl 176.0g
Agar 20.0g
$MgCl_2 \cdot 6H_2O$ 20.0g
HEPES (*N*-2-hydroxyethylpiperazine-*N*′-2-ethanesulfonic acid) buffer 11.9g
KNO_3 .... 5.0g
Yeast extract 5.0g
Hy-Case SF (Humko-Sheffield) 2.0g
KCl 2.0g
$CaCl_2 \cdot 2H_2O$ 0.1g

pH 6.7 ± 0.2 at 25°C

Preparation of Medium: Add components to distilled/deionized water and bring volume to 1.0L. Mix thoroughly. Gently heat and bring to boiling. Adjust pH to 6.7. Distribute into tubes or flasks. Autoclave for 15 min at 15 psi pressure–121°C. Pour into sterile Petri dishes or leave in tubes.

Use: For the anerobic cultivation and maintenance of *Haloferax (Halobacterium) denitrificans.*

Halobacterium halobium Defined Medium

Composition per liter:

NaCl 250.0g
$MgSO_4 \cdot 7H_2O$ 20.0g
L-Glutamic acid 1.3g
KCl 1.0g
Glycerol 1.0g
KCl 1.0g
L-Valine 1.0g
L-Lysine 0.85g
L-Leucine 0.8g
$CaCl_2 \cdot 7H_2O$ 0.71g
L-Serine 0.6g
Sodium citrate·$2H_2O$ 0.5g
L-Proline 0.5g
L-Threonine 0.5g
L-Alanine 0.43g
L-Arginine 0.4g
L-Methionine 0.37g
L-Phenylalanine 0.26g
L-Tyrosine 0.2g
KH_2PO_4 .... 0.15g
K_2HPO_4 .... 0.15g
KNO_3 .... 0.1g
KNO_3 .... 0.1g
Glycine 0.06g
L-Cysteine·HCl·H_2O 0.05g
L-Isoleucine 0.044g
$FeCl_2 \cdot 5H_2O$ 2.3mg
$ZnSO_4 \cdot 7H_2O$ 0.44mg
$CuSO_4 \cdot 5H_2O$ 0.005mg

Preparation of Medium: Add components to distilled/deionized water and bring volume to 1.0L. Mix thoroughly. Distribute into tubes or flasks. Autoclave for 15 min at 15 psi pressure–121°C.

Use: For the cultivation of *Halobacterium halobium.*

Halobacterium halobium/ Halobacterium salinarium Medium

Composition per liter:

NaCl	250.0g
$MgSO_4 \cdot 7H_2O$	20.0g
Pancreatic digest of casein	5.0g
Yeast extract	3.0g
Trisodium citrate·$2H_2O$	3.0g
KCl	2.0g
Trace metals solution	0.1mL

Trace Metals Solution:

Composition per liter:

$ZnSO_4 \cdot 7H_2O$	1.32g
$MnSO_4 \cdot H_2O$	0.34g
$Fe(NH_4)_2SO_4 \cdot 7H_2O$	0.78g
$CuSO_4 \cdot 5H_2O$	0.14g

Preparation of Trace Metals Solution: Add components to distilled/deionized water and bring volume to 1.0L. Mix thoroughly.

Preparation of Medium: Add components to distilled/deionized water and bring volume to 1.0L. Mix thoroughly. Distribute into tubes or flasks. Autoclave for 15 min at 15 psi pressure–121°C.

Use: For the cultivation of *Halobacterium halobium* and *Halobacterium salinarium.*

Halobacterium halobium/ Halobacterium salinarium Medium

Composition per liter:

NaCl	250.0g
$MgSO_4 \cdot 7H_2O$	20.0g
Trisodium citrate·$2H_2O$	3.0g
KCl	2.0g
Peptone	2.0g
Trace metals solution	0.1mL

Trace Metals Solution:

Composition per liter:

$ZnSO_4 \cdot 7H_2O$	1.32g
$Fe(NH_4)_2SO_4 \cdot 7H_2O$	0.78g
$MnSO_4 \cdot H_2O$	0.34g
$CuSO_4 \cdot 5H_2O$	0.14g

Preparation of Trace Metals Solution: Add components to distilled/deionized water and bring volume to 1.0L. Mix thoroughly.

Preparation of Medium: Add components to distilled/deionized water and bring volume to 1.0L. Mix thoroughly. Distribute into tubes or flasks. Autoclave for 15 min at 15 psi pressure–121°C.

Use: For the cultivation of *Halobacterium halobium* and *Halobacterium salinarium.*

Halobacterium/Halococcus Medium

Composition per liter:

Solution A	500.0mL
Salt solution	500.0mL

Solution A:

Composition per 500.0mL:

Skim milk	50.0g

Preparation of Solution A: Add 50.0g of skim milk to distilled/deionized water and bring volume to 500.0mL. Mix thoroughly.

Salt Solution:

Composition per 500.0mL:

$MgSO_4 \cdot 7H_2O$	10.0g
KNO_3	2.0g

Preparation of Salt Solution: Add components to distilled/deionized water and bring volume to 500.0mL. Mix thoroughly.

Preparation of Medium: Add components to distilled/deionized water and bring volume to 1.0L. Mix thoroughly. Distribute into tubes or flasks. Autoclave for 15 min at 15 psi pressure–121°C.

Use: For the cultivation of *Haloarcula vallismortis, Halobacterium cutirubrum, Halobacterium halobium, Halobacterium saccharovorum, Halobacterium salinarium,* and *Halococcus morrhuae.*

Halobacterium lacusprofundii Medium (DSMZ Medium 589)

Composition per liter:

NaCl	181.0g
$MgCl_2 \cdot 6H_2O$	75.0g
Na-succinate	10.0g
$MgSO_4 \cdot 7H_2O$	7.4g
KCl	7.4g
$CaCl_2 \cdot 2H_2O$	1.0g
Yeast extract	1.0g
Vitamin solution	1.0mL

pH 7.4 ± 0.2 at 25°C

Vitamin Solution:

Composition per 100.0mL:

Biotin	10.0mg
Vitamin B_{12}	10.0mg
Thiamine-HCl·$2H_2O$	10.0mg

Preparation of Vitamin Solution: Add components to distilled/deionized water and bring volume to 100.0mL. Mix thoroughly. Filter sterilize.

Preparation of Medium: Add components, except vitamin solution, to distilled/deionized water and bring volume to 999.0mL. Mix thoroughly. Autoclave for 15 min at 15 psi pressure–121°C. Cool to 25°C. Aseptically add 1.0mL vitamin solution. Mix thoroughly. Aseptically distribute into sterile tubes or flasks.

Use: For the cultivation of *Halorubrum lacusprofundii.*

Halobacterium Medium

Composition per liter:

Solution 1	500.0mL
Solution 2	500.0mL

pH 7.0 ± 0.2 at 25°C

Solution 1:

Composition per 500.0mL:

Yeast extract	10.0g
Pancreatic digest of casein	2.5g

Preparation of Solution 1: Add components to distilled/deionized water and bring volume to 500.0mL. Mix thoroughly. Gently heat and bring to boiling. Adjust pH to 7.0. Autoclave for 15 min at 15 psi pressure–121°C. Cool to 45°–50°C.

Solution 2:

Composition per 500.0mL:

NaCl	250.0g
$MgSO_4 \cdot 7H_2O$	10.0g

KCl	5.0g
$CaCl_2 \cdot 6H_2O$	0.2g

Preparation of Solution 2: Add components to distilled/deionized water and bring volume to 500.0mL. Mix thoroughly. Gently heat and bring to boiling. Autoclave for 15 min at 15 psi pressure–121°C. Cool to 45°–50°C.

Preparation of Medium: Aseptically combine sterile solution 1 and sterile solution 2. Mix thoroughly. Aseptically distribute into sterile tubes or flasks.

Use: For the cultivation of *Halobacterium salinarium*.

Halobacterium Medium (ATCC Medium 213)

Composition per liter:

Solution 1	500.0mL
Solution 2	500.0mL

pH 7.0 ± 0.2 at 25°C

Solution 1:

Composition per 500.0mL:

Yeast extract	10.0g
Pancreatic digest of casein	2.5g

Preparation of Solution 1: Add components to distilled/deionized water and bring volume to 500.0mL. Mix thoroughly. Gently heat and bring to boiling. Adjust pH to 7.0. Autoclave for 15 min at 15 psi pressure–121°C. Cool to 45°–50°C.

Solution 2:

Composition per 500.0mL:

NaCl	250.0g
Agar	20.0g
$MgSO_4 \cdot 7H_2O$	10.0g
KCl	5.0g
$CaCl_2 \cdot 6H_2O$	0.2g

Preparation of Solution 2: Add components to distilled/deionized water and bring volume to 500.0mL. Mix thoroughly. Gently heat and bring to boiling. Autoclave for 15 min at 15 psi pressure–121°C. Cool to 45°–50°C.

Preparation of Medium: Aseptically combine sterile solution 1 and sterile solution 2. Mix thoroughly. Aseptically pour into sterile Petri dishes or distribute into sterile tubes.

Use: For the cultivation of *Halobacterium salinarium*.

Halobacterium Medium (ATCC Medium 974)

Composition per 100.0mL:

Solution 1	75.0mL
Solution 2	25.0mL

pH 6.8 ± 0.2 at 25°C

Solution 1:

Composition per 75.0mL:

NaCl	12.5g
$MgCl_2 \cdot 6H_2O$	5.0g
K_2SO_4	0.5g
$CaCl_2 \cdot 6H_2O$	0.02g

Preparation of Solution 1: Add components to distilled/deionized water and bring volume to 75.0mL. Mix thoroughly. Adjust pH to 6.8. Autoclave for 15 min at 15 psi pressure–121°C. Cool to 45°–50°C.

Solution 2:

Composition per 25.0mL:

Agar	2.0g
Pancreatic digest of casein	0.5g
Yeast extract	0.5g

Preparation of Solution 2: Add components to distilled/deionized water and bring volume to 25.0mL. Mix thoroughly. Adjust pH to 6.8. Autoclave for 15 min at 15 psi pressure–121°C. Cool to 45°–50°C.

Preparation of Medium: Aseptically combine sterile solution 1 and sterile solution 2. Mix thoroughly. Pour into sterile Petri dishes or distribute into sterile tubes.

Use: For the cultivation and maintenance of *Haloferax volcanii*.

Halobacterium Medium (ATCC Medium 1176)

Composition per liter:

NaCl	156.0g
$MgSO_4 \cdot 7H_2O$	20.0g
$MgCl_2 \cdot 6H_2O$	13.0g
Yeast extract	5.0g
KCl	4.0g
$CaCl_2 \cdot 6H_2O$	1.0g
Glucose	1.0g
NaBr	0.5g
$NaHCO_3$	0.2g

pH 7.0 ± 0.2 at 25°C

Preparation of Medium: Add components to distilled/deionized water and bring volume to 1.0L. Mix thoroughly. Distribute into tubes or flasks. Autoclave for 15 min at 15 psi pressure–121°C.

Use: For the cultivation of *Haloferax mediterranei*.

Halobacterium Medium (ATCC Medium 1270)

Composition per liter:

NaCl	194.0g
$MgSO_4$	24.0g
$MgCl_2$	16.0g
KCl	5.0g
Yeast extract	5.0g
$CaCl_2$	1.0g
NaBr	0.5g
$NaHCO_3$	0.2g

pH 7.3 ± 0.2 at 25°C

Preparation of Medium: Add components to distilled/deionized water and bring volume to 1.0L. Mix thoroughly. Distribute into tubes or flasks. Autoclave for 15 min at 15 psi pressure–121°C.

Use: For the cultivation of *Haloarcula hispanica* and *Haloferax gibbonsii*.

Halobacterium pharaonis Medium

Composition per liter:

NaCl	250.0g
Agar	20.0g
Casamino acids	15.0g
Trisodium citrate·$2H_2O$	3.0g
Glutamic acid	2.5g

$MgSO_4·7H_2O$	2.5g
KCl	2.0g

pH 8.5 ± 0.2 at 25°C

Preparation of Medium: Add components to distilled/deionized water and bring volume to 1.0L. Mix thoroughly. Gently heat and bring to boiling. Adjust pH to 6.0. Autoclave for 15 min at 15 psi pressure–121°C. Cool to 50°C. Readjust pH to 8.5. Pour into sterile Petri dishes or distribute into sterile tubes.

Use: For the cultivation and maintenance of *Natronobacterium (Halobacterium) pharaonis.*

Halobacterium saccharovorum Medium

Composition per liter:

NaCl	250.0g
$MgCl·6H_2O$	20.0g
Glucose	10.0g
Casimino acids	5.0g
Yeast extract	2.5g
KCl	2.0g
$CaCl_2·7H_2O$	0.2g

pH 7.35 ± 0.2 at 25°C

Preparation of Medium: Add components to distilled/deionized water and bring volume to 1.0L. Mix thoroughly. Distribute into tubes or flasks. Autoclave for 15 min at 15 psi pressure–121°C.

Use: For the cultivation of *Halobacterium saccharovorum.*

Halobacterium soldomense Medium

Composition per liter:

$MgCl_2·6H_2O$	160.0g
NaCl	125.0g
K_2SO_4	5.0g
Soluble starch	2.0g
Peptone	1.0g
Yeast extract	1.0g
$CaCl_2·2H_2O$	0.13g

pH 7.0 ± 0.2 at 25°C

Preparation of Medium: Add components to distilled/deionized water and bring volume to 1.0L. Mix thoroughly. Distribute into tubes or flasks. Autoclave for 15 min at 15 psi pressure–121°C.

Use: For the cultivation of *Halobacterium soldomense.*

Halobacterium Starch Medium

Composition per liter:

$MgCl_2·6H_2O$	160.0g
NaCl	125.0g
K_2SO_4	5.0g
Soluble starch	2.0g
Peptone	1.0g
Yeast extract	1.0g
$CaCl_2·2H_2O$	0.13g

pH 7.0 ± 0.2 at 25°C

Preparation of Medium: Add components to distilled/deionized water and bring volume to 1.0L. Mix thoroughly. Distribute into tubes or flasks. Autoclave for 15 min at 15 psi pressure–121°C.

Use: For the cultivation and maintenance of *Haloarcula marismortui.*

Halobacterium volcanii Medium

Composition per 100.0mL:

NaCl	25.0g
$MgSO_4·7H_2O$	1.0g
KCl	0.5g
Glycine	0.2g
$CaCl_2·6H_2O$	0.02g
Yeast autolysate	1.0mL

pH 7.0 ± 0.2 at 25°C

Preparation of Medium: Add components to distilled/deionized water and bring volume to 100.0mL. Mix thoroughly. Adjust pH to 7.0. Distribute into tubes or flasks. Autoclave for 15 min at 15 psi pressure–121°C.

Use: For the specific enrichment of *Halobacterium volcanii.*

Halobacterium volcanii Medium

Composition per liter:

NaCl	125.0g
$MgCl_2·6H_2O$	50.0g
K_2SO_4	5.0g
Pancreatic digest of casein	5.0g
Yeast extract	5.0g
$CaCl_2·6H_2O$	0.2g

pH 6.8 ± 0.2 at 25°C

Preparation of Medium: Add components to distilled/deionized water and bring volume to 1.0L. Mix thoroughly. Gently heat until dissolved. Adjust pH to 6.8. Distribute into tubes or flasks. Autoclave for 15 min at 15 psi pressure–121°C.

Use: For the cultivation of *Halobacterium volcanii.*

Halobacteroides acetoethylicus Medium

Composition per liter:

NaCl	100.0g
Pancreatic digest of casein	10.0g
Yeast extract	3.0g
K_2HPO_4	1.5g
NH_4Cl	0.9g
KH_2PO_4	0.75g
$MgCl_2·6H_2O$	0.2g
Glucose solution	25.0mL
$Na_2S·9H_2O$ (10% solution)	10.0mL
Trace elements solution	9.0mL
Wolfe's Vitamin solution	5.0mL
Resazurin (0.025% solution)	4.0mL
$FeSO_4·7H_2O$ (10% solution)	0.03mL

pH 7.3 ± 0.2 at 25°C

Glucose Solution:

Composition per 100.0mL:

Glucose	20.0g

Preparation of Glucose Solution: Add glucose to distilled/deionized water and bring volume to 100.0mL. Mix thoroughly. Filter sterilize. Sparge with 100% N_2.

Trace Elements Solution:

Composition per liter:

Nitrilotriacetic acid	12.5g
NaCl	1.0g
$FeCl_3·4H_2O$	0.2g
$MnCl_2·4H_2O$	0.1g

$CaCl_2 \cdot 2H_2O$	0.1g
$ZnCl_2$	0.1g
$CuCl_2$	0.02g
Na_2SeO_3	0.02g
$CoCl_2 \cdot 6H_2O$	0.017g
H_3BO_3	0.01g
$Na_2MoO_4 \cdot 2H_2O$	0.01g

Preparation of Trace Elements Solution: Add nitrilotriacetic acid to 500.0mL of distilled/deionized water. Adjust pH to 6.5 with KOH. Add remaining components. Add distilled/deionized water to 1.0L.

Wolfe's Vitamin Solution:
Composition per liter:

Pyridoxine·HCl	10.0mg
Thiamine·HCl	5.0mg
Riboflavin	5.0mg
Nicotinic acid	5.0mg
Calcium DL-pantothenate	5.0mg
p-Aminobenzoic acid	5.0mg
Thioctic acid	5.0mg
Biotin	2.0mg
Folic acid	2.0mg
Cyanocobalamin	100.0μg

Preparation of Wolfe's Vitamin Solution: Add components to distilled/deionized water and bring volume to 1.0L. Mix thoroughly.

Preparation of Medium: Add components, except glucose solution, to distilled/deionized water and bring volume to 975.0mL. Mix thoroughly. Autoclave for 15 min at 15 psi pressure–121°C. While still hot, aseptically add 25.0mL of the sterile glucose solution under 97% N_2 + 3% H_2. Adjust pH to 7.3 if necessary. Aseptically and anaerobically distribute into tubes. Cap with rubber stoppers.

Use: For the cultivation and maintenance of *Halobacteroides acetoethylicus*.

Halobacteroides/Haloincola Medium

Composition per liter:

NaCl	100.0g
Peptone	5.0g
$CaCl_2$	0.33g
KCl	0.33g
KH_2PO_4	0.33g
$MgCl_2$	0.33g
NH_4Cl	0.33g
Resazurin	2.0mg
Glucose solution	30.0mL
$NaHCO_3$ solution	10.0mL
$Na_2S \cdot 9H_2O$ solution	10.0mL
Wolfe's vitamin solution	10.0mL
Trace elements solution SL-10	1.0mL

pH 7.5 ± 0.2 at 25°C

Glucose Solution:
Composition per 30.0mL:

D-Glucose	5.0g

Preparation of Glucose Solution: Add glucose to distilled/deionized water and bring volume to 30.0mL. Mix thoroughly. Filter sterilize. Sparge with 100% N_2.

$NaHCO_3$ Solution:
Composition per 10.0mL:

$NaHCO_3$	1.5g

Preparation of $NaHCO_3$ Solution: Add $NaHCO_3$ to distilled/deionized water and bring volume to 10.0mL. Mix thoroughly. Filter sterilize. Sparge with 100% N_2.

$Na_2S \cdot 9H_2O$ Solution:
Composition per 10.0mL:

$Na_2S \cdot 9H_2O$	0.5g

Preparation of $Na_2S \cdot 9H_2O$ Solution: Add $Na_2S \cdot 9H_2O$ to distilled/deionized water and bring volume to 10.0mL. Mix thoroughly. Sparge with 100% N_2. Autoclave for 15 min at 15 psi pressure–121°C.

Wolfe's Vitamin Solution:
Composition per liter:

Pyridoxine·HCl	10.0mg
p-Aminobenzoic acid	5.0mg
Lipoic acid	5.0mg
Nicotinic acid	5.0mg
Riboflavin	5.0mg
Thiamine·HCl	5.0mg
Calcium DL-pantothenate	5.0mg
Biotin	2.0mg
Folic acid	2.0mg
Vitamin B_{12}	0.1mg

Preparation of Wolfe's Vitamin Solution: Add components to distilled/deionized water and bring volume to 1.0L. Mix thoroughly. Filter sterilize.

Trace Elements Solution SL-10:
Composition per liter:

$FeCl_2 \cdot 4H_2O$	1.5g
$CoCl_2 \cdot 6H_2O$	190.0mg
$MnCl_2 \cdot 4H_2O$	100.0mg
$ZnCl_2$	70.0mg
$Na_2MoO_4 \cdot 2H_2O$	36.0mg
$NiCl_2 \cdot 6H_2O$	24.0mg
H_3BO_3	6.0mg
$CuCl_2 \cdot 2H_2O$	2.0mg
HCl (25% solution)	10.0mL

Preparation of Trace Elements Solution SL-10: Add $FeCl_2 \cdot 4H_2O$ to 10.0mL of HCl solution. Mix thoroughly. Add distilled/deionized water and bring volume to 1.0L. Add remaining components. Mix thoroughly. Sparge with 100% N_2. Autoclave for 15 min at 15 psi pressure–121°C.

Preparation of Medium: Prepare and dispense medium under 80% N_2 + 20% CO_2. Add components, except glucose solution, $NaHCO_3$ solution, $Na_2S \cdot 9H_2O$ solution, and Wolfe's vitamin solution, to distilled/deionized water and bring volume to 940.0mL Mix thoroughly. Gently heat and bring to boiling. Cool to room temperature while sparging with 80% N_2 + 20% CO_2. Autoclave for 15 min at 15 psi pressure–121°C. Aseptically and anaerobically add 30.0mL of sterile glucose solution, 10.0mL of sterile $NaHCO_3$ solution, 10.0mL of sterile $Na_2S \cdot 9H_2O$ solution, and 10.0mL of sterile Wolfe's vitamin solution. Mix thoroughly. Check that final pH is 7.5.

Use: For the cultivation of *Haloanaerobium (Haloincola) saccharolyticum.*

Halobacteroides/Haloincola Medium

Composition per liter:

NaCl	150.0g
Peptone	5.0g
$CaCl_2$	0.33g

KCl	0.33g
KH_2PO_4	0.33g
$MgCl_2$	0.33g
NH_4Cl	0.33g
Resazurin	2.0mg
Glucose solution	30.0mL
$NaHCO_3$ solution	10.0mL
$Na_2S \cdot 9H_2O$ solution	10.0mL
Wolfe's vitamin solution	10.0mL
Trace elements solution SL-10	1.0mL

pH 7.5 ± 0.2 at 25°C

Glucose Solution:

Composition per 30.0mL:

D-Glucose	5.0g

Preparation of Glucose Solution: Add glucose to distilled/deionized water and bring volume to 30.0mL. Mix thoroughly. Filter sterilize. Sparge with 100% N_2.

$NaHCO_3$ Solution:

Composition per 10.0mL:

$NaHCO_3$	1.5g

Preparation of $NaHCO_3$ Solution: Add $NaHCO_3$ to distilled/deionized water and bring volume to 10.0mL. Mix thoroughly. Filter sterilize. Sparge with 100% N_2.

$Na_2S \cdot 9H_2O$ Solution:

Composition per 10.0mL:

$Na_2S \cdot 9H_2O$	0.5g

Preparation of $Na_2S \cdot 9H_2O$ Solution: Add $Na_2S \cdot 9H_2O$ to distilled/deionized water and bring volume to 10.0mL. Mix thoroughly. Sparge with 100% N_2. Autoclave for 15 min at 15 psi pressure–121°C.

Wolfe's Vitamin Solution:

Composition per liter:

Pyridoxine·HCl	10.0mg
p-Aminobenzoic acid	5.0mg
Lipoic acid	5.0mg
Nicotinic acid	5.0mg
Riboflavin	5.0mg
Thiamine·HCl	5.0mg
Calcium DL-pantothenate	5.0mg
Biotin	2.0mg
Folic acid	2.0mg
Vitamin B_{12}	0.1mg

Preparation of Wolfe's Vitamin Solution: Add components to distilled/deionized water and bring volume to 1.0L. Mix thoroughly. Filter sterilize.

Trace Elements Solution SL-10:

Composition per liter:

$FeCl_2 \cdot 4H_2O$	1.5g
$CoCl_2 \cdot 6H_2O$	190.0mg
$MnCl_2 \cdot 4H_2O$	100.0mg
$ZnCl_2$	70.0mg
$Na_2MoO_4 \cdot 2H_2O$	36.0mg
$NiCl_2 \cdot 6H_2O$	24.0mg
H_3BO_3	6.0mg
$CuCl_2 \cdot 2H_2O$	2.0mg
HCl (25% solution)	10.0mL

Preparation of Trace Elements Solution SL-10: Add $FeCl_2 \cdot 4H_2O$ to 10.0mL of HCl solution. Mix thoroughly. Add distilled/deionized water and bring volume to 1.0L. Add remaining components. Mix thoroughly. Sparge with 100% N_2. Autoclave for 15 min at 15 psi pressure–121°C.

Preparation of Medium: Prepare and dispense medium under 80% N_2 + 20% CO_2. Add components, except glucose solution, $NaHCO_3$ solution, $Na_2S \cdot 9H_2O$ solution, and Wolfe's vitamin solution, to distilled/deionized water and bring volume to 940.0mL Mix thoroughly. Gently heat and bring to boiling. Cool to room temperature while sparging with 80% N_2 + 20% CO_2. Autoclave for 15 min at 15 psi pressure–121°C. Aseptically and anaerobically add 30.0mL of sterile glucose solution, 10.0mL of sterile $NaHCO_3$ solution, 10.0mL of sterile $Na_2S \cdot 9H_2O$ solution, and 10.0mL of sterile Wolfe's vitamin solution. Mix thoroughly. Check that final pH is 7.5.

Use: For the cultivation of *Haloabacteroides acunaris.*

Halobacteroides Medium

Composition per 990.0mL:

NaCl	88.0g
$MgCl_2 \cdot 6H_2O$	20.0g
$CaCl_2 \cdot 2H_2O$	7.4g
Yeast extract	5.0g
KCl	3.7g
L-Cysteine·HCl·H_2O	0.5g
Resazurin	1.0mg
Glucose (10% solution)	50.0mL
Sodium PIPES (piperazine-*N,N′*-bis-2-ethane sulfonate buffer, 1*M*, pH 6.8–7.0)	40.0mL

pH 6.8–7.0 at 25°C

Preparation of Medium: Filter sterilize glucose solution and PIPES buffer solution separately. Add remaining components—except glucose solution, PIPES buffer solution, and L-cysteine·HCl·H_2O—to distilled/deionized water and bring volume to 900.0mL. Mix thoroughly. Gently heat and bring to boiling under 100% N_2. Add L-cysteine·HCl·H_2O. Mix thoroughly. Autoclave for 15 min at 15 psi pressure–121°C. Cool to 45°–50°C. Aseptically add 50.0mL of sterile glucose solution and 40.0mL of sterile PIPES buffer solution. Mix thoroughly. Aseptically and anaerobically distribute into sterile tubes or flasks.

Use: For the cultivation and maintenance of *Halobacteroides halobius* and *Sporohalobacter marismortui.*

Halobacteroides Medium

Composition per liter:

NaCl	150.0g
Sucrose	5.0
$MgSO_4 \cdot 7H_2O$	4.0g
$NaHCO_3$	2.0g
$Na_2S \cdot 9H_2O$	0.5g
Yeast extract	0.5g
$CaCl_2 \cdot 2H_2O$	0.33g
KCl	0.33g
KH_2PO_4	0.33g
$MgCl_2 \cdot 6H_2O$	0.33g
NH_4Cl	0.33g
Resazurin	2.0mg
Wolfe's mineral solution	10.0mL
Wolfe's vitamin solution	10.0mL

pH 7.0 ± 0.2 at 25°C

Wolfe's Mineral Solution:

Composition per liter:

$MgSO_4 \cdot 7H_2O$	3.0g
Nitrilotriacetic acid	1.5g

NaCl	1.0g
$MnSO_4 \cdot 2H_2O$	0.5g
$CoCl_2 \cdot 6H_2O$	0.1g
$ZnSO_4 \cdot 7H_2O$	0.1g
$CaCl_2 \cdot 2H_2O$	0.1g
$FeSO_4 \cdot 7H_2O$	0.1g
$NiCl_2 \cdot 6H_2O$	0.025g
$KAl(SO_4)_2 \cdot 12H_2O$	0.02g
$CuSO_4 \cdot 5H_2O$	0.01g
H_3BO_3	0.01g
$Na_2MoO_4 \cdot 2H_2O$	0.01g
$Na_2SeO_3 \cdot 5H_2O$	0.3mg

Preparation of Wolfe's Mineral Solution: Add nitrilotriacetic acid to 500.0mL of distilled/deionized water. Adjust pH to 6.5 with KOH. Add remaining components one at a time. Add distilled/deionized water to 1.0L. Adjust pH to 6.8.

Wolfe's Vitamin Solution:
Composition per liter:

Pyridoxine·HCl	10.0mg
p-Aminobenzoic acid	5.0mg
Lipoic acid	5.0mg
Nicotinic acid	5.0mg
Riboflavin	5.0mg
Thiamine·HCl	5.0mg
Calcium DL-pantothenate	5.0mg
Biotin	2.0mg
Folic acid	2.0mg
Vitamin B_{12}	0.1mg

Preparation of Wolfe's Vitamin Solution: Add components to distilled/deionized water and bring volume to 1.0L. Mix thoroughly.

Preparation of Medium: Prepare and dispense medium under 80% N_2 + 10% CO_2 + 10% H_2. Add components, except $NaHCO_3$ and $Na_2S \cdot 9H_2O$, to distilled/deionized water and bring volume to 1.0L. Mix thoroughly. Gently heat and bring to boiling. Continue boiling for 3 min. Cool to room temperature while sparging with 80% N_2 + 10% CO_2 + 10% H_2. Add $NaHCO_3$ and $Na_2S \cdot 9H_2O$. Mix thoroughly. Adjust pH to 7.0. Anaerobically distribute into tubes or flasks. Autoclave for 15 min at 15 psi pressure–121°C.

Use: For the cultivation of *Haloanaerobacter lacunaris.*

Halobaculum gomorrense Medium (DSMZ Medium 823)

Composition per liter:

$MgCl_2 \cdot 6H_2O$	160.0g
NaCl	125.0g
K_2SO_4	5.0g
Starch	2.0g
Yeast extract	1.0g
Casamino acids	1.0g
$CaCl_2 \cdot 2H_2O$	0.1g

pH 7.0 ± 0.2 at 25°C

Preparation of Medium: Add components to distilled/deionized water and bring volume to 1.0L. Mix thoroughly. Distribute into tubes or flasks. Autoclave for 15 min at 15 psi pressure–121°C.

Use: For the cultivation of *Halobaculum gomorrense (Haloferax gomorrae).*

Halobius Medium

Composition per liter:

NaCl	116.0g
Agar	20.0g
$MgSO_4 \cdot 7H_2O$	20.0g
Yeast extract	10.0g
Vitamin-free casamino acids	7.5g
Sodium citrate	3.0g
KCl	2.0g
$FeCl_2$	0.023g

pH 6.2 ± 0.2 at 25°C

Preparation of Medium: Add components to distilled/deionized water and bring volume to 1.0L. Mix thoroughly. Gently heat and bring to boiling. Distribute into tubes or flasks. Autoclave for 15 min at 15 psi pressure–121°C. Pour into sterile Petri dishes or leave in tubes.

Use: For the cultivation and maintenance of *Micrococcus halobius.*

Halocella cellulolytica Medium

Composition per liter:

NaCl	150.0g
Cellobiose or microcrystalline cellulose	5.0g
Yeast extract	2.0g
$MgCl_2 \cdot 6H_2O$	3.6g
$CaCl_2$	0.33g
KCl	0.33g
KH_2PO_4	0.33g
NH_4Cl	0.33g
Resazurin	2.0mg
$NaHCO_3$ solution	10.0mL
$Na_2S \cdot 9H_2O$ solution	10.0mL
Wolfe's vitamin solution	10.0mL
Trace elements solution SL-10	1.0mL

pH 7.0 ± 0.2 at 25°C

$NaHCO_3$ Solution:
Composition per 10.0mL:

$NaHCO_3$	2.5g

Preparation of $NaHCO_3$ Solution: Add $NaHCO_3$ to distilled/deionized water and bring volume to 10.0mL. Mix thoroughly. Filter sterilize. Sparge with 100% N_2.

$Na_2S \cdot 9H_2O$ Solution:
Composition per 10.0mL:

$Na_2S \cdot 9H_2O$	0.5g

Preparation of $Na_2S \cdot 9H_2O$ Solution: Add $Na_2S \cdot 9H_2O$ to distilled/deionized water and bring volume to 10.0mL. Mix thoroughly. Sparge with 100% N_2. Autoclave for 15 min at 15 psi pressure–121°C.

Wolfe's Vitamin Solution:
Composition per liter:

Pyridoxine·HCl	10.0mg
p-Aminobenzoic acid	5.0mg
Lipoic acid	5.0mg
Nicotinic acid	5.0mg
Riboflavin	5.0mg
Thiamine·HCl	5.0mg
Calcium DL-pantothenate	5.0mg
Biotin	2.0mg
Folic acid	2.0mg
Vitamin B_{12}	0.1mg

Preparation of Wolfe's Vitamin Solution: Add components to distilled/deionized water and bring volume to 1.0L. Mix thoroughly. Filter sterilize.

Trace Elements Solution SL-10:

Composition per liter:

$FeCl_2 \cdot 4H_2O$	1.5g
$CoCl_2 \cdot 6H_2O$	190.0mg
$MnCl_2 \cdot 4H_2O$	100.0mg
$ZnCl_2$	70.0mg
$Na_2MoO_4 \cdot 2H_2O$	36.0mg
$NiCl_2 \cdot 6H_2O$	24.0mg
H_3BO_3	6.0mg
$CuCl_2 \cdot 2H_2O$	2.0mg
HCl (25% solution)	10.0mL

Preparation of Trace Elements Solution SL-10: Add $FeCl_2 \cdot 4H_2O$ to 10.0mL of HCl solution. Mix thoroughly. Add distilled/deionized water and bring volume to 1.0L. Add remaining components. Mix thoroughly. Sparge with 100% N_2. Autoclave for 15 min at 15 psi pressure–121°C.

Preparation of Medium: Prepare and dispense medium under 80% N_2 + 20% CO_2. Add components, except $NaHCO_3$ solution, $Na_2S \cdot 9H_2O$ solution, and Wolfe's vitamin solution, to distilled/deionized water and bring volume to 970.0mL. Mix thoroughly. Gently heat and bring to boiling. Cool to room temperature while sparging with 80% N_2 + 20% CO_2. Autoclave for 15 min at 15 psi pressure–121°C. Aseptically and anaerobically add 10.0mL of sterile $NaHCO_3$ solution, 10.0mL of sterile $Na_2S \cdot 9H_2O$ solution, and 10.0mL of sterile Wolfe's vitamin solution. Mix thoroughly. Check that final pH is 7.0.

Use: For the cultivation of *Halocella cellulolytica.*

Halococcus Agar

Composition per liter:

Solution A	500.0mL
Solution B	100.0mL
Solution C	400.0mL

pH 8.4 ± 0.2 at 25°C

Solution A:

Composition per 500.0mL:

Skim milk powder	50.0g

Preparation of Solution A: Add skim milk powder to distilled/deionized water and bring volume to 500.0mL. Autoclave for 15 min at 8 psi pressure–112°C. Cool to 50°–55°C.

Solution B:

Composition per 100.0mL:

NaCl	200.0g
KNO_3	2.0g
$MgSO_4 \cdot 7H_2O$	10.0mg
Ferric citrate	1.0μg

Preparation of Solution B: Add components to distilled/deionized water and bring volume to 100.0mL. Autoclave for 15 min at 15 psi pressure–121°C. Cool to 50°–55°C.

Solution C:

Composition per 400.0mL:

Agar	25.0g
Glycerol	10.0g
Neopeptone	5.0g

Preparation of Solution C: Add components to distilled/deionized water and bring volume to 400.0mL. Gently heat and bring to boiling. Autoclave for 15 min at 15 psi pressure–121°C. Cool to 50–55°C.

Preparation of Medium: Aseptically combine 100.0mL of sterile solution B with 400.0mL of sterile solution C. Mix thoroughly. Adjust pH to 8.4. Aseptically add 500.0mL of sterile solution A. Mix thoroughly. Pour into sterile Petri dishes or distribute into sterile tubes.

Use: For the cultivation and maintenance of *Halococcus morrhuae.*

Halococcus dombrowskii Medium (DSMZ Medium 954)

Composition per liter:

NaCl	200.0g
Agar	20.0g
$MgCl_2 \cdot 6H_2O$	20.0g
TRIS	12.1g
Casamino acids	5.0g
Yeast extract	5.0g
KCl	2.0g
$CaCl_2 \cdot 2H_2O$	0.2g

pH 7.4 ± 0.2 at 25°C

Preparation of Medium: Add components, except agar, to distilled/deionized water and bring volume to 1.0L. Mix thoroughly. Adjust pH to 7.4. Add agar. Gently heat and bring to boiling. Distribute into tubes or flasks. Autoclave for 15 min at 15 psi pressure–121°C. Pour into sterile Petri dishes or leave in tubes.

Use: For the cultivation and maintenance of *Halococcus dombrowskii.*

Halodurans Medium

Composition per liter:

NaCl	150.0g
Agar	20.0g
$MgSO_4 \cdot 7H_2O$	20.0g
Yeast extract	10.0g
Casamino acids	7.5g
Sodium citrate	3.0g
KCl	2.0g
$FeCl_2$	0.023g

pH 7.0 ± 0.2 at 25°C

Preparation of Medium: Add components to distilled/deionized water and bring volume to 1.0L. Mix thoroughly. Gently heat and bring to boiling. Distribute into tubes or flasks. Autoclave for 15 min at 15 psi pressure–121°C. Pour into sterile Petri dishes or leave in tubes.

Use: For the cultivation and maintenance of *Micrococcus varians.*

Haloferax mediterranei Medium

Composition per liter:

NaCl	195.0g
$MgSO_4 \cdot 7H_2O$	49.4g
$MgCl_2 \cdot 6H_2O$	34.6g
Yeast extract	5.0g
$CaCl_2 \cdot 2H_2O$	0.92g
NaBr	0.58g
KCl	0.5g
$NaHCO_3$	0.17g

pH 7.2 ± 0.2 at 25°C

Preparation of Medium: Add components to distilled/deionized water and bring volume to 1.0L. Mix thoroughly. Distribute into tubes or flasks. Autoclave for 15 min at 15 psi pressure–121°C.

Use: For the cultivation of *Haloferax mediterranei.*

Haloferax mediterranei Minimal Medium I

Composition per liter:

NaCl	156.0g
$MgSO_4 \cdot 7H_2O$	20.0g
$MgCl_2 \cdot 6H_2O$	13.0g
Glucose	10.0g
KCl	4.0g
NaH_4Cl	2.0g
$CaCl_2 \cdot 2H_2O$	1.0g
NaBr	0.58g
KH_2PO_4	0.5g
$NaHCO_3$	0.2g
$FeCl_2 \cdot 6H_2O$	0.005g

pH 7.2 ± 0.2 at 25°C

Preparation of Medium: Add components to distilled/deionized water and bring volume to 1.0L. Mix thoroughly. Distribute into tubes or flasks. Autoclave for 15 min at 15 psi pressure–121°C.

Use: For the cultivation of *Haloferax mediterranei.*

Haloferax mediterranei Minimal Medium II

Composition per liter:

NaCl	160.0g
$MgCl_2 \cdot 7H_2O$	20.0g
Sodium glutamate·H_2O	20.0g
Sucrose	10.0g
KCl	4.0g

pH 7.2 ± 0.2 at 25°C

Preparation of Medium: Add components to distilled/deionized water and bring volume to 1.0L. Mix thoroughly. Distribute into tubes or flasks. Autoclave for 15 min at 15 psi pressure–121°C.

Use: For the cultivation of *Haloferax mediterranei.*

Haloferax sulfurifontis Medium (DSMZ Medium 1018)

Composition per liter:

Casitone	3.0g
$MgSO_4 \cdot 7H_2O$	2.0g
$CaCl_2 \cdot 2H_2O$	0.07g
Trace elements solution SL-4	1.0mL
Vitamin solution	1.0mL

pH 7.1 ± 0.2 at 25°C

Vitamin Solution:

Composition per 10.0mL:

Vitamin B_{12}	5.0mg

Preparation of Vitamin Solution: Add vitamin B_{12} to distilled/deionized water and bring volume to 10.0mL. Mix thoroughly. Filter sterilize.

Trace Elements Solution SL-4:

Composition per liter:

EDTA	0.5g
$FeSO_4 \cdot 7H_2O$	0.2g
Trace elements solution SL-6	100.0mL

Trace Elements Solution SL-6:

Composition per liter:

$MnCl_2 \cdot 4H_2O$	0.5g
H_3BO_3	0.3g
$CoCl_2 \cdot 6H_2O$	0.2g
$ZnSO_4 \cdot 7H_2O$	0.1g
$Na_2MoO_4 \cdot 2H_2O$	0.03g
$NiCl_2 \cdot 6H_2O$	0.02g
$CuCl_2 \cdot 2H_2O$	0.01g

Preparation of Trace Elements Solution SL-6: Add components to distilled/deionized water and bring volume to 1.0L. Mix thoroughly. Autoclave for 15 min at 15 psi pressure–121°C.

Preparation of Trace Elements Solution SL-4: Add components to distilled/deionized water and bring volume to 1.0L. Mix thoroughly. Autoclave for 15 min at 15 psi pressure–121°C.

Preparation of Medium: Add components, except vitamin solution, to distilled/deionized water and bring volume to 1.0L. Mix thoroughly. Adjust pH to 7.1. Gently heat while stirring and bring to boiling. Autoclave for 15 min at 15 psi pressure–121°C. Cool to room temperature. Aseptically add 1.0mL vitamin solution. Aseptically distribute into sterile tubes or flasks.

Use: For the cultivation of *Haloferax sulfurifontis.*

Haloferax volcanii Low-Salt Medium

Composition per liter:

NaCl	125.0g
$MgCl_2 \cdot 6H_2O$	45.0g
$MgSO_4 \cdot 7H_2O$	10.0g
KCl	10.0g
$CaCl_2 \cdot 2H_2O$	10.0g
Pancreatic digest of casein	3.0g
Yeast extract	1.34g

pH 7.2 ± 0.2 at 25°C

Preparation of Medium: Add components to distilled/deionized water and bring volume to 1.0L. Mix thoroughly. Distribute into tubes or flasks. Autoclave for 15 min at 15 psi pressure–121°C.

Use: For the cultivation of *Haloferax volcanii.*

Haloferax volcanii Medium

Composition per liter:

NaCl	206.0g
$MgSO_4 \cdot 7H_2O$	37.0g
KCl	3.7g
Yeast tryptone solution	100.0mL
Tris[hydroxymethyl]aminomethane·HCl (1*M* solution, pH 7.2)	50.0mL
$CaCl_2 \cdot 2H_2O$ solution	5.0mL
$MnCl_2 \cdot 6H_2O$ solution	1.7mL

pH 7.2 ± 0.2 at 25°C

Yeast Tryptone Solution:

Composition per 100.0mL:

Tryptone	5.0g
Yeast extract	3.0g

Preparation of Yeast Tryptone Solution: Add components to distilled/deionized water and bring volume to 100.0mL. Mix thoroughly.

$CaCl_2 \cdot 2H_2O$ Solution:
Composition per 100.0mL:
$CaCl_2 \cdot 2H_2O$ 10.0g

Preparation of $CaCl_2 \cdot 2H_2O$ Solution: Add $CaCl_2 \cdot 2H_2O$ to distilled/deionized water and bring volume to 100.0mL. Mix thoroughly.

$MnCl_2 \cdot 6H_2O$ Solution:
Composition per liter:
$MnCl_2 \cdot 6H_2O$ 75.0mg

Preparation of $MnCl_2 \cdot 6H_2O$ Solution: Add $MnCl_2 \cdot 6H_2O$ to distilled/deionized water and bring volume to 1.0L. Mix thoroughly.

Preparation of Medium: Add components to distilled/deionized water and bring volume to 1.0L. Mix thoroughly. Distribute into tubes or flasks. Autoclave for 15 min at 15 psi pressure–121°C.

Use: For the cultivation of *Haloferax volcanii.*

Haloferax volcanii Minimal Medium

Composition per liter:
NaCl 206.0g
$MgSO_4 \cdot 7H_2O$ 36.9g
Glycerol 45.0mL
Sodium succinate solution 5.0mL
KCl solution 5.0mL
$CaCl_2 \cdot 2H_2O$ solution 5.0mL
NH_4Cl solution 5.0mL
$MnCl_2 \cdot 6H_2O$ solution 1.7mL
K_2HPO_4 solution 2.0mL
Trace elements solution 1.0mL
pH 7.2 ± 0.2 at 25°C

Sodium Succinate Solution:
Composition per 100.0mL:
Sodium succinate 10.0g

Preparation of Sodium Succinate Solution: Add sodium succinate to distilled/deionized water and bring volume to 100.0mL. Mix thoroughly.

KCl Solution:
Composition per 100.0mL:
KCl 7.45g

Preparation of KCl Solution: Add KCl to distilled/deionized water and bring volume to 100.0mL. Mix thoroughly.

$CaCl_2 \cdot 2H_2O$ Solution:
Composition per 100.0mL:
$CaCl_2 \cdot 2H_2O$ 10.0g

Preparation of $CaCl_2 \cdot 2H_2O$ Solution: Add $CaCl_2 \cdot 2H_2O$ to distilled/deionized water and bring volume to 100.0mL. Mix thoroughly. Filter sterilize.

NH_4Cl Solution:
Composition per 100.0mL:
NH_4Cl 5.35g

Preparation of NH_4Cl Solution: Add NH_4Cl to distilled/deionized water and bring volume to 100.0mL. Mix thoroughly.

$MnCl_2 \cdot 6H_2O$ Solution:
Composition per liter:
$MnCl_2 \cdot 6H_2O$ 75.0mg

Preparation of $MnCl_2 \cdot 6H_2O$ Solution: Add $MnCl_2 \cdot 6H_2O$ to distilled/deionized water and bring volume to 1.0L. Mix thoroughly.

K_2HPO_4 Solution:
Composition per 100.0mL:
K_2HPO_4 .... 8.7g

Preparation of K_2HPO_4 Solution: Add K_2HPO_4 to distilled/deionized water and bring volume to 100.0mL. Mix thoroughly. Adjust pH to 7.0.

Trace Elements Solution:
Composition per 100.0mL:
$FeSO_4$ 334.0mg
$MnCl_2$ 36.0mg
$ZnSO_4$ 44.0mg
$CuSO_4$ 5.0mg

Preparation of Trace Elements Solution: Add components to distilled/deionized water and bring volume to 100.0mL. Mix thoroughly.

Preparation of Medium: Add components, except $CaCl_2$ solution, to distilled/deionized water and bring volume to 995.0mL. Mix thoroughly. Adjust pH to 7.2. Autoclave for 15 min at 15 psi pressure–121°C. Aseptically add 5.0mL of sterile $CaCl_2$ solution. Mix thoroughly. Aseptically distribute into sterile tubes or flasks.

Use: For the cultivation of *Haloferax volcanii.*

Halomethanococcus Medium (*Methanohalophilus* Medium)

Composition per 1030.0mL:
Trimethylamine·HCl 2.5g
Na_2CO_3 .... 2.0g
$NaHCO_3$ 2.0g
Casamino acids 0.5g
L-Cysteine·HCl·H_2O 0.5g
$Na_2S \cdot 9H_2O$ solution 0.5g
NH_4Cl 0.5g
Pancreatic digest of casein 0.5g
Yeast extract 0.5g
K_2HPO_4 .... 0.2g
Ammonium-2-mercaptoethanesulfonate 1.0mg
Artificial brine 1.0L
Wolfe's vitamin solution 10.0mL
Wolfe's mineral solution 10.0mL
Volatile acids solution 10.0mL
pH 7.1 ± 0.2 at 25°C

$Na_2S \cdot 9H_2O$ Solution:
Composition per 10.0mL:
$Na_2S \cdot 9H_2O$ 2.0g

Preparation of $Na_2S \cdot 9H_2O$ Solution: Add $Na_2S \cdot 9H_2O$ to distilled/deionized water and bring volume to 10.0mL. Mix thoroughly. Filter sterilize.

Artificial Brine:
Composition per liter:
NaCl 80.7g
$MgCl_2 \cdot 6H_2O$ 35.1g
Na_2SO_4 .... 12.9g
KCl 5.7g
$CaCl_2$ 0.55g
$LiCl_2$ 0.13g
H_3BO_3 .... 0.12g

Preparation of Artificial Brine: Add components to distilled/deionized water and bring volume to 1.0L. Mix thoroughly.

Wolfe's Vitamin Solution:

Composition per liter:

Pyridoxine·HCl	0.01g
Thiamine·HCl	5.0mg
Riboflavin	5.0mg
Nicotinic acid	5.0mg
Calcium pantothenate	5.0mg
p-Aminobenzoic acid	5.0mg
Thioctic acid	5.0mg
Biotin	2.0mg
Folic acid	2.0mg
Cyanocobalamin	0.1mg

Preparation of Wolfe's Vitamin Solution: Add components to distilled/deionized water and bring volume to 1.0L. Mix thoroughly. Filter sterilize. Store at 4°C.

Wolfe's Mineral Solution:

Composition per liter:

$MgSO_4·7H_2O$	3.0g
Nitrilotriacetic acid	1.5g
NaCl	1.0g
$MnSO_4·H_2O$	0.5g
$FeSO_4·7H_2O$	0.1g
$CoCl_2·6H_2O$	0.1g
$CaCl_2$	0.1g
$ZnSO_4·7H_2O$	0.1g
$CuSO_4·5H_2O$	0.01g
$AlK(SO_4)_2·12H_2O$	0.01g
H_3BO_3	0.01g
$Na_2MoO_4·2H_2O$	0.01g

Preparation of Wolfe's Mineral Solution: Add nitrilotriacetic acid to 500.0mL of distilled/deionized water. Dissolve by adjusting pH to 6.5 with KOH. Add remaining components. Add distilled/deionized water to 1.0L.

Volatile Acids Solution:

Composition per liter:

α-Methylbutyric acid	0.5mL
Isobutyric acid	0.5mL
Isovaleric acid	0.5mL
Valeric acid	0.5mL

Preparation of Volatile Acids Solution: Add components to distilled/deionized water and bring volume to 1.0L. Mix thoroughly.

Preparation of Medium: Combine components, except the L-cysteine·HCl·H_2O, trimethylamine·HCl, and $Na_2S·9H_2O$ solution. Mix thoroughly. Gently heat and bring to boiling. Add L-cysteine·HCl·H_2O. Mix thoroughly. Cool in an ice-water bath under 80% N_2 + 20% CO_2. Add trimethylamine·HCl. Mix thoroughly. Adjust pH to 7.1. Aseptically and anaerobically distribute into tubes in 10.0mL volumes under 80% N_2 + 20% CO_2. Autoclave for 15 min at 15 psi pressure–121°C. Immediately prior to inoculation, aseptically add 0.25mL of sterile $Na_2S·9H_2O$ solution to each tube. Mix thoroughly.

Use: For the cultivation and maintenance of *Methanohalophilus mahii.*

Halomicrobium katesii Medium (DSMZ Medium 1164)

Composition per liter:

NaCl	195.0g
$MgSO_4·7H_2O$	49.83g
Proteose peptone	5.0g
Yeast extract	5.0g
KCl	5.0g
Glucose	1.0g
$CaCl_2·2H_2O$	0.916g
NaBr	0.35g
$NaHCO_3$	0.166g
$FeCl_3·6H_2O$, 0.5% soltuion	7 drops

pH 7.2 ± 0.2 at 25°C

Preparation of Medium: Add components to distilled/deionized water and bring volume to 1.0L. Mix thoroughly. Adjust pH to 7.2. Gently heat while stirring and bring to boiling. Distribute into tubes or flasks. Autoclave for 15 min at 15 psi pressure–121°C.

Use: For the cultivation of *Halomicrobium katesii.*

Halomonas desiderata Medium (DSMZ Medium 762)

Composition per liter:

Glucose	5.0g
KNO_3	2.0g
KH_2PO_4	1.0g
$MgCl_2·6H_2O$	0.2g
Carbonate solution	100.0mL

pH 9.5-10.0 at 25°C

Carbonate Solution

Composition per 100.0mL:

Na_2CO_3	5.4g
$NaHCO_3$	4.2g

Preparation of Carbonate Solution: Add components to distilled/deionized water and bring volume to 100.0mL. Mix thoroughly. Filter sterilize.

Preparation of Medium: Add components, except carbonate solution, to distilled/deionized water and bring volume to 900.0mL. Mix thoroughly. Autoclave for 15 min at 15 psi pressure–121°C. Cool to room temperature. Aseptically add 100.0mL sterile carbonate solution. Mix thoroughly. Aseptically distribute into sterile tubes or bottles.

Use: For the cultivation of *Halomonas desiderata.*

Halomonas desiderata Medium (LMG Medium 271)

Composition per liter:

Glucose	5.0g
KNO_3	2.0g
KH_2PO_4	1.0g
$MgCl_2·6H_2O$	0.2g
$Na_2CO_3/NaHCO_3$ solution	100.0mL

pH 9.5–10.0

$Na_2CO_3/NaHCO_3$ Solution:

Composition per 100.0mL:

Na_2CO_3	5.4g
$NaHCO_3$	4.2g

Preparation of $Na_2CO_3/NaHCO_3$ Solution: Add $NaHCO_3$ and Na_2CO_3 to 100.0mL of distilled/deionized water. Mix thoroughly. Filter sterilize.

Preparation of Medium: Add components, except Na_2CO_3/ $NaHCO_3$ solution, to distilled/deionized water and bring volume to 900.0mL. Mix thoroughly. Autoclave for 15 min at 15 psi pressure–121°C. Cool to 25°C. Aseptically add 100.0mL sterile Na_2CO_3/

$NaHCO_3$ solution. Pour into sterile Petri dishes or aseptically distribute into sterile tubes.

Use: For cultivation and maintenance of *Halomonas desiderata.*

Halomonas magadiensis Medium (DSMZ Medium 971)

Composition per liter:

Agar	20.0g
Glucose	10.0g
Peptone	5.0g
Yeast extract	5.0g
KH_2PO_4	1.0g
$MgSO_4 \cdot 7H_2O$	0.2g
NaCl solution	200.0mL
Na_2CO_3 solution	100.0mL

pH 10.0 ± 0.2 at 25°C

NaCl Solution:

Composition per 200.0mL:

NaCl	40.0g

Preparation of NaCl Solution: Add NaCl to distilled/deionized water and bring volume to 200.0mL. Mix thoroughly. Autoclave for 15 min at 15 psi pressure–121°C. Cool to 60°C.

Na_2CO_3 Solution:

Composition per 100.0mL:

Na_2CO_3	10.0g

Preparation of Na_2CO_3 Solution: Add Na_2CO_3 to distilled/deionized water and bring volume to 100.0mL. Mix thoroughly. Autoclave for 15 min at 15 psi pressure–121°C. Cool to 60°C.

Preparation of Medium: Add components, except NaCl solution and Na_2CO_3 solution, to distilled/deionized water and bring volume to 700.0mL. Mix thoroughly. Gently heat and bring to boiling. Autoclave for 15 min at 15 psi pressure–121°C. Cool to 60°C. Aseptically add 200.0mL NaCl solution and 100.0mL Na_2CO_3 solution. Mix thoroughly. Pour into sterile Petri dishes or distribute into sterile tubes.

Use: For the cultivation of *Halomonas magadiensis.*

Halomonas Medium

Composition per liter:

NaCl	80.0g
$MgSO_4 \cdot 7H_2O$	20.0g
Casamino acids with vitamins	7.5g
Proteose peptone No. 3	5.0g
Sodium citrate	3.0g
Yeast extract	1.0g
K_2HPO_4	0.5g
$Fe(NH_4)_2(SO_4)_2 \cdot 6H_2O$	0.05g

pH 7.0 ± 0.2 at 25°C

Preparation of Medium: Add components to distilled/deionized water and bring volume to 1.0L. Mix thoroughly. Adjust pH to 7.0 with KOH. Distribute into tubes or flasks. Autoclave for 15 min at 15 psi pressure–121°C.

Use: For the cultivation and maintenance of *Halomonas elongata.*

Halomonas pantelleriense Agar (DSMZ Medium 752)

Composition per liter:

NaCl	100.0g
Agar	20.0g
Yeast extract	10.0g
Na_2-citrate	3.0g
KCl	2.0g
$MgSO_4 \cdot 7H_2O$	1.0g
Na_2CO_3 solution	25.0mL
$MnCl_2$ solution	1.0mL
$FeSO_4$ solution	1.0mL

pH 9.0 ± 0.2 at 25°C

Na_2CO_3 Solution:

Composition per 25.0mL:

Na_2CO_3	2.5g

Preparation of Na_2CO_3 Solution: Add Na_2CO_3 to distilled/deionized water and bring volume to 25.0mL. Mix thoroughly.

$FeSO_4$ Solution:

Composition per 100.0mL:

$FeSO_4 \cdot 7H_2O$	5.0g

Preparation of $FeSO_4$ Solution: Add $FeSO_4 \cdot 7H_2O$ to distilled/deionized water and bring volume to 100.0mL. Mix thoroughly.

$MnCl_2$ Solution:

Composition per 100.0mL:

$MnCl_2 \cdot 4H_2O$	0.036g

Preparation of $MnCl_2$ Solution: Add $MnCl_2 \cdot 4H_2O$ to distilled/deionized water and bring volume to 100.0mL. Mix thoroughly.

Preparation of Medium: Add components, except Na_2CO_3 solution to distilled/deionized water and bring volume to 990.0L. Mix thoroughly. Adjust pH to 9.0. Autoclave for 15 min at 15 psi pressure–121°C. Aseptically add 10.0mL sterile Na_2CO_3 solution. Mix thoroughly. Pour into Petri dishes or distribute into sterile tubes.

Use: For the cultivation of *Halomonas pantelleriensis (Halomonas pantelleriense).*

Halomonas pantelleriense Medium (DSMZ Medium 752)

Composition per liter:

NaCl	100.0g
Yeast extract	10.0g
Na_2-citrate	3.0g
KCl	2.0g
$MgSO_4 \cdot 7H_2O$	1.0g
Na_2CO_3 solution	50.0mL
$MnCl_2$ solution	1.0mL
$FeSO_4$ solution	1.0mL

pH 9.0 ± 0.2 at 25°C

Na_2CO_3 Solution:

Composition per 50.0mL:

Na_2CO_3	2.5g

Preparation of Na_2CO_3 Solution: Add Na_2CO_3 to distilled/deionized water and bring volume to 50.0mL. Mix thoroughly. Filter sterilize.

$FeSO_4$ Solution:

Composition per 100.0mL:

$FeSO_4 \cdot 7H_2O$	5.0g

Preparation of $FeSO_4$ Solution: Add $FeSO_4 \cdot 7H_2O$ to distilled/deionized water and bring volume to 100.0mL. Mix thoroughly.

$MnCl_2$ Solution:

Composition per 100.0mL:

$MnCl_2 \cdot 4H_2O$	0.036g

Preparation of $MnCl_2$ Solution: Add $MnCl_2·4H_2O$ to distilled/deionized water and bring volume to 100.0mL. Mix thoroughly.

Preparation of Medium: Add components to distilled/deionized water and bring volume to 1.0L. Mix thoroughly. Adjust pH to 9.0. Distribute into tubes or flasks. Autoclave for 15 min at 15 psi pressure–121°C.

Use: For the cultivation of *Halomonas pantelleriensis (Halomonas pantelleriense).*

Halomonas pantelleriensis Medium (LMG Medium 268)

Composition per liter:

NaCl	100.0g
Agar	20.0g
Yeast extract	10.0g
Sodium citrate	3.0g
KCl	2.0g
$MgSO_4·7H_2O$	1.0g
Sodium carbonate solution	20-30mL
Manganese chloride solution	1.0mL
Iron sulfate solution	1.0mL

pH 9.0 ± 0.2 at 25°C

Manganese Chloride Solution:
Composition per 100.0mL:

$MnCl_2·4H_2O$	0.036g

Preparation of Manganese Chloride Solution: Add $MnCl_2·4H_2O$ to distilled/deionized water and bring volume to 100.0mL. Mix thoroughly.

Iron Sulfate Solution:
Composition per 10.0mL:

$FeSO_4·7H_2O$	0.5g

Preparation of Iron Sulfate Solution: Add $FeSO_4·7H_2O$ to distilled/deionized water and bring volume to 10.0mL. Mix thoroughly.

Sodium Carbonate Solution:
Composition per 100.0mL:

Na_2CO_3	10.0g

Preparation of Sodium Carbonate Solution: Add Na_2CO_3 to distilled/deionized water and bring volume to 100.0mL. Mix thoroughly.

Preparation of Medium: Add components, except sodium carbonate solution, to 950.0mL distilled/deionized water. Mix thoroughly. Gently heat and bring to boiling. Autoclave for 15 min at 15 psi pressure–121°C. Cool to 45°–50°C. Adjust pH to 9.0 by septically adding sterile sodium carbonate solution (20–30mL). Bring volume to 1.0L by adding sterile distilled/deionized water. Mix thoroughly. Pour into sterile Petri dishes or distribute into sterile tubes.

Use: For the cultivation of *Halomonas pantelleriensis.*

Halomonas subglaciescola Medium

Composition per liter:

Agar	25.0g
Peptone	10.0g
NaCl	10.0g
Beef extract	8.0g

Preparation of Medium: Add components to tap water and bring volume to 1.0L. Mix thoroughly. Gently heat and bring to boiling. Distribute into tubes or flasks. Autoclave for 15 min at 15 psi pressure–121°C. Pour into sterile Petri dishes or leave in tubes.

Use: For the cultivation and maintenance of *Halomonas subglaciescola.*

Halomonas subglaciescola Medium (Artificial Organic Lake Medium)

Composition per 1001.0mL:

NaCl	80.0g
$MgSO_4·7H_2O$	9.5g
Yeast extract	1.0g
KCl	0.5g
$CaCl_2·2H_2O$	0.2g
KNO_3	0.1g
$(NH_4)_2SO_4$	0.1g
Hutner's basal salts solution	20.0mL
Phosphate supplement	20.0mL
Vitamin solution	1.0mL

pH 8.0 ± 0.2 at 25°C

Hutner's Basal Salts Solution:
Composition per liter:

$MgSO_4·7H_2O$	29.7g
Nitrilotriacetic acid	10.0g
$CaCl_2·2H_2O$	3.335g
$FeSO_4·7H_2O$	99.0mg
$(NH_4)_6Mo_7O_{24}·4H_2O$	9.25mg
"Metals 44"	50.0mL

"Metals 44":
Composition per 100.0mL:

$ZnSO_4·7H_2O$	1.095g
$FeSO_4·7H_2O$	0.5g
Sodium EDTA	0.25g
$MnSO_4·H2O$	0.154g
$CuSO_4·5H_2O$	39.2mg
$Co(NO_3)_2·6H_2O$	24.8mg
$Na_2B_4O_7·10H_2O$	17.7mg

Preparation of "Metals 44": Add sodium EDTA to distilled/deionized water and bring volume to 90.0mL. Mix thoroughly. Add a few drops of concentrated H_2SO_4 to retard precipitation of heavy metal ions. Add remaining components. Mix thoroughly. Bring volume to 100.0mL with distilled/deionized water.

Preparation of Hutner's Basal Salts Solution: Add nitrilotriacetic acid to 500.0mL of distilled/deionized water. Adjust pH to 6.5 with KOH. Add remaining components. Add distilled/deionized water to 1.0L. Adjust pH to 6.8. Filter sterilize.

Phosphate Supplement:
Composition per 20.0mL:

K_2HPO_4	50.0mg
KH_2PO_4	50.0mg

Preparation of Phosphate Supplement: Add components to distilled/deionized water and bring volume to 20.0mL. Mix thoroughly. Filter sterilize.

Vitamin Solution:
Composition per liter:

Cyanocobalamin	10.0mg
Pyridoxine·HCl	10.0mg
Thiamine·HCl	10.0mg
Calcium DL-pantothenate	5.0mg
Nicotinamide	5.0mg

Biotin	2.0mg
Folic acid	2.0mg

Preparation of Vitamin Solution: Add components to distilled/deionized water and bring volume to 1.0L. Mix thoroughly. Filter sterilize.

Preparation of Medium: Add components, except Hutner's basal salts solution, phosphate supplement, and vitamin solution, to distilled/deionized water and bring volume to 960.0mL. Mix thoroughly. Adjust pH to 8.0. Autoclave for 15 min at 15 psi pressure–121°C. Aseptically add 20.0mL of sterile Hutner's basal salts solution, 20.0mL of sterile phosphate supplement, and 1.0mL of sterile vitamin solution. Mix thoroughly. Aseptically distribute into sterile tubes or flasks.

Use: For the cultivation of *Halomonas subglaciescola.*

Halonatronum saccharophilum Medium (DSMZ Medium 932)

Composition per liter:

Na_2CO_3	68.0g
NaCl	50.0g
$NaHCO_3$	38.0g
NH_4Cl	0.5g
KCl	0.2g
KH_2PO_4	0.2g
$MgCl_2$	0.1g
Resazurin	0.01g
Sucrose solution	50.0mL
$Na_2S·9H_2O$ solution	10.0mL
Yeast extract solution	10.0mL
Vitamin solution	10.0mL
Trace elements	1.0mL

pH 9.5–10.0 at 25°C

Sucrose Solution:

Composition per 50.0mL:

Sucrose	5.0g

Preparation of Sucrose Solution: Add sucrose to distilled/deionized water and bring volume to 50.0mL. Mix thoroughly. Sparge with 100% N_2. Autoclave for 15 min at 15 psi pressure–121°C. Cool to room temperature.

Yeast Extract Solution:

Composition per 10.0mL:

Yeast extract	0.2g

Preparation of Yeast Extract Solution: Add yeast extract to distilled/deionized water and bring volume to 10.0mL. Mix thoroughly. Sparge with 100% N_2. Autoclave under 100% N_2 for 15 min at 15 psi pressure–121°C. Cool to room temperature.

$Na_2S·9H_2O$ Solution:

Composition per 10.0mL:

$Na_2S·9H_2O$	0.7g

Preparation of $Na_2S·9H_2O$ Solution: Add $Na_2S·9H_2O$ to distilled/deionized water and bring volume to 10.0mL. Mix thoroughly. Autoclave under 100% N_2 for 15 min at 15 psi pressure–121°C. Cool to room temperature.

Trace Elements Solution:

Composition per liter:

$MgSO_4·7H_2O$	3.0g
Nitrilotriacetic acid	1.5g
NaCl	1.0g
$MnSO_4·2H_2O$	0.5g
$CoSO_4·7H_2O$	0.18g
$ZnSO_4·7H_2O$	0.18g
$CaCl_2·2H_2O$	0.1g
$FeSO_4·7H_2O$	0.1g
$NiCl_2·6H_2O$	0.025g
$KAl(SO_4)_2·12H_2O$	0.02g
H_3BO_3	0.01g
$Na_2MoO_4·4H_2O$	0.01g
$CuSO_4·5H_2O$	0.01g
$Na_2SeO_3·5H_2O$	0.3mg

Preparation of Trace Elements Solution: Add nitrilotriacetic acid to 500.0mL of distilled/deionized water. Dissolve by adjusting pH to 6.5 with KOH. Add remaining components. Add distilled/deionized water to 1.0L. Mix thoroughly.

Vitamin Solution:

Composition per liter:

Pyridoxine-HCl	10.0mg
Thiamine-HCl·$2H_2O$	5.0mg
Riboflavin	5.0mg
Nicotinic acid	5.0mg
D-Ca-pantothenate	5.0mg
p-Aminobenzoic acid	5.0mg
Lipoic acid	5.0mg
Biotin	2.0mg
Folic acid	2.0mg
Vitamin B_{12}	0.1mg

Preparation of Vitamin Solution: Add components to distilled/deionized water and bring volume to 1.0L. Mix thoroughly. Sparge with 80% H_2 + 20% CO_2. Filter sterilize.

Preparation of Medium: Prepare and dispense medium under 100% N_2 gas atmosphere. Add components, except $NaHCO_3$, NH_4Cl, Na_2CO_3, sucrose solution, $Na_2S·9H_2O$ solution, yeast extract solution, and vitamin solution, to distilled/deionized water and bring volume to 920.0mL. Mix thoroughly. Gently heat and bring to boiling. Boil for 5 min. Cool to room temperature while sparging with 100% N_2. Add solid $NaHCO_3$, NH_4Cl, and Na_2CO_3. Mix thoroughly. Distribute into anaerobe tubes or bottles. Autoclave for 15 min at 15 psi pressure–121°C. Aseptically and anaerobically add per liter of medium 50.0mL sucrose solution, 10.0mL yeast extract solution, 10.0mL $Na_2S·9H_2O$ solution, and 10.0mL vitamin solution. The final pH should be 9.5–10.0.

Use: For the cultivation of *Halonatronum saccharophilum.*

Halophile Agar

Composition per liter:

NaCl	156.0g
Agar	20.0g
$MgSO_4·7H_2O$	20.0g
$MgCl_2·6H_2O$	13.0g
Yeast extract	10.0g
KCl	4.0g
$CaCl_2·6H_2O$	1.0g
NaBr	0.5g
$NaHCO_3$	0.2g

pH 7.0 ± 0.2 at 25°C

Preparation of Medium: Add components to distilled/deionized water and bring volume to 1.0L. Mix thoroughly. Adjust pH to 7.0. Distribute into tubes or flasks. Autoclave for 15 min at 15 psi pressure–121°C.

Use: For the cultivation of *Haloarcula hispanica, Haloferax gibbonsii, Marinococcus halophilus, Planococcus* species, and *Sporosarcina halophila.*

Halophile Agar I

Composition per liter:

NaCl	250.0g
$MgSO_4 \cdot 7H_2O$	20.0g
Agar	15.0g
Yeast extract	10.0g
Casamino acids	7.5g
Trisodium citrate	3.0g
KCl	2.0g
$FeSO_4 \cdot 7H_2O$	0.05g
$MnSO_4 \cdot 4H_2O$	0.25mg

pH 7.4 ± 0.2 at 25°C

Preparation of Medium: Add components, except agar, to distilled/deionized water and bring volume to 1.0L. Mix thoroughly. Adjust pH to 7.4. Add agar. Gently heat and bring to boiling. Do not autoclave. Sterilize by steaming at 100°C for 30 min. Aseptically distribute into sterile tubes or flasks.

Use: For the cultivation and maintenance of *Halobacterium halobium, Halobacterium salinarium,* and *Halobacterium trapanicum.*

Halophile Medium

Composition per liter:

NaCl	30.0g
Agar	15.0g
Peptone	5.0g
Yeast extract	2.0g
Beef extract	1.0g

pH 7.4 ± 0.2 at 25°C

Preparation of Medium: Add components to distilled/deionized water and bring volume to 1.0L. Mix thoroughly. Gently heat and bring to boiling. Distribute into tubes or flasks. Autoclave for 15 min at 15 psi pressure–121°C. Pour into sterile Petri dishes or leave in tubes.

Use: For the cultivation and maintenance of *Bacillus tusciae, Chromobacterium maris-mortui, Flavobacterium tirrenicum, Flavobacterium uliginosum, Halomonas halmophila, Pseudomonas beijerinckii, Vibrio alginolyticus, Vibrio parahaemolyticus,* and *Vibrio proteolyticus.*

Halophile Medium

Composition per liter:

NaCl	100.0g
KCl	5.0g
$MgCl_2 \cdot 6H_2O$	5.0g
$MgSO_4 \cdot 7H_2O$	5.0g
NH_4Cl	5.0g
Peptone solution (15% solution)	30.0mL
Yeast extract solution (15% solution)	30.0mL
Ferric citrate solution (1% solution)	10.0mL
Trace elements solution	5.0mL

Trace Elements Solution:

Composition per liter:

$ZnSO_4 \cdot 7H_2O$	0.22g
$MgCl_2 \cdot 4H_2O$	0.18g
$CoCl_2 \cdot 6H_2O$	0.01g
$Na_2MoO_4 \cdot H_2O$	6.3mg
$CuSO_4 \cdot 5H_2O$	1.0mg

Preparation of Trace Elements Solution: Add components to distilled/deionized water and bring volume to 1.0L. Mix thoroughly.

Preparation of Medium: Add components to distilled/deionized water and bring volume to 1.0L. Mix thoroughly. Gently heat and bring to boiling. Distribute into tubes or flasks. Autoclave for 15 min at 15 psi pressure–121°C.

Use: For the cultivation of *Rhodospirillum salinarum.*

Halophile Medium III

Composition per liter:

NaCl	95.3g
$MgSO_4 \cdot 7H_2O$	81.3g
Glycerol	5.0g
NH_4Cl	1.0g
K_2HPO_4 solution	100.0mL
Wolfe's vitamin solution	10.0mL
Trace elements solution SL-10	1.0mL

pH 7.5 ± 0.2 at 25°C

Wolfe's Vitamin Solution:

Composition per liter:

Pyridoxine·HCl	10.0mg
Calcium D-(+)-pantothenate	5.0mg
Nicotinic acid	5.0mg
p-Aminobenzoic acid	5.0mg
Riboflavin	5.0mg
Thiamine·HCl	5.0mg
Thioctic acid	5.0mg
Biotin	2.0mg
Folic acid	2.0mg
Cyanocobalamin	100.0μg

Preparation of Wolfe's Vitamin Solution: Add components to distilled/deionized water and bring volume to 1.0L. Mix thoroughly. Filter sterilize.

Trace Elements Solution SL-10:

Composition per liter:

$FeCl_2 \cdot 4H_2O$	1.5g
$CoCl_2 \cdot 6H_2O$	190.0mg
$MnCl_2 \cdot 4H_2O$	100.0mg
$ZnCl_2$	70.0mg
$Na_2MoO_4 \cdot 2H_2O$	36.0mg
$NiCl_2 \cdot 6H_2O$	24.0mg
H_3BO_3	6.0mg
$CuCl_2 \cdot 2H_2O$	2.0mg
HCl (25% solution)	10.0mL

Preparation of Trace Elements Solution SL-10: Add $FeCl_2 \cdot 4H_2O$ to 10.0mL of HCl solution. Mix thoroughly. Add distilled/deionized water and bring volume to 1.0L. Add remaining components. Mix thoroughly. Sparge with 100% N_2. Autoclave for 15 min at 15 psi pressure–121°C.

K_2HPO_4 Solution:

Composition per 100.0mL:

K_2HPO_4	1.0g

Preparation of K_2HPO_4 Solution: Add K_2HPO_4 to distilled/deionized water and bring volume to 100.0mL. Mix thoroughly. Autoclave for 15 min at 15 psi pressure–121°C. Cool to 25°C.

Preparation of Medium: Add components, except K_2HPO_4 solution, Wolfe's vitamin solution, and trace elments solution SL-10, to distilled/deionized water and bring volume to 889.0mL. Mix thoroughly. Autoclave for 15 min at 15 psi pressure–121°C. Cool to 25°C. Aseptically add 100.0mL of sterile K_2HPO_4 solution, 10.0mL of sterile Wolfe's vitamin solution, and 1.0mL of sterile trace elements solution SL-10. Mix thoroughly. Aseptically adjust pH to 7.5. Aseptically distribute into sterile tubes or flasks.

Use: For the cultivation and maintenance of *Halovibrio variabilis.*

Halophilic Agar (HA)

Composition per liter:

NaCl	250.0g
$MgSO_4 \cdot 7H_2O$	25.0g
Agar	20.0g
Casamino acids	10.0g
Yeast extract	10.0g
Proteose peptone	5.0g
Trisodium citrate	3.0g
KCl	2.0g

pH 7.2 ± 0.2 at 25°C

Preparation of Medium: Combine the ingredients with distilled water and heat to boiling to dissolve completely. Autoclave at 121°C for 15 min.

Use: For the isolation and cultivation of halophilic microorganisms from foods, such as *Pseudomonas* species and *Flavobacterium* species from fish and salted foods.

Halophilic Broth (HB)

Composition per liter:

NaCl	250.0g
$MgSO_4 \cdot 7H_2O$	25.0g
Casamino acids	10.0g
Yeast extract	10.0g
Proteose peptone	5.0g
Trisodium citrate	3.0g
KCl	2.0g

pH 7.2 ± 0.2 at 25°C

Preparation of Medium: Add components to distilled/deionized water and bring volume to 1.0L. Mix thoroughly. Gently heat and bring to boiling. Distribute into tubes or flasks. Autoclave for 15 min at 15 psi pressure–121°C.

Use: For the isolation and cultivation of halophilic microorganisms from foods, such as *Pseudomonas* species and *Flavobacterium* species from fish and salted foods.

Halophilic *Chromatium* Medium

Composition per 1060.0mL:

NaCl	60.0g
KH_2PO_4	1.0g
NH_4Cl	1.0g
$MgCl_2 \cdot 6H_2O$	0.5g
Solution A	20.0mL
Solution B	20.0mL
Solution C	10.0mL
Solution D	10.0mL
Trace elements solution	1.0mL

pH 7.0 ± 0.2 at 25°C

Solution A:

Composition per 100.0mL:

$NaHCO_3$	10.0g

Preparation of Solution A: Add $NaHCO_3$ to distilled/deionized water and bring volume to 100.0mL. Mix thoroughly. Autoclave for 15 min at 15 psi pressure–121°C.

Solution B:

Composition per 100.0mL:

$Na_2S \cdot 9H_2O$	10.0g

Preparation of Solution B: Add $Na_2S \cdot 9H_2O$ to distilled/deionized water and bring volume to 100.0mL. Mix thoroughly. Autoclave for 15 min at 15 psi pressure–121°C.

Solution C:

Composition per 100.0mL:

$Na_2S_2O_3 \cdot 9H_2O$	10.0g

Preparation of Solution C: Add $Na_2S_2O_3 \cdot 9H_2O$ to distilled/deionized water and bring volume to 100.0mL. Mix thoroughly. Autoclave for 15 min at 15 psi pressure–121°C.

Solution D:

Composition per 100.0mL:

Sodium malate	10.0g

Preparation of Solution D: Add sodium malate to distilled/deionized water and bring volume to 100.0mL. Mix thoroughly. Autoclave for 15 min at 15 psi pressure–121°C.

Trace Elements Solution:

Composition per liter:

$FeCl_3 \cdot 6H_2O$	2.7g
H_3BO_3	0.1g
$ZnSO_4 \cdot 7H_2O$	0.1g
$Co(NO_3)_2 \cdot 6H_2O$	50.0mg
$CuSO_4 \cdot 5H_2O$	5.0mg
$MnCl_2 \cdot 6H_2O$	5.0mg

Preparation of Trace Elements Solution: Add components to distilled/deionized water and bring volume to 1.0L. Mix thoroughly.

Preparation of Medium: Add components, except solution A, solution B, solution C, and solution D, to distilled/deionized water and bring volume to 1.0L. Mix thoroughly. Bring pH to 7.0–7.2 with H_3PO_4. Distribute into tubes or flasks. Autoclave for 15 min at 15 psi pressure–121°C. Aseptically add 0.2mL of sterile solution A, 0.2mL of sterile solution B, 0.1mL of sterile solution C, and 0.1mL of sterile solution D for each 10.0mL of medium. Mix thoroughly. Use immediately.

Use: For the cultivation of halophilic *Chromatium* species.

Halophilic *Clostridium* Agar

Composition per liter:

L-Cysteine·HCl·H_2O	0.5g
Solution 1	1.0L
Solution 2	100.0mL

pH 6.2–7.0 at 25°C

Solution 1:

Composition per liter:

NaCl	105.0g
Agar	20.0g
KCl	7.5g
$CaCO_3$	5.0g
L-Glutamic acid	4.0g
Soluble starch	2.0g
Casamino acids	2.0g
Nutrient broth	2.0g
Yeast extract	2.0g
$FeSO_4 \cdot 7H_2O$	2.0mg

Resazurin 1.0mg
NaOH (2.5*N* solution) 12.5mL
Wolfe's vitamin solution 10.0mL
Wolfe's mineral solution 10.0mL

Preparation of Solution 1: Add components, except $CaCO_3$, to distilled/deionized water and bring volume to 1.0L. Mix thoroughly. Gently heat and bring to boiling. When all components have dissolved, add the $CaCO_3$. Mix thoroughly.

Wolfe's Vitamin Solution:
Composition per liter:
Pyridoxine·HCl 0.01g
Thiamine·HCl 5.0mg
Riboflavin 5.0mg
Nicotinic acid 5.0mg
Calcium pantothenate 5.0mg
p-Aminobenzoic acid 5.0mg
Thioctic acid 5.0mg
Biotin 2.0mg
Folic acid 2.0mg
Cyanocobalamin 0.1mg

Preparation of Wolfe's Vitamin Solution: Add components to distilled/deionized water and bring volume to 1.0L. Mix thoroughly.

Wolfe's Mineral Solution:
Composition per liter
$MgSO_4 \cdot 7H_2O$ 3.0g
Nitrilotriacetic acid 1.5g
NaCl 1.0g
$MnSO_4 \cdot H_2O$ 0.5g
$FeSO_4 \cdot 7H_2O$ 0.1g
$CoCl_2 \cdot 6H_2O$ 0.1g
$CaCl_2$ 0.1g
$ZnSO_4 \cdot 7H_2O$ 0.1g
$CuSO_4 \cdot 5H_2O$ 0.01g
$AlK(SO_4)_2 \cdot 12H_2O$ 0.01g
H_3BO_3 .... 0.01g
$Na_2MoO_4 \cdot 2H_2O$ 0.01g

Preparation of Wolfe's Mineral Solution: Add nitrilotriacetic acid to 500.0mL of distilled/deionized water. Dissolve by adjusting pH to 6.5 with KOH. Add remaining components. Add distilled/deionized water to 1.0L.

Solution 2:
Composition per 100.0mL:
$MgCl_2 \cdot 6H_2O$ 20.3g
$CaCl_2 \cdot 2H_2O$ 7.35g

Preparation of Solution 2: Add components to distilled/deionized water and bring volume to 100.0mL. Mix thoroughly. Gas with 100% N_2 for 20 min. Autoclave for 15 min at 15 psi pressure–121°C. Cool to 45°–50°C.

Preparation of Medium: Gently heat 1.0L of solution 1 and bring to boiling under 100% N_2. Add the L-cysteine·HCl·H_2O. Continue boiling until resazurin turns colorless, indicating reduction. The volume of solution 1 should be about 900.0mL. Anaerobically distribute into tubes in 9.0mL volumes under 100% N_2. Cap tubes with rubber stoppers. Place tubes in a press. Autoclave for 15 min at 15 psi pressure–121°C with fast exhaust. Cool to 50°C. Aseptically add 1.0mL of sterile solution 2 to each tube. In the presence of $CaCO_3$, the pH may be higher than 7.0. Do not adjust pH.

Use: For the cultivation and maintenance of *Sporohalobacter lortetii.*

Halophilic *Clostridium* Broth

Composition per liter:
L-Cysteine·HCl·H_2O 0.5g
Solution 1 .... 1.0L
Solution 2 .... 100.0mL

pH 6.2–7.0 at 25°C

Solution 1:
Composition per liter:
NaCl 105.0g
KCl 7.5g
L-Glutamic acid 4.0g
Casamino acids 2.0g
Nutrient broth 2.0g
Yeast extract 2.0g
$FeSO_4 \cdot 7H_2O$ 2.0mg
Resazurin 1.0mg
NaOH (2.5*N* solution) 12.5mL
Wolfe's vitamin solution 10.0mL
Wolfe's mineral solution 10.0mL

Preparation of Solution 1: Add components to distilled/deionized water and bring volume to 1.0L. Mix thoroughly.

Wolfe's Vitamin Solution:
Composition per liter:
Pyridoxine·HCl 0.01g
Thiamine·HCl 5.0mg
Riboflavin 5.0mg
Nicotinic acid 5.0mg
Calcium pantothenate 5.0mg
p-Aminobenzoic acid 5.0mg
Thioctic acid 5.0mg
Biotin 2.0mg
Folic acid 2.0mg
Cyanocobalamin 0.1mg

Preparation of Wolfe's Vitamin Solution: Add components to distilled/deionized water and bring volume to 1.0L. Mix thoroughly.

Wolfe's Mineral Solution:
Composition per liter:
$MgSO_4 \cdot 7H_2O$ 3.0g
Nitrilotriacetic acid 1.5g
NaCl 1.0g
$MnSO_4 \cdot H_2O$ 0.5g
$FeSO_4 \cdot 7H_2O$ 0.1g
$CoCl_2 \cdot 6H_2O$ 0.1g
$CaCl_2$ 0.1g
$ZnSO_4 \cdot 7H_2O$ 0.1g
$CuSO_4 \cdot 5H_2O$ 0.01g
$AlK(SO_4)_2 \cdot 12H_2O$ 0.01g
H_3BO_3 .... 0.01g
$Na_2MoO_4 \cdot 2H_2O$ 0.01g

Preparation of Wolfe's Mineral Solution: Add nitrilotriacetic acid to 500.0mL of distilled/deionized water. Dissolve by adjusting pH to 6.5 with KOH. Add remaining components. Add distilled/deionized water to 1.0L.

Solution 2:
Composition per 100.0mL:
$MgCl_2 \cdot 6H_2O$ 20.3g
$CaCl_2 \cdot 2H_2O$ 7.35g

Preparation of Solution 2: Add components to distilled/deionized water and bring volume to 100.0mL. Mix thoroughly. Gas with 100% N_2 for 20 min. Autoclave for 15 min at 15 psi pressure–121°C.

Preparation of Medium: Gently heat 1.0L of solution 1 and bring to boiling under 100% N_2. Add the L-cysteine·HCl·H_2O. Continue boiling until resazurin turns colorless, indicating reduction. The volume of solution 1 should be about 900.0mL. Anaerobically distribute into tubes in 9.0mL volumes under 100% N_2. Cap tubes with rubber stoppers. Place tubes in a press. Autoclave for 15 min at 15 psi pressure–121°C with fast exhaust. Cool to 25°C. Aseptically add 1.0mL of sterile solution 2 to each tube. Adjust pH to 6.2–7.0 if necessary with sterile O_2-free NaOH or HCl.

Use: For the cultivation and maintenance of *Sporohalobacter lortetii*.

Halophilic *Halobacterium* Medium

Composition per liter:

NaCl	200.0g
$MgSO_4 \cdot 7H_2O$	37.0g
$CaCl_2 \cdot 2H_2O$	0.7g
KCl	0.5g
$MnCl_2 \cdot 4H_2O$	0.05g
Yeast extract	100.0mL

pH 7.0 ± 0.2 at 25°C

Preparation of Medium: Add components to distilled/deionized water and bring volume to 1.0L. Mix thoroughly. Gently heat until dissolved. Adjust pH to 7.0. Distribute into tubes or flasks. Autoclave for 15 min at 15 psi pressure–121°C.

Use: For the cultivation of extremely halophilic *Halobacterium* species.

Halophilic HiVeg Agar

Composition per liter:

NaCl	250.0g
$MgSO_4$	25.0g
Agar	20.0g
Plant acid hydrolysate	10.0g
Yeast extract	10.0g
Plant peptone No. 3	5.0g
Na_3-citrate	3.0g
KCl	2.0g

pH 7.2 ± 0.2 at 25°C

Source: This medium is available as a premixed powder from HiMedia.

Preparation of Medium: Combine the ingredients with distilled water and heat to boiling to dissolve completely. Autoclave at 121°C for 15 min.

Use: For the isolation and cultivation of halophilic microorganisms from foods, such as *Pseudomonas* species and *Flavobacterium* species from fish and salted foods.

Halophilic HiVeg Broth

Composition per liter:

NaCl	250.0g
$MgSO_4$	25.0g
Plant acid hydrolysate	10.0g
Yeast extract	10.0g
Plant peptone No. 3	5.0g
KCl	2.0g
Na_3-citrate	3.0g

pH 7.2 ± 0.2 at 25°C

Source: This medium is available as a premixed powder from HiMedia.

Preparation of Medium: Add components to distilled/deionized water and bring volume to 1.0L. Mix thoroughly. Gently heat and bring to boiling. Distribute into tubes or flasks. Autoclave for 15 min at 15 psi pressure–121°C.

Use: For the isolation and cultivation of halophilic microorganisms from foods, such as *Pseudomonas* species and *Flavobacterium* species from fish and salted foods.

Halophilic Medium (DSMZ Medium 1125)

Composition per liter:

NaCl	180.0g
$MgCl_2 \cdot 6H_2O$	20.0g
HEPES (4-(2-hydroxyethyl)-1-piperazineethanesulfonic acid)	6.0g
K_2HPO_4	5.0g
NH_4Cl	0.5g
Sucrose	0.5g
$CaCl_2 \cdot 2H_2O$	0.1g
Yeast extract	0.1g
KH_2PO_4	0.05g

pH 6.5 ± 0.2 at 25°C

Preparation of Medium: Add components to distilled/deionized water and bring volume to 1.0L. Mix thoroughly. Adjust pH to 6.5. Gently heat while stirring and bring to boiling. Distribute into tubes or flasks. Autoclave for 15 min at 15 psi pressure–121°C.

Use: For the cultivation of *Haladaptatus paucihalophilus*.

Halophilic Methanotrophic Bacterium Medium (DSMZ Medium 1156)

Composition per liter:

NaCl	65.0g
$Na_2HPO_4 \cdot 2H_2O$	0.7g
NH_4Cl	0.5g
KH_2PO_4	0.3g
$MgSO_4 \cdot 7H_2O$	0.1g
$CaCl_2 \cdot 2H_2O$	0.01g
$FeSO_4 \cdot 7H_2O$	5.0mg
Methanol	5.0mL
Trace elements solution	1.0mL

pH 7.2 ± 0.2 at 25°C

Trace Elements Solution:

Composition per liter:

$ZnSO_4 \cdot 7H_2O$	0.44g
$CuSO_4 \cdot 5H_2O$	0.2g
$MnSO_4 \cdot 2H_2O$	0.17g
H_3BO_3	0.1g
$CoCl_2 \cdot 6H_2O$	0.08g
$Na_2MoO_4 \cdot 2H_2O$	0.06g

Preparation of Trace Elements Solution: Add components to distilled/deionized water and bring volume to 1.0L. Mix thoroughly. Autoclave for 15 min at 15 psi pressure–121°C.

Preparation of Medium: Add components to distilled/deionized water and bring volume to 1.0L. Mix thoroughly. Adjust pH to 7.6 with 10% NaOH. Gently heat while stirring and bring to boiling. Distribute 25.0mL aliquots into 120mL serum bottles closed with butyl rubber serum caps and aluminum crimps. Autoclave for 15 min at 15 psi pressure–121°C. The pH is critical and the final pH must be 6.5–7.5. Some insoluble phosphates may precipitate during medium preparation (the turbidity should not be more than OD 0.005). Instead of methanol a headspace with 20% methane, 2% CO_2, and 78% air can be used.

Use: For the cultivation of *Methylohalobius crimeensis.*

Halophilic Nutrient Agar (LMG Medium 220)

Composition per liter:

NaCl	60.0g
Agar	15.0g
Casein peptone, tryptic digest	10.0g
Yeast extract	5.0g
Glucose	5.0g

pH 7.3 ± 0.2 at 25°C

Preparation of Medium: Add components to tap water and bring volume to 1.0L. Mix thoroughly. Gently heat and bring to boiling. Distribute into tubes or flasks. Autoclave for 15 min at 15 psi pressure–121°C. Pour into sterile Petri dishes or leave in tubes.

Use: For the cultivation and maintenance of halophilic bacteria.

Halophilic Synthetic Medium

Composition per liter:

Glucose	0.1g
KNO_3	0.05g
$FePO_4$	0.01g
Artificial seawater	100.0mL

Artificial Seawater:
Composition per 100.0mL:

NaCl	2.4g
$MgCl_2·6H_2O$	1.1g
Na_2SO_4	0.4g
$CaCl_2·6H_2O$	0.2g
KCl	0.07g
$NaHCO_3$	0.02g
KBr	0.01g
$SrCl_2·6H_2O$	4.0mg
H_3BO_3	3.0mg
$Na_2SiO_3·9H_2O$	0.5mg
NaF	0.3mg

Preparation of Artificial Seawater: Add components to distilled/deionized water and bring volume to 100.0mL. Mix thoroughly.

Preparation of Medium: Add components to distilled/deionized water and bring volume to 1.0L. Mix thoroughly. Distribute into tubes or flasks. Autoclave for 15 min at 15 psi pressure–121°C.

Use: For the cultivation of halophilic bacteria.

Halopiger Medium (DSMZ Medium 1138)

Composition per liter:

NaCl	195.0g
$MgSO_4·7H_2O$	50.8g
$MgCl_2·6H_2O$	32.5g
Yeast extract	5.0g
KCl	5.0g
$CaCl_2·2H_2O$	0.8g
NaBr	0.6g
$NaHCO_3$	0.16g

pH 8.0 ± 0.2 at 25°C

Preparation of Medium: Add components to distilled/deionized water and bring volume to 1.0L. Mix thoroughly. Adjust pH to 8.0. Gently heat while stirring and bring to boiling. Distribute into tubes or flasks. Autoclave for 15 min at 15 psi pressure–121°C.

Use: For the cultivation of *Halopiger* spp.

Haloquadratum walsbyi Medium (DSMZ Medium 1091)

Composition per liter:

NaCl	195.0g
$MgSO_4·7H_2O$	50.0g
$MgCl_2·6H_2O$	35.0g
KCl	5.0g
$NaNO_3$	1.0g
Sodium pyruvate	1.0g
$CaCl_2·2H_2O$	0.5g
$NaHCO_3$	0.25g
KH_2PO_4	0.05g
Yeast extract	0.05g
NH_4Cl	0.03g
Tris buffer, 1*M*	20.0mL

pH 7.4 ± 0.2 at 25°C

Preparation of Medium: Add components to distilled/deionized water and bring volume to 1.0L. Mix thoroughly. Adjust pH to 7.4. Gently heat while stirring and bring to boiling. Distribute into tubes or flasks. Autoclave for 15 min at 15 psi pressure–121°C.

Use: For the cultivation of *Haloquadratum walsbyi.*

Halorhabdus utahensis Medium (DSMZ Medium 927)

Composition per 1003.25mL:

NaCl	270.0g
$MgSO_4·7H_2O$	20.0g
Tris-HCl	12.0g
KCl	5.0g
NH_4Cl	2.0g
Glucose	2.0g
Yeast extract	1.0g
NaBr	0.1g
Potassium phosphate solution	2.5mL
Calcium chloride solution	0.5mL
Iron chloride manganese chloride solution	0.25mL

pH 7.6 ± 0.2 at 25°C

Potassium Phosphate Solution:
Composition per 10.0mL:

KH_2PO_4	0.5g

Preparation of Potassium Phosphate Solution: Add KH_2PO_4 to distilled/deionized water and bring volume to 10.0mL. Mix thoroughly. Autoclave for 15 min at 15 psi pressure–121°C. Cool to room temperature.

Calcium Chloride Solution:
Composition per 10.0mL:

$CaCl_2·2H_2O$	1.0g

Preparation of Calcium Chloride Solution: Add $CaCl_2·2H_2O$ to distilled/deionized water and bring volume to 10.0mL. Mix thoroughly. Autoclave for 15 min at 15 psi pressure–121°C. Cool to room temperature.

Iron Chloride Manganese Chloride Solution:
Composition per 10.0mL:

$FeCl_2·4H_2O$	0.2g
$MnCl_2·4H_2O$	0.2g

Preparation of Iron Chloride Manganese Chloride Solution: Add components to distilled/deionized water and bring volume to 10.0mL. Mix thoroughly. Autoclave for 15 min at 15 psi pressure–121°C. Cool to room temperature.

Preparation of Medium: Add components, except potassium phosphate solution, calcium chloride solution, and iron chloride manganese chloride solution, to distilled/deionized water and bring volume to 1.0L. Mix thoroughly. Adjust pH to 7.6 with 5*M* NaOH. Autoclave for 15 min at 15 psi pressure–121°C. Cool to room temperature. Aseptically add 2.5mL phosphate solution, 0.5mL calcium chloride solution, and 0.25mL iron chloride manganese chloride solution. Mix thoroughly. Aseptically distribute to sterile tubes or flasks.

Use: For the cultivation of *Halorhabdus utahensis.*

Halorhodospira Medium (DSMZ Medium 999)

Composition per liter:

NaCl	100.0g
$MgCl_2·6H_2O$	2.0g
$MgSO_4·7H_2O$	1.0g
KH_2PO_4	1.0g
NH_4Cl	0.5g
$CaCl_2·2H_2O$	0.05g
$NaHCO_3$ solution	10.0mL
$Na_2S·9H_2O$ solution	7.5mL
Vitamin solution V7	1.0mL
Trace elements solution SL-12	1.0mL

pH 7.3 ± 0.2 at 25°C

$Na_2S·9H_2O$ Solution:
Composition per 10.0mL:

$Na_2S·9H_2O$	1.0g

Preparation of $Na_2S·9H_2O$ Solution: Add $Na_2S·9H_2O$ to distilled/deionized water and bring volume to 10.0mL. Mix thoroughly. Autoclave under 100% N_2 for 15 min at 15 psi pressure–121°C. Cool to room temperature.

$NaHCO_3$ Solution:
Composition per 10.0mL:

$NaHCO_3$	2.0g

Preparation of $NaHCO_3$ Solution: Add components to distilled/deionized water and bring volume to 10.0mL. Mix thoroughly. Sparge with 20% CO_2 + 80% H_2. Autoclave for 15 min at 15 psi pressure–121°C. Cool to room temperature.

Vitamin Solution V7:
Composition per liter:

Pyridoxine-HCl	50.0mg
Nicotinic acid	20.0mg
Vitamin B_{12}	20.0mg
Thiamine-HCl·$2H_2O$	10.0mg
p-Aminobenzoic acid	10.0mg
D-Ca-pantothenate	5.0mg
Biotin	2.0mg

Preparation of Vitamin Solution V7: Add components to distilled/deionized water and bring volume to 1.0L. Mix thoroughly. Filter sterilize.

Trace Elements Solution SL-12:
Composition per liter:

$FeSO_4·7H_2O$	1.1g
H_3BO_3	0.3g
$CoCl_2·6H_2O$	0.19g
$MnCl_2·2H_2O$	0.05g
$ZnCl_2$	42.0mg
$NiCl_2·6H_2O$	24.0mg
$Na_2MoO_4·4H_2O$	18.0mg
$CuCl_2·2H_2O$	2.0mg

Preparation of Trace Elements Solution SL12: Add components. to distilled/deionized water and bring volume to 1.0L. Mix thoroughly. Filter sterilize.

Preparation of Medium: Add components, except sulfide, bicarbonate, vitamin, and trace elements solutions, to distilled/deionized water and bring volume to 1.0L. Mix thoroughly. Adjust pH to 7.3. Gently heat while stirring and bring to boiling. Aseptically add sulfide, bicarbonate, vitamin, and trace elements solutions. Mix thoroughly. Distribute into screw-cap tubes, bottles, or flasks. Autoclave for 20 min at 15 psi pressure–121°C.

Use: For the cultivation of *Halorhodospira* spp.

Halorhodospira Medium with Succinate, Acetate, and Yeast Extract (DSMZ Medium 999)

Composition per liter:

NaCl	100.0g
$MgCl_2·6H_2O$	2.0g
$MgSO_4·7H_2O$	1.0g
KH_2PO_4	1.0g
NH_4Cl	0.5g
Yeast extract	0.5g
$CaCl_2·2H_2O$	0.05g
Na-succinate	0.27g
Na-acetate	0.17g
$NaHCO_3$ solution	10.0mL
$Na_2S·9H_2O$ solution	7.5mL
Vitamin solution V7	1.0mL
Trace elements solution SL-12	1.0mL

pH 7.3 ± 0.2 at 25°C

$Na_2S·9H_2O$ Solution:
Composition per 10.0mL:

$Na_2S·9H_2O$	1.0g

Preparation of $Na_2S·9H_2O$ Solution: Add $Na_2S·9H_2O$ to distilled/deionized water and bring volume to 10.0mL. Mix thoroughly. Autoclave under 100% N_2 for 15 min at 15 psi pressure–121°C. Cool to room temperature.

$NaHCO_3$ Solution:
Composition per 10.0mL:

$NaHCO_3$	2.0g

Preparation of $NaHCO_3$ Solution: Add components to distilled/deionized water and bring volume to 10.0mL. Mix thoroughly. Sparge with 20% CO_2 + 80% H_2. Autoclave for 15 min at 15 psi pressure–121°C. Cool to room temperature.

Vitamin Solution V7:

Composition per liter:

Pyridoxine-HCl	50.0mg
Nicotinic acid	20.0mg
Vitamin B_{12}	20.0mg
Thiamine-HCl·$2H_2O$	10.0mg
p-Aminobenzoic acid	10.0mg
D-Ca-pantothenate	5.0mg
Biotin	2.0mg

Preparation of Vitamin Solution V7: Add components to distilled/deionized water and bring volume to 1.0L. Mix thoroughly. Filter sterilize.

Trace Elements Solution SL-12:

Composition per liter:

$FeSO_4·7H_2O$	1.1g
H_3BO_3	0.3g
$CoCl_2·6H_2O$	0.19g
$MnCl_2·2H_2O$	0.05g
$ZnCl_2$	42.0mg
$NiCl_2·6H_2O$	24.0mg
$Na_2MoO_4·4H_2O$	18.0mg
$CuCl_2·2H_2O$	2.0mg

Preparation of Trace Elements Solution SL-12: Add components. to distilled/deionized water and bring volume to 1.0L. Mix thoroughly. Filter sterilize.

Preparation of Medium: Add components, except sulfide, bicarbonate, vitamin, and trace elements solutions, to distilled/deionized water and bring volume to 1.0L. Mix thoroughly. Adjust pH to 7.3. Gently heat while stirring and bring to boiling. Autoclave for 20 min at 15 psi pressure–121°C. Cool to room temperature. Aseptically add sulfide, bicarbonate, vitamin, and trace elements solutions. Mix thoroughly. Aseptically distribute into sterile screw-cap tubes, bottles, or flasks.

Use: For the improved growth during cultivation of pure cultures of *Halorhodospira* spp.

Halorubrum californiense Medium (DSMZ Medium 1194)

Composition per liter:

NaCl	156.0g
$MgSO_4·7H_2O$	40.6g
$MgCl_2·6H_2O$	26.0g
Yesat extract	10.0g
KCl	4.0g
$CaCl_2·2H_2O$	0.66g
NaBr	0.46g
$NaHCO_3$	0.12g

pH 7.2 ± 0.2 at 25°C

Preparation of Medium: Add components to distilled/deionized water and bring volume to 1.0L. Mix thoroughly. Adjust pH to 7.2. Gently heat while stirring and bring to boiling. Mix thoroughly. Distribute into tubes , bottles, or flasks. Autoclave for 15 min at 15 psi pressure–121°C.

Use: For the cultivation of *Halorubrum californiense.*

Halothermothrix orenii Medium (DSMZ Medium 761)

Composition per liter:

NaCl	100.0g
Glucose	10.0g
KCl	4.0g
$MgCl_2·6H_2O$	2.0g
NH_4Cl	1.0g
Na-acetate	1.0g
Trypticase™	0.5g
K_2HPO_4	0.3g
KH_2PO_4	0.3g
$CaCl_2·2H_2O$	0.2g
Resazurin	0.001g
$NaHCO_3$ solution	50.0mL
$Na_2S·9H_2O$ solution	10.0mL
Dithionite solution	5.0mL
Trace elements solution SL-6	1.0mL

pH 7.0 ± 0.2 at 25°C

Trace Elements Solution SL-6:

Composition per liter:

$MnCl_2·4H_2O$	0.5g
H_3BO_3	0.3g
$CoCl_2·6H_2O$	0.2g
$ZnSO_4·7H_2O$	0.1g
$Na_2MoO_4·2H_2O$	0.03g
$NiCl_2·6H_2O$	0.02g
$CuCl_2·2H_2O$	0.01g

Preparation of Trace Elements Solution SL-6: Add components to distilled/deionized water and bring volume to 1.0L. Mix thoroughly. Autoclave for 15 min at 15 psi pressure–121°C.

$Na_2S·9H_2O$ Solution:

Composition per 10.0mL:

$Na_2S·9H_2O$	0.2g

Preparation of $Na_2S·9H_2O$ Solution: Add $Na_2S·9H_2O$ to distilled/deionized water and bring volume to 10.0mL. Mix thoroughly. Sparge with 100% N_2. Autoclave for 15 min at 15 psi pressure–121°C. Neutralize to pH 7.0 with sterile HCl.

$NaHCO_3$ Solution:

Composition per 100.0mL:

$NaHCO_3$	10.0g

Preparation of $NaHCO_3$ Solution: Add $NaHCO_3$ to distilled/deionized water and bring volume to 100.0mL. Mix thoroughly. Sparge with 80% N_2 + 20% CO_2. Filter sterilize.

Dithionite Solution

Composition per 10.0mL:

Na-dithionite	2.0mg

Preparation of Dithionite Solution: Add Na-dithionite to distilled/deionized water and bring volume to 10.0mL. Mix thoroughly. Sparge with 100% N_2. Filter sterilize.

Preparation of Medium: Prepare and dispense medium under 80% N_2 + 20% CO_2 gas atmosphere. Add components, except $NaHCO_3$ solution, dithionite solution, and $Na_2S·9H_2O$ solution, to distilled/deionized water and bring volume to 935.0mL. Mix thoroughly. Adjust pH to 7.0. Gently heat and bring to boiling. Cool while sparging with 80% N_2 + 20% CO_2. Distribute into Hungate tubes under 80% N_2 + 20% CO_2. Autoclave for 15 min at 15 psi pressure–121°C. Cool to room temperature. Aseptically and anaerobically inject $NaHCO_3$ solution (0.5mL per

10mL medium), dithionite solution (0.05mL per 10mL medium), and $Na_2S·9H_2O$ solution (0.1mL per 10mL medium). Aseptically and anaerobically distribute into sterile tubes or bottles.

Use: For the cultivation of *Halothermothrix orenii.*

Halothiobacillus Medium (DSMZ Medium 864)

Composition per 1030.0mL:

NaCl	29.0g
$MgSO_4·7H_2O$	1.5g
$(NH_4)_2SO_4$	1.0g
KCl	0.7g
$CaCl_2·2H_2O$	0.42g
Bromthymol Blue	4.0mg
Na-thiosulfate solution	20.0mL
Phosphate solution	10.0mL
Trace elements solution	1.0mL

pH 6.5–7.0 at 25°C

Trace Elements Solution:
Composition per liter:

$FeCl_2·4H_2O$	1.5g
Na_2-EDTA	0.5g
$CoCl_2·6H_2O$	190.0mg
$MnCl_2·4H_2O$	100.0mg
$ZnCl_2$	70.0mg
$Na_2MoO_4·2H_2O$	36.0mg
$NiCl_2·6H_2O$	24.0mg
H_3BO_3	6.0mg
$CuCl_2·2H_2O$	2.0mg
HCl (25% solution)	10.0mL

Preparation of Trace Elements Solution: Add $FeCl_2·4H_2O$ to 10.0mL of HCl solution. Mix thoroughly. Add distilled/deionized water and bring volume to 1.0L. Add remaining components. Mix thoroughly. Adjust pH to 7.0. Sparge with 80% N_2 + 20% CO_2. Autoclave for 15 min at 15 psi pressure–121°C.

Na-thiosulfate Solution:
Composition per 20.0mL:

$Na_2S_2O_3·5H_2O$	5.0g

Preparation of Na-thiosulfate Solution: Add $Na_2S_2O_3·5H_2O$ to distilled/deionized water and bring volume to 20.0mL. Mix thoroughly. Sparge with 100% N_2. Filter sterilize.

Phosphate Solution:
Composition per 10.0mL:

K_2HPO_4	0.5g

Preparation of Phosphate Solution: Add K_2HPO_4 to distilled/deionized water and bring volume to 10.0mL. Mix thoroughly. Sparge with 100% N_2. Filter sterilize.

Preparation of Medium: Add components, except phosphate solution and Na-thiosulfate solution, to distilled/deionized water and bring volume to 1.0L. Mix thoroughly. Adjust pH to 6.7. Autoclave for 15 min at 15 psi pressure–121°C. Cool to room temperature. Aseptically add 20.0mL sterile Na-thiosulfate solution and 10.0mL sterile phosphate solution. Mix thoroughly. Aseptically distribute into sterile tubes or flasks.

Use: For the cultivation of *Halothiobacillus kellyi.*

Halovibrio Medium (DSMZ Medium 1176)

Composition per liter:

NaCl	120.0–240.0g
KH_2PO_4	3.0g
NH_4Cl	0.5g
Sodium acetate soltuion	10.0mL
HEPES solution	10.0mL
Yeast extract solution	2.0mL
Magnesium sulfate solution	2.0mL
Trace elements solution SL-6	1.0mL

pH 7.5 ± 0.2 at 25°C

Trace Elements Solution SL-6:
Composition per liter:

$MnCl_2·4H_2O$	0.5g
H_3BO_3	0.3g
$CoCl_2·6H_2O$	0.2g
$ZnSO_4·7H_2O$	0.1g
$Na_2MoO_4·2H_2O$	0.03g
$NiCl_2·6H_2O$	0.02g
$CuCl_2·2H_2O$	0.01g

Preparation of Trace Elements Solution SL-6: Add components to distilled/deionized water and bring volume to 1.0L. Mix thoroughly. Autoclave for 15 min at 15 psi pressure–121°C.

HEPES Solution:
Composition per 10.0mL:

HEPES	2.4g

Preparation of HEPES Solution: Add HEPES to distilled/deionized water and bring volume to 10.0mL. Mix thoroughly.

Sodium Acetate Solution:
Composition per 10.0mL:

Sodium acetate	1.64g

Preparation of Sodium Acetate Solution: Add sodium acetate to distilled/deionized water and bring volume to 10.0mL. Mix thoroughly. Autoclave for 15 min at 15 psi pressure–121°C.

Magnesium Sulfate Solution:
Composition per 10.0mL:

$MgSO_4·7H_2O$	2.5g

Preparation of Magnesium Sulfate Solution: Add $MgSO_4·7H_2O$ to distilled/deionized water and bring volume to 10.0mL. Mix thoroughly. Autoclave for 15 min at 15 psi pressure–121°C.

Yeast Extract Solution:
Composition per 10.0mL:

Yeast extract	1.0g

Preparation of Yeast Extract Solution: Add yeast extract to distilled/deionized water and bring volume to 10.0mL. Mix thoroughly. Autoclave for 15 min at 15 psi pressure–121°C.

Preparation of Medium: Add components, except yeast extract, magnesium sulfate, and sodium acetate solutions, to distilled/deionized water and bring volume to 970.0mL. Mix thoroughly. Adjust pH to 7.5. Gently heat while stirring and bring to boiling. Cool to room temperature. Dispense into sealed tubes, vials, or bottles. Autoclave for 15 min at 15 psi pressure–121°C. Cool to room temperature. Aseptically add yeast extract, magnesium sulfate, and sodium acetate solutions.

Use: For the cultivation of *Halovibrio* spp.

Halovibrio variabilis Medium

Composition per liter:

NaCl	95.0g
$MgSO_4 \cdot 7H_2O$	81.0g
Yeast extract	7.5g
Proteose peptone	2.5g
KCl	1.0g
Trace elements solution SL-4	10.0mL
Vitamin solution	10.0mL

pH 7.5 ± 0.2 at 25°C

Vitamin Solution:

Composition per liter:

Pyridoxine HCl	1.0mg
Lipoic acid	0.5mg
Nicotinic acid	0.5mg
p-Aminobenzoic acid	0.5mg
Pantothenic acid	0.5mg
Riboflavin	0.5mg
Thiamine HCl	0.5mg
Biotin	0.2mg
Folic acid	0.2mg
Cyanocobalamin	0.01mg

Preparation of Vitamin Solution: Add components to distilled/deionized water and bring volume to 1.0L. Mix thoroughly. Filter sterilize. Store at 5°C.

Trace Elements Solution SL-4:

Composition per 900.0mL:

EDTA	0.5g
$FeSO_4 \cdot 7H_2O$	0.2g
Trace elements solution SL-6	100.0mL

Trace Elements Solution SL-6:

Composition per liter:

H_3BO_3	0.3g
$CoCl_2 \cdot 6H_2O$	0.2g
$ZnSO_4 \cdot 7H_2O$	0.1g
$MnCl_2 \cdot 4H_2O$	0.03g
$CuCl_2 \cdot 2H_2O$	0.01g
$NiCl_2 \cdot 6H_2O$	0.02g
$Na_2MoO_4 \cdot 2H_2O$	0.03g

Preparation of Trace Elements Solution SL-6: Add components to distilled/deionized water and bring volume to 1.0L. Mix thoroughly.

Preparation of Trace Elements Solution SL-4: Add components to distilled/deionized water and bring volume to 1.0L. Mix thoroughly. Filter sterilize.

Preparation of Medium: Add components, except trace elements solution SL-4 and vitamin solution, to distilled/deionized water and bring volume to 980.0mL. Mix thoroughly. Autoclave for 15 min at 15 psi pressure–121°C. Aseptically add 10.0mL of sterile trace elements solution SL-4 and 10.0mL of sterile vitamin solution. Mix thoroughly. Aseptically distribute into sterile tubes or flasks.

Use: For the cultivation of *Halovibrio variabilis*.

Ham's F-10 Medium

Composition per liter:

NaCl	7.4g
$NaHCO_3$	1.2g
Glucose	1.1g
$NaH_2PO_4 \cdot H_2O$	0.29g
KCl	0.28g
L-Arginine·HCl	0.21g
L-Glutamine	0.15g
$MgSO_4 \cdot 7H_2O$	0.15g
Sodium pyruvate	0.11g
KH_2PO_4	0.08g
$CaCl_2 \cdot 2H_2O$	0.04g
L-Cystine·2HCl	0.04g
L-Histidine·HCl·H_2O	0.02g
L-Lysine·HCl	0.02g
L-Asparagine-H_2O	0.01g
L-Aspartic acid	0.01g
L-Glutamic acid	0.01g
L-Leucine	0.01g
L-Proline	0.01g
L-Serine	0.01g
L-Alanine	8.9mg
Glycine	7.5mg
D-Phenylalanine	5.0mg
L-Methionine	4.5mg
Hypoxanthine	4.1mg
L-Threonine	3.6mg
L-Valine	3.5mg
L-Isoleucine	2.6mg
L-Tyrosine	1.8mg
Vitamin B_{12}	1.4mg
Folic acid	1.3mg
Phenol Red	1.2mg
Thiamine·HCl	1.0mg
$FeSO_4 \cdot 7H_2O$	0.8mg
Choline chloride	0.7mg
D-Calcium pantothenate	0.7mg
Thymidine	0.7mg
Niacinamide	0.6mg
L-Tryptophan	0.6mg
Isoinositol	0.5mg
Riboflavin	0.4mg
Lipoic acid	0.2mg
Pyridoxine·HCl	0.2mg
$ZnSO_4 \cdot 7H_2O$	0.03mg
Biotin	0.02mg
$CuSO_4 \cdot 5H_2O$	3.0μg

pH 7.0 ± 0.2 at 25°C

Preparation of Medium: Add components to distilled/deionized water and bring volume to 1.0L. Mix thoroughly. Filter sterilize.

Use: For the growth of Y-1 cell cultures used in the mouse adrenal assay for heat-labile toxin of enterotoxigenic *Escherichia coli* and *Vibrio* species.

Hanahan's Broth (SOB Medium)

Composition per liter:

Casein enzymatic hydrolysate	20.0g
Yeast extract	5.0g
$MgSO_4$	2.4g
NaCl	0.5g
KCl	0.186g

pH 7.0 ± 0.2 at 25°C

Source: This medium is available as a premixed powder from SigmaAldrich.

Preparation of Medium: Add components to distilled/deionized water and bring volume to 1.0L. Mix thoroughly. Gently heat and bring to boiling. Distribute into tubes or flasks. Autoclave for 15 min at 15 psi pressure–121°C.

Use: For the cultivation of recombinant strains of *Escherichia coli.*

Hanahan's HiVeg Broth

Composition per liter:

Plant hydrolysate	20.0g
Yeast extract	5.0g
$MgSO_4$	2.4g
NaCl	0.5g
KCl	0.186g

pH 7.0± 0.2 at 25°C

Source: This medium is available as a premixed powder from HiMedia.

Preparation of Medium: Add components to distilled/deionized water and bring volume to 1.0L. Mix thoroughly. Gently heat and bring to boiling. Distribute into tubes or flasks. Autoclave for 15 min at 15 psi pressure–121°C.

Use: For the cultivation of recombinant strains of *Escherichia coli.*

Harpo's HTYE Trypticase Peptone Medium

Composition per liter:

Pancreatic digest of casein	5.0g
HEPES (*N*-[2-Hydroxyethyl]piperazine-*N*′-2-ethanesulfonic acid) buffer	4.0g
Yeast extract	2.0g

pH 6.8–7.0 at 25°C

Preparation of Medium: Add components to distilled/deionized water and bring volume to 1.0L. Mix thoroughly. Adjust pH to 7.0–7.2. Distribute into tubes or flasks. Autoclave for 15 min at 15 psi pressure–121°C.

Use: For the cultivation of *Cytophaga arvensicola* and *Flexibacter columnaris.*

Harpo's HTYEM Marine Medium

Pancreatic digest of casein	5.0g
HEPES	4.0g
Yeast extract	2.0g
Artificial seawater	1.0L

pH 7.5 ± 0.2 at 25°C

Artificial Seawater:

Composition per liter:

NaCl	27.5g
$MgSO_4 \cdot 7H_2O$	6.78g
$MgCl_2 \cdot 6H_2O$	5.38g
$CaCL_2 \cdot 2H_2O$	1.4g
KCl	0.72g
$NaHCO_3$	0.2g

Preparation of Artificial Seawater: Add components to distilled/deionized water and bring volume to 1.0L. Mix thoroughly.

Preparation of Medium: Add components to artificial seawater and bring volume to 1.0L. Mix thoroughly. Adjust pH to 7.5 with NaOH. Distribute into tubes or flasks. Autoclave for 15 min at 15 psi pressure–121°C.

Use: For the cultivation of *Cytophaga fermentans, Cytophaga latercula, Cytophaga uliginosa,* and *Microscilla aggregans.*

Harrold's Agar
see: **M 40 Y Agar**

Hartley's Digest Broth

Composition per 10.0L:

Ox heart	3000.0g
Pancreatin	50.0g
Na_2CO_3, anhydrous (0.8% solution)	5.0L
HCl, concentrated	80.0mL

pH 7.5 ± 0.2 at 25°C

Preparation of Medium: Finely mince the ox heart. Add the meat to 5.0L of distilled/deionized water. Gently heat and bring to 80°C. Add the 5.0L of Na_2CO_3 solution. Cool to 45°C. Add pancreatin and maintain at 45°C for 4 hr while stirring. Add the HCl and steam at 100°C for 30 min. Cool to room temperature. Adjust pH to 8.0 with 1*N* NaOH. Gently heat and bring to boiling. Continue boiling for 25 min. Filter while hot through Whatman #1 filter paper. Cool to room temperature. Adjust pH to 7.5. Autoclave for 15 min at 15 psi pressure–121°C.

Use: For the isolation and cultivation of *Actinobacillus lignieresii* from cattle.

Hartley's Digest HiVeg Broth

Composition per liter:

Plant hydrolysate	29.0g

pH 7.6 ± 0.2 at 25°C

Source: This medium is available as a premixed powder from HiMedia.

Preparation of Medium: Add components to distilled/deionized water and bring volume to 1.0L. Mix thoroughly. Gently heat and bring to boiling. Distribute into tubes or flasks. Autoclave for 15 min at 15 psi pressure–121°C.

Use: For the isolation and cultivation of *Actinobacillus lignieresii* from cattle.

Haskins Agar for *Tetrahymena*

Composition per liter:

Agar	16.0g
Dextrin	8.0g
Pancreatic digest of casein	5.0g
Yeast extract	5.0g
Liver concentrate	0.6g
Sodium acetate	0.6g

pH 7.2–7.4 at 25°C

Preparation of Medium: Add components to distilled/deionized water and bring volume to 1.0L. Mix thoroughly. Adjust pH to 7.2–7.4. Gently heat and bring to boiling. Distribute into tubes or flasks. Autoclave for 15 min at 15 psi pressure–121°C. Pour into sterile Petri dishes or leave in tubes. Overlay agar with a layer of sterile distilled water.

Use: For the cultivation of *Tetrahymena australis* and *Tetrahymena vorax.*

HAY
See: **Hay Extract Agar**

Hay Extract Agar (HAY)

Composition per liter:

Hay	50.0g
Agar	20.0g

pH 6.2 ± 0.2 at 25°C

Preparation of Medium: Add finely hashed hay to distilled/deionized water and bring volume to 1.0L. Mix thoroughly. Autoclave for 30 min at 15 psi pressure–121°C. Filter through Whatman filter paper. Adjust pH of filtrate to 6.2 with 5% potassium phosphate solution. Add agar. Bring volume to 1.0L wtih distilled/deionized water. Gently heat and bring to boiling. Distribute into tubes or flasks. Autoclave for 20 min at 10 psi pressure–115°C. Pour into sterile Petri dishes or leave in tubes.

Use: For the cultivation of *Alternaria porri, Botryotinia draytonii, Botryotinia fuckeliana, Botryotinia porri, Botryotinia narcissicola, Botryotinia polyblastis, Botrytis aclada, Botrytis hyacinthi, Botrytis cinerea, Cercospora beticola,* and *Sclerotinia sphaerosperma.*

Hay Extract Medium

See: **HE Medium**

Hay Infusion Agar

Composition per liter:

Hay, partially decomposed	50.0g
Agar	15.0g
K_2HPO_4	2.0g

pH 6.2 ± 0.3 at 25°C

Preparation of Medium: Add hay to distilled/deionized water and bring volume to 1.0L. Autoclave for 30 min at 15 psi pressure–121°C. Filter through paper and reserve filtrate. Add distilled/deionized water to filtrate and bring volume to 1.0L. Mix thoroughly. Add agar and K_2HPO_4. Mix thoroughly. Gently heat and bring to boiling. Adjust pH to 6.2. Distribute into tubes or flasks. Autoclave for 15 min at 15 psi pressure–121°C. Pour into sterile Petri dishes or leave in tubes.

Use: For the cultivation and maintenance of *Alternaria* species, *Caulochytrium protostelioides, Choanephora infundibulifera, Dipsacomyces acuminosporus, Eremascus albus, Eremascus fertilis, Eurotium chevalieri, Eurotium halophilicum, Eurotium herbariorum, Mortierella bisporalis, Protostelium irregularis,* and *Saksenaea vasiformis.*

Hayflick Medium

Composition per 107.5mL:

Mycoplasma broth base	70.0mL
Horse serum	20.0mL
Fresh yeast extract solution	10.0mL
Penicillin solution	5.0mL
Thallous acetate solution	2.5mL

pH 7.8 ± 0.2 at 25°C

Mycoplasma Broth Base:

Pancreatic digest of casein	7.0g
NaCl	5.0g
Beef extract	3.0g
Yeast extract	3.0g
Beef heart, solids from infusion	2.0g

pH 7.8 ± 0.2 at 25°C

Preparation of Mycoplasma Broth Base: Add components to distilled/deionized water and bring volume to 1.0L. Gently heat and bring to boiling. Mix thoroughly. Distribute into tubes or flasks. Autoclave for 15 min at 15 psi pressure–121°C.

Fresh Yeast Extract Solution:
Composition per 100.0mL:

Baker's yeast, live, pressed, starch-free	25.0g

Preparation of Fresh Yeast Extract Solution: Add live Baker's yeast to 100.0mL of distilled/deionized water. Autoclave for 90 min at 15 psi pressure–121°C. Allow to stand. Remove supernatant solution. Adjust pH to 6.6–6.8. Filter sterilize.

Penicillin Solution:
Composition per 5.0mL:

Penicillin	20,000U

Preparation of Penicillin Solution: Add penicillin to distilled/deionized water and bring volume to 5.0mL. Mix thoroughly. Filter sterilize.

Thallous Acetate Solution:
Composition per 10.0mL:

Thallous acetate	0.1g

Preparation of Thallous Acetate Solution: Add thallous acetate to distilled/deionized water and bring volume to 10.0mL. Mix thoroughly. Filter sterilize.

Use: For the cultivation of *Mycoplasma* species.

Hayflick Medium, Modified

Composition per 1212.0mL:

Beef heart, infusion from	500.0g
Tryptose	10.0g
Noble agar	9.6g
NaCl	5.0g
Horse serum, normal	200.0mL
Fresh yeast extract solution	100.0mL
Calf thymus DNA solution	12.0mL

pH 7.8 ± 0.2 at 25°C

Fresh Yeast Extract Solution:
Composition per 100.0mL:

Baker's yeast, live, pressed, starch-free	25.0g

Preparation of Fresh Yeast Extract Solution: Add the live Baker's yeast to 100.0mL of distilled/deionized water. Autoclave for 90 min at 15 psi pressure–121°C. Allow to stand. Remove supernatant solution. Adjust pH to 6.6–6.8. Filter sterilize.

Calf Thymus DNA Solution:
Composition per 20.0mL:

Calf thymus DNA	0.04g

Preparation of Calf Thymus DNA Solution: Add calf thymus DNA to distilled/deionized water and bring volume to 20.0mL. Mix thoroughly. Filter sterilize.

Preparation of Medium: Add components, except horse serum, fresh yeast extract solution, and calf thymus DNA solution, to distilled/deionized water and bring volume to 900.0mL. Mix thoroughly. Gently heat and bring to boiling. Autoclave for 15 min at 15 psi pressure–121°C. Cool to 45°–50°C. Aseptically add 200.0mL of sterile horse serum, 100.0mL of sterile fresh yeast extract solution, and 12.0mL of sterile calf thymus DNA solution. Mix thoroughly. Aseptically distribute into sterile tubes.

Use: For the cultivation and maintenance of *Mycoplasma mustelae.*

HB

See: **Halophilic Broth**

3HB

See: **Hydroxybutyrate Medium**

HBT Bilayer Medium (Human Blood Tween™ Bilayer Medium)

Composition per 1062.5mL:

Agar	13.5g
Pancreatic digest of casein	12.0g
Casein/meat peptone	10.0g
NaCl	5.0g
Peptic digest of animal tissue	5.0g
Beef extract	3.0g
Yeast extract	3.0g
Cornstarch	1.0g
Human blood, anticoagulated	25.0mL
Colistin solution	10.0mL
Nalidixic acid solution	10.0mL
Amphotericin B solution	10.0mL
Polysorbate 80 (Tween™ 80) solution	7.5mL

pH 7.3 ± 0.2 at 25°C

Source: This medium is available as a premixed powder from BD Diagnostic Systems.

Colistin Solution:

Composition per liter:

Colistin	0.01g

Preparation of Colistin: Add colistin to distilled/deionized water and bring volume to 10.0mL. Mix thoroughly. Filter sterilize.

Nalidixic Acid Solution:

Composition per liter:

Nalidixic acid	0.02g

Preparation of Nalidixic Acid Solution: Add nalidixic acid to distilled/deionized water and bring volume to 10.0mL. Mix thoroughly. Filter sterilize.

Amphotericin B Solution:

Composition per liter:

Amphotericin B	3.0mg

Preparation of Amphotericin B Solution: Add Amphotericin B to distilled/deionized water and bring volume to 10.0mL. Mix thoroughly. Filter sterilize.

Tween™ 80 Solution:

Composition per 100.0mL:

Tween™ 80	1.0mL

Preparation of Tween™ 80 Solution: Add Tween™ 80 to distilled/deionized water and bring volume to 100.0mL. Mix thoroughly. Adjust pH to 7.3. Filter sterilize.

Preparation of Medium: Add components, except amphotericin B solution, Tween™ 80, and human blood, to distilled/deionized water and bring volume to 1.0L. Mix thoroughly. Gently heat and bring to boiling. Divide the medium into two 500.0mL fractions. Autoclave both flasks of media for 15 min at 15 psi pressure–121°C. Cool to 45°–50°C. To one flask, aseptically add 5.0mL of sterile colistin solution, 5.0mL of sterile nalidixic acid solution, 5.0mL of sterile amphotericin B solution, and 3.75mL of Tween™ 80 solution. Mix thoroughly. Pour into sterile Petri dishes in 7.0mL volumes. Allow agar to harden. To remaining flask aseptically add 5.0mL of sterile colistin solution, 5.0mL of sterile nalidixic acid solution, 5.0mL of sterile amphotericin B solution, 3.75mL of sterile Tween™ 80 solution, and 25.0mL of sterile human blood. Mix thoroughly. Pour into the same Petri dishes that each contain 7.0mL of the agar medium without blood. The top layer should be approximately 14.0mL per plate.

Use: For the selective isolation, cultivation, and differentiation of *Gardnerella vaginalis* from clinical specimens.

HC Agar

See: **Hemorrhagic coli Agar**

HC Agar Base

Composition per liter:

Glucose	20.0g
Agar	15.0g
Yeast extract	5.0g
Na_2HPO_4	3.5g
KH_2PO_4	3.4g
Pancreatic digest of casein	2.5g
Peptic digest of animal tissue	2.5g
NH_4Cl	1.4g
Na_2CO_3	1.0g
Chloramphenicol	0.1g
$MgSO_4 \cdot 7H_2O$	0.06g
Polysorbate 80 (Tween™ 80) solution	20.0mL

pH 7.0 ± 0.2 at 25°C

Source: This medium is available as a premixed powder from BD Diagnostic Systems.

Preparation of Medium: Add components, except Tween™ 80, to distilled/deionized water and bring volume to 980.0mL. Mix thoroughly. Gently heat and bring to boiling. Add Tween™ 80. Mix thoroughly. Autoclave for 15 min at 15 psi pressure–121°C. Pour into sterile Petri dishes.

Use: For the cultivation and enumeration of molds in cosmetics and toiletries.

HD (1:10 Diluted) (DSMZ Medium 1124)

Composition per liter:

Casein peptone	0.5g
Glucose	0.1g
Yeast extract	0.25g

pH 5.0 ± 0.2 at 25°C

Preparation of Medium: Add components to distilled/deionized water and bring volume to 1.0L. Mix thoroughly. Adjust pH to 5.0. Gently heat while stirring and bring to boiling. Mix thoroughly. Distribute into tubes , bottles, or flasks. Autoclave for 15 min at 15 psi pressure–121°C.

Use: For the cultivation of *Edaphobacter modestus.*

HD Agar, 1:10 Diluted, Modified (DSMZ Medium 1135)

Composition per liter:

MES (2-[N-morpholino]ethane sulfonic acid)	1.95g
Casein peptone	0.5g
Yeast extract	0.25g

Glucose 0.1g
Agar solution 500.0mL

pH 5.5 ± 0.2 at 25°C

Agar Solution:
Composition per 10.0mL:
Agar 15.0g

Preparation of Agar Solution: Wash agar with double distilled water. Add agar to double distilled/deionized water and bring volume to 500.0mL. Adjsut pH to 5.5. Mix thoroughly. Autoclave for 15 min at 15 psi pressure–121°C. Cool to 50°C.

Preparation of Medium: Add components, except agar solution, to double distilled/deionized water and bring volume to 500.0mL. Mix thoroughly. Adjust pH to 5.5. Gently heat while stirring and bring to boiling. Mix thoroughly. Autoclave for 15 min at 15 psi–121°C. Cool to 50°C. Aseptically add 500.0mL agar solution. Pour into Petri dishes or aseptically distribute into tubes.

Use: For the cultivation of *Edaphobacter aggregans.*

HD- Medium, 1:10 Diluted, Modified (DSMZ Medium 1135)

Composition per liter:
MES (2-[N-Morpholino]ethane sulfonic acid) 1.95g
Casein peptone 0.5g
Yeast extract 0.25g
Glucose 0.1g

pH 5.5 ± 0.2 at 25°C

Preparation of Medium: Add components to double distilled/deionized water and bring volume to 1.0L. Mix thoroughly. Adjust pH to 5.5. Gently heat while stirring and bring to boiling. Mix thoroughly. Distribute into tubes , bottles, or flasks. Autoclave for 15 min at 15 psi pressure–121°C.

Use: For the cultivation of *Edaphobacter aggregans.*

HE Medium (Hay Extract Medium)

Composition per liter:
Agar 10.0g
Peptone 1.0g
Yeast extract 1.0g
Hay extract solution 500.0mL

pH 6.5 ± 0.2 at 25°C

Hay Extract Solution:
Composition per liter:
Hay, dried 50.0g

Preparation of Hay Extract Solution: Add dried barn hay to distilled/deionized water and bring volume to 1.0L. Mix thoroughly. Gently heat and bring to boiling. Filter through Whatman #40 filter paper.

Preparation of Medium: Add components to distilled/deionized water and bring volume to 1.0L. Mix thoroughly. Gently heat and bring to boiling. Distribute into tubes or flasks. Autoclave for 15 min at 15 psi pressure–121°C. Pour into sterile Petri dishes or leave in tubes.

Use: For the isolation and cultivation of *Spirochaeta aurantia.*

Heart Infusion Agar

Composition per liter:
Beef heart, infusion from 500.0g
Agar 15.0g
Tryptose 10.0g
NaCl 5.0g

pH 7.4 ± 0.2 at 25°C

Source: This medium is available as a premixed powder from BD Diagnostic Systems.

Preparation of Medium: Add components to distilled/deionized water and bring volume to 1.0L. Mix thoroughly. Gently heat and bring to boiling. Distribute into tubes or flasks. Autoclave for 15 min at 15 psi pressure–121°C. Pour into sterile Petri dishes or leave in tubes.

Use: For the isolation and cultivation of a wide variety of fastidious microorganisms. For the cultivation and maintenance of *Bacillus anthracis, Bacillus cereus, Bacillus mycoides, Serratia rubidaea, Staphylococcus aureus, Tsatumella ptyseos*, and *Vibrio vulnificus*. It can also be used as a base for the preparation of blood agar in determining hemolytic reactions. When using for blood agar, reduce volume to 950.0mL to allow for addition of 50.0mL of defibrinated horse blood.

Heart Infusion Agar (HIA) (BAM M159)

Composition per liter:
Proteose peptone 15.0g
Agar 12.0g
Yeast extract 5.0g
NaCl 5.0g
Liver digest 2.5g

pH 7.4 ± 0.2 at 25°C

Source: This medium is available as a premixed powder from BD Diagnostic Systems.

Preparation of Medium: Add components to distilled/deionized water and bring volume to 1.0L. Mix thoroughly. Gently heat and bring to boiling. Distribute into tubes or flasks. Autoclave for 15 min at 15 psi pressure–121°C. Pour into sterile Petri dishes or leave in tubes.

Use: For the isolation and cultivation of a wide variety of fastidious microorganisms. For the cultivation and maintenance of *Bacillus cereus, Staphylococcus aureus, Vibrio vulnificus,* and *Vibrio cholerae*. It can also be used as a base for the preparation of blood agar in determining hemolytic reactions. When using for blood agar, reduce volume to 950.0mL to allow for addition of 50.0mL of defibrinated horse blood. Blood is added aseptically after autoclaving.

Heart Infusion Agar (HIA) (BAM M60)

Composition per liter:
Agar 15.0g
Tryptose 10.0g
NaCl 5.0g
Beef heart, infusion from 500.0g 1.0L

pH 7.4 ± 0.2 at 25°C

Source: This medium is available as a premixed powder from BD Diagnostic Systems.

Preparation of Medium: Add components to distilled/deionized water and bring volume to 1.0L. Mix thoroughly. Gently heat and bring to boiling. Distribute into tubes or flasks. Autoclave for 15 min at 15 psi pressure–121°C. Pour into sterile Petri dishes or leave in tubes.

Use: For the isolation and cultivation of a wide variety of fastidious microorganisms. For the cultivation and maintenance of *Bacillus cereus, Staphylococcus aureus, Vibrio vulnificus,* and *Vibrio cholerae*.It can also be used as a base for the preparation of blood agar in determining hemolytic reactions. When using for blood agar, reduce volume to 950.0mL to allow for addition of 50.0mL of defibrinated horse blood. Blood is added aseptically after autoclaving.

Heart Infusion Agar with Glucose

Composition per liter:

Beef heart, infusion from	500.0g
Agar	15.0g
Tryptose	10.0g
NaCl	5.0g
Glucose	1.0g

pH 7.4 ± 0.2 at 25°C

Preparation of Medium: Add components to distilled/deionized water and bring volume to 1.0L. Mix thoroughly. Gently heat and bring to boiling. Distribute into tubes or flasks. Autoclave for 15 min at 15 psi pressure–121°C. Pour into sterile Petri dishes or leave in tubes.

Use: For the cultivation and maintenance of *Bacillus* species and *Pseudomonas* species.

Heart Infusion Agar with 0.1% Glucose

Composition per liter:

Beef heart, infusion from	500.0g
Agar	15.0g
Tryptose	10.0g
NaCl	5.0g
Glucose	1.0g

pH 7.4 ± 0.2 at 25°C

Source: This medium without glucose is available as a premixed powder from BD Diagnostic Systems.

Preparation of Medium: Add components to distilled/deionized water and bring volume to 1.0L. Mix thoroughly. Gently heat and bring to boiling. Distribute into tubes or flasks. Autoclave for 15 min at 15 psi pressure–121°C. Pour into sterile Petri dishes or leave in tubes.

Use: For the isolation and cultivation of *Bacillus circulans* and *Pseudomonas* sp.

Heart Infusion Agar, HiVeg

Composition per liter:

Agar	15.0g
Plant hydrolysate No. 1	10.0g
Plant infusion	10.0g
NaCl	5.0g

pH 7.4 ± 0.2 at 25°C

Source: This medium is available as a premixed powder from HiMedia.

Preparation of Medium: Add components to distilled/deionized water and bring volume to 1.0L. Mix thoroughly. Gently heat and bring to boiling. Distribute into tubes or flasks. Autoclave for 15 min at 15 psi pressure–121°C. Pour into sterile Petri dishes or leave in tubes.

Use: For the isolation and cultivation of a wide variety of fastidious microorganisms.

Heart Infusion Agar, HiVeg with Blood

Composition per liter:

Agar	15.0g
Plant hydrolysate No. 1	10.0g
Plant infusion	10.0g
NaCl	5.0g
Horse blood, defibrinated	50.0mL

pH 7.4 ± 0.2 at 25°C

Source: This medium, without blood, is available as a premixed powder from HiMedia.

Preparation of Medium: Add components, except blood, to distilled/deionized water and bring volume to 950.0mL. Mix thoroughly. Gently heat and bring to boiling. Distribute into tubes or flasks. Autoclave for 15 min at 15 psi pressure–121°C. Cool to 50°C. Aseptically add 50.0mL defibrinated blood. Mix thoroughly. Pour into sterile Petri dishes or leave in tubes.

Use: For the isolation and cultivation of a wide variety of fastidious microorganisms. For determining hemolytic reactions.

Heart Infusion Agar with Horse Serum and Fresh Yeast Extract

Composition per 930.0mL:

Heart infusion agar	720.0mL
Horse serum, unheated	200.0mL
Fresh yeast extract solution	10.0mL

pH 7.4 ± 0.2 at 25°C

Heart Infusion Agar:

Composition per liter:

Beef heart, infusion from	500.0g
Agar	15.0g
Tryptose	10.0g
NaCl	5.0g

Preparation of Heart Infusion Agar: Add components to distilled/deionized water and bring volume to 1.0L. Mix thoroughly. Gently heat and bring to boiling. Autoclave for 15 min at 15 psi pressure–121°C. Cool to 45°–50°C.

Fresh Yeast Extract Solution:

Composition per 100.0mL:

Baker's yeast, live, pressed, starch-free	25.0g

Preparation of Fresh Yeast Extract Solution: Add the live Baker's yeast to 100.0mL of distilled/deionized water. Autoclave for 90 min at 15 psi pressure–121°C. Allow to stand. Remove supernatant solution. Adjust pH to 6.6–6.8.

Preparation of Medium: To 720.0mL of sterile cooled heart infusion broth, aseptically add 200.0mL of horse serum and 10.0mL of fresh yeast extract solution. Mix thoroughly. Pour into sterile Petri dishes or distribute into sterile tubes.

Use: For the cultivation and maintenance of *Mycoplasma equigenitalium* and *Mycoplasma subdolum.*

Heart Infusion Agar (pH 7.6) with Inactivated Horse Serum (ATCC Medium 493)

Composition per liter:

Beef heart, infusion from	500.0g
Agar	15.0g
Tryptose	10.0g

NaCl	5.0g
Horse serum, inactivated	100.0mL

pH 7.6 ± 0.2 at 25°C

Preparation of Medium: Add components, except horse serum, to distilled/deionized water and bring volume to 900.0mL. Mix thoroughly. Gently heat and bring to boiling. Autoclave for 15 min at 15 psi pressure–121°C. Cool to 45°–50°C. Aseptically add sterile horse serum. Mix thoroughly. Pour into sterile Petri dishes or distribute into sterile tubes.

Use: For the cultivation and maintenance of *Corynebacterium* species.

Heart Infusion Agar with Inactivated Horse Serum

Composition per liter:

Beef heart, infusion from	500.0g
Agar	15.0g
Tryptose	10.0g
NaCl	5.0g
Horse serum, inactivated	100.0mL

pH 7.4 ± 0.2 at 25°C

Preparation of Medium: Add components, except horse serum, to distilled/deionized water and bring volume to 900.0mL. Mix thoroughly. Gently heat and bring to boiling. Autoclave for 15 min at 15 psi pressure–121°C. Cool to 45°–50°C. Aseptically add sterile horse serum. Mix thoroughly. Pour into sterile Petri dishes or distribute into sterile tubes.

Use: For the cultivation and maintenance of *Corynebacterium* species.

Heart Infusion Agar with Inactivated Horse Serum, Sodium Chloride, and Penicillin

Composition per liter:

Beef heart, infusion from	500.0g
NaCl	35.0g
Agar	15.0g
Tryptose	10.0g
Horse serum, inactivated	100.0mL
Penicillin solution	10.0mL

pH 7.4 ± 0.2 at 25°C

Penicillin Solution:

Composition per 10.0mL:

Penicillin	1,000,000U

Preparation of Penicillin Solution: Add penicillin to distilled/deionized water and bring volume to 10.0mL. Mix thoroughly. Filter sterilize.

Preparation of Medium: Add components, except penicillin solution and horse serum, to distilled/deionized water and bring volume to 890.0mL. Mix thoroughly. Gently heat and bring to boiling. Autoclave for 15 min at 15 psi pressure–121°C. Cool to 45°–50°C. Aseptically add 10.0mL of sterile penicillin solution and 100.0mL of sterile horse serum. Mix thoroughly. Pour into sterile Petri dishes or distribute into sterile tubes.

Use: For the cultivation and maintenance of *Corynebacterium* species.

Heart Infusion Agar with Sodium Chloride (HIA with NaCl) (BAM M60)

Composition per liter:

NaCl	20.0g
Agar	15.0g
Tryptose	10.0g
NaCl	20.0g
Beef heart, infusion from 500.0g	1.0L

pH 7.4 ± 0.2 at 25°C

Source: This medium is available as a premixed powder from BD Diagnostic Systems.

Preparation of Medium: Add components to distilled/deionized water and bring volume to 1.0L. Mix thoroughly. Gently heat and bring to boiling. Distribute into tubes or flasks. Autoclave for 15 min at 15 psi pressure–121°C. Pour into sterile Petri dishes or leave in tubes.

Use: For the isolation and cultivation of halophilic *Vibrio* spp.

Heart Infusion Agar with Rabbit Blood

Composition per liter:

Beef heart, infusion from	500.0g
Agar	15.0g
Tryptose	10.0g
NaCl	5.0g
Rabbit blood	50.0mL

pH 7.4 ± 0.2 at 25°C

Preparation of Medium: Add components, except rabbit blood, to distilled/deionized water and bring volume to 950.0mL. Mix thoroughly. Gently heat and bring to boiling. Autoclave for 15 min at 15 psi pressure–121°C. Cool to 45°–50°C. Aseptically add sterile rabbit blood. Mix thoroughly. Pour into sterile Petri dishes or distribute into sterile tubes.

Use: For the cultivation and maintenance of *Neisseria lactamica, Bartonella quintana, Bartonella elizabethae,* and *Bartonella henselae.*

Heart Infusion Agar with Yeast Extract

Composition per liter:

Beef heart, infusion from	500.0g
Agar	20.0g
Tryptose	10.0g
NaCl	5.0g
Yeast extract	5.0g

pH 7.4 ± 0.2 at 25°C

Preparation of Medium: Add components to distilled/deionized water and bring volume to 1.0L. Mix thoroughly. Gently heat and bring to boiling. Distribute into tubes or flasks. Autoclave for 15 min at 15 psi pressure–121°C. Pour into sterile Petri dishes or leave in tubes.

Use: For the cultivation and maintenance of *Moraxella nonliquefaciens.*

Heart Infusion Broth

Composition per liter:

Beef heart, infusion from	500.0g
Tryptose	10.0g
NaCl	5.0g

pH 7.4 ± 0.2 at 25°C

Source: This medium is available as a premixed powder from BD Diagnostic Systems.

Preparation of Medium: Add components to distilled/deionized water and bring volume to 1.0L. Mix thoroughly. Distribute into tubes or flasks. Autoclave for 15 min at 15 psi pressure–121°C.

Use: For the isolation and cultivation of a wide variety of fastidious microorganisms.

Heart Infusion Broth (HI) (BAM M60)

Composition per liter:

Agar	15.0g
Tryptose	10.0g
NaCl	5.0g
Beef heart, infusion from 500.0g	1.0L

pH 7.4 ± 0.2 at 25°C

Source: This medium without added NaCl is available as a premixed powder from BD Diagnostic Systems.

Preparation of Medium: Add components to distilled/deionized water and bring volume to 1.0L. Mix thoroughly. Distribute into tubes or flasks. Autoclave for 15 min at 15 psi pressure–121°C.

Use: For the isolation and cultivation of a wide variety of fastidious microorganisms. For the cultivation and maintenance of *Bacillus cereus, Staphylococcus aureus* and *Vibrio vulnificus,* and *Vibrio cholerae.* It can also be used as a base for the preparation of blood agar in determining hemolytic reactions. When using for blood broth, reduce volume to 950.0mL to allow for addition of 50.0mL of defibrinated horse blood. Blood is added aseptically after autoclaving.

Heart Infusion Broth with Additives for *Staphylococcus*

Composition per liter:

Beef heart, infusion from	500.0g
NaCl	30.0 g
Tryptose	10.0g
Horse serum, inactivated	100.0mL
Penicillin solution	10.0mL
Fresh yeast extract solution	5.0mL

pH 7.4 ± 0.2 at 25°C

Penicillin Solution:

Composition per 10.0mL:

Penicillin	1,000,000U

Preparation of Penicillin Solution: Add penicillin to distilled/deionized water and bring volume to 10.0mL. Mix thoroughly. Filter sterilize.

Fresh Yeast Extract Solution:

Composition per 100.0mL:

Baker's yeast, live, pressed, starch-free	10.0g

Preparation of Fresh Yeast Extract Solution: Add the live Baker's yeast to 100.0mL of distilled/deionized water. Autoclave for 90 min at 15 psi pressure–121°C. Allow to stand. Remove supernatant solution. Adjust pH to 6.6–6.8.

Preparation of Medium: Add components—except horse serum, fresh yeast extract solution, and penicillin solution—to distilled/deionized water and bring volume to 800.0mL. Mix thoroughly. Gently heat and bring to boiling. Autoclave for 15 min at 15 psi pressure–121°C. Cool to 45°–50°C. Aseptically add sterile horse serum. Mix thoroughly. Aseptically distribute into sterile tubes or flasks.

Use: For the cultivation of *Staphylococcus* species.

Heart Infusion Broth with Additives for *Streptobacillus*

Composition per liter:

Beef heart, infusion from	500.0g
Tryptose	10.0g
Peptone	10.0g
NaCl	5.0g
Glucose	0.5g
Horse serum, inactivated	200.0mL

pH 7.5 ± 0.2 at 25°C

Preparation of Medium: Add components, except horse serum, to distilled/deionized water and bring volume to 800.0mL. Mix thoroughly. Gently heat and bring to boiling. Autoclave for 15 min at 15 psi pressure–121°C. Cool to 45°–50°C. Aseptically add sterile horse serum. Mix thoroughly. Aseptically distribute into sterile tubes or flasks.

Use: For the cultivation and maintenance of *Streptobacillus moniliformis.*

Heart Infusion Broth with Glucose

Composition per liter:

Beef heart, infusion from	500.0g
Tryptose	10.0g
NaCl	5.0g
Glucose	1.0g

pH 7.4 ± 0.2 at 25°C

Preparation of Medium: Add components to distilled/deionized water and bring volume to 1.0L. Mix thoroughly. Distribute into tubes or flasks. Autoclave for 15 min at 15 psi pressure–121°C.

Use: For the cultivation and maintenance of *Arthrobacter* species, *Bacillus* species, and *Pseudomonas* species.

Heart Infusion Broth with 0.1% Glucose (ATCC Medium 544)

Composition per liter:

Beef heart, infusion from	500.0g
Tryptose	10.0g
NaCl	5.0g
Glucose	1.0g

pH 7.4 ± 0.2 at 25°C

Source: Heart infusion broth without glucose is available as a premixed powder from BD Diagnostic Systems.

Preparation of Medium: Add components to distilled/deionized water and bring volume to 1.0L. Mix thoroughly. Distribute into tubes or flasks. Autoclave for 15 min at 15 psi pressure–121°C.

Heart Infusion Broth with Glucose and Antibiotics

Composition per liter:

Beef heart, infusion from	500.0g
Tryptose	10.0g
NaCl	5.0g
Antibiotic inhibitor solution	10.0mL
Glucose solution	10.0mL

pH 7.4 ± 0.2 at 25°C

Antibiotic Inhibitor Solution:

Composition per 10.0mL:

Streptomycin sulfate	0.1g
Tetracycline·HCl	0.025g

Preparation of Antibiotic Inhibitor Solution: Add components to distilled/deionized water and bring volume to 10.0mL. Mix thoroughly. Filter sterilize.

Glucose Solution:

Composition per 10.0mL:

D-Glucose	1.0g

Preparation of Glucose Solution: Add D-glucose to distilled/deionized water and bring volume to 10.0mL. Mix thoroughly. Filter sterilize.

Preparation of Medium: Add components, except antibiotic inhibitor solution and glucose solution, to distilled/deionized water and bring volume to 980.0mL. Mix thoroughly. Gently heat and bring to boiling. Autoclave for 15 min at 15 psi pressure–121°C. Cool to 45°–50°C. Aseptically add sterile antibiotic inhibitor solution and glucose solution. Mix thoroughly. Aseptically distribute into sterile tubes or flasks.

Use: For the cultivation of *Escherichia coli.*

Heart Infusion Broth, HiVeg

Composition per liter:

Plant hydrolysate No. 1	10.0g
Plant infusion	10.0g
NaCl	5.0g

pH 7.4 ± 0.2 at 25°C

Source: This medium is available as a premixed powder from HiMedia.

Preparation of Medium: Add components to distilled/deionized water and bring volume to 1.0L. Mix thoroughly. Gently heat and bring to boiling. Distribute into tubes or flasks. Autoclave for 15 min at 15 psi pressure–121°C.

Use: For the isolation and cultivation of a wide variety of fastidious microorganisms.

Heart Infusion Broth with Horse Serum and Fresh Yeast Extract

Composition per 930.0mL:

Heart infusion broth	720.0mL
Horse serum, unheated	200.0mL
Fresh yeast extract solution	10.0mL

pH 7.4 ± 0.2 at 25°C

Heart Infusion Broth:

Composition per liter:

Beef heart, infusion from	500.0g
Tryptose	10.0g
NaCl	5.0g

Preparation of Heart Infusion Broth: Add components to distilled/deionized water and bring volume to 1.0L. Mix thoroughly. Autoclave for 15 min at 15 psi pressure–121°C. Cool to 45°–50°C.

Fresh Yeast Extract Solution:

Composition per 100.0mL:

Baker's yeast, live, pressed, starch-free	25.0g

Preparation of Fresh Yeast Extract Solution: Add the live Baker's yeast to 100.0mL of distilled/deionized water. Autoclave for 90 min at 15 psi pressure–121°C. Allow to stand. Remove supernatant solution. Adjust pH to 6.6–6.8.

Preparation of Medium: To 720.0mL of sterile cooled heart infusion broth, aseptically add 200.0mL of horse serum and 10.0mL of fresh yeast extract solution. Mix thoroughly. Aseptically distribute into sterile tubes or flasks.

Use: For the cultivation and maintenance of *Mycoplasma equigenitalium* and *Mycoplasma subdolum.*

Heart Infusion Broth with Horse Serum, Fresh Yeast Extract, and Penicillin

Composition per 940.0mL:

Heart infusion broth	720.0mL
Horse serum, unheated	200.0mL
Fresh yeast extract solution	10.0mL
Penicillin solution	10.0mL

pH 7.4 ± 0.2 at 25°C

Heart Infusion Broth:

Composition per liter:

Beef heart, infusion from	500.0g
Tryptose	10.0g
NaCl	5.0g

Preparation of Heart Infusion Broth: Add components to distilled/deionized water and bring volume to 1.0L. Mix thoroughly. Autoclave for 15 min at 15 psi pressure–121°C. Cool to 45°–50°C.

Fresh Yeast Extract Solution:

Composition per 100.0mL:

Baker's yeast, live, pressed, starch-free	25.0g

Preparation of Fresh Yeast Extract Solution: Add the live Baker's yeast to 100.0mL of distilled/deionized water. Autoclave for 90 min at 15 psi pressure–121°C. Allow to stand. Remove supernatant solution. Adjust pH to 6.6–6.8.

Penicillin Solution:

Composition per 10.0mL:

Penicillin	100,000U

Preparation of Penicillin Solution: Add penicillin to distilled/deionized water and bring volume to 10.0mL. Mix thoroughly. Filter sterilize.

Preparation of Medium: To 720.0mL of sterile cooled heart infusion broth, aseptically add 200.0mL of horse serum, 10.0mL of fresh yeast extract solution, and 10.0mL of sterile penicillin solution. Mix thoroughly. Aseptically distribute into sterile tubes or flasks.

Use: For the cultivation and maintenance of *Mycoplasma equigenitalium* and *Mycoplasma subdolum.*

Heart Infusion Broth with Human Serum and Fresh Yeast Extract

Composition per 930.0mL:

Heart infusion broth	720.0mL
Human serum, unheated	200.0mL
Fresh yeast extract solution	10.0mL

pH 7.4 ± 0.2 at 25°C

Heart Infusion Broth:

Composition per liter:

Beef heart, infusion from	500.0g
Tryptose	10.0g
NaCl	5.0g

Preparation of Heart Infusion Broth: Add components to distilled/deionized water and bring volume to 1.0L. Mix thoroughly. Autoclave for 15 min at 15 psi pressure–121°C. Cool to 45°–50°C.

Fresh Yeast Extract Solution:

Composition per 100.0mL:

Baker's yeast, live, pressed, starch-free	25.0g

Preparation of Fresh Yeast Extract Solution: Add the live Baker's yeast to 100.0mL of distilled/deionized water. Autoclave for 90 min at 15 psi pressure–121°C. Allow to stand. Remove supernatant solution. Adjust pH to 6.6–6.8.

Preparation of Medium: To 720.0mL of sterile, cooled heart infusion broth, aseptically add 200.0mL of human serum and 10.0mL of fresh yeast extract solution. Mix thoroughly. Aseptically distribute into sterile tubes or flasks.

Use: For the cultivation and maintenance of *Mycoplasma equigenitalium* and *Mycoplasma subdolum*.

Heart Infusion Broth with Inactivated Horse Serum

Composition per liter:

Beef heart, infusion from	500.0g
Tryptose	10.0g
NaCl	5.0g
Horse serum, inactivated	100.0mL

pH 7.6 ± 0.2 at 25°C

Preparation of Medium: Add components, except horse serum, to distilled/deionized water and bring volume to 900.0mL. Mix thoroughly. Gently heat and bring to boiling. Autoclave for 15 min at 15 psi pressure–121°C. Cool to 45°–50°C. Aseptically add sterile horse serum. Mix thoroughly. Aseptically distribute into sterile tubes or flasks.

Use: For the cultivation and maintenance of *Enterococcus faecium*.

Heart Infusion Broth with Inactivated Horse Serum and Fresh Yeast Extract

Composition per liter:

Beef heart, infusion from	500.0g
Tryptose	10.0g
NaCl	5.0g
Horse serum, inactivated	200.0mL
Fresh yeast extract solution	100.0mL

pH 7.5 ± 0.2 at 25°C

Source: This medium is available as a premixed powder from BD Diagnostic Systems.

Fresh Yeast Extract Solution:
Composition per 100.0mL:

Baker's yeast, live, pressed, starch-free	25.0g

Preparation of Fresh Yeast Extract Solution: Add the live Baker's yeast to 100.0mL of distilled/deionized water. Autoclave for 90 min at 15 psi pressure–121°C. Allow to stand. Remove supernatant solution. Adjust pH to 6.6–6.8.

Preparation of Medium: Add components, except horse serum and fresh yeast extract solution, to distilled/deionized water and bring volume to 700.0mL. Mix thoroughly. Gently heat and bring to boiling. Autoclave for 15 min at 15 psi pressure–121°C. Cool to 45°–50°C. Aseptically add sterile horse serum and fresh yeast extract solution. Mix thoroughly. Aseptically distribute into sterile tubes or flasks.

Use: For the cultivation and maintenance of *Acholeplasma* species, *Mycoplasma* species, and *Streptobacillus* species.

Heart Infusion Broth with Inactivated Horse Serum, Fresh Yeast Extract, and Sucrose

Composition per liter:

Beef heart, infusion from	500.0g
Sucrose	40.0g
Tryptose	10.0g
NaCl	5.0g
Horse serum, inactivated	200.0mL
Fresh yeast extract solution	100.0mL

pH 7.5 ± 0.2 at 25°C

Source: This medium is available as a premixed powder from BD Diagnostic Systems.

Fresh Yeast Extract Solution:
Composition per 100.0mL:

Baker's yeast, live, pressed, starch-free	25.0g

Preparation of Fresh Yeast Extract Solution: Add the live Baker's yeast to 100.0mL of distilled/deionized water. Autoclave for 90 min at 15 psi pressure–121°C. Allow to stand. Remove supernatant solution. Adjust pH to 6.6–6.8.

Preparation of Medium: Add components, except horse serum and fresh yeast extract solution, to distilled/deionized water and bring volume to 700.0mL. Mix thoroughly. Gently heat and bring to boiling. Autoclave for 15 min at 15 psi pressure–121°C. Cool to 45°–50°C. Aseptically add sterile horse serum and fresh yeast extract solution. Mix thoroughly. Aseptically distribute into sterile tubes or flasks.

Use: For the cultivation and maintenance of *Acholeplasma* species, *Mycoplasma* species, and *Streptobacillus* species.

Heart Infusion Broth (pH 7.5) with Inactivated Human Serum and Yeast Extract (ATCC Medium 245)

Composition per liter:

Heart infusion broth with yeast extract	800.0mL
Human serum, inactivated	200.0mL

pH 7.5 ± 0.2 at 25°C

Heart Infusion Broth with Yeast Extract:
Composition per liter:

Beef heart, infusion from	500.0g
Tryptose	10.0g
Yeast extract (Oxoid)	6.3g
NaCl	5.0g

Preparation of Heart Infusion Broth with Yeast Extract: Add components to distilled/deionized water and bring volume to 1.0L. Mix thoroughly. Adjust pH to 7.5. Autoclave for 15 min at 15 psi pressure–121°C. Cool to 25°.

Source: Heart infusion broth without yeast extract is available as a premixed powder from BD Diagnostic Systems.

Preparation of Medium: To 800.0mL of sterile cooled heart infusion broth with yeast extract, aseptically add 200.0mL of heat inactivated human serum. Mix thoroughly. Aseptically distribute into sterile tubes or flasks.

Use: For the cultivation and maintenance of *Corynebacterium pseudotuberculosis* and *Streptobacillus moniliformis*.

Heart Infusion Broth with Porcine Serum and Fresh Yeast Extract

Composition per liter:

Beef heart, infusion from	500.0g
Tryptose	10.0g
NaCl	5.0g
Porcine serum, inactivated	200.0mL
Fresh yeast extract (25% w/v solution)	100.0mL

pH 7.5 ± 0.2 at 25°C

Porcine Serum:

Composition per 200.0mL:

Porcine serum	200.0mL

Preparation of Porcine Serum: Adjust the pH of 200.0mL of porcine serum to 4.4 with sterile 1*N* HCl. Do not overshoot pH below 4.2. Allow serum to stand at 4°C for 18–20 hr. Adjust pH to 7.0 with sterile NaOH. Aseptically centrifuge at 9000 rpm for 20 min. Discard sediment. Filter supernatant solution through a 0.85μm filter. Filter sterilize through a 0.22μm filter. Store at –70°C.

Fresh Yeast Extract Solution:

Composition per 100.0mL:

Baker's yeast, live, pressed, starch-free	25.0g

Preparation of Fresh Yeast Extract Solution: Add the live Baker's yeast to 100.0mL of distilled/deionized water. Autoclave for 90 min at 15 psi pressure–121°C. Allow to stand. Remove supernatant solution. Adjust pH to 6.6–6.8.

Preparation of Medium: Add components, except porcine serum and fresh yeast extract solution, to distilled/deionized water and bring volume to 700.0mL. Mix thoroughly. Gently heat and bring to boiling. Autoclave for 15 min at 15 psi pressure–121°C. Cool to 45°–50°C. Aseptically add sterile porcine serum and fresh yeast extract solution. Mix thoroughly. Aseptically distribute into sterile tubes or flasks.

Use: For the cultivation of *Acholeplasma* species.

Heart Infusion Broth with Sodium Chloride (HI with NaCl) (BAM M60)

Composition per liter:

NaCl	20.0g
Tryptose	10.0g
NaCl	20.0g
Beef heart, infusion from 500.0g	1.0L

pH 7.4 ± 0.2 at 25°C

Source: This medium without added NaCl is available as a premixed powder from BD Diagnostic Systems.

Preparation of Medium: Add components to distilled/deionized water and bring volume to 1.0L. Mix thoroughly. Distribute into tubes or flasks. Autoclave for 15 min at 15 psi pressure–121°C.

Use: For the isolation and cultivation of halophilic *Vibrio* spp.

Heart Infusion Medium with Fetal Bovine Serum

Composition per liter:

Beef heart, infusion from	500.0g
Agar	15.0g
Tryptose	10.0g
NaCl	5.0g
Fetal bovine serum	100.0mL

pH 7.4 ± 0.2 at 25°C

Preparation of Medium: Add components, except fetal bovine serum, to distilled/deionized water and bring volume to 900.0mL. Mix thoroughly. Gently heat and bring to boiling. Autoclave for 15 min at 15 psi pressure–121°C. Cool to 45°–50°C. Aseptically add sterile fetal bovine serum. Aseptically distribute into sterile tubes or flasks.

Use: For the cultivation of *Haemophilus ducreyi*.

Heart Infusion Tyrosine Agar

Composition per liter:

Beef heart, infusion from	500.0g
Agar	15.0g
Tryptose	10.0g
NaCl	5.0g
L-Tyrosine	1.0g

pH 7.4 ± 0.2 at 25°C

Preparation of Medium: Add components to distilled/deionized water and bring volume to 1.0L. Mix thoroughly. Gently heat and bring to boiling. Distribute into tubes. Autoclave for 15 min at 15 psi pressure–121°C. Allow tubes to cool in a slanted position.

Use: For the cultivation and differentiation of *Bordetella parapertussis* based on browning of blood-free medium.

Hektoen Enteric Agar

Composition per liter:

Agar	13.5g
Lactose	12.0g
Peptic digest of animal tissue	12.0g
Sucrose	12.0g
Bile salts	9.0g
NaCl	5.0g
$Na_2S_2O_3$	5.0g
Yeast extract	3.0g
Salicin	2.0g
Ferric ammonium citrate	1.5g
Acid Fuchsin	0.1g
Bromthymol Blue	0.064g

pH 7.6 ± 0.2 at 25°C

Source: This medium is available as a premixed powder from BD Diagnostic Systems and Oxoid Unipath.

Caution: Acid Fuchsin is a potential carcinogen and care must be taken to avoid inhalation of the powdered dye and contact with the skin.

Preparation of Medium: Add components to distilled/deionized water and bring volume to 1.0L. Mix thoroughly. Gently heat while stirring until components are dissolved. Do not autoclave. Pour into sterile Petri dishes. Allow agar to solidify with the Petri dish covers partially off.

Use: For the isolation and cultivation of Gram-negative enteric microorganisms from a variety of clinical and nonclinical specimens based on lactose or sucrose fermentation and H_2S production. For the isolation and differentiation of *Salmonella* and *Shigella.* Bacteria that ferment lactose or sucrose appear as yellow to orange colonies. Bacteria that produce H_2S appear as colonies with black centers.

Hektoen Enteric Agar

Composition per liter:

Agar	15.0g
Proteose peptone	12.0g
Lactose	12.0g
Sucrose	12.0g
Bile salts	9.0g
NaCl	5.0g
$Na_2S_2O_3$	5.0g
Yeast extract	3.0g
Salicin	2.0g
Ferric ammonium citrate	1.5g

Acid Fuchsin 0.1g
Bromthymol Blue 0.065g

pH 7.5 ± 0.2 at 25°C

Source: This medium is available as a premixed powder from HiMedia.

Caution: Acid Fuchsin is a potential carcinogen and care must be taken to avoid inhalation of the powdered dye and contact with the skin.

Preparation of Medium: Add components to distilled/deionized water and bring volume to 1.0L. Mix thoroughly. Gently heat while stirring until components are dissolved. Do not autoclave. Pour into sterile Petri dishes. Allow agar to solidify with the Petri dish covers partially off.

Use: For the isolation and cultivation of Gram-negative enteric microorganisms from a variety of clinical and nonclinical specimens based on lactose or sucrose fermentation and H_2S production. For the isolation and differentiation of *Salmonella* and *Shigella*. Bacteria that ferment lactose or sucrose appear as yellow to orange colonies. Bacteria that produce H_2S appear as colonies with black centers.

Hektoen Enteric Agar, HiVeg

Composition per liter:

Plant peptone No. 3 .... 19.0g
Agar 15.0g
Lactose 12.0g
Sucrose 12.0g
NaCl 5.0g
$Na_2S_2O_3$ 5.0g
Yeast extract 3.0g
Synthetic detergent No. I 2.0g
Salicin 2.0g
Ferric ammonium citrate 1.5g
Acid Fuchsin 0.1g
Bromthymol Blue 0.065g

pH 7.5 ± 0.2 at 25°C

Source: This medium is available as a premixed powder from HiMedia.

Caution: Acid Fuchsin is a potential carcinogen and care must be taken to avoid inhalation of the powdered dye and contact with the skin.

Preparation of Medium: Add components to distilled/deionized water and bring volume to 1.0L. Mix thoroughly. Gently heat while stirring until components are dissolved. Do not autoclave. Pour into sterile Petri dishes. Allow agar to solidify with the Petri dish covers partially off.

Use: For the isolation and cultivation of Gram-negative enteric microorganisms from a variety of clinical and nonclinical specimens based on lactose or sucrose fermentation and H_2S production. For the isolation and differentiation of *Salmonella* and *Shigella*. Bacteria that ferment lactose or sucrose appear as yellow to orange colonies. Bacteria that produce H_2S appear as colonies with black centers.

Helicobacter Medium

Composition per 850.0mL:

Agar 15.0g
Beef extract 10.0g
Peptone 10.0g
NaCl 5.0g
Horse blood, laked 50.0mL
Antibiotic solution A 10.0mL
Antibiotic solution B 10.0mL

pH 7.3 ± 0.2 at 25°C

Antibiotic Solution A:
Composition per 10.0mL:

Vancomycin 1.0mg
Trimethoprim 5.0mg
Polymyxin B .... 250 U

Preparation of Antibiotic Solution A: Add components to distilled/deionized water and bring volume to 10.0mL. Mix thoroughly. Filter sterilize.

Antibiotic Solution B:
Composition per 10.0mL:

Amphotericin B 5.0mg
Dimethylformamide 2.0mL

Preparation of Antibiotic Solution B: Add components to distilled/deionized water and bring volume to 10.0mL. Mix thoroughly. Filter sterilize.

Preparation of Medium: Add components, except laked horse blood, antibiotic solution A, and antibiotic solution B, to distilled/deionized water and bring volume to 930.0mL. Mix thoroughly. Autoclave for 15 min at 15 psi pressure–121°C. Cool to 50°–55°C. Aseptically add 50.0mL of sterile laked horse blood, 10.0mL of sterile antibiotic solution A, and 10.0mL of sterile antibiotic solution B. Mix thoroughly. Pour into sterile Petri dishes or distribute into sterile tubes.

Use: For the cultivation of *Helicobacter muridarum* and *Helicobacter felis.*

Helicobacter pylori Isolation Agar

Composition per liter:

Agar 15.0g
Bitone 10.0g
Pancreatic digest of casein 5.0g
NaCl 5.0g
Peptic digest of animal tissue 5.0g
Tryptic digest of beef heart 3.0g
Cornstarch 1.0g
Horse blood, laked 35.0mL
Antibiotic inhibitor solution 10.0mL

pH 7.3 ± 0.2 at 25°C

Antibiotic Inhibitor Solution:
Composition per 10.0mL:

Vancomycin 0.01g
Amphotericin B 5.0mg
Cefsulodin 5.0mg
Trimethoprim lactate 5.0mg

Preparation of Antibiotic Inhibitor Solution: Add components to distilled/deionized water and bring volume to 10.0mL. Mix thoroughly. Filter sterilize.

Preparation of Medium: Add components, except horse blood and antibiotic inhibitor solution, to distilled/deionized water and bring volume to 955.0mL. Mix thoroughly. Gently heat and bring to boiling. Autoclave for 15 min at 15 psi pressure–121°C. Cool to 45°–50°C. Aseptically add sterile horse blood and sterile antibiotic inhibitor solution. Mix thoroughly. Pour into sterile Petri dishes or distribute into sterile tubes.

Use: For the isolation and cultivation of *Helicobacter pylori* from clinical specimens.

Helicobacter pylori Selective Medium

Composition per 1080.0mL:

Special peptone	23.0g
Agar	10.0g
NaCl	5.0g
Starch	1.0g
Horse blood, laked	70.0mL
Selective supplement solution	10.0mL

pH 7.3 ± 0.2 at 25°C

Source: This medium is available as a premixed powder from Oxoid Unipath.

Horse Blood, Laked:

Composition per 100.0mL:

Horse blood, fresh	100.0mL

Preparation of Horse Blood, Laked: Add blood to a sterile polypropylene bottle. Freeze overnight at –20°C. Thaw at 8°C. Refreeze at –20°C. Thaw again at 8°C.

Selective Supplement Solution:

Composition per 10.0mL:

Vancomycin	10.0mg
Trimethoprim	5.0mg
Cefsulodin	5.0mg
Amphotericin B	5.0mg

Preparation of Selective Supplement Solution: Add components to distilled/deionized water and bring volume to 10.0mL. Mix thoroughly. Filter sterilize.

Preparation of Medium: Add components, except selective supplement solution and laked horse blood, to distilled/deionized water and bring volume to 1.0L. Mix thoroughly. Gently heat while stirring and bring to boiling. Autoclave for 15 min at 15 psi pressure–121°C. Cool to 50°C. Aseptially add 10.0mL selective supplement solution and 70.0mL sterile laked horse blood. Mix thoroughly. Pour into sterile Petri dishes.

Use: For the isolation of *Helicobacter pylori* from clinical specimens. *H. pylori* forms discrete, translucent, and non-coalescent colonies.

Heliobacillus mobilis Medium

Composition per 966.0mL:

Yeast extract	10.0g
$MgSO_4$	0.1g
EDTA	2.0mg
Sodium pyruvate solution	100.0mL
Trace elements solution B	10.0mL
K_2HPO_4 solution	5.0mL
Trace elements solution A	1.0mL

pH 7.1 ± 0.2 at 25°C

Sodium Pyruvate Solution:

Composition per 100.0mL:

Sodium pyruvate	1.1g

Preparation of Sodium Pyruvate Solution: Add sodium pyruvate to distilled/deionized water and bring volume to 100.0mL. Mix thoroughly. Filter sterilize.

Trace Elements Solution B:

Composition per 100.0mL:

$CaCl_2 \cdot 2H_2O$	0.3g
Ferric ammonium citrate	0.2g

Preparation of Trace Elements Solution B: Add components to distilled/deionized water and bring volume to 100.0mL. Mix thoroughly. Filter sterilize.

K_2HPO_4 Solution:

Composition per 100.0mL:

K_2HPO_4	4.0g

Preparation of K_2HPO_4 Solution: Add K_2HPO_4 to distilled/deionized water and bring volume to 100.0mL. Mix thoroughly. Filter sterilize.

Trace Elements Solution A:

Composition per 100.0mL:

H_3BO_3	2.86g
$MnCl_2 \cdot 4H_2O$	1.81g
$Na_2MoO_4 \cdot 2H_2O$	0.39g
$ZnSO_4 \cdot 7H_2O$	0.222g
$CuSO_4 \cdot 5H_2O$	0.079g
$Co(NO_3)_2 \cdot 6H_2O$	49.4mg

Preparation of Trace Elements Solution A: Add components to distilled/deionized water and bring volume to 100.0mL. Mix thoroughly. Filter sterilize.

Preparation of Medium: Add components, except sodium pyruvate solution, trace elements solution B, K_2HPO_4 solution, and trace elements solution A, to distilled/deionized water and bring volume to 850.0mL. Mix thoroughly. Adjust pH to 7.1. Autoclave for 15 min at 15 psi pressure–121°C. Cool to room temperature. Aseptically add 100.0mL of sterile sodium pyruvate solution, 10.0mL of sterile trace elements solution B, 5.0mL of sterile K_2HPO_4 solution, and 1.0mL of sterile trace elements solution A. Mix thoroughly. Aseptically distibute into sterile tubes or flasks.

Use: For the cultivation and maintenance of *Heliobacillus mobilis.*

Heliobacterium chlorum Medium

Composition per liter:

Yeast extract	10.0g
K_2HPO_4	1.0g
$MgSO_4 \cdot 7H_2O$	1.0g
Sodium ascorbate	0.5g

pH 6.8 ± 0.2 at 25°C

Preparation of Medium: Add components, except sodium ascorbate, to distilled/deionized water and bring volume to 1.0L. Mix thoroughly. Gently heat and bring to boiling. Sparge with 100% N_2 and continue boiling for 3–4 min. Add sodium ascorbate and continue to sparge with 100% N_2. Adjust pH to 6.8. Under 100% N_2, immediately dispense 45.0mL of medium into 50.0mL screw-capped tubes fitted with rubber septa. Tighten screw caps. Autoclave for 15 min at 15 psi pressure–121°C.

Use: For the cultivation and maintenance of *Heliobacillus mobilis* and *Heliobacterium chlorum.*

Heliorestis Medium (DSMZ Medium 886)

Composition per liter:

Na-acetate	1.0g
$MgCl_2 \cdot 6H_2O$	0.6g
Yeast extract	0.5g
KH_2PO_4	0.5g
NaCl	0.5g
Resazurin	0.2g

$CaCl_2$ 0.1g
Vitamin B_{12} 20.0μg
Biotin 20.0μg
Solution A 50.0mL
Trace elements solution SL-6 .... 1.0mL

pH 9.0–9.5 at 25°C

Solution A :
Composition per 50.0mL:
Na_2CO_3 .... 2.5g
$NaHCO_3$ 2.5g
NH_4Cl 0.5g
$Na_2S{\cdot}9H_2O$ 0.4g

Preparation of Solution A: Add components to distilled/deionized water and bring volume to 50.0mL. Mix thoroughly. Sparge with 100% N_2. Autoclave for 15 min at 15 psi pressure–121°C. Cool to 25°C.

Trace Elements Solution SL-6:
Composition per liter:
$MnCl_2{\cdot}4H_2O$ 0.5g
H_3BO_3 .... 0.3g
$CoCl_2{\cdot}6H_2O$ 0.2g
$ZnSO_4{\cdot}7H_2O$ 0.1g
$Na_2MoO_4{\cdot}2H_2O$ 0.03g
$NiCl_2{\cdot}6H_2O$ 0.02g
$CuCl_2{\cdot}2H_2O$ 0.01g

Preparation of Trace Elements Solution SL-6: Add components to distilled/deionized water and bring volume to 1.0L. Mix thoroughly.

Preparation of Medium: Add components, except solution A, to distilled/deionized water and bring volume to 950.0mL. Mix thoroughly. Autoclave for 15 min at 15 psi pressure–121°C. Cool to 25°C. Aseptically add 50.0mL solution A. Mix thoroughly. Adjust pH to 9.0–9.5. Aseptically distribute to sterile tubes or flasks.

Use: For the cultivation of *Heliorestis baculata.*

Hemin Medium for *Mycobacterium*
See: Middlebrook 7H10 Agar with Middlebrook OADC Enrichment and Hemin

Hemmes Medium Base

Composition per liter:
Casein enzymatic hydrolysate 10.0g
Lactose 10.0g
Sucrose 10.0g
Agar 5.5g
NaCl 4.0g
Yeast extract 3.0g
Glucose 0.3g
$Na_2S_2O_3{\cdot}5H_2O$ 0.1g
$FeSO_4{\cdot}7H_2O$ 0.04g
Phenol Red 0.015g
Urea solution 5.0mL

pH 7.2 ± 0.2 at 25°C

Source: This medium is available from HiMedia.

Urea Solution:
Composition per 10.0mL:
Urea 4.0g

Preparation of Urea Solution: Add urea to distilled/deionized water and bring volume to 10.0mL. Mix thoroughly. Filter sterilize.

Preparation of Medium: Add components, except selective supplement solution, to distilled/deionized water and bring volume to 990.0mL. Mix thoroughly. Autoclave for 15 min at 15 psi pressure–121°C. Cool to 50°C. Aseptically add selective supplement solution. Mix thoroughly. Pour into Petri dishes or aseptically distribute into sterile tubes.

Use: For the detection of *Salmonella* spp. and *Shigella* spp. based upon 7 different metabolic reactions.

Hemo ID Quad Plate with Growth Factors (*Haemophilus* Identification Quadrant Plate with Growth Factors)

Composition per plate:
Quadrant I 5.0mL
Quadrant II 5.0mL
Quadrant III 5.0mL
Quadrant IV 5.0mL

Quadrant I:
Composition per 5.0mL:
Hemin 0.1mg
Brain heart infusion agar 5.0mL

Quadrant II:
Composition per 5.0mL:
Brain heart infusion agar 5.0mL
Supplement solution 0.05mL

Quadrant III:
Composition per 5.0mL:
Hemin 0.1mg
Brain heart infusion agar 5.0mL
Supplement solution 0.05mL

Quadrant IV:
Composition per 5.0mL:
Hemin 0.1mg
Brain heart infusion agar 5.0mL
Horse blood 0.25mL
Supplement solution 0.05mL

Source: The supplement solution (IsoVitaleX® enrichment) is available from BD Diagnostic Systems. This enrichment may be replaced by supplement VX from BD Diagnostic Systems.

Preparation of Quadrant Media: Sterilize Brain Heart Infusion Agar by autocalving for 15 min at 15 psi pressure–121°C. Cool to 45°–50°C. Add additional components as filter sterilized solutions. Mix and distribute as 5.0mL aliquots into quadrants.

Use: For the differentiation and presumptive identification of *Haemophilus* species. The Hemo ID Quad Plate is a four-sectored plate, each with a different medium.

Hemorrhagic Coli Agar (HC Agar)

Composition per liter:
Sorbitol 20.0g
Pancreatic digest of casein 20.0g
Agar 15.0g
NaCl 5.0g
Bile salts No. 3 .... 1.12g
Bromcresol Purple 0.015g

pH 7.2 ± 0.2 at 25°C

Preparation of Medium: Add components to distilled/deionized water and bring volume to 1.0L. Mix thoroughly. Gently heat and bring to boiling. Distribute into tubes or flasks. Autoclave for 15 min at 15 psi pressure–121°C. Pour into sterile Petri dishes.

Use: For the isolation and cultivation of enterohemorraghic *Escherichia coli* from food.

Hemorrhagic Coli Agar with MUG (HC Agar with MUG) (BAM M62)

Composition per liter:

Sorbitol	20.0g
Pancreatic digest of casein	20.0g
Agar	15.0g
NaCl	5.0g
Bile salts No. 3	1.12g
Bromcresol Purple	0.015g
MUG reagent	0.1g

pH 7.2 ± 0.2 at 25°C

Source: MUG reagent is available from Hach Company, Loveland, Colorado.

Preparation of Medium: Add components to distilled/deionized water and bring volume to 1.0L. Mix thoroughly. Gently heat and bring to boiling. Distribute into tubes or flasks. Autoclave for 15 min at 15 psi pressure–121°C. Pour into sterile Petri dishes.

Use: For the isolation and cultivation of enterohemorraghic *Escherichia coli* from food. For hemorrhagic colitis *E. coli* strains.

Heparin Medium

Composition per liter:

Agar	15.0g
Pancreatic digest of casein	3.5g
NaCl	1.0g
Pancreatic digest of soybean meal	0.6g
Glucose	0.5g
K_2HPO_4	0.5g
Heparin solution	10.0mL

pH 6.5 ± 0.2 at 25°C

Heparin Solution:

Composition per 10.0mL:

Heparin	0.02g

Preparation of Heparin Solution: Add heparin to distilled/deionized water and bring volume to 10.0mL. Mix thoroughly. Filter sterilize.

Preparation of Medium: Add components, except heparin solution, to distilled/deionized water and bring volume to 990.0mL. Mix thoroughly. Gently heat and bring to boiling. Autoclave for 15 min at 15 psi pressure–121°C. Cool to 45°–50°C. Aseptically add sterile heparin solution. Mix thoroughly. Pour into sterile Petri dishes or distribute into sterile tubes.

Use: For the cultivation of *Flavobacterium leparinum.*

Herbaspirillum Agar

Composition per liter:

Agar	15.0g
KH_2PO_4	4.0g
$MgSO_4 \cdot 7H_2O$	0.2g
K_2HPO_4	0.1g
NaCl	0.1g
Yeast extract	0.05g
$CaCl_2$	0.02g
$FeCl_2 \cdot 6H_2O$	0.01g
$NaMoO_4 \cdot 2H_2O$	2.0mg
Solution A	50.0mL

pH 7.0 ± 0.2 at 25°C

Solution A:

Composition per 50.0mL

Sodium malate	5.0g

Preparation of Solution A: Add sodium malate to distilled/deionized water and bring volume to 50.0mL. Mix thoroughly. Adjust pH to 7.0. Filter sterilize.

Preparation of Medium: Add components, except solution A, to distilled/deionized water and bring volume to 950.0mL. Mix thoroughly. Gently heat and bring to boiling. Autoclave for 15 min at 15 psi pressure–121°C. Aseptically add 50.0mL of sterile solution A. Mix thoroughly. Aseptically pour into sterile Petri dishes or distribute into sterile tubes.

Use: For the cultivation and maintenance of *Herbaspirillum seropedicae.*

Herellea Agar

Composition per liter:

Agar	16.0g
Pancreatic digest of casein	15.0g
Lactose	10.0g
Maltose	10.0g
Enzymatic digest of soybean meal	5.0g
NaCl	5.0g
Bile salts	1.25g
Bromcresol Purple	0.02g

pH 6.8 ± 0.2 at 25°C

Source: This medium is available as a premixed powder from BD Diagnostic Systems.

Preparation of Medium: Add components to distilled/deionized water and bring volume to 1.0L. Mix thoroughly. Gently heat and bring to boiling. Distribute into tubes or flasks. Autoclave for 15 min at 15 psi pressure–121°C. Pour into sterile Petri dishes or leave in tubes.

Use: For the isolation, cultivation, and differentiation of Gram-negative nonfermentative and fermentative bacteria. It is especially recommended for the differentiation of *Acinetobacter (Herellea)* species from *Neisseria gonorrhoeae* in urethral or vaginal specimens. Fermentative bacteria appear as yellow colonies surrounded by yellow zones. Nonfermentative bacteria, such as *Acinetobacter* species, appear as pale lavender colonies.

Herpetosiphon giganteus Medium

Composition per liter:

Pancreatic digest of casein	3.0g
Yeast extract	1.0g
$CaCl_2 \cdot 2H_2O$	0.5g

pH 7.2 ± 0.2 at 25°C

Preparation of Medium: Add components to distilled/deionized water and bring volume to 1.0L. Mix thoroughly. Adjust pH to 7.2. Distribute into tubes or flasks. Autoclave for 15 min at 15 psi pressure–121°C.

Use: For the cultivation of *Herpetosiphon giganteus.*

Hershey's Tris-Buffered Salts Medium

Composition per liter:

Tris(hydroxymethyl)amino-methane buffer (0.1*M* solution)	12.1g
NaCl	5.4g
KCl	3.0g
NH_4Cl	1.1g
$MgCl_2$	0.095g
KH_2PO_4	0.087g
Na_2SO_4	0.023g
$CaCl_2$	0.011g
$FeCl_3$	0.16mg
Glucose solution	100.0mL

pH 7.4 ± 0.2 at 25°C

Glucose Solution:

Composition per 100.0mL:

Glucose	2.0g

Preparation of Glucose Solution: Add glucose to distilled/deionized water and bring volume to 100.0mL. Mix thoroughly. Autoclave for 15 min at 15 psi pressure–121°C. Cool to 25°C.

Preparation of Medium: Add components, except glucose solution, to distilled/deionized water and bring volume to 900.0mL. Mix thoroughly. Gently heat and bring to boiling. Autoclave for 15 min at 15 psi pressure–121°C. Cool to 25°C. Aseptically add sterile glucose solution. Mix thoroughly. Aseptically distribute into sterile tubes or flasks.

Use: For the cultivation of a variety of heterotrophic microorganisms.

HESNW Medium

Composition per 1011.0mL:

Natural seawater	1.0L
Enrichment solution	10.0mL
Vitamin solution	1.0mL

Enrichment Solution:

Composition per liter:

$NaNO_3$	4.667g
$Na_2SiO_3 \cdot 9H_2O$	3.000g
Sodium glycerophosphate	0.667g
$EDTA \cdot 2H_2O$	0.553g
H_3BO_3	0.380g
$Fe(NH_4)_2(SO_4)_2 \cdot 6H_2O$	0.234g
$MnSO_4 \cdot 4H_2O$	0.054g
$FeCl_3 \cdot 6H_2O$	0.016g
$ZnSO_4 \cdot 7H_2O$	7.3mg
$CoSO_4 \cdot 7H_2O$	1.6mg

Preparation of Enrichment Solution: Add $Na_2SiO_3 \cdot 9H_2O$ to distilled/deionized water. Mix thoroughly. Neutralize $Na_2SiO_3 \cdot 9H_2O$ with 1*N* HCl. Add 500.0mL of distilled/deionized water. Mix thoroughly. Add remaining components and bring volume to 1.0L with distilled/deionized water. Mix thoroughly. Filter sterilize.

Vitamin Solution:

Composition per liter:

Thiamine	0.1g
Vitamin B_{12}	2.0mg
Biotin	1.0mg

Preparation of Vitamin Solution: Add components to distilled/deionized water and bring volume to 1.0L. Mix thoroughly. Filter sterilize.

Preparation of Medium: Allow natural seawater to age for 2 months. Filter sterilize. Aseptically add 10.0mL of sterile enrichment solution and 1.0mL of sterile vitamin solution. Mix thoroughly. Aseptically distribute into sterile tubes or flasks.

Use: For the cultivation of *Amphiprora hyalina, Chlamydomonas hedleyi, Chlamydomonas provasolii, Chlorella saccharophila, Chroomonas salina, Pavlova lutheri,* and *Trichosphaerium* species.

HESP1/SR1/TMC4/LUP Medium (DSMZ Medium 860)

Composition per 1090.0mL:

Yeast extract	1.0g
NH_4Cl	1.0g
NaCl	0.6g
$Cysteine\text{-}HCl \cdot H_2O$	0.5g
K_2HPO_4	0.3g
KH_2PO_4	0.3g
$MgCl_2 \cdot 6H_2O$	0.2g
$CaCl_2 \cdot 2H_2O$	0.1g
KCl	0.1g
Resazurin	0.5mg
$NaHCO_3$ solution	60.0mL
Substrate solution	20.0mL
$Na_2S \cdot 9H_2O$ solution	10.0mL
Trace elements solution SL-10	1.5mL

pH 7.0 ± 0.2 at 25°C

Trace Elements Solution SL-10:

Composition per liter:

$FeCl_2 \cdot 4H_2O$	1.5g
$CoCl_2 \cdot 6H_2O$	190.0mg
$MnCl_2 \cdot 4H_2O$	100.0mg
$ZnCl_2$	70.0mg
$Na_2MoO_4 \cdot 2H_2O$	36.0mg
$NiCl_2 \cdot 6H_2O$	24.0mg
H_3BO_3	6.0mg
$CuCl_2 \cdot 2H_2O$	2.0mg
HCl (25% solution)	10.0mL

Preparation of Trace Elements Solution SL-10: Add $FeCl_2 \cdot 4H_2O$ to 10.0mL of HCl solution. Mix thoroughly. Add distilled/deionized water and bring volume to 1.0L. Add remaining components. Mix thoroughly. Sparge with 100% N_2. Autoclave for 15 min at 15 psi pressure–121°C.

$NaHCO_3$ Solution:

Composition per 100.0mL:

$NaHCO_3$	10.0g

Preparation of $NaHCO_3$ Solution: Add $NaHCO_3$ to distilled/deionized water and bring volume to 100.0mL. Mix thoroughly. Sparge with 80% N_2 + 20% CO_2. Filter sterilize.

$Na_2S \cdot 9H_2O$ Solution:

Composition per 10.0mL:

$Na_2S \cdot 9H_2O$	0.3g

Preparation of $Na_2S \cdot 9H_2O$ Solution: Add $Na_2S \cdot 9H_2O$ to distilled/deionized water and bring volume to 10.0mL. Sparge with N_2. Autoclave for 15 min at 15 psi pressure–121°C. Cool to 25°C. Store anaerobically.

Substrate Solution:

Composition per 10.0mL:

Fructose ... 2.0g

Preparation of Substrate Solution: Add fructose to distilled/deionized water and bring volume to 10.0mL. Sparge with N_2. Filter sterilize.

Preparation of Medium: Prepare and dispense medium under 80% N_2 + 20% CO_2. Add components, except $NaHCO_3$ solution, $Na_2S \cdot 9H_2O$ solution, and substrate solution, to distilled/deionized water and bring volume to 1.0L. Mix thoroughly. Adjust pH to 7.0. Gently heat and bring to boiling. Boil for 10 min. Cool to room temperature while sparging with 80% N_2 + 20% CO_2. Adjust pH to 6.0. Dispense under 80% N_2 + 20% CO_2 into tubes or bottles. Autoclave for 15 min at 15 psi pressure–121°C. Cool to 25°C. Aseptically and anaerobically add 0.6mL sterile $NaHCO_3$ solution per 10.0mL medium, 0.1mL $Na_2S \cdot 9H_2O$ solution per 10.0mL medium, and 0.2mL substrate solution per 10.0mL medium. Final pH is 7.0.

Use: For the cultivation of *Clostridium xylanovorans* DSM 12503.

HESP1/SR1/TMC4/LUP Medium (DSMZ Medium 860)

Composition per 1090.0mL:

Yeast extract ... 1.0g
NH_4Cl ... 1.0g
NaCl ... 0.6g
Cysteine-HCl·H_2O ... 0.5g
K_2HPO_4 ... 0.3g
KH_2PO_4 ... 0.3g
$MgCl_2 \cdot 6H_2O$ ... 0.2g
$CaCl_2 \cdot 2H_2O$ ... 0.1g
KCl ... 0.1g
Resazurin ... 0.5mg
$NaHCO_3$ solution ... 60.0mL
Substrate solution ... 20.0mL
$Na_2S \cdot 9H_2O$ solution ... 10.0mL
Trace elements solution SL-10 ... 1.5mL

pH 7.0 ± 0.2 at 25°C

Trace Elements Solution SL-10:

Composition per liter:

$FeCl_2 \cdot 4H_2O$ ... 1.5g
$CoCl_2 \cdot 6H_2O$ ... 190.0mg
$MnCl_2 \cdot 4H_2O$ ... 100.0mg
$ZnCl_2$ ... 70.0mg
$Na_2MoO_4 \cdot 2H_2O$ ... 36.0mg
$NiCl_2 \cdot 6H_2O$ ... 24.0mg
H_3BO_3 ... 6.0mg
$CuCl_2 \cdot 2H_2O$ ... 2.0mg
HCl (25% solution) ... 10.0mL

Preparation of Trace Elements Solution SL-10: Add $FeCl_2 \cdot 4H_2O$ to 10.0mL of HCl solution. Mix thoroughly. Add distilled/deionized water and bring volume to 1.0L. Add remaining components. Mix thoroughly. Sparge with 100% N_2. Autoclave for 15 min at 15 psi pressure–121°C.

$NaHCO_3$ Solution:

Composition per 100.0mL:

$NaHCO_3$ ... 10.0g

Preparation of $NaHCO_3$ Solution: Add $NaHCO_3$ to distilled/deionized water and bring volume to 100.0mL. Mix thoroughly. Sparge with 80% N_2 + 20% CO_2. Filter sterilize.

$Na_2S \cdot 9H_2O$ Solution:

Composition per 10.0mL:

$Na_2S \cdot 9H_2O$ ... 0.3g

Preparation of $Na_2S \cdot 9H_2O$ Solution: Add $Na_2S \cdot 9H_2O$ to distilled/deionized water and bring volume to 10.0mL. Sparge with N_2. Autoclave for 15 min at 15 psi pressure–121°C. Cool to 25°C. Store anaerobically.

Substrate Solution:

Composition per 10.0mL:

Na-syringate ... 0.6g

Preparation of Substrate Solution: Add Na-syringate to distilled/deionized water and bring volume to 10.0mL. Sparge with N_2. Filter sterilize.

Preparation of Medium: Prepare and dispense medium under 80% N_2 + 20% CO_2. Add components, except $NaHCO_3$ solution, $Na_2S \cdot 9H_2O$ solution, and substrate solution, to distilled/deionized water and bring volume to 1.0L. Mix thoroughly. Adjust pH to 7.0. Gently heat and bring to boiling. Boil for 10 min. Cool to room temperature while sparging with 80% N_2 + 20% CO_2. Adjust pH to 6.0. Dispense under 80% N_2 + 20% CO_2 into tubes or bottles. Autoclave for 15 min at 15 psi pressure–121°C. Cool to 25°C. Aseptically and anaerobically add 0.6mL sterile $NaHCO_3$ solution per 10.0mL medium, 0.1mL $Na_2S \cdot 9H_2O$ solution per 10.0mL medium, and 0.2mL substrate solution per 10.0mL medium. Final pH is 7.0.

Use: For the cultivation of *Sporobacterium olearium (Clostridium sp.)* DSM 12504.

HESP1/SR1/TMC4/LUP Medium (DSMZ Medium 860)

Composition per 1090.0mL:

Yeast extract ... 1.0g
NH_4Cl ... 1.0g
NaCl ... 0.6g
Cysteine-HCl·H_2O ... 0.5g
K_2HPO_4 ... 0.3g
KH_2PO_4 ... 0.3g
$MgCl_2 \cdot 6H_2O$ ... 0.2g
$CaCl_2 \cdot 2H_2O$ ... 0.1g
KCl ... 0.1g
Resazurin ... 0.5mg
$NaHCO_3$ solution ... 60.0mL
Substrate solution ... 20.0mL
$Na_2S \cdot 9H_2O$ solution ... 10.0mL
Trace elements solution SL-10 ... 1.5mL

pH 7.0 ± 0.2 at 25°C

Trace Elements Solution SL-10:

Composition per liter:

$FeCl_2 \cdot 4H_2O$ ... 1.5g
$CoCl_2 \cdot 6H_2O$ ... 190.0mg
$MnCl_2 \cdot 4H_2O$ ... 100.0mg
$ZnCl_2$ ... 70.0mg
$Na_2MoO_4 \cdot 2H_2O$ ... 36.0mg
$NiCl_2 \cdot 6H_2O$ ... 24.0mg
H_3BO_3 ... 6.0mg
$CuCl_2 \cdot 2H_2O$ ... 2.0mg
HCl (25% solution) ... 10.0mL

Preparation of Trace Elements Solution SL-10: Add $FeCl_2 \cdot 4H_2O$ to 10.0mL of HCl solution. Mix thoroughly. Add distilled/deionized

water and bring volume to 1.0L. Add remaining components. Mix thoroughly. Sparge with 100% N_2. Autoclave for 15 min at 15 psi pressure–121°C.

$NaHCO_3$ Solution:
Composition per 100.0mL:

$NaHCO_3$ 10.0g

Preparation of $NaHCO_3$ Solution: Add $NaHCO_3$ to distilled/deionized water and bring volume to 100.0mL. Mix thoroughly. Sparge with 80% N_2 + 20% CO_2. Filter sterilize.

$Na_2S{\cdot}9H_2O$ Solution:
Composition per 10.0mL:

$Na_2S{\cdot}9H_2O$ 0.3g

Preparation of $Na_2S{\cdot}9H_2O$ Solution: Add $Na_2S{\cdot}9H_2O$ to distilled/deionized water and bring volume to 10.0mL. Sparge with N_2. Autoclave for 15 min at 15 psi pressure–121°C. Cool to 25°C. Store anaerobically.

Substrate Solution:
Composition per 10.0mL:

Casamino acids 0.5g

Preparation of Substrate Solution: Add casamino acids to distilled/deionized water and bring volume to 10.0mL. Sparge with N_2. Filter sterilize.

Preparation of Medium: Prepare and dispense medium under 80% N_2 + 20% CO_2. Add components, except $NaHCO_3$ solution, $Na_2S{\cdot}9H_2O$ solution, and substrate solution, to distilled/deionized water and bring volume to 1.0L. Mix thoroughly. Adjust pH to 7.0. Gently heat and bring to boiling. Boil for 10 min. Cool to room temperature while sparging with 80% N_2 + 20% CO_2. Adjust pH to 6.0. Dispense under 80% N_2 + 20% CO_2 into tubes or bottles. Autoclave for 15 min at 15 psi pressure–121°C. Cool to 25°C. Aseptically and anaerobically add 0.6mL sterile $NaHCO_3$ solution per 10.0mL medium, 0.1mL $Na_2S{\cdot}9H_2O$ solution per 10.0mL medium, and 0.2mL substrate solution per 10.0mL medium. Final pH is 7.0.

Use: For the cultivation of *Clostridium peptidivorans* DSM 12505.

HESP1/SR1/TMC4/LUP Medium (DSMZ Medium 860)

Composition per 1090.0mL:

Yeast extract 1.0g
NH_4Cl 1.0g
NaCl 0.6g
Cysteine-HCl·H_2O 0.5g
K_2HPO_4 .... 0.3g
KH_2PO_4 .... 0.3g
$MgCl_2{\cdot}6H_2O$ 0.2g
$CaCl_2{\cdot}2H_2O$ 0.1g
KCl 0.1g
Resazurin 0.5mg
$NaHCO_3$ solution 60.0mL
Substrate solution 20.0mL
$Na_2S{\cdot}9H_2O$ solution 10.0mL
Trace elements solution SL-10 .... 1.5mL

pH 7.0 ± 0.2 at 25°C

Trace Elements Solution SL-10:
Composition per liter:

$FeCl_2{\cdot}4H_2O$ 1.5g
$CoCl_2{\cdot}6H_2O$ 190.0mg
$MnCl_2{\cdot}4H_2O$ 100.0mg
$ZnCl_2$ 70.0mg
$Na_2MoO_4{\cdot}2H_2O$ 36.0mg
$NiCl_2{\cdot}6H_2O$ 24.0mg
H_3BO_3 .... 6.0mg
$CuCl_2{\cdot}2H_2O$ 2.0mg
HCl (25% solution) 10.0mL

Preparation of Trace Elements Solution SL-10: Add $FeCl_2{\cdot}4H_2O$ to 10.0mL of HCl solution. Mix thoroughly. Add distilled/deionized water and bring volume to 1.0L. Add remaining components. Mix thoroughly. Sparge with 100% N_2. Autoclave for 15 min at 15 psi pressure–121°C.

$NaHCO_3$ Solution:
Composition per 100.0mL:

$NaHCO_3$ 10.0g

Preparation of $NaHCO_3$ Solution: Add $NaHCO_3$ to distilled/deionized water and bring volume to 100.0mL. Mix thoroughly. Sparge with 80% N_2 + 20% CO_2. Filter sterilize.

$Na_2S{\cdot}9H_2O$ Solution:
Composition per 10.0mL:

$Na_2S{\cdot}9H_2O$ 0.3g

Preparation of $Na_2S{\cdot}9H_2O$ Solution: Add $Na_2S{\cdot}9H_2O$ to distilled/deionized water and bring volume to 10.0mL. Sparge with N_2. Autoclave for 15 min at 15 psi pressure–121°C. Cool to 25°C. Store anaerobically.

Substrate Solution:
Composition per 10.0mL:

Na-lactate 1.25g
$Na_2S_2O_3{\cdot}5H_2O$ 0.05g

Preparation of Substrate Solution: Add components to distilled/deionized water and bring volume to 10.0mL. Sparge with N_2. Filter sterilize.

Preparation of Medium: Prepare and dispense medium under 80% N_2 + 20% CO_2. Add components, except $NaHCO_3$ solution, $Na_2S{\cdot}9H_2O$ solution, and substrate solution, to distilled/deionized water and bring volume to 1.0L. Mix thoroughly. Adjust pH to 7.0. Gently heat and bring to boiling. Boil for 10 min. Cool to room temperature while sparging with 80% N_2 + 20% CO_2. Adjust pH to 6.0. Dispense under 80% N_2 + 20% CO_2 into tubes or bottles. Autoclave for 15 min at 15 psi pressure–121°C. Cool to 25°C. Aseptically and anaerobically add 0.6mL sterile $NaHCO_3$ solution per 10.0mL medium, 0.1mL $Na_2S{\cdot}9H_2O$ solution per 10.0mL medium, and 0.2mL substrate solution per 10.0mL medium. Final pH is 7.0.

Use: For the cultivation of *Desulfovibrio mexicanus* DSM 13116.

HESP1/SR1/TMC4/LUP Medium (DSMZ Medium 860)

Composition per 1090.0mL:

Yeast extract 1.0g
NH_4Cl 1.0g
NaCl 0.6g
Cysteine-HCl·H_2O 0.5g
K_2HPO_4 .... 0.3g
KH_2PO_4 .... 0.3g
$MgCl_2{\cdot}6H_2O$ 0.2g
$CaCl_2{\cdot}2H_2O$ 0.1g

KCl	0.1g
Resazurin	0.5mg
$NaHCO_3$ solution	60.0mL
Substrate solution	20.0mL
$Na_2S \cdot 9H_2O$ solution	10.0mL
Trace elements solution SL-10	1.5mL

pH 7.4 ± 0.2 at 25°C

Trace Elements Solution SL-10:
Composition per liter:

$FeCl_2 \cdot 4H_2O$	1.5g
$CoCl_2 \cdot 6H_2O$	190.0mg
$MnCl_2 \cdot 4H_2O$	100.0mg
$ZnCl_2$	70.0mg
$Na_2MoO_4 \cdot 2H_2O$	36.0mg
$NiCl_2 \cdot 6H_2O$	24.0mg
H_3BO_3	6.0mg
$CuCl_2 \cdot 2H_2O$	2.0mg
HCl (25% solution)	10.0mL

Preparation of Trace Elements Solution SL-10: Add $FeCl_2 \cdot 4H_2O$ to 10.0mL of HCl solution. Mix thoroughly. Add distilled/deionized water and bring volume to 1.0L. Add remaining components. Mix thoroughly. Sparge with 100% N_2. Autoclave for 15 min at 15 psi pressure–121°C.

$NaHCO_3$ Solution:
Composition per 100.0mL:

$NaHCO_3$	10.0g

Preparation of $NaHCO_3$ Solution: Add $NaHCO_3$ to distilled/deionized water and bring volume to 100.0mL. Mix thoroughly. Sparge with 80% N_2 + 20% CO_2. Filter sterilize.

$Na_2S \cdot 9H_2O$ Solution:
Composition per 10.0mL:

$Na_2S \cdot 9H_2O$	0.3g

Preparation of $Na_2S \cdot 9H_2O$ Solution: Add $Na_2S \cdot 9H_2O$ to distilled/deionized water and bring volume to 10.0mL. Sparge with N_2. Autoclave for 15 min at 15 psi pressure–121°C. Cool to 25°C. Store anaerobically.

Substrate Solution:
Composition per 10.0mL:

$Na_2S_2O_3 \cdot 5H_2O$	1.25g

Preparation of Substrate Solution: Add $Na_2S_2O_3 \cdot 5H_2O$ to distilled/deionized water and bring volume to 10.0mL. Sparge with N_2. Filter sterilize.

Preparation of Medium: Prepare and dispense medium under 80% N_2 + 20% CO_2. Add components, except $NaHCO_3$ solution, $Na_2S \cdot 9H_2O$ solution, and substrate solution, to distilled/deionized water and bring volume to 1.0L. Mix thoroughly. Adjust pH to 7.2. Gently heat and bring to boiling. Boil for 10 min. Cool to room temperature while sparging with 80% N_2 + 20% CO_2. Adjust pH to 6.0. Dispense under 80% N_2 + 20% CO_2 into tubes or bottles. Autoclave for 15 min at 15 psi pressure–121°C. Cool to 25°C. Aseptically and anaerobically add 0.6mL sterile $NaHCO_3$ solution per 10.0mL medium, 0.1mL $Na_2S \cdot 9H_2O$ solution per 10.0mL medium, and 0.2mL substrate solution per 10.0mL medium. Final pH is 7.4.

Use: For the cultivation of *Clostridium thiosulfatireducens* DSM 13105.

HESP1/SR1/TMC4/LUP Medium (DSMZ Medium 860)

Composition per 1090.0mL:

Yeast extract	1.0g
NH_4Cl	1.0g
NaCl	0.6g
Cysteine-HCl·H_2O	0.5g
K_2HPO_4	0.3g
KH_2PO_4	0.3g
$MgCl_2 \cdot 6H_2O$	0.2g
$CaCl_2 \cdot 2H_2O$	0.1g
KCl	0.1g
Resazurin	0.5mg
$NaHCO_3$ solution	60.0mL
Substrate solution	20.0mL
$Na_2S \cdot 9H_2O$ solution	10.0mL
Trace elements solution SL-10	1.5mL

pH 7.4 ± 0.2 at 25°C

Trace Elements Solution SL-10:
Composition per liter:

$FeCl_2 \cdot 4H_2O$	1.5g
$CoCl_2 \cdot 6H_2O$	190.0mg
$MnCl_2 \cdot 4H_2O$	100.0mg
$ZnCl_2$	70.0mg
$Na_2MoO_4 \cdot 2H_2O$	36.0mg
$NiCl_2 \cdot 6H_2O$	24.0mg
H_3BO_3	6.0mg
$CuCl_2 \cdot 2H_2O$	2.0mg
HCl (25% solution)	10.0mL

Preparation of Trace Elements Solution SL-10: Add $FeCl_2 \cdot 4H_2O$ to 10.0mL of HCl solution. Mix thoroughly. Add distilled/deionized water and bring volume to 1.0L. Add remaining components. Mix thoroughly. Sparge with 100% N_2. Autoclave for 15 min at 15 psi pressure–121°C.

$NaHCO_3$ Solution:
Composition per 100.0mL:

$NaHCO_3$	10.0g

Preparation of $NaHCO_3$ Solution: Add $NaHCO_3$ to distilled/deionized water and bring volume to 100.0mL. Mix thoroughly. Sparge with 80% N_2 + 20% CO_2. Filter sterilize.

$Na_2S \cdot 9H_2O$ Solution:
Composition per 10.0mL:

$Na_2S \cdot 9H_2O$	0.3g

Preparation of $Na_2S \cdot 9H_2O$ Solution: Add $Na_2S \cdot 9H_2O$ to distilled/deionized water and bring volume to 10.0mL. Sparge with N_2. Autoclave for 15 min at 15 psi pressure–121°C. Cool to 25°C. Store anaerobically.

Substrate Solution:
Composition per 10.0mL:

Glucose	1.0g

Preparation of Substrate Solution: Add glucose to distilled/deionized water and bring volume to 10.0mL. Sparge with N_2. Filter sterilize.

Preparation of Medium: Prepare and dispense medium under 80% N_2 + 20% CO_2. Add components, except $NaHCO_3$ solution, $Na_2S \cdot 9H_2O$ solution, and substrate solution, to distilled/deionized water and bring volume to 1.0L. Mix thoroughly. Adjust pH to 7.2. Gently heat and bring to boiling. Boil for 10 min. Cool to room temperature while sparging with 80% N_2 + 20% CO_2. Adjust pH to 6.0. Dispense under 80% N_2 + 20% CO_2 into tubes or bottles. Autoclave for 15 min at 15 psi pressure–121°C. Cool to 25°C. Aseptically and anaerobically add 0.6mL sterile $NaHCO_3$ solution per 10.0mL medium, 0.1mL $Na_2S \cdot 9H_2O$ solution per 10.0mL medium, and 0.2mL substrate solution per 10.0mL medium. Final pH is 7.4.

Use: For the cultivation of *Sporanaerobacter acetigenes* DSM 13106.

Heterotrophic Agar H3P (DSMZ Medium 428)

Composition per 1010.0mL:

Solution B	855.0mL
Solution A	50.0mL
Solution E	50.0mL
Solution D	30.0mL
Solution C	20.0mL
Standard vitamin solution	5.0mL

Solution A:

Composition per 50.0mL:

$Na_2HPO_4 \cdot 2H_2O$	2.9g
KH_2PO_4	2.3g

Preparation of Solution A: Add components to distilled/deionized water and bring volume to 50.0mL. Mix thoroughly. Autoclave for 15 min at 15 psi pressure–121°C. Cool to 50°C.

Solution B:

Composition per 855.0mL:

Agar	15.0g
NH_4Cl	1.0g
$MgSO_4 \cdot 7H_2O$	0.5g
$CaCl_2 \cdot 2H_2O$	0.010g
$MnCl_2 \cdot 4H_2O$	0.005g
$NaVO_2 \cdot H_2O$	0.005g
Trace elements solution SL-6	5.0mL

Preparation of Solution B: Add components to distilled/deionized water and bring volume to 855.0mL. Mix thoroughly. Autoclave for 15 min at 15 psi pressure–121°C. Cool to 50°C.

Trace Elements Solution SL-6:

Composition per liter:

$MnCl_2 \cdot 4H_2O$	0.5g
H_3BO_3	0.3g
$CoCl_2 \cdot 6H_2O$	0.2g
$ZnSO_4 \cdot 7H_2O$	0.1g
$Na_2MoO_4 \cdot 2H_2O$	0.03g
$NiCl_2 \cdot 6H_2O$	0.02g
$CuCl_2 \cdot 2H_2O$	0.01g

Preparation of Trace Elements Solution SL-6: Add components to distilled/deionized water and bring volume to 1.0L. Mix thoroughly. Autoclave for 15 min at 15 psi pressure–121°C. Cool to room temperature.

Solution C:

Composition per 20.0mL:

Ferric ammonium citrate	0.05g
Distilled water	20.0mL

Preparation of Solution C: Add ferric ammonium citrate to distilled/deionized water and bring volume to 20.0mL. Mix thoroughly. Autoclave for 15 min at 15 psi pressure–121°C. Cool to 50°C.

Solution D:

Composition per 30.0mL:

Yeast extract	1.0g
Na-acetate	1.0g
Na_2-succinate	1.0g
DL-Malate	1.0g

Preparation of Solution D: Add components to distilled/deionized water and bring volume to 30.0mL. Mix thoroughly. Adjust to pH 7.0. Autoclave for 15 min at 15 psi pressure–121°C. Cool to 50°C.

Solution E:

Composition per 50.0mL:

Na-lactate	1.0g
Na-pyruvate	1.0g
D-Mannitol	1.0g
D-Glucose	2.0g

Preparation of Solution E: Add components to distilled/deionized water and bring volume to 50.0mL. Mix thoroughly. Adjust to pH 7.0. Filter sterilize.

Standard Vitamin Solution:

Composition per 100.0mL:

Thiamine-HCl·$2H_2O$	50.0mg
Nicotinic acid	50.0mg
Pyridoxine-HCl	50.0mg
Ca-pantothenate	50.0mg
Riboflavin	10.0mg
Vitamin B_{12}	1.0mg
Folic acid	0.2mg
Biotin	0.1mg

Preparation of Standard Vitamin Solution: Add components to distilled/deionized water and bring volume to 100.0mL. Mix thoroughly. Filter sterilize.

Preparation of Medium: Aseptically mix solutions A, B, C, D, E, and standard vitamin solution. Pour into sterile Petri dishes or aseptically distribute to flasks or tubes.

Use: For the cultivation and maintenance of *Bacillus* spp., *Pseudomonas* spp., *Xanthomonas campestris* pvar. *campestris, Xanthobacter* sp., *Aquaspirillum arcticum, Herbaspirillum seropedicae, Sphingobium chlorophenolicum=Sphingomonas chlorophenolica,* and *Azoarcus* sp. Also H3P is a heterotrophic medium for growth, purity checking, and isolation of a broad spectrum of aerobic bacteria.

Heterotrophic Agar H3P

Composition per 1010.0mL:

Solution B	855.0mL
Solution A	50.0mL
Solution E	50.0mL
Solution C	20.0mL
Solution D	30.0mL
Standard vitamin solution	5.0mL

pH 6.8 ± 0.2 at 25°C

Solution A:

Composition per 50.0mL:

$Na_2HPO_4 \cdot 2H_2O$	2.9g
KH_2PO_4	2.3g

Preparation of Solution A: Add components to distilled/deionized water and bring volume to 50.0mL. Mix thoroughly. Autoclave for 15 min at 15 psi pressure–121°C. Cool to 50°C.

Solution B:

Composition per 855.0mL:

Agar	15.0g
NH_4Cl	1.0g
$MgSO_4 \cdot 7H_2O$	0.5g
$CaCl_2 \cdot 2H_2O$	0.01g
$MnCl_2 \cdot 4H_2O$	5.0mg
$NaVO_3 \cdot H_2O$	5.0mg
Trace elements solution SL-6	5.0mL

Trace Elements Solution SL-6:

Composition per liter:

$MnCl_2 \cdot 4H_2O$	0.5g
H_3BO_3	0.3g
$CoCl_2 \cdot 6H_2O$	0.2g
$ZnSO_4 \cdot 7H_2O$	0.1g
$Na_2MoO_4 \cdot 2H_2O$	0.03g
$NiCl_2 \cdot 6H_2O$	0.02g
$CuCl_2 \cdot 2H_2O$	0.01g

Preparation of Trace Elements Solution SL-6: Add components to distilled/deionized water and bring volume to 1.0L. Mix thoroughly.

Preparation of Solution B: Add components to distilled/deionized water and bring volume to 855.0mL. Gently heat and bring to boiling. Autoclave for 15 min at 15 psi pressure–121°C.

Solution C:

Composition per 20.0mL:

Ferric ammonium citrate	0.05g

Preparation of Solution C: Add ferric ammonium citrate to distilled/deionized water and bring volume to 20.0mL. Mix thoroughly. Autoclave for 15 min at 15 psi pressure–121°C. Cool to 50°C.

Solution D:

Composition per 30.0mL:

Disodium succinate	1.0g
DL-Malate	1.0g
Sodium acetate	1.0g
Yeast extract	1.0g

Preparation of Solution D: Add components to distilled/deionized water and bring volume to 30.0mL. Mix thoroughly. Adjust pH to 7.0. Autoclave for 15 min at 15 psi pressure–121°C. Cool to 50°C.

Solution E:

Composition per 50.0mL:

D-Glucose	2.0g
D-Mannitol	1.0g
Sodium lactate	1.0g
Sodium pyruvate	1.0g

Preparation of Solution E: Add components to distilled/deionized water and bring volume to 50.0mL. Mix thoroughly. Filter sterilize. Warm to 50°C.

Standard Vitamin Solution:

Composition per 100.0mL:

Calcium DL-pantothenate	50.0mg
Nicotinic acid	50.0mg
Pyridoxine HCl	50.0mg
Thiamine	50.0mg
Riboflavin	10.0mg
B_{12}	1.0mg
Folic acid	0.2mg
Biotin	0.1mg

Preparation of Standard Vitamin Solution: Add components to distilled/deionized water and bring volume to 100.0mL. Mix thoroughly. Filter sterilize. Warm to 50°C.

Preparation of Medium: Aseptically combine 50.0mL of sterile solution A, 855.0mL of sterile solution B, 20.0mL of sterile solution C, 30.0mL of sterile solution D, 50.0mL of sterile solution E, and 5.0mL of sterile standard vitamin solution. Mix thoroughly. Pour into sterile Petri dishes or distribute into sterile tubes.

Use: For the heterotrophic growth, isolation, and purity checking of a wide variety of aerobic bacteria, including *Ancylobacter* species, *Aquaspirillum* species, *Azospirillum* species, *Beijerinckia* species, *Derxia gummosa, Herbaspirillum seropedicae, Rhodococcus* species, *Xanthobacter* species, and *Xanthomonas campestris.*

Heterotrophic Broth H3P

Composition per 1010.0mL:

Solution B	855.0mL
Solution A	50.0mL
Solution E	50.0mL
Solution C	20.0mL
Solution D	30.0mL
Standard vitamin solution	5.0mL

pH 6.8 ± 0.2 at 25°C

Solution A:

Composition per 50.0mL:

$Na_2HPO_4 \cdot 2H_2O$	2.9g
KH_2PO_4	2.3g

Preparation of Solution A: Add components to distilled/deionized water and bring volume to 50.0mL. Mix thoroughly. Autoclave for 15 min at 15 psi pressure–121°C. Cool to 50°C.

Solution B:

Composition per 855.0mL:

NH_4Cl	1.0g
$MgSO_4 \cdot 7H_2O$	0.5g
$CaCl_2 \cdot 2H_2O$	0.01g
$MnCl_2 \cdot 4H_2O$	5.0mg
$NaVO_3 \cdot H_2O$	5.0mg
Trace elements solution SL-6	5.0mL

Trace Elements Solution SL-6:

Composition per liter:

$MnCl_2 \cdot 4H_2O$	0.5g
H_3BO_3	0.3g
$CoCl_2 \cdot 6H_2O$	0.2g
$ZnSO_4 \cdot 7H_2O$	0.1g
$Na_2MoO_4 \cdot 2H_2O$	0.03g
$NiCl_2 \cdot 6H_2O$	0.02g
$CuCl_2 \cdot 2H_2O$	0.01g

Preparation of Trace Elements Solution SL-6: Add components to distilled/deionized water and bring volume to 1.0L. Mix thoroughly.

Preparation of Solution B: Add components to distilled/deionized water and bring volume to 855.0mL. Autoclave for 15 min at 15 psi pressure–121°C.

Solution C:

Composition per 20.0mL:

Ferric ammonium citrate	0.05g

Preparation of Solution C: Add ferric ammonium citrate to distilled/deionized water and bring volume to 20.0mL. Mix thoroughly. Autoclave for 15 min at 15 psi pressure–121°C. Cool to 50°C.

Solution D:

Composition per 30.0mL:

Disodium succinate	1.0g
DL-Malate	1.0g
Sodium acetate	1.0g
Yeast extract	1.0g

Preparation of Solution D: Add components to distilled/deionized water and bring volume to 30.0mL. Mix thoroughly. Adjust pH to 7.0. Autoclave for 15 min at 15 psi pressure–121°C. Cool to 50°C.

Solution E:

Composition per 50.0mL:

D-Glucose	2.0g
D-Mannitol	1.0g
Sodium lactate	1.0g
Sodium pyruvate	1.0g

Preparation of Solution E: Add components to distilled/deionized water and bring volume to 50.0mL. Mix thoroughly. Filter sterilize. Warm to 50°C.

Standard Vitamin Solution:

Composition per 100.0mL:

Calcium DL-pantothenate	50.0mg
Nicotinic acid	50.0mg
Pyridoxine HCl	50.0mg
Thiamine	50.0mg
Riboflavin	10.0mg
B_{12}	1.0mg
Folic acid	0.2mg
Biotin	0.1mg

Preparation of Standard Vitamin Solution: Add components to distilled/deionized water and bring volume to 100.0mL. Mix thoroughly. Filter sterilize. Warm to 50°C.

Preparation of Medium: Aseptically combine 50.0mL of sterile solution A, 855.0mL of sterile solution B, 20.0mL of sterile solution C, 30.0mL of sterile solution D, 50.0mL of sterile solution E, and 5.0mL of sterile standard vitamin solution. Mix thoroughly. Aseptically distribute into sterile tubes.

Use: For the heterotrophic growth, isolation, and purity checking of a wide variety of aerobic bacteria, including *Ancylobacter* species, *Aquaspirillum* species, *Azospirillum* species, *Beijerinckia* species, *Derxia gummosa, Herbaspirillum seropedicae, Rhodococcus* species, *Xanthobacter* species, and *Xanthomonas campestris.*

Heterotrophic Hyperthermophilic Archaea Medium

Composition per liter:

Pancreatic digest of casein	3.0g
Glucose	3.0g
Yeast extract	3.0g
Artificial seawater	990.0mL
$Na_2S{\cdot}9H_2O$ solution	10.0mL

pH 7.2 ± 0.2 at 25°C

Artificial Seawater:

Composition per liter:

NaCl	20.0g
$MgSO_4{\cdot}7H_2O$	6.0g
$MgCl_2{\cdot}6H_2O$	3.0g
$(NH_4)_2SO_4$	1.0g
$NaHCO_3$	0.2g
$CaCl_2{\cdot}2H_2O$	0.3g
KCl	0.5g
KH_2PO_4	0.42g
NaBr	0.05g
$SrCl_2{\cdot}6H_2O$	0.02g
$Fe(NH_4)$ citrate	0.01g
Wolfe's mineral solution	5.0mL
Vitamin solution	5.0mL

Wolfe's Mineral Solution:

Composition per liter:

$MgSO_4{\cdot}7H_2O$	3.0g
Nitrilotriacetic acid	1.5g
NaCl	1.0g
$MnSO_4{\cdot}2H_2O$	0.5g
$CoCl_2{\cdot}6H_2O$	0.1g
$ZnSO_4{\cdot}7H_2O$	0.1g
$CaCl_2{\cdot}2H_2O$	0.1g
$FeSO_4{\cdot}7H_2O$	0.1g
$NiCl_2{\cdot}6H_2O$	0.025g
$KAl(SO_4)_2{\cdot}12H_2O$	0.02g
$CuSO_4{\cdot}5H_2O$	0.01g
H_3BO_3	0.01g
$Na_2MoO_4{\cdot}2H_2O$	0.01g
$Na_2SeO_3{\cdot}5H_2O$	0.3mg

Preparation of Wolfe's Mineral Solution: Add nitrilotriacetic acid to 500.0mL of distilled/deionized water. Adjust pH to 6.5 with KOH. Add remaining components one at a time. Add distilled/deionized water to 1.0L. Adjust pH to 6.8.

Vitamin Solution:

Composition per liter:

Niacin	10.0mg
Pantothenate	10.0mg
Lipoic acid	10.0mg
p-Aminobenzoic acid	10.0mg
Thiamine (B_1)	10.0mg
Riboflavin (B_2)	10.0mg
Pyridoxine (B_6)	10.0mg
Cobalamin (B_{12})	10.0mg
Biotin	4.0mg
Folic acid	4.0mg

Preparation of Vitamin Solution: Add components to distilled/deionized water and bring volume to 1.0L. Mix thoroughly.

Preparation of Artificial Seawater: Add components to distilled/deionized water and bring volume to 1.0L. Mix thoroughly.

$Na_2S{\cdot}9H_2O$ Solution:

Composition per 10.0mL:

$Na_2S{\cdot}9H_2O$	0.8g

Preparation of $Na_2S{\cdot}9H_2O$ Solution: Add $Na_2S{\cdot}9H_2O$ to distilled/deionized water and bring volume to 10.0mL. Mix thoroughly. Sparge with 100% N_2. Autoclave for 15 min at 15 psi pressure–121°C. Before use, neutralize to pH 7.0 with sterile HCl.

Preparation of Medium: Add components, except $Na_2S{\cdot}9H_2O$ solution, to artificial seawater and bring volume to 990.0mL. Mix thoroughly. Filter sterilize. Sparge with 100% N_2 for 20 min. Aseptically and anaerobically add 10.0mL of sterile $Na_2S{\cdot}9H_2O$ solution. Mix thoroughly. Adjust medium pH to 7.2 by adding sterile anaerobic HCl. Aseptically and anaerobically distribute into sterile tubes or bottles.

Use: For the cultivation of hyperthermophilic archaea.

Heterotrophic Medium H3P (DSMZ Medium 428)

Composition per 1010.0mL:

Solution B	855.0mL
Solution A	50.0mL
Solution E	50.0mL
Solution D	30.0mL
Solution C	20.0mL
Standard vitamin solution	5.0mL

Solution A:

Composition per 50.0mL:

$Na_2HPO_4 \cdot 2H_2O$	2.9g
KH_2PO_4	2.3g

Preparation of Solution A: Add components to distilled/deionized water and bring volume to 50.0mL. Mix thoroughly. Autoclave for 15 min at 15 psi pressure–121°C. Cool to 25°C.

Solution B:

Composition per 855.0mL:

NH_4Cl	1.0g
$MgSO_4 \cdot 7H_2O$	0.5g
$CaCl_2 \cdot 2H_2O$	0.010g
$MnCl_2 \cdot 4H_2O$	0.005g
$NaVO_2 \cdot H_2O$	0.005g
Trace elements solution SL-6	5.0mL

Preparation of Solution B: Add components to distilled/deionized water and bring volume to 855.0mL. Mix thoroughly. Autoclave for 15 min at 15 psi pressure–121°C. Cool to 25°C.

Trace Elements Solution SL-6:

Composition per liter:

$MnCl_2 \cdot 4H_2O$	0.5g
H_3BO_3	0.3g
$CoCl_2 \cdot 6H_2O$	0.2g
$ZnSO_4 \cdot 7H_2O$	0.1g
$Na_2MoO_4 \cdot 2H_2O$	0.03g
$NiCl_2 \cdot 6H_2O$	0.02g
$CuCl_2 \cdot 2H_2O$	0.01g

Preparation of Trace Elements Solution SL-6: Add components to distilled/deionized water and bring volume to 1.0L. Mix thoroughly. Autoclave for 15 min at 15 psi pressure–121°C. Cool to room temperature.

Solution C:

Composition per 20.0mL:

Ferric ammonium citrate	0.05g

Preparation of Solution C: Add ferric ammonium citrate to distilled/deionized water and bring volume to 20.0mL. Mix thoroughly. Autoclave for 15 min at 15 psi pressure–121°C. Cool to 25°C.

Solution D:

Composition per 30.0mL:

Yeast extract	1.0g
Na-acetate	1.0g
Na_2-succinate	1.0g
DL-Malate	1.0g

Preparation of Solution D: Add components to distilled/deionized water and bring volume to 30.0mL. Mix thoroughly. Adjust to pH 7.0. Autoclave for 15 min at 15 psi pressure–121°C. Cool to 25°C.

Solution E:

Composition per 50.0mL:

D-Glucose	2.0g
Na-lactate	1.0g
Na-pyruvate	1.0g
D-Mannitol	1.0g

Preparation of Solution E: Add components to distilled/deionized water and bring volume to 50.0mL. Mix thoroughly. Adjust to pH 7.0. Filter sterilize.

Standard Vitamin Solution:

Composition per 100.0mL:

Thiamine-HCl·$2H_2O$	50.0mg
Nicotinic acid	50.0mg
Pyridoxine-HCl	50.0mg
Ca-pantothenate	50.0mg
Riboflavin	10.0mg
Vitamin B_{12}	1.0mg
Folic acid	0.2mg
Biotin	0.1mg

Preparation of Standard Vitamin Solution: Add components to distilled/deionized water and bring volume to 100.0mL. Mix thoroughly. Filter sterilize.

Preparation of Medium: Aseptically mix solutions A, B, C, D, E, and standard vitamin solution. Aseptically distribute to flasks or tubes.

Use: For the cultivation and maintenance of *Bacillus* spp., *Pseudomonas* spp., *Xanthomonas campestris* pvar. *campestris, Xanthobacter* sp., *Aquaspirillum arcticum, Herbaspirillum seropedicae, Sphingobium chlorophenolicum=Sphingomonas chlorophenolica,* and *Azoarcus* sp. Also H3P is a heterotrophic medium for growth, purity checking, and isolation of a broad spectrum of aerobic bacteria.

Heterotrophic Medium for *Hydrogenomonas*

Composition per liter:

Agar	15.0g
Tryptose	5.0g
Cornstarch	2.0g
Sodium succinate·$6H_2O$	2.0g
Sodium glutamate	1.0g
Yeast extract	1.0g
Sodium citrate·$2H_2O$	0.5g
Sodium acetate·$3H_2O$	0.3g
KH_2PO_4	0.2g
$MgSO_4$	0.1g

pH 6.8–7.2 at 25°C

Preparation of Medium: Add components to distilled/deionized water and bring volume to 1.0L. Mix thoroughly. Gently heat and bring to boiling. Distribute into tubes or flasks. Autoclave for 15 min at 15 psi pressure–121°C. Pour into sterile Petri dishes or leave in tubes.

Use: For the heterotrophic cultivation of *Hydrogenomonas* species.

Heterotrophic Medium for Hydrogen-Oxidizing Bacteria

Composition per 1010.0mL:

Solution A	900.0mL
Solution C	100.0mL
Solution B	10.0mL

Solution A:

Composition per 900.0mL:

Noble agar	17.0g
$Na_2HPO_4 \cdot 12H_2O$	9.0g
KH_2PO_4	1.5g
NH_4Cl	1.0g
$MgSO_4 \cdot 7H_2O$	0.2g
Trace elements solution SL-6	1.0mL

Preparation of Solution A: Add components to distilled/deionized water and bring volume to 900.0mL. Mix thoroughly. Autoclave for 15 min at 15 psi pressure–121°C. Cool to 45°–50°C.

Trace Elements Solution SL-6:

Composition per liter:

H_3BO_3	0.3g
$CoCl_2 \cdot 6H_2O$	0.2g
$ZnSO_4 \cdot 7H_2O$	0.1g
$MnCl_2 \cdot 4H_2O$	0.03g
$Na_2MoO_4 \cdot H_2O$	0.03g
$NiCl_2 \cdot 6H_2O$	0.02g
$CuCl_2.2H_2O$	0.01g

Preparation of Trace Elements Solution SL-6: Add components to distilled/deionized water and bring volume to 1.0L. Mix thoroughly. Adjust pH to 3.4.

Solution B:

Composition per 10.0mL:

$CaCl_2 \cdot 2H_2O$	0.01g
Ferric ammonium citrate	5.0mg

Preparation of Solution B: Add components to distilled/deionized water and bring volume to 10.0mL. Mix thoroughly. Autoclave for 15 min at 15 psi pressure–121°C. Cool to 45°–50°C.

Solution C:

Composition per 100.0mL:

Sodium 3-hydroxybutyrate	2.0g

Preparation of Solution C: Add sodium 3-hydroxybutyrate to distilled/deionized water and bring volume to 100.0mL. Mix thoroughly. Filter sterilize. Warm to 45°–50°C.

Preparation of Medium: Aseptically combine 900.0mL of sterile solution A, 10.0mL of sterile solution B, and 100.0mL of sterile solution C. Mix thoroughly. Pour into sterile Petri dishes or distribute into sterile tubes.

Use: For the heterotrophic cultivation and maintenance of *Xanthobacter agilis.*

Heterotrophic *Nitrobacter* Medium (DSMZ Medium 756)

Composition per liter:

Yeast extract	1.5g
Peptone	1.5g
Na-pyruvate	0.55g
Stock solution	100.0mL
Trace elements solution	1.0mL

pH 7.4 ± 0.2 at 25°C

Stock Solution:

Composition per liter:

NaCl	5.0g
KH_2PO_4	1.5g
$MgSO_4 \cdot 7H_2O$	0.5g
$CaCO_3$	0.07g

Preparation of Stock Solution: Add components to distilled/deionized water and bring volume to 1.0L. Mix thoroughly.

Trace Elements Solution:

Composition per liter:

$FeSO_4 \cdot 7H_2O$	97.3mg
H_3BO_3	49.4mg
$ZnSO_4 \cdot 7H_2O$	43.1mg
$(NH_4)_6Mo_7O_{24} \cdot 4H_2O$	37.1mg
$MnSO_4 \cdot 2H_2O$	33.8mg
$CuSO_4 \cdot 5H_2O$	25.0mg

Preparation of Trace Elements Solution: Add components to distilled/deionized water and bring volume to 1.0L. Mix thoroughly.

Preparation of Medium: Add components to distilled/deionized water and bring volume to 1.0L. Mix thoroughly. Adjust pH to 7.4. Distribute into tubes or flasks. Autoclave for 15 min at 15 psi pressure–121°C.

Use: For the cultivation of *Nitrospira moscoviensis, Nitrobacter hamburgensis, Nitrobacter vulgaris,* and *Nitrobacter winogradskyi.*

Heterotrophic *Nitrobacter* Medium (LMG Medium 245)

Composition per liter:

Yeast extract	1.5g
Peptone	1.5g
Na-pyruvate	0.55g
Stock solution	100.0mL
Trace elements solution	1.0mL

pH 7.4 ± 0.2 at 25°C

Stock Solution:

Composition per liter:

NaCl	5.0g
KH_2PO_4	1.5g
$MgSO_4 \cdot 7H_2O$	0.5g
$CaCO_3$	0.07g

Preparation of Stock Solution: Add components to distilled/deionized water and bring volume to 1.0L. Mix thoroughly.

Trace Elements Solution:

Composition per liter:

$(NH_4)Mo7O_2$	437.1mg
$FeSO_4 \cdot 7H_2O$	97.3mg
$ZnSO_4 \cdot 7H_2O$	43.1mg
H_3BO_3	39.4mg
$MnSO_4 \cdot H_2O$	33.8mg
$CuSO_2 \cdot 5H_2O$	25.0mg

Preparation of Trace Elements Solution: Add components to distilled/deionized water and bring volume to 1.0L. Mix thoroughly.

Preparation of Medium: Add components to distilled/deionized water and bring volume to 1.0L. Mix thoroughly. Adjust pH to 7.4 with NaOH. Distribute into tubes or flasks. Autoclave for 15 min at 15 psi pressure–121°C.

Use: For the cultivation of heterotrophic *Nitrobactger* spp.

Heterotrophic Plate Count

See: **HPC Agar**

Hexamita Medium

Composition per liter:

TYGM-9 medium	250.0mL
Sonneborn's *Paramecium* medium	750.0mL

TYGM-9 Medium:

Composition per liter:

NaCl	7.5g
K_2HPO_4	2.8g
Casein digest	2.0g
Gastric mucin	2.0g
Yeast extract	1.0g
KH_2PO_4	0.4g
Bovine serum, heat inactivated	30.0mL
Rice starch solution	30.0mL
Tween™ solution	0.5mL

pH 7.4 ± 0.2 at 25°C

Tween™ Solution:

Composition per 100.0mL:

Tween™ 80	10.0mL

Preparation of Tween™ Solution: Add Tween™ 80 to absolute ethanol and bring volume to 100.0mL. Mix thoroughly. Filter sterilize.

Rice Starch Solution:

Composition per 100.0mL:

Rice starch	5.0g
Phosphate-buffered saline solution	100.0mL

Phosphate-Buffered Saline Solution:

Composition per liter:

NaCl	9.0g
$Na_2HPO_4 \cdot 7H_2O$	0.795g
KH_2PO_4	0.114g

Preparation of Phosphate-Buffered Saline Solution: Add components to distilled/deionized water and bring volume to 1.0L. Mix thoroughly. Adjust pH to 7.4. Autoclave for 15 min at 15 psi pressure–121°C. Cool to 25°C.

Preparation of Rice Starch Solution: Heat sterilize rice starch at 150°C for 2 hr. Aseptically add 100.0mL of sterile phosphate-buffered saline solution. Mix thoroughly. Use immediately.

Preparation of TYGM-9 Medium: Add components, except rice starch solution, Tween™ solution, and bovine serum, to distilled/deionized water and bring volume to 939.5mL. Mix thoroughly. Autoclave for 15 min at 15 psi pressure–121°C. Cool to 25°C. Aseptically add 30.0mL of sterile bovine serum, 30.0mL of sterile rice starch solution, and 0.5mL of sterile Tween™ solution. Mix thoroughly. Aseptically distribute into sterile, screw-capped tubes or flasks.

Sonneborn's *Paramecium* Medium:

Composition per liter:

Solution 1	1.0L
Klebsiella pneumoniae cultured on solution 2	variable

Solution 1:

Composition per liter:

Rye grass cerophyll	2.5g
Na_2HPO_4	0.5g

Source: Cerophyll can be obtained from Ward's Natural Science Establishment, Inc. Dairy Goat Nutrition distributes Grass Media Culture, which is equivalent. Cereal Leaf Product from Sigma Chemical is similar to cerophyll.

Preparation of Solution 1: Add cerophyll to distilled/deionized water and bring volume to 1.0L. Mix thoroughly. Gently heat and bring to boiling. Boil for 5 min. Filter through Whatman #1 filter paper. Add 0.5g of Na_2HPO_4. Bring volume to 1.0L with distilled/deionized water. Mix thoroughly. Distribute 10.0mL volumes into tubes. Autoclave for 15 min at 15 psi pressure–121°C.

Solution 2:

Composition per liter:

Agar	20.0g
Yeast extract	4.0g
Glucose	0.16g

Preparation of Solution 2: Add components to distilled/deionized water and bring volume to 1.0L. Mix thoroughly. Gently heat and bring to boiling. Distribute 5.0mL into tubes. Autoclave for 15 min at 15 psi pressure–121°C. Allow tubes to cool in a slanted position.

Preparation of Sonneborn's *Paramecium* Medium: Inoculate the surface of agar slants of solution 2 with a culture of *Klebsiella pneumoniae*. Incubate at 37°C for 24–48 hr. Scrape cells from the surface of the agar slants and add to 10.0mL of solution 1. Incubate at 30°C for 24 hr.

Preparation of Medium: Aseptically combine 3.0mL of sterile TYGM-9 medium with 9.0mL of Sonneborn's *Paramecium* medium in 16 × 125mm screw-capped test tubes.

Use: For the cultivation of *Hexamita inflata, Hexamita pusilla, Mastigamoeba invertens,* and *Trepomonas agilis*.

HHD Medium

Composition per liter:

Agar	20.0g
Pancreatic digest of casein	10.0g
Casamino acids	3.0g
Fructose	2.5g
KH_2PO_4	2.5g
Papaic digest of soybean meal	1.5g
Tween™ 80	1.0g
Yeast extract	1.0g
Bromcresol Green solution	20.0mL

pH 7.0 ± 0.2 at 25°C

Bromcresol Green Solution:

Composition per 30.0mL:

Bromcresol Green	0.1g
NaOH (0.01*N* solution)	30.0mL

Preparation of Bromcresol Green Solution: Add Bromcresol Green to 30.0mL of NaOH solution. Mix thoroughly. Filter sterilize.

Preparation of Medium: Add components to distilled/deionized water and bring volume to 1.0L. Mix thoroughly. Gently heat and bring to boiling. Adjust pH to 7.0. Distribute into tubes or flasks. Autoclave for 15 min at 15 psi pressure–121°C. Pour into sterile Petri dishes or leave in tubes.

Use: For the cultivation of *Salmonella* species from foods.

Hickey Tresner Agar

Composition per liter:

Agar	15.0g
Dextrin	10.0g
Pancreatic digest of casein	2.0g
Meat extract	1.0g

Yeast extract..1.0g
$CaCl_2$..2.0mg

pH 7.2 ± 0.2 at 25°C

Preparation of Medium: Add components to distilled/deionized water and bring volume to 1.0L. Mix thoroughly. Gently heat and bring to boiling. Distribute into tubes or flasks. Autoclave for 15 min at 15 psi pressure–121°C. Pour into sterile Petri dishes or leave in tubes.

Use: For the cultivation and maintenance of *Thermomonospora curvata.*

HiCrome™ Aureus Agar Base with Egg Tellurite (Aureus Agar Base, HiCrome™) (HiCrome™ *Staphylococcus aureus* Agar) (*Staphylococcus aureus* Agar, HiCrome)

Composition per liter:

Agar 20.0g
Casein enzymic hydrolysate 12.0g
Sodium pyruvate...10.0g
Beef extract...6.0g
LiCl 5.0g
Yeast extract...5.0g
Pancreatic digest of gelatin..3.0g
Chromogenic mixture 2.1g
Egg tellurite emulsion..50.0mL

pH 7.4 ± 0.2 at 25°C

Source: This medium is available as a premixed powder from HiMedia.

Caution: Lithium chloride is harmful. Avoid bodily contact and inhalation of vapors. On contact with skin, wash with plenty of water immediately.

Egg Yolk Tellurite Emulsion:

Composition per 100.0mL:

Sterile saline..64.0mL
Egg yolk 30.0mL
Sterile potassium tellurite solution, 3.5% 6.0mL

Source: This medium is available from Fluka, Sigma-Aldrich, and HiMedia.

Preparation of Medium: Add components, except egg yolk tellurite emulsion, to distilled/deionized water and bring volume to 950.0mL. Mix thoroughly. Gently heat and bring to boiling. Autoclave for 15 min at 15 psi pressure–121°C. Cool to 50°C. Aseptically add 50.0mL sterile egg yok tellurite emulsion. Mix thoroughly. Pour into sterile Petri dishes or distribute into sterile tubes.

Use: For the isolation and identification of staphylococci from environmental samples. For the isolation and enumeration of coagulase positive *Staphylococcus aureus*. Coagulase positive *S. aureus* gives brown-black colonies whereas *S. epidermidis* gives yellow, slightly brownish, colonies.

HiCrome™ *Bacillus* Agar wtih Polymyxin B (*Bacillus cereus* HiCrome™ Agar)

Composition per liter:

Agar 15.0g
Peptic digest of animal tissue...10.0g
D-Mannitol 10.0g
NaCl 10.0g
Chromogenic mixture 3.2g
Meat extract 1.0g
Phenol Red..0.025g
Polymyxin B solution 1.0mL

pH 7.1 ± 0.2 at 25°C

Source: This medium, without Polymyxin B solution, is available as a premixed powder from HiMedia.

Polymyxin B Solution:

Composition per 1.0mL:

Polymyxin B 1.0mg

Preparation of Polymyxin B Solution: Add Polymyxin B to distilled/deionized water and bring volume to 1.0mL. Mix thoroughly. Filter sterilize.

Preparation of Medium: Add components, except Polymyxin B solution, to distilled/deionized water and bring volume to 1.0L. Mix thoroughly and heat with frequent agitation until boiling. Boil until components are fully dissolved. Do not autoclave. Cool to 45°–50°C. Mix thoroughly. Aseptically add 1.0mL of Polymyxin B solution. Mix thoroughly. Pour into sterile Petri dishes.

Use: For the rapid identification of *Bacillus* species from a mixed culture by chromogenic method. For the enumeration of *Bacillus cereus* and *Bacillus thuringiensis* when present in large numbers in certain foodstuffs.

HiCrome™ *Candida* Agar

Composition per liter:

Agar 15.0g
Peptic digest of animal tissue 15.0g
Chromogenic mixture 11.22g
K_2HPO_4..1.0g
Chloramphenicol...0.5g

pH 6.9 ± 0.2 at 25°C

Source: This medium is available as a premixed powder from HiMedia.

Preparation of Medium: Add components to distilled/deionized water and bring volume to 1.0L. Mix thoroughly and heat with frequent agitation until boiling. Boil until components are fully dissolved. Do not autoclave. Cool to 45°–50°C. Pour into sterile Petri dishes.

Use: For the rapid isolation and identification of *Candida* species from mixed cultures.

HiCrome™ *Candida* Agar, HiVeg

Composition per liter:

Agar 15.0g
Plant peptone 15.0g
Chromogenic mixture 11.2g
K_2HPO_4..1.0g
Chloramphenicol...0.5g

pH 6.9 ± 0.2 at 25°C

Source: This medium is available as a premixed powder from HiMedia.

Preparation of Medium: Add components to distilled/deionized water and bring volume to 1.0L. Mix thoroughly and heat with frequent agitation until boiling. Boil until components are fully dissolved. Do not autoclave. Cool to 45°–50°C. Pour into sterile Petri dishes.

Use: For the rapid isolation and identification of *Candida* species from mixed cultures.

HiCrome™ *Candida* Differential Agar Base with *Candida* Selective Supplement

Composition per liter:

Agar	15.0g
Peptone, special	15.0g
Yeast extract	4.0g
Chromogenic mixture	7.220g
K_2HPO_4	1.0g
Chloramphenicol	0.5g
Gentamicin solution	1.0mL

pH 7.2 ± 0.2 at 25°C

Source: This medium is available as a premixed powder from HiMedia.

***Candida* Selective Supplement:**

Composition per 10.0mL:

Gentamicin	0.1g

Preparation of *Candida* Selective Supplement: Add gentamicin to distilled/deionized water and bring volume to 10.0mL. Mix thoroughly. Filter sterilize.

Preparation of Medium: Add components, except *Candida* selective supplement, to distilled/deionized water and bring volume to 990.0mL. Mix thoroughly. Gently heat and bring to boiling. Distribute into tubes or flasks. Do not autoclave. Do not overheat. Cool to 50°C. Aseptically add 10.0mL *Candida* selective supplement. Mix thoroughly. Pour into sterile Petri dishes or leave in tubes.

Use: For the rapid isolation and identification of *Candida* species from mixed cultures.

HiCrome™ *Candida* Differential Agar Base, Modified with *Candida* Selective Supplement

Composition per liter:

Agar	18.0g
Glucose	10.0g
Peptic digest of animal tissue	5.0g
Malt extract	3.0g
Yeast extract	3.0g
Chromogenic mixture	3.0g
Chloramphenicol	0.05g
Gentamicin solution	1.0mL

pH 7.2 ± 0.2 at 25°C

Source: This medium is available as a premixed powder from HiMedia.

***Candida* Selective Supplement:**

Composition per 10.0mL:

Gentamicin	0.1g

Preparation of *Candida* Selective Supplement: Add gentamicin to distilled/deionized water and bring volume to 10.0mL. Mix thoroughly. Filter sterilize.

Preparation of Medium: Add components, except *Candida* selective supplement, to distilled/deionized water and bring volume to 990.0mL. Mix thoroughly. Gently heat and bring to boiling. Distribute into tubes or flasks. Do not autoclave. Do not overheat. Cool to 50°C. Aseptically add 10.0mL *Candida* selective supplement. Mix thoroughly. Pour into sterile Petri dishes or leave in tubes.

Use: For the rapid isolation and identification of *Candida* species from mixed cultures.

HiCrome™ *Candida* HiVeg Agar Base, Modified wtih *Candida* Selective Supplement

Composition per liter:

Agar	18.0g
Glucose	10.0g
Plant peptone	5.0g
Malt extract	3.0g
Yeast extract	3.0g
Chromogenic mixture	3.0g
Chloramphenicol	0.05g
Gentamicn solution	1.0mL

pH 7.2 ± 0.2 at 25°C

Source: This medium is available as a premixed powder from HiMedia.

***Candida* Selective Supplement:**

Composition per 10.0mL:

Gentamicin	0.1g

Preparation of *Candida* Selective Supplement: Add gentamicin to distilled/deionized water and bring volume to 10.0mL. Mix thoroughly. Filter sterilize.

Preparation of Medium: Add components, except *Candida* selective supplement, to distilled/deionized water and bring volume to 990.0mL. Mix thoroughly. Gently heat and bring to boiling. Distribute into tubes or flasks. Do not autoclave. Do not overheat. Cool to 50°C. Aseptically add 10.0mL *Candida* selective supplement. Mix thoroughly. Pour into sterile Petri dishes or leave in tubes.

Use: For the rapid isolation and identification of *Candida* species from mixed cultures.

HiCrome™ Coliform Agar (Coliform Agar, HiCrome™)

Composition per liter:

Agar	12.0g
Sodium chloride	5.0g
Peptone, special	3.0g
K_2HPO_4	3.0g
KH_2PO_4	1.7g
Sodium pyruvate	1.0g
Tryptophan	1.0g
Chromogenic mixture	0.2g
Sodium lauryl sulphate	0.1g

pH 6.8 ± 0.2 at 25°C

Source: This medium is available as a premixed powder from HiMedia.

Preparation of Medium: Add components to distilled/deionized water and bring volume to 1.0L. Mix thoroughly. Gently heat while stirring and bring to boiling. Autoclave for 15 min at 15 psi pressure–121°C. Pour into sterile Petri dishes.

Use: For the simultaneous detection of *Escherichia coli* and total coliforms in water and food samples. Sodium lauryl sulphate inhibits Gram-positive organisms. The chromogenic mixture contains two chromogenic substrates as Salmon-GAL and X-glucuronide. The enzyme β-D-galactosidase produced by coliforms cleaves Salmon-GAL, resulting in the salmon to red coloration of coliform colonies. The enzyme β-D-glucuronidase produced by *E. coli* cleaves X-glucuronide. *Escherichia coli* forms dark blue to violet coloued colonies due to cleavage of both Salmon-GAL and X-glucuronide. The addition of tryptophan improves the indole reaction, thereby increasing detection reliability in combination with the two chromogens.

HiCrome™ Coliform Agar with Novobiocin (Coliform Agar with Novobiocin, HiCrome™)

Composition per liter:

Agar	12.0g
Sodium chloride	5.0g
Peptone, special	3.0g
K_2HPO_4	3.0g
KH_2PO_4	1.7g
Sodium pyruvate	1.0g
Tryptophan	1.0g
Chromogenic mixture	0.2g
Sodium lauryl sulphate	0.1g
Novobiocin	5.0mg

pH 6.8 ± 0.2 at 25°C

Source: This medium is available as a premixed powder from HiMedia.

Preparation of Medium: Add components to distilled/deionized water and bring volume to 1.0L. Mix thoroughly. Gently heat while stirring and bring to boiling. Autoclave for 15 min at 15 psi pressure–121°C. Pour into sterile Petri dishes.

Use: For the simultaneous detection of *Escherichia coli* and total coliforms in water and food samples when high numbers of Gram-positive bacteria may be present. Novobiocin and sodium lauryl sulphate inhibit Gram-positive organisms. The chromogenic mixture contains two chromogenic substrates as Salmon-GAL and X-glucuronide. The enzyme β-D-galactosidase produced by coliforms cleaves Salmon-GAL, resulting in the salmon to red coloration of coliform colonies. The enzyme β-D-glucuronidase produced by *E. coli*, cleaves X-glucuronide. *Escherichia coli* forms dark blue to violet colored colonies due to cleavage of both Salmon-GAL and X-glucuronide. The addition of tryptophan improves the indole reaction, thereby increasing detection reliability in combination with the two chromogens.

HiCrome™ Coliform Agar with SLS

Composition per liter:

Agar	12.0g
NaCl	5.0g
K_2HPO_4	3.0g
Peptone, special	3.0g
KH_2PO_4	1.7g
Sodium pyruvate	1.0g
Tryptophan	1.0g
Chromogenic mixture	0.2g
Sodium lauryl sulfate	0.1g

pH 7.1± 0.2 at 25°C

Source: This medium is available as a premixed powder from HiMedia.

Preparation of Medium: Add components to distilled/deionized water and bring volume to 1.0L. Mix thoroughly. Gently heat and bring to boiling. Distribute into tubes or flasks. Autoclave for 15 min at 15 psi pressure–121°C. Pour into sterile Petri dishes.

Use: For the simultaneous detection of *Escherichia coli* and total coliforms in water and food samples.

HiCrome™ Coliform Agar with SLS and Novobiocin

Composition per liter:

Agar	12.0g
NaCl	5.0g
K_2HPO_4	3.0g
Peptone, special	3.0g
KH_2PO_4	1.7g
Sodium pyruvate	1.0g
Tryptophan	1.0g
Chromogenic mixture	0.2g
Sodium lauryl sulfate	0.1g
Novobiocin	5.0mg

pH 7.1 ± 0.2 at 25°C

Source: This medium, wihout novobiocin, is available as a premixed powder from HiMedia.

Preparation of Medium: Add components to distilled/deionized water and bring volume to 1.0L. Mix thoroughly. Gently heat and bring to boiling. Distribute into tubes or flasks. Autoclave for 15 min at 15 psi pressure–121°C. Pour into sterile Petri dishes.

Use: For the simultaneous detection of *Escherichia coli* and total coliforms in water and food samples when a high number of Gram-positive accompanying bacteria are expected.

HiCrome™ Coliform HiVeg Agar wtih SLS

Composition per liter:

Agar	12.0g
NaCl	5.0g
K_2HPO_4	3.0g
Plant special peptone	3.0g
KH_2PO_4	1.7g
Sodium pyruvate	1.0g
Tryptophan	1.0g
Chromogenic mixture	0.2g
Sodium lauryl sulfate	0.1g

pH 7.1 ± 0.2 at 25°C

Source: This medium is available as a premixed powder from HiMedia.

Preparation of Medium: Add components to distilled/deionized water and bring volume to 1.0L. Mix thoroughly. Gently heat and bring to boiling. Distribute into tubes or flasks. Autoclave for 15 min at 15 psi pressure–121°C. Pour into sterile Petri dishes.

Use: For the simultaneous detection of *Escherichia coli* and total coliforms in water and food samples.

HiCrome™ *E. coli* Agar

Composition per liter:

Casein enzymic hydrolysate	14.0g
Agar	12.0g
Peptone, special	5.0g
NaCl	2.4g
Bile salts mixture	1.5g
Na_2HPO_4	1.0g
NaH_2PO_4	0.6g
X-Glucuronide	0.075g

pH 7.1 ± 0.2 at 25°C

Source: This medium is available as a premixed powder from HiMedia.

Preparation of Medium: Add components to distilled/deionized water and bring volume to 1.0L. Mix thoroughly. Gently heat and bring to boiling. Distribute into tubes or flasks. Autoclave for 15 min at 15 psi pressure–121°C. Pour into sterile Petri dishes.

Use: For the simultaneous detection of *Escherichia coli* in foods without further confirmation on membrane filter or by indole reagent.

HiCrome™ *E. coli* Agar

Composition per liter:

Casein enzymic hydrolysate	20.0g
Agar	15.0g
Bile salts mixture	1.50g
X-Glucuronide	7.5mg

pH 7.1 ± 0.2 at 25°C

Source: This medium is available as a premixed powder from HiMedia.

Preparation of Medium: Add components to distilled/deionized water and bring volume to 1.0L. Mix thoroughly. Gently heat and bring to boiling. Distribute into tubes or flasks. Autoclave for 15 min at 15 psi pressure–121°C. Pour into sterile Petri dishes.

Use: For the simultaneous detection of *Escherichia coli* in foods without further confirmation on membrane filter or by indole reagent.

HiCrome™ *E. coli* Agar A (*E. coli* Agar A, HiCrome™)

Composition per liter:

Casein enzymic hydrolysate	14.0g
Agar	12.0g
Peptone, special	5.0g
NaCl	2.4g
Bile salts mixture	1.5g
Na_2HPO_4	1.0g
NaH_2PO_4	0.6g
X-Glucuronide	0.075g

pH 7.2 ± 0.2 at 25°C

Source: This medium is available as a premixed powder from HiMedia.

Preparation of Medium: Add components to distilled/deionized water and bring volume to 1.0L. Mix thoroughly. Gently heat while stirring and bring to boiling. Autoclave for 15 min at 15 psi pressure–121°C. Cool to 50°C. Pour into sterile Petri dishes.

Use: For the detection and enumeration of *Escherichia coli* in foods without further confirmation on membrane filter or by indole reagent. The chromogenic agent X-glucuronide used in this medium helps to detect glucuronidase activity. *E. coli* cells absorb X-glucuronide and the intracellular glucuronidase splits the bond between the chromophore and the glucuronide. The released chromophore gives coloration to the colonies. Bile salts mixture inhibits Gram-positive organisms.

HiCrome™ *E. coli* Agar B (*E. coli* Agar B, HiCrome™)

Composition per liter:

Casein enzymic hydrolysate	20.0g
Agar	15.0g
Bile salts mixture	1.5g
X-Glucuronide	0.075g

pH 7.2 ± 0.2 at 25°C

Source: This medium is available as a premixed powder from HiMedia.

Preparation of Medium: Add components to distilled/deionized water and bring volume to 1.0L. Mix thoroughly. Gently heat while stirring and bring to boiling. Autoclave for 15 min at 15 psi pressure–121°C. Pour into sterile Petri dishes.

Use: For the detection and enumeration of *Escherichia coli* in foods without further confirmation on membrane filter or by indole reagent. The chromogenic agent X-glucuronide used in this medium helps to detect glucuronidase activity. *E. coli* cells absorb X-glucuronide and the intracellular glucuronidase splits the bond between the chromophore and the glucuronide. The released chromophore gives coloration to the colonies. Bile salts mixture inhibits Gram-positive organisms.

HiCrome™ *E. coli* Agar, HiVeg

Composition per liter:

Plant hydrolysate	14.0g
Agar	12.0g
Plant special peptone	5.0g
NaCl	2.4g
Synthetic detergent	1.5g
Na_2HPO_4	1.0g
NaH_2PO_4	0.6g
X-Glucuronide	0.075g

pH 7.1 ± 0.2 at 25°C

Source: This medium is available as a premixed powder from HiMedia.

Preparation of Medium: Add components to distilled/deionized water and bring volume to 1.0L. Mix thoroughly. Gently heat and bring to boiling. Distribute into tubes or flasks. Autoclave for 15 min at 15 psi pressure–121°C. Pour into sterile Petri dishes.

Use: For the simultaneous detection of *Escherichia coli* in foods without further confirmation on membrane filter or by indole reagent.

HiCrome™ EC O157:H7 Agar

Composition per liter:

Agar	15.0g
Casein enzymic hydrolysate	8.0g
Sorbitol	7.0g
Bile salts mixture	1.5g
Chromogenic mixture	0.25g
Sodium lauryl sulfate	0.1g

pH 7.1 ± 0.2 at 25°C

Source: This medium is available as a premixed powder from HiMedia.

Preparation of Medium: Add components to distilled/deionized water and bring volume to 1.0L. Mix thoroughly. Gently heat and bring to boiling. Distribute into tubes or flasks. Autoclave for 15 min at 15 psi pressure–121°C. Pour into sterile Petri dishes.

Use: For the isolation and differentiation of *Escherichia coli* O157:H7 from food and recommended for selective isolation and easy detection of *Escherichia coli* O157:H7.

HiCrome™ ECC Agar (ECC Agar, HiCrome™)

Composition per liter:

Chromogenic mixture	20.3g
Agar	15.0g

Peptone, special	5.0g
NaCl	5.0g
Na_2HPO_4	3.5g
Yeast extract	3.0g
Lactose	2.5g
KH_2PO_4	1.5g
Neutral Red	0.03g

pH 6.8 ± 0.2 at 25°C

Source: This medium is available as a premixed powder from HiMedia.

Preparation of Medium: Add components to distilled/deionized water and bring volume to 1.0L. Mix thoroughly. Gently heat while stirring and bring to boiling. Autoclave for 15 min at 15 psi pressure–121°C. Pour into sterile Petri dishes.

Use: For the presumptive identification of *Escherichia coli* and other coliforms in food and environmental samples. A differential medium for presumptive identification of *E. coli* and other coliforms in food samples. The chromogenic mixture contains two chromogens as X-glucuronide and Salmon-GAL. X-glucuronide is cleaved by the enzyme β-glucuronidase produced by *E. coli*. Salmon-GAL is cleaved by the enzyme galactosidase produced by the majority of coliforms, including *E. coli*.

HiCrome™ ECC HiVeg Agar

Composition per liter:

Chromogenic mixture	20.3g
Agar	15.0g
Plant special peptone	5.0g
NaCl	5.0g
Na_2HPO_4	3.5g
Yeast extract	3.0g
Lactose	2.5g
KH_2PO_4	1.5g
Neutral Red	0.03g

pH 7.2 ± 0.2 at 25°C

Source: This medium is available as a premixed powder from HiMedia.

Preparation of Medium: Add components to distilled/deionized water and bring volume to 1.0L. Mix thoroughly. Gently heat and bring to boiling. Distribute into tubes or flasks. Autoclave for 15 min at 15 psi pressure–121°C. Pour into sterile Petri dishes.

Use: For the presumptive identification of *Escherichia coli* and other coliforms in food and environmental samples.

HiCrome™ ECC Selective Agar (ECC Selective Agar, HiCrome™)

Composition per liter:

Agar	10.0g
Peptone, special	6.0g
Casein enzymic hydrolysate	3.3g
NaCl	2.0g
Sodium pyruvate	1.0g
Tryptophan	1.0g
Sorbitol	1.0g
Na_2HPO_4	1.0g
NaH_2PO_4	0.6g
Tergitol 7®	0.15g
Chromogenic mixture	0.43g

pH 6.8 ± 0.2 at 25°C

Source: This medium is available as a premixed powder from HiMedia.

Preparation of Medium: Add components to distilled/deionized water and bring volume to 1.0L. Mix thoroughly and heat with frequent agitation until components are completely dissolved (approximately 35 min). Do not autoclave. Cool to 45°–50°C. Pour into sterile Petri dishes. Medium may show haziness.

Use: For detection of *Escherichia coli* and coliforms in water and food samples.Tergitol inhibits Gram-positive as well as some Gram-negative bacteria other than coliforms. The chromogenic mixture contains two chromogenic substrates as Salmon-GAL and X-glucuronide. The enzyme b-D-galactosidase produced by coliforms cleaves Salmon-GAL, resulting in the salmon to red coloration of coliform colonies. The enzyme β-D-glucuronidase produced by *E. coli*, cleaves X-glucuronide. *E. coli* forms dark blue to violet colored colonies due to cleavage of both Salmon-GAL and X-glucuronide. The addition of tryptophan improves the indole reaction.

HiCrome™ ECC Selective Agar (ECC Selective Agar, HiCrome™)

Composition per liter:

Agar	10.0g
Peptone, special	6.0g
Casein enzymic hydrolysate	3.3g
NaCl	2.0g
Sodium pyruvate	1.0g
Tryptophan	1.0g
Sorbitol	1.0g
Na_2HPO_4	1.0g
NaH_2PO_4	0.6g
Tergitol 7®	0.15g
Chromogenic mixture	0.43g
Cefsulodin	5.0mg

pH 6.8 ± 0.2 at 25°C

Source: This medium is available as a premixed powder from HiMedia.

Preparation of Medium: Add components to distilled/deionized water and bring volume to 1.0L. Mix thoroughly and heat with frequent agitation until components are completely dissolved (approximately 35 min.). Do not autoclave. Cool to 45°–50°C. Pour into sterile Petri dishes. Medium may show haziness.

Use: For detection of *Escherichia coli* and coliforms in water and food samples using pour plate or streak plate methods.Tergitol inhibits Gram-positive as well as some Gram-negative bacteria other than coliforms. The cefsulodin inhibits *Pseudomonas* and *Aeromonas* species. The chromogenic mixture contains two chromogenic substrates as Salmon-GAL and X-glucuronide. The enzyme β-D-galactosidase produced by coliforms cleaves Salmon-GAL, resulting in the salmon to red coloration of coliform colonies. The enzyme β-D-glucuronidase produced by *E. coli*, cleaves X-glucuronide. *E. coli* forms dark blue to violet colored colonies due to cleavage of both Salmon-GAL and X-glucuronide. The addition of tryptophan improves the indole reaction.

HiCrome™ ECC Selective Agar (ECC Selective Agar, HiCrome™)

Composition per liter:

Agar	10.0g
Peptone, special	6.0g
Casein enzymic hydrolysate	3.3g

NaCl	2.0g
Sodium pyruvate	1.0g
Tryptophan	1.0g
Sorbitol	1.0g
Na_2HPO_4	1.0g
NaH_2PO_4	0.6g
Tergitol 7®	0.15g
Chromogenic mixture	0.43g
Cefsulodin	10.0mg

pH 6.8 ± 0.2 at 25°C

Source: This medium is available as a premixed powder from HiMedia.

Preparation of Medium: Add components to distilled/deionized water and bring volume to 1.0L. Mix thoroughly and heat with frequent agitation until components are completely dissolved (approximately 35 min). Do not autoclave. Cool to 45°–50°C. Pour into sterile Petri dishes. Medium may show haziness.

Use: For detection of *Escherichia coli* and coliforms in water and food samples using the membrane filter technique.Tergitol inhibits Gram-positive as well as some Gram-negative bacteria other than coliforms. The cefsulodin inhibits *Pseudomonas* and *Aeromonas* species. The chromogenic mixture contains two chromogenic substrates as Salmon-GAL and X-glucuronide. The enzyme β-D-galactosidase produced by coliforms cleaves Salmon-GAL, resulting in the salmon to red coloration of coliform colonies. The enzyme β-D-glucuronidase produced by *E. coli* cleaves X-glucuronide. *E. coli* forms dark blue to violet colored colonies due to cleavage of both Salmon-GAL and X-glucuronide. The addition of tryptophan improves the indole reaction.

HiCrome™ ECC Selective Agar Base

Composition per liter:

Agar	10.0g
Peptone, special	6.0g
Casein enzymic hydrolysate	3.3g
NaCl	2.0g
Na_2HPO_4	1.0g
Sodium pyruvate	1.0g
Sorbitol	1.0g
Tryptophan	1.0g
Sodium dihydrogen phosphate	0.6g
Chromogenic mixture	0.43g
Tergitol 7	0.15g

pH 7.0± 0.2 at 25°C

Source: This medium is available as a premixed powder from HiMedia.

Preparation of Medium: Add components to distilled/deionized water and bring volume to 1.0L. Mix thoroughly and heat with frequent agitation in boiling water bath or flowing steam. Boil until components are fully dissolved (approximately 35 min). Do not autoclave. Cool to 45°–50°C. Pour into sterile Petri dishes. Medium may show haziness. To inhibit *Pseudomonas* and *Aeromonas* species 5 mg cefsulodin may be added for surface and pour plate methods or 10 mg cefsulodin may be aseptically added to the medium for the membrane filter technique.

Use: For the detection of *Escherichia coli* and coliforms in water and food samples.

HiCrome™ ECC Selective Agar Base, HiVeg

Composition per liter:

Agar	10.0g
Plant special peptone	6.0g
Plant hydrolysate	3.3g
NaCl	2.0g
Sodium dihydrogen phosphate	0.6g
Na_2HPO_4	1.0g
Sodium pyruvate	1.0g
Sorbitol	1.0g
Tryptophan	1.0g
Chromogenic mixture	0.43g
Tergitol 7	0.15g

pH 7.0± 0.2 at 25°C

Source: This medium is available as a premixed powder from HiMedia.

Preparation of Medium: Add components to distilled/deionized water and bring volume to 1.0L. Mix thoroughly and heat with frequent agitation in boiling water bath or flowing steam. Boil until components are fully dissolved (approximately 35 min). Do not autoclave. Cool to 45°–50°C. Pour into sterile Petri dishes. Medium may show haziness. To inhibit *Pseudomonas* and *Aeromonas* species 5 mg cefsulodin may be added for surface and pour plate methods or 10 mg cefsulodin may be aseptically added to the medium for membrane filter technique.

Use: For the detection of *Escherichia coli* and coliforms in water and food samples.

HiCrome™ ECD Agar with MUG

Composition per liter:

Casein enzymic hydrolysate	20.0g
Agar	15.0g
Lactose	5.0g
NaCl	5.0g
K_2HPO_4	4.0g
Bile salts mixture	1.5g
KH_2PO_4	1.5g
L-Tryptophan	1.0g
Chromogenic substrate	0.1g
Fluorogenic substrate	0.07g

pH 7.0 ± 0.2 at 37°C

Source: This medium is available as a premixed powder from HiMedia.

Preparation of Medium: Add components to distilled/deionized water and bring volume to 1.0L. Mix thoroughly. Gently heat while stirring and bring to boiling. Autoclave for 15 min at 15 psi pressure–121°C. Cool to 50°C. Pour into sterile Petri dishes.

Use: For detection of *Escherichia coli* in a variety of specimens. Fluorescence in the UV and a positive indole test demonstrate the presence of *E. coli* in the colonies.

HiCrome™ ECD HiVeg Agar with MUG

Composition per liter:

Plant hydrolysate	20.0g
Agar	15.0g
NaCl	5.0g
Lactose	5.0g
K_2HPO_4	4.0g
KH_2PO_4	1.5g
Synthetic detergent	1.5g
Tryptophan	1.0g
Chromogenic substrate	0.1g
Fluorogenic substrate	0.07g

pH 7.0 ± 0.2 at 37°C

Source: This medium is available as a premixed powder from HiMedia.

Preparation of Medium: Add components to distilled/deionized water and bring volume to 1.0L. Mix thoroughly. Gently heat while stirring and bring to boiling. Autoclave for 15 min at 15 psi pressure–121°C. Cool to 50°C. Pour into sterile Petri dishes.

Use: For detection of *Escherichia coli* in a variety of specimens, including foods and water. Fluorescence in the UV and a positive indole test demonstrate the presence of *E. coli* in the colonies.

HiCrome™ Enrichment Broth Base for EC O157:H7

Composition per liter:

Casein enzymic hydrolysate	10.0g
Sorbitol	10.0g
Bile salts mixture	1.5g
Chromogenic mixture	1.3g

pH 7.1 ± 0.2 at 25°C

Source: This medium is available as a premixed powder from HiMedia.

Preparation of Medium: Add components to distilled/deionized water and bring volume to 1.0L. Mix thoroughly and heat with frequent agitation until boiling. Boil until components are fully dissolved. Do not autoclave. Cool to 45°–50°C.

Use: For the enrichment culture of *Escherichia coli* O157:H7 in foods and environmental samples

HiCrome™ *Enterobacter sakazakii* Agar

Composition per liter:

Casein enzymic hydrolysate	15.0g
Agar	15.0g
Chromogenic mixture	10.17g
Papaic digest of soybean meal	5.0g
NaCl	5.0g
$Na_2S_2O_3$	1.0g
Sodium deoxycholate	0.5g

pH 7.2 ± 0.2 at 25°C

Source: This medium is available as a premixed powder from HiMedia.

Preparation of Medium: Add components to distilled/deionized water and bring volume to 1.0L. Mix thoroughly. Gently heat and bring to boiling. Distribute into tubes or flasks. Autoclave for 15 min at 15 psi pressure–121°C. Pour into sterile Petri dishes.

Use: For the isolation and identification of *Enterobacter sakazakii* from food and environmental samples.

HiCrome™ *Enterobacter sakazakii* Agar, Modified

Composition per liter:

Agar	15.0g
Casein enzymatic hydrolysate	7.0g
NaCl	5.0g
Yeast extract	3.0g
Sodium deoxycholate	0.6g
Chromogenic substrate	0.15g
Crystal Violet	0.02g

pH 7.2 ± 0.2 at 25°C

Source: This medium is available as a premixed powder from HiMedia.

Preparation of Medium: Add components to distilled/deionized water and bring volume to 1.0L. Mix thoroughly. Gently heat and bring to boiling. Distribute into tubes or flasks. Autoclave for 15 min at 15 psi pressure–121°C. Pour into sterile Petri dishes.

Use: For the isolation and identification of *Enterobacter sakazakii* from food and environmental samples. Meets the formulation recommended by ISO Committee for the isolation and identification of *Enterobacter sakazakii* from milk and milk products.

HiCrome™ Enterococci Broth (Enterococci HiCrome™ Broth)

Composition per liter:

Peptone, special	10.0g
NaCl	5.0g
Polysorbate 80	2.0g
NaH_2PO_4	1.25g
NaN_3	0.3g
Chromogenic mixture	0.04g

pH 7.5 ± 0.2 at 25°C

Source: This medium is available as a premixed powder from HiMedia.

Caution: Sodium azide is toxic. It also has a tendency to form explosive metal azides with plumbing materials. It is advisable to use enough water to flush off the disposables.

Preparation of Medium: Add components to distilled/deionized water and bring volume to 1.0L. Mix thoroughly. Distribute into tubes or flasks. Autoclave for 15 min at 15 psi pressure–121°C.

Use: For the rapid and easy identification and differentiation of enterococci. It contains a chromogenic substrate which aids in the detection of enterococci, especially in water samples.

HiCrome™ Enterococci HiVeg Broth

Composition per liter:

Plant special peptone	10.0g
NaCl	5.0g
Polysorbate 80	2.0g
Na_2HPO_4	1.25g
NaN_3	0.3g
Chromogenic mixture	0.04g

pH 7.2 ± 0.2 at 25°C

Source: This medium is available as a premixed powder from HiMedia.

Caution: Sodium azide is toxic. It also has a tendency to form explosive metal azides with plumbing materials. It is advisable to use enough water to flush off the disposables.

Preparation of Medium: Add components to distilled/deionized water and bring volume to 1.0L. Mix thoroughly. Gently heat and bring to boiling. Distribute into tubes or flasks. Autoclave for 15 min at 15 psi pressure–121°C. Pour into sterile Petri dishes.

Use: For the identification and differentiation of Enterococci from water samples.

HiCrome™ *Enterococcus faecium* Agar Base

Composition per liter:

Peptone, special	23.0g
Agar	15.0g
Arabinose	10.0g

NaCl 5.0g
Cornstarch 1.0g
Chromogenic substrate 0.1g
Phenol Red 0.1g
pH 7.2 ± 0.2 at 25°C

Source: This medium is available as a premixed powder from HiMedia.

Preparation of Medium: Add components to distilled/deionized water and bring volume to 1.0L. Mix thoroughly. Gently heat and bring to boiling. Distribute into tubes or flasks. Autoclave for 15 min at 15 psi pressure–121°C. Pour into sterile Petri dishes.

Use: For the chromogenic identification of *Enterococcus faecium* from water and sewage samples.

HiCrome™ Improved *Salmonella* Agar

Composition per liter:
Agar 12.0g
Peptone, special 8.0g
Chromogenic mixture 3.25g
Yeast extract 2.0g
Sodium deoxycholate 1.0g
pH 7.1 ± 0.2 at 25°C

Source: This medium is available as a premixed powder from HiMedia.

Preparation of Medium: Add components to distilled/deionized water and bring volume to 1.0L. Mix thoroughly and heat with frequent agitation until boiling. Boil until components are fully dissolved. Do not autoclave. Cool to 45°–50°C. Mix well. Pour into sterile Petri dishes.

Use: For the improved selective and differential medium for *Salmonella* species.

HiCrome™ *Klebsiella* Selective Agar Base with Carbenicillin

Composition per liter:
Agar 15.0g
Peptone, special 12.0g
Yeast extract 7.0g
NaCl 5.0g
Bile salts mixture 1.5g
Chromogenic mixture 0.2g
Sodium lauryl sulfate 0.1g
Klebsiella selective supplement 2.0mL
pH 7.1 ± 0.2 at 25°C

Source: This medium, without *Klebsiella* selective supplement, is available as a premixed powder from HiMedia.

***Klebsiella* Selective Supplement:**
Composition per 2.0mL:
Carbenicillin 50.0mg

Preparation of *Klebsiella* Selective Supplement: Add components to distilled/deionized water and bring volume to 2.0mL. Mix thoroughly. Filter sterilize.

Preparation of Medium: Add components to distilled/deionized water and bring volume to 1.0L. Mix thoroughly and heat with frequent agitation until boiling. Boil until components are fully dissolved. Do not autoclave. Cool to 45°–50°C. Aseptically add *Klebsiella* selective supplement. Mix well. Pour into sterile Petri dishes.

Use: For the selective isolation and easy detection of *Klebsiella* species from water and sewage.

HiCrome™ *Listeria* Agar Base, Modified, with Moxalactam (*Listeria* HiCrome Agar Base, Modified)

Composition per liter:
Peptone, special 23.0g
Agar 13.0g
Rhamnose 10.0g
Chromogenic mixture 5.13g
LiCl 5.0g
Meat extract 5.0g
NaCl 5.0g
Yeast extract 1.0g
Phenol Red 0.12g
Moxolactam solution 10.0mL
pH 7.3 ± 0.2 at 25°C

Source: This medium is available as a premixed powder from HiMedia.

Caution: Lithium chloride is harmful. Avoid bodily contact and inhalation of vapors. On contact with skin, wash with plenty of water immediately.

Moxalactam Solution (*Listeria* Selective Supplement):
Composition per 10.0mL:
Moxolactam 0.2g

Preparation of Moxalactam Solution (*Listeria* Selective Supplement): Add moxalactam to distilled/deionized water and bring volume to 10.0mL. Mix thoroughly. Filter sterilize.

Preparation of Medium: Add components, except moxalactam solution, to distilled/deionized water and bring volume to 990.0mL. Mix thoroughly. Gently heat and bring to boiling. Distribute into tubes or flasks. Autoclave for 15 min at 15 psi pressure–121°C. Cool to 45°–50°C. Add 10.0mL moxalactam solution (*Listeria* selective supplement). Mix well. Pour into sterile Petri dishes.

Use: For the rapid and direct identification of *Listeria* species, specifically *Listeria monocytogenes*. A selective and differential agar medium recommended for rapid and direct identification of *Listeria* species, specifically *Listeria monocytogenes*.

HiCrome™ MacConkey-Sorbitol Agar (MacConkey-Sorbitol Agar, HiCrome)

Composition per liter:
Casein enzymic hydrolysate 17.0g
Agar 13.5g
Sorbitol 10.0g
NaCl 5.0g
Proteose peptone 3.0g
Bile salts mixture 1.5g
5-Bromo-4-chloro-3-indolyl-β-D-glucuronide sodium salt 0.1g
Neutral Red 0.03g
Crystal Violet 0.001g
pH 7.1 ± 0.3 at 25°C

Source: This medium is available as a premixed powder from HiMedia.

Preparation of Medium: Add components to distilled/deionized water and bring volume to 1.0L. Mix thoroughly. Gently heat while stirring and bring to boiling. Mix to completely dissolve components. Do not autoclave. Cool to 50°C. Pour into sterile Petri dishes.

Use: For the direct isolation and differentiation of *E. coli O157:H7* strains from foodstuffs. Recommended for selective isolation of *Escherichia coli* O157:H7 from food and animal feeds. The medium contains sorbitol instead of lactose. Enteropathogenic strains of *Escherichia coli* O157:H7 ferment lactose but do not ferment sorbitol and hence produce colorless colonies. Sorbitol fermenting strains of *Escherichia coli* produce pink-red colonies. The red color is due to production of acid from sorbitol, absorption of Neutral Red and a subsequent color change of the dye when the pH of the medium falls below 6.8. The chromogenic indicator is added to detect the presence of an enzyme β-D-glucuronidase. Strains of *Escherichia coli* possessing β-D-glucuronidase appear as blue colored colonies on the medium. Enteropathogenic strains of *Escherichia coli* O157 do not possess β-D-glucuronidase activity and thus produce colorless colonies. *Escherichia coli* fermenting sorbitol and possessing β-D-glucuronidase activity produce purple colored colonies. Most of the Gram-positive organisms are inhibited by Crystal Violet and bile salts.

HiCrome™ MacConkey-Sorbitol Agar with Tellurite-Cefixime Supplement (MacConkey-Sorbitol Agar with Tellurite-Cefixime Supplement)

Composition per 1004.0mL:

Casein enzymic hydrolysate	17.0g
Agar	13.5g
Sorbitol	10.0g
NaCl	5.0g
Proteose peptone	3.0g
Bile salts mixture	1.5g
5-Bromo-4-chloro-3-indolyl-β-D-glucuronide sodium salt	0.1g
Neutral Red	0.03g
Crystal Violet	0.001g
Tellurite-cefixime supplement	4.0mL

pH 7.1 ± 0.3 at 25°C

Source: This medium, without tellurite-cefixime supplement, is available as a premixed powder from HiMedia.

Tellurite-Cefixime Supplement:

Composition per 4.0mL:

K_2TeO_3	5.0mg
Cefixime	0.1mg

Preparation of Tellurite-Cefixime Supplement: Add components to 4.0mL of distilled/deionized water. Mix thoroughly. Filter sterilize.

Caution: Potassium tellurite is toxic.

Source: This medium is available from Fluka, Sigma-Aldrich. Tellurite-cefixime supplement is available from Oxoid Unipath.

Preparation of Medium: Add components, except tellurite-cefixime supplement, to distilled/deionized water and bring volume to 1.0L. Mix thoroughly. Gently heat while stirring and bring to boiling. Mix to completely dissolve components. Do not autoclave. Cool to 50°C. Add 4.0mL sterile tellurite-cefixime supplement. Mix thoroughly. Pour into sterile Petri dishes.

Use: For the direct isolation and differentiation of *E. coli O157:H7-* strains from foodstuffs. Recommended for selective isolation of *Escherichia coli* O157:H7 from food and animal feeding stuffs. The medium contains sorbitol instead of lactose. Enteropathogenic strains of *Escherichia coli* O157:H7 ferment lactose but do not ferment sorbitol and hence produce colorless colonies. Sorbitol fermenting strains of *Escherichia coli* produce pink-red colonies. The red color is due to production of acid from sorbitol, absorption of Neutral Red and a subsequent color change of the dye when the pH of the medium falls below 6.8. The chromogenic indicator is added to detect the presence of an enzyme β-D-glucuronidase. Strains of *Escherichia coli* possessing β-D-glucuronidase appear as blue colored colonies on the medium. Enteropathogenic strains of *Escherichia coli* O157 do not possess β-D-glucuronidase activity and thus produce colorless colonies. *Escherichia coli* fermenting sorbitol and possessing β-D-glucuronidase activity produce purple colored colonies. Most of the Gram-positive organisms are inhibited by Crystal Violet and bile salts. Addition of tellurite-cefixime supplement makes the medium selective. Potassium tellurite selects the serogroups O157 from other *E. coli* serogroups and inhibits *Aeromonas* species and *Providencia* species. Cefixime inhibits *Proteus* species.

HiCrome™ M-CP Agar Base (M-CP HiCrome™ Agar Base) (Membrane *Clostridium perfringens* HiCrome™ Agar Base)

Composition per liter:

Tryptose	30.0g
Yeast extract	20.0g
Agar	15.0g
Sucrose	5.0g
L-Cysteine·HCl·H_2O	1.0g
$MgSO_4 \cdot 7H_2O$	0.1g
$FeCl_3 \cdot 6H_2O$	0.09g
Indoxyl-β-D-glucoside	0.06g
Bromcresol Purple	0.04g

pH 7.6 ± 0.2 at 25°C

Source: This medium is available as a premixed powder from HiMedia.

Preparation of Medium: Add components to distilled/deionized water and bring volume to 1.0L. Mix thoroughly. Gently heat while stirring and bring to boiling. Autoclave for 15 min at 15 psi pressure–121°C. Pour into sterile Petri dishes.

Use: For the detection of *Clostridium perfringens*. Recommended by the Directive of the Council of the European Union 98/83/EC for isolation and enumeration of *C. perfringens* from water samples using the membrane filtration technique.

HiCrome™ MeReSa Agar with Methicillin

Composition per liter:

NaCl	40.0g
Agar	15.0g
Casein enzymic hydrolysate	13.0 g
Chromogenic mixture	5.3 g
Sodium pyruvate	5.0 g
Beef extract	2.5 g
Yeast extract	2.5 g
MRSA selective supplement	10.0mL

pH 7.0 ± 0.2 at 25°C

Source: This medium, without MRSA supplement, is available as a premixed powder from HiMedia.

MRSA Selective Supplement:
Composition per 10.0mL:

Methicillin	4.0mg

Preparation of MRSA Selective Supplement: Add methicillin to distilled/deionized water and bring volume to 10.0mL. Mix thoroughly. Filter sterilize.

Preparation of Medium: Add components, except MRSA supplement, to distilled/deionized water and bring volume to 990.0mL. Mix thoroughly. Gently heat and bring to boiling. Distribute into tubes or flasks. Autoclave for 15 min at 15 psi pressure–121°C. Cool to 50°C. Aseptically add 10.0mL MRSA supplement. Pour into sterile Petri dishes or leave in tubes.

Use: For the isolation and cultivation of methicillin-resistant *Staphylococcus aureus* (MRSA).

HiCrome™ MeReSa Agar with Oxacillin

Composition per liter:

NaCl	40.0g
Agar	15.0g
Casein enzymic hydrolysate	13.0 g
Chromogenic mixture	5.3 g
Sodium pyruvate	5.0 g
Beef extract	2.5 g
Yeast extract	2.5 g
MRSA selective supplement	10.0mL

pH 7.0 ± 0.2 at 25°C

Source: This medium, without MRSA supplement, is available as a premixed powder from HiMedia.

MRSA Selective Supplement:
Composition per 10.0mL:

Oxacillin	2.0mg

Preparation of MRSA Selective Supplement: Add oxacillin to distilled/deionized water and bring volume to 10.0mL. Mix thoroughly. Filter sterilize.

Preparation of Medium: Add components, except MRSA supplement, to distilled/deionized water and bring volume to 990.0mL. Mix thoroughly. Gently heat and bring to boiling. Distribute into tubes or flasks. Autoclave for 15 min at 15 psi pressure–121°C. Cool to 50°C. Aseptically add 10.0mL MRSA supplement. Pour into sterile Petri dishes or leave in tubes.

Use: For the isolation and cultivation of methicillin-resistant *Staphylococcus aureus* (MRSA).

HiCrome™ M-Lauryl Sulfate Agar

Composition per liter:

Peptic digest of animal tissue	40.0g
Lactose	30.0g
Agar	10.0g
Yeast extract	6.0g
Sodium lauryl sulfate	1.0g
Sodium pyruvate	0.5g
Chromogen	0.2g
Phenol Red	0.2g

pH 7.2 ± 0.2 at 25°C

Source: This medium is available as a premixed powder from HiMedia.

Preparation of Medium: Add components to distilled/deionized water and bring volume to 1.0L. Mix thoroughly. Gently heat and bring to boiling. Distribute into tubes or flasks. Autoclave for 15 min at 15 psi pressure–121°C. Pour into sterile Petri dishes.

Use: For the differentiation and enumeration of *Escherichia coli* and other coliforms by the membrane filtration method.

HiCrome™ MM Agar (HiCrome™ Miller and Mallinson Agar) (MM Agar HiCrome™)

Composition per liter:

Agar	15.0g
Lactose	10.0g
Peptic digest of animal tissue	10.0g
Chromogenic mixture	6.6g
D-Cellobiose	3.0g
Beef extract	2.0g
D-Trehalose	1.33g
D-Mannitol	1.2g

pH 7.6 ± 0.2 at 25°C

Source: This medium is available as a premixed powder from HiMedia.

Preparation of Medium: Add components to distilled/deionized water and bring volume to 1.0L. Mix thoroughly. Gently heat and bring to boiling. Distribute into tubes or flasks. Do not autoclave. Cool to 45°–50°C. Mix well. Pour into sterile Petri dishes.

Use: For the identification and differentiation of *Salmonella* and non-*Salmonella* like *Citrobacter* from water samples.

HiCrome™ MM HiVeg Agar

Composition per liter:

Agar	15.0g
Plant peptone	10.0g
Lactose	10.0g
Chromogenic mixture	6.6g
D-Cellobiose	3.0g
Plant extract	2.0g
D-Trehalose	1.33g
D-Mannitol	1.2g

pH 7.6 ± 0.2 at 25°C

Source: This medium is available as a premixed powder from HiMedia.

Preparation of Medium: Add components to distilled/deionized water and bring volume to 1.0L. Mix thoroughly. Gently heat and bring to boiling. Distribute into tubes or flasks. Do not autoclave. Cool to 45°–50°C. Mix well. Pour into sterile Petri dishes.

Use: For the identification and differentiation of *Salmonella* and non-*Salmonella* like *Citrobacter* from water samples.

HiCrome™ MS.O157 Agar (MS.O157 Agar HiCrome™)

Composition per liter:

Agar	12.0g
Peptone, special	10.0g
Sorbitol	4.0g

Bile salt mixture	1.0g
Chromogenic mixture	0.731g

pH 6.8 ± 0.2 at 25°C

Source: This medium is available from Fluka, Sigma-Aldrich.

Preparation of Medium: Add components to distilled/deionized water and bring volume to 1.0L. Mix thoroughly. Gently heat while stirring and bring to boiling. Do not autoclave. Cool to 50°C. Pour into sterile Petri dishes.

Use: For the simultaneous detection of *Escherichia coli, Escherichia coli O157:H7* and coliforms in water and food samples. *Escherichia coli O157:H7* gives colorless colonies because of non-fermentation of sorbitol and absence of β-glucuronidase activity, whereas other strains of *Escherichia coli* having β-glucuronidase activity and fermenting sorbitol appear as steel blue colored colonies. Some non *Escherichia coli* O157:H7 may have some colony color.

HiCrome™ MS.O157 Agar with Tellurite (MS.O157 Agar with Tellurite, HiCrome™)

Composition 1000.25mL:

Agar	12.0g
Peptone, special	10.0g
Sorbitol	4.0g
Bile salt mixture	1.0g
Chromogenic mixture	0.731g
Tellurite solution	0.25mL

pH 6.8 ± 0.2 at 25°C

Source: This medium is available as a premixed powder from HiMedia.

Tellurite Solution:

Composition per 10.0mL:

K_2TeO_3	0.1g

Preparation of Tellurite Solution: Add K_2TeO_3 to 10.0mL of distilled/deionized water. Mix thoroughly. Filter sterilize.

Caution: Potassium tellurite is toxic.

Preparation of Medium: Add components, exept tellurite solution, to distilled/deionized water and bring volume to 1.0L. Mix thoroughly. Gently heat while stirring and bring to boiling. Do not autoclave. Cool to 45°C. Aseptically add 0.25mL sterile tellurite solution. Mix thoroughly. Pour into sterile Petri dishes.

Use: For the simultaneous detection of *Escherichia coli, Escherichia coli O157:H7,* and coliforms in water and food samples. *Escherichia coli O157:H7* gives colorless colonies because of non-fermentation of sorbitol and absence of β-glucuronidase activity, whereas other strains of *Escherichia coli* having β-glucuronidase activity and fermenting sorbitol appear as steel blue colored colonies. Addition of tellurite makes the medium much more specific and selective.

HiCrome™ M-TEC Agar

Composition per liter:

Agar	15.0g
Lactose	10.0g
NaCl	7.5g
Proteose peptone	5.0g
K_2HPO_4	3.3g
Yeast extract	3.0g
KH_2PO_4	1.0g
Chromogen	0.5g
Sodium lauryl sulfate	0.2g
Sodium deoxycholate	0.1g

pH 7.3 ± 0.2 at 25°C

Source: This medium is available as a premixed powder from HiMedia.

Preparation of Medium: Add components to distilled/deionized water and bring volume to 1.0L. Mix thoroughly. Gently heat and bring to boiling. Distribute into tubes or flasks. Autoclave for 15 min at 15 psi pressure–121°C. Pour into sterile Petri dishes.

Use: For the differentiation and enumeration of thermotolerant *E. coli* from water by the membrane filtration method.

HiCrome Nickels and Leesment Medium

Composition per liter:

Casein enzymatic hydrolysate	18.0g
Agar	15.0g
Calcium lactate pentahydrate	8.0g
Tricalcium dicitrate tetrahydrate	6.65g
Yeast extract	4.5g
Glucose	4.5g
Lactose	4.5g
NaCl	3.6g
Gelatin	2.25g
Trisodium citrate dihydrate	1.8g
Carboxymethyl cellulose	0.4g
Chromogenic substrate (X-gal)	0.2g
Selective supplement solution	10.0mL

pH 6.65 ± 0.05 at 25°C

Source: This medium is available from HiMedia.

Selective Supplement Solution:

Composition per 10.0mL:

Vancomycin	0.2g

Preparation of Selective Supplement Solution: Add vancomycin to distilled/deionized water and bring volume to 10.0mL. Mix thoroughly. Filter sterilize.

Preparation of Medium: Add components, except selective supplement solution, to distilled/deionized water and bring volume to 990.0mL. Mix thoroughly. Autoclave for 15 min at 15 psi pressure–121°C. Cool to 50°C. Aseptically add selective supplement solution. Mix thoroughly. Pour into Petri dishes or aseptically distribute into sterile tubes.

Use: For the enumeration of citrate-fermenting lactic acid bacteria from milk.

HiCrome™ OGYE Agar Base (OGYE Agar Base HiCrome) (Oxytetracycline Glucose Yeast Extract Agar HiCrome)

Composition per liter:

Glucose	20.0g
Agar	12.0g
Yeast extract	4.0g
Chromogenic mixture	1.1g
Oxytetracycline selective supplement	10.0mL

pH 7.0 ± 0.2 at 25°C

Source: This medium is available as a premixed powder from HiMedia.

Oxytetracycline Selective Supplement:
Composition per 10.0mL:

Oxytetracycline	0.1g

Preparation of Oxytetracycline Selective Supplement: Add 0.1g oxytetracycline to 10.0mL of distilled/deionized water. Mix thoroughly. Filter sterilize.

Preparation of Medium: Add components, except oxytetracycline selective supplement, to distilled/deionized water and bring volume to 990.0mL. Mix thoroughly. Gently heat while stirring and bring to boiling. Autoclave for 15 min at 15 psi pressure–121°C. Cool to 50°C. Aseptically add 10.0mL sterile oxytetracycline selective supplement. Mix thoroughly. Pour into sterile Petri dishes.

Use: For the rapid isolation of yeasts and molds from milk and milk products. Oxytetracycline makes the medium more selective by inhibiting the growth of lactobacilli. *Aspergillus niger* appears as light blue colored colonies with black spores due to the presence of chromogenic mixture; *Candida albicans* shows green colored colonies and *Saccharomyces cerevisiae* gives colorless colonies.

HiCrome™ RajHans Medium (*Salmonella* Agar)

Composition per liter:

Agar	13.5g
Casein enzymic hydrolysate	8.0g
Chromogenic mixture	7.3g
NaCl	5.0g
Yeast extract	5.0g
Peptone	4.0g
Lactose	3.0g
Sodium deoxycholate	1.0g
Neutral Red	0.02g

pH 7.3 ± 0.2 at 25°C

Preparation of Medium: Add components to distilled/deionized water and bring volume to 1.0L. Mix thoroughly. Gently heat and bring to boiling. Distribute into tubes or flasks. Do not autoclave. Do not overheat. Cool to 50°C. Mix thoroughly. Pour into sterile Petri dishes or leave in tubes.

Use: For the identification and differentiation of *Salmonella* species from the members of Enterobacteriaceae, especially *Proteus* species.

HiCrome™ RajHans Medium, Modified (*Salmonella* Agar, Modified)

Composition per liter:

Agar	12.0g
Casein enzymic hydrolysate	8.0g
NaCl	5.0g
Yeast extract	5.0g
Chromogenic mixture	4.32g
Peptic digest of animal tissue	4.0g
Lactose	3.0g
Sodium deoxycholate	1.0g
Neutral Red	0.02g

pH 7.3 ± 0.2 at 25°C

Preparation of Medium: Add components to distilled/deionized water and bring volume to 1.0L. Mix thoroughly. Gently heat and bring to boiling. Distribute into tubes or flasks. Do not autoclave. Do not overheat. Cool to 50°C. Mix thoroughly. Pour into sterile Petri dishes or leave in tubes.

Use: For the identification and differentiation of *Salmonella* species from the members of Enterobacteriaceae, especially *Proteus* species.

HiCrome™ Rapid Coliform Broth (Coliform Rapid HiCrome™ Broth) (Rapid Coliform HiCrome™ Broth)

Composition per liter:

Peptone, special	5.0g
NaCl	5.0g
Na_2HPO_4	2.7g
KH_2PO_4	2.0g
Sorbitol	1.0g
Sodium lauryl sulfate	0.1g
IPTG	0.1g
Chromogenic substrate	0.08g
Fluorogenic substrate	0.05g

pH 6.8 ± 0.3 at 25°C

Source: This medium is available as a premixed powder from HiMedia.

Preparation of Medium: Add components to distilled/deionized water and bring volume to 1.0L. Mix thoroughly. Gently heat while stirring and bring to boiling. Autoclave for 15 min at 15 psi pressure–121°C. Pour into sterile Petri dishes.

Use: For the detection and confirmation of *Escherichia coli* and coliforms on the basis of enzyme substrate reaction from water samples, using a combination of chromogenic and fluorogenic substrate.

HiCrome™ Rapid Enterococci Agar (Enterococci Rapid HiCrome™ Agar) (Rapid Enterococci HiCrome™ Agar)

Composition per liter:

Agar	15.0g
Peptone special	10.0g
NaCl	5.0g
Polysorbate 80	2.0g
Na_2HPO_4	1.25g
NaN_3	0.3g
Chromogenic mixture	0.06g

pH 7.5 ± 0.2 at 25°C

Source: This medium is available as a premixed powder from HiMedia.

Caution: Sodium azide is toxic. It also has a tendency to form explosive metal azides with plumbing materials. It is advisable to use enough water to flush off the disposables.

Preparation of Medium: Add components to distilled/deionized water and bring volume to 1.0L. Mix thoroughly. Gently heat while stirring and bring to boiling. Autoclave for 15 min at 15 psi pressure–121°C. Pour into sterile Petri dishes.

Use: For the rapid and easy identification and differentiation of enterococci. It contains a chromogenic substrate, which aids in the detection of enterococci, especially from water samples.

HiCrome™ *Salmonella* Agar

Composition per liter:

Agar	13.0g
Peptic digest of animal tissue	6.0g
Chromogenic mixture	5.4g

Yeast extract	2.5g
Bile salts mixture	1.0g

pH 7.3 ± 0.2 at 25°C

Source: This medium is available as a premixed powder from HiMedia.

Preparation of Medium: Add components to distilled/deionized water and bring volume to 1.0L. Mix thoroughly. Gently heat and bring to boiling. Distribute into tubes or flasks. Do not autoclave. Cool to 45°–50°C. Mix well. Pour into sterile Petri dishes.

Use: For the simultaneous detection of *Escherichia coli* and *Salmonella* from food and water. For the identification, differentiation, and confirmation of enteric bacteria from specimens such as urine, water, or food which may contain large numbers of *Proteus* species as well as potentially pathogenic Gram-positive organisms.

HiCrome™ *Salmonella* Agar (*Salmonella* Agar, HiCrome™)

Composition per liter:

Agar	13.0g
Peptic digest of animal tissue	6.0g
Chromogenic mixture	5.4g
Yeast extract	2.5g
Chromogenic mix	1.5g
Bile salt mixture	1.0g

pH 7.7 ± 0.2 at 25°C

Source: This medium is available as a premixed powder from HiMedia.

Preparation of Medium: Add components to distilled/deionized water and bring volume to 1.0L. Mix thoroughly. Gently heat while stirring and bring to boiling. Mix to completely dissolve components. Do not autoclave. Cool to 50°C. Pour into sterile Petri dishes.

Use: A selective chromogenic medium used for the isolation and differentiation of *Salmonella* species from coliforms in food and water. *E. coli* and *Salmonella* are easily distinguishable due to the colony characteristics. *Salmonella* give light purple colonies with a halo. *E. coli* have a characteristic blue color. Other organisms give colorless colonies. The characteristic light purple and blue color is due to the chromogenic mixture. Chromogenic medium for detecting and identifying Enterobacteria, *Proteus* species, and other Gram-positive organisms.

HiCrome™ *Salmonella* Chromogen Agar (*Salmonella* Chromogen Agar, HiCrome™) (Rambach Equivalent Agar)

Composition per liter:

Agar	15.0g
Peptic digest of animal tissue	5.0g
NaCl	5.0g
Yeast extract	2.0g
Meat extract	1.0g
Na-deoxycholate	1.0g
Chromogenic mixture	1 vial

pH 7.3 ± 0.2 at 25°C

Source: This medium is available as a premixed powder from HiMedia.

Preparation of Medium: Add components to distilled/deionized water and bring volume to 1.0L. Mix thoroughly. Gently heat while stirring and bring to boiling. Mix to completely dissolve components. Do not autoclave. Mix and boil in 5 min sequences for 35–40 min. Cool to 50°C. Shake gently for 30–35 min. Pour into sterile Petri dishes.

Use: A selective medium used for the detectgion of *Salmonella* species. This medium exploits a novel phenotypic characteristic of *Salmonella* spp.: the formation of acid from propylene glycol. This characteristic may be used in combination with a chromogenic indicator of β-galactosidase to differentiate *Salmonella* spp. from *Proteus* spp. and the other members of the Enterobacteriaceae. Deoxycholate is included in the plate medium as an inhibitor of Gram-positive organisms. Non-typhi *Salmonella* spp. yield distinct, bright red colonies on this medium, allowing facilitated identification and unambiguous differentiation from *Proteus* spp. Coliforms produce blue-green to blue-violet colonies. Other Enterobacteriaceae and Gram-negative bacteria such as *Proteus, Shigella, Pseudomonas, Salmonella typhi,* and *S. paratyphi* A form colorless or yellow colonies.

HiCrome™ UTI Agar, HiVeg

Composition per liter:

Agar	15.0g
Plant peptone	15.0g
Chromogenic mixture	2.45g

pH 7.3 ± 0.2 at 25°C

Source: This medium is available as a premixed powder from HiMedia.

Preparation of Medium: Add components to distilled/deionized water and bring volume to 1.0L. Mix thoroughly. Gently heat and bring to boiling. Distribute into tubes or flasks. Autoclave for 15 min at 15 psi pressure–121°C. Pour into sterile Petri dishes.

Use: For the differentiation and enumeration of thermotolerant *E. coli* from water by the membrane filtration method. For the identification, differentiation, and confirmation of enteric bacteria from specimens such as urine, water, or food which may contain large numbers of *Proteus* species as well as potentially pathogenic Gram-positive organisms.

HiCrome™ UTI Agar, Modified

Composition per liter:

Peptic digest of animal tissue	18.0g
Agar	15.0g
Chromogenic mixture	12.44g
Beef extract	6.0g
Casein enzymic hydrolysate	4.0g

pH 7.2 ± 0.2 at 25°C

Source: This medium is available as a premixed powder from HiMedia.

Preparation of Medium: Add components to distilled/deionized water and bring volume to 1.0L. Mix thoroughly. Gently heat and bring to boiling. Distribute into tubes or flasks. Autoclave for 15 min at 15 psi pressure–121°C. Pour into sterile Petri dishes.

Use: For the differentiation and enumeration of thermotolerant *E. coli* from water by the membrane filtration method. For the identification, differentiation, and confirmation of enteric bacteria from specimens such as urine, water, or food which may contain large numbers of *Proteus* species as well as potentially pathogenic Gram-positive organisms.

HiCrome™ UTI Agar, Modified (UTI Agar, Modified HiCrome™)

Composition per liter:

Peptic digest of animal tissue	18.0g
Agar	15.0g
Chromogenic mixture	12.44g
Beef extract	4.0g
Casein enzymatic hydrolysate	4.0g

pH 7.2 ± 0.2 at 25°C

Source: This medium is available as a premixed powder from HiMedia.

Preparation of Medium: Add components to distilled/deionized water and bring volume to 1.0L. Mix thoroughly. Gently heat while stirring and bring to boiling. Mix to completely dissolve components. Do not autoclave. Cool to 50°C. Pour into sterile Petri dishes.

Use: A chromogenic medium used for detecting and identifying Enterobacteria, *Proteus* species, and other bacteria involved in urinary tract infections.

HiFluoro™ *Pseudomonas* Agar Base

Composition per liter:

Pancreatic digest of gelatin	18.0g
Agar	15.0g
K_2SO_4	10.0g
Fluorogenic mixture	2.05g
$MnCl_2$	1.4g
Cetrimide	0.3g
Glycerol	10.ml

pH 7.2 ± 0.2 at 25°C

Source: This medium, wihout glycerol, is available as a premixed powder from HiMedia.

Preparation of Medium: Add components to distilled/deionized water and bring volume to 1.0L. Mix thoroughly. Gently heat and bring to boiling. Distribute into tubes or flasks. Autoclave for 15 min at 15 psi pressure–121°C. Pour into sterile Petri dishes.

Use: For the selective isolation of *Pseudomonas aeruginosa* from clinical and nonclinical specimens by the fluorogenic method.

High Plate Count Agar

Composition per liter:

Agar	15.0g
Peptic digest of animal tissue	3.0g
Casein, soluble	0.5g
K_2HPO_4	0.2g
$MgSO_4 \cdot 7H_2O$	0.05g
$FeCl_3 \cdot 4H_2O$	1.0mg

pH 7.2 ± 0.2 at 25°C

Source: This medium is available from HiMedia.

Preparation of Medium: Add components to distilled/deionized water and bring volume to 1.0L. Mix thoroughly. Gently heat and bring to boiling. Distribute into tubes or flasks. Autoclave for 15 min at 15 psi pressure–121°C. Pour into sterile Petri dishes or leave in tubes.

Use: For obtaining higher colony counts by the spread plate, pour plate or membrane filter technique.

High Salt Nutrient Agar

Composition per liter:

NaCl	30.0g
Agar	15.0g
Peptic digest of animal tissue	5.0g
Meat extract	5.0g

pH 8.5 ± 0.2 at 25°C

Source: This medium is available from HiMedia.

Preparation of Medium: Add components to distilled/deionized water and bring volume to 1.0L. Mix thoroughly. Gently heat and bring to boiling. Distribute into tubes or flasks. Autoclave for 15 min at 15 psi pressure–121°C. Pour into sterile Petri dishes or leave in tubes.

Use: For the isolation and cultivation of salt-tolerant *Vibrio* species.

High Salt Peptone Yeast Extract Agar

Composition per liter:

NaCl	30.0g
Agar	15.0g
Peptic digest of animal tissue	10.0g
Yeast extract	6.0g
Meat extract	2.0g
Glucose	2.0g
L-Cysteine·HCl·H_2O	0.3g

pH 7.5 ± 0.2 at 25°C

Source: This medium is available from HiMedia.

Preparation of Medium: Add components to distilled/deionized water and bring volume to 1.0L. Mix thoroughly. Gently heat and bring to boiling. Distribute into tubes or flasks. Autoclave for 15 min at 15 psi pressure–121°C. Pour into sterile Petri dishes or leave in tubes.

Use: For the confirmation of *Vibrio* species.

Hippea Medium (DSMZ Medium 854)

Composition per 1010.0mL:

NaCl	25.0g
Sulfur, powdered	10.0g
Na-acetate	5.0g
MOPS [3-(*N*-morpholino) propane sulfonic acid]	3.0g
$Na_2S \cdot 9H_2O$	0.5g
NH_4Cl	0.33g
$CaCl_2 \cdot 2H_2O$	0.33g
$MgCl_2 \cdot 6H_2O$	0.33g
KCl	0.33g
KH_2PO_4	0.33g
Yeast extract	0.1g
Resazurin	0.5mg
Trace elements solution	10.0mL
Vitamin solution	10.0mL

pH 6.1 ± 0.2 at 25°C

Trace Elements Solution:

Composition per liter:

$MgSO_4 \cdot 7H_2O$	3.0g
Nitrilotriacetic acid	1.5g
NaCl	1.0g
$MnSO_4 \cdot 2H_2O$	0.5g
$CoSO_4 \cdot 7H_2O$	0.18g
$ZnSO_4 \cdot 7H_2O$	0.18g

$CaCl_2 \cdot 2H_2O$	0.1g
$FeSO_4 \cdot 7H_2O$	0.1g
$NiCl_2 \cdot 6H_2O$	0.025g
$KAl(SO_4)_2 \cdot 12H_2O$	0.02g
H_3BO_3	0.01g
$Na_2MoO_4 \cdot 4H_2O$	0.01g
$CuSO_4 \cdot 5H_2O$	0.01g
$Na_2SeO_3 \cdot 5H_2O$	0.3mg

Preparation of Trace Elements Solution: Add nitrilotriacetic acid to 500.0mL of distilled/deionized water. Dissolve by adjusting pH to 6.5 with KOH. Add remaining components. Add distilled/deionized water to 1.0L. Mix thoroughly.

Vitamin Solution:

Composition per liter:

Pyridoxine-HCl	10.0mg
Thiamine-HCl·2H$_2$O	5.0mg
Riboflavin	5.0mg
Nicotinic acid	5.0mg
D-Ca-pantothenate	5.0mg
p-Aminobenzoic acid	5.0mg
Lipoic acid	5.0mg
Biotin	2.0mg
Folic acid	2.0mg
Vitamin B_{12}	0.1mg

Preparation of Vitamin Solution: Add components to distilled/deionized water and bring volume to 1.0L. Mix thoroughly. Sparge with 80% H_2 + 20% CO_2. Filter sterilize.

Preparation of Medium: Add components, except vitamin solution, sulfur, and $Na_2S \cdot 9H_2O$, to distilled/deionized water and bring volume to 1.0L. Mix thoroughly. Adjust pH to 6.0. Sparge the medium with 80% N_2 + 20% CO_2 gas mixture for 30 min. Add $Na_2S \cdot 9H_2O$. Mix thoroughly. Readjust the pH to 6.0–6.2. Dispense medium under 80% N_2 + 20% CO_2 gas mixture into anaerobe tubes or bottles containing 100.0mg sulfur powder per 10mL medium. Autoclave 20 min at 110°C. Prior to use inject 0.1mL sterile vitamin solution per 10.0mL medium.

Use: For the cultivation of *Hippea maritima.*

Hippurate Broth

See: **Sodium Hippurate Broth**

Hippurate Hydrolysis Broth

Composition per liter:

Heart infusion powder	10.0g
Peptic digest of animal tissue	10.0g
Sodium hippurate	10.0g
NaCl	5.0g

pH 7.4 ± 0.2 at 25°C

Source: This medium is available from HiMedia.

Preparation of Medium: Add components to distilled/deionized water and bring volume to 1.0L. Mix thoroughly. Gently heat and bring to boiling. Distribute into tubes or flasks. Autoclave for 15 min at 15 psi pressure–121°C. Pour into sterile Petri dishes or leave in tubes.

Use: For the detection of hippurate-hydrolyzing bacteria.

Hirschia Medium

Composition per liter:

Pancreatic digest of casein	5.0g
HEPES	4.0g
Yeast extract	2.0g
Artificial seawater	250.0mL
Glucose solution	100.0mL

pH 7.4 ± 0.2 at 25°C

Glucose Solution:

Composition per 100.0mL:

Glucose	0.25g

Preparation of Glucose Solution: Add glucose to distilled/deionized water and bring volume to 100.0mL. Mix thoroughly. Filter sterilize.

Artificial Seawater:

Composition per liter:

NaCl	27.5g
$MgCl_2 \cdot 6H_2O$	5.38g
$MgSO_4 \cdot 7H_2O$	6.78g
KCl	0.72g
$NaHCO_3$	0.2g
$CaCL_2 \cdot 2H_2O$	1.4g

Preparation of Artificial Seawater: Add components to distilled/deionized water and bring volume to 1.0L. Mix thoroughly.

Preparation of Medium: Add components, except glucose solution, to distilled/deionized water and bring volume to 1.0L. Mix thoroughly. Adjust pH to 7.4. Autoclave for 15 min at 15 psi pressure–121°C. Cool to 25°C. Aseptically add 100.0mL of sterile glucose solution. Mix thoroughly. Aseptically distribute into sterile tubes or flasks.

Use: For the cultivation of *Hirschia baltica.*

Hi-Sensitivity Test Agar

Composition per liter:

Casein enzymic hydrolysate	11.0g
Agar	8.0g
NaCl	3.0g
Peptic digest of animal tissue	3.0g
Glucose	2.0g
Na_2HPO_4	2.0g
Sodium acetate	1.0g
Starch, soluble	1.0g
Magnesium glycerophosphate	0.2g
Calcium gluconate	0.1g
L-Cystine hydrochloride	0.02g
L-Tryptophan	0.02g
Adenine	0.01g
Guanine	0.01g
Uracil	0.01g
Xanthine	0.01g
Calcium pantothenate	3.0mg
Biotin	3.0mg
Nicotinamide	3.0mg
Pyridoxine hydrochloride	3.0mg
Manganese chloride	2.0mg
$ZnSO_4$	1.0mg
$CoSO_4$	1.0mg
$CuSO_4$	1.0mg
Cyanocobalamin	1.0mg
$FeSO_4$	1.0mg
Menadione	1.0mg
Thiamine hydrochloride	0.04mg

pH 7.2 ± 0.2 at 25°C

Source: This medium is available as a premixed powder from HiMedia.

Preparation of Medium: Add components to distilled/deionized water and bring volume to 1.0L. Mix thoroughly. Gently heat and bring to boiling. Distribute into tubes or flasks. Autoclave for 15 min at 15 psi pressure–121°C. Pour into sterile Petri dishes.

Use: For antimicrobial susceptibility tests.

Hi-Sensitivity Test Broth

Composition per liter:

Casein enzymic hydrolysate	11.0g
NaCl	3.0g
Peptic digest of animal tissue	3.0g
Glucose	2.0g
Na_2HPO_4	2.0g
Sodium acetate	1.0g
Starch, soluble	1.0g
Magnesium glycerophosphate	0.2g
Calcium gluconate	0.1g
L-Cystine hydrochloride	0.02g
L-Tryptophan	0.02g
Adenine	0.01g
Guanine	0.01g
Uracil	0.01g
Xanthine	0.01g
Calcium pantothenate	3.0mg
Biotin	3.0mg
Nicotinamide	3.0mg
Pyridoxine hydrochloride	3.0mg
Manganese chloride	2.0mg
$ZnSO_4$	1.0mg
$CoSO_4$	1.0mg
$CuSO_4$	1.0mg
Cyanocobalamin	1.0mg
$FeSO_4$	1.0mg
Menadione	1.0mg
Thiamine hydrochloride	0.04mg

pH 7.2 ± 0.2 at 25°C

Source: This medium is available as a premixed powder from HiMedia.

Preparation of Medium: Add components to distilled/deionized water and bring volume to 1.0L. Mix thoroughly. Gently heat and bring to boiling. Distribute into tubes or flasks. Autoclave for 15 min at 15 psi pressure–121°C.

Use: For antimicrobial susceptibility testing.

Hi-Sensitivity Test HiVeg Agar

Composition per liter:

Plant hydrolysate	11.0g
Agar	8.0g
NaCl	3.0g
Plant peptone	3.0g
Glucose	2.0g
Na_2HPO_4	2.0g
Sodium acetate	1.0g
Starch, soluble	1.0g
Magnesium glycerophosphate	0.2g
Calcium gluconate	0.1g
L-Cystine hydrochloride	0.02g
L-Tryptophan	0.02g
Adenine	0.01g
Guanine	0.01g
Uracil	0.01g
Xanthine	0.01g
Calcium pantothenate	3.0mg
Biotin	3.0mg
Nicotinamide	3.0mg
Pyridoxine hydrochloride	3.0mg
Manganese chloride	2.0mg
$ZnSO_4$	1.0mg
$CoSO_4$	1.0mg
$CuSO_4$	1.0mg
Cyanocobalamin	1.0mg
$FeSO_4$	1.0mg
Menadione	1.0mg
Thiamine hydrochloride	0.04mg

pH 7.4 ± 0.2 at 25°C

Source: This medium is available as a premixed powder from HiMedia.

Preparation of Medium: Add components to distilled/deionized water and bring volume to 1.0L. Mix thoroughly. Gently heat and bring to boiling. Distribute into tubes or flasks. Autoclave for 15 min at 15 psi pressure–121°C. Pour into sterile Petri dishes.

Use: For antimicrobial susceptibility tests.

Hi-Sensitivity Test HiVeg Broth

Composition per liter:

Plant hydrolysate	11.0g
NaCl	3.0g
Plant peptone	3.0g
Glucose	2.0g
Na_2HPO_4	2.0g
Sodium acetate	1.0g
Starch, soluble	1.0g
Magnesium glycerophosphate	0.2g
Calcium gluconate	0.1g
L-Cystine hydrochloride	0.02g
L-Tryptophan	0.02g
Adenine	0.01g
Guanine	0.01g
Uracil	0.01g
Xanthine	0.01g
Calcium pantothenate	3.0mg
Biotin	3.0mg
Nicotinamide	3.0mg
Pyridoxine hydrochloride	3.0mg
Manganese chloride	2.0mg
$ZnSO_4$	1.0mg
$CoSO_4$	1.0mg
$CuSO_4$	1.0mg
Cyanocobalamin	1.0mg
$FeSO_4$	1.0mg
Menadione	1.0mg
Thiamine hydrochloride	0.04mg

pH 7.4 ± 0.2 at 25°C

Source: This medium is available as a premixed powder from HiMedia.

Preparation of Medium: Add components to distilled/deionized water and bring volume to 1.0L. Mix thoroughly. Gently heat and bring

to boiling. Distribute into tubes or flasks. Autoclave for 15 min at 15 psi pressure–121°C.

Use: For antimicrobial susceptibility testing.

Hisitest Agar

Composition per liter:

Casein enzymic hydrolysate	11.0g
Agar	8.0g
Buffer salt	3.3g
Peptic digest of animal tissue	3.0g
NaCl	3.0g
Glucose	2.0g
Starch	1.0g
Nucleoside basis	0.02g
Thiamine	0.02mg

pH 7.2 ± 0.2 at 25°C

Source: This medium is available as a premixed powder from HiMedia.

Preparation of Medium: Add components to distilled/deionized water and bring volume to 1.0L. Mix thoroughly. Gently heat and bring to boiling. Distribute into tubes or flasks. Autoclave for 15 min at 15 psi pressure–121°C.

Use: For determination of antibiotic susceptibility of fastidious microorganisms.

Histidans Agar

Composition per liter:

Agar	20.0g
Glucose	10.0g
Yeast extract	10.0g
Na_2HPO_4	0.95g
KH_2PO_4	0.91g
$MgSO_4 \cdot 7H_2O$	0.5g

pH 7.0 ± 0.2 at 25°C

Preparation of Medium: Add components to distilled/deionized water and bring volume to 1.0L. Mix thoroughly. Gently heat and bring to boiling. Distribute into tubes or flasks. Autoclave for 15 min at 15 psi pressure–121°C. Pour into sterile Petri dishes or leave in tubes.

Use: For the cultivation of *Streptomyces* species.

Histoplasma capsulatum Agar

Composition per liter:

Agar	12.5g
Glucose	10.0g
Citric acid	10.0g
Potato starch	2.0g
α-Ketoglutaric acid	1.0g
L-Cystine·HCl·H_2O	1.0g
Glutathione, reduced	0.5g
L-Asparagine	0.1g
L-Tryptophan	0.02g
Solution 1	250.0mL
Solution 3	40.0mL
Solution 2	10.0mL
Solution 4	10.0mL
Solution 8	10.0mL
Solution 5	1.0mL
Solution 6	0.1mL
Solution 7	0.1mL

pH 6.5 ± 0.2 at 25°C

Solution 1:

Composition per liter:

KH_2PO_4	8.0g
$(NH_4)_2SO_4$	8.0g
$MgSO_4 \cdot 7H_2O$	0.86g
$CaCl_2$, anhydrous	0.08g
$ZnSO_4 \cdot 7H_2O$	0.05g

Preparation of Solution 1: Add components to distilled/deionized water and bring volume to 500.0mL. Mix thoroughly. Bring volume to 1.0L with distilled/deionized water. Store at 5°C.

Solution 2:

Composition per liter:

$FeSO_4 \cdot 7H_2O$	5.7g
$MnCl_2 \cdot 6H_2O$	0.8g
$NaMoO_4 \cdot 2H_2O$	0.15g
HCl, concentrated	1.0mL

Preparation of Solution 2: Add 1.0mL of concentrated HCl to 100.0mL of distilled water in a 1.0L volumetric flask. Dissolve each component completely in the sequence given. Bring volume to 1.0L with distilled/deionized water. Store at 5°C. Discard if red color or red precipitate appears.

Solution 3:

Composition per 100.0mL:

Casein, acid-hydrolyzed, vitamin-free	10.0g

Preparation of Solution 3: Add casein to distilled/deionized water and bring volume to 100.0mL.

Solution 4:

Composition per liter:

Calcium pantothenate	0.2g
Inositol	0.2g
Riboflavin	0.2g
Thiamine·HCl	0.2g
Nicotinamide	0.1g
Biotin	0.01g

Preparation of Solution 4: Add components to distilled/deionized water and bring volume to 1.0L. Mix thoroughly. Store at –20°C.

Solution 5:

Composition per 100.0mL:

Hemin	0.2g
NH_4OH, concentrated	0.3mL

Preparation of Solution 5: Add hemin to approximately 30.0mL of distilled/deionized water. Add NH_4OH. Mix thoroughly until dissolved. Bring volume to 100.0mL with distilled/deionized water. Store at 5° C.

Solution 6:

Composition per 10.0mL:

DL-Thioctic acid	0.01g
Ethanol (95% solution)	10.0mL

Preparation of Solution 6: Add DL-thioctic acid to 10.0mL of ethanol. Mix thoroughly. Store at –20°C.

Solution 7:

Composition per 10.0mL:

Coenzyme A	0.01g
$Na_2S \cdot 5H_2O$ (0.05% solution)	0.2mL

Preparation of Solution 7: Prepare $Na_2S{\cdot}5H_2O$ solution in freshly boiled distilled/deionized water. Add coenzyme A to 9.8mL of distilled/deionized water. Mix thoroughly. Add freshly prepared $Na_2S{\cdot}5H_2O$ solution. Mix thoroughly. Store the solution at –20°C.

Solution 8:

Composition per 100.0mL:

Oleic acid	0.1g

Preparation of Solution 8: Add oleic acid to 50.0mL of distilled/deionized water. Adjust pH to 9.0 with NaOH. Gently heat until dissolved. Bring volume to 100.0mL with distilled/deionized water. Store at 5°C.

Preparation of Medium: Add components—except agar, potato starch, and solution 8—to distilled/deionized water and bring volume to 400.0mL. Mix thoroughly. Adjust pH to 6.5 with 20% KOH solution. Filter sterilize. In a separate flask, add potato starch to 50.0mL of distilled/deionized water. Add the starch solution to 450.0mL of boiling distilled/deionized water. Add 10.0mL of solution 8 and the agar. Mix thoroughly. Autoclave for 15 min at 15 psi pressure–121°C. Cool to 70°C. Aseptically combine the two sterile solutions. Pour into sterile Petri dishes or distribute into sterile tubes.

Use: For the cultivation and maintenance of *Histoplasma capsulatum* in the yeast phase. For the cultivation of *Histoplasma duboisii, Blastomyces dermatitidis,* and *Sprotrichum schenckii.*

Histoplasma capsulatum Agar

Composition per liter:

Agar	15.0g
Glucose	10.0g
Potato starch	2.0g
α-Ketoglutaric acid	1.0g
L-Cystine·HCl·H_2O	1.0g
Glutathione, reduced	0.5g
L-Asparagine	0.1g
L-Tryptophan	0.02g
Solution 1	250.0mL
Solution 3	40.0mL
Solution 2	10.0mL
Solution 4	10.0mL
Solution 8	10.0mL
Solution 5	1.0mL
Solution 6	0.1mL
Solution 7	0.1mL

pH 6.5 ± 0.2 at 25°C

Solution 1:

Composition per liter:

KH_2PO_4	8.0g
$(NH_4)_2SO_4$	8.0g
$MgSO_4{\cdot}7H_2O$	0.86g
$CaCl_2$, anhydrous	0.08g

Preparation of Solution 1: Add components to distilled/deionized water and bring volume to 500.0mL. Mix thoroughly. Bring volume to 1.0L with distilled/deionized water. Store at 5°C.

Solution 2:

Composition per liter:

$FeSO_4{\cdot}7H_2O$	5.7g
$MnCl_2{\cdot}6H_2O$	0.8g
$NaMoO_4{\cdot}2H_2O$	0.15g
HCl, concentrated	1.0mL

Preparation of Solution 2: Add the 1.0mL of concentrated HCl to 100.0mL of distilled water in a 1.0L volumetric flask. Dissolve each component completely in the sequence given. Bring volume to 1.0L with distilled/deionized water. Store at 5°C. Discard if red color or red precipitate appears.

Solution 3:

Composition per 100.0mL:

Casein, acid-hydrolyzed, vitamin-free	10.0g

Preparation of Solution 3: Add casein to distilled/deionized water and bring volume to 100.0mL. Do not use enzymatically digested casein.

Solution 4:

Composition per liter:

Calcium pantothenate	0.2g
Inositol	0.2g
Riboflavin	0.2g
Thiamine·HCl	0.2g
Nicotinamide	0.1g
Biotin	0.01g

Preparation of Solution 4: Add components to distilled/deionized water and bring volume to 1.0L. Mix thoroughly. Store at –20°C.

Solution 5:

Composition per 100.0mL:

Hemin	0.2g
NH_4OH, concentrated	0.3mL

Preparation of Solution 5: Add hemin to approximately 30.0mL of distilled/deionized water. Add NH_4OH. Mix thoroughly until dissolved. Bring volume to 100.0mL with distilled/deionized water. Store at 5°C.

Solution 6:

Composition per 10.0mL:

DL-Thioctic acid	0.01g
Ethanol (95% solution)	10.0mL

Preparation of Solution 6: Add DL-thioctic acid to 10.0mL of ethanol. Mix thoroughly. Store solution at –20°C.

Solution 7:

Composition per 10.0mL:

Coenzyme A	0.01g
$Na_2S{\cdot}5H_2O$ (0.05% solution)	0.2mL

Preparation of Solution 7: Prepare $Na_2S{\cdot}5H_2O$ solution in freshly boiled distilled/deionized water. Add coenzyme A to 9.8mL of distilled/deionized water. Mix thoroughly. Add freshly prepared $Na_2S{\cdot}5H_2O$ solution. Mix thoroughly. Store the solution at –20°C.

Solution 8:

Composition per 100.0mL:

Oleic acid	0.1g

Preparation of Solution 8: Add oleic acid to 50.0mL of distilled/deionized water. Adjust pH to 9.0 with NaOH. Gently heat until dissolved. Bring volume to 100.0mL with distilled/deionized water. Store at 5°C.

Preparation of Medium: Add components—except agar, potato starch, and solution 8—to distilled/deionized water and bring volume to 400.0mL. Mix thoroughly. Adjust pH to 6.5 with 20% KOH solution. Filter sterilize. In a separate flask, add potato starch to 50.0mL of distilled/deionized water. Add the starch solution to 450.0mL of boiling distilled/deionized water. Add 10.0mL of solution 8 and the agar. Mix thoroughly. Autoclave for 15 min at 15 psi pressure–121°C. Cool to 70°C. Aseptically combine the two sterile solutions. Pour into sterile Petri dishes or distribute into sterile tubes.

Use: For the cultivation and maintenance of *Histoplasma capsulatum* in the mycelial phase.

Histoplasma capsulatum Broth

Composition per liter:

Glucose	10.0g
Citric acid	10.0g
α-Ketoglutaric acid	1.0g
L-Cystine·HCl·H_2O	1.0g
Potato starch	0.5g
Glutathione, reduced	0.5g
L-Asparagine	0.1g
L-Tryptophan	0.02g
Solution 1	250.0mL
Solution 3	40.0mL
Solution 2	10.0mL
Solution 4	10.0mL
Solution 5	1.0mL
Solution 8	1.0mL
Solution 6	0.1mL
Solution 7	0.1mL

pH 6.5 ± 0.2 at 25°C

Solution 1:

Composition per liter:

KH_2PO_4	8.0g
$(NH_4)_2SO_4$	8.0g
$MgSO_4·7H_2O$	0.86g
$CaCl_2$, anhydrous	0.08g
$ZnSO_4·7H_2O$	0.05g

Preparation of Solution 1: Add components to distilled/deionized water and bring volume to 500.0mL. Mix thoroughly. Bring volume to 1.0L with distilled/deionized water. Store at 5°C.

Solution 2:

Composition per liter:

$FeSO_4·7H_2O$	5.7g
$MnCl_2·6H_2O$	0.8g
$NaMoO_4·2H_2O$	0.15g
HCl, concentrated	1.0mL

Preparation of Solution 2: Add 1.0mL of concentrated HCl to 100.0mL of distilled water in a 1.0L volumetric flask. Dissolve each component completely in the sequence given. Bring volume to 1.0L with distilled/deionized water. Store at 5°C. Discard if red color or red precipitate appears.

Solution 3:

Composition per 100.0mL:

Casein, acid-hydrolyzed, vitamin-free	10.0g

Preparation of Solution 3: Add casein to distilled/deionized water and bring volume to 100.0mL. Do not use enzymatically digested casein.

Solution 4:

Composition per liter:

Calcium pantothenate	0.2g
Inositol	0.2g
Riboflavin	0.2g
Thiamine·HCl	0.2g
Nicotinamide	0.1g
Biotin	0.01g

Preparation of Solution 4: Add components to distilled/deionized water and bring volume to 1.0L. Mix thoroughly. Store at –20°C.

Solution 5:

Composition per 100.0mL:

Hemin	0.2g
NH_4OH, concentrated	0.3mL

Preparation of Solution 5: Add hemin to approximately 30.0mL of distilled/deionized water. Add NH_4OH. Mix thoroughly until dissolved. Bring volume to 100.0mL with distilled/deionized water. Store at 5°C.

Solution 6:

Composition per 10.0mL:

DL-Thioctic acid	0.01g
Ethanol (95% solution)	10.0mL

Preparation of Solution 6: Add DL-thioctic acid to 10.0mL of ethanol. Mix thoroughly. Store solution at –20°C.

Solution 7:

Composition per 10.0mL:

Coenzyme A	0.01g
$Na_2S·5H_2O$ (0.05% solution)	0.2mL

Preparation of Solution 7: Prepare $Na_2S·5H_2O$ solution in freshly boiled distilled/deionized water. Add coenzyme A to 9.8mL of distilled/deionized water. Mix thoroughly. Add freshly prepared $Na_2S·5H_2O$ solution. Mix thoroughly. Store the solution at –20°C.

Solution 8:

Composition per 100.0mL:

Oleic acid	0.1g

Preparation of Solution 8: Add oleic acid to 50.0mL of distilled/deionized water. Adjust pH to 9.0 with NaOH. Gently heat until dissolved. Bring volume to 100.0mL with distilled/deionized water. Store at 5°C.

Preparation of Medium: Add components—except potato starch and solution 8—to distilled/deionized water and bring volume to 400.0mL. Mix thoroughly. Adjust pH to 6.5 with 20% KOH solution. Filter sterilize. In a separate flask, add potato starch to 50.0mL of distilled/deionized water. Add the starch solution to 450.0mL of boiling distilled/deionized water. Add 1.0mL of solution 8. Mix thoroughly. Autoclave for 15 min at 15 psi pressure–121°C. Cool to 70°C. Aseptically combine the two sterile solutions. Pour into sterile Petri dishes or distribute into sterile tubes.

Use: For the cultivation of *Histoplasma capsulatum* in the yeast phase. For the cultivation of *Histoplasma duboisii, Blastomyces dermatitidis,* and *Sprotrichum schenckii.*

HiVeg Hydrolysate Agar with 2.5% Agar

Composition per liter:

Agar	25.0g
Plant hydrolysate	5.0g
Plant peptone	5.0g
NaCl	5.0g
Na_2HPO_4	2.5g
Plant infusion	1.5g
Yeast autolysate	1.5g
Glycerol	22.0mL

pH 7.8 ± 0.2 at 25°C

Source: This medium, without glycerol, is available as a premixed powder from HiMedia.

Preparation of Medium: Add components to distilled/deionized water and bring volume to 1.0L. Mix thoroughly. Gently heat and bring to boiling. Autoclave for 15 min at 15 psi pressure–121°C. Mix thoroughly. Pour into sterile Petri dishes or distribute into sterile tubes.

Use: For the cultivation of *Vibrio cholerae*. For the production of cholera vaccine.

HiVeg Magnesium Broth

Composition per liter:

Plant hydrolysate	10.0g
NaCl	5.0g
$MgSO_4$	0.94g

pH 7.0 ± 0.2 at 25°C

Source: This medium is available as a premixed powder from HiMedia.

Preparation of Medium: Add components to distilled/deionized water and bring volume to 1.0L. Mix thoroughly. Gently heat and bring to boiling. Autoclave for 15 min at 15 psi pressure–121°C. Mix thoroughly.

Use: For the cultivation of recombinant strains of *Escherichia coli*.

HiVeg Peptone Water

Composition per liter:

Plant peptone	10.0g
NaCl	5.0g

pH 7.2 ± 0.2 at 25°C

Source: This medium is available as a premixed powder from HiMedia.

Preparation of Medium: Add components to distilled/deionized water and bring volume to 1.0L. Mix thoroughly. Gently heat and bring to boiling. Autoclave for 15 min at 15 psi pressure–121°C. Mix thoroughly.

Use: For the cultivation of various bacteria.

HL Agar

Composition per plate:

Columbia agar base	10.0mL
Columbia blood top agar	5.0mL

pH 7.3 ± 0.2 at 25°C

Columbia Agar Base:

Composition per liter:

Agar	13.5g
Pancreatic digest of casein	12.0g
NaCl	5.0g
Peptic digest of animal tissue	5.0g
Beef extract	3.0g
Yeast extract	3.0g
Cornstarch	1.0g

Preparation of Columbia Agar Base: Add components to distilled/deionized water and bring volume to 1.0L. Mix thoroughly. Gently heat until boiling. Autoclave for 15 min at 15 psi pressure–121°C. Cool to 45°–50°C.

Columbia Blood Top Agar:

Composition per liter:

Agar	13.5g
Pancreatic digest of casein	12.0g
NaCl	5.0g
Peptic digest of animal tissue	5.0g
Beef extract	3.0g
Yeast extract	3.0g
Cornstarch	1.0g
Horse blood, defibrinated	50.0mL

Preparation of Columbia Blood Top Agar: Add components, except horse blood, to distilled/deionized water and bring volume to 950.0mL. Mix thoroughly. Gently heat until boiling. Autoclave for 15 min at 15 psi pressure–121°C. Cool to 45°–50°C. Aseptically add sterile horse blood. Mix thoroughly.

Preparation of Medium: Pour cooled, sterile Columbia agar base into sterile Petri dishes in 10.0mL volumes. Allow agar to solidify. Pour 5.0mL of cooled, sterile Columbia blood top agar over Columbia agar base that has solidified but is still warm.

Use: For the cultivation of *Listeria monocytogenes*.

HM Medium

Composition per liter:

NaCl	81.0g
Yeast extract	10.0g
$MgSO_4$	9.6g
$MgCl_2$	7.0g
Proteose peptone No. 3	5.0g
KCl	2.0g
Glucose	1.0g
$CaCl_2$	0.36g
$NaHCO_3$	60.0mg
NaBr	26.0mg

pH 7.1 ± 0.2 at 25°C

Preparation of Medium: Add components to distilled/deionized water and bring volume to 1.0L. Mix thoroughly. Adjust pH to 7.1. Distribute into tubes or flasks. Autoclave for 15 min at 15 psi pressure–121°C.

Use: For the cultivation of *Salinicoccus roseus* and *Salinicoccus hispanicus*.

HNS Agar (ATCC Medium 923)

Composition per liter:

Agar	15.0g
NaCl	9.6g
Heart infusion broth	990.0mL
Horse serum	10.0mL

pH 7.4 ± 0.2 at 25°C

Heart Infusion Broth:

Composition per 900.0mL:

Beef heart, infusion from	500.0g
Tryptose	10.0g
NaCl	5.0g

Preparation of Heart Infusion Broth: Add agar and NaCl to 990.0mL heart infusion broth. Mix thoroughly. Autoclave for 15 min at 15 psi pressure–121°C. Cool to 45°–50°C. Aseptically add 10.0mL sterile horse serum. Mix thoroughly. Pour into sterile Petri dishes or distribute into sterile tubes.

Use: For the cultivation and maintenance of *Corynebacterium* species.

HNW Medium (DSMZ Medium 997)

Composition per liter:

DMJ synthetic seawater	1.0L
Vitamin solution	10.0mL
$NaHCO_3$solution	10.0mL
$NaNO_3$solution	10.0mL
$Na_2S{\cdot}9H_2O$ solution	10.0mL
Tungstate solution	10.0mL

pH 7.2 ± 0.2 at 25°C

Tungstate Solution:

Composition per 10.0mL:

$Na_2WO_4{\cdot}2H_2O$	0.1mg

Preparation of Tungstate Solution: Add $Na_2WO_4{\cdot}2H_2O$ to distilled/deionized water and bring volume to 10.0mL. Mix thoroughly.

$NaHCO_3$ Solution:

Composition per 10.0mL:

$NaHCO_3$	1.0g

Preparation of $NaHCO_3$ Solution: Add $NaHCO_3$ to distilled/deionized water and bring volume to 10.0mL. Mix thoroughly. Sparge with 20% CO_2 + 80% H_2. Filter sterilize.

$NaNO_3$ Solution:

Composition per 10.0mL:

$NaNO_3$	1.0g

Preparation of $NaNO_3$ Solution: Add $NaNO_3$ to distilled/deionized water and bring volume to 10.0mL. Mix thoroughly. Sparge with 100% N_2. Filter sterilize.

Vitamin Solution:

Composition per liter:

Pyridoxine-HCl	10.0mg
Thiamine-HCl·$2H_2O$	5.0mg
Riboflavin	5.0mg
Nicotinic acid	5.0mg
D-Ca-pantothenate	5.0mg
p-Aminobenzoic acid	5.0mg
Lipoic acid	5.0mg
Biotin	2.0mg
Folic acid	2.0mg
Vitamin B_{12}	0.1mg

Preparation of Vitamin Solution: Add components to distilled/deionized water and bring volume to 1.0L. Mix thoroughly. Sparge with 80% H_2 + 20% CO_2. Filter sterilize.

$Na_2S{\cdot}9H_2O$ Solution:

Composition per 10.0mL:

$Na_2S{\cdot}9H_2O$	0.5g

Preparation of $Na_2S{\cdot}9H_2O$ Solution: Add $Na_2S{\cdot}9H_2O$ to distilled/deionized water and bring volume to 10.0mL. Mix thoroughly. Autoclave under 100% N_2 for 15 min at 15 psi pressure–121°C. Cool to room temperature. Adjust pH to 7.5.

DMJ Synthetic Seawater:

Composition per liter:

NaCl	30.0g
$MgCl_2{\cdot}6H_2O$	4.18g
$MgSO_4{\cdot}7H_2O$	3.4g
KCl	0.33g
NH_4Cl	0.25g
K_2HPO_4	0.14g
$CaCl_2{\cdot}2H_2O$	0.14g
$Fe(NH_4)_2(SO_4)_2{\cdot}6H_2O$	0.01g
$NiCl_2{\cdot}6H_2O$	0.5mg
$Na_2SeO_3{\cdot}5H_2O$	0.5mg
Trace elements solution SL-10	10.0mL

Trace Elements Solution SL-10:

Composition per liter:

$MgSO_4{\cdot}7H_2O$	3.0g
Nitrilotriacetic acid	1.5g
NaCl	1.0g
$MnSO_4{\cdot}2H_2O$	0.5g
$CoSO_4{\cdot}7H_2O$	0.18g
$ZnSO_4{\cdot}7H_2O$	0.18g
$CaCl_2{\cdot}2H_2O$	0.1g
$FeSO_4{\cdot}7H_2O$	0.1g
$NiCl_2{\cdot}6H_2O$	0.025g
$KAl(SO_4)_2{\cdot}12H_2O$	0.02g
H_3BO_3	0.01g
$Na_2MoO_4{\cdot}4H_2O$	0.01g
$CuSO_4{\cdot}5H_2O$	0.01g
$Na_2SeO_3{\cdot}5H_2O$	0.3mg

Preparation of Trace Elements Solution SL-10: Add nitrilotriacetic acid to 500.0mL of distilled/deionized water. Dissolve by adjusting pH to 6.5 with KOH. Add remaining components. Add distilled/deionized water to 1.0L. Mix thoroughly. Adjust pH to 7.0.

Preparation of DMJ Synthetic Seawater: Add components to distilled/deionized water and bring volume to 1.0L. Mix thoroughly. Autoclave for 15 min at 15 psi pressure–121°C. Cool to room temperature.

Preparation of Medium: Aseptically add 10.0mL each of vitamin solution, $NaHCO_3$solution, $NaNO_3$solution, $Na_2S{\cdot}9H_2O$ solution, and tungstate solution to 1.0L sterile DMJ synthetic seawater. Mix thoroughly. Distribute into tubes. Tightly seal the tubes with butyl rubber stoppers under a gas phase of 80% H_2 + 20% CO_2 (300 kPa).

Use: For the cultivation of *Persephonella hydrogeniphila* and *Hydrogenivirga caldilitoris.*

Hofer's Alkaline Medium

Composition per liter:

Agar	15.0g
Mannitol	10.0g
Yeast extract	1.0g
K_2HPO_4	0.5g
MgSO4	0.2g
NaCl	0.1g
Thymol Blue	0.016g

pH 11.0 ± 0.2 at 25°C

Source: This medium is available from HiMedia.

Preparation of Medium: Add components to distilled/deionized water and bring volume to 1.0L. Mix thoroughly. Gently heat and bring to boiling. Distribute into tubes or flasks. Autoclave for 15 min at 15 psi pressure–121°C. Pour into sterile Petri dishes or leave in tubes.

Use: For the selective isolation of *Agrobacterium* spp. from soil.

Hohn's Medium, Modified
See: **Steenken and Smith Agar**

HO-LE Trace Elements Solution

Composition per liter:

H_3BO_3	2.85g
$MnCl_2·4H_2O$	1.8g
Sodium tartrate	1.77g
$FeSO_4$	1.36g
$CoCl_2·6H_2O$	0.04g
$CuCl_2·2H_2O$	0.026g
$Na_2MoO_4·2H_2O$	0.025g
$ZnCl_2$	0.021g

Preparation of Medium: Add components to distilled/deionized water and bring volume to 1.0L. Mix thoroughly. Distribute into tubes or flasks. Autoclave for 15 min at 15 psi pressure–121°C.

Use: For use as an enrichment to other media that require trace minerals.

Hominis Agar
See: **H Agar**

Hominis Broth
See: **H Broth**

Horie Arabinose Ethyl Violet Broth (HAEB)

Composition per liter:

NaCl	30.0g
Peptone	5.0g
Beef extract	3.0g
Bromthymol Blue	0.03g
Ethyl Violet	1.0mg
Arabinose solution	100.0mL

pH 9.0 ± 0.2 at 25°C

Arabinose Solution:

Composition per 100.0mL:

Arabinose	5.0g

Preparation of Arabinose Solution: Add arabinose to distilled/deionized water and bring volume to 100.0mL. Mix thoroughly. Filter sterilize.

Preparation of Medium: Add components, except arabinose solution, to distilled/deionized water and bring volume to 900.0mL. Mix thoroughly. Gently heat and bring to boiling. Adjust pH to 9.0. Autoclave for 15 min at 15 psi pressure–121°C. Cool to 45°–50°C. Aseptically add sterile arabinose solution. Mix thoroughly. Aseptically distribute into sterile tubes or flasks.

Use: For the cultivation of *Vibrio* species from foods.

Horikoshi Alkaline Medium (DSMZ Medium 940)

Composition per liter:

Agar	15.0g
D-glucose	10.0g
Peptone	5.0g
Yeast extract	5.0g
Na_2CO_3	5.0g
KH_2PO_4	1.0g
$MgSO_4·7H_2O$	0.2g

pH 9.0 ± 0.2 at 25°C

Preparation of Medium: Add components to distilled/deionized water and bring volume to 1.0L. Mix thoroughly. Gently heat and bring to boiling. Distribute into tubes or flasks. Autoclave for 15 min at 15 psi pressure–121°C. Pour into sterile Petri dishes or leave in tubes.

Use: For the cultivation and maintenance of *Pannonibacter phragmitetus.*

Horikoshi-1 Medium (DSMZ Medium 1081)

Composition per liter:

Agar	15.0g
Glucose	10.0g
Polypeptone	5.0g
Yeast extract	5.0g
KH_2PO_4	1.0g
K_2HPO_4	1.0g
$MgSO_4·7H_2O$	0.2g
$NaCO_3$solution	100.0mL

pH 10.0 ± 0.2 at 25°C

$NaCO_3$ Solution:

Composition per 100.0mL:

$NaCO_3$	10.0g

Preparation of $NaCO_3$ Solution: Add $NaCO_3$ to distilled/deionized water and bring volume to 100.0mL. Mix thoroughly. Filter sterilize.

Preparation of Medium: Add components, except $NaCO_3$ solution, to double distilled/deionized water and bring volume to 900.0mL. Mix thoroughly. Adjust pH to 10.0. Gently heat while stirring and bring to boiling. Mix thoroughly. Autoclave for 15 min at 15 psi pressure–121°C. Cool to 50°C. Aseptically add 100.0mL $NaCO_3$ solution. Adjust pH to 10.0. Pour into Petri dishes or aseptically distribute into tubes.

Use: For the cultivation of "*Streptomyces sannurensis*" and *Salinicoccus alkaliphilus.*

Horikoshi-1 Medium with 10% Sodium Chloride (DSMZ Medium 1081a)

Composition per liter:

NaCl	100.0g
Agar	15.0g
Glucose	10.0g
Polypeptone	5.0g
Yeast extract	5.0g
KH_2PO_4	1.0g
K_2HPO_4	1.0g
$MgSO_4·7H_2O$	0.2g
$NaCO_3$solution	100.0mL

pH 10.0 ± 0.2 at 25°C

$NaCO_3$ Solution:

Composition per 100.0mL:

$NaCO_3$	10.0g

Preparation of $NaCO_3$ Solution: Add $NaCO_3$ to distilled/deionized water and bring volume to 100.0mL. Mix thoroughly. Filter sterilize.

Preparation of Medium: Add components, except $NaCO_3$ solution, to double distilled/deionized water and bring volume to 900.0mL. Mix thoroughly. Adjust pH to 10.0. Gently heat while stirring and bring to boiling. Mix thoroughly. Autoclave for 15 min at 15 psi pressure–121°C. Cool to 50°C. Aseptically add 100.0mL $NaCO_3$ solution. Adjust pH to 10.0. Pour into Petri dishes or aseptically distribute into tubes.

Use: For the cultivation of *Salinicoccus alkaliphilus*.

Horse Blood Agar

Composition per liter:

Beef heart, infusion from	500.0g
Agar	15.0g
Tryptose	10.0g
NaCl	5.0g
Horse blood, defibrinated	50.0mL

pH 6.8 ± 0.2 at 25°C

Preparation of Medium: Add components, except horse blood, to distilled/deionized water and bring volume to 950.0mL. Mix thoroughly. Gently heat and bring to boiling. Autoclave for 15 min at 15 psi pressure–121°C. Cool to 45°–50°C. Aseptically add sterile horse blood. Mix thoroughly. Pour into sterile Petri dishes or distribute into sterile tubes.

Use: For the cultivation and maintenance of *Yersinia pseudotuberculosis*.

Horse Serum Agar

Composition per liter:

Agar	15.0g
Pancreatic digest of gelatin	5.0g
Beef extract	3.0g
Horse serum	200.0mL

pH 6.8 ± 0.2 at 25°C

Preparation of Medium: Add components, except horse serum, to distilled/deionized water and bring volume to 800.0mL. Mix thoroughly. Gently heat and bring to boiling. Autoclave for 15 min at 15 psi pressure–121°C. Cool to 45°–50°C. Aseptically add sterile horse serum. Mix thoroughly. Pour into sterile Petri dishes or distribute into sterile tubes.

Use: For the cultivation and maintenance of *Pseudomonas aeruginosa* and *Streptobacillus moniliformis*.

Horse Serum Broth

Composition per liter:

Pancreatic digest of gelatin	5.0g
Beef extract	3.0g
Horse serum	200.0mL

pH 6.8 ± 0.2 at 25°C

Source: This medium is available as a premixed powder from BD Diagnostic Systems.

Preparation of Medium: Add components, except horse serum, to distilled/deionized water and bring volume to 800.0mL. Mix thoroughly. Gently heat and bring to boiling. Autoclave for 15 min at 15 psi pressure–121°C. Cool to 45°–50°C. Aseptically add sterile horse serum. Mix thoroughly. Aseptically distribute into sterile tubes or flasks.

Use: For the cultivation and maintenance of *Pseudomonas aeruginosa* and *Streptobacillus moniliformis*.

Hottinger Broth

Composition per liter:

Fish peptone	20.0g
Yeast extract	2.0g
Tryptophan	1.0g

pH 7.2 ± 0.2 at 25°C

Source: This medium is available as a premixed powder from HiMedia.

Preparation of Medium: Add components to distilled/deionized water and bring volume to 1.0L. Mix thoroughly. Gently heat and bring to boiling. Distribute into tubes or flasks. Autoclave for 15 min at 15 psi pressure–121°C.

Use: For cultivation of less fastidious microorganisms and determination of indole as per USSR State Pharmacopoeia.

Howardella Medium (DSMZ Medium 1085)

Composition per liter:

Casitone	20.0g
Yeast extract	5.0g
Na_2HPO_4	5.0g
$MgCl_2 \cdot 6H_2O$	1.1g
Urea	1.0g
Na-thioglycolate	0.75g
Resazurin	0.5mg
Urea solution	10.0mL

pH 7.4 ± 0.2 at 25°C

Urea Solution:

Composition per 10.0mL:

Urea	1.0g

Preparation of Urea Solution: Add urea to distilled/deionized water and bring volume to 10.0mL. Mix thoroughly. Sparge with 100% N_2. Filter sterilize.

Preparation of Medium: Add components, except thioglycolate and urea solution, to double distilled/deionized water and bring volume to 990.0mL. Cool to room temperature while sparging with 80% N_2 + 20% CO_2. Add thioglycolate. Mix thoroughly. Distribute into tubes or bottles under an atmosphere of 80% N_2 + 20% CO_2. Autoclave for 15 min at 15 psi pressure–121°C. Cool to room temperature. Aseptically add 10.0mL urea solution. Adjust pH to 7.4.

Use: For the cultivation of *Howardella* spp.

Hoyer's Medium

Composition per liter:

$(NH_4)_2SO_4$	1.0g
KH_2PO_4	0.9g
$MgSO_4 \cdot 7H_2O$	0.25g
K_2HPO_4	0.1g
$FeCl_3 \cdot 6H_2O$	0.02g
Ethanol solution	200.0mL

Ethanol Solution:

Composition per 200.0mL:

Ethanol	30.0mL

Preparation of Ethanol Solution: Add ethanol to distilled/deionized water and bring volume to 200.0mL. Mix thoroughly. Filter sterilize.

Preparation of Medium: Add components, except ethanol solution, to distilled/deionized water and bring volume to 800.0mL. Mix thoroughly. Distribute into tubes in 4.0mL volumes. Autoclave for 15 min at 15 psi pressure–121°C. Cool to 25°C. Aseptically add 1.0mL of sterile ethanol solution to each tube. Mix thoroughly.

Use: For the cultivation of *Acetobacter* species.

Hoyle Medium

Composition per 1060.0mL:

Agar	15.0g
Lab Lemco powder	10.0g
Peptone	10.0g
NaCl	5.0g
Horse blood, laked	50.0mL
Tellurite solution	10.0mL

pH 7.8 ± 0.2 at 25°C

Source: This medium is available as a premixed powder from Oxoid Unipath.

Horse Blood, Laked:

Composition per 50.0mL:

Horse blood, fresh	50.0mL

Preparation of Horse Blood, Laked: Add blood to a sterile polypropylene bottle. Freeze overnight at –20°C. Thaw at 8°C. Refreeze at –20°C. Thaw again at 8°C.

Tellurite Solution:

Composition per 100.0mL:

K_2TeO_3	3.5g

Preparation of Tellurite Solution: Add K_2TeO_3 to distilled/deionized water and bring volume to 100.0mL. Mix thoroughly. Filter sterilize.

Caution: Potassium tellurite is toxic.

Preparation of Medium: Add components, except laked horse blood and tellurite solution, to distilled/deionized water and bring volume to 1.0L. Mix thoroughly. Gently heat and bring to boiling. Autoclave for 15 min at 15 psi pressure–121°C. Cool to 50°C. Aseptically add 50.0mL sterile laked horse blood and 10.0mL sterile tellurite solution. Mix thoroughly. Pour into sterile Petri dishes or distribute into sterile tubes.

Use: For the isolation and differentiation of *Corynebacterium diphtheriae* strains. This medium permits very rapid growth of all types of *Corynebacterium diphtheriae*, so that diagnosis is possible after 18 hours' incubation.

Hoyle HiVeg Medium Base

Composition per liter:

Agar	15.0g
Plant extract	10.0g
Plant peptone	10.0g
NaCl	5.0g
Horse blood, laked	50.0mL
Tellurite solution	10.0mL

pH 7.8 ± 0.2 at 25°C

Source: This medium, without tellurite solution and laked blood, is available as a premixed powder from HiMedia.

Horse Blood, Laked:

Composition per 50.0mL:

Horse blood, fresh	50.0mL

Preparation of Horse Blood, Laked: Add blood to a sterile polypropylene bottle. Freeze overnight at –20°C. Thaw at 8°C. Refreeze at –20°C. Thaw again at 8°C.

Tellurite Solution:

Composition per 100.0mL:

K_2TeO_3	3.5g

Preparation of Tellurite Solution: Add K_2TeO_3 to distilled/deionized water and bring volume to 100.0mL. Mix thoroughly. Filter sterilize.

Caution: Potassium tellurite is toxic.

Preparation of Medium: Add components, except laked horse blood and tellurite solution, to distilled/deionized water and bring volume to 1.0L. Mix thoroughly. Gently heat and bring to boiling. Autoclave for 15 min at 15 psi pressure–121°C. Cool to 50°C. Aseptically add 50.0mL sterile laked horse blood and 10.0mL sterile tellurite solution. Mix thoroughly. Pour into sterile Petri dishes or distribute into sterile tubes.

Use: For the isolation and differentiation of *Corynebacterium diphtheriae* strains. This medium permits very rapid growth of all types of *Corynebacterium diphtheriae*, so that diagnosis is possible after 18 hours' incubation.

Hoyle Medium Base

Composition per liter:

Agar	15.0g
Beef extract	10.0g
Peptone	10.0g
NaCl	5.0g
Blood, laked	50.0mL
Tellurite solution	10.0mL

pH 7.8 ± 0.2 at 25°C

Source: This medium is available as a premixed powder from Oxoid Unipath.

Tellurite Solution:

Composition per 100.0mL:

K_2TeO_3	3.5g

Caution: Potassium tellurite is toxic.

Preparation of Tellurite Solution: Add K_2TeO_3 to distilled/deionized water and bring volume to 100.0mL. Mix thoroughly. Filter sterilize.

Horse Blood, Laked:

Composition per 50.0mL:

Horse blood, fresh	50.0mL

Preparation of Horse Blood, Laked: Add blood to a sterile polypropylene bottle. Freeze overnight at –20°C. Thaw at 8°C. Refreeze at –20°C. Thaw again at 8°C.

Preparation of Medium: Add components, except laked blood, to distilled/deionized water and bring volume to 940.0mL. Mix thoroughly. Gently heat and bring to boiling. Autoclave for 15 min at 15 psi pressure–121°C. Cool to 45°–50°C. Aseptically add sterile laked blood. Mix thoroughly. Pour into sterile Petri dishes or distribute into sterile tubes.

Use: For the isolation and differentiation of *Corynebacterium diphtheriae*.

HP 6 Agar

Composition per liter:

Agar	15.0g
Sodium glutaminate	10.0g
$MgSO_4 \cdot 7H_2O$	1.0g
Yeast extract	1.0g
Cyanocobalamin	0.5mg
Glucose solution	100.0mL

pH 7.2 ± 0.2 at 25°C

Glucose Solution:

Composition per 100.0mL:

D-Glucose	5.0g

Preparation of Glucose Solution: Add D-glucose to distilled/deionized water and bring volume to 100.0mL. Mix thoroughly. Autoclave for 15 min at 15 psi pressure–121°C. Cool to 25°C.

Preparation of Medium: Add components, except glucose solution, to distilled/deionized water and bring volume to 900.0mL. Mix thoroughly. Gently heat and bring to boiling. Autoclave for 15 min at 15 psi pressure–121°C. Cool to 45°–50°C. Aseptically add sterile glucose solution. Mix thoroughly. Pour into sterile Petri dishes or distribute into sterile tubes.

Use: For the isolation and cultivation of *Cytophaga* species, *Herpetosiphon* species, *Saprospira* species, and *Flexithrix* species.

HP 6 Agar Base

Composition per liter:

Plant hydrolysate	15.0g
Glucose	5.5g
Yeast extract	5.0g
NaCl	2.5g
Agar	1.0g
Sodium hydrosulphite	0.5g
Resazurin	1.0mg

pH 7.2 ± 0.2 at 25°C

Preparation of Medium: Add components to distilled/deionized water and bring volume to 1.0L. Mix thoroughly. Gently heat and bring to boiling. Autoclave for 15 min at 15 psi pressure–121°C. Mix thoroughly. Pour into sterile Petri dishes or distribute into sterile tubes.

Use: For the isolation and cultivation of *Cytophaga* species, *Herpetosiphon* species, *Saprospira* species, and *Flexithrix* species.

HP 6 Agar Base

Composition per liter:

Agar	15.0g
Sodium glutaminate	10.0g
$MgSO_4 \cdot 7H_2O$	1.0g
Yeast extract	1.0g
Cyanocobalamin	0.5mg
Glucose solution	100.0mL

Source: This medium is available from HiMedia.

Glucose Solution:

Composition per 100.0mL:

Glucose	5.0g

Preparation of Glucose Solution: Add glucose to distilled/deionized water and bring volume to 100.0mL. Mix thoroughly. Filter sterilize.

Preparation of Medium: Add components, except glucose solution, to distilled/deionized water and bring volume to 900.0mL. Mix thoroughly. Autoclave for 15 min at 15 psi pressure–121°C. Cool to 50°C. Aseptically add glucose solution. Mix thoroughly. Pour into Petri dishes or aseptically distribute into sterile tubes.

Use: For the isolation and cultivation of *Cytophaga, Herpetosiphon, Saprospira*, and *Flexithrix* species.

HP 74 Broth

Composition per liter:

Sodium glutaminate	10.0g
$MgSO_4 \cdot 7H_2O$	2.0g
Yeast extract	2.0g
Glucose solution	100.0mL
Phosphate buffer solution	20.0mL

pH 6.5 ± 0.2 at 25°C

Glucose Solution:

Composition per 100.0mL:

D-Glucose	10.0g

Preparation of Glucose Solution: Add D-glucose to distilled/deionized water and bring volume to 100.0mL. Mix thoroughly. Autoclave for 15 min at 15 psi pressure–121°C. Cool to 25°C.

Phosphate Buffer Solution:

Composition per 100.0mL:

K_2HPO_4	6.81g

Preparation of Phosphate Buffer Solution: Add K_2HPO_4 to distilled/deionized water and bring volume to 100.0mL. Mix thoroughly. Adjust pH to 6.5. Autoclave for 15 min at 15 psi pressure–121°C. Cool to 25°C.

Preparation of Medium: Add components, except glucose solution and phosphate buffer solution, to distilled/deionized water and bring volume to 880.0mL. Mix thoroughly. Gently heat and bring to boiling. Autoclave for 15 min at 15 psi pressure–121°C. Cool to 45°–50°C. Aseptically add 100.0mL of sterile glucose solution and 20.0mL of sterile phosphate buffer solution. Mix thoroughly. Aseptically distribute into sterile tubes or flasks.

Use: For the isolation and cultivation of *Cytophaga* species, *Herpetosiphon* species, *Saprospira* species, and *Flexithrix* species.

HP 101 Halophile Medium

Composition per liter:

NaCl	100.0g
Agar	20.0g
Peptone	10.0g
$MgSO_4 \cdot 7H_2O$	4.3g
$NaNO_3$	2.0g
Yeast extract	1.0g

pH 7.2 ± 0.2 at 25°C

Preparation of Medium: Add components to distilled/deionized water and bring volume to 1.0L. Mix thoroughly. Gently heat and bring to boiling. Distribute into tubes or flasks. Autoclave for 15 min at 15 psi pressure–121°C. Pour into sterile Petri dishes or leave in tubes.

Use: For the cultivation and maintenance of *Pseudomonas* species.

HP Medium

Composition per liter:

Pancreatic digest of soybean meal	20.0g
Beef extract	10.0g

Yeast extract	6.0g
Ammonium citrate	5.0g
Tween™ 80	0.5g
$MgSO_4 \cdot 7H_2O$	0.2g
$MnSO_4 \cdot 4H_2O$	0.05g
$FeSO_4 \cdot 7H_2O$	0.04g
Glucose solution	10.0mL
Tetracycline solution	10.0mL

Glucose Solution:
Composition per 100.0mL:

Glucose	10.0g

Preparation of Glucose Solution: Add glucose to distilled/deionized water and bring volume to 100.0mL. Mix thoroughly. Filter sterilize.

Tetracycline Solution:
Composition per 100.0mL:

Tetracycline	10.0g

Preparation of Tetracycline Solution: Add tetracycline to distilled/deionized water and bring volume to 100.0mL. Mix thoroughly. Filter sterilize.

Preparation of Medium: Add components, except glucose solution and tetracycline solution, to distilled/deionized water and bring volume to 990.0mL. Mix thoroughly. Gently heat and bring to boiling. Autoclave for 15 min at 15 psi pressure–121°C. Cool to 45°–50°C. Aseptically add sterile glucose solution and tetracycline solution. Mix thoroughly. Aseptically distribute into sterile tubes or flasks.

Use: For the cultivation and enumeration of *Leuconostoc* species.

HPC Agar
See: **NWRI Agar**

HPC Agar (Heterotrophic Plate Count Agar) (m-HPC Agar)

Composition per liter:

Gelatin	25.0g
Pancreatic digest of gelatin	20.0g
Agar	15.0g
Glycerol	10.0mL

pH 7.1 ± 0.2 at 25°C

Source: This medium is available from BD Diagnostic Systems.

Preparation of Medium: Add components, except glycerol, to distilled/deionized water and bring volume to 990.0mL. Mix thoroughly. Gently heat and bring to boiling. Add glycerol. Mix thoroughly. Autoclave for 15 min at 15 psi pressure–121°C. Cool to 45°–50°C. Pour into sterile Petri dishes.

Use: For the the cultivation and enumeration of microorganisms from potable water sources, swimming pools, and other water specimens by the membrane filter method and heterotrophic plate count technique.

HQGö1 Medium (DSMZ Medium 298a)

Composition per liter:

NaCl	1.0g
KCl	0.5g
$MgCl_2 \cdot 6H_2O$	0.4g
NH_4Cl	0.25g
KH_2PO_4	0.2g
$CaCl_2 \cdot 2H_2O$	0.15g
Resazurin	1.0mg
$NaHCO_3$ solution	10.0mL
Butanediol solution	10.0mL
$Na_2S \cdot 9H_2O$ solution	10.0mL
Vitamin solution	10.0mL
Gentisic acid solution	1.0mL
Trace elements solution SL-10	1.0mL

pH 7.2 ± 0.2 at 25°C

$Na_2S \cdot 9H_2O$ Solution:
Composition per 10.0mL:

$Na_2S \cdot 9H_2O$	0.36g

Preparation of $Na_2S \cdot 9H_2O$ Solution: Add $Na_2S \cdot 9H_2O$ to distilled/deionized water and bring volume to 10.0mL. Mix thoroughly. Autoclave under 100% N_2 for 15 min at 15 psi pressure–121°C. Cool to room temperature.

$NaHCO_3$ Solution:
Composition per 10.0mL:

$NaHCO_3$	2.5g

Preparation of $NaHCO_3$ Solution: Add $NaHCO_3$ to distilled/deionized water and bring volume to 10.0mL. Mix thoroughly. Sparge with 80% N_2 + 20% CO_2. Filter sterilize.

Vitamin Solution:
Composition per liter:

Pyridoxine-HCl	10.0mg
Thiamine-HCl·$2H_2O$	5.0mg
Riboflavin	5.0mg
Nicotinic acid	5.0mg
D-Ca-pantothenate	5.0mg
p-Aminobenzoic acid	5.0mg
Lipoic acid	5.0mg
Biotin	2.0mg
Folic acid	2.0mg
Vitamin B_{12}	0.1mg

Preparation of Vitamin Solution: Add components to distilled/deionized water and bring volume to 1.0L. Mix thoroughly. Sparge with 80% H_2 + 20% CO_2. Filter sterilize.

Butanediol Solution:
Composition per 10.0mL:

2,3 butanediol	0.9g

Preparation of Butanediol Solution: Add butanediol to distilled/deionized water and bring volume to 10.0mL. Mix thoroughly. Sparge with 100% N_2. Filter sterilize.

Gentisic Acid Solution:
Composition per 100.0mL:

Gentisic acid	3.08g

Preparation of Gentisic Acid Solution: Add gentisic acid to distilled/deionized water and bring volume to 100.0mL. Mix thoroughly. Adjust pH to 7.0. Sparge with 100% N_2. Filter sterilize.

Trace Elements Solution SL-10:
Composition per liter:

$FeCl_2 \cdot 4H_2O$	1.5g
$CoCl_2 \cdot 6H_2O$	190.0mg
$MnCl_2 \cdot 4H_2O$	100.0mg
$ZnCl_2$	70.0mg
$Na_2MoO_4 \cdot 2H_2O$	36.0mg

$NiCl_2 \cdot 6H_2O$	24.0mg
H_3BO_3	6.0mg
$CuCl_2 \cdot 2H_2O$	2.0mg
HCl (25% solution)	10.0mL

Preparation of Trace Elements Solution SL-10: Add $FeCl_2 \cdot 4H_2O$ to 10.0mL of HCl solution. Mix thoroughly. Add distilled/deionized water and bring volume to 1.0L. Add remaining components. Mix thoroughly. Sparge with 80% N_2 + 20% CO_2. Autoclave for 15 min at 15 psi pressure–121°C.

Preparation of Medium: Prepare and dispense medium under 80% N_2 + 20% CO_2 gas atmosphere. Add components, except $NaHCO_3$ solution, butanediol solution, $Na_2S \cdot 9H_2O$ solution, vitamin solution, gentisic acid solution, and trace elements solution SL-10, to distilled/deionized water and bring volume to 958.0mL. Mix thoroughly. Adjust pH to 7.2. Sparge with 80% N_2 + 20% CO_2. Autoclave for 15 min at 15 psi pressure–121°C. Aseptically and anaerobically add 10.0mL $NaHCO_3$ solution, 10.0mL butanediol solution, 10.0mL $Na_2S \cdot 9H_2O$ solution, 10.0mL vitamin solution, 1.0mL gentisic acid solution, and 1.0mL trace elements solution SL-10. Mix thoroughly. Aseptically and anaerobically distribute into sterile tubes or bottles. After inoculation, flush and repressurize the gas head space of culture bottles with sterile 80% N_2 + 20% CO_2 to 1 bar overpressure.

Use: For the cultivation of *Syntrophus gentianae*.

HR Antifungal Assay Medium Buffered with MOPS

Composition per liter:

MOPS (3-*N*-morpholino-propanesulfonic acid) buffer	34.53g
Glucose	10.0g
$(NH_4)_2SO_4$	2.5g
KH_2PO_4	1.0g
$NaHCO_3$	1.0g
Glutamine	0.58g
$MgSO_4 \cdot 7H_2O$	0.5g
$CaCl_2 \cdot 2H_2O$	0.1g
NaCl	0.1g
L-Lysine	0.07g
L-Isoleucine	0.05g
L-Leucine	0.05g
L-Threonine	0.05g
L-Valine	0.05g
L-Arginine	0.04g
L-Histidine	0.02g
L-Methionine	0.01g
L-Tryptophan	8.2mg
DL-Methionine	2.0mg
DL-Tryptophan	2.0mg
Inositol	2.0mg
L-Histidine·HCl	1.0mg
H_3BO_3	0.5mg
Calcium pantothenate	0.4mg
$MnSO_4 \cdot H_2O$	0.4mg
Niacin	0.4mg
Pyridoxine	0.4mg
Thiamine·HCl	0.4mg
$ZnSO_4 \cdot 7H_2O$	0.4mg
p-Aminobenzoic acid	0.2mg
$FeCl_3$	0.2mg
Riboflavin	0.2mg
Na_2MoO_3	0.2mg
KI	0.1mg
$CuSO_4 \cdot 5H_2O$	0.04mg
Biotin	2.0μg
Folic acid	2.0μg

pH 7.0 ± 0.2 at 25°C

Preparation of Medium: Add components, except $NaHCO_3$ and MOPS buffer, to distilled/deionized water and bring volume to 900.0mL. Mix thoroughly. Add $NaHCO_3$ and MOPS buffer. Mix thoroughly. Adjust pH to 7.0. Bring volume to 1.0L with distilled/deionized water. Filter sterilize.

Use: For testing the effectiveness of antifungal agents against clinical fungal isolates using the broth dilution susceptibility testing method.

HS HiVeg Medium

Plant hydrolysate	15.0g
Glucose	5.5g
Yeast extract	5.0g
Agar	1.0g
NaCl	2.5g
$Na_2S_2O_4$	0.5g
Resazurin	0.001g

pH 7.1 ± 0.2 at 25°C

Source: This medium is available as a premixed powder from HiMedia.

Preparation of Medium: Add components to distilled/deionized water and bring volume to 1.0L. Mix thoroughly. Gently heat and bring to boiling. Distribute into tubes or flasks. Autoclave for 15 min at 15 psi pressure–121°C.

Use: For cultivation of aerobic as well as anaerobic bacteria and sterility testing.

HS Medium

Casein enzymic hydrolysate	15.0g
Glucose	5.5g
Yeast extract	5.0g
Agar	1.0g
NaCl	2.5g
$Na_2S_2O_4$	0.5g
Resazurin	0.001g

pH 7.1 ± 0.2 at 25°C

Source: This medium is available as a premixed powder from HiMedia.

Preparation of Medium: Add components to distilled/deionized water and bring volume to 1.0L. Mix thoroughly. Gently heat and bring to boiling. Distribute into tubes or flasks. Autoclave for 15 min at 15 psi pressure–121°C.

Use: For cultivation of aerobic as well as anaerobic bacteria and sterility testing.

HTM

See: ***Haemophilus*** **Test Medium**

Hugh Leifson Glucose Broth

Composition per liter:

NaCl	30.0g
Glucose	10.0g
Agar	3.0g
Peptone	2.0g

Yeast extract 0.5g
Bromcresol Purple 0.015g

pH 7.4 ± 0.2 at 25°C

Preparation of Medium: Add components to distilled/deionized water and bring volume to 1.0L. Mix thoroughly. Gently heat while stirring and bring to boiling. Adjust pH to 7.4. Distibute into tubes or flasks. Autoclave for 15 min at 15 psi pressure–121°C.

Use: For the cultivation and differentiation of bacteria based on their ability to ferment glucose. Bacteria that ferment glucose turn the medium yellow.

Hugh Leifson Glucose HiVeg Medium

Composition per liter:

NaCl 30.0g
Glucose 10.0g
Agar 3.0g
Plant peptone 2.0g
Yeast extract 0.5g
Bromcresol Purple 0.015g

pH 7.4 ± 0.2 at 25°C

Source: This medium is available as a premixed powder from HiMedia.

Preparation of Medium: Add components to distilled/deionized water and bring volume to 1.0L. Mix thoroughly. Gently heat while stirring and bring to boiling. Adjust pH to 7.4. Distribute into tubes or flasks. Autoclave for 15 min at 15 psi pressure–121°C.

Use: For the cultivation and differentiation of bacteria based on their ability to ferment glucose. Bacteria that ferment glucose turn the medium yellow.

Hugh Leifson HiVeg Medium

Composition per liter:

Glucose 10.0g
NaCl 5.0g
Agar 2.0g
Plant peptone 2.0g
K_2HPO_4 .... 0.3g
Bromthymol Blue 0.05g

pH 6.8 ± 0.2 at 25°C

Source: This medium is available as a premixed powder from HiMedia.

Preparation of Medium: Add components to distilled/deionized water and bring volume to 1.0L. Mix thoroughly. Gently heat while stirring and bring to boiling. Adjust pH to 7.4. Distribute into tubes or flasks. Autoclave for 15 min at 15 psi pressure–121°C.

Use: For the cultivation and differentiation of bacteria based on their ability to ferment glucose. Bacteria that ferment glucose turn the medium yellow.

Hugh Leifson Oxidation-Fermentation Medium
See: **Oxidation-Fermentation Medium, Hugh-Leifson**

Human Blood Tween™ Bilayer Medium
See: **HBT Bilayer Medium**

Hungate's Habitat-Simulating Medium

Composition per 1140.2mL:

Rumen fluid 333.0mL
Mineral solution A 167.0mL
Mineral solution B 167.0mL
$NaHCO_3$ solution 53.0mL
L-Cysteine·HCl solution 10.6mL
Substrate solution 10.6mL
Resazurin solution 1.0mL

Mineral Solution A:

Composition per liter:

NaCl 6.0g
KH_2PO_4 .... 3.0g
$(NH_4)_2SO_4$ 3.0g
$CaCl_2$ 0.6g
$MgSO_4$ 0.6g

Preparation of Solution A: Add components to distilled/deionized water and bring volume to 1.0L. Mix thoroughly.

Mineral Solution B:

K_2HPO_4 .... 3.0

Preparation of Solution B: Add K_2HPO_4 to distilled/deionized water and bring volume to 1.0L. Mix thoroughly.

Resazurin Solution:

Composition per 100.0mL:

Resazurin 0.1g

Preparation of Resazurin Solution: Add resazurin to distilled/deionized water and bring volume to 100.0L. Mix thoroughly.

L-Cysteine·HCl Solution:

Composition per 100.0mL:

L-Cysteine·HCl 3.0g

Preparation of L-Cysteine·HCl Solution: Add L-cysteine·HCl to O_2-free distilled/deionized water and bring volume to 100.0L. Mix thoroughly. Gently heat and bring to boiling. Continue boiling for 2 min. Cool to 25°C under 100% N_2. Seal tube with a stopper that is wired in place. Autoclave for 15 min at 15 psi pressure–121°C. Cool to 25°C.

$NaHCO_3$ Solution:

Composition per 10.0mL:

$NaHCO_3$ 1.0g

Preparation of $NaHCO_3$ Solution: Add $NaHCO_3$ to O_2-free distilled/deionized water and bring volume to 10.0mL. Mix thoroughly. Filter sterilize. Gas with 100% CO_2 for 15 min.

Substrate Solution:

Composition per 100.0mL:

Sugar 10.0g

Preparation of Substrate Solution: Add sugar to O_2-free distilled/deionized water. Mix thoroughly. Gas with 100% N_2 for 15 min. Autoclave for 15 min at 15 psi pressure–121°C. Cool to 45°–50°C.

Preparation of Medium: Add 167.0mL of solution A, 167.0mL of solution B, and 1.0mL of resazurin solution to distilled/deionized water and bring volume to 733.0mL. Mix thoroughly. Gently heat and bring to boiling. Continue boiling until resazurin turns colorless, indicating reduction. Bring volume back to 733.0mL (some evaporation will have occurred) with O_2-free distilled/deionized water. Cool to 45°–50°C under O_2-free 100% CO_2. Anaerobically add rumen fluid. Anaerobically distribute into tubes in 10.0mL volumes. Cap with butyl rubber stoppers. Place tubes in a press. Autoclave for 15 min at 15 psi pressure–121°C. Cool to 25°C. Immediately prior to inoculation, aseptically and anaerobically add 0.1mL of sterile L-cysteine·HCl solution, 0.5mL of sterile $NaHCO_3$ solution, and 0.1mL of substrate solution per 10.0mL of medium in each tube.

Use: For the cultivation of *Bacteroides* species from rumens.

Hutner's Medium for *Euglena*

Composition per liter:

Agar, noble	12.0g
Pancreatic digest of peptone	0.6g
Yeast extract	0.4g
KH_2PO_4	0.02g
Potassium citrate·H_2O	0.04g
$MgSO_4·3H_2O$	0.02g
Thiamine	0.4mg
Vitamin B_{12}	0.5μg

Preparation of Medium: Add components to distilled/deionized water and bring volume to 1.0L. Mix thoroughly. Gently heat and bring to boiling. Distribute into tubes or flasks. Autoclave for 15 min at 15 psi pressure–121°C. Pour into sterile Petri dishes or leave in tubes.

Use: For the cultivation of *Astasia longa*, *Euglena gracilis*, *Polytoma* species, *Polytomella parva*, and *Polytomella caeca.*

Hutner's Medium for *Euglena*

Composition per liter:

Agar, noble	12.0g
Pancreatic digest of peptone	0.6g
Liver concentrate	0.2g
Potassium citrate·H_2O	0.04g
KH_2PO_4	0.02g
$MgSO_4·3H_2O$	0.02g

Preparation of Medium: Add components to distilled/deionized water and bring volume to 1.0L. Mix thoroughly. Gently heat and bring to boiling. Distribute into tubes or flasks. Autoclave for 15 min at 15 psi pressure–121°C. Pour into sterile Petri dishes or leave in tubes.

Use: For the cultivation of *Astasia longa*, *Euglena gracilis*, *Polytoma* species, *Polytomella parva*, and *Polytomella caeca.*

HY Agar for *Flavobacterium*

Composition per liter:

Agar	8.0g
Glutamic acid	5.0g
K_2HPO_4	0.1g
$MgSO_4·7H_2O$	0.1g

pH 7.3 ± 0.2 at 25°C

Preparation of Medium: Add components to distilled/deionized water and bring volume to 1.0L. Glutamic acid may be replaced by 1.0g of folic acid if desired. Mix thoroughly. Gently heat and bring to boiling. Distribute into tubes or flasks. Autoclave for 15 min at 15 psi pressure–121°C.

Use: For the cultivation of *Flavobacterium* species.

HY Medium for *Flavobacterium*

Composition per liter:

Glutamic acid	5.0g
K_2HPO_4	0.1g
$MgSO_4·7H_2O$	0.1g

pH 7.3 ± 0.2 at 25°C

Preparation of Medium: Add components to distilled/deionized water and bring volume to 1.0L. Glutamic acid may be replaced by 1.0g of folic acid if desired. Mix thoroughly. Gently heat and bring to boiling. Distribute into tubes or flasks. Autoclave for 15 min at 15 psi pressure–121°C.

Use: For the cultivation of *Flavobacterium* species.

HYA Agar

Composition per liter:

Agar	15.0g
Proteose peptone No. 3	10.0g
Beef extract	1.0g
Lactose solution	10.0mL
Galactose solution	10.0mL
Glucose solution	10.0mL

pH 6.8 ± 0.2 at 25°C

Lactose Solution:

Composition per 10.0mL:

Lactose	5.0g

Preparation of Lactose Solution: Add lactose to distilled/deionized water and bring volume to 10.0mL. Mix thoroughly. Filter sterilize.

Galactose Solution:

Composition per 10.0mL:

Galactose	2.5g

Preparation of Galactose Solution: Add galactose to distilled/deionized water and bring volume to 10.0mL. Mix thoroughly. Filter sterilize.

Glucose Solution:

Composition per 10.0mL:

Glucose	2.5g

Preparation of Glucose Solution: Add glucose to distilled/deionized water and bring volume to 10.0mL. Mix thoroughly. Filter sterilize.

Preparation of Medium: Add components—except lactose solution, galactose solution, and glucose solution—to distilled/deionized water and bring volume to 970.0mL. Mix thoroughly. Gently heat and bring to boiling. Adjust pH to 6.8. Autoclave for 15 min at 15 psi pressure–121°C. Cool to 45°–50°C. Aseptically add sterile lactose solution, galactose solution, and glucose solution. Mix thoroughly. Pour into sterile Petri dishes.

Use: For the cultivation of acidogenic microorganisms, especially *Lactobacillus bulgaricus* and *Streptococcus thermophilus,* from foods.

Hydrogen-Oxidizing Bacteria Medium

Composition per 1020.0mL:

Solution I	1.0L
Solution II	10.0mL
Solution III	10.0mL

Solution I:

Composition per liter:

$Na_2HPO_4·12H_2O$	9.0g
KH_2PO_4	1.5g
NH_4Cl	1.0g
$MgSO_4·7H_2O$	0.2g
Trace elements solution	1.0mL

Preparation of Solution I: Add components to distilled/deionized water and bring volume to 1.0L. Mix thoroughly. Gently heat until dissolved. Autoclave for 15 min at 15 psi pressure–121°C. Cool to 25°C.

Trace Elements Solution:
Composition per liter:

H_3BO_3	0.3g
$CoCl_2 \cdot 6H_2O$	0.2g
$ZnSO_4 \cdot 7H_2O$	0.1g
$MnCl_2 \cdot 4H_2O$	0.03g
$NaMoO_4 \cdot 2H_2O$	0.03g
$NiCl_2 \cdot 6H_2O$	0.02g
$CuCl_2 \cdot 2H_2O$	0.01g

Preparation of Trace Elements Solution: Add components to distilled/deionized water and bring volume to 1.0L. Mix thoroughly.

Solution II:
Composition per 100.0mL:

$CaCl_2 \cdot 2H_2O$	0.1g
Ferric ammonium citrate	0.05g

Preparation of Solution II: Add components to distilled/deionized water and bring volume to 100.0mL. Mix thoroughly. Autoclave for 15 min at 15 psi pressure–121°C. Cool to 25°C.

Solution III:
Composition per 100.0mL:

$NaHCO_3$	5.0g

Preparation of Solution III: Add $NaHCO_3$ to distilled/deionized water and bring volume to 100.0mL. Mix thoroughly. Filter sterilize.

Preparation of Medium: Aseptically combine 1.0L of cooled, sterile solution I, 10.0mL of cooled, sterile solution II, and 10.0mL of sterile solution III. Mix thoroughly. Aseptically distribute into sterile tubes or flasks.

Use: For the cultivation of hydrogen-oxidizing bacteria.

Hydrogen-oxidizing Medium (DSMZ Medium 1003)

Composition per liter:

$MgSO_4 \cdot 7H_2O$	7.0g
NaS_2O_3	2.0g
MES	1.95g
$MgCl_2 \cdot 6H_2O$	0.78g
KCl	0.48g
$CaCl_2 \cdot 2H_2O$	0.4g
Trace elements solution	10.0mL
Solution A	2.0mL
Solution B	1.5mL

pH 6.0 ± 0.2 at 25°C

Solution A:
Composition per liter:

NH_4Cl	100.0g
$MgCl_2 \cdot 6H_2O$	100.0g
$CaCl_2 \cdot 2H_2O$	40.0g

Preparation of Solution A: Add components to distilled/deionized water and bring volume to 1.0L. Mix thoroughly. Adjust pH to 4.0 with HCl.

Solution B:
Composition per liter:

$K_2HPO_4 \cdot 3H_2O$	200.0g

Preparation of Solution B: Add $K_2HPO_4 \cdot 3H_2O$ to distilled/deionized water and bring volume to 1.0L. Mix thoroughly.

Trace Elements Solution:
Composition per liter:

Na-EDTA	0.5g
$CoCl_2$	0.15g
$MnCl_2 \cdot 4H_2O$	0.1g
$FeSO_4 \cdot 7H_2O$	0.1g
$ZnCl_2$	0.1g
$AlCl_3 \cdot 6H_2O$	0.04g
$NaWoO_4 \cdot 2H_2O$	0.03g
$CuCl_2 \cdot 2H_2O$	0.02g
$NiSO_4 \cdot 6H_2O$	0.02g
H_3BO_3	0.01g
Na_2SeO_4	0.01g
$NaMoO_4 \cdot 2H_2O$	0.01g

Preparation of Trace Elements Solution: Add components to distilled/deionized water and bring volume to 1.0L. Mix thoroughly. Adjust the pH to 3.0.

Preparation of Medium: Sparge 1.0L of distilled/deionized water with 100% CO_2 to produce anaerobic water. Add components to 1.0L of the anaerobic water. Mix thoroughly. The pH should be 6.0. Sparge with 100% CO_2 for 20 min. Dispense 5.0mL aliquots into sealable culture tubes. Place stopper on culture tube and crimp tube cap onto stopper. Autoclave for 20 min at 15 psi pressure–121°C. Add 1.0mL of O_2 to each tube before inoculation. After inoculation pressurize the tubes with H_2 (138 KP).

Use: For the cultivation of *Sulfurihydrogenibium azorense.*

Hydrogenivirga okinawensis Medium (DSMZ Medium 1131)

Composition per liter:

Agar	20.0g
Mannitol	10.0g
Yeast extract	0.3g
K_2HPO_4	0.2g
$MgSO_4 \cdot 7H_2O$	0.2g
NaCl	0.05g

pH 7.0 ± 0.2 at 25°C

Preparation of Medium: Add components to distilled/deionized water and bring volume to 1.0L. Mix thoroughly. Adjust pH to 7.0. Distribute into tubes or flasks. Gently heat while stirring and bring to boiling. Mix thoroughly. Autoclave for 15 min at 15 psi pressure–121°C. Cool to 50°C. Pour into Petri dishes or leave in tubes.

Use: For the cultivation of *Hydrogenivirga okinawensis.*

Hydrogenobacter acidophilus Medium (DSMZ Medium 743)

Composition per liter:

Sulfur	5.0g
$(NH_4)_2SO_4$	1.0g
K_2HPO_4	1.0g
NaCl	1.0g
$MgSO_4 \cdot 7H_2O$	0.3g
$FeSO_4 \cdot 7H_2O$	1.0mg
$CaCl_2$	1.0mg
$NiSO_4 \cdot 6H_2O$	0.06mg
Trace elements solution	0.5mL

pH 3.0 ± 0.2 at 25°C

Trace Elements Solution:

Composition per liter:

$ZnSO_4·7H_2O$	28.0mg
MoO_3	4.0mg
H_3BO_3	4.0mg
$MnSO_4·5H_2O$	4.0mg
$CoCl_2·6H_2O$	4.0mg
$CuSO_4·5H_2O$	2.0mg

Preparation of Trace Elements Solution: Add components to distilled/deionized water and bring volume to 1.0L. Mix thoroughly. Sparge with 80% H_2 + 20% CO_2.

Preparation of Medium: Autoclave sulfur for 15 min at 9 psi pressure–113°C. Add components, except sulfur, to distilled/deionized water and bring volume to 1.0L. Mix thoroughly. Autoclave for 15 min at 15 psi pressure–121°C. Add 5.0g sterile sulfur. Mix thoroughly by swirling. Adjust pH to 3.0 with HCl. Aseptically distribute into sterile tubes or flasks.

Use: For the cultivation of *Hydrogenobaculum acidophilum=Hydrogenobacter acidophilus.*

Hydrogenobacter halophilus Medium (DSMZ Medium 744)

Composition per liter:

NaCl	29.3g
K_2HPO_4	2.5g
$(NH_4)_2SO_4$	2.0g
KH_2PO_4	0.5g
$MgSO_4·7H_2O$	0.2g
$CaCl_2$	10.0mg
$FeSO_4·7H_2O$	10.0mg
$NiSO_4·7H_2O$	0.6mg
Trace elements solution	0.5mL

pH 6.9 ± 0.2 at 25°C

Trace Elements Solution:

Composition per liter:

$ZnSO_4·7H_2O$	28.0mg
MoO_3	4.0mg
H_3BO_3	4.0mg
$MnSO_4·5H_2O$	4.0mg
$CoCl_2·6H_2O$	4.0mg
$CuSO_4·5H_2O$	2.0mg

Preparation of Trace Elements Solution: Add components to distilled/deionized water and bring volume to 1.0L. Mix thoroughly. Sparge with 80% H_2 + 20% CO_2.

Preparation of Medium: Add components to distilled/deionized water and bring volume to 1.0L. Mix thoroughly. Autoclave for 15 min at 15 psi pressure–121°C. Adjust pH to 6.9. Aseptically distribute into sterile tubes or flasks.

Use: For the cultivation of *Hydrogenovibrio marinus.*

Hydrogenobacter halophilus Medium

Composition per liter:

NaCl	29.3g
K_2HPO_4	2.5g
$(NH_4)_2SO_4$	2.0g
KH_2PO_4	0.5g
$CaCl_2·2H_2O$	0.25g
$MgSO_4·7H_2O$	0.2g
$FeSO_4·7H_2O$	10.0mg
$NiSO_4·7H_2O$	0.6mg
Trace elements solution	2.0mL

Trace Elements Solution:

Composition per liter:

$ZnSO_4·7H_2O$	7.0mg
MoO_3	1.0mg
H_3BO_3	1.0mg
$MnSO_4·H_2O$	1.0mg
$CoCl_2·6H_2O$	1.0mg
$CuSO_4·5H_2O$	0.5mg

Preparation of Trace Elements Solution: Add components to distilled/deionized water and bring volume to 1.0L. Mix thoroughly.

Preparation of Medium: Add components to distilled/deionized water and bring volume to 1.0L. Mix thoroughly. Distribute into tubes or flasks. Autoclave for 15 min at 15 psi pressure–121°C.

Use: For the cultivation of *Hydrogenovibrio marinus.*

Hydrogenobacter thermophilus Medium

Composition per liter:

Na_2HPO_4	4.5g
KH_2PO_4	1.5g
NH_4NO_3	1.0g
NaCl	1.0g
$MgSO_4·7H_2O$	0.2g
$CaCl_2$	10.0mg
$FeSO_4·7H_2O$	10.0mg
$NiSO_4·7H_2O$	0.06mg
Trace elements solution	2.0mL

Trace Elements Solution:

Composition per liter:

$ZnSO_4·7H_2O$	7.0mg
MoO_3	1.0mg
H_3BO_3	1.0mg
$MnSO_4·H_2O$	1.0mg
$CoCl_2·6H_2O$	1.0mg
$CuSO_4·5H_2O$	0.5mg

Preparation of Trace Elements Solution: Add components to distilled/deionized water and bring volume to 1.0L. Mix thoroughly.

Preparation of Medium: Add components to distilled/deionized water and bring volume to 1.0L. Mix thoroughly. Distribute into tubes or flasks. Autoclave for 15 min at 15 psi pressure–121°C.

Use: For the cultivation of *Calderobacterium hydrogenophilum* and *Hydrogenobacter thermophilus.*

Hydrogenothermus hirschii Medium (DSMZ Medium 783)

Composition per liter:

$MgSO_4·7H_2O$	7.0g
$MgCl_2·6H_2O$	5.5g
$NaHCO_3$	2.0g
KCl	0.65g
$CaCl_2·2H_2O$	0.5g
Sulfur, powdered	0.5g
NH_4Cl	0.15g
K_2HPO_4	0.15g
NaBr	0.1g

Trace elements solution 10.0mL
$CaCO_3$ solution 5.0mL

pH 7.0 ± 0.2 at 25°C

$CaCO_3$ Solution:
Composition per 10.0mL:
$CaCO_3$ 1.0g

Preparation of $CaCO_3$ Solution: Add $CaCO_3$ to 10.0mL of distilled/deionized water. Mix thoroughly. Autoclave for 15 min at 15 psi pressure–121°C. Cool to room temperature.

Trace Elements Solution:
Composition per liter:
$MgSO_4 \cdot 7H_2O$ 3.0g
Nitrilotriacetic acid 1.5g
NaCl 1.0g
$MnSO_4 \cdot 2H_2O$ 0.5g
$CoSO_4 \cdot 7H_2O$ 0.18g
$ZnSO_4 \cdot 7H_2O$ 0.18g
$CaCl_2 \cdot 2H_2O$ 0.1g
$FeSO_4 \cdot 7H_2O$ 0.1g
$NiCl_2 \cdot 6H_2O$ 0.025g
$KAl(SO_4)_2 \cdot 12H_2O$ 0.02g
H_3BO_3 .... 0.01g
$Na_2MoO_4 \cdot 4H_2O$ 0.01g
$CuSO_4 \cdot 5H_2O$ 0.01g
$Na_2SeO_3 \cdot 5H_2O$ 0.3mg

Preparation of Trace Elements Solution: Add nitrilotriacetic acid to 500.0mL of distilled/deionized water. Dissolve by adjusting pH to 6.5 with KOH. Add remaining components. Add distilled/deionized water to 1.0L. Mix thoroughly.

Preparation of Medium: Prepare the medium aerobically Add sulfur to 900.0mL distilled/deionized water. Dissolve sulfur using Ultra-Turrax dispersing instrument. Add remaining components. Bring volume to 1.0L with distilled/deionized water. Adjust pH to 7.0 using H_2SO_4. Fill 20.0mL medium into 100mL serum bottles. Seal with a rubber stopper. Change atmosphere to 80% H_2 + 20% CO_2 with an overpressure of two atmospheres. Autoclave for 20 min at 15 psi pressure–121°C. Cool to room temperature. Inject 20.0mL filter sterilized air and 0.1mL sterile $CaCO_3$ solution. Shake to mix.

Use: For the cultivation of *Hydrogenophilus hirschii (Hydrogenothermophilus hirschii).*

Hydroxybenzoate Agar

Composition per 1001.0mL:
Solution A 490.0mL
Solution D 500.0mL
Solution B 10.0mL
Solution C 1.0mL

pH 7.0 ± 0.2 at 25°C

Solution A:
Composition per 490.0mL:
4-Hydroxybenzoic acid 3.0g
$(NH_4)_2SO_4$ 3.0g
NaCl 2.5g
K_2HPO_4 .... 1.6g
Yeast extract 0.5g

Preparation of Solution A: Add components to distilled/deionized water and bring volume to 490.0mL. Mix thoroughly. Gently heat and bring to boiling. Autoclave for 15 min at 15 psi pressure–121°C. Cool to 45°–50°C.

Solution B:
Composition per 10.0mL:
$MgSO_4 \cdot 7H_2O$ 0.27g

Preparation of Solution B: Add $MgSO_4 \cdot 7H_2O$ to distilled/deionized water and bring volume to 10.0mL. Mix thoroughly. Autoclave for 15 min at 15 psi pressure–121°C. Cool to 45°–50°C.

Solution C:
Composition per 1.0mL:
$Fe(NH_4)_2(SO_4)_2 \cdot 6H_2O$ 0.05g

Preparation of Solution C: Add component to distilled/deionized water and bring volume to 1.0mL. Mix thoroughly. Filter sterilize. Prepare solution immediately before adding to solutions A and B.

Solution D:
Composition per 500.0mL:
Agar 14.0g

Preparation of Solution D: Add agar to distilled/deionized water and bring volume to 500.0mL. Mix thoroughly. Autoclave for 15 min at 15 psi pressure–121°C. Cool to 45°–50°C.

Preparation of Medium: Aseptically combine cooled sterile solution A, cooled sterile solution B, and cooled sterile solution D. Immediately add 1.0mL of freshly prepared sterile solution C. Adjust pH to 7.0 with 6*N* NaOH. Mix thoroughly. Pour into sterile Petri dishes or distribute into sterile tubes.

Use: For the cultivation and maintenance of *Comamonas testosteroni.*

Hydroxybenzoate Agar (*p*-Hydroxybenzoate Agar)

Composition per liter:
Agar 20.0g
$(NH_4)_2HPO_4$ 3.0g
p-hydroxybenzoic acid 3.0g
K_2HPO_4 .... 1.2g
NaCl 0.5g
$MgSO_4 \cdot 7H_2O$ 0.2g
$FeSO_4 \cdot 7H_2O$ 0.1g

pH 7.0 ± 0.2 at 25°C

Preparation of Medium: Add components to distilled/deionized water and bring volume to 1.0L. Mix thoroughly. Gently heat and bring to boiling. Distribute into tubes or flasks. Autoclave for 15 min at 15 psi pressure–121°C. Pour into sterile Petri dishes or leave in tubes.

Use: For the cultivation of *p*-hydroxybenzoate-utilizing bacteria.

Hydroxybenzoate Broth

Composition per 1001.0mL:
Solution A 990.0mL
Solution B 10.0mL
Solution C 1.0mL

pH 7.0 ± 0.2 at 25°C

Solution A:
Composition per 990.0mL:
4-Hydroxybenzoic acid 3.0g
$(NH_4)_2SO_4$ 3.0g
NaCl 2.5g

K_2HPO_4 .... 1.6g
Yeast extract 0.5g

Preparation of Solution A: Add components to distilled/deionized water and bring volume to 990.0mL. Mix thoroughly. Gently heat and bring to boiling. Autoclave for 15 min at 15 psi pressure–121°C. Cool to 45°–50°C.

Solution B:
Composition per 10.0mL:
$MgSO_4·7H_2O$ 0.27g

Preparation of Solution B: Add $MgSO_4·7H_2O$ to distilled/deionized water and bring volume to 10.0mL. Mix thoroughly. Autoclave for 15 min at 15 psi pressure–121°C. Cool to 45°–50°C.

Solution C:
Composition per 1.0mL:
$Fe(NH_4)_2(SO_4)_2·6H_2O$ 0.05g

Preparation of Solution C: Add component to distilled/deionized water and bring volume to 1.0mL. Mix thoroughly. Filter sterilize. Prepare solution immediately before adding to solutions A and B.

Preparation of Medium: Aseptically combine cooled sterile solution A and cooled sterile solution B. Immediately add 1.0mL of freshly prepared sterile solution C. Adjust pH to 7.0 with 6*N* NaOH. Mix thoroughly. Aseptically distribute into sterile tubes or flasks.

Use: For the cultivation and maintenance of *Comamonas testosteroni.*

Hydroxybenzoate Medium

Composition per liter:
Noble agar 20.0g
$(NH_4)_2HPO_4$ 3.0g
K_2HPO_4 .... 1.2g
NaCl 0.5g
$MgSO_4·7H_2O$ 0.2g
$FeSO_4·7H_2O$ 0.1g
p-Hydroxybenzoic acid solution 50.0mL
pH 7.0 ± 0.2 at 25°C

***p*-Hydroxybenzoic Acid Solution:**
Composition per 50.0mL:
p-Hydroxybenzoic acid 3.0g

Preparation of *p*-Hydroxybenzoic Acid Solution: Add *p*-hydroxybenzoic acid to distilled/deionized water and bring volume to 50.0mL. Mix thoroughly. Filter sterilize.

Preparation of Medium: Add components, except *p*-hydroxybenzoic acid solution, to distilled/deionized water and bring volume to 950.0mL. Mix thoroughly. Gently heat and bring to boiling. Adjust pH to 7.0 with 5*N* NaOH. Autoclave for 15 min at 15 psi pressure–121°C. Cool to 45°–50°C. Aseptically add sterile *p*-hydroxybenzoic acid solution. Mix thoroughly. Pour into sterile Petri dishes or distribute into sterile tubes.

Use: For the cultivation and maintenance of *Pseudomonas putida.*

Hydroxybenzoate Medium

Composition per 1002.0mL:
Solution A 920.0mL
Solution B 50.0mL
Solution E (Vitamin solution) 10.0mL
Solution F 10.0mL
Solution G 10.0mL
Solution C (Trace elements solution SL-10) 1.0mL
Solution D 1.0mL
pH 7.2–7.5 at 25°C

Solution A:
Composition per 920.0mL:
NaCl 1.0g
KCl 0.5g
$MgCl_2·6H_2O$ 0.4g
NH_4Cl 0.25g
KH_2PO_4 .... 0.2g
$CaCl_2·2H_2O$ 0.15g

Preparation of Solution A: Add components to distilled/deionized water and bring volume to 920.0mL. Mix thoroughly. Sparge with 80% N_2 + 20% CO_2. Autoclave for 15 min at 15 psi pressure–121°C.

Solution B:
Composition per 50.0mL:
$NaHCO_3$ 2.5g

Preparation of Solution B: Add $NaHCO_3$ to distilled/deionized water and bring volume to 50.0mL. Mix thoroughly. Sparge with 80% N_2 + 20% CO_2. Autoclave for 15 min at 15 psi pressure–121°C.

Solution C (Trace Elements Solution SL-10):
Composition per liter:
$FeCl_2·4H_2O$ 1.5g
$CoCl_2·6H_2O$ 190.0mg
$MnCl_2·4H_2O$ 100.0mg
$ZnCl_2$ 70.0mg
$Na_2MoO_4·2H_2O$ 36.0mg
$NiCl_2·6H_2O$ 24.0mg
H_3BO_3 .... 6.0mg
$CuCl_2·2H_2O$ 2.0mg
HCl (25% solution) 10.0mL

Preparation of Solution C (Trace Elements Solution SL-10): Add $FeCl_2·4H_2O$ to 10.0mL of HCl solution. Mix thoroughly. Add distilled/deionized water and bring volume to 1.0L. Add remaining components. Mix thoroughly. Gas under 80% N_2 + 20% CO_2. Autoclave for 15 min at 15 psi pressure–121°C.

Solution D:
Composition per 10.0mL:
NaOH 5.0mg
$Na_2WO_4·2H_2O$ 40.0μg
$Na_2SeO_3·5H_2O$ 30.0μg

Preparation of Solution D: Add components to distilled/deionized water and bring volume to 10.0mL. Mix thoroughly. Sparge with 80% N_2 + 20% CO_2. Autoclave for 15 min at 15 psi pressure–121°C.

Solution E (Vitamin Solution):
Composition per liter:
Pyridoxine·HCl 10.0mg
Calcium DL-pantothenate 5.0mg
Lipoic acid 5.0mg
Nicotinic acid 5.0mg
p-Aminobenzoic acid 5.0mg
Riboflavin 5.0mg
Thiamine·HCl 5.0mg
Biotin 2.0mg
Folic acid 2.0mg
Vitamin B_{12} 0.1mg

Preparation of Solution E (Vitamin Solution): Add components to distilled/deionized water and bring volume to 1.0L. Mix thoroughly.

Sparge with 80% N_2 + 20% CO_2. Autoclave for 15 min at 15 psi pressure–121°C.

Solution F:

Composition per 10.0mL:

Sodium dihydroxybenzoate	0.4g

Preparation of Solution F: Add sodium dihydroxybenzoate to distilled/deionized water and bring volume to 10.0mL. Mix thoroughly. Filter sterilize. Sparge with 80% N_2 + 20% CO_2.

Solution G:

Composition per 10.0mL:

$Na_2S \cdot 9H_2O$	0.36g

Preparation of Solution G: Add $Na_2S \cdot 9H_2O$ to distilled/deionized water and bring volume to 10.0mL. Mix thoroughly. Sparge with 80% N_2 + 20% CO_2. Autoclave for 15 min at 15 psi pressure–121°C.

Preparation of Medium: Prepare and dispense medium under 80% N_2 + 20% CO_2. Aseptically and anaerobically combine 920.0mL of sterile solution A, 50.0mL of sterile solution B, 1.0mL of sterile solution C, 1.0mL of sterile solution D, 10.0mL of sterile solution E, 10.0mL of sterile solution F, and 10.0mL of sterile solution G. Mix thoroughly. Aseptically and anaerobically distribute into sterile tubes or flasks.

Use: For the cultivation and maintenance of *Clostridium* species.

Hydroxybenzoic Acid Medium

Composition per liter:

Agar	15.0g
$K_2HPO_4 \cdot 3H_2O$	4.25g
NH_4Cl	2.0g
4-Hydroxybenzoic acid	1.0g
$NaH_2PO_4 \cdot H_2O$	1.0g
$MgSO_4 \cdot 7H_2O$	0.2g
Nitrilotriacetic acid	0.1g
$FeSO_4 \cdot 7H_2O$	0.012g
$MnSO_4 \cdot H_2O$	3.0mg
$ZnSO_4 \cdot 7H_2O$	3.0mg
$CoSO_4$	1.0mg

pH 7.2 ± 0.2 at 25°C

Preparation of Medium: Add 4-hydroxybenzoic acid and nitrilotriacetic acid to approximately 600.0mL of distilled/deionized water. Adjust pH to 8.0 with concentrated NaOH. Add remaining components. Mix thoroughly. Readjust pH to 7.2. Bring volume to 1.0L with distilled/deionized water. Distribute into tubes or flasks. Autoclave for 15 min at 15 psi pressure–121°C.

Use: For the cultivation and maintenance of *Bacillus* species.

Hydroxybutyrate Medium (3HB Medium) (LMG Medium 186)

Composition per liter:

Agar	20.0g
DL-3-hydroxybutyrate	3.0g
NH_4Cl	1.0g
$MgSO_4 \cdot 7H_2O$	0.5g
Ferric ammonium citrate	50.0mg
Yeast extract	50.0mg
Buffer solution	333.3mL

pH 6.8 ± 0.2 at 25°C

Buffer Solution:

KH_2PO_4	0.45g
$Na_2HPO_4 \cdot 12\ H_2O$	2.39g

Preparation of Buffer Solution: Add components to distilled/deionized water and bring volume to 1.0L. Mix thoroughly.

Preparation of Medium: Add components to distilled/deionized water and bring volume to 1.0L. Mix thoroughly. Gently heat and bring to boiling. Distribute into tubes or flasks. Autoclave for 15 min at 15 psi pressure–121°C. Pour into sterile Petri dishes or leave in tubes.

Use: For the cultivation and maintenance of *Paucimonas lemoignei.*

Hyperthermus butylicus Medium

Composition per 1010.0mL:

NaCl	17.0g
Pancreatic digest of casein	6.0g
Sulfur, powdered	6.0g
$MgSO_4 \cdot 7H_2O$	3.5g
$MgCl_2 \cdot 6H_2O$	2.75g
$NiCl_2 \cdot 6H_2O$	2.0g
Yeast extract	2.0g
$CaCl_2 \cdot 2H_2O$	0.75g
KH_2PO_4	0.5g
NH_4Cl	0.5g
KCl	0.325g
NaBr	0.05g
H_3BO_3	0.015g
$(NH_4)_2SO_4$	10.0mg
$SrCl_2 \cdot 6H_2O$	7.5mg
Citric acid	5.0mg
KI	2.5mg
Resazurin	1.0mg
Trace elements solution	10.0mL
$Na_2S \cdot 9H_2O$ solution	10.0mL

pH 6.0–6.5 at 25°C

Trace Elements Solution:

Composition per liter:

$MgSO_4 \cdot 7\ H_2O$	3.0g
Nitrilotriacetic acid	1.5 g
$CaCl_2 \cdot 2\ H_2O$	1.0g
NaCl	1.0g
$MnSO_4 \cdot 2\ H_2O$	0.5g
$CoSO_4 \cdot 7\ H_2O$	0.18g
$ZnSO_4 \cdot 7\ H_2O$	0.18g
$FeSO_4 \cdot 7\ H_2O$	0.1g
$NiCl_2 \cdot 6\ H_2O$	0.025g
$KAl(SO_4)_2 \cdot 12\ H_2O$	0.02g
$CuSO_4 \cdot 5\ H_2O$	0.01g
H_3BO_3	0.01g
$Na_2MoO_4 \cdot 2\ H_2O$	0.01g
$Na_2SeO_3 \cdot 5\ H_2O$	0.3mg

Preparation of Trace Elements Solution: Add nitrilotriacetic acid to 500.0mL of distilled/deionized water. Adjust pH to 6.5 with KOH. Add remaining components. Adjust pH to 7.0 with KOH. Add distilled/deionized water to 1.0L.

$Na_2S \cdot 9H_2O$ Solution:

Composition per 10.0mL:

$Na_2S \cdot 9H_2O$	0.5g

Preparation of $Na_2S \cdot 9H_2O$ Solution: Add $Na_2S \cdot 9H_2O$ to distilled/deionized water and bring volume to 10.0mL. Mix thoroughly.

Gas under 80% N_2 + 20% CO_2. Autoclave for 15 min at 15 psi pressure–121°C.

Preparation of Medium: Add components, except $Na_2S·9H_2O$ solution, to distilled/deionized water and bring volume to 1.0L. Mix thoroughly. Adjust pH to 6.0–6.5 with 6*N* H_2SO_4. Sparge wih 100% N_2. Sterilize by bringing to 90°C for 60 min on 3 consecutive days. Immediately prior to inoculation, add 10.0mL of sterile $Na_2S·9H_2O$ solution. Mix thoroughly.

Use: For the cultivation and maintenance of *Hyperthermus butylicus*.

Hyphomicrobium Enrichment Medium

Composition per 100.0mL:

KNO_3	0.04g
$Na_2HPO_4·7H_2O$	0.02g
$MgSO_4·7H_2O$	0.48mg
$FeCl_3·7H_2O$	0.02mg
$MnCl_2·4H_2O$	0.01mg

pH 7.2 ± 0.2 at 25°C

Preparation of Medium: Add components to distilled/deionized water and bring volume to 1.0L. Mix thoroughly. Adjust pH to 7.2. Distribute into tubes or flasks. Autoclave for 15 min at 15 psi pressure–121°C.

Use: For the cultivation and enrichment of *Hyphomicrobium* species.

Hyphomicrobium Medium

Composition per liter:

Agar	15.0g
Na_2HPO_4	2.13g
KH_2PO_4	1.36g
$MgSO_4·7H_2O$	0.2g
$CaCl_2·2H_2O$	9.95mg
$FeSO_4·7H_2O$	5.0mg
$MnSO_4·4H_2O$	2.5mg
$Na_2MoO_4·2H_2O$	2.5mg
Urea solution	30.0mL
Methanol	4.0mL

Urea Solution:

Composition per 100.0mL:

Urea	20.0g

Preparation of Urea Solution: Add urea to distilled/deionized water and bring volume to 100.0mL. Mix thoroughly. Filter sterilize.

Preparation of Medium: Filter sterilize methanol. Add components, except urea solution and methanol, to distilled/deionized water and bring volume to 966.0mL. Mix thoroughly. Gently heat and bring to boiling. Autoclave for 15 min at 15 psi pressure–121°C. Cool to 45°–50°C. Aseptically add sterile urea solution and sterile methanol. Mix thoroughly. Aseptically distribute into sterile tubes or bottles.

Use: For the cultivation of *Hyphomicrobium* species.

Hyphomicrobium Medium

Composition per liter:

Noble agar	18.0g
Na_2HPO_4	2.15g
KH_2PO_4	1.36g
$(NH_4)_2SO_4$	0.5g
$MgSO_4·7H_2O$	0.2g
Trace elements solution	5.0mL
Methylamine·HCl solution	20.0mL

pH 7.1 ± 0.1 at 25°C

Trace Elements Solution:

Composition per 100.0mL:

$CuCl_2$	0.15g
$FeSO_4·7H_2O$	0.1g
$Na_2MoO_4·2H_2O$	0.05g
$MnSO_4·H_2O$	0.035g

Preparation of Trace Elements Solution: Add components to distilled/deionized water and bring volume to 100.0mL. Mix thoroughly.

Methylamine·HCl Solution:

Composition per 20.0mL:

Methylamine·HCl	3.38g

Preparation of Methylamine·HCl Solution: Add methylamine·HCl to distilled/deionized water and bring volume to 20.0mL. Mix thoroughly. Filter sterilize.

Preparation of Medium: Add components, except methylamine·HCl solution, to distilled/deionized water and bring volume to 980.0mL. Mix thoroughly. Gently heat and bring to boiling. Autoclave for 15 min at 15 psi pressure–121°C. Cool to 45°–50°C. Aseptically add sterile methylamine·HCl solution. Mix thoroughly. Adjust pH to 7.1, if necessary. Pour into sterile Petri dishes or distribute into sterile tubes.

Use: For the cultivation and maintenance of *Hyphomicrobium aestuarii, Hyphomicrobium facilis, Hyphomicrobium hollandicum, Hyphomicrobium vulgare,* and *Hyphomicrobium zavarzinii.*

Hyphomicrobium Medium 337a

Composition per liter:

KH_2PO_4	1.3g
Na_2HPO_4	1.13g
$(NH_4)_2SO_4$	0.5g
$MgSO_4·7H_2O$	0.2g
$CaCl_2·2H_2O$	3.09mg
$FeSO_4·7H_2O$	2.0mg
$Na_2MoO_4·2H_2O$	1.0mg
$MnSO_4·4H_2O$	0.88mg

pH 7.2–7.5 at 25°C

Preparation of Medium: Add components to distilled/deionized water and bring volume to 1.0L. Mix thoroughly. Distribute into tubes or flasks. Autoclave for 15 min at 15 psi pressure–121°C.

Use: For the enrichment and cultivation of *Hyphomicrobium* species.

Hyphomicrobium methylovorum Medium

Composition per liter:

Agar	15.0g
$(NH_4)_2HPO_4$	3.0g
NaCl	1.0g
$MgSO_4·7H_2O$	0.2g
$FeSO_4·7H_2O$	10.0mg
$MnSO_4·2H_2O$	5.0mg
Tap water	1.0L
Methanol	10.0mL
Vitamin mixture	5.0mL

Vitamin Mixture:

Composition per liter:

Inositol	200.0mg
Choline	100.0mg
Calcium DL-pantothenate	40.0mg

Niacin	40.0mg
Pyridoxine·HCl	40.0mg
Riboflavin	40.0mg
p-Aminobenzoic acid	20.0mg
Thiamine·HCl	20.0mg
Biotin	0.2mg
Folic acid	0.2mg
Cyanocobalamin	2.0μg

Preparation of Vitamin Mixture: Add components to distilled/deionized water and bring volume to 1.0L. Mix thoroughly. Filter sterilize.

Preparation of Methanol: Filter sterilize 10.0mL of methanol.

Preparation of Medium: Add components, except methanol and vitamin mixture, to distilled/deionized water and bring volume to 985.0mL. Mix thoroughly. Autoclave for 15 min at 15 psi pressure–121°C. Aseptically add 10.0mL of sterile methanol and 5.0mL of sterile vitamin mixture. Mix thoroughly. Aseptically distribute into sterile tubes or flasks.

Use: For the cultivation of *Hyphomicrobium methylovorum.*

Hyphomicrobium Strain X Agar

Composition per liter:

Agar	15.0g
Methylamine·HCl	3.4g
K_2HPO_4	1.55g
$(NH_4)_2SO_4$	1.0g
$NaH_2PO_4 \cdot H_2O$	0.5g
$MgSO_4 \cdot 7H_2O$	0.2g
Trace elements solution	0.2mL

pH 7.2 ± 0.2 at 25°C

Trace Elements Solution:

Composition per liter:

Disodium EDTA	50.0g
$ZnSO_4 \cdot 7H_2O$	22.0g
$CaCl_2 \cdot 2H_2O$	5.54g
$MnCl_2 \cdot 4H_2O$	5.06g
$FeSO_4 \cdot 7H_2O$	5.0g
$CoCl_2 \cdot 6H_2O$	1.61g
$CuSO_4 \cdot 5H_2O$	1.57g
$(NH_4)_6Mo_7O_{24} \cdot 4H_2O$	1.1g

Preparation of Trace Elements Solution: Add components to distilled/deionized water and bring volume to 1.0L. Adjust pH to 7.0 with KOH.

Preparation of Medium: Add components to distilled/deionized water and bring volume to 1.0L. Mix thoroughly. Adjust pH to 7.2. Gently heat and bring to boiling. Distribute into tubes or flasks. Autoclave for 15 min at 15 psi pressure–121°C. Pour into sterile Petri dishes or leave in tubes.

Use: For the cultivation and maintenance of *Hyphomicrobium* species.

Hyphomicrobium Strain X Broth

Composition per liter:

K_2HPO_4	1.55g
$(NH_4)_2SO_4$	1.0g
Methylamine·HCl	0.7g
$NaH_2PO_4 \cdot H_2O$	0.5g
$MgSO_4 \cdot 7H_2O$	0.2g
Trace elements solution	0.2mL

pH 7.2 ± 0.2 at 25°C

Trace Elements Solution:

Composition per liter:

Disodium EDTA	50.0g
$ZnSO_4 \cdot 7H_2O$	22.0g
$CaCl_2 \cdot 2H_2O$	5.54g
$MnCl_2 \cdot 4H_2O$	5.06g
$FeSO_4 \cdot 7H_2O$	5.0g
$CoCl_2 \cdot 6H_2O$	1.61g
$CuSO_4 \cdot 5H_2O$	1.57g
$(NH_4)_6Mo_7O_{24} \cdot 4H_2O$	1.1g

Preparation of Trace Elements Solution: Add components to distilled/deionized water and bring volume to 1.0L. Adjust pH to 7.0 with KOH.

Preparation of Medium: Add components to distilled/deionized water and bring volume to 1.0L. Mix thoroughly. Adjust pH to 7.2. Gently heat and bring to boiling. Distribute into tubes or flasks. Autoclave for 15 min at 15 psi pressure–121°C. Pour into sterile Petri dishes or leave in tubes.

Use: For the cultivation and maintenance of *Hyphomicrobium* species.

Hyphomonas Enrichment Medium

Composition per liter:

Peptone	0.05g
Yeast extract	0.05g

Preparation of Medium: Add components to distilled/deionized water and bring volume to 1.0L. Mix thoroughly. Distribute into tubes or flasks. Autoclave for 15 min at 15 psi pressure–121°C.

Use: For the isolation and cultivation of *Hyphomonas* species.

Hyphomonas Medium

Composition per liter:

Pancreatic digest of casein	2.0g
$MgCl_2 \cdot 2H_2O$	2.0g
Yeast extract	1.0g

pH 7.5 ± 0.2 at 25°C

Preparation of Medium: Add components to distilled/deionized water and bring volume to 1.0L. Mix thoroughly. Adjust pH to 7.5 using indicator paper. Distribute into tubes or flasks. Autoclave for 15 min at 15 psi pressure–121°C.

Use: For the cultivation and maintenence of *Hyphomonas polymorpha.*

Hypoxanthine Agar

Composition per 1100.0mL:

Agar	15.0g
Peptone	5.0g
Hypoxanthine solution	5.0g
Beef extract	3.0g

pH 7.0 ± 0.1 at 25°C

Hypoxanthine Solution:

Composition per 100.0mL:

Hypoxanthine	5.0g

Preparation of Hypoxanthine Solution: Add hypoxanthine to distilled/deionized water and bring volume to 100.0mL. Mix thoroughly. Filter sterilize.

Preparation of Medium: Add components, except hypoxanthine solution, to distilled/deionized water and bring volume to 900.0mL. Mix thoroughly. Autoclave for 15 min at 15 psi pressure–121°C. Cool to 45°–50°C. Aseptically add 100.0mL of sterile hypoxanthine solution. Mix thoroughly. Pour into sterile 15mm × 100mm Petri dishes in 25.0mL volumes.

Use: For the cultivation and differentiation of bacteria based on hypoxanthine hydrolysis. Bacteria that hydrolyze hypoxanthine, such as *Streptomyces griseus*, appear with a clear zone under and around the colonies. *Nocardia asteroides* does not hydrolyze hypoxanthine.

IBB Agar

See: **Inositol Brilliant Green Bile Salts Agar**

Idiomarina Medium (DSMZ Medium 1016)

Composition per liter:

NaCl	30.0-60.0g
Glucose	10.0g
Proteose peptone	5.0g
Yeast extract	3.0g
Malt extract	3.0g

pH 7.0 ± 0.2 at 25°C

Preparation of Medium: Add componentsto distilled/deionized water and bring volume to 1.0L. Mix thoroughly. Adjust pH to 7.0. Distribute into tubes or flasks. Gently heat while stirring and bring to boiling. Mix thoroughly. Autoclave for 15 min at 15 psi pressure–121°C.

Use: For the cultivation of *Idiomarina* spp.

IE Medium

Composition per 1011.0mL:

Agar	15.0g
Peptone	5.0g
Yeast extract	1.0g
Basal salts solution	1.0L
Lactose solution	10.0g
Trace elements solution	1.0mL

Basal Salts Solution:

Composition per liter:

$MgSO_4$	0.5g
Phosphate solution	20.0mL
$(NH_4)SO_4$ (36% solution)	5.0mL

Phosphate Solution:

Composition per liter:

K_2HPO_4	95.0g
$NaH_2PO_4 \cdot 2H_2O$	78.0g

Preparation of Phosphate Solution: Add components to distilled/deionized water and bring volume to 1.0L. Mix thoroughly.

Preparation of Basal Salts Medium: Add components to distilled/deionized water and bring volume to 1.0L. Mix thoroughly. Adjust pH to 6.8-7.0.

Lactose Solution:

Composition per 10.0mL:

Lactose	2.5g

Preparation of Lactose Solution: Add lactose to distilled/deionized water and bring volume to 10.0mL. Mix thoroughly. Filter sterilize. Warm to 50°–55°C.

Trace Elements Solution:

Composition per liter:

Disodium EDTA	0.5g
$FeSO_4 \cdot 7H_2O$	0.2g
H_3BO_3	0.03g
$CoCl_2 \cdot 6H_2O$	0.02g
$ZnSO_4 \cdot 7H_2O$	0.01g
$MnCl_2 \cdot 4H_2O$	3.0mg
$Na_2MoO_4 \cdot 2H_2O$	3.0mg
$NiCl_2 \cdot 6H_2O$	2.0mg
$CaCl_2 \cdot 2H_2O$	1.0mg

Preparation of Trace Elements Solution: Add components to distilled/deionized water and bring volume to 1.0L. Mix thoroughly. Autoclave for 15 min at 15 psi pressure–121°C.

Preparation of Medium: Combine components, except lactose solution and trace elements solution. Mix thoroughly. Gently heat and bring to boiling. Autoclave for 15 min at 15 psi pressure–121°C. Cool to 50°–55°C. Aseptically add 10.0mL of sterile lactose solution and 1.0mL of sterile trace elements solution. Mix thoroughly. Pour into sterile Petri dishes or distribute into sterile tubes.

Use: For the cultivation of *Bacillus* species.

IFO Agar

Composition per liter:

Agar	20.0g
$(NH_4)_2HPO_4$	3.0g
NaCl	1.0g
$MgSO_4 \cdot 7H_2O$	0.2g
$FeSO_4 \cdot 7H_2O$	10.0mg
$MnSO_4 \cdot 6H_2O$	5.0mg
Riboflavin	0.02mg
Calcium pantothenate	0.02mg
Pyridoxine·HCl	0.02mg
Nicotinic acid	0.02mg
p-Aminobenzoic acid	0.01mg
Thiamine·HCl	0.01mg
Biotin	1.0μg
Methanol	10.0mL

pH 7.0 ± 0.2 at 25°C

Preparation of Medium: Add components, except agar and methanol, to distilled/deionized water and bring volume to 490.0mL. Mix thoroughly. Autoclave for 15 min at 15 psi pressure–121°C. Cool to 45°–50°C. In a separate flask, add agar to distilled/deionized water and bring volume to 500.0mL. Mix thoroughly. Gently heat and bring to boiling. Autoclave for 15 min at 15 psi pressure–121°C. Cool to 45°–50°C. Aseptically combine the two sterile solutions. Aseptically add 10.0mL of filter-sterilized methanol. Mix thoroughly. Adjust pH to 7.0. Pour into sterile Petri dishes or distribute into sterile tubes.

Use: For the cultivation and maintenance of *Hyphomicrobium methylovorum*.

IFO Broth

Composition per liter:

$(NH_4)_2HPO_4$	3.0g
NaCl	1.0g
$MgSO_4 \cdot 7H_2O$	0.2g
$FeSO_4 \cdot 7H_2O$	10.0mg
$MnSO_4 \cdot 6H_2O$	5.0mg
Riboflavin	20.0μg
Calcium pantothenate	20.0μg
Pyridoxine·HCl	20.0μg
Nicotinic acid	20.0μg

p-Aminobenzoic acid 10.0μg
Thiamine·HCl 10.0μg
Biotin 1.0μg
Methanol 10.0mL

pH 7.0 ± 0.2 at 25°C

Preparation of Medium: Add components, except methanol, to distilled/deionized water and bring volume to 990.0mL. Mix thoroughly. Autoclave for 15 min at 15 psi pressure–121°C. Aseptically add 10.0mL of filter-sterilized methanol. Mix thoroughly. Adjust pH to 7.0. Aseptically distribute into sterile tubes or flasks.

Use: For the cultivation and maintenance of *Hyphomicrobium methylovorum.*

IFO Medium 802

Composition per liter:
Polypeptone™ 10.0g
Yeast extract 2.0g
$MgSO_4 \cdot 7H_2O$ 1.0g

pH 7.0 ± 0.2 at 25°C

Preparation of Medium: Add components to distilled/deionized water and bring volume to 1.0L. Mix thoroughly. Adjust pH to 7.0. Distribute into tubes or flasks. Autoclave for 15 min at 15 psi pressure–121°C.

Use: For the cultivation of *Sphingomonas asaccharolytica, Sphingomonas pruni, Sphingomonas mali,* and *Sphingomonas rosa.*

Ignicoccus Medium (DSMZ Medium 897)

Composition per liter:
NaCl 13.65g
Sulfur, powdered 5.0g
$MgSO_4 \cdot 7H_2O$ 3.5g
$MgCl_2 \cdot 6H_2O$ 2.75g
Meat extract 1.0g
KH_2PO_4 .... 0.5g
$CaCl_2 \cdot 2H_2O$ 0.38g
KCl 0.33g
$(NH_4)_2SO_4$ 0.25g
NaBr 0.05g
H_3BO_3 .... 15.0mg
$SrCl_3 \cdot 6H_2O$ 7.50mg
KI 0.05mg
Resazurin 0.5mg
$Na_2S \cdot 9H_2O$ solution 10.0mL

pH 5.5 ± 0.2 at 25°C

$Na_2S \cdot 9H_2O$ Solution:
Composition per 10.0mL:
$Na_2S \cdot 9H_2O$ 0.2g

Preparation of $Na_2S \cdot 9H_2O$ Solution: Add $Na_2S \cdot 9H_2O$ to distilled/deionized water and bring volume to 10.0mL. Sparge with N_2. Autoclave for 15 min at 15 psi pressure–121°C. Cool to 25°C. Store anaerobically.

Preparation of Medium: Add components, except $Na_2S \cdot 9H_2O$ solution, to distilled/deionized water and bring volume to 990.0mL. Sparge medium with N_2 gas for 30–60 min. Mix thoroughly. Add 10.0mL $Na_2S \cdot 9H_2O$ solution. Mix thoroughly. Adjust pH to 5.5 with H_2SO_4. Distribute into tubes or bottles under 80% H_2 and 20% CO_2 gas mixture. Heat the vessels containing medium in boiling water for 1 hr before inoculation. After inoculation pressurize the vessels with 80% H_2 and 20% CO_2 gas mixture to 2 bar overpressure.

Use: For the cultivation of *Ignicoccus islandicus* and *Ignicoccus pacificus.*

Ignisphaera Medium (DSMZ Medium 1043)

Composition per liter:
Trypticase peptone 2.0g
Starch, soluble 2.0g
$(NH_4)_2SO_4$ 1.3g
$MgSO_4 \cdot 7H_2O$ 0.28g
KH_2PO_4 .... 0.28g
Yeast extract 0.1g
L-Cysteine 0.3g
$CaCl_2 \cdot 2H_2O$ 74.0mg
Resazurin 0.5mg
$FeCl_2 \cdot 6H_2O$ 0.5mg
Trace elements solution 10.0mL
$FeCl_2$ solution 10.0mL
$Na_2S \cdot 9H_2O$ solution 10.0mL

pH 6.5 ± 0.2 at 25°C

$FeCl_2$ Solution:
Composition per 10.0mL:
$FeCl_2 \cdot 6H_2O$ 0.5mg

Preparation of $FeCl_2$ Solution: Add $FeCl_2 \cdot 6H_2O$ to distilled/deionized water and bring volume to 10.0mL. Mix thoroughly. Sparge with 100% N_2. Filter sterilize.

$Na_2S \cdot 9H_2O$ Solution:
Composition per 10.0mL:
$Na_2S \cdot 9H_2O$ 0.3g

Preparation of $Na_2S \cdot 9H_2O$ Solution: Add $Na_2S \cdot 9H_2O$ to distilled/deionized water and bring volume to 10.0mL. Mix thoroughly. Autoclave under 100% N_2 for 15 min at 15 psi pressure–121°C. Cool to room temperature.

Trace Elements Solution:
Composition per liter:
$MgSO_4 \cdot 7H_2O$ 3.0g
Nitrilotriacetic acid 1.5g
NaCl 1.0g
$MnSO_4 \cdot 2H_2O$ 0.5g
$CoSO_4 \cdot 7H_2O$ 0.18g
$ZnSO_4 \cdot 7H_2O$ 0.18g
$CaCl_2 \cdot 2H_2O$ 0.1g
$FeSO_4 \cdot 7H_2O$ 0.1g
$NiCl_2 \cdot 6H_2O$ 0.025g
$KAl(SO_4)_2 \cdot 12H_2O$ 0.02g
H_3BO_3 .... 0.01g
$Na_2MoO_4 \cdot 4H_2O$ 0.01g
$CuSO_4 \cdot 5H_2O$ 0.01g
$Na_2SeO_3 \cdot 5H_2O$ 0.3mg

Preparation of Trace Elements Solution: Add nitrilotriacetic acid to 500.0mL of distilled/deionized water. Dissolve by adjusting pH to 6.5 with KOH. Add remaining components. Add distilled/deionized water to 1.0L. Mix thoroughly.

Preparation of Medium: Add components, except iron chloride and sulfide solutions, to distilled/deionized water and bring volume to 980.0mL. Mix thoroughly. Gently heat and bring to boiling. Boil for 1 min. Cool to room temperature while sparging with 100% N_2. Add $FeCl_3$ solution. Adjust pH to 6.3. Dispense under an atmosphere of 100% N_2 into suitable culture vessels (e.g., aliquots of 20mL medium into 50mL serum bottles). Autoclave for 15 min at 15 psi pressure–121°C. Cool to room temperature. Adjust pH to 6.5. Prior to inoculation aseptically and anoxically add sulfide solution.

Use: For the cultivation of *Ignisphaera* spp.

IGP Medium

See: **Intracellular Growth Phase Medium**

Ilyobacter Agar

Composition per liter:

NaCl	20.0g
Agar	15.0g
$MgCl_2 \cdot 6H_2O$	3.0g
KCl	0.5g
NH_4Cl	0.25g
KH_2PO_4	0.2g
$CaCl_2 \cdot 2H_2O$	0.15g
Resazurin	1.0mg
Sodium sulfide solution	10.0mL
Sodium L-tartrate solution	10.0mL
$NaHCO_3$ solution	10.0mL
Trace elements solution SL-7	1.0mL

pH 7.2 ± 0.2 at 25°C

Sodium Sulfide Solution:

Composition per 100.0mL:

$Na_2S \cdot 9H_2O$	3.6g

Preparation of Sodium Sulfide Solution: Add $Na_2S \cdot 9H_2O$ to distilled/deionized water and bring volume to 100.0mL. Mix thoroughly. Autoclave for 15 min at 15 psi pressure–121°C under N_2. Maintain under 100% N_2.

Sodium L-Tartrate Solution:

Composition per 10.0mL:

Sodium L-tartrate	2.0g

Preparation of Sodium L-Tartrate Solution: Add sodium L-tartrate to distilled/deionized water and bring volume to 10.0mL. Mix thoroughly. Filter sterilize. Maintain under 80% N_2 + 20% CO_2.

$NaHCO_3$ Solution:

Composition per 10.0mL:

$NaHCO_3$	2.5g

Preparation of $NaHCO_3$ Solution: Add $NaHCO_3$ to distilled/deionized water and bring volume to 10.0mL. Mix thoroughly. Filter sterilize. Maintain under 80% N_2 + 20% CO_2.

Trace Elements Solution SL-7:

Composition per liter:

$FeCl_2 \cdot 4H_2O$	1.5g
$CoCl_2 \cdot 6H_2O$	0.19g
$MnCl_2 \cdot 4H_2O$	0.1g
$ZnCl_2$	0.07g
H_3BO_3	0.062g
$Na_2MoO_4 \cdot 2H_2O$	0.036g
$NiCl_2 \cdot 6H_2O$	0.024g
$CuCl_2 \cdot 2H_2O$	0.017g
HCl (25% solution)	10.0mL

Preparation of Trace Elements Solution SL-7: Add the $FeCl_2 \cdot 4H_2O$ to the HCl. Add distilled/deionized water and bring volume to 1.0L. Add remaining components. Mix thoroughly. Filter sterilize. Maintain under 80% N_2 + 20% CO_2.

Preparation of Medium: Add components—except agar, sodium sulfide solution, sodium L-tartrate solution, $NaHCO_3$ solution, and trace elements solution SL-7—to distilled/deionized water and bring volume to 469.0mL. Mix thoroughly. Gently heat and bring to boiling. Autoclave for 15 min at 15 psi pressure–121°C. Cool to 45°–50°C. Maintain under 80% N_2 + 20% CO_2. Aseptically add 10.0mL of sodium L-tartrate solution, 10.0mL of $NaHCO_3$ solution, and 1.0mL of trace elements solution SL-7 under 80% N_2 + 20% CO_2. Mix thoroughly. In a separate flask, add agar to distilled/deionized water and bring volume to 500.0mL. Mix thoroughly. Gently heat and bring to boiling. Autoclave for 15 min at 15 psi pressure–121°C. Cool to 45°–50°C. Combine sterile agar and sterile basal medium. Adjust pH to 7.2. Aseptically add 10.0mL of sodium sulfide solution. Pour into sterile Petri dishes or distribute into sterile tubes. Maintain under 80% N_2 + 20% CO_2.

Use: For the cultivation and maintenance of *Ilyobacter tartaricus*.

Ilyobacter Broth

Composition per liter:

NaCl	20.0g
$MgCl_2 \cdot 6H_2O$	3.0g
KCl	0.5g
NH_4Cl	0.25g
KH_2PO_4	0.2g
$CaCl_2 \cdot 2H_2O$	0.15g
Resazurin	1.0mg
Sodium sulfide solution	10.0mL
Sodium L-tartrate solution	10.0mL
$NaHCO_3$ solution	10.0mL
Trace elements solution SL-7	1.0mL

pH 7.2 ± 0.2 at 25°C

Sodium Sulfide Solution:

Composition per 100.0mL:

$Na_2S \cdot 9H_2O$	3.6g

Preparation of Sodium Sulfide Solution: Add $Na_2S \cdot 9H_2O$ to distilled/deionized water and bring volume to 100.0mL. Mix thoroughly. Autoclave for 15 min at 15 psi pressure–121°C under 100% N_2. Maintain under 100% N_2.

Sodium L-Tartrate Solution:

Composition per 10.0mL:

Sodium L-tartrate	2.0g

Preparation of Sodium L-Tartrate Solution: Add sodium L-tartrate to distilled/deionized water and bring volume to 10.0mL. Mix thoroughly. Filter sterilize. Maintain under 80% N_2 + 20% CO_2.

$NaHCO_3$ Solution:

Composition per 10.0mL:

$NaHCO_3$	2.5g

Preparation of $NaHCO_3$ Solution: Add $NaHCO_3$ to distilled/deionized water and bring volume to 10.0mL. Mix thoroughly. Filter sterilize. Maintain under 80% N_2 + 20% CO_2.

Trace Elements Solution SL-7:

Composition per liter:

$FeCl_2 \cdot 4H_2O$	1.5g
$CoCl_2 \cdot 6H_2O$	0.19g
$MnCl_2 \cdot 4H_2O$	0.1g
$ZnCl_2$	0.07g

H_3BO_3	0.062g
$Na_2MoO_4 \cdot 2H_2O$	0.036g
$NiCl_2 \cdot 6H_2O$	0.024g
$CuCl_2 \cdot 2H_2O$	0.017g
HCl (25% solution)	10.0mL

Preparation of Trace Elements Solution SL-7: Add the $FeCl_2 \cdot 4H_2O$ to the HCl. Add distilled/deionized water and bring volume to 1.0L. Add remaining components. Mix thoroughly. Filter sterilize. Maintain under 80% N_2 + 20% CO_2.

Preparation of Medium: Add components—except sodium sulfide solution, sodium L-tartrate solution, $NaHCO_3$ solution, and trace elements solution SL-7—to distilled/deionized water and bring volume to 969.0mL. Mix thoroughly. Gently heat and bring to boiling. Autoclave for 15 min at 15 psi pressure–121°C. Cool to 45°–50°C. Maintain under 80% N_2 + 20% CO_2. Aseptically add 10.0mL of sodium L-tartrate solution, 10.0mL of $NaHCO_3$ solution, and 1.0mL of trace elements solution SL-7 under 80% N_2 + 20% CO_2. Mix thoroughly. Aseptically distribute into sterile tubes or flasks under 80% N_2 + 20% CO_2. Adjust pH to 7.2. At time of inoculation add sodium sulfide solution to a final concentration of 0.1%.

Use: For the cultivation and maintenance of *Ilyobacter tartaricus*.

Ilyobacter Medium

Composition per liter:

Crotonic acid	1.7g
NaCl	1.0g
Yeast extract	1.0g
$Na_2HPO_4 \cdot 12H_2O$	0.7g
KCl	0.5g
$MgCl_2 \cdot 6H_2O$	0.4g
NH_4Cl	0.3g
Na_2SO_4	0.1g
Sodium sulfide solution	10.0mL
$CaCl_2 \cdot 2H_2O$ (1.0%)	1.0mL
$FeCl_3$ (0.5%)	1.0mL
Modified SL-7 trace elements solution	1.0mL
Resazurin (0.1%)	1.0mL
Selenite-tungstate solution	1.0mL

pH 6.8–7.2 at 25°C

Sodium Sulfide Solution:

Composition per 100.0mL:

$Na_2S \cdot 9H_2O$	3.6g

Preparation of Sodium Sulfide Solution: Add $Na_2S \cdot 9H_2O$ to distilled/deionized water and bring volume to 100.0mL. Mix thoroughly. Autoclave for 15 min at 15 psi pressure–121°C under N_2. Maintain under 100% N_2.

Modified SL-7 Trace Elements Solution:

Composition per liter:

$CoCl_2 \cdot 6H_2O$	0.2g
$MnCl_2 \cdot 4H_2O$	0.1g
$ZnCl_2$	0.07g
H_3BO_3	0.06g
$Na_2MoO_4 \cdot 2H_2O$	0.04g
$CuCl_2 \cdot 2H_2O$	0.02g
$NiCl_2 \cdot 6H_2O$	0.02g
HCl (1*N*)	3.0mL

Preparation of Modified SL-7 Trace Elements Solution: Add components to distilled/deionized water and bring volume to 1.0L. Mix thoroughly. Filter sterilize. Maintain under 80% N_2 + 20% CO_2.

Selenite-Tungstate Solution:

Composition per liter:

NaOH	0.5g
$Na_2WO_4 \cdot 2H_2O$	4.0mg
$Na_2SeO_3 \cdot 5HO$	3.0mg

Preparation of Selenite-Tungstate Solution: Add components to distilled/deionized water and bring volume to 1.0L. Mix thoroughly. Filter sterilize. Maintain under 80% N_2 + 20% CO_2.

Preparation of Medium: Add components—except sodium sulfide solution, modified SL-7 trace elements solution, and selenite-tungstate solution—to distilled/deionized water and bring volume to 969.0mL. Mix thoroughly. Gently heat and bring to boiling. Autoclave for 15 min at 15 psi pressure–121°C. Adjust pH to 5.5. Cool to 45°–50°C under 100% N_2. Maintain under 100% N_2. Aseptically add 1.0mL of sterile modified SL-7 trace elements solution and 1.0mL of sterile selenite-tungstate solution under 100% N_2. Mix thoroughly. Aseptically distribute into sterile tubes or flasks under 100% N_2. At time of inoculation, add sodium sulfide solution to a final concentration of 1.0%. Maintain under 100% N_2.

Use: For the cultivation and maintenance of *Ilyobacter delafieldii*.

Ilyobacter polytropus Medium

Composition per 1011.0mL:

Solution A	1.0L
Solution B	10.0mL
Vitamin solution	1.0mL

pH 7.2–7.4 at 25°C

Solution A:

Composition per liter:

$NaHCO_3$	4.5g
Na_2SO_4	2.84g
Sodium 3-hydroxybutyrate	1.3g
NaCl	1.17g
Yeast extract	1.0g
$MgCl_2 \cdot 6H_2O$	0.4g
KCl	0.3g
NH_4Cl	0.27g
KH_2PO_4	0.2g
$CaCl_2 \cdot 2H_2O$	0.15g
Resazurin	0.5mg
Trace elements solution	1.0mL

Preparation of Solution A: Add components, except $NaHCO_3$ and vitamin solution, and bring volume to 1.0L. Mix thoroughly. Gently heat and bring to boiling. Continue boiling for 3 to 4 min. Allow to cool to room temperature while gassing under O_2-free 80% N_2 + 20% CO_2. Add $NaHCO_3$ and continue gassing with O_2-free 80% N_2 + 20% CO_2 until pH reaches 6.9–7.1. Seal the flask under 80% N_2 + 20% CO_2. Autoclave for 15 min at 15 psi pressure–121°C.

Trace Elements Solution:

Composition per liter:

$FeCl_2 \cdot 4H_2O$	1.5g
$CoCl_2 \cdot 6H_2O$	120.0mg
$MnCl_2 \cdot 4H_2O$	100.0mg
$ZnCl_2$	68.0mg
H_3BO_3	62.0mg
$Na_2MoO_4 \cdot 2H_2O$	24.0mg
$NiCl_2 \cdot 6H_2O$	24.0mg
$CuCl_2 \cdot 2H_2O$	17.0mg
HCl (25% solution)	10.0mL

Preparation of Trace Elements Solution: Add $FeCl_2 \cdot 4H_2O$ to 10.0mL of HCl solution. Mix thoroughly. Add distilled/deionized water and bring volume to 1.0L. Add remaining components. Mix thoroughly. Gas under 100% N_2.

Vitamin Solution:

Composition per 100.0mL:

Thiamine·HCl	10.0mg
p-Aminobenzoic acid	4.0mg
D(+)-Biotin	1.0mg

Preparation of Vitamin Solution: Add components to distilled/deionized water and bring volume to 100.0mL. Mix thoroughly. Filter sterilize. Gas under 100% N_2.

Solution B:

Composition per 10.0mL:

$Na_2S \cdot 9H_2O$	0.36g

Preparation of Solution B: Add $Na_2S \cdot 9H_2O$ to distilled/deionized water and bring volume to 10.0mL. Mix thoroughly. Gas under 100% N_2. Autoclave for 15 min at 15 psi pressure–121°C.

Preparation of Medium: To 1.0L of sterile solution A, add 10.0mL of sterile solution B and 1.0mL of sterile vitamin solution. Mix thoroughly. Adjust final pH to 7.2–7.4.

Use: For the cultivation and maintenance of *Ilyobacter polytropus*.

Ilyobacter tartaricus Medium

Composition per liter:

NaCl	20.0g
$MgCl_2 \cdot 6H_2O$	3.0g
KCl	0.5g
NH_4Cl	0.25g
KH_2PO_4	0.2g
$CaCl_2 \cdot 2H_2O$	0.15g
Resazurin	1.0mg
$NaHCO_3$ solution	20.0mL
Sodium L-tartrate solution	10.0mL
$Na_2S \cdot 9H_2O$ solution	10.0mL
Trace elements solution SL-10	1.0mL

pH 7.2 ± 0.2 at 25°C

$NaHCO_3$ Solution:

Composition per 20.0mL:

$NaHCO_3$	2.5g

Preparation of $NaHCO_3$ Solution: Add $NaHCO_3$ to distilled/deionized water and bring volume to 20.0mL. Mix thoroughly. Filter sterilize. Gas under 80% N_2 + 20% CO_2.

Sodium L-Tartrate Solution:

Composition per 10.0mL:

Sodium L-tartrate	2.0g

Preparation of Sodium L-Tartrate Solution: Add sodium L-tartrate to distilled/deionized water and bring volume to 10.0mL. Mix thoroughly. Filter sterilize. Gas under 80% N_2 + 20% CO_2.

$Na_2S \cdot 9H_2O$ Solution:

Composition per 10.0mL:

$Na_2S \cdot 9H_2O$	0.36g

Preparation of $Na_2S \cdot 9H_2O$ Solution: Add $Na_2S \cdot 9H_2O$ to distilled/deionized water and bring volume to 10.0mL. Mix thoroughly. Gas under 100% N_2. Autoclave for 15 min at 15 psi pressure–121°C.

Trace Elements Solution SL-10:

Composition per liter:

$FeCl_2 \cdot 4H_2O$	1.5g
$CoCl_2 \cdot 6H_2O$	190.0mg
$MnCl_2 \cdot 4H_2O$	100.0mg
$ZnCl_2$	70.0mg
$Na_2MoO_4 \cdot 2H_2O$	36.0mg
$NiCl_2 \cdot 6H_2O$	24.0mg
H_3BO_3	6.0mg
$CuCl_2 \cdot 2H_2O$	2.0mg
HCl (25% solution)	10.0mL

Preparation of Trace Elements Solution SL-10: Add $FeCl_2 \cdot 4H_2O$ to 10.0mL of HCl solution. Mix thoroughly. Add distilled/deionized water and bring volume to 1.0L. Add remaining components. Mix thoroughly. Gas under 100% N_2. Autoclave for 15 min at 15 psi pressure–121°C.

Preparation of Medium: Add components, except $NaHCO_3$ solution, sodium L-tartrate solution, and $Na_2S \cdot 9H_2O$ solution, to distilled/deionized water and bring volume to 960.0mL. Mix thoroughly. Gas under 80% N_2 + 20% CO_2. Autoclave for 15 min at 15 psi pressure–121°C. Aseptically and anaerobically add 20.0mL of sterile $NaHCO_3$ solution, 10.0mL of sterile sodium L-tartrate solution, and 10.0mL of sterile $Na_2S \cdot 9H_2O$ solution. Mix thoroughly. Distribute into sterile tubes or flasks.

Use: For the cultivation and maintenance of *Ilyobacter tartaricus*.

Ilyobacter tartaricus Medium

Composition per liter:

NaCl	20.0g
$MgCl_2 \cdot 6H_2O$	3.0g
KCl	0.5g
NH_4Cl	0.25g
KH_2PO_4	0.2g
$CaCl_2 \cdot 2H_2O$	0.15g
Resazurin	1.0mg
$NaHCO_3$ solution	20.0mL
Sodium L-tartrate solution	10.0mL
$Na_2S \cdot 9H_2O$ solution	10.0mL
Yeast extract solution	10.0mL
Trace elements solution SL-10	1.0mL

pH 7.2 ± 0.2 at 25°C

$NaHCO_3$ Solution:

Composition per 20.0mL:

$NaHCO_3$	2.5g

Preparation of $NaHCO_3$ Solution: Add $NaHCO_3$ to distilled/deionized water and bring volume to 20.0mL. Mix thoroughly. Filter sterilize. Gas under 80% N_2 + 20% CO_2.

Sodium L-Tartrate Solution:

Composition per 10.0mL:

Sodium L-tartrate	2.0g

Preparation of Sodium L-Tartrate Solution: Add sodium L-tartrate to distilled/deionized water and bring volume to 10.0mL. Mix thoroughly. Filter sterilize. Gas under 80% N_2 + 20% CO_2.

$Na_2S \cdot 9H_2O$ Solution:

Composition per 10.0mL:

$Na_2S \cdot 9H_2O$	0.36g

Preparation of $Na_2S·9H_2O$ Solution: Add $Na_2S·9H_2O$ to distilled/deionized water and bring volume to 10.0mL. Mix thoroughly. Gas under 100% N_2. Autoclave for 15 min at 15 psi pressure–121°C.

Yeast Extract Solution:
Composition per 10.0mL:

Yeast extract	1.0g

Preparation of Yeast Extract Solution: Add yeast extract to distilled/deionized water and bring volume to 10.0mL. Mix thoroughly. Gas under 100% N_2. Autoclave for 15 min at 15 psi pressure–121°C.

Trace Elements Solution SL-10:
Composition per liter:

$FeCl_2·4H_2O$	1.5g
$CoCl_2·6H_2O$	190.0mg
$MnCl_2·4H_2O$	100.0mg
$ZnCl_2$	70.0mg
$Na_2MoO_4·2H_2O$	36.0mg
$NiCl_2·6H_2O$	24.0mg
H_3BO_3	6.0mg
$CuCl_2·2H_2O$	2.0mg
HCl (25% solution)	10.0mL

Preparation of Trace Elements Solution SL-10: Add $FeCl_2·4H_2O$ to 10.0mL of HCl solution. Mix thoroughly. Add distilled/deionized water and bring volume to 1.0L. Add remaining components. Mix thoroughly. Gas under 100% N_2. Autoclave for 15 min at 15 psi pressure–121°C.

Preparation of Medium: Add components, except $NaHCO_3$ solution, sodium L-tartrate solution, yeast extract solution, and $Na_2S·9H_2O$ solution, to distilled/deionized water and bring volume to 950.0mL. Mix thoroughly. Gas under 80% N_2 + 20% CO_2. Autoclave for 15 min at 15 psi pressure–121°C. Aseptically and anaerobically add 20.0mL of sterile $NaHCO_3$ solution, 10.0mL of sterile sodium L-tartrate solution, 10.0mL of sterile yeast extract solution, and 10.0mL of sterile $Na_2S·9H_2O$ solution. Mix thoroughly. Distribute into sterile tubes or flasks.

Use: For the cultivation and maintenance of *Propionigenium modestum.*

IM
See: **Infection Medium**

Imhoff's Medium, Modified

Composition per liter:

NaCl	30.0g
$NaHCO_3$	3.0g
KH_2PO_4	1.0g
NH_4Cl	1.0g
Sodium acetate	1.0g
Na_2SO_4	0.7g
$MgCl_2·6H_2O$	0.5g
Sodium ascorbate	0.5g
$CaCl_2·2H_2O$	0.1g
Yeast extract	0.1g
Sodium sulfide solution	10.0mL
SLA trace elements solution	1.0mL
VA vitamin solution	1.0mL

pH 6.9–7.0 at 25°C

Sodium Sulfide Solution:
Composition per 100.0mL:

$Na_2S·9H_2O$	2.0g

Preparation of Sodium Sulfide Solution: Add $Na_2S·9H_2O$ to distilled/deionized water and bring volume to 100.0mL. Mix thoroughly. Autoclave for 15 min at 15 psi pressure–121°C under N_2. Maintain under 100% N_2.

SLA Trace Elements Solution:
Composition per liter:

$FeCl_2·4H_2O$	1.8g
H_3BO_3	0.5g
$CoCl_2·6H_2O$	0.25g
$ZnCl_2$	0.1g
$MnCl_2·4H_2O$	0.07g
$Na_2MoO_4·2H_2O$	0.03g
$CuCl_2·2H_2O$	0.01g
$Na_2SeO_3·5H_2O$	0.01g
$NiCl_2·6H_2O$	0.01g

Preparation of SLA Trace Elements Solution: Add components to distilled/deionized water and bring volume to 1.0L. Mix thoroughly. Adjust pH to 2–3. Filter sterilize.

VA Vitamin Solution:
Composition per 500.0mL:

Nicotinamide	0.17g
Thiamine·HCl	0.15g
p-Aminobenzoic acid	0.1g
Biotin	0.05g
Calcium pantothenate	0.05g
Pyridoxine·2HCl	0.05g
Cyanocobalamin	0.02g

Preparation of VA Vitamin Solution: Add components to distilled/deionized water and bring volume to 500.0mL. Mix thoroughly. Filter sterilize.

Preparation of Medium: Add components—except sodium sulfide solution, SLA trace elements solution, and VA vitamin solution—to distilled/deionized water and bring volume to 988.0mL. Mix thoroughly. Autoclave for 15 min at 15 psi pressure–121°C. Cool to 25°C. Aseptically add 1.0mL of sterile SLA trace elements solution and 1.0mL of sterile VA vitamin solution. Aseptically add 10.0mL of sterile sodium sulfide solution. Mix thoroughly. Aseptically distribute into sterile tubes or flasks.

Use: For the cultivation and maintenance of *Rhodobacter adriaticus* and *Rhodobacter sulfidophilus.*

Imidazole Utilization Medium

Composition per liter:

Imidazole	5.0g
KH_2PO_4	0.5g
$MgSO_4·7H_2O$	0.5g
$CaCl_2$	3.0mg
$FeSO_4·7H_2O$	3.0mg
Molybdenum solution	1.0mL
Trace elements solution	1.0mL

pH 6.0 ± 0.2 at 25°C

Molybdenum Solution:
Composition per 18.0mL:

$Na_2MoO_4·2H_2O$	0.5mg

Preparation of Molybdenum Solution: Add components to distilled/deionized water and bring volume to 18.0mL. Mix thoroughly. Filter sterilize.

Trace Elements Solution:

Composition per 18.0mL:

H_3BO_3	11.0mg
$MnCl_2 \cdot 4H_2O$	7.0mg
$Al_2(SO_4)_3 \cdot 18\ H_2O$	1.94mg
$Co(NO_3)_2 \cdot 6H_2O$	1.0mg
$CuSO_4 \cdot 5H_2O$	1.0mg
$NiSO_4 \cdot 6H_2O$	1.0mg
$ZnSO_4 \cdot H_2O$	0.62mg
KBr	0.5mg
KI	0.5mg
LiCl	0.5mg
$SnCl_2 \cdot 2H_2O$	0.5mg

Preparation of Trace Elements Solution: Add components to distilled/deionized water and bring volume to 18.0mL. Mix thoroughly. Filter sterilize.

Preparation of Medium: Add components, except molybdenum solution and trace elements solution, to distilled/deionized water and bring volume to 998.0mL. Mix thoroughly. Distribute into tubes or flasks. Autoclave for 15 min at 15 psi pressure–121°C. Cool to 25°C. Aseptically add 1.0mL of molybdenum solution and 1.0mL of trace elements solution. Mix thoroughly. Adjust pH to 6.0 with phosphoric acid. Mix thoroughly. Aseptically distribute into sterile tubes or flasks.

Use: For the cultivation and maintenance of *Pseudomonas* species.

Indole Medium

Composition per 200.0mL:

K_2HPO_4	3.13g
L-Tryptophan	1.0g
NaCl	1.0g
KH_2PO_4	0.27g

pH 7.2 ± 0.2 at 25°C

Preparation of Medium: Add components to distilled/deionized water and bring volume to 200.0mL. Mix thoroughly. Distribute into tubes or flasks. Autoclave for 15 min at 15 psi pressure–121°C.

Use: For the differentiation of microorganisms by means of indole production from the tryptophan test.

Indole Medium

Composition per liter:

Pancreatic digest of casein	20.0g

pH 7.3 ± 0.2 at 25°C

Preparation of Medium: Add pancreatic digest of casein to distilled/deionized water and bring volume to 1.0L. Mix thoroughly. Distribute into tubes or flasks. Autoclave for 15 min at 15 psi pressure–121°C.

Use: For the differentiation of microorganisms by means of the indole test.

Indole Medium, CDC (BAM M65)

Composition per liter:

Pancreatic digest of casein	20.0g

pH 7.3 ± 0.2 at 25°C

Preparation of Medium: Add pancreatic digest of casein to distilled/deionized water and bring volume to 1.0L. Mix thoroughly. Distribute into tubes or flasks. Autoclave for 15 min at 15 psi pressure–121°C.

Use: For the differentiation of microorganisms by means of the indole test.

Indole Nitrate HiVeg Medium (Tryptone Nitrate HiVeg Medium)

Composition per liter:

Plant hydrolysate	20.0g
Na_2HPO_4	2.0g
Agar	1.0g
Glucose	1.0g
Potassium nitrate	1.0g

pH 7.2 ± 0.2 at 25°C

Source: This medium is available as a premixed powder from HiMedia.

Preparation of Medium: Add components to distilled/deionized water and bring volume to 1.0L. Mix thoroughly. Gently heat and bring to boiling with frequent agitation. Distribute into tubes or flasks. Autoclave for 15 min at 15 psi pressure–121°C.

Use: For the identification of microorganisms by means of the nitrate reduction and indole tests.

Indole Nitrate Medium (Trypticase™ Nitrate Broth)

Composition per liter:

Pancreatic digest of casein	20.0g
Na_2HPO_4	2.0g
Agar	1.0g
Glucose	1.0g
KNO_3	1.0g

pH 7.2 ± 0.2 at 25°C

Source: This medium is available as a premixed powder from BD Diagnostic Systems.

Preparation of Medium: Add components to distilled/deionized water and bring volume to 1.0L. Mix thoroughly. Gently heat and bring to boiling with frequent agitation. Distribute into tubes or flasks. Autoclave for 15 min at 15 psi pressure–121°C.

Use: For the identification of microorganisms by means of the nitrate reduction and indole tests.

Infection Medium (IM)

Composition per 100.0mL:

Pancreatic digest of gelatin	0.05g
Bile salts No. 3	0.05g
Brain heart, solids from infusion	0.02g
Peptic digest of animal tissue	0.02g
NaCl	0.017g
Glucose	0.01g
Na_2HPO_4	8.0mg
Earle's balanced salts solution	80.0mL
Fetal bovine serum, heat inactivated (2 hr at 55°C)	20.0mL

pH 7.4 ± 0.2 at 25°C

Earle's Balanced Salts Solution:

Composition per liter:

NaCl	6.8g
$NaHCO_3$	2.2g

Glucose	1.0g
KCl	0.4g
$CaCl_2 \cdot 2H_2O$	0.265g
$MgSO_4 \cdot 7H_2O$	0.2g
$NaH_2PO_4 \cdot H_2O$	0.14g

Preparation of Earle's Balanced Salts Solution: Add components to distilled/deionized water and bring volume to 1.0L. Mix thoroughly. Filter sterilize.

Preparation of Medium: Combine components. Mix thoroughly. Filter sterilize. Store at 4°–10°C.

Use: For the screening of *Escherichia coli* for pathogenicity using the HeLa cell test for invasiveness.

Infusion Agar

See: **Blood Agar Base**

Infusion Broth

Composition per liter:

Pancreatic digest of casein	13.0g
NaCl	5.0g
Yeast extract	5.0g
Heart muscle, solids from infusion	2.0g

pH 7.4 ± 0.2 at 25°C

Source: This medium is available as a premixed powder from BD Diagnostic Systems.

Preparation of Medium: Add components to distilled/deionized water and bring volume to 1.0L. Mix thoroughly. Distribute into tubes or flasks. Autoclave for 15 min at 15 psi pressure–121°C.

Use: For the cultivation of a wide variety of microorganisms.

Infusion Cystine Agar Base, HiVeg

Composition per liter:

Agar	15.0g
Glucose	10.0g
Plant infusion	10.0g
Plant peptone No. 3	10.0g
NaCl	5.0g
L-Cystine	1.0g

pH 7.2 ± 0.2 at 25°C

Source: This medium is available as a premixed powder from HiMedia.

Preparation of Medium: Add components to distilled/deionized water and bring volume to1.0L. Mix thoroughly. Gently heat and bring to boiling with frequent agitation. Distribute into tubes or flasks. Autoclave for 15 min at 15 psi pressure–121°C. Pour into sterile Petri dishes.

Use: For the cultivation of Gram-negative cocci and other pathogenic organisms.

Infusion Cystine Agar Base, HiVeg with Hemoglobin

Composition per liter:

Agar	15.0g
Glucose	10.0g
Plant infusion	10.0g
Plant peptone No. 3	10.0g
NaCl	5.0g
L-Cystine	1.0g
Hemoglobin solution, 2%	100.0mL

pH 7.2 ± 0.2 at 25°C

Source: This medium, without hemoglobin solution, is available as a premixed powder from HiMedia.

Preparation of Medium: Add components, except hemoglobin solution, to distilled/deionized water and bring volume to 900.0mL. Mix thoroughly. Gently heat and bring to boiling with frequent agitation. Distribute into tubes or flasks. Autoclave for 15 min at 15 psi pressure–121°C. Cool to 50°C. Aseptically add 100.0mL of sterile hemoglobin solution. Mix thoroughly. Pour into sterile Petri dishes.

Use: For the cultivation of Gram-negative cocci and other pathogenic organisms.With added hemoglobin it is used for cultivation of *Francisella tularensis.*

Inhibitory Mold Agar

Composition per liter:

Agar	15.0g
Glucose	5.0g
Yeast extract	5.0g
Pancreatic digest of casein	3.0g
Na_2HPO_4	2.0g
Peptic digest of animal tissue	2.0g
Starch	2.0g
Dextrin	1.0g
$MgSO_4 \cdot 7H_2O$	0.8g
Chloramphenicol	0.125g
$FeSO_4$	0.04g
NaCl	0.04g
$MnSO_4$	0.16g

pH 6.7 ± 0.2 at 25°C

Source: This medium is available as a premixed powder from BD Diagnostic Systems.

Preparation of Medium: Add components to distilled/deionized water and bring volume to 1.0L. Mix thoroughly. Gently heat and bring to boiling with frequent agitation. Distribute into tubes or flasks. Autoclave for 15 min at 15 psi pressure–121°C. Pour into sterile Petri dishes or leave in tubes.

Use: For the isolation of pathogenic fungi.

Inorganic Salts-Maltose Medium (DSMZ Medium 754)

Composition per liter:

Yeast extract	4.0g
Peptone	2.0g
Inorganic salt solution	980.0mL
Maltose solution	20.0mL

Maltose Solution:

Composition per liter:

Maltose	5.0g

Preparation of Maltose Solution: Add maltose to 20.0mL distilled/deionized water. Mix thoroughly. Filter sterilize.

Inorganic Salt Solution:

Composition per liter:

$MgSO_4 \cdot 7H_2O$	49.37g
NaCl	43.8g
$CaCl_2 \cdot 2H_2O$	1.29g

Preparation of Inorganic Salt Solution: Add components in the order $CaCl_2{\cdot}2H_2O$, NaCl, $MgSO_4{\cdot}7H_2O$ to 900.0mL distilled/deionized water. After addition of each, mix thorougly to prevent precipitation. Add distilled/deionized water and bring volume to 1.0L. Mix thoroughly.

Preparation of Medium: Add components except maltose solution to 980.0mL inorganic salt solution. Mix thoroughly. Autoclave for 15 min at 15 psi pressure–121°C. Cool to 25°C. Aseptically add 20.0mL sterile maltose solution. Mix thoroughly. Aseptically distribute to sterile tubes or flasks.

Use: For the cultivation of *Spirochaeta halophila.*

Inorganic Salts Maltose Medium

Composition per liter:

Yeast extract	4.0g
Peptone	2.0g
Inorganic salts solution	980.0mL
Maltose solution	20.0mL

pH 7.5 ± 0.2 at 25°C

Inorganic Salts Solution:

Composition per liter:

$MgSO_4{\cdot}7H_2O$	49.37g
NaCl	43.8g
$CaCl_2{\cdot}2H_2O$	1.29g

Preparation of Inorganic Salts Solution: Add NaCl first and then other components to distilled/deionized water and bring volume to 1.0L. Mix thoroughly.

Maltose Solution:

Composition per 100.0mL:

Maltose	25.0g

Preparation of Maltose Solution: Add maltose to distilled/deionized water and bring volume to 100.0mL. Mix thoroughly. Filter sterilize.

Preparation of Medium: Add components, except maltose solution, to inorganic salts solution and bring volume to 980.0mL. Mix thoroughly. Adjust pH to 7.5 with KOH. Autoclave for 15 min at 15 psi pressure–121°C. Cool to 25°C. Aseptically add 20.0mL of sterile maltose solution. Mix thoroughly. Aseptically distribute into sterile tubes or flasks.

Use: For the cultivation and maintenance of *Spirochaeta halophila.*

Inorganic Salt Medium (Modified Raggios Medium)

Composition per liter:

Na_2CO_3	3.0g
KI	0.75g
$MgSO_4{\cdot}7H_2O$	0.7g
$CaCl_2{\cdot}2H_2O$	0.446g
KH_2PO_4	0.2g
Na_2SO_4	0.2g
KCl	0.165g
$MnSO_4{\cdot}2H_2O$	6.64mg
$ZnSO_4{\cdot}7H_2O$	2.67mg
$FeCl_3{\cdot}4H_2O$	2.5mg
H_3BO_3	1.5mg
$Na_2MoO_4{\cdot}2H_2O$	0.25mg
$CuSO_4{\cdot}5H_2O$	0.07mg

Source: This medium is available from HiMedia.

Preparation of Medium: Add components to distilled/deionized water and bring volume to 1.0L. Mix thoroughly. Gently heat and bring to boiling. Distribute into tubes or flasks. Autoclave for 15 min at 15 psi pressure–121°C. Pour into sterile Petri dishes or leave in tubes.

Use: For the cultivation of soil microorganisms such as *Rhizobium* spp.

Inorganic Salts Starch Agar

See: **ISP Medium 4**

Inositol Assay Medium

Composition per liter:

Glucose	100.0g
Potassium citrate	10.0g
Citric acid	2.0g
KH_2PO_4	1.1g
KCl	0.85g
L-Asparagine	0.8g
L-Glutamic acid	0.6g
L-Isoleucine	0.5g
L-Leucine	0.5g
L-Lysine	0.5g
L-Valine	0.5g
L-Arginine	0.48g
DL-Alanine	0.4g
DL-Threonine	0.4g
$CaCl_2$	0.25g
$MgSO_4{\cdot}7H_2O$	0.25g
DL-Aspartic acid	0.2g
DL-Phenylalanine	0.2g
Glycine	0.2g
L-Methionine	0.2g
L-Tyrosine	0.2g
L-Proline	0.2g
L-Histidine	0.124g
DL-Serine	0.1g
L-Cystine	0.1g
DL-Tryptophan	0.08g
$FeCl_3$	0.05g
$MnSO_4{\cdot}7H_2O$	0.05g
Calcium pantothenate	5.0mg
Pyridoxine·HCl	1.0mg
Thiamine·HCl	0.5mg
Biotin	0.01mg

pH 5.2 ± 0.2 at 25°C

Source: This medium is available as a premixed powder from BD Diagnostic Systems.

Preparation of Medium: Add components to distilled/deionized water and bring volume to 1.0L. Mix thoroughly. Distribute into tubes or flasks. Autoclave for 5 min at 15 psi pressure–121°C.

Use: For the microbiological assaying of inositol using *Saccharomyces uvarum* as the test organism.

Inositol Assay Medium KB

Composition per liter:

Glucose	40.0g
$(NH_4)_2SO_4$	4.0g
DL-Asparagine	4.0g
KH_2PO_4	3.0g
$MgSO_4{\cdot}7H_2O$	1.0g

$CaCl_2$	0.98g
$FeSO_4·7H_2O$	0.5mg
Calcium pantothenate	0.4mg
Nicotinic acid	0.4mg
Pyridoxine	0.4mg
Thiamine	0.4mg
H_3BO_3	0.2mg
KI	0.2mg
$CuSO_4·5H_2O$	0.09mg
$MnSO_4·7H_2O$	0.08mg
$ZnSO_4·7H_2O$	0.08mg
$(NH_4)_6Mo_7O_{24}·4H_2O$	0.04mg
Riboflavin	0.02mg
Biotin	0.4μg

pH 5.0 ± 0.2 at 25°C

Preparation of Medium: Add components to distilled/deionized water and bring volume to 1.0L. Mix thoroughly. Distribute into tubes or flasks. Autoclave for 5 min at 15 psi pressure–121°C.

Use: For the microbiological assaying of inositol using *Kloeckera apiculata* as the test organism.

Inositol Brilliant Green Bile Salts Agar (IBB Agar) (*Plesiomonas* Differential Agar)

Composition per liter:

Agar	15.0g
meso-Inositol	10.0g
Proteose peptone	10.0g
Bile salts No. 3	8.5g
Meat extract	5.0g
NaCL	5.0g
Neutral Red (2% solution)	1.25mL
Brilliant Green (0.1% solution)	0.33mL

pH 7.2 ± 0.1 at 25°C

Preparation of Medium: Add components to distilled/deionized water and bring volume to 1.0L. Mix thoroughly. Gently heat and bring to boiling. Distribute into tubes or flasks. Autoclave for 15 min at 15 psi pressure–121°C. Pour into sterile Petri dishes or leave in tubes.

Use: For the isolation of *Aeromonas* and *Plesiomonas* species.

Inositol Brilliant Green HiVeg Agar (*Plesiomonas* Differential HiVeg Agar)

Composition per liter:

Plant peptone No. 3	15.0g
Agar	13.5g
meso-Inositol	10.0g
Plant extract No. 1	6.5g
NaCl	5.0g
Synthetic detergent No. I	2.0g
Neutral Red	0.025g
Brilliant Green	0.33mg

pH 7.2 ± 0.1 at 25°C

Source: This medium is available as a premixed powder from HiMedia.

Preparation of Medium: Add components to distilled/deionized water and bring volume to 1.0L. Mix thoroughly. Gently heat and bring to boiling. Distribute into tubes or flasks. Autoclave for 15 min at 15 psi pressure–121°C. Pour into sterile Petri dishes or leave in tubes.

Use: For the isolation of *Aeromonas* and *Plesiomonas* species.

Inositol Gelatin Deeps

Composition per liter:

Gelatin	120.0g
Inositol	10.0g
Na_2HPO_4	5.0g
Yeast extract	5.0g
Phenol Red	0.05g

pH 7.4 ± 0.2 at 25°C

Preparation of Medium: Add components to distilled/deionized water and bring volume to 1.0L. Mix thoroughly. Gently heat and bring to boiling. Adjust pH to 7.4. Distribute into tubes in 5.0mL volumes. Autoclave for 15 min at 10 psi pressure–115°C.

Use: For the cultivation of *Plesiomonas shigelloides* from foods.

Inositol Urea Caffeic Acid Medium

Composition per liter:

Agar solution	900.0mL
Base solution	100.0mL

Agar Solution:

Composition per 900.0mL:

Agar	15.0g

Preparation of Agar Solution: Add agar to distilled/deionized water and bring volume to 900.0mL. Mix thoroughly. Gently heat and bring to boiling. Autoclave for 15 min at 15 psi pressure–121°C. Cool to 45°–50°C.

Base Solution:

Composition per 100.0mL:

Inositol	10.0g
Urea	5.0g
KH_2PO_4	1.0g
$MgSO_4·7H_2O$	0.5g
Caffeic acid	0.2g
NaCl	0.1g
$CaCl_2·2H_2O$	0.1g
Gentamicin sulfate	0.04g
H_3BO_3	0.5mg
$ZnSO_4·7H_2O$	0.4mg
$MnSO_4·4H_2O$	0.4mg
Thiamine·HCl	0.4mg
Pyroxidine·HCl	0.4mg
Niacin	0.4mg
Calcium pantothenate	0.4mg
p-Aminobenzoic acid	0.2mg
Riboflavin	0.2mg
$FeCl_3$	0.2mg
$Na_2MoO_4·4H_2O$	0.2mg
KI	0.1mg
$CuSO_4·5H_2O$	0.04mg
Folic acid	2.0μg
Biotin	2.0μg
Ferric citrate solution (1% solution)	1.0mL

Preparation of Base Solution: Add components, except urea, to distilled/deionized water and bring volume to 100.0mL. Mix thoroughly. Gently heat just until components are dissolved. Cool to 75°–80°C. Add urea. Mix thoroughly. Do not heat after addition of urea. Do not autoclave. Filter sterilize.

Preparation of Medium: Aseptically combine the cooled, sterile agar solution with the sterile base solution. Mix thoroughly. Pour into sterile Petri dishes.

Use: For the selective isolation and differentiation of *Cryptococcus* species based on inositol and urea utilization and pigment production from caffeic acid. On this medium, only *Cryptococcus* species utilize inositol as sole carbon source and urea as sole nitrogen source. *Cryptococcus neoformans* appears as dark brown colonies. Other *Cryptococcus* species are unpigmented.

International *Streptomyces* Project Medium 1
See: **ISP Medium 1**

International *Streptomyces* Project Medium 2
See: **ISP Medium 2**

International *Streptomyces* Project Medium 3
See: **ISP Medium 3**

International *Streptomyces* Project Medium 4
See: **ISP Medium 4**

International *Streptomyces* Project Medium 4 with Glucose
See: **ISP Medium 4 with Glucose**

International *Streptomyces* Project Medium 4 with Yeast Extract
See: **ISP Medium 4 with Yeast Extract**

International *Streptomyces* Project Medium 5
See: **ISP Medium 5**

International *Streptomyces* Project Medium 6
See: **ISP Medium 6**

International *Streptomyces* Project Medium 7
See: **Tyrosine Agar**

International *Streptomyces* Project Medium 8
See: **Nitrate Broth**

International *Streptomyces* Project Medium 9
See: **ISP Medium 9**

Intracellular Growth Phase Medium (IGP Medium)

Composition per 100.0mL:

Gentamicin sulfate	500.0mg
Lysozyme	30.0mg
Eagle MEM	72.0mL
Dulbecco's phosphate-buffered saline	20.0mL
Fetal bovine serum	8.0mL

pH 7.2–7.4 at 25°C

Eagle MEM:
Composition per liter:

NaCl	8.0g
Glucose	1.0g
KCl	0.4g
$CaCl_2·2H_2O$	0.14g
$MgSO_4·7H_2O$	0.1g
KH_2PO_4	0.06g
Na_2HPO_4	0.05g
L-Isoleucine	0.026g
L-Leucine	0.026g
L-Lysine	0.026g
L-Threonine	0.024g
L-Valine	0.0235g
L-Tyrosine	0.018g
L-Arginine	0.0174g
L-Phenylalanine	0.0165g
L-Cystine	0.012g
L-Histidine	8.0mg
L-Methionine	7.5mg
Phenol Red	5.0mg
L-Tryptophan	4.0mg
Inositol	1.8mg
Biotin	1.0mg
Folic acid	1.0mg
Calcium pantothenate	1.0mg
Choline chloride	1.0mg
Nicotinamide	1.0mg
Pyridoxal·HCl	1.0mg
Thiamine·HCl	1.0mg
Riboflavin	0.1mg

Preparation of Eagle MEM: Add components to distilled/deionized water and bring volume to 1.0L. Mix thoroughly.

Dulbecco's Phosphate-Buffered Saline:
Composition per liter:

NaCl	8.0g
$Na_2HPO_4·7H_2O$	2.16g
KCl	0.2g
KH_2PO_4	0.2g
$CaCl_2$	0.1g
$MnCl_2·6H_2O$	0.1g

Preparation of Dulbecco's Phosphate-Buffered Saline: Add components to distilled/deionized water and bring volume to 1.0L. Mix thoroughly.

Preparation of Medium: Combine components. Mix thoroughly. Filter sterilize. Aseptically distribute into sterile tubes or flasks.

Use: For the screening of *Escherichia coli* for pathogenicity using the HeLa cell test for invasiveness.

Ion Agar for *Ureaplasma*

Composition per 101.45mL:

HEPES (*N*-[2-hydroxyethyl]piperazine-*N′*-[2-ethanesulfonic acid]) buffer	1.19g
Ionagar No. 2	0.75g
Pancreatic digest of casein	0.7g
NaCl	0.5g
Beef extract	0.3g

Yeast extract....................0.3g
Beef heart, solids from infusion....................0.2g
Yeast extract....................0.1g
Horse serum, normal sterile....................10.0mL
Ampicillin solution....................1.0mL
Urea solution....................0.25mL
Nystatin solution....................0.1mL
Tripeptide solution....................0.1mL

pH 7.2 ± 0.2 at 25°C

Ampicillin Solution:

Composition per 10.0mL:

Ampicillin....................1.0g

Preparation of Ampicillin Solution: Add ampicillin to distilled/deionized water and bring volume to 10.0mL. Mix thoroughly. Filter sterilize.

Urea Solution:

Composition per 100.0mL:

Urea....................10.0g

Preparation of Urea Solution: Add urea to distilled/deionized water and bring volume to 100.0mL. Filter sterilize. Store at –20°C.

Nystatin Solution:

Composition per 1.0mL:

Nystatin....................50,000U

Preparation of Nystatin Solution: Add nystatin to distilled/deionized water and bring volume to 1.0mL. Filter sterilize.

Tripeptide Solution:

Composition per 10.0mL:

Glycyl-L-histidyl-L-lysine acetate....................0.2mg

Preparation of Tripeptide Solution: Add glycyl-L-histidyl-L-lysine acetate to distilled/deionized water and bring volume to 10.0mL. Mix thoroughly. Filter sterilize. Store at –20°C.

Preparation of Medium: Add components—except horse serum, ampicillin solution, urea solution, nystatin solution, and tripeptide solution—to distilled/deionized water and bring volume to 90.0mL. Mix thoroughly. Gently heat and bring to boiling. Autoclave for 15 min at 15 psi pressure–121°C. Cool to 45°–50°C. Aseptically add 10.0mL of sterile horse serum, 1.0mL of sterile ampicillin solution, 0.25mL of sterile urea solution, 0.1mL of sterile nystatin solution, and 0.1mL of sterile tripeptide solution. Mix thoroughly. Pour into sterile Petri dishes.

Use: For the cultivation of *Ureaplasma* species from clinical specimens.

Ionic Medium with Pipecolate

Composition per liter:

Agar....................30.0g
Ionic medium....................950.0mL
Pipecolic acid·HCl solution....................50.0mL

Ionic Medium:

Composition per liter:

K_2HPO_4....................4.1g
Na_2HPO_4....................3.34g
KH_2PO_4....................2.26g
NaH_2PO_4....................2.24g
Salt solution....................10.0mL

Preparation of Ionic Medium: Add components to distilled/deionized water and bring volume to 1.0L. Mix thoroughly.

Salt Solution:

Composition per liter:

$MgSO_4·7H_2O$....................14.8g
$FeSO_4·7H_2O$....................0.55g
$MnSO_4$....................0.045g
H_2SO_4, concentrated....................0.2mL

Preparation of Salt Solution: Add components to distilled/deionized water and bring volume to 1.0L. Mix thoroughly.

Pipecolic Acid·HCl Solution:

Composition per 100.0mL:

Pipecolic acid·HCl....................4.14g

Preparation of Pipecolic Acid·HCl Solution: Add pipecolic acid·HCl to distilled/deionized water and bring volume to 100.0mL. Mix thoroughly. Adjust pH to 7.0.

Preparation of Medium: Add agar to 950.0mL of ionic medium. Mix thoroughly. Gently heat and bring to boiling. Add 50.0mL of pipecolic acid·HCl. Mix thoroughly. Distribute into tubes or flasks. Autoclave for 15 min at 15 psi pressure–121°C. Pour into sterile Petri dishes or leave in tubes.

Use: For the cultivation and maintenance of *Pseudomonas putida*.

Irgasan® Ticarcillin Chlorate Broth (ITC Broth)

Composition per liter:

$MgCl_2·6H_20$....................60.0g
Pancreatic digest of casein....................10.0g
NaCl....................5.0g
$KClO_4$....................1.0g
Yeast extract....................1.0g
Malachite Green (0.2% solution)....................5.0mL
Irgasan solution....................1.0mL
Ticarcillin solution....................1.0mL

pH 7.6 ± 0.2 at 25°C

Irgasan Solution:

Composition per 10.0mL:

Irgasan (triclosan)....................1.0mg

Preparation of Irgasan Solution: Add Irgasan to distilled/deionized water and bring volume to 10.0mL. Mix thoroughly. Filter sterilize.

Ticarcillin Solution:

Composition per 10.0mL:

Ticarcillin....................1.0 mg

Preparation of Ticarcillin Solution: Add ticarcillin to distilled/deionized water and bring volume to 10.0mL. Mix thoroughly. Filter sterilize.

Preparation of Medium: Add components, except Irgasan solution and ticarcillin solution, to distilled/deionized water and bring volume to 998.0mL. Mix thoroughly. Autoclave for 15 min at 15 psi pressure–121°C. Cool to 45°–50°C. Adjust to pH 7.6. Aseptically add 1.0mL of Irgasan solution and 1.0mL of ticarcillin solution. Mix thoroughly. Aseptically distribute into sterile tubes or flasks.

Use: For the selective isolation and cultivation of *Yersinia* species.

Iron Agar, Lyngby (Lyngby Iron Agar)

Composition per liter:

Peptone	20.0g
Agar	12.0g
NaCl	5.0g
Beef extract	3.0g
Yeast extract	3.0g
L-Cysteine	0.6g
Ferric citrate	0.3g
$Na_2S_2O_3$	0.3g

pH 7.4 ± 0.2 at 25°C

Source: This medium is available as a premixed powder from Oxoid Unipath.

Preparation of Medium: Add components to distilled/deionized water and bring volume to 1.0L. Mix thoroughly. Gently heat and bring to boiling. Distribute into tubes or flasks. Autoclave for 15 min at 15 psi pressure–121°C. Pour into sterile Petri dishes or leave in tubes.

Use: For the cultivation and enumeration of H_2S-producing bacteria and total counts of heterotrophic bacteria from fish and fish products.

Iron Bacteria Isolation Medium

Composition per liter:

Agar	10.0g
$(NH_4)_2SO_4$	0.5g
Glucose	0.15g
$CaCO_3$	0.1g
K_2HPO_4	0.05g
$MgSO_4{\cdot}7H_2O$	0.05g
KCl	0.05g
$Ca(NO_3)_2$	0.01g
Vitamin solution	10.0mL

Vitamin Solution:
Composition per 10.0mL:

Thiamine	0.4mg
Cyanocobalamin	0.01mg

Preparation of Vitamin Solution: Add components to distilled/deionized water and bring volume to 10.0mL. Mix thoroughly. Filter sterilize.

Preparation of Medium: Add components, except vitamin solution, to distilled/deionized water and bring volume to 990.0mL. Mix thoroughly. Gently heat and bring to boiling. Autoclave for 15 min at 15 psi pressure–121°C. Cool to 45°–50°C. Aseptically add 10.0mL of vitamin solution. Mix thoroughly. Pour into sterile Petri dishes or distribute into sterile tubes.

Use: For the isolation of iron bacteria.

Iron Milk Medium

Composition per liter:

Iron filings	1.0g
Whole milk	1.0L

pH 6.8 ± 0.2 at 25°C

Preparation of Medium: Add iron filings, which may be small balls of steel wool, to whole milk and bring volume to 1.0L. Mix thoroughly. Distribute into tubes or flasks. Autoclave for 15 min at 15 psi pressure–121°C.

Use: For the cultivation of lactic acid bacteria. For the cultivation and differentiation of *Clostridium* species. The medium turns black if H_2S is produced. The medium turns red if acid is produced from milk carbohydrate fermentation. Acid and gas production is characteristic of *Clostridium perfringens* and *Clostridium butyricum*.

Iron Milk Medium, Modified

Composition per 1050.0mL:

Whole milk, fresh	1.0L
$FeSO_4{\cdot}7H_2O$	1.0g

Preparation of Medium: Add $FeSO_4{\cdot}7H_2O$ to distilled/deionized water and bring volume to 50.0mL. Add slowly to 1.0L of whole milk. Mix thoroughly. Distribute into tubes in 11.0mL volumes. Autoclave for 12 min at 13 psi pressure–118°C.

Use: For the cultivation and enumeration of *Clostridium perfringens* in foods.

Iron-Oxidizing Medium

Composition per liter:

$(NH_4)_2SO_4$	3.0g
K_2HPO_4	0.5g
$MgSO_4{\cdot}7H_2O$	0.5g
KCl	0.1g
$Ca(NO_3)_2$	0.01g
$FeSO_4{\cdot}7H_2O$ solution	300.0mL
H_2SO_4 (10*N*)	1.0mL

pH 3.0–3.6 at 25°C

$FeSO_4{\cdot}7H_2O$ Solution:
Composition per 300.0mL:

$FeSO_4{\cdot}7H_2O$	44.22g

Preparation of $FeSO_4{\cdot}7H_2O$ Solution: Add $FeSO_4{\cdot}7H_2O$ to distilled/deionized water and bring volume to 300.0mL. Mix thoroughly. Autoclave for 15 min at 15 psi pressure–121°C. Cool to 25°C.

Preparation of Medium: Add components, except $FeSO_4{\cdot}7H_2O$ solution, to distilled/deionized water and bring volume to 700.0mL. Mix thoroughly. Gently heat and bring to boiling. Autoclave for 15 min at 15 psi pressure–121°C. Cool to 25°C. Aseptically add 300.0mL of sterile $FeSO_4{\cdot}7H_2O$ solution. Mix thoroughly. Aseptically distribute into sterile tubes or flasks.

Use: For the enumeration, isolation, and cultivation of iron and sulfur bacteria, such as *Thiobacillus ferrooxidans*.

Iron Sulfite Agar

Composition per liter:

Agar	12.0g
Pancreatic digest of casein	10.0g
Ferric citrate	0.5g
$Na_2S{\cdot}9H_2O$	0.5g

pH 7.1 ± 0.2 at 25°C

Source: This medium is available as a premixed powder from Oxoid Unipath.

Preparation of Medium: Add components to distilled/deionized water and bring volume to 1.0L. Mix thoroughly. Gently heat and bring to boiling. Distribute into tubes or flasks. Autoclave for 15 min at 15 psi pressure–121°C. Mix thoroughly. Pour into sterile Petri dishes or leave in tubes.

Use: For the detection of thermophilic anaerobic organisms.

Iron Sulfite HiVeg Agar

Composition per liter:

Agar	15.0g
Plant hydrolysate	10.0g
Iron (III) citrate	0.5g
Na_2SO_3	0.5g

pH 7.1 ± 0.2 at 25°C

Source: This medium is available as a premixed powder from HiMedia.

Preparation of Medium: Add components to distilled/deionized water and bring volume to 1.0L. Mix thoroughly. Gently heat and bring to boiling. Distribute into tubes or flasks. Autoclave for 15 min at 15 psi pressure–121°C. Mix thoroughly. Pour into sterile Petri dishes or leave in tubes.

Use: For the detection of thermophilic anaerobic organisms.

Irradiated Tryptone Soya Broth

***See*: Cold Filterable Tryptone Soya Broth**

Irradiated Vegetable Peptone Broth

***See*: Cold Filterable Vegetable Peptone Broth**

Islam GBS Agar

See: **GBS Agar Base, Islam**

ISM Agar

Composition per liter:

$MgSO_4 \cdot 7H_2O$	49.2g
NaCl	43.5g
Agar	7.5g
Yeast extract	4.0g
Peptone	2.0g
$CaCl_2 \cdot 2H_2O$	1.5g
Maltose solution	100.0mL

Maltose Solution:

Composition per 100.0mL:

Maltose	5.0g

Preparation of Maltose Solution: Add maltose to distilled/deionized water and bring volume to 100.0mL. Mix thoroughly. Filter sterilize.

Preparation of Medium: Add components, except maltose solution, to distilled/deionized water and bring volume to 900.0mL. Mix thoroughly. Gently heat and bring to boiling. Autoclave for 15 min at 15 psi pressure–121°C. Cool to 45°–50°C. Aseptically add sterile maltose solution. Mix thoroughly. Pour into sterile Petri dishes or distribute into sterile tubes.

Use: For the cultivation and maintenance of *Spirochaeta halophila.*

Islams Medium Base for Group B Streptococci

Composition per liter:

Proteose peptone	23.0g
Agar	10.0g
Na_2HPO_4	5.749g
Starch, soluble	5.0g
KH_2PO_4	1.482g
Horse serum, sterile inactivated	50.0mL

pH 7.4 ± 0.2 at 25°C

Source: This medium is available from HiMedia.

Preparation of Medium: Add components, except horse serum, to distilled/deionized water and bring volume to 950.0mL. Mix thoroughly. Autoclave for 15 min at 10 psi pressure–115°C. Cool to 50°C. Aseptically add horse serum. Mix thoroughly. Pour into Petri dishes or aseptically distribute into sterile tubes.

Use: For the identification and cultivation of Group B streptococci from clinical specimens.

ISM Broth

Composition per liter:

$MgSO_4 \cdot 7H_2O$	49.2g
NaCl	43.5g
Yeast extract	4.0g
Peptone	2.0g
$CaCl_2 \cdot 2H_2O$	1.5g
Maltose solution	100.0mL

Maltose Solution:

Composition per 100.0mL:

Maltose	5.0g

Preparation of Maltose Solution: Add maltose to distilled/deionized water and bring volume to 100.0mL. Mix thoroughly. Filter sterilize.

Preparation of Medium: Add components, except maltose solution, to distilled/deionized water and bring volume to 900.0mL. Mix thoroughly. Gently heat and bring to boiling. Autoclave for 15 min at 15 psi pressure–121°C. Cool to 45°–50°C. Aseptically add sterile maltose solution. Mix thoroughly. Aseptically distribute into sterile tubes or flasks.

Use: For the cultivation of *Spirochaeta halophila.*

Iso-Sensitest Agar

Composition per liter:

Casein, hydrolyzed	11.0g
Agar	8.0g
Peptones	3.0g
NaCl	3.0g
Na_2HPO_4	2.0g
Glucose	2.0g
Sodium acetate	1.0g
Soluble starch	1.0g
Magnesium glycerophosphate	0.2g
Calcium gluconate	0.1g
L-Cysteine·HCl	0.02g
L-Tryptophan	0.02g
Adenine	0.01g
Guanine	0.01g
Xanthine	0.01g
Uracil	0.01g
Nicotinamide	3.0mg
Pantothenate	3.0mg
Pyridoxine	3.0mg
$MnCl_2 \cdot 4H_2O$	2.0mg
$CoSO_4$	1.0mg
$CuSO_4 \cdot 5H_2O$	1.0mg
$FeSO_4 \cdot 7H_2O$	1.0mg
Menadione	1.0mg
Cyanocobalamin	1.0mg
$ZnSO_4 \cdot 7H_2O$	1.0mg

Biotin	0.3mg
Thiamine	0.04mg

pH 7.4 ± 0.2 at 25°C

Preparation of Medium: Add components to distilled/deionized water and bring volume to 1.0L. Mix thoroughly. Gently heat and bring to boiling. Distribute into tubes or flasks. Autoclave for 15 min at 15 psi pressure–121°C. Pour into sterile Petri dishes or leave in tubes.

Use: For antimicrobial susceptibility testing.

Iso Sensitest Broth

Composition per liter:

Casein, hydrolyzed	11.0g
Peptones	3.0g
NaCl	3.0g
Glucose	2.0g
Na_2HPO_4	2.0g
Sodium acetate	1.0g
Soluble starch	1.0g
Magnesium glycerophosphate	0.2g
Calcium gluconate	0.1g
L-cysteine·HCl	0.02g
L-Tryptophan	0.02g
Adenine	0.01g
Guanine	0.01g
Xanthine	0.01g
Uracil	0.01g
Nicotinamide	3.0mg
Pantothenate	3.0mg
Pyridoxine	3.0mg
$MnCl_2·4H_2O$	2.0mg
$CoSO_4$	1.0mg
$CuSO_4·5H_2O$	1.0mg
$FeSO_4·7H_2O$	1.0mg
Menadione	1.0mg
Cyanocobalamin	1.0mg
$ZnSO_4·7H_2O$	1.0mg
Biotin	0.3mg
Thiamine	0.04mg

pH 7.4 ± 0.2 at 25°C

Preparation of Medium: Add components to distilled/deionized water and bring volume to 1.0L. Mix thoroughly. Gently heat and bring to boiling. Distribute into tubes or flasks. Autoclave for 15 min at 15 psi pressure–121°C. Pour into sterile Petri dishes or leave in tubes.

Use: For antimicrobial susceptibility testing.

Isoleucine Hydroxamate Medium

Composition per liter:

Agar	15.0g
K_2HPO_4	7.0g
Glucose	5.0g
KH_2PO_4	3.0g
L-Isoleucine hydroxamate	1.0g
$(NH_4)_2SO_4$	1.0g

pH 7.0 ± 0.2 at 25°C

Preparation of Medium: Add components to distilled/deionized water and bring volume to 1.0L. Mix thoroughly. Gently heat and bring to boiling. Distribute into tubes or flasks. Autoclave for 15 min at 15 psi pressure–121°C. Pour into sterile Petri dishes or leave in tubes.

Use: For the cultivation and maintenance of *Serratia marcescens*.

Isonema Medium

Composition per liter:

Pancreatic digest of casein	1.0g
Seawater	990.0mL
Horse serum, heat inactivated	10.0mL

Preparation of Medium: Add pancreatic digest of casein to seawater and bring volume to 990.0mL. Mix thoroughly. Autoclave for 15 min at 15 psi pressure–121°C. Cool to 25°C. Aseptically add 10.0mL of heat-inactivated horse serum. Mix thoroughly. Aseptically distribute into sterile tubes or flasks.

Use: For the cultivation of *Isonema papillatum*.

Isosphaera Agar

Composition per 1000.5mL:

Solution A	800.0mL
Solution B	100.0mL
Solution C	100.0mL
Vitamin solution	0.5mL

pH 7.6 ± 0.2 at 25°C

Solution A:

Composition per 800.0mL:

Agar, noble	15.0g
NaCl	0.25g
$(NH_4)_2SO_4$	0.125g
KCl	0.125g
$MgSO_4·7H_2O$	0.1g
$CaCl_2·2H_2O$	80.0mg
KH_2PO_4	75.0mg
$FeCl_3$	73.0μg
Trace elements solution SL-7	2.5mL

Trace Elements Solution SL-7:

Composition per liter:

$FeCl_2·4H_2O$	1.5g
$CoCl_2·6H_2O$	190.0mg
$MnCl_2·4H_2O$	100.0mg
$ZnCl_2$	70.0mg
H_3BO_3	62.0mg
$Na_2MoO_4·2H_2O$	36.0mg
$NiCl_2·6H_2O$	24.0mg
$CuCl_2·2H_2O$	17.0mg
HCl (25% solution)	10.0mL

Preparation of Trace Elements Solution SL-7: Add $FeCl_2·4H_2O$ to 10.0mL of HCl solution. Mix thoroughly. Add distilled/deionized water and bring volume to 1.0L. Add remaining components. Mix thoroughly.

Preparation of Solution A: Add agar to 400.0mL of distilled/deionized water. In another flask, add remaining components to 400.0mL of distilled/deionized water. Mix thoroughly. Adjust pH to 7.6 with NaOH. Remove any precipitate that forms by filtering through Whatman #1 filter paper. Autoclave both solutions separately for 15 min at 15 psi pressure–121°C. Aseptically combine the two sterile solutions. Cool to 50°–55°C.

Solution B:

Composition per 100.0mL:

$NaHCO_3$	0.42g

Preparation of Solution B: Add $NaHCO_3$ to distilled/deionized water and bring volume to 100.0mL. Mix thoroughly. Filter sterilize. Warm to 50°–55°C.

Solution C:

Composition per 100.5mL:

Glucose 0.25g
Casamino acids 0.25g

Preparation of Solution C: Add components to distilled/deionized water and bring volume to 100.0mL. Mix thoroughly. Filter sterilize. Warm to 50°–55°C.

Vitamin Solution:

Composition per 100.0mL:

Nicotinic acid 200.0mg
Thiamine HCl 200.0mg
p-Aminobenzoic acid 20.0mg
Biotin 2.0mg
Vitamin B_{12} 25.0μg

Preparation of Vitamin Solution: Add components to distilled/deionized water and bring volume to 100.0mL. Mix thoroughly. Filter sterilize. Warm to 50°–55°C.

Preparation of Medium: Aseptically combine 800.0mL of sterile solution A with 100.0mL of sterile solution B, 100.0mL of sterile solution C, and 0.5mL of sterile vitamin solution. Mix thoroughly. Pour into sterile Petri dishes or distribute into sterile tubes.

Use: For the cultivation of *Isosphaera pallida.*

Isosphaera Broth

Composition per 1000.5mL:

Solution A 800.0mL
Solution B 100.0mL
Solution C 100.0mL
Vitamin solution 0.5mL

pH 7.6 ± 0.2 at 25°C

Solution A:

Composition per 800.0mL:

NaCl 0.25g
$(NH_4)_2SO_4$ 0.125g
KCl 0.125g
$MgSO_4·7H_2O$ 0.1g
$CaCl_2·2H_2O$ 80.0mg
KH_2PO_4 .... 75.0mg
$FeCl_3$ 73.0μg
Trace elements solution SL-7 .... 2.5mL

Trace Elements Solution SL-7:

Composition per liter:

$FeCl_2·4H_2O$ 1.5g
$CoCl_2·6H_2O$ 190.0mg
$MnCl_2·4H_2O$ 100.0mg
$ZnCl_2$ 70.0mg
H_3BO_3 .... 62.0mg
$Na_2MoO_4·2H_2O$ 36.0mg
$NiCl_2·6H_2O$ 24.0mg
$CuCl_2·2H_2O$ 17.0mg
HCl (25% solution) 10.0mL

Preparation of Trace Elements Solution SL-7: Add $FeCl_2·4H_2O$ to 10.0mL of HCl solution. Mix thoroughly. Add distilled/deionized water and bring volume to 1.0L. Add remaining components. Mix thoroughly.

Preparation of Solution A: Add components to distilled/deionized water and bring volume to 800.0mL. Mix thoroughly. Adjust pH to 7.6 with NaOH. Remove any precipitate that forms by filtering through Whatman #1 filter paper. Autoclave for 15 min at 15 psi pressure–121°C.

Solution B:

Composition per 100.0mL:

$NaHCO_3$ 0.42g

Preparation of Solution B: Add $NaHCO_3$ to distilled/deionized water and bring volume to 100.0mL. Mix thoroughly. Filter sterilize.

Solution C:

Composition per 100.5mL:

Glucose 0.25g
Casamino acids 0.25g

Preparation of Solution C: Add components to distilled/deionized water and bring volume to 100.0mL. Mix thoroughly. Filter sterilize.

Vitamin Solution:

Composition per 100.0mL:

Nicotinic acid 200.0mg
Thiamine HCl 200.0mg
p-Aminobenzoic acid 20.0mg
Biotin 2.0mg
Vitamin B_{12} 25.0μg

Preparation of Vitamin Solution: Add components to distilled/deionized water and bring volume to 100.0mL. Mix thoroughly. Filter sterilize.

Preparation of Vitamin Solution: Aseptically combine 800.0mL of sterile solution A with 100.0mL of sterile solution B, 100.0mL of sterile solution C, and 0.5mL of sterile vitamin solution. Mix thoroughly. Aseptically distribute into sterile tubes or flasks.

Use: For the cultivation of *Isosphaera pallida.*

Isosphaera pallida Medium (DSMZ Medium 765)

Composition per 1005.0mL:

Solution A 900.0mL
$NaHCO_3$ solution 100.0mL
Vitamin solution 5.0mL

pH 7.9 ± 0.2 at 25°C

Solution A:

Composition per 900.0mL:

KCl 4.0g
$MgSO_4·7H_2O$ 2.0g
$(NH_4)_2SO_4$ 1.0g
NaCl 1.0g
KH_2PO_4 .... 0.3g
Glucose 0.25g
Casamino acids 0.25g
$CaCl_2·2H_2O$ 0.2g
$FeCl_3$ 0.292mg
Trace elements solution SL-7a 1.0mL

Trace Elements Solution SL-7a:

$CoCl_2·6H_2O$ 200.0mg
$MnCl_2·4H_2O$ 100.0mg
$ZnCl_2$ 70.0mg
H_3BO_3 .... 60.0mg
$Na_2MoO_4·4H_2O$ 40.0mg
$NiCl_2·6H_2O$ 20.0mg
$CuCl_2·2H_2O$ 20.0mg
HCl (25%) 1.0mL

Preparation of Trace Elements Solution SL-7a: Add components to distilled/deionized water and bring volume to 1.0L. Mix thoroughly.

Preparation of Solution A: Add components to distilled/deionized water and bring volume to 900.0mL. Mix thoroughly. Sparge with 95% air + 5% CO_2.

Vitamin Solution:

Composition per 100.0mL:

Nicotinic acid	20.0mg
Thiamine-HCl·$2H_2O$	10.0mg
p-Aminobenzoic acid	0.2mg
Vitamin B_{12}	0.1mg

Preparation of Vitamin Solution: Add components to distilled/deionized water and bring volume to 100.0mL. Mix thoroughly. Filter sterilize.

$NaHCO_3$ Solution:

Composition per 100.0mL:

$NaHCO_3$	4.2g

Preparation of $NaHCO_3$ Solution: Add $NaHCO_3$ to distilled/deionized water and bring volume to 100.0mL. Mix thoroughly. Sparge with 80% N_2 + 20% CO_2. Filter sterilize. Seal in serum bottle under 100% CO_2.

Preparation of Medium: Dispense under 95% air + 5% CO_2. Fill serum bottles so that the gas to liquid ratio is about 5:1 (v/v). Sparge with 95% air + 5% CO_2. Seal bottles. Autoclave for 15 min at 15 psi pressure–121°C. Cool to room temperature. Using a syringe, inject 10.0mL sterile $NaHCO_3$ solution/90.0mL solution A. Then inject 0.5 mL sterile vitamin solution per 100.0mL of the resulting medium.

Use: For the cultivation of *Isosphaera pallida=Isocystis pallida.*

IsoVitaleX® Enrichment

Composition per liter:

Glucose	100.0g
L-Cysteine·HCl	25.9g
L-Glutamine	10.0g
Adenine	1.0g
Thiamine pyrophosphate	0.1g
Vitamin B_{12}	0.1g
Guanine·HCl	0.03g
$Fe(NO_3)_3 \cdot 9H_2O$	0.02g
p-Aminobenzoic acid	0.013g
Thiamine·HCl	0.003g

Preparation of IsoVitaleX® Enrichment: Add components to distilled/deionized water and bring volume to 1.0L. Mix thoroughly. Filter sterilize.

Use: As a nutrient supplement for the isolation and cultivation of fastidious microorganisms.

ISP HiVeg Medium No. 1 (Tryptone Yeast Extract HiVeg Broth)

Composition per liter:

Plant hydrolysate	5.0g
Yeast extract	3.0g

pH 7.0–7.2 at 25°C

Source: This medium is available as a premixed powder from HiMedia.

Preparation of Medium: Add components to distilled/deionized water and bring volume to 1.0L. Mix thoroughly. Distribute into tubes or flasks. Autoclave for 15 min at 15 psi pressure–121°C.

Use: For the cultivation of *Streptomyces* species according to the International *Streptomyces* Project.

ISP HiVeg Medium No. 2 (Yeast Malt HiVeg Agar)

Composition per liter:

Agar	20.0g
Glucose	10.0g
Plant peptone	5.0g
Malt extract	3.0g
Yeast extract	3.0g

pH 7.3 ± 0.2 at 25°C

Source: This medium is available as a premixed powder from HiMedia.

Preparation of Medium: Add components to distilled/deionized water and bring volume to 1.0L. Mix thoroughly. Gently heat and bring to boiling. Distribute into tubes or flasks. Autoclave for 15 min at 15 psi pressure–121°C. Pour into sterile Petri dishes or leave in tubes.

Use: For the cultivation of *Streptomyces* species according to the International *Streptomyces* Project.

ISP HiVeg Medium No. 6 (Peptone Yeast Extract Iron HiVeg Agar)

Composition per liter:

Agar	15.0g
Plant peptone	15.0g
Plant peptone No. 3	5.0g
K_2HPO_4	1.0g
Yeast extract	1.0g
Ferric ammonium citrate	0.5g
$Na_2S_2O_3$	0.08g

pH 6.7 ± 0.2 at 25°C

Preparation of Medium: Add components to distilled/deionized water and bring volume to 1.0L. Mix thoroughly. Gently heat and bring to boiling. Distribute into tubes or flasks. Autoclave for 15 min at 15 psi pressure–121°C. Pour into sterile Petri dishes or leave in tubes.

Use: For the cultivation and maintenance of *Streptomyces* species.

ISP2 Medium (DSMZ Medium 987)

Composition per liter:

Agar	20.0g
Malt extract	10.0g
Yeast extract	4.0g
Glucose	4.0g

pH 7.2 ± 0.2 at 25°C

Preparation of Medium: Add components to distilled/deionized water and bring volume to 1.0L. Mix thoroughly. Adjust pH to 7.2. Distribute into tubes or flasks. Gently heat while stirring and bring to boiling. Mix thoroughly. Autoclave for 15 min at 15 psi pressure–121°C. Pour into Petri dishes or leave in tubes.

Use: For the cultivation of *Streptacidiphilus jiangxiensis.*

ISP 5 Medium
(DSMZ Medium 993)

Composition per liter:

Agar	20.0g
Glycerol	10.0g
L-asparagine, anhydrous	1.0g
K_2HPO_4, anhydrous	1.0g
Trace elements solution	1.0mL

pH 7.2 ± 0.2 at 25°C

Trace Elements Solution:

Composition per 100.0mL:

$FeSO_4 \cdot 7H_2O$	0.1g
$MnCl_2 \cdot 4H_2O$	0.1g
$ZnSO_4 \cdot 7H_2O$	0.1g

Preparation of Trace Elements Solution: Add components to distilled/deionized water and bring volume to 100.0mL. Mix thoroughly.

Preparation of Medium: Add components to distilled/deionized water and bring volume to 1.0L. Mix thoroughly. Adjust pH to 7.2. Distribute into tubes or flasks. Gently heat while stirring and bring to boiling. Mix thoroughly. Autoclave for 15 min at 15 psi pressure–121°C. Pour into Petri dishes or leave in tubes.

Use: For the cultivation of *Acrocarpospora macrocephala.*

ISP Medium 1
(International *Streptomyces* Project Medium 1)
(Tryptone Yeast Extract Broth)

Composition per liter:

Pancreatic digest of casein	5.0g
Yeast extract	3.0g

pH 7.0–7.2 at 25°C

Preparation of Medium: Add components to distilled/deionized water and bring volume to 1.0L. Mix thoroughly. Distribute into tubes or flasks. Autoclave for 15 min at 15 psi pressure–121°C.

Use: For the cultivation of *Streptomyces* species according to the International *Streptomyces* Project.

ISP Medium 2
(International *Streptomyces* Project Medium 2)
(Yeast Extract Malt Extract Agar)

Composition per liter:

Agar	20.0g
Malt extract	10.0g
Yeast extract	4.0g
Glucose	4.0g

pH 7.3 ± 0.2 at 25°C

Preparation of Medium: Add components to distilled/deionized water and bring volume to 1.0L. Mix thoroughly. Gently heat and bring to boiling. Distribute into tubes or flasks. Autoclave for 15 min at 15 psi pressure–121°C. Pour into sterile Petri dishes or leave in tubes.

Use: For the cultivation of *Streptomyces* species according to the International *Streptomyces* Project.

ISP Medium 2 with 5% Sodium Chloride

Composition per liter:

NaCl	50.0g
Agar	20.0g
Malt extract	10.0g
Glucose	4.0g
Yeast extract	4.0g

pH 7.3–7.5 at 25°C

Preparation of Medium: Add components to distilled/deionized water and bring volume to 1.0L. Mix thoroughly. Gently heat and bring to boiling. Distribute into tubes or flasks. Autoclave for 15 min at 15 psi pressure–121°C. Pour into sterile Petri dishes or leave in tubes.

Use: For the cultivation of *Streptomyces* species.

ISP Medium 3
(International *Streptomyces* Project Medium 3)
(Oatmeal Agar)

Composition per liter:

Oatmeal	20.0g
Agar	18.0g
Trace salts solution	1.0mL

Trace Salts Solution:

Composition per 100.0mL:

$FeSO_4 \cdot 7H_2O$	0.1g
$MnCl_2 \cdot 4H_2O$	0.1g
$ZnSO_4 \cdot 7H_2O$	0.1g

Preparation of Trace Salts Solution: Add components to distilled/deionized water and bring the volume to 100.0mL. Mix thoroughly. Filter sterilize.

Preparation of Medium: Add oatmeal to distilled/deionized water and bring volume to 1.0L. Mix thoroughly. Gently heat and bring to boiling. Steam for 20 min. Filter through cheesecloth. Add agar. Add sufficient distilled/deionized water to bring volume to 999.0mL. Gently heat and bring to boiling. Mix thoroughly. Distribute into tubes or flasks. Autoclave for 15 min at 15 psi pressure–121°C. Cool to 45°–50°C. Aseptically add 1.0mL of sterile trace salts solution. Mix thoroughly. Pour into sterile Petri dishes or distribute into sterile tubes.

Use: For the cultivation of *Streptomyces* species according to the International *Streptomyces* Project.

ISP Medium 4
(International *Streptomyces* Project Medium 4)
(Inorganic Salts Starch Agar)

Composition per liter:

Agar	20.0g
Soluble starch	10.0g
$CaCO_3$	2.0g
$(NH_4)_2SO_4$	2.0g
K_2HPO_4	1.0g
$MgSO_4 \cdot 7H_2O$	1.0g
NaCl	1.0g
$FeSO_4 \cdot 7H_2O$	1.0mg
$MnCl_2 \cdot 7H_2O$	1.0mg
$ZnSO_4 \cdot 7H_2O$	1.0mg

pH 7.2 ± 0.2 at 25°C

Preparation of Medium: Add components to distilled/deionized water and bring volume to 1.0L. Mix thoroughly. Gently heat and bring to boiling with frequent agitation. Distribute into tubes or flasks. Autoclave for 15 min at 15 psi pressure–121°C. Pour into sterile Petri dishes or leave in tubes.

Use: For characterizing *Streptomyces* species. For the cultivation and maintenance of *Actinomadura fastidiosa, Actinomadura roseoviolacea, Actinomadura* species, *Actinoplanes* species, *Amycolatopsis mediterranei, Kitasatosporia grisea, Kitasatosporia papulosa, Saccharomonospora internatus, Streptomyces* species, *Streptosporangium* species, *Saccharomonospora hirsuta,* and *Streptoverticillium* species.

ISP Medium 4 with Glucose (International *Streptomyces* Project Medium 4 with Glucose)

Composition per liter:

Agar	20.0g
Glucose	20.0g
Soluble starch	10.0g
$CaCO_3$	2.0g
$(NH_4)_2SO_4$	2.0g
K_2HPO_4	1.0g
$MgSO_4 \cdot 7H_2O$	1.0g
NaCl	1.0g
$FeSO_4 \cdot 7H_2O$	1.0mg
$MnCl_2 \cdot 7H_2O$	1.0mg
$ZnSO_4 \cdot 7H_2O$	1.0mg

pH 7.2 ± 0.2 at 25°C

Preparation of Medium: Add components to distilled/deionized water and bring volume to 1.0L. Mix thoroughly. Gently heat and bring to boiling with frequent agitation. Distribute into tubes or flasks. Autoclave for 15 min at 15 psi pressure–121°C. Pour into sterile Petri dishes with swirling or leave in tubes.

Use: For the cultivation and maintenance of *Streptomyces purpureus.*

ISP Medium 4 with Yeast Extract (International *Streptomyces* Project Medium 4 with Yeast Extract)

Composition per liter:

Agar	20.0g
Soluble starch	10.0g
$CaCO_3$	2.0g
$(NH_4)_2SO_4$	2.0g
K_2HPO_4	1.0g
$MgSO_4 \cdot 7H_2O$	1.0g
NaCl	1.0g
Yeast extract	1.0g
$FeSO_4 \cdot 7H_2O$	1.0mg
$MnCl_2 \cdot 7H_2O$	1.0mg
$ZnSO_4 \cdot 7H_2O$	1.0mg

pH 7.2 ± 0.2 at 25°C

Preparation of Medium: Add components to distilled/deionized water and bring volume to 1.0L. Mix thoroughly. Gently heat and bring to boiling with frequent agitation. Distribute into tubes or flasks. Autoclave for 15 min at 15 psi pressure–121°C. Pour into sterile Petri dishes with swirling or leave in tubes.

Use: For the cultivation and maintenance of *Thermomonospora mesouviformis.*

ISP Medium 5 (International *Streptomyces* Project Medium 5) (Glycerol Asparagine Agar)

Composition per liter:

Agar	20.0g
Glycerol	10.0g
L-Asparagine	1.0g
K_2HPO_4	1.0g
Trace salts solution	1.0mL

pH 7.4 ± 0.2 at 25°C

Trace Salts Solution:
Composition per 100.0mL:

$FeSO_4 \cdot 7H_2O$	0.1g
$MnCl_2 \cdot 4H_2O$	0.1g
$ZnSO_4 \cdot 7H_2O$	0.1g

Preparation of Trace Salts Solution: Add components to distilled/deionized water and bring the volume to 100.0mL. Mix thoroughly. Filter sterilize.

Preparation of Medium: Add components, except trace salts solution, to distilled/deionized water and bring volume to 999.0mL. Mix thoroughly. Gently heat and bring to boiling. Autoclave for 15 min at 15 psi pressure–121°C. Cool to 45°–50°C. Aseptically add 1.0mL of sterile trace salts solution. Mix thoroughly. Pour into sterile Petri dishes or distribute into sterile tubes.

Use: For the cultivation and maintenance of *Pseudonocardia* species and *Streptomyces peucetius.*

ISP Medium 6 (International *Streptomyces* Project Medium 6) (Peptone Yeast Extract Iron Agar)

Composition per liter:

Agar	15.0g
Peptone	15.0g
Proteose peptone	5.0g
K_2HPO_4	1.0g
Yeast extract	1.0g
Ferric ammonium citrate	0.5g
$Na_2S_2O_3$	0.08g

pH 7.2 ± 0.2 at 25°C

Preparation of Medium: Add components to distilled/deionized water and bring volume to 1.0L. Mix thoroughly. Gently heat and bring to boiling. Distribute into tubes or flasks. Autoclave for 15 min at 15 psi pressure–121°C. Pour into sterile Petri dishes or leave in tubes.

Use: For the cultivation and maintenance of *Streptomyces* species.

ISP Medium 7
See: **Tyrosine Agar**

ISP Medium 8
See: **Nitrate Broth**

ISP Medium 9 (International Streptomyces Project Medium 9)

Composition per liter:

$K_2HPO_4 \cdot 3H_2O$	5.65g
$(NH_4)_2SO_4$	2.64g
KH_2PO_4	2.38g
$MgSO_4 \cdot 7H_2O$	1.0g
Carbohydrate solution	100.0mL
Pridham and Gottlieb trace salts	1.0mL

pH 6.8–7.0 at 25°C

Carbohydrate Solution:
Composition per 100.0mL:

Carbohydrate	10.0g

Preparation of Carbohydrate Solution: Add carbohydrate to distilled/deionized water and bring volume to 100.0mL.Use glucose, arabinose, sucrose, xylose, inositol, mannitol, fructose, rhamnose, raffinose, or cellulose. Mix thoroughly. Filter sterilize.

Pridham And Gottlieb Trace Salts:

Composition per 100.0mL:

$MnCl_2 \cdot 7H_2O$	0.79g
$CuSO_4 \cdot 5H_2O$	0.64g
$ZnSO_4 \cdot 7H_2O$	0.15g
$FeSO_4 \cdot 7H_2O$	0.11g

Preparation of Pridham and Gottlieb Trace Salts: Add components to distilled/deionized water and bring volume to 100.0mL. Mix thoroughly.

Preparation of Medium: Add components, except carbohydrate solution, to distilled/deionized water and bring volume to 900.0mL. Mix thoroughly. Gently heat and bring to boiling with frequent agitation. Autoclave for 15 min at 15 psi pressure–121°C. Cool to 45°–50°C. Aseptically add sterile carbohydrate solution. Mix thoroughly. Aseptically distribute into sterile tubes or flasks.

Use: For the cultivation and differentiation of *Streptomyces purpureus* and other *Streptomyces* species based on carbohydrate utilization.

ISS1 Medium (DSMZ Medium 889)

Composition per liter:

NaCl	20.0g
$MgSO_4 \cdot 7H_2O$	7.0g
$MgCl_2 \cdot 6H_2O$	5.5g
$NaHCO_3$	2.0g
KCl	0.65g
$CaCl_2 \cdot 2H_2O$	0.5g
NH_4Cl	0.3g
K_2HPO_4	0.2g
Yeast extract	0.2g
NaBr	0.1g
Trace elements solution	10.0mL

pH 7.0 ± 0.2 at 25°C

Trace Elements Solution:

Composition per liter:

$MgSO_4 \cdot 7H_2O$	3.0g
Nitrilotriacetic acid	1.5g
NaCl	1.0g
$MnSO_4 \cdot 2H_2O$	0.5g
$CoSO_4 \cdot 7H_2O$	0.18g
$ZnSO_4 \cdot 7H_2O$	0.18g
$CaCl_2 \cdot 2H_2O$	0.1g
$FeSO_4 \cdot 7H_2O$	0.1g
$NiCl_2 \cdot 6H_2O$	0.025g
$KAl(SO_4)_2 \cdot 12H_2O$	0.02g
H_3BO_3	0.01g
$Na_2MoO_4 \cdot 4H_2O$	0.01g
$CuSO_4 \cdot 5H_2O$	0.01g
$Na_2SeO_3 \cdot 5H_2O$	0.3mg

Preparation of Trace Elements Solution: Add nitrilotriacetic acid to 500.0mL of distilled/deionized water. Dissolve by adjusting pH to 6.5 with KOH. Add remaining components. Add distilled/deionized water to 1.0L. Mix thoroughly.

Preparation of Medium: Add components to distilled/deionized water and bring volume to 1.0L. Mix thoroughly. Adjust pH to 7.0 using H_2SO_4. Distribute into anaerobe tubes or bottles. Autoclave for 15 min at 15 psi pressure–121°C. Provide an atmosphere of 78% H_2 + 20% CO_2 + 2% O_2.

Use: For the cultivation of unclassified bacterium DSM 12045.

ITC Broth

See: **Irgasan® Ticarcillin Chlorate Broth**

ITC HiVeg Broth Base with Ticarcillin and Potassium Chlorate (TTC HiVeg Broth Base)

Composition per liter:

$MnCl_2 \cdot 6H_2O$	60.0g
Plant hydrolysate	10.0g
NaCl	5.0g
Yeast extract	1.0g
Malachite Green	0.01g
Irgansan (Trichlosan)	1.0mg
Ticarcillin solution	10.0mL
Potassium chlorate solution	

pH 6.9 ± 0.2 at 25°C

Source: This medium, without ticarcillin and potassium chlorate, is available as a premixed powder from HiMedia.

Ticarcillin Solution:

Composition per 10.0mL:

Ticarcillin	1.0mg

Preparation of Ticarcillin Solution: Add ticarcillin to distilled/deionized water and bring volume to 10.0mL. Mix thoroughly. Filter sterilize.

Poassium Chlorate Solution:

Composition per 10.0mL:

$KClO_3$	1.0g

Preparation of Poassium Chlorate Solution: Add $KClO_3$ to distilled/deionized water and bring volume to 10.0mL. Mix thoroughly. Filter sterilize.

Caution: Potassium chlorate is a highly oxidizable agent and can cause explosions. Take proper precautions when handling.

Preparation of Medium: Add components, except ticarcillin and potassium chlorate solutions, to distilled/deionized water and bring volume to 980.0mL. Mix thoroughly. Gently heat and bring to boiling. Mix thoroughly. Autoclave for 15 min at 15 psi pressure–121°C. Cool to 45°–50°C. Aseptically add 10.0mL ticarcillin solution and 10.0mL potassium chlorate solution. Mix thoroughly. Distribute into tubes or flasks.

Use: For the selective isolation and cultivation of *Yersinia* species. For the selective enrichment and enumeration of *Yersinia enterocolitica*.

IUT Medium Base with Glycerol and Egg Emulsion

Composition per 1600.0mL:

L-Asparagine	3.6g
KH_2PO_4	2.46g
Magnesium citrate	0.6g
Malachite Green	0.4g
$MgSO_4 \cdot 7H_2O$	0.24g

Egg emulsion 1.0L
Glycerol 12.0mL

pH 7.0 ± 0.2 at 25°C

Source: This medium is available from HiMedia.

Egg Emulsion:
Composition per 1.0L:
Eggs .... Variable

Preparation of Egg Emulsion: Soak whole eggs with 1:100 dilution of saturated mercuric chloride solution for 1 min. Crack eggs. Beat to form emulsion. Avoiding the formation of air bubbles. Sterilize by inspissation at 85°C for 1 hr.

Preparation of Medium: Add glycerol to distilled/deionized water and bring volume to 600.0mL. Mix thoroughly. Add remaining components other than egg emulsion. Gently heat and bring to boiling. Autoclave for 15 min at 15 psi pressure–121°C. Cool to 50°C. Add sterile whole egg emulsion. Aseptically distribute into culture vessels.

Use: For the cultivation of *Mycobacterium tuberculosis.*

J Agar

Composition per liter:
Agar 20.0g
Yeast extract 15.0g
Pancreatic digest of casein 5.0g
K_2HPO_4 .... 3.0g
Glucose solution 10.0mL

pH 7.3–7.5 at 25°C

Glucose Solution:
Composition per 10.0mL:
Glucose 2.0g

Preparation of Glucose Solution: Add glucose to distilled/deionized water and bring volume to 10.0mL. Mix thoroughly. Filter sterilize.

Preparation of Medium: Add components, except glucose solution, to distilled/deionized water and bring volume to 990.0mL. Mix thoroughly. Gently heat and bring to boiling. Autoclave for 15 min at 15 psi pressure–121°C. Cool to 45°–50°C. Aseptically add sterile glucose solution. Mix thoroughly. Pour into sterile Petri dishes.

Use: For the cultivation of *Bacillus* species and *Sporolactobacillus* species.

J Broth

Composition per liter:
Yeast extract 15.0g
Pancreatic digest of casein 5.0g

pH 7.3–7.5 at 25°C

Preparation of Medium: Add components to distilled/deionized water and bring volume to 1.0L. Mix thoroughly. Adjust pH to 7.3–7.5. Distribute into tubes or flasks. Autoclave for 20 min at 15 psi pressure–121°C.

Use: For the cultivation of *Bacillus* species and *Sporolactobacillus* species for performing the Voges-Proskauer test.

JB Medium with Glucose

Composition per liter:
Yeast extract 15.0g
Pancreatic digest of casein 5.0g
K_2HPO_4 .... 3.0g
Glucose 2.0g

pH 7.3–7.5 at 25°C

Preparation of Medium: Add components to distilled/deionized water and bring volume to 1.0L. Mix thoroughly. Gently heat and bring to boiling. Distribute into tubes or flasks. Autoclave for 15 min at 15 psi pressure–121°C. Pour into sterile Petri dishes or leave in tubes.

Use: For the cultivation and maintenance of *Bacillus popilliae.*

JD1 Medium

Composition per liter:
Beef heart, solids from infusion 25.0g
Agar 15.0g
Peptone 5.0g
NaCl 2.5g
Bovine albumin 0.5g
Hemin chloride 0.04g

pH 7.8 ± 0.2 at 25°C

Preparation of Medium: Add components to distilled/deionized water and bring volume to 1.0L. Mix thoroughly. Gently heat and bring to boiling. Distribute into tubes or flasks. Autoclave for 15 min at 15 psi pressure–121°C. Pour into sterile Petri dishes or leave in tubes.

Use: For the isolation and cultivation of PD-ALS (Pierce's disease-almond leaf scorch) bacteria.

JD3 Medium

Composition per liter:
Pancreatic digest of casein 4.0g
Papaic digest of soybean meal 2.0g
Trisodium citrate 2.0g
K_2HPO_4 .... 1.5g
KH_2PO_4 .... 1.0g
$MgSO_4 \cdot 7H_2O$ 1.0g
Disodium succinate 0.01g
Bovine serum albumin solution 100.0mL
Hemin chloride solution 10.0mL

pH 7.0 ± 0.2 at 25°C

Bovine Serum Albumin Solution:
Composition per 100.0mL:
Bovine serum albumin fraction V 2.0g

Preparation of Bovine Serum Albumin Solution: Add 2.0g of bovine serum albumin fraction V to 100.0mL of distilled/deionized water. Mix thoroughly. Filter sterilize.

Hemin Chloride Solution:
Composition per 10.0mL:
Hemin chloride 0.01g

Preparation of Hemin Chloride Solution: Add 0.01g of hemin chloride to 10.0 mL of 0.5*N* NaOH. Mix thoroughly.

Preparation of Medium: Add components, except bovine serum albumin solution, to distilled/deionized water and bring volume to 900.0mL. Mix thoroughly. Adjust to pH 7.0. Autoclave for 20 min at 15 psi pressure–121°C. Cool to 25°C. Aseptically add 100.0mL of sterile bovine serum albumin solution. Mix thoroughly. Aseptically distribute into sterile tubes or flasks.

Use: For the cultivation of ATCC strain 33107.

Jensen's Medium

Composition per liter:

Sucrose	20.0g
Agar	15.0g
$CaCO_3$	2.0g
K_2HPO_4	1.0g
$MgSO_4 \cdot 7H_2O$	0.5g
NaCl	0.5g
$FeSO_4 \cdot 7H_2O$	0.1g
$Na_2MoO_4 \cdot 2H_2O$	5.0mg

Source: This medium is available from HiMedia.

Preparation of Medium: Add components to distilled/deionized water and bring volume to 1.0L. Mix thoroughly. Gently heat and bring to boiling. Distribute into tubes or flasks. Autoclave for 15 min at 15 psi pressure–121°C. Pour into sterile Petri dishes or leave in tubes.

Use: For the detection and cultivation of nitrogen fixing bacteria.

JO

Composition per liter:

Agar	20.0g
Sucrose	6.35g
$NaNO_3$	1.5g
Yeast extract	0.50g
K_2HPO_4	0.35g
$MgSO_4$	0.25g

Preparation of Medium: Add components to distilled/deionized water and bring volume to 1.0L. Mix thoroughly. Gently heat and bring to boiling. Distribute into tubes or flasks. Autoclave for 15 min at 15 psi pressure–121°C. Pour into sterile Petri dishes or leave in tubes.

Use: For the cultivation and maintenance of fungi with inhibition of *Mucor* species.

Johnson's Marine Medium

Composition per liter:

Peptone	5.0g
Yeast extract	1.0g
$Na_2S_2O_3$	0.3g
$FeSO_4 \cdot 7H_2O$	0.2g
Filtered, aged seawater	750.0mL

Preparation of Medium: Add components to distilled/deionized water and bring volume to 1.0L. Mix thoroughly. Distribute into tubes or flasks. Autoclave for 15 min at 15 psi pressure–121°C.

Use: For the cultivation of marine bacteria.

Jones–Kendrick Pertussis Transport Medium

Composition per liter:

Beef heart, solids from infusion	500.0g
Agar	20.0g
Soluble starch	10.0g
Tryptose	10.0g
NaCl	5.0g
Charcoal powder, activated	4.0g
Yeast extract	3.5g
Penicillin solution	10.0mL

pH 7.4 ± 0.2 at 25°C

Penicillin Solution:

Composition per 10.0mL:

Penicillin	300U

Preparation of Penicillin Solution: Add penicillin to distilled/deionized water and bring volume to 10.0mL. Mix thoroughly. Filter sterilize.

Preparation of Medium: Add components, except penicillin solution, starch, yeast extract, heart infusion, and agar, to water. Boil to dissolve. Add charcoal, mix well, and autoclave. Cool to 50°C, add penicillin, and dispense into small bottles as slants. Cool and seal tightly. Store at 5°C. Stable for 2 to 3 months.

Use: For the cultivation and transport of *Bordetella pertussis* between clinical isolation and laboratory cultivation.

Jordan's Tartrate Agar

Composition per liter:

Agar	15.0g
Pancreatic digest of casein	10.0g
Sodium potassium tartrate	10.0g
NaCl	5.0g
Phenol Red	0.024g

pH 7.7 ± 0.3 at 25°C

Source: This medium is available as a prepared medium in tubes from BD Diagnostic Systems.

Preparation of Medium: Add components to distilled/deionized water and bring volume to 1.0L. Mix thoroughly. Gently heat and bring to boiling. Adjust pH to 7.7. Distribute into tubes. Autoclave for 15 min at 15 psi pressure–121°C.

Use: For the differentiation and identification of members of the Enterobacteriaceae, especially *Salmonella* species, based upon the ability to utilize tartrate. Utilization of tartrate turns the medium yellow. *Salmonella enteritidis* utilizes tartrate. *Salmonella paratyphi* A does not utilize tartrate.

K101 *Flexibacter* Medium

Composition per liter:

Agar	10.0g
Casamino acids	1.0g
Glucose	1.0g
Tris(hydroxymethyl)aminomethane buffer	1.0g
$CaCl_2$	0.1g
KNO_3	0.1g
$MgSO_4 \cdot 7H_2O$	0.1g
Sodium glycerophosphate	0.1g
Thiamine·HCl	1.0mg
Cyanocobalamin	1.0μg
Trace elements solution HO-LE	1.0mL

pH 7.5 ± 0.2 at 25°C

Trace Elements Solution HO-LE:

Composition per liter:

H_3BO_3	2.85g
$MnCl_2 \cdot 4H_2O$	1.8g
Sodium tartrate	1.77g
$FeSO_4 \cdot 7H_2O$	1.36g
$CoCl_2 \cdot 6H_2O$	0.04g
$CuCl_2 \cdot 2H_2O$	0.027g
$Na_2MoO_4 \cdot 2H_2O$	0.025g
$ZnCl_2$	0.02g

Preparation of Trace Elements Solution HO-LE: Add components to distilled/deionized water and bring volume to 1.0L. Mix thoroughly. Filter sterilize.

Preparation of Medium: Add components, except trace elements solution HO-LE, to distilled/deionized water and bring volume to 999.0mL. Mix thoroughly. Gently heat and bring to boiling. Autoclave for 15 min at 15 psi pressure–121°C. Cool to 45°–50°C. Aseptically add 1.0mL of trace elements solution HO-LE. Mix thoroughly. Pour into sterile Petri dishes or distribute into sterile tubes.

Use: For the cultivation and maintenance of *Cytophaga* species, *Flexibacter* species, *Herpetosiphon geysericola*, and *Myxococcus fulvus*.

K7 Medium (DSMZ Medium 1199)

Composition per liter:

Agar	20.0g
Glucose	1.0g
Yeast extract	1.0g
Peptone	1.0g

pH 5.5 ± 0.2 at 25°C

Preparation of Medium: Add components to distilled/deionized water and bring volume to 1.0L. Mix thoroughly. Adjust pH to 5.5. Distribute into tubes or flasks. Gently heat while stirring and bring to boiling. Mix thoroughly. Autoclave for 15 min at 15 psi pressure–121°C. Pour into Petri dishes or leave in tubes.

Use: For the cultivation of *Novosphingobium acidiphilum*.

Kado's Agar

Composition per liter:

Agar	15.0g
Sucrose	10.0g
Pancreatic digest of casein	8.0g
Yeast extract	4.0g
K_2HPO_4	2.0g
$MgSO_4 \cdot 7H_2O$	30.0mg

Preparation of Medium: Add components to distilled/deionized water and bring volume to 1.0L. Mix thoroughly. Gently heat and bring to boiling. Distribute into tubes or flasks. Autoclave for 15 min at 15 psi pressure–121°C. Pour into sterile Petri dishes or leave in tubes.

Use: For the cultivation of *Lactococcus lactis* subspecies *hordniae*.

Kanamycin Esculin Azide Agar

Composition per liter:

Pancreatic digest of casein	20.0g
Agar	10.0g
NaCl	5.0g
Yeast extract	5.0g
Esculin	1.0g
Sodium citrate	1.0g
Ferric ammonium citrate	0.5g
NaN_3	0.15g
Kanamycin sulfate solution	10.0mL

pH 7.0 ± 0.2 at 25°C

Caution: Sodium azide is toxic. Azides also react with metals and disposal must be highly diluted.

Source: This medium is available as a premixed powder from Oxoid Unipath.

Kanamycin Sulfate Solution:
Composition per 10.0mL:

Kanamycin sulfate	20.0mg

Preparation of Kanamycin Sulfate Solution: Add kanamycin sulfate to distilled/deionized water and bring volume to 10.0mL. Mix thoroughly.

Preparation of Medium: Add components to distilled/deionized water and bring volume to 1.0L. Mix thoroughly. Gently heat and bring to boiling. Distribute into tubes or flasks. Autoclave for 15 min at 15 psi pressure–121°C. Pour into sterile Petri dishes or leave in tubes.

Use: For the isolation of enterococci from foods.

Kanamycin Esculin Azide Broth

Composition per liter:

Pancreatic digest of casein	20.0g
NaCl	5.0g
Yeast extract	5.0g
Esculin	1.0g
Sodium citrate	1.0g
Ferric ammonium citrate	0.5g
NaN_3	0.15g
Kanamycin sulfate solution	10.0mL

pH 7.0 ± 0.2 at 25°C

Caution: Sodium azide is toxic. Azides also react with metals and disposal must be highly diluted.

Source: This medium is available as a premixed powder from Oxoid Unipath.

Kanamycin Sulfate Solution:
Composition per 10.0mL:

Kanamycin sulfate	0.02g

Preparation of Kanamycin Sulfate Solution: Add kanamycin sulfate to distilled/deionized water and bring volume to 10.0mL. Mix thoroughly.

Preparation of Medium: Add components to distilled/deionized water and bring volume to 1.0L. Mix thoroughly. Distribute into tubes or flasks. Autoclave for 15 min at 15 psi pressure–121°C.

Use: For the isolation of enterococci from foods.

Kanamycin Esculin Azide HiVeg Agar

Composition per liter:

Plant hydrolysate	20.0g
Agar	12.0g
NaCl	5.0g
Yeast extract	5.0g
Esculin	1.0g
Sodium citrate	1.0g
Ferric ammonium citrate	0.5g
NaN_3	0.15g
Kanamycin sulfate	0.02g

pH 7.0 ± 0.2 at 25°C

Caution: Sodium azide is toxic. Azides also react with metals and disposal must be highly diluted.

Source: This medium is available as a premixed powder from HiMedia.

Preparation of Medium: Add components to distilled/deionized water and bring volume to 1.0L. Mix thoroughly. Distribute into tubes or flasks. Autoclave for 15 min at 15 psi pressure–121°C.

Use: For the isolation of enterococci from foods.

Kanamycin Esculin Azide HiVeg Agar Base with Kanamycin

Composition per liter:

Plant hydrolysate	20.0g
Agar	10.0g
Yeast extract	5.0g
NaCl	5.0g
Esculin	1.0g
Sodium citrate	1.0g
Ferric ammonium citrate	0.5g
NaN_3	0.15g
Kanamycin sulfate solution	10.0mL

pH 7.0 ± 0.2 at 25°C

Source: This medium, without kanamycin sulfate solution, is available as a premixed powder from HiMedia.

Caution: Sodium azide is toxic. Azides also react with metals and disposal must be highly diluted.

Kanamycin Sulfate Solution:

Composition per 10.0mL:

Kanamycin sulfate	0.02g

Preparation of Kanamycin Sulfate Solution: Add kanamycin sulfate to distilled/deionized water and bring volume to 10.0mL. Mix thoroughly.

Preparation of Medium: Add components to distilled/deionized water and bring volume to 1.0L. Mix thoroughly. Distribute into tubes or flasks. Autoclave for 15 min at 15 psi pressure–121°C. Pour into sterile Petri dishes or leave in tubes.

Use: For the isolation of enterococci from foods.

Kanamycin Esculin Azide HiVeg Broth

Composition per liter:

Plant hydrolysate	20.0g
NaCl	5.0g
Yeast extract	5.0g
Esculin	1.0g
Sodium citrate	1.0g
Ferric ammonium citrate	0.5g
NaN_3	0.15g
Kanamycin sulfate	0.02g

pH 7.0 ± 0.2 at 25°C

Caution: Sodium azide is toxic. Azides also react with metals and disposal must be highly diluted.

Source: This medium is available as a premixed powder from HiMedia.

Preparation of Medium: Add components to distilled/deionized water and bring volume to 1.0L. Mix thoroughly. Distribute into tubes or flasks. Autoclave for 15 min at 15 psi pressure–121°C.

Use: For the cultivation of enterococci from foods.

Kanamycin Esculin Azide HiVeg Broth Base with Kanamycin

Composition per liter:

Plant hydrolysate	20.0g
NaCl	5.0g
Yeast extract	5.0g
Esculin	1.0g
Sodium citrate	1.0g
Ferric ammonium citrate	0.5g
NaN_3	0.15g
Kanamycin sulfate solution	10.0mL

pH 7.0 ± 0.2 at 25°C

Caution: Sodium azide is toxic. Azides also react with metals and disposal must be highly diluted.

Source: This medium, without kanamycin sulfate solution, is available as a premixed powder from HiMedia.

Kanamycin Sulfate Solution:

Composition per 10.0mL:

Kanamycin sulfate	0.02g

Preparation of Kanamycin Sulfate Solution: Add kanamycin sulfate to distilled/deionized water and bring volume to 10.0mL. Mix thoroughly.

Preparation of Medium: Add components to distilled/deionized water and bring volume to 1.0L. Mix thoroughly. Distribute into tubes or flasks. Autoclave for 15 min at 15 psi pressure–121°C.

Use: For the cultivation of enterococci from foods.

Kanamycin L Broth Medium

Composition per liter:

Pancreatic digest of casein	10.0g
NaCl	5.0g
Yeast extract	5.0g
Glucose	1.0g
Kanamycin solution	10.0mL

Kanamycin Solution:

Composition per 10.0mL:

Kanamycin	50.0mg

Preparation of Kanamycin Solution: Add kanamycin to distilled/deionized water and bring volume to 10.0mL. Mix thoroughly. Filter sterilize.

Preparation of Medium: Add components, except kanamycin solution, to distilled/deionized water and bring volume to 990.0mL. Mix thoroughly. Bring pH to 7.0. Autoclave for 15 min at 15 psi pressure–121°C. Aseptically add 10.0mL of sterile kanamycin solution. Mix thoroughly. Aseptically distribute into sterile tubes or flasks.

Use: For the cultivation of *Escherichia coli*.

Kanamycin Luria Agar

Composition per liter:

Agar	15.0g
Pancreatic digest of casein	10.0g
Yeast extract	5.0g
NaCl	0.5g
Glucose solution	20.0mL
Kanamycin solution	10.0mL

Glucose Solution:

Composition per 100.0mL:

Glucose	10.0g

Preparation of Glucose Solution: Add glucose to distilled/deionized water and bring volume to 100.0mL. Mix thoroughly. Filter sterilize.

Kanamycin Solution:

Composition per 10.0mL:

Kanamycin	10.0mg

Preparation of Kanamycin Solution: Add kanamycin to distilled/deionized water and bring volume to 10.0mL. Mix thoroughly. Filter sterilize.

Preparation of Medium: Add components, except glucose solution and kanamycin solution, to distilled/deionized water and bring volume to 970.0mL. Mix thoroughly. Bring pH to 7.0. Autoclave for 15 min at 15 psi pressure–121°C. Aseptically add 20.0mL of sterile glucose solution and 10.0mL of sterile kanamycin solution. Mix thoroughly. Aseptically distribute into sterile tubes or flasks.

Use: For the cultivation of *Escherichia coli.*

Kanamycin Vancomycin Blood Agar (KVBA)

Composition per liter:

Agar	17.5g
Pancreatic digest of casein	15.0g
Papaic digest of soybean meal	5.0g
NaCl	5.0g
Kanamycin	0.1g
Sheep blood, defibrinated	50.0mL
Vancomycin solution	10.0mL
Vitamin K_1 solution	1.0mL

Vancomycin Solution:

Composition per 10.0mL:

Vancomycin	7.5mg

Preparation of Vancomycin Solution: Add vancomycin to distilled/deionized water and bring volume to 10.0mL. Mix thoroughly. Filter sterilize.

Vitamin K_1 Solution:

Composition per 100.0mL:

Vitamin K_1	1.0g

Preparation of Vitamin K_1 Solution: Add vitamin K_1 to 99.0mL of absolute ethanol. Mix thoroughly. Filter sterilize.

Preparation of Medium: Add components, except sheep blood, vancomycin solution, and vitamin K_1 solution, to distilled/deionized water and bring volume to 939.0mL. Mix thoroughly. Gently heat and bring to boiling. Autoclave for 15 min at 15 psi pressure–121°C. Cool to 45°–50°C. Aseptically add sheep blood, vancomycin solution, and 1.0mL vitamin K_1 solution. Mix thoroughly. Pour into sterile Petri dishes or distribute into sterile tubes.

Use: For the selective isolation of anaerobes, particularly *Bacteroides*, from clinical specimens.

Kanamycin Vancomycin Laked Blood Agar

Composition per liter:

Agar	17.5g
Pancreatic digest of casein	15.0g
Papaic digest of soybean meal	5.0g
NaCl	5.0g
Kanamycin	0.075g
Sheep blood, laked	50.0mL
Vancomycin solution	10.0mL
Vitamin K_1 solution	1.0mL

Vancomycin Solution:

Composition per 10.0mL:

Vancomycin	7.5mg

Preparation of Vancomycin Solution: Add vancomycin to distilled/deionized water and bring volume to 10.0mL. Mix thoroughly. Filter sterilize.

Vitamin K_1 Solution:

Composition per 100.0mL:

Vitamin K_1	1.0g

Preparation of Vitamin K_1 Solution: Add vitamin K_1 to 99.0mL of absolute ethanol. Mix thoroughly. Filter sterilize.

Preparation of Medium: The blood is laked (hemolyzed) by freezing whole blood overnight and then thawing. Add components, except sheep blood, vancomycin solution, and vitamin K_1 solution, to distilled/deionized water and bring volume to 939.0mL. Mix thoroughly. Gently heat and bring to boiling. Autoclave for 15 min at 15 psi pressure–121°C. Cool to 45°–50°C. Aseptically add sheep blood, vancomycin solution, and 1.0mL of vitamin K_1 solution. Mix thoroughly. Pour into sterile Petri dishes or distribute into sterile tubes.

Use: For isolation of the *Bacteroides melaninogenicus* group.

Karmali's *Campylobacter* Medium

See: ***Campylobacter*** **Selective Medium, Karmali's**

Kasai Medium

Composition per liter:

Pancreatic digest of casein	20.0g
Soluble starch	20.0g
L-Cysteine·HCl·H_2O	5.0g
K_2HPO_4	5.0g
NaCl	5.0g
Yeast extract	2.0g

Preparation of Medium: Add components to distilled/deionized water and bring volume to 1.0L. Mix thoroughly. Gently heat and bring to boiling. Distribute into tubes or flasks. Autoclave for 15 min at 15 psi pressure–121°C.

Use: For the isolation and cultivation of *Leptotrichia buccalis* from saliva and plaque.

KC Bottom Agar

Composition per liter:

Agar	10.0g
Pancreatic digest of casein	10.0g
KCl	2.5g
NaCl	2.5g
$CaCl_2$ solution	1.0mL

$CaCl_2$ Solution:

Composition per 10.0mL:

$CaCl_2·2H_2O$	1.47g

Preparation of $CaCl_2$ Solution: Add $CaCl_2·2H_2O$ to distilled/deionized water and bring volume to 10.0mL. Mix thoroughly. Filter sterilize.

Preparation of Medium: Add components, except $CaCl_2$ solution, to distilled/deionized water and bring volume to 999.0mL. Mix thoroughly. Gently heat and bring to boiling. Autoclave for 15 min at 15 psi pressure–121°C. Cool to 45°–50°C. Aseptically add 1.0mL of $CaCl_2$ solution. Mix thoroughly. Pour into sterile Petri dishes or distribute into sterile tubes.

Use: For the cultivation and maintenance of *Escherichia coli.*

KC Broth

Composition per liter:

Pancreatic digest of casein	10.0g
KCl	5.0g
$CaCl_2$ solution	0.5mL

$CaCl_2$ Solution:

Composition per 10.0mL:

$CaCl_2 \cdot 2H_2O$	1.47g

Preparation of $CaCl_2$ Solution: Add $CaCl_2 \cdot 2H_2O$ to distilled/deionized water and bring volume to 10.0mL. Mix thoroughly. Filter sterilize.

Preparation of Medium: Add components, except $CaCl_2$ solution, to distilled/deionized water and bring volume to 999.5mL. Mix thoroughly. Distribute into tubes or flasks. Autoclave for 15 min at 15 psi pressure–121°C. Cool to 25°C. Aseptically add 0.5mL of $CaCl_2$ solution. Mix thoroughly. Aseptically distribute into sterile tubes.

Use: For the cultivation of *Escherichia coli.*

KC Top Agar

Composition per liter:

Pancreatic digest of casein	10.0g
Agar	8.0g
NaCl	5.0g

Preparation of Medium: Add components to distilled/deionized water and bring volume to 1.0L. Mix thoroughly. Gently heat and bring to boiling. Distribute into tubes or flasks. Autoclave for 15 min at 15 psi pressure–121°C. Pour into sterile Petri dishes or leave in tubes.

Use: For the cultivation and maintenance of *Escherichia coli.*

KCN Broth

Composition per liter:

Na_2HPO_4	5.64g
NaCl	5.0g
Peptone	3.0g
KH_2PO_4	0.225g
KCN (0.5% solution)	15.0mL

pH 7.6 ± 0.2 at 25°C

Caution: Cyanide is toxic.

Preparation of Medium: Add components, except KCN solution, to distilled/deionized water and bring volume to 985.0mL. Mix thoroughly. Autoclave for 15 min at 15 psi pressure–121°C. Cool to 25°C. Aseptically add KCN solution. Mix thoroughly. Aseptically distribute into sterile tubes. Stopper immediately.

Use: For the differentiation of Enterobacteriaceae based upon growth in the presence of potassium cyanide.

KDM-2 Medium

Composition per liter:

Peptone	10.0g
L-Cysteine·HCl	1.0g
Yeast extract	0.5g
Calf serum	100.0mL

pH 6.5 ± 0.2 at 25°C

Preparation of Medium: Add components to distilled/deionized water and bring volume to 1.0L. Mix thoroughly. Distribute into tubes or flasks. Autoclave for 15 min at 15 psi pressure–121°C.

Use: For the cultivation of *Renibacterium salmoninarum.*

Keister's Modified TYI-S-33 Medium

Composition per liter:

Pancreatic digest of casein	20.0g
Glucose	10.0g
Yeast extract	10.0g
L-Cysteine·HCl	2.0g
NaCl	2.0g
K_2HPO_4	1.0g
Bovine bile	0.75g
KH_2PO_4	0.6g
Ascorbic acid	0.2g
Ferric ammonium citrate	22.8mg
Bovine serum, heat inactivated	100.0mL

Preparation of Medium: Add components, except bovine serum, to distilled/deionized water and bring volume to 900.0mL. Mix thoroughly. Autoclave for 15 min at 15 psi pressure–121°C. Aseptically add 1.0L of sterile, heat-inactivated bovine serum. Mix thoroughly. Aseptically distribute into sterile, screw-capped tubes or flasks.

Use: For the cultivation of *Giardia cati, Giardia intestinalis,* and *Hexamita* species.

Kelly Medium, Nonselective Modified

Composition per 1430.0mL:

HEPES buffer (*N*-2-hydroxyethylpiperazine-*N*-2-ethanesulfonic acid)	6.0g
Proteose peptone No. 2	5.0g
D-Glucose	3.0g
$NaHCO_3$	2.2g
Pancreatic digest of casein	1.0g
Yeast, autolyzed	1.0g
Sodium pyruvate	0.8g
Sodium citrate	0.7g
N-Acetylglucosamine	0.4g
$MgCl_2 \cdot 6H_2O$	0.3g
Gelatin solution	200.0mL
Bovine serum albumin solution	143.0mL
CMRL-1066 medium with glutamine, 10X	100.0mL
Rabbit serum, heat inactivated	86.0mL
Hemin solution	1.0mL

pH 7.2 ± 0.2 at 25°C

CMRL-1066 Medium with Glutamine, 10X:

Composition per liter:

NaCl	6.8g
$NaHCO_3$	2.2g
D-Glucose	1.0g
KCl	0.4g
L-Cysteine·HCl·H_2O	0.26g

$CaCl_2$, anhydrous	0.2g
$MgSO_4 \cdot 7H_2O$	0.2g
$NaH_2PO_4 \cdot H_2O$	0.14g
L-Glutamine	0.1g
Sodium acetate·$3H_2O$	0.083g
L-Glutamic acid	0.075g
L-Arginine·HCl	0.07g
L-Lysine·HCl	0.07g
L-Leucine	0.06g
Glycine	0.05g
Ascorbic acid	0.05g
L-Proline	0.04g
L-Tyrosine	0.04g
L-Aspartic acid	0.03g
L-Threonine	0.03g
L-Alanine	0.025g
L-Phenylalanine	0.025g
L-Serine	0.025g
L-Valine	0.025g
L-Cystine	0.02g
L-Histidine·HCl·H_2O	0.02g
L-Isoleucine	0.02g
Phenol Red	0.02g
L-Methionine	0.015g
Deoxyadenosine	0.01g
Deoxycytidine	0.01g
Deoxyguanosine	0.01g
Glutathione, reduced	0.01g
Thymidine	0.01g
Hydroxy-L-proline	0.01g
L-Tryptophan	0.01g
Nicotinamide adenine dinucleotide	7.0mg
Tween™ 80	5.0mg
Sodium glucuronate·H_2O	4.2mg
Coenzyme A	2.5mg
Cocarboxylase	1.0mg
Flavin adenine dinucleotide	1.0mg
Nicotinamide adenine dinucleotide phosphate	1.0mg
Uridine triphosphate	1.0mg
Choline chloride	0.5mg
Cholesterol	0.2mg
5-Methyldeoxycytidine	0.1mg
Inositol	0.05mg
p-Aminobenzoic acid	0.05mg
Niacin	0.025mg
Niacinamide	0.025mg
Pyridoxine	0.025mg
Pyridoxal·HCl	0.025mg
Biotin	0.01mg
D-Calcium pantothenate	0.01mg
Folic acid	0.01mg
Riboflavin	0.01mg
Thiamine·HCl	0.01mg

pH 7.2 ± 0.2 at 25°C

Source: This solution is available as a premixed powder from BD Diagnostics.

Preparation of CMRL-1066 Medium with Glutamine, 10X: Add components to distilled/deionized water and bring volume to 1.0L. Mix thoroughly. Adjust pH to 7.2. Filter sterilize.

Gelatin Solution:
Composition per 200.0mL:

Gelatin	14.0g

Preparation of Gelatin Solution: Add gelatin to distilled/deionized water and bring volume to 200.0mL. Mix thoroughly. Gently heat and bring to boiling. Autoclave for 15 min at 15 psi pressure–121°C. Cool to 50°C.

Hemin Solution:
Composition per 100.0mL:

Hemin	1.0g
NaOH (1*N* solution)	20.0mL

Preparation of Hemin Solution: Add hemin to 20.0mL of 1*N* NaOH solution. Mix thoroughly. Bring volume to 100.0mL with distilled/deionized water.

Bovine Serum Albumin Solution:
Composition per 200.0mL:

Bovine serum albumin	70.0g

Preparation of Bovine Serum Albumin Solution: Add bovine serum albumin to distilled/deionized water and bring volume to 200.0mL. Filter sterilize.

Preparation of Medium: Add components, except gelatin solution, bovine serum albumin solution, and rabbit serum, to distilled/deionized water and bring volume to 1001.0mL. Mix thoroughly. Bring pH to 7.6 with 5*N* NaOH. Filter sterilize. Aseptically add 200.0mL of sterile gelatin solution, 143.0mL of sterile bovine serum albumin, and 86.0mL of sterile heat-inactivated rabbit serum. Mix thoroughly. Aseptically dispense into sterile tubes or flasks.

Use: For the isolation of *Borrelia burgdorferi* and other spirochetes.

Kelly Medium, Selective Modified

Composition per 1270.0mL:

Bovine serum albumin fraction V	50.0g
HEPES buffer (*N*-2-hydroxyethylpiperazine-*N*-2-ethanesulfonic acid)	6.0g
Glucose	5.0g
Neopeptone	5.0g
$NaHCO_3$	2.2g
Sodium pyruvate	0.8g
Sodium citrate	0.7g
N-Acetylglucosamine	0.4g
Kanamycin	8.0mg
5-Fluorouracil	2.3mg
Gelatin solution	200.0mL
CMRL-1066 medium with glutamine, 10X	100.0mL
Rabbit serum, partially hemolyzed	70.0mL

pH 7.7 ± 0.2 at 25°C

Gelatin Solution:
Composition per 200.0mL:

Gelatin	14.0g

Preparation of Gelatin Solution: Add gelatin to distilled/deionized water and bring volume to 1.0L. Mix thoroughly. Gently heat and bring to boiling. Autoclave for 15 min at 15 psi pressure–121°C. Cool to 50°C.

CMRL-1066 Medium with Glutamine, 10X:
Composition per liter:

NaCl	6.8g
$NaHCO_3$	2.2g

D-Glucose	1.0g
KCl	0.4g
L-Cysteine·HCl·H_2O	0.26g
$CaCl_2$, anhydrous	0.2g
$MgSO_4·7H_2O$	0.2g
$NaH_2PO_4·H_2O$	0.14g
L-Glutamine	0.1g
Sodium acetate·$3H_2O$	0.083g
L-Glutamic acid	0.075g
L-Arginine·HCl	0.07g
L-Lysine·HCl	0.07g
L-Leucine	0.06g
Glycine	0.05g
Ascorbic acid	0.05g
L-Proline	0.04g
L-Tyrosine	0.04g
L-Aspartic acid	0.03g
L-Threonine	0.03g
L-Alanine	0.025g
L-Phenylalanine	0.025g
L-Serine	0.025g
L-Valine	0.025g
L-Cystine	0.02g
L-Histidine·HCl·H_2O	0.02g
L-Isoleucine	0.02g
Phenol Red	0.02g
L-Methionine	0.015g
Deoxyadenosine	0.01g
Deoxycytidine	0.01g
Deoxyguanosine	0.01g
Glutathione, reduced	0.01g
Thymidine	0.01g
Hydroxy-L-proline	0.01g
L-Tryptophan	0.01g
Nicotinamide adenine dinucleotide	7.0mg
Tween™ 80	5.0mg
Sodium glucuronate·H_2O	4.2mg
Coenzyme A	2.5mg
Cocarboxylase	1.0mg
Flavin adenine dinucleotide	1.0mg
Nicotinamide adenine dinucleotide phosphate	1.0mg
Uridine triphosphate	1.0mg
Choline chloride	0.5mg
Cholesterol	0.2mg
5-Methyldeoxycytidine	0.1mg
Inositol	0.05mg
p-Aminobenzoic acid	0.05mg
Niacin	0.025mg
Niacinamide	0.025mg
Pyridoxine	0.025mg
Pyridoxal·HCl	0.025mg
Biotin	0.01mg
D-Calcium pantothenate	0.01mg
Folic acid	0.01mg
Riboflavin	0.01mg
Thiamine·HCl	0.01mg

pH 7.2 ± 0.2 at 25°C

Source: This solution is available as a premixed powder from BD Diagnostics.

Preparation of CMRL-1066 Medium with Glutamine, 10X: Add components to distilled/deionized water and bring volume to 1.0L. Mix thoroughly. Adjust pH to 7.2. Filter sterilize.

Preparation of Medium: Add components, except gelatin solution, partially hemolyzed rabbit serum, kanamycin, and 5-fluorouracil, to distilled/deionized water and bring volume to 1.0L. Mix thoroughly. Bring pH to 7.6 with 5*N* NaOH. Filter sterilize. Aseptically add 200.0mL of sterile gelatin solution, 70.0mL of partially hemolyzed rabbit serum, 8.0mg of kanamycin, and 230.0mg of 5-fluorouracil. Mix thoroughly. Aseptically distribute into sterile tubes or flasks.

Use: For the isolation of *Borrelia burgdorferi*.

Kempler–McKay Agar

See: **KM Agar**

Kenknight and Munaier's Medium

Composition per liter:

Agar	15.0g
Glucose	1.0g
K_2HPO_4	0.1g
$NaNO_3$	0.1g
KCl	0.1g
$MgSO_4·7H_2O$	0.1g

Source: This medium is available from HiMedia.

Preparation of Medium: Add components to distilled/deionized water and bring volume to 1.0L. Mix thoroughly. Gently heat and bring to boiling. Distribute into tubes or flasks. Autoclave for 15 min at 15 psi pressure–121°C. Pour into sterile Petri dishes or leave in tubes.

Use: For isolating *Actinomyces* spp. from soil.

Kerosene Mineral Salts Medium

Composition per liter:

KH_2PO_4	1.0g
K_2HPO_4	1.0g
NH_4NO_3	1.0g
$MgSO_4·7H_2O$	0.2g
$CaCl_2$	0.02g
$FeCl_3$	0.05g
Kerosene	20.0mL

pH 6.9–7.0 at 25°C

Preparation of Medium: Add components, except kerosene, to distilled/deionized water and bring volume to 1.0L. Mix thoroughly. Adjust pH to 6.9–7.0 with dilute NaOH. Distribute into tubes in 10.0mL volumes or flasks in 100.0mL volumes. Autoclave for 15 min at 15 psi pressure–121°C. Overlay tubes with 0.2mL of kerosene per tube. Overlay flasks with 2.0mL of kerosene per flask.

Use: For the cultivation and maintenance of *Pseudomonas aeruginosa*.

Ketogluconate Broth

Composition per liter:

Potassium gluconate	20.0g
KH_2PO_4	5.4g
KNO_3	2.0g

pH 6.5 ± 0.2 at 25°C

Preparation of Medium: Add components to distilled/deionized water and bring volume to 1.0L. Mix thoroughly. Filter sterilize. Aseptically distribute into sterile tubes.

Use: For use in identifying bacteria that can utilize 2-ketogluconate.

Ketolactonate Broth

Composition per liter:

Agar	20.0g
Lactose	10.0g
Yeast extract	10.0g

Preparation of Medium: Add components to distilled/deionized water and bring volume to 1.0L. Mix thoroughly. Gently heat and bring to boiling. Distribute into tubes or flasks. Autoclave for 15 min at 15 psi pressure–121°C. Pour into sterile Petri dishes or leave in tubes.

Use: For use in the identification of agrobacteria and other bacteria based upon production of 3-ketogluconate. After incubation, Benedict's solution is added to the plates. Yellow zones around colonies indicate positive production of 3-ketogluconate.

KF Streptococcal Agar Base with Triphyenyltetrazolium Chloride

Composition per liter:

Agar	15.0g
Maltose	20.0g
Plant special peptone	10.0g
Sodium glycerophosphate	10.0g
Yeast extract	10.0g
NaCl	5.0g
Lactose	1.0g
NaN_3	0.4g
Bromcresol Purple	0.018g
2,3,5-Triphenyltetrazolium chloride solution	10.0mL

pH 7.2 ± 0.2 at 25°C

Caution: Sodium azide is toxic. Azides also react with metals and disposal must be highly diluted.

2,3,5-Triphenyltetrazolium Chloride Solution:

Composition per 10.0mL:

2,3,5-Triphenyltetrazolium chloride	0.1g

Preparation of 2,3,5-Triphenyltetrazolium Chloride Solution: Add 2,3,5-triphenyltetrazolium chloride to distilled/deionized water and bring volume to 10.0mL. Mix thoroughly. Filter sterilize.

Preparation of Medium: Add components, except 2,3,5-triphenyltetrazolium chloride solution, to distilled/deionized water and bring volume to 990.0mL. Mix thoroughly. Gently heat and bring to boiling. Autoclave for 10 min at 15 psi pressure–121°C. Cool to 45°–50°C. Aseptically add 2,3,5-triphenyltetrazolium chloride solution. Mix thoroughly. Pour into sterile Petri dishes or distribute into sterile tubes.

Use: For the detection and enumeration of fecal streptococci in waters and examination of feces and other materials.

KF Streptococcal HiVeg Agar Base

Composition per liter:

Agar	20.0g
Maltose	20.0g
Plant special peptone	10.0g
Sodium glycerophosphate	10.0g
Yeast extract	10.0g
NaCl	5.0g
Lactose	1.0g
NaN_3	0.4g
2,3,5-Triphenyltetrazolium chloride solution	10.0mL

pH 7.2 ± 0.2 at 25°C

Source: This medium, without 2,3,5-triphenyltetrazolium chloride solution, is available as a premixed powder from HiMedia.

Caution: Sodium azide is toxic. Azides also react with metals and disposal must be highly diluted.

2,3,5-Triphenyltetrazolium Chloride Solution:

Composition per 10.0mL:

2,3,5-Triphenyltetrazolium chloride	0.1g

Preparation of 2,3,5-Triphenyltetrazolium Chloride Solution: Add 2,3,5-triphenyltetrazolium chloride to distilled/deionized water and bring volume to 10.0mL. Mix thoroughly. Filter sterilize.

Preparation of Medium: Add components, except 2,3,5-triphenyltetrazolium chloride solution, to distilled/deionized water and bring volume to 990.0mL. Mix thoroughly. Gently heat and bring to boiling. Autoclave for 10 min at 15 psi pressure–121°C. Cool to 45°–50°C. Aseptically add 2,3,5-triphenyltetrazolium chloride solution. Mix thoroughly. Pour into sterile Petri dishes or distribute into sterile tubes.

Use: For the detection and enumeration of fecal streptococci in waters and examination of feces and other materials.

KF Streptococcal HiVeg Broth Base with BCP and Triphyenyltetrazolium Chloride

Composition per liter:

Maltose	20.0g
Proteose peptone	10.0g
Sodium glycerophosphate	10.0g
Yeast extract	10.0g
NaCl	5.0g
Lactose	1.0g
Na_2CO_3	0.636g
NaN_3	0.4g
Bromcresol Purple	0.018g
2,3,5-Triphenyltetrazolium chloride solution	10.0mL

pH 7.2 ± 0.2 at 25°C

Source: This medium, without 2,3,5-triphenyltetrazolium chloride solution, is available as a premixed powder from HiMedia.

Caution: Sodium azide is toxic. Azides also react with metals and disposal must be highly diluted.

2,3,5-Triphenyltetrazolium Chloride Solution:

Composition per 10.0mL:

2,3,5-Triphenyltetrazolium chloride	0.1g

Preparation of 2,3,5-Triphenyltetrazolium Chloride Solution: Add 2,3,5-triphenyltetrazolium chloride to distilled/deionized water and bring volume to 10.0mL. Mix thoroughly. Filter sterilize.

Preparation of Medium: Add components, except 2,3,5-triphenyltetrazolium chloride solution, to distilled/deionized water and bring volume to 990.0mL. Mix thoroughly. Gently heat and bring to boiling. Autoclave for 10 min at 15 psi pressure–121°C. Cool to 45°–50°C. Aseptically add 2,3,5-triphenyltetrazolium chloride solution. Mix thoroughly. Aseptically distribute into sterile tubes or flasks.

Use: For the selective cultivation of fecal streptococci.

KF Streptococcal HiVeg Broth Base with Triphyenyltetrazolium Chloride

Composition per liter:

Maltose	20.0g
Plant special peptone	10.0g
Sodium glycerophosphate	10.0g
Yeast extract	10.0g
NaCl	5.0g
Lactose	1.0g
Na_2CO_3	0.636g
NaN_3	0.4g
Phenol Red	0.018g
2,3,5-Triphenyltetrazolium chloride solution	10.0mL

pH 7.2 ± 0.2 at 25°C

Source: This medium, without 2,3,5-Triphenyltetrazolium chloride solution, is available as a premixed powder from HiMedia.

Caution: Sodium azide is toxic. Azides also react with metals and disposal must be highly diluted.

2,3,5-Triphenyltetrazolium Chloride Solution:
Composition per 10.0mL:

2,3,5-Triphenyltetrazolium chloride	0.1g

Preparation of 2,3,5-Triphenyltetrazolium Chloride Solution: Add 2,3,5-triphenyltetrazolium chloride to distilled/deionized water and bring volume to 10.0mL. Mix thoroughly. Filter sterilize.

Preparation of Medium: Add components, except 2,3,5-triphenyltetrazolium chloride solution, to distilled/deionized water and bring volume to 990.0mL. Mix thoroughly. Gently heat and bring to boiling. Autoclave for 10 min at 15 psi pressure–121°C. Cool to 45°–50°C. Aseptically add 2,3,5-triphenyltetrazolium chloride solution. Mix thoroughly. Aseptically distribute into sterile tubes or flasks.

Use: For the selective cultivation of fecal streptococci.

KF *Streptococcus* Agar

Composition per liter:

Agar	20.0g
Maltose	20.0g
Proteose peptone	10.0g
Sodium glycerophosphate	10.0g
Yeast extract	10.0g
NaCl	5.5g
Lactose	1.0g
NaN_3	0.4g
Bromcresol Purple	0.015g
2,3,5-Triphenyltetrazolium chloride solution	10.0mL

pH 7.2 ± 0.2 at 25°C

Caution: Sodium azide is toxic. Azides also react with metals and disposal must be highly diluted.

Source: This medium is available as a premixed powder from BD Diagnostic Systems and Oxoid Unipath.

2,3,5-Triphenyltetrazolium Chloride Solution:
Composition per 10.0mL:

2,3,5-Triphenyltetrazolium chloride	0.1g

Preparation of 2,3,5-Triphenyltetrazolium Chloride Solution: Add 2,3,5-triphenyltetrazolium chloride to distilled/deionized water and bring volume to 10.0mL. Mix thoroughly. Filter sterilize.

Preparation of Medium: Add components, except 2,3,5-triphenyltetrazolium chloride solution, to distilled/deionized water and bring volume to 990.0mL. Mix thoroughly. Gently heat and bring to boiling. Autoclave for 15 min at 15 psi pressure–121°C. Cool to 45°–50°C. Aseptically add 2,3,5-triphenyltetrazolium chloride solution. Mix thoroughly. Pour into sterile Petri dishes or distribute into sterile tubes.

Use: For the isolation and enumeration of enterococci.

KF *Streptococcus* Agar

Composition per liter:

Agar	20.0g
Maltose	20.0g
Sodium glycerophosphate	10.0g
Yeast extract	10.0g
NaCl	5.0g
Pancreatic digest of casein	5.0g
Peptic digest of animal tissue	5.0g
Lactose	1.0g
NaN_3	0.4g
2,3,5-Triphenyltetrazolium chloride solution	10.0mL

pH 7.2 ± 0.2 at 25°C

Caution: Sodium azide is toxic. Azides also react with metals and disposal must be highly diluted.

Source: This medium is available as a premixed powder from BD Diagnostic Systems.

2,3,5-Triphenyltetrazolium Chloride Solution:
Composition per 10.0mL:

2,3,5-Triphenyltetrazolium chloride	0.1g

Preparation of 2,3,5-Triphenyltetrazolium Chloride Solution: Add 2,3,5-triphenyltetrazolium chloride to distilled/deionized water and bring volume to 10.0mL. Mix thoroughly. Filter sterilize.

Preparation of Medium: Add components, except 2,3,5-triphenyltetrazolium chloride solution, to distilled/deionized water and bring volume to 990.0mL. Mix thoroughly. Gently heat and bring to boiling. Autoclave for 15 min at 15 psi pressure–121°C. Cool to 45°–50°C. Aseptically add 2,3,5-triphenyltetrazolium chloride solution. Mix thoroughly. Pour into sterile Petri dishes or distribute into sterile tubes.

Use: For the selective cultivation and enumeration of fecal streptococci.

KF *Streptococcus* Broth

Composition per liter:

Maltose	20.0g
Sodium glycerophosphate	10.0g
Yeast extract	10.0g
NaCl	5.0g
Pancreatic digest of casein	5.0g
Peptic digest of animal tissue	5.0g
Lactose	1.0g
Na_2CO_3	0.636g
NaN_3	0.4g
Phenol Red	0.018g
2,3,5-Triphenyltetrazolium chloride solution	10.0mL

pH 7.2 ± 0.2 at 25°C

Caution: Sodium azide is toxic. Azides also react with metals and disposal must be highly diluted.

Source: This medium is available as a premixed powder from BD Diagnostic Systems.

Preparation of Medium: Add components, except 2,3,5-triphenyltetrazolium chloride solution, to distilled/deionized water and bring volume to 990.0mL. Mix thoroughly. Gently heat and bring to boiling. Autoclave for 15 min at 15 psi pressure–121°C. Cool to 45°–50°C. Aseptically add 2,3,5-triphenyltetrazolium chloride solution. Mix thoroughly. Aseptically distribute into sterile tubes or flasks.

Use: For the selective cultivation of fecal streptococci.

KG Agar

See: **Kim-Goepfert Agar**

KG HiVeg Agar Base

Composition per liter:

Agar	18.0g
Plant peptone	1.0g
Yeast extract	0.5g
Phenol Red	0.025g
Egg yolk emulsion, 50%	100.0mL
Polymyxin B solution	1.0mL

pH 6.8 ± 0.2 at 25°C

Source: This medium, without egg yolk emulsion and polymyxin B solution, is available as a premixed powder from HiMedia.

Egg Yolk Emulsion, 50%:

Composition per 100.0mL:

Chicken egg yolks	11
Whole chicken egg	1
NaCl (0.9% solution)	50.0mL

Preparation of Egg Yolk Emulsion, 50%: Soak eggs with 1:100 dilution of saturated mercuric chloride solution for 1 min. Crack eggs and separate yolks from whites. Mix egg yolks with 1 chicken egg. Beat to form emulsion. Measure 50.0mL of egg yolk emulsion and add to 50.0mL of 0.9% NaCl solution. Mix thoroughly. Filter sterilize. Warm to 45°–50°C.

Polymyxin B Solution:

Composition per 5.0mL:

Polymyxin B sulfate	500,000U

Preparation of Polymyxin B Solution: Add polymyxin B sulfate to distilled/deionized water and bring volume to 5.0mL. Mix thoroughly. Filter sterilize.

Preparation of Medium: Add components, except egg yolk emulsion and polymyxin B solution, to distilled/deionized water and bring volume to 900.0mL. Mix thoroughly. Adjust pH to 6.8. Autoclave for 15 min at 15 psi pressure–121°C. Cool to 45°–50°C. Aseptically add 100.0mL of sterile egg yolk emulsion, 50%, and 1.0mL of sterile polymyxin B solution. Mix thoroughly. Pour into sterile Petri dishes. Allow plates to dry in the dark at 30°C for 24 hr before using.

Use: For the cultivation and differentiation of *Bacillus cereus.*

Kidney Bean Agar

Composition per liter:

Kidney beans, dry	30.0g
Agar	15.0g

Preparation of Medium: Add dry kidney beans to distilled/deionized water and bring volume to 1.0L. Autoclave for 30 min at 15 psi pressure–121°C. Filter solids through cheesecloth. Add agar to filtrate. Mix thoroughly. Bring volume to 1.0L with distilled/deionized water. Gently heat and bring to boiling. Distribute into tubes or flasks. Autoclave for 15 min at 15 psi pressure–121°C. Pour into sterile Petri dishes or leave in tubes.

Use: For the cultivation and maintenance of *Conidiobolus stromoideus, Phialophora gregata, Phytophthora cactorum, Phytophthora cryptogea, Phytophthora erythroseptica, Phytophthora fragariae, Phytophthora heveae,* and *Phytophthora syringae.*

Kievskaya Agar (Medium K)

Composition per liter:

Agar	15.0g
KNO_3	1.0g
KH_2PO_4	1.0g
K_2HPO_4	1.0g
NaCl	1.0g
$MgSO_4$	0.2g
$CaCl_2$	0.02g
$FeCl_3$	1.0mg

pH 6.8–7.0 at 25°C

Preparation of Medium: Add components to distilled/deionized water and bring volume to 1.0L. Mix thoroughly. Gently heat and bring to boiling. Distribute into tubes or flasks. Autoclave for 30 min at 3 psi pressure–105°C. Pour into sterile Petri dishes or leave in tubes. After inoculation, incubate in an atmosphere of natural gas.

Use: For the cultivation of *Rhodococcus luteus, Rhodococcus rhodochrous, Rhodococcus ruber, Rhodococcus* species, and other bacteria that can grow on natural gas.

Kievskaya Broth (Medium K)

Composition per liter:

KNO_3	1.0g
KH_2PO_4	1.0g
K_2HPO_4	1.0g
NaCl	1.0g
$MgSO_4$	0.2g
$CaCl_2$	0.02g
$FeCl_3$	1.0mg

pH 6.8–7.0 at 25°C

Preparation of Medium: Add components to distilled/deionized water and bring volume to 1.0L. Mix thoroughly. Distribute into tubes or flasks. Autoclave for 30 min at 15 psi pressure–121°C. After inoculation, sparge medium with natural gas.

Use: For the cultivation of *Rhodococcus luteus, Rhodococcus rhodochrous, Rhodococcus ruber, Rhodococcus* species, and other bacteria that can grow on natural gas.

Kievskaya Broth with *n*-Hexadecane

Composition per liter:

KNO_3	1.0g
KH_2PO_4	1.0g
K_2HPO_4	1.0g
NaCl	1.0g
$MgSO_4$	0.2g
$CaCl_2$	0.02g
$FeCl_3$	1.0mg
n-Hexadecane	10.0mL

pH 6.8–7.0 at 25°C

Preparation of Medium: Add components to distilled/deionized water and bring volume to 1.0L. Mix thoroughly. Distribute into tubes or flasks. Autoclave for 15 min at 15 psi pressure–121°C.

Use: For the cultivation of *Gordona terrae, Rhodococcus erythropolis, Rhodococcus luteus,* and *Rhodococcus maris.*

Kim-Goepfert Agar (KG Agar)

Composition per 1001.0mL:

Solution A	900.0mL
Egg yolk emulsion, 50%	100.0mL
Polymyxin B solution	1.0mL

pH 6.8 ± 0.2 at 25°C

Source: This medium is available as a premixed powder from BD Diagnostic Systems.

Solution A:

Composition per 900.0mL:

Agar	18.0g
Peptone	1.0g
Yeast extract	0.5g
Phenol Red	0.025g

Preparation of Solution A: Add components to distilled/deionized water and bring volume to 900.0mL. Mix thoroughly. Adjust pH to 6.8. Autoclave for 15 min at 15 psi pressure–121°C. Cool to 45°–50°C.

Egg Yolk Emulsion, 50%:

Composition per 100.0mL:

Chicken egg yolks	11
Whole chicken egg	1
NaCl (0.9% solution)	50.0mL

Preparation of Egg Yolk Emulsion, 50%: Soak eggs with 1:100 dilution of saturated mercuric chloride solution for 1 min. Crack eggs and separate yolks from whites. Mix egg yolks with 1 chicken egg. Beat to form emulsion. Measure 50.0mL of egg yolk emulsion and add to 50.0mL of 0.9% NaCl solution. Mix thoroughly. Filter sterilize. Warm to 45°–50°C.

Polymyxin B Solution:

Composition per 5.0mL:

Polymyxin B sulfate	500,000U

Preparation of Polymyxin B Solution: Add polymyxin B to distilled/deionized water and bring volume to 5.0mL. Mix thoroughly. Filter sterilize.

Preparation of Medium: To 900.0mL of cooled, sterile solution A, aseptically add 100.0mL of sterile egg yolk emulsion, 50%, and 1.0mL of sterile polymyxin B solution. Mix thoroughly. Pour into sterile Petri dishes. Allow plates to dry in the dark at 30°C for 24 hr before using.

Use: For the cultivation and differentiation of *Bacillus cereus.*

Kimmig's Agar

Composition per liter:

Agar	15.0g
Glucose	10.0g
Pancreatic digest of gelatin	9.5g
Beef extract	5.5g
NaCl	5.0g
Peptone	5.0g
Glycerol	5.0mL

pH 6.9 ± 0.2 at 35°C

Preparation of Medium: Add glycerol and then other components to distilled/deionized water and bring volume to 1.0L. Mix thoroughly. Gently heat and bring to boiling. Distribute into tubes or flasks. Autoclave for 15 min at 15 psi pressure–121°C. Pour into sterile Petri dishes or leave in tubes.

Use: For the assay of fungistatic agents. For agar dilution testing of antifungal agents. For the cultivation and preservation of various fungi.

Kimmig Fungi HiVeg Agar Base with Kimmig Supplement

Composition per liter:

Glucose	19.0g
Plant peptone	15.0g
Agar	15.0g
NaCl	1.0g
Cycloheximide	0.4g
Kimmig selective supplement	10.0mL
Glycerol	5.0mL

pH 6.9 ± 0.2 at 35°C

Source: This medium, without glycerol or Kimmig selective supplement, is available as a premixed powder from HiMedia.

Caution: Cycloheximide is toxic. Avoid skin contact or aerosol formation and inhalation.

Kimmig Selective Supplement:

Composition per 10.0mL:

Novobiocin	200.0mg
Colistin sulfate	80.0mg

Preparation of Kimmig Selective Supplement: Add components to distilled/deionized water and bring volume to 10.0mL. Mix thoroughly. Filter sterilize.

Preparation of Medium: Add glycerol and then other components, except Kimmig selective supplement, to distilled/deionized water and bring volume to 990.0L. Mix thoroughly. Gently heat and bring to boiling. Autoclave for 15 min at 15 psi pressure–121°C. Cool to 50°C. Aseptically add 10.0mL sterile Kimmig selective supplement. Mix thoroughly Pour into sterile Petri dishes or leave in tubes.

Use: For the assay of fungistatic agents. For agar dilution testing of antifungal agents. For the cultivation and preservation of various fungi.

Kimmig Fungi HiVeg Agar Base with George Kimmig Supplement

Composition per liter:

Glucose	19.0g
Plant peptone	15.0g
Agar	15.0g
NaCl	1.0g
Cycloheximide	0.4g
George Kimmig selective supplement	10.0mL
Glycerol	5.0mL

pH 6.9 ± 0.2 at 35°C

Source: This medium, without glycertol or George Kimmig selective supplement, is available as a premixed powder from HiMedia.

Caution: Cycloheximide is toxic. Avoid skin contact or aerosol formation and inhalation.

George Kimmig Selective Supplement:

Composition per 10.0mL:

Penicillin G	40,000U
Streptomycin	40,000U

Preparation of George Kimmig Selective Supplement: Add components to distilled/deionized water and bring volume to 10.0mL. Mix thoroughly. Filter sterilize.

Preparation of Medium: Add glycerol and then other components, except George Kimmig selective supplement, to distilled/deionized water and bring volume to 990.0L. Mix thoroughly. Gently heat and bring to boiling. Autoclave for 15 min at 15 psi pressure–121°C. Cool to 50°C. Aseptically add 10.0mL sterile George Kimmig selective supplement. Mix thoroughly Pour into sterile Petri dishes or leave in tubes.

Use: For the assay of fungistatic agents. For agar dilution testing of antifungal agents. For the cultivation and preservation of various fungi.

King's Medium A

Composition per liter:

Proteose peptone	20.0g
Agar	15.0g
Glycerol	10.0g
K_2SO_4	10.0g
$MgCl_2·6H_2O$	3.5g

pH 7.2–7.4 ± 0.2 at 25°C

Preparation of Medium: Add components to distilled/deionized water and bring volume to 1.0L. Mix thoroughly. Gently heat and bring to boiling. Distribute into tubes or flasks. Autoclave for 15 min at 15 psi pressure–121°C. Pour into sterile Petri dishes or leave in tubes.

Use: For the nonselective isolation, cultivation, and pigment production of *Pseudomonas.*

King's Medium B

Composition per liter:

Agar	20.0g
Proteose peptone No. 3	20.0g
K_2HPO_4, anhydrous	1.5g
$MgSO_4·7H_2O$	1.5g
Glycerol	15.0mL

pH 7.2 ± 0.2 at 25°C

Preparation of Medium: Add components to distilled/deionized water and bring volume to 1.0L. Mix thoroughly. Gently heat and bring to boiling. Distribute into tubes or flasks. Autoclave for 15 min at 15 psi pressure–121°C. Pour into sterile Petri dishes or leave in tubes.

Use: For the nonselective isolation, cultivation, and pigment production of *Pseudomonas* species.

King's Medium B

Composition per liter:

Proteose peptone No. 3	20.0g
Agar	15.0g
K_2HPO_4	1.5g
$MgSO_4·7H_2O$	1.5g
Glycerol	10.0mL

pH 7.2 ± 0.2 at 25°C

Preparation of Medium: Add components to distilled/deionized water and bring volume to 1.0L. Mix thoroughly. Gently heat and bring to boiling. Distribute into tubes or flasks. Autoclave for 15 min at 15 psi pressure–121°C. Pour into sterile Petri dishes or leave in tubes.

Use: For the cultivation and maintenance of *Pseudomonas glumae.*

King's OF Basal Medium (Oxidation-Fermentation Medium, King's) (King's OF Medium) (BAM M70)

Composition per liter:

Agar	3.0g
Pancreatic digest of casein	2.0g
Carbohydrate solution	100.0mL
Phenol Red (1.5% solution)	2.0mL

pH to 7.3 ± 0.2

Carbohydrate Solution:

Composition per 100.0mL:

Carbohydrate	10.0g

Preparation of Carbohydrate Solution: Add carbohydrate to distilled/deionized water and bring volume to 100.0mL. Mix thoroughly. Filter sterilize.

Preparation of Medium: Add components, except carbohydrate solution, to distilled/deionized water and bring volume to 900.0mL. Mix thoroughly. Gently heat and bring to boiling. Autoclave for 15 min at 15 psi pressure–121°C. Cool to 45°–50°C. Aseptically add 100.0mL of sterile carbohydrate solution. Mix thoroughly. Pour into sterile Petri dishes or distribute into sterile tubes.

Use: For differentiating bacteria based upon determining the oxidative and fermentative metabolism of carbohydrates. Bacteria that ferment the carbohydrate turn the medium yellow.

King's OF HiVeg Medium Base with Carbohydrate

Composition per liter:

Agar	0.3g
Plant hydrolysate	0.2g
Phenol Red	3.0mg
Carbohydrate solution	100.0mL

pH 6.9 ± 0.2 at 35°C

Source: This medium, without carbohydrate solution, is available as a premixed powder from HiMedia.

Carbohydrate Solution:

Composition per 100.0mL:

Carbohydrate	10.0g

Preparation of Carbohydrate Solution: Add carbohydrate to distilled/deionized water and bring volume to 100.0mL. Adonitol, arabinose, cellobiose, glucose, dulcitol, fructose, galactose, inositol, lactose, maltose, mannitol, raffinose, rhamnose, salicin, sorbitol, sucrose, trehalose, xylose, or other carbohydrates may be used. Mix thoroughly. Filter sterilize.

Preparation of Medium: Add components, except carbohydrate solution, to distilled/deionized water and bring volume to 900.0mL. Mix thoroughly. Gently heat and bring to boiling. Distribute into tubes or flasks. Autoclave for 15 min at 15 psi pressure–121°C. Cool to 50°C. Aseptically add 100.0mL of sterile carbohydrate solution. Mix thoroughly. Aseptically distribute into tubes.

Use: For studyling oxidation fermentation of carbohydrates by *Campylobacter* species.

King's OF Medium

See: **Oxidation-Fermentation Medium, King's**

Kirchner Enrichment Medium

Composition per liter:

$Na_2HPO_4 \cdot 12H_2O$	19.0g
Asparagine	5.0g
KH_2PO_4	2.5g
Sodium citrate	2.5g
$MgSO_4$	0.6g
Serum	100.0mL
Glycerol	20.0mL
Penicillin solution	10.0mL
Phenol Red (0.4% solution)	3.0mL

pH 7.4–7.6 at 25°C

Penicillin Solution:

Composition per 10.0mL:

Penicillin	100,000U

Preparation of Penicillin Solution: Add penicillin to distilled/deionized water and bring volume to 10.0mL. Mix thoroughly. Filter sterilize.

Preparation of Medium: Add components, except serum and penicillin solution, to distilled/deionized water and bring volume to 890.0mL. Mix thoroughly. Gently heat and bring to boiling. Autoclave for 15 min at 15 psi pressure–121°C. Cool to 45°–50°C. Aseptically add sterile serum and penicillin solution. Mix thoroughly. Aseptically distribute into sterile tubes or flasks.

Use: For the cultivation and enrichment of *Mycobacterium* species.

Kirchner Medium Base, Modified

Composition per liter:

L-Asparagine	5.0g
KH_2PO_4	4.0g
Na_2HPO_4	3.0g
Sodium citrate	2.5g
$MgSO_4 \cdot 7H_2O$	0.6g
Horse serum, sterile inactivated	100.0mL
Selective supplement solution	10.0mL

pH 7.4 ± 0.2 at 25°C

Source: This medium is available from HiMedia.

Selective Supplement Solution:

Composition per 10.0mL:

Peniclllin	100,000U
Amphotericin B	5.0mg

Preparation of Selective Supplement Solution: Add components to distilled/deionized water and bring volume to 10.0mL. Mix thoroughly. Filter sterilize.

Preparation of Medium: Add components, except horse serum and selective supplement solution, to distilled/deionized water and bring volume to 890.0mL. Mix thoroughly. Autoclave for 15 min at 15 psi pressure–121°C. Cool to 50°C. Aseptically add horse serum and selective supplement solution. Mix thoroughly. Pour into Petri dishes or aseptically distribute into sterile tubes.

Use: For the cultivation of *Mycobacterium tuberculosis.*

K-L Virulence Agar (Klebs-Loeffler Virulence Agar) (Elek Agar) (*Corynebacterium diphtheriae* Virulence Test Medium)

Composition per 1300.0mL:

K-L agar base	1.0L
Rabbit serum	200.0mL
K_2TeO_3 solution	100.0mL
K-L filter strips	100

pH 7.8 ± 0.2 at 25°C

Source: This medium is available as a premixed powder from BD Diagnostic Systems.

K-L Agar Base:

Composition per liter:

Meat peptone	20.0g
Agar	15.0g
NaCl	2.5g

Preparation of K-L Agar Base: Add components to distilled/deionized water and bring volume to 1.0L. Mix thoroughly. Gently heat and bring to boiling. Autoclave for 15 min at 15 psi pressure–121°C. Cool to 50°C.

K_2TeO_3 Solution:

Composition per 100.0mL:

K_2TeO_3	0.3g

Preparation of K_2TeO_3 Solution: Add K_2TeO_3 to distilled/deionized water and bring volume to 100.0mL. Mix thoroughly. Filter sterilize.

Caution: Potassium tellurite is toxic.

K-L Filter Strips:

Composition:

Whatman No. 3 filter paper	as needed
Diphtheria toxin solution	10.0mL

Preparation of K-L Strips: Cut Whatman No. 3 filter paper into 1.5cm × 7cm strips. Autoclave for 15 min at 15 psi pressure–121°C. Aseptically dip each strip into a sterile solution containing 1000U of purified diphtheria toxin/mL. Drain off excess liquid.

Preparation of Medium: Filter sterilize rabbit serum. To 1.0L of cooled, sterile K-L agar base, aseptically add sterile rabbit serum and sterile K_2TeO_3 solution. Mix thoroughly. Pour into sterile Petri dishes in 13.0mL volumes. Before the agar solidifies, aseptically add one K-L filter strip across the diameter of the plate. Allow the filter strip to sink to the bottom of the plate or press it down with sterile forceps. Allow the agar to solidify. Dry the surface of the plates by incubating at 35°C with lid of plate ajar for 2 hr.

Use: For *in vitro* toxigenicity testing of *Corynebacterium diphtheriae* by the agar diffusion technique. *Corynebacterium diphtheriae* that produce toxin form white precipitin lines at approximately 45° angles from the culture streak line.

K-L Virulence Agar (Klebs-Loeffler Virulence Agar)

Composition per 1250.0mL:

K-L agar base	1.0L
K-L enrichment	200.0mL
K_2TeO_3 solution	50.0mL
K-L filter strips	100

pH 7.8 ± 0.2 at 25°C

Source: This medium is available as a premixed powder from BD Diagnostic Systems.

K-L Agar Base:

Composition per liter:

Meat peptone	20.0g
Agar	15.0g
NaCl	2.5g

Preparation of K-L Agar Base: Add components to distilled/deionized water and bring volume to 1.0L. Mix thoroughly. Gently heat and bring to boiling. Autoclave for 15 min at 15 psi pressure–121°C. Cool to 50°C.

K-L Enrichment:
Composition per 200.0mL:

Casamino acids	4.0g
Glycerol	100.0mL
Tween™ 80	100.0mL

Preparation of K-L Enrichment: Combine components. Mix thoroughly. Filter sterilize.

K_2TeO_3 Solution:
Composition per 100.0mL:

K_2TeO_3	1.0g

Preparation of K_2TeO_3 Solution: Add K_2TeO_3 to distilled/deionized water and bring volume to 100.0mL. Mix thoroughly. Filter sterilize.

Caution: Potassium tellurite is toxic.

K-L Filter Strips:
Composition:

Diphtheria toxin solution	10.0mL
Whatman No. 3 filter paper	as needed

Preparation of K-L Strips: Cut Whatman #3 filter paper into 1.5cm × 7cm strips. Autoclave for 15 min at 15 psi pressure–121°C. Aseptically dip each strip into a sterile solution containing 1000U of purified diphtheria toxin/mL. Drain off excess liquid.

Preparation of Medium: To 1.0L of cooled, sterile K-L agar base, aseptically add sterile K-L enrichment and sterile K_2TeO_3 solution. Mix thoroughly. Pour into sterile Petri dishes in 13.0mL volumes. Before the agar solidifies, aseptically add one K-L filter strip across the diameter of the plate. Allow the filter strip to sink to the bottom of the plate or press it down with sterile forceps. Allow the agar to solidify. Dry the surface of the plates by incubating at 35°C with lid of plate ajar for 2 hr.

Use: For *in vitro* toxigenicity testing of *Corynebacterium diphtheriae* by the agar diffusion technique. *Corynebacterium diphtheriae* that produce toxin form white precipitin lines at approximately 45° angles from the culture streak line.

Kleb Agar
(m-Kleb Agar)

Composition per liter:

Agar	15.0g
Proteose peptone No. 3	10.0g
NaCl	5.0g
Adonitol	5.0g
Beef extract	1.0g
Aniline Blue	0.1g
Sodium lauryl sulfate	0.1g
Phenol Red	0.025g
Ethanol (95% solution)	20.0mL
Carbenicillin solution	10.0mL

pH 7.4 ± 0.2 at 25°C

Carbenicillin Solution:
Composition per 10.0mL:

Carbenicillin	0.05g

Preparation of Carbenicillin Solution: Add carbenicillin to distilled/deionized water and bring volume to 10.0mL. Mix thoroughly. Filter sterilize.

Preparation of Medium: Add components, except ethanol and carbenicillin solution, to distilled/deionized water and bring volume to 970.0mL. Mix thoroughly. Gently heat and bring to boiling. Autoclave for 15 min at 15 psi pressure–121°C. Cool to 45°–50°C. Aseptically add 20.0mL of ethanol and 10.0mL of carbenicillin solution. Mix thoroughly. Pour into sterile Petri dishes or distribute into sterile tubes.

Use: For the enumeration of bacteria from waters.

Klebs-Loeffler Virulence Agar
See: **K-L Virulence Agar**

Klebsiella Medium
(m-*Klebsiella* Medium)

Composition per 1041.0mL:

Agar	15.0g
Adonitol	4.0g
2X Salt solution	500.0mL
Uric acid solution	200.0mL
Phenol Red solution	10.0mL
Sodium taurocholate solution	30.0mL
Carbenicillin solution	1.0mL

2X Salt Solution:
Composition per liter:

KCl	8.0g
K_2HPO_4	3.0g
NaCl	2.0g
KH_2PO_4	1.0g
$MgSO_4 \cdot 7H_2O$	0.2g

Preparation of 2X Salt Solution: Add components to distilled/deionized water and bring volume to 1.0L. Mix thoroughly.

Uric Acid Solution:
Composition per 200.0mL:

Uric acid	0.3g

Preparation of Uric Acid Solution: Dissolve uric acid in a small volume of 1*N* NaOH. Bring volume to 200.0mL with distilled/deionized water. Adjust pH to 7.1 with 1*N* HCl. Filter sterilize.

Phenol Red Solution:
Composition per 10.0mL:

Phenol Red	0.1g

Preparation of Phenol Red Solution: Add Phenol Red to sterile distilled/deionized water and bring volume to 10.0mL. Mix thoroughly.

Sodium Taurocholate Solution:
Composition per 30.0mL:

Sodium taurocholate	0.4g

Preparation of Sodium Taurocholate Solution: Add sodium taurocholate to sterile distilled/deionized water and bring volume to 30.0mL. Mix thoroughly.

Carbenicillin Solution:
Composition per 1.0mL:

Carbenicillin	5.0mg

Preparation of Carbenicillin Solution: Add carbenicillin to distilled/deionized water and bring volume to 1.0mL. Mix thoroughly. Filter sterilize.

Preparation of Medium: Add adonitol and agar to 500.0mL of 2X salt solution. Bring volume to 800.0mL with distilled/deionized water. Mix thoroughly. Gently heat and bring to boiling. Autoclave for 15 min at 15 psi pressure–121°C. Cool to 45°–50°C. Aseptically add 200.0mL of uric acid solution, 30.0mL of sodium taurocholate solution, 10.0mL of Phenol Red solution, and 1.0mL of carbenicillin solution. Mix thoroughly. Pour into sterile Petri dishes or distribute into sterile tubes.

Use: For the enumeration of *Klebsiella* species by the membrane filter method.

Klebsiella Selective Agar

Composition per liter:

Agar	26.0g
DL–Phenylalanine	10.0g
L-Ornithine·HCl	10.0g
Raffinose	7.0g
Pancreatic digest of casein	2.5g
Yeast extract	2.5g
K_2HPO_4	2.0g
Phenol Red solution	10.0mL
Carbenicillin solution	10.0mL

pH 5.6 ± 0.2 at 25°C

Phenol Red Solution:

Composition per 10.0mL:

Phenol Red	0.5g

Preparation of Phenol Red Solution: Add Phenol Red to 50% ethanol and bring volume to 10.0mL. Mix thoroughly.

Carbenicillin Solution:

Composition per 1.0mL:

Carbenicillin	5.0mg

Preparation of Carbenicillin Solution: Add carbenicillin to distilled/deionized water and bring volume to 1.0mL. Mix thoroughly. Filter sterilize.

Preparation of Medium: Add components, except carbenicillin solution, to distilled/deionized water and bring volume to 990.0mL. Mix thoroughly. Gently heat and bring to boiling. Autoclave for 15 min at 15 psi pressure–121°C. Cool to 45°–50°C. Aseptically add 10.0mL carbenicillin solution. Mix thoroughly. Adjust pH to 5.6–5.7 with sterile 1*N* HCl. Pour into sterile Petri dishes or distribute into sterile tubes.

Use: For the isolation and identification of *Klebsiella pneumoniae* from clinical specimens.

Kligler Iron Agar

Composition per liter:

Peptone	20.0g
Agar	12.0g
Lactose	10.0g
NaCl	5.0g
Beef extract	3.0g
Yeast extract	3.0g
Glucose	1.0g
Ferric citrate	0.3g
$Na_2S_2O_3$	0.3g
Phenol Red	0.05g

pH 7.4 ± 0.2 at 25°C

Source: This medium is available as a premixed powder from BD Diagnostic Systems and Oxoid Unipath.

Preparation of Medium: Add components to distilled/deionized water and bring volume to 1.0L. Mix thoroughly. Gently heat and bring to boiling. Distribute into tubes. Autoclave for 15 min at 15 psi pressure–121°C. Pour into sterile Petri dishes or leave in tubes.

Use: For the differentiation and identification of Enterobacteriaceae based upon sugar fermentation and hydrogen sulfide production. Sugar fermentation is indicated by the medium turning yellow. H_2S production results in the medium turning black.

Kligler Iron Agar

Composition per liter:

Agar	15.0g
Lactose	10.0g
Pancreatic digest of casein	10.0g
Peptic digest of animal tissue	10.0g
NaCl	5.0g
Glucose	1.0g
Ferric ammonium citrate	0.5g
$Na_2S_2O_3$	0.5g
Phenol Red	0.025g

pH 7.4 ± 0.2 at 25°C

Source: This medium is available as a premixed powder from BD Diagnostic Systems.

Preparation of Medium: Add components to distilled/deionized water and bring volume to 1.0L. Mix thoroughly. Gently heat and bring to boiling. Distribute into tubes or flasks. Autoclave for 15 min at 15 psi pressure–121°C. Pour into sterile Petri dishes or leave in tubes.

Use: For the differentiation and identification of Enterobacteriaceae based upon sugar fermentation and hydrogen sulfide production. Sugar fermentation is indicated by the medium turning yellow. H_2S production results in the medium turning black.

Kligler Iron Agar (FDA M71)

Composition per liter:

Lactose	20.0g
Agar	15.0g
Pancreatic digest of casein	10.0g
Peptic digest of animal tissue	10.0g
NaCl	5.0g
Glucose	1.0g
Ferric ammonium citrate	0.5g
$Na_2S_2O_3$	0.5g
Phenol Red	0.025g

pH 7.4 ± 0.2 at 25°C

Preparation of Medium: Add components to distilled/deionized water and bring volume to 1.0L. Mix thoroughly. Gently heat and bring to boiling. Distribute into tubes or flasks. Autoclave for 15 min at 15 psi pressure–121°C. Pour into sterile Petri dishes or leave in tubes.

Use: For the differentiation and identification of Enterobacteriaceae based upon sugar fermentation and hydrogen sulfide production. Sugar fermentation is indicated by the medium turning yellow. H_2S production results in the medium turning black.

Kligler Iron Agar (BAM M71)

Composition per liter:

Lactose	20.0g
Polypeptone	20.0g

Agar	15.0g
Peptic digest of animal tissue	10.0g
NaCl	5.0g
Glucose	1.0g
Ferric ammonium citrate	0.5g
$Na_2S_2O_3$	0.5g
Phenol Red	0.025g

pH 7.4 ± 0.2 at 25°C

Preparation of Medium: Add components to distilled/deionized water and bring volume to 1.0L. Mix thoroughly. Gently heat and bring to boiling. Distribute into tubes or flasks. Autoclave for 15 min at 15 psi pressure–121°C. Pour into sterile Petri dishes or leave in tubes.

Use: For the differentiation and identification of Enterobacteriaceae based upon sugar fermentation and hydrogen sulfide production. Sugar fermentation is indicated by the medium turning yellow. H_2S production results in the medium turning black.

Kligler Iron Agar with Sodium Chloride (BAM M71)

Composition per liter:

NaCl	25.0g
Lactose	20.0g
Polypeptone	20.0g
Agar	15.0g
Peptic digest of animal tissue	10.0g
NaCl	5.0g
Glucose	1.0g
Ferric ammonium citrate	0.5g
$Na_2S_2O_3$	0.5g
Phenol Red	0.025g

pH 7.4 ± 0.2 at 25°C

Preparation of Medium: Add components to distilled/deionized water and bring volume to 1.0L. Mix thoroughly. Gently heat and bring to boiling. Distribute into tubes or flasks. Autoclave for 15 min at 15 psi pressure–121°C. Pour into sterile Petri dishes or leave in tubes.

Use: For the differentiation and identification of *Vibrio* spp. based upon sugar fermentation and hydrogen sulfide production. Sugar fermentation is indicated by the medium turning yellow. H_2S production results in the medium turning black.

Kligler Iron HiVeg Agar

Composition per liter:

Plant special peptone	15.0g
Lactose	10.0g
Agar	15.0g
Plant peptone No. 3	5.0g
NaCl	5.0g
Plant extract	3.0g
Yeast extract	3.0g
Glucose	1.0g
$Na_2S_2O_3$	0.3g
$FeSO_4$	0.2g
Phenol Red	0.024g

pH 7.4 ± 0.2 at 25°C

Source: This medium is available as a premixed powder from HiMedia.

Preparation of Medium: Add components to distilled/deionized water and bring volume to 1.0L. Mix thoroughly. Gently heat and bring to boiling. Distribute into tubes. Autoclave for 15 min at 15 psi pressure–121°C. Pour into sterile Petri dishes or leave in tubes.

Use: For the differentiation and identification of Enterobacteriaceae based upon sugar fermentation and hydrogen sulfide production. Sugar fermentation is indicated by the medium turning yellow. H_2S production results in the medium turning black.

Kligler Iron HiVeg Agar, Modified

Composition per liter:

Plant hydrolysate	20.0g
Agar	15.0g
Lactose	10.0g
NaCl	5.0g
Plant extract	3.0g
Yeast extract	3.0g
Glucose, anhydrous	1.0g
$Na_2S_2O_3 \cdot 5H_2O$	0.3g
$FeSO_4$	0.2g
Phenol Red	0.025g

pH 7.4 ± 0.2 at 25°C

Source: This medium is available as a premixed powder from HiMedia.

Preparation of Medium: Add components to distilled/deionized water and bring volume to 1.0L. Mix thoroughly. Gently heat and bring to boiling. Distribute into tubes. Autoclave for 15 min at 15 psi pressure–121°C. Pour into sterile Petri dishes or leave in tubes.

Use: For the differentiation and identification of Enterobacteriaceae based upon sugar fermentation and hydrogen sulfide production. Recommended for identification of *Yersinia enterocolitica.*

KM Agar (Kempler-McKay Agar)

Composition per liter:

Agar	15.0g
Milk, nonfat	10.0g
Glucose	5.0g
Milk protein hydrolysate	2.5g
Solution 1	10.0mL
Solution 2	10.0mL

pH 6.6 ± 0.2 at 25°C

Solution 1:

Composition per 100.0mL:

$K_3Fe(CN)_6$	10.0g

Caution: Cyanide is toxic.

Preparation of Solution 1: Add $K_3Fe(CN)_6$ to distilled/deionized water and bring volume to 100.0mL. Mix thoroughly. Filter sterilize.

Solution 2:

Composition per 40.0mL:

Ferric citrate	1.0g
Sodium citrate	1.0g

Preparation of Solution 2: Add components to distilled/deionized water and bring volume to 40.0mL. Mix thoroughly. Filter sterilize.

Preparation of Medium: Add components, except solution 1 and solution 2, to distilled/deionized water and bring volume to 980.0mL. Mix thoroughly. Gently heat and bring to boiling. Adjust pH to 6.6. Autoclave for 12 min at 10 psi pressure–115°C. Cool to 45°–50°C. Aseptically add sterile solution 1 and solution 2. Mix thoroughly. Pour into sterile Petri dishes. Allow plates to dry in the dark at 30°C for 24 hr before using.

Use: For the isolation and cultivation of acidogenic microorganisms from foods. For the differentiation of citrate-fermenting lactic bacteria, such as *Lactobacillus lactis* subspecies *diacetylactis*, from the noncitrate-fermenting *Lactobacillus lactis* subspecies *cremoris*.

Knisely Medium for *Bacillus anthracis*

Composition per liter:

Beef heart, solids from infusion	500.0g
Agar	15.0g
Pancreatic digest of casein	10.0g
NaCl	5.0g
EDTA	200.0mg
Lysozyme	40.0mg
Thallous acetate	40.0mg
Polymyxin	30,000U

Preparation of Medium: Add components, except EDTA, lysozyme, thallous acetate, and polymyxin, to distilled/deionized water and bring volume to 1.0mL. Mix thoroughly. Gently heat and bring to boiling. Adjust pH to 7.3. Autoclave for 15 min at 15 psi pressure–121°C. Cool to 45°–50°C. Aseptically add sterile EDTA, lysozyme, thallous acetate, and polymyxin. Mix thoroughly. Pour into sterile Petri dishes or distribute into sterile tubes.

Use: For the cultivation and maintenance of *Bacillus anthracis*.

Koch's K1 Medium

Composition per liter:

Glucose	1.8g
Peptone	0.6g
Yeast extract	0.4g

Preparation of Medium: Add components to distilled/deionized water and bring volume to 1.0L. Mix thoroughly. Distribute into tubes or flasks. Autoclave for 15 min at 15 psi pressure–121°C.

Use: For the cultivation of a variety of fungi.

Kohn Two Tube Medium No. 1 Base with Urea

Composition per liter:

Agar	16.0g
Peptic digest of animal tissue	15.0g
Mannitol	10.0g
Beef extract	2.0g
Yeast extract	2.0g
Glucose	1.0g
Phenol Red	0.05g
Urea solution	25.0mL

pH 7.3 ± 0.2 at 25°C

Source: This medium, without urea solution, is available as a premixed powder from HiMedia.

Urea Solution:

Composition per 100.0mL:

Urea	5.0g

Preparation of Urea Solution: Add urea to distilled/deionized water and bring volume to 100.0mL. Mix thoroughly. Filter sterilize.

Preparation of Medium: Add components to distilled/deionized water and bring volume to 1.0L. Mix thoroughly. Gently heat and bring to boiling. Distribute into tubes. Autoclave for 15 min at 10 psi pressure–115°C. Cool to 60°C. Aseptically add 25mL of sterile urea solution. Mix thoroughly. Allow to solidify as a slant in tubes with a genrous butt.

Use: For the identification of Enterobacteriaceae on the basis of glucose and mannitol fermentation and urease production.

Kohn Two Tube Medium No. 2

Composition per liter:

Casein enzymic hydrolysate	10.0g
Peptic digest of animal tissue	10.0g
Salicin	10.0g
Sucrose	10.0g
NaCl	5.0g
Agar	3.0g
Na_2HPO_4	0.09g
Bromthymol Blue	0.02g
$Na_2S_2O_3$	0.016g

pH 7.3 ± 0.2 at 25°C

Source: This medium is available as a premixed powder from HiMedia.

Preparation of Medium: Add components to distilled/deionized water and bring volume to 1.0L. Mix thoroughly. Gently heat and bring to boiling. Distribute into tubes. Autoclave for 15 min at 10 psi pressure–115°C. Allow to solidify in tubes in an upright position.

Use: For the identification of members of Enterobacteriaceae on the basis of sucrose and salicin fermentation, motility, H_2S production, and indole production.

Kohn Two Tube HiVeg Medium No. 1 Base with Urea

Composition per liter:

Agar	16.0g
Plant peptone	15.0g
Mannitol	10.0g
Plant extract	2.0g
Yeast extract	2.0g
Glucose	1.0g
Phenol Red	0.05g
Urea solution	25.0mL

pH 7.3 ± 0.2 at 25°C

Source: This medium, without urea solution, is available as a premixed powder from HiMedia.

Urea Solution:

Composition per 100.0mL:

Urea	5.0g

Preparation of Urea Solution: Add urea to distilled/deionized water and bring volume to 100.0mL. Mix thoroughly. Filter sterilize.

Preparation of Medium: Add components to distilled/deionized water and bring volume to 1.0L. Mix thoroughly. Gently heat and bring to boiling. Distribute into tubes. Autoclave for 15 min at 10 psi pressure–115°C. Cool to 60°C. Aseptically add 25.0mL of sterile urea solution. Mix htoroughly. Allow to solidify as a slant in tubes with a genrous butt.

Use: For the identification of Enterobacteriaceae on the basis of glucose and mannitol fermentation and urease production.

Kohn Two Tube HiVeg Medium No. 2

Composition per liter:

Plant hydrolysate	10.0g
Plant peptone	10.0g
Salicin	10.0g
Sucrose	10.0g
NaCl	5.0g
Agar	3.0g
Na_2HPO_4	0.09g
Bromthymol Blue	0.02g
$Na_2S_2O_3$	0.016g

pH 7.3 ± 0.2 at 25°C

Source: This medium is available as a premixed powder from HiMedia.

Preparation of Medium: Add components to distilled/deionized water and bring volume to 1.0L. Mix thoroughly. Gently heat and bring to boiling. Distribute into tubes. Autoclave for 15 min at 10 psi pressure–115°C. Allow to solidify in tubes in an upright position.

Use: For the identification of members of Enterobacteriaceae on the basis of sucrose and salicin fermentation, motility, H_2S production, and indole production.

KoKo Medium

Composition per 1020.0mL:

K_2HPO_4	1.6g
$NaH_2PO_4·2H_2O$	1.0g
Peptone, meat	1.0g
Pancreatic digest of casein	1.0g
Yeast extract	1.0g
NH_4Cl	0.5g
$MgSO_4·6H_2O$	0.16g
Resazurin	0.5mg
Glucose solution	100.0mL
$NaHCO_3$ solution	10.0mL
$CaCl_2·2H_2O$ solution	10.0mL
L-Cysteine·HCl·H_2O solution	10.0mL
$Na_2S·9H_2O$ solution	10.0mL
Trace elements solution SL-4	10.0mL
Wolfe's vitamin solution	10.0mL

pH 7.0 ± 0.2 at 25°C

Glucose Solution:

Composition per 100.0mL:

D-Glucose	5.0g

Preparation of Glucose Solution: Add glucose to distilled/deionized water and bring volume to 100.0mL. Mix thoroughly. Sparge with 100% N_2. Autoclave for 15 min at 15 psi pressure–121°C.

$NaHCO_3$ Solution:

Composition per 10.0mL:

$NaHCO_3$	1.0g

Preparation of $NaHCO_3$ Solution: Add $NaHCO_3$ to distilled/deionized water and bring volume to 10.0mL. Mix thoroughly. Sparge with 100% N_2. Autoclave for 15 min at 15 psi pressure–121°C.

$CaCl_2$ Solution:

Composition per 10.0mL:

$CaCl_2·2H_2O$	0.06g

Preparation of $CaCl_2$ Solution: Add $CaCl_2·2H_2O$ to distilled/deionized water and bring volume to 10.0mL. Mix thoroughly. Sparge with 100% N_2. Autoclave for 15 min at 15 psi pressure–121°C.

L-Cysteine·HCl·H_2O Solution:

Composition per 10.0mL:

L-Cysteine·HCl·H_2O	0.3g

Preparation of L-Cysteine·HCl·H_2O Solution: Add L-cysteine·HCl·H_2O to distilled/deionized water and bring volume to 10.0mL. Mix thoroughly. Sparge with 100% N_2. Autoclave for 15 min at 15 psi pressure–121°C.

$Na_2S·9H_2O$ Solution:

Composition per 10.0mL:

$Na_2S·9H_2O$	0.3g

Preparation of $Na_2S·9H_2O$ Solution: Add $Na_2S·9H_2O$ to distilled/deionized water and bring volume to 10.0mL. Mix thoroughly. Sparge with 100% N_2. Autoclave for 15 min at 15 psi pressure–121°C. Before use, neutralize to pH 7.0 with sterile HCl.

Trace Elements Solution SL-4:

Composition per liter:

EDTA	0.5g
$FeSO_4·7H_2O$	0.2g
Trace elements solution SL-6	100.0mL

Trace Elements Solution SL-6:

Composition per liter:

$MnCl_2·4H_2O$	0.5g
H_3BO_3	0.3g
$CoCl_2·6H_2O$	0.2g
$ZnSO_4·7H_2O$	0.1g
$Na_2MoO_4·2H_2O$	0.03g
$NiCl_2·6H_2O$	0.02g
$CuCl_2·2H_2O$	0.01g

Preparation of Trace Elements Solution SL-6: Add components to distilled/deionized water and bring volume to 1.0L. Mix thoroughly.

Preparation of Trace Elements Solution SL-4: Add components to distilled/deionized water and bring volume to 1.0L. Mix thoroughly. Filter sterilize.

Wolfe's Vitamin Solution:

Composition per liter:

Pyridoxine·HCl	10.0mg
p-Aminobenzoic acid	5.0mg
Lipoic acid	5.0mg
Nicotinic acid	5.0mg
Riboflavin	5.0mg
Thiamine·HCl	5.0mg
Calcium DL-pantothenate	5.0mg
Biotin	2.0mg
Folic acid	2.0mg
Vitamin B_{12}	0.1mg

Preparation of Wolfe's Vitamin Solution: Add components to distilled/deionized water and bring volume to 1.0L. Mix thoroughly. Sparge with 100% N_2. Filter sterilize.

Preparation of Medium: Prepare and dispense medium under 100% N_2. Add components, except glucose solution, $NaHCO_3$ solution, $CaCl_2·2H_2O$ solution, L-cysteine·HCl·H_2O solution, $Na_2S·9H_2O$ solution, and Wolfe's vitamin solution, to distilled/deionized water and bring volume to 850.0mL. Mix thoroughly. Adjust pH to 7.0. Sparge with 100% N_2. Autoclave for 15 min at 15 psi pressure–121°C. Aseptically and anaerobically add 100.0mL of sterile glucose solution, 10.0mL of sterile $NaHCO_3$ solution, 10.0mL of sterile $CaCl_2·2H_2O$ solution, 10.0mL of sterile L-cysteine·HCl·H_2O solution, 10.0mL of sterile $Na_2S·9H_2O$ solution, and 10.0mL of sterile Wolfe's vitamin solution. Mix thoroughly. The pH should be 7.0. A buffer solution of 1% MOPS from a 10% anaerobic solution at pH 6.9–7.0 may be added aseptically and anaerobically to enhance the buffer capacity of the medium. Aseptically and anaerobically distribute into sterile tubes or bottles.

Use: For the cultivation of *Thermoanaerobacter italicus.*

Korthof Medium

Composition per 1088.0mL:

NaCl	1.4g
$Na_2HPO_4 \cdot 2H_2O$	0.88g
Peptone	0.8g
KH_2PO_4	0.24g
$CaCl_2$	0.04g
KCl	0.04g
$NaHCO_3$	0.02g
Rabbit serum, inactivated	80.0mL
Rabbit hemoglobin solution	8.0mL

pH 7.2 ± 0.2 at 25°C

Rabbit Hemoglobin Solution:

Composition per 20.0mL:

Rabbit blood clot	10.0mL

Preparation of Rabbit Hemoglobin Solution: Add rabbit blood clot to 10.0mL of distilled/deionized water. Lyse the clot by freezing and thawing.

Preparation of Medium: Add components, except rabbit serum and rabbit hemoglobin solution, to distilled/deionized water and bring volume to 1.0L. Mix thoroughly. Gently heat and bring to boiling. Cool to 25°C. Filter through Whatman #1 filter paper. Distribute into flasks in 100.0mL volumes. Autoclave for 15 min at 15 psi pressure–121°C. Cool to 45°–50°C. Aseptically add 8.0mL of rabbit serum and 0.8mL of rabbit hemoglobin solution to each flask. Mix thoroughly.

Use: For the cultivation of *Leptospira* species.

Korthof Medium, Modified

Composition per liter:

NaCl	1.4g
$Na_2HPO_4 \cdot 2H_2O$	0.88g
Peptone	0.8g
KH_2PO_4	0.24g
$CaCl_2$	0.04g
KCl	0.04g
$NaHCO_3$	0.02g
Rabbit serum, heat inactivated at 56°C	100.0mL

pH 7.2–7.6 at 25°C

Preparation of Medium: Add components, except rabbit serum, to distilled/deionized water and bring volume to 900.0L. Mix thoroughly. Gently heat and bring to boiling. Boil for 20 min. Cool overnight at 4°C. Filter through Whatman #2 filter paper. Distribute into tubes or flasks. Autoclave for 15 min at 15 psi pressure–121°C. Cool to 50°–56°C. Aseptically add 100.0mL of rabbit serum. Mix thoroughly.

Use: For the cultivation of *Leptospira* species.

Koser Citrate Broth (BAM M72)

Composition per liter:

Sodium citrate	3.0g
$NaNH_4HPO_4 \cdot 4H_2O$	1.5g
KH_2PO_4	1.0g
$MgSO_4 \cdot 7H_2O$	0.2g

pH 6.7 ± 0.2 at 25°C

Source: This medium is available as a premixed powder from BD Diagnostic Systems.

Preparation of Medium: Add components to distilled/deionized water and bring volume to 1.0L. Mix thoroughly. Gently heat and bring to boiling. Distribute into tubes or flasks. Autoclave for 15 min at 15 psi pressure–121°C. Pour into sterile Petri dishes or leave in tubes.

Use: For the differentiation of *Escherichia coli* and *Enterobacter aerogenes* based on citrate utilization.

Koser Citrate Medium

Composition per liter:

Sodium citrate	3.0g
$NaNH_4HPO_4 \cdot 4H_2O$	1.5g
KH_2PO_4	1.0g
$MgSO_4 \cdot 7H_2O$	0.2g

pH 6.7 ± 0.2 at 25°C

Source: This medium is available as a premixed powder from BD Diagnostic Systems.

Preparation of Medium: Add components to distilled/deionized water and bring volume to 1.0L. Mix thoroughly. Gently heat and bring to boiling. Distribute into tubes or flasks. Autoclave for 15 min at 15 psi pressure–121°C. Pour into sterile Petri dishes or leave in tubes.

Use: For the differentiation of *Escherichia coli* and *Enterobacter aerogenes* based on citrate utilization.

Kosmachev's Medium

Composition per liter:

Agar	15.0g
$CaCO_3$	4.0g
KNO_3	1.0g
$(NH_4)_2SO_4$	1.0g
Na_2HPO_4	1.0g
$MgSO_4 \cdot 7H_2O$	0.5g
$FeSO_4 \cdot 7H_2O$	0.01g
Yeast autolysate (30% solution)	15.0mL

Preparation of Medium: Add components to distilled/deionized water and bring volume to 1.0L. Mix thoroughly. Gently heat and bring to boiling. Distribute into tubes or flasks. Autoclave for 15 min at 15 psi pressure–121°C. Pour into sterile Petri dishes or leave in tubes.

Use: For the isolation and cultivation of *Actinomadura* species, *Actinopolyspora* species, *Excellospora* species, and *Microspora* species.

KPL Medium

Composition per 1141.0mL:

Agar	15.0g
Galactose	10.0g
Glucose	10.0g
Lactic acid whey	1.0L
White table wine	140.0mL
Tween™ 80	1.0mL

pH 5.5 ± 0.2 at 25°C

Lactic Acid Whey:

Composition per liter:

Skim milk (10% solution)	1.0L

Preparation of Lactic Acid Whey: Adjust the pH of the skim milk to 5.5 with lactic acid. Gently heat and bring to boiling. Continue boiling for 30 min. Filter through Whatman #1 filter paper.

Preparation of Medium: Combine components, except white table wine. Mix thoroughly. Gently heat and bring to boiling. Adjust pH to 5.5. Autoclave for 15 min at 15 psi pressure–121°C. Cool to 45°–50°C. Filter sterilize white table wine. To cooled, sterile basal medium, aseptically add sterile white table wine. Mix thoroughly. Pour into sterile Petri dishes or distribute into sterile tubes.

Use: For the cultivation and maintenance of *Lactobacillus kefiranofaciens*.

Kracke Blood Culture HiVeg Medium

Composition per liter:

NaCl	49.0g
Glucose	10.0g
Plant peptone No. 3	10.0g
Na_2HPO_4	2.0g
Plant infusion	2.0g
Plant special infusion	1.0g
Sodium citrate	1.0g

pH 7.4 ± 0.2 at 25°C

Source: This medium is available as a premixed powder from HiMedia.

Preparation of Medium: Add components to distilled/deionized water and bring volume to 1.0L. Mix thoroughly. Gently heat and bring to boiling. Distribute into tubes or flasks. Autoclave for 15 min at 15 psi pressure–121°C.

Use: For the cultivation of pathogens in cases of bacteremia. The medium is inoculated with blood from a patient. Approximately 10-15mL of blood normally is inoculated into 50mL of medium.

Krainsky's Asparagine Agar

Composition per liter:

Agar	15.0g
Glucose	10.0g
K_2HPO_4	0.5g
L-Asparagine	0.5g

pH 7.0 ± 0.2 at 25°C

Preparation of Medium: Add components to distilled/deionized water and bring volume to 1.0L. Mix thoroughly. Gently heat and bring to boiling. Distribute into tubes or flasks. Autoclave for 15 min at 15 psi pressure–121°C. Pour into sterile Petri dishes or leave in tubes.

Use: For the cultivation and maintenance of *Streptomyces fragmentans*.

KRANEP Agar Base

Composition per liter:

KSCN	25.5g
Agar	18.3g
Sodium pyruvate	8.2g
Pancreatic digest of gelatin	6.1g
LiCl	5.1g
Mannitol	5.1g
Beef extract	3.7g
NaN_3	0.05g
Cycloheximide	0.041g
Egg yolk emulsion	100.0mL

pH 6.8 ± 0.2 at 25°C

Caution: Sodium azide is toxic. Azides also react with metals and disposal must be highly diluted.

Caution: Cycloheximide is toxic. Avoid skin contact or aerosol formation and inhalation.

Source: This medium is available as a premixed powder from Oxoid Unipath.

Egg Yolk Emulsion:
Composition:

Chicken egg yolks	11
Whole chicken egg	1

Preparation of Egg Yolk Emulsion: Soak eggs with 1:100 dilution of saturated mercuric chloride solution for 1 min. Crack eggs and separate yolks from whites. Mix egg yolks with 1 chicken egg.

Preparation of Medium: Add components, except egg yolk emulsion, to distilled/deionized water and bring volume to 900.0mL. Mix thoroughly. Gently heat and bring to boiling. Autoclave for 15 min at 15 psi pressure–121°C. Cool to 45°–50°C. Aseptically add 100.0mL of egg yolk emulsion. Mix thoroughly. Pour into sterile Petri dishes or distribute into sterile tubes.

Use: For the isolation and enumeration of staphylococci from foods.

KRANEP HiVeg Agar Base with Egg Yolk Emulsion

Composition per liter:

Potassium thiocyanate	25.5g
Agar	15.0g
Sodium pyruvate	8.2g
LiCl	5.1g
Mannitol	5.1g
Plant peptone	5.0g
NaCl	5.0g
Plant extract	1.5g
Yeast extract	1.5g
NaN_3	0.05g
Cycloheximide	0.041g
Egg yolk emulsion	100.0mL

pH 6.8 ± 0.2 at 25°C

Source: This medium, without egg yolk emulsion, is available as a premixed powder from HiMedia.

Caution: Sodium azide is toxic. Azides also react with metals and disposal must be highly diluted.

Caution: Lithium Chloride is harmful. Avoid bodily contact and inhalation of vapors. On contact with skin wash with plenty of water immediately.

Caution: Cycloheximide is toxic. Avoid skin contact or aerosol formation and inhalation.

Egg Yolk Emulsion:
Composition per liter:

Egg yolks	30.0mL
NaCl, 0.9% solution	70.0mL

Preparation of Egg Yolk Emulsion: Soak eggs with 1:100 dilution of saturated mercuric chloride solution for 1 min. Crack 11 eggs and separate yolks from whites. Mix egg yolks. Measure 30.0mL of egg yolk emulsion and add to 70.0mL of 0.9% sterile NaCl solution. Mix thoroughly. Warm to 45°–50°C.

Preparation of Medium: Add components, except egg yolk emulsion, to distilled/deionized water and bring volume to 900.0mL. Mix thoroughly. Gently heat and bring to boiling. Autoclave for 15 min at 15 psi pressure–121°C. Cool to 45°–50°C. Aseptically add 100.0mL of egg yolk emulsion. Mix thoroughly. Pour into sterile Petri dishes or distribute into sterile tubes.

Use: For the isolation and enumeration of staphylococci from foods.

Kreb's Yeast Lactate Medium

Composition per liter:

Yeast extract	10.0g
$Na_2HPO_4 \cdot 2H_2O$	3.0g
KH_2PO_4	1.0g
Sodium lactate solution	40.0mL

pH 7.0 ± 0.2 at 25°C

Sodium Lactate Solution:
Composition per 100.0mL:

Sodium lactate	70.0g

Preparation of Sodium Lactate Solution: Add sodium lactate to distilled/deionized water and bring volume to 100.0mL. Mix thoroughly. Filter sterilize.

Preparation of Medium: Add components, except sodium lactate solution, to distilled/deionized water and bring volume to 960.0mL. Mix thoroughly. Bring pH to 7.0. Autoclave for 15 min at 15 psi pressure–121°C. Aseptically add 40.0mL of sterile sodium lactate solution. Mix thoroughly. Aseptically distribute into sterile tubes or flasks.

Use: For the cultivation of *Propionibacterium acidipropionici, Propionibacterium freudenreichii, Propionibacterium intermedium, Propionibacterium jensenii, Propionibacterium* species, and *Propionibacterium thoenii.*

Kundrant Agar

Composition per liter:

Agar	10.0g
Meat peptone	7.8g
Casein peptone	7.8g
Starch	4.0g
Gelatin	4.0g
NaCl	3.0g
Yeast extract	2.8g
Glucose	1.0g
Bromcresol Purple	1.6mg

pH 6.8 ± 0.2 at 25°C

Source: This medium is available from HiMedia.

Preparation of Medium: Add components to distilled/deionized water and bring volume to 1.0L. Mix thoroughly. Gently heat and bring to boiling. Distribute into tubes or flasks. Autoclave for 15 min at 15 psi pressure–121°C.

Use: For the qualitative detection of residues from sulfonamides and other antimicrobics in animal-derived foods.

Kunkee Medium

Composition per 1000.88mL:

Pancreatic digest of casein	20.0g
Glucose	5.0g
Peptone	5.0g
Yeast extract	5.0g
Filtered tomato juice	250.0mL
Ethanol (96% solution)	176.0mL
Tween™ 80	0.5mL

pH 5.5 ± 0.2 at 25°C

Preparation of Medium: Add components, except ethanol, to distilled/deionized water and bring volume to 1.0L. Mix thoroughly. Distribute into tubes in 5.0mL volumes. Autoclave for 15 min at 15 psi pressure–121°C. Cool to room temperature. Aseptically add 0.88mL of ethanol solution to each tube containing 5.0mL of medium.

Use: For the cultivation of *Lactobacillus fructivorans.*

Kupferberg *Trichomonas* Base

Composition per liter:

Pancreatic digest of casein	20.0g
L-Cysteine·HCl·H_2O	1.5g
Agar	1.0g
Maltose	1.0g
Methylene Blue	3.0mg
Bovine serum	50.0mL

pH 6.0 ± 0.2 at 25°C

Source: This medium is available as a premixed powder from BD Diagnostic Systems.

Preparation of Medium: Add components, except bovine serum, to distilled/deionized water and bring volume to 950.0mL. Mix thoroughly. Gently heat and bring to boiling. Autoclave for 15 min at 15 psi pressure–121°C. Cool to 45°–50°C. Aseptically add 50.0mL of bovine serum. If desired, additional selectivity can be obtained by aseptically adding 250,000U of penicillin and 1.0g of streptomycin or 1.0g of chloramphenicol. Mix thoroughly. Pour into sterile Petri dishes or distribute into sterile tubes.

Use: For the cultivation of the *Trichomonas* species from clinical specimens.

Kupferberg *Trichomonas* Broth

Composition per liter:

Enzymatic digest of protein	20.0g
L-Cysteine·HCl·H_2O	1.5g
Agar	1.0g
Maltose	1.0g
Chloramphenicol	0.1g
Methylene Blue	3.0mg
Bovine serum	50.0mL

pH 6.0 ± 0.2 at 25°C

Source: This medium is available as a premixed powder from BD Diagnostic Systems.

Preparation of Medium: Add components, except bovine serum, to distilled/deionized water and bring volume to 950.0mL. Mix thoroughly. Gently heat and bring to boiling. Autoclave for 15 min at 15 psi pressure–121°C. Cool to 45°–50°C. Aseptically add bovine serum. If desired, additional selectivity can be obtained by aseptically adding 250,000U penicillin and 1.0g streptomycin or 1.0g chloramphenicol. Mix thoroughly. Pour into sterile Petri dishes or distribute into sterile tubes.

Use: For the cultivation of the *Trichomonas* species from clinical specimens.

Kupferberg *Trichomonas* HiVeg Broth Base with Serum and Selective Supplement (*Trichomonas* HiVeg Broth Base, Kupferberg)

Composition per liter:

Plant hydrolysate	20.0g
Agar	1.0g
L-Cysteine·HCl	1.5g
Maltose	1.0g
Methylene Blue	3.0mg
Bovine serum	50.0mL
Selective supplement	10.0mL

pH 6.0 ± 0.2 at 25°C

Source: This medium, without bovine serum and selective supplement, is available as a premixed powder from HiMedia.

Selective Supplement Solution:

Composition per 10.0mL:

Penicillin	250,000U

Preparation of Selective Supplement Solution: Add penicilliln to distilled/deionized water and bring volume to 10.0mL. Mix thoroughly. Filter sterilize.

Preparation of Medium: Add components, except bovine serum adn selective supplement, to distilled/deionized water and bring volume to 950.0mL. Mix thoroughly. Gently heat and bring to boiling. Autoclave for 15 min at 15 psi pressure–121°C. Cool to 45°–50°C. Aseptically add 50.0mL of bovine serum and 10.0mL selective supplement.

Use: For the cultivation of *Trichomonas* species from clinical specimens.

Kupferberg *Trichomonas* HiVeg Broth Base with Serum and Selective Supplement (*Trichomonas* HiVeg Broth Base, Kupferberg)

Composition per liter:

Plant hydrolysate	20.0g
Agar	1.0g
L-Cysteine·HCl	1.5g
Maltose	1.0g
Methylene Blue	3.0mg
Bovine serum	50.0mL
Selective supplement	10.0mL

pH 6.0 ± 0.2 at 25°C

Source: This medium, without bovine serum and selective supplement, is available as a premixed powder from HiMedia.

Selective Supplement Solution:

Composition per 10.0mL:

Streptomycin	1.0g

Preparation of Selective Supplement Solution: Add streptomycin to distilled/deionized water and bring volume to 10.0mL. Mix thoroughly. Filter sterilize.

Preparation of Medium: Add components, except bovine serum and selective supplement, to distilled/deionized water and bring volume to 950.0mL. Mix thoroughly. Gently heat and bring to boiling. Autoclave for 15 min at 15 psi pressure–121°C. Cool to 45°–50°C. Aseptically add 50.0mL of bovine serum and 10.0mL selective supplement.

Use: For the cultivation of *Trichomonas* species from clinical specimens.

Kupferberg *Trichomonas* HiVeg Broth Base with Serum and Selective Supplement (*Trichomonas* HiVeg Broth Base, Kupferberg)

Composition per liter:

Plant hydrolysate	20.0g
Agar	1.0g
L-Cysteine·HCl	1.5g
Maltose	1.0g
Methylene Blue	3.0mg
Bovine serum	50.0mL
Selective supplement	10.0mL

pH 6.0 ± 0.2 at 25°C

Source: This medium, without bovine serum and selective supplement, is available as a premixed powder from HiMedia.

Selective Supplement Solution:

Composition per 10.0mL:

Chloramphenicol	1.0g

Preparation of Selective Supplement Solution: Add chloramphenicol to distilled/deionized water and bring volume to 10.0mL. Mix thoroughly. Filter sterilize.

Preparation of Medium: Add components, except bovine serum adn selective supplement, to distilled/deionized water and bring volume to 950.0mL. Mix thoroughly. Gently heat and bring to boiling. Autoclave for 15 min at 15 psi pressure–121°C. Cool to 45°–50°C. Aseptically add 50.0mL of bovine serum and 10.0mL selective supplement.

Use: For the cultivation of *Trichomonas* species from clinical specimens.

Kushneria aurantia Medium (DSMZ Medium 1195)

Composition per liter:

NaCl	78.0g
$MgSO_4 \cdot 7H_2O$	20.3g
$MgCl_2 \cdot 6H_2O$	13.0g
Yeast extract	5.0g
KCl	2.0g
$CaCl_2 \cdot 2H_2O$	0.33g
NaBr	0.23g
$NaHCO_3$	0.06g

pH 7.3 ± 0.2 at 25°C

Preparation of Medium: Add components to distilled/deionized water and bring volume to 1.0L. Mix thoroughly. Adjust pH to 7.3. Distribute into tubes or flasks. Gently heat while stirring and bring to boiling. Mix thoroughly. Autoclave for 15 min at 15 psi pressure–121°C.

Use: For the cultivation of *Kushneria aurantia.*

KVBA

See: **Kanamycin Vancomycin Blood Agar**

KYE Agar

Composition per liter:

Agar	15.0g
$NaNO_3$	2.5g
KH_2PO_4	1.0g
Yeast extract	1.0g
$MgSO_4 \cdot 7H_2O$	0.3g

$CaCl_2 \cdot 6H_2O$ 0.15g
NaCl 0.1g
$FeCl_3$ 10.0mg

pH 10.0 ± 0.2 at 25°C

Preparation of Medium: Add components to distilled/deionized water and bring volume to 1.0L. Mix thoroughly. Gently heat and bring to boiling. Distribute into tubes or flasks. Autoclave for 15 min at 15 psi pressure–121°C. Pour into sterile Petri dishes or leave in tubes.

Use: For the cultivation of a variety of alkaliphilic bacteria.

L Agar (Luria Agar)

Composition per liter:

Agar 15.0g
Pancreatic digest of casein 10.0g
Yeast extract 5.0g
NaCl 0.5g
Glucose solution 20.0mL

pH 7.0 ± 0.2 at 25°C

Glucose Solution:
Composition per 100.0mL:

Glucose 10.0g

Preparation of Glucose Solution: Add glucose to distilled/deionized water and bring volume to 100.0mL. Mix thoroughly. Filter sterilize.

Preparation of Medium: Add components, except glucose solution, to distilled/deionized water and bring volume to 980.0mL. Mix thoroughly. Bring pH to 7.0. Gently heat and bring to boiling. Autoclave for 15 min at 15 psi pressure–121°C. Aseptically add 20.0mL of sterile glucose solution. Mix thoroughly. Pour into sterile Petri dishes or leave in tubes.

Use: For the cultivation of *Escherichia coli.*

L Broth (Luria Broth)

Composition per liter:

Pancreatic digest of casein 10.0g
NaCl 5.0g
Yeast extract 5.0g
Glucose 1.0g

pH 7.0 ± 0.2 at 25°C

Preparation of Medium: Add components to distilled/deionized water and bring volume to 1.0L. Mix thoroughly. Distribute into tubes or flasks. Autoclave for 15 min at 15 psi pressure–121°C.

Use: For the cultivation of *Escherichia coli.*

L Broth DAP Thymidine

Composition per liter:

Pancreatic digest of casein 10.0g
NaCl 5.0g
Yeast extract 5.0g
Diaminopimelic acid solution 10.0mL
Thymidine solution 10.0mL

pH 7.0 ± 0.2 at 25°C

Diaminopimelic Acid Solution:
Composition per 10.0mL:

Diaminopimelic acid 0.1g

Preparation of Diaminopimelic Acid Solution: Add diaminopimelic acid to distilled/deionized water and bring volume to 10.0mL. Mix thoroughly. Filter sterilize.

Thymidine Solution:
Composition per 10.0mL:

Thymidine 0.01g

Preparation of Thymidine Solution: Add thymidine to distilled/deionized water and bring volume to 10.0mL. Mix thoroughly. Filter sterilize.

Preparation of Medium: Add components, except diaminopimelic acid solution and thymidine solution, to distilled/deionized water and bring volume to 980.0mL. Mix thoroughly. Bring pH to 7.0. Autoclave for 15 min at 15 psi pressure–121°C. Aseptically add 10.0mL of sterile diaminopimelic acid solution and 10.0mL of sterile thymidine solution. Mix thoroughly. Aseptically distribute into sterile tubes or flasks.

Use: For the cultivation of *Escherichia coli.*

L Diphasic Blood Agar Medium (ATCC Medium 947)

Composition per 1150.0mL:

Blood agar, diphasic base medium 700.0mL
Rabbit blood, defibrinated 300.0mL
Locke's solution 150.0mL

pH 7.2–7.4 at 25°C

Blood Agar, Diphasic Base Medium:
Composition per 750.0mL:

Beef 25.0g
Agar 10.0g
Neopeptone 10.0g
NaCl 2.5g

Preparation of Blood Agar, Diphasic Base Medium: Trim beef to remove fat. Add 25.0g of lean beef to 250.0mL of distilled/deionized water. Gently heat and bring to boiling. Boil for 2–3 min. Filter through Whatman #2 filter paper. Add agar, neopeptone, and NaCl to filtrate. Bring volume to 750.0mL with distilled/deionized water. Mix thoroughly. Adjust pH to 7.2–7.4. Gently heat and bring to boiling. Autoclave for 15 min at 15 psi pressure–121°C. Cool to 50°–55°C.

Locke's Solution:
Composition per liter:

NaCl 8.0g
Glucose 2.5g
KH_2PO_4 .... 0.3g
$CaCl_2$ 0.2g
KCl 0.2g

Preparation of Locke's Solution: Add components to distilled/deionized water and bring volume to 1.0L. Mix thoroughly. Autoclave for 15 min at 15 psi pressure–121°C. Cool to 50°–55°C.

Preparation of Medium: Aseptically combine 700.0mL of sterile blood agar, diphasic base medium, with 300.0mL of sterile defibrinated rabbit blood warmed to 50°–55°C. Mix thoroughly. Aseptically distribute 5.0mL volumes into 16 × 125mm screw-capped test tubes. Allow to cool in a slanted position. Overlay the agar in each tube with 1.0mL of sterile Locke's solution.

Use: For the cultivation and maintenance of *Trypanosoma* species, *Leishmania donovani, Herpetomonas* species, and *Trypanosoma neveulemairei.*

L Diphasic Blood Agar Medium (ATCC Medium 1011)

Composition per 1150.0mL:

Blood agar, diphasic base medium	700.0mL
Rabbit blood, defibrinated	300.0mL
Locke's solution	150.0mL

pH 7.2–7.4 at 25°C

Blood Agar, Diphasic Base Medium:

Composition per 750.0mL:

Beef	25.0g
Agar	10.0g
Neopeptone	10.0g
NaCl	2.5g

Preparation of Blood Agar, Diphasic Base Medium: Trim beef to remove fat. Add 25.0g of lean beef to 250.0mL of distilled/deionized water. Gently heat and bring to boiling. Boil for 2–3 min. Filter through Whatman #2 filter paper. Add agar, neopeptone, and NaCl to filtrate. Bring volume to 750.0mL with distilled/deionized water. Mix thoroughly. Adjust pH to 7.2–7.4. Gently heat and bring to boiling. Autoclave for 15 min at 15 psi pressure–121°C. Cool to 50°–55°C.

Locke's Solution:

Composition per liter:

NaCl	8.0g
Glucose	2.5g
KH_2PO_4	0.3g
$CaCl_2$	0.2g
KCl	0.2g

Preparation of Locke's Solution: Add components to distilled/deionized water and bring volume to 1.0L. Mix thoroughly. Autoclave for 15 min at 15 psi pressure–121°C. Cool to 50°–55°C.

Preparation of Medium: Aseptically combine 700.0mL of sterile blood agar, diphasic base medium, with 300.0mL of sterile defibrinated rabbit blood warmed to 50°–55°C. Mix thoroughly. Aseptically distribute 5.0mL volumes into 16 × 125mm screw-capped test tubes. Allow to cool in a slanted position. Overlay the agar in each tube with 3.0mL of sterile Locke's solution.

Use: For the cultivation and maintenance of *Trypanosoma* species.

L and F Basal Salts, Modified with Heptadecane

Composition per liter:

NH_4Cl	2.0g
Na_2HPO_4	0.21g
$MgSO_4·7H_2O$	0.2g
NaH_2PO_4	0.09g
KCl	0.04g
$CaCl_2$	0.015g
$FeSO_4·7H_2O$	1.0mg
$ZnSO_4·7H_2O$	0.07mg
H_3BO_3	0.01mg
$MnSO_4·5H_2O$	0.01mg
MoO_3	0.01mg
$CuSO_4·5H_2O$	5.0μg
Heptadecane	2.0mL

pH 7.2 ± 0.2 at 25°C

Preparation of Medium: Add components, except heptadecane, to distilled/deionized water and bring volume to 1.0L. Mix thoroughly. Gently heat and bring to boiling. Distribute equally into four 250.0mL volumes. Autoclave for 15 min at 15 psi pressure–121°C. Cool to 60°C. Filter sterilize heptadecane. To one 250.0mL fraction of basal salts, aseptically add 0.5mL of sterile heptadecane. Pour mixture into a sterile blender. Homogenize slowly to mix heptadecane with basal salts and not to create excess bubbles. Rapidly distribute medium to sterile screw-capped tubes. Chill tubes quickly in an ice pack or in the refrigerator. Allow tubes to solidify in a slanted position.

Use: For the cultivation and maintenance of *Thermoleophilum album* and *Thermoleophilum minutum.*

L Medium (ATCC Medium 167)

Composition per liter:

Agar	20.0g
$NaNO_3$	2.0g
Na_2HPO_4	0.21g
$MgSO_4·7H_2O$	0.2g
NaH_2PO_4	0.09g
KCl	0.04g
$CaCl_2$	0.015g
$FeSO_4·7H_2O$	1.0mg
Salts solution	1.0mL

Salts Solution:

Composition per 100.0mL:

$ZnSO_4·7H_2O$	7.0mg
H_3BO_3	1.0mg
$MnSO_4·5H_2O$	1.0mg
MoO_3	1.0mg
$CuSO_4·5H_2O$	0.5mg

Preparation of Salts Solution: Add components to distilled/deionized water and bring volume to 100.0mL. Mix thoroughly.

Preparation of Medium: Add components to distilled/deionized water and bring volume to 1.0L. Mix thoroughly. Gently heat and bring to boiling. Distribute into tubes or flasks. Autoclave for 15 min at 15 psi pressure–121°C. Pour into sterile Petri dishes or leave in tubes.

Use: For the cultivation and maintenance of *Methylococcus capsulatus* and *Pseudomonas methanica.*

L Medium (ATCC Medium 1154)

Composition per liter:

Pancreatic digest of casein	10.0g
NaCl	5.0g
Yeast extract	5.0g

pH 7.0 ± 0.2 at 25°C

Preparation of Medium: Add components to distilled/deionized water and bring volume to 1.0L. Mix thoroughly. Adjust pH to 7.0. Distribute into tubes or flasks. Autoclave for 25 min at 15 psi pressure–121°C.

Use: For the cultivation and maintenance of *Escherichia coli.*

L Medium with Ampicillin

Composition per liter:

Pancreatic digest of casein	10.0g
NaCl	5.0g

Yeast extract 5.0g
Ampicillin solution 20.0mg
pH 7.0 ± 0.2 at 25°C

Ampicillin Solution:

Composition per 10.0mL:

Ampicillin 0.02g

Preparation of Ampicillin Solution: Add ampicillin to distilled/deionized water and bring volume to 10.0mL. Mix thoroughly. Filter sterilize.

Preparation of Medium: Add components to distilled/deionized water and bring volume to 1.0L. Mix thoroughly. Adjust pH to 7.0. Distribute into tubes or flasks. Autoclave for 15 min at 15 psi pressure–121°C.

Use: For the cultivation and maintenance of *Escherichia coli.*

L Medium with DAP and THY (L Medium with Diaminopimelic Acid and Thymidine)

Composition per liter:

Pancreatic digest of casein 10.0g
NaCl 5.0g
Yeast extract 5.0g
Glucose solution 10.0mL
Diaminopimelic acid solution 10.0mL
Thymidine solution 10.0mL
pH 7.0 ± 0.2 at 25°C

Glucose Solution:

Composition per 10.0mL:

D-Glucose 1.0g

Preparation of Glucose Solution: Add D-glucose to distilled/deionized water and bring volume to 10.0mL. Mix thoroughly. Filter sterilize.

Diaminopimelic Acid Solution:

Composition per 10.0mL:

DL-Diaminopimelic acid 0.1g

Preparation of Diaminopimelic Acid Solution: Add diaminopimelic acid to distilled/deionized water and bring volume to 10.0mL. Mix thoroughly. Filter sterilize.

Thymidine Solution:

Composition per 10.0mL:

Thymidine 0.02g

Preparation of Thymidine Solution: Add thymidine to distilled/deionized water and bring volume to 10.0mL. Mix thoroughly. Filter sterilize.

Preparation of Medium: Add components—except glucose solution, diaminopimelic acid solution, and thymidine solution—to distilled/deionized water and bring volume to 970.0mL. Mix thoroughly. Gently heat and bring to boiling. Adjust pH to 7.0. Autoclave for 15 min at 15 psi pressure–121°C. Cool to 45°–50°C. Aseptically add sterile glucose solution, diaminopimelic acid solution, and thymidine solution. Mix thoroughly. Aseptically distribute into sterile tubes or flasks.

Use: For the cultivation and maintenance of *Escherichia coli.*

L Medium with DAP, THY, and AMP (L Medium with Diaminopimelic Acid, Thymidine, and Ampicillin)

Composition per liter:

Pancreatic digest of casein 10.0g
NaCl 5.0g
Yeast extract 5.0g
Diaminopimelic acid solution 10.0mL
Thymidine solution 10.0mL
Ampicillin solution 10.0mL
pH 7.0 ± 0.2 at 25°C

Diaminopimelic Acid Solution:

Composition per 10.0mL:

DL-Diaminopimelic acid 0.1g

Preparation of Diaminopimelic Acid Solution: Add DL-diaminopimelic acid to distilled/deionized water and bring volume to 10.0mL. Mix thoroughly. Filter sterilize.

Thymidine Solution:

Composition per 10.0mL:

Thymidine 0.04g

Preparation of Thymidine Solution: Add thymidine to distilled/deionized water and bring volume to 10.0mL. Mix thoroughly. Filter sterilize.

Ampicillin Solution:

Composition per 10.0mL:

Ampicillin 0.02mg

Preparation of Ampicillin Solution: Add ampicillin to distilled/deionized water and bring volume to 10.0mL. Mix thoroughly. Filter sterilize.

Preparation of Medium: Add components—except ampicillin solution, diaminopimelic acid solution, and thymidine solution—to distilled/deionized water and bring volume to 970.0mL. Mix thoroughly. Gently heat and bring to boiling. Adjust pH to 7.0. Autoclave for 15 min at 15 psi pressure–121°C. Cool to 45°–50°C. Aseptically add sterile ampicillin solution, diaminopimelic acid solution, and thymidine solution. Mix thoroughly. Aseptically distribute into sterile tubes or flasks.

Use: For the cultivation and maintenance of *Escherichia coli.*

L Medium with Diaminopimelic Acid and Thymidine
See: **L Medium with DAP and THY**

L Medium with Diaminopimelic Acid, Thymidine, and Ampicillin
See: **L Medium with DAP, THY, and AMP**

L Medium with Methanol

Composition per liter:

Agar 20.0g
$NaNO_3$ 2.0g
Na_2HPO_4 .... 0.21g
$MgSO_4 \cdot 7H_2O$ 0.2g
NaH_2PO_4 .... 0.09g
KCl 0.04g
$CaCl_2$ 0.015g
$FeSO_4 \cdot 7H_2O$ 1.0mg
Methanol 20.0mL
Salts solution 1.0mL

Salts Solution:

Composition per 100.0mL:

$ZnSO_4 \cdot 7H_2O$	7.0mg
H_3BO_3	1.0mg
$MnSO_4 \cdot 5H_2O$	1.0mg
MoO_3	1.0mg
$CuSO_4 \cdot 5H_2O$	0.5mg

Preparation of Salts Solution: Add components to distilled/deionized water and bring volume to 100.0mL. Mix thoroughly.

Preparation of Medium: Add components, except methanol, to distilled/deionized water and bring volume to 980.0mL. Mix thoroughly. Gently heat and bring to boiling. Autoclave for 15 min at 15 psi pressure–121°C. Cool to 45°–50°C. Filter sterilize methanol. Aseptically add 20.0mL of sterile methanol to cooled, sterile basal medium. Mix thoroughly. Pour into sterile Petri dishes or distribute into sterile tubes.

Use: For the cultivation and maintenance of *Methylobacillus glycogenes*.

L Medium for *Salmonella*

Composition per liter:

Pancreatic digest of casein	10.0g
NaCl	5.0g
Yeast extract	5.0g
Glucose	1.0g

pH 7.2 ± 0.2 at 25°C

Preparation of Medium: Add components to distilled/deionized water and bring volume to 1.0L. Mix thoroughly. Adjust pH to 7.2. Distribute into tubes or flasks. Autoclave for 15 min at 15 psi pressure–121°C.

Use: For the cultivation and maintenance of *Salmonella choleraesuis*.

L Medium with Tetracycline

Composition per liter:

Pancreatic digest of casein	10.0g
NaCl	5.0g
Yeast extract	5.0g
Tetracycline solution	10.0mL

pH 7.0 ± 0.2 at 25°C

Tetracycline Solution:

Composition per 10.0mL:

Tetracycline	0.02mg

Preparation of Tetracycline Solution: Add tetracycline to distilled/deionized water and bring volume to 10.0mL. Mix thoroughly. Filter sterilize.

Preparation of Medium: Add components, except tetracycline solution, to distilled/deionized water and bring volume to 990.0mL. Mix thoroughly. Gently heat and bring to boiling. Adjust pH to 7.0. Autoclave for 15 min at 15 psi pressure–121°C. Cool to 45°–50°C. Aseptically add sterile tetracycline solution. Mix thoroughly. Aseptically distribute into sterile tubes or flasks.

Use: For the cultivation and maintenance of *Escherichia coli*.

L15 Medium, Modified Leibovitz

Composition per liter:

NaCl	8.0g
DL-Threonine	0.6g
Sodium pyruvate	0.6g
DL-Alanine	0.5g
L-Arginine, free base	0.5g
KCl	0.4g
L-Asparagine·H_2O	0.3g
L-Histidine, free base	0.3g
L-Glutamine	0.3g
L-Isoleucine	0.3g
L-Phenylalanine	0.3g
L-Tyrosine	0.3g
DL-Methionine	0.2g
DL-Valine	0.2g
Glycine	0.2g
L-Serine	0.2g
Na_2HPO_4, anhydrous	0.2g
$CaCl_2$, anhydrous	0.1g
L-Cysteine, free base	0.1g
L-Leucine·HCl	0.1g
$MgCl_2$, anhydrous	0.094g
D-Galactose	0.09g
KH_2PO_4	0.06g
L-Tryptophan	0.02g
Phenol Red	0.01g
i-Inositol	2.0mg
Choline chloride	1.0mg
D-Calcium pantothenate	1.0mg
Folic acid	1.0mg
Nicotinamide	1.0mg
Pyridoxine·HCl	1.0mg
Thiamine monophosphate·$2H_2O$	1.0mg
Riboflavin-5-phosphate	0.1mg

pH 7.5 ± 0.2 at 25°C

Preparation of Medium: Add components to distilled/deionized water and bring volume to 1.0L. Mix thoroughly. Filter sterilize. Store at 5°C.

Use: For the cultivation of oysters used for the growth of enteroviruses.

L. mono Confirmatory Agar Base (*Listeria monocytogenes* Confirmatory Agar, Base)

Composition per liter:

Special peptone	30.0g
Agar	12.0g
LiCl	10.0g
B.C. indicator	8.6g
Yeast extract	6.0g
NaCl	5.0g
α-Methyl-D-mannoside	3.0g
Na_2HPO_4, anhydrous	2.5g
Listeria mono enrichment supplement II	10.0mL
Listeria mono selective supplement I	10.0mL
Listeria mono selective supplement II	10.0mL

pH 7.2 ± 0.2 at 25°C

Source: This medium, without *Listeria mono* enrichment supplement II, *Listeria mono* selective supplement I, and *Listeria mono* selective supplement II, is available as a premixed powder from BioChemika.

Caution: LiCl is harmful. Avoid bodily contact and inhalation of vapors. On contact with skin wash with plenty of water immediately.

***Listeria mono* Enrichment Supplement II Solution:**
Composition per 10.0mL:

L-phosphatidylinositol	1.0g

Preparation of *Listeria mono* Enrichment Supplement II Solution: Add L-phosphatidylinositol to distilled/deionized water and bring volume to 10.0mL. Mix thoroughly. Filter sterilize.

***Listeria mono* Selective Supplement I Solution:**
Composition per 10.0mL:

Polymyxin B sulfate	76,700U

Preparation of *Listeria mono* Selective Supplement I Solution: Add polymyxin B sulfate to distilled/deionized water and bring volume to 10.0mL. Mix thoroughly. Filter sterilize.

***Listeria mono* Selective Supplement II Solution:**
Composition per 10.0mL:

Ceftazidime	20.0mg
Nalidixic acid, sodium salt	20.0mg
Amphotericin B	10.0mg

Preparation of *Listeria mono* Selective Supplement II Solution: Add components to distilled/deionized water and bring volume to 10.0mL. Mix thoroughly. Filter sterilize.

Preparation of Medium: Add components, except *Listeria mono* enrichment supplement II, *Listeria mono* selective supplement I, and *Listeria mono* selective supplement II, to distilled/deionized water and bring volume to 970.0mL. Mix thoroughly. Gently heat and bring to boiling. Autoclave for 15 min at 15 psi pressure–121°C. Cool to 45°–50°C. Aseptically add 10.0mL *Listeria mono* enrichment supplement II, 10.0mL *Listeria mono* selective supplement I, and 10.0mL *Listeria mono* selective supplement II. Mix thoroughly. Pour into sterile Petri dishes or distribute into sterile tubes.

Use: For the selective and differential isolation of *Listeria monocytogenes* from clinical and food specimens.

L. mono Confirmatory HiVeg Agar Base

Composition per liter:

Plant special peptone	30.0g
Agar	12.0g
LiCl	10.0g
B.C. indicator	8.6g
Yeast extract	6.0g
NaCl	5.0g
α-Methyl-D-mannoside	3.0g
Na_2HPO_4, anhydrous	2.5g
Listeria mono enrichment supplement II	10.0mL
Listeria mono selective supplement I	10.0mL
Listeria mono selective supplement II	10.0mL

pH 7.2 ± 0.2 at 25°C

Source: This medium, without *Listeria mono* enrichment supplement II, *Listeria mono* selective supplement I, and *Listeria mono* selective supplement II, is available as a premixed powder from HiMedia.

Caution: LiCl is harmful. Avoid bodily contact and inhalation of vapors. On contact with skin wash with plenty of water immediately.

***Listeria mono* Enrichment Supplement II Solution:**
Composition per 10.0mL:

L-phosphatidylinositol	1.0g

Preparation of *Listeria mono* Enrichment Supplement II Solution: Add L-phosphatidylinositol to distilled/deionized water and bring volume to 10.0mL. Mix thoroughly. Filter sterilize.

***Listeria mono* Selective Supplement I Solution:**
Composition per 10.0mL:

Polymyxin B sulfate	76,700U

Preparation of *Listeria mono* Selective Supplement I Solution: Add polymyxin B sulfate to distilled/deionized water and bring volume to 10.0mL. Mix thoroughly. Filter sterilize.

***Listeria mono* Selective Supplement II Solution:**
Composition per 10.0mL:

Ceftazidime	20.0mg
Nalidixic acid, sodium salt	20.0mg
Amphotericin B	10.0mg

Preparation of *Listeria mono* Selective Supplement II Solution: Add components to distilled/deionized water and bring volume to 10.0mL. Mix thoroughly. Filter sterilize.

Preparation of Medium: Add components, except *Listeria mono* enrichment supplement II, *Listeria mono* selective supplement I, and *Listeria mono* selective supplement II, to distilled/deionized water and bring volume to 970.0mL. Mix thoroughly. Gently heat and bring to boiling. Autoclave for 15 min at 15 psi pressure–121°C. Cool to 45°–50°C. Aseptically add 10.0mL *Listeria mono* enrichment supplement II, 10.0mL *Listeria mono* selective supplement I, and 10.0mL *Listeria mono* selective supplement II. Mix thoroughly. Pour into sterile Petri dishes or distribute into sterile tubes.

Use: For the selective and differential isolation of *Listeria monocytogenes* from clinical and food specimens.

L. mono Differential HiVeg Agar Base

Composition per liter:

Plant peptone No. 1	18.0g
Agar	15.0g
LiCl	10.0g
Yeast extract	10.0g
Plant hydrolysate	6.0g
NaCl	5.0g
Na_2HPO_4, anhydrous	2.5g
Glucose	2.0g
Sodium pyruvate	2.0g
Magnesium glycerophosphate	1.0g
$MgSO_4$	0.5g
Chromogenic substrate	0.05g
Listeria mono enrichment supplement II	10.0mL
Listeria mono selective supplement I	10.0mL
Listeria mono selective supplement II	10.0mL

pH 7.2 ± 0.2 at 25°C

Source: This medium, without *Listeria mono* enrichment supplement II, *Listeria mono* selective supplement I, and *Listeria mono* selective supplement II, is available as a premixed powder from HiMedia.

Caution: LiC lis harmful. Avoid bodily contact and inhalation of vapors. On contact with skin wash with plenty of water immediately.

***Listeria mono* Enrichment Supplement II Solution:**
Composition per 10.0mL:

L-phosphatidylinositol	1.0g

Preparation of *Listeria mono* Enrichment Supplement II Solution: Add L-phosphatidylinositol to distilled/deionized water and bring volume to 10.0mL. Mix thoroughly. Filter sterilize.

***Listeria mono* Selective Supplement I Solution:**

Composition per 10.0mL:

Polymyxin B sulfate	76,700U

Preparation of *Listeria mono* Selective Supplement I Solution: Add polymyxin B sulfate to distilled/deionized water and bring volume to 10.0mL. Mix thoroughly. Filter sterilize.

***Listeria mono* Selective Supplement II Solution:**

Composition per 10.0mL:

Ceftazidime	20.0mg
Nalidixic acid, sodium salt	20.0mg
Amphotericin B	10.0mg

Preparation of *Listeria mono* Selective Supplement II Solution: Add components to distilled/deionized water and bring volume to 10.0mL. Mix thoroughly. Filter sterilize.

Preparation of Medium: Add components, except *Listeria mono* enrichment supplement II, *Listeria mono* selective supplement I, and *Listeria mono* selective supplement II, to distilled/deionized water and bring volume to 970.0mL. Mix thoroughly. Gently heat and bring to boiling. Autoclave for 15 min at 15 psi pressure–121°C. Cool to 45°–50°C. Aseptically add 10.0mL *Listeria mono* enrichment supplement II, 10.0mL *Listeria mono* selective supplement I, and 10.0mL *Listeria mono* selective supplement II. Mix thoroughly. Pour into sterile Petri dishes or distribute into sterile tubes.

Use: For the selective and differential isolation of *Listeria monocytogenes* from clinical and food specimens.

L Salts for Thermophiles

See: **Mineral Salts for Thermophiles**

Lab-Lemco Agar

Composition per liter:

Agar	15.0g
Peptone	5.0g
Lab Llemco meat extract	3.0g

pH 7.4 ± 0.2 at 25°C

Preparation of Medium: Add components to distilled/deionized water and bring volume to 1.0L. Mix thoroughly. Gently heat and bring to boiling. Distribute into tubes or flasks. Autoclave for 15 min at 15 psi pressure–121°C. Pour into sterile Petri dishes or leave in tubes.

Use: For the cultivation and maintenance of a variety of heterotrophic microorganisms.

Lab-Lemco Broth

Composition per liter:

Peptone	5.0g
Lab Lemco meat extract	3.0g

pH 7.4 ± 0.2 at 25°C

Preparation of Medium: Add components to distilled/deionized water and bring volume to 1.0L. Mix thoroughly. Distribute into tubes or flasks. Autoclave for 15 min at 15 psi pressure–121°C.

Use: For the cultivation of a variety of heterotrophic microorganisms, including microorganisms from wastewater.

Lachica's Medium

See: **SA Agar, Modified**

Lachica's Medium Base

Composition per liter:

Beef heart, infusion from	500.0g
Agar	15.0g
Tryptose	10.0g
NaCl	5.0g
Amylose Azure	3.0g
Selective supplement solution	10.0mL

pH 7.4 ± 0.2 at 25°C

Source: This medium is available from HiMedia.

Selective Supplement Solution:

Composition per 10.0mL:

Ampicillin	10.0mg

Preparation of Selective Supplement Solution: Add ampicillin to distilled/deionized water and bring volume to 10.0mL. Mix thoroughly. Filter sterilize.

Preparation of Medium: Add components, except selective supplement solution, to distilled/deionized water and bring volume to 990.0mL. Mix thoroughly. Autoclave for 15 min at 15 psi pressure–121°C. Cool to 50°C. Aseptically add selective supplement solution. Mix thoroughly. Pour into Petri dishes or aseptically distribute into sterile tubes.

Use: For the isolation and cultivation of *Aeromonas hydrophila* from foods stored under different temperature conditions.

Lactate Agar

Composition per liter:

Yeast extract	3.0g
K_2HPO_4	2.8g
Agar	2.0g
Peptone	2.0g
KH_2PO_4	0.52g
Sodium lactate (60% solution)	10.0mL

pH 7.2 ± 0.2 at 25°C

Preparation of Medium: Add components to distilled/deionized water and bring volume to 1.0L. Mix thoroughly. Gently heat and bring to boiling. Adjust pH to 7.2. Distribute into tubes or flasks. Autoclave for 15 min at 15 psi pressure–121°C. Pour into sterile Petri dishes or leave in tubes.

Use: For the cultivation and maintenance of *Serpens flexibilis*.

Lactate Broth

Composition per liter:

Yeast extract	3.0g
K_2HPO_4	2.8g
Peptone	2.0g
KH_2PO_4	0.52g
Sodium lactate (60% solution)	10.0mL

pH 7.2 ± 0.2 at 25°C

Preparation of Medium: Add components to distilled/deionized water and bring volume to 1.0L. Mix thoroughly. Gently heat and bring to boiling. Adjust pH to 7.2. Distribute into tubes or flasks. Autoclave for 15 min at 15 psi pressure–121°C.

Use: For the cultivation and maintenance of *Serpens flexibilis*.

Lactate Seawater Minimal Medium

Composition per liter:

Tris(hydroxymethyl)aminomethane·HCl	7.88g
Sodium or potassium lactate	2.0g
NH_4Cl	1.0g
$K_2HPO_4·3H_2O$	0.075g
$FeSO_4·7H_2O$	0.028g
Artificial seawater	500.0mL

pH 7.5 ± 0.2 at 25°C

Artificial Seawater:

Composition per liter:

$MgSO_4·7H_2O$	24.7g
NaCl	23.4g
$CaCl_2·2H_2O$	2.9g
KCl	1.5g

Preparation of Artificial Seawater: Add components to distilled/deionized water and bring volume to 1.0L. Mix thoroughly.

Preparation of Medium: Add components to distilled/deionized water and bring volume to 1.0L. Mix thoroughly. Adjust pH to 7.5. Distribute into tubes or flasks. Autoclave for 15 min at 15 psi pressure–121°C.

Use: For the cultivation of ATCC strains 27134 and 27136.

Lactic Acid Bacteria Broth

Composition per liter:

Sodium acetate	12.0g
Pancreatic digest of casein	10.0g
Yeast autolysate	5.0g
Glucose	10.0g
Solution A	5.0mL
Solution B	5.0mL

pH 5.1 ± 0.2 at 25°C

Source: Yeast autolysate is available from Oxoid Unipath.

Solution A:

Composition per 100.0mL:

K_2HPO_4	10.0g
KH_2PO_4	10.0g

Preparation of Solution A: Add components to distilled/deionized water and bring volume to 100.0mL. Mix thoroughly.

Solution B:

Composition per 100.0mL:

$MgSO_4·7H_2O$	4.0g
$FeSO_4·7H_2O$	0.2g
$MnSO_4·H_2O$	0.2g
NaCl	0.2g

Preparation of Solution B: Add components to distilled/deionized water and bring volume to 100.0mL. Mix thoroughly.

Preparation of Medium: Add components to distilled/deionized water and bring volume to 1.0L. Mix thoroughly. Distribute into tubes or flasks. Autoclave for 15 min at 15 psi pressure–121°C.

Use: For the cultivation of *Lactobacillus malefermentans.*

Lactic Acid Bacteria Medium

Composition per liter:

Agar	15.0g
Glucose	10.0g
KH_2PO_4	5.0g
Peptone	5.0g
Sodium acetate·$3H_2O$	5.0g
Yeast extract	5.0g
Diammonium hydrogen citrate	2.0g
Tween™ 80	1.0g
Sorbic acid	0.5g
$MgSO_4·7H_2O$	0.5g
$MnSO_4·4H_2O$	0.2g
$FeSO_4·7H_2O$	0.05g

pH 5.3–5.4 at 25°C

Preparation of Medium: Add components to distilled/deionized water and bring volume to 1.0L. Mix thoroughly. Adjust pH to 5.3–5.4. Gently heat and bring to boiling. Distribute into tubes or flasks. Autoclave for 15 min at 15 psi pressure–121°C. Pour into sterile Petri dishes or leave in tubes.

Use: For the isolation and cultivation of lactic acid bacteria from wine.

Lactic Acid Bacteria Selective Agar Base

Composition per liter:

Casein enzymic hydrolysate	20.0g
Agar	17.0g
Maltose	10.0g
Yeast extract	5.0g
Fructose	5.0g
Glucose	5.0g
Potassium aspartate	2.5g
Potassium glutamate	2.5g
Betaine hydrochloride	2.0g
Diammonium hydrogen citrate	2.0g
$MgSO_4·7H_2O$	2.0g
KH_2PO_4	2.0g
Liver concentrate	1.0g
$MnSO_4·2H_2O$	0.66g
N-acetyl glucosamine	0.5g
Selective supplement solution	20.0mL

pH 5.4 ± 0.2 at 25°C

Source: This medium is available from HiMedia.

Selective Supplement Solution:

Composition per 20.0mL:

2-Phenylethanol	3.0g
Cycloheximide	10.0mg
Polysorbate 80	10.0mL

Preparation of Selective Supplement Solution: Add components to distilled/deionized water and bring volume to 20.0mL. Mix thoroughly. Filter sterilize.

Caution: Cycloheximide is toxic. Avoid skin contact or aerosol formation and inhalation.

Preparation of Medium: Add components, except selective supplement solution, to distilled/deionized water and bring volume to 980.0mL. Mix thoroughly. Autoclave for 15 min at 15 psi pressure–121°C. Cool to 50°C. Aseptically add selective supplement solution. Mix thoroughly. Pour into Petri dishes or aseptically distribute into sterile tubes.

Use: For the selective isolation of lactic acid bacteria from beer and brewing processes.

Lactic Agar

Composition per liter:

Casein enzymatic hydrolysate	20.0g
Agar	15.0g

Yeast extract	5.0g
Glucose	5.0g
Lactose	5.0g
Sucrose	5.0g
NaCl	4.0g
Gelatin	2.5g
Sodium acetate	1.5g
Ascorbic acid	0.5g

Source: This medium is available from HiMedia.

Preparation of Medium: Add components to distilled/deionized water and bring volume to 1.0L. Mix thoroughly. Gently heat and bring to boiling. Distribute into tubes or flasks. Autoclave for 15 min at 15 psi pressure–121°C. Pour into sterile Petri dishes or leave in tubes.

Use: For the enumeration and identification of lactic streptococci and lactobacilli by the pour plate method.

Lactic Agar for Yogurt Bacteria, Modified

Composition per liter:

Elliker agar	1.0L
Milk solution	70.0mL

pH 6.8 ± 0.1 at 25°C

Elliker Agar:

Composition per liter:

Pancreatic digest of casein	20.0g
Agar	15.0g
Yeast extract	10.0g
Gelatin	4.0g
Glucose	3.0g
Ascorbic acid	2.5g
Lactose	2.5g
NaCl	2.5g
Sodium acetate	2.5g
Sucrose	2.5g

Preparation of Elliker Agar: Add components to distilled/deionized water and bring volume to 1.0L. Mix thoroughly. Gently heat and bring to boiling. Autoclave for 15 min at 15 psi pressure–121°C. Cool to 45°–50°C.

Milk Solution:

Composition per 100.0mL:

Nonfat dry milk solids	11.0g

Preparation of Milk Solution: Add nonfat dry milk solids to distilled/deionized water and bring volume to 100.0mL. Mix thoroughly. Autoclave for 12 min at 15 psi pressure–121°C. Cool to 45°–50°C.

Preparation of Medium: Add 70.0mL of sterile milk solution to 1.0L of cooled, sterile Elliker agar. Mix thoroughly. Pour into sterile Petri dishes. Allow plates to dry at 28°–30°C for 18–24 hr.

Use: For the cultivation of acidogenic microorganisms, especially *Lactobacillus* species and lactic streptococci, from foods.

Lactic Bacteria Broth

Composition per liter:

Sodium acetate	12.0g
Glucose	10.0g
Pancreatic digest of casein	10.0g
Yeast autolysate	5.0g
Solution A	5.0mL
Solution B	5.0mL

pH 5.1–5.3 at 25°C

Solution A:

Composition per 100.0mL:

K_2HPO_4	10.0g
KH_2PO_4	10.0g

Preparation of Solution A: Add components to distilled/deionized water and bring volume to 100.0mL. Mix thoroughly.

Solution B:

Composition per 100.0mL:

$MgSO_4 \cdot 7H_2O$	4.0g
$FeSO_4$	0.2g
$MnSO_4 \cdot 5H_2O$	0.2g
NaCl	0.2g

Preparation of Solution B: Add components to distilled/deionized water and bring volume to 100.0mL. Mix thoroughly.

Preparation of Medium: Add components to distilled/deionized water and bring volume to 1.0L. Mix thoroughly. Distribute into tubes or flasks. Autoclave for 15 min at 15 psi pressure–121°C.

Use: For the cultivation and maintenance of *Lactobacillus buchneri, Lactobacillus delbrueckii,* and *Pediococcus damnosus.*

Lactic Bacteria Differential Agar

Composition per liter:

Agar	15.0g
Casein enzymic hydrolysate	10.0g
Casein acid hydrolysate	3.0g
Fructose	2.5g
KH_2PO_4	2.5g
Papaic digest of soybean meal	1.5g
Yeast extract	1.0g
Polysorbate 80	1.0g
Bromcresol Green	0.055g

pH 7.0 ± 0.2 at 25°C

Source: This medium, without polysorbate 80, is available from HiMedia.

Preparation of Medium: Add components to distilled/deionized water and bring volume to 1.0L. Mix thoroughly. Gently heat and bring to boiling. Distribute into tubes or flasks. Autoclave for 15 min at 15 psi pressure–121°C. Pour into sterile Petri dishes or leave in tubes.

Use: For the differentiation of homofermentative and heterofermentative lactic acid bacteria.

Lactic Bacteria Differential HiVeg Agar

Composition per liter:

Agar	15.0g
Plant hydrolysate	10.0g
Plant acid hydrolysate	3.0g
Fructose	2.5g
KH_2PO_4	2.5g
Papaic digest of soybean meal	1.5g
Yeast extract	1.0g
Bromcresol Green	0.055g
Polysorbate 80	1.0mL

pH 7.0 ± 0.2 at 25°C

Source: This medium, without polysorbate 80, is available as a premixed powder from HiMedia.

Preparation of Medium: Add components to distilled/deionized water and bring volume to 1.0L. Mix thoroughly. Adjust pH to 5.3–5.4.

Gently heat and bring to boiling. Distribute into tubes or flasks. Autoclave for 15 min at 15 psi pressure–121°C. Pour into sterile Petri dishes or leave in tubes.

Use: For the differentiation of homofermentative and heterofermentative lactic acid bacteria.

Lactic Bacteria Differential HiVeg Broth

Composition per liter:

Plant hydrolysate	10.0g
Plant acid hydrolysate	3.0g
Fructose	2.5g
KH_2PO_4	2.5g
Papaic digest of soybean meal	1.5g
Yeast extract	1.0g
Bromcresol Green	0.055g
Polysorbate 80	1.0mL

pH 7.0 ± 0.2 at 25°C

Source: This medium, without polysorbate 80, is available as a premixed powder from HiMedia.

Preparation of Medium: Add components to distilled/deionized water and bring volume to 1.0L. Mix thoroughly. Adjust pH to 5.3–5.4. Gently heat and bring to boiling. Distribute into tubes or flasks. Autoclave for 15 min at 15 psi pressure–121°C.

Use: For the differentiation of homofermentative and heterofermentative lactic acid bacteria.

Lactic HiVeg Agar

Composition per liter:

Plant hydrolysate	22.5g
Agar	15.0g
Glucose	5.0g
Lactose	5.0g
Sucrose	5.0g
Yeast extract	5.0g
NaCl	4.0g
Sodium acetate	1.5g
Ascorbic acid	0.5g

pH 6.0 ± 0.2 at 25°C

Source: This medium is available as a premixed powder from HiMedia.

Preparation of Medium: Add components to distilled/deionized water and bring volume to 1.0L. Mix thoroughly. Adjust pH to 5.3–5.4. Gently heat and bring to boiling. Distribute into tubes or flasks. Autoclave for 15 min at 15 psi pressure–121°C.

Use: For the enumeration and identification of lactic streptococci and lactobacilli by the pour plate method.

Lactic Streak Agar

Composition per liter:

Agar	15.0g
Sodium carboxymethylcellulose	10.0g
Calcium citrate	10.0g
Beef extract	5.0g
Papaic digest of soybean meal	5.0g
Polypeptone™	5.0g
Yeast extract	5.0g
Lactose	1.5g
L-Arginine·HCl	1.5g
Bromcresol Purple (0.1% solution)	20.0mL

pH 6.0 ± 0.1 at 25°C

Preparation of Medium: Add components—except Bromcresol Purple solution, calcium citrate, and sodium carboxymethylcellulose—to distilled/deionized water and bring volume to 800.0mL. Mix thoroughly. Gently heat and bring to boiling. In a blender, add the calcium citrate and sodium carboxymethylcellulose to 200.0mL of distilled/deionized water. Blend until mixed. Add the 200.0mL of citrate/carboxymethylcellulose solution to the hot agar solution. Mix thoroughly. Adjust pH to 6.0. Distribute into flasks in 100.0mL volumes. Autoclave for 10 min at 10 psi pressure–115°C. Cool to 45°–50°C. Aseptically add 2.0mL of sterile Bromcresol Purple to each flask. Mix thoroughly. Pour into sterile Petri dishes. Allow plates to dry at 37°C for 1 hr before use.

Use: For the differentiation of lactic streptococci. Bacteria that produce acid from lactose appear as yellow colonies.

Lactic Streak HiVeg Agar

Composition per liter:

Agar	15.0g
Calcium citrate	10.0g
Sodium carboxymethylcellulose	10.0g
Plant extract	5.0g
Plant peptone	5.0g
Papaic digest of soybean meal	5.0g
Yeast extract	5.0g
L-Arginine hydrochloride	1.5g
Lactose	1.5g
Bromcresol Purple	2.0mg

pH 6.0 ± 0.2 at 25°C

Source: This medium is available as a premixed powder from HiMedia.

Preparation of Medium: Add components to distilled/deionized water and bring volume to 1.0mL. Mix thoroughly. Gently heat and bring to boiling. Adjust pH to 6.0. Autoclave for 15 min at 15 psi pressure–121°C. Pour into sterile Petri dishes.

Use: For the differentiation of lactic streptococci. Bacteria that produce acid from lactose appear as yellow colonies.

Lactobacilli Agar, AOAC (Lactobacilli Agar, Association of Official Analytical Chemists)

Composition per liter:

Milk, peptonized	15.0g
Agar	10.0g
Glucose	10.0g
Yeast extract	5.0g
KH_2PO_4	2.0g
Sorbitan monooleate complex	1.0g
Tomato juice	100.0mL

pH 6.8 ± 0.2 at 25°C

Source: This medium is available as a premixed powder from BD Diagnostic Systems.

Preparation of Medium: Add components to distilled/deionized water and bring volume to 1.0L. Mix thoroughly. Gently heat and bring to boiling. Continue boiling for 2–3 min. Distribute into tubes in

10.0mL volumes. Autoclave for 15 min at 15 psi pressure–121°C. Allow tubes to cool in an upright position.

Use: For the cultivation and maintenance of stock cultures of *Lactobacillus casei* ATCC 7469, *Lactobacillus fermentum* ATCC 9338, *Lactobacillus leichmannii* ATCC 4797, and *Lactobacillus viridescens* ATCC 12706 used in the microbiological assay of vitamins.

Lactobacilli Broth, AOAC (Lactobacilli Broth, Association of Official Analytical Chemists)

Composition per liter:

Milk, peptonized	15.0g
Glucose	10.0g
Yeast extract	5.0g
KH_2PO_4	2.0g
Sorbitan monooleate complex	1.0g
Tomato juice	100.0mL

pH 6.8 ± 0.2 at 25°C

Source: This medium is available as a premixed powder from BD Diagnostic Systems.

Preparation of Medium: Add components to distilled/deionized water and bring volume to 1.0L. Mix thoroughly. Gently heat and bring to boiling. Continue boiling for 2–3 min. Distribute into tubes in 10.0mL volumes. Autoclave for 15 min at 15 psi pressure–121°C.

Use: For the cultivation and preparation of inocula of stock cultures of *Lactobacillus casei* ATCC 7469, *Lactobacillus fermentum* ATCC 9338, *Lactobacillus leichmannii* ATCC 4797, and *Lactobacillus viridescens* ATCC 12706 used in the microbiological assay of vitamins.

Lactobacilli deMan-Rogosa-Sharpe Broth

See: **Lactobacilli MRS Broth**

Lactobacilli HiVeg Broth (Elliker Broth, HiVeg)

Composition per liter:

Plant extract	20.0g
Yeast extract	10.0g
Gelatin	4.0g
Glucose	3.0g
Ascorbic acid	2.5g
Lactose	2.5g
NaCl	2.5g
Sodium acetate	2.5g
Sucrose	2.5g

pH 6.8 ± 0.1 at 25°C

Source: This medium is available as a premixed powder from HiMedia.

Preparation of Medium: Add components to distilled/deionized water and bring volume to 1.0mL. Mix thoroughly. Gently heat and bring to boiling. Autoclave for 15 min at 15 psi pressure–121°C.

Use: For the cultivation of *Lactobacillus* species and lactic streptococci, from foods.

Lactobacilli MRS Broth (Lactobacilli deMan-Rogosa-Sharpe Broth)

Composition per liter:

Glucose	20.0g
Beef extract	10.0g
Peptone	10.0g
Sodium acetate	5.0g
Yeast extract	5.0g
Ammonium citrate	2.0g
Na_2HPO_4	2.0g
Tween™ 80	1.0g
$MgSO_4 \cdot 7H_2O$	0.1g
$MnSO_4 \cdot 5H_2O$	0.05g

pH 6.5 ± 0.2 at 25°C

Source: This medium is available as a premixed powder from BD Diagnostic Systems.

Preparation of Medium: Add components to distilled/deionized water and bring volume to 1.0L. Mix thoroughly. Gently heat and bring to boiling. Distribute into tubes or flasks. Autoclave for 15 min at 15 psi pressure–121°C.

Use: For the cultivation of *Lactobacillus* species. Also used for the cultivation and maintenance of *Aerococcus viridans, Bifidobacterium coryneforme, Lactococcus plantarum, Leuconostoc* species, *Pectinatus cerevisiiphilus, Pediococcus* species, and *Sporolactobacillus inulinus.*

Lactobacilli MRS Broth with Mevalonic Acid

See: ***Pediococcus* Medium with Mevalonic Acid**

Lactobacilli MRS Broth

See: ***Pediococcus* Medium**

Lactobacilli MRS Broth with Cysteine

Composition per liter:

Glucose	20.0g
Beef extract	10.0g
Peptone	10.0g
Sodium acetate	5.0g
Yeast extract	5.0g
L-Cysteine	2.0g
Ammonium citrate	2.0g
Na_2HPO_4	2.0g
Tween™ 80	1.0g
$MgSO_4 \cdot 7H_2O$	0.1g
$MnSO_4 \cdot 5H_2O$	0.05g

pH 6.5 ± 0.2 at 25°C

Preparation of Medium: Add components to distilled/deionized water and bring volume to 1.0L. Mix thoroughly. Gently heat and bring to boiling. Distribute into tubes or flasks. Autoclave for 15 min at 15 psi pressure–121°C.

Use: For the cultivation and maintenance of *Lactobacillus* species, including *L. ruminis* and *L. vitulinus.*

Lactobacilli MRS Broth with 0.5% Cysteine (ATCC 1844)

Composition per liter:

Glucose	20.0g
Beef extract	10.0g
Peptone	10.0g
Sodium acetate	5.0g
Yeast extract	5.0g
L-Cysteine·HCl	5.0g
Ammonium citrate	2.0g

Na_2HPO_4	2.0g
Tween™ 80	1.0g
$MgSO_4 \cdot 7H_2O$	0.1g
$MnSO_4 \cdot 5H_2O$	0.05g

pH 6.5 ± 0.2 at 25°C

Preparation of Medium: Add components to distilled/deionized water and bring volume to 1.0L. Mix thoroughly. Gently heat and bring to boiling. Distribute into tubes or flasks. Autoclave for 15 min at 15 psi pressure–121°C.

Use: For the cultivation of *Lactobacillus sharpeae.*

Lactobacilli MRS Broth with Ethanol

Composition per liter:

Glucose	20.0g
Beef extract	10.0g
Peptone	10.0g
Sodium acetate	5.0g
Yeast extract	5.0g
L-Cysteine	2.0g
Ammonium citrate	2.0g
Na_2HPO_4	2.0g
Tween™ 80	1.0g
$MgSO_4 \cdot 7H_2O$	0.1g
$MnSO_4 \cdot 5H_2O$	0.05g
Ethanol (95% solution)	100.0mL

pH5.0 ± 0.2 at 25°C

Preparation of Medium: Add components, except ethanol, to distilled/deionized water and bring volume to 990.0mL. Mix thoroughly. Gently heat and bring to boiling. Autoclave for 15 min at 15 psi pressure–121°C. Cool to 25°C. Filter sterilize ethanol. Aseptically add 100.0mL of sterile ethanol. Mix thoroughly. Aseptically distribute into sterile tubes or flasks.

Use: For the cultivation and maintenance of *Lactobacillus* species.

Lactobacillus Agar 2

Composition per liter:

Glucose	20.0g
Sodium acetate	20.0g
Agar	15.0g
Tryptic digest of casein	10.0g
Yeast extract	5.0g
Meat extract	2.0g
K_2HPO_4	0.5g
KH_2PO_4	0.5g
$MgSO_4 \cdot 7H_2O$	200.0mg
DL-Mevalonic acid	30.0mg
$FeSO_4 \cdot 7H_2O$	10.0mg
$MnSO_4 \cdot H_2O$	7.5mg
Ethanol	40.0mL
Tween™ 80	1.0mL

pH 5.2 ± 0.2 at 25°C

Preparation of Medium: Add components to distilled/deionized water and bring volume to 1.0L. Mix thoroughly. Gently heat and bring to boiling. Distribute into tubes or flasks. Autoclave for 15 min at 15 psi pressure–121°C. Pour into sterile Petri dishes or leave in tubes.

Use: For the cultivation and maintenance of *Lactobacillus fructivorans* and *Lactobacillus homohiochii.*

Lactobacillus bifidus Medium

Composition per liter:

Lactose	70.0g
Sodium acetate, anhydrous	50.0g
Pancreatic digest of casein	10.0g
K_2HPO_4	5.0g
Tween™ 80	1.0g
Alanine	0.4g
L-Cystine	0.4g
Tryptophan	0.4g
Asparagine	0.2g
Adenine	0.02g
Guanine	0.02g
Uracil	0.02g
Xanthine	0.02g
Pyridoxine·HCl	2.4mg
Nicotinic acid	1.2mg
Calcium pantothenate	0.8mg
Riboflavin	0.4mg
Thiamine·HCl	0.4mg
p-Aminobenzoic acid	0.02mg
Folic acid	0.02mg
Biotin	8.0μg
Ascorbic acid solution	100.0mL
Human milk, skimmed	20.0mL
Salts B	10.0mL

pH 6.8 ± 0.2 at 25°C

Salts B:

Composition per 250.0mL:

$MgSO_4 \cdot 7H_2O$	10.0g
$FeSO_4 \cdot 7H_2O$	0.5g
NaCl	0.5g
$MnSO_4 \cdot 2H_2O$	0.337g

Preparation of Salts B: Add components to distilled/deionized water and bring volume to 250.0mL. Mix thoroughly.

Ascorbic Acid Solution:

Composition per 100.0mL:

Ascorbic acid	1.0g

Preparation of Ascorbic Acid Solution: Add ascorbic acid to distilled/deionized water and bring volume to 100.0mL. Mix thoroughly. Adjust pH to 6.5. Filter sterilize.

Preparation of Medium: Add components, except ascorbic acid solution and human milk, to distilled/deionized water and bring volume to 900.0mL. Mix thoroughly. Gently heat and bring to boiling. Autoclave for 15 min at 15 psi pressure–121°C. Cool to 45°–50°C. Aseptically add 100.0mL of sterile ascorbic acid solution. Mix thoroughly. This constitutes a double-strength medium. To prepare a single-strength medium, aseptically combine 500.0mL of sterile double-strength medium, 480.0mL of sterile distilled/deionized water, and 20.0mL of sterile human milk. Aseptically distribute into sterile tubes or flasks.

Use: For the cultivation and maintenance of *Bifidobacterium (Lactobacillus) bifidum.*

Lactobacillus bulgaricus Agar (LB Agar)

Composition per 900.0mL:

Agar	20.0g
Glucose	20.0g

Beef extract	10.0g
Pancreatic digest of casein	10.0g
Yeast extract	5.0g
K_2HPO_4	2.0g
Tween™ 80	1.0g
Acetate buffer	80.0mL
Tomato juice, filtered	40.0mL

pH 6.8 ± 0.2 at 25°C

Acetate Buffer:

Composition per liter:

Sodium acetate	113.55g
Acetic acid	10.0mL

Preparation of Acetate Buffer: Add components to distilled/deionized water and bring volume to 1.0L. Mix thoroughly.

Preparation of Medium: Add components, except acetate buffer, to distilled/deionized water and bring volume to 920.0mL. Mix thoroughly. Gently heat and bring to boiling. Adjust pH to 6.8. Add 80.0mL of acetate buffer. Mix thoroughly. Autoclave for 15 min at 15 psi pressure–121°C. Pour into sterile Petri dishes.

Use: For the isolation, cultivation, and enumeration of *Lactobacillus bulgaricus* from foods.

Lactobacillus bulgaricus HiVeg Agar Base with Tomato Juice and Acetate Buffer

Composition per liter:

Agar	20.0g
Glucose	20.0g
Plant extract	10.0g
Plant hydrolysate	10.0g
Yeast extract	5.0g
K_2HPO_4	2.0g
Tomato juice	2.0g
Polysorbate 80	1.0g
Acetate buffer	80.0mL
Tomato juice, filtered	40.0mL

pH 6.8 ± 0.2 at 25°C

Source: This medium, without tomato juice and acetate buffer, is available as a premixed powder from HiMedia.

Acetate Buffer:

Composition per liter:

Sodium acetate	113.55g
Acetic acid	10.0mL

Preparation of Acetate Buffer: Add components to distilled/deionized water and bring volume to 1.0L. Mix thoroughly.

Preparation of Medium: Add components, except acetate buffer, to distilled/deionized water and bring volume to 920.0mL. Mix thoroughly. Gently heat and bring to boiling. Adjust pH to 6.8. Add 80.0mL of acetate buffer. Mix thoroughly. Autoclave for 15 min at 15 psi pressure–121°C. Pour into sterile Petri dishes.

Use: For the isolation, cultivation, and enumeration of *Lactobacillus bulgaricus* from foods.

Lactobacillus Chloramphenicol Agar 1

Composition per liter:

Milk, peptonized	15.0g
Agar	10.0g
Glucose	10.0g
Yeast extract	5.0g
KH_2PO_4	2.0g
Sorbitan monooleate complex	1.0g
Chloramphenicol	0.3g
Tomato juice	100.0mL

pH 6.8 ± 0.2 at 25°C

Preparation of Medium: Add components to distilled/deionized water and bring volume to 1.0L. Mix thoroughly. Gently heat and bring to boiling. Continue boiling for 2–3 min. Distribute into tubes in 10.0mL volumes. Autoclave for 15 min at 15 psi pressure–121°C. Allow tubes to cool in an upright position.

Use: For the cultivation of *Lactobacillus casei.*

Lactobacillus Chloramphenicol Broth 1

Composition per liter:

Milk, peptonized	15.0g
Glucose	10.0g
Yeast extract	5.0g
KH_2PO_4	2.0g
Sorbitan monooleate complex	1.0g
Chloramphenicol	0.3g
Tomato juice	100.0mL

pH 6.8 ± 0.2 at 25°C

Preparation of Medium: Add components to distilled/deionized water and bring volume to 1.0L. Mix thoroughly. Gently heat and bring to boiling. Continue boiling for 2–3 min. Distribute into tubes in 10.0mL volumes. Autoclave for 15 min at 15 psi pressure–121°C.

Use: For the cultivation of *Lactobacillus casei.*

Lactobacillus Chloramphenicol Medium 2

Composition per liter:

Glucose	18.5g
Agar	13.5g
Pancreatic digest of gelatin	10.0g
Beef extract	8.0g
Yeast extract	4.0g
Sodium acetate	3.0g
K_2HPO_4	2.0g
Ammonium citrate	2.0g
Polysorbate 80	1.0g
$MgSO_4 \cdot 7H_2O$	0.2g
Chloramphenicol	0.1g
$MnSO_4 \cdot 4H_2O$	0.05g

pH 6.2 ± 0.2 at 25°C

Preparation of Medium: Add components to distilled/deionized water and bring volume to 1.0L. Mix thoroughly. Gently heat while stirring and bring to boiling. Distribute into tubes or flasks. Autoclave for 15 min at 15 psi pressure–121°C. Pour into sterile Petri dishes or leave in tubes.

Use: For the cultivation of *Lactobacillus casei* and *Pediococcus acidilactici.*

Lactobacillus Heteroferm Screen Agar (MRS, Modified)

Composition per liter:

Glucose	20.0g
Agar	15.0g
Proteose peptone No. 3	10.0g

Sodium acetate	5.0g
Yeast extract	5.0g
2-Phenylethyl alcohol	3.0g
Ammonium citrate	2.0g
K_2HPO_4	2.0g
$MgSO_4 \cdot 7H_2O$	0.1g
$MnSO_4 \cdot H_2O$	0.05g
Bromcresol Green	0.04g
Cycloheximide	4.0mg
Tween™ 80	1.0mL

pH 5.5 ± 0.01 at 25°C

Caution: Cycloheximide is toxic. Avoid skin contact or aerosol formation and inhalation.

Preparation of Medium: Add components to distilled/deionized water and bring volume to 1.0L. Mix thoroughly. Autoclave for 15 min at 15 psi pressure–121°C. Cool to 50°C. Adjust pH to 5.5 with glacial acetic acid. Pour into sterile Petri dishes.

Use: For the isolation and cultivation of *Lactobacillus* species from salad dressings.

Lactobacillus Heteroferm Screen Broth (MRS, Modified)

Composition per liter:

Glucose	20.0g
Proteose peptone No. 3	10.0g
Sodium acetate	5.0g
Yeast extract	5.0g
2-Phenylethyl alcohol	3.0g
Ammonium citrate	2.0g
K_2HPO_4	2.0g
$MgSO_4 \cdot 7H_2O$	0.1g
$MnSO_4 \cdot H_2O$	0.05g
Bromcresol Green	0.04g
Cycloheximide	4.0mg
Tween™ 80	1.0mL

pH 4.3 ± 0.01 at 25°C

Caution: Cycloheximide is toxic. Avoid skin contact or aerosol formation and inhalation.

Preparation of Medium: Add components to distilled/deionized water and bring volume to 1.0L. Mix thoroughly. Adjust pH to 4.3 with concentrated HCl. Distribute into test tubes that contain an inverted Durham tube. Autoclave for 15 min at 15 psi pressure–121°C.

Use: For the isolation and cultivation of *Lactobacillus* species from foods.

Lactobacillus homohiochii Medium

Composition per liter:

D-Glucose	20.0g
Tryptic digest of casein	10.0g
Yeast extract	5.0g
Meat extract	2.0g
K_2HPO_4	0.5g
KH_2PO_4	0.5g
$MgSO_4 \cdot 7H_2O$	0.2g
DL-Mevalonic acid	30.0mg
Sodium acetate	20.0mg
$FeSO_4 \cdot 7H_2O$	10.0mg
$MnSO_4 \cdot H_2O$	7.5mg
Ethanol	40.0mL
Tween™ 80	1.0mL

pH 5.2 ± 0.2 at 25°C

Preparation of Medium: Add components to distilled/deionized water and bring volume to 1.0L. Mix thoroughly. Adjust pH to 5.2. Distribute into tubes or flasks. Autoclave for 15 min at 15 psi pressure–121°C.

Use: For the cultivation and maintenance of *Lactobacillus homohiochii.*

Lactobacillus Medium (ATCC Medium 78)

Composition per liter:

Pancreatic digest of casein	20.0g
Agar	15.0g
Tryptose	5.0g
Yeast extract	5.0g
Glucose	3.0g
Lactose	2.0g
Liver extract concentrate	1.0g
Tween™ 80	0.05g
Tomato juice, filtered	200.0mL

pH 6.5 ± 0.2 at 25°C

Preparation of Medium: Add components to distilled/deionized water and bring volume to 1.0L. Mix thoroughly. Gently heat and bring to boiling. Adjust pH to 6.5. Distribute into tubes or flasks. Autoclave for 15 min at 15 psi pressure–121°C. Pour into sterile Petri dishes or leave in tubes.

Use: For the cultivation and maintenance of *Lactobacillus fermentum* and *Lactobacillus salivarius.*

Lactobacillus Medium (ATCC Medium 169)

Composition per liter:

Glucose	10.0g
Proteose peptone	7.5g
Yeast extract	7.5g
KH_2PO_4	2.0g
L-Cysteine·HCl·H_2O	1.0g
Tween™ 80	0.1g
Tomato juice, filtered	100.0mL

pH 7.0 ± 0.2 at 25°C

Preparation of Medium: Add components to distilled/deionized water and bring volume to 1.0L. Mix thoroughly. Gently heat and bring to boiling. Adjust pH to 7.0. Distribute into tubes or flasks. Autoclave for 15 min at 15 psi pressure–121°C. Pour into sterile Petri dishes or leave in tubes.

Use: For the cultivation and maintenance of *Lactobacillus casei.*

Lactobacillus Medium (ATCC Medium 1006)

Composition per liter:

Agar	20.0g
Pancreatic digest of casein	10.0g
Glucose	5.0g
Yeast extract	5.0g
KH_2PO_4	3.0g
K_2HPO_4	3.0g

Tryptose 3.0g
Sodium acetate 1.0g
L-Cysteine·HCl·H_2O 0.2g
Salt solution R 5.0mL
Tween™ 80 .... 1.0mL
pH 6.3 ± 0.2 at 25°C

Salt Solution R:

Composition per 100.0mL:

$MgSO_4·7H_2O$ 11.5g
$MnSO_4·2H_2O$ 2.4g
$FeSO_4·7H_2O$ 0.68g

Preparation of Salt Solution R: Add components to distilled/deionized water and bring volume to 100.0mL. Mix thoroughly. Store at 4°C.

Preparation of Medium: Add components to distilled/deionized water and bring volume to 1.0L. Mix thoroughly. Gently heat and bring to boiling. Distribute into tubes or flasks. Autoclave for 15 min at 15 psi pressure–121°C. Pour into sterile Petri dishes or leave in tubes.

Use: For the cultivation and maintenance of *Lactobacillus delbrueckii* and *Lactobacillus jensenii.*

Lactobacillus Medium

Composition per liter:

Glucose 20.0g
Sodium acetate 20.0g
Tryptone 10.0g
Yeast extract 5.0g
Beef extract 2.0g
K_2HPO_4 .... 0.5g
KH_2PO_4 .... 0.5g
$MgSO_4·7H_2O$ 200.0mg
$FeSO_4·7H_2O$ 10.0mg
$MnSO_4·H_2O$ 7.5mg
Ethanol 40.0mL
Mevalonic acid solution 10.0mL
Tween™ 80 .... 1.0mL
pH 5.2 ± 0.2 at 25°C

Mevalonic Acid Solution:

Composition per 10.0mL:

DL-Mevalonic acid 30.0mg

Preparation of Mevalonic Acid Solution: Add DL-mevalonic acid to distilled/deionized water and bring volume to 10.0mL. Mix thoroughly. Filter sterilize.

Preparation of Medium: Add components, except mevalonic acid solution and ethanol, to distilled/deionized water and bring volume to 950.0mL. Mix thoroughly. Autoclave for 15 min at 15 psi pressure–121°C. Aseptically add 40.0mL of filter-sterilized ethanol and 10.0mL of sterile mevalonic acid solution. Mix thoroughly. Aseptically distribute into sterile tubes or flasks.

Use: For the cultivation of *Lactobacillus homohiochii.*

Lactobacillus Medium (*Lactobacillus* Sake Medium)

Composition per liter:

Agar 13.0g
Yeast extract 5.0g
Liver extract concentrate 0.2g
Sake 700.0mL

Preparation of Medium: Add components to distilled/deionized water and bring volume to 1.0L. Mix thoroughly. Gently heat and bring to boiling. Distribute into tubes or flasks. Autoclave for 15 min at 15 psi pressure–121°C. Pour into sterile Petri dishes or leave in tubes.

Use: For the cultivation and maintenance of *Lactobacillus fructivorans* and *Lactobacillus homohiochii.*

Lactobacillus Medium II (DSMZ Medium 93)

Composition per liter:

Glucose 20.0g
Na-acetate 20.0g
Agar 15.0g
Pancreatic digest of casein 10.0g
Yeast extract 5.0g
Meat extract 2.0g
KH_2PO_4 .... 0.5g
K_2HPO_4 .... 0.5g
$MgSO_4·7H_2O$ 0.2g
DL-Mevalonic acid 30.0mg
$FeSO_4·7H_2O$ 10.0mg
$MnSO_4·2H_2O$ 7.5mg
Ethanol 40.0mL
Tween™ 80 .... 1.0mL
pH 5.2 ± 0.2 at 25°C

Preparation of Medium: Add components to distilled/deionized water and bring volume to 1.0L. Mix thoroughly. Gently heat and bring to boiling. Distribute into tubes or flasks. Autoclave for 15 min at 15 psi pressure–121°C. Pour into sterile Petri dishes or leave in tubes.

Use: For the cultivation and maintenance of *Lactobacillus fructivorans* and *Lactobacillus homohiochii.*

Lactobacillus Medium III

Composition per liter:

Pancreatic digest of casein 10.0g
Fructose 7.0g
Glucose 7.0g
Maltose 7.0g
Meat extract 5.0g
Sodium acetate·$3H_2O$ 5.0g
Yeast extract 5.0g
$K_2HPO_4·3H_2O$ 2.6g
Diammonium citrate 2.0g
Sodium gluconate 2.0g
L-Cysteine·HCl 0.5g
$MgSO_4·7H_2O$ 0.1g
$MnSO_4·4H_2O$ 0.05g
Tween™ 80 .... 1.0mL
pH 6.3 ± 0.2 at 25°C

Preparation of Medium: Add components to distilled/deionized water and bring volume to 1.0L. Mix thoroughly. Adjust pH to 6.3. Distribute into tubes or flasks. Autoclave for 15 min at 15 psi pressure–121°C.

Use: For the cultivation of *Lactobacillus* species.

Lactobacillus Medium IV (DSMZ Medium 859)

Composition per liter:

Tryptone 10.0g
Yeast extract 5.0g

Na-acetate·$3H_2O$	5.0g
Glucose	4.0g
Maltose	4.0g
Meat extract	3.0g
K_2HPO_4·$3H_2O$	2.6g
$(NH_4)_2$ citrate	2.0g
Cysteine-HCl·H_2O	0.5g
$MgSO_4$·$7H_2O$	0.1g
$MnSO_4$·$4H_2O$	0.05g
Tween™ 80	1.0mL

pH 6.3 ± 0.2 at 25°C

Preparation of Medium: Add components to distilled/deionized water and bring volume to 1.0L. Mix thoroughly. Distribute into tubes or flasks. Autoclave for 15 min at 15 psi pressure–121°C.

Use: For the cultivation of *Lactobacillus panis.*

Lactobacillus 8664 Medium

Composition per liter:

Maltose	20.0g
Peptone	10.0g
Yeast extract	10.0g
Glucose	5.0g

Preparation of Medium: Add components to distilled/deionized water and bring volume to 1.0L. Mix thoroughly. Distribute into tubes or flasks. Autoclave for 15 min at 15 psi pressure–121°C.

Use: For the cultivation of *Lactobacillus brevis*.

Lactobacillus MRS HiVeg Agar (MRS HiVeg Agar)

Composition per liter:

Glucose	20.0g
Agar	12.0g
Plant extract	10.0g
Plant peptone No. 3	10.0g
Sodium acetate	5.0g
Yeast extract	5.0g
Ammonium citrate	2.0g
K_2HPO_4	2.0g
Polysorbate 80	1.0g
$MgSO_4$	0.1g
$MnSO_4$	0.05g

pH 6.5 ± 0.2 at 25°C

Source: This medium is available as a premixed powder from HiMedia.

Preparation of Medium: Add components to distilled/deionized water and bring volume to 1.0mL. Mix thoroughly. Gently heat and bring to boiling. Adjust pH to 6.5. Autoclave for 15 min at 15 psi pressure–121°C. Pour into sterile Petri dishes.

Use: For the cultivation of *Lactobacillus* species.

Lactobacillus MRS HiVeg Broth (MRS HiVeg Broth)

Composition per liter:

Glucose	20.0g
Plant extract	10.0g
Plant peptone No. 3	10.0g
Sodium acetate	5.0g
Yeast extract	5.0g
Ammonium citrate	2.0g
K_2HPO_4	2.0g
Polysorbate 80	1.0g
$MgSO_4$	0.1g
$MnSO_4$	0.05g

pH 6.5 ± 0.2 at 25°C

Source: This medium is available as a premixed powder from HiMedia.

Preparation of Medium: Add components to distilled/deionized water and bring volume to 1.0L. Mix thoroughly. Gently heat and bring to boiling. Distribute into tubes or flasks. Autoclave for 15 min at 15 psi pressure–121°C.

Use: For the cultivation of *Lactobacillus* species.

Lactobacillus Orotic Acid Medium

Composition per liter:

Milk, peptonized	15.0g
Glucose	10.0g
Yeast extract	5.0g
KH_2PO_4	2.0g
Sorbitan monooleate complex	1.0g
Orotic acid	25.0mg
D-Pantothenic acid	0.2mg
Tomato juice	100.0mL
L-Cysteine·HCl·H_2O solution	7.5mL

pH 6.8 ± 0.2 at 25°C

L-Cysteine·HCl·H_2O Solution:

Composition per 10.0mL:

L-Cysteine·HCl·H_2O	0.15g

Preparation of L-Cysteine·HCl·H_2O Solution: Add L-cysteine·HCl·H_2O to distilled/deionized water and bring volume to 10.0mL. Mix thoroughly.

Preparation of Medium: Add components to distilled/deionized water and bring volume to 1.0L. Mix thoroughly. Gently heat and bring to boiling. Continue boiling for 2–3 min. Distribute into bottles in 20.0mL volumes. Add 0.15mL of L-cysteine·HCl·H_2O solution to each bottle containing 20.0mL of medium. Autoclave for 15 min at 15 psi pressure–121°C. Screw caps tightly to maintain reduced conditions.

Use: For the cultivation of *Lactobacillus helveticus*.

Lactobacillus rimae Medium

Composition per liter:

Yeast extract	10.0g
Peptone	5.0g
Pancreatic digest of casein	5.0g
Glucose	5.0g
$(NH_4)_2SO_4$	0.5g
L-Cysteine·HCl	0.5g
Resazurin	1.0mg
Mineral solution	40.0mL
Fatty acid mixture	3.1mL
Hemin solution	0.5mL
Vitamin K_1	0.2mL

pH 6.9 ± 0.2 at 25°C

Mineral Solution:

Composition per liter:

$NaHCO_3$	10.0g
NaCl	2.0g

KH_2PO_4	1.0g
K_2HPO_4	1.0g
$MgSO_4 \cdot 7H_2O$	0.48g
$CaCl_2 \cdot 2H_2O$	0.3g

Preparation of Mineral Solution: Add components to distilled/deionized water and bring volume to 1.0L. Mix thoroughly.

Fatty Acid Mixture:
Composition per 31.0mL:

Acetic acid	17.0mL
Propionic acid	6.0mL
n-Butyric acid	4.0mL
n-Valeric acid	1.0mL
iso-Valeric acid	1.0mL
iso-Butyric acid	1.0mL
DL-2-Methylbutyric acid	1.0mL

Preparation of Fatty Acid Mixture: Combine components. Mix thoroughly. Adjust pH to 7.5 with concentrated NaOH.

Hemin Solution:
Composition per 1.0mL:

Hemin	5.0mg
NaOH (1*N* solution)	1.0mL

Preparation of Hemin Solution: Add hemin to 1.0mL of NaOH solution. Mix thoroughly.

Preparation of Medium: Add components, except L-cysteine·HCl, hemin solution, and fatty acid mixture, to distilled/deionized water and bring volume to 1.0L. Mix thoroughly. Gently heat and bring to boiling. Continue boiling for 5 min. Cool to room temperature while sparging with 100% CO_2. Add L-cysteine·HCl, hemin solution, and fatty acid mixture. Adjust pH to 6.9 with 8*N* NaOH while continuing to sparge with 100% CO_2. After pH has been reached, sparge with 100% N_2. Anaerobically distribute into tubes or flasks. Autoclave for 15 min at 15 psi pressure–121°C.

Use: For the cultivation and maintenance of *Acetivibrio ethanolgignens, Lactobacillus rimae, Lactobacillus uli,* and *Sphaerotilus natans.*

Lactobacillus rimae Medium with Tween™

Composition per liter:

Yeast extract	10.0g
Peptone	5.0g
Pancreatic digest of casein	5.0g
Glucose	5.0g
$(NH_4)_2SO_4$	0.5g
L-Cysteine·HCl	0.5g
Tween™ 80	0.2g
Resazurin	1.0mg
Mineral solution	40.0mL
Fatty acid mixture	3.1mL
Hemin solution	0.5mL
Vitamin K_1	0.2mL

pH 6.9 ± 0.2 at 25°C

Mineral Solution:
Composition per liter:

$NaHCO_3$	10.0g
NaCl	2.0g
KH_2PO_4	1.0g
K_2HPO_4	1.0g
$MgSO_4 \cdot 7H_2O$	0.48g
$CaCl_2 \cdot 2H_2O$	0.3g

Preparation of Mineral Solution: Add components to distilled/deionized water and bring volume to 1.0L. Mix thoroughly.

Fatty Acid Mixture:
Composition per 31.0mL:

Acetic acid	17.0mL
Propionic acid	6.0mL
n-Butyric acid	4.0mL
n-Valeric acid	1.0mL
iso-Valeric acid	1.0mL
iso-Butyric acid	1.0mL
DL-2-Methylbutyric acid	1.0mL

Preparation of Fatty Acid Mixture: Combine components. Mix thoroughly. Adjust pH to 7.5 with concentrated NaOH.

Hemin Solution:
Composition per 1.0mL:

Hemin	5.0mg
NaOH (1*N* solution)	1.0mL

Preparation of Hemin Solution: Add hemin to 1.0mL of NaOH solution. Mix thoroughly.

Preparation of Medium: Add components, except L-cysteine·HCl, hemin solution, and fatty acid mixture, to distilled/deionized water and bring volume to 1.0L. Mix thoroughly. Gently heat and bring to boiling. Continue boiling for 5 min. Cool to room temperature while sparging with 100% CO_2. Add L-cysteine·HCl, hemin solution, and fatty acid mixture. Adjust pH to 6.9 with 8*N* NaOH while continuing to sparge with 100% CO_2. After pH has been reached, sparge with 100% N_2. Anaerobically distribute into tubes or flasks. Autoclave for 15 min at 15 psi pressure–121°C.

Use: For the cultivation and maintenance of *Lactobacillus rimae* and *Lactobacillus uli.*

Lactobacillus Selection Agar
See: **LBS™ Agar**

Lactobacillus Selection Agar Base

Composition per liter:

Sodium acetate	25.0g
Glucose	20.0g
Agar	15.0g
Casein enzymic hydrolysate	10.0g
KH_2PO_4	6.0g
Yeast extract	5.0g
Ammonium citrate	2.0g
Polysorbate 80	1.0g
$MgSO_4 \cdot 7H_2O$	0.575g
$MnSO_4 \cdot 2H_2O$	0.12g
$FeSO_3$	0.034g
Acetic acid, glacial	1.32mL

pH 5.5 ± 0.2 at 25°C

Source: This medium is available from HiMedia.

Preparation of Medium: Add acetic acid to distilled/deionized water and bring volume to 1.0L. Add remaining components. Mix thoroughly. Distribute into tubes or flasks. Gently heat and bring to boiling. Boil for 1–2 min. Do not autoclave unless storage is needed. If storage is necessary sterilize by autoclaving for 15 min at 15 psi pressure–121°C. Pour into sterile Petri dishes or leave in tubes.

Use: For the isolation and enumeration of lactobacilli from foods.

Lactobacillus Selection Broth

See: **LBS™ Broth**

Lactobacillus Selection HiVeg Agar Base with Acetic Acid and Polysorbate

Composition per liter:

Sodium acetate·3H_2O	25.0g
Glucose	20.0g
Agar	15.0g
Plant hydrolysate	10.0g
KH_2PO_4	6.0g
Yeast extract	5.0g
Ammonium citrate	2.0g
$MgSO_4$	0.575g
$FeSO_4$	0.034g
$MnSO_4$	0.12g
Acetic acid, glacial	1.32 mL
Polysorbate 80	1.0mL

pH 5.5 ± 0.2 at 25°C

Source: This medium, without polysorbate 80 or glacial acetic acid, is available as a premixed powder from HiMedia.

Preparation of Medium: Add components to distilled/deionized water and bring volume to 1.0L.Mix thoroughly. Gently heat and bring to boiling. Adjust pH to 5.5. Do not autoclave. Pour into sterile Petri dishes or distribute into sterile tubes.

Use: For the selective isolation, cultivation, and enumeration of lactobacilli from foods.

Lactobacillus Selection HiVeg Broth Base with Acetic Acid and Polysorbate

Composition per liter:

Sodium acetate	25.0g
Glucose	20.0g
Plant hydrolysate	10.0g
KH_2PO_4	6.0g
Yeast extract	5.0g
Polysorbate 80	1.0g
$MgSO_4$	0.575g
$MnSO_4$	0.12g
$FeSO_4$	0.034g
Acetic acid, glacial	1.32 mL
Polysorbate 80	1.0mL

pH 5.5 ± 0.2 at 25°C

Source: This medium, without polysorbate 80 or glacial acetic acid, is available as a premixed powder from HiMedia.

Preparation of Medium: Add components to distilled/deionized water and bring volume to 1.0L. Mix thoroughly. Gently heat and bring to boiling. Adjust pH to 5.5. Do not autoclave. Pour into sterile Petri dishes or distribute into sterile tubes.

Use: For the cultivation of lactobacilli.

Lactobacillus Selection Oxgall Agar (LBS™ Oxgall Agar)

Composition per liter:

Sodium acetate·3H_2O	25.0g
Glucose	20.0g
Agar	15.0g
Pancreatic digest of casein	10.0g
KH_2PO_4	6.0g
Yeast extract	5.0g
Ammonium citrate	2.0g
Oxgall	1.5g
Polysorbate 80	1.0g
$MgSO_4$	0.575g
$MnSO_4$	0.12g
$FeSO_4$	0.034g
Acetic acid, glacial	1.32mL

pH 5.5 ± 0.2 at 25°C

Preparation of Medium: Add components, except acetic acid, to distilled/deionized water and bring volume to 998.7mL. Mix thoroughly. Gently heat and bring to boiling. Add glacial acetic acid. Mix thoroughly. Gently heat while stirring and bring to 90°–100°C for 2–3 min. Do not autoclave. Pour into sterile Petri dishes or distribute into sterile tubes.

Use: For the selective isolation, cultivation, and enumeration of lactobacilli.

Lactobacillus Selection Oxgall Agar Base (LBS Oxgall Agar)

Composition per liter:

Sodium acetate	25.0g
Glucose	20.0g
Agar	15.0g
Casein enzymic hydrolysate	10.0g
KH_2PO_4	6.0g
Yeast extract	5.0g
Ammonium citrate	2.0g
Oxgall	1.5g
Polysorbate 80	1.0g
$MgSO_4 \cdot 7H_2O$	0.575g
$MnSO_4 \cdot 2H_2O$	0.12g
$FeSO_4 \cdot 7H_2O$	0.034g
Acetic acid, glacial	1.32mL

pH 5.4 ± 0.2 at 25°C

Source: This medium is available from HiMedia.

Preparation of Medium: Add acetic acid to distilled/deionized water and bring volume to 1.0L. Add remaining components. Mix thoroughly. Distribute into tubes or flasks. Gently heat and bring to boiling. Boil for 1–2 min. Do not autoclave unless storage is needed. If storage is necessary, sterilize by autoclaving for 15 min at 15 psi pressure–121°C. Pour into sterile Petri dishes or leave in tubes.

Use: For the selective isolation, cultivation, and enumeration of lactobacilli.

Lactobacillus-Streptococcus Differential Medium

See: **L-S Differential Medium**

Lactococcus piscium Medium

Composition per liter:

Glucose	10.0g
Peptone	10.0g
Beef extract	8.0g
NaCl	5.0g
Yeast extract	3.0g
KH_2PO_4	1.5g

$MgSO_4 \cdot 7H_2O$ 0.2g
$MnSO_4 \cdot 4H_2O$ 0.05g

pH 6.8 ± 0.2 at 25°C

Preparation of Medium: Add components to distilled/deionized water and bring volume to 1.0L. Mix thoroughly. Adjust pH to 6.8. Distribute into tubes or flasks. Autoclave for 15 min at 15 psi pressure–121°C.

Use: For the cultivation of *Lactococcus piscium.*

Lactose Blue Agar (B.T.B. Lactose Agar, Modified)

Composition per liter:

Lactose 15.5g
Agar 13.0g
NaCl 5.0g
Peptic digest of animal tissue 3.5g
Casein enzymic hydrolysate 3.5g
Bromthymol Blue 0.04g

pH 7.0 ± 0.2 at 25°C

Source: This medium is available as a premixed powder from HiMedia.

Preparation of Medium: Add components to distilled/deionized water and bring volume to 1.0mL. Mix thoroughly. Gently heat and bring to boiling. Autoclave for 15 min at 15 psi pressure–121°C. Pour into sterile Petri dishes.

Use: For the differentiation of lactose-fermenting and non-fermenting bacteria belonging to Enterobacteriaceae.

Lactose Blue HiVeg Agar (B.T.B. Lactose HiVeg Agar, Modified)

Composition per liter:

Lactose 15.5g
Agar 13.0g
NaCl 5.0g
Plant extract 3.5g
Plant hydrolysate 3.5g
Bromthymol Blue 0.04g

pH 7.0 ± 0.2 at 25°C

Source: This medium is available as a premixed powder from HiMedia.

Preparation of Medium: Add components to distilled/deionized water and bring volume to 1.0mL. Mix thoroughly. Gently heat and bring to boiling. Autoclave for 15 min at 15 psi pressure–121°C. Pour into sterile Petri dishes.

Use: For the differentiation of lactose-fermenting and non-fermenting bacteria belonging to Enterobacteriaceae.

Lactose Broth

Composition per liter:

Lactose 5.0g
Pancreatic digest of gelatin 5.0g
Beef extract 3.0g

pH 6.9 ± 0.2 at 25°C

Source: This medium is available as a premixed powder from BD Diagnostic Systems and Oxoid Unipath.

Preparation of Medium: Add components to distilled/deionized water and bring volume to 1.0L. Mix thoroughly. Distribute into tubes containing an inverted Durham tube in 10.0mL volumes. Autoclave for 12 min at 15 psi pressure–121°C. Cool broth quickly to 25°C. For testing water samples with 10.0mL volumes, prepare medium double strength.

Use: For the detection of lactose-fermenting, Gram-negative coliforms, as a preenrichment broth for *Salmonella* species, and in the study of lactose fermentation of bacteria in general.

Lactose Casein Hydrolysate Medium

Composition per liter:

Lactose 37.5g
Agar 15.0g
Casein hydrolysate 3.0g
KH_2PO_4 .... 1.0g
$MnSO_4$ 0.5g
Microelements solution 2.0mL

pH 6.0 ± 0.2 at 25°C

Microelements Solution:

Composition per liter:

$Fe(NO_3)_3 \cdot 9H_2O$ 723.5mg
$ZnSO_4 \cdot 7H_2O$ 439.8mg
$MnSO_4 \cdot 4H_2O$ 203.0mg
H_2SO_4 .... variable

Preparation of Microelements Solution: Add components, one at a time, to distilled/deionized water and bring volume to 1.0L. Mix thoroughly. Be sure one component is dissolved before adding the next. Add sulfuric acid to yield a clear solution.

Preparation of Medium: Add components to distilled/deionized water and bring volume to 1.0L. Mix thoroughly. Gently heat until boiling. Adjust pH to 6.0. Distribute into tubes or flasks. Autoclave for 15 min at 15 psi pressure–121°C. Pour into sterile Petri dishes or leave in tubes.

Use: For the cultivation and maintenance of *Drechslera catenaria.*

Lactose Distillers Solubles Medium

Composition per liter:

Lactose 20.0g
Distillers solubles 15.0g
Yeast, autolyzed 5.0g

pH 7.0 ± 0.2 at 25°C

Preparation of Medium: Add components to distilled/deionized water and bring volume to 1.0L. Mix thoroughly. Gently heat and bring to boiling. Distribute into tubes or flasks. Autoclave for 15 min at 15 psi pressure–121°C. Pour into sterile Petri dishes or leave in tubes.

Use: For the cultivation and maintenance of *Streptomyces avermitilis.*

Lactose Egg Yolk Milk Agar

Composition per 1206.0mL:

Lactose 12.0g
Agar 1.0g
Columbia blood agar base 800.0mL
Skim milk 150.0mL
Egg yolk emulsion, 50% 36.0mL
Inhibitor solution 20.0mL
Neutral Red (1% solution) 3.25mL

pH 7.0 ± 0.2 at 25°C

Columbia Blood Agar Base:

Composition per 800.0mL:

Agar	15.0g
Pantone	10.0g
Bitone	10.0g
NaCl	5.0g
Tryptic digest of beef heart	3.0g
Cornstarch	1.0g

Preparation of Columbia Blood Agar Base: Add components to distilled/deionized water and bring volume to 1.0L. Mix thoroughly. Gently heat until boiling.

Egg Yolk Emulsion, 50%:

Composition per 100.0mL:

Chicken egg yolks	11
Whole chicken egg	1
NaCl (0.9% solution)	50.0mL

Preparation of Egg Yolk Emulsion, 50%: Soak eggs with 1:100 dilution of saturated mercuric chloride solution for 1 min. Crack eggs and separate yolks from whites. Mix egg yolks with 1 chicken egg. Beat to form emulsion. Measure 50.0mL of egg yolk emulsion and add to 50.0mL of 0.9% NaCl solution. Mix thoroughly. Filter sterilize. Warm to 45°–50°C.

Inhibitor Solution:

Composition per 20.0mL:

Neomycin sulfate	0.18g
NaN_3	0.24g

Caution: Sodium azide is toxic. Azides also react with metals and disposal must be highly diluted.

Preparation of Inhibitor Solution: Add neomycin sulfate and NaN_3 to distilled/deionized water and bring volume to 20.0mL. Mix thoroughly. Filter sterilize.

Preparation of Medium: Combine Columbia blood agar base, lactose, agar, and Neutral Red and bring volume to 1.0L. Adjust pH to 7.0. Autoclave for 15 min at 15 psi pressure–121°C. Cool to 55°C. Filter sterilize skim milk. To 1.0L of cooled, sterile agar mixture, aseptically add 150.0mL of sterile skim milk, 36.0mL of sterile egg yolk emulsion, 50%, and 20.0mL of sterile inhibitor solution. Mix thoroughly. Pour into sterile Petri dishes or distribute into sterile tubes.

Use: For the cultivation and maintenance of *Clostridium* species.

Lactose Gelatin Medium

Composition per liter:

Gelatin	120.0g
Tryptose	15.0g
Lactose	10.0g
Yeast extract	10.0g
Phenol Red (0.5% solution)	10.0mL

pH 7.5 ± 0.2 at 25°C

Preparation of Medium: Add tryptose, yeast extract, and lactose to distilled/deionized water and bring volume to 400.0mL. Mix thoroughly. Add gelatin to distilled/deionized water and bring volume to 590.0mL. Gently heat gelatin solution while stirring and bring to 50°–60°C. Add Phenol Red. Mix the two solutions together. Distribute into tubes in 10.0mL volumes. Autoclave for 10 min at 15 psi pressure–121°C. If medium is not used in 8 hr, deoxygenate by heating to 50°–70°C for 2–3 hr prior to inoculation.

Use: For the cultivation of *Clostridium perfringens.*

Lactose Gelatin Medium, Modified

Composition per liter:

Gelatin	120.0g
Tryptose	15.0g
Yeast extract	10.0g
Lactose	10.0g
Na_2HPO_4	5.0g
Phenol Red	0.05g

pH 7.8 ± 0.2 at 25°C

Source: This medium is available from HiMedia.

Preparation of Medium: Add components to distilled/deionized water and bring volume to 1.0L. Mix thoroughly. Gently heat and bring to boiling. Distribute into screw-cap tubes or flasks. Autoclave for 15 min at 15 psi pressure–121°C. Just before use heat to remove oxygen and cool rapidly.

Use: For the cultivation of *Haemophilus* spp.

Lactose HiVeg Broth

Composition per liter:

Plant peptone	5.0g
Lactose	5.0g
Plant extract	3.0g

pH 7.0 ± 0.2 at 25°C

Source: This medium is available as a premixed powder from HiMedia.

Preparation of Medium: Add components to distilled/deionized water and bring volume to 1.0mL. Mix thoroughly. Gently heat and bring to boiling. Autoclave for 15 min at 15 psi pressure–121°C. Pour into sterile Petri dishes.

Use: For the detection of coliform bacteria in water, foods, and dairy products as per standard methods.

Lactose Lecithin Agar

Composition per liter:

Agar	15.0g
Casein enzymic hydrolysate	12.65g
Lactose	10.0g
Peptic digest of animal tissue	5.5g
NaCl	5.5g
Yeast extract	3.85g
Pancreatic digest of heart muscle	3.3g
Corn starch	1.1g
Egg lecithin	0.66g
L-Cysteine·HCl·H_2O	0.5g
NaN_3	0.2g
Neomycin sulfate	0.15g
$CaCl_2 \cdot 2H_2O$	0.05g
Bromcresol Purple	0.025g

pH 6.8 ± 0.2 at 25°C

Source: This medium is available from HiMedia.

Caution: Sodium azide has a tendency to form explosive metal azides with plumbing materials. It is advisable to use enough water to flush off the disposables.

Preparation of Medium: Add components to distilled/deionized water and bring volume to 1.0L. Mix thoroughly. Gently heat and bring to boiling. Distribute into tubes or flasks. Autoclave for 15 min at 15 psi pressure–121°C. Pour into sterile Petri dishes or leave in tubes.

Use: For the isolation and differentiation of histotoxic clostridia from clinical specimens.

Lactose Medium

Composition per liter:

Agar	15.0g
K_2HPO_4	0.5g
$(NH_4)_2SO_4$	0.5g
NaCl	50.0mg
Lactose solution	50.0mL

pH 7.2 ± 0.2 at 25°C

Lactose Solution:

Composition per 50.0mL:

Lactose	5.0g

Preparation of Lactose Solution: Add lactose to distilled/deionized water and bring volume to 50.0mL. Mix thoroughly. Autoclave for 15 min at 15 psi pressure–121°C.

Preparation of Medium: Add components, except lactose solution, to distilled/deionized water and bring volume to 950.0mL. Mix thoroughly. Adjust pH to 7.2. Autoclave for 15 min at 15 psi pressure–121°C. Aseptically add 50.0mL of sterile lactose solution. Mix thoroughly. Aseptically distribute into sterile tubes or flasks.

Use: For the cultivation of unidentified bacterium ATCC 51468.

Lactose Minimal Medium

Composition per liter:

Agar	20.0g
Lactose	15.0g
K_2HPO_4	5.0g
NH_4Cl	2.0g
NaCl	1.0g
$MgSO_4$	0.1g
Yeast extract	0.1g

Preparation of Medium: Add components to distilled/deionized water and bring volume to 1.0L. Mix thoroughly. Gently heat and bring to boiling. Distribute into tubes or flasks. Autoclave for 15 min at 15 psi pressure–121°C. Pour into sterile Petri dishes or leave in tubes.

Use: For the cultivation and maintenance of *Xanthomonas campestris*.

Lactose Peptone Agar

Composition per liter:

Agar	15.0g
Lactose	0.5g
Peptone	0.5g

Preparation of Medium: Add components to distilled/deionized water and bring volume to 1.0L. Mix thoroughly. Gently heat until boiling. Distribute into tubes or flasks. Autoclave for 15 min at 15 psi pressure–121°C. Pour into sterile Petri dishes or leave in tubes.

Use: For the cultivation and maintenance of *Acytostelium subglobosum*, numerous *Dictyostelium* species, *Didymium nigripes, Drechslera biseptata*, several *Mortierella* species, *Nodulisporium griseobrunneum, Penicillium chrysogenum, Polysphondylium pallidum, Polysphondylium violaceum, Pythium insidiosum*, and *Trichomycete* species.

Lactose Peptone Agar, Double Strength

Composition per liter:

Agar	18.0g
Lactose	1.0g
Peptone	1.0g

Preparation of Medium: Add components to distilled/deionized water and bring volume to 1.0L. Mix thoroughly. Gently heat until boiling. Distribute into tubes or flasks. Autoclave for 15 min at 15 psi pressure–121°C. Pour into sterile Petri dishes or leave in tubes.

Use: For the cultivation and maintenance of various fungi.

Lactose Peptone Agar, Half Strength

Composition per liter:

Agar	15.0g
Lactose	0.25g
Peptone	0.25g

Preparation of Medium: Add components to distilled/deionized water and bring volume to 1.0L. Mix thoroughly. Gently heat until boiling. Distribute into tubes or flasks. Autoclave for 15 min at 15 psi pressure–121°C. Pour into sterile Petri dishes or leave in tubes.

Use: For the cultivation and maintenance of various fungi.

Lactose Peptone Broth

Composition per liter:

Casein enzymic hydrolysate	17.0g
Lactose	10.0g
NaCl	5.0g
Papaic digest of soybean meal	3.0g
Bromcresol Purple	0.02g

pH 7.4 ± 0.2 at 25°C

Source: This medium is available from HiMedia.

Preparation of Medium: Add components to distilled/deionized water and bring volume to 1.0L. Mix thoroughly. Distribute into tubes with inverted Durham tubes. Autoclave for 15 min at 15 psi pressure–121°C.

Use: For the detection of coliform organisms in water.

Lactose Ricinoleate Broth

Composition per liter:

Lactose	10.0g
Peptone	5.0g
Sodium ricinoleate	1.0g

pH 7.6 ± 0.2 at 25°C

Preparation of Medium: Add components to distilled/deionized water and bring volume to 1.0L. Mix thoroughly. Distribute into tubes or flasks. Autoclave for 15 min at 15 psi pressure–121°C.

Use: For the selective cultivation of members of the Enterobacteriaceae.

Lactose Sulfite Broth Base

Composition per liter:

Lactose	10.0g
Casein enzymic hydrolysate	5.0g
Yeast extract	2.5g
NaCl	2.5g
L-Cysteine·HCl·H_2O	0.3g
Sodium metabisulfite solution	50.0mL
Ferric ammonium citrate	50.0mL

pH 7.1 ± 0.2 at 25°C

Source: This medium is available from HiMedia.

Selective Sodium Metabisulfite Solution:

Composition per 100.0mL:

Sodium metabisulfite	1.2g

Preparation of Sodium Metabisulfite Solution: Add components to distilled/deionized water and bring volume to 100.0mL. Mix thoroughly. Filter sterilize.

Selective Ferric Ammonium Citrate Solution:

Composition per 100.0mL:

Ferric ammonium citrate	1.0g

Preparation of Ferric Ammonium Citrate Solution: Add components to distilled/deionized water and bring volume to 100.0mL. Mix thoroughly. Filter sterilize.

Preparation of Medium: Add components, except sodium metabisulfite and ferric ammonium citrate solutions, to distilled/deionized water and bring volume to 999.0mL. Mix thoroughly. Autoclave for 15 min at 15 psi pressure–121°C. Cool to 50°C. Aseptically add sodium metabisulfite and ferric ammonium citrate solutions. Mix thoroughly. Pour into Petri dishes or aseptically distribute into sterile tubes.

Use: For the detection and enumeration of *Clostridium perfringens* in pharmaceutical products.

Lambda Broth

Composition per liter:

Pancreatic digest of casein	10.0g
NaCl	2.5g

Preparation of Medium: Add components to distilled/deionized water and bring volume to 1.0L. Mix thoroughly. Distribute into tubes or flasks. Autoclave for 25 min at 15 psi pressure–121°C.

Use: For the cultivation of *Escherichia coli* in the preparation of bacteriophage lysates.

Lambda Plates

Composition per liter:

Agar	10.0g
Pancreatic digest of casein	10.0g
NaCl	2.5g

Preparation of Medium: Add components to distilled/deionized water and bring volume to 1.0L. Mix thoroughly. Gently heat and bring to boiling. Distribute into tubes or flasks. Autoclave for 15 min at 15 psi pressure–121°C. Pour into sterile Petri dishes in 45.0mL volumes per plate.

Use: For use as a base agar to support the cultivation of *Escherichia coli* in the preparation of bacteriophages.

Lambda Top Agar

Composition per liter:

Pancreatic digest of casein	10.0g
Agar	7.0g
NaCl	2.5g

Preparation of Medium: Add components to distilled/deionized water and bring volume to 1.0L. Mix thoroughly. Gently heat and bring to boiling. Autoclave for 15 min at 15 psi pressure–121°C. Cool to 50°C. Distribute into flasks in 100.0mL volumes. Reautoclave for 15 min at 15 psi pressure–121°C. Store at 25°C.

Use: For use as a top agar for the distribution of bacteriophage or *Escherichia coli.*

Lange Medium

Composition per liter:

Agar	20.0g
Maltose	5.0g
$Ca(NO_3)_2$	0.5g
$MgSO_4$	0.5g
K_2HPO_4	0.25g
Peptone	0.1g
Horse dung extract	100.0mL

Horse Dung Extract:

Composition per liter:

Horse dung, fresh	100.0g

Preparation of Horse Dung Extract: Add fresh horse dung to distilled/deionized water and bring volume to 1.0L. Autoclave for 50 min at 15 psi pressure–121°C. Filter through Whatman #1 filter paper. Reserve filtrate.

Preparation of Medium: Add components to distilled/deionized water and bring volume to 1.0L. Mix thoroughly. Gently heat and bring to boiling. Distribute into tubes or flasks. Autoclave for 15 min at 15 psi pressure–121°C. Pour into sterile Petri dishes or leave in tubes.

Use: For the cultivation and maintenance of *Coprinus xanthothrix.*

Lash Serum Medium

Composition per liter:

Casamino acids	14.0g
NaCl	6.0g
Glucose	2.0g
Maltose	1.5g
Sodium lactate (60% solution)	0.5g
KCl	0.1g
$CaCl_2 \cdot 2H_2O$	0.1g
Serum solution	500.0mL

pH 5.8 ± 0.2 at 25°C

Serum Solution:

Composition per 500.0mL:

$NaHCO_3$	0.1g
Bovine serum	200.0mL

Preparation of Serum Solution: Add components to distilled/deionized water and bring volume to 500.0mL. Mix thoroughly. Filter sterilize.

Preparation of Medium: Add components, except serum solution, to distilled/deionized water and bring volume to 500.0mL. Mix thoroughly. Distribute into tubes in 5.0mL volumes. Autoclave for 15 min at 15 psi pressure–121°C. Cool to 25°C. Aseptically add 5.0mL of sterile serum solution to each tube. Mix thoroughly.

Use: For the cultivation of *Trichomonas vaginalis* from clinical specimens.

Lasseur Medium (LMG 170)

Composition per liter:

Glycerol	25.0g
Agar	15.0g
L-Asparagine	9.0g
$MgSO_4 \cdot 7H_2O$	5.0g
K_2HPO_4	2.5g
$CaCl_2 \cdot 2H_2O$	0.54g
$FeSO_4 \cdot 7H_2O$	0.1g

pH 6.7 ± 0.2 at 25°C

Preparation of Medium: Add K_2HPO_4 to 20.0mL of distilled/deionized water. Mix thoroughly. Add remaining components. Bring volume to 1.0L with distilled/deionized water. Gently heat and bring to boiling. Adjust pH to 6.7. Distribute into tubes or flasks. Autoclave for 15 min at 15 psi pressure–121°C. Pour into sterile Petri dishes or leave in tubes.

Use: For the cultivation and maintenance of *Pseudomonas fluorescens.*

Lauryl Sulfate Broth (m-Lauryl Sulfate Broth)

Composition per liter:

Peptone	39.0g
Lactose	30.0g
Yeast extract	6.0g
Sodium lauryl sulfate	1.0g
Phenol Red	0.2g

pH 7.4 ± 0.2 at 25°C

Source: This medium is available as a premixed powder from Oxoid Unipath.

Preparation of Medium: Add components to distilled/deionized water and bring volume to 1.0L. Mix thoroughly. Distribute into bottles or flasks. Autoclave for 15 min at 15 psi pressure–121°C.

Use: For the cultivation and enumeration of coliform bacteria, especially *Escherichia coli*, in water by the membrane filter method.

Lauryl Sulfate Broth (Lauryl Tryptose Broth)

Composition per liter:

Pancreatic digest of casein	20.0g
Lactose	5.0g
NaCl	5.0g
K_2HPO_4	2.75g
KH_2PO_4	2.75g
Sodium lauryl sulfate	0.1g

pH 6.8 ± 0.2 at 25°C

Source: This medium is available as a premixed powder from BD Diagnostic Systems.

Preparation of Medium: Add components to distilled/deionized water and bring volume to 1.0L. Mix thoroughly. Distribute into tubes containing an inverted Durham tube in 10.0mL volumes. Autoclave for 12 min at 15 psi pressure–121°C. Cool broth quickly to 25°C. For testing water samples with 10.0mL volumes, prepare medium double strength.

Use: For the detection of coliform bacteria in a variety of specimens. Also, for the enumeration of coliform bacteria by the multiple-tube fermentation technique.

Lauryl Sulfate Broth, Fluorocult (Fluorocult Lauryl Sulfate Broth)

Composition per liter:

Tryptose	20.0g
Lactose	5.0g
NaCl	5.0g
K_2HPO_4	2.75g
KH_2PO_4	2.75g
L-tryptophan	1.g
Sodium lauryl sulfate	0.1g
4-Methylumbelliferyl-ß-D-glucuronide	0.1g

pH 6.8 ± 0.2 at 25°C

Source: This medium is available from Merck.

Preparation of Medium: Add components to distilled/deionized water and bring volume to 1.0L. Mix thoroughly. Gently heat and bring to boiling. Cool. Distribute into test tubes containing inverted Durham tubes. Autoclave for 15 min at 15 psi pressure–121°C. The prepared broth is clear and yellowish-brown.

Use: For the detection of *E. coli* in milk. The medium complies with the German-DIN-Norm 10183 for the examination of milk, with the regulations acc. to § 35 LMBG (01.00/54) for the examination of food, and according to ISO/DIS 11886-2.2 (1994) for milk and milk products. The lauryl sulfate largely inhibits the growth of undesirable microbial flora. The presence of *E. coli* is indicated by fluorescence under a long wavelength UV lamp. A positive indole reaction and gas formation due to fermentation of lactose confirm the results.

Lauryl Sulfate Broth with MUG

Composition per liter:

Pancreatic digest of casein	20.0g
Lactose	5.0g
NaCl	5.0g
K_2HPO_4	2.75g
KH_2PO_4	2.75g
Sodium lauryl sulfate	0.1g
4-Methylumbelliferyl-β-D-glucuronide (MUG)	0.05g

pH 6.8 ± 0.2 at 25°C

Source: This medium is available as a premixed powder from BD Diagnostic Systems.

Preparation of Medium: Add components to distilled/deionized water and bring volume to 1.0L. Mix thoroughly. Distribute into tubes containing an inverted Durham tube in 10.0mL volumes. Autoclave for 12 min at 15 psi pressure–121°C. Cool broth quickly to 25°C. For testing water samples with 10.0mL volumes, prepare medium double strength.

Use: For the detection of *Escherichia coli* in water and food samples by a fluorogenic procedure.

Lauryl Sulfate HiVeg Broth (Lauryl Tryptose HiVeg Broth)

Composition per liter:

Plant hydrolysate No. 1	20.0g
NaCl	5.0g
Lactose	5.0g
K_2HPO_4	2.75g
KH_2PO_4	2.75g
Sodium lauryl sulfate	0.1g

pH 6.8 ± 0.2 at 25°C

Source: This medium is available as a premixed powder from HiMedia.

Preparation of Medium: Add components to distilled/deionized water and bring volume to 1.0L. Mix thoroughly. Distribute into bottles or flasks. Autoclave for 15 min at 15 psi pressure–121°C.

Use: For the cultivation and enumeration of coliform bacteria in water, wastewater, dairy products, and other foods.

Lauryl Tryptose Broth
See: **Lauryl Sulfate Broth**

Lauryl Tryptose Broth with MUG (Lauryl Sulfate Broth with MUG) (LST-MUG) (BAM M77)

Composition per liter:

Pancreatic digest of casein	20.0g
Lactose	5.0g
NaCl	5.0g
K_2HPO_4	2.75g
KH_2PO_4	2.75g
Sodium lauryl sulfate	0.1g
4-Methylumbelliferyl-β-D-glucuronide (MUG)	0.05g

pH 6.8 ± 0.2 at 25°C

Source: This medium is available as a premixed powder from BD Diagnostic Systems.

Preparation of Medium: Add components to distilled/deionized water and bring volume to 1.0L. Mix thoroughly. Distribute into tubes containing an inverted Durham tube in 10.0mL volumes. Autoclave for 12 min at 15 psi pressure–121°C. Cool broth quickly to 25°C. For testing water samples with 10.0mL volumes, prepare medium double strength.

Use: For the detection of *Escherichia coli* in water and food samples by a fluorogenic procedure.

Lauryl Tryptose Mannitol Broth with Tryptophan

Composition per liter:

Pancreatic digest of casein	20.0g
Lactose	5.0g
NaCl	5.0g
K_2HPO_4	2.75g
KH_2PO_4	2.75g
Sodium lauryl sulfate	0.1g
L-Tryptophan	0.2g

pH 6.8 ± 0.2 at 25°C

Source: This medium is available as a premixed powder from Oxoid Unipath.

Preparation of Medium: Add components to distilled/deionized water and bring volume to 1.0L. Mix thoroughly. Distribute into tubes containing an inverted Durham tube in 10.0mL volumes. Autoclave for 10 min at 10 psi pressure–115°C. Cool broth quickly to 25°C.

Use: For the detection of *Escherichia coli* in water samples.

LAVMm2 Medium

Composition per liter:

Lactalbumin hydrolysate	10.0g
Sodium acetate	5.0g
$MgCl_2 \cdot 6H_2O$	20.3mg
Nitrilotriacetic acid	19.1mg
$CaCl_2$	11.1mg
$FeSO_4$	0.152mg
Thiamine·HCl	0.05mg
Cupric acetate	0.04mg
Biotin	0.02mg

pH 8.0–8.1 at 25°C

Preparation of Medium: Add components to distilled/deionized water and bring volume to 1.0L. Mix thoroughly. Adjust pH to 7.5 with Na_2CO_3. Distribute into tubes or flasks. Autoclave for 15 min at 15 psi pressure–121°C. The pH should be 8.0–8.1 after autoclaving.

Use: For the cultivation of *Caryophanon latum*.

LB Agar

Composition per liter:

Agar	15.0g
Pancreatic digest of casein	10.0g
NaCl	5.0g
Yeast extract	5.0g
1*N* NaOH	1.0mL

pH 7.0 ± 0.2 at 25°C

Preparation of Medium: Add components to distilled/deionized water and bring volume to 1.0L. Mix thoroughly. Gently heat and bring to boiling. Adjust pH to 7.0. Distribute into tubes or flasks. Autoclave for 25 min at 15 psi pressure–121°C. Pour into sterile Petri dishes in 35–40.0mL volumes.

Use: For the cultivation of *Escherichia coli.*

LB Agar
See: Lactobacillus bulgaricus **Agar**

LB Broth, Modified

Composition per liter:

Pancreatic digest of casein	10.0g
NaCl	5.8g
Yeast extract	5.0g
NaCl solution	16.8mL
Glucose solution	10.0mL
$CaCl_2 \cdot 2H_2O$ solution	2.0mL
$MgCl_2$ solution	1.6mL

pH 7.0 ± 0.2 at 25°C

NaCl Solution:

Composition per 100.0mL:

NaCl	25.0g

Preparation of NaCl Solution: Add NaCl to distilled/deionized water and bring volume to 100.0mL. Mix thoroughly. Filter sterilize.

Glucose Solution:

Composition per 100.0mL:

D-Glucose	40.0g

Preparation of Glucose Solution: Add D-glucose to distilled/deionized water and bring volume to 100.0mL. Mix thoroughly. Filter sterilize.

$CaCl_2 \cdot 2H_2O$ Solution:

Composition per 10.0mL:

$CaCl_2 \cdot 2H_2O$	0.735g

Preparation of $CaCl_2 \cdot 2H_2O$ Solution: Add $CaCl_2 \cdot 2H_2O$ to distilled/deionized water and bring volume to 10.0mL. Mix thoroughly. Filter sterilize.

$MgCl_2$ Solution:

Composition per 10.0mL:

$MgCl_2$	0.95g

Preparation of $MgCl_2$ Solution: Add $MgCl_2$ to distilled/deionized water and bring volume to 10.0mL. Mix thoroughly. Filter sterilize.

Preparation of Medium: Add components—except NaCl solution, glucose solution, $CaCl_2 \cdot 2H_2O$ solution, and $MgCl_2$ solution—to distilled/deionized water and bring volume to 969.6mL. Mix thoroughly. Adjust pH to 7.0. Autoclave for 30 min at 15 psi pressure–121°C. Cool to 45°–50°C. Aseptically add 16.8mL of sterile NaCl solution, 10.0mL of sterile glucose solution, 2.0mL of sterile $CaCl_2 \cdot 2H_2O$ solution, and 1.6mL of sterile $MgCl_2$ solution. Mix thoroughly. Aseptically distribute into sterile tubes or flasks.

Use: For the cultivation and maintenance of *Escherichia coli.*

LB Medium (LB Broth, Miller) (ATCC Medium 1065) (ATCC Medium 1082)

Composition per liter:

NaCl	10.0g
Pancreatic digest of casein	10.0g
Yeast extract	5.0g

Source: This medium is available from BD Diagnostic Systems.

Preparation of Medium: Add components to distilled/deionized water and bring volume to 1.0L. Mix thoroughly. Distribute into tubes or flasks. Autoclave for 15 min at 15 psi pressure–121°C.

Use: For the cultivation and maintenance of *Bacillus subtilis, Corynebacterium glutamicum, Enterobacter cloacae, Erwinia uredovora, Escherichia coli, Klebsiella oxytoca,* and *Salmonella choleraesuis.*

LB Medium (Luria-Bertani Medium)

Composition per liter:

NaCl	10.0g
Pancreatic digest of casein	10.0g
Yeast extract	5.0g

pH 7.0 ± 0.2 at 25°C

Preparation of Medium: Add components to distilled/deionized water and bring volume to 1.0L. Mix thoroughly. Adjust pH to 7.0. Distribute into tubes or flasks. Autoclave for 15 min at 15 psi pressure–121°C.

Use: For the cultivation of *Bacillus subtilis, Daptobacter* species, and *Escherichia coli.*

LB Medium (Luria Broth) (Lenox Broth)

Composition per liter:

Pancreatic digest of casein	10.0g
NaCl	5.0g
Yeast extract	5.0g

pH 7.0 ± 0.2 at 25°C

Preparation of Medium: Add components to distilled/deionized water and bring volume to 1.0L. Mix thoroughly. Adjust pH to 7.0. Distribute into tubes or flasks. Autoclave for 25 min at 15 psi pressure–121°C.

Use: For the cultivation of *Escherichia coli.*

LB Medium with Ampicillin (ATCC Medium 1315)

Composition per liter:

NaCl	10.0g
Pancreatic digest of casein	10.0g
Yeast extract	5.0g
Ampicillin solution	10.0mL

pH 7.0 ± 0.2 at 25°C

Ampicillin Solution:
Composition per 10.0mL:

Ampicillin	0.1mg

Preparation of Ampicillin Solution: Add ampicillin to distilled/deionized water and bring volume to 10.0mL. Mix thoroughly. Filter sterilize.

Preparation of Medium: Add components, except ampicillin solution, to distilled/deionized water and bring volume to 990.0mL. Mix thoroughly. Adjust pH to 7.0. Autoclave for 15 min at 15 psi pressure–121°C. Cool to 45°–50°C. Aseptically add sterile ampicillin solution. Mix thoroughly. Aseptically distribute into sterile tubes or flasks.

Use: For the cultivation and maintenance of *Escherichia coli.*

LB Medium with Ampicillin (ATCC Medium 1364)

Composition per liter:

NaCl	10.0g
Pancreatic digest of casein	10.0g
Yeast extract	5.0g
Ampicillin solution	10.0mL

pH 7.0 ± 0.2 at 25°C

Ampicillin Solution:
Composition per 10.0mL:

Ampicillin	0.02mg

Preparation of Ampicillin Solution: Add ampicillin to distilled/deionized water and bring volume to 10.0mL. Mix thoroughly. Filter sterilize.

Preparation of Medium: Add components, except ampicillin solution, to distilled/deionized water and bring volume to 990.0mL. Mix thoroughly. Adjust pH to 7.0. Autoclave for 15 min at 15 psi pressure–121°C. Cool to 45°–50°C. Aseptically add sterile ampicillin solution. Mix thoroughly. Aseptically distribute into sterile tubes or flasks.

Use: For the cultivation and maintenance of *Escherichia coli.*

LB Medium with Chloramphenicol

Composition per liter:

NaCl	10.0g
Pancreatic digest of casein	10.0g
Yeast extract	5.0g
Chloramphenicol	0.01g

pH 7.0 ± 0.2 at 25°C

Preparation of Medium: Add components to distilled/deionized water and bring volume to 1.0L. Mix thoroughly. Adjust pH to 7.0. Distribute into tubes or flasks. Autoclave for 15 min at 15 psi pressure–121°C.

Use: For the cultivation and maintenance of *Escherichia coli.*

LB Medium with Glucose

Composition per liter:

NaCl	10.0g
Pancreatic digest of casein	10.0g
Glucose	5.0g
Yeast extract	5.0g

pH 7.0 ± 0.2 at 25°C

Preparation of Medium: Add components to distilled/deionized water and bring volume to 1.0L. Mix thoroughly. Adjust pH to 7.0. Distribute into tubes or flasks. Autoclave for 15 min at 15 psi pressure–121°C.

Use: For the cultivation and maintenance of *Escherichia coli.*

LB Medium with IPTG Medium

Composition per liter:

NaCl	10.0g
Pancreatic digest of casein	10.0g
Yeast extract	5.0g
IPTG solution	10.0mL

pH 7.0 ± 0.2 at 25°C

IPTG Solution:

Composition per 10.0mL:

IPTG (Isopropylthio-β-galactoside)	0.24g

Preparation of IPTG Solution: Add IPTG to distilled/deionized water and bring volume to 10.0mL. Mix thoroughly. Filter sterilize.

Preparation of Medium: Add components, except IPTG solution, to distilled/deionized water and bring volume to 990.0mL. Mix thoroughly. Adjust pH to 7.0. Autoclave for 15 min at 15 psi pressure–121°C. Aseptically add sterile IPTG solution. Mix thoroughly. Aseptically distribute into sterile tubes or flasks.

Use: For the cultivation and maintenance of *Escherichia coli.*

LB Medium with Kanamycin

Composition per liter:

Pancreatic digest of casein	10.0g
NaCl	10.0g
Yeast extract	5.0g
Kanamycin solution	50.0mL

Kanamycin Solution:

Composition per 50.0mL:

Kanamycin	50.0mg

Preparation of Kanamycin Solution: Add kanamycin to distilled/deionized water and bring volume to 50.0mL. Mix thoroughly. Filter sterilize. Warm to 50°C.

Preparation of Medium: Add components, except kanamycin solution, to distilled/deionized water and bring volume to 950.0mL. Mix thoroughly. Autoclave for 15 min at 15 psi pressure–121°C. Cool to 50–55°C. Aseptically add 50.0mL of sterile kanamycin solution. Mix thoroughly. Aseptically distribute into sterile tubes or flasks.

Use: For the cultivation of *Escherichia coli.*

LB Medium with 25mg of Kanamycin (ATCC Medium 1236)

Composition per liter:

NaCl	10.0g
Pancreatic digest of casein	10.0g
Yeast extract	5.0g
Kanamycin	0.025g

pH 7.0 ± 0.2 at 25°C

Preparation of Medium: Add components to distilled/deionized water and bring volume to 1.0L. Mix thoroughly. Adjust pH to 7.0. Distribute into tubes or flasks. Autoclave for 15 min at 15 psi pressure–121°C.

Use: For the cultivation and maintenance of *Escherichia coli.*

LB Medium with 50mg of Kanamycin

Composition per liter:

NaCl	10.0g
Pancreatic digest of casein	10.0g
Yeast extract	5.0g
Kanamycin	0.05g

pH 7.0 ± 0.2 at 25°C

Preparation of Medium: Add components to distilled/deionized water and bring volume to 1.0L. Mix thoroughly. Adjust pH to 7.0. Distribute into tubes or flasks. Autoclave for 15 min at 15 psi pressure–121°C.

Use: For the cultivation of *Escherichia coli.*

LB Medium with 100mg of Kanamycin (ATCC Medium 1468)

Composition per liter:

NaCl	10.0g
Pancreatic digest of casein	10.0g
Yeast extract	5.0g
Kanamycin	0.1g

pH 7.0 ± 0.2 at 25°C

Preparation of Medium: Add components to distilled/deionized water and bring volume to 1.0L. Mix thoroughly. Adjust pH to 7.0. Distribute into tubes or flasks. Autoclave for 15 min at 15 psi pressure–121°C.

Use: For the cultivation and maintenace of *Erwinia uredovora.*

LB Medium with Rifampicin

Composition per liter:

NaCl	10.0g
Pancreatic digest of casein	10.0g
Yeast extract	5.0g
Rifampicin	0.1g

pH 7.0 ± 0.2 at 25°C

Preparation of Medium: Add components to distilled/deionized water and bring volume to 1.0L. Mix thoroughly. Adjust pH to 7.0. Distribute into tubes or flasks. Autoclave for 15 min at 15 psi pressure–121°C.

Use: For the cultivation and maintenance of *Enterobacter cloacae.*

LB Medium with Tetracycline

Composition per liter:

NaCl	10.0g
Pancreatic digest of casein	10.0g
Yeast extract	5.0g
Tetracycline	0.02g

pH 7.0 ± 0.2 at 25°C

Preparation of Medium: Add components to distilled/deionized water and bring volume to 1.0L. Mix thoroughly. Adjust pH to 7.0. Distribute into tubes or flasks. Autoclave for 15 min at 15 psi pressure–121°C.

Use: For the cultivation and maintenance of *Escherichia coli.*

LB Medium with Tetracycline and Ampicillin (ATCC Medium 1226)

Composition per liter:

NaCl	10.0g
Pancreatic digest of casein	10.0g

Yeast extract 5.0g
Antibiotic solution 10.0mL
pH 7.0 ± 0.2 at 25°C

Antibiotic Solution:
Composition per 10.0mL:
Ampicillin 0.01g
Tetracycline 0.01g

Preparation of Antibiotic Solution: Add components to distilled/deionized water and bring volume to 10.0mL. Mix thoroughly. Filter sterilize.

Preparation of Medium: Add components, except antibiotic solution, to distilled/deionized water and bring volume to 990.0mL. Mix thoroughly. Adjust pH to 7.0. Autoclave for 15 min at 15 psi pressure–121°C. Cool to 45°–50°C. Aseptically add sterile antibiotic solution. Mix thoroughly. Aseptically distribute into sterile tubes or flasks.

Use: For the cultivation and maintenance of *Escherichia coli.*

LB Medium with Tetracycline and Ampicillin (ATCC Medium 1235)

Composition per liter:
NaCl 10.0g
Pancreatic digest of casein 10.0g
Yeast extract 5.0g
Antibiotic solution 10.0mL
pH 7.0 ± 0.2 at 25°C

Antibiotic Solution:
Composition per 10.0mL:
Ampicillin 0.01g
Tetracycline 5.0mg

Preparation of Antibiotic Solution: Add components to distilled/deionized water and bring volume to 10.0mL. Mix thoroughly. Filter sterilize.

Preparation of Medium: Add components, except antibiotic solution, to distilled/deionized water and bring volume to 990.0mL. Mix thoroughly. Adjust pH to 7.0. Autoclave for 15 min at 15 psi pressure–121°C. Cool to 45°–50°C. Aseptically add sterile antibiotic solution. Mix thoroughly. Aseptically distribute into sterile tubes or flasks.

Use: For the cultivation and maintenance of *Escherichia coli.*

LB Medium with Thiamine Monophosphate
See: **LB Medium with TMP**

LB Medium with Thiamine Pyrophosphate
See: **LB Medium with TPP**

LB Medium with TMP (LB Medium with Thiamine Monophosphate)

Composition per liter:
NaCl 10.0g
Pancreatic digest of casein 10.0g
Yeast extract 5.0g
Thiamine monophosphate 0.038mg
pH 7.0 ± 0.2 at 25°C

Preparation of Medium: Add components to distilled/deionized water and bring volume to 1.0L. Mix thoroughly. Adjust pH to 7.0. Distribute into tubes or flasks. Autoclave for 15 min at 15 psi pressure–121°C.

Use: For the cultivation and maintenance of *Escherichia coli.*

LB Medium with TPP (LB Medium with Thiamine Pyrophosphate)

Composition per liter:
NaCl 10.0g
Pancreatic digest of casein 10.0g
Yeast extract 5.0g
Thiamine pyrophosphate 0.046mg
pH 7.0 ± 0.2 at 25°C

Preparation of Medium: Add components to distilled/deionized water and bring volume to 1.0L. Mix thoroughly. Adjust pH to 7.0. Distribute into tubes or flasks. Autoclave for 15 min at 15 psi pressure–121°C.

Use: For the cultivation and maintenance of *Escherichia coli.*

LB Medium for X1776

Composition per liter:
NaCl 10.0g
Pancreatic digest of casein 10.0g
Yeast extract 5.0g
Glucose solution 10.0mL
Diaminopimelic acid solution 10.0mL
Thymidine solution 10.0mL
pH 7.0 ± 0.2 at 25°C

Glucose Solution:
Composition per 10.0mL:
D-Glucose 0.8g

Preparation of Glucose Solution: Add D-glucose to distilled/deionized water and bring volume to 10.0mL. Mix thoroughly. Filter sterilize.

Diaminopimelic Acid Solution:
Composition per 10.0mL:
DL-Diaminopimelic acid 0.1g

Preparation of Diaminopimelic Acid Solution: Add diaminopimelic acid to distilled/deionized water and bring volume to 10.0mL. Mix thoroughly. Filter sterilize.

Thymidine Solution:
Composition per 10.0mL:
Thymidine 0.02g

Preparation of Thymidine Solution: Add thymidine to distilled/deionized water and bring volume to 10.0mL. Mix thoroughly. Filter sterilize.

Preparation of Medium: Add components—except glucose solution, diaminopimelic acid solution, and thymidine solution—to distilled/deionized water and bring volume to 970.0mL. Mix thoroughly. Gently heat and bring to boiling. Autoclave for 15 min at 15 psi pressure–121°C. Cool to 45°–50°C. Aseptically add sterile glucose solution, diaminopimelic acid solution, and thymidine solution. Mix thoroughly. Aseptically distribute into sterile tubes or flasks.

Use: For the cultivation and maintenance of *Bacillus subtilis* and *Escherichia coli.*

LB Medium for X1776 with Tetracycline and Ampicillin

Composition per liter:
NaCl 10.0g
Pancreatic digest of casein 10.0g
Yeast extract 5.0g

Antibiotic solution 10.0mL
Glucose solution 10.0mL
Diaminopimelic acid solution 10.0mL
Thymidine solution 10.0mL
pH 7.0 ± 0.2 at 25°C

Antibiotic Solution:
Composition per 10.0mL:
Ampicillin 0.01g
Tetracycline 0.01g

Preparation of Antibiotic Solution: Add components to distilled/deionized water and bring volume to 10.0mL. Mix thoroughly. Filter sterilize.

Glucose Solution:
Composition per 10.0mL:
D-Glucose 0.8g

Preparation of Glucose Solution: Add D-glucose to distilled/deionized water and bring volume to 10.0mL. Mix thoroughly. Filter sterilize.

Diaminopimelic Acid Solution:
Composition per 10.0mL:
DL-Diaminopimelic acid 0.1g

Preparation of Diaminopimelic Acid Solution: Add diaminopimelic acid to distilled/deionized water and bring volume to 10.0mL. Mix thoroughly. Filter sterilize.

Thymidine Solution:
Composition per 10.0mL:
Thymidine 0.02g

Preparation of Thymidine Solution: Add thymidine to distilled/deionized water and bring volume to 10.0mL. Mix thoroughly. Filter sterilize.

Preparation of Medium: Add components—except glucose solution, diaminopimelic acid solution, and thymidine solution—to distilled/deionized water and bring volume to 960.0mL. Mix thoroughly. Gently heat and bring to boiling. Autoclave for 15 min at 15 psi pressure–121°C. Cool to 45°–50°C. Aseptically add sterile antibiotic solution, glucose solution, diaminopimelic acid solution, and thymidine solution. Mix thoroughly. Aseptically distribute into sterile tubes or flasks.

Use: For the cultivation and maintenance of *Bacillus subtilis* and *Escherichia coli.*

LB Modified Broth (ATCC Medium 1620)

Composition per 1030.4mL:
Tryptone 10.0g
NaCl 5.8g
Yeast extract 5.0g

NaCl Solution:
Composition per 100.0mL:
NaCl 20.0g

Preparation of NaCl Solution: Add NaCl to distilled/deionized water and bring volume to 100.0mL. Mix thoroughly. Autoclave for 15 min at 15 psi pressure–121°C.

$MgCl_2$ Solution:
Composition per 10.0mL:
$MgCl_2$ 0.9g

Preparation of $MgCl_2$ Solution: Add $MgCl_2$ to distilled/deionized water and bring volume to 10.0mL. Mix thoroughly. Autoclave for 15 min at 15 psi pressure–121°C.

$CaCl_2$ Solution:
Composition per 10.0mL:
$CaCl_2$ 0.5g

Preparation of $CaCl_2$ Solution: Add $CaCl_2$ to distilled/deionized water and bring volume to 10.0mL. Mix thoroughly. Autoclave for 15 min at 15 psi pressure–121°C.

Glucose Solution:
Composition per 100.0mL:
D-Glucose 40.0g

Preparation of Glucose Solution: Add glucose to distilled/deionized water and bring volume to 100.0mL. Mix thoroughly. Autoclave for 15 min at 15 psi pressure–121°C.

Preparation of Medium: Add components except salts and glucose solutions to distilled/deionized water and bring volume to 1.0L. Mix thoroughly. Autoclave for 30 min at 15 psi pressure–121°C. Cool to 45°–50°C. Aseptically add 16.8mL NaCl solution, 1.6 mL $MgCl_2$ solution, 2.0mL $CaCl_2$ solution, and 10.0mL glucose solution. Mix thoroughly.

Use: For the cultivation of *Escherichia coli* (Migula) Castellani and Chalmers.

LB Streptomycin Medium

Composition per liter:
NaCl 10.0g
Pancreatic digest of casein 10.0g
Yeast extract 5.0g
Streptomycin solution 10.0mL
pH 7.0 ± 0.2 at 25°C

Streptomycin Solution:
Composition per 10.0mL:
Streptomycin 0.2g

Preparation of Streptomycin Solution: Add streptomycin to distilled/deionized water and bring volume to 10.0mL. Mix thoroughly. Filter sterilize.

Preparation of Medium: Add components, except streptomycin solution, to distilled/deionized water and bring volume to 990.0mL. Mix thoroughly. Adjust pH to 7.0. Autoclave for 15 min at 15 psi pressure–121°C. Cool to 45°–50°C. Aseptically add sterile streptomycin solution. Mix thoroughly. Aseptically distribute into sterile tubes or flasks.

Use: For the cultivation of *Escherichia coli.*

LB Top Agar

Composition per liter:
Pancreatic digest of casein 10.0g
Agar 7.0g
NaCl 5.0g
Yeast extract 5.0g
pH 7.0 ± 0.2 at 25°C

Preparation of Medium: Add components to distilled/deionized water and bring volume to 1.0L. Mix thoroughly. Gently heat and bring to boiling. Adjust pH to 7.0. Autoclave for 15 min at 15 psi pressure–121°C. Cool to 50°C. Distribute into flasks in 100.0mL volumes. Reautoclave for 15 min at 15 psi pressure–121°C. Store at 25°C.

Use: For use as a top agar for the distribution of bacteriophage or *Escherichia coli*.

LBE Medium

Composition per liter:

NaCl	10.0g
Pancreatic digest of casein	10.0g
Yeast extract	5.0g
Glucose solution	10.0mL
50X medium E	4.0mL

pH 7.0 ± 0.2 at 25°C

Glucose Solution:

Composition per 100.0mL:

D-Glucose	20.0g

Preparation of Glucose Solution: Add D-glucose to distilled/deionized water and bring volume to 100.0mL. Mix thoroughly. Filter sterilize.

50X Medium E:

Composition per liter:

K_2HPO_4, anhydrous	500.0g
$Na(NH_4)HPO_4·4H_2O$	175.0g
Citric acid·H_2O	100.0g
$MgSO_4·7H_2O$	10.0g

Preparation of 50X Medium E: Add components to 670.0mL of distilled/deionized water in the following order: $MgSO_4·7H_2O$, citric acid·H_2O, K_2HPO_4, and $Na(NH_4)HPO_4·4H_2O$. Mix thoroughly. Bring volume to 1.0L with distilled/deionized water.

Preparation of Medium: Add components—except glucose solution and 50X medium E—to distilled/deionized water and bring volume to 986.0mL. Mix thoroughly. Autoclave for 15 min at 15 psi pressure–121°C. Cool to 45°–50°C. Aseptically add 10.0mL of sterile glucose solution and 4.0mL of sterile 50X medium E. Mix thoroughly. Aseptically distribute into sterile tubes or flasks.

Use: For the cultivation and maintenance of *Escherichia coli*.

LBS™ Agar (*Lactobacillus* Selection Agar)

Composition per liter:

Sodium acetate·$3H_2O$	25.0g
Glucose	20.0g
Agar	15.0g
Pancreatic digest of casein	10.0g
KH_2PO_4	6.0g
Yeast extract	5.0g
Ammonium citrate	2.0g
Polysorbate 80	1.0g
$MgSO_4$	0.575g
$FeSO_4$	0.034g
$MnSO_4$	0.12g
Acetic acid, glacial	1.32mL

pH 5.5 ± 0.2 at 25°C

Source: This medium is available as a premixed powder from BD Diagnostic Systems.

Preparation of Medium: Add components, except acetic acid, to distilled/deionized water and bring volume to 998.7mL. Mix thoroughly. Gently heat and bring to boiling. Add glacial acetic acid. Mix thoroughly. Gently heat while stirring and bring to 90°–100°C for 2–3 min. Do not autoclave. Pour into sterile Petri dishes or distribute into sterile tubes.

Use: For the selective isolation, cultivation, and enumeration of lactobacilli.

LBS™ Broth (*Lactobacillus* Selection Broth)

Composition per liter:

Sodium acetate·$3H_2O$	25.0g
Glucose	20.0g
Pancreatic digest of casein	10.0g
KH_2PO_4	6.0g
Yeast extract	6.0g
Ammonium citrate	2.0g
Polysorbate 80	1.0g
$MgSO_4$	0.575g
$FeSO_4$	0.034g
$MnSO_4$	0.12g
Acetic acid, glacial	1.32mL

pH 5.4 ± 0.2 at 25°C

Source: This medium is available as a premixed powder from BD Diagnostic Systems.

Preparation of Medium: Add components, except acetic acid, to distilled/deionized water and bring volume to 998.7mL. Mix thoroughly. Gently heat and bring to boiling. Add glacial acetic acid. Mix thoroughly. Gently heat while stirring and bring to 90°–100°C for 2–3 min. Do not autoclave. Aseptically distribute into sterile tubes.

Use: For the selective isolation and cultivation of lactobacilli.

LBS Oxgall Agar

See: ***Lactobacillus*** **Selection Oxgall Agar**

LC Broth

Composition per liter:

NaCl	10.0g
Pancreatic digest of casein	10.0g
Yeast extract	5.0g
Glucose	1.0g
$CaCl_2·2H_2O$ (1*M* solution)	5.0mL
$MgSO_4·7H_2O$ (1*M* solution)	5.0mL

pH 7.4 ± 0.2 at 25°C

$CaCl_2·2H_2O$ Solution:

Composition per 100.0mL:

$CaCl_2·2H_2O$	14.7g

Preparation of $CaCl_2·2H_2O$ Solution: Add $CaCl_2·2H_2O$ to distilled/deionized water and bring volume to 100.0mL. Mix thoroughly. Autoclave for 15 min at 15 psi pressure–121°C.

$MgSO_4·7H_2O$ Solution:

Composition per 100.0mL:

$MgSO_4·7H_2O$	24.65g

Preparation of $MgSO_4·7H_2O$ Solution: Add $MgSO_4·7H_2O$ to distilled/deionized water and bring volume to 100.0mL. Mix thoroughly. Autoclave for 15 min at 15 psi pressure–121°C.

Preparation of Medium: Add components, except $CaCl_2·2H_2O$ solution and $MgSO_4·7H_2O$ solution, to distilled/deionized water and bring volume to 990.0mL. Mix thoroughly. Adjust pH to 7.4. Autoclave for 15 min at 15 psi pressure–121°C. Cool to 50°C. Aseptically add 5.0mL of sterile $CaCl_2 \cdot 2H_2O$ solution and 5.0mL of sterile $MgSO_4 \cdot 7H_2O$ solution. Mix thoroughly. Aseptically distibute into sterile tubes or flasks.

Use: For the cultivation of *Escherichia coli.*

LD Agar
See: **Lombard-Dowell Agar**

LD Bile Agar
See: **Lombard-Dowell Bile Agar**

LD Broth
See: **Lombard-Dowell Broth**

LD Egg Yolk Agar
See: **Lombard-Dowell Egg Yolk Agar**

LD Esculin Agar
See: **Lombard-Dowell Esculin Agar**

LD Esculin HiVeg Agar (Lombard-Dowell Esculin Agar, HiVeg)

Composition per liter:

Agar	20.0g
Plant hydrolysate No. 1	5.0g
Yeast extract	5.0g
NaCl	2.5g
Esculin	1.0g
Ferric citrate	0.5g
L-Cystine	0.4g
L-Tryptophan	0.2g
$Fe_4(P_2O_7)_3 \cdot H_2O$	0.01g
Vitamin K_1	0.01g

pH 7.5 ± 0.2 at 25°C

Source: This medium is available as a premixed powder from HiMedia.

Preparation of Medium: Add components to distilled/deionized water and bring volume to 1.0L. Mix thoroughly. Distribute into tubes or flasks. Autoclave for 15 min at 15 psi pressure–121°C. Pour into sterile Petri plates.

Use: For the cultivation of a wide variety of anaerobic bacteria. For the differentiation of anaerobic bacteria based on esculin hydrolysis, H_2S production, and catalase production. Bacteria that hydrolyze esculin appear as colonies surrounded by a red-brown to dark brown zone. Bacteria that produce H_2S appear as black colonies.

LD Gelatin Agar
See: **Lombard-Dowell Gelatin Agar**

LD HiVeg Agar (Lombard-Dowell Agar, HiVeg)

Composition per liter:

Agar	20.0g
Plant hydrolysate	5.0g
Yeast extract	5.0g
NaCl	2.5g
L-Cystine	0.4g
L-Tryptophan	0.2g
Na_2SO_3	0.1g
Vitamin K_1	0.01g
$Fe_4(P_2O_7)_3 \cdot H_2O$	0.01g

pH 7.5 ± 0.2 at 25°C

Source: This medium is available as a premixed powder from HiMedia.

Preparation of Medium: Add components to distilled/deionized water and bring volume to 1.0L. Mix thoroughly. Distribute into tubes or flasks. Autoclave for 15 min at 15 psi pressure–121°C. Pour into sterile Petri plates.

Use: For the cultivation and identification of a variety of obligate anaerobic bacteria. For the cultivation of *Bacteroides* species, *Fusobacterium* species, *Clostridium* species, and nonspore-forming Gram-positive anaerobes.

Lead Acetate Agar

Composition per liter:

Agar	15.0g
Peptone	15.0g
Proteose peptone	5.0g
Glucose	1.0g
Lead acetate	0.2g
$Na_2S_2O_3$	0.08g

pH 6.6 ± 0.2 at 25°C

Preparation of Medium: Add components to distilled/deionized water and bring volume to 1.0L. Mix thoroughly. Gently heat and bring to boiling. Distribute into tubes or flasks. Autoclave for 15 min at 15 psi pressure–121°C. Pour into sterile Petri dishes or leave in tubes. Allow tubes to cool in a slanted position.

Use: For the cultivation and differentiation of Gram-negative coliform bacteria based on H_2S production. Bacteria that produce H_2S turn the medium brown.

LEB, FDA
*See: **Listeria*** **Enrichment Broth, FDA**

Lecithin Agar

Composition per liter:

Fraction B	500.0mL
Fraction A	450.0mL
Fraction C	50.0mL

pH 7.2 ± 0.2 at 25°C

Fraction A:

Composition per 500.0mL:

Agar	18.0g
Tryptone	10.0g
Yeast extract	5.0g
Glucose	5.0g

Preparation of Fraction A: Add components to distilled/deionized water and bring volume to 500.0mL. Mix thoroughly. Autoclave for 15 min at 15 psi pressure–121°C. Cool to 43°C.

Fraction B:

Composition per 450.0mL:

Crude soy lecithin	30.0g

Preparation of Fraction B: Add crude soy lethicin to distilled/deionized water and bring volume to 450.0mL. Mix thoroughly. Gently heat and bring to boiling. Swirl to form a viscous sol. Sonicate until homogeneous. Blending of unheated fraction A in a Waring blender for 2 min at high speed is also satisfactory. Autoclave for 15 min at 15 psi pressure–121°C. Cool to 43°C.

Fraction C:
Composition per 50.0mL:
$CaCl_2$ 0.6g

Preparation of Fraction C: Add $CaCl_2$ to distilled/deionized water and bring volume to 50.0mL. Mix thoroughly. Autoclave for 15 min at 15 psi pressure–121°C. Cool to 43°C.

Preparation of Medium: Combine fractions with gentle swirling. To prevent separation, immediately pour into sterile Petri plates.

Use: For the detection of microbial phospholipases.

Lecithin HiVeg Agar

Composition per liter:
Agar 20.5g
Plant hydrolysate 15.0g
Papaic digest of soybean meal 5.0g
Polysorbate 80 .... 5.0g
NaCl 5.0g
$Na_2S_2O_3$ 1.0g
L-Histidine 1.0g
Lecithin 0.7g
pH 7.3 ± 0.2 at 25°C

Source: This medium is available as a premixed powder from HiMedia.

Preparation of Medium: Add components to distilled/deionized water and bring volume to 1.0L. Mix thoroughly. Distribute into tubes or flasks. Autoclave for 15 min at 15 psi pressure–121°C. Pour into sterile Petri plates.

Use: For the detection of bacterial contamination of surfaces in unprotected and protected areas.

Lecithin Lactose Agar

Composition per liter:
Agar 15.0g
Pancreatic digest of casein 12.7g
Lactose 10.0g
NaCl 5.5g
Peptic digest of animal tissue 5.5g
Yeast extract 3.9g
Pancreatic digest of heart muscle 3.3g
Cornstarch 1.1g
Egg lecithin 0.66g
L-Cysteine·HCl·H_2O 0.5g
NaN_3 .... 0.2g
Neomycin sulfate 0.15g
$CaCl_2$ 0.05g
Bromcresol Purple 0.02g
pH 6.8 ± 0.2 at 25°C

Caution: Sodium azide is toxic. Azides also react with metals and disposal must be highly diluted.

Source: This medium is available as a prepared medium from BD Diagnostic Systems.

Preparation of Medium: Add components to distilled/deionized water and bring volume to 1.0L. Mix thoroughly. Gently heat and bring to boiling. Distribute into tubes or flasks. Autoclave for 15 min at 15 psi pressure–121°C. Pour into sterile Petri dishes.

Use: For the isolation, cultivation, and differentiation of histolytic clostridia from clinical specimens based on lecithinase production and lactose fermentation. It is especially useful for the differentiation of *Clostridium perfringens, Clostridium sordelli, Clostridium novyi, Clostridium septicum,* and *Clostridium histolyticum.* Bacteria that produce lecithinase appear as colonies surrounded by an opalescent zone. Bacteria that ferment lactose appear as colonies surrounded by a yellow zone.

Lecithin Lipase Anaerobic Agar

Composition per liter:
Pancreatic digest of casein 40.0g
Agar 25.0g
Yeast extract 5.0g
$Na_2HPO_4 \cdot 12H_2O$ 5.0g
Glucose 2.0g
NaCl 2.0g
KH_2PO_4 .... 1.0g
$MgSO_4 \cdot 7H_2O$ 0.1g
Egg yolk emulsion 100.0mL
pH 7.6 ± 0.2 at 25°C

Egg Yolk Emulsion:
Composition:
Chicken egg yolks 11
Whole chicken egg 1

Preparation of Egg Yolk Emulsion: Soak eggs with 1:100 dilution of saturated mercuric chloride solution for 1 min. Crack eggs and separate yolks from whites. Mix egg yolks with 1 chicken egg. Filter sterilize.

Preparation of Medium: Add components, except egg yolk emulsion, to distilled/deionized water and bring volume to 900.0mL. Mix thoroughly. Gently heat and bring to boiling. Autoclave for 15 min at 15 psi pressure–121°C. Cool to 45°–50°C. Aseptically add sterile egg yolk emulsion. Mix thoroughly. Pour into sterile Petri dishes or distribute into sterile tubes.

Use: For the isolation, cultivation, and differentiation of *Clostridium* species based on lecithinase production and lipase production. Bacteria that produce lecithinase appear as colonies surrounded by a zone of insoluble precipitate. Bacteria that produce lipase appear as colonies with a pearly iridescent sheen.

Lecithin Tween™ Medium (LT Medium)

Composition per liter:
Tween™ 80 .... 30.0g
Agar 15.0g
Pancreatic digest of casein 10.0g
Peptic digest of animal tissue 10.0g
NaCl 5.0g
Lecithin 5.0g
$Na_2S_2O_3 \cdot 5H_2O$ 5.0g
Glycerol 3.0g
Histidine, free base 1.0g
Glucose 1.0g
pH 7.5 ± 0.2 at 25°C

Antibiotic Solution:
Composition per 10.0mL:
5–Fluorocytosine 0.2g
Fosfomicin 0.1g
Ticarcillin 0.1g

Preparation of Antibiotic Solution: Add components to distilled/deionized water and bring volume to 10.0mL. Mix thoroughly. Filter sterilize.

Preparation of Medium: Add components, except antibiotic solution, to distilled/deionized water and bring volume to 990.0mL. Mix thoroughly. Gently heat and bring to boiling. Autoclave for 15 min at 15 psi pressure–121°C. Cool to 45°–50°C. Aseptically add sterile antibiotic solution. Mix thoroughly. Pour into sterile Petri dishes in 20.0mL volumes.

Use: For the isolation and cultivation of multiresistant lipophilic *Corynebacterium* species, especially *Corynebacterium* group JK found primarily in infections in immunocompromised hosts and patients with prosthetic valve endocarditis.

Lee's Agar

Composition per liter:

Agar	18.0g
Pancreatic digest of casein	10.0g
Yeast extract	10.0g
Lactose	5.0g
Sucrose	5.0g
$CaCO_3$	3.0g
K_2HPO_4	0.5g
Bromcresol Purple (0.2% solution)	10.0mL

pH 7.0 ± 0.2 at 25°C

Bromcresol Purple Solution:

Composition per 10.0mL:

Bromcresol Purple	0.02g

Preparation of Bromcresol Purple Solution: Add Bromcresol Purple to distilled/deionized water and bring volume to 10.0mL. Mix thoroughly. Filter sterilize.

Preparation of Medium: Add components, except Bromcresol Purple solution, to distilled/deionized water and bring volume to 990.0mL. Mix thoroughly. Adjust pH to 7.0. Gently heat and bring to boiling. Autoclave for 20 min at 15 psi pressure–121°C. Cool to 45°–50°C. Aseptically add sterile Bromcresol Purple solution. Mix thoroughly. Pour into sterile, chilled Petri dishes in 20–25mL volumes. Swirl flask while dispensing to evenly suspend $CaCO_3$. Dry plates at 30°C for 18–24 hr before use.

Use: For the isolation, cultivation, and enumeration of *Lactobacillus bulgaricus* from yogurt.

Lee's HiVeg Agar

Composition per liter:

Agar	18.0g
Plant hydrolysate	10.0g
Yeast extract	10.0g
Lactose	5.0g
Sucrose	5.0g
$CaCO_3$	3.0g
K_2HPO_4	0.5g
Bromcresol Purple	0.02g

pH 7.0 ± 0.2 at 25°C

Source: This medium is available as a premixed powder from HiMedia.

Preparation of Medium: Add components to distilled/deionized water and bring volume to 1.0L. Mix thoroughly. Adjust pH to 7.0. Gently heat and bring to boiling. Autoclave for 20 min at 15 psi pressure–121°C. Mix thoroughly. Pour into sterile, chilled Petri dishes in 20–25mL volumes. Swirl flask while dispensing to evenly suspend $CaCO_3$. Dry plates at 30°C for 18–24 hr before use.

Use: For the isolation, cultivation, and enumeration of *Lactobacillus bulgaricus* from yogurt.

Lee's Multidifferential Agar

Composition per liter:

Tomato juice, dessicated	20.0g
Peptonized milk	20.0g
Glucose	10.0g
Yeast extract	10.0g
Agar	15.0g
$CaCO_3$	5.0g
Calcium pantothenate	2.0g
Citric acid	1.1g
Polysorbate 80	0.5g
K_2HPO_4	0.5g
KH_2PO_4	0.5g
$MgSO_4 \cdot 7H_2O$	0.2g
$FeSO_4 \cdot 7H_2O$	0.01g
$MnSO_4 \cdot 7H_2O$	0.01g
NaCl	0.01g
Bromcresol Green	0.022g
Cycloheximide	7.0mg

pH 5.5 ± 0.2 at 25°C

Source: This medium is available as a premixed powder from HiMedia.

Caution: Cycloheximide is toxic. Avoid skin contact or aerosol formation and inhalation.

Preparation of Medium: Add components to distilled/deionized water and bring volume to 1.0L. Mix thoroughly. Distribute into tubes or flasks. Autoclave for 10 min at 15 psi pressure–121°C. Pour into sterile Petri plates while swirling to prevent calcium carbonate from settling. The medium will have a white precipitate of calcium carbonate.

Use: For the detection of most organisms commonly encountered in a brewery. Acid producing bacteria are identified by the development of a clear zone around the colonies. Further identification is facilitated by the characteristic color reactions.

Lee's Multidifferential HiVeg Agar

Composition per liter:

Tomato juice, dessicated	20.0g
Plant hydrolysate No. 3	20.0g
Agar	15.0g
Glucose	10.0g
Yeast extract	10.0g
$CaCO_3$	5.0g
Calcium pantothenate	2.0g
Citric acid	1.1g
Polysorbate 80	0.5g
K_2HPO_4	0.5g
KH_2PO_4	0.5g
$MgSO_4 \cdot 7H_2O$	0.2g
$FeSO_4 \cdot 7H_2O$	0.01g
$MnSO_4 \cdot 7H_2O$	0.01g
NaCl	0.01g
Bromcresol Green	0.022g
Cycloheximide	7.0mg

pH 6.7 ± 0.2 at 25°C

Source: This medium is available as a premixed powder from HiMedia.

Caution: Cycloheximide is toxic. Avoid skin contact or aerosol formation and inhalation.

Preparation of Medium: Add components to distilled/deionized water and bring volume to 1.0L. Mix thoroughly. Distribute into tubes or flasks. Pour into sterile Petri plates while swirling to prevent calcium carbonate from settling. The medium will have a white precipitate of calcium carbonate.

Use: For the detection of most organisms commonly encountered in a brewery. Acid producing bacteria are identified by the development of a clear zone around the colonies. Further identification is facilitated by the characteristic color reactions.

Legionella Agar Base (*Legionella* Medium) (BCYEα Agar, Modified)

Composition per liter:

Agar	17.0g
Yeast extract	10.0g
ACES buffer (*N*-2-acetamido-2-aminoethane sulfonic acid)	6.0g
Charcoal, activated	1.5g
KOH	1.5g
α-Ketoglutarate	1.0g
Legionella agar enrichment	10.0mL

pH 6.85–7.0 at 25°C

Source: This medium is available as a prepared medium from BD Diagnostic Systems.

***Legionella* Agar Enrichment:**

Composition per 10.0mL:

L-Cysteine·HCl·H_2O	0.4g
$Fe_4(P_2O_7)_3$	0.25g

Preparation of *Legionella* Agar Enrichment: Add components to distilled/deionized water and bring volume to 10.0mL. Mix thoroughly. Filter sterilize.

Preparation of Medium: Add components, except *Legionella* agar enrichment, to distilled/deionized water and bring volume to 990.0mL. Mix thoroughly. Gently heat to boiling. Autoclave for 15 min at 15 psi pressure–121°C. Cool to 50° C. Add 10.0mL of sterile *Legionella* agar enrichment. Adjust pH to 6.9 at 50°C by adding 4.0–4.5mL of 1.0*N* KOH. This is a critical step. Mix thoroughly. Pour into sterile Petri dishes in 20.0mL volumes. Swirl medium while pouring to keep charcoal in suspension.

Use: For the preparation of *Legionella* agars. For the isolation and cultivation of *Legionella* species from clinical and nonclinical materials.

Legionella pneumophila Medium (Charcoal Yeast Extract Diphasic Blood Culture Medium) (Diphasic Blood Culture Buffered Charcoal Yeast Extract Medium) (CYE-DBCM)

Composition per liter:

Agar phase	500.0mL
Broth phase	500.0mL

pH 6.9–7.0 at 25°C

Agar Phase:

Composition per 500.0mL:

Agar	17.0g
Charcoal, activated	2.0g

Preparation of Agar Phase: Add components to distilled/deionized water and bring volume to 500.0mL. Mix thoroughly. Gently heat and bring to boiling. Distribute in 20.0mL volumes into 125.0mL serum bottles with aluminum crimp seals and rubber stoppers. Autoclave for 20 min at 15 psi pressure–121°C. Cool to 50°C. Swirl medium to put charcoal in suspension. Allow agar to solidify so that a slant with a 6.0cm height is formed.

Broth Phase:

Composition per 500.0mL:

Yeast extract	20.0g
L-Cysteine·HCl·H_2O solution	0.4g
$Fe(NO_3)_3 \cdot 9H_2O$ solution	0.1g

Preparation of Broth Phase: Add yeast extract to distilled/deionized water and bring volume to 480.0mL. Mix thoroughly. Autoclave for 15 min at 15 psi pressure–121°C. Cool to 25°C. Aseptically add sterile L-cysteine·HCl·H_2O solution and $Fe(NO_3)_3 \cdot 9H_2O$ solution. Mix thoroughly. Adjust pH to 6.9 with 6.0mL of sterile 1*N* KOH.

L-Cysteine·HCl·H_2O Solution:

Composition per 10.0mL:

L-Cysteine·HCl·H_2O	0.04g

Preparation of L-Cysteine·HCl·H_2O Solution: Add L-cysteine·HCl·H_2O to distilled/deionized water and bring volume to 10.0mL. Mix thoroughly. Filter sterilize.

$Fe(NO_3)_3 \cdot 9H_2O$ Solution:

Composition per 10.0mL:

$Fe(NO_3)_3 \cdot 9H_2O$	0.04g

Preparation of $Fe(NO_3)_3 \cdot 9H_2O$ Solution: Add $Fe(NO_3)_3 \cdot 9H_2O$ to distilled/deionized water and bring volume to 10.0mL. Mix thoroughly. Filter sterilize.

Preparation of Medium: Add 20.0mL of sterile broth phase to 125.0mL serum bottles containing 20.0mL of solidified agar phase. Seal bottles by crimping metal caps over rubber stoppers.

Use: For the isolation and cultivation of *Legionella pneumophila* from blood cultures.

Legionella Selective Agar

Composition per liter:

Agar	15.0g
ACES (2-[(2-amino-2-oxoethyl)-amino]ethane sulfonic acid) buffer	10.0g
Yeast extract	10.0g
Charcoal, activated	2.0g
α-Ketoglutarate	1.0g
L-Cysteine·HCl·H_2O solution	10.0mL
$Fe_4(P_2O_7)_3$ solution	10.0mL
Antibiotic solution	10.0mL

pH 6.85–7.0 at 25°C

Source: This medium is available as a prepared medium from BD Diagnostic Systems.

L-Cysteine·HCl·H_2O Solution:

Composition per 10.0mL:

L-Cysteine·HCl·H_2O	0.4g

Preparation of L-Cysteine·HCl·H_2O Solution: Add L-cysteine·HCl·H_2O to distilled/deionized water and bring volume to 10.0mL. Mix thoroughly. Filter sterilize.

$Fe_4(P_2O_7)_3$ Solution:

Composition per 10.0mL:

$Fe_4(P_2O_7)_3$ 0.25g

Preparation of $Fe_4(P_2O_7)_3$ Solution: Add $Fe_4(P_2O_7)_3$ to distilled/deionized water and bring volume to 10.0mL. Mix thoroughly. Filter sterilize.

Antibiotic Solution:

Composition per 10.0mL:

Anisomycin 10.0mg
Colistin 3.75mg
Vancomycin 2.0mg

Preparation of Antibiotic Solution: Add components to distilled/deionized water and bring volume to 10.0mL. Mix thoroughly. Filter sterilize.

Preparation of Medium: Add components—except L-cysteine·HCl·H_2O, $Fe_4(P_2O_7)_3$, and antibiotic solutions—to distilled/deionized water and bring volume to 970.0mL. Mix thoroughly. Gently heat and bring to boiling. Autoclave for 15 min at 15 psi pressure–121°C. Cool to 45°–50°C. Aseptically add sterile L-cysteine·HCl·H_2O, $Fe_4(P_2O_7)_3$, and antibiotic solutions. Mix thoroughly. Pour into sterile Petri dishes. Swirl medium while pouring to keep charcoal in suspension.

Use: *Legionella* selective agar is used in qualitative procedures for the isolation of *Legionella* species from clinical and nonclinical specimens.

Legume Extract Agar

Composition per liter:

Alfalfa roots 35.0g
Agar 20.0g
Soybean meal 10.0g
Sucrose 10.0g
$CaCO_3$ 5.0g
Glucose 5.0g
K_2HPO_4 .... 1.0g
$MgSO_4·7H_2O$ 0.2g
$CaCl_2$ 0.1g
NaCl 0.1g
$FeCl_3$ 1.0mg

Preparation of Medium: Wash the alfalfa roots well and cut them up. Add 10.0g of soybean meal. Add three times the volume of distilled/deionized water. Steam for 1 hr. Let stand at 25°C overnight. Bring volume to 1.0L with distilled/deionized water. Filter through paper pulp. To the filtrate, add the K_2HPO_4, $CaCl_2$, $MgSO_4·7H_2O$, NaCl, $FeCl_3$, and agar. Autoclave for 20 min at 15 psi pressure–121°C. Cool to 45°–50°C. Add the $CaCO_3$, sucrose, and glucose. Mix thoroughly. Distribute into tubes or flasks. Autoclave for 20 min at 10 psi pressure–115°C.

Use: For the cultivation of *Rhizobium* species.

Leifson HiVeg Agar

Composition per liter:

Agar 12.0g
Lactose 10.0g
Plant extract No. 1 .... 6.5g
$Na_2S_2O_3$ 5.4g
Plant peptone No. 1 .... 5.0g
Synthetic detergent No. III 3.0g
Ferric citrate 1.0g
Neutral Red 0.02g

pH 7.5 ± 0.2 at 25°C

Source: This medium is available as a premixed powder from HiMedia.

Preparation of Medium: Add components to distilled/deionized water and bring volume to 1.0L. Mix thoroughly. Distribute into tubes or flasks. Gently bring to boiling. Do not autoclave. Pour into sterile Petri plates.

Use: For the isolation of *Salmonella* and *Shigella* species.

Leifson Medium

Composition per liter:

Agar 15.0g
Pancreatic digest of casein 2.0g
$MgCl_2$ 1.0g
Yeast extract 1.0g

pH 8.0 ± 0.2 at 25°C

Preparation of Medium: Add components to distilled/deionized water and bring volume to 1.0L. Mix thoroughly. Gently heat and bring to boiling. Adjust pH to 8.0. Distribute into tubes or flasks. Autoclave for 15 min at 15 psi pressure–121°C. Pour into sterile Petri dishes or leave in tubes.

Use: For the direct isolation and routine culturing of *Hyphomonas* species.

Leifson's Deoxycholate HiVeg Agar, Modified (Hugh Leifson Deoxycholate HiVeg Agar, Modified)

Composition per liter:

Agar 15.0g
Lactose 10.0g
Plant extract 5.0g
Plant peptone 5.0g
Sodium citrate 5.0g
$Na_2S_2O_3$ 5.0g
Synthetic detergent No. III 2.5g
Ferric citrate 1.0g
Neutral Red 0.025g

pH 7.0 ± 0.2 at 25°C

Source: This medium is available as a premixed powder from HiMedia.

Preparation of Medium: Add components to distilled/deionized water and bring volume to 1.0L. Mix thoroughly. Distribute into tubes or flasks. Gently bring to boiling. Do not autoclave. Pour into sterile Petri plates.

Use: For the selective isolation and differentiation of *Salmonella* and *Shigella* species.

Leishmania Medium

Composition per 100.0mL:

Sodium citrate 1.2g
NaCl 1.0g
Rabbit blood solution 10.0mL

Rabbit Blood Solution:

Composition per 10.0mL:

Rabbit blood, defibrinated 5.0mL

Preparation of Rabbit Blood Solution: Add 5.0mL of whole rabbit blood to 5.0mL of sterile distilled/deionized water. Freeze and thaw twice to lyse blood cells.

Preparation of Medium: Add components, except rabbit blood solution, to distilled/deionized water and bring volume to 90.0mL. Mix thoroughly. Autoclave for 15 min at 15 psi pressure–121°C. Aseptically add 10.0mL of sterile rabbit blood solution. Mix thoroughly. Aseptically distribute into sterile, screw-capped tubes or flasks.

Use: For the cultivation of *Leishmania donovani*, *Leishmania hertigi*, and *Leishmania tropica*.

LEMB Agar
See: **Levine EMB Agar**

Lenox Broth
See: **LB Medium**

Leonian's Agar

Composition per liter:

Agar	20.0g
Maltose	6.25g
Malt extract	6.25g
KH_2PO_4	1.25g
Peptone	0.625g
$MgSO_4 \cdot 7H_2O$	0.625g

Preparation of Medium: Add components to distilled/deionized water and bring volume to 1.0L. Mix thoroughly. Gently heat and bring to boiling. Distribute into tubes or flasks. Autoclave for 15 min at 15 psi pressure–121°C. Pour into sterile Petri dishes or leave in tubes.

Use: For the cultivation and maintenance of *Amorphotheca resinae, Apiosordaria rotula, Arachnotheca glomerata, Ascotricha erinacea, Auxarthron pseudauxarthron, Coniochaeta extramundana, Coniochaetidium ostreum, Coprinus velox, Cylindrocladium courataríae, Dicranidion fragile, Dicranidion gracilis, Eupenicillium brefeldianum, Isaria sulfurea, Linderina macrospora, Microthecium retisporum, Nigrospora sacchari, Orbicula parietina, Pectinotrichum llanense, Penicillium ochrochloron, Penicillium pinophilum, Pseudogymnoascus roseus, Thielavia terricola, Triangularia batistae, Tripospermum myrti,* and *Zopfiella pleuropora*.

Leptospira HiVeg Medium Base, Korthof, Modified with Rabbit Serum

Composition per liter:

NaCl	1.4g
Na_2HPO_4	0.88g
Plant peptone	0.8g
KH_2PO_4	0.24g
$CaCl_2$	0.04g
KCl	0.04g
$NaHCO_3$	0.02g
Rabbit serum, heat inactivated at 56°C	100.0mL

pH 7.2 ± 0.2 at 25°C

Source: This medium, without rabbit serum, is available as a premixed powder from HiMedia.

Preparation of Medium: Add components, except rabbit serum, to distilled/deionized water and bring volume to 900.0mL. Mix thoroughly. Gently heat and bring to boiling. Distribute into tubes or flasks. Autoclave for 15 min at 10 psi pressure–115°C. Cool to 50°–56°C. Aseptically add 100.0mL of rabbit serum. Mix thoroughly.

Use: For the cultivation of *Leptospira* species.

Leptospira HiVeg Medium Base, Korthof, Modified with Rabbit Serum and Hemoglobin

Composition per liter:

NaCl	1.4g
Na_2HPO_4	0.88g
Plant peptone	0.8g
KH_2PO_4	0.24g
$CaCl_2$	0.04g
KCl	0.04g
$NaHCO_3$	0.02g
Rabbit serum, heat inactivated at 56°C	80.0mL
Hemoglobin solution	8.0mL

pH 7.2 ± 0.2 at 25°C

Source: This medium, without rabbit serum, is available as a premixed powder from HiMedia.

Hemoglobin Solution:

Composition per 50.0mL:

Bovine hemoglobin	1.0g

Preparation of Hemoglobin Solution: Add bovine hemoglobin to distilled/deionized water and bring volume to 500.0mL. Mix thoroughly. Autoclave for 15 min at 15 psi pressure–121°C. Cool to 45°–50°C.

Preparation of Medium: Add components, except rabbit serum, and hemoglobin solution, to distilled/deionized water and bring volume to 1.0L. Mix thoroughly. Gently heat and bring to boiling. Distribute into tubes or flasks. Autoclave for 15 min at 10 psi pressure–115°C. Cool to 50°–56°C. Aseptically add 80.0mL of rabbit serum and 8.0mL of hemoglobin solution. Mix thoroughly. Aseptically dispense into tubes.

Use: For the cultivation of *Leptospira* species.

Leptospira Medium

Composition per liter:

$(NH_4)_2Fe(SO_4)_2 \cdot 6H_2O$	6.0g
NaH_2PO_4	0.53g
L-Asparagine	0.5g
Glycerol	0.2g
Tween™ 60	0.2g
$MgSO_4 \cdot 7H_2O$	0.15g
KH_2PO_4	0.069g
Tween™ 80	0.05g
EDTA	0.01g
$CaCO_3$	4.0mg
Thiamine·HCl	1.0mg
Vitamin B_{12}	1.0μg

pH 7.4–7.6 at 25°C

Preparation of Medium: Add components, except thiamine·HCl, to distilled/deionized water and bring volume to 990.0mL. Mix thoroughly. Gently heat and bring to boiling. Autoclave for 15 min at 15 psi pressure–121°C. Aseptically add 1.0mg of thiamine·HCl. Aseptically distribute into sterile tubes or flasks.

Use: For the cultivation of *Leptospira* species.

Leptospira Medium (DSMZ Medium 1113)

Composition per liter:

Agarose	1.5g
Na_2HPO_4	1.0g

NaCl	1.0g
KH_2PO_4	0.3g
NH_4Cl	0.25g
Thiamine	5.0mg
Leptospira enrichment EMJH (a solution of albumin, polysorbate 80 and additional growth factors available from BD Diagnostics	100.0mL

pH 7.5 ± 0.2 at 25°C

Preparation of Medium: Add agarose to distilled/deionized water and bring volume to 900.0mL. Mix thoroughly. Gently heat while stirring and bring to boiling. Boil with mixing until agarose dissolves. Add the other components, except *Leptospira* enrichment EMJH. Mix thoroughly. Autoclave for 15 min at 15 psi pressure–121°C. Cool to 50°C. Mix thoroughly. Warm the *Leptospira* enrichment to approximately 25°C and add to the warm medium. Mix thoroughly. Aseptically distribute into sterile tubes or flasks.

Use: For the cultivation of *Leptospira* spp.

Leptospira Medium, EMJH (*Leptospira* Medium, Ellinghausen-McCullough/ Johnson-Harris)

Composition per liter:

Na_2HPO_4	1.0g
NaCl	1.0g
KH_2PO_4	0.3g
NH_4Cl	0.25g
Thiamine	5.0mg
Rabbit serum	100.0mL

pH 7.5 ± 0.2 at 25°C

Source: This medium is available as a premixed powder from BD Diagnostic Systems.

Preparation of Medium: Add components, except rabbit serum, to distilled/deionized water and bring volume to 900.0mL. Mix thoroughly. Gently heat and bring to boiling. Autoclave for 15 min at 15 psi pressure–121°C. Cool to 25°C. Aseptically add sterile rabbit serum. Mix thoroughly. Aseptically distribute into sterile tubes or flasks.

Use: For the cultivation and maintenance of *Leptospira* species.

Leptospira Medium, Modified

Composition per liter:

Agar	1.5g
NaCl	0.5g
Peptone	0.3g
Beef extract	0.2g
Hemin solution	2.5mL
Sterile rabbit serum	100.0mL

pH 7.3 ± 0.1 at 25°C

Hemin Solution:

Composition per 100.0mL:

Hemin	0.05g
NaOH (1*N* solution)	1.0mL

Preparation of Hemin Solution: Add hemin to 1.0mL of 1*N* NaOH solution. Mix thoroughly. Bring volume to 100.0mL with distilled/deionized water. Autoclave for 15 min at 15 psi pressure–121°C. Cool to 45°–50°C.

Preparation of Medium: Add components, except hemin solution and rabbit serum, to distilled/deionized water and bring volume to 897.5mL. Mix thoroughly. Gently heat and bring to boiling. Adjust pH to 7.4. Autoclave for 15 min at 15 psi pressure–121°C. Cool to 45°–50°C. Aseptically add 2.5mL of sterile hemin solution and 100.0mL of sterile rabbit serum. Mix thoroughly. The pH of the medium should be 7.3. Store at 4°C for 24 hr. Inactivate medium at 56°C for 60 min. Aseptically distribute into sterile tubes or flasks.

Use: For the cultivation and maintenance of *Leptospira biflexa, Leptospira borgpetersenii, Leptospira interrogans, Leptospira meyeri, Leptospira noguchii, Leptospira santarosai,* and *Leptospira weili.*

Leptospira Protein-Free Medium (*Leptospira* PF Medium)

Composition per liter:

TES (*N*-Tris[hydroxymethyl]methyl-2-aminoethane sulfonic acid) buffer	1.2g
NaCl	0.9g
Sodium pyruvate	0.2g
CT-Tween™ 60	12.0mL
CT-Tween™ 40	3.0mL
$MgCl_2$-$CaCl_2$ solution	1.0mL
Cyanocobalamin (0.02% solution)	1.0mL
Glycerol (10% solution)	1.0mL
KH_2PO_4 (1% solution)	1.0mL
$MnSO_4 \cdot H_2O$ (0.1% solution)	1.0mL
$ZnSO_4$ (0.4% solution)	0.1mL

pH 7.6 ± 0.2 at 25°C

CT-Tween™ 60:

Composition per 200.0mL:

Charcoal, Norit A	40.0g
Tween™ 60	20.0g

Preparation of CT-Tween™ 60: Add Tween™ 60 to 200.0mL of distilled/deionized water. Mix thoroughly. While stirring, add charcoal. Stir mixture for 18 hr at 25°C. Allow charcoal to settle out of suspension for 18 hr at 4°C. Carefully decant the Tween™ solution off the sediment. Centrifuge the Tween™ solution at 10,000 × g for 1 hr. Decant supernatant solution. Pass Tween™ solution through a thin-channel ultrafiltration XM 100 membrane. Store stock solution at –20°C.

CT-Tween™ 40:

Composition per 200.0mL:

Charcoal, Norit A	40.0g
Tween™ 40	20.0g

Preparation of CT-Tween™ 40: Add Tween™ 40 to 200.0mL of distilled/deionized water. Mix thoroughly. While stirring, add charcoal. Stir mixture for 18 hr at 25°C. Allow charcoal to settle out of suspension for 18 hr at 4°C. Carefully decant the Tween™ solution off the sediment. Centrifuge the Tween™ solution at 10,000 × g for 1 hr. Decant supernatant solution. Pass Tween™ solution through a thin-channel ultrafiltration XM 100 membrane. Store stock solution at –20°C.

$MgCl_2$-$CaCl_2$ Solution:

Composition per 100.0mL:

$CaCl_2 \cdot 2H_2O$	1.5g
$MgCl_2 \cdot 6H_2O$	1.5g

Preparation of $MgCl_2$-$CaCl_2$ Solution: Add components to distilled/deionized water and bring volume to 100.0mL. Mix thoroughly.

Preparation of Medium: Add components to distilled/deionized water and bring volume to 1.0L. Mix thoroughly. Filter sterilize. Aseptically distribute into sterile tubes or flasks.

Use: For the cultivation of *Leptospira* species.

Leptospirillum ferrooxidans Medium

Composition per liter:

$FeSO_4 \cdot 7H_2O$	30.0g
$CaCl_2 \cdot 2H_2O$	0.147g
$(NH_4)_2SO_4$	0.13g
KH_2PO_4	27.0mg
$MgCl_2 \cdot 6H_2O$	25.0mg
Trace elements solution	1.0mL

pH 2.0 ± 0.2 at 25°C

Trace Elements Solution:

Composition per liter:

$CoCl_2 \cdot 6H_2O$	0.12g
$MnCl_2 \cdot 4H_2O$	0.1g
$Na_2MoO_4 \cdot 2H_2O$	85.2mg
$ZnCl_2$	70.0mg
H_3BO_3	31.0mg

Preparation of Trace Elements Solution: Add components to distilled/deionized water and bring volume to 1.0L. Mix thoroughly.

Preparation of Medium: Add H_2SO_4 to 900.0mL of distilled/deionized water and bring pH to 3.0. Add components. Mix thoroughly. Bring volume to 1.0L with distilled/deionized water. Mix thoroughly. Adjust pH to 2.0 with H_2SO_4. Distribute into tubes or flasks. Autoclave for 15 min at 15 psi pressure–121°C.

Use: For the cultivation of *Leptospirillum ferrooxidans.*

Leptospirillum HH Medium (DSMZ Medium 882)

Composition per 1001.0mL:

Solution A	950.0mL
Soultion B	50.0mL
Trace elements solution	1.0mL

pH 1.8 ± 0.2 at 25°C

Solution A:

$CaCl_2 \cdot 2H_2O$	147.0mg
$(NH_4)_2SO_4$	132.0mg
$MgCl_2 \cdot 6H_2O$	53.0mg
KH_2PO_4	27.0mg

Preparation of Solution A: Add components to distilled/deionized water and bring volume to 950.0mL. Mix thoroughly. Adjust pH to 1.8 with 10*N* H_2SO_4. Autoclave for 30 min at 112°C. Cool to room temperature.

Solution B:

$FeSO_4 \cdot 7H_2O$	20.0g
H_2SO_4, 0.25*N*	50.0mL

Preparation of Solution B: Add $FeSO_4 \cdot 7H_2O$ to 50.0mL 0.25*N* H_2SO_4. Mix thoroughly. The pH should be 1.2. Autoclave for 30 min at 112°C. Cool to room temperature.

Trace Elements Solution:

$ZnCl_2$	68.0mg
$CuCl_2 \cdot 2H_2O$	67.0mg
$CoCl_2 \cdot 6H_2O$	64.0mg
$MnCl_2 \cdot 2H_2O$	62.0mg
H_3BO_3	31.0mg
Na_2MoO_4	10.0mg

Preparation of Trace Elements Solution: Add components to distilled/deionized water and bring volume to 1.0L. Mix thoroughly. Adjust pH to 1.8 with 10*N* H_2SO_4. Autoclave for 15 min at 15 psi pressure–121°C. Cool to room temperature.

Preparation of Medium: Aseptically mix 950.0mL of solution A and 50.0mL solution B. Mix thoroughly. Aseptically add 1.0mL trace elements solution. Mix thoroughly. Adjust pH to 1.8.

Use: For the cultivation of *Acidithiobacillus ferrooxidans=Thiobacillus ferrooxidans* and *Leptospirillum* spp.

Leptothrix ochracea Medium

Composition per liter:

Agar	10.0g
Manganous acetate	0.1g
Manganese bicarbonate solution	100.0mL

pH 7.0 ± 0.2 at 25°C

Manganese Bicarbonate Solution:

Composition per 100.0mL:

$MnCO_3$	2.0g

Preparation of Manganese Bicarbonate Solution: Add $MnCO_3$ to distilled/deionized water and bring volume to 100.0mL. Mix thoroughly. Gas with 100% CO_2 for 20 min. Filter through Whatman #1 filter paper.

Preparation of Medium: Add components to distilled/deionized water and bring volume to 1.0L. Mix thoroughly. Gently heat and bring to boiling. Distribute into tubes or flasks. Autoclave for 15 min at 15 psi pressure–121°C. Pour into sterile Petri dishes or leave in tubes.

Use: For the cultivation of *Leptothrix ochracea.*

Leptothrix 2X PYG Medium

Composition per liter:

HEPES (*N*-2-Hhydroxyethyl piperazine-*N*′-2-ethanesulfonic acid) buffer	3.57g
$MgSO_4 \cdot 7H_2O$	0.6g
Glucose	0.5g
Peptone	0.5g
Yeast extract	0.5g
$CaCl_2 \cdot 2H_2O$	0.07g
$MnSO_4 \cdot H_2O$	0.017g

pH 7.3 ± 0.2 at 25°C

Preparation of Medium: Add components to distilled/deionized water and bring volume to 1.0L. Mix thoroughly. Adjust pH to 7.3. Distribute into tubes or flasks. Autoclave for 15 min at 15 psi pressure–121°C.

Use: For the cultivation and maintenance of *Leptothrix discophora.*

Leptothrix Strains Medium

Composition per liter:

Agar	7.5g
$MnCO_2$	2.0g
Beef extract	1.0g
$Fe(NH_4)_2(SO_4)_2$	0.15g
Sodium citrate	0.15g
Yeast extract	0.075g
Vitamin B_{12}	5.0μg

pH 7.0 ± 0.2 at 25°C

Preparation of Medium: Add components to distilled/deionized water and bring volume to 1.0L. Mix thoroughly. Distribute into tubes or flasks. Autoclave for 15 min at 15 psi pressure–121°C.

Use: For the cultivation and maintenance of *Leptothrix cholodnii, Leptothrix discophora,* and *Sphaerotilus natans*.

Leptothrix Strains Medium

Composition per liter:

Agar	12.0g
Peptone	5.0g
$MgSO_4 \cdot 7H_2O$	0.2g
Ferric ammonium citrate	0.15g
$CaCl_2$	0.05g
$FeCl_3 \cdot 6H_2O$	0.01g
$MnSO_4 \cdot H_2O$	0.01g

pH 7.0 ± 0.2 at 25°C

Preparation of Medium: Add components to distilled/deionized water and bring volume to 1.0L. Mix thoroughly. Gently heat and bring to boiling. Distribute into tubes or flasks. Autoclave for 15 min at 15 psi pressure–121°C. Pour into sterile Petri dishes or leave in tubes.

Use: For the isolation and cultivation of *Leptothrix* species.

Leptotrichia buccalis Medium

Composition per liter:

Agar	15.0g
Nutrient broth	8.0g
Yeast extract	2.0g
Glucose solution	10.0mL
L-Cysteine solution	10.0mL
Hemin solution	4.0mL

pH 7.2–7.6 at 25°C

Glucose Solution:

Composition per 10.0mL:

D-Glucose	2.0g

Preparation of Glucose Solution: Add glucose to distilled/deionized water and bring volume to 10.0mL. Mix thoroughly. Filter sterilize.

L-Cysteine Solution:

Composition per 10.0mL:

L-Cysteine·HCl·H_2O	1.0g

Preparation of L-Cysteine Solution: Add L-cysteine·HCl·H_2O to distilled/deionized water and bring volume to 10.0mL. Mix thoroughly. Filter sterilize.

Hemin Solution:

Composition per 10.0mL:

Hemin	2.5mg
Triethanolamine (50% solution)	10.0mL

Preparation of Hemin Solution: Add hemin to 10.0mL of triethanolamine solution. Mix thoroughly.

Preparation of Medium: Add components, except glucose solution, to distilled/deionized water and bring volume to 990.0mL. Mix thoroughly. Gently heat and bring to boiling. Autoclave for 15 min at 15 psi pressure–121°C. Cool to 45°–50°C. Aseptically add sterile glucose solution. Mix thoroughly. Pour into sterile Petri dishes or distribute into sterile tubes.

Use: For the cultivation and maintenance of *Leptotrichia buccalis*.

Leptotrichia Medium

Composition per liter:

Pancreatic digest of casein	10.0g
NaCl	5.0g
Peptone	5.0g
Yeast extract	3.0g
Na_2HPO_4	2.5g
L-Cysteine·HCl·H_2O	0.5g
Horse serum	100.0mL
Tomato decoction	50.0mL

pH 7.2–7.4 at 25°C

Tomato Decoction:

Composition per 100.0mL:

Tomatoes	50.0mL

Preparation of Tomato Decoction: Mince fresh tomatoes and measure 50.0mL. Add 50.0mL of tap water. Mix thoroughly. Gently heat and bring to boiling. Continue boiling for 10 min. Filter through Whatman #1 filter paper. Autoclave filtrate for 15 min at 15 psi pressure–121°C.

Preparation of Medium: Add components, except horse serum and tomato decoction, to distilled/deionized water and bring volume to 850.0mL. Mix thoroughly. Gently heat and bring to boiling. Adjust pH to 7.2–7.4. Autoclave for 15 min at 15 psi pressure–121°C. Cool to 25°C. Aseptically add sterile horse serum and tomato decoction. Mix thoroughly. Aseptically distribute into sterile tubes or flasks.

Use: For the cultivation and maintenance of *Leptotrichia buccalis*.

LES Endo Agar

See: **Endo Agar, LES**

Letheen Agar

Composition per liter:

Agar	15.0g
Tween™ 80	7.0g
Pancreatic digest of casein	5.0g
Beef extract	3.0g
Glucose	1.0g
Lecithin	1.0g

pH 7.0 ± 0.2 at 25°C

Source: This medium is available as a premixed powder from BD Diagnostic Systems.

Preparation of Medium: Add components to distilled/deionized water and bring volume to 1.0L. Mix thoroughly. Gently heat and bring to boiling. Distribute into tubes or flasks. Autoclave for 15 min at 15 psi pressure–121°C. Pour into sterile Petri dishes or leave in tubes.

Use: For determination of the antimicrobial activity of quaternary ammonium compounds.

Letheen Agar, Modified

Composition per liter:

Agar	20.0g
Thiotone	10.0g
Pancreatic digest of casein	10.0g
Tween™ 80	7.0g
NaCl	5.0g
Beef extract	3.0g
Yeast extract	2.0g
Glucose	1.0g
Lecithin	1.0g
$NaHSO_3$	0.1g

pH 7.2 ± 0.2 at 25°C

Preparation of Medium: Add components to distilled/deionized water and bring volume to 1.0L. Mix thoroughly. Gently heat and bring to boiling. Distribute into tubes or flasks. Autoclave for 15 min at 15 psi pressure–121°C. Pour into sterile Petri dishes.

Use: For determination of the antimicrobial activity of quaternary ammonium compounds.

Letheen Broth

Composition per liter:

Peptic digest of animal tissue	10.0g
Beef extract	5.0g
NaCl	5.0g
Tween™ 80	5.0g
Lecithin	0.7g

pH 7.0 ± 0.2 at 25°C

Source: This medium is available as a premixed powder from BD Diagnostic Systems.

Preparation of Medium: Add components to distilled/deionized water and bring volume to 1.0L. Mix thoroughly. Distribute into tubes or flasks. Autoclave for 15 min at 15 psi pressure–121°C.

Use: For determination of the antimicrobial activity of quaternary ammonium compounds.

Letheen Broth, Modified

Composition per liter:

Peptic digest of animal tissue	10.0g
Thiotone peptone	10.0g
Beef extract	5.0g
NaCl	5.0g
Tween™ 80	5.0g
Pancreatic digest of casein	5.0g
Yeast extract	2.0g
Lecithin	0.7g
$NaHSO_3$	0.1g

pH 7.2 ± 0.2 at 25°C

Preparation of Medium: Add components to distilled/deionized water and bring volume to 1.0L. Mix thoroughly. Distribute into screw-capped bottles in 90.0mL volumes. Autoclave for 15 min at 15 psi pressure–121°C.

Use: For determination of the antimicrobial activity of quaternary ammonium compounds.

Letheen HiVeg Agar

Composition per liter:

Agar	15.0g
Polysorbate 80	7.0g
Plant hydrolysate	5.0g
Plant extract	3.0g
Glucose	1.0g
Lecithin	1.0g

pH 7.0 ± 0.2 at 25°C

Source: This medium is available as a premixed powder from HiMedia.

Preparation of Medium: Add components to distilled/deionized water and bring volume to 1.0L. Mix thoroughly. Gently heat and bring to boiling. Distribute into tubes or flasks. Autoclave for 15 min at 15 psi pressure–121°C. Pour into sterile Petri dishes or leave in tubes.

Use: For the determination of the antimicrobial activity of quaternary ammonium compounds.

Letheen HiVeg Agar, Modified

Composition per liter:

Agar	20.0g
Plant peptone	20.0g
Plant extract	5.0g
Plant hydrolysate	5.0g
NaCl	5.0g
Polysorbate 80	5.0g
Yeast extract	2.0g
Lecithin	0.7g
$NaHSO_3$	0.1g

pH 7.0 ± 0.2 at 25°C

Source: This medium is available as a premixed powder from HiMedia.

Preparation of Medium: Add components to distilled/deionized water and bring volume to 1.0L. Mix thoroughly. Gently heat and bring to boiling. Distribute into tubes or flasks. Autoclave for 15 min at 15 psi pressure–121°C. Pour into sterile Petri dishes.

Use: For the screening of cosmetic products for microbial contamination.

Letheen HiVeg Broth, AOAC

Composition per liter:

Plant peptone	10.0g
Plant extract	5.0g
Polysorbate 80	5.0g
NaCl	5.0g
Lecithin	0.7g

pH 7.0 ± 0.2 at 25°C

Source: This medium is available as a premixed powder from HiMedia.

Preparation of Medium: Add components to distilled/deionized water and bring volume to 1.0L. Mix thoroughly. Gently heat and bring to boiling. Distribute into tubes or flasks. Autoclave for 15 min at 15 psi pressure–121°C.

Use: For the determination of the antimicrobial activity of quaternary ammonium compounds.

Letheen HiVeg Broth, Modified

Composition per liter:

Plant peptone	20.0g
Plant extract	5.0g
Plant hydrolysate	5.0g
NaCl	5.0g
Polysorbate 80	5.0g
Yeast extract	2.0g
Lecithin	0.7g
$NaHSO_3$	0.1g

pH 7.0 ± 0.2 at 25°C

Source: This medium is available as a premixed powder from HiMedia.

Preparation of Medium: Add components to distilled/deionized water and bring volume to 1.0L. Mix thoroughly. Gently heat and bring to boiling. Distribute into tubes or flasks. Autoclave for 15 min at 15 psi pressure–121°C.

Use: For the screening of cosmetic products for microbial contamination.

Leuconostoc Medium

Composition per liter:

$CaCO_3$	50.0g
Malt extract	50.0g
Agar	15.0g
NaCl	2.5g
Beef extract	1.0g
Polypeptone™	1.0g

Preparation of Medium: Add components to distilled/deionized water and bring volume to 1.0L. Mix thoroughly. Gently heat and bring to boiling. Distribute into tubes or flasks. Autoclave for 10 min at 15 psi pressure–121°C. Pour into sterile Petri dishes or leave in tubes.

Use: For the cultivation and maintenance of *Leuconostoc mesenteroides*.

Leuconostoc oenos Medium

Composition per liter:

Tryptic digest of casein	10.0g
Glucose	10.0g
Fructose	5.0g
Yeast extract	5.0g
Diammonium citrate	3.5g
L-Cysteine·HCl	0.5g
$MgSO_4 \cdot 7H_2O$	200.0mg
$MnSO_4 \cdot H_2O$	50.0mg
Tomato juice, filtered	100.0mL
Tween™ 80	1.0mL

pH 4.8 ± 0.2 at 25°C

Preparation of Medium: Add components to distilled/deionized water and bring volume to 1.0L. Mix thoroughly. Adjust pH to 4.8. Distribute into tubes or flasks. Autoclave for 15 min at 15 psi pressure–121°C.

Use: For the cultivation of *Leuconostoc oenos*.

Leuconostoc oenos Medium

Composition per liter:

Glucose	10.0g
Peptone	10.0g
Yeast extract	5.0g
$MnSO_4 \cdot 4H_2O$	0.1g
Tomato juice	250.0mL
L-Cysteine·HCl solution	10.0mL

pH 4.8 ± 0.2 at 25°C

L-Cysteine·HCl Solution:

Composition per 10.0mL:

L-Cysteine·HCl	0.5g

Preparation of L-Cysteine·HCl Solution: Add L-cysteine·HCl to distilled/deionized water and bring volume to 10.0mL. Mix thoroughly. Filter sterilize.

Preparation of Medium: Add components, except L-cysteine·HCl solution, to distilled/deionized water and bring volume to 990.0mL. Mix thoroughly. Gently heat and bring to boiling. Autoclave for 15 min at 15 psi pressure–121°C. Cool to 25°C. Aseptically add sterile L-cysteine·HCl solution. Mix thoroughly. Aseptically distribute into sterile tubes or flasks.

Use: For the cultivation of *Leuconostoc oenos*.

Leuconostoc oenos Medium

Composition per liter:

Glucose	10.0g
Tryptone	10.0g
Fructose	5.0g
Yeast extract	5.0g
Diammonium citrate	3.5g
L-Cysteine·HCl	0.5g
$So_4Mg \cdot 7H_2O$	0.2g
$MnSO_4 \cdot H_2O$	0.05g
Tomato juice, filtered	100.0mL
Tween™ 80	1.0mL

pH 4.8 ± 0.2 at 25°C

Preparation of Medium: Add components to distilled/deionized water and bring volume to 1.0L. Mix thoroughly. Adjust pH to 4.8. Distribute into tubes or flasks. Autoclave for 15 min at 15 psi pressure–121°C.

Use: For the cultivation of *Leuconostoc oenos.*

Leucothrix Medium

Composition per liter:

Pancreatic digest of casein	10.0g
Synthetic seawater	1000.0mL

pH 7.8 ± 0.2 at 25°C

Synthetic Seawater:

Composition per liter:

NaCl	24.0g
$MgCl_2 \cdot 6H_2O$	11.0g
Na_2SO_4	4.0g
$CaCl_2 \cdot 6H_2O$	2.0g
KCl	0.7g
KBr	0.1g
$SrCl_2 \cdot 6H_2O$	0.04g
H_3BO_3	0.03g
$NaSiO_3 \cdot 9H_2O$	5.0mg
NaF	3.0mg
NH_4NO_3	2.0mg
$FePO_4 \cdot 4H_2O$	1.0mg

Preparation of Synthetic Seawater: Add components to distilled/deionized water and bring volume to 1.0L. Mix thoroughly.

Preparation of Medium: Add 10.0g of pancreatic digest of casein to 1.0L of synthetic seawater. Mix thoroughly. Adjust pH to 7.8. Distribute into tubes or flasks. Autoclave for 15 min at 15 psi pressure–121°C.

Use: For the cultivation of *Leucothrix* species.

Leucothrix Medium (ATCC Medium 429)

Composition per liter:

NaCl	11.7g
Monosodium glutamate	10.0g
$MgCl_2 \cdot 6H_2O$	5.35g
Na_2SO_4	2.0g
$CaCl_2 \cdot 2H_2O$	0.75g
Tris(hydroxymethyl)aminomethane buffer	0.5g
KCl	0.35g
Na_2HPO_4	0.05g

pH 7.6 ± 0.2 at 25°C

Preparation of Medium: Add components to distilled/deionized water and bring volume to 1.0L. Mix thoroughly. Distribute into tubes or flasks. Autoclave for 15 min at 15 psi pressure–121°C.

Use: For the cultivation and maintenance of *Leucothrix mucor.*

Leucothrix Medium (ATCC Medium 430)

Composition per liter:

NaCl	11.7g
$MgCl_2 \cdot 6H_2O$	5.35g
Na_2SO_4	2.0g
$CaCl_2 \cdot 2H_2O$	0.75g
Pancreatic digest of casein	0.5g
Yeast extract	0.5g
Tris(hydroxymethyl)aminomethane buffer	0.5g
KCl	0.35g
Na_2HPO_4	0.05g

pH 7.6 ± 0.2 at 25°C

Preparation of Medium: Add components to distilled/deionized water and bring volume to 1.0L. Mix thoroughly. Distribute into tubes or flasks. Autoclave for 15 min at 15 psi pressure–121°C.

Use: For the cultivation and maintenance of *Leucothrix mucor.*

Leucothrix mucor Medium

Composition per liter:

NaCl	11.75g
Monosodium glutamate	10.0g
$MgCl_2 \cdot 6H_2O$	5.35g
Na_2SO_4	2.0g
Sodium lactate	2.0g
$CaCl_2 \cdot 6H_2O$	1.12g
Tris (hydroxymethyl)aminomethane buffer	0.5g
KCl	0.35g
Na_2HPO_4	0.05g

pH 7.6 ± 0.2 at 25°C

Preparation of Medium: Add components to distilled/deionized water and bring volume to 1.0L. Mix thoroughly. Distribute into tubes or flasks. Autoclave for 15 min at 15 psi pressure–121°C.

Use: For the cultivation of *Leucothrix mucor*.

Levine EMB Agar (Levine Eosin Methylene Blue Agar) (Eosin Methylene Blue Agar, Levine) (LEMB Agar)

Composition per liter:

Agar	15.0g
Lactose	10.0g
Peptone	10.0g
K_2HPO_4	2.0g
Eosin Y	0.4g
Methylene Blue	0.065mg

pH 7.1 ± 0.2 at 25°C

Source: This medium is available as a premixed powder from BD Diagnostic Systems.

Preparation of Medium: Add components to distilled/deionized water and bring volume to 1.0L. Mix thoroughly. Gently heat and bring to boiling. Distribute into tubes or flasks. Autoclave for 15 min at 15 psi pressure–121°C. Pour into sterile Petri dishes or leave in tubes.

Use: For the isolation, cultivation, and differentiation of Gram-negative enteric bacteria based on lactose fermentation. Bacteria that ferment lactose, especially the coliform bacterium *Escherichia coli,* appear as colonies with a green metallic sheen or blue-black to brown color. Bacteria that do not ferment lactose appear as colorless or transparent light purple colonies.

Levinthal's Agar

Composition per 105.0mL:

Nutrient agar, sterile	100.0mL
Rabbit blood or human blood, sterile	5.0mL

pH 6.8 ± 0.2 at 25°C

Nutrient Agar:

Composition per liter:

Agar	15.0g
Pancreatic digest of gelatin	5.0g
Beef extract	3.0g

Source: Nutrient agar is available as a premixed powder from BD Diagnostic Systems.

Preparation of Nutrient Agar: Add components to distilled/deionized water and bring volume to 1.0L. Mix thoroughly. Gently heat while stirring and bring to boiling. Distribute into tubes or flasks. Autoclave for 15 min at 15 psi pressure–121°C. Cool to 45°–50°C.

Preparation of Medium: To 100.0mL of cooled, sterile nutrient agar, aseptically add 5.0mL of human blood or rabbit blood. Mix thoroughly. Heat in a boiling water bath for 5 min. Allow the deposit to settle out of suspension. Pour the clear supernatant solution into sterile Petri dishes or distribute into sterile tubes.

Use: For the cultivation of *Haemophilus* species.

Levinthal's HiVeg Medium Base with Blood

Composition per liter:

Agar	20.0g
Plant extract	10.0g
Plant peptone	10.0g
NaCl	5.0g
Rabbit blood or human blood, sterile	50.0mL

pH 7.6 ± 0.2 at 25°C

Source: This medium, without blood, is available as a premixed powder from HiMedia.

Preparation of Medium: Add components, except blood, to distilled/deionized water and bring volume to 1.0L. Mix thoroughly. Gently heat while stirring and bring to boiling. Distribute into tubes or flasks. Autoclave for 15 min at 15 psi pressure–121°C. Cool to 45°–50°C. Aseptically add 50.0mL of human blood or rabbit blood. Mix thoroughly. Heat in a boiling water bath for 5 min. Allow the deposit to settle out of suspension. Pour the clear supernatant solution into sterile Petri dishes or distribute into sterile tubes.

Use: For the cultivation of *Haemophilus* species.

Levinthal's Medium Base

Composition per liter:

Agar	20.0g
Peptic digest of animal tissue	10.0g
Beef extract	10.0g

NaCl	5.0g
Rabbit or human blood	50.0mL

pH 7.6 ± 0.2 at 25°C

Source: This medium is available from HiMedia.

Preparation of Medium: Add components, except blood, to distilled/deionized water and bring volume to 950.0mL. Mix thoroughly. Autoclave for 15 min at 15 psi pressure–121°C. Cool to 50°C. Aseptically add blood. Mix thoroughly. Heat in boiling water bath. Allow the deposits to settle. Distribute clear supernatant into sterile tubes.

Use: For the cultivation of *Haemophilus* spp.

LGI Medium

See: ***Azospirillum amazonense*** **Medium**

LHET2 Medium

Composition per liter:

Solution A	500.0mL
Solution B	500.0mL

pH 2.5–3.0 at 25°C

Solution A:

Composition per 500.0mL:

$(NH_4)_2SO_4$	2.0g
K_2HPO_4	0.51g
$MgSO_4 \cdot 7H_2O$	0.5g
KCl	0.1g
Pancreatic digest of casein	0.06g
NaCl	0.02g
Papaic digest of soybean meal	0.01g

Preparation of Solution A: Add components to distilled/deionized water and bring volume to 500.0mL. Mix thoroughly. Gently heat and bring to boiling. Adjust pH to 2.5–3.0 with 1*N* H_2SO_4. Autoclave for 15 min at 15 psi pressure–121°C. Cool to 45°–50°C.

Solution B:

Composition per 500.0mL:

Agar	12.0g
Glucose	1.0g

Preparation of Solution B: Add components to distilled/deionized water and bring volume to 500.0mL. Mix thoroughly. Gently heat and bring to boiling. Autoclave for 15 min at 15 psi pressure–121°C. Cool to 45°–50°C.

Preparation of Medium: Aseptically combine sterile solution A and sterile solution B. Mix thoroughly. Pour into sterile Petri dishes or distribute into sterile tubes.

Use: For the cultivation and maintenance of *Acidiphilium cryptum*.

LHET2 Medium with Yeast Extract or Yeast Autolysate

Composition per liter:

Solution A	500.0mL
Solution B	500.0mL

pH 2.5–3.0 at 25°C

Solution A:

Composition per 500.0mL:

$(NH_4)_2SO_4$	2.0g
K_2HPO_4	0.51g
$MgSO_4 \cdot 7H_2O$	0.5g
KCl	0.1g
Yeast extract or yeast autolysate	0.1g
Pancreatic digest of casein	0.06g
NaCl	0.02g
Papaic digest of soybean meal	0.01g

Preparation of Solution A: Add components to distilled/deionized water and bring volume to 500.0mL. Mix thoroughly. Gently heat and bring to boiling. Adjust pH to 2.5–3.0 with 1*N* H_2SO_4. Autoclave for 15 min at 15 psi pressure–121°C. Cool to 45°–50°C.

Solution B:

Composition per 500.0mL:

Agar	12.0g
Glucose	1.0g

Preparation of Solution B: Add components to distilled/deionized water and bring volume to 500.0mL. Mix thoroughly. Gently heat and bring to boiling. Autoclave for 15 min at 15 psi pressure–121°C. Cool to 45°–50°C.

Preparation of Medium: Aseptically combine sterile solution A and sterile solution B. Mix thoroughly. Pour into sterile Petri dishes or distribute into sterile tubes.

Use: For the cultivation and maintenance of *Acidiphilium angustum, Acidiphilium facilis,* and *Acidiphilium rubrum.*

Lichen Fungi Medium

Composition per liter:

Agar	20.0g
Malt extract	20.0g
Yeast extract	2.0g

Preparation of Medium: Add components to distilled/deionized water and bring volume to 1.0L. Mix thoroughly. Gently heat until boiling. Distribute into tubes or flasks. Autoclave for 15 min at 15 psi pressure–121°C. Pour into sterile Petri dishes or leave in tubes.

Use: *For the cultivation of Acarospora fuscata, Acarospora smaragdula, Anaptychia cilaris, Anthracothecium albescens, Arthonia cinnabarina, Bacidia incompta, Baeomyces roseus, Buellia stillingiana, Caloplaca aurantiaca, Candelariella vitellina, Cetraria islandica, numerous Cladonia* species, *Dermatocarpon fluviatile, Graphis tenella, Lecanora cinerea, Lecanora dispersa, Lecanora rubina, Lecidea* species, *Microthelia albidella, Opegrapha lichenoides, Parmelia centrifuga, Parmelia conspersa, Phisica millegrana, Phisica stellaris, Porina sandwichensis, Pyrenula nitida, Ramalina americana, Sarcogyne simplex, Stereocaulon vulcani, Umbilicaria papulosa, Usnea florida, Xanthoria parietina,* and other fungi from lichen symbiotic relationships.

LICNR Broth (Lysine Iron Cystine Neutral Red Broth)

Composition per 500.0mL:

L-Lysine·HCl	10.0g
Mannitol	5.0g
Pancreatic digest of casein	5.0g
Yeast extract	3.0g
Glucose	1.0g
Salicin	1.0g
Ferric ammonium citrate	0.5g
L-Cystine	0.1g
$Na_2S_2O_3$	0.1g
Neutral Red	0.025g
Novobiocin solution	10.0mL

pH 6.2 ± 0.2 at 25°C

Novobiocin Solution:
Composition per 10.0mL:

Novobiocin	0.015g

Preparation of Novobiocin Solution: Add novobiocin to distilled/deionized water and bring volume to 10.0mL. Mix thoroughly. Filter sterilize.

Preparation of Medium: Add components, except novobiocin solution, to distilled/deionized water and bring volume to 990.0mL. Mix thoroughly. Gently heat and bring to boiling. Continue boiling for 2–3 min. Distribute into tubes in 10.0mL volumes. Autoclave for 15 min at 15 psi pressure–121°C. Cool to 45°–50°C. Aseptically add 0.1mL of sterile novobiocin solution to each tube. Mix thoroughly.

Use: For the rapid, presumptive detection of *Salmonella* in foods, food ingredients, and feed materials.

Lima Bean Agar (ATCC Medium 322)

Composition per liter:

Lima beans, infusion from 62.5g	8.0g
Agar	15.0g

pH 5.6 ± 0.2 at 25°C

Source: This medium is available as a premixed powder from BD Diagnostic Systems.

Preparation of Medium: Add components to distilled/deionized water and bring volume to 1.0L. Mix thoroughly. Gently heat and bring to boiling. Distribute into tubes or flasks. Autoclave for 15 min at 15 psi pressure–121°C. Pour into sterile Petri dishes or leave in tubes.

Use: For the cultivation of a variety of phytopathological fungi and other fungi.

Lima Bean Agar

Composition per liter:

Lima beans, solids from infusion	62.5g
Agar	15.0g

pH 5.6 ± 0.2 at 25°C

Preparation of Medium: Add components to distilled/deionized water and bring volume to 1.0L. Mix thoroughly. Gently heat and bring to boiling. Distribute into tubes or flasks. Autoclave for 15 min at 15 psi pressure–121°C. Pour into sterile Petri dishes or leave in tubes.

Use: For the cultivation of a variety of phytopathological fungi and other fungi.

Lima Bean Agar

Composition per liter:

Lima beans, frozen	250.0g
Agar	5.0g

pH 6.3 ± 0.3 at 25°C

Preparation of Medium: Add lima beans to distilled/deionized water and bring volume to 1.0L. Blend for 10 min. Add agar. Mix thoroughly. Gently heat and bring to boiling. Distribute into tubes or flasks. Autoclave for 15 min at 15 psi pressure–121°C. Pour into sterile Petri dishes or leave in tubes.

Use: For the cultivation and maintenance of numerous *Colletotrichum* species, *Coniothyrium fuckelii, Diheterospora chlamydosporia, Glomerella cingulata, Graphium fragrans, Monochaetia mali, Penicillium crustosum, Phleospora idahoensis, Phoma eupyrena, Phoma lingam, Phyllosticta sojaecola,* numerous *Phytophthora* species, *Polysphondylium violaceum, Pythium anandrum, Pythium irregulare, Pythium vexans, Scopulariopsis fimicola,* and *Sphaerostilbe repens.*

Limnobacter Medium (DSMZ Medium 919)

Composition per liter:

Yeast extract	0.5g
Proteose peptone	0.5g
Casamino acids	0.5g
Glucose	0.5g
Soluble starch	0.5g
Sodium pyruvate	0.3g
K_2HPO_4	0.3g
$MgSO_4 \cdot 7H_2O$	0.05g

pH 7.2 ± 0.2 at 25°C

Preparation of Medium: Add components to distilled/deionized water and bring volume to 1.0L. Mix thoroughly. Distribute into tubes or flasks. Autoclave for 15 min at 15 psi pressure–121°C.

Use: For the cultivation of *Limnobacter thiooxidans.*

Lindane Medium

Composition per liter:

Yeast extract	5.0g
γ-Hexachlorocyclohexane (Lindane)	0.1g

pH 7.0 ± 0.2 at 25°C

Preparation of Medium: Add components to distilled/deionized water and bring volume to 1.0L. Mix thoroughly. Distribute into tubes or flasks. Autoclave for 15 min at 15 psi pressure–121°C.

Use: For the cultivation of *Clostridium sphenoides.*

Lineola Agar

Composition per liter:

Mannitol	25.0g
Agar	15.0g
Yeast extract	5.0g
Peptone	3.0g

pH 6.8 ± 0.2 at 25°C

Preparation of Medium: Add components to distilled/deionized water and bring volume to 1.0L. Mix thoroughly. Gently heat and bring to boiling. Distribute into tubes or flasks. Autoclave for 15 min at 15 psi pressure–121°C. Pour into sterile Petri dishes or leave in tubes.

Use: For the cultivation and maintenance of *Bacillus macroides* and other *Bacillus* species.

Lipovitellin Salt Mannitol Agar

Composition per liter:

NaCl	75.0g
Egg yolk	20.0g
Agar	15.0g
D-Mannitol	10.0g
Polypeptone™	10.0g
Beef extract	1.0g
Phenol Red	0.025g

Preparation of Medium: Add components to distilled/deionized water and bring volume to 1.0L. Mix thoroughly. Gently heat and bring to boiling. Distribute into tubes or flasks. Autoclave for 15 min at 15 psi pressure–121°C. Pour into sterile Petri dishes or leave in tubes.

Use: For the detection of *Staphylococcus aureus* in swimming pool water based on lipovitellin-lipase activity and mannitol fermentation. *Staphylococcus aureus* and other bacteria with lipovitellin-lipase activity attack the egg yolk and appear as colonies surrounded by an opaque zone. Bacteria that ferment mannitol appear as colonies surrounded by a yellow zone.

Liquoid Broth

Composition per liter:

Beef heart, infusion from	250.0g
Calf brain, infusion from	200.0g
Proteose peptone	10.0g
NaCl	5.0g
Na_2HPO_4	2.5g
Glucose	2.0g
Sodium polyanethol sulfonate	0.5g

pH 7.4 ± 0.2 at 25°C

Source: This medium is available from HiMedia.

Preparation of Medium: Add components to distilled/deionized water and bring volume to 1.0L. Mix thoroughly. Gently heat and bring to boiling. Distribute into tubes or flasks. Autoclave for 15 min at 15 psi pressure–121°C.

Use: For the screening of blood specimens from suspected cases of bacteremia.

Listeria Enrichment Broth

Composition per liter:

Pancreatic digest of casein	17.0g
Yeast extract	6.0g
NaCl	5.0g
Papaic digest of soybean meal	3.0g
Glucose	2.5g
K_2HPO_4	2.5g
Cycloheximide	0.05g
Nalidixic acid	0.04g
Acriflavine·HCl	0.015g

pH 7.3 ± 0.2 at 25°C

Caution: Cycloheximide is toxic. Avoid skin contact or aerosol formation and inhalation.

Preparation of Medium: Add components to distilled/deionized water and bring volume to 1.0L. Mix thoroughly. Gently heat and bring to boiling. Distribute into tubes or flasks. Autoclave for 15 min at 15 psi pressure–121°C.

Use: For the isolation and cultivation of *Listeria monocytogenes* according to the FDA formula.

Listeria Enrichment Broth, FDA (LEB, FDA)

Composition per liter:

Soybean casein digest broth yeast extract	1.0L
Nalidixic acid solution	8.0mL
Cycloheximide solution	5.1mL
Acriflavin·HCl solution	3.0mL

pH 7.3 ± 0.2 at 25°C

Caution: Cycloheximide is toxic. Avoid skin contact or aerosol formation and inhalation.

Soybean Casein Digest Broth Yeast Extract:
Composition per liter:

Pancreatic digest of casein	17.0g
Yeast extract	6.0g
NaCl	5.0g
Papaic digest of soybean meal	3.0g
K_2HPO_4	2.5g
Glucose	2.5g

Source: This medium is available as a premixed powder from BD Diagnostic Systems.

Preparation of Soybean Casein Digest Broth Yeast Extract: Add components to distilled/deionized water and bring volume to 1.0L. Mix thoroughly. Autoclave for 15 min at 15 psi pressure–121°C.

Nalidixic Acid Solution:
Composition per 100.0mL:

Nalidixic acid	0.5g

Preparation of Nalidixic Acid Solution: Add nalidixic acid to distilled/deionized water and bring volume to 100.0mL. Mix thoroughly. Filter sterilize.

Cycloheximide Solution:
Composition per 100.0mL:

Cycloheximide	1.5g
Ethanol	40.0mL

Preparation of Cycloheximide Solution: Add cycloheximide to 40.0mL of ethanol. Mix thoroughly. Bring volume to 100.0mL with distilled/deionized water. Filter sterilize.

Caution: Cycloheximide is toxic. Avoid skin contact or aerosol formation and inhalation.

Acriflavin·HCl Solution:
Composition per 100.0mL:

Acriflavin·HCl solution	0.5g

Preparation of Acriflavin·HCl Solution: Add acriflavin·HCl solution to distilled/deionized water and bring volume to 100.0mL. Mix thoroughly. Filter sterilize.

Preparation of Medium: Add components—except nalidixic acid solution, acriflavin solution, and cycloheximide solution—to distilled/deionized water and bring volume to 990.0mL. Mix thoroughly. Gently heat and bring to boiling. Autoclave for 15 min at 15 psi pressure–121°C. Cool to 45°–50°C. Aseptically add 8.0mL of sterile nalidixic acid solution, 5.1mL of sterile cycloheximide solution, and 3.0mL of sterile acriflavin solution. Mix thoroughly. Pour into sterile Petri dishes or distribute into sterile tubes.

Use: For the isolation and enrichment of *Listeria monocytogenes* from foods.

Listeria Enrichment Broth I, USDA FSIS (*Listeria* Primary Selective Enrichment Broth, UVM I) (University of Vermont I *Listeria* Primary Selective Enrichment Broth)

Composition per liter:

NaCl	20.0g
Na_2HPO_4	12.0g
Beef extract	5.0g
Proteose peptone	5.0g
Pancreatic digest of casein	5.0g
Yeast extract	5.0g

KH_2PO_4	1.35g
Esculin	1.0g
Nalidixic acid solution	1.0mL
Acriflavine solution	1.0mL

pH 7.4 ± 0.2 at 25°C

Source: This medium is available as a premixed powder from Oxoid Unipath.

Nalidixic Acid Solution:

Composition per 10.0mL:

Nalidixic acid	0.2g
NaOH (0.1*M* solution)	10.0mL

Preparation of Nalidixic Acid Solution: Add nalidixic acid to 10.0mL of NaOH solution. Mix thoroughly. Filter sterilize.

Acriflavine Solution:

Composition per 10.0mL:

Acriflavine	0.12g

Preparation of Acriflavine Solution: Add acriflavine to distilled/deionized water and bring volume to 10.0mL. Mix thoroughly. Filter sterilize. Use freshly prepared solution.

Preparation of Medium: Add components, except nalidixic acid solution and acriflavine solution, to distilled/deionized water and bring volume to 998.0mL. Mix thoroughly. Gently heat and bring to boiling. Autoclave for 15 min at 15 psi pressure–121°C. Cool to 45°–50°C. Aseptically add 1.0mL of sterile nalidixic acid solution. Mix thoroughly. Store at 4°C. Immediately prior to use, aseptically add 1.0mL of sterile acriflavine solution. Mix thoroughly. Aseptically distribute into sterile tubes or flasks.

Use: For the selective isolation, cultivation, and enrichment of *Listeria monocytogenes* from food, milk, and dairy products.

Listeria Enrichment Broth II, USDA FSIS (*Listeria* Primary Selective Enrichment Broth, UVM II) (University of Vermont II *Listeria* Primary Selective Enrichment Broth

Composition per liter:

NaCl	20.0g
Na_2HPO_4	12.0g
Beef extract	5.0g
Protease peptone	5.0g
Pancreatic digest of casein	5.0g
Yeast extract	5.0g
KH_2PO_4	1.35g
Esculin	1.0g
Nalidixic acid solution	1.0mL
Acriflavine solution	1.0mL

pH 7.4 ± 0.2 at 25°C

Source: This medium is available as a premixed powder from Oxoid Unipath.

Nalidixic Acid Solution:

Composition per 10.0mL:

Nalidixic acid	0.2g
NaOH (0.1*M* solution)	10.0mL

Preparation of Nalidixic Acid Solution: Add nalidixic acid to 10.0mL of NaOH solution. Mix thoroughly. Filter sterilize.

Acriflavine Solution:

Composition per 10.0mL:

Acriflavine	0.25g

Preparation of Acriflavine Solution: Add acriflavine to distilled/deionized water and bring volume to 10.0mL. Mix thoroughly. Filter sterilize. Use freshly prepared solution.

Preparation of Medium: Add components, except nalidixic acid solution and acriflavine solution, to distilled/deionized water and bring volume to 998.0mL. Mix thoroughly. Gently heat and bring to boiling. Autoclave for 15 min at 15 psi pressure–121°C. Cool to 45°–50°C. Aseptically add 1.0mL of sterile nalidixic acid solution. Mix thoroughly. Store at 4°C. Immediately prior to use, aseptically add 1.0mL of sterile acriflavine solution. Mix thoroughly. Aseptically distribute into sterile tubes or flasks.

Use: For the selective isolation, cultivation, and enrichment of *Listeria monocytogenes* from food, milk, and dairy products.

Listeria Enrichment HiVeg Agar

Composition per liter:

Potassium thiocyanate	37.5g
Agar	13.0g
Plant hydrolysate	10.0g
Plant peptone	10.0g
NaCl	5.0g
Glucose	1.0g
Acriflavin hydrochloride (Trypaflavin)	0.01g
Thiaminium dichloride	5.0mg

pH 7.4 ± 0.2 at 25°C

Source: This medium is available as a premixed powder from HiMedia.

Preparation of Medium: Add components to distilled/deionized water and bring volume to 1.0L. Mix thoroughly. Gently heat and bring to boiling. Distribute into tubes or flasks. Autoclave for 15 min at 15 psi pressure–121°C. Pour into sterile Petri dishes.

Use: For the selective isolation of *Listeria monocytogenes* from clinical specimens.

Listeria Enrichment HiVeg Broth

Composition per liter:

Potassium thiocyanate	37.5g
Plant hydrolysate	10.0g
Plant peptone	10.0g
NaCl	5.0g
Glucose	1.0g
Acriflavin hydrochloride (Trypaflavin)	0.01g
Thiaminium dichloride	5.0mg

pH 7.4 ± 0.2 at 25°C

Source: This medium is available as a premixed powder from HiMedia.

Preparation of Medium: Add components to distilled/deionized water and bring volume to 1.0L. Mix thoroughly. Gently heat and bring to boiling. Distribute into tubes or flasks. Autoclave for 15 min at 15 psi pressure–121°C.

Use: For the selective enrichment of *Listeria monocytogenes* from clinical specimens.

Listeria Enrichment HiVeg Broth, Modified

Composition per liter:

NaCl	20.0g
Plant hydrolysate No. 1	10.0g
Na_2HPO_4	9.6g
Yeast extract	5.0g
Plant extract	5.0g
KH_2PO_4	1.35g
Esculin	1.0g
Nalidixic acid	0.02g
Acriflavin hydrochloride (Trypaflavin)	0.012g

pH 7.4 ± 0.2 at 25°C

Source: This medium is available as a premixed powder from HiMedia.

Preparation of Medium: Add components to distilled/deionized water and bring volume to 1.0L. Mix thoroughly. Gently heat and bring to boiling. Distribute into tubes or flasks. Autoclave for 15 min at 15 psi pressure–121°C.

Use: For the selective enrichment of *Listeria* species.

Listeria Enrichment HiVeg Medium Base with Acriflavine and Nalidixic Acid (UVM)

Composition per liter:

NaCl	20.0g
Na_2HPO_4	12.0g
Plant extract	5.0g
Plant hydrolysate	5.0g
Plant peptone No. 3	5.0g
Yeast extract	5.0g
KH_2PO_4	1.35g
Esculin	1.0g
Nalidixic acid solution	1.0mL
Acriflavine solution	1.0mL

pH 7.4 ± 0.2 at 25°C

Source: This medium, without nalidixic acid or acriflavine solutions, is available as a premixed powder from HiMedia.

Nalidixic Acid Solution:

Composition per 10.0mL:

Nalidixic acid	0.2g
NaOH (0.1*M* solution)	10.0mL

Preparation of Nalidixic Acid Solution: Add nalidixic acid to 10.0mL of NaOH solution. Mix thoroughly. Filter sterilize.

Acriflavine Solution:

Composition per 10.0mL:

Acriflavine	0.12g

Preparation of Acriflavine Solution: Add acriflavine to distilled/deionized water and bring volume to 10.0mL. Mix thoroughly. Filter sterilize. Use freshly prepared solution.

Preparation of Medium: Add components, except nalidixic acid solution and acriflavine solution, to distilled/deionized water and bring volume to 998.0mL. Mix thoroughly. Gently heat and bring to boiling. Autoclave for 15 min at 15 psi pressure–121°C. Cool to 45°–50°C. Aseptically add 1.0mL of sterile nalidixic acid solution. Mix thoroughly. Store at 4°C. Immediately prior to use, aseptically add 1.0mL of sterile acriflavine solution. Mix thoroughly. Aseptically distribute into sterile tubes or flasks.

Use: For the selective isolation, cultivation, and enrichment of *Listeria monocytogenes* from food, milk, and dairy products. For the selective isolation and cultivation of *Listeria monocytogenes* from clinical specimens.

Listeria Fermentation Broth

Composition per liter:

Proteose peptone No. 3	10.0g
NaCl	5.0g
Beef extract	1.0g
Bromcresol Purple	0.1g
Carbohydrate solution	10.0mL

Carbohydrate Solution:

Composition per 10.0mL:

Carbohydrate	5.0g

Preparation of Carbohydrate Solution: Add carbohydrate to distilled/deionized water and bring volume to 10.0mL. Mix thoroughly. Filter sterilize. Use glucose, salicin, rhamnose, dulcitol, or raffinose.

Preparation of Medium: Add components, except carbohydrate solution, to distilled/deionized water and bring volume to 990.0mL. Mix thoroughly. Gently heat and bring to boiling. Autoclave for 15 min at 15 psi pressure–121°C. Cool to 45°–50°C. Aseptically add sterile carbohydrate solution. Mix thoroughly. Aseptically distribute into sterile tubes or flasks.

Use: For the cultivation and differentiation of *Listeria* species based on the fermentation of glucose, salicin, rhamnose, dulcitol, and raffinose.

Listeria Identification HiVeg Agar Base with PALCAM Selective Supplement (PALCAM)

Composition per liter:

Plant peptone	23.0g
LiCl	15.0g
Agar	13.0g
Mannitol	10.0g
NaCl	5.0g
Starch	1.0g
Esculin	0.8g
Ammonium ferric citrate	0.5g
Glucose	0.5g
Phenol Red	0.08g
PALCAM selective supplement	10.0mL

pH 7.0 ± 0.2 at 25°C

Source: This medium, without PALCAM selective supplement, is available as a premixed powder from HiMedia.

Caution: LiCl is harmful. Avoid bodily contact and inhalation of vapors. On contact with skin wash with plenty of water immediately.

PALCAM Selective Supplement:

Composition per 10.0mL:

Ceftazidime	20.0mg
Polymyxin B	10.0mg
Acriflavine·HCl	5.0mg

Preparation of PALCAM Selective Supplement: Add components to distilled/deionized water and bring volume to 10.0mL. Mix thoroughly. Filter sterilize.

Preparation of Medium: Add components, except PALCAM selective supplement, to distilled/deionized water and bring volume to 990.0mL. Mix thoroughly. Gently heat and bring to boiling. Autoclave for 15 min at 15 psi pressure–121°C. Cool to 45°–50°C. Aseptically add sterile PALCAM selective supplement. Mix thoroughly. Pour into sterile Petri dishes.

Use: For the selective isolation and identification of *Listeria monocytogenes* and other *Listeria* species from foods.

Listeria Identification HiVeg Broth Base with PALCAM Selective Supplement (PALCAM)

Composition per liter:

Plant peptone	23.0g
LiCl	10.0g
D-Mannitol	5.0g
Yeast extract	5.0g
Polysorbate 80	2.0g
Soy lecithin	1.0g
Esculin	0.8g
Ammonium ferric citrate	0.5g
Phenol Red	0.08g
PALCAM selective supplement	10.0mL

pH 7.2 ± 0.2 at 25°C

Source: This medium, without PALCAM selective supplement, is available as a premixed powder from HiMedia.

Caution: LiCl is harmful. Avoid bodily contact and inhalation of vapors. On contact with skin wash with plenty of water immediately.

PALCAM Selective Supplement:

Composition per 10.0mL:

Ceftazidime	20.0mg
Polymyxin B	10.0mg
Acriflavine·HCl	5.0mg

Preparation of PALCAM Selective Supplement: Add components to distilled/deionized water and bring volume to 10.0mL. Mix thoroughly. Filter sterilize.

Preparation of Medium: Add components, except PALCAM selective supplement, to distilled/deionized water and bring volume to 990.0mL. Mix thoroughly. Gently heat and bring to boiling. Autoclave for 15 min at 15 psi pressure–121°C. Cool to 45°–50°C. Aseptically add sterile PALCAM selective supplement. Mix thoroughly.

Use: For the selective cultivation of f *Listeria monocytogenes* and other *Listeria* species.

Listeria Motility HiVeg Medium

Composition per liter:

Plant hydrolysate	20.0g
Plant peptone	6.1g
Agar	3.5g

pH 7.3 ± 0.2 at 25°C

Source: This medium is available as a premixed powder from HiMedia.

Preparation of Medium: Add components to distilled/deionized water and bring volume to 1.0L. Mix thoroughly. Gently heat and bring to boiling. Distribute into tubes or flasks. Autoclave for 15 min at 15 psi pressure–121°C. Pour into sterile Petri dishes.

Use: For testing motility of *Listeria monocytogenes*.

Listeria Motility Medium

Composition per liter:

Casein enzymatic hydrolysate	20.0g
Peptic digest of animal tissue	6.1g
Agar	3.5g

pH 7.3 ± 0.2 at 25°C

Source: This medium is available as a premixed powder from HiMedia.

Preparation of Medium: Add components to distilled/deionized water and bring volume to 1.0L. Mix thoroughly. Gently heat and bring to boiling. Distribute into tubes or flasks. Autoclave for 15 min at 15 psi pressure–121°C. Pour into sterile Petri dishes.

Use: For testing motility of *Listeria monocytogenes*.

Listeria Oxford HiVeg Medium Base with Antibiotic Inhibitor

Composition per liter:

Plant special peptone	23.0g
LiCl	15.0g
Agar	10.0g
NaCl	5.0g
Cornstarch	1.0g
Esculin	1.0g
Ammonium ferric citrate	0.5g
Antibiotic inhibitor	10.0mL

pH 7.0 ± 0.2 at 25°C

Source: This medium, without antibiotic inhibitor, is available as a premixed powder from HiMedia.

Caution: LiCl is harmful. Avoid bodily contact and inhalation of vapors. On contact with skin wash with plenty of water immediately.

Antibiotic Inhibitor:

Composition per 10.0mL:

Cycloheximide	0.4g
Colistin sulfate	0.02g
Fosfomycin	0.01g
Acriflavine	5.0mg
Cefotetan	2.0mg
Ethanol (50% solution)	10.0mL

Preparation of Antibiotic Inhibitor: Add antibiotics to 10.0mL of ethanol. Mix thoroughly. Filter sterilize.

Caution: Cycloheximide is toxic. Avoid skin contact or aerosol formation and inhalation.

Preparation of Medium: Add components, except antibiotic inhibitor, to distilled/deionized water and bring volume to 990.0mL. Mix thoroughly. Gently heat and bring to boiling. Autoclave for 15 min at 15 psi pressure–121°C. Cool to 45°–50°C. Aseptically add 10.0mL of sterile antibiotic inhibitor. Mix thoroughly. Pour into sterile Petri dishes or distribute into sterile tubes.

Use: For the isolation and cultivation of *Listeria monocytogenes* from specimens containing a mixed bacterial flora. For the isolation of *Listeria* species from pathological specimens.

Listeria Oxford Medium Base with Antibiotic Inhibitor

Composition per liter:

Peptone, special	23.0g
LiCl	15.0g

Agar	10.0g
NaCl	5.0g
Cornstarch	1.0g
Esculin	1.0g
Ammonium ferric citrate	0.5g
Antibiotic inhibitor	10.0mL

pH 7.0 ± 0.2 at 25°C

Source: This medium, without antibiotic inhibitor, is available as a premixed powder from HiMedia.

Caution: LiCl is harmful. Avoid bodily contact and inhalation of vapors. On contact with skin wash with plenty of water immediately.

Antibiotic Inhibitor:

Composition per 10.0mL:

Cycloheximide	0.4g
Colistin sulfate	0.02g
Fosfomycin	0.01g
Acriflavine	5.0mg
Cefotetan	2.0mg
Ethanol (50% solution)	10.0mL

Preparation of Antibiotic Inhibitor: Add antibiotics to 10.0mL of ethanol. Mix thoroughly. Filter sterilize.

Caution: Cycloheximide is toxic. Avoid skin contact or aerosol formation and inhalation.

Preparation of Medium: Add components, except antibiotic inhibitor, to distilled/deionized water and bring volume to 990.0mL. Mix thoroughly. Gently heat and bring to boiling. Autoclave for 15 min at 15 psi pressure–121°C. Cool to 45°–50°C. Aseptically add 10.0mL of sterile antibiotic inhibitor. Mix thoroughly. Pour into sterile Petri dishes or distribute into sterile tubes.

Use: For the isolation and cultivation of *Listeria monocytogenes* from specimens containing a mixed bacterial flora. For the isolation of *Listeria* species from pathological specimens.

Listeria Primary Selective Enrichment Broth, UVM I

See: ***Listeria* Enrichment Broth I, USDA FSIS**

Listeria Primary Selective Enrichment Broth, UVM II

See: ***Listeria* Enrichment Broth II, USDA FSIS**

Listeria Selective Agar, Modified Oxford

See: **Oxford Agar, Modified**

Listeria Selective Agar, Oxford

See: **Oxford Agar**

Listeria Selective HiVeg Agar

Composition per liter:

Potassium thiocyanate	37.5g
Agar	13.0g
Plant hydrolysate	10.0g
Plant peptone	10.0g
NaCl	5.0g
Glucose	1.0g
Nalidixic acid	0.04g
Acriflavin hydrochloride (Trypaflavin)	0.01g
Thiaminium dichloride	5.0mg

pH 7.4 ± 0.2 at 25°C

Source: This medium is available as a premixed powder from HiMedia.

Preparation of Medium: Add components to distilled/deionized water and bring volume to 1.0L. Mix thoroughly. Gently heat and bring to boiling. Autoclave for 15 min at 15 psi pressure–121°C. Cool to 45°–50°C. Pour into sterile Petri dishes or distribute into sterile tubes.

Use: For the selective isolation and cultivation of *Listeria* species from clinical specimens.

Listeria Selective HiVeg Broth Base with Antibiotic Inhibitor

Composition per liter:

Plant hydrolysate	17.0g
Yeast extract	6.0g
NaCl	5.0g
Papaic digest of soybean meal	3.0g
Glucose	2.5g
K_2HPO_4	2.5g
Antibiotic inhibitor	10.0mL

pH 7.3 ± 0.2 at 25°C

Source: This medium, without antibiotic inhibitor, is available as a premixed powder from HiMedia.

Antibiotic Inhibitor:

Composition per 10.0mL:

Cycloheximide	0.4g
Colistin sulfate	0.02g
Fosfomycin	0.01g
Acriflavine	5.0mg
Cefotetan	2.0mg
Ethanol (50% solution)	10.0mL

Preparation of Antibiotic Inhibitor: Add antibiotics to 10.0mL of ethanol. Mix thoroughly. Filter sterilize.

Caution: Cycloheximide is toxic. Avoid skin contact or aerosol formation and inhalation.

Preparation of Medium: Add components, except antibiotic inhibitor, to distilled/deionized water and bring volume to 990.0mL. Mix thoroughly. Gently heat and bring to boiling. Autoclave for 15 min at 15 psi pressure–121°C. Cool to 45°–50°C. Aseptically add 10.0mL of sterile antibiotic inhibitor. Mix thoroughly. Distribute into sterile tubes.

Use: For the cultivation of *Listeria monocytogenes* from specimens containing a mixed bacterial flora. For the cultivation of *Listeria* species from pathological specimens.

Listeria Transport Enrichment Medium

Composition per liter:

Sodium glycerophosphate	10.0g
Agar	2.0g
Sodium thioglycolate	1.0g
$CaCl_2$	0.1g
Nalidixic acid	0.04g
Acridine solution	2.0mL

pH 7.4 ± 0.2 at 25°C

Acridine Solution:

Composition per 10.0mL:

Acridine	0.04g

Preparation of Acridine Solution: Add acridine to distilled/deionized water and bring volume to 10.0mL. Mix thoroughly. Autoclave for 15 min at 15 psi pressure–121°C. Cool to 45°–50°C.

Preparation of Medium: Add components, except acridine solution, to distilled/deionized water and bring volume to 998.0mL. Mix thoroughly. Gently heat and bring to boiling. Autoclave for 15 min at 15 psi pressure–121°C. Cool to 45°–50°C. Aseptically add 2.0mL acridine solution. Mix thoroughly. Aseptically distribute into sterile tubes in 10.0mL volumes or fill bottles 4/5 full.

Use: For the maintenance—as a transport medium—and enrichment of *Listeria* species.

LIT Medium

Composition per liter:

Beef liver, infusion from	453.0g
Na_2HPO_4	8.0g
Pancreatic digest of casein	5.0g
Tryptose	5.0g
Glucose	1.0g
NaCl	1.0g
K_2HPO_4	0.5g
KCl	0.4g
Hemin	10.0mg
Fetal bovine serum, heat inactivated	100.0mL

pH 7.2 ± 0.2 at 25°C

Preparation of Medium: Add components to distilled/deionized water and bring volume to 1.0L. Mix thoroughly. Adjust pH to 7.2. Filter sterilize. Aseptically distribute into sterile tubes or flasks.

Use: For the cultivation of *Herpetomonas mariadeanei*, *Trypanoplasma borreli*, and *Trypanosoma cruzi*.

Lithium Chloride Phenylethanol Moxalactam Plating Agar
See: **LPM Agar**

Litmus Lactose Agar (LL Agar)

Composition per liter:

Agar	10.0g
Lactose	10.0g
Meat peptone	5.0g
Beef extract	3.0g
Litmus	1.0g

pH 7.0 ± 0.2 at 25°C

Preparation of Medium: Add components to distilled/deionized water and bring volume to 1.0L. Mix thoroughly. Gently heat and bring to boiling. Distribute into tubes or flasks. Autoclave for 15 min at 15 psi pressure–121°C. Pour into sterile Petri dishes or leave in tubes.

Use: For the maintenance of lactic acid bacteria and differentiation of bacteria on the basis of lactose fermentation. Bacteria that ferment lactose appear as red colonies and others as dark blue-purple colonies.

Litmus Lactose Agar with Crystal Violet (LLK Agar)

Composition per liter:

Agar	10.0g
Lactose	10.0g
Meat peptone	5.0g
Beef extract	3.0g
Litmus	1.0g
Crystal Violet	5.0mg

pH 7.0 ± 0.2 at 25°C

Preparation of Medium: Add components to distilled/deionized water and bring volume to 1.0L. Mix thoroughly. Gently heat and bring to boiling. Distribute into tubes or flasks. Autoclave for 15 min at 15 psi pressure–121°C. Pour into sterile Petri dishes or leave in tubes.

Use: For the maintenance of lactic acid bacteria and for the differentiation of several bacteria on the basis of lactose fermentation. Bacteria that ferment lactose appear as red colonies. Bacteria that do not ferment lactose appear as dark blue-purple colonies.

Litmus Lactose HiVeg Agar

Composition per liter:

Plant peptone	23.0g
Lactose	20.0g
Agar	15.0g
Plant extract	5.0g
NaCl	5.0g
Synthetic detergent No. V	2.0g
Litmus	0.5g

pH 7.4 ± 0.2 at 25°C

Source: This medium is available as a premixed powder from HiMedia.

Preparation of Medium: Add components to distilled/deionized water and bring volume to 1.0L. Mix thoroughly. Gently heat and bring to boiling. Distribute into tubes or flasks. Autoclave for 15 min at 15 psi pressure–121°C. Pour into sterile Petri dishes or leave in tubes.

Use: For the maintenance of lactic acid bacteria and differentiation of bacteria on the basis of lactose fermentation. Bacteria that ferment lactose appear as red colonies and others as dark blue-purple colonies.

Litmus Milk

Composition per liter:

Skim milk	100.0g
Azolitmin	0.5g
Na_2SO_3	0.5g

pH 6.5 ± 0.2 at 25°C

Source: This medium is available as a premixed powder from BD Diagnostic Systems and Oxoid Unipath.

Preparation of Medium: Add components to distilled/deionized water and bring volume to 1.0L. Mix thoroughly. Gently heat and bring to boiling. Distribute into tubes or flasks. Autoclave for 20 min at 10 psi pressure–115°C.

Use: For the maintenance of lactic acid bacteria and for the differentiation of several bacteria, especially *Clostridium* species, based on their action on milk. Bacteria that do not ferment carbohydrates, such as *Proteus vulgaris* or *Moraxella lacunata*, show no change in the azolitmin litmus indicator. Bacteria that ferment lactose or glucose with the production of gas, such as *Clostridium perfringens*, turn the medium pink and frothy. Bacteria that proteolytically degrade milk lactalbumin turn the medium blue. Bacteria that coagulate milk casein form a curd or clot. Bacteria that peptonize casein, such as *Pseudomonas aeruginosa*, show a dissolution of the clot.

Littman Oxgall Agar

Composition per liter:

Agar	20.0g
Oxgall	15.0g

Glucose 10.0g
Peptone 10.0g
Crystal Violet 0.01g
Streptomycin solution 10.0mL

pH 6.5 ± 0.2 at 25°C

Source: This medium is available as a premixed powder from BD Diagnostic Systems.

Streptomycin Solution:
Composition per 10.0mL:
Streptomycin 0.03g

Preparation of Streptomycin Solution: Add streptomycin to distilled/deionized water and bring volume to 10.0mL. Mix thoroughly. Filter sterilize.

Preparation of Medium: Add components, except streptomycin solution, to distilled/deionized water and bring volume to 990.0mL. Mix thoroughly. Gently heat and bring to boiling. Autoclave for 15 min at 15 psi pressure–121°C. Cool to 45°–50°C. Aseptically add sterile streptomycin solution. Mix thoroughly. Pour into sterile Petri dishes or distribute into sterile tubes. Allow tubes to cool in a slanted position.

Use: For the selective isolation and cultivation of fungi, especially dermatophytes.

Littman Oxgall HiVeg Agar Base with Streptomycin

Composition per liter:
Agar 20.0g
Plant peptone 20.0g
Glucose 10.0g
Synthetic detergent No. II 5.0g
Crystal Violet 0.01g
Streptomycin solution 10.0mL

pH 7.0 ± 0.2 at 25°C

Source: This medium, without streptomycin, is available as a premixed powder from HiMedia.

Streptomycin Solution:
Composition per 10.0mL:
Streptomycin 0.03g

Preparation of Streptomycin Solution: Add streptomycin to distilled/deionized water and bring volume to 10.0mL. Mix thoroughly. Filter sterilize.

Preparation of Medium: Add components, except streptomycin solution, to distilled/deionized water and bring volume to 990.0mL. Mix thoroughly. Gently heat and bring to boiling. Autoclave for 15 min at 15 psi pressure–121°C. Cool to 45°–50°C. Aseptically add sterile streptomycin solution. Mix thoroughly. Pour into sterile Petri dishes or distribute into sterile tubes. Allow tubes to cool in a slanted position.

Use: For the selective isolation and cultivation of fungi, especially dermatophytes.

Littman Oxgall HiVeg Broth Base with Streptomycin

Composition per liter:
Plant peptone 20.0g
Glucose 10.0g
Synthetic detergent No. II 5.0g
Crystal violet 0.01g
Streptomycin solution 10.0mL

pH 7.0 ± 0.2 at 25°C

Source: This medium, without streptomycin, is available as a premixed powder from HiMedia.

Streptomycin Solution:
Composition per 10.0mL:
Streptomycin 0.03g

Preparation of Streptomycin Solution: Add streptomycin to distilled/deionized water and bring volume to 10.0mL. Mix thoroughly. Filter sterilize.

Preparation of Medium: Add components, except streptomycin solution, to distilled/deionized water and bring volume to 990.0mL. Mix thoroughly. Gently heat and bring to boiling. Autoclave for 15 min at 15 psi pressure–121°C. Cool to 45°–50°C. Aseptically add sterile streptomycin solution. Mix thoroughly. Aseptically distribute into sterile tubes.

Use: For the selective cultivation of fungi, especially dermatophytes.

Liver Broth

Composition per liter:
Extracted liver tissue, minced 30.0g
Infusion from fresh liver 23.0g
Peptone 10.0g
K_2HPO_4 .... 1.0g
Agar overlay solution 200.0mL

pH 6.8 ± 0.2 at 25°C

Source: This medium is available as a premixed powder from Oxoid Unipath.

Agar Overlay Solution:
Composition per 200.0mL:
Agar, Oxoid Unipath No. 3 .... 4.0g

Preparation of Agar Overlay Solution: Add agar to distilled/deionized water and bring volume to 200.0mL. Mix thoroughly. Gently heat and bring to boiling. Autoclave for 15 min at 15 psi pressure–121°C. Cool to 45°–50°C.

Preparation of Medium: Add components, except agar, to distilled/deionized water and bring volume to 1.0L. Mix thoroughly. Gently heat and bring to boiling. Distribute into tubes to a depth of 50mm. Make sure that some liver particles are transferred to each tube. Autoclave for 20 min at 10 psi pressure–115°C. After inoculation, aseptically overlay the broth with 2.0mL of sterile, cooled agar solution per tube.

Use: For the isolation and cultivation of saccharolytic or putrefactive mesophilic and thermophilic anaerobic bacteria from foods.

Liver Broth

Composition per liter:
Beef liver, fresh 453.0g
Pancreatic digest of casein 10.0g
K_2HPO_4 .... 1.0g
Soluble starch 1.0g

pH 7.6 ± 0.2 at 25°C

Preparation of Medium: Remove the fat from fresh beef liver. Grind the liver. Add 1.0L of distilled/deionized water. Gently heat and bring to boiling. Continue boiling for 60 min. Adjust pH to 7.6. Filter through cheesecloth. Reserve meat. To filtrate, add pancreatic digest of casein, K_2HPO_4, and soluble starch. Bring volume to 1.0L with distilled/deionized water. Refilter solution. Add meat particles to test tubes to a depth of approximately 2cm. Distribute broth into tubes with meat particles in 15.0mL volumes. Autoclave for 20 min at 15 psi pressure–121°C.

Use: For the isolation and cultivation of anaerobic microorganisms, especially *Clostridium botulinum*, from foods.

Liver Broth with Sodium Chloride

Composition per liter:

Beef liver, fresh	453.0g
NaCl	150.0g
Pancreatic digest of casein	10.0g
K_2HPO_4	1.0g
Soluble starch	1.0g

pH 7.6 ± 0.2 at 25°C

Preparation of Medium: Remove the fat from fresh beef liver. Grind the liver. Add 1.0L of distilled/deionized water. Gently heat and bring to boiling. Continue boiling for 60 min. Adjust pH to 7.6. Filter through cheesecloth. Reserve meat. To filtrate, add pancreatic digest of casein, K_2HPO_4, NaCl, and soluble starch. Bring volume to 1.0L with distilled/deionized water. Refilter solution. Add meat particles to test tubes to a depth of approximately 2cm. Distribute broth into tubes with meat particles in 15.0mL volumes. Autoclave for 20 min at 15 psi pressure–121°C. After inoculation, overlay each tube with sterile, melted petroleum jelly.

Use: For the cultivation of obligate halophiles from brined and dry-salted vegetables.

Liver Infusion Agar

Composition per liter:

Beef liver, infusion from	500.0g
Agar	20.0g
Proteose peptone	10.0g
NaCl	5.0g

pH 6.9 ± 0.2 at 25°C

Source: This medium is available as a premixed powder from BD Diagnostic Systems.

Preparation of Medium: Add components to distilled/deionized water and bring volume to 1.0L. Mix thoroughly. Gently heat and bring to boiling. Distribute into tubes or flasks. Autoclave for 15 min at 15 psi pressure–121°C. Pour into sterile Petri dishes or leave in tubes.

Use: For the cultivation of *Brucella* species and other fastidious pathogenic bacteria.

Liver Infusion Agar, HiVeg

Composition per liter:

Agar	20.0g
Plant infusion No. 1	20.0g
Plant peptone No. 3	10.0g
NaCl	5.0g

pH 6.9 ± 0.2 at 25°C

Source: This medium is available as a premixed powder from HiMedia.

Preparation of Medium: Add components to distilled/deionized water and bring volume to 1.0L. Mix thoroughly. Gently heat and bring to boiling. Distribute into tubes or flasks. Autoclave for 15 min at 15 psi pressure–121°C. Pour into sterile Petri dishes or leave in tubes.

Use: For the cultivation of *Brucella* species and other fastidious pathogenic bacteria.

Liver Infusion Broth

Composition per liter:

Beef liver, infusion from	500.0g
Proteose peptone	10.0g
NaCl	5.0g

pH 6.9 ± 0.2 at 25°C

Source: This medium is available as a premixed powder from BD Diagnostic Systems.

Preparation of Medium: Add components to distilled/deionized water and bring volume to 1.0L. Mix thoroughly. Gently heat and bring to boiling. Distribute into tubes or flasks. Autoclave for 15 min at 15 psi pressure–121°C.

Use: For the cultivation of *Brucella* species and other fastidious pathogenic bacteria.

Liver Infusion Broth, HiVeg

Composition per liter:

Plant infusion No. 1	20.0g
Plant peptone No. 3	10.0g
NaCl	5.0g

pH 6.9 ± 0.2 at 25°C

Source: This medium is available as a premixed powder from HiMedia.

Preparation of Medium: Add components to distilled/deionized water and bring volume to 1.0L. Mix thoroughly. Gently heat and bring to boiling. Distribute into tubes or flasks. Autoclave for 15 min at 15 psi pressure–121°C.

Use: For the cultivation of *Brucella* species and other fastidious pathogenic bacteria.

Liver Infusion Sake Medium

Composition per liter:

Liver, fresh	400.0g
Agar	15.0g
Sake	400.0mL

Preparation of Medium: Finely mince liver and add to 600.0mL of distilled/deionized water. Gently heat and bring to boiling. Continue boiling for 5–10 min. Filter through Whatman #2 filter paper. To filtrate, add agar and sake. Bring volume to 1.0L with distilled/deionized water. Mix thoroughly. Gently heat and bring to boiling. Distribute into tubes or flasks. Autoclave for 25 min at 15 psi pressure–121°C. Pour into sterile Petri dishes or leave in tubes.

Use: For the cultivation and maintenance of *Lactobacillus fructivorans* and *Lactobacillus homohiochi.*

Liver Meat Infusion HiVeg Agar

Composition per liter:

Plant infusion No. 2	20.0g
Agar	11.0g
Na_2SO_3	1.2g
Glucose	0.75g
Starch	0.75g
Na_2CO_3	0.67g
Ferric ammonium citrate	0.5g

pH 7.6 ± 0.2 at 25°C

Source: This medium is available as a premixed powder from HiMedia.

Preparation of Medium: Add components to distilled/deionized water and bring volume to 1.0L. Mix thoroughly. Gently heat and bring to boiling. Distribute into tubes or flasks. Make sure that some liver and veal particles are transferred to each tube. Autoclave for 15 min at 15 psi pressure–121°C. Pour into sterile Petri dishes or leave in tubes.

Use: For the cultivation of a variety of anaerobic organisms.

Liver Veal Agar

Composition per liter:

Veal, infusion from	500.0g
Beef liver, infusion from	50.0g
Gelatin	20.0g
Proteose peptone	20.0g
Agar	15.0g
Soluble starch	10.0g
Glucose	5.0g
NaCl	5.0g
Casein	2.0g
$NaNO_3$	2.0g
Enzymatic digest of protein	1.3g
Pancreatic digest of casein	1.3g

pH 7.3 ± 0.2 at 25°C

Source: This medium is available as a premixed powder from BD Diagnostic Systems.

Preparation of Medium: Add components to distilled/deionized water and bring volume to 1.0L. Mix thoroughly. Gently heat and bring to boiling. Distribute into tubes or flasks. Make sure that some liver and veal particles are transferred to each tube. Autoclave for 15 min at 15 psi pressure–121°C. Pour into sterile Petri dishes or leave in tubes.

Use: For the cultivation of a variety of anaerobic organisms.

Liver Veal Egg Yolk Agar

Composition per 1080.0mL:

Liver veal agar	1.0L
Egg yolk emulsion, 50%	80.0mL

pH 7.3 ± 0.2 at 25°C

Liver Veal Agar:

Composition per liter:

Veal, infusion from	500.0g
Beef liver, infusion from	50.0g
Gelatin	20.0g
Proteose peptone	20.0g
Agar	15.0g
Soluble starch	10.0g
Glucose	5.0g
NaCl	5.0g
Casein	2.0g
$NaNO_3$	2.0g
Enzymatic digest of protein	1.3g
Pancreatic digest of casein	1.3g

Source: This medium is available as a premixed powder from BD Diagnostic Systems.

Preparation of Liver Veal Agar: Add components to distilled/deionized water and bring volume to 1.0L. Mix thoroughly. Gently heat and bring to boiling. Distribute into tubes or flasks. Make sure that some liver and veal particles are transferred to each tube. Autoclave for 15 min at 15 psi pressure–121°C. Cool to 50°C.

Egg Yolk Emulsion, 50%:

Composition per 100.0mL:

Chicken egg yolks	11
Whole chicken egg	1
NaCl (0.9% solution)	50.0mL

Preparation of Egg Yolk Emulsion, 50%: Soak eggs with 1:100 dilution of saturated mercuric chloride solution for 1 min. Crack eggs and separate yolks from whites. Mix egg yolks with 1 chicken egg. Beat to form emulsion. Measure 50.0mL of egg yolk emulsion and add to 50.0mL of 0.9% NaCl solution. Mix thoroughly. Filter sterilize. Warm to 45°–50°C.

Preparation of Medium: To 1.0L of cooled sterile liver veal agar, aseptically add 80.0mL of sterile egg yolk emulsion, 50%. Mix thoroughly. Pour into sterile Petri dishes. Dry plates at 35°C for 24 hr.

Use: For the cultivation of a variety of anaerobic organisms.

LL Agar

See: **Litmus Lactose Agar**

LLK Agar

See: **Litmus Lactose Agar with Crystal Violet**

LMX Broth Modified, Fluorocult (Fluorocult LMX Broth, Modified)

Composition per liter:

Tryptose	5.0
NaCl	5.0
Sorbitol	1.0
Tryptophan	1.0
K_2HPO_4	2.7g
KH_2PO_4	2.0g
Lauryl sulfate sodium salt	0.1g
1-Isopropyl-β-D-1-thio-galactopyranoside (IPTG)	0.1g
5-Bromo-4-chloro-3-indolyl-β-D-galactopyranoside (X-GAL)	0.08g
4-Methylumbelliferyl-β-D-glucuronide	0.05g

pH: 6.8 ± 0.2 at 25°C

Source: This medium is available from Merck.

Preparation of Medium: Add components to distilled/deionized water and bring volume to 1.0L. Mix thoroughly. Distribute into test tubes. Autoclave for 15 min at 15 psi pressure–121°C. The broth is clear and yellowish brown.

Use: For the simultaneous detection of total coliforms and *E. coli* in water, food, and diary products by the fluorogenic procedure. A color change of the broth from yellow to blue-green indicates the presence of coliforms. A blue fluorescence under long-wave UV light permits the rapid detection of *E.coli*. As tryptophan is added to the broth, the indole reaction is easily done by adding KOVACS reagent. The formation of a red ring additionally confirms the presence of *E. coli*.

Lobosphaera Medium

Composition per liter:

Polypeptone™	10.0g
Glycerol	7.5g
Yeast extract	2.0g
K_2HPO_4	1.0g
KH_2PO_4	1.0g
$MgSO_4·7H_2O$	0.5g

pH 6.5 ± 0.2 at 25°C

Preparation of Medium: Add components to distilled/deionized water and bring volume to 1.0L. Mix thoroughly. Adjust pH to 6.5. Filter sterilize. Aseptically distribute into sterile tubes or flasks.

Use: For the cultivation of *Lobosphaera* species.

Loefer's Medium

Composition per liter:

Proteose peptone	15.0g
Casitone	5.0g
Glucose	5.0g
NaCl	2.0g
Na_2HPO_4	1.0g
KH_2PO_4	1.0g
Yeast extract	0.5g
$MgCl_2$	0.3g
Asolectin solution	10.0mL

Asolectin Solution:

Asolectin	1.0g
Ethanol	100.0mL

Preparation of Asolectin Solution: Add 1.0g of asolectin to ethanol and bring volume to 100.0mL. Mix thoroughly.

Preparation of Medium: Add components to distilled/deionized water and bring volume to 1.0L. Mix thoroughly. Distribute into screw-capped tubes or flasks. Autoclave for 15 min at 15 psi pressure–121°C.

Use: For the cultivation of *Tetrahymena patula.*

Loeffler Blood Serum Medium

Composition per liter:

Beef blood serum	750.0mL
Dextrose broth	250.0mL

pH 7.1 ± 0.2 at 25°C

Source: This medium is available as a premixed powder from BD Diagnostic Systems.

Dextrose Broth:

Composition per liter:

Tryptose	10.0g
Glucose	5.0g
Sodium chloride	5.0g
Beef extract	3.0g

Preparation of Dextrose Broth: Add components to distilled/deionized water and bring volume to 1.0L. Mix thoroughly.

Preparation of Medium: Combine 750.0mL of beef blood serum with 250.0mL of dextrose broth. Mix thoroughly. Distribute into screw-capped tubes. Slant tubes in the autoclave. Close the autoclave door loosely. Autoclave for 10 min at 0 psi pressure–100°C. Close the autoclave door tightly. Autoclave for 15 min at 15 psi pressure–121°C.

Use: For the cultivation of *Corynebacterium diphtheriae.* For demonstration of pigment production and proteolysis by *Corynebacterium diphtheriae.*

Loeffler Blood Serum Medium

Composition per liter:

Beef blood serum	750.0mL
Dextrose broth	250.0mL

pH 7.1 ± 0.2 at 25°C

Dextrose Broth:

Composition per liter:

Enzymatic digest of protein	2.5g
Glucose	1.25g
NaCl	1.25g
Beef extract	0.75g

Preparation of Dextrose Broth: Add components to distilled/deionized water and bring volume to 1.0L. Mix thoroughly.

Preparation of Medium: Combine 750.0mL of beef blood serum with 250.0mL of dextrose broth. Mix thoroughly. Distribute into screw-capped tubes. Slant tubes in the autoclave. Close the autoclave door loosely. Autoclave for 10 min at 0 psi pressure–100°C. Close the autoclave door tightly. Autoclave for 15 min at 15 psi pressure–121°C.

Use: For the cultivation of *Corynebacterium diphtheriae.* For demonstration of pigment production and proteolysis by *Corynebacterium diphtheriae.*

Loeffler HiVeg Medium Base with Asolectin

Composition per liter:

Glucose	2.5g
Plant extract	2.5g
Plant special peptone	2.5g
NaCl	1.25g
Asolectin solution	10.0mL

pH 7.6 ± 0.2 at 25°C

Source: This medium, without asolectin, is available as a premixed powder from HiMedia.

Asolectin Solution:

Composition per 100.0mL:

Asolectin	1.0g
Ethanol	100.0mL

Preparation of Asolectin Solution: Add 1.0g of asolectin to ethanol and bring volume to 100.0mL. Mix thoroughly.

Preparation of Medium: Add components to distilled/deionized water and bring volume to 1.0L. Mix thoroughly. Distribute into screw-capped tubes or flasks. Autoclave for 15 min at 15 psi pressure–121°C.

Use: For the cultivation of *Tetrahymena patula.*

Loeffler Medium

Composition per liter:

Beef serum	70.0g
Egg, dried	7.5g
Heart muscle, solids from infusion	0.72g
Glucose	0.71g
Peptic digest of animal tissue	0.71g
NaCl	0.36g

pH 7.6 ± 0.2 at 25°C

Source: This medium is available as a premixed powder from BD Diagnostic Systems.

Preparation of Medium: Add components to distilled/deionized water and bring volume to 1.0L. Mix thoroughly. Distribute into screw-capped tubes. Slant tubes in the autoclave. Close the autoclave door loosely. Autoclave for 10 min at 0 psi pressure–100°C. Close the autoclave door tightly. Autoclave for 15 min at 15 psi pressure–121°C.

Use: For the cultivation of *Corynebacterium diphtheriae.* For demonstration of pigment production and proteolysis by *Corynebacterium* *diphtheriae*. For the cultivation and maintenance of *Moraxella lacunata*.

Loeffler Slant

Composition per liter:

Tryptose	5.0g
Glucose	1.0g
Beef serum	750.0mL

Preparation of Medium: Add components to distilled/deionized water and bring volume to 1.0L. Mix thoroughly. Distribute into screw-capped tubes. Slant tubes in the autoclave. Close the autoclave door loosely. Autoclave for 10 min at 0 psi pressure–100°C. Close the autoclave door tightly. Autoclave for 15 min at 15 psi pressure–121°C.

Use: For the cultivation of *Corynebacterium diphtheriae*. For demonstration of pigment production and proteolysis by *Corynebacterium diphtheriae*. For the cultivation and maintenance of *Moraxella lacunata*.

Loeffler Slant, Modified

Composition per liter:

Glucose	1.0g
Peptone	0.5g
Beef serum	300.0mL

pH 7.6 ± 0.2 at 25°C

Preparation of Medium: Add peptone and glucose to distilled/deionized water and bring volume to 100.0mL. Mix thoroughly. Add beef serum. Mix thoroughly. Adjust pH to 7.6. Distribute into screw-capped tubes in 3.0mL volumes. Slant tubes in the autoclave. Autoclave for 30 min at 0 psi pressure–100°C.

Use: For the cultivation of *Corynebacterium diphtheriae*. For demonstration of pigment production and proteolysis by *Corynebacterium diphtheriae*. For the cultivation and maintenance of *Moraxella lacunata*.

LOM Agar

See: **Lysine Ornithine Mannitol Agar**

Lombard–Dowell Agar (LD Agar)

Agar	20.0g
Pancreatic digest of casein	5.0g
Yeast extract	5.0g
NaCl	2.5g
L-Cystine	0.4g
L-Tryptophan	0.2g
Na_2SO_3	0.1g
Hemin	10.0mg
NaOH (1*N*)	5.0mL
Vitamin K_1 solution	1.0mL

pH 7.5 ± 0.2 at 25°C

Vitamin K_1 Solution:

Composition per 100.0mL:

Vitamin K_1	1.0g
Ethanol	99.0mL

Preparation of Vitamin K_1 Solution: Add vitamin K_1 to 99.0mL of absolute ethanol. Mix thoroughly.

Preparation of Medium: Add hemin and L-cystine to 5.0mL of NaOH. Mix thoroughly. Add remaining components. Bring volume to 1.0L with distilled/deionized water. Mix thoroughly. Gently heat and bring to boiling. Distribute into tubes or flasks. Autoclave for 15 min at 15 psi pressure–121°C. Pour into sterile Petri dishes.

Use: For the cultivation and identification of a variety of obligate anaerobic bacteria. For the cultivation of *Bacteroides species, Fusobacterium species, Clostridium* species, and nonspore-forming Gram-positive anaerobes.

Lombard-Dowell Bile Agar (LD Bile Agar)

Composition per liter:

Agar	20.0g
Oxgall	20.0g
Pancreatic digest of casein	5.0g
Yeast extract	5.0g
NaCl	2.5g
D-Glucose	1.0g
L-Cystine	0.4g
L-Tryptophan	0.2g
Na_2SO_3	0.1g
Hemin	10.0mg
Bile	200.0mL
NaOH (1*N*)	5.0mL
Vitamin K_1 solution	1.0mL

pH 7.5 ± 0.2 at 25°C

Vitamin K_1 Solution:

Composition per 100.0mL:

Vitamin K_1	1.0g
Ethanol	99.0mL

Preparation of Vitamin K_1 Solution: Add vitamin K_1 to 99.0mL of absolute ethanol. Mix thoroughly.

Preparation of Medium: Add hemin and L-cystine to 5.0mL of NaOH. Mix thoroughly. Add remaining components. Bring volume to 1.0L with distilled/deionized water. Mix thoroughly. Gently heat and bring to boiling. Distribute into tubes or flasks. Pour into sterile Petri dishes.

Use: For the cultivation and identification of a variety of obligate anaerobic bacteria in the presence of 20% bile.

Lombard-Dowell Broth (LD Broth)

Composition per liter:

Pancreatic digest of casein	5.0g
Yeast extract	5.0g
NaCl	2.5g
Agar	0.7g
L-Tryptophan	0.2g
Na_2SO_3	0.1g
NaOH (1*N*)	5.0mL
Hemin solution	1.0mL
Vitamin K_1 solution	1.0mL

pH 7.5 ± 0.2 at 25°C

Hemin Solution:

Composition per 100.0mL:

Hemin	1.0g
NaOH (1*N* solution)	20.0mL

Preparation of Hemin Solution: Add hemin to 20.0mL of 1*N* NaOH solution. Mix thoroughly. Bring volume to 100.0mL with distilled/deionized water.

Vitamin K_1 Solution:
Composition per 100.0mL:

Vitamin K_1	1.0g
Ethanol	99.0mL

Preparation of Vitamin K_1 Solution: Add vitamin K_1 to 99.0mL of absolute ethanol. Mix thoroughly.

Preparation of Medium: Add tryptophan to 5.0mL of NaOH. Mix thoroughly. Add remaining components. Bring volume to 1.0L with distilled/deionized water. Mix thoroughly. Gently heat and bring to boiling. Adjust pH to 7.5. Distribute into screw-capped tubes in 7.0mL volumes. Autoclave for 15 min at 15 psi pressure–121°C. Cool tubes, with caps loose, under 85% N_2 + 10% H_2 + 5% CO_2. Tighten caps.

Use: For the cultivation of a wide variety of anaerobic bacteria.

Lombard-Dowell Egg Yolk Agar (LD Egg Yolk Agar) (Egg Yolk Agar, Lombard-Dowell)

Composition per 9100.0mL:

$Na_2HPO_4 \cdot 12H_2O$	5.0g
Glucose	2.0g
LD Agar	9000.0mL
Egg yolk emulsion	100.0mL
$MgSO_4 \cdot 7H_2O$ (5% solution)	0.2mL

pH 7.5 ± 0.2 at 25°C

LD Agar:
Composition per liter:

Agar	20.0g
Pancreatic digest of casein	5.0g
Yeast extract	5.0g
NaCl	2.5g
L-Cystine	0.4g
L-Tryptophan	0.2g
Na_2SO_3	0.1g
Hemin	10.0mg
NaOH (1*N* NaOH)	5.0mL
Vitamin K_1 solution	1.0mL

Preparation of LD Agar: Add hemin and L-cystine to 5.0mL of NaOH. Mix thoroughly. Add remaining components. Mix thoroughly. Gently heat and bring to boiling.

Vitamin K_1 Solution:
Composition per 100.0mL:

Vitamin K_1	1.0g
Ethanol	99.0mL

Preparation of Vitamin K_1 Solution: Add vitamin K_1 to 99.0mL of absolute ethanol. Mix thoroughly.

Egg Yolk Emulsion:
Composition:

Chicken egg yolks	11
Whole chicken egg	1

Preparation of Egg Yolk Emulsion: Soak eggs with 1:100 dilution of saturated mercuric chloride solution for 1 min. Crack eggs and separate yolks from whites. Mix egg yolks with 1 chicken egg.

Preparation of Medium: Combine components, except egg yolk emulsion. Mix thoroughly. Autoclave for 15 min at 15 psi pressure–121°C. Cool to 45°–50°C. Aseptically add 100.0mL of egg yolk emulsion. Mix thoroughly. Pour into sterile Petri dishes.

Use: For the cultivation of a wide variety of anaerobic bacteria. For the differentiation of anaerobic bacteria based on lecithinase production, lipase production, and proteolytic ability. Bacteria that produce lecithinase appear as colonies surrounded by a zone of insoluble precipitate. Bacteria that produce lipase appear as colonies with a pearly iridescent sheen. Bacteria that produce proteolytic activity appear as colonies surrounded by a clear zone.

Lombard-Dowell Esculin Agar (LD Esculin Agar) (Esculin Agar, Lombard-Dowell)

Composition per liter:

Agar	20.0g
Pancreatic digest of casein	5.0g
Yeast extract	5.0g
NaCl	2.5g
Esculin	1.0g
Ferric citrate	0.5g
L-Cystine	0.4g
L-Tryptophan	0.2g
Hemin	10.0mg
NaOH (1*N*)	5.0mL
Vitamin K_1 solution	1.0mL

pH 7.5 ± 0.2 at 25°C

Vitamin K_1 Solution:
Composition per 100.0mL:

Vitamin K_1	1.0g
Ethanol	99.0mL

Preparation of Vitamin K_1 Solution: Add vitamin K_1 to 99.0mL of absolute ethanol. Mix thoroughly.

Preparation of Medium: Add hemin and L-cystine to 5.0mL of NaOH. Mix thoroughly. Add remaining components. Bring volume to 1.0L with distilled/deionized water. Mix thoroughly. Gently heat and bring to boiling. Distribute into tubes or flasks. Autoclave for 15 min at 15 psi pressure–121°C. Pour into sterile Petri dishes.

Use: For the cultivation of a wide variety of anaerobic bacteria. For the differentiation of anaerobic bacteria based on esculin hydrolysis, H_2S production, and catalase production. Bacteria that hydrolyze esculin appear as colonies surrounded by a red-brown to dark brown zone. Bacteria that produce H_2S appear as black colonies.

Lombard-Dowell Gelatin Agar (LD Gelatin Agar)

Composition per liter:

Agar	20.0g
Pancreatic digest of casein	5.0g
Yeast extract	5.0g
Gelatin	4.0g
NaCl	2.5g
Glucose	1.0g
L-Cystine	0.4g
L-Tryptophan	0.2g
Na_2SO_3	0.1g
Hemin	10.0mg
NaOH (1*N*)	5.0mL
Vitamin K_1 solution	1.0mL

pH 7.5 ± 0.2 at 25°C

Vitamin K_1 Solution:
Composition per 100.0mL:

Vitamin K_1	1.0g
Ethanol	99.0mL

Preparation of Vitamin K_1 Solution: Add vitamin K_1 to 99.0mL of absolute ethanol. Mix thoroughly.

Preparation of Medium: Add hemin and L-cystine to 5.0mL of NaOH. Mix thoroughly. Add remaining components, except agar and gelatin. Bring volume to 750.0mL with distilled/deionized water. Mix thoroughly. Gently heat and bring to boiling. In a separate flask, add gelatin to 100.0mL of cold distilled/deionized water. Gently heat and bring to 70°C. Add gelatin solution to the 750.0mL of basal medium. Mix thoroughly. Add agar. Bring volume to 1.0L with distilled/deionized water. Autoclave for 15 min at 15 psi pressure–121°C. Pour into sterile Petri dishes.

Use: For the cultivation of a wide variety of anaerobic bacteria. For the differentiation of anaerobic bacteria based on gelatinase production. After incubation of plates, gelatinase activity is determined by the addition of Frazier's reagent. Bacteria that hydrolyze gelatin appear as colonies surrounded by a clear zone.

Lombard-Dowell Neomycin Agar (Egg Yolk Agar with Neomycin)

Composition per 9100.0mL:

$Na_2HPO_4 \cdot 12H_2O$	5.0g
Glucose	2.0g
Neomycin sulfate	0.1g
LD Agar	9000.0mL
Egg yolk emulsion	100.0mL
$MgSO_4 \cdot 7H_20$ (5% solution)	0.2mL

pH 7.5 ± 0.2 at 25°C

LD Agar:
Composition per liter:

Agar	20.0g
Pancreatic digest of casein	5.0g
Yeast extract	5.0g
NaCl	2.5g
L-Cystine	0.4g
L-Tryptophan	0.2g
Na_2SO_3	0.1g
Hemin	10.0mg
NaOH (1*N* NaOH)	5.0mL
Vitamin K_1 solution	1.0mL

Preparation of LD Agar: Add hemin and L-cystine to 5.0mL of NaOH. Mix thoroughly. Add remaining components. Mix thoroughly. Gently heat and bring to boiling.

Vitamin K_1 Solution:
Composition per 100.0mL:

Vitamin K_1	1.0g
Ethanol	99.0mL

Preparation of Vitamin K_1 Solution: Add vitamin K_1 to 99.0mL of absolute ethanol. Mix thoroughly.

Egg Yolk Emulsion:
Composition:

Chicken egg yolks	11
Whole chicken egg	1

Preparation of Egg Yolk Emulsion: Soak eggs with 1:100 dilution of saturated mercuric chloride solution for 1 min. Crack eggs and separate yolks from whites. Mix egg yolks with 1 chicken egg.

Preparation of Medium: Combine components, except egg yolk emulsion and neomycin sulfate. Mix thoroughly. Autoclave for 15 min at 15 psi pressure–121°C. Cool to 45°–50°C. Aseptically add 100.0mL of egg yolk emulsion and neomycin sulfate. Mix thoroughly. Pour into sterile Petri dishes.

Use: For the selective cultivation of a wide variety of anaerobic bacteria. For the differentiation of anaerobic bacteria based on lecithinase production, lipase production, and proteolytic ability. Bacteria that produce lecithinase appear as colonies surrounded by a zone of insoluble precipitate. Bacteria that produce lipase appear as colonies with a pearly iridescent sheen. Bacteria that produce proteolytic activity appear as colonies surrounded by a clear zone.

Long-Term Preservation Medium

Composition per liter:

NaCl	30.0g
Peptone	10.0g
Agar	3.0g
Yeast extract	3.0g

Preparation of Medium: Add components to distilled/deionized water and bring volume to 1.0L. Mix thoroughly. Gently heat and bring to boiling. Distribute into screw-capped tubes in 4.0mL volumes. Autoclave for 15 min at 15 psi pressure–121°C.

Use: For the cultivation and maintenance of a wide variety of bacteria.

Low Iron YC Agar

Composition per 1033.0mL:

Solution 1	1.0L
Solution 4	30.0mL
Solution 2	2.0mL
Solution 3	1.0mL

pH 7.4 ± 0.2 at 25°C

Solution 1:
Composition per liter:

Yeast extract	20.0g
Noble agar	10.0g
Casamino acids	10.0g
KH_2PO_4	5.0g
$CaCl_2$	1.0g
Tryptophan	0.05g

Preparation of Solution 1: Add components to distilled/deionized water and bring volume to 1.0L. Mix thoroughly. Adjust pH to 7.4. Gently heat and bring to boiling. Filter through #40 ashless filter paper.

Solution 2:
Composition per 100.0mL:

$MgSO_4 \cdot 7H_2O$	22.5g
$CuSO_4 \cdot 5H_2O$	0.5g
$ZnSO_4 \cdot 5H_2O$	0.2g
β-Alanine	0.115g
Nicotinic Acid	0.115g
$MnCl_2 \cdot 4H_2O$	0.075g
Pimelic acid	7.5mg
HCl, concentrated	3.0mL

Preparation of Solution 2: Add components to distilled/deionized water and bring volume to 100.0mL. Mix thoroughly.

Solution 3:

Composition per 100.0mL:

L-Cystine	20.0g
HCl, concentrated	20.0mL

Preparation of Solution 3: Add components to distilled/deionized water and bring volume to 100.0mL. Mix thoroughly.

Solution 4:

Composition per 100.0mL:

Maltose	50.0g
$CaCl_2 \cdot 2H_2O$	0.5g

Preparation of Solution 4: Add components to distilled/deionized water and bring volume to 100.0mL. Mix thoroughly. Autoclave for 10 min at 11 psi pressure–116°C. Cool to 45°–50°C.

Preparation of Medium: To 1.0L of solution 1, add 2.0mL of solution 2 and 1.0mL of solution 3. Mix thoroughly. Autoclave for 15 min at 15 psi pressure–121°C. Cool to 45°–50°C. Aseptically add 30.0mL of sterile solution 4. Mix thoroughly. Pour into sterile Petri dishes or distribute into sterile tubes.

Use: For the cultivation and maintenance of *Corynebacterium diphtheriae.*

Low Iron YC Broth

Composition per 1033.0mL:

Solution 1	1.0L
Solution 4	30.0mL
Solution 2	2.0mL
Solution 3	1.0mL

pH 7.4 ± 0.2 at 25°C

Solution 1:

Composition per liter:

Yeast extract	20.0g
Casamino acids	10.0g
KH_2PO_4	5.0g
$CaCl_2 \cdot 2H_2O$	1.0g
Tryptophan	0.05g

Preparation of Solution 1: Add components to distilled/deionized water and bring volume to 1.0L. Mix thoroughly. Adjust pH to 7.4. Gently heat and bring to boiling. Filter through #40 ashless filter paper.

Solution 2:

Composition per 100.0mL:

$MgSO_4 \cdot 7H_2O$	22.5g
$CuSO_4 \cdot 5H_2O$	0.5g
$ZnSO_4 \cdot 5H_2O$	0.2g
β-Alanine	0.115g
Nicotinic acid	0.115g
$MnCl_2 \cdot 4H_2O$	0.075g
Pimelic acid	7.5mg
HCl, concentrated	3.0mL

Preparation of Solution 2: Add components to distilled/deionized water and bring volume to 100.0mL. Mix thoroughly.

Solution 3:

Composition per 100.0mL:

L-Cystine	20.0g
HCl, concentrated	20.0mL

Preparation of Solution 3: Add components to distilled/deionized water and bring volume to 100.0mL. Mix thoroughly.

Solution 4:

Composition per 100.0mL:

Maltose	50.0g
$CaCl_2 \cdot 2H_2O$	0.5g

Preparation of Solution 4: Add components to distilled/deionized water and bring volume to 100.0mL. Mix thoroughly. Autoclave for 10 min at 11 psi pressure–116°C. Cool to 25°C.

Preparation of Medium: To 1.0L of solution 1, add 2.0mL of solution 2 and 1.0mL of solution 3. Mix thoroughly. Autoclave for 15 min at 15 psi pressure–121°C. Cool to 25°C. Aseptically add 30.0mL of sterile solution 4. Mix thoroughly. Aseptically distribute into sterile tubes or flasks.

Use: For the cultivation and maintenance of *Corynebacterium diphtheriae.*

Low Phosphate Buffered Basal Medium, Modified

Composition per 1030.0mL:

Pectin	4.0g
NH_4Cl	1.0g
Na_2HPO_4	0.72g
KH_2PO_4	0.3g
$MgCl_2 \cdot 6H_2O$	0.2g
Reducing agent	20.0mL
Yeast extract solution	10.0mL
Trace minerals	10.0mL
Wolfe's vitamin solution	5.0mL
Resazurin (0.2% solution)	1.0mL
$FeSO_4 \cdot 7H_2O$ (2.5% solution)	0.03mL

pH 7.3 ± 0.1 at 25°C

Reducing Agent:

Composition per 20.0mL:

$Na_2S \cdot 9H_2O$	0.5g

Preparation of Reducing Agent: Add $Na_2S \cdot 9H_2O$ to distilled/deionized water and bring volume to 20.0mL. Mix thoroughly. Gas with 100% N_2 for 20 min. Cap with a rubber stopper. Autoclave for 45 min at 15 psi pressure–121°C. Use freshly prepared solution.

Yeast Extract Solution:

Composition per 10.0mL:

Yeast extract	1.0g

Preparation of Yeast Extract Solution: Add yeast extract to distilled/deionized water and bring volume to 10.0mL. Mix thoroughly. Autoclave for 45 min at 15 psi pressure–121°C. Cool to 25°C.

Trace Minerals:

Composition per liter:

Nitrilotriacetic acid	12.8g
NaCl	1.0g
$CoCl_2 \cdot 6H_2O$	0.16g
$CaCl_2 \cdot 2H_2O$	0.1g
$FeSO_4 \cdot 7H_2O$	0.1g
$MnCl_2 \cdot 4H_2O$	0.1g
$ZnCl_2$	0.1g
$NiSO_4 \cdot 6H_2O$	0.026g
$CuCl_2$	0.02g
Na_2SeO_3	0.02g
H_3BO_3	0.01g
$Na_2MoO_4 \cdot 2H_2O$	0.01g

Preparation of Trace Minerals: Add nitrilotriacetic acid to 500.0mL of distilled/deionized water. Dissolve by adjusting pH to 6.5

with KOH. Add remaining components. Add distilled/deionized water to 1.0L.

Wolfe's Vitamin Solution:

Composition per liter:

Pyridoxine·HCl	0.01g
Thiamine·HCl	5.0mg
Riboflavin	5.0mg
Nicotinic acid	5.0mg
Calcium pantothenate	5.0mg
p-Aminobenzoic acid	5.0mg
Thioctic acid	5.0mg
Biotin	2.0mg
Folic acid	2.0mg
Cyanocobalamin	0.1mg

Preparation of Wolfe's Vitamin Solution: Add components to distilled/deionized water and bring volume to 1.0L. Mix thoroughly.

Preparation of Medium: Add components, except yeast extract solution and reducing agent, to distilled/deionized water and bring volume to 1.0L. Mix thoroughly. Gently heat and bring to boiling. Cool under 90% N_2 + 10% CO_2. Anaerobically distribute into tubes in 6.0mL volumes. Autoclave for 45 min at 15 psi pressure–121°C. Aseptically add 0.06mL of sterile yeast extract solution to each tube. Mix thoroughly. Immediately prior to inoculation, aseptically add 0.12mL of sterile reducing agent to each tube. Mix thoroughly.

Use: For the cultivation and maintenance of *Clostridium thermosulfurogenes*.

Lowenstein-Gruft Medium

Composition per 1600.0mL:

Potato starch	30.0g
Asparagine	3.6g
KH_2PO_4	2.4g
Magnesium citrate	0.6g
Malachite Green	0.4g
$MgSO_4·7H_2O$	0.24g
Nalidixic acid	0.056g
Ribonucleic acid	0.08mg
Homogenized whole egg	1.0L
Glycerol	12.0mL
Penicillin	80,000U

Homogenized Whole Egg:

Composition per liter:

Whole eggs	18–24

Preparation of Homogenized Whole Egg: Use fresh eggs, less than 1 week old. Scrub the shells with soap. Let stand in a soap solution for 30 min. Rinse in running water. Soak eggs in 70% ethanol for 15 min. Break the eggs into a sterile container. Homogenize by shaking. Filter through four layers of sterile cheesecloth into a sterile graduated cylinder. Measure out 1.0L.

Preparation of Medium: Add glycerol to 600.0mL of distilled/deionized water. Mix thoroughly. Add remaining components, except fresh egg mixture. Mix thoroughly. Gently heat while stirring and bring to boiling. Autoclave for 15 min at 15 psi pressure–121°C. Cool to 50°C. Aseptically add 1.0L of homogenized whole egg. Mix thoroughly. Distribute into sterile screw-capped tubes. Place tubes in a slanted position. Inspissate at 85°C (moist heat) for 45 min.

Use: For the cultivation and differentiation of *Mycobacterium* species. *Mycobacterium tuberculosis* appears as granular, rough, dry colonies. *Mycobacterium kansasii* appears as smooth to rough photochromogenic colonies. *Mycobacterium gordonae* appears as smooth yellow-orange colonies. *Mycobacterium avium* appears as smooth, colorless colonies. *Mycobacterium smegmatis* appears as wrinkled, creamy white colonies.

Lowenstein-Jensen Medium

Composition per 1600.0mL:

Potato starch	30.0g
Asparagine	3.6g
KH_2PO_4	2.4g
Magnesium citrate	0.6g
Malachite Green	0.4g
$MgSO_4·7H_2O$	0.24g
Homogenized whole egg	1.0L
Glycerol	12.0mL

Source: This medium is available as a prepared medium from BD Diagnostic Systems and Oxoid Unipath.

Homogenized Whole Egg:

Composition per liter:

Whole eggs	18–24

Preparation of Homogenized Whole Egg: Use fresh eggs, less than 1 week old. Scrub the shells with soap. Let stand in a soap solution for 30 min. Rinse in running water. Soak eggs in 70% ethanol for 15 min. Break the eggs into a sterile container. Homogenize by shaking. Filter through four layers of sterile cheesecloth into a sterile graduated cylinder. Measure out 1.0L.

Preparation of Medium: Add glycerol to 600.0mL of distilled/deionized water. Mix thoroughly. Add remaining components, except fresh egg mixture. Mix thoroughly. Gently heat while stirring and bring to boiling. Autoclave for 15 min at 15 psi pressure–121°C. Cool to 50°C. Aseptically add 1.0L of homogenized whole egg. Mix thoroughly. Distribute into sterile screw-capped tubes. Place tubes in a slanted position. Inspissate at 85°C (moist heat) for 45 min.

Use: For the cultivation and differentiation of *Mycobacterium* species. *Mycobacterium tuberculosis* appears as granular, rough, dry colonies. *Mycobacterium kansasii* appears as smooth to rough photochromogenic colonies. *Mycobacterium gordonae* appears as smooth yellow-orange colonies. *Mycobacterium avium* appears as smooth, colorless colonies. *Mycobacterium smegmatis* appears as wrinkled, creamy white colonies. Also used for the cultivation and maintenance of *Gordona* species, *Nocardia* species, *Rhodococcus* species, and *Tsukamurella paurometabolum*.

Lowenstein-Jensen Medium with Sodium Chloride

Composition per 1600.0mL:

NaCl	80.0g
Potato starch	30.0g
Asparagine	3.6g
KH_2PO_4	2.4g
Magnesium citrate	0.6g
Malachite Green	0.4g
$MgSO_4·7H_2O$	0.24g
Homogenized whole egg	1.0L
Glycerol	12.0mL

Homogenized Whole Egg:

Composition per liter:

Whole eggs	18–24

Preparation of Homogenized Whole Egg: Use fresh eggs, less than 1 week old. Scrub the shells with soap. Let stand in a soap solution for 30 min. Rinse in running water. Soak eggs in 70% ethanol for 15 min. Break the eggs into a sterile container. Homogenize by shaking. Filter through four layers of sterile cheesecloth into a sterile graduated cylinder. Measure out 1.0L.

Preparation of Medium: Add glycerol to 600.0mL of distilled/deionized water. Mix thoroughly. Add remaining components, except fresh egg mixture. Mix thoroughly. Gently heat while stirring and bring to boiling. Autoclave for 15 min at 15 psi pressure–121°C. Cool to 50°C. Aseptically add 1.0L of homogenized whole egg. Mix thoroughly. Distribute into sterile screw-capped tubes. Place tubes in a slanted position. Inspissate at 85°C (moist heat) for 45 min.

Use: For the cultivation of *Mycobacterium smegmatis* and other salt-tolerant *Mycobacterium* species.

Lowenstein-Jensen Medium with Streptomycin

Composition per 1610.0mL:

Potato starch	30.0g
Asparagine	3.6g
KH_2PO_4	2.4g
Magnesium citrate	0.6g
Malachite Green	0.4g
$MgSO_4 \cdot 7H_2O$	0.24g
Homogenized whole egg	1.0L
Glycerol	12.0mL
Streptomycin solution	10.0mL

Homogenized Whole Egg:

Composition per liter:

Whole eggs	18–24

Preparation of Homogenized Whole Egg: Use fresh eggs, less than 1 week old. Scrub the shells with soap. Let stand in a soap solution for 30 min. Rinse in running water. Soak eggs in 70% ethanol for 15 min. Break the eggs into a sterile container. Homogenize by shaking. Filter through four layers of sterile cheesecloth into a sterile graduated cylinder. Measure out 1.0L.

Streptomycin Solution:

Composition per 10.0mL:

Streptomycin	0.1mg

Preparation of Streptomycin Solution: Add streptomycin to distilled/deionized water and bring volume to 10.0mL. Mix thoroughly. Filter sterilize.

Preparation of Medium: Add glycerol to 600.0mL of distilled/deionized water. Mix thoroughly. Add remaining components, except fresh egg mixture and streptomycin solution. Mix thoroughly. Gently heat while stirring and bring to boiling. Autoclave for 15 min at 15 psi pressure–121°C. Cool to 50°C. Aseptically add 1.0L of homogenized whole egg and 10.0mL of sterile streptomycin solution. Mix thoroughly. Distribute into sterile screw-capped tubes. Place tubes in a slanted position. Inspissate at 85°C (moist heat) for 45 min.

Use: For the cultivation and differentiation of *Mycobacterium* species. *Mycobacterium tuberculosis* appears as granular, rough, dry colonies. *Mycobacterium kansasii* appears as smooth to rough photochromogenic colonies. *Mycobacterium gordonae* appears as smooth yellow-orange colonies. *Mycobacterium avium* appears as smooth, colorless colonies. *Mycobacterium smegmatis* appears as wrinkled, creamy white colonies. Also used for the cultivation and maintenance of *Gordona* species, *Nocardia* species, *Rhodococcus* species, and *Tsukamurella paurometabolum*.

Lowenstein-Jensen Medium without Glycerol

Composition per 1600.0mL:

Potato starch	30.0g
Asparagine	3.6g
KH_2PO_4	2.4g
Magnesium citrate	0.6g
Malachite green	0.4g
$MgSO_4 \cdot 7H_2O$	0.24g
Homogenized whole egg	1.0L

Homogenized Whole Egg:

Composition per liter:

Whole eggs	18–24

Preparation of Homogenized Whole Egg: Use fresh eggs, less than 1 week old. Scrub the shells with soap. Let stand in a soap solution for 30 min. Rinse in running water. Soak eggs in 70% ethanol for 15 min. Break the eggs into a sterile container. Homogenize by shaking. Filter through four layers of sterile cheesecloth into a sterile graduated cylinder. Measure out 1.0L.

Preparation of Medium: Add components, except fresh egg mixture, to 600.0mL of distilled/deionized water. Mix thoroughly. Gently heat while stirring and bring to boiling. Autoclave for 15 min at 15 psi pressure–121°C. Cool to 50°C. Aseptically add 1.0L of homogenized whole egg. Mix thoroughly. Distribute into sterile screw-capped tubes. Place tubes in a slanted position. Inspissate at 85°C (moist heat) for 45 min.

Use: For the cultivation and maintenance of *Mycobacterium* species, especially *Mycobacterium bovis* and other species that are sensitive to glycerol.

LPBM Acido-Thermophile Medium

Composition per liter:

Agar	20.0g
Cellulose	5.0g
KH_2PO_4	1.0g
NH_4Cl	1.0g
Yeast extract	1.0g
Cellobiose	0.5g
$MgSO_4 \cdot 7H_2O$	0.2g
$Na_2HPO_4 \cdot 7H_2O$	0.1g
$CaCl_2 \cdot 2H_2O$	0.02g

pH 5.2 ± 0.2 at 25°C

Preparation of Medium: Add components, except cellulose and cellobiose, to distilled/deionized water and bring volume to 1.0L. Mix thoroughly. Adjust pH to 5.2 with H_3PO_4. Add cellulose and cellobiose. Mix thoroughly. Gently heat and bring to boiling. Distribute into tubes or flasks. Autoclave for 15 min at 15 psi pressure–121°C. Pour into sterile Petri dishes or leave in tubes.

Use: For the cultivation and maintenance of *Acidothermus cellulolyticus*.

LPM Agar (Lithium Chloride Phenylethanol Moxalactam Plating Agar)

Composition per liter:

Agar	15.0g
Glycine anhydride	10.0g

$LiCl_2$	5.0g
NaCl	5.0g
Pancreatic digest of casein	5.0g
Peptic digest of animal tissue	5.0g
Beef extract	3.0g
Phenylethyl alcohol	2.5g
Moxalactam solution	2.0mL

pH 7.3 ± 0.2 at 25°C

Source: This medium is available as a premixed powder from BD Diagnostic Systems.

Moxalactam Solution:

Composition per 10.0mL:

Moxalactam	0.1g

Preparation of Moxalactam Solution: Add moxalactam to distilled/deionized water and bring volume to 10.0mL. Mix thoroughly. Filter sterilize.

Preparation of Medium: Add components, except moxalactam solution, to distilled/deionized water and bring volume to 998.0mL. Mix thoroughly. Gently heat while stirring and bring to boiling. Autoclave for 12 min at 15 psi pressure–121°C. Cool to 45°–50°C. Aseptically add 2.0mL of sterile moxalactam solution. Mix thoroughly. Pour into sterile Petri dishes or distribute into sterile tubes.

Use: For the isolation and cultivation of *Listeria monocytogenes*.

LPM Agar (Lithium Phenylethanol Moxalactam Agar)

Composition per liter:

Agar	15.0g
Glycine anhydride	10.0g
Casein enzymic hydrolysate	5.0g
Peptic digest of animal tissue	5.0g
Beef extract	3.0g
LiCl	5.0g
NaCl	5.0g
Phenylethyl alcohol	2.5g
Selective supplement solution	10.0mL

pH 7.3 ± 0.2 at 25°C

Source: This medium is available from HiMedia.

Selective Supplement Solution:

Composition per 10.0mL:

Moxalactam	20.0mg

Preparation of Selective Supplement Solution: Add moxalactam to distilled/deionized water and bring volume to 10.0mL. Mix thoroughly. Filter sterilize.

Caution: Lithium chloride is harmful. Avoid bodily contact and inhalation of vapors. On contact with skin, wash with plenty of water immediately.

Preparation of Medium: Add components, except selective supplement solution, to distilled/deionized water and bring volume to 990.0mL. Mix thoroughly. Autoclave for 15 min at 15 psi pressure–121°C. Cool to 50°C. Aseptically add selective supplement solution. Mix thoroughly. Pour into Petri dishes or aseptically distribute into sterile tubes.

Use: For the isolation and cultivation of *Listeria monocytogenes* from food and dairy products.

LPM Agar with Esculin and Ferric Iron

Composition per liter:

Agar	15.0g
Glycine anhydride	10.0g
LiCl	5.0g
NaCl	5.0g
Pancreatic digest of casein	5.0g
Peptic digest of animal tissue	5.0g
Beef extract	3.0g
Phenylethyl alcohol	2.5g
Esculin	1.0g
Ferric ammonium citrate	0.5g
Moxalactam solution	2.0mL

pH 7.3 ± 0.2 at 25°C

Moxalactam Solution:

Composition per 10.0mL:

Moxalactam	0.1g

Preparation of Moxalactam Solution: Add moxalactam to distilled/deionized water and bring volume to 10.0mL. Mix thoroughly. Filter sterilize.

Preparation of Medium: Add components, except moxalactam solution, to distilled/deionized water and bring volume to 998.0mL. Mix thoroughly. Gently heat while stirring and bring to boiling. Autoclave for 12 min at 15 psi pressure–121°C. Cool to 45°–50°C. Aseptically add 2.0mL of sterile moxalactam solution. Mix thoroughly. Pour into sterile Petri dishes or distribute into sterile tubes.

Use: For the isolation and cultivation of *Listeria monocytogenes*.

LPM HiVeg Agar Base with Moxalactam

Composition per liter:

Agar	15.0g
Glycine anhydride	10.0g
Plant hydrolysate	5.0g
Plant peptone	5.0g
LiC	5.0g
NaCl	5.0g
Plant extract	3.0g
Phenylethyl alcohol	2.5g
Moxalactam solution	2.0mL

pH 7.3 ± 0.2 at 25°C

Source: This medium, without moxalactam, is available as a premixed powder from HiMedia.

Moxalactam Solution:

Composition per 10.0mL:

Moxalactam	0.1g

Preparation of Moxalactam Solution: Add moxalactam to distilled/deionized water and bring volume to 10.0mL. Mix thoroughly. Filter sterilize.

Preparation of Medium: Add components, except moxalactam solution, to distilled/deionized water and bring volume to 998.0mL. Mix thoroughly. Gently heat while stirring and bring to boiling. Autoclave for 12 min at 15 psi pressure–121°C. Cool to 45°–50°C. Aseptically add 2.0mL of sterile moxalactam solution. Mix thoroughly. Pour into sterile Petri dishes or distribute into sterile tubes.

Use: For the isolation and cultivation of *Listeria monocytogenes*.

LS Differential HiVeg Medium Base with Skim Milk and Triphenyltetrazolium Chloride

Composition per liter:

Glucose	20.0g
Agar	15.0g
Plant hydrolysate	10.0g
Plant extract	5.0g
Papaic digest of soybean meal	5.0g
NaCl	5.0g
Yeast extract	5.0g
L-Cysteine·HCl	0.3g
Skim milk solution	100.0mL
Triphenyltetrazolium chloride solution	10.0mL

pH 6.1 ± 0.2 at 25°C

Source: This medium, without skim milk and triphenyltetrazolium chloride solution, is available as a premixed powder from HiMedia.

Skim Milk Solution:

Composition per 100.0mL:

Skim milk, antibiotic free	10.0g

Preparation of Skim Milk Solution: Add skim milk to distilled/deionized water and bring volume to 100.0mL. Mix thoroughly. Autoclave for 5 min at 15 psi pressure–121°C. Cool to 50°C.

Triphenyltetrazolium Chloride Solution:

Composition per 10.0mL:

2,3,5-Triphenyltetrazolium chloride	0.2g

Preparation of Triphenyltetrazolium Chloride Solution: Add triphenyltetrazolium chloride to distilled/deionized water and bring volume to 10.0mL. Mix thoroughly. Filter sterilize. Warm to 50°C.

Preparation of Medium: Add components, except skim milk solution and triphenyltetrazolium chloride solution, to distilled/deionized water and bring volume to 890.0mL. Mix thoroughly. Gently heat and bring to boiling. Autoclave for 15 min at 15 psi pressure–121°C. Cool to 45°–50°C. Aseptically add sterile skim milk solution and sterile triphenyltetrazolium chloride solution. Mix thoroughly. Pour into sterile Petri dishes or distribute into sterile tubes.

Use: For the differentiation and enumeration of lactobacilli and streptococci in yogurt. *Lactobacillus* species appear as irregular, red colonies surrounded by a white, opaque zone. *Streptococcus* species appear as round, red colonies surrounded by a clear zone.

LS Differential Medium (*Lactobacillus Streptococcus* Differential Medium)

Composition per liter:

Glucose	20.0g
Agar	13.0g
Pancreatic digest of casein	10.0g
Beef extract	5.0g
NaCl	5.0g
Papaic digest of soybean meal	5.0g
Yeast extract	5.0g
L-Cysteine·HC1·H_2O	0.3g
Skim milk solution	100.0mL
Triphenyltetrazolium chloride solution	10.0mL

pH 6.1 ± 0.2 at 25°C

Source: This medium is available as a premixed powder from Oxoid Unipath.

Skim Milk Solution:

Composition per 100.0mL:

Skim milk, antibiotic free 10.0g

Preparation of Skim Milk Solution: Add skim milk to distilled/deionized water and bring volume to 100.0mL. Mix thoroughly. Autoclave for 5 min at 15 psi pressure–121°C. Cool to 50°C.

Triphenyltetrazolium Chloride Solution:

Composition per 10.0mL:

2,3,5-Triphenyltetrazolium chloride 0.2g

Preparation of Triphenyltetrazolium Chloride Solution: Add triphenyltetrazolium chloride to distilled/deionized water and bring volume to 10.0mL. Mix thoroughly. Filter sterilize. Warm to 50°C.

Preparation of Medium: Add components, except skim milk solution and triphenyltetrazolium chloride solution, to distilled/deionized water and bring volume to 890.0mL. Mix thoroughly. Gently heat and bring to boiling. Autoclave for 15 min at 15 psi pressure–121°C. Cool to 45°–50°C. Aseptically add sterile skim milk solution and sterile triphenyltetrazolium chloride solution. Mix thoroughly. Pour into sterile Petri dishes or distribute into sterile tubes.

Use: For the differentiation and enumeration of lactobacilli and streptococci in yogurt. *Lactobacillus* species appear as irregular, red colonies surrounded by a white, opaque zone. *Streptococcus* species appear as round, red colonies surrounded by a clear zone.

LST-MUG Broth

Composition per liter:

Tryptose	20.0g
Lactose	5.0g
K_2HPO_4	2.75g
KH_2PO_4	2.75g
NaCl	5.0g
L-Tryptophan	1.0g
Sodium lauryl sulfate	0.1g
4-Methylumbelliferyl-β-D-glucuronide	0.1g

pH 6.8 ± 0.2 at 37°C

Source: This medium is available from Fluka, Sigma-Aldrich.

Preparation of Medium: Add components to distilled/deionized water and bring volume to 1.0L. Mix thoroughly. Distribute into test tubes that contain an inverted Durham tube in 10.0mL volumes. Autoclave for 15 min at 15 psi pressure–121°C.

Use: For the detection of *E. coli* by the fluorogenic method. The presence of *E. coli* results in fluorescence in the UV. A positive indole test provides confirmation. β-D-glucoronidase, which is produced by *E. coli*, cleaves 4-methylumbelliferyl-β-D-glucuronide to 4-methylumbelliferone and glucuronide. The fluorogen 4-methylumbelliferone can be detected under a long wavelength UV lamp.

LT Medium

See: **Lecithin Tween™ Medium**

LTH Medium for *Thiothrix*

Composition per liter:

HEPES (*N*-[2-hydroxyethyl]piperazine-*N′*-2-ethanesulfonic acid) buffer	2.38.g
Sodium lactate	1.0g
$Na_2S_2O_3·5H_2O$	0.5g
$(NH_4)_2SO_4$	0.5g

K_2HPO_4	0.11g
$MgSO_4 \cdot 7H_2O$	0.1g
$CaCl_2 \cdot 2H_2O$	0.05g
KH_2PO_4	85.0mg
EDTA	3.0mg
$FeCl_3 \cdot 6H_2O$	2.0mg
Wolfe's vitamin solution	10.0mL

pH 7.3 ± 0.2 at 25°C

Wolfe's Vitamin Solution:

Composition per liter:

Pyridoxine·HCl	10.0mg
p-Aminobenzoic acid	5.0mg
Lipoic acid	5.0mg
Nicotinic acid	5.0mg
Riboflavin	5.0mg
Thiamine·HCl	5.0mg
Calcium DL-pantothenate	5.0mg
Biotin	2.0mg
Folic acid	2.0mg
Vitamin B_{12}	0.1mg

Preparation of Wolfe's Vitamin Solution: Add components to distilled/deionized water and bring volume to 1.0L. Mix thoroughly. Filter sterilize.

Preparation of Medium: Add components, except Wolfe's vitamin solution, to distilled/deionized water and bring volume to 990.0mL. Mix thoroughly. Adjust pH to 7.3 with NaOH. Autoclave for 15 min at 15 psi pressure–121°C. Cool to room temperature. Aseptically add 10.0mL of sterile Wolfe's vitamin solution. Mix thoroughly. Aseptically distribute into sterile tubes or flasks.

Use: For the cultivation of *Thiothrix* species.

LTH Medium for *Thiothrix* with Casitone

Composition per liter:

HEPES (*N*-[2-hydroxyethyl]piperazine-*N*′-2-ethanesulfonic acid) buffer	2.38g
Casitone	1.0g
Sodium lactate	1.0g
$Na_2S_2O_3 \cdot 5H_2O$	0.5g
$(NH_4)_2SO_4$	0.5g
K_2HPO_4	0.11g
$MgSO_4 \cdot 7H_2O$	0.1g
$CaCl_2 \cdot 2H_2O$	0.05g
KH_2PO_4	85.0mg
EDTA	3.0mg
$FeCl_3 \cdot 6H_2O$	2.0mg
Wolfe's vitamin solution	10.0mL

Wolfe's Vitamin Solution:

Composition per liter:

Pyridoxine·HCl	10.0mg
p-Aminobenzoic acid	5.0mg
Lipoic acid	5.0mg
Nicotinic acid	5.0mg
Riboflavin	5.0mg
Thiamine·HCl	5.0mg
Calcium DL-pantothenate	5.0mg
Biotin	2.0mg
Folic acid	2.0mg
Vitamin B_{12}	0.1mg

Preparation of Wolfe's Vitamin Solution: Add components to distilled/deionized water and bring volume to 1.0L. Mix thoroughly. Filter sterilize.

Preparation of Medium: Add components, except Wolfe's vitamin solution, to distilled/deionized water and bring volume to 990.0mL. Mix thoroughly. Adjust pH to 7.3 with NaOH. Autoclave for 15 min at 15 psi pressure–121°C. Aseptically add 10.0mL of sterile Wolfe's vitamin solution. Mix thoroughly. Aseptically distribute into sterile tubes or flasks.

Use: For the cultivation of unidentified bacterium ATCC 49750.

Luedemann Medium (DSMZ Medium 877)

Composition per 100.0mL:

Agar	1.5g
Malt extract broth	1.5g
Soluble starch	1.0g
Glucose	1.0g
Yeast extract	0.5g
NaCl	0.5g
$CaCO_3$	0.2g

pH 8.6 ± 0.2 at 25°C

Preparation of Medium: Add components to distilled/deionized water and bring volume to 1.0L. Mix thoroughly. Gently heat and bring to boiling. Distribute into tubes or flasks. Autoclave for 15 min at 15 psi pressure–121°C. Pour into sterile Petri dishes or leave in tubes.

Use: For the cultivation and maintenance of *Blastococcus saxobsidens* and *Blastococcus sp.*

Luminous Medium

Composition per liter:

NaCl	30.0g
Agar	20.0g
NH_4Cl	5.0g
Pancreatic digest of casein	5.0g
Yeast extract	5.0g
K_2HPO_4	3.9g
KH_2PO_4	2.1g
$CaCO_3$	1.0g
$MgSO_4 \cdot 7H_2O$	1.0g
KCl	0.75g
Tris buffer (1*M* solution, pH 7.5)	50.0mL
Glycerol	3.0mL

pH 7.2 ± 0.2 at 25°C

Preparation of Medium: Add components to distilled/deionized water and bring volume to 1.0L. Mix thoroughly. Gently heat and bring to boiling. Distribute into tubes or flasks. Autoclave for 15 min at 15 psi pressure–121°C. Pour into sterile Petri dishes or leave in tubes.

Use: For the cultivation and maintenance of *Alteromonas hanedai, Photobacterium* species, *Shewanella hanedai,* and *Vibrio* species.

LUP (Lupine Medium)

Composition per liter:

Agar	15.0g
Lupine stems	variable

Preparation of Medium: Add agar to distilled/deionized water and bring volume to 1.0L. Mix thoroughly. Gently heat and bring to boiling. Distribute 6.0mL volumes into tubes. Cut lupine stems into 8.0cm-long pieces. Add 2–3 lupine stems per tube. Autoclave for 15 min at 15 psi pressure–121°C. Allow tubes to cool in a slanted position.

Use: For the cultivation of *Botryosphaeria berengeriana* and *Mycosphaerella ligulicola.*

LuPhet1 Medium (DSMZ Medium 298e)

Composition per liter:

NaCl	1.0g
KCl	0.5g
$MgCl_2·6H_2O$	0.4g
NH_4Cl	0.25g
KH_2PO_4	0.2g
$CaCl_2·2H_2O$	0.15g
Resazurin	1.0mg
$NaHCO_3$ solution	10.0mL
Sodium lactate solution	10.0mL
$Na_2S·9H_2O$ solution	10.0mL
Yeast extract solution	10.0mL
Seven vitamin solution	1.0mL
Selenite-tungstate solution	1.0mL
Trace elements solution SL-10	1.0mL

pH 7.2 ± 0.2 at 25°C

$Na_2S·9H_2O$ Solution:
Composition per 10.0mL:

$Na_2S·9H_2O$	0.36g

Preparation of $Na_2S·9H_2O$ Solution: Add $Na_2S·9H_2O$ to distilled/deionized water and bring volume to 10.0mL. Mix thoroughly. Autoclave under 100% N_2 for 15 min at 15 psi pressure–121°C. Cool to room temperature.

$NaHCO_3$ Solution:
Composition per 10.0mL:

$NaHCO_3$	5.0g

Preparation of $NaHCO_3$ Solution: Add $NaHCO_3$ to distilled/deionized water and bring volume to 10.0mL. Mix thoroughly. Sparge with 80% N_2 + 20% CO_2. Filter sterilize.

Sodium Lactate Solution:
Composition per 10.0mL:

Sodium lactate	1.5g

Preparation of Sodium Lactate Solution: Add sodium lactate to distilled/deionized water and bring volume to 10.0mL. Mix thoroughly. Sparge with 100% N_2. Filter sterilize.

Yeast Extract Solution:
Composition per 10.0mL:

Yeast extract	1.0g

Preparation of Yeast Extract Solution: Add yeast extract to distilled/deionized water and bring volume to 10.0mL. Mix thoroughly. Sparge with 100% N_2. Filter sterilize.

Seven Vitamin Solution:
Composition per liter:

Pyridoxine hydrochloride	300.0mg
Thiamine-HCl·$2H_2O$	200.0mg
Nicotinic acid	200.0mg
Vitamin B_{12}	100.0mg
Calcium pantothenate	100.0mg
p-Aminobenzoic acid	80.0mg
D(+)-Biotin	20.0mg

Preparation of Seven Vitamin Solution: Add components to distilled/deionized water and bring volume to 1.0L. Sparge with 100% N_2. Mix thoroughly. Filter sterilize.

Trace Elements Solution SL-10:
Composition per liter:

$FeCl_2·4H_2O$	1.5g
$CoCl_2·6H_2O$	190.0mg
$MnCl_2·4H_2O$	100.0mg
$ZnCl_2$	70.0mg
$Na_2MoO_4·2H_2O$	36.0mg
$NiCl_2·6H_2O$	24.0mg
H_3BO_3	6.0mg
$CuCl_2·2H_2O$	2.0mg
HCl (25% solution)	10.0mL

Preparation of Trace Elements Solution SL-10: Add $FeCl_2·4H_2O$ to 10.0mL of HCl solution. Mix thoroughly. Add distilled/deionized water and bring volume to 1.0L. Add remaining components. Mix thoroughly. Sparge with 80% N_2 + 20% CO_2. Autoclave for 15 min at 15 psi pressure–121°C.

Selenite-Tungstate Solution
Composition per liter:

NaOH	0.5g
$Na_2WO_4·2H_2O$	4.0mg
$Na_2SeO_3·5H_2O$	3.0mg

Preparation of Selenite-Tungstate Solution: Add components to distilled/deionized water and bring volume to 1.0L. Mix thoroughly. Sparge with 100% N_2. Filter sterilize.

Preparation of Medium: Prepare and dispense medium under 80% N_2 + 20% CO_2 gas atmosphere. Add components, except $NaHCO_3$ solution, sodium lactate solution, $Na_2S·9H_2O$ solution, seven vitamin solution, selenite-tungstate solution, and trace elements solution SL-10, to distilled/deionized water and bring volume to 957.0mL. Mix thoroughly. Adjust pH to 7.2. Sparge with 80% N_2 + 20% CO_2. Autoclave for 15 min at 15 psi pressure–121°C. Aseptically and anaerobically add 10.0mL $NaHCO_3$ solution, 10.0mL $Na_2S·9H_2O$ solution, 10.0mL sodium lactate solution, 1.0mL seven vitamin solution, 1.0mL selenite- tungstate solution, and 1.0mL trace elements solution SL-10. Mix thoroughly. Aseptically and anaerobically distribute into sterile tubes or bottles. After inoculation, flush and repressurize the gas head space of culture bottles with sterile 80% N_2 + 20% CO_2 to 1 bar overpressure.

Use: For the cultivation of *Acetobacterium sp.*

Luria Agar
See: **L Agar**

Luria Agar Base, Miller

Composition per liter:

Agar	15.0g
Tryptone	10.0g
Yeast extract	5.0g
NaCl	0.5g

pH 7.0 ± 0.2 at 25°C

Preparation of Medium: Add components to distilled/deionized water and bring volume to 1.0L. Mix thoroughly. Gently heat and bring to boiling. Distribute into tubes or flasks. Autoclave for 15 min at 15 psi pressure–121°C. Pour into sterile Petri dishes or leave in tubes.

Use: For the cultivation and maintenance of bacteria for genetic and molecular studies.

Luria Bertani Agar, Miller (LB Agar, Miller)

Composition per liter:

Agar	15.0g
Tryptone	10.0g
NaCl	10.0g
Yeast extract	5.0g

pH 7.5 ± 0.2 at 25°C

Preparation of Medium: Add components to distilled/deionized water and bring volume to 1.0L. Mix thoroughly. Gently heat and bring to boiling. Distribute into tubes or flasks. Autoclave for 15 min at 15 psi pressure–121°C. Pour into sterile Petri dishes or leave in tubes.

Use: For the cultivation and maintenance of bacteria for genetic and molecular studies.

Luria Bertani HiVeg Agar, Miller

Composition per liter:

Agar	15.0g
Plant hydrolysate	10.0g
NaCl	10.0g
Yeast extract	5.0g

pH 7.5 ± 0.2 at 25°C

Source: This medium is available as a premixed powder from HiMedia.

Preparation of Medium: Add components to distilled/deionized water and bring volume to 1.0L. Mix thoroughly. Gently heat and bring to boiling. Distribute into tubes or flasks. Autoclave for 15 min at 15 psi pressure–121°C. Pour into sterile Petri dishes or leave in tubes.

Use: For the cultivation and maintenance of bacteria for genetic and molecular studies. For the cultivation and maintenance of recombinant strains of *Escherichia coli* for genetic studies.

Luria Bertani HiVeg Broth, Miller

Composition per liter:

Plant hydrolysate	10.0g
NaCl	10.0g
Yeast extract	5.0g

pH 7.5 ± 0.2 at 25°C

Source: This medium is available as a premixed powder from HiMedia.

Preparation of Medium: Add components to distilled/deionized water and bring volume to 1.0L. Mix thoroughly. Gently heat and bring to boiling. Distribute into tubes or flasks. Autoclave for 15 min at 15 psi pressure–121°C.

Use: For the cultivation of bacteria for genetic and molecular studies. For the cultivation and maintenance of recombinant strains of *Escherichia coli* for genetic studies.

Luria Bertani Medium

See: **LB Medium**

Luria Broth

See: **L Broth**
See: **LB Medium**

Luria Broth, HiVeg

Composition per liter:

Plant hydrolysate	10.0g
NaCl	5.0g
Yeast extract	5.0g

pH 7.0 ± 0.2 at 25°C

Source: This medium is available as a premixed powder from HiMedia.

Preparation of Medium: Add components to distilled/deionized water and bring volume to 1.0L. Mix thoroughly. Distribute into tubes or flasks. Autoclave for 15 min at 15 psi pressure–121°C.

Use: For the cultivation of *Escherichia coli.*

Luria HiVeg Agar

Composition per liter:

Agar	15.0g
Plant hydrolysate	10.0g
NaCl	5.0g
Yeast extract	5.0g

pH 7.0 ± 0.2 at 25°C

Source: This medium is available as a premixed powder from HiMedia.

Preparation of Medium: Add components to distilled/deionized water and bring volume to 1.0L. Mix thoroughly. Distribute into tubes or flasks. Autoclave for 15 min at 15 psi pressure–121°C. Pour into sterile Petri dishes.

Use: For the routine cultivation and estimation of not particularly fastidious microorganisms.

Luria HiVeg Agar with Glucose

Composition per liter:

Agar	15.0g
Plant hydrolysate	10.0g
NaCl	5.0g
Yeast extract	5.0g
Glucose solution	20.0mL

pH 7.0 ± 0.2 at 25°C

Source: This medium, without glucose solution, is available as a premixed powder from HiMedia.

Glucose Solution:

Composition per 100.0mL:

Glucose	10.0g

Preparation of Glucose Solution: Add glucose to distilled/deionized water and bring volume to 100.0mL. Mix thoroughly. Filter sterilize.

Preparation of Medium: Add components, except glucose solution, to distilled/deionized water and bring volume to 980.0mL. Mix thoroughly. Bring pH to 7.0. Gently heat and bring to boiling. Autoclave for 15 min at 15 psi pressure–121°C. Aseptically add 20.0mL of sterile glucose solution. Mix thoroughly. Pour into sterile Petri dishes or leave in tubes.

Use: For the cultivation of *Escherichia coli.*

LuTria3 Medium (DSMZ Medium 298d)

Composition per liter:

NaCl	1.0g
KCl	0.5g

$MgCl_2 \cdot 6H_2O$	0.4g
NH_4Cl	0.25g
KH_2PO_4	0.2g
$CaCl_2 \cdot 2H_2O$	0.15g
Resazurin	1.0mg
$NaHCO_3$ solution	10.0mL
Butanediol solution	10.0mL
$Na_2S \cdot 9H_2O$ solution	10.0mL
Seven vitamin solution	10.0mL
Triethanolamine hydrochloride solution	10.0mL
Trace elements solution SL-10	1.0mL

pH 7.2 ± 0.2 at 25°C

$Na_2S \cdot 9H_2O$ Solution:

Composition per 10.0mL:

$Na_2S \cdot 9H_2O$	0.36g

Preparation of $Na_2S \cdot 9H_2O$ Solution: Add $Na_2S \cdot 9H_2O$ to distilled/deionized water and bring volume to 10.0mL. Mix thoroughly. Autoclave under 100% N_2 for 15 min at 15 psi pressure–121°C. Cool to room temperature.

$NaHCO_3$ Solution:

Composition per 10.0mL:

$NaHCO_3$	2.5g

Preparation of $NaHCO_3$ Solution: Add $NaHCO_3$ to distilled/deionized water and bring volume to 10.0mL. Mix thoroughly. Sparge with 80% N_2 + 20% CO_2. Filter sterilize.

Butanediol Solution:

Composition per 10.0mL:

2,3 butanediol	0.9g

Preparation of Butanediol Solution: Add butanediol to distilled/deionized water and bring volume to 10.0mL. Mix thoroughly. Sparge with 100% N_2. Filter sterilize.

Seven Vitamin Solution:

Composition per liter:

Pyridoxine hydrochloride	300.0mg
Thiamine-HCl·$2H_2O$	200.0mg
Nicotinic acid	200.0mg
Vitamin B_{12}	100.0mg
Calcium pantothenate	100.0mg
p-Aminobenzoic acid	80.0mg
D(+)-Biotin	20.0mg

Preparation of Seven Vitamin Solution: Add components to distilled/deionized water and bring volume to 1.0L. Sparge with 100% N_2. Mix thoroughly. Filter sterilize.

Trace Elements Solution SL-10:

Composition per liter:

$FeCl_2 \cdot 4H_2O$	1.5g
$CoCl_2 \cdot 6H_2O$	190.0mg
$MnCl_2 \cdot 4H_2O$	100.0mg
$ZnCl_2$	70.0mg
$Na_2MoO_4 \cdot 2H_2O$	36.0mg
$NiCl_2 \cdot 6H_2O$	24.0mg
H_3BO_3	6.0mg
$CuCl_2 \cdot 2H_2O$	2.0mg
HCl (25% solution)	10.0mL

Preparation of Trace Elements Solution SL-10: Add $FeCl_2 \cdot 4H_2O$ to 10.0mL of HCl solution. Mix thoroughly. Add distilled/deionized water and bring volume to 1.0L. Add remaining components. Mix thoroughly. Sparge with 80% N_2 + 20% CO_2. Autoclave for 15 min at 15 psi pressure–121°C.

Triethanolamine Hydrochloride Solution:

Composition per 10.0mL:

Triethanolamine hydrochloride	1.4g

Preparation of Triethanolamine Hydrochloride Solution: Add triethanolamine hydrochloride to distilled/deionized water and bring volume to 10.0mL. Mix thoroughly. Adjust pH to 7.0. Sparge with 100% N_2. Autoclave under 100% N_2 for 15 min at 15 psi pressure–121°C. Cool to room temperature.

Preparation of Medium: Prepare and dispense medium under 80% N_2 + 20% CO_2 gas atmosphere. Add components, except $NaHCO_3$ solution, butanediol solution, $Na_2S \cdot 9H_2O$ solution, seven vitamin solution, triethanolamine hydrochloride solution, and trace elements solution SL-10, to distilled/deionized water and bring volume to 949.0mL. Mix thoroughly. Adjust pH to 7.2. Sparge with 80% N_2 + 20% CO_2. Autoclave for 15 min at 15 psi pressure–121°C. Aseptically and anaerobically add 10.0mL $NaHCO_3$ solution, 10.0mL butanediol solution, 10.0mL $Na_2S \cdot 9H_2O$ solution, 10.0mL seven vitamin solution, 10.0mL triethanolamine hydrochloride solution, and 1.0mL trace elements solution SL-10. Mix thoroughly. Aseptically and anaerobically distribute into sterile tubes or bottles. After inoculation, flush and repressurize the gas head space of culture bottles with sterile 80% N_2 + 20% CO_2 to 1 bar overpressure.

Use: For the cultivation of *Acetobacterium* sp.

LY Agar for *Filobasidium*

Composition per liter:

Agar	15.0g
Lactose	1.0g
Yeast extract	0.5g

Preparation of Medium: Add components to distilled/deionized water and bring volume to 1.0L. Mix thoroughly. Gently heat and bring to boiling. Distribute into tubes or flasks. Autoclave for 15 min at 15 psi pressure–121°C. Pour into sterile Petri dishes or leave in tubes.

Use: For the cultivation and maintenance of *Filobasidium floriforme*.

Lyngby Iron Agar

See: **Iron Agar, Lyngby**

Lysine Agar, Selective

Composition per liter:

Agar	15.0g
L-Lysine	10.0g
Peptone	5.0g
Glucose	3.5g
Yeast extract	3.0g
Bile salts mixture	1.5g
Sulfapyridine	0.3g
Bromcresol Purple	0.03g
Crystal Violet	0.001g

pH 6.8 ± 0.1 at 25°C

Source: This medium is available as a premixed powder from BD Diagnostic Systems.

Preparation of Medium: Add components to distilled/deionized water and bring volume to 1.0L. Mix thoroughly. Gently heat and bring to boiling. Distribute into tubes or flasks. Autoclave for 15 min at 15 psi pressure–121°C. Pour into sterile Petri dishes or leave in tubes.

Use: For the selective isolation and cultivation of *Salmonella* species from food by the hydrophobic grid membrane filter method.

Lysine Arginine Iron Agar

Composition per liter:

Agar	15.0g
L-Arginine	10.0g
L-Lysine	10.0g
Peptone	5.0g
Yeast extract	3.0g
Glucose	1.0g
Ferric ammonium citrate	0.5g
Sodium thiosulfate	0.04g
Bromcresol Purple	0.02g

pH 6.8 ± 0.2 at 25°C

Preparation of Medium: Add components to distilled/deionized water and bring volume to 1.0L. Mix thoroughly. Gently heat and bring to boiling. Adjust pH to 6.8. Distribute into screw-capped tubes in 5.0mL volumes. Autoclave for 12 min at 15 psi pressure–121°C. Allow tubes to cool in a slanted position.

Use: For the cultivation and differentiation of bacteria based on their ability to decarboxylate lysine, decarboxylate arginine, and produce H_2S. Bacteria that decarboxylate lysine or arginine turn the medium purple. Bacteria that produce H_2S appear as black colonies.

Lysine Arginine Iron HiVeg Agar

Composition per liter:

Agar	15.0g
L-Arginine	10.0g
L-Lysine	10.0g
Plant peptone	5.0g
Yeast extract	3.0g
Glucose	1.0g
Ferric ammonium citrate	0.5g
$Na_2S_2O_3$	0.04g
Bromcresol Purple	0.02g

pH 6.8 ± 0.2 at 25°C

Source: This medium is available as a premixed powder from HiMedia.

Preparation of Medium: Add components to distilled/deionized water and bring volume to 1.0L. Mix thoroughly. Gently heat and bring to boiling. Adjust pH to 6.8. Distribute into screw-capped tubes in 5.0mL volumes. Autoclave for 12 min at 15 psi pressure–121°C. Allow tubes to cool in a slanted position.

Use: For the cultivation and differentiation of bacteria based on their ability to decarboxylate lysine, decarboxylate arginine, and produce H_2S. Bacteria that decarboxylate lysine or arginine turn the medium purple.

Lysine Broth Falkow with Sodium Chloride (BAM M44)

Composition per liter:

L-Lysine	5.0g
Peptone or gelysate	5.0g
Yeast extract	3.0g
Glucose	1.0g
Bromcresol Purple	0.02g

pH 6.5 ± 0.2 at 25°C

Preparation of Medium: Add components to distilled/deionized water and bring volume to 1.0L. Mix thoroughly. Adjust pH so that is will be 6.5 ± 0.2 after sterilization. Distribute into 16 × 150mm screw-capped tubes in 5.0mL volumes. Autoclave medium with loosely capped tubes for 10 min at 15 psi pressure–121°C. Screw the caps on tightly for storage and after inoculation.

Use: For the cultivation and differentiation of *Vibrio* spp. based on their ability to decarboxylate the amino acid lysine. Bacteria that decarboxylate lysine turn the medium turbid purple.

Lysine Decarboxylase Broth, Falkow

Composition per liter:

Peptone	5.0g
L-Lysine	5.0g
Yeast extract	3.0g
Glucose	1.0g
Bromcresol Purple	0.02g

pH 6.5–6.8 at 25°C

Preparation of Medium: Add components to distilled/deionized water and bring volume to 1.0L. Mix thoroughly. Gently heat and bring to boiling. Adjust pH to 6.5–6.8. Distribute into tubes in 5.0mL volumes. Autoclave for 15 min at 15 psi pressure–121°C.

Use: For the cultivation and differentiation of bacteria, especially *Salmonella*, based on their ability to decarboxylate lysine. Bacteria that decarboxylate lysine turn the medium turbid purple.

Lysine Decarboxylase Broth, Taylor Modification

Composition per liter:

L-Lysine	5.0g
Yeast extract	3.0g
Glucose	1.0g
Bromcresol Purple	0.02g

pH 6.1 ± 0.2 at 25°C

Source: This medium is available as a premixed powder from Oxoid Unipath.

Preparation of Medium: Add components to distilled/deionized water and bring volume to 1.0L. Mix thoroughly. Gently heat and bring to boiling. Adjust pH to 6.1. Distribute into tubes in 5.0mL volumes. Autoclave for 15 min at 15 psi pressure–121°C.

Use: For the cultivation and differentiation of bacteria, especially *Salmonella*, based on their ability to decarboxylate lysine. Bacteria that decarboxylate lysine turn the medium turbid purple.

Lysine Decarboxylase Broth, Taylor Modification (Lysine Decarboxylase Broth)

Composition per liter:

L-Lysine	5.0g
Peptone	5.0g
Yeast extract	3.0g
Glucose	1.0g
Bromcresol Purple	0.02g

pH 6.8 ± 0.2 at 25°C

Source: This medium is available as a premixed powder from BD Diagnostic Systems.

Preparation of Medium: Add components to distilled/deionized water and bring volume to 1.0L. Mix thoroughly. Gently heat and bring

to boiling. Adjust pH to 6.1. Distribute into tubes in 5.0mL volumes. Autoclave for 15 min at 15 psi pressure–121°C.

Use: For the cultivation and differentiation of bacteria, especially *Salmonella*, based on their ability to decarboxylate lysine. Bacteria that decarboxylate lysine turn the medium turbid purple.

Lysine Decarboxylase HiVeg Broth

Composition per liter:

Plant peptone	5.0g
L-Lysine hydrochloride	5.0g
Yeast extract	3.0g
Glucose	1.0g
Bromcresol Purple	0.02g

pH 6.8 ± 0.2 at 25°C

Source: This medium is available as a premixed powder from HiMedia.

Preparation of Medium: Add components to distilled/deionized water and bring volume to 1.0L. Mix thoroughly. Gently heat and bring to boiling. Autoclave for 15 min at 15 psi pressure–121°C. Aseptically distribute into sterile tubes in 1.0mL volumes.

Use: For the cultivation and differentiation of Gram-negative, nonfermentative bacteria based on their ability to decarboxylate lysine. Bacteria that decarboxylate lysine turn the medium purple.

Lysine Decarboxylase Medium

Composition per liter:

Glucose	0.5g
KH_2PO_4	0.5g
L-Lysine·HCl	0.5g

pH 4.6 ± 0.2 at 25°C

Preparation of Medium: Add components to distilled/deionized water and bring volume to 1.0L. Mix thoroughly. Gently heat and bring to boiling. Adjust pH to 4.6. Autoclave for 15 min at 15 psi pressure–121°C. Aseptically distribute into sterile tubes in 1.0mL volumes.

Use: For the cultivation and differentiation of Gram-negative, nonfermentative bacteria based on their ability to decarboxylate lysine. Bacteria that decarboxylate lysine turn the medium turbid purple.

Lysine Iron Agar

Composition per liter:

Agar	13.5g
L-Lysine	10.0g
Pancreatic digest of gelatin	5.0g
Yeast extract	3.0g
Glucose	1.0g
Ferric ammonium citrate	0.5g
$Na_2S_2O_3 \cdot 5H_2O$	0.04g
Bromcresol Purple	0.02g

pH 6.7 ± 0.2 at 25°C

Source: This medium is available as a premixed powder from BD Diagnostic Systems and Oxoid Unipath.

Preparation of Medium: Add components to distilled/deionized water and bring volume to 1.0L. Mix thoroughly. Gently heat while stirring and bring to boiling. Distribute into tubes in 10.0mL volumes. Autoclave for 12 min at 15 psi pressure–121°C. Allow tubes to cool in a slanted position.

Use: For the cultivation and differentiation of members of the Enterobacteriaceae based on their ability to decarboxylate lysine and to form H_2S. Bacteria that decarboxylate lysine turn the medium purple. Bacteria that produce H_2S appear as black colonies.

Lysine Iron Cystine HiVeg Broth Base with Novobiocin

Composition per liter:

L-Lysine hydrochloride	10.0g
Plant hydrolysate	5.0g
Mannitol	5.0g
Yeast extract	3.0g
Glucose	1.0g
Salicin	1.0g
Ferric ammonium citrate	0.5g
$Na_2S_2O_3$	0.1g
L-Cystine	0.1g
Neutral Red	0.025g
Novobiocin solution	10.0mL

pH 6.2 ± 0.2 at 25°C

Source: This medium, without novobiocin solution, is available as a premixed powder from HiMedia.

Novobiocin Solution:

Composition per 10.0mL:

Novobiocin	0.015g

Preparation of Novobiocin Solution: Add novobiocin to distilled/deionized water and bring volume to 10.0mL. Mix thoroughly. Filter sterilize.

Preparation of Medium: Add components, except novobiocin solution, to distilled/deionized water and bring volume to 990.0mL. Mix thoroughly. Gently heat and bring to boiling. Continue boiling for 2–3 min. Distribute into tubes in 10.0mL volumes. Autoclave for 15 min at 15 psi pressure–121°C. Cool to 45°–50°C. Aseptically add 0.1mL of sterile novobiocin solution to each tube. Mix thoroughly.

Use: For the rapid, presumptive detection of *Salmonella* in foods, food ingredients, and feed materials.

Lysine Iron Cystine Neutral Red Broth

See: **LICNR Broth**

Lysine Iron HiVeg Agar

Composition per liter:

Agar	15.0g
L-Lysine	10.0g
Plant peptone	5.0g
Yeast extract	3.0g
Glucose	1.0g
Ferric ammonium citrate	0.5g
$Na_2S_2O_3$	0.04g
Bromcresol Purple	0.02g

pH 6.7 ± 0.2 at 25°C

Source: This medium is available as a premixed powder from HiMedia.

Preparation of Medium: Add components to distilled/deionized water and bring volume to 1.0L. Mix thoroughly. Gently heat while stirring and bring to boiling. Distribute into tubes in 10.0mL volumes. Autoclave for 12 min at 15 psi pressure–121°C. Allow tubes to cool in a slanted position.

Use: For the cultivation and differentiation of members of the Enterobacteriaceae based on their ability to decarboxylate lysine and to form H_2S. Bacteria that decarboxylate lysine turn the medium purple. Bacteria that produce H_2S appear as black colonies. For the differentiation of enteric organisms, especially *Salmonella* serotype Arizona.

Lysine Lactose HiVeg Broth

Composition per liter:

Lactose	10.0g
Plant peptone No. 2	5.0g
L-Lysine	5.0g
Yeast extract	3.0g
Glucose	1.0g
Bromcresol Purple	0.02g

pH 6.8 ± 0.2 at 25°C

Source: This medium is available as a premixed powder from HiMedia.

Preparation of Medium: Add components to distilled/deionized water and bring volume to 1.0L. Mix thoroughly. Gently heat and bring to boiling. Adjust pH to 6.8. Autoclave for 15 min at 15 psi pressure–121°C. Aseptically distribute into sterile tubes in 1.0mL volumes.

Use: For the determination of lysine decarboxylase activity of lactose nonfermenting members of Enterobacteriaceae, especially *Salmonella* species.

Lysine Medium

Composition per liter:

Glucose	44.5g
Agar	17.8g
KH_2PO_4	1.78g
Lysine	1.0g
$MgSO_4 \cdot 7H_2O$	0.89g
$CaCl_2 \cdot 2H_2O$	0.178g
NaCl	0.089g
Inositol	0.02g
Calcium pantothenate	2.0mg
Adenine	1.78mg
DL-Methionine	0.89mg
L-Histidine	0.89mg
DL-Tryptophan	0.89mg
Thiamine·HCl	0.4mg
Pyridoxine	0.4mg
Nicotinic acid	0.4mg
$FeSO_4 \cdot 7H_2O$	0.22mg
p-Aminobenzoic acid	0.2mg
Riboflavin	0.2mg
$MnSO_4 \cdot H_2O$	0.035mg
$ZnSO_4 \cdot 7H_2O$	0.035mg
$(NH_4)_2MoO_4 \cdot 4H_2O$	0.018mg
H_3BO_3	8.9μg
Biotin	2.0μg
Folic acid	1.0μg
Potassium lactate (50% solution)	10.0mL
Lactic acid	1.0mL

pH 4.8 ± 0.2 at 25°C

Source: This medium is available as a premixed powder from Oxoid Unipath.

Preparation of Medium: Add potassium lactate to distilled/deionized water and bring volume to 900.0mL. Mix thoroughly. Add remaining components, except lactic acid. Mix thoroughly. Gently heat while stirring and bring to boiling. Do not autoclave. Cool to 50°C. Adjust pH to 4.8 by adding 1.0mL of lactic acid. Pour into sterile Petri dishes. Dry the surface of the plates by incubation at 37°C for 24 hr.

Use: For the isolation, cultivation, and enumeration of wild yeasts encountered in brewing.

Lysine Ornithine Mannitol Agar (LOM Agar)

Composition per liter:

Agar	13.5g
L–Ornithine·HCl	6.5g
D–Mannitol	5.25g
L–Lysine·HCl	5.0g
NaCl	5.0g
Yeast extract	3.0g
Bromthymol Blue	0.3g
Vancomycin solution	10.0mL

pH 6.5 ± 0.2 at 25°C

Vancomycin Solution:

Composition per 10.0mL:

Vancomycin·HCl	0.03g

Preparation of Vancomycin Solution: Add vancomycin to distilled/deionized water and bring volume to 10.0mL. Mix thoroughly. Filter sterilize.

Preparation of Medium: Add components, except vancomycin solution, to distilled/deionized water and bring volume to 990.0mL. Mix thoroughly. Gently heat and bring to boiling. Autoclave for 15 min at 15 psi pressure–121°C. Cool to 45°–50°C. Aseptically add sterile vancomycin solution. Mix thoroughly. Pour into sterile Petri dishes or distribute into sterile tubes.

Use: For the cultivation and differentiation of Gram-negative bacilli based on their ability to decarboxylate lysine or ornithine and mannitol fermentation. Especially useful for the identification of *Enterobacter agglomerans*. Bacteria that ferment mannitol appear as dark yellow colonies. Bacteria that decarboxylate lysine or ornithine appear as green-yellow colonies.

Lysobacter deserti Medium (DSMZ Medium 1060)

Composition per liter:

Solution A	700.0mL
Solution B	300.0mL

pH 8.1 ± 0.2 at 25°C

Solution A:

Composition per 700.0mL:

Agar	15.0g
Glucose	10.0g
Peptone	5.0g
Yeast extract	5.0g
K_2HPO_4	1.0g
$MgSO_4 \cdot 7H_2O$	0.2g
$CaCl_2 \cdot 2H_2O$	40.0g

Preparation of Solution A: Add components to distilled/deionized water and bring volume to 700.0mL. Mix thoroughly. Autoclave for 15 min at 15 psi pressure–121°C. Cool to 50°C.

Solution B:

Composition per 300.0mL:

NaCl	20.0g
Na_2CO_3	1.0g

Preparation of Solution B: Add components to distilled/deionized water and bring volume to 300.0mL. Mix thoroughly. Autoclave for 15 min at 15 psi pressure–121°C. Cool to 50°C.

Preparation of Medium: Aseptically combine 700.0mL sterile solution A and 300.0mL sterile solution B. Mix thoroughly. Adjust pH to 8.1. Pour into Petri dishes or distribute into sterile tubes.

Use: For the cultivation of *Lysobacter deserti.*

Lysozyme Broth

Composition per 1005.0mL:

Basal glycerol broth	1.0L
Lysozyme solution	5.0mL

Basal Glycerol Broth:

Composition per liter:

Peptone	5.0g
Beef extract	3.0g
Glycerol	70.0mL

Preparation of Basal Glycerol Broth: Add components to distilled/deionized water and bring volume to 1.0L. Mix thoroughly. Distribute 500.0mL of the broth into screw-capped tubes in 5.0mL volumes. Autoclave the tubes and the flask with the remaining broth for 15 min at 15 psi pressure–121°C. Cool to 25°C.

Lysozyme Solution:

Composition per 100.0mL:

Lysozyme	0.1g
HCl (0.01*N* solution)	100.0mL

Preparation of Lysozyme Solution: Add lysozyme to 100.0mL of HCl solution. Mix thoroughly. Filter sterilize. Store for up to 1 week at 4°C.

Preparation of Medium: Add 5.0mL of the sterile lysozyme solution to 95.0mL of the cooled, sterile basal glycerol broth. Mix thoroughly. Aseptically distribute into sterile screw-capped tubes in 5.0mL volumes.

Use: For the cultivation and differentiation of *Nocardia asteroides, Streptomyces griseus,* and *Actinomadura madurae* based on sensitivity to lysozyme. *Nocardia asteroides* grows well in both the basal glycerol broth and the lysozyme broth. *Actinomadura madurae* and *Streptomyces griseus* grow well in the basal glycerol broth but not in the lysozyme broth.

Lysozyme Broth

Composition per 1010.0mL:

Nutrient broth	1.0L
Lysozyme solution	10.0mL

pH 6.9 ± 0.2 at 25°C

Nutrient Broth:

Composition per liter:

Pancreatic digest of gelatin	5.0g
Beef extract	3.0g

Source: Nutrient broth is available as a premixed powder from BD Diagnostic Systems.

Preparation of Nutrient Broth: Add components to distilled/deionized water and bring volume to 1.0L. Mix thoroughly. Distribute into bottles in 99.0mL volumes. Autoclave for 15 min at 15 psi pressure–121°C. Cool to 25°C.

Lysozyme Solution:

Composition per 100.0mL:

Lysozyme	0.1g

Preparation of Lysozyme Solution: Add lysozyme to distilled/deionized water and bring volume to 100.0mL. Mix thoroughly. Filter sterilize.

Preparation of Medium: Add 1.0mL of sterile lysozyme solution to 99.0mL of cooled, sterile nutrient broth. Mix thoroughly. Aseptically distribute into sterile tubes in 2.5mL volumes.

Use: For the cultivation and differentiation of *Bacillus cereus* in foods. *Bacillus cereus* is resistant to lysozyme and will grow in this medium.

M Broth

Composition per liter:

Pancreatic digest of casein	12.5g
K_2HPO_4	5.0g
NaCl	5.0g
Sodium citrate	5.0g
Yeast extract	5.0g
Mannose	2.0g
$MgSO_4 \cdot 7H_2O$	0.8g
Polysorbate 80	0.75g
$FeSO_4$	0.04g

pH 7.0 ± 0.22 at 25°C

Source: Available as a premixed powder from BD Diagnostic Systems.

Preparation of Medium: Add components to distilled/deionized water and bring volume to 1.0L. Mix thoroughly. Distribute into tubes or flasks. Autoclave for 15 min at 15 psi pressure–121°C.

Use: For the detection of *Salmonella* in dried foods and feeds.

M Medium

Composition per liter:

Beef	5.0g
Neopeptone	4.0g
NaCl	1.6g
Glucose	0.5g
$CaCl_2$	0.06g
KH_2PO_4	0.06g
KCl	0.04g
Rabbit blood solution	200.0mL

Rabbit Blood Solution:

Composition per liter:

Rabbit blood	500.0mL

Preparation of Rabbit Blood Solution: Add 500.0mL of whole rabbit blood to 500.0mL of sterile distilled/deionized water. Freeze and thaw twice to lyse blood cells.

Preparation of Medium: Trim beef to remove fat. Add 5.0g of lean beef to 200.0mL of distilled/deionized water. Gently heat and bring to boiling. Boil for 2–3 min. Filter through Whatman #2 filter paper. Add other components, except rabbit blood solution. Bring volume to 800.0mL with distilled/deionized water. Mix thoroughly. Adjust pH to 7.2 with 1*N* NaOH. Autoclave for 15 min at 15 psi pressure–121°C. Cool to 50°–55°C. Aseptically add 200.0mL of lysed rabbit blood solution. Mix thoroughly. Aseptically distribute into sterile tubes or flasks.

Use: For the cultivation of *Herpetomonas megaseliae*.

M1 Medium

Composition per liter:

L-Leucine	2.0g
L-Alanine	1.0g
L-Isoleucine	1.0g
L-Phenylalanine	1.0g
L-Proline	1.0g
L-Tryptophane	1.0g
L-Asparagine	0.5g
L-Lysine	0.5g
L-Methionine	0.5g
L-Tyrosine	0.4g
L-Valine	0.2g
L-Serine	0.2g
$MgSO_4 \cdot 7H_2O$	0.2g
NaCl	0.2g
KH_2PO_4	0.14g
L-Arginine	0.1g
L-Cysteine	0.1g
L-Glycine	0.1g
L-Histidine	0.1g
L-Threonine	0.1g
$CaCl_2$	2.0mg
$FeCl_3 \cdot 6H_2O$	2.0mg
Tris(hydroxymethyl)aminomethane buffer (0.01*M* solution, pH 7.6)	1.0L

pH 7.6 ± 0.2 at 25°C

Preparation of Medium: Add solid components to 1.0L of Tris buffer. Mix thoroughly. Filter sterilize. Aseptically distribute into tubes or flasks.

Use: For the cultivation of *Myxococcus xanthus*.

M1-*Nocardiopsis arabia* Medium (DSMZ Medium 1065)

Composition per liter:

NaCl	20.0g
Agar	18.0 g
Starch	10.0g
Yeast extract	4.0g
Peptone	2.0g
Seawater	1.0L

pH 7.2 ± 0.2 at 25°C

Preparation of Medium: Add components to seawater and bring volume to 1.0L. Mix thoroughly. Adjust pH to 7.2. Distribute into tubes or flasks. Gently heat while stirring and bring to boiling. Mix thoroughly. Autoclave for 15 min at 15 psi pressure–121°C. Pour into Petri dishes or leave in tubes.

Use: For the cultivation of *Nocardiopsis arabia*.

M1A Medium

Composition per 1001.0mL:

Sorbitol	23.3g
Peptone	6.0g
Sucrose	3.3g
Pancreatic digest of casein	3.3g
Beef heart infusion	2.0g
Glucose	1.3g
Yeast extract	1.0g
Fructose	0.3g
Phenol Red	20.0mg
Schneider's *Drosophila* medium	533.0mL
Fetal calf serum, heat inactivated	167.0mL
Fresh yeast extract solution	33.0mL
Penicillin solution	8.0mL

Schneider's *Drosophila* Medium:

Composition per liter:

$MgSO_4 \cdot 7H_2O$	3.7g
NaCl	2.1g
Yeast extract	2.0g
Trehalose	2.0g
D-Glucose	2.0g
L-Glutamine	1.8g
L-Lysine·HCl	1.7g
L-Proline	1.7g
KCl	1.6g
$Na_2HPO_4 \cdot 7H_2O$	1.3g
L-Glutamic acid	0.8g
L-Methionine	0.8g
$CaCl_2$, anhydrous	0.6g
KH_2PO_4	0.5g
β-Alanine	0.5g
L-Tyrosine	0.5g
L-Arginine	0.4g
L-Aspartic acid	0.4g
L-Histidine	0.4g
L-Threonine	0.4g
$NaHCO_3$	0.4g
Glycine	0.3g
L-Serine	0.3g
L-Valine	0.3g
L-Isoleucine	0.2g
L-Leucine	0.2g
L-Phenylalanine	0.2g
α-Ketoglutaric acid	0.2g
Fumaric acid	0.1g
Malic acid	0.1g
Succinic acid	0.1g
L-Cystine	0.1g
L-Tryptophan	0.1g
L-Cysteine	0.06g

Preparation of Schneider's *Drosophila* Medium: Add components to distilled/deionized water and bring volume to 1.0L. Mix thoroughly. Filter sterilize.

Penicillin Solution:

Composition per 10.0mL:

Penicillin	2,500,000U

Preparation of Penicillin Solution: Add penicillin to distilled/deionized water and bring volume to 10.0mL. Filter sterilize.

Fresh Yeast Extract Solution:

Composition per 100.0mL:

Baker's yeast, live, pressed, starch-free	25.0g

Preparation of Fresh Yeast Extract Solution: Add the live Baker's yeast to 100.0mL of distilled/deionized water. Autoclave for 90 min at 15 psi pressure–121°C. Allow to stand. Remove supernatant solution. Adjust pH to 6.6–6.8. Filter sterilize.

Preparation of Medium: Add components—except Schneider's *Drosophila* medium, fetal calf serum, fresh yeast extract solution, and penicillin solution— to distilled/deionized water and bring volume to 260.0mL. Mix thoroughly. Gently heat and bring to boiling. Autoclave for 15 min at 15 psi pressure–121°C. Cool to 45°–50°C. Aseptically add 533.0mL of sterile Schneider's *Drosophila* medium, 167.0mL of sterile fetal calf serum, 33.0mL of sterile fresh yeast extract solution, and 8.0mL of sterile penicillin solution. Mix thoroughly. Pour into sterile Petri dishes or distribute into sterile tubes.

Use: For the isolation and cultivation of *Spiroplasma* species that cause corn stunt.

M3 Agar

Composition per 1020.0mL:

Agar	18.0g
Na_2HPO_4	0.732g
KH_2PO_4	0.466g
NaCl	0.29g
Sodium propionate	0.2g
$MgSO_4 \cdot 7H_2O$	0.1g
$CaCO_3$	0.02g
KNO_3	0.01g
$FeSO_4 \cdot 7H_2O$	0.2mg
$ZnSO_4 \cdot 7H_2O$	0.18mg
$MnSO_4 \cdot 4H_2O$	0.02mg
Cycloheximide solution	10.0mL
Thiamine·HCl solution	10.0mL

pH 7.0 ± 0.2 at 25°C

Cycloheximide Solution:
Composition per 10.0mL:

Cycloheximide	0.05g

Preparation of Cycloheximide Solution: Add cycloheximide to distilled/deionized water and bring volume to 10.0mL. Mix thoroughly. Filter sterilize.

Caution: Cycloheximide is toxic. Avoid skin contact or aerosol formation and inhalation.

Thiamine·HCl Solution:
Composition per 10.0mL:

Thiamine·HCl	4.0mg

Preparation of Thiamine·HCl Solution: Add thiamine·HCl to distilled/deionized water and bring volume to 10.0mL. Mix thoroughly. Filter sterilize.

Preparation of Medium: Add components, except cycloheximide solution and thiamine·HCl solution, to distilled/deionized water and bring volume to 980.0mL. Mix thoroughly. Gently heat and bring to boiling. Autoclave for 15 min at 15 psi pressure–121°C. Cool to 45°–50°C. Aseptically add 10.0mL of sterile cycloheximide solution and 10.0mL of thiamine·HCl solution. Mix thoroughly. Pour into sterile Petri dishes or distribute into sterile tubes.

Use: For the selective isolation and cultivation of *Nocardia* species and *Rhodococcus* species.

M3 Agar Medium

Composition per liter:

Agar	18.0g
Na_2HPO_4	0.732g
KH_2PO_4	0.466g
NaCl	0.29g
Sodium propionate	0.2g
KNO_3	0.1g
$MgSO_4 \cdot 7H_2O$	0.1g
$CaCO_3$	0.02g
Thiamine·HCl	4.0mg
$FeSO_4 \cdot 7H_2O$	0.2mg
$ZnSO_4 \cdot 7H_2O$	0.18mg
$MnSO_4 \cdot 4H_2O$	0.02mg
Cycloheximide solution	10.0mL
Thiamine·HCl solution	10.0mL

pH 7.0 ± 0.2 at 25°C

Cycloheximide Solution:
Composition per 10.0mL:

Cycloheximide	0.04g

Preparation of Cycloheximide Solution: Add cycloheximide to distilled/deionized water and bring volume to 10.0mL. Mix thoroughly. Filter sterilize.

Caution: Cycloheximide is toxic. Avoid skin contact or aerosol formation and inhalation.

Thiamine·HCl Solution:
Composition per 10.0mL:

Thiamine·HCl	0.04g

Preparation of Thiamine·HCl Solution: Add thiamine·HCl to distilled/deionized water and bring volume to 10.0mL. Mix thoroughly. Filter sterilize.

Preparation of Medium: Add components, except cycloheximide solution and thiamine·HCl solution, to distilled/deionized water and bring volume to 980.0mL. Mix thoroughly. Gently heat and bring to boiling. Autoclave for 15 min at 15 psi pressure–121°C. Cool to 45°–50°C. Aseptically add sterile cycloheximide solution and thiamine·HCl solution. Mix thoroughly. Pour into sterile Petri dishes or distribute into sterile tubes.

Use: For the cultivation of *Micromonospora* species.

M3 Chytrid Agar

Composition per liter:

Agar	15.0g
Soluble starch	5.0g
Glucose	5.0g
Corn meal, solids from infusion	2.0g
Peptone	1.0g
Yeast extract	1.0g

Preparation of Medium: Add components to distilled/deionized water and bring volume to 1.0L. Mix thoroughly. Gently heat and bring to boiling. Distribute into tubes or flasks. Autoclave for 15 min at 15 psi pressure–121°C. Pour into sterile Petri dishes or leave in tubes.

Use: For the cultivation and maintenance of *Rhyzophydium* species.

M7 Medium

Composition per liter:

Yeast extract solution	200.0mL
Dialyzed fetal bovine serum	100.0mL
L-Methionine solution	30.0mL
Buffer solution	20.0mL
Glucose solution	20.0mL

Yeast Extract Solution:
Composition per liter:

Yeast extract	25.0g

Preparation of Yeast Extract Solution: Add yeast extract to distilled/deionized water and bring volume to 1.0L. Mix thoroughly. Autoclave for 15 min at 15 psi pressure–121°C. Cool to 25°C.

L-Methionine Solution:
Composition per liter:

L-Methionine	1.5g

Preparation of L-Methionine Solution: Add L-methionine to distilled/deionized water and bring volume to 1.0L. Mix thoroughly. Autoclave for 15 min at 15 psi pressure–121°C. Cool to 25°C.

Glucose Solution:
Composition per liter:

Glucose	270.0g

Preparation of Glucose Solution: Add glucose to distilled/deionized water and bring volume to 1.0L. Mix thoroughly. Autoclave for 15 min at 15 psi pressure–121°C. Cool to 25°C.

Buffer Solution:
Composition per liter:

Na_2HPO_4	25.0g
KH_2PO_4	18.1g

Preparation of Buffer Solution: Add components to distilled/deionized water and bring volume to 1.0L. Mix thoroughly. Autoclave for 15 min at 15 psi pressure–121°C. Cool to 25°C.

Dialyzed Fetal Bovine Serum:
Composition per 100.0mL:

Fetal bovine serum, heat inactivated	100.0mL

Preparation of Dialyzed Fetal Bovine Serum: Dialyze the heat-inactivated serum at 0°–4°C against 10 volumes of water. Clean the dialysis tubing before use by boiling in a solution of 0.37g/L of EDTA and rinsing with water. Change the water four times at 8–16 hr intervals. Centrifuge the dialyzed serum for 30 min at 35,000 x g and filter sterilize.

Preparation of Medium: Aseptically combine the sterile solutions. Mix thoroughly. Bring volume to 1.0L with sterile distilled/deionized water.

Use: For the cultivation of *Naegleria fowleri*, *Naegleria gruberi*, and *Nuclearia* species.

M9 Medium

Composition per liter:

Na_2HPO_4	6.0g
KH_2PO_4	3.0g
NH_4Cl	1.0g
NaCl	0.5g
Glucose solution	10.0mL
$MgSO_4·7H_2O$ solution	1.0mL
Thiamine·HCl solution	1.0mL
$CaCl_2$ solution	1.0mL

pH 7.0 ± 0.2 at 25°C

Glucose Solution:
Composition per 100.0mL:

D-Glucose	20.0g

Preparation of Glucose Solution: Add glucose to distilled/deionized water and bring volume to 1.0L. Mix thoroughly. Autoclave for 15 min at 15 psi pressure–121°C.

$MgSO_4·7H_2O$ Solution:
Composition per liter:

$MgSO_4·7H_2O$	246.5g

Preparation of $MgSO_4·7H_2O$ Solution: Add $MgSO_4·7H_2O$ to distilled/deionized water and bring volume to 1.0L. Mix thoroughly. Autoclave for 15 min at 15 psi pressure–121°C.

Thiamine·HCl Solution:
Composition per 10.0mL:

Thiamine·HCl	10.0mg

Preparation of Thiamine·HCl Solution: Add thiamine·HCl to distilled/deionized water and bring volume to 1.0L. Mix thoroughly. Filter sterilize.

$CaCl_2$ Solution:
Composition per liter:

$CaCl_2$ solution	14.7g

Preparation of $CaCl_2$ Solution: Add $CaCl_2$ solution to distilled/deionized water and bring volume to 1.0L. Mix thoroughly. Autoclave for 15 min at 15 psi pressure–121°C.

Preparation of Medium: Add components, except $MgSO_4·7H_2O$ solution, glucose solution, thiamine·HCl solution, and $CaCl_2$ solution, to distilled/deionized water and bring volume to 987.0mL. Mix thoroughly. Adjust pH to 7.0. Autoclave for 15 min at 15 psi pressure–121°C. Cool to room temperature. Aseptically add sterile $MgSO_4·7H_2O$ solution, sterile glucose solution, sterile thiamine·HCl solution, and sterile $CaCl_2$ solution. Mix thoroughly. Distribute into tubes or flasks.

Use: For the cultivation and maintenance of *Escherichia coli* and a variety of other bacteria.

M9 Medium with Arginine

Composition per 1013.0mL:

L-Proline	20.0mg
L-Arginine	20.0mg
10X M9 salts	100.0mL
Glucose solution	10.0mL
$CaCl_2·2H_2O$ solution	1.0mL
$MgSO_4$ solution	1.0mL
Thiamine·HCl solution	1.0mL

pH 7.4 ± 0.2 at 25°C

10X M9 Salts:
Composition per liter:

Na_2HPO_4	60.0g
KH_2PO_4	30.0g
NH_4Cl	10.0g
NaCl	5.0g

Preparation of 10X M9 Salts: Add components to distilled/deionized water and bring volume to 1.0L. Mix thoroughly. Adjust pH to 7.4.

Glucose Solution:
Composition per 100.0mL:

Glucose	20.0g

Preparation of Glucose Solution: Add glucose to distilled/deionized water and bring volume to 100.0mL. Mix thoroughly. Filter sterilize.

$CaCl_2·2H_2O$ Solution:
Composition per 100.0mL:

$CaCl_2·2H_2O$	1.47g

Preparation of $CaCl_2 \cdot 2H_2O$ Solution: Add $CaCl_2 \cdot 2H_2O$ to distilled/deionized water and bring volume to 100.0mL. Mix thoroughly. Autoclave for 15 min at 15 psi pressure–121°C.

$MgSO_4$ Solution:
Composition per 100.0mL:

$MgSO_4$	12.04g

Preparation of $MgSO_4$ Solution: Add $MgSO_4$ to distilled/deionized water and bring volume to 100.0mL. Mix thoroughly. Autoclave for 15 min at 15 psi pressure–121°C.

Thiamine·HCl Solution:
Composition per 10.0mL:

Thiamine·HCl	3.37g

Preparation of Thiamine·HCl Solution: Add thiamine·HCl to distilled/deionized water and bring volume to 10.0mL. Mix thoroughly. Filter sterilize.

Preparation of Medium: Add L-proline, L-arginine, and 10X M9 salts to distilled/deionized water and bring volume to 1.0L. Mix thoroughly. Autoclave for 15 min at 15 psi pressure–121°C. Aseptically add 10.0mL of sterile glucose solution, 1.0mL of sterile $CaCl_2 \cdot 2H_2O$ solution, 1.0mL of sterile $MgSO_4$ solution, and 1.0mL of sterile thiamine·HCl solution. Mix thoroughly. Aseptically distribute into sterile tubes or flasks.

Use: For the cultivation of *Escherichia coli.*

M9 Medium with Arginine (DSMZ Medium 450)

Composition per liter:

10X M9 salts	100.0mL
Glucose solution	10.0mL
Calcium chloride solution	10.0mL
Magnesium sulfate solution	10.0mL
Thiamine solution	1.0mL
Proline solution	1.0mL
Arginine solution	1.0mL

pH 7.4 ± 0.2 at 25°C

Glucose Solution:
Composition per 10.0mL:

D-Glucose	2.0g

Preparation of Glucose Solution: Add glucose to distilled/deionized water and bring volume to 10.0mL. Mix thoroughly. Filter sterilize.

Thiamine Solution:
Composition per 10.0mL:

Thiamine−$HCl \cdot 2H_2O$	3.7g

Preparation of Thiamine Solution: Add thiamine−$HCl \cdot 2H_2O$ to distilled/deionized water and bring volume to 10.0mL. Mix thoroughly. Filter sterilize.

Arginine Solution:
Composition per 10.0mL:

Arginine	0.2g

Preparation of Arginine Solution: Add arginine to distilled/deionized water and bring volume to 10.0mL. Mix thoroughly. Filter sterilize.

Proline Solution:
Composition per 10.0mL:

Proline	0.2g

Preparation of Proline Solution: Add proline to distilled/deionized water and bring volume to 10.0mL. Mix thoroughly. Filter sterilize.

Calcium Chloride Solution:
Composition per 10.0mL:

$CaCl_2$	0.1g

Preparation of Calcium Chloride Solution: Add $CaCl_2$ to distilled/deionized water and bring volume to 10.0mL. Mix thoroughly. Filter sterilize.

Magnesium Sulfate Solution:
Composition per 10.0mL:

$MgSO_4$	1.2g

Preparation of Magnesium Sulfate Solution: Add $MgSO_4$ to distilled/deionized water and bring volume to 10.0mL. Mix thoroughly. Filter sterilize.

10X M9 Salts Solution:
Composition per liter:

Na_2HPO_4	60.0g
KH_2PO_4	30.0g
NH_4Cl	10.0g
NaCl	5.0g
$MgSO_4$	1.2g

Preparation of 10X M9 Salts Solution: Add components to distilled/deionized water and bring volume to 1.0L. Mix thoroughly. Filter sterilize.

Preparation of Medium: Aseptically combine sterile component solutions. Mix thoroughly. Aseptically distribute into sterile tubes or flasks.

Use: For the cultivation of *Escherichia coli* arginine auxotrophs.

M9 Medium with Casamino Acids

Composition per liter:

Na_2HPO_4	6.0g
Casamino acids	5.0g
KH_2PO_4	3.0g
NH_4Cl	1.0g
NaCl	0.5g
Glucose solution	10.0mL
$MgSO_4 \cdot 7H_2O$ solution	1.0mL
Thiamine·HCl solution	1.0mL
$CaCl_2$ solution	1.0mL

pH 7.0 ± 0.2 at 25°C

Glucose solution:
Composition per 100.0mL:

D-Glucose	20.0g

Preparation of Glucose Solution: Add glucose to distilled/deionized water and bring volume to 1.0L. Mix thoroughly. Autoclave for 15 min at 15 psi pressure–121°C.

$MgSO_4 \cdot 7H_2O$ Solution:
Composition per liter:

$MgSO_4 \cdot 7H_2O$	246.5g

Preparation of $MgSO_4 \cdot 7H_2O$ Solution: Add $MgSO_4 \cdot 7H_2O$ to distilled/deionized water and bring volume to 1.0L. Mix thoroughly. Autoclave for 15 min at 15 psi pressure–121°C.

Thiamine·HCl Solution:
Composition per 10.0mL:

Thiamine·HCl	10.0mg

Preparation of Thiamine·HCl Solution: Add thiamine·HCl to distilled/deionized water and bring volume to 1.0L. Mix thoroughly. Filter sterilize.

$CaCl_2$ Solution:
Composition per liter:

$CaCl_2$ solution	14.7g

Preparation of $CaCl_2$ Solution: Add $CaCl_2$ solution to distilled/deionized water and bring volume to 1.0L. Mix thoroughly. Autoclave for 15 min at 15 psi pressure–121°C.

Preparation of Medium: Add components, except $MgSO_4 \cdot 7H_2O$ solution, glucose solution, thiamine·HCl solution, and $CaCl_2$ solution, to distilled/deionized water and bring volume to 987.0mL. Mix thoroughly. Adjust pH to 7.0. Autoclave for 15 min at 15 psi pressure–121°C. Cool to room temperature. Aseptically add sterile $MgSO_4 \cdot 7H_2O$ solution, sterile glucose solution, sterile thiamine·HCl solution, and sterile $CaCl_2$ solution. Mix thoroughly. Distribute into tubes or flasks.

Use: For the cultivation and maintenance of *Flavobacterium meningosepticum.*

M9 Medium with Tryptophan

Composition per 1013.0mL:

L-Proline	20.0mg
L-Tryptophan	20.0mg
10X M9 salts	100.0mL
Glucose solution	10.0mL
$CaCl_2 \cdot 2H_2O$ solution	1.0mL
$MgSO_4$ solution	1.0mL
Thiamine·HCl solution	1.0mL

pH 7.4 ± 0.2 at 25°C

10X M9 Salts:
Composition per liter:

Na_2HPO_4	60.0g
KH_2PO_4	30.0g
NH_4Cl	10.0g
NaCl	5.0g

Preparation of 10X M9 Salts: Add components to distilled/deionized water and bring volume to 1.0L. Mix thoroughly. Adjust pH to 7.4.

Glucose Solution:
Composition per 100.0mL:

Glucose	20.0g

Preparation of Glucose Solution: Add glucose to distilled/deionized water and bring volume to 100.0mL. Mix thoroughly. Filter sterilize.

$CaCl_2 \cdot 2H_2O$ Solution:
Composition per 100.0mL:

$CaCl_2 \cdot 2H_2O$	1.47g

Preparation of $CaCl_2 \cdot 2H_2O$ Solution: Add $CaCl_2 \cdot 2H_2O$ to distilled/deionized water and bring volume to 100.0mL. Mix thoroughly. Autoclave for 15 min at 15 psi pressure–121°C.

$MgSO_4$ Solution:
Composition per 100.0mL:

$MgSO_4$	12.04g

Preparation of $MgSO_4$ Solution: Add $MgSO_4$ to distilled/deionized water and bring volume to 100.0mL. Mix thoroughly. Autoclave for 15 min at 15 psi pressure–121°C.

Thiamine·HCl Solution:
Composition per 10.0mL:

Thiamine·HCl	3.37g

Preparation of Thiamine·HCl Solution: Add thiamine·HCl to distilled/deionized water and bring volume to 10.0mL. Mix thoroughly. Filter sterilize.

Preparation of Medium: Add L-proline, L-tryptophan, and 10X M9 salts to distilled/deionized water and bring volume to 1.0L. Mix thoroughly. Autoclave for 15 min at 15 psi pressure–121°C. Aseptically add 10.0mL of sterile glucose solution, 1.0mL of sterile $CaCl_2 \cdot 2H_2O$ solution, 1.0mL of sterile $MgSO_4$ solution, and 1.0mL of sterile thiamine·HCl solution. Mix thoroughly. Aseptically distribute into sterile tubes or flasks.

Use: For the cultivation of *Escherichia coli.*

M9 Medium with Tryptophan (DSMZ Medium 451)

Composition per liter:

10X M9 salts	100.0mL
Glucose solution	10.0mL
Calcium chloride solution	10.0mL
Magnesium sulfate solution	10.0mL
Thiamine solution	1.0mL
Proline solution	1.0mL
Tryptophan solution	1.0mL

pH 7.4 ± 0.2 at 25°C

Glucose Solution:
Composition per 10.0mL:

D-Glucose	2.0g

Preparation of Glucose Solution: Add glucose to distilled/deionized water and bring volume to 10.0mL. Mix thoroughly. Filter sterilize.

Thiamine Solution:
Composition per 10.0mL:

Thiamine−$HCl \cdot 2H_2O$	3.7g

Preparation of Thiamine Solution: Add thiamine−$HCl \cdot 2H_2O$ to distilled/deionized water and bring volume to 10.0mL. Mix thoroughly. Filter sterilize.

Tryptophan Solution:
Composition per 10.0mL:

Tryptophan	0.2g

Preparation of Tryptophan Solution: Add tryptophan to distilled/deionized water and bring volume to 10.0mL. Mix thoroughly. Filter sterilize.

Proline Solution:
Composition per 10.0mL:

Proline	0.2g

Preparation of Proline Solution: Add proline to distilled/deionized water and bring volume to 10.0mL. Mix thoroughly. Filter sterilize.

Calcium Chloride Solution:
Composition per 10.0mL:

$CaCl_2$	0.1g

Preparation of Calcium Chloride Solution: Add $CaCl_2$ to distilled/deionized water and bring volume to 10.0mL. Mix thoroughly. Filter sterilize.

Magnesium Sulfate Solution:
Composition per 10.0mL:

$MgSO_4$	1.2g

Preparation of Magnesium Sulfate Solution: Add $MgSO_4$ to distilled/deionized water and bring volume to 10.0mL. Mix thoroughly. Filter sterilize.

10X M9 Salts Solution:
Composition per liter:

Na_2HPO_4	60.0g
KH_2PO_4	30.0g
NH_4Cl	10.0g
NaCl	5.0g
$MgSO_4$	1.2g

Preparation of 10X M9 Salts Solution: Add components to distilled/deionized water and bring volume to 1.0L. Mix thoroughly. Filter sterilize.

Preparation of Medium: Aseptically combine sterile component solutions. Mix thoroughly. Aseptically distribute into sterile tubes or flasks.

Use: For the cultivation of *Escherichia coli* tryptophan auxotrophs.

M10 Broth

Composition per liter:

Pancreatic digest of casein	2.0g
Yeast extract	2.0g
Cellobiose	1.0g
Glucose	1.0g
Maltose	1.0g
Starch	1.0g
Resazurin	1.0mg
Mineral solution I	100.0mL
Mineral solution II	100.0mL
Na_2CO_3 solution	50.0mL
Hemin solution	10.0mL
L-Cysteine·HCl solution	10.0mL
Volatile fatty acid mixture	3.1mL

pH 6.8 ± 0.2 at 25°C

Mineral Solution I:
Composition per 100.0mL:

K_2HPO_4	0.2g

Preparation of Mineral Solution I: Add K_2HPO_4 to distilled/deionized water and bring volume to 100.0mL. Mix thoroughly.

Mineral Solution II:
Composition per 100.0mL:

NaCl	0.4g
$(NH_4)_2SO_4$	0.4g
KH_2PO_4	0.3g
$MgSO_4 \cdot 7H_2O$	0.09g
$CaCl_2$	0.05g

Preparation of Mineral Solution II: Add components to distilled/deionized water and bring volume to 100.0mL. Mix thoroughly.

Na_2CO_3 Solution:
Composition per 100.0mL:

Na_2CO_3	8.0g

Preparation of Na_2CO_3 Solution: Add Na_2CO_3 to distilled/deionized water and bring volume to 100.0mL. Mix thoroughly. Sparge with 100% CO_2. Autoclave for 15 min at 15 psi pressure–121°C.

Hemin Solution:
Composition per 100.0mL:

Hemin	1.0g
NaOH (1*N* solution)	10.0mL

Preparation of Hemin Solution: Add components to 100.0mL of distilled/deionized water. Mix thoroughly.

L-Cysteine·HCl Solution:
Composition per 10.0mL:

L-Cysteine·HCl	0.25g

Preparation of L-Cysteine·HCl Solution: Add L-cysteine·HCl to distilled/deionized water and bring volume to 10.0mL. Mix thoroughly. Autoclave under 100% N_2 for 15 min at 15 psi pressure–121°C.

Volatile Fatty Acid Mixture:
Composition per 31.0mL:

Acetic acid	17.0mL
Propionic acid	6.0mL
Butyric acid	4.0mL
DL-α-Methylbutyric acid	1.0mL
Isobutyric acid	1.0mL
Isovaleric acid	1.0mL
n-Valeric acid	1.0mL

Preparation of Volatile Fatty Acid Mixture: Combine components. Mix thoroughly. Store under 100% N_2.

Preparation of Medium: Prepare and dispense medium under 100% CO_2. Add components, except L-cysteine·HCl solution and Na_2CO_3 solution, to distilled/deionized water and bring volume to 930.0mL. Mix thoroughly. Sparge with 100% CO_2. Adjust pH to 6.8 with 1*N* NaOH. Distribute anaerobically in 9.3mL volumes into Hungate tubes. Autoclave for 15 min at 15 psi pressure–121°C. Aseptically and anaerobically add 0.2mL of sterile L-cysteine·HCl solution and 0.5mL of sterile Na_2CO_3 solution. Check that final pH is 6.8.

Use: For the cultivation of *Enterococcus* species, *Lactobacillus* species, *Streptococcus* species, and *Vagococcus fluvialis*.

M13 *Verrucomicrobium* Medium

Composition per liter:

Glucose	0.25g
Peptone	0.25g
Yeast extract	0.25g
Distilled water	670.0mL
Artificial seawater	250.0mL
Tris-HCl buffer, (0.1*M* solution, pH 7.5)	50.0mL
Modified Huntner's basal salts	20.0mL
Vitamin solution	10.0mL

pH 7.5 ± 0.2 at 25°C

Artificial Seawater:
Composition per liter:

NaCl	23.48g
$MgCl_2$	4.98g
Na_2SO_4	3.92g
$CaCl_2$	1.1g
KCl	0.66g
$NaHCO_3$	0.19g
H_3BO_3	0.026g
$SrCl_2$	0.024g
KBr	6.0mg
NaF	3.0mg

Preparation of Artificial Seawater: Add components to distilled/deionized water and bring volume to 1.0L. Mix thoroughly.

Modified Hutner's Basal Salts:

Composition per liter:

$MgSO_4 \cdot 7H_2O$	29.7g
Nitrilotriacetic acid	10.0g
$CaCl_2 \cdot 2H_2O$	3.34g
$FeSO_4 \cdot 7H_2O$	99.0mg
$(NH_4)_2MoO_4$	9.25mg
Metals "44"	50.0mL

Preparation of Modified Hutner's Basal Salts: Add nitrilotriacetic acid to 500.0mL of distilled/deionized water. Dissolve by adjusting pH to 6.5 with KOH. Add remaining components. Add distilled/deionized water to 1.0L.

Metals "44":

Composition per 100.0mL:

$ZnSO_4 \cdot 7H_2O$	1.1g
$FeSO_4 \cdot 7H_2O$	0.5g
EDTA	0.25g
$MnSO_4 \cdot 7H_2O$	0.154g
$CuSO_4 \cdot 5H_2O$	0.04g
$Co(NO_3)_2 \cdot 6H_2O$	0.025g
$Na_2B_4O_7 \cdot 10H_2O$	0.018g

Preparation of Metals "44": Add components to distilled/deionized water and bring volume to 100.0mL. Mix thoroughly.

Vitamin Solution:

Composition per liter:

D-Calcium pantothenate	5.0mg
Riboflavin	5.0mg
Thiamine·HCl	5.0mg
Biotin	2.0mg
Folic acid	2.0mg
Vitamin B_{12}	0.1mg

Preparation of Vitamin Solution: Add components to distilled/deionized water and bring volume to 1.0L. Mix thoroughly. Filter sterilize.

Preparation of Medium: Add components, except modified Hutner's basal salts, to distilled/deionized water and bring volume to 980.0mL. Mix thoroughly. Autoclave for 15 min at 15 psi pressure–121°C. Cool to room temperature. Aseptically add 20.0mL of sterile modified Hutner's basal salts. Mix thoroughly. Aseptically distribute into sterile tubes or flasks.

Use: For the cultivation and maintenance of *Verrucomicrobium spinosum*.

M14 Medium

Composition per liter:

Yeast extract	1.0g
D-Glucose	1.0g
Tris(hydroxymethyl)aminomethane	0.753g
Artificial seawater	250.0mL
Modified Hutner's basal salts	20.0mL

pH 7.5 ± 0.2 at 25°C

Artificial Seawater:

Composition per liter:

NaCl	23.48g
$MgCl_2$	4.98g
Na_2SO_4	3.92g
$CaCl_2$	1.1g
KCl	0.66g
$NaHCO_3$	0.19g
H_3BO_3	0.026g
$SrCl_2$	0.024g
KBr	6.0mg
NaF	3.0mg

Preparation of Artificial Seawater: Add components to distilled/deionized water and bring volume to 1.0L. Mix thoroughly.

Modified Hutner's Basal Salts:

Composition per liter:

$MgSO_4 \cdot 7H_2O$	29.7g
Nitrilotriacetic acid	10.0g
$CaCl_2 \cdot 2H_2O$	3.34g
$FeSO_4 \cdot 7H_2O$	99.0mg
$(NH_4)_2MoO_4$	9.25mg
Metals "44"	50.0mL

Preparation of Modified Hutner's Basal Salts: Add nitrilotriacetic acid to 500.0mL of distilled/deionized water. Dissolve by adjusting pH to 6.5 with KOH. Add remaining components. Add distilled/deionized water to 1.0L.

Metals "44":

Composition per 100.0mL:

$ZnSO_4 \cdot 7H_2O$	1.1g
$FeSO_4 \cdot 7H_2O$	0.5g
EDTA	0.25g
$MnSO_4 \cdot 7H_2O$	0.154g
$CuSO_4 \cdot 5H_2O$	0.04g
$Co(NO_3)_2 \cdot 6H_2O$	0.025g
$Na_2B_4O_7 \cdot 10H_2O$	0.018g

Preparation of Metals "44": Add components to distilled/deionized water and bring volume to 100.0mL. Mix thoroughly.

Preparation of Medium: Add components, except modified Hutner's basal salts, to distilled/deionized water and bring volume to 980.0mL. Mix thoroughly. Autoclave for 15 min at 15 psi pressure–121°C. Cool to room temperature. Aseptically add 20.0mL of sterile modified Hutner's basal salts. Mix thoroughly. Aseptically distribute into sterile tubes or flasks.

Use: For the cultivation and maintenance of *Pirellula marina*.

M16 Agar

Composition per liter:

Agar	10.0g
Beef extract	5.0g
Pancreatic digest of soybean meal	5.0g
Polypeptone™	5.0g
Sodium acetate·$3H_2O$	3.0g
Yeast extract	2.5g
Ascorbic acid	0.5g
Carbohydrate solution	50.0mL

pH 7.2 ± 0.2 at 25°C

Carbohydrate Solution:

Composition per 50.0mL:

Lactose or glucose	5.0g

Preparation of Carbohydrate Solution: Add lactose or glucose to distilled/deionized water and bring volume to 50.0mL. Mix thoroughly. Filter sterilize.

Preparation of Medium: Add components, except carbohydrate solution, to distilled/deionized water and bring volume to 950.0mL. Mix thoroughly. Gently heat and bring to boiling. Adjust pH to 7.2 with 2*N* NaOH. Autoclave for 15 min at 15 psi pressure–121°C. Cool to 45°–50°C. Aseptically add 50.0mL of sterile carbohydrate solution. Mix thoroughly. Pour into sterile Petri dishes or distribute into sterile tubes.

Use: For the cultivation of lactobacilli from cheddar cheese.

M17 Agar (LMG Medium 261)

Composition per liter:

Disodium β-glycerophosphate	19.0g
Agar	11.0g
Polypeptone™	5.0g
Beef extract	5.0g
Papaic digest of soybean meal	5.0g
Yeast extract	2.5g
Ascorbic acid	0.5g
Lactose solution	50.0mL
$MgSO_4 \cdot 7H_2O$ (1*M* solution)	1.0mL

pH 6.9 ± 0.2 at 25°C

Lactose Solution:

Composition per 100.0mL:

Lactose	10.0g

Preparation of Lactose Solution: Add lactose to distilled/deionized water and bring volume to 100.0mL. Mix thoroughly. Autoclave for 15 min at 15 psi pressure–121°C.

Preparation of Medium: Add components, except lactose solution, to distilled/deionized water and bring volume to 950.0mL. Mix thoroughly. Gently heat until boiling. Autoclave for 15 min at 15 psi pressure–121°C. Cool to 45°–50°C. Aseptically add 50.0mL of sterile lactose solution. Mix thoroughly. Pour into sterile Petri dishes or distribute into sterile tubes.

Use: For the cultivation of *Streptococcus thermophilus* and for the cultivation and maintenance of streptococci and their bacteriophages. Also used for the cultivation and maintenance of starter cultures for cheese and yogurt manufacture as well as detecting streptococcal mutants that are unable to ferment lactose

M17 Agar

Composition per liter:

Disodium β-glycerophosphate	19.0g
Agar	11.0g
Beef extract	5.0g
Papaic digest of soybean meal	5.0g
Yeast extract	2.5g
Ascorbic acid	0.5g
$MgSO_4 \cdot 7H_2O$	0.25g
Lactose solution	50.0mL

pH 6.9 ± 0.2 at 25°C

Source: This medium is available as a premixed powder from Oxoid Unipath.

Lactose Solution:

Composition per 100.0mL:

Lactose	10.0g

Preparation of Lactose Solution: Add lactose to distilled/deionized water and bring volume to 100.0mL. Mix thoroughly. Autoclave for 15 min at 15 psi pressure–121°C.

Preparation of Medium: Add components, except lactose solution, to distilled/deionized water and bring volume to 950.0mL. Mix thoroughly. Gently heat until boiling. Autoclave for 15 min at 15 psi pressure–121°C. Cool to 45°–50°C. Aseptically add 50.0mL of sterile lactose solution. Mix thoroughly. Pour into sterile Petri dishes or distribute into sterile tubes.

Use: For the cultivation and maintenance of streptococci and their bacteriophages. Also used for the cultivation and maintenance of starter cultures for cheese and yogurt manufacture as well as detecting streptococcal mutants which are unable to ferment lactose.

M17 Broth

Composition per liter:

Disodium β-glycerophosphate	19.0g
Beef extract	5.0g
Lactose	5.0g
Papaic digest of soybean meal	5.0g
Pancreatic digest of casein	2.5g
Peptic digest of animal tissue	2.5g
Yeast extract	2.5g
Ascorbic acid	0.5g
$MgSO_4 \cdot 7H_2O$	0.25g

pH 7.15 ± 0.05 at 25°C

Source: This medium is available as a premixed powder from BD Diagnostic Systems and Oxoid Unipath.

Preparation of Medium: Add components to distilled/deionized water and bring volume to 1.0L. Mix thoroughly. Distribute into tubes or flasks. Autoclave for 15 min at 15 psi pressure–121°C.

Use: For the cultivation and maintenance of streptococci and their bacteriophages. Also used for the cultivation and maintenance of starter cultures for cheese and yogurt manufacture as well as detecting streptococcal mutants that are unable to ferment lactose.

M17 HiVeg Agar Base with Disodium-β-glycerophosphate

Composition per liter:

Disodium-β-glycerophosphate	19.0g
Agar	10.0g
Plant extract	5.0g
Plant peptone	5.0g
Lactose	5.0g
Papaic digest of soybean meal	5.0g
Yeast extract	2.5g
Ascorbic acid	0.5g
$MgSO_4$	0.25g

pH 7.1 ± 0.2 at 25°C

Source: This medium, without disodium-β-glycerophosphate, is available as a premixed powder from HiMedia.

Preparation of Medium: Add components to distilled/deionized water and bring volume to 1.0L. Mix thoroughly. Gently heat until boiling. Autoclave for 15 min at 15 psi pressure–121°C. Pour into sterile Petri dishes or distribute into sterile tubes.

Use: For the cultivation and maintenance of streptococci and their bacteriophages. Also used for the cultivation and maintenance of starter cultures for cheese and yogurt manufacture as well as detecting streptococcal mutants which are unable to ferment lactose.

M17 Medium for *Filomicrobium fusiforme* (DSMZ Medium 768)

Composition per liter:

Na-acetate	1.0g
KNO_3	1.0g
Artificial seawater, concentrated	500.0mL
Hutner's salts solution	20.0mL
Vitamin solution	10.0mL

pH 7.2 ± 0.2 at 25°C

Hutner's Salts Solution:

Composition per liter:

$MgSO_4 \cdot 7H_2O$	29.7g
Nitrilotriacetic acid	10.0g
$CaCl_2 \cdot 2H_2O$	3.335g
$FeSO_4 \cdot 7H_2O$	99.0mg
$(NH_4)_6MoO_7O_{24} \cdot 4H_2O$	9.25mg
"Metals 44"	50.0mL

"Metals 44":

Composition per 100.0mL:

$ZnSO_4 \cdot 7H_2O$	1.095g
$FeSO_4 \cdot 7H_2O$	0.5g
Sodium EDTA	0.25g
$MnSO_4 \cdot H2O$	0.154g
$CuSO_4 \cdot 5H_2O$	39.2mg
$Co(NO_3)_2 \cdot 6H_2O$	24.8mg
$Na_2B_4O_7 \cdot 10H_2O$	17.7mg

Preparation of "Metals 44": Add sodium EDTA to distilled/deionized water and bring volume to 90.0mL. Mix thoroughly. Add a few drops of concentrated H_2SO_4 to retard precipitation of heavy metal ions. Add remaining components. Mix thoroughly. Bring volume to 100.0mL with distilled/deionized water.

Preparation of Hutner's Salts Solution: Add nitrilotriacetic acid to 500.0mL of distilled/deionized water. Adjust pH to 6.5 with KOH. Add remaining components. Add distilled/deionized water to 1.0L. Adjust pH to 6.8.

Artificial Seawater, Concentrated:

Composition per liter:

NaCl	70.43g
$MgCl_2 \cdot 6H_2O$	31.86g
Na_2SO_4	11.75g
$CaCl_2 \cdot 2H_2O$	4.35g
$NaHCO_3$	2.88g
KCl	1.99g
KBr	0.29g
H_3BO_3	0.08g

Preparation of Artificial Seawater: Add components to distilled/deionized water and bring volume to 1.0L. Mix thoroughly. Filter sterilize.

Vitamin Solution:

Composition per liter:

Riboflavin	5.0mg
Thiamine-HCl·$2H_2O$	5.0mg
Ca-pantothenate	5.0mg
Biotin	2.0mg
Folic acid	2.0mg
Vitamin B_{12}	0.1mg

Preparation of Vitamin Solution: Add components to distilled/deionized water and bring volume to 1.0L. Mix thoroughly. Filter sterilize.

Preparation of Medium: Add components, except artificial sea water and vitamin solution, to distilled/deionized water and bring volume to 490.0mL. Mix thoroughly. Adjust pH to 7.2. Autoclave for 15 min at 15 psi pressure–121°C. Cool to room temperature. Aseptically add 500.0mL artificial sea water and 10.0mL vitamin solution. Mix thoroughly. Aseptically and anaerobically distribute into sterile tubes or bottles.

Use: For the cultivation of *Filomicrobium fusiforme.*

M17 Medium for Lactic Streptococci (DSMZ Medium 449)

Composition per liter:

Na_2-ß-glycerophosphate	19.0g
Peptone from casein	5.0g
Soy peptone	5.0g
Peptone bacteriological	5.0g
Yeast extract	2.5g
Ascorbic acid	0.5g
$MgSO_4 \cdot 7H_2O$	0.25g
Lactose solution	10.0mL

pH 6.9 ± 0.2 at 25°C

Lactose Solution:

Composition per 10.0mL:

Lactose	5.0g

Preparation of Lactose Solution: Add lactose to distilled/deionized water and bring volume to 10.0mL. Mix thoroughly. Filter sterilize.

Preparation of Medium: Add components, except lactose solution, to distilled/deionized water and bring volume to 1.0L. Mix thoroughly. Autoclave for 15 min at 15 psi pressure–121°C. Cool to 25°C. Aseptically add 10.0mL sterile lactose solution. Aseptically distribute to sterile tubes or flasks.

Use: For the cultivation and maintenance of *Lactococcus lactis* subsp. *lactis=Streptococcus lactis.*

M17 Medium, Modified

Composition per 1001.2mL:

Disodium-ß-glycerophosphate	9.5g
Pancreatic digest of casein	5.0g
Meat peptone	5.0g
Papaic digest of soybean meal	5.0g
Yeast extract	2.5g
Ascorbic acid	0.5g
$MgSO_4 \cdot 7H_2O$	0.25g
Lactose solution	50.0mL
$CaCl_2$ solution	1.2mL

pH 7.15 ± 0.05 at 25°C

Lactose Solution:

Composition per 50.0mL:

Lactose	8.0g

Preparation of Lactose Solution: Add lactose to distilled/deionized water and bring volume to 50.0mL. Mix thoroughly. Filter sterilize.

$CaCl_2$ Solution:
Composition per 100.0mL:

$CaCl_2 \cdot 2H_2O$	14.7g

Preparation of $CaCl_2$ Solution: Add $CaCl_2 \cdot 2H_2O$ to distilled/deionized water and bring volume to 100.0mL. Mix thoroughly. Filter sterilize.

Preparation of Medium: Add components, except lactose solution and $CaCl_2$ solution, to distilled/deionized water and bring volume to 950.0mL. Mix thoroughly. Autoclave for 15 min at 15 psi pressure–121°C. Aseptically add 50.0mL of sterile lactose solution and 1.2mL of sterile $CaCl_2$ solution. Mix thoroughly. Aseptically adjust pH to 7.15 ± 0.05. Aseptically distribute into sterile tubes or flasks.

Use: For the cultivation of *Streptococcus thermophilus.*

M40 Y

See: **Medium for Osmophilic Fungi**

M40 Y Agar (Harrold's Agar)

Composition per liter:

Sucrose	400.0g
Agar	20.0g
Malt extract	20.0g
Yeast extract	5.0g

Preparation of Medium: Add components, except sucrose, to distilled/deionized water and bring volume to 400.0mL. Mix thoroughly. In a separate flask, add sucrose to distilled/deionized water and bring volume to 600.0mL. Mix thoroughly. Autoclave both solutions separately for 15 min at 15 psi pressure–121°C. Cool to 50°C. Combine the sterile solutions. Mix thoroughly. Pour into sterile Petri dishes or distribute into sterile tubes.

Use: For the cultivation and maintenance of *Ascosphaera osmophila, Aspergillus halophilicus, Aspergillus penicilloides, Aspergillus restictus, Aspergillus tonophilus, Eremascus albus, Eremascus fertilis, Eurotium halophilicum, Eurotium herbariorum, Geomyces pulvereus, Monascus bisporus, Monascus eremophilus, Oidiodendron sindenia, Penicillium isariiforme, Penicillium ochro-chloron, Penicillium pinophilum, Physalospora tucumanensis, Polypaecilum pisce, Saccharomyces cerevisiae, Trichophaea abundans, Trichophaea contradicta, Wallemia sebi, Wardomyces dimerus, Xeromyces bisporus,* and *Zygosaccharomyces rouxii.*

M56 Agar

Composition per liter:

Agar	15.0g
Na_2HPO_4	8.7g
KH_2PO_4	5.3g
D-Glucose	4.0g
$(NH_4)_2SO_4$	2.0g
$MgSO_4 \cdot 7H_2O$	0.1g
L-Histidine	0.05g
L-Leucine	0.05g
Uracil	0.03g
$Ca(NO_3)_2 \cdot 4H_2O$	5.0mg
$FeSO_4 \cdot 7H_2O$	5.0mg
$ZnSO_4 \cdot 7H_2O$	5.0mg

pH 7.0 ± 0.2 at 25°C

Preparation of Medium: Add components to distilled/deionized water and bring volume to 1.0L. Mix thoroughly. Gently heat until boiling. Autoclave for 15 min at 15 psi pressure–121°C. Pour into sterile Petri dishes or distribute into sterile tubes.

Use: For the cultivation and maintenance of *Escherichia coli.*

M56 Medium

Composition per liter:

Na_2HPO_4	8.7g
KH_2PO_4	5.3g
D-Glucose	4.0g
$(NH_4)_2SO_4$	2.0g
$MgSO_4 \cdot 7H_2O$	0.1g
L-Histidine	0.05g
L-Leucine	0.05g
Uracil	0.03g
$Ca(NO_3)_2 \cdot 4H_2O$	5.0mg
$FeSO_4 \cdot 7H_2O$	5.0mg
$ZnSO_4 \cdot 7H_2O$	5.0mg

pH 7.0 ± 0.2 at 25°C

Preparation of Medium: Add components to distilled/deionized water and bring volume to 1.0L. Mix thoroughly. Distribute into tubes or flasks. Autoclave for 15 min at 15 psi pressure–121°C.

Use: For the cultivation and maintenance of *Escherichia coli.*

M63 Medium, 5X

Composition per liter:

KH_2PO_4	68.0g
$(NH_4)_2SO_4$	10.0g
$FeSO_4 \cdot 7H_2O$	2.5mg
Carbohydrate solution	10.0mL
$MgSO_4 \cdot 7H_2O$ solution	1.0mL

pH 7.0 ± 0.2 at 25°C

Carbohydrate Solution:
Composition per 100.0mL:

Carbohydrate	20.0g

Preparation of Carbohydrate Solution: Add carbohydrate to distilled/deionized water and bring volume to 100.0mL. Glucose or glycerol may be used. Mix thoroughly. Filter sterilize.

$MgSO_4 \cdot 7H_2O$ Solution:
Composition per 100.0mL:

$MgSO_4 \cdot 7H_2O$	24.65g

Preparation of $MgSO_4 \cdot 7H_2O$ Solution: Add $MgSO_4 \cdot 7H_2O$ to distilled/deionized water and bring volume to 100.0mL. Mix thoroughly. Filter sterilize.

Preparation of Medium: Add components, except carbohydrate solution and $MgSO_4 \cdot 7H_2O$ solution, to distilled/deionized water and bring volume to 1.0L. Mix thoroughly. Gently heat and bring to boiling. Autoclave for 15 min at 15 psi pressure–121°C. Cool to 45°–50°C. To prepare medium for use (1×), aseptically dilute 200.0mL of 5× stock solution with 789.0mL of sterile distilled/deionized water. Aseptically add 10.0mL of sterile carbohydrate solution and 1.0mL of sterile $MgSO_4 \cdot 7H_2O$ solution. Mix thoroughly. Aseptically distribute into sterile tubes or flasks.

Use: For the cultivation of *Escherichia coli.*

MA Medium

Composition per 1002.0mL:

Peptone	10.0g
Pancreatic digest of casein	10.0g
Ribonucleic acid from *Torula* yeast	1.0g
Asolectin	0.2g
Artificial seawater	500.0mL
Vitamin solution	2.0mL

Preparation of Medium: Emulsify asolectin in warm, distilled water before adding remaining powdered ingredients. Adjust pH to 7.2. Add vitamin mix and artificial seawater; readjust to pH 7.2, if necessary. Dispense 5.0mL per 16 x 125mm screw-capped test tube and autoclave at 121°C for 15 min.

Artificial Seawater:

Composition per 500.0mL:

Aqua-Marin sea salts	20.8g

Source: Aqua-Marin sea salts are available from Aquatrol, Inc., Anaheim, CA.

Preparation of Artificial Seawater: Add Aqua-Marin sea salts to distilled/deionized water and bring volume to 500.0mL. Mix thoroughly.

Vitamin Solution:

Composition per 110.0mL:

Thiamine·HCl)	150.0mg
Calcium D-(+)-pantothenate	100.0mg
Folic acid	50.0mg
Nicotinamide	50.0mg
Pyridoxal·HCl	50.0mg
Riboflavin	50.0mg
DL-6-Thioctic acid	1.0mg
Biotin solution	10.0mL

Biotin Solution:

Composition per 10.0mL:

Biotin	0.01mg

Preparation of Biotin Solution: Add biotin to 10.0mL of absolute ethanol. Mix thoroughly.

Preparation of Vitamin Solution: Add components to distilled/deionized water and bring volume to 100.0mL. Mix thoroughly. For long-term storage, preserve under nitrogen at –20°C.

Preparation of Medium: Add components, except vitamin solution, to distilled/deionized water and bring volume to 1.0L. Mix thoroughly. Autoclave for 15 min at 15 psi pressure–121°C. Aseptically add 2.0mL of sterile vitamin solution. Mix thoroughly. Aseptically distribute into sterile tubes or flasks.

Use: For the cultivation of *Anophryoides soldoi, Metanophrys diminuta, Mesanophrys chesapeakensis, Miamiensis avidus, Parauronema acutum,* and *Paranophrys* species.

MA1

See: **Malt Agar 1%**

MA2

See: **Malt Agar 2%**

MA2 with Lupine Stems

See: **Malt Agar 2% with Lupine Stems**

MA4

See: **Malt Agar 4%**

MA4 with Lupine Stems

See: **Malt Agar 4% with Lupine Stems**

MA8

See: **Malt Agar 8%**

MAB1 Medium

Composition per 1003.0mL:

NaCl	20.0g
$MgCl_2 \cdot 6H_2O$	3.0g
Na_2SO_4	3.0g
KCl	0.5g
NH_4Cl	0.25g
Yeast extract	0.2g
KH_2PO_4	0.2g
Sodium benzoate	0.15g
$CaCl_2 \cdot 2H_2O$	0.15g
Resazurin	1.0mg
Wolfe's vitamin solution	20.0mL
$NaHCO_3$ solution	10.0mL
$Na_2S \cdot 9H_2O$ solution	10.0mL
Na_2SeO_3/Na_2WO_4 solution	1.0mL
Sodium dithionite solution	1.0mL
Trace elements solution SL-10	1.0mL

pH 7.2 ± 0.2 at 25°C

Wolfe's Vitamin Solution:

Composition per liter:

Pyridoxine·HCl	10.0mg
p-Aminobenzoic acid	5.0mg
Lipoic acid	5.0mg
Nicotinic acid	5.0mg
Riboflavin	5.0mg
Thiamine·HCl	5.0mg
Calcium DL-pantothenate	5.0mg
Biotin	2.0mg
Folic acid	2.0mg
Vitamin B_{12}	0.1mg

Preparation of Wolfe's Vitamin Solution: Add components to distilled/deionized water and bring volume to 1.0L. Mix thoroughly. Filter sterilize.

$NaHCO_3$ Solution:

Composition per 10.0mL:

$NaHCO_3$	2.5g

Preparation of $NaHCO_3$ Solution: Add $NaHCO_3$ to distilled/deionized water and bring volume to 10.0mL. Mix thoroughly. Filter sterilize. Sparge with 80% N_2 + 20% CO_2.

$Na_2S \cdot 9H_2O$ Solution:

Composition per 10.0mL:

$Na_2S \cdot 9H_2O$	0.3g

Preparation of $Na_2S \cdot 9H_2O$ Solution: Add $Na_2S \cdot 9H_2O$ to distilled/deionized water and bring volume to 10.0mL. Mix thoroughly. Sparge with 100% N_2. Autoclave for 15 min at 15 psi pressure–121°C.

Na_2SeO_3/Na_2WO_4 Solution:

Composition per liter:

NaOH	0.5g
$Na_2WO_4 \cdot 2H_2O$	4.0mg
$Na_2SeO_3 \cdot 5H_2O$	3.0mg

Preparation of Na_2SeO_3/Na_2WO_4 Solution: Add components to distilled/deionized water and bring volume to 1.0L. Mix thoroughly. Sparge with 100% N_2. Autoclave for 15 min at 15 psi pressure–121°C.

Sodium Dithionite Solution:

Composition per 10.0mL:

Sodium dithioninium	0.2g

Preparation of Sodium Dithionite Solution: Add sodium dithioninium to distilled/deionized water and bring volume to 10.0mL. Mix thoroughly. Sparge with 100% N_2. Autoclave for 15 min at 15 psi pressure–121°C.

Trace Elements Solution SL-10:

Composition per liter:

$FeCl_2·4H_2O$	1.5g
$CoCl_2·6H_2O$	190.0mg
$MnCl_2·4H_2O$	100.0mg
$ZnCl_2$	70.0mg
$Na_2MoO_4·2H_2O$	36.0mg
$NiCl_2·6H_2O$	24.0mg
H_3BO_3	6.0mg
$CuCl_2·2H_2O$	2.0mg
HCl (25% solution)	10.0mL

Preparation of Trace Elements Solution SL-10: Prepare and dispense under 80% N_2 + 20% CO_2. Add $FeCl_2·4H_2O$ to 10.0mL of HCl solution. Mix thoroughly. Add distilled/deionized water and bring volume to 1.0L. Add remaining components. Mix thoroughly. Sparge with 80% N_2 + 20% CO_2. Autoclave for 15 min at 15 psi pressure–121°C.

Preparation of Medium: Prepare medium and dispense under 80% N_2 + 20% CO_2. Add components, except Wolfe's vitamin solution, $NaHCO_3$ solution, sodium dithionite solution, $Na_2S·9H_2O$ solution, Na_2SeO_3/Na_2WO_4 solution, and trace elements solution SL-10, to distilled/deionized water and bring volume to 960.0mL. Mix thoroughly. Adjust pH to 7.2. Sparge with 80% N_2 + 20% CO_2. Autoclave for 15 min at 15 psi pressure–121°C. Aseptically and anaerobically add 20.0mL of Wolfe's vitamin solution, 10.0mL of sterile $NaHCO_3$ solution, 10.0mL of sterile $Na_2S·9H_2O$ solution, 1.0mL Na_2SeO_3/Na_2WO_4 solution, 1.0mL of sterile sodium dithionite solution, and 1.0mL of sterile trace elements solution SL-10. Mix thoroughly. Aseptically and anaerobically distribute into sterile tubes or flasks.

Use: For the cultivation of *Desulfotomaculum* species.

MACA with Maltose

Composition per liter:

Yeast extract	20.0g
Agar	10.0g
Maltose	10.0g
Glucose	10.0g
Proteose peptone No. 3	5.0g
KH_2PO_4	2.0g
Sorbitan monooleate complex	0.1g

pH 6.7 ± 0.2 at 25°C

Preparation of Medium: Add components to distilled/deionized water and bring volume to 1.0L. Mix thoroughly. Gently heat and bring to boiling. Distribute into tubes or flasks. Autoclave for 15 min at 15 psi pressure–121°C. Pour into sterile Petri dishes or leave in tubes.

Use: For the cultivation and maintenance of *Lactobacillus sanfrancisco.*

MacConkey Agar

Composition per liter:

Pancreatic digest of gelatin	17.0g
Agar	13.5g
Lactose	10.0g
NaCl	5.0g
Bile salts	1.5g
Pancreatic digest of casein	1.5g
Peptic digest of animal tissue	1.5g
Neutral Red	0.03g
Crystal Violet	1.0mg

pH 7.1 ± 0.2 at 25°C

Source: This medium is available as a premixed powder from BD Diagnostic Systems.

Preparation of Medium: Add components to distilled/deionized water and bring volume to 1.0L. Mix thoroughly. Gently heat while stirring until boiling. Autoclave for 15 min at 15 psi pressure–121°C. Pour into sterile Petri dishes or distribute into sterile tubes.

Use: For the selective isolation, cultivation, and differentiation of coliforms and enteric pathogens based on the ability to ferment lactose. Lactose-fermenting organisms appear as red to pink colonies. Lactose-nonfermenting organisms appear as colorless or transparent colonies.

MacConkey Agar

Composition per liter:

Peptone	20.0g
Agar	12.0g
Lactose	10.0g
Bile salts	5.0g
NaCl	5.0g
Neutral Red	0.075g

pH 7.4 ± 0.2 at 25°C

Source: This medium is available as a premixed powder from Oxoid Unipath.

Preparation of Medium: Add components to distilled/deionized water and bring volume to 1.0L. Mix thoroughly. Gently heat while stirring until boiling. Autoclave for 15 min at 15 psi pressure–121°C. Pour into sterile Petri dishes or distribute into sterile tubes.

Use: For the selective isolation, cultivation, and differentiation of coliforms and enteric pathogens based on the ability to ferment lactose. Lactose-fermenting organisms appear as red to pink colonies. Lactose-nonfermenting organisms appear as colorless or transparent colonies.

MacConkey Agar Base, HiVeg

Composition per liter:

Plant peptone	17.0g
Agar	13.5g
NaCl	5.0g
Plant peptone No. 3	3.0g
Synthetic detergent	1.5g
Neutral Red	0.03g
Crystal Violet	1.0mg

pH 7.1 ± 0.2 at 25°C

Source: This medium is available as a premixed powder from HiMedia.

Preparation of Medium: Add components to distilled/deionized water and bring volume to 1.0L. Mix thoroughly. Gently heat while stirring until boiling. Autoclave for 15 min at 15 psi pressure–121°C. Pour into sterile Petri dishes or distribute into sterile tubes.

Use: For the cultivation and differentiation of lactose-fermenting and nonfermenting Gram-negative bacteria. Lactose-fermenting organisms appear as red to pink colonies. Lactose-nonfermenting organisms appear as colorless or transparent colonies.

MacConkey Agar with 0.15% Bile Salts, Crystal Violet, and Sodium Chloride, HiVeg

Composition per liter:

Plant peptone No. 2	17.0g
Agar	15.0g
Lactose	10.0g
NaCl	5.0g
Plant hydrolysate	1.5g
Plant peptone	1.5g
Synthetic detergent	1.5g
Neutral Red	0.03g
Crystal Violet	1.0mg

pH 7.2 ± 0.2 at 25°C

Source: This medium is available as a premixed powder from HiMedia.

Preparation of Medium: Add components to distilled/deionized water and bring volume to 1.0L. Mix thoroughly. Gently heat while stirring until boiling. Autoclave for 15 min at 15 psi pressure–121°C. Pour into sterile Petri dishes or distribute into sterile tubes.

Use: For the selective isolation, cultivation, and differentiation of enteric pathogens, especially enterococci, in clinical specimens and in materials of sanitary importance.

MacConkey Agar, CS

Composition per liter:

Peptone	17.0g
Agar	13.5g
Lactose	10.0g
NaCl	5.0g
Proteose peptone	3.0g
Bile salts	1.5g
Neutral Red	0.03g
Crystal Violet	1.0mg

pH 7.1 ± 0.2 at 25°C

Source: This medium is available as a prepared medium from BD Diagnostic Systems.

Preparation of Medium: Add components to distilled/deionized water and bring volume to 1.0L. Mix thoroughly. Gently heat while stirring until boiling. Autoclave for 15 min at 15 psi pressure–121°C. Pour into sterile Petri dishes or distribute into sterile tubes.

Use: For the cultivation and differentiation of lactose-fermenting and lactose-nonfermenting Gram-negative bacteria while also controlling the swarming of *Proteus* species, if present. Lactose-fermenting organisms appear as red to pink colonies. Lactose-nonfermenting organisms appear as colorless or transparent colonies.

MacConkey Agar, Fluorocult (Fluorocult MacConkey Agar)

Composition per liter:

Peptone from casein	17.0g
Agar	13.5g
Lactose	10.0g
NaCl	5.0g
Peptone from meat	3.0g
Bile salt mixture	1.5g
4-Methylumbelliferyl-β-D-glucuronide	0.1g
Neutral Red	0.03g
Crystal Violet	0.001g

pH 7.1 ± 0.2 at 25°C

Source: This medium is available from Merck.

Preparation of Medium: Add components to distilled/deionized water and bring volume to 1.0L. Mix thoroughly. Autoclave for 15 min at 15 psi pressure–121°C. Cool to 45°–50°C. Pour into sterile Petri dishes. The plates are clear and red to red-brown.

Use: For the isolation of *Salmonella*, *Shigella,* and coliform bacteria, in particular *E. coli*, from various materials. The bile salts and Crystal Violet largely inhibit the growth of Gram-positive microbial flora. Lactose together with the pH indicator Neutral Red are used to detect lactose-positive colonies and *E. coli* can be seen among these because of fluorescence under UV light.

MacConkey Agar with Sorbitol
See: **Sorbitol MacConkey Agar**

MacConkey Agar without Crystal Violet

Composition per liter:

Agar	12.0g
Lactose	10.0g
Pancreatic digest of casein	10.0g
Peptic digest of animal tissue	10.0g
Bile salts	5.0g
NaCl	5.0g
Neutral Red	0.05g

pH 7.4 ± 0.2 at 25°C

Source: This medium is available as a premixed powder from BD Diagnostic Systems.

Preparation of Medium: Add components to distilled/deionized water and bring volume to 1.0L. Mix thoroughly. Gently heat while stirring until boiling. Autoclave for 15 min at 15 psi pressure–121°C. Pour into sterile Petri dishes or distribute into sterile tubes.

Use: For the detection of members of the Enterobacteriaceae and enterococci as well as some staphylococci. For the isolation and detection of coliforms and enteric pathogens from water and wastewater.

MacConkey Agar without Crystal Violet with Sodium Chloride and 0.5% Sodium Taurocholate, HiVeg

Composition per liter:

Agar	20.0g
Plant peptone	20.0g
Lactose	10.0g
Synthetic detergent No. V	5.0g
NaCl	5.0g
Neutral Red	0.04g

pH 7.2 ± 0.2 at 25°C

Source: This medium is available as a premixed powder from HiMedia.

Preparation of Medium: Add components to distilled/deionized water and bring volume to 1.0L. Mix thoroughly. Gently heat while stirring until boiling. Autoclave for 15 min at 15 psi pressure–121°C. Pour into sterile Petri dishes or distribute into sterile tubes.

Use: For the selective isolation, cultivation, and differentiation of *Vibrio* spp. in clinical specimens and in materials of sanitary importance.

MacConkey Agar without Crystal Violet and Sodium Chloride with 0.5% Sodium Taurocholate, HiVeg

Composition per liter:

Agar	20.0g
Plant peptone	20.0g
Lactose	10.0g
Synthetic detergent No. V	5.0g
Neutral Red	0.04g

pH 7.2 ± 0.2 at 25°C

Source: This medium, without NaCl, is available as a premixed powder from HiMedia.

Preparation of Medium: Add components to distilled/deionized water and bring volume to 1.0L. Mix thoroughly. Gently heat while stirring until boiling. Autoclave for 15 min at 15 psi pressure–121°C. Pour into sterile Petri dishes or distribute into sterile tubes.

Use: For the selective isolation, cultivation, and differentiation of enteric pathogens, especially enterococci, in clinical specimens and in materials of sanitary importance.

MacConkey Agar without Salt

Composition per liter:

Peptone	20.0g
Agar	12.0g
Lactose	10.0g
Bile salts	5.0g
Neutral Red	0.075g

pH 7.4 ± 0.2 at 25°C

Source: This medium is available as a premixed powder from BD Diagnostic Systems and Oxoid Unipath.

Preparation of Medium: Add components to distilled/deionized water and bring volume to 1.0L. Mix thoroughly. Gently heat while stirring until boiling. Autoclave for 15 min at 15 psi pressure–121°C. Pour into sterile Petri dishes or distribute into sterile tubes. Dry the surface of plates before inoculation.

Use: For the isolation and detection of coliforms and enteric pathogens from urine. Provides a low electrolyte medium on which most *Proteus* species will not swarm and therefore avoids overgrowth of the plate.

MacConkey Agar No. 2 (MacConkey II Agar)

Composition per liter:

Peptone	20.0g
Agar	15.0g
Lactose	10.0g
NaCl	5.0g
Bile salts No. 2	1.5g
Neutral Red	0.05g
Crystal Violet	1.0mg

pH 7.2 ± 0.2 at 25°C

Source: This medium is available as a premixed powder from BD Diagnostic Systems and Oxoid Unipath.

Preparation of Medium: Add components to distilled/deionized water and bring volume to 1.0L. Mix thoroughly. Gently heat while stirring until boiling. Autoclave for 15 min at 15 psi pressure–121°C. Pour into sterile Petri dishes or distribute into sterile tubes.

Use: For the selective isolation, cultivation, and differentiation of enteric pathogens, especially enterococci, in clinical specimens and in materials of sanitary importance.

MacConkey Agar No. 3

Composition per liter:

Peptone	20.0g
Agar	15.0g
Lactose	10.0g
NaCl	5.0g
Bile salts No. 3	1.5g
Neutral Red	0.03g
Crystal Violet	0.001g

pH 7.1 ± 0.2 at 25°C

Source: This medium is available as a premixed powder from Oxoid Unipath.

Preparation of Medium: Add components to distilled/deionized water and bring volume to 1.0L. Mix thoroughly. Gently heat while stirring until boiling. Autoclave for 15 min at 15 psi pressure–121°C. Pour into sterile Petri dishes or distribute into sterile tubes.

Use: For the selective isolation, cultivation, and differentiation of enteric pathogens, especially *Salmonella* and *Shigella*, in clinical specimens and in foods.

MacConkey Broth

Composition per liter:

Pancreatic digest of gelatin	20.0g
Lactose	10.0g
Oxgall	5.0g
Bromcresol Purple	0.02g

pH 7.3 ± 0.2 at 25°C

Source: This medium is available as a premixed powder from BD Diagnostic Systems.

Preparation of Medium: Add components to distilled/deionized water and bring volume to 1.0L. If testing 10.0mL samples, prepare medium double strength. Mix thoroughly. Gently heat while stirring until boiling. Distribute into test tubes containing inverted Durham tubes. Autoclave for 15 min at 15 psi pressure–121°C.

Use: For the selective isolation and cultivation of coliforms in milk and water.

MacConkey Broth

Composition per liter:

Peptone	20.0g
Lactose	10.0g
Bile salts	5.0g
NaCl	5.0g
Neutral Red	0.075g

pH 7.4 ± 0.2 at 25°C

Source: This medium is available as a premixed powder from Oxoid Unipath.

Preparation of Medium: Add components to distilled/deionized water and bring volume to 1.0L. If testing 10.0mL samples, prepare medium double strength. Mix thoroughly. Gently heat while stirring until boiling. Distribute into test tubes containing inverted Durham tubes. Autoclave for 15 min at 15 psi pressure–121°C.

Use: For the selective isolation and cultivation of coliforms in milk and water.

MacConkey Broth, Purple

Composition per liter:

Peptone	20.0g
Lactose	10.0g
Bile salts	5.0g
NaCl	5.0g
Bromcresol Purple	0.01g

pH 7.4 ± 0.2 at 25°C

Source: This medium is available as a premixed powder or tablets from Oxoid Unipath.

Preparation of Medium: Add components to distilled/deionized water and bring volume to 1.0L. If testing 10.0mL samples, prepare medium double strength. Mix thoroughly. Gently heat while stirring until boiling. Distribute into test tubes containing inverted Durham tubes. Autoclave for 15 min at 15 psi pressure–121°C.

Use: For the selective isolation and cultivation of coliforms in milk and water.

MacConkey Broth, Purple, with Bromcresol Purple, HiVeg

Composition per liter:

Plant special peptone	23.0g
Lactose	10.0g
NaCl	5.0g
Synthetic detergent No. V	2.0g
Bromcresol Purple	0.01g

pH 7.2 ± 0.2 at 25°C

Source: This medium is available as a premixed powder from HiMedia.

Preparation of Medium: Add components to distilled/deionized water and bring volume to 1.0L. Mix thoroughly. Gently heat while stirring until boiling. Autoclave for 15 min at 15 psi pressure–121°C.

Use: For the selective isolation, cultivation, and differentiation of enteric bacteria, especially coliforms.

MacConkey HiVeg Agar with Bromthymol Blue

Composition per liter:

Plant peptone	17.0g
Agar	15.0g
Lactose	10.0g
NaCl	5.0g
Plant peptone No. 3	3.0g
Synthetic detergent	1.5g
Bromthymol Blue	0.03g

pH 7.2 ± 0.2 at 25°C

Source: This medium is available as a premixed powder from HiMedia.

Preparation of Medium: Add components to distilled/deionized water and bring volume to 1.0L. Mix thoroughly. Gently heat while stirring until boiling. Autoclave for 15 min at 15 psi pressure–121°C. Pour into sterile Petri dishes or distribute into sterile tubes.

Use: For the selective isolation, cultivation, and differentiation of enteric bacteria.

MacConkey HiVeg Agar with Crystal Violet and Sodium Chloride

Composition per liter:

Plant peptone	20.0g
Agar	15.0g
Lactose	10.0g
NaCl	5.0g
Synthetic detergent	1.5g
Neutral Red	0.05g
Crystal Violet	1.0mg

pH 7.2 ± 0.2 at 25°C

Source: This medium is available as a premixed powder from HiMedia.

Preparation of Medium: Add components to distilled/deionized water and bring volume to 1.0L. Mix thoroughly. Gently heat while stirring until boiling. Autoclave for 15 min at 15 psi pressure–121°C. Pour into sterile Petri dishes or distribute into sterile tubes.

Use: For the selective isolation, cultivation, and differentiation of enteric bacteria. For the identification of Enterobacteriaceae in the presence of coliforms and lactose nonfermenters from water and sewage.

MacConkey HiVeg Agar with 1.35% Agar, Crystal Violet, and Sodium Chloride

Composition per liter:

Plant peptone No. 2	17.0g
Agar	13.5g
Lactose	10.0g
NaCl	5.0g
Plant hydrolysate	1.5g
Plant peptone	1.5g
Sodium acetate (anhydrous)	1.5g
Neutral Red	0.03g
Crystal Violet	1.0mg

pH 7.2 ± 0.2 at 25°C

Source: This medium is available as a premixed powder from HiMedia.

Preparation of Medium: Add components to distilled/deionized water and bring volume to 1.0L. Mix thoroughly. Gently heat while stirring until boiling. Autoclave for 15 min at 15 psi pressure–121°C. Pour into sterile Petri dishes or distribute into sterile tubes.

Use: For the selective isolation and differentiation of lactose-fermenting and lactose-nonfermenting enteric bacteria.

MacConkey HiVeg Agar without Crystal Violet and Sodium Chloride

Composition per liter:

Plant peptone	23.0g
Agar	12.0g
Lactose	10.0g
Synthetic detergent	2.0g
Neutral Red	0.075g

pH 7.2 ± 0.2 at 25°C

Source: This medium is available as a premixed powder from HiMedia.

Preparation of Medium: Add components to distilled/deionized water and bring volume to 1.0L. Mix thoroughly. Gently heat while stirring until boiling. Autoclave for 15 min at 15 psi pressure–121°C. Pour into sterile Petri dishes or distribute into sterile tubes.

Use: For the cultivation and differentiation of enteric bacteria, restricting swarming of *Proteus* species.

MacConkey HiVeg Agar, Modified

Composition per liter:

Plant peptone	17.0g
Agar	13.5g
Inositol	10.0g
NaCl	5.0g
Plant peptone No. 3	3.0g
Synthetic detergent	1.5g
Neutral Red	0.03g
Crystal Violet	1.0mg

pH 7.2 ± 0.2 at 25°C

Source: This medium is available as a premixed powder from HiMedia.

Preparation of Medium: Add components to distilled/deionized water and bring volume to 1.0L. Mix thoroughly. Gently heat while stirring until boiling. Autoclave for 15 min at 15 psi pressure–121°C. Pour into sterile Petri dishes or distribute into sterile tubes.

Use: For the isolation of *Klebsiella* species from water samples.

MacConkey HiVeg Broth (Double Strength) with Neutral Red

Composition per liter:

Plant peptone	46.0g
Lactose	20.0g
NaCl	10.0g
Synthetic detergent	4.0g
Neutral Red	0.15g

pH 7.2 ± 0.2 at 25°C

Source: This medium is available as a premixed powder from HiMedia.

Preparation of Medium: Add components to distilled/deionized water and bring volume to 1.0L. Mix thoroughly. Distribute into tubes or leave in flasks. Gently heat while stirring until boiling. Autoclave for 15 min at 15 psi pressure–121°C.

Use: For the primary isolation of coliforms from large samples such as water or wastewater.

MacConkey HiVeg Broth Purple with Bromo Cresol Purple

Composition per liter:

Plant special peptone	23.0g
Lactose	10.0g
NaCl	5.0g
Synthetic detergent No. V	2.0g
Bromcresol Purple	0.01g

pH 7.2 ± 0.2 at 25°C

Source: This medium is available as a premixed powder from HiMedia.

Preparation of Medium: Add components to distilled/deionized water and bring volume to 1.0L. Mix thoroughly. Gently heat while stirring until boiling. Autoclave for 15 min at 15 psi pressure–121°C. Pour into sterile Petri dishes or distribute into sterile tubes.

Use: For the presumptive identification of coliforms from water.

MacConkey Sorbitol HiVeg Agar (Sorbitol HiVeg Agar)

Composition per liter:

Plant peptone	17.0g
Agar	13.5g
D-Sorbitol	10.0g
NaCl	5.0g
Plant peptone No. 5	3.0g
Synthetic detergent No. I	1.5g
Neutral Red	0.03g
Crystal Violet	1.0mg

pH 7.2 ± 0.2 at 25°C

Source: This medium is available as a premixed powder from HiMedia.

Preparation of Medium: Add components to distilled/deionized water and bring volume to 1.0L. Mix thoroughly. Gently heat while stirring until boiling. Autoclave for 15 min at 15 psi pressure–121°C. Pour into sterile Petri dishes or distribute into sterile tubes.

Use: For the isolation and cultivation of pathogenic *Escherichia coli*.

M-*Aeromonas* Selective Agar Base, Havelaar

Composition per liter:

Agar	13.0g
Dextrin	11.4g
Tryptose	5.0g
NaCl	3.0g
KCl	2.0g
Yeast extract	2.0g
$MgSO_4 \cdot 7H_2O$	0.1g
Sodium deoxycholate	0.1g
Bromthymol Blue	0.08g
$FeCl_3 \cdot 6H_2O$	0.06g

pH 7.5 ± 0.2 at 25°C

Source: This medium is available from HiMedia.

Preparation of Medium: Add components to distilled/deionized water and bring volume to 1.0L. Mix thoroughly. Gently heat and bring to boiling. Distribute into tubes or flasks. Autoclave for 15 min at 15 psi pressure–121°C. Pour into sterile Petri dishes or leave in tubes.

Use: For the detection of *Aeromonas* species in water and other liquid samples by the membrane filter technique.

Magnesium Oxalate Agar (MOX Agar)

Composition per liter:

Pancreatic digest of casein	15.0g
Agar	15.0g
Papaic digest of soybean meal	5.0g
NaCl	5.0g
$MgCl_2 \cdot 6H_2O$	4.1g
Sodium oxalate	2.68g

pH 7.4–7.6 at 25°C

Preparation of Medium: Add components to distilled/deionized water and bring volume to 1.0L. Mix thoroughly. Gently heat and bring to boiling. Distribute into tubes or flasks. Autoclave for 15 min at 15 psi pressure–121°C. Pour into sterile Petri dishes or leave in tubes.

Use: For the cultivation of *Yersinia enterocolitica* from foods.

Magnetic *Spirillum* Growth Medium, Revised (MSGM, Revised)

Composition per liter:

Agar	1.3g
KH_2PO_4	0.68g
Tartaric acid	0.37g
Succinic acid	0.37g
$NaNO_3$	0.12g
Sodium acetate	0.05g
Ascorbic acid	0.035g
Wolfe's vitamin solution	10.0mL
Wolfe's mineral solution	5.0mL
Ferric quinate solution	2.0mL
Resazurin (0.1% solution)	0.45mL

pH 6.75 ± 0.2 at 25°C

Wolfe's Vitamin Solution:

Composition per liter:

Pyridoxine·HCl	10.0mg
Thiamine·HCl	5.0mg
Riboflavin	5.0mg
Nicotinic acid	5.0mg
Calcium pantothenate	5.0mg
p-Aminobenzoic acid	5.0mg
Thioctic acid	5.0mg
Biotin	2.0mg
Folic acid	2.0mg
Cyanocobalamin	100.0μg

Preparation of Wolfe's Vitamin Solution: Add components to distilled/deionized water and bring volume to 1.0L. Mix thoroughly.

Wolfe's Mineral Solution:

Composition per liter:

$MgSO_4 \cdot 7H_2O$	3.0g
Nitriloacetic acid	1.5g
NaCl	1.0g
$MnSO_4 \cdot H_2O$	0.5g
$FeSO_4 \cdot 7H_2O$	0.1g
$CoCl_2 \cdot 6H_2O$	0.1g
$CaCl_2$	0.1g
$ZnSO_4 \cdot 7H_2O$	0.1g
$CuSO_4 \cdot 5H_2O$	0.01g
$AlK(SO_4)_2 \cdot 12H_2O$	0.01g
H_3BO_3	0.01g
$Na_2MoO_4 \cdot 2H_2O$	0.01g

Preparation of Wolfe's Mineral Solution: Add nitrilotriacetic acid to 500.0mL of distilled/deionized water. Dissolve by adjusting pH to 6.5 with KOH. Add remaining components. Add distilled/deionized water to 1.0L.

Ferric Quinate Solution:

Composition per 100.0mL:

$FeCl_3$	0.27g
Quinic acid	0.19g

Preparation of Ferric Quinate Solution: Add components to distilled/deionized water and bring volume to 1.0L. Mix thoroughly. Autoclave for 15 min at 15 psi pressure–121°C.

Preparation of Medium: To 1.0L of distilled/deionized water add components in the following order: Wolfe's vitamin solution, Wolfe's mineral solution, ferric quinate solution, resazurin, KH_2PO_4, $NaNO_3$, ascorbic acid, tartaric acid, succinic acid, sodium acetate, and agar. Mix thoroughly after each addition. Adjust pH to 6.75 with NaOH. Autoclave for 15 min at 15 psi pressure–121°C. Aseptically distribute into sterile screw-capped tubes. Fill tubes to capacity with medium. Use a heavy inoculum in each tube and do not introduce a headspace of air. Screw down caps tightly.

Use: For the cultivation and maintenance of *Aquaspirillum magnetotacticum*.

Magnetospirillum Medium (DSMZ Medium 380)

Composition per liter:

KH_2PO_4	0.68g
L(+)-Tartaric acid	0.37g
Succinic acid	0.37g
$NaNO_3$	0.12g
Na-thioglycolate	0.05g
Na-acetate	0.05g
Resazurin	0.5mg
Vitamin solution	10.0mL
Trace elements solution	5.0mL
Ferric quinate solution	2.0mL

pH 6.8 ± 0.2 at 25°C

Trace Elements Solution:

Composition per liter:

$MgSO_4 \cdot 7H_2O$	3.0g
Nitrilotriacetic acid	1.5g
NaCl	1.0g
$MnSO_4 \cdot 2H_2O$	0.5g
$CoSO_4 \cdot 7H_2O$	0.18g
$ZnSO_4 \cdot 7H_2O$	0.18g
$CaCl_2 \cdot 2H_2O$	0.1g
$FeSO_4 \cdot 7H_2O$	0.1g
$NiCl_2 \cdot 6H_2O$	0.025g
$KAl(SO_4)_2 \cdot 12H_2O$	0.02g
H_3BO_3	0.01g
$Na_2MoO_4 \cdot 4H_2O$	0.01g
$CuSO_4 \cdot 5H_2O$	0.01g
$Na_2SeO_3 \cdot 5H_2O$	0.3mg

Preparation of Trace Elements Solution: Add nitrilotriacetic acid to 500.0mL of distilled/deionized water. Dissolve by adjusting pH to 6.5 with KOH. Add remaining components. Add distilled/deionized water to 1.0L. Mix thoroughly.

Vitamin Solution:

Composition per liter:

Pyridoxine-HCl	10.0mg
Thiamine-HCl·$2H_2O$	5.0mg
Riboflavin	5.0mg
Nicotinic acid	5.0mg
D-Ca-pantothenate	5.0mg
p-Aminobenzoic acid	5.0mg
Lipoic acid	5.0mg
Biotin	2.0mg
Folic acid	2.0mg
Vitamin B_{12}	0.1mg

Preparation of Vitamin Solution: Add components to distilled/deionized water and bring volume to 1.0L. Mix thoroughly. Sparge with 80% H_2 + 20% CO_2. Filter sterilize.

Ferric Quinate Solution:

Composition per 100.0mL:

$FeCl_3 \cdot 6H_2O$	0.45g
Quinic acid	0.19g

Preparation of Ferric Quinate Solution: Add components to distilled/deionized water and bring volume to 100.0mL. Sparge with N_2. Autoclave for 15 min at 15 psi pressure–121°C. Cool to 25°C.

Preparation of Medium: Add components, except vitamin solution and ferric quinate solution, to distilled/deionized water and bring volume to 988.0mL. Purge medium with N_2 gas for 10 min. Mix thoroughly. Autoclave for 15 min at 15 psi pressure–121°C. Cool to 25°C. Aseptically and anerobically add 10.0mL vitamin solution and 2.0mL ferric quinate solution. Mix thoroughly. Purge medium with N_2 gas for 10 min. Under the same atmosphere, anaerobically fill tubes to 1/3 of their volume and seal. Autoclave at 121°C for 15 min. Before inoculation, add sterile air (with hypodermic syringe through the rubber closure) to 1% O_2 concentration in the gas phase.

Use: For the cultivation of *Magnetospirillum magnetotacticum=Aquaspirillum magnetotacticum,* and *Magnetospirillum gryphiswaldense.*

Magnetospirillum 2 Medium

Composition per liter:

Sodium acetate	1.0g
K_2HPO_4	0.5g
Sodium thioglycolate	0.5g
NH_4Cl	0.1g
Yeast extract	0.1g
Ferric citrate	20.0μg

pH 6.8 ± 0.2 at 25°C

Preparation of Medium: Add components to distilled/deionized water and bring volume to 1.0L. Mix thoroughly. Distribute into screw-capped tubes in 5.0mL volumes. Autoclave for 15 min at 15 psi pressure–121°C. Allow medium to stand upright at room temperature for 2 to 3 days before inoculation. Do not shake.

Use: For the cultivation and maintenance of *Magnetospirillum gryphiswaldense.*

Magnetospirillum Semi-solid Medium (DSMZ Medium 380)

Composition per liter:

Agar	1.3g
KH_2PO_4	0.68g
L(+)-Tartaric acid	0.37g
Succinic acid	0.37g
$NaNO_3$	0.12g
Na-thioglycolate	0.05g
Na-acetate	0.05g
Resazurin	0.5mg
Vitamin solution	10.0mL
Trace elements solution	5.0mL
Ferric quinate solution	2.0mL

pH 6.8 ± 0.2 at 25°C

Trace Elements Solution:

Composition per liter:

$MgSO_4 \cdot 7H_2O$	3.0g
Nitrilotriacetic acid	1.5g
NaCl	1.0g
$MnSO_4 \cdot 2H_2O$	0.5g
$CoSO_4 \cdot 7H_2O$	0.18g
$ZnSO_4 \cdot 7H_2O$	0.18g
$CaCl_2 \cdot 2H_2O$	0.1g
$FeSO_4 \cdot 7H_2O$	0.1g
$NiCl_2 \cdot 6H_2O$	0.025g
$KAl(SO_4)_2 \cdot 12H_2O$	0.02g
H_3BO_3	0.01g
$Na_2MoO_4 \cdot 4H_2O$	0.01g
$CuSO_4 \cdot 5H_2O$	0.01g
$Na_2SeO_3 \cdot 5H_2O$	0.3mg

Preparation of Trace Elements Solution: Add nitrilotriacetic acid to 500.0mL of distilled/deionized water. Dissolve by adjusting pH to 6.5 with KOH. Add remaining components. Add distilled/deionized water to 1.0L. Mix thoroughly.

Vitamin Solution:

Composition per liter:

Pyridoxine-HCl	10.0mg
Thiamine-HCl·$2H_2O$	5.0mg
Riboflavin	5.0mg
Nicotinic acid	5.0mg
D-Ca-pantothenate	5.0mg
p-Aminobenzoic acid	5.0mg
Lipoic acid	5.0mg
Biotin	2.0mg
Folic acid	2.0mg
Vitamin B_{12}	0.1mg

Preparation of Vitamin Solution: Add components to distilled/deionized water and bring volume to 1.0L. Mix thoroughly. Sparge with 80% H_2 + 20% CO_2. Filter sterilize.

Ferric Quinate Solution:

Composition per 100.0mL:

$FeCl_3 \cdot 6H_2O$	0.45g
Quinic acid	0.19g

Preparation of Ferric Quinate Solution: Add components to distilled/deionized water and bring volume to 100.0mL. Sparge with N_2. Autoclave for 15 min at 15 psi pressure–121°C. Cool to 45°C.

Preparation of Medium: Add components, except vitamin solution and ferric quinate solution, to distilled/deionized water and bring volume to 988.0mL. Purge medium with N_2 gas for 10 min. Mix thoroughly. Autoclave for 15 min at 15 psi pressure–121°C. Cool to 45°C. Aseptically and anerobically add 10.0mL vitamin solution and 2.0mL ferric quinate solution. Mix thoroughly. Purge medium with N_2 gas for 10 min. Dispense 12mL of medium per 16 x 150mm anaerobe screw-cap tube under N_2 gas. Prior to inoculation, remove caps briefly under air, tighten the caps again, and wait several hours to establish oxygen gradients. The medium should be slightly pink in color. Strongly reduced conditions will not support growth of the organism. During growth O_2 will be consumed, resazurin decolorized, and the pH increased. Feed oxygen (by adding air) and succinic acid from sterile 0.05*M* solution (to maintain pH below 7.0). If higher densities of magnetic cell are wanted, ferric quinate also can be fed. For transfer use cell material which has been concentrated at the glass wall of the culture vessel by means of a magnetic rod attached outside.

Use: For the cultivation of *Magnetospirillum magnetotacticum (Aquaspirillum magnetotacticum)* and *Magnetospirillum gryphiswaldense.*

Maintenance HiVeg Medium for *B. subtilis* ATCC 6633

Composition per liter:

Agar	15.0g
Plant peptone	6.0g
Plant hydrolysate	4.0g
Yeast extract	3.0g
Plant extract	1.5g
Glucose	1.0g

pH 7.2 ± 0.2 at 25°C

Source: This medium is available as a premixed powder from HiMedia.

Preparation of Medium: Add components to distilled/deionized water and bring volume to 1.0L. Mix thoroughly. Gently heat while stirring until boiling. Autoclave for 15 min at 15 psi pressure–121°C. Pour into sterile Petri dishes or distribute into sterile tubes.

Use: For the cultivation of *Bacillus subtilis.*

Maintenance of L Antigen in *Neisseria*

Composition per liter:

Proteose peptone No. 3	20.0g
Agar	15.0g
Na_2HPO_4	5.0g
NaCl	5.0g
Glucose	0.5g
Rabbit blood, defibrinated	100.0mL

pH 7.4–7.6 at 25°C

Preparation of Medium: Add components, except rabbit blood, to distilled/deionized water and bring volume to 900.0mL. Mix thoroughly. Gently heat while stirring until boiling. Autoclave for 20 min at 15 psi pressure–121°C. Cool to 60°C. Aseptically add 100.0mL of sterile, defibrinated rabbit blood. Maintain at 75°C while shaking for 30 min. Pour into sterile Petri dishes or distribute into sterile tubes.

Use: For the cultivation and maintenance of *Neisseria gonorrhoeae.*

Maintenance SCY Medium

See: **SCY Medium**

Maintenance (SCY) HiVeg Medium

Composition per liter:

Agar	10.0g
Sucrose	1.0g
Plant hydrolysate	0.91g
Yeast extract	0.25g
NaCl	0.05g
Papaic digest of soybean meal	0.03g
K_2HPO_4	0.02g
Thiamine	0.4mg

pH 7.3 ± 0.2 at 25°C

Source: This medium is available as a premixed powder from HiMedia.

Preparation of Medium: Add components to distilled/deionized water and bring volume to 1.0L. Mix thoroughly. Filter sterilize.

Use: For the cultivation and maintenance of iron and sulfur bacteria.

Malachite Green Broth

Composition per liter:

Peptone	15.0g
Beef extract	9.0g
Malachite Green	0.01mg

pH 7.3 ± 0.2 at 25°C

Preparation of Medium: Add components to distilled/deionized water and bring volume to 1.0L. Mix thoroughly. Distribute into tubes or flasks. Autoclave for 15 min at 15 psi pressure–121°C.

Use: For the cultivation of *Pseudomonas aeruginosa.*

Maleate Medium for *Pseudomonas fluorescens*

Composition per liter:

Agar	15.0g
K_2HPO_4	1.13g
NH_4NO_3	1.0g
KH_2PO_4	0.48g
$MgSO_4 \cdot 7H_2O$	0.2g
Potassium maleate solution	8.61mL

pH 7.0 ± 0.2 at 25°C

Potassium Maleate Solution:

Composition per liter:

Maleic acid	116.07g
KOH (10*N* solution)	200.0mL

Preparation of Potassium Maleate Solution: Add maleic acid to distilled/deionized water and bring volume to 600.0mL. Slowly add KOH solution (generates heat). Bring volume to 1.0L with distilled/deionized water. Adjust pH to 7.0. Filter sterilize.

Preparation of Medium: Add components, except potassium maleate solution, to distilled/deionized water and bring volume to 991.4mL. Mix thoroughly. Gently heat while stirring until boiling. Autoclave for 20 min at 15 psi pressure–121°C. Cool to 50°C. Aseptically add 8.61mL of the potassium maleate solution. Mix thoroughly. Pour into sterile Petri dishes or distribute into sterile tubes.

Use: For the cultivation and maintenance of *Pseudomonas fluorescens* and *Mycoplasma pneumoniae.*

Maleate Medium for *Pseudomonas fluorescens* with Glucose and Phenol Red

Composition per liter:

Agar	15.0g
Glucose	10.0g
K_2HPO_4	1.13g
NH_4NO_3	1.0g
KH_2PO_4	0.48g
$MgSO_4 \cdot 7H_2O$	0.2g
Phenol Red	0.04g
Potassium maleate solution	8.61mL

pH 7.0 ± 0.2 at 25°C

Potassium Maleate Solution:

Composition per liter:

Maleic acid	116.07g
KOH (10*N* solution)	200.0mL

Preparation of Potassium Maleate Solution: Add maleic acid to distilled/deionized water and bring volume to 600.0mL. Slowly add KOH solution (generates heat). Bring volume to 1.0L with distilled/deionized water. Adjust pH to 7.0. Filter sterilize.

Preparation of Medium: Add components, except potassium maleate solution, to distilled/deionized water and bring volume to 991.4mL. Mix thoroughly. Gently heat while stirring until boiling. Autoclave for 20 min at 15 psi pressure–121°C. Cool to 50°C. Aseptically add 8.61mL of the potassium maleate solution. Mix thoroughly. Pour into sterile Petri dishes or distribute into sterile tubes.

Use: For the cultivation and maintenance of *Pseudomonas fluorescens* and *Mycoplasma pneumoniae.*

Malonate Broth

Composition per liter:

Sodium malonate	3.0g
NaCl	2.0g
$(NH_4)_2SO_4$	2.0g
K_2HPO_4	0.6g
KH_2PO_4	0.4g
Bromthymol Blue	0.025g

pH 6.7 ± 0.2 at 25°C

Source: This medium is available as a premixed powder from BD Diagnostic Systems.

Preparation of Medium: Add components to distilled/deionized water and bring volume to 1.0L. Mix thoroughly. Distribute into tubes or flasks. Autoclave for 15 min at 15 psi pressure–121°C. Avoid introduction of carbon and nitrogen from other sources.

Use: For the cultivation and differentiation of coliforms and other enteric organisms, particularly *Enterobacter* and *Escherichia,* based on their ability to utilize malonate as the sole carbon source and ammonium sulfate as the sole nitrogen source. Malonate-utilizing organisms turn the medium blue.

Malonate Broth, Ewing Modified

Composition per liter:

Sodium malonate	3.0g
NaCl	2.0g
$(NH_4)_2SO_4$	2.0g
Yeast extract	1.0g
Glucose	0.25g
K_2HPO_4	0.6g
KH_2PO_4	0.4g
Bromthymol Blue	0.025g

pH 6.7 ± 0.2 at 25°C

Source: This medium is available as a premixed powder from BD Diagnostic Systems.

Preparation of Medium: Add components to distilled/deionized water and bring volume to 1.0L. Mix thoroughly. Distribute into tubes or flasks. Autoclave for 15 min at 15 psi pressure–121°C.

Use: For the cultivation and differentiation of coliforms and other enteric organisms, particularly *Enterobacter* and *Escherichia,* based on their ability to utilize malonate as a carbon source and ammonium sulfate as a nitrogen source. The small amount of yeast extract and glucose encourages the growth of some organisms that may be distressed or fail to respond. Malonate-utilizing organisms turn the medium blue.

Malt Agar

Composition per liter:

Malt extract	30.0g
Agar	15.0g

pH 5.5 ± 0.2 at 25°C

Source: This medium is available as a premixed powder from BD Diagnostic Systems.

Preparation of Medium: Add components to distilled/deionized water and bring volume to 1.0L. Mix thoroughly. Gently heat while stirring until boiling. Distribute into tubes or flasks. Autoclave for 15 min at 15 psi pressure–118°C. Do not overheat or agar will not harden. Pour into sterile Petri dishes or distribute into sterile tubes.

Use: For the cultivation of yeasts and molds.

Malt Agar

Composition per liter:

Agar	20.0g
Malt extract	12.5g

Preparation of Medium: Add components to distilled/deionized water and bring volume to 1.0L. Mix thoroughly. Gently heat until boiling. Distribute into tubes or flasks. Autoclave for 15 min at 15 psi pressure–121°C. Pour into sterile Petri dishes or leave in tubes.

Use: For the cultivation and maintenance of fungi.

Malt Agar, 1/3 Strength (ATCC Medium 2365)

Composition per liter:

Agar	15.0g
Malt extract	10.0g

pH 5.5 ± 0.2 at 25°C

Source: This medium is available as a premixed powder from BD Diagnostic Systems.

Preparation of Medium: Add components to distilled/deionized water and bring volume to 1.0L. Mix thoroughly. Gently heat while stirring until boiling. Distribute into tubes or flasks. Autoclave for 15 min at 15 psi pressure–118°C. Do not overheat or agar will not harden. Pour into sterile Petri dishes or distribute into sterile tubes.

Use: For the cultivation of yeasts and molds.

Malt Agar, Blakeslee

Composition per liter:

Glucose	20.0g
Malt extract	20.0g
Agar	16.0g
Mycological peptone	1.0g

pH 6.8 ± 0.2 at 25°C

Preparation of Medium: Add components to distilled/deionized water and bring volume to 1.0L. Mix thoroughly. Gently heat and bring to boiling. Distribute into tubes or flasks. Autoclave for 15 min at 15 psi pressure–121°C. Pour into sterile Petri dishes or leave in tubes.

Use: For the cultivation and maintenance of *Acremonium chrysogenum, Agaricus bisporus, Agrocybe aegerita, Armillaria mellea, Aspergillus* species, *Aureobasidium pullulans, Basidiomycetes* species, *Candida* species, *Ceratocystis adiposa, Citeromyces matritensis, Cladosporium cucumerinum, Cochliobolus miyaheanus, Colletotrichum destructivum, Colletotrichum lindemuthianum, Cryptococcus albidus,* numerous other *Cryptococcus* species, *Debaryomyces polymorphus, Diaporthe magnusii, Diaporthephaseolorum, Drechslera spicifera, Fusarium graminearum, Geotrichum lactis,* other *Geotrichum* species, *Hanseniaspora uvarum, Hanseniaspora valbyensis, Hansenula subpelliculosa, Humicola* species, *Kloeckera apiculata, Kloeckera corticis, Kluyveromyces lodderi, Kluyveromyces marxianus, Metschnikowia pulcherrima, Monilinia fructigena,*

Myrothecium verrucaria, Myxozyma melibiosi, Nadsonia fulvescens, Neurospora crassa, Octosporomyces octosporus, Pachysolen tannophilus, Paecilomyces variotii, Penicillium aurantiogriseum, many other *Penicillium* species, *Pithoascus schumacheri, Rhizoctonia crocorum, Rhizomucor miehei, Rhodosporidium toruloides, Rhodotorula glutinis, Rhodotorula minuta, Rhodotorula rubra, Saccharomyces cerevisiae, Saccharomyces exiguus, Saccharomyces ludwigii, Saccharomyces capsularis, Saccharomyces fibuligera, Schizosaccharomyces pombe, Sporobolomyces pararoseus, Sporobolomyces salmonicolor, Sporopachydermia lactativora, Stachybotrys* species, *Talaromyces emersonii, Taphrina deformans, Torulaspora delbrueckii, Trichoderma harzianum, Trichoderma koningii, Trichoderma longibrachiatum, Trichoderma reesei, Trichoderma viride, Trichosporon beigelii, Wickerhamia fluorescens, Yarrowia lipolytica,* and *Zygosaccharomyces veronae.*

Malt Agar 1% (MA1)

Composition per liter:

Agar	15.0g
Malt extract	10.0g

Preparation of Medium: Add components to distilled/deionized water and bring volume to 1.0L. Mix thoroughly. Gently heat and bring to boiling. Distribute into tubes or flasks. Autoclave for 15 min at 15 psi pressure–121°C. Pour into sterile Petri dishes or leave in tubes.

Use: For the cultivation of *Acrodontium griseum, Acrodontium simplex, Acrophialophora fusispora, Acrothecium tenebrosum, Acrothecium capsic, Agaricus bisporus, Chaetocladium jonesii, Chaetomium globosum, Chaetomium cupreum, Chaetomium irregulare, Farlowiella carmichaeliana,* and many other filamentous fungi.

Malt Agar 2% (MA2)

Composition per liter:

Malt extract	20.0g
Agar	15.0g

Preparation of Medium: Add components to distilled/deionized water and bring volume to 1.0L. Mix thoroughly. Gently heat and bring to boiling. Distribute into tubes or flasks. Autoclave for 15 min at 15 psi pressure–121°C. Pour into sterile Petri dishes or leave in tubes.

Use: For the cultivation of *Acrophialophora fusispora, Agaricus augustus, Ceratocystis penicillata, Chaetocladium brefeldii,* and many other fungi.

Malt Agar 4% (MA4)

Composition per liter:

Malt extract	40.0g
Agar	15.0g

Preparation of Medium: Add components to distilled/deionized water and bring volume to 1.0L. Mix thoroughly. Gently heat and bring to boiling. Distribute into tubes or flasks. Autoclave for 15 min at 15 psi pressure–121°C. Pour into sterile Petri dishes or leave in tubes.

Use: For the cultivation of *Chaetomium globosum, Chaetomium indicum, Chaetomium funicola, Chaetomium megalocarpum, Chaetomium murorum, Chaetomium seminudum, Chaetomium pachypodioides, Chaetomium perlucidum, Chaetomium quadrangulatum, Chaetomium reflexum, Chaetomium subaffine, Chaetomium subspirilliferum, Chaetomium succineum, Chaetomium luknowense, Daedalea quercina, Mycosphaerella ligulicola, Polypaecilum pisce,* and many other fungi.

Malt Agar 8% (MA8)

Composition per liter:

Malt extract	80.0g
Agar	15.0g

Preparation of Medium: Add components to distilled/deionized water and bring volume to 1.0L. Mix thoroughly. Gently heat and bring to boiling. Distribute into tubes or flasks. Autoclave for 15 min at 15 psi pressure–121°C. Pour into sterile Petri dishes or leave in tubes.

Use: For the cultivation of *Moniliella pollinis.*

Malt Agar with 0.5% $CaCO_3$

Composition per liter:

Malt extract	20.0g
Glucose	20.0g
Agar	15.0g
$CaCO_3$	5.0g
Mycological peptone	1.0g

pH 7.2 ± 0.2 at 25°C

Preparation of Medium: Add components to distilled/deionized water and bring volume to 1.0L. Mix thoroughly. Gently heat and bring to boiling. Distribute into tubes or flasks. Autoclave for 15 min at 15 psi pressure–121°C. Pour into sterile Petri dishes or leave in tubes.

Use: For the cultivation and maintenance of *Brettanomyces anomalus, Brettanomyces bruxellensis, Brettanomyces claussenii, Brettanomyces lambicus, Dekkera bruxellensis,* and *Dekkera intermedia.*

Malt Agar with 2% Malt

Composition per liter:

Agar	20.0g
Malt extract	20.0g

pH 6.5 ± 0.2 at 25°C

Preparation of Medium: Add components to distilled/deionized water and bring volume to 1.0L. Mix thoroughly. Adjust pH to 6.5. Gently heat and bring to boiling. Distribute into tubes or flasks. Autoclave for 15 min at 15 psi pressure–121°C. Pour into sterile Petri dishes or leave in tubes.

Use: For the cultivation of a variety of yeasts and other fungi.

Malt Agar 2% with Lupine Stems (MA2 with Lupine Stems)

Composition per liter:

Malt extract	20.0g
Agar	15.0g
Lupine stems	variable

Preparation of Medium: Add components, except lupine stems, to distilled/deionized water and bring volume to 1.0L. Mix thoroughly. Gently heat and bring to boiling. Distribute 6.0mL volumes into tubes. Cut lupine stems into 8.0cm-long pieces. Add 2–3 lupine stems per tube. Autoclave for 15 min at 15 psi pressure–121°C. Allow tubes to cool in a slanted position.

Use: For the cultivation of *Chaetomium bostrychodes.*

Malt Agar 4% with Lupine Stems (MA4 with Lupine Stems)

Composition per liter:

Malt extract	40.0g
Agar	15.0g
Lupine stems	variable

Preparation of Medium: Add components, except lupine stems, to distilled/deionized water and bring volume to 1.0L. Mix thoroughly. Gently heat and bring to boiling. Distribute 6.0mL volumes into tubes. Cut lupine stems into 8.0cm-long pieces. Add 2–3 lupine stems per tube. Autoclave for 15 min at 15 psi pressure–121°C. Allow tubes to cool in a slanted position.

Use: For the cultivation of *Chaetomium indicum* and *Chaetospermum chaetosporum.*

Malt Dextrose Peptone Agar (MDPA)

Composition per liter:

Agar	25.0g
Malt extract	20.0g
Glucose	20.0g
Peptone	1.0g

Preparation of Medium: Add components to distilled/deionized water and bring volume to 1.0L. Mix thoroughly. Gently heat and bring to boiling. Distribute into tubes or flasks. Autoclave for 15 min at 15 psi pressure–121°C. Pour into sterile Petri dishes or leave in tubes.

Use: For the cultivation of *Aspergillus aeneus, Aspergillus candidus, Odontia uda, Oidiodendron chlamydosporicum, Oidiodendron cerealis, Oidiodendron flavum, Oidiodendron griseum, Oidiodendron periconioides, Oidiodendron tenuissimum, Penicillium spinulosum,* and other fungi.

Malt Dextrose 40% Peptone Agar (MDPA 40)

Composition per liter:

Glucose	400.0g
Agar	25.0g
Malt extract	20.0g
Peptone	1.0g

Preparation of Medium: Add components to distilled/deionized water and bring volume to 1.0L. Mix thoroughly. Gently heat and bring to boiling. Distribute into tubes or flasks. Autoclave for 15 min at 15 psi pressure–121°C. Pour into sterile Petri dishes or leave in tubes.

Use: For the cultivation of *Aspergillus penicilloides.*

Malt 4% Dextrose Peptone Yeast Agar (MDPYA4)

Composition per liter:

Malt extract	40.0g
Agar	20.0g
Glucose	20.0g
Peptone	1.0g
Yeast extract	1.0g

Preparation of Medium: Add components to distilled/deionized water and bring volume to 1.0L. Mix thoroughly. Gently heat and bring to boiling. Distribute into tubes or flasks. Autoclave for 15 min at 15 psi pressure–121°C. Pour into sterile Petri dishes or leave in tubes.

Use: For the cultivation of *Candida glabrata, Candida haemulonii, Candida lactis-condensi, Candida magnoliae, Candida nemodendra, Metschnikowia pulcherrima, Metschnikowia reukaufii, Phoma glomerata, Saccharomyces exiguus,* and *Trigonopsis variabilis.*

Malt 4% Dextrose Yeast Agar (MDYA4)

Composition per liter:

Malt extract	40.0g
Agar	20.0g
Glucose	20.0g
Yeast extract	1.0g

Preparation of Medium: Add components to distilled/deionized water and bring volume to 1.0L. Mix thoroughly. Gently heat and bring to boiling. Distribute into tubes or flasks. Autoclave for 15 min at 15 psi pressure–121°C. Pour into sterile Petri dishes or leave in tubes.

Use: For the cultivation of *Torulaspora delbrueckii.*

Malt Extract Agar

Composition per liter:

Malt extract	30.0g
Agar	15.0g
Peptone	5.0g

pH 7.0 ± 0.2 at 25°C

Preparation of Medium: Add components to distilled/deionized water and bring volume to 1.0L. Mix thoroughly. Gently heat and bring to boiling. Distribute into tubes or flasks. Autoclave for 20 min at 10 psi pressure–115°C. Pour into sterile Petri dishes or leave in tubes.

Use: For the cultivation of *Xanthomonas* species.

Malt Extract Agar (MEA)

Composition per liter:

Agar	20.0g
Glucose	20.0g
Malt extract	20.0g
Peptone	1.0g

pH 5.5 ± 0.2 at 25°C

Preparation of Medium: Add components to distilled/deionized water and bring volume to 1.0L. Mix thoroughly. Gently heat and bring to boiling. Distribute into tubes or flasks. Autoclave for 15 min at 15 psi pressure–121°C. Pour into sterile Petri dishes or leave in tubes.

Use: For the isolation, cultivation, and identification of heat-resistant filamentous fungi (molds) from foods.

Malt Extract Agar

Composition per liter:

Malt extract	30.0g
Agar	15.0g
Mycological peptone	5.0g

pH 5.4 ± 0.2 at 25°C

Source: This medium is available as a premixed powder from Oxoid Unipath.

Preparation of Medium: Add components to distilled/deionized water and bring volume to 1.0L. Mix thoroughly. Gently heat while stirring until boiling. Distribute into tubes or flasks. Autoclave for 10 min at 15 psi pressure–115°C. Do not overheat or agar will not harden. If a lower pH (3.5) is desired, cool medium to 55°C and aseptically add 100.0mL of sterile lactic acid. Pour into sterile Petri dishes or distribute into sterile tubes.

Use: For the detection, isolation, and enumeration of yeasts and molds. The addition of lactic acid suppresses bacterial growth.

Malt Extract Agar

Composition per liter:

Malt extract	30.0g
Agar	20.0g
Chlortetracycline solution	10.0mL

pH 5.5 ± 0.2 at 25°C

Chlortetracycline Solution:

Composition per 10.0mL:

Chlortetracycline	0.04mg

Preparation of Chlortetracycline Solution: Add chlortetracycline to distilled/deionized water and bring volume to 10.0mL. Mix thoroughly. Filter sterilize.

Preparation of Medium: Add components, except chlortetracycline solution, to distilled/deionized water and bring volume to 990.0mL. Mix thoroughly. Gently heat and bring to boiling. Autoclave for 15 min at 15 psi pressure–121°C. Cool to 45°–50°C. Aseptically add sterile chlortetracycline solution. Mix thoroughly. Pour into sterile Petri dishes in 20.0mL volumes.

Use: For the cultivation of yeasts and filamentous fungi (molds) from cosmetics.

Malt Extract Agar

Composition per liter:

Malt extract	20.0g
Agar	15.0g
Peptone	5.0g

Preparation of Medium: Add components to distilled/deionized water and bring volume to 1.0L. Mix thoroughly. Gently heat and bring to boiling. Distribute into tubes or flasks. Autoclave for 15 min at 15 psi pressure–121°C. Pour into sterile Petri dishes or leave in tubes.

Use: For the cultivation and maintenance of *Candida melibiosica, Cryptococcus curvatus, Kluyveromyces* species, *Metschnikowia hawaiiensis, Rhodosporidium paludigenum, Saccharomyces cerevisiae, Saccharomycodes ludwigii, Stephanoascus* species, and *Trichosporon nigrescens*.

Malt Extract Agar (ATCC Medium 109)

Composition per liter:

Agar	15.0g
Maltose	12.75g
Dextrin	2.75g
Glycerol	2.35g
Pancreatic digest of gelatin	0.78g

pH 4.6 ± 0.2 at 25°C

Source: This medium is available as a premixed powder from BD Diagnostic Systems.

Preparation of Medium: Add components to distilled/deionized water and bring volume to 1.0L. Mix thoroughly. Gently heat while stirring until boiling. Distribute into tubes or flasks. Autoclave for 15 min at 15 psi pressure–118°C. Do not overheat or agar will not harden. Pour into sterile Petri dishes or distribute into sterile tubes.

Use: For the cultivation and maintenance of yeasts, molds, and *Flavobacterium lucecoloratum*.

Malt Extract Agar (MEA)

Composition per liter:

Glucose	20.0g
Malt extract	20.0g
Agar	15.0g
Peptone	1.0g

Preparation of Medium: Add components to distilled/deionized water and bring volume to 1.0L. Mix thoroughly. Gently heat and bring to boiling. Distribute into tubes or flasks. Autoclave for 15 min at 15 psi pressure–121°C. Pour into sterile Petri dishes or leave in tubes.

Use: For the cultivation of *Arthrinium phaeospermum, Aspergillus fumigatus, Aspergillus clavatus, Penicillium verruculosum,* and *Penicillium spinulosum.*

Malt Extract Agar (BAM M93)

Composition per liter:

Malt extract	30.0g
Agar	20.0g

pH 5.5 ± 0.2 at 25°C

Preparation of Medium: Add components to distilled/deionized water and bring volume to 1.0L. Mix thoroughly. Gently heat and bring to boiling. Distribute into tubes or flasks. Autoclave for 15 min at 15 psi pressure–121°C. Pour into sterile Petri dishes or leave in tubes.

Use: For the isolation, cultivation, and identification of heat-resistant filamentous fungi (molds) from foods.

Malt Extract Agar, Blakeslee's

Composition per liter:

Malt extract	20.0g
Glucose	20.0g
Agar	20.0g
Peptone	1.0g

Preparation of Medium: Add components to distilled/deionized water and bring volume to 1.0L. Mix thoroughly. Gently heat and bring to boiling. Distribute into tubes or flasks. Autoclave for 15 min at 15 psi pressure–121°C. Pour into sterile Petri dishes or leave in tubes.

Use: For the cultivation and maintenance of a variety of yeasts, including *Candida versatilis, Cryptococcus elinovii, Kluyveromyces marxianus, Pichia* species, *Reniforma strues, Rhodotorula mucilaginosa, Saccharomyces cerevisiae, Saccharomycopsis fibuligera, Sporobolomyces roseus, Trichosporon nigrescens, Yarrowia lipolytica, Zygosaccharomyces rouxii,* and numerous filamentous fungi.

Malt Extract Agar, Half Strength (ATCC Medium 2418)

Composition per liter:

Agar	15.0g
Malt extract	10.0g
Peptone	2.5g

pH 5.5 ± 0.2 at 25°C

Preparation of Medium: Add components to distilled/deionized water and bring volume to 1.0L. Mix thoroughly. Gently heat while stirring until boiling. Distribute into tubes or flasks. Autoclave for 15 min at 15 psi pressure–118°C. Do not overheat or agar will not harden. Pour into sterile Petri dishes or distribute into sterile tubes.

Use: For the cultivation of yeasts and molds.

Malt Extract Agar for Yeasts and Molds (MEAYM) (BAM M182)

Composition per liter:

Agar	20.0g
Glucose	20.0g
Malt extract	20.0g
Peptone	1.0g

pH 5.4 ± 0.2 at 25°C

Preparation of Medium: Add components to distilled/deionized water and bring volume to 1.0L. Mix thoroughly. Gently heat and bring to boiling. Distribute into tubes or flasks. Autoclave for 15 min at 15 psi pressure–121°C. Pour into sterile Petri dishes or leave in tubes.

Use: For the isolation, cultivation, and identification of heat-resistant filamentous fungi (molds) from foods. Recommended for identification of *Aspergillus* spp. and *Penicillium* spp.

Malt Extract Broth

Composition per liter:

Malt extract	6.0g
Glucose	6.0g
Maltose	1.8g
Yeast extract	1.2g

pH 4.7 ± 0.2 at 25°C

Source: This medium is available as a premixed powder from BD Diagnostic Systems.

Preparation of Medium: Add components to distilled/deionized water and bring volume to 1.0L. Mix thoroughly. Distribute into tubes or flasks. Autoclave for 15 min at 15 psi pressure–121°C. Do not overheat—this results in darkening of the broth.

Use: For the isolation, cultivation, and enumeration of yeast and filamentous fungi (mold).

Malt Extract Broth

Composition per liter:

Malt extract	17.0g
Mycological peptone	3.0g

pH 5.4 ± 0.2 at 25°C

Source: This medium is available as a premixed powder from Oxoid Unipath.

Preparation of Medium: Add components to distilled/deionized water and bring volume to 1.0L. Mix thoroughly. Distribute into tubes or flasks. Autoclave for 10 min at 15 psi pressure–115°C.

Use: For the cultivation of molds and yeasts, especially for sterility testing.

Malt Extract Broth

Composition per liter:

Malt extract, purified solids	15.0g

pH 4.7 ± 0.2 at 25°C

Source: This medium is available as a premixed powder from BD Diagnostic Systems.

Preparation of Medium: Add components to distilled/deionized water and bring volume to 1.0L. Mix thoroughly. Distribute into tubes or flasks. Autoclave for 15 min at 15 psi pressure–118°C. Do not overheat.

Use: For the cultivation of yeasts and molds.

Malt Extract Charcoal Medium (DSMZ Medium 801)

Composition per liter:

Malt extract	30.0g
Agar	15.0g
Charcoal	3.0g

pH 6.5 ± 0.2 at 25°C

Preparation of Medium: Add components to distilled/deionized water and bring volume to 1.0L. Mix thoroughly. Gently heat and bring to boiling. Distribute into tubes or flasks. Autoclave for 10 min at 15 psi pressure–121°C. Pour into sterile Petri dishes or leave in tubes.

Use: For the cultivation and maintenance of *Boletus edulis, Xerocomus badius* DSM 4436, *Antrodia serialis,* and other filamentous fungi.

Malt Extract Glucose Agar (DSMZ Medium 735)

Composition per liter:

Agar	15.0g
Malt extract	10.0g
Glucose	5.0g

pH 7.1 ± 0.2 at 25°C

Preparation of Medium: Add components to distilled/deionized water and bring volume to 1.0L. Mix thoroughly. Gently heat and bring to boiling. Distribute into tubes or flasks. Autoclave for 15 min at 15 psi pressure–121°C. Pour into sterile Petri dishes or leave in tubes.

Use: For the cultivation and maintenance of *Morchella esculenta* spp., *Verpa digitaliformis, Disciotis venosa, Mitrophora semilibera, Gyromitra* spp., and *Mitrophora semilibera.*

Malt Extract HiVeg Agar Base

Composition per liter:

Malt extract	30.0g
Agar	15.0g
Plant peptone No. 4	5.0g

pH 5.4 ± 0.2 at 25°C

Source: This medium is available as a premixed powder from HiMedia.

Preparation of Medium: Add components to distilled/deionized water and bring volume to 1.0L. Mix thoroughly. Adjust pH to 5.4. Gently heat and bring to boiling. Distribute into tubes or flasks. Autoclave for 15 min at 15 psi pressure–121°C. Pour into sterile Petri dishes or leave in tubes.

Use: For the cultivation and maintenance of fungi, especially yeasts.

Malt Extract HiVeg Agar Base, Modified

Composition per liter:

Agar	15.0g
Maltose	12.75g
Dextrin	2.75g
Plant peptone	0.78g

pH 6.0 ± 0.2 at 25°C

Source: This medium is available as a premixed powder from HiMedia.

Preparation of Medium: Add components to distilled/deionized water and bring volume to 1.0L. Mix thoroughly. Adjust pH to 6.0. Gently heat and bring to boiling. Distribute into tubes or flasks. Autoclave for 15 min at 15 psi pressure–121°C. Pour into sterile Petri dishes or leave in tubes.

Use: For the cultivation of yeasts and molds.

Malt Extract HiVeg Broth Base

Composition per liter:

Malt extract	17.0g
Plant peptone No. 4	3.0g

pH 5.4 ± 0.2 at 25°C

Source: This medium is available as a premixed powder from HiMedia.

Preparation of Medium: Add components to distilled/deionized water and bring volume to 1.0L. Mix thoroughly. Adjust pH to 5.4. Gently heat and bring to boiling. Distribute into tubes or flasks. Autoclave for 15 min at 15 psi pressure–121°C.

Use: For the cultivation of yeasts and molds.

Malt Extract Peptone Agar

Composition per liter:

Malt extract	30.0g
Agar	15.0g
Soy peptone	3.0g

pH 5.6 ± 0.2 at 25°C

Preparation of Medium: Add components to distilled/deionized water and bring volume to 1.0L. Mix thoroughly. Adjust pH to 5.6. Gently heat and bring to boiling. Distribute into tubes or flasks. Autoclave for 10 min at 15 psi pressure–121°C. Pour into sterile Petri dishes or leave in tubes.

Use: For the cultivation and maintenance of *Basidiobolus ranarum* and *Sclerophoma pityophila.*

Malt Extract Peptone Agar

Composition per liter:

Malt extract	30.0g
Agar	15.0g
Papaic digest of soybean meal	3.0g

pH 6.8 ± 0.2 at 25°C

Preparation of Medium: Add components to distilled/deionized water and bring volume to 1.0L. Mix thoroughly. Gently heat and bring to boiling. Distribute into tubes or flasks. Autoclave for 15 min at 15 psi pressure–121°C. Pour into sterile Petri dishes or leave in tubes.

Use: For the cultivation and maintenance of *Ascobolus immersus, Aspergillus amstelodami,* and *Aspergillus clavatus.*

Malt Extract Yeast Extract 40% Glucose Agar (MY40G)

Composition per liter:

Glucose	400.0g
Agar	12.0g
Malt extract powder	12.0g
Yeast extract	3.0g

pH 5.5 ± 0.2 at 25°C

Preparation of Medium: Add components, except glucose, to 550.0mL of distilled/deionized water. Mix thoroughly. Gently heat and bring to boiling. Bring volume to 600.0mL with distilled/deionized water. While the solution is still hot, add the glucose all at once while stirring to avoid formation of lumps. Autoclave for 30 min at 0 psi pressure–100°C.

Use: For the isolation and cultivation of osmotolerant microorganisms from foods.

Malt and Peptone Medium

Composition per liter:

Agar	15.0g
Malt extract	10.0g
Peptone	5.0g
NaCl	1.0g

pH 6.5 ± 0.2 at 25°C

Preparation of Medium: Add components to distilled/deionized water and bring volume to 1.0L. Mix thoroughly. Distribute into tubes or flasks. Adjust pH to 6.5. Autoclave for 15 min at 15 psi pressure–121°C.

Use: For the cultivation and maintenance of *Flavobacterium* species.

Malt Peptone Yeast Extract Agar

See: **MPY Agar**

Malt Yeast Agar

Composition per liter:

Agar	20.0g
Glucose	10.0g
Peptone	5.0g
Malt extract	3.0g
Yeast extract	3.0g

pH 7.0 ± 0.2 at 25°C

Preparation of Medium: Add components to distilled/deionized water and bring volume to 1.0L. Mix thoroughly. Gently heat and bring to boiling. Distribute into tubes or flasks. Autoclave for 20 min at 15 psi pressure–121°C. Pour into sterile Petri dishes or leave in tubes.

Use: For the cultivation of *Arthrobacter viscosus.*

Malt Yeast Extract 50% Glucose Agar (MY50G) (ATCC Medium 2093)

Composition per liter:

Glucose	500.0g
Agar	10.0g
Malt extract	10.0g
Yeast extract	2.5g

pH 5.5 ± 0.2 at 25°C

Preparation of Medium: Add agar, yeast extract, and malt extract to distilled/deionized water and bring volume to 500mL. Mix thoroughly. Gently heat while stirring until boiling. Slowly add glucose while stirring to avoid lumps. Distribute into tubes or flasks. Autoclave for 15 min at 15 psi pressure–121°C. Do not overheat or agar will not harden. Pour into sterile Petri dishes or leave in tubes. Note: This agar hardens very slowly.

Use: For the cultivation of yeasts and molds.

Malt 2% Yeast Extract Agar (MYA2)

Composition per liter:

Agar	20.0g
Malt extract	20.0g
Yeast extract	1.0g

Preparation of Medium: Add components to distilled/deionized water and bring volume to 1.0L. Mix thoroughly. Gently heat and bring to boiling. Distribute into tubes or flasks. Autoclave for 15 min at 15 psi pressure–121°C. Pour into sterile Petri dishes or leave in tubes.

Use: For the cultivation of *Actinomucor elegans, Actinospora megalospora, Agaricus bisporus, Ceratocystis perfecta, Ceratocystis cana, Ceratocystis seticollis, Chaetomium trilaterale, Chaetomium indicum, Chaetomium seminudum, Chaetomium piluliferum, Cirrenalia macrocephala, Kluyveromyces species, Lepista inversa, Torula dematia, Trichoderma pseudokoningii*, and other fungi.

Malt 4% Yeast Extract Agar (MYA4)

Composition per liter:

Malt extract	40.0g
Agar	20.0g
Yeast extract	1.0g

Preparation of Medium: Add components to distilled/deionized water and bring volume to 1.0L. Mix thoroughly. Gently heat and bring to boiling. Distribute into tubes or flasks. Autoclave for 15 min at 15 psi pressure–121°C. Pour into sterile Petri dishes or leave in tubes.

Use: For the cultivation of *Arthrobotrys arthrobotryoides, Ascosphaera apis, Bettsia alvei, Ceratocystiopsis minuta, Chaetomium spinosum, Chaetomium piluliferum, Ciboriopsis simulata, Dactylella minuta, Dactylella rhombospora, Dactylella lysipaga, Eriopeziza caesia, Europhium clavigerum, Issatchenkia orientalis, Moniliella suaveolens,* and other fungi.

Malt 4% Yeast Extract Agar with Lupine Stems (MYA4 with Lupine Stems)

Composition per liter:

Malt extract	40.0g
Agar	20.0g
Yeast extract	1.0g
Lupine stems	variable

Preparation of Medium: Add components, except lupine stems, to distilled/deionized water and bring volume to 1.0L. Mix thoroughly. Gently heat and bring to boiling. Distribute 6.0mL volumes into tubes. Cut lupine stems into 8.0cm-long pieces. Add 2–3 lupine stems per tube. Autoclave for 15 min at 15 psi pressure–121°C. Allow tubes to cool in a slanted position.

Use: For the cultivation of *Ceratocystiopsis minuta-bicolor, Ceratocystiopsis retusi, Ceratocystis pilifera, Ceratocystis olivaceapini, Ceratocystis multiannulata, Ceratocystis nigra, Ceratocystis olivacea, Ceratocystis tremuloaurea, Chaetomium indicum,* and *Chaetospermum chaetosporum.*

Malt 8% Yeast Extract Agar (MYA8)

Composition per liter:

Malt extract	80.0g
Agar	20.0g
Yeast extract	1.0g

Preparation of Medium: Add components to distilled/deionized water and bring volume to 1.0L. Mix thoroughly. Gently heat and bring to boiling. Distribute into tubes or flasks. Autoclave for 15 min at 15 psi pressure–121°C. Pour into sterile Petri dishes or leave in tubes.

Use: For the cultivation of *Thielavia hyalocarpa.*

Maltea

Composition per liter:

Maltea	40.0g
Agar	22.0g
Yeast extract	3.0g

Preparation of Medium: Add components to distilled/deionized water and bring volume to 1.0L. Mix thoroughly. Gently heat and bring to boiling. Distribute into tubes or flasks. Autoclave for 15 min at 15 psi pressure–121°C. Pour into sterile Petri dishes or leave in tubes.

Use: For the cultivation of *Leucogyrophana mollusca.*

Maltose Peptone Yeast Extract Agar
See: **MPY Agar**

Maltose Peptone Yeast Extract Broth
See: **MPY Broth**

Maltose Peptone Yeast Extract Medium
See: **MPY Agar**

Manganese Acetate Agar

Composition per liter:

Agar, highly purified	10.0g
Manganous acetate	0.1g

pH 7.0 ± 0.2 at 25°C

Preparation of Medium: Add manganous acetate to distilled/deionized water and bring volume to 1.0L. Mix thoroughly. Adjust pH to 7.0. Add agar. Steam the medium to dissolve agar. Distribute into screw-capped tubes or bottles. Autoclave for 15 min at 15 psi pressure–121°C. Allow tubes or bottles to cool in a slanted position.

Use: For the cultivation of manganese-oxidizing bacteria.

Manganese Agar No. 1 (Mn Agar No. 1)

Composition per liter:

Agar	10.0g
$MnCO_3$	2.0g
Beef extract	1.0g
$Fe(NH_4)_2(SO_4)_2$	0.15g
Sodium citrate	0.15g
Yeast extract	0.075g
Cyanocobalamin solution	10.0mL

Cyanocobalamin Solution:

Composition per 10.0mL:

Cyanocobalamin	0.005mg

Preparation of Cyanocobalamin Solution: Add cyanocobalamin to distilled/deionized water and bring volume to 10.0mL. Mix thoroughly. Filter sterilize.

Preparation of Medium: Add components, except cyanocobalamin, to distilled/deionized water and bring volume to 990.0mL. Mix thoroughly. Autoclave for 15 min at 15 psi pressure–121°C. Cool to 45°–50°C. Aseptically add 10.0mL of the sterile cyanocobalamin solution. Mix thoroughly. Pour into sterile Petri dishes or distribute into sterile tubes.

Use: For the isolation and cultivation of iron and sulfur bacteria. Also used to differentiate *Leptothrix* (*Sphaerotilus*) *discophorus* from *Sphaerotilus natans*.

Manganese Agar No. 2 (Mn Agar No. 2)

Composition per liter:

Agar	15.0g
$MnSO_4 \cdot H_2O$	10.0mg

Preparation of Medium: Add components to distilled/deionized water and bring volume to 1.0L. Mix thoroughly. Gently heat and bring to boiling. Distribute into tubes or flasks. Autoclave for 15 min at 15 psi pressure–121°C. Pour into sterile Petri dishes or leave in tubes. Use freshly prepared solution.

Use: For the enumeration, enrichment, and isolation of iron and sulfur bacteria. For the isolation and cultivation of *Leptothrix* species from water.

Manganese Medium for *Pseudomonas* species

Composition per liter:

Noble agar	10.0g
$MnCO_3$	1.0g
$Fe(NH_4)_2(SO_4)_2 \cdot 6H_2O$	0.15g
Sodium citrate	0.15g
Yeast extract	0.075g
$Na_4P_2O_7 \cdot 10H_2O$	0.05g

pH 6.8 ± 0.2 at 25°C

Preparation of Medium: Add components to distilled/deionized water and bring volume to 1.0L. Mix thoroughly. Gently heat and bring to boiling. Distribute into tubes or flasks. Autoclave for 15 min at 15 psi pressure–121°C. Pour into sterile Petri dishes or leave in tubes.

Use: For the cultivation and maintenance of *Pseudomonas putida* and other *Pseudomonas* species.

Manganese Nutrient Agar

Composition per liter:

Agar	15.0g
Peptone	5.0g
Meat extract	3.0g
$MnSO_4 \cdot H_2O$	5.0mg

Preparation of Medium: Add components to distilled/deionized water and bring volume to 1.0L. Mix thoroughly. Gently heat and bring to boiling. Distribute into tubes or flasks. Autoclave for 15 min at 15 psi pressure–121°C. Pour into sterile Petri dishes or leave in tubes.

Use: For the cultivation and obtaining the sporulation of *Bacillus* species.

Manning Medium (DSMZ Medium 1023)

Composition per liter:

$(NH_4)_2SO_4$	6.0g
$MgSO_4 \cdot 7H_2O$	1.0g
KCl	0.2g
K_2HPO_4	0.2g
$Ca(NO_3)_2$	0.02g
Solution A	100.0mL
Solution B	10.0mL

pH 2.6 ± 0.1 at 25°C

Solution A:

Composition per 100.0mL:

$FeSO_4 \cdot 7H_2O$	33.4g

Preparation of Solution A: Add $FeSO_4 \cdot 7H_2O$ to distilled/deionized water and bring volume to 100.0mL. Mix thoroughly. Autoclave for 15 min at 15 psi pressure–121°C.

Solution B:

Composition per 10.0mL:

Yeast extract	0.2g

Preparation of Solution B: Add yeast extract to distilled/deionized water and bring volume to 10.0mL. Mix thoroughly. Autoclave for 15 min at 15 psi pressure–121°C.

Preparation of Medium: Add components, except solutions A and B, to distilled/deionized water and bring volume to 900.0mL. Mix thoroughly. Adjust pH to 2.6. Gently heat while stirring and bring to boiling. Mix thoroughly. Autoclave for 15 min at 15 psi pressure–121°C. Aseptically add solutions A and B. Adjust pH to 2.5–2.7 with 0.1*N* H_2SO_4.

Use: For the cultivation of iron-oxidizing and heterotrophic acidophilic bacteria from acid mine drainage.

Mannitol Agar

Composition per liter:

Mannitol	25.0g
Agar	15.0g
Yeast extract	5.0g
Peptone	3.0g

Preparation of Medium: Add components to distilled/deionized water and bring volume to 1.0L. Mix thoroughly. Gently heat and bring to boiling. Distribute into tubes or flasks. Autoclave for 15 min at 15 psi pressure–121°C. Pour into sterile Petri dishes or leave in tubes.

Use: For the cultivation and maintenance of *Acetobacter aceti, Acetobacter hansenii, Acetobacter pasteurianus, Frateuria aurantia, Gluconobacter asaii, Gluconobacter cerinus, Gluconobacter oxydans*, and other bacteria that can utilize mannitol as a carbon source.

Mannitol Agar with Prilion

Composition per liter:

D-Mannitol	15.0g
Agar	13.0g
Meat peptone	10.0g
Meat extract	7.0g
NaCl	3.0g
$Na_2H_2PO_4$	2.0g
Prilion	2.0g
Metachrome Yellow	1.875g
Water Blue	0.625g

pH 7.2 ± 0.2 at 25°C

Source: This medium is available from HiMedia.

Preparation of Medium: Add components to distilled/deionized water and bring volume to 1.0L. Mix thoroughly. Gently heat and bring to boiling. Distribute into tubes or flasks. Autoclave for 15 min at 15 psi pressure–121°C. Pour into sterile Petri dishes or leave in tubes.

Use: For the isolation and differentiation of *Salmonella* spp. from *Proteus* species.

Mannitol Egg Yolk Polymyxin Agar

Composition per liter:

Agar	15.0g
D-Mannitol	10.0g
Peptone	10.0g
NaCl	10.0g
Beef extract	1.0g
Phenol Red	0.025g
Egg yolk emulsion, 50%	50.0mL
Polymyxin B solution	10.0mL

pH 7.1 ± 0.1 at 25°C

Source: Available as a prepared medium from BD Diagnostic Systems.

Egg Yolk Emulsion, 50%:
Composition per 100.0mL:

Chicken egg yolks	11
Whole chicken egg	1
NaCl (0.9% solution)	80.0mL

Preparation of Egg Yolk Emulsion, 50%: Soak eggs with 1:100 dilution of saturated mercuric chloride solution for 1 min. Crack eggs and separate yolks from whites. Mix egg yolks with 1 chicken egg. Beat to form emulsion. Measure 50.0mL of egg yolk emulsion and add to 50.0mL of 0.9% NaCl solution. Mix thoroughly. Filter sterilize. Warm to 45°–50°C.

Polymyxin B Solution:
Composition per 10.0mL:

Polymyxin B	100,000U

Preparation of Polymyxin B Solution: Add polymyxin B to distilled/deionized water and bring volume to 10.0mL. Mix thoroughly. Filter sterilize.

Preparation of Medium: Add components—except egg yolk emulsion, 50%, and polymyxin B solution—to distilled/deionized water and bring volume to 940.0mL. Mix thoroughly. Gently heat and bring to boiling. Autoclave for 15 min at 15 psi pressure–121°C. Cool to 45°–50°C. Aseptically add 50.0mL of sterile egg yolk emulsion, 50%, and 10.0mL of sterile polymyxin B solution. Mix thoroughly. Pour into sterile Petri dishes.

Use: For the cultivation and enumeration of *Bacillus cereus* from foods.

Mannitol Lysine Crystal Violet Brilliant Green Agar
See: **MLCB Agar**

Mannitol Maltose Agar

Composition per liter:

NaCl	20.0g
Agar	13.0g
D-Mannitol	10.0g
Maltose	10.0g
Beef extract	5.0g
Papaic digest of soybean meal	5.0g
Polypeptone™	5.0g
Dye stock solution, 1000X	1.0mL

pH 7.8 ± 0.2 at 25°C

Dye Stock Solution, 1000X:
Composition per 100.0mL:

Bromthymol Blue	4.0g
Cresol Red	4.0g
Ethanol, 95%	100.0mL

Preparation of Dye Stock Solution, 1000X: Add Bromthymol Blue and Cresol Red to 100.0mL of ethanol. Mix thoroughly.

Preparation of Medium: Add components to distilled/deionized water and bring volume to 1.0L. Mix thoroughly. Adjust to pH 7.8. Gently heat and bring to boiling. Distribute into tubes or flasks. Autoclave for 15 min at 15 psi pressure–121°C. Pour into sterile Petri dishes or leave in tubes.

Use: For the cultivation of *Vibrio* species from foods.

Mannitol Motility Test HiVeg Medium

Composition per liter:

Plant peptone	20.0g
Agar	3.0g
Mannitol	2.0g
Potassium nitrate	1.0g
Phenol red	0.04g

pH 7.6 ± 0.2 at 25°C

Source: This medium is available as a premixed powder from HiMedia.

Preparation of Medium: Add components to distilled/deionized water and bring volume to 1.0L. Mix thoroughly. Gently heat while stirring and bring to boiling. Distribute into tubes or flasks. Autoclave for 15 min at 15 psi pressure–121°C. Pour into sterile Petri dishes or leave in tubes.

Use: For the selective isolation, cultivation, and enumeration of staphylococci from clinical and nonclinical specimens. Mannitol-utilizing organisms turn the medium yellow.

Mannitol Salt Agar

Composition per liter:

NaCl	75.0g
Agar	15.0g
D-Mannitol	10.0g
Pancreatic digest of casein	5.0g
Peptic digest of animal tissue	5.0g
Beef extract	1.0g
Phenol Red	0.025g

pH 7.4 ± 0.2 at 25°C

Source: This medium is available as a premixed powder from BD Diagnostic Systems and Oxoid Unipath.

Preparation of Medium: Add components to distilled/deionized water and bring volume to 1.0L. Mix thoroughly. Gently heat while stirring and bring to boiling. Distribute into tubes or flasks. Autoclave for 15 min at 15 psi pressure–121°C. Pour into sterile Petri dishes or leave in tubes.

Use: For the selective isolation, cultivation, and enumeration of staphylococci from clinical and nonclinical specimens. Mannitol-utilizing organisms turn the medium yellow.

Mannitol Salt Agar (BAM M97)

Composition per liter:

NaCl	75.0g
Agar	15.0g
D-Mannitol	10.0g
Polypeptone	10.0g

Beef extract 1.0g
Phenol Red 0.025g

pH 7.4 ± 0.2 at 25°C

Preparation of Medium: Add components to distilled/deionized water and bring volume to 1.0L. Mix thoroughly. Gently heat while stirring and bring to boiling. Distribute into tubes or flasks. Autoclave for 15 min at 15 psi pressure–121°C. Pour into sterile Petri dishes or leave in tubes.

Use: For the selective isolation, cultivation, and enumeration of staphylococci. Mannitol-utilizing organisms turn the medium yellow.

Mannitol Salt Agar with Egg Yolk Emulsion

Composition per liter:
NaCl 75.0g
Agar 15.0g
D-Mannitol 10.0g
Proteose peptone 10.0g
Beef extract 1.0g
Phenol Red 0.025g
Egg yolk emulsion 100.0mL

pH 7.4 ± 0.2 at 25°C

Source: This medium is available from HiMedia.

Egg Yolk Emulsion
Composition per 100.0mL:
Chicken egg yolks 11
Whole chicken egg 1
NaCl (0.9% solution) 50.0mL

Preparation of Egg Yolk Emulsion: Soak eggs with 1:100 dilution of saturated mercuric chloride solution for 1 min. Crack eggs and separate yolks from whites. Mix egg yolks with 1 chicken egg. Beat to form emulsion. Measure 50.0mL of egg yolk emulsion and add to 50.0mL of 0.9% NaCl solution. Mix thoroughly. Filter sterilize. Warm to 45°–50°C.

Preparation of Medium: Add components, except egg yolk emulsion, to distilled/deionized water and bring volume to 900.0mL. Mix thoroughly. Autoclave for 15 min at 15 psi pressure–121°C. Cool to 50°C. Aseptically add egg yolk emulsion. Mix thoroughly. Pour into Petri dishes or aseptically distribute into sterile tubes or flasks.

Use: For the selective isolation of pathogenic staphylococci.

Mannitol Salt Broth

Composition per liter:
NaCl 75.0g
Proteose peptone 10.0g
D-Mannitol 10.0g
Beef extract 1.0g
Phenol Red 0.025g

pH 7.4 ± 0.2 at 25°C

Source: This medium is available from HiMedia.

Preparation of Medium: Add components to distilled/deionized water and bring volume to 1.0L. Mix thoroughly. Gently heat and bring to boiling. Distribute into tubes or flasks. Autoclave for 15 min at 15 psi pressure–121°C. Pour into sterile Petri dishes or leave in tubes.

Use: For the selective isolation of presumptive pathogenic staphylococci.

Mannitol Salt Broth

Composition per liter:
NaCl 100.0g
Pancreatic digest of casein 17.0g
Papaic digest of soybean meal 3.0g
K_2HPO_4 .... 2.5g
D-Mannitol 2.5g
Phenol Red 0.025g

pH 7.4 ± 0.2 at 25°C

Source: This medium is available as a premixed powder from BD Diagnostic Systems.

Preparation of Medium: Add components to distilled/deionized water and bring volume to 1.0L. Mix thoroughly. Distribute into tubes or flasks. Autoclave for 15 min at 15 psi pressure–121°C.

Use: For the selective isolation and cultivation of staphylococci from foods and nonclinical specimens. Mannitol-utilizing organisms turn the medium yellow.

Mannitol Salt HiVeg Agar Base

Composition per liter:
NaCl 75.0g
Agar 15.0g
D-Mannitol 10.0g
Plant peptone No. 3 .... 10.0g
Plant extract 1.0g
Phenol Red 0.025g

pH 7.4 ± 0.2 at 25°C

Source: This medium is available as a premixed powder from HiMedia.

Preparation of Medium: Add components to distilled/deionized water and bring volume to 1.0L. Mix thoroughly. Gently heat while stirring and bring to boiling. Distribute into tubes or flasks. Autoclave for 15 min at 15 psi pressure–121°C. Pour into sterile Petri dishes or leave in tubes.

Use: For the selective isolation of pathogenic staphylococci.

Mannitol Salt HiVeg Broth

Composition per liter:
NaCl 75.0g
D-Mannitol 10.0g
Plant peptone No. 3 .... 10.0g
Plant extract 1.0g
Phenol Red 0.025g

pH 7.4 ± 0.2 at 25°C

Source: This medium is available as a premixed powder from HiMedia.

Preparation of Medium: Add components to distilled/deionized water and bring volume to 1.0L. Mix thoroughly. Gently heat while stirring and bring to boiling. Distribute into tubes or flasks. Autoclave for 15 min at 15 psi pressure–121°C.

Use: For the selective isolation of presumptive pathogenic staphylococci.

Mannitol Selenite Broth (Selenite Mannitol Broth)

Composition per liter:
NaH_2PO_4 .... 10.0g
Peptic digest of animal tissue 5.0g

Mannitol	4.0g
$NaHSeO_3$	4.0g

pH 7.1 ± 0.2 at 25°C

Source: This medium is available from HiMedia.

Caution: Sodium hydrogen selenite (sodium biselenite) is a very toxic, corrosive agent and causes teratogenicity; it should be handled with great care. If there is contact, wash immediately with lots of water.

Preparation of Medium: Add $NaHSeO_3$ to distilled/deionized water and bring volume to 1.0L. Mix thoroughly. Add remaining components. Mix thoroughly. Distribute into tubes or flasks. Gently heat and bring to boiling. Sterilize in a boiling water bath or free flowing steam for 10 min. Do not autoclave. Discard the prepared medium if a large amount of selenite is reduced (indicated by red precipitate at the bottom of the tube).

Use: For the selective enrichment of *Salmonella* spp. from clinical materials.

Mannitol Selenite Broth with Brilliant Green

Composition per liter:

Meat peptone	5.0g
Yeast extract	5.0g
Mannitol	5.0g
K_2HPO_4	4.35g
KH_2PO_4	3.4g
Sodium taurocholate	1.0g
Brilliant Green	0.005g
$Na_2SeO_3 \cdot 5H_2O$	1.0g

pH 7.0 ± 0.2 at 25°C

Source: This medium is available from HiMedia.

Caution: Sodium hydrogen selenite (sodium biselenite) is a very toxic, corrosive agent and causes teratogenicity; iit should be handled with great care. If there is contact, wash immediately with lots of water.

Preparation of Medium: Add $NaHSeO_3$ to distilled/deionized water and bring volume to 1.0L. Mix thoroughly. Add remaining components. Mix thoroughly. Distribute into tubes or flasks. Gently heat and bring to boiling. Sterilize in a boiling water bath or free flowing steam for 10 min. Do not autoclave. Discard the prepared medium if a large amount of selenite is reduced (indicated by red precipitate at the bottom of the tube).

Use: For the enrichment of *Salmonella* spp. from feces, foodstuffs, and other materials.

Mannitol Yeast Extract Medium (LMG 135)

Composition per liter:

Agar	20.0g
Mannitol	10.0g
Yeast extract	1.0g
KH_2PO_4	0.5g
NaCl	0.1g
$CaCl_2 \cdot 2H_2O$ solution	1.0mL
$FeCl_3 \cdot 6H_2O$ solution	1.0mL

$CaCl_2 \cdot 2H_2O$ Solution:
Composition per 10.0mL:

$CaCl_2 \cdot 2H_2O$	5.28mg

Preparation of $CaCl_2 \cdot 2H_2O$ Solution: Add $CaCl_2 \cdot 2H_2O$ to distilled/deionized water and bring volume to 10.0mL. Mix thoroughly.

$FeCl_3 \cdot 6H_2O$ Solution:
Composition per 10.0mL:

$FeCl_3 \cdot 6H_2O$	0.66mg

Preparation of $FeCl_3 \cdot 6H_2O$ Solution: Add $FeCl_3 \cdot 6H_2O$ to distilled/deionized water and bring volume to 10.0mL. Mix thoroughly.

Preparation of Medium: Add components to distilled/deionized water and bring volume to 1.0L. Mix thoroughly. Gently heat and bring to boiling. Distribute into tubes or flasks. Autoclave for 15 min at 15 psi pressure–121°C. Pour into sterile Petri dishes or leave in tubes.

Use: For the cultivation and maintenance of *Rhizobium* species and *Bradyrhizobium* species.

Mannitol-Yeast Extract-Peptone (MYP) (DSMZ Medium 1087)

Composition per liter:

D-Mannitol	25.0g
Agar	15.0g
Yeast extract	5.0g
Peptone	3.0g

pH 7.2 ± 0.2 at 25°C

Preparation of Medium: Add components to distilled/deionized water and bring volume to 1.0L. Mix thoroughly. Adjust pH to 7.2. Distribute into tubes or flasks. Gently heat while stirring and bring to boiling. Mix thoroughly. Autoclave for 15 min at 15 psi pressure–121°C. Pour into Petri dishes or leave in tubes.

Use: For the cultivation of *Acetobacter fabarum.*

Mannitol Yolk Polymyxin Agar (MYP Agar)

Composition per 110.0mL:

Agar	1.5g
NaCl	1.0g
Peptone	1.0g
D-Mannitol	1.0g
$(NH_4)_2PO_4$	0.1g
Meat extract	0.1g
Phenol Red	2.5mg
Egg yolk emulsion, 20%	10.0mL
Polymyxin B solution	1.0mL

pH 7.1 ± 0.2 at 25°C

Egg Yolk Emulsion, 20%:
Composition per 100.0mL:

Chicken egg yolks	11
Whole chicken egg	1
NaCl (0.9% solution)	80.0mL

Preparation of Egg Yolk Emulsion, 20%: Soak eggs with 1:100 dilution of saturated mercuric chloride solution for 1 min. Crack eggs and separate yolks from whites. Mix egg yolks with 1 chicken egg. Measure 20.0mL of egg yolk emulsion and add to 80.0mL of 0.9% NaCl solution. Mix thoroughly. Filter sterilize. Warm to 45°–50°C.

Polymyxin B Solution:
Composition per 1.0mL:

Polymyxin B	1.0mg

Preparation of Polymyxin B Solution: Add polymyxin B to distilled/deionized water and bring volume to 1.0mL. Mix thoroughly. Filter sterilize.

Preparation of Medium: Add components—except egg yolk emulsion, 20%, and polymyxin B solution—to distilled/deionized water and bring volume to 100.0mL. Mix thoroughly. Gently heat and bring to boiling. Autoclave for 15 min at 15 psi pressure–121°C. Cool to 45°–50°C. Aseptically add 10.0mL of sterile egg yolk emulsion, 20%, and 1.0mL of sterile polymyxin B solution. Mix thoroughly. Pour into sterile Petri dishes.

Use: For the cultivation and maintenance of *Bacillus cereus*.

Maricaulis Medium (DSMZ Medium 1025)

Composition per liter:

Sea salts, Sigma	30.0g
NH_4Cl	0.5g
Peptone yeast extract solution	20.0mL
Glucose solution	2.0mL
Riboflavin solution	5.0mL

pH 7.2 ± 0.2 at 25°C

Glucose Solution:

Composition per 10.0mL:

Glucose	5.0g

Preparation of Glucose Solution: Add glucose to distilled/deionized water and bring volume to 10.0mL. Mix thoroughly. Filter sterilize.

Riboflavin Solution:

Composition per 10.0mL:

Riboflavin	2.0mg

Preparation of Riboflavin Solution: Add riboflavin to distilled/deionized water and bring volume to 10.0mL. Mix thoroughly. Filter sterilize.

Peptone Yeast Extract Solution:

Composition per 100.0mL:

Peptone	10.0g
Yeast extract	5.0g

Preparation of Peptone Yeast Extract Solution: Add components to distilled/deionized water and bring volume to 100.0mL. Mix thoroughly. Autoclave for 15 min at 15 psi pressure–121°C.

Preparation of Medium: Add components, except riboflavin, glucose, and peptone yeast extract solutions, to distilled/deionized water and bring volume to 980.0mL. Mix thoroughly. Adjust pH to 7.2. Gently heat while stirring and bring to boiling. Mix thoroughly. Autoclave for 15 min at 15 psi pressure–121°C. Cool to room temperature. Aseptically add riboflavin, glucose, and peptone yeast extract solutions.

Use: For the cultivation of *Maricaulis* spp.

Marine Agar (DSMZ Medium 123)

Composition per liter:

Agar	15.0g
Tryptone	10.0g
Peptone	5.0g
Yeast extract	1.0g
Synthetic seawater	1.0L

pH 7.8 ± 0.2 at 25°C

Synthetic Seawater:

Composition per liter:

NaCl	24.0g
$MgCl_2 \cdot 6H_2O$	11.0g
Na_2SO_4	4.0g
$CaCl_2 \cdot 6H_2O$	2.0g
KCl	0.7g
KBr	0.1g
$SrCl_2 \cdot 6H_2O$	0.04g
H_3BO_3	0.03g
$NaSiO_3 \cdot 9H_2O$	5.0mg
NaF	3.0mg
NH_4NO_3	2.0mg
$Fe_3PO_4 \cdot 4H_2O$	1.0mg

Preparation of Synthetic Seawater: Add components to distilled water and bring volume to 1.0L. Mix thoroughly.

Preparation of Medium: Add agar, tryptone, peptone, and yeast extract to synthetic seawater and bring volume to 1.0L. Mix thoroughly. Adjust pH to 7.8. Gently heat and bring to boiling. Distribute into tubes or flasks. Autoclave for 15 min at 15 psi pressure–121°C. Pour into sterile Petri dishes or leave in tubes.

Use: For the cultivation and maintenance of *Halobacillus halophilus, Halomonas* spp., *Vibrio harveyi, Cobetia marina, and Ruegeria atlantica.*

Marine Agar 2216 (DSMZ Medium 604)

Composition per liter:

NaCl	19.45g
Agar	15.0g
$MgCl_2$	8.8g
Peptone	5.0g
Na_2SO_3	3.24g
$CaCl_2$	1.8g
Yeast extract	1.0g
KCl	0.55g
$NaHCO_3$	0.16g
Ferric citrate	0.1g
KBr	0.08g
$SrCl_2$	0.03g
H_3BO_3	0.02g
Na_2HPO_4	8.0mg
Na_2SiO_3	4.0mg
NaF	2.4mg
NH_4NO_3	1.6mg

pH 7.6 ± 0.2 at 25°C

Source: This medium is available as a premixed powder from BD Diagnostic Systems.

Preparation of Medium: Add components to distilled/deionized water and bring volume to 1.0L. Mix thoroughly. Gently heat while stirring and bring to boiling. Distribute into tubes or flasks. Autoclave for 15 min at 15 psi pressure–121°C. Pour into sterile Petri dishes or leave in tubes.

Use: For the cultivation and maintenance of *Hyphomonas* spp., *Oceanospirillum* spp., *Hyphomicrobium indicum, Psychroflexus gondwanensis=Flavobacterium gondwanense, Salegentibacter salegens=Flavobacterium salegens, Psychromonas antarctica, Sulfitobacter mediterraneus, Thalassomonas viridans, Vibrio* spp., *Marinospirillum minutulum=Oceanospirillum minutulum, Terasakiella pusilla=Oceanospirillum pusillum, Pseudoalteromonas atlantica=Alteromonas atlantica, Pseudomonas atlantica, Roseobacter* spp., *Erythrobacter longus, Pseudospirillum japonicum=Oceanospirillum japonicum, Marinobacter hydrocarbonoclasticus (Pseudomonas nautica), Psychrobacter* spp., and *Moritella japonica.* For the isolation, cultivation, and maintenance of a wide variety of heterotrophic marine bacteria.

Marine Agar with Biphenyl

Composition per liter:

NaCl	19.45g
Agar	15.0g
$MgCl_2$	8.8g
Peptone	5.0g
Na_2SO_3	3.24g
$CaCl_2$	1.8g
Yeast extract	1.0g
KCl	0.55g
$NaHCO_3$	0.16g
Ferric citrate	0.1g
KBr	0.08g
$SrCl_2$	0.03g
H_3BO_3	0.02g
Na_2HPO_4	8.0mg
Na_2SiO_3	4.0mg
NaF	2.4mg
NH_4NO_3	1.6mg
Biphenyl	1.0mg

pH 7.6 ± 0.2 at 25°C

Preparation of Medium: Add components, except biphenyl, to distilled/deionized water and bring volume to 1.0L. Mix thoroughly. Gently heat and bring to boiling. Distribute into tubes or flasks. Autoclave for 15 min at 15 psi pressure–121°C. Pour into sterile Petri dishes or leave in tubes. After agar solidifies, aseptically add a few crystals of biphenyl to each plate.

Use: For the cultivation and maintenance of biphenyl-utilizing marine bacteria, such as *Cycloclasticus pugetii.*

Marine Agar with Lambda Carrageenan

Composition per 1070.0mL:

Solution A	1.0L
Solution B	60.0mL
Solution C	10.0mL

pH 7.2 ± 0.2 at 25°C

Solution A:

Composition per liter:

NaCl	25.0g
Agar	15.0g
$MgSO_4·7H_2O$	5.0g
Casamino acids	2.5g
Lambda-carrageenan	2.5g
$NaNO_3$	2.0g
$CaCl_2·2H_2O$	0.2g
KCl	0.1g

Preparation of Solution A: Add components to distilled/deionized water and bring volume to 1.0L. Mix thoroughly. Gently heat and bring to boiling. Autoclave for 15 min at 15 psi pressure–121°C.

Solution B:

Composition per 100.0mL:

$Na_2HPO_4·2H_2O$	3.56g

Preparation of Solution B: Add $Na_2HPO_4·2H_2O$ to distilled/deionized water and bring volume to 100.0mL. Mix thoroughly. Autoclave for 15 min at 15 psi pressure–121°C.

Solution C:

Composition per 100.0mL:

$FeSO_4·7H_2O$	0.3g

Preparation of Solution C: Add $FeSO_4·7H_2O$ to distilled/deionized water and bring volume to 100.0mL. Mix thoroughly. Autoclave for 15 min at 15 psi pressure–121°C.

Preparation of Medium: Aseptically add 60.0mL of sterile solution B and 10.0mL of sterile solution C to 1.0L of sterile solution A. Mix thoroughly. Pour into sterile Petri dishes or distribute into sterile tubes.

Use: For the cultivation and maintenance of ATCC strain 43554.

Marine Agar with Kappa and Lambda Carrageenan

Composition per 1070.0mL:

Solution A	1.0L
Solution B	60.0mL
Solution C	10.0mL

pH 7.2 ± 0.2 at 25°C

Solution A:

Composition per liter:

NaCl	25.0g
Agar	15.0g
$MgSO_4·7H_2O$	5.0g
Casamino acids	2.5g
$NaNO_3$	2.0g
κ-Carrageenan	1.25g
λ-Carrageenan	1.25g
$CaCl_2·2H_2O$	0.2g
KCl	0.1g

Preparation of Solution A: Add components to distilled/deionized water and bring volume to 1.0L. Mix thoroughly. Gently heat and bring to boiling. Autoclave for 15 min at 15 psi pressure–121°C.

Solution B:

Composition per 100.0mL:

$Na_2HPO_4·2H_2O$	3.56g

Preparation of Solution B: Add $Na_2HPO_4·2H_2O$ to distilled/deionized water and bring volume to 100.0mL. Mix thoroughly. Autoclave for 15 min at 15 psi pressure–121°C.

Solution C:

Composition per 100.0mL:

$FeSO_4·7H_2O$	0.3g

Preparation of Solution C: Add $FeSO_4·7H_2O$ to distilled/deionized water and bring volume to 100.0mL. Mix thoroughly. Autoclave for 15 min at 15 psi pressure–121°C.

Preparation of Medium: Aseptically add 60.0mL of sterile solution B and 10.0mL of sterile solution C to 1.0L of sterile solution A. Mix thoroughly. Pour into sterile Petri dishes or distribute into sterile tubes.

Use: For the cultivation and maintenance of *Pseudomonas carrageenovora.*

Marine Agar with Naphthalene

Composition per liter:

NaCl	19.45g
Agar	15.0g
$MgCl_2$	8.8g

Peptone	5.0g
Na_2SO_3	3.24g
$CaCl_2$	1.8g
Yeast extract	1.0g
KCl	0.55g
$NaHCO_3$	0.16g
Ferric citrate	0.1g
KBr	0.08g
$SrCl_2$	0.03g
H_3BO_3	0.02g
Na_2HPO_4	8.0mg
Na_2SiO_3	4.0mg
NaF	2.4mg
NH_4NO_3	1.6mg
Naphthalene	1mg

pH 7.6 ± 0.2 at 25°C

Preparation of Medium: Add components, except naphthalene, to distilled/deionized water and bring volume to 1.0L. Mix thoroughly. Gently heat and bring to boiling. Distribute into tubes or flasks. Autoclave for 15 min at 15 psi pressure–121°C. Pour into sterile Petri dishes or leave in tubes. After agar solidifies, aseptically add a few crystals of naphthalene to each plate.

Use: For the cultivation and maintenance of naphthalene-utilizing marine bacteria

Marine Agar with Sulfur (ATCC Medium 1922)

Composition per liter:

NaCl	19.45g
Sulfur	10.0g
$MgCl_2$	8.8g
Peptone	5.0g
Na_2SO_3	3.24g
$CaCl_2$	1.8g
Yeast extract	1.0g
KCl	0.55g
$NaHCO_3$	0.16g
Ferric citrate	0.1g
KBr	0.08g
$SrCl_2$	0.03g
H_3BO_3	0.02g
Na_2HPO_4	8.0mg
Na_2SiO_3	4.0mg
NaF	2.4mg
NH_4NO_3	1.6mg

pH 7.6 ± 0.2 at 25°C

Preparation of Sulfur: Autoclave sulfur for 15 min at 0 psi pressure–100°C on 3 successive days.

Preparation of Medium: Prepare anaerobically under a gas phase of 80% N_2 + 10% CO_2 + 10% H_2. Add components, except sulfur, to distilled/deionized water and bring volume to 1.0L. Mix thoroughly. Gently heat while stirring and bring to boiling. Autoclave for 15 min at 15 psi pressure–121°C. Cool to 50–55°C. Aseptically add 10.0g of sterile sulfur. Mix thoroughly. Aseptically and anaerobically, under a gas phase of 80% N_2 + 10% CO_2 + 10% H_2, distribute into sterile tubes.

Use: For the cultivation and maintenance of *Thermococcus litoralis.*

Marine *Ameba* Medium

Composition per liter:

Agar	10.0g
Malt extract	0.1g
Yeast extract	0.1g
Artificial seawater	1.0L

Artificial Seawater:

Composition per liter:

NaCl	27.5g
$MgSO_4 \cdot 7H_2O$	6.78g
$MgCl_2 \cdot 6H_2O$	5.38g
KCl	0.72g
$NaHCO_3$	0.2g
$CaCL_2 \cdot 2H_2O$	1.4g

Preparation of Artificial Seawater: Add components to distilled/deionized water and bring volume to 1.0L. Mix thoroughly.

Preparation of Medium: Add components to artificial seawater and bring volume to 1.0L. Mix thoroughly. Gently heat and bring to boiling. Distribute into tubes or flasks. Autoclave for 15 min at 15 psi pressure–121°C. Pour into sterile Petri dishes or leave in tubes.

Use: For the cultivation of *Cochliopodium clarum, Heteramoeba clara, Lingulamoeba leei, Paramoeba pemaquidensis,* and *Vannella* species.

Marine Broth 2216 (LMG Medium 164)

Composition per liter:

NaCl	19.45g
$MgCl_2$	8.8g
Peptone	5.0g
Na_2SO_3	3.24g
$CaCl_2$	1.8g
Yeast extract	1.0g
KCl	0.55g
$NaHCO_3$	0.16g
Ferric citrate	0.1g
KBr	0.08g
$SrCl_2$	0.03g
H_3BO_3	0.02g
Na_2HPO_4	8.0mg
Na_2SiO_3	4.0mg
NaF	2.4mg
NH_4NO_3	1.6mg

pH 7.6 ± 0.2 at 25°C

Source: This medium is available as a premixed powder from BD Diagnostic Systems.

Preparation of Medium: Add components to distilled/deionized water and bring volume to 1.0L. Mix thoroughly. Gently heat while stirring and bring to boiling. Distribute into tubes or flasks. Autoclave for 15 min at 15 psi pressure–121°C.

Use: For the cultivation of *Vibrio liquefaciens* and for the isolation, cultivation, and maintenance of a wide variety of heterotrophic marine bacteria.

Marine Broth with Biphenyl

Composition per liter:

NaCl	19.45g
$MgCl_2$	8.8g

Peptone	5.0g
Na_2SO_3	3.24g
$CaCl_2$	1.8g
Yeast extract	1.0g
KCl	0.55g
$NaHCO_3$	0.16g
Ferric citrate	0.1g
KBr	0.08g
$SrCl_2$	0.03g
H_3BO_3	0.02g
Na_2HPO_4	8.0mg
Na_2SiO_3	4.0mg
NaF	2.4mg
NH_4NO_3	1.6mg
Biphenyl	1.0mg

pH 7.6 ± 0.2 at 25°C

Preparation of Medium: Add components, except biphenyl, to distilled/deionized water and bring volume to 1.0L. Mix thoroughly. Distribute into tubes or flasks. Autoclave for 15 min at 15 psi pressure–121°C. Aseptically add a few crytals of biphenyl to each tube or flask.

Use: For the cultivation of biphenyl-utilizing marine bacteria.

Marine Broth with Lambda Carrageenan

Composition per 1070.0mL:

Solution A	1.0L
Solution B	60.0mL
Solution C	10.0mL

pH 7.2 ± 0.2 at 25°C

Solution A:

Composition per liter:

NaCl	25.0g
$MgSO_4 \cdot 7H_2O$	5.0g
Casamino acids	2.5g
λ-Carrageenan	2.5g
$NaNO_3$	2.0g
$CaCl_2 \cdot 2H_2O$	0.2g
KCl	0.1g

Preparation of Solution A: Add components to distilled/deionized water and bring volume to 1.0L. Mix thoroughly. Gently heat and bring to boiling. Autoclave for 15 min at 15 psi pressure–121°C.

Solution B:

Composition per 100.0mL:

$Na_2HPO_4 \cdot 2H_2O$	3.56g

Preparation of Solution B: Add $Na_2HPO_4 \cdot 2H_2O$ to distilled/deionized water and bring volume to 100.0mL. Mix thoroughly. Autoclave for 15 min at 15 psi pressure–121°C.

Solution C:

Composition per 100.0mL:

$FeSO_4 \cdot 7H_2O$	0.3g

Preparation of Solution C: Add $FeSO_4 \cdot 7H_2O$ to distilled/deionized water and bring volume to 100.0mL. Mix thoroughly. Autoclave for 15 min at 15 psi pressure–121°C.

Preparation of Medium: Aseptically add 60.0mL of sterile solution B and 10.0mL of sterile solution C to 1.0L of sterile solution A. Mix thoroughly. Pour into sterile Petri dishes or distribute into sterile tubes.

Use: For the cultivation of ATCC strain 43554.

Marine Broth with Kappa and Lambda Carrageenan

Composition per 1070.0mL:

Solution A	1.0L
Solution B	60.0mL
Solution C	10.0mL

pH 7.2 ± 0.2 at 25°C

Solution A:

Composition per liter:

NaCl	25.0g
$MgSO_4 \cdot 7H_2O$	5.0g
Casamino acids	2.5g
$NaNO_3$	2.0g
κ-Carrageenan	1.25g
λ-Carrageenan	1.25g
$CaCl_2 \cdot 2H_2O$	0.2g
KCl	0.1g

Preparation of Solution A: Add components to distilled/deionized water and bring volume to 1.0L. Mix thoroughly. Gently heat and bring to boiling. Autoclave for 15 min at 15 psi pressure–121°C.

Solution B:

Composition per 100.0mL:

$Na_2HPO_4 \cdot 2H_2O$	3.56g

Preparation of Solution B: Add $Na_2HPO_4 \cdot 2H_2O$ to distilled/deionized water and bring volume to 100.0mL. Mix thoroughly. Autoclave for 15 min at 15 psi pressure–121°C.

Solution C:

Composition per 100.0mL:

$FeSO_4 \cdot 7H_2O$	0.3g

Preparation of Solution C: Add $FeSO_4 \cdot 7H_2O$ to distilled/deionized water and bring volume to 100.0mL. Mix thoroughly. Autoclave for 15 min at 15 psi pressure–121°C.

Preparation of Medium: Aseptically add 60.0mL of sterile solution B and 10.0mL of sterile solution C to 1.0L of sterile solution A. Mix thoroughly. Distribute into sterile tubes or flasks.

Use: For the cultivation and maintenance of *Pseudomonas carrageenovora.*

Marine Broth with Naphthalene

Composition per liter:

NaCl	19.45g
$MgCl_2$	8.8g
Peptone	5.0g
Na_2SO_3	3.24g
$CaCl_2$	1.8g
Yeast extract	1.0g
KCl	0.55g
$NaHCO_3$	0.16g
Ferric citrate	0.1g
KBr	0.08g
$SrCl_2$	0.03g
H_3BO_3	0.02g
Na_2HPO_4	8.0mg
Na_2SiO_3	4.0mg
NaF	2.4mg
NH_4NO_3	1.6mg
Naphthalene	1mg

pH 7.6 ± 0.2 at 25°C

Preparation of Medium: Add components, except naphthalene, to distilled/deionized water and bring volume to 1.0L. Mix thoroughly. Distribute into tubes or flasks. Autoclave for 15 min at 15 psi pressure–121°C. Aseptically add a few crytals of naphthalene to each tube or flask.

Use: For the cultivation of naphthalene-utilizing marine bacteria.

Marine Broth with Sulfur

Composition per liter:

NaCl	19.45g
Sulfur	10.0g
$MgCl_2$	8.8g
Peptone	5.0g
Na_2SO_3	3.24g
$CaCl_2$	1.8g
Yeast extract	1.0g
KCl	0.55g
$NaHCO_3$	0.16g
Ferric citrate	0.1g
KBr	0.08g
$SrCl_2$	0.03g
H_3BO_3	0.02g
Na_2HPO_4	8.0mg
Na_2SiO_3	4.0mg
NaF	2.4mg
NH_4NO_3	1.6mg

pH 7.6 ± 0.2 at 25°C

Preparation of Sulfur: Autoclave for 15 min at 0 psi pressure–100°C on three successive days.

Preparation of Medium: Prepare anaerobically under a gas phase of 80% N_2 + 10% CO_2 + 10% H_2. Add components, except sulfur, to distilled/deionized water and bring volume to 1.0L. Mix thoroughly. Gently heat while stirring and bring to boiling. Autoclave for 15 min at 15 psi pressure–121°C. Cool to 50°C. Aseptically add 10.0g of sulfur. Mix thoroughly. Aseptically and anaerobically, under a gas phase of 80% N_2 + 10% CO_2 + 10% H_2, distribute into sterile tubes.

Use: For the cultivation of *Thermococcus litoralis.*

Marine *Caulobacter* Medium

Composition per liter:

Proteose peptone	10.0g
Yeast extract	3.0g
Artificial seawater	1.0L

pH 7.2–7.4 at 25°C

Artificial Seawater:

Composition per liter:

Commercially available marine aquarium salts mixture	variable

Preparation of Artificial Seawater: Add commercially available marine aquarium salts mixture to distilled/deionized water and bring volume to 1.0L. Mix thoroughly.

Preparation of Medium: Combine components. Mix thoroughly. Distribute into tubes or flasks. Autoclave for 15 min at 15 psi pressure–121°C.

Use: For the cultivation of *Caulobacter halobacteroides* and *Caulobacter maris.*

Marine Chlorobiaceae Medium 2

Composition per 1051.0mL:

Solution 1	950.0mL
$Na_2S·9H_2O$ solution	60.0mL
$NaHCO_3$ solution	40.0mL
Vitamin B_{12} solution	1.0mL

pH 6.8 ± 0.2 at 25°C

Solution 1:

Composition per 950.0mL:

NaCl	20.0g
$MgSO_4·7H_2O$	3.0g
KH_2PO_4	1.0g
NH_4Cl	0.5g
$CaCl_2·2H_2O$	0.05g
Trace elements solution SL-8	1.0mL

Preparation of Solution 1: Add components to distilled/deionized water and bring volume to 950.0mL. Mix thoroughly. Autoclave for 15 min at 15 psi pressure–121°C. Cool to 45°–50°C.

Trace Elements Solution SL-8:

Composition per liter:

Disodium EDTA	5.2g
$FeCl_2·4H_2O$	1.5g
$CoCl_2·6H_2O$	0.19g
$MnCl_2·4H_2O$	0.1g
$ZnCl_2$	0.07g
H_3BO_3	0.06g
$NaMoO_4·2H_2O$	0.04g
$CuCl_2·2H_2O$	0.02g
$NiCl_2·6H_2O$	0.02g

Preparation of Trace Elements Solution SL-8: Add components to distilled/deionized water and bring volume to 1.0L. Mix thoroughly.

$Na_2S·9H_2O$ Solution:

Composition per 100.0mL:

$Na_2S·9H_2O$	5.0g

Preparation of $Na_2S·9H_2O$ Solution: Add $Na_2S·9H_2O$ to distilled/deionized water and bring volume to 100.0mL. Autoclave for 15 min at 15 psi pressure–121°C. Cool to 45°–50°C.

$NaHCO_3$ Solution:

Composition per 100.0mL:

$NaHCO_3$	5.0g

Preparation of $NaHCO_3$ Solution: Add $NaHCO_3$ to distilled/deionized water and bring volume to 100.0mL. Mix thoroughly. Filter sterilize.

Vitamin B_{12} Solution:

Composition per 100.0mL:

Vitamin B_{12}	2.0mg

Preparation of Vitamin B_{12} Solution: Add vitamin B_{12} to distilled/deionized water and bring volume to 100.0mL. Mix thoroughly. Filter sterilize.

Preparation of Medium: To 950.0mL of cooled, sterile solution 1, aseptically add 60.0mL of sterile $Na_2S·9H_2O$ solution, 40.0mL of sterile $NaHCO_3$ solution, and 1.0mL of sterile vitamin B_{12} solution. Mix thoroughly. Adjust pH to 6.8 with sterile H_2SO_4 or Na_2CO_3. Aseptically distribute into sterile 50.0mL or 100.0mL bottles with metal screw caps and rubber seals. Completely fill bottles with medium except for a pea-sized air bubble.

Use: For the isolation and cultivation of marine members of the Chlorobiaceae.

Marine Chromatiaceae Medium 2

Composition per 1051.0mL:

Solution 1	950.0mL
$Na_2S\cdot9H_2O$ solution	60.0mL
$NaHCO_3$ solution	40.0mL
Vitamin B_{12} solution	1.0mL

pH 7.3 ± 0.2 at 25°C

Solution 1:

Composition per 950.0mL:

NaCl	20.0g
$MgSO_4\cdot7H_2O$	3.0g
KH_2PO_4	1.0g
NH_4Cl	0.5g
$CaCl_2\cdot2H_2O$	0.05g
Trace elements solution SL-8	1.0mL

Preparation of Solution 1: Add components to distilled/deionized water and bring volume to 950.0mL. Mix thoroughly. Autoclave for 15 min at 15 psi pressure–121°C. Cool to 45°–50°C.

Trace Elements Solution SL-8:

Composition per liter:

Disodium EDTA	5.2g
$FeCl_2\cdot4H_2O$	1.5g
$CoCl_2\cdot6H_2O$	0.19g
$MnCl_2\cdot4H_2O$	0.1g
$ZnCl_2$	0.07g
H_3BO_3	0.06g
$NaMoO_4\cdot2H_2O$	0.04g
$CuCl_2\cdot2H_2O$	0.02g
$NiCl_2\cdot6H_2O$	0.02g

Preparation of Trace Elements Solution SL-8: Add components to distilled/deionized water and bring volume to 1.0L. Mix thoroughly.

$Na_2S\cdot9H_2O$ Solution:

Composition per 100.0mL:

$Na_2S\cdot9H_2O$	5.0g

Preparation of $Na_2S\cdot9H_2O$ Solution: Add $Na_2S\cdot9H_2O$ to distilled/deionized water and bring volume to 100.0mL. Autoclave for 15 min at 15 psi pressure–121°C. Cool to 45°–50°C.

$NaHCO_3$ Solution:

Composition per 100.0mL:

$NaHCO_3$	5.0g

Preparation of $NaHCO_3$ Solution: Add $NaHCO_3$ to distilled/deionized water and bring volume to 100.0mL. Mix thoroughly. Filter sterilize.

Vitamin B_{12} Solution:

Composition per 100.0mL:

Vitamin B_{12}	2.0mg

Preparation of Vitamin B_{12} Solution: Add vitamin B_{12} to distilled/deionized water and bring volume to 100.0mL. Mix thoroughly. Filter sterilize.

Preparation of Medium: To 950.0mL of cooled, sterile solution 1, aseptically add 60.0mL of sterile $Na_2S\cdot9H_2O$ solution, 40.0mL of sterile $NaHCO_3$ solution, and 1.0mL of sterile vitamin B_{12} solution. Mix thoroughly. Adjust pH to 7.3 with sterile H_2SO_4 or Na_2CO_3. Aseptically distribute into sterile 50.0mL or 100.0mL bottles with metal screw caps and rubber seals. Completely fill bottles with medium except for a pea-sized air bubble.

Use: For the isolation and cultivation of marine members of the Chromatiaceae.

Marine *Cytophaga* Agar

Composition per liter:

Agar	15.0g
Nutrient broth	8.0g
Yeast extract	5.0g
Salt solution	1.0L

Salt Solution:

Composition per liter:

NaCl	12.86g
$MgCl_2$	2.48g
KCl	0.75g
$CaCl_2$	0.56g
$Fe(SO_4)_2(NH_4)_2$	0.048g

Preparation of Salt Solution: Add components to distilled/deionized water and bring volume to 1.0L. Mix thoroughly.

Preparation of Medium: Add solid components to 1.0L of salt solution. Mix thoroughly. Gently heat while stirring and bring to boiling. Distribute into tubes or flasks. Autoclave for 15 min at 15 psi pressure–121°C. Pour into sterile Petri dishes or leave in tubes.

Use: For the cultivation and maintenance of *Cytophaga* species.

Marine *Cytophaga* Medium

Composition per liter:

NaCl	24.7g
Agar	15.0g
$MgSO_4\cdot7H_2O$	6.3g
$MgCl_2\cdot6H_2O$	4.6g
Tryptic digest of casein	1.0g
Yeast extract	1.0g
KCl	0.7g
$NaHCO_3$ solution	10.0mL
$CaCl_2\cdot2H_2O$ solution	10.0mL

pH 7.2 ± 0.2 at 25°C

$NaHCO_3$ Solution:

Composition per 10.0mL:

$NaHCO_3$	0.2g

Preparation of $NaHCO_3$ Solution: Add NaHCO3 to distilled/deionized water and bring volume to 10.0mL. Mix thoroughly. Autoclave for 15 min at 15 psi pressure–121°C.

$CaCl_2\cdot2H_2O$ Solution:

Composition per 10.0mL:

$CaCl_2\cdot2H_2O$	1.2g

Preparation of $CaCl_2\cdot2H_2O$ Solution: Add $CaCl_2\cdot2H_2O$ to distilled/deionized water and bring volume to 10.0mL. Mix thoroughly. Autoclave for 15 min at 15 psi pressure–121°C.

Preparation of Medium: Add components, except $NaHCO_3$ solution and $CaCl_2\cdot2H_2O$ solution, to distilled/deionized water and bring volume to 980.0mL. Mix thoroughly. Gently heat and bring to boiling. Autoclave for 15 min at 15 psi pressure–121°C. Cool to 50°–55°C. Aseptically add 10.0mL of sterile $NaHCO_3$ solution and 10.0mL of sterile $CaCl_2 \cdot 2H_2O$ solution. Mix thoroughly. Pour into sterile Petri dishes or distribute into sterile tubes.

Use: For the cultivation of *Cytophaga* species, *Flexibacter* species, *Microscilla* species, and *Saprospira grandis.*

Marine *Cytophaga* Medium A

Composition per liter:

Agar	15.0g
Pancreatic digest of casein	2.0g
Beef extract	0.5g
Yeast extract	0.5g
Sodium acetate	0.2g
Seawater	700.0mL

pH 7.2 ± 0.2 at 25°C

Preparation of Medium: Add components to distilled/deionized water and bring volume to 1.0L. Mix thoroughly. Gently heat and bring to boiling. Distribute into tubes or flasks. Autoclave for 15 min at 15 psi pressure–121°C. Pour into sterile Petri dishes or leave in tubes.

Use: For the cultivation of *Flexibacter maritimus.*

Marine *Cytophaga* Medium B

Composition per liter:

Agar	15.0g
Pancreatic digest of casein	2.0g
Beef extract	0.5g
Yeast extract	0.5g
Sodium acetate	0.2g
Seawater	500.0mL

pH 7.2 ± 0.2 at 25°C

Preparation of Medium: Add components to distilled/deionized water and bring volume to 1.0L. Mix thoroughly. Gently heat and bring to boiling. Distribute into tubes or flasks. Autoclave for 15 min at 15 psi pressure–121°C. Pour into sterile Petri dishes or leave in tubes.

Use: For the cultivation of *Vibrio ordalii.*

Marine *Cytophaga* Medium C

Composition per liter:

Agar	15.0g
Pancreatic digest of casein	2.0g
Beef extract	0.5g
Yeast extract	0.5g
Sodium acetate	0.2g

pH 7.2 ± 0.2 at 25°C

Preparation of Medium: Add components to seawater and bring volume to 1.0L. Mix thoroughly. Gently heat and bring to boiling. Distribute into tubes or flasks. Autoclave for 15 min at 15 psi pressure–121°C. Pour into sterile Petri dishes or leave in tubes.

Use: For the cultivation of *Cytophaga agarovorans, Cytophaga fermentans*, and *Cytophaga salmonicolor.*

Marine *Desulfovibrio* Medium

Composition per liter:

Solution A	980.0mL
Solution B	10.0mL
Solution C	10.0mL

pH 7.8 ± 0.2 at 25°C

Solution A:

Composition per 980.0mL:

NaCl	25.0g
DL-Sodium lactate	2.0g
$MgSO_4 \cdot 7H_2O$	2.0g
Na_2SO_4	1.0g
NH_4Cl	1.0g
Yeast extract	1.0g
K_2HPO_4	0.5g
$CaCl_2 \cdot 2H_2O$	0.1g
Resazurin	1.0mg

Preparation of Solution A: Add components to distilled/deionized water and bring volume to 980.0mL. Mix thoroughly. Gently heat and bring to boiling. Continue boiling for 3–4 min. Allow to cool to room temperature while gassing under 100% N_2.

Solution B:

Composition per 10.0mL:

$FeSO_4 \cdot 7H_2O$	0.5g

Preparation of Solution B: Add $FeSO_4 \cdot 7H_2O$ to distilled/deionized water and bring volume to 10.0mL. Mix thoroughly.

Solution C:

Composition per 10.0mL:

Ascorbic acid	0.1g
Sodium thioglycolate	0.1g

Preparation of Solution C: Add components to distilled/deionized water and bring volume to 10.0mL. Mix thoroughly.

Preparation of Medium: To 980.0mL of cooled solution A, anaerobically add 10.0mL of solution B and 10.0mL of solution C. Mix thoroughly. Adjust pH to 7.8 with NaOH. Distribute into tubes or flasks. During distribution, swirl the medium to keep the precipitate in suspension. Autoclave for 15 min at 15 psi pressure–121°C.

Use: For the cultivation and maintenance of *Desulfovibrio desulfuricans, Desulfovibrio salexigens,* and *Desulfovibrio vulgaris.*

Marine Flagellate Medium

Composition per 15.0mL:

Rice grains	2.0g
Seawater	15.0mL

Preparation of Medium: Autoclave rice grains for 15 min at 15 psi pressure–121°C. Add 2.0g of sterile rice grains to 15.0mL of filter-sterilized seawater. Aseptically distribute into T-25 tissue culture flasks.

Use: For the cultivation of *Acanthoecopsis unguiculata, Amastigomonas* species, *Bicosoeca vacillans, Bodo designis, Bodo variabilis, Caecitellus parvulus, Choanoeca perplexa, Codosiga gracilis, Diaphanoeca grandis, Entosiphon* species, *Goniomonas* species, *Procryptobia* species, *Pseudobodo tremulans, Rhynchomonas nasuta, Salpingoeca urceolata, Stephanoeca diplocostata,* and *Stephanopogon apogon.*

Marine Flagellate Medium with B-Vitamins

Composition per liter:

Seawater	990.0mL
Vitamin solution	10.0mL

Vitamin Solution:

Composition per 100.0mL:

Thiamine·HCl	0.15g
Calcium D-(+)-pantothenate	0.05g
Nicotinamide	0.05g

Pyridoxal·HCl 0.05g
Riboflavin 0.05g
Folic acid........ 0.025g
Pyridoxamine·HCl 0.025g
Biotin 12.5mg

Preparation of Vitamin Solution: Add components to distilled/deionized water and bring volume to 100.0mL. Mix thoroughly. Filter sterilize.

Preparation of Medium: Allow natural seawater to age for 2 months. Filter sterilize. Aseptically add 100.0mL of sterile vitamin solution. Mix thoroughly. Aseptically distribute into sterile tubes or flasks.

Use: For the cultivation of *Oikomonas* species.

Marine Glucose Trypticase™ Yeast Extract Agar (MGTY Agar)

Composition per liter:
Agar 8.0g
Glucose 2.0g
Pancreatic digest of casein........ 1.0g
Yeast extract........ 1.0g
L-Cysteine·HCl·H_2O........ 0.5g
Seawater........ 750.0mL
Tris-HCl buffer (5.0 m*M*, pH 7.5) 50.0mL
Resazurin (0.1% solution)........ 1.0mL
pH 7.5 ± 0.2 at 25°C

Preparation of Medium: Add components to distilled/deionized water and bring volume to 1.0L. Mix thoroughly. Gently heat while stirring and bring to boiling. Distribute into tubes or flasks under 97% N_2 + 3% H_2. Cap with rubber stoppers and place tubes in a press. Autoclave for 15 min at 15 psi pressure–121°C with fast exhaust.

Use: For the cultivation and maintenance of *Spirochaeta isovalerica*.

Marine Glucose Trypticase™ Yeast Extract Broth (MGTY Broth)

Composition per liter:
Glucose 2.0g
Pancreatic digest of casein........ 1.0g
Yeast extract........ 1.0g
L-Cysteine·HCl·H_2O........ 0.5g
Seawater........ 750.0mL
Tris-HCl buffer (5.0 m*M*, pH 7.5) 50.0mL
Resazurin (0.1% solution)........ 1.0mL
pH 7.5 ± 0.2 at 25°C

Preparation of Medium: Add components to distilled/deionized water and bring volume to 1.0L. Mix thoroughly. Gently heat while stirring and bring to boiling. Distribute into tubes or flasks under 97% N_2 + 3% H_2. Cap with rubber stoppers and place tubes in a press. Autoclave for 15 min at 15 psi pressure–121°C with fast exhaust.

Use: For the cultivation and maintenance of *Spirochaeta isovalerica*.

Marine *Methanogenium* Alcohol Medium

Composition per 1003.0mL:
NaCl........ 21.0g
$MgCl_2·6H_2O$........ 3.0g
NaCl 1.0g
KCl........ 0.5g
$MgCl_2·6H_2O$........ 0.5g
NH_4Cl 0.4g
Sodium acetate·$3H_2O$........ 0.4g
KH_2PO_4........ 0.2g
$CaCl_2·2H_2O$ 0.1g
$NaHCO_3$ solution........ 60.0mL
2-Propanol........ 5.0mL
$Na_2S·9H_2O$ solution........ 3.0mL
Cyanocobalamin solution 1.0mL
Selenite-molybdate-tungstate solution........ 1.0mL
Thiamine solution........ 1.0mL
Trace elements solution 1.0mL
Vitamin solution........ 1.0mL

Trace Elements Solution:
Composition per 100.0mL:
$FeSO_4·7H_2O$........ 1400.0mg
$ZnSO_4·7H_2O$ 145.0mg
$CoCl_2·6H_2O$ 120.0mg
$MnCl_2·4H_2O$ 100.0mg
$NiCl_2·6H_2O$........ 50.0mg
H_3BO_3........ 6.0mg
$CuSO_4·5H_2O$........ 3.0mg
HCl (25%,w/v)........ 8.0mL

Preparation of Trace Elements Solution: Add components to distilled/deionized water and bring volume to 1.0L. Mix thoroughly. Sparge with 100% N_2. Autoclave for 15 min at 15 psi pressure–121°C.

Selenite-Molybdate-Tungstate Solution:
Composition per liter:
NaOH........ 0.2g
$Na_2MoO_4·2H_2O$........ 40.0mg
$Na_2WO_4·2H_2O$........ 33.0mg
$Na_2SeO_3·2H_2O$........ 5.0mg

Preparation of Selenite-Molybdate-Tungstate Solution: Add components to distilled/deionized water and bring volume to 1.0L. Mix thoroughly. Sparge with 100% N_2. Autoclave for 15 min at 15 psi pressure–121°C.

$NaHCO_3$ Solution:
Composition per liter:
$NaHCO_3$........ 84.0g

Preparation of $NaHCO_3$ Solution: Add $NaHCO_3$ to distilled/deionized water and bring volume to 10.0mL. Mix thoroughly. Sparge with 80% N_2 + 20% CO_2. Autoclave for 15 min at 15 psi pressure–121°C.

$Na_2S·9H_2O$ Solution:
Composition per 100.0mL:
$Na_2S·9H_2O$........ 2.5g
NaOH........ 1 pellet

Preparation of $Na_2S·9H_2O$ Solution: Bring 100.0mL of distilled/deionized water to boiling. Cool to room temperature while sparging with 100%N_2. Dissolve 1 pellet of NaOH in the anaerobic water. Weigh out a little more than 2.5g of $Na_2S·9H_2O$. Briefly rinse the crystals in distilled/deionized water. Dry the crystals by blotting on paper towels or filter paper. Add 2.5g of washed $Na_2S·9H_2O$ crystals to 100.0mL of anaerobic NaOH solution. Distribute into serum bottles fitted with butyl rubber stoppers and aluminum seals. Do not grease stoppers. Pressurize to 60kPa with 100% N_2. Autoclave for 15 min at 15 psi pressure–121°C. Store at room temperature in an anaerobic chamber.

Preparation of 2-Propanol: Filter sterilize 10.0mL of 2-propanol. Sparge with 100% N_2.

Vitamin Solution:
Composition per liter:

Sodium 2-mercaptoethanesulfonate	0.25g
Pyridoxine·HCl	0.15g
Calcium pantothenate	0.1g
Nicotinic acid	0.1g
p-Aminobenzoic acid	40.0mg
Biotin	10.0mg
Potassium phosphate buffer (25m*M* solution, pH 7.0)	1.0L

Preparation of Vitamin Solution: Combine components. Mix thoroughly. Filter sterilize. Sparge with 100% N_2.

Thiamine Solution:
Composition per liter:

Thiamine·HCl	0.1g
Sodium phosphate buffer (0.1*M* solution, pH 3.6)	1.0L

Preparation of Thiamine Solution: Combine components. Mix thoroughly. Filter sterilize. Sparge with 100% N_2.

Cyanocobalamin Solution:
Composition per liter:

Cyanocobalamin	50.0mg

Preparation of Cyanocobalamin Solution: Add cyanocobalamin to distilled/deionized water and bring volume to 1.0L. Mix thoroughly. Filter sterilize. Sparge with 100% N_2.

Preparation of Medium: Prepare and dispense medium under 80% N_2 + 20% CO_2. Add components, except $NaHCO_3$ solution, 2-propanol, $Na_2S·9H_2O$ solution, cyanocobalamin solution, selenite-molybdate-tungstate solution, thiamine solution, trace elements solution, and vitamin solution, to distilled/deionized water and bring volume to 930.0mL. Mix thoroughly. Sparge with 80% N_2 + 20% CO_2. Autoclave for 15 min at 15 psi pressure–121°C. Aseptically and anaerobically add 60.0mL of sterile $NaHCO_3$ solution, 5.0mL of sterile 2-propanol, 3.0mL of sterile $Na_2S·9H_2O$ solution, 1.0mL of sterile cyanocobalamin solution, 1.0mL of sterile selenite-molybdate-tungstate solution, 1.0mL of sterile thiamine solution, 1.0mL of sterile trace elements solution, and 1.0mL of sterile vitamin solution. Mix thoroughly. Aseptically and anaerobically distribute into sterile tubes or bottles.

Use: For the cultivation of marine *Methanogenium* species.

Marine Methanol Medium

Composition per liter:

NaCl	20.0g
$(NH_4)_2SO_4$	2.0g
K_2HPO_4	2.0g
KH_2PO_4	1.0g
$MgSO_4·7H_2O$	0.3g
Methanol	10.0mL
Vitamin B_{12} solution	10.0mL
Trace metals solution	1.0mL

pH 7.0 ± 0.2 at 25°C

Vitamin B_{12} Solution:
Composition per 100.0mL:

Vitamin B_{12}	10.0μg

Preparation of Vitamin B_{12} Solution: Add the vitamin B_{12} to distilled/deionized water and bring volume to 100.0mL. Adjust pH to 5. Autoclave for 15 min at 15 psi pressure–121°C.

Trace Metals Solution:
Composition per liter:

$ZnSO_4·7H_2O$	1.4g
$MnSO_4·H_2O$	0.84g
$FeSO_4·7H_2O$	0.28g
$CuSO_4·5H_2O$	0.25g
$Na_2MoO_4·2H_2O$	0.24g
$CoCl_2·6H_2O$	0.24g
$CaCl_2·2H_2O$	0.15g

Preparation of Trace Metals Solution: Add components to distilled/deionized water and bring volume to 1.0L. Mix thoroughly.

Preparation of Medium: Add components, except vitamin B_{12} solution and methanol, to distilled/deionized water and bring volume to 980.0mL. Adjust pH to 7.0 with NaOH. Autoclave for 15 min at 15 psi pressure–121°C. Filter sterilize methanol. Aseptically add sterile vitamin B_{12} solution and filter-sterilized methanol. Distribute into sterile tubes or flasks.

Use: For the cultivation and maintenance of *Methylophaga thalassica*.

Marine Methylotroph Agar

Composition per 1003.0mL:

Agarose	12.0g
Bis (2-hydroxyethyl) aminotris (hydroxymethyl) methane	2.0g
KH_2PO_4	0.14g
Ferric ammonium citrate	0.06g
Methanol	2.0mL
Vitamin B_{12} solution	1.0mL

pH 7.4 ± 0.2 at 25°C

Vitamin B_{12} Solution:
Composition per 100.0mL:

Vitamin B_{12}	0.1mg

Preparation of Vitamin Solution: Add vitamin B_{12} to distilled/deionized water and bring volume to 100.0mL. Mix thoroughly. Filter sterilize. Store at 5°C.

Preparation of Medium: Add components, except methanol and vitamin B_{12} solution, to distilled/deionized water and bring volume to 1.0L. Mix thoroughly. Adjust pH to 7.4. Gently heat and bring to boiling. Autoclave for 15 min at 15 psi pressure–121°C. Cool to 50°C. Aseptically add 2.0mL of filter-sterilized methanol and 1.0mL of sterile vitamin B_{12} solution. Mix thoroughly. Pour into sterile Petri dishes or distribute into sterile tubes.

Use: For the cultivation of *Alteromonas* species, *Methylophaga marina, Methylophaga thalassica,* and *Methylophilus* species.

Marine Methylotroph Broth

Composition per 1003.0mL:

Bis (2-hydroxyethyl) aminotris (hydroxymethyl) methane	2.0g
KH_2PO_4	0.14g
Ferric ammonium citrate	0.06g
Methanol	2.0mL
Vitamin B_{12} solution	1.0mL

pH 7.4 ± 0.2 at 25°C

Vitamin B_{12} Solution:
Composition per 100.0mL:

Vitamin B_{12}	0.1mg

Preparation of Vitamin Solution: Add vitamin B_{12} to distilled/deionized water and bring volume to 100.0mL. Mix thoroughly. Filter sterilize. Store at 5°C.

Preparation of Medium: Add components, except methanol and vitamin B_{12} solution, to distilled/deionized water and bring volume to 1.0L. Mix thoroughly. Adjust pH to 7.4. Autoclave for 15 min at 15 psi pressure–121°C. Cool to 50°C. Aseptically add 2.0mL of filter-sterilized methanol and 1.0mL of sterile vitamin B_{12} solution. Mix thoroughly. Aseptically distribute into sterile tubes or flasks.

Use: For the cultivation of *Alteromonas* species, *Methylophaga marina, Methylophaga thalassica*, and *Methylophilus* species.

Marine Oxidation Fermentation HiVeg Medium (MOF HiVeg Medium)

Composition per liter:

NaCl	9.7g
$MnCl_2$	4.4g
Agar	3.0g
Na_2SO_4	1.6g
Plant hydrolysate	1.0g
$CaCl_2$	0.9g
$(NH_4)_2SO_4$	0.5g
Tris hydroxymethyl aminomethane	0.5g
KCl	0.275g
Yeast extract	0.1g
$NaHCO_3$	0.08g
KBr	0.04g
$SrCl_2$	0.017g
H_3BO_3	0.011g
Phenol Red	0.01g
Na_2HPO_4	4.0mg
Sodium silicate	2.0mg
NaFl	1.2mg
NH_4NO_3	0.8mg

pH 7.3 ± 0.2 at 25°C

Source: This medium is available as a premixed powder from HiMedia.

Preparation of Medium: Add components to distilled/deionized water and bring volume to 1.0L. Mix thoroughly. Gently heat while stirring and bring to boiling. Distribute into tubes or flasks. Autoclave for 15 min at 15 psi pressure–121°C.

Use: For the differentiation of marine bacteria on the basis of fermentative and oxidative metabolism of carbohydrates.

Marine Peptone Succinate Salts Medium (PSS Medium)

Composition per liter:

Peptone	10.0g
Succinic acid	1.0g
$(NH_4)_2SO_4$	1.0g
$MgSO_4 \cdot 7H_2O$	1.0g
$FeCl_3 \cdot 6H_2O$	2.0mg
$MnSO_4 \cdot H_2O$	2.0mg
Synthetic seawater	1.0L

pH 6.8 ± 0.2 at 25°C

Synthetic Seawater:

Composition per liter:

NaCl	27.5g
$MgCl_2$	5.0g
$MgSO_4 \cdot 7H_2O$	2.0g
KCl	1.0g
$CaCl_2$	0.5g
$FeSO_4$	1.0mg

Preparation of Synthetic Seawater: Add components to distilled/deionized water and bring volume to 1.0L. Mix thoroughly.

Preparation of Medium: Add components to 1.0L of synthetic seawater. Mix thoroughly. Gently heat while stirring and bring to boiling. Adjust pH to 6.8 with KOH. Distribute into tubes or flasks. Autoclave for 15 min at 15 psi pressure–121°C.

Use: For the cultivation and maintenance of *Oceanospirillum beijerinckii* and *Oceanospirillum multiglobuliferum*.

Marine Peptone Yeast Medium with Magnesium Sulfate

Composition per liter:

NaCl	20.0g
Peptone	10.0g
$MgSO_4 \cdot 7H_2O$	2.0g
$(NH_4)_2SO_4$	2.0g
Yeast extract	1.0g

pH 7.0 ± 0.2 at 25°C

Preparation of Medium: Add components to distilled/deionized water and bring volume to 1.0L. Mix thoroughly. Distribute into tubes or flasks. Autoclave for 15 min at 15 psi pressure–121°C.

Use: For the cultivation and maintenance of *Oceanospirillum pusillum*.

Marine *Pseudomonas* Medium

Composition per liter:

Agar	15.0g
Nutrient broth	8.0g
Yeast extract	5.0g
Salt solution	1.0L

Salt Solution:

Composition per liter:

NaCl	12.86g
$MgCl_2$	2.48g
KCl	0.75g
$CaCl_2$	0.56g
$Fe(SO_4)_2(NH_4)_2$	0.048g

Preparation of Salt Solution: Add components to distilled/deionized water and bring volume to 1.0L. Mix thoroughly.

Preparation of Medium: Add components to 1.0L of salt solution. Mix thoroughly. Gently heat and bring to boiling. Distribute into tubes or flasks. Autoclave for 15 min at 15 psi pressure–121°C. Pour into sterile Petri dishes or leave in tubes.

Use: For the cultivation and maintenance of *Alteromonas haloplanktis*.

Marine *Rhodococcus* Medium

Composition per liter:

Yeast extract	10.0g
Malt extract	4.0g
Glucose	4.0g
Seawater	750.0mL

Preparation of Medium: Add components to distilled/deionized water and bring volume to 1.0L. Mix thoroughly. Gently heat while stirring and bring to boiling. Distribute into tubes or flasks. Autoclave for 15 min at 15 psi pressure–121°C.

Use: For the cultivation and maintenance of *Rhodococcus marinonascens*.

Marine *Rhodopseudomonas* Medium

Composition per liter:

NaCl	30.4g
Yeast extract	1.0g
Disodium succinate	1.0g
KH_2PO_4	0.5g
$MgSO_4 \cdot 7H_2O$	0.4g
NH_4Cl	0.4g
$CaCl_2 \cdot 2H_2O$	0.05g
Ferric citrate (0.1% solution)	5.0mL
Trace elements solution SL-6	1.0mL
Ethanol	0.5mL

pH 6.8 ± 0.2 at 25°C

Trace Elements Solution SL-6:

Composition per liter:

H_3BO_3	0.3g
$CoCl_2 \cdot 6H_2O$	0.2g
$ZnSO_4 \cdot 7H_2O$	0.1g
$MnCl_2 \cdot 4H_2O$	0.03g
$Na_2MoO_4 \cdot H_2O$	0.03g
$NiCl_2 \cdot 6H_2O$	0.02g
$CuCl_2 \cdot 2H_2O$	0.01g

Preparation of Trace Elements Solution SL-6: Add components to distilled/deionized water and bring volume to 1.0L. Mix thoroughly. Adjust pH to 3.4.

Preparation of Medium: Add components to distilled/deionized water and bring volume to 1.0L. Mix thoroughly. Gently heat while stirring and bring to boiling. Distribute into tubes or flasks. Autoclave for 15 min at 15 psi pressure–121°C.

Use: For the cultivation and maintenance of *Rhodopseudomonas marina*.

Marine *Rhodopseudomonas* Medium

Composition per liter:

NaCl	30.0g
Peptone	2.5g
Yeast extract	2.5g

pH 7.0 ± 0.4 at 25°C

Preparation of Medium: Add components to distilled/deionized water and bring volume to 1.0L. Mix thoroughly. Distribute into screw-capped tubes. Autoclave for 15 min at 15 psi pressure–121°C. Before inoculating, loosen the screw caps, heat the medium to drive out O_2, and screw down the cap tightly.

Use: For the cultivation of *Rhodopseudomonas marina*.

Marine Salts Medium

Composition per liter:

NaCl	81.0g
Yeast extract	10.0g
$MgSO_4$	9.6g
$MgCl_2$	7.0g
Proteose peptone No.3	5.0g
KCl	2.0g
Glucose	1.0g
$CaCl_2$	0.36g
$NaHCO_3$	0.06g
NaBr	0.026g

pH 7.0 ± 0.2 at 25°C

Preparation of Medium: Add components to distilled/deionized water and bring volume to 1.0L. Mix thoroughly. Adjust pH to 7.0. Distribute into tubes or flasks. Autoclave for 15 min at 15 psi pressure–121°C.

Use: For the cultivation of marine bacteria.

Marine Spirochete Medium (DSMZ Medium 1008)

Composition per liter:

Trypticase peptone	2.0g
Yeast extract	1.0g
Na-thioglycolate	1.0g
Resazurin	0.5mg
Charcoal-filtered, natural seawater	800.0mL
Cellobiose solution	20.0mL

pH 7.5 ± 0.2 at 25°C

Cellobiose Solution:

Composition per 100.0mL:

Cellobiose	10.0g

Preparation of Cellobiose Solution: Add components to distilled/deionized water and bring volume to 100.0mL. Mix thoroughly. Sparge with 100% N_2. Filter sterilize.

Preparation of Medium: Add components, except thioglycolate and cellobiose solution, to seawater and bring volume to 800.0mL. Mix thoroughly. Bring volume to 980.0mL with distilled/deionized water. (Note: Bottled water from Biomaris GmbH can be used instead of filtered seawater.) Gently heat and bring to boiling. Boil for 3 min. Cool to room temperature while sparging with 100% N_2. Add the thioglycolate. Adjust pH to 7.5 with 10*N* NaOH. Dispense into tubes or bottles under 100% N_2. Autoclave for 15 min at 15 psi pressure–121°C. Cool to room temperature. Aseptically and anoxically add cellobiose solution.

Use: For the cultivation of *Spirochaeta bajacaliforniensis*.

Marine Salts Medium for *Sporosarcina halophila*

Composition per liter:

Marine salts mix	100.0g
Agar	20.0g
Yeast extract	10.0g
Proteose peptone No. 3	5.0g
Glucose	1.0g

pH 7.0 ± 0.2 at 25°C

Preparation of Medium: Add components to distilled/deionized water and bring volume to 1.0L. Mix thoroughly. Gently heat while stirring and bring to boiling. Distribute into tubes or flasks. Autoclave for 15 min at 15 psi pressure–121°C. Pour into sterile Petri dishes or leave in tubes.

Use: For the cultivation and maintenance of *Sporosarcina halophila*.

Marine Spirochete Medium

Composition per liter:

Cellobiose	2.0g
Peptone	2.0g
Yeast extract	1.0g
Sodium thioglycolate	1.0g
Seawater, charcoal filtered	800.0mL

pH 7.5 ± 0.2 at 25°C

Preparation of Medium: Add components, except sodium thioglycolate, to glass-distilled water and bring volume to 1.0L. Mix thoroughly. Bubble 100% N_2 into medium for 1.5 min. Add sodium thioglycolate. Adjust pH to 7.5 with 10*N* KOH. Distribute into tubes or flasks. Autoclave for 15 min at 15 psi pressure–121°C.

Use: For the cultivation and maintenance of *Spirochaeta bajacaliforniensis*.

Marine *Thermococcus* Medium (DSMZ Medium 760)

Composition per liter:

NaCl	19.45g
$MgCl_2$	8.8g
Sulfur	5.0g
Peptone	5.0g
Na_2SO_3	3.24g
$CaCl_2$	1.8g
Yeast extract	1.0g
KCl	0.55g
$NaHCO_3$	0.16g
Ferric citrate	0.1g
KBr	0.08g
$SrCl_2$	0.03g
H_3BO_3	0.02g
Na_2HPO_4	8.0mg
Na_2SiO_3	4.0mg
NaF	2.4mg
NH_4NO_3	1.6mg
$Na_2S{\cdot}9H_2O$ solution	0.5mL

pH 6.0 ± 0.2 at 25°C

$Na_2S{\cdot}9H_2O$ Solution:

Composition per 10.0mL:

$Na_2S{\cdot}9H_2O$	1.0g

Preparation of $Na_2S{\cdot}9H_2O$ Solution: Add $Na_2S{\cdot}9H_2O$ to distilled/deionized water and bring volume to 10.0mL. Mix thoroughly. Sparge with 100% N_2. Autoclave for 15 min at 15 psi pressure–121°C. Neutralize to pH 7.0 with sterile HCl.

Preparation of Medium: Add components, except sulfur and $Na_2S{\cdot}9H_2O$ solution, to distilled/deionized water and bring volume to 1.0L. Mix thoroughly. Filter through normal filter paper. An iron sediment will collect in the filter. Gently heat while stirring and bring to boiling. Boil for 5 min. Cool under an anaerobic gas mixture of N_2. Adjust pH to 6.0. Distribute the medium into Hungate tubes or serum bottles containing finely divided sulfur (0.5% w/v). Seal the tubes or bottles under the same anaerobic gas used when cooling the medium. Sterilize the medium at 100°C for 3 hr on 3 consecutive days. Reduce the medium by adding 10% neutralized $Na_2S{\cdot}9H_2O$ solution to a final concentration of 0.05%. The medium should not give a heavy black precipitate; if it does the iron sediment was not adequately removed by filtering in the initial stages and the medium should be made again, making sure that the iron is removed by filtering.

Use: For the cultivation and maintenance of *Thermococcus aegaeus* DSM 12767.

Marine *Thermococcus* Medium (DSMZ Medium 760)

Composition per liter:

NaCl	19.45g
$MgCl_2$	8.8g
Sulfur	5.0g
Peptone	5.0g
Na_2SO_3	3.24g
$CaCl_2$	1.8g
Yeast extract	1.0g
KCl	0.55g
$NaHCO_3$	0.16g
Ferric citrate	0.1g
KBr	0.08g
$SrCl_2$	0.03g
H_3BO_3	0.02g
Na_2HPO_4	8.0mg
Na_2SiO_3	4.0mg
NaF	2.4mg
NH_4NO_3	1.6mg
$Na_2S{\cdot}9H_2O$ solution	0.5mL

pH 6.5 ± 0.2 at 25°C

$Na_2S{\cdot}9H_2O$ Solution:

Composition per 10.0mL:

$Na_2S{\cdot}9H_2O$	1.0g

Preparation of $Na_2S{\cdot}9H_2O$ Solution: Add $Na_2S{\cdot}9H_2O$ to distilled/deionized water and bring volume to 10.0mL. Mix thoroughly. Sparge with 100% N_2. Autoclave for 15 min at 15 psi pressure–121°C. Neutralize to pH 7.0 with sterile HCl.

Preparation of Medium: Add components, except sulfur and $Na_2S{\cdot}9H_2O$ solution, to distilled/deionized water and bring volume to 1.0L. Mix thoroughly. Filter through normal filter paper. An iron sediment will collect in the filter. Gently heat while stirring and bring to boiling. Boil for 5 min. Cool under an anaerobic gas mixture of N_2. Adjust pH to 6.5. Distribute the medium into Hungate tubes or serum bottles containing finely divided sulfur (0.5% w/v). Seal the tubes or bottles under the same anaerobic gas used when cooling the medium. Sterilize the medium at 100°C for 3 hr on 3 consecutive days. Reduce the medium by adding 10% neutralized $Na_2S{\cdot}9H_2O$ solution to a final concentration of 0.05%. The medium should not give a heavy black precipitate; if it does the iron sediment was not adequately removed by filtering in the initial stages and the medium should be made again, making sure that the iron is removed by filtering.

Use: For the cultivation and maintenance of *Thermococcus pacificus* DSM 10394 and *Thermococcus gorgonarius* DSM 10395.

Marine *Thermococcus* Medium (DSMZ Medium 760)

Composition per liter:

NaCl	19.45g
$MgCl_2$	8.8g
Sulfur	5.0g
Peptone	5.0g
Na_2SO_3	3.24g
$CaCl_2$	1.8g
Yeast extract	1.0g

KCl	0.55g
$NaHCO_3$	0.16g
Ferric citrate	0.1g
KBr	0.08g
$SrCl_2$	0.03g
H_3BO_3	0.02g
Na_2HPO_4	8.0mg
Na_2SiO_3	4.0mg
NaF	2.4mg
NH_4NO_3	1.6mg
$Na_2S·9H_2O$ solution	0.5mL

pH 6.5 ± 0.2 at 25°C

$Na_2S·9H_2O$ Solution:
Composition per 10.0mL:

$Na_2S·9H_2O$	1.0g

Preparation of $Na_2S·9H_2O$ Solution: Add $Na_2S·9H_2O$ to distilled/deionized water and bring volume to 10.0mL. Mix thoroughly. Sparge with 100% N_2. Autoclave for 15 min at 15 psi pressure–121°C. Neutralize to pH 7.0 with sterile HCl.

Preparation of Medium: Add components, except sulfur and $Na_2S·9H_2O$ solution, to distilled/deionized water and bring volume to 1.0L. Mix thoroughly. Filter through normal filter paper. An iron sediment will collect in the filter. Gently heat while stirring and bring to boiling. Boil for 5 min. Cool under an anaerobic gas mixture of 80% N_2 + 20% CO_2. Adjust pH to 6.5. Distribute the medium into Hungate tubes or serum bottles containing finely divided sulfur (0.5% w/v). Seal the tubes or bottles under the same anaerobic gas used when cooling the medium. Sterilize the medium at 100°C for 3 hr on 3 consecutive days. Reduce the medium by adding 10% neutralized $Na_2S·9H_2O$ solution to a final concentration of 0.05%. The medium should not give a heavy black precipitate; if it does the iron sediment was not adequately removed by filtering in the initial stages and the medium should be made again, making sure that the iron is removed by filtering.

Use: For the cultivation and maintenance of *Thermococcus stetteri* DSM 5262.

Marine *Thermococcus* Medium (DSMZ Medium 760)

Composition per liter:

NaCl	19.45g
$MgCl_2$	8.8g
Sulfur	5.0g
Peptone	5.0g
Na_2SO_3	3.24g
$CaCl_2$	1.8g
Yeast extract	1.0g
KCl	0.55g
$NaHCO_3$	0.16g
Ferric citrate	0.1g
KBr	0.08g
$SrCl_2$	0.03g
H_3BO_3	0.02g
Na_2HPO_4	8.0mg
Na_2SiO_3	4.0mg
NaF	2.4mg
NH_4NO_3	1.6mg
$Na_2S·9H_2O$ solution	0.5mL

pH 5.8 ± 0.2 at 25°C

$Na_2S·9H_2O$ Solution:
Composition per 10.0mL:

$Na_2S·9H_2O$	1.0g

Preparation of $Na_2S·9H_2O$ Solution: Add $Na_2S·9H_2O$ to distilled/deionized water and bring volume to 10.0mL. Mix thoroughly. Sparge with 100% N_2. Autoclave for 15 min at 15 psi pressure–121°C. Neutralize to pH 7.0 with sterile HCl.

Preparation of Medium: Add components, except sulfur and $Na_2S·9H_2O$ solution, to distilled/deionized water and bring volume to 1.0L. Mix thoroughly. Filter through normal filter paper. An iron sediment will collect in the filter. Gently heat while stirring and bring to boiling. Boil for 5 min. Cool under an anaerobic gas mixture of N_2. Adjust pH to 5.8. Distribute the medium into Hungate tubes or serum bottles containing finely divided sulfur (0.5% w/v). Seal the tubes or bottles under the same anaerobic gas used when cooling the medium. Sterilize the medium at 100°C for 3 hr on 3 consecutive days. Reduce the medium by adding 10% neutralized $Na_2S·9H_2O$ solution to a final concentration of 0.05%. The medium should not give a heavy black precipitate; if it does the iron sediment was not adequately removed by filtering in the initial stages and the medium should be made again, making sure that the iron is removed by filtering.

Use: For the cultivation and maintenance of *Thermococcus celer* DSM 2476.

Marine *Thermococcus* Medium (DSMZ Medium 760)

Composition per liter:

NaCl	19.45g
$MgCl_2$	8.8g
Sulfur	5.0g
Peptone	5.0g
Na_2SO_3	3.24g
$CaCl_2$	1.8g
Yeast extract	1.0g
KCl	0.55g
$NaHCO_3$	0.16g
Ferric citrate	0.1g
KBr	0.08g
$SrCl_2$	0.03g
H_3BO_3	0.02g
Na_2HPO_4	8.0mg
Na_2SiO_3	4.0mg
NaF	2.4mg
NH_4NO_3	1.6mg
$Na_2S·9H_2O$ solution	0.5mL

pH 7.2 ± 0.2 at 25°C

$Na_2S·9H_2O$ Solution:
Composition per 10.0mL:

$Na_2S·9H_2O$	1.0g

Preparation of $Na_2S·9H_2O$ Solution: Add $Na_2S·9H_2O$ to distilled/deionized water and bring volume to 10.0mL. Mix thoroughly. Sparge with 100% N_2. Autoclave for 15 min at 15 psi pressure–121°C. Neutralize to pH 7.0 with sterile HCl.

Preparation of Medium: Add components, except sulfur and $Na_2S·9H_2O$ solution, to distilled/deionized water and bring volume to 1.0L. Mix thoroughly. Filter through normal filter paper. An iron sediment will collect in the filter. Gently heat while stirring and bring to boiling. Boil for 5 min. Cool under an anaerobic gas mixture of N_2. Adjust pH to 7.2. Distribute the medium into Hungate tubes or serum bottles containing finely divided sulfur (0.5% w/v). Seal the tubes or bottles under the same anaerobic gas used when cooling the medium. Sterilize the medium at 100°C for 3 hr on 3 consecutive days. Reduce the medium by adding 10% neutralized $Na_2S \cdot 9H_2O$ solution to a final concentration of 0.05%. The medium should not give a heavy black precipitate; if it does the iron sediment was not adequately removed by filtering in the initial stages and the medium should be made again, making sure that the iron is removed by filtering.

Use: For the cultivation and maintenance of *Thermococcus profundus* DSM 9503, *Thermococcus peptonophilus* DSM 10343, *Thermococcus guaymasensis* 11113, and *Thermococcus aggregans* DSM 12819.

Marine *Thermococcus* Medium (DSMZ Medium 760)

Composition per liter:

NaCl	19.45g
$MgCl_2$	8.8g
Sulfur	5.0g
Peptone	5.0g
Na_2SO_3	3.24g
$CaCl_2$	1.8g
Yeast extract	1.0g
KCl	0.55g
$NaHCO_3$	0.16g
Ferric citrate	0.1g
KBr	0.08g
$SrCl_2$	0.03g
H_3BO_3	0.02g
Na_2HPO_4	8.0mg
Na_2SiO_3	4.0mg
NaF	2.4mg
NH_4NO_3	1.6mg
$Na_2S \cdot 9H_2O$ solution	0.5mL

pH 7.5 ± 0.2 at 25°C

$Na_2S \cdot 9H_2O$ Solution:

Composition per 10.0mL:

$Na_2S \cdot 9H_2O$	1.0g

Preparation of $Na_2S \cdot 9H_2O$ Solution: Add $Na_2S \cdot 9H_2O$ to distilled/deionized water and bring volume to 10.0mL. Mix thoroughly. Sparge with 100% N_2. Autoclave for 15 min at 15 psi pressure–121°C. Neutralize to pH 7.0 with sterile HCl.

Preparation of Medium: Add components, except sulfur and $Na_2S \cdot 9H_2O$ solution, to distilled/deionized water and bring volume to 1.0L. Mix thoroughly. Filter through normal filter paper. An iron sediment will collect in the filter. Gently heat while stirring and bring to boiling. Boil for 5 min. Cool under an anaerobic gas mixture of N_2. Adjust pH to 7.5. Distribute the medium into Hungate tubes or serum bottles containing finely divided sulfur (0.5% w/v). Seal the tubes or bottles under the same anaerobic gas used when cooling the medium. Sterilize the medium at 100°C for 3 hr on 3 consecutive days. Reduce the medium by adding 10% neutralized $Na_2S \cdot 9H_2O$ solution to a final concentration of 0.05%. The medium should not give a heavy black precipitate; if it does the iron sediment was not adequately removed by filtering in the initial stages and the medium should be made again, making sure that the iron is removed by filtering.

Use: For the cultivation and maintenance of *Thermococcus litoralis* DSM 5473, *Thermococcus litoralis* 5474, *Thermococcus fumicolans* DSM 12820, and *Thermococcus sibiricus* DSM 12597.

Marinithermus hydrothermalis Medium (DSMZ Medium 973)

Composition per liter:

NaCl	30.0g
$MgCl_2 \cdot 6H_2O$	4.18g
$MgSO_4 \cdot 7H_2O$	3.4g
Yeast extract	1.0g
Tryptone	1.0g
KCl	0.33g
NH_4Cl	0.25g
K_2HPO_4	0.14g
$CaCl_2 \cdot 2H_2O$	0.14g
$Fe(NH_4)_2(SO_4)_2 \cdot 6H_2O$	10.0mg
$NiCl_2 \cdot 6H_2O$	0.5mg
$Na2Se_3 \cdot 5H_2O$	0.5mg
Trace elements solution	10.0mL

pH 7.0 ± 0.2 at 25°C

Trace Elements Solution:

Composition per liter:

$MgSO_4 \cdot 7H_2O$	3.0g
Nitrilotriacetic acid	1.5g
NaCl	1.0g
$MnSO_4 \cdot 2H_2O$	0.5g
$CoSO_4 \cdot 7H_2O$	0.18g
$ZnSO_4 \cdot 7H_2O$	0.18g
$CaCl_2 \cdot 2H_2O$	0.1g
$FeSO_4 \cdot 7H_2O$	0.1g
$NiCl_2 \cdot 6H_2O$	0.025g
$KAl(SO_4)_2 \cdot 12H_2O$	0.02g
H_3BO_3	0.01g
$Na_2MoO_4 \cdot 4H_2O$	0.01g
$CuSO_4 \cdot 5H_2O$	0.01g
$Na_2SeO_3 \cdot 5H_2O$	0.3mg

Preparation of Trace Elements Solution: Add nitrilotriacetic acid to 500.0mL of distilled/deionized water. Dissolve by adjusting pH to 6.5 with KOH. Add remaining components. Add distilled/deionized water to 1.0L. Mix thoroughly.

Preparation of Medium: Add components to distilled/deionized water and bring volume to 1.0L. Mix thoroughly. Distribute into tubes or flasks. Autoclave for 15 min at 15 psi pressure–121°C.

Use: For the cultivation of *Marinithermus hydrothermalis.*

Marinitoga Medium (DSMZ Medium 904)

Composition per 1045.0mL:

Sea salts	30.0g
PIPES	6.0g
Yeast extract	1.0g
Tryptone	1.0g
Resazurin	0.5mg
Glucose solution	25.0mL
$Na_2S \cdot 9H_2O$ solution	10.0mL
L-Cysteine solution	10.0mL

pH 7.0 ± 0.2 at 25°C

Glucose Solution:

Composition per 25.0mL:

Glucose	2.5g

Preparation of Glucose Solution: Add sucrose to distilled/deionized water and bring volume to 25.0mL. Mix thoroughly. Sparge with 100% N_2. Autoclave for 15 min at 15 psi pressure–121°C.

L-Cysteine Solution:

Composition per 10.0mL:

L-Cysteine·HCl·H_2O	0.5g

Preparation of L-Cysteine Solution: Add L-cysteine·HCl·H_2O to distilled/deionized water and bring volume to 10.0mL. Mix thoroughly. Sparge with 100% N_2. Autoclave for 15 min at 15 psi pressure–121°C.

$Na_2S·9H_2O$ Solution:

Composition per 10.0mL:

$Na_2S·9H_2O$	0.5g

Preparation of $Na_2S·9H_2O$ Solution: Add $Na_2S·9H_2O$ to distilled/deionized water and bring volume to 10.0mL. Mix thoroughly. Sparge with 100% N_2. Autoclave for 15 min at 15 psi pressure–121°C.

Preparation of Medium: Prepare and dispense medium under 100% N_2. Add components, except glucose solution, L-cysteine-HCl·H_2O solution, and $Na_2S·9H_2O$ solution, to distilled/deionized water and bring volume to 1.0L. Mix thoroughly. Adjust pH to 7.0. Distribute into anaerobe tubes or bottles. Autoclave for 15 min at 15 psi pressure–121°C. Aseptically and anaerobically add per liter, 50.0mL glucose solution, 10.0mL L-cysteine-HCl·H_2O solution, and 10.0mL $Na_2S·9H_2O$. Mix thoroughly. The final pH should be 7.0.

Use: For the cultivation of *Marinitoga camini* and *Caloranaerobacter azorensis.*

Marinitoga piezophila Medium (DSMZ Medium 945)

Composition per liter:

NaCl	30.0g
Yeast extract	5.0g
Trypticase™	5.0g
MES	1.95g
NH_4Cl	1.0g
Na-acetate	0.83g
K_2HPO_4	0.3g
KH_2PO_4	0.3g
$MgCl_2·6H_2O$	0.2g
$CaCl_2·2H_2O$	0.1g
KCl	0.1g
Resazurin	0.5mg
Maltose solution	100.0mL
$Na_2S·9H_2O$ solution	10.0mL
Cysteine solution	10.0mL

pH 6.0 ± 0.2 at 25°C

Maltose Solution:

Composition per 100.0mL:

Maltose	4.96g

Preparation of Maltose Solution: Add maltose to distilled/deionized water and bring volume to 100.0mL. Mix thoroughly. Sparge with 100% N_2. Filter sterilize.

Cysteine Solution:

Composition per 10.0mL:

L-Cysteine·HCl·H_2O	0.3g

Preparation of L-Cysteine Solution: Add L-cysteine·HCl·H_2O to distilled/deionized water and bring volume to 10.0mL. Mix thoroughly. Sparge with 100% N_2. Autoclave for 15 min at 15 psi pressure–121°C. Cool to 25°C.

$Na_2S·9H_2O$ Solution:

Composition per 10.0mL:

$Na_2S·9H_2O$	0.3g

Preparation of $Na_2S·9H_2O$ Solution: Add $Na_2S·9H_2O$ to distilled/deionized water and bring volume to 10.0mL. Sparge with N_2. Autoclave for 15 min at 15 psi pressure–121°C. Cool to 25°C. Store anaerobically.

Preparation of Medium: Add components, except maltose solution, $NaHCO_3$ solution, and $Na_2S·9H_2O$ solution, to 880.0mL distilled/deionized water. Mix thoroughly. Sparge for 30 min with 100% N_2. Adjust pH to 6.0 with concentrated NaOH. Distribute under 100% N_2 into anaerobic tubes or bottles. Autoclave for 15 min at 15 psi pressure–121°C. Cool to 25°C. Aseptically and anaerobically add 100.0mL sterile maltose solution, 10.0mL sterile $Na_2S·9H_2O$ solution, and 10.0mL sterile cysteine solution per liter medium. Mix thoroughly.

Use: For the cultivation of *Marinitoga piezophila.*

Marinitoga piezophila Medium (DSMZ Medium 945)

Composition per liter:

NaCl	30.0g
Sulfur	10.0g
Yeast extract	5.0g
Trypticase™	5.0g
MES	1.95g
NH_4Cl	1.0g
Na-acetate	0.83g
K_2HPO_4	0.3g
KH_2PO_4	0.3g
$MgCl_2·6H_2O$	0.2g
$CaCl_2·2H_2O$	0.1g
KCl	0.1g
Resazurin	0.5mg
$Na_2S·9H_2O$ solution	10.0mL
Cysteine solution	10.0mL

pH 6.0 ± 0.2 at 25°C

Cysteine Solution:

Composition per 10.0mL:

L-Cysteine·HCl·H_2O	0.3g

Preparation of L-Cysteine Solution: Add L-cysteine·HCl·H_2O to distilled/deionized water and bring volume to 10.0mL. Mix thoroughly. Sparge with 100% N_2. Autoclave for 15 min at 15 psi pressure–121°C. Cool to 25°C.

$Na_2S·9H_2O$ Solution:

Composition per 10.0mL:

$Na_2S·9H_2O$	0.3g

Preparation of $Na_2S·9H_2O$ Solution: Add $Na_2S·9H_2O$ to distilled/deionized water and bring volume to 10.0mL. Sparge with N_2. Autoclave for 15 min at 15 psi pressure–121°C. Cool to 25°C. Store anaerobically.

Preparation of Sulfur: Sterilize sulfur by steaming for 3 hr on each of 3 successive days.

Preparation of Medium: Add components, except sulfur, cysteine solution, and $Na_2S·9H_2O$ solution, to 980.0mL distilled/deionized water. Mix thoroughly. Sparge for 30 min with 100% N_2. Adjust pH to 6.0

with concentrated NaOH. Distribute under 100% N_2 into anaerobic tubes or bottles containing appropriate amounts of sterile sulfur (1g steam-sterilized sulfur per 100mL medium). Autoclave for 20 min at 110°C. Cool to room temperature. Aseptically and anaerobically add 10.0mL sterile $Na_2S{\cdot}9H_2O$ solution and 10.0mL sterile cysteine solution per liter medium. Mix thoroughly.

Use: For the cultivation of *Marinitoga piezophila.*

Marinobacter lutaoensis Medium (DSMZ Medium 1066)

Composition per liter:

Peptone	4.0g
Yeast extract	2.0g
NaCl	25.0g
$MgCl_2{\cdot}6H_2O$	2.0g

pH 7.0 ± 0.2 at 25°C

Preparation of Medium: Add components to distilled/deionized water and bring volume to 1.0L. Mix thoroughly. Adjust pH to 7.0. Distribute into tubes or flasks. Gently heat while stirring and bring to boiling. Mix thoroughly. Autoclave for 15 min at 15 psi pressure–121°C.

Use: For the cultivation of *Marinobacter lutaoensis.*

Marinobacter Medium (DSMZ Medium 941)

Composition per liter:

NaCl	6.0g
NH_4Cl	1.0g
Na-acetate	1.0g
$MgSO_4{\cdot}7H_2O$	0.2g
KCl	0.1g
KH_2PO_4	0.1g
Peptone	0.1g
$CaCl_2{\cdot}2H_2O$	0.04g
Trace elements solution SL-7	1.0mL
Vitamin solution, concentrated	1.0mL

pH 7.2 ± 0.2 at 25°C

Trace Elements Solution SL-7:

Composition per liter:

$FeCl_2{\cdot}7H_2O$	1.5g
$CoCl_2{\cdot}6H_2O$	190.0mg
$MnCl_2{\cdot}4H_2O$	100.0mg
$ZnCl_2$	70.0mg
$Na_2MoO_4{\cdot}2H_2O$	36.0mg
$NiCl_2{\cdot}6H_2O$	24.0mg
H_3BO_3	62.0mg
$CuCl_2{\cdot}2H_2O$	17.0mg
HCl (25% solution)	6.5mL

Preparation of Trace Elements Solution SL-7: Add $FeCl_2{\cdot}7H_2O$ to 10.0mL of HCl solution. Mix thoroughly. Add distilled/deionized water and bring volume to 1.0L. Add remaining components. Mix thoroughly. Sparge with 80% N_2 + 20% CO_2. Autoclave for 15 min at 15 psi pressure–121°C.

Vitamin Solution, Concentrated:

Composition per 100.0mL:

Pyridoxine-HCl	10.0mg
Thiamine-HCl·$2H_2O$	5.0mg
Riboflavin	5.0mg
Nicotinic acid	5.0mg
D-Ca-pantothenate	5.0mg
p-Aminobenzoic acid	5.0mg
Lipoic acid	5.0mg
Biotin	2.0mg
Folic acid	2.0mg
Vitamin B_{12}	0.1mg

Preparation of Vitamin Solution, Concentrated: Add components to distilled/deionized water and bring volume to 100.0mL. Mix thoroughly. Filter sterilize.

Preparation of Medium: Add components to distilled/deionized water and bring volume to 1.0L. Mix thoroughly. Distribute into tubes or flasks. Autoclave for 15 min at 15 psi pressure–121°C.

Use: For the cultivation of *Marionobacter* sp.

Marinobacter Medium (DSMZ Medium 970)

Composition per liter:

NaCl	11.7g
$MgSO_4$	7.85g
TRIS	6.0g
Yeast extract	5.0g
Peptone	5.0g
NH_4Cl	3.0g
$CaCl_2$	1.47g
KCl	0.74g

pH 7.8 ± 0.2 at 25°C

Preparation of Medium: Add components to distilled/deionized water and bring volume to 1.0L. Mix thoroughly. Distribute into tubes or flasks. Autoclave for 15 min at 15 psi pressure–121°C.

Use: For the cultivation of *Marinobacter* sp.

Marinococcus albus Agar (LMG Medium 212)

Composition per liter:

NaCl	81.0g
Agar	15.0g
Yeast	10.0g
$MgSO_4{\cdot}7H_2O$	9.6g
$MgCl_2{\cdot}6H_2O$	7.0g
Proteose peptone	5.0g
KCl	2.0g
Glucose	1.0g
$CaCl_2$	0.36g
NaB	226.0mg
$NaHCO_3$	60.0mg

pH 7.2 ± 0.2 at 25°C

Preparation of Medium: Add components to distilled/deionized water and bring volume to 1.0L. Mix thoroughly. Gently heat and bring to boiling. Distribute into tubes or flasks. Autoclave for 15 min at 15 psi pressure–121°C. Pour into sterile Petri dishes or leave in tubes.

Use: For the cultivation and maintenance of *Marinococcus albus.*

Marinococcus albus Medium

Composition per liter:

NaCl	81.0g
Yeast extract	10.0g
$MgSO_4{\cdot}7H_2O$	9.6g
$MgCl_2{\cdot}6H_2O$	7.0g
Proteose peptone No. 3	5.0g

KCl	2.0g
Glucose	1.0g
$CaCl_2$	0.36g
$NaHCO_3$	0.06g
NaBr	0.026g

pH 7.2 ± 0.2 at 25°C

Preparation of Medium: Add components to distilled/deionized water and bring volume to 1.0L. Mix thoroughly. Distribute into tubes or flasks. Autoclave for 15 min at 15 psi pressure–121°C.

Use: For the cultivation and maintenance of *Marinococcus albus*.

Marinomonas vaga Medium
See: **Nutrient Agar with 3% NaCl**

Marinomonas vaga Medium (DSMZ Medium 617)

Composition per liter:

NaCl	30.0g
Agar	15.0g
Beef extract	10.0g
Peptone	10.0g

pH 7.1 ± 0.2 at 25°C

Preparation of Medium: Add components to tap water and bring volume to 1.0L. Mix thoroughly. Gently heat and bring to boiling. Distribute into tubes or flasks. Autoclave for 15 min at 15 psi pressure–121°C. Pour into sterile Petri dishes or leave in tubes.

Use: For the cultivation and maintenance of *Marinomonas communis=Alteromonas communis* and *Marinomonas vaga=Alteromonas vaga.*

Martin-Lewis Agar

Composition per liter:

Agar	12.0g
Hemoglobin	10.0g
Pancreatic digest of casein	7.5g
Selected meat peptone	7.5g
NaCl	5.0g
K_2HPO_4	4.0g
Cornstarch	1.0g
KH_2PO_4	1.0g
Supplement solution	10.0mL
VCAT inhibitor	10.0mL

pH 7.2 ± 0.22 at 25°C

Source: Martin-Lewis agar is available as a prepared medium from BD Diagnostic Systems.

Supplement Solution:
Composition per liter:

Glucose	100.0g
L-Cysteine·HCl	25.9g
L-Glutamine	10.0g
L-Cystine	1.1g
Adenine	1.0g
Nicotinamide adenine dinucleotide	0.25g
Vitamin B_{12}	0.1g
Thiamine pyrophosphate	0.1g
Guanine·HCl	0.03g
$Fe(NO_3)_3 \cdot 6H_2O$	0.02g
p-Aminobenzoic acid	0.013g
Thiamine·HCl	3.0mg

Source: The supplement solution IsoVitaleX® enrichment is available from BD Diagnostic Systems. This enrichment may be replaced by supplement VX from BD Diagnostic Systems.

Preparation of Supplement Solution: Add components to distilled/deionized water and bring volume to 1.0L. Mix thoroughly. Filter sterilize.

VCAT Inhibitor:
Composition per 10.0mL:

Colistin	7.5mg
Trimethoprim lactate	5.0mg
Vancomycin	4.0mg
Anisomycin	0.02g

Preparation of VCAT Inhibitor: Add components to distilled/deionized water and bring volume to 10.0mL. Mix thoroughly. Filter sterilize.

Preparation of Medium: Add components, except supplement solution and VCAT inhibitor, to distilled/deionized water and bring volume to 980.0mL. Gently heat while stirring and bring to boiling. Autoclave for 15 min at 15 psi pressure–121°C. Cool to 45°–50°C. Aseptically add sterile supplement solution and sterile VCAT inhibitor. Mix thoroughly. Pour into sterile Petri dishes.

Use: For the isolation and cultivation of pathogenic *Neisseria* from specimens containing mixed flora of bacteria and fungi.

Martin-Lewis Agar, Enriched

Composition per liter:

Agar	12.0g
Pancreatic digest of casein	7.5g
Selected meat peptone	7.5g
NaCl	5.0g
K_2HPO_4	4.0g
Cornstarch	1.0g
KH_2PO_4	1.0g
Sarcina lutea suspension	20.0mL
Horse serum, inactivated	20.0mL
Supplement solution	10.0mL
PCAT inhibitor	10.0mL

pH 7.2 ± 0.22 at 25°C

Source: The supplement solution (IsoVitaleX® enrichment) is available from BD Diagnostic Systems. This enrichment may be replaced by supplement VX from BD Diagnostic Systems.

***Sarcina lutea* Suspension:**
Composition per 20.0mL:

Sarcina lutea FDA 1001	10^6–10^7 cells

Preparation of *Sarcina lutea* Suspension: Aseptically wash the growth of 24-hr cultures of *Sarcina lutea* FDA 1001 cells from Thayer-Martin plates with sterile soybean casein digest broth. Standardize the suspension by adding additional sterile tryptic soy broth to yield 40% light transmission at 530nm wavelength.

Soybean Casein Digest Broth:
Composition per liter:

Pancreatic digest of casein	17.0g
NaCl	5.0g
Papaic digest of soybean meal	3.0g
K_2HPO_4	2.5g
Glucose	2.5g

pH 7.3 ± 0.2 at 25°C

Preparation of Soybean Casein Digest Broth: Add components to distilled/deionized water and bring volume to 1.0L. Mix thoroughly. Distribute into tubes or flasks. Autoclave for 15 min at 15 psi pressure–121°C.

Supplement Solution:

Composition per liter:

Glucose	100.0g
L-Cysteine·HCl	25.9g
L-Glutamine	10.0g
L-Cystine	1.1g
Adenine	1.0g
Nicotinamide adenine dinucleotide	0.25g
Vitamin B_{12}	0.1g
Thiamine pyrophosphate	0.1g
Guanine·HCl	0.03g
$Fe(NO_3)_3 \cdot 6H_2O$	0.02g
p-Aminobenzoic acid	0.013g
Thiamine·HCl	3.0mg

Preparation of Supplement Solution: Add components to distilled/deionized water and bring volume to 1.0L. Mix thoroughly. Filter sterilize.

PCAT Inhibitor:

Composition per 10.0mL:

Anisomycin	0.02g
Colistin	7.5mg
Trimethoprim lactate	5.0mg
Penicillin G	25,000U

Preparation of PCAT Inhibitor: Add components to distilled/deionized water and bring volume to 10.0mL. Mix thoroughly. Filter sterilize.

Preparation of Medium: Add components—except *Sarcina lutea* suspension, horse serum, supplement solution, and PCAT inhibitor—to distilled/deionized water and bring volume to 940.0mL. Gently heat while stirring and bring to boiling. Autoclave for 15 min at 15 psi pressure–121°C. Cool to 45°–50°C. Aseptically add 20.0mL of sterile *Sarcina lutea* suspension, 20.0mL of sterile horse serum, 10.0mL of supplement solution, and 10.0mL of sterile PCAT inhibitor. Mix thoroughly. Pour into sterile Petri dishes.

Use: For the isolation and cultivation of pathogenic *Neisseria*, especially penicillinase-producing strains, from specimens containing mixed flora of bacteria and fungi.

Mating Agar

Composition per liter:

Agar	40.0g
Sucrose	10.0g
Xylose	2.0g
KH_2PO_4	1.0g
$MgSO_4$	0.5g
Yeast extract	0.5g
$CaCl_2$	0.1g
NaCl	0.1g
Biotin	5.0μg

pH 5.7 ± 0.2 at 25°C

Preparation of Medium: Add components to distilled/deionized water and bring volume to 1.0L. Mix thoroughly. Gently heat and bring to boiling. Adjust pH to 5.7. Distribute into tubes or flasks. Autoclave for 15 min at 15 psi pressure–121°C. Pour into sterile Petri dishes or leave in tubes.

Use: For the cultivation and maintenance of *Filobasidiella neoformans*.

Maximum Recovery Diluent

Composition per liter:

NaCl	8.5g
Peptone	1.0g

pH 7.0 ± 0.2 at 25°C

Source: This medium is available as a premixed powder from Oxoid Unipath.

Preparation of Medium: Add components to distilled/deionized water and bring volume to 1.0L. Mix thoroughly. Distribute into tubes or flasks. Autoclave for 15 min at 15 psi pressure–121°C.

Use: This diluent is a physiologically isotonic and protective medium for maximal recovery of microorganisms from a variety of sources.

M-Azide HiVeg Broth Base with Triphenyltetrazolium Chloride

Composition per liter:

Saccharose	100.0g
Plant hydrolysate No. 1	40.0g
Yeast extract	10.0g
K_2HPO_4	4.0g
Glucose	2.0g
NaN_3	0.4g
Triphenyltetrazolium chloride solution	5.0mL

pH 7.2 ± 0.2 at 25°C

Source: This medium is available as a premixed powder from HiMedia.

Caution: Sodium azide is toxic. Azides also react with metals and disposal must be highly diluted.

Triphenyltetrazolium Choride Solution:

Composition per 5.0mL:

Triphenyltetrazolium chloride	0.1g

Preparation of Triphenyltetrazolium Choride Solution: Add triphenyltetrazolium chloride to distilled/deionized water and bring volume to 5.0mL. Mix thoroughly. Filter sterilize.

Preparation of Medium: Add components, except triphenyltetrazolium chloride solution, to distilled/deionized water and bring volume to 1.0L.Mix thoroughly. Gently heat and bring to boiling. Distribute into tubes or flasks. Autoclave for 15 min at 15 psi pressure–121°C. Cool to 45°–50°C. Aseptically add 5.0mL triphenyltetrazolium chloride solution. Mix thoroughly. Aseptically distribute into tubes.

Use: For the detection and enrichment of fecal streptococci in water and sewage by the membrane filtration method.

MB Medium (DSMZ Medium 924)

Composition per liter:

NaCl	10.0g
$NaHCO_3$	4.0g
Yeast extract	2.0g
Trypticase™	2.0g
NH_4Cl	1.0g
$MgCl_2 \cdot 6H_2O$	1.0g
KCl	0.5g
$CaCl_2 \cdot 2H_2O$	0.4g

K_2HPO_4 .... 0.4g
Resazurin 0.5mg
Sodium formate solution 50.0mL
$Na_2S{\cdot}9H_2O$ solution 10.0mL
Cysteine-HCl·H_2O solution 10.0mL
Vitamin solution 10.0mL
Trace elements solution 10.0mL
Selenite-tungstate solution 1.0mL

pH 7.2 ± 0.2 at 25°C

Cysteine Solution:
Composition per 10.0mL:
L-Cysteine·HCl·H_2O 0.25g

Preparation of Cysteine Solution: Add L-cysteine·HCl·H_2O to distilled/deionized water and bring volume to 10.0mL. Mix thoroughly. Sparge with 100% N_2. Autoclave for 15 min at 15 psi pressure–121°C.

Sodium Formate Solution:
Composition per 50.0mL:
Na-formate 6.8g

Preparation of Sodium Formate Solution: Add sodium formate to distilled/deionized water and bring volume to 50.0mL. Mix thoroughly. Sparge with 100% N_2. Filter sterilize.

Selenite-Tungstate Solution
Composition per liter:
NaOH 0.5g
$Na_2WO_4{\cdot}2H_2O$ 4.0mg
$Na_2SeO_3{\cdot}5H_2O$ 3.0mg

Preparation of Selenite-Tungstate Solution: Add components to distilled/deionized water and bring volume to 1.0L. Mix thoroughly. Sparge with 100% N_2. Filter sterilize.

$Na_2S{\cdot}9H_2O$ Solution:
Composition per 10.0mL:
$Na_2S{\cdot}9H_2O$ 0.25g

Preparation of $Na_2S{\cdot}9H_2O$ Solution: Add $Na_2S{\cdot}9H_2O$ to distilled/deionized water and bring volume to 10.0mL. Sparge with N_2. Autoclave for 15 min at 15 psi pressure–121°C. Cool to 25°C. Store anaerobically.

Trace Elements Solution:
Composition per liter:
$MgSO_4{\cdot}7H_2O$ 3.0g
Nitrilotriacetic acid 1.5g
NaCl 1.0g
$MnSO_4{\cdot}2H_2O$ 0.5g
$CoSO_4{\cdot}7H_2O$ 0.18g
$ZnSO_4{\cdot}7H_2O$ 0.18g
$CaCl_2{\cdot}2H_2O$ 0.1g
$FeSO_4{\cdot}7H_2O$ 0.1g
$NiCl_2{\cdot}6H_2O$ 0.025g
$KAl(SO_4)_2{\cdot}12H_2O$ 0.02g
H_3BO_3 .... 0.01g
$Na_2MoO_4{\cdot}4H_2O$ 0.01g
$CuSO_4{\cdot}5H_2O$ 0.01g
$Na_2SeO_3{\cdot}5H_2O$ 0.3mg

Preparation of Trace Elements Solution: Add nitrilotriacetic acid to 500.0mL of distilled/deionized water. Dissolve by adjusting pH to 6.5 with KOH. Add remaining components. Add distilled/deionized water to 1.0L. Mix thoroughly.

Vitamin Solution:
Composition per liter:
Pyridoxine-HCl 10.0mg
Thiamine-HCl·$2H_2O$ 5.0mg
Riboflavin 5.0mg
Nicotinic acid 5.0mg
D-Ca-pantothenate 5.0mg
p-Aminobenzoic acid 5.0mg
Lipoic acid 5.0mg
Biotin 2.0mg
Folic acid 2.0mg
Vitamin B_{12} 0.1mg

Preparation of Vitamin Solution: Add components to distilled/deionized water and bring volume to 1.0L. Mix thoroughly. Sparge with 80% H_2 + 20% CO_2. Filter sterilize.

Preparation of Medium: Prepare and dispense medium under an oxygen-free 80% N_2 + 20% CO_2 gas mixture. Add components, except sodium formate solution, $NaHCO_3$, cysteine solution, and $Na_2S{\cdot}9H_2O$ solution, to distilled/deionized water and bring volume to 920.0mL. Mix thoroughly. Gently heat and bring to boiling. Boil for 3 min. Cool to 25°C while sparging with 80% N_2 + 20% CO_2. Add solid $NaHCO_3$. Mix thoroughly. Adjust pH to 6.8–7.0. Distribute into tubes or bottles. Autoclave for 15 min at 15 psi pressure–121°C. Aseptically and anaerobically add per liter 50.0mL sterile sodium formate solution, 10.0mL of sterile cysteine solution, and 10.0mL of sterile $Na_2S{\cdot}9H_2O$ solution. Mix thoroughly. The final pH should be 7.2.

Use: For the cultivation of *Methanocalculus taiwanensis, Methanococcus voltae (Methanococcus voltaei),* and *Methanofollis aquaemaris.*

MB Medium (DSMZ Medium 924)

Composition per liter:
NaCl 5.0g
$NaHCO_3$ 4.0g
Yeast extract 2.0g
Trypticase™ 2.0g
NH_4Cl 1.0g
$MgCl_2{\cdot}6H_2O$ 1.0g
KCl 0.5g
$CaCl_2{\cdot}2H_2O$ 0.4g
K_2HPO_4 .... 0.4g
Resazurin 0.5mg
Sodium formate solution 50.0mL
$Na_2S{\cdot}9H_2O$ solution 10.0mL
Cysteine-HCl·H_2O solution 10.0mL
Vitamin solution 10.0mL
Trace elements solution 10.0mL
Selenite-tungstate solution 1.0mL

pH 6.5 ± 0.2 at 25°C

Cysteine Solution:
Composition per 10.0mL:
L-Cysteine·HCl·H_2O 0.25g

Preparation of Cysteine Solution: Add L-cysteine·HCl·H_2O to distilled/deionized water and bring volume to 10.0mL. Mix thoroughly. Sparge with 100% N_2. Autoclave for 15 min at 15 psi pressure–121°C.

Sodium Formate Solution:
Composition per 50.0mL:
Na-formate 6.8g

Preparation of Sodium Formate Solution: Add sodium formate to distilled/deionized water and bring volume to 50.0mL. Mix thoroughly. Sparge with 100% N_2. Filter sterilize.

Selenite-Tungstate Solution
Composition per liter:

NaOH	0.5g
$Na_2WO_4 \cdot 2H_2O$	4.0mg
$Na_2SeO_3 \cdot 5H_2O$	3.0mg

Preparation of Selenite-Tungstate Solution: Add components to distilled/deionized water and bring volume to 1.0L. Mix thoroughly. Sparge with 100% N_2. Filter sterilize.

$Na_2S \cdot 9H_2O$ Solution:
Composition per 10.0mL:

$Na_2S \cdot 9H_2O$	0.25g

Preparation of $Na_2S \cdot 9H_2O$ Solution: Add $Na_2S \cdot 9H_2O$ to distilled/deionized water and bring volume to 10.0mL. Sparge with N_2. Autoclave for 15 min at 15 psi pressure–121°C. Cool to 25°C. Store anaerobically.

Trace Elements Solution:
Composition per liter:

$MgSO_4 \cdot 7H_2O$	3.0g
Nitrilotriacetic acid	1.5g
NaCl	1.0g
$MnSO_4 \cdot 2H_2O$	0.5g
$CoSO_4 \cdot 7H_2O$	0.18g
$ZnSO_4 \cdot 7H_2O$	0.18g
$CaCl_2 \cdot 2H_2O$	0.1g
$FeSO_4 \cdot 7H_2O$	0.1g
$NiCl_2 \cdot 6H_2O$	0.025g
$KAl(SO_4)_2 \cdot 12H_2O$	0.02g
H_3BO_3	0.01g
$Na_2MoO_4 \cdot 4H_2O$	0.01g
$CuSO_4 \cdot 5H_2O$	0.01g
$Na_2SeO_3 \cdot 5H_2O$	0.3mg

Preparation of Trace Elements Solution: Add nitrilotriacetic acid to 500.0mL of distilled/deionized water. Dissolve by adjusting pH to 6.5 with KOH. Add remaining components. Add distilled/deionized water to 1.0L. Mix thoroughly.

Vitamin Solution:
Composition per liter:

Pyridoxine-HCl	10.0mg
Thiamine-HCl·$2H_2O$	5.0mg
Riboflavin	5.0mg
Nicotinic acid	5.0mg
D-Ca-pantothenate	5.0mg
p-Aminobenzoic acid	5.0mg
Lipoic acid	5.0mg
Biotin	2.0mg
Folic acid	2.0mg
Vitamin B_{12}	0.1mg

Preparation of Vitamin Solution: Add components to distilled/deionized water and bring volume to 1.0L. Mix thoroughly. Sparge with 80% H_2 + 20% CO_2. Filter sterilize.

Preparation of Medium: Prepare and dispense medium under an oxygen-free 80% N_2 + 20% CO_2 gas mixture. Add components, except sodium formate solution, $NaHCO_3$, cysteine solution, and $Na_2S \cdot 9H_2O$ solution, to distilled/deionized water and bring volume to 920.0mL. Mix thoroughly. Gently heat and bring to boiling. Boil for 3 min. Cool to 25°C while sparging with 80% N_2 + 20% CO_2. Add solid $NaHCO_3$. Mix thoroughly. Adjust pH to 6.8–7.0. Distribute into tubes or bottles. Autoclave for 15 min at 15 psi pressure–121°C. Aseptically and anaerobically add per liter 50.0mL sterile sodium formate solution, 10.0mL of sterile cysteine solution, and 10.0mL of sterile $Na_2S \cdot 9H_2O$ solution. Mix thoroughly. The final pH should be 6.5.

Use: For the cultivation of *Methanofollis aquaemaris* DSM 14661.

MB Medium (DSMZ Medium 924)

Composition per liter:

NaCl	10.0g
$NaHCO_3$	4.0g
Yeast extract	2.0g
Trypticase™	2.0g
NH_4Cl	1.0g
$MgCl_2 \cdot 6H_2O$	1.0g
KCl	0.5g
$CaCl_2 \cdot 2H_2O$	0.4g
K_2HPO_4	0.4g
Resazurin	0.5mg
Sodium acetate solution	50.0mL
Sodium formate solution	50.0mL
$Na_2S \cdot 9H_2O$ solution	10.0mL
Cysteine-HCl·H_2O solution	10.0mL
Vitamin solution	10.0mL
Trace elements solution	10.0mL
Selenite-tungstate solution	1.0mL

pH 7.2 ± 0.2 at 25°C

Cysteine Solution:
Composition per 10.0mL:

L-Cysteine·HCl·H_2O	0.25g

Preparation of Cysteine Solution: Add L-cysteine·HCl·H_2O to distilled/deionized water and bring volume to 10.0mL. Mix thoroughly. Sparge with 100% N_2. Autoclave for 15 min at 15 psi pressure–121°C.

Sodium Formate Solution:
Composition per 50.0mL:

Na-formate	6.8g

Preparation of Sodium Formate Solution: Add sodium formate to distilled/deionized water and bring volume to 50.0mL. Mix thoroughly. Sparge with 100% N_2. Filter sterilize.

Sodium Acetate Solution:
Composition per 50.0mL:

Na-acetate	1.6g

Preparation of Sodium Acetate Solution: Add sodium acetate to distilled/deionized water and bring volume to 50.0mL. Mix thoroughly. Sparge with 100% N_2. Filter sterilize.

Selenite-Tungstate Solution
Composition per liter:

NaOH	0.5g
$Na_2WO_4 \cdot 2H_2O$	4.0mg
$Na_2SeO_3 \cdot 5H_2O$	3.0mg

Preparation of Selenite-Tungstate Solution: Add components to distilled/deionized water and bring volume to 1.0L. Mix thoroughly. Sparge with 100% N_2. Filter sterilize.

$Na_2S \cdot 9H_2O$ Solution:

Composition per 10.0mL:

$Na_2S \cdot 9H_2O$	0.25g

Preparation of $Na_2S \cdot 9H_2O$ Solution: Add $Na_2S \cdot 9H_2O$ to distilled/deionized water and bring volume to 10.0mL. Sparge with N_2. Autoclave for 15 min at 15 psi pressure–121°C. Cool to 25°C. Store anaerobically.

Trace Elements Solution:

Composition per liter:

$MgSO_4 \cdot 7H_2O$	3.0g
Nitrilotriacetic acid	1.5g
NaCl	1.0g
$MnSO_4 \cdot 2H_2O$	0.5g
$CoSO_4 \cdot 7H_2O$	0.18g
$ZnSO_4 \cdot 7H_2O$	0.18g
$CaCl_2 \cdot 2H_2O$	0.1g
$FeSO_4 \cdot 7H_2O$	0.1g
$NiCl_2 \cdot 6H_2O$	0.025g
$KAl(SO_4)_2 \cdot 12H_2O$	0.02g
H_3BO_3	0.01g
$Na_2MoO_4 \cdot 4H_2O$	0.01g
$CuSO_4 \cdot 5H_2O$	0.01g
$Na_2SeO_3 \cdot 5H_2O$	0.3mg

Preparation of Trace Elements Solution: Add nitrilotriacetic acid to 500.0mL of distilled/deionized water. Dissolve by adjusting pH to 6.5 with KOH. Add remaining components. Add distilled/deionized water to 1.0L. Mix thoroughly.

Vitamin Solution:

Composition per liter:

Pyridoxine-HCl	10.0mg
Thiamine-HCl·$2H_2O$	5.0mg
Riboflavin	5.0mg
Nicotinic acid	5.0mg
D-Ca-pantothenate	5.0mg
p-Aminobenzoic acid	5.0mg
Lipoic acid	5.0mg
Biotin	2.0mg
Folic acid	2.0mg
Vitamin B_{12}	0.1mg

Preparation of Vitamin Solution: Add components to distilled/deionized water and bring volume to 1.0L. Mix thoroughly. Sparge with 80% H_2 + 20% CO_2. Filter sterilize.

Preparation of Medium: Prepare and dispense medium under an oxygen-free 80% N_2 + 20% CO_2 gas mixture. Add components, except sodium acetate solution, sodium formate solution, $NaHCO_3$, cysteine solution, and $Na_2S \cdot 9H_2O$ solution, to distilled/deionized water and bring volume to 870.0mL. Mix thoroughly. Gently heat and bring to boiling. Boil for 3 min. Cool to 25°C while sparging with 80% N_2 + 20% CO_2. Add solid $NaHCO_3$. Mix thoroughly. Adjust pH to 6.8–7.0. Distribute into tubes or bottles. Autoclave for 15 min at 15 psi pressure–121°C. Aseptically and anaerobically add per liter 50.0mL sterile sodium acetate solution, 50.0mL sterile sodium formate solution, 10.0mL of sterile cysteine solution, and 10.0mL of sterile $Na_2S \cdot 9H_2O$ solution. Mix thoroughly. The final pH should be 7.2.

Use: For the cultivation of *Methanocalculus taiwanensis* DSM 14648.

MB Medium (DSMZ Medium 924)

Composition per liter:

NaCl	5.0g
$NaHCO_3$	4.0g
Yeast extract	2.0g
Trypticase™	2.0g
NH_4Cl	1.0g
$MgCl_2 \cdot 6H_2O$	1.0g
KCl	0.5g
$CaCl_2 \cdot 2H_2O$	0.4g
K_2HPO_4	0.4g
Resazurin	0.5mg
Sodium acetate solution	50.0mL
Sodium formate solution	50.0mL
$Na_2S \cdot 9H_2O$ solution	10.0mL
Cysteine-HCl·H_2O solution	10.0mL
Vitamin solution	10.0mL
Trace elements solution	10.0mL
Selenite-tungstate solution	1.0mL

pH 7.2 ± 0.2 at 25°C

Cysteine Solution:

Composition per 10.0mL:

L-Cysteine·HCl·H_2O	0.25g

Preparation of Cysteine Solution: Add L-cysteine·HCl·H_2O to distilled/deionized water and bring volume to 10.0mL. Mix thoroughly. Sparge with 100% N_2. Autoclave for 15 min at 15 psi pressure–121°C.

Sodium Formate Solution:

Composition per 50.0mL:

Na-formate	6.8g

Preparation of Sodium Formate Solution: Add sodium formate to distilled/deionized water and bring volume to 50.0mL. Mix thoroughly. Sparge with 100% N_2. Filter sterilize.

Sodium Acetate Solution:

Composition per 50.0mL:

Na-acetate	1.6g

Preparation of Sodium Acetate Solution: Add sodium acetate to distilled/deionized water and bring volume to 50.0mL. Mix thoroughly. Sparge with 100% N_2. Filter sterilize.

Selenite-Tungstate Solution

Composition per liter:

NaOH	0.5g
$Na_2WO_4 \cdot 2H_2O$	4.0mg
$Na_2SeO_3 \cdot 5H_2O$	3.0mg

Preparation of Selenite-Tungstate Solution: Add components to distilled/deionized water and bring volume to 1.0L. Mix thoroughly. Sparge with 100% N_2. Filter sterilize.

$Na_2S \cdot 9H_2O$ Solution:

Composition per 10.0mL:

$Na_2S \cdot 9H_2O$	0.25g

Preparation of $Na_2S \cdot 9H_2O$ Solution: Add $Na_2S \cdot 9H_2O$ to distilled/deionized water and bring volume to 10.0mL. Sparge with N_2. Autoclave for 15 min at 15 psi pressure–121°C. Cool to 25°C. Store anaerobically.

Trace Elements Solution:

Composition per liter:

$MgSO_4 \cdot 7H_2O$	3.0g
Nitrilotriacetic acid	1.5g
NaCl	1.0g
$MnSO_4 \cdot 2H_2O$	0.5g
$CoSO_4 \cdot 7H_2O$	0.18g
$ZnSO_4 \cdot 7H_2O$	0.18g
$CaCl_2 \cdot 2H_2O$	0.1g
$FeSO_4 \cdot 7H_2O$	0.1g
$NiCl_2 \cdot 6H_2O$	0.025g
$KAl(SO_4)_2 \cdot 12H_2O$	0.02g
H_3BO_3	0.01g
$Na_2MoO_4 \cdot 4H_2O$	0.01g
$CuSO_4 \cdot 5H_2O$	0.01g
$Na_2SeO_3 \cdot 5H_2O$	0.3mg

Preparation of Trace Elements Solution: Add nitrilotriacetic acid to 500.0mL of distilled/deionized water. Dissolve by adjusting pH to 6.5 with KOH. Add remaining components. Add distilled/deionized water to 1.0L. Mix thoroughly.

Vitamin Solution:

Composition per liter:

Pyridoxine-HCl	10.0mg
Thiamine-HCl·$2H_2O$	5.0mg
Riboflavin	5.0mg
Nicotinic acid	5.0mg
D-Ca-pantothenate	5.0mg
p-Aminobenzoic acid	5.0mg
Lipoic acid	5.0mg
Biotin	2.0mg
Folic acid	2.0mg
Vitamin B_{12}	0.1mg

Preparation of Vitamin Solution: Add components to distilled/deionized water and bring volume to 1.0L. Mix thoroughly. Sparge with 80% H_2 + 20% CO_2. Filter sterilize.

Preparation of Medium: Prepare and dispense medium under an oxygen-free 80% N_2 + 20% CO_2 gas mixture. Add components, except sodium acetate solution, sodium formate solution, $NaHCO_3$, cysteine solution, and $Na_2S \cdot 9H_2O$ solution, to distilled/deionized water and bring volume to 870.0mL. Mix thoroughly. Gently heat and bring to boiling. Boil for 3 min. Cool to 25°C while sparging with 80% N_2 + 20% CO_2. Add solid $NaHCO_3$. Mix thoroughly. Adjust pH to 6.8–7.0. Distribute into tubes or bottles. Autoclave for 15 min at 15 psi pressure–121°C. Aseptically and anaerobically add per liter 50.0mL sterile sodium acetate solution, 50.0mL sterile sodium formate solution, 10.0mL of sterile cysteine solution, and 10.0mL of sterile $Na_2S \cdot 9H_2O$ solution. Mix thoroughly. The final pH should be 7.2.

Use: For the cultivation of *Methanocalculus taiwanensis* DSM 14663.

M-BCG Yeast and Mold Agar

Composition per liter:

Glucose	50.0g
Agar	15.0g
Biopeptone	10.0g
Yeast extract	9.0g
$MgSO_4 \cdot 7H_2O$	2.1g
KH_2PO_4	2.0g
Diastase	0.05g
Thiamine hydrochloride	0.05g
Bromcresol Green	0.026g

pH 4.6 ± 0.2 at 25°C

Source: This medium is available from HiMedia.

Preparation of Medium: Add components to distilled/deionized water and bring volume to 1.0L. Mix thoroughly. Gently heat and bring to boiling. Distribute into tubes or flasks. Autoclave for 15 min at 15 psi pressure–121°C. Pour into sterile Petri dishes or leave in tubes.

Use: For the detection of fungi in routine analysis of beverages using the membrane filter technique.

m-Bismuth Sulfite Broth
See: **Bismuth Sulfite Broth**

M-Bismuth Sulfite Broth

Composition per liter:

Peptic digest of animal tissue	20.0g
Bismuth sulfite indicator	16.0g
Glucose	10.0g
Plant extract	10.0g
Na_2HPO_4	8.0g
$FeSO_4$	0.6g
Brilliant Green	0.05g

pH 7.7 ± 0.2 at 25°C

Source: This medium is available as a premixed powder from HiMedia.

Preparation of Medium: Add components to distilled/deionized water and bring volume to 1.0L. Mix thoroughly and heat with frequent agitation until boiling. Boil for 1 min. Do not autoclave. Cool to 45°–50°C. Mix to disperse the precipitate and aseptically distribute into sterile tubes or flasks. Use 2.0–2.2mL of medium for each membrane filter.

Use: For the selective isolation of *Salmonella typhi* and other enteric bacilli and for the detection of *Salmonella* by the membrane filter method.

M-Bismuth Sulfite HiVeg Broth

Composition per liter:

Plant peptone	20.0g
Bismuth sulfite indicator	16.0g
Glucose	10.0g
Plant extract	10.0g
Na_2HPO_4	8.0g
$FeSO_4$	0.6g
Brilliant Green	0.05g

pH 7.7 ± 0.2 at 25°C

Source: This medium is available as a premixed powder from HiMedia.

Preparation of Medium: Add components to distilled/deionized water and bring volume to 1.0L. Mix thoroughly and heat with frequent agitation until boiling. Boil for 1 min. Do not autoclave. Cool to 45°–50°C. Mix to disperse the precipitate and aseptically distribute into sterile tubes or flasks. Use 2.0–2.2mL of medium for each membrane filter.

Use: For the selective isolation of *Salmonella typhi* and other enteric bacilli and for the detection of *Salmonella* by the membrane filter method.

MBM Acetate Medium (Mineral Base Medium with Acetate)

Composition per liter:

Agar	16.0g
NaCl	5.0g
K_2HPO_4	1.0g
$NH_4H_2PO_4$	1.0g
Sodium acetate·$3H_2O$	1.0g
$MgSO_4 \cdot 7H_2O$	0.1g
Bromthymol Blue	0.01g

pH 6.5 ± 0.2 at 25°C

Preparation of Medium: Add components to distilled/deionized water and bring volume to 1.0L. Mix thoroughly. Adjust pH to 6.5. Gently heat and bring to boiling. Distribute into screw-capped tubes in 3.0mL volumes. Autoclave for 15 min at 15 psi pressure–121°C. Allow tubes to cool in a slanted position.

Use: For determining the nutritional independence of bacteria. Bacteria that are nutritionally independent turn the medium blue.

MBM Medium

See: **Methylene Blue Milk Medium**

MBM Medium (Modified) (DSMZ Medium 1020)

Composition per liter:

$NaNO_3$	0.2g
KH_2PO_4	0.2g
NH_4Cl	0.2g
$MgCl_2 \cdot 6H_2O$	0.4g
KCl	0.2g
$CaCl_2 \cdot 2H_2O$	0.1g
Resazurin	1.0mg
Thiosulfate solution	10.0mL
Trace elements solution SL-4	2.0mL

pH 7.0 ± 0.2 at 25°C

Thiosulfate Solution:

Composition per 10.0mL:

$Na_2S_2O_3$	2.5g

Preparation of Thiosulfate Solution: Add components to distilled/deionized water and bring volume to 10.0mL. Mix thoroughly. Sparge with 100% N_2. Filter sterilize.

Trace Elements Solution SL-4:

Composition per liter:

EDTA	0.5g
$FeSO_4 \cdot 7H_2O$	0.2g
Trace elements solution SL-6	100.0mL

Trace Elements Solution SL-6:

Composition per liter:

$MnCl_2 \cdot 4H_2O$	0.5g
H_3BO_3	0.3g
$CoCl_2 \cdot 6H_2O$	0.2g
$ZnSO_4 \cdot 7H_2O$	0.1g
$Na_2MoO_4 \cdot 2H_2O$	0.03g
$NiCl_2 \cdot 6H_2O$	0.02g
$CuCl_2 \cdot 2H_2O$	0.01g

Preparation of Trace Elements Solution SL-6: Add components to distilled/deionized water and bring volume to 1.0L. Mix thoroughly. Autoclave for 15 min at 15 psi pressure–121°C.

Preparation of Trace Elements Solution SL-4: Add components to distilled/deionized water and bring volume to 1.0L. Mix thoroughly. Sparge with 100% N_2. Filter sterilize.

Preparation of Medium: Add components, except thiosfulfate solution and trace elements SL-4 solution, to distilled/deionized water and bring volume to 990.0mL. Mix thoroughly. Sparge solution with 80% N_2 and 20% CO_2 gas mixture to make it anoxic. Distribute into culture vessels (e.g., 20mL in 120mL serum bottles) under a gas atmosphere of 80% N_2 and 20% CO_2. Autoclave for 15 min at 15 psi pressure–121°C. Cool to room temperature. Prior to inoculation, add the trace elements solution and thiosulfate solution to the medium. Adjust pH to 7.0. After inoculation, pressurize culture vessels to 0.5 bar overpressure with 100% H_2 gas.

Use: For the cultivation of *Sulfuricurvum kujiense.*

m-Brilliant Green Broth

See: **Brilliant Green Broth**

M-Brilliant Green Broth

Composition per liter:

Proteose peptone	20.0g
Lactose	20.0g
Saccharose	20.0g
NaCl	10.0g
Yeast extract	6.0g
Phenol Red	0.16g
Brilliant Green	0.025g

pH 6.9 ± 0.2 at 25°C

Source: This medium is available as a premixed powder from HiMedia.

Preparation of Medium: Add components to distilled/deionized water and bring volume to1.0L. Mix thoroughly. Distribute into tubes containing inverted Durham tubes, in 10.0mL amounts for testing 1.0mL or less of sample. Gently heat and bring to boiling. Do not autoclave. Cool the broth rapidly. Medium is sensitive to light.

Use: For the detection of coliform microorganisms in foods, dairy products, water, and wastewater, as well as in other materials of sanitary importance.

M-Brilliant Green HiVeg Broth

Composition per liter:

Plant peptone No. 3	20.0g
Lactose	20.0g
Saccharose	20.0g
NaCl	10.0g
Yeast extract	6.0g
Phenol Red	0.16g
Brilliant Green	0.025g

pH 6.9 ± 0.2 at 25°C

Source: This medium is available as a premixed powder from HiMedia.

Preparation of Medium: Add components to distilled/deionized water and bring volume to1.0L. Mix thoroughly. Distribute into tubes containing inverted Durham tubes, in 10.0mL amounts for testing 1.0mL or less of sample. Gently heat and bring to boiling. Do not autoclave. Cool the broth rapidly. Medium is sensitive to light.

Use: For the detection of coliform microorganisms in foods, dairy products, water, and wastewater, as well as in other materials of sanitary importance.

M-Broth, HiVeg

Composition per liter:

Plant hydrolysate	12.5g
K_2HPO_4	5.0g
NaCl	5.0g
Sodium citrate	5.0g
Yeast extract	5.0g
D-Mannose	2.0g
$MgSO_4$	0.8g
$MnCl_2$	0.14g
$FeSO_4$	0.04g
Polysorbate 80	0.75mL

pH 7.0 ± 0.22 at 25°C

Source: This medium, without polysorbate 80, is available as a premixed powder from HiMedia.

Preparation of Medium: Add components to distilled/deionized water and bring volume to 1.0L. Mix thoroughly. Distribute into tubes or flasks. Autoclave for 15 min at 15 psi pressure–121°C.

Use: For the detection of *Salmonella* in dried foods and feeds.

MCA

Composition per liter:

Carrots	200.0g
Agar	15.0g

Preparation of Medium: Peel and slice fresh carrots. Place carrots in a blender. Add 500.0mL of distilled/deionized water. Blend for 40 sec at high speed. Filter through four layers of cheesecloth. Squeeze out juice from the residue. Bring volume of filtrate to 1.0L with distilled/deionized water. Add agar. Mix thoroughly. Gently heat and bring to boiling. Distribute into tubes or flasks. Autoclave for 15 min at 15 psi pressure–121°C. Pour into sterile Petri dishes or leave in tubes.

Use: For the cultivation of *Phytophthora megasperma.*

McBride Agar, Modified

Composition per liter:

Agar	15.0g
Glycine anhydride	10.0g
Tryptose	10.0g
NaCl	5.0g
Beef extract	3.0g
Phenylethanol	2.5g
LiCl	0.5g
Cycloheximide solution	10.0mL

pH 7.3 ± 0.2 at 25°C

Cycloheximide Solution:
Composition per 10.0mL:

Cycloheximide	0.2g

Preparation of Cycloheximide Solution: Add cycloheximide to distilled/deionized water and bring volume to 10.0mL. Mix thoroughly. Filter sterilize.

Caution: Cycloheximide is toxic. Avoid skin contact or aerosol formation and inhalation.

Preparation of Medium: Add components, except cycloheximide solution, to distilled/deionized water and bring volume to 990.0mL. Mix thoroughly. Gently heat and bring to boiling. Autoclave for 15 min at 15 psi pressure–121°C. Cool to 45°–50°C. Aseptically add sterile cycloheximide solution. Mix thoroughly. Pour into sterile Petri dishes or distribute into sterile tubes.

Use: For the isolation of *Listeria monocytogenes* from dairy products.

McBride *Listeria* Agar

Composition per liter:

Agar	15.0g
Glycine	10.0g
Pancreatic digest of casein	5.0g
Peptic digest of animal tissue	5.0g
NaCl	5.0g
Beef extract	3.0g
Phenylethyl alcohol	2.5g
LiCl	0.5g

pH 7.3 ± 0.2 at 25°C

Source: This medium is available as a premixed powder from BD Diagnostic Systems.

Preparation of Medium: Add components to distilled/deionized water and bring volume to 1.0L. Mix thoroughly. Gently heat while stirring and bring to boiling. Distribute into tubes or flasks. Autoclave for 15 min at 15 psi pressure–121°C. Pour into sterile Petri dishes or leave in tubes.

Use: For the selective isolation of *Listeria monocytogenes* from clinical and nonclinical specimens containing mixed flora.

McBride *Listeria* HiVeg Agar Base with Blood and Selective Supplement

Composition per liter:

Agar	15.0g
Glycine anhydride	10.0g
Plant hydrolysate No. 1	10.0g
NaCl	5.0g
Plant extract	3.0g
Phenyl ethanol	2.5g
LiCl	0.5g
Sheep blood, defibrinated	50.0mL
Selective supplement solution	10.0mL

pH 7.3 ± 0.2 at 25°C

Source: This medium, without blood or selective supplement solution, is available as a premixed powder from HiMedia.

Caution: LiCl is harmful. Avoid bodily contact and inhalation of vapors. On contact with skin wash with plenty of water immediately.

Selective Supplement Solution:
Composition per 10.0mL:

Cycloheximide	0.2g

Preparation of Selective Supplement Solution: Add cycloheximide to distilled/deionized water and bring volume to 10.0mL. Mix thoroughly. Filter sterilize.

Caution: Cycloheximide is toxic. Avoid skin contact or aerosol formation and inhalation.

Preparation of Medium: Add components, except blood and selective supplement solution, to distilled/deionized water and bring volume to 940.0mL. Mix thoroughly. Gently heat and bring to boiling.

Distribute into tubes or flasks. Autoclave for 15 min at 15 psi pressure–121°C. Cool to 50°C. Aseptically add 50.0mL defibrinated blood and 10.0mL selective supplement solution. Mix thoroughly. Pour into sterile Petri dishes or leave in tubes.

Use: For the selective isolation of *Listeria monocytogenes* from clinical and nonclinical specimens containing mixed flora.

McBride *Listeria* HiVeg Agar Base, Modified, with Blood and Selective Supplement (Modified McBride *Listeria* HiVeg Agar Base)

Composition per liter:

Agar	15.0g
Glycine anhydride	10.0g
NaCl	5.0g
Plant extract	3.0g
Phenyl ethanol	2.5g
LiCl	0.5g
Sheep blood, defibrinated	50.0mL
Selective supplement solution	10.0mL

pH 7.3 ± 0.2 at 25°C

Source: This medium, without blood or selective supplement solution, is available as a premixed powder from HiMedia.

Caution: LiCl is harmful. Avoid bodily contact and inhalation of vapors. On contact with skin wash with plenty of water immediately.

Selective Supplement Solution:

Composition per 10.0mL:

Cycloheximide	0.2g

Preparation of Selective Supplement Solution: Add cycloheximide to distilled/deionized water and bring volume to 10.0mL. Mix thoroughly. Filter sterilize.

Caution: Cycloheximide is toxic. Avoid skin contact or aerosol formation and inhalation.

Preparation of Medium: Add components, except blood and selective supplement solution, to distilled/deionized water and bring volume to 940.0mL. Mix thoroughly. Gently heat and bring to boiling. Distribute into tubes or flasks. Autoclave for 15 min at 15 psi pressure–121°C. Cool to 50°C. Aseptically add 50.0mL defibrinated blood and 10.0mL selective supplement solution. Mix thoroughly. Pour into sterile Petri dishes or leave in tubes.

Use: For the selective isolation of *Listeria monocytogenes* from clinical and nonclinical specimens containing mixed flora.

McCarthy Agar

Composition per liter:

Cornstarch	10.0g
Naladixic acid	0.015g
Colistin	0.01g
GC agar base	1.0L

pH 7.2 ± 0.2 at 25°C

GC Agar Base:

Composition per liter:

Agar	10.0g
Pancreatic digest of casein	7.5g
Peptic digest of animal tissue	7.5g
NaCl	5.0g
K_2HPO_4	4.0g
Cornstarch	1.0g
KH_2PO_4	1.0g

Preparation of GC Agar Base: Add components to distilled/deionized water and bring volume to 1.0L. Mix thoroughly.

Preparation of Medium: To 1.0L of GC agar base, add the cornstarch. Gently heat while stirring to dissolve. Add the naladixic acid and colistin. Mix thoroughly. Distribute into tubes or flasks. Autoclave for 15 min at 15 psi pressure–121°C. Pour into sterile Petri dishes or leave in tubes.

Use: For the isolation and differentiation of *Gardnerella vaginalis* (*Haemophilus vaginalis, Corynebacterium vaginale*) from genitourinary specimens. Bacteria that can utilize starch appear as colonies surrounded by a clear zone.

McClung Carbon-Free Broth

Composition per liter:

$NaNO_3$	2.0g
K_2HPO_4	0.8g
$MgSO_4 \cdot 7H_2O$	0.5g
$FeCl_3$	0.01g
$MnCl_2 \cdot 4H_2O$	8.0mg
$ZnSO_4$	2.0mg

pH 7.2 ± 0.2 at 25°C

Preparation of Medium: Add components to distilled/deionized water and bring volume to 1.0L. Mix thoroughly. Gently heat without boiling until salts dissolve. Cool to 25°C. Adjust pH to 7.2. Filter sterilize.

Use: For use as a basal medium in determining the carbon assimilation capabilities of microorganisms.

McClung-Toabe Agar

Composition per liter:

Proteose peptone	40.0g
Agar	25.0g
Na_2HPO_4	5.0g
Glucose	2.0g
NaCl	2.0g
KH_2PO_4	1.0g
$MgSO_4 \cdot 7H_2O$	0.1g
Egg yolk emulsion, 50%	100.0mL

pH 7.3 ± 0.2 at 25°C

Source: This medium is available as a premixed powder from BD Diagnostic Systems.

Egg Yolk Emulsion, 50%:

Composition per 100.0mL:

Chicken egg yolks	11
Whole chicken egg	1
NaCl (0.9% solution)	50.0mL

Preparation of Egg Yolk Emulsion, 50%: Soak eggs with 1:100 dilution of saturated mercuric chloride solution for 1 min. Crack eggs and separate yolks from whites. Mix egg yolks with 1 chicken egg. Beat to form emulsion. Measure 50.0mL of egg yolk emulsion and add to 50.0mL of 0.9% NaCl solution. Mix thoroughly. Filter sterilize. Warm to 45°–50°C.

Preparation of Medium: Add components, except egg yolk emulsion, 50%, to distilled/deionized water and bring volume to 900.0mL. Mix thoroughly. Gently heat while stirring and bring to boiling. Autoclave for 15 min at 15 psi pressure–121°C. Cool to 50°–55°C. Aseptically add 100.0mL of sterile egg yolk emulsion, 50%. Mix thoroughly. Pour into sterile Petri dishes in 15.0mL volumes.

Use: For the isolation and cultivation of *Clostridium perfringens* in foods.

McClung-Toabe Agar, Modified

Composition per liter:

Proteose peptone No. 2	40.0g
Agar	20.0g
Na_2HPO_4	5.0g
Glucose	2.0g
NaCl	2.0g
KH_2PO_4	1.0g
$MgSO_4 \cdot 7H_2O$	0.1g
Egg yolk emulsion, 50%	100.0mL
Hemin solution	1.0mL

pH 7.6 ± 0.2 at 25°C

Egg Yolk Emulsion, 50%:

Composition per 100.0mL:

Chicken egg yolks	11
Whole chicken egg	1
NaCl (0.9% solution)	50.0mL

Preparation of Egg Yolk Emulsion, 50%: Soak eggs with 1:100 dilution of saturated mercuric chloride solution for 1 min. Crack eggs and separate yolks from whites. Mix egg yolks with 1 chicken egg. Beat to form emulsion. Measure 50.0mL of egg yolk emulsion and add to 50.0mL of 0.9% NaCl solution. Mix thoroughly. Filter sterilize. Warm to 45°–50°C.

Hemin Solution:

Composition per 100.0mL:

Hemin	0.5g
NaOH (1*N* solution)	20.0mL

Preparation of Hemin Solution: Add hemin to 20.0mL of 1*N* NaOH solution. Mix thoroughly. Bring volume to 100.0mL with distilled/deionized water.

Preparation of Medium: Add components, except egg yolk emulsion, 50%, to distilled/deionized water and bring volume to 900.0mL. Mix thoroughly. Gently heat while stirring and bring to boiling. Autoclave for 15 min at 15 psi pressure–121°C. Cool to 50°–55°C. Aseptically add 100.0mL of sterile egg yolk emulsion, 50%. Mix thoroughly. Pour into sterile Petri dishes in 20.0mL volumes.

Use: For the cultivation of a wide variety of anaerobic bacteria. For the differentiation of anaerobic bacteria based on lecithinase production and lipase production. Bacteria that produce lecithinase appear as colonies surrounded by a zone of insoluble precipitate. Bacteria that produce lipase appear as colonies with a pearly iridescent sheen.

McClung-Toabe Agar, Modified

Composition per liter:

Proteose peptone No. 2	40.0g
Agar	20.0g
Na_2HPO_4	5.0g
Glucose	2.0g
NaCl	2.0g
KH_2PO_4	1.0g
$MgSO_4 \cdot 7H_2O$	0.1g
Neomycin	0.1g
Egg yolk emulsion, 50%	100.0mL
Hemin solution	1.0mL

pH 7.6 ± 0.2 at 25°C

Egg Yolk Emulsion, 50%:

Composition per 100.0mL:

Chicken egg yolks	11
Whole chicken egg	1
NaCl (0.9% solution)	50.0mL

Preparation of Egg Yolk Emulsion, 50%: Soak eggs with 1:100 dilution of saturated mercuric chloride solution for 1 min. Crack eggs and separate yolks from whites. Mix egg yolks with 1 chicken egg. Beat to form emulsion. Measure 50.0mL of egg yolk emulsion and add to 50.0mL of 0.9% NaCl solution. Mix thoroughly. Filter sterilize. Warm to 45°–50°C.

Hemin Solution:

Composition per 100.0mL:

Hemin	0.5g
NaOH (1*N* solution)	20.0mL

Preparation of Hemin Solution: Add hemin to 20.0mL of 1*N* NaOH solution. Mix thoroughly. Bring volume to 100.0mL with distilled/deionized water.

Preparation of Medium: Add components, except egg yolk emulsion, 50%, to distilled/deionized water and bring volume to 900.0mL. Mix thoroughly. Gently heat while stirring and bring to boiling. Autoclave for 15 min at 15 psi pressure–121°C. Cool to 50°–55°C. Aseptically add 100.0mL of sterile egg yolk emulsion, 50%. Mix thoroughly. Pour into sterile Petri dishes in 20.0mL volumes.

Use: For the cultivation of *Clostridium* species. For the differentiation of *Clostridium* species based on lecithinase production and lipase production. Bacteria that produce lecithinase appear as colonies surrounded by a zone of insoluble precipitate. Bacteria that produce lipase appear as colonies with a pearly iridescent sheen.

McClung-Toabe Agar, Modified

Composition per liter:

Proteose peptone No. 2	20.0g
Agar	20.0g
Yeast extract	5.0g
Pancreatic digest of casein	5.0g
NaCl	5.0g
Sodium thioglycolate	1.0g
Egg yolk emulsion, 50%	80.0mL

pH 7.6 ± 0.2 at 25°C

Egg Yolk Emulsion, 50%:

Composition per 100.0mL:

Chicken egg yolks	11
Whole chicken egg	1
NaCl (0.9% solution)	50.0mL

Preparation of Egg Yolk Emulsion, 50%: Soak eggs with 1:100 dilution of saturated mercuric chloride solution for 1 min. Crack eggs and separate yolks from whites. Mix egg yolks with 1 chicken egg. Beat to form emulsion. Measure 50.0mL of egg yolk emulsion and add to 50.0mL of 0.9% NaCl solution. Mix thoroughly. Filter sterilize. Warm to 45°–50°C.

Preparation of Medium: Add components, except egg yolk emulsion, 50%, to distilled/deionized water and bring volume to 920.0mL. Mix thoroughly. Gently heat while stirring and bring to boiling. Autoclave for 15 min at 15 psi pressure–121°C. Cool to 50°–55°C. Aseptically add 80.0mL of sterile egg yolk emulsion, 50%. Mix thoroughly. Pour into sterile Petri dishes in 20.0mL volumes.

Use: For the cultivation of *Clostridium botulinum.*

McClung-Toabe Egg Yolk Agar, CDC Modified (CDC Modified McClung-Toabe Egg Yolk Agar)

Composition per liter:

Pancreatic digest of casein	40.0g
Agar	25.0g
$NaHPO_4$	5.0g
Yeast extract	5.0g
D-Glucose	2.0g
NaCl	2.0g
Egg yolk emulsion	100.0mL
$MgSO_4$ (5% solution)	0.2mL

pH 7.4 ± 0.2 at 25°C

Egg Yolk Emulsion:

Composition:

Chicken egg yolks	11
Whole chicken egg	1

Preparation of Egg Yolk Emulsion: Soak eggs with 1:100 dilution of saturated mercuric chloride solution for 1 min. Crack eggs. Separate yolks from whites for 11 eggs. Mix egg yolks with 1 chicken egg.

Preparation of Medium: Add components, except egg yolk emulsion, to distilled/deionized water and bring volume to 900.0mL. Mix thoroughly. Gently heat while stirring and bring to boiling. Autoclave for 15 min at 15 psi pressure–121°C. Cool to 60°C. Aseptically add 100.0mL of sterile egg yolk emulsion. Mix thoroughly. Pour into sterile Petri dishes in 20.0mL volumes.

Use: For the isolation, cultivation, and differentiation of anaerobic bacteria from foods. Bacteria that produce lecithinase appear as colonies surrounded by an insoluble opaque precipitate. Bacteria that produce lipase activity appear as colonies with a sheen or "pearly" surface. Bacteria that possess proteolytic activity appear as colonies surrounded by a clear zone.

McClung-Toabe Egg Yolk Agar, CDC Modified (CDC Modified McClung-Toabe Egg Yolk Agar)

Composition per liter:

Pancreatic digest of casein	40.0g
Agar	25.0g
Na_2HPO_4	5.0g
Yeast extract	5.0g
Glucose	2.0g
NaCl	2.0g
KH_2PO_4	1.0g
Egg yolk emulsion, 50%	100.0mL
$MgSO_4 \cdot 7H_2O$ (5% solution)	0.2mL

pH 7.3 ± 0.2 at 25°C

Egg Yolk Emulsion, 50%:

Composition per 100.0mL:

Chicken egg yolks	11
Whole chicken egg	1
NaCl (0.9% solution)	50.0mL

Preparation of Egg Yolk Emulsion, 50%: Soak eggs with 1:100 dilution of saturated mercuric chloride solution for 1 min. Crack eggs and separate yolks from whites. Mix egg yolks with 1 chicken egg. Beat to form emulsion. Measure 50.0mL of egg yolk emulsion and add to 50.0mL of 0.9% NaCl solution. Mix thoroughly. Filter sterilize. Warm to 45°–50°C.

Preparation of Medium: Add components—except egg yolk emulsion, 50%—to distilled/deionized water and bring volume to 900.0mL. Mix thoroughly. Gently heat while stirring and bring to boiling. Autoclave for 15 min at 15 psi pressure–121°C. Cool to 50°–55°C. Aseptically add 100.0mL of sterile egg yolk emulsion, 50%. Mix thoroughly. Pour into sterile Petri dishes in 15.0mL volumes.

Use: For the cultivation of a wide variety of anaerobic bacteria. For the differentiation of anaerobic bacteria based on lecithinase production, lipase production, and proteolytic ability. Bacteria that produce lecithinase appear as colonies surrounded by a zone of insoluble precipitate. Bacteria that produce lipase appear as colonies with a pearly iridescent sheen. Bacteria that produce proteolytic activity appear as colonies surrounded by a clear zone.

McClung Toabe HiVeg Agar Base wtih Egg Yolk

Composition per liter:

Plant peptone No. 3	40.0g
Agar	25.0g
Na_2HPO_4	5.0g
Glucose	2.0g
NaCl	2.0g
KH_2PO_4	1.0g
$MgSO_4$	0.1g
Egg yolk emulsion, 50%	100.0mL

pH 7.3 ± 0.2 at 25°C

Source: This medium, without egg yolk emulsion, is available as a premixed powder from HiMedia.

Egg Yolk Emulsion, 50%:

Composition per 100.0mL:

Chicken egg yolks	11
Whole chicken egg	1
NaCl (0.9% solution)	50.0mL

Preparation of Egg Yolk Emulsion, 50%: Soak eggs with 1:100 dilution of saturated mercuric chloride solution for 1 min. Crack eggs and separate yolks from whites. Mix egg yolks with 1 chicken egg. Beat to form emulsion. Measure 50.0mL of egg yolk emulsion and add to 50.0mL of 0.9% NaCl solution. Mix thoroughly. Filter sterilize. Warm to 45°–50°C.

Preparation of Medium: Add components, except egg yolk emulsion, 50%, to distilled/deionized water and bring volume to 900.0mL. Mix thoroughly. Gently heat while stirring and bring to boiling. Autoclave for 20 min at 15 psi pressure–121°C. Cool to 50°–55°C. Aseptically add 100.0mL of sterile egg yolk emulsion, 50%. Mix thoroughly. Pour into sterile Petri dishes in 15.0mL volumes.

Use: For the isolation and cultivation of *Clostridium perfringens* in foods.

M-CP Agar Base

Composition per liter:

Tryptose	30.0g
Yeast extract	20.0g
Agar	15.0g
Sucrose	5.0g
L-Cysteine·HCl·H_2O	1.0g
$MgSO_4 \cdot 7H_2O$	0.1g
$FeCl_3 \cdot 6H_2O$	0.09g
Indoxyl β-D-glucoside	0.06g

Bromcresol Purple 0.04g
Selective supplement solution B 20.0mL
Selective supplement solution A 10.0mL
pH 7.6 ± 0.2 at 25°C

Source: This medium is available from HiMedia.

Selective Supplement Solution A:
Composition per 10.0mL:
D-Cycloserine 400.0mg
Polymyxin B sulfate 25.0mg

Preparation of Selective Supplement Solution A: Add components to distilled/deionized water and bring volume to 10.0mL. Mix thoroughly. Filter sterilize.

Selective Supplement Solution B:
Composition per 20.0mL:
Phenolphthalein diphosphate 0.1g

Preparation of Selective Supplement Solution B: Add phenolphthalein diphosphate to distilled/deionized water and bring volume to 20.0mL. Mix thoroughly. Filter sterilize.

Preparation of Medium: Add components, except selective supplement solutions A and B, to distilled/deionized water and bring volume to 970.0mL. Mix thoroughly. Autoclave for 15 min at 15 psi pressure–121°C. Cool to 50°C. Aseptically add selective supplement solutions A and B. Mix thoroughly. Pour into Petri dishes or aseptically distribute into sterile tubes.

Use: Recommended by the Directive of the Council of the European Union 98/83/EC for the isolation and enumeration of *Clostridium perfringens* from water samples using the membrane filtration technique.

M-CP HiVeg Agar Base with Phenolphthalein Diphosphate

Composition per liter:
Plant hydrolysate No. 1 .... 30.0g
Agar 15.0g
L-Cysteine·HCl 1.0g
$MgSO_4 \cdot 7H_2O$ 0.1g
$FeCl_3 \cdot 6H_2O$ 0.09
Indoxyl β-D-glucoside 0.06
Bromcresol Purple 0.04
Sucrose 5.0g
Yeast extract 20.0g
Phenolphthalein diphosphate solution 10.0mL
pH 7.2 ± 0.2 at 25°C

Source: This medium, without phenolphthalein diphosphate, is available as a premixed powder from HiMedia.

Phenolphthalein Diphosphate Solution:
Composition per 10.0mL:
Phenolphthalein diphosphate 2.0g

Preparation of Phenolphthalein Diphosphate Solution: Add phenolphthalein diphosphate to distilled/deionized water and bring volume to 10.0mL. Mix thoroughly. Filter sterilize.

Preparation of Medium: Add components, except phenolphthalein diphosphate solution, to distilled/deionized water and bring volume to 990.0mL. Mix thoroughly. Gently heat while stirring until boiling. Autoclave for 15 min at 15 psi pressure–121°C. Cool to 45-50°C. Aseptially add 10.0mL of sterile phenolphthalein diphosphate solution. Mix thoroughly. Pour into sterile Petri dishes or distribute into sterile tubes.

Use: For the cultivation and identification of *Providencia stuartii.*

m-CP Medium

Composition per liter:
Tryptose 30.0g
Yeast extract 20.0g
Agar 15.0g
Sucrose 5.0g
L-Cysteine·HCl·H_2O 1.0g
$MgO_4 \cdot 7H_2O$ 0.1g
Bromcresol Purple 0.04g
Phenolphthalein biphosphate tetrazolium salt solution 10.0mL
Selective supplement solution 4.0mL
Indoxyl-β-D-glucoside solution 4.0mL
Ferric chloride solution 1.0mL
pH 7.6 ± 0.2 at 25°C

Selective Supplement Solution:
Composition per 4.0mL:
D-Cycloserine 0.4g
Polymyxin B sulfate 25.0mg

Preparation of Selective Supplement Solution: Add components to 4.0mL of distilled/deionized water. Mix thoroughly. Filter sterilize.

Phenolphthalein Biphosphate Tetrazolium Salt Solution:
Composition per 10.0mL:
Phenolphthalein biphosphate tetrazolium salt 25.0mg

Preparation of Phenolphthalein Biphosphate Tetrazolium Salt Solution: Add phenolphthalein biphosphate tetrazolium salt to 10.0mL of distilled/deionized water. Mix thoroughly. Filter sterilize.

Indoxyl-β-D-glucoside Solution:
Composition per 10.0mL:
Indoxyl-β-D-glucoside 0.45g

Preparation of Indoxyl-β-D-glucoside Solution: Add indoxyl-β-D-glucoside to 10.0mL of distilled/deionized water. Mix thoroughly. Filter sterilize.

Ferric Chloride Solution:
Composition per 4.0mL:
$FeCl_3 \cdot 6H_2O$ 30.0mg

Preparation of Ferric Chloride Solution: Add $FeCl_3 \cdot 6H_2O$ to 4.0mL of distilled/deionized water. Mix thoroughly. Filter sterilize.

Preparation of Medium: Add components, except selective supplement solution, phenolphthalein biphosphate tetrazolium salt solution, indoxyl-β-D-glucoside solution, and ferric chloride solution, to distilled/deionized water and bring volume to 981.0mL. Mix thoroughly. Autoclave for 15 min at 15 psi pressure–121°C. Cool to 45°–50°C. Aseptically add 4.0mL of selective supplement solution. Mix thoroughly. Aseptically add 10.0mL phenolphthalein biphosphate tetrazolium salt solution, 4.0mL indoxyl-β-D-glucoside solution, and 1.0mL ferric chloride solution. Mix thoroughly. Pour into sterile Petri dishes or aseptically distribute into tubes.

Use: A selective, chromogenic medium for the rapid identification and enumeration of *Clostridium perfringens* in water samples, including water used in food and beverage production.

MCP Medium
(Modified MacConkey Medium)
(MacConkey Phosphatase Medium)

Composition per liter:

Pancreatic digest of gelatin	17.0g
Agar	13.5g
Lactose	10.0g
NaCl	5.0g
Bile salts	1.5g
Pancreatic digest of casein	1.5g
Peptic digest of animal tissue	1.5g
Na_2HPO_4	0.6g
Glucose	0.2g
Methyl Blue	0.07g
Neutral Red	0.03g
Crystal Violet	1.0mg
Phenolphthalein diphosphate solution	10.0mL

pH 7.2 ± 0.2 at 25°C

Source: This medium, without phenolphthalein diphosphate, is available as a premixed powder from HiMedia.

Phenolphthalein Diphosphate Solution:

Composition per 10.0mL:

Phenolphthalein diphosphate	2.0g

Preparation of Phenolphthalein Diphosphate Solution: Add phenolphthalein diphosphate to distilled/deionized water and bring volume to 10.0mL. Mix thoroughly. Filter sterilize.

Preparation of Medium: Add components, except phenolphthalein diphosphate solution, to distilled/deionized water and bring volume to 990.0mL. Mix thoroughly. Gently heat while stirring until boiling. Autoclave for 15 min at 15 psi pressure–121°C. Cool to 45-50°C. Aseptially add 10.0mL of sterile phenolphthalein diphosphate solution. Mix thoroughly. Pour into sterile Petri dishes or distribute into sterile tubes.

Use: For the cultivation and identification of *Providencia stuartii.*

MD Medium

Composition per liter:

Agar	20.0g
L-Malic acid	20.0g
Pancreatic digest of casein	10.0g
D-Glucose	5.0g
Casamino acids	3.0g
Pancreatic digest of soybean meal	1.5g
Tween™ 80	1.0g
Yeast extract	1.0g
Bromcresol Green solution	20.0mL

pH 7.0 ± 0.2 at 25°C

Bromcresol Green Solution:

Composition per 30.0mL:

Bromcresol Green	0.1g
NaOH (0.01*N* solution)	30.0mL

Preparation of Bromcresol Green Solution: Add Bromcresol Green to 30.0mL of NaOH solution. Mix thoroughly. Filter sterilize.

Preparation of Medium: Add components to distilled/deionized water and bring volume to 1.0L. Mix thoroughly. Gently heat and bring to boiling. Distribute into tubes or flasks. Adjust pH to 7.0 with 10*N* KOH. Autoclave for 15 min at 15 psi pressure–121°C. Pour into sterile Petri dishes or leave in tubes.

Use: For the isolation and cultivation of *Salmonella* species from foods.

MD 1 Medium

Composition per liter:

Pancreatic digest of casein	3.0g
$MgSO_4 \cdot 7H_2O$	2.0g
$CaCl_2$	0.5g
Trace elements solution	1.0mL
Vitamin B_{12} solution	1.0mL

Trace Elements Solution:

Composition per liter:

EDTA	8.0g
$MnCl_2 \cdot 4H_2O$	0.1g
$CoCl_2$	0.02g
KBr	0.02g
$ZnCl_2$	0.02g
$CuSO_4$	0.01g
H_3BO_3	0.01g
$NaMoO_4 \cdot 2H_2O$	0.01g
$BaCl_2$	5.0mg
LiCl	5.0mg
$SnCl_2 \cdot 2H_2O$	5.0mg

Preparation of Trace Elements Solution: Add components to distilled/deionized water and bring volume to 1.0L. Mix thoroughly.

Vitamin B_{12} Solution:

Composition per 10.0mL:

Vitamin B_{12}	5.0mg

Preparation of Vitamin B_{12} Solution: Add vitamin B_{12} to distilled/deionized water and bring volume to 10.0mL. Mix thoroughly.

Preparation of Medium: Add components to distilled/deionized water and bring volume to 1.0L. Mix thoroughly. Distribute into tubes or flasks. Autoclave for 15 min at 15 psi pressure–121°C.

Use: For the cultivation of myxobacteria.

MD1- Medium
(DSMZ Medium 1118)

Composition per liter:

Casitone	3.0g
$MgSO_4 \cdot 7H_2O$	2.0g
$CaCl_2 \cdot 2H_2O$	0.7g
Vitamin solution	10.0mL
Trace elements solution SL-4	1.0mL

pH 7.1 ± 0.2 at 25°C

Trace Elements Solution SL-4:

Composition per liter:

EDTA	0.5g
$FeSO_4 \cdot 7H_2O$	0.2g
Trace elements solution SL-6	100.0mL

Trace Elements Solution SL-6:

Composition per liter:

$MnCl_2 \cdot 4H_2O$	0.5g
H_3BO_3	0.3g
$CoCl_2 \cdot 6H_2O$	0.2g
$ZnSO_4 \cdot 7H_2O$	0.1g
$Na_2MoO_4 \cdot 2H_2O$	0.03g

$NiCl_2 \cdot 6H_2O$ 0.02g
$CuCl_2 \cdot 2H_2O$ 0.01g

Preparation of Trace Elements Solution SL-6: Add components to distilled/deionized water and bring volume to 1.0L. Mix thoroughly. Autoclave for 15 min at 15 psi pressure–121°C.

Preparation of Trace Elements Solution SL-4: Add components to distilled/deionized water and bring volume to 1.0L. Mix thoroughly. Filter sterilize.

Vitamin Solution:
Composition per 10.0ml:
Vitamin B_{12} 0.5mg

Preparation of Vitamin Solution: Add components to distilled/deionized water and bring volume to 10.0mL. Mix thoroughly. Filter sterilize.

Preparation of Medium: Add components, except vitamin solution, to distilled/deionized water and bring volume to 990.0mL. Mix thoroughly. Adjust pH to 7.1. Distribute into tubes or flasks. Gently heat while stirring and bring to boiling. Mix thoroughly. Autoclave for 15 min at 15 psi pressure–121°C. Aseptically add 10.0mL sterile vitamin solution. Mix thorougly. Aseptically distribute into sterile tubes or flasks.

Use: For the cultivation of *Myxococcus xanthus.*

MDPA

See: **Malt Dextrose Peptone Agar**

MDPA with Calcium Carbonate

Composition per liter:
Agar 25.0g
Malt extract 20.0g
Glucose 20.0g
$CaCO_3$ 5.0g
Peptone 1.0g

Preparation of Medium: Add components to distilled/deionized water and bring volume to 1.0L. Mix thoroughly. Gently heat and bring to boiling. Distribute into tubes or flasks. Autoclave for 15 min at 15 psi pressure–121°C. Pour into sterile Petri dishes or leave in tubes.

Use: For the cultivation of *Dekkera* species.

MDPYA4

See: **Malt 4% Dextrose Peptone Yeast Agar**

MDYA4

See: **Malt 4% Dextrose Yeast Agar**

m-E Agar

See: **E Agar**

Me15% MH Agar (DSMZ Medium 582)

Composition per liter:
NaCl 121.5g
Agar 20.0g
$MgSO_4$ 14.4g
$MgCl_2$ 10.5g
Yeast extract 10.0g
Proteose peptone no. 3 .... 5.0g
KCl 3.0g
Glucose 1.0g
$CaCl_2$ 0.54g
NaBr 0.039g
$NaHCO_3$ solution 10.0mL

pH 7.5 ± 0.2 at 25°C

$NaHCO_3$ Solution:
Composition per 100.0mL:
$NaHCO_3$ 0.9g

Preparation of $NaHCO_3$ Solution: Add $NaHCO_3$ to distilled/deionized water and bring volume to 100.0mL. Mix thoroughly. Sparge with 80% N_2 + 20% CO_2. Autoclave for 15 min at 15 psi pressure–121°C. Cool to 50°C.

Preparation of Medium: Add components, except $NaHCO_3$ solution, to distilled/deionized water and bring volume to 990.0mL. Mix thoroughly. Autoclave for 15 min at 15 psi pressure–121°C. Cool to 50°C. Aseptically add 10.0mL $NaHCO_3$ solution. Mix thoroughly. Pour into Petri dishes or aseptically distribute into sterile tubes.

Use: For the cultivation of *Bacillus halophilus.*

Me15% MH Medium (DSMZ Medium 582)

Composition per liter:
NaCl 121.5g
$MgSO_4$ 14.4g
$MgCl_2$ 10.5g
Yeast extract 10.0g
Proteose peptone no. 3 .... 5.0g
KCl 3.0g
Glucose 1.0g
$CaCl_2$ 0.54g
NaBr 0.039g
$NaHCO_3$ solution 10.0mL

pH 7.5 ± 0.2 at 25°C

$NaHCO_3$ Solution:
Composition per 100.0mL:
$NaHCO_3$ 0.9g

Preparation of $NaHCO_3$ Solution: Add $NaHCO_3$ to distilled/deionized water and bring volume to 100.0mL. Mix thoroughly. Sparge with 80% N_2 + 20% CO_2. Autoclave for 15 min at 15 psi pressure–121°C. Cool to 25°C.

Preparation of Medium: Add components, except $NaHCO_3$ solution, to distilled/deionized water and bring volume to 990.0mL. Mix thoroughly. Autoclave for 15 min at 15 psi pressure–121°C. Cool to 25°C. Aseptically add 10.0mL $NaHCO_3$ solution. Mix thoroughly. Aseptically distribute into sterile tubes or flasks.

Use: For the cultivation of *Bacillus halophilus.*

MEA

See: **Malt Extract Agar**

Meat Extract with Peptone (Pepted Meat Broth)

Composition per liter:
NaCl 15.0g
Peptic digest of animal tissue 10.0g
Meat extract 3.0g

pH 7.2 ± 0.2 at 25°C

Source: This medium is available from HiMedia.

Preparation of Medium: Add components to distilled/deionized water and bring volume to 1.0L. Mix thoroughly. Gently heat and bring to boiling. Distribute into tubes or flasks. Autoclave for 15 min at 15 psi pressure–121°C. Pour into sterile Petri dishes or leave in tubes.

Use: For the cultivation and maintenance of *Alcaligenes* species.

Meat Extract with Peptone and 1.5% Salt

Composition per liter:

NaCl	15.0g
Peptone	10.0g
Meat extract	3.0g

Preparation of Medium: Add components to distilled/deionized water and bring volume to 1.0L. Mix thoroughly. Distribute into tubes or flasks. Autoclave for 15 min at 15 psi pressure–121°C.

Use: For the cultivation and maintenance of *Alcaligenes* species.

Meat Infusion Agar, HiVeg (Standard Infusion Agar, HiVeg)

Composition per liter:

Agar	25.0g
Plant infusion	10.0g
Plant peptone	10.0g
NaCl	5.0g

pH 7.5 ± 0.2 at 25°C

Source: This medium is available as a premixed powder from HiMedia.

Preparation of Medium: Add components to distilled/deionized water and bring volume to 1.0L. Mix thoroughly. Distribute into tubes or flasks. Autoclave for 15 min at 15 psi pressure–121°C.

Use: For the mass cultivation of organisms for vaccine or toxin production.

M-EC Test Agar

Composition per liter:

Agar	15.0g
Lactose	10.0g
NaCl	7.5g
Proteose peptone	5.0g
Dipotassium phosphate	3.3g
Yeast extract	3.0g
KH_2PO_4	1.0g
Sodium lauryl sulphate	0.2g
Sodium deoxycholate	0.1g
Bromcresol Purple	0.08g
Bromphenol Red	0.08g

pH 7.3 ± 0.2 at 25°C

Source: This medium is available from HiMedia.

Preparation of Medium: Add components to distilled/deionized water and bring volume to 1.0L. Mix thoroughly. Gently heat and bring to boiling. Distribute into tubes or flasks. Do not autoclave. Pour into sterile Petri dishes or leave in tubes.

Use: For the detection of *Escherichia coli* in water samples using the membrane filter technique.

MED IIa

Composition per liter:

Tris buffer stock solution	10.0mL
$CaCl_2$ (5.0% solution)	10.0mL
$MgSO_4 \cdot 7H_2O$ (3.33% solution)	1.0mL

pH 7.2 ± 0.2 at 25°C

Tris Buffer Stock Solution:

Composition per 500.0mL:

Tris(hydroxymethyl)aminomethane·HCl	35.01g
Tris(hydroxymethyl)aminomethane	3.35g

Preparation of Tris Buffer Stock Solution: Add components to distilled/deionized water and bring volume to 500.0mL. Mix thoroughly. Adjust pH to 7.2.

Preparation of Medium: Add components to distilled/deionized water and bring volume to 1.0L. Mix thoroughly. Distribute into tubes or flasks. Autoclave for 20 min at 15 psi pressure–121°C.

Use: For the cultivation and maintenance of *Vampirovibrio chlorellavorus*.

Medium A

Composition per liter:

D-Glucose	20.0g
Agar	20.0g
Yeast extract	10.0g
Biotin	1.0mg
Calcium pantothenate	1.0mg

pH 7.3 ± 0.2 at 25°C

Preparation of Medium: Add components, except biotin and calcium pantothenate, to distilled/deionized water and bring volume to 990.0mL. Mix thoroughly. Gently heat and bring to boiling. Autoclave for 15 min at 15 psi pressure–121°C. Cool to 45°–50°C. Add biotin and calcium pantothenate to distilled/deionized water and bring volume to 10.0mL. Mix thoroughly. Filter sterilize. Aseptically add the sterile biotin and calcium pantothenate solution to the cooled sterile basal medium. Mix thoroughly. Pour into sterile Petri dishes or distribute into sterile tubes.

Use: For the cultivation and maintenance of *Zymomonas mobilis*.

Medium A for Producing Lysates

Composition per liter:

Nutrient broth	8.0g
KCl	1.0g
$MgSO_4 \cdot 7H_2O$	0.25g
$MnCl_2$	1.25mg
$FeSO_4$ (1.0m*M* solution)	1.0mL
$Ca(NO_3)_2$ (1.0*M* solution)	1.0mL

pH 7.0–7.2 at 25°C

Preparation of Medium: Add components, except $FeSO_4$ and $Ca(NO_3)_2$, to distilled/deionized water and bring volume to 998.0mL. Mix thoroughly. Adjust pH to 7.0–7.2. Autoclave for 30 min at 15 psi pressure–115°C. Cool to 45°–50°C. Prepare 1.0m*M* $FeSO_4$ solution and 1.0*M* $Ca(NO_3)_2$ solution separately. Filter sterilize both solutions. Aseptically add the sterile $FeSO_4$ solution and sterile $Ca(NO_3)_2$ solution to the cooled sterile basal medium. Mix thoroughly. Distribute into sterile tubes or flasks.

Use: For the cultivation of microorganisms to be lysed.

Medium 2A

Composition per liter:

Arginine	10.0g
NaCl	5.0g
Agar	4.0g
Peptone	1.0g
$K_2HPO_4 \cdot 3H_2O$	0.3g
Phenol Red	0.01g

pH 7.2–7.4 at 25°C

Preparation of Medium: Add components to distilled/deionized water and bring volume to 1.0L. Mix thoroughly. Gently heat and bring to boiling. Distribute into tubes. Autoclave for 15 min at 15 psi pressure–121°C.

Use: For the cultivation and differentiation of *Pseudomonas* species based on their production of arginine dihydrolase activity.

Medium AS4

Composition per liter:

Sucrose	80.0g
PPLO broth without Crystal Violet	500.0mL
Horse serum	200.0mL
Phenol Red (0.5% solution)	5.0mL

pH 7.2 ± 0.2 at 25°C

PPLO Broth without Crystal Violet:

Composition per 500.0mL:

Beef heart, solids from infusion	11.53g
Peptone	2.33g
NaCl	1.15g

Source: PPLO broth without Crystal Violet is available as a premixed powder from BD Diagnostic Systems.

Preparation of PPLO Broth without Crystal Violet: Add components to distilled/deionized water and bring volume to 500.0mL. Mix thoroughly. Beef heart for infusion may be substituted; 100.0g of beef heart for infusion is equivalent to 500.0g of fresh heart tissue.

Preparation of Medium: Add components, except horse serum, to distilled/deionized water and bring volume to 800.0mL. Mix thoroughly. Adjust pH to 7.2. Autoclave for 10 min at 15 psi pressure–121°C. Cool to 45°–50°C. Aseptically add 200.0mL of noninactivated, sterile horse serum. Mix thoroughly. Aseptically distribute into sterile tubes or flasks.

Use: For the cultivation and maintenance of *Spiroplasma melliferum*.

Medium for *Acetivibrio cellulolyticus*
See: **BC Medium**

Medium for Aciduric, Thermophilic *Bacillus* Strains

Composition per liter:

Solution A	500.0mL
Solution B	500.0mL

pH 4.3 ± 0.2 at 25°C

Solution A:

Composition per 500.0mL:

KH_2PO_4	3.0g
Glucose	1.0g
Starch	1.0g
Tryptone	1.0g
Yeast extract	1.0g
$MgSO_4 \cdot 7H_2O$	0.5g
$CaCl_2 \cdot 2H_2O$	0.25g
$(NH_4)_2SO_4$	0.2g

Preparation of Solution A: Add components to distilled/deionized water and bring volume to 500.0mL. Mix thoroughly. Adjust pH to 4.3. Autoclave for 15 min at 15 psi pressure–121°C. Cool to 50°–55°C.

Solution B:

Composition per 500.0mL:

Agar	20.0g

Preparation of Solution B: Add agar to distilled/deionized water and bring volume to 500.0mL. Mix thoroughly. Autoclave for 15 min at 15 psi pressure–121°C. Cool to 50°–55°C.

Preparation of Medium: Aseptically combine 500.0mL of solution A and 500.0mL of solution B. Mix thoroughly. Pour into sterile Petri dishes or distribute into sterile tubes.

Use: For the cultivation of aciduric, thermophilic *Bacillus* strains.

Medium 2508-85-1 with Amino Acids

Composition per liter:

Agar	20.0g
Nutrient broth	8.0g
D-Glucose	5.0g
Polypeptone™	5.0g
Yeast extract	5.0g
L-Lysine	0.1g
L-Methionine	0.05g
Diaminopimelic acid	0.05g

Preparation of Medium: Add components to distilled/deionized water and bring volume to 1.0L. Mix thoroughly. Gently heat and bring to boiling. Distribute into tubes or flasks. Autoclave for 30 min at 15 psi pressure–121°C. Pour into sterile Petri dishes or leave in tubes.

Use: For the cultivation and maintenance of *Escherichia coli*.

Medium for Ammonia Oxidizers

Composition per liter:

$MgSO_4 \cdot 7H_2O$	0.2g
$(NH_4)_2SO_4$	0.13g
K_2HPO_4	0.09g
$CaCl_2 \cdot 2H_2O$	0.02g
Chelated iron	1.0mg
$MnCl_2 \cdot 4H_2O$	0.2mg
$Na_2MoO_4 \cdot 2H_2O$	0.1mg
$ZnSO_4 \cdot 7H_2O$	0.1mg
$CuSO_4 \cdot 5H_2O$	0.02mg
$CoCl_2 \cdot 6H_2O$	2.0μg

Preparation of Medium: Add components to distilled/deionized water and bring volume to 1.0L. Mix thoroughly. Distribute into tubes or flasks. Autoclave for 15 min at 15 psi pressure–121°C.

Use: For the isolation, cultivation, and enrichment of ammonia-oxidizing bacteria from soil.

Medium for Ammonia Oxidizers

Composition per liter:

$(NH_4)_2SO_4$	2.0g
$MgSO_4 \cdot 7H_2O$	0.2g
$CaCl_2 \cdot 2H_2O$	0.02g
K_2HPO_4	0.02g
Chelated iron	1.0mg

$MnCl_2·4H_2O$ 0.2mg
$Na_2MoO_4·2H_2O$ 0.1mg
$ZnSO_4·7H_2O$ 0.1mg
$CuSO_4·5H_2O$ 0.02mg
$CoCl_2·6H_2O$ 2.0μg

Preparation of Medium: Add components to distilled/deionized water and bring volume to 1.0L. Mix thoroughly. Distribute into tubes or flasks. Autoclave for 15 min at 15 psi pressure–121°C.

Use: For the isolation, cultivation, and enrichment of ammonia-oxidizing bacteria from soil.

Medium for Ammonia Oxidizers

Composition per liter:
K_2HPO_4 .... 0.5g
$(NH_4)_2SO_4$ 0.5g
Phenol Red 0.5g
$MgSO_4·7H_2O$ 0.05g
$CaCl_2·2H_2O$ 0.02g
NaCl 0.02g
$Na_2MoO_4·2H_2O$ 2.4μg
Metals "44" 1.0mL

Metals "44":
Composition per 100.0mL:
$ZnSO_4·7H_2O$ 1.1g
$FeSO_4·7H_2O$ 0.5g
EDTA 0.25g
$MnSO_4·7H_2O$ 0.154g
$CuSO_4·5H_2O$ 0.04g
$Co(NO_3)_2·6H_2O$ 0.025g
$Na_2B_4O_7·10H_2O$ 0.018g

Preparation of Metals "44": Add a few drops of H_2SO_4 to distilled/deionized water to inhibit precipitate formation. Add components to acidified distilled/deionized water and bring volume to 100.0mL. Mix thoroughly.

Preparation of Medium: Add components to distilled/deionized water and bring volume to 1.0L. Mix thoroughly. Distribute into tubes or flasks. Autoclave for 15 min at 15 psi pressure–121°C.

Use: For the isolation, cultivation, and enrichment of ammonia-oxidizing bacteria from soil.

Medium for Ammonia Oxidizers

Composition per liter:
$(NH_4)_2SO_4$ 0.5g
KH_2PO_4 .... 0.2g
$CaCl_2·2H_2O$ 0.04g
$MgSO_4·7H_2O$ 0.04g
Ferric citrate 0.5mg
Phenol Red 0.5mg

Preparation of Medium: Add components to distilled/deionized water and bring volume to 1.0L. Mix thoroughly. Distribute into tubes or flasks. Autoclave for 15 min at 15 psi pressure–121°C.

Use: For the isolation, cultivation, and enrichment of ammonia-oxidizing bacteria from soil.

Medium for Ammonia Oxidizers, Brackish

Composition per liter:
$CaCO_3$ 5.0g
NH_4Cl 0.5g
K_2HPO_4 .... 0.05g
Seawater 400.0mL

Preparation of Medium: Add components to distilled/deionized water and bring volume to 1.0L. Mix thoroughly. Distribute into tubes or flasks. Autoclave for 15 min at 15 psi pressure–121°C.

Use: For the isolation, cultivation, and enrichment of ammonia-oxidizing bacteria from brackish specimens.

Medium for Ammonia Oxidizers, Marine

Composition per liter:
$(NH_4)_2SO_4$ 1.32g
$MgSO_4·7H_2O$ 0.2g
Chelated iron 0.13g
K_2HPO_4 .... 0.11g
$CaCl_2·2H_2O$ 0.02g
$ZnSO_4·7H_2O$ 0.1mg
$CuSO_4·5H_2O$ 0.02mg
$CoCl_2·6H_2O$ 2.0μg
$MnCl_2·4H_2O$ 2.0μg
$Na_2MoO_4·2H_2O$ 1.0μg
Seawater 1.0L

Preparation of Medium: Combine components. Mix thoroughly. Distribute into tubes or flasks. Autoclave for 15 min at 15 psi pressure–121°C.

Use: For the isolation, cultivation, and enrichment of marine ammonia-oxidizing bacteria.

Medium for Ammonia-Oxidizing Bacteria

Composition per liter:
$(NH_4)_2SO_4$ 235.0mg
KH_2PO_4 .... 200.0mg
$CaCl_2·2H_2O$ 40.0mg
$MgSO_4·7H_2O$ 40.0mg
Iron-EDTA-Phenol Red solution 1.0mL
Na_2CO_3 solution .... variable

Na_2CO_3 Solution:
Composition per 100.0mL:
Na_2CO_3 .... 5.0g

Preparation of Na_2CO_3 Solution: Add Na_2CO_3 to distilled/deionized water and bring volume to 100.0mL. Mix thoroughly. Autoclave for 15 min at 15 psi pressure–121°C.

Iron-EDTA-Phenol Red Solution:
Composition per 100.0mL:
$FeSO_4·7H_2O$ 50.0mg
Sodium EDTA 50.0mg
Phenol Red 50.0mg

Preparation of Iron-EDTA-Phenol Red Solution: Add components to distilled/deionized water and bring volume to 100.0mL. Mix thoroughly.

Preparation of Medium: Add components, except Na_2CO_3 solution, to distilled/deionized water and bring volume to 1.0L. Mix thoroughly. Distribute into tubes or flasks. Autoclave for 15 min at 15 psi pressure–121°C. Add enough sterile Na_2CO_3 solution to turn the medium pale pink. During incubation and growth of bacteria, add additional sterile Na_2CO_3 solution to restore the pale pink color. Growth is complete when no further color change is observed.

Use: For the cultivation of *Nitrosolobus multiformis* and *Nitrosomonas europaea.*

Medium B for Sulfate Reducers (Postgate's Medium B for Sulfate Reducers)

Composition per liter:

Sodium lactate	3.5g
$MgSO_4 \cdot 7H_2O$	2.0g
NH_4Cl	1.0g
$CaSO_4$	1.0g
Yeast extract	1.0g
KH_2PO_4	0.5g
$FeSO_4 \cdot 7H_2O$	0.5g
Ascorbic acid	0.1g
Thioglycollic acid	0.1g

pH 7.0–7.5 at 25°C

Preparation of Medium: Add components, except ascorbic acid and thioglycollic acid, to tap water and bring volume to 1.0L. For marine bacteria, NaCl may be added or seawater used in place of tap water. Mix thoroughly. Adjust pH to 7.0–7.5. Thioglycolate and ascorbate should be added immediately prior to sterilization. Distribute into tubes or flasks. Autoclave for 15 min at 15 psi pressure–121°C.

Use: For the isolation, cultivation, and maintenance of *Desulfovibrio* species and *Desulfotomaculum* species. This medium turns black as a result of H_2S production due to bacterial growth.

Medium for *Bacillus schlegelii*

Composition per liter:

Agar	15.0g
$Na_2HPO_4 \cdot 2H_2O$	2.9g
KH_2PO_4	2.3g
NH_4Cl	1.0g
$MgSO_4 \cdot 7H_2O$	0.5g
$NaHCO_3$	0.5g
$CaCl_2 \cdot 2H_2O$	0.01g
$MnSO_4 \cdot H_2O$	10.0mg
Ferric ammonium citrate solution	20.0mL
Trace elements solution SL-6	5.0mL

pH 6.8 ± 0.2 at 25°C

Ferric Ammonium Citrate Solution:

Composition per 20.0mL:

Ferric ammonium citrate	0.05g

Preparation of Ferric Ammonium Citrate Solution: Add ferric ammonium citrate to distilled/deionized water and bring volume to 20.0mL. Mix thoroughly. Autoclave for 15 min at 15 psi pressure–121°C.

Trace Elements Solution SL-6:

Composition per liter:

$MnCl_2 \cdot 4H_2O$	0.5g
H_3BO_3	0.3g
$CoCl_2 \cdot 6H_2O$	0.2g
$ZnSO_4 \cdot 7H_2O$	0.1g
$Na_2MoO_4 \cdot 2H_2O$	0.03g
$NiCl_2 \cdot 6H_2O$	0.02g
$CuCl_2 \cdot 2H_2O$	0.01g

Preparation of Trace Elements Solution SL-6: Add components to distilled/deionized water and bring volume to 1.0L. Mix thoroughly.

Preparation of Medium: Add components, except ferric ammonium citrate solution, to distilled/deionized water and bring volume to 980.0mL. Mix thoroughly. Gently heat and bring to boiling. Autoclave for 15 min at 15 psi pressure–121°C. Aseptically add 20.0mL of sterile ferric ammonium citrate solution. Mix thoroughly. Pour into sterile Petri dishes or distribute into sterile tubes.

Use: For the chemolithotrophic growth of *Bacillus schlegelii.*

Medium for *Bacillus stearothermophilus*

Composition per 1001.0mL:

NH_4Cl	1.0g
K_2HPO_4	0.5g
Yeast extract	0.2g
Casamino acids	0.1g
$MgSO_4 \cdot 7H_2O$	0.02g
Phenol solution	100.0mL
Trace elements solution SL-4	1.0mL

pH 7.4 ± 0.2 at 25°C

Phenol Solution:

Composition per 100.0mL:

Phenol	0.47g

Preparation of Phenol Solution: Add phenol to distilled/deionized water and bring volume to 100.0mL. Mix thoroughly. Filter sterilize.

Trace Elements Solution SL-4:

Composition per liter:

EDTA	0.5g
$FeSO_4 \cdot 7H_2O$	0.2g
Trace elements solution SL-6	100.0mL

Trace Elements Solution SL-6:

Composition per liter:

$MnCl_2 \cdot 4H_2O$	0.5g
H_3BO_3	0.3g
$CoCl_2 \cdot 6H_2O$	0.2g
$ZnSO_4 \cdot 7H_2O$	0.1g
$Na_2MoO_4 \cdot 2H_2O$	0.03g
$NiCl_2 \cdot 6H_2O$	0.02g
$CuCl_2 \cdot 2H_2O$	0.01g

Preparation of Trace Elements Solution SL-6: Add components to distilled/deionized water and bring volume to 1.0L. Mix thoroughly. Autoclave for 15 min at 15 psi pressure–121°C.

Preparation of Trace Elements Solution SL-4: Add components to distilled/deionized water and bring volume to 1.0L. Mix thoroughly. Filter sterilize.

Preparation of Medium: Add components, except phenol solution and trace elements solution SL-4, to distilled/deionized water and bring volume to 900.0mL. Mix thoroughly. Autoclave for 15 min at 15 psi pressure–121°C. Aseptically add 100.0mL of sterile phenol solution and 1.0mL of sterile trace elements solution SL-4. Mix thoroughly. Aseptically distribute into sterile tubes or flasks.

Use: For the cultivation of *Bacillus stearothermophilus.*

Medium BG11 for Cyanobacteria
See: **BG11 Agar and BG11 Medium**

Medium BG11 for Marine Cyanobacteria
See: **BG11 Marine Agar and BG11 Marine Broth**

Medium with Biphenyl (DSMZ Medium 457d)

Composition per liter:

Na_2HPO_4	2.44g
KH_2PO_4	1.52g
$(NH_4)_2SO_4$	0.5g
$MgSO_4·7H_2O$	0.2g
$CaCl_2·2H_2O$	0.05g
Trace elements solution SL-4	10.0mL
Biphenyl solution	25.0mL

pH 6.9 ± 0.2 at 25°C

Trace Elements Solution SL-4:

Composition per liter:

EDTA	0.5g
$FeSO_4·7H_2O$	0.2g
Trace elements solution SL-6	100.0mL

Trace Elements Solution SL-6:

Composition per liter:

H_3BO_3	0.3g
$CoCl_2·6H_2O$	0.2g
$ZnSO_4·7H_2O$	0.1g
$MnCl_2·4H_2O$	0.03g
$Na_2MoO_4·H_2O$	0.03g
$NiCl_2·6H_2O$	0.02g
$CuCl_2·2H_2O$	0.01g

Preparation of Trace Elements Solution SL-6: Add components to distilled/deionized water and bring volume to 1.0L. Mix thoroughly. Adjust pH to 3.4.

Preparation of Trace Elements Solution SL-4: Add components to distilled/deionized water and bring volume to 1.0L. Mix thoroughly.

Biphenyl Solution:

Composition per liter:

Biphenyl	10.0g

Preparation of Biphenyl Solution: Add biphenyl to 1.0L ethanol. Mix thoroughly. Filter sterilize using a cellulose filter membrane.

Preparation of Medium: Add components, except biphenyl solution, to 1.0L distilled/deionized water. Adjust pH to 6.9. Autoclave for 15 min at 15 psi pressure–121°C. Cool to room temperature. Add an aliquot of the biphenyl solution to a sterile flask so that the final concentration will be 0.25g/L biphenyl, and let the ethanol evaporate. Aseptically add sterile medium to the crystal-layered flask.

Use: For the cultivation of biphenyl- utilizing bacteria.

Medium C for Sulfate Reducers (Postgate's Medium C for Sulfate Reducers)

Composition per liter:

Sodium lactate	6.0g
Na_2SO_4	4.5g
NH_4Cl	1.0g
Yeast extract	1.0g
KH_2PO_4	0.5g
Sodium citrate·$2H_2O$	0.3g
$CaCl_2·6H_2O$	0.06g
$MgSO_4·7H_2O$	0.06g
$FeSO_4·7H_2O$	0.004g

pH 7.5 ± 0.2 at 25°C

Preparation of Medium: Add components to distilled/deionized water and bring volume to 1.0L. For marine bacteria, NaCl may be added or sea water used in place of distilled/deionized water. Mix thoroughly. Adjust pH to 7.5. Distribute into tubes or flasks. Autoclave for 15 min at 15 psi pressure–121°C.

Use: For detection, culturing, and storage of *Desulfovibrio* species and many *Desulfotomaculum* species. This medium should be used when a clear culture medium is desired such as for chemostat culture. This medium may be cloudy after sterilization but usually clears on cooling. It turns black as a result of H_2S production due to bacterial growth.

Medium for *Campylobacter* DSM 806 (DSMZ Medium 121)

Composition per liter:

Na-aspartate	10.0g
$MgSO_4·7H_2O$	1.0g
Yeast extract	0.2g
$CaCl_2·2H_2O$	28.0mg
Resazurin	1.0mg
Cysteine phosphate solution	100.0mL
Trace elements solution SL-10	1.0mL

pH 7.0 ± 0.2 at 25°C

Trace Elements Solution SL-10:

Composition per liter:

$FeCl_2·4H_2O$	1.5g
$CoCl_2·6H_2O$	190.0mg
$MnCl_2·4H_2O$	100.0mg
$ZnCl_2$	70.0mg
$Na_2MoO_4·2H_2O$	36.0mg
$NiCl_2·6H_2O$	24.0mg
H_3BO_3	6.0mg
$CuCl_2·2H_2O$	2.0mg
HCl (25% solution)	10.0mL

Preparation of Trace Elements Solution SL-10: Add $FeCl_2·4H_2O$ to 10.0mL of HCl solution. Mix thoroughly. Add distilled/deionized water and bring volume to 1.0L. Add remaining components. Mix thoroughly. Gas under 80% N_2 + 20% CO_2.

Cysteine Phosphate Solution:

Composition per 100.0mL:

K_2HPO_4	0.75g
NaH_2PO_4	0.25g
Cysteine-HCl·H_2O	0.25g

Preparation of Cysteine Phosphate Solution: Add components to 100.0mL distilled/deionized water. Mix thoroughly. Gas under 80% N_2 + 20% CO_2. Autoclave for 15 min at 15 psi pressure–121°C.

Preparation of Medium: Add components, except cysteine phosphate solution, to distilled/deionized water and bring volume to 900.0mL. Mix thoroughly. Adjust pH to 7.0. Gently heat and bring to boiling. Autoclave for 15 min at 15 psi pressure–121°C. Cool to 25°C. Aseptically add 100.0mL of sterile cysteine phosphate solution. Mix thoroughly. Aseptically distribute into sterile tubes or flasks.

Use: For the cultivation of *Sulfurospirillum* sp.

Medium for Carbon Monoxide Oxidizers

Composition per liter:

Agar	12.0g
$Na_2HPO_4·12H_2O$	4.5g
NH_4Cl	1.5g
KH_2PO_4	0.75g
$MgSO_4·7H_2O$	0.2g

$CaCl_2 \cdot 2H_2O$ 0.03g
Ferric ammonium citrate 0.018g
Trace elements solution SL-6 .... 1.0mL
pH 7.0 ± 0.2 at 25°C

Trace Elements Solution SL-6:
Composition per liter:
$MnCl_2 \cdot 4H_2O$ 0.5g
H_3BO_3 .... 0.3g
$CoCl_2 \cdot 6H_2O$ 0.2g
$ZnSO_4 \cdot 7H_2O$ 0.1g
$Na_2MoO_4 \cdot 2H_2O$ 0.03g
$NiCl_2 \cdot 6H_2O$ 0.02g
$CuCl_2 \cdot 2H_2O$ 0.01g

Preparation of Trace Elements Solution SL-6: Add components to distilled/deionized water and bring volume to 1.0L. Mix thoroughly. Autoclave for 15 min at 15 psi pressure–121°C.

Preparation of Medium: Add components to distilled/deionized water and bring volume to 1.0L. Mix thoroughly. Gently heat and bring to boiling. Autoclave for 15 min at 15 psi pressure–121°C. Pour into sterile Petri dishes or distribute into sterile tubes. After inoculation, incubate in an atmosphere of 80% CO +10% O_2 + 10% N_2.

Use: For the chemoautotrophic cultivation and maintenance of *Alcaligenes* species, *Pseudomonas carboxydohydrogena,* and other *Pseudomonas* species.

Medium for Carbon Monoxide Oxidizers

Composition per liter:
Agar 12.0g
$Na_2HPO_4 \cdot 12H_2O$ 4.5g
Sodium acetate 3.0g
NH_4Cl 1.5g
KH_2PO_4 .... 0.75g
$MgSO_4 \cdot 7H_2O$ 0.2g
$CaCl_2 \cdot 2H_2O$ 0.03g
Ferric ammonium citrate 0.018g
Trace elements solution SL-6 .... 1.0mL
pH 7.0 ± 0.2 at 25°C

Trace Elements Solution SL-6:
Composition per liter:
$MnCl_2 \cdot 4H_2O$ 0.5g
H_3BO_3 .... 0.3g
$CoCl_2 \cdot 6H_2O$ 0.2g
$ZnSO_4 \cdot 7H_2O$ 0.1g
$Na_2MoO_4 \cdot 2H_2O$ 0.03g
$NiCl_2 \cdot 6H_2O$ 0.02g
$CuCl_2 \cdot 2H_2O$ 0.01g

Preparation of Trace Elements Solution SL-6: Add components to distilled/deionized water and bring volume to 1.0L. Mix thoroughly. Autoclave for 15 min at 15 psi pressure–121°C.

Preparation of Medium: Add components to distilled/deionized water and bring volume to 1.0L. Mix thoroughly. Gently heat and bring to boiling. Autoclave for 15 min at 15 psi pressure–121°C. Pour into sterile Petri dishes or distribute into sterile tubes. After inoculation incubate in air.

Use: For the chemoorganotrophic cultivation and maintenance of *Alcaligenes* species, *Pseudomonas carboxydohydrogena*, and other *Pseudomonas* species.

Medium for Chlorate Respirers (DSMZ Medium 908)

Composition per liter:
Solution A 1.0L
Solution B 10.0mL
Solution C 10.0mL
Vitamin solution 5.0mL
Trace elements solution SL-10 .... 1.0mL
pH 7.2 ± 0.2 at 25°C

Solution A:
Composition per liter:
$NaHCO_3$ 2.5g
Na-acetate 1.36g
$NaClO_3$ 1.0g
NaH_2PO_4 .... 0.6g
NH_4Cl 0.25g
KCl 0.1g

Preparation of Solution A: Add components to distilled/deionized water and bring volume to 1.0L. Mix thoroughly. Sparge with 80% N_2 + 20% CO_2. Autoclave for 15 min at 15 psi pressure–121°C. Cool to room temperature.

Solution B:
Composition per 10.0mL:
$MgSO_4$ 30.0mg
$CaCl_2 \cdot 2H_2O$ 10.0mg

Preparation of Solution B: Add components to distilled/deionized water and bring volume to 10.0mL. Mix thoroughly. Sparge with 80% N_2 + 20% CO_2. Autoclave for 15 min at 15 psi pressure–121°C. Cool to room temperature.

Solution C:
Na_2MoO_4 .... 25.0mg
$Na_2WO_4 \cdot 2H_2O$ 25.0mg

Preparation of Solution C: Add components to distilled/deionized water and bring volume to 10.0mL. Mix thoroughly. Sparge with 80% N_2 + 20% CO_2. Autoclave for 15 min at 15 psi pressure–121°C. Cool to room temperature.

Vitamin Solution:
Composition per liter:
Vitamin B_{12} 50.0mg
Pantothenic acid 50.0mg
Riboflavin 50.0mg
Alpha-lipoic acid 50.0mg
p-Aminobenzoic acid 50.0mg
Thiamine-HCl·$2H_2O$ 50.0mg
Nicotinic acid 25.0mg
Nicotinamide 25.0mg
Biotin 20.0mg
Folic acid 20.0mg
Pyridoxamine-HCl 10.0mg

Preparation of Vitamin Solution: Add components to distilled/deionized water and bring volume to 1.0L. Mix thoroughly. Filter sterilize.

Trace Elements Solution SL-10:
Composition per liter:
$FeCl_2 \cdot 4H_2O$ 1.5g
H_3BO_3 .... 300.0mg
$CoCl_2 \cdot 6H_2O$ 190.0mg
$MnCl_2 \cdot 4H_2O$ 100.0mg

$ZnCl_2$ 70.0mg
$Na_2MoO_4 \cdot 2H_2O$ 36.0mg
$NiCl_2 \cdot 6H_2O$ 24.0mg
$CuCl_2 \cdot 2H_2O$ 2.0mg
HCl (25% solution) 7.7mL

Preparation of Trace Elements Solution SL-10: Add $FeCl_2 \cdot 4H_2O$ to 10.0mL of HCl solution. Mix thoroughly. Add distilled/deionized water and bring volume to 1.0L. Add remaining components. Mix thoroughly. Sparge with 100% N_2. Autoclave for 15 min at 15 psi pressure–121°C. Cool to room temperature.

Preparation of Medium: Aseptically and anaerobically combine 1000.0mL solution A, 10.0mL solution B and 10.0mL solution C. Aseptically and anaerobically add 5.0mL vitamin solution and 1.0mL trace elements solution SL-10. Mix thoroughly. The pH should be 7.2.

Use: For the cultivation of *Dechloromonas agitata* and *Azospira oryzae.*

Medium with Chloroacrylic Acid (DSMZ Medium 457c)

Composition per liter:
Na_2HPO_4 .... 2.44g
KH_2PO_4 .... 1.52g
$(NH_4)_2SO_4$ 0.5g
$MgSO_4 \cdot 7H_2O$ 0.2g
$CaCl_2 \cdot 2H_2O$ 0.05g
Chloroacrylic acid solution 20.0mL
Trace elements solution SL-4 .... 10.0mL

pH 6.9 ± 0.2 at 25°C

Trace Elements Solution SL-4:
Composition per liter:
EDTA 0.5g
$FeSO_4 \cdot 7H_2O$ 0.2g
Trace elements solution SL-6 .... 100.0mL

Trace Elements Solution SL-6:
Composition per liter:
H_3BO_3 .... 0.3g
$CoCl_2 \cdot 6H_2O$ 0.2g
$ZnSO_4 \cdot 7H_2O$ 0.1g
$MnCl_2 \cdot 4H_2O$ 0.03g
$Na_2MoO_4 \cdot H_2O$ 0.03g
$NiCl_2 \cdot 6H_2O$ 0.02g
$CuCl_2 \cdot 2H_2O$ 0.01g

Preparation of Trace Elements Solution SL-6: Add components to distilled/deionized water and bring volume to 1.0L. Mix thoroughly. Adjust pH to 3.4.

Preparation of Trace Elements Solution SL-4: Add components to distilled/deionized water and bring volume to 1.0L. Mix thoroughly.

Chloroacrylic Acid Solution:
Composition per liter:
3-Chloroacrylic acid 4.0g

Preparation of Chloroacrylic Acid Solution: Add 3-chloroacrylic acid to distilled/deionized water and bring volume to 1.0L. Mix thoroughly. Adjust pH to 7.0. Filter sterilize.

Preparation of Medium: Add components, except chloroacrylic acid solution, to 1.0L distilled/deionized water. Adjust pH to 6.9. Autoclave for 15 min at 15 psi pressure–121°C. Cool to room temperature. Aseptically add 20.0mL sterile chloroacrylic acid solution. Mix thoroughly. Aseptically distribute to sterile tubes or flasks.

Use: For the cultivation of chloroacrylic acid-utilizing *Burkholderia* sp. *(Burkholderia cepacia), Rhodococcus erythropolis (Arthrobacter picolinophilus,* and *Nocardia* spp.

Medium for *Chlorobium ferrooxidans* (DSMZ Medium 29a)

Composition per 5.0L:
Solution A 4.0L
Solution B 860.0mL
Solution E 100.0mL
Solution F 30.0mL
Solution C 5.0mL
Solution D 5.0mL

pH 6.8 at 25°C

Solution A:
Composition per 4.0L:
$MgSO_4$ 2.5g
KH_2PO_4 .... 1.7g
NH_4Cl 1.7g
KCl 1.7g
$CaCl_2 \cdot 2H_2O$ 1.25g
Na-acetate 0.82g

Preparation of Solution A: Add components to 4.0L distilled water. Mix thoroughly. Autoclave for 45 min at 15 psi pressure–121°C in a 5-liter special bottle or flask with four openings at the top, together with a teflon-coated magnetic bar. In this 5-liter bottle, two openings are for tubes in the central, silicon rubber stopper; one is a short, gas-inlet tube with a sterile cotton filter, and the other is an outlet tube for medium, which reaches the bottom of the vessel at one end and has, at the other end, a silicon rubber tube with a pinch cock and a bell for aseptic dispensing of the medium into bottles. The other two openings have gas-tight screw caps; one of these openings is for the addition of sterile solutions and the other serves as a gas outlet. After autoclaving, cool solution A to room temperature under a N_2 atmosphere with a positive pressure of 0.05–0.1 atm (a manometer for low pressure will be required). Saturate the cold medium with CO_2 by magnetic stirring for 30 min under a CO_2 atmosphere of 0.05–0.1 atm.

Solution B:
Distilled water 860.0mL

Preparation of Solution B: Autoclave distilled water for 15 min at 15 psi pressure–121°C in a cotton-stoppered Erlenmeyer flask. Cool to room temperature under an atmosphere of N_2 in an anaerobic jar.

Solution C:
Composition per 100.0mL:
Vitamin B_{12} 2.0mg

Preparation of Solution C: Add vitamin B_{12} to distilled/deionized water and bring volume to 100.0mL. Mix thoroughly. Sparge under 100% N_2 gas for 3 min. Filter sterilize. Store under N_2 gas.

Solution D:
Composition per liter:
$FeCl_2 \cdot 4H_2O$ 1.5g
H_3BO_3 .... 300.0mg
$CoCl_2 \cdot 6H_2O$ 190.0mg
$MnCl_2 \cdot 4H_2O$ 100.0mg
$ZnCl_2$ 70.0mg
$Na_2MoO_4 \cdot 2H_2O$ 36.0mg

$NiCl_2·6H_2O$	24.0mg
$CuCl_2·2H_2O$	2.0mg
HCl (25% solution)	7.7mL

Preparation of Solution D: Add $FeCl_2·4H_2O$ to 10.0mL of HCl solution. Mix thoroughly. Add distilled/deionized water and bring volume to 1.0L. Add remaining components. Mix thoroughly. Sparge with 100% N_2. Autoclave for 15 min at 15 psi pressure–121°C.

Solution E:

Composition per 100.0mL:

$NaHCO_3$	4.2g

Preparation of Solution E: Add $NaHCO_3$ to distilled/deionized water and bring volume to 100.0mL. Mix thoroughly. Sparge with 100% CO_2 until saturated. Filter sterilize under 100% CO_2 into a sterile, gas-tight 100.0mL screw-capped bottle.

Solution F:

Composition per 100.0mL:

$FeSO_4$	25.0g

Preparation of Solution F: Add $FeSO_4$ to distilled/deionized water and bring volume to 100.0mL. Mix thoroughly. Sparge with 100% CO_2 until saturated. Filter sterilize under 100% CO_2 into a sterile, gas-tight 100.0mL screw-capped bottle.

Preparation of Medium: Add solutions B, C, D, and E to solution A through one of the screw-cap openings against a stream of either N_2 gas or, better, a mixture of 95% N_2 and 5% CO_2 while the medium is magnetically stirred. Adjust the pH of the medium with sterile HCl or Na_2CO_3 solution (2*M* solutions) to pH 6.8. Distribute the medium aseptically through the medium outlet tube into sterile, 100mL bottles (with metal caps and autoclavable rubber seals) using the positive gas pressure (0.05– 0.1 atm) of the N_2/CO_2 gas mixture: Leave a small air bubble in each bottle to meet possible pressure changes. The tightly sealed, screw-cap bottles can be stored for several weeks or months in the dark.

Use: For the cultivation of *Chlorobium ferrooxidans*.

Medium D

See: **Castenholz D Medium**

Medium D, Modified

See: **Castenholz D Medium, Modified**

Medium D for Sulfate Reducers (Postgate's Medium D for Sulfate Reducers)

Composition per liter:

Sodium pyruvate	3.5g
$MgCl_2·6H_2O$	1.6g
NH_4Cl	1.0g
Yeast extract	1.0g
KH_2PO_4	0.5g
$CaCl_2·2H_2O$	0.1g
$FeSO_4·7H_2O$	0.004g

pH 7.5 ± 0.2 at 25°C

Preparation of Medium: Add components to distilled/deionized water and bring volume to 1.0L. Malate or fumarate may also be used as a carbon source. For marine bacteria, NaCl may be added or seawater used in place of distilled/deionized water. Mix thoroughly. Adjust pH to 7.5. Filter sterilize. Aseptically distribute into sterile tubes or flasks.

Use: For the cultivation of *Desulfovibrio* species and *Desulfotomaculum* species that can grow in the absence of sulfate.

Medium D for Sulfate Reducers (Postgate's Medium D for Sulfate Reducers)

Composition per liter:

$MgCl_2·6H_2O$	1.6g
Choline chloride	1.0g
NH_4Cl	1.0g
Yeast extract	1.0g
KH_2PO_4	0.5g
$CaCl_2·2H_2O$	0.1g
$FeSO_4·7H_2O$	0.004g

pH 7.5 ± 0.2 at 25°C

Preparation of Medium: Add components to distilled/deionized water and bring volume to 1.0L. Malate or fumarate may also be used as a carbon source. For marine bacteria, NaCl may be added or seawater used in place of distilled/deionized water. Mix thoroughly. Adjust pH to 7.5. Filter sterilize. Aseptically distribute into sterile tubes or flasks.

Use: For the cultivation of *Desulfovibrio* species and *Desulfotomaculum* species that can grow in the absence of sulfate.

Medium D for *Thermus*

Composition per liter:

Pancreatic digest of casein	1.0g
Yeast extract	1.0g
$NaNO_3$	0.7g
KNO_3	0.1g
$MgSO_4·7H_2O$	0.1g
Na_2HPO_4	0.1g
Nitrilotriacetic acid	0.1g
$CaSO_4·2H_2O$	0.06g
NaCl	8.0mg
$MnSO_4·H_2O$	2.2mg
$ZnSO_4·7H_2O$	0.5mg
H_3BO_3	0.5mg
$FeCl_3$	0.28mg
$Na_2MoO_4·2H_2O$	0.03mg
$CuSO_4$	0.02mg

pH 8.2 ± 0.2 at 25°C

Preparation of Medium: Add nitrilotriacetic acid to 500.0mL of distilled/deionized water. Dissolve by adjusting pH to 6.5 with KOH. Add remaining components. Readjust pH to 8.2 with H_2SO_4 or KOH. Add distilled/deionized water to 1.0L. Distribute into tubes or flasks. Autoclave for 15 min at 15 psi pressure–121°C.

Use: For the cultivation of *Thermus* species.

Medium D for *Thermus*, Modified

Composition per liter:

Agar	25.0g
Pancreatic digest of casein	1.0g
Yeast extract	1.0g
Salt solution	100.0mL

pH 8.2 ± 0.2 at 25°C

Salt Solution:

Composition per liter:

$NaNO_3$	6.89g
$Na_2HPO_4·12H_2O$	2.8g

KNO_3 .... 1.03g
Nitrilotriacetic acid 1.0g
$MgSO_4 \cdot 7H_2O$ 1.0g
$CaSO_4 \cdot 2H_2O$ 0.6g
NaCl 0.08g
$FeCl_3 \cdot 6H_2O$ solution 10.0mL
Trace elements solution 10.0mL

$FeCl_3 \cdot 6H_2O$ Solution:
Composition per 100.0mL:
$FeCl_3 \cdot 6H_2O$ 47.0mg

Preparation of $FeCl_3 \cdot 6H_2O$ Solution: Add $FeCl_3 \cdot 6H_2O$ to distilled/deionized water and bring volume to 100.0mL. Mix thoroughly.

Trace Elements Solution:
Composition per liter:
$MnSO_4 \cdot 4H_2O$ 1.7g
$ZnSO_4 \cdot 7H_2O$ 0.5g
H_3BO_3 .... 0.5g
$CoCl_2 \cdot 6H_2O$ 46.0mg
$CuSO_4 \cdot 5H_2O$ 25.0mg
$Na_2MoO_4 \cdot 2H_2O$ 25.0mg
H_2SO_4 .... 0.5mL

Preparation of Trace Elements Solution: Add components to distilled/deionized water and bring volume to 1.0L. Mix thoroughly.

Preparation of Salt Solution: Add nitrilotriacetic acid to 500.0mL of distilled/deionized water. Dissolve by adjusting pH to 6.5 with KOH. Add remaining components. Readjust pH to 8.2 with H_2SO_4 or KOH. Add distilled/deionized water to 1.0L.

Preparation of Medium: Add components to distilled/deionized water and bring volume to 1.0L. Mix thoroughly. Gently heat and bring to boiling. Adjust pH to 8.2. Distribute into tubes or flasks. Autoclave for 15 min at 15 psi pressure–121°C. Pour into sterile Petri dishes or leave in tubes.

Use: For the cultivation of *Thermus aquaticus*.

Medium D2

Composition per liter:
Agar 15.0g
Glucose 10.0g
LiCl 5.0g
Pancreatic digest of casein 4.0g
Yeast extract 2.0g
Tris(hydroxymethyl)amino-methane·HCl buffer 1.2g
NH_4Cl 1.0g
$MgSO_4 \cdot 7H_2O$ 0.3g
Polymyxin sulfate solution 10.0mL
NaN_3 solution 10.0mL

pH 6.9 ± 0.2 at 25°C

Polymyxin Sulfate Solution:
Composition per 10.0mL:
Polymyxin sulfate 0.04g

Preparation of Polymyxin Sulfate Solution: Add polymyxin sulfate to distilled/deionized water and bring volume to 10.0mL. Mix thoroughly. Filter sterilize. Use freshly prepared solution.

NaN_3 Solution:
Composition per 10.0mL:
NaN_3 .... 2.0mg

Preparation of NaN_3 Solution: Add NaN_3 to distilled/deionized water and bring volume to 10.0mL. Mix thoroughly. Filter sterilize. Use freshly prepared solution.

Caution: Sodium azide is toxic. Azides also react with metals and disposal must be highly diluted.

Preparation of Medium: Add components, except polymyxin sulfate solution and NaN_3 solution, to distilled/deionized water and bring volume to 980.0mL. Mix thoroughly. Gently heat and bring to boiling. Autoclave for 15 min at 15 psi pressure–121°C. Cool to 45°–50°C. Aseptically add sterile polymyxin sulfate solution and NaN_3 solution. Mix thoroughly. Pour into sterile Petri dishes or distribute into sterile tubes.

Use: For the selective isolation and cultivation of *Corynebacterium* species.

Medium D4

Composition per liter:
Agar 15.0g
Sucrose 10.0g
NH_4Cl 5.0g
Na_2HPO_4, anhydrous 2.3g
Pancreatic digest of casein 1.0g
Sodium dodecyl sulfate 0.6g
Glycerol 10.0mL

Preparation of Medium: Add components to distilled/deionized water and bring volume to 1.0L. Mix thoroughly. Gently heat and bring to boiling. Distribute into tubes or flasks. Autoclave for 15 min at 15 psi pressure–121°C. Pour into sterile Petri dishes or leave in tubes.

Use: For the selective isolation and cultivation of *Pseudomonas syringae*.

Medium DG

See: **Castenholz DG Medium**

Medium DGN

See: **Castenholz DGN Medium**

Medium for DSM 14457 and DSM 14458 (DSMZ Medium 956)

Composition per liter:
KH_2PO_4 .... 2.0g
$(NH_4)_2SO_4$ 2.0g
NaCl 0.5g
$MgSO_4 \cdot 7H_2O$ 0.125g
$FeSO_4 \cdot 7H_2O$ 0.02g
Methanol solution 10.0mL

pH 7.2 ± 0.2 at 25°C

Methanol Solution:
Composition per 10.0mL:
Methanol 5.0mL

Preparation of Methanol Solution: Add methanol to distilled/deionized water and bring volume to 10.0mL. Mix thoroughly. Filter sterilize.

Preparation of Medium: Add components, except methanol solution, to distilled/deionized water and bring volume to 990.0mL. Mix thoroughly. Autoclave for 15 min at 15 psi pressure–121°C. Cool to room temperature. Aseptically add 10.0mL sterile methanol solution. Mix thoroughly. Aseptically distribute into sterile tubes or flasks.

Use: For the cultivation of *Methylobacterium lusitanum* and *Methylobacterium suomiense.*

Medium for DSM 14457 and DSM 14458 (DSMZ Medium 956)

Composition per liter:

KH_2PO_4	2.0g
$(NH_4)_2SO_4$	2.0g
NaCl	0.5g
$MgSO_4 \cdot 7H_2O$	0.125g
$FeSO_4 \cdot 7H_2O$	0.02g
Methylamine solution	10.0mL

pH 7.2 ± 0.2 at 25°C

Methylamine Solution:

Composition per 10.0mL:

Methylamine	3.0g

Preparation of Methylamine Solution: Add methylamine to distilled/deionized water and bring volume to 10.0mL. Mix thoroughly. Filter sterilize.

Preparation of Medium: Add components, except methylamine solution, to distilled/deionized water and bring volume to 990.0mL. Mix thoroughly. Autoclave for 15 min at 15 psi pressure–121°C. Cool to room temperature. Aseptically add 10.0mL sterile methylamine solution. Mix thoroughly. Aseptically distribute into sterile tubes or flasks.

Use: For the cultivation of *Methylobacterium lusitanum* and *Methylobacterium suomiense.*

Medium E for *Bacillus*

Composition per liter:

NaCl	50.0g
K_2HPO_4	10.6g
Sucrose	10.0g
KH_2PO_4	5.3g
$(NH_4)_2SO_4$	1.0g
$MgSO_4$	0.25g
Trace salts solution	10.0mL

Trace Salts Solution:

Composition per liter:

$MnSO_4 \cdot H_2O$	3.0g
Disodium EDTA	1.0g
$FeSO_4 \cdot 7H_2O$	0.1g
$CaCl_2 \cdot 2H_2O$	0.1g
$CoCl_2 \cdot 6H_2O$	0.1g
$ZnSO_4 \cdot 7H_2O$	0.1g
$CuSO_4 \cdot 5H_2O$	0.01g
$AlK(SO_4)_2 \cdot 12H_2O$	0.01g
H_3BO_3	0.01g
$Na_2MoO_4 \cdot 2H_2O$	0.01g

Preparation of Trace Salts Solution: Add components to distilled/deionized water and bring volume to 1.0L. Mix thoroughly.

Preparation of Medium: Add components to distilled/deionized water and bring volume to 1.0L. Mix thoroughly. Autoclave for 15 min at 15 psi pressure–121°C. Aseptically distribute into sterile tubes or flasks.

Use: For the cultivation and maintenance of *Bacillus* species.

Medium E for Sulfate Reducers (Postgate's Medium E for Sulfate Reducers)

Composition per liter:

Agar	15.0g
Sodium lactate	3.5g
$MgCl_2 \cdot 6H_2O$	2.0g
NH_4Cl	1.0g
Na_2SO_4	1.0g
$CaCl_2 \cdot 2H_2O$	1.0g
Yeast extract	1.0g
KH_2PO_4	0.5g
Ascorbic acid	0.1g
Thioglycollic acid	0.1g
$FeSO_4 \cdot 7H_2O$	0.004g

pH 7.6 ± 0.2 at 25°C

Preparation of Medium: Add components, except ascorbic acid and thioglycollic acid, to tap water and bring volume to 1.0L. For marine bacteria, NaCl may be added or seawater used in place of tap water. Mix thoroughly. Gently heat and bring to boiling. Adjust pH to 7.6. Thioglycolate and ascorbate should be added immediately prior to sterilization. Distribute into screw-capped tubes or flasks. Autoclave for 15 min at 15 psi pressure–121°C.

Use: For the cultivation and enumeration of *Desulfovibrio* species and *Desulfotomaculum* species as black colonies in deep agar cultures. Also used for the isolation of pure cultures of *Desulfovibrio* species and *Desulfotomaculum* species.

Medium E-2

Composition per liter:

$K_2HPO_4 \cdot 3H_2O$	7.5g
KH_2PO_4	3.7g
$NaNH_4HPO_4 \cdot 4H_2O$	3.5g
Tap water	1.0L
Thiamine solution	10.0mL
$MgSO_4 \cdot 7H_2O$ solution	10.0mL
n-Octane	variable

pH 7.0 ± 0.2 at 25°C

Thiamine Solution:

Composition per 10.0mL:

Thiamine	10.0mg

Preparation of Thiamine Solution: Add thiamine to distilled/deionized water and bring volume to 10.0mL. Mix thoroughly.

$MgSO_4 \cdot 7H_2O$ Solution:

Composition per 10.0mL:

$MgSO_4 \cdot 7H_2O$	0.246g

Preparation of $MgSO_4 \cdot 7H_2O$ Solution: Add $MgSO_4 \cdot 7H_2O$ to distilled/deionized water and bring volume to 10.0mL. Mix thoroughly.

Preparation of Medium: Add components, except octane, to tap water and bring volume to 1.0L. Mix thoroughly. Adjust pH to 7.0. Distribute into tubes or flasks. Autoclave for 15 min at 15 psi pressure–121°C. Inoculate tubes and place in a desiccator to which *n*-octane has been added and evaporated.

Use: For the cultivation of a recombinant strain of *Escherichia coli* that utilizes hydrocarbons.

Medium for *Ectothiorhodospira*

Composition per 1001.0mL:

Basal medium	800.0mL
Solution C	200.0mL
Vitamin solution B	1.0mL

Basal Medium:

Composition per 800.0mL:

NaCl	180.0g
Na_2SO_4	20.0g
Na_2CO_3	6.0g
$Na_2S \cdot 9H_2O$	1.0g
Sodium succinate	1.0g
NH_4Cl	0.8g
KH_2PO_4	0.5g
Yeast extract	0.5g
$MgCl_2 \cdot 6H_2O$	0.1g
$CaCl_2 \cdot 7H_2O$	0.05g
Trace elements solution A	1.0mL

pH 8.5 ± 0.2 at 25°C

Trace Elements Solution A:

Composition per liter:

$FeCl_2 \cdot 4H_2O$	1.8g
H_3BO_3	500.0mg
$CoCl_2 \cdot 6H_2O$	250.0mg
$ZnCl_2$	100.0mg
$MnCl_2 \cdot 4H_2O$	70.0mg
$Na_2MoO_4 \cdot 2H_2O$	30.0mg
$CuCl_2 \cdot 2H_2O$	10.0mg
$NiCl_2 \cdot 6H_2O$	10.0mg
$Na_2SeO_3 \cdot 5H_2O$	10.0mg

Preparation of Trace Elements Solution A: Add components to distilled/deionized water and bring volume to 990.0mL. Mix thoroughly. Adjust pH to 3 with 1*N* HCl. Bring volume to 1.0L with distilled/deionized water.

Preparation of Basal Solution: Add components to distilled/deionized water and bring volume to 800.0mL. Mix thoroughly. Adjust pH to 8.5. Distribute into screw-capped bottles. Autoclave for 15 min at 14 psi pressure–120°C.

Vitamin Solution B:

Composition per 100.0mL:

Nicotinamide	35.0mg
Thiamine dichloride	30.0mg
p-Aminobenzoic acid	20.0mg
Biotin	10.0mg
Calcium DL-pantothenate	10.0mg
Pyridoxal·HCl	10.0mg
Vitamin B_{12}	5.0mg

Preparation of Vitamin Solution B: Add components to distilled/deionized water and bring volume to 100.0mL. Mix thoroughly. Filter sterilize.

Solution C:

Composition per 200.0mL:

$NaHCO_3$	14.0g

Preparation of Solution C: Add $NaHCO_3$ to distilled/deionized water and bring volume to 200.0mL. Mix thoroughly. Filter sterilize.

Preparation of Medium: To 800.0mL of sterile basal solution, aseptically add 200.0mL of sterile solution C and 1.0mL of sterile vitamin solution B. Mix thoroughly.

Use: For the cultivation and maintenance of *Ectothiorhodospira halochloris*.

Medium with EDTA as Carbon and Nitrogen Source

Composition per liter:

Agar	15.0g
$MgSO_4 \cdot 7H_2O$	0.3g
Disodium ethylenediaminetetraacetate	0.25g
$CaCl_2 \cdot 2H_2O$	0.244g
Ferric ammonium citrate	0.05g
Phosphate solution	50.0mL
Trace elements solution SL-6	5.0mL
Schlegel's vitamin solution	5.0mL

pH 7.6 ± 0.4 at 25°C

Phosphate Solution:

Composition per 50.0mL:

$Na_2HPO_4 \cdot 2H_2O$	3.57g
KH_2PO_4	0.67g

Preparation of Phosphate Solution: Add components to distilled/deionized water and bring volume to 50.0mL. Mix thoroughly. Adjust pH to 7.6. Autoclave for 15 min at 15 psi pressure–121°C.

Trace Elements Solution SL-6:

Composition per liter:

$MnCl_2 \cdot 4H_2O$	0.5g
H_3BO_3	0.3g
$CoCl_2 \cdot 6H_2O$	0.2g
$ZnSO_4 \cdot 7H_2O$	0.1g
$Na_2MoO_4 \cdot 2H_2O$	0.03g
$NiCl_2 \cdot 6H_2O$	0.02g
$CuCl_2 \cdot 2H_2O$	0.01g

Preparation of Trace Elements Solution SL-6: Add components to distilled/deionized water and bring volume to 1.0L. Mix thoroughly.

Schlegel's Vitamin Solution:

Composition per 100.0mL:

Nicotinic acid	2.0g
Pyridoxamine	5.0mg
Cyanocobalamin	2.0mg
p-Aminobenzoate	1.0mg
Thiamine	1.0mg
Calcium DL-pantothenate	0.5mg
Biotin	0.2mg

Preparation of Schlegel's Vitamin Solution: Add components to distilled/deionized water and bring volume to 100.0mL. Filter sterilize.

Preparation of Medium: Add components, except phosphate solution and Schlegel's vitamin solution, to distilled/deionized water and bring volume to 945.0mL. Mix thoroughly. Gently heat and bring to boiling. Autoclave for 15 min at 15 psi pressure–121°C. Aseptically add 50.0mL of sterile phosphate solution and 5.0mL of sterile Schlegel's vitamin solution. Mix thoroughly. Pour into sterile Petri dishes or distribute into sterile tubes.

Use: For the cultivation of bacteria that can utilize EDTA as a carbon source.

Medium with EDTA as Carbon Source (DSMZ Medium 473)

Composition per liter:

Agar	15.0g
$MgSO_4 \cdot 7H_2O$	0.49g

Na_2-EDTA	0.2g
Ferric ammonium citrate	0.08g
$Ca(NO_3)_2·4H_2O$	0.02g
Phosphate solution	10.0mL
Trace elements solution SL-6	5.0mL
Vitamin solution	5.0mL

pH 7.5 ± 0.2 at 25°C

Phosphate Solution:
Composition per 10.0mL:

KH_2PO_4	0.272g

Preparation of Phosphate Solution: Add KH_2PO_4 to distilled/deionized water and bring volume to 10.0mL. Mix thoroughly. Autoclave for 15 min at 15 psi pressure–121°C. Cool to room temperature.

Vitamin Solution:
Composition per 100.0mL:

KH_2PO_4	0.272g
Biotin	0.08g
Folic acid	0.08g
Thiamin-HCl	0.08g

Preparation of Vitamin Solution: Add components to distilled/deionized water and bring volume to 100.0mL. Mix thoroughly. Filter sterilize.

Trace Elements Solution SL-6:
Composition per liter:

H_3BO_3	0.3g
$CoCl_2·6H_2O$	0.2g
$ZnSO_4·7H_2O$	0.1g
$MnCl_2·4H_2O$	0.03g
$Na_2MoO_4·H_2O$	0.03g
$NiCl_2·6H_2O$	0.02g
$CuCl_2·2H_2O$	0.01g

Preparation of Trace Elements Solution SL-6: Add components to distilled/deionized water and bring volume to 1.0L. Mix thoroughly. Adjust pH to 3.4.

Preparation of Medium: Add components, except vitamin solution and phosphate solution, to 985.0mL distilled/deionized water. Adjust pH to 7.5. Autoclave for 15 min at 15 psi pressure–121°C. Pour into sterile Petri dishes (20.0mL per Petri dish). Cool to room temperature. Aseptically add 10.0mL sterile phosphate solution and 5.0mL sterile vitamin solution. Mix thoroughly. Aseptically distribute to sterile tubes or flasks.

Use: For the cultivation of unclassified bacterium DSM6780.

Medium for *Erythrobacter longus* (DSMZ Medium 695)

Composition per liter:

Peptone	2.0g
Soytone	1.0g
Yeast extract	1.0g
Proteose peptone No.3	1.0g
Ferric citrate solution	2.0mL
Artificial seawater	700.0mL

pH 7.5 ± 0.2 at 25°C

Artificial Seawater:
Composition per liter:

NaCl	23.477g
$MgCl_2·6H_2O$	4.981g
Na_2SO_4	3.917g
$CaCl_2$	1.12g
KCl	664.0mg
$NaHCO_3$	192.0mg
H_3BO_3	26.0mg
$SrCl_2$	24.0mg
KBr	6.0mg
NaF	3.0mg

Preparation of Artificial Seawater: Add components to distilled/deionized water and bring volume to 1.0L. Mix thoroughly. Filter sterilize.

Ferric Citrate Solution:
Composition per 10.0mL:

Ferric citrate	0.5g

Preparation of Ferric Citrate Solution: Add ferric citrate to distilled/deionized water and bring volume to 10.0mL. Mix thoroughly.

Preparation of Medium: Add components, except artificial sea water, to distilled/deionized water and bring volume to 300.0mL. Mix thoroughly. Adjust pH to 7.5. Aseptically add 700.0mL artificial sea water. Mix thoroughly. Aseptically and anaerobically distribute into sterile tubes or bottles.

Use: For the cultivation of *Erythrobacter longus.*

Medium F

Composition per liter:

$MgSO_4·7H_2O$	0.5g
$(NH_4)_2SO_4$	0.15g
KCl	0.05g
KH_2PO_4	0.05g
$Ca(NO_3)_2$	0.01g
$FeSO_4·7H_2O$ solution	10.0mL

pH 3.5 ± 0.2 at 25°C

$FeSO_4·7H_2O$ Solution:
Composition per 10.0mL:

$FeSO_4·7H_2O$	1.0g

Preparation of $FeSO_4·7H_2O$ Solution: Add the $FeSO_4·7H_2O$ to distilled/deionized water and bring volume to 10.0mL. Mix thoroughly. Filter sterilize.

Preparation of Medium: Add components, except $FeSO_4·7H_2O$ solution, to tap water and bring volume to 990.0mL. Mix thoroughly. Gently heat until dissolved. Adjust pH to 3.5. Autoclave for 15 min at 15 psi pressure–121°C. Cool to 45°–50°C. Aseptically add 10.0mL of sterile $FeSO_4·7H_2O$ solution. Mix thoroughly. Aseptically distribute into sterile tubes or flasks.

Use: For the cultivation of *Thiobacillus* species.

Medium F for Sulfate Reducers (Postgate's Medium F for Sulfate Reducers)

Composition per liter:

Agar	12.0g
Pancreatic digest of casein	10.0g
Sodium lactate	3.5g
Ferrous citrate	0.5g
Na_2SO_3	0.5g
$MgSO_4·7H_2O$	0.2g
Ascorbic acid	0.1g
Sodium thioglycolate	0.1g

pH 7.1 ± 0.2 at 25°C

Preparation of Medium: Add components, except ascorbic acid and thioglycollic acid, to tap water and bring volume to 1.0L. For marine bacteria, NaCl may be added or seawater used in place of tap water. Mix thoroughly. Gently heat and bring to boiling. Adjust pH to 7.1. Thioglycolate and ascorbate should be added immediately prior to sterilization. Distribute into screw-capped tubes or flasks. Autoclave for 15 min at 15 psi pressure–121°C.

Use: For isolation and cultivation of *Desulfotomaculum nigrificans, Desulfovibrio* species, and other *Desulfotomaculum* species especially in food. These bacteria form black colonies in deep agar cultures.

Medium for Freshwater Flexibacteria

Composition per 1002.0mL:

Casamino acids	1.0g
$MgSO_4 \cdot 7H_2O$	1.0g
Tris (hydroxymethyl) amino methane	1.0g
$CaCl_2 \cdot 2H_2O$	0.1g
KNO_3	0.1g
Sodium glycerophosphate	0.1g
Thiamine	1.0mg
Cobalamine	1.0µg
Glucose solution	1.0mL
Trace elements solution	1.0mL

pH 7.5 ± 0.2 at 25°C

Glucose Solution:

Composition per 100.0mL:

Glucose	10.0g

Preparation of Glucose Solution: Add glucose to distilled/deionized water and bring volume to 100.0mL. Mix thoroughly. Filter sterilize.

Trace Elements Solution:

Composition per liter:

$ZnCl_2$	20.8g
H_3BO_3	2.85g
$MnCl_2 \cdot 4H_2O$	1.8g
Sodium tartrate	1.77g
$FeSO_4$	1.36g
$CoCl_2 \cdot 6H_2O$	40.4mg
$CuCl_2 \cdot 2H_2O$	26.9mg
$Na_2MoO_4 \cdot 2H_2O$	25.2mg

Preparation of Trace Elements Solution: Add components to distilled/deionized water and bring volume to 1.0L. Mix thoroughly. Autoclave for 15 min at 15 psi pressure–121°C.

Preparation of Medium: Add components, except glucose solution and trace elements solution, to distilled/deionized water and bring volume to 1.0L. Mix thoroughly. Autoclave for 15 min at 15 psi pressure–121°C. Aseptically add 1.0mL of sterile glucose solution and 1.0mL of sterile trace elements solution. Mix thoroughly. Aseptically distribute into sterile tubes or flasks.

Use: For the cultivation of *Cytophaga psychrophila, Flectobacillus major, Flexibacter aurantiacus, Flexibacter aurantiacus, Flexibacter elegans, Flexibacter flexilis, Flexibacter roseolus, Flexibacter ruber, Flexibacter sancti,* and *Herpetosiphon geysericola.*

Medium with Fluoranthene (DSMZ Medium 457b)

Composition per liter:

Na_2HPO_4	2.44g
KH_2PO_4	1.52g
$(NH_4)_2SO_4$	0.5g
$MgSO_4 \cdot 7H_2O$	0.2g
Tween 80	0.2g
Fluoranthene solution	50.0mL
$CaCl_2 \cdot 2H_2O$	0.05g
Trace elements solution SL-4	10.0mL
Fluoranthene solution	50.0mL

pH 6.9 ± 0.2 at 25°C

Trace Elements Solution SL-4:

Composition per liter:

EDTA	0.5g
$FeSO_4 \cdot 7H_2O$	0.2g
Trace elements solution SL-6	100.0mL

Trace Elements Solution SL-6:

Composition per liter:

H_3BO_3	0.3g
$CoCl_2 \cdot 6H_2O$	0.2g
$ZnSO_4 \cdot 7H_2O$	0.1g
$MnCl_2 \cdot 4H_2O$	0.03g
$Na_2MoO_4 \cdot H_2O$	0.03g
$NiCl_2 \cdot 6H_2O$	0.02g
$CuCl_2 \cdot 2H_2O$	0.01g

Preparation of Trace Elements Solution SL-6: Add components to distilled/deionized water and bring volume to 1.0L. Mix thoroughly. Adjust pH to 3.4.

Preparation of Trace Elements Solution SL-4: Add components to distilled/deionized water and bring volume to 1.0L. Mix thoroughly.

Fluoranthene Solution:

Composition per liter:

Fluoranthene	2.0g

Preparation of Fluoranthene Solution: Add fluoranthene to 1.0L acetone. Mix thoroughly. Filter sterilize using a cellulose filter membrane.

Preparation of Medium: Add components, except fluoranthene solution, to 1.0L distilled/deionized water. Adjust pH to 6.9. Autoclave for 15 min at 15 psi pressure–121°C. Cool to room temperature. Add an aliquot of the fluoranthene solution to a sterile flask so that the final concentration will be 0.1g/L fluoranthene, and let the acetone evaporate. Aseptically add sterile medium to the crystal-layered flask.

Use: For the cultivation of fluoranthene-utilizing *Pseudomonas frederiksbergensis Sphingomonas* sp. *(Pseudomonas paucimobilis),* and other bacteria.

Medium G for Sulfate Reducers (Postgate's Medium G for Sulfate Reducers)

Composition per 1015.2mL:

Solution 1	970.0mL
Solution 4	30.0mL
Solution 8A, 8B, 8C, 8D, or 8E	10.0mL
Solution 5	3.0mL
Solution 2	1.0mL
Solution 3	1.0mL
Solution 6	0.1mL
Solution 7	0.1mL

pH 7.2 ± 0.2 at 25°C

Solution 1:

Composition per 970.0mL:

Na_2SO_4	3.0g
NaCl	1.2g
$MgCl_2·6H_2O$	0.4g
KCl	0.3g
NH_4Cl	0.3g
KH_2PO_4	0.2g
$CaCl_2·2H_2O$	0.15g

Preparation of Solution 1: Add components to distilled/deionized water and bring volume to 970.0mL. Mix thoroughly. Adjust pH to 7.2 with 2*N* HCl. Autoclave for 15 min at 15 psi pressure–121°C. Cool to 25°C.

Solution 2:

Composition per 10.0mL:

NaOH	5.0mg
Na_2SeO_3	0.03mg

Preparation of Solution 2: Add NaOH and Na_2SeO_3 to distilled/deionized water and bring volume to 10.0mL. Mix thoroughly. Autoclave for 15 min at 15 psi pressure–121°C. Cool to 25°C.

Solution 3:

Composition per liter:

$FeCl_2·4H_2O$	1.5g
$CoCl_2·6H_2O$	0.12g
$MnCl_2·4H_2O$	0.1g
$ZnCl_2$	0.07g
H_3BO_3	0.06g
$NiCl_2·6H_2O$	0.025g
$NaMoO_4·2H_2O$	0.025g
$CuCl_2·2H_2O$	0.015g

Preparation of Solution 3: Add components to distilled/deionized water and bring volume to 1.0L. Mix thoroughly. Autoclave for 15 min at 15 psi pressure–121°C. Cool to 25°C.

Solution 4:

Composition per 30.0mL:

$NaHCO_3$	2.55g

Preparation of Solution 4: Add $NaHCO_3$ to distilled/deionized water and bring volume to 30.0mL. Mix thoroughly. Gas with 100% CO_2 for 10–15 min. Filter sterilize.

Solution 5:

Composition per 3.0mL:

$Na_2S·9H_2O$	0.36g

Preparation of Solution 5: Add $Na_2S·9H_2O$ to distilled/deionized water and bring volume to 3.0mL. Mix thoroughly. Gas with 100% N_2 for 5–10 min. Cap tube with a rubber stopper. Autoclave for 15 min at 15 psi pressure–121°C. Cool to 25°C.

Solution 6:

Composition per 100.0mL:

Thiamine·HCl	0.01g
Cyanocobalamin	5.0mg
p-Aminobenzoic acid	5.0mg
Biotin	1.0mg

Preparation of Solution 6: Add components to distilled/deionized water and bring volume to 100.0mL. Mix thoroughly. Filter sterilize.

Solution 7:

Composition per 100.0mL:

Succinic acid	0.6g
Isobutyric acid	0.5g
Valeric acid	0.5g
2-Methylbutyric acid	0.5g
3-Methylbutyric acid	0.5g
Caproic acid	0.2g

Preparation of Solution 7: Add components to distilled/deionized water and bring volume to 100.0mL. Mix thoroughly. Adjust pH to 9.0 with NaOH. Autoclave for 15 min at 15 psi pressure–121°C. Cool to 25°C.

Solution 8A:

Composition per 100.0mL:

Sodium acetate·$3H_2O$	20.0g

Solution 8B:

Composition per 100.0mL:

Propionic acid	7.0g

Solution 8C:

Composition per 100.0mL:

n-Butyric acid	8.0g

Solution 8D:

Composition per 100.0mL:

Benzoic acid	5.0g

Solution 8E:

Composition per 100.0mL:

n-Palmitic acid	5.0g

Preparation of Solutions 8A–E: Add the appropriate amount of component to distilled/deionized water and bring volume to 100.0mL. Mix thoroughly. Adjust pH to 9.0 with NaOH. Autoclave for 15 min at 15 psi pressure–121°C. Cool to 25°C.

Preparation of Medium: To 970.0mL of cooled, sterile solution 1, aseptically add 1.0mL of sterile solution 2, 1.0mL of sterile solution 3, 30.0mL of sterile solution 4, 3.0mL of sterile solution 5, 0.1mL of sterile solution 6, 0.1mL of sterile solution 7, and 10.0mL of sterile solution 8A, 8B, 8C, 8D, or 8E. Mix thoroughly. Aseptically distribute into sterile tubes or flasks.

Use: For the isolation and cultivation of *Desulfovibrio baarsii, Desulfovibrio sapovorans, Desulfobacter* species, *Desulfonema* species, *Desulfobulbus* species, and *Desulfotomaculum acetoxidans.*

Medium for Halophilic Archaea (DSMZ Medium 1184)

Composition per liter:

NaCl	195.0g
$MgSO_4·7H_2O$	50.8g
$MgCl_2·6H_2O$	32.5g
Yeast extract	5.0g
KCl	5.0g
$CaCl_2·2H_2O$	0.8g
NaBr	0.6g
$NaHCO_3$	0.16g

pH 6.7 ± 0.3 at 25°C

Preparation of Medium: Add components to distilled/deionized water and bring volume to 1.0L. Mix thoroughly. Adjust pH to 6.5–7.0. Distribute into tubes or flasks. Gently heat while stirring and bring to boiling. Mix thoroughly. Autoclave for 15 min at 15 psi pressure–121°C.

Use: For the cultivation of halophilic archaea. For the cultivation of *Pycnoporus cinnabarinus* and *Natrinema ejinorense.*

Medium for Halophilic Bacilli

Composition per liter:

NaCl	100.0g
Casamino acids	10.0g
Yeast extract	10.0g

pH 7.0 ± 0.2 at 25°C

Preparation of Medium: Add components to distilled/deionized water and bring volume to 1.0L. Mix thoroughly. Distribute into tubes or flasks. Autoclave for 15 min at 15 psi pressure–121°C.

Use: For the cultivation of halophilic *Bacillus* species.

Medium for Hydrocarbon-Degrading Bacteria

Composition per 1020.0mL:

NH_4Cl	0.5g
$MgSO_4 \cdot 7H_2O$	0.5g
NaCl	0.4g
Hydrocarbon	20.0mL
KH_2PO_4 solution	0.5mL
$Na_2HPO_4 \cdot H_2O$ solution	0.5mL

KH_2PO_4 Solution:

Composition per 100.0mL:

KH_2PO_4	10.0g

Preparation of KH_2PO_4 Solution: Add KH_2PO_4 to distilled/deionized water and bring volume to 100.0mL. Mix thoroughly. Autoclave for 15 min at 15 psi pressure–121°C. Cool to 25°C.

$Na_2HPO_4 \cdot H_2O$ Solution:

Composition per 100.0mL:

$Na_2HPO_4 \cdot H_2O$	10.0g

Preparation of $Na_2HPO_4 \cdot H_2O$ Solution: Add $Na_2HPO_4 \cdot H_2O$ to distilled/deionized water and bring volume to 100.0mL. Mix thoroughly. Autoclave for 15 min at 15 psi pressure–121°C. Cool to 25°C.

Preparation of Medium: Add components—except hydrocarbon, KH_2PO_4 solution, and $Na_2HPO_4 \cdot H_2O$ solution—to distilled/deionized water and bring volume to 999.0mL. Mix thoroughly. Gently heat and bring to boiling. Autoclave for 15 min at 15 psi pressure–121°C. Cool to 45°–50°C. Aseptically add 0.5mL of sterile KH_2PO_4 solution and 0.5mL of the sterile $Na_2HPO_4 \cdot H_2O$ solution. Mix thoroughly. Aseptically distribute into sterile tubes in 10.0mL volumes. Add 0.2mL of sterile hydrocarbon to each tube.

Use: For the cultivation and enumeration of hydrocarbon-degrading bacteria in fresh water.

Medium for Hydrocarbon-Degrading Bacteria (Naphthalene Mineral Salts Medium)

Composition per liter:

K_2HPO_4	1.0g
$(NH_4)_2SO_4$	1.0g
$MgSO_4 \cdot 7H_2O$	0.3g
$CaCl_2$	0.1g
$FeSO_4 \cdot 7H_2O$	0.02g
Naphthalene	2.0mL

pH 7.0 ± 0.2 at 25°C

Preparation of Medium: Add components, except naphthalene, to distilled/deionized water and bring volume to 998.0mL. Mix thoroughly. Gently heat and bring to boiling. Autoclave for 15 min at 15 psi pressure–121°C. Cool to 45°–50°C. Aseptically add 2.0mL of sterile naphthalene to 20.0mL of sterile basal salts. Ultrasonically homogenize the solution. Add the naphthalene–basal salts homogenate back to the remainder of the sterile basal salts medium. Mix thoroughly. Aseptically distribute into sterile tubes or flasks.

Use: For the cultivation and enrichment of hydrocarbon-degrading bacteria.

Medium K

See: **Kievskaya Broth**

Medium K (DSMZ Medium 1122)

Composition per liter:

Agar	20.0g
$(NH_4)_2SO_4$	2.0g
KH_2PO_4	2.0g
NaCl	0.5g
$MgSO_4 \cdot 7H_2O$	0.125g
$FeSO_4 \cdot 7H_2O$	0.002g
Methanol, sterilized by filtration	10.0mL

pH 7.2 ± 0.2 at 25°C

Preparation of Medium: Add components, except methanol, to distilled/deionized water and bring volume to 990.0mL. Mix thoroughly. Adjust pH to 7.2 0. Gently heat while stirring and bring to boiling. Mix thoroughly. Autoclave for 15 min at 15 psi pressure–121°C. Cool to 50°C. Aseptically add 10.0 sterile methanol. Mix thoroughly. Pour into Petri dishes or aseptically distribute into sterile tubes.

Use: For the cultivation of *Methylobacillus pratensis* and *Methylovorus mays.*

Medium for Lactobacilli (ATCC Medium 980)

Composition per liter:

Agar	20.0g
Peptone	12.5g
Glucose	11.0g
Sodium acetate	10.0g
Yeast extract	5.5g
KH_2PO_4	0.25g
K_2HPO_4	0.25g
$MgSO_4$	0.1g
$MnSO_4 \cdot 4H_2O$	0.05g
$FeSO_4 \cdot 7H_2O$	0.05g

pH 6.8 ± 0.2 at 25°C

Preparation of Medium: Add components to distilled/deionized water and bring volume to 1.0L. Mix thoroughly. Adjust pH to 6.8. Autoclave for 10 min at 15 psi pressure–120°C. Aseptically distribute into sterile tubes or flasks.

Use: For the cultivation and maintenance of *Pediococcus acidilactici* and *Bacillus* species.

Medium 4 m 1

Composition per liter:

Agar	15.0g
Peptone	3.0g

Pancreatic digest of casein	3.0g
Yeast extract	3.0g
Maltose	2.0g
Lactose	1.0g
Sodium dichromate solution	100.0mL

pH 7.0 ± 0.2 at 25°C

Sodium Dichromate Solution:

Composition per 100.0mL:

Sodium dichromate	0.05g

Preparation of Sodium Dichromate Solution: Add sodium dichromate to distilled/deionized water and bring volume to 100.0mL. Mix thoroughly. Autoclave for 15 min at 15 psi pressure–121°C. Cool to 45°–50°C.

Preparation of Medium: Add components, except sodium dichromate solution, to distilled/deionized water and bring volume to 900.0mL. Mix thoroughly. Gently heat and bring to boiling. Adjust pH to 7.2. Autoclave for 15 min at 15 psi pressure–121°C. Cool to 45°–50°C. Aseptically add sterile sodium dichromate solution. Mix thoroughly. Pour into sterile Petri dishes or distribute into sterile tubes.

Use: For the isolation and cultivation of *Corynebacterium sepedonicum.*

Medium M71

Composition per liter:

Agar	20.0g
Peptone	10.0g
Glucose	5.0g
H_3BO_3	1.0g
Pancreatic digest of casein	1.0g
Cycloheximide	0.05g
2,3,5-Triphenyltetrazolium·HCl solution	10.0mL

Caution: Cycloheximide is toxic. Avoid skin contact or aerosol formation and inhalation.

2,3,5-Triphenyltetrazolium·HCl Solution:

Composition per 10.0mL:

2,3,5-Triphenyltetrazolium·HCl	0.05g

Preparation of 2,3,5-Triphenyltetrazolium·HCl Solution: Add 2,3,5-triphenyltetrazolium·HCl to distilled/deionized water and bring volume to 10.0mL. Mix thoroughly. Autoclave for 15 min at 15 psi pressure–121°C.

Preparation of Medium: Add components, except 2,3,5-triphenyltetrazolium·HCl solution, to distilled/deionized water and bring volume to 990.0mL. Mix thoroughly. Gently heat and bring to boiling. Autoclave for 15 min at 15 psi pressure–121°C. Cool to 45°–50°C. Aseptically add 10.0mL of sterile 2,3,5-triphenyltetrazolium·HCl solution. Mix thoroughly. Pour into sterile Petri dishes.

Use: For the selective isolation and cultivation of *Pseudomonas syringae.*

Medium 523M

Composition per liter:

Agar	15.0g
Sucrose	10.0g
Casamino acids	2.0g
K_2HPO_4	2.0g
Yeast extract	2.0g
$MgSO_4·7H_2O$	0.3g

Preparation of Medium: Add components to distilled/deionized water and bring volume to 1.0L. Mix thoroughly. Gently heat and bring to boiling. Distribute into tubes or flasks. Autoclave for 15 min at 15 psi pressure–121°C. Pour into sterile Petri dishes or leave in tubes.

Use: For the cultivation of *Clavibacter toxicus.*

Medium for Marine Flexibacteria

Composition per 1001.0mL:

Pancreatic digest of casein	5.0g
Yeast extract	5.0g
Tris (hydroxymethyl) amino methane	1.0g
KNO_3	0.5g
Sodium glycerophosphate	0.1g
Trace elements solution	1.0mL

pH 7.0 ± 0.2 at 25°C

Trace Elements Solution:

Composition per liter:

$ZnCl_2$	20.8g
H_3BO_3	2.85g
$MnCl_2·4H_2O$	1.8g
Sodium tartrate	1.77g
$FeSO_4$	1.36g
$CoCl_2·6H_2O$	40.4mg
$CuCl_2·2H_2O$	26.9mg
$Na_2MoO_4·2H_2O$	25.2mg

Preparation of Trace Elements Solution: Add components to distilled/deionized water and bring volume to 1.0L. Mix thoroughly. Autoclave for 15 min at 15 psi pressure–121°C.

Preparation of Medium: Add components, except trace elements solution, to distilled/deionized water and bring volume to 1.0L. Mix thoroughly. Autoclave for 15 min at 15 psi pressure–121°C. Aseptically add 1.0mL of sterile trace elements solution. Mix thoroughly. Aseptically distribute into sterile tubes or flasks.

Use: For the cultivation of *Cytophaga aprica, Cytophaga diffluens, Cytophaga johnsonae, Cytophaga lytica, Cytophaga* species, *Flexibacter aggregans, Flexibacter aurantiacus, Flexibacter litoralis, Flexibacter tractuosus, Flexithrix dorotheae, Herpetosiphon cohaerens, Herpetosiphon nigricans, Herpetosiphon persicus, Microscilla arenaria, Microscilla furvescens, Microscilla marina, Microscilla sericea,* and *Saprospira grandis.*

Medium for Marine Methylotrophs (DSMZ Medium 750)

Composition per liter:

NaCl	25.0g
Agar	20.0g
Peptone	10.0g
Beef extract	7.0g
K_2HPO_4	1.0g
$(NH_4)_2SO_4$	1.0g
Methanol	10.0mL

pH 7.0 ± 0.2 at 25°C

Preparation of Medium: Add components, except methanol, to distilled/deionized water and bring volume to 990.0mL. Mix thoroughly. Gently heat and bring to boiling. Autoclave for 15 min at 15 psi pressure–121°C. Cool to 50°C. Aseptically add 10.0mL filter sterilized methanol. Mix thoroughly. Pour into sterile Petri dishes or distribute into sterile tubes.

Use: For the cultivation and maintenance of *Methylophaga marina* and *Methylophaga thalassica.*

Medium for *Methylobacterium podarium* (DSMZ Medium 1032)

Composition per liter:

Agar	15.0g
$Na_2HPO_4 \cdot 2H_2O$	7.9g
KH_2PO_4	1.5g
NH_4Cl	0.8g
$MgSO_4 \cdot 7H_2O$	0.1g
Methylamine solution	30.0mL
Trace metal solution (Kelly solution T)	10.0mL

pH 7.3 ± 0.2 at 25°C

Methylamine Solution:

Composition per 10.0ml:

Methylamine	0.5g

Preparation of Methylamine Solution: Add components to distilled/deionized water and bring volume to 10.0mL. Mix thoroughly. Filter sterilize.

Trace Metal Solution (Kelly Solution T):

Composition per liter:

EDTA	50.0g
NaOH	9.0g
$CaCl_2 \cdot 2H_2O$	7.34g
$FeSO_4 \cdot 7H_2O$	5.0g
$MnCl_2 \cdot 4H_2O$	2.5g
$ZnSO_4 \cdot 7H_2O$	1.0g
$CoCl_2 \cdot 6H_2O$	0.5g
$(NH_4)_2MoO_4$	0.5g
$CuSO_4 \cdot 5H_2O$	0.2g

Preparation of Trace Metal Solution: Add EDTA to 400.0mL distilled/deionized water. Add NaOH with constant mixing. This is best done in a 1–2L beaker on a magnetic stirrer. Add the other salts individually to about 30-40mL water to dissolve before adding to the EDTA-NaOH solution. Allow each component to mix thoroughly before adding the next component. Adjust pH to 6.0 using 1*M* NaOH (approximately 24.0mL). Bring volume to 1.0L with distilled/deionized water. Filter sterilize. Do not autoclave! Store in a dark bottle.

Preparation of Medium: Add components, except trace metal solution and methylamine solution, to distilled/deionized water and bring volume to 960.0mL. Mix thoroughly. Adjust pH to 7.4. Gently heat while stirring and bring to boiling. Mix thoroughly. Autoclave for 10 min at 105 psi pressure–115°C. Cool to 50°C. Aseptically add methylamine solution and trace metal solution. Mix thoroughly. Pour into Petri dishes or aseptically distribute into sterile tubes.

Use: For the cultivation of *Methylobacterium podarium.*

Medium N

Composition per liter:

Agar	20.0g
Glucose	20.0g
Yeast nitrogen base without amino acids	6.7g
Casamino acids, vitamin free	2.0g
Isoleucine	0.1g
Valine	0.1g
Deoxythymidine-5′-monophosphate solution	10.0mL

Deoxythymidine-5′-Monophosphate Solution:

Composition per 10.0mL:

Deoxythymidine-5′-monophosphate	15.0mg

Preparation of Deoxythymidine-5′-Monophosphate Solution: Add deoxythymidine-5′-monophosphate to distilled/deionized water and bring volume to 10.0mL. Mix thoroughly. Filter sterilize.

Preparation of Medium: Add components, except deoxythymidine-5′-monophosphate solution, to distilled/deionized water and bring volume to 990.0mL. Mix thoroughly. Gently heat and bring to boiling. Autoclave for 15 min at 15 psi pressure–121°C. Cool to 60°C. Aseptically add 10.0mL of sterile deoxythymidine-5′-monophosphate solution. Mix thoroughly. Pour into sterile Petri dishes or distribute into sterile tubes.

Use: For the cultivation and maintenance of *Saccharomyces cerevisiae.*

Medium N for Sulfate Reducers (Postgate's Medium N for Sulfate Reducers)

Composition per liter:

$(NH_4)_2SO_4$	7.0g
Sodium lactate	6.0g
NH_4Cl	1.0g
Yeast extract	1.0g
KH_2PO_4	0.5g
Sodium citrate·$2H_2O$	0.3g
$FeSO_4 \cdot 7H_2O$	0.1g
$CaCl_2 \cdot 6H_2O$	0.06g
$MgSO_4 \cdot 7H_2O$	0.06g

pH 7.5 ± 0.2 at 25°C

Preparation of Medium: Add components to distilled/deionized water and bring volume to 1.0L. For marine bacteria, NaCl may be added or seawater used in place of distilled/deionized water. Mix thoroughly. Adjust pH to 7.5. Distribute into tubes or flasks. Autoclave for 15 min at 15 psi pressure–121°C.

Use: For the detection, culturing, and storage of *Desulfovibrio* species and many *Desulfotomaculum* species. This medium should be used when a clear culture medium is desired such as for chemostat culture. This medium may be cloudy after sterilization but usually clears on cooling. It turns black as a result of H_2S production due to bacterial growth.

Medium ND

See: **Castenholz ND Medium**

Medium for Nitrite Oxidizers

Composition per liter:

$KHCO_3$	1.5g
KH_2PO_4	0.5g
K_2HPO_4	0.5g
KNO_2	0.3g
$MgSO_4 \cdot 7H_2O$	0.2g
NaCl	0.2g
$CaCl_2 \cdot 2H_2O$	0.01g
$FeSO_4 \cdot 7H_2O$	0.01g

Preparation of Medium: Add components to distilled/deionized water and bring volume to 1.0L. Mix thoroughly. Distribute into tubes or flasks. Autoclave for 15 min at 15 psi pressure–121°C.

Use: For the isolation, cultivation, and enrichment of nitrate-oxidizing bacteria.

Medium for Nitrite Oxidizers, Marine

Composition per liter:

$MgSO_4 \cdot 7H_2O$	0.1g
$NaNO_2$	0.07g
$CaCl_2 \cdot 2H_2O$	6.0mg
K_2HPO_4	1.74mg
Chelated iron	1.0mg
$MnCl_2 \cdot 4H_2O$	66.0μg
$Na_2MoO_4 \cdot 2H_2O$	30.0μg
$ZnSO_4 \cdot 7H_2O$	30.0μg
$CuSO_4 \cdot 5H_2O$	6.0μg
$CoCl_2 \cdot 6H_2O$	0.6μg
Seawater	700.0mL

Preparation of Medium: Add components to distilled/deionized water and bring volume to 1.0L. Mix thoroughly. Distribute into tubes or flasks. Autoclave for 15 min at 15 psi pressure–121°C.

Use: For the isolation, cultivation, and enrichment of marine nitrate-oxidizing bacteria.

Medium for Osmophilic Fungi (M 40 Y)

Composition per liter:

Sucrose	400.0g
Agar	20.0g
Malt extract	20.0g
Yeast extract	5.0g

Preparation of Medium: Add components to distilled/deionized water and bring volume to 1.0L. Mix thoroughly. Gently heat and bring to boiling. Distribute into tubes or flasks. Autoclave for 15 min at 15 psi pressure–121°C. Pour into sterile Petri dishes or leave in tubes.

Use: For the cultivation of osmophilic fungi.

Medium with Phenanthrene (DSMZ Medium 457b)

Composition per liter:

Na_2HPO_4	2.44g
KH_2PO_4	1.52g
$(NH_4)_2SO_4$	0.5g
$MgSO_4 \cdot 7H_2O$	0.2g
Tween 80	0.2g
$CaCl_2 \cdot 2H_2O$	0.05g
Phenanthrene solution	50.0mL
Trace elements solution SL-4	10.0mL

pH 6.9 ± 0.2 at 25°C

Trace Elements Solution SL-4:

Composition per liter:

EDTA	0.5g
$FeSO_4 \cdot 7H_2O$	0.2g
Trace elements solution SL-6	100.0mL

Preparation of Trace Elements Solution SL-4: Add components to distilled/deionized water and bring volume to 1.0L. Mix thoroughly.

Trace Elements Solution SL-6:

Composition per liter:

H_3BO_3	0.3g
$CoCl_2 \cdot 6H_2O$	0.2g
$ZnSO_4 \cdot 7H_2O$	0.1g
$MnCl_2 \cdot 4H_2O$	0.03g
$Na_2MoO_4 \cdot H_2O$	0.03g
$NiCl_2 \cdot 6H_2O$	0.02g
$CuCl_2 \cdot 2H_2O$	0.01g

Preparation of Trace Elements Solution SL-6: Add components to distilled/deionized water and bring volume to 1.0L. Mix thoroughly. Adjust pH to 3.4.

Phenanthrene Solution:

Composition per liter:

Phenanthrene	2.0g

Preparation of Phenanthrene Solution: Add phenanthrene to 1.0L acetone. Mix thoroughly. Filter sterilize using a cellulose filter membrane.

Preparation of Medium: Add components, except phenanthrene solution, to 1.0L distilled/deionized water. Adjust pH to 6.9. Autoclave for 15 min at 15 psi pressure–121°C. Cool to room temperature. Add an aliquot of the phenanthrene solution to a sterile flask so that the final concentration will be 0.1g/L phenanthrene, and let the acetone evaporate. Aseptically add sterile medium to the crystal-layered flask.

Use: For the cultivation of phenanthrene-utilizing *Sphingomonas* sp. *(Pseudomonas paucimobilis), Pseudomonas frederiksbergensis,* and other bacteria.

Medium with Polyhydroxybutyric Acid as Carbon Source (DSMZ Medium 474)

Composition per liter:

Agar	16.0g
Na_2HPO_4	2.44g
KH_2PO_4	1.52g
$(NH_4)_2SO_4$	0.5g
$MgSO_4 \cdot 7H_2O$	0.2g
$CaCl_2 \cdot 2H_2O$	0.05g
PHB solution	66.0mL
Trace elements solution SL-4	10.0mL

pH 6.9 ± 0.2 at 25°C

Trace Elements Solution SL-4:

Composition per liter:

EDTA	0.5g
$FeSO_4 \cdot 7H_2O$	0.2g
Trace elements solution SL-6	100.0mL

Trace Elements Solution SL-6:

Composition per liter:

H_3BO_3	0.3g
$CoCl_2 \cdot 6H_2O$	0.2g
$ZnSO_4 \cdot 7H_2O$	0.1g
$MnCl_2 \cdot 4H_2O$	0.03g
$Na_2MoO_4 \cdot H_2O$	0.03g
$NiCl_2 \cdot 6H_2O$	0.02g
$CuCl_2 \cdot 2H_2O$	0.01g

Preparation of Trace Elements Solution SL-6: Add components to distilled/deionized water and bring volume to 1.0L. Mix thoroughly. Adjust pH to 3.4.

Preparation of Trace Elements Solution SL-4: Add components to distilled/deionized water and bring volume to 1.0L. Mix thoroughly.

PHB Solution:

Composition per 100.0mL:

Poly-ß-hydroxybutyric acid (PHB)	3.0g

Preparation of PHB Solution: Add poly-ß-hydroxybutyric acid (PHB) to 100.0mL distilled/deionized water. Stir overnight. Sonicate until a white homogenous suspension is obtained. Autoclave for 5 min at 15 psi pressure–121°C. Cool to room temperature.

Preparation of Medium: Add components, except PHB solution, to 1.0L distilled/deionized water. Adjust pH to 6.9. Autoclave for 15 min at 15 psi pressure–121°C. Cool to 50°C. Use 500.0mL to prepare bottom layer of a double agar plate by aseptically pouring 10.0mL amounts into sterile Petri dishes. Allow to solidify. Warm the PHB solution to 50°C. Aseptically add 33 mL of sterile PHB solution to the remaining 500.0mL of the medium. Mix thoroughly. Pour the PHB containing agar as a top layer over the solidified base agar.

Use: For the cultivation of *Comamonas testosteroni.*

Medium for *Prosthecomicrobium* and *Ancalomicrobium*

Composition per liter:

Agar	15.0g
Peptone	0.1g
Hutner's mineral base solution	20.0mL
Vitamin solution	10.0mL

Hutner's Mineral Base Solution:

Composition per liter:

$MgSO_4·7H_2O$	29.7g
Nitrilotriacetic acid	10.0g
$CaCl_2·2H_2O$	3.34g
$FeSO_4·7H_2O$	0.1g
$(NH_4)_2MoO_4$	9.25mg
Metals "44"	50.0mL

Preparation of Hutner's Mineral Base Solution: Add nitrilotriacetic acid to 500.0mL of distilled/deionized water. Dissolve by adjusting pH to 6.5 with KOH. Add remaining components. Add distilled/deionized water to 1.0L.

Metals "44":

Composition per 100.0mL:

$ZnSO_4·7H_2O$	1.1g
$FeSO_4·7H_2O$	0.5g
EDTA	0.25g
$MnSO_4·7H_2O$	0.154g
$CuSO_4·5H_2O$	0.04g
$Co(NO_3)_2·6H_2O$	0.025g
$Na_2B_4O_7·10H_2O$	0.018g

Preparation of Metals "44": Add components to distilled/deionized water and bring volume to 100.0mL. Mix thoroughly.

Vitamin Solution:

Composition per liter:

Pyridoxine·HCl	0.01g
Calcium pantothenate	5.0mg
Nicotinamide	5.0mg
Riboflavin	5.0mg
Thiamine HCl	5.0mg
Biotin	2.0mg
Folic acid	2.0mg
Vitamin B_{12}	0.1mg

Preparation of Vitamin Solution: Add components to distilled/deionized water and bring volume to 1.0L. Mix thoroughly. Filter sterilize.

Preparation of Medium: Add components, except vitamin solution, to distilled/deionized water and bring volume to 990.0mL. Mix thoroughly. Gently heat and bring to boiling. Autoclave for 15 min at 15 psi pressure–121°C. Cool to 45°–50°C. Aseptically add sterile vitamin solution. Mix thoroughly. Pour into sterile Petri dishes or distribute into sterile tubes.

Use: For the isolation of *Prosthecomicrobium* species and *Ancalomicrobium* species.

Medium for *Prosthecomicrobium* and *Ancalomicrobium*

Composition per liter:

$(NH_4)_2SO_4$	0.25g
Glucose	0.25g
Na_2HPO_4	0.071g
Modified Hutner's basal salts	20.0mL
Vitamin solution	10.0mL

pH 7.2 ± 0.2 at 25°C

Modified Hutner's Basal Salts:

Composition per liter:

$MgSO_4·7H_2O$	29.7g
Nitrilotriacetic acid	10.0g
$CaCl_2·2H_2O$	3.34g
$FeSO_4·7H_2O$	0.1g
$(NH_4)_2MoO_4$	9.25mg
Metals "44"	50.0mL

Preparation of Modified Hutner's Basal Salts: Add nitrilotriacetic acid to 500.0mL of distilled/deionized water. Dissolve by adjusting pH to 6.5 with KOH. Add remaining components. Readjust pH to 7.2 with H_2SO_4 or KOH. Add distilled/deionized water to 1.0L. Store at 5°C.

Metals "44":

Composition per 100.0mL:

$ZnSO_4·7H_2O$	1.1g
$FeSO_4·7H_2O$	0.5g
EDTA	0.25g
$MnSO_4·7H_2O$	0.154g
$CuSO_4·5H_2O$	0.04g
$Co(NO_3)_2·6H_2O$	0.025g
$Na_2B_4O_7·10H_2O$	0.018g

Preparation of Metals "44": Add a few drops of H_2SO_4 to distilled/deionized water to inhibit precipitate formation. Add components to acidified distilled/deionized water and bring volume to 100.0mL. Mix thoroughly.

Vitamin Solution:

Composition per liter:

Thiamine·HCl	5.0mg
D-Calcium pantothenate	5.0mg
Riboflavin	5.0mg
Biotin	2.0mg
Folic acid	2.0mg
Vitamin B_{12}	0.1mg

Preparation of Vitamin Solution: Add components to distilled/deionized water and bring volume to 1.0L. Mix thoroughly. Filter sterilize.

Preparation of Medium: Add components, except vitamin solution, to distilled deionized water and bring volume to 990.0mL. Mix thoroughly. Autoclave for 15 min at 15 psi pressure–121°C. Cool to room temperature. Aseptically add 10.0mL of sterile vitamin solution. Mix thoroughly. Aseptically distribute into sterile tubes or flasks.

Use: For the cultivation and maintenance of *Prosthecomicrobium enhydrum*, *Prosthecomicrobium pneumaticum*, and *Ancalomicrobium* species.

Medium for *Prosthecomicrobium* and *Ancalomicrobium*, Modified

Composition per liter:

Agar	15.0g
Glucose	1.0g
$(NH_4)_2SO_4$	0.25g
Peptone	0.15g
Yeast extract	0.15g
Modified Hutner's basal salts	20.0mL
Vitamin solution	10.0mL

Modified Hutner's Basal Salts:

Composition per liter:

$MgSO_4 \cdot 7H_2O$	29.7g
Nitrilotriacetic acid	10.0g
$CaCl_2 \cdot 2H_2O$	3.34g
$FeSO_4 \cdot 7H_2O$	0.1g
$(NH_4)_2MoO_4$	9.25mg
Metals "44"	50.0mL

Preparation of Modified Hutner's Basal Salts: Add nitrilotriacetic acid to 500.0mL of distilled/deionized water. Dissolve by adjusting pH to 6.5 with KOH. Add remaining components. Readjust pH to 7.2 with H_2SO_4 or KOH. Add distilled/deionized water to 1.0L. Store at 5°C.

Metals "44":

Composition per 100.0mL:

$ZnSO_4 \cdot 7H_2O$	1.1g
$FeSO_4 \cdot 7H_2O$	0.5g
EDTA	0.25g
$MnSO_4 \cdot 7H_2O$	0.154g
$CuSO_4 \cdot 5H_2O$	0.04g
$Co(NO_3)_2 \cdot 6H_2O$	0.025g
$Na_2B_4O_7 \cdot 10H_2O$	0.018g

Preparation of Metals "44": Add a few drops of H_2SO_4 to distilled/deionized water to inhibit precipitate formation. Add components to acidified distilled/deionized water and bring volume to 100.0mL. Mix thoroughly.

Vitamin Solution:

Composition per liter:

Thiamine·HCl	5.0mg
D-Calcium pantothenate	5.0mg
Riboflavin	5.0mg
Biotin	2.0mg
Folic acid	2.0mg
Vitamin B_{12}	0.1mg

Preparation of Vitamin Solution: Add components to distilled/deionized water and bring volume to 1.0L. Mix thoroughly. Filter sterilize.

Preparation of Medium: Add components, except vitamin solution, to distilled deionized water and bring volume to 990.0mL. Mix thoroughly. Autoclave for 15 min at 15 psi pressure–121°C. Cool to room temperature. Aseptically add 10.0mL of sterile vitamin solution. Mix thoroughly. Aseptically distribute into sterile tubes or flasks.

Use: For the cultivation and maintenance of *Ancalomicrobium adetum*, *Prosthecomicrobium hirschii*, and *Prosthecomicrobium* species.

Medium for *Prosthecomicrobium* and *Ancalomicrobium* with Nicotinamide

Composition per liter:

$(NH_4)_2SO_4$	0.25g
Glucose	0.25g
Na_2HPO_4	0.071g
Modified Hutner's basal salts	20.0mL
Vitamin solution	10.0mL

pH 7.2 ± 0.2 at 25°C

Modified Hutner's Basal Salts:

Composition per liter:

$MgSO_4 \cdot 7H_2O$	29.7g
Nitrilotriacetic acid	10.0g
$CaCl_2 \cdot 2H_2O$	3.34g
$FeSO_4 \cdot 7H_2O$	0.1g
$(NH_4)_2MoO_4$	9.25mg
Metals "44"	50.0mL

Preparation of Modified Hutner's Basal Salts: Add nitrilotriacetic acid to 500.0mL of distilled/deionized water. Dissolve by adjusting pH to 6.5 with KOH. Add remaining components. Readjust pH to 7.2 with H_2SO_4 or KOH. Add distilled/deionized water to 1.0L. Store at 5°C.

Metals "44":

Composition per 100.0mL:

$ZnSO_4 \cdot 7H_2O$	1.1g
$FeSO_4 \cdot 7H_2O$	0.5g
EDTA	0.25g
$MnSO_4 \cdot 7H_2O$	0.154g
$CuSO_4 \cdot 5H_2O$	0.04g
$Co(NO_3)_2 \cdot 6H_2O$	0.025g
$Na_2B_4O_7 \cdot 10H_2O$	0.018g

Preparation of Metals "44": Add a few drops of H_2SO_4 to distilled/deionized water to inhibit precipitate formation. Add components to acidified distilled/deionized water and bring volume to 100.0mL. Mix thoroughly.

Vitamin Solution:

Composition per liter:

Thiamine·HCl	5.0mg
D-Calcium pantothenate	5.0mg
Riboflavin	5.0mg
Nicotinamide	5.0mg
Biotin	2.0mg
Folic acid	2.0mg
Vitamin B_{12}	0.1mg

Preparation of Vitamin Solution: Add components to distilled/deionized water and bring volume to 1.0L. Mix thoroughly. Filter sterilize.

Preparation of Medium: Add components, except vitamin solution, to distilled/deionized water and bring volume to 990.0mL. Mix thoroughly. Autoclave for 15 min at 15 psi pressure–121°C. Cool to room temperature. Aseptically add 10.0mL of sterile vitamin solution. Mix thoroughly. Aseptically distribute into sterile tubes or flasks.

Use: For the cultivation and maintenance of *Ancalomicrobium adetum* and *Prosthecomicrobium* species.

Medium R

Composition per liter:

$Na_2S_2O_3{\cdot}5H_2O$	5.0g
KNO_3	2.0g
$MgCl_2{\cdot}6H_2O$	0.5g
NH_4Cl	0.5g
KH_2PO_4 solution	10.0mL
$NaHCO_3$ solution	10.0mL
$FeSO_4{\cdot}7H_2O$ solution	10.0mL

pH 7.0 ± 0.2 at 25°C

KH_2PO_4 Solution:

Composition per 10.0mL:

KH_2PO_4	2.0g

Preparation of KH_2PO_4 Solution: Add KH_2PO_4 to distilled/deionized water and bring volume to 10.0mL. Mix thoroughly. Filter sterilize.

$NaHCO_3$ Solution:

Composition per 10.0mL:

$NaHCO_3$	1.0g

Preparation of $NaHCO_3$ Solution: Add the $NaHCO_3$ to distilled/deionized water and bring volume to 10.0mL. Mix thoroughly. Filter sterilize.

$FeSO_4{\cdot}7H_2O$ Solution:

Composition per 10.0mL:

$FeSO_4{\cdot}7H_2O$	10.0mg

Preparation of $FeSO_4{\cdot}7H_2O$ Solution: Add the $FeSO_4{\cdot}7H_2O$ to distilled/deionized water and bring volume to 10.0mL. Mix thoroughly. Filter sterilize.

Preparation of Medium: Add components—except KH_2PO_4 solution, $NaHCO_3$ solution, and $FeSO_4{\cdot}7H_2O$ solution—to tap water and bring volume to 970.0mL. Mix thoroughly. Gently heat until dissolved. Adjust pH to 7.0. Autoclave for 15 min at 15 psi pressure–121°C. Cool to 45°–50°C. Aseptically add 10.0mL of sterile KH_2PO_4 solution, 10.0mL of $NaHCO_3$ solution, and 10.0mL of $FeSO_4{\cdot}7H_2O$ solution. Mix thoroughly. Aseptically distribute into sterile tubes or flasks.

Use: For the cultivation of *Thiobacillus denitrificans.*

Medium for *Roseospira* (DSMZ Medium 998)

Composition per liter:

NaCl	20.0g
$MgCl_2{\cdot}6H_2O$	1.0g
$MgSO_4{\cdot}7H_2O$	0.25g
NH_4Cl	0.5g
Yeast extract	0.5g
KH_2PO_4	0.3g
$CaCl_2{\cdot}2H_2O$	0.05g
$NaHCO_3$ soltuion	10.0mL
Acetate solution	10.0mL
Succinate solution	10.0mL
Trace elements solution SL-12	1.0mL
Vitamin V7 solution	1.0mL

pH 6.9 ± 0.2 at 25°C

Acetate Solution:

Composition per 10.0mL:

Sodium acetate	0.41g

Preparation of Acetate Solution: Add sodium acetate to distilled/deionized water and bring volume to 10.0mL. Mix thoroughly. Sparge with 20% CO_2 + 80% N_2. Filter sterilize.

Succinate Solution:

Composition per 10.0mL:

Sodium succinate	0.85g

Preparation of Succinate Solution: Add sodium succinate to distilled/deionized water and bring volume to 10.0mL. Mix thoroughly. Sparge with 20% CO_2 + 80% N_2. Filter sterilize.

$NaHCO_3$ Solution:

Composition per 10.0mL:

$NaHCO_3$	1.5g

Preparation of $NaHCO_3$ Solution: Add $NaHCO_3$ to distilled/deionized water and bring volume to 10.0mL. Mix thoroughly. Sparge with 20% CO_2 + 80% N_2. Filter sterilize.

Vitamin Solution V7:

Composition per liter:

Pyridoxine-HCl	50.0mg
Nicotinic acid	20.0mg
Vitamin B_{12}	20.0mg
Thiamine-HCl·$2H_2O$	10.0mg
p-Aminobenzoic acid	10.0mg
D-Ca-pantothenate	5.0mg
Biotin	2.0mg

Preparation of Vitamin Solution V7: Add components to distilled/deionized water and bring volume to 1.0L. Mix thoroughly. Filter sterilize.

Trace Elements Solution SL-12:

Composition per liter:

$FeSO_4{\cdot}7H_2O$	1.1g
H_3BO_3	0.3g
$CoCl_2{\cdot}6H_2O$	0.19g
$MnCl_2{\cdot}2H_2O$	0.05g
$ZnCl_2$	42.0mg
$NiCl_2{\cdot}6H_2O$	24.0mg
$Na_2MoO_4{\cdot}4H_2O$	18.0mg
$CuCl_2{\cdot}2H_2O$	2.0mg

Preparation of Trace Elements Solution Sl-12: Add components to distilled/deionized water and bring volume to 1.0L. Mix thoroughly. Filter sterilize.

Preparation of Medium: Add components, except bicarbonate, vitamin, acetate, and succinate solutions, to distilled/deionized water and bring volume to 970.0mL. Mix thoroughly. Gently heat while stirring and bring to boiling. Boil for 1 min. Cool to room temperature while sparging with 90% N_2 + 10% CO_2 gas. Autoclave for 15 min at 15 psi pressure–121°C. Aseptically add bicarbonate and vitamin solutions. Mix thoroughly. Adjust pH to 6.9. Distribute into sterile 50mL screw-capped bottles. Add the organic acetate and succinate substrates.

Use: For the cultivation of *Roseospira* spp.

Medium for *Roseospira*

Composition per liter:

NaCl	20.0g
$MgCl_2{\cdot}6H_2O$	1.0g

$MgSO_4 \cdot 7H_2O$	0.25g
NH_4Cl	0.5g
Yeast extract	0.5g
KH_2PO_4	0.3g
$CaCl_2 \cdot 2H_2O$	0.05g
$NaHCO_3$ solution	10.0mL
Acetate solution	10.0mL
Succinate solution	10.0mL
$Na_2S \cdot 9H_2O$ solution	10.0mL
Trace elements solution SL-12	1.0mL
Vitamin V7 solution	1.0mL

pH 6.9 ± 0.2 at 25°C

Acetate Solution:

Composition per 10.0mL:

Sodium acetate	0.41g

Preparation of Acetate Solution: Add sodium acetate to distilled/deionized water and bring volume to 10.0mL. Mix thoroughly. Sparge with 20% CO_2 + 80% N_2. Filter sterilize.

Succinate Solution:

Composition per 10.0mL:

Sodium succinate	0.85g

Preparation of Succinate Solution: Add sodium succinate to distilled/deionized water and bring volume to 10.0mL. Mix thoroughly. Sparge with 20% CO_2 + 80% N_2. Filter sterilize.

$NaHCO_3$ Solution:

Composition per 10.0mL:

$NaHCO_3$	1.5g

Preparation of $NaHCO_3$ Solution: Add $NaHCO_3$ to distilled/deionized water and bring volume to 10.0mL. Mix thoroughly. Sparge with 20% CO_2 + 80% N_2. Filter sterilize.

Vitamin Solution V7:

Composition per liter:

Pyridoxine-HCl	50.0mg
Nicotinic acid	20.0mg
Vitamin B_{12}	20.0mg
Thiamine-HCl·$2H_2O$	10.0mg
p-Aminobenzoic acid	10.0mg
D-Ca-pantothenate	5.0mg
Biotin	2.0mg

Preparation of Vitamin Solution V7: Add components to distilled/deionized water and bring volume to 1.0L. Mix thoroughly. Filter sterilize.

Trace Elements Solution SL-12:

Composition per liter:

$FeSO_4 \cdot 7H_2O$	1.1g
H_3BO_3	0.3g
$CoCl_2 \cdot 6H_2O$	0.19g
$MnCl_2 \cdot 2H_2O$	0.05g
$ZnCl_2$	42.0mg
$NiCl_2 \cdot 6H_2O$	24.0mg
$Na_2MoO_4 \cdot 4H_2O$	18.0mg
$CuCl_2 \cdot 2H_2O$	2.0mg

Preparation of Trace Elements Solution Sl-12: Add components to distilled/deionized water and bring volume to 1.0L. Mix thoroughly. Filter sterilize.

$Na_2S \cdot 9H_2O$ Solution:

Composition per 10.0mL:

$Na_2S \cdot 9H_2O$	0.2g

Preparation of $Na_2S \cdot 9H_2O$ Solution: Add $Na_2S \cdot 9H_2O$ to distilled/deionized water and bring volume to 10.0mL. Mix thoroughly. Autoclave under 100% N_2 for 15 min at 15 psi pressure–121°C. Cool to room temperature.

Preparation of Medium: Add components, except sulfide, bicarbonate, vitamin, acetate, and succinate solutions, to distilled/deionized water and bring volume to 970.0mL. Mix thoroughly. Gently heat while stirring and bring to boiling. Boil for 1 min. Cool to room temperature while sparging with 90% N_2 + 10% CO_2 gas. Autoclave for 15 min at 15 psi pressure–121°C. Aseptically add bicarbonate and vitamin solutions. Mix thoroughly. Adjust pH to 6.9. Distribute into sterile 50mL screw-capped bottles. Add sulfide and organic acetate and succinate substrates.

Use: For the cultivation of *Roseospira navarrensis.*

Medium S

Composition per liter:

$Na_2S_2O_3 \cdot 5H_2O$	5.0g
$(NH_4)_2SO_4$	4.0g
KH_2PO_4	4.0g
$MgSO_4$	0.5g
$CaCl_2$	0.25g
$FeSO_4$	0.01g

Preparation of Medium: Add components to distilled/deionized water and bring volume to 1.0L. Mix thoroughly. Distribute into tubes or flasks. Autoclave for 15 min at 15 psi pressure–121°C.

Use: For the cultivation of *Thiobacillus* species.

Medium S

Composition per liter:

Glucose	10.0g
K_2HPO_4	4.0g
Peptone	4.0g
Yeast extract	4.0g
KH_2PO_4	2.0g
$MgSO_4 \cdot 7H_2O$	0.5g

pH 7.7 ± 0.2 at 25°C

Preparation of Medium: Add components to distilled/deionized water and bring volume to 1.0L. Mix thoroughly. Distribute into tubes or flasks. Autoclave for 15 min at 15 psi pressure–121°C.

Use: For the general cultivation of a wide variety of bacteria.

Medium SP 4

Composition per liter:

Pancreatic digest of casein	11.0g
Peptone	5.3g
Glucose	5.0g
NaCl	0.875g
Beef extract	0.525g
Yeast extract	0.525g
Beef heart, solids from infusion	0.35g
Fetal bovine serum, heat inactivated	170.0mL
Yeast extract solution	100.0mL
CMRL 1066, 10X solution	50.0mL
Fresh yeast extract solution	35.0mL

Phenol Red solution	20.0mL
Penicillin solution	10.0mL

pH 7.6 ± 0.2 at 25°C

Yeast Extract Solution:

Composition per 100.0mL:

Yeast extract	2.0g

Preparation of Yeast Extract Solution: Add yeast extract to distilled/deionized water and bring volume to 100.0mL. Mix thoroughly. Autoclave for 15 min at 15 psi pressure–121°C.

CMRL 1066, 10X Solution:

Composition per liter:

NaCl	6.8g
$NaHCO_3$	2.2g
D-Glucose	1.0g
KCl	0.4g
L-Cysteine·HCl·H_2O	0.26g
$CaCl_2$, anhydrous	0.2g
$MgSO_4 \cdot 7H_2O$	0.2g
$NaH_2PO_4 \cdot H_2O$	0.14g
L-Glutamine	0.1g
Sodium acetate·$3H_2O$	0.083g
L-Glutamic acid	0.075g
L-Arginine·HCl	0.07g
L-Lysine·HCl	0.07g
L-Leucine	0.06g
Glycine	0.05g
Ascorbic acid	0.05g
L-Proline	0.04g
L-Tyrosine	0.04g
L-Aspartic acid	0.03g
L-Threonine	0.03g
L-Alanine	0.025g
L-Phenylalanine	0.025g
L-Serine	0.025g
L-Valine	0.025g
L-Cystine	0.02g
L-Histidine·HCl·H_2O	0.02g
L-Isoleucine	0.02g
Phenol Red	0.02g
L-Methionine	0.015g
Deoxyadenosine	0.01g
Deoxycytidine	0.01g
Deoxyguanosine	0.01g
Glutathione, reduced	0.01g
Thymidine	0.01g
Hydroxy-L-proline	0.01g
L-Tryptophan	0.01g
Nicotinamide adenine dinucleotide	7.0mg
Tween™ 80	5.0mg
Sodium glucoronate·H_2O	4.2mg
Coenzyme A	2.5mg
Cocarboxylase	1.0mg
Flavin adenine dinucleotide	1.0mg
Nicotinamide adenine dinucleotide phosphate	1.0mg
Uridine triphosphate	1.0mg
Choline chloride	0.5mg
Cholesterol	0.2mg
5-Methyldeoxycytidine	0.1mg
Inositol	0.05mg
p-Aminobenzoic acid	0.05mg
Niacin	0.025mg
Niacinamide	0.025mg
Pyridoxine	0.025mg
Pyridoxal·HCl	0.025mg
Biotin	0.01mg
D-Calcium pantothenate	0.01mg
Folic acid	0.01mg
Riboflavin	0.01mg
Thiamine·HCl	0.01mg

Source: CMRL 1066, 10X medium is available as a premixed powder from BD Diagnostics.

Preparation of CMRL 1066, 10X Solution: Add components to distilled/deionized water and bring volume to 1.0L. Mix thoroughly. Adjust pH to 7.2. Filter sterilize.

Fresh Yeast Extract Solution:

Composition per 100.0mL:

Baker's yeast, live, pressed, starch-free	25.0g

Preparation of Fresh Yeast Extract Solution: Add the live Baker's yeast to 100.0mL of distilled/deionized water. Autoclave for 90 min at 15 psi pressure–121°C. Allow to stand. Remove supernatant solution. Adjust pH to 6.6–6.8. Filter sterilize.

Phenol Red Solution:

Composition per 100.0mL:

Phenol Red	0.01g

Preparation of Phenol Red Solution: Add Phenol Red to distilled/deionized water and bring volume to 10.0mL. Mix thoroughly. Filter sterilize.

Penicillin Solution:

Composition per 10.0mL:

Penicillin	1,000,000U

Preparation of Penicillin Solution: Add penicillin to distilled/deionized water and bring volume to 10.0mL. Filter sterilize.

Preparation of Medium: Add components—except fetal bovine serum, yeast extract solution, CMRL 1066, 10X solution, fresh yeast extract solution, Phenol Red solution, and penicillin solution—to distilled/deionized water and bring volume to 615.0mL. Mix thoroughly. Gently heat and bring to boiling. Autoclave for 15 min at 15 psi pressure–121°C. Cool to 45°–50°C. Aseptically add 170.0mL of sterile fetal bovine serum, 100.0mL of sterile yeast extract solution, 50.0mL of sterile CMRL 1066, 10X solution, 35.0mL of sterile fresh yeast extract solution, 20.0mL of sterile Phenol Red solution, and 10.0mL of sterile penicillin solution. Mix thoroughly. Aseptically distribute into sterile tubes or flasks.

Use: For the isolation and cultivation of *Spiroplasma* species from ticks.

Medium for Sulfate Reducers (ATCC Medium 1282)

Composition per 1050.0mL:

Modified Baar's medium for sulfate reducers	1020.0mL
Organic acid solution	10.0mL
Wolfe's vitamin solution	10.0mL
Wolfe's mineral solution	10.0mL

pH 7.5 ± 0.2 at 25°C

Modified Baar's Medium for Sulfate Reducers:
Composition per 1020.0mL:

Component I	400.0mL
Component III	400.0mL
Component II	200.0mL
$Fe(NH_4)_2(SO_4)_2$ (5% solution)	20.0mL

Component I:
Composition per 400.0mL:

Sodium citrate	5.0g
$MgSO_4$	2.0g
$CaSO_4$	1.0g
NH_4Cl	1.0g

Preparation of Component I: Add components to distilled/deionized water and bring volume to 400.0mL. Mix thoroughly. Adjust pH to 7.5. Autoclave for 15 min at 15 psi pressure–121°C.

Component II:
Composition per 200.0mL:

K_2HPO_4	0.5g

Preparation of Component II: Add K_2HPO_4 to distilled/deionized water and bring volume to 200.0mL. Mix thoroughly. Adjust pH to 7.5. Autoclave for 15 min at 15 psi pressure–121°C.

Component III:
Composition per 400.0mL:

Sodium lactate	3.5g
Yeast extract	1.0g

Preparation of Component III: Add components to distilled/deionized water and bring volume to 400.0mL. Mix thoroughly. Adjust pH to 7.5. Autoclave for 15 min at 15 psi pressure–121°C.

Preparation of Modified Baar's Medium for Sulfate Reducers: Aseptically combine the three sterile solutions, except the $Fe(NH_4)_2(SO_4)_2$ solution. Mix thoroughly. Distribute 5.0mL volumes into tubes under 97% N_2 + 3% H_2. Add medium to tubes while still warm to exclude as much O_2 as possible. Prepare a 5% solution of ferrous ammonium sulfate, $Fe(NH_4)_2(SO_4)_2$. Sterilize by filtration. Add 0.2mL of sterile $Fe(NH_4)_2(SO_4)_2$ solution to 10.0mL of medium immediately prior to inoculation.

Organic Acid Solution:
Composition per 100.0mL:

Butyric acid	5.18mL
Caproic acid	2.4mL
Octanoic acid	1.25mL

Preparation of Organic Acid Solution: Add components to distilled/deionized water and bring volume to 75.0mL. Adjust pH to 7.0 with 5*N* NaOH. Bring volume to 100.0mL with distilled/deionized water. Filter sterilize.

Wolfe's Vitamin Solution:
Composition per liter:

Pyridoxine·HCl	10.0mg
Thiamine·HCl	5.0mg
Riboflavin	5.0mg
Nicotinic acid	5.0mg
Calcium pantothenate	5.0mg
p-Aminobenzoic acid	5.0mg
Thioctic acid	5.0mg
Biotin	2.0mg
Folic acid	2.0mg
Cyanocobalamin	100.0μg

Preparation of Wolfe's Vitamin Solution: Add components to distilled/deionized water and bring volume to 1.0L. Mix thoroughly. Filter sterilize.

Wolfe's Mineral Solution:
Composition per liter:

$MgSO_4 \cdot 7H_2O$	3.0g
Nitrilotriacetic acid	1.5g
NaCl	1.0g
$MnSO_4 \cdot H_2O$	0.5g
$FeSO_4 \cdot 7H_2O$	0.1g
$CoCl_2 \cdot 6H_2O$	0.1g
$CaCl_2$	0.1g
$ZnSO_4 \cdot 7H_2O$	0.1g
$CuSO_4 \cdot 5H_2O$	0.01g
$AlK(SO_4)_2 \cdot 12H_2O$	0.01g
H_3BO_3	0.01g
$Na_2MoO_4 \cdot 2H_2O$	0.01g

Preparation of Wolfe's Mineral Solution: Add nitrilotriacetic acid to 500.0mL of distilled/deionized water. Dissolve by adjusting pH to 6.5 with KOH. Add remaining components. Add distilled/deionized water to 1.0L. Filter sterilize.

Preparation of Medium: To each test tube containing 10.0mL of modified Baar's medium for sulfate reducers, aseptically add 0.1mL of sterile organic acid solution, 0.1mL of sterile Wolfe's vitamin solution, and 0.1mL of sterile Wolfe's mineral solution immediately prior to inoculation.

Use: For the cultivation and maintenance of *Desulfotomaculum thermobenzoicum* and *Desulfovibrio sapovorans.*

Medium for Sulfate Reducers (Postgate's Medium for Sulfate Reducers) (ATCC Medium 1283)

Composition per liter:

Part A	869.0mL
Part C	100.0mL
Part D	10.0mL
Part E	10.0mL
Part F	10.0mL
Part B	1.0mL

pH 7.7 ± 0.2 at 25°C

Part A:
Composition per 869.0mL:

Na_2SO_4	3.0g
NaCl	1.0g
KCl	0.5g
$MgCl_2 \cdot 6H_2O$	0.4g
NH_4Cl	0.3g
KH_2PO_4	0.2g
$CaCl_2 \cdot 2H_2O$	0.15g

Preparation of Part A: Add components to distilled/deionized water and bring volume to 869.0mL. Mix thoroughly. Prepare and autoclave part A under 90% N_2 + 10% CO_2. Autoclave for 15 min at 15 psi pressure–121°C. Cool to room temperature.

Part B:
Composition per liter:

$FeCl_2 \cdot 4H_2O$	1.5g
$CoCl_2 \cdot 6H_2O$	0.19g
$MnCl_2 \cdot 4H_2O$	0.1g

$ZnCl_2$ 0.07g
H_3BO_3 .... 0.06g
$Na_2MoO_4·2H_2O$ 0.04g
$NiCl_2·6H_2O$ 0.02g
$CuCl_2·2H_2O$ 0.02g
HCl, 25% 10.0mL

Preparation of Part B: Add the $FeCl_2·4H_2O$ to the HCl. Add distilled/deionized water and bring volume to 1.0L. Add remaining components. Mix thoroughly. Autoclave under 100% N_2 for 15 min at 15 psi pressure–121°C. Cool to room temperature.

Part C:
Composition per 100.0mL:
$NaHCO_3$ 5.0g

Preparation of Part C: Add the $NaHCO_3$ to distilled/deionized water and bring volume to 100.0mL. Mix thoroughly. Filter sterilize. Gas with 90% N_2 + 10% CO_2 to remove residual O_2.

Part D:
Composition per 10.0mL:
Sodium butyrate 0.7g
Sodium caproate 0.3g
Sodium octanoate 0.15g

Preparation of Part D: Add components to distilled/deionized water and bring volume to 10.0mL. Mix thoroughly. Autoclave under 100% N_2 for 15 min at 15 psi pressure–121°C. Cool to room temperature.

Part E:
Composition per 10.0mL:
Yeast extract 1.0g
Thiamine·HCl 100.0μg
p-Aminobenzoic acid 40.0μg
D(+)-Biotin 10.0μg

Preparation of Part E: Add components to distilled/deionized water and bring volume to 10.0mL. Mix thoroughly. Autoclave under 100% N_2 for 15 min at 15 psi pressure–121°C. Cool to room temperature.

Part F:
Composition per 10.0mL:
$Na_2S·9H_2O$ 0.4g

Preparation of Part F: Add $Na_2S·9H_2O$ to distilled/deionized water and bring volume to 10.0mL. Mix thoroughly. Autoclave under 100% N_2 for 15 min at 15 psi pressure–121°C. Cool to room temperature.

Preparation of Medium: To 869.0mL of sterile cooled part A, aseptically add the remaining sterile solutions in the following order: part B, part C, part D, part E, and part F. Mix thoroughly. Adjust pH to 7.7. Anaerobically distribute under 80% N_2 + 20% CO_2 into sterile tubes or flasks.

Use: For the cultivation and maintenance of *Desulfovibrio baarsii* and *Desulfovibrio sapovorans*.

Medium for Thermophilic Actinomycetes

Composition per liter:
Agar 20.0g
Soluble starch 10.0g
Maize extract 5.0g
NaCl 5.0g
Peptone 5.0g
$CaCl_2$ 0.5g

pH 6.5 ± 0.2 at 25°C

Preparation of Medium: Add components to distilled/deionized water and bring volume to 1.0L. Mix thoroughly. Gently heat and bring to boiling. Autoclave for 15 min at 15 psi pressure–121°C. Pour into sterile Petri dishes or distribute into sterile tubes.

Use: For the cultivation of thermophilic actinomycetes.

Medium for *Treponema pectinovorum*

Composition per liter:
Polypeptone™ 5.0g
Heart infusion broth 5.0g
Yeast extract 5.0g
NaCl 5.0g
K_2HPO_4 .... 2.0g
$(NH_4)_2SO_4$ 2.0g
Agar 1.0g
Pectin 0.8g
L-Cysteine·HCl·H_2O 0.68g
Rumen fluid 500.0mL
Resazurin (25.0 mg/100.0mL water) 4.0mL

pH 7.0–7.2 at 25°C

Preparation of Medium: Add components to distilled/deionized water and bring volume to 1.0L. Prepare and distribute anaerobically under 90% N_2 + 10% CO_2. Mix thoroughly. Adjust pH to 7.0–7.2. Distribute into screw-capped tubes or flasks. Autoclave for 15 min at 15 psi pressure–121°C.

Use: For the cultivation and maintenance of *Treponema pectinovorum.*

Medium for *Ureaplasma*
See: **B Broth**

Medium VTY

Composition per 100.0mL:
Peptone 1.0g
Noble agar 0.7g
Yeast extract 0.5g
L-Cysteine·HCl·H_2O 0.1g
Salts A 20.0mL
Salts B 20.0mL
Glucose solution 5.0mL
$NaHCO_3$ (5% solution) 1.0mL
Hemin solution 1.0mL
Volatile fatty acid solution 0.31mL
Resazurin (0.1% solution) 0.1mL

pH 7.2 ± 0.2 at 25°C

$NaHCO_3$ Solution:
Composition per 10.0mL:
$NaHCO_3$ 0.5g

Preparation of $NaHCO_3$ Solution: Add $NaHCO_3$ to distilled/deionized water and bring volume to 10.0mL. Mix thoroughly. Sparge with 100% N_2. Filter sterilize.

Glucose Solution:
Composition per 10.0mL:
Glucose 1.0g

Preparation of Glucose Solution: Add glucose to distilled/deionized water and bring volume to 10.0mL. Mix thoroughly. Filter sterilize.

Salts A:

Composition per liter:

$CaCl_2·2H_2O$	0.6g
$MgSO_4$	0.45g

Preparation of Salts A: Add components to distilled/deionized water and bring volume to 1.0L. Mix thoroughly.

Salts B:

Composition per liter:

NaCl	4.5g
$(NH_4)_2SO_4$	4.5g
Potassium phosphate buffer (0.05*M*, pH 7.4)	1.0L

Preparation of Salts B: Add NaCl and $(NH_4)_2SO_4$ to 1.0L of 0.05*M* potassium phosphate buffer, pH 7.4. Mix thoroughly.

Hemin Solution:

Composition per liter:

Hemin	0.5g
NaOH (0.01*N* solution)	1.0mL

Preparation of Hemin Solution: Add hemin to 1.0mL of 0.01*N* NaOH solution. Mix thoroughly.

Volatile Fatty Acid Solution:

Composition per 31.0mL:

Acetic acid	17.0mL
Propionic acid	6.0mL
n-Butyric acid	4.0mL
n-Valeric acid	1.0mL
Isovaleric acid	1.0mL
Isobutyric acid	1.0mL
DL-α-Methylbutyric acid	1.0mL

Preparation of Volatile Fatty Acid Solution: Combine components. Mix thoroughly.

Preparation of Medium: Add components, except glucose and $NaHCO_3$ solutions, to distilled/deionized water and bring volume to 94.0mL. Mix thoroughly. Adjust pH to 7.2. Gently heat and gas with 95% N_2 + 5% CO_2 until reduced. Anaerobically distribute into tubes or flasks. Cap with rubber stoppers. Autoclave for 20 min at 15 psi pressure–121°C. Cool to 50°C. Filter sterilize glucose solution and $NaHCO_3$ solution separately. Aseptically and anaerobically add sterile glucose solution and sterile $NaHCO_3$ solution to cooled, sterile basal medium.

Use: For the cultivation and maintenance of *Roseburia cecicola.*

Megasphaera Medium

Composition per liter:

Yeast extract	4.0g
K_2HPO_4	3.2g
KH_2PO_4	1.6g
Agar	1.0g
NH_4Cl	0.5g
Sodium thioglycolate	0.45g
$CaCl_2$	0.2g
$MgCl_2$	0.2g
Sodium lactate (60% solution)	16.0mL

pH 7.0 ± 0.2 at 25°C

Preparation of Medium: Add components to distilled/deionized water and bring volume to 1.0L. Mix thoroughly. Gently heat and bring to boiling. Distribute into tubes or flasks. Autoclave for 15 min at 15 psi pressure–121°C.

Use: For the cultivation and maintenance of *Megasphaera elsdenii.*

Mehlman's Maintenance HiVeg Medium

Composition per liter:

Plant peptone No. 3	15.0g
Yeast extract	7.5g
K_2HPO_4	5.0g
Plant hydrolysate	5.0g
Agar	3.0g
$(NH_4)_2SO_4$	1.5g
Starch, soluble	1.0g
Neutral Red	0.02g

pH 7.3 ± 0.22 at 25°C

Source: This medium is available as a premixed powder from HiMedia.

Preparation of Medium: Add components to distilled/deionized water and bring volume to 1.0L. Mix thoroughly. Distribute into tubes or flasks. Autoclave for 15 min at 15 psi pressure–121°C.

Use: For the cultivation and maintenance of *Campylobacter* spp.

Melin–Norkrans Medium (MN)

Composition per 1001.2mL:

Agar	15.0g
Glucose	10.0g
Malt extract	2.8g
KH_2PO_4	0.5g
$(NH_4)_2HPO_4$	0.25g
$MgSO_4·7H_2O$	0.15g
$CaCl_2$	0.05g
NaCl	0.025g
Thiamine	0.1mg
Biotine	0.005mg
Oligo solution	1.66mL
$FeCl_3$ solution	1.2mL

$FeCl_3$ Solution:

Composition per 10.0mL:

$FeCl_3$	1.0g

Preparation of $FeCl_3$ Solution: Add $FeCl_3$ to distilled/deionized water and bring volume to 10.0mL. Mix thoroughly.

Oligo Solution:

Composition per 1.66mL:

Lilly and Barnett solution	1.0mL
Hoagland 1% solution	0.66mL

Hoagland Solution:

Composition per 100.0mL:

$Fe(NO_3)_3·9H_2OH_3BO_3$	2.86g
$MnCl_2$	1.81g
ZnSO4·$7H_2O$	0.22g
$CuSO_4·5H_2O$	0.08g
$H_2MoO_4·H_20$	0.01g

Preparation of Hoagland Solution: Add components to distilled/deionized water and bring volume to 100.0mL. Mix thoroughly.

Lilly and Barnett Solution:

Composition per 100.0mL:

$Fe(NO_3)_3 \cdot 9H_2O$	723.5mg
$ZnSO_4 \cdot 7H_2O$	439.8mg

Preparation of Lilly and Barnett Solution: Add components to distilled/deionized water and bring volume to 100.0mL. Mix thoroughly.

Preparation of Oligo Solution: Combine 1.0mL of Lilly and Barnett solution and 0.66mL of 1% Hoagland solution.

Preparation of Medium: Add components to distilled/deionized water and bring volume to 1.0L. Mix thoroughly. Gently heat and bring to boiling. Distribute into tubes or flasks. Autoclave for 15 min at 15 psi pressure–121°C. Pour into sterile Petri dishes or leave in tubes.

Use: For the cultivation of *Agaricus xanthoderma, Agaricus macrosporus, Antrodia serialis, Armillaria mellea, Auricularia fuscosuccinea, Boletinellus merulioides, Boletus leucophaeus, Cephaliophora irregularis, Circinella umbellata, Kuehneromyces mutabilis, Laccaria laccata, Lentinus tigrinus, Lenzites betulina, Leucogyrophana mollusca, Lycoperdon foetidum, Macrolepiota rhacodes, Macrolepiota procera, Pholiota lenta, Phoma exigua,* and many other fungi.

Melissococcus pluton Medium

Composition per liter:

Glucose	10.0g
Neopeptone	5.0g
Peptone	2.5g
Yeast extract	2.5g
Soluble starch	2.0g
Pancreatic digest of casein	2.0g
L-Cysteine·HCl·H_2O	0.25g
Phosphate buffer (1*M*, pH 6.7)	50.0mL

pH 7.2 ± 0.2 at 25°C

Preparation of Medium: Add components to distilled/deionized water and bring volume to 1.0L. Mix thoroughly. Adjust pH to 7.2. Gently heat and bring to boiling. Distribute into tubes or flasks that have been flushed with 90% N_2 + 10% CO_2. Cap with butyl rubber stoppers. Place tubes in a press. Autoclave for 15 min at 15 psi pressure–121°C.

Use: For the cultivation and maintenance of *Melissococcus pluton.*

Melissococcus pluton Medium

Composition per liter:

Agar	20.0g
KH_2PO_4	13.5g
Glucose	10.0g
Peptone	10.0g
Soluble starch	10.0g

pH 6.6 ± 0.2 at 25°C

Preparation of Medium: Add components to distilled/deionized water and bring volume to 1.0L. Mix thoroughly. Gently heat and bring to boiling. Adjust pH to 6.6 with KOH. Distribute into tubes or flasks. Autoclave for 15 min at 15 psi pressure–121°C. Pour into sterile Petri dishes or leave in tubes.

Use: For the cultivation and maintenance of *Melissococcus pluton.*

Melissococcus plutonius Agar (LMG Medium 110)

Composition per liter:

Agar	20.0g
Glucose	10.0g
Neopeptone	5.0g
Peptone	2.5g
Yeast extract	2.5g
Starch, soluble	2.0g
Trypticase™	2.0g
L-Cysteine·HCl	0.25g
Phosphate buffer solution	50.0mL

Phosphate Buffer Solution:

Composition per liter:

KH_2PO_4	4.5g
$Na_2HPO_4 \cdot 2H_2O$	5.8g

Preparation of Phosphate Buffer Solution: Add components to distilled/deionized water and bring volume to 1.0L. Mix thoroughly. Adjust pH to 6.7.

Preparation of Medium: Add components to distilled/deionized water and bring volume to 1.0L. Mix thoroughly. Adjust pH to 7.2. Anaerobically distribute into tubes sparged with a gas mixture of 100% N_2 + 10% CO_2. Immediately plug with butyl rubber stoppers. Autoclave for 15 min at 15 psi pressure–121°C.

Use: For the cultivation of *Melissococcus plutonius.*

Membrane *Clostridium perfringens* Medium (m-CP Medium)

Composition per 1040.0mL:

Tryptose	30.0g
Yeast extract	20.0g
Agar	15.0g
Sucrose	5.0g
L-Cysteine·HCl·H_2O	1.0g
$MgSO_4 \cdot 7H_2O$	0.1g
Bromcresol Purple	0.04g
Phenolphthalein solution	20.0mL
Indoxyl-β-D-glucoside solution	8.0mL
Selective supplement solution	8.0mL
Ferric chloride solution	4.0mL

pH 7.6 ± 0.2 at 25°C

Source: This medium is available as a premixed powder from Oxoid Unipath.

Selective Supplement Solution:

Composition per 8.0mL:

D-Cycloserine	0.8g
Polymyxin B sulfate	50.0mg

Preparation of Selective Supplement Solution: Add components to distilled/deionized water and bring volume to 8.0mL. Mix thoroughly. Filter sterilize.

Indoxyl-β-D-glucoside Solution:

Composition per 10.0mL:

Indoxyl-β-D-glucoside	75.0mg

Preparation of Indoxyl-β-D-glucoside Solution: Add indoxyl-β-D-glucoside to distilled/deionized water and bring volume to 10.0mL. Mix thoroughly. Filter sterilize.

Ferric Chloride Solution:
Composition per 10.0mL:
$FeCl_3 \cdot 6H_2O$ 0.45g

Preparation of Ferric Chloride Solution: Add ferric chloride to distilled/deionized water and bring volume to 10.0mL. Mix thoroughly. Filter sterilize.

Phenolphthalein Solution:
Composition per 20.0mL:
Phenolphthalein biphosphate tetrasodium salt 0.15g

Preparation of Phenolphthalein Solution: Add phenolphthalein biphosphate tetrasodium salt to distilled/deionized water and bring volume to 20.0mL. Mix thoroughly. Filter sterilize.

Preparation of Medium: Add components, except selective supplement solution, phenolphthalein solution, ferric chloride solution, and indoxyl-β-D-glucoside solution to distilled/deionized water and bring volume to 1.0L. Mix thoroughly. Distribute into tubes or flasks. Autoclave for 15 min at 15 psi pressure–121°C. Cool to 50°C. Aseptically add 8.0mL selective supplement solution, 20.0mL phenolphthalein solution, 4.0mL ferric chloride solution, and 8.0mL indoxyl-β-D-glucoside solution. Mix thoroughly. Pour into sterile Petri dishes or aseptically distribute into sterile tubes.

Use: For the rapid isolation and presumptive identification of *Clostridium perfringens* from food and water samples. A selective and chromogenic medium for the presumptive identification of *Clostridium perfringens,* especially from water samples. Recommended in European Council Directive 98/83/EC for testing the quality of water intended for human consumption. *C. perfringens* colonies have a characteristic opaque yellow appearance. Most other *Clostridium* spp. will appear as either purple colonies, due to the lack of sucrose fermentation, or blue/green colonies where the organism is still cleaving indoxyl-β-D-glucoside and also fermenting sucrose.

Membrane Lactose Glucuronide Agar (MLGA)

Composition per liter:
Peptone 40.0g
Lactose 30.0g
Agar 10.0g
Yeast extract 6.0g
Sodium lauryl sulfate 1.0g
Sodium pyruvate 0.5g
5-Bromo-4-chloro-3-indoxyl-β-D-glucuronic acid 0.2g
Phenol Red 0.2g
pH 7.4 ± 0.2 at 25°C

Source: This medium is available from Oxoid Unipath.

Preparation of Medium: Add components to distilled/deionized water and bring volume to 1.0L. Mix thoroughly. Autoclave for 15 min at 15 psi pressure–121°C. Cool to 45°–50°C. Pour into sterile Petri dishes.

Use: For the direct enumeration of *E. coli* and coliforms in foods by the membrane filtration method. The chromogenic substrate 5-bromo-4-chloro-3-indoxyl-β-D-glucuronic acid (BCIG) is cleaved by the enzyme β-glucuronidase and produces a blue chromophore that builds up within the bacterial cells. In addition, the incorporation of Phenol Red detects lactose fermentation and results in yellow colonies when acid is produced. Since coliform colonies are lactose positive, they will appear yellow on this medium and as *E. coli* colonies are both lactose and β-glucuronidase positive, they will appear green.

Membrane Lauryl Sulfate Broth

Composition per liter:
Peptone 39.0g
Lactose 30.0g
Yeast extract 6.0g
Sodium lauryl sulfate 1.0g
Phenol Red 0.2g
pH 7.4 ± 0.2 at 25°C

Preparation of Medium: Add components to distilled/deionized water and bring volume to 1.0L. Mix thoroughly. Distribute into tubes or flasks. Autoclave for 15 min at 15 psi pressure–121°C.

Use: For the enumeration of coliform organisms and *Escherichia coli* in water.

m-Endo Agar, LES
See: **Endo Agar, LES**

m-Endo Broth
See: **Endo Broth**

M-Endo HiVeg Agar LES

Composition per liter:
Agar 15.0g
Lactose 9.4g
Plant hydrolysate No. 1 .... 7.5g
NaCl 3.7g
Plant hydrolysate 3.7g
Plant peptone 3.7g
K_2HPO_4 .... 3.3g
Na_2SO_3 .... 1.6g
Yeast extract 1.2g
KH_2PO_4 .... 1.0g
Basic Fuchsin 0.8g
Synthetic detergent No. III 0.1g
Sodium lauryl sulfate 0.05g
pH 7.2 ± 0.2 at 25°C

Source: This medium is available as a premixed powder from Hi-Media.

Caution: Basic Fuchsin is a potential carcinogen and care must be taken to avoid inhalation of the powdered dye and contact with the skin.

Preparation of Medium: Add ethanol to approximately 900.0mL of distilled/deionized water. Add remaining components. Bring volume to 1.0L with distilled/deionized water. Mix thoroughly. Gently heat and bring to boiling. Autoclave for 15 min at 15 psi pressure–121°C. Pour into sterile 60mm Petri dishes in 4.0mL volumes. Protect from the light.

Use: For the cultivation and enumeration of coliform bacteria by the membrane filter method.

M-Endo HiVeg Broth

Composition per liter:
Lactose 25.0g
Plant peptone 20.0g
K_2HPO_4 .... 7.0g
Yeast extract 6.0g

Na_2SO_3	2.5g
Basic Fuchsin	1.0g

pH 7.2 ± 0.1 at 25°C

Source: This medium is available as a premixed powder from HiMedia.

Caution: Basic Fuchsin is a potential carcinogen and care must be taken to avoid inhalation of the powdered dye and contact with the skin.

Preparation of Medium: Add ethanol to approximately 900.0mL of distilled/deionized water. Add remaining components. Bring volume to 1.0L with distilled/deionized water. Mix thoroughly. Gently heat and bring to boiling. Rapidly cool broth below 45°C. Do not autoclave. Use 1.8–2.0mL for each filter pad. Protect from the light. Prepare broth freshly.

Use: For the cultivation and enumeration of coliform bacteria from water by the membrane filter method.

M-Endo HiVeg Broth MF (MF Endo HiVeg Medium) (M-Coliform HiVeg Broth)

Composition per liter:

Lactose	12.5g
Plant hydrolysate No. 1	10.0g
Plant hydrolysate	5.0g
Plant special peptone	5.0g
NaCl	5.0g
K_2HPO_4	4.375g
Na_2SO_3	2.1g
Yeast extract	1.5g
KH_2PO_4	1.375g
Basic Fuchsin	1.05g
Synthetic detergent No. III	0.1g
Sodium lauryl sulfate	0.05g

pH 7.2 ± 0.1 at 25°C

Source: This medium is available as a premixed powder from HiMedia.

Caution: Basic Fuchsin is a potential carcinogen and care must be taken to avoid inhalation of the powdered dye and contact with the skin.

Preparation of Medium: Add ethanol to approximately 900.0mL of distilled/deionized water. Add remaining components. Bring volume to 1.0L with distilled/deionized water. Mix thoroughly. Gently heat and bring to boiling. Rapidly cool broth below 45°C. Do not autoclave. Use 1.8–2.0mL for each filter pad. Protect from the light. Prepare broth freshly.

Use: For the cultivation and enumeration of coliform bacteria from water by the membrane filter method.

Meniscus glaucopis Agar

Composition per liter:

Agar	15.0g
$CaCO_3$	10.0g
Maltose	5.0g
Yeast extract	1.0g
KH_2PO_4	0.5g
NaCl	0.4g
NH_4Cl	0.4g
Sodium thioglycolate	0.3g
$MgSO_4 \cdot 7H_2O$	0.2g
$CaCl_2 \cdot H_2O$	0.01g
$FeSO_4 \cdot 7H_2O$	1.0mg
Resazurin (0.025% solution)	4.0mL
Trace elements solution SL-6	1.0mL
Vitamin solution	10.0mL

pH 7.3 ± 0.2 at 25°C

Trace Elements Solution SL-6:
Composition per liter:

H_3BO_3	0.3g
$CoCl_2 \cdot 6H_2O$	0.2g
$ZnSO_4 \cdot 7H_2O$	0.1g
$MnCl_2 \cdot 4H_2O$	0.03g
$Na_2MoO_4 \cdot H_2O$	0.03g
$NiCl_2 \cdot 6H_2O$	0.02g
$CuCl_2 \cdot 2H_2O$	0.01g

Preparation of Trace Elements Solution SL-6: Add components to distilled/deionized water and bring volume to 1.0L. Mix thoroughly. Adjust pH to 3.4.

Vitamin Solution:
Composition per liter:

Vitamin B_{12}	2.8mg
Thiamine·HCl	0.28mg

Preparation of Vitamin Solution: Add components to distilled/deionized water and bring volume to 10.0mL. Mix thoroughly. Filter sterilize.

Preparation of Medium: Add components, except vitamin solution, to distilled/deionized water and bring volume to 990.0mL. Mix thoroughly. Adjust pH to 7.3 with 10% Na_2CO_3. Gently heat and bring to boiling. Continue boiling until resazurin changes color. Cool to 50°C. Distribute into tubes in 7.0mL volumes under O_2-free 97% N_2 + 3% H_2. Cap with rubber stoppers under O_2-free 97% N_2 + 3% H_2. Place tubes in a press. Autoclave for 15 min at 15 psi pressure–121°C using fast exhaust. Cool to 50°C. Aseptically add 0.25mL of sterile vitamin solution to each tube.

Use: For the cultivation and maintenance of *Meniscus glaucopis*.

Meniscus glaucopis Broth

Composition per liter:

Maltose	5.0g
Agar	3.0g
Yeast extract	1.0g
KH_2PO_4	0.5g
NaCl	0.4g
NH_4Cl	0.4g
Sodium thioglycolate	0.3g
$MgSO_4 \cdot 7H_2O$	0.2g
$CaCl_2 \cdot H_2O$	0.01g
$FeSO_4 \cdot 7H_2O$	1.0mg
Resazurin (0.025% solution)	4.0mL
Trace elements solution SL-6	1.0mL
Vitamin solution	10.0mL

pH 7.3 ± 0.2 at 25°C

Trace Elements Solution SL-6:
Composition per liter:

H_3BO_3	0.3g
$CoCl_2 \cdot 6H_2O$	0.2g
$ZnSO_4 \cdot 7H_2O$	0.1g
$MnCl_2 \cdot 4H_2O$	0.03g

$Na_2MoO_4 \cdot H_2O$	0.03g
$NiCl_2 \cdot 6H_2O$	0.02g
$CuCl_2 \cdot 2H_2O$	0.01g

Preparation of Trace Elements Solution SL-6: Add components to distilled/deionized water and bring volume to 1.0L. Mix thoroughly. Adjust pH to 3.4.

Vitamin Solution:
Composition per liter:

Vitamin B_{12}	2.8mg
Thiamine·HCl	0.28mg

Preparation of Vitamin Solution: Add components to distilled/deionized water and bring volume to 10.0mL. Mix thoroughly. Filter sterilize.

Preparation of Medium: Add components, except vitamin solution, to distilled/deionized water and bring volume to 990.0mL. Mix thoroughly. Adjust pH to 7.3 with 10% Na_2CO_3. Gently heat and bring to boiling. Continue boiling until resazurin changes color. Cool to 50°C. Distribute into tubes in 7.0mL volumes under O_2-free 97% N_2 + 3% H_2. Cap with rubber stoppers under O_2-free 97% N_2 + 3% H_2. Place tubes in a press. Autoclave for 15 min at 15 psi pressure–121°C using fast exhaust. Cool to 50°C. Aseptically add 0.25mL of sterile vitamin solution to each tube.

Use: For the cultivation and maintenance of *Meniscus glaucopis*.

M-Enrichment Broth

Composition per liter:

Proteose peptone	40.0g
Yeast extract	6.0g
NaCl	5.0g
K_2HPO_4	3.0g

pH 7.0 ± 0.2 at 25°C

Source: This medium is available from HiMedia.

Preparation of Medium: Add components to distilled/deionized water and bring volume to 1.0L. Mix thoroughly. Gently heat and bring to boiling. Distribute into tubes or flasks. Autoclave for 15 min at 15 psi pressure–121°C. Pour into sterile Petri dishes or leave in tubes.

Use: For the preliminary enrichment of organisms on membrane filter prior to using selective media.

m-*Enterococcus* Agar
See: ***Enterococcus*** **Agar**

M-*Enterococcus* Agar Base with Polysorbate 80 and Sodium Carbonate

Composition per liter:

Agar	10.0g
Casein enzymic hydrolysate	15.0g
Papaic digest of soybean meal	5.0g
Yeast extract	5.0g
K_2HPO_4	4.0g
Glucose	2.0g
NaN_3	0.4g
Triphenyl tetrazolium chloride	0.1g
Sodium carbonate solution	2.0mL
Polysorbate 80	0.5mL

pH 7.2 ± 0.2 at 25°C

Source: This medium is available from HiMedia.

Caution: Sodium azide has a tendency to form explosive metal azides with plumbing materials. It is advisable to use enough water to flush off the disposables.

Sodium Carbonate Solution:
Composition per 10.0mL:

Na_2CO_3	1.0g

Preparation of Sodium Carbonate Solution: Add Na_2CO_3 to distilled/deionized water and bring volume to 10.0mL. Mix thoroughly. Filter sterilize.

Preparation of Medium: Add components, except polysorbate 80 and sodium carbonate solution, to distilled/deionized water and bring volume to 997.5mL. Mix thoroughly. Gently heat to dissolve components. Do not autoclave. Cool to 50°C. Add polysorbate 80 and sodium carbonate solution. Mix thoroughly. Pour into Petri dishes or aseptically distribute into sterile tubes.

Use: For the selective isolation and enumeration of enterococci from water, sewage, food, or other materials.

M-*Enterococcus* Agar Base, Modified

Composition per liter:

Yeast extract	30.0g
Pancreatic digest of gelatin	10.0g
Agar	15.0g
NaCl	15.0g
Esculin	1.0g
Nalidixic acid	0.25g
NaN_3	0.15g
Cycloheximide	0.05g
Selective supplement solution	15.0mL

pH 7.1 ± 0.2 at 25°C

Source: This medium is available from HiMedia.

Caution: Sodium azide has a tendency to form explosive metal azides with plumbing materials. It is advisable to use enough water to flush off the disposables.

Caution: Cycloheximide is toxic. Avoid skin contact or aerosol formation and inhalation.

Selective Supplement Solution:
Composition per 20.0mL:

2,3,5-Triphenyl tetrazolium chloride	0.2g

Preparation of Selective Supplement Solution: Add 2,3,5-triphenyl tetrazolium chloride to distilled/deionized water and bring volume to 20.0mL. Mix thoroughly. Filter sterilize.

Preparation of Medium: Add components, except selective supplement solution, to distilled/deionized water and bring volume to 985.0mL. Mix thoroughly. Autoclave for 15 min at 15 psi pressure–121°C. Cool to 50°C. Aseptically add selective supplement solution. Mix thoroughly. Pour into Petri dishes or aseptically distribute into sterile tubes.

Use: For the recovery of enterococci in water samples using the membrane filter technique.

M-*Enterococcus* HiVeg Agar Base

Composition per liter:

Plant hydrolysate	15.0g
Agar	10.0g
Papaic digest of soybean meal	5.0g
Yeast extract	5.0g

KH_2PO_4	4.0g
Glucose	2.0g
NaN_3	0.4g
Triphenyl tetrazolium chloride	0.1g

pH 7.2 ± 0.2 at 25°C

Source: This medium is available as a premixed powder from HiMedia.

Caution: Sodium azide is toxic. Azides also react with metals and disposal must be highly diluted.

Preparation of Medium: Add components to distilled/deionized water and bring volume to 1.0L. Mix thoroughly. Gently heat and bring to boiling. Cool to 45°–50°C. Do not autoclave. Pour into sterile Petri dishes.

Use: For the isolation, cultivation, and enumeration of entercocci in water, sewage, and feces by the membrane filter method. For the direct plating of specimens for the detection and enumeration of fecal streptococci.

MeReSa Agar Base with Methicillin (Methicillin-Resistant *Staphylococcus aureus* Agar)

Composition per liter:

Agar	20.0g
Casein enzymic hydrolysate	10.0g
Glycine	10.0g
Mannitol	10.0g
NaCl	10.0g
Sodium pyruvate	10.0g
LiCl	5.0g
Beef extract	5.0g
Indicator mix	0.13g
MRSA selective supplement	10.0mL

pH 7.0 ± 0.2 at 25°C

Source: This medium, without MRSA selective supplement, is available as a premixed powder from HiMedia.

Caution: Lithium chloride is harmful. Avoid bodily contact and inhalation of vapors. On contact with skin wash with plenty of water immediately.

MRSA Selective Supplement:

Composition per 10.0mL:

Methicillin	4.0mg

Preparation of MRSA Selective Supplement: Add methicillin to distilled/deionized water and bring volume to 10.0mL. Mix thoroughly. Filter sterilize.

Preparation of Medium: Add components, except MRSA selective supplement, to distilled/deionized water and bring volume to 990.0mL. Mix thoroughly. Gently heat and bring to boiling. Distribute into tubes or flasks. Autoclave for 15 min at 15 psi pressure–121°C. Cool to 50°C. Aseptically add 10.0mL MRSA selective supplement. Pour into sterile Petri dishes or leave in tubes.

Use: For the isolation and cultivation of methicillin-resistant *Staphylococcus aureus* (MRSA).

MeReSa Agar Base with Oxacillin (Methicillin-Resistant *Staphylococcus aureus* Agar)

Composition per liter:

Agar	20.0g
Casein enzymic hydrolysate	10.0g
Glycine	10.0g
Mannitol	10.0g
NaCl	10.0g
Sodium pyruvate	10.0g
LiCl	5.0g
Beef extract	5.0g
Indicator mix	0.13g
MRSA selective supplement	10.0mL

pH 7.0 ± 0.2 at 25°C

Source: This medium, without MRSA supplement solution, is available as a premixed powder from HiMedia.

Caution: Lithium chloride is harmful. Avoid bodily contact and inhalation of vapors. On contact with skin wash with plenty of water immediately.

MRSA Selective Supplement:

Composition per 10.0mL:

Oxacillin	2.0mg

Preparation of MRSA Selective Supplement: Add oxacillin to distilled/deionized water and bring volume to 10.0mL. Mix thoroughly. Filter sterilize.

Preparation of Medium: Add components, except MRSA selective supplement, to distilled/deionized water and bring volume to 990.0mL. Mix thoroughly. Gently heat and bring to boiling. Distribute into tubes or flasks. Autoclave for 15 min at 15 psi pressure–121°C. Cool to 50°C. Aseptically add 10.0mL MRSA selective supplement. Pour into sterile Petri dishes or leave in tubes.

Use: For the isolation and cultivation of methicillin-resistant *Staphylococcus aureus* (MRSA).

MES Agar

See: **U Agar Plates**

Metal Acetate Agar

Composition per liter:

Agar	15.0g
Sodium acetate	2.0g
Beijerinck's solution	50.0mL
Phosphate buffer solution	50.0mL
Trace elements solution	1.0mL

pH 6.8 ± 0.2 at 25°C

Beijerinck's Solution:

Composition per liter:

NH_4Cl	10.0g
$MgSO_4 \cdot 7H_2O$	0.4g
$CaCl_2 \cdot 2H_2O$	0.2g

Preparation of Beijerinck's Solution: Add $CaCl_2 \cdot 2H_2O$ to distilled/deionized water and bring volume to 500.0mL. Mix thoroughly. Add the remaining components to distilled/deionized water and bring volume to 500.0mL in a separate flask. Combine the two solutions.

Phosphate Buffer Solution:

Composition per liter:

K_2HPO_4	28.8g
KH_2PO_4	14.4g

Preparation of Phosphate Buffer Solution: Add components to distilled/deionized water and bring volume to 1.0L. Mix thoroughly. Adjust pH to 6.8. Autoclave for 15 min at 15 psi pressure–121°C.

Trace Elements Solution:
Composition per liter:

EDTA	50.0g
H_3BO_3 solution	200.0mL
$ZnSO_4 \cdot 7H_2O$ solution	100.0mL
$CoCl_2 \cdot 6H_2O$ solution	50.0mL
$CuSO_4 \cdot 5H_2O$ solution	50.0mL
$FeSO_4 \cdot 7H_2O$ solution	50.0mL
$MnCl_2 \cdot 4H_2O$ solution	50.0mL
$(NH_4)_6Mo_7O_{24} \cdot 4H_2O$ solution	50.0mL

H_3BO_3 Solution:
Composition per 200.0mL:

H_3BO_3	11.4g

Preparation of H_3BO_3 Solution: Add H_3BO_3 to distilled/deionized water and bring volume to 200.0mL. Mix thoroughly.

$ZnSO_4 \cdot 7H_2O$ Solution:
Composition per 100.0mL:

$ZnSO_4 \cdot 7H_2O$	22.0g

Preparation of $ZnSO_4 \cdot 7H_2O$ Solution: Add $ZnSO_4 \cdot 7H_2O$ to distilled/deionized water and bring volume to 100.0mL. Mix thoroughly.

$MnCl_2 \cdot 4H_2O$ Solution:
Composition per 50.0mL:

$MnCl_2 \cdot 4H_2O$	5.06g

Preparation of $MnCl_2 \cdot 4H_2O$ Solution: Add $MnCl_2 \cdot 4H_2O$ to distilled/deionized water and bring volume to 50.0mL. Mix thoroughly.

$FeSO_4 \cdot 7H_2O$ Solution:
Composition per 50.0mL:

$FeSO_4 \cdot 7H_2O$	4.99g

Preparation of $FeSO_4 \cdot 7H_2O$ Solution: Add $FeSO_4 \cdot 7H_2O$ to distilled/deionized water and bring volume to 50.0mL. Mix thoroughly.

$CoCl_2 \cdot 6H_2O$ Solution:
Composition per 50.0mL:

$CoCl_2 \cdot 6H_2O$	1.61g

Preparation of $CoCl_2 \cdot 6H_2O$ Solution: Add $CoCl_2 \cdot 6H_2O$ to distilled/deionized water and bring volume to 50.0mL. Mix thoroughly.

$CuSO_4 \cdot 5H_2O$ Solution:
Composition per 50.0mL:

$CuSO_4 \cdot 5H_2O$	1.57g

Preparation of $CuSO_4 \cdot 5H_2O$ Solution: Add $CuSO_4 \cdot 5H_2O$ to distilled/deionized water and bring volume to 50.0mL. Mix thoroughly.

$(NH_4)_6Mo_7O_{24} \cdot 4H_2O$ Solution:
Composition per 50.0mL:

$(NH_4)_6Mo_7O_{24} \cdot 4H_2O$	1.1g

Preparation of $(NH_4)_6Mo_7O_{24} \cdot 4H_2O$ Solution: Add 1.1 g of $(NH_4)_6Mo_7O_{24} \cdot 4H_2O$ to distilled/deionized water and bring volume to 50.0mL. Mix thoroughly.

Preparation of Trace Elements Solution: Add EDTA to distilled/deionized water and bring volume to 250.0mL. Mix thoroughly. Gently heat and bring to boiling. Continue boiling until dissolved. Add 200.0mL of H_3BO_3 solution, 100.0mL of $ZnSO_4 \cdot 7H_2O$ solution, 50.0mL of $MnCl_2 \cdot 4H_2O$ solution, 50.0mL of $FeSO_4 \cdot 7H_2O$ solution, 50.0mL of $CoCl_2 \cdot 6H_2O$ solution, 50.0mL of $CuSO_4 \cdot 5H_2O$ solution, and 50.0mL of $(NH_4)_6Mo_7O_{24} \cdot 4H_2O$ solution. Gently heat and bring to boiling. Cool to 70°C. Adjust pH to 6.8 with hot (70°C) 20% KOH solution. Add distilled/deionized water and bring volume to 1.0L. Allow solution to stand in a 2.0L cotton-stoppered flask at room temperature until the solution turns purple (approximately 2 weeks). Filter using two layers of Whatman #1 filter paper. Filter until clear.

Preparation of Medium: Add components, except phosphate buffer solution, to distilled/deionized water and bring volume to 950.0mL. Mix thoroughly. Gently heat and bring to boiling. Autoclave for 15 min at 15 psi pressure–121°C. Cool to 50°–55°C. Aseptically add 50.0mL of sterile phosphate solution. Mix thoroughly. Pour into sterile Petri dishes or distribute into sterile tubes.

Use: For the cultivation and maintenance of *Chlamydomonas reinhardtii*.

Metal Acetate Broth

Composition per liter:

Sodium acetate	2.0g
Beijerinck's solution	50.0mL
Phosphate buffer solution	50.0mL
Trace elements solution	1.0mL

pH 6.8 ± 0.2 at 25°C

Beijerinck's Solution:
Composition per liter:

NH_4Cl	10.0g
$MgSO_4 \cdot 7H_2O$	0.4g
$CaCl_2 \cdot 2H_2O$	0.2g

Preparation of Beijerinck's Solution: Add $CaCl_2 \cdot 2H_2O$ to distilled/deionized water and bring volume to 500.0mL. Mix thoroughly. Add the remaining components to distilled/deionized water and bring volume to 500.0mL in a separate flask. Combine the two solutions.

Phosphate Buffer Solution:
Composition per liter:

K_2HPO_4	28.8g
KH_2PO_4	14.4g

Preparation of Phosphate Buffer Solution: Add components to distilled/deionized water and bring volume to 1.0L. Mix thoroughly. Adjust pH to 6.8. Autoclave for 15 min at 15 psi pressure–121°C.

Trace Elements Solution:
Composition per liter:

EDTA	50.0g
H_3BO_3 solution	200.0mL
$ZnSO_4 \cdot 7H_2O$ solution	100.0mL
$CoCl_2 \cdot 6H_2O$ solution	50.0mL
$CuSO_4 \cdot 5H_2O$ solution	50.0mL
$FeSO_4 \cdot 7H_2O$ solution	50.0mL
$MnCl_2 \cdot 4H_2O$ solution	50.0mL
$(NH_4)_6Mo_7O_{24} \cdot 4H_2O$ solution	50.0mL

H_3BO_3 Solution:
Composition per 200.0mL:

H_3BO_3	11.4g

Preparation of H_3BO_3 Solution: Add H_3BO_3 to distilled/deionized water and bring volume to 200.0mL. Mix thoroughly.

$ZnSO_4 \cdot 7H_2O$ Solution:
Composition per 100.0mL:

$ZnSO_4 \cdot 7H_2O$	22.0g

Preparation of $ZnSO_4 \cdot 7H_2O$ Solution: Add $ZnSO_4 \cdot 7H_2O$ to distilled/deionized water and bring volume to 100.0mL. Mix thoroughly.

$MnCl_2·4H_2O$ Solution:
Composition per 50.0mL:

$MnCl_2·4H_2O$	5.06g

Preparation of $MnCl_2·4H_2O$ Solution: Add $MnCl_2·4H_2O$ to distilled/deionized water and bring volume to 50.0mL. Mix thoroughly.

$FeSO_4·7H_2O$ Solution:
Composition per 50.0mL:

$FeSO_4·7H_2O$	4.99g

Preparation of $FeSO_4·7H_2O$ Solution: Add $FeSO_4·7H_2O$ to distilled/deionized water and bring volume to 50.0mL. Mix thoroughly.

$CoCl_2·6H_2O$ Solution:
Composition per 50.0mL:

$CoCl_2·6H_2O$	1.61g

Preparation of $CoCl_2·6H_2O$ Solution: Add $CoCl_2·6H_2O$ to distilled/deionized water and bring volume to 50.0mL. Mix thoroughly.

$CuSO_4·5H_2O$ Solution:
Composition per 50.0mL:

$CuSO_4·5H_2O$	1.57g

Preparation of $CuSO_4.5H_2O$ Solution: Add $CuSO_4·5H_2O$ to distilled/deionized water and bring volume to 50.0mL. Mix thoroughly.

$(NH_4)_6Mo_7O_{24}·4H_2O$ Solution:
Composition per 50.0mL:

$(NH_4)_6Mo_7O_{24}·4H_2O$	1.1g

Preparation of $(NH_4)_6Mo_7O_{24}·4H_2O$ Solution: Add 1.1 g of $(NH_4)_6Mo_7O_{24}·4H_2O$ to distilled/deionized water and bring volume to 50.0mL. Mix thoroughly.

Preparation of Trace Elements Solution: Add EDTA to distilled/deionized water and bring volume to 250.0mL. Mix thoroughly. Gently heat and bring to boiling. Continue boiling until dissolved. Add 200.0mL of H_3BO_3 solution, 100.0mL of $ZnSO_4·7H_2O$ solution, 50.0mL of $MnCl_2·4H_2O$ solution, 50.0mL of $FeSO_4·7H_2O$ solution, 50.0mL of $CoCl_2·6H_2O$ solution, 50.0mL of $CuSO_4·5H_2O$ solution, and 50.0mL of $(NH_4)_6Mo_7O_{24}·4H_2O$ solution. Gently heat and bring to boiling. Cool to 70°C. Adjust pH to 6.8 with hot (70°C) 20% KOH solution (approximately 80.0–90.0mL). Add distilled/deionized water and bring volume to 1.0L. Allow solution to stand in a 2.0L cotton-stoppered flask at room temperature until the solution turns purple (approximately 2 weeks). Filter using two layers of Whatman #1 filter paper. Filter until clear. Store at 4°C or at –20°C.

Preparation of Medium: Add components, except phosphate buffer solution, to distilled/deionized water and bring volume to 950.0mL. Mix thoroughly. Autoclave for 15 min at 15 psi pressure–121°C. Aseptically add 50.0mL of sterile phosphate solution. Mix thoroughly. Aseptically distribute into sterile tubes or flasks.

Use: For the cultivation of *Chlamydomonas reinhardtii*.

Metal Acetate Yeast Broth with Arginine

Composition per liter:

Yeast extract	4.0g
Sodium acetate	2.0g
Arginine	0.1g
Beijerinck's solution	50.0mL
Phosphate buffer solution	50.0mL
Trace elements solution	1.0mL

pH 6.8 ± 0.2 at 25°C

Beijerinck's Solution:
Composition per liter:

NH_4Cl	10.0g
$MgSO_4·7H_2O$	0.4g
$CaCl_2·2H_2O$	0.2g

Preparation of Beijerinck's Solution: Add $CaCl_2·2H_2O$ to distilled/deionized water and bring volume to 500.0mL. Mix thoroughly. Add the remaining components to distilled/deionized water and bring volume to 500.0mL in a separate flask. Combine the two solutions.

Phosphate Buffer Solution:
Composition per liter:

K_2HPO_4	28.8g
KH_2PO_4	14.4g

Preparation of Phosphate Buffer Solution: Add components to distilled/deionized water and bring volume to 1.0L. Mix thoroughly. Adjust pH to 6.8. Autoclave for 15 min at 15 psi pressure–121°C.

Trace Elements Solution:
Composition per liter:

EDTA	50.0g
H_3BO_3 solution	200.0mL
$ZnSO_4·7H_2O$ solution	100.0mL
$CoCl_2·6H_2O$ solution	50.0mL
$CuSO_4·5H_2O$ solution	50.0mL
$FeSO_4·7H_2O$ solution	50.0mL
$MnCl_2·4H_2O$ solution	50.0mL
$(NH_4)_6Mo_7O_{24}·4H_2O$ solution	50.0mL

H_3BO_3 Solution:
Composition per 200.0mL:

H_3BO_3	11.4g

Preparation of H_3BO_3 Solution: Add H_3BO_3 to distilled/deionized water and bring volume to 200.0mL. Mix thoroughly.

$ZnSO_4·7H_2O$ Solution:
Composition per 100.0mL:

$ZnSO_4·7H_2O$	22.0g

Preparation of $ZnSO_4·7H_2O$ Solution: Add $ZnSO_4·7H_2O$ to distilled/deionized water and bring volume to 100.0mL. Mix thoroughly.

$MnCl_2·4H_2O$ Solution:
Composition per 50.0mL:

$MnCl_2·4H_2O$	5.06g

Preparation of $MnCl_2·4H_2O$ Solution: Add $MnCl_2·4H_2O$ to distilled/deionized water and bring volume to 50.0mL. Mix thoroughly.

$FeSO_4·7H_2O$ Solution:
Composition per 50.0mL:

$FeSO_4·7H_2O$	4.99g

Preparation of $FeSO_4·7H_2O$ Solution: Add $FeSO_4·7H_2O$ to distilled/deionized water and bring volume to 50.0mL. Mix thoroughly.

$CoCl_2·6H_2O$ Solution:
Composition per 50.0mL:

$CoCl_2·6H_2O$	1.61g

Preparation of $CoCl_2·6H_2O$ Solution: Add $CoCl_2·6H_2O$ to distilled/deionized water and bring volume to 50.0mL. Mix thoroughly.

$CuSO_4·5H_2O$ Solution:
Composition per 50.0mL:

$CuSO_4·5H_2O$	1.57g

Preparation of $CuSO_4.5H_2O$ Solution: Add $CuSO_4 \cdot 5H_2O$ to distilled/deionized water and bring volume to 50.0mL. Mix thoroughly.

$(NH_4)_6Mo_7O_{24} \cdot 4H_2O$ Solution:
Composition per 50.0mL:
$(NH_4)_6Mo_7O_{24} \cdot 4H_2O$ 1.1g

Preparation of $(NH_4)_6Mo_7O_{24} \cdot 4H_2O$ Solution: Add 1.1 g of $(NH_4)_6Mo_7O_{24} \cdot 4H_2O$ to distilled/deionized water and bring volume to 50.0mL. Mix thoroughly.

Preparation of Trace Elements Solution: Add EDTA to distilled/deionized water and bring volume to 250.0mL. Mix thoroughly. Gently heat and bring to boiling. Continue boiling until dissolved. Add 200.0mL of H_3BO_3 solution, 100.0mL of $ZnSO_4 \cdot 7H_2O$ solution, 50.0mL of $MnCl_2 \cdot 4H_2O$ solution, 50.0mL of $FeSO_4 \cdot 7H_2O$ solution, 50.0mL of $CoCl_2 \cdot 6H_2O$ solution, 50.0mL of $CuSO_4 \cdot 5H_2O$ solution, and 50.0mL of $(NH_4)_6Mo_7O_{24} \cdot 4H_2O$ solution. Gently heat and bring to boiling. Cool to 70°C. Adjust pH to 6.8 with hot (70°C) 20% KOH solution (approximately 80.0–90.0mL). Add distilled/deionized water and bring volume to 1.0L. Allow solution to stand in a 2.0L cotton-stoppered flask at room temperature until the solution turns purple (approximately 2 weeks). Filter using two layers of Whatman #1 filter paper. Filter until clear. Store at 4°C or –20°C.

Preparation of Medium: Add components, except phosphate buffer solution, to distilled/deionized water and bring volume to 950.0mL. Mix thoroughly. Autoclave for 15 min at 15 psi pressure–121°C. Aseptically add 50.0mL of sterile phosphate solution. Mix thoroughly. Aseptically distribute into sterile tubes or flasks.

Use: For the cultivation of *Chlamydomonas reinhardtii*.

Metallogenium Cultivation Broth

Composition per liter:
Gum arabic 20.0g
$MnCO_3$ 0.5g

$MnCO_3$:
Composition per 100.0mL:
$MnCl_2$ 20.0g
$NaHCO_3$ (25% solution) 25.0mL

Preparation of $MnCO_3$: Add $MnCl_2$ to distilled/deionized water and bring volume to 100.0mL. Mix thoroughly. Add $NaHCO_3$ solution. Filter through Whatman #1 filter paper. Save the $MnCO_3$ precipitate. Wash and store under distilled/deionized water.

Preparation of Medium: Add components to distilled/deionized water and bring volume to 1.0L. Mix thoroughly. Distribute into tubes or flasks. Autoclave for 15 min at 15 psi pressure–121°C.

Use: For the cultivation of *Metallogenium* species.

Metallogenium Cultivation Broth

Composition per liter:
Starch, hydrolyzed 20.0g
$MnCO_3$ 0.5g

$MnCO_3$:
Composition per 100.0mL:
$MnCl_2$ 20.0g
$NaHCO_3$ (25% solution) 25.0mL

Preparation of $MnCO_3$: Add $MnCl_2$ to distilled/deionized water and bring volume to 100.0mL. Mix thoroughly. Add $NaHCO_3$ solution. Filter through Whatman #1 filter paper. Save the $MnCO_3$ precipitate. Wash and store under distilled/deionized water.

Preparation of Medium: Hydrolyze starch with HCl. Add components to distilled/deionized water and bring volume to 1.0L. Mix thoroughly. Distribute into tubes or flasks. Autoclave for 15 min at 15 psi pressure–121°C.

Use: For the cultivation of *Metallogenium* species.

Metallogenium Isolation Agar

Composition per liter:
Agar 15.0g
Manganese acetate 0.1g

Preparation of Medium: Add components to distilled/deionized water and bring volume to 1.0L. Mix thoroughly. Gently heat and bring to boiling. Distribute into tubes or flasks. Autoclave for 15 min at 15 psi pressure–121°C. Pour into sterile Petri dishes or leave in tubes.

Use: For the isolation and cultivation of *Metallogenium* species.

Metallogenium Medium

Composition per liter:
$MnCO_3$ 2.0g
Starch, hydrolyzed 1.0g
DNA 0.01g
Catalase 5.0mg
Mycoplasma broth base 100.0mL
Yeast extract, ultrafiltrate 100.0mL
Horse serum 10.0mL

***Mycoplasma* Broth Base:**
Composition per liter:
Pancreatic digest of casein 7.0g
NaCl 5.0g
Beef extract 3.0g
Yeast extract 3.0g
Beef heart, solids from infusion 2.0g

Preparation of *Mycoplasma* Broth Base: Add components to distilled/deionized water and bring volume to 1.0L. Mix thoroughly. Autoclave for 15 min at 15 psi pressure–121°C. Cool to 25°C.

$MnCO_3$:
Composition per 100.0mL:
$MnCl_2$ 20.0g
$NaHCO_3$ (25% solution) 25.0mL

Preparation of $MnCO_3$: Add $MnCl_2$ to distilled/deionized water and bring volume to 100.0mL. Mix thoroughly. Add $NaHCO_3$ solution. Filter through Whatman #1 filter paper. Save the $MnCO_3$ precipitate. Wash and store under distilled/deionized water.

Preparation of Medium: Add $MnCO_3$, hydrolyzed starch, and DNA to 25.0mL of distilled/deionized water. Mix thoroughly. Autoclave for 15 min at 15 psi pressure–121°C. Cool to 45°–50°C. Aseptically add 100.0mL of sterile *Mycoplasma* broth base, 100.0mL of ultrafiltrate of yeast extract, 10.0mL of horse serum, and 5.0mg of catalase. Mix thoroughly. Aseptically distribute into sterile tubes or flasks.

Use: For the cultivation of *Metallogenium* species.

Metallosphaera Medium

Composition per liter:
$(NH_4)_2SO_4$ 1.3g
Yeast extract 1.0g
KH_2PO_4 .... 0.28g
$MgSO_4 \cdot 7H_2O$ 0.25g

$CaCl_2 \cdot 2H_2O$ 0.07g
$FeCl_3 \cdot 6H_2O$ 0.02g
$Na_2B_4 \cdot 10H_2O$ 4.5mg
$MnCl_2 \cdot 4H_2O$ 1.8mg
$ZnSO_4 \cdot 7H_2O$ 0.22mg
$CuCl_2 \cdot 2H_2O$ 0.05mg
$Na_2MoO_4 \cdot 2H_2O$ 0.03mg
$VOSO_4 \cdot 2H_2O$ 0.03mg
$CoSO_4$ 0.01mg

Preparation of Medium: Add components to distilled/deionized water and bring volume to 1.0L. Mix thoroughly. Adjust pH to 2.0 using 10*N* H_2SO_4. Distribute into tubes or flasks. Autoclave for 15 min at 15 psi pressure–121°C.

Use: For the cultivation of *Metallosphaera sedula*.

Methanobacillus Medium

Composition per liter:
KH_2PO_4 .... 9.0g
K_2HPO_4 .... 6.0g
NH_4Cl 5.0g
$MgCl_2$ 1.0g
$CaCl_2$ 0.01g
$FeSO_4 \cdot 7H_2O$ 0.01g
Ethanol 10.0mL

pH 7.4 ± 0.2 at 25°C

Preparation of Medium: Filter sterilize ethanol. Add components, except ethanol, to tap water and bring volume to 990.0mL. Mix thoroughly. Gently heat until dissolved. Autoclave for 20 min at 10psi pressure–115°C. Cool to 45°–50°C. Aseptically add sterile ethanol. Mix thoroughly. Aseptically distribute into sterile tubes or flasks.

Use: For the selective isolation and cultivation of *Methanobacillus* species from mixed cultures.

Methanobacteria Medium

Composition per liter:
Mineral solution 2 .... 50.0mL
Sodium carbonate solution 50.0mL
Mineral solution 1 .... 25.0mL
L-Cysteine-sulfide reducing agent 20.0mL
Wolfe's mineral solution 10.0mL
Wolfe's vitamin solution 10.0mL
Resazurin (0.025% solution) 4.0mL

pH 7.2 ± 0.2 at 25°C

Mineral Solution 1:
Composition per liter:
K_2HPO_4 .... 6.0g

Preparation of Medium: Add K_2HPO_4 to distilled/deionized water and bring volume to 1.0L. Mix thoroughly.

Mineral Solution 2:
Composition per liter:
NaCl 12.0g
KH_2PO_4 .... 6.0g
$(NH_4)_2SO_4$ 6.0g
$MgSO_4 \cdot 7H_2O$ 2.4g
$CaCl_2 \cdot 2H_2O$ 1.6g

Preparation of Mineral Solution 2: Add components to distilled/deionized water and bring volume to 1.0L. Mix thoroughly.

Sodium Carbonate Solution:
Composition per 100.0mL:
Na_2CO_3 .... 8.0g

Preparation of Sodium Carbonate Solution: Add Na_2CO_3 to distilled/deionized water and bring volume to 100.0mL. Mix thoroughly.

L-Cysteine-Sulfide Reducing Agent:
Composition per 20.0mL:
L-Cysteine·HCl·H_2O 0.3g
$Na_2S \cdot 9H_2O$ 0.3g

Preparation of L-Cysteine-Sulfide Reducing Agent: Add L-cysteine·HCl·H_2O to 10.0mL of distilled/deionized water. Mix thoroughly. In a separate tube, add $Na_2S \cdot 9H_2O$ to 10.0mL of distilled/deionized water. Mix thoroughly. Gas both solutions with 100% N_2 and cap tubes. Autoclave both solutions for 15 min at 15 psi pressure–121°C using fast exhaust. Cool to 50°C. Aseptically combine the two solutions under 100% N_2.

Wolfe's Mineral Solution:
Composition per liter
$MgSO_4 \cdot 7H_2O$ 3.0g
Nitrilotriacetic acid 1.5g
NaCl 1.0g
$MnSO_4 \cdot H_2O$ 0.5g
$FeSO_4 \cdot 7H_2O$ 0.1g
$CoCl_2 \cdot 6H_2O$ 0.1g
$CaCl_2$ 0.1g
$ZnSO_4 \cdot 7H_2O$ 0.1g
$CuSO_4 \cdot 5H_2O$ 0.01g
$AlK(SO_4)_2 \cdot 12H_2O$ 0.01g
H_3BO_3 .... 0.01g
$Na_2MoO_4 \cdot 2H_2O$ 0.01g

Preparation of Wolfe's Mineral Solution: Add nitrilotriacetic acid to 500.0mL of distilled/deionized water. Dissolve by adjusting pH to 6.5 with KOH. Add remaining components. Add distilled/deionized water and bring volume to 1.0L.

Wolfe's Vitamin Solution:
Composition per liter:
Pyridoxine·HCl 10.0mg
Thiamine·HCl 5.0mg
Riboflavin 5.0mg
Nicotinic acid 5.0mg
Calcium pantothenate 5.0mg
p-Aminobenzoic acid 5.0mg
Thioctic acid 5.0mg
Biotin 2.0mg
Folic acid 2.0mg
Cyanocobalamin 100.0μg

Preparation of Wolfe's Vitamin Solution: Add components to distilled/deionized water and bring volume to 1.0L. Mix thoroughly. Filter sterilize.

Preparation of Medium: Add components, except vitamin solution and L-cysteine-sulfide reducing agent, to distilled/deionized water and bring volume to 970.0mL. Mix thoroughly. Autoclave for 15 min at 15 psi pressure–121°C. Cool under 80% N_2 + 20% CO_2. Aseptically add the sterile vitamin solution and then the sterile L-cysteine-sulfide reducing agent. Adjust the pH to 7.2. Distribute aseptically and anaerobically into sterile tubes.

Use: For the cultivation and maintenance of *Acetogenium kivui, Methanobacterium formicicum, Methanobacterium thermoautotrophicum,* and *Methanobrevibacter arboriphilicus.*

Methanobacteria Medium with Glucose and Yeast Extract

Composition per liter:

Glucose	5.0g
Yeast extract	2.0g
Mineral solution 2	50.0mL
Sodium carbonate solution	50.0mL
Mineral solution 1	25.0mL
L-Cysteine-sulfide reducing agent	20.0mL
Wolfe's mineral solution	10.0mL
Wolfe's vitamin solution	10.0mL
Resazurin (0.025% solution)	4.0mL

pH 7.2 ± 0.2 at 25°C

Mineral Solution 1:

Composition per liter:

K_2HPO_4	6.0g

Preparation of Mineral Solution 1: Add K_2HPO_4 to distilled/deionized water and bring volume to 1.0L. Mix thoroughly.

Mineral Solution 2:

Composition per liter:

NaCl	12.0g
KH_2PO_4	6.0g
$(NH_4)_2SO_4$	6.0g
$MgSO_4 \cdot 7H_2O$	2.4g
$CaCl_2 \cdot 2H_2O$	1.6g

Preparation of Mineral Solution 2: Add components to distilled/deionized water and bring volume to 1.0L. Mix thoroughly.

Sodium Carbonate Solution:

Composition per 100.0mL:

Na_2CO_3	8.0g

Preparation of Sodium Carbonate Solution: Add Na_2CO_3 to distilled/deionized water and bring volume to 100.0mL. Mix thoroughly.

L-Cysteine-Sulfide Reducing Agent:

Composition per 20.0mL:

L-Cysteine·HCl·H_2O	0.3g
$Na_2S \cdot 9H_2O$	0.3g

Preparation of L-Cysteine-Sulfide Reducing Agent: Add L-cysteine·HCl·H_2O to 10.0mL of distilled/deionized water. Mix thoroughly. In a separate tube, add $Na_2S \cdot 9H_2O$ to 10.0mL of distilled/deionized water. Mix thoroughly. Gas both solutions with 100% N_2 and cap tubes. Autoclave both solutions for 15 min at 15 psi pressure–121°C using fast exhaust. Cool to 50°C. Aseptically combine the two solutions under 100% N_2.

Wolfe's Mineral Solution:

Composition per liter

$MgSO_4 \cdot 7H_2O$	3.0g
Nitrilotriacetic acid	1.5g
NaCl	1.0g
$MnSO_4 \cdot H_2O$	0.5g
$FeSO_4 \cdot 7H_2O$	0.1g
$CoCl_2 \cdot 6H_2O$	0.1g
$CaCl_2$	0.1g
$ZnSO_4 \cdot 7H_2O$	0.1g
$CuSO_4 \cdot 5H_2O$	0.01g
$AlK(SO_4)_2 \cdot 12H_2O$	0.01g
H_3BO_3	0.01g
$Na_2MoO_4 \cdot 2H_2O$	0.01g

Preparation of Wolfe's Mineral Solution: Add nitrilotriacetic acid to 500.0mL of distilled/deionized water. Dissolve by adjusting pH to 6.5 with KOH. Add remaining components. Add distilled/deionized water to 1.0L.

Wolfe's Vitamin Solution:

Composition per liter:

Pyridoxine·HCl	10.0mg
Thiamine·HCl	5.0mg
Riboflavin	5.0mg
Nicotinic acid	5.0mg
Calcium pantothenate	5.0mg
p-Aminobenzoic acid	5.0mg
Thioctic acid	5.0mg
Biotin	2.0mg
Folic acid	2.0mg
Cyanocobalamin	100.0μg

Preparation of Wolfe's Vitamin Solution: Add components to distilled/deionized water and bring volume to 1.0L. Mix thoroughly. Filter sterilize.

Preparation of Medium: Add components, except vitamin solution and L-cysteine-sulfide reducing agent, to distilled/deionized water and bring volume to 970.0mL. Mix thoroughly. Autoclave for 15 min at 15 psi pressure–121°C. Cool under 80% N_2 + 20% CO_2. Aseptically add the sterile vitamin solution and then the sterile L-cysteine-sulfide reducing agent. Adjust the pH to 7.2. Distribute aseptically and anaerobically into sterile tubes.

Use: For the cultivation and maintenance of *Clostridium saccharolyticum, Clostridium thermoaceticum,* and *Clostridium thermohydrosulfuricum.*

Methanobacteria Medium with Xylose, Yeast Extract, and Tryptone

Composition per liter:

Pancreatic digest of casein	10.0g
Xylose	5.0g
Yeast extract	3.0g
Mineral solution 2	50.0mL
Sodium carbonate solution	50.0mL
Mineral solution 1	25.0mL
L-Cysteine-sulfide reducing agent	20.0mL
Wolfe's mineral solution	10.0mL
Wolfe's vitamin solution	10.0mL
Resazurin (0.025% solution)	4.0mL

pH 7.2 ± 0.2 at 25°C

Mineral Solution 1:

Composition per liter:

K_2HPO_4	6.0g

Preparation of Medium: Add K_2HPO_4 to distilled/deionized water and bring volume to 1.0L. Mix thoroughly.

Mineral Solution 2:

Composition per liter:

NaCl	12.0g
KH_2PO_4	6.0g

$(NH_4)_2SO_4$ 6.0g
$MgSO_4 \cdot 7H_2O$ 2.4g
$CaCl_2 \cdot 2H_2O$ 1.6g

Preparation of Mineral Solution 2: Add components to distilled/deionized water and bring volume to 1.0L. Mix thoroughly.

Sodium Carbonate Solution:
Composition per 100.0mL:
Na_2CO_3 .... 8.0g

Preparation of Sodium Carbonate Solution: Add Na_2CO_3 to distilled/deionized water and bring volume to 100.0mL. Mix thoroughly.

L-Cysteine-Sulfide Reducing Agent:
Composition per 20.0mL:
L-Cysteine·HCl·H_2O 0.3g
$Na_2S \cdot 9H_2O$ 0.3g

Preparation of L-Cysteine-Sulfide Reducing Agent: Add L-cysteine·HCl·H_2O to 10.0mL of distilled/deionized water. Mix thoroughly. In a separate tube, add $Na_2S \cdot 9H_2O$ to 10.0mL of distilled/deionized water. Mix thoroughly. Gas both solutions with 100% N_2 and cap tubes. Autoclave both solutions for 15 min at 15 psi pressure–121°C using fast exhaust. Cool to 50°C. Aseptically combine the two solutions under 100% N_2.

Wolfe's Mineral Solution:
Composition per liter:
$MgSO_4 \cdot 7H_2O$ 3.0g
Nitrilotriacetic acid 1.5g
NaCl 1.0g
$MnSO_4 \cdot H_2O$ 0.5g
$FeSO_4 \cdot 7H_2O$ 0.1g
$CoCl_2 \cdot 6H_2O$ 0.1g
$CaCl_2$ 0.1g
$ZnSO_4 \cdot 7H_2O$ 0.1g
$CuSO_4 \cdot 5H_2O$ 0.01g
$AlK(SO_4)_2 \cdot 12H_2O$ 0.01g
H_3BO_3 .... 0.01g
$Na_2MoO_4 \cdot 2H_2O$ 0.01g

Preparation of Wolfe's Mineral Solution: Add nitrilotriacetic acid to 500.0mL of distilled/deionized water. Dissolve by adjusting pH to 6.5 with KOH. Add remaining components. Add distilled/deionized water to 1.0L.

Wolfe's Vitamin Solution:
Composition per liter:
Pyridoxine·HCl 10.0mg
Thiamine·HCl 5.0mg
Riboflavin 5.0mg
Nicotinic acid 5.0mg
Calcium pantothenate 5.0mg
p-Aminobenzoic acid 5.0mg
Thioctic acid 5.0mg
Biotin 2.0mg
Folic acid 2.0mg
Cyanocobalamin 100.0μg

Preparation of Wolfe's Vitamin Solution: Add components to distilled/deionized water and bring volume to 1.0L. Mix thoroughly. Filter sterilize.

Preparation of Medium: Add components, except vitamin solution and L-cysteine-sulfide reducing agent, to distilled/deionized water and bring volume to 970.0mL. Mix thoroughly. Autoclave for 15 min at 15 psi pressure–121°C. Cool under 80% N_2 + 20% CO_2. Aseptically add the sterile vitamin solution and then the sterile L-cysteine-sulfide reducing agent. Adjust the pH to 7.2. Distribute aseptically and anaerobically into sterile tubes.

Use: For the cultivation and maintenance of *Thermobacteroides acetoethylicus*.

Methanobacteria Medium with Yeast Extract, Sodium Acetate, and Methanol

Composition per liter:
Glucose 5.0g
Sodium acetate 4.1g
Yeast extract 2.0g
Mineral solution 2 .... 50.0mL
Sodium carbonate solution 50.0mL
Mineral solution 1 .... 25.0mL
L-cysteine-sulfide reducing agent 20.0mL
Wolfe's mineral solution 10.0mL
Wolfe's vitamin solution 10.0mL
Methanol 4.0mL
Resazurin (0.025% solution) 4.0mL
pH 7.2 ± 0.2 at 25°C

Mineral Solution 1:
Composition per liter:
K_2HPO_4 .... 6.0g

Preparation of Mineral Solution 1: Add K_2HPO_4 to distilled/deionized water and bring volume to 1.0L. Mix thoroughly.

Mineral Solution 2:
Composition per liter:
NaCl 12.0g
KH_2PO_4 .... 6.0g
$(NH_4)_2SO_4$ 6.0g
$MgSO_4 \cdot 7H_2O$ 2.4g
$CaCl_2 \cdot 2H_2O$ 1.6g

Preparation of Mineral Solution 2: Add components to distilled/deionized water and bring volume to 1.0L. Mix thoroughly.

Sodium Carbonate Solution:
Composition per 100.0mL:
Na_2CO_3 .... 8.0g

Preparation of Sodium Carbonate Solution: Add Na_2CO_3 to distilled/deionized water and bring volume to 100.0mL. Mix thoroughly.

L-Cysteine-Sulfide Reducing Agent:
Composition per 20.0mL:
L-Cysteine·HCl·H_2O 300.0mg
$Na_2S \cdot 9H_2O$ 300.0mg

Preparation of L-Cysteine-Sulfide Reducing Agent: Add L-cysteine·HCl·H_2O to 10.0mL of distilled/deionized water. Mix thoroughly. In a separate tube, add $Na_2S \cdot 9H_2O$ to 10.0mL of distilled/deionized water. Mix thoroughly. Gas both solutions with 100% N_2 and cap tubes. Autoclave both solutions for 15 min at 15 psi pressure–121°C using fast exhaust. Cool to 50°C. Aseptically combine the two solutions under 100% N_2.

Wolfe's Mineral Solution:
Composition per liter:
$MgSO_4 \cdot 7H_2O$ 3.0g
Nitrilotriacetic acid 1.5g

NaCl	1.0g
$MnSO_4·H_2O$	0.5g
$FeSO_4·7H_2O$	0.1g
$CoCl_2·6H_2O$	0.1g
$CaCl_2$	0.1g
$ZnSO_4·7H_2O$	0.1g
$CuSO_4·5H_2O$	0.01g
$AlK(SO_4)_2·12H_2O$	0.01g
H_3BO_3	0.01g
$Na_2MoO_4·2H_2O$	0.01g

Preparation of Wolfe's Mineral Solution: Add nitrilotriacetic acid to 500.0mL of distilled/deionized water. Dissolve by adjusting pH to 6.5 with KOH. Add remaining components. Add distilled/deionized water to 1.0L.

Wolfe's Vitamin Solution:
Composition per liter:

Pyridoxine·HCl	10.0mg
Thiamine·HCl	5.0mg
Riboflavin	5.0mg
Nicotinic acid	5.0mg
Calcium pantothenate	5.0mg
p-Aminobenzoic acid	5.0mg
Thioctic acid	5.0mg
Biotin	2.0mg
Folic acid	2.0mg
Cyanocobalamin	100.0μg

Preparation of Wolfe's Vitamin Solution: Add components to distilled/deionized water and bring volume to 1.0L. Mix thoroughly. Filter sterilize.

Preparation of Medium: Add components, except vitamin solution, L-cysteine-sulfide reducing agent, and methanol, to distilled/deionized water and bring volume to 970.0mL. Mix thoroughly. Autoclave for 15 min at 15 psi pressure–121°C. Cool under 80% N_2 + 20% CO_2. Filter sterilize methanol. Aseptically add 4.0mL of sterile methanol to cooled, sterile basal medium. Aseptically add the sterile vitamin solution and then the sterile L-cysteine-sulfide reducing agent. Adjust the pH to 7.2. Distribute aseptically and anaerobically into sterile tubes.

Use: For the cultivation and maintenance of *Butyribacterium methylotrophicum.*

Methanobacterium alcaliphilum Medium

Composition per liter:

$NaHCO_3$	10.0g
Yeast extract	2.0g
Peptone	2.0g
NH_4Cl	1.0g
L-Cysteine·HCl·H_2O	0.5g
K_2HPO_4	0.4g
$MgCl_2·6H_2O$	0.1g
$CaCl_2$	0.02g
Resazurin	1.0mg
Salt solution	5.0mL

pH 8.4 ± 0.2 at 25°C

Salt Solution:
Composition per 100.0mL:

Sodium EDTA·$2H_2O$	0.1g
$CoCl_2·6H_2O$	0.03g
$MnCl_2·4H_2O$	0.02g
$ZnCl_2$	0.02g
$AlCl_3·6H_2O$	8.0mg
$CuCl_2·2H_2O$	4.0mg
$NiSO_4·6H_2O$	4.0mg
Na_2SeO_3	2.7mg
$FeSO_4·7H_2O$	2.0mg
H_3BO_3	2.0mg
$NaMoO_4·2H_2O$	2.0mg

Preparation of Salt Solution: Add components to distilled/deionized water and bring volume to 100.0mL. Mix thoroughly.

Preparation of Medium: Add components, except $NaHCO_3$, yeast extract, peptone, and L-cysteine·HCl·H_2O, to distilled/deionized water and bring volume to 990.0mL. Gently heat and bring to boiling under O_2-free 100% N_2. Continue boiling until resazurin becomes pale, indicating partial reduction. Add the yeast extract, peptone, and L-cysteine·HCl·H_2O and continue boiling under O_2-free 100% N_2 until resazurin becomes colorless, indicating complete reduction. Cool to room temperature under O_2-free 100% N_2. Add $NaHCO_3$ to 10.0mL of distilled/deionized water. Mix thoroughly. Gas with O_2-free 100% N_2 in a sealed tube. Add reduced $NaHCO_3$ solution to cooled reduced medium. Distribute anaerobically into tubes. Cap with butyl rubber stoppers and secure with closures. Autoclave for 15 min at 15 psi pressure–121°C with fast exhaust.

Use: For the cultivation and maintenance of *Methanobacterium alcaliphilum.*

Methanobacterium Enrichment Medium

Composition per liter:

$CaCO_3$	100.0g
K_2HPO_4	5.0g
$(NH_4)_2SO_4$	0.3g
$MgSO_47H_2O$	0.1g
$FeSO_4·7H_2O$	0.02g
Na_2CO_3 solution	10.0mL
$Na_2S·9H_2O$ solution	10.0mL
Ethanol	10.0mL
Yeast autolysate	5.0mL

pH 7.2 ± 0.2 at 25°C

Na_2CO_3 Solution:
Composition per 10.0mL:

$NaHCO_3$	0.5g

Preparation of Na_2CO_3 Solution: Add Na_2CO_3 to distilled/deionized water and bring volume to 10.0mL. Mix thoroughly. Filter sterilize.

$Na_2S·9H_2O$ Solution:
Composition per 10.0mL:

$Na_2S·9H_2O$	0.1g

Preparation of $Na_2S·9H_2O$ Solution: Add $Na_2S·9H_2O$ to distilled/deionized water and bring volume to 10.0mL. Mix thoroughly. Filter sterilize.

Preparation of Medium: Filter sterilize ethanol. Add components—except ethanol, Na_2CO_3 solution, and $Na_2S·9H_2O$ solution—to distilled/deionized water and bring volume to 970.0mL. Mix thoroughly. Gently heat and bring to boiling. Autoclave for 15 min at 15 psi pressure–121°C. Cool to 45°–50°C. Aseptically add sterile ethanol, Na_2CO_3 solution, and $Na_2S·9H_2O$ solution. Mix thoroughly. Aseptically distribute into sterile tubes or flasks.

Use: For the cultivation and enrichment of *Methanobacterium* species.

Methanobacterium espanolae Medium

Composition per 1020.0mL:

Sodium acetate·$3H_2O$	2.5g
Na_2CO_3	0.5g
$(NH_4)_2SO_4$	0.45g
K_2HPO_4	0.29g
KH_2PO_4	0.18g
$MgSO_4 \cdot 7H_2O$	0.12g
$CaCl_2 \cdot 2H_2O$	0.06g
NaCl	0.05g
L-Cysteine·HCl	0.25g
$Na_2S \cdot 9H_2O$	0.25g
Resazurin	1.0mg
Trace elements solution	10.0mL
Vitamin solution	10.0mL

pH 5.5 ± 0.2 at 25°C

Trace Elements Solution:

Composition per liter:

$MgSO_4 \cdot 7H_2O$	3.0g
Nitrilotriacetic acid	1.5g
NaCl	1.0g
$MnSO_4 \cdot 2H_2O$	0.5g
$CoSO_4 \cdot 7H_2O$	0.18g
$ZnSO_4 \cdot 7H_2O$	0.18g
$CaCl_2 \cdot 2H_2O$	0.1g
$FeSO_4 \cdot 7H_2O$	0.1g
$NiCl_2 \cdot 6H_2O$	0.025g
$KAl(SO_4)_2 \cdot 12H_2O$	0.02g
$CuSO_4 \cdot 5H_2O$	0.01g
H_3BO_3	0.01g
$Na_2MoO_4 \cdot 2H_2O$	0.01g
$Na_2SeO_3 \cdot 5H_2O$	0.3mg

Preparation of Trace Elements Solution: Add nitrilotriacetic acid to 500.0mL of distilled/deionized water. Adjust pH to 6.5 with KOH. Add remaining components. Add distilled/deionized water to 1.0L.

Vitamin Solution:

Composition per liter:

Pyridoxine·HCl	10.0mg
Calcium DL-pantothenate	5.0mg
Lipoic acid	5.0mg
Nicotinic acid	5.0mg
p-Aminobenzoic acid	5.0mg
Riboflavin	5.0mg
Thiamine·HCl	5.0mg
Biotin	2.0mg
Folic acid	2.0mg
Vitamin B_{12}	0.1mg

Preparation of Vitamin Solution: Add components to distilled/deionized water and bring volume to 1.0L. Adjust pH to 7.0. Mix thoroughly. Sparge with 80% N_2 + 20% CO_2.

Preparation of Medium: Prepare and dispense the medium anaerobically with 80% N_2 and 20% CO_2. Add components to distilled/deionized water and bring volume to 1.0L. Mix thoroughly. Adjust pH to 5.5–6.0. Anaerobically distribute into tubes or flasks. Autoclave for 15 min at 15 psi pressure–121°C. Check the pH of the medium after autoclaving.

Use: For the cultivation of *Methanobacterium espanolae.*

Methanobacterium formicicum Medium

Composition per liter:

$NaCH_3CO_2$	2.0g
$(NH_4)_2SO_4$	1.48g
NaCl	0.45g
L-Cysteine·HCl·H_2O	0.27g
$CaCl_2 \cdot 2H_2O$	0.06g
$Fe(NH_4)(SO_4)_2$	0.06g
$MgSO_4$	45.0mg
Na_2MoO_4	24.0mg
K_2HPO_4	21.0mg
KH_2PO_4	21.0mg
Resazurin	1.0mg
Na_2SeO_3	0.2mg
$Na_2S \cdot 9H_2O$ solution	1.0mL

pH 6.8 ± 0.2 at 25°C

$Na_2S \cdot 9H_2O$ Solution:

Composition per 100.0mL:

$Na_2S \cdot 9H_2O$	25.0g

Preparation of $Na_2S \cdot 9H_2O$ Solution: Bring 100.0mL of distilled/deionized water to boiling. Cool to room temperature while sparging with 100%N_2. Add $Na_2S \cdot 9H_2O$ to the 100.0mL of anaerobic water. Mix thoroughly. Distribute into serum bottles fitted with butyl rubber stoppers and aluminum seals. Do not grease stoppers. Autoclave for 20 min at 15 psi pressure–121°C.

Preparation of Medium: Add components, except $Na_2S \cdot 9H_2O$ solution, to distilled/deionized water and bring volume to 1.0L. Mix thoroughly. Adjust pH to 6.8 with concentrated HCl. Distribute into tubes or flasks. Autoclave for 15 min at 15 psi pressure–121°C. Cool medium while sparging with 100% N_2. Prior to inoculation, aseptically and anaerobically add 1.0mL of sterile $Na_2S \cdot 9H_2O$ solution per liter of medium. Repeat the addition of 1.0mL of sterile $Na_2S \cdot 9H_2O$ solution per liter of medium every 48 hr during growth.

Use: For the cultivation of *Methanobacterium formicicum.*

Methanobacterium Mass-Culturing Medium

Composition per liter:

NH_4Cl	2.0g
NaCl	0.6g
KH_2PO_4	0.42g
K_2HPO_4	0.23g
Na_2CO_3	0.16g
$MgCl_2 \cdot 2H_2O$	0.04g
$CaCl_2 \cdot 2H_2O$	0.03g
Resazurin	0.001g
$Na_2S \cdot 9H_2O$ solution	10.0mL
Trace elements solution	10.0mL

$Na_2S \cdot 9H_2O$ Solution:

Composition per 100.0mL:

NaOH	1 pellet
$Na_2S \cdot 9H_2O$	2.5g

Preparation of $Na_2S \cdot 9H_2O$ Solution: Bring 100.0mL of distilled/deionized water to boiling. Cool to room temperature while sparging with 100%N_2. Dissolve 1 pellet of NaOH in the anaerobic water. Weigh out a little more than 2.5g of $Na_2S \cdot 9H_2O$. Briefly rinse the crystals in distilled/deionized water. Dry the crystals by blotting on paper towels or filter paper. Add 2.5g of washed $Na_2S \cdot 9H_2O$ crystals to 100.0mL of anaerobic NaOH solution. Distribute into serum bottles fitted with butyl rubber stoppers and aluminum seals. Do not grease stoppers. Pressurize to 60kPa with 100% N_2. Autoclave for 15 min at 15 psi pressure–121°C. Store at room temperature in an anaerobic chamber.

Trace Elements Solution:
Composition per liter:

Sodium nitrilotriacetate	1.67g
$CoCl_2 \cdot 6H_2O$	0.18g
$FeCl_2 \cdot 4H_2O$	0.14g
$MnCl_2 \cdot 4H_2O$	0.09g
$NiCl_2 \cdot 6H_2O$	0.09g
$ZnCl_2$	0.09g
$CaCl_2$	0.06g
$Na_2MoO_4 \cdot 2H_2O$	0.046g
$CuSO_4$	0.045g

Preparation of Trace Elements Solution: Add sodium nitrilotriacetate to 500.0mL of distilled/deionized water. Adjust pH to 6.5. Add remaining components. Add distilled/deionized water to 1.0L. Adjust pH to 7.0.

Preparation of Medium: Prepare and dispense medium under 80% H_2 + 20% CO_2. Add components, except Na_2CO_3 and $Na_2S \cdot 9H_2O$ solution, to distilled/deionized water and bring volume to 990.0mL. Mix thoroughly. Gently heat and bring to boiling. Continue boiling for 3 min. Cool to room temperature while sparging with 80% H_2 + 20% CO_2. Add Na_2CO_3. Mix thoroughly. Anaerobically distribute 9.9mL volumes into anaerobic tubes. Autoclave for 15 min at 15 psi pressure–121°C. Aseptically and anaerobically add 0.1mL of sterile $Na_2S \cdot 9H_2O$ solution to each tube. Mix thoroughly.

Use: For the cultivation of *Methanobacterium* species.

Methanobacterium Medium

Composition per 1010.0mL:

$NaHCO_3$	4.0g
Sodium formate	2.0g
Sodium acetate	1.0g
Yeast extract	1.0g
L-Cysteine·HCl	0.5g
KH_2PO_4	0.5g
$Na_2S \cdot 9H_2O$	0.5g
$MgSO_4 \cdot 7H_2O$	0.4g
NaCl	0.4g
NH_4Cl	0.4g
$CaCl_2 \cdot 2H_2O$	0.05g
$FeSO_4 \cdot 7H_2O$	2.0mg
Resazurin	1.0mg
Sludge fluid	50.0mL
Fatty acid mixture	20.0mL
Trace elements solution SL-10	1.0mL

pH 6.7 ± 0.2 at 25°C

Sludge Fluid:
Composition per 100.0mL:

Yeast extract	0.4g
Sludge	100.0mL

Preparation of Sludge Fluid: To 100.0mL of sludge from an anaerobic digester, add 0.4g of yeast extract. Sparge with 100% N_2 for a few minutes. Incubate at 37°C for 24 hr. Centrifuge the sludge at 13,000 × g for 15 min. Decant the clear supernatant solution. Sparge with 100% N_2 for a few minutes. Store in screw-capped bottles at room temperature in the dark.

Fatty Acid Mixture:
Composition per 20.0mL:

α-Methylbutyric acid	0.5g
Isobutyric acid	0.5g
Isovaleric acid	0.5g
Valeric acid	0.5g

Preparation of Fatty Acid Mixture: Add components to distilled/deionized water and bring volume to 20.0mL. Mix thoroughly. Adjust pH to 7.5 with concentrated NaOH.

Trace Elements Solution SL-10:
Composition per liter:

$FeCl_2 \cdot 4H_2O$	1.5g
$CoCl_2 \cdot 6H_2O$	190.0mg
$MnCl_2 \cdot 4H_2O$	100.0mg
$ZnCl_2$	70.0mg
$Na_2MoO_4 \cdot 2H_2O$	36.0mg
$NiCl_2 \cdot 6H_2O$	24.0mg
H_3BO_3	6.0mg
$CuCl_2 \cdot 2H_2O$	2.0mg
HCl (25% solution)	10.0mL

Preparation of Trace Elements Solution SL-10: Add $FeCl_2 \cdot 4H_2O$ to 10.0mL of HCl solution. Mix thoroughly. Add distilled/deionized water and bring volume to 1.0L. Add remaining components. Mix thoroughly. Autoclave for 15 min at 15 psi pressure–121°C.

Preparation of Medium: Prepare and dispense medium anaerobically under 80% H_2 + 20% CO_2. Add components to distilled/deionized water and bring volume to 1010.0mL. Mix thoroughly. Sparge with 80% H_2 + 20% CO_2. Autoclave for 15 min at 15 psi pressure–121°C.

Use: For the cultivation and maintenance of *Methanobacterium* species.

Methanobacterium ruminantium Medium

Composition per liter:

$NaHCO_3$	6.0g
NaCl	2.0g
L-Cysteine·HCl·H_2O	1.0g
$K_2HPO_4 \cdot 3H_2O$	1.0g
KH_2PO_4	1.0g
NH_4Cl	1.0g
$CaCl_2 \cdot 2H_2O$	0.1g
$MgSO_4 \cdot 7H_2O$	0.1g
Resazurin	1.0mg
Rumen fluid	300.0mL
$Na_2S \cdot 9H_2O$ solution	10.0mL

6.8 ± 0.2 at 25°C

$Na_2S \cdot 9H_2O$ Solution:
Composition per 10.0mL:

$Na_2S \cdot 9H_2O$	0.25g

Preparation of $Na_2S \cdot 9H_2O$ Solution: Add $Na_2S \cdot 9H_2O$ to distilled/deionized water and bring volume to 10.0mL. Mix thoroughly. Autoclave for 15 min at 15 psi pressure–121°C. Cool to 25°C.

Preparation of Medium: Prepare and distribute medium anaerobically under 80% H_2 + 20% CO_2. Add components, except rumen fluid and $Na_2S \cdot 9H_2O$ solution, to distilled/deionized water and bring volume to 690.0mL. Mix thoroughly. Gently heat and bring to boiling. Continue boiling until resazurin turns colorless, indicating reduction. Autoclave for 15 min at 15 psi pressure–121°C. Cool to 25°C. Aseptically add 10.0mL of sterile $Na_2S \cdot 9H_2O$ solution and 300.0mL of sterile rumen fluid. Mix thoroughly. Aseptically and anaerobically distribute into sterile tubes or flasks.

Use: For the cultivation of *Methanobacterium ruminantium.*

Methanobacterium subterraneum Medium (DSMZ Medium 814)

Composition per liter:

Na-formate	3.2g
Na-acetate	1.0g
Tryptone	1.0g
K_2HPO_4	0.5g
NaCl	0.45g
NH_4Cl	0.4g
$MgCl_2·6H_2O$	0.03g
$FeSO_4·7H_2O$	0.003g
Resazurin	0.5mg
Trace elements solution	10.0mL
Cysteine solution	10.0mL
$Na_2S·9H_2O$ solution	10.0mL
$NaHCO_3$ solution	10.0mL
Seven vitamin solution	1.0mL

pH 8.3 ± 0.2 at 25°C

$NaHCO_3$ Solution:

Composition per 10.0mL:

$NaHCO_3$	2.0g

Preparation of $NaHCO_3$ Solution: Add $NaHCO_3$ to distilled/deionized water and bring volume to 10.0mL. Mix thoroughly. Sparge with 100% N_2. Autoclave for 15 min at 15 psi pressure–121°C. Cool to 25°C. Must be prepared freshly.

Cysteine Solution:

Composition per 10.0mL:

L-Cysteine-HCl·H_2O	0.25g

Preparation of Cysteine Solution: Add L-cysteine·HCl·H_2O to distilled/deionized water and bring volume to 10.0mL. Mix thoroughly. Sparge with 100% N_2. Autoclave for 15 min at 15 psi pressure–121°C.

$Na_2S·9H_2O$ Solution:

Composition per 10.0mL:

$Na_2S·9H_2O$	0.25g

Preparation of $Na_2S·9H_2O$ Solution: Add $Na_2S·9H_2O$ to distilled/deionized water and bring volume to 10.0mL. Sparge with N_2. Autoclave for 15 min at 15 psi pressure–121°C. Cool to 25°C. Store anaerobically.

Trace Elements Solution:

Composition per liter:

$MgSO_4·7H_2O$	3.0g
Nitrilotriacetic acid	1.5g
NaCl	1.0g
$MnSO_4·2H_2O$	0.5g
$CoSO_4·7H_2O$	0.18g
$ZnSO_4·7H_2O$	0.18g
$CaCl_2·2H_2O$	0.1g
$FeSO_4·7H_2O$	0.1g
$NiCl_2·6H_2O$	0.025g
$KAl(SO_4)_2·12H_2O$	0.02g
H_3BO_3	0.01g
$Na_2MoO_4·4H_2O$	0.01g
$CuSO_4·5H_2O$	0.01g
$Na_2SeO_3·5H_2O$	0.3mg

Preparation of Trace Elements Solution: Add nitrilotriacetic acid to 500.0mL of distilled/deionized water. Dissolve by adjusting pH to 6.5 with KOH. Add remaining components. Add distilled/deionized water to 1.0L. Mix thoroughly.

Seven Vitamin Solution:

Composition per liter:

Pyridoxine hydrochloride	300.0mg
Thiamine-HCl·$2H_2O$	200.0mg
Nicotinic acid	200.0mg
Vitamin B_{12}	100.0mg
Calcium pantothenate	100.0mg
p-Aminobenzoic acid	80.0mg
D(+)-Biotin	20.0mg

Preparation of Seven Vitamin Solution: Add components to distilled/deionized water and bring volume to 1.0L. Sparge with 100% N_2. Mix thoroughly. Filter sterilize.

Preparation of Medium: Prepare and dispense medium under 100% N_2. Add components, except $Na_2S·9H_2O$ solution, seven vitamin solution, cysteine solution, and $NaHCO_3$ solution, to distilled/deionized water and bring volume to 969.0mL. Mix thoroughly. Flush medium with N_2 for 5 min. Adjust medium pH to 8.3. Autoclave for 15 min at 15 psi pressure–121°C. Aseptically and anaerobically add 1.0mL seven vitamin mix, 10.0mL cysteine solution, 10.0mL $Na_2S·9H_2O$, and 10.0mL $NaHCO_3$ solution. Mix thoroughly. Adjust pH to 8.3. Aseptically and anaerobically distribute into sterile tubes or bottles.

Use: For the cultivation of *Methanobacterium subterraneum.*

Methanobacterium thermoautotrophicum Marburg Medium

Composition per liter:

L-Cysteine·HCl·H_2O	12.5g
$Na_2S·9H_2O$	12.5g
KH_2PO_4	6.8g
NH_4Cl	2.1g
Na_2CO_3 (1*M* solution)	32.0mL
Trace elements solution	10.0mL
Resazurin (0.2% solution)	0.3mL

Trace Elements Solution:

Composition per liter:

$MgCl_2·6H_2O$	4.0g
Nitrilotriacetic acid	3.0g
$FeCl_2·4H_2O$	1.0g
$NiCl_2·6H_2O$	0.12g
$CoCl_2·6H_2O$	0.02g
$NaMoO_4·2H_2O$	0.02g

Preparation of Trace Elements Solution: Add nitrilotriacetic acid to 500.0mL of distilled/deionized water. Adjust pH to 6.5 with NaOH. Add remaining components. Mix thoroughly. Add distilled/deionized water to 1.0L. Adjust pH to 7.0.

Preparation of Medium: Prepare and dispense medium under 80% H_2 + 20% CO_2. Add components, except L-cysteine·HCl·H_2O and $Na_2S·9H_2O$, to distilled/deionized water and bring volume to 1.0L. Mix thoroughly. Gently heat and bring to boiling. Continue boiling for 3 min. Cool to room temperature while sparging with 80% H_2 + 20% CO_2. Add L-cysteine·HCl·H_2O and $Na_2S·9H_2O$. Mix thoroughly. Anaerobically distribute into serum bottles fitted with butyl rubber stoppers and aluminum seals. Do not grease stoppers. Autoclave for 15 min at 15 psi pressure–121°C.

Use: For the cultivation of *Methanobacterium thermoautotrophicum.*

Methanobacterium thermoautotrophicum Medium (DSMZ Medium 131)

Composition per liter:

Na_2CO_3	4.0g
$(NH_4)_2SO_4$	1.5g
NaCl	0.6g
KH_2PO_4	0.3g
K_2HPO_4	0.15g
$MgSO_4 \cdot 7H_2O$	0.12g
$CaCl_2 \cdot 2H_2O$	0.08g
$FeSO_4 \cdot 7H_2O$	4.0mg
Resazurin	1.0mg
Vitamin solution	10.0mL
Trace elements solution	10.0mL
L-Cysteine solution	10.0mL
$Na_2S \cdot 9H_2O$ solution	10.0mL

pH 7.2 ± 0.2 at 25°C

Vitamin Solution:

Composition per liter:

Pyridoxine-HCl	10.0mg
Thiamine-HCl·2H_2O	5.0mg
Riboflavin	5.0mg
Nicotinic acid	5.0mg
Ca-pantothenate	5.0mg
Biotin	2.0mg
Folic acid	2.0mg
p-Aminobenzoic acid	1.0mg
Vitamin B_{12}	0.01mg

Preparation of Vitamin Solution: Add components to distilled/deionized water and bring volume to 1.0L. Mix thoroughly. Sparge with 80% H_2 + 20% CO_2. Filter sterilize.

Trace Elements Solution:

Composition per liter:

$MgSO_4 \cdot 7H_2O$	6.2g
NaCl	1.0g
Na_2-EDTA	0.64g
$MnSO_4 \cdot 4H_2O$	0.55g
$ZnSO_4 \cdot 7H_2O$	0.18g
$CoCl_2 \cdot 6H_2O$	0.17g
$CaCl_2 \cdot 2H_2O$	0.13g
$FeSO_4 \cdot 7H_2O$	0.1g
$CuSO_4$	0.05g
$NiCl_2 \cdot 6H_2O$	0.025g
$KAl(SO_4)_2 \cdot 12H_2O$	0.018g
$Na_2MoO_4 \cdot 4H_2O$	0.011g
H_3BO_3	0.01g

Preparation of Trace Elements Solution: Add Na_2-EDTA to 500.0mL distilled/deionized water. Mix thoroughly. Add other components and bring volume to 1.0L with distilled/deionized water. Mix thoroughly. Sparge with 80% H_2 + 20% CO_2. Autoclave for 15 min at 15 psi pressure–121°C.

$Na_2S \cdot 9H_2O$ Solution:

Composition per 10.0mL:

$Na_2S \cdot 9H_2O$	1.5g

Preparation of $Na_2S \cdot 9H_2O$ Solution: Add $Na_2S \cdot 9H_2O$ to distilled/deionized water and bring volume to 10.0mL. Sparge with N_2. Autoclave for 15 min at 15 psi pressure–121°C. Cool to 25°C. Store anaerobically.

L-Cysteine Solution:

Composition per 10.0mL:

L-Cysteine·HCl·H_2O	1.5g

Preparation of L-Cysteine Solution: Add L-cysteine·HCl·H_2O to distilled/deionized water and bring volume to 10.0mL. Mix thoroughly. Sparge with 100% N_2. Autoclave for 15 min at 15 psi pressure–121°C.

Preparation of Medium: Prepare and dispense medium under an oxygen-free 80% H_2 + 20% CO_2 gas mixture. Add components, except vitamin solution, L-cysteine solution, and $Na_2S \cdot 9H_2O$ solution, to distilled/deionized water and bring volume to 1.0L. Mix thoroughly. Sparge with 80% H_2 + 20% CO_2. Autoclave for 15 min at 15 psi pressure–121°C. Cool to 25°C while sparging with 80% H_2 + 20% CO_2. Aseptically and anaerobically add 10.0mL vitamin solution, 10.0mL of sterile L-cysteine solution, and 10.0mL of sterile $Na_2S \cdot 9H_2O$ solution. Mix thoroughly. Aseptically and anaerobically distribute into sterile tubes or flasks. Alternately the medium can be distributed to tubes under anaerobic conditions and autoclaved in tubes prior to addition of substrate solution, vitamin solution, and $Na_2S \cdot 9H_2O$ solution. Appropriate amounts of these solutions can then be added to each tube by syringes to yield the desired concentrations.

Use: For the cultivation of *Ectothiorhodospira shaposhnikovii* and *Methanothermobacter thermautotrophicus.*

Methanobacterium thermoautotrophicum Medium

Composition per liter:

Na_2CO_3	4.0g
$(NH_4)_2SO_4$	1.5g
NaCl	0.6g
KH_2PO_4	0.3g
K_2HPO_4	0.15g
$MgSO_4 \cdot 7H_2O$	0.12g
$CaCl_2 \cdot 2H_2O$	0.08g
$FeSO_4 \cdot 7H_2O$	4.0mg
Resazurin	1.0mg
Trace elements solution	10.0mL
Vitamin solution	10.0mL
L-Cysteine·HCl solution	10.0mL
$Na_2S \cdot 9H_2O$ solution	10.0mL

pH 7.2 ± 0.2 at 25°C

Trace Elements Solution:

Composition per liter:

NaCl	1.0g
Disodium EDTA	0.64g
$MnSO_4 \cdot 4H_2O$	0.55g
$MgSO_4 \cdot 7H_2O$	0.2g
$ZnSO_4 \cdot 7H_2O$	0.18g
$CoCl_2 \cdot 6H_2O$	0.17g
$CuSO_4$	0.15g
$CaCl_2 \cdot 2H_2O$	0.13g
$FeSO_4 \cdot 7H_2O$	0.1g
$NiCl_2 \cdot H_2O$	0.025g
$KAl(SO_4)_2 \cdot 12H_2O$	0.018g
$Na_2MoO_4 \cdot 2H_2O$	0.011g
H_3BO_3	0.01g

Preparation of Trace Elements Solution: Add disodium EDTA to 500.0mL of distilled/deionized water. Mix thoroughly to dissolve. Add remaining components. Bring volume to 1.0L with distilled/deionized water.

Vitamin Solution:

Composition per liter:

Pyridoxine·HCl	10.0mg
Calcium DL-pantothenate	5.0mg
Nicotinic acid	5.0mg
Riboflavin	5.0mg
Thiamine·HCl	5.0mg
Biotin	2.0mg
Folic acid	2.0mg
p-Aminobenzoic acid	1.0mg
Cyanocobalamin	0.01mg

Preparation of Vitamin Solution: Add components to distilled/deionized water and bring volume to 1.0L. Adjust pH to 7.0. Mix thoroughly.

L-Cysteine·HCl Solution:

Composition per 10.0mL:

L-Cysteine·HCl	1.5g

Preparation of L-Cysteine·HCl Solution: Add L-cysteine·HCl to distilled/deionized water and bring volume to 10.0mL. Mix thoroughly. Autoclave under 80% N_2 + 20% CO_2 for 15 min at 15 psi pressure–121°C.

$Na_2S·9H_2O$ Solution:

Composition per 10.0mL:

$Na_2S·9H_2O$	1.5g

Preparation of $Na_2S·9H_2O$ Solution: Add $Na_2S·9H_2O$ to distilled/deionized water and bring volume to 10.0mL. Mix thoroughly. Sparge with 80% N_2 + 20% CO_2. Autoclave for 15 min at 15 psi pressure–121°C.

Preparation of Medium: Prepare and dispense medium under 80% H_2 + 20% CO_2. Add components, except L-cysteine·HCl solution and $Na_2S·9H_2O$ solution, to distilled/deionized water and bring volume to 980.0mL. Mix thoroughly. Anaerobically distribute into tubes or flasks fitted with butyl rubber stoppers. Autoclave for 15 min at 15 psi pressure–121°C. Anaerobically add 10.0mL of sterile L-cysteine·HCl solution and 10.0mL of sterile $Na_2S·9H_2O$ solution to each liter of medium or, using a syringe, inject the appropriate amount of sterile L-cysteine·HCl solution and sterile $Na_2S·9H_2O$ solution into individual tubes containing medium.

Use: For the cultivation and maintenance of *Methanobacterium thermoautotrophicus* and *Methanoculleus oldenburgensis*.

Methanobacterium thermoautotrophicum Medium, Taylor and Pirt

Composition per liter:

Na_2CO_3	4.0g
$(NH_4)_2SO_4$	3.0g
NaCl	1.2g
KH_2PO_4	0.6g
L-Cysteine·HCl·H_2O	0.5g
K_2HPO_4	0.3g
Nitrilotriacetic acid	0.03g
$CoCl_2$	0.02g
$CaCl_2$	0.01g
$FeSO_4$	0.01g
$MgSO_4$	0.01g
$MnSO_4$	0.01g
$ZnSO_4$	2.0mg
Resazurin	1.0mg
$AlK(SO_4)_2$	0.2mg
$CuSO_4$	0.2mg
H_3BO_3	0.2mg
Na_2MoO_4	0.2mg
$Na_2S·9H_2O$ solution	10.0mL

pH 7.2 ± 0.2 at 25°C

$Na_2S·9H_2O$ Solution:

Composition per 10.0mL:

$Na_2S·9H_2O$	0.5g

Preparation of $Na_2S·9H_2O$ Solution: Add $Na_2S·9H_2O$ to distilled/deionized water and bring volume to 10.0mL. Mix thoroughly. Autoclave for 15 min at 15 psi pressure–121°C. Cool to 25°C.

Preparation of Medium: Prepare and distribute medium anaerobically under 80% H_2 + 20% CO_2. Add components, except $Na_2S·9H_2O$ solution, to distilled/deionized water and bring volume to 990.0mL. Mix thoroughly. Gently heat and bring to boiling. Continue boiling until resazurin turns colorless, indicating reduction. Autoclave for 15 min at 15 psi pressure–121°C. Cool to 25°C. Aseptically add 10.0mL of sterile $Na_2S·9H_2O$ solution. Mix thoroughly. Aseptically and anaerobically distribute into sterile tubes or flasks.

Use: For the cultivation of *Methanobacterium thermoautotrophicum*.

Methanobacterium thermoautotrophicum MS Medium

Composition per liter:

$NaHCO_3$	8.4g
Pancreatic digest of casein	2.0g
Yeast extract	2.0g
$MgCl_2·6H_2O$	1.0g
NH_4Cl	1.0g
K_2HPO_4	0.84g
$CaCl_2·2H_2O$	0.4g
Reazurin	0.001g
Trace minerals solution	1.0mL

pH 7.1 ± 0.2 at 25°C

Trace Minerals Solution:

Composition per liter:

Disodium EDTA·$2H_2O$	0.5g
$ZnSO_4·2H_2O$	0.21g
$CoCl_2·6H_2O$	0.15g
$AlK(SO_4)_2$	0.14g
$FeSO_4·7H_2O$	0.1g
$MnSO_4·H_2O$	0.1g
$Na_2WO_4·2H_2O$	0.03g
$CuCl_2·2H_2O$	0.02g
$NiCl_2·6H_2O$	0.02g
H_3BO_3	0.01g
$Na_2MoO_2·2H_2O$	0.01g
Na_2SeO_4	0.01g

Preparation of Trace Minerals Solution: Add components, except $MnSO_4·H_2O$, to distilled/deionized water and bring volume to 1.0L. Mix thoroughly. Adjust pH to 7.1. Add $MnSO_4·H_2O$. Mix thoroughly.

Preparation of Medium: Prepare and dispense medium under 80% H_2 + 20% CO_2. Add components to distilled/deionized water and bring volume to 1.0L. Mix thoroughly. Gently heat and bring to boiling. Continue boiling for 3 min. Cool to room temperature while sparging with 80% H_2 + 20% CO_2. Adjust pH to 7.1 with 6*N* HCl. Anaerobically distribute into serum bottles fitted with butyl rubber stoppers and aluminum seals. Do not grease stoppers. Autoclave for 15 min at 15 psi pressure–121°C. Store at room temperature in an anaerobic chamber.

Use: For the cultivation of *Methanobacterium thermoautotrophicum.*

Methanobacterium Transduction Agar

Composition per liter:

Agar solution	500.0mL
Basal salts solution	100.0mL
Na_2CO_3 solution	24.0mL
Titanium (III) citrate solution	16.0mL
Trace elements solution	10.0mL
L-Cysteine/Na_2S solution	10.0mL

Agar Solution:

Composition per 500.0mL:

Agar	15.0g

Preparation of Agar Solution: Add agar to distilled/deionized water and bring volume to 500.0mL. Mix thoroughly. Gently heat and bring to boiling. Sparge with 100% N_2. Autoclave for 15 min at 15 psi pressure–121°C. Cool to 60°C while sparging with 80% N_2 + 20% CO_2.

Basal Salts Solution:

Composition per liter:

KH_2PO_4	68.0g
NH_4Cl	21.2g
Resazurin	10.0mg

Preparation of Basal Salts Solution: Add components to distilled/deionized water and bring volume to 1.0L. Mix thoroughly. Sparge with 100% N_2.

Trace Elements Solution:

Composition per liter:

Nitrilotriacetic acid	9.55g
$MgCl_2 \cdot 7H_2O$	4.06g
$FeCl_2 \cdot 7H_2O$	0.98g
$NaWO_4 \cdot 6H_2O$	0.264g
$NiCl_2 \cdot 6H_2O$	0.118g
$Na_2Se_2O_3$	0.1g
$NaMoO_4 \cdot 2H_2O$	0.024g
$CoCl_2 \cdot 6H_2O$	0.023g

Preparation of Trace Elements Solution: Add nitrilotriacetic acid to 500.0mL of distilled/deionized water. Adjust pH to 7.0 with KOH. Add remaining components one at a time. Add distilled/deionized water to 1.0L. Adjust pH to 7.0. Sparge with 100% N_2.

Na_2CO_3 Solution:

Composition per liter:

Na_2CO_3	10.6g

Preparation of Na_2CO_3 Solution: Add Na_2CO_3 to distilled/deionized water and bring volume to 100.0mL. Mix thoroughly. Sparge with 80% N_2 + 20% CO_2.

L-Cysteine/Na_2S Solution:

Composition per 50.0mL:

L-Cysteine·HCl	5.0g
$Na_2S \cdot 9H_2O$	5.0g

Preparation of L-Cysteine/Na_2S Solution: Gently heat and bring 50.0mL of distilled/deionized water to boiling. Add L-cysteine·HCl and $Na_2S \cdot 9H_2O$. Mix thoroughly. Cool while sparging with 100% N_2. Bring volume to 50.0mL with distilled/deionized water. Sparge with 100% N_2. Adjust pH to 10.0 with 3*N* NaOH.

Titanium (III) Citrate Solution:

Composition per 55.0mL:

Sodium citrate·$9H_2O$	2.9g
Titanium (III) chloride	0.75g

Preparation of Titanium (III) Citrate Solution: Add components to distilled/deionized water and bring volume to 55.0mL. Mix thoroughly. Filter sterilize. Sparge with 100% N_2.

Preparation of Medium: Prepare and dispense medium under 80% N_2 + 20% CO_2. Add components, except agar solution and titanium citrate solution, to distilled/deionized water and bring volume to 484.0mL. Mix thoroughly. Sparge with 80% N_2 + 20% CO_2. Autoclave for 15 min at 15 psi pressure–121°C. Cool to 50°–55°C. Aseptically and anaerobically add 55.0mL of sterile titanium citrate solution to 500.0mL of sterile agar solution. Add the agar/titanium/citrate solution to the rest of the medium. Mix thoroughly. Aseptically and anaerobically, under an atmosphere of 80% H_2 + 20% CO_2, distribute into sterile Petri dishes, sterile tubes, or sterile bottles.

Use: For the cultivation of *Methanobacterium* species for transduction.

Methanobacterium Transduction Medium

Composition per liter:

Basal salts solution	100.0mL
Na_2CO_3 solution	24.0mL
Trace elements solution	10.0mL
L-Cysteine/Na_2S solution	10.0mL

Basal Salts Solution:

Composition per liter:

KH_2PO_4	68.0g
NH_4Cl	21.2g
Resazurin	10.0mg

Preparation of Basal Salts Solution: Add components to distilled/deionized water and bring volume to 1.0L. Mix thoroughly. Sparge with 100% N_2.

Trace Elements Solution:

Composition per liter:

Nitrilotriacetic acid	9.55g
$MgCl_2 \cdot 7H_2O$	4.06g
$FeCl_2 \cdot 7H_2O$	0.98g
$NaWO_4 \cdot 6H_2O$	0.264g
$NiCl_2 \cdot 6H_2O$	0.118g
$Na_2Se_2O_3$	0.1g
$NaMoO_4 \cdot 2H_2O$	0.024g
$CoCl_2 \cdot 6H_2O$	0.023g

Preparation of Trace Elements Solution: Add nitrilotriacetic acid to 500.0mL of distilled/deionized water. Adjust pH to 7.0 with KOH. Add remaining components one at a time. Add distilled/deionized water to 1.0L. Adjust pH to 7.0. Sparge with 100% N_2.

Na_2CO_3 Solution:

Composition per liter:

Na_2CO_3	10.6g

Preparation of Na_2CO_3 Solution: Add Na_2CO_3 to distilled/deionized water and bring volume to 100.0mL. Mix thoroughly. Sparge with 80% N_2 + 20% CO_2.

L-Cysteine/Na_2S Solution:

Composition per 50.0mL:

L-Cysteine·HCl	5.0g
$Na_2S·9H_2O$	5.0g

Preparation of L-Cysteine/Na_2S Solution: Gently heat and bring 50.0mL of distilled/deionized water to boiling. Add L-cysteine·HCl and $Na_2S·9H_2O$. Mix thoroughly. Cool while sparging with 100% N_2. Bring volume to 50.0mL with distilled/deionized water. Sparge with 100% N_2. Adjust pH to 10.0 with 3*N* NaOH.

Preparation of Medium: Prepare and dispense medium under 80% N_2 + 20% CO_2. Add components to distilled/deionized water and bring volume to 1.0L. Mix thoroughly. Sparge with 80% N_2 + 20% CO_2. Anaerobically distribute into tubes or bottles. Autoclave for 15 min at 15 psi pressure–121°C. Prior to inoculation, pressurize to 200kPa with an atmosphere of 80% H_2 + 20% CO_2.

Use: For the cultivation of *Methanobacterium* species for transduction.

Methanobacterium wolfei Medium

Composition per liter:

Na_2CO_3	4.0g
Sodium acetate·$3H_2O$	1.0g
L-Cysteine·HCl	0.5g
NaCl	0.5g
$Na_2S·9H_2O$	0.5g
K_2HPO_4	0.26g
KH_2PO_4	0.26g
$(NH_4)_2SO_4$	0.26g
$MgSO_4·7H_2O$	0.1g
$CaCl_2·2H_2O$	0.07g
$NaWO_4$	2.6mg
$FeSO_4$	2.0mg
Resazurin	1.0mg
Vitamin solution	10.0mL
Trace elements solution	10.0mL

pH 7.2 ± 0.2 at 25°C

Trace Elements Solution:

Composition per liter:

Nitrilotriacetic acid	9.5g
$MgCl_2$	4.06g
$FeCl_2$	0.99g
$NiCl_2·6H_2O$	0.12g
Na_2MoO_4	0.024g
$CoCl_2·6H_2O$	0.023g

Preparation of Trace Elements Solution: Add nitrilotriacetic acid to 500.0mL of distilled/deionized water. Adjust pH to 6.5 with KOH. Add remaining components one at a time. Add distilled/deionized water to 1.0L. Mix thoroughly. Sparge with 100% N_2. Adjust pH to 7.0 with 10*N* NaOH.

Vitamin Solution:

Composition per liter:

Calcium DL-pantothenate	5.0g
Vitamin B_{12}	0.1g
Pyridoxine·HCl	10.0mg
p-Aminobenzoic acid	5.0mg
Lipoic acid	5.0mg
Nicotinic acid	5.0mg
Riboflavin	5.0mg
Thiamine·HCl	5.0mg
Biotin	2.0mg
Folic acid	2.0mg

Preparation of Vitamin Solution: Add components to distilled/deionized water and bring volume to 1.0L. Mix thoroughly. Sparge with 100% N_2.

Preparation of Medium: Prepare and dispense medium under 80% H_2 + 20% CO_2. Add components to distilled/deionized water and bring volume to 1.0L. Mix thoroughly. Adjust pH to 7.2. Sparge with 80% H_2 + 20% CO_2. Autoclave for 15 min at 15 psi pressure–121°C. Aseptically and anaerobically distribute into sterile tubes or bottles.

Use: For the cultivation of *Methanobacterium wolfei*.

Methanobacterium II Medium (DSMZ Medium 825)

Composition per 1050.0mL:

Yeast extract	1.0g
NH_4Cl	1.0g
NaCl	0.6g
Cysteine-HCl·H_2O	0.5g
Sodium acetate	0.5g
K_2HPO_4	0.3g
KH_2PO_4	0.3g
$MgCl_2·6H_2O$	0.2g
$CaCl_2·2H_2O$	0.1g
KCl	0.1g
Resazurin	0.5mg
$NaHCO_3$ solution	40.0mL
Trace elements solution	10.0mL
Vitamin solution	10.0mL
$Na_2S·9H_2O$ solution	10.0mL

pH 7.0 ± 0.2 at 25°C

Trace Elements Solution:

Composition per liter:

$MgSO_4·7H_2O$	3.0g
Nitrilotriacetic acid	1.5g
NaCl	1.0g
$MnSO_4·2H_2O$	0.5g
$CoSO_4·7H_2O$	0.18g
$ZnSO_4·7H_2O$	0.18g
$CaCl_2·2H_2O$	0.1g
$FeSO_4·7H_2O$	0.1g
$NiCl_2·6H_2O$	0.025g
$KAl(SO_4)_2·12H_2O$	0.02g
H_3BO_3	0.01g
$Na_2MoO_4·4H_2O$	0.01g
$CuSO_4·5H_2O$	0.01g
$Na_2SeO_3·5H_2O$	0.3mg

Preparation of Trace Elements Solution: Add nitrilotriacetic acid to 500.0mL of distilled/deionized water. Dissolve by adjusting pH to 6.5 with KOH. Add remaining components. Add distilled/deionized water to 1.0L. Mix thoroughly.

Vitamin Solution:

Composition per liter:

Pyridoxine-HCl	10.0mg
Thiamine-HCl·$2H_2O$	5.0mg
Riboflavin	5.0mg
Nicotinic acid	5.0mg
D-Ca-pantothenate	5.0mg
p-Aminobenzoic acid	5.0mg

Lipoic acid 5.0mg
Biotin 2.0mg
Folic acid 2.0mg
Vitamin B_{12} 0.1mg

Preparation of Vitamin Solution: Add components to distilled/deionized water and bring volume to 1.0L. Mix thoroughly. Sparge with 80% H_2 + 20% CO_2. Filter sterilize.

$Na_2S·9H_2O$ Solution:
Composition per 10.0mL:
$Na_2S·9H_2O$ 0.3g

Preparation of $Na_2S·9H_2O$ Solution: Add $Na_2S·9H_2O$ to distilled/deionized water and bring volume to 10.0mL. Sparge with N_2. Autoclave for 15 min at 15 psi pressure–121°C. Cool to 25°C. Store anaerobically.

$NaHCO_3$ Solution:
Composition per 100.0mL:
$NaHCO_3$ 10.0g

Preparation of $NaHCO_3$ Solution: Add $NaHCO_3$ to distilled/deionized water and bring volume to 100.0mL. Mix thoroughly. Autoclave for 15 min at 15 psi pressure–121°C. Cool to 25°C. Must be prepared freshly.

Preparation of Medium: Prepare and dispense medium under 80% H_2 + 20% CO_2 gas mixture. Add components, except $NaHCO_3$ solution and $Na_2S·9H_2O$ solution, to 1.0L distilled/deionized water. Mix thoroughly. Sparge with 80% H_2 + 20% CO_2. Dispense 5.0mL aliquots into Hungate tubes under 80% H_2 + 20% CO_2 gas mixture. Autoclave for 15 min at 15 psi pressure–121°C. Prior to use inject, for each 5.0mL medium, 0.2mL sterile $NaHCO_3$ solution and 0.05mL sterile $Na_2S·9H_2O$ solution. The pH should be 7.0. After inoculation, pressurize culture vessels to 2 bar 80% H_2 + 20% CO_2 overpressure.

Use: For the cultivation of *Methanobacterium formicicum, Methanosarcina barkeri, Methanosarcina mazei=Methanococcus mazei (Methanosarcina frisia), Methanobacterium bryantii, and Methanobacterium oryzae (Methanobacterium* sp.*).*

Methanobacterium II Medium (DSMZ Medium 825)

Composition per 1060.0mL:
Yeast extract 1.0g
NH_4Cl 1.0g
NaCl 0.6g
Cysteine-HCl·H_2O 0.5g
Sodium acetate 0.5g
K_2HPO_4 ..... 0.3g
KH_2PO_4 ..... 0.3g
$MgCl_2·6H_2O$ 0.2g
$CaCl_2·2H_2O$ 0.1g
KCl 0.1g
Resazurin 0.5mg
$NaHCO_3$ solution 40.0mL
Trace elements solution 10.0mL
Vitamin solution 10.0mL
$Na_2S·9H_2O$ solution 10.0mL
Methanol solution 10.0mL

pH 7.0 ± 0.2 at 25°C

Methanol Solution:
Composition per 100.0mL:
Methanol 15.0mL

Preparation of Methanol Solution: Add methanol to distilled/deionized water and bring volume to 100.0mL. Mix thoroughly. Sparge with 100% N_2. Autoclave for 15 min at 15 psi pressure–121°C.

Trace Elements Solution:
Composition per liter:
$MgSO_4·7H_2O$ 3.0g
Nitrilotriacetic acid 1.5g
NaCl 1.0g
$MnSO_4·2H_2O$ 0.5g
$CoSO_4·7H_2O$ 0.18g
$ZnSO_4·7H_2O$ 0.18g
$CaCl_2·2H_2O$ 0.1g
$FeSO_4·7H_2O$ 0.1g
$NiCl_2·6H_2O$ 0.025g
$KAl(SO_4)_2·12H_2O$ 0.02g
H_3BO_3 ..... 0.01g
$Na_2MoO_4·4H_2O$ 0.01g
$CuSO_4·5H_2O$ 0.01g
$Na_2SeO_3·5H_2O$ 0.3mg

Preparation of Trace Elements Solution: Add nitrilotriacetic acid to 500.0mL of distilled/deionized water. Dissolve by adjusting pH to 6.5 with KOH. Add remaining components. Add distilled/deionized water to 1.0L. Mix thoroughly.

Vitamin Solution:
Composition per liter:
Pyridoxine-HCl 10.0mg
Thiamine-HCl·$2H_2O$ 5.0mg
Riboflavin 5.0mg
Nicotinic acid 5.0mg
D-Ca-pantothenate 5.0mg
p-Aminobenzoic acid 5.0mg
Lipoic acid 5.0mg
Biotin 2.0mg
Folic acid 2.0mg
Vitamin B_{12} 0.1mg

Preparation of Vitamin Solution: Add components to distilled/deionized water and bring volume to 1.0L. Mix thoroughly. Sparge with 80% H_2 + 20% CO_2. Filter sterilize.

$Na_2S·9H_2O$ Solution:
Composition per 10.0mL:
$Na_2S·9H_2O$ 0.3g

Preparation of $Na_2S·9H_2O$ Solution: Add $Na_2S·9H_2O$ to distilled/deionized water and bring volume to 10.0mL. Sparge with N_2. Autoclave for 15 min at 15 psi pressure–121°C. Cool to 25°C. Store anaerobically.

$NaHCO_3$ Solution:
Composition per 100.0mL:
$NaHCO_3$ 10.0g

Preparation of $NaHCO_3$ Solution: Add $NaHCO_3$ to distilled/deionized water and bring volume to 100.0mL. Mix thoroughly. Autoclave for 15 min at 15 psi pressure–121°C. Cool to 25°C. Must be prepared freshly.

Preparation of Medium: Prepare and dispense medium under 80% N_2 + 20% CO_2 gas mixture. Add components, except methanol solution, $NaHCO_3$ solution, and $Na_2S·9H_2O$ solution, to 1.0L distilled/deionized water. Mix thoroughly. Sparge with 80% N_2 + 20% CO_2. Dispense 5.0mL aliquots into Hungate tubes under 80% H_2 + 20% CO_2 gas mixture. Autoclave for 15 min at 15 psi pressure–121°C. Prior to use inject, for each 5.0mL medium, 0.2mL sterile $NaHCO_3$ solution, 0.05mL sterile $Na_2S·9H_2O$ solution, and 0.05mL sterile methanol solution. The pH should be 7.0.

Use: For the cultivation of *Methanosarcina barkeri* DSM 10131 *and Methanosarcina mazei=Methanococcus mazei (Methanosarcina frisia)* DSM 10132.

Methanobacterium II Medium (DSMZ Medium 825)

Composition per 1065.0mL:

Yeast extract	1.0g
NH_4Cl	1.0g
NaCl	0.6g
Cysteine-HCl·H_2O	0.5g
Sodium acetate	0.5g
K_2HPO_4	0.3g
KH_2PO_4	0.3g
$MgCl_2·6H_2O$	0.2g
$CaCl_2·2H_2O$	0.1g
KCl	0.1g
Resazurin	0.5mg
$NaHCO_3$ solution	40.0mL
Na-formate solution	15.0mL
Trace elements solution	10.0mL
Vitamin solution	10.0mL
$Na_2S·9H_2O$ solution	10.0mL

pH 7.0 ± 0.2 at 25°C

Na-Formate Solution:
Composition per 10.0mL:

Na-formate	5.0g

Preparation of Na-Formate Solution: Add Na-formate to distilled/deionized water and bring volume to 10.0mL. Mix thoroughly. Sparge with 100% N_2. Autoclave for 15 min at 15 psi pressure–121°C.

Trace Elements Solution:
Composition per liter:

$MgSO_4·7H_2O$	3.0g
Nitrilotriacetic acid	1.5g
NaCl	1.0g
$MnSO_4·2H_2O$	0.5g
$CoSO_4·7H_2O$	0.18g
$ZnSO_4·7H_2O$	0.18g
$CaCl_2·2H_2O$	0.1g
$FeSO_4·7H_2O$	0.1g
$NiCl_2·6H_2O$	0.025g
$KAl(SO_4)_2·12H_2O$	0.02g
H_3BO_3	0.01g
$Na_2MoO_4·4H_2O$	0.01g
$CuSO_4·5H_2O$	0.01g
$Na_2SeO_3·5H_2O$	0.3mg

Preparation of Trace Elements Solution: Add nitrilotriacetic acid to 500.0mL of distilled/deionized water. Dissolve by adjusting pH to 6.5 with KOH. Add remaining components. Add distilled/deionized water to 1.0L. Mix thoroughly.

Vitamin Solution:
Composition per liter:

Pyridoxine-HCl	10.0mg
Thiamine-HCl·$2H_2O$	5.0mg
Riboflavin	5.0mg
Nicotinic acid	5.0mg
D-Ca-pantothenate	5.0mg
p-Aminobenzoic acid	5.0mg
Lipoic acid	5.0mg
Biotin	2.0mg
Folic acid	2.0mg
Vitamin B_{12}	0.1mg

Preparation of Vitamin Solution: Add components to distilled/deionized water and bring volume to 1.0L. Mix thoroughly. Sparge with 80% H_2 + 20% CO_2. Filter sterilize.

$Na_2S·9H_2O$ Solution:
Composition per 10.0mL:

$Na_2S·9H_2O$	0.3g

Preparation of $Na_2S·9H_2O$ Solution: Add $Na_2S·9H_2O$ to distilled/deionized water and bring volume to 10.0mL. Sparge with N_2. Autoclave for 15 min at 15 psi pressure–121°C. Cool to 25°C. Store anaerobically.

$NaHCO_3$ Solution:
Composition per 100.0mL:

$NaHCO_3$	10.0g

Preparation of $NaHCO_3$ Solution: Add $NaHCO_3$ to distilled/deionized water and bring volume to 100.0mL. Mix thoroughly. Autoclave for 15 min at 15 psi pressure–121°C. Cool to 25°C. Must be prepared freshly.

Preparation of Medium: Prepare and dispense medium under 80% N_2 + 20% CO_2 gas mixture. Add components, except Na-formate solution, $NaHCO_3$ solution, and $Na_2S·9H_2O$ solution, to 1.0L distilled/deionized water. Mix thoroughly. Sparge with 80% N_2 + 20% CO_2. Dispense 5.0mL aliquots into Hungate tubes under 80% H_2 + 20% CO_2 gas mixture. Autoclave for 15 min at 15 psi pressure–121°C. Prior to use inject, for each 5.0mL medium, 0.2mL sterile $NaHCO_3$ solution, 0.05mL sterile $Na_2S·9H_2O$ solution, and 0.075mL sterile Na-formate solution. The pH should be 7.0.

Use: For the cultivation of *Methanobacterium formicicum* DSM 10111.

Methanobacterium II Medium (DSMZ Medium 825)

Composition per 1075.0mL:

Yeast extract	1.0g
NH_4Cl	1.0g
NaCl	0.6g
Cysteine-HCl·H_2O	0.5g
Sodium acetate	0.5g
K_2HPO_4	0.3g
KH_2PO_4	0.3g
$MgCl_2·6H_2O$	0.2g
$CaCl_2·2H_2O$	0.1g
KCl	0.1g
Resazurin	0.5mg
$NaHCO_3$ solution	40.0mL
Na-formate solution	15.0mL
Trace elements solution	10.0mL
Vitamin solution	10.0mL
$Na_2S·9H_2O$ solution	10.0mL
Trypticase™ solution	10.0mL

pH 7.0 ± 0.2 at 25°C

Trypticase™ Solution:
Composition per 10.0mL:

Trypticase™	1.0g

Preparation of Trypticase™ Solution: Add Trypticase™ to distilled/deionized water and bring volume to 10.0mL. Mix thoroughly. Autoclave under 100% N_2 for 15 min at 15 psi pressure–121°C. Cool to room temperature.

Na-Formate Solution:
Composition per 10.0mL:

Na-formate 5.0g

Preparation of Na-Formate Solution: Add Na-formate to distilled/deionized water and bring volume to 10.0mL. Mix thoroughly. Sparge with 100% N_2. Autoclave for 15 min at 15 psi pressure–121°C.

Trace Elements Solution:
Composition per liter:

$MgSO_4 \cdot 7H_2O$	3.0g
Nitrilotriacetic acid	1.5g
NaCl	1.0g
$MnSO_4 \cdot 2H_2O$	0.5g
$CoSO_4 \cdot 7H_2O$	0.18g
$ZnSO_4 \cdot 7H_2O$	0.18g
$CaCl_2 \cdot 2H_2O$	0.1g
$FeSO_4 \cdot 7H_2O$	0.1g
$NiCl_2 \cdot 6H_2O$	0.025g
$KAl(SO_4)_2 \cdot 12H_2O$	0.02g
H_3BO_3	0.01g
$Na_2MoO_4 \cdot 4H_2O$	0.01g
$CuSO_4 \cdot 5H_2O$	0.01g
$Na_2SeO_3 \cdot 5H_2O$	0.3mg

Preparation of Trace Elements Solution: Add nitrilotriacetic acid to 500.0mL of distilled/deionized water. Dissolve by adjusting pH to 6.5 with KOH. Add remaining components. Add distilled/deionized water to 1.0L. Mix thoroughly.

Vitamin Solution:
Composition per liter:

Pyridoxine-HCl	10.0mg
Thiamine-HCl·$2H_2O$	5.0mg
Riboflavin	5.0mg
Nicotinic acid	5.0mg
D-Ca-pantothenate	5.0mg
p-Aminobenzoic acid	5.0mg
Lipoic acid	5.0mg
Biotin	2.0mg
Folic acid	2.0mg
Vitamin B_{12}	0.1mg

Preparation of Vitamin Solution: Add components to distilled/deionized water and bring volume to 1.0L. Mix thoroughly. Sparge with 80% H_2 + 20% CO_2. Filter sterilize.

$Na_2S \cdot 9H_2O$ Solution:
Composition per 10.0mL:

$Na_2S \cdot 9H_2O$ 0.3g

Preparation of $Na_2S \cdot 9H_2O$ Solution: Add $Na_2S \cdot 9H_2O$ to distilled/deionized water and bring volume to 10.0mL. Sparge with N_2. Autoclave for 15 min at 15 psi pressure–121°C. Cool to 25°C. Store anaerobically.

$NaHCO_3$ Solution:
Composition per 100.0mL:

$NaHCO_3$ 10.0g

Preparation of $NaHCO_3$ Solution: Add $NaHCO_3$ to distilled/deionized water and bring volume to 100.0mL. Mix thoroughly. Autoclave for 15 min at 15 psi pressure–121°C. Cool to 25°C. Must be prepared freshly.

Preparation of Medium: Prepare and dispense medium under 80% N_2 + 20% CO_2 gas mixture. Add components, except Trypticase™ solution, Na-formate solution, $NaHCO_3$ solution, and $Na_2S \cdot 9H_2O$ solution, to 1.0L distilled/deionized water. Mix thoroughly. Sparge with 80% N_2 + 20% CO_2. Dispense 5.0mL aliquots into Hungate tubes under 80% H_2 + 20% CO_2 gas mixture. Autoclave for 15 min at 15 psi pressure–121°C. Prior to use inject, for each 5.0mL medium, 0.2mL sterile $NaHCO_3$ solution, 0.05mL sterile Trypticase™ solution, 0.05mL sterile $Na_2S \cdot 9H_2O$ solution, and 0.075mL sterile Na-formate solution. The pH should be 7.0.

Use: For the cultivation of *Methanobacterium oryzae* DSM 11106.

Methanobrevibacter curvatus Medium (DSMZ Medium 734)

Composition per 1010.0mL:

NaCl	1.0g
KCl	0.5g
Casamino acids	0.5g
Yeast extract	0.5g
$MgCl_2 \cdot 6H_2O$	0.4g
NH_4Cl	0.3g
KH_2PO_4	0.2g
Na_2SO_4	0.15g
$CaCl_2 \cdot 2H_2O$	0.1g
Resazurin	0.5mg
$NaHCO_3$ solution	40.0mL
Dithionite solution	10.0mL
Trace elements solution SL-10	1.0mL
Selenite-tungstate solution	1.0mL
Seven vitamin solution	1.0mL

pH 7.4 ± 0.2 at 25°C

$NaHCO_3$ Solution:
Composition per 40.0mL:

$NaHCO_3$ 5.8g

Preparation of $NaHCO_3$ Solution: Add $NaHCO_3$ to distilled/deionized water and bring volume to 40.0mL. Mix thoroughly. Autoclave for 15 min at 15 psi pressure–121°C. Cool to 25°C. Must be prepared freshly.

Selenite-Tungstate Solution
Composition per liter:

NaOH	0.5g
$Na_2WO_4 \cdot 2H_2O$	4.0mg
$Na_2SeO_3 \cdot 5H_2O$	3.0mg

Preparation of Selenite-Tungstate Solution: Add components to distilled/deionized water and bring volume to 1.0L. Mix thoroughly. Sparge with 100% N_2. Filter sterilize.

Dithionite Solution
Composition per 10.0mL:

Na-dithionite 2.0mg

Preparation of Dithionite Solution: Add Na-dithionite to distilled/deionized water and bring volume to 10.0mL. Mix thoroughly. Sparge with 100% N_2. Filter sterilize.

Trace Elements Solution SL-10:
Composition per liter:

$FeCl_2 \cdot 4H_2O$	1.5g
$CoCl_2 \cdot 6H_2O$	190.0mg

$MnCl_2 \cdot 4H_2O$ 100.0mg
$ZnCl_2$ 70.0mg
$Na_2MoO_4 \cdot 2H_2O$ 36.0mg
$NiCl_2 \cdot 6H_2O$ 24.0mg
H_3BO_3 .... 6.0mg
$CuCl_2 \cdot 2H_2O$ 2.0mg
HCl (25% solution) 10.0mL

Preparation of Trace Elements Solution SL-10: Add $FeCl_2 \cdot 4H_2O$ to 10.0mL of HCl solution. Mix thoroughly. Add distilled/deionized water and bring volume to 1.0L. Add remaining components. Mix thoroughly. Sparge with 80% N_2 + 20% CO_2. Autoclave for 15 min at 15 psi pressure–121°C.

Seven Vitamin Solution:
Composition per liter:
Pyridoxine hydrochloride 300.0mg
Thiamine-HCl·$2H_2O$ 200.0mg
Nicotinic acid 200.0mg
Vitamin B_{12} 100.0mg
Calcium pantothenate 100.0mg
p-Aminobenzoic acid 80.0mg
D(+)-Biotin 20.0mg

Preparation of Seven Vitamin Solution: Add components to distilled/deionized water and bring volume to 1.0L. Sparge with 100% N_2. Mix thoroughly. Filter sterilize.

Preparation of Medium: Prepare and dispense medium under 80% H_2 + 20% CO_2 gas atmosphere. Add components, except $NaHCO_3$ solution, seven vitamin solution, selenite-tungstate solution, and trace elements solution SL-10, to distilled/deionized water and bring volume to 947.0mL. Mix thoroughly. Adjust pH to 7.6. Sparge with 80% H_2 + 20% CO_2. Autoclave for 15 min at 15 psi pressure–121°C. Aseptically and anaerobically add 40.0mL $NaHCO_3$ solution, 1.0mL selenite-tungstate solution, 1.0mL seven vitamin solution, and 1.0mL trace elements solution SL-10. Mix thoroughly. Aseptically and anaerobically distribute into sterile tubes or bottles. Adjust pH to 7.6. Prior to inoculation add dithionite solution (0.1mL per 10mL medium) as reductant.

Use: For the cultivation of *Methanobrevibacter curvatus.*

Methanobrevibacter curvatus Medium (DSMZ Medium 734)

Composition per 1022.0mL:
NaCl 1.0g
KCl 0.5g
Casamino acids 0.5g
Yeast extract 0.5g
$MgCl_2 \cdot 6H_2O$ 0.4g
NH_4Cl 0.3g
KH_2PO_4 .... 0.2g
Na_2SO_4 .... 0.15g
$CaCl_2 \cdot 2H_2O$ 0.1g
Resazurin 0.5mg
Rumen fluid, bovine, clarified 400mL
$NaHCO_3$ solution 40.0mL
Dithionite solution 10.0mL
MOPS buffer 10.0mL
Nutrient supplement solution 2.0mL
Trace elements solution SL-10 .... 1.0mL
Selenite-tungstate solution 1.0mL
Seven vitamin solution 1.0mL
pH 7.2 ± 0.2 at 25°C

$NaHCO_3$ Solution:
Composition per 40.0mL:
$NaHCO_3$ 5.8g

Preparation of $NaHCO_3$ Solution: Add $NaHCO_3$ to distilled/deionized water and bring volume to 40.0mL. Mix thoroughly. Autoclave for 15 min at 15 psi pressure–121°C. Cool to 25°C. Must be prepared freshly.

Selenite-Tungstate Solution:
Composition per liter:
NaOH 0.5g
$Na_2WO_4 \cdot 2H_2O$ 4.0mg
$Na_2SeO_3 \cdot 5H_2O$ 3.0mg

Preparation of Selenite-Tungstate Solution: Add components to distilled/deionized water and bring volume to 1.0L. Mix thoroughly. Sparge with 100% N_2. Filter sterilize.

Dithionite Solution:
Composition per 10.0mL:
Na-dithionite 2.0mg

Preparation of Dithionite Solution: Add Na-dithionite to distilled/deionized water and bring volume to 10.0mL. Mix thoroughly. Sparge with 100% N_2. Filter sterilize.

Trace Elements Solution SL-10:
Composition per liter:
$FeCl_2 \cdot 4H_2O$ 1.5g
$CoCl_2 \cdot 6H_2O$ 190.0mg
$MnCl_2 \cdot 4H_2O$ 100.0mg
$ZnCl_2$ 70.0mg
$Na_2MoO_4 \cdot 2H_2O$ 36.0mg
$NiCl_2 \cdot 6H_2O$ 24.0mg
H_3BO_3 .... 6.0mg
$CuCl_2 \cdot 2H_2O$ 2.0mg
HCl (25% solution) 10.0mL

Preparation of Trace Elements Solution SL-10: Add $FeCl_2 \cdot 4H_2O$ to 10.0mL of HCl solution. Mix thoroughly. Add distilled/deionized water and bring volume to 1.0L. Add remaining components. Mix thoroughly. Sparge with 80% N_2 + 20% CO_2. Autoclave for 15 min at 15 psi pressure–121°C.

Seven Vitamin Solution:
Composition per liter:
Pyridoxine hydrochloride 300.0mg
Thiamine-HCl·$2H_2O$ 200.0mg
Nicotinic acid 200.0mg
Vitamin B_{12} 100.0mg
Calcium pantothenate 100.0mg
p-Aminobenzoic acid 80.0mg
D(+)-Biotin 20.0mg

Preparation of Seven Vitamin Solution: Add components to distilled/deionized water and bring volume to 1.0L. Sparge with 100% N_2. Mix thoroughly. Filter sterilize.

Nutrient Supplement Solution:
Composition per liter:
Pancreatic digest of gelatin 5.0g
Beef extract 3.0g

Preparation of Nutrient Supplement Solution: Add components to distilled/deionized water and bring volume to 1.0L. Mix thoroughly. Sparge with 100% N_2. Autoclave for 15 min at 15 psi pressure–121°C. Cool to room tempterature.

MOPS Buffer:
Composition per 10.0mL:
MOPS [3-(*N*-morpholino) propane sulfonic acid] 2.1g
Na-acetate 0.3g
EDTA 0.1g

Preparation of MOPS Buffer: Add components to distilled/deionized water and bring volume to 10.0mL. Mix thoroughly. Sparge with 100% N_2. Adjust to pH 7.2. Filter sterilize.

Preparation of Medium: Prepare and dispense medium under 80% H_2 + 20% CO_2 gas atmosphere. Add components, except clarified bovine rumen fluid, MOPS buffer, nutrient supplement solution, $NaHCO_3$ solution, seven vitamin solution, selenite-tungstate solution, and trace elements solution SL-10, to distilled/deionized water and bring volume to 547.0mL. Mix thoroughly. Adjust pH to 7.2. Sparge with 80% H_2 + 20% CO_2. Autoclave for 15 min at 15 psi pressure–121°C. Aseptically and anaerobically add 400.0mL sterile clarified bovine rumen fluid, 10.0mL MOPS buffer, 2.0mL nutrient supplement solution, 40.0mL $NaHCO_3$ solution, 1.0mL selenite-tungstate solution, 1.0mL seven vitamin solution, and 1.0mL trace elements solution SL-10. Mix thoroughly. Aseptically and anaerobically distribute into sterile tubes or bottles. Adjust pH to 7.2. Prior to inoculation add dithionite solution (0.1mL per 10mL medium) as reductant.

Use: For the cultivation of *Methanobrevibacter curvatus* DSM 11111 (strain RFM-2).

Methanobrevibacter curvatus Medium (DSMZ Medium 734)

Composition per 1020.0mL:
NaCl 1.0g
KCl 0.5g
Casamino acids 0.5g
Yeast extract 0.5g
$MgCl_2 \cdot 6H_2O$ 0.4g
NH_4Cl 0.3g
KH_2PO_4 .... 0.2g
Na_2SO_4 .... 0.15g
$CaCl_2 \cdot 2H_2O$ 0.1g
Resazurin 0.5mg
Rumen fluid, bovine, clarified 200mL
$NaHCO_3$ solution 40.0mL
Dithionite solution 10.0mL
MOPS buffer 10.0mL
Trace elements solution SL-10 .... 1.0mL
Selenite-tungstate solution 1.0mL
Seven vitamin solution 1.0mL
pH 7.7 ± 0.2 at 25°C

$NaHCO_3$ Solution:
Composition per 40.0mL:
$NaHCO_3$ 5.8g

Preparation of $NaHCO_3$ Solution: Add $NaHCO_3$ to distilled/deionized water and bring volume to 40.0mL. Mix thoroughly. Autoclave for 15 min at 15 psi pressure–121°C. Cool to 25°C. Must be prepared freshly.

Selenite-Tungstate Solution:
Composition per liter:
NaOH 0.5g
$Na_2WO_4 \cdot 2H_2O$ 4.0mg
$Na_2SeO_3 \cdot 5H_2O$ 3.0mg

Preparation of Selenite-Tungstate Solution: Add components to distilled/deionized water and bring volume to 1.0L. Mix thoroughly. Sparge with 100% N_2. Filter sterilize.

Dithionite Solution:
Composition per 10.0mL:
Na-dithionite 2.0mg

Preparation of Dithionite Solution: Add Na-dithionite to distilled/deionized water and bring volume to 10.0mL. Mix thoroughly. Sparge with 100% N_2. Filter sterilize.

Trace Elements Solution SL-10:
Composition per liter:
$FeCl_2 \cdot 4H_2O$ 1.5g
$CoCl_2 \cdot 6H_2O$ 190.0mg
$MnCl_2 \cdot 4H_2O$ 100.0mg
$ZnCl_2$ 70.0mg
$Na_2MoO_4 \cdot 2H_2O$ 36.0mg
$NiCl_2 \cdot 6H_2O$ 24.0mg
H_3BO_3 .... 6.0mg
$CuCl_2 \cdot 2H_2O$ 2.0mg
HCl (25% solution) 10.0mL

Preparation of Trace Elements Solution SL-10: Add $FeCl_2 \cdot 4H_2O$ to 10.0mL of HCl solution. Mix thoroughly. Add distilled/deionized water and bring volume to 1.0L. Add remaining components. Mix thoroughly. Sparge with 80% N_2 + 20% CO_2. Autoclave for 15 min at 15 psi pressure–121°C.

Seven Vitamin Solution:
Composition per liter:
Pyridoxine hydrochloride 300.0mg
Thiamine-HCl·$2H_2O$ 200.0mg
Nicotinic acid 200.0mg
Vitamin B_{12} 100.0mg
Calcium pantothenate 100.0mg
p-Aminobenzoic acid 80.0mg
D(+)-Biotin 20.0mg

Preparation of Seven Vitamin Solution: Add components to distilled/deionized water and bring volume to 1.0L. Sparge with 100% N_2. Mix thoroughly. Filter sterilize.

MOPS Buffer:
Composition per 10.0mL:
MOPS [3-(*N*-morpholino) propane sulfonic acid] 2.1g
Na-acetate 0.3g
EDTA 0.1g

Preparation of MOPS Buffer: Add components to distilled/deionized water and bring volume to 10.0mL. Mix thoroughly. Sparge with 100% N_2. Adjust to pH 7.7. Filter sterilize.

Preparation of Medium: Prepare and dispense medium under 80% H_2 + 20% CO_2 gas atmosphere. Add components, except clarified bovine rumen fluid, MOPS buffer, $NaHCO_3$ solution, seven vitamin solution, selenite-tungstate solution, and trace elements solution SL-10, to distilled/deionized water and bring volume to 747.0mL. Mix thoroughly. Adjust pH to 7.7. Sparge with 80% H_2 + 20% CO_2. Autoclave for 15 min at 15 psi pressure–121°C. Aseptically and anaerobically add 200.0mL sterile clarified bovine rumen fluid, 10.0mL MOPS buffer, 240.0mL $NaHCO_3$ solution, 1.0mL selenite-tungstate solution, 1.0mL seven vitamin solution, and 1.0mL trace elements solution SL-10. Mix thoroughly. Aseptically and anaerobically distribute into sterile tubes or bottles. Adjust pH to 7.7. Prior to inoculation add dithionite solution (0.1mL per 10mL medium) as reductant.

Use: For the cultivation of *Methanobrevibacter curvatus* DSM 11111 DSM 11139 (strain RFM-1)

Methanocalculus halotolerans Medium (DSMZ Medium 905)

Composition per 1010.0mL:

NaCl	50.0g
NH_4Cl	1.0g
Na-acetate	0.5g
Yeast extract	0.5g
Trypticase™	0.5g
K_2HPO_4	0.3g
KH_2PO_4	0.3g
KCl	0.17g
Resazurin	0.5mg
Magnesium chloride solution	30.0mL
$NaHCO_3$ solution	20.0mL
Trace elements solution	10.0mL
Calcium chloride solution	10.0mL
$Na_2S{\cdot}9H_2O$ solution	10.0mL
L-Cysteine solution	10.0mL

pH 7.2-7.6 at 25°C

Magnesium Chloride Solution:

Composition per 30.0mL:

$MgCl_2{\cdot}6H_2O$	3.2g

Preparation of Magnesium Chloride Solution: Add $MgCl_2{\cdot}6H_2O$ to distilled/deionized water and bring volume to 30.0mL. Mix thoroughly. Sparge with 100% N_2. Autoclave for 15 min at 15 psi pressure–121°C.

Calcium Chloride Solution:

Composition per 10.0mL:

$CaCl_2{\cdot}2H_2O$	0.6g

Preparation of Calcium Chloride Solution: Add $CaCl_2{\cdot}2H_2O$ to distilled/deionized water and bring volume to 10.0mL. Mix thoroughly. Sparge with 100% N_2. Autoclave for 15 min at 15 psi pressure–121°C.

L-Cysteine Solution:

Composition per 10.0mL:

L-Cysteine·HCl·H_2O	0.5g

Preparation of L-Cysteine Solution: Add L-cysteine·HCl·H_2O to distilled/deionized water and bring volume to 10.0mL. Mix thoroughly. Sparge with 100% N_2. Autoclave for 15 min at 15 psi pressure–121°C.

$Na_2S{\cdot}9H_2O$ Solution:

Composition per 10.0mL:

$Na_2S{\cdot}9H_2O$	0.3g

Preparation of $Na_2S{\cdot}9H_2O$ Solution: Add $Na_2S{\cdot}9H_2O$ to distilled/deionized water and bring volume to 10.0mL. Mix thoroughly. Sparge with 100% N_2. Autoclave for 15 min at 15 psi pressure–121°C.

$NaHCO_3$ Solution:

Composition per 20.0mL:

$NaHCO_3$	2.0g

Preparation of $NaHCO_3$ Solution: Add $NaHCO_3$ to distilled/deionized water and bring volume to 20.0mL. Mix thoroughly. Sparge with 80% N_2 + 20% CO_2. Autoclave for 15 min at 15 psi pressure–121°C. Cool to 25°C. Must be prepared freshly.

Trace Elements Solution:

Composition per liter:

$MgSO_4{\cdot}7H_2O$	3.0g
Nitrilotriacetic acid	1.5g
NaCl	1.0g
$MnSO_4{\cdot}2H_2O$	0.5g
$CoSO_4{\cdot}7H_2O$	0.18g
$ZnSO_4{\cdot}7H_2O$	0.18g
$CaCl_2{\cdot}2H_2O$	0.1g
$FeSO_4{\cdot}7H_2O$	0.1g
$NiCl_2{\cdot}6H_2O$	0.025g
$KAl(SO_4)_2{\cdot}12H_2O$	0.02g
H_3BO_3	0.01g
$Na_2MoO_4{\cdot}4H_2O$	0.01g
$CuSO_4{\cdot}5H_2O$	0.01g
$Na_2SeO_3{\cdot}5H_2O$	0.3mg

Preparation of Trace Elements Solution: Add nitrilotriacetic acid to 500.0mL of distilled/deionized water. Dissolve by adjusting pH to 6.5 with KOH. Add remaining components. Add distilled/deionized water to 1.0L. Mix thoroughly.

Preparation of Medium: Prepare and dispense medium under sparge with 80% N_2 + 20% CO_2. Add components, except $NaHCO_3$ solution, magnesium chloride solution, calcium chloride solution, L-cysteine-HCl·H_2O solution, and $Na_2S{\cdot}9H_2O$ solution, to distilled/deionized water and bring volume to 950.0mL. Mix thoroughly. Gently heat and bring to boiling. Boil for 3 min. Cool to room temperature while sparging with 80% N_2 + 20% CO_2. Add 10.0mL L-cysteine-HCl·H_2O solution and 20.0mL $NaHCO_3$ solution. Mix thoroughly. Adjust pH to 7.5. Distribute into anaerobic tubes or bottles. Autoclave for 15 min at 15 psi pressure–121°C. Aseptically and anaerobically add per liter, 30.0mL magnesium chloride solution, 10.0mL calcium chloride solution, and 10.0mL $Na_2S{\cdot}9H_2O$. Mix thoroughly. After inoculation pressurize vessels with 80% H_2 + 20% CO_2 gas mixture to 1 bar overpressure and add sulfide from a sterile, anaerobic stock solution. The final pH of the medium should be 7.2–7.6.

Use: For the cultivation of *Methanocalculus halotolerans.*

Methanocalculus pumilus Medium (DSMZ Medium 892)

Composition per 1080.0mL:

NaCl	10.0g
Yeast extract	2.0g
Trypticase™	2.0g
NH_4Cl	0.9g
K_2HPO_4	0.4g
$MgCl_2{\cdot}6H_2O$	0.36g
Resazurin	0.5mg
Na_2CO_3 solution	50.0mL
$Na_2S{\cdot}9H_2O$ solution	15.0mL
L-Cysteine-HCl·H_2O solution	15.0mL
Vitamin solution	10.0mL
Trace elements solution	10.0mL

L-Cysteine Solution:

Composition per 15.0mL:

L-Cysteine·HCl·H_2O	0.5g

Preparation of L-Cysteine Solution: Add L-cysteine·HCl·H_2O to distilled/deionized water and bring volume to 15.0mL. Mix thoroughly. Sparge with 100% N_2. Autoclave for 15 min at 15 psi pressure–121°C.

Na_2CO_3 Solution:
Composition per 50.0mL:

Na_2CO_3	5.0g

Preparation of Na_2CO_3 Solution: Add Na_2CO_3 to distilled/deionized water and bring volume to 10.0mL. Mix thoroughly. Sparge with 80% N_2 + 20% CO_2. Filter sterilize.

$Na_2S{\cdot}9H_2O$ Solution:
Composition per 15mL:

$Na_2S{\cdot}9H_2O$	0.5g

Preparation of $Na_2S{\cdot}9H_2O$ Solution: Add $Na_2S{\cdot}9H_2O$ to distilled/deionized water and bring volume to 15.0mL. Mix thoroughly. Autoclave under 100% N_2 for 15 min at 15 psi pressure–121°C. Cool to room temperature.

Trace Elements Solution:
Composition per liter:

$MgSO_4{\cdot}7H_2O$	3.0g
Nitrilotriacetic acid	1.5g
NaCl	1.0g
$MnSO_4{\cdot}2H_2O$	0.5g
$CoSO_4{\cdot}7H_2O$	0.18g
$ZnSO_4{\cdot}7H_2O$	0.18g
$CaCl_2{\cdot}2H_2O$	0.1g
$FeSO_4{\cdot}7H_2O$	0.1g
$NiCl_2{\cdot}6H_2O$	0.025g
$KAl(SO_4)_2{\cdot}12H_2O$	0.02g
H_3BO_3	0.01g
$Na_2MoO_4{\cdot}4H_2O$	0.01g
$CuSO_4{\cdot}5H_2O$	0.01g
$Na_2SeO_3{\cdot}5H_2O$	0.3mg

Preparation of Trace Elements Solution: Add nitrilotriacetic acid to 500.0mL of distilled/deionized water. Dissolve by adjusting pH to 6.5 with KOH. Add remaining components. Add distilled/deionized water to 1.0L. Mix thoroughly.

Vitamin Solution:
Composition per liter:

Pyridoxine-HCl	10.0mg
Thiamine-HCl·2H_2O	5.0mg
Riboflavin	5.0mg
Nicotinic acid	5.0mg
D-Ca-pantothenate	5.0mg
p-Aminobenzoic acid	5.0mg
Lipoic acid	5.0mg
Biotin	2.0mg
Folic acid	2.0mg
Vitamin B_{12}	0.1mg

Preparation of Vitamin Solution: Add components to distilled/deionized water and bring volume to 1.0L. Mix thoroughly. Sparge with 80% H_2 + 20% CO_2. Filter sterilize.

Preparation of Medium: Add components, except Na_2CO_3 solution, $Na_2S{\cdot}9H_2O$ solution, and L-cysteine solution, to distilled/deionized water and bring volume to 1.0L. Mix thoroughly. Sparge with 100% N_2 for 30 min. Autoclave for 15 min at 15 psi pressure–121°C. Aseptically distribute 5.0mL aliquots into Hungate tubes under 100% N_2. Aseptically and anaerobically add per 5.0mL medium 0.25mL Na_2CO_3 solution, 0.075mL $Na_2S{\cdot}9H_2O$ solution, and 0.075mL L-cysteine solution. Mix thoroughly. Replace N_2 atmosphere with atmosphere of 80% H_2 + 20% CO_2. Repeat atmosphere replacement several times with overpressurization. The initial pH of 9.0 will decrease over a 30 min period to 7.3–7.5. After inoculation use atmosphere of 80% H_2 + 20% CO_2 to 1.5 bar overpressure.

Use: For the cultivation of *Methanocalculus pumilus.*

Methanococcal Complex Medium
See: ***Methanococcus* McC Medium**

Methanococcoides Medium

Composition per liter:

NaCl	18.0g
$NaHCO_3$	5.0g
$MgCl_2{\cdot}6H_2O$	4.0g
$MgSO_4{\cdot}7H_2O$	3.45g
Trimethylamine·HCl	3.0g
Trypticase™	2.0g
Yeast extract	2.0g
Sodium acetate	1.0g
L-Cysteine·HCl	0.5g
$Na_2S{\cdot}9H_2O$	0.5g
KCl	0.335g
NH_4Cl	0.25g
$CaCl_2{\cdot}2H_2O$	0.14g
K_2HPO_4	0.14g
$Fe(NH_4)_2(SO_4)_2{\cdot}7H_2O$	2.0mg
Resazurin	1.0mg
Trace elements solution	10.0mL
Vitamin solution	10.0mL

pH 7.0 ± 0.2 at 25°C

Trace Elements Solution:
Composition per liter:

$MgSO_4{\cdot}7H_2O$	3.0g
Nitrilotriacetic acid	1.5g
NaCl	1.0g
$MnSO_4{\cdot}2H_2O$	0.5g
$CoSO_4{\cdot}7H_2O$	0.18g
$ZnSO_4{\cdot}7H_2O$	0.18g
$CaCl_2{\cdot}2H_2O$	0.1g
$FeSO_4{\cdot}7H_2O$	0.1g
$NiCl_2{\cdot}6H_2O$	0.025g
$KAl(SO_4)_2{\cdot}12H_2O$	0.02g
$CuSO_4{\cdot}5H_2O$	0.01g
H_3BO_3	0.01g
$Na_2MoO_4{\cdot}2H_2O$	0.01g
$Na_2SeO_3{\cdot}5H_2O$	0.3mg

Preparation of Trace Elements Solution: Add nitrilotriacetic acid to 500.0mL of distilled/deionized water. Adjust pH to 6.5 with KOH. Add remaining components. Adjust pH to 7.0. Add distilled/deionized water to 1.0L.

Vitamin Solution:
Composition per liter:

Pyridoxine·HCl	10.0mg
Calcium DL-pantothenate	5.0mg
Lipoic acid	5.0mg
Nicotinic acid	5.0mg
p-Aminobenzoic acid	5.0mg
Riboflavin	5.0mg
Thiamine·HCl	5.0mg
Biotin	2.0mg
Folic acid	2.0mg
Vitamin B_{12}	0.1mg

Preparation of Vitamin Solution: Add components to distilled/deionized water and bring volume to 1.0L. Adjust pH to 7.0. Mix thoroughly.

Preparation of Medium: Prepare the medium anaerobically under 80% H_2 + 20% CO_2. Add components to distilled/deionized water and bring volume to 1.0L. Mix thoroughly. Sparge with 80% H_2 + 20% CO_2. Autoclave for 15 min at 15 psi pressure–121°C.

Use: For the cultivation and maintenance of *Methanococcoides methylutens*.

Methanococcus deltae Medium

Composition per liter:

NaCl	35.0g
$NaHCO_3$	5.0g
$MgCl_2·6H_2O$	4.0g
NH_4Cl	2.7g
Sodium acetate	2.5g
L-Cysteine·HCl	0.3g
K_2HPO_4	0.3g
KH_2PO_4	0.3g
$Na_2S·9H_2O$	0.3g
$MgSO_4·7H_2O$	0.13g
Resazurin	1.0mg
$(NH_4)_2SO_4$	0.3mg
Trace elements solution	10.0mL
Vitamin solution	10.0mL
L-Cysteine·HCl solution	10.0mL
$Na_2S·9H_2O$ solution	10.0mL

pH 6.9 ± 0.2 at 25°C

Trace Elements Solution:

Composition per liter:

$MgSO_4·7H_2O$	3.0g
Nitrilotriacetic acid	1.5g
NaCl	1.0g
$MnSO_4·2H_2O$	0.5g
$CoSO_4·7H_2O$	0.18g
$ZnSO_4·7H_2O$	0.18g
$CaCl_2·2H_2O$	0.1g
$FeSO_4·7H_2O$	0.1g
$NiCl_2·6H_2O$	0.025g
$KAl(SO_4)_2·12H_2O$	0.02g
$CuSO_4·5H_2O$	0.01g
H_3BO_3	0.01g
$Na_2MoO_4·2H_2O$	0.01g
$Na_2SeO_3·5H_2O$	0.3mg

Preparation of Trace Elements Solution: Add nitrilotriacetic acid to 500.0mL of distilled/deionized water. Adjust pH to 6.5 with KOH. Add remaining components. Adjust pH to 7.0. Add distilled/deionized water to 1.0L.

Vitamin Solution:

Composition per liter:

Pyridoxine·HCl	10.0mg
Calcium DL-pantothenate	5.0mg
Lipoic acid	5.0mg
Nicotinic acid	5.0mg
p-Aminobenzoic acid	5.0mg
Riboflavin	5.0mg
Thiamine·HCl	5.0mg
Biotin	2.0mg
Folic acid	2.0mg
Vitamin B_{12}	0.1mg

Preparation of Vitamin Solution: Add components to distilled/deionized water and bring volume to 1.0L. Adjust pH to 7.0. Mix thoroughly.

L-Cysteine·HCl Solution:

Composition per 10.0mL:

L-Cysteine·HCl	0.3g

Preparation of L-Cysteine·HCl Solution: Add L-cysteine·HCl to distilled/deionized water and bring volume to 10.0mL. Mix thoroughly. Autoclave under 100% N_2 for 15 min at 15 psi pressure–121°C.

$Na_2S·9H_2O$ Solution:

Composition per 10.0mL:

$Na_2S·9H_2O$	0.3g

Preparation of $Na_2S·9H_2O$ Solution: Add $Na_2S·9H_2O$ to distilled/deionized water and bring volume to 10.0mL. Mix thoroughly. Sparge with 100% N_2. Autoclave for 15 min at 15 psi pressure–121°C.

Preparation of Medium: Prepare and dispense medium under 80% H_2 + 20% CO_2. Add components, except L-cysteine·HCl solution and $Na_2S·9H_2O$ solution, to distilled/deionized water and bring volume to 980.0mL. Mix thoroughly. Anaerobically distribute into tubes or flasks fitted with butyl rubber stoppers. Autoclave for 15 min at 15 psi pressure–121°C. Anaerobically add 10.0mL of sterile L-cysteine·HCl solution and 10.0mL of sterile $Na_2S·9H_2O$ solution to each liter of medium or, using a syringe, inject the appropriate amount of sterile L-cysteine·HCl solution and sterile $Na_2S·9H_2O$ solution into individual tubes containing medium.

Use: For the cultivation and maintenance of *Methanococcus deltae.*

Methanococcus jannaschii Medium

Composition per liter:

NaCl	30.0g
$MgSO_4·7H_2O$	3.40g
$MgCl_2·2H_2O$	2.7g
$NaHCO_3$	1.0g
$Na_2S·9H_2O$	0.5g
KCl	0.33g
NH_4Cl	0.25g
$CaCl_2·2H_2O$	0.14g
K_2HPO_4	0.14g
$Fe(NH_4)_2(SO_4)_2·6H_2O$	0.01g
Resazurin	1.0mg
$Na_2SeO_3·5H_2O$	0.5mg
$NiCl_2·6H_2O$	0.5mg
Trace elements solution	10.0mL
Vitamin solution	10.0mL
L-Cysteine·HCl solution	10.0mL
$Na_2S·9H_2O$ solution	10.0mL

pH 6.0 ± 0.2 at 25°C

Trace Elements Solution:

Composition per liter:

$MgSO_4·7H_2O$	3.0g
Nitrilotriacetic acid	1.5g
NaCl	1.0g
$MnSO_4·2H_2O$	0.5g
$CoSO_4·7H_2O$	0.18g
$ZnSO_4·7H_2O$	0.18g
$CaCl_2·2H_2O$	0.1g

$FeSO_4 \cdot 7H_2O$	0.1g
$NiCl_2 \cdot 6H_2O$	0.025g
$KAl(SO_4)_2 \cdot 12H_2O$	0.02g
$CuSO_4 \cdot 5H_2O$	0.01g
H_3BO_3	0.01g
$Na_2MoO_4 \cdot 2H_2O$	0.01g
$Na_2SeO_3 \cdot 5H_2O$	0.3mg

Preparation of Trace Elements Solution: Add nitrilotriacetic acid to 500.0mL of distilled/deionized water. Adjust pH to 6.5 with KOH. Add remaining components. Adjust pH to 7.0. Add distilled/deionized water to 1.0L.

Vitamin Solution:

Composition per liter:

Pyridoxine·HCl	10.0mg
Calcium DL-pantothenate	5.0mg
Lipoic acid	5.0mg
Nicotinic acid	5.0mg
p-Aminobenzoic acid	5.0mg
Riboflavin	5.0mg
Thiamine·HCl	5.0mg
Biotin	2.0mg
Folic acid	2.0mg
Vitamin B_{12}	0.1mg

Preparation of Vitamin Solution: Add components to distilled/deionized water and bring volume to 1.0L. Adjust pH to 7.0. Mix thoroughly.

L-Cysteine·HCl Solution:

Composition per 10.0mL:

L-Cysteine·HCl	0.5g

Preparation of L-Cysteine·HCl Solution: Add L-cysteine·HCl to distilled/deionized water and bring volume to 10.0mL. Mix thoroughly. Autoclave under 100% N_2 for 15 min at 15 psi pressure–121°C.

$Na_2S \cdot 9H_2O$ Solution:

Composition per 10.0mL:

$Na_2S \cdot 9H_2O$	0.3g

Preparation of $Na_2S \cdot 9H_2O$ Solution: Add $Na_2S \cdot 9H_2O$ to distilled/deionized water and bring volume to 10.0mL. Mix thoroughly. Sparge with 100% N_2. Autoclave for 15 min at 15 psi pressure–121°C.

Preparation of Medium: Prepare and dispense medium under 80% H_2 + 20% CO_2. Add components, except L-cysteine·HCl solution and $Na_2S \cdot 9H_2O$ solution, to distilled/deionized water and bring volume to 980.0mL. Mix thoroughly. Anaerobically distribute into tubes or flasks. Autoclave for 15 min at 15 psi pressure–121°C. Anaerobically add 10.0mL of sterile L-cysteine·HCl solution and 10.0mL of sterile $Na_2S \cdot 9H_2O$ solution to each liter of medium or, using a syringe, inject the appropriate amount of sterile L-cysteine·HCl solution and sterile $Na_2S \cdot 9H_2O$ solution into individual tubes containing medium.

Use: For the cultivation and maintenance of *Methanococcus* species.

Methanococcus jannaschii Medium

Composition per 1020.0mL:

PIPES (piperazine-*N,N′*-bis[2-ethanesulfonic acid]) buffer	15.12g
$MgCl_2 \cdot 6H_2O$	4.3g
$MgSO_4 \cdot 7H_2O$	3.4g
NaCl	3.0g
NH_4Cl	0.25g
K_2HPO_4	0.14g
$CaCl_2 \cdot 2H_2O$	0.14g
KCl	0.33g
Minerals solution	10.0mL
$Na_2S_2O_3$ solution	10.0mL
β-Mercaptoethanol solution	10.0mL
SeO_2 solution	1.0mL

pH 7.0 ± 0.2 at 25°C

Minerals Solution:

Composition per liter:

Nitrilotriacetic acid	4.5g
$FeCl_2 \cdot 4H_2O$	0.4g
$CoCl_2 \cdot 2H_2O$	0.17g
$MnCl_2 \cdot 4H_2O$	0.1g
$ZnCl_2$	0.1g
$NaMoO_4 \cdot 6H_2O$	36.0mg
$CaCl_2 \cdot H_2O$	27.0mg

Preparation of Minerals Solution: Add components to distilled/deionized water and bring volume to 1.0L. Mix thoroughly.

$Na_2S_2O_3$ Solution:

Composition per 10.0mL:

$Na_2S_2O_3 \cdot 5H_2O$	0.63g

Preparation of $Na_2S_2O_3$ Solution: Add $Na_2S_2O_3 \cdot 5H_2O$ to distilled/deionized water and bring volume to 10.0mL. Mix thoroughly. Sparge with 100% N_2. Autoclave for 15 min at 15 psi pressure–121°C.

β-Mercaptoethanol Solution:

Composition per 10.0mL:

β-Mercaptoethanol	0.39g

Preparation of β-Mercaptoethanol Solution: Add β-mercaptoethanol to distilled/deionized water and bring volume to 10.0mL. Mix thoroughly. Sparge with 100% N_2. Autoclave for 15 min at 15 psi pressure–121°C.

SeO_2 Solution:

Composition per 100.0mL:

SeO_2	0.011g

Preparation of SeO_2 Solution: Add SeO_2 to distilled/deionized water and bring volume to 100.0mL. Mix thoroughly.

Preparation of Medium: Prepare and dispense medium under 100% N_2. Add components, except $Na_2S_2O_3$ solution and β-mercaptoethanol solution, to distilled/deionized water and bring volume to 1.0L. Mix thoroughly. Gently heat and bring to boiling. Continue boiling for 3 min. Cool to room temperature while sparging with 100% N_2. Adjust pH to 7.0 with KOH. Anaerobically distribute 100.0mL volumes into anaerobic bottles. Autoclave for 15 min at 15 psi pressure–121°C. Aseptically and anaerobically add 1.0mL of sterile $Na_2S_2O_3$ solution and 1.0mL of sterile β-mercaptoethanol solution to each bottle. Mix thoroughly. Prior to inoculation, flush each bottle with 80% H_2 + 20% CO_2.

Use: For the cultivation of *Methanococcus jannaschii.*

Methanococcus McC Medium

Composition per 1100.0mL:

$NaHCO_3$	5.0g
Yeast extract	2.0g
L-Cysteine·HCl·H_2O	0.5g
General salts solution	500.0mL
NaCl solution	75.0mL
$Na_2S \cdot 9H_2O$ solution	20.0mL
K_2HPO_4 solution	10.0mL

Trace minerals solution 10.0mL
Sodium acetate solution 10.0mL
Iron stock solution 5.0mL
Resazurin solution 1.0mL

General Salts Solution:
Composition per liter:

$MgSO_4 \cdot 7H_2O$	6.9g
$MgCl_2 \cdot 6H_2O$	5.5g
NH_4Cl	1.0g
KCl	0.67g
$CaCl_2 \cdot 2H_2O$	0.28g

Preparation of General Salts Solution: Add components to distilled/deionized water and bring volume to 1.0L. Mix thoroughly.

NaCl Solution:
Composition per 100.0mL:
NaCl 29.3g

Preparation of NaCl Solution: Add NaCl to distilled/deionized water and bring volume to 100.0mL. Mix thoroughly.

$Na_2S \cdot 9H_2O$ Solution:
Composition per 100.0mL:

NaOH	1 pellet
$Na_2S \cdot 9H_2O$	2.5g

Preparation of $Na_2S \cdot 9H_2O$ Solution: Bring 100.0mL of distilled/deionized water to boiling. Cool to room temperature while sparging with 100%N_2. Dissolve 1 pellet of NaOH in the anaerobic water. Weigh out a little more than 2.5g of $Na_2S \cdot 9H_2O$. Briefly rinse the crystals in distilled/deionized water. Dry the crystals by blotting on paper towels or filter paper. Add 2.5g of washed $Na_2S \cdot 9H_2O$ crystals to 100.0mL of anaerobic NaOH solution. Distribute into serum bottles fitted with butyl rubber stoppers and aluminum seals. Do not grease stoppers. Pressurize to 60kPa with 100% N_2. Autoclave for 15 min at 15 psi pressure–121°C. Store at room temperature in an anaerobic chamber.

K_2HPO_4 Solution:
Composition per 100.0mL:
K_2HPO_4 .... 1.4g

Preparation of K_2HPO_4 Solution: Add K_2HPO_4 to distilled/deionized water and bring volume to 100.0mL. Mix thoroughly.

Trace Minerals Solution:
Composition per liter:

Nitrilotriacetic acid	1.5g
$Na_2WO_4 \cdot 2H_2O$	1.0g
$Fe(NH_4)_2(SO_4)_2 \cdot 6H_2O$	0.2g
Na_2SeO_3	0.2g
$Na_2MoO_4 \cdot 2H_2O$	0.1g
$Mn_4 \cdot 2H_2O$	0.1g
$Zn_4 \cdot 7H_2O$	0.1g
$NiCl_2 \cdot 7H_2O$	0.025g
$CuSO_4 \cdot 5H_2O$	0.01g

Preparation of Trace Minerals Solution: Add nitrilotriacetic acid to 500.0mL of distilled/deionized water. Adjust pH to 6.5 with KOH. Add remaining components. Add distilled/deionized water to 1.0L. Adjust pH to 7.0.

Sodium Acetate Solution:
Composition per 100.0mL:
Sodium acetate·$3H_2O$ 13.6g

Preparation of Sodium Acetate Solution: Add sodium acetate·$3H_2O$ to distilled/deionized water and bring volume to 100.0mL. Mix thoroughly.

Iron Stock Solution:
Composition per 100.0mL:
$Fe(NH_4)_2(SO_4)_2 \cdot 6H_2O$ 0.2g

Preparation of Iron Stock Solution: Add $Fe(NH_4)_2(SO_4)_2 \cdot 6H_2O$ to 5.0mL of distilled H_2O containing 2 drops of concentrated HCl. Mix thoroughly. When the $Fe(NH_4)_2(SO_4)_2 \cdot 6H_2O$ has dissolved, bring the volume to 100.0mL with distilled/deionized water.

Resazurin Solution:
Composition per 10.0mL:
Resazurin 10.0mg

Preparation of Resazurin Solution: Add resazurin to distilled/deionized water and bring volume to 10.0mL. Mix thoroughly.

Preparation of Medium: Prepare and dispense medium under 80% H_2 + 20% CO_2. Add components, except $NaHCO_3$ and $Na_2S \cdot 9H_2O$ solution, to distilled/deionized water and bring volume to 1080.0mL. Mix thoroughly. Gently heat and bring to boiling. Continue boiling for 3 min. Cool to room temperature while sparging with 80% H_2 + 20% CO_2. Add $NaHCO_3$. Mix thoroughly. Anaerobically distribute 9.8mL volumes into anaerobic tubes. Autoclave for 15 min at 15 psi pressure–121°C. Aseptically and anaerobically add 0.2mL of sterile $Na_2S \cdot 9H_2O$ solution to each tube. Mix thoroughly.

Use: For the cultivation of *Methanococcus* species.

Methanococcus McN Medium

Composition per 1100.0mL:

$NaHCO_3$	5.0g
L-Cysteine·HCl·H_2O	0.5g
General salts solution	500.0mL
NaCl solution	75.0mL
$Na_2S \cdot 9H_2O$ solution	20.0mL
K_2HPO_4 solution	10.0mL
Trace minerals solution	10.0mL
Iron stock solution	5.0mL
Resazurin solution	1.0mL

General Salts Solution:
Composition per liter:

$MgSO_4 \cdot 7H_2O$	6.9g
$MgCl_2 \cdot 6H_2O$	5.5g
NH_4Cl	1.0g
KCl	0.67g
$CaCl_2 \cdot 2H_2O$	0.28g

Preparation of General Salts Solution: Add components to distilled/deionized water and bring volume to 1.0L. Mix thoroughly.

NaCl Solution:
Composition per 100.0mL:
NaCl 29.3g

Preparation of NaCl Solution: Add NaCl to distilled/deionized water and bring volume to 100.0mL. Mix thoroughly.

$Na_2S \cdot 9H_2O$ Solution:
Composition per 100.0mL:

$Na_2S \cdot 9H_2O$	2.5g
NaOH	1 pellet

Preparation of $Na_2S \cdot 9H_2O$ Solution: Bring 100.0mL of distilled/deionized water to boiling. Cool to room temperature while sparging with 100%N_2. Dissolve 1 pellet of NaOH in the anaerobic water. Weigh out a little more than 2.5g of $Na_2S{\cdot}9H_2O$. Briefly rinse the crystals in distilled/deionized water. Dry the crystals by blotting on paper towels or filter paper. Add 2.5g of washed $Na_2S{\cdot}9H_2O$ crystals to 100.0mL of anaerobic NaOH solution. Distribute into serum bottles fitted with butyl rubber stoppers and aluminum seals. Do not grease stoppers. Pressurize to 60kPa with 100% N_2. Autoclave for 15 min at 15 psi pressure–121°C. Store at room temperature in an anaerobic chamber.

K_2HPO_4 Solution:

Composition per 100.0mL:

K_2HPO_4	1.4g

Preparation of K_2HPO_4 Solution: Add K_2HPO_4 to distilled/deionized water and bring volume to 100.0mL. Mix thoroughly.

Trace Minerals Solution:

Composition per liter:

Nitrilotriacetic acid	1.5g
$Na_2WO_4{\cdot}2H_2O$	1.0g
$Fe(NH_4)_2(SO_4)_2{\cdot}6H_2O$	0.2g
Na_2SeO_3	0.2g
$Na_2MoO_4{\cdot}2H_2O$	0.1g
$Mn_4{\cdot}2H_2O$	0.1g
$Zn_4{\cdot}7H_2O$	0.1g
$NiCl_2{\cdot}7H_2O$	0.025g
$CuSO_4{\cdot}5H_2O$	0.01g

Preparation of Trace Minerals Solution: Add nitrilotriacetic acid to 500.0mL of distilled/deionized water. Adjust pH to 6.5 with KOH. Add remaining components. Add distilled/deionized water to 1.0L. Adjust pH to 7.0.

Iron Stock Solution:

Composition per 100.0mL:

$Fe(NH_4)_2(SO_4)_2{\cdot}6H_2O$	0.2g

Preparation of Iron Stock Solution: Add $Fe(NH_4)_2(SO_4)_2{\cdot}6H_2O$ to 5.0mL of distilled H_2O containing 2 drops of concentrated HCl. Mix thoroughly. When the $Fe(NH_4)_2(SO_4)_2{\cdot}6H_2O$ has dissolved, bring the volume to 100.0mL with distilled/deionized water.

Resazurin Solution:

Composition per 10.0mL:

Resazurin	10.0mg

Preparation of Resazurin Solution: Add resazurin to distilled/deionized water and bring volume to 10.0mL. Mix thoroughly.

Preparation of Medium: Prepare and dispense medium under 80% H_2 + 20% CO_2. Add components, except $NaHCO_3$ and $Na_2S{\cdot}9H_2O$ solution, to distilled/deionized water and bring volume to 1080.0mL. Mix thoroughly. Gently heat and bring to boiling. Continue boiling for 3 min. Cool to room temperature while sparging with 80% H_2 + 20% CO_2. Add $NaHCO_3$. Mix thoroughly. Anaerobically distribute 9.8mL volumes into anaerobic tubes. Autoclave for 15 min at 15 psi pressure–121°C. Aseptically and anaerobically add 0.2mL of sterile $Na_2S{\cdot}9H_2O$ solution to each tube. Mix thoroughly.

Use: For the cultivation of *Methanococcus* species.

Methanococcus McNail Medium

Composition per 1100.0mL:

$NaHCO_3$	5.0g
L-Leucine	1.0g
L-Isoleucine	0.5g
L-Cysteine·HCl·H_2O	0.5g
General salts solution	500.0mL
NaCl solution	75.0mL
$Na_2S{\cdot}9H_2O$ solution	20.0mL
K_2HPO_4 solution	10.0mL
Trace minerals solution	10.0mL
Sodium acetate solution	10.0mL
Pantoyllactone solution	10.0mL
Iron stock solution	5.0mL
Resazurin solution	1.0mL

General Salts Solution:

Composition per liter:

$MgSO_4{\cdot}7H_2O$	6.9g
$MgCl_2{\cdot}6H_2O$	5.5g
NH_4Cl	1.0g
KCl	0.67g
$CaCl_2{\cdot}2H_2O$	0.28g

Preparation of General Salts Solution: Add components to distilled/deionized water and bring volume to 1.0L. Mix thoroughly.

NaCl Solution:

Composition per 100.0mL:

NaCl	29.3g

Preparation of NaCl Solution: Add NaCl to distilled/deionized water and bring volume to 100.0mL. Mix thoroughly.

$Na_2S{\cdot}9H_2O$ Solution:

Composition per 100.0mL:

NaOH	1 pellet
$Na_2S{\cdot}9H_2O$	2.5g

Preparation of $Na_2S{\cdot}9H_2O$ Solution: Bring 100.0mL of distilled/deionized water to boiling. Cool to room temperature while sparging with 100%N_2. Dissolve 1 pellet of NaOH in the anaerobic water. Weigh out a little more than 2.5g of $Na_2S{\cdot}9H_2O$. Briefly rinse the crystals in distilled/deionized water. Dry the crystals by blotting on paper towels or filter paper. Add 2.5g of washed $Na_2S{\cdot}9H_2O$ crystals to 100.0mL of anaerobic NaOH solution. Distribute into serum bottles fitted with butyl rubber stoppers and aluminum seals. Do not grease stoppers. Pressurize to 60kPa with 100% N_2. Autoclave for 15 min at 15 psi pressure–121°C. Store at room temperature in an anaerobic chamber.

K_2HPO_4 Solution:

Composition per 100.0mL:

K_2HPO_4	1.4g

Preparation of K_2HPO_4 Solution: Add K_2HPO_4 to distilled/deionized water and bring volume to 100.0mL. Mix thoroughly.

Trace Minerals Solution:

Composition per liter:

Nitrilotriacetic acid	1.5g
$Na_2WO_4{\cdot}2H_2O$	1.0g
$Fe(NH_4)_2(SO_4)_2{\cdot}6H_2O$	0.2g
Na_2SeO_3	0.2g
$Na_2MoO_4{\cdot}2H_2O$	0.1g
$Mn_4{\cdot}2H_2O$	0.1g
$Zn_4{\cdot}7H_2O$	0.1g
$NiCl_2{\cdot}7H_2O$	0.025g
$CuSO_4{\cdot}5H_2O$	0.01g

Preparation of Trace Minerals Solution: Add nitrilotriacetic acid to 500.0mL of distilled/deionized water. Adjust pH to 6.5 with KOH. Add remaining components. Add distilled/deionized water to 1.0L. Adjust pH to 7.0.

Sodium Acetate Solution:

Composition per 100.0mL:

Sodium acetate·$3H_2O$	13.6g

Preparation of Sodium Acetate Solution: Add 13.6g of sodium acetate·$3H_2O$ to distilled/deionized water and bring volume to 100.0mL. Mix thoroughly.

Pantoyllactone Solution:

Composition per 100.0mL:

Pantoyllactone	0.013g

Preparation of Pantoyllactone Solution: Add pantoyllactone to distilled/deionized water and bring volume to 100.0mL. Mix thoroughly.

Iron Stock Solution:

Composition per 100.0mL:

$Fe(NH_4)_2(SO_4)_2 \cdot 6H_2O$	0.2g

Preparation of Iron Stock Solution: Add $Fe(NH_4)_2(SO_4)_2 \cdot 6H_2O$ to 5.0mL of distilled H_2O containing 2 drops of concentrated HCl. Mix thoroughly. When the $Fe(NH_4)_2(SO_4)_2 \cdot 6H_2O$ has dissolved, bring the volume to 100.0mL with distilled/deionized water.

Resazurin Solution:

Composition per 10.0mL:

Resazurin	10.0mg

Preparation of Resazurin Solution: Add resazurin to distilled/deionized water and bring volume to 10.0mL. Mix thoroughly.

Preparation of Medium: Prepare and dispense medium under 80% H_2 + 20% CO_2. Add components, except $NaHCO_3$ and $Na_2S \cdot 9H_2O$ solution, to distilled/deionized water and bring volume to 1080.0mL. Mix thoroughly. Gently heat and bring to boiling. Continue boiling for 3 min. Cool to room temperature while sparging with 80% H_2 + 20% CO_2. Add $NaHCO_3$. Mix thoroughly. Anaerobically distribute 9.8mL volumes into anaerobic tubes. Autoclave for 15 min at 15 psi pressure–121°C. Aseptically and anaerobically add 0.2mL of sterile $Na_2S \cdot 9H_2O$ solution to each tube. Mix thoroughly.

Use: For the cultivation of *Methanococcus* species.

Methanococcus Medium

Composition per liter:

NaCl	18.0g
$Mg_2SO_4 \cdot 7H_2O$	3.45g
$MgCl_2 \cdot 2H_2O$	2.75g
Pancreatic digest of casein	2.0g
Yeast extract	2.0g
Sodium acetate	1.0g
L-Cysteine·HCl	0.5g
$Na_2S \cdot 9H_2O$	0.5g
NH_4HCO_3	0.5g
KCl	0.335g
NH_4Cl	0.225g
$CaCl_2 \cdot 2H_2O$	0.14g
KH_2PO_4	0.14g
Calcium DL-pantothenate	5.0mg
Na_2SeO_3	2.0mg
$FeSO_4 \cdot 2H_2O$	1.0mg
Resazurin	1.0mg
Trace minerals stock solution	10mL

pH 6.5 ± 0.2 at 25°C

Trace Minerals Stock Solution:

Composition per liter:

Nitrilotriacetic acid	1.5g
$MnSO_4 \cdot H_2O$	0.5g
$CoCl_2$	0.1g
$ZnSO_4$	0.1g
$AIK(SO_4)_2 \cdot 12H_2O$	0.01g
$CuSO_4 \cdot 5H_2O$	0.01g
H_3BO_3	0.01g
$Na_2MoO_4 \cdot 2H_2O$	0.01g
$NiCl_2$	0.01g

Preparation of Trace Minerals Stock Solution: Add nitrilotriacetic acid to 500.0mL of distilled/deionized water. Adjust pH to 6.5 with KOH. Add remaining components one at a time. Add distilled/deionized water to 1.0L. Adjust pH to 6.8.

Preparation of Medium: Prepare medium anaerobically under 80% N_2 + 20% CO_2. Add components, except Na_2CO_3, L-cysteine·HCl, and $Na_2S \cdot 9H_2O$, to distilled/deionized water and bring volume to 1.0L. Mix thoroughly. Gently heat and bring to boiling. Cool while sparging with 80% N_2 + 20% CO_2. Add Na_2CO_3, L-cysteine·HCl, and $Na_2S \cdot 9H_2O$. Dispense into tubes, bottles, or flasks under an atmosphere of 80% H_2 + 20% CO_2. Seal with butyl rubber stoppers secured with aluminum crimp seals. Autoclave for 15 min at 15 psi pressure–121°C. Cool to room temperature. Pressurize the head space to 69kPA with 80% H_2 + 20% CO_2.

Use: For the cultivation of *Methanococcus* species.

Methanococcus sp. Medium (DSMZ Medium 141a)

Composition per 1040.0mL:

NaCl	18.0g
$MgCl_2 \cdot 6H_2O$	4.0g
$MgSO_4 \cdot 7H_2O$	3.45g
Na-acetate	1.0g
KCl	0.335g
NH_4Cl	0.25g
$CaCl_2 \cdot 2H_2O$	0.14g
K_2HPO_4	0.14g
$Fe(NH_4)_2(SO_4)_2 \cdot 7H_2O$	2.0mg
Resazurin	1.0mg
Trace elements solution	10.0mL
Vitamin solution	10.0mL
L-Cysteine solution	10.0mL
$Na_2S \cdot 9H_2O$ solution	10.0mL

pH 7.2 ± 0.2 at 25°C

Trace Elements Solution:

Composition per liter:

$MgSO_4 \cdot 7H_2O$	3.0g
Nitrilotriacetic acid	1.5g
NaCl	1.0g
$MnSO_4 \cdot 2H_2O$	0.5g
$CoSO_4 \cdot 7H_2O$	0.18g
$ZnSO_4 \cdot 7H_2O$	0.18g
$CaCl_2 \cdot 2H_2O$	0.1g
$FeSO_4 \cdot 7H_2O$	0.1g
$NiCl_2 \cdot 6H_2O$	0.025g
$KAl(SO_4)_2 \cdot 12H_2O$	0.02g
H_3BO_3	0.01g
$Na_2MoO_4 \cdot 4H_2O$	0.01g

$CuSO_4 \cdot 5H_2O$ 0.01g
$Na_2SeO_3 \cdot 5H_2O$ 0.3mg

Preparation of Trace Elements Solution: Add nitrilotriacetic acid to 500.0mL of distilled/deionized water. Dissolve by adjusting pH to 6.5 with KOH. Add remaining components. Add distilled/deionized water to 1.0L. Mix thoroughly.

Vitamin Solution:
Composition per liter:
Pyridoxine-HCl 10.0mg
Thiamine-HCl·$2H_2O$ 5.0mg
Riboflavin 5.0mg
Nicotinic acid 5.0mg
D-Ca-pantothenate 5.0mg
p-Aminobenzoic acid 5.0mg
Lipoic acid 5.0mg
Biotin 2.0mg
Folic acid 2.0mg
Vitamin B_{12} 0.1mg

Preparation of Vitamin Solution: Add components to distilled/deionized water and bring volume to 1.0L. Mix thoroughly. Sparge with 80% H_2 + 20% CO_2. Filter sterilize.

$Na_2S \cdot 9H_2O$ Solution:
Composition per 10.0mL:
$Na_2S \cdot 9H_2O$ 0.5g

Preparation of $Na_2S \cdot 9H_2O$ Solution: Add $Na_2S \cdot 9H_2O$ to distilled/deionized water and bring volume to 10.0mL. Sparge with N_2. Autoclave for 15 min at 15 psi pressure–121°C. Cool to 25°C. Store anaerobically.

L-Cysteine Solution:
Composition per 10.0mL:
L-Cysteine·HCl·H_2O 0.5g

Preparation of L-Cysteine Solution: Add L-cysteine·HCl·H_2O to distilled/deionized water and bring volume to 10.0mL. Mix thoroughly. Sparge with 100% N_2. Autoclave for 15 min at 15 psi pressure–121°C.

Preparation of Medium: Prepare and dispense medium under an oxygen-free 80% H_2 + 20% CO_2 gas mixture. Add components, except vitamin solution, L-cysteine solution, and $Na_2S \cdot 9H_2O$ solution, to distilled/deionized water and bring volume to 1.0L. Mix thoroughly. Sparge with 80% H_2 + 20% CO_2. Autoclave for 15 min at 15 psi pressure–121°C. Cool to 25°C while sparging with 80% H_2 + 20% CO_2. Aseptically and anaerobically add 10.0mL vitamin solution, 10.0mL of sterile L-cysteine solution, and 10.0mL of sterile $Na_2S \cdot 9H_2O$ solution. Mix thoroughly. Aseptically and anaerobically distribute into sterile tubes or flasks. Alternately the medium can be distributed to tubes under anaerobic conditions and autoclaved in tubes prior to addition of substrate solution, vitamin solution, and $Na_2S \cdot 9H_2O$ solution. Appropriate amounts of these solutions can then be added to each tube by syringes to yield the desired concentrations. After inoculation, pressurize culture vessels to 2 bar 80% H_2 + 20% CO_2 overpressure.

Use: For the cultivation of *Methanococcus* sp.

Methanococcus vannielii Medium

Composition per 1020.0mL:
Solution A 500.0mL
Inorganic salts solution 500.0mL
$Na_2S \cdot 9H_2O$ solution 10.0mL
Na_2CO_3 solution 10.0mL

Solution A:
Composition per 500.0mL:
Sodium formate 10.0g
Phenol Red 3.0mg
Methylene Blue 2.0mg

Preparation of Solution A: Add components to distilled/deionized water and bring volume to 500.0mL. Mix thoroughly. Autoclave for 15 min at 15 psi pressure–121°C. Cool to 25°C.

Inorganic Salts Solution:
Composition per 500.0mL:
$K_2HPO_4 \cdot 3H_2O$ 1.45g
NH_4Cl 1.0g
KH_2PO_4 .... 0.75g
$MgCl_2 \cdot 6H_2O$ 0.2g
Nitrilotriacetic acid 0.04g
$CaCl_2 \cdot 2H_2O$ 0.02g
$FeCl_2 \cdot 4H_2O$ 3.6mg
$CoCl_2 \cdot 6H_2O$ 1.5mg
$MnCl_2$·4H2O 0.9mg
$ZnCl_2$ 0.9mg
H_3BO_2 .... 0.17mg
$Na_2MoO_4 \cdot 2H_2O$ 0.09mg

Preparation of Inorganic Salts Solution: Add nitrilotriacetic acid to 250.0mL of distilled/deionized water. Dissolve by adjusting pH to 6.5 with KOH. Add remaining components. Readjust pH to 7.2 with H_2SO_4 or KOH. Add distilled/deionized water to 500.0mL. Filter sterilize.

$Na_2S \cdot 9H_2O$ Solution:
Composition per 10.0mL:
$Na_2S \cdot 9H_2O$ 0.3g

Preparation of $Na_2S \cdot 9H_2O$ Solution: Add $Na_2S \cdot 9H_2O$ to distilled/deionized water and bring volume to 10.0mL. Mix thoroughly. Autoclave for 15 min at 15 psi pressure–121°C. Cool to 25°C.

Na_2CO_3 Solution:
Composition per 10.0mL:
Na_2CO_3 .... 2.5g

Preparation of Na_2CO_3 Solution: Add Na_2CO_3 to distilled/deionized water and bring volume to 10.0mL. Mix thoroughly. Autoclave for 15 min at 15 psi pressure–121°C. Cool to 25°C.

Preparation of Medium: Prepare and distribute medium anaerobically under 80% N_2 + 20% CO_2. Aseptically and anaerobically combine 500.0mL of sterile inorganic salts solution, 500.0mL of sterile solution A, 10.0mL of sterile $Na_2S \cdot 9H_2O$ solution, and 10.0mL of sterile Na_2CO_3 solution. Mix thoroughly. Aseptically and anaerobically distribute into sterile tubes or flasks.

Use: For the isolation and cultivation of *Methanococcus vannielii* from marine mud.

Methanococcus vannielii Medium

Composition per liter:
Sodium formate 15.0g
K_2HPO_4 .... 3.48g
$CoCl_2 \cdot 6H_2O$ 2.38g
NH_4Cl 1.0g
L-Cysteine·HCl·H_2O 0.3g
$MgSO_4 \cdot 7H_2O$ 0.2g
$CaCl_2 \cdot 2H_2O$ 0.01g
$FeSO_4 \cdot 7H_2O$ 0.01g

$MnSO_4 \cdot H_2O$	7.5mg
$Na_2MoO_4 \cdot 2H_2O$	7.5mg
Na_2SeO_3	1.7mg
$Na_2S \cdot 9H_2O$ solution	10.0mL

$Na_2S \cdot 9H_2O$ Solution:
Composition per 10.0mL:

$Na_2S \cdot 9H_2O$	0.15g

Preparation of $Na_2S \cdot 9H_2O$ Solution: Add $Na_2S \cdot 9H_2O$ to distilled/deionized water and bring volume to 10.0mL. Mix thoroughly. Autoclave for 15 min at 15 psi pressure–121°C. Cool to 25°C.

Preparation of Medium: Prepare and distribute medium anaerobically under 100% N_2. Add components, except $Na_2S \cdot 9H_2O$ solution, to distilled/deionized water and bring volume to 990.0mL. Mix thoroughly. Gently heat and bring to boiling. Continue boiling until resazurin turns colorless, indicating reduction. Autoclave for 15 min at 15 psi pressure–121°C. Cool to 25°C. Aseptically add 10.0mL of sterile $Na_2S \cdot 9H_2O$ solution. Mix thoroughly. Aseptically and anaerobically distribute into sterile tubes or flasks.

Use: For the cultivation of *Methanococcus vannielii*.

Methanococcus voltae BD Medium

Composition 1003.0mL:

L-Leucine	50.0g
NaCl	18.0g
Sodium panthothenate	5.0g
$MgSO_4 \cdot 7H_2O$	3.48g
$MgCl_2 \cdot 6H_2O$	2.75g
Sodium acetate·$3H_2O$	1.0g
KCl	0.34g
NH_4Cl	0.26g
$CaCl_2 \cdot 2H_2O$	0.14g
K_2PO_4	0.14g
L-Isoleucine	0.1g
L-Cysteine/Na_2S solution	17.5mL
Trace minerals solution	10.0mL
Vitamin solution	10.0mL
Na_2CO_3 solution	6.0mL
$FeSO_4 \cdot 7H_2O$ solution	4.5mL

Trace Minerals Solution:
Composition per liter:

Nitrilotriacetic acid	1.5g
$MnSO_4 \cdot 7H_2O$	0.5g
$NiCl_2 \cdot 6H_2O$	0.12g
$CoCl_2 \cdot 6H_2O$	0.1g
$FeSO_4 \cdot 7H_2O$	0.1g
Resazurin	0.1g
$ZnSO_4 \cdot 7H_2O$	0.1g
$AlK(SO_4)_2 \cdot 5H_2O$	0.01g
$CuSO_4 \cdot 5H_2O$	0.01g
H_3BO_3	0.01g

Preparation of Trace Minerals Solution: Add nitrilotriacetic acid to 500.0mL of distilled/deionized water. Adjust pH to 6.5 with KOH. Add remaining components one at a time. Add distilled/deionized water to 1.0L. Adjust pH to 6.8.

Vitamin Solution:
Composition per liter:

Pyridoxine·HCl	10.0mg
p-Aminobenzoic acid	5.0mg
Lipoic acid	5.0mg
Nicotinic acid	5.0mg
Riboflavin	5.0mg
Thiamine·HCl	5.0mg
Biotin	2.0mg
Folic acid	2.0mg
Vitamin B_{12}	0.5mg

Preparation of Vitamin Solution: Add components to distilled/deionized water and bring volume to 1.0L. Mix thoroughly. Sparge with 100% N_2. Filter sterilize.

L-Cysteine/Na_2S Solution:
Composition per liter:

L-Cysteine·HCl	1.25g
$Na_2S \cdot 9H_2O$	1.25g

Preparation of L-Cysteine/Na_2S Solution: Add L-cysteine·HCl and $Na_2S \cdot 9H_2O$ to distilled/deionized water and bring volume to 10.0mL. Mix thoroughly. Sparge with 100% N_2. Autoclave for 15 min at 15 psi pressure–121°C. Before use, neutralize to pH 7.0 with sterile HCl.

Na_2CO_3 Solution:
Composition per 100.0mL:

Na_2CO_3	1.0g

Preparation of Na_2CO_3 Solution: Add Na_2CO_3 to distilled/deionized water and bring volume to 100.0mL. Mix thoroughly. Sparge with 80% N_2 + 20% CO_2. Autoclave for 15 min at 15 psi pressure–121°C.

$FeSO_4 \cdot 7H_2O$ Solution:
Composition per 100.0mL:

$FeSO_4 \cdot 7H_2O$	0.1g

Preparation of $FeSO_4 \cdot 7H_2O$ Solution: Add $FeSO_4 \cdot 7H_2O$ to distilled/deionized water and bring volume to 100.0mL. Mix thoroughly. Filter sterilize. Sparge with 100% N_2.

Preparation of Medium: Prepare and dispense medium under 80% H_2 + 20% CO_2. Add components, except L-cysteine/Na_2S solution, Na_2CO_3 solution, and $FeSO_4 \cdot 7H_2O$ solution, to distilled/deionized water and bring volume to 975.0mL. Mix thoroughly. Gently heat and bring to boiling while sparging with 80% H_2 + 20% CO_2. Cool to room temperature while sparging with 80% H_2 + 20% CO_2. Add 17.5mL of L-cysteine/$Na_2S \cdot 9H_2O$ solution and 6.0mL of Na_2CO_3 solution. Anaerobically dispense into tubes or bottles in 10.0mL aliquots under an atmosphere of 80% H_2 + 20% CO_2. Seal with butyl rubber stoppers secured with aluminum crimp seals. Autoclave for 15 min at 15 psi pressure–121°C. Immediately prior to inoculation, aseptically and anaerobically add 0.05mL of sterile $FeSO_4 \cdot 7H_2O$ solution to each tube.

Use: For the cultivation of *Methanococcus voltae*.

Methanocorpusculum Medium (DSMZ Medium 279)

Composition per liter:

Na-acetate	4.0g
$NaHCO_3$	4.0g
Na-formate	2.0g
Yeast extract	1.0g
KH_2PO_4	0.5g
$MgSO_4 \cdot 7H_2O$	0.4g
NaCl	0.4g
NH_4Cl	0.4g
$CaCl_2 \cdot 2H_2O$	0.05g
$FeSO_4 \cdot 7H_2O$	0.002g

Resazurin	0.001g
$NiCl_2 \cdot 6H_2O$	24.0mg
Sludge fluid	50.0mL
Fatty acid mixture	20.0mL
L-Cysteine solution	10.0mL
$Na_2S \cdot 9H_2O$	10.0mL
Trace elements solution SL-10	1.0mL

pH 6.7–7.0

Sludge Fluid:

Composition per 500.0mL:

Yeast extract	2.0g
Sludge	500.0mL

Preparation of Sludge Fluid: Add yeast extract to sludge from an anaerobic digester. Gas with nitrogen gas for a few minutes. Incubate at 37°C for 24 hrs. Centrifuge the sludge at 13,000 x g. Autoclave for 15 min at 15 psi pressure–121°C. Place the resulting, clear supernatant in screw-capped vessels under nitrogen gas. The sludge fluid can be stored at room temperature in the dark.

Fatty Acid Mixture:

Composition per 20.0mL:

Valeric acid	0.5g
Isovaleric acid	0.5g
α-Methylbutyric acid	0.5g
Isobutyric acid	0.5g
Distilled water	20.0mL

Preparation of Fatty Acid Mixture: Add components to 20.0mL distilled/deionized water. Mix thoroughly.

Trace Elements Solution SL-10:

Composition per liter:

$FeCl_2 \cdot 4H_2O$	1.5g
$CoCl_2 \cdot 6H_2O$	190.0mg
$MnCl_2 \cdot 4H_2O$	100.0mg
$ZnCl_2$	70.0mg
$Na_2MoO_4 \cdot 2H_2O$	36.0mg
$NiCl_2 \cdot 6H_2O$	24.0mg
H_3BO_3	6.0mg
$CuCl_2 \cdot 2H_2O$	2.0mg
HCl (25% solution)	10.0mL

Preparation of Trace Elements Solution SL-10: Add $FeCl_2 \cdot 4H_2O$ to 10.0mL of HCl solution. Mix thoroughly. Add distilled/deionized water and bring volume to 1.0L. Add remaining components. Mix thoroughly. Sparge with 100% N_2. Autoclave for 15 min at 15 psi pressure–121°C.

$Na_2S \cdot 9H_2O$ Solution:

Composition per 10.0mL:

$Na_2S \cdot 9H_2O$	0.5g

Preparation of $Na_2S \cdot 9H_2O$ Solution: Add $Na_2S \cdot 9H_2O$ to distilled/deionized water and bring volume to 10.0mL. Mix thoroughly. Sparge with 100% N_2. Autoclave for 15 min at 15 psi pressure–121°C. Before use, neutralize to pH 7.0 with sterile HCl.

L-Cysteine Solution:

Composition per 10.0mL:

L-Cysteine·HCl·H_2O	0.5g

Preparation of L-Cysteine Solution: Add L-cysteine·HCl·H_2O to distilled/deionized water and bring volume to 10.0mL. Mix thoroughly. Sparge with 100% N_2. Autoclave for 15 min at 15 psi pressure–121°C.

Preparation of Medium: Prepare and dispense medium under 80% H_2 + 20% CO_2 gas atmosphere. Add components, except L-cysteine solution, $Na_2S \cdot 9H_2O$ solution, and trace elements solution SL-10, to distilled/deionized water and bring volume to 920.0mL. Mix thoroughly. Adjust pH to 6.8. Sparge with 80% H_2 + 20% CO_2. Autoclave for 15 min at 15 psi pressure–121°C. Aseptically and anaerobically add 10.0mL L-cysteine solution, 10.0mL $Na_2S \cdot 9H_2O$ solution, and 1.0mL trace elements solution SL-10. Mix thoroughly. Aseptically and anaerobically distribute into sterile tubes or bottles. After inoculation, flush and repressurize the gas head space of culture bottles with sterile 80% H_2 + 20% CO_2 to 1 bar overpressure. Alternately, the medium without L-cysteine solution, $Na_2S \cdot 9H_2O$ solution, and trace elements solution SL-10 can be distributed to tubes anaerobically prior to autoclaving. After autoclaving in tubes the appropriate volumes of the individual solutions can be injected through the stoppers so that the final concentrations of the medium are achieved.

Use: For the cultivation of *Methanococcoides* spp. and *Methanolobus bombayensis.*

Methanoculleus olentangyi Medium

Composition per liter:

$NaHCO_3$	5.0g
NH_4Cl	2.7g
Sodium acetate	2.5g
NaCl	0.61g
K_2HPO_4	0.3g
KH_2PO_4	0.3g
$MgSO_4 \cdot 7H_2O$	0.13g
Resazurin	1.0mg
$(NH_4)_2SO_4$	0.3mg
Trace elements solution	10.0mL
Vitamin solution	10.0mL
L-Cysteine·HCl solution	10.0mL
$Na_2S \cdot 9H_2O$ solution	10.0mL

pH 6.9 ± 0.2 at 25°C

Trace Elements Solution:

Composition per liter:

$MgSO_4 \cdot 7H_2O$	3.0g
Nitrilotriacetic acid	1.5g
NaCl	1.0g
$MnSO_4 \cdot 2H_2O$	0.5g
$CoSO_4 \cdot 7H_2O$	0.18g
$ZnSO_4 \cdot 7H_2O$	0.18g
$CaCl_2 \cdot 2H_2O$	0.1g
$FeSO_4 \cdot 7H_2O$	0.1g
$KAl(SO_4)_2 \cdot 12H_2O$	0.02g
$CuSO_4 \cdot 5H_2O$	0.01g
H_3BO_3	0.01g
$Na_2MoO_4 \cdot 2H_2O$	0.01g
$NiCl_2 \cdot 6H_2O$	0.025g
$Na_2SeO_3 \cdot 5H_2O$	0.3mg

Preparation of Trace Elements Solution: Add nitrilotriacetic acid to 500.0mL of distilled/deionized water. Adjust pH to 6.5 with KOH. Add remaining components. Add distilled/deionized water to 1.0L.

Vitamin Solution:

Composition per liter:

Pyridoxine·HCl	10.0mg
Calcium DL-pantothenate	5.0mg
Lipoic acid	5.0mg

Nicotinic acid	5.0mg
p-Aminobenzoic acid	5.0mg
Riboflavin	5.0mg
Thiamine·HCl	5.0mg
Biotin	2.0mg
Folic acid	2.0mg
Vitamin B_{12}	0.1mg

Preparation of Vitamin Solution: Add components to distilled/deionized water and bring volume to 1.0L. Mix thoroughly.

L-Cysteine·HCl Solution:
Composition per 10.0mL:

L-Cysteine·HCl	0.3g

Preparation of L-Cysteine·HCl Solution: Add L-cysteine·HCl to distilled/deionized water and bring volume to 10.0mL. Mix thoroughly. Autoclave under 100% N_2 for 15 min at 15 psi pressure–121°C.

$Na_2S·9H_2O$ Solution:
Composition per 10.0mL:

$Na_2S·9H_2O$	0.3g

Preparation of $Na_2S·9H_2O$ Solution: Add $Na_2S·9H_2O$ to distilled/deionized water and bring volume to 10.0mL. Mix thoroughly. Sparge with 100% N_2. Autoclave for 15 min at 15 psi pressure–121°C.

Preparation of Medium: Prepare and dispense medium under 80% H_2 + 20% CO_2. Add components, except L-cysteine·HCl solution and $Na_2S·9H_2O$ solution, to distilled/deionized water and bring volume to 980.0mL. Mix thoroughly. Anaerobically distribute into tubes or flasks. Autoclave for 15 min at 15 psi pressure–121°C. Anaerobically add 10.0mL of sterile L-cysteine·HCl solution and 10.0mL of sterile $Na_2S·9H_2O$ solution to each liter of medium or, using a syringe, inject the appropriate amount of sterile L-cysteine·HCl solution and sterile $Na_2S·9H_2O$ solution into individual tubes containing medium.

Use: For the cultivation and maintenance of *Methanoculleus olentangyi.*

Methanogen Enrichment Medium, Barker

Composition per liter:

$CaCO_3$	20.0g
NH_4Cl	1.0g
$K_2HPO_4·3H_2O$	0.4g
$MgCl_2·6H_2O$	0.1g
Methanol	20.0mL

pH 7.0 ± 0.2 at 25°C

Preparation of Medium: Add components, except methanol and $CaCO_3$, to distilled/deionized water and bring volume to 1.0L. Mix thoroughly. Gently heat and bring to boiling. Autoclave for 15 min at 15 psi pressure–121°C. Cool to 25°C. Aseptically add filter-sterilized methanol solution. Mix thoroughly. Add 1.0g of $CaCO_3$ to each of 50.0mL screw-capped bottles. Autoclave for 15 min at 15 psi pressure–121°C. Cool to 25°C. Fill each bottle to capacity with enrichment medium.

Use: For the cultivation of methanogenic bacteria.

Methanogen Medium

Composition per 106.0mL:

$CaCO_3$	10.0g
Calcium acetate	2.0g
NH_4Cl	0.1g
$K_2HPO_4·3H_2O$	0.04g
$MgCl_2·6H_2O$	0.01g
$Na_2S·9H_2O$ solution	3.0mL
Na_2CO_3 solution	3.0mL

$Na_2S·9H_2O$ Solution:
Composition per 10.0mL:

$Na_2S·9H_2O$	0.1g

Preparation of $Na_2S·9H_2O$ Solution: Add $Na_2S·9H_2O$ to distilled/deionized water and bring volume to 10.0mL. Mix thoroughly. Autoclave for 15 min at 15 psi pressure–121°C. Cool to 25°C.

Na_2CO_3 Solution:
Composition per 10.0mL:

Na_2CO_3	0.5g

Preparation of Na_2CO_3 Solution: Add Na_2CO_3 to distilled/deionized water and bring volume to 10.0mL. Mix thoroughly. Autoclave for 15 min at 15 psi pressure–121°C. Cool to 25°C.

Preparation of Medium: Prepare and distribute medium anaerobically under 100% N_2. Add components, except $Na_2S·9H_2O$ solution and Na_2CO_3 solution, to distilled/deionized water and bring volume to 100.0mL. Mix thoroughly. Autoclave for 15 min at 15 psi pressure–121°C. Cool to 25°C. Aseptically add 3.0mL of sterile $Na_2S·9H_2O$ solution and 3.0mL of sterile Na_2CO_3 solution. Mix thoroughly. Aseptically and anaerobically distribute into sterile tubes or flasks.

Use: For the cultivation and enrichment of acetate-utilizing methanogenic bacteria.

Methanogen Medium, Zeikus

Composition per 1010.0mL:

Inorganic salts solution	500.0mL
Vitamin solution	500.0mL
$Na_2S·9H_2O$ solution	10.0mL

pH 7.0 ± 0.2 at 25°C

Inorganic Salts Solution:
Composition per 500.0mL:

$K_2HPO_4·3H_2O$	1.45g
NH_4Cl	1.0g
KH_2PO_4	0.75g
$MgCl_2·6H_2O$	0.2g
Nitrilotriacetic acid	0.04g
$CaCl_2·2H_2O$	0.02g
$FeCl_2·4H_2O$	3.6mg
$CoCl_2·6H_2O$	1.5mg
$MnCl_2$·4H2O	0.9mg
$ZnCl_2$	0.9mg
H_3BO_2	0.17mg
$Na_2MoO_4·2H_2O$	.09mg

Preparation of Inorganic Salts Solution: Add nitrilotriacetic acid to 250.0mL of distilled/deionized water. Dissolve by adjusting pH to 6.5 with KOH. Add remaining components. Readjust pH to 7.2 with H_2SO_4 or KOH. Add distilled/deionized water to 500.0mL. Filter sterilize.

Vitamin Solution:
Composition per 500.0mL:

Pyridoxine·HCl	1.0mg
p-Aminobenzoic acid	0.5mg
Ca-D-pantothenate	0.5mg
Nicotinic acid	0.5mg
Riboflavin	0.5mg
Thiamine·HCl	0.5mg
Thioctic acid	0.5mg

Biotin 0.2mg
Folic acid 0.2mg
Vitamin B_{12} 0.01mg

Preparation of Vitamin Solution: Add components to distilled/deionized water and bring volume to 500.0mL. Mix thoroughly. Filter sterilize.

$Na_2S·9H_2O$ Solution:
Composition per 10.0mL:
$Na_2S·9H_2O$ 0.3g

Preparation of $Na_2S·9H_2O$ Solution: Add $Na_2S·9H_2O$ to distilled/deionized water and bring volume to 10.0mL. Mix thoroughly. Autoclave for 15 min at 15 psi pressure–121°C. Cool to 25°C.

Preparation of Medium: Prepare and distribute medium anaerobically under 95% N_2 + 5% CO_2. Aseptically and anaerobically combine 500.0mL of sterile inorganic salts solution, 500.0mL of sterile vitamin solution, and 10.0mL of sterile $Na_2S·9H_2O$ solution. Mix thoroughly. Aseptically and anaerobically distribute into sterile tubes or flasks.

Use: For the cultivation of methanogenic bacteria.

Methanogenium aggregans Medium (DSMZ Medium 321)

Composition per liter:
Na-formate 5.0g
Na_2CO_3 .... 1.5g
Na-acetate 1.0g
Trypticase™ 1.0g
Yeast extract 1.0g
NH_4Cl 1.0g
$K_2HPO_4·3H_2O$ 0.4g
$MgCl_2·6H_2O$ 0.4g
Resazurin 1.0mg
Mineral solution 50.0mL
Cysteine solution 10.0mL
$Na_2S·9H_2O$ solution 10.0mL
Trace elements solution 10.0mL
Sludge fluid 5.0mL
pH 6.8 ± 0.2 at 25°C

Mineral Solution:
Composition per liter:
NaCl 12.0g
KH_2PO_4 .... 6.0g
$(NH_4)_2SO_4$ 6.0g
$MgSO_4·7H_2O$ 2.6g
$CaCl_2·2H_2O$ 0.16g

Preparation of Mineral Solution: Add components to distilled/deionized water and bring volume to 1.0L. Mix thoroughly.

Trace Elements Solution:
Composition per liter:
$MgSO_4·7H_2O$ 3.0g
Nitrilotriacetic acid 1.5g
NaCl 1.0g
$MnSO_4·2H_2O$ 0.5g
$CoSO_4·7H_2O$ 0.18g
$ZnSO_4·7H_2O$ 0.18g
$CaCl_2·2H_2O$ 0.1g
$FeSO_4·7H_2O$ 0.1g
$NiCl_2·6H_2O$ 0.025g
$KAl(SO_4)_2·12H_2O$ 0.02g
H_3BO_3 .... 0.01g
$Na_2MoO_4·4H_2O$ 0.01g
$CuSO_4·5H_2O$ 0.01g
$Na_2SeO_3·5H_2O$ 0.3mg

Preparation of Trace Elements Solution: Add nitrilotriacetic acid to 500.0mL of distilled/deionized water. Dissolve by adjusting pH to 6.5 with KOH. Add remaining components. Add distilled/deionized water to 1.0L. Mix thoroughly.

Cysteine Solution:
Composition per 10.0mL:
L-Cysteine·HCl·H_2O 0.2g

Preparation of Cysteine Solution: Add L-cysteine·HCl·H_2O to distilled/deionized water and bring volume to 10.0mL. Mix thoroughly. Sparge with 100% N_2. Autoclave for 15 min at 15 psi pressure–121°C.

$Na_2S·9H_2O$ Solution:
Composition per 10.0mL:
$Na_2S·9H_2O$ 0.2g

Preparation of $Na_2S·9H_2O$ Solution: Add $Na_2S·9H_2O$ to distilled/deionized water and bring volume to 10.0mL. Mix thoroughly. Autoclave under 100% N_2 for 15 min at 15 psi pressure–121°C. Cool to room temperature.

Sludge Fluid:
Composition per 500.0mL:
Yeast extract 2.0g
Sludge 500.0mL

Preparation of Sludge Fluid: Add yeast extract to sludge from an anaerobic digester. Gas with nitrogen gas for a few minutes. Incubate at 37°C for 24 hours. Centrifuge the sludge at 13,000 x g. Autoclave for 15 min at 15 psi pressure–121°C. Place the resulting, clear supernatant in screw-capped vessels under nitrogen gas. The sludge fluid can be stored at room temperature in the dark.

Preparation of Medium: Prepare and dispense medium under an oxygen-free 80% H_2 + 20% CO_2 gas mixture. Add components, except sludge fluid, cysteine solution, and $Na_2S·9H_2O$ solution, to distilled/deionized water and bring volume to 975.0mL. Mix thoroughly. Sparge with 80% H_2 + 20% CO_2. Autoclave for 15 min at 15 psi pressure–121°C. Cool to 25°C while sparging with 80% H_2 + 20% CO_2. Aseptically and anaerobically add 5.0mL sterile sludge fluid, 10.0mL of sterile cysteine solution, and 10.0mL of sterile $Na_2S·9H_2O$ solution. Mix thoroughly. Aseptically and anaerobically distribute into sterile tubes or flasks.

Use: For the cultivation of *Methanocorpusculum aggregans* (*Methanogenium aggregans*).

Methanogenium aggregans Medium

Composition per liter:
Sodium formate 5.0g
NH_4Cl 1.0g
Sodium acetate 1.0g
Trypticase™ 1.0g
Yeast extract 1.0g
$K_2HPO_4·3H_2O$ 0.4g
$MgCl_2·6H_2O$ 0.4g
Resazurin 1.0mg
Mineral solution 50.0mL
Trace elements solution 10.0mL
L-Cysteine·HCl solution 10.0mL

$Na_2S{\cdot}9H_2O$ solution....................10.0mL
Na_2CO_3 solution....................10.0mL
Sludge fluid....................5.0mL

pH 6.8 ± 0.2 at 25°C

Mineral Solution:

Composition per liter:

NaCl....................12.0g
KH_2PO_4....................6.0g
$(NH_4)_2SO_4$....................6.0g
$MgSO_4{\cdot}7H_2O$....................2.6g
$CaCl_2{\cdot}2H_2O$....................0.16g

Preparation of Mineral Solution: Add components to distilled/deionized water and bring volume to 1.0L. Mix thoroughly.

Trace Elements Solution:

Composition per liter:

$MgSO_4{\cdot}7H_2O$....................3.0g
Nitrilotriacetic acid....................1.5g
NaCl....................1.0g
$MnSO_4{\cdot}2H_2O$....................0.5g
$CoSO_4{\cdot}7H_2O$....................0.18g
$ZnSO_4{\cdot}7H_2O$....................0.18g
$CaCl_2{\cdot}2H_2O$....................0.1g
$FeSO_4{\cdot}7H_2O$....................0.1g
$NiCl_2{\cdot}6H_2O$....................0.025g
$KAl(SO_4)_2{\cdot}12H_2O$....................0.02g
$CuSO_4{\cdot}5H_2O$....................0.01g
H_3BO_3....................0.01g
$Na_2MoO_4{\cdot}2H_2O$....................0.01g
$Na_2SeO_3{\cdot}5H_2O$....................0.3mg

Preparation of Trace Elements Solution: Add nitrilotriacetic acid to 500.0mL of distilled/deionized water. Adjust pH to 6.5 with KOH. Add remaining components. Adjust pH to 7.0. Add distilled/deionized water to 1.0L.

L-Cysteine·HCl Solution:

Composition per 10.0mL:

L-Cysteine·HCl....................0.2g

Preparation of L-Cysteine·HCl Solution: Add L-cysteine·HCl to distilled/deionized water and bring volume to 10.0mL. Mix thoroughly. Autoclave under 100% N_2 for 15 min at 15 psi pressure–121°C.

$Na_2S{\cdot}9H_2O$ Solution:

Composition per 10.0mL:

$Na_2S{\cdot}9H_2O$....................0.2g

Preparation of $Na_2S{\cdot}9H_2O$ Solution: Add $Na_2S{\cdot}9H_2O$ to distilled/deionized water and bring volume to 10.0mL. Mix thoroughly. Sparge with 100% N_2. Autoclave for 15 min at 15 psi pressure–121°C.

Na_2CO_3 Solution:

Composition per 10.0mL:

Na_2CO_3....................1.5g

Preparation of Na_2CO_3 Solution: Add Na_2CO_3 to distilled/deionized water and bring volume to 10.0mL. Mix thoroughly. Sparge with 100% N_2. Autoclave for 15 min at 15 psi pressure–121°C.

Sludge Fluid:

Composition per 100.0mL:

Yeast extract....................0.4g
Sludge....................100.0mL

Preparation of Sludge Fluid: To 100.0mL of sludge from an anaerobic digester, add 0.4g of yeast extract. Sparge with 100% N_2 for a few minutes. Incubate at 37°C for 24 hr. Centrifuge the sludge at 13,000 × g for 15 min. Decant the clear supernatant solution. Sparge with 100% N_2 for a few minutes. Store in screw-capped bottles at room temperature in the dark.

Preparation of Medium: Prepare and dispense medium under 80% H_2 + 20% CO_2. Add components, except L-cysteine·HCl solution, $Na_2S{\cdot}9H_2O$ solution, and Na_2CO_3 solution, to distilled/deionized water and bring volume to 970.0mL. Mix thoroughly. Adjust pH to 6.8. Anaerobically distribute into tubes or flasks fitted with butyl rubber stoppers. Autoclave for 15 min at 15 psi pressure–121°C. Anaerobically add 10.0mL of sterile L-cysteine·HCl solution, 10.0mL of sterile $Na_2S{\cdot}9H_2O$ solution, and 10.0mL of sterile Na_2CO_3 solution to each liter of medium or, using a syringe, inject the appropriate amount of sterile L-cysteine·HCl solution, sterile $Na_2S{\cdot}9H_2O$ solution, and sterile Na_2CO_3 solution into individual tubes containing medium.

Use: For the cultivation and maintenance of *Methanocorpusculum aggregans*.

Methanogenium Alcohol Medium

Composition per 1003.0mL:

NaCl....................1.0g
KCl....................0.5g
$MgCl_2{\cdot}6H_2O$....................0.5g
NH_4Cl....................0.4g
Sodium acetate·$3H_2O$....................0.4g
KH_2PO_4....................0.2g
$CaCl_2{\cdot}2H_2O$....................0.1g
$NaHCO_3$ solution....................60.0mL
2-Propanol....................5.0mL
$Na_2S{\cdot}9H_2O$ solution....................3.0mL
Cyanocobalamin solution....................1.0mL
Selenite-molybdate-tungstate solution....................1.0mL
Thiamine solution....................1.0mL
Trace elements solution....................1.0mL
Vitamin solution....................1.0mL

Trace Elements Solution:

Composition per 100.0mL:

$FeSO_4{\cdot}7H_2O$....................1400.0mg
$ZnSO_4{\cdot}7H_2O$....................145.0mg
$CoCl_2{\cdot}6H_2O$....................120.0mg
$MnCl_2{\cdot}4H_2O$....................100.0mg
$NiCl_2{\cdot}6H_2O$....................50.0mg
H_3BO_3....................6.0mg
$CuSO_4{\cdot}5H_2O$....................3.0mg
HCl (25%,w/v)....................8.0mL

Preparation of Trace Elements Solution: Add components to distilled/deionized water and bring volume to 1.0L. Mix thoroughly. Sparge with 100% N_2. Autoclave for 15 min at 15 psi pressure–121°C.

Selenite-Molybdate-Tungstate Solution:

Composition per liter:

NaOH....................0.2g
$Na_2MoO_4{\cdot}2H_2O$....................40.0mg
$Na_2WO_4{\cdot}2H_2O$....................33.0mg
$Na_2SeO_3{\cdot}2H_2O$....................5.0mg

Preparation of Selenite-Molybdate-Tungstate Solution: Add components to distilled/deionized water and bring volume to 1.0L. Mix thoroughly. Sparge with 100% N_2. Autoclave for 15 min at 15 psi pressure–121°C.

$NaHCO_3$ Solution:
Composition per liter:

$NaHCO_3$	84.0g

Preparation of $NaHCO_3$ Solution: Add $NaHCO_3$ to distilled/deionized water and bring volume to 10.0mL. Mix thoroughly. Sparge with 80% N_2 + 20% CO_2. Autoclave for 15 min at 15 psi pressure–121°C.

$Na_2S{\cdot}9H_2O$ Solution:
Composition per 100.0mL:

$Na_2S{\cdot}9H_2O$	2.5g
NaOH	1 pellet

Preparation of $Na_2S{\cdot}9H_2O$ Solution: Bring 100.0mL of distilled/deionized water to boiling. Cool to room temperature while sparging with 100%N_2. Dissolve 1 pellet of NaOH in the anaerobic water. Weigh out a little more than 2.5g of $Na_2S{\cdot}9H_2O$. Briefly rinse the crystals in distilled/deionized water. Dry the crystals by blotting on paper towels or filter paper. Add 2.5g of washed $Na_2S{\cdot}9H_2O$ crystals to 100.0mL of anaerobic NaOH solution. Distribute into serum bottles fitted with butyl rubber stoppers and aluminum seals. Do not grease stoppers. Pressurize to 60kPa with 100% N_2. Autoclave for 15 min at 15 psi pressure–121°C. Store at room temperature in an anaerobic chamber.

Preparation of 2-Propanol: Filter sterilize 10.0mL of 2-propanol. Sparge with 100% N_2.

Vitamin Solution:
Composition per liter:

Sodium 2-mercaptoethanesulfonate	0.25g
Pyridoxine·HCl	0.15g
Calcium pantothenate	0.1g
Nicotinic acid	0.1g
p-Aminobenzoic acid	40.0mg
Biotin	10.0mg
Potassium phosphate buffer (25m*M* solution, pH 7.0)	1.0L

Preparation of Vitamin Solution: Combine components. Mix thoroughly. Filter sterilize. Sparge with 100% N_2.

Thiamine Solution:
Composition per liter:

Thiamine·HCl	0.1g
Sodium phosphate buffer (0.1*M* solution, pH 3.6)	1.0L

Preparation of Thiamine Solution: Combine components. Mix thoroughly. Filter sterilize. Sparge with 100% N_2.

Cyanocobalamin Solution:
Composition per liter:

Cyanocobalamin	50.0mg

Preparation of Cyanocobalamin Solution: Add cyanocobalamin to distilled/deionized water and bring volume to 1.0L. Mix thoroughly. Filter sterilize. Sparge with 100% N_2.

Preparation of Medium: Prepare and dispense medium under 80% N_2 + 20% CO_2. Add components, except $NaHCO_3$ solution, 2-propanol, $Na_2S{\cdot}9H_2O$ solution, cyanocobalamin solution, selenite-molybdate-tungstate solution, thiamine solution, trace elements solution, and vitamin solution, to distilled/deionized water and bring volume to 930.0mL. Mix thoroughly. Sparge with 80% N_2 + 20% CO_2. Autoclave for 15 min at 15 psi pressure–121°C. Aseptically and anaerobically add 60.0mL of sterile $NaHCO_3$ solution, 5.0mL of sterile 2-propanol, 3.0mL of sterile $Na_2S{\cdot}9H_2O$ solution, 1.0mL of sterile cyanocobalamin solution, 1.0mL of sterile selenite-molybdate-tungstate solution, 1.0mL of sterile thiamine solution, 1.0mL of sterile trace elements solution, and 1.0mL of sterile vitamin solution. Mix thoroughly. Aseptically and anaerobically distribute into sterile tubes or bottles.

Use: For the cultivation of *Methanogenium* species.

Methanogenium bourgense Medium (DSMZ Medium 322)

Composition per liter:

Na-formate	5.0g
Na_2CO_3	1.5g
Na-acetate	1.0g
Trypticase™ peptone	1.0g
Yeast extract	1.0g
NH_4Cl	1.0g
$K_2HPO_4{\cdot}3H_2O$	0.4g
$MgCl_2{\cdot}6H_2O$	0.1g
Resazurin	1.0mg
Cysteine solution	10.0mL
$Na_2S{\cdot}9H_2O$ solution	10.0mL

pH 6.0–7.0 at 25°C

Cysteine Solution:
Composition per 10.0mL:

L-Cysteine·HCl·H_2O	0.5g

Preparation of Cysteine Solution: Add L-cysteine·HCl·H_2O to distilled/deionized water and bring volume to 10.0mL. Mix thoroughly. Sparge with 100% N_2. Autoclave for 15 min at 15 psi pressure–121°C.

$Na_2S{\cdot}9H_2O$ Solution:
Composition per 10.0mL:

$Na_2S{\cdot}9H_2O$	0.2g

Preparation of $Na_2S{\cdot}9H_2O$ Solution: Add $Na_2S{\cdot}9H_2O$ to distilled/deionized water and bring volume to 10.0mL. Mix thoroughly. Autoclave under 100% N_2 for 15 min at 15 psi pressure–121°C. Cool to room temperature.

Preparation of Medium: Prepare and dispense medium under an oxygen-free 80% H_2 + 20% CO_2 gas mixture. Add components, except cysteine solution and $Na_2S{\cdot}9H_2O$ solution, to distilled/deionized water and bring volume to 980.0mL. Mix thoroughly. Sparge with 80% H_2 + 20% CO_2. Autoclave for 15 min at 15 psi pressure–121°C. Cool to 25°C while sparging with 80% H_2 + 20% CO_2. Aseptically and anaerobically add 10.0mL of sterile cysteine solution and 10.0mL of sterile $Na_2S{\cdot}9H_2O$ solution. Mix thoroughly. Aseptically and anaerobically distribute into sterile tubes or flasks.

Use: For the cultivation of *Methanoculleus bourgensis=Methanogenium bourgense.*

Methanogenium bourgense Medium

Composition per liter:

Sodium formate	5.0g
Sodium acetate	1.0g
NH_4Cl	1.0g
Trypticase™	1.0g
Yeast extract	1.0g
L-Cysteine·HCl	0.5g
$K_2HPO_4{\cdot}3H_2O$	0.4g
$MgCl_2{\cdot}6H_2O$	0.1g
Resazurin	1.0mg

Na_2CO_3 solution....10.0mL
$Na_2S·9H_2O$ solution....10.0mL

pH 6.7 ± 0.2 at 25°C

Na_2CO_3 Solution:
Composition per 10.0mL:
Na_2CO_3....1.5g

Preparation of Na_2CO_3 Solution: Add Na_2CO_3 to distilled/deionized water and bring volume to 10.0mL. Mix thoroughly. Sparge with 100% N_2. Autoclave for 15 min at 15 psi pressure–121°C.

$Na_2S·9H_2O$ Solution:
Composition per 10.0mL:
$Na_2S·9H_2O$....0.2g

Preparation of $Na_2S·9H_2O$ Solution: Add $Na_2S·9H_2O$ to distilled/deionized water and bring volume to 10.0mL. Mix thoroughly. Sparge with 100% N_2. Autoclave for 15 min at 15 psi pressure–121°C.

Preparation of Medium: Prepare and dispense medium under 80% H_2 + 20% CO_2. Add components, except Na_2CO_3 solution and $Na_2S·9H_2O$ solution, to distilled/deionized water and bring volume to 980.0mL. Mix thoroughly. Adjust pH to 67. Anaerobically distribute into tubes or flasks fitted with butyl rubber stoppers. Autoclave for 15 min at 15 psi pressure–121°C. Anaerobically add 10.0mL of sterile Na_2CO_3 solution and 10.0mL of sterile $Na_2S·9H_2O$ solution to each liter of medium or, using a syringe, inject the appropriate amount of sterile Na_2CO_3 solution and sterile $Na_2S·9H_2O$ solution into individual tubes containing medium.

Use: For the cultivation and maintenance of *Methanogenium bourgense.*

Methanogenium CV Medium

Composition per liter:
NaCl....18.0g
Isopropanol....7.5g
$NaHCO_3$....5.0g
$MgCl_2·6H_2O$....4.0g
$MgSO_4·7H_2O$....3.45g
Trypticase™....2.0g
Yeast extract....2.0g
Sodium acetate....1.0g
KCl....0.335g
NH_4Cl....0.25g
$CaCl_2·2H_2O$....0.14g
K_2HPO_4....0.14g
L-Cysteine·HCl....0.5g
$Na_2S·9H_2O$....0.4g
$Fe(NH_4)_2(SO_4)_2·7H_2O$....2.0mg
Resazurin....1.0mg
$Na_2WO_4·2H_2O$....0.03mg
Trace elements solution....10.0mL
Vitamin solution....10.0mL

pH 6.8–7.0 at 25°C

Trace Elements Solution:
Composition per liter:
$MgSO_4·7H_2O$....3.0g
Nitrilotriacetic acid....1.5g
NaCl....1.0g
$MnSO_4·2H_2O$....0.5g
$CoSO_4·7H_2O$....0.18g
$ZnSO_4·7H_2O$....0.18g
$CaCl_2·2H_2O$....0.1g
$FeSO_4·7H_2O$....0.1g
$NiCl_2·6H_2O$....0.025g
$KAl(SO_4)_2·12H_2O$....0.02g
$CuSO_4·5H_2O$....0.01g
H_3BO_3....0.01g
$Na_2MoO_4·2H_2O$....0.01g
$Na_2SeO_3·5H_2O$....0.3mg

Preparation of Trace Elements Solution: Add nitrilotriacetic acid to 500.0mL of distilled/deionized water. Adjust pH to 6.5 with KOH. Add remaining components. Adjust pH to 7.0. Add distilled/deionized water to 1.0L.

Vitamin Solution:
Composition per liter:
Pyridoxine·HCl....10.0mg
Calcium DL-pantothenate....5.0mg
Nicotinic acid....5.0mg
p-Aminobenzoic acid....5.0mg
Lipoic acid....5.0mg
Riboflavin....5.0mg
Thiamine·HCl....5.0mg
Biotin....2.0mg
Folic acid....2.0mg
Vitamin B_{12}....0.1mg

Preparation of Vitamin Solution: Add components to distilled/deionized water and bring volume to 1.0L. Mix thoroughly. Filter sterilize. Sparge with 80% H_2 + 20% CO_2.

Preparation of Medium: Prepare and dispense medium under 80% H_2 + 20% CO_2. Add components, except vitamin solution, to distilled/deionized water and bring volume to 990.0mL. Mix thoroughly. Adjust pH to 6.8–7.0. Anaerobically distribute into tubes or flasks fitted with butyl rubber stoppers. Autoclave for 15 min at 15 psi pressure–121°C. Anaerobically add 10.0mL of sterile vitamin solution to each liter of medium or, using a syringe, inject the appropriate amount of sterile vitamin solution into individual tubes containing medium.

Use: For the cultivation and maintenance of *Methanogenium organophilum.*

Methanogenium Medium

Composition per liter:
NaCl....18.0g
$NaHCO_3$....5.0g
$MgSO_4·7H_2O$....3.45g
$MgCl_2·7H_2O$....2.75g
Yeast extract....2.0g
Pancreatic digest of casein....2.0g
Sodium acetate....1.0g
L-Cysteine·HCl·H_2O....0.5g
$Na_2S·9H_2O$....0.5g
KCl....0.335g
NH_4Cl....0.25g
$CaCl_2·2H_2O$....0.14g
K_2HPO_4....0.14g
$Fe(NH_4)_2(SO_4)_2·7H_2O$....2.0mg
Resazurin....1.0mg
Trace elements solution SL-6....10.0mL
Wolfe's vitamin solution....10.0mL

pH 6.8 ± 0.2 at 25°C

Trace Elements Solution SL-6:

Composition per liter:

H_3BO_3	0.3g
$CoCl_2 \cdot 6H_2O$	0.2g
$ZnSO_4 \cdot 7H_2O$	0.1g
$MnCl_2 \cdot 4H_2O$	0.03g
$Na_2MoO_4 \cdot H_2O$	0.03g
$NiCl_2 \cdot 6H_2O$	0.02g
$CuCl_2 \cdot 2H_2O$	0.01g

Preparation of Trace Elements Solution SL-6: Add components to distilled/deionized water and bring volume to 1.0L. Mix thoroughly. Adjust pH to 3.4.

Wolfe's Vitamin Solution:

Composition per liter:

Pyridoxine·HCl	10.0mg
Thiamine·HCl	5.0mg
Riboflavin	5.0mg
Nicotinic acid	5.0mg
Calcium pantothenate	5.0mg
p-Aminobenzoic acid	5.0mg
Thioctic acid	5.0mg
Biotin	2.0mg
Folic acid	2.0mg
Cyanocobalamin	100.0μg

Preparation of Wolfe's Vitamin Solution: Add components to distilled/deionized water and bring volume to 1.0L. Mix thoroughly. Filter sterilize.

Preparation of Medium: Prepare and dispense medium under 80% N_2 + 20% CO_2. Add components, except Wolfe's vitamin solution, to distilled/deionized water and bring volume to 990.0mL. Mix thoroughly. Adjust pH to 6.8. Autoclave for 15 min at 15 psi pressure–121°C. Cool under 80% N_2 + 20% CO_2. Aseptically add sterile Wolfe's vitamin solution. Aseptically and anaerobically distribute into sterile tubes or flasks.

Use: For the cultivation and maintenance of *Methanococcus frisius, Methanococcus maripaludis, Methanococcus thermolithotrophicus,* and *Methanoplanus limicola.*

Methanogenium Medium

Composition per liter:

$NaHCO_3$	4.0g
Sodium acetate	4.0g
Sodium formate	2.0g
Yeast extract	1.0g
L-Cysteine·HCl	0.5g
KH_2PO_4	0.5g
$Na_2S \cdot 9H_2O$	0.5g
$MgSO_4 \cdot 7H_2O$	0.4g
NaCl	0.4g
NH_4Cl	0.4g
$CaCl_2 \cdot 2H_2O$	0.05g
$NiCl_2 \cdot 6H_2O$	24.0mg
$FeSO_4 \cdot 7H_2O$	2.0mg
Resazurin	1.0mg
Sludge fluid	50.0mL
Fatty acid mixture	20.0mL
Trace elements solution SL-10	1.0mL

pH 6.7 ± 0.2 at 25°C

Sludge Fluid:

Composition per 100.0mL:

Sludge	100.0mL
Yeast extract	0.4g

Preparation of Sludge Fluid: To 100.0mL of sludge from an anaerobic digester, add 0.4g of yeast extract. Sparge with 100% N_2 for a few minutes. Incubate at 37°C for 24 hr. Centrifuge the sludge at 13,000 × g for 15 min. Decant the clear supernatant solution. Sparge with 100% N_2 for a few minutes. Store in screw-capped bottles at room temperature in the dark.

Fatty Acid Mixture:

Composition per 20.0mL:

α-Methylbutyric acid	0.5g
Isobutyric acid	0.5g
Isovaleric acid	0.5g
Valeric acid	0.5g

Preparation of Fatty Acid Mixture: Add components to distilled/deionized water and bring volume to 20.0mL. Mix thoroughly. Adjust pH to 7.5 with concentrated NaOH.

Trace Elements Solution SL-10:

Composition per liter:

$FeCl_2 \cdot 4H_2O$	1.5g
$CoCl_2 \cdot 6H_2O$	190.0mg
$MnCl_2 \cdot 4H_2O$	100.0mg
$ZnCl_2$	70.0mg
$Na_2MoO_4 \cdot 2H_2O$	36.0mg
$NiCl_2 \cdot 6H_2O$	24.0mg
H_3BO_3	6.0mg
$CuCl_2 \cdot 2H_2O$	2.0mg
HCl (25% solution)	10.0mL

Preparation of Trace Elements Solution SL-10: Add $FeCl_2 \cdot 4H_2O$ to 10.0mL of HCl solution. Mix thoroughly. Add distilled/deionized water and bring volume to 1.0L. Add remaining components. Mix thoroughly. Sparge with 100% N_2. Autoclave for 15 min at 15 psi pressure–121°C.

Preparation of Medium: Prepare and dispense medium anaerobically under 80% H_2 + 20% CO_2. Add components to distilled/deionized water and bring volume to 1.0L. Mix thoroughly. Sparge with 80% H_2 + 20% CO_2. Autoclave for 15 min at 15 psi pressure–121°C.

Use: For the cultivation and maintenance of *Methanobacterium* species, *Methanococcus* species, *Methanocorpusculum* species, *Methanoculleus* species, *Methanogenium* species, *Methanoplanus* species, and *Thermotoga* species.

Methanogenium olentangyi Medium (DSMZ Medium 287)

Composition per liter:

$NaHCO_3$	5.0g
NH_4Cl	2.7g
Na-acetate	2.5g
NaCl	0.61g
K_2HPO_4	0.3g
KH_2PO_4	0.3g
$(NH_4)_2SO_4$	0.3g
$CaCl_2 \cdot H_2O$	0.14g
$MgSO_4 \cdot 7H_2O$	0.13g
Resazurin	1.0mg
Trace elements solution	10.0mL

Vitamin solution 10.0mL
L-Cysteine-HCl·H_2O solution 10.0mL
$Na_2S·9H_2O$ solution 10.0mL
pH 6.9 ± 0.2 at 25°C

Trace Elements Solution:
Composition per liter:
$MgSO_4·7H_2O$ 3.0g
Nitrilotriacetic acid 1.5g
NaCl 1.0g
$MnSO_4·2H_2O$ 0.5g
$CoSO_4·7H_2O$ 0.18g
$ZnSO_4·7H_2O$ 0.18g
$FeSO_4·7H_2O$ 0.1g
$CaCl_2·2H_2O$ 0.1g
$NiCl_2·6H_2O$ 0.025g
$KAl(SO_4)_2·12H_2O$ 0.02g
$CuSO_4·5H_2O$ 0.01g
H_3BO_3 .... 0.01g
$Na_2MoO_4·4H_2O$ 0.01g
$Na_2SeO_3·5H_2O$ 0.3mg

Preparation of Trace Elements Solution: Add nitrilotriacetic acid to 500.0mL of distilled/deionized water. Adjust pH to 6.5 with KOH. Add remaining components. Add distilled/deionized water to 1.0L. Adjust pH to 7.0 with KOH.

Vitamin Solution:
Composition per liter:
Pyridoxine-HCl 10.0mg
Thiamine-HCl·$2H_2O$ 5.0mg
Riboflavin 5.0mg
Nicotinic acid 5.0mg
D-Ca-pantothenate 5.0mg
p-Aminobenzoic acid 5.0mg
Lipoic acid 5.0mg
Biotin 2.0mg
Folic acid 2.0mg
Vitamin B_{12} 0.1mg

Preparation of Vitamin Solution: Add components to distilled/deionized water and bring volume to 1.0L. Mix thoroughly. Filter sterilize.

L-Cysteine-HCl·H_2O Solution:
Composition per 10.0mL:
L-Cysteine-HCl·H_2O 0.3g

Preparation of L-Cysteine-HCl·H_2O Solution: Add L-cysteine·HCl·H_2O to distilled/deionized water and bring volume to 10.0mL. Mix thoroughly. Sparge with 100% N_2. Autoclave for 15 min at 15 psi pressure–121°C. Cool to room temperature.

$Na_2S·9H_2O$ Solution:
Composition per 10.0mL:
$Na_2S·9H_2O$ 0.3g

Preparation of $Na_2S·9H_2O$ Solution: Add $Na_2S·9H_2O$ to distilled/deionized water and bring volume to 10.0mL. Mix thoroughly. Autoclave under 100% N_2 for 15 min at 15 psi pressure–121°C. Cool to room temperature.

Preparation of Medium: Prepare and dispense medium under 80% H_2 + 20% CO_2 gas atmosphere. Add components, except L-cysteine-HCl·H_2O solution, $Na_2S·9H_2O$ solution, and vitamin solution, to distilled/deionized water and bring volume to 970.0mL. Mix thoroughly. Adjust pH to 6.9. Sparge with 80% H_2 + 20% CO_2. Autoclave for 15 min at 15 psi pressure–121°C. Aseptically and anaerobically add 10.0mL L-cysteine-HCl·H_2O solution, 10.0mL $Na_2S·9H_2O$ solution, and 10.0mL vitamin solution. Mix thoroughly. Aseptically and anaerobically distribute into sterile tubes or bottles. After inoculation, flush and repressurize the gas head space of culture bottles with sterile 80% H_2 + 20% CO_2 to 1 bar overpressure.

Use: For the cultivation of *Methanoculleus olentangyi=Methanogenium olentangyi.*

Methanogenium species Medium

Composition per liter:
NaCl 5.0g
$NaHCO_3$ 4.0g
Sodium formate 2.0g
Sodium acetate 1.0g
Yeast extract 1.0g
L-Cysteine·HCl 0.5g
KH_2PO_4 .... 0.5g
$Na_2S·9H_2O$ 0.5g
$MgSO_4·7H_2O$ 0.4g
NH_4Cl 0.4g
$CaCl_2·2H_2O$ 0.05g
$FeSO_4·7H_2O$ 2.0mg
Resazurin 1.0mg
$NiCl_2·6H_2O$ 2.5mg
Trace elements solution SL-10 .... 1.0mL
pH 6.7 ± 0.2 at 25°C

Trace Elements Solution SL-10:
Composition per liter:
$FeCl_2·4H_2O$ 1.5g
$CoCl_2·6H_2O$ 190.0mg
$MnCl_2·4H_2O$ 100.0mg
$ZnCl_2$ 70.0mg
$Na_2MoO_4·2H_2O$ 36.0mg
$NiCl_2·6H_2O$ 24.0mg
H_3BO_3 .... 6.0mg
$CuCl_2·2H_2O$ 2.0mg
HCl (25% solution) 10.0mL

Preparation of Trace Elements Solution SL-10: Add $FeCl_2·4H_2O$ to 10.0mL of HCl solution. Mix thoroughly. Add distilled/deionized water and bring volume to 1.0L. Add remaining components. Mix thoroughly. Autoclave for 15 min at 15 psi pressure–121°C.

Preparation of Medium: Prepare and dispense medium anaerobically under 80% H_2 + 20% CO_2. Add components to distilled/deionized water and bring volume to 1.0L. Mix thoroughly. Sparge with 80% H_2 + 20% CO_2. Autoclave for 15 min at 15 psi pressure–121°C.

Use: For the cultivation and maintenance of *Methanoculleus marisnigri* and *Methanogenium* species.

Methanogenium thermophilicum Medium

Composition per liter:
NaCl 11.7g
Pancreatic digest of casein 6.0g
Sodium formate 5.0g
$MgCl_2·6H_2O$ 4.0g
$NaHCO_3$ 4.0g
Yeast extract 2.0g
L-Cysteine·HCl 0.5g
$Na_2S·9H_2O$ 0.25g
K_2HPO_4 .... 0.14g

KH_2PO_4	0.14g
$CaCl_2 \cdot 2H_2O$	0.075g
Resazurin	1.0mg
Vitamin solution	10.0mL

Vitamin Solution:
Composition per liter:

Pyridoxine·HCl	10.0mg
Calcium DL-pantothenate	5.0mg
Nicotinic acid	5.0mg
p-Aminobenzoic acid	5.0mg
Lipoic acid	5.0mg
Riboflavin	5.0mg
Thiamine·HCl	5.0mg
Biotin	2.0mg
Folic acid	2.0mg
Vitamin B_{12}	0.1mg

Preparation of Vitamin Solution: Add components to distilled/deionized water and bring volume to 1.0L. Mix thoroughly. Filter sterilize. Sparge with 80% H_2 + 20% CO_2.

Preparation of Medium: Add components, except L-cysteine·HCl and $Na_2S \cdot 9H_2O$, to distilled/deionized water and bring volume to 1.0L. Mix thoroughly. Bubble the solution with a stream of oxygen-free 80% N_2 + 20% CO_2 for 20 min. Adjust the pH to 7.2 with 2*M* KOH. Add the L-cysteine·HCl and $Na_2S \cdot 9H_2O$. Mix thoroughly. Tube the medium under oxygen-free 80% N_2 + 20% CO_2 in either serum tubes or bottles. Autoclave for 15 min at 15 psi pressure–121°C. After sterilization, the medium may form a precipitate. Allow it to cool and mix thoroughly to bring the precipitate back into solution.

Use: For the cultivation and maintenance of *Methanogenium thermophilicum.*

Methanohalobium Medium

Composition per liter:

NaCl	250.0g
$MgSO_4 \cdot 7H_2O$	4.0g
$CaCl_2 \cdot 2H_2O$	0.33g
KCl	0.33g
KH_2PO_4	0.33g
$MgCl_2 \cdot 6H_2O$	0.33g
NH_4Cl	0.33g
Resazurin	1.0mg
$NaHCO_3$ solution	100.0mL
Trimethylamine·HCl solution	20.0mL
$Na_2S \cdot 9H_2O$ solution	10.0mL
Yeast extract solution	10.0mL
Vitamin solution	10.0mL
Trace elements solution SL-10	10.0mL
Na_2CO_3 solution	variable

pH 7.4 ± 0.2 at 25°C

$NaHCO_3$ Solution:
Composition per 10.0mL:

$NaHCO_3$	5.0g

Preparation of $NaHCO_3$ Solution: Add $NaHCO_3$ to distilled/deionized water and bring volume to 10.0mL. Mix thoroughly. Sparge with 80% N_2 + 20% CO_2. Autoclave for 15 min at 15 psi pressure–121°C.

Trimethylamine·HCl Solution:
Composition per 20.0mL:

Trimethylamine·HCl	0.2g

Preparation of Trimethylamine·HCl Solution: Add trimethylamine·HCl to distilled/deionized water and bring volume to 20.0mL. Mix thoroughly. Sparge with 100% N_2. Autoclave for 15 min at 15 psi pressure–121°C.

$Na_2S \cdot 9H_2O$ Solution:
Composition per 10.0mL:

$Na_2S \cdot 9H_2O$	0.5g

Preparation of $Na_2S \cdot 9H_2O$ Solution: Add $Na_2S \cdot 9H_2O$ to distilled/deionized water and bring volume to 10.0mL. Mix thoroughly. Sparge with 100% N_2. Autoclave for 15 min at 15 psi pressure–121°C.

Yeast Extract Solution:
Composition per 5.0mL:

Yeast extract	50.0mg

Preparation of Yeast Extract Solution: Add yeast extract to distilled/deionized water and bring volume to 5.0mL. Mix thoroughly. Sparge with 100% N_2. Autoclave for 15 min at 15 psi pressure–121°C.

Vitamin Solution:
Composition per liter:

Pyridoxine·HCl	10.0mg
Calcium DL-pantothenate	5.0mg
Nicotinic acid	5.0mg
p-Aminobenzoic acid	5.0mg
Lipoic acid	5.0mg
Riboflavin	5.0mg
Thiamine·HCl	5.0mg
Biotin	2.0mg
Folic acid	2.0mg
Vitamin B_{12}	0.1mg

Preparation of Vitamin Solution: Add components to distilled/deionized water and bring volume to 1.0L. Mix thoroughly. Filter sterilize. Sparge with 80% H_2 + 20% CO_2.

Trace Elements Solution SL-10:
Composition per liter:

$FeCl_2 \cdot 4H_2O$	1.5g
$CoCl_2 \cdot 6H_2O$	190.0mg
$MnCl_2 \cdot 4H_2O$	100.0mg
$ZnCl_2$	70.0mg
$Na_2MoO_4 \cdot 2H_2O$	36.0mg
$NiCl_2 \cdot 6H_2O$	24.0mg
H_3BO_3	6.0mg
$CuCl_2 \cdot 2H_2O$	2.0mg
HCl (25% solution)	10.0mL

Preparation of Trace Elements Solution SL-10: Add $FeCl_2 \cdot 4H_2O$ to 10.0mL of HCl solution. Mix thoroughly. Add distilled/deionized water and bring volume to 1.0L. Add remaining components. Mix thoroughly. Sparge with 100% N_2. Autoclave for 15 min at 15 psi pressure–121°C.

Na_2CO_3 Solution:
Composition per 10.0mL:

Na_2CO_3	0.5g

Preparation of Na_2CO_3 Solution: Add Na_2CO_3 to distilled/deionized water and bring volume to 10.0mL. Mix thoroughly. Sparge with 100% N_2. Autoclave for 15 min at 15 psi pressure–121°C.

Preparation of Medium: Prepare and dispense medium under 80% H_2 + 20% CO_2. Add components, except $NaHCO_3$ solution, trimethylamine·HCl solution, $Na_2S \cdot 9H_2O$ solution, yeast extract solution, vitamin solution, trace elements solution SL-10, and Na_2CO_3 solution, to distilled/deionized water and bring volume to 840.0mL. Mix thoroughly. Autoclave for 15 min at 15 psi pressure–121°C. Anaerobically add 100.0mL of sterile $NaHCO_3$ solution, 20.0mL of sterile trimethylamine·HCl solution, 10.0mL of sterile $Na_2S·9H_2O$ solution, 10.0mL of sterile yeast extract solution, 10.0mL of sterile vitamin solution, and 10.0mL of sterile trace elements solution SL-10. Mix thoroughly. Add a sufficient quantity of sterile Na_2CO_3 solution to bring the pH to 7.4.

Use: For the cultivation and maintenance of *Methanohalobium evestigatus*.

Methanohalophilus euhalobius Medium

Composition per 1010.0mL:

NaCl	60.0g
$MgCl_2·6H_2O$	2.4g
Yeast extract	2.0g
KCl	1.0g
Disodium EDTA	0.5g
NH_4Cl	0.5g
KH_2PO_4	0.4g
Resazurin	0.5mg
Trimethylamine solution	50.0mL
$NaHCO_3$ solution	40.0mL
$CaCl_2$ solution	20.0mL
$Na_2S·9H_2O$ solution	15.0mL
L-Cysteine·HCl solution	15.0mL
Wolfe's mineral solution	10.0mL
Wolfe's vitamin solution	5.0mL

pH 6.8–7.0 at 25°C

Trimethylamine Solution:

Composition per 50.0mL:

Trimethylamine·HCl	10.0g

Preparation of Trimethylamine Solution: Add trimethylamine·HCl to distilled/deionized water and bring volume to 50.0mL. Mix thoroughly. Sparge with 100% N_2. Autoclave for 15 min at 15 psi pressure–121°C.

$NaHCO_3$ Solution:

Composition per 40.0mL:

$NaHCO_3$	4.0g

Preparation of $NaHCO_3$ Solution: Add $NaHCO_3$ to distilled/deionized water and bring volume to 40.0mL. Mix thoroughly. Sparge with 80% N_2 + 20% CO_2. Autoclave for 15 min at 15 psi pressure–121°C.

$CaCl_2$ Solution:

Composition per 20.0mL:

$CaCl_2·2H_2O$	2.0g

Preparation of $CaCl_2$ Solution: Add $CaCl_2·2H_2O$ to distilled/deionized water and bring volume to 20.0mL. Mix thoroughly. Sparge with 80% N_2 + 20% CO_2 gas mixture. Autoclave for 15 min at 15 psi pressure–121°C.

$Na_2S·9H_2O$ Solution:

Composition per 20.0mL:

$Na_2S·9H_2O$	0.6g

Preparation of $Na_2S·9H_2O$ Solution: Add $Na_2S·9H_2O$ to distilled/deionized water and bring volume to 20.0mL. Mix thoroughly. Sparge with 100% N_2. Autoclave for 15 min at 15 psi pressure–121°C. Before use, neutralize to pH 7.0 with sterile HCl.

L-Cysteine·HCl Solution:

Composition per 20.0mL:

L-Cysteine·HCl·H_2O	0.6g

Preparation of L-Cysteine·HCl Solution: Add L-cysteine·HCl·H_2O to distilled/deionized water and bring volume to 20.0mL. Mix thoroughly. Sparge with 100% N_2. Autoclave for 15 min at 15 psi pressure–121°C.

Wolfe's Mineral Solution:

Composition per liter:

$MgSO_4·7H_2O$	3.0g
Nitrilotriacetic acid	1.5g
NaCl	1.0g
$MnSO_4·2H_2O$	0.5g
$CoCl_2·6H_2O$	0.1g
$ZnSO_4·7H_2O$	0.1g
$CaCl_2·2H_2O$	0.1g
$FeSO_4·7H_2O$	0.1g
$NiCl_2·6H_2O$	0.025g
$KAl(SO_4)_2·12H_2O$	0.02g
$CuSO_4·5H_2O$	0.01g
H_3BO_3	0.01g
$Na_2MoO_4·2H_2O$	0.01g
$Na_2SeO_3·5H_2O$	0.3mg

Preparation of Wolfe's Mineral Solution: Add nitrilotriacetic acid to 500.0mL of distilled/deionized water. Adjust pH to 6.5 with KOH. Add remaining components. Add distilled/deionized water to 1.0L. Adjust pH to 6.8.

Wolfe's Vitamin Solution:

Composition per liter:

Pyridoxine·HCl	10.0mg
p-Aminobenzoic acid	5.0mg
Lipoic acid	5.0mg
Nicotinic acid	5.0mg
Riboflavin	5.0mg
Thiamine·HCl	5.0mg
Calcium DL-pantothenate	5.0mg
Biotin	2.0mg
Folic acid	2.0mg
Vitamin B_{12}	0.1mg

Preparation of Wolfe's Vitamin Solution: Add components to distilled/deionized water and bring volume to 1.0L. Mix thoroughly. Filter sterilize.

Preparation of Medium: Prepare and dispense medium under 80% N_2 + 20% CO_2 gas mixture. Add components, except trimethylamine solution, $NaHCO_3$ solution, $CaCl_2$ solution solution, L-cysteine·HCl solution, and $Na_2S·9H_2O$ solution, to distilled/deionized water and bring volume to 860.0mL. Mix thoroughly. Adjust pH to 6.8–7.0. Sparge with 80% N_2 + 20% CO_2 gas mixture. Autoclave for 15 min at 15 psi pressure–121°C. Aseptically and anaerobically add 50.0mL of sterile trimethylamine solution, 40.0mL of sterile $NaHCO_3$ solution, 20.0mL of sterile $CaCl_2$ solution, 15.0mL of sterile L-cysteine·HCl solution, and 15.0mL of sterile $Na_2S·9H_2O$ solution. Mix thoroughly. Aseptically and anaerobically distribute into sterile tubes or bottles.

Use: For the cultivation of *Methanohalophilus euhalobius*.

Methanohalophilus halophilus Medium

Composition per 1011.0mL:

NaCl	70.0g
KCl	0.33g

KH_2PO_4	0.33g
NH_4Cl	0.33g
Yeast extract	0.05g
K_2SO_4	0.01g
Resazurin	1.0mg
Methylamine·HCl solution	20.0mL
$MgCl_2/CaCl_2$ solution	10.0mL
$NaHCO_3$ solution	10.0mL
Vitamin solution	10.0mL
L-Cysteine·HCl solution	10.0mL
$Na_2S\cdot 9H_2O$ solution	10.0mL
Trace elements solution SL-10	1.0mL

pH 7.2 ± 0.2 at 25°C

$MgCl_2/CaCl_2$ Solution:
Composition per 10.0mL:

$MgCl_2\cdot 6H_2O$	0.33g
$CaCl_2\cdot 2H_2O$	0.33g

Preparation of $MgCl_2/CaCl_2$ Solution: Add components to distilled/deionized water and bring volume to 10.0mL. Mix thoroughly. Sparge with 100% N_2. Autoclave for 15 min at 15 psi pressure–121°C.

$NaHCO_3$ Solution:
Composition per 10.0mL:

$NaHCO_3$	5.0g

Preparation of $NaHCO_3$ Solution: Add $NaHCO_3$ to distilled/deionized water and bring volume to 10.0mL. Mix thoroughly. Sparge with 80% N_2 + 20% CO_2. Autoclave for 15 min at 15 psi pressure–121°C.

Vitamin Solution:
Composition per liter:

Pyridoxine·HCl	10.0mg
Calcium DL-pantothenate	5.0mg
Nicotinic acid	5.0mg
p-Aminobenzoic acid	5.0mg
Riboflavin	5.0mg
Thiamine·HCl	5.0mg
Biotin	2.0mg
Folic acid	2.0mg
Vitamin B_{12}	0.1mg

Preparation of Vitamin Solution: Add components to distilled/deionized water and bring volume to 1.0L. Mix thoroughly. Filter sterilize. Sparge with 80% H_2 + 20% CO_2.

Methylamine·HCl Solution:
Composition per 20.0mL:

Methylamine·HCl	5.0g

Preparation of Methylamine·HCl Solution: Add methylamine·HCl to distilled/deionized water and bring volume to 20.0mL. Mix thoroughly. Sparge with 100% N_2. Autoclave for 15 min at 15 psi pressure–121°C.

L-Cysteine·HCl Solution:
Composition per 10.0mL:

L-Cysteine·HCl	0.2g

Preparation of L-Cysteine·HCl Solution: Add L-cysteine·HCl to distilled/deionized water and bring volume to 10.0mL. Mix thoroughly. Autoclave under 100% N_2 for 15 min at 15 psi pressure–121°C.

$Na_2S\cdot 9H_2O$ Solution:
Composition per 10.0mL:

$Na_2S\cdot 9H_2O$	0.2g

Preparation of $Na_2S\cdot 9H_2O$ Solution: Add $Na_2S\cdot 9H_2O$ to distilled/deionized water and bring volume to 10.0mL. Mix thoroughly. Sparge with 100% N_2. Autoclave for 15 min at 15 psi pressure–121°C.

Trace Elements Solution SL-10:
Composition per liter:

$FeCl_2\cdot 4H_2O$	1.5g
$CoCl_2\cdot 6H_2O$	190.0mg
$MnCl_2\cdot 4H_2O$	100.0mg
$ZnCl_2$	70.0mg
$Na_2MoO_4\cdot 2H_2O$	36.0mg
$NiCl_2\cdot 6H_2O$	24.0mg
H_3BO_3	6.0mg
$CuCl_2\cdot 2H_2O$	2.0mg
HCl (25% solution)	10.0mL

Preparation of Trace Elements Solution SL-10: Add $FeCl_2\cdot 4H_2O$ to 10.0mL of HCl solution. Mix thoroughly. Add distilled/deionized water and bring volume to 1.0L. Add remaining components. Mix thoroughly. Sparge with 100% N_2. Autoclave for 15 min at 15 psi pressure–121°C.

Preparation of Medium: Prepare and dispense medium under 80% H_2 + 20% CO_2. Add components, except $MgCl_2/CaCl_2$ solution, $NaHCO_3$ solution, vitamin solution, methylamine·HCl solution, L-cysteine·HCl solution, and $Na_2S\cdot 9H_2O$ solution, to distilled/deionized water and bring volume to 840.0mL. Mix thoroughly. Adjust pH to 6.9–7.0. Sparge with 80% H_2 + 20% CO_2. Autoclave for 15 min at 15 psi pressure–121°C. Anaerobically add 10.0mL of sterile $MgCl_2/CaCl_2$ solution, 10.0mL of sterile $NaHCO_3$ solution, 10.0mL of sterile vitamin solution, 20.0mL of sterile methylamine·HCl solution, 10.0mL of sterile L-cysteine·HCl solution, and 10.0mL of sterile $Na_2S\cdot 9H_2O$ solution. Mix thoroughly.

Use: For the cultivation and maintenance of *Methanohalophilus halophilus*.

Methanohalophilus mahii Medium

Composition per 1010.0mL:

NaCl	87.0g
$MgCl_2\cdot 6H_2O$	6.0g
$NaHCO_3$	4.0g
Trypticase™	2.0g
Yeast extract	2.0g
KCl	1.5g
NH_4Cl	1.0g
$CaCl_2\cdot 2H_2O$	0.4g
$K_2HPO_4\cdot 3H_2O$	0.4g
Coenzyme M (mercaptoethane sulfonic acid)	0.2g
Resazurin	1.0mg
Trace elements solution	10.0mL
$Na_2S\cdot 9H_2O$ solution	10.0mL
Methanol	4.0mL

pH 7.0 ± 0.2 at 25°C

Trace Elements Solution:
Composition per liter:

$MgSO_4\cdot 7H_2O$	3.0g
Nitrilotriacetic acid	1.5g
NaCl	1.0g
$MnSO_4\cdot 2H_2O$	0.5g
$FeSO_4\cdot 7H_2O$	0.1g
$CoSO_4\cdot 7H_2O$	0.18g
$ZnSO_4\cdot 7H_2O$	0.18g

$CaCl_2·2H_2O$ 0.1g
$KAl(SO_4)_2·12H_2O$ 0.02g
$CuSO_4·5H_2O$ 0.01g
H_3BO_3 .... 0.01g
$Na_2MoO_4·2H_2O$ 0.01g
$NiCl_2·6H_2O$ 0.025g
$Na_2SeO_3·5H_2O$ 0.3mg

Preparation of Trace Elements Solution: Add nitrilotriacetic acid to 500.0mL of distilled/deionized water. Adjust pH to 6.5 with KOH. Add remaining components. Adjust pH to 7.0. Add distilled/deionized water to 1.0L.

$Na_2S·9H_2O$ Solution:
Composition per 10.0mL:
$Na_2S·9H_2O$ 0.25g

Preparation of $Na_2S·9H_2O$ Solution: Add $Na_2S·9H_2O$ to distilled/deionized water and bring volume to 10.0mL. Mix thoroughly. Sparge with 100% N_2. Autoclave for 15 min at 15 psi pressure–121°C.

Preparation of Medium: Prepare and dispense medium under 80% N_2 + 20% CO_2. Add components, except $Na_2S·9H_2O$ solution, to distilled/deionized water and bring volume to 1.0L. Mix thoroughly. Sparge with 80% N_2 + 20% CO_2. Autoclave for 15 min at 15 psi pressure–121°C. Aseptically and anaerobically add 10.0mL of sterile $Na_2S·9H_2O$ solution. Mix thoroughly.

Use: For the cultivation and maintenance of *Methanohalophilus mahii*.

Methanohalophilus Medium
See: ***Halomethanococcus*** **Medium**

Methanohalophilus oregonense Medium

Composition per 1010.0mL:
NaCl 29.0g
Trimethylamine·HCl 2.0g
Trypticase™ 2.0g
Yeast extract 2.0g
$MgCl_2·6H_2O$ 1.7g
KCl 1.5g
NH_4Cl 1.0g
Resazurin 1.0g
Coenzyme M (mercaptoethane sulfonic acid) 0.5g
$K_2HPO_4·3H_2O$ 0.4g
$Na_2S·9H_2O$ solution 10.0mL
Trace elements solution 10.0mL
Na_2CO_3 solution .... variable

pH 8.5 ± 0.2 at 25°C

Trace Elements Solution:
Composition per liter:
$MgSO_4·7H_2O$ 3.0g
Nitrilotriacetic acid 1.5g
NaCl 1.0g
$MnSO_4·2H_2O$ 0.5g
$CoSO_4·7H_2O$ 0.18g
$ZnSO_4·7H_2O$ 0.18g
$CaCl_2·2H_2O$ 0.1g
$FeSO_4·7H_2O$ 0.1g
$KAl(SO_4)_2·12H_2O$ 0.02g
$CuSO_4·5H_2O$ 0.01g
H_3BO_3 .... 0.01g
$Na_2MoO_4·2H_2O$ 0.01g
$NiCl_2·6H_2O$ 0.025g
$Na_2SeO_3·5H_2O$ 0.3mg

Preparation of Trace Elements Solution: Add nitrilotriacetic acid to 500.0mL of distilled/deionized water. Adjust pH to 6.5 with KOH. Add remaining components. Adjust pH to 7.0. Add distilled/deionized water to 1.0L.

$Na_2S·9H_2O$ Solution:
Composition per 10.0mL:
$Na_2S·9H_2O$ 0.3g

Preparation of $Na_2S·9H_2O$ Solution: Add $Na_2S·9H_2O$ to distilled/deionized water and bring volume to 10.0mL. Mix thoroughly. Sparge with 100% N_2. Autoclave for 15 min at 15 psi pressure–121°C.

Na_2CO_3 Solution:
Composition per 10.0mL:
Na_2CO_3 .... 1.0g

Preparation of Na_2CO_3 Solution: Add Na_2CO_3 to distilled/deionized water and bring volume to 10.0mL. Mix thoroughly. Sparge with 100% N_2. Autoclave for 15 min at 15 psi pressure–121°C.

Preparation of Medium: Prepare and dispense medium under 80% N_2 + 20% CO_2. Add components, except $Na_2S·9H_2O$ solution, to distilled/deionized water and bring volume to 1.0L. Mix thoroughly. Sparge with 80% N_2 + 20% CO_2. Autoclave for 15 min at 15 psi pressure–121°C. Aseptically and anaerobically add 10.0mL of sterile $Na_2S·9H_2O$ solution. Mix thoroughly. Adjust pH to 8.5 with a sufficient quantity of sterile Na_2CO_3 solution (approximately 0.1mL per 10.0mL of medium).

Use: For the cultivation and maintenance of *Methanohalophilus oregonense*.

Methanohalophilus zhilinae Medium

Composition per liter:
NaCl 40.0g
$MgCl_2·6H_2O$ 3.5g
$MgSO_4·7H_2O$ 3.0g
Trypticase™ 2.0g
Yeast extract 2.0g
KCl 1.0g
NH_4Cl 1.0g
L-Cysteine·HCl 0.5g
K_2HPO_4 .... 0.4g
Resazurin 1.0mg
$Na_2SeO_3·5H_2O$ 0.1mg
Trimethylamine·HCl solution 20.0mL
$NaHCO_3$ solution 10.0mL
Na_2CO_3 solution 10.0mL
$Na_2S·9H_2O$ solution 10.0mL
Trace elements solution 5.0mL

pH 9.2 ± 0.2 at 25°C

Trimethylamine·HCl Solution:
Composition per 20.0mL:
Trimethylamine·HCl 2.0g

Preparation of Trimethylamine·HCl Solution: Add trimethylamine·HCl to distilled/deionized water and bring volume to 20.0mL. Mix thoroughly. Sparge with 100% N_2. Autoclave for 15 min at 15 psi pressure–121°C.

$NaHCO_3$ Solution:
Composition per 10.0mL:
$NaHCO_3$ 0.5g

Preparation of $NaHCO_3$ Solution: Add $NaHCO_3$ to distilled/deionized water and bring volume to 10.0mL. Mix thoroughly. Sparge with 100% N_2. Autoclave for 15 min at 15 psi pressure–121°C.

Na_2CO_3 Solution:
Composition per 10.0mL:
Na_2CO_3 .... 2.0g

Preparation of Na_2CO_3 Solution: Add Na_2CO_3 to distilled/deionized water and bring volume to 10.0mL. Mix thoroughly. Sparge with 100% N_2. Autoclave for 15 min at 15 psi pressure–121°C.

$Na_2S{\cdot}9H_2O$ Solution:
Composition per 10.0mL:
$Na_2S{\cdot}9H_2O$ 0.25g

Preparation of $Na_2S{\cdot}9H_2O$ Solution: Add $Na_2S{\cdot}9H_2O$ to distilled/deionized water and bring volume to 10.0mL. Mix thoroughly. Sparge with 100% N_2. Autoclave for 15 min at 15 psi pressure–121°C.

Trace Elements Solution:
Composition per liter:

$MgSO_4{\cdot}7H_2O$	3.0g
Nitrilotriacetic acid	1.5g
NaCl	1.0g
$MnSO_4{\cdot}2H_2O$	0.5g
$CoSO_4{\cdot}7H_2O$	0.18g
$ZnSO_4{\cdot}7H_2O$	0.18g
$CaCl_2{\cdot}2H_2O$	0.1g
$FeSO_4{\cdot}7H_2O$	0.1g
$KAl(SO_4)_2{\cdot}12H_2O$	0.02g
$CuSO_4{\cdot}5H_2O$	0.01g
H_3BO_3	0.01g
$Na_2MoO_4{\cdot}2H_2O$	0.01g
$NiCl_2{\cdot}6H_2O$	0.025g
$Na_2SeO_3{\cdot}5H_2O$	0.3mg

Preparation of Trace Elements Solution: Add nitrilotriacetic acid to 500.0mL of distilled/deionized water. Adjust pH to 6.5 with KOH. Add remaining components. Adjust pH to 7.0. Add distilled/deionized water to 1.0L.

Preparation of Medium: Prepare and dispense medium under 100% N_2. Add components, except trimethylamine·HCl solution, $NaHCO_3$ solution, Na_2CO_3 solution, L-cysteine·HCl, and $Na_2S{\cdot}9H_2O$ solution, to distilled/deionized water and bring volume to 950.0mL. Mix thoroughly. Gently heat and bring to boiling. Continue boiling for 5 min. Allow to cool to room temperature while sparging with 100% N_2. Add L-cysteine·HCl. Distribute medium into tubes and seal. Autoclave for 15 min at 15 psi pressure–121°C. Anaerobically add 20.0mL of sterile trimethylamine·HCl solution, 10.0mL of sterile $NaHCO_3$ solution, 10.0mL of sterile Na_2CO_3 solution, and 10.0mL of sterile $Na_2S{\cdot}9H_2O$ solution. Mix thoroughly. Check pH.

Use: For the cultivation and maintenance of *Methanohalophilus zhilinae.*

Methanol Agar (LMG Medium 72)

Composition per liter:

Agar	15.0g
K_2HPO_4	1.2g
KH_2PO_4	0.62g
$(NH_4)_2SO_4$	0.5g
$MgSO_4{\cdot}7H_2O$	0.2g
NaCl	0.1g
$CaCl_2{\cdot}2H_2O$	34.0mg
$FeCl_3{\cdot}6H_2O$	1.0mg
Methanol	10.0mL
Trace elements solution	1.0mL

pH 7.0 ± 0.2 at 25°C

Trace Elements Solution:
Composition per liter:

$ZnSO_4{\cdot}7H_2O$	70.0mg
$Na_2MoO_4{\cdot}2H_2O$	10.0mg
H_3BO_3	10.0mg
$MnSO_4{\cdot}H_2O$	7.0mg
$CuSO_4{\cdot}5H_2O$	5.0mg
$CoCl_2{\cdot}6H_2O$	5.0mg

Preparation of Trace Elements Solution: Add components to distilled/deionized water and bring volume to 1.0L. Mix thoroughly.

Preparation of Medium: Add components, except methanol, to 990.0mL distilled/deionized water. Mix thoroughly. Gently heat and bring to boiling. Autoclave for 15 min at 15 psi pressure–121°C. Cool to 45°C. Aseptically add 10.0mL sterile methanol. Mix thoroughly. Pour into sterile Petri dishes or distribute into sterile tubes.

Use: For the cultivation of *Methylobacterium* spp., *Methylobacillus glycogens,* and *Methylophilus methylotrophus.*

Methanol Ammonium Salts Medium

Composition per liter:

$MgSO_4{\cdot}7H_2O$	1.0g
NH_4Cl	0.5g
Na_2HPO_4	0.33g
KH_2PO_4	0.26g
$CaCl_2$	0.2g
Ferrous EDTA	5.0mg
$Na_2MoO_4{\cdot}2H_2O$	2.0mg
$FeSO_4{\cdot}7H_2O$	500.0μg
$ZnSO_4{\cdot}7H_2O$	400.0μg
EDTA	250.0μg
$CoCl_2{\cdot}6H_2O$	50.0μg
$MnCl_2{\cdot}4H_2O$	20.0μg
H_3BO_4	15.0μg
$NiCl_2{\cdot}6H_2O$	10.0μg
Methanol	5.0mL

pH 6.8 ± 0.2 at 25°C

Preparation of Medium: Add Na_2HPO_4 and KH_2PO_4 to distilled/deionized water and bring volume to 100.0mL. Mix thoroughly. In a separate container, add remaining components, except methanol, to distilled/deionized water and bring volume to 895.0mL. Mix thoroughly. Autoclave both solutions for 15 min at 15 psi pressure–121°C. Cool to room temperature. Filter sterilize methanol. Aseptically add the sterile phosphate solution and the sterile methanol to the cooled, sterile basal medium. Mix thoroughly. Aseptically distribute into sterile tubes or flasks.

Use: For the maintenance and cultivation of *Methylomonas methylotrophus.*

Methanol Medium (ATCC Medium 436)

Composition per liter:

Agar	15.0g
K_2HPO_4	7.0g
$(NH_4)_2SO_4$	3.0g
KH_2PO_4	2.0g
$MgSO_4 \cdot 7H_2O$	0.5g
Yeast extract	0.2g
$FeSO_4 \cdot 7H_2O$	0.01g
$MnSO_4 \cdot H_2O$	8.0mg
Biotin	0.2μg
Thiamine·HCl	0.2μg
Methanol	10.0mL

pH 7.0 ± 0.2 at 25°C

Preparation of Medium: Add components, except methanol, to distilled/deionized water and bring volume to 990.0mL. Mix thoroughly. Gently heat and bring to boiling. Autoclave for 15 min at 15 psi pressure–121°C. Cool to 50°–55°C. Filter sterilize methanol. Aseptically add the sterile methanol to the cooled, sterile basal medium. Mix thoroughly. Aseptically distribute into sterile tubes or flasks.

Use: For the cultivation and maintenance of *Ancylobacter* species, *Methanomonas methylovora*, and *Methylobacterium* species.

Methanol Medium (ATCC Medium 1096)

Composition per liter:

NH_4NO_3	0.75g
$FeCl_3$	0.743g
Methanol	0.45g
$MgSO_4$	0.09g
KH_2PO_4	0.044g
$Na_2MoO_4 \cdot 2H_2O$	0.1mg

pH 7.0 ± 0.2 at 25°C

Preparation of Medium: Prepare and dispense medium under 97% N_2 + 3% H_2. Add components, except methanol, to distilled/deionized water and bring volume to 999.0mL. Mix thoroughly. Gently heat and bring to boiling. Autoclave for 15 min at 15 psi pressure–121°C. Cool to 50°–55°C. Filter sterilize methanol. Aseptically add the sterile methanol to the cooled, sterile basal medium. Mix thoroughly. Aseptically distribute into sterile tubes or flasks.

Use: For the cultivation and maintenance of *Ancylobacter* species, *Methylobacterium* species, and *Methanomonas methylovora.*

Methanol Medium for *Achromobacter*

Composition per liter:

NH_4Cl	5.0g
KH_2PO_4	2.0g
NaCl	0.5g
$MgSO_4$	0.2g
Yeast extract	0.2g
$FeSO_4$	2.0mg
$MnCl_2$	2.0mg
Methanol	20.0mL

pH 7.0 ± 0.2 at 25°C

Preparation of Medium: Add components, except methanol, to distilled/deionized water and bring volume to 980.0mL. Mix thoroughly. Autoclave for 15 min at 15 psi pressure–121°C. Cool to 50°–55°C. Filter sterilize methanol. Aseptically add the sterile methanol to the cooled, sterile basal medium. Mix thoroughly. Aseptically distribute into sterile tubes or flasks.

Use: For the cultivation and maintenance of *Achromobacter methanolophila, Methylobacterium rhodesianum, Pseudomonas insueta,* and *Pseudomonas polysaccharogenes*.

Methanol Medium with 1% Peptone

Composition per liter:

Agar	15.0g
Peptone	10.0g
K_2HPO_4	7.0g
$(NH_4)_2SO_4$	3.0g
KH_2PO_4	2.0g
$MgSO_4 \cdot 7H_2O$	0.5g
Yeast extract	0.2g
$FeSO_4 \cdot 7H_2O$	0.01g
$MnSO_4 \cdot H_2O$	8.0mg
Biotin	0.2μg
Thiamine·HCl	0.2μg
Methanol	10.0mL

pH 7.0 ± 0.2 at 25°C

Preparation of Medium: Add components, except methanol, to distilled/deionized water and bring volume to 990.0mL. Mix thoroughly. Autoclave for 15 min at 15 psi pressure–121°C. Cool to 50°–55°C. Filter sterilize methanol. Aseptically add the sterile methanol to the cooled, sterile basal medium. Mix thoroughly. Aseptically distribute into sterile tubes or flasks.

Use: For the cultivation and maintenance of *Methylobacterium* species.

Methanol Mineral Salts Medium

Composition per liter:

Agar	20.0g
$(NH_4)_2SO_4$	2.0g
NH_4Cl	2.0g
$(NH_4)_2HPO_4$	2.0g
Yeast extract	2.0g
KH_2PO_4	1.0g
K_2HPO_4	1.0g
$MgSO_4 \cdot 7H_2O$	0.5g
$Fe_2SO_4 \cdot 7H_2O$	0.01g
$CaCl_2 \cdot 2H_2O$	0.01g
Methanol	10.0mL

pH 7.0 ± 0.2 at 25°C

Preparation of Medium: Add components, except methanol, to distilled/deionized water and bring volume to 990.0mL. Mix thoroughly. Gently heat and bring to boiling. Autoclave for 15 min at 15 psi pressure–121°C. Cool to 50°–55°C. Filter sterilize methanol. Aseptically add the sterile methanol to the cooled, sterile basal medium. Mix thoroughly. Aseptically distribute into sterile Petri dishes or sterile tubes.

Use: For the cultivation and maintenance of *Pseudomonas viscogena.*

Methanol Salts Medium

Composition per liter:

Agar, noble	20.0g
K_2HPO_4	1.2g
KH_2PO_4	0.62g
$(NH_4)_2SO_4$	0.5g
$MgSO_4 \cdot 7H_2O$	0.2g

NaCl	0.1g
$CaCl_2 \cdot 6H_2O$	0.05g
$ZnSO_4 \cdot 7H_2O$	70.0μg
H_3BO_3	10.0μg
$MnSO_4 \cdot 5H_2O$	10.0μg
$Na_2MoO_4 \cdot 2H_2O$	10.0μg
$CoCl_2 \cdot 6H_2O$	5.0μg
$CuSO_4 \cdot 5H_2O$	5.0μg
$FeCl_3 \cdot 6H_2O$	1.0mg
Methanol	1.0mL

pH 7.0 ± 0.2 at 25°C

Preparation of Medium: Add components, except methanol, to distilled/deionized water and bring volume to 999.0mL. Mix thoroughly. Gently heat and bring to boiling. Autoclave for 15 min at 15 psi pressure–121°C. Cool to 50°C. Aseptically add 1.0mL of filter-sterilized methanol. Mix thoroughly. Pour into sterile Petri dishes or distribute into sterile tubes.

Use: For the cultivation of *Hyphomicrobium* species, *Methylobacillus glycogenes, Methylobacterium extorquens, Methylobacterium fujisawaense, Methylobacterium mesophilicum, Methylobacterium organophilum, Methylobacterium radiotolerans, Methylobacterium rhodesianum, Methylobacterium rhodinum, Methylobacterium* species, *Methylobacterium zatmanii, Methylomonas* species, *Methylophilus methylotrophus, Methylovorus glucosotrophus, Paracoccus* species, *Protaminobacter thiaminophaga*, and *Pseudomonas insueta.*

Methanol Urea Mineral Salts Medium

Composition per liter:

Na_2HPO_4	2.13g
KH_2PO_4	1.36g
$(NH_4)_2SO_4$	0.5g
$MgSO_4 \cdot 7H_2O$	0.2g
$CaCl_2 \cdot 2H_2O$	0.01g
$FeSO_4 \cdot 7H_2O$	5.0mg
$MnSO_4 \cdot 5H_2O$	2.5mg
$NaMoO_4 \cdot 2H_2O$	2.5mg
Urea solution	30.0mL
Methanol	5.0mL

Urea Solution:

Composition per 50.0mL:

Urea	10.0g

Preparation of Urea Solution: Add urea to distilled/deionized water and bring volume to 50.0mL. Mix thoroughly. Filter sterilize.

Preparation of Methanol: Filter sterilize 5.0mL of methanol using a teflon filter.

Preparation of Medium: Add components, except urea solution and methanol, to distilled/deionized water and bring volume to 965.0mL. Mix thoroughly. Autoclave for 15 min at 15 psi pressure–121°C. Aseptically add 30.0mL of sterile urea solution and 5.0mL of sterile methanol. Mix thoroughly. Aseptically distribute into sterile tubes or flasks.

Use: For the cultivation of *Hyphomicrobium vulgare.*

Methanol-Utilizing Bacteria Medium B

Composition per liter:

Na_2HPO_4	3.0g
$(NH_4)_2SO_4$	3.0g
KH_2PO_4	1.4g
$MgSO_4 \cdot 7H_2O$	0.2g
$CaCl_2 \cdot 2H_2O$	30.0mg
Ferric citrate	30.0mg
$MnCl_2 \cdot 4H_2O$	5.0mg
$ZnSO_4 \cdot 7H_2O$	5.0mg
$CuSO_4 \cdot 5H_2O$	0.5mg
Thiamine·HCl	0.4mg
Methanol	10.0mL

pH 7.1 ± 0.2 at 25°C

Preparation of Medium: Add components to distilled/deionized water and bring volume to 1.0L. Mix thoroughly. Adjust pH to 7.1. Distribute into tubes or flasks. Autoclave for 15 min at 15 psi pressure–121°C.

Use: For the cultivation of methanol-utilizing bacteria.

Methanol-Utilizing Bacteria Medium B

Composition per liter:

Na_2HPO_4	3.0g
$(NH_4)_2SO_4$	3.0g
KH_2PO_4	1.4g
Yeast extract	0.2g
$MgSO_4 \cdot 7H_2O$	0.2g
$CaCl_2 \cdot 2H_2O$	30.0mg
Ferric citrate	30.0mg
$MnCl_2 \cdot 4H_2O$	5.0mg
$ZnSO_4 \cdot 7H_2O$	5.0mg
$CuSO_4 \cdot 5H_2O$	0.5mg
Methanol	10.0mL

pH 7.1 ± 0.2 at 25°C

Preparation of Medium: Add components to distilled/deionized water and bring volume to 1.0L. Mix thoroughly. Adjust pH to 7.1. Distribute into tubes or flasks. Autoclave for 15 min at 15 psi pressure–121°C.

Use: For the cultivation of methanol-utilizing bacteria.

Methanol-Utilizing Bacteria Medium D

Composition per liter:

Na_2HPO_4	3.0g
$(NH_4)_2SO_4$	3.0g
KH_2PO_4	1.4g
$MgSO_4 \cdot 7H_2O$	0.2g
$CaCl_2 \cdot 2H_2O$	30.0mg
Ferric citrate	30.0mg
$MnCl_2 \cdot 4H_2O$	5.0mg
$ZnSO_4 \cdot 7H_2O$	5.0mg
$CuSO_4 \cdot 5H_2O$	0.5mg
Thiamine·HCl	0.4mg
Methanol	10.0mL

pH 9.0 ± 0.2 at 25°C

Preparation of Medium: Add components to distilled/deionized water and bring volume to 1.0L. Mix thoroughly. Autoclave for 15 min at 15 psi pressure–121°C. Aseptically adjust pH to 9.0 with filter-sterilized 10% $NaCO_3$ solution. Mix thoroughly. Aseptically distribute into sterile tubes or flasks.

Use: For the cultivation of *Paracoccus alcaliphilus.*

Methanol-Utilizing Bacteria Medium D

Composition per liter:

Na_2HPO_4	3.0g
$(NH_4)_2SO_4$	3.0g
KH_2PO_4	1.4g
Yeast extract	0.2g
$MgSO_4 \cdot 7H_2O$	0.2g
$CaCl_2 \cdot 2H_2O$	30.0mg
Ferric citrate	30.0mg
$MnCl_2 \cdot 4H_2O$	5.0mg
$ZnSO_4 \cdot 7H_2O$	5.0mg
$CuSO_4 \cdot 5H_2O$	0.5mg
Methanol	10.0mL

pH 9.0 ± 0.2 at 25°C

Preparation of Medium: Add components to distilled/deionized water and bring volume to 1.0L. Mix thoroughly. Autoclave for 15 min at 15 psi pressure–121°C. Aseptically adjust pH to 9.0 with filter-sterilized 10% $NaCO_3$ solution. Mix thoroughly. Aseptically distribute into sterile tubes or flasks.

Use: For the cultivation of *Paracoccus alcaliphilus*.

Methanol-Utilizing Bacteria Medium E

Composition per 1010.0mL:

NaCl	30.0g
Na_2HPO_4	3.0g
$(NH_4)_2SO_4$	3.0g
KH_2PO_4	1.4g
$MgSO_4 \cdot 7H_2O$	0.2g
$CaCl_2 \cdot 2H_2O$	30.0mg
Ferric citrate	30.0mg
$MnCl_2 \cdot 4H_2O$	5.0mg
$ZnSO_4 \cdot 7H_2O$	5.0mg
$CuSO_4 \cdot 5H_2O$	0.5mg
Thiamine·HCl	0.4mg
Methanol	10.0mL
Vitamin B_{12}	10.0μg

pH 9.0 ± 0.2 at 25°C

Preparation of Medium: Add components to distilled/deionized water and bring volume to 1.0L. Mix thoroughly. Autoclave for 15 min at 15 psi pressure–121°C. Aseptically adjust pH to 9.0 with filter-sterilized 10% $NaCO_3$ solution. Mix thoroughly. Aseptically distribute into sterile tubes or flasks.

Use: For the cultivation of *Methylophaga marina* and *Methylophaga thalassica.*

Methanol-Utilizing Bacteria Medium E

Composition per 1010.0mL:

NaCl	30.0g
Na_2HPO_4	3.0g
$(NH_4)_2SO_4$	3.0g
KH_2PO_4	1.4g
Yeast extract	0.2g
$MgSO_4 \cdot 7H_2O$	0.2g
$CaCl_2 \cdot 2H_2O$	30.0mg
Ferric citrate	30.0mg
$MnCl_2 \cdot 4H_2O$	5.0mg
$ZnSO_4 \cdot 7H_2O$	5.0mg
$CuSO_4 \cdot 5H_2O$	0.5mg
Methanol	10.0mL
Vitamin B_{12}	10.0μg

pH 9.0 ± 0.2 at 25°C

Preparation of Medium: Add components to distilled/deionized water and bring volume to 1.0L. Mix thoroughly. Autoclave for 15 min at 15 psi pressure–121°C. Aseptically adjust pH to 9.0 with filter-sterilized 10% $NaCO_3$ solution. Mix thoroughly. Aseptically distribute into sterile tubes or flasks.

Use: For the cultivation of *Methylophaga marina* and *Methylophaga thalassica.*

Methanolobus Medium

Composition per liter:

NaCl	18.0g
$NaHCO_3$	5.0g
$MgSO_4 \cdot 7H_2O$	3.45g
$MgCl_2 \cdot 6H_2O$	2.75g
L-Cysteine·HCl·H_2O	0.5g
$Na_2S \cdot 9H_2O$	0.5g
KCl	0.335g
NH_4Cl	0.25g
$CaCl_2 \cdot 2H_2O$	0.14g
K_2HPO_4	0.14g
$Fe(NH_4)_2(SO_4)_2 \cdot 6H_2O$	2.0mg
Resazurin	1.0mg
Wolfe's mineral solution	10.0mL
Wolfe's vitamin solution	10.0mL
Methanol	5.0mL

pH 6.5 ± 0.2 at 25°C

Wolfe's Mineral Solution:

Composition per liter:

$MgSO_4 \cdot 7H_2O$	3.0g
Nitrilotriacetic acid	1.5g
NaCl	1.0g
$MnSO_4 \cdot H_2O$	0.5g
$FeSO_4 \cdot 7H_2O$	0.1g
$CoCl_2 \cdot 6H_2O$	0.1g
$CaCl_2$	0.1g
$ZnSO_4 \cdot 7H_2O$	0.1g
$CuSO_4 \cdot 5H_2O$	0.01g
$AlK(SO_4)_2 \cdot 12H_2O$	0.01g
H_3BO_3	0.01g
$Na_2MoO_4 \cdot 2H_2O$	0.01g

Preparation of Wolfe's Mineral Solution: Add nitrilotriacetic acid to 500.0mL of distilled/deionized water. Dissolve by adjusting pH to 6.5 with KOH. Add remaining components. Add distilled/deionized water to 1.0L.

Wolfe's Vitamin Solution:

Composition per liter:

Pyridoxine·HCl	10.0mg
Thiamine·HCl	5.0mg
Riboflavin	5.0mg
Nicotinic acid	5.0mg
Calcium pantothenate	5.0mg
p-Aminobenzoic acid	5.0mg
Thioctic acid	5.0mg
Biotin	2.0mg
Folic acid	2.0mg
Cyanocobalamin	100.0μg

Preparation of Wolfe's Vitamin Solution: Add components to distilled/deionized water and bring volume to 1.0L. Mix thoroughly. Filter sterilize.

Preparation of Medium: Prepare and dispense medium under 80% N_2 + 20% CO_2. Add components, except methanol and Wolfe's vitamin solution, to distilled/deionized water and bring volume to 985.0mL. Mix thoroughly. Autoclave for 15 min at 15 psi pressure–121°C. Cool under 80% N_2 + 20% CO_2. Aseptically add sterile Wolfe's vitamin solution and sterile methanol. Adjust pH to 6.5. Aseptically and anaerobically distribute into sterile tubes or flasks.

Use: For the cultivation and maintenance of *Methanolobus siciliae* and *Methanolobus tindarius*.

Methanolobus 2 Medium

Composition per liter:

NaCl	18.0g
$NaHCO_3$	5.0g
$MgCl_2·6H_2O$	4.0g
$MgSO_4·7H_2O$	3.45g
Trypticase™	2.0g
Yeast extract	2.0g
Sodium acetate	1.0g
L-Cysteine·HCl	0.5g
$Na_2S·9H_2O$	0.4g
KCl	0.335g
NH_4Cl	0.25g
$CaCl_2·2H_2O$	0.14g
K_2HPO_4	0.14g
$Fe(NH_4)_2(SO_4)_2·7H_2O$	2.0mg
Resazurin	1.0mg
$Na_2WoO_4·2H_2O$	0.03mg
Trace elements solution	10.0mL
Vitamin solution	10.0mL
Methanol solution	5.0mL

pH 6.8–7.0 at 25°C

Trace Elements Solution:

Composition per liter:

$MgSO_4·7H_2O$	3.0g
Nitrilotriacetic acid	1.5g
NaCl	1.0g
$MnSO_4·2H_2O$	0.5g
$CoSO_4·7H_2O$	0.18g
$ZnSO_4·7H_2O$	0.18g
$CaCl_2·2H_2O$	0.1g
$FeSO_4·7H_2O$	0.1g
$NiCl_2·6H_2O$	0.025g
$KAl(SO_4)_2·12H_2O$	0.02g
$CuSO_4·5H_2O$	0.01g
H_3BO_3	0.01g
$Na_2MoO_4·2H_2O$	0.01g
$Na_2SeO_3·5H_2O$	0.3mg

Preparation of Trace Elements Solution: Add nitrilotriacetic acid to 500.0mL of distilled/deionized water. Adjust pH to 6.5 with KOH. Add remaining components. Adjust pH to 7.0. Add distilled/deionized water to 1.0L.

Vitamin Solution:

Composition per liter:

Pyridoxine·HCl	10.0mg
Calcium DL-pantothenate	5.0mg
Nicotinic acid	5.0mg
p-Aminobenzoic acid	5.0mg
Lipoic acid	5.0mg
Riboflavin	5.0mg
Thiamine·HCl	5.0mg
Biotin	2.0mg
Folic acid	2.0mg
Vitamin B_{12}	0.1mg

Preparation of Vitamin Solution: Add components to distilled/deionized water and bring volume to 1.0L. Mix thoroughly. Filter sterilize. Sparge with 80% H_2 + 20% CO_2.

Methanol Solution:

Composition per 5.0mL:

Methanol	5.0mL

Preparation of Methanol Solution: Sparge 5.0mL of methanol with 100% N_2. Autoclave for 15 min at 15 psi pressure–121°C.

Preparation of Medium: Prepare and dispense medium under 80% N_2 + 20% CO_2. Add components, except vitamin solution, to distilled/deionized water and bring volume to 980.0mL. Mix thoroughly. Adjust pH to 6.9–7.0. Anaerobically distribute into tubes or flasks fitted with butyl rubber stoppers. Autoclave for 15 min at 15 psi pressure–121°C. Anaerobically add 10.0mL of sterile vitamin solution and 10.0mL of sterile methanol to each liter of medium or, using a syringe, inject the appropriate amount of sterile vitamin solution and sterile methanol into individual tubes containing medium.

Use: For the cultivation and maintenance of *Methanolobus siciliae*, *Methanolobus vulcani*, and *Methanosarcina* species.

Methanolobus taylorii Medium (DSMZ Medium 490a)

Composition per liter:

NaCl	29.0g
Yeast extract	2.0g
Trypticase™ peptone	2.0g
$MgCl_2·6H_2O$	1.7g
KCl	1.5g
NH_4Cl	1.0g
Resazurin	1.0g
Mercaptoethanesulfonic acid (coenzyme M)	0.5g
$K_2HPO_4·3H_2O$	0.4g
Trace elements solution	10.0mL
Trimethylamine·HCl solution	10.0mL
$Na_2S·9H_2O$ solution	10.0mL
$NaHCO_3$ solution	10.0mL

pH 8.5 ± 0.2 at 25°C

Trace Elements Solution:

Composition per liter:

$MgSO_4·7H_2O$	3.0g
Nitrilotriacetic acid	1.5g
NaCl	1.0g
$MnSO_4·2H_2O$	0.5g
$CoSO_4·7H_2O$	0.18g
$ZnSO_4·7H_2O$	0.18g
$CaCl_2·2H_2O$	0.1g
$FeSO_4·7H_2O$	0.1g
$NiCl_2·6H_2O$	0.025g
$KAl(SO_4)_2·12H_2O$	0.02g
H_3BO_3	0.01g
$Na_2MoO_4·4H_2O$	0.01g

$CuSO_4 \cdot 5H_2O$ 0.01g
$Na_2SeO_3 \cdot 5H_2O$ 0.3mg

Preparation of Trace Elements Solution: Add nitrilotriacetic acid to 500.0mL of distilled/deionized water. Dissolve by adjusting pH to 6.5 with KOH. Add remaining components. Add distilled/deionized water to 1.0L. Mix thoroughly.

$Na_2S \cdot 9H_2O$ Solution:
Composition per 10.0mL:
$Na_2S \cdot 9H_2O$ 0.3g

Preparation of $Na_2S \cdot 9H_2O$ Solution: Add $Na_2S \cdot 9H_2O$ to distilled/deionized water and bring volume to 10.0mL. Sparge with N_2. Autoclave for 15 min at 15 psi pressure–121°C. Cool to 25°C. Store anaerobically.

Trimethylamine·HCl Solution:
Composition per 10.0mL:
Trimethylamine·HCl 2.0g

Preparation of Trimethylamine·HCl Solution: Add trimethylamine·HCl to distilled/deionized water and bring volume to 10.0mL. Mix thoroughly. Autoclave for 15 min at 15 psi pressure–121°C. Cool to 25°C.

$NaHCO_3$ Solution:
Composition per 10.0mL:
$NaHCO_3$ 2.0g

Preparation of $NaHCO_3$ Solution: Add $NaHCO_3$ to distilled/deionized water and bring volume to 10.0mL. Mix thoroughly. Autoclave for 15 min at 15 psi pressure–121°C. Cool to 25°C. Should be prepared freshly.

Preparation of Medium: Prepare and dispense medium under an oxygen-free 100% N_2 gas mixture. Add components, except trimethylamine·HCl solution, $NaHCO_3$ solution, and $Na_2S \cdot 9H_2O$ solution, to 970.0mL distilled/deionized water. Mix thoroughly. Sparge for 20 min with 100% N_2. Adjust pH to 8.0. Autoclave for 15 min at 15 psi pressure–121°C. Cool to 25°C. Aseptically and anaerobically add 10.0mL sterile trimethylamine·HCl solution, and 10.0mL $Na_2S \cdot 9H_2O$ solution. Adjust pH to 8.5 using sterile $NaHCO_3$ solution, approximately 10.0mL per liter medium. Aseptically and anaerobically distribute to sterile tubes or flasks under 100% N_2.

Use: For the cultivation of *Methanolobus taylorii.*

Methanomicrobium Medium

Composition per liter:
$NaHCO_3$ 2.0g
Yeast extract 1.0g
Pancreatic digest of casein 1.0g
NaCl 0.6g
L-Cysteine·HCl·H_2O 0.5g
$Na_2S \cdot 9H_2O$ 0.5g
K_2HPO_4 .... 0.3g
KH_2PO_4 .... 0.3g
$(NH_4)_2SO_4$ 0.3g
$MgSO_4 \cdot 7H_2O$ 0.13g
$CaCl_2 \cdot 2H_2O$ 8.0mg
$FeSO_4 \cdot 7H_2O$ 2.0mg
Rumen fluid, clarified 300.0mL
Fatty acid mixture 20.0mL
Wolfe's mineral solution 10.0mL
Wolfe's vitamin solution 10.0mL
Resazurin (0.1% solution) 1.0mL
pH 6.5 ± 0.2 at 25°C

Fatty Acid Mixture:
Composition per liter:
Valeric acid 0.7mL
Isovaleric acid 0.7mL
α-Methylbutyric acid 0.5mL
Isobutyric acid 0.5mL

Preparation of Fatty Acid Mixture: Add components to distilled/deionized water and bring volume to 1.0L. Mix thoroughly.

Wolfe's Mineral Solution:
Composition per liter:
$MgSO_4 \cdot 7H_2O$ 3.0g
Nitrilotriacetic acid 1.5g
NaCl 1.0g
$MnSO_4 \cdot H_2O$ 0.5g
$FeSO_4 \cdot 7H_2O$ 0.1g
$CoCl_2 \cdot 6H_2O$ 0.1g
$CaCl_2$ 0.1g
$ZnSO_4 \cdot 7H_2O$ 0.1g
$CuSO_4 \cdot 5H_2O$ 0.01g
$AlK(SO_4)_2 \cdot 12H_2O$ 0.01g
H_3BO_3 .... 0.01g
$Na_2MoO_4 \cdot 2H_2O$ 0.01g

Preparation of Wolfe's Mineral Solution: Add nitrilotriacetic acid to 500.0mL of distilled/deionized water. Dissolve by adjusting pH to 6.5 with KOH. Add remaining components. Add distilled/deionized water to 1.0L.

Wolfe's Vitamin Solution:
Composition per liter:
Pyridoxine·HCl 10.0mg
Thiamine·HCl 5.0mg
Riboflavin 5.0mg
Nicotinic acid 5.0mg
Calcium pantothenate 5.0mg
p-Aminobenzoic acid 5.0mg
Thioctic acid 5.0mg
Biotin 2.0mg
Folic acid 2.0mg
Cyanocobalamin 100.0μg

Preparation of Wolfe's Vitamin Solution: Add components to distilled/deionized water and bring volume to 1.0L. Mix thoroughly.

Preparation of Medium: Prepare and dispense medium under 80% N_2 + 20% CO_2. Add components to distilled/deionized water and bring volume to 1.0L. Mix thoroughly. Adjust pH to 6.5. Distribute into tubes or flasks under 80% N_2 + 20% CO_2. Cap with rubber stoppers. Autoclave for 15 min at 15 psi pressure–121°C.

Use: For the cultivation and maintenance of *Methanomicrobium mobile.*

Methanomicrobium mobile Medium (DSMZ Medium 161)

Composition per liter:
$NaHCO_3$ 2.0g
Yeast extract 1.0g
Trypticase™ 1.0g
NaCl 0.6g

Cysteine-HCl·H_2O	0.5g
$Na_2S·9H_2O$	0.5g
K_2HPO_4	0.3g
KH_2PO_4	0.3g
$(NH_4)_2SO_4$	0.3g
$MgSO_4·7H_2O$	0.13g
$CaCl_2·2H_2O$	0.008g
$FeSO_4·7H_2O$	0.002g
Resazurin	0.001g
Rumen fluid, clarified	300.0mL
Fatty acid mixture	20.0mL
Trace elements solution	10.0mL
Vitamin solution	10.0mL

pH 6.5 ± 0.2 at 25°C

Trace Elements Solution:
Composition per liter:

$MgSO_4·7H_2O$	3.0g
Nitrilotriacetic acid	1.5g
NaCl	1.0g
$MnSO_4·2H_2O$	0.5g
$CoSO_4·7H_2O$	0.18g
$ZnSO_4·7H_2O$	0.18g
$CaCl_2·2H_2O$	0.1g
$FeSO_4·7H_2O$	0.1g
$NiCl_2·6H_2O$	0.025g
$KAl(SO_4)_2·12H_2O$	0.02g
H_3BO_3	0.01g
$Na_2MoO_4·4H_2O$	0.01g
$CuSO_4·5H_2O$	0.01g
$Na_2SeO_3·5H_2O$	0.3mg

Preparation of Trace Elements Solution: Add nitrilotriacetic acid to 500.0mL of distilled/deionized water. Dissolve by adjusting pH to 6.5 with KOH. Add remaining components. Add distilled/deionized water to 1.0L. Mix thoroughly.

Vitamin Solution:
Composition per liter:

Pyridoxine-HCl	10.0mg
Thiamine-HCl·$2H_2O$	5.0mg
Riboflavin	5.0mg
Nicotinic acid	5.0mg
D-Ca-pantothenate	5.0mg
p-Aminobenzoic acid	5.0mg
Lipoic acid	5.0mg
Biotin	2.0mg
Folic acid	2.0mg
Vitamin B_{12}	0.1mg

Preparation of Vitamin Solution: Add components to distilled/deionized water and bring volume to 1.0L. Mix thoroughly. Sparge with 80% H_2 + 20% CO_2. Filter sterilize.

Fatty Acid Mixture:
Composition per 20.0mL:

Valeric acid	0.5g
Isovaleric acid	0.5g
α-Methylbutyric acid	0.5g
Isobutyric acid	0.5g
Distilled water	20.0mL

Preparation of Fatty Acid Mixture: Add components to 20.0mL distilled/deionized water. Mix thoroughly.

Preparation of Medium: Prepare and dispense medium under an oxygen-free 80% H_2 + 20% CO_2 gas mixture. Add components, except vitamin solution, to 990.0mL distilled/deionized water. Mix thoroughly. Sparge with 80% H_2 + 20% CO_2. Adjust pH to 6.5 with concentrated NaOH. Autoclave for 15 min at 15 psi pressure–121°C. Cool to 25°C while sparging with 80% H_2 + 20% CO_2. Aseptically and anaerobically add 10.0mL sterile vitamin solution. Mix thoroughly. Aseptically and anaerobically distribute into sterile tubes or flasks. Alternately the medium can be distributed to tubes under anaerobic conditions and autoclaved in tubes prior to addition of vitamin solution. Appropriate amounts of the vitamin solution can then be added to each tube by syringes. After inoculation, pressurize culture vessels to 2 bar 80% H_2 + 20% CO_2 overpressure.

Use: For the cultivation of *Methanomicrobium mobile.*

Methanomicrobium paynteri Medium

Composition per liter:

3-(*N*-morpholino) propane sulfonic acid (MOPS buffer)	20.93g
NaCl	6.31g
$NaHCO_3$	5.0g
Sodium acetate·$3H_2O$	4.14g
$MgSO_4·7H_2O$	3.40g
$MgCl_2·2H_2O$	2.75g
NH_4Cl	1.5g
KCl	0.34g
$CaCl_2·2H_2O$	0.14g
K_2HPO_4	0.14g
$Fe(NH_4)_2(SO_4)_2·6H_2O$	2.0mg
Resazurin	1.0mg
Trace elements solution	10.0mL
Vitamin solution	10.0mL
L-Cysteine·HCl solution	10.0mL
$Na_2S·9H_2O$ solution	10.0mL

pH 7.0 ± 0.2 at 25°C

Trace Elements Solution:
Composition per liter:

$MgSO_4·7H_2O$	3.0g
NaCl	1.0g
Nitrilotriacetic acid	1.5g
$MnSO_4·2H_2O$	0.5g
$CoSO_4·7H_2O$	0.18g
$ZnSO_4·7H_2O$	0.18g
$CaCl_2·2H_2O$	0.1g
$FeSO_4·7H_2O$	0.1g
$KAl(SO_4)_2·12H_2O$	0.02g
$CuSO_4·5H_2O$	0.01g
H_3BO_3	0.01g
$Na_2MoO_4·2H_2O$	0.01g
$NiCl_2·6H_2O$	0.025g
$Na_2SeO_3·5H_2O$	0.3mg

Preparation of Trace Elements Solution: Add nitrilotriacetic acid to 500.0mL of distilled/deionized water. Adjust pH to 6.5 with KOH. Add remaining components. Adjust pH to 7.0. Add distilled/deionized water to 1.0L.

Vitamin Solution:
Composition per liter:

Pyridoxine·HCl	10.0mg
Calcium DL-pantothenate	5.0mg
Lipoic acid	5.0mg

Nicotinic acid....................5.0mg
p-Aminobenzoic acid....................5.0mg
Riboflavin....................5.0mg
Thiamine·HCl....................5.0mg
Biotin....................2.0mg
Folic acid....................2.0mg
Vitamin B_{12}....................0.1mg

Preparation of Vitamin Solution: Add components to distilled/deionized water and bring volume to 1.0L. Mix thoroughly. Filter sterilize. Sparge with 80% H_2 + 20% CO_2.

L-Cysteine·HCl Solution:
Composition per 10.0mL:
L-Cysteine·HCl....................0.5g

Preparation of L-Cysteine·HCl Solution: Add L-cysteine·HCl to distilled/deionized water and bring volume to 10.0mL. Mix thoroughly. Autoclave under 100% N_2 for 15 min at 15 psi pressure–121°C.

$Na_2S \cdot 9H_2O$ Solution:
Composition per 10.0mL:
$Na_2S \cdot 9H_2O$....................0.5g

Preparation of $Na_2S \cdot 9H_2O$ Solution: Add $Na_2S \cdot 9H_2O$ to distilled/deionized water and bring volume to 10.0mL. Mix thoroughly. Sparge with 100% N_2. Autoclave for 15 min at 15 psi pressure–121°C.

Preparation of Medium: Prepare and dispense medium under 80% H_2 + 20% CO_2. Add components, except MOPS buffer, L-cysteine·HCl solution, and $Na_2S \cdot 9H_2O$ solution, to distilled/deionized water and bring volume to 890.0mL. Mix thoroughly. Sparge with O_2-free 80% H_2 + 20% CO_2. In a separate flask, add MOPS buffer to distilled/deionized water and bring volume to 90.0mL. Adjust pH to 7.0 with 2*M* KOH. Add the 90.0mL of MOPS solution to the 890.0mL of medium and continue sparging with O_2-free 80% H_2 + 20% CO_2. Anaerobically distribute into tubes or flasks fitted with butyl rubber stoppers. Autoclave for 15 min at 15 psi pressure–121°C. Anaerobically add 10.0mL of sterile L-cysteine·HCl solution and 10.0mL of sterile $Na_2S \cdot 9H_2O$ solution to each liter of medium or, using a syringe, inject the appropriate amount of sterile L-cysteine·HCl solution and sterile $Na_2S \cdot 9H_2O$ solution into individual tubes containing medium.

Use: For the cultivation and maintenance of *Methanomicrobium paynteri*.

Methanomicrococcus Medium (DSMZ Medium 120b)

Composition 1112.0mL:
NaCl....................2.25g
Yeast extract....................2.0g
Casitone....................2.0g
NH_4Cl....................0.5g
$MgSO_4 \cdot 7H_2O$....................0.5g
K_2HPO_4....................0.348g
$CaCl_2 \cdot 2H_2O$....................0.25g
KH_2PO_4....................0.227g
$FeSO_4 \cdot 7H_2O$....................0.002g
Resazurin....................0.001g
$NaHCO_3$ solution....................40.0mL
Methanol solution....................20.0mL
L-Cysteine-HCl·H_2O solution....................15.0mL
$Na_2S \cdot 9H_2O$ solution....................15.0mL
Na-acetate solution....................10.0mL
Vitamin solution....................10.0mL
Coenzyme M solution....................1.0mL
Trace elements solution SL-10....................1.0mL
pH 7.2 ± 0.2 at 25°C

Vitamin Solution:
Composition per liter:
Pyridoxamine-HCl....................10.0mg
Pantothenic acid....................5.0mg
Riboflavin....................5.0mg
Alpha-lipoic acid....................5.0mg
p-Aminobenzoic acid....................5.0mg
Thiamine-HCl·$2H_2O$....................5.0mg
Nicotinic acid....................5.0mg
Biotin....................2.0mg
Folic acid....................2.0mg
Vitamin B_{12}....................0.1mg

Preparation of Vitamin Solution: Add components to distilled/deionized water and bring volume to 1.0L. Mix thoroughly. Sparge with 80% N_2 + 20% CO_2. Filter sterilize.

Trace Elements Solution SL-10:
Composition per liter:
$FeCl_2 \cdot 4H_2O$....................1.5g
$CoCl_2 \cdot 6H_2O$....................190.0mg
$MnCl_2 \cdot 4H_2O$....................100.0mg
$ZnCl_2$....................70.0mg
$Na_2MoO_4 \cdot 2H_2O$....................36.0mg
$NiCl_2 \cdot 6H_2O$....................24.0mg
H_3BO_3....................6.0mg
$CuCl_2 \cdot 2H_2O$....................2.0mg
HCl (25% solution)....................10.0mL

Preparation of Trace Elements Solution SL-10: Add $FeCl_2 \cdot 4H_2O$ to 10.0mL of HCl solution. Mix thoroughly. Add distilled/deionized water and bring volume to 1.0L. Add remaining components. Mix thoroughly. Sparge with 80% N_2 + 20% CO_2. Autoclave for 15 min at 15 psi pressure–121°C.

$NaHCO_3$ Solution:
Composition per 100.0mL:
$NaHCO_3$....................5.0g

Preparation of $NaHCO_3$ Solution: Add $NaHCO_3$ to distilled/deionized water and bring volume to 100.0mL. Mix thoroughly. Sparge with 80% N_2 + 20% CO_2. Autoclave for 15 min at 15 psi pressure–121°C.

$Na_2S \cdot 9H_2O$ Solution:
Composition per 100.0mL:
$Na_2S \cdot 9H_2O$....................3.0g

Preparation of $Na_2S \cdot 9H_2O$ Solution: Add $Na_2S \cdot 9H_2O$ to distilled/deionized water and bring volume to 100.0mL. Mix thoroughly. Sparge with 100% N_2. Autoclave for 15 min at 15 psi pressure–121°C. Before use, neutralize to pH 7.0 with sterile HCl.

Methanol Solution:
Composition per 100.0mL:
Methanol....................50.0mL

Preparation of Methanol Solution: Add methanol to distilled/deionized water and bring volume to 100.0mL. Mix thoroughly. Sparge with 100% N_2. Autoclave for 15 min at 15 psi pressure–121°C.

Na-Acetate Solution:
Composition per 100.0mL:
Na-acetate....................25.0g

Preparation of Na-Acetate Solution: Add Na-acetate to distilled/deionized water and bring volume to 100.0mL. Mix thoroughly. Sparge with 100% N_2. Autoclave for 15 min at 15 psi pressure–121°C.

Coenzyme M Solution:
Composition per 10.0mL:

Coenzyme M	0.1g

Preparation of Coenzyme M Solution: Add coenzyme M to distilled/deionized water and bring volume to 10.0mL. Mix thoroughly. Sparge with 100% N_2. Autoclave for 15 min at 15 psi pressure–121°C.

L-Cysteine Solution:
Composition per 100.0mL:

L-Cysteine·HCl·H_2O	3.0g

Preparation of L-Cysteine Solution: Add L-cysteine·HCl·H_2O to distilled/deionized water and bring volume to 100.0mL. Mix thoroughly. Sparge with 100% N_2. Autoclave for 15 min at 15 psi pressure–121°C.

Preparation of Medium: Prepare and dispense medium under 80% N_2 + 20% CO_2 gas atmosphere. Add components, except vitamin solution, $NaHCO_3$ solution, methanol solution, L-cysteine-HCl·H_2O solution, $Na_2S·9H_2O$ solution, Na-acetate solution, coenzyme M solution, and trace elements solution SL-10, to distilled/deionized water and bring volume to 1.0L. Mix thoroughly. Adjust pH to 7.2. Sparge with 80% N_2 + 20% CO_2. Autoclave for 15 min at 15 psi pressure–121°C. Aseptically and anaerobically add 10.0mL vitamin solution, 40.0mL $NaHCO_3$ solution, 20.0mL methanol solution, 15.0mL L-cysteine-HCl·H_2O solution, 15.0mL $Na_2S·9H_2O$ solution, 10.0mL Na-acetate solution, 1.0mL coenzyme M solution, and 1.0mL trace elements solution SL-10. Mix thoroughly. Aseptically and anaerobically distribute into sterile tubes or bottles. After inoculation, flush and repressurize the gas head space of culture bottles with sterile 80% H_2 + 20% CO_2 to 1 bar overpressure.

Use: For the cultivation of *Methanomicrococcus blatticola.*

Methanomonas Autotrophic Medium

Composition per liter:

$NaNO_3$	2.0g
Na_2HPO_4	0.21g
$MgSO_4·7H_2O$	0.2g
NaH_2PO_4	0.09g
KCl	0.04g
$CaCl_2$	0.015g
$FeSO_4·7H_2O$	1.0mg
$ZnSO_4·7H_2O$	0.3mg
$CuSO_4·5H_2O$	0.2mg
H_3BO_3	0.06mg
$MnSO_4·H_2O$	0.03mg
MoO_3	0.015mg

Preparation of Medium: Add components to distilled/deionized water and bring volume to 1.0L. Mix thoroughly. Gently heat until dissolved. Distribute into tubes or flasks. Autoclave for 15 min at 15 psi pressure–121°C.

Use: For the autotrophic cultivation of *Methanomonas* species.

Methanopyrus Medium

Composition per liter:

NaCl	11.80g
$MgCl_2·6H_2O$	4.50g
$MgSO_4·7H_2O$	1.75g
Na_2SO_4	0.81g
$CaCl_2·2H_2O$	0.78g
KCl	0.30g
KH_2PO_4	0.09g
$K_2HPO_4·3H_2O$	0.07g
$Na_2WO_4·2H_2O$	2.0mg
$(NH_4)_2Fe(SO_4)_2·6H_2O$	2.0mg
$(NH_4)_2Ni(SO_4)_2$	2.0mg
Resazurin	0.2mg
Marine trace elements	10.0mL
Trace elements solution	10.0mL
Vitamin solution	10.0mL
$NaHCO_3$ solution	10.0mL
$Na_2S·9H_2O$ solution	10.0mL

pH 6.5 ± 0.2 at 25°C

Marine Trace Elements:
Composition per liter:

NaBr	4.0g
$SrCl_2·6H_2O$	1.8g
H_3BO_3	1.3g
Sodium silicate	100.0mg
NaF	60.0mg
KNO_3	40.0mg
KI	1.25mg
$Na_2HPO_4·3H_2O$	0.25mg

Preparation of Marine Trace Elements: Add components to distilled/deionized water and bring volume to 1.0L. Mix thoroughly.

Trace Elements Solution:
Composition per liter:

$MgSO_4·7H_2O$	3.0g
Nitrilotriacetic acid	1.5g
NaCl	1.0g
$MnSO_4·2H_2O$	0.5g
$CoSO_4·7H_2O$	0.18g
$ZnSO_4·7H_2O$	0.18g
$CaCl_2·2H_2O$	0.1g
$FeSO_4·7H_2O$	0.1g
$NiCl_2·6H_2O$	0.025g
$KAl(SO_4)_2·12H_2O$	0.02g
$CuSO_4·5H_2O$	0.01g
H_3BO_3	0.01g
$Na_2MoO_4·2H_2O$	0.01g
$Na_2SeO_3·5H_2O$	0.3mg

Preparation of Trace Elements Solution: Add nitrilotriacetic acid to 500.0mL of distilled/deionized water. Adjust pH to 6.5 with KOH. Add remaining components. Adjust pH to 7.0. Add distilled/deionized water to 1.0L.

Vitamin Solution:
Composition per liter:

Pyridoxine·HCl	10.0mg
Calcium DL-pantothenate	5.0mg
Lipoic acid	5.0mg
Nicotinic acid	5.0mg
p-Aminobenzoic acid	5.0mg
Riboflavin	5.0mg
Thiamine·HCl	5.0mg
Biotin	2.0mg
Folic acid	2.0mg
Vitamin B_{12}	0.1mg

Preparation of Vitamin Solution: Add components to distilled/deionized water and bring volume to 1.0L. Mix thoroughly. Filter sterilize. Sparge with 80% H_2 + 20% CO_2.

$NaHCO_3$ Solution:

Composition per 10.0mL:

$NaHCO_3$	0.5g

Preparation of $NaHCO_3$ Solution: Add $NaHCO_3$ to distilled/deionized water and bring volume to 10.0mL. Mix thoroughly. Sparge with 80% N_2 + 20% CO_2. Autoclave for 15 min at 15 psi pressure–121°C.

$Na_2S·9H_2O$ Solution:

Composition per 10.0mL:

$Na_2S·9H_2O$	0.5g

Preparation of $Na_2S·9H_2O$ Solution: Add $Na_2S·9H_2O$ to distilled/deionized water and bring volume to 10.0mL. Mix thoroughly. Sparge with 100% N_2. Autoclave for 15 min at 15 psi pressure–121°C.

Preparation of Medium: Prepare and dispense medium under 80% H_2 + 20% CO_2. Add components, except vitamin solution, $NaHCO_3$ solution, and $Na_2S·9H_2O$ solution, to distilled/deionized water and bring volume to 970.0mL. Mix thoroughly. Anaerobically distribute into tubes or flasks fitted with butyl rubber stoppers. Autoclave for 15 min at 15 psi pressure–121°C. Anaerobically add 10.0mL of sterile vitamin solution, 10.0mL of sterile $NaHCO_3$ solution, and 10.0mL of sterile $Na_2S·9H_2O$ solution to each liter of medium or, using a syringe, inject the appropriate amount of sterile vitamin solution, sterile $NaHCO_3$ solution, and sterile $Na_2S·9H_2O$ solution into individual tubes containing medium. Check that final pH of medium is 6.5.

Use: For the cultivation and maintenance of *Methanopyrus kandleri*.

Methanosarcina acetivorans Medium

Composition per liter:

NaCl	23.4g
Agar	10.0g
$MgSO_4$	6.3g
Na_2CO_3	5.0g
Trimethylamine·HCl	3.0g
Yeast extract	1.0g
NH_4Cl	1.0g
KCl	0.8g
Na_2HPO_4	0.6g
L-Cysteine·HCl·H_2O	0.25g
$Na_2S·9H_2O$	0.25g
$CaCl_2·2H_2O$	0.14g
Resazurin	1.0mg
Wolfe's mineral solution	10.0mL

pH 7.2 ± 0.2 at 25°C

Wolfe's Mineral Solution:

Composition per liter:

$MgSO_4·7H_2O$	3.0g
Nitrilotriacetic acid	1.5g
NaCl	1.0g
$MnSO_4·H_2O$	0.5g
$FeSO_4·7H_2O$	0.1g
$CoCl_2·6H_2O$	0.1g
$CaCl_2$	0.1g
$ZnSO_4·7H_2O$	0.1g
$CuSO_4·5H_2O$	0.01g
$AlK(SO_4)_2·12H_2O$	0.01g
H_3BO_3	0.01g
$Na_2MoO_4·2H_2O$	0.01g

Preparation of Wolfe's Mineral Solution: Add nitrilotriacetic acid to 500.0mL of distilled/deionized water. Dissolve by adjusting pH to 6.5 with KOH. Add remaining components. Add distilled/deionized water to 1.0L.

Preparation of Medium: Add components, except $Na_2S·9H_2O$, to glass-distilled water and bring volume to 990.0mL. Mix thoroughly. Methanol or methylamine·HCl may be substituted for the trimethylamine·HCl at a concentration of 50 m*M*. Heat gently and bring to boiling. Adjust pH to 7.2 with 6*N* HCl. Autoclave for 5 min at 10 psi pressure–115°C. Cool to 50°C under 80% N_2 + 20% CO_2. If a large precipitate is present, add a small amount of HCl and mix thoroughly. Add $Na_2S·9H_2O$. Mix thoroughly. Distribute into tubes under 80% N_2 + 20% CO_2. Cap with butyl rubber stoppers. Autoclave for 15 min at 15 psi pressure–121°C. A precipitate will form but resolubilizes as the medium cools. Invert tubes as they are cooling to facilitate resolubilization. Allow tubes to cool in a slanted position.

Use: For the cultivation and maintenance of *Methanococcoides methylutens* and *Methanosarcina acetivorans*.

Methanosarcina acetivorans Medium

Composition per 1010.0mL:

NaCl	23.4g
$MgSO_4·7H_2O$	9.45g
Na_2CO_3	5.0g
Yeast extract	1.0g
NH_4Cl	1.0g
$CaCl_2·2H_2O$	0.14g
KCl	0.1g
Na_2HPO_4	0.6g
L-Cysteine·HCl·H_2O	0.025g
$Na_2S·9H_2O$	0.025g
Resazurin	1.0mg
Trace elements solution	10.0mL
Methanol	5.0mL

pH 6.8 ± 0.2 at 25°C

Trace Elements Solution:

Composition per liter:

Nitrilotriacetic acid	12.8g
$FeCl_3·6H_2O$	1.35g
NaCl	1.0g
$NiCl_2·6H_2O$	0.12g
$MnCl_2·4H_2O$	0.1g
$CaCl_2·2H_2O$	0.1g
$ZnCl_2$	0.1g
$Na_2SeO_3·5H_2O$	0.026g
$CuCl_2·2H_2O$	0.025g
$CoCl_2·6H_2O$	0.024g
$Na_2MoO_4·2H_2O$	0.024g
H_3BO_3	0.01g

Preparation of Trace Elements Solution: Add nitrilotriacetic acid to approximately 500.0mL of distilled/deionized water. Dissolve by adding KOH and adjust pH to 6.5. Add remaining components. Bring volume to 1.0L with additional distilled/deionized water. Adjust pH to 7.0 with KOH.

Preparation of Medium: Prepare and dispense medium under 80% N_2 + 20% CO_2. Add components to distilled/deionized water and bring volume to 1010.0mL. Mix thoroughly. Adjust pH to 7.0 with 1*N* HCl.

Sparge under 80% N_2 + 20% CO_2 for 3–4 min. Autoclave for 15 min at 15 psi pressure–121°C.

Use: For the cultivation and maintenance of *Methanosarcina acetivorans*.

Methanosarcina barkeri Medium

Composition per liter:

$NaHCO_3$	2.5g
NaCl	0.46g
Yeast extract	0.24g
KH_2PO_4	0.23g
K_2HPO_4	0.23g
$(NH_4)_2SO_4$	0.23g
$MgCl_2 \cdot 6H_2O$	0.09g
$CaCl_2 \cdot 2H_2O$	0.06g
$NiCl_2 \cdot 6H_2O$	2.0mg
Methanol	10.0mL
Trace elements solution	10.0mL
Vitamin solution	10.0mL
L-Cysteine·HCl·H_2O solution	10.0mL
$Na_2S \cdot 9H_2O$ solution	10.0mL
Resazurin (0.01% solution)	1.0mL

Preparation of Methanol: Filter sterilize 10.0mL of methanol.

Trace Elements Solution:

Composition per liter:

$MgSO_4 \cdot 5H_2O$	3.0g
Nitrilotriacetic acid	1.5g
NaCl	1.0g
$MnSO_4 \cdot 2H_2O$	0.5g
$NaS_4 \cdot SeO_3 \cdot 5H_2O$	0.3g
$NiCl_2 \cdot 6H_2O$	0.25g
$CoSO_4 \cdot 7H_2O$	0.18g
$ZnSO_4 \cdot 7H_2O$	0.18g
$CaCl_2 \cdot 7H_2O$	0.1g
$FeSO_4 \cdot 7H_2O$	0.1g
$KAI(SO_4)_2 \cdot 12H_2O$	0.02g
$CuSO_4 \cdot 5H_2O$	0.01g
H_3BO_3	0.01g
$Na_2MoO_4 \cdot 2H_2O$	0.01g

Preparation of Trace Elements Solution: Add nitrilotriacetic acid to 500.0mL of distilled/deionized water. Adjust pH to 6.5 with KOH. Add remaining components. Mix thoroughly. Add distilled/deionized water to 1.0L. Adjust pH to 6.8.

Vitamin Solution:

Composition per liter:

Calcium DL-pantothenate	5.0g
Vitamin B_{12}	0.1g
Pyridoxine·HCl	10.0mg
p-Aminobenzoic acid	5.0mg
Lipoic acid	5.0mg
Nicotinic acid	5.0mg
Riboflavin	5.0mg
Thiamine·HCl	5.0mg
Biotin	2.0mg
Folic acid	2.0mg

Preparation of Vitamin Solution: Add components to distilled/deionized water and bring volume to 1.0L. Mix thoroughly.

L-Cysteine·HCl·H_2O Solution:

Composition per 100.0mL:

L-Cysteine·HCl·H_2O	2.5g

Preparation of L-Cysteine·HCl·H_2O Solution: Bring 100.0mL of distilled/deionized water to boiling. Cool to room temperature while sparging with 100% N_2. Add L-cysteine·HCl·H_2O to the 100.0mL of anaerobic water. Distribute into serum bottles fitted with butyl rubber stoppers and aluminum seals. Do not grease stoppers. Autoclave for 20 min at 15 psi pressure–121°C.

$Na_2S \cdot 9H_2O$ Solution:

Composition per 100.0mL:

NaOH	1 pellet
$Na_2S \cdot 9H_2O$	2.5g

Preparation of $Na_2S \cdot 9H_2O$ Solution: Bring 100.0mL of distilled/deionized water to boiling. Cool to room temperature while sparging with 100%N_2. Dissolve 1 pellet of NaOH in the anaerobic water. Weigh out a little more than 2.5g of $Na_2S \cdot 9H_2O$. Briefly rinse the crystals in distilled/deionized water. Dry the crystals by blotting on paper towels or filter paper. Weigh out 2.5g of washed $Na_2S \cdot 9H_2O$ crystals. Add to the 100.0mL of anaerobic NaOH solution. Distribute into serum bottles fitted with butyl rubber stoppers and aluminum seals. Do not grease stoppers. Pressurize to 60kPa with 100% N_2. Autoclave for 15 min at 15 psi pressure–121°C. Store at room temperature in an anaerobic chamber.

Preparation of Medium: Prepare and dispense medium under 80% N_2 + 20% CO_2. Add components, except methanol, L-cysteine·HCl·H_2O solution, and $Na_2S \cdot 9H_2O$ solution, to distilled/deionized water and bring volume to 970.0mL. Mix thoroughly. Gently heat and bring to boiling. Continue boiling for 3 min. Cool to room temperature while sparging with 80% H_2 + 20% CO_2. Anaerobically distribute 9.7mL volumes into anaerobic tubes. Autoclave for 20 min at 15 psi pressure–121°C. Aseptically and anaerobically add 0.1mL of sterile methanol, 0.1mL of sterile L-cysteine·HCl·H_2O solution, and 0.1mL of sterile $Na_2S \cdot 9H_2O$ solution to each tube. Mix thoroughly.

Use: For the cultivation of *Methanosarcina barkeri*.

Methanosarcina BCYT Medium (DSMZ Medium 318)

Composition per liter:

$KHCO_3$	2.0g
NH_4Cl	1.0g
NaCl	0.6g
Yeast extract	0.5g
Trypticase™	0.5g
KH_2PO_4	0.3g
$MgCl_2 \cdot 6H_2O$	0.1g
$CaCl_2 \cdot 2H_2O$	0.08g
Resazurin	1.0mg
Cysteine solution	10.0mL
$Na_2S \cdot 9H_2O$ solution	10.0mL
Trace elements solution	10.0mL
Vitamin solution	10.0mL
Methanol	5.0mL

pH 6.8 ± 0.2 at 25°C

Vitamin Solution:

Composition per liter:

Pyridoxine-HCl	10.0mg
Thiamine-HCl·$2H_2O$	5.0mg
Riboflavin	5.0mg
Nicotinic acid	5.0mg
D-Ca-pantothenate	5.0mg
p-Aminobenzoic acid	5.0mg
Lipoic acid	5.0mg

Biotin	2.0mg
Folic acid	2.0mg
Vitamin B_{12}	0.1mg

Preparation of Vitamin Solution: Add components to distilled/deionized water and bring volume to 1.0L. Mix thoroughly. Sparge with 80% H_2 + 20% CO_2. Filter sterilize.

Trace Elements Solution:
Composition per liter:

$MgSO_4 \cdot 7H_2O$	3.0g
Nitrilotriacetic acid	1.5g
NaCl	1.0g
$MnSO_4 \cdot 2H_2O$	0.5g
$CoSO_4 \cdot 7H_2O$	0.18g
$ZnSO_4 \cdot 7H_2O$	0.18g
$CaCl_2 \cdot 2H_2O$	0.1g
$FeSO_4 \cdot 7H_2O$	0.1g
$NiCl_2 \cdot 6H_2O$	0.025g
$KAl(SO_4)_2 \cdot 12H_2O$	0.02g
H_3BO_3	0.01g
$Na_2MoO_4 \cdot 4H_2O$	0.01g
$CuSO_4 \cdot 5H_2O$	0.01g
$Na_2SeO_3 \cdot 5H_2O$	0.3mg

Preparation of Trace Elements Solution: Add nitrilotriacetic acid to 500.0mL of distilled/deionized water. Dissolve by adjusting pH to 6.5 with KOH. Add remaining components. Add distilled/deionized water to 1.0L. Mix thoroughly.

$Na_2S \cdot 9H_2O$ Solution:
Composition per 10.0mL:

$Na_2S \cdot 9H_2O$	0.3g

Preparation of $Na_2S \cdot 9H_2O$ Solution: Add $Na_2S \cdot 9H_2O$ to distilled/deionized water and bring volume to 10.0mL. Mix thoroughly. Autoclave under 100% N_2 for 15 min at 15 psi pressure–121°C. Cool to room temperature.

Cysteine Solution:
Composition per 10.0mL:

L-Cysteine·HCl·H_2O	0.3g

Preparation of Cysteine Solution: Add L-cysteine·HCl·H_2O to distilled/deionized water and bring volume to 10.0mL. Mix thoroughly. Sparge with 100% N_2. Autoclave for 15 min at 15 psi pressure–121°C.

Preparation of Medium: Add components, except methanol, vitamin solution, cysteine solution, and $Na_2S \cdot 9H_2O$ solution, to distilled/deionized water and bring volume to 985.0mL. Mix thoroughly. Sparge with 100% CO_2. Autoclave for 15 min at 15 psi pressure–121°C. Cool to 25°C while sparging with 100% CO_2. Aseptically and anaerobically add 5.0mL filter sterilized methanol, 10.0mL vitamin solution, 10.0mL cysteine solution, and 10.0mL of sterile $Na_2S \cdot 9H_2O$ solution. Mix thoroughly. Aseptically and anaerobically distribute into sterile tubes or flasks.

Use: For the cultivation of *Methanosarcina* spp.

Methanosarcina DPB Medium

Composition per 1001.0mL:

Sodium acetate·$3H_2O$	4.1g
NH_4Cl	1.4g
K_2HPO_4	1.3g
KH_2PO_4	1.3g
$MgSO_4$	0.5g
NaCl	0.5g
L-Cysteine·HCl·H_2O	0.27g
$CaCl_2 \cdot 2H_2O$	0.06g
$Fe(NH_4)_2(SO_4)_2$	0.01g
Antifoam C	10.0mL
Trace elements solution	10.0mL
Vitamin solution	10.0mL
$Na_2S \cdot 9H_2O$ solution	1.0mL

pH 6.8 ± 0.2 at 25°C

Source: Antifoam C is available from Sigma Chemical Co.

Trace Elements Solution:
Composition per liter:

$MgSO_4 \cdot 5H_2O$	3.0g
Nitrilotriacetic acid	1.5g
NaCl	1.0g
$MnSO_4 \cdot 2H_2O$	0.5g
$NaS_4 \cdot SeO_3 \cdot 5H_2O$	0.3g
$NiCl_2 \cdot 6H_2O$	0.25g
$CoSO_4 \cdot 7H_2O$	0.18g
$ZnSO_4 \cdot 7H_2O$	0.18g
$CaCl_2 \cdot 7H_2O$	0.1g
$FeSO_4 \cdot 7H_2O$	0.1g
$KAl(SO_4)_2 \cdot 12H_2O$	0.02g
$CuSO_4 \cdot 5H_2O$	0.01g
H_3BO_3	0.01g
$Na_2MoO_4 \cdot 2H_2O$	0.01g

Preparation of Trace Elements Solution: Add nitrilotriacetic acid to 500.0mL of distilled/deionized water. Adjust pH to 6.5 with KOH. Add remaining components. Mix thoroughly. Add distilled/deionized water to 1.0L. Adjust pH to 6.8.

Vitamin Solution:
Composition per liter:

Calcium DL-pantothenate	5.0g
Vitamin B_{12}	0.1g
Pyridoxine·HCl	10.0mg
p-Aminobenzoic acid	5.0mg
Lipoic acid	5.0mg
Nicotinic acid	5.0mg
Riboflavin	5.0mg
Thiamine·HCl	5.0mg
Biotin	2.0mg
Folic acid	2.0mg

Preparation of Vitamin Solution: Add components to distilled/deionized water and bring volume to 1.0L. Mix thoroughly.

$Na_2S \cdot 9H_2O$ Solution:
Composition per 100.0mL:

$Na_2S \cdot 9H_2O$	25.0g

Preparation of $Na_2S \cdot 9H_2O$ Solution: Bring 100.0mL of distilled/deionized water to boiling. Cool to room temperature while sparging with 100%N_2. Add $Na_2S \cdot 9H_2O$ to the 100.0mL of anaerobic water. Mix thoroughly. Distribute into serum bottles fitted with butyl rubber stoppers and aluminum seals. Do not grease stoppers. Autoclave for 20 min at 15 psi pressure–121°C.

Preparation of Medium: Add components, except $Na_2S \cdot 9H_2O$ solution, to distilled/deionized water and bring volume to 1.0L. Mix thoroughly. Adjust pH to 6.8 with concentrated HCl. Distribute into tubes or flasks. Autoclave for 15 min at 15 psi pressure–121°C. Cool medium while sparging with 100% N_2. Prior to inoculation, aseptically and anaerobically add 1.0mL of sterile $Na_2S \cdot 9H_2O$ solution per liter of medium. Repeat the addition of 1.0mL of sterile $Na_2S \cdot 9H_2O$ solution per liter of medium every 48 hr during growth.

Use: For the cultivation of *Methanosarcina* species.

Methanosarcina frisia Medium

Composition per liter:

NaCl	18.0g
$NaHCO_3$	5.0g
$MgCl_2 \cdot 6H_2O$	4.0g
$MgSO_4 \cdot 7H_2O$	3.45g
Trypticase™	2.0g
Yeast extract	2.0g
Sodium acetate	1.0g
KCl	0.335g
NH_4Cl	0.25g
$CaCl_2 \cdot 2H_2O$	0.14g
K_2HPO_4	0.14g
L-Cysteine·HCl	0.5g
$Na_2S \cdot 9H_2O$	0.5g
$Fe(NH_4)_2(SO_4)_2 \cdot 7H_2O$	2.0mg
Resazurin	1.0mg
Trace elements solution	10.0mL
Vitamin solution	10.0mL
Methanol solution	2.0mL

pH 6.8–7.0 at 25°C

Trace Elements Solution:

Composition per liter:

$MgSO_4 \cdot 7H_2O$	3.0g
Nitrilotriacetic acid	1.5g
NaCl	1.0g
$MnSO_4 \cdot 2H_2O$	0.5g
$CoSO_4 \cdot 7H_2O$	0.18g
$ZnSO_4 \cdot 7H_2O$	0.18g
$CaCl_2 \cdot 2H_2O$	0.1g
$FeSO_4 \cdot 7H_2O$	0.1g
$KAl(SO_4)_2 \cdot 12H_2O$	0.02g
$CuSO_4 \cdot 5H_2O$	0.01g
H_3BO_3	0.01g
$Na_2MoO_4 \cdot 2H_2O$	0.01g
$NiCl_2 \cdot 6H_2O$	0.025g
$Na_2SeO_3 \cdot 5H_2O$	0.3mg

Preparation of Trace Elements Solution: Add nitrilotriacetic acid to 500.0mL of distilled/deionized water. Adjust pH to 6.5 with KOH. Add remaining components. Adjust pH to 7.0. Add distilled/deionized water to 1.0L.

Vitamin Solution:

Composition per liter:

Pyridoxine·HCl	10.0mg
Calcium DL-pantothenate	5.0mg
Lipoic acid	5.0mg
Nicotinic acid	5.0mg
p-Aminobenzoic acid	5.0mg
Riboflavin	5.0mg
Thiamine·HCl	5.0mg
Biotin	2.0mg
Folic acid	2.0mg
Vitamin B_{12}	0.1mg

Preparation of Vitamin Solution: Add components to distilled/deionized water and bring volume to 1.0L. Mix thoroughly. Sparge with 80% H_2 + 20% CO_2.

Methanol Solution:

Composition per 10.0mL:

Methanol	10.0mL

Preparation of Methanol Solution: Sparge 10.0mL of methanol with 100% N_2. Autoclave for 15 min at 15 psi pressure–121°C.

Preparation of Medium: Prepare and dispense medium under 80% N_2 + 20% CO_2. Add components, except methanol solution, to distilled/deionized water and bring volume to 1.0L. Mix thoroughly. Anaerobically distribute into tubes or flasks fitted with butyl rubber stoppers. Autoclave for 15 min at 15 psi pressure–121°C. Anaerobically add 10.0mL of sterile methanol solution to each liter of medium or, using a syringe, inject the appropriate amount of sterile methanol solution into individual tubes containing medium.

Use: For the cultivation and maintenance of *Methanosarcina frisia*.

Methanosarcina Mass-Culturing Medium

Composition per liter:

NaCl	0.58g
NH_4Cl	0.53g
K_2HPO_4	0.44g
KH_2PO_4	0.35g
$MgCl_2 \cdot 6H_2O$	0.04g
$CaCl_2 \cdot 2H_2O$	0.03g
Resazurin	0.001g
Trace elements solution	10.0mL
Vitamin solution	10.0mL
Methanol	8.0mL
$Na_2S \cdot 9H_2O$ solution	8.0mL

Trace Elements Solution:

Composition per liter:

Sodium nitrilotriacetate	1.67g
$CoCl_2 \cdot 6H_2O$	0.18g
$FeCl_2 \cdot 4H_2O$	0.14g
$MnCl_2 \cdot 4H_2O$	0.09g
$NiCl_2 \cdot 6H_2O$	0.09g
$ZnCl_2$	0.09g
$CaCl_2$	0.06g
$Na_2MoO_4 \cdot 2H_2O$	0.046g
$CuSO_4$	0.045g

Preparation of Trace Elements Solution: Add sodium nitrilotriacetate to 500.0mL of distilled/deionized water. Adjust pH to 6.5. Add remaining components. Add distilled/deionized water to 1.0L. Adjust pH to 7.0.

Vitamin Solution:

Composition per liter:

Calcium DL-pantothenate	5.0g
Vitamin B_{12}	0.1g
Pyridoxine·HCl	10.0mg
p-Aminobenzoic acid	5.0mg
Lipoic acid	5.0mg
Nicotinic acid	5.0mg
Riboflavin	5.0mg
Thiamine·HCl	5.0mg
Biotin	2.0mg
Folic acid	2.0mg

Preparation of Vitamin Solution: Add components to distilled/deionized water and bring volume to 1.0L. Mix thoroughly. Filter sterilize.

Preparation of Methanol: Filter sterilize 10.0mL of methanol.

$Na_2S{\cdot}9H_2O$ Solution:

Composition per 100.0mL:

NaOH	1 pellet
$Na_2S{\cdot}9H_2O$	2.5g

Preparation of $Na_2S{\cdot}9H_2O$ Solution: Bring 100.0mL of distilled/deionized water to boiling. Cool to room temperature while sparging with 100%N_2. Dissolve 1 pellet of NaOH in the anaerobic water. Weigh out a little more than 2.5g of $Na_2S{\cdot}9H_2O$. Briefly rinse the crystals in distilled/deionized water. Dry the crystals by blotting on paper towels or filter paper. Add 2.5g of washed $Na_2S{\cdot}9H_2O$ crystals to 100.0mL of anaerobic NaOH solution. Distribute into serum bottles fitted with butyl rubber stoppers and aluminum seals. Do not grease stoppers. Pressurize to 60kPa with 100% N_2. Autoclave for 15 min at 15 psi pressure–121°C. Store at room temperature in an anaerobic chamber.

Preparation of Medium: Prepare and dispense medium under 80% H_2 + 20% CO_2. Add components, except methanol, vitamin solution, and $Na_2S{\cdot}9H_2O$ solution, to distilled/deionized water and bring volume to 974.0mL. Mix thoroughly. Gently heat and bring to boiling. Continue boiling for 3 min. Cool to room temperature while sparging with 80% H_2 + 20% CO_2. Anaerobically distribute 9.7mL volumes into anaerobic tubes. Autoclave for 15 min at 15 psi pressure–121°C. Aseptically and anaerobically add 0.1mL of sterile vitamin solution, 0.08mL of sterile methanol, and 0.08mL of sterile $Na_2S{\cdot}9H_2O$ solution to each tube. Mix thoroughly.

Use: For the cultivation of *Methanosarcina* species.

Methanosarcina mazei Alpha Basal Medium

Composition per liter:

$NaHCO_3$	4.4g
Pancreatic digest of casein	2.0g
Yeast extract	2.0g
NH_4Cl	1.0g
$Na_2S{\cdot}6H_2O$	0.5g
K_2HPO_4	0.4g
Sodium acetate·$3H_2O$	0.27g
$MgCl_2{\cdot}6H_2O$	0.08g
$CaCl_2{\cdot}2H_2O$	0.04g
$CoCl_2{\cdot}6H_2O$	1.5mg
$FeSO_4{\cdot}7H_2O$	1.0mg
$MnCl_2{\cdot}4H_2O$	1.0mg
Resazurin	1.0mg
H_3BO_4	0.1mg
$NaMoO_4{\cdot}2H_2O$	0.1mg
$ZnCl_2$	0.1mg

pH 7.2 ± 0.2 at 25°C

Preparation of Medium: Prepare and dispense medium under 70% N_2 + 30% CO_2. Add components to distilled/deionized water and bring volume to 1.0L. Mix thoroughly. Adjust pH to 7.2. Sparge with 70% N_2 + 30% CO_2. Autoclave for 15 min at 15 psi pressure–121°C. Aseptically and anaerobically distribute into sterile tubes or bottles.

Use: For the cultivation of *Methanosarcina mazei.*

Methanosarcina mazei Medium (DSMZ Medium 120)

Composition 1112.0mL:

NaCl	2.25g
Yeast extract	2.0g
Casitone	2.0g
$NaHCO_3$	0.85g
NH_4Cl	0.5g
$MgSO_4{\cdot}7H_2O$	0.5g
K_2HPO_4	0.348g
$CaCl_2{\cdot}2H_2O$	0.25g
KH_2PO_4	0.227g
$FeSO_4{\cdot}7H_2O$	0.002g
Resazurin	0.001g
Methanol solution	20.0mL
L-Cysteine-HCl·H_2O solution	15.0mL
$Na_2S{\cdot}9H_2O$ solution	15.0mL
Na-acetate solution	10.0mL
Vitamin solution	10.0mL
$NaHCO_3$ solution	10.0mL
Trace elements solution SL-10	1.0mL

pH 6.5–6.8 at 25°C

Vitamin Solution:

Composition per liter:

Pyridoxamine-HCl	10.0mg
Pantothenic acid	5.0mg
Riboflavin	5.0mg
Alpha-lipoic acid	5.0mg
p-Aminobenzoic acid	5.0mg
Thiamine-HCl·$2H_2O$	5.0mg
Nicotinic acid	5.0mg
Biotin	2.0mg
Folic acid	2.0mg
Vitamin B_{12}	0.1mg

Preparation of Vitamin Solution: Add components to distilled/deionized water and bring volume to 1.0L. Mix thoroughly. Sparge with 80% N_2 + 20% CO_2. Filter sterilize.

Trace Elements Solution SL-10:

Composition per liter:

$FeCl_2{\cdot}4H_2O$	1.5g
$CoCl_2{\cdot}6H_2O$	190.0mg
$MnCl_2{\cdot}4H_2O$	100.0mg
$ZnCl_2$	70.0mg
$Na_2MoO_4{\cdot}2H_2O$	36.0mg
$NiCl_2{\cdot}6H_2O$	24.0mg
H_3BO_3	6.0mg
$CuCl_2{\cdot}2H_2O$	2.0mg
HCl (25% solution)	10.0mL

Preparation of Trace Elements Solution SL-10: Add $FeCl_2{\cdot}4H_2O$ to 10.0mL of HCl solution. Mix thoroughly. Add distilled/deionized water and bring volume to 1.0L. Add remaining components. Mix thoroughly. Sparge with 80% N_2 + 20% CO_2. Autoclave for 15 min at 15 psi pressure–121°C.

$NaHCO_3$ Solution:

Composition per 100.0mL:

$NaHCO_3$	5.0g

Preparation of $NaHCO_3$ Solution: Add $NaHCO_3$ to distilled/deionized water and bring volume to 100.0mL. Mix thoroughly. Sparge with 80% N_2 + 20% CO_2. Autoclave for 15 min at 15 psi pressure–121°C.

$Na_2S·9H_2O$ Solution:
Composition per 100.0mL:

$Na_2S·9H_2O$	3.0g

Preparation of $Na_2S·9H_2O$ Solution: Add $Na_2S·9H_2O$ to distilled/deionized water and bring volume to 100.0mL. Mix thoroughly. Sparge with 100% N_2. Autoclave for 15 min at 15 psi pressure–121°C. Before use, neutralize to pH 7.0 with sterile HCl.

Methanol Solution:
Composition per 100.0mL:

Methanol	50.0mL

Preparation of Methanol Solution: Add methanol to distilled/deionized water and bring volume to 100.0mL. Mix thoroughly. Sparge with 100% N_2. Autoclave for 15 min at 15 psi pressure–121°C.

Na-Acetate Solution:
Composition per 100.0mL:

Na-acetate	25.0g

Preparation of Na-Acetate Solution: Add Na-acetate to distilled/deionized water and bring volume to 100.0mL. Mix thoroughly. Sparge with 100% N_2. Autoclave for 15 min at 15 psi pressure–121°C.

L-Cysteine Solution:
Composition per 100.0mL:

L-Cysteine·HCl·H_2O	3.0g

Preparation of L-Cysteine Solution: Add L-cysteine·HCl·H_2O to distilled/deionized water and bring volume to 100.0mL. Mix thoroughly. Sparge with 100% N_2. Autoclave for 15 min at 15 psi pressure–121°C.

Preparation of Medium: Prepare and dispense medium under 80% N_2 + 20% CO_2 gas atmosphere. Add components, except vitamin solution, $NaHCO_3$ solution, methanol solution, L-cysteine-HCl·H_2O solution, $Na_2S·9H_2O$ solution, Na-acetate solution, and trace elements solution SL-10, to distilled/deionized water and bring volume to 1.0L. Mix thoroughly. Adjust pH to 6.5. Sparge with 80% N_2 + 20% CO_2. Autoclave for 15 min at 15 psi pressure–121°C. Aseptically and anaerobically add 10.0mL vitamin solution, 10.0mL $NaHCO_3$ solution, 20.0mL methanol solution, 15.0mL L-cysteine-HCl·H_2O solution, 15.0mL $Na_2S·9H_2O$ solution, 10.0mL Na-acetate solution, and 1.0mL trace elements solution SL-10. Mix thoroughly. Aseptically and anaerobically distribute into sterile tubes or bottles. After inoculation, flush and repressurize the gas head space of culture bottles with sterile 80% H_2 + 20% CO_2 to 1 bar overpressure.

Use: For the cultivation of *Mathanosarcina mazei.*

Methanosarcina Medium

Composition per liter:

Agar	20.0g
NaCl	2.25g
Yeast extract	2.0g
Pancreatic digest of casein	2.0g
NH_4Cl	0.5g
$MgSO_4·7H_2O$	0.5g
K_2HPO_4	0.35g
$CaCl_2·2H_2O$	0.25g
KH_2PO_4	0.23g
$FeSO_4·7H_2O$	2.0mg
$NaHCO_3$ solution	20.0mL
L-Cysteine-sulfide reducing agent	20.0mL
Wolfe's vitamin solution	10.0mL
Methanol	10.0mL
Resazurin (0.025% solution)	4.0mL
Trace elements solution SL-6	3.0mL

pH 6.8 ± 0.2 at 25°C

$NaHCO_3$ Solution:
Composition per 20.0mL:

$NaHCO_3$	850.0mg

Preparation of $NaHCO_3$ Solution: Add $NaHCO_3$ to distilled/deionized water and bring volume to 20.0mL. Mix thoroughly. Filter sterilize. Gas with 100% CO_2 for 20 min.

L-Cysteine-Sulfide Reducing Agent:
Composition per 20.0mL:

L-Cysteine·HCl·H_2O	0.3g
$Na_2S·9H_2O$	0.3g

Preparation of L-Cysteine-Sulfide Reducing Agent: Add L-cysteine·HCl·H_2O to 10.0mL of distilled/deionized water. Mix thoroughly. In a separate tube, add $Na_2S·9H_2O$ to 10.0mL of distilled/deionized water. Mix thoroughly. Gas both solutions with 100% N_2 and cap tubes. Autoclave both solutions for 15 min at 15 psi pressure–121°C using fast exhaust. Cool to 50°C. Aseptically combine the two solutions under 100% N_2.

Wolfe's Vitamin Solution:
Composition per liter:

Pyridoxine·HCl	10.0mg
Thiamine·HCl	5.0mg
Riboflavin	5.0mg
Nicotinic acid	5.0mg
Calcium pantothenate	5.0mg
p-Aminobenzoic acid	5.0mg
Thioctic acid	5.0mg
Biotin	2.0mg
Folic acid	2.0mg
Cyanocobalamin	100.0μg

Preparation of Wolfe's Vitamin Solution: Add components to distilled/deionized water and bring volume to 1.0L. Mix thoroughly. Filter sterilize.

Trace Elements Solution SL-6:
Composition per liter:

H_3BO_3	0.3g
$CoCl_2·6H_2O$	0.2g
$ZnSO_4·7H_2O$	0.1g
$MnCl_2·4H_2O$	0.03g
$Na_2MoO_4·H_2O$	0.03g
$NiCl_2·6H_2O$	0.02g
$CuCl_2·2H_2O$	0.01g

Preparation of Trace Elements Solution SL-6: Add components to distilled/deionized water and bring volume to 1.0L. Mix thoroughly. Adjust pH to 3.4.

Preparation of Medium: Add components, except $NaHCO_3$ solution, L-cysteine-sulfide reducing agent, Wolfe's vitamin solution, and methanol, to distilled/deionized water and bring volume to 940.0mL. Mix thoroughly. Autoclave for 15 min at 15 psi pressure–121°C. Cool under 80% N_2 + 20% CO_2. Aseptically and anaerobically add the sterile $NaHCO_3$ solution, the sterile L-cysteine-sulfide reducing agent, the sterile Wolfe's vitamin solution, and filter-sterilized methanol. Mix thoroughly. Adjust pH to 6.8. Aseptically and anaerobically distribute into sterile tubes or flasks.

Use: For the cultivation and maintenance of *Bifidobacterium asteroides, Methanosarcina barkeri, Methanosarcina* species, and *Methanosarcina vacuolata.*

Methanosarcina Medium (BCYT)

Composition per liter:

$KHCO_3$	2.0g
NH_4Cl	1.0g
NaCl	0.6g
Pancreatic digest of casein	0.5g
Yeast extract	0.5g
KH_2PO_4	0.3g
$MgCl_2·6H_2O$	0.1g
$CaCl_2·2H_2O$	0.08g
Resazurin	1.0mg
Trace elements solution	10.0mL
Vitamin solution	10.0mL
L-Cysteine·HCl·H_2O solution	10.0mL
$Na_2S·9H_2O$ solution	10.0mL
Methanol	5.0mL

pH 6.8 ± 0.2 at 25°C

Trace Elements Solution:

Composition per liter:

Nitrilotriacetic acid	12.8g
$FeCl_3·6H_2O$	1.35g
NaCl	1.0g
$NiCl_2·6H_2O$	0.12g
$MnCl_2·4H_2O$	0.1g
$CaCl_2·2H_2O$	0.1g
$ZnCl_2$	0.1g
$Na_2SeO_3·5H_2O$	0.026g
$CuCl_2·2H_2O$	0.025g
$CoCl_2·6H_2O$	0.024g
$Na_2MoO_4·2H_2O$	0.024g
H_3BO_3	0.01g

Preparation of Trace Elements Solution: Add nitrilotriacetic acid to approximately 500.0mL of distilled/deionized water. Dissolve by adding KOH and adjust pH to 6.5. Add remaining components. Bring volume to 1.0L with additional distilled/deionized water. Adjust pH to 7.0 with KOH.

Vitamin Solution:

Composition per liter:

Pyridoxine·HCl	10.0mg
Calcium DL-pantothenate	5.0mg
Lipoic acid	5.0mg
Nicotinic acid	5.0mg
p-Aminobenzoic acid	5.0mg
Riboflavin	5.0mg
Thiamine·HCl	5.0mg
Biotin	2.0mg
Folic acid	2.0mg
Vitamin B_{12}	0.1mg

Preparation of Vitamin Solution: Add components to distilled/deionized water and bring volume to 1.0L. Mix thoroughly. Filter sterilize. Sparge with 80% N_2 + 20% CO_2.

L-Cysteine·HCl·H_2O Solution:

Composition per 10.0mL:

L-Cysteine·HCl·H_2O	0.25g

Preparation of L-Cysteine·HCl·H_2O Solution: Add L-cysteine·HCl·H_2O to distilled/deionized water and bring volume to 10.0mL. Mix thoroughly. Autoclave under 80% N_2 + 20% CO_2 for 15 min at 15 psi pressure–121°C.

$Na_2S·9H_2O$ Solution:

Composition per 10.0mL:

$Na_2S·9H_2O$	0.3g

Preparation of $Na_2S·9H_2O$ Solution: Add $Na_2S·9H_2O$ to distilled/deionized water and bring volume to 10.0mL. Mix thoroughly. Sparge with 80% N_2 + 20% CO_2. Autoclave for 15 min at 15 psi pressure–121°C.

Preparation of Medium: Add components, except vitamin solution, L-cysteine·HCl·H_2O solution, and $Na_2S·9H_2O$ solution, to distilled/deionized water and bring volume to 960.0mL. Mix thoroughly. Sparge under 80% N_2 + 20% CO_2 for 3–4 min. Autoclave for 15 min at 15 psi pressure–121°C. Aseptically and anaerobically add 20.0mL of sterile vitamin solution, 10.0mL of sterile L-cysteine·HCl·H_2O solution, and 10.0mL of sterile $Na_2S·9H_2O$ solution. Mix thoroughly. Aseptically and anaerobically distribute into sterile screw-cappped bottles under 80% N_2 + 20% CO_2.

Use: For the cultivation and maintenance of *Methanosarcina mazei* and *Methanosarcina thermophila.*

Methanosarcina Medium

Composition per liter:

NaCl	2.25g
Pancreatic digest of casein	2.0g
Yeast extract	2.0g
$NaHCO_3$	0.85g
$MgSO_4·7H_2O$	0.5g
NH_4Cl	0.5g
K_2HPO_4	0.348g
$CaCl_2·2H_2O$	0.25g
KH_2PO_4	0.227g
$FeSO_4·7H_2O$	2.0mg
Resazurin	1.0mg
Methanol solution	10.0mL
Vitamin solution	10.0mL
L-Cysteine·HCl·H_2O solution	10.0mL
$Na_2S·9H_2O$ solution	10.0mL
Trace elements solution SL-10	1.0mL

pH 6.5–6.8 at 25°C

Methanol Solution:

Composition per 10.0mL:

Methanol	5.0mL

Preparation of Methanol Solution: Add methanol to distilled/deionized water and bring volume to 10.0mL. Sparge with 100% N_2. Autoclave for 15 min at 15 psi pressure–121°C.

Vitamin Solution:

Composition per liter:

Pyridoxine·HCl	10.0mg
Calcium DL-pantothenate	5.0mg
Lipoic acid	5.0mg
Nicotinic acid	5.0mg
p-Aminobenzoic acid	5.0mg

Riboflavin	5.0mg
Thiamine·HCl	5.0mg
Biotin	2.0mg
Folic acid	2.0mg
Vitamin B_{12}	0.1mg

Preparation of Vitamin Solution: Add components to distilled/deionized water and bring volume to 1.0L. Mix thoroughly. Filter sterilize. Sparge with 80% N_2 + 20% CO_2.

L-Cysteine·HCl·H_2O Solution:

Composition per 10.0mL:

L-Cysteine·HCl·H_2O	0.3g

Preparation of L-Cysteine·HCl·H_2O Solution: Add L-cysteine·HCl·H_2O to distilled/deionized water and bring volume to 10.0mL. Mix thoroughly. Autoclave under 100% N_2 for 15 min at 15 psi pressure–121°C.

$Na_2S·9H_2O$ Solution:

Composition per 10.0mL:

$Na_2S·9H_2O$	0.3g

Preparation of $Na_2S·9H_2O$ Solution: Add $Na_2S·9H_2O$ to distilled/deionized water and bring volume to 10.0mL. Mix thoroughly. Sparge with 100% N_2. Autoclave for 15 min at 15 psi pressure–121°C.

Trace Elements Solution SL-10:

Composition per liter:

$FeCl_2·4H_2O$	1.5g
$CoCl_2·6H_2O$	190.0mg
$MnCl_2·4H_2O$	100.0mg
$ZnCl_2$	70.0mg
$Na_2MoO_4·2H_2O$	36.0mg
$NiCl_2·6H_2O$	24.0mg
H_3BO_3	6.0mg
$CuCl_2·2H_2O$	2.0mg
HCl (25% solution)	10.0mL

Preparation of Trace Elements Solution SL-10: Add $FeCl_2·4H_2O$ to 10.0mL of HCl solution. Mix thoroughly. Add distilled/deionized water and bring volume to 1.0L. Add remaining components. Mix thoroughly. Sparge with 100% N_2.

Preparation of Medium: Add components, except methanol solution, vitamin solution, L-cysteine·HCl·H_2O solution, and $Na_2S·9H_2O$ solution, to distilled/deionized water and bring volume to 960.0mL. Mix thoroughly. Sparge under 80% N_2 + 20% CO_2 for 3–4 min. Autoclave for 15 min at 15 psi pressure–121°C. Aseptically and anaerobically add 10.0mL of sterile methanol, 10.0mL of sterile vitamin solution, 10.0mL of sterile L-cysteine·HCl·H_2O solution, and 10.0mL of sterile $Na_2S·9H_2O$ solution. Mix thoroughly. Aseptically and anaerobically distribute into sterile screw-cappped bottles under 80% N_2 + 20% CO_2.

Use: For the cultivation and maintenance of *Methanosarcina* species.

Methanosarcina MP Medium

Composition per 1015.0mL:

NH_4Cl	1.0g
Yeast extract	1.0g
L-Cysteine·HCl	0.5g
K_2HPO_4	0.4g
$MgCl_2·6H_2O$	0.2g
Resazurin	1.0mg
Mineral solution	50.0mL
Trace elements solution	10.0mL
Na_2CO_3 solution	10.0mL
$Na_2S·9H_2O$ solution	10.0mL
Methanol	5.0mL

pH 6.8 ± 0.2 at 25°C

Mineral Solution:

Composition per liter:

NaCl	12.0g
KH_2PO_4	6.0g
$(NH_4)_2SO_4$	6.0g
$MgSO_4·7H_2O$	2.6g
$CaCl_2·2H_2O$	0.16g

Preparation of Mineral Solution: Add components to distilled/deionized water and bring volume to 1.0L. Mix thoroughly.

Trace Elements Solution:

Composition per liter:

$MgSO_4·7H_2O$	3.0g
Nitrilotriacetic acid	1.5g
NaCl	1.0g
$MnSO_4·2H_2O$	0.5g
$CoSO_4·7H_2O$	0.18g
$ZnSO_4·7H_2O$	0.18g
$CaCl_2·2H_2O$	0.1g
$FeSO_4·7H_2O$	0.1g
$NiCl_2·6H_2O$	0.025g
$KAl(SO_4)_2·12H_2O$	0.02g
$CuSO_4·5H_2O$	0.01g
H_3BO_3	0.01g
$Na_2MoO_4·2H_2O$	0.01g
$Na_2SeO_3·5H_2O$	0.3mg

Preparation of Trace Elements Solution: Add nitrilotriacetic acid to 500.0mL of distilled/deionized water. Adjust pH to 6.5 with KOH. Add remaining components. Adjust pH to 7.0. Add distilled/deionized water to 1.0L.

Na_2CO_3 Solution:

Composition per 10.0mL:

Na_2CO_3	1.0g

Preparation of Na_2CO_3 Solution: Add Na_2CO_3 to distilled/deionized water and bring volume to 10.0mL. Mix thoroughly. Sparge with 100% N_2. Autoclave for 15 min at 15 psi pressure–121°C.

$Na_2S·9H_2O$ Solution:

Composition per 10.0mL:

$Na_2S·9H_2O$	0.25g

Preparation of $Na_2S·9H_2O$ Solution: Add $Na_2S·9H_2O$ to distilled/deionized water and bring volume to 10.0mL. Mix thoroughly. Sparge with 100% N_2. Autoclave for 15 min at 15 psi pressure–121°C.

Preparation of Medium: Prepare and dispense medium under 80% N_2 + 20% CO_2. Add components, except L-cysteine·HCl, Na_2CO_3 solution, and $Na_2S·9H_2O$ solution, to distilled/deionized water and bring volume to 980.0mL. Mix thoroughly. Gently heat and bring to boiling. Continue boiling for 5 min. Cool to room temperature while sparging with 80% N_2 + 20% CO_2. Add L-cysteine·HCl. Mix thoroughly. Adjust pH to 6.8–7.0. Autoclave for 15 min at 15 psi pressure–121°C. Aseptically and anaerobically add 10.0mL of sterile Na_2CO_3 solution and 10.0mL of sterile $Na_2S·9H_2O$ solution. Mix thoroughly.

Use: For the cultivation and maintenance of *Methanosarcina* species.

Methanosarcina Nitrogen-Fixing Medium

Composition per 1005.0mL:

NaH_2PO_4	1.38g
K_2HPO_4	0.34g
$MgCl_2 \cdot 2H_2O$	0.1g
Yeast extract	0.1g
$CaCl_2 \cdot 2H_2O$	0.05g
Resazurin	0.001g
Trace elements solution	10.0mL
Methanol	2.0mL
$Na_2S \cdot 9H_2O$ solution	2.0mL
$NaHCO_3$ solution	1.0mL

pH 7.0 ± 0.2 at 25°C

Trace Elements Solution:

Composition per liter:

Nitrilotriacetic acid	4.5g
$FeCl_2 \cdot 7H_2O$	0.4g
H_3BO_3	0.19g
$CoCl_2 \cdot 6H_2O$	0.17g
$MnCl_2 \cdot 4H_2O$	0.1g
$ZnCl_2$	0.1g
$NiCl_2 \cdot 6H_2O$	0.02g
$Na_2MoO_4 \cdot 6H_2O$	0.01g

Preparation of Trace Elements Solution: Add components to distilled/deionized water and bring volume to 1.0L. Mix thoroughly. Sparge with 100% N_2. Adjust pH to 7.0 with 10*N* NaOH. Autoclave for 15 min at 15 psi pressure–121°C.

Preparation of Methanol: Sparge 2-propanol with 100% N_2. Filter sterilize.

$Na_2S \cdot 9H_2O$ Solution:

Composition per 10.0mL:

$Na_2S \cdot 9H_2O$	2.0g
NaOH	0.1g

Preparation of $Na_2S \cdot 9H_2O$ Solution: Add $Na_2S \cdot 9H_2O$ and NaOH to distilled/deionized water and bring volume to 10.0mL. Mix thoroughly. Sparge with 100% N_2. Autoclave for 15 min at 15 psi pressure–121°C. Before use, neutralize to pH 7.0 with sterile HCl.

$NaHCO_3$ Solution:

Composition per 10.0mL:

$NaHCO_3$	1.0g

Preparation of $NaHCO_3$ Solution: Add $NaHCO_3$ to distilled/deionized water and bring volume to 10.0mL. Mix thoroughly. Sparge with 80% N_2 + 20% CO_2. Autoclave for 15 min at 15 psi pressure–121°C.

Preparation of Medium: Prepare and dispense medium under 100% N_2. Add components, except methanol, $Na_2S \cdot 9H_2O$ solution, and $NaHCO_3$ solution, to distilled/deionized water and bring volume to 1.0L. Mix thoroughly. Adjust pH to 7.5. Sparge with 100% N_2. Autoclave for 15 min at 15 psi pressure–121°C. Aseptically and anaerobically add 2.0mL of sterile methanol, 2.0mL of sterile $Na_2S \cdot 9H_2O$ solution, and 2.0mL of sterile $NaHCO_3$ solution. Mix thoroughly. Adjust pH to 7.0 by adding sterile anaerobic 10*N* NaOH. Aseptically and anaerobically distribute into sterile tubes or bottles.

Use: For the cultivation of nitrogen-fixing *Methanosarcina* species.

Methanosarcina semesiae Medium

Composition per 1214.0mL:

Solution A	950.0mL
$NaHCO_3$ solution	50.0mL
Solution E	30.0mL
$Na_2S \cdot 9H_2O$ solution	15.0mL
Solution D	10.0mL
Solution F	10.0mL
Solution G	10.0mL
Trace elements solution	10.0mL
Vitamin solution	10.0mL
Methanol solution	5.0mL
Solution B	1.0mL
Solution C	1.0mL
Seven vitamin solution	1.0mL
Dithionite solution	1.0mL

pH 7.2 ± 0.2 at 25°C

Solution A:

Composition per 950.0mL:

KH_2PO_4	1.4g
NH_4Cl	0.5g
$MgCl_2 \cdot 6H_2O$	0.2g
$CaCl_2 \cdot 2H_2O$	0.15g
Resazurin	1.0mg

Preparation of Solution A: Add components to distilled/deionized water and bring volume to 950.0mL. Mix thoroughly. Sparge with 80% N_2 + 20% CO_2. Filter sterilize.

Solution B:

Composition per liter:

$FeCl_2 \cdot 4H_2O$	1.5g
$CoCl_2 \cdot 6H_2O$	190.0mg
$MnCl_2 \cdot 4H_2O$	100.0mg
$ZnCl_2$	70.0mg
$Na_2MoO_4 \cdot 2H_2O$	36.0mg
$NiCl_2 \cdot 6H_2O$	24.0mg
H_3BO_3	6.0mg
$CuCl_2 \cdot 2H_2O$	2.0mg
HCl (25% solution)	10.0mL

Preparation of Solution B: Add $FeCl_2 \cdot 4H_2O$ to 10.0mL of HCl solution. Mix thoroughly. Add distilled/deionized water and bring volume to 1.0L. Add remaining components. Mix thoroughly. Sparge with 80% N_2 + 20% CO_2. Autoclave for 15 min at 15 psi pressure–121°C.

Solution C:

Composition per liter:

NaOH	0.5g
$Na_2WO_4 \cdot 2H_2O$	4.0mg
$Na_2SeO_3 \cdot 5H_2O$	3.0mg

Preparation of Solution C: Add components to distilled/deionized water and bring volume to 1.0L. Mix thoroughly. Sparge with 100% N_2. Filter sterilize.

Solution D:

Composition per liter:

Pyridoxine-HCl	10.0mg
Thiamine-HCl·$2H_2O$	5.0mg
Riboflavin	5.0mg
Nicotinic acid	5.0mg
D-Ca-pantothenate	5.0mg
p-Aminobenzoic acid	5.0mg
Lipoic acid	5.0mg
Biotin	2.0mg
Folic acid	2.0mg
Vitamin B_{12}	0.1mg

Preparation of Solution D: Add components to distilled/deionized water and bring volume to 1.0L. Mix thoroughly. Sparge with 80% H_2 + 20% CO_2. Filter sterilize.

Solution E:

Composition per 100.0mL:

$NaHCO_3$	5.0g

Preparation of Solution E: Add $NaHCO_3$ to distilled/deionized water and bring volume to 150.0mL. Mix thoroughly. Sparge with 80% N_2 + 20% CO_2. Filter sterilize.

Solution F:

Composition per 10.0mL:

Na_2-maleate	1.6g

Preparation of Solution F: Add Na_2-maleate to distilled/deionized water and bring volume to 150.0mL. Mix thoroughly. Sparge with 80% N_2 + 20% CO_2. Filter sterilize.

Solution G:

Composition per 10.0mL:

$Na_2S·9H_2O$	0.25g

Preparation of Solution G: Add $Na_2S·9H_2O$ to distilled/deionized water and bring volume to 10.0mL. Mix thoroughly. Autoclave under 100% N_2 for 15 min at 15 psi pressure–121°C. Cool to room temperature.

$NaHCO_3$ Solution:

Composition per 100.0mL:

$NaHCO_3$	5.0g

Preparation of $NaHCO_3$ Solution: Add $NaHCO_3$ to distilled/deionized water and bring volume to 100.0mL. Mix thoroughly. Sparge with 100% N_2. Autoclave for 15 min at 15 psi pressure–121°C. Cool to 25°C.

Trace Elements Solution:

Composition per liter:

$MgSO_4·7H_2O$	3.0g
Nitrilotriacetic acid	1.5g
NaCl	1.0g
$MnSO_4·2H_2O$	0.5g
$CoSO_4·7H_2O$	0.18g
$ZnSO_4·7H_2O$	0.18g
$CaCl_2·2H_2O$	0.1g
$FeSO_4·7H_2O$	0.1g
$NiCl_2·6H_2O$	0.025g
$KAl(SO_4)_2·12H_2O$	0.02g
H_3BO_3	0.01g
$Na_2MoO_4·4H_2O$	0.01g
$CuSO_4·5H_2O$	0.01g
$Na_2SeO_3·5H_2O$	0.3mg

Preparation of Trace Elements Solution: Add nitrilotriacetic acid to 500.0mL of distilled/deionized water. Dissolve by adjusting pH to 6.5 with KOH. Add remaining components. Add distilled/deionized water to 1.0L. Mix thoroughly.

Vitamin Solution:

Composition per liter:

Pyridoxine-HCl	10.0mg
Thiamine-HCl·$2H_2O$	5.0mg
Riboflavin	5.0mg
Nicotinic acid	5.0mg
D-Ca-pantothenate	5.0mg
p-Aminobenzoic acid	5.0mg
Lipoic acid	5.0mg
Biotin	2.0mg
Folic acid	2.0mg
Vitamin B_{12}	0.1mg

Preparation of Vitamin Solution: Add components to distilled/deionized water and bring volume to 1.0L. Mix thoroughly. Sparge with 80% H_2 + 20% CO_2. Filter sterilize.

Seven Vitamin Solution:

Composition per liter:

Pyridoxine hydrochloride	300.0mg
Thiamine-HCl·$2H_2O$	200.0mg
Nicotinic acid	200.0mg
Vitamin B_{12}	100.0mg
Calcium pantothenate	100.0mg
p-Aminobenzoic acid	80.0mg
D(+)-Biotin	20.0mg

Preparation of Seven Vitamin Solution: Add components to distilled/deionized water and bring volume to 1.0L. Sparge with 100% N_2. Mix thoroughly. Filter sterilize.

Methanol Solution:

Composition per 100.0mL:

Methanol	50.0mL

Preparation of Methanol Solution: Add methanol to distilled/deionized water and bring volume to 100.0mL. Mix thoroughly. Sparge with 100% N_2. Autoclave for 15 min at 15 psi pressure–121°C.

$Na_2S·9H_2O$ Solution:

Composition per 20.0mL:

$Na_2S·9H_2O$	0.6g

Preparation of $Na_2S·9H_2O$ Solution: Add $Na_2S·9H_2O$ to distilled/deionized water and bring volume to 20.0mL. Mix thoroughly. Autoclave under 100% N_2 for 15 min at 15 psi pressure–121°C. Cool to room temperature.

Dithionite Solution

Composition per 10.0mL:

Na-dithionite	0.25g

Preparation of Dithionite Solution: Add Na-dithionite to distilled/deionized water and bring volume to 10.0mL. Mix thoroughly. Sparge with 100% N_2. Filter sterilize.

Preparation of Medium: Prepare and dispense medium under 80% N_2 + 20% CO_2. Combine component solutions A-G; do not add $NaHCO_3$ solution, $Na_2S·9H_2O$ solution, trace elements solution, vitamin solution, methanol solution, seven vitamin solution, and dithionite solution. Mix thoroughly. Flush medium with 80% N_2 + 20% CO_2 for 5 min. Distribute into serum bottles. Autoclave for 15 min at 15 psi pressure–121°C. Cool to 25°C. For every 10.0mL medium aseptically and anaerobically add from sterile anaerobic solution: 0.5mL $NaHCO_3$ solution, 0.1mL trace elements solution, 0.1mL vitamin solution, 0.01mL seven vitamin solution, 0.01mL dithionite solution, 0.15mL $Na_2S·9H_2O$ solution, and 0.05mL methanol solution. Final pH should be 7.0.

Use: For the cultivation of *Methanosarcina semesiae.*

Methanosarcina Thermophilic Medium (DSMZ Medium 164)

Composition per 1051.0mL:

NaCl	2.25g
Yeast extract	2.0g

Casitone	2.0g
NH_4Cl	0.5g
$MgSO_4·7H_2O$	0.5g
K_2HPO_4	0.348g
$CaCl_2·2H_2O$	0.25g
KH_2PO_4	0.227g
$FeSO_4·7H_2O$	0.002g
Resazurin	0.001g
Rumen fluid, clarified	50.0mL
Methanol solution	10.0mL
Vitamin solution	10.0mL
$NaHCO_3$ solution	10.0mL
L-Cysteine solution	10.0mL
$Na_2S·9H_2O$ solution	10.0mL
Trace elements solution SL-10	1.0mL

pH 6.5–6.8 at 25°C

Vitamin Solution:

Composition per liter:

Pyridoxine-HCl	10.0mg
Thiamine-HCl·$2H_2O$	5.0mg
Riboflavin	5.0mg
Nicotinic acid	5.0mg
D-Ca-pantothenate	5.0mg
p-Aminobenzoic acid	5.0mg
Lipoic acid	5.0mg
Biotin	2.0mg
Folic acid	2.0mg
Vitamin B_{12}	0.1mg

Preparation of Vitamin Solution: Add components to distilled/deionized water and bring volume to 1.0L. Mix thoroughly. Sparge with 80% N_2 + 20% CO_2. Filter sterilize.

Methanol Solution:

Composition per 100.0mL:

Methanol	50.0mL

Preparation of Methanol Solution: Add methanol to distilled/deionized water and bring volume to 100.0mL. Mix thoroughly. Sparge with 100% N_2. Autoclave for 15 min at 15 psi pressure–121°C.

$Na_2S·9H_2O$ Solution:

Composition per 100.0mL:

$Na_2S·9H_2O$	3.0g

Preparation of $Na_2S·9H_2O$ Solution: Add $Na_2S·9H_2O$ to distilled/deionized water and bring volume to 100.0mL. Mix thoroughly. Sparge with 100% N_2. Autoclave for 15 min at 15 psi pressure–121°C. Before use, neutralize to pH 7.0 with sterile HCl.

$NaHCO_3$ Solution:

Composition per 100.0mL:

$NaHCO_3$	8.5g

Preparation of $NaHCO_3$ Solution: Add $NaHCO_3$ to distilled/deionized water and bring volume to 100.0mL. Mix thoroughly. Autoclave for 15 min at 15 psi pressure–121°C. Cool to 25°C. Must be prepared freshly.

Trace Elements Solution SL-10:

Composition per liter:

$FeCl_2·4H_2O$	1.5g
H_3BO_3	300.0mg
$CoCl_2·6H_2O$	190.0mg
$MnCl_2·4H_2O$	100.0mg
$ZnCl_2$	70.0mg
$Na_2MoO_4·2H_2O$	36.0mg
$NiCl_2·6H_2O$	24.0mg
$CuCl_2·2H_2O$	2.0mg
HCl (25% solution)	10.0mL

Preparation of Trace Elements Solution SL-10: Add $FeCl_2·4H_2O$ to 10.0mL of HCl solution. Mix thoroughly. Add distilled/deionized water and bring volume to 1.0L. Add remaining components. Mix thoroughly. Sparge with 100% N_2. Autoclave for 15 min at 15 psi pressure–121°C.

L-Cysteine Solution:

Composition per 10.0mL:

L-Cysteine·HCl·H_2O	0.3g

Preparation of L-Cysteine Solution: Add L-cysteine·HCl·H_2O to distilled/deionized water and bring volume to 10.0mL. Mix thoroughly. Sparge with 100% N_2. Autoclave for 15 min at 15 psi pressure–121°C.

Preparation of Medium: Prepare and dispense medium under 80% N_2 + 20% CO_2 gas atmosphere. Add components, except vitamin solution, $NaHCO_3$ solution, methanol solution, L-cysteine-HCl·H_2O solution, $Na_2S·9H_2O$ solution, and trace elements solution SL-10, to distilled/deionized water and bring volume to 1.0L. Mix thoroughly. Adjust pH to 6.5. Sparge with 80% N_2 + 20% CO_2. Autoclave for 15 min at 15 psi pressure–121°C. Aseptically and anaerobically add 10.0mL vitamin solution, 10.0mL $NaHCO_3$ solution, 10.0mL methanol solution, 10.0mL L-cysteine-HCl·H_2O solution, 10.0mL $Na_2S·9H_2O$ solution, and 1.0mL trace elements solution SL-10. Mix thoroughly. Aseptically and anaerobically distribute into sterile tubes or bottles. After inoculation, flush and repressurize the gas head space of culture bottles with sterile 80% H_2 + 20% CO_2 to 1 bar overpressure. Alternately, the medium without vitamin solution, $NaHCO_3$ solution, methanol solution, L-cysteine-HCl·H_2O solution, $Na_2S·9H_2O$ solution, and trace elements solution SL-10 can be distributed to tubes anaerobically prior to autoclaving. After autoclaving in tubes the appropriate volumes of the individual solutions can be injected through the stoppers so that the final concentrations of the medium are achieved.

Use: For the cultivation of *Methanosarcina thermophila.*

Methanosarcina Thermophilic Medium (DSMZ Medium 164)

Composition per 1051.0mL:

NaCl	2.25g
Yeast extract	2.0g
Casitone	2.0g
NH_4Cl	0.5g
$MgSO_4·7H_2O$	0.5g
K_2HPO_4	0.348g
$CaCl_2·2H_2O$	0.25g
KH_2PO_4	0.227g
$FeSO_4·7H_2O$	0.002g
Resazurin	0.001g
Sludge fluid	50.0mL
Methanol solution	10.0mL
Vitamin solution	10.0mL
$NaHCO_3$ solution	10.0mL
L-Cysteine solution	10.0mL
$Na_2S·9H_2O$ solution	10.0mL
Trace elements solution SL-10	1.0mL

pH 6.5–6.8 at 25°C

Sludge Fluid:

Composition per 500.0mL:

Yeast extract	2.0g
Sludge	500.0mL

Preparation of Sludge Fluid: Add yeast extract to sludge from an anaerobic digester. Gas with nitrogen gas for a few minutes. Incubate at 37°C for 24 hr. Centrifuge the sludge at 13,000 x g. Autoclave for 15 min at 15 psi pressure–121°C. Place the resulting, clear supernatant in screw-capped vessels under nitrogen gas. The sludge fluid can be stored at room temperature in the dark.

Vitamin Solution:

Composition per liter:

Pyridoxine-HCl	10.0mg
Thiamine-HCl·$2H_2O$	5.0mg
Riboflavin	5.0mg
Nicotinic acid	5.0mg
D-Ca-pantothenate	5.0mg
p-Aminobenzoic acid	5.0mg
Lipoic acid	5.0mg
Biotin	2.0mg
Folic acid	2.0mg
Vitamin B_{12}	0.1mg

Preparation of Vitamin Solution: Add components to distilled/deionized water and bring volume to 1.0L. Mix thoroughly. Sparge with 80% N_2 + 20% CO_2. Filter sterilize.

Methanol Solution:

Composition per 100.0mL:

Methanol	50.0mL

Preparation of Methanol Solution: Add methanol to distilled/deionized water and bring volume to 100.0mL. Mix thoroughly. Sparge with 100% N_2. Autoclave for 15 min at 15 psi pressure–121°C.

$Na_2S·9H_2O$ Solution:

Composition per 100.0mL:

$Na_2S·9H_2O$	3.0g

Preparation of $Na_2S·9H_2O$ Solution: Add $Na_2S·9H_2O$ to distilled/deionized water and bring volume to 100.0mL. Mix thoroughly. Sparge with 100% N_2. Autoclave for 15 min at 15 psi pressure–121°C. Before use, neutralize to pH 7.0 with sterile HCl.

$NaHCO_3$ Solution:

Composition per 100.0mL:

$NaHCO_3$	8.5g

Preparation of $NaHCO_3$ Solution: Add $NaHCO_3$ to distilled/deionized water and bring volume to 100.0mL. Mix thoroughly. Autoclave for 15 min at 15 psi pressure–121°C. Cool to 25°C. Must be prepared freshly.

Trace Elements Solution SL-10:

Composition per liter:

$FeCl_2·4H_2O$	1.5g
H_3BO_3	300.0mg
$CoCl_2·6H_2O$	190.0mg
$MnCl_2·4H_2O$	100.0mg
$ZnCl_2$	70.0mg
$Na_2MoO_4·2H_2O$	36.0mg
$NiCl_2·6H_2O$	24.0mg
$CuCl_2·2H_2O$	2.0mg
HCl (25% solution)	7.7mL

Preparation of Trace Elements Solution SL-10: Add 1.5g of $FeCl_2·4H_2O$ to 10.0mL of HCl solution. Mix thoroughly. Add distilled/deionized water and bring volume to 1.0L. Add remaining components. Mix thoroughly. Sparge with 100% N_2. Autoclave for 15 min at 15 psi pressure–121°C.

L-Cysteine Solution:

Composition per 10.0mL:

L-Cysteine·HCl·H_2O	0.3g

Preparation of L-Cysteine Solution: Add L-cysteine·HCl·H_2O to distilled/deionized water and bring volume to 10.0mL. Mix thoroughly. Sparge with 100% N_2. Autoclave for 15 min at 15 psi pressure–121°C.

Preparation of Medium: Prepare and dispense medium under 80% N_2 + 20% CO_2 gas atmosphere. Add components, except vitamin solution, $NaHCO_3$ solution, methanol solution, L-cysteine-HCl·H_2O solution, $Na_2S·9H_2O$ solution, and trace elements solution SL-10, to distilled/deionized water and bring volume to 1.0L. Mix thoroughly. Adjust pH to 6.5. Sparge with 80% N_2 + 20% CO_2. Autoclave for 15 min at 15 psi pressure–121°C. Aseptically and anaerobically add 10.0mL vitamin solution, 10.0mL $NaHCO_3$ solution, 10.0mL methanol solution, 10.0mL L-cysteine-HCl·H_2O solution, 10.0mL $Na_2S·9H_2O$ solution, and 1.0mL trace elements solution SL-10. Mix thoroughly. Aseptically and anaerobically distribute into sterile tubes or bottles. After inoculation, flush and repressurize the gas head space of culture bottles with sterile 80% H_2 + 20% CO_2 to 1 bar overpressure. Alternately, the medium without vitamin solution, $NaHCO_3$ solution, methanol solution, L-cysteine-HCl·H_2O solution, $Na_2S·9H_2O$ solution, and trace elements solution SL-10 can be distributed to tubes anaerobically prior to autoclaving. After autoclaving in tubes the appropriate volumes of the individual solutions can be injected through the stoppers so that the final concentrations of the medium are achieved.

Use: For the cultivation of *Methanosarcina thermophila.*

Methanosphaera Medium 1

Composition per liter:

KH_2PO_4	2.8g
Trypticase™	2.0g
Yeast extract	2.0g
NH_4Cl	1.0g
K_2HPO_4	0.6g
NaCl	0.6g
Sodium acetate	0.5g
Sodium formate	0.5g
$(NH_4)_2SO_4$	0.3g
$MgSO_4·7H_2O$	0.15g
$CaCl_2·2H_2O$	0.076g
$FeSO_4·7H_2O$	3.0mg
Na_2SeO_4	1.9mg
Resazurin	1.0mg
$NiCl_2·6H_2O$	0.7mg
Rumen fluid, clarified	100.0mL
Trace elements solution	10.0mL
L-Cysteine·HCl·H_2O solution	10.0mL
$Na_2S·9H_2O$ solution	10.0mL
Methanol solution	5.0mL
Vitamin solution	1.0mL

pH 6.5 ± 0.2 at 25°C

Trace Elements Solution:

Composition per liter:

$MgSO_4 \cdot 7H_2O$	3.0g
Nitrilotriacetic acid	1.5g
NaCl	1.0g
$MnSO_4 \cdot 2H_2O$	0.5g
$CoSO_4 \cdot 7H_2O$	0.18g
$ZnSO_4 \cdot 7H_2O$	0.18g
$CaCl_2 \cdot 2H_2O$	0.1g
$FeSO_4 \cdot 7H_2O$	0.1g
$NiCl_2 \cdot 6H_2O$	0.025g
$KAl(SO_4)_2 \cdot 12H_2O$	0.02g
$CuSO_4 \cdot 5H_2O$	0.01g
H_3BO_3	0.01g
$Na_2MoO_4 \cdot 2H_2O$	0.01g
$Na_2SeO_3 \cdot 5H_2O$	0.3mg

Preparation of Trace Elements Solution: Add nitrilotriacetic acid to 500.0mL of distilled/deionized water. Adjust pH to 6.5 with KOH. Add remaining components. Adjust pH to 7.0. Add distilled/deionized water to 1.0L.

L-Cysteine·HCl·H_2O Solution:

Composition per 10.0mL:

L-Cysteine·HCl·H_2O	0.875g

Preparation of L-Cysteine·HCl·H_2O Solution: Add L-cysteine·HCl·H_2O to distilled/deionized water and bring volume to 10.0mL. Mix thoroughly. Autoclave under 80% N_2 + 20% CO_2 for 15 min at 15 psi pressure–121°C.

$Na_2S \cdot 9H_2O$ Solution:

Composition per 10.0mL:

$Na_2S \cdot 9H_2O$	0.375g

Preparation of $Na_2S \cdot 9H_2O$ Solution: Add $Na_2S \cdot 9H_2O$ to distilled/deionized water and bring volume to 10.0mL. Mix thoroughly. Sparge with 80% N_2 + 20% CO_2. Autoclave for 15 min at 15 psi pressure–121°C.

Methanol Solution:

Composition per 10.0mL:

Methanol	10.0mL

Preparation of Methanol Solution: Sparge 10.0mL of methanol with 100% N_2. Autoclave for 15 min at 15 psi pressure–121°C.

Vitamin Solution:

Composition per 10.0mL:

Calcium-D-pantothenate	20.0mg
Nicotinamide	20.0mg
Pyridoxine·HCl	20.0mg
Riboflavin	20.0mg
Thiamine·HCl	20.0mg
Biotin	10.0mg
p-Aminobenzoic acid	1.0mg
Folic acid	0.5mg
Vitamin B_{12}	0.2mg

Preparation of Vitamin Solution: Add components to distilled/deionized water and bring volume to 1.0L. Mix thoroughly. Filter sterilize. Sparge with 80% N_2 + 20% CO_2.

Preparation of Medium: Add components, except methanol solution, L-cysteine·HCl·H_2O solution, and $Na_2S \cdot 9H_2O$ solution, to distilled/deionized water and bring volume to 975.0mL. Mix thoroughly. Gently heat and bring to boiling. Continue boiling for 5 min. Cool to room temperature while sparging with 100% N_2. Autoclave for 15 min at 15 psi pressure–121°C. Aseptically and anaerobically add 10.0mL of sterile L-cysteine·HCl·H_2O solution, 10.0mL of sterile $Na_2S \cdot 9H_2O$ solution, and 5.0mL of sterile methanol solution. Mix thoroughly. Aseptically and anaerobically distribute into sterile screw-cappped bottles under 100% N_2.

Use: For the cultivation and maintenance of *Methanosphaera stadtmanae.*

Methanosphaera Medium II

Composition per 1005.5mL:

$NaHCO_3$	3.0g
Sodium acetate·$3H_2O$	3.0g
Trypticase™	1.0g
Yeast extract	1.0g
$Fe(NH_4)_2(SO_4)_2 \cdot 7H_2O$	2.0mg
Resazurin	1.0mg
$NiCl_2 \cdot 6H_2O$	0.2mg
$Na_2SeO_3 \cdot 5H_2O$ (1m*M*)	0.1mg
Mineral solution 1	40.0mL
Mineral solution 2	40.0mL
Trace elements solution	10.0mL
Vitamin solution	10.0mL
L-Cysteine·HCl·H_2O solution	10.0mL
$Na_2S \cdot 9H_2O$ solution	10.0mL
Methanol solution	5.0mL

pH 7.0 ± 0.2 at 25°C

Mineral Solution 1:

Composition per liter:

K_2HPO_4	6.0g

Preparation of Mineral Solution 1: Add K_2HPO_4 to distilled/deionized water and bring volume to 1.0L. Mix thoroughly.

Mineral Solution 2:

Composition per liter:

NaCl	12.0g
KH_2PO_4	6.0g
$(NH_4)_2SO_4$	6.0g
$MgSO_4 \cdot 7H_2O$	2.6g
$CaCl_2 \cdot 2H_2O$	0.16g

Preparation of Mineral Solution 2: Add components to distilled/deionized water and bring volume to 1.0L. Mix thoroughly.

Trace Elements Solution:

Composition per liter:

$MgSO_4 \cdot 7H_2O$	3.0g
Nitrilotriacetic acid	1.5g
NaCl	1.0g
$MnSO_4 \cdot 2H_2O$	0.5g
$CoSO_4 \cdot 7H_2O$	0.18g
$ZnSO_4 \cdot 7H_2O$	0.18g
$CaCl_2 \cdot 2H_2O$	0.1g
$FeSO_4 \cdot 7H_2O$	0.1g
$NiCl_2 \cdot 6H_2O$	0.025g
$KAl(SO_4)_2 \cdot 12H_2O$	0.02g
$CuSO_4 \cdot 5H_2O$	0.01g
H_3BO_3	0.01g
$Na_2MoO_4 \cdot 2H_2O$	0.01g
$Na_2SeO_3 \cdot 5H_2O$	0.3mg

Preparation of Trace Elements Solution: Add nitrilotriacetic acid to 500.0mL of distilled/deionized water. Adjust pH to 6.5 with KOH.

Add remaining components. Adjust pH to 7.0. Add distilled/deionized water to 1.0L.

Vitamin Solution:

Composition per liter:

Pyridoxine·HCl	10.0mg
Calcium DL-pantothenate	5.0mg
Lipoic acid	5.0mg
Nicotinic acid	5.0mg
p-Aminobenzoic acid	5.0mg
Riboflavin	5.0mg
Thiamine·HCl	5.0mg
Biotin	2.0mg
Folic acid	2.0mg
Vitamin B_{12}	0.1mg

Preparation of Vitamin Solution: Add components to distilled/deionized water and bring volume to 1.0L. Mix thoroughly. Sparge with 80% N_2 + 20% CO_2.

L-Cysteine·HCl·H_2O Solution:

Composition per 10.0mL:

L-Cysteine·HCl·H_2O	0.3g

Preparation of L-Cysteine·HCl·H_2O Solution: Add L-cysteine·HCl·H_2O to distilled/deionized water and bring volume to 10.0mL. Mix thoroughly. Autoclave under 80% N_2 + 20% CO_2 for 15 min at 15 psi pressure–121°C.

$Na_2S·9H_2O$ Solution:

Composition per 10.0mL:

$Na_2S·9H_2O$	0.3g

Preparation of $Na_2S·9H_2O$ Solution: Add $Na_2S·9H_2O$ to distilled/deionized water and bring volume to 10.0mL. Mix thoroughly. Sparge with 80% N_2 + 20% CO_2. Autoclave for 15 min at 15 psi pressure–121°C.

Methanol Solution:

Composition per 10.0mL:

Methanol	10.0mL

Preparation of Methanol Solution: Sparge 10.0mL of methanol with 100% N_2. Autoclave for 15 min at 15 psi pressure–121°C.

Preparation of Medium: Add components, except L-cysteine·HCl·H_2O solution, $Na_2S·9H_2O$ solution, and methanol solution, to distilled/deionized water and bring volume to 975.0mL. Mix thoroughly. Gently heat and bring to boiling. Continue boiling for 5 min. Cool to room temperature while sparging with 100% N_2. Autoclave for 15 min at 15 psi pressure–121°C. Aseptically and anaerobically add 10.0mL of sterile L-cysteine·HCl·H_2O solution, 10.0mL of sterile $Na_2S·9H_2O$ solution, and 5.0mL of sterile methanol solution. Mix thoroughly. Aseptically and anaerobically distribute into sterile screw-cappped bottles under 100% N_2.

Use: For the cultivation and maintenance of *Methanosphaera cuniculi*.

Methanosphaerula Medium (DSMZ Medium 1094)

Composition per liter:

Solution A	1.0L
Solution B	1.0mL
Solution C	12.55mL
Solution D	20.0mL
Solution E	10.0mL
Solution F	10.0mL
Solution G	10.0mL
Solution H	10.0mL

pH 6.9 ± 0.2 at 25°C

Solution A:

Composition per liter:

NH_4 Cl	26.8mg
KH_2PO_4	13.6mg
KCl	1.5mg

Preparation of Solution A: Add components to distilled/deionized water and bring volume to 1.0L. Mix thoroughly.

Solution B:

Composition per liter:

Na_2EDTA	37.229g
$KAl(SO_4)_2·12H_2O$	3.446g
$CaCl_2·2H_2O$	2.336g
$MgSO_4·7H_2O$	1.556g
$FeCl_2·4H_2O$	1.344g
$CoCl_2·6H_2O$	0.24g
$ZnCl_2$	0.075g
$MnSO_4·2H_2O$	0.026g
$NiCl_2·6H_2O$	0.024g
$NaMoO_4·2H_2O$	0.024g
H_3BO_3	0.019g
$CuSO_4$	0.009g

Preparation of Solution B: Add components to distilled/deionized water and bring volume to 1.0L. Mix thoroughly.

Solution C:

Composition per 12.55mL:

TRIS-HCl, 1.0*M*, pH 8.0	7.2mL
Sodium nitrilotriacetic acid, 0.5M	4.8mL
$TiCl_3$, 15% in HCl	0.55mL

Preparation of Solution C: Combine solutions. Mix thoroughly. Filter sterilize. Store under N_2 gas atmosphere.

Solution D:

Composition per 20.0mL:

MES	3.905g
NaOH	0.8g

Preparation of Solution D: Add components to distilled/deionized water and bring volume to 20.0mL. Mix thoroughly. Adjust to pH 7.5 with NaOH. Filter sterilize. Store under N_2 gas atmosphere.

Solution E:

Composition per liter:

Pyridoxine-HCl	10.0mg
Thiamine-HCl·$2H_2O$	5.0mg
Riboflavin	5.0mg
Nicotinic acid	5.0mg
D-Ca-pantothenate	5.0mg
p-Aminobenzoic acid	5.0mg
Lipoic acid	5.0mg
Biotin	2.0mg
Folic acid	2.0mg
Vitamin B_{12}	0.1mg

Preparation of Solution E: Add components to distilled/deionized water and bring volume to 1.0L. Mix thoroughly. Sparge with 100% N_2. Filter sterilize. Store under N_2 gas atmosphere.

Solution F:

Composition per 10.0mL:

2-Mercaptoethanesulfonic acid (coenzyme M) 0.082g

Preparation of Solution F: Add components to distilled/deionized water and bring volume to 10.0mL. Mix thoroughly. Sparge with N_2 gas. Autoclave for 15 min at 15 psi pressure–121°C. Store under N_2 gas atmosphere.

Solution G:

Composition per 10.0mL:

Na-acetate 0.68g

Preparation of Solution G: Add Na-acetate to distilled/deionized water and bring volume to 10.0mL. Mix thoroughly. Sparge with N_2 gas. Autoclave for 15 min at 15 psi pressure–121°C. Store under N_2 gas atmosphere.

Solution H:

Composition per 10.0mL:

$Na_2S·9H_2O$ 0.96mg

Preparation of Solution H: Add $Na_2S·9H_2O$ to distilled/deionized water and bring volume to 10.0mL. Mix thoroughly. Sparge with N_2 gas. Filter sterilize. Store in sealed vial under N_2 gas atmosphere.

Preparation of Medium: Add 1.0mL of solution B to 1.0L of solution A. Gently heat and bring to boiling. Boil for 1 min. Cool to room temperature under 80% N_2 + 20% CO_2 gas mixture. Dispense under same gas atmosphere into culture vessels. Autoclave for 15 min at 15 psi pressure–121°C. Aseptically add solutions C to H in the sequence indicated and let medium equilibrate overnight. After inoculation pressurize vials to 1 bar overpressure with 80% H_2 + 20% CO_2 gas mixture. The final pH should be 5.7.

Use: For the cultivation of *Methanosphaerula* spp.

Methanospirillum hungatei JMA Medium

Composition per liter:

Sodium acetate 2.5g
NH_4Cl 1.9g
Mineral solution 1 .... 75.0mL
Mineral solution 2 .... 75.0mL
Wolfe's mineral solution 10.0mL
Vitamin solution 10.0mL
L-Cysteine·HCl·H_2O solution 10.0mL
$Na_2S·9H_2O$ solution 10.0mL
$FeSO_4·6H_2O$ solution 6.0mL
Na_2CO_3 solution 4.8mL
Resazurin 1.0mL
$NiCl_2·6H_2O$ solution 0.1mL

$FeSO_4·6H_2O$ Solution:

Composition per 100.0mL:

$FeSO_4·6H_2O$ 0.2g

Preparation of $FeSO_4·6H_2O$ Solution: Add $FeSO_4·6H_2O$ to distilled/deionized water and bring volume to 100.0mL. Mix thoroughly.

$NiCl_2·6H_2O$ Solution:

Composition per liter:

$NiCl_2·6H_2O$ 1.2g

Preparation of $NiCl_2·6H_2O$ Solution: Add $NiCl_2·6H_2O$ to distilled/deionized water and bring volume to 1.0L. Mix thoroughly.

Mineral Solution 1:

Composition per liter:

K_2HPO_4 .... 39.0g

Preparation of Mineral Solution 1: Add K_2HPO_4 to distilled/deionized water and bring volume to 1.0L. Mix thoroughly.

Mineral Solution 2:

Composition per liter:

KH_2PO_4 .... 24.0g
$(NH_4)_2SO_4$ 6.0g
$MgSO_4·7H_2O$ 1.2g
$CaCl_2·2H_2O$ 0.79g
NaCl 0.59g

Preparation of Mineral Solution 2: Add components to distilled/deionized water and bring volume to 1.0L. Mix thoroughly.

Wolfe's Mineral Solution:

Composition per liter:

$MgSO_4·7H_2O$ 3.0g
Nitrilotriacetic acid 1.5g
NaCl 1.0g
$MnSO_4·2H_2O$ 0.5g
$CoCl_2·6H_2O$ 0.1g
$ZnSO_4·7H_2O$ 0.1g
$CaCl_2·2H_2O$ 0.1g
$FeSO_4·7H_2O$ 0.1g
$NiCl_2·6H_2O$ 0.025g
$KAl(SO_4)_2·12H_2O$ 0.02g
$CuSO_4·5H_2O$ 0.01g
H_3BO_3 .... 0.01g
$Na_2MoO_4·2H_2O$ 0.01g
$Na_2SeO_3·5H_2O$ 0.3mg

Preparation of Wolfe's Mineral Solution: Add nitrilotriacetic acid to 500.0mL of distilled/deionized water. Adjust pH to 6.5 with KOH. Add remaining components one at a time. Add distilled/deionized water to 1.0L. Adjust pH to 6.8.

Vitamin Solution:

Composition per liter:

Calcium DL-pantothenate 5.0g
Vitamin B_{12} 0.1g
Pyridoxine·HCl 10.0mg
p-Aminobenzoic acid 5.0mg
Lipoic acid 5.0mg
Nicotinic acid 5.0mg
Riboflavin 5.0mg
Thiamine·HCl 5.0mg
Biotin 2.0mg
Folic acid 2.0mg
Vitamin B_{12} 0.4mg

Preparation of Vitamin Solution: Add components to distilled/deionized water and bring volume to 1.0L. Mix thoroughly. Sparge with 100% N_2. Filter sterilize.

Na_2CO_3 Solution:

Composition per liter:

Na_2CO_3 .... 0.84g

Preparation of Na_2CO_3 Solution: Add Na_2CO_3 to distilled/deionized water and bring volume to 10.0mL. Mix thoroughly. Sparge with 80% N_2 + 20% CO_2. Autoclave for 15 min at 15 psi pressure–121°C.

$Na_2S·9H_2O$ Solution:

Composition per 10.0mL:

$Na_2S·9H_2O$ 0.2g

Preparation of $Na_2S·9H_2O$ Solution: Add $Na_2S·9H_2O$ to distilled/deionized water and bring volume to 10.0mL. Mix thoroughly. Sparge with 100% N_2. Autoclave for 15 min at 15 psi pressure–121°C. Before use, neutralize to pH 7.0 with sterile HCl.

L-Cysteine·HCl·H_2O Solution:

Composition per 10.0mL:

L-Cysteine·HCl·H_2O	0.22g

Preparation of L-Cysteine·HCl·H_2O Solution: Add L-cysteine·HCl·H_2O to distilled/deionized water and bring volume to 10.0mL. Mix thoroughly. Sparge with 100% N_2. Autoclave for 15 min at 15 psi pressure–121°C.

Preparation of Medium: Prepare and dispense medium under 80% N_2 + 20% CO_2. Add components, except Na_2CO_3 solution, L-cysteine·HCl solution, and $Na_2S·9H_2O$ solution, to distilled/deionized water and bring volume to 975.2mL. Mix thoroughly. Gently heat and bring to boiling. Cool while sparging with 80% N_2 + 20% CO_2. Autoclave for 15 min at 15 psi pressure–121°C. Aseptically and anaerobically add 4.8mL of sterile Na_2CO_3 soltuion, 10.0mL of sterile L-cysteine·HCl solution, and 10.0mL of sterile $Na_2S·9H_2O$ solution. Mix thoroughly. Aseptically and anaerobically distribute into sterile tubes or bottles.

Use: For the cultivation of *Methanospirillum hungatei.*

Methanospirillum hungatei Medium

Composition per 100.0mL:

Na_2CO_3	0.4g
Sodium formate	0.2g
Pancreatic digest of casein	0.2g
Yeast extract	0.2g
NaCl	0.05g
L-Cysteine·HCl·H_2O	0.03g
K_2HPO_4	0.02g
KH_2PO_4	0.02g
$(NH_4)_2SO_4$	0.02g
$MgSO_4·7H_2O$	9.0mg
$CaCl_2·2H_2O$	6.0mg
Resazurin	0.1mg
$Na_2S·9H_2O$ solution	10.0mL
Vitamin solution	1.0mL
Trace metal solution	1.0mL

pH 7.0 ± 0.2 at 25°C

$Na_2S·9H_2O$ Solution:

Composition per 10.0.mL:

$Na_2S·9H_2O$	0.03g

Preparation of $Na_2S·9H_2O$ Solution: Add $Na_2S·9H_2O$ to distilled/deionized water and bring volume to 10.0mL. Mix thoroughly. Autoclave for 15 min at 15 psi pressure–121°C. Cool to 25°C.

Vitamin Solution:

Composition per 1000.0mL:

Pyridoxine·HCl	1.0mg
p-Aminobenzoic acid	0.5mg
Calcium-D-pantothenate	0.5mg
Nicotinic acid	0.5mg
Riboflavin	0.5mg
Thiamine·HCl	0.5mg
Thioctic acid	0.5mg
Biotin	0.2mg
Folic acid	0.2mg
Vitamin B_{12}	0.01mg

Preparation of Vitamin Solution: Add components to distilled/deionized water and bring volume to 1.0L. Mix thoroughly. Filter sterilize.

Trace Metal Solution:

Composition per liter:

$K_2HPO_4·3H_2O$	9.0g
K_2HPO_4	6.0g
NH_4Cl	5.0g
$MgCl_2·6H_2O$	1.0g
$CaCl_2·2H_2O$	0.01g

Preparation of Trace Metal Solution: Add components to distilled/deionized water and bring volume to 1.0L. Mix thoroughly.

Preparation of Medium: Prepare and distribute medium anaerobically under 80% H_2 + 20% CO_2. Add components, except $Na_2S·9H_2O$ solution, to distilled/deionized water and bring volume to 90.0mL. Mix thoroughly. Gently heat and bring to boiling. Continue boiling until resazurin turns colorless, indicating reduction. Autoclave for 15 min at 15 psi pressure–121°C. Cool to 25°C. Aseptically add 10.0mL of sterile $Na_2S·9H_2O$ solution. Mix thoroughly. Aseptically and anaerobically distribute into sterile tubes or flasks.

Use: For the cultivation of *Methanospirillum hungatei.*

Methanospirillum hungatei SAM Medium

Composition per liter:

Na_2CO_3	2.63
Sodium acetate·$3H_2O$	2.5g
$(NH_4)_2SO_4$	0.45g
K_2HPO_4	0.3g
KH_2PO_4	0.18g
$MgSO_4·9H_2O$	0.12g
$CaCl_2·2H_2O$	0.06g
NaCl	0.05g
Na_2CO_3 solution	20.0mL
Trace minerals solution	10.0mL
Vitamin solution	10.0mL
L-Cysteine/Na_2S solution	10.0mL
$FeSO_4·7H_2O$ solution	5.0mL
Resazurin solution	0.2mL

pH 7.1 ± 0.2 at 25°C

Na_2CO_3 Solution:

Composition per liter:

Na_2CO_3	100.0g

Preparation of Na_2CO_3 Solution: Bring 1.0L of distilled/deionized water to boiling. Cool to room temperature while sparging with 80% H_2 + 20% CO_2. Add Na_2CO_3. Distribute into serum bottles fitted with butyl rubber stoppers and aluminum seals. Do not grease stoppers. Autoclave for 15 min at 15 psi pressure–121°C. Store at room temperature in an anaerobic chamber.

Trace Minerals Solution:

Composition per liter:

Nitrolotriacetic acid	1.5g
$MnSO_4·2H_2O$	0.5g
$Na_2MoO_4·2H_2O$	0.24g
$Na_2WO_4·2H_2O$	0.165g
Na_2SeO_3	0.15g
$CoCl_2·6H_2O$	0.1g
$NiCl_2·6H_2O$	0.1g
$ZnSO_4·7H_2O$	0.1g
$AlK(SO_4)_2·12H_2O$	0.01g

$CuSO_4 \cdot 5H_2O$ 0.01g
H_3BO_3 .... 0.01g

Preparation of Trace Minerals Solution: Add nitrilotriacetic acid to 500.0mL of distilled/deionized water. Adjust pH to 6.5 with KOH. Add remaining components. Mix thoroughly. Add distilled/deionized water to 1.0L. Adjust pH to 7.0.

Vitamin Solution:
Composition per liter:
Vitamin B_{12} 0.1g
Pyridoxine·HCl 10.0mg
p-Aminobenzoic acid 5.0mg
Lipoic acid 5.0mg
Nicotinic acid 5.0mg
Riboflavin 5.0mg
Thiamine·HCl 5.0mg
Biotin 2.0mg
Folic acid 2.0mg
Vitamin B_{12} 0.4mg

Preparation of Vitamin Solution: Add components to distilled/deionized water and bring volume to 1.0L. Mix thoroughly.

L-Cysteine/Na_2S Solution:
Composition per liter:
L-Cysteine·HCl 1.25g
$Na_2S \cdot 7H_2O$ 1.25g

Preparation of L-Cysteine/Na_2S Solution: Gently heat and bring 75.0mL of distilled/deionized water to boiling. Add L-cysteine·HCl. Adjust pH to 10.0 with 3*N* NaOH. Add $Na_2S \cdot 9H_2O$. Mix thoroughly. Gently heat and bring to boiling. Cool to room temperature while sparging with 100% N_2. Bring volume to 100.0mL with distilled/deionized water. Gently heat and bring to boiling. Cool to 60°C while sparging with 100% N_2. Anaerobically distribute into serum bottles. Autoclave for 15 min at 15 psi pressure–121°C.

$FeSO_4 \cdot 6H_2O$ Solution:
Composition per 100.0mL:
$FeSO_4 \cdot 6H_2O$ 0.2g

Preparation of $FeSO_4 \cdot 6H_2O$ Solution: Add $FeSO_4 \cdot 6H_2O$ to distilled/deionized water and bring volume to 100.0mL. Mix thoroughly.

Resazurin Solution:
Composition per 10.0mL:
Resazurin 10.0mg

Preparation of Resazurin Solution: Add resazurin to distilled/deionized water and bring volume to 10.0mL. Mix thoroughly.

Preparation of Medium: Prepare and dispense medium under 80% H_2 + 20% CO_2. Add components, except Na_2CO_3, Na_2CO_3 solution, and L-cysteine/Na_2S solution, to distilled/deionized water and bring volume to 1.0L. Mix thoroughly. Adjust pH to 7.0. Gently heat and bring to boiling. Continue boiling for 3 min. Cool to 60°C while sparging with 80% H_2 + 20% CO_2. Add Na_2CO_3 and L-cysteine/Na_2S solution. Mix thoroughly. Anaerobically distribute 10.0mL volumes into anaerobic tubes. Autoclave for 20 min at 15 psi pressure–121°C. Aseptically and anaerobically add 0.2mL of sterile Na_2CO_3 solution to each tube. Mix thoroughly. Adjust pH to 7.1.

Use: For the cultivation of *Methanospirillum hungatei.*

Methanospirillum SK Medium

Composition per liter:
NaCl 18.0g
Isopropanol 7.5g
$NaHCO_3$ 5.0g
$MgCl_2 \cdot 6H_2O$ 4.0g
$MgSO_4 \cdot 7H_2O$ 3.45g
Trypticase™ 2.0g
Yeast extract 2.0g
Sodium acetate 1.0g
L-Cysteine·HCl 0.5g
KCl 0.335g
NH_4Cl 0.25g
$CaCl_2 \cdot 2H_2O$ 0.14g
K_2HPO_4 .... 0.14g
$Fe(NH_4)_2(SO_4)_2 \cdot 7H_2O$ 2.0mg
Resazurin 1.0mg
Trace elements solution 10.0mL
Vitamin solution 10.0mL
$Na_2S \cdot 9H_2O$ solution 10.0mL
pH 7.0–7.4 at 25°C

Trace Elements Solution:
Composition per liter:
$MgSO_4 \cdot 7H_2O$ 3.0g
Nitrilotriacetic acid 1.5g
NaCl 1.0g
$MnSO_4 \cdot 2H_2O$ 0.5g
$CoSO_4 \cdot 7H_2O$ 0.18g
$ZnSO_4 \cdot 7H_2O$ 0.18g
$CaCl_2 \cdot 2H_2O$ 0.1g
$FeSO_4 \cdot 7H_2O$ 0.1g
$NiCl_2 \cdot 6H_2O$ 0.025g
$KAl(SO_4)_2 \cdot 12H_2O$ 0.02g
$CuSO_4 \cdot 5H_2O$ 0.01g
H_3BO_3 .... 0.01g
$Na_2MoO_4 \cdot 2H_2O$ 0.01g
$Na_2SeO_3 \cdot 5H_2O$ 0.3mg

Preparation of Trace Elements Solution: Add nitrilotriacetic acid to approximately 500.0mL of distilled/deionized water. Dissolve by adding KOH and adjust pH to 6.5. Add remaining components. Bring volume to 1.0L with additional distilled/deionized water. Adjust pH to 7.0 with KOH.

Vitamin Solution:
Composition per liter:
Pyridoxine·HCl 10.0mg
Calcium DL-pantothenate 5.0mg
Lipoic acid 5.0mg
Nicotinic acid 5.0mg
p-Aminobenzoic acid 5.0mg
Riboflavin 5.0mg
Thiamine·HCl 5.0mg
Biotin 2.0mg
Folic acid 2.0mg
Vitamin B_{12} 0.1mg

Preparation of Vitamin Solution: Add components to distilled/deionized water and bring volume to 1.0L. Mix thoroughly. Sparge with 80% N_2 + 20% CO_2.

$Na_2S \cdot 9H_2O$ Solution:
Composition per 10.0mL:
$Na_2S \cdot 9H_2O$ 0.4g

Preparation of $Na_2S \cdot 9H_2O$ Solution: Add $Na_2S \cdot 9H_2O$ to distilled/deionized water and bring volume to 10.0mL. Mix thoroughly.

Sparge with 80% N_2 + 20% CO_2. Autoclave for 15 min at 15 psi pressure–121°C.

Preparation of Medium: Add components, except $Na_2S{\cdot}9H_2O$ solution, to distilled/deionized water and bring volume to 990.0mL. Mix thoroughly. Sparge with 80% N_2 + 20% CO_2. Autoclave for 15 min at 15 psi pressure–121°C. Aseptically and anaerobically add 10.0mL of sterile $Na_2S{\cdot}9H_2O$ solution. Mix thoroughly.

Use: For the cultivation and maintenance of *Methanospirillum hungatei*.

Methanothermobacter/Methanobacterium Transduction Agar (DSMZ Medium 863)

Composition per 1016.0mL:

Agar	15.0g
Basal salts solution	100.0mL
Sodium carbonate solution	24.0mL
Titanium (III) citrate solution	16.0mL
Trace elements solution	10.0mL
Cysteine solution	5.0mL
NaS solution	5.0mL

pH 7.0 ± 0.2 at 25°C

Sodium Carbonate Solution:

Composition per 100.0mL:

Na_2CO_3	10.6g

Preparation of Sodium Carbonate Solution: Add Na_2CO_3 to distilled/deionized water and bring volume to 100.0mL. Mix thoroughly. Sparge with 100% N_2. Autoclave for 15 min at 15 psi pressure–121°C. Cool to 25°C. Store anaerobically.

Basal Salts Solution:

Composition per liter:

KH_2PO_4	68.0g
NH_4Cl	21.2g
Resazurin	10.0mg

Preparation of Basal Salts Solution: Add components to distilled/deionized water and bring volume to 1.0L. Mix thoroughly. Sparge with 100% N_2.

Trace Elements Solution:

Composition per liter:

Nitrilotriacetic acid	9.55g
$MgCl_2{\cdot}6H_2O$	4.06g
$FeCl_2{\cdot}4H_2O$	0.98g
$NaWO_4{\cdot}6H_2O$	0.264g
$NiCl_2{\cdot}6H_2O$	0.118g
$Na_2Se_2O_3$	0.1g
$Na_2MoO_4{\cdot}4H_2O$	0.024g
$CoCl_2{\cdot}6H_2O$	0.023g

Preparation of Trace Elements Solution: Add nitrilotriacetic acid to 500.0mL of distilled/deionized water. Dissolve by adjusting pH to 7.0 with KOH. Add remaining components. Add distilled/deionized water to 1.0L. Mix thoroughly.

Cysteine Solution:

Composition per 10.0mL:

L-Cysteine·HCl·H_2O	1.0g

Preparation of Cysteine Solution: Add L-cysteine·HCl·H_2O to distilled/deionized water and bring volume to 10.0mL. Mix thoroughly. Adjust pH to 10.0. Sparge with 100% N_2. Store anaerobically.

NaS Solution:

Composition per 10.0mL:

$Na_2S{\cdot}9H_2O$	1.0g

Preparation of NaS Solution: Add $Na_2S{\cdot}9H_2O$ to distilled/deionized water and bring volume to 10.0mL. Mix thoroughly. Adjust pH to 10.0. Sparge with N_2. Store anaerobically.

Titanium (III) Citrate Solution:

Sodium citrate, 0.2*M*	50.0mL
Titanium(III) chloride, 15%	5.0mL

Preparation of Titanium (III) Citrate Solution: Mix solutions. Neutralize with sodium carbonate solution. Sparge with N_2. Store anaerobically.

Preparation of Medium: Prepare and dispense medium under 80% N_2 + 20% CO_2 gas atmosphere. Add components, except titanium citrate solution, to distilled/deionized water and bring volume to 1.0L. Mix thoroughly. Gently heat and bring to boiling. Autoclave for 15 min at 15 psi pressure–121°C. Cool to 60°C. Add 16.0mL titanium citrate solution. Mix thoroughly. Pour into sterile Petri dishes or distribute to sterile tubes. Store anaerobically.

Use: For the transduction of *Methanobacter* spp. and *Methanobacterium* spp.

Methanothermobacter/Methanobacterium Transduction Medium (DSMZ Medium 863)

Composition per liter:

Basal salts solution	100.0mL
Sodium carbonate solution	24.0mL
Trace elements solution	10.0mL
Cysteine solution	5.0mL
NaS solution	5.0mL

pH 7.0 ± 0.2 at 25°C

Sodium Carbonate Solution:

Composition per 100.0mL:

Na_2CO_3	10.6g

Preparation of Sodium Carbonate Solution: Add Na_2CO_3 to distilled/deionized water and bring volume to 100.0mL. Mix thoroughly. Sparge with 100% N_2. Autoclave for 15 min at 15 psi pressure–121°C. Cool to 25°C. Store anaerobically.

Basal Salts Solution:

KH_2PO_4	68.0g
NH_4Cl	21.2g
Resazurin	10.0mg

Preparation of Basal Salts Solution: Add components to distilled/deionized water and bring volume to 1.0L. Mix thoroughly. Sparge with 100% N_2.

Trace Elements Solution:

Nitrilotriacetic acid	9.55g
$MgCl_2{\cdot}6H_2O$	4.06g
$FeCl_2{\cdot}4H_2O$	0.98g
$NaWO_4{\cdot}6H_2O$	0.264g
$NiCl_2{\cdot}6H_2O$	0.118g
$Na_2Se_2O_3$	0.1g
$Na_2MoO_4{\cdot}4H_2O$	0.024g
$CoCl_2{\cdot}6H_2O$	0.023g

Preparation of Trace Elements Solution: Add nitrilotriacetic acid to 500.0mL of distilled/deionized water. Dissolve by adjusting pH to 7.0 with KOH. Add remaining components. Add distilled/deionized water to 1.0L. Mix thoroughly.

Cysteine Solution:

Composition per 10.0mL:

L-Cysteine·HCl·H_2O 1.0g

Preparation of Cysteine Solution: Add L-cysteine·HCl·H_2O to distilled/deionized water and bring volume to 10.0mL. Mix thoroughly. Adjust pH to 10.0. Sparge with 100% N_2. Store anaerobically.

NaS Solution:

Composition per 10.0mL:

$Na_2S·9H_2O$ 1.0g

Preparation of NaS Solution: Add $Na_2S·9H_2O$ to distilled/deionized water and bring volume to 10.0mL. Mix thoroughly. Adjust pH to 10.0. Sparge with N_2. Store anaerobically.

Preparation of Medium: Prepare and dispense medium under 80% N_2 + 20% CO_2 gas atmosphere. Add components to distilled/deionized water and bring volume to 1.0L. Mix thoroughly. Anaerobically distribute 10.0mL aliquots into serum vials under 80% N_2 + 20% CO_2 gas atmosphere. Seal vials. Autoclave for 15 min at 15 psi pressure–121°C. Cool to room temperature. Pressurize to 2 bar with 80% N_2 + 20% CO_2 gas atmosphere.

Use: For the transduction of *Methanobacter* spp. and *Methanobacterium* spp.

Methanothermus fervidus Medium

Composition per liter:

Na_2SO_4 .... 3.4g
$NaHCO_3$ 2.0g
Tryticase™ 2.0g
Yeast extract 2.0g
L-Cysteine·HCl 0.5g
$Na_2S·9H_2O$ 0.5g
$FeSO_4·7H_2O$ 2.0mg
$Ni(NH_4)_2(SO_4)_2$ 2.0mg
Resazurin 1.0mg
Mineral solution 1 .... 37.5mL
Mineral solution 2 .... 37.5mL
Trace elements solution 10.0mL
Vitamin solution 10.0mL

pH 6.5 ± 0.2 at 25°C

Mineral Solution 1:

Composition per liter:

K_2HPO_4 .... 6.0g

Preparation of Mineral Solution 1: Add K_2HPO_4 to distilled/deionized water and bring volume to 1.0L. Mix thoroughly.

Mineral Solution 2:

Composition per liter:

NaCl 12.0g
K_2HPO_4 .... 6.0g
$(NH_4)_2SO_4$ 6.0g
$MgSO_4·7H_2O$ 2.4g
$CaCl_2·2H_2O$ 1.6g

Preparation of Mineral Solution 2: Add components to distilled/deionized water and bring volume to 1.0L. Mix thoroughly.

Trace Elements Solution:

Composition per liter:

$MgSO_4·7H_2O$ 3.0g
Nitrilotriacetic acid 1.5g
NaCl 1.0g
$MnSO_4·2H_2O$ 0.5g
$CoSO_4·7H_2O$ 0.18g
$ZnSO_4·7H_2O$ 0.18g
$CaCl_2·2H_2O$ 0.1g
$FeSO_4·7H_2O$ 0.1g
$NiCl_2·6H_2O$ 0.025g
$KAl(SO_4)_2·12H_2O$ 0.02g
$CuSO_4·5H_2O$ 0.01g
H_3BO_3 .... 0.01g
$Na_2MoO_4·2H_2O$ 0.01g
$Na_2SeO_3·5H_2O$ 0.3mg

Preparation of Trace Elements Solution: Add nitrilotriacetic acid to approximately 500.0mL of distilled/deionized water. Dissolve by adding KOH and adjust pH to 6.5. Add remaining components. Bring volume to 1.0L with additional distilled/deionized water. Adjust pH to 7.0 with KOH.

Vitamin Solution:

Composition per liter:

Pyridoxine·HCl 10.0mg
Calcium DL-pantothenate 5.0mg
Lipoic acid 5.0mg
Nicotinic acid 5.0mg
p-Aminobenzoic acid 5.0mg
Riboflavin 5.0mg
Thiamine·HCl 5.0mg
Biotin 2.0mg
Folic acid 2.0mg
Vitamin B_{12} 0.1mg

Preparation of Vitamin Solution: Add components to distilled/deionized water and bring volume to 1.0L. Mix thoroughly. Sparge with 80% N_2 + 20% CO_2.

Preparation of Medium: Add components, except $NaHCO_3$ and $Na_2S·9H_2O$, to distilled/deionized water and bring volume to 1.0L. Adjust pH to 6.5 with 10*N* H_2SO_4. Add $NaHCO_3$ and sparge with 80% N_2 + 20% CO_2 for 15 min. Add $Na_2S·9H_2O$. Mix thoroughly. Anaerobically distribute 20.0mL of medium into 100mL alkali-rich soda lime glass bottles. Pressurize to 2 bar with 80% H_2 + 20% CO_2. Autoclave for 15 min at 15 psi pressure–121°C.

Use: For the cultivation and maintenance of *Methanothermus fervidus*.

Methanothermus fervidus Medium

Composition per liter:

Na_2SO_4 .... 3.4g
$NaHCO_3$ 2.0g
L-Cysteine·HCl 0.5g
$Na_2S·9H_2O$ 0.5g
$FeSO_4·7H_2O$ 2.0mg
$Ni(NH_4)_2(SO_4)_2$ 2.0mg
Resazurin 1.0mg
Mineral solution 1 .... 37.5mL
Mineral solution 2 .... 37.5mL
Trace elements solution 10.0mL

pH 6.5 ± 0.2 at 25°C

Mineral Solution 1:

Composition per liter:

K_2HPO_4 .... 6.0g

Preparation of Mineral Solution 1: Add K_2HPO_4 to distilled/deionized water and bring volume to 1.0L. Mix thoroughly.

Mineral Solution 2:

Composition per liter:

NaCl	12.0g
K_2HPO_4	6.0g
$(NH_4)_2SO_4$	6.0g
$MgSO_4 \cdot 7H_2O$	2.4g
$CaCl_2 \cdot 2H_2O$	1.6g

Preparation of Mineral Solution 2: Add components to distilled/deionized water and bring volume to 1.0L. Mix thoroughly.

Trace Elements Solution:

Composition per liter:

$MgSO_4 \cdot 7H_2O$	3.0g
Nitrilotriacetic acid	1.5g
NaCl	1.0g
$MnSO_4 \cdot 2H_2O$	0.5g
$CoSO_4 \cdot 7H_2O$	0.18g
$ZnSO_4 \cdot 7H_2O$	0.18g
$CaCl_2 \cdot 2H_2O$	0.1g
$FeSO_4 \cdot 7H_2O$	0.1g
$NiCl_2 \cdot 6H_2O$	0.025g
$KAl(SO_4)_2 \cdot 12H_2O$	0.02g
$CuSO_4 \cdot 5H_2O$	0.01g
H_3BO_3	0.01g
$Na_2MoO_4 \cdot 2H_2O$	0.01g
$Na_2SeO_3 \cdot 5H_2O$	0.3mg

Preparation of Trace Elements Solution: Add nitrilotriacetic acid to approximately 500.0mL of distilled/deionized water. Dissolve by adding KOH and adjust pH to 6.5. Add remaining components. Bring volume to 1.0L with additional distilled/deionized water. Adjust pH to 7.0 with KOH.

Preparation of Medium: Add components, except $NaHCO_3$ and $Na_2S \cdot 9H_2O$, to distilled/deionized water and bring volume to 1.0L. Adjust pH to 6.5 with 10*N* H_2SO_4. Add $NaHCO_3$ and sparge with 80% N_2 + 20% CO_2 for 15 min. Add $Na_2S \cdot 9H_2O$. Mix thoroughly. Anaerobically distribute 20.0mL of medium into 100mL alkali-rich soda lime glass bottles. Pressurize to 2 bar with 80% H_2 + 20% CO_2. Autoclave for 15 min at 15 psi pressure–121°C.

Use: For the cultivation and maintenance of *Methanothermus sociabilis*.

Methanothrix Medium

Composition per liter:

Sodium acetate	6.8g
$KHCO_3$	4.0g
NH_4Cl	1.0g
NaCl	0.6g
KH_2PO_4	0.3g
$MgCl_2 \cdot 6H_2O$	0.1g
$CaCl_2 \cdot 2H_2O$	0.08g
Resazurin	1.0mg
Trace elements solution	10.0mL
Vitamin solution	10.0mL
L-Cysteine·HCl·H_2O solution	10.0mL
$Na_2S \cdot 9H_2O$ solution	10.0mL

pH 7.0 ± 0.2 at 25°C

Trace Elements Solution:

Composition per liter:

Nitrilotriacetic acid	12.8g
$FeCl_3 \cdot 6H_2O$	1.35g
NaCl	1.0g
$NiCl_2 \cdot 6H_2O$	0.12g
$CaCl_2 \cdot 2H_2O$	0.10g
$MnCl_2 \cdot 4H_2O$	0.10g
$ZnCl_2$	0.10g
$Na_2SeO_3 \cdot 5H_2O$	0.026g
$CuCl_2 \cdot 2H_2O$	0.025g
$CoCl_2 \cdot 6H_2O$	0.024g
$Na_2MoO_4 \cdot 2H_2O$	0.024g
H_3BO_3	0.01g

Preparation of Trace Elements Solution: Add nitrilotriacetic acid to approximately 500.0mL of distilled/deionized water. Dissolve by adding KOH and adjust pH to 6.5. Add remaining components. Bring volume to 1.0L with additional distilled/deionized water. Adjust pH to 7.0 with KOH.

Vitamin Solution:

Composition per liter:

Pyridoxine·HCl	10.0mg
Calcium DL-pantothenate	5.0mg
Lipoic acid	5.0mg
Nicotinic acid	5.0mg
p-Aminobenzoic acid	5.0mg
Riboflavin	5.0mg
Thiamine·HCl	5.0mg
Biotin	2.0mg
Folic acid	2.0mg
Vitamin B_{12}	0.1mg

Preparation of Vitamin Solution: Add components to distilled/deionized water and bring volume to 1.0L. Mix thoroughly. Filter sterilize. Sparge with 80% N_2 + 20% CO_2.

L-Cysteine·HCl·H_2O Solution:

Composition per 10.0mL:

L-Cysteine·HCl·H_2O	0.3g

Preparation of L-Cysteine·HCl·H_2O Solution: Add L-cysteine·HCl·H_2O to distilled/deionized water and bring volume to 10.0mL. Mix thoroughly. Autoclave under 80% N_2 + 20% CO_2 for 15 min at 15 psi pressure–121°C.

$Na_2S \cdot 9H_2O$ Solution:

Composition per 10.0mL:

$Na_2S \cdot 9H_2O$	0.3g

Preparation of $Na_2S \cdot 9H_2O$ Solution: Add $Na_2S \cdot 9H_2O$ to distilled/deionized water and bring volume to 10.0mL. Mix thoroughly. Sparge with 80% N_2 + 20% CO_2. Autoclave for 15 min at 15 psi pressure–121°C.

Preparation of Medium: Add components, except vitamin solution, L-cysteine·HCl·H_2O solution, and $Na_2S \cdot 9H_2O$ solution, to distilled/deionized water and bring volume to 970.0mL. Mix thoroughly. Autoclave for 15 min at 15 psi pressure–121°C. Aseptically and anaerobically add 10.0mL of sterile vitamin solution, 10.0mL of sterile L-cysteine·HCl·H_2O solution, and 10.0mL of sterile $Na_2S \cdot 9H_2O$ solution. Mix thoroughly. Check that final pH is 7.0. Use 20% inoculum.

Use: For the cultivation and maintenance of *Methanosaeta concilii*.

Methermicoccus Medium (DSMZ Medium 1084)

Composition per liter:

NaCl	24.0g
$MgCl_2 \cdot 6H_2O$	10.2g
Yeast extract	2.0g
KCl	0.34g
NH_4Cl	0.25g
K_2HPO_4	0.2g
Resazurin	0.5mg
Cysteine solution	10.0mL
Sulfide solution	10.0mL
Bicarbonate solution	10.0mL
Coenzyme M solution	10.0mL
Methanol	8.0mL
Sludge fluid	5.0mL

pH 9.5 ± 0.2 at 25°C

Methanol:

Composition per 10.0mL:

Methanol	10.0mL

Preparation of Methanol: Sparge 10.0mL of methanol with 100% N_2. Filter sterilize.

Coenzyme M Solution:

Composition per 10.0mL:

Mercaptoethanesulfonic acid (coenzyme M)	2.5g

Preparation of Coenzyme M: Add coenzyme M to distilled/deionized water and bring volume to 10.0mL. Mix thoroughly. Sparge with 100% N_2. Filter sterilize.

B icarbonate Solution:

Composition per 10.0mL:

$NaHCO_3$	2.5g

Preparation of Bicarbonate Solution: Add $NaHCO_3$ to distilled/deionized water and bring volume to 10.0mL. Mix thoroughly. Sparge with 20% CO_2 + 80% H_2. Filter sterilize.

Sulfide Solution:

Composition per 10.0mL:

$Na_2S \cdot 9H_2O$	0.5g

Preparation of Sulfide Solution: Add $Na_2S \cdot 9H_2O$ to distilled/deionized water and bring volume to 10.0mL. Mix thoroughly. Autoclave under 100% N_2 for 15 min at 15 psi pressure–121°C. Cool to room temperature.

Cysteine Solution:

Composition per 10.0mL:

L-Cysteine-$HCl \cdot 2H_2O$	0.3g

Preparation of Cysteine Solution: Add L-cysteine-$HCl \cdot 2H_2O$ to distilled/deionized water and bring volume to 10.0mL. Mix thoroughly. Sparge with 100% N_2. Filter sterilize.

Sludge Fluid:

Composition per 500.0mL:

Yeast extract	2.0g
Sludge	500.0mL

Preparation of Sludge Fluid: Add yeast extract to sludge from an anaerobic digester. Gas with nitrogen gas for a few minutes. Incubate at 37°C for 24 hr. Centrifuge the sludge at 13,000 x g. Autoclave for 15 min at 15 psi pressure–121°C. Place the resulting, clear supernatant in screw-capped vessels under nitrogen gas. The sludge fluid can be stored at room temperature in the dark.

Preparation of Medium: Add components, except bicarbonate, sludge fluid, methanol, coenzyme M, cysteine, and sulfide solutions, to distilled/deionized water and bring volume to 947.0mL. Gently heat and bring to boiling. Boil medium for 1 min. Cool to room temperature under 80% N_2 + 20% CO_2 gas mixture. Dispense under same gas atmosphere into culture vessels. Autoclave for 15 min at 15 psi pressure–121°C. Cool to room temperature under 80% N_2 + 20% CO_2 gas mixture. Aseptically add coenzyme M, methanol, sludge fluid, cysteine, and sulfide from sterile anoxic stock solutions. Adjust pH to 6.0–6.5.

Use: For the cultivation of *Methermicoccus* spp.

Methyl Red Voges-Proskauer Broth

See: **MRVP Broth**

Methyl Red Voges-Proskauer Medium

See: **MRVP Medium**

Methylamine Salts Medium

Composition per liter:

Agar	15.0g
Methylamine·HCl	6.75g
K_2HPO_4	2.12g
KH_2PO_4	1.0g
Solution A	5.0mL
Solution B	1.0mL

pH 7.0 ± 0.2 at 25°C

Solution A:

Composition per 100.0mL:

$MgSO_4 \cdot 7H_2O$	2.0g
$CaCl_2 \cdot 2H_2O$	0.2g
$FeSO_4 \cdot 7H_2O$	0.2g

Preparation of Solution A: Add components to distilled/deionized water and bring volume to 100.0mL. Mix thoroughly.

Solution B:

Composition per 100.0mL:

$MnSO_4 \cdot 7H_2O$	0.05g
$Na_2MoO_4 \cdot 2H_2O$	0.05g

Preparation of Solution B: Add components to distilled/deionized water and bring volume to 100.0mL. Mix thoroughly.

Preparation of Medium: Add components to distilled/deionized water and bring volume to 1.0L. Mix thoroughly. Gently heat and bring to boiling. Distribute into tubes or flasks. Autoclave for 15 min at 15 psi pressure–121°C. Pour into sterile Petri dishes or leave in tubes.

Use: For the cultivation and maintenance of *Methylobacterium extorquens* and *Pseudomonas* species.

Methylene Blue Milk Medium (MBM Medium)

Composition per liter:

Skim milk, dehydrated	100.0g
Methylene Blue	10.0g

pH 6.4 ± 0.2 at 25°C

Preparation of Medium: Add components to distilled/deionized water and bring volume to 1.0L. Mix thoroughly. Distribute into tubes or flasks. Autoclave for 20 min at 10 psi pressure–115°C.

Use: For the cultivation and differentiation of group D streptococci (enterococci) from other *Streptococcus* species.

Methylobacterium Agar

Composition per liter:

Agar	12.0g
KNO_3	1.0g
Na_2HPO_4	0.23g
$MgSO_4 \cdot 7H_2O$	0.2g
NaH_2PO_4	0.07g
$CaCl_2 \cdot 2H_2O$	0.02g
$FeSO_4 \cdot 7H_2O$	1.0mg
$ZnSO_4 \cdot 7H_2O$	70.0μg
H_3BO_3	10.0μg
$MnSO_4 \cdot 5H_2O$	10.0μg
MoO_3	10.0μg
$CuSO_4 \cdot 5H_2O$	5.0μg
Methanol, filter sterilized	5.0mL

pH 6.8 ± 0.2 at 25°C

Preparation of Medium: Add components, except methanol, to distilled/deionized water and bring volume to 995.0mL. Mix thoroughly. Gently heat and bring to boiling. Autoclave for 15 min at 15 psi pressure–121°C. Cool to 50°–55°C. Aseptically add 5.0mL of filter-sterilized methanol. Mix thoroughly. Pour into sterile Petri dishes or distribute into sterile tubes.

Use: For the cultivation and maintenance of *Methylobacterium organophilum*.

Methylobacterium Medium (DSMZ Medium 125)

Composition per liter:

Agar	12.0g
KNO_3	1.0g
Na_2HPO_4	0.23g
$MgSO_4 \cdot 7H_2O$	0.2g
NaH_2PO_4	0.07g
$CaCl_2 \cdot 2H_2O$	0.02g
$FeSO_4 \cdot 7H_2O$	1.0mg
$ZnSO_4 \cdot 7H_2O$	70.0μg
MoO_3	10.0μg
H_3BO_3	10.0μg
$MnSO_4 \cdot 5H_2O$	10.0μg
$CuSO_4 \cdot 5H_2O$	5.0μg
Methanol	15.0mL

pH 4.0–5.4 at 25°C

Preparation of Medium: Add components to distilled/deionized water and bring volume to 1.0L. Mix thoroughly. Adjust pH to 6.8. Gently heat and bring to boiling. Distribute into tubes or flasks. Autoclave for 15 min at 15 psi pressure–121°C. Pour into sterile Petri dishes or leave in tubes.

Use: For the cultivation and maintenance of *Acidomonas methanolica* and *Methanomonas methylovora.*

Methylobacterium Medium (DSMZ Medium 125)

Composition per liter:

Agar	12.0g
KNO_3	1.0g
Na_2HPO_4	0.23g
$MgSO_4 \cdot 7H_2O$	0.2g
NaH_2PO_4	0.07g
$CaCl_2 \cdot 2H_2O$	0.02g
$FeSO_4 \cdot 7H_2O$	1.0mg
$ZnSO_4 \cdot 7H_2O$	70.0μg
MoO_3	10.0μg
H_3BO_3	10.0μg
$MnSO_4 \cdot 5H_2O$	10.0μg
$CuSO_4 \cdot 5H_2O$	5.0μg
Methanol	5.0mL

pH 6.8 ± 0.2 at 25°C

Preparation of Medium: Add components to distilled/deionized water and bring volume to 1.0L. Mix thoroughly. Adjust pH to 6.8. Gently heat and bring to boiling. Distribute into tubes or flasks. Autoclave for 15 min at 15 psi pressure–121°C. Pour into sterile Petri dishes or leave in tubes.

Use: For the cultivation and maintenance of *Methylobacterium organophilum.*

Methylobacterium Medium (DSMZ Medium 125)

Composition per liter:

KNO_3	1.0g
Na_2HPO_4	0.23g
$MgSO_4 \cdot 7H_2O$	0.2g
NaH_2PO_4	0.07g
$CaCl_2 \cdot 2H_2O$	0.02g
$FeSO_4 \cdot 7H_2O$	1.0mg
$ZnSO_4 \cdot 7H_2O$	70.0μg
MoO_3	10.0μg
H_3BO_3	10.0μg
$MnSO_4 \cdot 5H_2O$	10.0μg
$CuSO_4 \cdot 5H_2O$	5.0μg
Methane/air mixture (4:1)	variable

pH 6.8 ± 0.2 at 25°C

Preparation of Medium: Add components to distilled/deionized water and bring volume to 1.0L. Mix thoroughly. Adjust pH to 6.8. Distribute into tubes or flasks. Autoclave for 15 min at 15 psi pressure–121°C. Sparge and incubate under methane/air mixture (4:1).

Use: For the cultivation and maintenance of *Methylobacterium organophilum.*

Methylobacterium Medium

Composition per liter:

Agar	15.0g
K_2HPO_4	1.2g
KH_2PO_4	0.62g
$(NH_4)_2SO_4$	0.5g
$MgSO_4 \cdot 7H_2O$	0.2g
NaCl	0.1g
$CaCl_2 \cdot 2H_2O$	34.0mg
$FeCl_3 \cdot H_2O$	1.0mg
Trace elements solution	1.0mL
Methanol, filter sterilized	10.0mL

pH 7.0 ± 0.2 at 25°C

Trace Elements Solution:

Composition per liter:

$ZnSO_4 \cdot 7H_2O$	70.0mg
H_3BO_3	10.0mg

$Na_2MoO_4 \cdot 2H_2O$ 10.0mg
$MnSO_4 \cdot H_2O$ 7.0mg
$CoCl_2 \cdot H_2O$ 5.0mg
$CuSO_4 \cdot 5H_2O$ 5.0mg

Preparation of Medium: Add components, except methanol, to distilled/deionized water and bring volume to 990.0mL. Mix thoroughly. Gently heat and bring to boiling. Autoclave for 15 min at 15 psi pressure–121°C. Cool to 50°–55°C. Aseptically add 10.0mL of sterile methanol. Mix thoroughly. Pour into sterile Petri dishes or distribute into sterile tubes.

Use: For the cultivation and maintenance of *Methylobacterium* species.

Methylobacterium thiocyanatum Medium (DSMZ Medium 805)

Composition per liter:
$Na_2HPO_4 \cdot 2H_2O$ 7.9g
Glucose 4.5g
K_2HPO_4 .... 1.5g
KSCN 0.25g
$MgSO_4 \cdot 7H_2O$ 0.1g
Iron sulfate solution 1.0mL
pH 7.1 ± 0.1 at 25°C

Iron Sulfate Solution:
Composition per 10.0mL:
$FeSO_4 \cdot 7H_2O$ 0.2g

Preparation of Iron Sulfate Solution: Add $FeSO_4 \cdot 7H_2O$ to distilled/deionized water and bring volume to 10.0mL. Mix thoroughly.

Preparation of Medium: Add components to distilled/deionized water and bring to 1.0L. Mix thoroughly. Distribute into tubes or flasks. Autoclave for 10 min at 10 psi pressure–115°C.

Use: For the cultivation of *Methylobacterium thiocyanatum.*

Methylocapsa acidophila Medium (DSMZ Medium 922)

Composition per liter:
KH_2PO_4 .... 100.0mg
$MgSO_4 \cdot 7H_2O$ 50.0mg
$CaCl_2 \cdot 2H_2O$ 10.0mg
Trace elements 1.0mL
pH 4.5-5.8 at 25°C

Trace Elements:
Composition per liter:
EDTA 5.0g
$FeSO_4 \cdot 7H_2O$ 2.0g
$CoCl_2 \cdot 6H_2O$ 0.2g
$CuCl_2 \cdot 5H_2O$ 0.1g
$ZnSO_4 \cdot 7H_2O$ 0.1g
Na_2MoO_4 .... 0.03g
$NiCl_2 \cdot 6H_2O$ 0.02g

Preparation of Trace Elements: Add components to distilled/deionized water and bring volume to 1.0L. Mix thoroughly.

Preparation of Medium: Add components to distilled/deionized water and bring volume to 1.0L. Mix thoroughly. Distribute into tubes or flasks. Autoclave for 15 min at 15 psi pressure–121°C. The final pH should be 4.5–5.8. The medium is fairly weakly buffered so the pH should be checked before and after autoclaving. Incubate under atmosphere of 10–30% methane.

Use: For the cultivation of *Methylocapsa acidiphila.*

Methylocapsa acidophila Medium (DSMZ Medium 922)

Composition per liter:
KNO_3 .... 100.0mg
KH_2PO_4 .... 100.0mg
$MgSO_4 \cdot 7H_2O$ 50.0mg
$CaCl_2 \cdot 2H_2O$ 10.0mg
Trace elements 1.0mL
pH 4.5–5.8 at 25°C

Trace Elements:
Composition per liter:
EDTA 5.0g
$FeSO_4 \cdot 7H_2O$ 2.0g
$CoCl_2 \cdot 6H_2O$ 0.2g
$CuCl_2 \cdot 5H_2O$ 0.1g
$ZnSO_4 \cdot 7H_2O$ 0.1g
Na_2MoO_4 .... 0.03g
$NiCl_2 \cdot 6H_2O$ 0.02g

Preparation of Trace Elements: Add components to distilled/deionized water and bring volume to 1.0L. Mix thoroughly.

Preparation of Medium: Add components to distilled/deionized water and bring volume to 1.0L. Mix thoroughly. Distribute into tubes or flasks. Autoclave for 15 min at 15 psi pressure–121°C. The final pH should be 4.5–5.8. The medium is fairly weakly buffered so the pH should be checked before and after autoclaving. Incubate under atmosphere of 10–30% methane.

Use: For the cultivation and enhanced growth of *Methylocapsa acidiphila.*

Methylocella silverstris Medium (DSMZ Medium 1181)

Composition per liter:
$Na_2HPO_4 \cdot 12H_2O$ 0.717g
$KH_2PO_4 \cdot 2H_2O$ 0.272g
$MgSO_4 \cdot 7H_2O$ 0.2g
$NaNO_3$ 0.2g
$CaCl_2 \cdot 2H_2O$ 6.0mg
Fe(III)NH_4-EDTA 4.0mg
Phosphate buffer solution 5.0mL
Trace elements solution 0.5mL
pH 5.8 ± 0.2 at 25°C

Phosphate Buffer Solution:
Composition per liter:
$NaH_2PO_4 \cdot 2H_2O$ 28.71g
$Na_2HPO_4 \cdot 12H_2O$ 5.73g

Preparation of Phosphate Buffer Solution: Add components to distilled/deionized water and bring volume to 1.0L. Filter sterilize

Trace Elements Solution:
Composition per liter:
Composition per liter:
H_3BO_3 .... 30.0mg
$CaCl_2 \cdot 2H_2O$ 20.0mg
$ZnSO_4 \cdot 2H_2O$ 10.0mg

$MnCl_2 \cdot 4H_2O$	3.0mg
$Na_2MoO_4 \cdot 2H_2O$	3.0mg
$CuCl_2 \cdot 6H_2O$	1.0mg

Preparation of Trace Elements Solution: Add components to distilled/deionized water and bring volume to 1.0L. Mix thoroughly. Filter sterilize.

Preparation of Medium: Add components, except phosphate buffer solution, to distilled/deionized water and bring volume to 995.0mL. Gently heat and bring to boiling. Autoclave for 15 min at 15 psi pressure–121°C. Cool to room temperature. Aseptically add 5.0mL sterile phosphate buffer solution. Adjust pH to 5.8.

Use: For the cultivation of *Methylocella silverstris.*

Methylococcus Medium

Composition per liter:

Agar	8.0g
$NaNO_3$ (20% solution)	10.0mL
L-F salts solution	10.0mL
Sodium-potassium phosphate buffer	6.5mL

pH 7.1 ± 0.2 at 25°C

Sodium-Potassium Phosphate Buffer:
Composition per liter:

KH_2PO_4	136.0g
NaOH	28.8g

Preparation of Sodium-Potassium Phosphate Buffer: Add components to distilled/deionized water and bring volume to 1.0L. Mix thoroughly. Adjust pH to 7.1.

L-F Salts Solution:
Composition per liter:

$MgSO_4 \cdot 7H_2O$ (10% solution)	200.0mL
$CaCl_2 \cdot 2H_2O$ (10% solution)	20.0mL
$FeSO_4$ (10% solution)	10.0mL
$ZnSO_4 \cdot 7H_2O$ (1% solution)	4.9mL
H_3BO_3 (1% solution)	0.6mL
$MnSO_4 \cdot H_2O$ (1% solution)	0.27mL
$CuSO_4 \cdot 5H_2O$ (1% solution)	0.2mL

Preparation of L-F Salts Solution: Filter sterilize $FeSO_4$ solution immediately prior to use. Add all components to distilled/deionized water and bring volume to 1.0L. Mix thoroughly.

Preparation of Medium: Add components to distilled/deionized water and bring volume to 1.0L. Mix thoroughly. Adjust pH to 7.1. Autoclave for 15 min at 15 psi pressure–121°C. Pour into sterile Petri dishes or leave in tubes.

Use: For the cultivation and maintenance of *Methylococcus* species.

Methylohalomonas Medium (DSMZ Medium 1171)

Composition per liter:

NaCl	120.0g
K_2HPO_4	2.0g
NH_4Cl	0.5g
Magnesium sulfate solution	10.0mL
Thiosulfate solution	10.0mL
Bicarbonate solution	10.0mL
Vitamin solution	10.0mL
Methanol	2.0mL
Trace elements solution SL-6	1.0mL

pH 7.5 ± 0.2 at 25°C

Trace Elements Solution SL-6:
Composition per liter:

$MnCl_2 \cdot 4H_2O$	0.5g
H_3BO_3	0.3g
$CoCl_2 \cdot 6H_2O$	0.2g
$ZnSO_4 \cdot 7H_2O$	0.1g
$Na_2MoO_4 \cdot 2H_2O$	0.03g
$NiCl_2 \cdot 6H_2O$	0.02g
$CuCl_2 \cdot 2H_2O$	0.01g

Preparation of Trace Elements Solution SL-6: Add components to distilled/deionized water and bring volume to 1.0L. Mix thoroughly. Autoclave for 15 min at 15 psi pressure–121°C.

Magnesium Sulfate Solution:
Composition per 10.0mL:

$MgSO_4 \cdot 7H_2O$	2.0g

Preparation of Magnesium Sulfate Solution: Add $MgSO_4 \cdot 7H_2O$ to distilled/deionized water and bring volume to 10.0mL. Mix thoroughly. Autoclave for 15 min at 15 psi pressure–121°C. Cool to room temperature.

Vitamin Solution:
Composition per 10.0mL:

Vitamin B_{12}	0.1mg

Preparation of Vitamin Solution: Add vitamin B_{12} to distilled/deionized water and bring volume to 10.0mL. Mix thoroughly. Filter sterilize.

Methanol:
Composition per 10.0mL:

Methanol	10.0mL

Preparation of Methanol: Autoclave for 15 min at 15 psi pressure–121°C. Cool to room temperature.

Bicarbonate Solution:
Composition per 10.0mL:

$NaHCO_3$	11.0g

Preparation of Bicarbonate Solution: Add $NaHCO_3$ to distilled/deionized water and bring volume to 10.0mL. Mix thoroughly. Autoclave for 15 min at 15 psi pressure–121°C. Cool to room temperature.

Thiosulfate Solution:
Composition per 10.0mL:

$NaS_2O_3 \cdot 5H_2O$	18.0g

Preparation of Thiosulfate Solution: Add $NaS_2O_3 \cdot 5H_2O$ to distilled/deionized water and bring volume to 10.0mL. Mix thoroughly. Autoclave for 15 min at 15 psi pressure–121°C. Cool to room temperature.

Preparation of Medium: Add components, except methanol, vitamin, bicarbonate, thiosulfate, and magnesium sulfate solutions, to distilled/deionized water and bring volume to 960.0mL. Mix thoroughly. Distribute into closed vessels with a medium to headspace ratio of 1:5 to 1:10. Autoclave for 15 min at 15 psi pressure–121°C. Cool to room temperature. Aseptically add the methanol, vitamin, bicarbonate, thiosulfate, and magnesium sulfate solutions.

Use: For the cultivation of *Methylohalomonas* spp.

Methylohalomonas Medium with Acetate (DSMZ Medium 1171)

Composition per liter:

NaCl	120.0g
K_2HPO_4	2.0g
NH_4Cl	0.5g
Magnesium sulfate solution	10.0mL
Thiosulfate solution	10.0mL
Bicarbonate solution	10.0mL
Acetate solution	10.0mL
Vitamin solution	10.0mL
Trace elements solution SL-6	1.0mL

pH 7.5 ± 0.2 at 25°C

Trace Elements Solution SL-6:

Composition per liter:

$MnCl_2·4H_2O$	0.5g
H_3BO_3	0.3g
$CoCl_2·6H_2O$	0.2g
$ZnSO_4·7H_2O$	0.1g
$Na_2MoO_4·2H_2O$	0.03g
$NiCl_2·6H_2O$	0.02g
$CuCl_2·2H_2O$	0.01g

Preparation of Trace Elements Solution SL-6: Add components to distilled/deionized water and bring volume to 1.0L. Mix thoroughly. Autoclave for 15 min at 15 psi pressure–121°C.

Magnesium Sulfate Solution:

Composition per 10.0mL:

$MgSO_4·7H_2O$	2.0g

Preparation of Magnesium Sulfate Solution: Add $MgSO_4·7H_2O$ to distilled/deionized water and bring volume to 10.0mL. Mix thoroughly. Autoclave for 15 min at 15 psi pressure–121°C. Cool to room temperature.

Vitamin Solution:

Composition per 10.0mL:

Vitamin B_{12}	0.1mg

Preparation of Vitamin Solution: Add vitamin B_{12} to distilled/deionized water and bring volume to 10.0mL. Mix thoroughly. Filter sterilize.

Aceate Solution:

Composition per 10.0mL:

Sodium acetate	0.2g

Preparation of Acetate Solution: Add sodium acetate to distilled/deionized water and bring volume to 10.0mL. Mix thoroughly. Autoclave for 15 min at 15 psi pressure–121°C. Cool to room temperature.

B icarbonate Solution:

Composition per 10.0mL:

$NaHCO_3$	11.0g

Preparation of Bicarbonate Solution: Add $NaHCO_3$ to distilled/deionized water and bring volume to 10.0mL. Mix thoroughly. Autoclave for 15 min at 15 psi pressure–121°C. Cool to room temperature.

Thiosulfate Solution:

Composition per 10.0mL:

$NaS_2O_3·5H_2O$	18.0g

Preparation of Thiosulfate Solution: Add $NaS_2O_3·5H_2O$ to distilled/deionized water and bring volume to 10.0mL. Mix thoroughly. Autoclave for 15 min at 15 psi pressure–121°C. Cool to room temperature.

Preparation of Medium: Add components, except acetate, vitamin, bicarbonate, thiosulfate, and magnesium sulfate solutions, to distilled/deionized water and bring volume to 950.0mL. Mix thoroughly. Distribute into closed vessels with a medium to headspace ratio of 1:5 to 1:10. Autoclave for 15 min at 15 psi pressure–121°C. Cool to room temperature. Aseptically add the acetate, vitamin, bicarbonate, thiosulfate, and magnesium sulfate solutions.

Use: For the cultivation with faster growth rates of *Methylohalomonas* spp.

Methylomicrobium Medium (DSMZ Medium 1180)

Composition per liter:

NaCl	30.0g
$MgSO_4·7H_2O$	0.2g
$CaCl_2·2H_2O$	0.02g
KNO_3	1.0g
Bicarbonate solution	50.0mL
Phosphate buffer solution	20.0mL
Carbonate solution	5.0mL
Trace elements solution	1.0mL

pH 8.7 ± 0.2 at 25°C

Phosphate Buffer Solution:

Composition per liter:

$Na_2HPO_4·12H_2O$	30.0g
KH_2PO_4	14.0g

Preparation of Phosphate Buffer Solution: Add components to distilled/deionized water and bring volume to 1.0L. Filter sterilize

Bicarbonate Solution:

Composition per 100.0mL:

$NaHCO_3$	7.2g

Preparation of Bicarbonate Solution: Add $NaHCO_3$ to distilled/deionized water and bring volume to 100.0mL. Mix thoroughly. Autoclave for 15 min at 15 psi pressure–121°C. Cool to room temperature.

Carbonate Solution:

Composition per 10.0mL:

Na_2CO_3	1.1g

Preparation of Carbonate Solution: Add Na_2CO_3 to distilled/deionized water and bring volume to 10.0mL. Mix thoroughly. Autoclave for 15 min at 15 psi pressure–121°C. Cool to room temperature.

Trace Elements Solution:

Composition per liter:

EDTA	5.0g
$FeSO_4·7H_2O$	2.0g
$CoCl_2·6H_2O$	0.2g
$CuCl_2·2H_2O$	0.1g
$ZnSO_4·7H_2O$	0.1g
$MnCl_2·4H_2O$	0.03g
H_3BO_3	0.03g
$Na_2MoO_4·2H_2O$	0.03g
$NiCl_2·6H_2O$	0.02g

Preparation of Trace Elements Solution: Add components to distilled/deionized water and bring volume to 1.0L. Mix thoroughly. Autoclave for 15 min at 15 psi pressure–121°C.

Preparation of Medium: Add components, except carbonate, bicarbonate, and phosphate buffer solutions, to distilled/deionized water and bring volume to 925.0mL. Mix thoroughly. Autoclave for 15 min at 15 psi pressure–121°C. Cool to room temperature. Aseptically add the carbonate, bicarbonate, and phosphate buffer solutions. Aseptically distribute into culture vessels.

Use: For the cultivation with faster growth rates of *Methylomicrobium* spp.

Methylonatrum Medium (DSMZ Medium 1170)

Composition per liter:

NaCl	18.0g
K_2HPO_4	1.0g
Magnesium sulfate solution	10.0mL
Thiosulfate solution	10.0mL
Carbonate solution	10.0mL
Vitamin solution	10.0mL
Methanol	2.0mL
Trace elements solution SL-6	1.0mL

pH 10.0 ± 0.2 at 25°C

Trace Elements Solution SL-6:

Composition per liter:

$MnCl_2 \cdot 4H_2O$	0.5g
H_3BO_3	0.3g
$CoCl_2 \cdot 6H_2O$	0.2g
$ZnSO_4 \cdot 7H_2O$	0.1g
$Na_2MoO_4 \cdot 2H_2O$	0.03g
$NiCl_2 \cdot 6H_2O$	0.02g
$CuCl_2 \cdot 2H_2O$	0.01g

Preparation of Trace Elements Solution SL-6: Add components to distilled/deionized water and bring volume to 1.0L. Mix thoroughly. Autoclave for 15 min at 15 psi pressure–121°C.

Magnesium Sulfate Solution:

Composition per 10.0mL:

$MgSO_4 \cdot 7H_2O$	2.0g

Preparation of Magnesium Sulfate Solution: Add $MgSO_4 \cdot 7H_2O$ to distilled/deionized water and bring volume to 10.0mL. Mix thoroughly. Autoclave for 15 min at 15 psi pressure–121°C. Cool to room temperature.

Vitamin Solution:

Composition per 10.0mL:

Vitamin B_{12}	0.1mg

Preparation of Vitamin Solution: Add vitamin B_{12} to distilled/deionized water and bring volume to 10.0mL. Mix thoroughly. Filter sterilize.

Methanol:

Composition per 10.0mL:

Methanol	10.0mL

Preparation of Methanol: Autoclave for 15 min at 15 psi pressure–121°C. Cool to room temperature.

Carbonate Solution:

Composition per 10.0mL:

$NaHCO_3$	11.0g

Preparation of Carbonate Solution: Add $NaHCO_3$ to distilled/deionized water and bring volume to 10.0mL. Mix thoroughly. Autoclave for 15 min at 15 psi pressure–121°C. Cool to room temperature.

Thiosulfate Solution:

Composition per 10.0mL:

$NaS_2O_3 \cdot 5H_2O$	18.0g

Preparation of Thiosulfate Solution: Add $NaS_2O_3 \cdot 5H_2O$ to distilled/deionized water and bring volume to 10.0mL. Mix thoroughly. Autoclave for 15 min at 15 psi pressure–121°C. Cool to room temperature.

Preparation of Medium: Add components, except methanol, vitamin, carbonate, thiosulfate, and magnesium sulfate solutions, to distilled/deionized water and bring volume to 960.0mL. Mix thoroughly. Distribute into closed vessels with a medium to headspace ratio of 1:5 to 1:10. Autoclave for 15 min at 15 psi pressure–121°C. Cool to room temperature. Aseptically add the methanol, vitamin, carbonate, thiosulfate, and magnesium sulfate solutions.

Use: For the cultivation with faster growth rates of *Methylonatrum* spp.

Methylonatrum Medium with Acetate (DSMZ Medium 1170)

Composition per liter:

NaCl	18.0g
K_2HPO_4	1.0g
Magnesium sulfate solution	10.0mL
Thiosulfate solution	10.0mL
Carbonate solution	10.0mL
Acetate solution	10.0mL
Vitamin solution	10.0mL
Trace elements solution SL-6	1.0mL

pH 10.0 ± 0.2 at 25°C

Trace Elements Solution SL-6:

Composition per liter:

$MnCl_2 \cdot 4H_2O$	0.5g
H_3BO_3	0.3g
$CoCl_2 \cdot 6H_2O$	0.2g
$ZnSO_4 \cdot 7H_2O$	0.1g
$Na_2MoO_4 \cdot 2H_2O$	0.03g
$NiCl_2 \cdot 6H_2O$	0.02g
$CuCl_2 \cdot 2H_2O$	0.01g

Preparation of Trace Elements Solution SL-6: Add components to distilled/deionized water and bring volume to 1.0L. Mix thoroughly. Autoclave for 15 min at 15 psi pressure–121°C.

Magnesium Sulfate Solution:

Composition per 10.0mL:

$MgSO_4 \cdot 7H_2O$	2.0g

Preparation of Magnesium Sulfate Solution: Add $MgSO_4 \cdot 7H_2O$ to distilled/deionized water and bring volume to 10.0mL. Mix thoroughly. Autoclave for 15 min at 15 psi pressure–121°C. Cool to room temperature.

Vitamin Solution:

Composition per 10.0mL:

Vitamin B_{12}	0.1mg

Preparation of Vitamin Solution: Add vitamin B_{12} to distilled/deionized water and bring volume to 10.0mL. Mix thoroughly. Filter sterilize.

Carbonate Solution:

Composition per 10.0mL:

Na_2CO_3	11.0g

Preparation of Carbonate Solution: Add Na_2CO_3 to distilled/deionized water and bring volume to 10.0mL. Mix thoroughly. Autoclave for 15 min at 15 psi pressure–121°C. Cool to room temperature.

Thiosulfate Solution:

Composition per 10.0mL:

$Na_2S_2O_3 \cdot 5H_2O$ 18.0g

Preparation of Thiosulfate Solution: Add $Na_2S_2O_3 \cdot 5H_2O$ to distilled/deionized water and bring volume to 10.0mL. Mix thoroughly. Autoclave for 15 min at 15 psi pressure–121°C. Cool to room temperature.

Acetate Solution:

Composition per 10.0mL:

Sodium acetate 0.2g

Preparation of Acetate Solution: Add sodium acetate to distilled/deionized water and bring volume to 10.0mL. Mix thoroughly. Autoclave for 15 min at 15 psi pressure–121°C. Cool to room temperature.

Preparation of Medium: Add components, except acetate, vitamin, carbonate, thiosulfate, and magnesium sulfate solutions, to distilled/deionized water and bring volume to 960.0mL. Mix thoroughly. Distribute into closed vessels with a medium to headspace ratio of 1:5 to 1:10. Autoclave for 15 min at 15 psi pressure–121°C. Cool to room temperature. Aseptically add the acetate, vitamin, carbonate, thiosulfate, and magnesium sulfate solutions.

Use: For the cultivation with faster growth rates of *Methylonatrum* spp.

Methylophaga Agar

Composition per 103.0mL:

Agar solution 50.0mL
Mineral base, 2X 50.0mL
Solution T 2.0mL
Vitamin B_{12} solution 1.0mL
Methanol 0.3mL

pH 7.3 ± 0.2 at 25°C

Agar Solution:

Composition per 500.0mL:

Agar 15.0g

Preparation of Agar Solution: Add agar to distilled/deionized water and bring volume to 500.0mL. Mix thoroughly. Autoclave for 15 min at 15 psi pressure–121°C. Cool to 50°C.

Mineral Base, 2X:

Composition per 500.0mL:

NaCl 24.0g
$MgCl_2 \cdot 6H_2O$ 3.0g
$MgSO_4 \cdot 7H_2O$ 2.0g
$CaCl_2 \cdot 2H_2O$ 1.0g
KCl 0.5g
Bis-Tris buffer (bis[2-hydroxyethyl]amino-tris[hydroxymethyl]-methane) 0.5g
Wolfe's mineral solution 10.0mL

Preparation of Mineral Base, 2X: Add components to distilled/deionized water and bring volume to 500.0mL. Mix thoroughly. Adjust pH to 7.3. Autoclave for 15 min at 15 psi pressure–121°C. Cool to 50°C.

Wolfe's Mineral Solution:

Composition per liter:

$MgSO_4 \cdot 7H_2O$ 3.0g
Nitrilotriacetic acid 1.5g
NaCl 1.0g
$MnSO_4 \cdot H_2O$ 0.5g
$FeSO_4 \cdot 7H_2O$ 0.1g
$CoCl_2 \cdot 6H_2O$ 0.1g
$CaCl_2$ 0.1g
$ZnSO_4 \cdot 7H_2O$ 0.1g
$CuSO_4 \cdot 5H_2O$ 0.01g
$AlK(SO_4)_2 \cdot 12H_2O$ 0.01g
H_3BO_3 .... 0.01g
$Na_2MoO_4 \cdot 2H_2O$ 0.01g

Preparation of Wolfe's Mineral Solution: Add nitrilotriacetic acid to 500.0mL of distilled/deionized water. Dissolve by adjusting pH to 6.5 with KOH. Add remaining components. Add distilled/deionized water to 1.0L.

Solution T:

Composition per 100.0mL:

NH_4Cl 10.0g
Bis-Tris buffer (bis[2-hydroxyethyl]amino-tris[hydroxymethyl]-methane) 10.0g
KH_2PO_4 .... 0.7g
Ferric ammonium citrate 0.3g

Preparation of Solution T: Add components to distilled/deionized water and bring volume to 100.0mL. Mix thoroughly. Adjust pH to 7.3. Autoclave for 15 min at 15 psi pressure–121°C.

Vitamin B_{12} Solution:

Composition per 10.0mL:

Vitamin B_{12} 1.0μg

Preparation of Vitamin B_{12} Solution: Add vitamin B_{12} to 10.0mL of distilled/deionized water. Mix thoroughly. Filter sterilize.

Preparation of Medium: Aseptically mix 50.0mL of the sterile agar solution with 50.0mL of the sterile mineral base, 2X. Aseptically combine sterile solution T and sterile vitamin B_{12} solution with the sterile mineral base. Filter sterilize methanol and add to basal medium. Pour into sterile Petri dishes or distribute into sterile tubes.

Use: For the cultivation and maintenance of *Methylophaga marina.*

Methylophaga alcalica Agar (DSMZ Medium 976)

Composition per liter:

NaCl 30.0g
Agar 20.0g
KH_2PO_4 .... 1.0g
KNO_3 .... 1.0g
$MgSO_4 \cdot 7H_2O$ 0.22g
Na_2CO_3 solution 50.0mL
Methanol solution 50.0mL
Trace elements solution 1.0mL

pH 9.5 ± 0.2 at 25°C

Methanol Solution:

Composition per 50.0mL:

Methanol 10.0mL

Preparation of Methanol Solution: Add methanol to distilled/deionized water and bring volume to 50.0mL. Mix thoroughly. Autoclave for 15 min at 15 psi pressure–121°C.

Na_2CO_3 Solution:

Composition per 50.0mL:

Na_2CO_3	5.0g

Preparation of Na_2CO_3 Solution: Add Na_2CO_3 to distilled/deionized water and bring volume to 50.0mL. Mix thoroughly. Autoclave for 15 min at 15 psi pressure–121°C.

Trace Elements Solution:

Composition per liter:

Ferric citrate	30.0mg
$CaCl_2 \cdot 2H_2O$	30.0mg
$MgCl_2 \cdot 4H_2O$	5.0mg
$ZnSO_4 \cdot 7H_2O$	5.0mg
$CuSO_4 \cdot 5H_2O$	0.5mg

Preparation of Trace Elements Solution: Add components to distilled/deionized water and bring volume to 1.0L. Mix thoroughly.

Preparation of Medium: Add components, except methanol solution and Na_2CO_3 solution, to distilled/deionized water and bring volume to 900.0mL. Mix thoroughly. Gently heat and bring to boiling. Autoclave for 15 min at 15 psi pressure–121°C. Cool to 55°C. Aseptically add 50.0mL warm sterile Na_2CO_3 solution and 50.0mL warm sterile methanol solution. Mix thoroughly. Pour into Petri dishes or aseptically distribute into sterile tubes.

Use: For the cultivation and maintenance of *Methylophaga alcalica.*

Methylophaga alcalica Medium (DSMZ Medium 976)

Composition per liter:

NaCl	30.0g
KH_2PO_4	1.0g
KNO_3	1.0g
$MgSO_4 \cdot 7H_2O$	0.22g
Na_2CO_3 solution	50.0mL
Methanol solution	50.0mL
Trace elements solution	1.0mL

pH 9.5 ± 0.2 at 25°C

Methanol Solution:

Composition per 50.0mL:

Methanol	10.0mL

Preparation of Methanol Solution: Add methanol to distilled/deionized water and bring volume to 50.0mL. Mix thoroughly. Autoclave for 15 min at 15 psi pressure–121°C.

Na_2CO_3 Solution:

Composition per 50.0mL:

Na_2CO_3	5.0g

Preparation of Na_2CO_3 Solution: Add Na_2CO_3 to distilled/deionized water and bring volume to 50.0mL. Mix thoroughly. Autoclave for 15 min at 15 psi pressure–121°C.

Trace Elements Solution:

Composition per liter:

Ferric citrate	30.0mg
$CaCl_2 \cdot 2H_2O$	30.0mg
$MgCl_2 \cdot 4H_2O$	5.0mg
$ZnSO_4 \cdot 7H_2O$	5.0mg
$CuSO_4 \cdot 5H_2O$	0.5mg

Preparation of Trace Elements Solution: Add components to distilled/deionized water and bring volume to 1.0L. Mix thoroughly.

Preparation of Medium: Add components, except methanol solution and Na_2CO_3 solution, to distilled/deionized water and bring volume to 900.0mL. Mix thoroughly. Autoclave for 15 min at 15 psi pressure–121°C. Cool to 25°C. Aseptically add 50.0mL sterile Na_2CO_3 solution and 50.0mL sterile methanol solution. Mix thoroughly. Aseptically distribute into sterile tubes or flasks.

Use: For the cultivation of *Methylophaga alcalica.*

Methylophaga Broth

Composition per 103.0mL:

Mineral base	100.0mL
Solution T	2.0mL
Vitamin B_{12} solution	1.0mL
Methanol	0.3mL

pH 7.3 ± 0.2 at 25°C

Mineral Base:

Composition per liter:

NaCl	24.0g
$MgCl_2 \cdot 6H_2O$	3.0g
$MgSO_4 \cdot 7H_2O$	2.0g
$CaCl_2 \cdot 2H_2O$	1.0g
KCl	0.5g
Bis-Tris buffer (bis[2-hydroxyethyl]amino-tris[hydroxymethyl]-methane)	0.5g
Wolfe's mineral solution	10.0mL

Preparation of Mineral Base: Add components to distilled/deionized water and bring volume to 1.0L. Mix thoroughly. Adjust pH to 7.3. Autoclave for 15 min at 15 psi pressure–121°C.

Wolfe's Mineral Solution:

Composition per liter:

$MgSO_4 \cdot 7H_2O$	3.0g
Nitrilotriacetic acid	1.5g
NaCl	1.0g
$MnSO_4 \cdot H_2O$	0.5g
$FeSO_4 \cdot 7H_2O$	0.1g
$CoCl_2 \cdot 6H_2O$	0.1g
$CaCl_2$	0.1g
$ZnSO_4 \cdot 7H_2O$	0.1g
$CuSO_4 \cdot 5H_2O$	0.01g
$AlK(SO_4)_2 \cdot 12H_2O$	0.01g
H_3BO_3	0.01g
$Na_2MoO_4 \cdot 2H_2O$	0.01g

Preparation of Wolfe's Mineral Solution: Add nitrilotriacetic acid to 500.0mL of distilled/deionized water. Dissolve by adjusting pH to 6.5 with KOH. Add remaining components. Add distilled/deionized water to 1.0L.

Solution T:

Composition per 100.0mL:

NH_4Cl	10.0g
Bis-Tris buffer (bis[2-hydroxyethyl]amino-tris[hydroxymethyl]-methane)	10.0g
KH_2PO_4	0.7g
Ferric ammonium citrate	0.3g

Preparation of Solution T: Add components to distilled/deionized water and bring volume to 100.0mL. Mix thoroughly. Adjust pH to 7.3. Autoclave for 15 min at 15 psi pressure–121°C.

Vitamin B_{12} Solution:
Composition per 10.0mL:
Vitamin B_{12} 1.0μg

Preparation of Vitamin B_{12} Solution: Add vitamin B_{12} to 10.0mL of distilled/deionized water. Mix thoroughly. Filter sterilize.

Preparation of Medium: Aseptically combine sterile solution T and sterile vitamin B_{12} solution with the sterile mineral base. Filter sterilize methanol and add to basal medium. Aseptically distribute into sterile tubes or sterile flasks.

Use: For the cultivation and maintenance of *Methylophaga marina*.

Methylophaga Medium

Composition per liter:
NaCl 25.0g
Agar 20.0g
Peptone 10.0g
Beef extract 7.0g
K_2HPO_4 .... 1.0g
$(NH_4)_2SO_4$ 1.0g
Methanol, filter sterilized 10.0mL
pH 7.0 ± 0.2 at 25°C

Preparation of Medium: Add components, except methanol, to distilled/deionized water and bring volume to 990.0mL. Mix thoroughly. Gently heat and bring to boiling. Autoclave for 15 min at 15 psi pressure–121°C. Cool to 50°–55°C. Aseptically add 10.0mL of sterile methanol. Mix thoroughly. Pour into sterile Petri dishes or distribute into sterile tubes.

Use: For the cultivation and maintenance of *Methylophaga marina* and *Methylophaga thalassica*.

Methylophaga sulfidovorans Medium (DSMZ Medium 951)

Composition per liter:
NaCl 15.0g
Na_2CO_3 .... 2.0g
$MgSO_4 \cdot 7H_2O$ 1.0g
$(NH_4)_2SO_4$ 0.5g
$CaCl_2 \cdot 6H_2O$ 0.33g
KCl 0.2g
KH_2PO_4 .... 0.02g
DMS (Dimethylsulphide) 62.0mg
$FeSO_4 \cdot 7H_2O$ 1.0mg
Trace elements solution SL-10 .... 1.0mL
Vitamin solution 1.0mL
pH 7.5 ± 0.3 at 25°C

Trace Elements Solution SL-10:
Composition per liter:
$FeCl_2 \cdot 4H_2O$ 1.5g
$CoCl_2 \cdot 6H_2O$ 190.0mg
$MnCl_2 \cdot 4H_2O$ 100.0mg
$ZnCl_2$ 70.0mg
$Na_2MoO_4 \cdot 2H_2O$ 36.0mg
$NiCl_2 \cdot 6H_2O$ 24.0mg
H_3BO_3 .... 6.0mg
$CuCl_2 \cdot 2H_2O$ 2.0mg
HCl (25% solution) 10.0mL

Preparation of Trace Elements Solution SL-10: Add $FeCl_2 \cdot 4H_2O$ to 10.0mL of HCl solution. Mix thoroughly. Add distilled/deionized water and bring volume to 1.0L. Add remaining components. Mix thoroughly. Sparge with 80% N_2 + 20% CO_2.

Vitamin Solution:
Composition per liter:
Pyridoxine-HCl 500.0mg
Nicotinic acid 200.0mg
Thiamine 100.0mg
p-Aminobenzoic acid 100.0mg
Pantothenate 50.0mg
Biotin 20.0mg
Riboflavin 10.0mg
Vitamin B_{12} 10.0mg

Preparation of Vitamin Solution: Add components to distilled/deionized water and bring volume to 1.0L. Mix thoroughly. Sparge with 80% H_2 + 20% CO_2.

Preparation of Medium: Add components to distilled/deionized water and bring volume to 1.0L. Mix thoroughly. Adjust pH to 7.5. Distribute into tubes or bottles. Autoclave for 15 min at 15 psi pressure–121°C.

Use: For the cultivation of *Methylophaga sulfidovorans*.

Methylophaga thalassica Agar (LMG Medium 73)

Composition per liter:
NaCl 25.0g
Agar 20.0g
Peptone 10.0g
Lab Lemco beef extract 7.0g
K_2HPO_4 .... 1.0g
$(NH_4)_2SO_4$ 1.0g
pH 7.0 ± 0.2 at 25°C

Preparation of Medium: Add components, except methanol, to 990.0mL distilled/deionized water. Mix thoroughly. Gently heat and bring to boiling. Autoclave for 15 min at 15 psi pressure–121°C. Cool to 45°C. Aseptically add 10.0mL sterile methanol. Mix thoroughly. Pour into sterile Petri dishes or distribute into sterile tubes.

Use: For the cultivation of *Methylophaga thalassica*.

Methylosarcina quisquillarum/ Methylosarcina fibrata Medium (DSMZ Medium 921)

Composition per 1012.1mL:
Solution 1 .... 100.0mL
Phosphate buffer 10.0mL
Solution 3 .... 1.0mL
Trace elements 1.0mL
Solution 2 .... 0.1mL
pH 7.0 ± 0.2 at 25°C

Solution 1 (10X NMS Salts):
Composition per liter:
KNO_3 .... 10.0g
$MgSO_4 \cdot 6H_2O$ 10.0g
$CaCl_2 \cdot 2H_2O$ 2.0g

Preparation of Solution 1 (10X NMS Salts): Add components to 700.0mL distilled/deionized water. Mix thoroughly. Bring volume to 1.0L with distilled/deionized water. Mix thoroughly.

Solution 2 (Fe EDTA):

Composition per liter:

Fe EDTA	3.8g

Preparation of Solution 2 (Fe EDTA): Add Fe EDTA to distilled/deionized water and bring volume to 1.0L. Mix thoroughly.

Solution 3 (Sodium Molybdate):

Composition per liter:

$Na_2MoO_4 \cdot 4H_2O$	0.26g

Preparation of Solution 3 (Sodium Molybdate): Add $Na_2MoO_4 \cdot 4H_2O$ to distilled/deionized water and bring volume to 1.0L. Mix thoroughly.

Trace Elements:

Composition per 100.0mL:

$CuSO_4 \cdot 5H_2O$	100.0mg
$FeSO_4 \cdot 7H_2O$	50.0mg
$ZnSO_4 \cdot 7H_2O$	40.0mg
EDTA disodium salt	25.0mg
$CoCl_2 \cdot 6H_2O$	5.0mg
$MnCl_2 \cdot 4H_2O$	2.0mg
H_3BO_3	1.5mg
$NiCl_2 \cdot 6H_2O$	1.0mg

Preparation of Trace Elements: Add components to distilled/deionized water and bring volume to 100.0mL. Mix thoroughly.

Phosphate Buffer:

Composition per liter:

$Na_2HPO_4 \cdot 2H_2O$	71.6g
KH_2PO_4	26.0g

Preparation of Phosphate Buffer: Add components to 800.0mL distilled/deionized water. Mix thoroughly. Adjust pH to 6.8. Bring volume to 1.0L. Mix thoroughly. Autoclave for 15 min at 15 psi pressure–121°C. Cool to 55°C.

Preparation of Medium: Add 100.0mL solution 1 to distilled/deionized water and bring volume to 1.0L. Mix thoroughly. Add 1.0mL of solution 3, 1.0mL of the trace elements, and 0.1mL of solution 2. Autoclave for 15 min at 15 psi pressure–121°C. Cool to 55°C. Aseptically add 10.0mL phosphate buffer. Mix thoroughly. Aseptically distribute to sterile tubes or bottles.

Use: For the cultivation of *Methylosarcina fibrata* and *Methylosarcina quisquiliarum.*

Methylotrophic *Arthrobacter Hyphomicrobium* Medium (DSMZ Medium 939)

Composition per liter:

$Na_2HPO_4 \cdot 2H_2O$	7.9g
Dimethylsulfone	1.9g
KH_2PO_4	1.5g
NH_4Cl	0.8g
$MgSO_4 \cdot 7H_2O$	0.1g
Trace elements solution	10.0mL

pH 7.2-7.5 at 25°C

Trace Elements Solution:

Composition per liter:

EDTA disodium salt	50.0g
NaOH	9.0g
$CaCl_2 \cdot 2H_2O$	7.34g
$FeSO_4 \cdot 7H_2O$	5.0g
$MnCl_2 \cdot 4H_2O$	2.5g
$ZnSO_4 \cdot 7H_2O$	1.0g
$CoCl_2 \cdot 6H_2O$	0.5g
$NH_4(MoO_4)$	0.5g
$CuSO_4 \cdot 5H_2O$	0.2g

Preparation of Trace Elements Solution: Add Na_2-EDTA to 400.0mL distilled/deionized water. Mix thoroughly. Add 9.0g NaOH. Mix thoroughly. Individually dissolve each of the other components in 40.0mL distilled/deionized water. Add each of the other dissolved components to the EDTA solution and bring volume to 1.0L with distilled/deionized water. Mix thoroughly. Adjust the pH to 6.0 with 1*M* NaOH.

Preparation of Medium: Add components to distilled/deionized water and bring volume to 1.0L. Mix thoroughly. Distribute into tubes or flasks. Autoclave for 15 min at 15 psi pressure–121°C.

Use: For the cultivation of *Hyphomicrobium sulfonivorans.*

Methylotrophic *Arthrobacter Hyphomicrobium* Medium (DSMZ Medium 939)

Composition per liter:

$Na_2HPO_4 \cdot 2H_2O$	7.9g
KH_2PO_4	1.5g
NH_4Cl	0.8g
$MgSO_4 \cdot 7H_2O$	0.1g
Trace elements solution	10.0mL
Methanol	10.0mL

pH 7.2–7.5 at 25°C

Trace Elements Solution:

Composition per liter:

EDTA disodium salt	50.0g
NaOH	9.0g
$CaCl_2 \cdot 2H_2O$	7.34g
$FeSO_4 \cdot 7H_2O$	5.0g
$MnCl_2 \cdot 4H_2O$	2.5g
$ZnSO_4 \cdot 7H_2O$	1.0g
$CoCl_2 \cdot 6H_2O$	0.5g
$NH_4(MoO_4)$	0.5g
$CuSO_4 \cdot 5H_2O$	0.2g

Preparation of Trace Elements Solution: Add Na_2-EDTA to 400.0mL distilled/deionized water. Mix thoroughly. Add 9.0g NaOH. Mix thoroughly. Individually dissolve each of the other components in 40.0mL distilled/deionized water. Add each of the other dissolved components to the EDTA solution and bring volume to 1.0L with distilled/deionized water. Mix thoroughly. Adjust the pH to 6.0 with 1*M* NaOH.

Preparation of Medium: Add components to distilled/deionized water and bring volume to 1.0L. Mix thoroughly. Distribute into tubes or flasks. Autoclave for 15 min at 15 psi pressure–121°C.

Use: For the cultivation of *Hyphomicrobium sulfonivorans.*

Methylotrophic *Arthrobacter Hyphomicrobium* Medium (DSMZ Medium 939)

Composition per liter:

$Na_2HPO_4 \cdot 2H_2O$	7.9g
KH_2PO_4	1.5g
Methylamine	1.0g
NH_4Cl	0.8g
$MgSO_4 \cdot 7H_2O$	0.1g
Trace elements solution	10.0mL

pH 7.2–7.5 at 25°C

Trace Elements Solution:

Composition per liter:

EDTA disodium salt	50.0g
NaOH	9.0g
$CaCl_2 \cdot 2H_2O$	7.34g
$FeSO_4 \cdot 7H_2O$	5.0g
$MnCl_2 \cdot 4H_2O$	2.5g
$ZnSO_4 \cdot 7H_2O$	1.0g
$CoCl_2 \cdot 6H_2O$	0.5g
$NH_4(MoO_4)$	0.5g
$CuSO_4 \cdot 5H_2O$	0.2g

Preparation of Trace Elements Solution: Add Na_2-EDTA to 400.0mL distilled/deionized water. Mix thoroughly. Add 9.0g NaOH. Mix thoroughly. Individually dissolve each of the other components in 40.0mL distilled/deionized water. Add each of the other dissolved components to the EDTA solution and bring volume to 1.0L with distilled/deionized water. Mix thoroughly. Adjust the pH to 6.0 with 1*M* NaOH.

Preparation of Medium: Add components to distilled/deionized water and bring volume to 1.0L. Mix thoroughly. Distribute into tubes or flasks. Autoclave for 15 min at 15 psi pressure–121°C.

Use: For the cultivation of *Hyphomicrobium sulfonivorans.*

Methylotrophic *Arthrobacter Hyphomicrobium* Medium (DSMZ Medium 939)

Composition per liter:

$Na_2HPO_4 \cdot 2H_2O$	7.9g
KH_2PO_4	1.5g
Glucosel	1.0g
NH_4Cl	0.8g
$MgSO_4 \cdot 7H_2O$	0.1g
Trace elements solution	10.0mL

pH 7.2–7.5 at 25°C

Trace Elements Solution:

Composition per liter:

EDTA disodium salt	50.0g
NaOH	9.0g
$CaCl_2 \cdot 2H_2O$	7.34g
$FeSO_4 \cdot 7H_2O$	5.0g
$MnCl_2 \cdot 4H_2O$	2.5g
$ZnSO_4 \cdot 7H_2O$	1.0g
$CoCl_2 \cdot 6H_2O$	0.5g
$NH_4(MoO_4)$	0.5g
$CuSO_4 \cdot 5H_2O$	0.2g

Preparation of Trace Elements Solution: Add Na_2-EDTA to 400.0mL distilled/deionized water. Mix thoroughly. Add 9.0g NaOH. Mix thoroughly. Individually dissolve each of the other components in 40.0mL distilled/deionized water. Add each of the other dissolved components to the EDTA solution and bring volume to 1.0L with distilled/deionized water. Mix thoroughly. Adjust the pH to 6.0 with 1*M* NaOH.

Preparation of Medium: Add components to distilled/deionized water and bring volume to 1.0L. Mix thoroughly. Distribute into tubes or flasks. Autoclave for 15 min at 15 psi pressure–121°C.

Use: For the cultivation of *Hyphomicrobium sulfonivorans.*

Methylotrophic *Arthrobacter Hyphomicrobium* Medium (DSMZ Medium 939)

Composition per liter:

$Na_2HPO_4 \cdot 2H_2O$	7.9g
KH_2PO_4	1.5g
Fructose	1.0g
NH_4Cl	0.8g
$MgSO_4 \cdot 7H_2O$	0.1g
Trace elements solution	10.0mL

pH 7.2–7.5 at 25°C

Trace Elements Solution:

Composition per liter:

EDTA disodium salt	50.0g
NaOH	9.0g
$CaCl_2 \cdot 2H_2O$	7.34g
$FeSO_4 \cdot 7H_2O$	5.0g
$MnCl_2 \cdot 4H_2O$	2.5g
$ZnSO_4 \cdot 7H_2O$	1.0g
$CoCl_2 \cdot 6H_2O$	0.5g
$NH_4(MoO_4)$	0.5g
$CuSO_4 \cdot 5H_2O$	0.2g

Preparation of Trace Elements Solution: Add Na_2-EDTA to 400.0mL distilled/deionized water. Mix thoroughly. Add 9.0g NaOH. Mix thoroughly. Individually dissolve each of the other components in 40.0mL distilled/deionized water. Add each of the other dissolved components to the EDTA solution and bring volume to 1.0L with distilled/deionized water. Mix thoroughly. Adjust the pH to 6.0 with 1*M* NaOH.

Preparation of Medium: Add components to distilled/deionized water and bring volume to 1.0L. Mix thoroughly. Distribute into tubes or flasks. Autoclave for 15 min at 15 psi pressure–121°C.

Use: For the cultivation of *Hyphomicrobium sulfonivorans.*

Methylpyridine Medium

Composition per 1002.0mL:

K_2HPO_4	0.61g
KH_2PO_4	0.39g
KCl	0.25g
Yeast extract	0.1g
Wolfe's mineral solution	10.0mL
2-Methylpyridine	1.0mL

Wolfe's Mineral Solution:

Composition per liter:

$MgSO_4 \cdot 7H_2O$	3.0g
Nitrilotriacetic acid	1.5g
NaCl	1.0g
$MnSO_4 \cdot 2H_2O$	0.5g
$CoCl_2 \cdot 6H_2O$	0.1g
$ZnSO_4 \cdot 7H_2O$	0.1g
$CaCl_2 \cdot 2H_2O$	0.1g
$FeSO_4 \cdot 7H_2O$	0.1g
$NiCl_2 \cdot 6H_2O$	0.025g
$KAl(SO_4)_2 \cdot 12H_2O$	0.02g
$CuSO_4 \cdot 5H_2O$	0.01g
H_3BO_3	0.01g
$Na_2MoO_4 \cdot 2H_2O$	0.01g
$Na_2SeO_3 \cdot 5H_2O$	0.3mg

Preparation of Wolfe's Mineral Solution: Add nitrilotriacetic acid to 500.0mL of distilled/deionized water. Adjust pH to 6.5 with

KOH. Add remaining components one at a time. Add distilled/deionized water to 1.0L. Adjust pH to 6.8.

Preparation of Medium: Add components, except 2-methylpyridine, to distilled/deionized water and bring volume to 1.0L. Mix thoroughly. Autoclave for 15 min at 15 psi pressure–121°C. Cool to room temperature. In a fume hood, aseptically add 1.0mL of 2-methylpyridine. Mix thoroughly. Aseptically distribute into sterile tubes or flasks. Use polyurethane foam closures to eliminate odors caused by volatilization of 2-methylpyridine.

Use: For the cultivation of *Arthrobacter* species.

M-FC Agar

See: **FC Agar**

M-FC Broth

See: **FC Broth**

M-FC HiVeg Agar Base with Rosalic Acid

Composition per liter:

Agar	15.0g
Lactose	12.5g
Plant hydrolysate No. 1	10.0g
Plant peptone No. 3	5.0g
NaCl	5.0g
Yeast extract	3.0g
Synthetic detergent No. I	1.5g
Aniline Blue	0.1g
Rosolic acid solution	10.0mL

pH 7.4 ± 0.2 at 25°C

Source: This medium, without rosolic acid, is available as a premixed powder from HiMedia.

Rosolic Acid Solution:

Composition per 100.0mL:

Rosolic acid	1.0g

Preparation of Rosolic Acid Solution: Add rosolic acid to 0.2*N* NaOH and bring volume to 100.0L. Mix thoroughly.

Preparation of Medium: Add 10.0mL rosolic acid solution to 950.0mL of distilled/deionized water. Mix thoroughly. Add other components and bring volume to 1.0L with distilled/deionized water. Mix thoroughly. Gently heat and bring to boiling with frequent mixing. Do not autoclave. Pour into sterile Petri dishes or leave in tubes.

Use: For the detection and enumeration of fecal coliforms using the membrane filter technique.

M-FC HiVeg Agar Base, Modified with Rosalic Acid

Composition per liter:

Agar	15.0g
Plant hydrolysate No. 1	10.0g
Inositol	10.0g
Plant peptone No. 3	5.0g
NaCl	5.0g
Yeast extract	3.0g
Synthetic detergent No. I	1.5g
Aniline blue	0.1g
Rosolic acid solution	10.0mL

pH 7.4 ± 0.2 at 25°C

Source: This medium, without rosolic acid, is available as a premixed powder from HiMedia.

Rosolic Acid Solution:

Composition per 100.0mL:

Rosolic acid	1.0g

Preparation of Rosolic Acid Solution: Add rosolic acid to 0.2*N* NaOH and bring volume to 100.0L. Mix thoroughly.

Preparation of Medium: Add 10.0mL rosolic acid solution to 950.0mL of distilled/deionized water. Mix thoroughly. Add other components and bring volume to 1.0L with distilled/deionized water. Mix thoroughly. Gently heat and bring to boiling with frequent mixing. Do not autoclave. Pour into sterile Petri dishes or leave in tubes.

Use: For the detection and enumeration of fecal coliforms using membrane filter technique.

M-FC HiVeg Broth Base with Rosalic Acid

Composition per liter:

Lactose	12.5g
Plant hydrolysate No. 1	10.0g
Plant peptone No. 3	5.0g
NaCl	5.0g
Yeast extract	3.0g
Synthetic detergent No. I	1.5g
Aniline blue	0.1g
Rosolic acid solution	10.0mL

pH 7.4 ± 0.2 at 25°C

Source: This medium, without rosolic acid, is available as a premixed powder from HiMedia.

Rosolic Acid Solution:

Composition per 100.0mL:

Rosolic acid	1.0g

Preparation of Rosolic Acid Solution: Add rosolic acid to 0.2*N* NaOH and bring volume to 100.0L. Mix thoroughly.

Preparation of Medium: Add 10.0mL of rosolic acid solution to 950.0mL of distilled/deionized water. Mix thoroughly. Add other components and bring volume to 1.0L with distilled/deionized water. Mix thoroughly. Gently heat and bring to boiling with frequent mixing. Do not autoclave. Pour into sterile Petri dishes or leave in tubes.

Use: For the cultivation of fecal coliform bacteria from waters and the enumeration of coliform bacteria using the membrane filtration method.

M-FC Agar

Composition per liter:

Agar	15.0g
Tryptose	10.0g
Inositol	10.0g
Proteose peptone	5.0g
NaCl	5.0g
Yeast extract	3.0g
Bile salts mixture	1.5g
Aniline Blue	0.1g
Selective supplement solution	10.0mL
Rosolic acid solution	10.0mL

pH 7.4 ± 0.2 at 25°C

Source: This medium is available from HiMedia.

Rosolic Acid Solution:

Composition per 10.0mL:

Rosolic acid	0.1g

Preparation of Rosolic Acid Solution: Add rosoloic acid to distilled/deionized water and bring volume to 10.0mL. Mix thoroughly. Filter sterilize.

Selective Supplement Solution:
Composition per 10.0mL:

Carbenicillin	0.05g

Preparation of Selective Supplement Solution: Add carbenicillin to distilled/deionized water and bring volume to 10.0mL. Mix thoroughly. Filter sterilize.

Preparation of Medium: Add components, except selective supplement solution, to distilled/deionized water and bring volume to 990.0mL. Mix thoroughly. Autoclave for 15 min at 15 psi pressure–121°C. Cool to 50°C. Aseptically add selective supplement solution. Mix thoroughly. Pour into Petri dishes or aseptically distribute into sterile tubes.

Use: For the rapid enumeration of *Klebsiella* using the membrane filter technique.

M-FC Agar Base

Composition per liter:

Agar	15.0g
Lactose	12.5g
Tryptose	10.0g
Proteose peptone	5.0g
NaCl	5.0g
Yeast extract	3.0g
Bile salts mixture	1.5g
Aniline Blue	0.1g
Rosolic acid solution	10.0mL

pH 7.4 ± 0.2 at 25°C

Source: This medium is available from HiMedia.

Rosolic Acid Solution:
Composition per 10.0mL:

Rosolic acid	0.1g

Preparation of Rosolic Acid Solution: Add rosolic acidto distilled/deionized water and bring volume to 10.0mL. Mix thoroughly. Filter sterilize.

Preparation of Medium: Add components to distilled/deionized water and bring volume to 1.0L. Mix thoroughly. Gently heat to dissolve components. Do not autoclave. Cool to 50°C. Pour into Petri dishes or aseptically distribute into sterile tubes.

Use: For the detection and enumeration of fecal coliforms using the membrane filter technique at 44.5°C.

m-Fecal Coliform Agar
See: **FC Agar**

m-Fecal Coliform Agar, Modified
See: **Fecal Coliform Agar, Modified**

m-Fecal Coliform Broth
See: **FC Broth**

MG Medium

Composition per liter:

NaCl	100.0g
$MgSO_4 \cdot 7H_2O$	3.45g
$MgCl_2 \cdot 6H_2O$	2.75g
Sodium acetate	1.0g
KCl	0.335g
NH_4Cl	0.25g
$CaCl_2 \cdot 2H_2O$	0.14g
$K_2HPO_4 \cdot 3H_2O$	0.14g
Resazurin	1.0mg
$NaHCO_3$ solution	80.0mL
Trimethylamine·HCl solution	20.0mL
Na_2CO_3 solution	10.0mL
Trace elements solution	10.0mL
Vitamin solution	10.0mL
L-Cysteine·HCl solution	10.0mL
$Na_2S \cdot 9H_2O$ solution	10.0mL

pH 6.9 ± 0.2 at 25°C

$NaHCO_3$ Solution:
Composition per 100.0mL:

$NaHCO_3$	5.0g

Preparation of $NaHCO_3$ Solution: Add $NaHCO_3$ to distilled/deionized water and bring volume to 100.0mL. Mix thoroughly. Autoclave for 15 min at 15 psi pressure–121°C.

Trimethylamine·HCl Solution:
Composition per 20.0mL:

Trimethylamine·HCl	5.0g

Preparation of Trimethylamine·HCl Solution: Add trimethylamine·HCl to distilled/deionized water and bring volume to 20.0mL. Mix thoroughly. Sparge with 100% N_2. Autoclave for 15 min at 15 psi pressure–121°C.

Na_2CO_3 Solution:
Composition per 10.0mL:

Na_2CO_3	0.5g

Preparation of Na_2CO_3 Solution: Add Na_2CO_3 to distilled/deionized water and bring volume to 10.0mL. Mix thoroughly. Sparge with 100% N_2. Autoclave for 15 min at 15 psi pressure–121°C.

Trace Elements Solution:
Composition per liter:

$MgSO_4 \cdot 7H_2O$	3.0g
Nitrilotriacetic acid	1.5g
NaCl	1.0g
$MnSO_4 \cdot 2H_2O$	0.5g
$CoSO_4 \cdot 7H_2O$	0.18g
$ZnSO_4 \cdot 7H_2O$	0.18g
$CaCl_2 \cdot 2H_2O$	0.1g
$FeSO_4 \cdot 7H_2O$	0.1g
$NiCl_2 \cdot 6H_2O$	0.025g
$KAl(SO_4)_2 \cdot 12H_2O$	0.02g
$CuSO_4 \cdot 5H_2O$	0.01g
H_3BO_3	0.01g
$Na_2MoO_4 \cdot 2H_2O$	0.01g
$Na_2SeO_3 \cdot 5H_2O$	0.3mg

Preparation of Trace Elements Solution: Add nitrilotriacetic acid to approximately 500.0mL distilled/deionized water. Dissolve by adding KOH and adjust pH to 6.5. Add remaining components. Bring volume to 1.0L with additional distilled/deionized water. Adjust pH to 7.0 with KOH.

Vitamin Solution:
Composition per liter:

Pyridoxine·HCl	10.0mg
Calcium DL-pantothenate	5.0mg
Lipoic acid	5.0mg

Nicotinic acid	5.0mg
p-Aminobenzoic acid	5.0mg
Riboflavin	5.0mg
Thiamine·HCl	5.0mg
Biotin	2.0mg
Folic acid	2.0mg
Vitamin B_{12}	0.1mg

Preparation of Vitamin Solution: Add components to distilled/deionized water and bring volume to 1.0L. Mix thoroughly. Filter sterilize. Sparge with 100% N_2.

L-Cysteine·HCl Solution:
Composition per 10.0mL:

L-Cysteine·HCl	0.5g

Preparation of L-Cysteine·HCl Solution: Add L-cysteine·HCl to distilled/deionized water and bring volume to 10.0mL. Mix thoroughly. Autoclave under 100% N_2 for 15 min at 15 psi pressure–121°C.

$Na_2S·9H_2O$ Solution:
Composition per 10.0mL:

$Na_2S·9H_2O$	0.5g

Preparation of $Na_2S·9H_2O$ Solution: Add $Na_2S·9H_2O$ to distilled/deionized water and bring volume to 10.0mL. Mix thoroughly. Sparge with 100% N_2. Autoclave for 15 min at 15 psi pressure–121°C.

Preparation of Medium: Add components, except $NaHCO_3$ solution, trimethylamine·HCl solution, Na_2CO_3 solution, vitamin solution, L-cysteine·HCl solution, and $Na_2S·9H_2O$ solution, to distilled/deionized water and bring volume to 860.0mL. Mix thoroughly. Sparge with 100% N_2 for 20 min. Then sparge with 80% N_2 + 20% CO_2 for 10 min. Anaerobically distribute into tubes or bottles. Autoclave for 15 min at 15 psi pressure–121°C. Aseptically and anaerobically add 80.0mL of sterile $NaHCO_3$ solution, 20.0mL of sterile trimethylamine·HCl solution, 10.0mL of sterile Na_2CO_3 solution, 10.0mL of sterile vitamin solution, 10.0mL of sterile L-cysteine·HCl solution, and 10.0mL of sterile $Na_2S·9H_2O$ solution. Mix thoroughly.

Use: For the cultivation and maintenance of *Methanohalobium* species, *Methanohalophilus halophilus*, and *Methanohalophilus* species.

MG Medium

Composition per liter:

NaCl	150.0g
$MgSO_4·7H_2O$	3.45g
$MgCl_2·6H_2O$	2.75g
Sodium acetate	1.0g
KCl	0.335g
NH_4Cl	0.25g
$CaCl_2·2H_2O$	0.14g
$K_2HPO_4·3H_2O$	0.14g
Resazurin	1.0mg
$NaHCO_3$ solution	80.0mL
Trimethylamine·HCl solution	20.0mL
Na_2CO_3 solution	10.0mL
Trace elements solution	10.0mL
Vitamin solution	10.0mL
L-Cysteine·HCl solution	10.0mL
$Na_2S·9H_2O$ solution	10.0mL

pH 6.9 ± 0.2 at 25°C

$NaHCO_3$ Solution:
Composition per 100.0mL:

$NaHCO_3$	5.0g

Preparation of $NaHCO_3$ Solution: Add $NaHCO_3$ to distilled/deionized water and bring volume to 100.0mL. Mix thoroughly. Autoclave for 15 min at 15 psi pressure–121°C.

Trimethylamine·HCl Solution:
Composition per 20.0mL:

Trimethylamine·HCl	5.0g

Preparation of Trimethylamine·HCl Solution: Add trimethylamine·HCl to distilled/deionized water and bring volume to 20.0mL. Mix thoroughly. Sparge with 100% N_2. Autoclave for 15 min at 15 psi pressure–121°C.

Na_2CO_3 Solution:
Composition per 10.0mL:

Na_2CO_3	0.5g

Preparation of Na_2CO_3 Solution: Add Na_2CO_3 to distilled/deionized water and bring volume to 10.0mL. Mix thoroughly. Sparge with 100% N_2. Autoclave for 15 min at 15 psi pressure–121°C.

Trace Elements Solution:
Composition per liter:

$MgSO_4·7H_2O$	3.0g
Nitrilotriacetic acid	1.5g
NaCl	1.0g
$MnSO_4·2H_2O$	0.5g
$CoSO_4·7H_2O$	0.18g
$ZnSO_4·7H_2O$	0.18g
$CaCl_2·2H_2O$	0.1g
$FeSO_4·7H_2O$	0.1g
$NiCl_2·6H_2O$	0.025g
$KAl(SO_4)_2·12H_2O$	0.02g
$CuSO_4·5H_2O$	0.01g
H_3BO_3	0.01g
$Na_2MoO_4·2H_2O$	0.01g
$Na_2SeO_3·5H_2O$	0.3mg

Preparation of Trace Elements Solution: Add nitrilotriacetic acid to approximately 500.0mL of distilled/deionized water. Dissolve by adding KOH and adjust pH to 6.5. Add remaining components. Bring volume to 1.0L with additional distilled/deionized water. Adjust pH to 7.0 with KOH.

Vitamin Solution:
Composition per liter:

Pyridoxine·HCl	10.0mg
Calcium DL-pantothenate	5.0mg
Lipoic acid	5.0mg
Nicotinic acid	5.0mg
p-Aminobenzoic acid	5.0mg
Riboflavin	5.0mg
Thiamine·HCl	5.0mg
Biotin	2.0mg
Folic acid	2.0mg
Vitamin B_{12}	0.1mg

Preparation of Vitamin Solution: Add components to distilled/deionized water and bring volume to 1.0L. Mix thoroughly. Filter sterilize. Sparge with 100% N_2.

L-Cysteine·HCl Solution:
Composition per 10.0mL:

L-Cysteine·HCl	0.5g

Preparation of L-Cysteine·HCl Solution: Add L-cysteine·HCl to distilled/deionized water and bring volume to 10.0mL. Mix thoroughly. Autoclave under 100% N_2 for 15 min at 15 psi pressure–121°C.

$Na_2S·9H_2O$ Solution:
Composition per 10.0mL:
$Na_2S·9H_2O$ 0.5g

Preparation of $Na_2S·9H_2O$ Solution: Add $Na_2S·9H_2O$ to distilled/deionized water and bring volume to 10.0mL. Mix thoroughly. Sparge with 100% N_2. Autoclave for 15 min at 15 psi pressure–121°C.

Preparation of Medium: Add components, except $NaHCO_3$ solution, trimethylamine·HCl solution, Na_2CO_3 solution, vitamin solution, L-cysteine·HCl solution, and $Na_2S·9H_2O$ solution, to distilled/deionized water and bring volume to 860.0mL. Mix thoroughly. Sparge with 100% N_2 for 20 min. Then sparge with 80% N_2 + 20% CO_2 for 10 min. Anaerobically distribute into tubes or bottles. Autoclave for 15 min at 15 psi pressure–121°C. Aseptically and anaerobically add 80.0mL of sterile $NaHCO_3$ solution, 20.0mL of sterile trimethylamine·HCl solution, 10.0mL of sterile Na_2CO_3 solution, 10.0mL of sterile vitamin solution, 10.0mL of sterile L-cysteine·HCl solution, and 10.0mL of sterile $Na_2S·9H_2O$ solution. Mix thoroughly.

Use: For the cultivation and maintenance of *Methanohalophilus* species.

MGA Agar

Composition per liter:
Agar 20.0g
Glucose 2.0g
L-Asparagine 1.0g
K_2HPO_4 .... 0.5g
$MgSO_4·7H_2O$ 0.5g
Trace salts solution 1.0mL
pH 7.4 ± 0.2 at 25°C

Trace Salts Solution:
Composition per liter:
$FeSO_4·7H_2O$ 10.0g
$CuSO_4·5H_2O$ 1.0g
$MnSO_4·7H_2O$ 1.0g
$ZnSO_4·7H_2O$ 1.0g

Preparation of Trace Salts Solution: Add components to distilled/deionized water and bring volume to 1.0L. Mix thoroughly. Adjust pH to 8.0.

Preparation of Medium: Add components to distilled/deionized water and bring volume to 1.0L. Mix thoroughly. Gently heat and bring to boiling. Distribute into tubes or flasks. Autoclave for 15 min at 15 psi pressure–121°C. Pour into sterile Petri dishes or leave in tubes.

Use: For the isolation and cultivation of *Actinomadura* species, *Actinopolyspora* species, *Excellospora* species, and *Microspora* species.

m-Green Yeast and Mold Broth
See: **Green Yeast and Mold Broth**

MGTY Agar
See: **Marine Glucose Trypticase™ Yeast Extract Agar**

MGTY Broth
See: **Marine Glucose Trypticase™ Yeast Extract Broth**

MH Agar 15% (LMG Medium 258)

Composition per liter:
NaCl 121.5g
Agar 20.0g
$MgSO_4·7H_2O$ 14.4g
$MgCl_2$ 10.5g
Yeast extract 10.0g
Proteose peptone no.3 .... 5.0g
KCl 3.0g
Glucose 1.0g
$CaCl_2$ 0.54g
NaBr 39.0mg
$NaHCO_3$ solution 10.0mL
pH 7.5 ± 0.2 at 25°C

$NaHCO_3$ Solution:
Composition per 10.0mL:
$NaHCO_3$ 0.09g

Preparation of $NaHCO_3$ Solution: Add $NaHCO_3$ to 10.0mL of distilled/deionized water. Mix thoroughly. Filter sterilize.

Preparation of Medium: Add components, except $NaHCO_3$ solution, to distilled/deionized water and bring volume to 990.0mL. Mix thoroughly. Autoclave for 15 min at 15 psi pressure–121°C. Aseptically add 10.0mL sterile $NaHCO_3$ solution. Pour into sterile Petri dishes or aseptically distribute into sterile tubes.

Use: For cultivation and maintenance of *Bacillus halophilus.*

MH IH Agar

Composition per liter:
Solution A 490.0mL
Solution B 490.0mL
Supplement solution 20.0mL
pH 6.9 ± 0.2 at 25°C

Solution A:
Composition per 490.0mL:
Beef infusion 300.0g
Acid hydrolysate of casein 17.5g
Agar 17.0g
Starch 1.5g

Preparation of Solution A: Add components to distilled/deionized water and bring volume to 490.0mL. Mix thoroughly. Gently heat and bring to boiling. Autoclave for 15 min at 15 psi pressure–121°C. Cool to 45°–50°C.

Solution B:
Composition per 490.0mL:
Hemoglobin 10.0g

Preparation of Solution B: Add hemoglobin to distilled/deionized water and bring volume to 490.0mL. Mix thoroughly. Gently heat and bring to boiling. Autoclave for 15 min at 15 psi pressure–121°C. Cool to 45°–50°C.

Supplement Solution:
Composition per liter:
Glucose 100.0g
L-Cysteine·HCl 25.9g
L-Glutamine 10.0g
L-Cystine 1.1g
Adenine 1.0g
Nicotinamide adenine dinucleotide 0.25g
Vitamin B_{12} 0.1g
Thiamine pyrophosphate 0.1g
Guanine·HCl 0.03g
$Fe(NO_3)_3·6H_2O$ 0.02g

p-Aminobenzoic acid 0.013g
Thiamine·HCl 3.0mg

Source: The supplement solution IsoVitaleX® enrichment is available from BD Diagnostic Systems. This enrichment may be replaced by supplement VX from BD Diagnostic Systems.

Preparation of Supplement Solution: Add components to distilled/deionized water and bring volume to 1.0L. Mix thoroughly. Filter sterilize.

Preparation of Medium: Aseptically combine cooled, sterile solution A and cooled, sterile solution B. Mix thoroughly. Adjust pH to 6.9 with sterile 1*N* HCl or sterile 1*N* KOH. Aseptically add 20.0mL of sterile supplement solution. Pour into sterile Petri dishes or distribute into sterile tubes.

Use: For the cultivation and differentiation of *Legionella* species.

MH Medium

Composition per liter:
NaCl 60.7g
Agar 20.0g
$MgCl_2 \cdot 6H_2O$ 15.0g
Yeast extract 10.0g
$MgSO_4 \cdot 7H_2O$ 7.4g
Proteose peptone No. 3 .... 5.0g
KCl 1.5g
Glucose 1.0g
$CaCl_2$ 0.27g
$NaHCO_3$ 0.45g
NaBr 0.19g

Preparation of Medium: Add components to distilled/deionized water and bring volume to 1.0L. Mix thoroughly. Gently heat and bring to boiling. Distribute into tubes or flasks. Autoclave for 15 min at 15 psi pressure–121°C. Pour into sterile Petri dishes or leave in tubes.

Use: For the cultivation and maintenance of *Deleya salina* and *Volcaniella eurihalina.*

MH Medium

Composition per liter:
NaCl 60.7g
$MgCl_2 \cdot 6H_2O$ 15.0g
Yeast extract 10.0g
$MgSO_4 \cdot 7H_2O$ 7.4g
Proteose peptone No. 3 .... 5.0g
KCl 1.5g
Glucose 1.0g
$CaCl_2$ 0.27g
$NaHCO_3$ 0.045g
NaBr 0.019g

pH 7.2 ± 0.2 at 25°C

Preparation of Medium: Add components to distilled/deionized water and bring volume to 1.0L. Mix thoroughly. Adjust pH to 7.2. Distribute into tubes or flasks. Autoclave for 15 min at 15 psi pressure–121°C.

Use: For the cultivation of *Halomonas eurihalina.*

MH Medium, 2%

Composition per liter:
KH_2PO_4 .... 2.0g
$(NH_4)_2SO_4$ 2.0g
NaCl 0.5g
$MgSO_4 \cdot 7H_2O$ 0.025g
$FeSO_4 \cdot 7H_2O$ 2.0mg
Glucose solution 20.0mL
Methanol, filter sterilized 20.0mL

pH 7.0–7.5 at 25°C

Glucose Solution:
Composition per 20.0mL:
Glucose 1.5g

Preparation of Glucose Solution: Add glucose to distilled/deionized water and bring volume to 20.0mL. Mix thoroughly. Filter sterilize.

Preparation of Medium: Add components, except glucose solution and methanol, to distilled/deionized water and bring volume to 960.0mL. Mix thoroughly. Autoclave for 15 min at 15 psi pressure–121°C. Aseptically add 20.0mL of sterile glucose solution and 20.0mL of filter-sterilized methanol. Mix thoroughly. Aseptically distribute into sterile tubes or flasks.

Use: For the cultivation of *Methylobacillus fructoseoxidans* and *Methylophilus glucoseoxidans.*

MH Medium, 10% (LMG Medium 270)

Composition per liter:
NaCl 81.0g
Agar 15.0g
Yeast extract 10.0g
$MgSO_4 \cdot 7H_2O$ 9.6g
$MgCl_2 \cdot 6H_2O$ 7.0g
Proteose peptone No.3 .... 5.0g
KCl 2.0g
Glucose 1.0g
$CaCl_2$ 0.54g
NaBr 26.0mg
$NaHCO_3$ solution 10.0mL

pH 7.5 ± 0.2 at 25°C

$NaHCO_3$ Solution:
Composition per 10.0mL:
$NaHCO_3$ 0.06g

Preparation of $NaHCO_3$ Solution: Add $NaHCO_3$ to 10.0mL of distilled/deionized water. Mix thoroughly. Filter sterilize.

Preparation of Medium: Add components, except $NaHCO_3$ solution, to distilled/deionized water and bring volume to 990.0mL. Mix thoroughly. Autoclave for 15 min at 15 psi pressure–121°C. Cool to 45°C. Aseptically add 10.0mL sterile $NaHCO_3$ solution. Pour into sterile Petri dishes or aseptically distribute into sterile tubes.

Use: For cultivation and maintenance of *Chromohalobacter israelensis* and *Chromohalobacter canadensis.*

MH Medium, 10%

Composition per liter:
NaCl 81.0g
Yeast extract 10.0g
$MgSO_4 \cdot 7H_2O$ 9.6g
$MgCl_2 \cdot 6H_2O$ 7.0g
Proteose peptone No. 3 .... 5.0g
KCl 2.0g
Glucose 1.0g

$CaCl_2$ 0.36g
NaBr 0.026g
$NaHCO_3$ solution 10.0mL

pH 7.5 ± 0.2 at 25°C

$NaHCO_3$ Solution:
Composition per 10.0mL:
$NaHCO_3$ 0.06g

Preparation of $NaHCO_3$ Solution: Add $NaHCO_3$ to distilled/deionized water and bring volume to 10.0mL. Mix thoroughly. Filter sterilize.

Preparation of Medium: Add components, except $NaHCO_3$ solution, to distilled/deionized water and bring volume to 990.0mL. Mix thoroughly. Autoclave for 15 min at 15 psi pressure–121°C. Aseptically add 10.0mL of sterile $NaHCO_3$ solution. Mix thoroughly. Aseptically distribute into sterile tubes or flasks.

Use: For the cultivation of *Zygomonas mobilis, Salinicoccus hispanicus, Salinicoccus roseus,* and *Pseudomonas beijerinckii.*

MH Medium 15% (LMG Medium 258)

Composition per liter:
NaCl 121.5g
$MgSO_4 \cdot 7H_2O$ 14.4g
$MgCl_2$ 10.5g
Yeast extract 10.0g
Proteose peptone No.3 .... 5.0g
KCl 3.0g
Glucose 1.0g
$CaCl_2$ 0.54g
NaBr 39.0mg
$NaHCO_3$ solution 10.0mL

pH 7.5 ± 0.2 at 25°C

$NaHCO_3$ Solution:
Composition per 10.0mL:
$NaHCO_3$ 0.09g

Preparation of $NaHCO_3$ Solution: Add $NaHCO_3$ to 10.0mL of distilled/deionized water. Mix thoroughly. Filter sterilize.

Preparation of Medium: Add components, except $NaHCO_3$ solution, to distilled/deionized water and bring volume to 990.0mL. Mix thoroughly. Autoclave for 15 min at 15 psi pressure–121°C. Aseptically add 10.0mL sterile $NaHCO_3$ solution. Aseptically distribute into sterile tubes or flasks.

Use: For cultivation and maintenance of *Bacillus halophilus.*

MH Salts

Composition per liter:
NaCl 120.5g
$MgCl_2 \cdot 6H_2O$ 22.4g
Agar 20.0g
$MgSO_4$ 14.4g
Yeast extract 10.0g
Proteose peptone No. 3 .... 5.0g
KCl 3.0g
Glucose 1.0g
$CaCl_2$ 0.54g
$NaHCO_3$ 0.09g
NaBr 0.039g

pH 7.5 ± 0.2 at 25°C

Preparation of Medium: Add components to distilled/deionized water and bring volume to 1.0L. Mix thoroughly. Adjust pH to 7.5. Gently heat and bring to boiling. Distribute into tubes or flasks. Autoclave for 15 min at 15 psi pressure–121°C. Pour into sterile Petri dishes or leave in tubes.

Use: For the cultivation and maintenance of *Bacillus halophilus.*

MIB with Maltose

Composition per liter:
Yeast extract 20.0g
Maltose 10.0g
Glucose 10.0g
Proteose peptone No. 3 .... 5.0g
KH_2PO_4 .... 2.0g
Sorbitan monooleate complex 0.1g

pH 6.7 ± 0.2 at 25°C

Preparation of Medium: Add components to distilled/deionized water and bring volume to 1.0L. Mix thoroughly. Distribute into tubes or flasks. Autoclave for 15 min at 15 psi pressure–121°C.

Use: For the cultivation of *Lactobacillus sanfrancisco.*

Micro Assay Culture Agar

Composition per liter:
Yeast extract 20.0g
Agar 10.0g
Glucose 10.0g
Proteose peptone No. 3 .... 5.0g
KH_2PO_4 .... 2.0g
Sorbitan monooleate complex 0.1g

pH 6.7 ± 0.2 at 25°C

Source: This medium is available as a premixed powder from BD Diagnostic Systems.

Preparation of Medium: Add components to distilled/deionized water and bring volume to 1.0L. Mix thoroughly. Gently heat and bring to boiling. Distribute into tubes in 10.0mL volumes. Autoclave for 15 min at 15 psi pressure–121°C. Just prior to solidification of the agar, disperse precipitate by gently twirling tube.

Use: For carrying stock cultures of lactobacilli and other test microorganisms used in microbiological assays. For the general cultivation of lactobacilli.

Micro Assay Culture Agar

Composition per liter:
Agar 10.0g
Glucose 10.0g
Proteose peptone No. 3 .... 5.0g
Yeast extract 5.0g
KH_2PO_4 .... 2.0g
Polysorbate 80 .... 0.1g

pH 7.2 ± 0.2 at 25°C

Source: This medium is available as a premixed powder from BD Diagnostic Systems.

Preparation of Medium: Add components to distilled/deionized water and bring volume to 1.0L. Mix thoroughly. Gently heat and bring to boiling. Distribute into tubes or flasks. Autoclave for 15 min at 15 psi pressure–121°C. Pour into sterile Petri dishes or leave in tubes.

Use: For the cultivation of lactobacilli and other microorganisms used in microbiological assays.

Micro Assay Culture Broth

Composition per liter:

Glucose	10.0g
Proteose Peptone No. 3	5.0g
Yeast Extract	5.0g
KH_2PO_4	2.0g
Polysorbate 80	0.1g

pH 7.2 ± 0.2 at 25°C

Source: This medium is available as a premixed powder from BD Diagnostic Systems.

Preparation of Medium: Add components to distilled/deionized water and bring volume to 1.0L. Mix thoroughly. Gently heat and bring to boiling. Distribute into tubes or flasks. Autoclave for 15 min at 15 psi pressure–121°C.

Use: For the cultivation of lactobacilli and other microorganisms used in microbiological assays.

Micro Inoculum Broth

Composition per liter:

Yeast extract	20.0g
Glucose	10.0g
Proteose peptone No. 3	5.0g
KH_2PO_4	2.0g
Sorbitan monooleate complex	0.1g

pH 6.7 ± 0.2 at 25°C

Source: This medium is available as a premixed powder from BD Diagnostic Systems.

Preparation of Medium: Add components to distilled/deionized water and bring volume to 1.0L. Mix thoroughly. Distribute into tubes in 10.0mL volumes. Autoclave for 15 min at 15 psi pressure–121°C.

Use: For the cultivation of lactobacilli used in microbiological assays. It is of particular value in the preparation of the inoculum for these tests.

Micro Vitamin Test Culture HiVeg Agar

Composition per liter:

Yeast extract	20.0g
Agar	15.0g
Glucose	10.0g
Plant peptone	5.0g
KH_2PO_4	2.0g
Polysorbate 80	0.1g

pH 6.7 ± 0.2 at 25°C

Source: This medium is available as a premixed powder from HiMedia.

Preparation of Medium: Add components to distilled/deionized water and bring volume to 1.0L. Mix thoroughly. Gently heat and bring to boiling. Distribute into tubes or flasks. Autoclave for 15 min at 15 psi pressure–121°C. Pour into sterile Petri dishes or leave in tubes.

Use: For the cultivation of lactobacilli and other microorganisms used in microbiological assays.

Micro Vitamin Test Inoculum HiVeg Broth

Composition per liter:

Yeast extract	20.0g
Glucose	10.0g
Plant peptone No. 3	5.0g
KH_2PO_4	2.0g
Polysorbate 80	0.1g

pH 6.7 ± 0.1 at 25°C

Source: This medium is available as a premixed powder from HiMedia.

Preparation of Medium: Add components to distilled/deionized water and bring volume to 1.0L. Mix thoroughly. Gently heat and bring to boiling. Distribute into tubes or flasks. Autoclave for 15 min at 15 psi pressure–121°C.

Use: For the cultivation of lactobacilli and other microorganisms used in microbiological assays.

Microbacterium Medium

Composition per liter:

Glucose	10.0g
KH_2PO_4	5.0g
K_2HPO_4	5.0g
Potassium aspartate	5.0g
$(NH_4)_2SO_4$	2.0g
$MgSO_4 \cdot 7H_2O$	0.5g
Calcium pantothenate	0.2g
β-Mercaptopurine	0.1g
$FeSO_4 \cdot 7H_2O$	0.01g
Thiamine·HCl	0.01g
Biotin	0.1mg

pH 7.0 ± 0.2 at 25°C

Preparation of Medium: Add components to distilled/deionized water and bring volume to 1.0L. Mix thoroughly. Distribute into tubes or flasks. Autoclave for 10 min at 15 psi pressure–121°C.

Use: For the cultivation and maintenance of *Microbacterium* species.

Microbial Content Test Agar

Composition per liter:

Agar	15.0g
Pancreatic digest of casein	15.0g
NaCl	5.0g
Tween™ 80	5.0g
Enzymatic hydrolysate of soybean meal	5.0g
Lecithin	0.7g

pH 7.3 ± 0.2 at 25°C

Source: This medium is available as a premixed powder from BD Diagnostic Systems.

Preparation of Medium: Add components to distilled/deionized water and bring volume to 1.0L. Mix thoroughly. Gently heat and bring to boiling. Boil for 1–2 min. Distribute into tubes or flasks. Autoclave for 15 min at 15 psi pressure–121°C. Pour into sterile Petri dishes or leave in tubes.

Use: For use in the microbial content test of water-soluble cosmetic products. Also used for determining the efficiency of sanitization of containers, equipment, and environmental surfaces.

Microbial Content Test HiVeg Agar (Tryptone Soy HiVeg Agar with Lecithin and Tween 80)

Composition per liter:

Agar	15.0g
Plant hydrolysate	15.0g

Papaic digest of soybean meal	5.0g
NaCl	5.0g
Lecithin	0.7g
Polysorbate 80	5.0mL

pH 7.3 ± 0.2 at 25°C

Source: This medium, without polysorbate 80, is available as a premixed powder from HiMedia.

Preparation of Medium: Add components to distilled/deionized water and bring volume to 1.0L. Mix thoroughly. Gently heat and bring to boiling. Boil for 1–2 min. Distribute into tubes or flasks. Autoclave for 15 min at 15 psi pressure–121°C. Pour into sterile Petri dishes or leave in tubes.

Use: For use in the microbial content test of water-soluble cosmetic products. Also used for determining the efficiency of sanitization of containers, equipment, and environmental surfaces.

Microcella alkaliphila Medium (DSMZ Medium 1063)

Composition per liter:

Pancreatic digest of casein	17.0g
NaCl	5.0g
Papaic digest of soybean meal	3.0g
Yeast extract	3.0g
K_2HPO_4	2.5g
Glucose	2.5g
Na-sesquicarbonate solution	100.0mL
Mineral salt solution	50.0mL
Vitamin solution	5.0mL
Trace elements solution SL-10	1.0mL

pH 9.5 ± 0.2 at 25°C

Sodium Sesquicarbonate Solution:

Composition per 100.0mL:

Na_2CO_3, anhydrous	10.6g
$NaHCO_3$	8.42g

Preparation of Sodium Sesquicarbonate Solution: Add components to distilled/deionized water and bring volume to 100.0mL. Mix thoroughly. Filter sterilize.

Mineral Solution:

Composition per liter:

KH_2PO_4	10.0g
$MgCl_2·6H_2O$	6.6g
NaCl	8.0g
NH_4Cl	8.0g
$CaCl_2·2H_2O$	1.0g

Preparation of Mineral Solution: Add components to distilled/deionized water and bring volume to 1.0L. Mix thoroughly.

Vitamin Solution:

Composition per liter:

Pyridoxine·HCl	6.2mg
Nicotinic acid	2.5mg
Thiamine·HCl	1.25mg
p-Aminobenzoic acid	1.25mg
Pantothenic acid	0.62mg
Biotin	0.25mg

Preparation of Vitamin Solution: Add components to distilled/deionized water and bring volume to 1.0L. Mix thoroughly.

Trace Elements Solution SL-10:

Composition per liter:

$MgSO_4·7H_2O$	3.0g
Nitrilotriacetic acid	1.5g
NaCl	1.0g
$MnSO_4·2H_2O$	0.5g
$CoSO_4·7H_2O$	0.18g
$ZnSO_4·7H_2O$	0.18g
$CaCl_2·2H_2O$	0.1g
$FeSO_4·7H_2O$	0.1g
$NiCl_2·6H_2O$	0.025g
$KAl(SO_4)_2·12H_2O$	0.02g
H_3BO_3	0.01g
$Na_2MoO_4·4H_2O$	0.01g
$CuSO_4·5H_2O$	0.01g
$Na_2SeO_3·5H_2O$	0.3mg

Preparation of Trace Elements Solution SL-10: Add nitrilotriacetic acid to 500.0mL of distilled/deionized water. Dissolve by adjusting pH to 6.5 with KOH. Add remaining components. Add distilled/deionized water to 1.0L. Mix thoroughly. Adjust pH to 7.0.

Preparation of Medium: Add components, except sodium sesquicarbonate solution, to distilled/deionized water and bring volume to 900.0mL. Mix thoroughly. Autoclave for 15 min at 15 psi pressure–121°C. Cool to room temperature. Aseptically add sterile 1*M* sodium sesquicarbonate solution to achieve a final pH of 9.5. Aseptically dispense into tubes, flasks, or bottles.

Use: For the cultivation of *Microcella alkaliphila*.

Micrococcus Medium

Composition per liter:

Agar	15.0g
Peptone	5.0g
Yeast extract	3.0g
Beef extract	1.5g
Glucose	1.0g

pH 7.4 ± 0.2 at 25°C

Preparation of Medium: Add components to distilled/deionized water and bring volume to 1.0L. Mix thoroughly. Gently heat and bring to boiling. Distribute into tubes or flasks. Autoclave for 15 min at 15 psi pressure–121°C. Pour into sterile Petri dishes or leave in tubes.

Use: For the cultivation and maintenance of *Staphylococcus aureus* and *Micrococcus* species.

Micrococcus Medium, FDA

Composition per liter:

Agar	15.0g
Proteose peptone	10.0g
Beef extract	5.0g
NaCl	5.0g

pH 7.2 ± 0.2 at 25°C

Preparation of Medium: Add components to distilled/deionized water and bring volume to 1.0L. Mix thoroughly. Gently heat and bring to boiling. Distribute into tubes or flasks. Autoclave for 15 min at 15 psi pressure–121°C. Pour into sterile Petri dishes or leave in tubes.

Use: For the cultivation and maintenance of *Staphylococcus aureus* and *Micrococcus* species.

Micrococcus/Sarcina Medium

Composition per liter:

Agar	16.0g
Pancreatic digest of casein	5.0g
Sodium succinate·$6H_2O$	2.0g
Starch	2.0g
Yeast autolysate	1.0g
Sodium citrate·$2H_2O$	0.5g
Sodium acetate·$3H_2O$	0.3g
K_2HPO_4	0.2g

pH 7.0 ± 0.2 at 25°C

Preparation of Medium: Add components to distilled/deionized water and bring volume to 1.0L. Mix thoroughly. Gently heat and bring to boiling. Distribute into tubes or flasks. Autoclave for 15 min at 15 psi pressure–121°C. Pour into sterile Petri dishes or leave in tubes.

Use: For the cultivation and maintenance of *Micrococcus luteus* and *Sarcina* species.

Microcyclus eburneus Medium

Composition per liter:

K_2HPO_4	7.0g
$(NH_4)SO_4$	3.0g
KH_2PO_4	2.0g
$MgSO_4 \cdot 7H_2O$	0.5g
Yeast extract	0.2g
Thiamine·HCl	0.2mg
Biotin	0.02mg
$FeSO_4 \cdot 7H_2O$	2.0μg
$MnSO_4 \cdot 4H_2O$	2.0μg

Preparation of Medium: Add components to distilled/deionized water and bring volume to 1.0L. Mix thoroughly. Distribute into tubes or flasks. Autoclave for 15 min at 15 psi pressure–121°C.

Use: For the cultivation of *Microcyclus eburneus.*

Microcyclus major Medium

Composition per liter:

Glucose	1.0g
Peptone	1.0g
KNO_3	0.1g
K_2HPO_4	0.07g
$MgSO_4 \cdot 7H_2O$	0.03g
Trace elements solution	1.0mL

Trace Elements Solution:

Composition per liter:

Disodium EDTA	10.0g
$FeSO_4 \cdot 7H_20$	9.3g
$NaBO_3 \cdot 4H_2O$	2.6g
$MnCl_2 \cdot 4H_2O$	1.8g
$CaCl_2$	1.2g
$(NH_4)_6Mo_7O_{24} \cdot 4H_2O$	1.0g
$ZnSO_4 \cdot 7H_2O$	0.2g
$CuSO_4 \cdot 5H_2O$	0.08g
$Co(NO_3)_2 \cdot H_2O$	0.02g

Preparation of Trace Elements Solution: Add components to distilled/deionized water and bring volume to 1.0L. Mix thoroughly.

Preparation of Medium: Add components to distilled/deionized water and bring volume to 1.0L. Mix thoroughly. Distribute into tubes or flasks. Autoclave for 15 min at 15 psi pressure–121°C.

Use: For the cultivation of *Microcyclus major.*

Microcyclus marinus Medium

Composition per liter:

NaCl	23.5g
$MgCl_2$	5.0g
Na_2SO_4	4.0g
$CaCl_2 \cdot 2H_2O$	1.5g
KCl	0.7g
$NaHCO_3$	0.2g

Preparation of Medium: Add components to distilled/deionized water and bring volume to 1.0L. Mix thoroughly. Distribute into tubes or flasks. Autoclave for 15 min at 15 psi pressure–121°C.

Use: For the cultivation of *Microcyclus marinus.*

Microcyclus Medium

Composition per liter:

Agar	15.0g
Glucose	5.0g
Peptone	5.0g
Yeast extract	5.0g

pH 6.8 ± 0.2 at 25°C

Preparation of Medium: Add components to distilled/deionized water and bring volume to 1.0L. Mix thoroughly. Gently heat and bring to boiling. Distribute into tubes or flasks. Autoclave for 15 min at 15 psi pressure–121°C. Pour into sterile Petri dishes or leave in tubes.

Use: For the cultivation and maintenance of *Flectobacillus major* and *Microcyclus* species.

Microcyclus/Spirosoma Medium

Composition per liter:

Agar	15.0g
Glucose	1.0g
Peptone	1.0g
Yeast extract	1.0g

pH 6.8–7.0 at 25°C

Preparation of Medium: Add components to distilled/deionized water and bring volume to 1.0L. Mix thoroughly. Gently heat and bring to boiling. Distribute into tubes or flasks. Autoclave for 15 min at 15 psi pressure–121°C. Pour into sterile Petri dishes or leave in tubes.

Use: For the cultivation and maintenance of *Spirosoma linguale* and *Microcyclus* species.

Microlunatus Medium (DSMZ Medium 776)

Composition per liter:

Glucose	0.5g
Peptone	0.5g
Yeast extract	0.5g
Na-glutamate	0.5g
KH_2PO_4	0.5g
$(NH_4)_2SO_4$	0.1g
$MgSO_4 \cdot 7H_2O$	0.1g

pH 7.0 ± 0.2 at 25°C

Preparation of Medium: Add components to distilled/deionized water and bring volume to 1.0L. Mix thoroughly. Adjust pH to 7.0.

Distribute into tubes or flasks. Autoclave for 15 min at 15 psi pressure–121°C.

Use: For the cultivation of *Microlunatus phosphovorus* and *Kineosphaera limosa.*

Micromonospora megalomicea Agar

Composition per liter:

Soluble starch	20.0g
Agar	15.0g
Glucose	10.0g
Pancreatic digest of casein	5.0g
Yeast extract	5.0g
$CaCO_3$	1.0g

Preparation of Medium: Add components to distilled/deionized water and bring volume to 1.0L. Mix thoroughly. Gently heat and bring to boiling. Autoclave for 15 min at 15 psi pressure–121°C. Pour into sterile Petri dishes or distribute into sterile tubes.

Use: For the cultivation and maintenance of *Micromonospora rhodorangea* and *Micromonospora rosaria.*

Micromonospora Starch Agar

Composition per liter:

Starch, soluble	20.0g
Agar	15.0g
Glucose	10.0g
N-Z amine type A	5.0g
Yeast extract	5.0g
$CaCO_3$	1.0g

pH 7.2 ± 0.2 at 25°C

Preparation of Medium: Add components to distilled/deionized water and bring volume to 1.0L. Mix thoroughly. Adjust pH to 7.2. Gently heat and bring to boiling. Distribute into tubes or flasks. Autoclave for 15 min at 15 psi pressure–121°C. Pour into sterile Petri dishes or leave in tubes.

Use: For the cultivation and maintenance of *Ampullariella campanulata, Micromonospora aurantiaca, Micromonospora brunnea, Micromonospora chalcea, Micromonospora halophytica, Micromonospora inositola, Micromonospora melanosporea, Micromonospora purpurea, Micromonospora purpureochromogenes, Micromonospora rhodorangea,* and *Micromonospora glauca.*

Microvirgula Medium (DSMZ Medium 957)

Composition per liter:

Na-succinate	1.5g
KNO_3	1.5g
$(NH_4)_2SO_4$	1.0g
KH_2PO_4	0.82g
K_2HPO_4	0.7g
$MgSO_4 \cdot 7H_2O$	0.5g
Yeast extract	0.25g

pH 7.0 ± 0.2 at 25°C

Preparation of Medium: Add components to distilled/deionized water and bring volume to 1.0L. Mix thoroughly. Sparge with 100% N_2. Distribute into tubes or bottles. Autoclave for 15 min at 15 psi pressure–121°C. The final pH should be 7.0.

Use: For the cultivation of *Microvirgula aerodenitrificans.*

Middlebrook 13A Medium

Composition per 112.5mL:

Casein hydrolysate	0.1g
Tween™ 80	0.02g
Sodium polyanetholesulfonate	0.025g
Middlebrook 7H9 broth	100.0mL
Middlebrook 13A enrichment	12.5mL
Catalase	36,000U
^{14}C-substrate	125μCi (185kBq)

pH 6.6 ± 0.2 at 25°C

Middlebrook 7H9 Broth:

Composition per liter:

Na_2HPO_4	2.5g
KH_2PO_4	1.0g
Monosodium glutamate	0.5g
$(NH_4)_2SO_4$	0.5g
Sodium citrate	0.1g
$MgSO_4 \cdot 7H_2O$	0.05g
Ferric ammonium citrate	0.04g
$CuSO_4 \cdot 5H_2O$	1.0mg
Pyridoxine	1.0mg
$ZnSO_4 \cdot 7H_2O$	1.0mg
Biotin	0.5mg
$CaCl_2 \cdot 2H_2O$	0.5mg
Glycerol	2.0mL

Preparation of Middlebrook 7H9 Broth: Add components to distilled/deionized water and bring volume to 1.0L. Mix thoroughly.

Middlebrook 13A Enrichment:

Composition per 20.0mL:

Bovine serum albumin	3.0g

Preparation of Middlebrook 13A Enrichment: Add bovine serum albumin to distilled/deionized water and bring volume to 20.0mL. Mix thoroughly. Filter sterilize.

Preparation of Medium: To 100.0mL of Middlebrook 7H9 broth, add remaining components, except Middlebrook 13A enrichment. Mix thoroughly. Filter sterilize. Aseptically distribute into bottles in 4.0mL volumes. Prior to inoculation, aseptically add 0.5mL of Middlebrook 13A enrichment to each bottle. Mix thoroughly.

Use: For the cultivation of *Mycobacterium* species from the blood of patients suspected of having mycobacteremia.

Middlebrook ADC Enrichment (Middlebrook Albumin Dextrose Catalase Enrichment)

Composition per 100.0mL:

Bovine albumin fraction V	5.0g
Glucose	2.0g
Catalase	0.003g

Source: This medium is available as a prepared enrichment from BD Diagnostic Systems.

Preparation of Enrichment: Add components to distilled/deionized water and bring volume to 100.0mL. Mix thoroughly. Filter sterilize.

Use: For use as a supplement to other Middlebrook media for the isolation, cultivation, and maintenance of *Mycobacterium* species. Also used as a supplement to other Middlebrook media for determining the antimicrobial susceptibility of mycobacteria.

Middlebrook 7H9 Broth with Middlebrook ADC Enrichment

Composition per liter:

Na_2HPO_4	2.5g
KH_2PO_4	1.0g
Monosodium glutamate	0.5g
$(NH_4)_2SO_4$	0.5g
Sodium citrate	0.1g
$MgSO_4 \cdot 7H_2O$	0.05g
Ferric ammonium citrate	0.04g
$CuSO_4 \cdot 5H_2O$	1.0mg
Pyridoxine	1.0mg
$ZnSO_4 \cdot 7H_2O$	1.0mg
Biotin	0.5mg
$CaCl_2 \cdot 2H_2O$	0.5mg
Middlebrook ADC enrichment	100.0mL
Glycerol	2.0mL

pH 6.6 ± 0.2 at 25°C

Source: This medium is available as a premixed powder from BD Diagnostic Systems.

Middlebrook ADC Enrichment:

Composition per 100.0mL:

Bovine albumin fraction V	5.0g
Glucose	2.0g
Catalase	3.0mg

Source: This enrichment is available as a prepared enrichment from BD Diagnostic Systems.

Preparation of Middlebrook ADC Enrichment: Add components to distilled/deionized water and bring volume to 100.0mL. Mix thoroughly. Filter sterilize.

Preparation of Medium: Add glycerol to 900.0mL of distilled/deionized water and add remaining components, except Middlebrook ADC enrichment. Mix thoroughly. Gently heat and bring to boiling. Autoclave for 15 min at 15 psi pressure–121°C. Cool to 50°–55°C. Aseptically add 100.0mL of sterile Middlebrook ADC enrichment. Mix thoroughly. Distribute into sterile tubes or flasks.

Use: For the isolation, cultivation, and maintenance of *Mycobacterium* species, including *Mycobacterium tuberculosis.* Also used for determining the antimicrobial susceptibility of mycobacteria.

Middlebrook 7H9 Broth with Middlebrook OADC Enrichment

Composition per liter:

Na_2HPO_4	2.5g
KH_2PO_4	1.0g
Monosodium glutamate	0.5g
$(NH_4)_2SO_4$	0.5g
Sodium citrate	0.1g
$MgSO_4 \cdot 7H_2O$	0.05g
Ferric ammonium citrate	0.04g
$CuSO_4 \cdot 5H_2O$	1.0mg
Pyridoxine	1.0mg
$ZnSO_4 \cdot 7H_2O$	1.0mg
Biotin	0.5mg
$CaCl_2 \cdot 2H_2O$	0.5mg
Middlebrook OADC enrichment	100.0mL
Glycerol	2.0mL

pH 6.6 ± 0.2 at 25°C

Source: This medium is available as a premixed powder from BD Diagnostic Systems.

Middlebrook OADC Enrichment:

Composition per 100.0mL:

Bovine albumin fraction V	5.0g
Glucose	2.0g
NaCl	0.85g
Oleic acid	0.05g
Catalase	4.0mg

Source: This enrichment is available as a prepared enrichment from BD Diagnostic Systems.

Preparation of Middlebrook OADC Enrichment: Add components to distilled/deionized water and bring volume to 100.0mL. Mix thoroughly. Filter sterilize.

Preparation of Medium: Add glycerol to 900.0mL of distilled/deionized water and add remaining components, except Middlebrook OADC enrichment. Mix thoroughly. Gently heat and bring to boiling. Autoclave for 15 min at 15 psi pressure–121°C. Cool to 50°–55°C. Aseptically add 100.0mL of sterile Middlebrook OADC enrichment. Mix thoroughly. Distribute into sterile tubes or flasks.

Use: For the isolation, cultivation, and maintenance of *Mycobacterium* species, including *Mycobacterium tuberculosis.* Also used for determining the antimicrobial susceptibility of mycobacteria.

Middlebrook 7H9 Broth with Middlebrook OADC Enrichment and Triton™ WR 1339

Composition per liter:

Na_2HPO_4	2.5g
KH_2PO_4	1.0g
Monosodium glutamate	0.5g
$(NH_4)_2SO_4$	0.5g
Sodium citrate	0.1g
$MgSO_4 \cdot 7H_2O$	0.05g
Ferric ammonium citrate	0.04g
$CuSO_4 \cdot 5H_2O$	1.0mg
Pyridoxine	1.0mg
$ZnSO_4 \cdot 7H_2O$	1.0mg
Biotin	0.5mg
$CaCl_2 \cdot 2H_2O$	0.5mg
Middlebrook OADC enrichment with Triton™ WR 1339	100.0mL
Glycerol	2.0mL

pH 6.6 ± 0.2 at 25°C

Source: This medium is available as a premixed powder from BD Diagnostic Systems.

Middlebrook OADC Enrichment with Triton™ WR 1339:

Composition per 100.0mL:

Bovine albumin fraction V	5.0g
Glucose	2.0g
NaCl	0.85g
Triton™ WR 1339	0.25g
Oleic acid	0.05g
Catalase	4.0mg

Source: This enrichment is available as a prepared enrichment from BD Diagnostic Systems.

Preparation of Middlebrook OADC Enrichment with Triton™ WR 1339: Add components to distilled/deionized water and bring volume to 100.0mL. Mix thoroughly. Filter sterilize.

Preparation of Medium: Add glycerol to 900.0mL of distilled/deionized water and add remaining components, except Middlebrook OADC enrichment with Triton™ WR 1339. Mix thoroughly. Gently heat and bring to boiling. Autoclave for 15 min at 15 psi pressure–121°C. Cool to 50°–55°C. Aseptically add 100.0mL of sterile Middlebrook OADC enrichment with Triton™ WR 1339. Mix thoroughly. Distribute into sterile tubes or flasks.

Use: For the isolation, cultivation, and maintenance of *Mycobacterium* species, including *Mycobacterium tuberculosis*. Also used for determining the antimicrobial susceptibility of mycobacteria.

Middlebrook 7H9 Broth, Supplemented

Composition per liter:

Na_2HPO_4	2.5g
KH_2PO_4	1.0g
Monosodium glutamate	0.5g
$(NH_4)_2SO_4$	0.5g
Tween™ 80	0.5g
Sodium citrate	0.1g
$MgSO_4 \cdot 7H_2O$	0.05g
Ferric ammonium citrate	0.04g
Mycobactin J	2.0mg
$CuSO_4 \cdot 5H_2O$	1.0mg
Pyridoxine	1.0mg
$ZnSO_4 \cdot 7H_2O$	1.0mg
Biotin	0.5mg
$CaCl_2 \cdot 2H_2O$	0.5mg
Dubos oleic albumin complex	100.0mL
Glycerol	2.0mL

pH 6.6 ± 0.2 at 25°C

Source: Mycobactin J is available from Allied Laboratories, Inc.

Dubos Oleic Albumin Complex:

Composition per 100.0mL:

Bovine serum albumin, fraction V	5.0g
Oleic acid, sodium salt	0.05g
NaCl (0.85% solution)	100.0mL

Preparation of Dubos Oleic Albumin Complex: Add bovine serum albumin and oleic acid to 100.0mL of NaCl solution. Mix thoroughly. Filter sterilize.

Preparation of Medium: Add components, except Dubos oleic albumin complex, to distilled/deionized water and bring volume to 900.0mL. Mix thoroughly. Gently heat and bring to boiling. Autoclave for 15 min at 15 psi pressure–121°C. Cool to 45°–50°C. Aseptically add sterile Dubos oleic albumin complex. Mix thoroughly. Pour into sterile Petri dishes or distribute into sterile tubes.

Use: For the cultivation and maintenance of *Mycobacterium avium*.

Middlebrook 7H10 Agar with Glycerol

See: **Middlebrook 7H10 Agar with Middlebrook ADC Enrichment**

Middlebrook 7H10 Agar with Middlebrook ADC Enrichment

Composition per liter:

Agar	15.0g
Na_2HPO_4	1.5g
KH_2PO_4	1.5g
$(NH_4)_2SO_4$	0.5g
L-Glutamic acid	0.5g
Sodium citrate	0.4g
Ferric ammonium citrate	0.04g
$MgSO_4 \cdot 7H_2O$	0.025g
$ZnSO_4 \cdot 7H_2O$	1.0mg
$CuSO_4 \cdot 5H_2O$	1.0mg
Pyridoxine	1.0mg
Biotin	0.5mg
$CaCl_2 \cdot 2H_2O$	0.5mg
Malachite Green	0.25mg
Middlebrook ADC enrichment	100.0mL
Glycerol	5.0mL

pH 6.6 ± 0.2 at 25°C

Middlebrook ADC Enrichment:

Composition per 100.0mL:

Bovine albumin fraction V	5.0g
Glucose	2.0g
Catalase	0.003g

Source: The medium and enrichment are available as a prepared enrichment from BD Diagnostic Systems.

Preparation of Middlebrook ADC Enrichment: Add components to distilled/deionized water and bring volume to 100.0mL. Mix thoroughly. Filter sterilize.

Preparation of Medium: Add glycerol to 900.0mL of distilled/deionized water and add remaining components, except Middlebrook ADC enrichment. Mix thoroughly. Gently heat and bring to boiling. Autoclave for 15 min at 15 psi pressure–121°C. Cool to 50°–55°C. Aseptically add 100.0mL of sterile Middlebrook ADC enrichment. Mix thoroughly. Pour into sterile Petri dishes or distribute into sterile tubes.

Use: For the isolation, cultivation, and maintenance of *Mycobacterium* species, including *Mycobacterium tuberculosis*. Also used for determining the antimicrobial susceptibility of mycobacteria.

Middlebrook 7H10 Agar with Middlebrook OADC Enrichment (Middlebrook and Cohn 7H10 Agar)

Composition per liter:

Agar	15.0g
Na_2HPO_4	1.5g
KH_2PO_4	1.5g
$(NH_4)_2SO_4$	0.5g
L-Glutamic acid	0.5g
Sodium citrate	0.4g
Ferric ammonium citrate	0.04g
$MgSO_4 \cdot 7H_2O$	0.025g
$ZnSO_4 \cdot 7H_2O$	1.0mg
$CuSO_4 \cdot 5H_2O$	1.0mg
Pyridoxine	1.0mg
Biotin	0.5mg
$CaCl_2 \cdot 2H_2O$	0.5mg
Malachite Green	0.25mg
Middlebrook OADC enrichment	100.0mL
Glycerol	5.0mL

pH 6.6 ± 0.2 at 25°C

Source: This medium is available as a premixed powder from BD Diagnostic Systems.

Middlebrook OADC Enrichment:

Composition per 100.0mL:

Bovine albumin fraction V	5.0g
Glucose	2.0g
NaCl	0.85g
Oleic acid	0.05g
Catalase	4.0mg

Source: This enrichment is available as a prepared enrichment from BD Diagnostic Systems.

Preparation of Middlebrook OADC Enrichment: Add components to distilled/deionized water and bring volume to 100.0mL. Mix thoroughly. Filter sterilize.

Preparation of Medium: Add glycerol to 900.0mL of distilled/deionized water and add remaining components, except Middlebrook OADC enrichment. Mix thoroughly. Gently heat and bring to boiling. Autoclave for 15 min at 15 psi pressure–121°C. Cool to 50°–55°C. Aseptically add 100.0mL of sterile Middlebrook OADC enrichment. Mix thoroughly. Pour into sterile Petri dishes or distribute into sterile tubes.

Use: For the isolation, cultivation, and maintenance of *Mycobacterium* species, including *Mycobacterium tuberculosis*. Also used for determining the antimicrobial susceptibility of mycobacteria.

Middlebrook 7H10 Agar with Middlebrook OADC Enrichment and Hemin (Hemin Medium for *Mycobacterium*)

Composition per liter:

Agar	15.0g
Na_2HPO_4	1.5g
KH_2PO_4	1.5g
$(NH_4)_2SO_4$	0.5g
L-Glutamic acid	0.5g
Sodium citrate	0.4g
Ferric ammonium citrate	0.04g
$MgSO_4 \cdot 7H_2O$	0.025g
$ZnSO_4 \cdot 7H_2O$	1.0mg
$CuSO_4 \cdot 5H_2O$	1.0mg
Pyridoxine	1.0mg
Biotin	0.5mg
$CaCl_2 \cdot 2H_2O$	0.5mg
Malachite Green	0.25mg
Middlebrook OADC enrichment	100.0mL
Glycerol	5.0mL
Hemin solution	3.9mL

pH 6.6 ± 0.2 at 25°C

Source: This medium is available as a premixed powder from BD Diagnostic Systems.

Middlebrook OADC Enrichment:

Composition per 100.0mL:

Bovine albumin fraction V	5.0g
Glucose	2.0g
NaCl	0.85g
Oleic acid	0.05g
Catalase	4.0mg

Preparation of Middlebrook OADC Enrichment: Add components to distilled/deionized water and bring volume to 100.0mL. Mix thoroughly. Filter sterilize.

Hemin Solution:

Composition per 100.0mL:

Hemin	1.0g
NaOH (1*N* solution)	20.0mL

Preparation of Hemin Solution: Add hemin to 20.0mL of 1*N* NaOH solution. Mix thoroughly. Bring volume to 100.0mL with distilled/deionized water.

Preparation of Medium: Add glycerol to 891.1mL of distilled/deionized water and add remaining components, except Middlebrook OADC enrichment. Mix thoroughly. Gently heat and bring to boiling. Autoclave for 15 min at 15 psi pressure–121°C. Cool to 50°–55°C. Aseptically add 100.0mL of sterile Middlebrook OADC enrichment. Mix thoroughly. Pour into sterile Petri dishes or distribute into sterile tubes.

Use: For the isolation, cultivation, and maintenance of *Mycobacterium* species, including *Mycobacterium tuberculosis*. For the cultivation and maintenance of *Mycobacterium haemophilum*. Also used for determining the antimicrobial susceptibility of mycobacteria.

Middlebrook 7H10 Agar with Middlebrook OADC Enrichment and Triton™ WR 1339

Composition per liter:

Agar	15.0g
Na_2HPO_4	1.5g
KH_2PO_4	1.5g
$(NH_4)_2SO_4$	0.5g
L-Glutamic acid	0.5g
Sodium citrate	0.4g
Ferric ammonium citrate	0.04g
$MgSO_4 \cdot 7H_2O$	0.025g
$ZnSO_4 \cdot 7H_2O$	1.0mg
$CuSO_4 \cdot 5H_2O$	1.0mg
Pyridoxine	1.0mg
Biotin	0.5mg
$CaCl_2 \cdot 2H_2O$	0.5mg
Malachite Green	0.25mg
Middlebrook OADC enrichment with Triton™ WR 1339	100.0mL
Glycerol	5.0mL

pH 6.6 ± 0.2 at 25°C

Source: This medium is available as a premixed powder from BD Diagnostic Systems.

Middlebrook OADC Enrichment with Triton™ WR 1339:

Composition per 100.0mL:

Bovine albumin fraction V	5.0g
Glucose	2.0g
NaCl	0.85g
Triton™ WR 1339	0.25g
Oleic acid	0.05g
Catalase	4.0mg

Source: This enrichment is available as a prepared enrichment from BD Diagnostic Systems.

Preparation of Middlebrook OADC Enrichment with Triton™ WR 1339: Add components to distilled/deionized water and bring volume to 100.0mL. Mix thoroughly. Filter sterilize.

Preparation of Medium: Add glycerol to 900.0mL of distilled/deionized water and add remaining components, except Middlebrook OADC enrichment with Triton™ WR 1339. Mix thoroughly. Gently heat and bring to boiling. Autoclave for 15 min at 15 psi pressure–

121°C. Cool to 50°–55°C. Aseptically add 100.0mL of sterile Middlebrook OADC enrichment with Triton™ WR 1339. Mix thoroughly. Pour into sterile Petri dishes or distribute into sterile tubes.

Use: For the isolation, cultivation, and maintenance of *Mycobacterium* species, including *Mycobacterium tuberculosis*. Also used for determining the antimicrobial susceptibility of mycobacteria.

Middlebrook 7H10 Agar with Streptomycin

Composition per liter:

Agar	15.0g
Na_2HPO_4	1.5g
KH_2PO_4	1.5g
$(NH_4)_2SO_4$	0.5g
L-Glutamic acid	0.5g
Sodium citrate	0.4g
Ferric ammonium citrate	0.04g
$MgSO_4 \cdot 7H_2O$	0.025g
$ZnSO_4 \cdot 7H_2O$	1.0mg
$CuSO_4 \cdot 5H_2O$	1.0mg
Pyridoxine	1.0mg
Biotin	0.5mg
$CaCl_2 \cdot 2H_2O$	0.5mg
Malachite Green	0.25mg
Glycerol	5.0mL
Streptomycin	100.0mg

pH 6.6 ± 0.2 at 25°C

Source: This medium is available as a premixed powder from BD Diagnostic Systems.

Preparation of Medium: Add glycerol to 1.0L of distilled/deionized water and add remaining components. Mix thoroughly. Gently heat and bring to boiling. Autoclave for 15 min at 15 psi pressure–121°C. Cool to 50°–55°C. Aseptically add streptomycin. Mix thoroughly. Pour into sterile Petri dishes or distribute into sterile tubes.

Use: For the isolation, cultivation, and maintenance of *Mycobacterium kansasii*.

Middlebrook 7H11 Agar, Selective

Composition per liter:

Agar	15.0g
Na_2HPO_4	1.5g
KH_2PO_4	1.5g
Pancreatic digest of casein	1.0g
$(NH_4)_2SO_4$	0.5g
L-Glutamic acid	0.5g
Sodium citrate	0.4g
$MgSO_4 \cdot 7H_2O$	0.05g
Ferric ammonium citrate	0.04g
Pyridoxine	1.0mg
$ZnSO_4 \cdot 7H_2O$	1.0mg
$CuSO_4 \cdot 5H_2O$	1.0mg
$CaCl_2 \cdot 2H_2O$	0.5mg
Malachite Green	0.25mg
D-Biotin	0.5μg
Middlebrook OADC enrichment	100.0mL
Antibiotic solution	10.0mL
Glycerol	5.0mL

pH 6.6 ± 0.2 at 25°C

Middlebrook OADC Enrichment:

Composition per 100.0mL:

Bovine albumin fraction V	5.0g
Glucose	2.0g
NaCl	0.85g
Oleic acid	0.05g
Catalase	4.0mg

Source: This enrichment is available as a prepared enrichment from BD Diagnostic Systems.

Preparation of Middlebrook OADC Enrichment: Add components to distilled/deionized water and bring volume to 100.0mL. Mix thoroughly. Filter sterilize.

Antibiotic Solution:

Composition per 10.0mL:

Carbenicillin	0.05mg
Trimethoprim lactate	0.02mg
Amphotericin B	0.01mg
Polymyxin B	200,000U

Preparation of Antibiotic Solution: Add components to distilled/deionized water and bring volume to 10.0mL. Mix thoroughly. Filter sterilize.

Preparation of Medium: Add glycerol to 890.0mL of distilled/deionized water and add remaining components, except Middlebrook OADC enrichment and antibiotic solution. Mix thoroughly. Gently heat and bring to boiling. Autoclave for 15 min at 15 psi pressure–121°C. Cool to 50°–55°C. Aseptically add 100.0mL of sterile Middlebrook OADC enrichment and 10.0mL of sterile antibiotic solution. Mix thoroughly. Pour into sterile Petri dishes or distribute into sterile tubes.

Use: For the selective isolation and cultivation of pathogenic mycobacteria from specimens potentially contaminated with bacteria and fungi.

Middlebrook 7H11 Agar with Middlebrook ADC Enrichment (Mycobacteria 7H11 Agar with Middlebrook ADC Enrichment)

Composition per liter:

Agar	15.0g
Na_2HPO_4	1.5g
KH_2PO_4	1.5g
Pancreatic digest of casein	1.0g
$(NH_4)_2SO_4$	0.5g
L-Glutamic acid	0.5g
Sodium citrate	0.4g
$MgSO_4 \cdot 7H_2O$	0.05g
Ferric ammonium citrate	0.04g
Pyridoxine	1.0mg
Malachite Green	0.25mg
D-Biotin	0.5μg
Middlebrook ADC enrichment	100.0mL
Glycerol	5.0mL

pH 6.6 ± 0.2 at 25°C

Source: This medium is available as a premixed powder from BD Diagnostic Systems.

Middlebrook ADC Enrichment:

Composition per 100.0mL:

Bovine albumin fraction V	5.0g

Glucose 2.0g
Catalase 0.003g

Source: This enrichment is available as a prepared enrichment from BD Diagnostic Systems.

Preparation of Middlebrook ADC Enrichment: Add components to distilled/deionized water and bring volume to 100.0mL. Mix thoroughly. Filter sterilize.

Preparation of Medium: Add glycerol to 900.0mL of distilled/deionized water and add remaining components, except Middlebrook ADC enrichment. Mix thoroughly. Gently heat and bring to boiling. Autoclave for 15 min at 15 psi pressure–121°C. Cool to 50°–55°C. Aseptically add 100.0mL of sterile Middlebrook ADC enrichment. Mix thoroughly. Pour into sterile Petri dishes or distribute into sterile tubes.

Use: For the cultivation of drug-resistant (isoniazid [INH]) strains of *Mycobacterium tuberculosis*. For the cultivation of particularly fastidious strains of tubercle bacilli that occur following treatment of tuberculosis patients with secondary antitubercular drugs. Generally, these strains fail to grow on 7H10 medium.

Middlebrook 7H11 Agar with Middlebrook OADC Enrichment (Mycobacteria 7H11 Agar with Middlebrook OADC Enrichment)

Composition per liter:

Agar 15.0g
Na_2HPO_4 .... 1.5g
KH_2PO_4 .... 1.5g
Pancreatic digest of casein 1.0g
$(NH_4)_2SO_4$ 0.5g
L-Glutamic acid 0.5g
Sodium citrate 0.4g
$MgSO_4 \cdot 7H_2O$ 0.05g
Ferric ammonium citrate 0.04g
Pyridoxine 1.0mg
Malachite Green 0.25mg
D-Biotin 0.5μg
Middlebrook OADC enrichment 100.0mL
Glycerol 5.0mL

pH 6.6 ± 0.2 at 25°C

Source: This medium is available as a premixed powder from BD Diagnostic Systems.

Middlebrook OADC Enrichment:

Composition per 100.0mL:

Bovine albumin fraction V 5.0g
Glucose 2.0g
NaCl 0.85g
Oleic acid 0.05g
Catalase 4.0mg

Source: This enrichment is available as a prepared enrichment from BD Diagnostic Systems.

Preparation of Middlebrook OADC Enrichment: Add components to distilled/deionized water and bring volume to 100.0mL. Mix thoroughly. Filter sterilize.

Preparation of Medium: Add glycerol to 900.0mL of distilled/deionized water and add remaining components, except Middlebrook OADC enrichment. Mix thoroughly. Gently heat and bring to boiling. Autoclave for 15 min at 15 psi pressure–121°C. Cool to 50°–55°C. Aseptically add 100.0mL of sterile Middlebrook OADC enrichment. Mix thoroughly. Pour into sterile Petri dishes or distribute into sterile tubes.

Use: For the cultivation of drug-resistant (isoniazid [INH]) strains of *Mycobacterium tuberculosis*. For the cultivation of particularly fastidious strains of tubercle bacilli that occur following treatment of tuberculosis patients with secondary antitubercular drugs. Generally, these strains fail to grow on 7H10 medium.

Middlebrook 7H11 Agar with Middlebrook OADC Enrichment and Triton™ WR 1339 (Mycobacteria 7H11 Agar with Middlebrook OADC Enrichment and Triton™ WR 1339)

Composition per liter:

Agar 15.0g
Na_2HPO_4 .... 1.5g
KH_2PO_4 .... 1.5g
Pancreatic digest of casein 1.0g
$(NH_4)_2SO_4$ 0.5g
L-Glutamic acid 0.5g
Sodium citrate 0.4g
$MgSO_4 \cdot 7H_2O$ 0.05g
Ferric ammonium citrate 0.04g
Pyridoxine 1.0mg
Malachite Green 0.25mg
D-Biotin 0.5μg
Middlebrook OADC enrichment
with Triton™ WR 1339 .... 100.0mL
Glycerol 5.0mL

pH 6.6 ± 0.2 at 25°C

Source: This medium is available as a premixed powder from BD Diagnostic Systems.

Middlebrook OADC Enrichment with Triton™ WR 1339:

Composition per 100.0mL:

Bovine albumin fraction V 5.0g
Glucose 2.0g
NaCl 0.85g
Triton™ WR 1339 .... 0.25g
Oleic acid 0.05g
Catalase 4.0mg

Source: This enrichment is available as a prepared enrichment from BD Diagnostic Systems.

Preparation of Middlebrook OADC Enrichment with Triton™ WR 1339: Add components to distilled/deionized water and bring volume to 100.0mL. Mix thoroughly. Filter sterilize.

Preparation of Medium: Add glycerol to 900.0mL of distilled/deionized water and add remaining components, except Middlebrook OADC enrichment with Triton™ WR-1339. Mix thoroughly. Gently heat and bring to boiling. Autoclave for 15 min at 15 psi pressure–121°C. Cool to 50°–55°C. Aseptically add 100.0mL of sterile Middlebrook OADC enrichment with Triton™ WR-1339. Mix thoroughly. Pour into sterile Petri dishes or distribute into sterile tubes.

Use: For the cultivation of drug-resistant (isoniazid [INH]) strains of *Mycobacterium tuberculosis*. For the cultivation of particularly fastidious strains of tubercle bacilli that occur following treatment of tuberculosis patients with secondary antitubercular drugs. Generally, these strains fail to grow on 7H10 medium.

Middlebrook 7H11 HiVeg Agar Base with Middlebrook ADC Enrichment

Composition per liter:

Agar	15.0g
Na_2HPO_4	1.5g
KH_2PO_4	1.5g
Plant hydrolysate	1.0g
L-Glutamic acid	0.5g
$(NH_4)_2SO_4$	0.5g
Sodium citrate	0.4g
$MgSO_4$	0.05g
Ferric ammonium citrate	0.04g
Pyridoxine	1.0mg
Malachite green	1.0mg
Biotin	0.5mg
Middlebrook ADC enrichment	100.0mL
Glycerol	5.0mL

pH 6.6 ± 0.2 at 25°C

Source: This medium, without glycerol and Middlebrook ADC enrichment, is available as a premixed powder from HiMedia.

Middlebrook ADC Enrichment:

Composition per 100.0mL:

Bovine albumin fraction V	5.0g
Glucose	2.0g
Catalase	0.003g

Preparation of Middlebrook ADC Enrichment: Add components to distilled/deionized water and bring volume to 100.0mL. Mix thoroughly. Filter sterilize.

Preparation of Medium: Add glycerol to 900.0mL of distilled/deionized water and add remaining components, except Middlebrook ADC enrichment. Mix thoroughly. Gently heat and bring to boiling. Autoclave for 15 min at 15 psi pressure–121°C. Cool to 50°–55°C. Aseptically add 100.0mL of sterile Middlebrook ADC enrichment. Mix thoroughly. Pour into sterile Petri dishes or distribute into sterile tubes.

Use: For the cultivation of drug-resistant (isoniazid [INH]) strains of *Mycobacterium tuberculosis*. For the cultivation of particularly fastidious strains of tubercle bacilli that occur following treatment of tuberculosis patients with secondary antitubercular drugs.

Middlebrook 7H11 HiVeg Agar Base with Middlebrook OADC Enrichment

Composition per liter:

Agar	15.0g
Na_2HPO_4	1.5g
KH_2PO_4	1.5g
Plant hydrolysate	1.0g
L-Glutamic acid	0.5g
$(NH_4)_2SO_4$	0.5g
Sodium citrate	0.4g
$MgSO_4$	0.05g
Ferric ammonium citrate	0.04g
Pyridoxine	1.0mg
Malachite green	1.0mg
Biotin	0.5mg
Middlebrook OADC enrichment	100.0mL
Glycerol	5.0mL

pH 6.6 ± 0.2 at 25°C

Source: This medium, without glycerol and Middlebrook OADC enrichment, is available as a premixed powder from HiMedia.

Middlebrook OADC Enrichment:

Composition per 100.0mL:

Bovine albumin fraction V	5.0g
Glucose	2.0g
NaCl	0.85g
Oleic acid	0.05g
Catalase	4.0mg

Preparation of Middlebrook OADC Enrichment: Add components to distilled/deionized water and bring volume to 100.0mL. Mix thoroughly. Filter sterilize.

Preparation of Medium: Add glycerol to 900.0mL of distilled/deionized water and add remaining components, except Middlebrook OADC enrichment. Mix thoroughly. Gently heat and bring to boiling. Autoclave for 15 min at 15 psi pressure–121°C. Cool to 50°–55°C. Aseptically add 100.0mL of sterile Middlebrook OADC enrichment. Mix thoroughly. Pour into sterile Petri dishes or distribute into sterile tubes.

Use: For the cultivation of drug-resistant (isoniazid [INH]) strains of *Mycobacterium tuberculosis*. For the cultivation of particularly fastidious strains of tubercle bacilli that occur following treatment of tuberculosis patients with secondary antitubercular drugs.

Middlebrook 7H12 Medium

Composition per 102.5mL:

Bovine serum albumin	0.5g
Casein hydrolysate	0.1g
Catalase	4800U
^{14}C-Palmitic acid	100μCi
Middlebrook 7H9 broth	100.0mL
Antibiotic solution	2.5mL

pH 6.8 ± 0.1 at 25°C

Middlebrook 7H9 Broth:

Composition per liter:

Na_2HPO_4	2.5g
KH_2PO_4	1.0g
Monosodium glutamate	0.5g
$(NH_4)_2SO_4$	0.5g
Sodium citrate	0.1g
$MgSO_4 \cdot 7H_2O$	0.05g
Ferric ammonium citrate	0.04g
$CuSO_4 \cdot 5H_2O$	1.0mg
Pyridoxine	1.0mg
$ZnSO_4 \cdot 7H_2O$	1.0mg
Biotin	0.5mg
$CaCl_2 \cdot 2H_2O$	0.5mg
Glycerol	2.0mL

Preparation of Middlebrook 7H9 Broth: Add components to distilled/deionized water and bring volume to 1.0L. Mix thoroughly.

Antibiotic Solution:

Composition per 5.0mL:

Nalidixic acid	0.2g
Azlocillin	0.1g
Amphotericin B	0.05g

Trimethoprim 0.05g
Polymyxin B 500,000U

Preparation of Antibiotic Solution: Add components to distilled/deionized water and bring volume to 5.0mL. Mix thoroughly. Filter sterilize.

Preparation of Medium: To 100.0mL of Middlebrook 7H9 broth, add remaining components, except antibiotic solution. Mix thoroughly. Filter sterilize. Aseptically distribute into bottles in 4.0mL volumes. Prior to inoculation, aseptically add 0.1mL of antibiotic solution to each bottle. Mix thoroughly.

Use: For the cultivation of *Mycobacterium* species from the blood of patients suspected of having mycobacteremia.

Middlebrook and Cohn 7H10 Agar
See: **Middlebrook 7H10 Agar with Middlebrook OADC Enrichment**

Middlebrook Medium (DSMZ Medium 645)

Composition per liter:

Agar	15.0g
Na_2HPO_4	1.5g
KH_2PO_4	1.5g
$(NH_4)_2SO_4$	0.5g
L-Glutamic acid	0.5g
Sodium citrate	0.4g
Ferric ammonium citrate	0.04g
$MgSO_4 \cdot 7H_2O$	0.025g
$ZnSO_4 \cdot 7H_2O$	1.0mg
$CuSO_4 \cdot 5H_2O$	1.0mg
Pyridoxine	1.0mg
Biotin	0.5mg
$CaCl_2 \cdot 2H_2O$	0.5mg
Malachite Green	0.25mg
Middlebrook OADC enrichment	100.0mL
Glycerol	5.0mL

pH 6.6 ± 0.2 at 25°C

Source: This medium is available as a premixed powder from BD Diagnostic Systems.

Middlebrook OADC Enrichment:

Composition per 100.0mL:

Bovine albumin fraction V	5.0g
Glucose	2.0g
Catalase	0.003g
Distilled water	100.0mL

Source: This enrichment is available as a prepared enrichment from BD Diagnostic Systems.

Preparation of Middlebrook OADC Enrichment: Add components to distilled/deionized water and bring volume to 100.0mL. Mix thoroughly. Filter sterilize.

Preparation of Medium: Add glycerol to 900.0mL of distilled/deionized water and add remaining components, except Middlebrook OADC enrichment. Mix thoroughly. Gently heat and bring to boiling. Autoclave for 15 min at 15 psi pressure–121°C. Cool to 50°–55°C. Aseptically add 100.0mL of sterile Middlebrook OADC enrichment. Mix thoroughly. Pour into sterile Petri dishes or distribute into sterile tubes.

Use: For the isolation, cultivation, and maintenance of *Mycobacterium* species.

Middlebrook Medium with Mycobactin (DSMZ Medium 780)

Composition per liter:

Agar	15.0g
Na_2HPO_4	1.5g
KH_2PO_4	1.5g
$(NH_4)_2SO_4$	0.5g
L-Glutamic acid	0.5g
Sodium citrate	0.4g
Ferric ammonium citrate	0.04g
$MgSO_4 \cdot 7H_2O$	0.025g
Mycobactin J	2.0mg
$ZnSO_4 \cdot 7H_2O$	1.0mg
$CuSO_4 \cdot 5H_2O$	1.0mg
Pyridoxine	1.0mg
Biotin	0.5mg
$CaCl_2 \cdot 2H_2O$	0.5mg
Malachite Green	0.25mg
Middlebrook ADC enrichment	100.0mL
Glycerol	5.0mL

pH 6.6 ± 0.2 at 25°C

Source: Mycobactin J is available from Allied Laboratories, Inc.

Middlebrook ADC Enrichment:

Composition per 100.0mL:

Bovine albumin fraction V	5.0g
Glucose	2.0g
Catalase	0.003g
Distilled water	100.0mL

Source: This medium and enrichment is available from BD Diagnostic Systems.

Preparation of Middlebrook ADC Enrichment: Add components to distilled/deionized water and bring volume to 100.0mL. Mix thoroughly. Filter sterilize.

Preparation of Medium: Add glycerol to 900.0mL of distilled/deionized water and add remaining components, except Middlebrook ADC enrichment and mycobactin. Mix thoroughly. Gently heat and bring to boiling. Autoclave for 15 min at 15 psi pressure–121°C. Cool to 50°–55°C. Aseptically add 100.0mL of sterile Middlebrook ADC enrichment and mycobactin. The mycobactin is dissolved in 2.0mL ethanol. Be sure to add all of the mycobactin; wash with additional 2.0mL ethanol if needed. Mix thoroughly. Pour into sterile Petri dishes or distribute into sterile tubes.

Use: For the cultivation of *Mycobacterium avium* subsp. *paratuberculosis.*

Middlebrook OADC Enrichment (Middlebrook Oleic Albumin Dextrose Catalase Enrichment)

Composition per 100.0mL:

Bovine albumin fraction V	5.0g
Glucose	2.0g
NaCl	0.85g
Oleic acid	0.05g
Catalase	4.0mg

Source: This enrichment is available as a prepared enrichment from BD Diagnostic Systems.

Preparation of Enrichment: Add components to distilled/deionized water and bring volume to 100.0mL. Mix thoroughly. Filter sterilize.

Use: For use as a supplement to other Middlebrook media for the isolation, cultivation, and maintenance of *Mycobacterium* species. Also used as a supplement to other Middlebrook media for determining the antimicrobial susceptibility of mycobacteria.

Middlebrook OADC Enrichment with Triton™ WR 1339 (Middlebrook Oleic Albumin Dextrose Catalase Enrichment with Triton™ WR 1339)

Composition per 100.0mL:

Bovine albumin fraction V	5.0g
Glucose	2.0g
NaCl	0.85g
Triton™ WR 1339	0.25g
Oleic acid	0.05g
Catalase	4.0mg

Source: This enrichment is available as a prepared enrichment from BD Diagnostic Systems.

Preparation of Enrichment: Add components to distilled/deionized water and bring volume to 100.0mL. Mix thoroughly. Filter sterilize.

Use: For use as a supplement to other Middlebrook media for the isolation, cultivation, and maintenance of *Mycobacterium* species. Also used as a supplement to other Middlebrook media for determining the antimicrobial susceptibility of mycobacteria.

Middlebrook Oleic Albumin Dextrose Catalase Enrichment

See: **Middlebrook OADC Enrichment**

Middlebrook Oleic Albumin Dextrose Catalase Enrichment with Triton™ WR 1339

See: **Middlebrook OADC Enrichment with Triton™ WR 1339**

MIL Medium (Motility Indole Lysine Medium)

Composition per liter:

Peptone	10.0g
Pancreatic digest of casein	10.0g
L-Lysine·HCl	10.0g
Yeast extract	3.0g
Agar	2.0g
Dextrose	1.0g
Ferric ammonium citrate	0.5g
Bromcresol Purple	0.02g

pH 6.6 ± 0.2 at 25°C

Source: This medium is available as a premixed powder and prepared medium from BD Diagnostic Systems.

Preparation of Medium: Add components to distilled/deionized water and bring volume to 1.0L. Mix thoroughly. Gently heat and bring to boiling. Distribute into tubes in 5.0mL volumes. Autoclave for 15 min at 15 psi pressure–121°C.

Use: For the cultivation and differentiation of members of the Enterobacteriaceae on the basis of motility, lysine decarboxylase activity, lysine deaminase activity, and indole production.

Milk Agar

See: **Skim Milk Agar**

Milk Agar

Composition per liter:

Agar	15.0g
Peptone	5.0g
Yeast extract	3.0g
Milk (solids or	1.0g
fresh milk)	10.0mL

pH 7.2 ± 0.2 at 25°C

Source: This medium is available as a premixed powder from Oxoid Unipath.

Preparation of Medium: Add components to distilled/deionized water and bring volume to 1.0L. Mix thoroughly. Gently heat and bring to boiling. Distribute into tubes or flasks. Autoclave for 15 min at 15 psi pressure–121°C.

Use: For the cultivation of microorganisms from dairy and water samples.

Milk Agar

Composition per liter:

Mixture A	500.0mL
Mixture B	500.0mL

Mixture A:

Composition per 500.0mL:

Instant nonfat milk	100.0g

Preparation of Mixture A: Add instant nonfat milk to distilled/deionized water and bring volume to 500.0mL. Mix thoroughly. Autoclave for 15 min at 15 psi pressure–121°C. Cool rapidly to 55°C.

Mixture B:

Composition per 500.0mL:

Agar	15.0g
Nutrient broth	12.5g
NaCl	2.5g

Preparation of Mixture B: Add components to distilled/deionized water and bring volume to 500.0mL. Mix thoroughly. Gently heat and bring to boiling. Autoclave for 15 min at 15 psi pressure–121°C. Cool rapidly to 55°C.

Preparation of Medium: Aseptically combine cooled, sterile mixture A with cooled, sterile mixture B. Mix thoroughly. Pour into sterile Petri dishes in 20.0mL volumes.

Use: For the cultivation and estimation of the numbers of *Pseudomonas aeruginosa* in water by the membrane filter method.

Milk HiVeg Agar

Composition per liter:

Agar	15.0g
Plant peptone	5.0g

Yeast extract 3.0g
Milk solids 1.0g

pH 7.2 ± 0.2 at 25°C

Source: This medium is available as a premixed powder from HiMedia.

Preparation of Medium: Add components to distilled/deionized water and bring volume to 1.0L. Mix thoroughly. Gently heat and bring to boiling. Distribute into tubes or flasks. Autoclave for 15 min at 15 psi pressure–121°C. Pour into sterile Petri dishes or leave in tubes.

Use: For the cultivation of microorganisms from dairy and water samples.

Milk HiVeg Agar (Brown and Scott Modified)

Composition per liter:

Instant nonfat milk 100.0g
Agar 15.0g
Plant peptone 5.0g
NaCl 5.0g
Plant extract 1.5g
Yeast extract 1.5g

pH 7.2 ± 0.2 at 25°C

Source: This medium is available as a premixed powder from HiMedia.

Preparation of Medium: Add components, except milk, to distilled/deionized water and bring volume to 500.0mL. Mix thoroughly. Gently heat and bring to boiling. Autoclave for 15 min at 15 psi pressure–121°C. Cool to 50°–55°C. Separately add nonfat milk to distilled/deionized water and bring volume to 500.0mL. Mix thoroughly. Gently heat and bring to boiling. Autoclave for 5 min at 15 psi pressure–121°C. Cool to 50°–55°C. Aseptically combine the two sterile solutions. Mix thoroughly. Pour into sterile Petri dishes or leave in tubes.

Use: For the confirmation of *Pseudomonas aeruginosa* in swimming pool waters.

Milk Protein Hydrolysate Agar
See: **MPH Agar**

Milk Salt HiVeg Agar Base

Composition per liter:

NaCl 65.0g
Agar 15.0g
Plant extract 3.0g
Plant peptone 5.0g

pH 7.2 ± 0.2 at 25°C

Source: This medium is available as a premixed powder from HiMedia.

Preparation of Medium: Add components to distilled/deionized water and bring volume to 1.0L. Mix thoroughly. Gently heat and bring to boiling. Distribute into tubes or flasks. Autoclave for 15 min at 15 psi pressure–121°C. Pour into sterile Petri dishes or leave in tubes.

Use: For the cultivation of halophilic microorganisms.

Miller Luria Bertani HiVeg Agar (Luria Bertani HiVeg Agar, Miller)

Composition per liter:

Agar 15.0g
NaCl 10.0g
Plant peptones 10.0g
Yeast extract 5.0g

pH 7.0 ± 0.2 at 25°C

Source: This medium is available as a premixed powder from HiMedia.

Preparation of Medium: Add components to distilled/deionized water and bring volume to 1.0L. Mix thoroughly. Adjust pH to 7.0. Distribute into tubes or flasks. Autoclave for 15 min at 15 psi pressure–121°C. Pour into sterile Petri dishes or leave in tubes.

Use: For the cultivation of *Bacillus subtilis* and *Escherichia coli.*

Miller Luria Bertani HiVeg Broth (Luria Bertani HiVeg Broth, Miller)

Composition per liter:

NaCl 10.0g
Plant peptones 10.0g
Yeast extract 5.0g

pH 7.0 ± 0.2 at 25°C

Source: This medium is available as a premixed powder from HiMedia.

Preparation of Medium: Add components to distilled/deionized water and bring volume to 1.0L. Mix thoroughly. Adjust pH to 7.0. Distribute into tubes or flasks. Autoclave for 15 min at 15 psi pressure–121°C.

Use: For the cultivation of *Bacillus subtilis* and *Escherichia coli.*

Mineral Agar

Composition per liter:

NH_4Cl 0.5g
$Na_2HPO_4 \cdot 7H_2O$ 670.0mg
KH_2PO_4 .... 340.0mg
$MgSO_4 \cdot 7H_2O$ 112.0mg
$CaCl_2$ 14.0mg
$ZnSO_4 \cdot 7H_2O$ 5.0mg
$Na_2MoO_4 \cdot 2H_2O$ 2.5mg
$FeCl_3$ 0.13mg
1,4-Dichlorobenzene .... variable

pH 7.0 ± 0.2 at 25°C

Preparation of Medium: Add components, except 1,4-dichlorobenzene, to distilled/deionized water and bring volume to 1.0L. Mix thoroughly. Gently heat and bring to boiling. Distribute into tubes or flasks. Autoclave for 15 min at 15 psi pressure–121°C. Pour into sterile Petri dishes or leave in tubes. After inoculation, place Petri dishes or tubes into a desiccator. Add a few crystals of 1,4-dichlorobenzene to the desiccator.

Use: For the cultivation of dichlorobenzene-degrading *Pseudomonas* species.

Mineral Base E for Autotrophic Growth

Composition per liter:

Noble agar 15.0g
K_2HPO_4 .... 1.2g
KH_2PO_4 .... 0.624g
$(NH_4)_2SO_4$ 0.5g
NaCl 0.1g
$CaCl_2 \cdot 6H_2O$ solution 10.0mL
$MgSO_4 \cdot 7H_2O$ solution 10.0mL

Mineral solution	1.0mL
p-Aminobenzoic acid solution	1.0mL

$CaCl_2·6H_2O$ Solution:

Composition per liter:

$CaCl_2·6H_2O$	5.0g

Preparation of $CaCl_2·6H_2O$ Solution: Add $CaCl_2·6H_2O$ to distilled/deionized water and bring volume to 1.0L. Mix thoroughly. Autoclave for 15 min at 15 psi pressure–121°C.

$MgSO_4·7H_2O$ Solution:

Composition per liter:

$MgSO_4·7H_2O$	20.0g

Preparation of $MgSO_4·7H_2O$ Solution: Add $MgSO_4·7H_2O$ to distilled/deionized water and bring volume to 1.0L. Mix thoroughly. Autoclave for 15 min at 15 psi pressure–121°C.

***p*-Aminobenzoic Acid Solution:**

Composition per 10.0.mL:

p-Aminobenzoic acid	100.0mg

Preparation of *p*-Aminobenzoic Acid Solution: Add *p*-aminobenzoic acid to distilled/deionized water and bring volume to 10.0mL. Mix thoroughly. Autoclave for 15 min at 15 psi pressure–121°C.

Mineral Solution:

Composition per 1000.0mL:

Disodium EDTA	1.58g
ZnSO4·7H2O	0.7g
$MnSO_4·4H_2O$	0.18g
$FeSO_4·7H_2O$	0.16g
$CoCl_2·6H_2O$	0.052g
$Na_2MoO_4·2H_2O$	0.047g
$CuSO_4·5H_2O$	0.047g

Preparation of Medium: Add components, except $CaCl_2·6H_2O$ solution, $MgSO_4·7H_2O$ solution, and *p*-aminobenzoic acid solution, to distilled/deionized water and bring volume to 979.0mL. Mix thoroughly. Autoclave for 15 min at 15 psi pressure–121°C. Cool to 50°C. Aseptically add in the following order: 10.0mL of sterile $CaCl_2·6H_2O$ solution, 10.0mL of sterile $MgSO_4·7H_2O$ solution, and 1.0mL of sterile *p*-aminobenzoic acid solution. Mix thoroughly. Aseptically distribute into sterile tubes or flasks. Incubate inoculated tubes in 50% CO_2.

Use: For the autotrophic cultivation and maintenance of *Pseudomonas thermocarboxydovorans*.

Mineral Base E for Heterotrophic Growth

Composition per liter:

Noble agar	15.0g
K_2HPO_4	1.2g
KH_2PO_4	0.624g
$(NH_4)_2SO_4$	0.5g
NaCl	0.1g
$CaCl_2·6H_2O$ solution	10.0mL
$MgSO_4·7H_2O$ solution	10.0mL
Sodium pyruvate solution	10.0mL
Mineral solution	1.0mL
p-Aminobenzoic acid solution	1.0mL

$CaCl_2·6H_2O$ Solution:

Composition per liter:

$CaCl_2·6H_2O$	5.0g

Preparation of $CaCl_2·6H_2O$ Solution: Add $CaCl_2·6H_2O$ to distilled/deionized water and bring volume to 1.0L. Mix thoroughly. Autoclave for 15 min at 15 psi pressure–121°C.

$MgSO_4·7H_2O$ Solution:

Composition per liter:

$MgSO_4·7H_2O$	20.0g

Preparation of $MgSO_4·7H_2O$ Solution: Add $MgSO_4·7H_2O$ to distilled/deionized water and bring volume to 1.0L. Mix thoroughly. Autoclave for 15 min at 15 psi pressure–121°C.

Sodium Pyruvate Solution:

Composition per 10.0mL:

Sodium pyruvate	2.0g

Preparation of Sodium Pyruvate Solution: Add sodium pyruvate to distilled/deionized water and bring volume to 10.0mL. Mix thoroughly. Filter sterilize.

***p*-Aminobenzoic Acid Solution:**

Composition per 10.0mL:

p-Aminobenzoic acid	100.0mg

Preparation of *p*-Aminobenzoic Acid Solution: Add *p*-aminobenzoic acid to distilled/deionized water and bring volume to 10.0mL. Mix thoroughly. Autoclave for 15 min at 15 psi pressure–121°C.

Mineral Solution:

Composition per 100.0mL:

Disodium EDTA	1.58g
ZnSO4·7H2O	0.7g
$MnSO_4·4H_2O$	0.18g
$FeSO_4·7H_2O$	0.16g
$CoCl_2·6H_2O$	0.052g
$Na_2MoO_4·2H_2O$	0.047g
$CuSO_4·5H_2O$	0.047g

Preparation of Medium: Add components, except $CaCl_2·6H_2O$ solution, $MgSO_4·7H_2O$ solution, sodium pyruvate solution, and *p*-aminobenzoic acid solution, to distilled/deionized water and bring volume to 969.0mL. Mix thoroughly. Autoclave for 15 min at 15 psi pressure–121°C. Cool to 45°–50°C. Aseptically add in the following order: 10.0mL of the sterile $CaCl_2·6H_2O$ solution, 10.0mL of the sterile $MgSO_4·7H_2O$ solution, 10.0mL of sterile sodium pyruvate solution, and 1.0mL of sterile *p*-aminobenzoic acid solution. Mix thoroughly. Aseptically distribute into sterile tubes or flasks.

Use: For the heterotrophic cultivation and maintenance of *Pseudomonas thermocarboxydovorans*.

Mineral Base Medium with Acetate

See: **MBM Acetate Medium**

Mineral Lactate Medium

Composition per liter:

$K_2HPO_4·3H_2O$	1.13g
NaCl	1.0g
NH_4Cl	1.0g
KH_2PO_4	0.88g
$MgSO_4·7H_2O$	0.5g
Sodium lactate	0.5g
$CaCl_2·2H_2O$	5.0mg
Trace elements solution	1.2mL

pH 7.0 ± 0.2 at 25°C

Trace Elements Solution:
Composition per liter:

Disodium EDTA	50.0g
$ZnSO_4 \cdot 7H_2O$	22.0g
$CaCl_2 \cdot 2H_2O$	5.54g
$MnCl_2 \cdot 4H_2O$	5.06g
$FeSO_4 \cdot 7H_2O$	5.0g
$CoCl_2 \cdot 6H_2O$	1.61g
$CuSO_4 \cdot 5H_2O$	1.57g
$(NH_4)_6Mo_7O_{24} \cdot 4H_2O$	1.1g

Preparation of Trace Elements Solution: Add components to distilled/deionized water and bring volume to 1.0L. Mix thoroughly. Adjust pH to 7.0 with KOH.

Preparation of Medium: Add components to distilled/deionized water and bring volume to 1.0L. Mix thoroughly. Adjust pH to 7.0. Distribute into tubes or flasks. Autoclave for 15 min at 15 psi pressure–121°C.

Use: For the cultivation and maintenance of *Pseudomonas* species and *Spirillum* species.

Mineral Medium

Composition per liter:

Yeast extract	2.0g
Mineral base 5X	200.0mL
Trace elements solution SL-6	1.0mL
Thiamine·HCl	3.0μg
Biotin	0.2μg

pH 6.8 ± 0.2 at 25°C

Mineral Base 5X:
Composition per liter:

NaCl	5.0g
NH_4Cl	2.0g
KH_2PO_4	1.35g
$MgSO_4 \cdot 7H_2O$	1.0g
K_2HPO_4	0.87g
$CaCl_2$	0.05g
$FeCl_3 \cdot 6H_2O$	1.25mg

Preparation of Mineral Base 5X: Add components to distilled/deionized water and bring volume to 1.0L. Mix thoroughly.

Trace Elements Solution SL-6:
Composition per liter:

H_3BO_3	0.3g
$CoCl_2 \cdot 6H_2O$	0.2g
$ZnSO_4 \cdot 7H_2O$	0.1g
$MnCl_2 \cdot 4H_2O$	0.03g
$Na_2MoO_4 \cdot H_2O$	0.03g
$NiCl_2 \cdot 6H_2O$	0.02g
$CuCl_2 \cdot 2H_2O$	0.01g

Preparation of Trace Elements Solution SL-6: Add components to distilled/deionized water and bring volume to 1.0L. Mix thoroughly. Adjust pH to 3.4.

Preparation of Medium: Add components to distilled/deionized water and bring volume to 1.0L. Mix thoroughly. Adjust pH to 6.8. Distribute into tubes or flasks. Autoclave for 15 min at 15 psi pressure–121°C.

Use: For the cultivation of *Arthrobacter* species.

Mineral Medium

Composition per liter:

NH_4Cl	0.5g
Yeast extract	0.2g
1,4-Dichlorobenzene	0.1g
$Na_2HPO_4 \cdot 7H_2O$	670.0mg
KH_2PO_4	340.0mg
$MgSO_4 \cdot 7H_2O$	112.0mg
$CaCl_2$	14.0mg
$ZnSO_4 \cdot 7H_2O$	5.0mg
$Na_2MoO_4 \cdot 2H_2O$	2.5mg
$FeCl_3$	0.13mg

pH 7.0 ± 0.2 at 25°C

Preparation of Medium: Add components to distilled/deionized water and bring volume to 1.0L. Mix thoroughly. Distribute into tubes or flasks. Autoclave for 15 min at 15 psi pressure–121°C.

Use: For the cultivation of dichlorobenzene-degrading *Pseudomonas* species.

Mineral Medium (DSMZ Medium 994)

Composition per liter:

D-Glucose	3.6g
KH_2PO_4	1.0g
NH_4Cl	0.6g
NaCl	0.4g
$MgSO_4 \cdot 7H_2O$	0.4g
$CaCl_2 \cdot 2H_2O$	0.5mg
Vitamin solution	5.0mL
Trace elements solution SL-4	1.0mL

pH 6.8 ± 0.2 at 25°C

Vitamin Solution:
Composition per liter:

Thiamine-HCl·$2H_2O$	50.0mg
Riboflavin	50.0mg
Vitamin B_{12}	50.0mg
D-Ca-pantothenate	50.0mg
p-Aminobenzoic acid	50.0mg
Lipoic acid	50.0mg
Nicotinic acid	25.0mg
Niconinamide	25.0mg
Biotin	20.0mg
Folic acid	20.0mg
Pyridoxine-HCl	10.0mg

Preparation of Vitamin Solution: Add components to distilled/deionized water and bring volume to 1.0L. Mix thoroughly for several hours. Filter sterilize.

Trace Elements Solution SL-4:
Composition per liter:

EDTA	0.5g
$FeSO_4 \cdot 7H_2O$	0.2g
Trace elements solution SL-6	100.0mL

Trace Elements Solution SL-6:
Composition per liter:

$MnCl_2 \cdot 4H_2O$	0.5g
H_3BO_3	0.3g
$CoCl_2 \cdot 6H_2O$	0.2g
$ZnSO_4 \cdot 7H_2O$	0.1g
$Na_2MoO_4 \cdot 2H_2O$	0.03g

$NiCl_2·6H_2O$ 0.02g
$CuCl_2·2H_2O$ 0.01g

Preparation of Trace Elements Solution SL-6: Add components to distilled/deionized water and bring volume to 1.0L. Mix thoroughly. Autoclave for 15 min at 15 psi pressure–121°C.

Preparation of Trace Elements Solution SL-4: Add components to distilled/deionized water and bring volume to 1.0L. Mix thoroughly. Filter sterilize.

Preparation of Medium: Add components, except vitamin and trace elements solutions, to distilled/deionized water and bring volume to 994.0mL. Mix thoroughly. Autoclave for 15 min at 15 psi pressure–121°C. Cool to room temperature. Aseptically add sterile trace elements and vitamin solutions. Mix thoroughly. Adjust pH to 6.8. Aseptically dispense into tubes, flasks, or bottles.

Use: For the cultivation of *Stenotrophomonas maltophilia.*

Mineral Medium (DSMZ Medium 1007)

Composition per liter:
KNO_3 .... 0.25g
KH_2PO_4 .... 0.1g
$MgSO_4·7H_2O$ 0.05g
$CaCl_2·2H_2O$ 0.01g
Methanol 75.0mL
Trace elements solution 1.0mL
pH 5.7 ± 0.2 at 25°C

Trace Elements Solution:
Composition per liter:
EDTA 5.0g
$FeSO_4·7H_2O$ 2.0g
$CoCl_2·6H_2O$ 0.2g
$CuCl_2·2H_2O$ 0.1g
$ZnSO_4·7H_2O$ 0.1g
$MnCl_2·4H_2O$ 0.03g
H_3BO_3 .... 0.03g
$Na_2MoO_4·2H_2O$ 0.03g
$NiCl_2·6H_2O$ 0.02g

Preparation of Trace Elements Solution: Add components to distilled/deionized water and bring volume to 1.0L. Mix thoroughly. Autoclave for 15 min at 15 psi pressure–121°C.

Preparation of Medium: Add components, except trace elements solution, to distilled/deionized water and bring volume to 999.0mL. Mix thoroughly. Autoclave for 15 min at 15 psi pressure–121°C. Cool to room temperature. Aseptically add sterile trace elements solution. Mix thoroughly. Adjust pH to 5.7. Aseptically dispense into tubes, flasks, or bottles. Note: Methane can be substituted for methanol for the growth of some strains.

Use: For the cultivation of *Methylocella tundrae* and *Methylocystis heyeri.*

Mineral Medium A

Composition per liter:
$(NH_4)_2SO_4$ 1.0g
K_2HPO_4 .... 1.0g

Preparation of Medium: Add components to tap water and bring volume to 1.0L. Mix thoroughly. Distribute into tubes or flasks. Autoclave for 15 min at 15 psi pressure–121°C.

Use: For the cultivation of *Saccharobacterium ovale.*

Mineral Medium with 3-Aminophenol (DSMZ Medium 465f)

Composition per liter:
$Na_2HPO_4·2H_2O$ 3.5g
KH_2PO_4 .... 1.0g
$(NH_4)_2SO_4$ 0.5g
$MgCl_2·6H_2O$ 0.1g
$Ca(NO_3)_2·4H_2O$ 0.05g
Aminophenol solution 10.0mL
Trace elements solution SL-4 .... 1.0mL
pH 7.25 ± 0.2 at 25°C

Trace Elements Solution SL-4:
Composition per liter:
EDTA 0.5g
$FeSO_4·7H_2O$ 0.2g
Trace elements solution SL-6 .... 100.0mL

Trace Elements Solution SL-6:
Composition per liter:
H_3BO_3 .... 0.3g
$CoCl_2·6H_2O$ 0.2g
$ZnSO_4·7H_2O$ 0.1g
$MnCl_2·4H_2O$ 0.03g
$Na_2MoO_4·H_2O$ 0.03g
$NiCl_2·6H_2O$ 0.02g
$CuCl_2·2H_2O$ 0.01g

Preparation of Trace Elements Solution SL-6: Add components to distilled/deionized water and bring volume to 1.0L. Mix thoroughly. Adjust pH to 3.4.

Preparation of Trace Elements Solution SL-4: Add components to distilled/deionized water and bring volume to 1.0L. Mix thoroughly.

Aminophenol Solution:
Composition per 100.0mL:
3-Aminophenol 1.0g

Preparation of Aminophenol Solution: Add 100.0mL boiling water to 1.0g aminophenol crystals. Stir the solution to mix thoroughly. Cool to room temperature. Sterilize by filtration.

Preparation of Medium: Add components, except aminophenol solution, to 990.0mL distilled/deionized water. Adjust pH to 7.25. Autoclave for 15 min at 15 psi pressure–121°C. Aseptically add 10.0mL aminophenol solution. Mix thoroughly. Aseptically distribute to sterile tubes or flasks.

Use: For the cultivation of *Arthrobacter* sp. and other aminophenol-utilizing bacteria.

Mineral Medium with Antipyrin

Composition per liter:
Antipyrin 1.0g
$Na_2HPO_4·12H_2O$ 0.7g
$(NH_4)_2HPO_4$ 0.7g
KH_2PO_4 .... 0.3g
$(NH_4)H_2PO_4$ 0.3g
$MgSO_4·7H_2O$ 0.25g
$(NH_4)_2SO_4$ 0.1g
$CaCl_2·6H_2O$ 0.05g
H_3BO_3 .... 0.5mg
$MnSO_4·4H_2O$ 0.4mg
$ZnSO_4·7H_2O$ 0.4mg
$FeCl_3·6H_2O$ 0.2mg

$(NH_4)_6Mo_7O_{24}·4H_2O$ 0.2mg
KI 0.1mg
$CuSO_4·5H_2O$ 0.04mg
Vitamin solution 20.0mL

pH 6.8–7.0 at 25°C

Vitamin Solution:

Composition per 20.0mL:

Biotin 0.1mg
Vitamin B_{12} 0.03mg

Preparation of Vitamin Solution: Add biotin and vitamin B_{12} to 20.0mL of distilled/deionized water. Mix thoroughly. Filter sterilize.

Preparation of Medium: Add components, except vitamin solution, to distilled/deionized water and bring volume to 980.0mL. Mix thoroughly. Adjust pH to 6.8–7.0 with 1*N* NaOH. Autoclave for 20 min at 15 psi pressure–121°C. Cool to 45°–50°C. Aseptically add the sterile vitamin solution. Mix thoroughly. Distribute into sterile tubes or flasks.

Use: For the cultivation and maintenance of *Phenylobacterium immobile.*

Mineral Medium with Atrazine (DSMZ Medium 465i)

Composition per liter:

$Na_2HPO_4·2H_2O$ 3.5g
Na-citrate 1.0g
KH_2PO_4 .... 1.0g
$MgCl_2·6H_2O$ 0.1g
$CaCl_2$ 0.05g
Atrazine solution 10.0mL
Trace elements solution SL-4 .... 1.0mL

pH 7.25 ± 0.2 at 25°C

Trace Elements Solution SL-4:

Composition per liter:

EDTA 0.5g
$FeSO_4·7H_2O$ 0.2g
Trace elements solution SL-6 .... 100.0mL

Preparation of Trace Elements Solution SL-4: Add components to distilled/deionized water and bring volume to 1.0L. Mix thoroughly.

Trace Elements Solution SL-6:

Composition per liter:

H_3BO_3 .... 0.3g
$CoCl_2·6H_2O$ 0.2g
$ZnSO_4·7H_2O$ 0.1g
$MnCl_2·4H_2O$ 0.03g
$Na_2MoO_4·H_2O$ 0.03g
$NiCl_2·6H_2O$ 0.02g
$CuCl_2·2H_2O$ 0.01g

Preparation of Trace Elements Solution SL-6: Add components to distilled/deionized water and bring volume to 1.0L. Mix thoroughly. Adjust pH to 3.4.

Atrazine Solution:

Composition per 10.0mL:

Atrazine 100mg

Preparation of Atrazine Solution: Add (2-chloro-4(ethylamino)-6-(isopropylamino)-1,3,5-triazine) to 10.0mL methanol. Mix thoroughly. Shortly sonicate to reduce particle size.

Preparation of Medium: Add components, except atrazine solution, to 990.0mLL distilled/deionized water. Adjust pH to 7.25. Autoclave for 15 min at 15 psi pressure–121°C. Cool to room temperature. Aseptically add 10.0mL atazine solution. Mix thoroughly. Aseptically distribute to sterile tubes or flasks.

Use: For the cultivation of *Pseudomonas* sp. and other atrazine-utilizing bacteria.

Mineral Medium with Benzylcyanide (DSMZ Medium 465d)

Composition per liter:

$Na_2HPO_4·2H_2O$ 3.5g
KH_2PO_4 .... 1.0g
$MgCl_2·6H_2O$ 0.1g
$Ca(NO_3)_2·4H_2O$ 0.05g
Benzylcyanide solution 10.0mL
Glucose solution 10.0mL
Trace elements solution SL-4 .... 1.0mL

pH 7.25 ± 0.2 at 25°C

Trace Elements Solution SL-4:

Composition per liter:

EDTA 0.5g
$FeSO_4·7H_2O$ 0.2g
Trace elements solution SL-6 .... 100.0mL

Trace Elements Solution SL-6:

Composition per liter:

H_3BO_3 .... 0.3g
$CoCl_2·6H_2O$ 0.2g
$ZnSO_4·7H_2O$ 0.1g
$MnCl_2·4H_2O$ 0.03g
$Na_2MoO_4·H_2O$ 0.03g
$NiCl_2·6H_2O$ 0.02g
$CuCl_2·2H_2O$ 0.01g

Preparation of Trace Elements Solution SL-6: Add components to distilled/deionized water and bring volume to 1.0L. Mix thoroughly. Adjust pH to 3.4.

Preparation of Trace Elements Solution SL-4: Add components to distilled/deionized water and bring volume to 1.0L. Mix thoroughly.

Glucose Solution:

Composition per 10.0mL:

Glucose 1.8g

Preparation of Glucose Solution: Add glucose to 10.0mL distilled/deionized water. Mix thoroughly. Filter sterilize.

Benzylcyanide Solution:

Composition per 100.0mL:

Benzylcyanide 0.12g

Preparation of Benzylcyanide Solution: Add benzylcyanide to 10.0mL distilled/deionized water. Mix thoroughly. Do not sterilize.

Preparation of Medium: Add components, except benzylcyanide solution and glucose solution, to 980.0mL distilled/deionized water. Adjust pH to 7.25. Autoclave for 15 min at 15 psi pressure–121°C. Cool to room temperature. Aseptically add 10.0mL glucose solution and 10.0mL benzylcyanide solution to the medium. Mix thoroughly. Aseptically distribute the medium to sterile tubes or flasks.

Use: For the cultivation of *Pseudomonas* sp., *Rhodococcus erythropolis,* and other benzylcyanide-utilizing bacteria.

Mineral Medium, Brunner

Composition per liter:

Na_2HPO_4	2.44g
KH_2PO_4	1.52g
$(NH_4)_2SO_4$	0.5g
$MgSO_4 \cdot 7H_2O$	0.2g
$CaCl_2 \cdot 2H_2O$	0.05g
Trace elements solution SL-4	10.0mL

pH 6.9 ± 0.2 at 25°C

Trace Elements Solution SL-4:

Composition per liter:

EDTA	0.5g
$FeSO_4 \cdot 7H_2O$	0.2g
Trace elements solution SL-6	100.0mL

Trace Elements Solution SL-6:

Composition per liter:

$MnCl_2 \cdot 4H_2O$	0.5g
H_3BO_3	0.3g
$CoCl_2 \cdot 6H_2O$	0.2g
$ZnSO_4 \cdot 7H_2O$	0.1g
$Na_2MoO_4 \cdot 2H_2O$	0.03g
$NiCl_2 \cdot 6H_2O$	0.02g
$CuCl_2 \cdot 2H_2O$	0.01g

Preparation of Trace Elements Solution SL-6: Add components to distilled/deionized water and bring volume to 1.0L. Mix thoroughly.

Preparation of Trace Elements Solution SL-4: Add components to distilled/deionized water and bring volume to 1.0L. Mix thoroughly.

Preparation of Medium: Add components to distilled/deionized water and bring volume to 1.0L. Mix thoroughly. Distribute into tubes or flasks. Autoclave for 15 min at 15 psi pressure–121°C.

Use: For the cultivation and maintenance of *Alcaligenes* species, *Bacillus benzoevorans*, *Bacillus gordonae*, *Comamonas acidovorans*, *Hyphomicrobium* species, *Moraxella* species, *Nocardia* species, *Pseudomonas* species, *Rhodococcus* species, *Sphingomonas* species, and *Xanthobacter* species.

Mineral Medium with Camphor

Composition per liter:

$Na_2HPO_4 \cdot 12H_2O$	9.0g
Ferric ammonium citrate	5.0g
$MnSO_4 \cdot H_2O$	3.0g
NH_4Cl	2.0g
KH_2PO_4	1.5g
$MgSO_4 \cdot 7H_2O$	0.2g
$ZnSO_4 \cdot 7H_2O$	0.2g
Titriplex I	10.0mg
$CoSO_4$	10.0μg
Camphor crumbs	variable

pH 7.0 ± 0.2 at 25°C

Preparation of Medium: Add components, except camphor crumbs, to distilled/deionized water and bring volume to 1.0L. Mix thoroughly. Distribute into tubes or flasks. Autoclave for 15 min at 15 psi pressure–121°C. Inoculate tubes or flasks and place in a dessicator jar in which crumbs of camphor will be evaporated.

Use: For the cultivation of bacteria that can utilize camphor as sole carbon source.

Mineral Medium for Chemolithotrophic Growth

Composition per 985.0mL:

Agar	15.0g
$Na_2HPO_4 \cdot 2H_2O$	2.9g
KH_2PO_4	2.3g
NH_4Cl	1.0g
$MgSO_4 \cdot 7H_2O$	0.5g
$NaHCO_3$	0.5g
$CaCl_2 \cdot 2H_2O$	0.01g
Ferric ammonium citrate solution	20.0mL
Trace elements solution SL-6	5.0mL

pH 6.8 ± 0.2 at 25°C

Ferric Ammonium Citrate Solution:

Composition per 20.0mL:

Ferric ammonium citrate	0.05g

Preparation of Ferric Ammonium Citrate Solution: Add ferric ammonium citrate to distilled/deionized water and bring volume to 20.0mL. Mix thoroughly. Autoclave for 15 min at 15 psi pressure–121°C.

Trace Elements Solution SL-6:

Composition per liter:

$MnCl_2 \cdot 4H_2O$	0.5g
H_3BO_3	0.3g
$CoCl_2 \cdot 6H_2O$	0.2g
$ZnSO_4 \cdot 7H_2O$	0.1g
$Na_2MoO_4 \cdot 2H_2O$	0.03g
$NiCl_2 \cdot 6H_2O$	0.02g
$CuCl_2 \cdot 2H_2O$	0.01g

Preparation of Trace Elements Solution SL-6: Add components to distilled/deionized water and bring volume to 1.0L. Mix thoroughly.

Preparation of Medium: Add components, except ferric ammonium citrate solution, to distilled/deionized water and bring volume to 980.0mL. Mix thoroughly. Gently heat and bring to boiling. Autoclave for 15 min at 15 psi pressure–121°C. Cool to 50°–55°C. Aseptically add 20.0mL of sterile ferric ammonium citrate solution. Mix thoroughly. Pour into sterile Petri dishes or distribute into sterile tubes.

Use: For the chemolithotrophic growth and cultivation of a wide variety of bacteria.

Mineral Medium with Chloridazon

Composition per liter:

Chloridazon	1.0g
$Na_2HPO_4 \cdot 12H_2O$	0.7g
$(NH_4)_2HPO_4$	0.7g
KH_2PO_4	0.3g
$(NH_4)H_2PO_4$	0.3g
$MgSO_4 \cdot 7H_2O$	0.25g
$(NH_4)_2SO_4$	0.1g
$CaCl_2 \cdot 6H_2O$	0.05g
H_3BO_3	0.5mg
$MnSO_4 \cdot 4H_2O$	0.4mg
$ZnSO_4 \cdot 7H_2O$	0.4mg
$FeCl_3 \cdot 6H_2O$	0.2mg
$(NH_4)_6Mo_7O_{24} \cdot 4H_2O$	0.2mg
KI	0.1mg
$CuSO_4 \cdot 5H_2O$	0.04mg
Vitamin solution	20.0mL

pH 6.8–7.0 at 25°C

Vitamin Solution:
Composition per 20.0mL:

Biotin	0.1mg
Vitamin B_{12}	0.03mg

Preparation of Vitamin Solution: Add biotin and vitamin B_{12} to 20.0mL of distilled/deionized water. Mix thoroughly. Filter sterilize.

Preparation of Medium: Add components, except vitamin solution, to distilled/deionized water and bring volume to 980.0mL. Mix thoroughly. Adjust pH to 6.8–7.0 with 1*N* NaOH. Autoclave for 20 min at 15 psi pressure–121°C. Cool to 45°–50°C. Aseptically add the sterile vitamin solution. Mix thoroughly. Distribute into sterile tubes or flasks.

Use: For the cultivation and maintenance of *Phenylobacterium immobile*.

Mineral Medium with 2-Chlorobenzoate (DSMZ Medium 457a)

Composition per liter:

Na_2HPO_4	2.44g
KH_2PO_4	1.52g
$(NH_4)_2SO_4$	0.5g
$MgSO_4 \cdot 7H_2O$	0.2g
Tween 80	0.2g
$CaCl_2 \cdot 2H_2O$	0.05g
Trace elements solution SL-4	10.0mL
Chlorobenzoate solution	10.0mL

pH 7.4 ± 0.2 at 25°C

Trace Elements Solution SL-4:
Composition per liter:

EDTA	0.5g
$FeSO_4 \cdot 7H_2O$	0.2g
Trace elements solution SL-6	100.0mL

Trace Elements Solution SL-6:
Composition per liter:

H_3BO_3	0.3g
$CoCl_2 \cdot 6H_2O$	0.2g
$ZnSO_4 \cdot 7H_2O$	0.1g
$MnCl_2 \cdot 4H_2O$	0.03g
$Na_2MoO_4 \cdot H_2O$	0.03g
$NiCl_2 \cdot 6H_2O$	0.02g
$CuCl_2 \cdot 2H_2O$	0.01g

Preparation of Trace Elements Solution SL-6: Add components to distilled/deionized water and bring volume to 1.0L. Mix thoroughly. Adjust pH to 3.4.

Preparation of Trace Elements Solution SL-4: Add components to distilled/deionized water and bring volume to 1.0L. Mix thoroughly.

Chlorobenzoate Solution:
Composition per liter:

2-Chlorobenzoic acid	78.3g

Preparation of Chlorobenzoate Solution: Add 2-chlorobenzoic acid to distilled/deionized water and bring volume to 1.0L. Mix thoroughly. Slowly add concentrated NaOH to adjust pH to 7.4. Filter sterilize.

Preparation of Medium: Add components, except chlorobenzoate solution, to 990.0mL distilled/deionized water. Adjust pH to 7.4. Autoclave for 15 min at 15 psi pressure–121°C. Cool to room temperature. Aseptically add 10.0mL chlorobenzoate solution. Mix thoroughly. Aseptically distribute to sterile tubes or flasks.

Use: For the cultivation of chlorobenzoate-utilizing bacteria.

Mineral Medium with Crude Oil

Composition per 100.0mL:

K_2HPO_4	0.45g
$(NH_4)_2SO_4$	0.1g
$MgSO_4 \cdot 7H_2O$	0.02g
NaCl	0.01g
$CaCl_2$	0.01g
$FeCl_3$	0.002g
Crude oil	5.0mL

pH 7.2 ± 0.3 at 25°C

Preparation of Medium: Add components, except crude oil, to distilled/deionized water and bring volume to 1.0L. Mix thoroughly. Autoclave for 15 min at 15 psi pressure–121°C. Cool to 50°C. Aseptically add 5.0mL of filter-sterilized crude oil. Mix thoroughly. Aseptically distribute into sterile tubes or flasks.

Use: For the cultivation of *Acinetobacter baumannii*.

Mineral Medium with Cyanuric Acid as Nitrogen Source (DSMZ Medium 465g)

Composition per liter:

$Na_2HPO_4 \cdot 2H_2O$	3.5g
KH_2PO_4	1.0g
$MgCl_2 \cdot 6H_2O$	0.1g
$Ca(NO_3)_2 \cdot 4H_2O$	0.05g
Cyanuric acid solution	10.0mL
Vitamin solution	10.0mL
Glycerol solution	10.0mL
Trace elements solution SL-4	1.0mL

pH 7.25 ± 0.2 at 25°C

Trace Elements Solution SL-4:
Composition per liter:

EDTA	0.5g
$FeSO_4 \cdot 7H_2O$	0.2g
Trace elements solution SL-6	100.0mL

Trace Elements Solution SL-6:
Composition per liter:

H_3BO_3	0.3g
$CoCl_2 \cdot 6H_2O$	0.2g
$ZnSO_4 \cdot 7H_2O$	0.1g
$MnCl_2 \cdot 4H_2O$	0.03g
$Na_2MoO_4 \cdot H_2O$	0.03g
$NiCl_2 \cdot 6H_2O$	0.02g
$CuCl_2 \cdot 2H_2O$	0.01g

Preparation of Trace Elements Solution SL-6: Add components to distilled/deionized water and bring volume to 1.0L. Mix thoroughly. Adjust pH to 3.4.

Preparation of Trace Elements Solution SL-4: Add components to distilled/deionized water and bring volume to 1.0L. Mix thoroughly.

Vitamin Solution:
Composition per liter:

Vitamin B_{12}	50.0mg
Pantothenic acid	50.0mg
Riboflavin	50.0mg
Alpha-lipoic acid	50.0mg
p-Aminobenzoic acid	50.0mg
Thiamine-HCl·2H₂O	50.0mg
Nicotinic acid	25.0mg

Nicotinamide	25.0mg
Biotin	20.0mg
Folic acid	20.0mg
Pyridoxamine-HCl	10.0mg

Preparation of Vitamin Solution: Add components to distilled/deionized water and bring volume to 1.0L. Mix thoroughly. Filter sterilize.

Cyanuric Acid Solution:

Composition per 10.0mL:

Cyanuric acid	645.0mg

Preparation of Cyanuric Acid Solution: Add cyanuric acid to 10.0mL distilled/deionized water. Mix thoroughly. Adjust pH to 7.0. Filter sterilize.

Glycerol Solution:

Composition per 10.0mL:

Glycerol	5.5g

Preparation of Glycerol Solution: Add glycerol to distilled/deionized water. Mix thoroughly. Filter sterilize.

Preparation of Medium: Add components, except cyanuric acid solution, glycerol solution, and vitamin solution, to 1.0L distilled/deionized water. Adjust pH to 7.25. Autoclave for 15 min at 15 psi pressure–121°C. Aseptically add 10.0mL cyanuric acid solution, 10.0mL glycerol solution, and 10.0mL vitamin solution. Mix thoroughly. Aseptically distribute to sterile tubes or flasks.

Use: For the cultivation of *Gordonia rubripertincta=Rhodococcus rubropertinctus* and other cyanuric acid-utilizing bacteria.

Mineral Medium with Dichlorobenzoate

Composition per liter:

Na_2HPO_4	2.78g
KH_2PO_4	2.78g
$(NH_4)_2SO_4$	1.0g
Hutner's mineral base	20.0mL
2,4-Dichlorobenzoate solution	10.0mL

pH 6.8 ± 0.2 at 25°C

Hutner's Mineral Base:

Composition per liter:

$MgSO_4 \cdot 7H_2O$	29.7g
Nitrilotriacetic acid	10.0g
$CaCl_2 \cdot 2H_2O$	3.34g
$FeSO_4 \cdot 7H_2O$	99.0mg
$(NH_4)_2MoO_4$	9.25mg
Metals "44"	50.0mL

Preparation of Hutner's Mineral Base: Add nitrilotriacetic acid to 500.0mL of distilled/deionized water. Dissolve by adjusting pH to 6.5 with KOH. Add remaining components. Readjust pH to 7.2 with H_2SO_4 or KOH. Add distilled/deionized water to 1.0L.

Metals "44":

Composition per 100.0mL:

$ZnSO_4 \cdot 7H_2O$	1.1g
$FeSO_4 \cdot 7H_2O$	0.5g
EDTA	0.25g
$MnSO_4 \cdot 7H_2O$	0.154g
$CuSO_4 \cdot 5H_2O$	0.04g
$Co(NO_3)_2 \cdot 6H_2O$	0.025g
$Na_2B_4O_7 \cdot 10H_2O$	0.018g

Preparation of Metals "44": Add a few drops of H_2SO_4 to distilled/deionized water to inhibit precipitate formation. Add components to acidified distilled/deionized water and bring volume to 100.0mL. Mix thoroughly.

2,4-Dichlorobenzoate Solution:

Composition per 10.0mL:

2,4-Dichlorobenzoate	5.0mg

Preparation of 2,4-Dichlorobenzoate Solution: Add 2,4-dichlorobenzoate to 10.0mL of distilled/deionized water. Mix thoroughly. Filter sterilize.

Preparation of Medium: Add components, except 2,4-dichlorobenzoate solution, to distilled/deionized water and bring volume to 990.0mL. Mix thoroughly. Adjust pH to 6.8 with 1*N* KOH. Autoclave for 15 min at 15 psi pressure–121°C. Cool to 45°–50°C. Aseptically add the sterile 2,4-dichlorobenzoate solution. Mix thoroughly. Distribute into sterile tubes or flasks.

Use: For the cultivation and maintenance of *Actinomyces viscosus*.

Mineral Medium with Dichloromethane (DSMZ Medium 465c)

Composition per liter:

$Na_2HPO_4 \cdot 2H_2O$	3.5g
KH_2PO_4	1.0g
$(NH_4)_2SO_4$	0.5g
$MgCl_2 \cdot 6H_2O$	0.1g
$Ca(NO_3)_2 \cdot 4H_2O$	0.05g
Bromthymol blue	50.0mg
Methanol	10.0mL
Trace elements solution SL-4	1.0mL
Dichloromethane	variable

pH 7.25 ± 0.2 at 25°C

Trace Elements Solution SL-4:

Composition per liter:

EDTA	0.5g
$FeSO_4 \cdot 7H_2O$	0.2g
Trace elements solution SL-6	100.0mL

Trace Elements Solution SL-6:

Composition per liter:

H_3BO_3	0.3g
$CoCl_2 \cdot 6H_2O$	0.2g
$ZnSO_4 \cdot 7H_2O$	0.1g
$MnCl_2 \cdot 4H_2O$	0.03g
$Na_2MoO_4 \cdot H_2O$	0.03g
$NiCl_2 \cdot 6H_2O$	0.02g
$CuCl_2 \cdot 2H_2O$	0.01g

Preparation of Trace Elements Solution SL-6: Add components to distilled/deionized water and bring volume to 1.0L. Mix thoroughly. Adjust pH to 3.4.

Preparation of Trace Elements Solution SL-4: Add components to distilled/deionized water and bring volume to 1.0L. Mix thoroughly.

Preparation of Medium: Add components, except methanol and dichloromethane, to 1.0L distilled/deionized water. Adjust pH to 7.25. Autoclave for 15 min at 15 psi pressure–121°C. Cool to room temperature. Inoculate the medium and supply methanol via incubation atmosphere (up to 10.0mL methanol per liter medium). Subsequently feed small amounts of dichloromethane (toxic to bacteria at 2 mmol per liter or less) via incubation atmosphere. Readjust pH of the culture with sterile 1*M* NaOH as necessary

Use: For the cultivation of *Methylophilus leisingeri, Rhizobium radiobacter,* and other dichloromethane-utilizing bacteria.

Mineral Medium with 2,4-Dichlorophenoxyacetic Acid

Composition per liter:

$Na_2HPO_4 \cdot 12H_2O$	9.0g
Ferric ammonium citrate	5.0g
$MnSO_4 \cdot H_2O$	3.0g
NH_4Cl	2.0g
KH_2PO_4	1.5g
$MgSO_4 \cdot 7H_2O$	0.2g
$ZnSO_4 \cdot 7H_2O$	0.2g
Titriplex I	10.0mg
$CoSO_4$	10.0μg
2,4-Dichlorophenoxyacetic acid solution	10.0mL

pH 7.0 ± 0.2 at 25°C

2,4-Dichlorophenoxyacetic Acid Solution:

Composition per 10.0mL:

2,4-Dichlorophenoxyacetic acid	0.5g

Preparation of 2,4-Dichlorophenoxyacetic Acid Solution: Add 2,4-dichlorophenoxyacetic acid to distilled/deionized water and bring volume to 10.0mL. Mix thoroughly. Filter sterilize.

Preparation of Medium: Add components, except 2,4-dichlorophenoxyacetic acid solution, to distilled/deionized water and bring volume to 990.0mL. Mix thoroughly. Autoclave for 15 min at 15 psi pressure–121°C. Aseptically add 10.0mL of sterile 2,4-dichlorophenoxyacetic acid solution. Mix thoroughly. Aseptically distribute into sterile tubes or flasks.

Use: For the cultivation of bacteria that can utilize 2,4-dichlorophenoxyacetic acid as sole carbon source.

Mineral Medium with 2,4-Dichlorotoluene

Composition per liter:

NH_4Cl	0.5g
Yeast extract	0.1g
2,4-Dichlorotoluene	0.1g
$Na_2HPO_4 \cdot 7H_2O$	670.0mg
KH_2PO_4	340.0mg
$MgSO_4 \cdot 7H_2O$	112.0mg
$CaCl_2$	14.0mg
$ZnSO_4 \cdot 7H_2O$	5.0mg
$Na_2MoO_4 \cdot 2H_2O$	2.5mg
$FeCl_3$	0.13mg

pH 7.0 ± 0.2 at 25°C

Preparation of Medium: Add components to distilled/deionized water and bring volume to 1.0L. Mix thoroughly. Distribute into tubes or flasks. Autoclave for 15 min at 15 psi pressure–121°C.

Use: For the cultivation of 2,4-dichlorotoluene-degrading *Pseudomonas* species.

Mineral Medium with Glucose

Composition per liter:

Agar	20.0g
Na_2HPO_4	4.8g
KH_2PO_4	4.4g
NH_4Cl	1.0g
$MgSO_4 \cdot 7H_2O$	0.5g
Solution B	10.0mL
Solution A	5.0mL

pH 6.8 ± 0.2 at 25°C

Solution A:

Composition per 100.0mL:

Ferric ammonium citrate	1.0g
$CaCl_2$	0.1g

Preparation of Solution A: Add ferric ammonium citrate and $CaCl_2$ to distilled/deionized water and bring volume to 100.0mL. Mix thoroughly. Filter sterilize.

Solution B:

Composition per 100.0mL:

Glucose	10.0g

Preparation of Solution B: Add glucose to distilled/deionized water and bring volume to 100.0mL. Mix thoroughly. Filter sterilize.

Preparation of Medium: Add components, except solution A and solution B, to distilled/deionized water and bring volume to 985.0mL. Mix thoroughly. Gently heat and bring to boiling. Autoclave for 15 min at 15 psi pressure–121°C. Cool to 50°C. Aseptically add sterile solution A and sterile solution B. Mix thoroughly. Pour into sterile Petri dishes or distribute into sterile tubes.

Use: For the cultivation and maintenance of *Alcaligenes latus.*

Mineral Medium H-3

Composition per liter:

Agar (if needed)	20.0g
$Na_2HPO_4 \cdot 2H_2O$	2.9g
KH_2PO_4	2.3g
NH_4Cl	1.0g
$MgSO_4 \cdot 7H_2O$	0.5g
$NaHCO_3$	0.5g
$CaCl_2 \cdot 2H_2O$	0.01g
Ferric ammonium citrate solution	20.0mL
Trace elements solution	5.0mL

Ferric Ammonium Citrate Solution:

Composition per 20.0mL:

Ferric ammonium citrate	0.05g

Preparation of Ferric Ammonium Citrate Solution: Add ferric ammonium citrate to distilled/deionized water and bring volume to 20.0mL. Mix thoroughly. Autoclave for 15 min at 15 psi pressure–121°C.

Trace Elements Solution:

Composition per liter:

H_3BO_3	0.3g
$CoCl_3 \cdot 6H_2O$	0.2g
$ZnSO_4 \cdot 7H_2O$	0.1g
$MnCl_2 \cdot 4H_2O$	0.03g
$Na_2MoO_4 \cdot 2H_2O$	0.03g
$CuCl_2 \cdot 2H_2O$	0.01g
$NiCl_2 \cdot 6H_2O$	0.002g

Preparation of Trace Elements Solution: Add components to distilled/deionized water and bring volume to 1.0L. Mix thoroughly.

Preparation of Medium: Add components, except ferric ammonium citrate solution, to distilled/deionized water and bring volume to 980.0mL. Mix thoroughly. Gently heat and bring to boiling. Autoclave for 15 min at 15 psi pressure–121°C. Cool to 50°–55°C. Aseptically add 20.0mL of sterile ferric ammonium citrate solution. Mix thoroughly. Pour into sterile Petri dishes or distribute into sterile tubes. Incubate cultures in 60% H_2 + 28% N_2 + 10% CO_2 + 2% O_2.

Use: For the chemolithotrophic growth of *Alcaligenes eutrophus.*

Mineral Medium for Hydrogen Bacteria

Composition per liter:

Agar	15.0g
$Na_2HPO_4 \cdot 2H_2O$	2.9g
KH_2PO_4	2.3g
NH_4Cl	1.0g
$MgSO_4 \cdot 7H_2O$	0.5g
$NaHCO_3$	0.5g
$CaCl_2 \cdot 2H_2O$	0.01g
Ferric ammonium citrate solution	20.0mL

Ferric Ammonium Citrate Solution:
Composition per 20.0mL:

Ferric ammonium citrate	0.05g

Preparation of Ferric Ammonium Citrate Solution: Add ferric ammonium citrate to 20.0mL of distilled/deionized water. Filter sterilize.

Preparation of Medium: Add components, except ferric ammonium citrate solution, to distilled/deionized water and bring volume to 980.0mL. Mix thoroughly. Gently heat and bring to boiling. Autoclave for 15 min at 15 psi pressure–121°C. Cool to 50°C. Aseptically add the sterile ferric ammonium citrate solution. Mix thoroughly. Pour into sterile Petri dishes or distribute into sterile tubes. Incubate inoculated medium at 30°C under 60% H_2 + 25% N_2 + 10% CO_2 + 5% O_2.

Use: For the cultivation and maintenance of *Alcaligenes eutrophus, Hydrogenophaga flava,* and *Hydrogenophaga pseudoflava.*

Mineral Medium with 2-Hydroxybiphenyl (DSMZ Medium 465a)

Composition per1010.0mL:

$Na_2HPO_4 \cdot 2H_2O$	3.5g
KH_2PO_4	1.0g
$(NH_4)_2SO_4$	0.5g
$MgCl_2 \cdot 6H_2O$	0.1g
$Ca(NO_3)_2 \cdot 4H_2O$	0.05g
Hydroxybiphenyl solution	10.0mL
Trace elements solution SL-4	1.0mL

pH 7.25 ± 0.2 at 25°C

Trace Elements Solution SL-4:
Composition per liter:

EDTA	0.5g
$FeSO_4 \cdot 7H_2O$	0.2g
Trace elements solution SL-6	100.0mL

Trace Elements Solution SL-6:
Composition per liter:

H_3BO_3	0.3g
$CoCl_2 \cdot 6H_2O$	0.2g
$ZnSO_4 \cdot 7H_2O$	0.1g
$MnCl_2 \cdot 4H_2O$	0.03g
$Na_2MoO_4 \cdot H_2O$	0.03g
$NiCl_2 \cdot 6H_2O$	0.02g
$CuCl_2 \cdot 2H_2O$	0.01g

Preparation of Trace Elements Solution SL-6: Add components to distilled/deionized water and bring volume to 1.0L. Mix thoroughly. Adjust pH to 3.4.

Preparation of Trace Elements Solution SL-4: Add components to distilled/deionized water and bring volume to 1.0L. Mix thoroughly.

Hydroxybiphenyl Solution:
Composition per 100.0mL:

2-Hydroxybiphenyl	5.0g

Preparation of Hydroxybiphenyl Solution: Add 2-hydroxybiphenyl to 100.0mL ethanol. Mix thoroughly. Filter sterilize.

Preparation of Medium: Add components, except hydroxybiphenyl solution, to 1.0L distilled/deionized water. Adjust pH to 7.25. Autoclave for 15 min at 15 psi pressure–121°C. Aseptically add hydroxybiphenyl solution to a sterile culture vessel so that the final concentration of 2-hydroxybiphenyl will be 0.5g/L medium. Let the ethanol evaporate under sterile conditions. Aseptically add the liquid medium to the culture vessel to achieve the appropriate concentration of biphenyl.

Use: For the cultivation of *Pseudomonas* sp. and other hydroxybiphenyl-utilizing bacteria.

Mineral Medium M9

Composition per 1013.0mL:

Proline	20.0mg
10X M9 salts	100.0mL
Glucose solution	10.0mL
$CaCl_2 \cdot 2H_2O$ solution	1.0mL
$MgSO_4$ solution	1.0mL
Thiamine·HCl solution	1.0mL

pH 7.4 ± 0.2 at 25°C

10X M9 Salts:
Composition per liter:

Na_2HPO_4	60.0g
KH_2PO_4	30.0g
NH_4Cl	10.0g
NaCl	5.0g

Preparation of 10X M9 Salts: Add components to distilled/deionized water and bring volume to 1.0L. Mix thoroughly. Adjust pH to 7.4.

Glucose Solution:
Composition per 100.0mL:

Glucose	20.0g

Preparation of Glucose Solution: Add glucose to distilled/deionized water and bring volume to 100.0mL. Mix thoroughly. Filter sterilize.

$CaCl_2 \cdot 2H_2O$ Solution:
Composition per 100.0mL:

$CaCl_2 \cdot 2H_2O$	1.47g

Preparation of $CaCl_2 \cdot 2H_2O$ Solution: Add $CaCl_2 \cdot 2H_2O$ to distilled/deionized water and bring volume to 100.0mL. Mix thoroughly. Autoclave for 15 min at 15 psi pressure–121°C.

$MgSO_4$ Solution:
Composition per 100.0mL:

$MgSO_4$	12.04g

Preparation of $MgSO_4$ Solution: Add $MgSO_4$ to distilled/deionized water and bring volume to 100.0mL. Mix thoroughly. Autoclave for 15 min at 15 psi pressure–121°C.

Thiamine·HCl Solution:
Composition per 10.0mL:

Thiamine·HCl	3.37g

Preparation of Thiamine·HCl Solution: Add thiamine·HCl to distilled/deionized water and bring volume to 10.0mL. Mix thoroughly. Filter sterilize.

Preparation of Medium: Add proline and 10X M9 salts to distilled/deionized water and bring volume to 1.0L. Mix thoroughly. Autoclave for 15 min at 15 psi pressure–121°C. Aseptically add 10.0mL of sterile glucose solution, 1.0mL of sterile $CaCl_2·2H_2O$ solution, 1.0mL of sterile $MgSO_4$ solution, and 1.0mL of sterile thiamine·HCl solution. Mix thoroughly. Aseptically distribute into sterile tubes or flasks.

Use: For the cultivation of *Escherichia coli* JM strains.

Mineral Medium with 6-Methylnicotinate (DSMZ Medium 465h)

Composition per liter:

$Na_2HPO_4·2H_2O$	3.5g
KH_2PO_4	1.0g
$(NH_4)_2SO_4$	0.5g
$MgCl_2·6H_2O$	0.1g
$Ca(NO_3)_2·4H_2O$	0.05g
Nicotinate solution	10.0mL
Selenite-tungstate solution	1.0mL
Vitamin solution	1.0mL
Trace elements solution SL-4	1.0mL

pH 7.25 ± 0.2 at 25°C

Trace Elements Solution SL-4:

Composition per liter:

EDTA	0.5g
$FeSO_4·7H_2O$	0.2g
Trace elements solution SL-6	100.0mL

Trace Elements Solution SL-6:

Composition per liter:

H_3BO_3	0.3g
$CoCl_2·6H_2O$	0.2g
$ZnSO_4·7H_2O$	0.1g
$MnCl_2·4H_2O$	0.03g
$Na_2MoO_4·H_2O$	0.03g
$NiCl_2·6H_2O$	0.02g
$CuCl_2·2H_2O$	0.01g

Preparation of Trace Elements Solution SL-6: Add components to distilled/deionized water and bring volume to 1.0L. Mix thoroughly. Adjust pH to 3.4.

Preparation of Trace Elements Solution SL-4: Add components to distilled/deionized water and bring volume to 1.0L. Mix thoroughly.

Vitamin Solution:

Composition per liter:

Vitamin B_{12}	50.0mg
Pantothenic acid	50.0mg
Riboflavin	50.0mg
Alpha-lipoic acid	50.0mg
p-Aminobenzoic acid	50.0mg
Thiamine-HCl·$2H_2O$	50.0mg
Nicotinic acid	25.0mg
Nicotinamide	25.0mg
Biotin	20.0mg
Folic acid	20.0mg
Pyridoxamine-HCl	10.0mg

Preparation of Vitamin Solution: Add components to distilled/deionized water and bring volume to 1.0L. Mix thoroughly. Filter sterilize.

Selenite-Tungstate Solution

Composition per liter:

NaOH	0.5g
$Na_2WO_4·2H_2O$	4.0mg
$Na_2SeO_3·5H_2O$	3.0mg

Preparation of Selenite-Tungstate Solution: Add components to distilled/deionized water and bring volume to 1.0L. Mix thoroughly. Sparge with 100% N_2. Filter sterilize.

Nicotinate Solution:

Composition per 10.0mL:

6-Methylnicotinate	5.0g

Preparation of Nicotinate Solution: Add nicotinate to 10.0mL distilled/deionized wate. Mix thoroughly. Autoclave for 15 min at 15 psi pressure–121°C. Cool to room temperature.

Preparation of Medium: Add components, except nicotinate solution, selenite-tungstate solution, and vitamin solution, to 9898.0mL distilled/deionized water. Adjust pH to 7.25. Autoclave for 15 min at 15 psi pressure–121°C. Cool to room temperature. Aseptically add 10.0mL nicotinate solution, 1.0mL selenite-tungstate solution, and 1.0mL vitamin solution. Mix thoroughly. Aseptically distribute to sterile tubes or flasks.

Use: For the cultivation of *Paenibacillus* sp. and other nicotinate-utilizing bacteria.

Mineral Medium (N_4)

Composition per liter:

HEPES buffer	11.92g
NaCl	0.5g
$(NH_4)_2SO_4$	0.5g
KH_2PO_4	0.2g
$MgSO_4·7H_2O$	0.04g
$CaCl_2·2H_2O$	0.02g
$FeSO_4$/EDTA solution	10.0mL
Phenol Red (0.04%)	1.0mL

$FeSO_4$/EDTA Solution:

Composition per 100.0mL:

EDTA	0.019g
$FeSO_4·7H_2O$	0.018g

Preparation of $FeSO_4$/EDTA Solution: Add components to distilled/deionized water and bring volume to 100.0mL. Mix thoroughly.

Preparation of Medium: Add components to distilled/deionized water and bring volume to 1.0L. Mix thoroughly. Distribute into tubes or flasks. Autoclave for 15 min at 15 psi pressure–121°C.

Use: For the cultivation of *Bacillus* species.

Mineral Medium, Nagel and Andreesen (DSM Medium 461)

Composition per liter:

$MgSO_4·7H_2O$	0.5g
NH_4Cl	0.3g
NaCl	0.05g
$CaCl_2$	0.01g
$MnSO_4$	0.01g
Phosphate solution	50.0mL

Vitamin solution 5.0mL
Trace elements solution SL-10 .... 1.0mL

pH 7.5 ± 0.2 at 25°C

Phosphate Solution:
Composition per 50.0mL:
$Na_2HPO_4 \cdot 2H_2O$ 1.45g
KH_2PO_4 .... 0.25g

Preparation of Phosphate Solution: Add components to distilled/deionized water and bring volume to 50.0mL. Mix thoroughly. Adjust pH to 7.5. Autoclave for 15 min at 15 psi pressure–121°C.

Vitamin Solution:
Composition per liter:
Folic acid 20.0g
α-Lipoic acid 50.0mg
p-Aminobenzoic acid 50.0mg
Pantothenic acid 50.0mg
Riboflavin 50.0mg
Thamine·HCl 50.0mg
Vitamin B_{12} 50.0mg
Nicotinamide 25.0mg
Biotin 20.0mg
Nicotinic acid 20.0mg
Pyridoxamine·HCl 10.0mg

Preparation of Vitamin Solution: Add components to distilled/deionized water and bring volume to 1.0L Mix thoroughly. Stir for 2 hr. Filter sterilize.

Trace Elements Solution SL-10:
Composition per liter:
$FeCl_2 \cdot 4H_2O$ 1.5g
$CoCl_2 \cdot 6H_2O$ 190.0mg
$MnCl_2 \cdot 4H_2O$ 100.0mg
$ZnCl_2$ 70.0mg
$Na_2MoO_4 \cdot 2H_2O$ 36.0mg
$NiCl_2 \cdot 6H_2O$ 24.0mg
H_3BO_3 .... 6.0mg
$CuCl_2 \cdot 2H_2O$ 2.0mg
HCl (25% solution) 10.0mL

Preparation of Trace Elements Solution SL-10: Add $FeCl_2 \cdot 4H_2O$ to 10.0mL of HCl solution. Mix thoroughly. Add distilled/deionized water and bring volume to 1.0L. Add remaining components. Mix thoroughly. Autoclave for 15 min at 15 psi pressure–121°C.

Preparation of Medium: Add components, except phosphate solution, vitamin solution, and trace elements solution SL-10, to distilled/deionized water and bring volume to 944.0mL. Mix thoroughly. Autoclave for 15 min at 15 psi pressure–121°C. Aseptically add 50.0mL of sterile phosphate solution, 5.0mL of sterile vitamin solution, and 1.0mL of sterile trace elements solution SL-10. Mix thoroughly. Aseptically distribute into sterile tubes or flasks.

Use: For the cultivation of *Arthrobacter* species, *Mycobacterium* species, *Paracoccus denitrificans*, *Pseudomonas putida*, and *Rhodococcus* species.

Mineral Medium with Naphthalene

Composition per liter:
$Na_2HPO_4 \cdot 12H_2O$ 9.0g
Ferric ammonium citrate 5.0g
$MnSO_4 \cdot H_2O$ 3.0g
NH_4Cl 2.0g
KH_2PO_4 .... 1.5g
$MgSO_4 \cdot 7H_2O$ 0.2g
$ZnSO_4 \cdot 7H_2O$ 0.2g
Titriplex I 10.0mg
$CoSO_4$ 10.0μg
Naphthalene crumbs .... variable

pH 7.0 ± 0.2 at 25°C

Preparation of Medium: Add components, except naphthalene crumbs, to distilled/deionized water and bring volume to 1.0L. Mix thoroughly. Distribute into tubes or flasks. Autoclave for 15 min at 15 psi pressure–121°C. Inoculate tubes or flasks and place in a dessicator jar in which crumbs of naphthalene will be evaporated.

Use: For the cultivation of bacteria that can utilize naphthalene as sole carbon source.

Mineral Medium with Nicotinic Acid

Composition per liter:
$Na_2HPO_4 \cdot 12H_2O$ 9.0g
Ferric ammonium citrate 5.0g
$MnSO_4 \cdot H_2O$ 3.0g
NH_4Cl 2.0g
KH_2PO_4 .... 1.5g
$MgSO_4 \cdot 7H_2O$ 0.2g
$ZnSO_4 \cdot 7H_2O$ 0.2g
Titriplex I 10.0mg
$CoSO_4$ 10.0μg
Nicotinic acid solution 10.0mL

pH 7.0 ± 0.2 at 25°C

Nicotinic Acid Solution:
Composition per 10.0mL:
Nicotinic acid 0.5g

Preparation of Nicotinic Acid Solution: Add nicotinic acid to distilled/deionized water and bring volume to 10.0mL. Mix thoroughly. Filter sterilize.

Preparation of Medium: Add components, except nicotinic acid solution, to distilled/deionized water and bring volume to 990.0mL. Mix thoroughly. Autoclave for 15 min at 15 psi pressure–121°C. Aseptically add 10.0mL of sterile nicotinic acid solution. Mix thoroughly. Aseptically distribute into sterile tubes or flasks.

Use: For the cultivation of bacteria that can utilize nicotinic acid as sole carbon source.

Mineral Medium with *o*-Nitrophenol (DSMZ Medium 461c)

Composition per liter:
$MgSO_4 \cdot 7H_2O$ 0.5g
NH_4Cl 0.3g
NaCl 0.05g
$CaCl_2$ 0.01g
$MnSO_4$ 0.01g
o-Nitrophenol solution 200.0mL
Phosphate solution 100.0mL
Trace elements solution SL-10 .... 1.0mL

pH 7.2 ± 0.2 at 25°C

Trace Elements Solution SL-10:
Composition per liter:
$FeCl_2 \cdot 4H_2O$ 1.5g
H_3BO_3 .... 300.0mg
$CoCl_2 \cdot 6H_2O$ 190.0mg
$MnCl_2 \cdot 4H_2O$ 100.0mg

$ZnCl_2$ 70.0mg
$Na_2MoO_4·2H_2O$ 36.0mg
$NiCl_2·6H_2O$ 24.0mg
$CuCl_2·2H_2O$ 2.0mg
HCl (25% solution) 7.7mL

Preparation of Trace Elements Solution SL-10: Add $FeCl_2·4H_2O$ to 10.0mL of HCl solution. Mix thoroughly. Add distilled/deionized water and bring volume to 1.0L. Add remaining components. Mix thoroughly. Sparge with 100% N_2. Autoclave for 15 min at 15 psi pressure–121°C. Cool to room temperature.

Phosphate Solution:
Composition per 100.0mL:
$Na_2HPO_4·2H_2O$ 1.45g
KH_2PO_4 .... 0.25g

Preparation of Phosphate Solution: Add components to distilled/deionized water and bring volume to 100.0mL. Mix thoroughly. Autoclave for 15 min at 15 psi pressure–121°C. Cool to room temperature.

***o*-Nitrophenol Solution:**
Composition per liter:
o-Nitrophenol 0.5g

Preparation of *o*-Nitrophenol Solution: Dissolve *o*-nitrophenol in phosphate buffer (50 m*M*, pH 7.5) and bring volume to 1.0L. Sterilize by filtration.

Preparation of Medium: Add components, except phosphate solution and *o*-nitrophenol solution, to distilled/deionized water and bring volume to 700.0mL. Mix thoroughly. Autoclave for 15 min at 15 psi pressure–121°C. Cool to room temperature. Aseptically add 100.0mL phosphate solution and 200.0mL *o*-nitrophenol solution. Mix thoroughly. Aseptically distribute into sterile tubes or flasks.

Use: For the cultivation of *o*-nitrophenol-utilizing bacteria.

Mineral Medium with PCP (DSMZ Medium 465b)

Composition per liter:
$Na_2HPO_4·2H_2O$ 3.5g
KH_2PO_4 .... 1.0g
$(NH_4)_2SO_4$ 0.5g
$MgCl_2·6H_2O$ 0.1g
$Ca(NO_3)_2·4H_2O$ 0.05g
Pentachlorophenol solution 100.0mL
Trace elements solution SL-4 .... 1.0mL
pH 7.25 ± 0.2 at 25°C

Trace Elements Solution SL-4:
Composition per liter:
EDTA 0.5g
$FeSO_4·7H_2O$ 0.2g
Trace elements solution SL-6 .... 100.0mL

Trace Elements Solution SL-6:
Composition per liter:
H_3BO_3 .... 0.3g
$CoCl_2·6H_2O$ 0.2g
$ZnSO_4·7H_2O$ 0.1g
$MnCl_2·4H_2O$ 0.03g
$Na_2MoO_4·H_2O$ 0.03g
$NiCl_2·6H_2O$ 0.02g
$CuCl_2·2H_2O$ 0.01g

Preparation of Trace Elements Solution SL-6: Add components to distilled/deionized water and bring volume to 1.0L. Mix thoroughly. Adjust pH to 3.4.

Preparation of Trace Elements Solution SL-4: Add components to distilled/deionized water and bring volume to 1.0L. Mix thoroughly.

Pentachlorophenol Solution:
Composition per liter:
Pentachlorophenol 1.0g

Preparation of Pentachlorophenol Solution: Add pentachlorophenol to 1.0L 0.1*M* NaOH. Mix thoroughly. Filter sterilize.

Preparation of Medium: Add components, except pentachlorophenol solution, to 900.0mL distilled/deionized water. Adjust pH to 7.25. Autoclave for 15 min at 15 psi pressure–121°C. Aseptically add 100.0mL sterile pentachlorophenol solution. Mix thoroughly. Aseptically distribute to sterile tubes or flasks.

Use: For the cultivation of *Sphingobium chlorophenolicum=Sphingomonas chlorophenolica* and other pentachlorophenol-utilizing bacteria.

Mineral Medium, pH 7.25

Composition per liter:
$Na_2HPO_4·2H_2O$ 3.5g
KH_2PO_4 .... 1.0g
$(NH_4)_2SO_4$ 0.5g
$MgCl_2·6H_2O$ 0.1g
$Ca(NO_3)_2·4H_2O$ 0.05g
Trace elements solution SL-4 .... 1.0mL
pH 7.25 ± 0.2 at 25°C

Trace Elements Solution SL-4:
Composition per liter:
EDTA 0.5g
$FeSO_4·7H_2O$ 0.2g
Trace elements solution SL-6 .... 100.0mL

Trace Elements Solution SL-6:
Composition per liter:
$MnCl_2·4H_2O$ 0.5g
H_3BO_3 .... 0.3g
$CoCl_2·6H_2O$ 0.2g
$ZnSO_4·7H_2O$ 0.1g
$Na_2MoO_4·2H_2O$ 0.03g
$NiCl_2·6H_2O$ 0.02g
$CuCl_2·2H_2O$ 0.01g

Preparation of Trace Elements Solution SL-6: Add components to distilled/deionized water and bring volume to 1.0L. Mix thoroughly.

Preparation of Trace Elements Solution SL-4: Add components to distilled/deionized water and bring volume to 1.0L. Mix thoroughly.

Preparation of Medium: Add components to distilled/deionized water and bring volume to 1.0L. Mix thoroughly. Distribute into tubes or flasks. Autoclave for 15 min at 15 psi pressure–121°C.

Use: For the cultivation of *Azotobacter* species, *Pseudomonas* species, and *Sphingomonas* species.

Mineral Medium with Phenol

Composition per liter:
Phenol 1.0g
K_2HPO_4 .... 1.0g
NH_4NO_3 .... 1.0g
$(NH_4)_2SO_4$ 0.5g

$MgSO_4$ 0.5g
KH_2PO_4 .... 0.5g
NaCl 0.5g
$CaCl_2$ 0.02g
$FeSO_4$ 0.02g
Wolfe's mineral solution 10.0mL

Wolfe's Mineral Solution:
Composition per liter:
$MgSO_4·7H_2O$ 3.0g
Nitrilotriacetic acid 1.5g
NaCl 1.0g
$MnSO_4·H_2O$ 0.5g
$FeSO_4·7H_2O$ 0.1g
$CoCl_2·6H_2O$ 0.1g
$CaCl_2$ 0.1g
$ZnSO_4·7H_2O$ 0.1g
$CuSO_4·5H_2O$ 0.01g
$AlK(SO_4)_2·12H_2O$ 0.01g
H_3BO_3 .... 0.01g
$Na_2MoO_4·2H_2O$ 0.01g

Preparation of Wolfe's Mineral Solution: Add nitrilotriacetic acid to 500.0mL of distilled/deionized water. Dissolve by adjusting pH to 6.5 with KOH. Add remaining components. Add distilled/deionized water to 1.0L.

Preparation of Medium: Add components, except phenol, to distilled/deionized water and bring volume to 1.0L. Mix thoroughly. Autoclave for 15 min at 15 psi pressure–121°C. Aseptically add the phenol. Mix thoroughly. Distribute into sterile tubes or flasks.

Use: For the cultivation and maintenance of *Pseudomonas putida.*

Mineral Medium with Phenylacetate

Composition per liter:
$Na_2HPO_4·12H_2O$ 9.0g
Ferric ammonium citrate 5.0g
$MnSO_4·H_2O$ 3.0g
NH_4Cl 2.0g
KH_2PO_4 .... 1.5g
$MgSO_4·7H_2O$ 0.2g
$ZnSO_4·7H_2O$ 0.2g
Titriplex I 10.0mg
$CoSO_4$ 10.0μg
Phenylacetic acid solution 10.0mL
pH 7.0 ± 0.2 at 25°C

Phenylacetic Acid Solution:
Composition per 10.0mL:
Phenylacetic acid 0.5g

Preparation of Phenylacetic Acid Solution: Add phenylacetic acid to distilled/deionized water and bring volume to 10.0mL. Mix thoroughly. Filter sterilize.

Preparation of Medium: Add components, except phenylacetic acid solution, to distilled/deionized water and bring volume to 990.0mL. Mix thoroughly. Autoclave for 15 min at 15 psi pressure–121°C. Aseptically add 10.0mL of sterile phenylacetic acid solution. Mix thoroughly. Aseptically distribute into sterile tubes or flasks.

Use: For the cultivation of bacteria that can utilize phenylacetic acid as sole carbon source.

Mineral Medium with Pyrrolic Acid (DSMZ Medium 461b)

Composition per liter:
$MgSO_4·7H_2O$ 0.5g
NH_4Cl 0.3g
NaCl 0.05g
$CaCl_2$ 0.01g
$MnSO_4$ 0.01g
Phosphate solution 100.0mL
Pyrrolic acid solution 10.0mL
Trace elements solution SL-10 .... 1.0mL
pH 7.2 ± 0.2 at 25°C

Trace Elements Solution SL-10:
Composition per liter:
$FeCl_2·4H_2O$ 1.5g
H_3BO_3 .... 300.0mg
$CoCl_2·6H_2O$ 190.0mg
$MnCl_2·4H_2O$ 100.0mg
$ZnCl_2$ 70.0mg
$Na_2MoO_4·2H_2O$ 36.0mg
$NiCl_2·6H_2O$ 24.0mg
$CuCl_2·2H_2O$ 2.0mg
HCl (25% solution) 7.7mL

Preparation of Trace Elements Solution SL-10: Add $FeCl_2·4H_2O$ to 10.0mL of HCl solution. Mix thoroughly. Add distilled/deionized water and bring volume to 1.0L. Add remaining components. Mix thoroughly. Sparge with 100% N_2. Autoclave for 15 min at 15 psi pressure–121°C. Cool to room temperature.

Phosphate Solution:
Composition per 100.0mL:
$Na_2HPO_4·2H_2O$ 1.45g
KH_2PO_4 .... 0.25g

Preparation of Phosphate Solution: Add components to distilled/deionized water and bring volume to 100.0mL. Mix thoroughly. Autoclave for 15 min at 15 psi pressure–121°C. Cool to room temperature.

Pyrrolic Acid Solution:
Composition per liter:
Pyrrolic acid 10.0g

Preparation of Pyrrolic Acid Solution: Dissolve pyrrolic acid in phosphate buffer (50 m*M*, pH 7.5) and bring volume to 1.0L. Sterilize by filtration.

Preparation of Medium: Add components, except phosphate solution and pyrrolic acid solution, to distilled/deionized water and bring volume to 890.0mL. Mix thoroughly. Autoclave for 15 min at 15 psi pressure–121°C. Cool to room temperature. Aseptically add 100.0mL phosphate solution and 10.0mL pyrrolic acid solution. Mix thoroughly. Aseptically distribute into sterile tubes or flasks

Use: For the cultivation of pyrrolic acid-utilizing bacteria.

Mineral Medium with Quinoline (DSMZ Medium 461a)

Composition per liter:
$MgSO_4·7H_2O$ 0.5g
NH_4Cl 0.3g
NaCl 0.05g
$CaCl_2$ 0.01g
$MnSO_4$ 0.01g

Phosphate solution	100.0mL
Quinoline emulsion	100.0mL
Vitamin solution	5.0mL
Trace elements solution SL-10	1.0mL

pH 7.2 ± 0.2 at 25°C

Trace Elements Solution SL-10:

Composition per liter:

$FeCl_2 \cdot 4H_2O$	1.5g
H_3BO_3	300.0mg
$CoCl_2 \cdot 6H_2O$	190.0mg
$MnCl_2 \cdot 4H_2O$	100.0mg
$ZnCl_2$	70.0mg
$Na_2MoO_4 \cdot 2H_2O$	36.0mg
$NiCl_2 \cdot 6H_2O$	24.0mg
$CuCl_2 \cdot 2H_2O$	2.0mg
HCl (25% solution)	7.7mL

Preparation of Trace Elements Solution SL-10: Add $FeCl_2 \cdot 4H_2O$ to 10.0mL of HCl solution. Mix thoroughly. Add distilled/deionized water and bring volume to 1.0L. Add remaining components. Mix thoroughly. Sparge with 100% N_2. Autoclave for 15 min at 15 psi pressure–121°C. Cool to room temperature.

Vitamin Solution:

Composition per liter:

Vitamin B_{12}	50.0mg
Pantothenic acid	50.0mg
Riboflavin	50.0mg
Alpha-lipoic acid	50.0mg
p-Aminobenzoic acid	50.0mg
Thiamine-HCl·$2H_2O$	50.0mg
Nicotinic acid	25.0mg
Nicotinamide	25.0mg
Biotin	20.0mg
Folic acid	20.0mg
Pyridoxamine-HCl	10.0mg

Preparation of Vitamin Solution: Add components to distilled/deionized water and bring volume to 1.0L. Mix thoroughly. Filter sterilize.

Phosphate Solution:

Composition per 100.0mL:

$Na_2HPO_4 \cdot 2H_2O$	1.45g
KH_2PO_4	0.25g

Preparation of Phosphate Solution: Add components to distilled/deionized water and bring volume to 100.0mL. Mix thoroughly. Autoclave for 15 min at 15 psi pressure–121°C. Cool to room temperature.

Quinoline Emulsion:

Composition per liter:

Quinoline	3.0g

Preparation of Quinoline Emulsion: Prepare an emulsion of quinoline in 50 m*M* phosphate buffer by stirring or ultrasonication. Add quinoline acid to phosphate buffer (50 m*M*, pH 7.5) and bring volume to 1.0L. Stir vigorously or sonicate to form emulsion. Autoclave for 15 min at 15 psi pressure–121°C in a gas tight vessel. Cool to room temperature.

Preparation of Medium: Add components, except phosphate solution, vitamin solution, and quinoline emulsion to distilled/deionized water and bring volume to 795.0mL. Mix thoroughly. Autoclave for 15 min at 15 psi pressure–121°C. Cool to room temperature. Aseptically add 100.0mL phosphate solution, 5.0mL vitamin solution, and 100.0mL quinoline emulsion. Mix thoroughly. Aseptically distribute into sterile tubes or flasks

Use: For the cultivation of *Rhodococcus rhodochrous (Rhodococcus roseus)* and other quinoline-utilizing bacteria.

Mineral Medium S with 1% Sucrose

Composition per liter:

Sucrose	10.0g
$NH_4H_2PO_4$	1.5g
$MgSO_4 \cdot 7H_2O$	0.2g
$CaCl_2$	0.1g
KCl	0.1g
$FeCl_3$	0.005g
NaOH (1*N* solution)	10.0mL

Preparation of Medium: Add components to distilled/deionized water and bring volume to 1.0L. Mix thoroughly. Distribute into tubes or flasks. Autoclave for 15 min at 15 psi pressure–121°C.

Use: For the cultivation of *Saccharobacterium acuminatum.*

Mineral Medium with Salicylate

Composition per liter:

$Na_2HPO_4 \cdot 12H_2O$	9.0g
Ferric ammonium citrate	5.0g
$MnSO_4 \cdot H_2O$	3.0g
NH_4Cl	2.0g
KH_2PO_4	1.5g
$MgSO_4 \cdot 7H_2O$	0.2g
$ZnSO_4 \cdot 7H_2O$	0.2g
Titriplex I	10.0mg
$CoSO_4$	10.0μg
Salicylic acid solution	10.0mL

pH 7.0 ± 0.2 at 25°C

Salicylic Acid Solution:

Composition per 10.0mL:

Salicylic acid	0.5g

Preparation of Salicylic Acid Solution: Add salicylic acid to distilled/deionized water and bring volume to 10.0mL. Mix thoroughly. Filter sterilize.

Preparation of Medium: Add components, except salicylic acid solution, to distilled/deionized water and bring volume to 990.0mL. Mix thoroughly. Autoclave for 15 min at 15 psi pressure–121°C. Aseptically add 10.0mL of sterile salicylic acid solution. Mix thoroughly. Aseptically distribute into sterile tubes or flasks.

Use: For the cultivation of bacteria that can utilize salicylic acid as sole carbon source.

Mineral Medium with Santonin

Composition per liter:

K_2HPO_4	6.3g
α-Santonin	4.0g
KH_2PO_4	1.82g
NH_4NO_3	1.0g
$CaCl_2 \cdot 2H_2O$	0.75g
$MgSO_4 \cdot 7H_2O$	0.1g
$FeSO_4 \cdot 7H_2O$	0.06g

$MnSO_4$ (anhydrous) 600.0μg
$Na_2MoO_4 \cdot 2H_2O$ 600.0μg

pH 7.0 ± 0.2 at 25°C

Preparation of Medium: Add components to distilled/deionized water and bring volume to 1.0L. Mix thoroughly. Autoclave for 15 min at 15 psi pressure–121°C. Distribute into sterile tubes or flasks.

Use: For the cultivation and maintenance of *Pseudomonas* species.

Mineral Medium with Sulfobenzoic Acid (DSMZ Medium 465e)

Composition per liter:

$Na_2HPO_4 \cdot 2H_2O$ 3.5g
KH_2PO_4 .... 1.0g
$(NH_4)_2SO_4$ 0.5g
$MgCl_2 \cdot 6H_2O$ 0.1g
$Ca(NO_3)_2 \cdot 4H_2O$ 0.05g
2-Sulfobenzoic acid 50.0mL
Vitamin solution 2.5mL
Trace elements solution SL-4 .... 1.0mL

pH 7.25 ± 0.2 at 25°C

Trace Elements Solution SL-4:
Composition per liter:

EDTA 0.5g
$FeSO_4 \cdot 7H_2O$ 0.2g
Trace elements solution SL-6 .... 100.0mL

Trace Elements Solution SL-6:
Composition per liter:

H_3BO_3 .... 0.3g
$CoCl_2 \cdot 6H_2O$ 0.2g
$ZnSO_4 \cdot 7H_2O$ 0.1g
$MnCl_2 \cdot 4H_2O$ 0.03g
$Na_2MoO_4 \cdot H_2O$ 0.03g
$NiCl_2 \cdot 6H_2O$ 0.02g
$CuCl_2 \cdot 2H_2O$ 0.01g

Preparation of Trace Elements Solution SL-6: Add components to distilled/deionized water and bring volume to 1.0L. Mix thoroughly. Adjust pH to 3.4.

Preparation of Trace Elements Solution SL-4: Add components to distilled/deionized water and bring volume to 1.0L. Mix thoroughly.

Vitamin Solution:
Composition per 100.0mL:

Pyridoxamine 5.0mg
Vitamin B_{12} 2.0mg
Nicotinic acid 2.0mg
Thiamine-HCl·$2H_2O$ 1.0mg
p-Aminobenzoate 1.0mg
Ca-pantothenate 0.5mg
Biotin 0.2mg

Preparation of Vitamin Solution: Add components to distilled/deionized water and bring volume to 100.0mL. Mix thoroughly. Filter sterilize.

2-Sulfobenzoic Acid Solution:
Composition per 100.0mL:

2-Sulfobenzoic acid 2.0g

Preparation of 2-Sulfobenzoic Acid Solution: Add 2-sulfobenzoic acid to 100.0mL distilled/deionized water. Mix thoroughly. Filter sterilize.

Preparation of Medium: Add components, except vitamin solution and 2-sulfobenzoic acid solution, to 947.5mL distilled/deionized water. Adjust pH to 7.25. Autoclave for 15 min at 15 psi pressure–121°C. Cool to room temperature. Aseptically add 2-sulfobenzoic acid solution and vitamin solution. Mix thoroughly. Aseptically distribute to sterile tubes or flasks.

Use: For the cultivation of sulfobenzoic acid-utilizing bacteria.

Mineral Medium with Toluene

Composition per liter:

$Na_2HPO_4 \cdot 12H_2O$ 9.0g
Ferric ammonium citrate 5.0g
$MnSO_4 \cdot H_2O$ 3.0g
NH_4Cl 2.0g
KH_2PO_4 .... 1.5g
$MgSO_4 \cdot 7H_2O$ 0.2g
$ZnSO_4 \cdot 7H_2O$ 0.2g
Titriplex I 10.0mg
$CoSO_4$ 10.0μg
Toluene .... 1 drop

pH 7.0 ± 0.2 at 25°C

Preparation of Medium: Add components, except toluene, to distilled/deionized water and bring volume to 1.0L. Mix thoroughly. Distribute into tubes or flasks. Autoclave for 15 min at 15 psi pressure–121°C. Inoculate tubes or flasks and place in a dessicator jar in which 1 drop of toluene will be evaporated.

Use: For the cultivation of bacteria that can utilize toluene as sole carbon source.

Mineral Medium with Vitamins

Composition per 1002.5mL:

Mineral medium 1000.0mL
Schlegel's vitamin solution 2.5mL

Mineral Medium:
Composition per 1010.0mL:

Na_2HPO_4 .... 2.44g
KH_2PO_4 .... 1.52g
$(NH_4)_2SO_4$ 0.5g
$MgSO_4 \cdot 7H_2O$ 0.2g
$CaCl_2 \cdot 2H_2O$ 0.05g
Trace elements solution SL-4 .... 10.0mL

pH 6.9 ± 0.2 at 25°C

Trace Elements Solution SL-4:
Composition per liter:

EDTA 0.5g
$FeSO_4 \cdot 7H_2O$ 0.2g
Trace elements solution SL-6 .... 100.0mL

Trace Elements Solution SL-6:
Composition per liter:

$MnCl_2 \cdot 4H_2O$ 0.5g
H_3BO_3 .... 0.3g
$CoCl_2 \cdot 6H_2O$ 0.2g
$ZnSO_4 \cdot 7H_2O$ 0.1g
$Na_2MoO_4 \cdot 2H_2O$ 0.03g
$NiCl_2 \cdot 6H_2O$ 0.02g
$CuCl_2 \cdot 2H_2O$ 0.01g

Preparation of Trace Elements Solution SL-6: Add components to distilled/deionized water and bring volume to 1.0L. Mix thoroughly.

Preparation of Trace Elements Solution SL-4: Add components to distilled/deionized water and bring volume to 1.0L. Mix thoroughly.

Schlegel's Vitamin Solution:

Composition per 100.0mL:

Nicotinic acid	2.0g
Pyridoxamine	5.0mg
Cyanocobalamin	2.0mg
p-Aminobenzoate	1.0mg
Thiamine	1.0mg
Calcium DL-pantothenate	0.5mg
Biotin	0.2mg

Preparation of Schlegel's Vitamin Solution: Add components to distilled/deionized water and bring volume to 100.0mL. Filter sterilize.

Preparation of Medium: Add components, except Schlegel's vitamin solution, to distilled/deionized water and bring volume to 1.0L. Mix thoroughly. Autoclave for 15 min at 15 psi pressure–121°C. Aseptically add 2.5mL of sterile Schlegel's vitamin solution. Mix thoroughly. Aseptically distribute into sterile tubes or flasks.

Use: For the cultivation of *Pseudomonas chlororaphis* and *Rhodococcus* species.

Mineral Pectin 5 Medium

See: **MP 5 Medium**

Mineral Pectin 7 Medium

See: **MP 7 Medium**

Mineral Salts Agar

Composition per liter:

Agar	15.0g
$NaNO_3$	2.0g
K_2HPO_4	1.2g
$MgSO_4$	0.5g
KCl	0.5g
KH_2PO_4	0.14g
Yeast extract	0.02g
$Fe_2(SO_4)_3 \cdot H_2O$	0.01g

pH 7.2 ± 0.2 at 25°C

Preparation of Medium: Add components to distilled/deionized water and bring volume to 1.0L. Mix thoroughly. Adjust pH to 7.2. Gently heat and bring to boiling. Distribute into tubes. Autoclave for 15 min at 15 psi pressure–121°C. Allow tubes to cool in a slanted position. Add a strip of sterile filter paper onto cooled slant. Inoculate organisms on filter paper.

Use: For the cultivation and maintenance of *Cytophaga aurantiaca* and *Sporocytophaga myxococcoides*.

Mineral Salts Enrichment Medium

Composition per liter:

KH_2PO_4	1.36g
$(NH_4)_2SO_4$	0.5g
$MgSO_4 \cdot 7H_2O$	0.2g
$CaCl_2 \cdot 2H_2O$	0.01g
$FeSO_4 \cdot 7H_2O$	5.0mg
$MnSO_4 \cdot 7H_2O$	2.5mg
$Na_2MoO_4 \cdot 2H_2O$	2.5mg
Na_2HPO_4	2.13mg

pH 7.2 ± 0.2 at 25°C

Preparation of Medium: Add components to distilled/deionized water and bring volume to 1.0L. Mix thoroughly. Distribute into tubes or flasks. Autoclave for 15 min at 15 psi pressure–121°C.

Use: For the enrichment and cultivation of *Caulobacter* species.

Mineral Salts with Butane

Composition per liter of tap water:

$(NH_4)_2HPO_4$	8.0g
$Na_2HPO_4 \cdot 12H_2O$	2.5g
KH_2PO_4	2.0g
$MgSO_4 \cdot 7H_2O$	0.5g
Yeast extract	100.0mg
$CaCl_2 \cdot 2H_2O$	60.0mg
$FeSO_4 \cdot 7H_2O$	30.0mg
$MnCl_2 \cdot 4H_2O$	60.0μg
$CuSO_4 \cdot 5H_2O$	15.0μg

pH 7.1 ± 0.2 at 25°C

Preparation of Medium: Add components to distilled/deionized water and bring volume to 1.0L. Mix thoroughly. Distribute into tubes or flasks. Autoclave for 15 min at 15 psi pressure–121°C. Incubate inoculated medium in 88% air + 7% *n*-butane + 5% CO_2.

Use: For the cultivation and maintenance of *Pseudomonas butanovora*.

Mineral Salts Medium

Composition per liter:

Na_2HPO_4	4.0g
KH_2PO_4	1.5g
NH_4Cl	1.0g
$MgSO_4 \cdot 7H_2O$	0.2g
Ferric ammonium citrate	5.0mg
Modified Hoagland trace elements solution	1.0mL

pH 7.0 ± 0.2 at 25°C

Modified Hoagland Trace Elements Solution:

Composition per 3.6L:

H_3BO_3	11.0g
$MnCl_2 \cdot 4H_2O$	7.0g
$AlCl_3$	1.0g
$CoCl_2$	1.0g
$CuCl_2$	1.0g
KI	1.0g
$NiCl_2$	1.0g
$ZnCl_2$	1.0g
$BaCl_2$	0.5g
KBr	0.5g
LiCl	0.5g
Na_2MoO_4	0.5g
$SeCl_4$	0.5g
$SnCl_2 \cdot 2H_2O$	0.5g
$NaVO_3 \cdot H_2O$	0.1g

Preparation of Modified Hoagland Trace Elements Solution: Prepare each component as a separate solution. Dissolve each salt in approximately 100.0mL of distilled/deionized water. Adjust the pH of each solution to below 7.0. Combine all the salt solutions and bring the volume to 3.6L with distilled/deionized water. Adjust the pH to 3–4. A yellow precipitate may form after mixing. After a few days, it will turn into a fine white precipitate. Mix the solution thoroughly before using.

Preparation of Medium: Add components to distilled/deionized water and bring volume to 1.0L. Mix thoroughly. Distribute into tubes or flasks. Autoclave for 15 min at 15 psi pressure–121°C.

Use: For the cultivation and maintenance of *Rhodococcus rhodochrous*.

Mineral Salts Medium with Methanol

Composition per liter:

$NaNH_4HPO_4 \cdot 4H_2O$	1.74g
$NaH_2PO_4 \cdot H_2O$	0.54g
$MgSO_4 \cdot 7H_2O$	0.2g
KCl	0.04g
$FeSO_4 \cdot 7H_2O$	5.0mg
Methanol	5.0mL
Trace minerals solution	1.0mL

pH 7.2 ± 0.2 at 25°C

Trace Minerals Solution:

Composition per liter:

H_3BO_3	2.86g
$MnCl_2 \cdot 4H_2O$	1.81g
$ZnSO_4 \cdot 7H_2O$	0.22g
$CuSO_4 \cdot 5H_2O$	0.08g
$CoCl_2 \cdot 6H_2O$	0.06g
$Na_2MoO_4 \cdot 2H_2O$	25.0mg

Preparation of Trace Minerals Solution: Add components to distilled/deionized water and bring volume to 1.0L. Mix thoroughly.

Preparation of Medium: Add components, except methanol, to distilled/deionized water and bring volume to 1.0L. Mix thoroughly. Distribute into tubes or flasks. Autoclave for 15 min at 15 psi pressure–121°C. Cool to 50°C. Filter sterilize methanol. Aseptically add sterile methanol to cooled, sterile basal medium.

Use: For the cultivation and maintenance of *Rhodococcus rhodochrous*.

Mineral Salts Medium with Methanol and Yeast Extract

Composition per liter:

$NaNH_4HPO_4 \cdot 4H_2O$	1.74g
$NaH_2PO_4 \cdot H_2O$	0.54g
$MgSO_4 \cdot 7H_2O$	0.2g
Yeast extract	0.2g
KCl	0.04g
$FeSO_4 \cdot 7H_2O$	5.0mg
Methanol	5.0mL
Trace minerals solution	1.0mL

pH 7.2 ± 0.2 at 25°C

Trace Minerals Solution:

Composition per liter:

H_3BO_3	2.86g
$MnCl_2 \cdot 4H_2O$	1.81g
$ZnSO_4 \cdot 7H_2O$	0.22g
$CuSO_4 \cdot 5H_2O$	0.08g
$CoCl_2 \cdot 6H_2O$	0.06g
$Na_2MoO_4 \cdot 2H_2O$	25.0mg

Preparation of Trace Minerals Solution: Add components to distilled/deionized water and bring volume to 1.0L. Mix thoroughly.

Preparation of Medium: Add components, except methanol, to distilled/deionized water and bring volume to 1.0L. Mix thoroughly. Autoclave for 15 min at 15 psi pressure–121°C. Cool to 50°C. Filter sterilize methanol. Aseptically add sterile methanol to cooled, sterile basal medium. Aseptically distribute into sterile tubes or flasks.

Use: For the cultivation and maintenance of *Pseudomonas* species.

Mineral Salts Medium for Thermophiles

Composition per liter:

$NaNO_3$	0.25g
NH_4Cl	0.25g
Na_2HPO_4	210.0mg
$MgSO_4 \cdot 7H_2O$	200.0mg
NaH_2PO_4	90.0mg
KCl	40.0mg
$CaCl_2$	15.0mg
$FeSO_4$	1.0mg
Trace minerals solution	10.0mL
n-Heptadecane	1.0mL

Trace Minerals Solution:

Composition per liter:

$ZnSO_4 \cdot 7H_2O$	7.0mg
H_3BO_4	1.0mg
MoO_3	1.0mg
$CuSO_4 \cdot 5H_2O$	500.0μg
$CoSO_4 \cdot 7H_2O$	18.0μg
$MnSO_4 \cdot 5H_2O$	7.0μg

Preparation of Trace Minerals Solution: Add components to distilled/deionized water and bring volume to 1.0L. Mix thoroughly.

Preparation of Medium: Add components to distilled/deionized water and bring volume to 1.0L. Mix thoroughly. Distribute into tubes or flasks. Autoclave for 15 min at 15 psi pressure–121°C.

Use: For the cultivation of *Bacillus thermoleovorans*.

Mineral Salts Peptonized Milk Agar (SPMA)

Composition per liter:

Agar	15.0g
Milk, peptonized	1.0g
Mineral solution	100.0mL

Mineral Solution:

Composition per 100.0mL:

$MgSO_4 \cdot 7H_2O$	0.5g
$CaCl_2$	0.25g
K_2HPO_4	0.25g
$(NH_4)_2SO_4$	0.1g
$FeCl_3 \cdot 6H_2O$	0.01g
$MnCl_2$	0.1mg

Preparation of Mineral Solution: Add components to distilled/deionized water and bring volume to 100.0mL. Mix thoroughly. Filter sterilize.

Preparation of Medium: Add components, except mineral solution, to distilled/deionized water and bring volume to 900.0mL. Mix thoroughly. Gently heat and bring to boiling. Autoclave for 15 min at 15 psi pressure–121°C. Cool to 45°–50°C. Aseptically add 100.0mL of sterile mineral solution. Mix thoroughly. Pour into sterile Petri dishes or distribute into sterile tubes.

Use: For the cultivation of freshwater *Myxobacterium* species.

Mineral Salts for Thermophiles (L Salts for Thermophiles)

Composition per liter:

$NaNO_3$	0.25g
NH_4Cl	0.25g
Na_2HPO_4	0.21g
$MgSO_4·7H_2O$	0.2g
NaH_2PO_4	0.09g
KCl	0.04g
$CaCl_2$	0.02g
$FeSO_4$	1.0mg
Trace minerals solution	10.0mL
n-Heptadecane	1.0mL

Trace Minerals Solution:
Composition per liter:

$ZnSO_4·7H_2O$	7.0mg
H_3BO_4	1.0mg
MoO_3	1.0mg
$CuSO_4·5H_2O$	500.0μg
$CoSO_4·7H_2O$	18.0μg
$MnSO_4·5H_2O$	7.0μg

Preparation of Trace Minerals Solution: Add components to distilled/deionized water and bring volume to 1.0L. Mix thoroughly.

Preparation of Medium: Add components, except *n*-heptadecane, to distilled/deionized water and bring volume to 1.0L. Mix thoroughly. Autoclave for 15 min at 15 psi pressure–121°C. Aseptically add the *n*-heptadecane. Mix thoroughly. Distribute into sterile tubes or flasks.

Use: For the cultivation and maintenance of *Bacillus thermoleovorans*.

Minerals Modified Glutamate Agar

Composition per liter:

Agar	15.0g
Lactose	10.0g
Sodium glutamate	6.35g
NH_4Cl	2.5g
K_2HPO_4	0.9g
Sodium formate	0.25g
$MgSO_4·7H_2O$	0.1g
L-Aspartic acid	0.024g
L-Arginine	0.02g
L-Cystine	0.02g
Bromcresol Purple	0.01g
$CaCl_2·2H_2O$	0.01g
Ferric ammonium citrate	0.01g
Nicotinic acid	1.0mg
Pantothenic acid	1.0mg
Thiamine	1.0mg

pH 6.7 ± 0.2 at 25°C

Preparation of Medium: Add components to distilled/deionized water and bring volume to 1.0L. Mix thoroughly. Gently heat and bring to boiling. Distribute into tubes or flasks. Autoclave for 10 min at 11 psi pressure–116°C. Pour into sterile Petri dishes in 20.0mL volumes.

Use: For the cultivation of coliform bacteria from foods.

Minerals Modified Medium

Composition per liter:

Lactose	20.0g
Sodium glutamate	12.7g
NH_4Cl	5.0g
K_2HPO_4	1.8g
Sodium formate	0.5g
$MgSO_4·7H_2O$	0.2g
L-Aspartic acid	0.048g
L-Cystine	0.04g
L-Arginine	0.04g
Ferric ammonium citrate	0.02g
$CaCl_2·2H_2O$	0.02g
Bromcresol Purple	0.02g
Thiamine	2.0mg
Nicotinic acid	2.0mg
Pantothenic acid	2.0mg

pH 6.7 ± 0.2 at 25°C

Source: This medium is available as a premixed powder from Oxoid Unipath.

Preparation of Medium: Add NH_4Cl to distilled/deionized water and bring volume to 800.0mL. Add remaining components and bring volume to 1.0L. Mix thoroughly. Adjust pH to 6.7. Distribute into tubes or flasks. Autoclave for 10 min at 10 psi pressure–116°C. Check pH after autoclaving. This medium is double strength.

Use: For the enumeration of coliform bacteria in water.

Minimal Agar I

Composition per liter:

Agar	20.0g
Glucose	20.0g
$(NH_4)_2SO_4$	5.0g
KH_2PO_4	1.0g
$MgSO_4·7H_2O$	0.5g
NaCl	0.1g
$CaCl_2·2H_2O$	0.1g
Inositol	2.0mg
KI	1.0mg
H_3BO_3	0.5mg
$ZnSO_4·7H_2O$	0.4mg
$MnSO_4·4H_2O$	0.4mg
Thiamine·HCl	0.4mg
Pyroxidine·HCl	0.4mg
Niacin	0.4mg
Calcium pantothenate	0.4mg
p-Aminobenzoic acid	0.2mg
Riboflavin	0.2mg
$FeCl_3$	0.2mg
$Na_2MoO_4·4H_2O$	0.2mg
$CuSO_4·5H_2O$	0.04mg
Folic acid	2.0μg
Biotin	2.0μg
Growth supplement solution	10.0mL

pH 5.5 ± 0.2 at 25°C

Growth Supplement Solution:
Composition per 10.0mL:

L-Tryptophan	20.0mg
Uracil	20.0mg
L-Histidine·HCl	20.0mg

Preparation of Growth Supplement Solution: Add components to distilled/deionized water and bring volume to 10.0mL. Mix thoroughly. Filter sterilize.

Preparation of Medium: Add components, except growth supplement solution, to distilled/deionized water and bring volume to 990.0mL. Mix thoroughly. Gently heat and bring to boiling. Autoclave for 15 min at 15 psi pressure–121°C. Cool to 50°–55°C. Aseptically add 10.0mL of sterile growth supplement solution. Mix thoroughly. Pour into sterile Petri dishes or distribute into sterile tubes.

Use: For the cultivation of *Saccharomyces cerevisiae.*

Minimal Agar II

Composition per liter:

Agar	20.0g
Glucose	20.0g
$(NH_4)_2SO_4$	5.0g
KH_2PO_4	1.0g
$MgSO_4 \cdot 7H_2O$	0.5g
NaCl	0.1g
$CaCl_2 \cdot 2H_2O$	0.1g
Inositol	2.0mg
KI	1.0mg
H_3BO_3	0.5mg
$ZnSO_4 \cdot 7H_2O$	0.4mg
$MnSO_4 \cdot 4H_2O$	0.4mg
Thiamine·HCl	0.4mg
Pyroxidine·HCl	0.4mg
Niacin	0.4mg
Calcium pantothenate	0.4mg
p-Aminobenzoic acid	0.2mg
Riboflavin	0.2mg
$FeCl_3$	0.2mg
$Na_2MoO_4 \cdot 4H_2O$	0.2mg
$CuSO_4 \cdot 5H_2O$	0.04mg
Folic acid	2.0μg
Biotin	2.0μg
Growth supplement solution	10.0mL

pH 5.5 ± 0.2 at 25°C

Growth Supplement Solution:
Composition per 10.0mL:

Adenine sulfate	20.0mg
L-Arginine·HCl	20.0mg
L-Histidine·HCl	20.0mg

Preparation of Growth Supplement Solution: Add components to distilled/deionized water and bring volume to 10.0mL. Mix thoroughly. Filter sterilize.

Preparation of Medium: Add components, except growth supplement solution, to distilled/deionized water and bring volume to 990.0mL. Mix thoroughly. Gently heat and bring to boiling. Autoclave for 15 min at 15 psi pressure–121°C. Cool to 50°–55°C. Aseptically add 10.0mL of sterile growth supplement solution. Mix thoroughly. Pour into sterile Petri dishes or distribute into sterile tubes.

Use: For the cultivation of *Saccharomyces cerevisiae.*

Minimal Agar III

Composition per liter:

Agar	20.0g
Glucose	20.0g
$(NH_4)_2SO_4$	5.0g
KH_2PO_4	1.0g
$MgSO_4 \cdot 7H_2O$	0.5g
NaCl	0.1g
$CaCl_2 \cdot 2H_2O$	0.1g
Inositol	2.0mg
KI	1.0mg
H_3BO_3	0.5mg
$ZnSO_4 \cdot 7H_2O$	0.4mg
$MnSO_4 \cdot 4H_2O$	0.4mg
Thiamine·HCl	0.4mg
Pyroxidine·HCl	0.4mg
Niacin	0.4mg
Calcium pantothenate	0.4mg
p-Aminobenzoic acid	0.2mg
Riboflavin	0.2mg
$FeCl_3$	0.2mg
$Na_2MoO_4 \cdot 4H_2O$	0.2mg
$CuSO_4 \cdot 5H_2O$	0.04mg
Folic acid	2.0μg
Biotin	2.0μg
Growth supplement solution	10.0mL

pH 5.5 ± 0.2 at 25°C

Growth Supplement Solution:
Composition per 10.0mL:

L-Leucine	30.0mg
L-Tryptophan	20.0gm
Uracil	20.0mg

Preparation of Growth Supplement Solution: Add components to distilled/deionized water and bring volume to 10.0mL. Mix thoroughly. Filter sterilize.

Preparation of Medium: Add components, except growth supplement solution, to distilled/deionized water and bring volume to 990.0mL. Mix thoroughly. Gently heat and bring to boiling. Autoclave for 15 min at 15 psi pressure–121°C. Cool to 50°–55°C. Aseptically add 10.0mL of sterile growth supplement solution. Mix thoroughly. Pour into sterile Petri dishes or distribute into sterile tubes.

Use: For the cultivation of *Saccharomyces cerevisiae.*

Minimal Agar, Davis

Composition per liter:

Agar	15.0g
K_2HPO_4	7.0g
KH_2PO_4	2.0g
$(NH_4)_2SO_4$	1.0g
Glucose	1.0g
Sodium citrate	0.5g
$MgSO_4 \cdot 7H_2O$	0.1g

pH 7.0 ± 0.2 at 25°C

Source: This medium is available as a premixed powder from BD Diagnostic Systems.

Preparation of Medium: Add components to cold distilled/deionized water and bring volume to 1.0L. Mix thoroughly. Gently heat and bring to boiling. Distribute into tubes or flasks. Autoclave for 15 min at 15 psi pressure–121°C. Pour into sterile Petri dishes or leave in tubes.

Use: For the isolation, cultivation, and characterization of nutritional mutants of *Escherichia coli.*

Minimal Broth I

Composition per liter:

Glucose	20.0g
$(NH_4)_2SO_4$	5.0g
KH_2PO_4	1.0g
$MgSO_4 \cdot 7H_2O$	0.5g
NaCl	0.1g
$CaCl_2 \cdot 2H_2O$	0.1g
Inositol	2.0mg
KI	1.0mg
H_3BO_3	0.5mg
$ZnSO_4 \cdot 7H_2O$	0.4mg
$MnSO_4 \cdot 4H_2O$	0.4mg
Thiamine·HCl	0.4mg
Pyroxidine·HCl	0.4mg
Niacin	0.4mg
Calcium pantothenate	0.4mg
p-Aminobenzoic acid	0.2mg
Riboflavin	0.2mg
$FeCl_3$	0.2mg
$Na_2MoO_4 \cdot 4H_2O$	0.2mg
$CuSO_4 \cdot 5H_2O$	0.04mg
Folic acid	2.0μg
Biotin	2.0μg
Growth supplement solution	10.0mL

pH 5.5 ± 0.2 at 25°C

Growth Supplement Solution:
Composition per 10.0mL:

L-Tryptophan	20.0mg
Uracil	20.0mg
L-Histidine·HCl	20.0mg

Preparation of Growth Supplement Solution: Add components to distilled/deionized water and bring volume to 10.0mL. Mix thoroughly. Filter sterilize.

Preparation of Medium: Add components, except growth supplement solution, to distilled/deionized water and bring volume to 990.0mL. Mix thoroughly. Autoclave for 15 min at 15 psi pressure–121°C. Aseptically add 10.0mL of sterile growth supplement solution. Mix thoroughly. Aseptically distribute into sterile tubes or flasks.

Use: For the cultivation of *Saccharomyces cerevisiae.*

Minimal Broth II

Composition per liter:

Glucose	20.0g
$(NH_4)_2SO_4$	5.0g
KH_2PO_4	1.0g
$MgSO_4 \cdot 7H_2O$	0.5g
NaCl	0.1g
$CaCl_2 \cdot 2H_2O$	0.1g
Inositol	2.0mg
KI	1.0mg
H_3BO_3	0.5mg
$ZnSO_4 \cdot 7H_2O$	0.4mg
$MnSO_4 \cdot 4H_2O$	0.4mg
Thiamine·HCl	0.4mg
Pyroxidine·HCl	0.4mg
Niacin	0.4mg
Calcium pantothenate	0.4mg
p-Aminobenzoic acid	0.2mg
Riboflavin	0.2mg
$FeCl_3$	0.2mg
$Na_2MoO_4 \cdot 4H_2O$	0.2mg
$CuSO_4 \cdot 5H_2O$	0.04mg
Folic acid	2.0μg
Biotin	2.0μg
Growth supplement solution	10.0mL

pH 5.5 ± 0.2 at 25°C

Growth Supplement Solution:
Composition per 10.0mL:

Adenine sulfate	20.0mg
L-Arginine·HCl	20.0mg
L-Histidine·HCl	20.0mg

Preparation of Growth Supplement Solution: Add components to distilled/deionized water and bring volume to 10.0mL. Mix thoroughly. Filter sterilize.

Preparation of Medium: Add components, except growth supplement solution, to distilled/deionized water and bring volume to 990.0mL. Mix thoroughly. Autoclave for 15 min at 15 psi pressure–121°C. Aseptically add 10.0mL of sterile growth supplement solution. Mix thoroughly. Aseptically distribute into sterile tubes or flasks.

Use: For the cultivation of *Saccharomyces cerevisiae.*

Minimal Broth III

Composition per liter:

Agar	20.0g
Glucose	20.0g
$(NH_4)_2SO_4$	5.0g
KH_2PO_4	1.0g
$MgSO_4 \cdot 7H_2O$	0.5g
NaCl	0.1g
$CaCl_2 \cdot 2H_2O$	0.1g
Inositol	2.0mg
KI	1.0mg
H_3BO_3	0.5mg
$ZnSO_4 \cdot 7H_2O$	0.4mg
$MnSO_4 \cdot 4H_2O$	0.4mg
Thiamine·HCl	0.4mg
Pyroxidine·HCl	0.4mg
Niacin	0.4mg
Calcium pantothenate	0.4mg
p-Aminobenzoic acid	0.2mg
Riboflavin	0.2mg
$FeCl_3$	0.2mg
$Na_2MoO_4 \cdot 4H_2O$	0.2mg
$CuSO_4 \cdot 5H_2O$	0.04mg
Folic acid	2.0μg
Biotin	2.0μg
Growth supplement solution	10.0mL

pH 5.5 ± 0.2 at 25°C

Growth Supplement Solution:
Composition per 10.0mL:

L-Leucine	30.0mg
L-Tryptophan	20.0mg
Uracil	20.0mg

Preparation of Growth Supplement Solution: Add components to distilled/deionized water and bring volume to 10.0mL. Mix thoroughly. Filter sterilize.

Preparation of Medium: Add components, except growth supplement solution, to distilled/deionized water and bring volume to

990.0mL. Mix thoroughly. Autoclave for 15 min at 15 psi pressure–121°C. Aseptically add 10.0mL of sterile growth supplement solution. Mix thoroughly. Aseptically distribute into sterile tubes or flasks.

Use: For the cultivation of *Saccharomyces cerevisiae.*

Minimal Broth, Davis

Composition per liter:

K_2HPO_4	7.0g
KH_2PO_4	2.0g
$(NH_4)_2SO_4$	1.0g
Glucose	1.0g
Sodium citrate	0.5g
$MgSO_4 \cdot 7H_2O$	0.1g

pH 7.0 ± 0.2 at 25°C

Source: This medium is available as a premixed powder from BD Diagnostic Systems.

Preparation of Medium: Add components to cold distilled/deionized water and bring volume to 1.0L. Mix thoroughly. Distribute into tubes or flasks. Autoclave for 15 min at 15 psi pressure–121°C.

Use: For the isolation, cultivation, and characterization of nutritional mutants of *Escherichia coli.* Also recommended for the isolation and characterization of nutritional mutants from wild-type strains of *Bacillus subtilis* when used in conjunction with minimal agar Davis and antibiotic medium 3.

Minimal F-Top Agar

Composition per liter:

NaCl	8.0g
Agar	4.5g
K_2HPO_4	2.1g
KH_2PO_4	0.6g
$(NH_4)_2SO_4$	0.3g
Glucose	0.3g
Sodium citrate	0.15g
$MgSO_4 \cdot 7H_2O$	0.03g

pH 7.0 ± 0.2 at 25°C

Preparation of Minimal Agar: Add components to cold distilled/deionized water and bring volume to 1.0L. Mix thoroughly. Gently heat and bring to boiling. Distribute into tubes or flasks. Autoclave for 15 min at 15 psi pressure–121°C. Pour into sterile Petri dishes or leave in tubes.

Use: For the distribution of bacteriophage or bacterial cells evenly in a thin layer over the surface of a plate.

Minimal Lactate Medium

See: **ML Medium**

Minimal Medium for Denitrifying Bacteria

Composition per liter:

Solution A	980.0mL
Solution B	10.0mL
Solution C	10.0mL

Solution A:

Composition per 980.0mL:

KNO_3	5.0g
Carbon source	4.0g
$(NH_4)_2SO_4$	1.0g
$K_2HPO_4 \cdot 3H_2O$	0.87g
KH_2PO_4	0.54g

Preparation of Solution A: Add components to distilled/deionized water and bring volume to 1.0L. Mix thoroughly. Autoclave for 15 min at 15 psi pressure–121°C. Cool to 25°C.

Solution B:

Composition per 100.0mL:

$MgSO_4 \cdot 7H_2O$	2.0g

Preparation of Solution B: Add $MgSO_4 \cdot 7H_2O$ to distilled/deionized water and bring volume to 100.0mL. Mix thoroughly. Autoclave for 15 min at 15 psi pressure–121°C. Cool to 25°C.

Solution C:

Composition per 100.0mL:

$CaCl_2 \cdot 2H_2O$	0.2g
$FeSO_4 \cdot 7H_2O$	0.1g
$MnSO_4 \cdot H_2O$	0.05g
$CuSO_4 \cdot 5H_2O$	0.01g
$Na_2MoO_4 \cdot 2H_2O$	0.01g
HCl (0.1*N* solution)	100.0mL

Preparation of Solution C: Combine components. Mix thoroughly. Autoclave for 15 min at 15 psi pressure–121°C. Cool to 25°C.

Preparation of Medium: Aseptically combine 980.0mL of cooled sterile solution A, 10.0mL of cooled sterile solution B, and 10.0mL of cooled sterile solution C. Mix thoroughly. Aseptically distribute into sterile tubes or flasks.

Use: For the isolation and cultivation of denitrifying bacteria.

Minimal Medium for *Penicillium* Interspecific Hybrids

Composition per liter:

Glucose	40.0g
$NaNO_3$	3.0g
KH_2PO_4	1.0g
KCl	0.5g
$MgSO_4 \cdot 7H_2O$	0.5g
$FeSO_4 \cdot 7H_2O$	0.01g

pH 6.0 ± 0.2 at 25°C

Preparation of Medium: Add components to distilled/deionized water and bring volume to 1.0L. Mix thoroughly. Adjust pH to 6.0. Distribute into tubes or flasks. Autoclave for 15 min at 15 psi pressure–121°C.

Use: For the cultivation of genetic variants of *Penicillium* species.

Minimum Essential Medium with Bicarbonate, Serum, and Antibiotics

Composition per 1100.0mL:

Minimum essential medium	950.0mL
Fetal bovine serum, heat inactivated	100.0mL
$NaHCO_3$ solution	40.0mL
Penicillin-streptomycin solution	10.0mL

pH 7.2 ± 0.2 at 25°C

Minimum Essential Medium (MEM):

Composition per liter:

Inorganic salt solution	400.0mL
Other component solution	400.0mL
Amino acid solution	100.0mL
Vitamin solution	100.0mL

Inorganic Salt Solution:
Composition per 400.0mL:

NaCl	6.8g
KCl	0.4g
$CaCl_2$, anhydrous	0.2g
$NaH_2PO_4 \cdot H_2O$	0.14g
$MgSO_4$, anhydrous	97.67mg

Preparation of Inorganic Salt Solution: Add components to distilled/deionized water and bring volume to 400.0mL. Mix thoroughly. Autoclave for 15 min at 15 psi pressure–121°C. Cool to 25°C.

Other Component Solution:
Composition per 400.0mL:

D-Glucose	1.0g
Phenol Red	10.0mg

Preparation of Other Component Solution: Add components to distilled/deionized water and bring volume to 400.0mL. Mix thoroughly. Autoclave for 15 min at 15 psi pressure–121°C. Cool to 25°C.

Amino Acid Solution:
Composition per 100.0mL:

L-Glutamine	292.0mg
L-Arginine·HCl	126.1mg
L-Lysine·HCl	72.5mg
L-Isoleucine	52.0mg
L-Leucine	52.0mg
L-Tyrosine, disodium salt	52.0mg
L-Threonine	48.0mg
L-Valine	46.0mg
L-Histidine·HCl·H_2O	42.0mg
L-Phenylalanine	32.0mg
L-Cysteine·2HCl	31.0mg
L-Methionine	15.0mg
L-Glutamic acid	14.7mg
L-Aspartic acid	13.3mg
L-Asparagine·H_2O	13.2mg
L-Proline	11.5mg
L-Serine	10.5mg
L-Tryptophan	10.0mg
L-Alanine	8.9mg
Glycine	7.5mg

Preparation of Amino Acid Solution: Add components to distilled/deionized water and bring volume to 100.0mL. Mix thoroughly. Filter sterilize.

Vitamin Solution:
Composition per 100.0mL:

i-Inositol	2.0mg
D-Ca pantothenate	1.0mg
Choline chloride	1.0mg
Folic acid	1.0mg
Niacinamide	1.0mg
Pyridoxal·HCl	1.0mg
Thiamine·HCl	1.0mg
Riboflavin	0.1mg

Preparation of Vitamin Solution: Add components to distilled/deionized water and bring volume to 100.0mL. Mix thoroughly. Filter sterilize.

Preparation of Minimum Essential Medium (MEM): Aseptically combine 400.0mL of sterile inorganic salt solution, 400.0mL of sterile other component solution, 100.0mL of sterile amino acid solution, and 100.0mL of sterile vitamin solution.

$NaHCO_3$ Solution:
Composition per 40.0mL:

$NaHCO_3$	2.0g

Preparation of $NaHCO_3$ Solution: Add $NaHCO_3$ to distilled/deionized water and bring volume to 40.0mL. Mix thoroughly. Filter sterilize.

Penicillin-Streptomycin Solution
Composition per 10.0mL:

Penicillin	0.01g
Streptomycin	0.01g

Preparation of Penicillin-Streptomycin Solution: Add components to distilled/deionized water and bring volume to 10.0mL. Mix thoroughly. Filter sterilize.

Preparation of Medium: Aseptically combine 950.0mL of sterile minimum essential medium, 100.0mL of sterile heat inactivated fetal bovine serum, 40.0mL of sterile $NaHCO_3$ solution, and 10.0mL of sterile penicillin-streptomycin solution. Adjust pH to 7.2 with humidified 10% CO_2 in 90% air.

Use: For the cultivation of *Encephalitozoon cuniculi, Encephalitozoon hellem, Encephalitozoon intestinalis, Naegleria fowleri,* and *Nosema corneum.*

Minimum Salts with HiVeg Acid Hydrolysate

Composition per liter:

Na_2HPO_4	6.8g
Glucose	4.0g
Plant acid hydrolysate	4.0g
KH_2PO_4	3.0g
NH_4Cl	1.0g
NaCl	0.5g
$MgSO_4$	0.24g

pH 6.8 ± 0.2 at 25°C

Source: This medium is available as a premixed powder from HiMedia.

Preparation of Medium: Add components to distilled/deionized water and bring volume to 1.0L. Mix thoroughly. Heat gently and bring to boiling. Distribute into tubes or flasks. Autoclave for 15 min at 15 psi pressure–121°C.

Use: For the cultivation of *Escherichia coli* strains used for genetic and molecular studies.

MIO HiVeg Medium (Motility Indole Ornithine HiVeg Medium)

Composition per liter:

Plant hydrolysate	10.0g
Plant peptone	10.0g
L-Ornithine hydrochloride	5.0g
Yeast extract	3.0g
Agar	2.0g
Glucose	1.0g
Bromcresol Purple	0.02g

pH 6.6 ± 0.2 at 25°C

Source: This medium is available as a premixed powder from HiMedia.

Preparation of Medium: Add components to distilled/deionized water and bring to 1.0L. Mix thoroughly. Gently heat and bring to boiling. Distribute into tubes or flasks. Autoclave for 15 min at 15 psi pressure–121°C.

Use: For the differentiation of Gram-negative enteric bacteria based on their motility, indole production, and ornithine decarboxylase activity.

MIO Medium

See: **Motility Indole Ornithine Medium**

Mist Agar

Composition per liter:

Cow manure, dry	50.0g
Agar	15.0g

Preparation of Medium: Add cow manure to 1.0L of tap water. Boil for 1 hr. Filter through cheesecloth. Filter through paper. Add agar to filtrate and bring volume to 1.0L with tap water. Gently heat and bring to boiling. Distribute into tubes or flasks. Autoclave for 15 min at 15 psi pressure–121°C. Pour into sterile Petri dishes or leave in tubes.

Use: For the cultivation and maintenance of *Streptomyces fragmentosporus.*

Mitis Salivarius Agar

Composition per liter:

Sucrose	50.0g
Agar	15.0g
Enzymatic digest of protein	10.0g
Proteose peptone	10.0g
K_2HPO_4	4.0g
Dextrose	1.0g
Trypan Blue	0.08g
Crystal Violet	0.8mg
Na_2TeO_3 solution	1.0mL

pH 7.0 ± 0.2 at 25°C

Source: This medium is available as a premixed powder from BD Diagnostic Systems.

Na_2TeO_3 Solution:

Composition per 10.0mL:

Na_2TeO_3	0.1g

Preparation of Na_2TeO_3 Solution: Add Na_2TeO_3 to 10.0mL of distilled/deionized water. Mix thoroughly. Filter sterilize.

Caution: Potassium tellurite is toxic.

Preparation of Medium: Add components to distilled/deionized water and bring volume to 999.0mL. Mix thoroughly. Gently heat and bring to boiling. Autoclave for 15 min at 15 psi pressure–121°C. Cool medium to 50°–55°C. Aseptically add 1.0mL of the sterile Na_2TeO_3 solution to the cooled basal medium. Mix thoroughly. Pour into sterile Petri dishes or distribute into sterile tubes.

Use: For the selective isolation of *Streptococcus mitis, Streptococcus salivarius,* and other viridans streptococci and enterococci.

Mitis Salivarius HiVeg Agar Base with Tellurite

Composition per liter:

Sucrose	50.0g
Agar	15.0g
Plant hydrolysate	15.0g
Plant peptone	5.0g
K_2HPO_4	4.0g
Glucose	1.0g
Trypan Blue	0.075g
Crystal Violet	0.8mg
Na_2TeO_3 solution	1.0mL

pH 7.0 ± 0.2 at 25°C

Source: This medium, without tellurite, is available as a premixed powder from HiMedia.

Na_2TeO_3 Solution:

Composition per 10.0mL:

Na_2TeO_3	0.1g

Preparation of Na_2TeO_3 Solution: Add Na_2TeO_3 to 10.0mL of distilled/deionized water. Mix thoroughly. Filter sterilize.

Caution: Potassium tellurite is toxic.

Preparation of Medium: Add components to distilled/deionized water and bring volume to 999.0mL. Mix thoroughly. Gently heat and bring to boiling. Autoclave for 15 min at 15 psi pressure–121°C. Cool medium to 50°–55°C. Aseptically add 1.0mL of the sterile Na_2TeO_3 solution to the cooled basal medium. Mix thoroughly. Pour into sterile Petri dishes or distribute into sterile tubes.

Use: For the selective isolation of *Streptococcus mitis, Streptococcus salivarius,* and other viridans streptococci and enterococci.

Mitsuokella dentalis Medium

Composition per 1003.3mL:

Yeast extract	10.0g
Beef extract	5.0g
Glucose	5.0g
Trypticase™	5.0g
L-Cysteine·HCl	0.5g
$(NH_4)_2SO_4$	0.5g
Resazurin	1.0mg
Bovine serum	50.0mL
Mineral solution	40.0mL
Hemin solution	1.0mL
Vitamin K_1 solution	0.2mL

pH 6.9 ± 0.2 at 25°C

Bovine Serum:

Composition per 50.0mL:

Bovine serum	50.0mL

Preparation of Bovine Serum: Incubate 50.0mL of bovine serum in a GasPak™ container overnight to make anaerobic.

Mineral Solution:

Composition per liter:

$NaHCO_3$	10.0g
NaCl	2.0g
K_2HPO_4	1.0g
KH_2PO_4	1.0g
$MgSO_4·7H_2O$	0.48g
$CaCl_2·2H_2O$	0.3g

Preparation of Mineral Solution: Add components to distilled/deionized water and bring volume to 1.0L. Mix thoroughly.

Hemin Solution:

Composition per 100.0mL:

Hemin	5.0mg
NaOH (0.002% solution)	1.0mL

Preparation of Hemin Solution: Add hemin to 1.0mL of NaOH solution. Mix thoroughly.

Vitamin K_1 Solution:
Composition per 100.0mL:
Vitamin K_1 .... 1.09g

Preparation of Vitamin K_1 Solution: Add vitamin K_1 to 99.0mL of absolute ethanol. Store in the dark at 4°C.

Preparation of Medium: Add components, except L-cysteine·HCl and bovine serum, to distilled/deionized water and bring volume to 950.0mL. Mix thoroughly. Gently heat and bring to boiling. Continue boiling for 5 min. Cool to room temperature while sparging with 100% CO_2. Add L-cysteine·HCl. Mix thoroughly. Adjust pH to 6.0 with 8*N* NaOH. After pH has been reached, change sparging gas to 100% N_2. Anaerobically distribute into bottles under 100% N_2. Autoclave for 15 min at 15 psi pressure–121°C. Aseptically and anaerobically add 50.0mL of bovine serum. Mix thoroughly.

Use: For the cultivation and maintenance of *Mitsuokella dentalis*.

Mixed Cereal Agar

Composition per liter:
Gerber™ mixed cereal (oats, wheat, corn, barley) 50.0g
Agar 15.0g
Sucrose 15.0g
Thiamine·HCl 5.0mg

Preparation of Medium: Add components to distilled/deionized water and bring volume to 1.0L. Mix thoroughly. Gently heat and bring to boiling. Distribute into tubes or flasks. Autoclave for 15 min at 15 psi pressure–121°C. Pour into sterile Petri dishes or leave in tubes.

Use: For the cultivation and maintenance of *Tilletia caries*.

Mixotrophic *Nitrobacter* Medium (DSMZ Medium 756a)

Composition per liter:
$NaNO_2$ 2.0g
Yeast extract 1.5g
Peptone 1.5g
Na-pyruvate 0.55g
Stock solution 100.0mL
Trace elements solution 1.0mL
pH 7.4 ± 0.2 at 25°C

Stock Solution:
Composition per liter:
NaCl 5.0g
KH_2PO_4 .... 1.5g
$MgSO_4·7H_2O$ 0.5g
$CaCO_3$ 0.07g

Preparation of Stock Solution: Add components to distilled/deionized water and bring volume to 1.0L. Mix thoroughly.

Trace Elements Solution:
Composition per liter:
$FeSO_4·7H_2O$ 97.3mg
H_3BO_3 .... 49.4mg
$ZnSO_4·7H_2O$ 43.1mg
$(NH_4)_6Mo_7O_{24}·4H_2O$ 37.1mg
$MnSO_4·2H_2O$ 33.8mg
$CuSO_4·5H_2O$ 25.0mg

Preparation of Trace Elements Solution: Add components to distilled/deionized water and bring volume to 1.0L. Mix thoroughly.

Preparation of Medium: Add components to distilled/deionized water and bring volume to 1.0L. Mix thoroughly. Adjust pH to 7.4. Distribute into tubes or flasks. Autoclave for 15 min at 15 psi pressure–121°C.

Use: For the cultivation of *Nitrobacter vulgaris*.

Mixotrophic *Nitrobacter* Medium (LMG Medium 246)

Composition per liter:
$NaNO_2$ 2.0g
Yeast extract 1.50g
Peptone 1.50g
Na-pyruvate 0.55g
Stock solution 100.0mL
Trace elements solution 1.0mL
pH 7.4 ± 0.2 at 25°C

Stock Solution:
Composition per liter:
NaCl 5.0g
KH_2PO_4 .... 1.5g
$MgSO_4·7H_2O$ 0.5g
$CaCO_3$ 0.07g

Preparation of Stock Solution: Add components to distilled/deionized water and bring volume to 1.0L. Mix thoroughly.

Trace Elements Solution:
Composition per liter:
$(NH_4)Mo7O_2$ 437.10mg
$FeSO_4·7H_2O$ 97.30mg
$ZnSO_4·7H_2O$ 43.10mg
H_3BO_3 .... 39.40mg
$MnSO_4·H_2O$ 33.80mg
$CuSO_2·5H_2O$ 25.00mg

Preparation of Trace Elements Solution: Add components to distilled/deionized water and bring volume to 1.0L. Mix thoroughly.

Preparation of Medium: Add components to distilled/deionized water and bring volume to 1.0L. Mix thoroughly. Adjust pH to 7.4 with NaOH. Distribute into tubes or flasks. Autoclave for 15 min at 15 psi pressure–121°C.

Use: For the cultivation of mixotrophic *Nitrobacter* spp.

Mixotrophic *Nitrobacter* Medium, 10% (DSMZ Medium 756b)

Composition per liter:
$NaNO_2$ 2.0g
Yeast extract 0.15g
Peptone 0.15g
Na-pyruvate 0.055g
Stock solution 100.0mL
Trace elements solution 1.0mL
pH 8.6 ± 0.2 at 25°C

Stock Solution:
Composition per liter:
NaCl 5.0g
KH_2PO_4 .... 1.5g

$MgSO_4 \cdot 7H_2O$ 0.5g
$CaCO_3$ 0.07g

Preparation of Stock Solution: Add components to distilled/deionized water and bring volume to 1.0L. Mix thoroughly.

Trace Elements Solution:

Composition per liter:

$FeSO_4 \cdot 7H_2O$ 97.3mg
H_3BO_3 .... 49.4mg
$ZnSO_4 \cdot 7H_2O$ 43.1mg
$(NH_4)_6Mo_7O_{24} \cdot 4H_2O$ 37.1mg
$MnSO_4 \cdot 2H_2O$ 33.8mg
$CuSO_4 \cdot 5H_2O$ 25.0mg

Preparation of Trace Elements Solution: Add components to distilled/deionized water and bring volume to 1.0L. Mix thoroughly.

Preparation of Medium: Add components to distilled/deionized water and bring volume to 1.0L. Mix thoroughly. Adjust pH to 8.6. Distribute into tubes or flasks. Autoclave for 15 min at 15 psi pressure–121°C.

Use: For the cultivation of *Nitrobacter hamburgensis* and *Nitrobacter vulgaris.*

MJ Medium (DSMZ Medium 1011)

Composition per liter:

NaCl 30.0g
$MgCl_2 \cdot 6H_2O$ 4.18g
$MgSO_4 \cdot 7H_2O$ 3.4g
KCl 0.33g
NH_4Cl 0.25g
KH_2PO_4 .... 0.14g
$CaCl_2 \cdot 2H_2O$ 0.014g
$Fe(NH_4)_2(SO_4)_2 \cdot 6H_2O$ 0.01g
$NiCl_2 \cdot 6H_2O$ 0.5mg
$Na_2SeO_3 \cdot 5H_2O$ 0.5mg
Trace elements solution 10.0mL
Vitamin solution 10.0mL
Thiosulfate solution 10.0mL
Bicarbonate solution 10.0mL

pH 6.7 ± 0.2 at 25°C

Bicarbonate Solution:

Composition per 10.0mL:

$NaHCO_3$ 1.5g

Preparation of Bicarbonate Solution: Add components to distilled/deionized water and bring volume to 10.0mL. Mix thoroughly. Filter sterilize.

Thiosulfate Solution:

Composition per 10.0mL:

$NaS_2O_3 \cdot 5H_2O$ 1.5g

Preparation of Thiosulfate Solution: Add components to distilled/deionized water and bring volume to 10.0mL. Mix thoroughly. Autoclave for 15 min at 15 psi pressure–121°C. Cool to room temperature.

Trace Elements Solution:

Composition per liter:

$MgSO_4 \cdot 7H_2O$ 3.0g
Nitrilotriacetic acid 1.5g
NaCl 1.0g
$MnSO_4 \cdot 2H_2O$ 0.5g
$CoSO_4 \cdot 7H_2O$ 0.18g
$ZnSO_4 \cdot 7H_2O$ 0.18g
$CaCl_2 \cdot 2H_2O$ 0.1g
$FeSO_4 \cdot 7H_2O$ 0.1g
$NiCl_2 \cdot 6H_2O$ 0.025g
$KAl(SO_4)_2 \cdot 12H_2O$ 0.02g
H_3BO_3 .... 0.01g
$Na_2MoO_4 \cdot 4H_2O$ 0.01g
$CuSO_4 \cdot 5H_2O$ 0.01g
$Na_2SeO_3 \cdot 5H_2O$ 0.3mg

Preparation of Trace Elements Solution: Add nitrilotriacetic acid to 500.0mL of distilled/deionized water. Dissolve by adjusting pH to 6.5 with KOH. Add remaining components. Add distilled/deionized water to 1.0L. Mix thoroughly.

Vitamin Solution:

Composition per liter:

Pyridoxine-HCl 10.0mg
Thiamine-HCl·$2H_2O$ 5.0mg
Riboflavin 5.0mg
Nicotinic acid 5.0mg
D-Ca-pantothenate 5.0mg
p-Aminobenzoic acid 5.0mg
Lipoic acid 5.0mg
Biotin 2.0mg
Folic acid 2.0mg
Vitamin B_{12} 0.1mg

Preparation of Vitamin Solution: Add components to distilled/deionized water and bring volume to 1.0L. Mix thoroughly. Sparge with 80% H_2 + 20% CO_2. Filter sterilize.

Preparation of Medium: Add components, except thiosulfate, bicarbonate, and vitamin solutions, to distilled/deionized water and bring volume to 970.0mL. Mix thoroughly. Gently heat and bring to boiling. Boil for 1 min. Cool to room temperature while sparging with a gas mixture of 80% N_2 + 20% CO_2. Dispense under the same atmosphere into culture vessels. Fill up to a volume of 20% volume of vessel. Autoclave for 15 min at 15 psi pressure–121°C. Cool to room temperature under an atmosphere of 0% N_2 + 20% CO_2. Aseptically add sterile thiosulfate, bicarbonate, and vitamin solutions. Mix thoroughly. Adjust pH to 6.7. Aseptically dispense into tubes, flasks, or bottles. After inoculation pressurize vessels to 0.5 bar overpressure with 80% N_2 + 20% CO_2 gas mixture. Add sterile air in an amount that is equivalent to a volume of 20% of the headspace. After inoculation reduce medium with 10–20mg sodium dithionite per liter medium, added from a 5% solution freshly prepared under N_2 and filter sterilized. Pressurize vessels to 2 bar overpressure with 80% H_2 and 20% CO_2.

Use: For the cultivation of *Sulfurimonas autotrophica, Thiomicrospira thermophila*, and *Desulfothermus okinawensis.*

MJ Medium (DSMZ Medium 1011)

Composition per liter:

NaCl 30.0g
$MgCl_2 \cdot 6H_2O$ 4.18g
$MgSO_4 \cdot 7H_2O$ 3.4g
KCl 0.33g
NH_4Cl 0.25g
KH_2PO_4 .... 0.14g
$CaCl_2 \cdot 2H_2O$ 0.014g
$Fe(NH_4)_2(SO_4)_2 \cdot 6H_2O$ 0.01g

$NiCl_2 \cdot 6H_2O$ 0.5mg
$Na_2SeO_3 \cdot 5H_2O$ 0.5mg
Trace elements solution 10.0mL
Vitamin solution 10.0mL
Thiosulfate solution 10.0mL
Bicarbonate solution 10.0mL
Pyruvate solution 10.0mL
Yeast extract solution 10.0mL
Lactate soltuion 10.0mL

pH 6.7 ± 0.2 at 25°C

Pyruvate Solution:
Composition per 10.0mL:
Sodium pyruvate 0.5g

Preparation of Pyruvate Solution: Add sodium pyruvate to distilled/deionized water and bring volume to 10.0mL. Mix thoroughly. Filter sterilize.

Lactate Solution:
Composition per 10.0mL:
Sodium lactate 0.5g

Preparation of Lactate Solution: Add sodium lactate to distilled/deionized water and bring volume to 10.0mL. Mix thoroughly. Filter sterilize.

Yeast Extract Solution:
Composition per 10.0mL:
Yeast extract 0.1g

Preparation of Yeast Extract Solution: Add yeast extract to distilled/deionized water and bring volume to 10.0mL. Mix thoroughly. Filter sterilize.

Bicarbonate Solution:
Composition per 10.0mL:
$NaHCO_3$ 1.5g

Preparation of Bicarbonate Solution: Add $NaHCO_3$ to distilled/deionized water and bring volume to 10.0mL. Mix thoroughly. Filter sterilize.

Thiosulfate Solution:
Composition per 10.0mL:
$NaS_2O_3 \cdot 5H_2O$ 1.5g

Preparation of Thiosulfate Solution: Add $NaS_2O_3 \cdot 5H_2O$ to distilled/deionized water and bring volume to 10.0mL. Mix thoroughly. Autoclave for 15 min at 15 psi pressure–121°C. Cool to room temperature.

Trace Elements Solution:
Composition per liter:
$MgSO_4 \cdot 7H_2O$ 3.0g
Nitrilotriacetic acid 1.5g
NaCl 1.0g
$MnSO_4 \cdot 2H_2O$ 0.5g
$CoSO_4 \cdot 7H_2O$ 0.18g
$ZnSO_4 \cdot 7H_2O$ 0.18g
$CaCl_2 \cdot 2H_2O$ 0.1g
$FeSO_4 \cdot 7H_2O$ 0.1g
$NiCl_2 \cdot 6H_2O$ 0.025g
$KAl(SO_4)_2 \cdot 12H_2O$ 0.02g
H_3BO_3 .... 0.01g
$Na_2MoO_4 \cdot 4H_2O$ 0.01g
$CuSO_4 \cdot 5H_2O$ 0.01g
$Na_2SeO_3 \cdot 5H_2O$ 0.3mg

Preparation of Trace Elements Solution: Add nitrilotriacetic acid to 500.0mL of distilled/deionized water. Dissolve by adjusting pH to 6.5 with KOH. Add remaining components. Add distilled/deionized water to 1.0L. Mix thoroughly.

Vitamin Solution:
Composition per liter:
Pyridoxine-HCl 10.0mg
Thiamine-HCl·$2H_2O$ 5.0mg
Riboflavin 5.0mg
Nicotinic acid 5.0mg
D-Ca-pantothenate 5.0mg
p-Aminobenzoic acid 5.0mg
Lipoic acid 5.0mg
Biotin 2.0mg
Folic acid 2.0mg
Vitamin B_{12} 0.1mg

Preparation of Vitamin Solution: Add components to distilled/deionized water and bring volume to 1.0L. Mix thoroughly. Sparge with 80% H_2 + 20% CO_2. Filter sterilize.

Preparation of Medium: Add components, except lactate, pyruvate, yeast extract, thiosulfate, bicarbonate, and vitamin solutions, to distilled/deionized water and bring volume to 970.0mL. Mix thoroughly. Gently heat and bring to boiling. Boil for 1 min. Cool to room temperature while sparging with a gas mixture of 80% H_2 + 20% CO_2. Dispense under the same atmosphere into culture vessels. Fill up to a volume of 20% volume of vessel. Autoclave for 15 min at 15 psi pressure–121°C. Cool to room temperature under an atmosphere of 80% H_2 + 20% CO_2. Aseptically add sterile lactate, pyruvate, yeast extract, thiosulfate, bicarbonate, and vitamin solutions. Mix thoroughly. Adjust pH to 6.7. Aseptically dispense into tubes, flasks, or bottles. After inoculation pressurize vessels to 0.5 bar overpressure with 80% H_2 + 20% CO_2 gas mixture. Add sterile air in an amount that is equivalent to a volume of 20% of the headspace. After inoculation reduce medium with 10–20 mg sodium dithionite per liter medium, added from a 5% solution freshly prepared under N_2 and filter sterilized. Pressurize vessels to 2 bar overpressure with 80% H_2 and 20% CO_2.

Use: For the cultivation of *Desulfothermus okinawensis.*

MJ Medium for *Thiobacter subterraneus* (DSMZ Medium 1011a)

Composition per liter:
$MnCl_2 \cdot 6H_2O$ 4.18g
NaCl 3.0g
$MgSO_4 \cdot 7H_2O$ 0.34g
KCl 0.33g
NH_4Cl 0.25g
KH_2PO_4 .... 0.14g
$CaCl_2 \cdot 2H_2O$ 0.014g
$Fe(NH_4)_2(SO_4)_2 \cdot 6H_2O$ 0.01g
$NiCl_2 \cdot 6H_2O$ 0.5mg
$Na_2SeO_3 \cdot 5H_2O$ 0.5mg
Trace elements solution 10.0mL
Vitamin solution 10.0mL
Thiosulfate solution 10.0mL
Bicarbonate solution 10.0mL

pH 6.7 ± 0.2 at 25°C

Bicarbonate Solution:
Composition per 10.0mL:
$NaHCO_3$ 1.5g

Preparation of Bicarbonate Solution: Add $NaHCO_3$ to distilled/deionized water and bring volume to 10.0mL. Mix thoroughly. Filter sterilize.

Thiosulfate Solution:

Composition per 10.0mL:

$NaS_2O_3 \cdot 5H_2O$	1.5g

Preparation of Thiosulfate Solution: Add $NaS_2O_3 \cdot 5H_2O$ to distilled/deionized water and bring volume to 10.0mL. Mix thoroughly. Autoclave for 15 min at 15 psi pressure–121°C. Cool to room temperature.

Trace Elements Solution:

Composition per liter:

$MgSO_4 \cdot 7H_2O$	3.0g
Nitrilotriacetic acid	1.5g
NaCl	1.0g
$MnSO_4 \cdot 2H_2O$	0.5g
$CoSO_4 \cdot 7H_2O$	0.18g
$ZnSO_4 \cdot 7H_2O$	0.18g
$CaCl_2 \cdot 2H_2O$	0.1g
$FeSO_4 \cdot 7H_2O$	0.1g
$NiCl_2 \cdot 6H_2O$	0.025g
$KAl(SO_4)_2 \cdot 12H_2O$	0.02g
H_3BO_3	0.01g
$Na_2MoO_4 \cdot 4H_2O$	0.01g
$CuSO_4 \cdot 5H_2O$	0.01g
$Na_2SeO_3 \cdot 5H_2O$	0.3mg

Preparation of Trace Elements Solution: Add nitrilotriacetic acid to 500.0mL of distilled/deionized water. Dissolve by adjusting pH to 6.5 with KOH. Add remaining components. Add distilled/deionized water to 1.0L. Mix thoroughly.

Vitamin Solution:

Composition per liter:

Pyridoxine-HCl	10.0mg
Thiamine-HCl·$2H_2O$	5.0mg
Riboflavin	5.0mg
Nicotinic acid	5.0mg
D-Ca-pantothenate	5.0mg
p-Aminobenzoic acid	5.0mg
Lipoic acid	5.0mg
Biotin	2.0mg
Folic acid	2.0mg
Vitamin B_{12}	0.1mg

Preparation of Vitamin Solution: Add components to distilled/deionized water and bring volume to 1.0L. Mix thoroughly. Sparge with 80% H_2 + 20% CO_2. Filter sterilize.

Preparation of Medium: Add components, except thiosulfate, bicarbonate, and vitamin solutions, to distilled/deionized water and bring volume to 970.0mL. Mix thoroughly. Gently heat and bring to boiling. Boil for 1 min. Cool to room temperature while sparging with a gas mixture of 80% N_2 + 20% CO_2. Dispense under the same atmosphere into culture vessels. Fill up to a volume of 20% volume of vessel. Autoclave for 15 min at 15 psi pressure–121°C. Cool to room temperature under an atmosphere of 80% N_2 + 20% CO_2. Aseptically add sterile thiosulfate, bicarbonate, and vitamin solutions. Mix thoroughly. Adjust pH to 6.7. Aseptically dispense into tubes, flasks, or bottles. After inoculation pressurize vessels to 0.5 bar overpressure with 80% N_2 + 20% CO_2 gas mixture. Add sterile air in an amount that is equivalent to a volume of 50% of the headspace. After inoculation reduce medium with 10–20mg sodium dithionite per liter medium, added from a 5% solution freshly prepared under N_2 and filter sterilized. Pressurize vessels to 2 bar overpressure with 80% H_2 and 20% CO_2.

Use: For the cultivation of *Thiobacter subterraneus*.

MJANHOX-NO_3 Medium with Supplement (DSMZ Medium 1000)

Composition per liter:

NaCl	3.0g
$MgCl_2 \cdot 6H_2O$	0.418g
Na_2SiO_3	0.5g
NH_4Cl	0.4g
$MgSO_4 \cdot 7H_2O$	0.34g
KH_2PO_4	0.14g
KCl	0.033
$CaCl_2 \cdot 2H_2O$	0.014g
$Fe_2(SO_4)_3 \cdot 7H_2O$	0.005g
$NiCl_2 \cdot 6H_2O$	0.005mg
$Na_2SeO_3 \cdot 5H_2O$	0.005mg
Bicarbonate solution	10.0mL
Nitrate solution	10.0mL
Thiosulfate solution	10.0mL
Trace elements solution	1.0mL

pH 7.7 ± 0.2 at 25°C

Thiosulfate Solution:

Composition per 10.0mL:

$Na_2S_2O_3 \cdot 5H_2O$	2.4g

Preparation of Thiosulfate Solution: Add $Na_2S_2O_3 \cdot 5H_2O$ to distilled/deionized water and bring volume to 10.0mL. Mix thoroughly. Sparge with 100% N_2. Filter sterilize.

Bicarbonate Solution:

Composition per 10.0mL:

$NaHCO_3$	1.5g

Preparation of Bicarbonate Solution: Add $NaHCO_3$ to distilled/deionized water and bring volume to 10.0mL. Mix thoroughly. Filter sterilize.

Nitrate Solution:

Composition per 10.0mL:

$NaNO_3$	2.0g

Preparation of Nitrate Solution: Add $NaNO_3$ to distilled/deionized water and bring volume to 10.0mL. Mix thoroughly. Filter sterilize.

Trace Elements Solution:

Composition per liter:

Nitrilotriacetic acid	1.5g
$MnSO_4 \cdot 2H_2O$	0.5g
$CoSO_4 \cdot 7H_2O$	0.5g
$ZnSO_4 \cdot 7H_2O$	0.18g
$CuSO_4 \cdot 5H_2O$	0.01g
$KAl(SO_4)_2 \cdot 12H_2O$	0.02g
H_3BO_3	0.01g
$Na_2MoO_4 \cdot 4H_2O$	0.01g

Preparation of Trace Elements Solution: Add nitrilotriacetic acid to 500.0mL of distilled/deionized water. Dissolve by adjusting pH to 6.5 with KOH. Add remaining components. Add distilled/deionized water to 1.0L. Mix thoroughly.

Preparation of Medium: Add components, except thiosulfate solution, bicarbonate solution, nitrate solution, and vitamin solution, to distilled/deionized water and bring volume to 970.0mL. Mix thoroughly. Adjust the pH to 7.7. Autoclave for 15 min at 15 psi pressure–121°C. Cool to room temperature. Aseptically add thiosulfate solution, bicarbonate solution, nitrate solution, and vitamin solution. Sparge with a gas mixture of 80% H_2 + 20% CO_2 for 5 min. Compress gas mixture into gas phase (> 80% volume of the tube or bottle) at 3 atm.

Use: For the cultivation of *Sulfurihydrogenibium subterraneum.*

m-Kleb Agar
See: **Kleb Agar**

m-*Klebsiella* Medium
See: ***Klebsiella*** **Medium**

ML Medium (Minimal Lactate Medium)

Composition per liter:

Sodium lactate	5.0g
$MgSO_4 \cdot 7H_2O$	2.0g
NH_4Cl	1.0g
Na_2SO_4	1.0g
Yeast extract	1.0g
K_2HPO_4	0.5g
L-Cysteine	0.5g
$CaCl_2 \cdot 6H_2O$	0.1g
Resazurin	1.0mg
$NaHCO_3$ solution	25.0mL
$FeSO_4 \cdot 7H_2O$ solution	25.0mL

pH 6.8 ± 0.2 at 25°C

$NaHCO_3$ Solution:

Composition per 25.0mL:

$NaHCO_3$	4.0g

Preparation of $NaHCO_3$ Solution: Add $NaHCO_3$ to distilled/deionized water and bring volume to 25.0mL. Mix thoroughly. Filter sterilize. Gas with O_2-free 97% N_2 + 3% H_2. Cap with a rubber stopper.

$FeSO_4 \cdot 7H_2O$ Solution:

Composition per 25.0mL:

$FeSO_4 \cdot 7H_2O$	4.0mg

Preparation of $FeSO_4 \cdot 7H_2O$ Solution: Add $FeSO_4 \cdot 7H_2O$ to distilled/deionized water and bring volume to 25.0mL. Mix thoroughly. Filter sterilize. Gas with O_2-free 97% N_2 + 3% H_2. Cap with a rubber stopper.

Preparation of Medium: Add components, except $NaHCO_3$ solution and $FeSO_4 \cdot 7H_2O$ solution, to distilled/deionized water and bring volume to 1.0L. Gently heat and bring to boiling under O_2-free 97% N_2 + 3% H_2. Adjust pH to 6.8. Continue boiling until resazurin becomes colorless, indicating reduction. Distribute anaerobically under O_2-free 97% N_2 + 3% H_2 into tubes in 10.0mL volumes. Cap with rubber stoppers. Place tubes in a press. Autoclave for 15 min at 15 psi pressure–121°C. Cool to room temperature. Prior to inoculation, add 0.25mL of sterile $NaHCO_3$ solution and 0.25mL of sterile $FeSO_4 \cdot 7H_2O$ solution to each test tube containing 10.0mL of sterile basal medium.

Use: For the cultivation and maintenance of *Desulfovibrio* species.

m-Lauryl Sulfate Broth
See: **Lauryl Sulfate Broth**

M-Lauryl Sulfate HiVeg Broth

Composition per liter:

Plant special peptone	39.0g
Lactose	30.0g
Yeast extract	6.0g
Sodium lauryl sulfate	1.0g
Phenol Red	0.2g

pH 7.4 ± 0.2 at 25°C

Source: This medium is available as a premixed powder from HiMedia.

Preparation of Medium: Add components to distilled/deionized water and bring volume to 1.0L. Mix thoroughly. Distribute into bottles or flasks. Autoclave for 15 min at 15 psi pressure–121°C.

Use: For the cultivation and enumeration of coliform bacteria, especially *Escherichia coli,* in water by the membrane filter method.

MLCB Agar (Mannitol Lysine Crystal Violet-Brilliant Green Agar)

Composition per liter:

Agar	15.0g
Peptone	10.0g
Yeast extract	5.0g
L-Lysine·HCl	5.0g
NaCl	4.0g
$Na_2S_2O_3$	4.0g
Mannitol	3.0g
Beef extract	2.0g
Ferric ammonium citrate	1.0g
Crystal Violet	0.01g
Brilliant Green	12.5mg

pH 6.8 ± 0.1 at 25°C

Source: This medium is available as a premixed powder from Oxoid Unipath.

Preparation of Medium: Add components to distilled/deionized water and bring volume to 1.0L. Mix thoroughly. Gently heat while stirring and bring to boiling. Do not autoclave. Cool to 50°C. Pour into sterile Petri dishes in 20.0mL volumes.

Use: For the selective isolation and cultivation of *Salmonella* species from fecal material and foods.

m-LES, Endo Agar
See: **Endo Agar, LES**

MM10 Agar (Modified Medium 10 Agar)

Composition per liter:

Base	954.0mL
Dithiothreitol solution	20.0mL
Sheep blood	20.0mL
Na_2CO_3 solution	5.0mL
Menadione solution	1.0mL

pH 7.2 ± 0.2 at 25°C

Base:

Composition per 954.0mL:

Agar	15.0g
Casein peptone	2.0g
Glucose	1.0g
Sodium formate	1.0g

KNO_3 .... 0.5g
Yeast extract 0.5g
Hemin 0.01g
Mineral salt solution 1 .... 38.0mL
Mineral salt solution 2 .... 38.0mL
Sodium lactate solution 4.0mL

Preparation of Base: Add components to distilled/deionized water and bring volume to 954.0mL. Mix thoroughly. Gently heat and bring to boiling. Autoclave for 15 min at 15 psi pressure–121°C. Cool to 45°–50°C.

Mineral Salt Solution 1:
Composition per 100.0mL:
K_2HPO_4 .... 0.6g

Preparation of Mineral Salt Solution 1: Add K_2HPO_4 to distilled/deionized water and bring volume to 100.0mL. Mix thoroughly.

Mineral Salt Solution 2:
Composition per 100.0mL:
NaCl 1.2g
$(NH_4)_2SO_4$ 1.2g
KH_2PO_4 .... 0.6g
$CaCl_2$ 0.12g

Preparation of Mineral Salt Solution 2: Add components to distilled/deionized water and bring volume to 100.0mL. Mix thoroughly.

Sodium Lactate Solution:
Composition per 100.0mL:
Sodium lactate 60.0g

Preparation of Sodium Lactate Solution: Add sodium lactate to distilled/deionized water and bring volume to 100.0mL. Mix thoroughly.

Dithiothreitol Solution:
Composition per 100.0mL:
Dithiothreitol 1.0g

Preparation of Dithiothreitol Solution: Add dithiothreitol to distilled/deionized water and bring volume to 100.0mL. Mix thoroughly. Filter sterilize.

Menadione Solution:
Composition per 100.0mL:
Vitamin K_1 (phytomenadione) 0.05g
Ethanol (95% solution) 100.0mL

Preparation of Menadione Solution: Add vitamin K_1 to 100.0mL of ethanol. Mix thoroughly. Filter sterilize.

Na_2CO_3 Solution:
Composition per 100.0mL:
Na_2CO_3 .... 8.0g

Preparation of Na_2CO_3 Solution: Add Na_2CO_3 to distilled/deionized water and bring volume to 100.0mL. Mix thoroughly. Filter sterilize.

Preparation of Medium: To 954.0mL of sterile cooled base, aseptically add 20.0mL of sterile dithiothreitol solution, 20.0mL of sterile, defibrinated sheep blood, 5.0mL of sterile Na_2CO_3 solution, and 1.0mL of sterile menadione solution. Mix thoroughly. Pour into sterile Petri dishes or distribute into sterile screw-capped tubes.

Use: For the isolation and quantitation of plaque bacteria, especially *Streptococcus mutans, Streptococcus sanguis,* and *Streptococcus salivarius.*

MMA Salts Medium

Composition per liter:
Agar, noble 20.0g
K_2HPO_4 .... 1.2g
KH_2PO_4 .... 0.62g
$(NH_4)_2SO_4$ 0.5g
$MgSO_4·7H_2O$ 0.2g
NaCl 0.1g
$CaCl_2·6H_2O$ 0.05g
$ZnSO_4·7H_2O$ 70.0μg
H_3BO_3 .... 10.0μg
$MnSO_4·5H_2O$ 10.0μg
$Na_2MoO_4·2H_2O$ 10.0μg
$CoCl_2·6H_2O$ 5.0μg
$CuSO_4·5H_2O$ 5.0μg
$FeCl_3·6H_2O$ 1.0mg
Monomethylamine solution 10.0mL
pH 7.0 ± 0.2 at 25°C

Monomethylamine Solution:
Composition per 10.0mL:
Monomethylamine 1.0g

Preparation of Monomethylamine Solution: Add monomethylamine to distilled/deionized water and bring volume to 10.0mL. Mix thoroughly. Filter sterilize.

Preparation of Medium: Add components, except monomethylamine solution, to distilled/deionized water and bring volume to 990.0mL. Mix thoroughly. Gently heat and bring to boiling. Autoclave for 15 min at 15 psi pressure–121°C. Cool to 50°C. Aseptically add 10.0mL of sterile methylamine solution. Mix thoroughly. Pour into sterile Petri dishes or distribute into sterile tubes.

Use: For the cultivation of *Hyphomicrobium* species and *Methylophilus methylotrophus*.

MMB Medium

Composition per liter:
Glucose 1.0g
$(NH_4)_2SO_4$ 0.25g
Peptone 0.15g
Yeast extract 0.15g
Glucose solution 20.0mL
Hutner's basal salts solution 20.0mL
Vitamin solution 10.0mL
pH 7.2 ± 0.2 at 25°C

Glucose Solution:
Composition per 20.0mL:
Glucose 1.5g

Preparation of Glucose Solution: Add glucose to distilled/deionized water and bring volume to 20.0mL. Mix thoroughly. Filter sterilize.

Hutner's Basal Salts Solution:
Composition per liter:
$MgSO_4·7H_2O$ 29.7g
Nitrilotriacetic acid 10.0g
$CaCl_2·2H_2O$ 3.335g
$FeSO_4·7H_2O$ 99.0mg
$(NH_4)_6MoO_7O_{24}·4H_2O$ 9.25mg
"Metals 44" 50.0mL

"Metals 44":
Composition per 100.0mL:

$ZnSO_4 \cdot 7H_2O$	1.095g
$FeSO_4 \cdot 7H_2O$	0.5g
Sodium EDTA	0.25g
$MnSO_4 \cdot H2O$	0.154g
$CuSO_4 \cdot 5H_2O$	39.2mg
$Co(NO_3)_2 \cdot 6H_2O$	24.8mg
$Na_2B_4O_7 \cdot 10H_2O$	17.7mg

Preparation of "Metals 44": Add sodium EDTA to distilled/deionized water and bring volume to 90.0mL. Mix thoroughly. Add a few drops of concentrated H_2SO_4 to retard precipitation of heavy metal ions. Add remaining components. Mix thoroughly. Bring volume to 100.0mL with distilled/deionized water.

Preparation of Hutner's Basal Salts Solution: Add nitrilotriacetic acid to 500.0mL of distilled/deionized water. Adjust pH to 6.5 with KOH. Add remaining components. Add distilled/deionized water to 1.0L. Adjust pH to 6.8.

Vitamin Solution:
Composition per 100.0mL:

Vitamin B_{12}	0.01mg
Calcium DL-pantothenate	0.5mg
Nicotinamide	0.5mg
Pyridoxine·HCl	0.5mg
Riboflavin	0.5mg
Thiamine·HCl	0.5mg
Biotin	0.2mg
Folic acid	0.2mg

Preparation of Vitamin Solution: Add components to distilled/deionized water and bring volume to 100.0mL. Mix thoroughly. Filter sterilize.

Preparation of Medium: Add components, except glucose solution and vitamin solution, to distilled/deionized water and bring volume to 970.0mL. Mix thoroughly. Autoclave for 15 min at 15 psi pressure–121°C. Aseptically add 20.0mL of sterile glucose solution and 10.0mL of sterile vitamin solution. Mix thoroughly. Aseptically distribute into sterile tubes or flasks.

Use: For the cultivation of *Angulomicrobium tetraedrale, Labrys monachus, Prosthecomicrobium polyspheroidum,* and *Aquabacter spiritensis.*

MMJS Medium (Modified) (DSMZ Medium 1121)

Composition per liter:

NaCl	20.0g
$MgSO_4 \cdot 7H_2O$	4.0g
Sulfur, elemental	3.0g
$MgCl_2 \cdot 6H_2O$	3.0g
$Na_2S_2O_3 \cdot 5H_2O$	2.0g
NH_4Cl	1.25g
$CaCl_2 \cdot 2H_2O$	0.8g
KCl	0.33g
K_2HPO_4	0.09g
KH_2PO_4	0.07g
$Fe_2(SO_4)_3 \cdot 7H_2O$	0.01g
$Na_2SeO_3 \cdot 5H_2O$	1.0mg
H_2WO_4	1.0mg
$NiCl_2 \cdot 6H_2O$	1.0mg
Bicarbonate solution	10.0mL
Trace elements solution	10.0mL
Vitamin solution	1.0mL

pH 6.8 ± 0.2 at 25°C

Trace Elements Solution:
Composition per liter:

Nitrilotriacetic acid	1.5g
$MnSO_4 \cdot 2H_2O$	0.5g
$CoSO_4 \cdot 7H_2O$	0.5g
$ZnSO_4 \cdot 7H_2O$	0.18g
$CaCl_2 \cdot 2H_2O$	0.1g
$FeSO_4 \cdot 7H_2O$	0.1g
$NiCl_2 \cdot 6H_2O$	0.025g
$KAl(SO_4)_2 \cdot 12H_2O$	0.02g
$CuSO_4 \cdot 5H_2O$	0.01g
H_3BO_3	0.01g
NaBr	0.01g
$SrCl_2 \cdot 6H_2O$	0.01g
KI	0.01g
$Na_2MoO_4 \cdot 4H_2O$	1.0mg
$Na_2SeO_3 \cdot 5H_2O$	0.3mg

Preparation of Trace Elements Solution: Add nitrilotriacetic acid to 500.0mL of distilled/deionized water. Dissolve by adjusting pH to 6.5 with KOH. Add remaining components. Add distilled/deionized water to 1.0L. Mix thoroughly.

Vitamin Solution:
Composition per liter:

Pyridoxine-HCl	10.0mg
Thiamine-HCl·$2H_2O$	5.0mg
Riboflavin	5.0mg
Nicotinic acid	5.0mg
D-Ca-pantothenate	5.0mg
p-Aminobenzoic acid	5.0mg
Lipoic acid	5.0mg
Biotin	2.0mg
Folic acid	2.0mg
Vitamin B_{12}	0.1mg

Preparation of Vitamin Solution: Add components to distilled/deionized water and bring volume to 1.0L. Mix thoroughly. Sparge with 80% H_2 + 20% CO_2. Filter sterilize.

Bicarbonate Solution:
Composition per 10.0mL:

$NaHCO_3$	2.0g

Preparation of Bicarbonate Solution: Add $NaHCO_3$ to distilled/deionized water and bring volume to 10.0mL. Mix thoroughly. Filter sterilize.

Preparation of Medium: Add components, except trace elements solution, bicarbonate solution, and vitamin solution, to distilled/deionized water and bring volume to 980.0mL. Mix thoroughly. Adjust the pH to 6.8. Autoclave for 15 min at 15 psi pressure–121°C. Cool to room temperature. Aseptically add the trace elements solution, bicarbonate solution, and vitamin solution. Sparge wtih a gas mixture of 80% N_2 + 20% CO_2 for 5 min. Compress gas mixture of 79% N_2 + 20% CO_2 + 1% O_2 into gas phase (> 80% volume of the tube or bottle) at 2 atm.

Use: For the cultivation *Sulfurivirga caldicuralii.*

MMN Agar

Composition per liter:

Agar	15.0g
D-Glucose	10.0g

Malt extract	3.0g
KH_2PO_4	0.5g
Ammonium tartrate	0.25g
$MgSO_4 \cdot 7H_2O$	0.15g
$CaCl_2$	0.05g
NaCl	0.025g
Thiamine·HCl	0.1mg
$FeCl_3$ solution	1.2mL

$FeCl_3$ Solution:
Composition per 100.0mL:

$FeCl_3$	1.0g

Preparation of $FeCl_3$ Solution: Add $FeCl_3$ to distilled/deionized water and bring volume to 100.0mL. Mix thoroughly.

Preparation of Medium: Add components to distilled/deionized water and bring volume to 1.0L. Mix thoroughly. Gently heat until boiling. Distribute into tubes or flasks. Autoclave for 15 min at 15 psi pressure–121°C. Pour into sterile Petri dishes or leave in tubes.

Use: For the cultivation and maintenance of *Cenococcum geophilum, Cortinarius* species, *Gyrodon lividus, Hebeloma crustuliniforme, Hebeloma pusillum, Hygrophorous purpurascens, Hygrophorus russula, Laccaria bicolor, Laccaria laccata, Lyophyllum fumosum, Lyophyllum shimeji, Macrolepiota rhacodes, Obolarina dryophila, Paxillus atromentosus, Phaeolepiota aurea, Pisolithus tinctoruis, Rhizopogon colossus, Rhizopogon ellenae,* many *Rhizopogon* species, *Sarcodon aspratu, Scleroderma albidum, Scleroderma aurantium,* many *Suillus* species, *Tricholoma flavovirens,* and many *Tricholoma* species.

MMS Medium for *Thermotoga neapolitana*

Composition per liter:

NaCl	6.93g
Starch	5.0g
$MgSO_4 \cdot 7H_2O$	1.75g
$MgCl_2 \cdot 6H_2O$	1.38g
KH_2PO_4	0.5g
$Na_2S \cdot 9H_2O$	0.5g
$CaCl_2$	0.38g
KCl	0.16g
NaBr	25.0mg
H_3BO_3	7.5mg
$SrCl_2 \cdot 6H_2O$	3.8mg
$(NH_4)_2Ni(SO_4)_2$	2.0mg
Resazurin	1.0mg
KI	0.025mg
Wolfe's mineral solution	15.0mL

pH 6.5 ± 0.2 at 25°C

Wolfe's Mineral Solution:
Composition per liter:

$MgSO_4 \cdot 7H_2O$	3.0g
Nitrilotriacetic acid	1.5g
NaCl	1.0g
$MnSO_4 \cdot H_2O$	0.5g
$FeSO_4 \cdot 7H_2O$	0.1g
$CoCl_2 \cdot 6H_2O$	0.1g
$CaCl_2$	0.1g
$ZnSO_4 \cdot 7H_2O$	0.1g
$CuSO_4 \cdot 5H_2O$	0.01g
$AlK(SO_4)_2 \cdot 12H_2O$	0.01g
H_3BO_3	0.01g
$Na_2MoO_4 \cdot 2H_2O$	0.01g

Preparation of Wolfe's Mineral Solution: Add nitrilotriacetic acid to 500.0mL of distilled/deionized water. Dissolve by adjusting pH to 6.5 with KOH. Add remaining components. Add distilled/deionized water to 1.0L.

Preparation of Medium: Prepare and dispense medium under 80% N_2 and 20% CO_2. Add components to distilled/deionized water and bring volume to 1.0L. Mix thoroughly. Adjust pH to 6.5 with H_2SO_4. Distribute into tubes or flasks. Autoclave for 15 min at 15 psi pressure–121°C.

Use: For the cultivation and maintenance of *Thermotoga neapolitana.*

MN
See: **Melin–Norkrans Medium**

Mn Agar No. 1
See: **Manganese Agar No. 1**

Mn Agar No. 2
See: **Manganese Agar No. 2**

Mn HiVeg Agar Base with Manganese

Composition per liter:

Agar	12.0g
KH_2PO_4	2.0g
$MnCO_3$	2.0g
Plant extract	1.0g
$Fe(NH_4)_2SO_4$	0.15g
Sodium citrate	0.15g
Yeast extract	0.075g
Cyanocobalamin solution	10.0mL

pH 7.0 ± 0.2 at 25°C

Cyanocobalamin Solution:
Composition per 10.0mL:

Cyanocobalamin	5.0mg

Preparation of Cyanocobalamin: Add cyanocarbalamin to distilled/deionized water and bring volume to 10.0mL. Mix thoroughly. Filter sterilize.

Preparation of Medium: Add components, except cyanocarbalamin solution, to distilled/deionized water and bring volume to 990.0mL. Mix thoroughly. Gently heat and bring to boiling. Distribute into tubes or flasks. Autoclave for 15 min at 15 psi pressure–121°C. Cool to 50°C. Aseptically add 10.0mL sterile cyanocarbolamin solution. Mix thoroughly. Pour into sterile Petri dishes or aseptically distribute into tubes.

Use: For the cultivation of *Leptothrix* spp. and detection of *Leptothrix* by its ability to oxidize manganous ions.

MN Marine Medium

Composition per liter:

Noble agar	10.0g
$NaNO_3$	0.75g
$MgSO_4 \cdot 7H_2O$	0.04g
$CaCl_2 \cdot 2H_2O$	0.02g
$K_2HPO_4 \cdot 3H_2O$	0.02g
Na_2CO_3	0.02g
Citric acid	3.0mg
Ferric ammonium citrate	3.0mg

Disodium potassium EDTA	0.5mg
Trace metal mix A-5	1.0mL

pH 8.5 ± 0.2 at 25°C

A-5 Trace Metal Mix:

Composition per liter:

H_3BO_3	2.86g
$MnCl_2 \cdot 4H_2O$	1.81g
$ZnSO_4 \cdot 7H_2O$	0.222g
$CuSO_4 \cdot 5H_2O$	0.079g
$Na_2MoO_4 \cdot 2H_2O$	0.039g
$Co(NO_3)_2 \cdot 6H_2O$	0.049g

Preparation of A-5 Trace Metal Mix: Add components to distilled/deionized water and bring volume to 1.0L. Mix thoroughly.

Preparation of Medium: Add components to 750.0mL of seawater and bring volume to 1.0L with glass-distilled water. Mix thoroughly. Gently heat and bring to boiling. Autoclave for 15 min at 15 psi pressure–121°C. After autoclaving, adjust pH to 8.5 with KOH.

Use: For the cultivation and maintenance of marine cyanobacteria.

MN Marine Medium with Vitamin B_{12}

Composition per liter:

Noble agar	10.0g
$NaNO_3$	0.75g
$MgSO_4 \cdot 7H_2O$	0.04g
$CaCl_2 \cdot 2H_2O$	0.02g
$K_2HPO_4 \cdot 3H_2O$	0.02g
Na_2CO_3	0.02g
Citric acid	3.0mg
Ferric ammonium citrate	3.0mg
Disodium potassium EDTA	0.5mg
Vitamin B_{12}	20.0μg
Trace metal mix A-5	1.0mL

pH 8.5 ± 0.2 at 25°C

A-5 Trace Metal Mix:

Composition per liter:

H_3BO_3	2.86g
$MnCl_2 \cdot 4H_2O$	1.81g
$ZnSO_4 \cdot 7H_2O$	0.222g
$CuSO_4 \cdot 5H_2O$	0.079g
$Na_2MoO_4 \cdot 2H_2O$	0.039g
$Co(NO_3)_2 \cdot 6H_2O$	0.049g

Preparation of A-5 Trace Metal Mix: Add components to distilled/deionized water and bring volume to 1.0L. Mix thoroughly.

Preparation of Medium: Add components to 750.0mL of seawater and bring volume to 1.0L with glass-distilled water. Mix thoroughly. Gently heat and bring to boiling. Autoclave for 15 min at 15 psi pressure–121°C. After autoclaving, adjust pH to 8.5 with KOH.

Use: For the cultivation and maintenance of *Dermocarpa* species, *Dermocarpella* species, *Myxosarcina* species, *Phormidium* species, *Pleurocapsa* species, *Synechococcus* species, *Synechocystis* species, and *Xenococcus* species.

Mobiluncus Agar (LMG Medium 117)

Composition per liter:

Special peptone	23.0g
Agar No. 1	10.0g
Starch, soluble	10.0g
NaCl	5.0g
Resazurin	10.0μg
Rabbit serum	20.0mL

pH 7.1–7.5

Source: Special peptone and Agar No. 1 are available from Oxoid Unipath.

Preparation of Medium: Add components, except rabbit serum, to distilled/deionized water and bring volume to 990.0mL. Mix thoroughly. Gently heat and bring to boiling. Autoclave for 15 min at 15 psi pressure–121°C. Cool to 45°–50°C. Aseptically add 20.0mL sterile rabbit serum. Mix thoroughly. Pour into sterile Petri dishes or distribute into sterile tubes.

Use: For the cultivation of *Mobiluncus curtisii* subsp. *holmesii* and *Mobiluncus mulieris*.

Moderate Halophilic Medium (HM)

Composition per liter:

NaCl	81.0g
Agar	20.0g
Yeast extract	10.0g
$MgSO_4 \cdot 7H_2O$	9.6g
$MgCl_2 \cdot 6H_2O$	7.0g
Proteose peptone No. 3	5.0g
KCl	2.0g
Glucose	1.0g
$CaCl_2 \cdot 2H_2O$	0.36g
$NaHCO_3$	0.06g
NaBr	0.026g

pH 7.2 ± 0.2 at 25°C

Preparation of Medium: Add components to distilled/deionized water and bring volume to 1.0L. Mix thoroughly. Gently heat and bring to boiling. Distribute into tubes or flasks. Autoclave for 15 min at 15 psi pressure–121°C. Pour into sterile Petri dishes or leave in tubes.

Use: For the cultivation and maintenance of *Bacillus halophilus, Halococcus saccharolyticus, Marinococcus albus, Marinococcus halophilus,* and *Marinococcus hispanicus.*

Modified AEA Sporulation Medium Base

Composition 1070.0mL:

Biopeptone	10.0g
Yeast extract	10.0g
Na_2HPO_4	4.36g
Ammonium acetate	1.5g
KH_2PO_4	0.25g
$MgSO_4 \cdot 7H_2O$	0.2g
Raffinose solution	40.0mL
Carbonate solution	10.0mL
Cobalt chloride solution	10.0mL
Sodium ascorbate solution	10.0mL

pH 7.8 ± 0.2 at 25°C

Source: This medium is available from HiMedia.

Raffinose Solution:

Composition per 40.0mL:

Raffinose	4.0g

Preparation of Raffinose Solution: Add raffinose to distilled/deionized water and bring volume to 40.0mL. Mix thoroughly. Filter sterilize.

Carbonate Solution:

Composition per 10.0mL:

Na_2CO_3 .... 0.7g

Preparation of Carbonate Solution: Add Na_2CO_3 to distilled/deionized water and bring volume to 10.0mL. Mix thoroughly. Filter sterilize.

Cobalt Cloride Solution:

Composition per 10.0mL:

$CoCl_2$ 0.032g

Preparation of Cobalt Chloride Solution: Add $CoCl_2$ to distilled/deionized water and bring volume to 10.0mL. Mix thoroughly. Filter sterilize.

Sodium Ascorbate Solution:

Composition per 10.0mL:

Sodium ascorbate 0.15g

Preparation of Sodium Ascorbate Solution: Add sodium ascorbate to distilled/deionized water and bring volume to 10.0mL. Mix thoroughly. Filter sterilize.

Preparation of Medium: Add components, except raffinose, carbonate, cobalt chloride, and sodium ascorbate solutions, to distilled/deionized water and bring volume to 1.0L. Mix thoroughly. Distribute 15.0mL aliquots into screw capped tubes. Autoclave for 15 min at 15 psi pressure–121°C. Cool to 50°C. Add 0.6L raffinose solution and 0.2mL each of the carbonate and cobalt chloride solutions dropwise to the medium in each of the tubes. Just before using, steam the medium for 10 min. Cool to room temperature. Aseptically add 0.2mL of freshly prepared sodium ascorbate solution to each tube of the medium.

Use: For the early sporulation of *Clostridium perfringens* from foods.

Modified Biebl and Pfennig's Medium (DSMZ Medium 1069)

Composition per liter:

NaCl 20.0g
Malate/pyruvate 3.0g
$MgSO_4 \cdot 7H_2O$ 2.0g
KH_2PO_4 .... 0.5g
Yeast extract 0.4g
NH_4Cl 0.34g
$CaCl_2 \cdot 2H_2O$ 0.15g
Ferric citrate solution 5.0mL
Trace elements solution SL-7 .... 1.0mL
Vitamin solution 1.0mL

pH 6.8 ± 0.2 at 25°C

Vitamin Solution:

Composition per 10.0ml:

Vitamin B_{12} 0.1mg

Preparation of Vitamin Solution: Add vitamin B_{12} to distilled/deionized water and bring volume to 10.0mL. Mix thoroughly. Filter sterilize.

Ferric Citrate Solution:

Composition per 10.0mL:

Ferric citrate 0.01g

Preparation of Ferric Citrate Solution: Add ferric citrate to distilled/deionized water and bring volume to 10.0mL. Mix thoroughly. Filter sterilize.

Trace Elements Solution SL-7:

Composition per 1001.0mL:

$CoCl_2 \cdot 6H_2O$ 200.0mg
$MnCl_2 \cdot 4H_2O$ 100.0mg
$ZnCl_2$ 70.0mg
H_3BO_3 .... 60.0mg
$Na_2MoO_4 \cdot 2H_2O$ 40.0mg
$CuCl_2 \cdot 2H_2O$ 20.0mg
$NiCl_2 \cdot 6H_2O$ 20.0mg
HCl (25%) 1.0mL

Preparation of Trace Elements Solution SL-7: Add components to distilled/deionized water and bring volume to 1.0L. Mix thoroughly.

Preparation of Medium: Add components, except vitamin solution, to distilled/deionized water and bring volume to 1.0L. Mix thoroughly. Adjust the pH to 6.8. Autoclave for 15 min at 15 psi pressure–121°C. Cool to room temperature. Aseptically add the vitamin solution. Mix thoroughly. Aseptically dispense into culture vessels.

Use: For the cultivation of *Rhodovulum imhoffii.*

Modified Bile Esculin Azide Agar

Composition per liter:

Casein enzymic hydrolysate 17.0g
Agar 13.5g
Oxgall 10.0g
Yeast extract 5.0g
NaCl 5.0g
Peptic digest of animal tissue 3.0g
Sodium citrate 1.0g
Esculin 1.0g
Ferric ammonium citrate 0.5g
NaN_3 .... 0.25g

pH 7.1 ± 0.2 at 25°C

Source: This medium is available from HiMedia.

Caution: Sodium azide has a tendency to form explosive metal azides with plumbing materials. It is advisable to use enough water to flush off the disposables.

Preparation of Medium: Add components to distilled/deionized water and bring volume to 1.0L. Mix thoroughly. Gently heat and bring to boiling. Distribute into tubes or flasks. Autoclave for 15 min at 15 psi pressure–121°C. Pour into sterile Petri dishes or leave in tubes.

Use: For the selective isolation and enumeration of group D streptococci.

Modified Buffered Charcoal HiVeg Agar Base with Cysteine

Composition per liter:

Agar 17.0g
ACES buffer 10.0g
Plant peptone No. 3 .... 10.0g
Charcoal, activated 2.0g
α-Ketoglutarate monopotassium salt 1.0g
L-Cysteine solution 4.0mL

pH 6.9 ± 0.2 at 25°C

Source: This medium, without L-cysteine solution, is available as a premixed powder from HiMedia.

L-Cysteine Solution:

Composition per 10.0mL:

L-cysteine·HCl·H_2O 0.4g

L-Cysteine Solution: Add L-cysteine·HCl·H_2O to distilled/deionized water and bring volume to 10.0mL. Mix thoroughly. Filter sterilize.

Preparation of Medium: Add components, except L-cysteine solution, to distilled/deionized water and bring volume to 1.0L. Mix thoroughly. Adjust medium to pH 6.9 with 1*N* KOH. Heat gently and bring to boiling for 1 min. Autoclave for 15 min at 15 psi pressure–121°C. Cool to 50°–55°C. Aseptically add 4.0mL of L-cysteine solution. Mix thoroughly. Pour into sterile Petri dishes with constant agitation to keep charcoal in suspension.

Use: For the isolation, cultivation, and maintenance of *Legionella pneumophila* and other *Legionella* species from environmental and clinical specimens.

Modified *Campylobacter* Blood-Free Selective Agar Base (Modified *Campylobacter* Charcoal Differential Agar) (Modified CCDA) (BAM M30a)

Composition per 1012.0mL:

Agar	12.0g
Beef extract	10.0g
Peptone	10.0g
NaCl	5.0g
Charcoal	4.0g
Casein hydrolysate	3.0g
Yeast extract	2.0g
Sodium deoxycholate	1.0g
$FeSO_4$	0.25g
Sodium pyruvate	0.25g
Cefoperazone solution	4.0mL
Amphotericin B solution	4.0mL
Rifampicin solution	4.0mL

pH 7.4 ± 0.2 at 25°C

Cefoperazone Solution:

Composition per 10.0mL:

Cefoperazone	0.037g

Preparation of Cefoperazone Solution: Add cefoperazone to distilled/deionized water and bring volume to 10.0mL. Mix thoroughly. Filter sterilize.

Rifampicin Solution:

Composition per 100.0mL:

Rifampicin	0.25g
Ethanol, absolute	50.0mL

Preparation of Rifampicin Solution: Add rifampicin to 50.0mL of ethanol. Mix thoroughly. Bring volume to 100.0mL with distilled/deionized water. Filter sterilize.

Amphotericin B Solution:

Composition per 10.0mL:

Amphotericin B	0.005g

Preparation of Amphotericin B Solution: Add amphotericin B to distilled/deionized water and bring volume to 10.0mL. Mix thoroughly. Filter sterilize. Can be stored for 1 year at –20°C.

Preparation of Medium: Add components, except cefoperazone solution, amphotericin B solution, and rifampicin solution, to distilled/deionized water and bring volume to 1.0L. Mix thoroughly. Gently heat and bring to boiling. Autoclave for 15 min at 15 psi pressure–121°C. Cool to 45°–50°C. Aseptically add 10.0mL of sterile cefoperazone solution, 10.0mL of sterile amphotericin B solution, and 10.0mL of sterile rifampicin solution. Mix thoroughly. Pour into sterile Petri dishes or distribute into sterile tubes.

Use: For the cultivation of *Campylobacter* species. For the recovery of injured *Campylobacter* spp. from foods.

Modified CM + YE Agar

See: **CM plus YE Agar, Modified**

Modified CPLM HiVeg Medium Base with Horse Serum (*Trichomonas* Modified CPLM HiVeg Medium Base)

Composition per liter:

Plant peptone	32.0g
Liver digest	20.0g
L-Cysteine·HCl	2.4g
Maltose	1.6g
Ringer's salt solution, 1/4X	1.0L
Horse serum	100.0mL

pH 6.0 ± 0.2 at 25°C

Ringer's Salt Solution, 1/4X:

Composition per 400.0mL:

NaCl	9.0g
KCl	0.042g
$CaCl_2$	0.024g

Preparation of Ringer's Salt Solution, 1/4X: Add components to distilled/deionized water and bring volume to 400.0mL. Mix thoroughly.

Preparation of Medium: Add components, except horse serum, to 1.0L Ringer's salt solution, 1/4X. Mix thoroughly. Adjust pH to 6.0. Gently heat and bring to boiling. Autoclave for 10 min at 15 psi pressure–121°C. Cool to 25°C. Aseptically add 100.0mL of sterile, heat-inactivated horse serum. Mix thoroughly. Aseptically distribute into sterile, screw-capped tubes or flasks.

Use: For the cultivation of *Trichomonas vaginalis*.

Modified CPLM HiVeg Medium Base with Horse Serum, Penicillin, Streptomycin, and Nystatin (*Trichomonas* Modified CPLM HiVeg Medium Base)

Composition per liter:

Plant peptone	32.0g
Liver digest	20.0g
L-Cysteine·HCl	2.4g
Maltose	1.6g
Ringer's salt solution, 1/4X	1.0L
Horse serum	100.0mL
Peniclllin-streptomycin solution	10.0mL
Nystatin solution	10.0mL

pH 6.0 ± 0.2 at 25°C

Ringer's Salt Solution, 1/4X:

Composition per 400.0mL:

NaCl	9.0g
KCl	0.042g
$CaCl_2$	0.024g

Preparation of Ringer's Salt Solution, 1/4X: Add components to distilled/deionized water and bring volume to 400.0mL. Mix thoroughly.

Penicllin-Streptomycin Solution:
Composition per 10.0mL:

Streptomycin	0.1g
Penicillin	1,000,000U

Preparation of Penicllin-Streptomycin Solution: Add components to distilled/deionized water and bring volume to 10.0mL. Mix thoroughly. Filter sterilize

Nystatin Solution:
Composition per 10.0mL:

Nystatin	50,000U

Preparation of Nystatin Solution: Add nystatin to distilled/deionized water and bring volume to 10.0mL. Mix thoroughly. Filter sterilize

Preparation of Medium: Add components, except horse serum, penicillin-streptomycin solution, and nystatin solution, to 1.0L of Ringer's salt solution, 1/4X. Mix thoroughly. Adjust pH to 6.0. Gently heat and bring to boiling. Autoclave for 10 min at 15 psi pressure–121°C. Cool to 25°C. Aseptically add 100.0mL of sterile, heat-inactivated horse serum, 10.0mL sterile penicllin-streptomycin solution, and 10.0mL sterile nystatin solution. . Mix thoroughly. Aseptically distribute into sterile, screw-capped tubes or flasks.

Use: For the selective cultivation of *Trichomonas vaginalis*.

Modified Differential Clostridial Broth

Composition per liter:

Meat extract	8.0g
Casein enzymic hydrolysate	5.0g
Meat peptone	5.0g
Sodium acetate	5.0g
Yeast extract	1.0g
Starch	1.0g
Glucose	1.0g
L-Cysteine hydrochloride	0.5g
$NaHSO_3$	0.5g
Ammonium ferric citrate	0.5g
Resazurin	0.002g

pH 7.2 ± 0.2 at 25°C

Source: This medium is available from HiMedia.

Preparation of Medium: Add components to distilled/deionized water and bring volume to 1.0L. Mix thoroughly. Gently heat and bring to boiling. Distribute into tubes or flasks. Autoclave for 15 min at 15 psi pressure–121°C.

Use: For the detection of *Clostridium* spp. from foods by the MPN technique.

Modified Duncan Strong HiVeg Medium (DS HiVeg Medium)

Composition per liter:

Plant peptone No. 3	15.0g
Na_2HPO_4	10.0g
Raffinose	4.0g
Yeast extract	4.0g
Na-thioglycollate	1.0g

pH 7.8 ± 0.2 at 25°C

Source: This medium is available as a premixed powder from HiMedia.

Preparation of Medium: Add components to distilled/deionized water and bring volume to 1.0L. Mix thoroughly. Gently heat and bring to boiling. Distribute into tubes or flasks. Autoclave for 15 min at 15 psi pressure–121°C. Adjust pH to 7.8 with filter-sterilized 0.66*M* Na_2CO_3. Pour into sterile Petri dishes or leave in tubes.

Use: For the cultivation and induction of sporulation of *Clostridium perfringens.*

Modified Duncan Strong Medium (DS Medium)

Composition per liter:

Proteose peptone	15.0g
Na_2HPO_4	10.0g
Raffinose	4.0g
Yeast extract	4.0g
Na-thioglycollate	1.0g

pH 7.8 ± 0.2 at 25°C

Source: This medium is available as a premixed powder from HiMedia.

Preparation of Medium: Add components to distilled/deionized water and bring volume to 1.0L. Mix thoroughly. Gently heat and bring to boiling. Distribute into tubes or flasks. Autoclave for 15 min at 15 psi pressure–121°C. Adjust pH to 7.8 with filter-sterilized 0.66*M* Na_2CO_3. Pour into sterile Petri dishes or leave in tubes.

Use: For the cultivation and induction of sporulation of *Clostridium perfringens.*

Modified Fungal Agar Base (Modified Inhibitory Mold Agar)

Composition per liter:

Glucose	20.0g
Agar	15.0g
Casein enzymic hydrolysate	2.5g
Peptic digest of animal tissue	2.5g
Yeast extract	5.0g
Na_2HPO_4	3.5g
KH_2PO_4	3.4g
NH_4Cl	1.4g
$NaCO_3$	1.0g
Chloramphenicol	0.1g
$MgSO_4 \cdot 7H_2O$	0.06g
Polysorbate 80	20.0mL

pH 7.0 ± 0.2 at 25°C

Source: This medium is available from HiMedia.

Preparation of Medium: Add components, except polysorbate 80, to distilled/deionized water and bring volume to 980.0mL. Mix thoroughly. Gently heat and bring to boiling. Add polysorbate 80. Distribute into tubes or flasks. Autoclave for 15 min at 15 psi pressure–121°C. Mix thoroughly. Pour into Petri dishes or aseptically distribute into sterile tubes.

Use: For the detection and enumeration of molds in cosmetics and toiletries.

Modified Fungal HiVeg Agar Base (Modified Inhibitory Mold HiVeg Agar Base)

Composition per liter:

Glucose	20.0g
Agar	15.0g

Yeast extract 5.0g
Na_2HPO_4 .... 3.5g
KH_2PO_4 .... 3.4g
Plant hydrolysate 2.5g
Plant peptone 2.5g
NH_4Cl 1.4g
Na_2CO_3 .... 1.0g
Chloramphenicol 0.1g
$MgSO_4$ 0.06g
Polysorbate 80 .... 20.0mL

pH 7.0 ± 0.2 at 25°C

Source: This medium, without polysorbate 80, is available as a premixed powder from HiMedia.

Preparation of Medium: Add components to distilled/deionized water and bring volume to 1.0L. Mix thoroughly. Gently heat and bring to boiling with frequent agitation. Distribute into tubes or flasks. Autoclave for 15 min at 15 psi pressure–121°C. Pour into sterile Petri dishes or leave in tubes.

Use: For the isolation of pathogenic fungi.

Modified FWM Medium (DSMZ Medium 503)

Composition per 1013.0mL:

Solution A 940.0mL
Solution E 50.0mL
Solution F 10.0mL
Solution G 10.0mL
Solution B 1.0mL
Solution C 1.0mL
Solution D 1.0mL

pH 7.3 ± 0.2 at 25°C

Solution A:

Composition per 940.0mL:

NaCl 1.0g
KCl 0.5g
$MgCl_2 \cdot 6H_2O$ 0.4g
NH_4Cl 0.25g
KH_2PO_4 .... 0.2g
$CaCl_2 \cdot 2H_2O$ 0.15g
Resazurin 0.5mg

Preparation of Solution A: Prepare under 80% N_2 + 20% CO_2 gas atmosphere. Add components to distilled/deionized water and bring volume to 940.0mL. Mix thoroughly. Adjust pH to 7.2. Sparge with 80% N_2 + 20% CO_2. Autoclave for 15 min at 15 psi pressure–121°C. Cool to 25°C.

Solution B:

Composition per liter:

$FeCl_2 \cdot 4H_2O$ 1.5g
$CoCl_2 \cdot 6H_2O$ 190.0mg
$MnCl_2 \cdot 4H_2O$ 100.0mg
$ZnCl_2$ 70.0mg
$Na_2MoO_4 \cdot 2H_2O$ 36.0mg
$NiCl_2 \cdot 6H_2O$ 24.0mg
H_3BO_3 .... 6.0mg
$CuCl_2 \cdot 2H_2O$ 2.0mg
HCl (25% solution) 10.0mL

Preparation of Solution B: Add $FeCl_2 \cdot 4H_2O$ to 10.0mL of HCl solution. Mix thoroughly. Add distilled/deionized water and bring volume to 1.0L. Add remaining components. Mix thoroughly. Sparge with 80% N_2 + 20% CO_2. Autoclave for 15 min at 15 psi pressure–121°C. Cool to 25°C.

Solution C:

Composition per liter:

Pyridoxine hydrochloride 300.0mg
Thiamine-HCl·$2H_2O$ 200.0mg
Nicotinic acid 200.0mg
Vitamin B_{12} 100.0mg
Calcium pantothenate 100.0mg
p-Aminobenzoic acid 80.0mg
D(+)-Biotin 20.0mg

Preparation of Solution C: Add components to distilled/deionized water and bring volume to 1.0L. Sparge with 100% N_2. Mix thoroughly. Filter sterilize.

Solution D:

Composition per liter:

NaOH 0.5g
$Na_2WO_4 \cdot 2H_2O$ 4.0mg
$Na_2SeO_3 \cdot 5H_2O$ 3.0mg

Preparation of Solution D: Add components to distilled/deionized water and bring volume to 1.0L. Mix thoroughly. Sparge with 100% N_2. Filter sterilize.

Solution E:

Composition per 100.0mL:

$NaHCO_3$ 5.0g

Preparation of Solution E: Add $NaHCO_3$ to distilled/deionized water and bring volume to 100.0mL. Mix thoroughly. Sparge with 100% N_2 gas mixture. Autoclave for 15 min at 15 psi pressure–121°C. Cool to 25°C.

Solution F:

Composition per 10.0mL:

Na-(D,L)-3-hydroxybutyrate 1.5g

Preparation of Solution F: Add Na-(D,L)-3-hydroxybutyrate to distilled/deionized water and bring volume to 10.0mL. Mix thoroughly. Sparge with 100% N_2 gas mixture. Autoclave for 15 min at 15 psi pressure–121°C. Cool to 25°C.

Solution G:

Composition per 10.0mL:

$Na_2S \cdot 9H_2O$ 0.125g

Preparation of Solution G: Add $Na_2S \cdot 9H_2O$ to distilled/deionized water and bring volume to 10.0mL. Mix thoroughly. Autoclave under 100% N_2 for 15 min at 15 psi pressure–121°C. Cool to 25°C.

Preparation of Medium: Prepare and dispense medium under 80% N_2 + 20% CO_2 gas atmosphere. Sequentially add 1.0mL solution B, 1.0mL solution C, 1.0mL solution D, 50.0mL solution E, 10.0mL solution F, and 10.0mL solution G, to 940.0mL solution A. Distribute anaerobically under 80% N_2 + 20% CO_2 into appropriate vessels. The pH should be 7.2–7.4.

Use: For the cultivation of *Clostridium homopropionicum* DSM 5847.

Modified FWM Medium (DSMZ Medium 503)

Composition per 1013.0mL:

Solution A 940.0mL
Solution E 50.0mL
Solution F 10.0mL

Solution G	10.0mL
Solution B	1.0mL
Solution C	1.0mL
Solution D	1.0mL

pH 7.3 ± 0.2 at 25°C

Solution A:

Composition per 940.0mL:

NaCl	1.0g
KCl	0.5g
$MgCl_2 \cdot 6H_2O$	0.4g
NH_4Cl	0.25g
KH_2PO_4	0.2g
$CaCl_2 \cdot 2H_2O$	0.15g
Resazurin	0.5mg

Preparation of Solution A: Prepare under 80% N_2 + 20% CO_2 gas atmosphere. Add components to distilled/deionized water and bring volume to 940.0mL. Mix thoroughly. Adjust pH to 7.2. Sparge with 80% N_2 + 20% CO_2. Autoclave for 15 min at 15 psi pressure–121°C. Cool to 25°C.

Solution B:

Composition per liter:

$FeCl_2 \cdot 4H_2O$	1.5g
$CoCl_2 \cdot 6H_2O$	190.0mg
$MnCl_2 \cdot 4H_2O$	100.0mg
$ZnCl_2$	70.0mg
$Na_2MoO_4 \cdot 2H_2O$	36.0mg
$NiCl_2 \cdot 6H_2O$	24.0mg
H_3BO_3	6.0mg
$CuCl_2 \cdot 2H_2O$	2.0mg
HCl (25% solution)	10.0mL

Preparation of Solution B: Add $FeCl_2 \cdot 4H_2O$ to 10.0mL of HCl solution. Mix thoroughly. Add distilled/deionized water and bring volume to 1.0L. Add remaining components. Mix thoroughly. Sparge with 80% N_2 + 20% CO_2. Autoclave for 15 min at 15 psi pressure–121°C. Cool to 25°C.

Solution C:

Composition per liter:

Pyridoxine hydrochloride	300.0mg
Thiamine-HCl·$2H_2O$	200.0mg
Nicotinic acid	200.0mg
Vitamin B_{12}	100.0mg
Calcium pantothenate	100.0mg
p-Aminobenzoic acid	80.0mg
D(+)-Biotin	20.0mg

Preparation of Solution C: Add components to distilled/deionized water and bring volume to 1.0L. Sparge with 100% N_2. Mix thoroughly. Filter sterilize.

Solution D:

Composition per liter:

NaOH	0.5g
$Na_2WO_4 \cdot 2H_2O$	4.0mg
$Na_2SeO_3 \cdot 5H_2O$	3.0mg

Preparation of Solution D: Add components to distilled/deionized water and bring volume to 1.0L. Mix thoroughly. Sparge with 100% N_2. Filter sterilize.

Solution E:

Composition per 100.0mL:

$NaHCO_3$	5.0g

Preparation of Solution E: Add $NaHCO_3$ to distilled/deionized water and bring volume to 100.0mL. Mix thoroughly. Sparge with 100% N_2 gas mixture. Autoclave for 15 min at 15 psi pressure–121°C. Cool to 25°C.

Solution F:

Composition per 10.0mL:

Fructose	2.0g

Preparation of Solution F: Add fructose to distilled/deionized water and bring volume to 10.0mL. Mix thoroughly. Sparge with 100% N_2 gas mixture. Autoclave for 15 min at 15 psi pressure–121°C. Cool to 25°C.

Solution G:

Composition per 10.0mL:

$Na_2S \cdot 9H_2O$	0.125g

Preparation of Solution G: Add $Na_2S \cdot 9H_2O$ to distilled/deionized water and bring volume to 10.0mL. Mix thoroughly. Autoclave under 100% N_2 for 15 min at 15 psi pressure–121°C. Cool to 25°C.

Preparation of Medium: Prepare and dispense medium under 80% N_2 + 20% CO_2 gas atmosphere. Sequentially add 1.0mL solution B, 1.0mL solution C, 1.0mL solution D, 50.0mL solution E, 10.0mL solution F, and 10.0mL solution G to 940.0mL solution A. Distribute anaerobically under 80% N_2 + 20% CO_2 into appropriate vessels. The pH should be 7.2–7.4.

Use: For the cultivation of *Clostridium homopropionicum* DSM 5847.

Modified FWM Medium (DSMZ Medium 503)

Composition per 1013.0mL:

Solution A	940.0mL
Solution E	50.0mL
Solution F	10.0mL
Solution G	10.0mL
Solution B	1.0mL
Solution C	1.0mL
Solution D	1.0mL

pH 7.3 ± 0.2 at 25°C

Solution A:

Composition per 940.0mL:

NaCl	1.0g
KCl	0.5g
$MgCl_2 \cdot 6H_2O$	0.4g
NH_4Cl	0.25g
KH_2PO_4	0.2g
$CaCl_2 \cdot 2H_2O$	0.15g
Resazurin	0.5mg

Preparation of Solution A: Prepare under 80% N_2 + 20% CO_2 gas atmosphere. Add components to distilled/deionized water and bring volume to 940.0mL. Mix thoroughly. Adjust pH to 7.2. Sparge with 80% N_2 + 20% CO_2. Autoclave for 15 min at 15 psi pressure–121°C. Cool to 25°C.

Solution B:

Composition per liter:

$FeCl_2 \cdot 4H_2O$	1.5g
$CoCl_2 \cdot 6H_2O$	190.0mg
$MnCl_2 \cdot 4H_2O$	100.0mg
$ZnCl_2$	70.0mg
$Na_2MoO_4 \cdot 2H_2O$	36.0mg
$NiCl_2 \cdot 6H_2O$	24.0mg

H_3BO_3 .... 6.0mg
$CuCl_2 \cdot 2H_2O$ 2.0mg
HCl (25% solution) 10.0mL

Preparation of Solution B: Add $FeCl_2 \cdot 4H_2O$ to 10.0mL of HCl solution. Mix thoroughly. Add distilled/deionized water and bring volume to 1.0L. Add remaining components. Mix thoroughly. Sparge with 80% N_2 + 20% CO_2. Autoclave for 15 min at 15 psi pressure–121°C. Cool to 25°C.

Solution C:

Composition per liter:

Pyridoxine hydrochloride 300.0mg
Thiamine-HCl·$2H_2O$ 200.0mg
Nicotinic acid 200.0mg
Vitamin B_{12} 100.0mg
Calcium pantothenate 100.0mg
p-Aminobenzoic acid 80.0mg
D(+)-Biotin 20.0mg

Preparation of Solution C: Add components to distilled/deionized water and bring volume to 1.0L. Sparge with 100% N_2. Mix thoroughly. Filter sterilize.

Solution D:

Composition per liter:

NaOH 0.5g
$Na_2WO_4 \cdot 2H_2O$ 4.0mg
$Na_2SeO_3 \cdot 5H_2O$ 3.0mg

Preparation of Solution D: Add components to distilled/deionized water and bring volume to 1.0L. Mix thoroughly. Sparge with 100% N_2. Filter sterilize.

Solution E:

Composition per 100.0mL:

$NaHCO_3$ 5.0g

Preparation of Solution E: Add $NaHCO_3$ to distilled/deionized water and bring volume to 100.0mL. Mix thoroughly. Sparge with 100% N_2 gas mixture. Autoclave for 15 min at 15 psi pressure–121°C. Cool to 25°C.

Solution F:

Composition per 10.0mL:

Na_2-succinate 2.5g

Preparation of Solution F: Add Na_2-succinate to distilled/deionized water and bring volume to 10.0mL. Mix thoroughly. Sparge with 100% N_2 gas mixture. Autoclave for 15 min at 15 psi pressure–121°C. Cool to 25°C.

Solution G:

Composition per 10.0mL:

$Na_2S \cdot 9H_2O$ 0.125g

Preparation of Solution G: Add $Na_2S \cdot 9H_2O$ to distilled/deionized water and bring volume to 10.0mL. Mix thoroughly. Autoclave under 100% N_2 for 15 min at 15 psi pressure–121°C. Cool to 25°C.

Preparation of Medium: Prepare and dispense medium under 80% N_2 + 20% CO_2 gas atmosphere. Sequentially add 1.0mL solution B, 1.0mL solution C, 1.0mL solution D, 50.0mL solution E, 10.0mL solution F, and 10.0mL solution G to 940.0mL solution A. Distribute anaerobically under 80% N_2 + 20% CO_2 into appropriate vessels. The pH should be 7.2–7.4.

Use: For the cultivation of unidentified bacterium DSM 5849.

Modified FWM Medium (DSMZ Medium 503)

Composition per 1013.0mL:

Solution A 940.0mL
Solution E 50.0mL
Solution F 10.0mL
Solution G 10.0mL
Solution B 1.0mL
Solution C 1.0mL
Solution D 1.0mL

pH 7.3 ± 0.2 at 25°C

Solution A:

Composition per 940.0mL:

NaCl 1.0g
KCl 0.5g
$MgCl_2 \cdot 6H_2O$ 0.4g
NH_4Cl 0.25g
KH_2PO_4 .... 0.2g
$CaCl_2 \cdot 2H_2O$ 0.15g
Resazurin 0.5mg

Preparation of Solution A: Prepare under 80% N_2 + 20% CO_2 gas atmosphere. Add components to distilled/deionized water and bring volume to 940.0mL. Mix thoroughly. Adjust pH to 7.2. Sparge with 80% N_2 + 20% CO_2. Autoclave for 15 min at 15 psi pressure–121°C. Cool to 25°C.

Solution B:

Composition per liter:

$FeCl_2 \cdot 4H_2O$ 1.5g
$CoCl_2 \cdot 6H_2O$ 190.0mg
$MnCl_2 \cdot 4H_2O$ 100.0mg
$ZnCl_2$ 70.0mg
$Na_2MoO_4 \cdot 2H_2O$ 36.0mg
$NiCl_2 \cdot 6H_2O$ 24.0mg
H_3BO_3 .... 6.0mg
$CuCl_2 \cdot 2H_2O$ 2.0mg
HCl (25% solution) 10.0mL

Preparation of Solution B: Add $FeCl_2 \cdot 4H_2O$ to 10.0mL of HCl solution. Mix thoroughly. Add distilled/deionized water and bring volume to 1.0L. Add remaining components. Mix thoroughly. Sparge with 80% N_2 + 20% CO_2. Autoclave for 15 min at 15 psi pressure–121°C. Cool to 25°C.

Solution C:

Composition per liter:

Pyridoxine hydrochloride 300.0mg
Thiamine-HCl·$2H_2O$ 200.0mg
Nicotinic acid 200.0mg
Vitamin B_{12} 100.0mg
Calcium pantothenate 100.0mg
p-Aminobenzoic acid 80.0mg
D(+)-Biotin 20.0mg

Preparation of Solution C: Add components to distilled/deionized water and bring volume to 1.0L. Sparge with 100% N_2. Mix thoroughly. Filter sterilize.

Solution D:

Composition per liter:

NaOH 0.5g
$Na_2WO_4 \cdot 2H_2O$ 4.0mg
$Na_2SeO_3 \cdot 5H_2O$ 3.0mg

Preparation of Solution D: Add components to distilled/deionized water and bring volume to 1.0L. Mix thoroughly. Sparge with 100% N_2. Filter sterilize.

Solution E:

Composition per 100.0mL:

$NaHCO_3$ 5.0g

Preparation of Solution E: Add $NaHCO_3$ to distilled/deionized water and bring volume to 100.0mL. Mix thoroughly. Sparge with 100% N_2 gas mixture. Autoclave for 15 min at 15 psi pressure–121°C. Cool to 25°C.

Solution F:

Composition per 10.0mL:

Na-(D/L)-3-hydroxybutyrate 1.5g

Preparation of Solution F: Add Na-(D/L)-3-hydroxybutyrate to distilled/deionized water and bring volume to 10.0mL. Mix thoroughly. Sparge with 100% N_2 gas mixture. Autoclave for 15 min at 15 psi pressure–121°C. Cool to 25°C.

Solution G:

Composition per 10.0mL:

$Na_2S{\cdot}9H_2O$ 0.125g

Preparation of Solution G: Add $Na_2S{\cdot}9H_2O$ to distilled/deionized water and bring volume to 10.0mL. Mix thoroughly. Autoclave under 100% N_2 for 15 min at 15 psi pressure–121°C. Cool to 25°C.

Preparation of Medium: Prepare and dispense medium under 80% N_2 + 20% CO_2 gas atmosphere. Sequentially add 1.0mL solution B, 1.0mL solution C, 1.0mL solution D, 50.0mL solution E, 10.0mL solution F, and 10.0mL solution G to 940.0mL solution A. Distribute anaerobically under 80% N_2 + 20% CO_2 into appropriate vessels. The pH should be 7.2–7.4.

Use: For the cultivation of *Desulfococcus biacutus* DSM 5651.

Modified FWM Medium (DSMZ Medium 503)

Composition per 1013.0mL:

Solution A 940.0mL
Solution E 50.0mL
Solution F 10.0mL
Solution G 10.0mL
Solution B 1.0mL
Solution C 1.0mL
Solution D 1.0mL

pH 7.3 ± 0.2 at 25°C

Solution A:

Composition per 940.0mL:

NaCl 1.0g
KCl 0.5g
$MgCl_2{\cdot}6H_2O$ 0.4g
NH_4Cl 0.25g
KH_2PO_4 .... 0.2g
$CaCl_2{\cdot}2H_2O$ 0.15g
Resazurin 0.5mg

Preparation of Solution A: Prepare under 80% N_2 + 20% CO_2 gas atmosphere. Add components to distilled/deionized water and bring volume to 940.0mL. Mix thoroughly. Adjust pH to 7.2. Sparge with 80% N_2 + 20% CO_2. Autoclave for 15 min at 15 psi pressure–121°C. Cool to 25°C.

Solution B:

Composition per liter:

$FeCl_2{\cdot}4H_2O$ 1.5g
$CoCl_2{\cdot}6H_2O$ 190.0mg
$MnCl_2{\cdot}4H_2O$ 100.0mg
$ZnCl_2$ 70.0mg
$Na_2MoO_4{\cdot}2H_2O$ 36.0mg
$NiCl_2{\cdot}6H_2O$ 24.0mg
H_3BO_3 .... 6.0mg
$CuCl_2{\cdot}2H_2O$ 2.0mg
HCl (25% solution) 10.0mL

Preparation of Solution B: Add $FeCl_2{\cdot}4H_2O$ to 10.0mL of HCl solution. Mix thoroughly. Add distilled/deionized water and bring volume to 1.0L. Add remaining components. Mix thoroughly. Sparge with 80% N_2 + 20% CO_2. Autoclave for 15 min at 15 psi pressure–121°C. Cool to 25°C.

Solution C:

Composition per liter:

Pyridoxine hydrochloride 300.0mg
Thiamine-HCl·$2H_2O$ 200.0mg
Nicotinic acid 200.0mg
Vitamin B_{12} 100.0mg
Calcium pantothenate 100.0mg
p-Aminobenzoic acid 80.0mg
D(+)-Biotin 20.0mg

Preparation of Solution C: Add components to distilled/deionized water and bring volume to 1.0L. Sparge with 100% N_2. Mix thoroughly. Filter sterilize.

Solution D:

Composition per liter:

NaOH 0.5g
$Na_2WO_4{\cdot}2H_2O$ 4.0mg
$Na_2SeO_3{\cdot}5H_2O$ 3.0mg

Preparation of Solution D: Add components to distilled/deionized water and bring volume to 1.0L. Mix thoroughly. Sparge with 100% N_2. Filter sterilize.

Solution E:

Composition per 100.0mL:

$NaHCO_3$ 5.0g

Preparation of Solution E: Add $NaHCO_3$ to distilled/deionized water and bring volume to 100.0mL. Mix thoroughly. Sparge with 100% N_2 gas mixture. Autoclave for 15 min at 15 psi pressure–121°C. Cool to 25°C.

Solution F:

Composition per 10.0mL:

Xylan 2.0g

Preparation of Solution F: Add xylan to distilled/deionized water and bring volume to 10.0mL. Mix thoroughly. Sparge with 100% N_2 gas mixture. Autoclave for 15 min at 15 psi pressure–121°C. Cool to 25°C.

Solution G:

Composition per 10.0mL:

$Na_2S{\cdot}9H_2O$ 0.125g

Preparation of Solution G: Add $Na_2S{\cdot}9H_2O$ to distilled/deionized water and bring volume to 10.0mL. Mix thoroughly. Autoclave under 100% N_2 for 15 min at 15 psi pressure–121°C. Cool to 25°C.

Preparation of Medium: Prepare and dispense medium under 80% N_2 + 20% CO_2 gas atmosphere. Sequentially add 1.0mL solution B, 1.0mL solution C, 1.0mL solution D, 50.0mL solution E, 10.0mL solution F, and 10.0mL solution G to 940.0mL solution A. Distribute anaerobically under 80% N_2 + 20% CO_2 into appropriate vessels. The pH should be 7.2–7.4.

Use: For the cultivation of *Cytophaga xylanolytica* DSM 6779.

Modified FWM Medium (DSMZ Medium 503)

Composition per 1013.0mL:

Solution A	940.0mL
Solution E	50.0mL
Solution F	10.0mL
Solution G	10.0mL
Solution B	1.0mL
Solution C	1.0mL
Solution D	1.0mL

pH 7.3 ± 0.2 at 25°C

Solution A:

Composition per 940.0mL:

NaCl	1.0g
KCl	0.5g
$MgCl_2 \cdot 6H_2O$	0.4g
NH_4Cl	0.25g
KH_2PO_4	0.2g
$CaCl_2 \cdot 2H_2O$	0.15g
Resazurin	0.5mg

Preparation of Solution A: Prepare under 80% N_2 + 20% CO_2 gas atmosphere. Add components to distilled/deionized water and bring volume to 940.0mL. Mix thoroughly. Adjust pH to 7.2. Sparge with 80% N_2 + 20% CO_2. Autoclave for 15 min at 15 psi pressure–121°C. Cool to 25°C.

Solution B:

Composition per liter:

$FeCl_2 \cdot 4H_2O$	1.5g
$CoCl_2 \cdot 6H_2O$	190.0mg
$MnCl_2 \cdot 4H_2O$	100.0mg
$ZnCl_2$	70.0mg
$Na_2MoO_4 \cdot 2H_2O$	36.0mg
$NiCl_2 \cdot 6H_2O$	24.0mg
H_3BO_3	6.0mg
$CuCl_2 \cdot 2H_2O$	2.0mg
HCl (25% solution)	10.0mL

Preparation of Solution B: Add $FeCl_2 \cdot 4H_2O$ to 10.0mL of HCl solution. Mix thoroughly. Add distilled/deionized water and bring volume to 1.0L. Add remaining components. Mix thoroughly. Sparge with 80% N_2 + 20% CO_2. Autoclave for 15 min at 15 psi pressure–121°C. Cool to 25°C.

Solution C:

Composition per liter:

Pyridoxine hydrochloride	300.0mg
Thiamine-HCl·$2H_2O$	200.0mg
Nicotinic acid	200.0mg
Vitamin B_{12}	100.0mg
Calcium pantothenate	100.0mg
p-Aminobenzoic acid	80.0mg
D(+)-Biotin	20.0mg

Preparation of Solution C: Add components to distilled/deionized water and bring volume to 1.0L. Sparge with 100% N_2. Mix thoroughly. Filter sterilize.

Solution D:

Composition per liter:

NaOH	0.5g
$Na_2WO_4 \cdot 2H_2O$	4.0mg
$Na_2SeO_3 \cdot 5H_2O$	3.0mg

Preparation of Solution D: Add components to distilled/deionized water and bring volume to 1.0L. Mix thoroughly. Sparge with 100% N_2. Filter sterilize.

Solution E:

Composition per 100.0mL:

$NaHCO_3$	5.0g

Preparation of Solution E: Add $NaHCO_3$ to distilled/deionized water and bring volume to 100.0mL. Mix thoroughly. Sparge with 100% N_2 gas mixture. Autoclave for 15 min at 15 psi pressure–121°C. Cool to 25°C.

Solution F:

Composition per 10.0mL:

Xylose	2.0g

Preparation of Solution F: Add xylose to distilled/deionized water and bring volume to 10.0mL. Mix thoroughly. Sparge with 100% N_2 gas mixture. Autoclave for 15 min at 15 psi pressure–121°C. Cool to 25°C.

Solution G:

Composition per 10.0mL:

$Na_2S \cdot 9H_2O$	0.125g

Preparation of Solution G: Add $Na_2S \cdot 9H_2O$ to distilled/deionized water and bring volume to 10.0mL. Mix thoroughly. Autoclave under 100% N_2 for 15 min at 15 psi pressure–121°C. Cool to 25°C.

Preparation of Medium: Prepare and dispense medium under 80% N_2 + 20% CO_2 gas atmosphere. Sequentially add 1.0mL solution B, 1.0mL solution C, 1.0mL solution D, 50.0mL solution E, 10.0mL solution F, and 10.0mL solution G to 940.0mL solution A. Distribute anaerobically under 80% N_2 + 20% CO_2 into appropriate vessels. The pH should be 7.2–7.4.

Use: For the cultivation of *Cytophaga xylanolytica* DSM 6779.

Modified FWM Medium (DSMZ Medium 503)

Composition per 1013.0mL:

Solution A	940.0mL
Solution E	50.0mL
Solution F	10.0mL
Solution G	10.0mL
Solution B	1.0mL
Solution C	1.0mL
Solution D	1.0mL

pH 7.3 ± 0.2 at 25°C

Solution A:

Composition per 940.0mL:

NaCl	1.0g
KCl	0.5g
$MgCl_2 \cdot 6H_2O$	0.4g
NH_4Cl	0.25g
KH_2PO_4	0.2g

$CaCl_2 \cdot 2H_2O$ 0.15g
Resazurin 0.5mg

Preparation of Solution A: Prepare under 80% N_2 + 20% CO_2 gas atmosphere. Add components to distilled/deionized water and bring volume to 940.0mL. Mix thoroughly. Adjust pH to 7.2. Sparge with 80% N_2 + 20% CO_2. Autoclave for 15 min at 15 psi pressure–121°C. Cool to 25°C.

Solution B:
Composition per liter:
$FeCl_2 \cdot 4H_2O$ 1.5g
$CoCl_2 \cdot 6H_2O$ 190.0mg
$MnCl_2 \cdot 4H_2O$ 100.0mg
$ZnCl_2$ 70.0mg
$Na_2MoO_4 \cdot 2H_2O$ 36.0mg
$NiCl_2 \cdot 6H_2O$ 24.0mg
H_3BO_3 .... 6.0mg
$CuCl_2 \cdot 2H_2O$ 2.0mg
HCl (25% solution) 10.0mL

Preparation of Solution B: Add $FeCl_2 \cdot 4H_2O$ to 10.0mL of HCl solution. Mix thoroughly. Add distilled/deionized water and bring volume to 1.0L. Add remaining components. Mix thoroughly. Sparge with 80% N_2 + 20% CO_2. Autoclave for 15 min at 15 psi pressure–121°C. Cool to 25°C.

Solution C:
Composition per liter:
Pyridoxine hydrochloride 300.0mg
Thiamine-HCl·$2H_2O$ 200.0mg
Nicotinic acid 200.0mg
Vitamin B_{12} 100.0mg
Calcium pantothenate 100.0mg
p-Aminobenzoic acid 80.0mg
D(+)-Biotin 20.0mg

Preparation of Solution C: Add components to distilled/deionized water and bring volume to 1.0L. Sparge with 100% N_2. Mix thoroughly. Filter sterilize.

Solution D:
Composition per liter:
NaOH 0.5g
$Na_2WO_4 \cdot 2H_2O$ 4.0mg
$Na_2SeO_3 \cdot 5H_2O$ 3.0mg

Preparation of Solution D: Add components to distilled/deionized water and bring volume to 1.0L. Mix thoroughly. Sparge with 100% N_2. Filter sterilize.

Solution E:
Composition per 100.0mL:
$NaHCO_3$ 5.0g

Preparation of Solution E: Add $NaHCO_3$ to distilled/deionized water and bring volume to 100.0mL. Mix thoroughly. Sparge with 100% N_2 gas mixture. Autoclave for 15 min at 15 psi pressure–121°C. Cool to 25°C.

Solution F:
Composition per 10.0mL:
Pivalic acid 1.0g
$NaNO_3$ 0.85g

Preparation of Solution F: Add components to distilled/deionized water and bring volume to 10.0mL. Mix thoroughly. Sparge with 100% N_2 gas mixture. Autoclave for 15 min at 15 psi pressure–121°C. Cool to 25°C.

Solution G:
Composition per 10.0mL:
$Na_2S \cdot 9H_2O$ 0.125g

Preparation of Solution G: Add $Na_2S \cdot 9H_2O$ to distilled/deionized water and bring volume to 10.0mL. Mix thoroughly. Autoclave under 100% N_2 for 15 min at 15 psi pressure–121°C. Cool to 25°C.

Preparation of Medium: Prepare and dispense medium under 80% N_2 + 20% CO_2 gas atmosphere. Sequentially add 1.0mL solution B, 1.0mL solution C, 1.0mL solution D, 50.0mL solution E, 10.0mL solution F, and 10.0mL solution G to 940.0mL solution A. Distribute anaerobically under 80% N_2 + 20% CO_2 into appropriate vessels. The pH should be 7.2–7.4.

Use: For the cultivation of *Thavera pivalivorans* DSM 14691.

Modified FWM Medium (DSMZ Medium 503)

Composition per 1013.0mL:
Solution A 940.0mL
Solution E 50.0mL
Solution F 10.0mL
Solution G 10.0mL
Solution B 1.0mL
Solution C 1.0mL
Solution D 1.0mL

pH 7.3 ± 0.2 at 25°C

Solution A:
Composition per 940.0mL:
NaCl 1.0g
KCl 0.5g
$MgCl_2 \cdot 6H_2O$ 0.4g
NH_4Cl 0.25g
KH_2PO_4 .... 0.2g
$CaCl_2 \cdot 2H_2O$ 0.15g
Resazurin 0.5mg

Preparation of Solution A: Prepare under 80% N_2 + 20% CO_2 gas atmosphere. Add components to distilled/deionized water and bring volume to 940.0mL. Mix thoroughly. Adjust pH to 7.2. Sparge with 80% N_2 + 20% CO_2. Autoclave for 15 min at 15 psi pressure–121°C. Cool to 25°C.

Solution B:
Composition per liter:
$FeCl_2 \cdot 4H_2O$ 1.5g
$CoCl_2 \cdot 6H_2O$ 190.0mg
$MnCl_2 \cdot 4H_2O$ 100.0mg
$ZnCl_2$ 70.0mg
$Na_2MoO_4 \cdot 2H_2O$ 36.0mg
$NiCl_2 \cdot 6H_2O$ 24.0mg
H_3BO_3 .... 6.0mg
$CuCl_2 \cdot 2H_2O$ 2.0mg
HCl (25% solution) 10.0mL

Preparation of Solution B: Add $FeCl_2 \cdot 4H_2O$ to 10.0mL of HCl solution. Mix thoroughly. Add distilled/deionized water and bring volume to 1.0L. Add remaining components. Mix thoroughly. Sparge with 80% N_2 + 20% CO_2. Autoclave for 15 min at 15 psi pressure–121°C. Cool to 25°C.

Solution C:

Composition per liter:

Pyridoxine hydrochloride	300.0mg
Thiamine-HCl·$2H_2O$	200.0mg
Nicotinic acid	200.0mg
Vitamin B_{12}	100.0mg
Calcium pantothenate	100.0mg
p-Aminobenzoic acid	80.0mg
D(+)-Biotin	20.0mg

Preparation of Solution C: Add components to distilled/deionized water and bring volume to 1.0L. Sparge with 100% N_2. Mix thoroughly. Filter sterilize.

Solution D:

Composition per liter:

NaOH	0.5g
$Na_2WO_4·2H_2O$	4.0mg
$Na_2SeO_3·5H_2O$	3.0mg

Preparation of Solution D: Add components to distilled/deionized water and bring volume to 1.0L. Mix thoroughly. Sparge with 100% N_2. Filter sterilize.

Solution E:

Composition per 100.0mL:

$NaHCO_3$	5.0g

Preparation of Solution E: Add $NaHCO_3$ to distilled/deionized water and bring volume to 100.0mL. Mix thoroughly. Sparge with 100% N_2 gas mixture. Autoclave for 15 min at 15 psi pressure–121°C. Cool to 25°C.

Solution F:

Composition per 10.0mL:

Glucose	5.0g
Yeast extract	2.0g

Preparation of Solution F: Add components to distilled/deionized water and bring volume to 10.0mL. Mix thoroughly. Sparge with 100% N_2 gas mixture. Autoclave for 15 min at 15 psi pressure–121°C. Cool to 25°C.

Solution G:

Composition per 10.0mL:

$Na_2S·9H_2O$	0.125g

Preparation of Solution G: Add $Na_2S·9H_2O$ to distilled/deionized water and bring volume to 10.0mL. Mix thoroughly. Autoclave under 100% N_2 for 15 min at 15 psi pressure–121°C. Cool to 25°C.

Preparation of Medium: Prepare and dispense medium under 80% N_2 + 20% CO_2 gas atmosphere. Sequentially add 1.0mL solution B, 1.0mL solution C, 1.0mL solution D, 50.0mL solution E, 10.0mL solution F, and 10.0mL solution G to 940.0mL solution A. Distribute anaerobically under 80% N_2 + 20% CO_2 into appropriate vessels. The pH should be 7.2–7.4.

Use: For the cultivation of *Opitutus* sp. DSM 14424.

Modified FWM Medium (DSMZ Medium 503)

Composition per 1013.0mL:

Solution A	940.0mL
Solution E	50.0mL
Solution F	10.0mL
Solution G	10.0mL
Solution B	1.0mL
Solution C	1.0mL
Solution D	1.0mL

pH 7.3 ± 0.2 at 25°C

Solution A:

Composition per 940.0mL:

NaCl	1.0g
KCl	0.5g
$MgCl_2·6H_2O$	0.4g
NH_4Cl	0.25g
KH_2PO_4	0.2g
$CaCl_2·2H_2O$	0.15g
Resazurin	0.5mg

Preparation of Solution A: Prepare under 80% N_2 + 20% CO_2 gas atmosphere. Add components to distilled/deionized water and bring volume to 940.0mL. Mix thoroughly. Adjust pH to 7.2. Sparge with 80% N_2 + 20% CO_2. Autoclave for 15 min at 15 psi pressure–121°C. Cool to 25°C.

Solution B:

Composition per liter:

$FeCl_2·4H_2O$	1.5g
$CoCl_2·6H_2O$	190.0mg
$MnCl_2·4H_2O$	100.0mg
$ZnCl_2$	70.0mg
$Na_2MoO_4·2H_2O$	36.0mg
$NiCl_2·6H_2O$	24.0mg
H_3BO_3	6.0mg
$CuCl_2·2H_2O$	2.0mg
HCl (25% solution)	10.0mL

Preparation of Solution B: Add $FeCl_2·4H_2O$ to 10.0mL of HCl solution. Mix thoroughly. Add distilled/deionized water and bring volume to 1.0L. Add remaining components. Mix thoroughly. Sparge with 80% N_2 + 20% CO_2. Autoclave for 15 min at 15 psi pressure–121°C. Cool to 25°C.

Solution C:

Composition per liter:

Pyridoxine hydrochloride	300.0mg
Thiamine-HCl·$2H_2O$	200.0mg
Nicotinic acid	200.0mg
Vitamin B_{12}	100.0mg
Calcium pantothenate	100.0mg
p-Aminobenzoic acid	80.0mg
D(+)-Biotin	20.0mg

Preparation of Solution C: Add components to distilled/deionized water and bring volume to 1.0L. Sparge with 100% N_2. Mix thoroughly. Filter sterilize.

Solution D:

Composition per liter:

NaOH	0.5g
$Na_2WO_4·2H_2O$	4.0mg
$Na_2SeO_3·5H_2O$	3.0mg

Preparation of Solution D: Add components to distilled/deionized water and bring volume to 1.0L. Mix thoroughly. Sparge with 100% N_2. Filter sterilize.

Solution E:

Composition per 100.0mL:

$NaHCO_3$	5.0g

Preparation of Solution E: Add $NaHCO_3$ to distilled/deionized water and bring volume to 100.0mL. Mix thoroughly. Sparge with

100% N_2 gas mixture. Autoclave for 15 min at 15 psi pressure–121°C. Cool to 25°C.

Solution F:

Composition per 10.0mL:

Na₂-fumarate	3.2g
Na-acetate	0.8g

Preparation of Solution F: Add components to distilled/deionized water and bring volume to 10.0mL. Mix thoroughly. Sparge with 100% N_2 gas mixture. Autoclave for 15 min at 15 psi pressure–121°C. Cool to 25°C.

Solution G:

Composition per 10.0mL:

$Na_2S·9H_2O$	0.125g

Preparation of Solution G: Add $Na_2S·9H_2O$ to distilled/deionized water and bring volume to 10.0mL. Mix thoroughly. Autoclave under 100% N_2 for 15 min at 15 psi pressure–121°C. Cool to 25°C.

Preparation of Medium: Prepare and dispense medium under 80% N_2 + 20% CO_2 gas atmosphere. Sequentially add 1.0mL solution B, 1.0mL solution C, 1.0mL solution D, 50.0mL solution E, 10.0mL solution F, and 10.0mL solution G to 940.0mL solution A. Distribute anaerobically under 80% N_2 + 20% CO_2 into appropriate vessels. The pH should be 7.2–7.4.

Use: For the cultivation of *Geovibrio thiophilus* DSM 11263.

Modified FWM Medium (DSMZ Medium 503)

Composition per 1013.0mL:

Solution A	940.0mL
Solution E	50.0mL
Solution F	10.0mL
Solution G	10.0mL
Solution B	1.0mL
Solution C	1.0mL
Solution D	1.0mL

pH 7.3 ± 0.2 at 25°C

Solution A:

Composition per 940.0mL:

NaCl	1.0g
KCl	0.5g
$MgCl_2·6H_2O$	0.4g
NH_4Cl	0.25g
KH_2PO_4	0.2g
$CaCl_2·2H_2O$	0.15g
Resazurin	0.5mg

Preparation of Solution A: Prepare under 80% N_2 + 20% CO_2 gas atmosphere. Add components to distilled/deionized water and bring volume to 940.0mL. Mix thoroughly. Adjust pH to 7.2. Sparge with 80% N_2 + 20% CO_2. Autoclave for 15 min at 15 psi pressure–121°C. Cool to 25°C.

Solution B:

Composition per liter:

$FeCl_2·4H_2O$	1.5g
$CoCl_2·6H_2O$	190.0mg
$MnCl_2·4H_2O$	100.0mg
$ZnCl_2$	70.0mg
$Na_2MoO_4·2H_2O$	36.0mg
$NiCl_2·6H_2O$	24.0mg
H_3BO_3	6.0mg
$CuCl_2·2H_2O$	2.0mg
HCl (25% solution)	10.0mL

Preparation of Solution B: Add $FeCl_2·4H_2O$ to 10.0mL of HCl solution. Mix thoroughly. Add distilled/deionized water and bring volume to 1.0L. Add remaining components. Mix thoroughly. Sparge with 80% N_2 + 20% CO_2. Autoclave for 15 min at 15 psi pressure–121°C. Cool to 25°C.

Solution C:

Composition per liter:

Pyridoxine hydrochloride	300.0mg
Thiamine-HCl·2H_2O	200.0mg
Nicotinic acid	200.0mg
Vitamin B_{12}	100.0mg
Calcium pantothenate	100.0mg
p-Aminobenzoic acid	80.0mg
D(+)-Biotin	20.0mg

Preparation of Solution C: Add components to distilled/deionized water and bring volume to 1.0L. Sparge with 100% N_2. Mix thoroughly. Filter sterilize.

Solution D:

Composition per liter:

NaOH	0.5g
$Na_2WO_4·2H_2O$	4.0mg
$Na_2SeO_3·5H_2O$	3.0mg

Preparation of Solution D: Add components to distilled/deionized water and bring volume to 1.0L. Mix thoroughly. Sparge with 100% N_2. Filter sterilize.

Solution E:

Composition per 100.0mL:

$NaHCO_3$	5.0g

Preparation of Solution E: Add $NaHCO_3$ to distilled/deionized water and bring volume to 100.0mL. Mix thoroughly. Sparge with 100% N_2 gas mixture. Autoclave for 15 min at 15 psi pressure–121°C. Cool to 25°C.

Solution F:

Composition per 10.0mL:

Na-glycolate	2.0g
Yeast extract	1.0g

Preparation of Solution F: Add components to distilled/deionized water and bring volume to 10.0mL. Mix thoroughly. Sparge with 100% N_2 gas mixture. Autoclave for 15 min at 15 psi pressure–121°C. Cool to 25°C.

Solution G:

Composition per 10.0mL:

$Na_2S·9H_2O$	0.125g

Preparation of Solution G: Add $Na_2S·9H_2O$ to distilled/deionized water and bring volume to 10.0mL. Mix thoroughly. Autoclave under 100% N_2 for 15 min at 15 psi pressure–121°C. Cool to 25°C.

Preparation of Medium: Prepare and dispense medium under 80% N_2 + 20% CO_2 gas atmosphere. Sequentially add 1.0mL solution B, 1.0mL solution C, 1.0mL solution D, 50.0mL solution E, 10.0mL solution F, and 10.0mL solution G to 940.0mL solution A. Distribute anaerobically under 80% N_2 + 20% CO_2 into appropriate vessels. The pH should be 7.2–7.4.

Use: For the cultivation of *Clostridium* sp. DSM 11261.

Modified FWM Medium (DSMZ Medium 503)

Composition per 1013.0mL:

Solution A	940.0mL
Solution E	50.0mL
Solution F	10.0mL
Solution G	10.0mL
Solution B	1.0mL
Solution C	1.0mL
Solution D	1.0mL

pH 7.3 ± 0.2 at 25°C

Solution A:

Composition per 940.0mL:

NaCl	1.0g
KCl	0.5g
$MgCl_2 \cdot 6H_2O$	0.4g
NH_4Cl	0.25g
KH_2PO_4	0.2g
$CaCl_2 \cdot 2H_2O$	0.15g
Resazurin	0.5mg

Preparation of Solution A: Prepare under 80% N_2 + 20% CO_2 gas atmosphere. Add components to distilled/deionized water and bring volume to 940.0mL. Mix thoroughly. Adjust pH to 7.2. Sparge with 80% N_2 + 20% CO_2. Autoclave for 15 min at 15 psi pressure–121°C. Cool to 25°C.

Solution B:

Composition per liter:

$FeCl_2 \cdot 4H_2O$	1.5g
$CoCl_2 \cdot 6H_2O$	190.0mg
$MnCl_2 \cdot 4H_2O$	100.0mg
$ZnCl_2$	70.0mg
$Na_2MoO_4 \cdot 2H_2O$	36.0mg
$NiCl_2 \cdot 6H_2O$	24.0mg
H_3BO_3	6.0mg
$CuCl_2 \cdot 2H_2O$	2.0mg
HCl (25% solution)	10.0mL

Preparation of Solution B: Add $FeCl_2 \cdot 4H_2O$ to 10.0mL of HCl solution. Mix thoroughly. Add distilled/deionized water and bring volume to 1.0L. Add remaining components. Mix thoroughly. Sparge with 80% N_2 + 20% CO_2. Autoclave for 15 min at 15 psi pressure–121°C. Cool to 25°C.

Solution C:

Composition per liter:

Pyridoxine hydrochloride	300.0mg
Thiamine-HCl·$2H_2O$	200.0mg
Nicotinic acid	200.0mg
Vitamin B_{12}	100.0mg
Calcium pantothenate	100.0mg
p-Aminobenzoic acid	80.0mg
D(+)-Biotin	20.0mg

Preparation of Solution C: Add components to distilled/deionized water and bring volume to 1.0L. Sparge with 100% N_2. Mix thoroughly. Filter sterilize.

Solution D:

Composition per liter:

NaOH	0.5g
$Na_2WO_4 \cdot 2H_2O$	4.0mg
$Na_2SeO_3 \cdot 5H_2O$	3.0mg

Preparation of Solution D: Add components to distilled/deionized water and bring volume to 1.0L. Mix thoroughly. Sparge with 100% N_2. Filter sterilize.

Solution E:

Composition per 100.0mL:

$NaHCO_3$	5.0g

Preparation of Solution E: Add $NaHCO_3$ to distilled/deionized water and bring volume to 100.0mL. Mix thoroughly. Sparge with 100% N_2 gas mixture. Autoclave for 15 min at 15 psi pressure–121°C. Cool to 25°C.

Solution F:

Composition per 10.0mL:

Na-3-hydroxybutyrate	3.0g
Yeast extract	1.0g

Preparation of Solution F: Add components to distilled/deionized water and bring volume to 10.0mL. Mix thoroughly. Sparge with 100% N_2 gas mixture. Autoclave for 15 min at 15 psi pressure–121°C. Cool to 25°C.

Solution G:

Composition per 10.0mL:

$Na_2S \cdot 9H_2O$	0.125g

Preparation of Solution G: Add $Na_2S \cdot 9H_2O$ to distilled/deionized water and bring volume to 10.0mL. Mix thoroughly. Autoclave under 100% N_2 for 15 min at 15 psi pressure–121°C. Cool to 25°C.

Preparation of Medium: Prepare and dispense medium under 80% N_2 + 20% CO_2 gas atmosphere. Sequentially add 1.0mL solution B, 1.0mL solution C, 1.0mL solution D, 50.0mL solution E, 10.0mL solution F, and 10.0mL solution G to 940.0mL solution A. Distribute anaerobically under 80% N_2 + 20% CO_2 into appropriate vessels. The pH should be 7.2–7.4.

Use: For the cultivation of unclassified bacterium DSM 11262.

Modified FWM Medium (DSMZ Medium 503)

Composition per 1023.0mL:

Solution A	940.0mL
Solution E	50.0mL
Solution F	10.0mL
Solution G	20.0mL
Solution B	1.0mL
Solution C	1.0mL
Solution D	1.0mL

pH 7.3 ± 0.2 at 25°C

Solution A:

Composition per 940.0mL:

Na_2SO_4	2.8 g
NaCl	1.0g
KCl	0.5g
$MgCl_2 \cdot 6H_2O$	0.4g
NH_4Cl	0.25g
KH_2PO_4	0.2g
$CaCl_2 \cdot 2H_2O$	0.15g
Resazurin	0.5mg

Preparation of Solution A: Prepare under 80% N_2 + 20% CO_2 gas atmosphere. Add components to distilled/deionized water and bring volume to 940.0mL. Mix thoroughly. Adjust pH to 7.2. Sparge with 80% N_2

+ 20% CO_2. Autoclave for 15 min at 15 psi pressure–121°C. Cool to 25°C.

Solution B:

Composition per liter:

$FeCl_2·4H_2O$	1.5g
$CoCl_2·6H_2O$	190.0mg
$MnCl_2·4H_2O$	100.0mg
$ZnCl_2$	70.0mg
$Na_2MoO_4·2H_2O$	36.0mg
$NiCl_2·6H_2O$	24.0mg
H_3BO_3	6.0mg
$CuCl_2·2H_2O$	2.0mg
HCl (25% solution)	10.0mL

Preparation of Solution B: Add $FeCl_2·4H_2O$ to 10.0mL of HCl solution. Mix thoroughly. Add distilled/deionized water and bring volume to 1.0L. Add remaining components. Mix thoroughly. Sparge with 80% N_2 + 20% CO_2. Autoclave for 15 min at 15 psi pressure–121°C. Cool to 25°C.

Solution C:

Composition per liter:

Pyridoxine hydrochloride	300.0mg
Thiamine-HCl·$2H_2O$	200.0mg
Nicotinic acid	200.0mg
Vitamin B_{12}	100.0mg
Calcium pantothenate	100.0mg
p-Aminobenzoic acid	80.0mg
D(+)-Biotin	20.0mg

Preparation of Solution C: Add components to distilled/deionized water and bring volume to 1.0L. Sparge with 100% N_2. Mix thoroughly. Filter sterilize.

Solution D:

Composition per liter:

NaOH	0.5g
$Na_2WO_4·2H_2O$	4.0mg
$Na_2SeO_3·5H_2O$	3.0mg

Preparation of Solution D: Add components to distilled/deionized water and bring volume to 1.0L. Mix thoroughly. Sparge with 100% N_2. Filter sterilize.

Solution E:

Composition per 100.0mL:

$NaHCO_3$	5.0g

Preparation of Solution E: Add $NaHCO_3$ to distilled/deionized water and bring volume to 100.0mL. Mix thoroughly. Sparge with 100% N_2 gas mixture. Autoclave for 15 min at 15 psi pressure–121°C. Cool to 25°C.

Solution F:

Composition per 10.0mL:

Syringic acid	0.6 g

Preparation of Solution F: Add syringic acid to distilled/deionized water and bring volume to 10.0mL. Mix thoroughly. Adjust pH to 7.0. Sparge with 100% N_2 gas mixture. Autoclave for 15 min at 15 psi pressure–121°C. Cool to 25°C.

Solution G:

Composition per 20.0mL:

$Na_2S·9H_2O$	0.25g

Preparation of Solution G: Add $Na_2S·9H_2O$ to distilled/deionized water and bring volume to 20.0mL. Mix thoroughly. Autoclave under 100% N_2 for 15 min at 15 psi pressure–121°C. Cool to 25°C.

Preparation of Medium: Prepare and dispense medium under 80% N_2 + 20% CO_2 gas atmosphere. Sequentially add 1.0mL solution B, 1.0mL solution C, 1.0mL solution D, 50.0mL solution E, 10.0mL solution F, and 20.0mL solution G to 940.0mL solution A. Distribute anaerobically under 80% N_2 + 20% CO_2 into appropriate vessels. The pH should be 7.2–7.4. Prior to inoculation the completed medium should equilibrate overnight.

Use: For the cultivation of *Parasporobacterium paucivorans* DSM 15970.

Modified FWM Medium (DSMZ Medium 503)

Composition per 1023.0mL:

Solution A	940.0mL
Solution E	50.0mL
Solution F	10.0mL
Solution G	10.0mL
Solution H	10.0mL
Solution B	1.0mL
Solution C	1.0mL
Solution D	1.0mL

pH 6.5–7.0 at 25°C

Solution A:

Composition per 940.0mL:

NaCl	1.0g
KCl	0.5g
$MgCl_2·6H_2O$	0.4g
NH_4Cl	0.25g
KH_2PO_4	0.2g
$CaCl_2·2H_2O$	0.15g
Resazurin	0.5mg

Preparation of Solution A: Prepare under 80% N_2 + 20% CO_2 gas atmosphere. Add components to distilled/deionized water and bring volume to 940.0mL. Mix thoroughly. Adjust pH to 7.2. Sparge with 80% N_2 + 20% CO_2. Autoclave for 15 min at 15 psi pressure–121°C. Cool to 25°C.

Solution B:

Composition per liter:

$FeCl_2·4H_2O$	1.5g
$CoCl_2·6H_2O$	190.0mg
$MnCl_2·4H_2O$	100.0mg
$ZnCl_2$	70.0mg
$Na_2MoO_4·2H_2O$	36.0mg
$NiCl_2·6H_2O$	24.0mg
H_3BO_3	6.0mg
$CuCl_2·2H_2O$	2.0mg
HCl (25% solution)	10.0mL

Preparation of Solution B: Add $FeCl_2·4H_2O$ to 10.0mL of HCl solution. Mix thoroughly. Add distilled/deionized water and bring volume to 1.0L. Add remaining components. Mix thoroughly. Sparge with 80% N_2 + 20% CO_2. Autoclave for 15 min at 15 psi pressure–121°C. Cool to 25°C.

Solution C:

Composition per liter:

Pyridoxine hydrochloride	300.0mg
Thiamine-HCl·$2H_2O$	200.0mg

Nicotinic acid 200.0mg
Vitamin B_{12} 100.0mg
Calcium pantothenate 100.0mg
p-Aminobenzoic acid 80.0mg
D(+)-Biotin 20.0mg

Preparation of Solution C: Add components to distilled/deionized water and bring volume to 1.0L. Sparge with 100% N_2. Mix thoroughly. Filter sterilize.

Solution D:

Composition per liter:

NaOH 0.5g
$Na_2WO_4 \cdot 2H_2O$ 4.0mg
$Na_2SeO_3 \cdot 5H_2O$ 3.0mg

Preparation of Solution D: Add components to distilled/deionized water and bring volume to 1.0L. Mix thoroughly. Sparge with 100% N_2. Filter sterilize.

Solution E:

Composition per 100.0mL:

$NaHCO_3$ 5.0g

Preparation of Solution E: Add $NaHCO_3$ to distilled/deionized water and bring volume to 100.0mL. Mix thoroughly. Sparge with 100% N_2 gas mixture. Autoclave for 15 min at 15 psi pressure–121°C. Cool to 25°C.

Solution F:

Composition per 10.0mL:

Trimethylamine hydrochloride 2.0g
Methanol 0.6g

Preparation of Solution F: Add components to distilled/deionized water and bring volume to 10.0mL. Mix thoroughly. Sparge with 100% N_2 gas mixture. Autoclave for 15 min at 15 psi pressure–121°C. Cool to 25°C.

Solution G:

Composition per 10.0mL:

$Na_2S \cdot 9H_2O$ 0.125g

Preparation of Solution G: Add $Na_2S \cdot 9H_2O$ to distilled/deionized water and bring volume to 10.0mL. Mix thoroughly. Autoclave under 100% N_2 for 15 min at 15 psi pressure–121°C. Cool to 25°C.

Solution H:

Composition per 10.0mL:

$FeSO_4 \cdot 6H_2O$ 0.02g

Preparation of Solution H: Add $FeSO_4 \cdot 6H_2O$ to distilled/deionized water and bring volume to 10.0mL. Mix thoroughly. Sparge with 100% N_2 gas mixture. Autoclave for 15 min at 15 psi pressure–121°C. Cool to 25°C.

Preparation of Medium: Prepare and dispense medium under 80% N_2 + 20% CO_2 gas atmosphere. Sequentially add 1.0mL solution B, 1.0mL solution C, 1.0mL solution D, 50.0mL solution E, 10.0mL solution F, and 10.0mL solution G to 940.0mL solution A. Distribute anaerobically under 80% N_2 + 20% CO_2 into appropriate vessels. The pH should be 6.5–7.0. After inoculation aseptically inject 0.1mL of solution H for each 10.0mL of medium.

Use: For the cultivation of *Methanomethylovorans hollandica* DSM 15978.

Modified Gorodkowa Agar

Composition per liter:

Agar 20.0g
Peptic digest of animal tissue 10.0g
NaCl 5.0g
Glucose 1.0g

pH 6.8 ± 0.2 at 25°C

Source: This medium is available as a premixed powder from HiMedia.

Preparation of Medium: Add components to distilled/deionized water and bring volume to 1.0L. Mix thoroughly. Gently heat and bring to boiling with frequent agitation. Distribute into tubes or flasks. Autoclave for 15 min at 15 psi pressure–121°C. Pour into sterile Petri dishes or leave in tubes.

Use: For promoting sporulation of yeasts.

Modified ISP5 Medium (DSMZ Medium 993b)

Composition per liter:

KCl 100.0
Agar 20.0g
Glycerol 10.0g
Yeast extract 5.0g
L-asparagine, anhydrous 1.0g
K_2HPO_4, anhydrous 1.0g
Trace elements solution 1.0mL

pH 7.2 ± 0.2 at 25°C

Trace Elements Solution:

Composition per 100.0mL:

$FeSO_4 \cdot 7H_2O$ 0.1g
$MnCl_2 \cdot 4H_2O$ 0.1g
$ZnSO_4 \cdot 7H_2O$ 0.1g

Preparation of Trace Elements Solution: Add components to distilled/deionized water and bring volume to 100.0mL. Mix thoroughly.

Preparation of Medium: Add components to distilled/deionized water and bring volume to 1.0L. Mix thoroughly. Adjust pH to 7.2. Distribute into tubes or flasks. Gently heat while stirring and bring to boiling. Mix thoroughly. Autoclave for 15 min at 15 psi pressure–121°C. Pour into Petri dishes or leave in tubes.

Use: For the cultivation of *Nesterenkonia xinjiangensis* and *Microbacterium halotolerans*.

Modified ISP5 Medium with Sodium Chloride (DSMZ Medium 993c)

Composition per liter:

NaCl 100.0g
Agar 20.0g
Glycerol 10.0g
L-asparagine, anhydrous 1.0g
K_2HPO_4, anhydrous 1.0g
Trace elements solution 1.0mL

pH 7.2 ± 0.2 at 25°C

Trace Elements Solution:

Composition per 100.0mL:

$FeSO_4 \cdot 7H_2O$ 0.1g
$MnCl_2 \cdot 4H_2O$ 0.1g
$ZnSO_4 \cdot 7H_2O$ 0.1g

Preparation of Trace Elements Solution: Add components to distilled/deionized water and bring volume to 100.0mL. Mix thoroughly.

Preparation of Medium: Add components to distilled/deionized water and bring volume to 1.0L. Mix thoroughly. Adjust pH to 7.2. Distribute into tubes or flasks. Gently heat while stirring and bring to boiling. Mix thoroughly. Autoclave for 15 min at 15 psi pressure–121°C. Pour into Petri dishes or leave in tubes.

Use: For the cultivation of *Nesterenkonia lutea*.

Modified Letheen HiVeg Agar (Letheen HiVeg Agar, Modified)

Composition per liter:

Agar	20.0g
Plant peptone	20.0g
Plant extract	5.0g
Plant hydrolysate	5.0g
NaCl	5.0g
Polysorbate 80	5.0g
Yeast extract	2.0g
Lecithin	0.7g
$NaHSO_3$	0.1g

pH 7.2 ± 0.2 at 25°C

Source: This medium is available as a premixed powder from HiMedia.

Preparation of Medium: Add components to distilled/deionized water and bring volume to 1.0L. Mix thoroughly. Gently heat and bring to boiling. Distribute into tubes or flasks. Autoclave for 15 min at 15 psi pressure–121°C. Pour into sterile Petri dishes.

Use: For the screening cosmetic products for microbial contamination.

Modified Letheen HiVeg Broth (Letheen HiVeg Broth, Modified)

Composition per liter:

Plant peptone	20.0g
Plant extract	5.0g
Plant hydrolysate	5.0g
NaCl	5.0g
Polysorbate 80	5.0g
Yeast extract	2.0g
Lecithin	0.7g
$NaHSO_3$	0.1g

pH 7.0 ± 0.2 at 25°C

Source: This medium is available as a premixed powder from HiMedia.

Preparation of Medium: Add components to distilled/deionized water and bring volume to 1.0L. Mix thoroughly. Gently heat and bring to boiling. Distribute into tubes or flasks. Autoclave for 15 min at 15 psi pressure–121°C.

Use: For the screening of cosmetic products for microbial contamination.

Modified McBride *Listeria* HiVeg Agar Base with Blood and Cycloheximide

Composition per liter:

Agar	15.0g
Glycine anhydride	10.0g
Plant hydrolysate	5.0g
Plant peptone	5.0g
NaCl	5.0g
Plant extract	3.0g
Phenyl ethanol	2.5g
LiCl	0.5g
Sheep blood, defibrinated	50.0mL
Selective supplement solution	10.0mL

pH 7.3 ± 0.2 at 25°C

Source: This medium, without blood or selective supplement solution, is available as a premixed powder from HiMedia.

Caution: LiCl is harmful. Avoid bodily contact and inhalation of vapors. On contact with skin wash with plenty of water immediately.

Caution: Cycloheximide is toxic. Avoid skin contact or aerosol formation and inhalation.

Selective Supplement Solution:
Composition per 10.0mL:

Cycloheximide	0.2g

Preparation of Selective Supplement Solution: Add cycloheximide to distilled/deionized water and bring volume to 10.0mL. Mix thoroughly. Filter sterilize.

Preparation of Medium: Add components, except blood and selective solution, to distilled/deionized water and bring volume to 940.0mL. Mix thoroughly. Gently heat and bring to boiling. Distribute into tubes or flasks. Autoclave for 15 min at 15 psi pressure–121°C. Cool to 50°C. Aseptically add 50.0mL defibrinated blood and 10.0mL selective supplement solution. Mix thoroughly. Pour into sterile Petri dishes or leave in tubes.

Use: For the selective isolation of *Listeria monocytogenes* from clinical and nonclinical specimens containing mixed flora.

Modified Marine Broth (Modified Medium 514) (DSMZ Medium 1173)

Composition per liter:

NaCl	39.45g
Peptone	10.0g
$MgCl_2$	8.8g
Yeast extract	5.0g
Na_2SO_3	3.24g
$CaCl_2$	1.8g
KCl	0.55g
$NaHCO_3$	0.16g
Ferric citrate	0.1g
KBr	0.08g
$SrCl_2$	0.03g
H_3BO_3	0.02g
Na_2HPO_4	8.0mg
Na_2SiO_3	4.0mg
NaF	2.4mg
NH_4NO_3	1.6mg

pH 7.6 ± 0.2 at 25°C

Preparation of Medium: Add components to distilled/deionized water and bring volume to 1.0L. Mix thoroughly. Gently heat while stirring and bring to boiling. Distribute into tubes or flasks. Autoclave for 15 min at 15 psi pressure–121°C.

Use: For the cultivation of *Shewanella atlantica*.

Modified MYP Agar Base

Composition per liter:

Agar	12.0g
Peptic digest of animal tissue	10.0g
D-Mannitol	10.0g
NaCl	10.0g
Meat extract	1.0g
Phenol Red	0.025g
Egg yolk emulsion	100.0mL
Selective supplement solution	10.0mL

pH 7.1 ± 0.2 at 25°C

Source: This medium is available from HiMedia.

Egg Yolk Emulsion
Composition per 100.0mL:

Chicken egg yolks	11
Whole chicken egg	1
NaCl (0.9% solution)	50.0mL

Preparation of Egg Yolk Emulsion: Soak eggs with 1:100 dilution of saturated mercuric chloride solution for 1 min. Crack eggs and separate yolks from whites. Mix egg yolks with 1 chicken egg. Beat to form emulsion. Measure 50.0mL of egg yolk emulsion and add to 50.0mL of 0.9% NaCl solution. Mix thoroughly. Filter sterilize. Warm to 45°–50°C.

Selective Supplement Solution:
Composition per 10.0mL:

Polymyxin B	1000U

Preparation of Selective Supplement Solution: Add polymyxin B to distilled/deionized water and bring volume to 10.0mL. Mix thoroughly. Filter sterilize.

Preparation of Medium: Add components, except egg yolk emulsion and selective supplement solution, to distilled/deionized water and bring volume to 890.0mL. Mix thoroughly. Autoclave for 15 min at 15 psi pressure–121°C. Cool to 50°C. Aseptically add egg yolk emulsion and selective supplement solution. Mix thoroughly. Pour into Petri dishes or aseptically distribute into sterile tubes.

Use: For the isolation and identification of *Bacillus* species and pathogenic staphylococci.

Modified MYP HiVeg Agar Base with Egg Yok and Polymyxin B

Composition per liter:

Agar	12.0g
D-Mannitol	10.0g
Plant peptone	10.0g
NaCl	10.0g
Plant extract No. 1	1.0g
Phenol red	0.025g
Egg yolk emulsion, 20%	10.0mL
Polymyxin B solution	1.0mL

pH 7.1 ± 0.2 at 25°C

Source: This medium, without egg yolk emulsion and polymyxin B solution, is available as a premixed powder from HiMedia.

Egg Yolk Emulsion, 20%:
Composition per 100.0mL:

Chicken egg yolks	11
Whole chicken egg	1
NaCl (0.9% solution)	80.0mL

Preparation of Egg Yolk Emulsion, 20%: Soak eggs with 1:100 dilution of saturated mercuric chloride solution for 1 min. Crack eggs and separate yolks from whites. Mix egg yolks with 1 chicken egg. Measure 20.0mL of egg yolk emulsion and add to 80.0mL of 0.9% NaCl solution. Mix thoroughly. Filter sterilize. Warm to 45°–50°C.

Polymyxin B Solution:
Composition per 1.0mL:

Polymyxin B	1.0mg

Preparation of Polymyxin B Solution: Add polymyxin B to distilled/deionized water and bring volume to 1.0mL. Mix thoroughly. Filter sterilize.

Preparation of Medium: Add components—except egg yolk emulsion, 20%, and polymyxin B solution—to distilled/deionized water and bring volume to 989.0mL. Mix thoroughly. Gently heat and bring to boiling. Autoclave for 15 min at 15 psi pressure–121°C. Cool to 45°–50°C. Aseptically add 10.0mL of sterile egg yolk emulsion, 20%, and 1.0mL of sterile polymyxin B solution. Mix thoroughly. Pour into sterile Petri dishes.

Use: For the cultivation and maintenance of *Bacillus cereus*. For the isolation and identification of *Bacillus* species and pathogenic staphylococci.

Modified Oxford *Listeria* Selective Agar (Oxford Agar, Modified) (*Listeria* Selective Agar, Modified Oxford) (MOX Agar) (BAM M103a)

Composition per 1002.0mL:

Special peptone	23.0g
LiCl	15.0g
Agar	12.0g
NaCl	5.0g
Cornstarch	1.0g
Esculin	1.0g
Ferric ammonium citrate	0.5g
Buffered colistin methane sulfonate solution	1.0mL
Buffered moxalactam solution	2.0mL

pH 7.2 ± 0.2 at 25°C

Buffered Colistin Methane Sulfonate Solution:
Composition per 100.0mL:

Colistin methane sulfonate	1.0g
Potassium phosphate buffer, 0.1*M*. pH 6.0	100.0mL

Preparation of Buffered Colistin Methane Sulfonate Solution: Add colistin methane sulfonate to 100.0mL 0.1*M* potassium phosphate buffer. Mix thoroughly. Adjust pH to 6.0. Filter sterilize. Store at –20°C.

Buffered Moxalactam Solution:
Composition per 100.0mL:

Sodium or ammonium moxalactam	1.0g

Preparation of Buffered Moxalactam Solution: Add sodium or ammonium moxalactam to distilled/deionized water and bring volume to 100.0mL. Mix thoroughly. Filter sterilize. Store at –20°C.

Preparation of Medium: Gradually add components, except buffered moxalactam solution, to distilled/deionized water and bring volume to 1.0L. Mix thoroughly. Adjust pH to 7.2. Gently heat and bring to boiling. Autoclave for 10 min at 15 psi pressure–121°C. Cool quickly to 46°C. Aseptically add 2.0mL of sterile moxalactam solution. Mix thoroughly. Pour into sterile Petri dishes or distribute into sterile tubes.

Use: For the isolation and cultivation of *Listeria monocytogenes* from specimens containing a mixed bacterial flora.

Modified PYNFH Medium

Composition per liter:

Peptone	10.0g
Yeast extract	10.0g
Yeast nucleic acid	1.0g
Folic acid	15.0mg
Hemin	1.0mg
Fetal bovine serum, heat inactivated	100.0mL
Buffer solution	20.0mL

pH 6.5 ± 0.2 at 25°C

Buffer Solution:

Composition per liter:

Na_2HPO_4	25.0g
KH_2PO_4	18.1g

Preparation of Buffer Solution: Add components to distilled/deionized water and bring volume to 1.0L. Mix thoroughly. Adjust pH to 6.5. Autoclave for 15 min at 15 psi pressure–121°C. Cool to 25°C.

Preparation of Medium: Add components, except buffer solution and fetal bovine serum, to distilled/deionized water and bring volume to 880.0mL. Mix thoroughly. Autoclave for 15 min at 15 psi pressure–121°C. Cool to 25°C. Aseptically add 20.0mL of sterile buffer solution and 100.0mL of sterile, heat-inactivated fetal bovine serum. Mix thoroughly. Aseptically distribute into sterile tubes or flasks.

Use: For the cultivation of *Dexiostoma campyla, Hartmannella vermiformis, Naegleria australiensis, Naegleria fowleri, Naegleria gruberi, Naegleria jadini, Phytomonas davidi, Tetrahymena* species, *Vahlkampfia avara,* and *Willaertia magna.*

Modified Rappaport Vassiliadis HiVeg Medium

Composition per liter:

$MgCl_2 \cdot 6H_2O$	40.0g
NaCl	8.0g
Papaic digest of soybean meal	5.0g
KH_2PO_4	1.6
Malachite Green	0.04g

pH 5.2 ± 0.2 at 25°C

Source: This medium is available as a premixed powder from HiMedia.

Preparation of Medium: Add components to distilled/deionized water and bring volume to 1.0L. Mix thoroughly. Gently heat and bring to boiling. Distribute into tubes or flasks. Autoclave for 15 min at 10 psi pressure–115°C.

Use: For the selective enrichment of *Salmonella* species from food and environmental specimens.

Modified Rogosa HiVeg Agar (M16 HiVeg Agar)

Composition per liter:

Agar	10.0g
Glucose	5.0g
Plant extract	5.0g
Plant hydrolysate No. 1	5.0g
Papaic digest of soybean meal	5.0g
Sodium acetate	3.0g
Yeast extract	2.5g
Ascorbic acid	0.5g

pH 7.2 ± 0.2 at 25°C

Source: This medium is available as a premixed powder from HiMedia.

Preparation of Medium: Add components to distilled/deionized water and bring volume to 1.0L. Mix thoroughly. Gently heat and bring to boiling. Adjust pH to 7.2 with 2*N* NaOH. Autoclave for 15 min at 15 psi pressure–121°C. Pour into sterile Petri dishes or distribute into sterile tubes.

Use: For the cultivation of lactobacilli from cheese. For the cultivation and enumeration of microorganisms encountered in the dairy industry.

Modified Salt Broth

See: **Salt Broth, Modified**

Modified Semisolid Rappaport Vassiliadis Medium (MSRV Medium)

Composition per liter:

$MgCl_2$, anhydrous	10.93g
NaCl	7.34g
Casein hydrolysate	4.59g
Tryptose	4.59g
Agar	2.7g
KH_2PO_4	1.47g
Malachite Green oxalate	.0.037g
Novobiocin solution	10.0mL

pH 5.2 ± 0.2 at 25°C

Source: This medium is available as a premixed powder from Oxoid Unipath.

Novobiocin Solution:

Composition per 10.0mL:

Novobiocin	0.02g

Preparation of Novobiocin Solution: Add novobiocin to 10.0mL of distilled/deionized water. Mix thoroughly. Filter sterilize.

Preparation of Medium: Add components, except novobiocin solution, to distilled/deionized water and bring volume to 990.0mL. Mix thoroughly. Gently heat to boiling. Do not autoclave. Cool to 45°–50°C. Aseptically add 10.0mL of sterile novobiocin solution. Mix thoroughly. Pour into sterile Petri dishes. Air-dry plates for at least 1 hr.

Use: For the isolation and cultivation of motile *Salmonella* species from food and environmental samples.

Modified Shieh Agar (LMG Medium 215)

Composition per liter:

Agar	15.0g
Peptone	5.0g
Yeast extract	1.0g
$MgSO_4 \cdot 7H_2O$	0.3g
K_2HPO_4	0.1g
KH_2PO_4	50.0mg
$NaHCO_3$	50.0mg
Na-acetate	10.0mg
$BaCl_2 \cdot H_2O$	10.0mg

$CaCl_2 \cdot 2H_2O$ .... 6.7 mg
$FeSO_4 \cdot 7H_2O$ 1.0mg

pH 7.3 ± 0.2 at 25°C

Preparation of Medium: Add components to distilled/deionized water and bring volume to 1.0L. Mix thoroughly. Gently heat and bring to boiling. Distribute into tubes or flasks. Autoclave for 15 min at 15 psi pressure–121°C. Pour into sterile Petri dishes or leave in tubes.

Use: For the cultivation and maintenance of *Flavobacterium* spp., *Flexibacter* spp., *Chitinophaga pinensis,* and *Flectobacillus major.*

Modified Skim Milk HiVeg Agar

Composition per liter:

Agar 15.0g
Plant hydrolysate 5.0g
Yeast extract 2.5g
Skim milk powder 1.0g
Glucose monohydrate 1.0g

pH 7.0 ± 0.1 at 25°C

Source: This medium is available as a premixed powder from HiMedia.

Preparation of Medium: Add components to distilled/deionized water and bring volume to 1.0L. Mix thoroughly. Gently heat and bring to boiling. Distribute into tubes or flasks. Autoclave for 15 min at 15 psi pressure–121°C. Pour into sterile Petri dishes or aseptically distribute into sterile tubes.

Use: For the cultivation and differentiation of bacteria based on proteolytic activity. For the cultivation and enumeration of lactic streptococci used in manufacturing of cheddar cheese.

Modified Soyabean Bile Broth Base

Composition per liter:

Casein enzymic hydrolysate 17.0g
NaCl 5.0g
K_2HPO_4 .... 4.0g
Papaic digest of soybean meal 3.0g
D-Glucose 2.5g
Bile salts mixture 1.5g
Selective supplement solution 10.0mL

pH 7.3 ± 0.2 at 25°C

Source: This medium is available from HiMedia.

Selective Supplement Solution:
Composition per 10.0mL:

Novobiocin 10.0mg

Preparation of Selective Supplement Solution: Add novobiocin to distilled/deionized water and bring volume to 10.0mL. Mix thoroughly. Filter sterilize.

Preparation of Medium: Add components, except selective supplement solution, to distilled/deionized water and bring volume to 990.0mL. Mix thoroughly. Autoclave for 15 min at 15 psi pressure–121°C. Cool to 50°C. Aseptically add selective supplement solution. Mix thoroughly. Pour into Petri dishes or aseptically distribute into sterile tubes.

Use: For the detection of *Escherichia coli* O157:H7 from food.

Modified Soybean HiVeg Agar

Composition per liter:

Plant hydrolysate 17.0g
Agar 15.0g
NaCl 5.0g
Papaic digest of soyabean meal 3.0g
K_2HPO_4 .... 2.5g
Glucose 2.5g
Polysorbate 80 .... 10.0mL

pH 7.3 ± 0.2 at 25°C

Source: This medium, without polysorbate 80, is available as a premixed powder from HiMedia.

Preparation of Medium: Add components to distilled/deionized water and bring volume to 1.0L. Mix thoroughly. Adjust pH to 7.3. Distribute into tubes or flasks. Autoclave for 15 min at 15 psi pressure–121°C.

Use: For the microbiological assay of polymyxin B, colistin sulfate, and sodium colistimethate.

Modified TH Agar (DSMZ Medium 1061)

Composition per liter:

Agar 15.0g
Tryptone 5.0g
NaCl 5.0g
Yeast extract 3.0g
$MnSO_4 \cdot 2H_2O$ 0.01g
Trace elements solution SL-6 .... 1.0mL

pH 7.5 ± 0.2 at 25°C

Trace Elements Solution SL-6:
Composition per liter:

$MnCl_2 \cdot 4H_2O$ 0.5g
H_3BO_3 .... 0.3g
$CoCl_2 \cdot 6H_2O$ 0.2g
$ZnSO_4 \cdot 7H_2O$ 0.1g
$Na_2MoO_4 \cdot 2H_2O$ 0.03g
$NiCl_2 \cdot 6H_2O$ 0.02g
$CuCl_2 \cdot 2H_2O$ 0.01g

Preparation of Trace Elements Solution SL-6: Add components to distilled/deionized water and bring volume to 1.0L. Mix thoroughly. Autoclave for 15 min at 15 psi pressure–121°C.

Preparation of Medium: Add components to distilled/deionized water and bring volume to 1.0L. Mix thoroughly. Adjust pH to 7.5. Gently heat while stirring and bring to boiling. Distribute into tubes or flasks. Autoclave for 15 min at 15 psi pressure–121°C. Pour into Petri dishes or leave in tubes.

Use: For the cultivation of *Anoxybacillus voinovskiensis.*

Modified Thayer-Martin Agar
See: **Thayer-Martin Agar, Modified**

Modified *Thermus* Medium (DSMZ Medium 630)

Composition per liter:

Agar 28.0g
Yeast extract 2.5g
Tryptone 2.5g
$MgCl_2 \cdot 6H_2O$ 200.0mg
Nitrilotriacetic acid 100.0mg
$CaSO_4 \cdot 2H_2O$ 40.0mg
Phosphate solution 100.0mL

Ferric citrate solution......0.5mL
Trace elements solution......0.5mL

pH 7.2 ± 0.2 at 25°C

Phosphate Solution:
Composition per liter:
$Na_2HPO_4 \cdot 12H_2O$......43.0g
KH_2PO_4......5.44g

Preparation of Phosphate Solution: Add components to distilled/deionized water and bring volume to 1.0L. Mix thoroughly. Adjust pH to 7.2. Autoclave for 15 min at 15 psi pressure–121°C. Cool to 50°C.

Ferric Citrate Solution:
Composition per liter:
Ferric citrate......2.5g

Preparation of Ferric Citrate Solution: Add ferric citrate to distilled/deionized water and bring volume to 1.0L. Mix thoroughly.

Trace Elements Solution:
Composition per liter:
Nitrilotriacetic acid......12.8g
$FeCl_2 \cdot 4H_2O$......1.0g
$MnCl_2 \cdot 4H_2O$......0.5g
$CoCl_2 \cdot 4H_2O$......0.3g
$Na_2MoO_4 \cdot 4H_2O$......50.0mg
$CuCl_2 \cdot 2H_2O$......50.0mg
$NiCl_2 \cdot 6H_2O$......20.0mg
H_3BO_3......20.0mg

Preparation of Trace Elements Solution: Add nitrilotriacetic acid to 500.0mL of distilled/deionized water. Dissolve by adjusting pH to 7.0 with KOH. Add remaining components. Add distilled/deionized water to 1.0L. Mix thoroughly. Adjust pH to 6.8.

Preparation of Medium: Add components, except phosphate solution, to distilled/deionized water and bring volume to 900.0mL. Mix thoroughly. Autoclave for 15 min at 15 psi pressure–121°C. Cool to 50°C. Aseptically add 100.0mL phosphate solution. Mix thoroughly. Adjust pH to 7.2. Pour into Petri dishes or aseptically distribute into sterile tubes.

Use: For the cultivation and maintenance of *Thermus* sp., *Rhodothermus marinus,* and *Albidovulum inexpectatum.*

Modified V.P. HiVeg Broth

Composition per liter:
Plant peptone No. 3......7.0g
Glucose......5.0g
NaCl......5.0g

pH 6.9 ± 0.2 at 25°C

Source: This medium is available as a premixed powder from HiMedia.

Preparation of Medium: Add components to distilled/deionized water and bring volume to 1.0L. Mix thoroughly. Adjust pH to 6.9. Distribute into tubes. Autoclave for 15 min at 15 psi pressure–121°C.

Use: For the cultivation and differentiation of bacteria based on their ability to produce acetoin.

Modified VWM Medium (DSMZ Medium 503a)

Composition per 1013.4mL:
Solution A......940.0mL
Solution E......50.0mL
Solution F......10.0mL
Solution G......10.0mL
Solution B......1.0mL
Solution C......1.0mL
Solution D......1.0mL
Solution H......0.4mL

pH 7.3 ± 0.2 at 25°C

Solution A:
Composition per 940.0mL:
NaCl......1.0g
KCl......0.5g
$MgCl_2 \cdot 6H_2O$......0.4g
NH_4Cl......0.25g
KH_2PO_4......0.2g
$CaCl_2 \cdot 2H_2O$......0.15g
Resazurin......0.5mg

Preparation of Solution A: Prepare under 80% N_2 + 20% CO_2 gas atmosphere. Add components to distilled/deionized water and bring volume to 940.0mL. Mix thoroughly. Adjust pH to 7.2. Sparge with 80% N_2 + 20% CO_2. Autoclave for 15 min at 15 psi pressure–121°C. Cool to 25°C.

Solution B:
Composition per liter:
$FeCl_2 \cdot 4H_2O$......1.5g
$CoCl_2 \cdot 6H_2O$......190.0mg
$MnCl_2 \cdot 4H_2O$......100.0mg
$ZnCl_2$......70.0mg
$Na_2MoO_4 \cdot 2H_2O$......36.0mg
$NiCl_2 \cdot 6H_2O$......24.0mg
H_3BO_3......6.0mg
$CuCl_2 \cdot 2H_2O$......2.0mg
HCl (25% solution)......10.0mL

Preparation of Solution B: Add $FeCl_2 \cdot 4H_2O$ to 10.0mL of HCl solution. Mix thoroughly. Add distilled/deionized water and bring volume to 1.0L. Add remaining components. Mix thoroughly. Sparge with 80% N_2 + 20% CO_2. Autoclave for 15 min at 15 psi pressure–121°C. Cool to 25°C.

Solution C:
Composition per liter:
Pyridoxine hydrochloride......300.0mg
Thiamine-HCl·$2H_2O$......200.0mg
Nicotinic acid......200.0mg
Vitamin B_{12}......100.0mg
Calcium pantothenate......100.0mg
p-Aminobenzoic acid......80.0mg
D(+)-Biotin......20.0mg

Preparation of Solution C: Add components to distilled/deionized water and bring volume to 1.0L. Sparge with 100% N_2. Mix thoroughly. Filter sterilize.

Solution D:
Composition per liter:
NaOH......0.5g
$Na_2WO_4 \cdot 2H_2O$......4.0mg
$Na_2SeO_3 \cdot 5H_2O$......3.0mg

Preparation of Solution D: Add components to distilled/deionized water and bring volume to 1.0L. Mix thoroughly. Sparge with 100% N_2. Filter sterilize.

Solution E:

Composition per 100.0mL:

$NaHCO_3$ 5.0g

Preparation of Solution E: Add $NaHCO_3$ to distilled/deionized water and bring volume to 100.0mL. Mix thoroughly. Sparge with 100% N_2 gas mixture. Autoclave for 15 min at 15 psi pressure–121°C. Cool to 25°C.

Solution F:

Composition per 10.0mL:

Taurine 2.5g

Preparation of Solution F: Add taurine to distilled/deionized water and bring volume to 100.0mL. Mix thoroughly. Sparge with 100% N_2 gas mixture. Autoclave for 15 min at 15 psi pressure–121°C. Cool to 25°C.

Solution G:

Composition per 10.0mL:

$Na_2S{\cdot}9H_2O$ 0.125g

Preparation of Solution G: Add $Na_2S{\cdot}9H_2O$ to distilled/deionized water and bring volume to 10.0mL. Mix thoroughly. Autoclave under 100% N_2 for 15 min at 15 psi pressure–121°C. Cool to 25°C.

Solution H:

Composition per 10.0mL:

Na-dithionite 0.5g

Preparation of Solution H: Add Na-dithionite to distilled/deionized water and bring volume to 10.0mL. Mix thoroughly. Autoclave under 100% N_2 for 15 min at 15 psi pressure–121°C. Cool to 25°C.

Preparation of Medium: Prepare and dispense medium under 80% N_2 + 20% CO_2 gas atmosphere. Sequentially add 1.0mL solution B, 1.0mL solution C, 1.0mL solution D, 50.0mL solution E, 10.0mL solution F, 10.0mL solution G, and 0.4mL solution H to 940.0mL solution A. Distribute anaerobically under 80% N_2 + 20% CO_2 into appropriate vessels. The pH should be 7.2–7.4.

Use: For the cultivation of *Desulfonispora thiosulfatigenes*.

Moeller Decarboxylase Broth

Composition per liter:

Amino acid 10.0g
Peptic digest of animal tissue 5.0g
Beef extract 5.0g
Glucose 0.5g
Bromcresol Purple 0.01g
Cresol Red 5.0mg
Pyridoxal 5.0mg

pH 6.0 ± 0.2 at 25°C

Source: This medium is available as a premixed powder from BD Diagnostic Systems.

Preparation of Medium: Add components to distilled/deionized water and bring volume to 1.0L. Use L-lysine, L-arginine, or L-ornithine. Mix thoroughly. Gently heat until dissolved. Distribute into screw-capped tubes in 5.0mL volumes. Autoclave for 15 min at 15 psi pressure–121°C. A slight precipitate may form in the ornithine broth.

Use: For the differentiation of Gram-negative enteric bacteria based on the production of arginine dihydrolase, lysine decarboxylase, or ornithine decarboxylase.

Moeller Decarboxylase HiVeg Broth Base (Decarboxylase Broth Base, Moeller)

Composition per liter:

Plant extract 5.0g
Plant peptone 5.0g
Glucose 0.5g
Bromcresol Purple 0.01g
Pyridoxal 5.0mg
Cresol Red 5.0mg

pH 6.0 ± 0.2 at 25°C

Source: This medium is available as a premixed powder from HiMedia.

Preparation of Medium: Add components to distilled/deionized water and bring volume to 1.0L. Mix thoroughly. Gently heat until dissolved. Distribute into screw-capped tubes in 5.0mL volumes. Autoclave for 15 min at 15 psi pressure–121°C.

Use: With the addition of amino acid solutions, this medium is used for the differentiation of Gram-negative enteric bacteria based on the production of amino acid decarboxylation reactions.

Moeller Decarboxylase HiVeg Broth Arginine HCl (Decarboxylase Broth Base, Moeller with Arginine)

Composition per liter:

L-Arginine hydrochloride 10.0g
Plant extract 5.0g
Plant peptone 5.0g
Glucose 0.5g
Bromcresol Purple 0.01g
Pyridoxal 5.0mg
Cresol Red 5.0mg

pH 6.0 ± 0.2 at 25°C

Source: This medium is available as a premixed powder from HiMedia.

Preparation of Medium: Add components to distilled/deionized water and bring volume to 1.0L. Mix thoroughly. Gently heat until dissolved. Distribute into screw-capped tubes in 5.0mL volumes. Autoclave for 15 min at 15 psi pressure–121°C.

Use: For the differentiation of Gram-negative enteric bacteria based on the production of arginine dihydrolase.

Moeller Decarboxylase HiVeg Broth with Lysine HCl (Decarboxylase Broth Base, Moeller with Lysine)

Composition per liter:

L-Lysine hydrochloride 10.0g
Plant extract 5.0g
Plant peptone 5.0g
Glucose 0.5g
Bromcresol Purple 0.01g
Pyridoxal 5.0mg
Cresol Red 5.0mg

pH 6.0 ± 0.2 at 25°C

Source: This medium is available as a premixed powder from HiMedia.

Preparation of Medium: Add components to distilled/deionized water and bring volume to 1.0L. Mix thoroughly. Gently heat until dissolved. Distribute into screw-capped tubes in 5.0mL volumes. Autoclave for 15 min at 15 psi pressure–121°C.

Use: For the differentiation of Gram-negative enteric bacteria based on the production of lysine decarboxylase.

Moeller Decarboxylase HiVeg Broth with Ornithine HCl

Composition per liter:

L-Ornithine hydrochloride	10.0g
Plant extract	5.0g
Plant peptone	5.0g
Glucose	0.5g
Bromcresol Purple	0.01g
Pyridoxal	5.0mg
Cresol Red	5.0mg

pH 6.0 ± 0.2 at 25°C

Source: This medium is available as a premixed powder from HiMedia.

Preparation of Medium: Add components to distilled/deionized water and bring volume to 1.0L. Mix thoroughly. Gently heat until dissolved. Distribute into screw-capped tubes in 5.0mL volumes. Autoclave for 15 min at 15 psi pressure–121°C.

Use: For the differentiation of Gram-negative enteric bacteria based on the production of ornithine decarboxylase.

Moeller KCN Broth Base

Composition per liter:

Na_2HPO_4	5.64g
NaCl	5.0g
Pancreatic digest of casein	1.5g
Peptic digest of animal tissue	1.5g
KH_2PO_4	0.225g
KCN solution	0.15mL

pH 7.6 ± 0.2 at 25°C

Source: This medium is available as a premixed powder from BD Diagnostic Systems.

KCN Solution:

Composition per 100.0mL:

KCN	0.5g

Preparation of KCN Solution: Add KCN to 100.0mL of cold distilled/deionized water. Mix thoroughly and cap. Do not mouth pipette.

Caution: Cyanide is toxic.

Preparation of Medium: Add components, except KCN solution, to distilled/deionized water and bring volume to 1.0L. Mix thoroughly. Autoclave for 15 min at 15 psi pressure–121°C. Cool to room temperature. Prior to use, add 0.15mL of KCN solution. Mix thoroughly. Aseptically distribute into sterile tubes.

Use: For the differentiation of Gram-negative enteric bacteria on the basis of their ability to grow in the presence of cyanide.

MOF HiVeg Medium with Carbohydrate

Composition per liter:

NaCl	9.7g
$MnCl_2$	4.4g
Agar	3.0g
Na_2SO_4	1.6g
Plant hydrolysate	1.0g
$CaCl_2$	0.9g
$(NH_4)_2SO_4$	0.5g
Tris hydroxymethyl aminomethane	0.5g
KCl	0.275g
Yeast extract	0.1g
$NaHCO_3$	0.08g
KBr	0.04g
$SrCl_2$	0.017g
H_3BO_3	0.011g
Phenol red	0.01g
Na_2HPO_4	4.0mg
Sodium silicate	2.0mg
NaFl	1.2mg
NH_4NO_3	0.8mg
Carbohydrate solution	100.0mL

pH 7.0 ± 0.2 at 25°C

Source: This medium is available as a premixed powder from HiMedia.

Carbohydrate Solution:

Composition per 100.0mL:

Carbohydrate	10.0g

Preparation of Carbohydrate Solution: Add carbohydrate to distilled/deionized water and bring volume to 100.0mL. Adonitol, arabinose, cellobiose, glucose, dulcitol, fructose, galactose, inositol, lactose, maltose, mannitol, raffinose, rhamnose, salicin, sorbitol, sucrose, trehalose, xylose, or other carbohydrates may be used. Mix thoroughly. Filter sterilize.

Preparation of Medium: Add components, except carbohydrate solution, to distilled/deionized water and bring volume to 900.0mL. Mix thoroughly. Gently heat until dissolved. Autoclave for 15 min at 15 psi pressure–121°C. Cool to 50°C. Aseptically add 100.0mL sterile carbohydrate soltuion. Mix thoroughly. Aseptically distribute into screw-capped tubes in 5.0mL volumes.

Use: For the differentiation of marine bacteria on the basis of fermentative and oxidative metabolism of carbohydrates.

Molybdate Agar

Composition per101.5mL:

Base	100.0mL
Phosphomolybdic acid solution	1.5mL

pH 5.3 ± 0.2 at 25°C

Base:

Composition per liter:

Sucrose	40.0g
Agar	15.0g
Meat peptone	10.0g

Preparation of Base: Add components to distilled/deionized water and bring volume to 1.0L. Mix thoroughly. Adjust pH to 7.6. Gently heat and bring to boiling. Autoclave for 15 min at 15 psi pressure–121°C. Cool to 45°–50°C.

Phosphomolybdic Acid Solution:

Composition per 100.0mL:

$P_2O_5 \cdot 20MoO_3$	12.5g

Preparation of Base: Add $P_2O_5 \cdot 20MoO_3$ (phospho-12-molybdic acid, 12–molybdophosphoric acid, or PMA) to sterile distilled/deionized water. Mix thoroughly. Do not adjust pH.

Preparation of Medium: To 100.0mL of cooled sterile base, add 1.5mL of phosphomolybdic acid solution. Mix thoroughly. Pour into sterile Petri dishes or distribute into sterile tubes.

Use: For the isolation and presumptive identification of yeast, especially *Candida* species. *Candida albicans* appears as smooth, medium olive colonies with medium olive bottoms. *Candida stellatoidea* appears as shiny, light gray colonies with light gray bottoms. *Candida tropicalis* appears as smooth, shiny, dark blue/gray colonies with dark blue/gray bottoms. *Candida krusei* appears as smooth, dull white colonies with white bottoms. *Saccharomyces cerevisiae* appears as smooth, shiny light blue/dark blue colonies with dark blue/green bottoms.

Monsur Agar
(Taurocholate Tellurite Gelatin Agar)

Composition per liter:

Gelatin	30.0g
Agar	15.0g
Casein peptone	10.0g
NaCl	10.0g
Sodium taurocholate	5.0g
$Na_2CO_3 \cdot H_2O$	1.0g
K_2TeO_3 solution	10.0mL

pH 8.5 ± 0.2 at 25°C

K_2TeO_3 Solution:

Composition per 10.0mL:

K_2TeO_3	0.02g

Preparation of K_2TeO_3 Solution: Add K_2TeO_3 to 10.0mL of distilled/deionized water. Mix thoroughly. Filter sterilize.

Caution: Potassium tellurite is toxic.

Preparation of Medium: Add components, except K_2TeO_3 solution, to distilled/deionized water and bring volume to 990.0mL. Mix thoroughly. Gently heat and bring to boiling. Autoclave for 15 min at 15 psi pressure–121°C. Cool to 45°–50°C. Add 10.0mL of sterile K_2TeO_3 solution. Mix thoroughly. Pour into sterile Petri dishes or distribute into sterile tubes.

Use: For the isolation of *Vibrio cholerae* from fecal specimens.

Moorella glycerini Medium
(DSMZ Medium 793)

Composition per liter:

$NaHCO_3$	10.0g
Yeast extract	0.5g
KH_2PO_4	0.33g
NH_4Cl	0.33g
KCl	0.33g
$MgCl_2 \cdot 6H_2O$	0.33g
$CaCl_2 \cdot 2H_2O$	0.33g
Resazurin	0.5mg
Vitamin solution	10.0mL
$Na_2S \cdot 9H_2O$ solution	10.0mL
Glycerol, 87%	3.0mL
Trace elements solution SL-10	1.0mL
Selenite-tungstate solution	1.0mL

pH 6.7± 0.2 at 25°C

$Na_2S \cdot 9H_2O$ Solution:

Composition per 10.0mL:

$Na_2S \cdot 9H_2O$	0.5g

Preparation of $Na_2S \cdot 9H_2O$ Solution: Add $Na_2S \cdot 9H_2O$ to distilled/deionized water and bring volume to 10.0mL. Mix thoroughly. Sparge with 100% N_2. Autoclave for 15 min at 15 psi pressure–121°C.

Selenite-Tungstate Solution:

Composition per liter:

NaOH	0.5g
$Na_2WO_4 \cdot 2H_2O$	4.0mg
$Na_2SeO_3 \cdot 5H_2O$	3.0mg

Preparation of Selenite-Tungstate Solution: Add components to distilled/deionized water and bring volume to 1.0L. Mix thoroughly. Sparge with 100% N_2. Filter sterilize.

Vitamin Solution:

Composition per liter:

Pyridoxine-HCl	10.0mg
Thiamine-HCl·$2H_2O$	5.0mg
Riboflavin	5.0mg
Nicotinic acid	5.0mg
D-Ca-pantothenate	5.0mg
p-Aminobenzoic acid	5.0mg
Lipoic acid	5.0mg
Biotin	2.0mg
Folic acid	2.0mg
Vitamin B_{12}	0.1mg

Preparation of Vitamin Solution: Add components to distilled/deionized water and bring volume to 1.0L. Mix thoroughly. Sparge with 80% H_2 + 20% CO_2. Filter sterilize.

Trace Elements Solution SL-10:

Composition per liter:

$FeCl_2 \cdot 4H_2O$	1.5g
$CoCl_2 \cdot 6H_2O$	190.0mg
$MnCl_2 \cdot 4H_2O$	100.0mg
$ZnCl_2$	70.0mg
$Na_2MoO_4 \cdot 2H_2O$	36.0mg
$NiCl_2 \cdot 6H_2O$	24.0mg
H_3BO_3	6.0mg
$CuCl_2 \cdot 2H_2O$	2.0mg
HCl (25% solution)	10.0mL

Preparation of Trace Elements Solution SL-10: Add $FeCl_2 \cdot 4H_2O$ to 10.0mL of HCl solution. Mix thoroughly. Add distilled/deionized water and bring volume to 1.0L. Add remaining components. Mix thoroughly. Sparge with 100% N_2. Autoclave for 15 min at 15 psi pressure–121°C.

Preparation of Medium: Prepare and dispense medium under 100% CO_2 gas atmosphere. Add components, except $NaHCO_3$, $Na_2S \cdot 9H_2O$ solution, and vitamin solution, to distilled/deionized water and bring volume to 980.0mL. Mix thoroughly. Gently heat and bring to boiling. Boil for 5 min. Cool to room temperature while sparging with 100% CO_2. Add solid $NaHCO_3$. Adjust pH to 6.7 by adding NaOH and equilibrating with 100% CO_2. Distribute into tubes or bottles. Autoclave for 20 min at 15 psi pressure–121°C. Aseptically and anaerobically add 10.0mL $Na_2S \cdot 9H_2O$ solution and vitamin solution, 0.1mL per 10mL medium. Mix thoroughly. Aseptically and anaerobically distribute into sterile tubes or bottles.

Use: For the cultivation of *Moorella glycerini.*

Moraxella Medium (LMG Medium 204)

Composition per liter:

Special peptone	23.0g
Agar	15.0g
Glucose	5.0g
NaCl	5.0g
Soluble starch	1.0g
Cysteine hydrochloride	0.3g
Sheep blood, sterile defibrinated	50.0mL

pH 7.1 ± 0.2 at 25°C

Source: Special peptone is available from Oxoid Unipath.

Preparation of Medium: Add components, except sheep blood, to 950.0mL distilled/deionized water and bring volume to 1.0L. Mix thoroughly. Gently heat and bring to boiling. Autoclave for 15 min at 15 psi pressure–121°C. Cool to 45°–50°C. Aseptically add 50.0mL sterile sheep blood. Mix thoroughly. Pour into sterile Petri dishes or distribute into sterile tubes.

Use: For the cultivation and maintenance of *Moraxella osloensis, Moraxella atlantae,* and *Cellulomonas hominis.*

Motility GI Medium

Composition per liter:

Gelatin	53.4g
Heart infusion broth	25.0g
Agar	3.0g

pH 7.2 ± 0.2 at 25°C

Source: This medium is available as a premixed powder from BD Diagnostic Systems.

Preparation of Medium: Add components to cold distilled/deionized water and bring to 1.0L. Mix thoroughly. Gently heat and bring to boiling. Distribute into tubes or flasks. Autoclave for 15 min at 15 psi pressure–121°C. Pour into sterile Petri dishes in 20.0mL volumes or leave in tubes.

Use: For demonstrating the motility of microorganisms and for separating organisms in their motile phase.

Motility Indole Lysine Medium
See: **MIL Medium**

Motility-Indole-Lysine HiVeg Medium (MIL HiVeg Medium)

Composition per liter:

Plant hydrolysate	10.0g
Plant peptone	10.0g
L-Lysine hydrochloride	10.0g
Yeast extract	3.0g
Agar	2.0g
Glucose	1.0g
Ferric ammonium citrate	0.5g
Bromcresol Purple	0.02g

pH 6.6 ± 0.2 at 25°C

Source: This medium is available as a premixed powder from HiMedia.

Preparation of Medium: Add components to distilled/deionized water and bring volume to 1.0L. Mix thoroughly. Gently heat and bring to boiling. Distribute into tubes in 5.0mL volumes. Autoclave for 15 min at 15 psi pressure–121°C.

Use: For the cultivation and differentiation of members of the Enterobacteriaceae on the basis of motility, lysine decarboxylase activity, lysine deaminase activity, and indole production.

Motility Indole Ornithine Medium (MIO Medium)

Composition per liter:

Pancreatic digest of gelatin	10.0g
Pancreatic digest of casein	9.5g
L-Ornithine·HCl	5.0g
Yeast extract	3.0g
Agar	2.0g
Glucose	1.5g
Bromcresol Purple	0.02g

pH 6.6 ± 0.2 at 25°C

Source: This medium is available as a premixed powder from BD Diagnostic Systems.

Preparation of Medium: Add components to distilled/deionized water and bring to 1.0L. Mix thoroughly. Gently heat and bring to boiling. Distribute into tubes or flasks. Autoclave for 15 min at 15 psi pressure–121°C.

Use: For the differentiation of Gram-negative enteric bacteria based on their motility, indole production, and ornithine decarboxylase activity.

Motility Indole Ornithine Medium (MIO Medium) (BAM M99)

Composition per liter:

Tryptone	10.0g
Peptone	10.0g
L-Ornithine·HCl	5.0g
Yeast extract	3.0g
Agar	2.0g
Glucose	1.0g
Bromcresol Purple	0.02g

pH 6.6 ± 0.2 at 25°C

Source: This medium is available as a premixed powder from BD Diagnostic Systems.

Preparation of Medium: Add components to distilled/deionized water and bring to 1.0L. Mix thoroughly. Gently heat and bring to boiling. Distribute into tubes or flasks. Autoclave for 15 min at 15 psi pressure–121°C.

Use: For the differentiation of Gram-negative enteric bacteria based on their motility, indole production, and ornithine decarboxylase activity.

Motility Medium

Composition per liter:

Pancreatic digest of casein	10.0g
Glucose	5.0g
Agar	3.0g
Na_2HPO_4	2.5g
Yeast extract	2.5g

pH 7.4 ± 0.2 at 25°C

Preparation of Medium: Add components to distilled/deionized water and bring volume to 1.0L. Mix thoroughly. Gently heat and bring

to boiling. Distribute into tubes in 2.0mL volumes. Autoclave for 15 min at 15 psi pressure–121°C. Allow tubes to stand at 25°C for 2 days prior to inoculation.

Use: For the cultivation and observation of motility of *Bacillus cereus*.

Motility Medium S

Composition per liter:

Beef heart, solids from infusion	500.0g
Gelatin	30.0g
Enzymatic hydrolyzate of protein	10.0g
NaCl	5.0g
K_2HPO_4	2.0g
KNO_3	2.0g
Agar	1.0g
2,3,5-Triphenyltetrazolium chloride solution	10.0mL

pH 7.2 ± 0.2 at 25°C

2,3,5-Triphenyltetrazolium Chloride Solution:

Composition per 10.0mL:

2,3,5-Triphenyltetrazolium chloride	0.1g

Preparation of 2,3,5-Triphenyltetrazolium Chloride Solution: Add 2,3,5-triphenyltetrazolium chloride to distilled/deionized water and bring volume to 10.0mL. Mix thoroughly. Filter sterilize.

Preparation of Medium: Add components, except 2,3,5-triphenyltetrazolium chloride solution, to distilled/deionized water and bring volume to 990.0mL. Mix thoroughly. Gently heat while stirring and bring to boiling. Autoclave for 15 min at 15 psi pressure–121°C. Cool to 60°C. Aseptically add 10.0mL of the sterile 2,3,5-triphenyltetrazolium chloride solution. Mix thoroughly. Aseptically distribute into sterile tubes. Keep at 4°–8°C until used.

Use: For the determination of bacterial motility.

Motility Nitrate Agar

Composition per liter:

Beef heart, solids from infusion	100.0g
Tryptose	12.0g
Agar	3.0g
NaCl	1.0g
KNO_3	1.0g
Glucose	0.5g

pH 7.4 ± 0.2 at 25°C

Preparation of Medium: Add components to distilled/deionized water and bring volume to 1.0L. Mix thoroughly. Gently heat and bring to boiling. Distribute into tubes in 4.0mL volumes. Autoclave for 15 min at 15 psi pressure–121°C.

Use: For the cultivation and observation of motility and nitrate reduction in a variety of Gram-negative bacteria.

Motility Nitrate HiVeg Medium, Buffered

Composition per liter:

Galactose	5.0g
Plant extract	3.0g
Plant peptone	5.0g
KNO_3	5.0g
Agar	3.0g
Na_2HPO_4	2.5g
Glycerol	5.0mL

pH 7.4 ± 0.2 at 25°C

Source: This medium, without glycerol, is available as a premixed powder from HiMedia.

Preparation of Medium: Add glycerol followed by other components to distilled/deionized water and bring volume to 1.0L. Mix thoroughly. Gently heat and bring to boiling. Distribute into tubes in 4.0mL volumes. Autoclave for 15 min at 15 psi pressure–121°C.

Use: For the cultivation and observation of motility and nitrate reduction in a variety of Gram-negative bacteria.

Motility Nitrate Medium (FDA M101)

Composition per liter:

Beef heart, solids from infusion	100.0g
Tryptose	12.0g
Agar	3.0g
NaCl	1.0g
KNO_3, nitrite free	1.0g

pH 7.4 ± 0.2 at 25°C

Preparation of Medium: Add components to distilled/deionized water and bring volume to 1.0L. Mix thoroughly. Gently heat and bring to boiling. Distribute into screw-capped tubes in 4.0mL volumes. Autoclave for 15 min at 15 psi pressure–121°C.

Use: For the cultivation and differentiation of Gram-negative nonfermentative bacteria from cosmetics based on their motility and their ability to reduce nitrate to nitrite.

Motility Nitrate Medium

Composition per liter:

Beef heart, solids from infusion	100.0g
Tryptose	12.0g
Agar	3.0g
NaCl	1.0g
KNO_3	1.0g

Preparation of Medium: Add components to distilled/deionized water and bring volume to 1.0L. Mix thoroughly. Gently heat and bring to boiling. Distribute into tubes in 4.0mL volumes. Autoclave for 15 min at 15 psi pressure–121°C.

Use: For the cultivation and observation of motility and nitrate reduction in a variety of Gram-negative nonfermentative bacteria isolated from cosmetics.

Motility Nitrate Medium (BAM M101)

Composition per liter:

Tryptose	10.0g
Agar	3.0g
Beef heart, infusion from 500g	2.0g
Tryptose	2.0g
NaCl	1.0g
KNO_3	1.0g
Glucose	0.5g

pH 7.2 ± 0.2 at 25°C

Source: This medium is available as a premixed powder from BD Diagnostic Systems.

Preparation of Medium: Add components to distilled/deionized water and bring volume to 1.0L. Mix thoroughly. Gently heat with agitation to dissolve agar. Distribute into screw-cap tubes. Autoclave for 15 min at 15 psi pressure–121°C.

Use: For the detection of motility in Gram-negative nonfermentative bacteria.

Motility Nitrate Medium, Buffered

Composition per liter:

Peptone	5.0g
Galactose	5.0g
Agar	3.0g
Beef extract	3.0g
Na_2HPO_4	2.5g
KNO_3	1.0g
Glycerin	5.0mL

pH 7.3 ± 0.1 at 25°C

Preparation of Medium: Add components, except agar, to distilled/deionized water and bring volume to 1.0L. Mix thoroughly. Gently heat until dissolved. Add agar. Gently heat until boiling. Distribute into tubes in 11.0mL volumes. Autoclave for 15 min at 15 psi pressure–121°C. If not used within 4 hr, heat tubes to 100°C for 10 min.

Use: For the cultivation and observation of the motility of *Clostridium perfringens*.

Motility Sulfide Medium

Composition per liter:

Gelatin	80.0g
Proteose peptone	10.0g
NaCl	5.0g
Agar	4.0g
Beef extract	3.0g
Sodium citrate	2.0g
L-Cystine	0.2g
Ferrous ammonium citrate	0.2g

pH 7.3 ± 0.2 at 25°C

Source: This medium is available as a premixed powder from BD Diagnostic Systems.

Preparation of Medium: Add components to distilled/deionized water and bring volume to 1.0L. Mix thoroughly. Gently heat while stirring and bring to boiling. Distribute into tubes in 4–5.0mL volumes. Autoclave for 15 min at 10 psi pressure–116°C.

Use: For the determination of bacterial motility and the ability of bacteria to produce H_2S from L-cystine. For the differentiation of Gram-negative bacteria of the Enterobacteriaceae.

Motility Test HiVeg Medium

Composition per liter:

Plant hydrolysate No. 1	10.0g
Agar	5.0g
NaCl	5.0g

pH 7.4 ± 0.2 at 25°C

Source: This medium is available as a premixed powder from HiMedia.

Preparation of Medium: Add components to distilled/deionized water and bring volume to 1.0L. Mix thoroughly. Gently heat and bring to boiling. Distribute into tubes in 4.0mL volumes. Autoclave for 15 min at 15 psi pressure–121°C.

Use: For the detection of bacterial motility.

Motility Test HiVeg Medium (Edwards and Ewing HiVeg Medium)

Composition per liter:

Plant peptone	10.0g
NaCl	5.0g
Agar	4.0g
Plant extract	3.0g

pH 7.2 ± 0.2 at 25°C

Source: This medium is available as a premixed powder from HiMedia.

Preparation of Medium: Add components to distilled/deionized water and bring volume to 1.0L. Mix thoroughly. Gently heat and bring to boiling. Distribute into tubes in 4.0mL volumes. Autoclave for 15 min at 15 psi pressure–121°C.

Use: For testing motility of enteric bacteria.

Motility Test HiVeg Medium with Triphenyltetrazolium Chloride (Edwards and Ewing HiVeg Medium)

Composition per liter:

Plant peptone	10.0g
NaCl	5.0g
Agar	4.0g
Plant extract	3.0g
Triphenyltetrazolium chloride solution	5.0mL

pH 7.2 ± 0.2 at 25°C

Source: This medium, without triphenyltetrazolium chloride solution, is available as a premixed powder from HiMedia.

Triphenyltetrazolium Choride Solution:

Composition per 5.0mL:

Triphenyltetrazolium chloride	0.1g

Preparation of Triphenyltetrazolium Choride Solution: Add triphenyltetrazolium chloride to distilled/deionized water and bring volume to 5.0mL. Mix thoroughly. Filter sterilize.

Preparation of Medium: Add components, except triphenyltetrazolium chloride solution, to distilled/deionized water and bring volume to 1.0L.Mix thoroughly. Gently heat and bring to boiling. Distribute into tubes or flasks. Autoclave for 15 min at 15 psi pressure–121°C. Cool to 45°–50°C. Aseptically add 5.0mL triphenyltetrazolium chloride solution. Mix thoroughly. Aseptically distribute into tubes.

Motility Test and Maintenance Medium

Composition per liter:

Peptone	10.0g
NaCl	5.0g
Agar	4.0g
Beef extract	3.0g
2,3,5-Triphenyltetrazolium chloride	0.05g

Preparation of Medium: Add components to distilled/deionized water and bring volume to 1.0L. Mix thoroughly. Distribute into screw-capped tubes in 8.0mL volumes. Autoclave for 15 min at 15 psi pressure–121°C.

Use: For the cultivation, maintenance, and observation of the motility of *Listeria monocytogenes*.

Motility Test and Maintenance Medium

Composition per liter:

Agar	9.0g
Tryptose	8.0g
NaCl	5.0g
Pancreatic digest of gelatin	2.5g
Beef extract	1.5g

pH 7.2 ± 0.1 at 25°C

Preparation of Medium: Add components to distilled/deionized water and bring volume to 1.0L. Mix thoroughly. Gently heat and bring to boiling. Distribute into tubes in 7.0mL volumes. Autoclave for 15 min at 15 psi pressure–121°C. Cool to 45°–50°C. Pass the cooled tubes into an anaerobic chamber containing 85% N_2 + 10% H_2 + 5% CO_2.

Use: For the cultivation, maintenance, and observation of the motility in a variety of anaerobic bacteria.

Motility Test and Maintenance Medium

Composition per liter:

Peptone	10.0g
NaCl	5.0g
Agar	4.0g
Beef extract	3.0g

pH 7.4 ± 0.1 at 25°C

Preparation of Medium: Add components to distilled/deionized water and bring volume to 1.0L. Mix thoroughly. Distribute into screw-capped tubes in 8.0mL volumes. Autoclave for 15 min at 15 psi pressure–121°C.

Use: For the cultivation, maintenance, and observation of motility in members of the Enterobacteriaceae.

Motility Test and Maintenance Medium, Gilardi

Composition per liter:

Pancreatic digest of casein	10.0g
NaCl	5.0g
Agar	3.0g
Yeast extract	3.0g

pH 7.2 ± 0.1 at 25°C

Preparation of Medium: Add components to distilled/deionized water and bring volume to 1.0L. Mix thoroughly. Distribute into screw-capped tubes in 3.5mL volumes. Autoclave for 15 min at 15 psi pressure–121°C.

Use: For the cultivation, maintenance, and observation of motility in nonfermenting Gram–negative bacteria.

Motility Test and Maintenance Medium, Tatum

Composition per liter:

Tryptose	8.0g
NaCl	5.0g
Agar	4.0g
Pancreatic digest of gelatin	2.5g
Beef extract	1.5g

pH 6.9 ± 0.2 at 25°C

Preparation of Medium: Add components to distilled/deionized water and bring volume to 1.0L. Mix thoroughly. Distribute into screw-capped tubes in 8.0mL volumes. Autoclave for 15 min at 15 psi pressure–121°C.

Use: For the cultivation, maintenance, and observation of motility in nonfermenting Gram–negative bacteria.

Motility Test Medium

Composition per liter:

Pancreatic digest of gelatin	10.0g
NaCl	5.0g
Agar	4.0g
Beef extract	3.0g
Triphenyltetrazolium chloride solution	5.0mL

pH 7.3 ± 0.2 at 25°C

Source: This medium is available as a premixed powder from BD Diagnostic Systems.

2,3,5-Triphenyltetrazolium Chloride Solution:
Composition per 10.0mL:

2,3,5-Triphenyltetrazolium chloride	0.1g

Preparation of 2,3,5-Triphenyltetrazolium Chloride Solution: Add 2,3,5-triphenyltetrazolium chloride to distilled/deionized water and bring volume to 10.0mL. Mix thoroughly. Filter sterilize.

Preparation of Medium: Add components to distilled/deionized water and bring volume to 995.0mL. Mix thoroughly. Gently heat while stirring and bring to boiling. Autoclave for 15 min at 15 psi pressure–121°C. Cool to 45°–50°C. Aseptically add 5.0mL of sterile 2,3,5-triphenyltetrazolium chloride solution. Mix thoroughly. Aseptically distribute into sterile tubes.

Use: For detection of the motility of Gram-negative enteric bacteria.

Motility Test Medium

Composition per liter:

Tryptose	10.0g
NaCl	5.0g
Agar	5.0g

pH 7.2 ± 0.2 at 25°C

Source: This medium is available as a premixed powder from BD Diagnostic Systems.

Preparation of Medium: Add components to distilled/deionized water and bring volume to 1.0L. Mix thoroughly. Gently heat while stirring and bring to boiling. Distribute into tubes in 4–5.0mL volumes. Autoclave for 15 min at 15 psi pressure–121°C. Cool tubes quickly in an upright position.

Use: For the determination of bacterial motility.

Motility Test Medium, Semisolid

Composition per liter:

Peptone	10.0g
NaCl	5.0g
Agar	4.0g
Beef extract	3.0g

pH 7.4 ± 0.2 at 25°C

Preparation of Medium: Add components to distilled/deionized water and bring volume to 1.0L. Mix thoroughly. Gently heat while stirring and bring to boiling. Distribute into screw-capped tubes in 8.0mL or 20.0mL volumes. Autoclave for 15 min at 15 psi pressure–121°C. Pour into sterile Petri dishes in 20.0mL volumes or leave in tubes.

Use: For the cultivation and observation of motility in a variety of bacteria, especially *Salmonella* species.

Motility Test Medium, Semisolid with Sodium Chloride (BAM M103)

Composition per liter:

NaCl	25.0g
Peptone	10.0g
Agar	4.0g
Beef extract	3.0g

pH 7.4 ± 0.2 at 25°C

Preparation of Medium: Add components to distilled/deionized water and bring volume to 1.0L. Mix thoroughly. Gently heat while stirring and bring to boiling. Distribute into screw-capped tubes in 8.0mL volumes. Autoclave for 15 min at 15 psi pressure–121°C. Leave in tubes.

Use: For the cultivation and observation of motility in a variety of bacteria, especially *Salmonella* spp. and *Listeria* spp. For *Listeria* spp. keep tubes tightly screw capped and sealed with parafilm. For *Salmonella* spp. use Petri plates prepared on the same day as use.

Use: For the cultivation of yeasts and fungi.

MOX Agar

See: **Magnesium Oxalate Agar**
See: **Oxford Agar, Modified**

MOX HiVeg Agar

Composition per liter:

Agar	15.0g
Plant hydrolysate	15.0g
Papaic digest of soybean meal	5.0g
NaCl	5.0g
$MnCl_2$	4.067g
Sodium oxalate	2.68g

pH 7.5 ± 0.2 at 25°C

Source: This medium is available as a premixed powder from HiMedia.

Preparation of Medium: Add components to distilled/deionized water and bring volume to 1.0L. Mix thoroughly. Gently heat and bring to boiling. Distribute into tubes or flasks. Autoclave for 15 min at 15 psi pressure–121°C. Pour into sterile Petri dishes or leave in tubes.

Use: For the cultivation of *Yersinia enterocolitica* from foods.

MP Agar

Composition per liter:

Agar	15.0g
Sodium acetate	0.1g
Basal medium	1.0L
Sodium sulfide solution	3.0mL

pH 7.0–7.5 at 25°C

Basal Medium:

Composition per liter:

$CaSO_4 \cdot 2H_2O$ (saturated solution)	20.0mL
NH_4Cl (4% solution)	5.0mL
Trace elements solution	5.0mL
$MgSO_4 \cdot 7H_2O$ (1% solution)	1.0mL
K_2HPO_4 (1% solution)	1.0mL

Preparation of Basal Medium: Add components to distilled/deionized water and bring volume to 1.0L. Mix thoroughly.

Trace Elements Solution:

Composition per liter:

Ferrous EDTA solution	20.0mL
$ZnSO_4 \cdot 7H_2O$ (0.1% solution)	10.0mL
$MnSO_4 \cdot 4H_2O$ (0.02% solution)	10.0mL
$CuSO_4 \cdot 5H_2O$ (0.00005% solution)	10.0mL
H_3BO_3 (0.1% solution)	10.0mL
$Co(NO_3)_2$ or $CoCl_2 \cdot 6H_2O$ (0.01% solution)	10.0mL
$Na_2MoO_4 \cdot 2H_2O$ (0.01% solution)	10.0mL

Preparation of Trace Elements Solution: Add components to distilled/deionized water and bring volume to 1.0L. Mix thoroughly.

Ferrous EDTA Solution:

Composition per 100.0mL:

$FeSO_4 \cdot 7H_2O$	7.0g
EDTA	2.0g
HCl, concentrated	1.0mL

Preparation of Ferrous EDTA Solution: Add components to distilled/deionized water and bring volume to 100.0mL. Mix thoroughly.

Sodium Sulfide Solution:

Composition per 10.0mL:

$Na_2S \cdot 9H_20$	1.0g

Preparation of Sodium Sulfide Solution: Add $Na_2S \cdot 9H_20$ to distilled/deionized water and bring volume to 10.0mL. Mix thoroughly. Autoclave for 15 min at 15 psi pressure–121°C. Prepare freshly.

Preparation of Medium: Add sodium acetate and agar to 1.0L of basal medium. Mix thoroughly. Adjust pH to 7.0–7.5. Gently heat and bring to boiling. Autoclave for 15 min at 15 psi pressure–121°C. Cool to 45°–50°C. Aseptically add 3.0mL of sterile sodium sulfide solution immediately prior to dispensing. Mix thoroughly. Pour into sterile Petri dishes or distribute into sterile screw-capped tubes.

Use: For the isolation and cultivation of *Beggiatoa* species and myxotrophic strains of *Thiothrix* species from water and environmental sources.

MP 5 Medium (Mineral Pectin 5 Medium)

Composition per liter:

Agar solution	500.0mL
Basal medium	250.0mL
Mineral solution	250.0mL

pH 5.0–6.0 at 25°C

Agar Solution:

Composition per 500.0mL:

Agar	15.0g

Preparation of Agar Solution: Add agar to distilled/deionized water and bring volume to 500.0mL. Mix thoroughly. Gently heat and bring to boiling. Autoclave for 15 min at 15 psi pressure–121°C. Cool to 45°–50°C.

Basal Medium:

Composition per 250.0mL:

Na_2HPO_4	6.0g
Pectin, citrus or apple	5.0g

KH_2PO_4 .... 4.0g
NH_4SO_4 .... 2.0g
Yeast extract 1.0g

Preparation of Basal Medium: Add components to distilled/deionized water and bring volume to 250.0mL. Mix thoroughly. Gently heat and bring to boiling.

Mineral Solution:
Composition per 250.0mL:

$FeSO_4$ (0.1% solution) 1.0mL
$MgSO_4·7H_2O$ (20% solution) 1.0mL
$CaCl_2·2H_2O$ (0.1% solution) 1.0mL
H_3BO_3 (0.001% solution) 1.0mL
$MnSO_4·H_2O$ (0.001% solution) 1.0mL
$ZnSO_4·7H_2O$ (0.007% solution 1.0mL
$CuSO_4·5H_2O$ (0.005% solution) 1.0mL
MoO_3 (0.001% solution) 1.0mL

Preparation of Mineral Solution: Add components to distilled/deionized water and bring volume to 250.0mL. Mix thoroughly.

Preparation of Medium: Combine 250.0mL of basal medium and 250.0mL of mineral solution. Mix thoroughly. Adjust pH to 5.0–6.0 with 1*N* HCl. Autoclave the basal medium-mineral solution and agar solution separately for 15 min at 15 psi pressure–121°C. Cool to 45°–50°C. Aseptically combine the two sterile solutions. Mix thoroughly. Pour immediately into sterile Petri dishes to prevent hydrolysis of the agar.

Use: For the cultivation of microorganisms that produce polygalactanase.

MP 7 Medium
(Mineral Pectin 7 Medium)

Composition per liter:

Basal medium 500.0mL
Mineral solution 500.0mL

pH 7.2 ± 0.2 at 25°C

Basal Medium:
Composition per 500.0mL:

Agar 15.0g
Na_2HPO_4 .... 6.0g
Pectin (citrus or apple) 5.0g
KH_2PO_4 .... 4.0g
NH_4SO_4 .... 2.0g
Yeast extract 1.0g

Preparation of Basal Medium: Add components to distilled/deionized water and bring volume to 500.0mL. Mix thoroughly. Gently heat and bring to boiling.

Mineral Solution:
Composition per 500.0mL:

$FeSO_4$ (0.1% solution) 1.0mL
$MgSO_4·7H_2O$ (20% solution) 1.0mL
$CaCl_2·2H_2O$ (0.1% solution) 1.0mL
H_3BO_3 (0.001% solution) 1.0mL
$MnSO_4·H_2O$ (0.001% solution) 1.0mL
$ZnSO_4·7H_2O$ (0.007% solution 1.0mL
$CuSO_4·5H_2O$ (0.005% solution 1.0mL
MoO_3 (0.001% solution) 1.0mL

Preparation of Mineral Solution: Add components to distilled/deionized water and bring volume to 500.0mL. Mix thoroughly.

Preparation of Medium: Combine 500.0mL of basal medium and 500.0mL of mineral solution. Mix thoroughly. Adjust pH to 7.2. Autoclave for 15 min at 15 psi pressure–121°C. Cool to 50°C. Pour into sterile Petri dishes.

Use: For the cultivation of microorganisms that produce pectate lyase.

m-PA Agar
See: **PA Agar**

m-PA-C Agar
See: **PA-C Agar**

MPH Agar
(Milk Protein Hydrolysate Agar)

Composition per liter:

Agar 15.0g
Casein hydrolysate 9.0g
Glucose 1.0g

pH 7.0 ± 0.2 at 25°C

Source: This medium is available as a premixed powder from BD Diagnostic Systems.

Preparation of Medium: Add components to distilled/deionized water and bring volume to 1.0L. Mix thoroughly. Gently heat while stirring and bring to boiling. Autoclave for 15 min at 15 psi pressure–121°C. Aseptically distribute into sterile tubes. Cool to 43°–45°C before using.

Use: For use in the enumeration of bacteria in water and dairy products.

MPHM Medium

Composition per 1275.0mL:

McCoy's 5A medium, modified 1.0L
Peptone solution 250.0mL
Hemin solution 25.0mL

Source: McCoy's 5A medium, modified, is available from Gibco.

McCoys 5A Medium, Modified:
Composition per liter:

Inorganic salt solution 400.0mL
Other component solution 400.0mL
Amino acid solution 100.0mL
Vitamin solution 100.0mL

Inorganic Salt Solution:
Composition per 400.0mL:

NaCl 6.46g
$NaHCO_3$ 2.2g
$NaH_2PO_4·H_2O$ 0.58g
KCl 0.4g
$CaCl_2$, anhydrous 0.1g
$MgSO_4$, anhydrous 97.67mg

Preparation of Inorganic Salt Solution: Add components to distilled/deionized water and bring volume to 400.0mL. Mix thoroughly. Autoclave for 15 min at 15 psi pressure–121°C. Cool to 25°C.

Other Component Solution:
Composition per 400.0mL:

D-Glucose 3.0g
Peptone 0.6g

Phenol Red 0.1g
Glutathione, reduced 0.5mg

Preparation of Other Component Solution: Add components to distilled/deionized water and bring volume to 400.0mL. Mix thoroughly. Autoclave for 15 min at 15 psi pressure–121°C. Cool to 25°C.

Amino Acid Solution:
Composition per 100.0mL:
L-Glutamine 219.2mg
L-Asparagine 45.0mg
L-Arginine·HCl 42.1mg
L-Isoleucine 39.3mg
L-Leucine 39.3mg
L-Lysine·HCl 36.5mg
L-Cysteine 31.5mg
L-Serine 26.3mg
L-Tyrosine·2Na·$2H_2O$ 26.2mg
L-Glutamic acid 22.1mg
L-Histidine·HCl·H_2O 21.0mg
L-Aspartic acid 20.0mg
L-Hydroxyproline 19.7mg
L-Threonine 17.9mg
L-Valine 17.6mg
L-Proline 17.3mg
L-Phenylalanine 16.5mg
L-Methionine 15.0mg
L-Alanine 13.9mg
Glycine 7.5mg
L-Tryptophan 3.1mg

Preparation of Amino Acid Solution: Add components to distilled/deionized water and bring volume to 100.0mL. Mix thoroughly. Filter sterilize.

Vitamin Solution:
Composition per 100.0mL:
i-Inositol 36.0mg
Folic acid 10.0mg
Choline chloride 5.0mg
Vitamin B_{12} 2.0mg
p-Aminobenzoic acid 1.0mg
Ascorbic acid 0.5mg
Niacinamide 0.5mg
Nicotinic acid 0.5mg
Pyridoxal·HCl 0.5mg
Pyridoxine·HCl 0.5mg
Biotin 0.2mg
D-Ca pantothenate 0.2mg
Riboflavin 0.2mg
Thiamine·HCl 0.2mg

Preparation of Vitamin Solution: Add components to distilled/deionized water and bring volume to 100.0mL. Mix thoroughly. Filter sterilize.

Preparation of McCoys 5A Medium, Modified: Aseptically combine 400.0mL of sterile inorganic salt solution, 400.0mL of sterile other component solution, 100.0mL of sterile amino acid solution, and 100.0mL of sterile vitamin solution.

Peptone Solution:
Composition per 300.0mL:
Proteose peptone No. 3 .... 7.5g

Preparation of Peptone Solution: Add components to distilled/deionized water and bring volume to 300.0mL. Mix thoroughly. Autoclave for 15 min at 15 psi pressure–121°C. Cool to 25°C.

Hemin Solution:
Composition per 100.0mL:
NaCl 0.8g
Na_2HPO_4 .... 0.12g
Hemin 0.05g
KCl 0.04g
$MgCl_2·6H_2O$ 0.03g
$NaH_2PO_4·H_2O$ 0.02g
$CaCl_2$ 0.011g

Preparation of Hemin Solution: Add NaCl, KCl, $NaH_2PO_4·H_2O$, and Na_2HPO_4 to distilled/deionized water and bring volume to 50.0mL. Mix thoroughly. Separately add the $CaCl_2$ and $MgCl_2·6H_2O$ to distilled/deionized water and bring volume to 50.0mL. Mix thoroughly. Combine the two solutions. Mix thoroughly. Add 0.05g of hemin. Mix thoroughly. Adjust to pH 9.0 with 10*N* NaOH to dissolve the hemin. Mix thoroughly. Filter sterilize.

Preparation of Medium: Aseptically combine 1.0L of sterile modified McCoy's medium 5A, 250.0mL of sterile peptone solution, and 25.0mL of sterile hemin solution. Aseptically distribute into sterile screw-capped tubes or flasks.

Use: For the cultivation of *Trypanosoma theileri*.

m-Plate Count Broth
See: **Plate Count Broth**

MPOB Medium

Composition per 1012.0mL:
Disodium fumarate 3.2g
$Na_2HPO_4·2H_2O$ 0.53g
KH_2PO_4 .... 0.41g
NaCl 0.3g
NH_4Cl 0.3g
Yeast extract 0.2g
$CaCl_2·2H_2O$ 0.11g
$MgCl_2·6H_2O$ 0.10g
Resazurin 0.5mg
$NaHCO_3$ solution 20.0mL
$Na_2S·9H_2O$ solution 10.0mL
Wolfe's vitamin solution 10.0mL
Selenite-tungstate solution 1.0mL
Trace elements solution SL-10 .... 1.0mL
pH 7.0–7.2 at 25°C

$NaHCO_3$ Solution:
Composition per 20.0mL:
$NaHCO_3$ 4.0g

Preparation of $NaHCO_3$ Solution: Add $NaHCO_3$ to distilled/deionized water and bring volume to 20.0mL. Mix thoroughly. Sparge with 80% N_2 + 20% CO_2. Autoclave for 15 min at 15 psi pressure–121°C.

$Na_2S·9H_2O$ Solution:
Composition per 10.0mL:
$Na_2S·9H_2O$ 0.5g

Preparation of $Na_2S·9H_2O$ Solution: Add $Na_2S·9H_2O$ to distilled/deionized water and bring volume to 10.0mL. Mix thoroughly. Sparge with 100% N_2. Autoclave for 15 min at 15 psi pressure–121°C.

Wolfe's Vitamin Solution:
Composition per liter:

Pyridoxine·HCl	10.0mg
p-Aminobenzoic acid	5.0mg
Lipoic acid	5.0mg
Nicotinic acid	5.0mg
Riboflavin	5.0mg
Thiamine·HCl	5.0mg
Calcium DL-pantothenate	5.0mg
Biotin	2.0mg
Folic acid	2.0mg
Vitamin B_{12}	0.1mg

Preparation of Wolfe's Vitamin Solution: Add components to distilled/deionized water and bring volume to 1.0L. Mix thoroughly. Filter sterilize.

Selenite-tungstate Solution:
Composition per liter:

NaOH	0.5g
$Na_2WO_4 \cdot 2H_2O$	4.0mg
$Na_2SeO_3 \cdot 5H_2O$	3.0mg

Preparation of Selenite-tungstate Solution: Add components to distilled/deionized water and bring volume to 1.0L. Mix thoroughly. Sparge with 100% N_2. Autoclave for 15 min at 15 psi pressure–121°C.

Trace Elements Solution SL-10:
Composition per liter:

$FeCl_2 \cdot 4H_2O$	1.5g
$CoCl_2 \cdot 6H_2O$	190.0mg
$MnCl_2 \cdot 4H_2O$	100.0mg
$ZnCl_2$	70.0mg
$Na_2MoO_4 \cdot 2H_2O$	36.0mg
$NiCl_2 \cdot 6H_2O$	24.0mg
H_3BO_3	6.0mg
$CuCl_2 \cdot 2H_2O$	2.0mg
HCl (25% solution)	10.0mL

Preparation of Trace Elements Solution SL-10: Add $FeCl_2 \cdot 4H_2O$ to 10.0mL of HCl solution. Mix thoroughly. Add distilled/deionized water and bring volume to 1.0L. Add remaining components. Mix thoroughly. Sparge with 100% N_2. Autoclave for 15 min at 15 psi pressure–121°C.

Preparation of Medium: Prepare medium under 80% N_2 + 20% CO_2. Add components, except $NaHCO_3$ solution and $Na_2S \cdot 9H_2O$ solution, to distilled/deionized water and bring volume to 970.0mL. Mix thoroughly. Sparge with 80% N_2 + 20% CO_2. Autoclave for 15 min at 15 psi pressure–121°C. Aseptically and anaerobically add 20.0mL of sterile $NaHCO_3$ solution and 10.0mL of sterile $Na_2S \cdot 9H_2O$ solution. Mix thoroughly. Aseptically and anaerobically distribute into sterile tubes or flasks. After inoculation, bring culture bottles to 0.7 bar 80% N_2 + 20% CO_2 overpressure.

Use: For the cultivation of *Syntrophobacter* species.

m-*Pseudomonas aeruginosa* Agar

See: ***Pseudomonas aeruginosa*** **Agar**

MPSS Broth

Composition per liter:

Peptone	5.0g
$MgSO_4 \cdot 7H_2O$	1.0g
$(NH_4)_2SO_4$	1.0g
Succinic acid	1.0g
$FeCl_3 \cdot 6H_2O$ (0.2% solution)	1.0mL
$MnSO_4 \cdot H_2O$ (0.2% solution)	1.0mL

pH 6.8 ± 0.2 at 25°C

Preparation of Medium: Add components to distilled/deionized water and bring volume to 1.0L. Mix thoroughly. Distribute into tubes or flasks. Autoclave for 15 min at 15 psi pressure–121°C.

Use: For the cultivation of *Spirillum volutans*.

MPY Agar (Malt Peptone Yeast Extract Agar) (ATCC Medium 582)

Composition per liter:

Agar	15.0g
Malt extract	5.0g
Xylose	2.0g
Fructose	2.0g
Lactose	2.0g
Peptone	1.0g
Yeast extract	1.0g

pH 7.0 ± 0.2 at 25°C

Preparation of Medium: Add components to distilled/deionized water and bring volume to 1.0L. Mix thoroughly. Gently heat and bring to boiling. Distribute into tubes or flasks. Autoclave for 15 min at 15 psi pressure–121°C. Pour into sterile Petri dishes or leave in tubes.

Use: For the cultivation and maintenance of *Streptomyces naniwaensis*.

MPY Agar (Maltose Peptone Yeast Extract Medium) (ATCC Medium 518)

Composition per liter:

Agar	10.0g
Maltose	2.0g
Peptone	2.0g
Yeast extract	1.0g
Potassium phosphate buffer (1*M*, pH 7.5)	10.0mL

pH 7.5 ± 0.2 at 25°C

Preparation of Medium: Add components, except potassium phosphate buffer, to distilled/deionized water and bring volume to 990.0mL. Mix thoroughly. Gently heat and bring to boiling. Autoclave for 15 min at 15 psi pressure–121°C. Cool to 45°–50°C. Filter sterilize potassium phosphate buffer. Aseptically add sterile potassium phosphate buffer to sterile, cooled basal medium. Distribute into sterile tubes or flasks.

Use: For the cultivation and maintenance of *Spirochaeta aurantia*.

MPY Broth (Maltose Peptone Yeast Extract Broth)

Composition per liter:

Maltose	2.0g
Peptone	2.0g
Yeast extract	1.0g
Potassium phosphate buffer (1M, pH 7.5)	10.0mL

pH 7.5 ± 0.2 at 25°C

Preparation of Medium: Add components, except potassium phosphate buffer, to distilled/deionized water and bring volume to

990.0mL. Mix thoroughly. Autoclave for 15 min at 15 psi pressure–121°C. Filter sterilize potassium phosphate buffer. Aseptically add sterile potassium phosphate buffer to sterile cooled basal medium. Distribute into sterile tubes or flasks.

Use: For the cultivation and maintenance of *Spirochaeta aurantia*.

MPYG Medium

See: **Peptone Yeast Extract Glucose Medium, Modified**

MRS Agar (DeMan, Rogosa, Sharpe Agar)

Composition per liter:

Glucose	20.0g
Peptone	10.0g
Agar	10.0g
Beef extract	8.0g
Sodium acetate·3H_2O	5.0g
Yeast extract	4.0g
K_2HPO_4	2.0g
Triammonium citrate	2.0g
$MgSO_4$·7H_2O	0.2g
$MnSO_4$·4H_2O	0.05g
Sorbitan monooleate	1.0mL

pH 6.2 ± 0.2 at 25°C

Source: This medium is available as a premixed powder from Oxoid Unipath.

Preparation of Medium: Add components to distilled/deionized water and bring volume to 1.0L. Mix thoroughly. Gently heat and bring to boiling. Distribute into tubes or flasks. Autoclave for 15 min at 15 psi pressure–121°C. Pour into sterile Petri dishes or leave in tubes.

Use: For the cultivation of lactic acid bacteria.

MRS Agar (DeMan, Rogosa, Sharpe Agar)

Composition per liter:

Glucose	18.5g
Agar	13.5g
Pancreatic digest of gelatin	10.0g
Beef extract	8.0g
Yeast extract	4.0g
Sodium acetate	3.0g
K_2HPO_4	2.0g
Ammonium citrate	2.0g
Polysorbate 80	1.0g
$MgSO_4$·7H_2O	0.2g
$MnSO_4$·4H_2O	0.05g

pH 6.2 ± 0.2 at 25°C

Source: This medium is available as a premixed powder from BD Diagnostic Systems.

Preparation of Medium: Add components to distilled/deionized water and bring volume to 1.0L. Mix thoroughly. Gently heat while stirring and bring to boiling. Distribute into tubes or flasks. Autoclave for 15 min at 15 psi pressure–121°C. Pour into sterile Petri dishes or leave in tubes.

Use: For the isolation and cultivation of *Lactobacillus* species from clinical specimens, foods, and dairy products.

MRS Agar with Cysteine (LMG Medium 118)

Composition per liter:

Glucose	20.0g
Peptone	10.0g
Agar	12.0g
Beef extract	5.0g
Sodium acetate·3H_2O	5.0g
Yeast extract	5.0g
KH_2PO_4	2.0g
Diammonium hydrogen citrate	2.0g
$MgSO_4$·7H_2O	0.2g
$MnSO_4$·4H_2O	38.0mg
L-Cysteine solution	10.0mL
Tween™ 80	1.0mL

pH 6.5 ± 0.2 at 25°C

L-Cysteine Solution:

Composition per 10.0mL:

L-Cysteine	0.5g

Preparation of L-Cysteine Solution: Add 0.1g of L-cysteine to distilled/deionized water and bring volume to 10.0mL. Mix thoroughly. Filter sterilize.

Preparation of Medium: Add components, except L-cysteine solution, to distilled/deionized water and bring volume to 990.0mL. Mix thoroughly. Adjust pH to 6.5. Gently heat and bring to boiling. Distribute into tubes or flasks. Autoclave for 15 min at 15 psi pressure–121°C. Cool to 50°C. Aseptically add 10.0mL of sterile L-cysteine solution. Mix thoroughly. Aseptically pour into sterile Petri dishes or distribute into sterile tubes.

Use: For the isolation and cultivation of *Leuconostoc mesenteroides* and other lactic acid bacteria.

MRS Agar with 0.5% Cysteine (LMG Medium 131)

Composition per liter:

Glucose	20.0g
Agar	12.0g
Peptone	10.0g
Beef extract	5.0g
Sodium acetate·3H_2O	5.0g
Yeast extract	5.0g
KH_2PO_4	2.0g
Diammonium hydrogen citrate	2.0g
$MgSO_4$·7H_2O	0.2g
$MnSO_4$·4H_2O	38.0mg
L-Cysteine·HCl solution	100.0mL
Tween™ 80	1.0mL

pH 6.5 ± 0.2 at 25°C

L-Cysteine·HCl Solution:

Composition per 100.0mL:

L-Cysteine·HCl	5.0g

Preparation of L-Cysteine·HCl Solution: Add 5.0g of L-cysteine·HCl to distilled/deionized water and bring volume to 10.0mL. Mix thoroughly. Filter sterilize.

Preparation of Medium: Add components, except L-cysteine·HCl solution, to distilled/deionized water and bring volume to 900.0mL. Mix thoroughly. Adjust pH to 6.5. Gently heat and bring to boiling. Distribute into tubes or flasks. Autoclave for 15 min at 15 psi pressure–

121°C. Cool to 50°C. Aseptically add 100.0mL of sterile L-cysteine·HCl solution. Mix thoroughly. Aseptically pour into sterile Petri dishes or distribute into sterile tubes.

Use: For the isolation and cultivation of *Lactobacillus ruminis* and *Lactobacillus vitulinus.*

MRS Agar with Ethanol (LMG Medium 130)

Composition per liter:

Glucose	20.0g
Agar	12.0g
Peptone	10.0g
Beef extract	5.0g
Sodium acetate·3H_2O	5.0g
Yeast extract	5.0g
KH_2PO_4	2.0g
Diammonium hydrogen citrate	2.0g
$MgSO_4$·7H_2O	0.2g
$MnSO_4$·4H_2O	38.0mg
Ethanol	100.0mL
Tween™ 80	1.0mL

pH 5.0± 0.2 at 25°C

Preparation of Medium: Add components, except ethanol, to distilled/deionized water and bring volume to 900.0mL. Mix thoroughly. Autoclave for 15 min at 15 psi pressure–121°C. Cool to 50°C. Aseptically add 100.0mL of sterile ethanol. Mix thoroughly. Aseptically distribute into sterile tubes or pour into sterile Petri dishes.

Use: For the cultivation of *Lactobacillus acetotolerans.*

MRS Agar with Tomato Juice, pH 5.2 (LMG Medium 248)

Composition per liter:

Glucose	20.0g
Peptone	10.0g
Agar	12.0g
Beef extract	5.0g
Sodium acetate·3H_2O	5.0g
Yeast extract	5.0g
KH_2PO_4	2.0g
Diammonium hydrogen citrate	2.0g
$MgSO_4$·7H_2O	0.2g
$MnSO_4$·4H_2O	38.0mg
Tomato juice	100.0mL
Tween™ 80	1.0mL

pH 5.2 ± 0.2 at 25°C

Preparation of Medium: Add components to distilled/deionized water and bring volume to 1.0L. Mix thoroughly. Adjust pH to 5.2 Gently heat and bring to boiling. Distribute into tubes or flasks. Autoclave for 15 min at 15 psi pressure–121°C. Pour into sterile Petri dishes or leave in tubes.

Use: For the isolation and cultivation of *Lactobacillus kunkeei.*

MRS Agar, Half Strength (LMG Medium 281)

Composition per liter:

Glucose	10.0g
Agar	6.0g
Peptone	5.0g
Beef extract	2.5g
Sodium acetate·3H_2O	2.5g
Yeast extract	2.5g
KH_2PO_4	1.0g
Diammonium hydrogen citrate	1.0g
$MgSO_4$·7H_2O	0.1g
$MnSO_4$·4H_2O	19.0mg
Tween™ 80	0.5mL

pH 6.5± 0.2 at 25°C

Preparation of Medium: Add components to distilled/deionized water and bring volume to 1.0L. Mix thoroughly. Distribute into tubes or flasks. Autoclave for 15 min at 15 psi pressure–121°C.

Use: For the cultivation of *Lactobacillus acidipiscis* and *Weissella thailandensis.*

MRS Broth

Composition per liter:

Glucose	20.0g
Peptone	10.0g
Beef extract	5.0g
Sodium acetate·3H_2O	5.0g
Yeast extract	5.0g
K_2HPO_4	2.0g
Diammonium hydrogen citrate	2.0g
$MgSO_4$·7H_2O	0.2g
$MnSO_4$·4H_2O	38.0mg
Sorbitan monooleate	1.0mL

pH 6.2 ± 0.2 at 25°C

Preparation of Medium: Add components to distilled/deionized water and bring volume to 1.0L. Mix thoroughly. Gently heat and bring to boiling. Distribute into tubes or flasks. Autoclave for 15 min at 15 psi pressure–121°C.

Use: For the cultivation of *Leuconostoc pseudomesenteroides, Leuconostoc carnosum,* and *Lactobacillus crispatus.*

MRS Broth (DeMan, Rogosa, Sharpe Broth)

Composition per liter:

Glucose	20.0g
Peptone	10.0g
Beef extract	8.0g
Sodium acetate·3H_2O	5.0g
Yeast extract	4.0g
K_2HPO_4	2.0g
Triammonium citrate	2.0g
$MgSO_4$·7H_2O	0.2g
$MnSO_4$·4H_2O	0.05g
Sorbitan monooleate	1.0mL

pH 6.2 ± 0.2 at 25°C

Source: This medium is available as a premixed powder from Oxoid Unipath.

Preparation of Medium: Add components to distilled/deionized water and bring volume to 1.0L. Mix thoroughly. Gently heat and bring to boiling. Distribute into tubes or flasks. Autoclave for 15 min at 15 psi pressure–121°C.

Use: For the cultivation of lactic acid bacteria.

MRS Broth (DeMan, Rogosa, Sharpe Broth)

Composition per liter:

Glucose	18.5g
Pancreatic digest of gelatin	10.0g
Beef extract	8.0g
Yeast extract	4.0g
Sodium acetate	3.0g
K_2HPO_4	2.0g
Ammonium citrate	2.0g
Polysorbate 80	1.0g
$MgSO_4 \cdot 7H_2O$	0.2g
$MnSO_4 \cdot 4H_2O$	0.05g

pH 6.2 ± 0.2 at 25°C

Source: This medium is available as a premixed powder from BD Diagnostic Systems.

Preparation of Medium: Add components to distilled/deionized water and bring volume to 1.0L. Mix thoroughly. Distribute into tubes or flasks. Autoclave for 15 min at 15 psi pressure–121°C.

Use: For the isolation and cultivation of *Lactobacillus* species from clinical specimens, foods, and dairy products.

MRS Broth with $CaCO_3$ (LMG Medium 166)

Composition per liter:

$CaCO_3$	30.0g
Glucose	20.0g
Peptone	10.0g
Beef extract	5.0g
Sodium acetate·$3H_2O$	5.0g
Yeast extract	5.0g
K_2HPO_4	2.0g
Diammonium hydrogen citrate	2.0g
$MgSO_4 \cdot 7H_2O$	0.2g
$MnSO_4 \cdot 4H_2O$	38.0mg
Sorbitan monooleate	1.0mL

pH 6.2 ± 0.2 at 25°C

Preparation of Medium: Add components to distilled/deionized water and bring volume to 1.0L. Mix thoroughly. Gently heat and bring to boiling. Distribute into tubes or flasks. Autoclave for 15 min at 15 psi pressure–121°C.

Use: For the cultivation of *Sporolactobacillus inulinus.*

MRS Broth with Cysteine

Composition per liter:

Glucose	18.5g
Pancreatic digest of gelatin	10.0g
Beef extract	8.0g
Yeast extract	4.0g
Sodium acetate	3.0g
K_2HPO_4	2.0g
Ammonium citrate	2.0g
Polysorbate 80	1.0g
$MgSO_4 \cdot 7H_2O$	0.2g
$MnSO_4 \cdot 4H_2O$	0.05g
L-Cysteine solution	0.5mL

pH 6.2 ± 0.2 at 25°C

L-Cysteine Solution:

Composition per 10.0mL:

L-Cysteine	0.1g

Preparation of L-Cysteine Solution: Add 0.1g of L-cysteine to distilled/deionized water and bring volume to 10.0mL. Mix thoroughly. Filter sterilize.

Preparation of Medium: Add components, except L-cysteine solution, to distilled/deionized water and bring volume to 999.5mL. Mix thoroughly. Adjust pH to 6.2. Autoclave for 15 min at 15 psi pressure–121°C. Aseptically add 0.5mL of sterile L-cysteine solution. Mix thoroughly. Aseptically distribute into sterile tubes or flasks.

Use: For the isolation and cultivation of *Leuconostoc mesenteroides*.

MRS Broth with Ethanol (LMG Medium 240)

Composition per liter:

Glucose	20.0g
Peptone	10.0g
Beef extract	5.0g
Sodium acetate·$3H_2O$	5.0g
Yeast extract	5.0g
KH_2PO_4	2.0g
Diammonium hydrogen citrate	2.0g
$MgSO_4 \cdot 7H_2O$	0.2g
$MnSO_4 \cdot 4H_2O$	38.0mg
Ethanol	100.0mL
Tween™ 80	1.0mL

pH 5.0± 0.2 at 25°C

Preparation of Medium: Add components, except ethanol, to distilled/deionized water and bring volume to 900.0mL. Mix thoroughly. Autoclave for 15 min at 15 psi pressure–121°C. Cool to 50°C. Aseptically add 100.0mL of sterile ethanol. Mix thoroughly. Aseptically distribute into sterile tubes or sterile flasks.

Use: For the cultivation of *Lactobacillus acetotolerans.*

MRS Broth with Ethanol

Composition per liter:

Glucose	20.0g
Peptone	10.0g
Beef extract	8.0g
Sodium acetate·$3H_2O$	5.0g
Yeast extract	4.0g
K_2HPO_4	2.0g
Triammonium citrate	2.0g
$MgSO_4 \cdot 7H_2O$	0.2g
$MnSO_4 \cdot 4H_2O$	0.05g
Sorbitan monooleate	1.0mL
Ethanol (95% solution), filter sterilized	100.0mL

pH 5.0 ± 0.2 at 25°C

Preparation of Medium: Add components, except ethanol, to distilled/deionized water and bring volume to 900.0mL. Mix thoroughly. Autoclave for 15 min at 15 psi pressure–121°C. Aseptically add 100.0mL of sterile ethanol. Mix thoroughly. Aseptically distribute into sterile tubes or flasks.

Use: For the cultivation of *Lactobacillus acetotolerans.*

MRS Chalk

Composition per liter:

$CaCO_3$	30.0g
Glucose	20.0g
Beef extract	10.0g
Peptone	10.0g
Yeast extract	5.0g
K_2HPO_4	2.0g
Triammonium citrate	2.0g
$MgSO_4 \cdot 7H_2O$	0.2g
$MnSO_4 \cdot 4H_2O$	0.05g
Tween™ 80™	1.0mL

pH 6.2 ± 0.4 at 25°C

Preparation of Medium: Add components to distilled/deionized water and bring volume to 1.0L. Mix thoroughly. Adjust pH to 6.2. Distribute into tubes or flasks. Autoclave for 15 min at 15 psi pressure–121°C.

Use: For the cultivation of *Sporolactobacillus inulinus* and other *Sporolactobacillus* species.

MRS with Fermented Wort

Composition per liter:

Solution A	800.0mL
Solution B	200.0mL

pH 6.2 ± 0.2 at 25°C

Solution A:

Composition per liter:

Glucose	20.0g
Peptone	10.0g
Agar	10.0g
Beef extract	8.0g
Sodium acetate·$3H_2O$	5.0g
Yeast extract	4.0g
K_2HPO_4	2.0g
Triammonium citrate	2.0g
$MgSO_4 \cdot 7H_2O$	0.2g
$MnSO_4 \cdot 4H_2O$	0.05g
Sorbitan monooleate	1.0mL

Preparation of Solution A: Add components to distilled/deionized water and bring volume to 1.0L. Mix thoroughly. Gently heat and bring to boiling. Autoclave for 15 min at 15 psi pressure–121°C. Cool to 50°–55°C.

Solution B:

Composition per liter:

Malt extract	15.0g
Maltose	12.75g
Dextrin	2.75g
Glycerol	2.35g
K_2HPO_4	1.0g
NH_4Cl	1.0g
Pancreatic digest of gelatin	0.78g

Preparation of Solution B: Add components to distilled/deionized water and bring volume to 1.0L. Mix thoroughly. Gently heat and bring to boiling. Autoclave for 15 min at 15 psi pressure–121°C. Cool to room temperature. Inoculate with *Saccharomyces cerevisiae* to allow wort to ferment. Incubate for 24–48 hr. Allow to cool to room temperature. Filter through Whatman filter paper to remove solids. Filter sterilize.

Preparation of Medium: Aseptically combine 800.0mL of sterile solution A and 200.0mL of sterile solution B. Mix thoroughly. Aseptically distribute into sterile tubes or flasks.

Use: For the cultivation of yeasts and fungi.

MRS Fructose Medium

Composition per liter:

Beef extract	10.0g
Fructose	10.0g
Peptone	10.0g
Yeast extract	5.0g
K_2HPO_4	2.0g
Triammonium citrate	2.0g
$MgSO_4 \cdot 7H_2O$	0.2g
$MnSO_4 \cdot 4H_2O$	0.05g
Tween™ 80	1.0mL

pH 6.2 ± 0.4 at 25°C

Preparation of Medium: Add components to distilled/deionized water and bring volume to 1.0L. Mix thoroughly. Adjust pH to 6.2. Distribute into tubes or flasks. Autoclave for 15 min at 15 psi pressure–121°C.

Use: For the cultivation of *Carnobacterium divergens*.

MRS HiVeg Agar
(*Lactobacillus* MRS HiVeg Agar)

Composition per liter:

Glucose	20.0g
Peptone	10.0g
Agar	10.0g
Plant extract	8.0g
Sodium acetate·$3H_2O$	5.0g
Yeast extract	4.0g
K_2HPO_4	2.0g
Triammonium citrate	2.0g
$MgSO_4 \cdot 7H_2O$	0.2g
$MnSO_4 \cdot 4H_2O$	0.05g

pH 6.2 ± 0.2 at 25°C

Source: This medium is available as a premixed powder from HiMedia.

Preparation of Medium: Add components to distilled/deionized water and bring volume to 1.0L. Mix thoroughly. Gently heat and bring to boiling. Distribute into tubes or flasks. Autoclave for 15 min at 15 psi pressure–121°C. Pour into sterile Petri dishes or leave in tubes.

Use: For the cultivation of lactic acid bacteria.

MRS HiVeg Agar, Modified
(Lactobacilli Heteroferm Screen HiVeg Agar)

Composition per liter:

Glucose	20.0g
Agar	15.0g
Plant peptone No. 3	10.0g
Sodium acetate	5.0g
Yeast extract	5.0g
2-Phenylethyl alcohol	3.0g
Ammonium citrate	2.0g
K_2HPO_4	2.0g
$MgSO_4$	0.1g
$MnSO_4$	0.05g

Bromcresol Green	0.04g
Cycloheximide	4.0mg

pH 6.2 ± 0.2 at 25°C

Source: This medium is available as a premixed powder from HiMedia.

Caution: Cycloheximide is toxic. Avoid skin contact or aerosol formation and inhalation.

Preparation of Medium: Add components to distilled/deionized water and bring volume to 1.0L. Mix thoroughly. Gently heat and bring to boiling. Distribute into tubes or flasks. Autoclave for 15 min at 15 psi pressure–121°C. Pour into sterile Petri dishes or leave in tubes.

Use: For the isolation and cultivation of *Lactobacillus* species from salad dressings.

MRS HiVeg Broth (*Lactobacillus* MRS HiVeg Broth)

Composition per liter:

Glucose	20.0g
Peptone	10.0g
Plant extract	8.0g
Sodium acetate·3H_2O	5.0g
Yeast extract	4.0g
K_2HPO_4	2.0g
Triammonium citrate	2.0g
$MgSO_4$·7H_2O	0.2g
$MnSO_4$·4H_2O	0.05g

pH 6.2 ± 0.2 at 25°C

Source: This medium is available as a premixed powder from HiMedia.

Preparation of Medium: Add components to distilled/deionized water and bring volume to 1.0L. Mix thoroughly. Gently heat and bring to boiling. Distribute into tubes or flasks. Autoclave for 15 min at 15 psi pressure–121°C.

Use: For the cultivation of lactic acid bacteria.

MRS HiVeg Broth, Modified (*Lactobacillus* Heteroferm Screen HiVeg Broth)

Composition per liter:

Glucose	20.0g
Plant peptone no. 3	10.0g
Sodium acetate	5.0g
Yeast extract	5.0g
2-Phenylethyl alcohol	3.0g
Ammonium citrate	2.0g
K_2HPO_4	2.0g
$MgSO_4$	0.1g
$MnSO_4$	0.05g
Bromcresol Green	0.04g
Cycloheximide	4.0mg

pH 6.2 ± 0.2 at 25°C

Source: This medium is available as a premixed powder from HiMedia.

Caution: Cycloheximide is toxic. Avoid skin contact or aerosol formation and inhalation.

Preparation of Medium: Add components to distilled/deionized water and bring volume to 1.0L. Mix thoroughly. Gently heat and bring to boiling. Distribute into tubes or flasks. Autoclave for 15 min at 15 psi pressure–121°C.

Use: For the isolation and cultivation of *Lactobacillus* species from foods.

MRS Medium (DSMZ Medium 11)

Composition per liter:

Glucose	20.0g
Casein peptone, tryptic digest	10.0g
Meat extract	10.0g
Yeast extract	5.0g
Na-acetate	5.0g
K_2HPO_4	2.0g
$(NH_4)_2$ citrate	2.0g
$MgSO_4$·7H_2O	0.2g
$MnSO_4$·2H_2O	0.05g
Tween™ 80	1.0mL

pH 6.2–6.5

Preparation of Medium: Add components to distilled/deionized water and bring volume to 1.0L. Mix thoroughly. Adjust pH to 6.2 – 6.5. Gently heat and bring to boiling. Distribute into tubes or flasks. Autoclave for 15 min at 15 psi pressure–121°C.

Use: For the cultivation and maintenance of *Lactobacillus* spp., *Leuconostoc mesenteroides*, and *Pediococcus pentosaceus*.

MRS Medium, Modified

Composition per liter:

Agar	15.0g
Pancreatic digest of casein	10.0g
Beef extract	5.0g
Sodium acetate·3H_2O	5.0g
Yeast extract	5.0g
K_2HPO_4·3H_2O	2.6g
Ammonium citrate	2.0g
Tween™ 80	1.0g
L-Cysteine·HCl·H_2O	0.5g
$MgSO_4$·7H_2O	0.1g
$MnSO_4$·4H_2O	50.0mg
Carbohydrate solution	100.0mL

pH 6.3 ± 0.2 at 25°C

Carbohydrate Solution:
Composition per 100.0mL:

Fructose	7.0g
Glucose	7.0g
Maltose	7.0g
Sodium gluconate	2.0g

Preparation of Carbohydrate Solution: Add components to distilled/deionized water and bring volume to 100.0mL. Mix thoroughly. Filter sterilize.

Preparation of Medium: Add components, except carbohydrate solution, to distilled/deionized water and bring volume to 900.0mL. Mix thoroughly. Gently heat and bring to boiling. Adjust pH to 6.3. Autoclave for 15 min at 15 psi pressure–121°C. Cool to 50°–55°C. Aseptically add 100.0mL of sterile carbohydrate solution. Mix thoroughly. Pour into sterile Petri dishes or distribute into sterile tubes.

Use: For the cultivation of *Lactobacillus pontis*.

MRS Medium, pH 5.5

Composition per liter:

Glucose	20.0g
Tryptic digest of casein	10.0g
Meat extract	10.0g
Sodium acetate	5.0g
Yeast extract	5.0g
Diammonium citrate	2.0g
K_2HPO_4	2.0g
Tween™ 80	1.0g
$MgSO_4·7H_2O$	0.2g
$MnSO_4·H_2O$	50.0mg

pH 5.5 ± 0.2 at 25°C

Preparation of Medium: Add components to distilled/deionized water and bring volume to 1.0L. Mix thoroughly. Adjust pH to 5.5. Distribute into tubes or flasks. Autoclave for 15 min at 15 psi pressure–121°C.

Use: For the cultivation of *Lactobacillus kefiranofaciens.*

MRS, Modified

See: ***Lactobacillus*** **Heteroferm Screen Broth**

MRS Medium with L-Cysteine

Composition per liter:

Glucose	20.0g
Peptone	10.0g
Agar	10.0g
Beef extract	8.0g
L-Cysteine·HCl	5.0g
Sodium acetate·$3H_2O$	5.0g
Yeast extract	4.0g
K_2HPO_4	2.0g
Triammonium citrate	2.0g
$MgSO_4·7H_2O$	0.2g
$MnSO_4·4H_2O$	0.05g
Sorbitan monooleate	1.0mL

pH 6.2 ± 0.2 at 25°C

Preparation of Medium: Add components to distilled/deionized water and bring volume to 1.0L. Mix thoroughly. Gently heat and bring to boiling. Distribute into tubes or flasks. Autoclave for 15 min at 15 psi pressure–121°C. Pour into sterile Petri dishes or leave in tubes.

Use: For the cultivation of *Lactobacillus ruminis.*

MRS Medium with L-Cysteine

Composition per liter:

Glucose	20.0g
Tryptic digest of casein	10.0g
Meat extract	10.0g
Sodium acetate	5.0g
Yeast extract	5.0g
L-Cysteine·HCl	5.0g
Diammonium citrate	2.0g
K_2HPO_4	2.0g
$MgSO_4·7H_2O$	2.0g
Tween™ 80	1.0g
$MnSO_4·H_2O$	0.05g

Preparation of Medium: Add components to distilled/deionized water and bring volume to 1.0L. Mix thoroughly. Distribute into tubes or flasks. Autoclave for 15 min at 15 psi pressure–121°C.

Use: For the cultivation of *Lactobacillus* species, *Pectinatus* species, *Selenomonas lacticifex, Zymophilus paucivorans*, and *Zymophilus raffinosivorans*.

MRS Salts

Composition per liter:

NaCl	100.0g
Glucose	20.0g
Beef extract	10.0g
Peptone	10.0g
Yeast extract	5.0g
K_2HPO_4	2.0g
Triammonium citrate	2.0g
$MgSO_4·7H_2O$	0.2g
$MnSO_4·4H_2O$	0.05g
Tween™ 80	1.0mL

pH 6.2 ± 0.4 at 25°C

Preparation of Medium: Add components to distilled/deionized water and bring volume to 1.0L. Mix thoroughly. Adjust pH to 6.2. Distribute into tubes or flasks. Autoclave for 15 min at 15 psi pressure–121°C.

Use: For the cultivation of *Tetragenococcus halophila.*

MRSASelect™

Composition per liter:

Proprietary

Source: Available from BioRad

Preparation of Medium: Prepared plates.

Use: For the rapid screening of nasal specimens for MRSA (methicillin-resistant *Staphylococcus aureus*).

MRVP Broth (Methyl Red- Voges-Proskauer Broth)

Composition per liter:

Glucose	5.0g
KH_2PO_4	5.0g
Pancreatic digest of casein	3.5g
Peptic digest of animal tissue	3.5g

pH 6.9 ± 0.2 at 25°C

Source: Available as a premixed powder from BD Diagnostic Systems and as a prepared medium from BD Diagnostic Systems.

Preparation of Medium: Add components to distilled/deionized water and bring volume to 1.0L. Mix thoroughly. Distribute into tubes or flasks. Autoclave for 15 min at 15 psi pressure–121°C.

Use: For the differentiation of bacteria based on acid production (Methyl Red test) and acetoin production (Voges-Proskauer reaction).

MRVP Broth (Methyl Red-Voges-Proskauer Broth) (BAM M104 Medium 1)

Composition per liter:

Buffered peptone-water powder	7.0g
Glucose	5.0g
KH_2PO_4	5.0g

pH 7.0 ± 0.2 at 25°C

Source: Buffered peptone-water powder is available from BD Diagnostic Systems.

Preparation of Medium: Add components to distilled/deionized water and bring volume to 1.0L. Mix thoroughly. Distribute into tubes or flasks. Autoclave for 15 min at 15 psi pressure–121°C.

Use: For the differentiation of bacteria based on acid production (Methyl Red test) and acetoin production (Voges-Proskauer reaction).

MRVP Broth with Sodium Chloride (Methyl Red-Voges-Proskauer Broth with NaCl) (BAM M104 Medium 1)

Composition per liter:

NaCl	30.0g
Buffered peptone-water powder	7.0g
Glucose	5.0g
KH_2PO_4	5.0g

pH 7.0 ± 0.2 at 25°C

Source: Buffered peptone-water powder is available from BD Diagnostic Systems.

Preparation of Medium: Add components to distilled/deionized water and bring volume to 1.0L. Mix thoroughly. Distribute into tubes or flasks. Autoclave for 15 min at 15 psi pressure–121°C.

Use: For the differentiation of halophilic *Vibrio* spp. based on acid production (Methyl Red test) and acetoin production (Voges-Proskauer reaction).

MRVP Broth with Sodium Chloride (Methyl Red-Voges-Proskauer Broth with NaCl) (BAM M104 Medium 2)

Composition per liter:

NaCl	30.0g
Glucose	5.0g
KH_2PO_4	5.0g
Pancreatic digest of casein	3.5g
Peptic digest of animal tissue	3.5g

pH 6.9 ± 0.2 at 25°C

Source: Available as a premixed powder from BD Diagnostic Systems and as a prepared medium from BD Diagnostic Systems.

Preparation of Medium: Add components to distilled/deionized water and bring volume to 1.0L. Mix thoroughly. Distribute into tubes or flasks. Autoclave for 15 min at 15 psi pressure–121°C.

Use: For the differentiation of halophilic *Vibrio* spp. based on acid production (Methyl Red test) and acetoin production (Voges-Proskauer reaction).

MRVP Broth with Sodium Chloride (Methyl Red-Voges-Proskauer Broth with NaCl) (Methyl Red-Voges-Proskauer Medium)

Composition per liter:

NaCl	30.0g
Glucose	5.0g
Peptone	5.0g
Phosphate buffer	5.0g

pH 7.5 ± 0.2 at 25°C

Source: This medium is available as a premixed powder from Oxoid Unipath.

Preparation of Medium: Add components to distilled/deionized water and bring volume to 1.0L. Mix thoroughly. Distribute into tubes or flasks. Autoclave for 15 min at 15 psi pressure–121°C.

Use: For the differentiation of halophilic *Vibrio* spp. based on acid production (Methyl Red test) and acetoin production (Voges-Proskauer reaction).

MRVP Medium (Methyl Red-Voges-Proskauer Medium)

Composition per liter:

Glucose	5.0g
Peptone	5.0g
Phosphate buffer	5.0g

pH 7.5 ± 0.2 at 25°C

Source: This medium is available as a premixed powder from Oxoid Unipath.

Preparation of Medium: Add components to distilled/deionized water and bring volume to 1.0L. Mix thoroughly. Distribute into tubes or flasks. Autoclave for 15 min at 15 psi pressure–121°C.

Use: For the differentiation of bacteria based on acid production (Methyl Red test) and acetoin production (Voges-Proskauer reaction).

MS Agar

Composition per liter:

Agar	15.0g
Peptone	1.0g
Yeast extract	1.0g
Glucose	1.0g

pH 6.8–7.2 at 25°C

Preparation of Medium: Add components to distilled/deionized water and bring volume to 1.0L. Mix thoroughly. Gently heat and bring to boiling. Distribute into tubes or flasks. Autoclave for 15 min at 15 psi pressure–121°C. Pour into sterile Petri dishes or leave in tubes.

Use: For the cultivation of *Runella slithyformis*.

MS 1 Agar

Composition per liter:

Agar	15.0g
Seawater	1.0L

Preparation of Medium: Add agar to 1.0L of natural seawater. Mix thoroughly. Gently heat and bring to boiling. Distribute into tubes or flasks. Autoclave for 15 min at 15 psi pressure–121°C. Pour into sterile Petri dishes or leave in tubes.

Use: For the isolation and cultivation of *Cytophaga* species, *Herpetosiphon* species, *Saprospira* species, and *Flexithrix* species.

MS 3 Agar

Composition per liter:

Agar	15.0g
$(NH_4)_2SO_4$	1.0g
Seawater	1.0L

Preparation of Medium: Add agar to 500.0mL of natural seawater. Mix thoroughly. Gently heat and bring to boiling. In a separate flask, add $(NH_4)_2SO_4$ to 500.0mL of natural seawater. Mix thoroughly. Autoclave both solutions separately for 15 min at 15 psi pressure–121°C. Aseptically combine the two sterile solutions. Pour into sterile Petri dishes or distribute into sterile tubes.

Use: For the isolation and cultivation of *Cytophaga* species, *Herpetosiphon* species, *Saprospira* species, and *Flexithrix* species.

MS 4 Agar

Composition per liter:

Agar	15.0g
Glucose	2.0g
$(NH_4)_2SO_4$	1.0g
Seawater	1.0L

Preparation of Medium: Add agar to 500.0mL of natural seawater. Mix thoroughly. Gently heat and bring to boiling. Add $(NH_4)_2SO_4$ to 250.0mL of natural seawater. Mix thoroughly. Add glucose to 250.0mL of natural seawater. Mix thoroughly. Autoclave the three solutions separately for 15 min at 15 psi pressure–121°C. Aseptically combine the three sterile solutions. Pour into sterile Petri dishes or distribute into sterile tubes.

Use: For the isolation and cultivation of *Cytophaga* species, *Herpetosiphon* species, *Saprospira* species, and *Flexithrix* species.

MS Medium (DSMZ Medium 670)

Composition per liter:

$(NH_4)_2SO_4$	2.0g
$MgSO_4 \cdot 7H_2O$	0.25g
K_2HPO_4	0.1g
KCl	0.1g
$FeSO_4$ solution	50.0mL

pH 2.2 ± 0.2 at 25°C

$FeSO_4$ Solution:

Composition per 100.0mL:

$FeSO_4 \cdot 7H_2O$	40.0g

Preparation of $FeSO_4$ Solution: Add $FeSO_4 \cdot 7H_2O$ to 0.2*N* sulfuric acid and bring volume to 100.0mL. Mix thoroughly. Autoclave under 100% N_2 for 15 min at 15 psi pressure–121°C. Cool to room temperature.

Preparation of Medium: Add components, except $FeSO_4$ solution, to distilled/deionized water and bring volume to 950.0mL. Mix thoroughly. Adjust pH to 2.2 with 4*N* sulfuric acid. Autoclave for 15 min at 15 psi pressure–121°C. Cool to 25°C. Aseptically add 50.0mL sterile $FeSO_4$ solution. Mix thoroughly. Distribute into tubes or flasks.

Use: For the cultivation and maintenance of *Acidithiobacillus ferrooxidans* DSM2392, DSM9464, and DSM9465.

MS Medium (DSMZ Medium 670)

Composition per liter:

Sulfur	5.0g
$(NH_4)_2SO_4$	2.0g
$MgSO_4 \cdot 7H_2O$	0.25g
K_2HPO_4	0.1g
KCl	0.1g

pH 3.5 ± 0.2 at 25°C

Preparation of Medium: Sulfur is sterilized by steaming for 3 hr on each of 3 successive days. Add components, except sulfur, to distilled/deionized water and bring volume to 1.0L. Mix thoroughly. Adjust pH to 3.5 with 1*N* sulfuric acid. Autoclave for 15 min at 15 psi pressure–121°C. Cool to 25°C. Aseptically add 5.0g sterile sulfur. Mix thoroughly. Distribute into tubes or flasks.

Use: For the cultivation and maintenance of *Acidithiobacillus* DSM9463.

MS Medium (DSMZ Medium 670)

Composition per liter:

Sulfur	5.0g
$(NH_4)_2SO_4$	2.0g
$MgSO_4 \cdot 7H_2O$	0.25g
K_2HPO_4	0.1g
KCl	0.1g
Yeast extract solution	10.0mL

pH 3.5 ± 0.2 at 25°C

Yeast Extract Solution:

Composition per 10.0mL:

Yeast extract	0.2g

Preparation of Yeast Extract Solution: Add yeast extract to distilled/deionized water and bring volume to 10.0mL. Mix thoroughly. Autoclave under 100% N_2 for 15 min at 15 psi pressure–121°C. Cool to room temperature.

Preparation of Medium: Sulfur is sterilized by steaming for 3 hours on each of 3 successive days. Add components, except sulfur and yeast extract solution, to distilled/deionized water and bring volume to 990.0mL. Mix thoroughly. Adjust pH to 3.5 with 1*N* sulfuric acid. Autoclave for 15 min at 15 psi pressure–121°C. Cool to 25°C. Aseptically add 5.0g sterile sulfur and 10.0mL sterile yeast extract solution. Mix thoroughly. Distribute into tubes or flasks.

Use: For the cultivation and maintenance of *Acidithiobacillus caldus* DSM9466.

MS Medium (DSMZ Medium 670)

Composition per liter:

$(NH_4)_2SO_4$	2.0g
$MgSO_4 \cdot 7H_2O$	0.25g
K_2HPO_4	0.1g
KCl	0.1g
Glucose solution	20.0mL
Yeast extract solution	10.0mL

pH 3.0 ± 0.2 at 25°C

Yeast Extract Solution:

Composition per 10.0mL:

Yeast extract	0.1g

Preparation of Yeast Extract Solution: Add yeast extract to distilled/deionized water and bring volume to 10.0mL. Mix thoroughly. Autoclave under 100% N_2 for 15 min at 15 psi pressure–121°C. Cool to room temperature.

Glucose Solution:

Composition per 20.0mL:

Glucose	2.0g

Preparation of Glucose Solution: Add glucose to distilled/deionized water and bring volume to 20.0mL. Mix thoroughly. Filter sterilize.

Preparation of Medium: Add components, except yeast extract solution and glucose solution, to distilled/deionized water and bring volume to 970.0mL. Mix thoroughly. Adjust pH to 3.0 with 2*N* sulfuric acid. Autoclave for 15 min at 15 psi pressure–121°C. Cool to 25°C.

Aseptically add 20.0mL sterile glucose solution and 10.0mL sterile yeast extract solution. Mix thoroughly. Distribute into tubes or flasks.

Use: For the cultivation and maintenance of *Acidiphillum cryptum* DSM9467.

MS Medium (DSMZ Medium 670)

Composition per liter:

$(NH_4)_2SO_4$	2.0g
$MgSO_4 \cdot 7H_2O$	0.25g
K_2HPO_4	0.1g
KCl	0.1g
$FeSO_4$ solution	50.0mL

pH 1.6 ± 0.2 at 25°C

$FeSO_4$ Solution:

Composition per 100.0mL:

$FeSO_4 \cdot 7H_2O$	40.0g

Preparation of $FeSO_4$ Solution: Add $FeSO_4 \cdot 7H_2O$ to 0.2*N* sulfuric acid and bring volume to 100.0mL. Mix thoroughly. Autoclave under 100% N_2 for 15 min at 15 psi pressure–121°C. Cool to room temperature.

Preparation of Medium: Add components, except $FeSO_4$ solution, to distilled/deionized water and bring volume to 950.0mL. Mix thoroughly. Adjust pH to 1.6 with 4*N* sulfuric acid. Autoclave for 15 min at 15 psi pressure–121°C. Cool to 25°C. Aseptically add 50.0mL sterile $FeSO_4$ solution. Mix thoroughly. Distribute into tubes or flasks.

Use: For the cultivation and maintenance of *Leptospirillum* sp. DSM9468.

MS Medium for *Acidiphilum cryptum*

Composition per liter:

$(NH_4)_2SO_4$	2.0g
$MgSO_4 \cdot 7H_2O$	0.25g
K_2HPO_4	0.1g
KCl	0.1g
Glucose solution	20.0mL
Yeast extract solution	10.0mL

pH 3.0 ± 0.2 at 25°C

Glucose Solution:

Composition per 100.0mL:

D-Glucose	10.0g

Preparation of Glucose Solution: Add glucose to distilled/deionized water and bring volume to 100.0mL. Mix thoroughly. Filter sterilize.

Yeast Extract Solution:

Composition per 10.0mL:

Yeast extract	1.0g

Preparation of Yeast Extract Solution: Add yeast extract to distilled/deionized water and bring volume to 10.0mL. Mix thoroughly. Autoclave for 15 min at 15 psi pressure–121°C.

Preparation of Medium: Add components, except glucose solution and yeast extract solution, to distilled/deionized water and bring volume to 970.0mL. Mix thoroughly. Adjust pH to 3.0 with 2*N* H_2SO_4. Autoclave for 15 min at 15 psi pressure–121°C. Aseptically add 20.0mL of sterile glucose solution and 10.0mL of sterile yeast extract solution. Mix thoroughly. Aseptically distribute into sterile tubes or flasks.

Use: For the cultivation of *Acidiphilium cryptum.*

MS Medium for *Leptospirillum* species

Composition per liter:

$(NH_4)_2SO_4$	2.0g
$MgSO_4 \cdot 7H_2O$	0.25g
K_2HPO_4	0.1g
KCl	0.1g
$FeSO_4$ solution	50.0mL

pH 1.6 ± 0.2 at 25°C

$FeSO_4$ Solution:

Composition per 10.0mL:

$FeSO_4 \cdot 7H_2O$	4.0g

Preparation of $FeSO_4$ Solution: Add $FeSO_4 \cdot 7H_2O$ to distilled/deionized water and bring volume to 10.0mL. Mix thoroughly. Sparge with 100% N_2. Filter sterilize.

Preparation of Medium: Add components, except $FeSO_4$ solution, to distilled/deionized water and bring volume to 950.0mL. Mix thoroughly. Adjust pH to 1.6 with 4*N* H_2SO_4. Autoclave for 15 min at 15 psi pressure–121°C. Aseptically add 50.0mL of sterile $FeSO_4$ solution. Mix thoroughly. Aseptically distribute into sterile tubes or flasks.

Use: For the cultivation of *Leptospirillum* species.

MS Medium for Methanogens

Composition per 408.8mL:

Agar	8.0g
$NaHCO_3$	2.4g
L-Cysteine-sulfide reducing agent	16.0mL
Mineral solution 1	15.0mL
Mineral solution 2	15.0mL
Sodium formate (20% solution)	6.0mL
Yeast extract-soybean casein solution	4.0mL
Sodium acetate (25% solution)	4.0mL
Wolfe's vitamin solution	4.0mL
Wolfe's mineral solution	4.0mL
$FeSO_4 \cdot 7H_2O$ (0.2% solution)	0.4mL
Resazurin (0.1% solution)	0.4mL

pH 7.0 ± 0.2 at 25°C

L-Cysteine-Sulfide Reducing Agent:

Composition per 400.0mL:

L-Cysteine·HCl·H_2O	5.0g
Na_2S (12.5% solution)	40.0mL
NaOH (1*N* solution)	30.0mL

Preparation of L-Cysteine-Sulfide Reducing Agent: Add distilled/deionized water to a 500.0mL round-bottomed flask. Add freshly prepared NaOH solution. Gently heat and bring to boiling under 100% N_2. Remove gassing probe. Add L-cysteine·HCl·H_2O. Add freshly prepared Na_2S solution. Renew gassing for several minutes. Cap with rubber stoppers. Distribute into 8.0mL/18.0mm Hungate tubes.

Mineral Solution 1:

Composition per liter:

K_2HPO_4	6.0g

Preparation of Mineral Solution 1: Add K_2HPO_4 to distilled/deionized water and bring volume to 1.0L. Mix thoroughly.

Mineral Solution 2:

Composition per liter:

NaCl	12.0g
KH_2PO_4	6.0g

$(NH_4)_2SO_4$	6.0g
$MgSO_4 \cdot 7H_2O$	2.6g
$CaCl_2 \cdot 2H_2O$	0.16g

Preparation of Mineral Solution 2: Add components to distilled/deionized water and bring volume to 1.0L. Mix thoroughly.

Yeast Extract-Soybean Casein Solution:

Composition per 100.0mL:

Yeast extract	20.0g
Pancreatic digest of casein	20.0g

Preparation of Yeast Extract-Soybean Casein Solution: Add components to distilled/deionized water and bring volume to 100.0mL. Mix thoroughly.

Wolfe's Mineral Solution:

Composition per liter:

$MgSO_4 \cdot 7H_2O$	3.0g
Nitrilotriacetic acid	1.5g
NaCl	1.0g
$MnSO_4 \cdot H_2O$	0.5g
$FeSO_4 \cdot 7H_2O$	0.1g
$CoCl_2 \cdot 6H_2O$	0.1g
$CaCl_2$	0.1g
$ZnSO_4 \cdot 7H_2O$	0.1g
$CuSO_4 \cdot 5H_2O$	0.01g
$AlK(SO_4)_2 \cdot 12H_2O$	0.01g
H_3BO_3	0.01g
$Na_2MoO_4 \cdot 2H_2O$	0.01g

Preparation of Wolfe's Mineral Solution: Add nitrilotriacetic acid to 500.0mL of distilled/deionized water. Dissolve by adjusting pH to 6.5 with KOH. Add remaining components. Add distilled/deionized water to 1.0L.

Wolfe's Vitamin Solution:

Composition per liter:

Pyridoxine·HCl	10.0mg
Thiamine·HCl	5.0mg
Riboflavin	5.0mg
Nicotinic acid	5.0mg
Calcium pantothenate	5.0mg
p-Aminobenzoic acid	5.0mg
Thioctic acid	5.0mg
Biotin	2.0mg
Folic acid	2.0mg
Cyanocobalamin	100.0μg

Preparation of Wolfe's Vitamin Solution: Add components to distilled/deionized water and bring volume to 1.0L. Mix thoroughly.

Preparation of Medium: Add components to distilled/deionized water and bring volume to 408.8mL. Gently heat and bring to boiling under 80% N_2 + 20% CO_2. Continue boiling until resazrin turns colorless, indicating reduction. Adjust pH to 7.0. Anaerobically distribute into tubes under 80% N_2 + 20% CO_2. Cap with rubber stoppers and aluminum crimp closures. Autoclave for 15 min at 15 psi pressure–121°C.

Use: For the cultivation and maintenance of *Methanobacterium thermoautotrophicum, Methanobacterium wolfei, Methanobrevibacter smithii, Methanogenium bourgense,* and *Methanogenium* species.

MS Medium (Modified) (DSMZ Medium 1145)

Composition per liter:

$MgSO_4 \cdot 7H_2O$	7.0g
Sulfur, elemental	5.0g
$NaS_2O_3 \cdot 5H_2O$	2.0g
$MgCl_2 \cdot 6H_2O$	0.8g
KCl	0.48g
$CaCl_2 \cdot 2H_2O$	0.4g
MS Buffer	200.0mL
Solution A	20.0mL
Solution D	10.0mL
Solution B	1.5mL

pH 6.6 ± 0.2 at 25°C

MS Buffer Solution:

Composition per liter:

NaOH	4.0g

Preparation of MS Buffer Solution: Prepare anaerobic water by sparging with double distilled water with constant gassing with 100% N_2. Add NaOH to double distilled/deionized anaerobic water and bring volume to 1.0L. Mix thoroughly. Sparge with CO_2 to saturate. Filter sterilize.

Solution A:

Composition per liter:

NH_4Cl	100.0g
$MgCl_2 \cdot 6H_2O$	100.0g
$CaCl_2 \cdot 2H_2O$	40.0g

Preparation of Solution A: Prepare anaerobic water by sparging with double distilled water with constant gassing with 100% N_2. Add components to double distilled/deionized anaerobic water and bring volume to 1.0L. Mix thoroughly. Adjust pH 4.0 with HCl.

Solution B:

Composition per liter:

$K_2HPO_4 \cdot 3H_2O$	200.0g

Preparation of Solution B: Prepare anaerobic water by sparging with double distilled water with constant gassing with 100% N_2. Add $K_2HPO_4 \cdot 3H_2O$ to double distilled/deionized anaerobic water and bring volume to 1.0L. Mix thoroughly.

Solution D:

Composition per liter:

Na_2-EDTA	0.5g
$CoCl_2 \cdot 6H_2O$	150.0mg
$MnCl_2 \cdot 4H_2O$	100.0mg
$FeSO_4 \cdot 7H_2O$	100.0mg
$ZnCL_2$	100.0mg
$AlCl_3 \cdot 6H_2O$	40.0mg
$Na_2WO_4 \cdot 2H_2O$	30.0mg
$CuCl_2 \cdot 2H_2O$	20.0mg
$NiSO_4 \cdot 6H_2O$	20.0mg
$Na_2MoO_4 \cdot 2H_2O$	10.0mg
H_2SeO_3	10.0mg
H_3BO_3	10.0mg

Preparation of Solution D: Prepare anaerobic water by sparging with double distilled water with constant gassing with 100% N_2. Add components to double distilled/deionized anaerobic water and bring volume to 1.0L. Mix thoroughly. Adjust pH to 3.0.

Preparation of Medium: Prepare anaerobic water by sparging with double distilled water with constant gassing with 100% N_2. Add components, except sulfur, to double distilled/deionized anaerobic water and bring volume to 1.0L. Adjust pH to 6.6. Autoclave for 15 min at 15 psi pressure–121°C. Sterilize sulfur separately in screw-capped tubes by steaming in a water bath for 3 hr on each of 3 successive days. Aseptically add the sterilized sulfur to the medium. Mix thoroughly.

Use: For the cultivation of *Sulfurihydrogenibium kristjanssonii.*

MS Medium for *Thiobacillus caldus*

Composition per liter:

Sulfur, powdered	5.0g
$(NH_4)_2SO_4$	2.0g
$MgSO_4·7H_2O$	0.25g
K_2HPO_4	0.1g
KCl	0.1g
Yeast extract solution	10.0mL

pH 3.5 ± 0.2 at 25°C

Preparation of Sulfur: Sterilize powdered elemental sulfur by steaming for 3 hr at 0 psi pressure–100°C on 3 successive days.

Yeast Extract Solution:

Composition per 10.0mL:

Yeast extract	0.2g

Preparation of Yeast Extract Solution: Add yeast extract to distilled/deionized water and bring volume to 10.0mL. Mix thoroughly. Autoclave for 15 min at 15 psi pressure–121°C.

Preparation of Medium: Add components, except elemental sulfur and yeast extract solution, to distilled/deionized water and bring volume to 990.0mL. Mix thoroughly. Adjust pH to 3.5 with 1*N* sulfuric acid. Autoclave for 15 min at 15 psi pressure–121°C. Aseptically add 5.0g of sterile elemental sulfur and 10.0mL of sterile yeast extract solution. Mix thoroughly. Aseptically distribute into sterile tubes or flasks.

Use: For the cultivation of *Thiobacillus caldus.*

MS Medium for *Thiobacillus ferrooxidans*

Composition per liter:

$(NH_4)_2SO_4$	2.0g
$MgSO_4·7H_2O$	0.25g
K_2HPO_4	0.1g
KCl	0.1g
$FeSO_4$ solution	50.0mL

pH 2.2 ± 0.2 at 25°C

$FeSO_4$ Solution:

Composition per 10.0mL:

$FeSO_4·7H_2O$	4.0g

Preparation of $FeSO_4$ Solution: Add $FeSO_4·7H_2O$ to distilled/deionized water and bring volume to 10.0mL. Mix thoroughly. Sparge with 100% N_2. Filter sterilize.

Preparation of Medium: Add components, except $FeSO_4$ solution, to distilled/deionized water and bring volume to 950.0mL. Mix thoroughly. Adjust pH to 2.2 with 4*N* H_2SO_4. Autoclave for 15 min at 15 psi pressure–121°C. Aseptically add 50.0mL of sterile $FeSO_4$ solution. Mix thoroughly. Aseptically distribute into sterile tubes or flasks.

Use: For the cultivation of *Thiobacillus ferrooxidans.*

MS Medium for *Thiobacillus thiooxidans*

Composition per liter:

Sulfur, powdered	5.0g
$(NH_4)_2SO_4$	2.0g
$MgSO_4·7H_2O$	0.25g
K_2HPO_4	0.1g
KCl	0.1g

pH 3.5 ± 0.2 at 25°C

Preparation of Sulfur: Sterilize powdered elemental sulfur by steaming for 3 hr at 0 psi pressure–100°C on three successive days.

Preparation of Medium: Add components, except elemental sulfur, to distilled/deionized water and bring volume to 1.0L. Mix thoroughly. Adjust pH to 3.5 with 1*N* sulfuric acid. Autoclave for 15 min at 15 psi pressure–121°C. Aseptically add 5.0g of sterile elemental sulfur. Mix thoroughly. Aseptically distribute into sterile tubes or flasks.

Use: For the cultivation of *Thiobacillus thiooxidans.*

MSA-Fe Medium

Composition per liter:

NaCl	5.8g
Pancreatic digest of casein	2.0g
Yeast extract	2.0g
$MgCl_2·6H_2O$	1.0g
NH_4Cl	1.0g
Mercaptoethanesulfonic acid	0.5g
$K_2HPO_4·3H_2O$	0.4g
Resazurin	0.5mg
Trace elements solution	10.0mL

pH 7.5 ± 0.2 at 25°C

Trace Elements Solution:

Composition per 100.0mL:

Disodium EDTA·$2H_2O$	50.0mg
$CoCl_2·6H_2O$	15.0mg
$FeSO_4·7H_2O$	10.0mg
$MnCl_2·4H_2O$	10.0mg
$ZnCl_2$	10.0mg
$AlCl_3·6H_2O$	4.0mg
$Na_2WO_4·2H_2O$	3.0mg
$CuCl_2·2H_2O$	2.0mg
$NiSO_4·6H_2O$	2.0mg
H_2SeO_3	1.0mg
H_3BO_3	1.0mg
$Na_2MoO_4·2H_2O$	1.0mg

Preparation of Trace Elements Solution: Add components to distilled/deionized water and bring volume to 100.0mL. Mix thoroughly.

Preparation of Medium: Prepare and dispense medium under 100% N_2. Add components to distilled/deionized water and bring volume to 1.0L. Mix thoroughly. Adjust pH to 7.5 with 2*N* NaOH. Sparge with 100% N_2 for 30 min. Anaerobically distribute into tubes. Autoclave for 15 min at 15 psi pressure–121°C.

Use: For the cultivation of *Bacillus infermus.*

m-Seven Hour Fecal Coliform Agar

See: **Seven-Hour Fecal Coliform Agar**

M-Slanetz *Enterococcus* Broth Base with Triphenyltetrazolium Chloride

Composition per liter:

Sucrose	100.0g
Casein enzymic hydrolysate	25.0g
Peptic digest of animal tissue	15.0g
Yeast extract	10.0g

K_2HPO_4 .... 4.0g
Glucose 2.0g
NaN_3 .... 0.4g
Triphenyltetrazolium chloride solution 5.0mL
pH 7.2 ± 0.2 at 25°C

Source: This medium, without triphenyltetrazolium chloride solution, is available as a premixed powder from HiMedia.

Caution: Sodium azide is toxic. Azides also react with metals and disposal must be highly diluted.

Triphenyltetrazolium Choride Solution:
Composition per 5.0mL:
Triphenyltetrazolium chloride 0.1g

Preparation of Triphenyltetrazolium Choride Solution: Add triphenyltetrazolium chloride to distilled/deionized water and bring volume to 5.0mL. Mix thoroughly. Filter sterilize.

Preparation of Medium: Add components, except triphenyltetrazolium chloride solution, to distilled/deionized water and bring volume to 1.0L.Mix thoroughly. Gently heat and bring to boiling. Distribute into tubes or flasks. Autoclave for 15 min at 15 psi pressure–121°C. Cool to 45°–50°C. Aseptically add 5.0mL triphenyltetrazolium chloride solution. Mix thoroughly.

Use: For the isolation, cultivation, and enumeration of entercocci in water, sewage, and feces by the membrane filter method. For the direct plating of specimens for the detection and enumeration of fecal streptococci. For the isolation and detection of enterococci using the membrane filter technique.

M-Slanetz *Enterococcus* HiVeg Broth Base with Triphenyltetrazolium Chloride

Composition per liter:
Sucrose 100.0g
Plant hydrolysate 25.0g
Plant peptone 15.0g
Yeast extract 10.0g
K_2HPO_4 .... 4.0g
Glucose 2.0g
NaN_3 .... 0.4g
Triphenyltetrazolium chloride solution 5.0mL
pH 7.2 ± 0.2 at 25°C

Source: This medium, without triphenyltetrazolium chloride solution, is available as a premixed powder from HiMedia.

Caution: Sodium azide is toxic. Azides also react with metals and disposal must be highly diluted.

Triphenyltetrazolium Choride Solution:
Composition per 5.0mL:
Triphenyltetrazolium chloride 0.1g

Preparation of Triphenyltetrazolium Choride Solution: Add triphenyltetrazolium chloride to distilled/deionized water and bring volume to 5.0mL. Mix thoroughly. Filter sterilize.

Preparation of Medium: Add components, except triphenyltetrazolium chloride solution, to distilled/deionized water and bring volume to 1.0L.Mix thoroughly. Gently heat and bring to boiling. Distribute into tubes or flasks. Autoclave for 15 min at 15 psi pressure–121°C. Cool to 45°–50°C. Aseptically add 5.0mL triphenyltetrazolium chloride solution. Mix thoroughly.

Use: For the cultivation, and enumeration of entercocci in water, sewage, and feces by the membrane filter method. For the isolation and detection of enterococci using the membrane filter technique.

m-Sporulation Agar
See: **Sporulation Agar**

MSRV Medium
See: **Modified Semisolid Rappaport Vassiliadis Medium**

m-ST Holding Medium
See: **ST Holding Medium**

m-Standard Methods
See: **Standard Methods Broth**

m-*Staphylococcus* Broth
See: ***Staphylococcus* Broth**

M-*Staphylococcus* HiVeg Broth

Composition per liter:
NaCl 75.0g
Plant hydrolysate 10.0g
Mannitol 10.0g
K_2HPO_4 .... 5.0g
Yeast extract 2.5g
Lactose 2.0g
NaN_3 .... 0.049g
pH 7.0 ± 0.2 at 25°C

Source: This medium is available as a premixed powder from HiMedia.

Caution: Sodium azide is toxic. Azides also react with metals and disposal must be highly diluted.

Preparation of Medium: Add components to distilled/deionized water and bring volume to 1.0L. Mix thoroughly. Distribute into tubes or flasks. Autoclave for 15 min at 15 psi pressure–121°C.

Use: For the cultivation and enumeration of pathogenic and enterotoxigenic staphylococci by the membrane filter method.

MSL86 Medium (DSMZ Medium 1068)

Composition per liter:
NaCl 10.0g
$MgSO_4 \cdot 7H_2O$ 2.0g
NH_4Cl 1.0g
Na_2SO_4 .... 1.0g
Yeast extract 0.5g
$CaCl_2 \cdot 2H_2O$ 0.1g
Resazurin 0.5mg
Vitamin solution 10.0mL
Sulfide solution 10.0mL
Cysteine solution 10.0mL
Lactate solution 10.0mL
Trace element solution SL-10 .... 1.0mL
pH 7.5 ± 0.2 at 25°C

Sulfide Solution:
Composition per 10.0mL:
$Na_2S \cdot 9H_2O$ 0.25g

Preparation of Sulfide Solution: Add $Na_2S{\cdot}9H_2O$ to distilled/deionized water and bring volume to 10.0mL. Mix thoroughly. Autoclave under 100% N_2 for 15 min at 15 psi pressure–121°C. Cool to room temperature.

Cysteine Solution:

Composition per 10.0mL:

L-Cysteine-HCl·$2H_2O$	0.25g

Preparation of Cysteine Solution: Add L-cysteine-HCl·$2H_2O$ to to distilled/deionized water and bring volume to 10.0mL. Mix thoroughly. Sparge with 100% N_2. Filter sterilize.

Lactate Solution:

Composition per 10.0mL:

Sodium lactate	2.5g

Preparation of Lactate Solution: Add sodium lactate to distilled/deionized water and bring volume to 10.0mL. Mix thoroughly. Filter sterilize.

Trace Elements Solution SL-10:

Composition per liter:

$MgSO_4{\cdot}7H_2O$	3.0g
Nitrilotriacetic acid	1.5g
NaCl	1.0g
$MnSO_4{\cdot}2H_2O$	0.5g
$CoSO_4{\cdot}7H_2O$	0.18g
$ZnSO_4{\cdot}7H_2O$	0.18g
$CaCl_2{\cdot}2H_2O$	0.1g
$FeSO_4{\cdot}7H_2O$	0.1g
$NiCl_2{\cdot}6H_2O$	0.025g
$KAl(SO_4)_2{\cdot}12H_2O$	0.02g
H_3BO_3	0.01g
$Na_2MoO_4{\cdot}4H_2O$	0.01g
$CuSO_4{\cdot}5H_2O$	0.01g
$Na_2SeO_3{\cdot}5H_2O$	0.3mg

Preparation of Trace Elements Solution SL-10: Add nitrilotriacetic acid to 500.0mL of distilled/deionized water. Dissolve by adjusting pH to 6.5 with KOH. Add remaining components. Add distilled/deionized water to 1.0L. Mix thoroughly. Adjust pH to 7.0.

Vitamin Solution:

Composition per liter:

Pyridoxine-HCl	10.0mg
Thiamine-HCl·$2H_2O$	5.0mg
Riboflavin	5.0mg
Nicotinic acid	5.0mg
D-Ca-pantothenate	5.0mg
p-Aminobenzoic acid	5.0mg
Lipoic acid	5.0mg
Biotin	2.0mg
Folic acid	2.0mg
Vitamin B_{12}	0.1mg

Preparation of Vitamin Solution: Add components to distilled/deionized water and bring volume to 1.0L. Mix thoroughly. Sparge with 80% H_2 + 20% CO_2. Filter sterilize.

Preparation of Medium: Add components, except lactate, cysteine, sulfide, and vitamin solutions, to distilled/deionized water and bring volume to 960.0mL. Mix thoroughly. Sparge wtih 100% N_2 gas for at least 45 min. Dispense the medium under same gas atmosphere in culture vessels. Autoclave for 15 min at 15 psi pressure–121°C. Cool to room temperature. Aseptically add the lactate, cysteine, sulfide, and vitamin solutions. Adjust pH of the completed medium to 7.5.

Use: For the cultivation of *Desulfopila aestuarii.*

MSV AcS Agar

Composition per liter:

$Na_2S{\cdot}9H_2O$	0.187g
Sodium acetate	0.15g
MSV agar	1.0L

pH 7.2–7.5 at 25°C

MSV Agar:

Composition per liter:

Agar	12.0g
$(NH_4)_2SO_4$	0.5g
K_2HPO_4	0.11g
KH_2PO_4	0.085g
$MgSO_4{\cdot}7H_2O$	0.05g
$CaCl_2{\cdot}2H_20$	0.05g
EDTA	3.0mg
$FeCl_3{\cdot}H_2O$	2.0mg
Vitamin mix	1.0mL

Preparation of MSV Agar: Add components to distilled/deionized water and bring volume to 1.0L. Mix thoroughly.

Vitamin Mix:

Composition per 100.0mL:

Calcium pantothenate	0.01g
Niacin	0.01g
Pyridoxine	0.01g
p-Aminobenzoic acid	0.01g
Cocarboxylase	0.01g
Inositol	0.01g
Thiamine	0.01g
Riboflavin	0.01g
Biotin	0.5mg
Cyanocobalamin	0.5mg
Folic acid	0.5mg

Preparation of Vitamin Mix: Add components to distilled/deionized water and bring volume to 100.0mL. Mix thoroughly.

Preparation of Medium: To 1.0L of MSV agar add sodium acetate and $Na_2S{\cdot}9H_2O$. Adjust pH to 7.2–7.5. Gently heat to boiling. Distribute into tubes or flasks. Autoclave for 15 min at 15 psi pressure–121°C. Pour into sterile Petri dishes or leave in tubes.

Use: For the isolation, cultivation, and enrichment of heterotrophic strains of *Thiothrix* species from water and environmental sources.

MSV Agar

Composition per liter:

Agar	12.0g
$(NH_4)_2SO_4$	0.5g
K_2HPO_4	0.11g
KH_2PO_4	0.085g
$MgSO_4{\cdot}7H_2O$	0.05g
$CaCl_2{\cdot}2H_20$	0.05g
EDTA	3.0mg
$FeCl_3{\cdot}H_2O$	2.0mg
Vitamin mix	1.0mL

pH 7.2–7.5 at 25°C

Vitamin Mix:

Composition per 100.0mL:

Calcium pantothenate	0.01g
Niacin	0.01g
Pyridoxine	0.01g
p-Aminobenzoic acid	0.01g
Cocarboxylase	0.01g
Inositol	0.01g
Thiamine	0.01g
Riboflavin	0.01g
Biotin	0.5mg
Cyanocobalamin	0.5mg
Folic acid	0.5mg

Preparation of Vitamin Mix: Add components to distilled/deionized water and bring volume to 100.0mL. Mix thoroughly.

Preparation of Medium: Add components to distilled/deionized water and bring volume to 1.0L. Mix thoroughly. Adjust pH to 7.2–7.5. Gently heat to boiling. Distribute into tubes or flasks. Autoclave for 15 min at 15 psi pressure–121°C. Pour into sterile Petri dishes or leave in tubes.

Use: For the isolation, cultivation, and enrichment of heterotrophic strains of *Thiothrix* species from water and environmental sources.

MSV Broth

Composition per liter:

$(NH_4)_2SO_4$	0.5g
K_2HPO_4	0.11g
KH_2PO_4	0.085g
$MgSO_4 \cdot 7H_2O$	0.05g
$CaCl_2 \cdot 2H_20$	0.05g
EDTA	3.0mg
$FeCl_3 \cdot H_2O$	2.0mg
Vitamin mix	1.0mL

pH 7.2–7.5 at 25°C

Vitamin Mix:

Composition per 100.0mL:

Calcium pantothenate	0.01g
Niacin	0.01g
Pyridoxine	0.01g
p-Aminobenzoic acid	0.01g
Cocarboxylase	0.01g
Inositol	0.01g
Thiamine	0.01g
Riboflavin	0.01g
Biotin	0.5mg
Cyanocobalamin	0.5mg
Folic acid	0.5mg

Preparation of Vitamin Mix: Add components to distilled/deionized water and bring volume to 100.0mL. Mix thoroughly.

Preparation of Medium: Add components to distilled/deionized water and bring volume to 1.0L. Mix thoroughly. Adjust pH to 7.2–7.5. Distribute into tubes or flasks. Autoclave for 15 min at 15 psi pressure–121°C.

Use: For the isolation, cultivation, and enrichment of heterotrophic strains of *Thiothrix* species from water and environmental sources.

MSV GS Agar

Composition per liter:

$Na_2S \cdot 9H_2O$	0.187g
Glucose	0.15g
MSV agar	1.0L

pH 7.2–7.5 at 25°C

MSV Agar:

Composition per liter:

Agar	12.0g
$(NH_4)_2SO_4$	0.5g
K_2HPO_4	0.11g
KH_2PO_4	0.085g
$MgSO_4 \cdot 7H_2O$	0.05g
$CaCl_2 \cdot 2H_20$	0.05g
EDTA	3.0mg
$FeCl_3 \cdot H_2O$	2.0mg
Vitamin mix	1.0mL

Preparation of MSV Agar: Add components to distilled/deionized water and bring volume to 1.0L. Mix thoroughly.

Vitamin Mix:

Composition per 100.0mL:

Calcium pantothenate	0.01g
Niacin	0.01g
Pyridoxine	0.01g
p-Aminobenzoic acid	0.01g
Cocarboxylase	0.01g
Inositol	0.01g
Thiamine	0.01g
Riboflavin	0.01g
Biotin	0.5mg
Cyanocobalamin	0.5mg
Folic acid	0.5mg

Preparation of Vitamin Mix: Add components to distilled/deionized water and bring volume to 100.0mL. Mix thoroughly.

Preparation of Medium: To 1.0L of MSV Agar add glucose and $Na_2S \cdot 9H_2O$. Adjust pH to 7.2–7.5. Gently heat to boiling. Distribute into tubes or flasks. Autoclave for 15 min at 15 psi pressure–121°C. Pour into sterile Petri dishes or leave in tubes.

Use: For the isolation, cultivation, and enrichment of heterotrophic strains of *Thiothrix* species from water and environmental sources.

MSV I Agar

Composition per liter:

Glucose	0.15g
MSV agar	1.0L

pH 7.2–7.5 at 25°C

MSV Agar:

Composition per liter:

Agar	12.0g
$(NH_4)_2SO_4$	0.5g
K_2HPO_4	0.11g
KH_2PO_4	0.085g
$MgSO_4 \cdot 7H_2O$	0.05g
$CaCl_2 \cdot 2H_20$	0.05g
EDTA	3.0mg
$FeCl_3 \cdot H_2O$	2.0mg
Vitamin mix	1.0mL

Preparation of MSV Agar: Add components to distilled/deionized water and bring volume to 1.0L. Mix thoroughly.

Vitamin Mix:

Composition per 100.0mL:

Calcium pantothenate	0.01g
Niacin	0.01g
Pyridoxine	0.01g
p-Aminobenzoic acid	0.01g
Cocarboxylase	0.01g
Inositol	0.01g
Thiamine	0.01g
Riboflavin	0.01g
Biotin	0.5mg
Cyanocobalamin	0.5mg
Folic acid	0.5mg

Preparation of Vitamin Mix: Add components to distilled/deionized water and bring volume to 100.0mL. Mix thoroughly.

Preparation of Medium: To 1.0L of MSV agar, add glucose. Adjust pH to 7.2–7.5. Gently heat to boiling. Distribute into tubes or flasks. Autoclave for 15 min at 15 psi pressure–121°C. Pour into sterile Petri dishes or leave in tubes.

Use: For the isolation, cultivation, and enrichment of heterotrophic strains of *Thiothrix* species from water and environmental sources.

MSV LT Agar

Composition per liter:

Sodium lactate	0.5g
$Na_2S_2O_3$	0.5g
MSV agar	1.0L

pH 7.2–7.5 at 25°C

MSV Agar:

Composition per liter:

Agar	12.0g
$(NH_4)_2SO_4$	0.5g
K_2HPO_4	0.11g
KH_2PO_4	0.085g
$MgSO_4 \cdot 7H_2O$	0.05g
$CaCl_2 \cdot 2H_20$	0.05g
EDTA	3.0mg
$FeCl_3 \cdot H_2O$	2.0mg
Vitamin mix	1.0mL

Preparation of MSV Agar: Add components to distilled/deionized water and bring volume to 1.0L. Mix thoroughly.

Vitamin Mix:

Composition per 100.0mL:

Calcium pantothenate	0.01g
Niacin	0.01g
Pyridoxine	0.01g
p-Aminobenzoic acid	0.01g
Cocarboxylase	0.01g
Inositol	0.01g
Thiamine	0.01g
Riboflavin	0.01g
Biotin	0.5mg
Cyanocobalamin	0.5mg
Folic acid	0.5mg

Preparation of Vitamin Mix: Add components to distilled/deionized water and bring volume to 100.0mL. Mix thoroughly.

Preparation of Medium: To 1.0L of MSV agar, add sodium lactate and $Na_2S_2O_3$. Adjust pH to 7.2–7.5. Gently heat to boiling. Distribute into tubes or flasks. Autoclave for 15 min at 15 psi pressure–121°C. Pour into sterile Petri dishes or leave in tubes.

Use: For the isolation, cultivation, and enrichment of heterotrophic strains of *Thiothrix* species from water and environmental sources.

MSV S Agar

Composition per liter:

$Na_2S \cdot 9H_2O$	0.187g
MSV agar	1.0L

pH 7.2–7.5 at 25°C

MSV Agar:

Composition per liter:

Agar	12.0g
$(NH_4)_2SO_4$	0.5g
K_2HPO_4	0.11g
KH_2PO_4	0.085g
$MgSO_4 \cdot 7H_2O$	0.05g
$CaCl_2 \cdot 2H_20$	0.05g
EDTA	3.0mg
$FeCl_3 \cdot H_2O$	2.0mg
Vitamin mix	1.0mL

Preparation of MSV Agar: Add components to distilled/deionized water and bring volume to 1.0L. Mix thoroughly.

Vitamin Mix:

Composition per 100.0mL:

Calcium pantothenate	0.01g
Niacin	0.01g
Pyridoxine	0.01g
p-Aminobenzoic acid	0.01g
Cocarboxylase	0.01g
Inositol	0.01g
Thiamine	0.01g
Riboflavin	0.01g
Biotin	0.5mg
Cyanocobalamin	0.5mg
Folic acid	0.5mg

Preparation of Vitamin Mix: Add components to distilled/deionized water and bring volume to 100.0mL. Mix thoroughly.

Preparation of Medium: To 1.0L of MSV agar, add $Na_2S \cdot 9H_2O$. Adjust pH to 7.2–7.5. Gently heat to boiling. Distribute into tubes or flasks. Autoclave for 15 min at 15 psi pressure–121°C. Pour into sterile Petri dishes or leave in tubes.

Use: For the isolation, cultivation, and enrichment of heterotrophic strains of *Thiothrix* species from water and environmental sources.

MSV SS Agar

Composition per liter:

$Na_2S \cdot 9H_2O$	0.187g
Sucrose	0.15g
MSV agar	1.0L

pH 7.2–7.5 at 25°C

MSV Agar:

Composition per liter:

Agar	12.0g
$(NH_4)_2SO_4$	0.5g
K_2HPO_4	0.11g

KH_2PO_4 .... 0.085g
$MgSO_4 \cdot 7H_2O$ 0.05g
$CaCl_2 \cdot 2H_20$ 0.05g
EDTA 3.0mg
$FeCl_3 \cdot H_2O$ 2.0mg
Vitamin mix 1.0mL

Preparation of MSV Agar: Add components to distilled/deionized water and bring volume to 1.0L. Mix thoroughly.

Vitamin Mix:
Composition per 100.0mL:
Calcium pantothenate 0.01g
Niacin 0.01g
Pyridoxine 0.01g
p-Aminobenzoic acid 0.01g
Cocarboxylase 0.01g
Inositol 0.01g
Thiamine 0.01g
Riboflavin 0.01g
Biotin 0.5mg
Cyanocobalamin 0.5mg
Folic acid 0.5mg

Preparation of Vitamin Mix: Add components to distilled/deionized water and bring volume to 100.0mL. Mix thoroughly.

Preparation of Medium: To 1.0L of MSV agar, add $Na_2S \cdot 9H_2O$ and sucrose. Adjust pH to 7.2–7.5. Gently heat to boiling. Distribute into tubes or flasks. Autoclave for 15 min at 15 psi pressure–121°C. Pour into sterile Petri dishes or leave in tubes.

Use: For the isolation, cultivation, and enrichment of heterotrophic strains of *Thiothrix* species from water and environmental sources.

MSV SUC Agar

Composition per liter:
Sodium succinate 0.15g
MSV agar 1.0L

pH 7.2–7.5 at 25°C

MSV Agar:
Composition per liter:
Agar 12.0g
$(NH_4)_2SO_4$ 0.5g
K_2HPO_4 .... 0.11g
KH_2PO_4 .... 0.085g
$MgSO_4 \cdot 7H_2O$ 0.05g
$CaCl_2 \cdot 2H_20$ 0.05g
EDTA 3.0mg
$FeCl_3 \cdot H_2O$ 2.0mg
Vitamin mix 1.0mL

Preparation of MSV Agar: Add components to distilled/deionized water and bring volume to 1.0L. Mix thoroughly.

Vitamin Mix:
Composition per 100.0mL:
Calcium pantothenate 0.01g
Niacin 0.01g
Pyridoxine 0.01g
p-Aminobenzoic acid 0.01g
Cocarboxylase 0.01g
Inositol 0.01g
Thiamine 0.01g
Riboflavin 0.01g
Biotin 0.5mg
Cyanocobalamin 0.5mg
Folic acid 0.5mg

Preparation of Vitamin Mix: Add components to distilled/deionized water and bring volume to 100.0mL. Mix thoroughly.

Preparation of Medium: To 1.0L of MSV agar, add sodium succinate. Adjust pH to 7.2–7.5. Gently heat to boiling. Distribute into tubes or flasks. Autoclave for 15 min at 15 psi pressure–121°C. Pour into sterile Petri dishes or leave in tubes.

Use: For the isolation, cultivation, and enrichment of heterotrophic strains of *Thiothrix* species from water and environmental sources.

MSVP Agar

Composition per 984.0mL:
Agar, noble 15.0g
HEPES (*N*-[2-hydroxyethyl]piperazine-*N′*-2-ethanesulfonic acid) buffer 2.383g
$(NH_4)_2SO_4$ 0.24g
$CaCl_2 \cdot 2H_2O$ 0.06g
$MgSO_4 \cdot 7H_2O$ 0.06g
Na_2HPO_4 .... 0.03g
KH_2PO_4 .... 0.02g
$FeSO_4$ (10m*M* solution) 1.0mL
Sodium pyruvate solution 5.0mL
Vitamin solution 1.0mL

pH 7.2 ± 0.2 at 25°C

Sodium Pyruvate Solution:
Composition per 50.0mL:
Sodium pyruvate 10.0g

Preparation of Sodium Pyruvate Solution: Add sodium pyruvate to distilled/deionized water and bring volume to 50.0mL. Mix thoroughly. Filter sterilize.

Vitamin Solution:
Composition per liter:
Pyridoxine·HCl 100.0mg
p-Aminobenzoic acid 50.0mg
D-(+)-Calcium pantothenate 50.0mg
Nicotinic acid 50.0mg
Riboflavin 50.0mg
Thiamine·HCl 50.0mg
Biotin 20.0mg
Folic acid 20.0mg
Vitamin B_{12} 1.0mg

Preparation of Vitamin Solution: Add components to distilled/deionized watcr and bring volume to 1.0L. Mix thoroughly. Filter sterilize.

Preparation of Medium: Add components, except sodium pyruvate solution and vitamin solution, to distilled/deionized water and bring volume to 994.0mL. Mix thoroughly. Gently heat and bring to boiling. Adjust pH to 7.2. Autoclave for 15 min at 15 psi pressure–121°C. Cool to 50°–55°C. Aseptically add 5.0mL of sterile sodium pyruvate solution and 1.0mL of sterile vitamin solution. Mix thoroughly. Pour into sterile Petri dishes or distribute into sterile tubes.

Use: For the cultivation of *Leptothrix discophora.*

m-T7 Agar Base
See: **T7 Agar Base**

m-TEC Agar
See: **TEC Agar**

m-Teepol Broth, Enriched
See: **Teepol Broth, Enriched**

m-Tetrathionate Broth
See: **Tetrathionate Broth**

M-Tetrathionate HiVeg Broth Base with Iodine

Composition per liter:

$Na_2S_2O_3$	30.0g
Plant peptone No. 3	5.0g
Synthetic detergent	1.0g
Iodine solution	20.0mL

Source: This medium, without iodine solution, is available as a premixed powder from HiMedia.

Iodine Solution:

Composition per 20.0mL:

Iodine	6.0g
KI	5.0g

Preparation of Iodine Solution: Add iodine and KI to distilled/deionized water and bring volume to 20.0mL. Mix thoroughly.

Preparation of Medium: Add components, except iodine solution, to distilled/deionized water and bring volume to 980.0mL. Mix thoroughly. Gently heat and bring to boiling. Do not autoclave. Cool to 40°C. Add 20.0mL of iodine solution. Mix thoroughly. Distribute into tubes in 10.0mL volumes. Use medium the same day it is prepared.

Use: For the selective isolation and enrichment of *Salmonella typhi* and other salmonellae from fecal specimens, sewage, and other specimens.

m-TGE Broth
See: **TGE Broth**

MTM II
See: **Thayer-Martin Agar, Modified**

MTP4 Medium

Composition per 1001.0mL:

Solution A	870.0mL
Solution C	100.0mL
Solution D (Vitamin solution)	10.0mL
Solution E	10.0mL
Solution B (Trace elements soluiton SL-10)	1.0mL
Methanol	1.0mL
Methanethiol gas	1–2.0mL

pH 7.1–7.4 at 25°C

Solution A:

Composition per 870.0mL:

NaCl	21.0g
$MgCl_2·6H_2O$	3.1g
KCl	0.5g
NH_4Cl	0.3g
KH_2PO_4	0.2g
$CaCl_2·2H_2O$	0.15g
Resazurin	1.0mg

Preparation of Solution A: Add components to distilled/deionized water and bring volume to 870.0mL. Mix thoroughly. Gently heat and bring to boiling. Continue boiling for 3-4 min. Allow to cool to room temperature while gassing under 80% N_2 + 20% CO_2. Continue gassing until pH reaches below 6.0. Seal the flask under 80% N_2 + 20% CO_2. Autoclave for 15 min at 15 psi pressure–121°C.

Solution B (Trace Elements Solution SL-10):

Composition per liter:

$FeCl_2·4H_2O$	1.5g
$CoCl_2·6H_2O$	190.0mg
$MnCl_2·4H_2O$	100.0mg
$ZnCl_2$	70.0mg
$Na_2MoO_4·2H_2O$	36.0mg
$NiCl_2·6H_2O$	24.0mg
H_3BO_3	6.0mg
$CuCl_2·2H_2O$	2.0mg
HCl (25% solution)	10.0mL

Preparation of Solution B (Trace Elements Solution SL-10): Add $FeCl_2·4H_2O$ to 10.0mL of HCl solution. Mix thoroughly. Add distilled/deionized water and bring volume to 1.0L. Add remaining components. Mix thoroughly. Gas under 100% N_2. Autoclave for 15 min at 15 psi pressure–121°C.

Solution C:

Composition per 100.0mL:

$NaHCO_3$	5.0g

Preparation of Solution C: Add $NaHCO_3$ to distilled/deionized water and bring volume to 100.0mL. Mix thoroughly. Filter sterilize. Gas under 80% N_2 + 20% CO_2.

Solution D (Vitamin Solution):

Composition per liter:

Pyridoxine·HCl	10.0mg
Calcium DL-pantothenate	5.0mg
Lipoic acid	5.0mg
Nicotinic acid	5.0mg
p-Aminobenzoic acid	5.0mg
Riboflavin	5.0mg
Thiamine·HCl	5.0mg
Biotin	2.0mg
Folic acid	2.0mg
Vitamin B_{12}	0.1mg

Preparation of Solution D (Vitamin Solution): Add components to distilled/deionized water and bring volume to 1.0L. Mix thoroughly. Gas under 100% N_2. Autoclave for 15 min at 15 psi pressure–121°C.

Solution E:

Composition per 10.0mL:

$Na_2S·9H_2O$	0.4g

Preparation of Solution E: Add $Na_2S·9H_2O$ to distilled/deionized water and bring volume to 10.0mL. Mix thoroughly. Gas under 100% N_2. Autoclave for 15 min at 15 psi pressure–121°C.

Preparation of Medium: Aseptically and anaerobically combine solution A with solution B, solution C, solution D, and solution E, in that order. Mix thoroughly. Anaerobically distribute into sterile tubes or flasks under 80% N_2 + 20% CO_2. Prior to inoculation, aseptically and anaerobically add 1.0mL of filter-sterilized methanol and 1.0-2.0mL of methanethiol gas to each liter of medium. Addition of 10-20mg of sodium dithionite per liter (e.g., from a 5% solution, freshly prepared under N_2 and filter-sterilized) may stimulate growth at the beginning.

Use: For the cultivation and maintenance of *Methanosarcina* species.

m-TT Broth
See: **Tetrathionate Broth**

Mucate Broth

Composition per liter:

Mucic acid	10.0g
Peptone	10.0g
Bromthymol Blue	0.024g

pH 7.4 ± 0.1 at 25°C.

Preparation of Medium: Add components to distilled/deionized water and bring volume to 1.0L. Mix thoroughly. Add 5*N* NaOH while stirring until mucic acid dissolves. Distribute into screw-capped tubes in 5.0mL volumes. Autoclave for 10 min at 15 psi pressure–121°C.

Use: For the isolation and cultivation of enterovirulent *Escherichia coli* and *Shigella* species.

Mucate Control Broth

Composition per liter:

Peptone	10.0g
Bromthymol Blue	0.024g

pH 7.4 ± 0.1 at 25°C

Preparation of Medium: Add components to distilled/deionized water and bring volume to 1.0L. Mix thoroughly. Distribute into screw-capped tubes in 5.0mL volumes. Autoclave for 10 min at 15 psi pressure–121°C.

Use: For the isolation and cultivation of enterovirulent *Escherichia coli* and *Shigella* species.

Mucate Control HiVeg Broth

Composition per liter:

Plant peptone	10.0g
Bromthymol Blue	0.024g

pH 7.4 ± 0.2 at 25°C

Source: This medium is available as a premixed powder from HiMedia.

Preparation of Medium: Add components to distilled/deionized water and bring volume to 1.0L. Mix thoroughly. Distribute into screw-capped tubes in 5.0mL volumes. Autoclave for 10 min at 15 psi pressure–121°C.

Use: For the isolation and cultivation of enterovirulent *Escherichia coli* and *Shigella* species.

Mucate HiVeg Broth

Composition per liter:

Plant peptone	10.0g
Mucic acid	10.0g
Bromthymol Blue	0.024g

pH 7.4 ± 0.2 at 25°C.

Source: This medium is available as a premixed powder from HiMedia.

Preparation of Medium: Add components to distilled/deionized water and bring volume to 1.0L. Mix thoroughly. Add 5*N* NaOH while stirring until mucic acid dissolves. Distribute into screw-capped tubes in 5.0mL volumes. Autoclave for 10 min at 15 psi pressure–121°C.

Use: For the isolation and cultivation of enterovirulent *Escherichia coli* and *Shigella* species.

MUD SF Broth Base

Composition per liter:

Tryptose	40.0g
KH_2PO_4	10.0g
D-Galactose	2.0g
Tween 80 (polysorbate 80)	1.5g
4-Methylumbelliferyl-β-D-glucoside (MUD)	0.15g
Selective supplement solution A	10.0mL
Selective supplement solution B	1.0mL

pH 7.5 ± 0.2 at 25°C

Source: This medium is available from HiMedia.

Selective Supplement Solution A:

Composition per 10.0mL:

Thallium acetate	2.0g
Nalidixic acid	0.25g

Preparation of Selective Supplement Solution A: Add components to distilled/deionized water and bring volume to 10.0mL. Mix thoroughly. Filter sterilize.

Selective Supplement Solution B:

Composition per 20.0mL:

2,3,5,Triphenyltetrazolium chloride	0.2g

Preparation of Selective Supplement Solution B: Add 0.2g of 2,3,5,triphenyltetrazolium chloride to distilled/deionized water and bring volume to 20.0mL. Mix thoroughly. Filter sterilize.

Preparation of Medium: Add components, except selective supplement solutions A and B, to distilled/deionized water and bring volume to 989.0mL. Mix thoroughly. Autoclave for 15 min at 15 psi pressure–121°C. Cool to 50°C. Aseptically add selective supplement solutions A and B. Mix thoroughly. Aseptically distribute into sterile tubes or flasks.

Use: For the detection and enumeration of intestinal enterococci in surface and wastewater in accordance with ISO committee under ISO 7899-1:1998.

Mueller-Hinton Agar

Composition per liter:

Beef infusion	300.0g
Acid hydrolysate of casein	17.5g
Agar	17.0g
Starch	1.5g

pH 7.4 ± 0.2 at 25°C

Source: This medium is available as a premixed powder from BD Diagnostic Systems and Oxoid Unipath.

Preparation of Medium: Add components to distilled/deionized water and bring to 1.0L. Mix thoroughly. Gently heat and bring to boiling. Distribute into tubes or flasks. Autoclave for 15 min at 15 psi pressure–121°C. Pour into sterile Petri dishes or leave in tubes.

Use: For the isolation of pathogenic *Neisseria* species. For antimicrobial susceptibility testing of a variety of bacterial species. For the cultivation and maintenance of *Moraxella osloensis* and *Neisseria meningitidis*.

Mueller-Hinton Agar with Horse Blood (LMG Medium 49)

Composition per liter:

Beef, infusion from	300.0g
Acid hydrolysate of casein	17.5g
Agar	17.0g
Starch	1.5g
Horse blood, sterile defibrinated	50.0mL

pH 7.4 ± 0.2 at 25°C

Source: This medium without horse blood is available as a premixed powder from BD Diagnostic Systems and Oxoid Unipath.

Preparation of Medium: Add components, except horse blood, to 950.0mL distilled/deionized water. Mix thoroughly. Gently heat and bring to boiling. Autoclave for 15 min at 15 psi pressure–121°C. Cool to 45°–50°C. Aseptically add 50.0mL sterile horse blood. Mix thoroughly. Pour into sterile Petri dishes or distribute into sterile tubes.

Use: For the cultivation and maintenance of *Campylobacter* spp., *Arcobacter* spp., *Helicobacter* spp., *Moraxella lincolnii,* and other bacteria.

Mueller-Hinton Agar with IsoVitaleX® and Hemoglobin

Composition per liter:

Component A	490.0mL
Component B	490.0mL
IsoVitaleX® enrichment	20.0mL

pH 6.9 ± 0.2 at 25°C

Component A:

Composition per 490.0mL:

Beef infusion	300.0g
Acid hydrolysate of casein	17.5g
Agar	17.0g
Starch	1.5g

Preparation of Component A: Add components to distilled/deionized water and bring to 490.0mL. Mix thoroughly. Gently heat and bring to boiling. Autoclave for 15 min at 15 psi pressure–121°C. Cool to 45°–50°C.

Component B:

Composition per 490.0mL:

Hemoglobin	10.0g

Preparation of Component B: Add hemoglobin to distilled/deionized water and bring to 490.0mL. Mix thoroughly. Gently heat and bring to boiling. Autoclave for 15 min at 15 psi pressure–121°C. Cool to 45°–50°C.

IsoVitaleX® Enrichment:

Composition per liter:

Glucose	100.0g
L-Cysteine·HCl	25.9g
L-Glutamine	10.0g
L-Cystine	1.1g
Adenine	1.0g
Nicotinamide adenine dinucleotide	0.25g
Vitamin B_{12}	0.1g
Thiamine pyrophosphate	0.1g
Guanine·HCl	0.03g
$Fe(NO_3)_3 \cdot 6H_2O$	0.02g
p-Aminobenzoic acid	0.013g
Thiamine·HCl	3.0mg

Source: The supplement solution IsoVitaleX® enrichment is available from BD Diagnostic Systems. This enrichment may be replaced by supplement VX from BD Diagnostic Systems.

Preparation of IsoVitaleX® Enrichment: Add components to distilled/deionized water and bring volume to 1.0L. Mix thoroughly. Filter sterilize. Warm to 45°–50°C.

Preparation of Medium: Aseptically combine 490.0mL of component A, 490.0mL of component B, and 20.0mL of IsoVitaleX® enrichment. Mix thoroughly. Adjust pH to 6.9. Pour into sterile Petri dishes in 20.0mL volumes.

Use: For the isolation and cultivation of *Legionella pneumophila.*

Mueller-Hinton Agar with Sodium Chloride (BAM M107)

Composition per liter:

Beef infusion from	300.0g
NaCl	30.0g
Acid hydrolysate of casein	17.5g
Agar	17.0g
Starch	1.5g

pH 7.3 ± 0.2 at 25°C

Source: This medium without NaCl is available as a premixed powder from BD Diagnostics and Oxoid Unipath.

Preparation of Medium: Add components to distilled/deionized water and bring to 1.0L. Mix thoroughly. Gently heat and bring to boiling. Distribute into tubes or flasks. Autoclave for 15 min at 11 psi pressure–116°C. Pour into sterile Petri dishes or leave in tubes.

Use: For antimicrobial susceptibility testing of a variety of halophilic *Vibrio* spp.

Mueller-Hinton Broth

Composition per liter:

Acid hydrolysate of casein	17.5g
Beef extract	3.0g
Starch	1.5g

pH 7.3 ± 0.1 at 25°C

Source: This medium is available as a premixed powder from BD Diagnostic Systems and Oxoid Unipath.

Preparation of Medium: Add components to distilled/deionized water and bring to 1.0L. Mix thoroughly. Gently heat and bring to boiling. Distribute into tubes or flasks. Autoclave for 10 min at 10 psi pressure–115°C. Do not overheat.

Use: For the cultivation of a wide variety of microorganisms. For antimicrobial susceptibility testing.

Mueller-Hinton Chocolate Agar

Composition per liter:

Beef infusion	300.0g
Acid hydrolysate of casein	17.5g
Agar	17.0g
Starch	1.5g
Sheep blood	50.0mL

pH 7.4 ± 0.2 at 25°C

Preparation of Medium: Add components, except sheep blood, to distilled/deionized water and bring volume to 950.0mL. Mix thoroughly. Gently heat and bring to boiling. Autoclave for 15 min at 15 psi pressure–121°C. Cool to 45°–50°C. Aseptically add sterile sheep blood. Mix thoroughly. Gently heat to 70°C for 10 min. Pour into sterile Petri dishes or distribute into sterile tubes.

Use: For the cultivation and maintenance of *Neisseria gonorrhoeae* and *Neisseria meningitidis*. For antimicrobial susceptibility testing of fastidious microorganisms.

Mueller-Hinton II Agar

Composition per liter:

Acid hydrolysate of casein	17.5g
Agar	17.0g
Beef extract	2.0g
Starch	1.5g

pH 7.3 ± 0.1 at 25°C

Source: This medium is available as a premixed powder from BD Diagnostic Systems.

Preparation of Medium: Add components to distilled/deionized water and bring to 1.0L. Mix thoroughly. Gently heat and bring to boiling. Distribute into tubes or flasks. Autoclave for 15 min at 15 psi pressure–121°C. Pour into sterile Petri dishes or leave in tubes.

Use: For antimicrobial disc diffusion susceptibility testing by the Bauer-Kirby method of a variety of bacteria. This medium supplemented with 5% sheep blood is recommended for use in antimicrobial susceptibility testing of *Streptococcus pneumoniae* and *Haemophilus influenzae.*

Mueller-Hinton HiVeg Agar

Composition per liter:

Plant infusion	300.0g
Plant acid hydrolysate	17.5g
Agar	17.0g
Starch	1.5g

pH 7.3 ± 0.2 at 25°C

Source: This medium is available as a premixed powder from HiMedia.

Preparation of Medium: Add components to distilled/deionized water and bring to 1.0L. Mix thoroughly. Gently heat and bring to boiling. Distribute into tubes or flasks. Autoclave for 15 min at 15 psi pressure–121°C. Pour into sterile Petri dishes or leave in tubes.

Use: For the cultivation of *Neisseria* species and for determination of susceptibility of microorganisms to antimicrobial agents.

Mueller-Hinton HiVeg Agar No. 2

Composition per liter:

Plant hydrolysate	17.5g
Agar	17.0g
Plant infusion	2.0g
Starch, soluble	1.5g

pH 7.3 ± 0.2 at 25°C

Source: This medium is available as a premixed powder from HiMedia.

Preparation of Medium: Add components to distilled/deionized water and bring to 1.0L. Mix thoroughly. Gently heat and bring to boiling. Distribute into tubes or flasks. Autoclave for 15 min at 15 psi pressure–121°C. Pour into sterile Petri dishes or leave in tubes.

Use: For the isolation of pathogenic *Neisseria* species. For testing susceptibility of common and rapidly growing bacteria using antimicrobial discs by the Bauer-Kirby method.

Mueller-Hinton HiVeg Broth

Composition per liter:

Plant acid hydrolysate	17.5g
Plant infusion	2.0g
Starch	1.5g

pH 7.3 ± 0.1 at 25°C

Source: This medium is available as a premixed powder from HiMedia.

Preparation of Medium: Add components to distilled/deionized water and bring to 1.0L. Mix thoroughly. Gently heat and bring to boiling. Distribute into tubes or flasks. Autoclave for 10 min at 10 psi pressure–115°C. Do not overheat.

Use: For the cultivation of a wide variety of microorganisms. For antimicrobial susceptibility testing.

Mueller-Hinton Medium with Garden Soil

Composition per liter:

Garden soil, sterile	300.0g
Beef infusion	300.0g
Acid hydrolysate of casein	17.5g
Agar	17.0g
Starch	1.5g

pH 7.4 ± 0.2 at 25°C

Preparation of Medium: Add components to distilled/deionized water and bring volume to 1.0L. Mix thoroughly. Gently heat and bring to boiling. Distribute into tubes or flasks. Autoclave for 15 min at 15 psi pressure–121°C. Pour into sterile Petri dishes or leave in tubes. Swirl flask while pouring to disperse soil.

Use: For the cultivation and maintenance of *Chromobacterium violaceum.*

Mueller-Hinton Medium with Rabbit Serum

Composition per liter:

Beef infusion	300.0g
Acid hydrolysate of casein	17.5g
Agar	17.0g
Starch	1.5g
Rabbit serum	100.0mL

pH 7.4 ± 0.2 at 25°C

Preparation of Medium: Add components, except rabbit serum, to distilled/deionized water and bring volume to 900.0mL. Mix thoroughly. Gently heat and bring to boiling. Autoclave for 15 min at 15 psi pressure–121°C. Cool to 45°–50°C. Aseptically add sterile rabbit serum. Pour into sterile Petri dishes or distribute into sterile tubes.

Use: For the cultivation and maintenance of *Corynebacterium* species.

Mueller-Kauffmann Tetrathionate Broth

Composition per 1028.0mL:

$Na_2S_2O_3$	40.7g
$CaCO_3$	25.0g
Pancreatic digest of casein	7.0g
Ox bile	4.75g
Soya peptone	2.3g
NaCl	2.3g
Iodine solution	19.0mL
Brilliant Green solution	9.5mL

Iodine Solution:

Composition per 100.0mL:

Iodine	20.0g
KI	25.0g

Preparation of Iodine Solution: Add the KI to approximately 5.0mL of distilled/deionized water. Mix thoroughly. Add the iodine. Gently heat to dissolve. Bring volume to 100.0mL with distilled/deionized water. Filter sterilize.

Brilliant Green Solution:

Composition per 100.0mL:

Brilliant Green	0.1g

Preparation of Brilliant Green Solution: Add the Brilliant Green to distilled/deionized water and bring volume to 100.0mL. Mix thoroughly. Gently heat while stirring and bring to boiling. Continue boiling for 30 min while stirring until dye has dissolved. Filter sterilize. Store protected from light.

Preparation of Medium: Add components, except iodine solution and Brilliant Green solution, to distilled/deionized water and bring volume to 1.0L. Mix thoroughly. Gently heat and bring to boiling. Do not autoclave. Cool to 45°C. Prior to use, add 19.0mL of iodine solution and 9.5mL of Brilliant Green solution. Mix thoroughly. Aseptically distribute into sterile tubes or flasks.

Use: For the isolation and cultivation of *Salmonella* species from specimens with a mixed flora.

Mueller-Kauffman Tetrathionate HiVeg Broth Base with Iodine and Brilliant Green

Composition per liter:

$Na_2S_2O_3$	40.7g
$CaCO_3$	25.0g
Plant hydrolysate	9.75
Papaic digest of soybean meal	2.3g
NaCl	2.3g
Synthetic detergent No. II	2.0g
Iodine solution	19.0mL
Brilliant Green solution	9.5mL

pH 7.2 ± 0.2 at 25°C

Source: This medium, without iodine and Brilliant Green, is available as a premixed powder from HiMedia.

Iodine Solution:

Composition per 100.0mL:

Iodine	20.0g
KI	25.0g

Preparation of Iodine Solution: Add the KI to approximately 5.0mL of distilled/deionized water. Mix thoroughly. Add the iodine. Gently heat to dissolve. Bring volume to 100.0mL with distilled/deionized water. Filter sterilize.

Brilliant Green Solution:

Composition per 100.0mL:

Brilliant Green	0.1g

Preparation of Brilliant Green Solution: Add the Brilliant Green to distilled/deionized water and bring volume to 100.0mL. Mix thoroughly. Gently heat while stirring and bring to boiling. Continue boiling for 30 min while stirring until dye has dissolved. Filter sterilize. Store protected from light.

Preparation of Medium: Add components, except iodine solution and Brilliant Green solution, to distilled/deionized water and bring volume to 1.0L. Mix thoroughly. Gently heat and bring to boiling. Do not autoclave. Cool to 45°C. Prior to use, add 19.0mL of iodine solution and 9.5mL of Brilliant Green solution. Mix thoroughly. Aseptically distribute into sterile tubes or flasks.

Use: For the isolation and cultivation of *Salmonella* species from specimens with a mixed flora.

Mueller-Tellurite Medium

Composition per liter:

Casamino acids	20.0g
Agar	20.0g
Casein	5.0g
KH_2PO_4	0.3g
$MgSO_4 \cdot 7H_2O$	0.1g
L-Tryptophan	0.05g
Mueller-tellurite serum	25.0mL

pH 7.4 ± 0.1 at 25°C

Mueller-Tellurite Serum:

Composition per 100.0mL:

K_2TeO_3 solution	0.4g
Calcium pantothenate	0.2mg
Horse or beef serum, sterile	50.0mL
Sodium lactate solution	40.0mL
Ethyl alcohol	10.0mL

Preparation of Mueller-Tellurite Serum: Add calcium pantothenate to 1.0mL of distilled/deionized water. Autoclave for 15 min at 15 psi pressure–121°C. Add K_2TeO_3 to 1.0mL of sterile distilled/deionized water. To 40.0mL of cooled, sterile sodium lactate solution, add filter-sterilized ethanol, sterile calcium pantothenate solution, sterile serum, and K_2TeO_3 solution. Mix thoroughly.

Sodium Lactate Solution:

Composition per 100.0mL:

Lactic acid (85% solution)	50.0mL
Phenol Red solution (0.2g in 50% ethanol)	0.1mL

Preparation of Sodium Lactate Solution: Add lactic acid to distilled/deionized water and bring volume to 100.0mL. Add 0.1mL of Phenol Red solution. Add enough 40% NaOH solution to adjust pH to 7.0. Gently heat and bring to boiling for 5 min. Add more NaOH solution to retain red color, if necessary. Autoclave for 15 min at 15 psi pressure–121°C. Cool to 50°C.

Caution: Potassium tellurite is toxic.

Preparation of Medium: Add components to distilled/deionized water and bring volume to 975.0mL. Gently heat and bring to boiling. Autoclave for 15 min at 15 psi pressure–121°C. Cool quickly to 50°C. Aseptically add 25.0mL of Mueller-Tellurite serum. Mix thoroughly. Distribute into sterile Petri dishes. Allow the surface of the plates to dry by partially removing the covers during solidification.

Use: For the isolation, cultivation, and differentiation of *Corynebacterium diphtheriae*.

MUG Bromcresol Purple Broth with Lactose

Composition per liter:

Casein enzymic hydrolysate	17.0g
Lactose	10.0g
NaCl	5.0g
Papaic digest of soybean meal	3.0g
Tryptophan	1.0g
Bromcresol Purple	0.02g
4-Methylumbelliferyl-β-D-glucuronide (MUG)	0.01g

pH 7.0 ± 0.2 at 25°C

Source: This medium is available from HiMedia.

Preparation of Medium: Add components to distilled/deionized water and bring volume to 1.0L. Mix thoroughly. Gently heat and bring to boiling. Distribute into tubes with inverted Durham tubes. Autoclave for 20 min at 10 psi pressure–115°C.

Use: For the identification of *Escherichia coli* and coliform bacteria from water samples by a fluorogenic assay method.

MUG EC O157 Agar

Composition per liter:

Casein peptone	20.0g
Agar	13.0g
Sorbitol	10.0g
NaCl	5.0g
Meat extract	2.0g
$Na_2S_2O_3$	2.0g
Sodium deoxycholate	1.12g
Yeast extract	1.0g
Ferric ammonium citrate	0.5g
4-Methylumbellifery-lβ-D-glucuronide (MUG)	0.1g
Bromthymol Blue	0.025g

pH 7.4 ± 0.2 at 25°C

Source: This medium is available from HiMedia.

Preparation of Medium: Add components to distilled/deionized water and bring volume to 1.0L. Mix thoroughly. Gently heat and bring to boiling. Distribute into tubes or flasks. Autoclave for 15 min at 15 psi pressure–121°C. Pour into sterile Petri dishes or leave in tubes.

Use: For the isolation and differentiation of enterohemorrhagic *Escherichia coli* O157:H7 from foodstuffs, water, and clinical samples by a fluorogenic method.

MUG EC O157 Agar, Modified

Composition per liter:

Peptic digest of animal tissue	20.0g
Sorbitol	20.0g
Agar	12.0g
NaCl	5.0g
Bile salts	1.12g
4-Methylumbelliferyl-β-D-glucuronide (MUG)	0.05g
Bromcresol Purple	0.01g

pH 7.2 ± 0.2 at 25°C

Source: This medium is available from HiMedia.

Preparation of Medium: Add components to distilled/deionized water and bring volume to 1.0L. Mix thoroughly. Gently heat and bring to boiling. Distribute into tubes or flasks. Autoclave for 15 min at 15 psi pressure–121°C. Pour into sterile Petri dishes or leave in tubes.

Use: For the isolation and differentiation of enterohemorrhagic *Escherichia coli* O157:H7 from foodstuffs, water, and clinical samples by a fluorogenic method.

MUG EC Broth

Composition per liter:

Casein enzymatic hydrolysate	20.0g
Lactose	5.0g
NaCl	5.0g
K_2HPO_4	4.0g
KH_2PO_4	1.5g
Bile salts mixture	1.5g
4-Methylumbelliferyl-β-D-glucuronide (MUG)	0.05g

pH 6.9 ± 0.2 at 25°C

Source: This medium is available as a premixed powder from HiMedia.

Preparation of Medium: Add components to distilled/deionized water and bring volume to 1.0L. Mix thoroughly. Distribute into test tubes that contain an inverted Durham tube in 10.0mL volumes. Autoclave for 15 min at 15 psi pressure–121°C.

Use: For the detection of *Escherichia coli* in water and food samples by a fluorogenic procedure.

MUG EC Broth, Modified

Composition per liter:

Casein enzymic hydrolysate	40.0g
Salicin	1.0g
Triton X-100	1.0g
4-Methylumbelliferyl-β-D-glucuronide (MUG)	0.1g

pH 6.9 ± 0.2 at 25°C

Source: This medium is available from HiMedia.

Preparation of Medium: Add components to distilled/deionized water and bring volume to 1.0L. Mix thoroughly. Gently heat and bring to boiling. Distribute into tubes or flasks. Autoclave for 15 min at 15 psi pressure–121°C.

Use: For the detection and enumeration of *Escherichia coli* in surface and waste water by the miniaturized method (MPN) in accordance with the ISO committee under ISO 9308-3:1998.

MUG EC HiVeg Broth

Composition per liter:

Plant hydrolysate	20.0g
Lactose	5.0g
NaCl	5.0g
K_2HPO_4	4.0g
KH_2PO_4	1.5g
Synthetic detergent No. I	1.5g
4-Methylumbelliferyl-β-D-glucuronide (MUG)	0.05g

pH 6.9 ± 0.2 at 25°C

Source: This medium is available as a premixed powder from HiMedia.

Preparation of Medium: Add components to distilled/deionized water and bring volume to 1.0L. Mix thoroughly. Distribute into test tubes that contain an inverted Durham tube in 10.0mL volumes. Autoclave for 15 min at 15 psi pressure–121°C.

Use: For the detection of *Escherichia coli* in water and food samples by a fluorogenic procedure.

MUG Lauryl Sulfate Broth

Composition per liter:

Casein enzymic hydrolysate	20.0g
Lactose	5.0g
NaCl	5.0g
K_2HPO_4	2.75g
KH_2PO_4	2.75g
Sodium lauryl sulfate	0.1g
4-Methylumbelliferyl-β-D-glucuronide (MUG)	0.05g

pH 6.8 ± 0.2 at 25°C

Source: This medium is available from HiMedia.

Preparation of Medium: Add components to distilled/deionized water and bring volume to 1.0L. Mix thoroughly. Gently heat and bring to boiling. Distribute into tubes or flasks. Autoclave for 15 min at 15 psi pressure–121°C.

Use: For the detection of coliform bacteria in water and food specimens by a fluorogenic procedure.

MUG Lauryl Sulfate Broth, Modified

Composition per liter:

Casein enzymatic hydrolysate	20.0g
Lactose	5.0g
NaCl	5.0g
K_2HPO_4	2.75g
KH_2PO_4	2.75g
Sodium lauryl sulfate	0.1g
4-Methylumbelliferyl-β-D-glucuronide (MUG)	0.05g

pH 6.8 ± 0.2 at 25°C

Source: This medium is available as a premixed powder from HiMedia.

Preparation of Medium: Add components to distilled/deionized water and bring volume to 1.0L. Mix thoroughly. Distribute into tubes containing an inverted Durham tube in 10.0mL volumes. Autoclave for 12 min at 15 psi pressure–121°C. Cool broth quickly to 25°C. For testing water samples with 10.0mL volumes, prepare medium double strength.

Use: For the detection of *Escherichia coli* in water and food samples by a fluorogenic procedure.

MUG Lauryl Sulfate HiVeg Broth, Modified

Composition per liter:

Plant hydrolysate	20.0g
Lactose	5.0g
NaCl	5.0g
K_2HPO_4	2.75g
KH_2PO_4	2.75g
Sodium lauryl sulfate	0.1g
4-Methylumbelliferyl-β-D-glucuronide (MUG)	0.05g

pH 6.8 ± 0.2 at 25°C

Source: This medium is available as a premixed powder from HiMedia.

Preparation of Medium: Add components to distilled/deionized water and bring volume to 1.0L. Mix thoroughly. Distribute into tubes containing an inverted Durham tube in 10.0mL volumes. Autoclave for 12 min at 15 psi pressure–121°C. Cool broth quickly to 25°C. For testing water samples with 10.0mL volumes, prepare medium double strength.

Use: For the detection of *Escherichia coli* in water and food samples by a fluorogenic procedure.

MUG MacConkey HiVeg Agar

Composition per liter:

Plant peptone	20.0g
Agar	15.0g
Lactose	10.0g
NaCl	5.0g
Synthetic detergent No. I	1.5g
4-Methylumbelliferyl-β-D-glucuronide (MUG)	0.1g
Neutral Red	0.03g
Crystal Violet	1.0mg

pH 7.4 ± 0.2 at 25°C

Source: This medium is available as a premixed powder from HiMedia.

Preparation of Medium: Add components to distilled/deionized water and bring volume to 1.0L. Mix thoroughly. Gently heat while stirring until boiling. Autoclave for 15 min at 15 psi pressure–121°C. Pour into sterile Petri dishes or distribute into sterile tubes.

Use: For the selective isolation, cultivation, and differentiation of coliforms and enteric pathogens based on the ability to ferment lactose by a fluorogenic procedure.

MUG Nutrient Agar

Composition per liter:

Agar	15.0g
Peptic digest of animal tissue	5.0g
NaCl	5.0g
Beef extract	1.5g
Yeast extract	1.5g
4-Methylumbelliferyl-β-D-glucuronide (MUG)	0.1g

pH 7.4 ± 0.2 at 25°C

Source: This medium is available from HiMedia.

Preparation of Medium: Add components to distilled/deionized water and bring volume to 1.0L. Mix thoroughly. Gently heat and bring to boiling. Distribute into tubes or flasks. Autoclave for 15 min at 15 psi pressure–121°C. Pour into sterile Petri dishes or leave in tubes.

Use: For the detection of *Escherichia coli* in water and food samples by a fluorogenic procedure.

MUG Plate Count Agar

Composition per liter:

Agar	15.0g
Casein enzymic hydrolysate	5.0g
Yeast extract	2.5g
Glucose	1.0g
4-Methylumbelliferyl-β-D-glucuronide (MUG)	0.1g

pH 7.4 ± 0.2 at 25°C

Source: This medium is available from HiMedia.

Preparation of Medium: Add components to distilled/deionized water and bring volume to 1.0L. Mix thoroughly. Gently heat and bring to boiling. Distribute into tubes or flasks. Autoclave for 15 min at 15 psi pressure–121°C. Pour into sterile Petri dishes or leave in tubes.

Use: For the enumeration of bacteria in a variety of samples by a fluorogenic procedure.

MUG Sorbitol Agar

Composition per liter:

Peptic digest of animal tissue	17.0g
Agar	13.5g
D-Sorbitol	10.0g
NaCl	5.0g
Proteose peptone	3.0g
Bile salts mixture	1.5g
4-Methylumbelliferyl-β-D-glucuronide (MUG)	0.1g
Neutral Red	0.03g
Crystal Violet	0.001g

pH 7.1 ± 0.2 at 25°C

Source: This medium is available from HiMedia.

Preparation of Medium: Add components to distilled/deionized water and bring volume to 1.0L. Mix thoroughly. Gently heat and bring

to boiling. Distribute into tubes or flasks. Autoclave for 15 min at 15 psi pressure–121°C. Pour into sterile Petri dishes or leave in tubes.

Use: For the isolation and identification of enteropathogenic *Escherichia coli* associated with infant diarrhea by fluorogenic method.

MUG Tryptone Soy Agar

Composition per liter:

Agar	15.0g
Casein enzymic hydrolysate	15.0g
Papaic digest of soybean meal	5.0g
NaCl	5.0g
4-Methylumbelliferyl-β-D-glucuronide (MUG)	0.1g

pH 7.3 ± 0.2 at 25°C

Source: This medium is available from HiMedia.

Preparation of Medium: Add components to distilled/deionized water and bring volume to 1.0L. Mix thoroughly. Gently heat and bring to boiling. Distribute into tubes or flasks. Autoclave for 15 min at 15 psi pressure–121°C. Pour into sterile Petri dishes or leave in tubes.

Use: For the cultivation of fastidious and non-fastidious microorganisms by fluorogenic method.

MUG Violet Red Agar

Composition per liter:

Agar	15.0g
Lactose	10.0g
Peptic digest of animal tissue	7.0g
NaCl	5.0g
Yeast extract	3.0g
Synthetic detergent No. I	1.5g
4-Methylumbelliferyl-β-D-glucuronide (MUG)	0.1g
Neutral Red	0.03g
Crystal Violet	2.0mg

pH 7.4 ± 0.2 at 25°C

Source: This medium is available as a premixed powder from HiMedia.

Preparation of Medium: Add components to distilled/deionized water and bring volume to 1.0L. Mix thoroughly. Gently heat while stirring and bring to boiling. Distribute into tubes or flasks. Do not autoclave. Pour immediately into sterile Petri dishes or leave in tubes.

Use: For the differentiation of *Escherichia coli* from dairy products and other foods by a fluorogenic procedure based on their ability to produce β-glucuronidase.

MUG Violet Red HiVeg Agar

Composition per liter:

Agar	15.0g
Lactose	10.0g
Plant peptone	7.0g
NaCl	5.0g
Yeast extract	3.0g
Synthetic detergent No. I	1.5g
4-Methylumbelliferyl-β-D-glucuronide (MUG)	0.1g
Neutral Red	0.03g
Crystal Violet	2.0mg

pH 7.4 ± 0.2 at 25°C

Source: This medium is available as a premixed powder from HiMedia.

Preparation of Medium: Add components to distilled/deionized water and bring volume to 1.0L. Mix thoroughly. Gently heat while stirring and bring to boiling. Distribute into tubes or flasks. Do not autoclave. Pour immediately into sterile Petri dishes or leave in tubes.

Use: For the differentiation of *Escherichia coli* from dairy products and other foods by a fluorogenic procedure based on their ability to produce β-glucuronidase.

MV Medium

Composition per 1001.0mL:

Na_2SO_4	2.0g
$MgSO_4 \cdot 7H_2O$	1.0g
$Na_2S_2O_35H_2O$	1.0g
NH_4Cl	1.0g
Yeast extract	1.0g
KH_2PO_4	0.5g
$CaCl_2 \cdot 2H_2O$	0.1g
Resazurin	0.5mg
Wolfe's vitamin solution	10.0mL
Sodium malate solution	10.0mL
Sodium pyruvate solution	10.0mL
$NaHCO_3$ solution	10.0mL
$Na_2S \cdot 9H_2O$ solution	10.0mL
Trace elements solution SL-10	1.0mL

pH 7.0–7.2 at 25°C

Sodium Malate Solution:

Composition per 100.0mL:

Sodium malate	1.0g

Preparation of Sodium Malate Solution: Add sodium malate to distilled/deionized water and bring volume to 10.0mL. Mix thoroughly. Sparge with 100% N_2. Autoclave for 15 min at 15 psi pressure–121°C.

Sodium Pyruvate Solution:

Composition per 100.0mL:

Sodium pyruvate	1.0g

Preparation of Sodium Pyruvate Solution: Add sodium pyruvate to distilled/deionized water and bring volume to 10.0mL. Mix thoroughly. Sparge with 100% N_2. Autoclave for 15 min at 15 psi pressure–121°C.

$NaHCO_3$ Solution:

Composition per 10.0mL:

$NaHCO_3$	2.0g

Preparation of $NaHCO_3$ Solution: Add $NaHCO_3$ to distilled/deionized water and bring volume to 10.0mL. Mix thoroughly. Sparge with 80% N_2 + 20% CO_2. Autoclave for 15 min at 15 psi pressure–121°C.

$Na_2S \cdot 9H_2O$ Solution:

Composition per 10.0mL:

$Na_2S \cdot 9H_2O$	0.75μg

Preparation of $Na_2S \cdot 9H_2O$ Solution: Add $Na_2S \cdot 9H_2O$ to distilled/deionized water and bring volume to 10.0mL. Mix thoroughly. Sparge with 100% N_2. Autoclave for 15 min at 15 psi pressure–121°C. Before use, neutralize to pH 7.0 with sterile HCl.

Wolfe's Vitamin Solution:

Composition per liter:

Pyridoxine·HCl	10.0mg
p-Aminobenzoic acid	5.0mg
Lipoic acid	5.0mg

Nicotinic acid	5.0mg
Riboflavin	5.0mg
Thiamine·HCl	5.0mg
Calcium DL-pantothenate	5.0mg
Biotin	2.0mg
Folic acid	2.0mg
Vitamin B_{12}	0.1mg

Preparation of Wolfe's Vitamin Solution: Add components to distilled/deionized water and bring volume to 1.0L. Mix thoroughly. Sparge with 100% N_2. Filter sterilize.

Trace Elements Solution SL-10:

Composition per liter:

$FeCl_2 \cdot 4H_2O$	1.5g
$CoCl_2 \cdot 6H_2O$	190.0mg
$MnCl_2 \cdot 4H_2O$	100.0mg
$ZnCl_2$	70.0mg
$Na_2MoO_4 \cdot 2H_2O$	36.0mg
$NiCl_2 \cdot 6H_2O$	24.0mg
H_3BO_3	6.0mg
$CuCl_2 \cdot 2H_2O$	2.0mg
HCl (25% solution)	10.0mL

Preparation of Trace Elements Solution SL-10: Add $FeCl_2 \cdot 4H_2O$ to 10.0mL of HCl solution. Mix thoroughly. Add distilled/deionized water and bring volume to 1.0L. Add remaining components. Mix thoroughly. Sparge with 100% N_2. Autoclave for 15 min at 15 psi pressure–121°C.

Preparation of Medium: Prepare and dispense medium under 100% N_2. Add components, except Wolfe's vitamin solution, sodium malate solution, sodium pyruvate solution, $NaHCO_3$ solution, $Na_2S \cdot 9H_2O$ solution, and trace elements solution SL-10, to distilled/deionized water and bring volume to 950.0mL. Mix thoroughly. Adjust pH to 7.0–7.2. Gently heat and bring to boiling. Cool while sparging with 100% N_2. Autoclave for 15 min at 15 psi pressure–121°C. Aseptically and anaerobically add 10.0mL of sterile Wolfe's vitamin solution, 10.0mL of sterile sodium malate solution, 10.0mL of sterile sodium pyruvate solution, 10.0mL of sterile $NaHCO_3$ solution, 10.0mL of sterile $Na_2S \cdot 9H_2O$ solution, and 1.0mL of sterile trace elements solution SL-10. Mix thoroughly. Aseptically and anaerobically distribute into sterile tubes or bottles.

Use: For the cultivation of *Desulfotomaculum orientis.*

MVL Medium

Composition per liter:

Agar	15.0g
Pancreatic digest of casein	10.0g
Yeast extract	5.0g
Beef extract	2.0g
Glucose	2.0g
Mineral solution 1	75.0mL
Mineral solution 2	75.0mL
Na_2CO_3 (8.0% solution)	50.0mL
Hemin (0.07% solution)	10.0mL
L-Cysteine·HCl·H_2O (3.0% solution)	10.0mL
Resazurin (0.1% solution)	1.0mL

pH 7.0 ± 0.2 at 25°C

Mineral Solution 1:

Composition per 100.0mL:

K_2HPO_4	0.6g

Preparation of Mineral Solution 1: Add K_2HPO_4 to distilled/deionized water and bring volume to 100.0mL. Mix thoroughly.

Mineral Solution 2:

Composition per 100.0mL:

$(NH_4)_2SO_4$	1.2g
NaCl	1.2g
KH_2PO_4	0.6g
$MgSO_4 \cdot 7H_2O$	0.12g
$CaCl_2 \cdot 2H_2O$	0.12g

Preparation of Mineral Solution 2: Add components to distilled/deionized water and bring volume to 100.0mL. Mix thoroughly.

Preparation of Medium: Prepare and dispense medium under 100% CO_2. Add components to distilled/deionized water and bring volume to 1.0L. Mix thoroughly. Adjust pH to 7.0. Anaerobically distribute into tubes. Autoclave for 15 min at 15 psi pressure–121°C.

Use: For the cultivation of *Tonsillophilus suis*.

MVTY Medium

Composition per liter:

Noble agar	7.0g
Yeast extract	2.0g
Pancreatic digest of casein	2.0g
L-Cysteine·HCl·H_2O	1.0g
Salts A	200.0mL
Salts B	200.0mL
$NaHCO_3$ solution	100.0mL
Glucose (10% solution)	20.0mL
Resazurin (0.1% solution)	1.0mL
n-Butyric acid	0.4mL
Isobutyric acid	0.2mL
DL-2-Methylbutyric acid	0.2mL
n-Valeric acid	0.2mL
Isovaleric acid	0.2mL

pH 6.9 ± 0.2 at 25°C

Salts A:

Composition per liter:

$CaCl_2$	0.45g
$MgSO_4$	0.45g

Preparation of Salts A: Add components to distilled/deionized water and bring volume to 1.0L. Mix thoroughly.

Salts B:

Composition per liter:

NaCl	4.5g
$(NH_4)_2SO_4$	4.5g
KH_2PO_4	2.25g
K_2HPO_4	2.25g

Preparation of Salts B: Add components to distilled/deionized water and bring volume to 1.0L. Mix thoroughly.

$NaHCO_3$ Solution:

Composition per 100.0mL:

$NaHCO_3$	5.0g

Preparation of $NaHCO_3$ Solution: Add $NaHCO_3$ to distilled/deionized water and bring volume to 100.0mL. Mix thoroughly. Filter sterilize.

Preparation of Medium: Separately autoclave the glucose solution, for 15 min at 15 psi pressure–121°C. Prepare and dispense the basal salts A and B under 100% CO_2. Add the agar, yeast extract, pancreatic digest of casein, L-cysteine·HCl·H_2O, salts A, salts B, and resazurin to distilled/deionized water and bring volume to 478.0mL. Mix thoroughly. Adjust pH to 7.0. Autoclave for 15 min at 15 psi pressure–121°C. Cool to 50°C. Add filter sterilized *n*-butyric acid, isobutyric acid, DL-2-methylbutyric acid, *n*-valeric acid, isovaleric acid, sterile $NaHCO_3$ solution, and sterile glucose solution. Mix thoroughly. Aseptically and anaerobically distribute into sterile tubes.

Use: For the cultivation and maintenance of *Treponema saccharophilum.*

MWY Medium (Wadowsky and Yee Medium, Modified)

Composition per liter:

Agar	13.0g
Yeast extract	10.0g
Glycine	3.0g
ACES buffer (2-[(2-amino-2-oxoethyl)-amino]-ethane sulfonic acid)	2.0g
Charcoal, activated	2.0g
α-Ketoglutarate	0.2g
$Fe_4(P_2O_7)_3·9H_2O$	0.05g
Bromcresol Purple	0.01g
Bromcresol Blue	0.01g
Antibiotic inhibitor	10.0mL
L-Cysteine·HCl·H_2O solution	10.0mL

pH 6.9 ± 0.2 at 25°C

Antibiotic Inhibitor:
Composition per 10.0mL:

Anisomycin	0.16g
Cefamandole	4.0mg
Vancomycin	1.0mg
Polymyxin B	130,000U

Preparation of Antibiotic Inhibitor: Add components to distilled/deionized water and bring volume to 10.0mL. Mix thoroughly. Filter sterilize.

L-Cysteine·HCl·H_2O Solution:
Composition per 10.0mL:

L-Cysteine·HCl·H_2O	0.08g

Preparation of L-Cysteine·HCl·H_2O Solution: Add L-cysteine·HCl·H_2O to distilled/deionized water and bring volume to 10.0mL. Mix thoroughly. Filter sterilize.

Preparation of Medium: Add components, except L-cysteine solution and antibiotic inhibitor, to distilled/deionized water and bring volume to 980.0mL. Mix thoroughly. Adjust medium to pH 6.9 with 1*N* KOH. Heat gently and bring to boiling for 1 min. Autoclave for 15 min at 15 psi pressure–121°C. Cool to 50°–55°C. Add 10.0mL of the sterile L-cysteine·HCl·H_2O solution and 10.0mL of the sterile antibiotic solution. Mix thoroughly. Pour into sterile Petri dishes with constant agitation to keep charcoal in suspension.

Use: For the selective isolation and cultivation of *Legionella pneumophila* and other *Legionella* species.

MY Agar

Composition per liter:

Agar	15.0g
Sodium acetate	0.1g
Yeast extract	0.1g
Pancreatic digest of gelatin	0.06g
Beef extract	0.04g
Basal medium	1.0L
Sodium sulfide solution	3.0mL

pH 7.0–7.5 at 25°C

Basal Medium:
Composition per liter:

$CaSO_4·2H_2O$ (saturated solution)	20.0mL
NH_4Cl (4% solution)	5.0mL
Trace elements solution	5.0mL
K_2HPO_4 (1% solution)	1.0mL
$MgSO_4·7H_2O$ (1% solution)	1.0mL

Preparation of Basal Medium: Add components to distilled/deionized water and bring volume to 1.0L. Mix thoroughly.

Trace Elements Solution:
Composition per liter:

Ferrous EDTA solution	20.0mL
$ZnSO_4·7H_2O$ (0.1% solution)	10.0mL
$MnSO_4·4H_2O$ (0.02% solution)	10.0mL
$CuSO_4·5H_2O$ (0.00005% solution)	10.0mL
H_3BO_3 (0.1% solution)	10.0mL
$Co(NO_3)_2$ or $CoCl_2·6H_2O$ (0.01% solution)	10.0mL
$Na_2MoO_4·2H_2O$ (0.01% solution)	10.0mL

Preparation of Trace Elements Solution: Add components to distilled/deionized water and bring volume to 1.0L. Mix thoroughly.

Ferrous EDTA Solution:
Composition per 100.0mL:

$FeSO_4·7H_2O$	7.0g
EDTA	2.0g
HCl, concentrated	1.0mL

Preparation of Ferrous EDTA Solution: Add components to distilled/deionized water and bring volume to 100.0mL. Mix thoroughly.

Sodium Sulfide Solution:
Composition per 10.0mL:

$Na_2S·9H_20$	1.0g

Preparation of Sodium Sulfide Solution: Add $Na_2S·9H_20$ to distilled/deionized water and bring volume to 10.0mL. Mix thoroughly. Autoclave for 15 min at 15 psi pressure–121°C. Prepare freshly.

Preparation of Medium: Add sodium acetate, beef extract, pancreatic digest of gelatin, yeast extract, and agar to 1.0L of basal medium. Mix thoroughly. Adjust pH to 7.0–7.5. Gently heat and bring to boiling. Autoclave for 15 min at 15 psi pressure–121°C. Cool to 45°–50°C. Aseptically add 3.0mL of sterile sodium sulfide solution immediately prior to dispensing. Mix thoroughly. Pour into sterile Petri dishes or distribute into sterile screw-capped tubes.

Use: For the isolation and cultivation of *Beggiatoa* species and myxotrophic strains of *Thiothrix* species from water and environmental sources.

MY20 Agar

Composition per liter:

Glucose	200.0g
Agar	20.0g
Peptone	5.0g
Yeast extract	3.0g
Malt extract	3.0g

Preparation of Medium: Add components to distilled/deionized water and bring volume to 1.0L. Mix thoroughly. Gently heat and bring to boiling. Distribute into tubes or flasks. Autoclave for 15 min at 15 psi pressure–121°C. Pour into sterile Petri dishes or leave in tubes.

Use: For the cultivation of a variety of osmophilic heterotrophic bacteria and fungi.

MY40 Agar

Composition per liter:

Sucrose	400.0g
Agar	20.0g
Malt extract	20.0g
Yeast extract	5.0g

Preparation of Medium: Add components to distilled/deionized water and bring volume to 1.0L. Mix thoroughly. Gently heat and bring to boiling. Distribute into tubes or flasks. Autoclave for 15 min at 15 psi pressure–121°C. Pour into sterile Petri dishes or leave in tubes.

Use: For the cultivation of a variety of osmophilic heterotrophic bacteria and fungi.

MY 40G Agar

Composition per liter:

Glucose	400.0g
Agar	12.0g
Malt extract	12.0g
Yeast extract	3.0g

pH 5.5 ± 0.2 at 25°C

Source: This medium is available from HiMedia.

Preparation of Medium: Add components to distilled/deionized water and bring volume to 1.0L. Mix thoroughly. Gently heat and bring to boiling. Steam for 30 min. Distribute into tubes or flasks. Do not autoclave. Pour into sterile Petri dishes or leave in tubes.

Use: For the cultivation of osmophilic microorganisms.

MY60 Agar

Composition per liter:

Sucrose	600.0g
Agar	20.0g
Malt extract	20.0g
Yeast extract	5.0g

Preparation of Medium: Add components to distilled/deionized water and bring volume to 1.0L. Mix thoroughly. Gently heat and bring to boiling. Distribute into tubes or flasks. Autoclave for 15 min at 15 psi pressure–121°C. Pour into sterile Petri dishes or leave in tubes.

Use: For the cultivation of a variety of osmophilic heterotrophic bacteria and fungi.

MY Medium (DSMZ Medium 753)

Composition per liter:

Maltose	20.0g
Yeast extract	10.0g
NaCl	5.0g
KH_2PO_4	2.0g

pH 6.0 ± 0.2 at 25°C

Preparation of Medium: Add components to distilled/deionized water and bring volume to 1.0L. Mix thoroughly. Adjust pH to 6.0. Distribute into tubes or flasks. Autoclave for 15 min at 15 psi pressure–121°C.

Use: For the cultivation of *Zymobacter palmae.*

MY Medium

Composition per liter:

Maltose	20.0g
Yeast extract	10.0g
NaCl	5.0g
KH_2PO_4	2.0g

Preparation of Medium: Add components to distilled/deionized water and bring volume to 1.0L. Mix thoroughly. Adjust pH to 6.0. Distribute into tubes or flasks. Autoclave for 15 min at 15 psi pressure–121°C.

Use: For the cultivation of *Zymobacter palmae.*

MY10-12 Medium (DSMZ Medium 982)

Composition per liter:

Glucose	120.0g
NaCl	100.0g
Malt extract	20.0g
Agar	20.0g
Yeast extract	5.0g

pH 6.7 ± 0.2 at 25°C

Preparation of Medium: Add components to distilled/deionized water and bring volume to 1.0L. Mix thoroughly. Distribute into tubes or flasks. Sterilize by steaming for 30 min. Overheating will cause softening. Pour into sterile Petri dishes or leave in tubes.

Use: For the cultivation and maintenance of osmophilic yeasts and fungi.

MY10-12 Medium (DSMZ Medium 982)

Composition per liter:

Glucose	120.0g
NaCl	100.0g
Agar	20.0g
Malt extract	20.0g
Yeast extract	5.0g

pH 6.5 ± 0.2 at 25°C

Preparation of Medium: Add components to distilled/deionized water and bring volume to 1.0L. Mix thoroughly. Distribute into tubes or flasks. Steam for 30 min at 0 psi pressure–10°C. Overheating will cause softening. Final aw is 0.88.

Use: For the cultivation of *Aspergillus vitricolae.*

MYA2

See: **Malt 2% Yeast Extract Agar**

MYA4

See: **Malt 4% Yeast Extract Agar**

MYA4 with 40% Sucrose

Composition per liter:

Sucrose	400.0g
Agar	20.0g
Yeast extract	1.0g

Preparation of Medium: Add components to distilled/deionized water and bring volume to 1.0L. Mix thoroughly. Gently heat and bring to boiling. Distribute into tubes or flasks. Autoclave for 15 min at 15 psi pressure–121°C. Pour into sterile Petri dishes or leave in tubes.

Use: For the cultivation of *Ascosphaera apis* and *Moniliella pollinis.*

MYA8

See: **Malt 8% Yeast Extract Agar**

MYCA Medium

Composition per liter:

Yeast extract 3.0g
Casamino acids 2.0g
Sodium DL-malate 1.0g
$(NH_4)_2SO_4$ 0.5g

pH 6.8 ± 0.2 at 25°C

Preparation of Medium: Add components to distilled/deionized water and bring volume to 1.0L. Mix thoroughly. Distribute into tubes or flasks. Autoclave for 15 min at 15 psi pressure–121°C.

Use: For the cultivation of *Rhodoferax fermentans.*

Mycin Assay Agar

Composition per liter:

Glucose 15.0g
Peptone 5.0g
Beef extract 3.0g

pH 7.9 ± 0.1 at 25°C

Preparation of Medium: Add components to distilled/deionized water and bring volume to 1.0L. Mix thoroughly. Distribute into tubes or flasks. Autoclave for 15 min at 15 psi pressure–121°C.

Use: For the microbiological assay of antibiotics in pharmaceutical products, body fluids, feeds, and other sample materials.

Mycobacteria 7H11 Agar with Middlebrook ADC Enrichment

See: **Middlebrook 7H11 Agar with Middlebrook ADC Enrichment**

Mycobacteria 7H11 Agar with Middlebrook OADC Enrichment

See: **Middlebrook 7H11 Agar with Middlebrook ADC Enrichment**

Mycobacteria 7H11 Agar with Middlebrook OADC Enrichment and Triton™ WR 1339

See: **Middlebrook 7H11 Agar with Middlebrook OADC Enrichment and Triton™ WR 1339**

Mycobacterium Agar

Composition per liter:

Agar 12.0g
$Na_2HPO_4 \cdot 12H_2O$ 2.5g
Tryptic digest of casein 2.0g
Proteose peptone No.3 .... 2.0g
Yeast extract 2.0g
Sodium citrate 1.5g
KH_2PO_4 .... 1.0g
$MgSO_4 \cdot 7H_2O$ 0.6g
Glycerol 50.0mL

pH 7.0 ± 0.2 at 25°C

Preparation of Medium: Add components to distilled/deionized water and bring volume to 1.0L. Mix thoroughly. Adjust pH to 7.0. Gently heat and bring to boiling. Distribute into tubes or flasks. Autoclave for 15 min at 15 psi pressure–121°C. Pour into sterile Petri dishes or leave in tubes.

Use: For the cultivation and maintenance of *Actinomyces israelii*, *Mycobacterium* species, and *Nocardia farcinica.*

Mycobacterium intracellulare Agar

Composition per liter:

Glycerol 20.0g
Urea 20.0g
Agar 15.0g
Peptone 5.0g
Meat extract 3.0g

pH 7.0 ± 0.2 at 25°C

Preparation of Medium: Add components to distilled/deionized water and bring volume to 1.0L. Mix thoroughly. Adjust pH to 7.0. Gently heat and bring to boiling. Distribute into tubes or flasks. Autoclave for 15 min at 15 psi pressure–121°C. Pour into sterile Petri dishes or leave in tubes.

Use: For the cultivation and maintenance of *Mycobacterium phlei.*

Mycobacterium Medium

Composition per liter:

Noble agar 15.0g
$(NH_4)_2SO_4$ 1.0g
Na_2HPO_4 .... 0.5g
KH_2PO_4 .... 0.5g
$MgSO_4$ 0.2g
$FeSO_4 \cdot 7H_2O$ 5.0mg
$MnSO_4$ 2.0mg
Liquid paraffin 5.0mL

Preparation of Medium: Add components, except agar, to distilled/deionized water and bring volume to 1.0L. Homogenize in a blender. Add agar. Gently heat and bring to boiling. Distribute into tubes or flasks. Autoclave for 15 min at 15 psi pressure–121°C. Pour into sterile Petri dishes or leave in tubes.

Use: For the cultivation and maintenance of *Mycobacterium paraffinicum.*

Mycobacterium smegmatis Medium

Composition per liter:

Agar 15.0g
Peptone 10.0g
Beef extract 4.0g
NaCl 5.0g

pH 7.2 ± 0.2 at 25°C

Preparation of Medium: Add components to distilled/deionized water and bring volume to 1.0L. Mix thoroughly. Gently heat and bring to boiling. Distribute into tubes or flasks. Autoclave for 15 min at 15 psi pressure–121°C. Pour into sterile Petri dishes or leave in tubes.

Use: For the cultivation of *Mycobacterium smegmatis.*

Mycobacterium Yeast Extract Medium

Composition per liter:

Agar	15.0g
Pancreatic digest of casein	5.0g
Yeast extract	2.5g
Glucose	1.0g

Preparation of Medium: Add components to distilled/deionized water and bring volume to 1.0L. Mix thoroughly. Gently heat and bring to boiling. Distribute into tubes or flasks. Autoclave for 15 min at 15 psi pressure–121°C. Pour into sterile Petri dishes or leave in tubes.

Use: For the cultivation and maintenance of *Mycobacterium* species and *Rhodococcus* species.

Mycobactin Medium

Serum Agar Medium:

Composition per liter:

Noble agar	15.0g
Casamino acids	2.5g
Na_2HPO_4, anhydrous	2.5g
Sodium citrate	1.5g
KH_2PO_4	1.0g
$MgSO_4 \cdot 7H_2O$	0.6g
Asparagine	0.3g
Crude mycobactin	0.16g
Chloramphenicol	0.05g
Primaricine (myprozine)	0.05g
Penicillin	100,000U
Bovine serum, 56°C-inactivated	200.0mL
Tween™ 80 (1% solution)	50.0mL
Glycerol	25.0mL

pH 7.2 ± 0.2 at 25°C

Preparation of Crude Mycobactin: Grow *Mycobacterium phlei* in 600.0mL of mycobactin production broth for 2 weeks at 37°C. Autoclave the culture for 15 min at 15 psi pressure–121°C. Filter the cells and wash with distilled/deionized water. Dry cells under $CaCl_2$. Treat 100.0g of dried culture with three successive acetone extractions—500.0mL of acetone for 30 min in a 1.0L flask fitted with a reflux condenser. Evaporate the acetone to dryness. Extract the residue in a Soxhlet apparatus with petroleum ether for 18–20 hr at 40°–60°C. A hard red residue will remain. Dissolve the residue in warm absolute ethanol. Centrifuge for 30 min at 2250 rpm to remove debris. Evaporate the supernatant to dryness. Grind the residue to a powder of crude mycobactin.

Source: Purified mycobactin is available from Allied Labs, Inc., 2520 Hunt St., Ames, IA 50010.

Mycobactin Production Broth:

Composition per 600.0mL:

Solution B	500.0mL
Solution A	100.0mL

Preparation of Mycobactin Production Broth: Aseptically mix cooled sterile solution A and solution B.

Solution A:

Composition per 100.0mL:

L-Asparagine	5.0g
Na_2HPO_4	2.0g
KH_2PO_4	1.0g
Glycerol	30.0mL

Preparation of Solution A: Add components to distilled/deionized water and bring volume to 100.0mL. Mix thoroughly. Autoclave for 15 min at 15 psi pressure–121°C. Cool to 45°–50°C.

Solution B:

Composition per 500.0mL:

Glucose	10.0g
$MgSO_4 \cdot 7H_2O$	0.2g

Preparation of Solution B: Add components to distilled/deionized water and bring volume to 500.0mL. Mix thoroughly. Autoclave for 15 min at 15 psi pressure–121°C. Cool to 45°–50°C.

Preparation of Medium: Add components, except penicillin, chloramphenicol, primaricine, and bovine serum, to distilled/deionized water and bring volume to 800.0mL. Mix thoroughly. Gently heat with a minimum of heat. Autoclave for 15 min at 10 psi pressure–116°C. Cool to 50°C. Aseptically add penicillin, chloramphenicol, primaricine, and sterile bovine serum. Mix thoroughly. Adjust pH to 7.2. Distribute into sterile tubes or flasks.

Use: For the cultivation and maintenance of *Mycobacterium avium* and *Mycobacterium paratuberculosis*.

Mycobactosel™ Agar

Composition per liter:

Agar	13.5g
Bovine albumin fraction V	5.0g
Glucose	2.0g
Na_2HPO_4	1.5g
KH_2PO_4	1.5g
Pancreatic digest of casein	1.0g
NaCl	0.85g
$(NH_4)_2SO_4$	0.5g
Monosodium glutamate	0.5g
Sodium citrate	0.4g
$MgSO_4 \cdot 7H_2O$	0.05g
Ferric ammonium citrate	0.04g
Pyridoxine	1.0mg
$ZnSO_4 \cdot 7H_2O$	1.0mg
$CuSO_4 \cdot 5H_2O$	1.0mg
Biotin	0.5mg
$CaCl_2 \cdot H_2O$	0.5mg
Malachite Green	0.25mg
Antibiotic solution	10.0mL
Catalase solution	10.0mL
Glycerol	5.0mL
Oleic acid	0.06mL

pH 6.6 ± 0.2 at 25°C

Source: This medium is available as a prepared medium from BD Diagnostic Systems.

Antibiotic Solution:

Composition per 10.0mL:

Cycloheximide	0.36g
Nalidixic acid	0.02g
Lincomycin	2.0mg

Preparation of Antibiotic Solution: Add components to distilled/deionized water and bring volume to 10.0mL. Mix thoroughly. Filter sterilize.

Caution: Cycloheximide is toxic. Avoid skin contact or aerosol formation and inhalation.

Catalase Solution:

Composition per 10.0mL:

Catalase	3.0mg

Preparation of Catalase Solution: Add catalase to distilled/deionized water and bring volume to 10.0mL. Mix thoroughly. Filter sterilize.

Preparation of Medium: Add components, except antibiotic solution and catalase solution, to distilled/deionized water and bring volume to 980.0mL. Mix thoroughly. Gently heat and bring to boiling. Autoclave for 15 min at 15 psi pressure–121°C. Cool to 45°–50°C. Aseptically add sterile antibiotic solution and sterile catalase solution. Mix thoroughly. Pour into sterile Petri dishes or distribute into sterile tubes.

Use: For the selective isolation of mycobacteria from specimens containing mixed flora.

Mycobactosel™ L-J Medium

Composition per liter:

Potato flour	30.0g
L-Asparagine	3.6g
KH_2PO_4, anhydrous	2.5g
Sodium citrate	0.6g
$MgSO_4 \cdot 7H_2O$	0.24g
Homogenized whole egg	1.0L
Malachite Green solution	20.0mL
Glycerol	12.0mL
Antibiotic solution	10.0mL

pH 7.0 ± 0.2 at 25°C

Source: This medium is available as a prepared medium from BD Diagnostic Systems.

Homogenized Whole Egg:

Composition per liter:

Whole eggs	18–24

Preparation of Whole Egg: Use fresh eggs, less than 1 week old. Scrub the shells with soap. Let stand in a soap solution for 30 min. Rinse in running water. Soak eggs in 70% ethanol for 15 min. Break the eggs into a sterile container. Homogenize by shaking. Filter through four layers of sterile cheesecloth into a sterile graduated cylinder. Measure out 1.0L.

Malachite Green Solution:

Composition per 20.0mL:

Malachite Green	0.4g

Preparation of Malachite Green Solution: Add Malachite Green to sterile distilled/deionized water and bring volume to 20.0mL in a sterile container. Mix thoroughly.

Antibiotic Solution:

Composition per 10.0mL:

Cycloheximide	0.64g
Nalidixic acid	0.056g
Lincomycin	3.2mg

Preparation of Antibiotic Solution: Add components to distilled/deionized water and bring volume to 10.0mL. Mix thoroughly. Filter sterilize.

Caution: Cycloheximide is toxic. Avoid skin contact or aerosol formation and inhalation.

Preparation of Medium: Add components—except whole egg, Malachite Green solution, and antibiotic solution—to distilled/deionized water and bring volume to 600.0mL. Mix thoroughly. Autoclave for 30 min at 15 psi pressure–121°C. Cool to room temperature. Add the homogenized whole egg, Malachite Green solution, and antibiotic solution. Distribute into sterile tubes in 8.0mL volumes. Coagulate medium in a slanted position at 85°C (moist heat) for 50 min.

Use: For the isolation and cultivation of *Mycobacterium* species from clinical specimens.

Mycobiotic Agar (Cycloheximide Chloramphenicol Agar)

Composition per liter:

Agar	15.0g
Enzymatic hydrolysate of soybean meal	10.0g
Glucose	10.0g
Cycloheximide	0.5g
Chloramphenicol	0.05g

pH 6.5 ± 0.2 at 25°C

Source: This medium is available as a premixed powder from BD Diagnostic Systems.

Caution: Cycloheximide is toxic. Avoid skin contact or aerosol formation and inhalation.

Preparation of Medium: Add components to distilled/deionized water and bring volume to 1.0L. Mix thoroughly. Gently heat and bring to boiling. Distribute into tubes or flasks. Autoclave for 15 min at 15 psi pressure–121°C. Cool tubes quickly in a slanted position.

Use: For the selective isolation and cultivation of pathogenic fungi.

Mycological Agar

Composition per liter:

Agar	15.0g
Enzymatic hydrolysate of soybean meal	10.0g
Glucose	10.0g

pH 7.0 ± 0.2 at 25°C

Source: This medium is available as a premixed powder from BD Diagnostic Systems.

Preparation of Medium: Add components to distilled/deionized water and bring volume to 1.0L. Mix thoroughly. Gently heat and bring to boiling. Distribute into tubes or flasks. Autoclave for 15 min at 15 psi pressure–121°C.

Use: For the selective isolation, cultivation, and maintenance of pathogenic fungi.

Mycological Agar with Low pH

Composition per liter:

Agar	15.0g
Enzymatic hydrolysate of soybean meal	10.0g
Glucose	10.0g

pH 4.8 ± 0.2 at 25°C

Source: This medium is available as a premixed powder from BD Diagnostic Systems.

Preparation of Medium: Add components to distilled/deionized water and bring volume to 1.0L. Mix thoroughly. Gently heat and bring to boiling. Distribute into tubes or flasks. Autoclave for 15 min at 15 psi pressure–121°C.

Use: For the selective isolation, cultivation, and maintenance of pathogenic fungi.

Mycological Broth

Composition per liter:

Glucose	40.0g
Enzymatic hydrolysate of soybean meal	10.0g

pH 7.0 ± 0.2 at 25°C

Source: This medium is available as a premixed powder from BD Diagnostic Systems.

Preparation of Medium: Add components to distilled/deionized water and bring volume to 1.0L. Mix thoroughly. Gently heat and bring to boiling. Distribute into tubes or flasks. Autoclave for 15 min at 15 psi pressure–121°C.

Use: For the cultivation of fungi.

Mycological Broth with Low pH

Composition per liter:

Glucose	40.0g
Enzymatic hydrolysate of soybean meal	10.0g

pH 4.8 ± 0.2 at 25°C

Source: This medium is available as a premixed powder from BD Diagnostic Systems.

Preparation of Medium: Add components to distilled/deionized water and bring volume to 1.0L. Mix thoroughly. Gently heat and bring to boiling. Adjust pH to 4.8. Distribute into tubes or flasks. Autoclave for 15 min at 15 psi pressure–121°C.

Use: For the cultivation of saprophytic species of yeasts and molds. It is also suitable for cultivation of aciduric bacteria.

Mycophil™ Agar

Composition per liter:

Agar	16.0g
Papaic digest of soybean meal	10.0g
Glucose	10.0g

pH 7.0 ± 0.2 at 25°C

Source: This medium is available as a premixed powder from BD Diagnostic Systems.

Preparation of Medium: Add components to distilled/deionized water and bring volume to 1.0L. Mix thoroughly. Gently heat and bring to boiling. Distribute into tubes or flasks. Autoclave for 15 min at 15 psi pressure–121°C. Pour into sterile Petri dishes or distribute into sterile tubes.

Use: For the cultivation, maintenance, and enumeration of fungi. For the demonstration of pigment production in fungal species. Also used for the cultivation and maintenance of *Bacillus* species.

Mycophil™ Agar with Low pH

Composition per liter:

Agar	18.0g
Papaic digest of soybean meal	10.0g
Glucose	10.0g

pH 4.7 ± 0.2 at 25°C

Source: This medium is available as a premixed powder from BD Diagnostic Systems.

Preparation of Medium: Add components to distilled/deionized water and bring volume to 1.0L. Mix thoroughly. Gently heat and bring to boiling. Distribute into tubes or flasks. Autoclave for 15 min at 15 psi pressure–121°C. Adjust pH to 4.7 by adding approximately 10.0mL of sterile 10% lactic acid. Mix thoroughly. Pour into sterile Petri dishes or distribute into sterile tubes.

Use: For the isolation and enumeration of yeasts and molds.

Mycophil™ Broth

Composition per liter:

Glucose	40.0g
Pancreatic digest of casein	5.0g
Peptic digest of animal tissue	5.0g

pH 7.0 ± 0.2 at 25°C

Source: This medium is available as a premixed powder from BD Diagnostic Systems.

Preparation of Medium: Add components to distilled/deionized water and bring volume to 1.0L. Mix thoroughly. Gently heat and bring to boiling. Distribute into tubes or flasks. Autoclave for 15 min at 15 psi pressure–118°C. Do not overheat.

Use: For the isolation and cultivation of a wide variety of fungi.

Mycoplasma Agar

Composition per liter:

Basal medium	700.0mL
Horse serum	200.0mL
Fresh yeast extract solution	100.0mL

pH 7.5–7.8 at 25°C

Basal Medium:

Composition per 700.0mL:

Sorbitol	50.0g
Beef heart, solids from infusion	16.2g
Noble agar	13.0g
Peptone	3.26g
NaCl	1.62g
Fructose	1.0g
Glucose	1.0g
Sucrose	1.0g
Pancreatic digest of casein	1.0g

Preparation of Basal Medium: Add components to distilled/deionized water and bring volume to 700.0mL. Mix thoroughly. Adjust pH to 7.5–7.8. Autoclave for 15 min at 15 psi pressure–121°C. Cool to 50°C.

Fresh Yeast Extract Solution:

Composition per 100.0mL:

Baker's yeast, live, pressed, starch-free	25.0g

Preparation of Fresh Yeast Extract Solution: Add the live Baker's yeast to 100.0mL of distilled/deionized water. Autoclave for 90 min at 15 psi pressure–121°C. Allow to stand. Remove supernatant solution. Adjust pH to 6.6–6.8.

Preparation of Medium: Filter sterilize horse serum and fresh yeast extract solution. Aseptically add to cooled, sterile basal medium. Mix thoroughly. Aseptically distribute into sterile tubes or flasks.

Use: For the cultivation and maintenance of *Mycoplasma mycoides, Spiroplasma apis, Spiroplasma citri,* and *Spiroplasma melliferum.*

Mycoplasma Agar

Composition per 1004.0mL:

Noble agar	8.0g
Arginine	1.0g
Glucose	1.0g

L-Cysteine·HCl·H_2O	1.0g
Mycoplasma broth base	850.0mL
Horse serum, not inactivated	100.0mL
Fresh yeast extract (25% solution)	50.0mL
Phenol Red (1.0% solution)	2.0mL
DNA calf thymus solution	2.0mL

pH 7.8 ± 0.2 at 25°C

***Mycoplasma* Broth Base:**
Composition per 850.0mL:

Pancreatic digest of casein	7.0g
NaCl	5.0g
Beef extract	3.0g
Yeast extract	3.0g
Beef heart, solids from infusion	2.0g

Preparation of *Mycoplasma* Broth Base: Add components to distilled/deionized water and bring volume to 850.0mL. Add the 8.0g of Noble agar. Mix thoroughly. Gently heat and bring to boiling. Autoclave for 15 min at 15 psi pressure–121°C. Cool to 50°C.

Fresh Yeast Extract Solution:
Composition per 100.0mL:

Baker's yeast, live, pressed, starch-free	25.0g

Preparation of Fresh Yeast Extract Solution: Add the live Baker's yeast to 100.0mL of distilled/deionized water. Autoclave for 90 min at 15 psi pressure–121°C. Allow to stand. Remove supernatant solution. Adjust pH to 6.6–6.8.

DNA Calf Thymus Solution:
Composition per 10.0mL:

DNA calf thymus	1.0g

Preparation of DNA Calf Thymus Solution: Add DNA calf thymus to distilled/deionized water and bring volume to 10.0mL. Mix thoroughly. Filter sterilize.

Preparation of Medium: Combine components, except *Mycoplasma* broth base and DNA calf thymus solution, and mix thoroughly. Filter sterilize through a 0.2μm membrane. Add sterile solution to 850.0mL of cooled, sterile *Mycoplasma* broth base. Aseptically add 2.0mL of sterile DNA calf thymus solution. Mix thoroughly. Pour into sterile Petri dishes or distribute into sterile tubes.

Use: For the cultivation and maintenance of *Mycoplasma lipophilum* and *Mycoplasma* species.

Mycoplasma Agar (ATCC Medium 555)

Composition per 103.0mL:

Noble agar	0.7g
Hartley's digest broth	30.0mL
Pig serum	20.0mL
Enzymatic hydrolysate of lactalbumin	10.0mL
Hanks' balanced salt solution, 10X	4.0mL
Fresh yeast extract solution	2.0mL
Phenol Red (0.25% solution)	1.0mL

pH 7.4 ± 0.2 at 25°C

Hartley's Digest Broth:
Composition per 10.0L:

Ox heart	3,000.0g
Pancreatin	50.0g
Na_2CO_3, anhydrous (0.8% solution)	5.0L
HCl, concentrated	80.0mL

pH 7.5 ± 0.2 at 25°C

Preparation of Hartley's Digest Broth: Finely mince the ox heart. Add the meat to 5.0L of distilled/deionized water. Gently heat and bring to 80°C. Add the 5.0L of Na_2CO_3 solution. Cool to 45°C. Add pancreatin and maintain at 45°C for 4 hr while stirring. Add the HCl and steam at 100°C for 30 min. Cool to room temperature. Adjust pH to 8.0 with 1*N* NaOH. Gently heat and bring to boiling. Continue boiling for 25 min. Filter while hot. Cool to room temperature. Adjust pH to 7.5. Autoclave for 15 min at 15 psi pressure–121°C.

Pig Serum:
Composition per 100.0mL:

Pig serum	100.0mL

Preparation of Pig Serum: Adjust pH of pig serum to 4.4 with sterile 1*N* HCl. Do not let pH go below 4.2. Let serum stand at 4°C for 18-20 hr. Adjust pH to 7.0 with sterile 1*N* NaOH. Centrifuge at 9000 rpm for 20 min. Discard pellet. Filter supernatant solution through a 0.2μm membrane. Store at –70°C.

Fresh Yeast Extract Solution:
Composition per 100.0mL:

Baker's yeast, live, pressed, starch-free	25.0g

Preparation of Fresh Yeast Extract Solution: Add the live Baker's yeast to 100.0mL of distilled/deionized water. Autoclave for 90 min at 15 psi pressure–121°C. Allow to stand. Remove supernatant solution. Adjust pH to 6.6–6.8.

Enzymatic Hydrolysate of Lactalbumin:
Composition per 100.0mL:

Enzymatic hydrolysate of lactalbumin	5.0g

Preparation of Enzymatic Hydrolysate of Lactalbumin: Add enzymatic hydrolysate of lactalbumin to 100.0mL of phosphate buffered saline, 1X, pH 7.0.

Phosphate Buffered Saline Solution, 1X:
Composition per liter:

NaCl	8.0g
$Na_2HPO_4 \cdot 7H_2O$	2.16g
KCl	0.2g
KH_2PO_4	0.2g
$MgCl_2 \cdot 6H_2O$	0.1g
$CaCl_2$	0.1g

Phosphate Buffered Saline Solution, 1X: Add components to distilled/deionized water and bring volume to 1.0L. Mix thoroughly.

Hanks' Balanced Salt Solution, 10X:
Composition per liter:

NaCl	80.0g
Glucose	10.0g
KCl	4.0g
$CaCl_2$	1.4g
$MgCl_2 \cdot 6H_2O$	1.0g
$MgSO_4 \cdot 7H_2O$	1.0g
$Na_2HPO_4 \cdot 7H_2O$	0.9g
KH_2PO_4	0.6g

Preparation of Hanks' Balanced Salt Solution, 10X: Add components to distilled/deionized water and bring volume to 1.0L. Mix thoroughly.

Preparation of Medium: Combine components, except agar, in the following order: Hanks' balanced salt solution, 10X, Phenol Red, Hartley's digest broth, pig serum, enzymatic hydrolysate of lactalbumin, and fresh yeast extract solution. Mix thoroughly. Adjust pH to 7.4 with 1*N* NaOH. Filter sterilize through a 0.2μm membrane. Add 0.7g of Noble agar to 36.0mL of distilled/deionized water. Autoclave for 15 min at 15 psi pressure–121°C. Cool to 56°C. Warm basal medium to 56°C. Aseptically combine the two solutions. Pour into sterile Petri dishes or distribute into sterile tubes.

Use: For the cultivation of *Mycoplasma* species.

Mycoplasma Agar (ATCC Medium 1435)

Composition per 930.0mL:

Agar, noninhibitory to mycoplasmas	10.0g
Glucose	1.0g
Nicotinamide adenine dinucleotide	0.1g
PPLO broth without Crystal Violet	680.0mL
Swine serum (56°C, 30 min)	150.0mL
Yeast extract (25% w/v solution)	100.0mL

pH 7.8 ± 0.2 at 25°C

PPLO Broth without Crystal Violet:

Composition per 680.0mL:

Beef heart, infusion from	11.3g
Peptone	2.28g
NaCl	1.13g

Source: PPLO broth without Crystal Violet is available as a premixed powder from BD Diagnostic Systems.

Preparation of PPLO Broth without Crystal Violet: Add components and agar to distilled/deionized water and bring volume to 680.0mL. Mix thoroughly. Gently heat and bring to boiling. Autoclave for 15 min at 15 psi pressure–121°C. Cool to 45°–50°C.

Preparation of Medium: Mix glucose, nicotinamide adenine dinucleotide, swine serum, and fresh yeast extract solution. Mix thoroughly. Heat to 56°C. Add to cooled, sterile PPLO broth without Crystal Violet. Mix thoroughly. Pour into sterile Petri dishes or distribute into sterile tubes.

Use: For the cultivation and maintenance of *Mycoplasma anseris* and *Mycoplasma lipofaciens*.

Mycoplasma Agar Base (PPLO Agar Base)

Composition per 1300.0mL:

Agar	14.0g
Pancreatic digest of casein	7.0g
NaCl	5.0g
Beef extract	3.0g
Yeast extract	3.0g
Beef heart, solids from infusion	2.0g
Horse serum	260.0mL
Fresh yeast extract solution	65.0mL

pH 7.8 ± 0.2 at 25°C

Source: This medium is available as a premixed powder from BD Diagnostic Systems.

Fresh Yeast Extract Solution:

Composition per 100.0mL:

Baker's yeast, live, pressed, starch-free	25.0g

Preparation of Fresh Yeast Extract Solution: Add the live Baker's yeast to 100.0mL of distilled/deionized water. Autoclave for 90 min at 15 psi pressure–121°C. Allow to stand. Remove supernatant solution. Adjust pH to 6.6–6.8.

Preparation of Medium: Add components, except horse serum and fresh yeast extract, to distilled/deionized water and bring volume to 1.0L. Mix thoroughly. Gently heat and bring to boiling. Distribute into tubes or flasks. Autoclave for 15 min at 15 psi pressure–121°C. Cool to 50°C. To each 75.0mL of cooled, sterile basal medium, add 20.0mL of sterile horse serum and 5.0mL of special yeast extract. Mix thoroughly. Pour into sterile Petri dishes or distribute into sterile tubes.

Use: For the preparation of media for the cultivation of *Mycoplasma*.

Mycoplasma Agar with Increased Selectivity

Composition per 1300.0mL:

Agar	14.0g
Pancreatic digest of casein	7.0g
NaCl	5.0g
Beef extract	3.0g
Yeast extract	3.0g
Beef heart, solids from infusion	2.0g
Thallous acetate	0.7g
Penicillin	70,000U
Horse serum	260.0mL
Fresh yeast extract solution	65.0mL

pH 7.8 ± 0.2 at 25°C

Caution: Thallous acetate is a poison.

Fresh Yeast Extract Solution:

Composition per 100.0mL:

Baker's yeast, live, pressed, starch-free	25.0g

Preparation of Fresh Yeast Extract Solution: Add the live Baker's yeast to 100.0mL of distilled/deionized water. Autoclave for 90 min at 15 psi pressure–121°C. Allow to stand. Remove supernatant solution. Adjust pH to 6.6–6.8.

Preparation of Medium: Add components—except horse serum, special yeast extract, thallous acetate, and penicillin—to distilled/deionized water and bring volume to 1.0L. Mix thoroughly. Gently heat and bring to boiling. Distribute into tubes or flasks. Autoclave for 15 min at 15 psi pressure–121°C. Cool to 50°C. To each 70.0mL of cooled, sterile basal medium, add 20.0mL of sterile horse serum, 10.0mL of special yeast extract, 50.0mg of thallous acetate, and 5000U of penicillin. Mix thoroughly. Pour into sterile Petri dishes or distribute into sterile tubes.

Use: For the preparation of media for the cultivation of *Mycoplasma* species.

Mycoplasma Agar with Supplement G

Composition per liter:

Agar	10.0g
Bacteriological peptone	10.0g
Beef extract	10.0g
NaCl	5.0g
Special mineral supplement	0.5g
Mycoplasma supplement G	250.0mL

pH 7.8 ± 0.2 at 25°C

Source: This medium is available as a premixed powder from Oxoid Unipath.

***Mycoplasma* Supplement G:**

Composition per 20.0mL:

Thallous acetate	25.0mg
Horse serum	20.0mL

Yeast extract (25% solution)	10.0mL
Penicillin	20,000U

Preparation of *Mycoplasma* Supplement G: Add components to distilled/deionized water and bring volume to 20.0mL. Mix thoroughly. Filter sterilize.

Caution: Thallous acetate is a poison.

Preparation of Medium: Add components, except *Mycoplasma* supplement G, to distilled/deionized water and bring volume to 1.0L. Mix thoroughly. Gently heat and bring to boiling. Distribute into flasks in 80.0mL volumes. Autoclave for 15 min at 15 psi pressure–121°C. Cool to 50°C. Aseptically add 20.0mL of sterile *Mycoplasma* supplement G to each 80.0mL of basal medium. Mix thoroughly.

Use: For the growth of *Mycoplasma* species.

Mycoplasma Agar with Supplement P

Composition per liter:

Agar	10.0g
Bacteriological peptone	10.0g
Beef extract	10.0g
NaCl	5.0g
Special mineral supplement	0.5g
Mycoplasma supplement P	250.0mL

pH 7.8 ± 0.2 at 25°C

Source: This medium is available as a premixed powder from Oxoid Unipath.

***Mycoplasma* Supplement P:**
Composition per 20.0mL:

Glucose	0.3g
Mycoplasma broth base	0.145g
Thallous acetate	8.0mg
Phenol Red	1.2mg
Methylene Blue chloride	0.3mg
Penicillin	12,000U
Horse serum	6.0mL
Fresh yeast extract (25% solution)	3.0mL

Preparation of *Mycoplasma* Supplement P: Add components to distilled/deionized water and bring volume to 20.0mL. Mix thoroughly. Filter sterilize.

Caution: Thallous acetate is a poison.

Fresh Yeast Extract Solution:
Composition per 100.0mL:

Baker's yeast, live, pressed, starch-free	25.0g

Preparation of Fresh Yeast Extract Solution: Add the live Baker's yeast to 100.0mL of distilled/deionized water. Autoclave for 90 min at 15 psi pressure–121°C. Allow to stand. Remove supernatant solution. Adjust pH to 6.6–6.8.

Preparation of Medium: Add components, except *Mycoplasma* supplement P, to distilled/deionized water and bring volume to 1.0L. Mix thoroughly. Gently heat and bring to boiling. Distribute into bottles in 1.0mL volumes. Autoclave for 15 min at 15 psi pressure–121°C. Cool to room temperature. Aseptically add 2.0mL of sterile *Mycoplasma* supplement P to each bottle.

Use: For the growth of *Mycoplasma* species.

Mycoplasma Broth

Composition per liter:

Beef heart, infusion from 50g	6.0g
NaCl	5.0g
Mycoploasma supplement solution	300.0mL

pH 7.8 ± 0.2 at 25°C

Source: This medium is available as a premixed powder from BD Diagnostic Systems.

***Mycoplasma* Supplement Solution:**
Composition per 300.0mL:

Horse serum, desiccated	16.0g
Yeast extract	0.1g

Preparation of *Mycoplasma* Supplement Solution: Add components to distilled/deionized water and bring volume to 300.0mL. Mix thoroughly. Filter sterilize

Preparation of Medium: Add components, except *Mycoplasma* supplement solution, to distilled/deionized water and bring volume to 700.0mL. Mix thoroughly. Gently heat and bring to boiling. Boil for 1 min. Autoclave for 15 min at 15 psi pressure–121°C. Cool to 45°–50°C. Aseptically add sterile *Mycoplasma* supplement solution. Mix thoroughly.

Use: For the cultivation of *Mycoplasma* species.

Mycoplasma Broth

Composition per liter:

Pancreatic digest of casein	7.0g
NaCl	5.0g
Beef extract	3.0g
Yeast extract	3.0g
Beef heart, infusion from, solids	2.0g
Mycoploasma supplement solution	300.0mL

pH 7.8 ± 0.2 at 25°C

Source: This medium is available as a premixed powder from BD Diagnostic Systems.

***Mycoplasma* Supplement Solution:**
Composition per 300.0mL:

Horse serum	200.0mL
Yeast extract (fresh autolysate)	100.0mL
Thallium acetate	50.0 mg

Preparation of *Mycoplasma* Supplement Solution: Combine components. Mix thoroughly. Filter sterilize.

Preparation of Medium: Add components, except *Mycoplasma* supplement solution, to distilled/deionized water and bring volume to 700.0mL. Mix thoroughly. Gently heat and bring to boiling. Boil for 1 min. Autoclave for 15 min at 15 psi pressure–121°C. Cool to 45°–50°C. Aseptically add sterile *Mycoplasma* supplement solution. Mix thoroughly.

Use: For the cultivation of *Mycoplasma* species.

Mycoplasma Broth

Composition per liter:

Basal medium	700.0mL
Horse serum	200.0mL
Fresh yeast extract solution	100.0mL

pH 7.5–7.8 at 25°C

Basal Medium:

Composition per 700.0mL:

Sorbitol 50.0g
Beef heart, solids from infusion 16.2g
Peptone 3.26g
NaCl 1.62g
Fructose 1.0g
Glucose 1.0g
Sucrose 1.0g
Pancreatic digest of casein 1.0g

Preparation of Basal Medium: Add components to distilled/deionized water and bring volume to 700.0mL. Mix thoroughly. Adjust pH to 7.5–7.8. Autoclave for 15 min at 15 psi pressure–121°C. Cool to 50°C.

Fresh Yeast Extract Solution:

Composition per 100.0mL:

Baker's yeast, live, pressed, starch-free 25.0g

Preparation of Fresh Yeast Extract Solution: Add the live Baker's yeast to 100.0mL of distilled/deionized water. Autoclave for 90 min at 15 psi pressure–121°C. Allow to stand. Remove supernatant solution. Adjust pH to 6.6–6.8.

Preparation of Medium: Filter sterilize horse serum and fresh yeast extract solution. Aseptically add to cooled, sterile basal medium. Mix thoroughly. Aseptically distribute into sterile tubes or flasks.

Use: For the cultivation and maintenance of *Mycoplasma mycoides*, *Spiroplasma apis*, *Spiroplasma citri*, and *Spiroplasma melliferum*.

Mycoplasma Broth

Composition per 950.0mL:

Glucose 1.0g
Nicotinamide adenine dinucleotide 0.1g
PPLO broth without Crystal Violet 680.0mL
Swine serum (56°C, 30 min) 150.0mL
Fresh yeast extract solution 100.0mL
Phenol Red (0.1% w/v solution) 20.0mL

pH 7.8 ± 0.2 at 25°C

PPLO Broth without Crystal Violet:

Composition per 680.0mL:

Beef heart, solids from infusion 11.3g
Peptone 2.28g
NaCl 1.13g

Source: PPLO broth without Crystal Violet is available as a premixed powder from BD Diagnostic Systems.

Preparation of PPLO Broth without Crystal Violet: Add components to distilled/deionized water and bring volume to 680.0mL. Autoclave for 15 min at 15 psi pressure–121°C. Cool to 56°C.

Fresh Yeast Extract Solution:

Composition per 100.0mL:

Baker's yeast, live, pressed, starch-free 25.0g

Preparation of Fresh Yeast Extract Solution: Add the live Baker's yeast to 100.0mL of distilled/deionized water. Autoclave for 90 min at 15 psi pressure–121°C. Allow to stand. Remove supernatant solution. Adjust pH to 6.6–6.8.

Preparation of Medium: Mix glucose, nicotinamide adenine dinucleotide, swine serum, fresh yeast extract solution, and Phenol Red. Mix thoroughly. Heat to 56°C. Add to cooled, sterile PPLO broth without Crystal Violet. Mix thoroughly. Aseptically distribute into sterile tubes or flasks.

Use: For the cultivation and maintenance of *Mycoplasma anseris* and *Mycoplasma lipofaciens*.

Mycoplasma Broth (ATCC Medium 555)

Composition per 103.0mL:

Hartley's digest broth 30.0mL
Pig serum 20.0mL
Enzymatic hydrolysate of lactalbumin 10.0mL
Hanks' balanced salt solution, 10X 4.0mL
Fresh yeast extract solution 2.0mL
Phenol Red (0.25% solution) 1.0mL

pH 7.4 ± 0.2 at 25°C

Hartley's Digest Broth:

Composition per 10.0L:

Ox heart 3,000.0g
Pancreatin 50.0g
Na_2CO_3, anhydrous (0.8% solution) 5.0L
HCl, concentrated 80.0mL

pH 7.5 ± 0.2 at 25°C

Preparation of Hartley's Digest Broth: Finely mince the ox heart. Add the meat to 5.0L of distilled/deionized water. Gently heat and bring to 80°C. Add the 5.0L of Na_2CO_3 solution. Cool to 45°C. Add pancreatin and maintain at 45°C for 4 hr while stirring. Add the HCl and steam at 100°C for 30 min. Cool to room temperature. Adjust pH to 8.0 with 1*N* NaOH. Gently heat and bring to boiling. Continue boiling for 25 min. Filter while hot. Cool to room temperature. Adjust pH to 7.5. Autoclave for 15 min at 15 psi pressure–121°C.

Pig Serum:

Composition per 100.0mL:

Pig serum 100.0mL

Preparation of Pig Serum: Adjust pH of pig serum to 4.4 with sterile 1*N* HCl. Do not let pH go below 4.2. Let serum stand at 4°C for 18-20 hr. Adjust pH to 7.0 with sterile 1*N* NaOH. Centrifuge at 9000 rpm for 20 min. Discard pellet. Filter supernatant solution through a 0.2μm membrane. Store at –70°C.

Fresh Yeast Extract Solution:

Composition per 100.0mL:

Baker's yeast, live, pressed, starch-free 25.0g

Preparation of Fresh Yeast Extract Solution: Add the live Baker's yeast to 100.0mL of distilled/deionized water. Autoclave for 90 min at 15 psi pressure–121°C. Allow to stand. Remove supernatant solution. Adjust pH to 6.6–6.8.

Enzymatic Hydrolysate of Lactalbumin:

Composition per 100.0mL:

Enzymatic hydrolysate of lactalbumin 5.0g

Preparation of Enzymatic Hydrolysate of Lactalbumin: Add enzymatic hydrolysate of lactalbumin to 100.0mL of phosphate buffered saline, 1X, pH 7.0.

Phosphate Buffered Saline Solution, 1X:

Composition per liter:

NaCl 8.0g
$Na_2HPO_4 \cdot 7H_2O$ 2.16g
KCl 0.2g
KH_2PO_4 .... 0.2g

$MgCl_2·6H_2O$ 0.1g
$CaCl_2$ 0.1g

Hanks' Balanced Salt Solution, 10X:
Composition per liter:

NaCl	80.0g
Glucose	10.0g
KCl	4.0g
$CaCl_2$	1.4g
$MgCl_2·6H_2O$	1.0g
$MgSO_4·7H_2O$	1.0g
$Na_2HPO_4·7H_2O$	0.9g
KH_2PO_4	0.6g

Preparation of Hanks' Balanced Salt Solution, 10X: Add components to distilled/deionized water and bring volume to 1.0L. Mix thoroughly.

Preparation of Medium: Combine components in the following order: Hanks' balanced salt solution, 10X, Phenol Red, Hartley's digest broth, pig serum, enzymatic hydrolysate of lactalbumin, and fresh yeast extract solution. Mix thoroughly. Add 36.0mL of distilled/deionized water. Adjust pH to 7.4 with 1*N* NaOH. Filter sterilize through a 0.2μm membrane. Store at 4°C for up to 3 weeks.

Use: For the cultivation of *Mycoplasma* species.

Mycoplasma Broth Base (PPLO Broth Base without Crystal Violet)

Composition per liter:

Pancreatic digest of casein	7.0g
NaCl	5.0g
Beef extract	3.0g
Yeast extract	3.0g
Beef heart, solids from infusion	2.0g

pH 7.8 ± 0.2 at 25°C

Source: This medium is available as a premixed powder from BD Diagnostic Systems.

Preparation of Medium: Add components to distilled/deionized water and bring volume to 1.0L. Gently heat and bring to boiling. Mix thoroughly. Distribute into tubes or flasks. Autoclave for 15 min at 15 psi pressure–121°C.

Use: Used as a basal medium that should be enriched for the isolation and cultivation of *Mycoplasma* species.

Mycoplasma Broth Base, Frey with Horse Serum

Composition per 1100.0mL:

Pancreatic digest of casein	7.5g
Papaic digest of soybean meal	2.5g
KCl	0.4g
$MgSO_4$	0.2g
Na_2PO_4	1.6g
KH_2PO_4	0.1g
Horse serum	100.0mL

pH 7.7 ± 0.2 at 25°C

Source: This medium is available as a premixed powder from BD Diagnostic Systems.

Preparation of Medium: Add components, except horse serum, to distilled/deionized water and bring volume to 1.0L. Mix thoroughly. Gently heat and bring to boiling. Autoclave for 15 min at 15 psi pressure–121°C. Cool to 50°C. Add sterile, heat-inactivated horse serum. Mix thoroughly. Aseptically distribute into sterile tubes.

Use: For the cultivation of avian mycoplasmas.

Mycoplasma Broth Base, Frey with Horse Serum

Composition per 1100.0mL:

Pancreatic digest	7.5g
Yeast extract	5.0g
NaCl	5.0g
Papaic digest of soybean meal	2.5g
Na_2HPO_4	1.6g
KCl	0.4g
$MgSO_4·7H_2O$	0.2g
KH_2PO_4	0.1g
Horse serum	100.0mL

pH 7.7 ± 0.2 at 25°C

Preparation of Medium: Add components, except horse serum, to distilled/deionized water and bring volume to 1.0L. Mix thoroughly. Gently heat and bring to boiling. Autoclave for 15 min at 15 psi pressure–121°C. Cool to 50°C. Add sterile, heat-inactivated horse serum. Mix thoroughly. Aseptically distribute into sterile tubes.

Use: For the cultivation of avian mycoplasmas.

Mycoplasma Broth, Supplemented

Composition per liter:

Pancreatic digest of casein	7.0g
NaCl	5.0g
Beef extract	3.0g
Yeast extract	3.0g
Beef heart, solids from infusion	2.0g
Horse serum	260.0mL
Fresh yeast extract solution	65.0mL

pH 7.8 ± 0.2 at 25°C

Fresh Yeast Extract Solution:
Composition per 100.0mL:

Baker's yeast, live, pressed, starch-free 25.0g

Preparation of Fresh Yeast Extract Solution: Add the live Baker's yeast to 100.0mL of distilled/deionized water. Autoclave for 90 min at 15 psi pressure–121°C. Allow to stand. Remove supernatant solution. Adjust pH to 6.6–6.8.

Preparation of Medium: Add components, except horse serum and fresh yeast extract solution, to distilled/deionized water and bring volume to 1.0L. Mix thoroughly. Gently heat and bring to boiling. Distribute into tubes or flasks. Autoclave for 15 min at 15 psi pressure–121°C. Cool to 50°C. To each 75.0mL of cooled, sterile basal medium, add 20.0mL of sterile horse serum and 5.0mL of fresh yeast extract solution. Mix thoroughly. Aseptically distribute into sterile tubes.

Use: For the isolation and cultivation of *Mycoplasma* species.

Mycoplasma Broth with Supplement G

Composition per liter:

Bacteriological peptone	10.0g
Beef extract	10.0g
NaCl	5.0g
Special mineral supplement, Oxoid Unipath	0.5g
Mycoplasma supplement G	250.0mL

pH 7.8 ± 0.2 at 25°C

Source: This medium is available as a premixed powder from Oxoid Unipath.

***Mycoplasma* Supplement G:**

Composition per 20.0mL:

Thallous acetate	25.0mg
Horse serum	20.0mL
Yeast extract (25% solution)	10.0mL
Penicillin	20,000U

Preparation of *Mycoplasma* Supplement G: Add components to distilled/deionized water and bring volume to 20.0mL. Mix thoroughly. Filter sterilize.

Caution: Thallous acetate is a poison.

Preparation of Medium: Add components, except *Mycoplasma* supplement G, to distilled/deionized water and bring volume to 1.0L. Mix thoroughly. Gently heat and bring to boiling. Distribute into flasks in 80.0mL volumes. Autoclave for 15 min at 15 psi pressure–121°C. Cool to 50°C. Aseptically add 20.0mL of sterile *Mycoplasma* supplement G to each 80.0mL of basal medium. Mix thoroughly.

Use: For the growth of *Mycoplasma* species.

Mycoplasma Broth with Supplement P

Composition per liter:

Bacteriological peptone	10.0g
Beef extract	10.0g
NaCl	5.0g
Special mineral supplement, Oxoid Unipath	0.5g
Mycoplasma supplement P	250.0mL

pH 7.8 ± 0.2 at 25°C

Source: This medium is available as a premixed powder from Oxoid Unipath.

***Mycoplasma* Supplement P:**

Composition per 20.0mL:

Glucose	0.3g
Mycoplasma broth base	0.145g
Thallous acetate	8.0mg
Phenol Red	1.2mg
Methylene Blue chloride	0.3mg
Penicillin	12,000U
Horse serum	6.0mL
Yeast extract (25% solution)	3.0mL

Preparation of *Mycoplasma* Supplement P: Add components to distilled/deionized water and bring volume to 20.0mL. Mix thoroughly. Filter sterilize.

Caution: Thallous acetate is a poison.

Preparation of Medium: Add components, except *Mycoplasma* supplement P, to distilled/deionized water and bring volume to 1.0L. Mix thoroughly. Gently heat and bring to boiling. Distribute into bottles in 1.0mL volumes. Autoclave for 15 min at 15 psi pressure–121°C. Cool to room temperature. Aseptically add 2.0mL of sterile *Mycoplasma* supplement P to each bottle.

Use: For the cultivation of *Mycoplasma* species.

Mycoplasma Broth with 10% Swine Serum

Composition per liter:

Pancreatic digest of casein	5.6g
NaCl	4.0g
Yeast extract	2.6g
Beef extract	2.4g
Beef heart, solids from infusion	1.6g
Swine serum, heat inactivated	100.0mL
Fresh yeast extract solution	100.0mL

pH 7.8 ± 0.2 at 25°C

Fresh Yeast Extract Solution:

Composition per 100.0mL:

Baker's yeast, live, pressed, starch-free	25.0g

Preparation of Fresh Yeast Extract Solution: Add the live Baker's yeast to 100.0mL of distilled/deionized water. Autoclave for 90 min at 15 psi pressure–121°C. Allow to stand. Remove supernatant solution. Adjust pH to 6.6–6.8.

Preparation of Medium: Add components, except swine serum and fresh yeast extract solution, to distilled/deionized water and bring volume to 800.0mL. Mix thoroughly. Gently heat and bring to boiling. Autoclave for 15 min at 15 psi pressure–121°C. Cool to 50°C. Add sterile swine serum and fresh yeast extract solution. Mix thoroughly. Aseptically distribute into sterile tubes.

Use: For the cultivation and maintenance of *Mycoplasma columbinum* and *Mycoplasma columborale*.

Mycoplasma HiVeg Agar Base with Horse Serum and Yeast Extract (PPLO HiVeg Agar Base)

Composition per liter:

Agar	15.0g
Plant peptone	10.0g
Plant infusion	6.0g
NaCl	5.0g
Horse serum	260.0mL
Fresh yeast extract solution	65.0mL

pH 7.8 ± 0.2 at 25°C

Source: This medium, without horse serum and yeast extract solution, is available as a premixed powder from HiMedia.

Fresh Yeast Extract Solution:

Composition per 100.0mL:

Baker's yeast, live, pressed, starch-free	25.0g

Preparation of Fresh Yeast Extract Solution: Add the live Baker's yeast to 100.0mL of distilled/deionized water. Autoclave for 90 min at 15 psi pressure–121°C. Allow to stand. Remove supernatant solution. Adjust pH to 6.6–6.8.

Preparation of Medium: Add components, except horse serum and fresh yeast extract solution, to distilled/deionized water and bring volume to 1.0L. Mix thoroughly. Gently heat and bring to boiling. Distribute into tubes or flasks. Autoclave for 15 min at 15 psi pressure–121°C. Cool to 50°C. To each 75.0mL of cooled, sterile basal medium, add 20.0mL of sterile horse serum and 5.0mL of special yeast extract solution. Mix thoroughly. Pour into sterile Petri dishes or distribute into sterile tubes.

Use: For the preparation of media for the cultivation of *Mycoplasma.*

Mycoplasma HiVeg Broth Base with Crystal Violet and Tellurite (PPLO HiVeg Broth Base with CV)

Composition per liter:

Plant peptone	10.0g
Plant infusion	6.0g
NaCl	5.0g
Crystal Violet	0.01g

Chapman tellurite solution	2.85mL
Ascitic fluid	250.0mL

pH 7.8 ± 0.2 at 25°C

Source: This medium, without tellurite, is available as a premixed powder from HiMedia.

Chapman Tellurite Solution:

Composition per 100.0mL:

K_2TeO_3	1.0g

Preparation of Chapman Tellurite Solution: Add K_2TeO_3 to distilled/deionized water and bring volume to 100.0mL. Mix thoroughly. Filter sterilize.

Caution: Potassium tellurite is toxic.

Preparation of Medium: Add components, except ascitic fluid and Chapman tellurite solution, to distilled/deionized water and bring volume to 747.15mL. Mix thoroughly. Autoclave for 15 min at 15 psi pressure–121°C. Cool to less than 37°C. Aseptically add sterile ascitic fluid and 2.85mL of Chapman tellurite solution. Mix thoroughly. Aseptically distribute into sterile tubes or flasks.

Use: For the isolation of *Mycoplasma* species from clinical specimens.

Mycoplasma HiVeg Broth Base without Crystal Violet and with Ascitic Fluid (PPLO HiVeg Broth Base without CV)

Composition per liter:

Plant peptone	10.0g
Plant infusion	6.0g
NaCl	5.0g
Ascitic fluid	250.0mL

pH 7.8 ± 0.2 at 25°C

Source: This medium, without ascitic fluid, is available as a premixed powder from HiMedia.

Preparation of Medium: Add components, except ascitic fluid, to distilled/deionized water and bring volume to 750.0mL. Mix thoroughly. Autoclave for 15 min at 15 psi pressure–121°C. Cool to less than 37°C. Aseptically add sterile ascitic fluid. If desired, 0.5g of thallium acetate or 100,000U of penicillin may be added for a more selective medium. Mix thoroughly. Aseptically distribute into sterile tubes or flasks.

Use: For the enrichment of pleuro-pneumonia-like organisms (PPLOs) and *Mycoplasma* species from clinical specimens.

Mycoplasma Horse Serum Broth (ATCC Medium 1959)

Mycoplasma broth base	660.0mL
Horse serum	200.0mL
Fresh yeast extract solution	100.0mL
Phenol Red (0.1%)	20.0mL
Glucose solution	10.0mL
NaOH (1*N* solution)	6.25mL
Arginine solution	5.0mL

***Mycoplasma* Broth Base:**

Composition per liter:

Pancreatic digest of casein	7.0g
NaCl	5.0g
Beef extract	3.0g
Yeast extract	3.0g
Beef heart, solids from infusion	2.0g

Preparation of *Mycoplasma* Broth Base: Add components to distilled/deionized water and bring volume to 1.0L. Mix thoroughly.

Fresh Yeast Extract Solution:

Composition per 100.0mL:

Baker's yeast, live, pressed, starch-free	25.0g

Preparation of Fresh Yeast Extract Solution: Add the live Baker's yeast to 100.0mL of distilled/deionized water. Autoclave for 90 min at 15 psi pressure–121°C. Allow to stand. Remove supernatant solution. Adjust pH to 6.6–6.8.

Glucose Solution:

Composition per 10.0mL:

Glucose	5.0g

Preparation of Glucose Solution: Add glucose to 10.0mL of distilled/deionized water. Mix thoroughly. Filter sterilize.

Arginine Solution:

Composition per 10.0mL:

L-Arginine	4.2g

Preparation of Arginine Solution: Add arginine to 10.0mL of distilled/deionized water. Mix thoroughly. Filter sterilize.

Preparation of Medium: Combine 660.0mL *Mycoplasma* broth base, 20.0mL Phenol Red, and 6.25mL 1*N* NaOH. Mix thoroughly. Autoclave for 15 min at 15 psi pressure–121°C. Cool to 25°C. Aseptically add 5.0mL sterile arginine solution, 100.0mL sterile fresh yeast extract solution, 10.0mL sterile glucose solution, and 200.0mL filter sterilized horse serum. Mix thoroughly. Aseptically distribute into sterile tubes or flasks.

Use: For the preparation of media for the cultivation of *Mycoplasma* spp.

Mycoplasma Liquid Medium

Composition per 1004.0mL:

Arginine	1.0g
Glucose	1.0g
L-Cysteine·HCl·H_2O	1.0g
Mycoplasma broth base	850.0mL
Horse serum, not inactivated	100.0mL
Fresh yeast extract (25% solution)	50.0mL
Phenol Red (1.0% solution)	2.0mL
DNA calf thymus solution	2.0mL

pH 7.8 ± 0.2 at 25°C

***Mycoplasma* Broth Base:**

Composition per 850.0mL:

Pancreatic digest of casein	7.0g
NaCl	5.0g
Beef extract	3.0g
Yeast extract	3.0g
Beef heart, solids from infusion	2.0g

Preparation of *Mycoplasma* Broth Base: Add components to distilled/deionized water and bring volume to 850.0mL. Mix thoroughly. Gently heat and bring to boiling. Autoclave for 15 min at 15 psi pressure–121°C. Cool to 50°C.

Fresh Yeast Extract Solution:

Composition per 100.0mL:

Baker's yeast, live, pressed, starch-free	25.0g

Preparation of Fresh Yeast Extract Solution: Add the live Baker's yeast to 100.0mL of distilled/deionized water. Autoclave for 90 min at 15 psi pressure–121°C. Allow to stand. Remove supernatant solution. Adjust pH to 6.6–6.8.

DNA Calf Thymus Solution:

Composition per 10.0mL:

DNA calf thymus	1.0g

Preparation of DNA Calf Thymus Solution: Add DNA calf thymus to distilled/deionized water and bring volume to 10.0mL. Mix thoroughly. Filter sterilize.

Preparation of Medium: Combine components, except *Mycoplasma* broth base and DNA calf thymus solution, and mix thoroughly. Filter sterilize through a 0.2μm membrane. Add sterile solution to 850.0mL of cooled, sterile *Mycoplasma* broth base. Aseptically add 2.0mL of sterile DNA calf thymus solution. Mix thoroughly. Aseptically distribute into sterile tubes or flasks.

Use: For the cultivation and maintenance of *Mycoplasma lipophilum* and *Mycoplasma* species.

Mycoplasma Medium

Composition per liter:

Heart infusion broth	25.0g
Mucin, bacteriological grade	5.0g
Agar, purified (optional)	7.0g
Hemoglobin	2.0g
Turkey serum, sterile inactivated	100.0mL

pH 7.8 ± 0.2 at 25°C

Preparation of Medium: Add components, except turkey serum and agar, to 850.0mL distilled water. Adjust pH to 7.8. Heat mixture at 93°–95°C for 30 min in a water bath. Restore to original volume. Add 0.5% diatomaceous earth. Mix thoroughly. Filter through Whatman GFA (glass fiber paper) in Buchner filter. Clarify using 0.45μm Millipore filter. Add 15% inactivated turkey serum. Sterilize using S3 (0.1μm) Seitz filter. Use positive pressure. For solid medium: Prepare 42.5mL of double-strength broth and 42.5mL of distilled water containing 0.7g of purified agar. Sterilize the solutions separately and combine aseptically at 56°C with 15.0mL of sterile inactivated turkey serum for a final volume of 150.0mL.

Use: For the cultivation and maintenance of *Mycoplasma hyosynoviae.*

Mycoplasma Medium (CIP Medium 89) (DSMZ Medium 1080)

Composition per liter:

Pancreatic digest of casein	7.0g
NaCl	5.0g
Beef extract	3.0g
Yeast extract	3.0g
Beef heart, infusion from (solids)	2.0g
Selective supplement solution	210.0mL
Yeast extract solution	100.0mL
Phenol Red solution	20.0mL

Yeast Extract Solution:

Composition per 100.0mL:

Yeast extract	25.0g

Preparation of Yeast Extract Solution: Add yeast extract to distilled/deionized water and bring volume to 100.0mL. Mix thoroughly. Adjust pH to 7.0. Autoclave for 15 min at 15 psi pressure–121°C.

Phenol Red Solution:

Composition per 100.0mL:

Phenol Red	1.0g

Preparation of Phenol Red Solution: Add Phenol Red to distilled/deionized water and bring volume to 100.0mL. Mix thoroughly. Adjust pH to 7.0. Autoclave for 15 min at 15 psi pressure–121°C.

Selective Supplement Solution:

Composition per 210.0mL:

Ampicillin	1.0g
Horse serum	200.0mL
Arginine solution	10.0mL

Arginine Solution:

Composition per 100.0mL:

L-Arginine	50.0g

Preparation of Arginine Solution: Add L-arginine to distilled/deionized water and bring volume to 100.0mL. Mix thoroughly. Adjust pH to 7.0.

Preparation of Selective Supplement Solution: Add ampicillin to 10.0mL arginine solution. Add horse serum. Mix thoroughly. Filter sterilize.

Preparation of Medium: Add components, except yeast extract, Phenol Red, and selective supplement solutions, to distilled/deionized water and bring volume to 670.0mL. Mix thoroughly. Adjust pH to 7.2. Distribute into tubes or flasks. Gently heat while stirring and bring to boiling. Autoclave for 15 min at 15 psi pressure–121°C. Cool to room temperature. Aseptically add yeast extract, Phenol Red, and selective supplement solutions. Mix thoroughly. Aseptically distribute into tubes or flasks.

Use: For the cultivation of *Mycoplasma* spp.

Mycoplasma Medium, Revised

Composition per 1030.0mL:

Noble agar	10.0g
Distilled water	360.0mL
Heart infusion broth	300.0mL
Pig serum, heat inactivated	200.0mL
Enzymatic hydrolysate of lactalbumin	100.0mL
Hanks' balanced salt solution, 10X	40.0mL
Fresh yeast extract solution	20.0mL
Phenol Red (0.25% solution)	10.0mL

Heart Infusion Broth:

Composition per liter:

Beef heart, infusion from	500.0g
Tryptose	10.0g
NaCl	5.0g

Preparation of Heart Infusion Broth: Add components to distilled/deionized water and bring volume to 1.0L. Mix thoroughly. Gently heat and bring to boiling.

Enzymatic Hydrolysate of Lactalbumin:

Composition per 100.0mL:

Enzymatic hydrolysate of lactalbumin	5.0g
Phosphate buffered saline, 1X, pH7.0	100.0mL

Preparation of Enzymatic Hydrolysate of Lactalbumin: Add enzymatic hydrolysate of lactalbumin to 100.0mL of phosphate buffered saline, 1X, pH 7.0.

Phosphate Buffered Saline Solution, 1X:
Composition per liter:

NaCl	8.0g
$Na_2HPO_4 \cdot 7H_2O$	2.16
KCl	0.2g
KH_2PO_4	0.2g
$MgCl_2 \cdot 6H_2O$	0.1g
$CaCl_2$	0.1g

Preparation of Phosphate Buffered Saline Solution, 1X: Add components to distilled/deionized water and bring volume to 1.0L. Mix thoroughly.

Hanks' Balanced Salt Solution, 10X:
Composition per liter:

Na_2Cl	80.0g
Glucose	10.0g
KCl	4.0g
$CaCl_2$	1.4g
$MgCl_2 \cdot 6H_2O$	1.0g
$MgSO_4 \cdot 7H_2O$	1.0g
$Na_2HPO_4 \cdot 7H_2O$	0.9g
KH_2PO_4	0.6g

Preparation of Hanks' Balanced Salt Solution, 10X: Add components to distilled/deionized water and bring volume to 1.0L. Mix thoroughly.

Fresh Yeast Extract Solution:
Composition per 100.0mL:

Baker's yeast, live, pressed, starch-free	25.0g

Preparation of Fresh Yeast Extract Solution: Add the live Baker's yeast to 100.0mL of distilled/deionized water. Autoclave for 90 min at 15 psi pressure–121°C. Allow to stand. Remove supernatant solution. Adjust pH to 6.6–6.8.

Preparation of Medium: Add agar and heart infusion broth to distilled/deionized water and bring volume to 660.0mL. Mix thoroughly. Adjust pH to 7.4. Autoclave for 15 min at 15 psi pressure–121°C. Aseptically add pig serum, enzymatic hydrolysate of lactalbumin, Hanks' balanced salt solution, 10X, fresh yeast extract solution, and Phenol Red solution. Mix thoroughly. Aseptically distribute into sterile tubes.

Use: For the cultivation and maintenance of *Mycoplasma dispar*, *Mycoplasma flocculare*, and *Mycoplasma hyopneumoniae*.

Mycoplasma pneumoniae Isolation Medium

Composition per 1200.0mL:

Beef heart for infusion	50.0g
Peptone	10.0g
NaCl	5.0g
Water	900.0mL
Yeast extract solution	100.0mL
α-Gamma horse serum, unheated	200.0mL

pH 7.6–7.8 at 25°C

Yeast Extract Solution:
Composition per 10.0mL:

Yeast, active, dry, Baker's	250.0g

Preparation of Yeast Extract Solution: Add yeast to 1.0L of distilled/deionized water. Mix thoroughly. Gently heat and bring to boiling. Filter through Whatman #2 filter paper. Adjust the pH of the filtrate to 8.0 with NaOH. Distribute into tubes in 10.0mL volumes. Autoclave for 15 min at 15 psi pressure–121°C. Store at –20°C.

Preparation of Medium: Add components, except yeast extract solution and α-gamma horse serum, to distilled/deionized water and bring volume to 990.0mL. Mix thoroughly. Gently heat and bring to boiling. Autoclave for 15 min at 15 psi pressure–121°C. Cool to 45°–50°C. Aseptically add sterile yeast extract solution and α-gamma horse serum. Mix thoroughly. Aseptically distribute into sterile tubes.

Use: For the isolation and cultivation of *Mycoplasma pneumoniae*.

Mycoplasmal Agar

Composition per liter:

Papaic digest of soy meal	20.0g
Agarose	10.0g
NaCl	5.0g
Phenol Red (2% solution)	1.0mL

pH 7.3 ± 0.2 at 25°C

Preparation of Medium: Add components, except agarose, to distilled/deionized water and bring volume to 1.0L. Mix thoroughly. Adjust pH to 7.3 with 1*N* NaOH. Add agarose. Mix thoroughly. Gently heat and bring to boiling. Distribute into tubes or flasks. Autoclave for 15 min at 15 psi pressure–121°C. Pour into sterile Petri dishes or leave in tubes.

Use: For the isolation and cultivation of human mycoplasmas and ureaplasmas.

Mycorrhiza Medium

Composition per liter:

Agar	15.0g
Glucose	4.0g
Ammonium tartrate	1.0g
Malt extract	1.0g
KH_2PO_4	0.2g
$MgSO_4 \cdot 7H_2O$	0.1g
$CaCl_2 \cdot 2H_2O$	26.0mg
NaCl	20.0mg
Inositol	10.0mg
$ZnSO_4 \cdot 7H_2O$	0.88mg
$MnSO_4 \cdot 4H_2O$	0.81mg
$FeCl_3 \cdot 6H_2O$	0.8mg
Nicotinamide	100.0μg
p-Aminobenzoic acid	100.0μg
Pantothenic acid	100.0μg
Pyridoxine	100.0μg
Thiamine	100.0μg
Biotin	25.0μg
Riboflavin	25.0μg

Preparation of Medium: Add components to distilled/deionized water and bring volume to 1.0L. Mix thoroughly. Gently heat until boiling. Distribute into tubes or flasks. Autoclave for 10 min at 15 psi pressure–121°C. Pour into sterile Petri dishes or leave in tubes.

Use: For the cultivation and maintenance of *Thelephora terrestris*.

Mycosel™ Agar
(Cycloheximide Chloramphenicol Agar)

Composition per liter:

Agar	15.5g
Papaic digest of soybean meal	10.0g
Glucose	10.0g

Cycloheximide 0.4g
Chloramphenicol 0.05g

pH 6.9 ± 0.2 at 25°C

Caution: Cycloheximide is toxic. Avoid skin contact or aerosol formation and inhalation.

Preparation of Medium: Add components to distilled/deionized water and bring volume to 1.0L. Mix thoroughly. Gently heat while stirring and bring to boiling. Autoclave for 15 min at 14 psi pressure–118°C. Avoid overheating. Pour into sterile Petri dishes or distribute into sterile tubes.

Use: For the selective isolation of pathogenic fungi from specimens with other fungi and bacteria.

MYCT Medium (DSMZ Medium 972)

Composition per liter:

KH_2PO_4 ..... 13.6g
Cyclomaltoheptaose (ß-cyclodextrin) 7.0g
$(NH_4)_2SO_4$ 4.0g
Yeast extract 3.0g
Casein hydrolysate 3.0g
Tween™ 80 ..... 1.0g
$MgCl_2$ 0.2g
Sodium citrate 0.25g
$FeSO_4 \cdot 7H_2O$ 25.0mg
$MnSO_4 \cdot 4H_2O$ 25.0mg

pH 6.5 ± 0.2 at 25°C

Preparation of Medium: Add components to distilled/deionized water and bring volume to 1.0L. Mix thoroughly. Distribute into tubes or flasks. Autoclave for 15 min at 15 psi pressure–121°C.

Use: For the cultivation of *Lactobacillus* sp.

Mykorrhiza Agar

Composition per liter:

Agar 15.0g
Malt extract 8.0g
Glucose 7.0g
Casein hydrolysate 1.0g
Yeast extract 1.0g
Asparagine 0.5g
KH_2PO_4 ..... 0.5g
$MgSO_4 \cdot 7H_2O$ 0.5g

Preparation of Medium: Add components to distilled/deionized water and bring volume to 1.0L. Mix thoroughly. Gently heat and bring to boiling. Distribute into tubes or flasks. Autoclave for 15 min at 15 psi pressure–121°C. Pour into sterile Petri dishes or leave in tubes.

Use: For the cultivation and maintenance of *Klebsiella pneumoniae.*

MYP Agar

See: **Mannitol Yolk Polymyxin Agar**

MYP Agar Base, HiVeg with Egg Yolk and Polymyxin B (Phenol Red Egg Yolk Polymyxin Agar Base, HiVeg)

Composition per liter:

Agar 15.0g
D-Mannitol 10.0g
Plant peptone 10.0g
NaCl 10.0g
Plant extract No. 1 ..... 1.0g
Phenol Red 0.025g
Egg yolk emulsion, 20% 10.0mL
Polymyxin B solution 1.0mL

pH 7.1 ± 0.2 at 25°C

Source: This medium, without egg yolk and polymyxin B, is available as a premixed powder from HiMedia.

Egg Yolk Emulsion, 20%:

Composition per 100.0mL:

Chicken egg yolks 11
Whole chicken egg 1
NaCl (0.9% solution) 80.0mL

Preparation of Egg Yolk Emulsion, 20%: Soak eggs with 1:100 dilution of saturated mercuric chloride solution for 1 min. Crack eggs and separate yolks from whites. Mix egg yolks with 1 chicken egg. Measure 20.0mL of egg yolk emulsion and add to 80.0mL of 0.9% NaCl solution. Mix thoroughly. Filter sterilize. Warm to 45°–50°C.

Polymyxin B Solution:

Composition per 1.0mL:

Polymyxin B 1.0mg

Preparation of Polymyxin B Solution: Add polymyxin B to distilled/deionized water and bring volume to 1.0mL. Mix thoroughly. Filter sterilize.

Preparation of Medium: Add components—except egg yolk emulsion, 20%, and polymyxin B solution—to distilled/deionized water and bring volume to 100.0mL. Mix thoroughly. Gently heat and bring to boiling. Autoclave for 15 min at 15 psi pressure–121°C. Cool to 45°–50°C. Aseptically add 10.0mL of sterile egg yolk emulsion, 20%, and 1.0mL of sterile polymyxin B solution. Mix thoroughly. Pour into sterile Petri dishes.

Mysorens Medium

Composition per liter:

Peptone 10.0g
Meat extract 10.0g
Yeast extract 5.0g
NaCl 3.0g

Preparation of Medium: Add components to distilled/deionized water and bring volume to 1.0L. Mix thoroughly. Distribute into tubes or flasks. Autoclave for 15 min at 15 psi pressure–121°C.

Use: For the cultivation and maintenance of *Arthrobacter mysorens*.

MYX Agar (DSMZ Medium 729)

Composition per liter:

Na_2-glutamate 5.0g
Yeast extract 1.0g
$MgSO_4 \cdot 7H_2O$ 1.0g
Glucose solution 10.0mL

pH 7.2 ± 0.2 at 25°C

Glucose Solution:

Composition per10.0mL:

Glucose 2.0g

Preparation of Glucose Solution: Add glucose to distilled/deionized water and bring volume to 10.0mL. Mix thoroughly. Filter sterilize.

Preparation of Medium: Add components, except glucose solution, to distilled/deionized water and bring volume to 990.0mL. Mix thoroughly. Adjust pH to 7.2. Aseptically add 10.0mL glucose solution. Mix thoroughly. Aseptically distribute into sterile tubes or bottles.

Use: For the cultivation of *Taxeobacter* spp.

Myxobacteria Medium

Composition per liter:

Agar	15.0g
Skim milk powder	5.0g
Yeast extract	0.5g

Preparation of Medium: Add components to distilled/deionized water and bring volume to 1.0L. Mix thoroughly. Do not adjust pH. Gently heat and bring to boiling. Distribute into tubes or flasks. Autoclave for 15 min at 15 psi pressure–121°C.

Use: For the cultivation and maintenance of *Archangium primigenium, Chondrococcus macrosporus*, and *Myxococcus coralloides*.

Myxococcus flavescens Medium

Composition per liter:

Agar	15.0g
Soluble starch	5.0g
Casitone	2.5g
Galactose	1.0g
Raffinose	1.0g
Sucrose	1.0g
Yeast extract	1.0g
$MgSO_4 \cdot 7H_2O$	0.5g
K_2HPO_4	0.25g

pH 6.0 ± 0.2 at 25°C

Preparation of Medium: Add components to distilled/deionized water and bring volume to 1.0L. Mix thoroughly. Adjust pH to 6.0. Gently heat and bring to boiling. Distribute into tubes or flasks. Autoclave for 15 min at 15 psi pressure–121°C. Pour into sterile Petri dishes or leave in tubes.

Use: For the cultivation and maintenance of *Myxococcus flavescens.*

Myxococcus Medium

Composition per liter:

Agar	12.0g
Pancreatic digest of casein	1.0g
Meat extract	1.0g
Glucose solution	50.0mL

pH 7.2 ± 0.2 at 25°C

Glucose Solution:

Composition per 50.0mL:

Glucose	1.0g

Preparation of Glucose Solution: Add glucose to distilled/deionized water and bring volume to 50.0mL. Mix thoroughly. Autoclave for 15 min at 15 psi pressure–121°C. Cool to 25°C.

Preparation of Medium: Add components, except glucose solution, to distilled/deionized water and bring volume to 950.0mL. Mix thoroughly. Adjust pH to 7.2. Gently heat and bring to boiling. Autoclave for 15 min at 15 psi pressure–121°C. Cool to 45°–50°C. Aseptically add sterile glucose solution. Mix thoroughly. Pour into sterile Petri dishes or distribute into sterile tubes or bottles. Allow tubes or bottles to cool in a slanted position.

Use: For the cultivation of *Myxococcus* species.

Myxococcus xanthus Medium

Composition per liter:

Agar	20.0g
Pancratic digest of casein	10.0g
$MgSO_4 \cdot 7H_2O$	0.5g
K_2HPO_4	0.148g
KH_2PO_4	0.017g

pH 7.6 ± 0.2 at 25°C

Preparation of Medium: Add components to distilled/deionized water and bring volume to 1.0L. Mix thoroughly. Gently heat and bring to boiling. Distribute into tubes or flasks. Autoclave for 15 min at 15 psi pressure–121°C. Pour into sterile Petri dishes or leave in tubes.

Use: For the cultivation and maintenance of *Myxococcus xanthus*.

N plus C Medium

Composition per liter:

Pancreatic digest of casein	10.0g
Glucose	10.0g
Citric acid·H_2O	4.04g
KH_2PO_4	2.0g
Yeast extract	1.5g
$CaCl_2 \cdot 2H_2O$	0.6g
$MgSO_4 \cdot 7H_2O$	0.6g
$FeCl_2 \cdot 4H_2O$	0.06g
$ZnSO_4 \cdot 7H_2O$	0.034g
Hemin solution	1.0mL

pH 4.6 ± 0.2 at 25°C

Hemin Solution:

Composition per 100.0mL:

NaOH	1.0g
Hemin	250.0mg

Preparation of Hemin Solution: Add components to distilled/deionized water and bring volume to 100.0mL. Mix thoroughly. Autoclave for 15 min at 15 psi pressure–121°C.

Preparation of Medium: Add components, except hemin solution, to distilled/deionized water and bring volume to 1.0L. Mix thoroughly. Adjust pH to 4.6. Distribute into tubes or flasks. Autoclave for 15 min at 15 psi pressure–121°C. Prior to inoculation, add 0.1mL of hemin solution per 100.0mL of medium.

Use: For the cultivation and maintenance of *Physarum polycephalum.*

N DeVogel Medium (Vogel N Medium)

Composition per liter:

Sucrose	15.0g
KH_2PO_4	5.0g
Trisodium citrate·$2H_2O$	3.0g
NH_4NO_3	2.0g
$MgSO_4 \cdot 7H_2O$	0.2g
$CaCl_2 \cdot H_2O$ solution	20.0mL
Biotin solution	5.0mL
Trace elements solution	5.0mL

$CaCl_2 \cdot H_2O$ Solution:

Composition per 20.0mL:

$CaCl_2 \cdot H_2O$	0.1g

Preparation of $CaCl_2 \cdot H_2O$ Solution: Add $CaCl_2 \cdot H_2O$ to distilled/deionized water and bring volume to 20.0mL. Mix thoroughly.

Biotin Solution:

Composition per 100.0mL:

Biotin	5.0mg

Preparation of Biotin Solution: Add biotin to 50% ethanol and bring volume to 100.0mL. Mix thoroughly. Filter sterilize. Store at 5°C.

Trace Elements Solution:

Composition per 100.0mL:

Citric acid·H_2O	5.0g
$ZnSO_4 \cdot 7H_2O$	5.0g
$Fe(NH_4)_2(SO_4)_2 \cdot 6H_2O$	1.0g
$CuSO_4 \cdot 5H_2O$	0.25g
H_3BO_3, anhydrous	0.05g
$MnSO_4 \cdot H_2O$	0.05g
$Na_2MoO_4 \cdot 2H_2O$	0.05g

Preparation of Trace Elements Solutions: Add components successively to distilled/deionized water and bring volume to 100.0mL. Mix thoroughly after addition of each component. Filter sterilize. Add 2–3.0mL of chloroform as a preservative. Store at 25°C.

Preparation of Medium: Add components, except biotin solution and trace elements solution, to distilled/deionized water and bring volume to 990.0mL. Mix thoroughly. Autoclave for 15 min at 15 psi pressure–121°C. Aseptically add 5.0mL of sterile biotin solution and 5.0mL of sterile trace elements solution. Mix thoroughly. Aseptically distribute into sterile tubes or flasks.

Use: For the cultivation and maintenance of *Neurospora crassa.*

NAG Medium (DSMZ Medium 366)

Composition per liter:

Agar	15.0g
Glucose	10.0g
Peptone	5.0g
Meat extract	3.0g

pH 7.0 ± 0.2 at 25°C

Preparation of Medium: Add components to distilled/deionized water and bring volume to 1.0L. Mix thoroughly. Gently heat and bring to boiling. Distribute into tubes or flasks. Autoclave for 15 min at 15 psi pressure–121°C. Do not overheat. Pour into sterile Petri dishes or leave in tubes.

Use: For the isolation and cultivation of *Xanthomonas fragariae*.

Nakayama Glucose Agar

Composition per 1001.0mL:

Yeast extract	15.0g
Agar	10.0g
Glucose	10.0g
Peptone	10.0g
Solution A	10.0mL
Solution B	10.0mL
Solution C	1.0mL

pH 6.8 ± 0.2 at 25°C

Solution A:

Composition per 100.0mL:

K_2HPO_4	0.5g
KH_2PO_4	0.5g

Preparation of Solution A: Add components to distilled/deionized water and bring volume to 100.0mL. Mix thoroughly. Autoclave for 15 min at 15 psi pressure–121°C.

Solution B:

Composition per 100.0mL:

$MgSO_4 \cdot 7H_2O$	3.0g
$MnSO_4 \cdot 5H_2O$	0.1g
NaCl	0.1g
$CuSO_4 \cdot 5H_2O$	0.01g

Preparation of Solution B: Add components to distilled/deionized water and bring volume to 100.0mL. Mix thoroughly. Autoclave for 15 min at 15 psi pressure–121°C.

Solution C:

Composition per 100.0mL:

Trisodium citrate	2.0g
$FeSO_4 \cdot 7H_2O$	0.1g

Preparation of Solution C: Add components to distilled/deionized water and bring volume to 100.0mL. Mix thoroughly. Autoclave for 15 min at 15 psi pressure–121°C.

Preparation of Medium: Add components, except solution A, solution B, and solution C, to distilled/deionized water and bring volume to 980.0mL. Mix thoroughly. Gently heat and bring to boiling. Autoclave for 15 min at 15 psi pressure–121°C. Cool to 50°–55°C. Aseptically add 10.0mL of sterile solution A, 10.0mL of sterile solution B, and 1.0mL of sterile solution C. Mix thoroughly. Pour into sterile Petri dishes or distribute into sterile tubes.

Use: For the cultivation and maintenance of *Bacillus laevolacticus*.

NAM Medium

Composition per liter:

Pancreatic digest of casein	15.0g
Agar	15.0g
Papaic digest of soybean meal	5.0g
NaCl	5.0g
Sheep blood, defibrinated	50.0mL
Hemin solution	10.0mL
N-Acetyl muramic acid (NAM) solution	1.0mL

pH 7.3 ± 0.2 at 25°C

Hemin Solution:

Composition per 100.0mL:

Hemin	0.050g
NaOH (1*N* solution)	1.0mL

Preparation of Hemin Solution: Add hemin to NaOH solution. Mix thoroughly. Adjust volume to 100.0mL with distilled/deionized water. Autoclave for 15 min at 15 psi pressure–121°C. Cool to 45°–50°C.

***N*-Acetyl Muramic Acid (NAM) Solution:**

Composition per 10.0mL:

N-Acetyl muramic acid	100.0mg

Preparation of *N*-Acetyl Muramic Acid (NAM) Solution: Add *N*-acetyl muramic acid to distilled/deionized water and bring volume to 10.0mL. Filter sterilize.

Preparation of Medium: Add components, except sheep blood, hemin solution, and *N*-acetyl muramic acid (NAM) solution, to distilled/deionized water and bring volume to 49.0mL. Mix thoroughly. Autoclave for 15 min at 15 psi pressure–121°C. Cool to 50°–55°C. Aseptically add 50.0mL of sterile sheep blood, 10.0mL of sterile hemin solution, and 1.0mL of sterile *N*-acetyl muramic acid (NAM) solution.

Mix thoroughly. Aseptically and anaerobically distribute into sterile tubes under a gas phase of 80% N_2 + 10% CO_2 + 10% H_2.

Use: For the cultivation of *Bacteroides forsythus.*

NAMn

See: ***Nutrient Agar with Manganese***

NANAT Agar (Nalidixic Acid Novobiocin Actidione Tellurite Agar)

Composition per liter:

Pancreatic digest of casein	17.0g
Agar	15.0g
NaCl	5.0g
Tween™ 80	5.0g
Papaic digest of soybean meal	3.0g
K_2HPO_4	2.5g
Glucose	2.5g
Yeast extract	1.0g
Tellurite solution	10.0mL
Antibiotic solution	10.0mL

pH 7.2 ± 0.2 at 25°C

Tellurite Solution:

Composition per 100.0mL:

K_2TeO_3	0.05g

Preparation of Tellurite Solution: Add K_2TeO_3 to distilled/deionized water and bring volume to 100.0mL. Mix thoroughly. Filter sterilize.

Antibiotic Solution:

Composition per 10.0mL:

Actidione (cycloheximide)	0.04g
Polymyxin B (optional)	0.03g
Novobiocin	0.025g
Nalidixic acid	0.02g

Preparation of Antibiotic Solution: Add components to distilled/deionized water and bring volume to 10.0mL. Mix thoroughly. Filter sterilize.

Caution: Potassium tellurite is toxic.

Caution: Cycloheximide is toxic. Avoid skin contact or aerosol formation and inhalation.

Preparation of Medium: Add components, except tellurite solution and antibiotic solution, to distilled/deionized water and bring volume to 980.0mL. Mix thoroughly. Gently heat and bring to boiling. Autoclave for 15 min at 15 psi pressure–121°C. Cool to 45°–50°C. Aseptically add sterile tellurite solution and antibiotic solution. Mix thoroughly. Pour into sterile Petri dishes or distribute into sterile tubes.

Use: For the isolation and cultivation of *Rhodococcus (Corynebacterium) equi* from animal feces, especially from horses and swine. The addition of polymyxin B inhibits the growth of *Pseudomonas aeruginosa* which may interfere with the isolation of *Rhodococcus equi.*

Nannocystis Agar

Composition per liter:

Agar	15.0g
$CaCl_2 \cdot 2H_2O$	1.0g

pH 7.2 ± 0.2 at 25°C

Preparation of Medium: Add components to distilled/deionized water and bring volume to 1.0L. Mix thoroughly. Gently heat and bring to boiling. Autoclave for 15 min at 15 psi pressure–121°C. Pour into sterile Petri dishes. After the agar has solidified, overlay the surface with 0.5mL of a suspension of dead (autoclaved) *Escherichia coli* cells.

Use: For the cultivation and maintenance of *Nannocystis* species.

Naphthalene Medium

Composition per liter:

NH_4NO_3	2.5g
$Na_2HPO_4 \cdot 2H_2O$	1.0g
Naphthalene	0.64g
$MgSO_4 \cdot 7H_2O$	0.5g
$Fe(SO_4)_3 \cdot 5H_2O$	0.01g
$Co(NO_3)_2 \cdot 6H_2O$	5.0mg
$CaCl_2 \cdot 2H_2O$	1.0mg
KH_2PO_4	0.5mg
$MnSO_4 \cdot 2H_2O$	0.1mg
$(NH_4)_6Mo_7O_{24} \cdot 4H_2O$	0.1mg

Preparation of Medium: Add components to distilled/deionized water and bring volume to 1.0L. Mix thoroughly. Distribute into tubes or flasks. Autoclave for 15 min at 15 psi pressure–121°C.

Use: For the cultivation of *Pseudomonas alcaligenes.*

Naphthalene Mineral Salts Medium

See: **Medium for Hydrocarbon-Degrading Bacteria**

Naphthalene Sulfonic Acid Medium

Composition per 1004.0mL:

$Na_2HPO_4 \cdot 2H_2O$	3.5g
KH_2PO_4	1.0g
NH_4Cl	0.31g
$MgCl_2 \cdot 6H_2O$	0.1g
$Ca(NO_3)_2 \cdot 4H_2O$	0.05g
Solution A	100.0mL
Solution B	3.0mL
Trace elements solution SL-4	1.0mL

pH 7.0 ± 0.2 at 25°C

Solution A:

Composition per liter:

Glucose	3.0g
Glycerol	3.0g
Sodium succinate	3.0g

Preparation of Solution A: Add components to distilled/deionized water and bring volume to 1.0L. Mix thoroughly. Filter sterilize.

Solution B:

Composition per liter:

Naphthalene sulfonic acid	2.3g

Preparation of Solution B: Add naphthalene sulfonic acid to distilled/deionized water and bring volume to 1.0L. Mix thoroughly. Filter sterilize.

Trace Elements Solution SL-4:

Composition per liter:

EDTA	0.5g
$FeSO_4 \cdot 7H_2O$	0.2g
Trace elements solution SL-6	100.0mL

Preparation of Trace Elements Solution SL-4: Add components to distilled/deionized water and bring volume to 1.0L. Mix thoroughly. Filter sterilize.

Trace Elements Solution SL-6:

Composition per liter:

$MnCl_2 \cdot 4H_2O$	0.5g
H_3BO_3	0.3g
$CoCl_2 \cdot 6H_2O$	0.2g
$ZnSO_4 \cdot 7H_2O$	0.1g
$Na_2MoO_4 \cdot 2H_2O$	0.03g
$NiCl_2 \cdot 6H_2O$	0.02g
$CuCl_2 \cdot 2H_2O$	0.01g

Preparation of Trace Elements Solution SL-6: Add components to distilled/deionized water and bring volume to 1.0L. Mix thoroughly.

Preparation of Medium: Add components, except solution A, solution B, and trace elements solution SL-4, to distilled/deionized water and bring volume to 900.0mL. Mix thoroughly. Autoclave for 15 min at 15 psi pressure–121°C. Aseptically add 100.0mL of sterile solution A, 3.0mL of sterile solution B, and 1.0mL of sterile trace elements solution SL-4. Mix thoroughly. Aseptically distribute into sterile tubes or flasks.

Use: For the cultivation of *Pseudomonas putida.*

Natranaerobius Medium (DSMZ Medium 1095)

Composition per liter:

NaCl	100.0g
Yeast extract	10.0g
Tryptone	10.0g
NH_4Cl	0.5g
KH_2PO_4	0.2g
$MgCl_2 \cdot 6H_2O$	0.1g
Sucrose solution	10.0mL
Vitamin solution	10.0mL
Cysteine solution	10.0mL
Bicarbonate solution	10.0mL
Carbonate solution	10.0mL
Trace elements solution SL-10	1.0mL

pH 8.5 ± 0.2 at 25°C

Vitamin Solution:

Composition per liter:

Pyridoxine-HCl	10.0mg
Thiamine-HCl·$2H_2O$	5.0mg
Riboflavin	5.0mg
Nicotinic acid	5.0mg
D-Ca-pantothenate	5.0mg
p-Aminobenzoic acid	5.0mg
Lipoic acid	5.0mg
Biotin	2.0mg
Folic acid	2.0mg
Vitamin B_{12}	0.1mg

Preparation of Vitamin Solution: Add components to distilled/deionized water and bring volume to 1.0L. Mix thoroughly. Sparge with 80% H_2 + 20% CO_2. Filter sterilize.

Trace Elements Solution SL-10:

Composition per liter:

$MgSO_4 \cdot 7H_2O$	3.0g
Nitrilotriacetic acid	1.5g
NaCl	1.0g
$MnSO_4 \cdot 2H_2O$	0.5g
$CoSO_4 \cdot 7H_2O$	0.18g
$ZnSO_4 \cdot 7H_2O$	0.18g
$CaCl_2 \cdot 2H_2O$	0.1g
$FeSO_4 \cdot 7H_2O$	0.1g
$NiCl_2 \cdot 6H_2O$	0.025g
$KAl(SO_4)_2 \cdot 12H_2O$	0.02g
H_3BO_3	0.01g
$Na_2MoO_4 \cdot 4H_2O$	0.01g
$CuSO_4 \cdot 5H_2O$	0.01g
$Na_2SeO_3 \cdot 5H_2O$	0.3mg

Preparation of Trace Elements Solution SL-10: Add nitrilotriacetic acid to 500.0mL of distilled/deionized water. Dissolve by adjusting pH to 6.5 with KOH. Add remaining components. Add distilled/deionized water to 1.0L. Mix thoroughly. Adjust pH to 7.0.

Sucrose Solution:

Composition per 10.0mL:

Sucrose	5.0g

Preparation of Sucrose Solution: Add sucrose to distilled/deionized water and bring volume to 10.0mL. Mix thoroughly. Filter sterilize.

Carbonate Solution:

Composition per 10.0mL:

Na_2CO_3	68.0g

Preparation of Carbonate Solution: Add Na_2CO_3 to distilled/deionized water and bring volume to 10.0mL. Mix thoroughly. Filter sterilize.

Bicarbonate Solution:

Composition per 10.0mL:

$NaHCO_3$	38.0g

Preparation of Bicarbonate Solution: Add $NaHCO_3$ to distilled/deionized water and bring volume to 10.0mL. Mix thoroughly. Filter sterilize.

Cysteine Solution:

Composition per 10.0mL:

L-Cysteine-HCl·$2H_2O$	0.07g

Preparation of Cysteine Solution: Add L-Cysteine-HCl·$2H_2O$ to to distilled/deionized water and bring volume to 10.0mL. Mix thoroughly. Sparge with 100% N_2. Filter sterilize.

Preparation of Medium: Add components, except carbonate, bicarbonate, cysteine, sucrose, and vitamin solutions, to distilled/deionized water and bring volume to 950.0mL. Mix thoroughly. Sparge wtih 100% N_2 gas. Gently heat and bring to boiling. Boil for 1 min. Cool to room temperature while sparging with 100% N_2. Add cysteine, carbonate, and bicarbonate solutions. Adjust pH to 8.5. Autoclave for 15 min at 15 psi pressure–121°C. Cool to room temperature. Aseptically add the vitamins and sucrose solutions. Aseptically and anoxically dispense into cuture vessels.

Use: For the cultivation of *Natranaerobius* spp.

Natroniella Medium (DSMZ Medium 784)

Composition per liter:

Na_2CO_3	68.3g
NaCl	15.7g
NH_4Cl	1.0g
KCl	0.2g
KH_2PO_4	0.2g
Yeast extract	0.2g
$MgCl_2 \cdot 6H_2O$	0.1g

Resazurin	0.5mg
$NaHCO_3$ solution	100.0mL
Vitamin solution	10.0mL
$Na_2S \cdot 9H_2O$ solution	10.0mL
Ethanol solution	10.0mL
Trace elements solution SL-11	1.0mL

pH 9.7–10.0 at 25°C

$Na_2S \cdot 9H_2O$ Solution:

Composition per 10.0mL:

$Na_2S \cdot 9H_2O$	1.0g

Preparation of $Na_2S \cdot 9H_2O$ Solution: Add $Na_2S \cdot 9H_2O$ to distilled/deionized water and bring volume to 10.0mL. Mix thoroughly. Autoclave under 100% N_2 for 15 min at 15 psi pressure–121°C. Cool to room temperature.

$NaHCO_3$ Solution:

Composition per 100.0mL:

$NaHCO_3$	38.3g

Preparation of $NaHCO_3$ Solution: Add $NaHCO_3$ to distilled/deionized water and bring volume to 100.0mL. Mix thoroughly. Sparge with 80% N_2 + 20% CO_2. Filter sterilize.

Ethanol Solution:

Composition per 10.0mL:

Ethanol	5.0mL

Preparation of Ethanol Solution: Add ethanol to distilled/deionized water and bring volume to 10.0mL. Mix thoroughly. Sparge with 100% N_2. Filter sterilize.

Vitamin Solution:

Composition per liter:

Pyridoxine-HCl	10.0mg
Thiamine-HCl·$2H_2O$	5.0mg
Riboflavin	5.0mg
Nicotinic acid	5.0mg
D-Ca-pantothenate	5.0mg
p-Aminobenzoic acid	5.0mg
Lipoic acid	5.0mg
Biotin	2.0mg
Folic acid	2.0mg
Vitamin B_{12}	0.1mg

Preparation of Vitamin Solution: Add components to distilled/deionized water and bring volume to 1.0L. Mix thoroughly. Sparge with 80% H_2 + 20% CO_2. Filter sterilize.

Trace Elements Solution SL-4:

Composition per liter:

EDTA	0.5g
$FeSO_4 \cdot 7H_2O$	0.2g
Trace elements solution SL-6	100.0mL

Trace Elements Solution SL-6:

Composition per liter:

H_3BO_3	0.3g
$CoCl_2 \cdot 6H_2O$	0.2g
$ZnSO_4 \cdot 7H_2O$	0.1g
$MnCl_2 \cdot 4H_2O$	0.03g
$Na_2MoO_4 \cdot H_2O$	0.03g
$NiCl_2 \cdot 6H_2O$	0.02g
$CuCl_2 \cdot 2H_2O$	0.01g

Preparation of Trace Elements Solution SL-6: Add components to distilled/deionized water and bring volume to 1.0L. Mix thoroughly. Adjust pH to 3.4.

Preparation of Trace Elements Solution SL-4: Add components to distilled/deionized water and bring volume to 1.0L. Mix thoroughly.

Preparation of Medium: Prepare and dispense medium under 100% N_2. Add components, except vitamin solution, $NaHCO_3$ solution, ethanol solution, and $Na_2S \cdot 9H_2O$ solution, to distilled/deionized water and bring volume to 870.0mL. Mix thoroughly. Gently heat and bring to boiling. Cool to room temperature while sparging with 100% N_2. Autoclave for 15 min at 15 psi pressure–121°C. Cool to 25°C. Aseptically and anaerobically add 10.0mL sterile vitamin solution, 100.0mL of sterile $NaHCO_3$ solution, 10.0mL sterile ethanol solution, and 10.0mL of sterile $Na_2S \cdot 9H_2O$ solution. Mix thoroughly. Adjust pH to 9.7–10.0. Aseptically and anaerobically distribute into sterile tubes or flasks.

Use: For the cultivation of *Natroniella acetigena (Acetohalobium* sp.*)*.

Natroniella Medium (DSMZ Medium 784)

Composition per liter:

Na_2CO_3	68.3g
NaCl	15.7g
NH_4Cl	1.0g
KCl	0.2g
KH_2PO_4	0.2g
Yeast extract	0.2g
$MgCl_2 \cdot 6H_2O$	0.1g
Resazurin	0.5mg
$NaHCO_3$ solution	100.0mL
Vitamin solution	10.0mL
$Na_2S \cdot 9H_2O$ solution	10.0mL
Ethanol solution	10.0mL
Trace elements solution SL-4	1.0mL

pH 9.7–10.0 at 25°C

$Na_2S \cdot 9H_2O$ Solution:

Composition per 10.0mL:

$Na_2S \cdot 9H_2O$	1.0g

Preparation of $Na_2S \cdot 9H_2O$ Solution: Add $Na_2S \cdot 9H_2O$ to distilled/deionized water and bring volume to 10.0mL. Mix thoroughly. Autoclave under 100% N_2 for 15 min at 15 psi pressure–121°C. Cool to room temperature.

$NaHCO_3$ Solution:

Composition per 100.0mL:

$NaHCO_3$	38.3g

Preparation of $NaHCO_3$ Solution: Add $NaHCO_3$ to distilled/deionized water and bring volume to 100.0mL. Mix thoroughly. Sparge with 80% N_2 + 20% CO_2. Filter sterilize.

Ethanol Solution:

Composition per 10.0mL:

Ethanol	5.0mL

Preparation of Ethanol Solution: Add ethanol to distilled/deionized water and bring volume to 10.0mL. Mix thoroughly. Sparge with 100% N_2. Filter sterilize.

Vitamin Solution:

Composition per liter:

Pyridoxine-HCl	10.0mg
Thiamine-HCl·$2H_2O$	5.0mg

Riboflavin 5.0mg
Nicotinic acid 5.0mg
D-Ca-pantothenate 5.0mg
p-Aminobenzoic acid 5.0mg
Lipoic acid 5.0mg
Biotin 2.0mg
Folic acid 2.0mg
Vitamin B_{12} 0.1mg

Preparation of Vitamin Solution: Add components to distilled/deionized water and bring volume to 1.0L. Mix thoroughly. Sparge with 80% H_2 + 20% CO_2. Filter sterilize.

Trace Elements Solution SL-11:
Composition per liter:
$FeCl_2·4H_2O$ 1.5g
$CoCl_2·6H_2O$ 190.0mg
$MnCl_2·4H_2O$ 100.0mg
$ZnCl_2$ 70.0mg
$Na_2MoO_4·H_2O$ 36.0mg
$NiCl_2·6H_2O$ 24.0mg
H_3BO_3 .... 6.0mg
Na_2-EDTA 5.2g
$CuCl_2·2H_2O$ 2.0mg

Preparation of Trace Elements Solution SL-11: Add components to distilled/deionized water and bring volume to 1.0L. Mix thoroughly. Adjust pH to 6.0.

Preparation of Medium: Prepare and dispense medium under 100% N_2. Add components, except vitamin solution, $NaHCO_3$ solution, ethanol, and $Na_2S·9H_2O$ solution, to distilled/deionized water and bring volume to 870.0mL. Mix thoroughly. Gently heat and bring to boiling. Cool to room temperature while sparging with 100% N_2. Autoclave for 15 min at 15 psi pressure–121°C. Cool to 25°C. Aseptically and anaerobically add 10.0mL sterile vitamin solution, 100.0mL of sterile $NaHCO_3$ solution, 10.0mL sterile ethanol solution, and 10.0mL of sterile $Na_2S·9H_2O$ solution. Mix thoroughly. Adjust pH to 9.7–10.0. Aseptically and anaerobically distribute into sterile tubes or flasks.

Use: For the cultivation of *Natroniella acetigena (Acetohalobium* sp.*)*.

Natronincola histidinovorans Medium (DSMZ Medium 930)

Composition per liter:
NaCl 80.0g
Na_2CO_3 .... 6.83g
$NaHCO_3$ 3.83g
NH_4Cl 1.0g
KCl 0.2g
KH_2PO_4 .... 0.2g
$MgCl_2·6H_2O$ 0.1g
Resazurin 0.01g
Histidine solution 50.0mL
$Na_2S·9H_2O$ solution 10.0mL
Yeast extract solution 10.0mL
Vitamin solution 2.0mL
Trace elements solution 1.0mL
pH 8.9 ± 0.2 at 25°C

Histidine Solution:
Composition per 50.0mL:
Histidine 5.0g

Preparation of Histidine Solution: Add histidine to distilled/deionized water and bring volume to 50.0mL. Mix thoroughly. Sparge with 100% N_2. Autoclave for 15 min at 15 psi pressure–121°C. Cool to room temperature.

Yeast Extract Solution:
Composition per 10.0mL:
Yeast extract 0.2g

Preparation of Yeast Extract Solution: Add yeast extract to distilled/deionized water and bring volume to 10.0mL. Mix thoroughly. Sparge with 100% N_2. Autoclave under 100% N_2 for 15 min at 15 psi pressure–121°C. Cool to room temperature.

$Na_2S·9H_2O$ Solution:
Composition per 10.0mL:
$Na_2S·9H_2O$ 0.2g

Preparation of $Na_2S·9H_2O$ Solution: Add $Na_2S·9H_2O$ to distilled/deionized water and bring volume to 10.0mL. Mix thoroughly. Autoclave under 100% N_2 for 15 min at 15 psi pressure–121°C. Cool to room temperature.

Trace Elements Solution:
Composition per liter:
$MgSO_4·7H_2O$ 3.0g
Nitrilotriacetic acid 1.5g
NaCl 1.0g
$MnSO_4·2H_2O$ 0.5g
$CoSO_4·7H_2O$ 0.18g
$ZnSO_4·7H_2O$ 0.18g
$CaCl_2·2H_2O$ 0.1g
$FeSO_4·7H_2O$ 0.1g
$NiCl_2·6H_2O$ 0.025g
$KAl(SO_4)_2·12H_2O$ 0.02g
H_3BO_3 .... 0.01g
$Na_2MoO_4·4H_2O$ 0.01g
$CuSO_4·5H_2O$ 0.01g
$Na_2SeO_3·5H_2O$ 0.3mg

Preparation of Trace Elements Solution: Add nitrilotriacetic acid to 500.0mL of distilled/deionized water. Dissolve by adjusting pH to 6.5 with KOH. Add remaining components. Add distilled/deionized water to 1.0L. Mix thoroughly.

Vitamin Solution:
Composition per liter:
Pyridoxine-HCl 10.0mg
Thiamine-HCl·$2H_2O$ 5.0mg
Riboflavin 5.0mg
Nicotinic acid 5.0mg
D-Ca-pantothenate 5.0mg
p-Aminobenzoic acid 5.0mg
Lipoic acid 5.0mg
Biotin 2.0mg
Folic acid 2.0mg
Vitamin B_{12} 0.1mg

Preparation of Vitamin Solution: Add components to distilled/deionized water and bring volume to 1.0L. Mix thoroughly. Sparge with 80% H_2 + 20% CO_2. Filter sterilize.

Preparation of Medium: Prepare and dispense medium under 100% N_2 gas atmosphere. Add components, except $NaHCO_3$, NH_4Cl, Na_2CO_3, histidine solution, $Na_2S·9H_2O$ solution, yeast extract solution, and vitamin solution, to distilled/deionized water and bring volume to 928.0mL. Mix thoroughly. Gently heat and bring to boiling. Boil for 5 min. Cool to room temperature while sparging with 100% N_2. Add solid $NaHCO_3$, NH_4Cl, and Na_2CO_3. Mix thoroughly. Distribute into anaerobe tubes or bottles. Autoclave for 15 min at 15 psi pressure–121°C. Aseptically and anaerobically add per liter of medium 50.0mL histidine solution, 10.0mL yeast extract solution, 10.0mL $Na_2S{\cdot}9H_2O$ solution, and 2.0mL vitamin solution. The final pH should be 8.9.

Use: For the cultivation of *Natronincola histidinovorans.*

Natronobacteria Medium

Composition per liter:

NaCl	200.0g
Agar	20.0g
Yeast extract	5.0g
Casamino acids	5.0g
KH_2PO_4	1.0g
KCl	1.0g
NH_4Cl	1.0g
Sodium glutamate	1.0g
$MgSO_4{\cdot}7H_2O$	0.24g
$CaSO_4{\cdot}2H_2O$	0.17g
Na_2CO_3 solution	100.0mL
Trace elements solution SL-6	1.0mL

pH 9.0 ± 0.2 at 25°C

Na_2CO_3 Solution:
Composition per 100.0mL:

Na_2CO_3	5.0g

Preparation of Na_2CO_3 Solution: Add Na_2CO_3 to distilled/deionized water and bring volume to 100.0mL. Mix thoroughly. Autoclave for 15 min at 15 psi pressure–121°C. Cool to 50°C.

Trace Elements Solution SL-6:
Composition per liter:

H_3BO_3	0.3g
$CoCl_2{\cdot}6H_2O$	0.2g
$ZnSO_4{\cdot}7H_2O$	0.1g
$MnCl_2{\cdot}4H_2O$	0.03g
$Na_2MoO_4{\cdot}H_2O$	0.03g
$NiCl_2{\cdot}6H_2O$	0.02g
$CuCl_2{\cdot}2H_2O$	0.01g

Preparation of Trace Elements Solution SL-6: Add components to distilled/deionized water and bring volume to 1.0L. Mix thoroughly. Adjust pH to 3.4.

Preparation of Medium: Add components, except Na_2CO_3 solution, to distilled/deionized water and bring volume to 900.0mL. Mix thoroughly. Gently heat and bring to boiling. Autoclave for 15 min at 15 psi pressure–121°C. Cool to 45°–50°C. Aseptically add sterile Na_2CO_3 solution. Mix thoroughly. Adjust pH to 9.0, if necessary. Pour into sterile Petri dishes or distribute into sterile tubes.

Use: For the cultivation and maintenance of *Natronobacterium gregoryi, Natronobacterium magadii, Natronobacterium pharaonis,* and *Natronococcus occultus.*

Natronobacterium Medium

Composition per liter:

NaCl	250.0g
Agar	20.0g
Casamino acids	15.0g
Trisodium citrate·$2H_2O$	3.0g
Glutamic acid	2.5g
$MgSO_4{\cdot}7H_2O$	2.5g
KCl	2.0g
Na_2CO_3 solution	variable

pH 8.5 ± 0.2 at 25°C

Na_2CO_3 Solution:
Composition per 100.0mL:

Na_2CO_3	5.0g

Preparation of Na_2CO_3 Solution: Add Na_2CO_3 to distilled/deionized water and bring volume to 100.0mL. Mix thoroughly. Filter sterilize.

Preparation of Medium: Add components, except salt and Na_2CO_3 solution, to distilled/deionized water and bring volume to 850.0mL. Mix thoroughly. Gently heat and bring to boiling. Add salt. Mix thoroughly. Bring pH to 7.0. Autoclave for 15 min at 15 psi pressure–121°C. Cool to 50°–55°C. Adjust pH to 8.5 with sterile Na_2CO_3 solution.

Use: For the cultivation and maintenance of *Natronobacterium magadii, Natronobacterium pharaonis,* and *Natronococcus occultus.*

Natronobacterium pharaonis Medium

Composition per liter:

NaCl	250.0g
Casamino acids	15.0g
Sodium citrate	3.0g
Glutamic acid	2.5g
$MgSO_4{\cdot}7H_2O$	2.0g
KCl	2.0g

pH 8.5 ± 0.2 at 25°C

Preparation of Medium: Add components to distilled/deionized water and bring volume to 1.0L. Mix thoroughly. Adjust pH to 8.5. Distribute into tubes or flasks. Autoclave for 15 min at 15 psi pressure–121°C.

Use: For the cultivation of *Natronobacterium pharaonis.*

Nautilia Medium (DSMZ Medium 946)

Composition per liter:

Sulfur	10.0g
NH_4Cl	0.33g
KH_2PO_4	0.33g
Resazurin	0.5mg
Synthetic seawater, concentrated	500.0mL
$NaHCO_3$ solution	50.0mL
$Na_2S{\cdot}9H_2O$ solution	20.0mL
Sodium formate solution	15.0mL
Trace elements solution	10.0mL
Vitamin solution	10.0mL
Selenite-tungstate solution	1.0mL

pH 6.8 ± 0.2 at 25°C

Synthetic Seawater, Concentrated:
Composition per 500.0mL:

NaCl	55.4g
$MgSO_4{\cdot}7H_2O$	14.0g
$MgCl_2{\cdot}6H_2O$	11.0g
$CaCl_2{\cdot}2H_2O$	1.5g
KCl	1.3g
NaBr	0.2g
H_3BO_3	0.06g
$SrCl_2{\cdot}6H_2O$	0.03g

Na_3-citrate 20.0mg
KI 0.1mg

Preparation of Synthetic Seawater, Concentrated: Add components to distilled/deionized water and bring volume to 500.0mL. Mix thoroughly.

Preparation of Vitamin Solution: Add components to distilled/deionized water and bring volume to 1.0L. Mix thoroughly. Sparge with 80% H_2 + 20% CO_2. Filter sterilize.

Trace Elements Solution:
Composition per liter:
$MgSO_4 \cdot 7H_2O$ 3.0g
Nitrilotriacetic acid 1.5g
NaCl 1.0g
$MnSO_4 \cdot 2H_2O$ 0.5g
$CoSO_4 \cdot 7H_2O$ 0.18g
$ZnSO_4 \cdot 7H_2O$ 0.18g
$CaCl_2 \cdot 2H_2O$ 0.1g
$FeSO_4 \cdot 7H_2O$ 0.1g
$NiCl_2 \cdot 6H_2O$ 0.025g
$KAl(SO_4)_2 \cdot 12H_2O$ 0.02g
H_3BO_3 .... 0.01g
$Na_2MoO_4 \cdot 4H_2O$ 0.01g
$CuSO_4 \cdot 5H_2O$ 0.01g
$Na_2SeO_3 \cdot 5H_2O$ 0.3mg

Preparation of Trace Elements Solution: Add nitrilotriacetic acid to 500.0mL of distilled/deionized water. Dissolve by adjusting pH to 6.5 with KOH. Add remaining components. Add distilled/deionized water to 1.0L. Mix thoroughly.

Vitamin Solution:
Composition per liter:
Pyridoxine-HCl 10.0mg
Thiamine-HCl·$2H_2O$ 5.0mg
Riboflavin 5.0mg
Nicotinic acid 5.0mg
D-Ca-pantothenate 5.0mg
p-Aminobenzoic acid 5.0mg
Lipoic acid 5.0mg
Biotin 2.0mg
Folic acid 2.0mg
Vitamin B_{12} 0.1mg

Preparation of Vitamin Solution: Add components to distilled/deionized water and bring volume to 1.0L. Mix thoroughly. Sparge with 80% H_2 + 20% CO_2. Filter sterilize.

Sodium Formate Solution:
Composition per 50.0mL:
Na-formate 10.0g

Preparation of Sodium Formate Solution: Add sodium formate to distilled/deionized water and bring volume to 50.0mL. Mix thoroughly. Sparge with 100% N_2. Filter sterilize.

$Na_2S \cdot 9H_2O$ Solution:
Composition per 20.0mL:
$Na_2S \cdot 9H_2O$ 0.6g

Preparation of $Na_2S \cdot 9H_2O$ Solution: Add $Na_2S \cdot 9H_2O$ to distilled/deionized water and bring volume to 20.0mL. Mix thoroughly. Autoclave under 100% N_2 for 15 min at 15 psi pressure–121°C. Cool to room temperature.

$NaHCO_3$ Solution:
Composition per 100.0mL:
$NaHCO_3$ 5.0g

Preparation of $NaHCO_3$ Solution: Add $NaHCO_3$ to distilled/deionized water and bring volume to 100.0mL. Mix thoroughly. Sparge with 80% N_2 + 20% CO_2. Filter sterilize.

Selenite-Tungstate Solution
Composition per liter:
NaOH 0.5g
$Na_2WO_4 \cdot 2H_2O$ 4.0mg
$Na_2SeO_3 \cdot 5H_2O$ 3.0mg

Preparation of Selenite-Tungstate Solution: Add components to distilled/deionized water and bring volume to 1.0L. Mix thoroughly. Sparge with 100% N_2. Filter sterilize.

Preparation of Sulfur: Sterilize sulfur by steaming for 3 h on each of 3 successive days.

Preparation of Medium: Prepare and dispense medium under 80% N_2 + 20% CO_2 gas atmosphere. Add components, except sulfur, $NaHCO_3$ solution, sodium formate solution, $Na_2S \cdot 9H_2O$ solution, vitamin solution, selenite-tungstate solution, and trace elements solution, to distilled/deionized water and bring volume to 894.0mL. Mix thoroughly. Adjust pH to 6.8. Sparge with 80% N_2 + 20% CO_2. Autoclave for 15 min at 15 psi pressure–121°C. Aseptically and anaerobically add 10.0g steam sterilized sulfur, 50.0mL $NaHCO_3$ solution, 15.0mL sodium formate solution, 20.0mL $Na_2S \cdot 9H_2O$ solution, 10.0mL vitamin solution, 1.0mL selenite-tungstate solution, and 10.0mL trace elements solution. Mix thoroughly. Adjust pH to 6.8. Aseptically and anaerobically distribute into sterile tubes or bottles.

Use: For the cultivation of *Caldithrix abyssi* and *Nautilia lithotrophica.*

Nautilia Medium (DSMZ Medium 946)

Composition per liter:
NH_4Cl 0.33g
KH_2PO_4 .... 0.33g
Resazurin 0.5mg
Synthetic seawater, concentrated 500.0mL
$NaHCO_3$ solution 50.0mL
$Na_2S \cdot 9H_2O$ solution 20.0mL
Trace elements solution 10.0mL
Vitamin solution 10.0mL
Selenite-tungstate solution 1.0mL
pH 6.8 ± 0.2 at 25°C

Synthetic Seawater, Concentrated:
NaCl 55.4g
$MgSO_4 \cdot 7H_2O$ 14.0g
$MgCl_2 \cdot 6H_2O$ 11.0g
$CaCl_2 \cdot 2H_2O$ 1.5g
KCl 1.3g
NaBr 0.2g
H_3BO_3 .... 0.06g
$SrCl_2 \cdot 6H_2O$ 0.03g
Na_3-citrate 20.0mg
KI 0.1mg

Trace Elements Solution:
Composition per liter:
$MgSO_4 \cdot 7H_2O$ 3.0g
Nitrilotriacetic acid 1.5g
NaCl 1.0g

$MnSO_4 \cdot 2H_2O$	0.5g
$CoSO_4 \cdot 7H_2O$	0.18g
$ZnSO_4 \cdot 7H_2O$	0.18g
$CaCl_2 \cdot 2H_2O$	0.1g
$FeSO_4 \cdot 7H_2O$	0.1g
$NiCl_2 \cdot 6H_2O$	0.025g
$KAl(SO_4)_2 \cdot 12H_2O$	0.02g
H_3BO_3	0.01g
$Na_2MoO_4 \cdot 4H_2O$	0.01g
$CuSO_4 \cdot 5H_2O$	0.01g
$Na_2SeO_3 \cdot 5H_2O$	0.3mg

Preparation of Trace Elements Solution: Add nitrilotriacetic acid to 500.0mL of distilled/deionized water. Dissolve by adjusting pH to 6.5 with KOH. Add remaining components. Add distilled/deionized water to 1.0L. Mix thoroughly.

Vitamin Solution:

Composition per liter:

Pyridoxine-HCl	10.0mg
Thiamine-HCl·$2H_2O$	5.0mg
Riboflavin	5.0mg
Nicotinic acid	5.0mg
D-Ca-pantothenate	5.0mg
p-Aminobenzoic acid	5.0mg
Lipoic acid	5.0mg
Biotin	2.0mg
Folic acid	2.0mg
Vitamin B_{12}	0.1mg

Preparation of Vitamin Solution: Add components to distilled/deionized water and bring volume to 1.0L. Mix thoroughly. Sparge with 80% H_2 + 20% CO_2. Filter sterilize.

$Na_2S \cdot 9H_2O$ Solution:

Composition per 20.0mL:

$Na_2S \cdot 9H_2O$	0.6g

Preparation of $Na_2S \cdot 9H_2O$ Solution: Add $Na_2S \cdot 9H_2O$ to distilled/deionized water and bring volume to 20.0mL. Mix thoroughly. Autoclave under 100% N_2 for 15 min at 15 psi pressure–121°C. Cool to room temperature.

$NaHCO_3$ Solution:

Composition per 100.0mL:

$NaHCO_3$	5.0g

Preparation of $NaHCO_3$ Solution: Add $NaHCO_3$ to distilled/deionized water and bring volume to 100.0mL. Mix thoroughly. Sparge with 80% N_2 + 20% CO_2. Filter sterilize.

Selenite-Tungstate Solution

Composition per liter:

NaOH	0.5g
$Na_2WO_4 \cdot 2H_2O$	4.0mg
$Na_2SeO_3 \cdot 5H_2O$	3.0mg

Preparation of Selenite-Tungstate Solution: Add components to distilled/deionized water and bring volume to 1.0L. Mix thoroughly. Sparge with 100% N_2. Filter sterilize.

Yeast Extract Solution:

Composition per 20.0mL:

Yeast extract	4.0g

Preparation of Yeast Extract Solution: Add yeast extract to distilled/deionized water and bring volume to 20.0mL. Mix thoroughly. Sparge with 100% N_2. Autoclave under 100% N_2 for 15 min at 15 psi pressure–121°C. Cool to room temperature.

Preparation of Medium: Prepare and dispense medium under 80% N_2 + 20% CO_2 gas atmosphere. Add components, except $NaHCO_3$ solution, yeast extract solution, $Na_2S \cdot 9H_2O$ solution, vitamin solution, selenite-tungstate solution, and trace elements solution, to distilled/deionized water and bring volume to 894.0mL. Mix thoroughly. Adjust pH to 6.8. Sparge with 80% N_2 + 20% CO_2. Autoclave for 15 min at 15 psi pressure–121°C. Aseptically and anaerobically add 50.0mL $NaHCO_3$ solution, 15.0mL yeast extract solution, 20.0mL $Na_2S \cdot 9H_2O$ solution, 10.0mL vitamin solution, 1.0mL selenite-tungstate solution, and 10.0mL trace elements solution. Mix thoroughly. Adjust pH to 6.8. Aseptically and anaerobically distribute into sterile tubes or bottles.

Use: For the cultivation of *Caldithrix abyssi* DSM 13497.

NBA Medium

Composition per liter:

Pancreatic digest of gelatin	5.0g
Casamino acids	5.0g
Beef extract	3.0g
Yeast extract	1.0g

pH 6.8 ± 0.2 at 25°C

Preparation of Medium: Add components to distilled/deionized water and bring volume to 1.0L. Mix thoroughly. Distribute into tubes or flasks. Autoclave for 15 min at 15 psi pressure–121°C.

Use: For the cultivation of *Bdellovibrio* species.

NBY Medium (Nutrient Broth Yeast Extract Medium)

Composition per liter:

Nutrient broth, dehydrated	8.0g
Yeast extract	2.0g
K_2HPO_4	2.0g
KH_2PO_4	0.5g
Glucose solution	50.0mL
$MgSO_4 \cdot 7H_2O$ (1*M* solution)	1.0mL

Glucose Solution:

Composition per 50.0mL:

D-Glucose	5.0g

Preparation of Glucose Solution: Add glucose to distilled/deionized water and bring volume to 50.0mL. Mix thoroughly. Filter sterilize.

Preparation of Medium: Add components, except glucose solution, to distilled/deionized water and bring volume to 950.0mL. Mix thoroughly. Gently heat and bring to boiling. Autoclave for 15 min at 15 psi pressure–121°C. Cool to 45°–50°C. Aseptically add sterile glucose solution. Mix thoroughly. Pour into sterile Petri dishes or distribute into sterile tubes.

Use: For the cultivation and maintenance of *Curtobacterium flaccumfaciens* and *Pseudomonas syringae*.

NBY Medium (Nutrient Broth Yeast Extract Medium) (ATCC Medium 763)

Composition per liter:

Agar	15.0g
Nutrient broth	8.0g

Yeast extract 2.0g
K_2HPO_4 .... 2.0g
KH_2PO_4 .... 0.5g
Glucose solution 50.0mL
$MgSO_4$ solution 50.0mL

Glucose Solution:
Composition per 50.0mL:
D-Glucose 5.0g

Preparation of Glucose Solution: Add glucose to distilled/deionized water and bring volume to 50.0mL. Mix thoroughly. Filter sterilize.

$MgSO_4$ Solution:
Composition per 50.0mL:
$MgSO_4 \cdot 7H_2O$ 0.25g

Preparation of $MgSO_4$ Solution: Add the solid $MgSO_4 \cdot 7H_2O$ to distilled/deionized water and bring volume to 50.0mL. Mix thoroughly. Autoclave for 15 min at 15 psi pressure–121°C. Cool to 50°C.

Preparation of Medium: Add components, except glucose solution and $MgSO_4$ solution, to distilled/deionized water and bring volume to 900.0mL. Mix thoroughly. Gently heat and bring to boiling. Autoclave for 25 min at 15 psi pressure–121°C. Cool to 45°–50°C. Aseptically add sterile glucose solution and $MgSO_4$ solution. Mix thoroughly. Pour into sterile Petri dishes or distribute into sterile tubes.

Use: For the cultivation and maintenance of *Bacillus sphaericus*.

Neisseria Medium

Composition per liter:
Biosate 10.0g
Polypeptone™ 10.0g
NaCl 5.0g
Myosate 3.0g
Agar 1.5g
Phenol Red 0.017g
Carbohydrate solution 50.0mL
pH 7.4–7.6 at 25°C

Carbohydrate Solution:
Composition per 50.0mL:
Carbohydrate 10.0g

Preparation of Carbohydrate Solution: Add glucose, sucrose, or maltose to distilled/deionized water and bring volume to 50.0mL. Mix thoroughly. Filter sterilize.

Preparation of Medium: Add components, except carbohydrate solution, to distilled/deionized water and bring volume to 950.0mL. Mix thoroughly. Gently heat and bring to boiling. Autoclave for 15 min at 15 psi pressure–121°C. Cool to 45°–50°C. Aseptically add sterile carbohydrate solution. Mix thoroughly. Pour into sterile Petri dishes or distribute into sterile tubes.

Use: For the cultivation of *Neisseria* species.

Neisseria meningitidis Medium

Composition per liter:
Beef infusion 300.0g
Acid hydrolysate of casein 17.5g
Agar 17.0g
Starch 1.5g
Antibiotic solution 10.0mL
pH 7.4 ± 0.2 at 25°C

Antibiotic Solution:
Composition per 10.0mL:
Vancomycin 3.0mg
Colistin 7.5mg
Nystatin 12,500U

Preparation of Antibiotic Solution: Add components to distilled/deionized water and bring volume to 10.0mL. Mix thoroughly. Filter sterilize.

Preparation of Medium: Add components, except antibiotic solution, to distilled/deionized water and bring volume to 990.0mL. Mix thoroughly. Gently heat and bring to boiling. Autoclave for 15 min at 15 psi pressure–121°C. Cool to 45°–50°C. Aseptically add sterile antibiotic solution. Mix thoroughly. Pour into sterile Petri dishes or distribute into sterile tubes.

Use: For the selective isolation and cultivation of *Neisseria meningitidis*.

Nelson Culture Medium for *Naegleria*

Composition per 100.0mL:
Glucose 0.17g
Panmede 0.17g
Na_2HPO_4 .... 14.2mg
KH_2PO_4 .... 13.6mg
NaCl 12.0mg
$CaCl_2 \cdot 2H_2O$ 0.4mg
$MgSO_4 \cdot 7H_2O$ 0.4mg
Bovine serum, heat-inactivated fetal 10.0mL

Source: Panmede is available from Paines and Byrne Ltd., Greenford, England, and Harrisons and Crosfield, Inc., Bronxville, NY.

Preparation of Medium: Add components, except bovine serum, to distilled/deionized water and bring volume to 90.0mL. Mix thoroughly. Autoclave for 15 min at 15 psi pressure–121°C. Cool to 25°C. Aseptically add 10.0mL of sterile, heat-inactivated fetal bovine serum. Mix thoroughly. Aseptically distribute into sterile tubes or flasks. Use immediately.

Use: For the cultivation of *Naegleria fowleri* and *Paratetramitus jugosus*.

Nelson Culture Medium for *Naegleria*

Composition per 100.0mL:
Glucose 0.17g
Liver infusion 0.17g
Na_2HPO_4 .... 14.2mg
KH_2PO_4 .... 13.6mg
NaCl 12.0mg
$CaCl_2 \cdot 2H_2O$ 0.4mg
$MgSO_4 \cdot 7H_2O$ 0.4mg
Bovine serum, heat-inactivated fetal 10.0mL

Preparation of Medium: Add components, except bovine serum, to distilled/deionized water and bring volume to 90.0mL. Mix thoroughly. Autoclave for 15 min at 15 psi pressure–121°C. Cool to 25°C. Aseptically add 10.0mL of sterile, heat-inactivated fetal bovine serum. Mix thoroughly. Aseptically distribute into sterile tubes or flasks. Use immediately.

Use: For the cultivation of *Naegleria fowleri* and *Paratetramitus jugosus*.

Nelson Medium for *Naegleria fowleri*

Composition per liter:

Glucose	1.0g
Ox liver digest	1.0g
Page's amoeba saline	1.0L
Fetal calf serum, inactivated	20.0mL

Page's Amoeba Saline:
Composition per liter:

Na_2HPO_4	0.142g
KH_2PO_4	0.136g
NaCl	0.12g
$MgSO_4 \cdot 7H_2O$	4.0mg
$CaCl_2 \cdot 2H_2O$	4.0mg

Preparation of Page's Amoeba Saline: Add components to distilled/deionized water and bring volume to 10.0mL. Mix thoroughly.

Preparation of Medium: Add the glucose and ox liver digest to 1.0L of Page's amoeba saline. Mix thoroughly. Distribute into screw-capped tubes in 10.0mL volumes. Autoclave for 15 min at 15 psi pressure–121°C. Cool to 25°C. Aseptically add 0.2mL of sterile fetal calf serum to each tube. Mix thoroughly.

Use: For the cultivation of *Naegleria fowleri.*

Neocallimastix Medium (ANO2 Fungus II)

Composition per liter:

$NaHCO_3$	6.6g
Cellobiose	2.0g
Glucose	2.0g
Maltose	2.0g
Starch	2.0g
PIPES (piperazine-*N,N′*-bis[2-ethanesulfonic acid]) buffer	1.0g
Trypticase™ peptone	1.0g
Yeast extract	1.0g
L-Cysteine·HCl·H_2O	0.5g
KH_2PO_4	0.5g
$MgCl_2 \cdot 6H_2O$	0.4g
NaCl	0.4g
NH_4Cl	0.4g
$Na_2S \cdot 9H_2O$	0.1g
$CaCl_2 \cdot 2H_2O$	50.0mg
Resazurin	1.0mg
Rumen fluid	100.0mL
Wolfe's vitamin solution	10.0mL
Trace elements solution SL-10	1.0mL

pH 6.9 ± 0.2 at 25°C

Wolfe's Vitamin Solution:
Composition per liter:

Pyridoxine·HCl	10.0mg
p-Aminobenzoic acid	5.0mg
Lipoic acid	5.0mg
Nicotinic acid	5.0mg
Riboflavin	5.0mg
Thiamine·HCl	5.0mg
Calcium DL-pantothenate	5.0mg
Biotin	2.0mg
Folic acid	2.0mg
Vitamin B_{12}	0.1mg

Preparation of Wolfe's Vitamin Solution: Add components to distilled/deionized water and bring volume to 1.0L. Mix thoroughly.

Trace Elements Solution SL-10:
Composition per liter:

$FeCl_2 \cdot 4H_2O$	1.5g
$CoCl_2 \cdot 6H_2O$	190.0mg
$MnCl_2 \cdot 4H_2O$	100.0mg
$ZnCl_2$	70.0mg
$Na_2MoO_4 \cdot 2H_2O$	36.0mg
$NiCl_2 \cdot 6H_2O$	24.0mg
H_3BO_3	6.0mg
$CuCl_2 \cdot 2H_2O$	2.0mg
HCl (25% solution)	10.0mL

Preparation of Trace Elements Solution SL-10: Add $FeCl_2 \cdot 4H_2O$ to 10.0mL of HCl solution. Mix thoroughly. Add distilled/deionized water and bring volume to 1.0L. Add remaining components. Mix thoroughly.

Preparation of Medium: Prepare and dispense medium under 100% CO_2. Add components to distilled/deionized water and bring volume to 1.0L. Mix thoroughly. Gently heat and bring to boiling. Continue boiling for 3 min. Cool to room temperature while sparging with 100% CO_2. Anaerobically distribute into tubes. Autoclave for 15 min at 15 psi pressure–121°C. Adjust pH to 6.9.

Use: For the cultivation of *Neocallimastix frontalis.*

Neomycin Agar

Composition per liter:

Agar	15.0g
Peptone	6.0g
Pancreatic digest of casein	4.0g
Yeast extract	3.0g
Beef extract	1.5g
Glucose	1.0g
Neomycin solution	10.0mL

pH 7.0 ± 0.2 at 25°C

Neomycin Solution:
Composition per 10.0mL:

Neomycin sulfate	1.0g

Preparation of Neomycin Solution: Add neomycin sulfate to distilled/deionized water and bring volume to 10.0mL. Mix thoroughly. Filter sterilize.

Preparation of Medium: Add components, except neomycin solution, to distilled/deionized water and bring volume to 990.0mL. Mix thoroughly. Gently heat and bring to boiling. Autoclave for 15 min at 15 psi pressure–121°C. Cool to 45°–50°C. Aseptically add sterile neomycin solution. Mix thoroughly. Pour into sterile Petri dishes or distribute into sterile tubes.

Use: For the cultivation and maintenance of *Micrococcus luteus.*

Neomycin Agar, Modified

Composition per liter:

Agar	15.0g
Peptone	6.0g
Pancreatic digest of casein	4.0g
Yeast extract	3.0g
Beef extract	1.5g
Glucose	1.0g

Methanol 20.0mL
Neomycin solution 10.0mL
pH 7.0 ± 0.2 at 25°C

Neomycin Solution:
Composition per 10.0mL:
Neomycin sulfate 1.0g

Preparation of Neomycin Solution: Add neomycin sulfate to distilled/deionized water and bring volume to 10.0mL. Mix thoroughly. Filter sterilize.

Preparation of Medium: Add components, except methanol and neomycin solution, to distilled/deionized water and bring volume to 970.0mL. Mix thoroughly. Gently heat and bring to boiling. Autoclave for 15 min at 15 psi pressure–121°C. Cool to 45°–50°C. Filter sterilize methanol. To cooled, sterile basal medium, aseptically add sterile methanol and sterile neomycin solution. Mix thoroughly. Pour into sterile Petri dishes or distribute into sterile tubes.

Use: For the cultivation and maintenance of *Bordetella bronchiseptica.*

Neomycin Assay Agar

See: **Antibiotic Medium 11**

Neomycin Blood Agar

Composition per liter:
Pancreatic digest of casein 14.5g
Agar 14.0g
Papaic digest of soybean meal 5.0g
NaCl 5.0g
Growth factors 1.5g
Sheep blood, defibrinated 50.0mL
Neomycin solution 10.0mL
pH 7.3 ± 0.2 at 25°C

Source: This medium is available as a premixed powder from BD Diagnostic Systems.

Neomycin Solution:
Composition per 10.0mL:
Neomycin sulfate 0.03g

Preparation of Neomycin Solution: Add neomycin sulfate to distilled/deionized water and bring volume to 10.0mL. Mix thoroughly. Filter sterilize.

Preparation of Medium: Add components, except sheep blood and neomycin solution, to distilled/deionized water and bring volume to 940.0mL. Mix thoroughly. Gently heat and bring to boiling. Autoclave for 15 min at 15 psi pressure–121°C. Cool to 45°–50°C. Aseptically add sterile sheep blood and sterile neomycin solution. Mix thoroughly. Pour into sterile Petri dishes or distribute into sterile tubes.

Use: For the isolation and cultivation of group A streptococci (*Streptococcus pyogenes)* and group B streptococci (*Streptococcus agalactiae*) from throat cultures and other clinical specimens.

Neomycin Luria Agar

Composition per liter:
Agar 15.0g
Pancreatic digest of casein 10.0g
Yeast extract 5.0g
NaCl 0.5g
Glucose solution 20.0mL
Neomycin solution 10.0mL
pH 7.0 ± 0.1 at 25°C

Glucose Solution:
Composition per 100.0mL:
Glucose 10.0g

Preparation of Glucose Solution: Add glucose to distilled/deionized water and bring volume to 100.0mL. Mix thoroughly. Filter sterilize.

Neomycin Solution:
Composition per 10.0mL:
Neomycin sulfate 12.0mg

Preparation of Neomycin Solution: Add neomycin sulfate to distilled/deionized water and bring volume to 10.0mL. Mix thoroughly. Filter sterilize.

Preparation of Medium: Add components, except glucose solution and neomycin solution, to distilled/deionized water and bring volume to 970.0mL. Mix thoroughly. Gently heat and bring to boiling. Autoclave for 15 min at 15 psi pressure–121°C. Aseptically add 20.0mL of sterile glucose solution and 10.0mL of sterile neomycin solution. Mix thoroughly. Pour into sterile Petri dishes or distribute into sterile tubes.

Use: For the cultivation of *Escherichia coli.*

Neomycin Medium No. 1

Composition per liter:
Agar 15.0g
Peptone 5.0g
NaCl 5.0g
Yeast extract 2.0g
Beef extract 1.0g
Sucrose solution 20.0mL
Neomycin solution 10.0mL
pH 7.0 ± 0.1 at 25°C

Sucrose Solution:
Composition per 100.0mL:
Sucrose 2.5g

Preparation of Sucrose Solution: Add sucrose to distilled/deionized water and bring volume to 100.0mL. Mix thoroughly. Filter sterilize.

Neomycin Solution:
Composition per 10.0mL:
Neomycin sulfate 500.0mg

Preparation of Neomycin Solution: Add neomycin sulfate to distilled/deionized water and bring volume to 10.0mL. Mix thoroughly. Filter sterilize.

Preparation of Medium: Add components, except sucrose solution and neomycin solution, to distilled/deionized water and bring volume to 970.0mL. Mix thoroughly. Gently heat and bring to boiling. Autoclave for 15 min at 15 psi pressure–121°C. Aseptically add 20.0mL of sterile sucrose solution and 10.0mL of sterile neomycin solution. Mix thoroughly. Pour into sterile Petri dishes or distribute into sterile tubes.

Use: For the cultivation of *Micrococcus luteus* and *Pseudomonas aeruginosa.*

Neopeptone Glucose Agar

Composition per liter:

Agar	20.0g
Glucose	10.0g
Neopeptone	5.0g

pH 6.5 ± 0.2 at 25°C

Preparation of Medium: Add components to distilled/deionized water and bring volume to 1.0L. Mix thoroughly. Gently heat and bring to boiling. Distribute into tubes or flasks. Autoclave for 15 min at 15 psi pressure–121°C. Adjust pH to 6.5. Pour into sterile Petri dishes or leave in tubes.

Use: For the maintenance of stock cultures of a variety of microorganisms.

Neopeptone Glucose Rose Bengal Aureomycin® Agar

Composition per liter:

Agar	20.0g
Neopeptone	5.0g
Glucose	1.0g
Tetracycline solution	5.0mL
Rose Bengal solution	3.5mL

pH 6.5 ± 0.2 at 25°C

Tetracycline Solution:

Composition per 150.0mL:

Tetracycline	1.0g

Preparation of Tetracycline Solution: Add tetracycline to distilled/deionized water and bring volume to 150.0mL. Mix thoroughly. Filter sterilize.

Rose Bengal Solution:

Composition per 100.0mL:

Rose Bengal	1.0g

Preparation of Rose Bengal Solution: Add Rose Bengal to distilled/deionized water and bring volume to 100.0mL. Mix thoroughly. Filter sterilize.

Preparation of Medium: Add components, except tetracycline solution, to distilled/deionized water and bring volume to 995.0mL. Mix thoroughly. Gently heat and bring to boiling. Autoclave for 15 min at 15 psi pressure–121°C. Cool to 45°–50°C. Aseptically add 5.0mL of sterile tetracycline solution. Mix thoroughly. Pour into sterile Petri dishes or distribute into sterile tubes.

Use: For the isolation and cultivation of a wide variety of fungal species.

Neopeptone Infusion Agar

Composition per liter:

Beef heart, infusion from	500.0g
Neopeptone	20.0g
Agar	20.0g
NaCl	5.0g
Sheep blood, defibrinated	50.0mL

pH 7.4 ± 0.2 at 25°C

Preparation of Medium: Add components, except sheep blood, to distilled/deionized water and bring volume to 950.0mL. Mix thoroughly. Gently heat and bring to boiling. Autoclave for 15 min at 15 psi pressure–121°C. Cool to 45°–50°C. Aseptically add sterile sheep blood. Mix thoroughly. Pour into sterile Petri dishes or distribute into sterile tubes.

Use: For the cultivation of a wide variety of fastidious microorganisms.

Nesterenkonia Modified Medium (DSMZ Medium 993a)

Composition per liter:

$MgCl_2 \cdot 6H_2O$	100.0g
Agar	20.0g
Glycerol	10.0g
Yeast extract	5.0g
L-Asparagine, anhydrous	1.0g
K_2HPO_4, anhydrous	1.0g
Trace elements solution	1.0mL

pH 7.2 ± 0.2 at 25°C

Trace Elements Solution:

Composition per 100.0mL:

$FeSO_4 \cdot 7H_2O$	0.1g
$MnCl_2 \cdot 4H_2O$	0.1g
$ZnSO_4 \cdot 7H_2O$	0.1g

Preparation of Trace Elements Solution: Add components to distilled/deionized water and bring volume to 100.0mL. Mix thoroughly.

Preparation of Medium: Add components to distilled/deionized water and bring volume to 1.0L. Mix thoroughly. Adjust pH to 7.2. Distribute into tubes or flasks. Gently heat while stirring and bring to boiling. Mix thoroughly. Autoclave for 15 min at 15 psi pressure–121°C. Pour into Petri dishes or leave in tubes.

Use: For the cultivation of *Nesterenkonia* spp.

Neurospora Culture Agar

Composition per liter:

Maltose	40.0g
Agar	15.0g
Proteose peptone No. 3	5.0g
Yeast extract	5.0g

pH 6.7 ± 0.2 at 25°C

Source: This medium is available as a premixed powder from BD Diagnostic Systems.

Preparation of Medium: Add components to distilled/deionized water and bring volume to 1.0L. Mix thoroughly. Gently heat while stirring and bring to boiling. Distribute into tubes in 8.0mL volumes. Autoclave for 15 min at 15 psi pressure–121°C. Allow tubes to cool in a slanted position.

Use: For the cultivation of *Neurospora intermedia* used in the microbiological assay of pyridoxine. Also used for the cultivation of other fungi.

Neurospora Medium

Composition per liter:

Agar	15.0g
Glucose	5.0g
Malt syrup, spray dried	5.0g
Sucrose	5.0g
Yeast extract	2.5g
Vitamin solution	10.0mL
Casein, hydrolyzed	5.0mL

Vitamin Solution:

Composition per liter:

Ribonucleic acid, alkali hydrolyzed 0.5g
Inositol 0.4g
Choline 0.2g
Nicotinamide 0.2g
Pantothenic acid 0.2g
Thiamine 0.1g
p-Aminobenzoic acid 0.05g
Pyridoxine 0.05g
Riboflavin 0.05g
Folic acid 4.0μg

Preparation of Vitamin Solution: Add components to distilled/deionized water and bring volume to 1.0L. Mix thoroughly. Filter sterilize.

Preparation of Medium: Add components, except vitamin solution, to distilled/deionized water and bring volume to 990.0mL. Mix thoroughly. Gently heat and bring to boiling. Autoclave for 15 min at 15 psi pressure–121°C. Cool to 45°–50°C. Aseptically add 10.0mL of sterile vitamin solution. Mix thoroughly. Pour into sterile Petri dishes or distribute into sterile tubes.

Use: For the cultivation of *Neurospora* species on complete medium.

Neurospora Medium

Composition per liter:

Sucrose 15.0g
Ammonium tartrate 5.0g
KH_2PO_4 .... 1.0g
NH_4NO_3 .... 1.0g
$MgSO_4 \cdot 7H_2O$ 0.5g
$CaCl_2$ 0.1g
NaCl 0.1g
$ZnCl_2$ 2.0mg
$FeCl_3$ 0.2mg
Cu Cl_2 .... 0.1mg
$MnCl_2$ 0.02mg
$Na_2MoO_4 \cdot 2H_2O$ 0.02mg
H_3BO_3 .... 0.01mg
Biotin 1.0μg

Preparation of Medium: Add components to distilled/deionized water and bring volume to 1.0L. Mix thoroughly. Distribute into tubes or flasks. Autoclave for 15 min at 15 psi pressure–121°C.

Use: For the cultivation of *Neurospora* species on minimal medium.

Neurospora Minimal Medium

Composition per liter:

Sucrose 20.0g
Ammonium tartrate 5.0g
KH_2PO_4 .... 1.0g
$NaNO_3$ 1.0g
$MgSO_4 \cdot 7H_2O$ 0.5g
$CaCl_2 \cdot 2H_2O$ 0.1g
NaCl 0.1g
$ZnSO_4 \cdot 7H_2O$ 5.5mg
$FeSO_4 \cdot 7H_2O$ 0.54mg
$CuSO_4 \cdot 5H_2O$ 0.39mg
$MnSO_4 \cdot 4H_2O$ 0.063mg
H_3BO_3 .... 0.057mg
$Na_2MoO_4 \cdot 2H_2O$ 0.05mg
Biotin 5.0μg

pH 7.3 ± 0.2 at 25°C

Preparation of Medium: Add components to distilled/deionized water and bring volume to 1.0L. Mix thoroughly. Gently heat and bring to boiling. Distribute into tubes or flasks. Autoclave for 15 min at 15 psi pressure–121°C.

Use: For the cultivation of *Neurospora* species and the detection of *Neurospora* mutants.

Neurospora Minimal Medium

Composition per liter:

Sucrose 20.0g
Agar 20.0g
KH_2PO_4 .... 5.0g
Trisodium citrate·$2H_2O$ 2.5g
NH_4NO_3 .... 2.0g
$MgSO_4 \cdot 7H_2O$ 0.2g
$CaCl_2 \cdot 2H_2O$ 0.1g
Sucrose-agar solution 200.0mL
Biotin solution 1.0mL
Trace elements solution 0.1mL

pH 5.8 ± 0.2 at 25°C

Sucrose-Agar Solution:

Composition per 200.0mL:

Sucrose 20.0g
Agar 20.0g

Preparation of Sucrose-Agar Solution: Add components to distilled/deionized water and bring volume to 200.0mL. Mix thoroughly. Gently heat and bring to boiling. Autoclave for 15 min at 15 psi pressure–121°C. Cool to 45°–50°C.

Biotin Solution:

Composition per 10.0mL:

Biotin 50.0μg

Preparation of Biotin Solution: Add biotin to distilled/deionized water and bring volume to 10.0mL. Mix thoroughly. Filter sterilize.

Trace Elements Solution:

Composition per 100.0mL

Citric acid·H_2O 5.0g
$ZnSO_4 \cdot 7H_2O$ 5.0g
$Fe(NH_4)_2(SO_4)_2 \cdot 6H_2O$ 1.0g
$CuSO_4 \cdot 5H_2O$ 0.25g
$MnSO_4 \cdot H_2O$ 0.05g
H_3BO_3 .... 0.05g
$Na_2MoO_4 \cdot 2H_2O$ 0.05g

Preparation of Trace Elements Solution: Add components one at a time to distilled/deionized water and bring volume to 100.0mL. Mix thoroughly. Filter sterilize. Add 1.0mL of chloroform as a preservative. Store at room temperature.

Preparation of Medium: Add trisodium citrate·$2H_2O$, KH_2PO_4, NH_4NO_3, $MgSO_4 \cdot 7H_2O$, and $CaCl_2 \cdot 2H_2O$ to distilled/deionized water and bring volume to 800.0mL. Make sure that one component is dissolved completely before adding the next one. This is conveniently done in a Fernbach flask on a shaker. Mix thoroughly. Filter sterilize. Warm solution to 45°–50°C. Add 200.0mL of sterile sucrose-agar solution, 1.0mL of sterile biotin solution, and 0.1mL of sterile trace elements solution. Mix thoroughly. Pour into sterile Petri dishes or distribute into sterile tubes.

Use: For the cultivation and maintenance of *Cochliobolus sativus, Fusarium solani, Neurospora crassa,* and *Neurospora sitophila.*

Neutral Red Broth
See: **LICNR Broth**

Nevskia Medium (DSMZ Medium 828)

Composition per liter:

Na-lactate	0.56g
$MgSO_4 \cdot 7H_2O$	0.05g
$CaCl_2 \cdot 2H_2O$	0.014g
Potassium phosphate buffer, 1*M*, pH 7	27.5mL
Trace elements solution SL-9	0.5mL
Seven vitamin solution	0.5mL

pH 7.1 ± 0.2 at 25°C

Trace Elements Solution SL-9:

Composition per liter:

Nitrilotriacetic acid	12.8g
$FeCl_2 \cdot 4H_2O$	3.5g
$CoCl_2 \cdot 6H_2O$	190.0mg
$MnCl_2 \cdot 4H_2O$	100.0mg
$ZnCl_2$	70.0mg
$Na_2MoO_4 \cdot 2H_2O$	36.0mg
$NiCl_2 \cdot 6H_2O$	24.0mg
H_3BO_3	6.0mg
$CuCl_2 \cdot 2H_2O$	2.0mg
HCl (25% solution)	10.0mL

Preparation of Trace Elements Solution SL-9: Add $FeCl_2 \cdot 4H_2O$ to 10.0mL of HCl solution. Mix thoroughly. Add distilled/deionized water and bring volume to 1.0L. Add remaining components. Mix thoroughly. Adjust pH to 6.0. Sparge with 80% N_2 + 20% CO_2. Autoclave for 15 min at 15 psi pressure–121°C.

Seven Vitamin Solution:

Composition per liter:

Pyridoxine hydrochloride	300.0mg
Thiamine-HCl·$2H_2O$	200.0mg
Nicotinic acid	200.0mg
Vitamin B_{12}	100.0mg
Calcium pantothenate	100.0mg
p-Aminobenzoic acid	80.0mg
D(+)-Biotin	20.0mg

Preparation of Seven Vitamin Solution: Add components to distilled/deionized water and bring volume to 1.0L. Sparge with 100% N_2. Mix thoroughly. Filter sterilize.

Preparation of Medium: Add components, except seven vitamin solution, to distilled/deionized water and bring volume to 999.5mL. Mix thoroughly. Adjust pH to 7.1. Autoclave for 15 min at 15 psi pressure–121°C. Cool to room temperature. Aseptically add 0.5mL seven vitamin solution. Mix thoroughly. Aseptically distribute into sterile tubes or flasks.

Use: For the cultivation of *Nevskia ramosa.*

New York City Medium

Composition per liter:

NYC basal medium	640.0mL
Horse blood cells	200.0mL
Horse plasma, citrated	120.0mL
Yeast dialysate	25.0mL
Glucose solution	10.0mL
Antibiotic VCNT solution	5.0mL

pH 7.4 ± 0.2 at 25°C

NYC Basal Medium:

Composition per 640.0mL:

Solution 1	400.0mL
Solution 3	200.0mL
Solution 2	40.0mL

Preparation of NYC Basal Medium: Combine solution 1, solution 2, and solution 3. Mix thoroughly. Autoclave for 15 min at 15 psi pressure–121°C. Cool to 45°–50°C.

Solution 1:

Composition per 400.0mL:

Agar	20.0g

Preparation of Solution 1: Add agar to distilled/deionized water and bring volume to 400.0mL. Mix thoroughly. Melt agar in autoclave for 10 min at 0 psi pressure–100°C. Cool to 45°–50°C.

Solution 2:

Composition per 40.0mL:

Cornstarch	1.0g

Preparation of Solution 2: Add cornstarch to distilled/deionized water and bring volume to 40.0mL. Mix thoroughly. Warm to 45°–50°C.

Solution 3:

Composition per 200.0mL:

Proteose peptone No. 3	15.0g
NaCl	5.0g
K_2HPO_4	4.0g
KH_2PO_4	1.0g

Preparation of Solution 3: Add components to distilled/deionized water and bring volume to 1.0L. Mix thoroughly. Gently heat and bring to boiling. Cool to 45°–50°C.

Horse Blood Cells:

Composition per 200.0mL:

Horse blood cells, sedimented	6.0mL

Preparation of Horse Blood Cells: Cow blood may be used instead of horse blood but do not use sheep blood. Use cells freshly packed by sedimentation. Do not pack by centrifugation. Aseptically add 6.0mL of sedimented blood cells to 200.0mL of sterile distilled/deionized water. Mix thoroughly.

Horse Plasma, Citrated:

Composition per 6.0L:

Horse blood	5400.0mL
Citrate solution	600.0mL

Preparation of Horse Plasma, Citrated: Place 600.0mL of sterile citrate solution into a receiving bottle. Draw horse blood to the 6.0L mark. Allow cells to sediment out. Aseptically remove plasma.

Citrate Solution:

Composition per liter:

Sodium citrate	150.0g
NaCl	81.13g

Preparation of Citrate Solution: Add components to distilled/deionized water and bring volume to 1.0L. Mix thoroughly. Filter sterilize.

Glucose Solution:

Composition per 10.0mL:

D-Glucose	5.0g

Preparation of Glucose Solution: Add D-glucose to distilled/deionized water and bring volume to 10.0mL. Mix thoroughly. Autoclave for 10 min at 10 psi pressure–115°C. Cool to 45°–50°C.

Yeast Dialysate:

Composition per 2500.0mL:

Baker's yeast, fresh	908.0g

Preparation of Yeast Dialysate: Add fresh baker's yeast to 2500.0mL of distilled/deionized water. Mix thoroughly. Autoclave for 10 min at 15 psi pressure–121°C. Cool to 25°C. Put into dialysis tubing. Dialyze against 2.0L of distilled/deionized water for 48 hr at 4°C.

Antibiotic VCNT Solution:

Composition per 5.0mL:

Colistin	7.5mg
Trimethorprim lactate	3.0mg
Vancomycin·HCl	2.0mg
Nystatin	12.5U

Preparation of Antibiotic Solution: Add components to distilled/deionized water and bring volume to 5.0mL. Mix thoroughly. Filter sterilize.

Preparation of Medium: Have all solutions prepared and at 45°–50°C. Aseptically combine components. Mix thoroughly. Pour into sterile Petri dishes.

Use: For the isolation and cultivation of pathogenic *Neisseria* species. Used as a transport medium for urogenital and other clinical specimens. For the isolation and presumptive identification of Mycoplasmatales, including large-colony species (*Mycoplasma pneumoniae*) and T–mycoplasmas from urogenital specimens.

New York City Medium, Modified

Composition per liter:

NYC basal medium	840.0mL
α-Gamma horse serum (Flow Labs)	120.0mL
Yeast dialysate	25.0mL
Glucose solution	10.0mL
Antibiotic LCNT solution	5.0mL

pH 7.4 ± 0.2 at 25°C

NYC Basal Medium:

Composition per 840.0mL:

Horse blood	5400.0mL
Solution 1	600.0mL
Solution 3	200.0mL
Solution 2	40.0mL

Preparation of NYC Basal Medium: Combine solution 1, solution 2, and solution 3. Mix thoroughly. Autoclave for 15 min at 15 psi pressure–121°C. Cool to 45°–50°C.

Solution 1:

Composition per 600.0mL:

Agar	20.0g

Preparation of Solution 1: Add agar to distilled/deionized water and bring volume to 600.0mL. Mix thoroughly. Melt agar in autoclave for 10 min at 0 psi pressure–100°C. Cool to 45°–50°C.

Solution 2:

Composition per 40.0mL:

Cornstarch	1.0g

Preparation of Solution 2: Add cornstarch to distilled/deionized water and bring volume to 40.0mL. Mix thoroughly. Warm to 45°–50°C.

Solution 3:

Composition per 200.0mL:

Proteose peptone No. 3	15.0g
NaCl	5.0g
K_2HPO_4	4.0g
KH_2PO_4	1.0g

Preparation of Solution 3: Add components to distilled/deionized water and bring volume to 200.0mL. Mix thoroughly. Gently heat and bring to boiling. Cool to 45°–50°C.

Glucose Solution:

Composition per 10.0mL:

D-Glucose	5.0g

Preparation of Glucose Solution: Add glucose to distilled/deionized water and bring volume to 10.0mL. Mix thoroughly. Autoclave for 10 min at 10 psi pressure–115°C. Cool to 45°–50°C.

Yeast Dialysate:

Composition per 2500.0mL:

Baker's yeast, fresh	908.0g

Preparation of Yeast Dialysate: Add fresh Baker's yeast to 2500.0mL of distilled/deionized water. Mix thoroughly. Autoclave for 10 min at 15 psi pressure–121°C. Cool to 25°C. Put into dialysis tubing. Dialyze against 2.0L of distilled/deionized water for 48 hr at 4°C.

Antibiotic LCNT Solution:

Composition per 5.0mL:

Colistin	7.5mg
Lincomycin·HCl	4.0mg
Trimethorprim lactate	3.0mg
Nystatin	12.5U

Preparation of Antibiotic LCNT Solution: Add the components to distilled/deionized water and bring volume to 5.0mL. Mix thoroughly. Filter sterilize the solution.

Preparation of Medium: Have all solutions prepared and at 45°–50°C. Aseptically combine components. Mix thoroughly. Pour into sterile Petri dishes.

Use: For the isolation and cultivation of pathogenic *Neisseria* species. Used as a transport medium for urogenital and other clinical specimens. For the isolation and presumptive identification of Mycoplasmatales, including large-colony species (*Mycoplasma pneumoniae*) and T–mycoplasmas from urogenital specimens.

Niacin Assay HiVeg Medium

Composition per liter:

Glucose	40.0g
Sodium acetate	20.0g
Plant acid hydrolysate, vitamin free	12.0g
KH_2PO_4	1.0g
K_2HPO_4	1.0g
L-Cystine	0.4g
$MgSO_4$	0.4g
DL-Tryptophan	0.2g
Adenine sulfate	0.02g
$FeSO_4$	0.02g
Guanine hydrochloride	0.02g
$MnSO_4$	0.02g
Uracil	0.02g
NaCl	0.02g
Pyridoxine hydrochloride	0.4mg
Riboflavin (Vitamin B_2)	0.4mg

Calcium pantothenate 0.2mg
Thiamine hydrochloride 0.2mg
p-Aminobenzoic acid (PABA) 0.1mg
Biotin 0.08mg

pH 6.7 ± 0.2 at 25°C

Source: This medium is available as a premixed powder from HiMedia.

Preparation of Medium: Add components to distilled/deionized water and bring volume to 1.0L. Mix thoroughly. Gently heat and bring to boiling. Continue boiling for 2–3 min. Distribute into tubes in 5.0mL volumes. Add standard solution or test solutions to each tube. Adjust the volume of each tube to 10.0mL with distilled/deionized water. Autoclave for 10 min at 15 psi pressure–121°C.

Use: For the microbiological assay of nicotinic acid or nicotinamide (niacin) using *Lactobacillus plantarum* as the test organism.

Niacin Assay Medium

Composition per liter:

Glucose 40.0g
Sodium acetate 20.0g
Vitamin assay casamino acids 12.0g
K_2HPO_4 .... 1.0g
KH_2PO_4 .... 1.0g
L-Cystine 0.4g
$MgSO_4 \cdot 7H_2O$ 0.4g
DL-Tryptophan 0.2g
Adenine sulfate 0.02g
$FeSO_4 \cdot 5H_2O$ 0.02g
Guanine·HCl 0.02g
$MnSO_4 \cdot H_2O$ 0.02g
NaCl 0.02g
Uracil 0.02g
Pyridoxine·HCl 0.4mg
Riboflavin 0.4mg
Calcium pantothenate 0.2mg
Thiamine·HCl 0.2mg
p-Aminobenzoic acid 0.1mg
Biotin 0.8μg

pH 6.7 ± 0.2 at 25°C

Source: This medium is available as a premixed powder from BD Diagnostic Systems.

Preparation of Medium: Add components to distilled/deionized water and bring volume to 1.0L. Mix thoroughly. Gently heat and bring to boiling. Continue boiling for 2–3 min. Distribute into tubes in 5.0mL volumes. Add standard solution or test solutions to each tube. Adjust the volume of each tube to 10.0mL with distilled/deionized water. Autoclave for 10 min at 15 psi pressure–121°C.

Use: For the microbiological assay of nicotinic acid or nicotinamide (niacin) using *Lactobacillus plantarum* as the test organism.

Nickels and Leesment Agar, Modified

Composition per liter:

Part 1 .... 750.0mL
Part 2 .... 100.0mL
Part 3 .... 100.0mL
Part 4 .... 50.0mL

pH 6.65 ± 0.2 at 25°C

Part 1:

Composition per 750.0mL:

Pancreatic digest of casein 20.0g
Lactose 10.0g
Yeast extract 5.0g
NaCl 4.0g
Gelatin 2.5g
Sodium citrate 2.0g

Preparation of Part 1: Add components to distilled/deionized water and bring volume to 750.0mL. Mix thoroughly. Adjust pH to 6.65.

Part 2:

Composition per 500.0mL:

Nonfat dry milk 50.0g

Preparation of Part 2: Add nonfat dry milk to distilled/deionized water and bring volume to 500.0mL. Inoculate with *Lactobacillus bulgaricus*. Incubate at 20°C for 24 hr. Centrifuge at 5000 rpm for 10 min to separate the curd. Collect the supernatant solution. Autoclave for 15 min at 15 psi pressure–121°C. Store at 4°C.

Part 3:

Composition per 100.0mL:

Calcium citrate 13.3g
Carboxymethylcellulose 0.8g

Preparation of Part 3: Combine the calcium citrate and carboxymethylcellulose in a mortar and grind until a fine powder. Add powder to 100.0mL of hot distilled/deionized water. Mix thoroughly. Filter through cheesecloth.

Part 4:

Composition per 50.0mL:

Calcium lactate 8.0g

Preparation of Part 4: Add calcium lactate to distilled/deionized water and bring volume to 50.0mL. Mix thoroughly. Gently heat until dissolved.

Preparation of Medium: Prepare each of the four parts separately. Autoclave each part for 15 min at 15 psi pressure–121°C. Aseptically combine part 1, part 2, part 3, and part 4. Mix thoroughly. Pour into sterile Petri dishes. Swirl flask while dispensing medium.

Use: For the isolation and cultivation of acid-producing microorganisms from foods.

Nickerson Medium

See: **BiGGY Agar**

Nicotinic Acid Medium (DSMZ Medium 152)

Composition per liter:

Yeast extract 10.0g
Nicotinic acid 5.0g
$Cysteine\text{-}HCl \cdot H_2O$ 0.5g
$NaHCO_3$ 0.4g
NaCl 80.0mg
KH_2PO_4 .... 40.0mg
K_2HPO_4 .... 40.0mg
$CaCl_2 \cdot 2H_2O$ 8.0mg
$MgSO_4 \cdot 7H_2O$ 8.0mg
Distilled water 1000.0mL

pH 8.2 ± 0.2 at 25°C

Preparation of Medium: Add nicotinic acid to distilled/deionized water and bring volume to 1.0L. Gently heat and bring to boiling. Boil with mixing until nicotinic acid is fully dissolved. Cool to 25°C. Add remaining components. Readjust volume to 1.0L with distilled/deionized water. Mix thoroughly. Adjust pH to 8.2. Distribute into tubes or flasks under 100% nitrogen atmosphere. Autoclave for 15 min at 15 psi pressure–121°C.

Use: For the cultivation and maintenance of *Eubacterium barkeri.*

Niger Seed Agar
See: **Bird Seed Agar**

Niger Seed Salts Agar with Yeast Extract

Composition per liter:

Niger seeds	50.0g
Agar	20.0g
Glucose	1.0g
Yeast extract	1.0g
KH_2PO_4	1.0g
$MgSO_4 \cdot 7H_2O$	0.5g

Preparation of Medium: Add Niger seeds to 250.0mL of distilled/deionized water in a blender container. Soak the seeds at 60°–70°C for 1.5 hr. Blend. Filter through Whatman #1 filter paper. Reserve filtrate. Combine filtrate with remaining components. Add distilled/deionized water and bring volume to 1.0L. Mix thoroughly. Gently heat and bring to boiling. Distribute into tubes or flasks. Autoclave for 15 min at 15 psi pressure–121°C. Pour into sterile Petri dishes or leave in tubes.

Use: For the cultivation and maintenance of *Arthroderma otae* and *Arthroderma vanbreuseghemii.*

NIH Agar

Composition per liter:

Pancreatic digest of casein	15.0g
Agar	15.0g
Glucose	5.5g
Yeast extract	5.0g
NaCl	2.5g
L-Cystine	0.05g

pH 7.1 ± 0.2 at 25°C

Source: This medium is available as a premixed powder from BD Diagnostic Systems.

Preparation of Medium: Add components to distilled/deionized water and bring volume to 1.0L. Mix thoroughly. Gently heat while stirring and bring to boiling. Distribute into tubes or flasks. Autoclave for 15 min at 15 psi pressure–121°C. Pour into sterile Petri dishes or leave in tubes.

Use: For the cultivation and maintenance of microorganisms isolated from sterility testing of biological products. Also used as a solid medium for sterility testing.

NIH Thioglycolate Broth

Composition per liter:

Pancreatic digest of casein	15.0g
Glucose	5.5g
Yeast extract	5.0g
NaCl	2.5g
L-Cystine	0.5g
Sodium thioglycolate	0.5g

pH 7.1 ± 0.2 at 25°C

Source: This medium is available as a premixed powder from BD Diagnostic Systems.

Preparation of Medium: Add components to distilled/deionized water and bring volume to 1.0L. Mix thoroughly. Gently heat while stirring and bring to boiling. Distribute into tubes or flasks. Autoclave for 15 min at 15 psi pressure–121°C.

Use: For sterility testing of biological products that are turbid or otherwise cannot be cultivated in fluid thioglycolate broth because of its viscosity.

Nine K Medium (9K Medium)

Composition per liter:

$FeSO_4 \cdot 7H_2O$	50.0g
$(NH_4)_2SO_4$	3.0g
$Ca(NO_3)_2$	1.0g
K_2HPO_4	0.5g
$MgSO_4 \cdot 7H_2O$	0.5g
KCl	0.1g
H_2SO_4, 10*N*	1.0mL

pH 3.0 ± 0.2 at 25°C

Preparation of Medium: Add components to distilled/deionized water and bring volume to 1.0L. Mix thoroughly. Adjust pH to 3.0. Distribute into tubes or flasks. Autoclave for 15 min at 15 psi pressure–121°C.

Use: For the cultivation of *Thiobacillus ferrooxidans.*

Nitrate Agar

Composition per liter:

Agar	12.0g
Peptone	5.0g
Beef extract	3.0g
KNO_3	1.0g

pH 6.8 ± 0.2 at 25°C

Preparation of Medium: Add components to distilled/deionized water and bring volume to 1.0L. Mix thoroughly. Gently heat and bring to boiling. Distribute into tubes. Autoclave for 15 min at 15 psi pressure–121°C. Allow tubes to cool in a slanted position.

Use: For the differentiation of aerobic and facultative Gram-negative microorganisms based on their ability to reduce nitrate. Test for nitrates with sulfanilic acid and α-naphthylamine reagents. Bacteria that reduce nitrate to nitrite turn the reagents red or pink.

Nitrate Assimilation Medium, Auxanographic Method for Yeast Identification

Composition per liter:

Noble agar	20.0g
Glucose	10.0g
KH_2PO_4	1.0g
$MgSO_4 \cdot 7H_2O$	0.5g
NaCl	0.1g
$CaCl_2 \cdot 2H_2O$	0.1g
DL-Methionine	0.02g
DL-Tryptophan	0.02g
L-Histidine·HCl	0.01g

Inositol	2.0mg
H_3BO_3	0.5mg
$ZnSO_4 \cdot 7H_2O$	0.4mg
$MnSO_4 \cdot 4H_2O$	0.4mg
Thiamine·HCl	0.4mg
Pyridoxine	0.4mg
Niacin	0.4mg
Calcium pantothenate	0.4mg
p-Aminobenzoic acid	0.2mg
Riboflavin	0.2mg
$FeCl_3$	0.2mg
$Na_2MoO_4 \cdot 4H_2O$	0.2mg
KI	0.1mg
$CuSO_4 \cdot 5H_2O$	0.04mg
Folic Acid	2.0μg
Biotin	2.0μg

pH 4.5 ± 0.2 at 25°C

Preparation of Medium: Add components to distilled/deionized water and bring volume to 1.0L. Mix thoroughly. Gently heat and bring to boiling. Distribute into screw-capped tubes in 20.0mL volumes. Autoclave for 15 min at 15 psi pressure–121°C.

Use: For nitrate assimilation tests by the auxanographic method.

Nitrate Broth (International Streptomyces Project Medium 8) (ISP Medium 8) (ATCC Medium 872)

Composition per liter:

Peptone	5.0g
Beef extract	3.0g
KNO_3	1.0g

pH 7.0 ± 0.2 at 25°C

Source: This medium is available as a premixed powder from BD Diagnostic Systems.

Preparation of Medium: Add components to distilled/deionized water and bring volume to 1.0L. Mix thoroughly. Distribute into tubes or flasks. Autoclave for 15 min at 15 psi pressure–121°C.

Use: For the differentiation of aerobic and facultative Gram-negative microorganisms based on their ability to reduce nitrate. Test for nitrates with sulfanilic acid and α-naphthylamine reagents. Bacteria that reduce nitrate to nitrite turn the reagents red or pink.

Nitrate Broth

Composition per liter:

Pancreatic digest of gelatin	20.0g
KNO_3	2.0g

pH 7.2 ± 0.2 at 25°C

Source: This medium is available as a premixed powder from BD Diagnostic Systems.

Preparation of Medium: Add components to distilled/deionized water and bring volume to 1.0L. Mix thoroughly. Distribute into tubes or flasks. Autoclave for 15 min at 15 psi pressure–121°C.

Use: For the differentiation of aerobic and facultative Gram-negative microorganisms based on their ability to reduce nitrate. Test for nitrates with sulfanilic acid and α-naphthylamine reagents. Bacteria that reduce nitrate to nitrite turn the reagents red or pink.

Nitrate Broth, *Campylobacter*

Composition per liter:

Beef heart, solids from infusion	500.0g
Tryptose	10.0g
NaCl	5.0g
KNO_3	2.0g

pH 7.0 ± 0.2 at 25°C

Preparation of Medium: Add components to distilled/deionized water and bring volume to 1.0L. Mix thoroughly. Adjust pH to 7.0. Distribute 4.0mL volumes into test tubes that contain inverted Durham tubes. Autoclave for 15 min at 15 psi pressure–121°C.

Use: For the differentiation of *Campylobacter* species based on their ability to reduce nitrate.

Nitrate Broth, Enriched

Composition per liter:

Pancreatic digest of casein	13.0g
NaCl	5.0g
Yeast extract	5.0g
Heart muscle, solids from infusion	2.0g
KNO_3	2.0g

pH 7.3 ± 0.2 at 25°C

Preparation of Medium: Add components to distilled/deionized water and bring volume to 1.0L. Mix thoroughly. Distribute into test tubes that contain an inverted Durham tube. Autoclave for 15 min at 15 psi pressure–121°C.

Use: For the differentiation of aerobic and facultative Gram-negative microorganisms based on their ability to reduce nitrate to nitrite or form N_2 gas. Test for nitrates with sulfanilic acid and α-naphthylamine reagents. Bacteria that reduce nitrate to nitrite turn the reagents red or pink.

Nitrate HiVeg Agar

Composition per liter:

Agar	12.0g
Plant peptone	5.0g
Plant extract	3.0g
KNO_3	1.0g

pH 6.8 ± 0.2 at 25°C

Source: This medium is available as a premixed powder from HiMedia.

Preparation of Medium: Add components to distilled/deionized water and bring volume to 1.0L. Mix thoroughly. Gently heat and bring to boiling. Distribute into tubes. Autoclave for 15 min at 15 psi pressure–121°C. Allow tubes to cool in a slanted position.

Use: For the detection of nitrate reduction by bacteria.

Nitrate HiVeg Broth

Composition per liter:

Plant peptone	5.0g
Plant extract	3.0g
KNO_3	1.0g

pH 7.0 ± 0.2 at 25°C

Source: This medium is available as a premixed powder from HiMedia.

Preparation of Medium: Add components to distilled/deionized water and bring volume to 1.0L. Mix thoroughly. Distribute into tubes or flasks. Autoclave for 15 min at 15 psi pressure–121°C.

Use: For the differentiation of aerobic and facultative Gram-negative microorganisms based on their ability to reduce nitrate. Test for ntrates with sulfanilic acid and α-naphthylamine reagents. Bacteria that reduce nitrate to nitrite turn the reagents red or pink.

Nitrate Liquid Medium

Composition per liter:

Solution A	500.0mL
Solution B	250.0mL
Solution C	250.0mL

Solution A:

Composition per 500.0mL:

Mannitol	10.0g
KNO_3	0.6g
$Na_2HPO_4 \cdot 12H_2O$	0.45g
Na_2SO_4	0.03g

Preparation of Solution A: Add components to distilled/deionized water and bring volume to 500.0mL. Mix thoroughly. Autoclave for 15 min at 15 psi pressure–121°C. Cool to 25°C.

Solution B:

Composition per 250.0mL:

$MgSO_4 \cdot 7H_2O$	0.12g
$CaCl_2 \cdot 6H_2O$	0.1g
$FeCl_3 \cdot 6H_2O$	0.01g

Preparation of Solution B: Add components to distilled/deionized water and bring volume to 250.0mL. Mix thoroughly. Autoclave for 15 min at 15 psi pressure–121°C. Cool to 25°C.

Solution C:

Composition per 250.0mL:

Calcium pantothenate	0.5mg
Thiamine·HCl	0.1mg
Biotin	0.5μg

Preparation of Solution C: Add components to distilled/deionized water and bring volume to 250.0mL. Mix thoroughly. Filter sterilize.

Preparation of Medium: Aseptically combine 500.0mL of cooled, sterile solution A, 250.0mL of cooled, sterile solution B, and 250.0mL of sterile solution C. Mix thoroughly. Aseptically distribute into sterile tubes or flasks.

Use: For the isolation and cultivation of *Rhizobium* species.

Nitrate Methanol Medium

Composition per liter:

$NaNO_3$	5.0g
K_2HPO_4	2.0g
NaCl	1.0g
$MgSO_4 \cdot 7H_2O$	0.02g
$Na_2MoO_4 \cdot H_2O$	1.0mg
Riboflavin	0.2mg
Calcium pantothenate	0.2mg
Pyridoxine·HCl	0.2mg
Nicotinic acid	0.2mg
Thiamine·HCl	0.1mg
p-Aminobenzoic acid	0.1mg
Biotin	0.01mg
Methanol	10.0mL

pH 7.0 ± 0.2 at 25°C

Preparation of Medium: Add components, except methanol, to distilled/deionized water and bring volume to 990.0mL. Mix thoroughly. Autoclave for 15 min at 15 psi pressure–121°C. Cool to 45°–50°C. Filter sterilize methanol. Aseptically add sterile methanol to cooled sterile medium. Mix thoroughly. Aseptically distribute into sterile tubes or flasks.

Use: For the cultivation and maintenance of *Methylobacterium rhodinum*.

Nitrate Mineral Salts Medium (NMS Medium)

Composition per liter:

Noble agar	12.5g
$MgSO_4 \cdot 7H_2O$	1.0g
KNO_3	1.0g
$Na_2HPO_4 \cdot 12H_2O$	0.717g
KH_2PO_4	0.272g
$CaCl_2 \cdot 6H_2O$	0.2g
Ferric ammonium EDTA	4.0mg
Trace elements solution	0.5mL

pH 6.8 ± 0.2 at 25°C

Trace Elements Solution:

Composition per liter:

Disodium EDTA	0.5g
$FeSO_4 \cdot 7H_2O$	0.2g
H_3BO_3	0.03g
$CoCl_2 \cdot 6H_2O$	0.02g
$ZnSO_4 \cdot 7H_2O$	0.01g
$MnCl_2 \cdot 4H_2O$	3.0mg
$Na_2MoO_4 \cdot 2H_2O$	3.0mg
$NiCl_2 \cdot 6H_2O$	2.0mg
$CaCl_2 \cdot 2H_2O$	1.0mg

Preparation of Trace Elements Solution: Add components to distilled/deionized water and bring volume to 1.0L. Mix thoroughly.

Preparation of Medium: Add components to distilled/deionized water and bring volume to 1.0L. Mix thoroughly. Gently heat and bring to boiling. Adjust pH to 6.8. Distribute into tubes or flasks. Autoclave for 15 min at 15 psi pressure–121°C. Pour into sterile Petri dishes or leave in tubes.

Use: For the cultivation and maintenance of *Methylobacterium* species, *Methylococcus capsulatus*, *Methylomonas agile*, and *Methylomonas methanica*.

Nitrate Mineral Salts Medium with Methanol (NMS Medium with Methanol)

Composition per liter:

Noble agar	12.5g
$MgSO_4 \cdot 7H_2O$	1.0g
KNO_3	1.0g
$Na_2HPO_4 \cdot 12H_2O$	0.717g
KH_2PO_4	0.272g
$CaCl_2 \cdot 6H_2O$	0.2g
Ferric ammonium EDTA	4.0mg

Trace elements solution 0.5mL
Methanol 1.0mL

pH 6.8 ± 0.2 at 25°C

Trace Elements Solution:

Composition per liter:

Disodium EDTA 0.5g
$FeSO_4 \cdot 7H_2O$....................0.2g
H_3BO_3 ....0.03g
$CoCl_2 \cdot 6H_2O$ 0.02g
$ZnSO_4 \cdot 7H_2O$ 0.01g
$MnCl_2 \cdot 4H_2O$....................3.0mg
$Na_2MoO_4 \cdot 2H_2O$ 3.0mg
$NiCl_2 \cdot 6H_2O$ 2.0mg
$CaCl_2 \cdot 2H_2O$....................1.0mg

Preparation of Trace Elements Solution: Add components to distilled/deionized water and bring volume to 1.0L. Mix thoroughly.

Preparation of Medium: Add components, except methanol, to distilled/deionized water and bring volume to 999.0mL. Mix thoroughly. Gently heat and bring to boiling. Adjust pH to 6.8. Distribute into tubes or flasks. Autoclave for 15 min at 15 psi pressure–121°C. Cool to 45°–50°C. Filter sterilize methanol. Aseptically add sterile methanol to cooled sterile medium. Mix thoroughly. Pour into sterile Petri dishes or leave in tubes.

Use: For the cultivation and maintenance of *Methylobacterium fujisawaense, Methylobacterium* species, and *Methylomonas clara.*

Nitrate Reduction Broth

Composition per liter:

Pancreatic digest of gelatin....................5.0g
Beef extract....................3.0g
KNO_3 ....1.0g

pH 6.9 ± 0.2 at 25°C

Preparation of Medium: Add components to distilled/deionized water and bring volume to 1.0L. Mix thoroughly. Distribute into test tubes that contain an inverted Durham tube. Autoclave for 15 min at 15 psi pressure–121°C.

Use: For the differentiation of members of the Pseudomonadaceae based on their ability to reduce nitrate to nitrite or form N_2 gas. Test for nitrates with sulfanilic acid and α-naphthylamine reagents. Bacteria that reduce nitrate to nitrite turn the reagents red or pink.

Nitrate Reduction Broth

Composition per liter:

Pancreatic digest of casein....................13.0g
NaCl 5.0g
Yeast extract....................5.0g
Heart muscle, solids from infusion....................2.0g
KNO_3 or $NaNO_3$ 2.0g

pH 7.4 ± 0.2 at 25°C

Preparation of Medium: Add components to distilled/deionized water and bring volume to 1.0L. Mix thoroughly. Distribute into test tubes that contain an inverted Durham tube. Autoclave for 15 min at 15 psi pressure–121°C.

Use: For the differentiation of a variety of Gram-negative bacteria based on their ability to reduce nitrate to nitrite or form N_2 gas. Test for nitrates with sulfanilic acid and α-naphthylamine reagents. Bacteria that reduce nitrate to nitrite turn the reagents red or pink.

Nitrate Reduction Broth

Composition per liter:

Pancreatic digest of casein....................13.0g
NaCl....................5.0g
Yeast extract....................5.0g
Heart muscle, solids from infusion....................2.0g
KNO_3 or $NaNO_3$ 2.0g

pH 7.4 ± 0.2 at 25°C

Preparation of Medium: Add components to distilled/deionized water and bring volume to 1.0L. Mix thoroughly. Distribute into test tubes that contain an inverted Durham tube. Autoclave for 15 min at 15 psi pressure–121°C.

Use: For the differentiation of a variety of nonfermenting Gram-negative bacteria based on their ability to reduce nitrate to nitrite or form N_2 gas. Test for nitrates with sulfanilic acid and α-naphthylamine reagents. Bacteria that reduce nitrate to nitrite turn the reagents red or pink.

Nitrate Reduction Broth, Clark

Composition per liter:

Peptone 20.0g
KNO_3 or $NaNO_3$ 2.0g

Preparation of Medium: Add components to distilled/deionized water and bring volume to 1.0L. Mix thoroughly. Distribute into test tubes that contain an inverted Durham tube. Autoclave for 15 min at 15 psi pressure–121°C.

Use: For the differentiation of a variety of Gram-negative bacteria based on their ability to reduce nitrate to nitrite or form N_2 gas. Test for nitrates with sulfanilic acid and α-naphthylamine reagents. Bacteria that reduce nitrate to nitrite turn the reagents red or pink.

Nitrate Reduction Broth for *Pseudomonas* and Related Genera

Composition per liter:

Peptone 5.0g
NaCl....................5.0g
Yeast extract....................2.0g
Beef extract....................1.0g
$NaNO_3$ 0.1g

pH 7.4 ± 0.2 at 25°C

Preparation of Medium: Add components to distilled/deionized water and bring volume to 1.0L. Mix thoroughly. Distribute into test tubes that contain an inverted Durham tube. Autoclave for 15 min at 15 psi pressure–121°C.

Use: For the differentiation of members of the Pseudomonadaceae based on their ability to reduce nitrate to nitrite or form N_2 gas. Test for nitrates with sulfanilic acid and α-naphthylamine reagents. Bacteria that reduce nitrate to nitrite turn the reagents red or pink.

Nitratiruptor and *Nitratifactor* Medium (DSMZ Medium 1024)

Composition per liter:

Sulfur, elemental....................3.0g
$Na_2S_2O_3 \cdot 5H_2O$ 1.0g
$NaNO_3$ 1.0g
Bicarbonate solution....................10.0mL

Vitamin solution	10.0mL
DMJ synthetic seawater	1.0L

pH 7.0 ± 0.2 at 25°C

Bicarbonate Solution:
Composition per 10.0mL:

$NaHCO_3$	1.0g

Preparation of Bicarbonate Solution: Add $NaHCO_3$ to distilled/deionized water and bring volume to 10.0mL. Mix thoroughly. Filter sterilize.

DMJ Synthetic Seawater:
Composition per liter:

NaCl	30.0g
$MgCl_2 \cdot 6H_2O$	4.18g
$MgSO_4 \cdot 7H_2O$	3.4g
KCl	0.33g
NH_4Cl	0.25g
K_2HPO_4	0.14g
$CaCl_2 \cdot 2H_2O$	0.14g
$Fe(NH_4)_2(SO_4)_2 \cdot 6H_2O$	0.01g
$NiCl_2 \cdot 6H_2O$	0.5mg
$Na_2SeO_3 \cdot 5H_2O$	0.5mg
Trace elements solution SL-10	10.0mL

Trace Elements Solution SL-10:
Composition per liter:

$MgSO_4 \cdot 7H_2O$	3.0g
Nitrilotriacetic acid	1.5g
NaCl	1.0g
$MnSO_4 \cdot 2H_2O$	0.5g
$CoSO_4 \cdot 7H_2O$	0.18g
$ZnSO_4 \cdot 7H_2O$	0.18g
$CaCl_2 \cdot 2H_2O$	0.1g
$FeSO_4 \cdot 7H_2O$	0.1g
$NiCl_2 \cdot 6H_2O$	0.025g
$KAl(SO_4)_2 \cdot 12H_2O$	0.02g
H_3BO_3	0.01g
$Na_2MoO_4 \cdot 4H_2O$	0.01g
$CuSO_4 \cdot 5H_2O$	0.01g
$Na_2SeO_3 \cdot 5H_2O$	0.3mg

Preparation of Trace Elements Solution SL-10: Add nitrilotriacetic acid to 500.0mL of distilled/deionized water. Dissolve by adjusting pH to 6.5 with KOH. Add remaining components. Add distilled/deionized water to 1.0L. Mix thoroughly. Adjust pH to 7.0.

Preparation of DMJ Synthetic Seawater: Add components to distilled/deionized water and bring volume to 1.0L. Mix thoroughly. Autoclave for 15 min at 15 psi pressure–121°C. Cool to room temperature.

Vitamin Solution:
Composition per liter:

Pyridoxine-HCl	10.0mg
Thiamine-HCl·$2H_2O$	5.0mg
Riboflavin	5.0mg
Nicotinic acid	5.0mg
D-Ca-pantothenate	5.0mg
p-Aminobenzoic acid	5.0mg
Lipoic acid	5.0mg
Biotin	2.0mg
Folic acid	2.0mg
Vitamin B_{12}	0.1mg

Preparation of Vitamin Solution: Add components to distilled/deionized water and bring volume to 1.0L. Mix thoroughly. Sparge with 80% H_2 + 20% CO_2. Filter sterilize.

Preparation of Medium: Add components, except sulfur, bicarbonate solution, and vitamin solution to seawater and bring volume to 1.0L. Dispense into serum bottles. Autoclave for 15 min at 15 psi pressure–121°C under an atmosphere of air. Sterilize sulfur separately in screw-capped tubes by steaming in a water bath for 3 hr on each of 3 successive days. Aseptically add the sterilized sulfur, bicarbonate, and vitamin solutions. Mix thoroughly. Sparge with 80% H_2 + 20% CO_2. Seal the serum tubes with butyl rubber stoppers. Increase the 80% H_2 + 20% CO_2 gas phase pressure to 300 kPa.

Use: For the cultivation of *Nitratiruptor* spp. and *Nitratifactor* spp.

Nitriliruptor alkaliphilus Medium (DSMZ Medium 1105)

Composition per liter:

Na_2CO_3	22.0g
Na_2HCO_3	8.0g
NaCl	6.0g
K_2HPO_4	0.5g
Isobutyroamide solution	10.0mL
Trace elements solution	1.0mL
Magnesium sulfate solution	1.0mL
Vitamin solution	1.0mL
Thiosulfate solution	0.01mL

pH 9.5 ± 0.5 at 25°C

Isobutyroamide Solution:
Composition per 10.0ml:

Isobutyroamide	0.87g

Preparation of Isobutyroamide Solution: Add isobutyroamide to distilled/deionized water and bring volume to 10.0mL. Mix thoroughly. Filter sterilize.

Vitamin Solution:
Composition per 10.0ml:

Vitamin B_{12}	1.0mg

Preparation of Vitamin Solution: Add vitamin B_{12} to distilled/deionized water and bring volume to 10.0mL. Mix thoroughly. Filter sterilize.

Magnesium Sulfate Solution:
Composition per 10.0mL:

$MgSO_4 \cdot 7H_2O$	2.0g

Preparation of Magnesium Sulfate Solution: Add $MgSO_4 \cdot 7H_2O$ to distilled/deionized water and bring volume to 10.0mL. Mix thoroughly. Autoclave for 15 min at 15 psi pressure–121°C. Cool to room temperature.

Thiosulfate Solution:
Composition per 10.0mL:

$Na_2S_2O_3 \cdot 5H_2O$	1.6g

Preparation of Thiosulfate Solution: Add $Na_2S_2O_3 \cdot 5H_2O$ to distilled/deionized water and bring volume to 10.0mL. Mix thoroughly. Sparge with 100% N_2. Adjust to pH 10.0. Filter sterilize.

Trace Elements Solution:
Composition per liter:

EDTA	5.0g
$FeSO_4 \cdot 7H_2O$	2.0g
H_3BO_3	0.01g

$ZnSO_4 \cdot 7H_2O$	0.3g
$CoCl_2 \cdot 6H_2O$	0.2g
$MnCl_2 \cdot 4H_2O$	0.03g
$NiCl_2 \cdot 2H_2O$	0.02g
$NaMoO_4 \cdot 2H_2O$	0.02g
$CuCl_2$	0.01g

Preparation of Trace Elements Solution: Add components to distilled/deionized water and bring volume to 1.0L. Mix thoroughly. Adjust pH to 3-4.

Preparation of Medium: Add components, except vitamin, trace elements, thiosulfate, isobutyroamide, and magnesium sulfate solutions, to distilled/deionized water and bring volume to 987.0mL. Mix thoroughly. Dispense into closed bottles. Autoclave for 15 min at 15 psi pressure–121°C. Cool to room temperature. After cooling, there may be some precipitate on the bottom. Decant into sterile bottle to eliminate precipitate. Aseptically add vitamin, trace elements, thiosulfate, isobutyroamide, and magnesium sulfate solutions. Adjust pH to 9.5. Aseptically dispense into culture vessels.

Use: For the cultivation of *Nitriliruptor alkaliphilus*.

Nitrilotriacetate Medium

Composition per 1002.0mL:

$MgSO_4 \cdot 7H_2O$	1.0g
Nitrilotriacetate	1.0g
$Na_2HPO_4 \cdot 2H_2O$	0.41g
KH_2PO_4	0.26g
$CaCl_2 \cdot 2H_2O$	0.2g
Trace elements solution	1.0mL
Vitamin solution	1.0mL

pH 6.5 ± 0.2 at 25°C

Trace Elements Solution:

Composition per liter:

$FeCl_2 \cdot 4H_2O$	1.5g
$CoCl_2 \cdot 6H_2O$	120.0mg
$MnCl_2 \cdot 4H_2O$	100.0mg
$ZnCl_2$	68.0mg
H_3BO_3	62.0mg
$Na_2MoO_4 \cdot 2H_2O$	24.0mg
$NiCl_2 \cdot 6H_2O$	24.0mg
$CuCl_2 \cdot 2H_2O$	17.0mg
HCl (0.05*M* solution)	1.0L

Preparation of Trace Elements Solution: Add $FeCl_2 \cdot 4H_2O$ to 1.0L of HCl solution. Mix thoroughly. Add distilled/deionized water and bring volume to 1.0L. Add remaining components. Mix thoroughly.

Vitamin Solution:

Composition per liter:

Folic acid	20.0g
α-Lipoic acid	50.0mg
p-Aminobenzoic acid	50.0mg
Pantothenic acid	50.0mg
Riboflavin	50.0mg
Thamine·HCl	50.0mg
Vitamin B_{12}	50.0mg
Nicotinamide	25.0mg
Biotin	20.0mg
Nicotinic acid	20.0mg
Pyridoxamine·HCl	10.0mg

Preparation of Vitamin Solution: Add components to distilled/deionized water and bring volume to 1.0L. Stir for 2-3 hr. Filter sterilize.

Preparation of Medium: Add components to distilled/deionized water and bring volume to 1.0L. Mix thoroughly. Distribute into tubes or flasks. Autoclave for 15 min at 15 psi pressure–121°C.

Use: For the cultivation of *Chelatobacter heintzii* and *Chelatococcus saccharophobus*.

Nitrincola Medium (DSMZ Medium 1174)

Composition per liter:

NaCl	17.5g
Na-acetate	10.0g
$Na_2B_4O_7$	4.0g
NH_4Cl	0.5g
K_2HPO_4	0.25g
Yeast extract	0.10g

pH 7.0 ± 0.2 at 25°C

Preparation of Medium: Add components to distilled/deionized water and bring volume to 1.0L. Mix thoroughly. Adjust pH to 9.0. Autoclave for 15 min at 15 psi pressure–121°C. Cool to room temperature.

Use: For the cultivation of *Nitrincola* spp.

Nitrobacter agilis Medium

Composition per liter:

$CaCO_3$	10.0g
NaCl	0.3g
Na_2CO_3	0.25g
KNO_2	0.17g
K_2HPO_4	0.14g
$MgSO_4 \cdot 7H_2O$	0.14g
$FeSO_4 \cdot 7H_2O$	0.03g
$MnSO_4 \cdot 4H_2O$	0.01g
Biotin solution	10.0mL

Biotin Solution:

Composition per 10.0mL:

Biotin	0.15g

Preparation of Biotin Solution: Add biotin to distilled/deionized water and bring volume to 10.0mL. Mix thoroughly. Filter sterilize.

Preparation of Medium: Add Na_2CO_3 to distilled/deionized water and bring volume to 200.0mL. Mix thoroughly. In a separate flask, add the remaining components, except the biotin solution, to distilled/deionized water and bring volume to 790.0mL. Autoclave the Na_2CO_3 solution and salts solution separately for 15 min at 15 psi pressure–121°C. Cool to 25°C. Aseptically combine the sterile Na_2CO_3 solution, sterile salts solution, and sterile biotin solution. Mix thoroughly. Aseptically distribute into sterile tubes or flasks.

Use: For the cultivation of *Nitrobacter agilis*.

Nitrobacter Medium 203

Composition per liter:

Solution C	1.0mL
Solution A	0.5mL
Solution B	0.5mL
Solution D	0.5mL

Solution E 0.5mL
Solution F 0.2mL

Solution A:

Composition per 100.0mL:

$CaCl_2$ 2.0g

Preparation of Solution A: Add $CaCl_2$ to distilled/deionized water and bring volume to 100.0mL. Mix thoroughly.

Solution B:

Composition per 100.0mL:

$MgSO_4 \cdot 7H_2O$ 20.0g

Preparation of Solution B: Add $MgSO_4 \cdot 7H_2O$ to distilled/deionized water and bring volume to 100.0mL. Mix thoroughly.

Solution C:

Composition per 100.0mL:

Chelated iron (Sequestrene) 0.1g

Preparation of Solution C: Add chelated iron to distilled/deionized water and bring volume to 100.0mL. Mix thoroughly.

Solution D:

Composition per liter:

$MnCl_2 \cdot 4H_2O$ 0.2g
$Na_2MoO_4 \cdot 2H_2O$ 0.1g
$ZnSO_4 \cdot 7H_2O$ 0.1g
$CuSO_4 \cdot 5H_2O$ 0.02g
$CoCl_2 \cdot 6H_2O$ 2.0mg

Preparation of Solution D: Add components to distilled/deionized water and bring volume to 1.0L. Mix thoroughly.

Solution E:

Composition per 100.0mL:

$NaNO_2$ 41.4g

Preparation of Solution E: Add $NaNO_2$ to distilled/deionized water and bring volume to 100.0mL. Mix thoroughly.

Solution F:

Composition per 100.0mL:

K_2HPO_4 .... 1.74g

Preparation of Solution F: Add K_2HPO_4 to distilled/deionized water and bring volume to 100.0mL. Mix thoroughly.

Preparation of Medium: Add the appropriate volumes of solutions A–F to distilled/deionized water and bring volume to 1.0L. Mix thoroughly. Distribute into tubes or flasks. Autoclave for 15 min at 15 psi pressure–121°C.

Use: For the cultivation and maintenance of *Nitrobacter* species and *Nitrobacter winogradskyi*.

Nitrobacter Medium 204

Composition per liter:

Seawater 700.0mL
Solution C 1.0mL
Solution A 0.5mL
Solution B 0.5mL
Solution D 0.5mL
Solution E 0.5mL
Solution F 0.2mL

Solution A:

Composition per 100.0mL:

$CaCl_2$ 2.0g

Preparation of Solution A: Add $CaCl_2$ to distilled/deionized water and bring volume to 100.0mL. Mix thoroughly.

Solution B:

Composition per 100.0mL:

$MgSO_4 \cdot 7H_2O$ 20.0g

Preparation of Solution B: Add $MgSO_4 \cdot 7H_2O$ to distilled/deionized water and bring volume to 100.0mL. Mix thoroughly.

Solution C:

Composition per 100.0mL:

Chelated iron (Sequestrene) 0.1g

Preparation of Solution C: Add chelated iron to distilled/deionized water and bring volume to 100.0mL. Mix thoroughly.

Solution D:

Composition per liter:

$MnCl_2 \cdot 4H_2O$ 0.2g
$Na_2MoO_4 \cdot 2H_2O$ 0.1g
$ZnSO_4 \cdot 7H_2O$ 0.1g
$CuSO_4 \cdot 5H_2O$ 0.02g
$CoCl_2 \cdot 6H_2O$ 2.0mg

Preparation of Solution D: Add components to distilled/deionized water and bring volume to 1.0L. Mix thoroughly.

Solution E:

Composition per 100.0mL:

$NaNO_2$ 41.4g

Preparation of Solution E: Add $NaNO_2$ to distilled/deionized water and bring volume to 100.0mL. Mix thoroughly.

Solution F:

Composition per 100.0mL:

K_2HPO_4 .... 1.74g

Preparation of Solution F: Add K_2HPO_4 to distilled/deionized water and bring volume to 100.0mL. Mix thoroughly.

Preparation of Medium: Add the appropriate volumes of solutions A–F and seawater to distilled/deionized water and bring volume to 1.0L. Mix thoroughly. Distribute into tubes or flasks. Autoclave for 15 min at 15 psi pressure–121°C.

Use: For the cultivation and maintenance of *Nitrococcus mobilis*.

Nitrobacter Medium B

Composition per liter:

$NaNO_2$ 1.0g
K_2HPO_4 .... 0.5g
$MgSO_4$ 0.5g
NaCl 0.3g
$Fe_2(SO_4)_3$ 5.0mg
$MnSO_4$ 2.0mg
Marble chips .... as needed

pH 7.5 ± 0.2 at 25°C

Preparation of Medium: Add components, except marble chips, to distilled/deionized water and bring volume to 1.0L. Mix thoroughly. Autoclave for 15 min at 15 psi pressure–121°C. Cool to 25°C. Wash marble chips in distilled/deionized water. Put a few chips into test tubes. Autoclave for 60 min at 15 psi pressure–121°C. Cool to 25°C. Aseptically distribute cooled sterile medium into test tubes to cover marble chips.

Use: For the cultivation of *Nitrobacter* species.

Nitrococcus Medium

Composition per 1004.0mL:

$NaNO_2$ solution	1.0mL
K_2HPO_4 solution	1.0mL
$NaHCO_3$ solution	1.0mL
Chelated metals solution	1.0mL

pH 7.5 ± 0.1 at 25°C

$NaNO_2$ Solution:

Composition per 100.0mL:

$NaNO_2$	10.0g

Preparation of $NaNO_2$ Solution: Add $NaNO_2$ to distilled/deionized water and bring volume to 100.0mL. Mix thoroughly. Filter sterilize.

K_2HPO_4 Solution:

Composition per 100.0mL:

K_2HPO_4	2.5g

Preparation of K_2HPO_4 Solution: Add K_2HPO_4 to distilled/deionized water and bring volume to 100.0mL. Mix thoroughly. Filter sterilize.

$NaHCO_3$ Solution:

Composition per 100.0mL:

$NaHCO_3$	5.0g

Preparation of $NaHCO_3$ Solution: Add $NaHCO_3$ to distilled/deionized water and bring volume to 100.0mL. Mix thoroughly. Filter sterilize.

Chelated Metals Solution:

Composition per liter:

EDTA	6.0g
$FeCl_3 \cdot 6H_2O$	1.0g
$MnSO_4 \cdot H_2O$	0.6g
$ZnSO_4 \cdot 7H_2O$	0.3g
$Na_2MoO_4 \cdot 2H_2O$	0.15g
$CoCl_2 \cdot 6H_2O$	4.0mg
$CuSO_4 \cdot 5H_2O$	4.0mg

Preparation of Chelated Metals Solution: Add components to distilled/deionized water and bring volume to 1.0L. Mix thoroughly. Filter sterilize.

Preparation of Medium: Adjust pH of 1.0L of seawater to pH 7.5 with NaOH. Add 1.0mL of chelated metals solution to the seawater. Mix thoroughly. Autoclave for 15 min at 15 psi pressure–121°C. Cool to 50°C. Aseptically add 1.0mL each of sterile $NaNO_2$, K_2HPO_4, and $NaHCO_3$ solutions. Mix thoroughly. Aseptically distribute into sterile tubes or flasks.

Use: For the cultivation of *Nitrococcus* species.

Nitrogen-Fixing Hydrocarbon Oxidizers Medium

Composition per liter:

Na_2HPO_4	0.3g
KH_2PO_4	0.2g
$MgSO_4 \cdot 7H_2O$	0.1g
$FeSO_4 \cdot 7H_2O$	5.0mg
$Na_2MoO_4 \cdot 2H_2O$	2.0mg

Preparation of Medium: Add components to distilled/deionized water and bring volume to 1.0L. Mix thoroughly. Distribute into tubes or flasks. Autoclave for 15 min at 15 psi pressure–121°C.

Use: For the cultivation and enrichment of nitrogen-fixing hydrocarbon-oxidizing bacteria.

Nitrogen-Fixing Marine Medium

Composition per liter:

Noble agar	10.0g
$MgSO_4 \cdot 7H_2O$	0.04g
$CaCl_2 \cdot 2H_2O$	0.02g
$K_2HPO_4 \cdot 3H_2O$	0.02g
Na_2CO_3	0.02g
Citric acid	3.0mg
Ferric ammonium citrate	3.0mg
Disodium potassium EDTA	0.5mg
Seawater	750.0mL
Trace metals A-5 mix	1.0mL

pH 8.5 ± 0.2 at 25°C

Trace Metals A-5 Mix:

Composition per liter:

H_3BO_3	2.86g
$MnCl_2 \cdot 4H_2O$	1.81g
$ZnSO_4 \cdot 7H_2O$	0.222g
$CuSO_4 \cdot 5H_2O$	0.079g
$Co(NO_3)_2 \cdot 6H_2O$	0.05g
$Na_2MoO_4 \cdot 2H_2O$	0.039g

Preparation of Trace Metals A-5 Mix: Add components to distilled/deionized water and bring volume to 1.0L. Mix thoroughly.

Preparation of Medium: Add components to glass-distilled water and bring volume to 1.0L. Mix thoroughly. Gently heat and bring to boiling. Autoclave for 15 min at 15 psi pressure–121°C. Adjust pH to 8.5 with KOH. Pour into sterile Petri dishes or distribute into sterile tubes.

Use: For the cultivation and maintenance of *Anabaena* species.

Nitrogen-Free Agar

Composition per liter:

Agar	15.0g
$CaCO_3$	1.0g
K_2HPO_4	1.0g
$MgSO_4 \cdot 7H_2O$	0.2g
NaCl	0.2g
$FeSO_4 \cdot 7H_2O$	0.1g
$Na_2MoO_4 \cdot 2H_2O$	5.0mg
Glucose solution	50.0mL

pH 7.0 ± 0.2 at 25°C

Glucose Solution:

Composition per 50.0mL:

Glucose	10.0g

Preparation of Glucose Solution: Add glucose to distilled/deionized water and bring volume to 50.0mL. Mix thoroughly. Filter sterilize.

Preparation of Medium: Add components, except glucose solution and agar, to distilled/deionized water and bring volume to 950.0mL. Mix thoroughly. Add agar. Gently heat and bring to boiling. Autoclave for 15 min at 15 psi pressure–121°C. Cool to 50°–55°C. Aseptically add 50.0mL of sterile glucose solution. Mix thoroughly. Pour into sterile Petri dishes or distribute into sterile tubes.

Use: For the cultivation of *Azomonas agilis, Azomonas insignis, Azomonas macrocytogenes, Azorhizophilus paspali, Azotobacter beijerinckii, Azotobacter chroococcum, Azotobacter vinelandii, Beijerinckia acida, Beijerinckia fluminensis, Beijerinckia indica, Beijerinckia mobilis,* and *Derxia gummosa.*

Nitrogen-Free Agar (Norris Agar)

Composition per liter:

Agar	15.0g
$CaCO_3$	1.0g
K_2HPO_4	1.0g
$MgSO_4 \cdot 7H_2O$	0.2g
NaCl	0.2g
$FeSO_4 \cdot 7H_2O$	0.1g
$Na_2MoO_4 \cdot 2H_2O$	5.0mg
Glucose solution	50.0mL

Glucose Solution:

Composition per 100.0mL:

Glucose	20.0g

Preparation of Glucose Solution: Add 20.0g of glucose to distilled/deionized water and bring volume to 100.0mL. Mix thoroughly. Filter sterilize. Warm to 50°C.

Preparation of Medium: Add components, except glucose solution, to distilled/deionized water and bring volume to 950.0mL. Mix thoroughly. Adjust pH to 7.2. Gently heat and bring to boiling. Autoclave for 15 min at 15 psi pressure–121°C. Cool to 50°–55°C. Aseptically add 50.0mL of sterile glucose solution. Mix thoroughly. Pour into sterile Petri dishes or distribute into sterile tubes.

Use: For the cultivation and maintenance of *Azomonas agilis, Azotobacter chroococcum,* and *Azotobacter vinelandii.*

Nitrogen-Free Medium for *Pseudomonas stutzeri*

Composition per liter:

Disodium DL-malate	6.6g
K_2HPO_4	0.5g
$MgSO_4 \cdot 7H_2O$	0.2g
Yeast extract	0.2g
NaCl	0.1g
$FeCl_3 \cdot 6H_2O$	15.0mg

pH 7.0 ± 0.2 at 25°C

Preparation of Medium: Add components to distilled/deionized water and bring volume to 1.0L. Mix thoroughly. Distribute into tubes or flasks. Autoclave for 15 min at 15 psi pressure–121°C.

Use: For the cultivation of *Pseudomonas stutzeri.*

Nitrogen-Free Mineral Agar for *Derxia*

Composition per liter:

Agar	15.0g
Glucose	10.0g
K_2HPO_4	0.5g
$MgSO_4 \cdot 7H_2O$	0.25g
NaCl	0.25g
$CaCl_2$	0.1g
$FeSO_4 \cdot 7H_2O$	0.1g
$Na_2MoO_4 \cdot 2H_2O$	5.0mg

pH 6.9 ± 0.2 at 25°C

Preparation of Medium: Add components to distilled/deionized water and bring volume to 1.0L. Mix thoroughly. Adjust pH to 6.9. Gently heat and bring to boiling. Distribute into tubes or flasks. Autoclave for 15 min at 15 psi pressure–121°C. Pour into sterile Petri dishes or leave in tubes.

Use: For the cultivation and maintenance of *Derxia gummosa.*

Nitrogen-Free Mineral Medium for *Beijerinckia*

Composition per liter:

Glucose	20.0g
KH_2PO_4	0.8g
$MgSO_4 \cdot 7H_2O$	0.5g
K_2HPO_4	0.2g
$FeCl_3 \cdot 6H_2O$	0.1g
$CaCl_2 \cdot 2H_2O$	0.05g
$Na_2MoO_4 \cdot 2H_2O$	5.0mg

Preparation of Medium: Add components to distilled/deionized water and bring volume to 1.0L. Mix thoroughly. Distribute into tubes or flasks. Autoclave for 15 min at 15 psi pressure–121°C.

Use: For the cultivation of *Beijerinckia indica.*

Nitrosococcus Medium

Composition per liter:

$(NH_4)_2SO_4$	1.32g
$MgSO_4 \cdot 7H_2O$	0.38g
$CaCl_2 \cdot 2H_2O$	0.02g
K_2HPO_4	8.7mg
Chelated iron	1.0mg
$MnCl_2 \cdot 4H_2O$	0.2mg
$Na_2MoO_4 \cdot 2H_2O$	0.1mg
$ZnSO_4 \cdot 7H_2O$	0.1mg
$CoCl_2 \cdot 6H_2O$	2.0μg
Phenol Red (0.04% solution)	3.25mL

pH 7.5–7.8 at 25°C

Preparation of Medium: Add components to filtered seawater and bring volume to 1.0L. Mix thoroughly. Adjust pH to 7.5–7.8 with 1*N* HCl. Distribute into tubes. Autoclave for 15 min at 15 psi pressure–121°C.

Use: For the cultivation of *Nitrosococcus oceanus.*

Nitrosococcus oceanus Medium

Composition per 1001.0mL:

Phenol Red	5.0g
$NH_4 \cdot Cl$	0.635g
$MgSO_4 \cdot 7H_2O$	0.357g
K_2HPO_4	43.0mg
$CaCl_2 \cdot H_2O$	20.0mg
Chelated metals solution	1.0mL

pH 7.5 ± 0.2 at 25°C

Chelated Metals Solution:

Composition per liter:

EDTA	6.0g
$FeCl_3 \cdot 6H_2O$	1.0g
$MnSO_4 \cdot H_2O$	0.6g
$ZnSO_4 \cdot 7H_2O$	0.3g
$Na_2MoO_4 \cdot 2H_2O$	0.15g
$CoCl_2 \cdot 6H_2O$	4.0mg
$CuSO_4 \cdot 5H_2O$	4.0mg

Preparation of Chelated Metals Solution: Add components to distilled/deionized water and bring volume to 1.0L. Mix thoroughly. Filter sterilize.

Preparation of Medium: Add components, except chelated metals solution, to filtered seawater and bring volume to 1.0L. Mix thoroughly. Adjust pH to 7.5 with sterile 0.1*M* K_2CO_3. Autoclave for 15 min at 15 psi pressure–121°C. Aseptically add 1.0mL of sterile chelated metals solution. Mix thoroughly. Aseptically distribute into sterile tubes or flasks.

Use: For the cultivation of *Nitrosococcus oceanus*.

Nitrosolobus Medium (ATCC Medium 438)

Composition per liter:

$(NH_4)_2SO_4$	1.65g
$MgSO_4 \cdot 7H_2O$	0.2g
K_2HPO_4	0.087g
$CaCl_2 \cdot 2H_2O$	0.02g
Phenol Red	5.0mg
Disodium EDTA	1.0mg
$MnCl_2 \cdot 4H_2O$	0.2mg
$Na_2MoO_4 \cdot 2H_2O$	0.1mg
$ZnSO_4 \cdot 7H_2O$	0.1mg
$CuSO_4 \cdot 5H_2O$	0.02mg
$CoCl_2 \cdot 6H_2O$	2.0μg

pH 7.5 ± 0.2 at 25°C

Preparation of Medium: Add components to distilled/deionized water and bring volume to 1.0L. Mix thoroughly. Adjust pH to 7.5 with 0.1*M* K_2CO_3. Distribute into tubes or flasks. Autoclave for 15 min at 15 psi pressure–121°C.

Use: For the cultivation and maintenance of *Nitrosolobus multiformis*.

Nitrosolobus Medium (ATCC Medium 929)

Composition per liter:

$(NH_4)_2SO_4$	1.32g
$MgSO_4 \cdot 7H_2O$	0.38g
K_2HPO_4	0.087g
$CaCl_2 \cdot 2H_2O$	0.02g
Chelated iron	1.0mg
$MnCl_2 \cdot 4H_2O$	0.2mg
$Na_2MoO_4 \cdot 2H_2O$	0.1mg
$ZnSO_4 \cdot 7H_2O$	0.1mg
$CoCl_2 \cdot 6H_2O$	2.0μg
Phenol Red (0.5% solution)	0.25mL

pH 7.5 ± 0.2 at 25°C

Preparation of Medium: Add components to distilled/deionized water and bring volume to 1.0L. Mix thoroughly. Adjust pH to 7.5 with 0.1*M* K_2CO_3. Distribute into tubes or flasks. Autoclave for 15 min at 15 psi pressure–121°C.

Use: For the cultivation and maintenance of *Nitrosolobus multiformis*.

Nitrosomonas europaea Medium

Composition per liter:

$(NH_4)_2SO_4$	1.7g
$MgSO_4 \cdot 7H_2O$	0.2g
$CaCl_2 \cdot 2H_2O$	0.02g
K_2HPO_4	0.015g
Ferric EDTA	1.0mg
Trace elements solution	1.0mL

pH 7.5 ± 0.2 at 25°C

Trace Elements Solution:

Composition per 100.0mL:

$MnCl_2 \cdot 4H_2O$	0.02g
$Na_2MoO_4 \cdot 2H_2O$	0.01g
$ZnSO_4 \cdot 7H_2O$	0.01g
$CuSO_4 \cdot 5H_2O$	2.0mg
$CoCl_2 \cdot 6H_2O$	0.2mg

Preparation of Trace Elements Solution: Add components to distilled/deionized water and bring volume to 1.0L. Mix thoroughly.

Preparation of Medium: Add components to distilled/deionized water and bring volume to 1.0L. Mix thoroughly. Adjust pH to 7.5 with K_2CO_3. Distribute into tubes or flasks. Autoclave for 15 min at 15 psi pressure–121°C. After inoculation, maintain pH at 7.5–7.8 with sterile 50% K_2CO_3 solution.

Use: For the cultivation and maintenance of *Nitrosomonas europaea*.

Nitrosomonas Medium

Composition per liter:

$(NH_4)_2SO_4$	3.0g
K_2HPO_4	0.5g
$MgSO_4 \cdot 7H_2O$	0.05g
$CaCl_2 \cdot 2H_2O$	4.0mg
Cresol Red (0.0005% solution)	25.0mL
Ferric EDTA solution	0.1mL

pH 8.2–8.4 at 25°C

Ferric EDTA Solution:

Composition per 100.0mL:

$FeSO_4 \cdot 7H_2O$	0.5g
Disodium EDTA	0.14g
H_2SO_4, concentrated	0.05mL

Preparation of Ferric EDTA Solution: Add components to distilled/deionized water and bring volume to 100.0mL. Mix thoroughly.

Preparation of Medium: Add $CaCl_2 \cdot 2H_2O$ and $MgSO_4 \cdot 7H_2O$ to distilled/deionized water and bring volume to 500.0mL. Mix thoroughly. In a separate flask, add remaining components to distilled/deionized water and bring volume to 500.0mL. Mix thoroughly. Autoclave both solutions separately for 15 min at 15 psi pressure–121°C. Cool to 25°C. Aseptically combine the two sterile solutions. Mix thoroughly. Aseptically distribute into sterile tubes or flasks. After inoculation, maintain pH at 8.2–8.4 with sterile 50% K_2CO_3 solution.

Use: For the cultivation and maintenance of *Nitrosomonas europaea*.

Nitrospira moscoviensis Medium (DSMZ Medium 756d)

Composition per liter:

$NaNO_2$	0.5g
Stock solution	100.0mL
Trace elements solution	1.0mL

pH 8.6 ± 0.2 at 25°C

Stock Solution:

Composition per liter:

NaCl	5.0g
KH_2PO_4	1.5g
$MgSO_4 \cdot 7H_2O$	0.5g
$CaCO_3$	0.07g

Preparation of Stock Solution: Add components to distilled/deionized water and bring volume to 1.0L. Mix thoroughly.

Trace Elements Solution:

Composition per liter:

$FeSO_4 \cdot 7H_2O$	97.3mg
H_3BO_3	49.4mg

$ZnSO_4 \cdot 7H_2O$	43.1mg
$(NH_4)_6Mo_7O_{24} \cdot 4H_2O$	37.1mg
$MnSO_4 \cdot 2H_2O$	33.8mg
$CuSO_4 \cdot 5H_2O$	25.0mg

Preparation of Trace Elements Solution: Add components to distilled/deionized water and bring volume to 1.0L. Mix thoroughly.

Preparation of Medium: Add components to distilled/deionized water and bring volume to 1.0L. Mix thoroughly. Adjust pH to 8.6. Distribute into tubes or flasks. Autoclave for 15 min at 15 psi pressure–121°C. Allow to stand for 2–3 days so that pH adjusts itself to 7.4–7.6.

Use: For the cultivation of *Nitrospira moscoviensis.*

NL 333-Agar Medium (DSMZ Medium 984)

Composition per liter:

Agar-Agar	20.0g
Starch, soluble	10.0g
Malt extract	10.0g
Glucose	5.0g
Yeast extract	3.0g
Casein peptone	3.0g
NH_4NO_3	3.0g
$CaCO_3$	2.0g

pH 7.2 ± 0.2 at 25°C

Preparation of Medium: Add components to distilled/deionized water and bring volume to 1.0L. Mix thoroughly. Adjust pH to 7.2. Distribute into tubes or flasks. Gently heat while stirring and bring to boiling. Mix thoroughly. Autoclave for 15 min at 15 psi pressure–121°C. Pour into Petri dishes or leave in tubes.

Use: For the cultivation of a *Micromonospora* spp.

NMS Medium

See: **Nitrate Mineral Salts Medium**

NMS Medium

Composition per 1000.5mL:

Purified agar	12.5g
KNO_3	1.0g
$MgSO_4 \cdot 7H_2O$	1.0g
$Na_2HPO_4 \cdot 12H_2O$	0.717g
KH_2PO_4	0.272g
$CaCl_2 \cdot 6H_2O$	0.2g
Ferric ammonium EDTA	4.0mg
Trace elements solution	0.5mL

pH 6.8 ± 0.2 at 25°C

Trace Elements Solution:

Composition per liter:

Disodium EDTA	0.5g
$FeSO_4 \cdot 7H_2O$	0.2g
H_3BO_3	30.0mg
$CoCl_2 \cdot 6H_2O$	20.0mg
$ZnSO_4 \cdot 7H_2O$	10.0mg
$MnCl_2 \cdot 4H_2O$	3.0mg
$Na_2MoO_4 \cdot 2H_2O$	3.0mg
$NiCl_2 \cdot 6H_2O$	2.0mg
$CaCl_2 \cdot 2H_2O$	1.0mg

Preparation of Trace Elements Solution: Add components to distilled/deionized water and bring volume to 1.0L. Mix thoroughly. Autoclave for 15 min at 15 psi pressure–121°C.

Preparation of Medium: Add components, except trace elements solution, to distilled/deionized water and bring volume to 1.0L. Mix thoroughly. Adjust pH to 6.8. Autoclave for 15 min at 15 psi pressure–121°C. Aseptically add 0.5mL of sterile trace elements solution. Mix thoroughly. Aseptically distribute into sterile tubes or flasks.

Use: For the cultivation of *Methylomonas clara.*

NMS Medium with Methanol

See: **Nitrate Mineral Salts Medium with Methanol**

NMS Medium for Methanotrophs (DSMZ Medium 1179)

Composition per liter:

Agar, purified	12.5g
$MgSO_4 \cdot 7H_2O$	1.0g
$Na_2HPO_4 \cdot 12H_2O$	0.717g
K_2HPO_4	0.272g
$CaCl_2 \cdot 2H_2O$	0.2g
Fe(III)NH_4-EDTA	4.0mg
KNO_3	1.0g
Trace elements solution	0.5mL

pH 6.8 ± 0.2 at 25°C

Trace Elements Solution:

Composition per liter:

Disodium EDTA	0.5g
$FeSO_4 \cdot 7H_2O$	0.2g
H_3BO_3	0.03g
$CoCl_2 \cdot 6H_2O$	0.02g
$ZnSO_4 \cdot 7H_2O$	0.01g
$MnCl_2 \cdot 4H_2O$	3.0mg
$Na_2MoO_4 \cdot 2H_2O$	3.0mg
$NiCl_2 \cdot 6H_2O$	2.0mg
$CaCl_2 \cdot 2H_2O$	1.0mg

Preparation of Trace Elements Solution: Add components to distilled/deionized water and bring volume to 1.0L. Mix thoroughly.

Preparation of Medium: Add components to distilled/deionized water and bring volume to 1.0L. Mix thoroughly. Adjust pH to 6.8. Distribute into tubes or flasks. Gently heat while stirring and bring to boiling. Mix thoroughly. Autoclave for 15 min at 15 psi pressure–121°C. Pour into Petri dishes or leave in tubes.

Use: For the cultivation of *Methylocystis hirsuta.*

NNN Medium (Novy, MacNeal, and Nicole Medium)

Composition per liter:

Agar	7.0g
NaCl	3.0g
Rabbit blood, defibrinated	150.0mL

Preparation of Medium: Add components, except rabbit blood, to distilled/deionized water and bring volume to 850.0mL. Mix thoroughly. Gently heat and bring to boiling. Autoclave for 15 min at 15 psi pressure–121°C. Cool to 50°C. Aseptically add sterile rabbit blood. Mix thoroughly. Aseptically distribute into sterile tubes in 5.0mL volumes. Allow tubes to cool in a slanted position at 4°C.

Use: For the cultivation and maintenance of *Leishmania* species and *Trypanosoma cruzi.*

Nocardia histidans Medium

Composition per liter:

Agar	20.0g
Yeast extract	10.0g
Glucose	10.0g
Na_2HPO_4	0.95g
KH_2PO_4	0.91g
$MgSO_4 \cdot 7H_2O$	0.5g

pH 7.0 ± 0.2 at 25°C

Preparation of Medium: Add components to distilled/deionized water and bring volume to 1.0L. Mix thoroughly. Gently heat and bring to boiling. Distribute into tubes or flasks. Autoclave for 15 min at 15 psi pressure–121°C. Pour into sterile Petri dishes or leave in tubes.

Use: For the cultivation and maintenance of *Nocardia histidans* and *Streptomyces* species.

Nocardia Medium

Composition per liter:

Agar	20.0g
Peptone	10.0g
Beef extract	5.0g
NaCl	2.5g

Preparation of Medium: Add components to distilled/deionized water and bring volume to 1.0L. Mix thoroughly. Gently heat and bring to boiling. Distribute into tubes or flasks. Autoclave for 15 min at 15 psi pressure–121°C. Pour into sterile Petri dishes or leave in tubes.

Use: For the cultivation and maintenance of *Rhodococcus globerulus* and *Nocardia* species.

Nocardia Medium 1

Composition per 1010.0mL:

Agar	12.0g
Proteose peptone	10.0g
Veal infusion solids	10.0g
NaCl	3.0g
Na_2HPO_4	2.0g
Glucose	2.0g
Sodium acetate	1.0g
Adenine sulfate	0.01g
Guanine·HCl	0.01g
Uracil	0.01g
Xanthine	0.01g
Thiamine	0.02mg
Additives solution	10.0mL

pH 7.4 ± 0.2 at 25°C

Additives Solution:

Composition per 10.0mL:

Actidione (cycloheximide)	0.05mg
Mycostatin	0.05mg
Dimethylchlortetracycline·HCl	5.0μg

Preparation of Additives Solution: Add components to distilled/deionized water and bring volume to 10.0mL. Mix thoroughly. Filter sterilize.

Caution: Cycloheximide is toxic. Avoid skin contact or aerosol formation and inhalation.

Preparation of Medium: Add components, except additives solution, to distilled/deionized water and bring volume to 990.0mL. Mix thoroughly. Gently heat and bring to boiling. Autoclave for 15 min at 15 psi pressure–121°C. Cool to 45°–50°C. Aseptically add additives solution. Mix thoroughly. Pour into sterile Petri dishes or distribute into sterile tubes.

Use: For the isolation and cultivation of *Nocardia.*

Nocardia Medium 2

Composition per 1010.0mL:

Agar	12.0g
Proteose peptone	10.0g
Veal infusion solids	10.0g
NaCl	3.0g
Na_2HPO_4	2.0g
Glucose	2.0g
Sodium acetate	1.0g
Adenine sulfate	0.01g
Guanine·HCl	0.01g
Uracil	0.01g
Xanthine	0.01g
Thiamine	0.02mg
Additives solution	10.0mL

pH 7.4 ± 0.2 at 25°C

Additives Solution:

Composition per 10.0mL:

Actidione (cycloheximide)	0.05mg
Mycostatin	0.05mg
Methacycline·HCl	0.01mg

Preparation of Additives Solution: Add components to distilled/deionized water and bring volume to 10.0mL. Mix thoroughly. Filter sterilize.

Caution: Cycloheximide is toxic. Avoid skin contact or aerosol formation and inhalation.

Preparation of Medium: Add components, except additives solution, to distilled/deionized water and bring volume to 990.0mL. Mix thoroughly. Gently heat and bring to boiling. Autoclave for 15 min at 15 psi pressure–121°C. Cool to 45°–50°C. Aseptically add additives solution. Mix thoroughly. Pour into sterile Petri dishes or distribute into sterile tubes.

Use: For the isolation and cultivation of *Nocardia.*

Nocardia Medium 3

Composition per 1010.0mL:

Agar	12.0g
Proteose peptone	10.0g
Veal infusion solids	10.0g
NaCl	3.0g
Na_2HPO_4	2.0g
Glucose	2.0g
Sodium acetate	1.0g
Adenine sulfate	0.01g
Guanine·HCl	0.01g
Uracil	0.01g
Xanthine	0.01g
Thiamine	0.02mg
Actidione	0.05mg
Mycostatin	0.05mg
Chlortetracycline·HCl	0.045mg

Demethylchlortetracycline·HCl	5.0μg
Additives solution	10.0mL

pH 7.4 ± 0.2 at 25°C

Additives Solution:

Composition per 10.0mL:

Actidione (cycloheximide)	0.05mg
Mycostatin	0.05mg
Dimethylchlortetracycline·HCl	5.0μg

Preparation of Additives Solution: Add components to distilled/deionized water and bring volume to 10.0mL. Mix thoroughly. Filter sterilize.

Caution: Cycloheximide is toxic. Avoid skin contact or aerosol formation and inhalation.

Preparation of Medium: Add components, except additives solution, to distilled/deionized water and bring volume to 990.0mL. Mix thoroughly. Gently heat and bring to boiling. Autoclave for 15 min at 15 psi pressure–121°C. Cool to 45°–50°C. Aseptically add additives solution. Mix thoroughly. Pour into sterile Petri dishes or distribute into sterile tubes.

Use: For the isolation and cultivation of *Nocardia*.

Nocardia Medium 4

Composition per 1010.0mL:

Agar	12.0g
Proteose peptone	10.0g
Veal infusion solids	10.0g
NaCl	3.0g
Na_2HPO_4	2.0g
Glucose	2.0g
Sodium acetate	1.0g
Adenine sulfate	0.01g
Guanine·HCl	0.01g
Uracil	0.01g
Xanthine	0.01g
Thiamine	0.02mg
Additives solution	10.0mL

pH 7.4 ± 0.2 at 25°C

Additives Solution:

Composition per 10.0mL:

Actidione (cycloheximide)	0.05mg
Mycostatin	0.05mg
Chlortetracycline·HCl	0.045mg
Methacycline·HCl	0.01mg

Preparation of Additives Solution: Add components to distilled/deionized water and bring volume to 10.0mL. Mix thoroughly. Filter sterilize.

Caution: Cycloheximide is toxic. Avoid skin contact or aerosol formation and inhalation.

Preparation of Medium: Add components, except additives solution, to distilled/deionized water and bring volume to 990.0mL. Mix thoroughly. Gently heat and bring to boiling. Autoclave for 15 min at 15 psi pressure–121°C. Cool to 45°–50°C. Aseptically add sadditives solution. Mix thoroughly. Pour into sterile Petri dishes or distribute into sterile tubes.

Use: For the isolation and cultivation of *Nocardia* species.

Nonfat Dry Milk, Reconstituted

Composition per liter:

Milk, nonfat dry	100.0g

pH 6.8 ± 0.2 at 25°C

Preparation of Medium: Add 100.0g of nonfat dry milk to distilled/deionized water and bring volume to 1.0L. Mix thoroughly. Distribute into tubes or flasks. Autoclave for 15 min at 15 psi pressure–121°C.

Use: For the cultivation of *Salmonella* species and monkey kidney cells in tissue culture.

Nonnutrient Agar

Composition per liter:

Agar	15.0g

Preparation of Medium: Add agar to distilled/deionized water and bring volume to 1.0L. Mix thoroughly. Gently heat and bring to boiling. Distribute into tubes or flasks. Autoclave for 15 min at 15 psi pressure–121°C. Pour into sterile Petri dishes or leave in tubes.

Use: For the cultivation of *Naegleria lovaniensis*.

Nonnutrient Agar Plates

Composition per liter:

Agar	15.0g
Page's amoeba saline	1.0L

Page's Amoeba Saline:

Composition per liter:

Na_2HPO_4	0.142g
KH_2PO_4	0.136g
NaCl	0.12g
$MgSO_4 \cdot 7H_2O$	4.0mg
$CaCl_2 \cdot 2H_2O$	4.0mg

Preparation of Page's Amoeba Saline: Add components to distilled/deionized water and bring volume to 1.0mL. Mix thoroughly.

Preparation of Medium: Add agar to 1.0L of Page's amoeba saline. Mix thoroughly. Gently heat and bring to boiling. Autoclave for 15 min at 15 psi pressure–121°C. Cool to 60°C. Pour into sterile Petri dishes in 20.0mL volumes. Store at 4°C for up to 3 months.

Use: For the isolation and cultivation of pathogenic free-living amoebae.

Norris Agar

See: **Nitrogen-Free Agar**

NOS Medium, Modified

Composition per 100.67mL:

Basal medium	94.0mL
$NaHCO_3$ solution	2.67mL
TPP/VFA mixture	2.0mL
Rabbit serum, heat inactivated	2.0mL

pH 7.4 ± 0.2 at 25°C

Basal Medium:

Composition per 94.0mL:

Pancreatic digest of casein	1.0g
Pancreatic digest of gelatin	0.48g
Yeast extract	0.25g
Brain heart, solids from infusion	0.2g
Peptic digest of animal tissue	0.2g

D-Glucose 0.2g
NaCl 0.17g
Glucose 0.1g
L-Cysteine·HCl·H_2O 0.1g
Na_2HPO_4 .... 0.085g
Sodium thioglycolate 0.05g
L-Asparagine 0.025g
Resazurin (0.1% w/v solution) 0.1mL

Preparation of Basal Medium: Add components to distilled/deionized water and bring volume to 94.0mL. Mix thoroughly. Gently heat and bring to boiling. Gas under O_2-free 85% N_2 + 10% CO_2 + 5% H_2. Stopper and wire flask closed. Autoclave for 20 min at 15 psi pressure–121°C. Cool to 45°–50°C.

$NaHCO_3$ Solution:
Composition per 10.0mL:
$NaHCO_3$ 0.75g

Preparation of $NaHCO_3$ Solution: Add the $NaHCO_3$ to distilled/deionized water and bring volume to 10.0mL. Mix thoroughly. Filter sterilize.

TPP/VFA Mixture:
Composition per 10.9mL:
Thiamine pyrophosphate (0.2% solution) 1.5mL
VFA solution 1.0mL

Preparation of TPP/VFA Mixture: Add components to distilled/deionized water and bring volume to 10.9mL. Mix thoroughly. Filter sterilize. Store at –20°C.

VFA Solution:
Composition per 100.0mL:
NaOH (0.1*N* solution) 98.0mL
Isobutyric acid 0.5mL
2-Methylbutyric acid 0.5mL
Isovaleric acid 0.5mL
Valeric acid 0.5mL

Preparation of VFA Solution: Add volatile fatty acids to 98.0mL of NaOH solution. Mix thoroughly. Filter sterilize. Store at 4°C.

Preparation of Medium: Open the flask containing 94.0mL of cooled sterile basal medium while flushing with O_2-free 85% N_2 + 10% CO_2 + 5% H_2. Aseptically add sterile $NaHCO_3$ solution, sterile TPP/VFA mixture, and filter-sterilized rabbit serum. Mix thoroughly.

Use: For the cultivation and maintenance of *Treponema vincentii* and other *Treponema* species.

NOS Spirochete Medium

Composition per 1045.0mL:
Basal medium 1.0L
$NaHCO_3$ (10% solution) 20.0mL
Rabbit serum, heat inactivated 20.0mL
Thiamine pyrophosphate (0.2% solution) 3.0mL
VFA solution 2.0mL

pH 7.4 ± 0.2 at 25°C

Basal Medium:
Composition per liter:
Pancreatic digest of casein 10.0g
Pancreatic digest of gelatin 4.85g
Noble agar 3.0g
Yeast extract 2.5g
Brain heart, solids from infusion 2.0g
Peptic digest of animal tissue 2.0g
Glucose 2.0g
NaCl 1.65g
Glucose 1.0g
L-Cysteine·HCl·H_2O 1.0g
Na_2HPO_4 .... 0.85g
Sodium thioglycolate 0.5g
L-Asparagine 0.25g

Preparation of Basal Medium: Add components to distilled/deionized water and bring volume to 1.0L. Mix thoroughly. Gently heat and bring to boiling. Gas under O_2-free 80% N_2 + 10% CO_2 + 10% H_2. Stopper and wire flask closed. Autoclave for 20 min at 15 psi pressure–121°C. Cool to 45°–50°C.

VFA Solution:
Composition per 100.0mL:
KOH (0.1*N* solution) 98.0mL
Isobutyric acid 0.5mL
2-Methylbutyric acid 0.5mL
Isovaleric acid 0.5mL
Valeric acid 0.5mL

Preparation of VFA Solution: Add volatile fatty acids to 98.0mL of KOH solution. Mix thoroughly. Filter sterilize. Store at 4°C.

Preparation of Medium: Combine 20.0mL of $NaHCO_3$ solution, 20.0mL of rabbit serum, 3.0mL of thiamine pyrophosphate solution, and 2.0mL of VFA solution. Mix thoroughly. Filter sterilize. Open the flask containing 1.0L of cooled, sterile basal medium while flushing with O_2-free 85% N_2 + 10% CO_2 + 5% H_2. Aseptically add the filter-sterilized mixture. Mix thoroughly. Aseptically and anaerobically distribute into sterile tubes or flasks.

Use: For the cultivation and maintenance of *Treponema denticola* and *Treponema socranskii.*

Novobiocin Agar

Composition per liter:
Agar 15.0g
Peptone 5.0g
NaCl 5.0g
Yeast extract 2.0g
Beef extract 1.0g
Novobiocin solution 10.0mL

Novobiocin Solution:
Composition per 10.0mL:
Novobiocin 10.0mg

Preparation of Novobiocin Solution: Add novobiocin to distilled/deionized water and bring volume to 10.0mL. Mix thoroughly. Filter sterilize.

Preparation of Medium: Add components, except novobiocin solution, to distilled/deionized water and bring volume to 990.0mL. Mix thoroughly. Gently heat and bring to boiling. Autoclave for 15 min at 15 psi pressure–121°C. Aseptically add 10.0mL of sterile novobiocin solution. Mix thoroughly. Pour into sterile Petri dishes or distribute into sterile tubes.

Use: For the cultivation of *Staphylococcus aureus.*

NPB Medium (DSMZ Medium 995)

Composition per liter:

Tryptone peptone	10.0g
D-Glucose	5.0g
Yeast extract	2.0g
$MgSO_4·7H_2O$	1.0g
K_2HPO_4	1.0g
KH_2PO_4	0.5g

pH 7.0 ± 0.2 at 25°C

Preparation of Medium: Add components to distilled/deionized water and bring volume to 1.0L. Mix thoroughly. Adjust pH to 7.0. Distribute into tubes or flasks. Gently heat while stirring and bring to boiling. Mix thoroughly. Autoclave for 15 min at 15 psi pressure–121°C.

Use: For the cultivation of *Catellibacterium nectariphilum.*

NSMP, Modified

Composition per liter:

Casamino acids	5.0g
Glucose	2.0g
KH_2PO_4	0.86g
Sodium citrate	0.6g
K_2HPO_4	0.55g
$MgCl_2·6H_2O$	0.43g
$CaCl_2$	0.1g
$MnCl_2·4H_2O$	0.016g
$ZnCl_2$	7.0mg
$FeCl_3$	3.0mg

pH 6.5 ± 0.2 at 25°C

Preparation of Medium: Add components to distilled/deionized water and bring volume to 1.0L. Mix thoroughly. Distribute into tubes or flasks. Autoclave for 15 min at 15 psi pressure–121°C.

Use: For the cultivation and maintenance of *Bacillus thuringiensis*.

NTYG

Composition per liter:

Glucose	10.0g
Pancreatic digest of casein	5.0g
Yeast extract	5.0g
Fetal bovine serum, dialyzed	20.0mL
Sheep blood, defibrinated	10.0mL

Fetal Bovine Serum, Dialyzed:
Composition per 100.0mL:

Fetal bovine serum, heat inactivated	100.0mL

Preparation of Fetal Bovine Serum, Dialyzed: Dialyze the heat-inactivated serum at 0–4°C against 10 volumes of distilled/deionized water. Clean the dialysis tubing before use by boiling in 1.0L of a 0.037% EDTA solution. Rinse the tubing with distilled/deionized water. Change the water four times at 8–16 hr intervals. Centrifuge the dialyzed serum for 30 min at 35,000X g. Filter sterilize.

Preparation of Medium: Add components, except dialyzed fetal bovine serum and sheep blood, to distilled/deionized water and bring volume to 975.0mL. Mix thoroughly. Autoclave for 15 min at 15 psi pressure–121°C. Aseptically add 20.0mL of sterile, dialyzed fetal bovine serum, and 5.0mL of sterile sheep blood. Mix thoroughly. Aseptically distribute into sterile screw-capped tubes or flasks.

Use: For the cultivation of *Naegleria lovaniensis*.

Nutrient Agar (LMG Medium 160)

Composition per liter:

Agar	3.0g
Lab-Lemco beef extract	1.0g
Peptone	1.0g
NaCl	0.5g

pH 7.3 ± 0.2 at 25°C

Preparation of Medium: Add components to distilled/deionized water and bring volume to 1.0L. Mix thoroughly. Distribute into tubes or flasks. Autoclave for 15 min at 15 psi pressure–121°C.

Use: For the cultivation of heterotrophic bacteria.

Nutrient Agar

Composition per liter:

Agar	15.0g
Peptone	5.0g
NaCl	5.0g
Yeast extract	2.0g
Beef extract	1.0g

pH 7.4 ± 0.2 at 25°C

Source: This medium is available as a premixed powder from Oxoid Unipath.

Preparation of Medium: Add components to distilled/deionized water and bring volume to 1.0L. Mix thoroughly. Gently heat and bring to boiling. Distribute into tubes or flasks. Autoclave for 15 min at 15 psi pressure–121°C. Pour into sterile Petri dishes or leave in tubes.

Use: For the cultivation and maintenance of a wide variety of microorganisms.

Nutrient Agar (ATCC Medium 3) (BAM M113)

Composition per liter:

Agar	15.0g
Pancreatic digest of gelatin	5.0g
Beef extract	3.0g

pH 6.8 ± 0.2 at 25°C

Source: This medium is available as a premixed powder from BD Diagnostic Systems.

Preparation of Medium: Add components to distilled/deionized water and bring volume to 1.0L. Mix thoroughly. Gently heat while stirring and bring to boiling. Distribute into tubes or flasks. Autoclave for 15 min at 15 psi pressure–121°C. Pour into sterile Petri dishes or leave in tubes.

Use: For the cultivation of a wide variety of bacteria and for the enumeration of organisms in water, sewage, feces, and other materials. For the cultivation of *Bacillus cereus*.

Nutrient Agar (Oxoid CM3) (LMG Medium 1)

Composition per liter:

Agar	15.0g
Peptone	5.0g
NaCl	5.0g

Yeast extract	2.0g
Lab-Lemco beef extract	1.0g

pH 7.4 ± 0.2 at 25°C

Preparation of Medium: Add components to distilled/deionized water and bring volume to 1.0L. Mix thoroughly. Gently heat and bring to boiling. Distribute into tubes or flasks. Autoclave for 15 min at 15 psi pressure–121°C. Pour into sterile Petri dishes or leave in tubes.

Use: For the cultivation and maintenance of *Pseudomonas* spp., *Acidovorax* spp., *Ralstonia* spp., *Delftia acidovorans*, *Burkholderia* spp., *Comamonas testosteroni*, *Microbacterium flavescens*, and other bacteria.

Nutrient Agar, 1.5% (ATCC Medium 105)

Composition per liter:

Agar	15.0g
NaCl	8.0g
Pancreatic digest of gelatin	5.0g
Beef extract	3.0g

pH 7.3 ± 0.2 at 25°C

Source: This medium is available as a premixed powder from BD Diagnostic Systems.

Preparation of Medium: Add components to distilled/deionized water and bring volume to 1.0L. Mix thoroughly. Gently heat while stirring and bring to boiling. Distribute into tubes or flasks. Autoclave for 15 min at 15 psi pressure–121°C. Pour into sterile Petri dishes or leave in tubes.

Use: For the cultivation and maintenance of a variety of nonfastidious bacteria.

Nutrient Agar, Alkaline (LMG Medium 53)

Composition per liter:

Agar	15.0g
Peptone	5.0g
NaCl	5.0g
Yeast extract	2.0g
Lab-Lemco beef extract	1.0g

pH 9.5–10.0 at 25°C

Preparation of Medium: Add components to distilled/deionized water and bring volume to 1.0L. Mix thoroughly. Adjust pH to 9.5–10.0 with sterile Na_2CO_3 solution. Gently heat and bring to boiling. Distribute into tubes or flasks. Autoclave for 15 min at 15 psi pressure–121°C. Pour into sterile Petri dishes or leave in tubes.

Use: For the cultivation and maintenance of *Bacillus alcalophilus* and *Bacillus cohnii.*

Nutrient Agar, Buffered

Composition per liter:

Agar	15.0g
Peptone	5.0g
NaCl	5.0g
$Na_2HPO_4 \cdot 12H_2O$	2.39g
Yeast extract	2.0g
Beef extract	1.0g
KH_2PO_4	0.45g

pH 6.8 ± 0.2 at 25°C

Preparation of Medium: Add components to distilled/deionized water and bring volume to 1.0L. Mix thoroughly. Gently heat and bring to boiling. Adjust pH to 6.8. Distribute into tubes or flasks. Autoclave for 15 min at 15 psi pressure–121°C. Pour into sterile Petri dishes or leave in tubes.

Use: For the cultivation and maintenance of *Acidovorax avenae, Acidovorax avenae, Acidovorax delafieldii, Acidovorax facilis, Acidovorax konjaci, Acidovorax temperans, Aminobacter aminovorans, Chryseomonas luteola, Comamonas acidovorans, Comamonas testosteroni, Flavimonas oryzihabitans, Flavobacterium breve, Flavobacterium mizutaii, Hydrogenoflava palleronii, Hydrogenophaga flava, Hydrogenophaga pseudoflava, Hydrogenophaga taeniospiralis,* numerous *Pseudomonas* species, *Sphingobacterium multivorum, Sphingobacterium spiritivorum, Weeksella virosa, Weeksella zoohelcum,* and *Moraxella atlantae.*

Nutrient Agar with Formate, Fumarate, and Horse Blood (LMG Medium 250)

Composition per liter:

Agar	15.0g
Peptone	5.0g
NaCl	5.0g
Fumaric acid	3.0g
Sodium formate	2.0g
Yeast extract	2.0g
Lab-Lemco beef extract	1.0g
Horse blood, sterile defibrinated	50.0mL

pH 7.2 ± 0.2 at 25°C

Preparation of Medium: Add components, except horse blood, to distilled/deionized water and bring volume to 950.0mL. Mix thoroughly. Adjust pH to 7.2. Gently heat and bring to boiling. Autoclave for 15 min at 15 psi pressure–121°C. Cool to 45°C. Aseptically add 50.0mL sterile defibrinated horse blood. Mix thoroughly. Pour into sterile Petri dishes or distribute into sterile tubes.

Use: For the cultivation and maintenance of *Campylobacter rectus* and *Campylobacter gracilis.*

Nutrient Agar, Half Strength

Composition per liter:

Agar	15.0g
Peptone	2.5g
NaCl	2.5g
Yeast extract	1.0g
Beef extract	0.5g

pH 7.0 ± 0.2 at 25°C

Preparation of Medium: Add components to distilled/deionized water and bring volume to 1.0L. Mix thoroughly. Gently heat and bring to boiling. Distribute into tubes or flasks. Autoclave for 15 min at 15 psi pressure–121°C. Pour into sterile Petri dishes or leave in tubes.

Use: For the cultivation of *Chromobactertium* species and *Thermomonospora chromogena.*

Nutrient Agar 1.5%, HiVeg

Composition per liter:

Agar	15.0g
NaCl	8.0g

Plant peptone.. 5.0g
Plant extract 3.0g

pH 7.4 ± 0.2 at 25°C

Source: This medium is available as a premixed powder from HiMedia.

Preparation of Medium: Add components to distilled/deionized water and bring volume to 1.0L. Mix thoroughly. Gently heat and bring to boiling. Distribute into tubes or flasks. Autoclave for 15 min at 15 psi pressure–121°C. Pour into sterile Petri dishes or leave in tubes.

Use: For the cultivation and maintenance of a wide variety of microorganisms.

Nutrient Agar 1.5%, HiVeg with Ascitic Fluid

Composition per liter:

Agar 15.0g
NaCl 8.0g
Plant peptone.. 5.0g
Plant extract 3.0g
Ascitic fluid..250.0mL

pH 7.4 ± 0.2 at 25°C

Source: This medium, without ascitic fluid, is available as a premixed powder from HiMedia.

Preparation of Medium: Add components, except ascitic fluid, to distilled/deionized water and bring volume to 750.0mL. Mix thoroughly. Autoclave for 15 min at 15 psi pressure–121°C. Cool to less than 37°C. Aseptically add sterile ascitic fluid. If desired, 0.5g of thallium acetate or 100,000U of penicillin may be added for a more selective medium. Mix thoroughly. Aseptically distribute into sterile tubes or flasks.

Use: For the enrichment of pleuro-pneumonia-like organisms (PPLOs) and *Mycoplasma* species from clinical specimens.

Nutrient Agar with Maltose

Composition per liter:

Agar 15.0g
Maltose.. 10.0g
Pancreatic digest of gelatin 5.0g
Beef extract 3.0g

pH 6.8 ± 0.2 at 25°C

Source: Nutrient agar is available as a premixed powder from BD Diagnostic Systems.

Preparation of Medium: Add components to distilled/deionized water and bring volume to 1.0L. Mix thoroughly. Gently heat and bring to boiling. Distribute into tubes or flasks. Autoclave for 15 min at 15 psi pressure–121°C. Pour into sterile Petri dishes or leave in tubes.

Use: For the cultivation and maintenance of *Nocardia corynebacteroides.*

Nutrient Agar No. 2

Composition per liter:

Agar 20.0g
Meat extract 10.0g
Peptone.. 10.0g
NaCl 5.0g

pH 7.0 ± 0.2 at 25°C

Preparation of Medium: Add components to distilled/deionized water and bring volume to 1.0L. Mix thoroughly. Gently heat and bring to boiling. Distribute into tubes or flasks. Autoclave for 15 min at 15 psi pressure–121°C. Pour into sterile Petri dishes or leave in tubes.

Use: For the cultivation and maintenance of a wide variety of bacteria.

Nutrient Agar No. 2, Diluted 1/10 (DSMZ Medium 952)

Composition per liter:

Agar 20.0g
Peptone 1.0g
Meat extract 1.0g
NaCl 0.5g

pH 7.1 ± 0.2 at 25°C

Preparation of Medium: Add components to distilled/deionized water and bring volume to 1.0L. Mix thoroughly. Adjust pH to 7.1. Gently heat and bring to boiling. Distribute into tubes or flasks. Autoclave for 15 min at 15 psi pressure–121°C. Pour into sterile Petri dishes or leave in tubes.

Use: For the cultivation and maintenance of *Agromonas oligotrophica.*

Nutrient Agar No. 2, Diluted 1/10

Composition per liter:

Agar 20.0g
Meat extract 1.0g
Peptone 1.0g
NaCl 0.5g

pH 7.0 ± 0.2 at 25°C

Preparation of Medium: Add components to distilled/deionized water and bring volume to 1.0L. Mix thoroughly. Gently heat and bring to boiling. Distribute into tubes or flasks. Autoclave for 15 min at 15 psi pressure–121°C. Pour into sterile Petri dishes or leave in tubes.

Use: For the cultivation and maintenance of *Agromonas oligotrophica* and *Kurthia gibsonii.*

Nutrient Agar No. 2 Diluted 1/100 (DSMZ Medium 952)

Composition per liter:

Agar 20.0g
Peptone 0.1g
Meat extract 0.1g
NaCl 50.0mg

pH 7.1 ± 0.2 at 25°C

Preparation of Medium: Add components to distilled/deionized water and bring volume to 1.0L. Mix thoroughly. Adjust pH to 7.1. Gently heat and bring to boiling. Distribute into tubes or flasks. Autoclave for 15 min at 15 psi pressure–121°C. Pour into sterile Petri dishes or leave in tubes.

Use: For the cultivation and maintenance of *Bacillus pumilus.*

Nutrient Agar No. 2 Diluted 1/100

Composition per liter:

Agar 20.0g
Meat extract 0.1g
Peptone 0.1g
NaCl 50.0mg

pH 7.0 ± 0.2 at 25°C

Preparation of Medium: Add components to distilled/deionized water and bring volume to 1.0L. Mix thoroughly. Adjust pH to 7.0.

Gently heat and bring to boiling. Distribute into tubes or flasks. Autoclave for 15 min at 15 psi pressure–121°C. Pour into sterile Petri dishes or leave in tubes.

Use: For the cultivation and maintenance of *Agromonas oligotrophica.*

Nutrient Agar No. 2, HiVeg

Composition per liter:

Agar	15.0g
Plant extract	10.0g
Plant peptone	10.0g
NaCl	5.0g

pH 7.4 ± 0.2 at 25°C

Source: This medium is available as a premixed powder from HiMedia.

Preparation of Medium: Add components to distilled/deionized water and bring volume to 1.0L. Mix thoroughly. Gently heat and bring to boiling. Distribute into tubes or flasks. Autoclave for 15 min at 15 psi pressure–121°C. Pour into sterile Petri dishes or leave in tubes.

Use: For the cultivation and maintenance of a wide variety of microorganisms.

Nutrient Agar Oxoid CM3 with Phosphate (DSMZ Medium 605a)

Composition per liter:

Agar	15.0g
Peptone	5.0g
NaCl	5.0g
$Na_2HPO_4 \cdot 12H_2O$	2.39g
Yeast extract	2.0g
Lab-Lemco beef extract	1.0g
KH_2PO_4	0.45g

pH 6.8 ± 0.2 at 25°C

Preparation of Medium: Add components to distilled/deionized water and bring volume to 1.0L. Mix thoroughly. Gently heat and bring to boiling. Distribute into tubes or flasks. Autoclave for 15 min at 15 psi pressure–121°C. Pour into sterile Petri dishes or leave in tubes.

Use: For the cultivation and maintenance of *Devosia riboflavina, Brevundimonas diminuta,* and *Herbaspirillum rubrisubalbicans.*

Nutrient Agar with 1% Peptone, HiVeg

Composition per liter:

Agar	15.0g
Plant peptone	10.0g
Plant extract	5.0g
NaCl	5.0g

pH 7.4 ± 0.2 at 25°C

Source: This medium is available as a premixed powder from HiMedia.

Preparation of Medium: Add components to distilled/deionized water and bring volume to 1.0L. Mix thoroughly. Gently heat and bring to boiling. Distribute into tubes or flasks. Autoclave for 15 min at 15 psi pressure–121°C. Pour into sterile Petri dishes or leave in tubes.

Use: For the cultivation and maintenance of a wide variety of microorganisms.

Nutrient Agar, pH 5.0 (Oxoid Nutrient Agar No. 1, pH 5.0)

Composition per liter:

Agar	15.0g
Peptone	5.0g
NaCl	5.0g
Yeast extract	2.0g
Beef extract	1.0g

pH 5.0 ± 0.2 at 25°C

Source: This medium is available as a premixed powder from Oxoid Unipath.

Preparation of Medium: Add components to distilled/deionized water and bring volume to 1.0L. Mix thoroughly. Gently heat and bring to boiling. Adjust pH to 5.0. Distribute into tubes or flasks. Autoclave for 15 min at 15 psi pressure–121°C. Pour into sterile Petri dishes or leave in tubes.

Use: For the cultivation and maintenance of *Pseudomonas phenazinium.*

Nutrient Agar pH 6.0

Composition per liter:

Agar	15.0g
Peptone	5.0g
Beef extract	3.0g

pH 6.0 ± 0.2 at 25°C

Source: This medium is available as a premixed powder from BD Diagnostic Systems.

Preparation of Medium: Add components to distilled/deionized water and bring volume to 1.0L. Mix thoroughly. Gently heat and bring to boiling. Adjust pH to 6.0. Distribute into tubes or flasks. Autoclave for 15 min at 15 psi pressure–121°C. Pour into sterile Petri dishes or leave in tubes.

Use: For the cultivation of microorganisms that prefer a slightly acid nutrient agar.

Nutrient Agar pH 6.0 with 0.8% NaCl, HiVeg

Composition per liter:

Agar	15.0g
NaCl	8.0g
Plant peptone	5.0g
Plant extract	3.0g

pH 6.0± 0.2 at 25°C

Source: This medium is available as a premixed powder from HiMedia.

Preparation of Medium: Add components to distilled/deionized water and bring volume to 1.0L. Mix thoroughly. Gently heat and bring to boiling. Distribute into tubes or flasks. Autoclave for 15 min at 15 psi pressure–121°C. Pour into sterile Petri dishes or leave in tubes.

Use: For the cultivation of microorganisms that prefer a slightly acid nutrient agar.

Nutrient Agar pH 6.8 (LMG Medium 2)

Composition per liter:

Agar	15.0g
Peptone	5.0g
NaCl	5.0g
$Na_2HPO_4 \cdot 12\ H_2O$	2.39g
Yeast extract	2.0g

Lab-Lemco beef extract	1.0g
KH_2PO_4	0.45g

pH 6.8 ± 0.2 at 25°C

Preparation of Medium: Add components to distilled/deionized water and bring volume to 1.0L. Mix thoroughly. Adjust pH to 6.8. Distribute into tubes or flasks. Autoclave for 15 min at 15 psi pressure–121°C.

Use: For the cultivation and maintenance of *Acidovorax facilis, Pseudomonas geniculata, Arthrobacter siderocapsulatus, Delftia acidovorans, Pseudomonas* spp., and various other heterotrophic bacteria.

Nutrient Agar, pH 6.8, HiVeg

Composition per liter:

Agar	15.0g
Plant peptone	5.0g
Plant extract	3.0g

pH 6.8 ± 0.2 at 25°C

Source: This medium is available as a premixed powder from HiMedia.

Preparation of Medium: Add components to distilled/deionized water and bring volume to 1.0L. Mix thoroughly. Gently heat and bring to boiling. Distribute into tubes or flasks. Autoclave for 15 min at 15 psi pressure–121°C. Pour into sterile Petri dishes or leave in tubes.

Use: For the cultivation and maintenance of a wide variety of microorganisms.

Nutrient Agar pH 6.8, HiVeg with Blood

Composition per liter:

Agar	15.0g
Plant peptone	5.0g
Plant extract	3.0g
Sheep blood, defibrinated	50.0mL

pH 6.8 ± 0.2 at 25°C

Source: This medium, without blood, is available as a premixed powder from HiMedia.

Preparation of Medium: Add components, except sheep blood, to distilled/deionized water and bring volume to 950.0mL. Mix thoroughly. Heat with frequent agitation and boil for 1 min to completely dissolve. Autoclave for 15 min at 15 psi pressure–121°C. Cool to 45°–50°C. Aseptically add 50.0mL of sterile, defibrinated sheep blood. Mix thoroughly and pour into sterile Petri dishes.

Use: For the cultivation of streptococci and other fastidious microorganisms.

Nutrient Agar pH 6.8 with Horse Blood (LMG Medium 234)

Composition per liter:

Agar	15.0g
Peptone	5.0g
NaCl	5.0g
$Na_2HPO_4 \cdot 12\ H_2O$	2.39g
Yeast extract	2.0g
Lab-Lemco beef extract	1.0g
KH_2PO_4	0.45g
Horse blood, sterile defibrinated	50.0mL

pH 6.8 ± 0.2 at 25°C

Preparation of Medium: Add components, except horse blood, to distilled/deionized water and bring volume to 950.0mL. Mix thoroughly. Adjust pH to 6.8. Autoclave for 15 min at 15 psi pressure–121°C. Cool to 45°–50°C. Aseptically add 50.0mL sterile horse blood. Mix thoroughly. Pour into sterile Petri dishes or distribute into sterile tubes.

Use: For the cultivation and maintenance of *Neisseria weaveri*.

Nutrient Agar pH 8.0

Composition per liter:

Agar	15.0g
Pancreatic digest of gelatin	5.0g
Beef extract	3.0g

pH 8.0 ± 0.2 at 25°C.

Preparation of Medium: Add components to distilled/deionized water and bring volume to 1.0L. Mix thoroughly. Gently heat and bring to boiling. Adjust pH to 8.0. Distribute into tubes or flasks. Autoclave for 15 min at 15 psi pressure–121°C. Pour into sterile Petri dishes or leave in tubes. Allow tubes to cool in a slanted position.

Use: For the cultivation and maintenance of *Bacillus alcalophilus*.

Nutrient Agar pH 9.0 (ATCC Medium 2029)

Composition per liter:

Agar	15.0g
Pancreatic digest of gelatin	5.0g
Beef extract	3.0g

pH 9.0 ± 0.1 at 25°C

Source: This medium is available as a premixed powder from BD Diagnostic Systems.

Preparation of Medium: Add components to distilled/deionized water and bring volume to 1.0L. Mix thoroughly. Adjust pH to 9.0. Gently heat while stirring and bring to boiling. Distribute into tubes or flasks. Autoclave for 15 min at 15 psi pressure–121°C. Pour into sterile Petri dishes or leave in tubes.

Use: For the cultivation of alkilophilic bacteria.

Nutrient Agar pH 9.5

Composition per liter:

Agar	15.0g
Pancreatic digest of gelatin	5.0g
Beef extract	3.0g
Na_2CO_3 solution	variable

pH 9.5–10.0 at 25°C

Na_2CO_3 Solution:

Composition per 100.0mL:

Na_2CO_3	5.0g

Preparation of Na_2CO_3 Solution: Add Na_2CO_3 to distilled/deionized water and bring volume to 100.0mL. Mix thoroughly. Autoclave for 15 min at 15 psi pressure–121°C. Cool to 50°C.

Preparation of Medium: Add components, except Na_2CO_3 solution, to distilled/deionized water and bring volume to 1.0L. Mix thoroughly. Gently heat and bring to boiling. Autoclave for 15 min at 15 psi pressure–121°C. Cool to 45°–50°C. Adjust pH to 9.5 with sterile Na_2CO_3 solution. Mix thoroughly. Pour into sterile Petri dishes or distribute into sterile tubes. Allow tubes to cool in a slanted position.

Use: For the cultivation and maintenance of *Bacillus alcalophilus*.

Nutrient Agar pH 10.0 (ATCC Medium 2030)

Composition per liter:

Agar	15.0g
Pancreatic digest of gelatin	5.0g
Beef extract	3.0g

pH 10.0 ± 0.1 at 25°C

Source: This medium is available as a premixed powder from BD Diagnostic Systems.

Preparation of Medium: Add components to distilled/deionized water and bring volume to 1.0L. Mix thoroughly. Adjust pH to 10.0. Gently heat while stirring and bring to boiling. Distribute into tubes or flasks. Autoclave for 15 min at 15 psi pressure–121°C. Pour into sterile Petri dishes or leave in tubes.

Use: For the cultivation of alkilophilic bacteria.

Nutrient Agar with Cysteine (ATCC Medium 812)

Composition per liter:

Agar	15.0g
Pancreatic digest of gelatin	5.0g
Beef extract	3.0g
L-Cysteine	0.1g

pH 6.8 ± 0.2 at 25°C

Source: Nutrient agar is available as a premixed powder from BD Diagnostic Systems.

Preparation of Medium: Add components to distilled/deionized water and bring volume to 1.0L. Mix thoroughly. Gently heat and bring to boiling. Distribute into tubes or flasks. Autoclave for 15 min at 15 psi pressure–121°C. Pour into sterile Petri dishes or leave in tubes.

Use: For the cultivation and maintenance of *Salmonella choleraesuis.*

Nutrient Agar with Dihydrostreptomycin

Composition per liter:

Agar	15.0g
Pancreatic digest of gelatin	5.0g
Beef extract	3.0g
Dihydrostreptomycin solution	10.0mL

pH 6.8 ± 0.2 at 25°C

Source: Nutrient agar is available as a premixed powder from BD Diagnostic Systems.

Dihydrostreptomycin Solution:

Composition per 10.0mL:

Dihydrostreptomycin	0.625g

Preparation of Dihydrostreptomycin Solution: Add dihydrostreptomycin to distilled/deionized water and bring volume to 10.0mL. Mix thoroughly. Filter sterilize.

Preparation of Medium: Add components, except dihydrostreptomycin solution, to distilled/deionized water and bring volume to 990.0mL. Mix thoroughly. Gently heat and bring to boiling. Autoclave for 15 min at 15 psi pressure–121°C. Cool to 45°–50°C. Aseptically add sterile dihydrostreptomycin solution. Mix thoroughly. Pour into sterile Petri dishes or distribute into sterile tubes.

Use: For the cultivation and maintenance of *Escherichia coli, Micrococcus luteus, Shigella* species, and *Vibrio cholerae.*

Nutrient Agar with Erythromycin

Composition per liter:

Agar	15.0g
Peptone	5.0g
Beef extract	3.0g
Erythromycin solution	10.0mL

pH 6.8 ± 0.2 at 25°C.

Erythromycin Solution:

Composition per 10.0mL:

Erythromycin	0.6g

Preparation of Erythromycin Solution: Add erythromycin to distilled/deionized water and bring volume to 10.0mL. Mix thoroughly. Filter sterilize.

Preparation of Medium: Add components, except erythromycin solution, to distilled/deionized water and bring volume to 990.0mL. Mix thoroughly. Gently heat and bring to boiling. Autoclave for 15 min at 15 psi pressure–121°C. Cool to 45°–50°C. Aseptically add sterile erythromycin solution. Mix thoroughly. Pour into sterile Petri dishes or distribute into sterile tubes.

Use: For the cultivation and maintenance of *Micrococcus luteus.*

Nutrient Agar with Ethanolamine

Composition per liter:

Agar	15.0g
Pancreatic digest of gelatin	5.0g
Beef extract	3.0g
Ethanolamine	2.0mL

pH 6.8 ± 0.2 at 25°C

Source: Nutrient agar is available as a premixed powder from BD Diagnostic Systems.

Preparation of Medium: Add components to distilled/deionized water and bring volume to 1.0L. Mix thoroughly. Gently heat and bring to boiling. Distribute into tubes or flasks. Autoclave for 15 min at 15 psi pressure–121°C. Pour into sterile Petri dishes or leave in tubes.

Use: For the cultivation and maintenance of *Flavobacterium tirrenicum* and *Pseudomonas* species.

Nutrient Agar with Ethylene Glycol

Composition per liter:

Agar	15.0g
Pancreatic digest of gelatin	5.0g
Beef extract	3.0g
Ethylene glycol	2.0mL

pH 6.8 ± 0.2 at 25°C

Source: Nutrient agar is available as a premixed powder from BD Diagnostic Systems.

Preparation of Medium: Add components to distilled/deionized water and bring volume to 1.0L. Mix thoroughly. Gently heat and bring to boiling. Distribute into tubes or flasks. Autoclave for 15 min at 15 psi pressure–121°C. Pour into sterile Petri dishes or leave in tubes.

Use: For the cultivation and maintenance of *Pseudomonas putida.*

Nutrient Agar with Glucose

Composition per liter:

Agar	15.0g
Pancreatic digest of gelatin	5.0g

Beef extract 3.0g
Glucose 10.0g

pH 6.8 ± 0.2 at 25°C

Source: Nutrient agar is available as a premixed powder from BD Diagnostic Systems.

Preparation of Medium: Add components to distilled/deionized water and bring volume to 1.0L. Mix thoroughly. Gently heat and bring to boiling. Distribute into tubes or flasks. Autoclave for 15 min at 15 psi pressure–121°C. Pour into sterile Petri dishes or leave in tubes.

Use: For the cultivation and maintenance of *Amycolata saturnea, Arthrobacter* species, *Corynebacterium* species, *Curtobacterium flaccumfaciens, Deinococcus radiodurans, Escherichia coli, Hafnia alvei, Micrococcus aurantiacus, Myxomicrobium multiplex, Nocardia petroleophila, Nocardia* species, *Pseudomonas* species, *Rhodococcus rhodochrous, Streptomyces piedadensis,* and *Xanthomonas* species.

Nutrient Agar with Horse Serum (LMG Medium 38)

Composition per liter:

Agar 15.0g
Peptone 5.0g
NaCl 5.0g
Yeast extract 2.0g
Lab-Lemco beef extract 1.0g
Horse serum 100.0mL

pH 7.4 ± 0.2 at 25°C

Preparation of Medium: Add components, except horse serum, to distilled/deionized water and bring volume to 1.0L. Mix thoroughly. Gently heat and bring to boiling. Autoclave for 15 min at 15 psi pressure–121°C. Cool medium to 45–50°C. Aseptically add 100.0mL of sterile horse serum. Mix thoroughly. Pour into sterile Petri dishes or distribute into sterile tubes.

Use: For the cultivation of fastidious bacteria.

Nutrient Agar with Horse Serum

Composition per liter:

Agar 15.0g
Pancreatic digest of gelatin 5.0g
Beef extract 3.0g
Horse serum 100.0mL

pH 6.8 ± 0.2 at 25°C

Source: Nutrient agar is available as a premixed powder from BD Diagnostic Systems.

Preparation of Medium: Add components, except horse serum, to distilled/deionized water and bring volume to 900.0mL. Mix thoroughly. Gently heat and bring to boiling. Autoclave for 15 min at 15 psi pressure–121°C. Cool to 45°–50°C. Aseptically add sterile horse serum. Mix thoroughly. Pour into sterile Petri dishes or distribute into sterile tubes.

Use: For the cultivation and maintenance of *Alysiella filiformis* and *Simonsiella crassa.*

Nutrient Agar with 10% Horse Serum (DSMZ Medium 302)

Composition per liter:

Agar 15.0g
Pancreatic digest of gelatin 5.0g
Beef extract 3.0g
Horse serum 100.0mL

pH 7.0 ± 0.2 at 25°C

Preparation of Medium: Add components, except horse serum, to distilled/deionized water and bring volume to 900.0mL. Mix thoroughly. Gently heat and bring to boiling. Autoclave for 15 min at 15 psi pressure–121°C. Cool to 25°C. Aseptically add 100.0mL filter sterilized horse serum. Mix thoroughly. Pour into sterile Petri dishes or aseptically distrubute into tubes.

Use: For the isolation and cultivation of *Simonsiella crassa.*

Nutrient Agar with Manganese (NAMn)

Composition per liter:

Agar 15.0g
Pancreatic digest of gelatin 5.0g
Beef extract 3.0g
Manganese sulfate solution 1.0mL

pH 6.8 ± 0.2 at 25°C

Manganese Sulfate Solution:

Composition per 100.0mL:

$MnSO_4 \cdot H_2O$ 3.08g

Preparation of Manganese Sulfate Solution: Add $MnSO_4 \cdot H_2O$ to distilled/deionized water and bring volume to 100.0mL. Mix thoroughly.

Preparation of Medium: Add components to distilled/deionized water and bring volume to 1.0L. Mix thoroughly. Gently heat and bring to boiling. Distribute into tubes or flasks. Autoclave for 15 min at 15 psi pressure–121°C. Pour into sterile Petri dishes or leave in tubes.

Use: For the cultivation of *Bacillus* species from canned foods. For enhanced spore production by *Bacillus* species.

Nutrient Agar with Manganese Sulfate (ATCC Medium 1859)

Composition per liter:

Agar 15.0g
Pancreatic digest of gelatin 5.0g
Beef extract 3.0g
Manganese sulfate solution 1.0mL

pH 6.8 ± 0.2 at 25°C

Manganese Sulfate Solution:

Composition per 100.0mL:

$MnSO_4 \cdot H_2O$ 0.5g

Preparation of Manganese Sulfate Solution: Add $MnSO_4 \cdot H_2O$ to distilled/deionized water and bring volume to 100.0mL. Mix thoroughly.

Preparation of Medium: Add components to distilled/deionized water and bring volume to 1.0L. Mix thoroughly. Gently heat and bring to boiling. Distribute into tubes or flasks. Autoclave for 15 min at 15 psi pressure–121°C. Pour into sterile Petri dishes or leave in tubes.

Use: For the isolation and cultivation of *Paenibacillus chondroitinus* and *Paenibacillus alginolyticus.*

Nutrient Agar with 1% Methanol (ATCC Medium 620)

Composition per liter:

Agar 15.0g
Pancreatic digest of gelatin 5.0g

Beef extract ... 3.0g
Methanol ... 10.0mL

pH 6.8 ± 0.2 at 25°C

Source: Nutrient agar is available as a premixed powder from BD Diagnostic Systems.

Preparation of Medium: Filter sterilize methanol. Add components, except methanol, to distilled/deionized water and bring volume to 990.0mL. Mix thoroughly. Gently heat and bring to boiling. Autoclave for 15 min at 15 psi pressure–121°C. Cool to 45°–50°C. Aseptically add sterile methanol. Mix thoroughly. Pour into sterile Petri dishes or distribute into sterile tubes.

Use: For the cultivation and maintenance of *Bacillus* species, *Methylomonas clara*, and *Pseudomonas methanolica.*

Nutrient Agar with 2% Methanol (ATCC Medium 628)

Composition per liter:

Agar ... 15.0g
Pancreatic digest of gelatin ... 5.0g
Beef extract ... 3.0g
Methanol ... 20.0mL

pH 6.8 ± 0.2 at 25°C

Source: Nutrient agar is available as a premixed powder from BD Diagnostic Systems.

Preparation of Medium: Filter sterilize methanol. Add components, except methanol, to distilled/deionized water and bring volume to 980.0mL. Mix thoroughly. Gently heat and bring to boiling. Autoclave for 15 min at 15 psi pressure–121°C. Cool to 45°–50°C. Aseptically add sterile methanol. Mix thoroughly. Pour into sterile Petri dishes or distribute into sterile tubes.

Use: For the cultivation and maintenance of *Pseudomonas* species.

Nutrient Agar with 0.5% Sodium Chloride

Composition per liter:

Agar ... 15.0g
NaCl ... 5.0g
Pancreatic digest of gelatin ... 5.0g
Beef extract ... 3.0g

pH 6.8 ± 0.2 at 25°C

Source: Nutrient agar is available as a premixed powder from BD Diagnostic Systems.

Preparation of Medium: Add components to distilled/deionized water and bring volume to 1.0L. Mix thoroughly. Gently heat and bring to boiling. Distribute into tubes or flasks. Autoclave for 15 min at 15 psi pressure–121°C. Pour into sterile Petri dishes or leave in tubes.

Use: For the cultivation and maintenance of *Agrobacterium tumefaciens*, *Escherichia coli*, *Pseudomonas aeruginosa*, *Salmonella choleraesuis*, *Shigella dysenteriae*, *Shigella flexneri*, *Vibrio* species, and *Yersinia* species.

Nutrient Agar with 0.5% Sodium Chloride, pH 9.5-10.0 (LMG Medium 253)

Composition per liter:

Agar ... 15.0g
NaCl ... 10.0g
Peptone ... 5.0g
Yeast extract ... 2.0g
Lab-Lemco beef extract ... 1.0g

pH 9.5–10.0

Preparation of Medium: Add components to distilled/deionized water and bring volume to 1.0L. Mix thoroughly. Gently heat and bring to boiling. Distribute into tubes or flasks. Autoclave for 15 min at 15 psi pressure–121°C. Cool to 45°-50°C. Adjust pH to 9.5-10.0 with sterile Na_2CO_3 solution. Pour into sterile Petri dishes or leave in tubes.

Use: For the cultivation and maintenance of *Bacillus* spp.

Nutrient Agar with 1.5% Sodium Chloride

Composition per liter:

NaCl ... 15.0g
Agar ... 15.0g
Pancreatic digest of gelatin ... 5.0g
Beef extract ... 3.0g

pH 6.8 ± 0.2 at 25°C

Source: Nutrient agar is available as a premixed powder from BD Diagnostic Systems.

Preparation of Medium: Add components to distilled/deionized water and bring volume to 1.0L. Mix thoroughly. Gently heat and bring to boiling. Distribute into tubes or flasks. Autoclave for 15 min at 15 psi pressure–121°C. Pour into sterile Petri dishes or leave in tubes.

Use: For the cultivation and maintenance of *Photobacterium leiognathi, Pseudomonas fluorescens,* and *Vibrio natriegens.*

Nutrient Agar with 2% Sodium Chloride

Composition per liter:

NaCl ... 20.0g
Agar ... 15.0g
Pancreatic digest of gelatin ... 5.0g
Beef extract ... 3.0g

pH 6.8 ± 0.2 at 25°C

Preparation of Medium: Add components to distilled/deionized water and bring volume to 1.0L. Mix thoroughly. Gently heat and bring to boiling. Distribute into tubes or flasks. Autoclave for 15 min at 15 psi pressure–121°C. Pour into sterile Petri dishes or leave in tubes.

Use: For the cultivation and maintenance of *Vibrio alginolyticus, Vibrio bivalvii, Vibrio mediterranei, Vibrio natriegens, Vibrio ordali, Vibrio orientalis, Vibrio parahaemolyticus, Vibrio pelagigius,* and *Vibrio vulnificus.*

Nutrient Agar with 3% Sodium Chloride (LMG Medium 257)

Composition per liter:

NaCl ... 35.0g
Agar ... 15.0g
Peptone ... 5.0g
Yeast extract ... 2.0g
Lab-Lemco beef extract ... 1.0g

pH 7.4 ± 0.2 at 25°C

Preparation of Medium: Add components to distilled/deionized water and bring volume to 1.0L. Mix thoroughly. Gently heat and bring to boiling. Distribute into tubes or flasks. Autoclave for 15 min at 15 psi pressure–121°C. Cool to 45°–50°C. Adjust pH to 9.5–10.0 with sterile Na_2CO_3 solution. Pour into sterile Petri dishes or leave in tubes.

Use: For the cultivation and maintenance of *Bacillus cirroflagellosus.*

Nutrient Agar with 3% Sodium Chloride

Composition per liter:

NaCl	30.0g
Agar	15.0g
Pancreatic digest of gelatin	5.0g
Beef extract	3.0g

pH 6.8 ± 0.2 at 25°C

Source: Nutrient agar is available as a premixed powder from BD Diagnostic Systems.

Preparation of Medium: Add components to distilled/deionized water and bring volume to 1.0L. Mix thoroughly. Gently heat and bring to boiling. Distribute into tubes or flasks. Autoclave for 15 min at 15 psi pressure–121°C. Pour into sterile Petri dishes or leave in tubes.

Use: For the cultivation and maintenance of *Bacillus* species, *Alteromonas nigrifaciens*, *Halococcus* species, *Planococcus citreus*, *Pseudomonas beijerinckii*, *Staphylococcus* species, *Streptococcus pyogenes*, and *Vibrio* species.

Nutrient Agar with 5% Sodium Chloride, pH 9.5–10.0 (LMG Medium 254)

Composition per liter:

NaCl	55.0g
Agar	15.0g
Peptone	5.0g
Yeast extract	2.0g
Lab-Lemco beef extract	1.0g

pH 9.5–10.0

Preparation of Medium: Add components to distilled/deionized water and bring volume to 1.0L. Mix thoroughly. Gently heat and bring to boiling. Distribute into tubes or flasks. Autoclave for 15 min at 15 psi pressure–121°C. Cool to 45°–50°C. Adjust pH to 9.5–10.0 with sterile Na_2CO_3 solution. Pour into sterile Petri dishes or leave in tubes.

Use: For the cultivation and maintenance of *Bacillus haloalkaliphilus*.

Nutrient Agar with 10% Sodium Chloride

Composition per liter:

NaCl	100.0g
Agar	15.0g
Pancreatic digest of gelatin	5.0g
Beef extract	3.0g

pH 6.8 ± 0.2 at 25°C

Source: Nutrient agar is available as a premixed powder from BD Diagnostic Systems.

Preparation of Medium: Add components to distilled/deionized water and bring volume to 1.0L. Mix thoroughly. Gently heat and bring to boiling. Distribute into tubes or flasks. Autoclave for 15 min at 15 psi pressure–121°C. Pour into sterile Petri dishes or leave in tubes.

Use: For the cultivation and maintenance of *Paracoccus halodenitrificans* and *Micrococcus* species.

Nutrient Agar with 10% Sodium Chloride and Maltose

Composition per liter:

NaCl	100.0g
Agar	15.0g
Maltose	10.0g
Pancreatic digest of gelatin	5.0g
Beef extract	3.0g

pH 6.8 ± 0.2 at 25°C

Source: Nutrient agar is available as a premixed powder from BD Diagnostic Systems.

Preparation of Medium: Add components to distilled/deionized water and bring volume to 1.0L. Mix thoroughly. Gently heat and bring to boiling. Distribute into tubes or flasks. Autoclave for 15 min at 15 psi pressure–121°C. Pour into sterile Petri dishes or leave in tubes.

Use: For the cultivation and maintenance of *Paracoccus halodenitrificans* and *Micrococcus* species.

Nutrient Agar with 0.5% Sodium Chloride and Sodium Citrate

Composition per liter:

Agar	15.0g
NaCl	5.0g
Pancreatic digest of gelatin	5.0g
Beef extract	3.0g
Sodium citrate	2.94g

pH 6.8 ± 0.2 at 25°C

Source: Nutrient agar is available as a premixed powder from BD Diagnostic Systems.

Preparation of Medium: Add components to distilled/deionized water and bring volume to 1.0L. Mix thoroughly. Gently heat and bring to boiling. Distribute into tubes or flasks. Autoclave for 15 min at 15 psi pressure–121°C. Pour into sterile Petri dishes or leave in tubes.

Use: For the cultivation and maintenance of *Escherichia coli*.

Nutrient Agar with Nalidixic Acid (ATCC Medium 2089)

Composition per liter:

Agar	15.0g
Pancreatic digest of gelatin	5.0g
NaCl	5.0g
Beef extract	3.0g
Nalidixic acid solution	10.0mL

pH 6.8 ± 0.2 at 25°C

Nalidixic Acid Solution:

Composition per 10.0mL:

Nalidixic acid	0.1g

Preparation of Nalidixic Acid Solution: Add nalidixic acid to distilled/deionized water and bring volume to 10.0mL. Mix thoroughly. Filter sterilize.

Preparation of Medium: Add components, except nalidixic acid solution, to distilled/deionized water and bring volume to 990.0mL. Mix thoroughly. Gently heat while stirring and bring to boiling. Autoclave for 15 min at 15 psi pressure–121°C. Cool to 45°–50°C. Aseptically add sterile nalidixic acid solution. Mix thoroughly. Pour into sterile Petri dishes or distribute into sterile tubes.

Use: For the cultivation of nalidixic acid-resistant bacteria.

Nutrient Agar with Phosphate

See: **Nutrient Agar, Buffered**

Nutrient Agar with Phytone™

Composition per liter:

Agar	15.0g
Phytone™	10.0g
Pancreatic digest of gelatin	5.0g
Beef extract	3.0g

pH 6.8 ± 0.2 at 25°C

Preparation of Medium: Add components to distilled/deionized water and bring volume to 1.0L. Mix thoroughly. Gently heat while stirring and bring to boiling. Distribute into tubes or flasks. Autoclave for 15 min at 15 psi pressure–121°C. Pour into sterile Petri dishes or leave in tubes.

Use: For the cultivation of a wide variety of bacteria.

Nutrient Agar with Potato Starch

Composition per liter:

Agar	15.0g
Potato starch	10.0g
Pancreatic digest of gelatin	5.0g
Beef extract	3.0g

pH 6.8 ± 0.2 at 25°C

Source: Nutrient agar is available as a premixed powder from BD Diagnostic Systems.

Preparation of Medium: Add components to distilled/deionized water and bring volume to 1.0L. Mix thoroughly. Gently heat while stirring and bring to boiling. Distribute into tubes or flasks. Autoclave for 15 min at 15 psi pressure–121°C. Pour into sterile Petri dishes or leave in tubes.

Use: For the cultivation and maintenance of *Bacillus polymyxa* and *Bacillus subtilis.*

Nutrient Agar with Soil Extract

Composition per liter:

Agar	15.0g
Pancreatic digest of gelatin	5.0g
Beef extract	3.0g
Soil extract	250.0mL

pH 6.8 ± 0.2 at 25°C

Source: Nutrient agar is available as a premixed powder from BD Diagnostic Systems.

Soil Extract:
Composition per 300.0mL:

African Violet soil	115.5g
Na_2CO_3	0.3g

Preparation of Soil Extract: Add components to tap water and bring volume to 300.0mL. Autoclave for 60 min at 15 psi pressure–121°C. Filter through Whatman filter paper.

Preparation of Medium: Add components to distilled/deionized water and bring volume to 1.0L. Mix thoroughly. Gently heat and bring to boiling. Distribute into tubes or flasks. Autoclave for 15 min at 15 psi pressure–121°C. Pour into sterile Petri dishes.

Use: For the cultivation and maintenance of *Auerobacterium* species, *Bacillus* species, and *Saccharomonospora viridis.*

Nutrient Agar with Streptomycin

Composition per liter:

Agar	15.0g
Peptone	5.0g
Beef extract	3.0g
Streptomycin solution	10.0mL

pH 6.8 ± 0.2 at 25°C

Streptomycin Solution:
Composition per 10.0mL:

Streptomycin	0.02g

Preparation of Streptomycin Solution: Add streptomycin to distilled/deionized water and bring volume to 10.0mL. Mix thoroughly. Filter sterilize.

Preparation of Medium: Add components, except streptomycin solution, to distilled/deionized water and bring volume to 990.0mL. Mix thoroughly. Gently heat and bring to boiling. Autoclave for 15 min at 15 psi pressure–121°C. Cool to 45°–50°C. Aseptically add sterile streptomycin solution. Mix thoroughly. Pour into sterile Petri dishes or distribute into sterile tubes.

Use: For the cultivation and maintenance of *Corynebacterium glutamicum* and *Corynebacterium herculis.*

Nutrient Agar with Sucrose

Composition per liter:

Sucrose	20.0g
Agar	15.0g
Pancreatic digest of gelatin	5.0g
Beef extract	3.0g

pH 6.8 ± 0.2 at 25°C

Source: Nutrient agar is available as a premixed powder from BD Diagnostic Systems.

Preparation of Medium: Add components to distilled/deionized water and bring volume to 1.0L. Mix thoroughly. Gently heat and bring to boiling. Distribute into tubes or flasks. Autoclave for 15 min at 15 psi pressure–121°C. Pour into sterile Petri dishes or leave in tubes.

Use: For the cultivation and maintenance of *Pseudomonas* species.

Nutrient Agar with 2% Sucrose (ATCC Medium 1297)

Composition per liter:

Sucrose	20.0 g
Agar	15.0g
Pancreatic digest of gelatin	5.0g
Beef extract	3.0g

pH 6.8 ± 0.2 at 25°C

Preparation of Medium: Add components to distilled/deionized water and bring volume to 1.0L. Mix thoroughly. Gently heat and bring to boiling. Distribute into tubes or flasks. Autoclave for 15 min at 15 psi pressure–121°C. Pour into sterile Petri dishes or leave in tubes.

Use: For the cultivation of osmophilic bacteria

Nutrient Agar with Tetracycline

Composition per liter:

Agar	15.0g
Pancreatic digest of gelatin	5.0g
Beef extract	3.0g
Tetracycline solution	10.0mL

pH 6.8 ± 0.2 at 25°C

Source: Nutrient agar is available as a premixed powder from BD Diagnostic Systems.

Tetracycline Solution:

Composition per 10.0mL:

Tetracycline 0.025g

Preparation of Tetracycline Solution: Add tetracycline to distilled/deionized water and bring volume to 10.0mL. Mix thoroughly. Filter sterilize.

Preparation of Medium: Add components, except tetracycline solution, to distilled/deionized water and bring volume to 990.0mL. Mix thoroughly. Gently heat and bring to boiling. Autoclave for 15 min at 15 psi pressure–121°C. Cool to 45°–50°C. Aseptically add sterile tetracycline solution. Mix thoroughly. Pour into sterile Petri dishes or distribute into sterile tubes.

Use: For the cultivation and maintenance of *Salmonella choleraesuis.*

Nutrient Agar with Uracil

Composition per liter:

Agar 15.0g
Pancreatic digest of gelatin 5.0g
Beef extract 3.0g
Uracil 0.01g

pH 6.8 ± 0.2 at 25°C

Source: Nutrient agar is available as a premixed powder from BD Diagnostic Systems.

Preparation of Medium: Add components to distilled/deionized water and bring volume to 1.0L. Mix thoroughly. Gently heat and bring to boiling. Distribute into tubes or flasks. Autoclave for 15 min at 15 psi pressure–121°C. Pour into sterile Petri dishes or leave in tubes.

Use: For the cultivation and maintenance of *Escherichia coli.*

Nutrient Agar with 5% Urea

Composition per liter:

Agar 15.0g
Pancreatic digest of gelatin 5.0g
Beef extract 3.0g
Urea solution 100.0mL

pH 6.8 ± 0.2 at 25°C

Urea Solution:

Composition per 100.0mL:

Urea 50.0g

Preparation of Urea Solution: Add urea to distilled/deionized water and bring volume to 100.0mL. Mix thoroughly. Filter sterilize.

Preparation of Medium: Add components, except urea solution, to distilled/deionized water and bring volume to 900.0mL. Mix thoroughly. Gently heat and bring to boiling. Autoclave for 15 min at 15 psi pressure–121°C. Cool to 45°–50°C. Aseptically add 100.0mL of sterile urea solution. Mix thoroughly. Pour into sterile Petri dishes or aseptically distribute into sterile tubes.

Use: For the cultivation and maintenance of *Aspergillus* species.

Nutrient Agar with Ureum (LMG Medium 139)

Composition per liter:

Ureum 20.0g
Agar 15.0g
Peptone 5.0g
NaCl 5.0g
Yeast extract 2.0g
Lab-Lemco beef extract 1.0g

pH 7.4 ± 0.2 at 25°C

Preparation of Medium: Add components to distilled/deionized water and bring volume to 1.0L. Mix thoroughly. Gently heat and bring to boiling. Distribute into tubes or flasks. Autoclave for 15 min at 15 psi pressure–121°C. Pour into sterile Petri dishes or leave in tubes.

Use: For the cultivation and maintenance of *Sporosarcina pasteurii.*

Nutrient Agar with V-8™ Juice

Composition per liter:

Agar 15.0g
Pancreatic digest of gelatin 5.0g
Beef extract 3.0g
V-8™ juice 200.0mL

pH 6.8 ± 0.2 at 25°C

Source: Nutrient agar is available as a premixed powder from BD Diagnostic Systems.

Preparation of Medium: Add components to distilled/deionized water and bring volume to 1.0L. Mix thoroughly. Gently heat and bring to boiling. Distribute into tubes or flasks. Autoclave for 15 min at 15 psi pressure–121°C. Pour into sterile Petri dishes or leave in tubes.

Use: For the cultivation and maintenance of *Pseudomonas tolaasii.*

Nutrient Agar with Yeast Extract

Composition per liter:

Yeast extract 20.0g
Agar 15.0g
Pancreatic digest of gelatin 5.0g
Beef extract 3.0g

pH 6.8 ± 0.2 at 25°C

Source: Nutrient agar is available as a premixed powder from BD Diagnostic Systems.

Preparation of Medium: Add components to distilled/deionized water and bring volume to 1.0L. Mix thoroughly. Gently heat and bring to boiling. Distribute into tubes or flasks. Autoclave for 15 min at 15 psi pressure–121°C. Pour into sterile Petri dishes or leave in tubes.

Use: For the cultivation and maintenance of *Bacillus anthracis* and *Comamonas testosteroni.*

Nutrient Broth

Composition per liter:

Peptone 5.0g
NaCl 5.0g
Yeast extract 2.0g
Beef extract 1.0g

pH 7.4 ± 0.2 at 25°C

Source: This medium is available as a premixed powder from Oxoid Unipath.

Preparation of Medium: Add components to distilled/deionized water and bring volume to 1.0L. Mix thoroughly. Distribute into tubes or flasks. Autoclave for 15 min at 15 psi pressure–121°C.

Use: For the cultivation of a wide variety of nonfastidious microorganisms.

Nutrient Broth (DSMZ Medium 948)

Composition per liter:

Peptone	5.0g
NaCl	5.0g
Yeast extract	2.0g
Lab-Lemco beef extract	1.0g

pH 7.2 ± 0.2 at 25°C

Preparation of Medium: Add components to distilled/deionized water and bring volume to 1.0L. Mix thoroughly. Distribute into tubes or flasks. Autoclave for 15 min at 15 psi pressure–121°C.

Use: For the cultivation of *Psychrobacter urativorans.*

Nutrient Broth

Composition per liter:

Pancreatic digest of gelatin	5.0g
Beef extract	3.0g

pH 6.9 ± 0.2 at 25°C

Source: This medium is available as a premixed powder from BD Diagnostic Systems.

Preparation of Medium: Add components to distilled/deionized water and bring volume to 1.0L. Mix thoroughly. Distribute into tubes or flasks. Autoclave for 15 min at 15 psi pressure–121°C.

Use: For the cultivation of a wide variety of nonfastidious microorganisms.

Nutrient Broth with Bovine Serum

Composition per liter:

NaCl	10.0g
Pancreatic digest of gelatin	5.0g
Beef extract	3.0g
Bovine serum	100.0mL

pH 6.9 ± 0.2 at 25°C

Preparation of Medium: Add components, except bovine serum, to distilled/deionized water and bring volume to 900.0mL. Mix thoroughly. Autoclave for 15 min at 15 psi pressure–121°C. Cool to 45°–50°C. Aseptically add sterile bovine serum. Mix thoroughly. Aseptically distribute into sterile tubes or flasks.

Use: For the cultivation and maintenance of *Pseudomonas anguilliseptica.*

Nutrient Broth, Diluted 1:100

Nutrient Broth:

Composition per liter:

Pancreatic digest of gelatin	5.0g
Beef extract	3.0g

pH 6.8 ± 0.2 at 25°C

Preparation of Nutrient Broth: Add components to distilled/deionized water and bring volume to 1.0L. Mix thoroughly.

Preparation of Medium: Add 10.0mL of nutrient broth to distilled/deionized water and bring volume to 1.0L. Mix thoroughly. Distribute into tubes or flasks. Autoclave for 15 min at 15 psi pressure–121°C.

Use: For the cultivation and maintenance of *Agromonas oligotrophica.*

Nutrient Broth Glycerol Medium

Composition per liter:

Glycerol	100.0g
Pancreatic digest of gelatin	5.0g
Beef extract	3.0g

pH 7.2–7.4 ± 0.2 at 25°C

Preparation of Medium: Add components to distilled/deionized water and bring volume to 1.0L. Mix thoroughly. Distribute into tubes or flasks. Autoclave for 15 min at 15 psi pressure–121°C.

Use: For the cultivation of *Pseudomonas* species.

Nutrient Broth with Manganese Sulfate

Composition per liter:

$MnSO_4$	5.0g
Pancreatic digest of gelatin	5.0g
Beef extract	3.0g

pH 6.9 ± 0.2 at 25°C

Preparation of Medium: Add components to distilled/deionized water and bring volume to 1.0L. Mix thoroughly. Distribute into tubes or flasks. Autoclave for 15 min at 15 psi pressure–121°C.

Use: For the cultivation of *Bacillus chondroitinus* and *Bacillus alginolyticus.*

Nutrient Broth with Rifampicin

Composition per liter:

Peptone	5.0g
NaCl	5.0g
Yeast extract	2.0g
Beef extract	1.0g
Rifampicin solution	10.0mL

pH 7.4 ± 0.2 at 25°C

Rifampicin Solution:

Composition per 10.0mL:

Rifampicin	10.0mg

Preparation of Rifampicin Solution: Add rifampicin to distilled/deionized water and bring volume to 10.0mL. Mix thoroughly. Filter sterilize.

Preparation of Medium: Add components, except rifampicin solution, to distilled/deionized water and bring volume to 990.0mL. Mix thoroughly. Adjust pH to 7.4. Autoclave for 15 min at 15 psi pressure–121°C. Cool to room temperature. Aseptically add 10.0mL of sterile rifampicin solution. Mix thoroughly. Aseptically distribute into sterile tubes or flasks.

Use: For the cultivation of *Agrobacterium tumefaciens.*

Nutrient Broth with 6% Sodium Chloride

Composition per liter:

NaCl	60.0g
Pancreatic digest of gelatin	5.0g
Beef extract	3.0g

pH 6.8 ± 0.2 at 25°C

Preparation of Medium: Add components to distilled/deionized water and bring volume to 1.0L. Mix thoroughly. Distribute into tubes or flasks. Autoclave for 15 min at 15 psi pressure–121°C.

Use: For the cultivation of organisms in water, sewage, feces, and other materials. For the cultivation and maintenance of *Paracoccus halodenitrificans.*

Nutrient Broth Sodium Chloride Thymine Medium

Composition per liter:

Pancreatic digest of gelatin	5.0g
NaCl	5.0g
Beef extract	3.0g
Thymine	0.03g

pH 7.2 ± 0.2 at 25°C

Preparation of Medium: Add components to distilled/deionized water and bring volume to 1.0L. Mix thoroughly. Distribute into tubes or flasks. Autoclave for 15 min at 15 psi pressure–121°C.

Use: For the cultivation and maintenance of *Salmonella choleraesuis*.

Nutrient Broth Salts Medium

Composition per liter:

Pancreatic digest of gelatin	5.0g
Beef extract	3.0g
NaCl	3.0g
$MgSO_4 \cdot 7H_2O$	0.2g
$CaCl_2 \cdot 2H_2O$	0.15g
$MnSO_4 \cdot H_2O$	0.05g

pH 6.8 ± 0.2 at 25°C

Preparation of Medium: Add components to distilled/deionized water and bring volume to 1.0L. Mix thoroughly. Distribute into tubes or flasks. Autoclave for 15 min at 15 psi pressure–121°C.

Use: For the cultivation and maintenance of *Bacillus subtilis*.

Nutrient Broth, Standard II

Composition per liter:

Special peptone	8.6g
NaCl	6.4g

pH 7.5 ± 0.1 at 37°C

Preparation of Medium: Add components to distilled/deionized water and bring volume to 1.0L. Mix thoroughly. Distribute into tubes or flasks. Autoclave for 15 min at 15 psi pressure–121°C.

Use: For the cultivation of a variety of fastidious and nonfastidious microorganisms.

Nutrient Broth, 1/2 Strength

Composition per liter:

Pancreatic digest of gelatin	2.5g
Beef extract	1.5g

pH 6.9 ± 0.2 at 25°C

Preparation of Medium: Add components to distilled/deionized water and bring volume to 1.0L. Mix thoroughly. Distribute into tubes or flasks. Autoclave for 15 min at 15 psi pressure–121°C.

Use: For the cultivation of *Cytophaga allerginae, Cytophaga allerginae*, and *Thermomonospora chromogena*.

Nutrient Broth with Streptomycin

Composition per liter:

Peptone	5.0g
Beef extract	3.0g
Streptomycin solution	10.0mL

pH 6.8 ± 0.2 at 25°C

Streptomycin Solution:

Composition per 10.0mL:

Streptomycin	0.02g

Preparation of Streptomycin Solution: Add streptomycin to distilled/deionized water and bring volume to 10.0mL. Mix thoroughly. Filter sterilize.

Preparation of Medium: Add components, except streptomycin solution, to distilled/deionized water and bring volume to 990.0mL. Mix thoroughly. Gently heat and bring to boiling. Autoclave for 15 min at 15 psi pressure–121°C. Cool to 45°–50°C. Aseptically add sterile streptomycin solution. Mix thoroughly. Pour into sterile Petri dishes or distribute into sterile tubes.

Use: For the cultivation of *Corynebacterium glutamicum* and *Corynebacterium herculis*.

Nutrient Broth No. 2

Composition per liter:

Beef extract	10.0g
Peptone	10.0g
NaCl	5.0g

pH 7.5 ± 0.2 at 25°C

Source: This medium is available as a premixed powder from Oxoid Unipath.

Preparation of Medium: Add components to distilled/deionized water and bring volume to 1.0L. Mix thoroughly. Distribute into tubes or flasks. Autoclave for 15 min at 15 psi pressure–121°C.

Use: For the cultivation of a variety of fastidious and nonfastidious microorganisms.

Nutrient Broth Yeast Extract Medium

See: **NBY Medium**

Nutrient Broth Yeast Extract Medium

See: **NY Medium**

Nutrient Gelatin

Composition per liter:

Gelatin	120.0g
Pancreatic digest of gelatin	5.0g
Beef extract	3.0g

pH 6.8 ± 0.2 at 25°C

Source: This medium is available as a premixed powder from BD Diagnostic Systems and Oxoid Unipath.

Preparation of Medium: Add components to distilled/deionized water and bring volume to 1.0L. Mix thoroughly. Gently heat while stirring to 50°C. Distribute into tubes. Autoclave for 15 min at 15 psi pressure–121°C.

Use: For the cultivation and differentiation of bacteria based on their ability to liquefy gelatin.

Nutrient Gelatin

Composition per liter:

Beef heart, solids from infusion	500.0g
Gelatin	120.0g
Tryptose	10.0g
NaCl	5.0g

pH 7.4 ± 0.2 at 25°C

Preparation of Medium: Add components to distilled/deionized water and bring volume to 1.0L. Mix thoroughly. Gently heat and bring

to boiling. Distribute into screw-capped tubes in 4.0mL volumes. Autoclave for 15 min at 15 psi pressure–121°C.

Use: For the cultivation of Gram-negative, nonfermentative bacteria from foods.

Nutrient Gelatin, CDC (BAM M115)

Composition per liter:

Gelatin	120.0g
Pancreatic digest of casein	13.0g
NaCl	5.0g
Yeast extract	5.0g
Heart muscle, solids from infusion	2.0g

pH 7.4 ± 0.2 at 25°C

Preparation of Medium: Add components to distilled/deionized water and bring volume to 1.0L. Mix thoroughly. Gently heat and bring to boiling. Cool to 55°C. Adjust pH to 7.4. Distribute into screw-capped tubes in 4.0mL volumes. Autoclave for 15 min at 15 psi pressure–121°C.

Use: For the cultivation of Gram-negative, nonfermentative bacteria from foods.

Nutrient Glucose Agar (LMG Medium 51)

Composition per liter:

Agar	15.0g
Glucose	10.0g
Peptone	5.0g
NaCl	5.0g
Yeast extract	2.0g
Lab-Lemco beef extract	1.0g

pH 7.4 ± 0.2 at 25°C

Preparation of Medium: Add components to distilled/deionized water and bring volume to 1.0L. Mix thoroughly. Gently heat and bring to boiling. Distribute into tubes or flasks. Autoclave for 15 min at 15 psi pressure–121°C. Pour into sterile Petri dishes or leave in tubes.

Use: For the cultivation and maintenance of *Paenibacillus* spp., *Pseudomonas amygdali, Bacillus circulans, Brevibacillus laterosporus, Brevibacillus invocatus,* and other bacteria.

Nutrient HiVeg Agar

Composition per liter:

Agar	15.0g
Plant peptone	5.0g
NaCl	5.0g
Plant extract	1.5g
Yeast extract	1.5g

pH 7.4 ± 0.2 at 25°C

Source: This medium is available as a premixed powder from HiMedia.

Preparation of Medium: Add components to distilled/deionized water and bring volume to 1.0L. Mix thoroughly. Gently heat and bring to boiling. Distribute into tubes or flasks. Autoclave for 15 min at 15 psi pressure–121°C. Pour into sterile Petri dishes or leave in tubes.

Use: For the cultivation and maintenance of a wide variety of microorganisms. This medium may be used for blood culture work with the addition of 5–10% sterile defibrinated blood.

Nutrient HiVeg Agar, 1.5%

Composition per liter:

Agar	15.0g
Plant peptone	5.0g
NaCl	5.0g
Plant extract	3.0g

pH 7.4 ± 0.2 at 25°C

Source: This medium is available as a premixed powder from HiMedia.

Preparation of Medium: Add components to distilled/deionized water and bring volume to 1.0L. Mix thoroughly. Gently heat and bring to boiling. Distribute into tubes or flasks. Autoclave for 15 min at 15 psi pressure–121°C. Pour into sterile Petri dishes or leave in tubes.This medium may be used for blood culture work with the addition of 5–10% sterile defibrinated blood.

Use: For the cultivation and maintenance of a wide variety of microorganisms.

Preparation of Medium: Add components to distilled/deionized water and bring volume to 1.0L. Mix thoroughly. Gently heat and bring to boiling. Distribute into tubes. Autoclave for 15 min at 15 psi pressure–121°C.

Use: For the cultivation of a variety of microorganisms.

Nutrient HiVeg Agar pH 6.0 with 0.8% Sodium Chloride

Composition per liter:

Agar	15.0g
NaCl	8.0g
Plant peptone	5.0g
Plant extract	3.0g

pH 6.0 ± 0.2 at 25°C

Source: This medium is available as a premixed powder from HiMedia.

Preparation of Medium: Add components to distilled/deionized water and bring volume to 1.0L. Mix thoroughly. Gently heat and bring to boiling. Distribute into tubes or flasks. Autoclave for 15 min at 15 psi pressure–121°C. Pour into sterile Petri dishes or leave in tubes.

Use: For the cultivation and maintenance of a wide variety of microorganisms. For the microbiological analysis of water and cultivation of bacteria requiring slightly acidic pH.

Nutrient HiVeg Agar with Manganese

Composition per liter:

Agar	15.0g
Plant peptone No. 2	5.0g
Plant extract	3.0g
$MnSO_4$	0.03g

pH 6.8 ± 0.2 at 25°C

Source: This medium is available as a premixed powder from HiMedia.

Preparation of Medium: Add components to distilled/deionized water and bring volume to 1.0L. Mix thoroughly. Gently heat and bring to boiling. Distribute into tubes or flasks. Autoclave for 15 min at 15 psi pressure–121°C. Pour into sterile Petri dishes or leave in tubes.

Use: For the cultivation of *Bacillus* species from canned foods. For enhanced spore production by *Bacillus* species.

Nutrient HiVeg Agar for Oxidase

Composition per liter:

Agar	15.0g
NaCl	5.0g
Plant extract no. 1	1.0g
Plant peptone	1.0g

pH 7.2 ± 0.2 at 25°C

Source: This medium is available as a premixed powder from HiMedia.

Preparation of Medium: Add components to distilled/deionized water and bring volume to 1.0L. Mix thoroughly. Gently heat and bring to boiling. Distribute into tubes or flasks. Autoclave for 15 min at 15 psi pressure–121°C. Pour into sterile Petri dishes or leave in tubes.

Use: For the confirmation of presence of oxidase in microorganisms in water.

Nutrient HiVeg Agar with 1% Peptone

Composition per liter:

Agar	15.0g
Plant peptone	10.0g
NaCl	5.0g
Plant extract	5.0g

pH 7.4 ± 0.2 at 25°C

Source: This medium is available as a premixed powder from HiMedia.

Preparation of Medium: Add components to distilled/deionized water and bring volume to 1.0L. Mix thoroughly. Gently heat and bring to boiling. Distribute into tubes or flasks. Autoclave for 15 min at 15 psi pressure–121°C. Pour into sterile Petri dishes or leave in tubes.This medium may be used for blood culture work with the addition of 5–10% sterile defibrinated blood.

Use: For the cultivation and maintenance of a wide variety of microorganisms.

Nutrient HiVeg Agar No. 2

Composition per liter:

Agar	15.0g
Plant peptone	10.0g
Plant extract	10.0g
NaCl	5.0g

pH 7.2 ± 0.2 at 25°C

Source: This medium is available as a premixed powder from HiMedia.

Preparation of Medium: Add components to distilled/deionized water and bring volume to 1.0L. Mix thoroughly. Gently heat and bring to boiling. Distribute into tubes or flasks. Autoclave for 15 min at 15 psi pressure–121°C. Pour into sterile Petri dishes or leave in tubes.

Use: For the cultivation and maintenance of a wide variety of microorganisms.

Nutrient HiVeg Broth

Composition per liter:

Plant peptone	5.0g
NaCl	5.0g
Plant extract	1.5g
Yeast extract	1.5g

pH 7.4 ± 0.2 at 25°C

Source: This medium is available as a premixed powder from HiMedia.

Preparation of Medium: Add components to distilled/deionized water and bring volume to 1.0L. Mix thoroughly. Gently heat and bring to boiling. Distribute into tubes or flasks. Autoclave for 15 min at 15 psi pressure–121°C.

Use: For the cultivation and maintenance of a wide variety of microorganisms.

Nutrient HiVeg Broth with 1% Peptone

Composition per liter:

Plant extract	10.0g
Plant peptone	10.0g
NaCl	5.0g

pH 7.2 ± 0.2 at 25°C

Source: This medium is available as a premixed powder from HiMedia.

Preparation of Medium: Add components to distilled/deionized water and bring volume to 1.0L. Mix thoroughly. Gently heat and bring to boiling. Distribute into tubes or flasks. Autoclave for 15 min at 15 psi pressure–121°C.

Use: For the cultivation of a wide variety of microorganisms.

Nutrient HiVeg Broth, pH 6.9 without Sodium Chloride

Composition per liter:

Plant peptone	5.0g
Plant extract	3.0g

pH 6.9 ± 0.2 at 25°C

Source: This medium is available as a premixed powder from HiMedia.

Preparation of Medium: Add components to distilled/deionized water and bring volume to 1.0L. Mix thoroughly. Gently heat and bring to boiling. Distribute into tubes or flasks. Autoclave for 15 min at 15 psi pressure–121°C.

Use: For the cultivation of a wide variety of microorganisms.

Nutrient Soil Extract Agar (LMG Medium 30)

Composition per liter:

Agar	15.0g
Yeast extract	2.0g
Peptone	5.0g
NaCl	5.0g
Lab-Lemco beef extract	1.0g
Soil extract	1.0L

pH 7.4 ± 0.2 at 25°C

Soil Extract:

Composition per liter:

Soil	1.0kg
$CaCO_3$	2.0g

Preparation of Soil Extract: Add 1.0kg soil to 1.0L tap water. Autoclave for 30 min at 15 psi pressure–121°C. Add 2.0g $CaCO_3$. Filter through Whatman filter paper. Bring volume to 1.0L with tap water. Mix thoroughly.

Preparation of Medium: Add components to 1.0L soil extract. Mix thoroughly. Gently heat and bring to boiling. Distribute into tubes or flasks. Autoclave for 15 min at 15 psi pressure–121°C. Pour into sterile Petri dishes or leave in tubes.

Use: For the cultivation and maintenance of soil bacteria.

Nutrient Yeast Glucose Medium

Composition per liter:

Glucose	10.0g
Pancreatic digest of gelatin	5.0g
Yeast extract	5.0g
Beef extract	3.0g

pH 6.8 ± 0.2 at 25°C

Preparation of Medium: Add components to distilled/deionized water and bring volume to 1.0L. Mix thoroughly. Distribute into tubes or flasks. Autoclave for 15 min at 15 psi pressure–121°C.

Use: For the cultivation and maintenance of *Erwinia amylovora.*

NWRI Agar (HPC Agar)

Composition per liter:

Agar	15.0g
Peptone	3.0g
Soluble casein	0.5g
K_2HPO_4	0.2g
$MgSO_4$	0.05g
$FeCl_3$	1.0mg

pH 7.2 ± 0.2 at 25°C

Preparation of Medium: Add components to distilled/deionized water and bring volume to 1.0L. Mix thoroughly. Gently heat and bring to boiling. Adjust pH to 7.2. Distribute into tubes or flasks. Autoclave for 15 min at 15 psi pressure–121°C. Pour into sterile Petri dishes.

Use: For estimation of the number of live heterotrophic bacteria in water using the heterotrophic plate count technique.

NY Medium (Nutrient Broth Yeast Extract Medium)

Composition per liter:

NaCl	8.0g
Peptone	5.0g
Yeast extract	5.0g
Beef extract	3.0g

pH 7.2–7.4 at 25°C

Preparation of Medium: Add components to distilled/deionized water and bring volume to 1.0L. Mix thoroughly. Distribute into tubes or flasks. Autoclave for 15 min at 15 psi pressure–121°C.

Use: For the cultivation of *Pseudomonas* species.

Nystatin Assay Agar

See: **Antibiotic Medium 19**

N-Z Amine A™ Glycerol Agar

Composition per liter:

Agar	15.0g
N-Z Amine A™	5.0g
Beef extract	1.0g
Glycerol	70.0mL

pH 6.5–7.0 at 25°C

Preparation of Medium: Add components to distilled/deionized water and bring volume to 1.0L. Mix thoroughly. Gently heat and bring to boiling. Distribute into tubes or flasks. Autoclave for 15 min at 15 psi pressure–121°C. Pour into sterile Petri dishes or leave in tubes.

Use: For the isolation and cultivation of *Actinomadura* species, *Actinopolyspora* species, *Excellospora* species, and *Microspora* species.

N-Z Amine A™ Medium

Composition per liter:

N-Z Amine A™	5.0g
Beef extract	1.0g
Glycerol	80.0mL

Preparation of Medium: Add components to tap water and bring volume to 1.0L. Mix thoroughly. Distribute into tubes or flasks. Autoclave for 15 min at 15 psi pressure–121°C.

Use: For the cultivation and maintenance of *Nocardia brevicatena.*

N-Z Amine A™ Medium with Soluble Starch and Glucose

Composition per liter:

Soluble starch	20.0g
Agar	15.0g
Glucose	10.0g
Yeast extract	5.0g
N-Z-Amine A™	5.0g
$CaCO_3$	1.0g

Preparation of Medium: Add components to distilled/deionized water and bring volume to 1.0L. Mix thoroughly. Gently heat and bring to boiling. Distribute into tubes or flasks. Autoclave for 15 min at 15 psi pressure–121°C. Pour into sterile Petri dishes or leave in tubes.

Use: For the cultivation and maintenance of *Actinomadura* species, *Actinoplanes* species, *Amycolatopsis fastidiosa, Catenuloplanes japonicus, Dactylosporangium* species, *Geodermatophilus obscurus, Glycomyces* species, *Kitasatosporia mediocidica, Micromonospora* species, *Saccharomonospora caesia, Saccharothrix aerocolonigenes, Streptomyces* species, *and Streptosporangium* species.

N-Z Amine Medium (DSMZ Medium 554)

Composition per liter:

Starch, soluble	20.0g
Agar	15.0g
Glucose	10.0g
Yeast extract	5.0g
N-Z-Amine	5.0g
$CaCO_3$	1.0g

pH 7.2 ± 0.2 at 25°C

Preparation of Medium: Add components to distilled/deionized water and bring volume to 1.0L. Mix thoroughly. Distribute into tubes or flasks. Autoclave for 15 min at 15 psi pressure–121°C. Pour into sterile Petri dishes or leave in tubes.

Use: For the cultivation and maintenance of *Streptomyces nojiriensis, Streptomyces pulveraceus, Streptomyces capillispiralis, Saccharothrix mutabilis* subsp. *mutabilis=Nocardiopsis mutabilis, Couchioplanes* *caeruleus* subsp. *caeruleus=Actinoplanes caeruleus, Micromonospora inositola, Dactylosporangium vinaceum, Frankia* sp., *Streptosporangium koreanum, Streptosporangium roseum* subsp. *incarnatum, Actinomadura fibrosa, Micromonospora* sp., *Amycolatopsis kentuckyensis,* and *Amycolatopsis pretoriensis.*

NZC Broth

Composition per liter:

N-Z Amine A™	10.0g
NaCl	5.0g
$MgCl_2 \cdot 6H_2O$	2.0g
Casamino acids solution	5.0mL

Casamino Acids Solution:

Composition per 10.0mL:

Casamino acids	2.0g

Preparation of Casamino Acids Solution: Add casamino acids to distilled/deionized water and bring volume to 10.0mL. Mix thoroughly. Filter sterilize.

Preparation of Medium: Add components, except casamino acids solution, to distilled/deionized water and bring volume to 990.0mL. Mix thoroughly. Gently heat and bring to boiling. Autoclave for 15 min at 15 psi pressure–121°C. Cool to 45°–50°C. Aseptically add 5.0mL of sterile casamino acids solution. Mix thoroughly. Aseptically distribute into sterile tubes or flasks.

Use: For the cultivation of *Escherichia coli.*

NZCYM Medium

Composition per liter:

Pancreatic digest of casein	10.0g
NaCl	5.0g
Yeast extract	5.0g
Maltose	2.0g
$MgSO_4 \cdot 7H_2O$	2.0g
Casamino acids	1.0g

pH 7.5 ± 0.2 at 25°C

Preparation of Medium: Add components to distilled/deionized water and bring volume to 1.0L. Mix thoroughly. Distribute into tubes or flasks. Autoclave for 15 min at 15 psi pressure–121°C.

Use: For the cultivation of *Escherichia coli* and *Pseudomonas* species.

NZM Broth

Composition per liter:

Enzymatic digest of casein	10.0g
NaCl	5 .0g
$MgSO_4 \cdot 7H_2O$	2 .0g

pH 7.0 ± 0.2 at 25°C

Source: This medium is available from Acumedia, Neogen Corp.

Preparation of Medium: Add components to distilled/deionized water and bring volume to 1.0L. Mix thoroughly. Distribute into tubes or flasks. Autoclave for 15 min at 15 psi pressure–121°C.

Use: For the cultivation of recombinant strains of *Escherichia coli.*

NZY Agar

Composition per liter:

Agar	20.0g
N-Z Amine A™	10.0g
NaCl	5.0g
Yeast extract	5.0g
$MgCl_2 \cdot 6H_2O$	2.0g

Preparation of Medium: Add components to distilled/deionized water and bring volume to 1.0L. Mix thoroughly. Gently heat and bring to boiling. Distribute into tubes. Autoclave for 15 min at 15 psi pressure–121°C. Allow tubes to cool in a slanted position.

Use: For the cultivation and maintenance of a variety of microorganisms.

NZY Agar

Composition per liter:

Agar	11.0g
N-Z Amine A™	10.0g
NaCl	5.0g
Yeast extract	5.0g
$MgCl_2 \cdot 6H_2O$	2.0g

Preparation of Medium: Add components to distilled/deionized water and bring volume to 1.0L. Mix thoroughly. Gently heat and bring to boiling. Distribute into tubes or flasks. Autoclave for 15 min at 15 psi pressure–121°C. Pour into sterile Petri dishes.

Use: For the cultivation and enumeration of a variety of microorganisms.

NZY Broth

Composition per liter:

N-Z-Amine A™	10.0g
NaCl	5.0g
Yeast extract	5.0g
$MgCl_2 \cdot 6H_2O$	2.0g

Preparation of Medium: Add components to distilled/deionized water and bring volume to 1.0L. Mix thoroughly. Gently heat and bring to boiling. Autoclave for 15 min at 15 psi pressure–121°C.

Use: For the cultivation and enumeration of a variety of microorganisms

NZYM Broth

Composition per liter:

Enzymatic digest of casein	10.0g
NaCl	5 .0g
Yeast extract	5.0g
Casamino acids	1.0g
$MgSO_4 \cdot 7H_2O$	0.98g

pH 7.0 ± 0.2 at 25°C

Source: This medium is available from Acumedia, Neogen Corp.

Preparation of Medium: Add components to distilled/deionized water and bring volume to 1.0L. Mix thoroughly. Distribute into tubes or flasks. Autoclave for 15 min at 15 psi pressure–121°C.

Use: For the cultivation of recombinant strains of *Escherichia coli.*

O157:H7(+) Plating Medium

Composition per liter:

Agar	15.0g
Sorbitol	12.0g
Salicin	10.0g
Inositol	10.0g
Peptone	10.0g
Adonitol	8.0g
NaCl	5.0g
Tryptone	5.0g
Proteose peptone	3.0g

Bile salts No. 3 .... 1.25g
Indoxyl-β-D-galactopyranoside 0.12g
5-Bromo-4-chloro-3-indoxyl-β-D-galactopyranoside 0.12g
Phenol Red 0.1g
Isopropyl-β-D-thiogalactopyranoside 0.1g
Novobiocin solution 1.0mL
Tellurite solution 1.0mL

pH 6.8 ± 0.2 at 25°C

Source: This medium is available as a premixed powder from BIOSYNTH International, Inc.

Novobiocin Solution:

Composition per 10.0mL:

Novobiocin 0.1g

Preparation of Novobiocin Solution: Add novobiocin to distilled/deionized water and bring volume to 10.0mL. Mix thoroughly. Filter sterilize.

Tellurite Solution:

Composition per 10.0mL:

K_2TeO_3 .... 0.01g

Preparation of Tellurite Solution: Add K_2TeO_3 to distilled/deionized water and bring volume to 10.0mL. Mix thoroughly. Filter sterilize.

Caution: Potassium tellurite is toxic.

Preparation of Medium: Add components, except novobiocin solution and tellurite solution, to distilled/deionized water and bring volume to 998.0mL. Mix thoroughly. Gently heat while stirring and bring to boiling. Autoclave for 15 min at 15 psi pressure–121°C. Cool to 50°C. Aseptially add 1.0mL sterile novobiocin solution and 1.0 sterile tellurite solution. Mix thouroughly. Pour into sterile Petri dishes.

Use: For the detection of *Escherichia coli* O157:H7. *E.coli* O157:H7 grow with blue-black colonies and *E.coli* non-O157 with green-yellow colonies.

O157:H7 ID Agar

Composition per liter:

Proprietary

Source: This medium is available from bioMérieux.

Use: A new chromogenic medium for the detection of *Escherichia coli* O157:H7. In accordance with USDA, BAM and ISO 16654 standards, O157:H7 ID Agar allows the detection and presumptive identification of *E. coli* O157:H7.

OA

See: **Oat Meal Agar**

Oak Wilt Fungus HiVeg Agar

Composition per liter:

Malt extract 22.0g
Agar 15.0g
Plant peptone No. 4 .... 8.0g
Synthetic detergent No. II 5.0g

pH 5.7 ± 0.2 at 25°C

Source: This medium is available as a premixed powder from HiMedia.

Preparation of Medium: Add components to distilled/deionized water and bring volume to 1.0L. Mix thoroughly. Distribute into tubes or flasks. Autoclave for 15 min at 15 psi pressure–121°C. Pour into sterile Petri dishes or leave in tubes.

Use: For the cultivation of oak wilt fungi.

Oat Flake Medium (DSMZ Medium 189)

Composition per liter:

Oat flakes 30.0g
Agar 15.0g

pH 7.2 ± 0.2 at 25°C

Preparation of Medium: Add rolled oats to 500.0mL distilled/deionized water. Gently heat and bring to boiling. Boil for 10 min. Filter. Add agar and bring volume to 1.0L with distilled/deionized water. Mix thoroughly. Gently heat and bring to boiling. Mix by shaking. Distribute into tubes or flasks. Autoclave for 20 min at 15 psi pressure–121°C. Mix by shaking. Pour into sterile Petri dishes or leave in tubes.

Use: For the cultivation and maintenance of actinomycetes and fungi.

Oat Flakes Agar

Composition per liter:

Oat flakes 30.0g
Agar 15.0g

Preparation of Medium: Add oat flakes to distilled/deionized water and bring volume to 990.0mL. Mix thoroughly. Gently heat and bring to boiling. Continue boiling for 10 min. Add agar. Autoclave for 20 min at 15 psi pressure–121°C. Pour into sterile Petri dishes or distribute into sterile tubes. Swirl flask while decanting medium.

Use: For the cultivation and maintenance of *Actinoplanes* species and *Streptomycoides glaucoflavus*.

Oat Flakes Agar with Yeast Extract (DSMZ Medium 248)

Composition per liter:

Oat flakes 20.0g
Agar 12.0g
Yeast extract 2.0g
Trace elements solution 1.0mL

pH 7.2 ± 0.2 at 25°C

Trace Elements Solution:

Composition per 100.0mL:

$FeSO_4 \cdot 7H_2O$ 0.1g
$MnCl_2 \cdot 4H_2O$ 0.1g
$ZnSO_4 \cdot 7H_2O$ 0.1g

Preparation of Trace Elements Solution: Add components to distilled/deionized water and bring volume to 100.0mL. Mix thoroughly.

Preparation of Medium: Add oat flakes to 500.0mL distilled/deionized water. Gently heat and bring to boiling. Boil for 20 min. Filter through cheese cloth. Add agar and yeast extract to the filtrate and bring volume to 1.0L with distilled/deionized water. Mix thoroughly. Add 1.0mL trace elements solution. Gently heat and bring to boiling. Distribute into tubes or flasks. Autoclave for 15 min at 15 psi pressure–121°C. Pour into sterile Petri dishes or leave in tubes.

Use: For the cultivation and maintenance of *Streptomyces purpureus=Kitasatoa purpurea* and *Thermoactinomyces putidus*.

Oat Flakes Agar with Yeast Extract

Composition per liter:

Oat flakes 20.0g
Agar 12.0g

Yeast extract 2.0g
Trace elements solution SL-6 .... 1.0mL

pH 7.2 ± 0.2 at 25°C

Trace Elements Solution SL-6:
Composition per liter:

$MnCl_2 \cdot 4H_2O$	0.5g
H_3BO_3	0.3g
$CoCl_2 \cdot 6H_2O$	0.2g
$ZnSO_4 \cdot 7H_2O$	0.1g
$Na_2MoO_4 \cdot 2H_2O$	0.03g
$NiCl_2 \cdot 6H_2O$	0.02g
$CuCl_2 \cdot 2H_2O$	0.01g

Preparation of Trace Elements Solution SL-6: Add components to distilled/deionized water and bring volume to 1.0L. Mix thoroughly.

Preparation of Medium: Add oat flakes to distilled/deionized water and bring volume to 990.0mL. Mix thoroughly. Gently heat and bring to boiling. Continue boiling for 10 min. Filter through cheesecloth. Add remaining components to filtrate and bring volume to 1.0L with distilled/deionized water. Autoclave for 15 min at 15 psi pressure–121°C. Pour into sterile Petri dishes or distribute into sterile tubes.

Use: For the cultivation of actinomycetes.

Oatmeal Agar
See: **ISP Medium 3**

Oatmeal Agar

Composition per liter:

Agar	15.0g
Oat flakes	10.0g
Oatmeal	10.0g

pH 7.0 ± 0.2 at 25°C

Preparation of Medium: Add components to distilled/deionized water and bring volume to 1.0L. Mix thoroughly. Gently heat and bring to boiling. Distribute into tubes or flasks. Autoclave for 15 min at 15 psi pressure–121°C. Pour into sterile Petri dishes or leave in tubes.

Use: For the cultivation and maintenance of *Actinomadura* species, *Actinoplanes* species, *Geodermatophilus obscurus, Glycomyces harbinensis, Microbispora rosea, Micromonospora* species, *Nocardioides* species, *Nocardioides luteus, Promicromonospora citrea, Sporichthya polymorpha, Streptomyces* species, *Streptosporangium amethystogenes, Streptosporangium roseum, Streptosporangium violaceochromogenes, Thermoactinomyces putidus,* and *Thermoactinomyces vulgaris.*

Oatmeal Agar

Composition per 750.0mL:

Oatmeal, instant for babies	40.0g
Agar	5.0g

Preparation of Medium: Add agar to distilled/deionized water and bring volume to 500.0mL. Mix thoroughly. Gently heat and bring to boiling. Add instant oatmeal for babies to distilled/deionized water and bring volume to 250.0mL. Mix thoroughly. Add the oatmeal solution to the melted agar solution. Mix thoroughly. Autoclave for 15 min at 15 psi pressure–121°C. Pour into sterile Petri dishes or aseptically distribute into sterile tubes.

Use: For the cultivation of fungi.

Oatmeal Agar (ATCC 551)

Composition per liter:

Oatmeal	60.0g
Agar	12.5g

pH 6.0 ± 0.2 at 25°C

Source: This medium is available as a premixed powder from BD Diagnostic Systems.

Preparation of Medium: Add components to distilled/deionized water and bring volume to 1.0L. Mix thoroughly. Gently heat and bring to boiling. Distribute into tubes or flasks. Autoclave for 15 min at 15 psi pressure–121°C. Pour into sterile Petri dishes or leave in tubes.

Use: For cultivation of fungi and actinomycetes, particularly for macrospore formation.

Oatmeal Agar (OA)

Composition per liter:

Agar	12.0g
Oatmeal, rolled oats	10.0g
Glycerin	5.0mL
Lactic acid	0.2mL

Preparation of Medium: Add components to distilled/deionized water and bring volume to 1.0L. Mix thoroughly. Gently heat and bring to boiling. Distribute into tubes or flasks. Autoclave for 15 min at 15 psi pressure–121°C. Pour into sterile Petri dishes or leave in tubes.

Use: For the cultivation of many filamentous fungi.

Oatmeal Agar A (ISP-3) (DSMZ Medium 609)

Composition per liter:

Oatmeal (Quaker White Oats)	20.0g
Agar	18.0g
Trace salts solution	1.0mL

pH 7.0 ± 0.2 at 25°C

Trace Salts Solution:
Composition per 100.0mL:

$FeSO_4 \cdot 7H_2O$	0.1g
$MnCl_2 \cdot 4H_2O$	0.1g
$ZnSO_4 \cdot 7H_2O$	0.1g

Preparation of Trace Salts Solution: Add components to 100.0mL distilled/deionized water. Mix thoroughly.

Preparation of Medium: Add oats to 1.0L distilled/deionized water. Adjust pH to 7.2. Gently heat and bring to boiling. Cook or steam for 20 min. Filter through cheesecloth. Add 18.0g agar. Add 1.0mL trace salts solution. Bring volume to 1.0L with distilled/deionized water. Gently heat and bring to boiling. Autoclave for 15 min at 15 psi pressure–121°C. Pour into Petri dishes or distribute to sterile tubes.

Use: For the cultivation of various fungi.

Oatmeal Agar B

Composition per liter:

Agar	15.0g
Oatmeal, powdered	3.0g
K_2HPO_4	0.5g

KNO_3 .... 0.2g
$MgSO_4 \cdot 7H_2O$ 0.2g

pH 7.0 ± 0.2 at 25°C

Preparation of Medium: Add components to distilled/deionized water and bring volume to 1.0L. Mix thoroughly. Gently heat and bring to boiling. Distribute into tubes or flasks. Autoclave for 15 min at 15 psi pressure–121°C. Pour into sterile Petri dishes or leave in tubes.

Use: For the cultivation of *Actinomadura viridis, Actinoplanes awajinensis, Actinoplanes brasiliensis, Actinoplanes consettensis, Actinoplanes derwentensis, Actinoplanes durhamensis, Actinoplanes humidus, Actinoplanes palleronii, Actinoplanes pyriformis, Amycolatopsis orientalis, Excellospora viridilutea, Kineosporia aurantiaca, Microbispora bispora, Microbispora rosea, Microtetraspora fastidiosa, Microtetraspora flexuosa, Microtetraspora fusca, Microtetraspora glauca, Microtetraspora polychroma, Microtetraspora pusilla, Nocardioides albus, Nocardioides fulvus, Nocardiopsis streptosporus, Planobispora longispora, Planomonospora parontospora, Streptomyces* species, *Streptomycoides glaucoflavus, Streptosporangium album, Thermomonospora curvata,* and *Thermomonospora mesophila.*

Oatmeal Agar A (ISP-3) (DSMZ Medium 609)

Composition per liter:
Oatmeal (Quaker White Oats) 20.0g
Agar 18.0g
Trace salts solution 1.0mL

pH 7.0 ± 0.2 at 25°C

Trace Salts Solution:
Composition per 100.0mL:
$FeSO_4 \cdot 7H_2O$ 0.1g
$MnCl_2 \cdot 4H_2O$ 0.1g
$ZnSO_4 \cdot 7H_2O$ 0.1g

Preparation of Trace Salts Solution: Add components to 100.0mL distilled/deionized water. Mix thoroughly.

Preparation of Medium: Add oats to 1.0L distilled/deionized water. Adjust pH to 7.2. Gently heat and bring to boiling. Cook or steam for 20 min. Filter through cheesecloth. Add 18.0g agar. Add 1.0mL trace salts solution. Bring volume to 1.0L with distilled/deionized water. Gently heat and bring to boiling. Autoclave for 15 min at 15 psi pressure–121°C. Pour into Petri dishes or distribute to sterile tubes.

Use: For the cultivation of various fungi.

Oatmeal Agar B

Composition per liter:
Agar 15.0g
Oatmeal, powdered 3.0g
K_2HPO_4 .... 0.5g
KNO_3 .... 0.2g
$MgSO_4 \cdot 7H_2O$ 0.2g

pH 7.0 ± 0.2 at 25°C

Preparation of Medium: Add components to distilled/deionized water and bring volume to 1.0L. Mix thoroughly. Gently heat and bring to boiling. Distribute into tubes or flasks. Autoclave for 15 min at 15 psi pressure–121°C. Pour into sterile Petri dishes or leave in tubes.

Use: For the cultivation of *Actinomadura viridis, Actinoplanes awajinensis, Actinoplanes brasiliensis, Actinoplanes consettensis, Actinoplanes derwentensis, Actinoplanes durhamensis, Actinoplanes humidus, Actinoplanes palleronii, Actinoplanes pyriformis, Amycolatopsis orientalis, Excellospora viridilutea, Kineosporia aurantiaca, Microbispora bispora, Microbispora rosea, Microtetraspora fastidiosa, Microtetraspora flexuosa, Microtetraspora fusca, Microtetraspora glauca, Microtetraspora polychroma, Microtetraspora pusilla, Nocardioides albus, Nocardioides fulvus, Nocardiopsis streptosporus, Planobispora longispora, Planomonospora parontospora, Streptomyces* species, *Streptomycoides glaucoflavus, Streptosporangium album, Thermomonospora curvata,* and *Thermomonospora mesophila.*

Oatmeal Agar with Lupine Stems

Composition per liter:
Oatmeal 30.0g
Agar 20.0g
Lupine stems .... variable

Preparation of Medium: Add oatmeal to 1.0L of distilled/deionized water. Gently heat and bring to boiling. Continue boiling for 20 min. Strain through muslin. Bring volume of filtrate to 1.0L with distilled/deionized water. Add 20.0g of agar. Gently heat and bring to boiling. Distribute into tubes. Autoclave for 15 min at 15 psi pressure–121°C. Pour into sterile Petri dishes or leave in tubes. Cut green lupine stems into 1.0cm pieces. Place 6 to 9 lupine stem pieces in tubes with 3.0mL of distilled/deionized water. Autoclave for 15 min at 15 psi pressure–121°C. Melt agar tubes by gently heating. Cool to 50°C. Aseptically add 1 sterile lupine stem to each tube. Allow to solidify in a slanted position.

Use: For the cultivation and maintenance of *Ascochyta fabae, Ascochyta pinodes, Ascochyta pisi, Ascochyta punctata, Ascochyta viciae-pannonicae, Ascochyta viciae-villosae*, and *Colletotrichum orbiculare.*

Oatmeal Agar with Lupine Stems (OA-LUP)

Composition per liter:
Agar 12.0g
Oatmeal, rolled oats 10.0g
Glycerin 5.0mL
Lactic acid 0.2mL
Lupine stems .... variable

Preparation of Medium: Add components, except lupine stems, to distilled/deionized water and bring volume to 1.0L. Mix thoroughly. Gently heat and bring to boiling. Distribute 6.0mL volumes into tubes. Cut lupine stems into 8.0cm long pieces. Add 2–3 lupine stems per tube. Autoclave for 15 min at 15 psi pressure–121°C. Allow tubes to cool in a slanted position.

Use: For the cultivation of *Ceratocystis brunnea, Ceratocystis major,* and *Ceratocystis radicicola.*

Oatmeal Nitrate Agar

Composition per liter:
Agar 15.0g
Oatmeal 3.0g
K_2HPO_4 .... 0.5g
KNO_3 .... 0.2g
$MgSO_4$ 0.2g

pH 7.0 ± 0.2 at 25°C

Preparation of Medium: Add components to distilled/deionized water and bring volume to 1.0L. Mix thoroughly. Gently heat and bring to boiling. Distribute into tubes or flasks. Autoclave for 15 min at 15 psi pressure–121°C. Pour into sterile Petri dishes or leave in tubes.

Use: For the cultivation and maintenance of *Microtetraspora flexuosa.*

Oatmeal Salts Agar

Composition per liter:

Agar	18.0g
Oatmeal, baby (Beechnut™)	10.0g
$MgSO_4$	1.0g
KH_2PO_4	1.0g
$NaNO_3$	1.0g

pH 5.6 ± 0.2 at 25°C

Preparation of Medium: Add components to distilled/deionized water and bring volume to 1.0L. Mix thoroughly. Gently heat and bring to boiling. Adjust pH to 5.6. Distribute into tubes or flasks. Autoclave for 15 min at 15 psi pressure–121°C. Pour into sterile Petri dishes or leave in tubes.

Use: For the cultivation and maintenance of fungi.

Oatmeal Soy Peptone Medium

Composition per liter:

Agar	20.0g
Oatmeal	20.0g
Glucose	2.0g
Soy peptone	2.0g

Preparation of Medium: Add components to distilled/deionized water and bring volume to 1.0L. Mix thoroughly. Gently heat and bring to boiling. Distribute into tubes or flasks. Autoclave for 15 min at 15 psi pressure–121°C. Pour into sterile Petri dishes or leave in tubes.

Use: For the cultivation and maintenance of *Streptomyces metachromogenes* and *Actinosporangium* species.

OAY Agar

Composition per liter:

Oatmeal	20.0g
Agar	18.0g
Yeast extract	1.0g
Trace salts solution	1.0mL

Trace Salts Solution:

Composition per 100.0mL:

$FeSO_4·7H_2O$	0.1g
$MnCl_2·4H_2O$	0.1g
$ZnSO_4·7H_2O$	0.1g

Preparation of Trace Salts Solution: Add components to distilled/deionized water and bring the volume to 100.0mL. Mix thoroughly. Filter sterilize.

Preparation of Medium: Add oatmeal to distilled/deionized water and bring volume to 1.0L. Mix thoroughly. Gently heat and bring to boiling. Steam for 20 min. Filter through cheesecloth. Add agar. Add sufficient distilled/deionized water to bring volume to 999.0mL. Gently heat and bring to boiling. Mix thoroughly. Distribute into tubes or flasks. Autoclave for 15 min at 15 psi pressure–121°C. Cool to 45°–50°C. Aseptically add 1.0mL of sterile trace salts solution. Mix thoroughly. Pour into sterile Petri dishes or distribute into sterile tubes.

Use: For the cultivation of *Actinomadura citrea, Actinomadura coerulea, Actinomadura cremea, Actinomadura kijaniata, Actinomadura rubrobrunea, Actinomadura vinacea, Actinomadura viridis, Actinoplanes* species, *Glycomyces harbinensis, Microbispora rosea, Micromonospora* species, *Microstreptospora cinerea, Microtetraspora africana, Microtetraspora fastidiosa, Microtetraspora ferruginea, Microtetraspora roseola, Microtetraspora rosevolacea, Microtetraspora rubra, Microtetraspora salmonea, Saccharothrix coeruleofusca, Saccharothrix longispora, Spirillospora rubra, Streptomyces* species, *Streptosporangium pseudovulgare, Streptosporangium sibiricum,* and *Thermoactinomyces vulgaris.*

Oceanithermus Medium (DSMZ Medium 1149)

Composition per liter:

NaCl	20.0g
Yeast extract	2.0g
K_2HPO_4	0.78g
KH_2PO_4	0.75g
NH_4Cl	0.53g
$MgCl_2·6H_2O$	0.36g
$CaCl_2·2H_2O$	0.33g
$NaNO_3$	0.2g
Na_2SO_4	0.14g
Bicarbonate solution	100.0mL
Trace elements solution	10.0mL
Vitamin solution	10.0mL

Bicarbonate Solution:

Composition per 100.0mL:

$NaHCO_3$	5.0g

Preparation of Bicarbonate Solution: Add $NaHCO_3$ to distilled/deionized water and bring volume to 100.0mL. Mix thoroughly. Adjust pH to 6.5. Filter sterilize.

Trace Elements Solution:

Composition per liter:

Nitrilotriacetic acid	12.8g
$FeCl_3·6H_2O$	1.35g
NaCl	1.0g
$NiCl_2·6H_2O$	0.12g
$MnCl_2·4H_2O$	0.1g
$CaCl_2·2H_2O$	0.1g
$ZnCl_2$	0.1g
$Na_2SeO_3·5H_2O$	0.026g
$CuCl_2·2H_2O$	0.025g
$CoCl_2·6H_2O$	0.024g
$Na_2MoO_4·2H_2O$	0.024g
H_3BO_3	0.01g

Preparation of Trace Elements Solution: Add nitrilotriacetic acid to 500.0mL of distilled/deionized water. Dissolve by adjusting pH to 6.5 with KOH. Add remaining components. Add distilled/deionized water to 1.0L. Mix thoroughly. Adjust pH to 7.0.

Vitamin Solution:

Composition per liter:

Pyridoxine-HCl	10.0mg
Thiamine-HCl·$2H_2O$	5.0mg
Riboflavin	5.0mg
Nicotinic acid	5.0mg
D-Ca-pantothenate	5.0mg
Biotin	2.0mg
Folic acid	2.0mg

Preparation of Vitamin Solution: Add components to distilled/deionized water and bring volume to 1.0L. Mix thoroughly. Filter sterilize.

Preparation of Medium: Add components, except bicarbonate solution and vitamin solution, to distilled/deionized water and bring volume to 890.0mL. Mix thoroughly. Adjust pH to 7.0. Gently heat while stirring and bring to boiling. Autoclave for 15 min at 15 psi pressure–121°C. Cool to room temperature. Aseptically add bicarbonate and vitamin solutions. Mix thoroughly. Aseptically distribute into tubes or flasks.

Use: For the cultivation of *Oceanithermus* spp.

Oceanithermus profundus Medium (DSMZ Medium 975)

Composition per liter:

NaCl	30.0g
HEPES	2.38g
NH_4Cl	0.33g
KCl	0.33g
Potassium nitrate solution	10.0mL
Calcium chloride solution	10.0mL
Magnesium chloride solution	10.0mL
Yeast extract solution	10.0mL
Sucrose solution	10.0mL
Tryptone solution	10.0mL
Vitamin solution	1.0mL
Trace elements solution	1.0mL

pH 7.0–7.5 at 25°C

Potassium Nitrate Solution:
Composition per 10.0mL:

KNO_3	0.33g

Preparation of Potassium Nitrate Solution: Add KNO_3 to distilled/deionized water and bring volume to 10.0mL. Mix thoroughly. Autoclave for 15 min at 15 psi pressure–121°C. Cool to room temperature.

Magnesium Chloride Solution:
Composition per 10.0mL:

$MgCl_2{\cdot}6H_2O$	0.33g

Preparation of Magnesium Chloride Solution: Add 0.33g of $MgCl_2{\cdot}6H_2O$ to distilled/deionized water and bring volume to 10.0mL. Mix thoroughly. Autoclave for 15 min at 15 psi pressure–121°C. Cool to room temperature.

Calcium Chloride Solution:
Composition per 10.0mL:

$CaCl_2{\cdot}2H_2O$	0.33g

Preparation of Calcium Chloride Solution: Add $CaCl_2{\cdot}2H_2O$ to distilled/deionized water and bring volume to 10.0mL. Mix thoroughly. Autoclave for 15 min at 15 psi pressure–121°C. Cool to room temperature.

Yeast Extract Solution:
Composition per 10.0mL:

Yeast extract	0.2g

Preparation of Yeast Extract Solution: Add yeast extract to distilled/deionized water and bring volume to 10.0mL. Mix thoroughly. Autoclave for 15 min at 15 psi pressure–121°C. Cool to 25°C.

Sucrose Solution:
Composition per 10.0mL:

Sucrose	2.0g

Preparation of Sucrose Solution: Add sucrose to distilled/deionized water and bring volume to 10.0mL. Mix thoroughly. Autoclave for 15 min at 15 psi pressure–121°C. Cool to room temperature.

Tryptone Solution:
Composition per 10.0mL:

Tryptone	1.0g

Preparation of Tryptone Solution: Add tryptone to distilled/deionized water and bring volume to 10.0mL. Mix thoroughly. Autoclave for 15 min at 15 psi pressure–121°C. Cool to room temperature.

Trace Elements Solution:
Composition per liter:

$MgSO_4{\cdot}7H_2O$	3.0g
Nitrilotriacetic acid	1.5g
NaCl	1.0g
$MnSO_4{\cdot}2H_2O$	0.5g
$CoSO_4{\cdot}7H_2O$	0.18g
$ZnSO_4{\cdot}7H_2O$	0.18g
$CaCl_2{\cdot}2H_2O$	0.1g
$FeSO_4{\cdot}7H_2O$	0.1g
$NiCl_2{\cdot}6H_2O$	0.025g
$KAl(SO_4)_2{\cdot}12H_2O$	0.02g
H_3BO_3	0.01g
$Na_2MoO_4{\cdot}4H_2O$	0.01g
$CuSO_4{\cdot}5H_2O$	0.01g
$Na_2SeO_3{\cdot}5H_2O$	0.3mg

Preparation of Trace Elements Solution: Add nitrilotriacetic acid to 500.0mL of distilled/deionized water. Dissolve by adjusting pH to 6.5 with KOH. Add remaining components. Add distilled/deionized water to 1.0L. Mix thoroughly. Sparge with 100% N_2. Autoclave for 15 min at 15 psi pressure–121°C. Cool to room temperature.

Vitamin Solution:
Composition per liter:

Pyridoxine-HCl	10.0mg
Thiamine-HCl·$2H_2O$	5.0mg
Riboflavin	5.0mg
Nicotinic acid	5.0mg
D-Ca-pantothenate	5.0mg
p-Aminobenzoic acid	5.0mg
Lipoic acid	5.0mg
Biotin	2.0mg
Folic acid	2.0mg
Vitamin B_{12}	0.1mg

Preparation of Vitamin Solution: Add components to distilled/deionized water and bring volume to 1.0L. Mix thoroughly. Sparge with 80% H_2 + 20% CO_2. Filter sterilize.

Preparation of Medium: Prepare and dispense medium under 100% N_2 gas atmosphere. Add components, except yeast extract solution, sucrose solution, calcium chloride solution, magnesium chloride solution, potassium nitrate solution, tryptone solution, vitamin solution, and trace elements solution, to distilled/deionized water and bring volume to 938.0mL. Mix thoroughly. Distribute into anaerobe tubes or bottles. Autoclave for 15 min at 15 psi pressure–121°C. Aseptically and anaerobically add per liter, 10.0mL sucrose solution, 10.0mL yeast extract solution, 10.0mL potassium nitrate solution, 10.0mL tryptone solution, 10.0mL magnesium chloride solution, 10.0mL calcium chloride solution, 1.0mL vitamin solution, and 1.0mL trace elements solution. Mix thoroughly. Adjust pH to 7.0–7.5

Use: For the cultivation of *Oceanithermus profundus.*

Ochromonas Medium

Composition per liter:

Glucose	20.0g
Vitamin-free casamino acids	10.0g
Diammonium hydrogen citrate	1.6g
KH_2PO_4	0.6g
DL-Methionine	0.4g
$MgSO_4 \cdot 7H_2O$	0.4g
$CaCO_3$	0.3g
$ZnSO_4 \cdot 7H_2O$	0.22g
L-Cystine	0.2g
DL-Tryptophan	0.2g
$MnSO_4 \cdot H_2O$	0.123g
EDTA	0.1g
$Na_2MoO_4 \cdot 2H_2O$	0.1g
$FeSO_4 \cdot 7H_2O$	20.0mg
Inositol	20.0mg
$CoSO_4 \cdot 7H_2O$	6.0mg
Choline chloride	4.0mg
Thiamine	4.0mg
PAB	2.0mg
H_3BO_3	1.2mg
$CuSO_4 \cdot 5H_2O$	800.0μg
Biotin	20.0μg
KI	20.0μg
Tween™ 80	2.0mL

pH 5.8 ± 0.2 at 25°C

Preparation of Medium: Add components to distilled/deionized water and bring volume to 1.0L. Mix thoroughly. Adjust pH to 5.8. Gently heat and bring to boiling. Continue boiling for 2–3 min. Distribute 5.0mL volumes into 6 × 125mm screw-capped test tubes. Swirl the medium while dispensing to evenly distribute the fine precipitate which forms. Autoclave for 10 min at 15 psi pressure–121°C. Do not overheat.

Use: For the cultivation of *Ochromonas malhamensis*.

Ochromonas Medium

Composition per liter:

Glucose	20.0g
Vitamin-free casamino acids	10.0g
Diammonium hydrogen citrate	1.6g
KH_2PO_4	0.6g
$MgSO_4 \cdot 7H_2O$	0.4g
$CaCO_3$	0.3g
$ZnSO_4 \cdot 7H_2O$	0.22g
L-Cystine	0.2g
DL-Tryptophan	0.2g
$MnSO_4 \cdot H_2O$	0.123g
EDTA	0.1g
$Na_2MoO_4 \cdot 2H_2O$	0.1g
DL-Methionine	0.4g
$FeSO_4 \cdot 7H_2O$	20.0mg
Inositol	20.0mg
$CoSO_4 \cdot 7H_2O$	6.0mg
Choline chloride	4.0mg
Thiamine	4.0mg
PAB	2.0mg
H_3BO_3	1.2mg
Vitamin B_{12}	0.4ng
$CuSO_4 \cdot 5H_2O$	800.0μg
Biotin	20.0μg
KI	20.0μg
Tween™ 80	2.0mL

pH 5.8 ± 0.2 at 25°C

Preparation of Medium: Add components to distilled/deionized water and bring volume to 1.0L. Mix thoroughly. Adjust pH to 5.8. Gently heat and bring to boiling. Continue boiling for 2–3 min. Distribute 5.0mL volumes into 6 × 125mm screw-capped test tubes. Swirl the medium while dispensing to evenly distribute the fine precipitate which forms. Autoclave for 10 min at 15 psi pressure–121°C. Do not overheat.

Use: For the maintenance of *Ochromonas malhamensis*.

Oddoux Medium

Composition per liter:

Agar	13.0g
Malt extract	8.0g
Glucose	7.0g
Casein hydrolysate	1.0g
Asparagine	0.5g
KH_2PO_4	0.5g
$MgSO_4 \cdot 7H_2O$	0.5g
Vitamin solution	10.0mL

Vitamin Solution:

Composition per 10.0mL:

Calcium pantothenate	0.5mg
Nicotinamide	0.5mg
Pyridoxine HCl	0.5mg
Riboflavin	0.5mg
Thiamine	0.5mg

Preparation of Vitamin Solution: Add components to distilled/deionized water and bring volume to 10.0mL. Mix thoroughly. Filter sterilize.

Preparation of Medium: Add components, except vitamin solution, to distilled/deionized water and bring volume to 990.0mL. Mix thoroughly. Gently heat and bring to boiling. Autoclave for 15 min at 15 psi pressure–121°C. Aseptically add 10.0mL of sterile vitamin solution. Mix thoroughly. Pour into sterile Petri dishes or distribute into sterile tubes.

Use: For the cultivation and maintenance of *Boletus granulatus, Boletus luteus, Lactarius deliciosus,* and *Tricholoma equestre.*

OF Basal HiVeg Medium with Carbohydrate

Composition per liter:

NaCl	5.0g
Agar	2.0g
Plant hydrolysate	2.0g
K_2HPO_4	0.3g
Bromthymol Blue	0.08g
Carbohydrate solution	100.0mL

pH 6.8 ± 0.2 at 25°C

Source: This medium, without carbohydrate solution, is available as a premixed powder from HiMedia.

Carbohydrate Solution:

Composition per 100.0mL:

Carbohydrate	10.0g

Preparation of Carbohydrate Solution: Add carbohydrate to distilled/deionized water and bring volume to 100.0mL. Mix thoroughly. Filter sterilize.

Preparation of Medium: Add components, except carbohydrate solution, to distilled/deionized water and bring volume to 1.0L. Mix thoroughly. Gently heat and bring to boiling. Distribute into tubes in 3.0mL volumes. Autoclave for 15 min at 15 psi pressure–121°C. Cool to 45°–50°C. Aseptically add 0.3mL of sterile carbohydrate solution to each tube. Mix thoroughly.

Use: For the cultivation and differentiation of a variety of microorganisms based on their ability to ferment a specific carbohydrate. Bacteria that ferment the specific carbohydrate turn the medium yellow.

OF Glucose Medium, Semisolid
See: **Oxidative Fermentative Glucose Medium, Semisolid**

OF Glucose Medium, Semisolid with Sodium Chloride
See: **Oxidative Fermentative Glucose Medium, Semisolid, with Sodium Chloride**

OF Test Medium
See: **Oxidative Fermentative Test Medium**

Ogawa Egg Medium

Composition per liter:

Chicken eggs, whole	200.0mL
Sodium glutamate-KH_2PO_4 solution	100.0mL
Glycerol	6.0mL
Malachite Green (2.0% solution)	6.0mL

pH 6.8 ± 0.2 at 25°C

Sodium Glutamate-KH_2PO_4 Solution:

Composition per 100.0mL:

Sodium glutamate	1.0g
KH_2PO_4	1.0g

Preparation of Sodium Glutamate-KH_2PO_4 Solution: Add components to distilled/deionized water and bring volume to 100.0mL. Mix thoroughly. Filter sterilize.

Preparation of Medium: Soak eggs with 1:100 dilution of saturated mercuric chloride solution for 1 min. Aseptically break eggs into a sterile graduated cylinder. Homogenize eggs. Add remaining components. Mix thoroughly. Aseptically distribute into sterile tubes in 10.0mL volumes. Inspissate at 90°C (moist heat) for 60 min.

Use: For the selective isolation and cultivation of *Nocardia* and *Rhodococci* species.

Ogawa TB Medium

Composition per 300.0mL:

KH_2PO_4	1.0g
Homogenized whole egg	200.0mL
Glycerol	6.0mL
Malachite Green (2% solution)	6.0mL

pH 6.5 ± 0.2 at 25°C

Homogenized Whole Egg:

Composition per liter:

Whole eggs	18–24

Preparation of Homogenized Whole Egg: Use fresh eggs, less than 1 week old. Scrub the shells with soap. Let stand in a soap solution for 30 min. Rinse in running water. Soak eggs in 70% ethanol for 15 min. Break the eggs into a sterile container. Homogenize by shaking. Filter through four layers of sterile cheesecloth into a sterile graduated cylinder. Measure out 1.0L.

Preparation of Medium: Add components, except homogenized whole egg, to distilled/deionized water and bring volume to 100.0mL. Mix thoroughly. Autoclave for 15 min at 15 psi pressure–121°C. Cool to 45°–50°C. Aseptically add 200.0mL of sterile homogenized whole egg. Mix thoroughly. Aseptically distribute into sterile screw-capped tubes in 7.0mL volumes. Inspissate at 85°–90°C (moist heat) for 60 min.

Use: For the isolation and cultivation of *Mycobacterium* species, except for *Mycobacterium leprae.*

OGYE Agar
See: **Oxytetracycline Glucose Yeast Extract Agar**

Oil Agar Medium

Composition per liter:

Agar, purified	20.0g
NaCl	10.0g
Oil powder	10.0g
NH_4NO_3	1.0g
$MgSO_4$	0.5g
Amphotericin B solution	10.0mL
K_2HPO_4 solution	7.0mL
KH_2PO_4 solution	3.0mL
$FeCl_3$	0.1mL

Oil Powder:

Composition per 10.0g:

Hydrocarbon	10.0g
Silica gel	10.0g
Diethyl ether	30.0mL

Preparation of Oil Powder: Add 10.0g of hydrocarbon to 30.0mL of diethyl ether. Mix thoroughly. Add 10.0g of silica gel. Allow ether to evaporate.

Amphotericin B Solution:

Composition per 10.0mL:

Amphotericin B	0.01g

Preparation of Amphotericin B Solution: Add amphotericin B to distilled/deionized water and bring volume to 10.0mL. Mix thoroughly. Filter sterilize.

K_2HPO_4 Solution:

Composition per 100.0mL:

K_2HPO_4	10.0g

Preparation of K_2HPO_4 Solution: Add K_2HPO_4 to distilled/deionized water and bring volume to 100.0mL. Mix thoroughly. Autoclave for 15 min at 15 psi pressure–121°C. Cool to 25°C.

KH_2PO_4 Solution:

Composition per 100.0mL:

KH_2PO_4	10.0g

Preparation of KH_2PO_4 Solution: Add KH_2PO_4 to distilled/deionized water and bring volume to 100.0mL. Mix thoroughly. Autoclave for 15 min at 15 psi pressure–121°C. Cool to 25°C.

Preparation of Medium: Add components—except amphotericin B solution, K_2HPO_4 solution, and KH_2PO_4 solution—to distilled/deionized water and bring volume to 980.0mL. Mix thoroughly. Gently heat and bring to boiling. Autoclave for 15 min at 15 psi pressure–121°C. Cool to 45°–50°C. Aseptically add 10.0mL of sterile amphotericin B solution, 7.0mL of sterile K_2HPO_4 solution, and 3.0mL of sterile KH_2PO_4 solution. Mix thoroughly. Pour into sterile Petri dishes or distribute into sterile tubes.

Use: For the cultivation and enumeration of hydrocarbon-utilizing bacteria by direct plating of estuarine water and sediment samples.

Oleic Albumin Complex

Composition per liter:

Bovine albumin fraction V	50.0g
NaCl	8.5g
Oleic acid	0.6mL

Preparation of Medium: Add components to distilled/deionized water and bring volume to 1.0L. Mix thoroughly. Filter sterilize.

Use: For use in media employed for the cultivation of mycobacteria.

OM-2 Medium (DSMZ Medium 782)

Composition per liter:

NH_4-oxalate	15.0g
$NaHCO_3$	10.0g
$NaH_2PO_4·2H_2O$	10.0g
$NaS_2O_3·5H_2O$	1.0g
NaCl	0.7g
KCl	0.57g
$MgCl_2·6H_2O$	0.1g
$CaCl_2·2H_2O$	0.01g

pH 6.9 ± 0.2 at 25°C

Preparation of Medium: Add NH_4-oxalate, $NaH_2PO_4·2H_2O$, and $NaHCO_3$ to distilled/deionized water and bring volume to 1.0L. Mix thoroughly after adding each until dissolved. Add remaining components. Mix thoroughly. Distribute into tubes or flasks. Autoclave for 20 min at 15 psi pressure–121°C.

Use: For the cultivation of *Ammoniphilus oxalivorans* and *Ammoniphilus oxalaticus.*

ONE Broth-*Listeria* (Oxoid Novel Enrichment (ONE) Broth-*Listeria*)

Composition per 1005.0mL:

Peptone	28.0g
Salt mix	10.0g
Carbohydrate mix	6.0g
ONE Broth-*Listeria* selective supplement (proprietary)	5.0mL

pH 7.4 ± 0.2 at 25°C

Source: This medium is available as a premixed powder from Oxoid Unipath.

Preparation of Medium: Add components, except ONE Broth-*Listeria* selective supplement, to distilled/deionized water and bring volume to 1.0L. Mix thoroughly. Gently heat while stirring and bring to boiling. Autoclave for 15 min at 15 psi pressure–121°C. Cool to 50°C. Aseptically add 5.0mL ONE Broth-*Listeria* selective supplement. Mix thoroughly. Pour into sterile Petri dishes.

Use: For the selective enrichment of *Listeria* spp. from food and environmental samples. A selective enrichment broth for *Listeria* species from food samples in 24 hr.

Önöz *Salmonella* Agar

Composition per liter:

Agar	15.0g
Sucrose	13.0g
Lactose	11.5g
Trisodium citrate–5,5–hydrate	9.3g
Meat peptone	6.8g
Beef extract	6.0g
L-Phenylalanine	5.0g
$Na_2S_2O_3·5H_2O$	4.25g
Bile salt mixture	3.825g
Yeast extract	3.0g
$Na_2HPO_4·2H_2O$	1.0g
Ferric citrate	0.5g
Metachrome Yellow	0.47g
$MgSO_4·7H_2O$	0.4g
Aniline Blue	0.25g
Neutral Red	0.022g
Brilliant Green	0.00166g

pH 7.1 ± 0.2 at 25°C

Preparation of Medium: Add components to distilled/deionized water and bring volume to 1.0L. Mix thoroughly. Gently heat and bring to boiling. Distribute into tubes or flasks. Autoclave for 15 min at 15 psi pressure–121°C. Pour into sterile Petri dishes or leave in tubes.

Use: For the isolation and cultivation of *Salmonella* from feces.

ONPG Broth

Composition per liter:

Peptone water	750.0mL
ONPG solution	250.0mL

pH 7.2–7.4 at 25°C

ONPG Solution:

Composition per 250.0mL:

ONPG (*o*-nitrophenyl-β-D-galactopyranoside)	1.5g
Sodium phosphate buffer (0.01*M*, pH 7.5)	250.0mL

Preparation of ONPG Solution: Add ONPG to 250.0mL of sodium phosphate buffer. Mix thoroughly. Filter sterilize.

Peptone Water:

Composition per 750.0mL:

Peptone	7.5g
NaCl	3.75g

Preparation of Peptone Water: Add components to distilled/deionized water and bring volume to 750.0mL. Mix thoroughly. Gently heat and bring to boiling. Adjust pH to 8.0–8.4. Continue boiling for 10 min. Filter through Whatman #1 filter paper. Readjust pH of filtrate to 7.2–7.4. Autoclave for 20 min at 10 psi pressure–115°C. Cool to 25°C.

Preparation of Medium: Aseptically combine the sterile ONPG solution with the cooled, sterile peptone water. Mix thoroughly. Aseptically distribute into tubes in 2.5–3.0mL volumes. Store at 4°C for up to 1 month.

Use: For the differentiation of a variety of Gram-negative bacteria based on production of β-galactosidase. For the differentiation of lactose-delayed bacteria from lactose-negative bacteria. For the differentiation of *Pseudomonas cepacia* (positive) and *Pseudomonas maltophila* (positive) from other *Pseudomonas* species (negative). Bacteria that produce β-galactosidase turn the medium yellow.

ONR7a Medium (DSMZ Medium 950)

Composition per liter:

Solution 1	800.0mL
Solution 2	150.0mL
Solution 3	50.0mL

Solution 1:

Composition per 800.0mL:

NaCl	22.79g
Na_2SO_4	3.98g
TAPSO	1.3g
KCl	0.72g
NH_4Cl	0.27g
NaBr	83.0mg
$NaHCO_3$	31.0mg
H_3BO_3	27.0mg
NaF	2.60mg
$Na_2HPO_4 \cdot 7H_2O$	89.0mg

pH 7.6 ± 0.2 at 25°C

Preparation of Solution 1: Add components to distilled/deionized water and bring volume to 800.0mL. Mix thoroughly. Adjust pH to 7.6. with NaOH. Autoclave for 15 min at 15 psi pressure–121°C. Cool to 25°C.

Solution 2:

Composition per 150.0mL:

$MgCl_2 \cdot 6H_2O$	11.18g
$CaCl_2 \cdot 2H_2O$	1.46g
$SrCl_2 \cdot 6H_2O$	24.0mg

Preparation of Solution 2: Add components to distilled/deionized water and bring volume to 150.0mL. Mix thoroughly. Autoclave for 15 min at 15 psi pressure–121°C. Cool to 25°C.

Solution 3:

Composition per 50.0mL:

$FeCl_2 \cdot 4H_2O$	2.0mg

Preparation of Solution 3: Add $FeCl_2 \cdot 4H_2O$ to distilled/deionized water and bring volume to 50.0mL. Mix thoroughly. Autoclave for 15 min at 15 psi pressure–121°C. Cool to 25°C.

Preparation of Medium: Aseptically combine 800.0mL sterile solution 1, 150.0mL sterile solution 2, and 50.0mL sterile solution 3. Mix thoroughly. Distribute into sterile tubes or flasks.

Use: For the cultivation of *Oleiphilus messinensis.*

ONR7a Medium (DSMZ Medium 983)

Composition per liter:

NaCl	22.79g
Agar	15.0g
Na_2SO_4	3.98g
$CaCl_2 \cdot 2H_2O$	1.46g
TAPSO	1.3g
$MgCl_2 \cdot 6H_2O$	1.18g
KCl	0.72g
NH_4Cl	0.27g
Na_2HPO_4	89.0mg
NaBr	83.0mg
$NaHCO_3$	31.0mg
H_3BO_3	27.0mg
$SrCl_2 \cdot 6H_2O$	24.0mg
NaF	2.6mg
$FeCl_2 \cdot 4H_2O$	2.0mg
Tetradecane	10.0mL

pH 7.0 ± 0.2 at 25°C

Preparation of Medium: Add components, except tetradecane, to distilled/deionized water and bring volume to 1.0L. Mix thoroughly. Adjust pH to 7.0. Distribute into tubes or flasks. Gently heat while stirring and bring to boiling. Mix thoroughly. Autoclave for 15 min at 15 psi pressure–121°C. Pour into Petri dishes. Invert after agar solidifies. Add tetradecane to the lid of the Petri dish.

Use: For the cultivation of *Oleispira antarctica.*

OR Indicator Agar (Oxidation-Reduction Indicator Agar)

Composition per liter:

Agar	15.0g
Sodium glycerol phosphate	10.0g
Sodium thioglycolate	1.7g
$CaCl_2 \cdot 2H_2O$	0.1g
Methylene Blue	6.0mg

Preparation of Medium: Add components to distilled/deionized water and bring volume to 1.0L. Mix thoroughly. Gently heat and bring to boiling. Distribute into tubes or flasks. Autoclave for 15 min at 15 psi pressure–121°C. Pour into sterile Petri dishes or leave in tubes.

Use: For use as an indicator of oxygen-free conditions in anaerobic culture chambers.

Oral *Fusobacterium* Medium

Composition per liter:

Agar	15.0g
Proteose peptone	10.0g
Na_2HPO_4	5.0g
Glucose	5.0g
Beef extract	3.0g
Soluble starch	2.0g
$NaNO_3$	1.0g
Yeast extract	1.0g
L-Cysteine·HCl·H_2O	0.5g
Ethyl Violet solution	10.0mL
Bacitracin solution	10.0mL

pH 7.6 ± 0.2 at 25°C

Ethyl Violet Solution:

Composition per 10.0mL:

Ethyl Violet	0.04g

Preparation of Ethyl Violet Solution: Add Ethyl Violet to distilled/deionized water and bring volume to 10.0mL. Mix thoroughly. Filter sterilize.

Bacitracin Solution:

Composition per 10.0mL:

Bacitracin	1.0mg

Preparation of Bacitracin Solution: Add bacitracin to distilled/deionized water and bring volume to 10.0mL. Mix thoroughly. Filter sterilize.

Preparation of Medium: Add components, except Ethyl Violet solution and bacitracin solution, to distilled/deionized water and bring volume to 980.0mL. Mix thoroughly. Gently heat and bring to boiling. Autoclave for 15 min at 15 psi pressure–121°C. Cool to 45°–50°C. Aseptically add sterile Ethyl Violet solution and bacitracin solution. Mix thoroughly. Pour into sterile Petri dishes or distribute into sterile tubes.

Use: For the selective isolation and cultivation of oral *Fusobacterium* species, especially *Fusobacterium nucleatum.*

Oral *Treponema* Medium

Composition per 1250.0mL:

Veal heart, fresh ground	1.0Kg
Thiopeptone	20.0g
NaCl	10.0g
Ionagar No. 2	2.0g
Glutathione (1% solution)	100.0mL
Rabbit serum or ascitic fluid	100.0mL
Eggs, whole fresh	2

pH 6.8–7.0 at 25°C

Preparation of Medium: Add agar to 50.0mL of distilled/deionized water. Gently heat and bring to boiling. Autoclave for 15 min at 15 psi pressure–121°C. Cool to 45°–50°C. In a separate flask, add finely ground veal heart to 1.0L of distilled/deionized water. Add remaining components, except glutathione, rabbit serum, and agar. Gently heat and bring to 70°C. Adjust pH to 7.4. Gently heat and bring to 100°C. Continue heating at 100°C for 2 hr. Skim off fat. Filter through glass wool. Adjust pH to 7.6. Gently heat and bring to 100°C. Maintain at 100°C for 30 min. Store at 4°C for 18 hr. Sterilize in an Arnold sterilizer for 30 min at 100°C on 2 consecutive days. Cool to 45°–50°C. Aseptically add rabbit serum, glutathione solution, and sterile cooled agar solution. Mix thoroughly. Aseptically distribute into sterile tubes or flasks.

Use: For the isolation and cultivation of *Treponema denticola* and *Treponema oralis*.

Orange Serum Agar

Composition per liter:

Agar	17.0g
Pancreatic digest of casein	10.0g
Glucose	4.0g
Yeast extract	3.0g
K_2HPO_4	2.5g
Orange serum	200.0mL

pH 5.5 ± 0.2 at 25°C

Source: This medium is available as a premixed powder from BD Diagnostic Systems.

Preparation of Medium: Add components to distilled/deionized water and bring volume to 1.0L. Mix thoroughly. Gently heat and bring to boiling. Distribute into tubes or flasks. Autoclave for 15 min at 15 psi pressure–121°C. Pour into sterile Petri dishes or leave in tubes.

Use: For the cultivation and enumeration of microorganisms associated with the spoilage of citrus products. For the cultivation of lactobacilli, other aciduric microorganisms, and pathogenic fungi.

Orange Serum Agar

Composition per liter:

Agar	14.0g
Pancreatic digest of casein	10.0g
Glucose	4.0g
Orange serum, equivalent solids	3.5g
Yeast extract	3.0g
K_2HPO_4	2.5g

pH 5.5 ± 0.2 at 25°C

Source: This medium is available as a premixed powder from Oxoid Unipath.

Preparation of Medium: Add components to distilled/deionized water and bring volume to 1.0L. Mix thoroughly. Gently heat and bring to boiling. Distribute into tubes or flasks. Autoclave for 15 min at 15 psi pressure–121°C. Pour into sterile Petri dishes or leave in tubes.

Use: For the isolation and enumeration of spoilage organisms from citrus products.

Orange Serum Agar

Composition per liter:

Agar	15.5g
Orange serum	10.0g
Pancreatic digest of casein	10.0g
Glucose	4.0g
Yeast extract	3.0g
K_2HPO_4	2.5g

pH 5.5 ± 0.2 at 25°C

Source: This medium is available as a premixed powder from BD Diagnostic Systems.

Preparation of Medium: Add components to distilled/deionized water and bring volume to 1.0L. Mix thoroughly. Gently heat and bring to boiling. Distribute into tubes or flasks. Autoclave for 10 min at 15 psi pressure–121°C. Pour into sterile Petri dishes or leave in tubes.

Use: For the enumeration and cultivation of microorganisms from citrus juice and other products. For the cultivation of lactobacilli, pathogenic fungi, and other aciduric microorganisms.

Orange Serum Broth Concentrate 10X

Composition per liter:

Pancreatic digest of casein	10.0g
Glucose	4.0g
Yeast extract	3.0g
K_2HPO_4	2.5g
Orange serum concentrate	100.0mL

pH 5.6 ± 0.2 at 25°C

Source: This medium is available as a premixed solution from BD Diagnostic Systems.

Preparation of Medium: Add components to distilled/deionized water and bring volume to 1.0L. Mix thoroughly. Distribute into tubes or flasks. Autoclave for 15 min at 15 psi pressure–121°C.

Use: For the cultivation and enumeration of microorganisms associated with the spoilage of citrus products. For the cultivation of lactobacilli, other aciduric microorganisms, and pathogenic fungi.

Orange Serum HiVeg Agar

Composition per liter:

Agar	17.0g
Plant hydrolysate	10.0g
Orange serum (solids from 200mL)	9.0g
Glucose	4.0g
Yeast extract	3.0g
K_2HPO_4	2.5g

pH 5.5 ± 0.2 at 25°C

Source: This medium is available as a premixed powder from HiMedia.

Preparation of Medium: Add components to distilled/deionized water and bring volume to 1.0L. Mix thoroughly. Gently heat and bring to boiling. Distribute into tubes or flasks. Autoclave for 15 min at 15 psi pressure–121°C. Pour into sterile Petri dishes or leave in tubes.

Use: For the cultivation and enumeration of microorganisms associated with the spoilage of citrus products. For the cultivation of lactobacilli, other aciduric microorganisms, and pathogenic fungi.

Orange Serum HiVeg Broth

Composition per liter:

Plant hydrolysate	10.0g
Orange serum (solids from 200mL)	9.0g
Glucose	4.0g
Yeast extract	3.0g
K_2HPO_4	2.5g

pH 5.5 ± 0.2 at 25°C

Source: This medium is available as a premixed powder from HiMedia.

Preparation of Medium: Add components to distilled/deionized water and bring volume to 1.0L. Mix thoroughly. Gently heat and bring to boiling. Distribute into tubes or flasks. Autoclave for 15 min at 15 psi pressure–121°C.

Use: For the cultivation of lactobacilli, other aciduric microorganisms, and pathogenic fungi.

Organic Acid Medium KP (Organic Acid Medium, Kauffmann and Petersen)

Composition per liter:

Gelatin	10.0g
Bromthymol Blue	0.024g
Organic acid solution	100.0mL

pH 7.4 ± 0.2 at 25°C

Organic Acid Solution:

Composition per 100.0mL:

Organic acid	10.0g

Preparation of Organic Acid Solution: Add organic acid to distilled/deionized water and bring volume to 100.0mL. Sodium potassium D tartrate, sodium citrate, or mucic acid may be used. Mix thoroughly.

Preparation of Medium: Add components, except organic acid solution, to distilled/deionized water and bring volume to 900.0mL. Mix thoroughly. Gently heat and bring to boiling. Autoclave for 15 min at 15 psi pressure–121°C. Cool to 45°–50°C. Aseptically add sterile organic acid solution. Mix thoroughly. Aseptically distribute into sterile tubes or flasks.

Use: For the cultivation and differentiation of members of the Enterobacteriaceae based on their ability to utilize different organic acids as carbon source. Bacteria that utilize tartrate, citrate, or mucate turn the medium yellow.

Organic Medium 79

Composition per liter:

Glucose	10.0g
Peptone	10.0g
NaCl	6.0g
Pancreatic digest of casein	2.0g
Yeast extract	2.0g

pH 7.8 ± 0.2 at 25°C

Preparation of Medium: Add components to tap water and bring volume to 1.0L. Mix thoroughly. Distribute into tubes or flasks. Autoclave for 15 min at 15 psi pressure–121°C.

Use: For the cultivation of *Acinetobacter* species, *Actinomadura aurantiaca, Actinomadura madurae, Actinomadura pelletieri, Actinomadura vinacea, Amycolatopsis mediterranei, Amycolatopsis orientalis, Amycolatopsis sulphurea, Bacillus cereus, Bacillus laterosporus, Bacillus licheniformis, Bacillus stearothermophilus, Cellulomonas cellulans, Cellulomonas turbata, Citrobacter freundii, Citrobacter koseri, Comamonas acidovorans, Corynebacterium pseudodiphtheriticum, Corynebacterium xerosis, Enterobacter aerogenes, Escherichia coli, Gordona bronchialis, Gordona rubropertinctus, Gordona sputi, Gordona terrae, Microbispora rosea, Microstreptospora cinerea, Mycobacterium fortuitum, Mycobacterium kansasii, Mycobacterium vaccae, Nocardia amarae, Nocardia asteroides, Nocardia brasiliensis, Nocardia carnea, Nocardia farcinica, Nocardia globerula, Nocardia otitidiscaviarum, Nocardia petroleophila, Nocardia* species, *Nocardia transvalensis, Nocardia vaccinii, Nocardioides* species, *Oerskovia* species, *Promicromonospora citrea, Promicromonospora enterophila, Proteus mirabilis, Proteus rettgeri, Proteus vulgaris, Pseudomonas aeruginosa, Pseudomonas boreopolis, Pseudomonas perlurida, Pseudomonas putida, Pseudomonas stutzeri, Rhodococcus coprophilus, Rhodococcus equi, Rhodococcus erythropolis, Rhodococcus maris, Rhodococcus rhodnii, Rhodococcus rhodochrous, Rhodococcus ruber, Saccharomonospora viridis, Saccharopolyspora hirsuta, Saccharopolyspora rectivirgula, Saccharothrix aerocolonigenes, Staphylococcus aureus, Thermoactinomyces glaucus,* and *Tsukamurella paurometabolum.*

Ornithine Broth (BAM M44)

Composition per liter:

L-Ornithine	5.0g
Peptone or gelysate	5.0g
Yeast extract	3.0g
Glucose	1.0g
Bromcresol Purple	0.02g

pH 6.5 ± 0.2 at 25°C

Preparation of Medium: Add components to distilled/deionized water and bring volume to 1.0L. Mix thoroughly. Adjust pH so that it will be 6.5 ± 0.2 after sterilization. Distribute into 16 × 150mm screw-capped tubes in 5.0mL volumes. Autoclave medium with loosely capped tubes for 10 min at 15 psi pressure–121°C. Screw the caps on tightly for storage and after inoculation.

Use: For the cultivation and differentiation of bacteria based on their ability to decarboxylate the amino acid ornithine. Bacteria that decarboxylate ornithine turn the medium turbid purple.

Ornithine Broth with Sodium Chloride (BAM M44)

Composition per liter:

L-Ornithine	5.0g
Peptone or gelysate	5.0g
Yeast extract	3.0g
Glucose	1.0g
Bromcresol Purple	0.02g

pH 6.5 ± 0.2 at 25°C

Preparation of Medium: Add components to distilled/deionized water and bring volume to 1.0L. Mix thoroughly. Adjust pH so that it will be 6.5 ± 0.2 after sterilization. Distribute into 16 × 150mm screw-capped tubes in 5.0mL volumes. Autoclave medium with loosely capped tubes for 10 min at 15 psi pressure–121°C. Screw the caps on tightly for storage and after inoculation.

Use: For the cultivation and differentiation of *Vibrio* spp. based on their ability to decarboxylate the amino acid ornithine. Bacteria that decarboxylate ornithine turn the medium turbid purple.

Orotic Acid Medium

Composition per liter:

K_2HPO_4	6.95g
Tryptone	5.0g
Sodium orotate	2.5g
KH_2PO_4	1.36g
Sodium thioglycolate	0.5g
Yeast extract	0.5g
Riboflavin	15.0mg
Resazurin	1.0mg

pH 7.5 ± 0.2 at 25°C

Preparation of Medium: Add components to distilled/deionized water and bring volume to 1.0L. Mix thoroughly. Sparge with 100% N_2. Anaerobically distribute into tubes or flasks. Autoclave for 15 min at 15 psi pressure–121°C.

Use: For the cultivation and maintenance of *Clostridium oroticum*.

OS Medium (DSMZ Medium 887)

Composition per liter:

$NaHCO_3$	1.0g
$Na_2S_2O_3 \cdot 5H_2O$	1.0g
NaCl	256.0mg
Na_2SO_4	23.0mg
KCl	15.0mg
H_3BO_3	10.3mg
KH_2PO_4	1.7mg
$CaCl_2 \cdot 2H_2O$	0.8mg
$NaNO_3$	0.3mg
$FeCl_3 \cdot 6H_2O$	0.1mg
$MnSO_4 \cdot 4H_2O$	0.06mg
Trace elements solution	10.0mL

pH 7.0 ± 0.2 at 25°C

Trace Elements Solution:

Composition per liter:

$MgSO_4 \cdot 7H_2O$	3.0g
Nitrilotriacetic acid	1.5g
NaCl	1.0g
$MnSO_4 \cdot 2H_2O$	0.5g
$CoSO_4 \cdot 7H_2O$	0.18g
$ZnSO_4 \cdot 7H_2O$	0.18g
$CaCl_2 \cdot 2H_2O$	0.1g
$FeSO_4 \cdot 7H_2O$	0.1g
$NiCl_2 \cdot 6H_2O$	0.025g
$KAl(SO_4)_2 \cdot 12H_2O$	0.02g
H_3BO_3	0.01g
$Na_2MoO_4 \cdot 4H_2O$	0.01g
$CuSO_4 \cdot 5H_2O$	0.01g
$Na_2SeO_3 \cdot 5H_2O$	0.3mg

Preparation of Trace Elements Solution: Add nitrilotriacetic acid to 500.0mL of distilled/deionized water. Dissolve by adjusting pH to 6.5 with KOH. Add remaining components. Add distilled/deionized water to 1.0L. Mix thoroughly.

Preparation of Medium: Add components to distilled/deionized water and bring volume to 1.0L. Mix thoroughly. Adjust pH to 7.0 with Na_2SO_4. Autoclave for 15 min at 15 psi pressure–121°C. Distribute 10mL medium into 28mL serum bottles and seal with a rubber stopper. Gas with atmosphere of 94% N_2 + 3% H_2 + 3% O_2 at 300kPa.

Use: For the cultivation and maintenance of *Thermocrinis ruber*.

OS Solid Medium (DSMZ Medium 887)

Composition per liter:

Gelrite	15.0g
$NaHCO_3$	1.0g
$Na_2S_2O_3 \cdot 5H_2O$	1.0g
NaCl	256.0mg
Na_2SO_4	23.0mg
KCl	15.0mg
H_3BO_3	10.3mg
KH_2PO_4	1.7mg
$CaCl_2 \cdot 2H_2O$	0.8mg
$NaNO_3$	0.3mg
$FeCl_3 \cdot 6H_2O$	0.1mg
$MnSO_4 \cdot 4H_2O$	0.06mg
Trace elements solution	10.0mL

pH 7.0 ± 0.2 at 25°C

Trace Elements Solution:

Composition per liter:

$MgSO_4 \cdot 7H_2O$	3.0g
Nitrilotriacetic acid	1.5g
NaCl	1.0g
$MnSO_4 \cdot 2H_2O$	0.5g
$CoSO_4 \cdot 7H_2O$	0.18g
$ZnSO_4 \cdot 7H_2O$	0.18g
$CaCl_2 \cdot 2H_2O$	0.1g
$FeSO_4 \cdot 7H_2O$	0.1g
$NiCl_2 \cdot 6H_2O$	0.025g
$KAl(SO_4)_2 \cdot 12H_2O$	0.02g
H_3BO_3	0.01g
$Na_2MoO_4 \cdot 4H_2O$	0.01g
$CuSO_4 \cdot 5H_2O$	0.01g
$Na_2SeO_3 \cdot 5H_2O$	0.3mg

Preparation of Trace Elements Solution: Add nitrilotriacetic acid to 500.0mL of distilled/deionized water. Dissolve by adjusting pH to 6.5 with KOH. Add remaining components. Add distilled/deionized water to 1.0L. Mix thoroughly.

Preparation of Medium: Add components to distilled/deionized water and bring volume to 1.0L. Mix thoroughly. Adjust pH to 7.0 with Na_2SO_4. Autoclave for 15 min at 15 psi pressure–121°C. Pour into Petri dishes. Incubate under atmosphere of 94% N_2 + 3% H_2 + 3% O_2.

Use: For the cultivation and maintenance of *Thermocrinis ruber*.

Osmophilic Agar

Composition per liter:

Glucose	50.0g
Agar	15.0g
Yeast extract	3.0g

pH 7.0 ± 0.2 at 25°C

Preparation of Medium: Add components to distilled/deionized water and bring volume to 1.0L. Mix thoroughly. Gently heat and bring to boiling. Distribute into tubes or flasks. Autoclave for 15 min at 15 psi pressure–121°C. Pour into sterile Petri dishes or leave in tubes.

Use: For the cultivation and maintenance of *Zygosaccharomyces rouxii.*

Osmophilic Fungi Medium (M 40 Y) (DSMZ Medium 187)

Composition per liter:

Sucrose	400.0g
Malt extract	20.0g
Agar	20.0g
Yeast extract	5.0g

pH 6.6 ± 0.2 at 25°C

Preparation of Medium: Add components to distilled/deionized water and bring volume to 1.0L. Mix thoroughly. Gently heat and bring to boiling. Distribute into tubes or flasks. Autoclave for 15 min at 15 psi pressure–121°C. Pour into sterile Petri dishes or leave in tubes.

Use: For the cultivation and maintenance of *Eremascus fertilis, Eurotium glabrum, Eurotium tonophilum, Zygosaccharomyces rouxii, Eurotium intermedium, Eremascus albus, Eurotium tonophilum, Chrysosporium xerophilum, Wallemia sebi, Eurotium chevalieri, Eurotium rubrum, Zygosaccharomyces bisporus, Zygosaccharomyces bailii,* and *Zygosaccharomyces rouxii.*

OSrt Broth (ATCC Medium 2340)

Composition per liter:

Yeast extract	2.0 g
Glycerol	10.0 mL

Preparation of Medium: Add components to distilled/deionized water and bring volume to 1.0L. Mix thoroughly. Distribute into flasks or tubes. Autoclave for 15 min at 15 psi pressure–121°C.

Use: For the cultivation and maintenance of *Geodermatophilus obscurus* subspecies *utahensis*.

OTI Medium, Modified

Composition per 1100.0mL:

Beef heart, solids from infusion	100.0g
NaCl	6.0g
Polypeptone™	5.0g
Yeast extract	5.0g
K_2HPO_4	2.0g
Tryptose	2.0g
Agar	1.6g
Pectin	0.8g
Glucose	0.8g
Starch	0.8g
Sucrose	0.8g
Maltose	0.8g
Sodium pyruvate	0.8g
Ribose	0.8g
L-Cysteine·HCl·H_2O	0.68g
$MgSO_4 \cdot 7H_2O$	0.1g
Rumen fluid, clarified	500.0 mL
Rabbit serum, inactivated	50.0mL
Thiamine pyrophosphate solution	50.0mL

pH 7.0 ± 0.2 at 25°C

Thiamine Pyrophosphate Solution:
Composition per 50.0mL:

Thiamine pyrophosphate	7.5mg

Preparation of Thiamine Pyrophosphate Solution: Add thiamine pyrophosphate to distilled/deionized water and bring volume to 50.0mL. Mix thoroughly. Filter sterilize.

Preparation of Medium: Prepare and dispense medium under O_2-free 100% N_2. Add components, except rabbit serum and thiamine pyrophosphate solution, to distilled/deionized water and bring volume to 1.0L. Mix thoroughly. Gently heat and bring to boiling. Adjust pH to 7.0. Anaerobically distribute into tubes in 10.0mL volumes. Autoclave for 15 min at 15 psi pressure–121°C. Cool to 45°–50°C. Aseptically add 0.5 mL of sterile rabbit solution and 0.5mL of sterile thiamine pyrophosphate solution to each tube. Mix thoroughly.

Use: For the cultivation and maintenance of *Treponema socranskii*.

Ottow's Agar Medium

Composition per liter:

Agar	13.0g
Peptone	7.5g
Beef extract	5.0g
NaCl	5.0g
Casamino acids	2.5g
Yeast extract	2.5g
Glucose	1.0g

Preparation of Medium: Add components to tap water and bring volume to 1.0L. Mix thoroughly. Gently heat and bring to boiling. Distribute into tubes or flasks. Autoclave for 15 min at 15 psi pressure–121°C. Pour into sterile Petri dishes or leave in tubes.

Use: For the cultivation and maintenance of *Bacillus pallidus*.

Ottow's Medium

Composition per liter:

Peptone	7.5g
Meat extract	5.0g
NaCl	5.0g
Casamino acids	2.5g
Yeast extract	2.5g
Glucose	1.0g

pH 8.5 ± 0.2 at 25°C

Preparation of Medium: Add components to tap water and bring volume to 1.0L. Mix thoroughly. Distribute into tubes or flasks. Autoclave for 15 min at 15 psi pressure–121°C.

Use: For the cultivation of *Bacillus pallidus*, *Bacillus thermocloacae*, and *Sphaerobacter thermophilus*.

Oxacillin Resistance Screening Agar Base

Composition per liter:

NaCl	55.0g
Agar	12.5g
Peptic digest of animal tissue	11.8g
Mannitol	10.0g
Yeast extract	9.0g
LiCl	5.0g
Aniline Blue	0.2g

pH 7.2 ± 0.2 at 25°C

Source: This medium is available from HiMedia.

Caution: Lithium chloride is harmful. Avoid bodily contact and inhalation of vapors. On contact with skin, wash with plenty of water immediately.

Selective Supplement Solution:
Composition per 10.0mL:

Oxacillin	2.0mg
Polymyxin B	50,000U

Preparation of Selective Supplement Solution: Add components to distilled/deionized water and bring volume to 10.0mL. Mix thoroughly. Filter sterilize.

Preparation of Medium: Add components, except selective supplement solution, to distilled/deionized water and bring volume to 990.0mL. Mix thoroughly. Autoclave for 15 min at 15 psi pressure–121°C. Cool to 50°C. Aseptically add selective supplement solution. Mix thoroughly. Pour into Petri dishes or aseptically distribute into sterile tubes.

Use: For the screening of oxacillin-resistant microorganisms.

Oxalate Maintenance Medium

Composition per liter:

Pancreatic digest of casein	10.0g
Sodium oxalate	5.0g
Na_2CO_3	4.0g
Yeast extract	1.0g
Sodium acetate	0.82g
$(NH_4)_2SO_4$	0.5g
L-Cysteine·HCl·H_2O	0.5g
K_2HPO_4	0.25g
KH_2PO_4	0.25g
$MgSO_4$·$7H_2O$	0.025g
Resazurin	0.001g

pH 6.8 ± 0.2 at 25°C

Preparation of Medium: Add components, except Na_2CO_3 and L-cysteine·HCl·H_2O, to distilled/deionized water and bring volume to 1.0L. Mix thoroughly. Adjust pH to 6.8. Gently heat and bring to boiling. Cool under O_2-free 100% CO_2. Add Na_2CO_3 and L-cysteine·HCl·H_2O. Mix thoroughly. Anaerobically distribute into tubes. Cap with rubber stoppers. Place tubes in a press. Autoclave for 15 min at 15 psi pressure–121°C with fast exhaust.

Use: For the cultivation and maintenance of *Oxalobacter formigenes*.

Oxalate Medium

Basal Medium:
Composition per liter:

K_2HPO_4	4.4g
KH_2PO_4	3.4g
Potasssium oxalate	2.0g
$(NH_4)_2SO_4$	0.5g
$MgSO_4$·$7H_2O$	0.2g
$FeCl_3$	0.015g
Phenol Red (0.4% solution)	5.0mL
Mineral stock solution	1.0mL

Mineral Stock Solution:
Composition per liter:

$ZnSO_4$·$7H_2O$	11.0g
$MnSO_4$·H_2O	5.0g
Na_2MoO_4·$2H_2O$	2.0g
$CoSO_4$	0.05g
H_3BO_3	0.05g
$CuSO_4$·$5H_2O$	7.0mg

Preparation of Mineral Stock Solution: Add components to distilled/deionized water and bring volume to 1.0L. Mix thoroughly.

Preparation of Medium: Add components to distilled/deionized water and bring volume to 1.0L. Mix thoroughly. Distribute into tubes or flasks. Autoclave for 15 min at 15 psi pressure–121°C.

Use: For the isolation and cultivation of oxalate-decomposing *Alcaligenes* species.

Oxalate Medium, Modified

Composition per liter:

Pancreatic digest of casein	10.0g
Sodium oxalate	5.0g
Na_2CO_3	4.0g
Yeast extract	1.0g
Sodium acetate	0.82g
$(NH_4)_2SO_4$	0.5g
L-Cysteine·HCl·H_2O	0.5g
K_2HPO_4	0.25g
KH_2PO_4	0.25g
$MgSO_4$·$7H_2O$	0.025g
Resazurin	0.001g
Trace elements solution	20.0mL

pH 7.0 ± 0.2 at 25°C

Trace Elements Solution:
Composition per liter:

H_3BO_3	0.03g
$CoCl_2$·$6H_2O$	0.02g
$ZnSO_4$·$7H_2O$	0.01g
$MnCl_2$·$4H_2O$	3.0mg
Na_2MoO_4·$2H_2O$	3.0mg
$NiCl_2$·$6H_2O$	2.0mg
$CuCl_2$·$6H_2O$	1.0mg

Preparation of Trace Elements Solution: Add components to distilled/deionized water and bring volume to 1.0L. Mix thoroughly.

Preparation of Medium: Add components, except Na_2CO_3 and L-cysteine·HCl·H_2O, to distilled/deionized water and bring volume to 1.0L. Mix thoroughly. Adjust pH to 7.0. Gently heat and bring to boiling. Cool under O_2-free 100% CO_2. Add Na_2CO_3 and L-cysteine·HCl·H_2O. Mix thoroughly. Anaerobically distribute into tubes. Cap with rubber stoppers. Place tubes in a press. Autoclave for 15 min at 15 psi pressure–121°C with fast exhaust.

Use: For the cultivation and maintenance of *Oxalobacter formigenes*.

Oxalate Utilization Medium

Composition per liter:

Agar	12.0g
Potassium oxalate	1.0g
NaCl	1.0g
$(NH_4)_2HPO_4$	1.0g
KH_2PO_4	0.5g
$MgSO_4$·$7H_2O$	0.2g
$CaCl_2$ solution	80.0mL

pH 7.0 ± 0.2 at 25°C

$CaCl_2$ Solution:
Composition per 100.0mL:

$CaCl_2$	1.47g

Preparation of $CaCl_2$ Solution: Add $CaCl_2$ to distilled/deionized water and bring volume to 100.0mL. Mix thoroughly. Gently heat until dissolved. Filter sterilize.

Preparation of Medium: Add components, except $CaCl_2$ solution, to distilled/deionized water and bring volume to 920.0mL. Mix thoroughly. Gently heat and bring to boiling. Distribute into flasks in 92.0mL volumes. Autoclave for 15 min at 15 psi pressure–121°C. Cool to 45°–50°C. Aseptically add 8.0mL of sterile $CaCl_2$ solution to each flask. A fine precipitate of calcium oxalate will form. Mix thoroughly. Pour into sterile Petri dishes. Swirl flask while dispensing agar to disperse precipitate evenly.

Use: For the cultivation and differentiation of streptomycetes based on oxalate utilization. Bacteria that utilize oxalate turn the medium dark blue.

Oxalobacter Medium

Composition per 1001.0mL:

Sodium oxalate	10.0g
Na_2CO_3	4.0g
Yeast extract	1.0g
Sodium acetate	0.82g
L-Cysteine·HCl	0.5g
$(NH_4)_2SO_4$	0.5g
K_2HPO_4	0.25g
KH_2PO_4	0.25g
$MgSO_4 \cdot 7H_2O$	0.025g
Resazurin	1.0mg
Trace elements solution SL-10	1.0mL

pH 6.8 ± 0.2 at 25°C

Trace Elements Solution SL-10:
Composition per liter:

$FeCl_2 \cdot 4H_2O$	1.5g
$CoCl_2 \cdot 6H_2O$	190.0mg
$MnCl_2 \cdot 4H_2O$	100.0mg
$ZnCl_2$	70.0mg
$Na_2MoO_4 \cdot 2H_2O$	36.0mg
$NiCl_2 \cdot 6H_2O$	24.0mg
H_3BO_3	6.0mg
$CuCl_2 \cdot 2H_2O$	2.0mg
HCl (25% solution)	10.0mL

Preparation of Trace Elements Solution SL-10: Add $FeCl_2 \cdot 4H_2O$ to 10.0mL of HCl solution. Mix thoroughly. Add distilled/deionized water and bring volume to 1.0L. Add remaining components. Mix thoroughly. Sparge with 100% N_2. Autoclave for 15 min at 15 psi pressure–121°C.

Preparation of Medium: Add components to distilled/deionized water and bring volume to 1.0L. Mix thoroughly. Sparge with 100% CO_2. Anaerobically distribute into tubes or flasks. Autoclave for 15 min at 15 psi pressure–121°C.

Use: For the cultivation of *Aureobacterium testaceum* and *Oxalobacter formigenes*.

Oxford Agar (*Listeria* Selective Agar, Oxford)

Composition per liter:

Special peptone	23.0g
LiCl	15.0g
Agar	10.0g
NaCl	5.0g
Cornstarch	1.0g
Esculin	1.0g
Ferric ammonium citrate	0.5g
Antibiotic inhibitor	10.0mL

pH 7.0 ± 0.2 at 25°C

Source: This medium is available as a premixed powder from Oxoid Unipath.

Caution: Lithium chloride is harmful. Avoid bodily contact and inhalation of vapors. On contact with skin, wash with plenty of water immediately.

Antibiotic Inhibitor:
Composition per 10.0mL:

Cycloheximide	0.4g
Colistin sulfate	0.02g
Fosfomycin	0.01g
Acriflavine	5.0mg
Cefotetan	2.0mg
Ethanol (50% solution)	10.0mL

Preparation of Antibiotic Inhibitor: Add antibiotics to 10.0mL of ethanol. Mix thoroughly. Filter sterilize.

Caution: Cycloheximide is toxic. Avoid skin contact or aerosol formation and inhalation.

Preparation of Medium: Add components, except antibiotic inhibitor, to distilled/deionized water and bring volume to 990.0mL. Mix thoroughly. Gently heat and bring to boiling. Autoclave for 15 min at 15 psi pressure–121°C. Cool to 45°–50°C. Aseptically add 10.0mL of sterile antibiotic inhibitor. Mix thoroughly. Pour into sterile Petri dishes or distribute into sterile tubes.

Use: For the isolation and cultivation of *Listeria monocytogenes* from specimens containing a mixed bacterial flora.

Oxford Agar, Modified (*Listeria* Selective Agar, Modified Oxford) (MOX Agar)

Composition per liter:

Special peptone	23.0g
LiCl	15.0g
Agar	12.0g
NaCl	5.0g
Cornstarch	1.0g
Esculin	1.0g
Ferric ammonium citrate	0.5g
Antibiotic inhibitor	10.0mL

pH 7.0 ± 0.2 at 25°C

Caution: Lithium chloride is harmful. Avoid bodily contact and inhalation of vapors. On contact with skin, wash with plenty of water immediately.

Antibiotic Inhibitor:
Composition per 10.0mL:

Moxalactam	0.015g
Colistin sulfate	0.01g

Preparation of Antibiotic Inhibitor: Add components to distilled/deionized water and bring volume to 10.0mL. Mix thoroughly. Filter sterilize.

Preparation of Medium: Add components, except antibiotic inhibitor, to distilled/deionized water and bring volume to 990.0mL. Mix thoroughly. Gently heat and bring to boiling. Autoclave for 10 min at

15 psi pressure–121°C. Cool to 45°–50°C. Aseptically add 10.0mL of sterile antibiotic inhibitor. Mix thoroughly. Pour into sterile Petri dishes or distribute into sterile tubes.

Use: For the isolation and cultivation of *Listeria monocytogenes* from specimens containing a mixed bacterial flora.

Oxford Medium (BAM M118)

Composition per liter:

Special peptone	23.0g
LiCl	15.0g
Agar	10.0g
NaCl	5.0g
Cornstarch	1.0g
Esculin	1.0g
Ferric ammonium citrate	0.5g
Antibiotic inhibitor	10.0mL

pH 7.0 ± 0.2 at 25°C

Source: This medium is available as a premixed powder from Oxoid Unipath.

Caution: Lithium chloride is harmful. Avoid bodily contact and inhalation of vapors. On contact with skin, wash with plenty of water immediately.

Supplement Mix:

Composition per 10.0mL:

Cycloheximide	0.4g
Colistin sulfate	0.02g
Fosfomycin	0.01g
Acriflavine	5.0mg
Cefotetan	2.0mg
Ethanol (50% solution)	10.0mL

Preparation of Supplement Mix: Add components to 10.0mL of ethanol. Mix thoroughly. Filter sterilize.

Caution: Cycloheximide is toxic. Avoid skin contact or aerosol formation and inhalation.

Preparation of Medium: Add components, except supplement mix, to distilled/deionized water and bring volume to 990.0mL. Mix thoroughly. Gently heat and bring to boiling. Autoclave for 15 min at 15 psi pressure–121°C. Cool to 50°C. Aseptically add 10.0mL of sterile supplement mix. Mix thoroughly. Pour into sterile Petri dishes or distribute into sterile tubes.

Use: For the isolation and cultivation of *Listeria monocytogenes* from specimens containing a mixed bacterial flora.

Oxidation-Fermentation Medium (OF Medium)

Composition per liter:

NaCl	5.0g
Agar	2.5g
Pancreatic digest of casein	2.0g
K_2HPO_4	0.3g
Bromthymol Blue	0.03g
Carbohydrate solution	100.0mL

pH 6.8 ± 0.1 at 25°C

Source: This medium is available as a premixed powder from BD Diagnostic Systems.

Carbohydrate Solution:

Composition per 100.0mL:

Carbohydrate	10.0g

Preparation of Carbohydrate Solution: Add carbohydrate to distilled/deionized water and bring volume to 100.0mL. Mix thoroughly. Filter sterilize.

Preparation of Medium: Add components, except carbohydrate solution, to distilled/deionized water and bring volume to 900.0mL. Mix thoroughly. Gently heat and bring to boiling. Autoclave for 15 min at 15 psi pressure–121°C. Cool to 45°–50°C. Aseptically add 100.0mL of sterile carbohydrate solution. Mix thoroughly. Pour into sterile Petri dishes or distribute into sterile tubes.

Use: For differentiating Gram-negative bacteria based upon determining the oxidative and fermentative metabolism of carbohydrates.

Oxidation-Fermentation Medium, Hugh-Leifson's (Hugh-Leifson's Oxidation Fermentation Medium)

Composition per liter:

NaCl	5.0g
Agar	3.0g
Peptone	2.0g
K_2HPO_4	0.3g
Carbohydrate solution	100.0mL
Bromthymol Blue solution (0.2%)	15.0mL

pH 7.1 ± 0.2 at 25°C

Carbohydrate Solution:

Composition per 100.0mL:

Carbohydrate	10.0g

Preparation of Carbohydrate Solution: Add carbohydrate to distilled/deionized water and bring volume to 100.0mL. Mix thoroughly. Filter sterilize.

Preparation of Medium: Add components, except carbohydrate solution, to distilled/deionized water and bring volume to 900.0mL. Mix thoroughly. Gently heat and bring to boiling. Autoclave for 15 min at 15 psi pressure–121°C. Cool to 45°–50°C. Aseptically add 100.0mL of sterile carbohydrate solution. Mix thoroughly. Pour into sterile Petri dishes or distribute into sterile tubes.

Use: For differentiating Gram-negative bacteria, such as *Vibrio* species, based upon determining the oxidative and fermentative metabolism of carbohydrates. Bacteria that ferment the carbohydrate turn the medium yellow.

Oxidation-Fermentation Medium, King's (King's OF Medium)

Composition per liter:

Agar	3.0g
Pancreatic digest of casein	2.0g
Carbohydrate solution	100.0mL
Phenol Red (1.5% solution)	2.0mL
Carbohydrate solution	100.0mL

pH to 7.3 ± 0.2

Preparation of Medium: Add components, except carbohydrate solution, to distilled/deionized water and bring volume to 900.0mL. Mix thoroughly. Gently heat and bring to boiling. Autoclave for 15 min at 15 psi pressure–121°C. Cool to 45°–50°C. Aseptically add 100.0mL of sterile carbohydrate solution. Mix thoroughly. Pour into sterile Petri dishes or distribute into sterile tubes.

Carbohydrate Solution:

Composition per 100.0mL:

Carbohydrate.. 10.0g

Preparation of Carbohydrate Solution: Add carbohydrate to distilled/deionized water and bring volume to 100.0mL. Mix thoroughly. Filter sterilize.

Preparation of Medium: Add components, except carbohydrate solution, to distilled/deionized water and bring volume to 900.0mL. Mix thoroughly. Gently heat and bring to boiling. Autoclave for 15 min at 15 psi pressure–121°C. Cool to 45°–50°C. Aseptically add 100.0mL of sterile carbohydrate solution. Mix thoroughly. Pour into sterile Petri dishes or distribute into sterile tubes.

Use: For differentiating bacteria based upon determining the oxidative and fermentative metabolism of carbohydrates. Bacteria that ferment the carbohydrate turn the medium yellow.

Oxidation-Reduction Indicator Agar
See: **OR Indicator Agar**

Oxidative-Fermentative Medium (OF Medium)

Composition per liter:

NaCl.. 5.0g
Agar.. 2.0g
Pancreatic digest of casein.. 2.0g
K_2HPO_4.. 0.3g
Bromthymol Blue.. 0.08g
Carbohydrate solution.. 100.0mL

pH 6.8 ± 0.1 at 25°C

Source: This medium is available as a premixed powder from BD Diagnostic Systems and Oxoid Unipath.

Carbohydrate Solution:

Composition per 100.0mL:

Carbohydrate.. 10.0g

Preparation of Carbohydrate Solution: Add carbohydrate to distilled/deionized water and bring volume to 100.0mL. Mix thoroughly. Filter sterilize.

Preparation of Medium: Add components, except carbohydrate solution, to distilled/deionized water and bring volume to 900.0mL. Mix thoroughly. Gently heat and bring to boiling. Autoclave for 15 min at 15 psi pressure–121°C. Cool to 45°–50°C. Aseptically add 100.0mL of sterile carbohydrate solution. Mix thoroughly. Pour into sterile Petri dishes or distribute into sterile tubes.

Use: For differentiating bacteria based upon determining the oxidative and fermentative metabolism of carbohydrates. Bacteria that ferment the carbohydrate turn the medium yellow.

Oxidative-Fermentative Glucose Medium, Semisolid (OF Glucose Medium, Semisolid)

Composition per liter:

Glucose.. 10.0g
NaCl.. 5.0g
Agar.. 2.0g
Pancreatic digest of casein.. 2.0g
K_2HPO_4.. 0.3g
Bromthymol Blue.. 0.08g

pH 6.8 ± 0.2 at 25°C

Preparation of Medium: Add components to distilled/deionized water and bring volume to 1.0L. Mix thoroughly. Gently heat and bring to boiling. Distribute into tubes or flasks. Autoclave for 15 min at 15 psi pressure–121°C. Pour into sterile Petri dishes or leave in tubes.

Use: For differentiating Gram-negative bacteria based upon determining the oxidative and fermentative metabolism of glucose. Bacteria that ferment glucose turn the medium yellow.

Oxidative-Fermentative Glucose Medium, Semisolid, with Sodium Chloride (OF Glucose Medium, Semisolid with NaCl)

Composition per liter:

NaCl.. 20.0g
Glucose.. 10.0g
Agar.. 2.0g
Pancreatic digest of casein.. 2.0g
K_2HPO_4.. 0.3g
Bromthymol Blue.. 0.08g

pH 6.8 ± 0.2 at 25°C

Preparation of Medium: Add components to distilled/deionized water and bring volume to 1.0L. Mix thoroughly. Gently heat and bring to boiling. Distribute into tubes or flasks. Autoclave for 15 min at 15 psi pressure–121°C. Pour into sterile Petri dishes or leave in tubes.

Use: For differentiating halophilic *Vibrio* species based upon determining the oxidative and fermentative metabolism of glucose. Bacteria that ferment glucose turn the medium yellow.

Oxidative-Fermentative Test Medium (OF Test Medium)

Composition per liter:

NaCl.. 5.0g
Agar.. 3.0g
Peptone.. 2.0g
K_2HPO_4.. 0.3g
Bromthymol Blue.. 0.03g
Carbohydrate solution.. 100.0mL

Carbohydrate Solution:

Composition per 100.0mL:

Carbohydrate.. 10.0g

Preparation of Carbohydrate Solution: Add carbohydrate to distilled/deionized water and bring volume to 100.0mL. Mix thoroughly. Filter sterilize.

Preparation of Medium: Add components, except carbohydrate solution, to distilled/deionized water and bring volume to 1.0L. Mix thoroughly. Gently heat and bring to boiling. Distribute into tubes in 3.0mL volumes. Autoclave for 15 min at 15 psi pressure–121°C. Cool to 45°–50°C. Aseptically add 0.3mL of sterile carbohydrate solution to each tube. Mix thoroughly.

Use: For the cultivation and differentiation of a variety of microorganisms based on their ability to ferment a specific carbohydrate. Bacteria that ferment the specific carbohydrate turn the medium yellow.

Oxytetra Glucose Yeast Agar Base (OGYE Agar Base)

Composition per liter:

Glucose.. 20.0g
Agar.. 12.0g

Yeast extract 5.0g
Oxytetracycline solution 10.0mL

pH 7.0 ± 0.2 at 25°C

Source: This medium, without oxytetracycline solution, is available as a premixed powder from HiMedia.

Oxytetracycline Solution:
Composition per 10.0mL:
Oxytetracycline 0.1g
Tris(hydroxymethyl) aminomethane buffer (0.1*M*, pH 7.0) 10.0mL

Preparation of Oxytetracycline Solution: Add oxytetracycline to 10.0mL of Tris buffer. Mix thoroughly. Filter sterilize.

Preparation of Medium: Add components, except oxytetracycline solution, to distilled/deionized water and bring volume to 990.0mL. Mix thoroughly. Gently heat and bring to boiling. Autoclave for 10 min at 10 psi pressure–115°C. Cool to 45°–50°C. Aseptically add sterile oxytetracycline solution. Mix thoroughly. Pour into sterile Petri dishes or distribute into sterile tubes.

Use: For the isolation, enumeration, and cultivation of yeasts and other fungi from foods.

Oxytetra Glucose Yeast Agar Base with Biotin

Composition per liter:
Glucose 20.0g
Agar 12.0g
Yeast extract 5.0g
Biotin 0.0001g
Selective supplement solution 10.0mL

pH 7.0 ± 0.2 at 25°C

Source: This medium is available from HiMedia.

Selective Supplement Solution:
Composition per 10.0mL:
Oxytetracycline 50.0mg

Preparation of Selective Supplement Solution: Add oxytetracycline to distilled/deionized water and bring volume to 10.0mL. Mix thoroughly. Filter sterilize.

Preparation of Medium: Add components, except selective supplement solution, to distilled/deionized water and bring volume to 990.0mL. Mix thoroughly. Autoclave for 15 min at 15 psi pressure–121°C. Cool to 50°C. Aseptically add selective supplement solution. Mix thoroughly. Pour into Petri dishes or aseptically distribute into sterile tubes.

Use: For the isolation and enumeration of yeasts and molds from foodstuffs.

Oxytetracycline Glucose Yeast Extract Agar (OGYE Agar)

Composition per liter:
Glucose 20.0g
Agar 12.0g
Yeast extract 5.0g
Biotin 0.1mg
Oxytetracycline solution 10.0mL

pH 7.0 ± 0.2 at 25°C

Source: This medium is available as a premixed powder from Oxoid Unipath.

Oxytetracycline Solution:
Composition per 10.0mL:
Oxytetracycline 0.1g
Tris(hydroxymethyl)
aminomethane buffer (0.1*M*, pH 7.0) 10.0mL

Preparation of Oxytetracycline Solution: Add oxytetracycline to 10.0mL of Tris buffer. Mix thoroughly. Filter sterilize.

Preparation of Medium: Add components, except oxytetracycline solution, to distilled/deionized water and bring volume to 990.0mL. Mix thoroughly. Gently heat and bring to boiling. Autoclave for 10 min at 10 psi pressure–115°C. Cool to 45°–50°C. Aseptically add sterile oxytetracycline solution. Mix thoroughly. Pour into sterile Petri dishes or distribute into sterile tubes.

Use: For the isolation, enumeration, and cultivation of yeasts and other fungi from foods.

Oxytetracycline Glucose Yeast Extract Agar (OGYE Agar)

Composition per liter:
Glucose 20.0g
Agar 12.0g
Yeast extract 5.0g
Oxytetracycline solution 10.0mL

pH 7.0 ± 0.2 at 25°C

Source: This medium is available as a premixed powder from BD Diagnostic Systems.

Oxytetracycline Solution:
Composition per 10.0mL:
Oxytetracycline 0.1g

Preparation of Oxytetracycline Solution: Add oxytetracycline to distilled/deionized water and bring volume to 10.0mL. Mix thoroughly. Filter sterilize.

Preparation of Medium: Add components, except oxytetracycline solution, to distilled/deionized water and bring volume to 990.0mL. Mix thoroughly. Gently heat and bring to boiling. Autoclave for 15 min at 15 psi pressure–121°C. Cool to 45°–50°C. Aseptically add sterile oxytetracycline solution. Mix thoroughly. Pour into sterile Petri dishes or distribute into sterile tubes.

Use: For the isolation, enumeration, and cultivation of yeasts and other fungi from foods.

OZR Medium

Composition per liter:
Agar 15.0g
Pancreatic digest of casein 1.0g
Yeast extract 1.0g

Preparation of Medium: Add components to seawater and bring volume to 1.0L. Mix thoroughly. Gently heat and bring to boiling. Distribute into tubes or flasks. Autoclave for 15 min at 15 psi pressure–121°C. Pour into sterile Petri dishes or leave in tubes.

Use: For the cultivation and maintenance of *Leucothrix mucor*.

P Agar

Composition per liter:
Agar 15.0g
Peptone 10.0g
NaCl 5.0g

Yeast extract 5.0g
Glucose 1.0g

pH 7.5 ± 0.2 at 25°C

Preparation of Medium: Add components to distilled/deionized water and bring volume to 1.0L. Mix thoroughly. Gently heat and bring to boiling. Distribute into tubes or flasks. Autoclave for 15 min at 15 psi pressure–121°C. Pour into sterile Petri dishes or leave in tubes.

Use: For the cultivation of *Staphylococcus* species.

P-2 Medium

Composition per liter:

Yeast extract 5.0g
K_2HPO_4 .... 3.0g
KH_2PO_4 .... 2.0g
NH_4Cl 2.0g
L-Cysteine·HCl 0.5g
$Na_2S \cdot 9H_2O$ 0.5g
$MgCl_2 \cdot 6H_2O$ 0.2g
$CaCl_2 \cdot 2H_2O$ 0.05g
Resazurin 0.5mg
Glucose solution 5.0g

pH 7.0–7.2 at 25°C

Glucose Solution:
Composition per 100.0mL:

D-Glucose 5.0g

Preparation of Glucose Solution: Add glucose to distilled/deionized water and bring volume to 100.0mL. Mix thoroughly. Sparge with 100% N_2. Autoclave for 15 min at 15 psi pressure–121°C.

Preparation of Medium: Prepare and dispense medium under 100% N_2. Add components, except glucose solution, to distilled/deionized water and bring volume to 900.0mL. Mix thoroughly. Adjust pH to 7.0–7.2. Sparge with 100% N_2. Autoclave for 15 min at 15 psi pressure–121°C. Aseptically and anaerobically add 100.0mL of sterile glucose solution. Mix thoroughly. Aseptically and anaerobically distribute into sterile tubes or bottles.

Use: For the cultivation of *Clostridium uzonii, Thermoanaerobium crenophilum,* and *Thermoanaerobium olidum.*

PA Agar

See: ***Pseudomonas aeruginosa*** **Agar**

PA Broth

See: **Presence Absence Broth**

PA HiVeg Broth

Composition per liter:

Plant hydrolysate No. 1 .... 9.83g
Lactose 7.46g
Plant peptone 5.0g
Plant extract 3.0g
NaCl 2.46g
KH_2PO_4 .... 1.35g
K_2HPO_4 .... 1.35g
Sodium lauryl sulfate 0.05g
Bromcresol Purple 0.8.5mg

pH 6.8 ± 0.2 at 25°C

Source: This medium is available as a premixed powder from HiMedia.

Preparation of Medium: Add components to distilled/deionized water and bring volume to 1.0L. Mix thoroughly. Adjust pH to 6.5. Autoclave for 15 min at 15 psi pressure–121°C. Cool to 55°–60°C. Readjust pH to 7.1. Mix thoroughly. Pour into 50mm × 12mm Petri dishes in 3.0mL volumes.

Use: For the cultivation and estimation of numbers of *Pseudomonas aeruginosa* in water by the membrane filter method. For the detection of presence and absence of coliform bacteria in water from treatment plants or distribution systems.

Pablum Cereal Agar

Composition per liter:

Pablum cereal, precooked 100.0g
Agar 18.0g
Chloramphenicol 0.05g

Preparation of Medium: Add components to distilled/deionized water and bring volume to 1.0L. Mix thoroughly. Gently heat and bring to boiling. Distribute into tubes or flasks. Autoclave for 15 min at 15 psi pressure–121°C. Pour into sterile Petri dishes or leave in tubes.

Use: For the cultivation of dematiaceous fungi and stimulation of spore formation.

PA-C Agar (mPA-C Agar)

Composition per liter:

Agar 12.0g
L-Lysine·HCl 5.0g
NaCl 5.0g
$Na_2S_2O_3$ 5.0g
Yeast extract 2.0g
$MgSO_4 \cdot 7H_2O$ 1.5g
Lactose 1.25g
Sucrose 1.25g
Xylose 1.25g
Ferric ammonium citrate 0.8g
Phenol Red 0.08g
Nalidixic acid 0.037g
Kanamycin 8.0mg

pH 7.2 ± 0.1 at 25°C

Source: This medium is available as a premixed powder from BD Diagnostic Systems.

Preparation of Medium: Add components to distilled/deionized water and bring volume to 1.0L. Mix thoroughly. Gently heat and bring to boiling. Distribute into tubes or flasks. Autoclave for 15 min at 15 psi pressure–121°C. Pour into sterile Petri dishes or leave in tubes.

Use: For the selective recovery and enumeration of *Pseudomonas aeruginosa* from water samples.

Packer's Agar

See: **Azide Blood Agar with Crystal Violet**

Pagano Levin Agar

Composition per liter:

Glucose 40.0g
Agar 15.0g
Peptone 10.0g
Yeast extract 1.0g

Neomycin	0.5g
2,3,5-Triphenyltetrazolium chloride	0.1g

pH 6.0 ± 0.1 at 25°C

Source: This medium is available as a premixed powder from BD Diagnostic Systems.

Preparation of Medium: Add components, except neomycin and 2,3,5-triphenyltetrazolium chloride, to distilled/deionized water and bring volume to 1.0L. Mix thoroughly. Gently heat and bring to boiling. Autoclave for 15 min at 15 psi pressure–121°C. Cool to 45°–50°C. Aseptically add neomycin and 2,3,5-triphenyltetrazolium chloride. Mix thoroughly. Pour into sterile Petri dishes or distribute into sterile tubes. Allow tubes to cool in a slanted position.

Use: For the isolation, cultivation, and differentiation of *Candida* species. *Candida albicans* appears as smooth, shiny, cream-light pink colonies.

Pages Balanced Salt Solution (PBS)

Composition per liter:

Solution A	500.0mL
Solution B	500.0mL

Solution A:

Composition per 500.0mL:

Na_2HPO_4	2.84g
KH_2PO_4	2.72g

Preparation of Solution A: Add components to distilled/deionized water and bring volume to 500.0mL. Mix thoroughly. Autoclave for 20 min at 15 psi pressure–121°C. Cool to 25°C.

Solution B:

Composition per 500.0mL:

NaCl	0.24g
$CaCl_2 \cdot 2H_2O$	8.0mg
$MgSO_4 \cdot 7H_2O$	8.0mg

Preparation of Solution B: Add components to distilled/deionized water and bring volume to 500.0mL. Mix thoroughly. Autoclave for 20 min at 15 psi pressure–121°C. Cool to 25°C.

Preparation of Medium: Aseptically combine component solutions. Mix thoroughly. Aseptically distribute into sterile tubes or flasks.

Use: For the cultivation of *Tokophrya lemnarum*.

Pai Medium

Composition per liter:

Homogenized whole egg	666.0mL
NaCl (0.85% solution)	334.0mL

pH 6.75 ± 0.2 at 25°C

Homogenized Whole Egg:

Composition per liter:

Whole eggs	18-24

Preparation of Homogenized Whole Egg: Use fresh eggs, less than 1 week old. Scrub the shells with soap. Let stand in a soap solution for 30 min. Rinse in running water. Soak eggs in 70% ethanol for 15 min. Break the eggs into a sterile container. Homogenize by shaking. Filter through four layers of sterile cheesecloth into a sterile graduated cylinder. Measure out 1.0L.

Preparation of Medium: Combine components. Mix thoroughly. Aseptically distribute into sterile tubes. Inspissate tubes in a slanted position at 80°–90°C (moist heat) for 30 min.

Use: For the maintenance of stock cultures of *Salmonella typhi* and other *Salmonella* species.

Pai Medium

Composition per 1120.0mL:

Glucose	5.0g
Homogenized whole egg	1.0L
Glycerol	120.0mL

pH 6.75 ± 0.2 at 25°C

Homogenized Whole Egg:

Composition per liter:

Whole eggs	18–24

Preparation of Homogenized Whole Egg: Use fresh eggs, less than 1 week old. Scrub the shells with soap. Let stand in a soap solution for 30 min. Rinse in running water. Soak eggs in 70% ethanol for 15 min. Break the eggs into a sterile container. Homogenize by shaking. Filter through four layers of sterile cheesecloth into a sterile graduated cylinder. Measure out 1.0L.

Preparation of Medium: Combine components. Mix thoroughly. Aseptically distribute into sterile tubes. Inspissate tubes in a slanted position at 80°–90°C (moist heat) for 30 min.

Use: For the isolation and cultivation of *Corynebacterium* spp.

PALCAM Agar (Polymyxin Acriflavine Lithium Chloride Ceftazidime Esculin Mannitol Agar)

Composition per liter:

Peptone	23.0g
$LiCl_2$	15.0g
Agar	10.0g
Mannitol	10.0g
NaCl	5.0g
Yeast extract	3.0g
Starch	1.0g
Esculin	0.8g
Ferric ammonium citrate	0.5g
Glucose	0.5g
Phenol Red	0.08g
PALCAM selective supplement	10.0mL

pH 7.2 ± 0.2 at 25°C

Source: This medium is available as a premixed powder from Oxoid Unipath.

PALCAM Selective Supplement:

Composition per 10.0mL:

Ceftazidime	20.0mg
Polymyxin B	10.0mg
Acriflavine·HCl	5.0mg

Preparation of PALCAM Selective Supplement: Add components to distilled/deionized water and bring volume to 10.0mL. Mix thoroughly. Filter sterilize.

Preparation of Medium: Add components, except PALCAM selective supplement, to distilled/deionized water and bring volume to 990.0mL. Mix thoroughly. Gently heat and bring to boiling. Autoclave for 15 min at 15 psi pressure–121°C. Cool to 45°–50°C. Aseptically add sterile PALCAM selective supplement. Mix thoroughly.

Use: For the selective isolation, cultivation, and differentiation of *Listeria monocytogenes* and other *Listeria* species from foods.

PALCAM Agar with Egg Yolk Emulsion (Polymyxin Acriflavine Lithium Chloride Ceftazidime Esculin Mannitol Agar with Egg Yolk Emulsion)

Composition per liter:

Peptone	23.0g
$LiCl_2$	15.0g
Agar	10.0g
Mannitol	10.0g
NaCl	5.0g
Yeast extract	3.0g
Starch	1.0g
Esculin	0.8g
Ferric ammonium citrate	0.5g
Glucose	0.5g
Phenol Red	0.08g
Egg yolk emulsion	25.0mL
PALCAM selective supplement	10.0mL

pH 7.2 ± 0.2 at 25°C

Source: This medium is available as a premixed powder from Oxoid Unipath.

PALCAM Selective Supplement:

Composition per 10.0mL:

Ceftazidime	20.0mg
Polymyxin B	10.0mg
Acriflavine·HCl	5.0mg

Preparation of PALCAM Selective Supplement: Add components to distilled/deionized water and bring volume to 10.0mL. Mix thoroughly. Filter sterilize.

Egg Yolk Emulsion:

Composition:

Chicken egg yolks	11
Whole chicken egg	1

Preparation of Egg Yolk Emulsion: Soak eggs with 1:100 dilution of saturated mercuric chloride solution for 1 min. Crack eggs and separate yolks from whites. Mix egg yolks with 1 chicken egg.

Preparation of Medium: Add components, except PALCAM selective supplement and egg yolk emulsion, to distilled/deionized water and bring volume to 990.0mL. Mix thoroughly. Gently heat and bring to boiling. Autoclave for 15 min at 15 psi pressure–121°C. Cool to 45°–50°C. Aseptically add sterile PALCAM selective supplement and egg yolk emulsion. Mix thoroughly. Pour into sterile Petri dishes or distribute into sterile tubes.

Use: For the selective isolation, cultivation, and differentiation of *Listeria monocytogenes* and other *Listeria* species from foods The addition of egg yolk emulsion aids in the recovery of damaged *Listeria.*

PALCAM Listeria Selective Agar

See: **PALCAM Agar**

Palleroni and Doudoroff Mineral Base Agar, Modified

Composition per liter:

Agar	15.0g
$Na_2HPO_4 \cdot 12H_2O$	6.0g
KH_2PO_4	2.4g
$NH_4 \cdot Cl$	1.0g
$MgSO_4 \cdot 7H_2O$	0.5g
$CaCl_2 \cdot 6H_2O$	0.01g
$FeCl_3 \cdot 6H_2O$	0.01g

pH 6.8 ± 0.2 at 25°C

Preparation of Medium: Add components to distilled/deionized water and bring volume to 1.0L. Mix thoroughly. Gently heat and bring to boiling. Distribute into tubes or flasks. Autoclave for 15 min at 15 psi pressure–121°C. Pour into sterile Petri dishes or leave in tubes.

Use: For the cultivation of *Alcaligenes eutrophus, Alcaligenes latus,* and *Alcaligenes xylosoxydans.*

Pantothenate Assay HiVeg Medium

Composition per liter:

Glucose	40.0g
Sodium acetate	20.0g
Plant acid hydrolysate	10.0g
K_2HPO_4	1.0g
KH_2PO_4	1.0g
L-Cystine	0.4g
$MgSO_4$	0.4g
DL-Tryptophan	0.2g
Adenine sulfate	0.02g
$FeSO_4$	0.02g
Guanine hydrochloride	0.02g
$MnSO_4$	0.02g
NaCl	0.02g
Uracil	0.02g
Pyridoxine	8.0mg
Riboflavin	4.0mg
Thiamine hydrochloride	2.0mg
p-Aminobenzoic acid (PABA)	2.0mg
Niacin	1.0mg
Biotin	0.8μg

pH 6.7 ± 0.1 at 25°C

Source: This medium is available as a premixed powder from HiMedia.

Preparation of Medium: Add components to distilled/deionized water and bring volume to 1.0L. Mix thoroughly. Gently heat and bring to boiling. Continue boiling for 2–3 min. Distribute into tubes in 5.0mL volumes. Add standard solution or test solution to each tube. Adjust the volume of each tube to 10.0mL with distilled/deionized water. Autoclave for 15 min at 15 psi pressure–121°C.

Use: For the microbiological assaying of pantothenic acid and its salts using *Lactobacillus plantarum* as the test organism.

Pantothenate Assay Medium

Composition per 100.0mL:

Glucose	40.0g
Sodium acetate	20.0g
Vitamin assay casamino acids	10.0g
K_2HPO_4	1.0g
KH_2PO_4	1.0g
L-Cystine	0.4g
$MgSO_4 \cdot 7H_2O$	0.4g
DL-Tryptophan	0.2g
Adenine sulfate	0.02g
$FeSO_4 \cdot 7H_2O$	0.02g
Guanine·HCl	0.02g
$MnSO_4 \cdot H_2O$	0.02g

NaCl 0.02g
Uracil 0.02g
Niacin 1.0mg
Pyridoxine 0.8mg
Riboflavin 0.4mg
p-Aminobenzoic acid 0.2mg
Thiamine·HCl 0.2mg
Biotin 0.8μg

pH 6.7 ± 0.1 at 25°C

Source: This medium is available as a premixed powder from BD Diagnostic Systems.

Preparation of Medium: Add components to distilled/deionized water and bring volume to 100.0mL. Mix thoroughly. Gently heat and bring to boiling. Continue boiling for 2–3 min. Distribute into tubes in 5.0mL volumes. Add standard solution or test solution to each tube. Adjust the volume of each tube to 10.0mL with distilled/deionized water. Autoclave for 15 min at 15 psi pressure–121°C.

Use: For the microbiological assaying of pantothenic acid and its salts using *Lactobacillus plantarum* as the test organism.

Pantothenate Assay Medium

Composition per liter:

Glucose 38.0g
Sodium acetate 20.0g
Vitamin-free casamino acids 10.0g
K_2HPO_4 .... 3.0g
$(NH_4)_2SO_4$ 2.0g
NaCl 1.0g
$MgSO_4·7H_2O$ 0.4g
L-Tryptophan 0.1g
$MnSO_4·H_2O$ 0.026g
Xanthine 0.02g
Adenine 0.02g
Guanine 0.02g
Uracil 0.02g
$(NH_4)_2SO_4·FeSO_4·6H_2O$ 0.02g
Niacin 5.0mg
Pyridoxine·HCl 4.0mg
Riboflavin 2.0mg
Thiamine·HCl 1.0mg
p-Aminobenzoic acid 1.0mg
Biotin 0.05mg

Source: This medium is available as a premixed powder from BD Diagnostic Systems.

Preparation of Medium: Add components to distilled/deionized water and bring volume to 1.0L. Mix thoroughly. Gently heat and bring to boiling. Continue boiling for 2–3 min. Distribute into tubes in 5.0mL volumes. Add standard solution or test solution to each tube. Adjust the volume of each tube to 10.0mL with distilled/deionized water. Autoclave for 15 min at 15 psi pressure–121°C.

Use: For determination of the pantothenate content of pharmaceutical products and other materials using *Lactobacillus plantarum* as the test organism.

Pantothenate Culture Agar, USP

Composition per liter:

Yeast extract 20.0g
Agar 15.0g
Glucose 5.0g
Sodium acetate 5.0g

pH 6.8 ± 0.2 at 25°C

Preparation of Medium: Add components to distilled/deionized water and bring volume to 1.0L. Mix thoroughly. Gently heat and bring to boiling. Distribute into tubes or flasks in 10.0mL volumes. Autoclave for 15 min at 15 psi pressure–121°C. Pour into sterile Petri dishes or leave in tubes.

Use: For the maintenance of *Lactobacillus plantarum* used in the microbiological assay of pantothenic acid or pantothenate. Also used for the cultivation of other *Lactobacillus* species.

Pantothenate Inoculum HiVeg Broth

Composition per liter:

Plant hydrolysate No. 4 .... 15.0g
Glucose 10.0g
Tomato juice (100 ml) 5.0g
Yeast extract 5.0g
KH_2PO_4 .... 2.0g
Polysorbate 80 .... 1.0g

pH 6.8 ± 0.1 at 25°C

Source: This medium is available as a premixed powder from HiMedia.

Preparation of Medium: Add components to distilled/deionized water and bring volume to 100.0mL. Mix thoroughly. Gently heat and bring to boiling. Continue boiling for 2–3 min. Distribute into tubes in 5.0mL volumes. Add standard solution or test solution to each tube. Adjust the volume of each tube to 10.0mL with distilled/deionized water. Autoclave for 15 min at 15 psi pressure–121°C.

Use: For the microbiological assaying of pantothenic acid and its salts using *Lactobacillus plantarum* as the test organism.

Pantothenate Medium, AOAC USP

Composition per liter:

Glucose 40.0g
Sodium acetate 20.0g
Vitamin assay casamino acids 10.0g
K_2HPO_4 .... 1.0g
KH_2PO_4 .... 1.0g
L-Cystine 0.4g
$MgSO_4·7H_2O$ 0.4g
L-Tryptophan 0.1g
Sorbitan monooleate complex 0.1g
Adenine sulfate 0.02g
$FeSO_4·7H_2O$ 0.02g
Guanine·HCl 0.02g
$MnSO_4·H_2O$ 0.02g
NaCl 0.02g
Uracil 0.02g
Nicotinic acid 1.0mg
Pyridoxine·HCl 0.8mg
Riboflavin 0.4mg
p-Aminobenzoic acid 0.2mg
Thiamine·HCl 0.2mg
Biotin 0.8μg

pH 6.7 ± 0.1 at 25°C

Source: This medium is available as a premixed powder from BD Diagnostic Systems.

Preparation of Medium: Add components to distilled/deionized water and bring volume to 100.0mL. Mix thoroughly. Gently heat and bring to boiling. Continue boiling for 2–3 min. Distribute into tubes in 5.0mL volumes. Add standard solution or test solution to each tube. Adjust the volume of each tube to 10.0mL with distilled/deionized water. Autoclave for 15 min at 15 psi pressure–121°C.

Use: For the determination of pantothenic acid and its salts using *Lactobacillus plantarum* as the test organism.

Panthenol Assay Medium

Composition per 950.0mL:

Panthenol supplement	475.0mL
Base medium	225.0mL

pH 6.0 ± 0.2 at 25°C

Source: This medium is available as a premixed powder from BD Diagnostic Systems.

Base Medium:

Composition per 900.0mL:

Glucose	15.0g
Pancreatic digest of casein, charcoal treated	10.0g
Sodium citrate	2.0g
Vitamin assay casamino acids	2.0g
K_2HPO_4	1.0g
KH_2PO_4	1.0g
$MgSO_4 \cdot 7H_2O$	0.8g
L-Tryptophan	0.2g
$MnSO_4 \cdot H_2O$	0.16g
L-Cystine	0.15g
$FeSO_4 \cdot 7H_2O$	0.04g
Liver concentrate	0.04g
NaCl	0.04g
Adenine sulfate	0.01g
Guanine·HCl	0.01g
Uracil	0.01g
p-Aminobenzoic acid	2.0mg
β-Alanine	2.0mg
Nicotinic acid	2.0mg
Pyridoxine·HCl	2.0mg
Riboflavin	2.0mg
Thiamine·HCl	2.0mg
Folic acid	0.02mg
Biotin	0.016mg

Preparation of Base Medium: Add components to distilled/deionized water and bring volume to 900.0mL. Mix thoroughly. Gently heat and bring to boiling. Mix thoroughly.

Panthenol Supplement:

Composition per 100.0mL:

Glycerol	33.0g
Sorbitan monooleate complex	2.0g
Lactic acid	0.68g

Preparation of Panthenol Supplement: Add components to distilled/deionized water and bring volume to 100.0mL. Mix thoroughly. Filter sterilize.

Preparation of Medium: Distribute base medium into 50.0mL flasks in 4.5mL volumes. Add standard solution or test solution to each flask. Adjust the volume in each flask to 9.5mL with distilled/deionized water. Autoclave for 10 min at 15 psi pressure–121°C. Cool to 25°C. Aseptically add 0.5mL of panthenol supplement to each flask.

Use: For the microbiological assaying of panthenol using *Gluconobacter oxydans* subspecies *suboxydans* as the test organism.

Papillibacter Medium (DSMZ Medium 872)

Composition per 1070.0mL:

Yeast extract	5.0g
Trypticase™	5.0g
NaCl	1.0g
Cysteine-HCl·H_2O	0.5g
KCl	0.5g
$MgCl_2 \cdot 6H_2O$	0.4g
NH_4Cl	0.3g
K_2HPO_4	0.2g
$CaCl_2 \cdot 2H_2O$	0.1g
Resazurin	0.5mg
$NaHCO_3$ solution	50.0mL
$Na_2S \cdot 9H_2O$ solution	10.0mL
Cinnamate solution	10.0mL
Trace elements solution SL-10	1.0mL

pH 7.0–7.5 at 25°C

Cinnamate Solution:

Composition per 10.0mL:

Na-trans-cinnamate	0.8g

Preparation of Cinnamate Solution: Add Na-trans-cinnamate to distilled/deionized water and bring volume to 10.0mL. Sparge with N_2. Filter sterilie.

$Na_2S \cdot 9H_2O$ Solution:

Composition per 10.0mL:

$Na_2S \cdot 9H_2O$	0.3g

Preparation of $Na_2S \cdot 9H_2O$ Solution: Add $Na_2S \cdot 9H_2O$ to distilled/deionized water and bring volume to 10.0mL. Sparge with N_2. Autoclave for 15 min at 15 psi pressure–121°C. Cool to 25°C. Store anaerobically.

$NaHCO_3$ Solution:

Composition per 100.0mL:

$NaHCO_3$	10.0g

Preparation of $NaHCO_3$ Solution: Add $NaHCO_3$ to distilled/deionized water and bring volume to 100.0mL. Mix thoroughly. Autoclave for 15 min at 15 psi pressure–121°C. Cool to 25°C. Must be prepared freshly.

Trace Elements Solution SL-10:

Composition per liter:

$FeCl_2 \cdot 4H_2O$	1.5g
H_3BO_3	300.0mg
$CoCl_2 \cdot 6H_2O$	190.0mg
$MnCl_2 \cdot 4H_2O$	100.0mg
$ZnCl_2$	70.0mg
$Na_2MoO_4 \cdot 2H_2O$	36.0mg
$NiCl_2 \cdot 6H_2O$	24.0mg
$CuCl_2 \cdot 2H_2O$	2.0mg
HCl (25% solution)	7.7mL

Preparation of Trace Elements Solution SL-10: Add 1.5g of $FeCl_2 \cdot 4H_2O$ to 10.0mL of HCl solution. Mix thoroughly. Add distilled/deionized water and bring volume to 1.0L. Add remaining components. Mix thoroughly. Sparge with 100% N_2. Autoclave for 15 min at 15 psi pressure–121°C. Cool to room temperature.

Preparation of Medium: Prepare and dispense medium under 80% N_2 + 20% CO_2 gas atmosphere. Add components, except cysteine, $NaHCO_3$ solution, $Na_2S·9H_2O$ solution, and cinnamate solution, to distilled/deionized water and bring volume to 1.0L. Mix thoroughly. Gently heat and bring to boiling. Boil for 10 min. Cool to room temperature while sparging with 80% N_2 + 20% CO_2. Add 0.5g cysteine-HCl·H_2O. Mix thoroughly. Adjust pH to 7.0. Distribute into anaerobe tubes. Autoclave for 15 min at 15 psi pressure–121°C. For every 10.0mL medium inject 0.5mL sterile $NaHCO_3$ solution, 0.1mL $Na_2S·9H_2O$ solution, and 0.1mL cinnamate solution. Mix thoroughly. Final pH is 7.0–7.5.

Use: For the cultivation of *Papillibacter cinnamivorans.*

Paracoccus alcaliphilus Medium (DSMZ Medium 772)

Composition per liter:

$(NH_4)_2SO_4$	3.0g
Na_2HPO_4	3.0g
Yeast extract	2.0g
KH_2PO_4	1.4g
$MgSO_4·7H_2O$	0.2g
Fe-citrate	30.0mg
$CaCl_2·2H_2O$	30.0mg
$MnCl_2·4H_2O$	5.0mg
$ZnSO_4·7H_2O$	5.0mg
$CuSO_4·2H_2O$	0.5mg
Methanol	10.0mL
$NaHCO_3$ solution	variable

pH 9.0 ± 0.2 at 25°C

$NaHCO_3$ Solution:

Composition per 50.0mL:

$NaHCO_3$	5.0g

Preparation of $NaHCO_3$ Solution: Add $NaHCO_3$ to distilled/deionized water and bring volume to 50.0mL. Mix thoroughly. Sparge with 80% N_2 + 20% CO_2. Filter sterilize.

Preparation of Medium: Add components, except $NaHCO_3$ solution and methanol, to distilled/deionized water and bring volume to 990.0mL. Mix thoroughly. Autoclave for 15 min at 15 psi pressure–121°C. Filter sterilize 10.0mL of methanol. Aseptically add the 10.0mL filter sterilized methanol. Mix thoroughly. Adjust pH to 9.0 using the sterile $NaHCO_3$ solution.

Use: For the cultivation of *Paracoccus alcaliphilus.*

Paracoccus alcaliphilus Medium (DSMZ Medium 772)

Composition per liter:

$(NH_4)_2SO_4$	3.0g
Na_2HPO_4	3.0g
KH_2PO_4	1.4g
$MgSO_4·7H_2O$	0.2g
Fe-citrate	30.0mg
$CaCl_2·2H_2O$	30.0mg
$MnCl_2·4H_2O$	5.0mg
$ZnSO_4·7H_2O$	5.0mg
$CuSO_4·2H_2O$	0.5mg
Thiamine-HCl·$2H_2O$	0.4mg
Methanol	10.0mL
$NaHCO_3$ solution	variable

pH 9.0 ± 0.2 at 25°C

$NaHCO_3$ Solution:

Composition per 50.0mL:

$NaHCO_3$	5.0g

Preparation of $NaHCO_3$ Solution: Add $NaHCO_3$ to distilled/deionized water and bring volume to 50.0mL. Mix thoroughly. Sparge with 80% N_2 + 20% CO_2. Filter sterilize.

Preparation of Medium: Add components, except $NaHCO_3$ solution and methanol, to distilled/deionized water and bring volume to 990.0mL. Mix thoroughly. Autoclave for 15 min at 15 psi pressure–121°C. Aseptically add 10.0mL filter sterilized methanol. Adjust pH to 9.0 using the sterile $NaHCO_3$ solution.

Use: For the cultivation of *Paracoccus alcaliphilus.*

Paracoccus aminophilus *Paracoccus aminovorans* Medium (DSMZ Medium 774)

Composition per liter:

Agar	20.0g
Peptone	5.0g
Yeast extract	5.0g
Glucose	5.0g

pH 7.0 ± 0.2 at 25°C

Preparation of Medium: Add components to distilled/deionized water and bring volume to 1.0L. Mix thoroughly. Gently heat and bring to boiling. Distribute into tubes or flasks. Autoclave for 15 min at 15 psi pressure–121°C. Pour into sterile Petri dishes or leave in tubes.

Use: For the cultivation and maintenance of *Paracoccus aminovorans* and *Paracoccus aminophilus.*

Paracoccus halodenitrificans Agar

Composition per liter:

NaCl	60.0g
Agar	15.0g
Peptone	5.0g
NaCl	5.0g
Yeast extract	4.0g
Beef extract	1.0g

pH 7.2 ± 0.2 at 25°C

Preparation of Medium: Add components to distilled/deionized water and bring volume to 1.0L. Mix thoroughly. Gently heat and bring to boiling. Distribute into tubes or flasks. Autoclave for 15 min at 15 psi pressure–121°C. Pour into sterile Petri dishes or leave in tubes.

Use: For the cultivation and maintenance of *Deleya aquamarina, Deleya halophila, Deleya venusta, Halomonas halmophila, Marinococcus communis, Micrococcus halobius, Micrococcus varians,* and *Paracoccus halodenitrificans.*

Paracoccus halodenitrificans Agar

Composition per liter:

NaCl	60.0g
Urea	20.0g
Agar	15.0g
Peptone	5.0g
Meat extract	3.0g

pH 7.0 ± 0.2 at 25°C

Preparation of Medium: Add components to distilled/deionized water and bring volume to 1.0L. Mix thoroughly. Adjust pH to 7.0.

Gently heat and bring to boiling. Distribute into tubes or flasks. Autoclave for 15 min at 15 psi pressure–121°C. Pour into sterile Petri dishes or leave in tubes.

Use: For the cultivation of *Paracoccus halodenitrificans*.

Paracoccus kocurii Medium (DSMZ Medium 773)

Composition per liter:

$NaH_2PO_4 \cdot 2H_2O$	1.71g
K_2HPO_4	1.41g
$(NH_4)_2SO_4$	1.0g
Hutner's salt solution	20.0mL
Tetramethyl ammonium chloride solution	10.0mL
Thiamine hydrochloride solution	10.0mL

pH 6.9 ± 0.2 at 25°C

Tetramethyl Ammonium Chloride Solution:

Composition per 10.0mL:

Tetramethyl ammonium chloride	1.0g

Preparation of Tetramethyl Ammonium Chloride Solution: Add tetramethyl ammonium chloride to distilled/deionized water and bring volume to 10.0mL. Mix thoroughly. Filter sterilize.

Thiamine Hydrochloride Solution:

Composition per 10.0mL:

Thiamine-HCl·$2H_2O$	0.5mg

Preparation of Thiamine Hydrochloride Solution: Add thiamine-HCl·$2H_2O$ to distilled/deionized water and bring volume to 10.0mL. Mix thoroughly. Filter sterilize.

Hutner's Salt Solution:

Composition per liter:

$MgSO_4 \cdot 7H_2O$	29.7g
Nitrilotriacetic acid	10.0g
$CaCl_2 \cdot 2H_2O$	3.335g
$FeSO_4 \cdot 7H_2O$	99.0mg
$(NH_4)_6MoO_7O_{24} \cdot 4H_2O$	9.25mg
"Metals 44"	50.0mL

"Metals 44":

Composition per 100.0mL:

$ZnSO_4 \cdot 7H_2O$	1.095g
$FeSO_4 \cdot 7H_2O$	0.5g
Sodium EDTA	0.25g
$MnSO_4$·H2O	0.154g
$CuSO_4 \cdot 5H_2O$	39.2mg
$Co(NO_3)_2 \cdot 6H_2O$	24.8mg
$Na_2B_4O_7 \cdot 10H_2O$	17.7mg

Preparation of "Metals 44": Add sodium EDTA to distilled/deionized water and bring volume to 90.0mL. Mix thoroughly. Add a few drops of concentrated H_2SO_4 to retard precipitation of heavy metal ions. Add remaining components. Mix thoroughly. Bring volume to 100.0mL with distilled/deionized water.

Preparation of Hutner's Salt Solution: Add nitrilotriacetic acid to 500.0mL of distilled/deionized water. Adjust pH to 6.5 with KOH. Add remaining components. Add distilled/deionized water to 1.0L. Adjust pH to 6.8.

Preparation of Medium: Add components, except tetramethyl ammonium chloride solution and thiamine hydrochloride solution, to distilled/deionized water and bring volume to 980.0L. Mix thoroughly. Autoclave for 15 min at 15 psi pressure–121°C. Cool to room temperature. Aseptically add 10.0mL sterile tetramethyl ammonium chloride solution and 10.0mL sterile thiamine hydrochloride solution. Mix thoroughly. Adjust pH to 6.9.

Use: For the cultivation of *Paracoccus kocurii.*

Paraffin Agar

Composition per liter:

Agar	15.0g
K_2HPO_4	6.0g
NH_4NO_3	4.0g
KH_2PO_4	2.0g
Paraffin, liquid	1.0g
$ZnSO_4 \cdot 7H_2O$	0.049g
$MnCl_2 \cdot 4H_2O$	0.046g
$FeSO_4 \cdot 7H_2O$	5.4mg
$CuSO_4 \cdot 5H_2O$	2.5mg
$Na_2B_4O_7 \cdot 10H_2O$	0.94mg
$(NH_4)_6Mo_7O_{24} \cdot 4H_2O$	0.2mg

Preparation of Medium: Add components to distilled/deionized water and bring volume to 1.0L. Mix thoroughly. Gently heat and bring to boiling. Distribute into tubes or flasks. Autoclave for 15 min at 15 psi pressure–121°C. Pour into sterile Petri dishes or leave in tubes.

Use: For the selective isolation and cultivation of streptomycetes.

Paraffin Medium with McClung Carbon-Free Broth

Composition per liter:

Paraffin pellets	1lb
McClung carbon-free broth	1.0L

pH 7.2 ± 0.2 at 25°C

McClung Carbon-Free Broth:

Composition per liter:

$NaNO_3$	2.0g
K_2HPO_4	0.8g
$MgSO_4 \cdot 7H_2O$	0.5g
$FeCl_3$	0.01g
$MnCl_2 \cdot 4H_2O$	0.008g
$ZnSO_4$	0.002g

Preparation of McClung Carbon-Free Broth: Add components to distilled/deionized water and bring volume to 1.0L. Mix thoroughly. Gently heat (low heat for 15–30 min) until salts dissolve. Cool to 25°C. Adjust pH to 7.2, if necessary. Filter sterilize.

Preparation of Medium: Fill glass screw-capped tubes 60% full with paraffin pellets. Place on slanted rack and autoclave for 15 min at 121°C. Let tubes cool in a slanted position. Add 2.5mL of sterile McClung carbon-free broth to each paraffin slant. Tighten screw caps. Sterility is tested by the addition of 2.5mL of sterile Trypticase™ soy broth and sample to each slant.

Use: For use in the sterility testing of various specimens.

Paramecium Medium

Composition per liter:

Solution C	500.0mL
Solution A	10.0mL
Solution B	1.0mL

Solution A:

Composition per liter:

Thiamine·HCl	1.5g
Calcium pantothenate	1.0g

Nicotinamide	0.5g
Pyridoxal·HCl	0.5g
Riboflavin	0.5g
Folic acid	0.5g
α-Lipoic acid	0.1g
Biotin	0.1mg

Preparation of Solution A: Add components to distilled/deionized water and bring volume to 1.0L. Mix thoroughly. Distribute while stirring into screw-capped tubes in 10.0mL volumes. Store at –20°C. Thaw as needed.

Solution B:
Composition per 100.0mL:

TEM-4T (Hachmeister, Pittsburgh)	10.0g
Stigmasterol	0.5g
Ethanol, absolute	100.0mL

Preparation of Solution B: Add TEM-4T and stigmasterol to 100.0mL of hot ethanol. Mix thoroughly. Store at 4°C.

Solution C:
Composition per 500.0mL:

Proteose peptone	10.0g
Pancreatic digest of casein	5.0g
Ribonucleic acid	1.0g
$MgSO_4 \cdot 7H_2O$	0.5g

Preparation of Solution C: Add components to distilled/deionized water and bring volume to 500.0mL. Mix thoroughly.

Preparation of Medium: Combine 500.0mL of solution C, 10.0mL of solution A, and 1.0mL of solution B. Mix thoroughly. Bring volume to 1.0L with distilled/deionized water. Adjust pH to 7.0–7.2 with 0.1*N* NaOH. Autoclave for 20 min at 15 psi pressure–121°C. Cool to 45°–50°C. Aseptically distribute into sterile tubes or flasks.

Use: For the cultivation of *Paramecium* species to be used as host cells by bacterial symbionts.

Park and Sanders Enrichment Broth

Composition per 1010.0mL:

Basal medium	1.0L
Supplement A	5.0mL
Supplement B	5.0mL

pH 7.0 ± 0.2 at 25°C

Basal Medium:
Composition per liter:

Pancreatic digest of casein	10.0g
Peptic digest of animal tissue	10.0g
NaCl	5.0g
Yeast extract	2.0g
Glucose	1.0g
Sodium pyruvate	0.25g
$NaHSO_3$	0.1g
Horse blood, lysed	50.0mL

Preparation of Basal Medium: Add components, except horse blood, to distilled/deionized water and bring volume to 950.0mL. Mix thoroughly. Gently heat and bring to boiling. Autoclave for 15 min at 15 psi pressure–121°C. Cool to 25°C. Aseptically add sterile horse blood. Mix thoroughly.

Supplement A:
Composition per 5.0mL:

Vancomycin	0.01g
Trimethoprim lactate	0.01g

Preparation of Supplement A: Add components to distilled/deionized water and bring volume to 5.0mL. Mix thoroughly. Filter sterilize.

Supplement B:
Composition per 5.0mL:

Cefoperazone	0.032g
Cycloheximide	0.1g

Preparation of Supplement B: Add components to distilled/deionized water and bring volume to 5.0mL. Mix thoroughly. Filter sterilize.

Caution: Cycloheximide is toxic. Avoid skin contact or aerosol formation and inhalation.

Preparation of Medium: To 1.0L of cooled, sterile basal medium, aseptically add 5.0mL of sterile supplement A. Mix thoroughly. Aseptically distribute into flasks in 100.0mL volumes. Inoculate medium with food samples. Incubate at 31°–32°C for 4 hr to recover and resuscitate injured cells. Aseptically add 0.5mL of supplement B to each 100.0mL of medium. Incubate cultures at 37°C for 2 hr.

Use: For the cultivation and enrichment of *Campylobacter* species from foods.

Park and Sanders Enrichment HiVeg Broth Base with Horse Blood and Selective Antibiotics

Composition per liter:

Plant hydrolysate	10.0g
Plant peptone	10.0g
NaCl	5.0g
Yeast extract	2.0g
Glucose	1.0g
Sodium pyruvate	0.25g
Sodium biselenite	0.1g
Horse blood, lysed	50.0mL
Supplement A	5.0mL
Supplement B	5.0mL

pH 7.0 ± 0.2 at 25°C

Source: This medium, without horse blood and selective antibiotic supplements, is available as a premixed powder from HiMedia.

Supplement A:
Composition per 5.0mL:

Vancomycin	0.01g
Trimethoprim lactate	0.01g

Preparation of Supplement A: Add components to distilled/deionized water and bring volume to 5.0mL. Mix thoroughly. Filter sterilize.

Supplement B:
Composition per 5.0mL:

Cefoperazone	0.032g
Cycloheximide	0.1g

Preparation of Supplement B: Add components to distilled/deionized water and bring volume to 5.0mL. Mix thoroughly. Filter sterilize.

Caution: Cycloheximide is toxic. Avoid skin contact or aerosol formation and inhalation.

Preparation of Medium: Add components, except horse blood, supplement A and supplement B, to distilled/deionized water and bring volume to 940.0mL. Mix thoroughly. Gently heat and bring to boiling. Autoclave for 15 min at 15 psi pressure–121°C. Cool to 25°C. Aseptically add 50.0mL sterile horse blood. Mix thoroughly. Aseptically add 5.0mL of sterile supplement A. Mix thoroughly. Aseptically distribute into flasks in 100.0mL volumes. Inoculate medium with food samples. Incubate at 31°–32°C for 4 hr to recover and resuscitate injured cells. Aseptically add 0.5mL of supplement B to each 100.0mL of medium. Incubate cultures at 37°C for 2 hr.

Use: For the cultivation and enrichment of *Campylobacter* species from foods.

Paromomycin Vancomycin Blood Agar
See: PV Blood Agar

Pasteurella haemolytica Selective Medium

Composition per 1010.0mL:

Tryptose agar with peptic digest of blood	1.0L
Antibiotic solution	10.0mL

pH 7.2 ± 0.2 at 25°C

Tryptose Agar with Peptic Digest of Blood:
Composition per liter:

Agar	15.0g
Pancreatic digest of casein	10.0g
Peptic digest of animal tissue	10.0g
NaCl	5.0g
Glucose	1.0g
Peptic digest of blood	50.0mL

Preparation of Tryptose Agar with Peptic Digest of Blood: Add components to distilled/deionized water and bring volume to 950.0mL. Mix thoroughly. Gently heat and bring to boiling. Autoclave for 15 min at 15 psi pressure–121°C. Cool to 45°–50°C. Aseptically add peptic digest of blood. Mix thoroughly.

Antibiotic Solution:
Composition per 10.0mL:

Actidione (cycloheximide)	0.1g
Novobiocin	2.0mg
Neomycin	1.5mg

Preparation of Antibiotic Solution: Add components to distilled/deionized water and bring volume to 10.0mL. Mix thoroughly. Filter sterilize.

Caution: Cycloheximide is toxic. Avoid skin contact or aerosol formation and inhalation.

Preparation of Medium: To 1.0L of cooled, sterile tryptose agar with peptic digest of blood, aseptically add 10.0mL of sterile antibiotic solution. Mix thoroughly. Pour into sterile Petri dishes or distribute into sterile tubes.

Use: For the selective cultivation of *Pasteurella haemolytica*.

Pasteurella multocida Selective Medium

Composition per 1020.0mL:

Tryptose agar with peptic digest of blood	1.0L
Antibiotic solution	10.0mL
K_2TeO_3 solution	10.0mL

pH 7.2 ± 0.2 at 25°C

Tryptose Agar with Peptic Digest of Blood:
Composition per liter:

Agar	15.0g
Pancreatic digest of casein	10.0g
Peptic digest of animal tissue	10.0g
NaCl	5.0g
Glucose	1.0g
Peptic digest of blood	50.0mL

Preparation of Tryptose Agar with Peptic Digest of Blood: Add components to distilled/deionized water and bring volume to 950.0mL. Mix thoroughly. Gently heat and bring to boiling. Autoclave for 15 min at 15 psi pressure–121°C. Cool to 45°–50°C. Aseptically add peptic digest of blood. Mix thoroughly.

Antibiotic Solution:
Composition per 10.0mL:

Actidione (cycloheximide)	0.1g
Novobiocin	0.01g
Erythrocin	5.0mg

Preparation of Antibiotic Solution: Add components to distilled/deionized water and bring volume to 10.0mL. Mix thoroughly. Filter sterilize.

Caution: Cycloheximide is toxic. Avoid skin contact or aerosol formation and inhalation.

K_2TeO_3 Solution:
Composition per 10.0mL:

K_2TeO_3	5.0mg

Preparation of K_2TeO_3 Solution: Add K_2TeO_3 to distilled/deionized water and bring volume to 10.0mL. Mix thoroughly. Filter sterilize.

Caution: Potassium tellurite is toxic.

Preparation of Medium: To 1.0L of cooled, sterile tryptose agar with peptic digest of blood, aseptically add 10.0mL of sterile antibiotic solution and 10.0mL of sterile K_2TeO_3 solution. Mix thoroughly. Pour into sterile Petri dishes or distribute into sterile tubes.

Use: For the selective cultivation of *Pasteurella multocida*.

Payne, Seghal, and Gibbons Medium (DSMZ Medium 1160)

Composition per liter:

NaCl	250.0g
$MgSO_4 \cdot 7H_2O$	20.0g
Yeast extract	10.0g
Casamino acids	7.5g
Trisodium citrate	3.0g
KCl	2.0g
$FeCl_2 \cdot 4H_2O$	36.0mg
$MnCl_2 \cdot 4H_2O$	0.36mg

pH 7.4 ± 0.2 at 25°C

Preparation of Medium: Add components to distilled/deionized water and bring volume to 1.0L. Mix thoroughly. Adjust pH to 7.4. Distribute into tubes or flasks. Autoclave for 15 min at 15 psi pressure–121°C.

Use: For the cultivation of *Haloarcula californiae, Haloarcula hispanica, Haloarcula japonica, Haloarcula marismortui, Haloarcula sinaiiensis, Haloarcula vallismortis, Halobacterium cutirubrum, Halobacterium distributum, Halobacterium lacusprofundi, Halobacterium saccharovorum, Haloferax gibbonsii, Haloferax mediterranei,Halobacterium salinarium, Halobacterium simoncinii, Halobacterium* species, *Halobacterium trapanicum, Halobacterium volcanii, Halococcus morrhuae, Halococcus saccharolyticus, Halococcus* species, *Halococcus turkmenicus, Haloferax denitrificans, Halomonas elongata,* and *Salinicoccus roseus.* For the cultivation of *Natrinema pallidum.*

PB90-1 Medium
(DSMZ Medium 298g)

Composition per liter:

NaCl	1.0g
KCl	0.5g
$MgCl_2 \cdot 6H_2O$	0.4g
NH_4Cl	0.25g
KH_2PO_4	0.2g
$CaCl_2 \cdot 2H_2O$	0.15g
Resazurin	1.0mg
$NaHCO_3$ solution	10.0mL
Butanediol solution	10.0mL
$Na_2S \cdot 9H_2O$ solution	10.0mL
Vitamin solution	10.0mL
Seven vitamin solution	10.0mL
Glucose solution	10.0mL
Trace elements solution SL-10	1.0mL

pH 7.2 ± 0.2 at 25°C

$Na_2S \cdot 9H_2O$ Solution:
Composition per 10.0mL:

$Na_2S \cdot 9H_2O$	0.36g

Preparation of $Na_2S \cdot 9H_2O$ Solution: Add $Na_2S \cdot 9H_2O$ to distilled/deionized water and bring volume to 10.0mL. Mix thoroughly. Autoclave under 100% N_2 for 15 min at 15 psi pressure–121°C. Cool to room temperature.

$NaHCO_3$ Solution:
Composition per 10.0mL:

$NaHCO_3$	2.5g

Preparation of $NaHCO_3$ Solution: Add $NaHCO_3$ to distilled/deionized water and bring volume to 10.0mL. Mix thoroughly. Sparge with 80% N_2 + 20% CO_2. Filter sterilize.

Butanediol Solution:
Composition per 10.0mL:

2,3 butanediol	0.9g

Preparation of Butanediol Solution: Add butanediol to distilled/deionized water and bring volume to 10.0mL. Mix thoroughly. Sparge with 100% N_2. Filter sterilize.

Trace Elements Solution SL-10:
Composition per liter:

$FeCl_2 \cdot 4H_2O$	1.5g
$CoCl_2 \cdot 6H_2O$	190.0mg
$MnCl_2 \cdot 4H_2O$	100.0mg
$ZnCl_2$	70.0mg
$Na_2MoO_4 \cdot 2H_2O$	36.0mg
$NiCl_2 \cdot 6H_2O$	24.0mg
H_3BO_3	6.0mg
$CuCl_2 \cdot 2H_2O$	2.0mg
HCl (25% solution)	10.0mL

Preparation of Trace Elements Solution SL-10: Add $FeCl_2 \cdot 4H_2O$ to 10.0mL of HCl solution. Mix thoroughly. Add distilled/deionized water and bring volume to 1.0L. Add remaining components. Mix thoroughly. Sparge with 80% N_2 + 20% CO_2. Autoclave for 15 min at 15 psi pressure–121°C.

Vitamin Solution:
Composition per liter:

Pyridoxine-HCl	10.0mg
Thiamine-HCl·$2H_2O$	5.0mg
Riboflavin	5.0mg
Nicotinic acid	5.0mg
D-Ca-pantothenate	5.0mg
p-Aminobenzoic acid	5.0mg
Lipoic acid	5.0mg
Biotin	2.0mg
Folic acid	2.0mg
Vitamin B_{12}	0.1mg

Preparation of Vitamin Solution: Add components to distilled/deionized water and bring volume to 1.0L. Mix thoroughly. Sparge with 80% H_2 + 20% CO_2. Filter sterilize.

Seven Vitamin Solution:
Composition per liter:

Pyridoxine hydrochloride	300.0mg
Thiamine-HCl·$2H_2O$	200.0mg
Nicotinic acid	200.0mg
Vitamin B_{12}	100.0mg
Calcium pantothenate	100.0mg
p-Aminobenzoic acid	80.0mg
D(+)-Biotin	20.0mg

Preparation of Seven Vitamin Solution: Add components to distilled/deionized water and bring volume to 1.0L. Sparge with 100% N_2. Mix thoroughly. Filter sterilize.

Glucose Solution:
Composition per 10.0mL:

Glucose	0.7g

Preparation of Glucose Solution: Add glucose to distilled/deionized water and bring volume to 10.0mL. Mix thoroughly. Sparge with 100% N_2. Filter sterilize.

Preparation of Medium: Prepare and dispense medium under 80% N_2 + 20% CO_2 gas atmosphere. Add components, except $NaHCO_3$ solution, butanediol solution, $Na_2S \cdot 9H_2O$ solution, vitamin solution, seven vitamin solution, glucose solution, and trace elements solution SL-10, to distilled/deionized water and bring volume to 939.0mL. Mix thoroughly. Adjust pH to 7.2. Sparge with 80% N_2 + 20% CO_2. Autoclave for 15 min at 15 psi pressure–121°C. Aseptically and anaerobically add 10.0mL $NaHCO_3$ solution, 10.0mL butanediol solution, 10.0mL $Na_2S \cdot 9H_2O$ solution, 10.0mL vitamin solution, 10.0mL seven vitamin solution, 10.0mL glucose solution, and 1.0mL trace elements solution SL-10. Mix thoroughly. Aseptically and anaerobically distribute into sterile tubes or bottles. After inoculation, flush and repressurize the gas head space of culture bottles with sterile 80% N_2 + 20% CO_2 to 1 bar overpressure.

Use: For the cultivation of *Opitutus terrae.*

PBS

See: **Pages Balanced Salt Solution**

PCA

Composition per liter:

Potatoes	300.0g
Carrots	25.0g
Agar	15.0g

Preparation of Medium: Slice potatoes with skin. Peel and slice carrots. Place 300.0g of potatoes and 25.0g of carrots in 1.0L of distilled/deionized water. Gently heat and bring to boiling. Allow to boil for 20 min. Filter through Whatman filer paper. Add agar to filtrate. Bring volume to 1.0 L with distilled/deionized water. Mix thoroughly. Gently heat and bring to boiling. Distribute into tubes or flasks. Autoclave for 15 min at 15 psi pressure–121°C. Pour into sterile Petri dishes or leave in tubes.

Use: For the cultivation of *Cochliobolus ravenelii, Humicola fuscoatra, Humicola grisea, Pyrenophora tritici-repentis,* and *Verticillium fungicola.*

PCA+T+T

Composition per liter:

Potatoes	300.0g
Carrots	25.0g
Agar	15.0g
β-sitosterol	30.0mg
Tryptophan	20.0mg

Preparation of Medium: Slice potatoes with skin. Peel and slice carrots. Place 300.0g of potatoes and 25.0g of carrots in 1.0L of distilled/deionized water. Gently heat and bring to boiling. Allow to boil for 20 min. Filter through Whatman filer paper. Add agar and other components to filtrate. Bring volume to 1.0L with distilled/deionized water. Mix thoroughly. Gently heat and bring to boiling. Distribute into tubes or flasks. Autoclave for 15 min at 15 psi pressure–121°C. Pour into sterile Petri dishes or leave in tubes.

Use: For the cultivation of *Phytophthora nicotianae, Phytophthora citrophthora,* and *Phytophthora cactorum.*

PD2 Medium

Composition per liter:

Agar	15.0g
Pancreatic digest of casein	4.0g
Papaic digest of soybean meal	2.0g
K_2HPO_4	1.5g
Disodium succinate	1.0g
KH_2PO_4	1.0g
$MgSO_4 \cdot 7H_2O$	1.0g
Trisodium citrate	1.0g
Bovine serum albumin solution	10.0mL
Hemin chloride solution	10.0mL

pH 7.0 ± 0.2 at 25°C

Bovine Serum Albumin Solution:

Composition per 10.0mL:

Bovine serum albumin	2.0g

Preparation of Bovine Serum Albumin Solution: Add bovine serum albumin to distilled/deionized water and bring volume to 10.0mL. Mix thoroughly. Filter sterilize.

Hemin Chloride Solution:

Composition per 100.0mL:

Hemin chloride	0.1g
NaOH (0.05*N* solution)	100.0mL

Preparation of Hemin Chloride Solution: Add hemin chloride to 100.0mL of NaOH solution. Mix thoroughly.

Preparation of Medium: Add components, except bovine serum albumin solution, to distilled/deionized water and bring volume to 990.0mL. Mix thoroughly. Gently heat and bring to boiling. Adjust pH to 7.0. Autoclave for 15 min at 15 psi pressure–121°C. Cool to 45°–50°C. Aseptically add sterile bovine serum albumin solution. Mix thoroughly. Pour into sterile Petri dishes or distribute into sterile tubes.

Use: For the isolation and cultivation of PD-ALS (Pierce's disease-almond leaf scorch) bacteria.

PD3 Agar (LMG Medium 200)

Composition per liter:

Agar	15.0g
Pancreatic digest of casein	4.0g
Papaic digest of soybean meal	2.0g
Potato starch	2.0g
K_2HPO_4	1.5g
Trisodium citrate	1.0g
Disodium succinate	1.0g
$MgSO_4 \cdot 7H_2O$	1.0g
KH_2PO_4	1.0g
Hemin chloride solution	10.0mL

pH 7.0 ± 0.2 at 25°C

Hemin Chloride Solution:

Composition per 100.0mL:

Hemin chloride	0.1g
NaOH (0.05*N* solution)	100.0mL

Preparation of Hemin Chloride Solution: Add hemin chloride to 100.0mL of NaOH solution. Mix thoroughly.

Preparation of Medium: Add components to distilled/deionized water and bring volume to 1.0L. Mix thoroughly. Gently heat and bring to boiling. Adjust pH to 7.0. Distribute into tubes or flasks. Autoclave for 15 min at 15 psi pressure–121°C. Pour into sterile Petri dishes or leave into tubes.

Use: For the isolation and cultivation of PD-ALS (Pierce's disease-almond leaf scorch) bacteria.

PDA Agar

See: **Potato Dextrose Agar**

PDA and Yeast Medium

See: **Potato Dextrose Agar and Yeast Medium**

PDA-LUP

Composition per liter:

Potatoes	500.0g
Glucose	20.0g
Agar	15.0g
Lupine stems	variable

Preparation of Medium: Slice potatoes with skin. Place 500.0g of potatoes in 1.0L of distilled/deionized water. Gently heat to 60°C. Allow to steep at 60°C for 60 min. Filter through cheesecloth. Add agar and glucose to filtrate. Bring volume to 1.0L with distilled/deionized water. Mix thoroughly. Gently heat and bring to boiling. Distribute 6.0mL volumes into tubes. Cut lupine stems into 8.0cm-long pieces. Add 2–3 lupine stems per tube. Autoclave for 15 min at 15 psi pressure–121°C. Allow tubes to cool in a slanted position.

Use: For the cultivation of *Ceratocystis adiposa, Ceratocystis coerulescens,* and *Ceratocystis fimbriata.*

PDAmb Agar

See: **Potato Dextrose Agar with Methionine and Biotin**

PDM-114 Medium

Composition per liter:

Base solution	770.0mL
Bovine serum, heat inactivated	100.0mL
Solution A	50.0mL

Solution B 50.0mL
Special 107 vitamin mix 30.0mL

pH 6.8 ± 0.2 at 25°C

Base Solution:

Composition per 770.0mL:

Pancreatic digest of casein 20.0g
NaCl 2.0g
L-Cysteine·HCl 1.0g
Ascorbic acid 0.2g
Ammonium chloride 0.16g
Adenine 73.0mg
Guanosine 59.0mg
Adenosine 5′-monophosphate 52.0mg
Uracil 37.0mg
Cytosine 30.0mg
Ferric ammonium citrate 22.8mg
Adenosine 5′-diphosphate 16.6mg
Adenosine 5′-triphosphate 7.6mg

Preparation of Base Solution: Add components to distilled/deionized water and bring volume to 770.0mL. Mix thoroughly. Adjust pH to 6.8 with NaOH. Autoclave for 15 min at 15 psi pressure–121°C. Cool to room temperature.

Solution A:

Composition per 50.0.0mL:

Glucose 10.0g

Preparation of Solution A: Add glucose to distilled/deionized water and bring volume to 50.0mL. Mix thoroughly. Filter sterilize.

Solution B:

Composition per 50.0.0mL:

K_2HPO_4 ....1.0g
KH_2PO_4 ....0.6g

Preparation of Solution B: Add components to distilled/deionized water and bring volume to 50.0mL. Mix thoroughly. Autoclave for 15 min at 15 psi pressure–121°C. Cool to room temperature.

Special 107 Vitamin Mix:

Composition per 120.0.0mL:

Solution 4 vitamins 100.0mL
Solution 2 ....1.2mL
Solution 1 ....0.4mL
Solution 3 ....0.4mL

Solution 1:

Composition per 100.0mL:

Absolute ethanol 100.0mL
DL-6,8-Thioctic acid, oxidized 100.0mg

Preparation of Solution 1: Add DL-6,8-thioctic acid to absolute ethanol and bring volume to 100.0mL. Mix thoroughly. Filter sterilize.

Solution 2:

Composition per 100.0mL:

Vitamin B_{12} 40.0mg

Preparation of Solution 2: Add vitamin B_{12} to distilled/deionized water and bring volume to 100.0mL. Mix thoroughly. Filter sterilize.

Solution 3:

Tween™ 80 ....50.0g
Absolute ethanol 100.0mL

Preparation of Solution 3: Add Tween™ 80 to absolute ethanol and bring volume to 100.0mL. Mix thoroughly. Filter sterilize.

Solution 4 Vitamins:

Composition per liter:

Vitamin B_{12} 10.0mg
Calcium D-(+)-pantothenate 2.3mg
Choline chloride 1.25mg
Riboflavin 0.7mg
Vitamin A, crystallized alcohol 0.25mg
Calciferol (vitamin D_2) 0.25mg
i-Inositol 0.125mg
p-Aminobenzoic acid 0.125mg
Niacin 0.0625mg
Niacinamide 0.0625mg
Pyridoxal·HCl 0.0625mg
Pyridoxine·HCl 0.0625mg
α-Tocopherol phosphate, disodium salt 0.025mg
Biotin 0.025mg
Folic acid 0.025mg
Menadione (vitamin K_3) 0.025mg
Thiamine·HCl 0.025mg

Preparation of Solution 4 Vitamins: Add components to distilled/deionized water and bring volume to 1.0L. Mix thoroughly. Filter sterilize.

Preparation of Special 107 Vitamin Mix: Aseptically combine 0.4mL of sterile solution 1, 1.2mL of sterile solution 2, 0.4mL of sterile solution 3, and 100.0mL of sterile solution 4 vitamins. Bring volume to 120.0mL with sterile distilled/deionized water.

Preparation of Medium: Aseptically combine 770.0mL of sterile base solution, 100.0mL of heat-inactivated bovine serum, 50.0mL of sterile solution A, 50.0mL of sterile solution B, and 30.0mL of sterile special 107 vitamin mix. Aseptically distribute into sterile tubes or flasks.

Use: For the cultivation of *Entamoeba histolytica.*

PDY Agar

See: **Potato Dextrose Yeast Agar**

PDY Agar

See: **Potato Dextrose Yeast Agar**

PE-2 HiVeg Medium

Composition per liter:

Plant peptone 20.0g
Yeast extract 3.0g
Bromcresol Purple 0.04g
Alaska seed peas ....variable

pH 6.8 ± 0.2 at 25°C

Source: This medium is available as a premixed powder from HiMedia.

Preparation of Medium: Add components, except alaska seed peas, to distilled/deionized water and bring volume to 1.0L. Mix thoroughly. Gently heat and bring to boiling. Distribute into screw-cap tubes. Add 8–10 untreated Alaska seed peas per tube and let the tubes stand for 1 hr to permit hydration. Autoclave for 15 min at 15 psi pressure–121°C.

Use: For the cultivation of *Clostridium botulinum* from foods.

PE2 Medium

Composition per liter:

Peptone 20.0g
Yeast extract 3.0g

Alaska seed peas 416–520
Bromcresol Purple solution 2.0mL

Bromcresol Purple Solution:
Composition per 100.0mL:

Bromcresol Purple 2.0g
Ethanol 10.0mL

Preparation of Bromcresol Purple Solution: Add Bromcresol Purple to 10.0mL of ethanol. Mix thoroughly. Bring volume to 100.0mL with distilled/deionized water. Filter sterilize.

Preparation of Medium: Add components, except Alaska seed peas, to distilled/deionized water and bring volume to 1.0L. Mix thoroughly. Gently heat until dissolved. Add 8–10 Alaska seed peas to each of 18 × 150 mm screw-capped tubes. Distribute the broth into each tube in 19.0mL volumes. Allow tubes to stand for 1 hr. Autoclave for 15 min at 15 psi pressure–121°C.

Use: For the cultivation of *Clostridium botulinum* from foods.

Pea Agar Medium

Composition per 3.0L:

Dry peas 360.0g
Agar 45.0g

Preparation of Medium: Add dry peas to 2.2L of distilled/deionized water. Autoclave for 2 hr at 15 psi pressure–121°C. Filter pea mash through cheesecloth. Reserve filtrate. Bring volume of filtrate to 3.0L with distilled/deionized water. Mix thoroughly. Add agar. Gently heat and bring to boiling. Distribute into tubes or flasks. Autoclave for 15 min at 15 psi pressure–121°C. Pour into sterile Petri dishes or leave in tubes. Allow tubes to cool in a slanted position.

Use: For the cultivation and maintenance of *Ascochyta pinodes, Ascochyta pisi,* and *Phoma medicaginis.*

Pectin Agar

Composition per plate:

Base agar 15.0mL
Pectin gel 7.0mL

Base Agar:
Composition per liter:

Agar 15.0g
K_2HPO_4 .... 5.0g
$CaCl_2 \cdot 2H_2O$ 3.0g
NH_4Cl 1.0g
$MgSO_4 \cdot 7H_2O$ 0.2g
Tris(hydroxymethyl)amino–
methane buffer (1*M*, pH 8.0) 100.0mL
Trace elements solution 1.0mL

Preparation of Base Agar: Add components to distilled/deionized water and bring volume to 1.0L. Mix thoroughly. Gently heat and bring to boiling. Autoclave for 15 min at 15 psi pressure–121°C. Cool to 45°–50°C.

Trace Elements Solution:
Composition per liter:

Disodium EDTA 8.0g
$MnCl_2 \cdot 4H_2O$ 0.1g
$CoCl_2 \cdot 6H_2O$ 0.02g
$ZnCl_2$ 0.02g
KBr 0.02g
KI 0.02g
$CuSO_4$ 0.01g
$Na_2MoO_4 \cdot 2H_2O$ 0.01g
H_3BO_3 .... 0.01g
LiCl 5.0mg
$SnCl_2 \cdot 2H_2O$ 5.0mg

Preparation of Trace Elements Solution: Add components to distilled/deionized water and bring volume to 1.0L. Mix thoroughly.

Pectin Gel:
Composition per liter:

Pectin, low esterified 20.0g

Preparation of Pectin Gel: Add pectin to 70°C distilled/deionized water and bring volume to 1.0L. Autoclave for 10 min at 7 psi pressure–110°C. Cool to 45°–50°C.

Preparation of Medium: Pour cooled, sterile base agar into sterile Petri dishes in 15.0mL volumes. Allow agar to solidify. Overlay each plate with 7.0mL of cooled, sterile pectin gel. The pectin may take 5 hr to form a gel.

Use: For the isolation and cultivation of *Cytophaga* species, *Herpetosiphon* species, *Saprospira* species, and *Flexithrix* species.

Pectin Medium

Composition per liter:

Pectin 30.0g
Yeast extract 5.0g
Bromthymol Blue (0.1% solution) 1.0mL
$CaCl_2 \cdot 2H_2O$ (10.0% solution) 0.6mL

pH 7.3 ± 0.2 at 25°C

Preparation of Medium: Add 100.0mL of distilled/deionized water to a 2.0L flask and place on a magnetic stirrer with no heat. While stirring, slowly add $CaCl_2 \cdot 2H_2O$, Bromthymol Blue solution, yeast extract, and pectin. Add slowly to ensure each particle is wetted. Gently heat and bring to almost boiling. Adjust pH to 7.3 with 1*N* NaOH. Do not overshoot pH 7.3.

Use: For the isolation and presumptive identification of *Yersinia enterocolitica* and *Yersinia pseudotuberculosis* from other fermenting Gram-negative bacilli such as *Enterocolitica agglomerans* which is often confused with *Yersinia. Enterocolitica agglomerans* does not produce pectinase. *Yersinia enterocolitica* is strongly positive for pectinase activity. *Yersinia pseudotuberculosis* is weakly positive for pectinase activity. *Yersinia pestis* is negative for pectinase activity. Also used for the differentiation of *Klebsiella oxytoca*, which is pectinase positive.

Pectobacterium carotovorum Medium (LMG Medium 172)

Composition per liter:

Glucose 5.0g
Peptone 5.0g
Yeast extract 3.0g

pH 7.2 ± 0.2 at 25°C

Preparation of Medium: Add components to distilled/deionized water and bring volume to 1.0L. Mix thoroughly. Distribute into tubes or flasks. Autoclave for 15 min at 15 psi pressure–121°C.

Use: For the cultivation of *Pectobacterium carotovorum* subsp. *odoriferum.*

Pediococccus cereviseae and *Aerococcus viridans* Medium

Composition per liter:

Pancreatic digest of casein	12.5g
Glucose	10.0g
Yeast extract	7.5g
K_2HPO_4	5.0g
NaCl	5.0g
Sodium citrate	5.0g
$MgSO_4$	0.8g
Tween™ 80	0.2g
$MnCl_2$	0.14g
$FeSO_4$	0.04g
Thiamine·HCl	0.1mg

pH 6.7 ± 0.2 at 25°C

Preparation of Medium: Add components to distilled/deionized water and bring volume to 1.0L. Mix thoroughly. Gently heat until dissolved. Distribute into tubes or flasks. Autoclave for 15 min at 15 psi pressure–121°C.

Use: For the cultivation of *Pediocococcus cereviseae* and *Aerococcus viridans*.

Pediococcus damnosus Medium

Composition per liter:

Glucose	20.0g
Tryptic digest of casein	10.0g
Meat extract	10.0g
Sodium acetate	5.0g
Yeast extract	5.0g
K_2HPO_4	2.0g
Diammonium citrate	2.0g
Tween™ 80	1.0g
$MgSO_4 \cdot 7H_2O$	0.2g
$MnSO_4 \cdot H_2O$	0.05g

pH 5.2 ± 0.2 at 25°C

Preparation of Medium: Add components to distilled/deionized water and bring volume to 1.0L. Mix thoroughly. Adjust pH to 5.2. Distribute into tubes or flasks. Autoclave for 15 min at 15 psi pressure–121°C.

Use: For the cultivation of *Lactobacillus acetotolerans, Lactobacillus lindneri*, and *Pediococcus damnosus*.

Pediococcus halophilus Medium

Composition per liter:

NaCl	65.0g
Glucose	20.0g
Pancreatic digest of casein	10.0g
Meat extract	10.0g
Sodium acetate	5.0g
Yeast extract	5.0g
K_2HPO_4	2.0g
Diammonium citrate	2.0g
Tween™ 80	1.0g
$MgSO_4 \cdot 7H_2O$	0.2g
$MnSO_4 \cdot H_2O$	0.05g

Preparation of Medium: Add components to distilled/deionized water and bring volume to 1.0L. Mix thoroughly. Distribute into tubes or flasks. Autoclave for 15 min at 15 psi pressure–121°C.

Use: For the cultivation of *Tetragenococcus halophilus*.

Pediococcus Medium (Lactobacilli MRS Broth)

Composition per liter:

Glucose	20.0g
Beef extract	10.0g
Peptone	10.0g
Sodium acetate	5.0g
Yeast extract	5.0g
Ammonium citrate	2.0g
Na_2HPO_4	2.0g
Tween™ 80	1.0g
$MgSO_4 \cdot 7H_2O$	0.1g
$MnSO_4 \cdot 5H_2O$	0.05g

pH 5.2 ± 0.2 at 25°C

Source: This medium is available from BD Diagnostic Systems.

Preparation of Medium: Add components to distilled/deionized water and bring volume to 1.0L. Mix thoroughly. Gently heat and bring to boiling. Distribute into tubes or flasks. Autoclave for 15 min at 15 psi pressure–121°C.

Use: For the cultivation and maintenance of *Pediococcus damnosus*.

Pediococcus Medium with Mevalonic Acid (Lactobacilli MRS Broth with Melvalonic Acid)

Composition per liter:

Mevalonic acid	30.0g
Glucose	20.0g
Beef extract	10.0g
Peptone	10.0g
Sodium acetate	5.0g
Yeast extract	5.0g
Ammonium citrate	2.0g
Na_2HPO_4	2.0g
Tween™ 80	1.0g
$MgSO_4 \cdot 7H_2O$	0.1g
$MnSO_4 \cdot 5H_2O$	0.05g

pH 5.2 ± 0.2 at 25°C

Preparation of Medium: Add components to distilled/deionized water and bring volume to 1.0L. Mix thoroughly. Gently heat and bring to boiling. Distribute into tubes or flasks. Autoclave for 15 min at 15 psi pressure–121°C.

Use: For the cultivation and maintenance of *Pediococcus damnosus*.

Pedomicrobium PSM Medium

Composition per liter:

Yeast extract	0.5g
Sodium acetate solution	10.0mL
Vitamin solution	10.0mL
Trace elements solution	1.0mL

pH 7.1 ± 0.2 at 25°C

Sodium Acetate Solution:

Composition per 10.0mL:

Sodium acetate·$3H_2O$	1.36g

Preparation of Sodium Acetate Solution: Add sodium acetate to distilled/deionized water and bring volume to 10.0mL. Mix thoroughly. Filter sterilize.

Vitamin Solution:

Composition per liter:

Pyridoxine·HCl	10.0mg
Thiamine·HCl	5.0mg
Calcium pantothenate	5.0mg
Riboflavin	5.0mg
Nicotinic acid	5.0mg
p-Aminobenzoic acid	5.0mg
D-Biotin	2.0mg
Folic acid	2.0mg
Cyanocobalamin	0.1mg

Preparation of Vitamin Solution: Add components to distilled/deionized water and bring volume to 1.0L. Mix thoroughly. Filter sterilize.

Trace Elements Solution:

Composition per 50.0mL:

$ZnSO_4 \cdot 7H_2O$	0.55g
$FeSO_4 \cdot 7H_2O$	0.25g
EDTA	0.125g
$MnSO_4 \cdot H_2O$	0.077g
$CuSO_4 \cdot 5H_2O$	0.02g
$Co(NO_3)_2 \cdot 6H_2O$	0.012g
$Na_2B_4O_7 \cdot 10H_2O$	8.85mg

Preparation of Trace Elements Solution: Add components to distilled/deionized water and bring volume to 50.0mL. Mix thoroughly. Filter sterilize.

Preparation of Medium: Add components, except sodium acetate solution and vitamin solution, to distilled/deionized water and bring volume to 980.0mL. Mix thoroughly. Adjust pH to 7.1 with KOH. Autoclave for 15 min at 15 psi pressure–121°C. Cool to 25°C. Aseptically add sterile sodium acetate solution and 10.0mL of sterile vitamin solution. Mix thoroughly. Aseptically distribute into sterile tubes or flasks.

Use: For the cultivation and maintenance of *Pedomicrobium americanum* and *Pedomicrobium ferrugineum*.

Pedomicrobium PSM Medium with Ribose

Composition per liter:

Peptone	0.25g
Yeast extract	0.25g
Ribose solution	10.0mL
Vitamin solution	10.0mL
Trace elements solution	1.0mL

pH 7.1 ± 0.2 at 25°C

Ribose Solution:

Composition per 100.0mL:

D-Ribose	10.0g

Preparation of Ribose Solution: Add ribose to distilled/deionized water and bring volume to 100.0mL. Mix thoroughly. Filter sterilize.

Vitamin Solution:

Composition per liter:

Pyridoxine·HCl	10.0mg
Thiamine·HCl	5.0mg
Calcium pantothenate	5.0mg
Riboflavin	5.0mg
Nicotinic acid	5.0mg
p-Aminobenzoic acid	5.0mg
D-Biotin	2.0mg
Folic acid	2.0mg
Cyanocobalamin	0.1mg

Preparation of Vitamin Solution: Add components to distilled/deionized water and bring volume to 1.0L. Mix thoroughly. Filter sterilize.

Trace Elements Solution:

Composition per 50.0mL:

$ZnSO_4 \cdot 7H_2O$	0.55g
$FeSO_4 \cdot 7H_2O$	0.25g
EDTA	0.125g
$MnSO_4 \cdot H_2O$	0.077g
$CuSO_4 \cdot 5H_2O$	0.02g
$Co(NO_3)_2 \cdot 6H_2O$	0.012g
$Na_2B_4O_7 \cdot 10H_2O$	8.85mg

Preparation of Trace Elements Solution: Add components to distilled/deionized water and bring volume to 50.0mL. Mix thoroughly. Filter sterilize.

Preparation of Medium: Add components, except ribose solution and vitamin solution, to distilled/deionized water and bring volume to 980.0mL. Mix thoroughly. Adjust pH to 7.1 with KOH. Autoclave for 15 min at 15 psi pressure–121°C. Cool to 25°C. Aseptically add 10.0mL of sterile ribose solution and 10.0mL of sterile vitamin solution. Mix thoroughly. Aseptically distribute into sterile tubes or flasks.

Use: For the cultivation and maintenance of *Pedomicrobium americanum* and *Pedomicrobium ferrugineum*.

Pedomicrobium PYVM Medium

Composition per liter:

Peptone	0.25g
Yeast extract	0.25g
Trace elements solution	20.0mL
Malate solution	10.0mL
Vitamin solution	10.0mL

pH 7.1 ± 0.2 at 25°C

Malate Solution:

Composition per 100.0mL:

Malic acid	1.36g

Preparation of Malate Solution: Add malic acid to distilled/deionized water and bring volume to 100.0mL. Adjust pH to 7.0 with concentrated NaOH. Mix thoroughly. Filter sterilize.

Vitamin Solution:

Composition per liter:

Pyridoxine·HCl	10.0mg
Thiamine·HCl	5.0mg
Calcium pantothenate	5.0mg
Riboflavin	5.0mg
Nicotinic acid	5.0mg
p-Aminobenzoic acid	5.0mg
D-Biotin	2.0mg
Folic acid	2.0mg
Cyanocobalamin	0.1mg

Preparation of Vitamin Solution: Add components to distilled/deionized water and bring volume to 1.0L. Mix thoroughly. Filter sterilize.

Trace Elements Solution:

Composition per 50.0mL:

$ZnSO_4 \cdot 7H_2O$	0.55g
$FeSO_4 \cdot 7H_2O$	0.25g
EDTA	0.125g
$MnSO_4 \cdot H_2O$	0.077g

$CuSO_4 \cdot 5H_2O$ 0.02g
$Co(NO_3)_2 \cdot 6H_2O$ 0.012g
$Na_2B_4O_7 \cdot 10H_2O$ 8.85mg

Preparation of Trace Elements Solution: Add components to distilled/deionized water and bring volume to 50.0mL. Mix thoroughly. Filter sterilize.

Preparation of Medium: Add components, except malate solution and vitamin solution, to distilled/deionized water and bring volume to 980.0mL. Mix thoroughly. Adjust pH to 7.1 with KOH. Autoclave for 15 min at 15 psi pressure–121°C. Cool to 25°C. Aseptically add 10.0mL of sterile malate solution and 10.0mL of sterile vitamin solution. Mix thoroughly. Aseptically distribute into sterile tubes or flasks.

Use: For the cultivation of *Pedomicrobium* species.

Pedomicrobium PYVM Medium

Composition per liter:
Peptone 0.25g
Yeast extract 0.25g
Modified Hutner's basal salts 20.0mL
DL-Malate solution 10.0mL
Vitamin solution 10.0mL
pH 7.5 ± 0.2 at 25°C

Modified Hutner's Basal Salts:
Composition per liter:
$MgSO_4 \cdot 7H_2O$ 29.7g
Nitrilotriacetic acid 10.0g
$CaCl_2 \cdot 2H_2O$ 3.34g
$FeSO_4 \cdot 7H_2O$ 0.1g
$(NH_4)_2MoO_4$ 9.25mg
Metals "44" 50.0mL

Preparation of Modified Hutner's Basal Salts: Add nitrilotriacetic acid to 500.0mL of distilled/deionized water. Dissolve by adjusting pH to 6.5 with KOH. Add remaining components. Readjust pH to 7.2 with H_2SO_4 or KOH. Add distilled/deionized water to 1.0L. Store at 5°C.

Metals "44":
Composition per 100.0 mL:
$ZnSO_4 \cdot 7H_2O$ 1.1g
$FeSO_4 \cdot 7H_2O$ 0.5g
EDTA 0.25g
$MnSO_4 \cdot 7H_2O$ 0.154g
$CuSO_4 \cdot 5H_2O$ 0.04g
$Co(NO_3)_2 \cdot 6H_2O$ 0.025g
$Na_2B_4O_7 \cdot 10H_2O$ 0.018g

Preparation of Metals "44": Add a few drops of H_2SO_4 to distilled/deionized water to inhibit precipitate formation. Add components to acidified distilled/deionized water and bring volume to 100.0mL. Mix thoroughly.

DL-Malate Solution:
Composition per 100.0mL:
DL-Malic acid 20.0g

Preparation of DL-Malate Solution: Add malic acid to distilled/deionized water and bring volume to 80.0mL. Mix thoroughly. Adjust pH to 7.5 with NaOH. Bring volume to 100.0mL with distilled/deionized water. Filter sterilize.

Vitamin Solution:
Composition per liter:
Pyridoxine·HCl 10.0mg
Riboflavin 5.0mg
Nicotinamide 5.0mg
p-Aminobenzoic acid 5.0mg
Thiamine·HCl 5.0mg
Calcium D-(+)-pantothenate 5.0mg
D-(+)-Biotin 2.0mg
Folic acid 2.0mg
Cyanocobalamine 0.1mg

Preparation of Vitamin Solution: Add components to distilled/deionized water and bring volume to 1.0L. Mix thoroughly. Filter sterilize.

Preparation of Medium: Add components, except DL-malate solution and vitamin solution, to distilled/deionized water and bring volume to 980.0mL. Mix thoroughly. Gently heat and bring to boiling. Autoclave for 15 min at 15 psi pressure–121°C. Aseptically add 10.0mL of sterile DL-malate solution and 10.0mL of sterile vitamin solution. Mix thoroughly. Distribute into sterile tubes or flasks.

Use: For the cultivation of *Pedomicrobium americanum.*

Pelobacter acetylenicus Medium

Composition per liter:
NaCl 20.0g
$MgCl_2 \cdot 6H_2O$ 3.0g
KCl 0.5g
NH_4Cl 0.25g
KH_2PO_4 .... 0.2g
$CaCl_2 \cdot 2H_2O$ 0.15g
Resazurin 1.0mg
$NaHCO_3$ solution 10.0mL
$Na_2S \cdot 9H_2O$ solution 10.0mL
Acetoin solution 10.0mL
Trace elements solution SL-10 .... 1.0mL
pH 7.2 ± 0.2 at 25°C

$NaHCO_3$ Solution:
Composition per 10.0mL:
$NaHCO_3$ 2.5g

Preparation of $NaHCO_3$ Solution: Add $NaHCO_3$ to distilled/deionized water and bring volume to 10.0mL. Mix thoroughly. Sparge with 80% N_2 + 20% CO_2. Autoclave for 15 min at 15 psi pressure–121°C.

$Na_2S \cdot 9H_2O$ Solution:
Composition per 10.0mL:
$Na_2S \cdot 9H_2O$ 0.36g

Preparation of $Na_2S \cdot 9H_2O$ Solution: Add $Na_2S \cdot 9H_2O$ to distilled/deionized water and bring volume to 10.0mL. Mix thoroughly. Sparge with 100% N_2. Autoclave for 15 min at 15 psi pressure–121°C.

Acetoin Solution:
Composition per 10.0mL:
Acetoin 1.0g

Preparation of Acetoin Solution: Add acetoin to distilled/deionized water and bring volume to 10.0mL. Mix thoroughly. Sparge with 100% N_2. Autoclave for 15 min at 15 psi pressure–121°C.

Trace Elements Solution SL-10:
Composition per liter:

$FeCl_2 \cdot 4H_2O$	1.5g
$CoCl_2 \cdot 6H_2O$	0.19g
$MnCl_2 \cdot 4H_2O$	100.0mg
$ZnCl_2$	70.0mg
$Na_2MoO_4 \cdot 2H_2O$	36.0mg
$NiCl_2 \cdot 6H_2O$	24.0mg
H_3BO_3	6.0mg
$CuCl_2 \cdot 2H_2O$	2.0mg
HCl (25% solution)	10.0mL

Preparation of Trace Elements Solution SL-10: Add $FeCl_2 \cdot 4H_2O$ to 10.0mL of HCl solution. Mix thoroughly. Add distilled/deionized water and bring volume to 1.0L. Add remaining components. Mix thoroughly.

Preparation of Medium: Prepare and dispense medium under 80% N_2 + 20% CO_2. Add components, except $NaHCO_3$ solution, $Na_2S \cdot 9H_2O$ solution, and acetoin solution, to distilled/deionized water and bring volume to 970.0mL. Mix thoroughly. Sparge with 80% N_2 + 20% CO_2. Autoclave for 15 min at 15 psi pressure–121°C. Aseptically and anaerobically add 10.0mL of sterile $NaHCO_3$ solution, 10.0mL of sterile $Na_2S \cdot 9H_2O$ solution, and 10.0mL of sterile acetoin solution or, using a syringe, inject the appropriate amount of sterile $NaHCO_3$ solution, sterile $Na_2S \cdot 9H_2O$ solution, and sterile acetoin solution into individual tubes containing medium.

Use: For the cultivation and maintenance of *Pelobacter acetylenicus*.

Pelobacter acidigallici Medium

Composition per 1001.0mL:

NaCl	20.0g
$MgCl_2 \cdot 6H_2O$	3.0g
KCl	0.5g
NH_4Cl	0.25g
KH_2PO_4	0.2g
$CaCl_2 \cdot 2H_2O$	0.15g
Resazurin	1.0mg
$NaHCO_3$ solution	10.0mL
$Na_2S \cdot 9H_2O$ solution	10.0mL
Gallic acid solution	10.0mL
Trace elements solution SL-10	1.0mL

pH 7.2 ± 0.2 at 25°C

$NaHCO_3$ Solution:
Composition per 10.0mL:

$NaHCO_3$	2.5g

Preparation of $NaHCO_3$ Solution: Add $NaHCO_3$ to distilled/deionized water and bring volume to 10.0mL. Mix thoroughly. Filter sterilize. Sparge with 80% N_2 + 20% CO_2.

$Na_2S \cdot 9H_2O$ Solution:
Composition per 10.0mL:

$Na_2S \cdot 9H_2O$	0.36g

Preparation of $Na_2S \cdot 9H_2O$ Solution: Add $Na_2S \cdot 9H_2O$ to distilled/deionized water and bring volume to 10.0mL. Mix thoroughly. Sparge with 100% N_2. Autoclave for 15 min at 15 psi pressure–121°C.

Gallic Acid Solution:
Composition per 10.0mL:

Gallic acid	1.0g

Preparation of Gallic Acid Solution: Add gallic acid to distilled/deionized water and bring volume to 10.0mL. Mix thoroughly. Filter sterilize. Sparge with 80% N_2 + 20% CO_2.

Trace Elements Solution SL-10:
Composition per liter:

$FeCl_2 \cdot 4H_2O$	1.5g
$CoCl_2 \cdot 6H_2O$	0.19g
$MnCl_2 \cdot 4H_2O$	100.0mg
$ZnCl_2$	70.0mg
$Na_2MoO_4 \cdot 2H_2O$	36.0mg
$NiCl_2 \cdot 6H_2O$	24.0mg
H_3BO_3	6.0mg
$CuCl_2 \cdot 2H_2O$	2.0mg
HCl (25% solution)	10.0mL

Preparation of Trace Elements Solution SL-10: Add $FeCl_2 \cdot 4H_2O$ to 10.0mL of HCl solution. Mix thoroughly. Add distilled/deionized water and bring volume to 1.0L. Add remaining components. Mix thoroughly.

Preparation of Medium: Prepare and dispense medium under 80% H_2 + 20% CO_2. Add components, except $NaHCO_3$ solution, $Na_2S \cdot 9H_2O$ solution, gallic acid solution, and trace elements solution SL-10, to distilled/deionized water and bring volume to 970.0mL. Mix thoroughly. Sparge with 80% N_2 + 20% CO_2. Autoclave for 15 min at 15 psi pressure–121°C. Aseptically and anaerobically add 10.0mL of sterile $NaHCO_3$ solution, 10.0mL of sterile $Na_2S \cdot 9H_2O$ solution, 10.0mL of sterile gallic acid solution, and 1.0mL of sterile trace elements solution SL-10 or, using a syringe, inject the appropriate amount of sterile $NaHCO_3$ solution, sterile $Na_2S \cdot 9H_2O$ solution, sterile gallic acid solution, and sterile trace elements solution SL-10 into individual tubes containing medium.

Use: For the cultivation and maintenance of *Pelobacter acidigallici*.

Pelobacter carbinolicus Medium

Composition per 1001.0mL:

NaCl	20.0g
$MgCl_2 \cdot 6H_2O$	3.0g
KCl	0.5g
NH_4Cl	0.25g
KH_2PO_4	0.2g
$CaCl_2 \cdot 2H_2O$	0.15g
Resazurin	1.0mg
$NaHCO_3$ solution	10.0mL
$Na_2S \cdot 9H_2O$ solution	10.0mL
2,3-Butanediol solution	10.0mL
Trace elements solution SL-10	1.0mL

pH 7.2 ± 0.2 at 25°C

$NaHCO_3$ Solution:
Composition per 10.0mL:

$NaHCO_3$	2.5g

Preparation of $NaHCO_3$ Solution: Add $NaHCO_3$ to distilled/deionized water and bring volume to 10.0mL. Mix thoroughly. Filter sterilize. Sparge with 80% N_2 + 20% CO_2.

$Na_2S \cdot 9H_2O$ Solution:
Composition per 10.0mL:

$Na_2S \cdot 9H_2O$	0.36g

Preparation of $Na_2S \cdot 9H_2O$ Solution: Add $Na_2S \cdot 9H_2O$ to distilled/deionized water and bring volume to 10.0mL. Mix thoroughly. Sparge with 100% N_2. Autoclave for 15 min at 15 psi pressure–121°C.

2,3-Butanediol Solution:
Composition per 10.0mL:

2,3-Butanediol	0.68g

Preparation of 2,3-Butanediol Solution: Add 2,3-butanediol to distilled/deionized water and bring volume to 10.0mL. Mix thoroughly. Filter sterilize. Sparge with 80% N_2 + 20% CO_2.

Trace Elements Solution SL-10:
Composition per liter:

$FeCl_2 \cdot 4H_2O$	1.5g
$CoCl_2 \cdot 6H_2O$	0.19g
$MnCl_2 \cdot 4H_2O$	100.0mg
$ZnCl_2$	70.0mg
$Na_2MoO_4 \cdot 2H_2O$	36.0mg
$NiCl_2 \cdot 6H_2O$	24.0mg
H_3BO_3	6.0mg
$CuCl_2 \cdot 2H_2O$	2.0mg
HCl (25% solution)	10.0mL

Preparation of Trace Elements Solution SL-10: Add $FeCl_2 \cdot 4H_2O$ to 10.0mL of HCl solution. Mix thoroughly. Add distilled/deionized water and bring volume to 1.0L. Add remaining components. Mix thoroughly.

Preparation of Medium: Prepare and dispense medium under 80% H_2 + 20% CO_2. Add components, except $NaHCO_3$ solution, $Na_2S \cdot 9H_2O$ solution, 2,3-butanediol solution, and trace elements solution SL-10, to distilled/deionized water and bring volume to 970.0mL. Mix thoroughly. Sparge with 80% N_2 + 20% CO_2. Autoclave for 15 min at 15 psi pressure–121°C. Aseptically and anaerobically add 10.0mL of sterile $NaHCO_3$ solution, 10.0mL of sterile $Na_2S \cdot 9H_2O$ solution, 10.0mL of sterile 2,3-butanediol solution, and 1.0mL of sterile trace elements solution SL-10 or, using a syringe, inject the appropriate amount of sterile $NaHCO_3$ solution, sterile $Na_2S \cdot 9H_2O$ solution, sterile 2,3-butanediol solution, and sterile trace elements solution SL-10 into individual tubes containing medium.

Use: For the cultivation and maintenance of *Pelobacter carbinolicus*.

Pelobacter Medium

Composition per liter:

$KHCO_3$	4.5g
NH_4Cl	1.0g
NaCl	0.6g
Trypticase™	0.5g
Yeast extract	0.5g
KH_2PO_4	0.3g
$MgCl_2 \cdot 6H_2O$	0.1g
$CaCl_2 \cdot 2H_2O$	0.08g
Resazurin	1.0mg
Trace elements solution	10.0mL
Vitamin solution	10.0mL
Sodium gallate solution	10.0mL
L-Cysteine·HCl·H_2O solution	10.0mL
$Na_2S \cdot 9H_2O$ solution	10.0mL

pH 7.3 ± 0.2 at 25°C

Trace Elements Solution:
Composition per liter:

Nitrilotriacetic acid	12.8g
$FeCl_3 \cdot 6H_2O$	1.35g
NaCl	1.0g
$NiCl_2 \cdot 6H_2O$	0.12g
$CaCl_2 \cdot 2H_2O$	0.10g
$MnCl_2 \cdot 4H_2O$	0.10g
$ZnCl_2$	0.10g
$Na_2SeO_3 \cdot 5H_2O$	0.026g
$CuCl_2 \cdot 2H_2O$	0.025g
$CoCl_2 \cdot 6H_2O$	0.024g
$Na_2MoO_4 \cdot 2H_2O$	0.024g
H_3BO_3	0.01g

Preparation of Trace Elements Solution: Add nitrilotriacetic acid to 500.0mL of distilled/deionized water. Adjust pH to 6.5 with KOH. Add remaining components. Adjust pH to 7.0. Add distilled/deionized water to 1.0L.

Vitamin Solution:
Composition per liter:

Biotin	2.0mg
Folic acid	2.0mg
Pyridoxine·HCl	10.0mg
Thiamine·HCl	5.0mg
Riboflavin	5.0mg
Nicotinic acid	5.0mg
Calcium DL-pantothenate	5.0mg
Vitamin B_{12}	0.1mg
p-Aminobenzoic acid	5.0mg
Lipoic acid	5.0mg

Preparation of Vitamin Solution: Add components to distilled/deionized water and bring volume to 1.0L. Mix thoroughly. Filter sterilize. Sparge with 80% N_2 + 20% CO_2.

Sodium Gallate Solution:
Composition per 10.0mL:

Gallic acid	1.88g
NaOH (1*M* solution)	variable

Preparation of Sodium Gallate Solution: Add gallic acid to distilled/deionized water and bring volume to 8.0mL. Mix thoroughly. Add sufficient NaOH solution to bring pH to 7.3. Filter sterilize. Sparge with 80% N_2 + 20% CO_2.

L-Cysteine·HCl·H_2O Solution:
Composition per 10.0mL:

L-Cysteine·HCl·H_2O	0.3g

Preparation of L-Cysteine·HCl·H_2O Solution: Add L-cysteine·HCl·H_2O to distilled/deionized water and bring volume to 10.0mL. Mix thoroughly. Sparge with 100% N_2. Autoclave for 15 min at 15 psi pressure–121°C.

$Na_2S \cdot 9H_2O$ Solution:
Composition per 10.0mL:

$Na_2S \cdot 9H_2O$	0.3g

Preparation of $Na_2S \cdot 9H_2O$ Solution: Add $Na_2S \cdot 9H_2O$ to distilled/deionized water and bring volume to 10.0mL. Mix thoroughly. Sparge with 100% N_2. Autoclave for 15 min at 15 psi pressure–121°C.

Preparation of Medium: Prepare and dispense medium under 80% N_2 + 20% CO_2. Add components, except sodium gallate solution, $NaHCO_3$ solution, $Na_2S \cdot 9H_2O$ solution, and vitamin solution, to distilled/deionized water and bring volume to 960.0mL. Mix thoroughly. Sparge with 80% N_2 + 20% CO_2. Autoclave for 15 min at 15 psi pressure–121°C. Aseptically and anaerobically add 10.0mL of sterile $NaHCO_3$ solution, 10.0mL of sterile $Na_2S \cdot 9H_2O$ solution, 10.0mL of sterile sodium gallate solution, and 10.0mL of sterile vitamin solution or, using a syringe, inject the appropriate amount of sterile $NaHCO_3$ solution, sterile $Na_2S \cdot 9H_2O$ solution, sterile sodium gallate solution, and sterile vitamin solution into individual tubes containing medium.

Use: For the cultivation and maintenance of *Pelobacter acidigallici*.

Pelobacter Medium with Gallic Acid

Composition per liter:

Solution A	950.0mL
Solution B	25.0mL
Solution C	25.0mL

pH 7.2 ± 0.2 at 25°C

Solution A:

Composition per liter:

NaCl	20.0g
$MgCl_2 \cdot 6H_2O$	3.65g
$NaHCO_3$	2.5g
KCl	0.5g
NH_4Cl	0.25g
KH_2PO_4	0.2g
$CaCl_2 \cdot 2H_2O$	0.15g
Resazurin	0.5mg
Modified Wolfe's mineral solution	10.0mL
Wolfe's vitamin solution	10.0mL

Preparation of Solution A: Prepare and dispense solution under 80% N_2 + 20% CO_2. Add components, except $NaHCO_3$, to distilled/deionized water and bring volume to 950.0mL. Mix thoroughly. Gently heat and bring to boiling. Continue boiling for 3 min. Cool to room temperature while sparging with 80% N_2 + 20% CO_2. Add $NaHCO_3$. Mix thoroughly. Anaerobically distribute 9.5mL volumes into anaerobic tubes. Autoclave for 15 min at 15 psi pressure–121°C.

Modified Wolfe's Mineral Solution:

Composition per liter:

$MgSO_4 \cdot 7H_2O$	3.0g
Nitrilotriacetic acid	1.5g
NaCl	1.0g
$MnSO_4 \cdot H_2O$	0.5g
$CaCl_2$	0.1g
$CoCl_2 \cdot 6H_2O$	0.1g
$FeSO_4 \cdot 7H_2O$	0.1g
$ZnSO_4 \cdot 7H_2O$	0.1g
$AlK(SO_4)_2 \cdot 12H_2O$	0.01g
$CuSO_4 \cdot 5H_2O$	0.01g
H_3BO_3	0.01g
$Na_2MoO_4 \cdot 2H_2O$	0.01g
Na_2SeO_3	0.01g
$NaWO_4 \cdot 2H_2O$	0.01g
$NiCl_2 \cdot 6H_2O$	0.01g

Preparation of Modified Wolfe's Mineral Solution: Add nitrilotriacetic acid to 500.0mL of distilled/deionized water. Adjust pH to 6.5 with KOH. Add remaining components one at a time. Add distilled/deionized water to 1.0L. Adjust pH to 6.8.

Wolfe's Vitamin Solution:

Composition per liter:

Pyridoxine·HCl	10.0mg
p-Aminobenzoic acid	5.0mg
Lipoic acid	5.0mg
Nicotinic acid	5.0mg
Riboflavin	5.0mg
Thiamine·HCl	5.0mg
Calcium DL-pantothenate	5.0mg
Biotin	2.0mg
Folic acid	2.0mg
Vitamin B_{12}	0.1mg

Preparation of Wolfe's Vitamin Solution: Add components to distilled/deionized water and bring volume to 1.0L. Mix thoroughly.

Solution B:

Composition per 25.0mL:

Gallic acid	0.85g

Preparation of Solution B: Prepare solution B immediately prior to use. Add gallic acid to distilled/deionized water and bring volume to 25.0mL. Mix thoroughly. Rapidly adjust pH to 6.5. Filter sterilize. Sparge with 100% N_2.

Solution C:

Composition per 25.0mL:

$Na_2S \cdot 9H_2O$	0.4g

Preparation of Solution C: Add $Na_2S \cdot 9H_2O$ to distilled/deionized water and bring volume to 25.0mL. Mix thoroughly. Sparge with 100% N_2. Autoclave for 15 min at 15 psi pressure–121°C.

Preparation of Medium: Aseptically and anaerobically add 0.25mL of sterile solution B and 0.25mL of sterile solution C to each tube containing 9.5mL of sterile solution A. Mix thoroughly. Adjust pH to 7.2.

Use: For the cultivation of *Pelobacter acidigallici* and *Pelobacter massiliensis*.

Pelobacter propionicus Medium (DSMZ Medium 298)

Composition per liter:

NaCl	1.0g
KCl	0.5g
$MgCl_2 \cdot 6H_2O$	0.4g
NH_4Cl	0.25g
KH_2PO_4	0.2g
$CaCl_2 \cdot 2H_2O$	0.15g
Resazurin	1.0mg
$NaHCO_3$ solution	10.0mL
Butanediol solution	10.0mL
$Na_2S \cdot 9H_2O$ solution	10.0mL
Trace elements solution SL-10	1.0mL

pH 7.2 ± 0.2 at 25°C

$Na_2S \cdot 9H_2O$ Solution:

Composition per 10.0mL:

$Na_2S \cdot 9H_2O$	0.36g

Preparation of $Na_2S \cdot 9H_2O$ Solution: Add $Na_2S \cdot 9H_2O$ to distilled/deionized water and bring volume to 10.0mL. Mix thoroughly. Autoclave under 100% N_2 for 15 min at 15 psi pressure–121°C. Cool to room temperature.

$NaHCO_3$ Solution:

Composition per 10.0mL:

$NaHCO_3$	2.5g

Preparation of $NaHCO_3$ Solution: Add $NaHCO_3$ to distilled/deionized water and bring volume to 10.0mL. Mix thoroughly. Sparge with 80% N_2 + 20% CO_2. Filter sterilize.

Butanediol Solution:

Composition per 10.0mL:

2,3-Butanediol	0.9g

Preparation of Butanediol Solution: Add butanediol to distilled/deionized water and bring volume to 10.0mL. Mix thoroughly. Sparge with 100% N_2. Filter sterilize.

Trace Elements Solution SL-10:

Composition per liter:

$FeCl_2·4H_2O$	1.5g
$CoCl_2·6H_2O$	190.0mg
$MnCl_2·4H_2O$	100.0mg
$ZnCl_2$	70.0mg
$Na_2MoO_4·2H_2O$	36.0mg
$NiCl_2·6H_2O$	24.0mg
H_3BO_3	6.0mg
$CuCl_2·2H_2O$	2.0mg
HCl (25% solution)	10.0mL

Preparation of Trace Elements Solution SL-10: Add $FeCl_2·4H_2O$ to 10.0mL of HCl solution. Mix thoroughly. Add distilled/deionized water and bring volume to 1.0L. Add remaining components. Mix thoroughly. Sparge with 80% N_2 + 20% CO_2. Autoclave for 15 min at 15 psi pressure–121°C.

Preparation of Medium: Prepare and dispense medium under 80% N_2 + 20% CO_2 gas atmosphere. Add components, except $NaHCO_3$ solution, butanediol solution, $Na_2S·9H_2O$ solution, and trace elements solution SL-10, to distilled/deionized water and bring volume to 969.0mL. Mix thoroughly. Adjust pH to 7.2. Sparge with 80% N_2 + 20% CO_2. Autoclave for 15 min at 15 psi pressure–121°C. Aseptically and anaerobically add 10.0mL $NaHCO_3$ solution, 10.0mL butanediol solution, 10.0mL $Na_2S·9H_2O$ solution, and 1.0mL trace elements solution SL-10. Mix thoroughly. Aseptically and anaerobically distribute into sterile tubes or bottles. After inoculation, flush and repressurize the gas head space of culture bottles with sterile 80% N_2 + 20% CO_2 to 1 bar overpressure.

Use: For the cultivation of *Pelobacter propionicus.*

Pelobacter propionicus Medium

Composition per liter:

NaCl	1.0g
KCl	0.5g
$MgCl_2·6H_2O$	0.4g
NH_4Cl	0.25g
KH_2PO_4	0.2g
$CaCl_2·2H_2O$	0.15g
Resazurin	1.0mg
2,3-Butanediol solution	50.0mL
$NaHCO_3$ solution	20.0mL
$Na_2S·9H_2O$ solution	10.0mL
Trace elements solution SL-10	1.0mL

pH 7.2 ± 0.2 at 25°C

Trace Elements Solution SL-10:

Composition per liter:

$FeCl_2·4H_2O$	1.5g
$CoCl_2·6H_2O$	190.0mg
$MnCl_2·4H_2O$	100.0mg
$ZnCl_2$	70.0mg
$Na_2MoO_4·2H_2O$	36.0mg
$NiCl_2·6H_2O$	24.0mg
H_3BO_3	6.0mg
$CuCl_2·2H_2O$	2.0mg
HCl (25% solution)	10.0mL

Preparation of Trace Elements Solution SL-10: Prepare and dispense under 80% N_2 + 20% CO_2. Add $FeCl_2·4H_2O$ to 10.0mL of HCl solution. Mix thoroughly. Add distilled/deionized water and bring volume to 1.0L. Add remaining components. Mix thoroughly. Gas under 80% N_2 + 20% CO_2. Autoclave for 15 min at 15 psi pressure–121°C.

2,3-Butanediol Solution:

Composition per 50.0mL:

2,3-Butanediol	0.9g

Preparation of 2,3-Butanediol Solution: Add 2,3-butanediol to distilled/deionized water and bring volume to 50.0mL. Mix thoroughly. Filter sterilize. Gas under 80% N_2 + 20% CO_2.

$NaHCO_3$ Solution:

Composition per 20.0mL:

$NaHCO_3$	2.5g

Preparation of $NaHCO_3$ Solution: Add $NaHCO_3$ to distilled/deionized water and bring volume to 20.0mL. Mix thoroughly. Filter sterilize. Gas under 80% N_2 + 20% CO_2.

$Na_2S·9H_2O$ Solution:

Composition per 10.0mL:

$Na_2S·9H_2O$	0.36g

Preparation of $Na_2S·9H_2O$ Solution: Add $Na_2S·9H_2O$ to distilled/deionized water and bring volume to 10.0mL. Mix thoroughly. Gas under 100% N_2. Autoclave for 15 min at 15 psi pressure–121°C.

Preparation of Medium: Prepare medium and dispense under 80% N_2 + 20% CO_2. Add components, except 2,3-butanediol solution, $NaHCO_3$ solution, and $Na_2S·9H_2O$ solution, to distilled/deionized water and bring volume to 920.0mL. Mix thoroughly. Sparge with 80% N_2 + 20% CO_2. Autoclave for 15 min at 15 psi pressure–121°C. Aseptically and anaerobically add 50.0mL of sterile 2,3-butanediol solution, 20.0mL of sterile $NaHCO_3$ solution, and 10.0mL of sterile $Na_2S·9H_2O$ solution. Mix thoroughly. Aseptically and anaerobically distribute into sterile tubes or flasks.

Use: For the cultivation and maintenance of *Pelobacter propionicus*.

Pelobacter venetianus Marine Medium (DSMZ Medium 296)

Composition per liter:

NaCl	20.0g
$MgCl_2·6H_2O$	3.0g
KCl	0.5g
NH_4Cl	0.25g
KH_2PO_4	0.2g
$CaCl_2·2H_2O$	0.15g
Resazurin	1.0mg
$NaHCO_3$ solution	10.0mL
Polyethylene glycol solution	10.0mL
$Na_2S·9H_2O$ solution	10.0mL
Trace elements solution SL-10	1.0mL

pH 7.2 ± 0.2 at 25°C

$Na_2S·9H_2O$ Solution:

Composition per 10.0mL:

$Na_2S·9H_2O$	0.36g

Preparation of $Na_2S·9H_2O$ Solution: Add $Na_2S·9H_2O$ to distilled/deionized water and bring volume to 10.0mL. Mix thoroughly. Autoclave under 100% N_2 for 15 min at 15 psi pressure–121°C. Cool to room temperature.

$NaHCO_3$ Solution:

Composition per 10.0mL:

$NaHCO_3$	2.5g

Preparation of $NaHCO_3$ Solution: Add $NaHCO_3$ to distilled/deionized water and bring volume to 10.0mL. Mix thoroughly. Sparge with 80% N_2 + 20% CO_2. Filter sterilize.

Polyethylene Glycol Solution:

Composition per 10.0mL:

Polyethylene glycol (molecular weight 106–20000)	1.0g

Preparation of Polyethylene Glycol Solution: Add polyethylene glycol to distilled/deionized water and bring volume to 10.0mL. Mix thoroughly. Sparge with 100% N_2. Filter sterilize.

Trace Elements Solution SL-10:

Composition per liter:

$FeCl_2·4H_2O$	1.5g
$CoCl_2·6H_2O$	190.0mg
$MnCl_2·4H_2O$	100.0mg
$ZnCl_2$	70.0mg
$Na_2MoO_4·2H_2O$	36.0mg
$NiCl_2·6H_2O$	24.0mg
H_3BO_3	6.0mg
$CuCl_2·2H_2O$	2.0mg
HCl (25% solution)	10.0mL

Preparation of Trace Elements Solution SL-10: Add $FeCl_2·4H_2O$ to 10.0mL of HCl solution. Mix thoroughly. Add distilled/deionized water and bring volume to 1.0L. Add remaining components. Mix thoroughly. Sparge with 80% N_2 + 20% CO_2. Filter sterilize.

Preparation of Medium: Prepare and dispense medium under 80% N_2 + 20% CO_2 gas atmosphere. Add components, except $NaHCO_3$ solution, polyethylene glycol solution, $Na_2S·9H_2O$ solution, and trace elements solution SL-10, to distilled/deionized water and bring volume to 969.0mL. Mix thoroughly. Adjust pH to 7.2. Sparge with 80% N_2 + 20% CO_2. Autoclave for 15 min at 15 psi pressure–121°C. Aseptically and anaerobically add 10.0mL $NaHCO_3$ solution, 10.0mL polyethylene glycol solution, 10.0mL $Na_2S·9H_2O$ solution, and 1.0mL trace elements solution SL-10. Mix thoroughly. Aseptically and anaerobically distribute into sterile tubes or bottles. After inoculation, flush and repressurize the gas head space of culture bottles with sterile 80% N_2 + 20% CO_2 to 1 bar overpressure.

Use: For the cultivation of *Pelobacter venetianus.*

Pelobacter venetianus Medium

Composition per liter:

NaCl	1.0g
KCl	0.5g
$MgCl_2·6H_2O$	0.4g
NH_4Cl	0.25g
KH_2PO_4	0.2g
$CaCl_2·2H_2O$	0.15g
Resazurin	1.0mg
Polyethylene glycol solution	50.0mL
$NaHCO_3$ solution	20.0mL
$Na_2S·9H_2O$ solution	10.0mL
Trace elements solution SL-10	1.0mL

pH 7.2 ± 0.2 at 25°C

Trace Elements Solution SL-10:

Composition per liter:

$FeCl_2·4H_2O$	1.5g
$CoCl_2·6H_2O$	190.0mg
$MnCl_2·4H_2O$	100.0mg
$ZnCl_2$	70.0mg
$Na_2MoO_4·2H_2O$	36.0mg
$NiCl_2·6H_2O$	24.0mg
H_3BO_3	6.0mg
$CuCl_2·2H_2O$	2.0mg
HCl (25% solution)	10.0mL

Preparation of Trace Elements Solution SL-10: Prepare and dispense under 80% N_2 + 20% CO_2. Add $FeCl_2·4H_2O$ to 10.0mL of HCl solution. Mix thoroughly. Add distilled/deionized water and bring volume to 1.0L. Add remaining components. Mix thoroughly. Gas under 80% N_2 + 20% CO_2. Autoclave for 15 min at 15 psi pressure–121°C.

Polyethylene Glycol Solution:

Composition per 10.0mL:

Polyethylene glycol	1.0g

Preparation of Polyethylene Glycol Solution: Add polyethylene glycol to distilled/deionized water and bring volume to 10.0mL. Mix thoroughly. Filter sterilize. Sparge with 80% N_2 + 20% CO_2.

$NaHCO_3$ Solution:

Composition per 20.0mL:

$NaHCO_3$	2.5g

Preparation of $NaHCO_3$ Solution: Add $NaHCO_3$ to distilled/deionized water and bring volume to 20.0mL. Mix thoroughly. Filter sterilize. Gas under 80% N_2 + 20% CO_2.

$Na_2S·9H_2O$ Solution:

Composition per 10.0mL:

$Na_2S·9H_2O$	0.36g

Preparation of $Na_2S·9H_2O$ Solution: Add $Na_2S·9H_2O$ to distilled/deionized water and bring volume to 10.0mL. Mix thoroughly. Gas under 100% N_2. Autoclave for 15 min at 15 psi pressure–121°C.

Preparation of Medium: Prepare medium and dispense under 80% N_2 + 20% CO_2. Add components, except polyethylene glycol solution, $NaHCO_3$ solution, and $Na_2S·9H_2O$ solution, to distilled/deionized water and bring volume to 970.0mL. Mix thoroughly. Sparge with 80% N_2 + 20% CO_2. Autoclave for 15 min at 15 psi pressure–121°C. Aseptically and anaerobically add 50.0mL of sterile polyethylene glycol solution, 20.0mL of sterile $NaHCO_3$ solution, and 10.0mL of sterile $Na_2S·9H_2O$ solution. Mix thoroughly. Aseptically and anaerobically distribute into sterile tubes or flasks.

Use: For the cultivation and maintenance of freshwater strains of *Pelobacter venetianus.*

Pelobacter venetianus Medium

Composition per liter:

NaCl	20.0g
$MgCl_2·6H_2O$	3.0g
KCl	0.5g
NH_4Cl	0.25g
KH_2PO_4	0.2g
$CaCl_2·2H_2O$	0.15g
Resazurin	1.0mg
$NaHCO_3$ solution	10.0mL
$Na_2S·9H_2O$ solution	10.0mL
Polyethylene glycol solution	10.0mL
Trace elements solution SL-10	1.0mL

pH 7.2 ± 0.2 at 25°C

$NaHCO_3$ Solution:

Composition per 10.0mL:

$NaHCO_3$	2.5g

Preparation of $NaHCO_3$ Solution: Add $NaHCO_3$ to distilled/deionized water and bring volume to 10.0mL. Mix thoroughly. Filter sterilize. Gas under 80% N_2 + 20% CO_2.

$Na_2S·9H_2O$ Solution:
Composition per 10.0mL:

$Na_2S·9H_2O$	0.36g

Preparation of $Na_2S·9H_2O$ Solution: Add $Na_2S·9H_2O$ to distilled/deionized water and bring volume to 10.0mL. Mix thoroughly. Gas under 100% N_2. Autoclave for 15 min at 15 psi pressure–121°C.

Polyethylene Glycol Solution:
Composition per 10.0mL:

Polyethylene glycol	1.0g

Preparation of Polyethylene Glycol Solution: Add polyethylene glycol to distilled/deionized water and bring volume to 10.0mL. Mix thoroughly. Filter sterilize. Sparge with 80% N_2 + 20% CO_2.

Trace Elements Solution SL-10:
Composition per liter:

$FeCl_2·4H_2O$	1.5g
$CoCl_2·6H_2O$	190.0mg
$MnCl_2·4H_2O$	100.0mg
$ZnCl_2$	70.0mg
$Na_2MoO_4·2H_2O$	36.0mg
$NiCl_2·6H_2O$	24.0mg
H_3BO_3	6.0mg
$CuCl_2·2H_2O$	2.0mg
HCl (25% solution)	10.0mL

Preparation of Trace Elements Solution SL-10: Prepare and dispense under 80% N_2 + 20% CO_2. Add $FeCl_2·4H_2O$ to 10.0mL of HCl solution. Mix thoroughly. Add distilled/deionized water and bring volume to 1.0L. Add remaining components. Mix thoroughly. Gas under 80% N_2 + 20% CO_2. Autoclave for 15 min at 15 psi pressure–121°C.

Preparation of Medium: Prepare medium and dispense under 80% N_2 + 20% CO_2. Add components, except $NaHCO_3$ solution, $Na_2S·9H_2O$ solution, polyethylene glycol solution, and trace elements solution SL-10, to distilled/deionized water and bring volume to 960.0mL. Mix thoroughly. Sparge with 80% N_2 + 20% CO_2. Autoclave for 15 min at 15 psi pressure–121°C. Aseptically and anaerobically add 10.0mL of sterile $NaHCO_3$ solution, 10.0mL of sterile $Na_2S·9H_2O$ solution, 10.0mL of sterile polyethylene glycol solution, and 10.0mL of sterile trace elements solution SL-10. Mix thoroughly. Aseptically and anaerobically distribute into sterile tubes or flasks.

Use: For the cultivation and maintenance of marine strains of *Pelobacter venetianus*.

Pelotomaculum Medium (DSMZ Medium 960)

Composition per liter:

$NaHCO_3$	2.5g
NH_4Cl	0.54g
$MgCl_2·6H_2O$	0.2g
$CaCl_2·2H_2O$	0.15g
KH_2PO_4	0.14g
Yeast extract	0.1g
Resazurin	0.5mg
Na-pyruvate solution	20.0mL
Cysteine solution	10.0mL
$Na_2S·9H_2O$ solution	10.0mL
Vitamin solution	5.0mL
Trace elements solution	1.0mL
Selenite-tungstate solution	1.0mL

pH 7.0 ± 0.2 at 25°C

Na-Pyruvate Solution:
Composition per 20.0mL:

Na-pyruvate	2.2g

Preparation of Na-Pyruvate Solution: Add Na-pyruvate to distilled/deionized water and bring volume to 20.0mL. Mix thoroughly. Autoclave under 100% N_2 for 15 min at 15 psi pressure–121°C. Cool to room temperature.

$Na_2S·9H_2O$ Solution:
Composition per 10.0mL:

$Na_2S·9H_2O$	0.3g

Preparation of $Na_2S·9H_2O$ Solution: Add $Na_2S·9H_2O$ to distilled/deionized water and bring volume to 10.0mL. Mix thoroughly. Sparge with 100% N_2. Autoclave for 15 min at 15 psi pressure–121°C. Cool to room temperature.

Cysteine Solution:
Composition per 10.0mL:

L-Cysteine·HCl·H_2O	0.3g

Preparation of Cysteine Solution: Add L-cysteine·HCl·H_2O to distilled/deionized water and bring volume to 10.0mL. Mix thoroughly. Sparge with 100% N_2. Autoclave for 15 min at 15 psi pressure–121°C. Cool to room temperature.

Vitamin Solution:
Composition per liter:

Pyridoxine-HCl	10.0mg
Thiamine-HCl·$2H_2O$	5.0mg
Riboflavin	5.0mg
Nicotinic acid	5.0mg
D-Ca-pantothenate	5.0mg
p-Aminobenzoic acid	5.0mg
Lipoic acid	5.0mg
Biotin	2.0mg
Folic acid	2.0mg
Vitamin B_{12}	0.1mg

Preparation of Vitamin Solution: Add components to distilled/deionized water and bring volume to 1.0L. Mix thoroughly. Sparge with 80% H_2 + 20% CO_2. Filter sterilize.

Trace Elements Solution:
Composition per liter:

$MgSO_4·7H_2O$	3.0g
Nitrilotriacetic acid	1.5g
NaCl	1.0g
$MnSO_4·2H_2O$	0.5g
$CoSO_4·7H_2O$	0.18g
$ZnSO_4·7H_2O$	0.18g
$CaCl_2·2H_2O$	0.1g
$FeSO_4·7H_2O$	0.1g
$NiCl_2·6H_2O$	0.025g
$KAl(SO_4)_2·12H_2O$	0.02g
H_3BO_3	0.01g
$Na_2MoO_4·4H_2O$	0.01g
$CuSO_4·5H_2O$	0.01g
$Na_2SeO_3·5H_2O$	0.3mg

Preparation of Trace Elements Solution: Add nitrilotriacetic acid to 500.0mL of distilled/deionized water. Dissolve by adjusting pH to

6.5 with KOH. Add remaining components. Add distilled/deionized water to 1.0L. Mix thoroughly.

Selenite-Tungstate Solution

Composition per liter:

NaOH	0.5g
$Na_2WO_4 \cdot 2H_2O$	4.0mg
$Na_2SeO_3 \cdot 5H_2O$	3.0mg

Preparation of Selenite-Tungstate Solution: Add components to distilled/deionized water and bring volume to 1.0L. Mix thoroughly. Sparge with 100% N_2. Filter sterilize.

Preparation of Medium: Add components, except Na-pyruvate solution, cysteine solution, and $Na_2S \cdot 9H_2O$ solution, to distilled/deionized water and bring volume to 960.0mL. Mix thoroughly. Sparge with 80% N_2 + 20% CO_2. Equilibrate with this gas mixture to reach pH 7.0. Distribute into anaerobe tubes or bottles. Autoclave for 15 min at 15 psi pressure–121°C. Aseptically and anaerobically add per liter of medium 20.0mL sterile Na-pyruvate solution, 10.0mL sterile cysteine solution, and 10.0mL of sterile $Na_2S \cdot 9H_2O$ solution. Mix thoroughly.

Use: For the cultivation of *Pelotomaculum thermopropionicum.*

Pelotomaculum Medium (DSMZ Medium 960)

Composition per liter:

$NaHCO_3$	2.5g
NH_4Cl	0.54g
$MgCl_2 \cdot 6H_2O$	0.2g
$CaCl_2 \cdot 2H_2O$	0.15g
KH_2PO_4	0.14g
Yeast extract	0.1g
Resazurin	0.5mg
Ethanol solution	20.0mL
Cysteine solution	10.0mL
$Na_2S \cdot 9H_2O$ solution	10.0mL
Vitamin solution	5.0mL
Trace elements solution	1.0mL
Selenite-tungstate solution	1.0mL

pH 7.0 ± 0.2 at 25°C

Ethanol Solution:

Composition per 20.0mL:

Ethanol	0.92mL

Preparation of Ethanol Solution: Add ethanol to distilled/deionized water and bring volume to 20.0mL. Mix thoroughly. Autoclave under 100% N_2 for 15 min at 15 psi pressure–121°C. Cool to room temperature.

$Na_2S \cdot 9H_2O$ Solution:

Composition per 10.0mL:

$Na_2S \cdot 9H_2O$	0.3g

Preparation of $Na_2S \cdot 9H_2O$ Solution: Add $Na_2S \cdot 9H_2O$ to distilled/deionized water and bring volume to 10.0mL. Mix thoroughly. Sparge with 100% N_2. Autoclave for 15 min at 15 psi pressure–121°C. Cool to room temperature.

Cysteine Solution:

Composition per 10.0mL:

L-Cysteine·HCl·H_2O	0.3g

Preparation of Cysteine Solution: Add L-cysteine·HCl·H_2O to distilled/deionized water and bring volume to 10.0mL. Mix thoroughly. Sparge with 100% N_2. Autoclave for 15 min at 15 psi pressure–121°C. Cool to room temperature.

Vitamin Solution:

Composition per liter:

Pyridoxine-HCl	10.0mg
Thiamine-HCl·$2H_2O$	5.0mg
Riboflavin	5.0mg
Nicotinic acid	5.0mg
D-Ca-pantothenate	5.0mg
p-Aminobenzoic acid	5.0mg
Lipoic acid	5.0mg
Biotin	2.0mg
Folic acid	2.0mg
Vitamin B_{12}	0.1mg

Preparation of Vitamin Solution: Add components to distilled/deionized water and bring volume to 1.0L. Mix thoroughly. Sparge with 80% H_2 + 20% CO_2. Filter sterilize.

Trace Elements Solution:

Composition per liter:

$MgSO_4 \cdot 7H_2O$	3.0g
Nitrilotriacetic acid	1.5g
NaCl	1.0g
$MnSO_4 \cdot 2H_2O$	0.5g
$CoSO_4 \cdot 7H_2O$	0.18g
$ZnSO_4 \cdot 7H_2O$	0.18g
$CaCl_2 \cdot 2H_2O$	0.1g
$FeSO_4 \cdot 7H_2O$	0.1g
$NiCl_2 \cdot 6H_2O$	0.025g
$KAl(SO_4)_2 \cdot 12H_2O$	0.02g
H_3BO_3	0.01g
$Na_2MoO_4 \cdot 4H_2O$	0.01g
$CuSO_4 \cdot 5H_2O$	0.01g
$Na_2SeO_3 \cdot 5H_2O$	0.3mg

Preparation of Trace Elements Solution: Add nitrilotriacetic acid to 500.0mL of distilled/deionized water. Dissolve by adjusting pH to 6.5 with KOH. Add remaining components. Add distilled/deionized water to 1.0L. Mix thoroughly.

Selenite-Tungstate Solution

Composition per liter:

NaOH	0.5g
$Na_2WO_4 \cdot 2H_2O$	4.0mg
$Na_2SeO_3 \cdot 5H_2O$	3.0mg

Preparation of Selenite-Tungstate Solution: Add components to distilled/deionized water and bring volume to 1.0L. Mix thoroughly. Sparge with 100% N_2. Filter sterilize.

Preparation of Medium: Add components, except ethanol solution, cysteine solution, and $Na_2S \cdot 9H_2O$ solution, to distilled/deionized water and bring volume to 960.0mL. Mix thoroughly. Sparge with 80% N_2 + 20% CO_2. Equilibrate with this gas mixture to reach pH 7.0. Distribute into anaerobe tubes or bottles. Autoclave for 15 min at 15 psi pressure–121°C. Aseptically and anaerobically add per liter of medium 20.0mL sterile ethanol solution, 10.0mL sterile cysteine solution, and 10.0mL of sterile $Na_2S \cdot 9H_2O$ solution. Mix thoroughly.

Use: For the cultivation of *Pelotomaculum thermopropionicum* DSM 13752.

PEM

See: **Pre-Enrichment Medium**

PEMBA
See: **Polymyxin Pyruvate Egg Yolk Mannitol Bromthymol Blue Agar**

Penassay Base Agar
See: **Antibiotic Medium 2**

Penassay Broth
See: **Antibiotic Medium 3**

Penassay Broth with Chloramphenicol

Composition per liter:

Peptone	5.0g
K_2HPO_4	3.68g
NaCl	3.5g
Beef extract	1.5g
Yeast extract	1.5g
KH_2PO_4	1.32g
Glucose	1.0g
Chloramphenicol	10.0mg

pH 7.0 ± 0.05 at 25°C

Preparation of Medium: Add components to distilled/deionized water and bring volume to 1.0L. Mix thoroughly. Gently heat and bring to boiling. Distribute into tubes or flasks. Autoclave for 15 min at 15 psi pressure–121°C.

Use: For the cultivation and maintenance of *Bacillus subtilis*.

Penassay Broth with Magnesium

Composition per liter:

Pancreatic digest of gelatin	5.0g
K_2HPO_4	3.68g
NaCl	3.5g
Beef extract	1.5g
Yeast extract	1.5g
KH_2PO_4	1.32g
Glucose	1.0g
$MgCl_2$	0.095g

pH 7.0 ± 0.05 at 25°C

Preparation of Medium: Add components to distilled/deionized water and bring volume to 1.0L. Mix thoroughly. Gently heat and bring to boiling. Distribute into tubes or flasks. Autoclave for 15 minutes at 15 psi–121°C.

Use: For the cultivation and maintenance of *Bacillus subtilis*.

Penassay G-THY Medium (Penassay Glucose Thymine Medium)

Composition per liter:

Glucose	21.0g
Pancreatic digest of gelatin	5.0g
K_2HPO_4	3.68g
NaCl	3.5g
Beef extract	1.5g
Yeast extract	1.5g
KH_2PO_4	1.32g
Thymine	0.05g

pH 7.0 ± 0.05 at 25°C

Preparation of Medium: Add components to distilled/deionized water and bring volume to 1.0L. Mix thoroughly. Gently heat and bring to boiling. Distribute into tubes or flasks. Autoclave for 15 min at 15 psi pressure–121°C.

Use: For the cultivation and maintenance of *Bacillus subtilis*.

Penassay Seed Agar
See: **Antibiotic Medium 1**

Penicillinase-Producing *Neisseria gonorrhoeae* Medium
See: **PPNG Selective Medium**

Pentachloronitrobenzene Rose Bengal Yeast Extract Sucrose Agar (PRYES Agar)

Composition per liter:

Sucrose	150.0g
Agar	20.0g
Yeast extract	20.0g
Pentachloronitrobenzene (PCNB)	0.1g
Chloramphenicol	0.05g
Chlortetracycline·HCl	0.05g
Rose Bengal	0.025g

pH 5.6 ± 0.2 at 25°C

Preparation of Medium: Add components, except chloramphenicol and chlortetracycline, to distilled/deionized water and bring volume to 1.0L. Mix thoroughly. Adjust pH to 5.6 with tartaric acid. Autoclave for 15 min at 15 psi pressure–121°C. Cool to 45°–50°C. Aseptically add chloramphenicol and chlortetracycline. Mix thoroughly. Pour into sterile Petri dishes.

Use: For the cultivation and differentiation of nephrotoxin-producing strains of *Penicillium viridicatum* and related species isolated from foods. Colonies exhibiting a violet brown pigment on the reverse are counted as potential ochratoxin- and citrinin-producing strains of *Penicillium viridicatum.* Colonies exhibiting a yellow reverse and obverse are counted as potential xanthomegnin- and viomellein-producing strains of *Penicillium viridicatum* and *Penicillium aurantiogriseum* (*Penicillium cyclopium*).

Pentachlorophenol Medium

Composition per liter:

NH_4NO_3	2.5g
$Na_2HPO_4·2H_2O$	1.0g
$MgSO_4·7H_2O$	0.5g
$Fe(SO_4)_3·5H_2O$	0.01g
$Co(NO_3)_2·6H_2O$	0.005g
$CaCl_2·2H_2O$	1.0mg
Pentachlorophenol	1.0mg
KH_2PO_4	0.5mg
$MnSO_4·2H_2O$	0.1mg
$(NH_4)_6Mo_7O_{24}·4H_2O$	0.1mg

Preparation of Medium: Add components to distilled/deionized water and bring volume to 1.0L. Mix thoroughly. Distribute into tubes or flasks. Autoclave for 15 min at 15 psi pressure–121°C.

Use: For the cultivation of *Flavobacterium* species.

Pentachlorophenol Medium

Composition per 1007.0mL:

Sodium glutamate	4.0g
K_2HPO_4	0.65g
$NaNO_3$	0.5g

KH_2PO_4	0.19g
$MgSO_4·7H_2O$	0.1g
Pentachlorophenol solution	5.0mL
$FeSO_4$ solution	2.0mL

pH 7.3 ± 0.1 at 25°C

Pentachlorophenol Solution:
Composition per 100.0mL:

Pentachlorophenol	1.0g
NaOH (0.5*N* solution)	100.0mL

Preparation of Pentachlorophenol Solution: Add pentachlorophenol to 100.0mL of NaOH solution. Mix thoroughly. Filter sterilize.

$FeSO_4$ Solution:
Composition per 100.0mL:

$FeSO_4$	2.5g

Preparation of $FeSO_4$ Solution: Add $FeSO_4$ to distilled/deionized water and bring volume to 100.0mL. Mix thoroughly. Filter sterilize.

Preparation of Medium: Add components, except pentachlorophenol solution and $FeSO_4$ solution, to distilled/deionized water and bring volume to 1.0L. Mix thoroughly. Adjust pH to 7.3–7.4. Autoclave for 15 min at 15 psi pressure–121°C. Aseptically add 2.0mL of sterile $FeSO_4$ solution. Mix thoroughly. Aseptically distribute into sterile flasks. Inoculate flasks and place on a shaker at 200 rpm at 25°–30°C. Monitor growth with a spectrophotometer at 560 nm. When absorbance at 560 nm (A_{560}) is 0.5, add 5.0mL of sterile pentachlorophenol solution per liter of medium.

Use: For the cultivation of *Pseudomonas mendocina.*

PEP Medium

Composition per liter:

Agar	10.0g
Peptone	5.0g
Yeast extract	0.5g
K_2HPO_4	0.1g

Preparation of Medium: Add components to distilled/deionized water and bring volume to 1.0L. Mix thoroughly. Gently heat and bring to boiling. Distribute into tubes or flasks. Autoclave for 15 min at 15 psi pressure–121°C. Pour into sterile Petri dishes or leave in tubes.

Use: For the isolation and cultivation of *Spirochaeta aurantia.*

Pept Carb Soluble Starch Agar (Peptone Carbonate Starch Agar)

Composition per liter:

Solution A	900.0mL
Solution B	100.0mL

Solution A:
Composition per 900.0mL:

Soluble starch	20.0g
Agar	15.0g
Peptone	5.0g
Yeast extract	5.0g
K_2HPO_4	1.0g
$MgSO_4·7H_2O$	0.2g

Preparation of Solution A: Add components to distilled/deionized water and bring volume to 900.0mL. Mix thoroughly. Gently heat and bring to boiling. Autoclave for 15 min at 10 psi pressure–115°C. Cool to 45°–50°C.

Solution B:
Composition per 100.0mL:

Na_2CO_3	10.0g

Preparation of Solution B: Add Na_2CO_3 to distilled/deionized water and bring volume to 100.0mL. Mix thoroughly. Autoclave for 15 min at 10 psi pressure–115°C. Cool to 45°–50°C.

Preparation of Medium: Aseptically combine cooled sterile solution A with cooled sterile solution B. Mix thoroughly. Aseptically distribute into sterile tubes or flasks.

Use: For the cultivation and maintenance of *Bacillus* species.

Peptococcus glycinophilus Medium (DSMZ Medium 228)

Composition per liter:

Yeast extract	10.0g
NaCl	8.0g
Trypticase™ peptone	5.0g
Peptone	5.0g
Beef extract	5.0g
Glycine	3.0g
K_2HPO_4	2.0g
L-Cysteine·HCl	0.5g
Resazurin	1.0mg
Salt solution	40.0mL
Hemin solution	10.0mL
Tween™ 80	1.0mL
Vitamin K_1 solution	0.2mL

pH 7.2± 0.2 at 25°C

Salt Solution:
Composition per liter:

$NaHCO_3$	10.0g
NaCl	2.0g
K_2HPO_4	1.0g
KH_2PO_4	1.0g
$MgSO_4·7H_2O$	0.5g
$CaCl_2·2H_2O$	0.25g

Preparation of Salt Solution: Add components to distilled/deionized water and bring volume to 1.0L. Mix thoroughly.

Vitamin K_1 Solution:
Composition per 20.0mL:

Vitamin K_1	0.1g

Preparation of Vitamin K_1 Solution: Add vitamin K_1 to 20.0mL of 95% ethanol. Mix thoroughly. Store refrigerated in a brown bottle.

Hemin Solution:
Composition per 20.0mL:

Hemin	50.0mg
NaOH (1*N* solution)	20.0mL

Preparation of Hemin Solution: Add hemin to 1.0mL of 1*N* NaOH solution. Mix thoroughly. Bring volume to 20.0mL with distilled/deionized water. Store refrigerated.

Preparation of Medium: Add components, except L-cysteine·HCl, vitamin K_1 solution, and hemin solution, to distilled/deionized water and bring volume to 990.0mL. Mix thoroughly. Gently heat and bring to boiling. Sparge with CO_2. Add L-cysteine·HCl. Cool to 25°C. Aseptically add 10.0mL hemin solution and 0.2mL vitamin K_1 solution. Mix thoroughly. Adjust pH to 7.2 using 8*N* NaOH. Distribute into tubes or flasks under N_2. Autoclave for 15 min at 15 psi pressure–121°C.

Use: For the cultivation of *Micromonas micros=Peptostreptococcus micros.*

Peptococcus glycinophilus Medium

Composition per liter:

Pancreatic digest of casein	5.0g
Yeast extract	5.0g
Glycine	3.0g
Agar	2.0g
Reducing agent	20.0mL
Potassium phosphate buffer (1*M*, pH 7.1)	5.0mL
Salts B	1.0mL

Reducing Agent:

Composition per 100.0mL:

$NaHCO_3$	5.0g
$Na_2S_2O_4$	1.0g

Preparation of Reducing Agent: Add components to distilled/deionized water and bring volume to 100.0mL. Mix thoroughly.

Salts B:

Composition per 100.0mL:

$MgSO_4 \cdot 7H_2O$	20.0g
$FeSO_4 \cdot 7H_2O$	1.0g
$MnSO_4 \cdot H_2O$	0.5g

Preparation of Salts B: Add components to distilled/deionized water and bring volume to 100.0mL. Mix thoroughly.

Preparation of Medium: Add components to distilled/deionized water and bring volume to 1.0L. Mix thoroughly. Gently heat and bring to boiling. Distribute into tubes or flasks. Autoclave for 15 min at 15 psi pressure–121°C. Pour into sterile Petri dishes or leave in tubes.

Use: For the cultivation and maintenance of *Peptostreptococcus micros (Peptococcus glycinophilus).*

Peptococcus Medium

Composition per liter:

Casein peptone	10.0g
Beef extract	3.0g
Yeast extract	3.0g
Glucose	2.0g
L-Cysteine·HCl	0.5g
Salt solution	40.0mL
Tween™ 80	1.0mL

pH 7.2 ± 0.2 at 25°C

Salt Solution:

Composition per liter:

$NaHCO_3$	10.0g
NaCl	2.0g
K_2HPO_4	1.0g
KH_2PO_4	1.0g
$CaCl_2$	0.2g
$MgSO_4 \cdot 7H_2O$	0.2g

Preparation of Salt Solution: Add components to distilled/deionized water and bring volume to 1.0L. Mix thoroughly.

Preparation of Medium: Add components to distilled/deionized water and bring volume to 1.0L. Mix thoroughly. Adjust pH to 7.2. Distribute into tubes or flasks. Autoclave for 15 min at 15 psi pressure–121°C.

Use: For the cultivation of *Gemella morbillorum, Megasphaera elsdenii,* and *Streptococcus pleomorphus.*

Peptone Broth

Composition per liter:

Peptone	5.0g

pH 7.2 ± 0.2 at 25°C

Preparation of Medium: Add peptone to distilled/deionized water and bring volume to 1.0L. Mix thoroughly. Distribute into tubes that contain a 3-inch strip of Whatman #1 filter paper. Add enough broth to cover about two thirds of the filter paper. Autoclave for 15 min at 15 psi pressure–121°C.

Use: For the cultivation and maintenance of *Cellvibrio gilvus* and *Pseudomonas* species.

Peptone Carbonate Starch Agar

See: **Pept Carb Soluble Starch Agar**

Peptone Cholic Acid Recovery

Composition per liter:

Meat extract	10.0g
Peptone	10.0g
Cholic acid	10.0g
NaCl	5.0g
NaOH	1.0g

pH 6.8 ± 0.2 at 25°C

Preparation of Medium: Add components to distilled/deionized water and bring volume to 1.0L. Mix thoroughly. Gently heat and bring to boiling. Adjust pH to 6.8. Distribute into tubes or flasks. Autoclave for 15 min at 15 psi pressure–121°C.

Use: For the cultivation and maintenance of *Arthrobacter* species and *Corynebacterium* species.

Peptone Corn Agar

Composition per liter:

Agar	16.0g
Corn steep liquor	5.0g
NaCl	5.0g
Peptone	5.0g
$CaCl_2 \cdot 2H_2O$	0.5g

pH 7.2 ± 0.2 at 25°C

Preparation of Medium: Add components to distilled/deionized water and bring volume to 1.0L. Mix thoroughly. Gently heat and bring to boiling. Distribute into tubes or flasks. Autoclave for 15 min at 15 psi pressure–121°C. Pour into sterile Petri dishes or leave in tubes.

Use: For the cultivation and maintenance of *Actinomadura rubrobrunea, Pseudonocardia thermophila, Saccharopolyspora hordei, Thermoactinomyces candidus, Thermoactinomyces intermedius, Thermoactinomyces putidus, Saccharopolyspora rectivirgula, Streptomyces macrosporeus, Streptomyces rimosus, Thermoactinomyces dichotomicus, Thermoactinomyces thalpophilus,* and *Thermoactinomyces vulgaris.*

Peptone, Czapek's

Composition per liter:

Agar	15.0g
Sucrose	15.0g
Peptone	5.0g
$NaNO_3$	3.0g

K_2HPO_4 .... 1.0g
KCl 0.5g
$MgSO_4$ 0.5g
$FeSO_4$ 0.01g

Preparation of Medium: Add components to distilled/deionized water and bring volume to 1.0L. Mix thoroughly. Gently heat and bring to boiling. Distribute into tubes or flasks. Autoclave for 15 min at 15 psi pressure–121°C. Pour into sterile Petri dishes or leave in tubes.

Use: For the cultivation and maintenance of *Pilimelia anulata, Pilimelia terevasa,* and other *Pilimelia* species.

Peptone Fumarate Sulfate Medium

See: **PFS Medium**

Peptone Glucose Liver Extract Medium

See: **PGLE Medium**

Peptone Glucose Salt Agar

See: **PGS Agar**

Peptone Glucose Yeast Extract Agar

See: **PGY Agar**

Peptone Glycerol Phosphate Broth

See: **PGP Broth**

Peptone Iron Agar

Composition per liter:
Agar 15.0g
Peptone 15.0g
Proteose peptone 5.0g
Sodium glycerophosphate 1.0g
Ferric ammonium citrate 0.5g
$Na_2S_2O_3$ 0.08g

pH 6.7 ± 0.2 at 25°C

Source: This medium is available as a premixed powder from BD Diagnostic Systems.

Preparation of Medium: Add components to distilled/deionized water and bring volume to 1.0L. Mix thoroughly. Gently heat and bring to boiling. Distribute into tubes. Autoclave for 15 min at 15 psi pressure–121°C. Allow tubes to cool in an upright position.

Use: For the cultivation and differentiation of microorganisms based on their ability to produce H_2S. Microorganisms that produce H_2S turn the medium black.

Peptone Iron HiVeg Agar

Composition per liter:
Agar 15.0g
Plant peptone 15.0g
Plant peptone No. 3 .... 5.0g
Sodium glycerophosphate 1.0g
Ferric ammonium citrate 0.5g
$Na_2S_2O_3$ 0.08g

pH 6.7 ± 0.2 at 25°C

Source: This medium is available as a premixed powder from HiMedia.

Preparation of Medium: Add components to distilled/deionized water and bring volume to 1.0L. Mix thoroughly. Gently heat and bring to boiling. Distribute into tubes. Autoclave for 15 min at 15 psi pressure–121°C. Allow tubes to cool in an upright position.

Use: For the cultivation and differentiation of microorganisms based on their ability to produce H_2S. Microorganisms that produce H_2S turn the medium black.

Peptone Meat Extract Glycerol Agar

Composition per liter:
Agar 12.0g
Proteose peptone No. 3 .... 5.0g
Meat extract 3.0g
Glycerol 20.0mL

pH 7.0 ± 0.2 at 25°C

Preparation of Medium: Add components to distilled/deionized water and bring volume to 1.0L. Mix thoroughly. Gently heat and bring to boiling. Distribute into tubes or flasks. Autoclave for 15 min at 15 psi pressure–121°C. Pour into sterile Petri dishes or leave in tubes.

Use: For the cultivation and maintenance of *Mycobacterium intracellulare, Mycobacterium kansasii, Mycobacterium terrae, Mycobacterium vaccae, Nocardia vaccinii,* and *Rhodococcus* species.

Peptone Meat Extract Soil Extract Agar (PFE Agar)

Composition per liter:
Agar 12.0g
Proteose peptone No. 3 .... 5.0g
Meat extract 3.0g
Tap water 850.0mL
Soil extract 150.0mL
Glycerol 20.0mL

pH 7.0 ± 0.2 at 25°C

Soil Extract:
Composition per liter:
Garden soil, air dried 400.0g

Preparation of Soil Extract: Pass 400.0g of air-dried garden soil through a coarse sieve. Add soil to 960.0mL of tap water. Mix thoroughly. Autoclave for 60 min at 15 psi pressure–121°C. Cool to room temperature. Allow residue to settle. Decant supernatant solution. Filter through Whatman filter paper. Distribute into bottles in 200.0mL volumes. Autoclave for 15 min at 15 psi pressure–121°C. Store at room temperature until clear.

Preparation of Medium: Add components to tap water and bring volume to 1.0L. Mix thoroughly. Gently heat and bring to boiling. Distribute into tubes or flasks. Autoclave for 15 min at 15 psi pressure–121°C. Pour into sterile Petri dishes or leave in tubes.

Use: For the cultivation and maintenance of *Mycobacterium avium, Mycobacterium gastri, Mycobacterium kansasii, Mycobacterium marinum, Mycobacterium scrofulaceum,* and *Mycobacterium terrae.*

Peptone Medium

Composition per liter:
Peptone 10.0g

Preparation of Medium: Add peptone to distilled/deionized water and bring volume to 1.0L. Mix thoroughly. Distribute into tubes or flasks. Autoclave for 15 min at 15 psi pressure–121°C.

Use: For the cultivation and maintenance of *Escherichia coli.*

Peptone Recovery Broth

Composition per liter:

Meat extract 10.0g
Peptone 10.0g
NaCl 5.0g

pH 6.8 ± 0.2 at 25°C

Preparation of Medium: Add components to distilled/deionized water and bring volume to 1.0L. Mix thoroughly. Adjust pH to 6.8. Distribute into tubes or flasks. Autoclave for 15 min at 15 psi pressure–121°C.

Use: For the cultivation of *Brevibacterium* species.

Peptone-Reduced Agar

Composition per liter:

Agar 12.5g
Peptone 10.0g
Beef extract 5.0g
NaCl 3.0g

Preparation of Medium: Add components to distilled/deionized water and bring volume to 1.0L. Mix thoroughly. Gently heat and bring to boiling. Distribute into tubes or flasks. Autoclave for 15 min at 15 psi pressure–121°C. Pour into sterile Petri dishes or leave in tubes.

Use: For the cultivation of *Arthrobacter* species.

Peptone Sodium Cholate

Composition per liter:

Meat extract 10.0g
Peptone 10.0g
NaCl 5.0g
Sodium cholate 5.0g

pH 6.8 ± 0.2 at 25°C

Preparation of Medium: Add components to distilled/deionized water and bring volume to 1.0L. Mix thoroughly. Adjust pH to 6.8. Distribute into tubes or flasks. Autoclave for 15 min at 15 psi pressure–121°C.

Use: For the cultivation and maintenance of *Anthrobacter* species.

Peptone Sorbitol Bile Broth

Composition per liter:

Sorbitol 10.0g
Na_2HPO_4 .... 8.23g
NaCl 5.0g
Peptone 5.0g
Bile salts No. 3 .... 1.5g
NaH_2PO_4 .... 1.2g

pH 7.6 ± 0.2 at 25°C

Preparation of Medium: Add components to distilled/deionized water and bring volume to 1.0L. Mix thoroughly. Distribute into bottles in 100.0mL volumes. Autoclave for 15 min at 15 psi pressure–121°C.

Use: For the enrichment and cultivation of *Yersinia* species.

Peptone Sorbitol HiVeg Broth

Composition per liter:

Sorbitol 10.0g
Na_2HPO_4 .... 8.23g
Plant peptone 5.0g
NaCl 5.0g
Synthetic detergent No. I 1.5g
NaH_2PO_4 .... 1.2g

pH 7.6 ± 0.2 at 25°C

Source: This medium is available as a premixed powder from HiMedia.

Preparation of Medium: Add components to distilled/deionized water and bring volume to 1.0L. Mix thoroughly. Distribute into bottles in 100.0mL volumes. Autoclave for 15 min at 15 psi pressure–121°C.

Use: For the enrichment and cultivation of *Yersinia* species.

Peptone Starch Carbonate Medium

Composition per liter:

Agar 15.0g
Soluble starch 10.0g
Peptone 5.0g
Yeast extract 5.0g
K_2HPO_4 .... 1.0g
$MgSO_4 \cdot 7H_2O$ 0.2g
Na_2CO_3 solution 100.0mL

Na_2CO_3 Solution:

Composition per 100.0mL:

Na_2CO_3 .... 10.0g

Preparation of Na_2CO_3 Solution: Add Na_2CO_3 to distilled/deionized water and bring volume to 100.0mL. Mix thoroughly. Autoclave for 15 min at 15 psi pressure–121°C. Cool to 45°–50°C.

Preparation of Medium: Add components, except Na_2CO_3 solution, to distilled/deionized water and bring volume to 990.0mL. Mix thoroughly. Gently heat and bring to boiling. Autoclave for 15 min at 15 psi pressure–121°C. Cool to 45°–50°C. Aseptically add sterile Na_2CO_3 solution. Mix thoroughly. Pour into sterile Petri dishes or distribute into sterile tubes.

Use: For the cultivation and maintenance of *Bacillus alcalophilus* and other *Bacillus* species.

Peptone Starch Dextrose Agar (PSD Agar) (Dunkelberg Agar)

Composition per liter:

Proteose peptone No. 3 .... 20.0g
Agar 15.0g
Soluble starch 10.0g
Glucose 2.0g
Na_2HPO_4 .... 1.0g
NaH_2PO_4 .... 1.0g

pH 6.8 ± 0.2 at 25°C

Preparation of Medium: Add starch to approximately 100.0mL of cold distilled/deionized water. Mix thoroughly. Add starch solution to 400.0mL of boiling distilled/deionized water. Add remaining components. Mix thoroughly. Bring volume to 1.0L with distilled/deionized water. Autoclave for 12 min at 8 psi pressure–112°C. Pour into sterile Petri dishes or distribute into screw-capped tubes.

Use: For the selective isolation and cultivation of *Gardnerella vaginalis*.

Peptone Succinate Agar

Composition per liter:

Agar 15.0g
Peptone 5.0g

Succinic acid	1.68g
$MgSO_4·7H_2O$	1.0g
$(NH_4)_2SO_4$	1.0g
$FeCl_3·6H_2O$	2.0mg
$MnSO_4·H_2O$	2.0mg

pH 7.0 ± 0.2 at 25°C

Preparation of Medium: Add components to distilled/deionized water and bring volume to 1.0L. Mix thoroughly. Gently heat and bring to boiling. Distribute into tubes or flasks. Autoclave for 15 min at 15 psi pressure–121°C. Pour into sterile Petri dishes or leave in tubes.

Use: For the cultivation and maintenance of *Aquaspirillum bengal, Aquaspirillum dispar,* and *Spirillum volutans*.

Peptone Succinate Agar

Composition per liter:

Peptone	5.0g
Succinic acid	1.68g
Agar	1.5g
$MgSO_4·7H_2O$	1.0g
NH_4SO_4	1.0g
$FeCl_3·6H_2O$	2.0mg
$MnSO_4·H_2O$	2.0mg

pH 7.0 ± 0.2 at 25°C

Preparation of Medium: Add components to distilled/deionized water and bring volume to 1.0L. Mix thoroughly. Adjust pH to 7.0. Distribute into tubes or flasks. Autoclave for 15 min at 15 psi pressure–121°C.

Use: For the cultivation of *Aquaspirillum serpens.*

Peptone Succinate Agar in Seawater

Composition per liter:

Peptone	5.0g
Succinic acid	1.68g
Agar	1.5g
$MgSO_4·7H_2O$	1.0g
$(NH_4)_2SO_4$	1.0g
$FeCl_3·6H_2O$	2.0mg
$MnSO_4·H_2O$	2.0mg
Seawater	1.0L

pH 7.0 ± 0.2 at 25°C

Preparation of Medium: Combine components. Mix thoroughly. Gently heat and bring to boiling. Distribute into tubes or flasks. Autoclave for 15 min at 15 psi pressure–121°C.

Use: For the cultivation and maintenance of *Oceanospirillum maris*.

Peptone Succinate Salts Broth (PSS Broth)

Composition per 100.0mL:

Peptone	1.0g
$MgSO_4·7H_2O$	0.1g
$(NH_4)_2SO_4$	0.1g
Succinic acid	0.1g
$FeCl_3·6H_2O$	0.2mg
$MnSO_4·H_2O$	0.2mg

pH 6.8 ± 0.2 at 25°C

Preparation of Medium: Add components to distilled/deionized water and bring volume to 1.0L. Mix thoroughly. Adjust pH to 6.8 with KOH. Distribute into tubes or flasks. Autoclave for 15 min at 15 psi pressure–121°C.

Use: For the cultivation of *Spirillum* species.

Peptone Succinate Salts Medium (PSS Medium)

Composition per liter:

Peptone	10.0g
$MgSO_4·7H_2O$	1.0g
$(NH_4)_2SO_4$	1.0g
Succinic acid	1.0g
$FeCl_3·6H_2O$	2.0mg
$MnSO_4·H_2O$	2.0mg
Synthetic seawater	1.0L

pH 6.8 ± 0.2 at 25°C

Synthetic Seawater:

Composition per liter:

NaCl	27.5g
$MgCl_2$	5.0g
$MgSO_4$	2.0g
KCl	1.0g
$CaCl_2$	0.5g
$FeSO_4$	1.0μg

Preparation of Synthetic Seawater: Add components to distilled/deionized water and bring volume to 1.0L. Mix thoroughly.

Preparation of Medium: Add solid components to 1.0L synthetic seawater. Mix thoroughly. Adjust pH to 6.8 with 2*N* KOH. Distribute into tubes or flasks. Autoclave for 15 min at 15 psi pressure–121°C.

Use: For the cultivation and maintenance of *Aquaspirillum anulus*.

Peptone Succinate Salts in Seawater

Composition per liter:

Peptone	10.0g
$MgSO_4·7H_2O$	1.0g
$(NH_4)_2SO_4$	1.0g
Succinic acid	1.0g
$FeCl_3·6H_2O$	2.0mg
$MnSO_4·H_2O$	2.0mg
Synthetic seawater	1.0L

pH 6.8 ± 0.2 at 25°C

Synthetic Seawater:

Composition per liter:

NaCl	27.5g
$MgCl_2$	5.0g
$MgSO_4$	2.0g
KCl	1.0g
$CaCl_2$	0.5g
$FeSO_4$	1.0μg

Preparation of Synthetic Seawater: Add components to distilled/deionized water and bring volume to 1.0L. Mix thoroughly.

Preparation of Medium: Add solid components to 1.0L synthetic seawater. Mix thoroughly. Adjust pH to 6.8 with 2*N* KOH. Distribute into tubes or flasks. Autoclave for 15 min at 15 psi pressure–121°C.

Use: For the cultivation and maintenance of *Oceanospirillum maris*.

Peptone Sucrose Broth

Composition per liter:

Sucrose	20.0g
Peptone	10.0g

Preparation of Medium: Add components to distilled/deionized water and bring volume to 1.0L. Mix thoroughly. Distribute into tubes or flasks. Autoclave for 15 min at 15 psi pressure–121°C.

Use: For the cultivation and maintenance of *Xanthomonas campestris*.

Peptone Water

Composition per liter:

Peptone	10.0g
NaCl	5.0g

pH 7.2 ± 0.2 at 25°C

Source: This medium is available as a premixed powder from BD Diagnostic Systems and Oxoid Unipath.

Preparation of Medium: Add components to distilled/deionized water and bring volume to 1.0L. Mix thoroughly. Distribute into tubes or flasks. Autoclave for 15 min at 15 psi pressure–121°C.

Use: For the cultivation of nonfastidious microorganisms, for carbohydrate fermentation tests, and for performing the indole test.

Peptone Water with Andrade's Indicator

Composition per liter:

Peptone	10.0g
NaCl	5.0g
Andrade's indicator	100.0mL
Carbohydrate solution	20.0mL

pH 7.4 ± 0.2 at 25°C

Source: This medium is available as a premixed powder from Oxoid Unipath.

Andrade's Indicator:

Composition per 100.0mL:

NaOH (1*N* solution)	16.0mL
Acid Fuchsin	0.1 g

Preparation of Andrade's Indicator: Add Acid Fuchsin to NaOH solution and bring volume to 100.0mL with distilled/deionized water.

Caution: Acid Fuchsin is a potential carcinogen and care must be taken to avoid inhalation of the powdered dye and contact with the skin.

Carbohydrate Solution:

Composition per 20.0mL:

Carbohydrate	5.0–10.0g

Preparation of Carbohydrate Solution: Add carbohydrate to distilled/deionized water and bring volume to 20.0mL. Mix thoroughly. Filter sterilize.

Preparation of Medium: Add components, except carbohydrate solution, to distilled/deionized water and bring volume to 980.0mL. Mix thoroughly. Adjust pH to 7.4 if necessary. Distribute into tubes containing an inverted Durham tube. Fill each tube with 9.8mL of medium. Autoclave for 15 min at 15 psi pressure–121°C. Aseptically add 0.2mL of sterile carbohydrate solution to each tube.

Use: For use in carbohydrate fermentation tests. Fermentation is determined by the production of acid—broth turns pink—and formation of gas—bubble trapped in Durham tube.

Peptone Water, HiVeg

Composition per liter:

Plant peptone	10.0g
NaCl	5.0g

pH 7.2 ± 0.2 at 25°C

Source: This medium is available as a premixed powder from HiMedia.

Preparation of Medium: Add components to distilled/deionized water and bring volume to 1.0L. Mix thoroughly. Distribute into tubes or flasks. Autoclave for 15 min at 15 psi pressure–121°C.

Use: For the cultivation of nonfastidious microorganisms, for carbohydrate fermentation tests, and for performing the indole test. Note: When sterile solutions are to be added after autoclaving, reduce the volume of water for reconstitution by an equal amount. When used for carbohydrate fermentation studies add inverted Durham tubes into the final containers.

Peptone Yeast Extract Agar (ATCC Medium 526)

Composition per liter:

Agar	15.0g
Peptone	10.0g
Yeast extract	3.0g

Preparation of Medium: Add components to distilled/deionized water and bring volume to 1.0L. Mix thoroughly. Gently heat and bring to boiling. Distribute into tubes or flasks. Autoclave for 15 min at 15 psi pressure–121°C. Pour into sterile Petri dishes or leave in tubes.

Use: For the cultivation and maintenance of *Bdellovibrio bacteriovorus* and *Bdellovibrio stolpii*.

Peptone Yeast Extract Agar (ATCC Medium 1093)

Composition per liter:

Agar	15.0g
Peptone	0.5g
Yeast extract	0.5g

Preparation of Medium: Add components to distilled/deionized water and bring volume to 1.0L. Mix thoroughly. Gently heat and bring to boiling. Distribute into tubes or flasks. Autoclave for 15 min at 15 psi pressure–121°C. Pour into sterile Petri dishes or leave in tubes.

Use: For the cultivation and maintenance of *Angiococcus disciformis*.

Peptone Yeast Extract Agar (ATCC 135)

Composition per liter:

Agar	15.0g
Peptone	10.0g
Yeast extract	0.5g

Preparation of Medium: Add components to distilled/deionized water and bring volume to 1.0L. Mix thoroughly. Gently heat and bring to boiling. Distribute into tubes or flasks. Autoclave for 15 min at 15 psi pressure–121°C. Pour into sterile Petri dishes or leave in tubes.

Use: For the cultivation of *Arthrobacter* species.

Peptone Yeast Extract Agar (ATCC 1815)

Composition per liter:

Agar	20.0g
Glucose	5.0g
Peptone	5.0g
Yeast extract	3.0g
K_2HPO_4	0.2g

pH 7.0 ± 0.2 at 25°C

Preparation of Medium: Add components to tap water and bring volume to 1.0L. Mix thoroughly. Gently heat and bring to boiling. Adjust pH to 7.0. Distribute into tubes or flasks. Autoclave for 15 min at 15 psi pressure–121°C. Pour into sterile Petri dishes or leave in tubes.

Use: For the cultivation of *Glycomyces tenuis.*

Peptone Yeast Extract Carboxymethyl Cellulose Medium

See: **PY CMC Medium**

Peptone Yeast Extract Glucose Agar

Composition per liter:

Agar	15.0g
Glucose	10.0g
Peptone	5.0g
Yeast extract	5.0g

pH 7.0 ± 0.2 at 25°C

Preparation of Medium: Add components to distilled/deionized water and bring volume to 1.0L. Mix thoroughly. Gently heat and bring to boiling. Distribute into tubes or flasks. Autoclave for 15 min at 15 psi pressure–121°C. Pour into sterile Petri dishes or leave in tubes.

Use: For the cultivation and maintenance of *Alcaligenes latus, Clavibacter iranicum, Clavibacter michiganense, Clavibacter rathayi, Clavibacter tritici, Curtobacterium flaccumfaciens, Erwinia amylovora, Erwinia mallotivora, Erwinia nigrifluens, Erwinia quercina, Erwinia rubrifaciens, Erwinia salicis, Gordona bronchialis, Gordona terrae, Rhodococcus fascians,* and *Acinetobacter baumannii.*

Peptone Yeast Extract Glucose Agar with Casein

Composition per liter:

Agar	15.0g
Glucose	10.0g
Peptone	5.0g
Yeast extract	5.0g
Casein hydrolysate	0.1g

pH 7.0 ± 0.2 at 25°C

Preparation of Medium: Add components to distilled/deionized water and bring volume to 1.0L. Mix thoroughly. Gently heat and bring to boiling. Distribute into tubes or flasks. Autoclave for 15 min at 15 psi pressure–121°C. Pour into sterile Petri dishes or leave in tubes.

Use: For the cultivation and maintenance of *Clavibacter michiganense.*

Peptone Yeast Extract Glucose Broth

See: **PYG Broth**

Peptone Yeast Extract Glucose Maltose Medium

See: **PYGM Medium**

Peptone Yeast Extract Glucose Medium

See: **PYG Medium**

Peptone Yeast Extract Glucose Medium for *Spirillum*

See: **PYG Medium for *Spirillum***

Peptone Yeast Extract Glucose Medium, Modified (MPYG Medium)

Composition per 950.0mL:

Peptone	10.0g
Yeast extract	10.0g
Glucose	5.0g
L-Cysteine·HCl·H_2O	0.5g
$(NH_4)_2SO_4$	0.5g
Salt solution	40.0mL
Vitamin K-heme solution	10.0mL
Resazurin (0.025% solution)	4.0mL
Volatile fatty acid solution	3.1mL

pH 7.0 ± 0.2 at 25°C

Salt Solution:

Composition per liter:

$NaHCO_3$	10.0g
NaCl	2.0g
K_2HPO_4	1.0g
KH_2PO_4	1.0g
$CaCl_2$, anhydrous	0.2g
$MgSO_4$	0.2g

Preparation of Salt Solution: Add $CaCl_2$ and $MgSO_4$ to 300.0mL of distilled/deionized water. Mix thoroughly until dissolved. Bring volume to 800.0mL with distilled/deionized water. Add remaining components while stirring. Bring volume to 1.0L. Mix thoroughly. Store at 4°C.

Vitamin K-Heme Solution:

Composition per liter:

Part A	100.0mL
Part B	1.0mL

Preparation of Vitamin K-Heme Solution: Aseptically add 1.0mL of sterile part B to 100.0mL of cooled sterile part A. Mix thoroughly.

Part A:

Composition per 100.0mL:

Hemin	50.0mg
NaOH (1*N* solution)	1.0mL

Preparation of Part A: Add hemin to NaOH solution and bring volume to 100.0mL with distilled/deionized water. Mix thoroughly. Autoclave for 15 min at 15 psi pressure–121°C. Cool to 45°–50°C.

Part B:

Composition per 30.0mL:

Menadione (vitamin K_3)	100.0mg
Ethanol (95% solution)	30.0mL

Preparation of Part B: Add menadione to ethanol. Mix thoroughly. Filter sterilize.

Volatile Fatty Acid Solution:

Composition per 31.0mL:

Acetic acid	17.0mL
Propionic acid	6.0mL
n-Butyric acid	4.0mL
n-Valeric acid	1.0mL
Isovaleric acid	1.0mL

Isobutyric acid 1.0mL
DL-α-Methyl butyric acid 1.0mL

Preparation of Volatile Fatty Acid Solution: Combine components. Mix thoroughly.

Preparation of Medium: Add components—except vitamin K-heme solution, L-cysteine·HCl·H_2O, and volatile fatty acid solution—to distilled/deionized water and bring volume to 936.9mL. Gently heat and bring to boiling under 97% N_2 + 3% H_2. Continue boiling until resazurin turns colorless, indicating reduction. Cool to 45°–50°C. Add vitamin K-heme solution, L-cysteine·HCl·H_2O, and volatile fatty acid solution. Adjust pH to 7.0. Distribute into tubes under 97% N_2 + 3% H_2. Cap with rubber stoppers. Place tubes in a press. Autoclave for 15 min at 15 psi pressure–121°C with fast exhaust.

Use: For the cultivation and maintenance of *Acetivibrio ethanolgignens, Butyrivibrio fibrisolvens, Lachnospira multipara,* and *Succinivibrio dextrinosolvens.*

Peptone Yeast Extract Glucose Salt Medium
See: **PYEX Glucose Salt Medium**

Peptone Yeast Extract Glucose Vitamin Marine Medium
See: **PYGV Marine Medium**

Peptone Yeast Extract Glucose Vitamin Medium
See: **PYGV Medium**

Peptone Yeast Extract Inositol Medium
See: **PY Inositol Medium**

Peptone Yeast Extract Iron Agar
See: **ISP Medium 6**

Peptone Yeast Extract Medium (ATCC Medium 828)

Composition per liter:
Peptone 20.0g
Yeast extract 1.5g

pH 7.0 ± 0.2 at 25°C

Preparation of Medium: Add components to distilled/deionized water and bring volume to 1.0L. Mix thoroughly. Gently heat and bring to boiling. Distribute into tubes or flasks. Autoclave for 15 min at 15 psi pressure–121°C.

Use: For the cultivation and maintenance of *Acinetobacter lwoffii.*

Peptone Yeast Extract Medium (ATCC Medium 1366)

Composition per liter:
Peptone 10.0g
NaCl 5.0g
Yeast extract 5.0g

pH 7.2 ± 0.2 at 25°C

Preparation of Medium: Add components to distilled/deionized water and bring volume to 1.0L. Mix thoroughly. Gently heat and bring to boiling. Adjust pH to 7.2. Distribute into tubes or flasks. Autoclave for 15 min at 15 psi pressure–121°C.

Use: For the cultivation and maintenance of *Xenorhabdus nematophilus.*

Peptone Yeast Extract Medium (PY Medium) (ATCC Medium 1524)

Composition per 950.0mL:
Yeast extract 10.0g
Peptone 5.0g
Pancreatic digest of casein 5.0g
L-Cysteine·HCl·H_2O 0.5g
Salt solution 40.0mL
Hemin solution 10.0mL
Resazurin (0.025% solution) 4.0mL
Vitamin K_1 solution 0.2mL

pH 7.0 ± 0.2 at 25°C

Salt Solution:
Composition per liter:
$NaHCO_3$ 10.0g
NaCl 2.0g
K_2HPO_4 .... 1.0g
KH_2PO_4 .... 1.0g
$CaCl_2$, anhydrous 0.2g
$MgSO_4$ 0.2g

Preparation of Salt Solution: Add $CaCl_2$ and $MgSO_4$ to 300.0mL of distilled/deionized water. Mix thoroughly until dissolved. Bring volume to 800.0mL with distilled/deionized water. Add remaining components while stirring. Bring volume to 1.0L. Mix thoroughly. Store at 4°C.

Hemin Solution:
Composition per 100.0mL:
Hemin 0.05g
NaOH (1*N* solution) 1.0mL

Preparation of Hemin Solution: Add hemin to NaOH solution and bring volume to 100.0mL with distilled/deionized water. Mix thoroughly. Autoclave for 15 min at 15 psi pressure–121°C. Cool to 45°–50°C.

Vitamin K_1 Solution:
Composition per 30.0mL:
Vitamin K_1 .... 0.15g
Ethanol (95% solution) 30.0mL

Preparation of Vitamin K_1 Solution: Add vitamin K_1 to ethanol. Mix thoroughly. Filter sterilize.

Preparation of Medium: Add components—except vitamin K_1 solution, hemin solution, and L-cysteine·HCl·H_2O, to distilled/deionized water and bring volume to 939.8mL. Gently heat and bring to boiling under 80% N_2 + 10% H_2 + 10% CO_2. Continue boiling until resazurin turns colorless, indicating reduction. Cool to 45°–50°C. Add vitamin K_1 solution, hemin solution, and L-cysteine·HCl·H_2O. Adjust pH to 7.0. Distribute into tubes under 80% N_2 + 10% H_2 + 10% CO_2. Cap with rubber stoppers. Place tubes in a press. Autoclave for 15 min at 15 psi pressure–121°C with fast exhaust.

Use: For the cultivation and maintenance of *Megasphaera cerevisiae* and *Clostridium* species.

Peptone Yeast Extract 1% Medium
See: **PY 1% Medium**

Peptone Yeast Extract Medium with Fructose
See: **PY Medium with Fructose**

Peptone Yeast Extract Medium with Glucose
See: **PY Medium with Glucose**

Peptone Yeast Extract Salt Agar
See: **PYS Agar**

Peptone Yeast Extract Salt Medium
See: **PY Salt Medium**

Peptone Yeast Glutamate Medium

Composition per liter:

Peptone	20.0g
Yeast extract	10.0g
Monosodium glutamate	4.0g
Sodium thioglycolate	1.0g

pH 7.2 ± 0.2 at 25°C

Preparation of Medium: Add components to distilled/deionized water and bring volume to 1.0L. Mix thoroughly. Distribute into tubes or flasks. Autoclave for 15 min at 15 psi pressure–121°C.

Use: For the cultivation of *Peptococcus aerogenes* and a variety of other bacteria.

Peptone Yeast Medium with Magnesium Sulfate (DSMZ Medium 790)

Composition per liter:

Peptone	10.0g
Yeast extract, dehydrated	1.0g
$MgSO_4 \cdot 7H_2O$	2.0g
$(NH_4)_2SO_4$	2.0g

pH 7.0 ± 0.2 at 25°C

Preparation of Medium: Add components to distilled/deionized water and bring volume to 1.0L. Mix thoroughly. Distribute into tubes or flasks. Autoclave for 15 min at 15 psi pressure–121°C.

Use: For the cultivation of *Aquaspirillum psychrophilum* and *Aquaspirillum peregrinum* subsp. *integrum.*

Peptone Yeast Medium with $MgSO_4$

Composition per liter:

Peptone	10.0g
$MgSO_4 \cdot 7H_2O$	2.0g
$(NH_4)_2SO_4$	2.0g
Yeast extract	1.0g

pH 7.0 ± 0.2 at 25°C

Preparation of Medium: Add components to distilled/deionized water and bring volume to 1.0L. Mix thoroughly. Distribute into tubes or flasks. Autoclave for 15 min at 15 psi pressure–121°C.

Use: For the cultivation and maintenance of *Aquaspirillum itersonii, Aquaspirillum peregrinum,* and *Aquaspirillum psychrophilum.*

Peptone Yeast Trypticase™ Agar (ATCC Medium 118)

Composition per liter:

Agar	15.0g
Peptone	6.0g
Trypticase™ (pancreatic digest of casein)	4.0g
Yeast extract	3.0g
Beef extract	1.5g
Glucose	1.0g

pH 7.0 ± 0.2 at 25°C

Preparation of Medium: Add components to distilled/deionized water and bring volume to 1.0L. Mix thoroughly. Gently heat and bring to boiling. Distribute into tubes or flasks. Autoclave for 15 min at 15 psi pressure–121°C. Pour into sterile Petri dishes or leave in tubes.

Use: For the cultivation of a variety of heterotrophic bacteria.

Peptonized Milk Agar (PMA Medium)

Composition per liter:

Agar	15.0g
Milk, peptonized	1.0g

Preparation of Medium: Add components to distilled/deionized water and bring volume to 1.0L. Mix thoroughly. Gently heat and bring to boiling. Distribute into tubes or flasks. Autoclave for 15 min at 15 psi pressure–121°C. Pour into sterile Petri dishes or leave in tubes.

Use: For the cultivation of freshwater *Myxobacterium* species.

Perfringens Agar, OPSP
See: ***Clostridium perfringens*** **Agar, OPSP**

Perfringens HiVeg Agar Base (O.P.S.P.) with Antibiotics

Composition per liter:

Agar	15.0g
Plant hydrolysate	15.0g
Plant extract No. 2	7.0g
Papaic digest of soybean meal	5.0g
Yeast extract	5.0g
Tris buffer	1.5g
Ferric ammonium citrate	1.0g
$Na_2S_2O_5$	1.0g
Antibiotic inhibitor	10.0mL

pH 7.3 ± 0.2 at 25°C

Source: This medium, without antibiotic inhibitor, is available as a premixed powder from HiMedia.

Antibiotic Inhibitor:

Composition per 10.0mL:

Sodium sulfadiazine	0.1g
Oleandomycin phosphate	0.5mg
Polymyxin B	10,000U

Preparation of Antibiotic Inhibitor: Add components to distilled/deionized water and bring volume to 10.0mL. Mix thoroughly. Filter sterilize.

Preparation of Medium: Add components, except antibiotic inhibitor, to distilled/deionized water and bring volume to 990.0mL. Mix thoroughly. Gently heat and bring to boiling. Autoclave for 15 min at 15 psi pressure–121°C. Cool to 45°–50°C. Aseptically add sterile antibiotic inhibitor. Mix thoroughly. Pour into sterile Petri dishes or distribute into sterile tubes.

Use: For the presumptive identification and enumeration of *Clostridium perfringens* in foods.

Perfringens HiVeg Agar Base with Egg Yolk and Antibiotics (T.S.C./S.F.P. HiVeg Agar Base)

Composition per liter:

Agar	15.0g
Plant hydrolysate No. 1	15.0g
Papaic digest of soybean meal	5.0g
Plant extract	5.0g
Yeast extract	5.0g
$Na_2S_2O_5$	1.0g
Ferric ammonium citrate	1.0g
Egg yolk emulsion	25.0mL
Perfringens SFP supplement	4.0mL
Perfringens TSC supplement	4.0mL

pH 7.6 ± 0.2 at 25°C

Source: This medium, without egg yolk emulsion, perfringens SFP supplement, and perfringens TSC supplement, is available as a premixed powder from HiMedia.

Egg Yolk Emulsion:

Composition per 100.0mL:

Chicken egg yolks	9
Whole chicken egg	1
NaCl (0.9% solution)	25.0mL

Preparation of Egg Yolk Emulsion: Soak eggs with 1:100 dilution of saturated mercuric chloride solution for 1 min. Crack eggs and separate yolks from whites. Mix egg yolks with 1 chicken egg. Beat to form emulsion. Measure 50.0mL of egg yolk emulsion and add to 50.0mL of 0.9% NaCl solution. Mix thoroughly. Filter sterilize. Warm to 45°–50°C.

Perfringens SFP Supplement:

Composition per 10.0mL:

Kanamycin sulfate	30.0mg
Polymyxin B	75,000U

Preparation of Perfringens SFP Supplement: Add components to distilled/deionized water and bring volume to 10.0mL. Mix thoroughly. Filter sterilize.

Perfringens TSC Supplement:

Composition per 10.0mL:

D-Cycloserine	1.0g

Preparation of Perfringens TSC Supplement: Add D-cycloserine to distilled/deionized water and bring volume to 10.0mL. Mix thoroughly. Filter sterilize.

Preparation of Medium: Add components, except perfringens SFP supplement, egg yolk emulsion, and perfringens TSC supplement, to distilled/deionized water and bring volume to 975mL. Mix thoroughly. Gently heat and bring to boiling. Distribute into tubes or flasks. Autoclave for 15 min at 15 psi pressure–121°C. Cool to 55°C. Aseptically add 25.0mL egg yolk emulsion, 4.0mL perfringens SFP supplement, and 4.00mL perfringens TSC supplement. Mix thoroughly. Pour into sterile Petri dishes or distribute into sterile tubes.

Use: For the cultivation, enumeration, and presumptive identification of *Clostridium perfringens* from foods.

Perkinsus Agar Medium (ATCC Medium 2289)

Composition per liter:

Modified Perkinsus Medium	485.0mL
Agar Medium	485.0mL
Fetal bovine serum, heat inactivated	20.0mL
Lipid concentrate	10.0mL

Modified Perkinsus Medium

Composition per liter:

HEPES	11.92g
Nutrient mix F-12 Ham	10.8g
Dulbecco's modified Eagle's medium	8.4g
L-Glutamine	0.29g
$NaHCO_3$	1.3g
SASW 2X solution	860.0mL
JPL carbohydrate solution	20.0mL
Phenol Red (0.5%) solution	1.0mL

SASW 2X Solution:

Composition per liter:

Seawater, synthetic basal mixture	36.4g

Preparation of 2X SASW Solution: Add seawater synthetic basal mixture to distilled/deionized water and bring volume to 1.0L. Mix thoroughly.

JLP Carbohydrate Solution:

Composition per 100.0mL:

Glucose	5.0g
Galactose	1.0g
Trehalose	1.0g

Preparation of JLP Carbohydrate Solution: Add components to distilled/deionized water and bring volume to 100.0mL. Mix thoroughly.

Phenol Red Solution:

Composition per 10.0mL:

Phenol Red	0.05g

Preparation of Phenol Red Solution: Add Phenol Red to distilled/deionized water and bring volume to 10.0mL.

Preparation of Modified Perkinsus Medium: Add components to distilled/deionized water and bring volume to 1.0L. Mix thoroughly. Filter sterilize

Agar Medium:

Composition per liter:

Agar	30.0g

Preparation of Agar Medium: Add agar to distilled/deionized water and bring volume to 1.0L. Mix thoroughly. Gently heat while stirring and bring to boiling. Autoclave for 15 min at 15 psi pressure–121°C. Cool to 45°–50°C.

Lipid Concentrate:

Composition per liter:

Pluronic™ F68	10.0g
Tween™ 80	2.5g
Cod liver oil	1.0g
Cholesterol	0.45g
DL-α-Tocopherol acetate	0.2g

Preparation of Lipid Concentrate: Add components to distilled/deionized water and bring volume to 1.0L. Mix thoroughly. Filter sterilize.

Preparation of Medium: Warm Modified Perkinsus Medium to 50°C and combine with Agar Medium at 50°C. Mix thoroughly and maintain at 50°C. Aseptically add heat-inactivated fetal bovine serum and lipid concentrate. Mix thoroughly. Aliquot in 20mL amounts to Petri dishes and allow to solidify.

Use: For the cultivation of *Perkinsus marinus, P. andrewsi, P. chesapeaki,* and *P. atlanticus.*

Perkinsus Medium

Composition per liter:

NaCl	9.0g
$NaHCO_3$	2.1g
Glucose	1.1g
$NaH_2PO_4 \cdot H_2O$	0.29g
KCl	0.38g
L-Arginine·HCl	0.21g
L-Glutamine	0.30g
$MgSO_4 \cdot 7H_2O$	0.17g
Sodium pyruvate	0.11g
KH_2PO_4	0.08g
$CaCl_2 \cdot 2H_2O$	0.09g
L-Cystine·2HCl	0.04g
L-Lysine·HCl	0.04g
L-Leucine	0.2g
L-Isoleucine	0.02g
L-Histidine·HCl·H_2O	0.02g
L-Arginine	0.02g
L-Threonine	0.02g
L-Valine	0.02g
L-Tyrosine	0.02g
L-Methionine	0.02mg
L-Cystine	0.016g
L-Phenylalanine	0.014g
L-Serine	0.01g
L-Asparagine·H_2O	0.01g
L-Aspartic Acid	0.01g
L-Glutamic acid	0.01g
L-Histidine	0.01g
L-Proline	0.01g
L-Glycine	0.01g
L-Alanine	8.9mg
D-Phenylalanine	5.0mg
L-Methionine	4.5mg
Hypoxanthine	4.1mg
L-Tryptophan	4.6mg
L-Threonine	3.6mg
L-Valine	3.5mg
L-Tyrosine	1.8mg
Vitamin B_{12}	1.4mg
Folic acid	2.3mg
Phenol Red	1.2mg
Thiamine·HCl	2.0mg
$FeSO_4 \cdot 7H_2O$	0.8mg
Choline chloride	1.7mg
Calcium DL-pantothenate	1.7mg
Thymidine	0.7mg
Niacinamide	1.6mg
Pyridoxal	1.0mg
Inositol	0.7mg
Riboflavin	0.5mg
Lipoic acid	0.2mg
Pyridoxine·HCl	0.2mg
$ZnSO_4 \cdot 7H_2O$	0.03mg
$FeNO_3 \cdot 7H_2O$	0.025mg
Biotin	0.02mg
$CuSO_4 \cdot 5H_2O$	3.0μg
SASW solution	910.0mL
HEPES (*N*-[2-hydroxyethyl] piperazine-*N′*-2-ethanesulfonic acid) buffer (1.0*M* solution)	25.0mL
Fetal bovine serum, heat inactivated	20.0mL
JLP carbohydrate solution	10.0mL
Lipid concentrate (100X)	10.0mL
$NaHCO_3$ solution	8.6mL
L-Glutamine solution	5.0mL
Phenol Red solution	0.5mL

SASW Solution:

Composition per liter:

Seawater, synthetic basal mixture	18.2g

Preparation of SASW Solution: Add seawater synthetic basal mixture to distilled/deionized water and bring volume to 1.0L. Mix thoroughly.

JLP Carbohydrate Solution:

Composition per 100.0mL:

Glucose	5.0g
Galactose	1.0g
Trehalose	1.0g

Preparation of JLP Carbohydrate Solution: Add components to distilled/deionized water and bring volume to 100.0mL. Mix thoroughly.

Phenol Red Solution:

Composition per 10.0mL:

Phenol Red	0.05g

Preparation of Phenol Red Solution: Add Phenol Red to distilled/deionized water and bring volume to 10.0mL.

Glutamine Solution:

Composition per 10.0mL:

L-Glutamine	0.29g

Preparation of Glutamine Solution: Add L-glutamine to distilled/deionized water and bring volume to 10.0mL.

$NaHCO_3$ Solution:

Composition per 10.0mL:

$NaHCO_3$	0.75g

Preparation of $NaHCO_3$ Solution: Add $NaHCO_3$ to distilled/deionized water and bring volume to 10.0mL.

Lipid Concentrate:

Composition per liter:

Pluronic™ F68	10.0g
Tween™ 80	2.5g
Cod liver oil	1.0g
Cholesterol	0.45g
DL-α-Tocopherol acetate	0.2g

Preparation of Lipid Concentrate: Add components to distilled/deionized water and bring volume to 1.0L. Mix thoroughly. Filter sterilize.

Preparation of Medium: Add all components, except lipid concentrate and fetal bovine serum, to distilled/deionized water and bring volume to 1.0L. Mix thoroughly. Filter sterilize. Aseptically add 20.0mL of sterile fetal bovine serum and 10.0mL sterile lipid concentrate. Mix thoroughly. Aseptically distribute into sterile tubes or flasks. Use immediately.

Use: For the cultivation of *Perkinsus marinus, P. andrewsi, P. chesapeaki,* and *P. atlanticus.*

Persephonella Medium (DSMZ Medium 996)

Composition per liter:

NaCl	29.0g
$MgSO_4 \cdot 7H_2O$	7.0g
NaOH	2.0g
$Na_2S_2O_3$	2.0g
$MgCl_2 \cdot 6H_2O$	1.36g
KCl	0.5g
$CaCl_2 \cdot 2H_2O$	0.4g
K_2HPO_4	0.3g
NH_4Cl	0.2g
Trace elements solution	10.0mL

pH 6.0 ± 0.2 at 25°C

Trace Elements Solution:

Composition per liter:

$Na\text{-}EDTA \cdot 2H_2O$	0.5g
$CoCl_2 \cdot 6H_2O$	0.15g
$MnCl_2 \cdot 4H_2O$	0.1g
$FeSO_4 \cdot 7H_2O$	0.1g
$ZnCl_2$	0.1g
$AlCl_3 \cdot 6H_2O$	40.0mg
$Na_2O_4W \cdot 6H_2O$	30.0mg
CuCl	20.0mg
$Ni_2SO_4 \cdot 6H_2O$	20.0mg
Se-acide	10.0mg
H_3BO_3	10.0mg
$Na_2MoO_4 \cdot 2H_2O$	10.0mg

Preparation of Trace Elements Solution: Add components to distilled/deionized water and bring volume to 1.0L. Adjust pH to 3.0. Mix thoroughly.

Preparation of Medium: Prepare anaerobic distilled/deionized water by sparging with 100% CO_2. Add components to distilled/deionized anaerobic water and bring volume to 1.0L. Adjust pH to 6.0 with H_2SO_4. Autoclave for 15 min at 15 psi pressure–121°C. Dispense under a CO_2 atmosphere into Bellco tubes (5mL medium per 27mL tube). Stopper with butyl stoppers. Cap and crimp closures. Autoclave for 15 min at 15 psi pressure–121°C. After autoclaving a white precipitate might be present; this precipitate can be redissolved by shaking the medium.It can take up to an hour before all precipitate is dissolved. Add 3.8% O_2 to each tube (1mL of O_2 per 27mL tube). After inoculation pressurize the tubes with H_2 to 20psi (or 138kPa).

Use: For the cultivation of *Persephonella* spp.

Petragnani Medium

Composition per 2398.0mL:

Skim milk	100.0g
Potato flour	36.4g
L-Asparagine	5.1g
Pancreatic digest of casein	5.1g
Malachite Green	1.2g
Whole egg	1277.0mL
Egg yolk	121.0mL
Glycerol	60.0mL

pH 7.0 ± 0.2 at 25°C

Source: This medium is available as a prepared medium from BD Diagnostic Systems.

Preparation of Medium: Add components—except whole egg, egg yolk, and glycerol—to distilled/deionized water and bring volume to 940.0mL. Mix thoroughly. Add glycerol. Gently heat while stirring and bring to boiling. Autoclave for 15 min at 15 psi pressure–121°C. Cool to 45°–50°C. Scrub the eggshells with soap. Let stand in a soap solution for 30 min. Rinse in running water. Soak eggs in 70% ethanol for 15 min. Break the eggs into a sterile container. Homogenize by shaking. Filter through four layers of sterile cheesecloth into a sterile graduated cylinder. Measure out 1277.0mL. Add separated egg yolks to another sterile container. Measure out 121.0mL. Aseptically add homogenized whole egg and egg yolk to cooled sterile basal medium. Mix thoroughly. Aseptically distribute into sterile tubes. Inspissate at 85°–90°C (moist heat) for 45 min.

Use: For the isolation and cultivation of *Mycobacterium* species from clinical specimens. For the cultivation and maintenance of *Mycobacterium smegmatis*.

Petragnani Medium

Composition per 2285.0mL:

Potato	500.0g
Potato flour	36.0g
Malachite Green	1.2g
Whole egg	1200.0mL
Whole milk	900.0mL
Egg yolk	115.0mL
Glycerol	70.0mL

pH 7.2 ± 0.2 at 25°C

Source: This medium is available as a prepared medium from BD Diagnostic Systems.

Preparation of Medium: Peel and dice potato. Add potato to 500.0mL of distilled/deionized water. Gently heat and bring to boiling. Continue boiling for 30 min. Filter solids through two layers of cheesecloth. Combine potato solids with remaining components, except whole egg, egg yolk, and glycerol. Mix thoroughly. Add glycerol. Gently heat while stirring and bring to boiling. Autoclave for 15 min at 15 psi pressure–121°C. Cool to 45°–50°C. Scrub the eggshells with soap. Let stand in a soap solution for 30 min. Rinse in running water. Soak eggs in 70% ethanol for 15 min. Break the eggs into a sterile container. Homogenize by shaking. Filter through four layers of sterile cheesecloth into a sterile graduated cylinder. Measure out 1200.0mL. Add separated egg yolks to another sterile container. Measure out 115.0mL. Aseptically add homogenized whole egg and egg yolk to cooled sterile basal medium. Mix thoroughly. Aseptically distribute into sterile tubes. Inspissate at 85°–90°C (moist heat) for 45 min.

Use: For the isolation and cultivation of *Mycobacterium* species from clinical specimens. For the cultivation and maintenance of *Mycobacterium smegmatis*.

Petrotoga Medium

Composition per liter:

NaCl	18.0g
$MgSO_4 \cdot 7H_2O$	3.45g
$MgCl_2 \cdot 7H_2O$	2.75g
$NaHCO_3$	1.0g
L-Cysteine·HCl·H_2O	0.5g
KCl	0.335g
NH_4Cl	0.25g
$CaCl_2 \cdot 2H_2O$	0.14g
K_2HPO_4	0.14g
$Fe(NH_4)_2(SO_4)_2 \cdot 7H_2O$	2.0mg
Resazurin	1.0mg
Glucose solution	50.0mL

Trace elements solution SL-6 ....10.0mL
Pancreatic digest of casein solution 10.0mL
Yeast extract solution 10.0mL
$Na_2S·9H_2O$ solution 10.0mL
Wolfe's vitamin solution 10.0mL
pH 6.5–6.7 at 25°C

Glucose Solution:
Composition per 50.0mL:
D-Glucose....5.0g

Preparation of Glucose Solution: Add glucose to distilled/deionized water and bring volume to 50.0mL. Mix thoroughly. Sparge with 100% N_2. Autoclave for 15 min at 15 psi pressure–121°C.

Trace Elements Solution SL-6:
Composition per liter:
H_3BO_3 ....0.3g
$CoCl_2·6H_2O$ 0.2g
$ZnSO_4·7H_2O$ 0.1g
$MnCl_2·4H_2O$....0.03g
$Na_2MoO_4·H_2O$ 0.03g
$NiCl_2·6H_2O$ 0.02g
$CuCl_2·2H_2O$....0.01g

Preparation of Trace Elements Solution SL-6: Add components to distilled/deionized water and bring volume to 1.0L. Mix thoroughly. Adjust pH to 3.4.

Pancreatic Digest of Casein Solution:
Composition per 10.0mL:
Pancreatic digest of casein....1.0g

Preparation of Pancreatic Digest of Casein Solution: Add pancreatic digest of casein to distilled/deionized water and bring volume to 10.0mL. Mix thoroughly. Sparge with 100% N_2. Autoclave for 15 min at 15 psi pressure–121°C.

Yeast Extract Solution:
Composition per 10.0mL:
Yeast extract....1.0g

Preparation of Yeast Extract Solution: Add yeast extract to distilled/deionized water and bring volume to 10.0mL Mix thoroughly. Sparge with 100% N_2. Autoclave for 15 min at 15 psi pressure–121°C.

$Na_2S·9H_2O$ Solution:
Composition per 10.0mL:
$Na_2S·9H_2O$....0.5g

Preparation of $Na_2S·9H_2O$ Solution: Add $Na_2S·9H_2O$ to distilled/deionized water and bring volume to 10.0mL. Mix thoroughly. Sparge with 100% N_2. Autoclave for 15 min at 15 psi pressure–121°C.

Wolfe's Vitamin Solution:
Composition per liter:
Pyridoxine·HCl 10.0mg
Thiamine·HCl 5.0mg
Riboflavin 5.0mg
Nicotinic acid....5.0mg
Calcium pantothenate 5.0mg
p-Aminobenzoic acid....5.0mg
Thioctic acid 5.0mg
Biotin 2.0mg
Folic acid....2.0mg
Cyanocobalamin 100.0μg

Preparation of Wolfe's Vitamin Solution: Add components to distilled/deionized water and bring volume to 1.0L. Mix thoroughly. Filter sterilize. Sparge with 100% N_2.

Preparation of Medium: Prepare and dispense medium under 80% N_2 + 20% CO_2. Add components, except $NaHCO_3$, glucose solution, pancreatic digest of casein solution, yeast extract solution, $Na_2S·9H_2O$ solution, and Wolfe's vitamin solution, to distilled/deionized water and bring volume to 910.0mL. Mix thoroughly. Adjust pH to 6.5–6.7. Gently heat and bring to boiling. Continue boiling for 3 min. Cool to room temperature while sparging with 80% N_2 + 20% CO_2. Add $NaHCO_3$. Adjust pH to 6.5–6.7. Anaerobically distribute 9.1mL volumes into anaerobic tubes. Autoclave for 15 min at 15 psi pressure–121°C. Aseptically add 0.5mL of sterile glucose solution, 0.1mL of sterile pancreatic digest of casein solution, 0.1mL of sterile yeast extract solution, 0.1mL of sterile $Na_2S·9H_2O$ solution, and 0.1mL of sterile Wolfe's vitamin solution to each tube. Mix thoroughly.

Use: For the cultivation of *Petrotoga miotherma.*

Petrotoga Medium (ATCC 1881)

Composition per liter:
NaCl....20.0g
Sodium PIPES (piperazine-*N,N′*-bis[2-ethanesulfonic acid]) buffer 5.24g
Pancreatic digest of casein....5.0g
Yeast extract....2.0g
Resazurin 1.0g
Soluble starch....1.0g
L-Cysteine·HCl·H_2O 0.5g
pH 7.2 ± 0.2 at 25°C

Preparation of Medium: Add components, except L-cysteine·HCl·H_2O, to distilled/deionized water and bring volume to 1.0L. Mix thoroughly. Adjust pH to 7.4. Gently heat and bring to boiling. Continue boiling for 3 min. Cool to room temperature while sparging with 100% N_2. Add L-cysteine·HCl·H_2O. Mix thoroughly. Anaerobically distribute into tubes. Autoclave for 15 min at 15 psi pressure–121°C.

Use: For the cultivation of *Petrotoga miotherma.*

PFE Agar
See: **Peptone Meat Extract Soil Extract Agar**

Pfennig's Medium I, Modified for Marine Purple Sulfur Bacteria (DSMZ Medium 28)

Composition per 5.0L:
Solution A....4.0L
Solution B 860.0mL
Solution E 100.0mL
Solution F....20.0mL
Solution C 5.0mL
Solution D....5.0mL
pH 7.3 at 25°C

Solution A:
Composition per 4.0L:
NaCl....100.0g
$MgSO_4$....15.0g
KH_2PO_4....1.7g
NH_4Cl 1.7g

KCl 1.7g
$CaCl_2 \cdot 2H_2O$ 1.25g

Preparation of Solution A: Add components to 4.0L distilled water. Mix thoroughly. Autoclave for 45 min at 15 psi pressure–121°C in 5-liter special bottle or flask with four openings at the top, together with a teflon-coated magnetic bar. In this 5-liter bottle, two openings for tubes are in the central, silicon rubber stopper: a short, gas-inlet tube with a sterile cotton filter and an outlet tube for medium, which reaches the bottom of the vessel at one end and has, at the other end, a silicon rubber tube with a pinch cock and a bell for aseptic dispensing of the medium into bottles. The other two openings have gas-tight screw caps; one of these openings is for the addition of sterile solutions and the other serves as a gas outlet. After autoclaving, cool to room temperature under a N_2 atmosphere with a positive pressure of 0.05–0.1 atm (a manometer for low pressure will be required). Saturate the cold medium with CO_2 by magnetic stirring for 30 min under a CO_2 atmosphere of 0.05– 0.1 atm.

Solution B:

Distilled water 860.0mL

Preparation of Solution B: Autoclave distilled water for 15 min at 15 psi pressure–121°C in a cotton-stoppered Erlenmeyer flask. Cool to room temperature under an atmosphere of N_2 in an anaerobic jar.

Solution C:

Composition per 100.0mL:

Vitamin B_{12} 2.0mg

Preparation of Solution C: Add vitamin B_{12} to distilled/deionized water and bring volume to 100.0mL. Mix thoroughly. Sparge under 100% N_2 gas for 3 min. Filter sterilize. Store under N_2 gas.

Solution D:

Composition per liter:

Disodium ethylendiamine-tetraacetate
(Disodium EDTA) 3.0g
$FeSO_4 \cdot 7H_2O$ 1.1g
H_3BO_3 .... 0.3g
$CoCl_2 \cdot 6H_2O$ 0.19g
$MnCl_2 \cdot 2H_2O$ 50.0mg
$ZnCl_2$ 42.0mg
$NiCl_2 \cdot 6H_2O$ 24.0mg
$Na_2MoO_4 \cdot 2H_2O$ 18.0mg
$CuCl_2 \cdot 2H_2O$ 2.0mg

Preparation of Solution D: Add components to distilled/deionized water and bring volume to 1.0L. Mix thoroughly. Autoclave for 15 min at 15 psi pressure–121°C.

Solution E:

Composition per 100.0mL:

$NaHCO_3$ 7.5g

Preparation of Solution E: Add $NaHCO_3$ to distilled/deionized water and bring volume to 100.0mL. Mix thoroughly. Sparge with 100% CO_2 until saturated. Filter sterilize under 100% CO_2 into a sterile, gas-tight 100.0mL screw-capped bottle.

Solution F:

Composition per 100.0mL:

$Na_2S \cdot 9H_2O$ 10.0g

Preparation of Solution F: Add $Na_2S \cdot 9H_2O$ to distilled/deionized water in a 250.0mL screw-capped bottle fitted with a butyl rubber septum and bring volume to 100.0mL. Mix thoroughly. Sparge under 100% N_2 gas for 3 min. Autoclave for 15 min at 15 psi pressure–121°C. Cool to room temperature.

Preparation of Medium: Add solutions B, C, D, E, and F to solution A through one of the screw-cap openings against a stream of either N_2 gas or, better, a mixture of 95% N_2 and 5% CO_2 while the medium is magnetically stirred. Adjust the pH of the medium with sterile HCl or Na_2CO_3 solution (2*M* solution) to pH 7.3. Distribute the medium aseptically through the medium outlet tube into sterile, 100mL bottles (with metal caps and autoclavable rubber seals) using the positive gas pressure (0.05–0.1 atm) of the N_2/CO_2 gas mixture. Leave a small air bubble in each bottle to meet possible pressure changes. The tightly sealed, screw-cap bottles can be stored for several weeks or months in the dark. During the first 24 hr, the iron of the medium precipitates in the form of black flocks. No other sediment should arise in the otherwise clear medium. Incubate in the light using a tungsten lamp. Feed periodically with neutralized solution of sodium sulfide to replenish sulfide and with other supplement solutions.

Use: For the cultivation of marine purple sulfur bacteria.

Pfennig's Medium 1, Modified for Marine Purple Sulfur Bacteria (DSMZ Medium 28)

Composition per 4990.0mL:

Solution A 4000.0mL
Solution B 860.0mL
Solution E 100.0mL
Solution F 20.0mL
Solution C (Vitamin B_{12} solution) 5.0mL
Solution D (Trace elements solution SL-12B) 5.0mL
Neutralized sulfide solution .... As needed
pH 7.3 ± 0.2 at 25°C

Solution A:

Composition per 4000.0mL:

NaCl 100.0g
$MgSO_4 \cdot 7H_2O$ 15.0g
$CaCl_2 \cdot 2H_2O$ 1.25g
KH_2PO_4 .... 1.7g
NH_4Cl 1.7g
KCl 1.7g

Preparation of Solution A: Add components to distilled/deionized water and bring volume to 4.0L. Mix thoroughly. Adjust pH to 6.0. Dispense into a 5L flask with four openings at the top (two openings are in a central silicon rubber stopper and two openings are gas-tight screw caps). Add a teflon-coated magnetic stir bar to the flask. Autoclave for 45 min at 15 psi pressure–121°C. Cool to room temperature under 100% N_2 at 0.05–0.1 atm pressure (use a manometer to measure low pressure).

Solution B:

Composition per 860.0mL:

Distilled/deionized water 860.0mL

Preparation of Solution B: Add 860.0mL of distilled/deionized water to a cotton-stoppered flask. Autoclave for 20 min at 15 psi pressure–121°C. Cool to room temperature under 100% N_2 in an anaerobic jar.

Solution C (Vitamin B_{12} Solution):

Composition per 5.0mL:

Vitamin B_{12} 1.0mg

Preparation of Solution C (Vitamin B_{12} Solution): Add vitamin B_{12} to distilled/deionized water and bring volume to 5.0mL. Mix thoroughly. Filter sterilize.

Solution D (Trace Elements Solution SL-12B):

Composition per liter:

Disodium ethylendiamine-tetraacetate (Na_2-EDTA)	3.0g
$FeSO_4 \cdot 7H_2O$	1.1g
H_3BO_3	0.3g
$CoCl_2 \cdot 6H_2O$	0.19g
$MnCl_2 \cdot 2H_2O$	50.0mg
$ZnCl_2$	42.0mg
$NiCl_2 \cdot 6H_2O$	24.0mg
$Na_2MoO_4 \cdot 2H_2O$	18.0mg
$CuCl_2 \cdot 2H_2O$	2.0mg

Preparation of Solution D (Trace Elements Solution SL-12B): Add components to distilled/deionized water and bring volume to 1.0L. Mix thoroughly. Autoclave for 15 min at 15 psi pressure–121°C.

Solution E:

Composition per 100.0mL:

$NaHCO_3$	7.5g

Preparation of Solution E: Add $NaHCO_3$ to distilled/deionized water and bring volume to 100.0mL. Mix thoroughly. Sparge with 100% CO_2 until saturated. Filter sterilize under 100% CO_2 into a sterile, gas-tight 100.0mL screw-capped bottle.

Solution F:

Composition per 100.0mL:

$Na_2S \cdot 9H_2O$	10.0g

Preparation of Solution F: Add $Na_2S \cdot 9H_2O$ to distilled/deionized water and bring volume to 100.0mL. Mix thoroughly. Dispense into a screw-capped bottle. Sparge with 100% N_2 for 3–4 min. Autoclave for 15 min at 15 psi pressure–121°C.

Neutralized Sulfide Solution:

Composition per 100.0mL:

$Na_2S \cdot 9H_2O$	1.5g

Preparation of Neutralized Sulfide Solution: Add $Na_2S \cdot 9H_2O$ to distilled/deionized water in a 250.0mL screw-capped bottle fitted with a butyl rubber septum and bring volume to 100.0mL. Add a magnetic stir bar. Mix thoroughly. Sparge under 100% N_2 gas for 3 min. Autoclave for 15 min at 15 psi pressure–121°C. Cool to room temperature. Adjust pH to about 7.3 with sterile 2*M* H_2SO_4. Do not open the bottle to add H_2SO_4; use a sterile syringe. Stir the solution continuously to avoid precipitation of elemental sulfur. The final solution should be clear and yellow in color.

Preparation of Medium: Saturate cooled solution A under 100% CO_2 at 0.05–0.1 atm pressure for 30 min with magnetic stirring. Add 860.0mL of solution B, 5.0mL of solution C, 5.0mL of solution D, 100.0mL of solution E, and 20.0mL of solution F through one of the screw-cappped openings under 95% N_2 and 5% CO_2 with magnetic stirring. Adjust pH to 7.3 with sterile 2*M* HCl or sterile 2*M* Na_2CO_3 solution. Aseptically and anaerobically distribute the medium through the medium outlet tube into sterile 100.0mL bottles under 95% N_2 and 5% CO_2 at 0.05–0.1 atm pressure. Leave a small gas bubble in each bottle to accommodate pressure changes. After 24 hr, the iron in the medium will precipitate out of solution as black flocs. No other sediment should arise in the otherwise clear medium. Incubate in the light using a tungsten lamp. Feed periodically with neutralized solution of sodium sulfide to replenish sulfide and with other supplement solutions.

Use: For the cultivation of purple sulfur bacteria. For the cultivation of purple sulfur bacteria.For the cultivation and maintenance of *Amoebobacter pendens, Amoebobacter roseus, Chromatium minus, Chromatium minutissimum, Chromatium okenii, Chromatium* species, *Chromatium vinosum, Chromatium violascens, Chromatium weissei, Lamprocystis roseopersicina,* and *Streptomyces venezuelae.*

Pfennig's Medium 1 with 1% Salt

Composition per 4990.0mL:

Solution A	4000.0mL
Solution B	860.0mL
Solution E	100.0mL
Solution F	20.0mL
Solution C (Vitamin B_{12} solution)	5.0mL
Solution D (Trace elements solution SL-12B)	5.0mL

pH 7.3 ± 0.2 at 25°C

Solution A:

Composition per 4000.0mL:

NaCl	50.0g
$CaCl_2 \cdot 2H_2O$	1.25g
KH_2PO_4	1.7g
NH_4Cl	1.7g
KCl	1.7g
$MgSO_4$	2.5g

Preparation of Solution A: Add components to distilled/deionized water and bring volume to 4.0L. Mix thoroughly. Adjust pH to 6.0. Dispense into a 5L flask with four openings at the top (two openings are in a central silicon rubber stopper and two openings are gas-tight screw caps). Add a teflon-coated magnetic stir bar to the flask. Autoclave for 45 min at 15 psi pressure–121°C. Cool to room temperature under 100% N_2 at 0.05–0.1 atm pressure (use a manometer to measure low pressure).

Solution B:

Composition per 860.0mL:

Distilled/deionized water	860.0mL

Preparation of Solution B: Add 860.0mL of distilled/deionized water to a cotton-stoppered flask. Autoclave for 20 min at 15 psi pressure–121°C. Cool to room temperature under 100% N_2 in an anaerobic jar.

Solution C (Vitamin B_{12} Solution):

Composition per 5.0mL:

Vitamin B_{12}	1.0mg

Preparation of Solution C (Vitamin B_{12} Solution): Add vitamin B_{12} to distilled/deionized water and bring volume to 5.0mL. Mix thoroughly. Filter sterilize.

Solution D (Trace Elements Solution SL-12B):

Composition per liter:

Disodium ethylendiamine-tetraacetate (Na_2-EDTA)	3.0g
$FeSO_4 \cdot 7H_2O$	1.1g
H_3BO_3	0.3g
$CoCl_2 \cdot 6H_2O$	0.19g
$MnCl_2 \cdot 2H_2O$	50.0mg
$ZnCl_2$	42.0mg
$NiCl_2 \cdot 6H_2O$	24.0mg
$Na_2MoO_4 \cdot 2H_2O$	18.0mg
$CuCl_2 \cdot 2H_2O$	2.0mg

Preparation of Solution D (Trace Elements Solution SL-12B): Add components to distilled/deionized water and bring volume to 1.0L. Mix thoroughly. Autoclave for 15 min at 15 psi pressure–121°C.

Solution E:

Composition per 100.0mL:

$NaHCO_3$ 7.5g

Preparation of Solution E: Add $NaHCO_3$ to distilled/deionized water and bring volume to 100.0mL. Mix thoroughly. Sparge with 100% CO_2 until saturated. Filter sterilize under 100% CO_2 into a sterile, gas-tight 100.0mL screw-capped bottle.

Solution F:

Composition per 100.0mL:

$Na_2S{\cdot}9H_2O$ 10.0g

Preparation of Solution F: Add $Na_2S{\cdot}9H_2O$ to distilled/deionized water and bring volume to 100.0mL. Mix thoroughly. Dispense into a screw-capped bottle. Sparge with 100% N_2 for 3–4 min. Autoclave for 15 min at 15 psi pressure–121°C.

Preparation of Medium: Saturate cooled solution A under 100% CO_2 at 0.05–0.1 atm pressure for 30 min with magnetic stirring. Add 860.0mL of solution B, 5.0mL of solution C, 5.0mL of solution D, 100.0mL of solution E, and 20.0mL of solution F through one of the screw-cappped openings under 95% N_2 and 5% CO_2 with magnetic stirring. Adjust pH to 7.3 with sterile 2*M* HCl or sterile 2*M* Na_2CO_3 solution. Aseptically and anaerobically distribute the medium through the medium outlet tube into sterile 100.0mL bottles under 95% N_2 and 5% CO_2 at 0.05–0.1 atm pressure. Leave a small gas bubble in each bottle to accommodate pressure changes. After 24 hr, the iron in the medium will precipitate out of solution as black flocs.

Use: For the cultivation and maintenance of *Ectothiorhodospira mobilis*.

Pfennig's Medium 1 with 3% Salt

Composition per 4990.0mL:

Solution A 4000.0mL
Solution B 860.0mL
Solution E 100.0mL
Solution F 20.0mL
Solution C (Vitamin B_{12} solution) 5.0mL
Solution D (Trace elements solution SL-12B) 5.0mL
pH 7.3 ± 0.2 at 25°C

Solution A:

Composition per 4000.0mL:

NaCl 150.0g
$CaCl_2{\cdot}2H_2O$ 1.25g
KH_2PO_4 .... 1.7g
NH_4Cl 1.7g
KCl 1.7g
$MgSO_4$ 2.5g

Preparation of Solution A: Add components to distilled/deionized water and bring volume to 4.0L. Mix thoroughly. Adjust pH to 6.0. Dispense into a 5-liter flask with four openings at the top (two openings are in a central silicon rubber stopper and two openings are gas-tight screw caps). Add a teflon-coated magnetic stir bar to the flask. Autoclave for 45 min at 15 psi pressure–121°C. Cool to room temperature under 100% N_2 at 0.05–0.1 atm pressure (use a manometer to measure low pressure).

Solution B:

Composition per 860.0mL:

Distilled/deionized water 860.0mL

Preparation of Solution B: Add 860.0mL of distilled/deionized water to a cotton-stoppered flask. Autoclave for 20 min at 15 psi pressure–121°C. Cool to room temperature under 100% N_2 in an anaerobic jar.

Solution C (Vitamin B_{12} Solution):

Composition per 5.0mL:

Vitamin B_{12} 1.0mg

Preparation of Solution C (Vitamin B_{12} Solution): Add vitamin B_{12} to distilled/deionized water and bring volume to 5.0mL. Mix thoroughly. Filter sterilize.

Solution D (Trace Elements Solution SL-12B):

Composition per liter:

Disodium ethylendiamine-tetraacetate (Na_2-EDTA) 3.0g
$FeSO_4{\cdot}7H_2O$ 1.1g
H_3BO_3 .... 0.3g
$CoCl_2{\cdot}6H_2O$ 0.19g
$MnCl_2{\cdot}2H_2O$ 50.0mg
$ZnCl_2$ 42.0mg
$NiCl_2{\cdot}6H_2O$ 24.0mg
$Na_2MoO_4{\cdot}2H_2O$ 18.0mg
$CuCl_2{\cdot}2H_2O$ 2.0mg

Preparation of Solution D (Trace Elements Solution SL-12B): Add components to distilled/deionized water and bring volume to 1.0L. Mix thoroughly. Autoclave for 15 min at 15 psi pressure–121°C.

Solution E:

Composition per 100.0mL:

$NaHCO_3$ 7.5g

Preparation of Solution E: Add $NaHCO_3$ to distilled/deionized water and bring volume to 100.0mL. Mix thoroughly. Sparge with 100% CO_2 until saturated. Filter sterilize under 100% CO_2 into a sterile, gas-tight 100.0mL screw-capped bottle.

Solution F:

Composition per 100.0mL:

$Na_2S{\cdot}9H_2O$ 10.0g

Preparation of Solution F: Add $Na_2S{\cdot}9H_2O$ to distilled/deionized water and bring volume to 100.0mL. Mix thoroughly. Dispense into a screw-capped bottle. Sparge with 100% N_2 for 3–4 min. Autoclave for 15 min at 15 psi pressure–121°C.

Preparation of Medium: Saturate cooled solution A under 100% CO_2 at 0.05–0.1 atm pressure for 30 min with magnetic stirring. Add 860.0mL of solution B, 5.0mL of solution C, 5.0mL of solution D, 100.0mL of solution E, and 20.0mL of solution F through one of the screw-cappped openings under 95% N_2 and 5% CO_2 with magnetic stirring. Adjust pH to 7.3 with sterile 2*M* HCl or sterile 2*M* Na_2CO_3 solution. Aseptically and anaerobically distribute the medium through the medium outlet tube into sterile 100.0mL bottles under 95% N_2 and 5% CO_2 at 0.05–0.1 atm pressure. Leave a small gas bubble in each bottle to accommodate pressure changes. After 24 hr, the iron in the medium will precipitate out of solution as black flocs.

Use: For the cultivation and maintenance of *Chromatium gracile, Thiocystis gelatinosa,* and *Thiocystis violacea.*

Pfennig's Medium 1 with Yeast Extract

Composition per 4990.0mL:

Solution A	4000.0mL
Solution B	860.0mL
Solution E	100.0mL
Solution F	20.0mL
Solution C (Vitamin B_{12} solution)	5.0mL
Solution D (Trace elements solution SL-12B)	5.0mL

pH 7.3 ± 0.2 at 25°C

Solution A:

Composition per 4000.0mL:

$CaCl_2·2H_2O$	1.25g
KH_2PO_4	1.7g
NH_4Cl	1.7g
KCl	1.7g
$MgSO_4$	2.5g
Yeast extract	2.5g

Preparation of Solution A: Add components to distilled/deionized water and bring volume to 4.0L. Mix thoroughly. Adjust pH to 6.0. Dispense into a 5L flask with four openings at the top (two openings are in a central silicon rubber stopper and two openings are gas-tight screw caps). Add a teflon-coated magnetic stir bar to the flask. Autoclave for 45 min at 15 psi pressure–121°C. Cool to room temperature under 100% N_2 at 0.05–0.1 atm pressure.

Solution B:

Composition per 860.0mL:

Distilled/deionized water	860.0mL

Preparation of Solution B: Add 860.0mL of distilled/deionized water to a cotton-stoppered flask. Autoclave for 20 min at 15 psi pressure–121°C. Cool to room temperature under 100% N_2 in an anaerobic jar.

Solution C (Vitamin B_{12} Solution):

Composition per 5.0mL:

Vitamin B_{12}	1.0mg

Preparation of Solution C (Vitamin B_{12} Solution): Add vitamin B_{12} to distilled/deionized water and bring volume to 5.0mL. Mix thoroughly. Filter sterilize.

Solution D (Trace Elements Solution SL-12B):

Composition per liter:

Disodium ethylendiamine-tetraacetate (Na_2-EDTA)	3.0g
$FeSO_4·7H_2O$	1.1g
H_3BO_3	0.3g
$CoCl_2·6H_2O$	0.19g
$MnCl_2·2H_2O$	50.0mg
$ZnCl_2$	42.0mg
$NiCl_2·6H_2O$	24.0mg
$Na_2MoO_4·2H_2O$	18.0mg
$CuCl_2·2H_2O$	2.0mg

Preparation of Solution D (Trace Elements Solution SL-12B): Add components to distilled/deionized water and bring volume to 1.0L. Mix thoroughly. Autoclave for 15 min at 15 psi pressure–121°C.

Solution E:

Composition per 100.0mL:

$NaHCO_3$	7.5g

Preparation of Solution E: Add $NaHCO_3$ to distilled/deionized water and bring volume to 100.0mL. Mix thoroughly. Sparge with 100% CO_2 until saturated. Filter sterilize under 100% CO_2 into a sterile, gas-tight 100.0mL screw-capped bottle.

Solution F:

Composition per 100.0mL:

$Na_2S·9H_2O$	10.0g

Preparation of Solution F: Add $Na_2S·9H_2O$ to distilled/deionized water and bring volume to 100.0mL. Mix thoroughly. Dispense into a screw-capped bottle. Sparge with 100% N_2 for 3–4 min. Autoclave for 15 min at 15 psi pressure–121°C.

Preparation of Medium: Saturate cooled solution A under 100% CO_2 at 0.05–0.1 atm pressure for 30 min with magnetic stirring. Add 860.0mL of solution B, 5.0mL of solution C, 5.0mL of solution D, 100.0mL of solution E, and 20.0mL of solution F through one of the screw-cappped openings under 95% N_2 and 5% CO_2 with magnetic stirring. Adjust pH to 7.3 with sterile 2*M* HCl or sterile 2*M* Na_2CO_3 solution. Aseptically and anaerobically distribute the medium through the medium outlet tube into sterile 100.0mL bottles under 95% N_2 and 5% CO_2 at 0.05–0.1 atm pressure. Leave a small gas bubble in each bottle to accommodate pressure changes. After 24 hr, the iron in the medium will precipitate out of solution as black flocs.

Use: For the cultivation and maintenance of *Amoebobacter pendens*.

Pfennig's Medium 2, Modified for Green Sulfur Bacteria

Composition per 5000.0mL:

Solution A	4000.0mL
Solution B	860.0mL
Solution E	100.0mL
Solution F	30.0mL
Solution C (Vitamin B_{12} solution)	5.0mL
Solution D (Trace elements solution SL-10B)	5.0mL

pH 6.8 ± 0.2 at 25°C

Solution A:

Composition per 4000.0mL:

$CaCl_2·2H_2O$	1.25g
KH_2PO_4	1.7g
NH_4Cl	1.7g
KCl	1.7g
$MgSO_4$	2.5g

Preparation of Solution A: Add components to distilled/deionized water and bring volume to 4.0L. Mix thoroughly. Adjust pH to 6.0. Dispense into a 5L flask with four openings at the top (two openings are in a central silicon rubber stopper and two openings are gas-tight screw caps). Add a teflon-coated magnetic stir bar to the flask. Autoclave for 45 min at 15 psi pressure–121°C. Cool to room temperature under 100% N_2 at 0.05–0.1 atm pressure (use a manometer to measure low pressure).

Solution B:

Composition per 860.0mL:

Distilled/deionized water	860.0mL

Preparation of Solution B: Add 860.0mL of distilled/deionized water to a cotton-stoppered flask. Autoclave for 20 min at 15 psi pressure–121°C. Cool to room temperature under 100% N_2 in an anaerobic jar.

Solution C (Vitamin B_{12} Solution):

Composition per 5.0mL:

Vitamin B_{12}	1.0mg

Preparation of Solution C (Vitamin B_{12} Solution): Add vitamin B_{12} to distilled/deionized water and bring volume to 5.0mL. Mix thoroughly. Filter sterilize.

Solution D (Trace Elements Solution SL-10B):

Composition per liter:

$FeCl_2 \cdot 4H_2O$	1.5g
H_3BO_3	300.0mg
$CoCl_2 \cdot 6H_2O$	190.0mg
$MnCl_2 \cdot 4H_2O$	100.0mg
$ZnCl_2$	70.0mg
$Na_2MoO_4 \cdot 2H_2O$	36.0mg
$NiCl_2 \cdot 6H_2O$	24.0mg
$CuCl_2 \cdot 2H_2O$	2.0mg
HCl (25% solution)	10.0mL

Preparation of Solution D (Trace Elements Solution SL-10B): Add $FeCl_2 \cdot 4H_2O$ to 10.0mL of HCl solution. Mix thoroughly. Add distilled/deionized water and bring volume to 1.0L. Add remaining components. Mix thoroughly.

Solution E:

Composition per 100.0mL:

$NaHCO_3$	7.5g

Preparation of Solution E: Add $NaHCO_3$ to distilled/deionized water and bring volume to 100.0mL. Mix thoroughly. Sparge with 100% CO_2 until saturated. Filter sterilize under 100% CO_2 into a sterile, gas-tight 100.0mL screw-capped bottle.

Solution F:

Composition per 100.0mL:

$Na_2S \cdot 9H_2O$	10.0g

Preparation of Solution F: Add $Na_2S \cdot 9H_2O$ to distilled/deionized water and bring volume to 100.0mL. Mix thoroughly. Dispense into a screw-capped bottle. Sparge with 100% N_2 for 3–4 min. Autoclave for 15 min at 15 psi pressure–121°C.

Preparation of Medium: Saturate cooled solution A under 100% CO_2 at 0.05–0.1 atm pressure for 30 min with magnetic stirring. Add 860.0mL of solution B, 5.0mL of solution C, 5.0mL of solution D, 100.0mL of solution E, and 20.0mL of solution F through one of the screw-cappped openings under 95% N_2 and 5% CO_2 with magnetic stirring. Adjust pH to 6.8 with sterile 2*M* HCl or sterile 2*M* Na_2CO_3 solution. Aseptically and anaerobically distribute the medium through the medium outlet tube into sterile 100.0mL bottles under 95% N_2 and 5% CO_2 at 0.05–0.1 atm pressure. Leave a small gas bubble in each bottle to accommodate pressure changes. After 24 hr, the iron in the medium will precipitate out of solution as black flocs.

Use: For the cultivation and maintenance of *Chlorobium limicola* and *Chlorobium phaeovibrioides*.

Pfennig's Medium 2 Modified for Green Sulfur Bacteria (DSMZ Medium 29)

Composition per 5.0L:

Solution A	4.0L
Solution B	860.0mL
Solution E	100.0mL
Solution F	30.0mL
Solution C	5.0mL
Solution D	5.0mL
Neutralized sulfide solution	As needed

pH 6.8 at 25°C

Solution A:

Composition per 4.0L:

$MgSO_4$	2.5g
KH_2PO_4	1.7g
NH_4Cl	1.7g
KCl	1.7g
$CaCl_2 \cdot 2H_2O$	1.25g

Preparation of Solution A: Add components to 4.0L distilled water. Mix thoroughly. Autoclave for 45 min at 15 psi pressure–121°C in 5-liter special bottle or flask with four openings at the top, together with a teflon-coated magnetic bar. In this 5-liter bottle, two openings for tubes are in the central, silicon rubber stopper: a short, gas-inlet tube with a sterile cotton filter; and an outlet tube for medium, which reaches the bottom of the vessel at one end and has, at the other end, a silicon rubber tube with a pinch cock and a bell for aseptic dispensing of the medium into bottles. The other two openings have gas-tight screw caps; one of these openings is for the addition of sterile solutions and the other serves as a gas outlet. After autoclaving, cool to room temperature under a N_2 atmosphere with a positive pressure of 0.05–0.1 atm (a manometer for low pressure will be required). Saturate the cold medium with CO_2 by magnetic stirring for 30 min under a CO_2 atmosphere of 0.05–0.1 atm.

Solution B:

Distilled water	860.0mL

Preparation of Solution B: Autoclave distilled water for 15 min at 15 psi pressure–121°C in a cotton-stoppered Erlenmeyer flask. Cool to room temperature under an atmosphere of N_2 in an anaerobic jar.

Solution C:

Composition per 100.0mL:

Vitamin B_{12}	2.0mg

Preparation of Solution C: Add vitamin B_{12} to distilled/deionized water and bring volume to 100.0mL. Mix thoroughly. Sparge under 100% N_2 gas for 3 min. Filter sterilize. Store under N_2 gas.

Solution D:

Composition per liter:

$FeCl_2 \cdot 4H_2O$	1.5g
$CoCl_2 \cdot 6H_2O$	190.0mg
$MnCl_2 \cdot 4H_2O$	100.0mg
$ZnCl_2$	70.0mg
$Na_2MoO_4 \cdot 2H_2O$	36.0mg
$NiCl_2 \cdot 6H_2O$	24.0mg
H_3BO_3	300.0mg
$CuCl_2 \cdot 2H_2O$	2.0mg
HCl (25% solution)	7.7mL

Preparation of Solution D: Add $FeCl_2 \cdot 4H_2O$ to 10.0mL of HCl solution. Mix thoroughly. Add distilled/deionized water and bring volume to 1.0L. Add remaining components. Mix thoroughly. Sparge with 100% N_2. Autoclave for 15 min at 15 psi pressure–121°C.

Solution E:

Composition per 100.0mL:

$NaHCO_3$	7.5g

Preparation of Solution E: Add $NaHCO_3$ to distilled/deionized water and bring volume to 100.0mL. Mix thoroughly. Sparge with 100% CO_2 until saturated. Filter sterilize under 100% CO_2 into a sterile, gas-tight 100.0mL screw-capped bottle.

Solution F:

Composition per 100.0mL:

$Na_2S \cdot 9H_2O$	10.0g

Preparation of Solution F: Add $Na_2S \cdot 9H_2O$ to distilled/deionized water in a 250.0mL screw-capped bottle fitted with a butyl rubber septum and bring volume to 100.0mL. Mix thoroughly. Sparge under

100% N_2 gas for 3 min. Autoclave for 15 min at 15 psi pressure–121°C. Cool to room temperature.

Neutralized Sulfide Solution:
Composition per 100.0mL:

$Na_2S·9H_2O$	1.5g

Preparation of Neutralized Sulfide Solution: Add $Na_2S·9H_2O$ to distilled/deionized water in a 250.0mL screw-capped bottle fitted with a butyl rubber septum and bring volume to 100.0mL. Add a magnetic stir bar. Mix thoroughly. Sparge under 100% N_2 gas for 3 min. Autoclave for 15 min at 15 psi pressure–121°C. Cool to room temperature. Adjust pH to about 6.8 with sterile 2*M* H_2SO_4. Do not open the bottle to add H_2SO_4; use a sterile syringe. Stir the solution continuously to avoid precipitation of elemental sulfur. The final solution should be clear and yellow in color.

Preparation of Medium: Add solution B, C, D, E, and F to solution A through one of the screw-cap openings against a stream of either N_2 gas or, better, a mixture of 95% N_2 and 5% CO_2 while the medium is magnetically stirred. Adjust the pH of the medium with sterile HCl or Na_2CO_3 solution (2*M* solution) to pH 6.8. Distribute the medium aseptically through the medium outlet tube into sterile, 100mL bottles (with metal caps and autoclavable rubber seals) using the positive gas pressure (0.05–0.1 atm) of the N_2/CO_2 gas mixture. Leave a small air bubble in each bottle to meet possible pressure changes. The tightly sealed, screw-cap bottles can be stored for several weeks or months in the dark. During the first 24 hr, the iron of the medium precipitates in the form of black flocs. No other sediment should arise in the otherwise clear medium. Incubate in the light using a tungsten lamp. Feed periodically with neutralized solution of sodium sulfide to replenish sulfide and with other supplement solutions.

Use: For the cultivation of green sulfur bacteria.

Pfennig's Medium 2 with Salt

Composition per 5000.0mL:

Solution A	4000.0mL
Solution B	860.0mL
Solution E	100.0mL
Solution F	30.0mL
Solution C (Vitamin B_{12} solution)	5.0mL
Solution D (Trace elements solution SL-10B)	5.0mL

pH 6.8 ± 0.2 at 25°C

Solution A:
Composition per 4000.0mL:

NaCl	50.0g
$CaCl_2·2H_2O$	1.25g
KH_2PO_4	1.7g
NH_4Cl	1.7g
KCl	1.7g
$MgSO_4$	2.5g

Preparation of Solution A: Add components to distilled/deionized water and bring volume to 4.0L. Mix thoroughly. Adjust pH to 6.0. Dispense into a 5L flask with four openings at the top (two openings are in a central silicon rubber stopper and two openings are gas-tight screw caps). Add a teflon-coated magnetic stir bar to the flask. Autoclave for 45 min at 15 psi pressure–121°C. Cool to room temperature under 100% N_2 at 0.05–0.1 atm pressure (use a manometer to measure low pressure).

Solution B:
Composition per 860.0mL:

Distilled/deionized water	860.0mL

Preparation of Solution B: Add 860.0mL of distilled/deionized water to a cotton-stoppered flask. Autoclave for 20 min at 15 psi pressure–121°C. Cool to room temperature under 100% N_2 in an anaerobic jar.

Solution C (Vitamin B_{12} Solution):
Composition per 5.0mL:

Vitamin B_{12}	1.0mg

Preparation of Solution C (Vitamin B_{12} Solution): Add vitamin B_{12} to distilled/deionized water and bring volume to 5.0mL. Mix thoroughly. Filter sterilize.

Solution D (Trace Elements Solution SL-10B):
Composition per liter:

$FeCl_2·4H_2O$	1.5g
$CoCl_2·6H_2O$	0.19g
$MnCl_2·4H_2O$	100.0mg
$ZnCl_2$	70.0mg
$Na_2MoO_4·2H_2O$	36.0mg
$NiCl_2·6H_2O$	24.0mg
H_3BO_3	6.0mg
$CuCl_2·2H_2O$	2.0mg
HCl (25% solution)	10.0mL

Preparation of Solution D (Trace Elements Solution SL-10B): Add $FeCl_2·4H_2O$ to 10.0mL of HCl solution. Mix thoroughly. Add distilled/deionized water and bring volume to 1.0L. Add remaining components. Mix thoroughly.

Solution E:
Composition per 100.0mL:

$NaHCO_3$	7.5g

Preparation of Solution E: Add $NaHCO_3$ to distilled/deionized water and bring volume to 100.0mL. Mix thoroughly. Sparge with 100% CO_2 until saturated. Filter sterilize under 100% CO_2 into a sterile, gas-tight 100.0mL screw-capped bottle.

Solution F:
Composition per 100.0mL:

$Na_2S·9H_2O$	10.0g

Preparation of Solution F: Add $Na_2S·9H_2O$ to distilled/deionized water and bring volume to 100.0mL. Mix thoroughly. Dispense into a screw-capped bottle. Sparge with 100% $N_2$for 3–4 min. Autoclave for 15 min at 15 psi pressure–121°C.

Preparation of Medium: Saturate cooled solution A under 100% CO_2 at 0.05–0.1 atm pressure for 30 min with magnetic stirring. Add 860.0mL of solution B, 5.0mL of solution C, 5.0mL of solution D, 100.0mL of solution E, and 20.0mL of solution F through one of the screw-cappped openings under 95% N_2 and 5% CO_2 with magnetic stirring. Adjust pH to 6.8 with sterile 2*M* HCl or sterile 2*M* Na_2CO_3 solution. Aseptically and anaerobically distribute the medium through the medium outlet tube into sterile 100.0mL bottles under 95% N_2 and 5% CO_2 at 0.05–0.1 atm pressure. Leave a small gas bubble in each bottle to accommodate pressure changes. After 24 hr, the iron in the medium will precipitate out of solution as black flocs.

Use: For the cultivation and maintenance of *Chlorobium phaeobacteroides, Chlorobium vibrioforme, Pelodictyon luteolum, Pelodictyon phaeum,* and *Prosthecochloris aestuarii.*

Pfizer Selective *Enterococcus* Agar (PSE Agar)

Peptone C	17.0g
Agar	15.0g

Bile 10.0g
NaCl 5.0g
Yeast extract 5.0g
Peptone B 3.0g
Esculin 1.0g
Sodium citrate 1.0g
Ferric ammonium citrate 0.5g
NaN_3 .... 0.25g

pH 7.1 ± 0.2 at 25°C

Caution: Sodium azide is toxic. Azides also react with metals and disposal must be highly diluted.

Preparation of Medium: Add components to distilled/deionized water and bring volume to 1.0L. Mix thoroughly. Gently heat and bring to boiling. Distribute into tubes or flasks. Autoclave for 15 min at 15 psi pressure–121°C. Pour into sterile Petri dishes or leave in tubes.

Use: For the selective isolation, cultivation, and enumeration of *Enterococcus* species by the multiple tube technique.

Pfizer Selective *Enterococcus* HiVeg Agar

Composition per liter:

Plant hydrolysate 21.0g
Agar 15.0g
Plant peptone 6.0g
NaCl 5.0g
Yeast extract 5.0g
Synthetic detergent 3.0g
Esculin 1.0g
Sodium citrate 1.0g
Ferric ammonium citrate 0.5g
NaN_3 .... 0.25g

pH 7.1 ± 0.2 at 25°C

Source: This medium is available as a premixed powder from HiMedia.

Caution: Sodium azide is toxic. Azides also react with metals and disposal must be highly diluted.

Preparation of Medium: Add components to distilled/deionized water and bring volume to 1.0L. Mix thoroughly. Gently heat and bring to boiling. Distribute into tubes or flasks. Autoclave for 15 min at 15 psi pressure–121°C. Pour into sterile Petri dishes or leave in tubes.

Use: For the selective isolation, cultivation, and enumeration of *Enterococcus* species by the multiple tube technique.

Pfizer TB Medium Base with Glycerol, Egg Yolk, Glucose, and Malachite Green

Composition per liter:

Agar 15.0g
Potato starch 15.0g
Casein acid hydrolysate 10.0g
K_2HPO_4 .... 3.5g
Beef extract 3.0g
L-Asparagine 3.0g
Citric acid 0.1g
Ferric ammonium citrate 0.1g
$MgSO_4$ 0.015g
Egg yolk emulsion 100.0mL
Glycerol 40.0mL
Malachite Green solution 13.0mL
Glucose solution 1.0mL

pH 7.0 ± 0.2 at 35°C

Source: This medium, without glycerol, glucose solution, Malachite Green solution, and egg yolk emulsion, is available as a premixed powder from HiMedia.

Egg Yolk Emulsion:

Composition per 100.0mL:

Chicken egg yolks 9
Whole chicken egg 1
NaCl (0.9% solution) 25.0mL

Preparation of Egg Yolk Emulsion: Soak eggs with 1:100 dilution of saturated mercuric chloride solution for 1 min. Crack eggs and separate yolks from whites. Mix egg yolks with 1 chicken egg. Beat to form emulsion. Measure 50.0mL of egg yolk emulsion and add to 50.0mL of 0.9% NaCl solution. Mix thoroughly. Filter sterilize. Warm to 45°–50°C.

Glucose Solution:

Composition per 100.0mL:

Glucose 20.0g

Preparation of Glucose Solution: Add glucose to distilled/deionized water and bring volume to 100.0mL. Mix thoroughly. Filter sterilize.

Malachite Green Solution:

Composition per 20.0mL:

Malachite Green 0.2g

Preparation of Malachite Green Solution: Add Malachite Green to distilled/deionized water and bring volume to 100.0mL. Mix thoroughly. Filter sterilize.

Preparation of Medium: Add glycerol and the other components, except glucose solution, egg yolk emulsion, and Malachite Green solution, to distilled/deionized water and bring volume to 1.0L. Mix thoroughly. Gently heat and bring to boiling. Distribute into tubes or flasks. Autoclave for 15 min at 12 psi pressure–118°C. Cool to 55°C. Aseptically add 100.0mL egg yolk emulsion, 1.0mL glucose solution, and 13.0mL Malachite Green solution. Mix thoroughly. Aseptically dispense into sterile tubes. Allow to solidity as slants.

Use: For the cultivation of *Mycobacterium tuberculosis*.

PFS Medium (Peptone Fumarate Sulfate Medium)

Composition per liter:

Peptone 10.0g
Fumaric acid 2.0g
$(NH_4)_2SO_4$ 1.0g
$MgSO_4 \cdot 7H_2O$ 0.5g
$FeCl_3 \cdot 6H_2O$ 0.2mg
$MnSO_4 \cdot H_2O$ 0.2mg

pH 7.0 ± 0.2 at 25°C

Preparation of Medium: Add components to distilled/deionized water and bring volume to 1.0L. Mix thoroughly. Adjust pH to 7.0 with KOH. Distribute into tubes or flasks. Autoclave for 15 min at 15 psi pressure–121°C.

Use: For the cultivation and maintenance of *Aquaspirillum fasciculus*.

PGA
(Potato Glucose Agar)

Composition per liter:

Potatoes	500.0g
Glucose	20.0g
Agar	15.0g
Yeast extract	5.0g

Preparation of Medium: Slice potatoes with skin. Place 500.0g of potatoes in 1.0L of distilled/deionized water. Gently heat and bring to boiling. Allow to boil for 20 min. Filter through Whatman filter paper. Add agar and other components to filtrate. Bring volume to 1.0L with distilled/deionized water. Mix thoroughly. Gently heat and bring to boiling. Distribute into tubes or flasks. Autoclave for 15 min at 15 psi pressure–121°C. Pour into sterile Petri dishes or leave in tubes.

Use: For the cultivation of numerous filamentous fungi.

PGLE Medium
(Peptone Glucose Liver Extract Medium)

Composition per liter:

Agar	25.0g
Glucose	20.0g
Peptone	5.0g
Yeast extract	5.0g
K_2HPO_4	1.0g
Liver extract	0.5g

pH 9.0 ± 0.2 at 25°C

Preparation of Medium: Add components to distilled/deionized water and bring volume to 1.0L. Mix thoroughly. Gently heat and bring to boiling. Distribute into tubes or flasks. Autoclave for 15 min at 15 psi pressure–121°C. Pour into sterile Petri dishes or leave in tubes.

Use: For the cultivation and maintenance of *Enterobacter cloacae*.

PGP Broth
(Peptone Glycerol Phosphate Broth)

Composition per liter:

Peptone	5.0g
K_2HPO_4	2.0g
Glycerol	10.0mL

Preparation of Medium: Add components to distilled/deionized water and bring volume to 1.0L. Mix thoroughly. Distribute into tubes or flasks. Autoclave for 15 min at 15 psi pressure–121°C.

Use: For the cultivation and maintenance of *Serratia marcescens*.

PGS Agar
(Peptone Glucose Salt Agar)

Composition per liter:

Agar	15.0g
Glucose	10.0g
NaCl	10.0g
Peptone	10.0g

pH 7.2 ± 0.2 at 25°C

Preparation of Medium: Add components to distilled/deionized water and bring volume to 1.0L. Mix thoroughly. Gently heat and bring to boiling. Distribute into tubes or flasks. Autoclave for 15 min at 15 psi pressure–121°C. Pour into sterile Petri dishes or leave in tubes.

Use: For the cultivation and maintenance of *Rhodococcus australis*.

PGT Medium

Composition per liter:

Casamino acids	30.0g
L-Glutamic acid	0.5g
$MgSO_4 \cdot 7H_2O$	0.45g
Maltose	0.2g
L-Cystine	0.2g
DL-Tryptophan	0.1g
Solution 3	100.0mL
Solution 2	2.0mL
Calcium pantothenate (0.1% solution)	0.5mL

pH 6.8 ± 0.2 at 25°C

Solution 3:

Composition per 500.0mL:

Maltose	200.0g
$CaCl_2$	1.5g
Calcium pantothenate (0.1% solution)	3.0mL
$FeSO_4$ (1% in 1*N* HCl)	0.2mL

Preparation of Solution 3: Add components to distilled/deionized water and bring volume to 500.0mL. Mix thoroughly. Autoclave for 15 min at 7 psi pressure–111°C. Cool to 45°–50°C.

Solution 2:

Composition per 100.0mL:

β-Alanine	0.115g
Nicotinic acid	0.115g
$CuSO_4 \cdot 5H_2O$	0.05g
$ZnSO_4 \cdot 7H_2O$	0.045g
$MnCl_2 \cdot 4H_2O$	0.015g
Pimelic acid	7.5mg
HCl, concentrated	3.0mL

Preparation of Solution 2: Add components to distilled/deionized water and bring volume to 100.0mL. Mix thoroughly.

Preparation of Medium: Add components, except solution 3, to distilled/deionized water and bring volume to 900.0mL. Mix thoroughly. Adjust pH to 6.8 with 50% KOH. Gently heat and bring to boiling. Autoclave for 15 min at 15 psi pressure–121°C. Cool to 45°–50°C. Aseptically add sterile solution 3. Mix thoroughly. Aseptically distribute into sterile tubes or flasks.

Use: For the cultivation of *Corynebacterium diphtheriae*.

PGY Agar
(Peptone Glucose Yeast Extract Agar)

Composition per liter:

Agar	15.0g
Peptone	10.0g
Yeast extract	5.0g
Glucose	1.0g

Preparation of Medium: Add components to distilled/deionized water and bring volume to 1.0L. Mix thoroughly. Gently heat and bring to boiling. Distribute into tubes or flasks. Autoclave for 15 min at 15 psi pressure–121°C. Pour into sterile Petri dishes or leave in tubes.

Use: For the cultivation and maintenance of *Micrococcus luteus*.

PH Medium
(DSMZ Medium 1077)

Composition per 1118.0mL:

Pancreatic digest of casein	7.0g
NaCl	5.0g

Beef extract	3.0g
Yeast extract	3.0g
Beef heart, solids from infusion	2.0g
Supplement solution	200.0mL

pH 7.8 ± 0.2 at 25°C

Supplement Solution:

Composition per liter:

Horse serum, inactivated	187.0mL
Yeast extract solution	93.6mL
Fish sperm DNA solution, filter sterilized	18.7mL

Yeast Extract Solution:

Composition per 100.0mL:

Yeast extract	25.0g

Preparation of Yeast Extract Solution: Add yeast extract to distilled/deionized water and bring volume to 100.0mL. Mix thoroughly.

Preparation of Supplement Solution: Aseptically add components to sterile distilled/deionized water and bring volume to 318.0mL. Mix thoroughly.

Preparation of Medium: Add components, except supplement solution, to distilled/deionized water and bring volume to 800.0L. Mix thoroughly. Adjust pH to 6.5. Autoclave for 15 min at 15 psi pressure–121°C. Cool to 25°C. Aseptically add 318.0mL of supplement solution. Adjust pH to 7.8. Mix thoroughly. Aseptically distribute into culture vessels.

Use: For the cultivation of *Mycoplasma orale.*

PHB Delafield Agar (LMG Medium 123)

Composition per liter:

Agar	20.0g
NH_4Cl	1.0g
$MgSO_4 \cdot 7H_2O$	0.5g
Casamino acids	0.1g
Ferric ammonium citrate	0.05g
Yeast extract	0.05g
Phosphate buffer solution	1.0L
Solution A/B	264.0mL

pH 6.8 ± 0.2 at 25°C

Phosphate Buffer Solution:

Composition per liter:

KH_2PO_4	4.5g
$Na_2HPO_4 \cdot 2H_2O$	5.8g

Preparation of Phosphate Buffer Solution: Add components to distilled/deionized water and bring volume to 1.0L. Mix thoroughly. Adjust pH to 6.8.

Solution A/B:

Composition per 700.0mL:

Solution A	140.0mL
Solution B	560.0mL

Solution A:

Composition per 140.0mL:

Poly-β-hydroxybutyrate, powdered	1.75g

Preparation of Solution A: Add poly-β-hydroxybutyrate to distilled/deionized water and bring volume to 140.0mL. Mix thoroughly. Sonicate for 30 min. Autoclave for 15 min at 15 psi pressure–121°C. Sonicate again for 30 min in the sterile bottle. Warm to 50°C.

Solution B:

Composition per 560.0mL:

Agar	14.0g

Preparation of Solution B: Add agar to distilled/deionized water and bring volume to 560.0mL. Mix thoroughly. Gently heat and bring to boiling. Autoclave for 15 min at 15 psi pressure–121°C. Cool to 50°C.

Preparation of Solution A/B: Aseptically combine 140.0mL of solution A with 560.0mL of solution B. Warm to 50°C.

Preparation of Medium: Combine components except solution A/B. Mix thoroughly. Gently heat and bring to boiling. Autoclave for 15 min at 15 psi pressure–121°C. Pour into sterile Petri dishes. Allow agar to solidify. Bring the Petri dishes to 65°C. Aseptically add 4.0mL of sterile solution A/B onto the surface of each agar plate.

Use: For the cultivation and maintenance of *Pseudomonas lemoignei* Delafield.

PHB Medium

See: **Poly-β-Hydroxybutyrate Medium**

PHB/Pyruvate Medium (DSMZ Medium 193a)

Composition per 991.0mL:

Solution A	870.0mL
Solution C	100.0mL
Solution D	10.0mL
Solution E (Vitamin solution)	10.0mL
Solution B (Trace elements solution SL-10)	1.0mL

pH 7.1–7.4 at 25°C

Solution A:

Composition per 870.0mL:

Na-pyruvate	5.0g
NaCl	7.0g
Na_2SO_4	3.0g
$MgCl_2 \cdot 6H_2O$	1.3g
Poly-hydroxybutyrate (PHB)	1.0g
KH_2PO_4	0.2g
NH_4Cl	0.3g
KCl	0.5g
$CaCl_2 \cdot 2H_2O$	0.15g
Resazurin	1.0mg

Preparation of Solution A: Add components to distilled/deionized water and bring volume to 870.0mL Mix thoroughly.

Solution B (Trace Elements Solution SL-10):

Composition per liter:

$FeCl_2 \cdot 4H_2O$	1.5g
$CoCl_2 \cdot 6H_2O$	190.0mg
$MnCl_2 \cdot 4H_2O$	100.0mg
$ZnCl_2$	70.0mg
$Na_2MoO_4 \cdot 2H_2O$	36.0mg
$NiCl_2 \cdot 6H_2O$	24.0mg
H_3BO_3	6.0mg
$CuCl_2 \cdot 2H_2O$	2.0mg
HCl (25% solution)	10.0mL

Preparation of Solution B (Trace Elements Solution SL-10): Add $FeCl_2 \cdot 4H_2O$ to 10.0mL of HCl solution. Mix thoroughly. Add distilled/deionized water and bring volume to 1.0L. Add remaining components. Mix thoroughly. Sparge with 100% N_2. Autoclave for 15 min at 15 psi pressure–121°C.

Solution C:

Composition per 100.0mL:

$NaHCO_3$ 5.0g

Preparation of Solution C: Add $NaHCO_3$ to distilled/deionized water and bring volume to 100.0mL. Mix thoroughly. Filter sterilize. Flush with 80% N_2 + 20% CO_2 to remove dissolved oxygen.

Solution D:

Composition per 10.0mL:

Na-acetate·$3H_2O$ 2.5g

Preparation of Solution D: Add Na-acetate·$3H_2O$ to distilled/deionized water and bring volume to 10.0mL. Mix thoroughly. Sparge with 100% N_2. Autoclave for 15 min at 15 psi pressure–121°C.

Solution E (Vitamin Solution):

Composition per liter:

Pyridoxine-HCl 10.0mg
Thiamine-HCl·$2H_2O$ 5.0mg
Riboflavin 5.0mg
Nicotinic acid 5.0mg
D-Ca-pantothenate 5.0mg
p-Aminobenzoic acid 5.0mg
Lipoic acid 5.0mg
Biotin 2.0mg
Folic acid 2.0mg
Vitamin B_{12} 0.10mg

Solution E (Vitamin Solution): Add components to distilled/deionized water and bring volume to 1.0L. Mix thoroughly. Sparge with 100% N_2. Autoclave for 15 min at 15 psi pressure–121°C.

Solution F:

Composition per 10.0mL:

$Na_2S \cdot 9H_2O$ 0.4g

Preparation of Solution F: Add $Na_2S \cdot 9H_2O$ to distilled/deionized water and bring volume to 10.0mL. Mix thoroughly. Sparge with 100% N_2. Autoclave for 15 min at 15 psi pressure–121°C.

Preparation of Medium: Gently heat solution A and bring to boiling. Boil solution A for a few minutes. Cool to room temperature. Gas with 80% N_2 + 20% CO_2 gas mixture to reach a pH below 6. Autoclave for 15 min at 15 psi pressure–121°C. Cool to room temperature. Sequentially add 1.0mL solution B, 100.0mL solution C, 10.0mL solution D, 10.0mL solution E, and 10.0mL solution F. Distribute anaerobically under 80% N_2 + 20% CO_2 into appropriate vessels. Addition of 10–20mg sodium dithionite per liter from a 5% (w/v) solution, freshly prepared under N_2 and filter sterilized, may stimulate growth. After inoculation, pressurize vessels up to 1 bar $H_2 + CO_2$ (80% + 20%) overpressure.

Use: For the cultivation of unclassified bacterium DSM 6754.

PHC Medium
(Polyhexamethylene Carbonate Medium)
(DSMZ Medium 885)

K_2HPO_4 .... 1.6g
Polyhexamethylene carbonate, emulsified 1.0g
$(NH_4)_2SO_4$ 1.0g
YKH_2PO_4 .... 0.2g
Yeast extract 0.1g
Surfactant, Plysurf A210G 60.0mg
$CaCl_2 \cdot 2H_2O$ 20.0mg
NaCl 10.0mg
$FeSO_4 \cdot 7H_2O$ 10.0mg
$Na_2MoO_4 \cdot 4H_2O$ 0.5mg
$Na_2WO_4 \cdot 2H_2O$ 0.5mg
$MnSO_4$ 0.5mg

pH 7.0 ± 0.2 at 25°C

Preparation of Medium: Add components to distilled/deionized water and bring volume to 1.0L. Mix thoroughly. Distribute into tubes or flasks. Autoclave for 15 min at 15 psi pressure–121°C.

Use: For the cultivation of *Roseateles depolymerans.*

Phenethyl Alcohol Agar
(Phenylethanol Agar)
(Phenylethyl Alcohol Agar)

Composition per liter:

Agar 15.0g
Pancreatic digest of casein 15.0g
NaCl 5.0g
Papaic digest of soybean meal 5.0g
β-Phenethyl alcohol 2.5g
Blood 50.0mL

pH 7.3 ± 0.2 at 25°C

Source: This medium is available as a premixed powder from BD Diagnostic Systems.

Preparation of Medium: Add components, except blood, to distilled/deionized water and bring volume to 950.0mL. Mix thoroughly. Gently heat and bring to boiling. Autoclave for 15 min at 13 psi pressure–118°C. Cool to 45°–50°C. Aseptically add sterile defibrinated blood. Mix thoroughly. Pour into sterile Petri dishes or distribute into sterile tubes.

Use: For the selective isolation of Gram-positive bacteria, particularly Gram-positive cocci, from specimens with a mixed flora. Do not use for the observation of hemolytic reactions.

Phenol Nutrient-Supplemented Broth

Composition per 1000.05mL:

$(NH_4)_2SO_4$ 2.0g
Na_2HPO_4 .... 1.0g
Nutrient broth 20.0mL
Phenol (90% solution) 1.05mL

pH 6.8 ± 0.2 at 25°C

Nutrient Broth:

Composition per liter:

Peptone 5.0g
Beef extract 3.0g

Preparation of Nutrient Broth: Add components to tap water and bring volume to 1.0L. Mix thoroughly. Gently heat and bring to boiling.

Preparation of Medium: Add components, except phenol, to distilled/deionized water and bring volume to 1.0L. Mix thoroughly. Autoclave for 15 min at 15 psi pressure–121°C. Cool to 45°–50°C. Aseptically add phenol. Mix thoroughly.

Use: For the cultivation of ATCC strain 1413.

Phenol Red Adonitol Broth

Composition per liter:

Proteose peptone 10.0g
NaCl 5.0g

Adonitol 5.0g
Beef extract 1.0g
Phenol Red 0.018g

pH 7.4 ± 0.2 at 25°C

Source: This medium is available from HiMedia.

Preparation of Medium: Add components to distilled/deionized water and bring volume to 1.0L. Mix thoroughly. Distribute into tubes with inverted Durham tubes. Autoclave for 15 min at 15 psi pressure–121°C.

Use: For the detection of adonitol fermenting bacteria.

Phenol Red Agar

Composition per liter:

Agar 15.0g
Pancreatic digest of casein 10.0g
NaCl 5.0g
Phenol Red 0.018g
Carbohydrate solution 20.0mL

pH 7.4 ± 0.2 at 25°C

Source: This medium is available as a premixed powder from BD Diagnostic Systems.

Carbohydrate Solution:
Composition per 20.0mL:

Carbohydrate 5.0–10.0g

Preparation of Carbohydrate Solution: Add carbohydrate to distilled/deionized water and bring volume to 20.0mL. Mix thoroughly. Filter sterilize.

Preparation of Medium: Add components, except carbohydrate solution, to distilled/deionized water and bring volume to 980.0mL. Mix thoroughly. Adjust pH to 7.4 if necessary. Autoclave for 15 min at 15 psi pressure–121°C. Cool to 45°–50°C. Aseptically add 20.0mL of sterile carbohydrate solution. Pour into sterile Petri dishes or distribute into sterile tubes. Allow tubes to cool in a slanted position.

Use: For the determination of fermentation reactions. Bacteria that can ferment the added carbohydrate turn the medium yellow.

Phenol Red Agar

Composition per liter:

Agar 15.0g
Proteose peptone No. 3 .... 10.0g
NaCl 5.0g
Beef extract 1.0g
Phenol Red 0.025g
Carbohydrate solution 20.0mL

pH 7.4 ± 0.2 at 25°C

Source: This medium is available as a premixed powder from BD Diagnostic Systems.

Carbohydrate Solution:
Composition per 20.0mL:

Carbohydrate 5.0–10.0g

Preparation of Carbohydrate Solution: Add carbohydrate to distilled/deionized water and bring volume to 20.0mL. Mix thoroughly. Filter sterilize.

Preparation of Medium: Add components, except carbohydrate solution, to distilled/deionized water and bring volume to 980.0mL. Mix thoroughly. Adjust pH to 7.4 if necessary. Autoclave for 15 min at 15 psi pressure–121°C. Cool to 45°–50°C. Aseptically add 20.0mL of sterile carbohydrate solution. Pour into sterile Petri dishes or distribute into sterile tubes. Allow tubes to cool in a slanted position.

Use: For the determination of fermentation reactions. Bacteria that can ferment the added carbohydrate turn the medium yellow.

Phenol Red Arabinose Broth

Composition per liter:

Proteose peptone 10.0g
NaCl 5.0g
Arabinose 5.0g
Beef extract 1.0g
Phenol Red 0.018g

pH 7.4 ± 0.2 at 25°C

Source: This medium is available from HiMedia.

Preparation of Medium: Add components to distilled/deionized water and bring volume to 1.0L. Mix thoroughly. Distribute into tubes with inverted Durham tubes. Autoclave for 15 min at 15 psi pressure–121°C.

Use: For the detection of arabinose-fermenting bacteria.

Phenol Red Broth

Composition per liter:

Pancreatic digest of casein 10.0g
NaCl 5.0g
Phenol Red 0.018g
Carbohydrate solution 20.0mL

pH 7.4 ± 0.2 at 25°C

Source: This medium is available as a premixed powder from BD Diagnostic Systems.

Carbohydrate Solution:
Composition per 20.0mL:

Carbohydrate 5.0–10.0g

Preparation of Carbohydrate Solution: Add carbohydrate to distilled/deionized water and bring volume to 20.0mL. Mix thoroughly. Filter sterilize.

Preparation of Medium: Add components, except carbohydrate solution, to distilled/deionized water and bring volume to 980.0mL. Mix thoroughly. Adjust pH to 7.4 if necessary. Distribute into tubes containing an inverted Durham tube. Fill each tube with 9.8mL of medium. Autoclave for 15 min at 13 psi pressure–118°C. Cool to 45°–50°C. Aseptically add 0.2mL of sterile carbohydrate solution to each tube.

Use: For the determination of fermentation reactions in the differentiation of microorganisms. Fermentation is determined by the production of acid—broth turns yellow—and formation of gas—bubble trapped in Durham tube.

Phenol Red Broth Base with Plant Extract No. 1

Composition per liter:

Plant hydrolysate 10.0g
NaCl 5.0g
Plant extract No. 1 .... 1.0g
Phenol Red 0.018g

pH 7.4 ± 0.2 at 25°C

Source: This medium, without carbohydrate, is available as a premixed powder from HiMedia.

Carbohydrate Solution:

Composition per 20.0mL:

Carbohydrate	5.0–10.0g

Preparation of Carbohydrate Solution: Add carbohydrate to distilled/deionized water and bring volume to 20.0mL. Mix thoroughly. Filter sterilize.

Preparation of Medium: Add components, except carbohydrate solution, to distilled/deionized water and bring volume to 980.0mL. Mix thoroughly. Adjust pH to 7.4 if necessary. Autoclave for 15 min at 15 psi pressure–121°C. Cool to 45°–50°C. Aseptically add 20.0mL of sterile carbohydrate solution. Aseptically distribute into sterile tubes or flasks.

Use: For the determination of fermentation reactions. Bacteria that can ferment the added carbohydrate turn the medium yellow.

Phenol Red Carbohydrate Broth (BAM M121)

Composition per liter:

Pancreatic digest of casein	10.0g
NaCl	5.0g
Beef extract	1.0g
Carbohydrate solution	200.0mL
Phenol Red solution	7.2mL

pH 7.4 ± 0.2 at 25°C

Source: This medium is available as a premixed powder from BD Diagnostics.

Phenol Red Solution:

Composition per 10.0mL:

Phenol Red	0.025g

Preparation of Phenol Red Solution: Add Phenol Red to distilled/deionized water and bring volume to 10.0mL. Mix thoroughly.

Carbohydrate Solution:

Composition per 200.0mL:

Carbohydrate	5.0–10.0g

Preparation of Carbohydrate Solution: Add carbohydrate (5.0g dulcitol, 10.0g lactose, or 10.0g sucrose) to distilled/deionized water and bring volume to 200.0mL. Mix thoroughly. Filter sterilize.

Preparation of Medium: Add components, except carbohydrate solution, to distilled/deionized water and bring volume to 800.0mL. Mix thoroughly. Adjust pH to 7.4 if necessary. Distribute into tubes containing an inverted Durham tube. Fill each tube with 8.0mL of medium. Autoclave for 15 min at 13 psi pressure–118°C. Cool to 45°–50°C. Aseptically add 2.0mL of sterile carbohydrate solution to each tube.

Use: For the determination of fermentation reactions in the differentiation of microorganisms, such as *Salmonella* spp. Fermentation is determined by the production of acid—broth turns yellow—and formation of gas—bubble trapped in Durham tube.

Phenol Red Carbohydrate Broth with Sodium Chloride (BAM M121)

Composition per liter:

NaCl	30.0g
Pancreatic digest of casein	10.0g
Beef extract	1.0g
Carbohydrate solution	200.0mL
Phenol Red solution	7.2mL

pH 7.4 ± 0.2 at 25°C

Source: This medium, without added NaCl, is available as a premixed powder from BD Diagnostics.

Phenol Red Solution:

Composition per 10.0mL:

Phenol Red	0.025g

Preparation of Phenol Red Solution: Add Phenol Red to distilled/deionized water and bring volume to 10.0mL. Mix thoroughly.

Carbohydrate Solution:

Composition per 200.0mL:

Carbohydrate	5.0–10.0g

Preparation of Carbohydrate Solution: Add carbohydrate (5.0g dulcitol, 10.0g lactose, or 10.0g sucrose) to distilled/deionized water and bring volume to 200.0mL. Mix thoroughly. Filter sterilize.

Preparation of Medium: Add components, except carbohydrate solution, to distilled/deionized water and bring volume to 800.0mL. Mix thoroughly. Adjust pH to 7.4 if necessary. Distribute into tubes containing an inverted Durham tube. Fill each tube with 8.0mL of medium. Autoclave for 15 min at 13 psi pressure–118°C. Cool to 45°–50°C. Aseptically add 2.0mL of sterile carbohydrate solution to each tube.

Use: For the determination of fermentation reactions in the differentiation of halophilic *Vibrio* spp. Fermentation is determined by the production of acid—broth turns yellow—and formation of gas—bubble trapped in Durham tube.

Phenol Red Dextrose Broth

Composition per liter:

Proteose peptone	10.0g
NaCl	5.0g
Glucose	5.0g
Beef extract	1.0g
Phenol Red	0.018g

pH 7.4 ± 0.2 at 25°C

Source: This medium is available from HiMedia.

Preparation of Medium: Add components to distilled/deionized water and bring volume to 1.0L. Mix thoroughly. Distribute into tubes with inverted Durham tubes. Autoclave for 15 min at 15 psi pressure–121°C.

Use: For glucose fermentation studies of microorganisms.

Phenol Red Dulcitol HiVeg Broth

Composition per liter:

Plant peptone No. 3	10.0g
Dulcitol	5.0g
NaCl	5.0g
Plant extract	1.0g
Phenol Red	0.018g

pH 7.4 ± 0.2 at 25°C

Source: This medium is available as a premixed powder from HiMedia.

Preparation of Medium: Add components to distilled/deionized water and bring volume to 1.0L. Mix thoroughly. Gently heat and bring to boiling. Distribute into tubes or flasks. Autoclave for 15 min at 15 psi pressure–121°C.

Use: For the determination of the ability of a microorganism to ferment dulcitol. Fermentation is determined by the production of acid—medium turns yellow.

Phenol Red Galactose Broth

Composition per liter:

Proteose peptone	10.0g
NaCl	5.0g
Galactose	5.0g
Beef extract	1.0g
Phenol Red	0.018g

pH 7.4 ± 0.2 at 25°C

Source: This medium is available from HiMedia.

Preparation of Medium: Add components to distilled/deionized water and bring volume to 1.0L. Mix thoroughly. Distribute into tubes with inverted Durham tubes. Autoclave for 15 min at 15 psi pressure–121°C.

Use: For galactose fermentation studies of microorganisms.

Phenol Red Glucose Broth

Composition per liter:

Pancreatic digest of casein	10.0g
Glucose	5.0g
NaCl	5.0g
Phenol Red	0.018g

pH 7.3 ± 0.2 at 25°C

Source: This medium is available as a premixed powder from BD Diagnostic Systems.

Preparation of Medium: Add components to distilled/deionized water and bring volume to 1.0L. Mix thoroughly. Adjust pH to 7.3 if necessary. Distribute into tubes containing an inverted Durham tube. Fill each tube with 10.0mL of medium. Autoclave for 15 min at 13 psi pressure–118°C.

Use: For determination of the ability of a microorganism to ferment glucose. Fermentation is determined by the production of acid—broth turns yellow—and formation of gas—bubble trapped in Durham tube.

Phenol Red Glucose Broth (BAM M122)

Composition per liter:

Proteose peptone No. 3	10.0g
Glucose	5.0g
NaCl	5.0g
Beef extract	1.0g
Phenol Red solution	7.2mL

pH 7.4 ± 0.2 at 25°C

Source: This medium is available as a premixed powder from BD Diagnostics.

Phenol Red Solution:

Composition per 10.0mL:

Phenol Red	0.025g

Preparation of Phenol Red Solution: Add Phenol Red to distilled/deionized water and bring volume to 10.0mL. Mix thoroughly.

Preparation of Medium: Add components to distilled/deionized water and bring volume to 1.0L. Mix thoroughly. Adjust pH to 7.4 if necessary. Distribute into tubes containing an inverted Durham tube. Fill each tube with 10.0mL of medium. Autoclave for 10 min at 13 psi pressure–118°C.

Use: For determination of the ability of a microorganism to ferment glucose. Fermentation is determined by the production of acid—broth turns yellow—and formation of gas—bubble trapped in Durham tube.

Phenol Red Glucose Broth, HiVeg

Composition per liter:

Plant peptone No. 3	10.0g
Glucose	5.0g
NaCl	5.0g
Plant extract	1.0g
Phenol Red	0.018g

pH 7.3 ± 0.2 at 25°C

Source: This medium is available as a premixed powder from HiMedia.

Preparation of Medium: Add components to distilled/deionized water and bring volume to 1.0L. Mix thoroughly. Adjust pH to 7.3 if necessary. Distribute into tubes containing an inverted Durham tube. Fill each tube with 10.0mL of medium. Autoclave for 15 min at 13 psi pressure–118°C.

Use: For determination of the ability of a microorganism to ferment glucose. Fermentation is determined by the production of acid—broth turns yellow—and formation of gas—bubble trapped in Durham tube.

Phenol Red Glucose Broth with Sodium Chloride (BAM M122)

Composition per liter:

NaCl	30.0g
Proteose peptone No. 3	10.0g
Glucose	5.0g
Beef extract	1.0g
Phenol Red solution	7.2mL

pH 7.4 ± 0.2 at 25°C

Source: This medium, without added NaCl, is available as a premixed powder from BD Diagnostics.

Phenol Red Solution:

Composition per 10.0mL:

Phenol Red	0.025g

Preparation of Phenol Red Solution: Add Phenol Red to distilled/deionized water and bring volume to 10.0mL. Mix thoroughly.

Preparation of Medium: Add components to distilled/deionized water and bring volume to 1.0L. Mix thoroughly. Adjust pH to 7.4 if necessary. Distribute into tubes containing an inverted Durham tube. Fill each tube with 10.0mL of medium. Autoclave for 10 min at 13 psi pressure–118°C.

Use: For determination of the ability of halophilic *Vibrio* spp. to ferment glucose. Fermentation is determined by the production of acid—broth turns yellow—and formation of gas—bubble trapped in Durham tube.

Phenol Red Glucose HiVeg Agar

Composition per liter:

Agar	15.0g
Plant peptone No. 3	10.0g
Glucose	10.0g
NaCl	5.0g

Plant extract 1.0g
Phenol Red 0.025g

pH 7.4 ± 0.2 at 25°C

Source: This medium is available as a premixed powder from HiMedia.

Preparation of Medium: Add components to distilled/deionized water and bring volume to 1.0L. Mix thoroughly. Gently heat and bring to boiling. Distribute into tubes or flasks. Autoclave for 15 min at 13 psi pressure–118°C. Pour into sterile Petri dishes or leave in tubes. Allow tubes to cool in a slanted position.

Use: For the determination of the ability of a microorganism to ferment glucose. Fermentation is determined by the production of acid—medium turns yellow.

Phenol Red HiVeg Agar Base with Carbohydrate

Composition per liter:

Agar 15.0g
Plant peptone No. 3 .... 10.0g
NaCl 5.0g
Plant extract 1.0g
Phenol Red 0.025g
Carbohydrate solution 20.0mL

pH 7.4 ± 0.2 at 25°C

Source: This medium, without carbohydrate, is available as a premixed powder from HiMedia.

Carbohydrate Solution:

Composition per 20.0mL:

Carbohydrate 5.0–10.0g

Preparation of Carbohydrate Solution: Add carbohydrate to distilled/deionized water and bring volume to 20.0mL. Mix thoroughly. Filter sterilize.

Preparation of Medium: Add components, except carbohydrate solution, to distilled/deionized water and bring volume to 980.0mL. Mix thoroughly. Adjust pH to 7.4 if necessary. Autoclave for 15 min at 15 psi pressure–121°C. Cool to 45°–50°C. Aseptically add 20.0mL of sterile carbohydrate solution. Pour into sterile Petri dishes or distribute into sterile tubes. Allow tubes to cool in a slanted position.

Use: For the determination of fermentation reactions. Bacteria that can ferment the added carbohydrate turn the medium yellow.

Phenol Red HiVeg Broth Base

Composition per liter:

Plant peptone No. 3 .... 10.0g
NaCl 5.0g
Plant extract 1.0g
Phenol Red 0.018g
Carbohydrate solution 20.0mL

pH 7.4 ± 0.2 at 25°C

Source: This medium, without carbohydrate, is available as a premixed powder from HiMedia.

Carbohydrate Solution:

Composition per 20.0mL:

Carbohydrate 5.0–10.0g

Preparation of Carbohydrate Solution: Add carbohydrate to distilled/deionized water and bring volume to 20.0mL. Mix thoroughly. Filter sterilize.

Preparation of Medium: Add components, except carbohydrate solution, to distilled/deionized water and bring volume to 980.0mL. Mix thoroughly. Adjust pH to 7.4 if necessary. Autoclave for 15 min at 15 psi pressure–121°C. Cool to 45°–50°C. Aseptically add 20.0mL of sterile carbohydrate solution. Aseptically distribute into sterile tubes or flasks.

Use: For the determination of fermentation reactions. Bacteria that can ferment the added carbohydrate turn the medium yellow.

Phenol Red Inositol Broth

Composition per liter:

Proteose peptone 10.0g
NaCl 5.0g
Inositol 5.0g
Beef extract 1.0g
Phenol Red 0.018g

pH 7.4 ± 0.2 at 25°C

Source: This medium is available from HiMedia.

Preparation of Medium: Add components to distilled/deionized water and bring volume to 1.0L. Mix thoroughly. Distribute into tubes with inverted Durham tubes. Autoclave for 15 min at 15 psi pressure–121°C.

Use: For inositol fermentation studies of microorganisms.

Phenol Red Lactose Agar

Composition per liter:

Agar 15.0g
Lactose 10.0g
Proteose peptone No. 3 .... 10.0g
NaCl 5.0g
Beef extract 1.0g
Phenol Red 25.0mg

pH 7.4 ± 0.2 at 25°C

Source: This medium is available as a premixed powder from BD Diagnostic Systems.

Preparation of Medium: Add components to distilled/deionized water and bring volume to 1.0L. Mix thoroughly. Gently heat and bring to boiling. Distribute into tubes or flasks. Autoclave for 15 min at 13 psi pressure–118°C. Pour into sterile Petri dishes or leave in tubes. Allow tubes to cool in a slanted position.

Use: For the determination of the ability of a microorganism to ferment lactose. Fermentation is determined by the production of acid—medium turns yellow.

Phenol Red Lactose Broth

Composition per liter:

Pancreatic digest of casein 10.0g
Lactose 5.0g
NaCl 5.0g
Phenol Red 0.018g

pH 7.3 ± 0.2 at 25°C

Source: This medium is available as a premixed powder from BD Diagnostic Systems.

Preparation of Medium: Add components to distilled/deionized water and bring volume to 1.0L. Mix thoroughly. Adjust pH to 7.3 if necessary. Distribute into tubes containing an inverted Durham tube.

Fill each tube with 10.0mL of medium. Autoclave for 15 min at 13 psi pressure–118°C.

Use: For determination of the ability of a microorganism to ferment lactose. Fermentation is determined by the production of acid—broth turns yellow—and formation of gas—bubble trapped in Durham tube.

Phenol Red Lactose Broth

Composition per liter:

Proteose peptone	10.0g
NaCl	5.0g
Lactose	5.0g
Beef extract	1.0g
Phenol Red	0.018g

pH 7.4 ± 0.2 at 25°C

Source: This medium is available from HiMedia.

Preparation of Medium: Add components to distilled/deionized water and bring volume to 1.0L. Mix thoroughly. Distribute into tubes with inverted Durham tubes. Autoclave for 15 min at 15 psi pressure–121°C.

Use: For the detection of lactose-fermenting bacteria.

Phenol Red Lactose Broth

Composition per liter:

Peptic digest of animal tissue	10.0g
Lactose	10.0g
NaCl	5.0g
Beef extract	1.0g
Phenol Red	0.018g

pH 7.4 ± 0.2 at 25°C

Source: This medium is available from HiMedia.

Preparation of Medium: Add components to distilled/deionized water and bring volume to 1.0L. Mix thoroughly. Distribute into tubes with inverted Durham tubes. Autoclave for 15 min at 15 psi pressure–121°C.

Use: For lactose fermentation studies in accordance with the ISO Committee under the specification ISO 9308-1:1990.

Phenol Red Lactose HiVeg Agar

Composition per liter:

Agar	15.0g
Plant peptone No. 3	10.0g
Lactose	10.0g
NaCl	5.0g
Plant extract	1.0g
Phenol Red	0.025g

pH 7.4 ± 0.2 at 25°C

Source: This medium is available as a premixed powder from Hi-Media.

Preparation of Medium: Add components to distilled/deionized water and bring volume to 1.0L. Mix thoroughly. Gently heat and bring to boiling. Distribute into tubes or flasks. Autoclave for 15 min at 13 psi pressure–118°C. Pour into sterile Petri dishes or leave in tubes. Allow tubes to cool in a slanted position.

Use: For the determination of the ability of a microorganism to ferment lactose. Fermentation is determined by the production of acid—medium turns yellow.

Phenol Red Lactose HiVeg Broth

Composition per liter:

Plant peptone No. 3	10.0g
Lactose	5.0g
NaCl	5.0g
Plant extract	1.0g
Phenol Red	0.018g

pH 7.4 ± 0.2 at 25°C

Source: This medium is available as a premixed powder from Hi-Media.

Preparation of Medium: Add components to distilled/deionized water and bring volume to 1.0L. Mix thoroughly. Gently heat and bring to boiling. Distribute into tubes or flasks. Autoclave for 15 min at 13 psi pressure–118°C.

Use: For the determination of the ability of a microorganism to ferment lactose. Fermentation is determined by the production of acid—medium turns yellow.

Phenol Red Maltose Broth

Composition per liter:

Proteose peptone	10.0g
NaCl	5.0g
Maltose	5.0g
Beef extract	1.0g
Phenol Red	0.018g

pH 7.4 ± 0.2 at 25°C

Source: This medium is available from HiMedia.

Preparation of Medium: Add components to distilled/deionized water and bring volume to 1.0L. Mix thoroughly. Distribute into tubes with inverted Durham tubes. Autoclave for 15 min at 15 psi pressure–121°C.

Use: For maltose fermentation studies of microorganisms.

Phenol Red Maltose HiVeg Agar

Composition per liter:

Agar	15.0g
Plant peptone No. 3	10.0g
Maltose	10.0g
NaCl	5.0g
Plant extract	1.0g
Phenol Red	0.025g

pH 7.4 ± 0.2 at 25°C

Source: This medium is available as a premixed powder from Hi-Media.

Preparation of Medium: Add components to distilled/deionized water and bring volume to 1.0L. Mix thoroughly. Gently heat and bring to boiling. Distribute into tubes or flasks. Autoclave for 15 min at 13 psi pressure–118°C. Pour into sterile Petri dishes or leave in tubes. Allow tubes to cool in a slanted position.

Use: For the determination of the ability of a microorganism to ferment maltose. Fermentation is determined by the production of acid—medium turns yellow.

Phenol Red Maltose HiVeg Broth

Composition per liter:

Plant peptone No. 3	10.0g
Maltose	5.0g

NaCl...5.0g
Plant extract 1.0g
Phenol Red...0.018g

pH 7.4 ± 0.2 at 25°C

Source: This medium is available as a premixed powder from HiMedia.

Preparation of Medium: Add components to distilled/deionized water and bring volume to 1.0L. Mix thoroughly. Gently heat and bring to boiling. Distribute into tubes or flasks. Autoclave for 15 min at 13 psi pressure–118°C.

Use: For the determination of the ability of a microorganism to ferment maltose. Fermentation is determined by the production of acid—medium turns yellow.

Phenol Red Mannitol Agar

Composition per liter:

Agar 15.0g
Mannitol..10.0g
Proteose peptone No. 3 ....10.0g
NaCl...5.0g
Beef extract...1.0g
Phenol Red...25.0mg

pH 7.4 ± 0.2 at 25°C

Source: This medium is available as a premixed powder from BD Diagnostic Systems.

Preparation of Medium: Add components to distilled/deionized water and bring volume to 1.0L. Mix thoroughly. Gently heat and bring to boiling. Distribute into tubes or flasks. Autoclave for 15 min at 13 psi pressure–118°C. Pour into sterile Petri dishes or leave in tubes. Allow tubes to cool in a slanted position.

Use: For determination of the ability of a microorganism to ferment mannitol. Fermentation is determined by the production of acid—medium turns yellow.

Phenol Red Mannitol Broth

Composition per liter:

Pancreatic digest of casein 10.0g
D-Mannitol 5.0g
NaCl...5.0g
Phenol Red...0.018g

pH 7.3 ± 0.2 at 25°C

Source: This medium is available as a premixed powder from BD Diagnostic Systems.

Preparation of Medium: Add components to distilled/deionized water and bring volume to 1.0L. Mix thoroughly. Adjust pH to 7.3 if necessary. Distribute into tubes containing an inverted Durham tube. Fill each tube with 10.0mL of medium. Autoclave for 15 min at 13 psi pressure–118°C.

Use: For determination of the ability of a microorganism to ferment mannitol. Fermentation is determined by the production of acid—broth turns yellow—and formation of gas—bubble trapped in Durham tube.

Phenol Red Mannitol HiVeg Agar

Composition per liter:

Agar 15.0g
Plant peptone No. 3..10.0g
Mannitol..10.0g
NaCl...5.0g
Plant extract 1.0g
Phenol Red...0.025g

pH 7.4 ± 0.2 at 25°C

Source: This medium is available as a premixed powder from HiMedia.

Preparation of Medium: Add components to distilled/deionized water and bring volume to 1.0L. Mix thoroughly. Gently heat and bring to boiling. Distribute into tubes or flasks. Autoclave for 15 min at 13 psi pressure–118°C. Pour into sterile Petri dishes or leave in tubes. Allow tubes to cool in a slanted position.

Use: For the determination of the ability of a microorganism to ferment mannitol. Fermentation is determined by the production of acid—medium turns yellow.

Phenol Red Mannitol HiVeg Broth

Composition per liter:

Plant peptone No. 3..10.0g
Mannitol..5.0g
NaCl...5.0g
Plant extract 1.0g
Phenol Red...0.018g

pH 7.4 ± 0.2 at 25°C

Source: This medium is available as a premixed powder from HiMedia.

Preparation of Medium: Add components to distilled/deionized water and bring volume to 1.0L. Mix thoroughly. Gently heat and bring to boiling. Distribute into tubes or flasks. Autoclave for 15 min at 13 psi pressure–118°C.

Use: For the determination of the ability of a microorganism to ferment mannitol. Fermentation is determined by the production of acid—medium turns yellow.

Phenol Red Raffinose Broth

Composition per liter:

Proteose peptone...10.0g
NaCl...5.0g
Raffinose...5.0g
Beef extract...1.0g
Phenol Red...0.018g

pH 7.4 ± 0.2 at 25°C

Source: This medium is available from HiMedia.

Preparation of Medium: Add components to distilled/deionized water and bring volume to 1.0L. Mix thoroughly. Distribute into tubes with inverted Durham tubes. Autoclave for 15 min at 15 psi pressure–121°C.

Use: For raffinose fermentation studies of microorganisms.

Phenol Red Rhamnose Broth

Composition per liter:

Proteose peptone...10.0g
NaCl...5.0g
Rhamnose 5.0g
Beef extract...1.0g
Phenol Red...0.018g

pH 7.4 ± 0.2 at 25°C

Source: This medium is available from HiMedia.

Preparation of Medium: Add components to distilled/deionized water and bring volume to 1.0L. Mix thoroughly. Distribute into tubes with inverted Durham tubes. Autoclave for 15 min at 15 psi pressure–121°C.

Use: For rhamnose fermentation studies of microorganisms.

Phenol Red Salicin Broth

Composition per liter:

Proteose peptone	10.0g
NaCl	5.0g
Salicin	5.0g
Beef extract	1.0g
Phenol Red	0.018g

pH 7.4 ± 0.2 at 25°C

Source: This medium is available from HiMedia.

Preparation of Medium: Add components to distilled/deionized water and bring volume to 1.0L. Mix thoroughly. Distribute into tubes with inverted Durham tubes. Autoclave for 15 min at 15 psi pressure–121°C.

Use: For salicin fermentation studies of microorganisms.

Phenol Red Sorbitol Broth

Composition per liter:

Proteose peptone	10.0g
NaCl	5.0g
Sorbitol	5.0g
Beef extract	1.0g
Phenol Red	0.018g

pH 7.4 ± 0.2 at 25°C

Source: This medium is available from HiMedia.

Preparation of Medium: Add components to distilled/deionized water and bring volume to 1.0L. Mix thoroughly. Distribute into tubes with inverted Durham tubes. Autoclave for 15 min at 15 psi pressure–121°C.

Use: For sorbitol fermentation studies of microorganisms.

Phenol Red Starch Broth

Composition per liter:

Proteose peptone	10.0g
NaCl	5.0g
Starch	5.0g
Beef extract	1.0g
Phenol Red	0.018g

pH 7.4 ± 0.2 at 25°C

Source: This medium is available from HiMedia.

Preparation of Medium: Add components to distilled/deionized water and bring volume to 1.0L. Mix thoroughly. Distribute into tubes with inverted Durham tubes. Autoclave for 15 min at 15 psi pressure–121°C.

Use: For starch fermentation studies of microorganisms.

Phenol Red Sucrose Broth

Composition per liter:

Pancreatic digest of casein	10.0g
NaCl	5.0g
Sucrose	5.0g
Phenol Red	0.018g

pH 7.3 ± 0.2 at 25°C

Source: This medium is available as a premixed powder from BD Diagnostic Systems.

Preparation of Medium: Add components to distilled/deionized water and bring volume to 1.0L. Mix thoroughly. Adjust pH to 7.3 if necessary. Distribute into tubes containing an inverted Durham tube. Fill each tube with 10.0mL of medium. Autoclave for 15 min at 13 psi pressure–118°C.

Use: For determination of the ability of a microorganism to ferment sucrose. Fermentation is determined by the production of acid—broth turns yellow—and formation of gas—bubble trapped in Durham tube.

Phenol Red Sucrose Broth

Composition per liter:

Proteose peptone	10.0g
NaCl	5.0g
Sucrose	5.0g
Beef extract	1.0g
Phenol Red	0.018g

pH 7.4 ± 0.2 at 25°C

Source: This medium is available from HiMedia.

Preparation of Medium: Add components to distilled/deionized water and bring volume to 1.0L. Mix thoroughly. Distribute into tubes with inverted Durham tubes. Autoclave for 15 min at 15 psi pressure–121°C.

Use: For sucrose fermentation studies of microorganisms.

Phenol Red Sucrose HiVeg Agar

Composition per liter:

Agar	15.0g
Plant peptone No. 3	10.0g
Sucrose	10.0g
NaCl	5.0g
Plant extract	1.0g
Phenol Red	0.025g

pH 7.4 ± 0.2 at 25°C

Source: This medium is available as a premixed powder from HiMedia.

Preparation of Medium: Add components to distilled/deionized water and bring volume to 1.0L. Mix thoroughly. Gently heat and bring to boiling. Distribute into tubes or flasks. Autoclave for 15 min at 13 psi pressure–118°C. Pour into sterile Petri dishes or leave in tubes. Allow tubes to cool in a slanted position.

Use: For the determination of the ability of a microorganism to ferment sucrose. Fermentation is determined by the production of acid—medium turns yellow.

Phenol Red Sucrose HiVeg Broth

Composition per liter:

Plant peptone No. 3	10.0g
NaCl	5.0g
Sucrose	5.0g
Plant extract	1.0g
Phenol Red	0.018g

pH 7.4 ± 0.2 at 25°C

Source: This medium is available as a premixed powder from HiMedia.

Preparation of Medium: Add components to distilled/deionized water and bring volume to 1.0L. Mix thoroughly. Gently heat and bring to boiling. Distribute into tubes or flasks. Autoclave for 15 min at 13 psi pressure–118°C.

Use: For the determination of the ability of a microorganism to ferment sucrose. Fermentation is determined by the production of acid—medium turns yellow.

Phenol Red Tartrate Agar

Composition per liter:

Agar	15.0g
Peptone	10.0g
Potassium tartrate	10.0g
NaCl	5.0g
Phenol Red	0.024g

pH 7.6 ± 0.2 at 25°C

Source: This medium is available as a premixed powder from BD Diagnostic Systems.

Preparation of Medium: Add components to cold distilled/deionized water and bring volume to 1.0L. Mix thoroughly. Gently heat and bring to boiling. Distribute into tubes or flasks. Autoclave for 15 min at 13 psi pressure–118°C. Pour into sterile Petri dishes or leave in tubes. Allow tubes to cool in an upright position.

Use: For the differentiation of Gram-negative bacteria of the intestinal groups, particularly members of the *Salmonella* (paratyphoid) group based on their ability to ferment tartrate.

Phenol Red Tartrate Broth

Composition per liter:

Pancreatic digest of casein	10.0g
Potassium tartrate	10.0g
Agar	5.0g
NaCl	5.0g
Phenol Red	0.024g

pH 7.6 ± 0.2 at 25°C

Preparation of Medium: Add components to distilled/deionized water and bring volume to 1.0L. Mix thoroughly. Distribute into tubes or flasks. Autoclave for 15 min at 15 psi pressure–121°C.

Use: For the differentiation of Gram-negative bacteria of the intestinal groups, particularly members of the *Salmonella* (paratyphoid) group based on their ability to ferment tartrate.

Phenol Red Tartrate HiVeg Agar

Composition per liter:

Agar	15.0g
Plant peptone	10.0g
Sodium potassium tartrate	10.0g
NaCl	5.0g
Phenol Red	0.024g

pH 7.6 ± 0.2 at 25°C

Source: This medium is available as a premixed powder from HiMedia.

Preparation of Medium: Add components to cold distilled/deionized water and bring volume to 1.0L. Mix thoroughly. Gently heat and bring to boiling. Distribute into tubes or flasks. Autoclave for 15 min at 13 psi pressure–118°C. Pour into sterile Petri dishes or leave in tubes. Allow tubes to cool in an upright position.

Use: For the differentiation of Gram-negative bacteria of the intestinal groups, particularly members of the *Salmonella* (paratyphoid) group based on their ability to ferment tartrate.

Phenol Red Trehalose Broth

Composition per liter:

Proteose peptone	10.0g
NaCl	5.0g
Trehalose	5.0g
Beef extract	1.0g
Phenol Red	0.018g

pH 7.4 ± 0.2 at 25°C

Source: This medium is available from HiMedia.

Preparation of Medium: Add components to distilled/deionized water and bring volume to 1.0L. Mix thoroughly. Distribute into tubes with inverted Durham tubes. Autoclave for 15 min at 15 psi pressure–121°C.

Use: For trehalose fermentation studies of microorganisms.

Phenol Red Xylose Broth

Composition per liter:

Proteose peptone	10.0g
NaCl	5.0g
Xylose	5.0g
Beef extract	1.0g
Phenol Red	0.018g

pH 7.4 ± 0.2 at 25°C

Source: This medium is available from HiMedia.

Preparation of Medium: Add components to distilled/deionized water and bring volume to 1.0L. Mix thoroughly. Distribute into tubes with inverted Durham tubes. Autoclave for 15 min at 15 psi pressure–121°C.

Use: For xylose fermentation studies of microorganisms.

Phenolphthalein Phosphate HiVeg Agar

Composition per liter:

Agar	15.0g
Plant peptone	5.0g
NaCl	5.0g
Plant extract	3.0g
Sodium phenolphthalein phosphate	0.012g

pH 7.4 ± 0.2 at 25°C

Source: This medium is available as a premixed powder from HiMedia.

Preparation of Medium: Add components to distilled/deionized water and bring volume to 1.0L. Mix thoroughly. Gently heat and bring to boiling. Distribute into tubes or flasks. Autoclave for 15 min at 15 psi pressure–21°C.

Use: For the identification of phosphatase-positive *Staphylococcus aureus*.

Phenylalanine Agar (Phenylalanine Deaminase Medium)

Composition per liter:

Agar	12.0g
NaCl	5.0g
Yeast extract	3.0g

DL-Phenylalanine 2.0g
Na_2HPO_4 .... 1.0g

pH 7.3 ± 0.2 at 25°C

Source: This medium is available as a premixed powder from BD Diagnostic Systems.

Preparation of Medium: Add components to distilled/deionized water and bring volume to 1.0L. Mix thoroughly. Gently heat while stirring and bring to boiling. Distribute into tubes or flasks. Autoclave for 10 min at 15 psi pressure–121°C. Pour into sterile Petri dishes or leave in tubes.

Use: For the differentiation of enteric Gram-negative bacilli on the basis of their ability to produce phenylpyruvic acid from phenylalanine. After appropriate incubation of bacteria, ferric chloride reagent is added on the agar. Formation of a green color in 1–5 min indicates the production of phenylpyruvic acid.

Phenylalanine Malonate Broth

Composition per liter:

Sodium malonate 3.0g
DL-Phenylalanine 2.0g
NaCl 2.0g
$(NH_4)_2SO_4$ 2.0g
Yeast extract 1.0g
K_2HPO_4 .... 0.6g
KH_2PO_4 .... 0.4g
Bromthymol Blue 0.025g

pH 7.3 ± 0.2 at 25°C

Source: This medium is available as a premixed powder from BD Diagnostic Systems.

Preparation of Medium: Add components to distilled/deionized water and bring volume to 1.0L. Mix thoroughly. Distribute into tubes or flasks. Autoclave for 10 min at 10 psi pressure–115°C.

Use: For the differentiation of Gram-negative enteric bacilli on the basis of malonate utilization and formation of pyruvic acid from phenylalanine.

Phenylethanol Agar

Composition per liter:

Agar 15.0g
Tryptose 10.0g
NaCl 5.0g
Beef extract 3.0g
Phenylethanol 2.5g

pH 7.3 ± 0.2 at 25°C

Source: This medium is available as a premixed powder from BD Diagnostic Systems.

Preparation of Medium: Add components to distilled/deionized water and bring volume to 1.0L. Mix thoroughly. Gently heat and bring to boiling. Distribute into tubes or flasks. Autoclave for 15 min at 15 psi pressure–121°C. Pour into sterile Petri dishes or leave in tubes.

Use: For the isolation of staphylococci and streptococci from specimens containing a mixed flora.

Phenylethanol Blood Agar

Composition per liter:

Agar 15.0g
Tryptose 10.0g
NaCl 5.0g
Beef extract 3.0g
Phenylethanol 2.5g
Blood, defibrinated 50.0mL

pH 7.3 ± 0.2 at 25°C

Preparation of Medium: Add components, except blood, to distilled/deionized water and bring volume to 950.0mL. Mix thoroughly. Gently heat and bring to boiling. Autoclave for 15 min at 13 psi pressure–118°C. Cool to 45°–50°C. Aseptically add sterile defibrinated blood. Mix thoroughly. Pour into sterile Petri dishes or distribute into sterile tubes.

Use: For the isolation of staphylococci and streptococci from specimens containing a mixed flora.

Phenylethyl Alcohol Agar

See: **Phenethyl Alcohol Agar**

Phenylethyl Alcohol HiVeg Agar

Composition per liter:

Agar 15.0g
Plant hydrolysate 15.0g
Papaic digest of soybean meal 5.0g
NaCl 5.0g
Phenylethyl alcohol 2.5g

pH 7.3 ± 0.2 at 25°C

Source: This medium is available as a premixed powder from HiMedia.

Preparation of Medium: Add components to distilled/deionized water and bring volume to 1.0L. Mix thoroughly. Gently heat and bring to boiling. Distribute into tubes or flasks. Autoclave for 15 min at 15 psi pressure–121°C. Pour into sterile Petri dishes or leave in tubes.

Use: For the isolation of staphylococci and streptococci from specimens containing a mixed flora.

Phenylketonuria Test Agar

See: **PKU Test Agar**

Phenylobacterium Agar

Composition per liter:

Agar 20.0g
Chloridazon herbicide 2.0g
Peptone 1.0g
Yeast extract 1.0g
$Na_2HPO_4 \cdot 12H_2O$ 0.7g
$(NH_4)_2HPO_4$ 0.7g
KH_2PO_4 .... 0.3g
$NH_4H_2PO_4$ 0.3g
$MgSO_4 \cdot 7H_2O$ 0.25g
$(NH_4)_2SO_4$ 0.1g
$CaCl_2 \cdot 6H_2O$ 0.05g
H_3BO_3 .... 0.5mg
$MnSO_4 \cdot 4H_2O$ 0.4mg
$ZnSO_4 \cdot 7H_2O$ 0.4mg
$FeCl_3 \cdot 6H_2O$ 0.2mg
$(NH_4)_6Mo_7O_{24} \cdot 4H_2O$ 0.2mg
Biotin 0.1mg
KJ 0.1mg
$CuSO_4 \cdot 5H_2O$ 0.04mg
Vitamin B_{12} 0.03mg

pH 6.8–7.0 at 25°C

Preparation of Medium: Add components to distilled/deionized water and bring volume to 1.0L. Mix thoroughly. Gently heat and bring to boiling. Distribute into tubes or flasks. Autoclave for 15 min at 15 psi pressure–121°C. Pour into sterile Petri dishes or leave in tubes.

Use: For the cultivation and maintenance of *Phenylobacterium immobile.*

Phenylobacterium Medium

Composition per liter:

Agar	15.0g
Antipyrin	1.0g
$Na_2HPO_4 \cdot 12H_2O$	0.7g
$(NH_4)_2HPO_4$	0.7g
KH_2PO_4	0.3g
$NH_4H_2PO_4$	0.3g
$MgSO_4 \cdot 7H_2O$	0.25g
$FeCl_3 \cdot 6H_2O$	0.2g
$(NH_4)_2SO_4$	0.1g
$CaCl_2 \cdot 6H_2O$	50.0mg
H_3BO_3	0.5mg
$MnSO_4 \cdot 4H_2O$	0.4mg
$ZnSO_4 \cdot 7H_2O$	0.4mg
$(NH_4)_2MoO_4$	0.2mg
Biotin	0.1mg
$CuSO_4 \cdot 5H_2O$	0.1mg
KI	0.1mg
Vitamin B_{12}	30.0μg

pH 6.8 ± 0.2 at 25°C

Preparation of Medium: Add components to distilled/deionized water and bring volume to 1.0L. Mix thoroughly. Gently heat and bring to boiling. Distribute into tubes or flasks. Autoclave for 15 min at 15 psi pressure–121°C. Pour into sterile Petri dishes or leave in tubes.

Use: For the cultivation of *Phenylobacterium immobile.*

Phenylobacterium Medium

Composition per liter:

Antipyrine	1.0g
$Na_2HPO_4 \cdot 12H_2O$	0.7g
$(NH_4)_2HPO_4$	0.7g
KH_2PO_4	0.3g
$NH_4H_2PO_4$	0.3g
$(NH_4)_2SO_4$	0.1g
$MgSO_4 \cdot 7H_2O$	0.25g
$CaCl_2 \cdot 6H_2O$	0.05g
H_3BO_3	0.5mg
$MnSO_4 \cdot 4H_2O$	0.4mg
$ZnSO_4 \cdot 7H_2O$	0.4mg
$FeCl_3 \cdot 6H_2O$	0.2mg
$(NH_4)_2MoO_4$	0.2mg
Biotin	0.1mg
KI	0.1mg
$CuSO_4 \cdot 5H_2O$	0.04mg
Vitamin B_{12}	0.03mg

Preparation of Medium: Add components to distilled/deionized water and bring volume to 1.0L. Mix thoroughly. Distribute into tubes or flasks. Autoclave for 15 min at 15 psi pressure–121°C.

Use: For the cultivation and maintenance of *Phenylobacterium immobile.*

Phosphate Mineral Salts Medium with Octane

Composition per liter:

$(NH_4)_2HPO_4$	10.0g
K_2HPO_4	5.0g
Na_2SO_4	0.5g
Octane	10.0mL

Preparation of Medium: Add components, except octane, to tap water and bring volume to 990.0mL. Mix thoroughly. Autoclave for 15 min at 15 psi pressure–121°C. Prior to inoculation, filter sterilize octane. Aseptically add sterile octane to sterile medium. Aseptically distribute into sterile tubes or flasks.

Use: For the cultivation and maintenance of *Pseudomonas oleovorans.*

Photobacterium Agar (DSMZ Medium 32)

Composition per liter:

Seawater aquarium salt	33.0g
Agar	20.0g
Tris	6.0g
Yeast extract	5.0g
Tryptone	5.0g
NH_4Cl	5.0g
Glycerol	3.0g
$CaCO_3$	1.0g

pH 7.2–7.5 at 25°C

Preparation of Medium: Add components to distilled/deionized water and bring volume to 1.0L. Mix thoroughly. Adjust pH to 7.2–7.5. Gently heat and bring to boiling. Distribute into tubes or flasks. Autoclave for 15 min at 15 psi pressure–121°C. Pour into sterile Petri dishes or leave in tubes.

Use: For the cultivation and maintenance of *Vibrio fischeri.*

Photobacterium Agar

Composition per liter:

Seawater aquarium salt	33.0g
Agar	20.0g
Tris(hydroxymethyl)aminomethane	6.0g
NH_4Cl	5.0g
Pancreatic digest of casein	5.0g
Yeast extract	5.0g
Glycerol	3.0g
$CaCO_3$	1.0g

pH 7.2 ± 0.2 at 25°C

Preparation of Medium: Add components to distilled/deionized water and bring volume to 1.0L. Mix thoroughly. Gently heat and bring to boiling. Distribute into tubes or flasks. Autoclave for 15 min at 15 psi pressure–121°C. Pour into sterile Petri dishes or leave in tubes.

Use: For the cultivation and maintenance of *Acetobacter pasteurianus.*

Photobacterium Broth (DSMZ Medium 608)

Composition per liter:

NaCl	30.0g
Na-glycerol phosphate	23.5g
Tryptone	5.0g
KH_2PO_4	3.0g

Yeast extract.......2.5g
$CaCO_3$.......1.0g
NH_4Cl.......0.3g
Mg_2SO_4.......0.3g
$FeCl_2$.......0.01g

pH 7.0 ± 0.2 at 25°C

Preparation of Medium: Add components to distilled/deionized water and bring volume to 1.0L. Mix thoroughly. Adjust pH to 7.0. Distribute into tubes or flasks. Autoclave for 15 min at 15 psi pressure–121°C.

Use: For the cultivation of *Shewanella hanedai=Alteromonas hanedai.*

Photobacterium Broth

Composition per liter:

NaCl.......30.0g
Sodium glycerol phosphate.......23.5g
Pancreatic digest of casein.......5.0g
KH_2PO_4.......3.0g
Yeast extract.......2.5g
$CaCO_3$.......1.0g
NH_4Cl.......0.3g
$MgSO_4 \cdot 7H_2O$.......0.3g
$FeCl_3$.......0.01g

pH 7.0 ± 0.2 at 25°C

Source: This medium is available as a premixed powder from BD Diagnostic Systems.

Preparation of Medium: Add components to distilled/deionized water and bring volume to 1.0L. Mix thoroughly. Distribute into tubes or flasks to form a shallow layer of medium. Autoclave for 15 min at 15 psi pressure–121°C.

Use: For the cultivation and demonstration of luminescence by photobacteria. For the cultivation and maintenance of *Alteromonas hanedai, Photobacterium phosphoreum, Shewanella hanedai, Vibrio fischeri, Vibrio harveyi,* and other *Vibrio* species.

Photobacterium Broth

Composition per liter:

Seawater aquarium salt.......33.0g
Tris(hydroxymethyl)aminomethane.......6.0g
NH_4Cl.......5.0g
Pancreatic digest of casein.......5.0g
Yeast extract.......5.0g
Glycerol.......3.0g

pH 7.2 ± 0.2 at 25°C

Preparation of Medium: Add components to distilled/deionized water and bring volume to 1.0L. Mix thoroughly. Distribute into tubes or flasks. Autoclave for 15 min at 15 psi pressure–121°C.

Use: For the cultivation and maintenance of *Acetobacter pasteurianus.*

Photobacterium HiVeg Broth

Composition per liter:

NaCl.......30.0g
Sodium glycerophosphate.......23.5g
Plant hydrolysate.......5.0g
KH_2PO_4.......3.0g
Yeast extract.......2.5g
$CaCO_3$.......1.0g
NH_4Cl.......0.3g
$MgSO_4$.......0.3g
$FeCl_3$.......0.01g

pH 7.0 ± 0.2 at 25°C

Source: This medium is available as a premixed powder from HiMedia.

Preparation of Medium: Add components to distilled/deionized water and bring volume to 1.0L. Mix thoroughly. Distribute into tubes or flasks to form a shallow layer of medium. Autoclave for 15 min at 15 psi pressure–121°C.

Use: For the cultivation and demonstration of luminescence by photobacteria. For the cultivation and maintenance of *Vibrio* species.

Photobacterium Medium (DSMZ Medium 32)

Composition per liter:

Seawater aquarium salt.......33.0g
Tris.......6.0g
Yeast extract.......5.0g
Tryptone.......5.0g
NH_4Cl.......5.0g
Glycerol.......3.0g

pH 7.2–7.5 at 25°C

Preparation of Medium: Add components to distilled/deionized water and bring volume to 1.0L. Mix thoroughly. Adjust pH to 7.2–7.5. Gently heat and bring to boiling. Distribute into tubes or flasks. Autoclave for 15 min at 15 psi pressure–121°C.

Use: For the cultivation and maintenance of *Vibrio fischeri.*

Photobacterium MPY Medium

Composition per liter:

NaCl.......28.2g
$MgSO_4 \cdot 7H_2O$.......6.9g
$MgCl_2 \cdot 6H_2O$.......5.5g
Peptone.......5.0g
Yeast extract.......3.0g
$CaCl_2 \cdot 2H_2O$.......1.5g
KCl.......0.7g

pH 7.4 ± 0.2 at 25°C

Preparation of Medium: Add components to distilled/deionized water and bring volume to 1.0L. Mix thoroughly. Distribute into tubes or flasks. Autoclave for 15 min at 15 psi pressure–121°C.

Use: For the cultivation and maintenance of *Photobacterium leiognathi.*

Phthalic Acid Medium

Composition per liter:

Solution 1.......400.0mL
Solution 2.......400.0mL
Potassium hydrogen phthalate solution.......200.0mL

pH 6.8 ± 0.2 at 25°C

Solution 1:

Composition per 400.0mL:

KH_2PO_4.......9.1g
$(NH_4)_2SO_4$.......1.2g

Preparation of Solution 1: Add components to distilled/deionized water and bring volume to 400.0mL. Mix thoroughly. Adjust pH to 6.8

with KOH. Autoclave for 15 min at 15 psi pressure–121°C. Cool to 25°C.

Solution 2:

Composition per 400.0mL:

$MgSO_4 \cdot 7H_2O$	0.4g
$FeSO_4 \cdot 7H_2O$	0.01g

Preparation of Solution 2: Add components to distilled/deionized water and bring volume to 400.0mL. Mix thoroughly. Adjust pH to 6.8 with KOH. Autoclave for 15 min at 15 psi pressure–121°C. Cool to 25°C.

Potassium Hydrogen Phthalate Solution:

Composition per 200.0mL:

Potassium hydrogen phthalate	1.0g

Preparation of Potassium Hydrogen Phthalate Solution: Add potassium hydrogen phthalate to distilled/deionized water and bring volume to 200.0mL. Mix thoroughly. Adjust pH to 6.8 with KOH. Autoclave for 15 min at 15 psi pressure–121°C. Cool to 25°C.

Preparation of Medium: Aseptically combine the three sterile solutions. Mix thoroughly. Aseptically distribute into sterile tubes or flasks.

Use: For the cultivation and maintenance of *Pseudomonas cepacia.*

PHYG Medium

Composition per 110.1mL:

Beef heart, solids from infusion	10.0g
Polypeptone™	2.0g
Gelatin	1.0g
Glucose	1.0g
Yeast extract	1.0g
$NaHCO_3$	0.5g
Tryptose	0.2g
Agar	0.16g
NaCl	0.1g
L-Cysteine·HCl·H_2O	0.09g
$(NH_4)_2SO_4$	0.05g
Resazurin	0.16mg
Salts solution	50.0mL
Rabbit serum, inactivated	10.0mL
Thiamine pyrophosphate solution	0.1mL

pH 7.2–7.5 at 25°C

Salts Solution:

Composition per 400.0mL:

K_2HPO_4	0.9g
NaCl	0.8g
KH_2PO_4	0.4g
$MnCl_2 \cdot 4H_2O$	0.16g
$MgSO_4$	0.08g

Preparation of Salts Solution: Add components to distilled/deionized water and bring volume to 400.0mL. Mix thoroughly.

Thiamine Pyrophosphate Solution:

Composition per 10.0mL:

Thiamine pyrophosphate	0.05g

Preparation of Thiamine Pyrophosphate Solution: Add thiamine pyrophosphate to distilled/deionized water and bring volume to 10.0mL. Mix thoroughly. Filter sterilize.

Preparation of Medium: Add components, except rabbit serum and thiamine pyrophosphate solution, to distilled/deionized water and bring volume to 100.0mL. Mix thoroughly. Gently heat and bring to boiling. Autoclave for 15 min at 15 psi pressure–121°C. Cool to 45°–50°C. Aseptically add 10.0mL of sterile rabbit serum and 0.1mL of sterile thiamine pyrophosphate solution. Mix thoroughly. Pour into sterile Petri dishes or distribute into sterile tubes.

Use: For the cultivation of treponemes.

Phymatotrichum Medium

Composition per liter:

Glucose	40.0g
Agar	20.0g
NH_4NO_3	1.19g
$MgSO_4 \cdot 7H_2O$	0.75g
KCl	0.15g
K_2HPO_4 (0.1*M* solution)	28.1mL
KH_2PO_4 (0.1*M* solution)	18.8mL
$Fe_2(SO_4)_3 \cdot 2H_2O$ (0.23*M* solution)	1.0mL
$MnSO_4 \cdot H_2O$ (0.045*M* solution)	1.0mL
$ZnSO_4 \cdot 7H_2O$ (0.38*M* solution)	1.0mL

Preparation of Medium: Add components to distilled/deionized water and bring volume to 1.0L. Mix thoroughly. Gently heat and bring to boiling. Distribute into tubes or flasks. Autoclave for 15 min at 15 psi pressure–121°C. Pour into sterile Petri dishes or leave in tubes.

Use: For the cultivation and maintenance of *Phymatotrichum omnivorum.*

Phytone™ Yeast Extract Agar

Composition per liter:

Glucose	40.0g
Agar	17.0g
Papaic digest of soybean meal	10.0g
Yeast extract	5.0g
Chloramphenicol	0.05g
Streptomycin	0.03g

pH 6.6 ± 0.2 at 25°C

Source: This medium is available as a premixed powder from BD Diagnostic Systems.

Preparation of Medium: Add components to distilled/deionized water and bring volume to 1.0L. Mix thoroughly. Gently heat and bring to boiling. Distribute into tubes or flasks. Autoclave for 15 min at 15 psi pressure–121°C. Pour into sterile Petri dishes or leave in tubes.

Use: For the selective isolation of dermatophytes, particularly *Trichophyton verrucosum,* and other pathogenic fungi, from clinical specimens.

Picrophilus Medium

Composition per 1020.0mL:

$(NH_4)_2SO_4$	1.3g
KH_2PO_4	0.28g
$MgSO_4 \cdot 7H_2O$	0.25g
$CaCl_2 \cdot 2H_2O$	0.07g
$FeCl_3 \cdot 6H_2O$	0.02g
$Na_2B_4 \cdot 10H_2O$	4.5mg
$MnCl_2 \cdot 4H_2O$	1.8mg
$ZnSO_4 \cdot 7H_2O$	0.22mg
$CuCl_2 \cdot 2H_2O$	0.05mg
$Na_2MoO_4 \cdot 2H_2O$	0.03mg
$VOSO_4 \cdot 2H_2O$	0.03mg
$CoSO_4$	0.01mg

Sulfuric acid solution 1.0L
Yeast extract solution 20.0mL

pH 1.0 ± 0.2 at 25°C

Sulfuric Acid Solution:
Composition per liter:
H_2SO_4 (0.5*M* solution) 300.0mL

Preparation of Sulfuric Acid Solution: Add 300.0mL of 0.5*M* H_2SO_4 solution to 700.0mL of distilled/deionized water. Mix thoroughly.

Yeast Extract Solution:
Composition per 20.0mL:
Yeast extract 2.0g

Preparation of Yeast Extract Solution: Add yeast extract to distilled/deionized water and bring volume to 20.0mL. Mix thoroughly. Autoclave for 15 min at 15 psi pressure–121°C.

Preparation of Medium: Combine components, except yeast extract solution. Mix thoroughly. Autoclave for 15 min at 15 psi pressure–121°C. Aseptically add 20.0mL of sterile yeast extract solution. Mix thoroughly. Aseptically distribute into sterile tubes or flasks.

Use: For the cultivation of *Picrophilus oshimae* and *Picrophilus torridus*.

Pike Streptococcal Broth

Composition per liter:
Pancreatic digest of casein 10.0g
Tryptose 10.0g
Yeast extract 10.0g
Glucose 0.2g
NaN_3 .... 0.065g
Crystal Violet 2.0mg
Rabbit blood, defibrinated 50.0mL

pH 7.4 ± 0.2 at 25°C

Caution: Sodium azide is toxic. Azides also react with metals and disposal must be highly diluted.

Preparation of Medium: Add components, except rabbit blood, to distilled/deionized water and bring volume to 950.0mL. Mix thoroughly. Gently heat and bring to boiling. Distribute into flasks in 100.0mL volumes. Autoclave for 15 min at 15 psi pressure–121°C. Cool to 45°–50°C. Aseptically add 5.0mL of sterile rabbit blood to each flask. Mix thoroughly.

Use: For the isolation and enrichment of hemolytic streptococci from throat swabs and other clinical specimens. After incubation of bacteria for 18–24 hr in this medium, they may be isolated by streaking the culture onto blood agar plates.

Pike Streptococcal HiVeg Broth Base with Blood

Composition per liter:
Yeast extract 10.0g
Plant hydrolysate 10.0g
Plant hydrolysate No. 1 .... 10.0g
Glucose 0.2g
NaN_3 .... 0.065g
Crystal Violet 2.0mg
Rabbit blood, defibrinated 50.0mL

pH 7.4 ± 0.2 at 25°C

Caution: Sodium azide is toxic. Azides also react with metals and disposal must be highly diluted.

Preparation of Medium: Add components, except rabbit blood, to distilled/deionized water and bring volume to 950.0mL. Mix thoroughly. Gently heat and bring to boiling. Distribute into flasks in 100.0mL volumes. Autoclave for 15 min at 15 psi pressure–121°C. Cool to 45°–50°C. Aseptically add 5.0mL of sterile rabbit blood to each flask. Mix thoroughly.

Use: For the isolation and enrichment of hemolytic streptococci from throat swabs and other clinical specimens. After incubation of bacteria for 18–24 hr in this medium, they may be isolated by streaking the culture onto blood agar plates.

Pikovskayas Agar

Composition per liter:
Agar 15.0g
Glucose 10.0g
$Ca_3(PO_4)_2$ 5.0g
Yeast extract 0.5g
$(NH_4)_2SO_4$ 0.5g
KCl 0.2g
$MgSO_4 \cdot 7H_2O$ 0.1g
$MnSO_4 \cdot 2H_2O$ 0.1mg
$FeSO_4$ 0.1mg

Source: This medium is available from HiMedia.

Preparation of Medium: Add components to distilled/deionized water and bring volume to 1.0L. Mix thoroughly. Gently heat and bring to boiling. Distribute into tubes or flasks. Autoclave for 15 min at 15 psi pressure–121°C. Pour into sterile Petri dishes or leave in tubes.

Use: For the detection of phosphate-solubilizing microorganisms.

Pilobolus Agar

Composition per liter:
Agar 15.0g
Sodium acetate 10.0g
Yeast extract 2.0g
K_2HPO_4 .... 1.0g
$(NH_4)_2SO_4$ 0.66g
$MgSO_4 \cdot 7H_2O$ 0.5g
Thiamine 10.0mg
Hemin 10.0mg
NaOH (0.1*N* solution) 37.5mL

pH 8.0 ± 0.2 at 25°C

Preparation of Medium: Add hemin to NaOH solution and mix thoroughly to dissolve. Add remaining components and bring volume to 1.0L with distilled/deionized water. Mix thoroughly. Gently heat and bring to boiling. Distribute into tubes or flasks. Autoclave for 15 min at 15 psi pressure–121°C. Pour into sterile Petri dishes or leave in tubes.

Use: For the cultivation and maintenance of *Pilobolus* species.

Pilobolus Medium (DSMZ Medium 192)

Composition per liter:
Agar 15.0g
Na-acetate 10.0g
Yeast extract 2.0g
K_2HPO_4 .... 1.0g
$(NH_4)_2SO_4$ 0.66g
$MgSO_4 \cdot 7H_2O$ 0.5g

Thiamine-HCl·$2H_2O$ 10.0mg
Hemin solution .... 37.5 mL

pH 8.0 ± 0.2 at 25°C

Hemin Solution:

Composition per 10.0mL:

Hemin 10.0mg

Preparation of Hemin Solution: Add hemin to 37.5 mL 0.1*N* NaOH. Mix thoroughly.

Preparation of Medium: Add components to distilled/deionized water and bring volume to 1.0L. Mix thoroughly. Adjust pH to 8.0. Gently heat and bring to boiling. Distribute into tubes or flasks. Autoclave for 15 min at 15 psi pressure–121°C. Pour into sterile Petri dishes or leave in tubes.

Use: For the cultivation and maintenance of *Pilobolus sphaerosporus.*

PIM

See: ***Prototheca*** **Isolation Agar**

Pine and Drouhet's *Histoplasma* Yeast Phase Medium

Composition per liter:

Casein hydrolysate-vitamin-glucose base 500.0mL
Agar starch base 500.0mL

Casein Hydrolysate-Vitamin-Glucose Base:

Composition per 500.0mL:

Glucose 10.0g
Citric acid 10.0g
L-Cysteine·HCl 1.0g
L-Asparagine 1.0g
Glutathione, reduced 0.5g
L-Tryptophan 0.02g
Solution I 250.0mL
Solution II 10.0mL
Solution III 40.0mL
Solution IV 10.0mL
Solution V 1.0mL
Solution VI 0.1mL
Solution VII 0.1mL

Solution I:

Composition per liter:

KH_2PO_4 .... 8.0g
$(NH_4)_2SO_4$ 8.0g
$MgSO_4 \cdot 7H_2O$ 0.86g
$CaCl_2$ 0.08g

Preparation of Solution I: Add components to distilled/deionized water and bring volume to 1.0L. Mix thoroughly. Store at 5°C.

Solution II:

Composition per liter:

$FeSO_4 \cdot 7H_2O$ 5.7g
$MnCl_2 \cdot 6H_2O$ 0.8g
$Na_2MoO_4 \cdot 2H_2O$ 0.15g
HCl, concentrated 1.0mL

Preparation of Solution II: Add the concentrated HCl to 100.0mL of distilled/deionized water. Add $FeSO_4 \cdot 7H_2O$, $MnCl_2 \cdot 6H_2O$, and $Na_2MoO_4 \cdot 2H_2O$, in that order. Make sure each salt is completely dissolved before adding the next. Mix thoroughly. Bring volume to 1.0L with distilled/deionized water. Mix thoroughly. Store at 5°C. Solution is good until a red precipitate forms.

Solution III:

Composition per 100.0mL:

Acid hydrolysate of casein, vitamin free 10.0g

Preparation of Solution III: Add acid hydrolysate of casein to distilled/deionized water and bring volume to 100.0mL. Mix thoroughly. Gently heat until dissolved.

Solution IV:

Composition per liter:

Inositol 0.2g
Thiamine·HCl 0.2g
Calcium pantothenate 0.2g
Riboflavin 0.2g
Nicotinamide 0.1g
Biotin 10.0mg

Preparation of Solution IV: Add components to distilled/deionized water and bring volume to 1.0L. Mix thoroughly. Store at 5°C. Suspension is good for 6 months to 1 year.

Solution V:

Composition per 100.0mL:

Hemin 0.2g
NH_4OH, concentrated solution .... 2–3 drops

Preparation of Solution V: Add hemin to 10.0mL of distilled/deionized water. Dissolve by adding a few drops of concentrated NH_4OH. Mix thoroughly. Bring volume to 100.0mL with distilled/deionized water. Store at 5°C.

Solution VI:

Composition per 100.0mL:

DL-Thioctic acid 10.0mg
Ethanol (95% solution) 10.0mL

Preparation of Solution VI: Add thioctic acid to 10.0mL of 95% ethanol. Mix thoroughly. Store at –20°C.

Solution VII:

Composition per 10.0mL:

Coenzyme A 10.0mg
$Na_2S \cdot 5H_2O$ (0.05% solution in freshly boiled, distilled/deionized H_2O) .... 2 drops

Preparation of Solution VII: Add coenzyme A to distilled/deionized water and bring volume to 10.0mL. Mix thoroughly. Add 2 drops of 0.05% $Na_2S \cdot 5H_2O$ solution. Store at –20°C.

Preparation of Casein Hydrolysate-Vitamin-Glucose Base: Add components to distilled/deionized water and bring volume to 450.0mL. Mix thoroughly. Adjust pH to 6.5 with 20% KOH solution. Bring volume to 500.0mL with distilled/deionized water. Mix thoroughly. Filter sterilize.

Agar Starch Base:

Composition per 500.0mL:

Agar 12.5g
Potato starch, insoluble 2.0g
Solution VIII 10.0mL

Solution VIII:

Composition per 100.0mL:

Oleic acid 0.1g

Preparation of Solution VIII: Add oleic acid to distilled/deionized water and bring volume to 50.0mL. Mix thoroughly. Adjust pH to

7.0 with NaOH. Bring volume to 100.0mL with distilled/deionized water.

Preparation of Agar Starch Base: Add potato starch to distilled/deionized water and bring volume to 50.0mL. Mix thoroughly. Pour into 440.0mL of boiling, distilled/deionized water. Add 10.0mL of solution VIII and the agar. Mix thoroughly. Autoclave for 20 min at 15 psi pressure–121°C.

Preparation of Medium: Aseptically combine 500.0mL of filter-sterilized casein hydrolysate-vitamin-glucose base with 500.0mL of hot agar starch base. Mix thoroughly. Heat nearly to boiling while stirring constantly. Aseptically distribute into sterile tubes. Allow tubes to cool in a slanted position.

Use: For the cultivation and maintenance of *Blastomyces dermatitidis* and *Histoplasma capsulatum* in the yeast phase.

Pirellula marina Medium M14

Composition per liter:

Glucose	1.0g
Yeast extract	1.0g
Artificial seawater	680.0mL
Tris·HCl buffer, (0.1*M* solution, pH 7.5)	50.0mL
Hutner's basal salts	20.0mL
Vitamin solution	10.0mL

pH 7.5 ± 0.2 at 25°C

Artificial Seawater:

Composition per liter:

NaCl	23.47g
$MgCl_2 \cdot 6H_2O$	10.64g
Na_2SO_4	3.92g
$CaCl_2$	1.1g
KCl	664.0mg
$NaHCO_3$	192.0mg
KBr	96.0mg
H_3BO_3	26.0mg
$SrCl_2$	24.0mg
NaF	3.0mg

Preparation of Artificial Seawater: Add components to distilled/deionized water and bring volume to 1.0L. Mix thoroughly.

Vitamin Solution:

Composition per liter:

Nicotinamide	9.0mg
Calcium DL-pantothenate	5.0mg
Riboflavin	5.0mg
Thiamine·HCl	5.0mg
Biotin	2.0mg
Folic acid	2.0mg
Cyanocobalamin	0.1mg

Preparation of Vitamin Solution: Add components to distilled/deionized water and bring volume to 1.0L. Mix thoroughly. Filter sterilize.

Preparation of Medium: Add components, except vitamin solution, to distilled/deionized water and bring volume to 990.0mL. Mix thoroughly. Autoclave for 15 min at 15 psi pressure–121°C. Aseptically add 10.0mL of sterile vitamin solution. Mix thoroughly. Aseptically distribute into sterile tubes or flasks.

Use: For the cultivation of *Pirellula marina.*

Pisu Medium

Composition per 1512.0mL:

Agar base	960.0mL
Horse serum, sterile	360.0mL
L-Cystine solution	180.0mL
Lead acetate solution	12.0mL

pH 6.8 ± 0.2 at 25°C

Agar Base:

Composition per liter:

Proteose peptone No. 3	20.0g
Agar	7.0g
NaCl	5.0g
Meat extract	4.0g

Preparation for Agar Base: Add components to distilled/deionized water and bring volume to 1.0L. Mix thoroughly. Dissolve in steam for 15 min at 0 psi pressure–100°C. Cool to 45°–50°C. Adjust pH to 7.5. Filter sterilize. Distribute into 200.0mL Erlenmeyer flasks in 80.0mL volumes. Autoclave for 60 min at 0 psi pressure–100°C. Cool to 60°C.

L-Cystine Solution:

Composition per 200.0mL:

L-Cystine	2.0g

Preparation of L-Cystine Solution: Add L-cystine to distilled/deionized water and bring volume to 200.0mL. Mix thoroughly. Filter sterilize.

Lead Acetate Solution:

Composition per 100.0mL:

Lead acetate	10.0g

Preparation of Lead Acetate Solution: Add lead acetate to distilled/deionized water and bring volume to 100.0mL. Mix thoroughly. Filter sterilize.

Preparation of Medium: To each flask containing 80.0mL of cooled, sterile agar base, aseptically add 30.0mL of sterile horse serum, 15.0mL of sterile L-cystine solution, and 1.0mL of sterile lead acetate solution. Mix thoroughly. Aseptically distribute into small sterile tubes in 2.0–3.0mL volumes.

Use: For the cultivation and differentiation of bacteria based on their ability to produce cystinase. Cystinase-producing bacteria turn the medium black.

Pityrosporum Agar

Composition per liter:

Malt extract	40.0g
Ox bile, dessicated	20.0g
Agar	15.0g
Tween™ 40	10.0g
Glycerol monooleate	2.5g

Preparation of Medium: Add components to distilled/deionized water and bring volume to 1.0L. Mix thoroughly. Gently heat and bring to boiling. Distribute into tubes or flasks. Autoclave for 15 min at 15 psi pressure–121°C. Pour into sterile Petri dishes or leave in tubes.

Use: For the cultivation and maintenance of *Pityrosporum* species.

Pityrosporum Medium

Composition per liter:

Malt extract	30.0g
Agar	22.5g

Ox bile, dessicated 20.0g
Tween™ 40 .... 10.0g
Peptone 7.5g
Glycerol monooleate 2.5g

Preparation of Medium: Add components to distilled/deionized water and bring volume to 1.0L. Mix thoroughly. Gently heat and bring to boiling. Distribute into tubes or flasks. Autoclave for 15 min at 15 psi pressure–121°C. Pour into sterile Petri dishes or leave in tubes.

Use: For the cultivation and maintenance of a variety of yeasts including *Malassezia furfur, Malassezia pachydermatis, Pityrosporum ovale*, and other *Pityrosporum* species.

Pityrosporum Medium II

Composition per liter:
Glucose 40.0g
Agar 18.0g
Neopeptone 10.0g
Glycerol monostearate 2.5g
Yeast extract 0.1g
Olive oil 20.0mL
Tween™ 80 .... 2.0mL

pH 5.6 ± 0.2 at 25°C

Preparation of Medium: Add components to distilled/deionized water and bring volume to 1.0L. Mix thoroughly. Gently heat and bring to boiling. Adjust pH to 5.6. Distribute into tubes or flasks. Autoclave for 15 min at 15 psi pressure–121°C. Pour into sterile Petri dishes or leave in tubes.

Use: For the cultivation and maintenance of *Malassezia furfur*.

PKU Test Agar (Phenylketonuria Test Agar)

Composition per liter:
Agar 15.0g
K_2HPO_4 .... 15.0g
Glucose 10.0g
KH_2PO_4 .... 5.0g
$(NH_4)Cl$ 2.5g
$(NH_4)NO_3$ 0.5g
Asparagine 0.5g
DL-Alanine 0.5g
L-Glutamic acid 0.5g
Na_2SO_4 .... 0.5g
$MgSO_4 \cdot 7H_2O$ 0.05g
$FeCl_3$ 5.0mg
$MnCl_2 \cdot 4H_2O$ 5.0mg
β-2-Thienylalanine 3.3mg
$CaCl_2 \cdot 2H_2O$ 2.5mg
Bacillus subtilis spore suspension 10.0mL

pH 7.0 ± 0.2 at 25°C

Source: This medium is available as a premixed powder from BD Diagnostic Systems.

Preparation of Medium: Add components to distilled/deionized water and bring volume to 1.0L. Mix thoroughly. Gently heat and bring to boiling. Continue boiling for 5 min. Do not autoclave. Cool to 50°C. Add 10.0mL of a suspension of *Bacillus subtilis* ATCC 6633 spores. Mix thoroughly. Pour into sterile Petri dishes or other containers.

Use: For the determination of phenylalanine concentrations in serum or urine. Used in the Guthrie-modified bacterial-inhibition assay procedure for screening newborn infants for phenylketonuria (PKU).

PKU Test Agar (Phenylketonuria Test Agar)

Composition per liter:
K_2HPO_4 .... 15.0g
Agar 13.5g
Glucose 5.0g
KH_2PO_4 .... 5.0g
$(NH_4)Cl$ 2.5g
L-Asparagine 0.5g
L-Glutamic acid 0.5g
Na_2SO_4 .... 0.5g
$(NH_4)NO_3$ 0.5g
L-Alanine 0.25g
$MgSO_4 \cdot 7H_2O$ 0.05g
$FeCl_3$ 0.005g
$MnSO_4$ 0.005g
$CaCl_2 \cdot 2H_2O$ 2.5mg
β-2-Thienylalanine solution 1.0mL

pH 6.9 ± 0.2 at 25°C

Source: This medium is available as a premixed powder from BD Diagnostic Systems.

β-2-Thienylalanine Solution:
Composition per 100.0mL:
β-2-Thienylalanine 0.33g

Preparation of β-2-Thienylalanine Solution: Add β-2-thienylalanine to distilled/deionized water and bring volume to 100.0mL. Mix thoroughly. Filter sterilize.

Preparation of Medium: Add components, except β-2-thienylalanine solution, to distilled/deionized water and bring volume to 1.0L. Mix thoroughly. Gently heat and bring to boiling. Continue boiling for 5 min. Do not autoclave. Cool to 50°C. Add 10.0mL of a suspension of *Bacillus subtilis* ATCC 6633 spores and 1.0mL of sterile β-2-thienylalanine solution. Mix thoroughly. Pour into sterile Petri dishes or other containers.

Use: For the determination of phenylalanine concentrations in serum or urine. Used in the Guthrie-modified bacterial-inhibition assay procedure for screening newborn infants for phenylketonuria (PKU).

PL Agar

Composition per liter:
Agar 15.0g
Mannitol 7.5g
L-Arabinose 5.0g
Peptone 5.0g
NaCl 5.0g
Lysine 2.0g
Yeast extract 2.0g
Bile salts No. 2 .... 1.0g
Inositol 1.0g
Phenol Red 0.08g

pH 7.4 ± 0.2 at 25°C

Preparation of Medium: Add components to distilled/deionized water and bring volume to 1.0L. Mix thoroughly. Gently heat and bring to boiling. Adjust pH to 7.4. Distribute into tubes or flasks. Autoclave

for 15 min at 10 psi pressure–115°C. Pour into sterile Petri dishes or leave in tubes.

Use: For the isolation and cultivation of *Pleisomonas shigelloides* from foods.

PL HiVeg Agar

Composition per liter:

Agar	15.0g
Mannitol	7.5g
Plant peptone	5.0g
L-Arabinose	5.0g
NaCl	5.0g
Yeast extract	2.0g
Lysine	2.0g
Inositol	1.0g
Synthetic detergent	1.0g
Phenol Red	0.08g

pH 7.4 ± 0.2 at 25°C

Source: This medium is available as a premixed powder from HiMedia.

Preparation of Medium: Add components to distilled/deionized water and bring volume to 1.0L. Mix thoroughly. Gently heat and bring to boiling. Adjust pH to 7.4. Distribute into tubes or flasks. Autoclave for 15 min at 10 psi pressure–115°C. Pour into sterile Petri dishes or leave in tubes.

Use: For the isolation and cultivation of *Pleisomonas shigelloides* from foods.

Plaice Medium

Composition per liter:

Fresh plaice, minced	200.0g
NaCl	120.0g
Peptone	20.0g

pH 7.3 ± 0.2 at 25°C

Preparation of Medium: Soak plaice in 4.0L of water and allow to stand for 2 hr. Boil for 1 hr and filter. Add peptone and NaCl. Adjust pH. Boil for a few minutes and filter. Mix thoroughly. Distribute into tubes or flasks. Autoclave for 15 min at 15 psi pressure–121°C.

Use: For the cultivation of bioluminescent bacteria.

Planctomyces Medium

Composition per liter:

Glucose	1.0g
$(NH_4)_2SO_4$	0.25g
Peptone	0.15g
Yeast extract	0.15g
Seawater, aged filtered	1.0L

Preparation of Medium: Add components to distilled/deionized water and bring volume to 1.0L. Mix thoroughly. Distribute into tubes or flasks. Autoclave for 15 min at 15 psi pressure–121°C.

Use: For the cultivation of *Planctomyces* species.

Plant *Mycoplasma* Agar

Composition per 291.2mL:

Schneider's *Drosophila* medium	160.0mL
Solution 1	70.0mL
Fetal calf serum	50.0mL
Fresh yeast extract solution	10.0mL
Phenol Red (0.5% solution)	1.2mL

pH 7.4 ± 0.2 at 25°C

Schneider's *Drosophila* Medium:
Composition per liter:

$MgSO_4 \cdot 7H_2O$	3.7g
NaCl	2.1g
Yeast extract	2.0g
Trehalose	2.0g
D-Glucose	2.0g
L-Glutamine	1.8g
L-Lysine·HCl	1.7g
L-Proline	1.7g
KCl	1.6g
$Na_2HPO_4 \cdot 7H_2O$	1.3g
L-Glutamic acid	0.8g
L-Methionine	0.8g
$CaCl_2$, anhydrous	0.6g
KH_2PO_4	0.5g
β-Alanine	0.5g
L-Tyrosine	0.5g
L-Arginine	0.4g
L-Aspartic acid	0.4g
L-Histidine	0.4g
L-Threonine	0.4g
$NaHCO_3$	0.4g
Glycine	0.3g
L-Serine	0.3g
L-Valine	0.3g
L-Isoleucine	0.2g
L-Leucine	0.2g
L-Phenylalanine	0.2g
α-Ketoglutaric acid	0.2g
Fumaric acid	0.1g
Malic acid	0.1g
Succinic acid	0.1g
L-Cystine	0.1g
L-Tryptophan	0.1g
L-Cysteine	0.06g

Preparation of Schneider's *Drosophila* Medium: Add components to distilled/deionized water and bring volume to 1.0L. Mix thoroughly. Filter sterilize.

Solution 1:
Composition per 70.0mL:

Sorbitol	7.0g
Noble agar	5.0g
Beef heart, solids from infusion	5.0g
Peptone	1.8g
Pancreatic digest of casein	1.0g
Sucrose	1.0g
NaCl	0.5g
D-Fructose	0.1g
D-Glucose	0.1g

Preparation of Solution 1: Add components to distilled/deionized water and bring volume to 70.0mL. Mix thoroughly. Adjust pH to 7.8 with 1*N* NaOH. Autoclave for 15 min at 15 psi pressure–121°C. Cool to 50°C.

Fresh Yeast Extract Solution:
Composition per 100.0mL:

Baker's yeast, live, pressed, starch-free	25.0g

Preparation of Fresh Yeast Extract Solution: Add the live Baker's yeast to 100.0mL of distilled/deionized water. Autoclave for 90 min at 15 psi pressure–121°C. Allow to stand. Remove supernatant solution. Adjust pH to 6.6–6.8.

Preparation of Medium: Bring fetal calf serum and Phenol Red solution to 56°C. Rapidly bring Schneider's *Drosophila* medium to 37°C. Rapidly combine the components. Mix thoroughly. Pour into sterile Petri dishes or distribute into sterile tubes or flasks.

Use: For the cultivation and maintenance of *Spiroplasma floricola, Spiroplasma kunkelii, Spiroplasma melliferum,* and *Spiroplasma* species.

Plant *Mycoplasma* Broth

Composition per 291.2mL:

Schneider's *Drosophila* medium	160.0mL
Solution 1	70.0mL
Fetal calf serum	50.0mL
Fresh yeast extract solution	10.0mL
Phenol Red (0.5% solution)	1.2mL

pH 7.4 ± 0.2 at 25°C

Schneider's *Drosophila* Medium:

Composition per liter:

$MgSO_4 \cdot 7H_2O$	3.7g
NaCl	2.1g
Trehalose	2.0g
Yeast extract	2.0g
D-Glucose	2.0g
L-Glutamine	1.8g
L-Lysine·HCl	1.7g
L-Proline	1.7g
KCl	1.6g
$Na_2HPO_4 \cdot 7H_2O$	1.3g
L-Glutamic acid	0.8g
L-Methionine	0.8g
$CaCl_2$, anhydrous	0.6g
KH_2PO_4	0.5g
β-Alanine	0.5g
L-Tyrosine	0.5g
L-Arginine	0.4g
L-Aspartic acid	0.4g
L-Histidine	0.4g
L-Threonine	0.4g
$NaHCO_3$	0.4g
Glycine	0.3g
L-Serine	0.3g
L-Valine	0.3g
L-Isoleucine	0.2g
L-Leucine	0.2g
L-Phenylalanine	0.2g
α-Ketoglutaric acid	0.2g
Fumaric acid	0.1g
Malic acid	0.1g
Succinic acid	0.1g
L-Cystine	0.1g
L-Tryptophan	0.1g
L-Cysteine	0.06g

Preparation of Schneider's *Drosophila* Medium: Add components to distilled/deionized water and bring volume to 1.0L. Mix thoroughly. Filter sterilize.

Solution 1:

Composition per 70.0mL:

Sorbitol	7.0g
Beef heart, solids from infusion	5.0g
Peptone	1.8g
Pancreatic digest of casein	1.0g
Sucrose	1.0g
NaCl	0.5g
D-Fructose	0.1g
D-Glucose	0.1g

Preparation of Solution 1: Add components to distilled/deionized water and bring volume to 70.0mL. Mix thoroughly. Adjust pH to 7.8 with 1*N* NaOH. Autoclave for 15 min at 15 psi pressure–121°C. Cool to 25°C.

Fresh Yeast Extract Solution:

Composition per 100.0mL:

Baker's yeast, live, pressed, starch-free	25.0g

Preparation of Fresh Yeast Extract Solution: Add the live Baker's yeast to 100.0mL of distilled/deionized water. Autoclave for 90 min at 15 psi pressure–121°C. Allow to stand. Remove supernatant solution. Adjust pH to 6.6–6.8.

Preparation of Medium: Bring fetal calf serum and Phenol Red solution to 56°C. Rapidly bring Schneider's *Drosophila* medium to 37°C. Rapidly combine the components. Mix thoroughly. Aseptically distribute into sterile tubes or flasks.

Use: For the cultivation and maintenance of *Spiroplasma floricola, Spiroplasma kunkelii, Spiroplasma melliferum,* and *Spiroplasma* species.

Plate Count Agar

Composition per liter:

Agar	15.0g
Pancreatic digest of casein	5.0g
Yeast extract	2.5g
Glucose	1.0g

pH 7.0 ± 0.2 at 25°C

Preparation of Medium: Add components to distilled/deionized water and bring volume to 1.0L. Mix thoroughly. Gently heat and bring to boiling. Distribute into tubes or flasks. Autoclave for 15 min at 15 psi pressure–121°C. Pour into sterile Petri dishes or leave in tubes.

Use: For the cultivation and maintenance of *Chelatobacter heintzii, Chelatococcus saccharophobus*, and *Pseudomonas* species.

Plate Count Agar (Tryptone Glucose Yeast Agar)

Composition per liter:

Agar	9.0g
Pancreatic digest of casein	5.0g
Yeast extract	2.5g
Glucose	1.0g

pH 7.0 ± 0.2 at 25°C

Source: This medium is available as a premixed powder from Oxoid Unipath.

Preparation of Medium: Add components to distilled/deionized water and bring volume to 1.0L. Mix thoroughly. Gently heat while stirring and bring to boiling. Distribute into tubes or flasks. Autoclave

for 15 min at 15 psi pressure–121°C. Pour into sterile Petri dishes or leave in tubes.

Use: For the enumeration of viable bacteria in milk and dairy products. Also used for the estimation of the number of live heterotrophic bacteria in water. For the cultivation and maintenance of *Brevibacterium casei, Brevibacterium epidermidis,* and *Methylobacterium mesophilicum.*

Plate Count Agar (ATCC Medium 1048)

Composition per liter:

Agar	15.0g
Pancreatic digest of casein	5.0g
Yeast extract	2.5g
Glucose	1.0g

pH 7.0 ± 0.2 at 25°C

Source: This medium is available as a premixed powder from BD Diagnostic Systems and Oxoid Unipath.

Preparation of Medium: Add components to distilled/deionized water and bring volume to 1.0L. Mix thoroughly. Gently heat and bring to boiling. Distribute into tubes or flasks. Autoclave for 15 min at 15 psi pressure–121°C. Pour into sterile Petri dishes or leave in tubes.

Use: For the enumeration of bacteria in milk, water, food, and dairy products.

Plate Count Agar with Antibiotic

Composition per liter:

Agar	9.0g
Pancreatic digest of casein	5.0g
Yeast extract	2.5g
Glucose	1.0g
Chloramphenicol	0.1g

pH 7.0 ± 0.2 at 25°C

Preparation of Medium: Add components to distilled/deionized water and bring volume to 1.0L. Mix thoroughly. Gently heat and bring to boiling. Distribute into tubes or flasks. Autoclave for 15 min at 15 psi pressure–121°C. Pour into sterile Petri dishes or leave in tubes.

Use: For the cultivation of yeasts from foods.

Plate Count Agar with Antibiotic-Free Skim Milk

Composition per liter:

Agar	10.0g
Pancreatic digest of casein	5.0g
Yeast extract	2.5g
Skim milk, antibiotic free	1.0g
Glucose	1.0g

pH 6.9 ± 0.1 at 25°C

Source: This medium is available as a premixed powder from Oxoid Unipath.

Preparation of Medium: Add components to distilled/deionized water and bring volume to 1.0L. Mix thoroughly. Gently heat while stirring and bring to boiling. Distribute into tubes or flasks. Autoclave for 15 min at 15 psi pressure–121°C. Pour into sterile Petri dishes or leave in tubes.

Use: For the enumeration of viable bacteria in milk and dairy products.

Plate Count Agar, HiVeg (Standard Methods Agar, HiVeg)

Composition per liter:

Agar	15.0g
Plant hydrolysate	5.0g
Yeast extract	2.5g
Glucose	1.0g

pH 7.0 ± 0.2 at 25°C

Source: This medium is available as a premixed powder from Oxoid Unipath.

Preparation of Medium: Add components to distilled/deionized water and bring volume to 1.0L. Mix thoroughly. Gently heat while stirring and bring to boiling. Distribute into tubes or flasks. Autoclave for 15 min at 15 psi pressure–121°C. Pour into sterile Petri dishes or leave in tubes.

Use: For the enumeration of viable bacteria in milk and dairy products. Also used for the estimation of the number of live heterotrophic bacteria in water.

Plate Count Agar, Modified

Composition per liter:

Pancreatic digest of casein	20.0g
Yeast extract	20.0g
Agar	10.0g
Glucose	4.0g

pH 7.0 ± 0.1 at 25°C

Preparation of Medium: Add components to distilled/deionized water and bring volume to 1.0L. Mix thoroughly. Gently heat and bring to boiling. Distribute into bottles. Autoclave for 15 min at 15 psi pressure–121°C.

Use: For the cultivation and enumeration of microorganisms from food by the plate count method.

Plate Count Agar, Special

Composition per liter:

Agar	30.1g
Pancreatic digest of casein	6.13g
Yeast extract	3.06g
Glucose	1.23g

pH 7.0 ± 0.2 at 25°C

Source: This medium is available from HiMedia.

Preparation of Medium: Add components to distilled/deionized water and bring volume to 1.0L. Mix thoroughly. Gently heat while stirring and bring to boiling. Distribute into tubes or flasks. Autoclave for 20 min at 15 psi pressure–121°C. Pour into sterile Petri dishes or leave in tubes.

Use: For the enumeration of viable bacteria in raw milk, milk, and other dairy products.

Plate Count Broth (m-Plate Count Broth)

Composition per liter:

Yeast extract	5.0g
Glucose	2.0g

pH 7.0 ± 0.2 at 25°C

Source: This medium is available as a premixed powder from BD Diagnostic Systems.

Preparation of Medium: Add components to distilled/deionized water and bring volume to 1.0L. Mix thoroughly. Distribute into tubes or flasks. Autoclave for 15 min at 15 psi pressure–121°C.

Use: For the determination of bacterial counts by the membrane filter method.

Plate Count HiVeg Agar

Composition per liter:

Agar	9.0g
Plant hydrolysate	5.0g
Yeast extract	2.5g
Glucose	1.0g

pH 7.0 ± 0.2 at 25°C

Source: This medium is available as a premixed powder from Hi-Media.

Preparation of Medium: Add components to distilled/deionized water and bring volume to 1.0L. Mix thoroughly. Gently heat while stirring and bring to boiling. Distribute into tubes or flasks. Autoclave for 15 min at 15 psi pressure–121°C. Pour into sterile Petri dishes or leave in tubes.

Use: For the enumeration of viable bacteria in milk and dairy products by the pour plate technique.

Plate Count MUG Agar

Composition per liter:

Agar	14.0g
Casein peptone	5.0g
Yeast extract	2.5g
D(+)-Glucose	1.0g
Tryptophan	1.0g
4-Methylumbelliferyl-β-D-glucuronide	0.07g

pH 7.0 ± 0.2 at 37°C

Source: This medium is available from Fluka, Sigma-Aldrich.

Preparation of Medium: Add components to distilled/deionized water and bring volume to 1.0L. Mix thoroughly. Gently heat while stirring and bring to boiling. Autoclave for 15 min at 15 psi pressure–121°C. Cool to 50°C. Pour into sterile Petri dishes.

Use: For the determination of bacterial counts in milk, dairy products, water, and other material. *E. coli* can be identified by fluorescence in the UV and verified by means of a positive indole test.

Plates with Fluoranthene (DSMZ Medium 462a)

Composition per 1052.0mL:

Agar	15.0g
Na_2HPO_4	2.44g
KH_2PO_4	1.52g
$(NH_4)_2SO_4$	0.5g
$MgSO_4 \cdot 7H_2O$	0.2g
$CaCl_2 \cdot 2H_2O$	0.05g
Fluoranthene solution	50.0mL
Trace elements solution SL-4	10.0mL
Vitamin solution	2.5mL

pH 7.1 ± 0.2 at 25°C

Trace Elements Solution SL-4:

Composition per liter:

EDTA	0.5g
$FeSO_4 \cdot 7H_2O$	0.2g
Trace elements solution SL-6	100.0mL

Trace Elements Solution SL-6:

Composition per liter:

H_3BO_3	0.3g
$CoCl_2 \cdot 6H_2O$	0.2g
$ZnSO_4 \cdot 7H_2O$	0.1g
$MnCl_2 \cdot 4H_2O$	0.03g
$Na_2MoO_4 \cdot H_2O$	0.03g
$NiCl_2 \cdot 6H_2O$	0.02g
$CuCl_2 \cdot 2H_2O$	0.01g

Preparation of Trace Elements Solution SL-6: Add components to distilled/deionized water and bring volume to 1.0L. Mix thoroughly. Adjust pH to 3.4.

Preparation of Trace Elements Solution SL-4: Add components to distilled/deionized water and bring volume to 1.0L. Mix thoroughly.

Vitamin Solution:

Composition per liter:

Pyridoxamine	5.0mg
Vitamin B_{12}	2.0mg
Nicotinic acid	2.0mg
p-Aminobenzoate	1.0mg
Thiamine–HCl·$2H_2O$	1.0mg
Ca-pantothenate	0.5mg
Biotin	0.2mg

Preparation of Vitamin Solution: Add components to distilled/deionized water and bring volume to 1.0L. Mix thoroughly. Filter sterilize.

Fluoranthene Solution:

Composition per 100.0mL:

Fluoranthene	10.0g

Preparation of Fluoranthene Solution: Add fluoranthene to 100.0mL acetone. Mix thoroughly. Filter sterilize.

Preparation of Medium: Add components, except vitamin solution and fluoranthene solution, to 1.0L distilled/deionized water. Adjust pH to 7.1. Gently heat and bring to boil. Autoclave for 15 min at 15 psi pressure–121°C. Cool to 45°C. Aseptically add 2.5mL vitamin solution. Mix thoroughly. Pour into sterile Petri dishes (20.0mL per Petri dish). Allow plates to dry for 48 hr. Aseptically spread 1.0mL of fluoranthene solution onto the surface of the dried plates. Allow the acetone to evaporate under sterile conditions.

Use: For the cultivation of *Mycobacterium gilvum, Sphingomonas* sp. *(Sphingomonas paucimobilis),* unclassified bacterium *(Gordona* sp.*), Mycobacterium* sp., *Sphingomonas* sp., and other biphenyl-utilizing bacteria.

Plesiomonas Differential Agar

See: **Inositol Brilliant Green Bile Salts Agar**

PLET Agar

Composition per liter:

Beef heart, solids from infusion	500.0g
Agar	15.0g
Tryptose	10.0g
NaCl	5.0g
Ethylenediamine tetracetic acid (EDTA)	0.3g
Thallous acetate	0.04g
Antibiotic inhibitor	10.0mL

Antibiotic Inhibitor:

Composition per 10.0mL:

Lysozyme	300,000U
Polymyxin	30,000U

Preparation of Antibiotic Inhibitor: Add components to distilled/deionized water and bring volume to 10.0mL. Mix thoroughly. Filter sterilize.

Preparation of Medium: Add components, except antibiotic inhibitor, to distilled/deionized water and bring volume to 990.0mL. Mix thoroughly. Gently heat and bring to boiling. Autoclave for 15 min at 15 psi pressure–121°C. Cool to 50°C. Aseptically add sterile antibiotic inhibitor. Mix thoroughly. Pour into sterile Petri dishes or distribute into sterile tubes.

Use: For the selective isolation and cultivation of *Bacillus anthracis*.

Pluton Medium

Composition per 1010.0mL:

Glucose	10.0g
Malt extract	5.0g
Neopeptone	5.0g
Peptone	2.5g
Yeast extract	2.5g
Starch	2.0g
Trypticase™	2.0g
Potassium phosphate buffer (1*M* solution, pH 7.2)	50.0mL
L-Cysteine·HCl solution	10.0mL

pH 7.2 ± 0.2 at 25°C

L-Cysteine·HCl Solution:

Composition per 10.0mL:

L-Cysteine·HCl	0.25g

Preparation of L-Cysteine·HCl Solution: Add L-cysteine·HCl to distilled/deionized water and bring volume to 10.0mL. Mix thoroughly. Sparge with 100% N_2. Autoclave for 15 min at 15 psi pressure–121°C.

Preparation of Medium: Add components, except L-cysteine·HCl solution, to distilled/deionized water and bring volume to 990.0mL. Mix thoroughly. Distribute 10.0mL volumes into tubes. Autoclave for 15 min at 15 psi pressure–121°C. Aseptically add 0.1mL of sterile L-cysteine·HCl solution to each tube. Mix thoroughly.

Use: For the cultivation of *Melissococcus pluton.*

PM Indicator Agar

Composition per liter:

Agar	15.0g
Glucose	5.25g
Peptone	5.0g
Beef extract	3.0g
Pancreatic digest of casein	1.7g
Sorbitan monooleate complex	1.0g
NaCl	0.5g
Papaic digest of soybean meal	0.3g
K_2HPO_4	0.25g
Bromcresol Purple	0.06g

pH 7.8 ± 0.2 at 25°C

Source: This medium is available as a premixed powder from BD Diagnostic Systems.

Preparation of Medium: Add components to distilled/deionized water and bring volume to 1.0L. Mix thoroughly. Gently heat and bring to boiling. Do not autoclave. Inoculate medium with a suspension of *Bacillus stearothermophilus* spores. Pour into 100mm flat-bottomed Petri dishes in 6.0mL volumes. Use medium immediately.

Use: For the rapid detection of trace amounts of penicillin in milk using the AOAC *Bacillus stearothermophilus* Qualitative Disc Method II.

PMA Medium

See: **Peptonized Milk Agar**

PMP Broth

Composition per liter:

Na_2HPO_4	7.9g
D-Mannitol	2.5g
Peptone	2.5g
NaH_2PO_4	1.1g

pH 7.6 ± 0.2 at 25°C

Preparation of Medium: Add components to distilled/deionized water and bring volume to 1.0L. Mix thoroughly. Distribute into tubes or flasks. Autoclave for 15 min at 15 psi pressure–121°C.

Use: For the enrichment and cultivation of *Yersinia pseudotuberculosis* from food.

PmTG Agar

Composition per liter:

Agar	10.0g
Glucose	5.0g
Peptonized milk	1.0g
Pancreatic digest of casein	1.0g

Preparation of Medium: Add components to distilled/deionized water and bring volume to 1.0L. Mix thoroughly. Gently heat and bring to boiling. Distribute into tubes or flasks. Autoclave for 15 min at 15 psi pressure–121°C. Pour into sterile Petri dishes or leave in tubes.

Use: For the cultivation and maintenance of *Monoblepharis polymorpha*.

PMY Medium

Composition per liter:

Agar	15.0g
Glucose	10.0g
NaCl	5.0g
Polypeptone™	5.0g
Beef extract	2.0g
Yeast extract	1.0g
$MgSO_4.7H_2O$	0.5g

pH 7.0 ± 0.2 at 25°C

Preparation of Medium: Add components to distilled/deionized water and bring volume to 1.0L. Mix thoroughly. Gently heat and bring to boiling. Distribute into tubes or flasks. Autoclave for 15 min at 15 psi pressure–121°C. Pour into sterile Petri dishes or leave in tubes.

Use: For the cultivation and maintenance of *Xanthomonas campestris*.

PMYA II Medium

Composition per liter:

Agar	15.0g
Milk, peptonized	1.0g
Yeast extract	0.2g
Sodium acetate	0.02g

Preparation of Medium: Add components to distilled/deionized water and bring volume to 1.0L. Mix thoroughly. Gently heat and bring

to boiling. Distribute into tubes or flasks. Autoclave for 15 min at 15 psi pressure–121°C. Pour into sterile Petri dishes or leave in tubes.

Use: For the isolation and cultivation of *Flavobacterium* species.

Poly-β-Hydroxybutyrate Medium (PHB Medium)

Composition per liter:

Part A 900.0mL
Part B 100.0mL

pH 7.2 ± 0.2 at 25°C

Part A:

Composition per 900.0mL:

$K_2HPO_4 \cdot 3H_2O$ 0.6g
KH_2PO_4 .... 0.2g
$MgSO_4 \cdot 7H_2O$ 0.2g
$(NH_4)_2SO_4$ 0.2g

Preparation of Medium: Add components to distilled/deionized water and bring volume to 900.0mL. Mix thoroughly. Adjust pH to 7.2. Autoclave for 15 min at 15 psi pressure–121°C. Cool to 25°C.

Part B:

Composition per 100.0mL:

Glucose 10.0g

Preparation of Medium: Add glucose to distilled/deionized water and bring volume to 100.0mL. Mix thoroughly. Autoclave for 15 min at 15 psi pressure–121°C. Cool to 25°C.

Preparation of Medium: Aseptically combine 900.0mL of cooled, sterile part A and 100.0mL of cooled, sterile part B. Mix thoroughly. Aseptically distribute into sterile tubes or flasks.

Use: For the cultivation and differentiation of *Pseudomonas* species based on their ability to produce intracellular poly-β-hydroxybutyrate. Production of poly-β-hydroxybutyrate is determined by staining cells with Sudan Black B.

Polyangium Agar

Composition per liter:

Agar 10.0g
K_2HPO_4 .... 1.0g
KNO_3 .... 1.0g
$MgSO_4 \cdot 7H_2O$ 0.3g
$CaCl_2 \cdot 2H_2O$ 0.13g
$FeCl_3 \cdot 6H_2O$ 32.0mg
Filter paper strips .... variable

pH 7.2 ± 0.2 at 25°C

Preparation of Medium: Add components to distilled/deionized water and bring volume to 1.0L. Mix thoroughly. Gently heat and bring to boiling. Distribute into tubes or flasks. Autoclave for 15 min at 15 psi pressure–121°C. Pour into sterile Petri dishes or leave in tubes. Allow tubes to cool in a slanted position. Aseptically add a strip of sterile filter paper to the surface of solidified slants. For Petri dishes, add 4–6 strips of sterile filter paper to the surface of the agar in each plate.

Use: For the cultivation and maintenance of *Polyangium cellulosum.*

Polymyxin Acriflavine LiCl Ceftazidime Esculin Mannitol Agar

See: **PALCAM Agar**

Polymyxin Base Agar

See: **Antibiotic Medium 9**

Polymyxin Pyruvate Egg Yolk Mannitol Bromthymol Blue Agar (PEMBA)

Composition per 110.0mL:

Agar 1.8g
D-Mannitol 1.0g
Sodium pyruvate 1.0g
Na_2HPO_4 .... 0.25g
NaCl 0.2g
Peptone 0.1g
KH_2PO_4 .... 0.025g
Bromthymol Blue 0.01g
$MgSO_4 \cdot 7H_2O$ 0.01g
Antibiotic inhibitor 5.0mL
Egg yolk emulsion (20% solution) 5.0mL

pH 7.4 ± 0.2 at 25°C

Antibiotic Inhibitor:

Composition per 5.0mL:

Polymyxin B 10,000U

Preparation of Antibiotic Inhibitor: Add components to distilled/deionized water and bring volume to 5.0mL. Mix thoroughly. Filter sterilize.

Egg Yolk Emulsion, 20% Solution:

Composition per 100.0mL:

Chicken egg yolks 11
Whole chicken egg 1
NaCl (0.9% solution) 80.0mL

Preparation of Egg Yolk Emulsion, 20% Solution: Soak eggs with 1:100 dilution of saturated mercuric chloride solution for 1 min. Crack eggs and separate yolks from whites. Mix egg yolks with 1 chicken egg. Measure 20.0mL of egg yolk emulsion and add to 80.0mL of 0.9% NaCl solution. Mix thoroughly. Filter sterilize. Warm to 45°–50°C.

Preparation of Medium: Add components, except antibiotic inhibitor and egg yolk emulsion, 20%, to distilled/deionized water and bring volume to 100.0mL. Mix thoroughly. Gently heat and bring to boiling. Autoclave for 15 min at 15 psi pressure–121°C. Cool to 45°–50°C. Aseptically add sterile antibiotic inhibitor and egg yolk emulsion, 20%. Mix thoroughly. Pour into sterile Petri dishes.

Use: For the cultivation of *Bacillus cereus.*

Polymyxin Seed Agar

See: **Antibiotic Medium 10**

Polymyxin *Staphylococcus* Medium

Composition per liter:

Agar 15.0g
Pancreatic digest of gelatin 5.0g
Beef extract 3.0g
Lecithin 0.7g
Polymyxin 0.075g
Tween™ 80 .... 10.2mL

pH 6.8 ± 0.2 at 25°C

Preparation of Medium: Add components to distilled/deionized water and bring volume to 1.0L. Mix thoroughly. Gently heat and bring to boiling. Distribute into tubes or flasks. Autoclave for 15 min at 15 psi pressure–121°C. Pour into sterile Petri dishes.

Use: For the selective isolation and cultivation of pathogenic, coagulase-positive *Staphylococcus aureus*. *Proteus* species will grow on this medium but appear as translucent colonies.

Polypectate Gel Medium

Composition per liter:

Sodium polypectate	70.0g
K_2HPO_4	5.0g
Peptone	5.0g
KH_2PO_4	1.0g
$CaCl_2 \cdot 2H_2O$	0.6g

pH 7.0 ± 0.2 at 25°C

Preparation of Medium: Add components, except sodium polypectate, to 500.0mL of boiling water. Put into a blender and mix on low speed to dissolve. Add the sodium polypectate slowly with the blender on low speed to minimize air bubbles. Adjust pH to 7.0. Bring volume to 1.0L with distilled/deionized water. Autoclave for 15 min at 15 psi pressure–121°C. Cool to 45°–50°C. Pour into sterile Petri dishes.

Use: For the cultivation of microoganisms and detection of pectate-degrading enzymes.

Polysorbate 80 Agar (Tween™ 80 Agar)

Composition per 1010.0mL:

Agar	15.0g
Peptone	10.0g
Tween™ 80	10.0mL

pH 7.2 ± 0.2 at 25°C

Preparation of Medium: Add components, except Tween™ 80, to distilled/deionized water and bring volume to 1.0L. Mix thoroughly. Gently heat and bring to boiling. Autoclave for 20 min at 15 psi pressure–121°C. Cool to 45°–50°C. Add 10.0mL Tween™ 80. Mix thoroughly. Pour into sterile Petri dishes.

Use: For the cultivation of a variety of microorganisms.

Polysorbate 80 HiVeg Agar

Composition per liter:

Agar	15.0g
Plant peptone	10.0g
Polysorbate 80	10.0g

pH 7.2 ± 0.2 at 25°C

Source: This medium is available as a premixed powder from HiMedia.

Preparation of Medium: Add components, except Tween™ 80, to distilled/deionized water and bring volume to 1.0L. Mix thoroughly. Gently heat and bring to boiling. Autoclave for 20 min at 15 psi pressure–121°C. Cool to 45°–50°C. Pour into sterile Petri dishes.

Use: For the cultivation of a variety of microorganisms.

Pontecorvo's *Aspergillus* Medium

Composition per liter:

Glucose	30.0g
$NaNO_3$	6.0g
KH_2PO_4	1.52g
KCl	0.52g
$MgSO_4 \cdot 7H_2O$	0.52g
$ZnSO_4$	0.001g
$FeCl_3$	0.85mg
Biotin	40.0μg
Trace metals solution	1.0mL

pH 6.2 ± 0.2 at 25°C

Trace Metals Solution:

Composition per liter:

$CuSO_4 \cdot 5H_2O$	0.25g
$Na_2B_4O_7 \cdot 10H_2O$	0.1g
$MnCl_2 \cdot 4H_2O$	50.0mg
$Na_2MoO_4 \cdot 2H_2O$	50.0mg

Preparation of Trace Metals Solution: Add components to distilled/deionized water and bring volume to 1.0L. Mix thoroughly.

Preparation of Medium: Add components to distilled/deionized water and bring volume to 1.0L. Mix thoroughly. Gently heat and bring to boiling. Adjust pH to 6.2 with KOH. Distribute into tubes or flasks. Autoclave for 15 min at 15 psi pressure–121°C.

Use: For the cultivation of *Aspergillus awamori.*

Pontecorvo's *Aspergillus* Medium with 0.05% Yeast Extract

Composition per liter:

Glucose	30.0g
$NaNO_3$	6.0g
KH_2PO_4	1.52g
KCl	0.52g
$MgSO_4 \cdot 7H_2O$	0.52g
Yeast extract	0.5g
$ZnSO_4$	0.001g
$FeCl_3$	0.85mg
Biotin	40.0μg
Trace metals solution	1.0mL

pH 6.2 ± 0.2 at 25°C

Trace Metals Solution:

Composition per liter:

$CuSO_4 \cdot 5H_2O$	0.25g
$Na_2B_4O_7 \cdot 10H_2O$	0.1g
$MnCl_2 \cdot 4H_2O$	50.0mg
$Na_2MoO_4 \cdot 2H_2O$	50.0mg

Preparation of Trace Metals Solution: Add components to distilled/deionized water and bring volume to 1.0L. Mix thoroughly.

Preparation of Medium: Add components to distilled/deionized water and bring volume to 1.0L. Mix thoroughly. Gently heat and bring to boiling. Adjust pH to 6.2 with KOH. Distribute into tubes or flasks. Autoclave for 15 min at 15 psi pressure–121°C.

Use: For the cultivation of *Aspergillus nidulans.*

Porcine Heart Agar

Composition per liter:

Porcine heart, infusion from	375.0g
Agar	15.0g
Papaic digest of soybean meal	6.5g
Glucose	5.0g
NaCl	5.0g
Proteose peptone No. 3	5.0g
Yeast extract	3.5g
Sheep blood, defibrinated	50.0mL

pH 7.2 ± 0.2 at 25°C

Source: This medium is available as a premixed powder from BD Diagnostic Systems.

Preparation of Medium: Add components, except sheep blood, to distilled/deionized water and bring volume to 950.0mL. Mix thoroughly. Gently heat and bring to boiling. Autoclave for 15 min at 15 psi pressure–121°C. Cool to 45°–50°C. Aseptically add sterile sheep blood. Mix thoroughly. Pour into sterile Petri dishes.

Use: For determination of the sensitivity of microorganisms using the disc plate technique.

Pork Plasma Fibrinogen Overlay Agar

Composition per plate:

Baird-Parker agar, modified	15.0mL
Pork plasma fibrinogen overlay agar	8.0mL

pH 7.0 ± 0.1 at 25°C

Baird-Parker Agar, Modified:

Composition per liter:

Agar	17.0g
Glycine	12.0g
Sodium pyruvate	10.0g
Pancreatic digest of casein	10.0g
Beef extract	5.0g
LiCl	5.0g
Yeast extract	1.0g
Chapman tellurite solution	1.0mL

pH 7.0 ± 0.2 at 25°C

Chapman Tellurite Solution:

Composition per 100.0mL:

K_2TeO_3	1.0g

Preparation of Chapman Tellurite Solution: Add K_2TeO_3 to distilled/deionized water and bring volume to 100.0mL. Mix thoroughly. Filter sterilize.

Preparation of Baird-Parker Agar, Modified: Add components, except Chapman tellurite solution, to distilled/deionized water and bring volume to 999.0mL. Mix thoroughly. Gently heat and bring to boiling. Autoclave for 15 min at 15 psi pressure–121°C. Cool to 45–50°C. Aseptically add 1.0mL of sterile Chapman tellurite solution. Mix thoroughly but gently.

Pork Plasma Fibrinogen Overlay Agar:

Composition per 100.5mL:

Agar solution	50.0mL
Bovine fibrinogen solution	47.5mL
Pork plasma	2.5mL
Trypsin inhibitor solution	0.5mL

Agar Solution:

Composition per 50.0mL:

Agar	0.7g

Preparation of Agar Solution: Add agar to distilled/deionized water and bring volume to 50.0mL. Mix thoroughly. Autoclave for 15 min at 15 psi pressure–121°C. Cool to 45°–50°C.

Bovine Fibrinogen Solution:

Composition per 50.0mL:

Bovine fibrinogen, fraction I	0.4g
Sodium phosphate buffer (0.05*M* solution, pH 7.0)	50.0mL

Preparation of Bovine Fibrinogen Solution: Grind bovine fibrinogen in a mortar to a fine powder. Add bovine fibrinogen to 50.0mL of sodium phosphate buffer. Mix thoroughly on a magnetic stirrer for 30 min. Filter through Whatman #1 filter paper. Filter sterilize.

Pork Plasma:

Composition per 10.0mL:

Pork plasma-EDTA	10.0mL

Preparation of Pork Plasma: Filter sterilize fresh or rehydrated commercial pork plasma-EDTA.

Trypsin Inhibitor Solution:

Composition per 5.0mL:

Trypsin inhibitor	0.015g
Sodium phosphate buffer (0.05*M* solution, pH 7.0)	5.0mL

Preparation of Trypsin Inhibitor Solution: Add trypsin inhibitor to 5.0mL of sodium phosphate buffer. Mix thoroughly. Filter sterilize.

Preparation of Pork Plasma Fibrinogen Overlay Agar: Aseptically combine 50.0mL of cooled, sterile agar solution, 47.5mL of sterile bovine fibrinogen solution, 2.5mL of sterile pork plasma, and 0.5mL of trypsin inhibitor solution. Mix thoroughly. Maintain at 45°–50°C but use within 1 hr.

Caution: Potassium tellurite is toxic.

Preparation of Medium: Pour cooled, sterile Baird-Parker agar, modified, into sterile Petri dishes in 15.0mL volumes. Allow agar to solidify. Overlay each plate with 8.0mL of sterile pork plasma fibrinogen overlay agar.

Use: For the cultivation of *Staphylococcus aureus* from foods.

Porphyrobacter tepidarius Medium (DSMZ Medium 791)

Composition per liter:

Sodium glutamate	0.5g
Sodium succinate	0.5g
Sodium acetate	0.5g
Yeast extract	0.5g
Casamino acids	0.5g
$(NH_4)_2SO_4$	0.5g
Sodium thiosulfate	0.5g
Basal salts solution	5.0mL
Phosphate solution	5.0mL
Vitamin solution	1.0mL

pH 7.5 ± 0.2 at 25°C

Basal Salts Solution:

Composition per liter:

$MgSO_4·7H_2O$	24.65g
NaCl	23.4g
Tri-sodium EDTA	4.53g
$CaCl_2·2H_2O$	2.94g
$FeSO_4·7H_2O$	1.11g
$MnSO_4·4H_2O$	112.0mg
H_3BO_3	31.0mg
$ZnSO_4·7H_2O$	29.0mg
$Co(NO_3)_2·6H_2O$	29.0mg
$CuSO_4·5H_2O$	25.4mg
$Na_2MoO_4·4H_2O$	24.0mg

Preparation of Basal Salts Solution: Add components to distilled/deionized water and bring volume to 1.0L. Mix thoroughly.

Phosphate Solution:

Composition per liter:

KH_2PO_4	75.0g
K_2HPO_4	78.0g

Preparation of Phosphate Solution: Add components to distilled/deionized water and bring volume to 1.0L. Mix thoroughly.

Vitamin Solution

Composition per 100.0mL:

Tri-sodium EDTA	200.0mg
Nicotinic acid	100.0mg
Thiamine	100.0mg
p-Aminobenzoic acid	50.0mg
Calcium pantothenate	50.0mg
Pyridoxine hydrochloride	50.0mg
Folic acid	50.0mg
Biotin	5.0mg
Vitamin B_{12}	1.0mg

Preparation of Vitamin Solution: Add components to distilled/deionized water and bring volume to 100.0mL. Mix thoroughly. Filter sterilize.

Preparation of Medium: Add components, except vitamin solution, to distilled/deionized water and bring volume to 999.0mL. Mix thoroughly. Autoclave for 15 min at 15 psi pressure–121°C. Cool to 25°C. Aseptically add 1.0mL sterile vitamin solution. Mix thoroughly. Aseptically distribute into sterile tubes or flasks.

Use: For the cultivation of *Porphyrobacter tepidarius, Porphyrobacter cryptus,* and *Porphyrobacter* sp.

Postgate's Medium

Composition per liter:

$MgSO_4 \cdot 7H_2O$	2.0g
$CaSO_4$	1.0g
NH_4Cl	1.0g
Yeast extract	1.0g
K_2HPO_4	0.5g
Sodium lactate (70% solution)	3.5mL

Preparation of Medium: Add components, except $FeSO_4 \cdot 7H_2O$, ascorbic acid, and thioglycollic acid, to distilled/deionized water and bring volume to 1.0L. Mix thoroughly. Sparge with 80% N_2 + 20% CO_2 for 10–15 min. Add $FeSO_4 \cdot 7H_2O$, ascorbic acid, and thioglycollic acid. Mix thoroughly. Continue to sparge with 80% N_2 + 20% CO_2 and adjust pH to 7.4. Anaerobically distribute into tubes or flasks. Autoclave for 10 min at 10 psi pressure–115°C.

Use: For the cultivation of *Desulfobulbus proprionicus, Desulfotomaculum nigrificans, Desulfotomaculum orientis, Desulfotomaculum ruminis, Desulfovibrio africanus, Desulfovibrio desulfuricans, Desulfovibrio gigas, Desulfovibrio multispirans, Desulfovibrio* species, and *Desulfovibrio vulgaris.*

Postgate's Medium for Sulfate Reducers
See: **Medium for Sulfate Reducers**

Postgate's Medium B for Sulfate Reducers
See: **Medium B for Sulfate Reducers**

Postgate's Medium C for Sulfate Reducers
See: **Medium C for Sulfate Reducers**

Postgate's Medium D for Sulfate Reducers
See: **Medium D for Sulfate Reducers**

Postgate's Medium E for Sulfate Reducers
See: **Medium E for Sulfate Reducers**

Postgate's Medium F for Sulfate Reducers
See: **Medium F for Sulfate Reducers**

Postgate's Medium G for Sulfate Reducers
See: **Medium G for Sulfate Reducers**

Postgate's Medium N for Sulfate Reducers
See: **Medium N for Sulfate Reducers**

Potassium Cyanide Broth

Composition per liter:

Na_2HPO_4	5.64g
NaCl	5.0g
Proteose peptone No. 3	3.0g
KH_2PO_4	0.225g
KCN solution	15.0mL

pH 7.6 ± 0.2 at 25°C

KCN Solution:

Composition per 100.0mL:

KCN	0.5g

Preparation of KCN Solution: Add KCN to distilled/deionized water and bring volume to 100.0mL. Mix thoroughly.

Caution: Cyanide is toxic.

Preparation of Medium: Add components, except KCN solution, to distilled/deionized water and bring volume to 985.0mL. Mix thoroughly. Gently heat and bring to boiling. Autoclave for 15 min at 15 psi pressure–121°C. Cool to 25°C. Aseptically add 15.0mL of KCN solution. Mix thoroughly. Distribute into sterile screw-capped tubes or flasks in 1.0–1.5mL volumes. Close caps tightly.

Use: For the cultivation and differentiation of urease-negative, Gram-negative enteric bacteria. *Salmonella* species and *Shigella* species are nonmotile in this medium. *Proteus* species are motile in this medium.

Potassium Cyanide HiVeg Broth Base with KCN

Composition per liter:

Na_2HPO_4	5.64g
NaCl	5.0g
Plant peptone No. 3	3.0g
KH_2PO_4	0.225g
KCN solution	15.0mL

pH 7.6 ± 0.2 at 25°C

Source: This medium, without potassium cyanide solution, is available as a premixed powder from HiMedia.

KCN Solution:

Composition per 100.0mL:

KCN	0.5g

Preparation of KCN Solution: Add KCN to distilled/deionized water and bring volume to 100.0mL. Mix thoroughly.

Caution: Cyanide is toxic.

Preparation of Medium: Add components, except KCN solution, to distilled/deionized water and bring volume to 985.0mL. Mix thoroughly. Gently heat and bring to boiling. Autoclave for 15 min at 15 psi pressure–121°C. Cool to 25°C. Aseptically add 15.0mL of KCN solution. Mix thoroughly. Distribute into sterile screw-capped tubes or flasks in 1.0–1.5mL volumes. Close caps tightly.

Use: For the cultivation and differentiation of urease-negative, Gram-negative enteric bacteria. *Salmonella* species and *Shigella* species are nonmotile in this medium. *Proteus* species are motile in this medium.

Potassium Tellurite Agar

Composition per liter:

Beef heart, solids from infusion	500.0g
Agar	15.0g
Tryptose	10.0g
NaCl	5.0g
Blood, defibrinated	50.0mL
K_2TeO_3 solution	20.0mL

pH 6.0 ± 0.2 at 25°C

K_2TeO_3 Solution:

Composition per 20.0mL:

K_2TeO_3	0.5g

Preparation of K_2TeO_3 Solution: Add K_2TeO_3 to distilled/deionized water and bring volume to 20.0mL. Mix thoroughly. Filter sterilize.

Caution: Potassium tellurite is toxic.

Preparation of Medium: Add components, except K_2TeO_3 solution, to distilled/deionized water and bring volume to 930.0mL. Mix thoroughly. Gently heat and bring to boiling. Autoclave for 15 min at 15 psi pressure–121°C. Cool to 45°–50°C. Aseptically add sterile K_2TeO_3 solution and 50.0mL of blood. Rabbit or sheep blood may be used. Mix thoroughly. Pour into sterile Petri dishes or distribute into sterile tubes. Allow tubes to cool in a slanted position.

Use: For the cultivation and differentiation of *Enterococcus faecalis*. *Enterococcus faecalis* appears as black colonies.

Potato Agar

Composition per liter:

Potatoes	200.0g
Agar	20.0g

Preparation of Medium: Dice potatoes and place in 1.0L of tap water. Gently heat and bring to boiling. Boil potatoes until thoroughly cooked. Filter through cheesecloth. Bring volume of filtrate to 1.0L with tap water. Add 20.0g of agar. Gently heat and bring to boiling. Distribute into tubes or flasks. Autoclave for 15 min at 15 psi pressure–121°C. Pour into sterile Petri dishes or leave in tubes.

Use: For the cultivation and maintenance of *Cercospora medicaginis, Metarhizium anisopliae, Metarhizium flavoviride, Phoma chrysanthemi, Phoma eupyrena, Phoma exigua, Phoma foveata, Phoma lingam, Phoma macrostoma, Phoma pinodella, Phoma putaminum, Polyscytalum pustulans, Sclerotium cepivorum, Ulocladium atrum, Ulocladium botrytis, Ulocladium chartarum, Ulacladium corsortiale*, and *Ulocladium curcurbitae*.

Potato Agar with Cereal Culms

Composition per liter:

Potatoes	200.0g
Agar	20.0g
Cereal culms	variable

Preparation of Medium: Peel and dice potatoes. Add potatoes to 1.0L of tap water. Gently heat and bring to boiling. Boil potatoes until thoroughly cooked. Filter through cheesecloth. Bring volume of filtrate to 1.0L with tap water. Add 20.0g of agar. Gently heat and bring to boiling. Distribute into tubes. Autoclave for 15 min at 15 psi pressure–121°C. Cut cereal culms into 1.0cm pieces. Place 6 to 9 cereal culm pieces in tubes with 3.0mL of distilled/deionized water. Autoclave for 15 min at 15 psi pressure–121°C. Melt agar tubes by gently heating. Cool to 50°C. Aseptically add 2 to 3 sterile cereal culm pieces to each tube. Allow to solidify in a slanted position.

Use: For the cultivation of fungi.

Potato Carrot Agar

Composition per liter:

Potatoes	20.0g
Carrots	20.0g
Agar	20.0g

Preparation of Medium: Wash and peel potatoes and carrots. Grate potatoes and carrots and place in 1.0L of tap water. Gently heat and bring to boiling. Boil for 30 min. Filter through cheesecloth. Bring volume of filtrate to 1.0L with tap water. Add 20.0g of agar. Gently heat and bring to boiling. Distribute into tubes or flasks. Autoclave for 20 min at 15 psi pressure–121°C. Pour into sterile Petri dishes or leave in tubes.

Use: For the cultivation and maintenance of *Alternaria alternata, Alternaria brassicae, Alternaria citri, Alternaria dianthi, Alternaria dianthicola, Alternaria radicina, Alternaria solani, Alternaria tenuissima, Chalaropsis thielavioides, Cladosporium fulvum, Drechslera bicolor, Drechslera cynodontis, Drechslera graminea, Drechslera rostrata, Drechslera sorokiniana, Drechslera spicifera, Drechslera teres, Drechslera verticillata, Drechslera victoriae, Embellisia allii, Glomerella cingulata, Helminthosporium carbonum, Helminthosporium solini, Monilinia fructicola*, and *Trichurus spiralis*.

Potato Carrot Agar

Composition per liter:

Potato, infusion from	250.0g
Carrot, infusion from	200.0g
Agar	15.0g

pH 6.5 ± 0.2 at 25°C

Source: This medium is available as a premixed powder from HiMedia.

Preparation of Medium: Add components to distilled/deionized water and bring volume to 1.0L. Mix thoroughly. Gently heat and bring to boiling. Distribute into tubes or flasks. Autoclave for 15 min at 15 psi pressure–121°C. Pour into sterile Petri dishes or leave in tubes.

Use: For the cultivation and maintenance of fungi and other aciduric microorganisms. For the reproduction of *Pyronema domesticum* and for the cultivation and maintenance of *Actinoplanes awajinensis, Actinoplanes nirasakiensis, Amorphosphorangium auranticolor, Streptomyces flaveus*, and *Thermoactinomyces vulgaris*.

Potato Carrot Agar, Diluted 1/10

Composition per liter:

Potato, peeled and cut 30.0g
Agar 15.0g
Carrot, peeled and cut 2.5g

Preparation of Medium: Boil potatoes and carrots until cooked and filter through cheesecloth. Add components to distilled/deionized water and bring volume to 1.0L. Mix thoroughly. Gently heat and bring to boiling. Distribute into tubes or flasks. Autoclave for 15 min at 15 psi pressure–121°C. Pour into sterile Petri dishes or leave in tubes.

Use: For the cultivation and maintenance of *Actinoplanes kinshanensis, Actinoplanes nipponensis, Actinoplanes pyriformis, Actinoplanes* species, *Actinoplanes utahensis, Microbispora rosea, Planobispora longispora, Planomonospora venezuelensis, Spirillospora albida,* and *Streptosporangium album.*

Potato Carrot Agar with Manganese

Composition per liter:

Agar 20.0g
Potatoes 20.0g
Carrots 20.0g
Manganese solution 10.0mL

pH 6.5 ± 0.2 at 25°C

Manganese Solution:

Composition per 100.0mL:

$MnCl_2 \cdot 4H_2O$ 15.0g

Preparation of Manganese Solution: Add $MnCl_2 \cdot 4H_2O$ to to distilled/deionized water and bring volume to 100.0mL. Mix thoroughly. Filter sterilize.

Preparation of Medium: Wash and peel potatoes and carrots. Grate potatoes and carrots and place in 1.0L of tap water. Gently heat and bring to boiling. Boil for 30 min. Filter through cheesecloth. Bring volume of filtrate to 900.0mL with tap water. Add 20.0g of agar. Gently heat and bring to boiling. Distribute into flasks. Autoclave for 20 min at 15 psi pressure–121°C. Cool to 55°C. Aseptically add 100.0mL of sterile manganese solution. Mix thoroughly. Pour into sterile Petri dishes or distribute into tubes.

Use: For the selective cultivation and maintenance of *Alternaria, Epicoccum*, and *Phoma* species. The additon of manganese inhibits the growth of other fungi.

Potato Carrot Broth

Composition per liter:

Potatoes 20.0g
Carrots 20.0g

Preparation of Medium: Wash and peel potatoes and carrots. Grate potatoes and carrots and place in 1.0L of tap water. Gently heat and bring to boiling. Boil for 30 min. Filter through cheesecloth. Bring volume of filtrate to 1.0L with tap water. Gently heat and bring to boiling. Distribute into tubes or flasks. Autoclave for 20 min at 15 psi pressure–121°C.

Use: For the cultivation of *Alternaria species* and a variety of other fungi.

Potato Carrot Broth

Composition per liter:

Potato, infusion from 250.0g
Carrot, infusion from 200.0g

pH 6.5 ± 0.2 at 25°C

Preparation of Medium: Add components to distilled/deionized water and bring volume to 1.0L. Mix thoroughly. Gently heat and bring to boiling. Distribute into tubes or flasks. Autoclave for 15 min at 15 psi pressure–121°C.

Use: For the cultivation of *Alternaria* and a variety of other fungi.

Potato Carrot Medium

Composition per liter of tap water:

Potatoes, sliced with skin 300.0g
Carrots, peeled and sliced 25.0g
Agar 15.0g

Preparation of Medium: Slice potatoes with the skin on. Peel carrots and slice. Add potatoes and carrots to approximately 700.0mL of tap water. Gently heat and bring to boiling. Continue boiling for 20 min. Filter through cheesecloth. Bring volume of filtrate to 1.0L with distilled/deionized water. Add agar. Gently heat and bring to boiling. Distribute into tubes or flasks. Autoclave for 20 min at 15 psi pressure–121°C. Pour into sterile Petri dishes or leave in tubes.

Use: For the cultivation and maintenance of *Actinoplanes awajinensis, Actinoplanes nirasakinensis, Amorphosporangium auranticolor, Streptomyces flaveus,* and *Thermoactinomyces vulgaris*.

Potato Dextrose Adenine Agar (ATCC Medium 2204)

Composition per liter:

Glucose 20.0g
Agar 15.0g
Potatoes, infusion from 500.0mL
Adenine solution 50.0mL

pH 5.6 ± 0.2 at 25°C

Potato Infusion:

Composition per 500.0mL:

Potatoes 300.0g

Preparation of Potato Infusion: Peel and dice potatoes. Add 500.0mL of distilled/deionized water. Gently heat and bring to boiling. Continue boiling for 30 min. Filter through cheesecloth. Reserve filtrate.

Adenine Solution:

Composition per 50.0mL:

Adenine 0.1g

Preparation of Adenine Solution: Add adenine to distilled/deionized water and bring volume to 50.0mL. Mix thoroughly. Filter sterilize.

Preparation of Medium: Add components, except adenine solution, to distilled/deionized water and bring volume to 950.0L. Mix thoroughly. Gently heat and bring to boiling. Autoclave for 15 min at 15 psi pressure–121°C. Cool to 45°–50°C. Aseptically add 50.0mL of sterile adenine solution. Mix thoroughly. Pour into sterile Petri dishes or distriubte to sterile tubes.

Use: For the cultivation of yeasts and molds.

Potato Dextrose Agar

Composition per liter:

Glucose	20.0g
Agar	15.0g
Potato, infusion from	4.0g
Tartaric acid solution	14.0mL

pH 5.6 ± 0.2 at 25°C

Source: This medium is available as a premixed powder from BD Diagnostic Systems and Oxoid Unipath.

Tartaric Acid Solution:

Composition per 50.0mL:

Tartaric acid	5.0g

Preparation of Tartaric Acid Solution: Add tartaric acid to distilled/deionized water and bring volume to 50.0mL. Mix thoroughly. Filter sterilize.

Preparation of Medium: Add components to distilled/deionized water and bring volume to 986.0mL. Mix thoroughly. Gently heat and bring to boiling. Distribute into tubes or flasks. Autoclave for 15 min at 15 psi pressure–121°C. Cool to 45°–50°C. Aseptically add 14.0mL of sterile tartaric acid solution. Mix thoroughly. If medium is to be used for the enumeration of yeasts and molds in butter, adjust pH to 3.5. Pour into sterile Petri dishes or distribute into sterile tubes.

Use: For the cultivation and enumeration of yeasts and molds. For the enumeration of yeasts and molds in butter by the plate count method.

Potato Dextrose Agar

Composition per liter:

Agar	20.0g
Glucose	20.0g
Potato infusion	200.0mL

pH 5.6 ± 0.2 at 25°C

Potato Infusion:

Composition per 10.0mL:

Potatoes, unpeeled and sliced	200.0g

Preparation of Potato Infusion: Add potato slices to 1.0L of distilled/deionized water. Gently heat and bring to boiling. Continue boiling for 30 min. Filter through cheesecloth. Reserve filtrate.

Preparation of Medium: Add components to distilled/deionized water and bring volume to 1.0L. Mix thoroughly. Gently heat and bring to boiling. Distribute into tubes or flasks. Autoclave for 15 min at 15 psi pressure–121°C. Pour into sterile Petri dishes or leave in tubes.

Use: For thc cultivation and enumeration of yeasts and filamentous fungi (molds) from foods.

Potato Dextrose Agar

Composition per liter:

Potatoes	300.0g
Glucose	20.0g
Agar	15.0g

Preparation of Medium: Dice potatoes and place in 500.0mL of boiling water for 30 min. Strain through cheesecloth. Adjust volume to 1.0L with distilled/deionized water. Mix thoroughly. Add agar. Gently heat and bring to boiling. Add 20.0g of glucose. Mix thoroughly. Distribute into tubes or flasks. Autoclave for 15 min at 15 psi pressure–121°C. Pour into sterile Petri dishes or leave in tubes.

Use: For the cultivation and maintenance of numerous fungi such as *Cephaleuros virescens.*

Potato Dextrose Agar (PDA Agar)

Composition per liter:

Glucose	20.0g
Agar	15.0g
Potatoes, infusion from	1.0L

Potatoes, Infusion From:

Composition per liter:

Potatoes	300.0g

Preparation of Potatoes, Infusion From: Peel and dice potatoes. Add 500.0mL of distilled/deionized water. Gently heat and bring to boiling. Continue boiling for 30 min. Filter through cheesecloth. Bring volume of filtrate to 1.0L.

Preparation of Medium: To 1.0L of potato infusion, add glucose and agar. Mix thoroughly. Gently heat and bring to boiling. Distribute into tubes or flasks. Autoclave for 15 min at 15 psi pressure–121°C. Pour into sterile Petri dishes or leave in tubes.

Use: For the cultivation and maintenance of *Bacillus megaterium, Bacillus subtilis, Pseudomonas lindbergii, Pseudomonas syringae, Streptomyces testaceus,* and *Xanthomonas campestris.*

Potato Dextrose Agar (PDA Agar) (ATCC Medium 336)

Composition per liter:

Glucose	20.0g
Agar	15.0g
Potatoes, infusion from	500.0mL

pH 5.6 ± 0.2 at 25°C

Source: This medium is available as a premixed powder from BD Diagnostic Systems.

Potato Infusion:

Composition per 500.0mL:

Potatoes	300.0g

Preparation of Potato Infusion: Peel and dice potatoes. Add 500.0mL of distilled/deionized water. Gently heat and bring to boiling. Continue boiling for 30 min. Filter through cheesecloth. Reserve filtrate.

Preparation of Medium: Add components to distilled/deionized water and bring volume to 1.0L. Mix thoroughly. Gently heat and bring to boiling. Distribute into tubes or flasks. Autoclave for 15 min at 15 psi pressure–121°C. Pour into sterile Petri dishes or leave in tubes.

Use: For the cultivation of yeasts and molds from dairy products and other foods. Also used to induce sporulation in many fungi.

Potato Dextrose Agar (BAM M127)

Composition per liter:

Agar	20.0g
Glucose	20.0g
Potato infusion	1.0L

pH 5.6 ± 0.2 at 25°C

Potato Infusion:

Composition per liter:

Potatoes, unpeeled and sliced	200.0g

Preparation of Potato Infusion: Add unpeeled potato slices to 1.0L of distilled/deionized water. Gently heat and bring to boiling. Continue boiling for 30 min. Filter through cheesecloth. Reserve filtrate. Bring volume to 1.0L with distilled/deionized water.

Preparation of Medium: Add agar and glucose to 1.0L potato infusion. Mix thoroughly. Gently heat and bring to boiling. Distribute into tubes or flasks. Autoclave for 15 min at 15 psi pressure–121°C. Pour into sterile Petri dishes or leave in tubes.

Use: For the cultivation and enumeration of yeasts and filamentous fungi (molds) from foods.

Potato Dextrose Agar with Antibiotics

Composition per liter:

Glucose	20.0g
Agar	15.0g
Potato, infusion from	4.0g
Antibiotic solution	20.0mL

Antibiotic Solution:

Composition per 100.0mL:

Chlortetracycline·HCl	0.5g
Chloramphenicol	0.5g

Preparation of Antibiotic Solution: Add components to distilled/deionized water and bring volume to 100.0mL. Mix thoroughly. Filter sterilize.

Preparation of Medium: Add components, except antibiotic solution, to distilled/deionized water and bring volume to 980.0mL. Mix thoroughly. Gently heat and bring to boiling. Autoclave for 15 min at 15 psi pressure–121°C. Cool to 45°–50°C. Aseptically add 20.0mL of sterile antibiotic solution. Mix thoroughly. Pour into sterile Petri dishes.

Use: For the cultivation of fungi from foods.

Potato Dextrose Agar with Gentamicin (ATCC Medium 2286)

Composition per liter:

Glucose	20.0g
Agar	15.0g
Potatoes, infusion from	500.0mL
Gentamicin solution	10.0mL

pH 5.6 ± 0.2 at 25°C

Source: This medium is available as a premixed powder from BD Diagnostic Systems.

Potato Infusion:

Composition per 500.0mL:

Potatoes	300.0g

Preparation of Potato Infusion: Peel and dice potatoes. Add 500.0mL of distilled/deionized water. Gently heat and bring to boiling. Continue boiling for 30 min. Filter through cheesecloth. Reserve filtrate.

Gentamicin Solution:

Composition per 10.0mL:

Gentamycin	100.0mg

Preparation of Gentamicin Solution: Add gentamycin to distilled/deionized water and bring volume to 10.0mL. Mix thoroughly. Filter sterilize.

Preparation of Medium: Add components, except gentamicin, to distilled/deionized water and bring volume to 990.0L. Mix thoroughly. Gently heat and bring to boiling. Autoclave for 15 min at 15 psi pressure–121°C. Cool to 45°–50°C. Aseptically add 10.0mL of sterile gentamicin solution. Mix thoroughly. Pour into sterile Petri dishes or sterile tubes.

Use: For the cultivation of yeasts and molds from dairy products and other foods.

Potato Dextrose Agar with 2% Glucose and 60% Sucrose

Composition per liter:

Sucrose	600.0g
Glucose	20.0g
Agar	15.0g
Potato, solids from infusion	4.0g

Preparation of Medium: Add components to distilled/deionized water and bring volume to 1.0L. Mix thoroughly. Gently heat and bring to boiling. Distribute into tubes or flasks. Autoclave for 15 min at 15 psi pressure–121°C.

Use: For the isolation and cultivation of *Saccharomyces rouxii* from chocolate syrup.

Potato Dextrose Agar with Methionine and Biotin (PDAmb Agar)

Composition per liter:

Glucose	20.0g
Agar	15.0g
DL-Methionine	100.0mg
Biotin	1.0mg
Potato infusion	500.0mL

pH 5.6 ± 0.2 at 25°C

Potato Infusion:

Composition per 500.0mL:

Potatoes	300.0g

Preparation of Potato Infusion: Peel and dice potatoes. Add 500.0mL of distilled/deionized water. Gently heat and bring to boiling. Continue boiling for 30 min. Filter through cheesecloth. Reserve filtrate.

Preparation of Medium: Add components to distilled/deionized water and bring volume to 1.0L. Mix thoroughly. Gently heat and bring to boiling. Distribute into tubes or flasks. Autoclave for 15 min at 15 psi pressure–121°C. Pour into sterile Petri dishes or leave in tubes.

Use: For the cultivation of *Cryphonectria gyrosa* and *Cryphonectria parasitica.*

Potato Dextrose Agar, pH 5.0 (ATCC Medium 368)

Composition per liter:

Glucose	20.0g
Agar	15.0g
Potatoes, infusion from	500.0mL

pH 5.0 ± 0.2 at 25°C

Potato Infusion:

Composition per 500.0mL:

Potatoes	300.0g

Preparation of Potato Infusion: Peel and dice potatoes. Add 500.0mL of distilled/deionized water. Gently heat and bring to boiling.

Continue boiling for 30 min. Filter through cheesecloth. Reserve filtrate.

Preparation of Medium: Add components to distilled/deionized water and bring volume to 1.0L. Mix thoroughly. Adjust pH to 5.0. Gently heat and bring to boiling. Distribute into tubes or flasks. Autoclave for 15 min at 15 psi pressure–121°C. Pour into sterile Petri dishes or leave in tubes.

Use: For the cultivation of yeasts and molds from dairy products and other foods.

Potato Dextrose Agar with 7.5% Sodium Chloride

Composition per liter:

NaCl	75.0g
Glucose	20.0g
Agar	15.0g
Potato infusion	500.0mL

pH 5.6 ± 0.2 at 25°C

Potato Infusion:

Composition per 500.0mL:

Potatoes	300.0g

Preparation of Potato Infusion: Peel and dice potatoes. Add 500.0mL of distilled/deionized water. Gently heat and bring to boiling. Continue boiling for 30 min. Filter through cheesecloth. Reserve filtrate.

Preparation of Medium: Add components to distilled/deionized water and bring volume to 1.0L. Mix thoroughly. Gently heat and bring to boiling. Distribute into tubes or flasks. Autoclave for 15 min at 15 psi pressure–121°C. Pour into sterile Petri dishes or leave in tubes.

Use: For the cultivation and maintenance of *Aspergillus caesiellus, Aspergillus gracilis, Aspergillus mangini, Aspergillus montevidensis, Aspergillus niveus, Aspergillus penicilloides, Aspergillus repens, Aspergillus ruber, Aspergillus sulphureus, Aspergillus versicolor,* and *Eurotium chevalierii.*

Potato Dextrose Agar, 1/3 Strength (ATCC Medium 2369)

Composition per liter:

Glucose	6.67g
Agar	15.0g
Potatoes, infusion from	333.0mL

pH 5.6 ± 0.2 at 25°C

Potato Infusion:

Composition per liter:

Potatoes	300.0g

Preparation of Potato Infusion: Peel and dice potatoes. Add 500.0mL of distilled/deionized water. Gently heat and bring to boiling. Continue boiling for 30 min. Filter through cheesecloth. Bring volume to 1.0L. Reserve filtrate.

Preparation of Medium: Add components to distilled/deionized water and bring volume to 1.0L. Mix thoroughly. Gently heat and bring to boiling. Distribute into tubes or flasks. Autoclave for 15 min at 15 psi pressure–121°C. Pour into sterile Petri dishes or leave in tubes.

Use: For the cultivation of yeasts and molds.

Potato Dextrose Agar, 1/4 Strength (ATCC Medium 2218)

Composition per liter:

Glucose	20.0g
Agar	15.0g
Potatoes, infusion from	125.0mL

pH 5.6 ± 0.2 at 25°C

Potato Infusion:

Composition per 500.0mL:

Potatoes	300.0g

Preparation of Potato Infusion: Peel and dice potatoes. Add 500.0mL of distilled/deionized water. Gently heat and bring to boiling. Continue boiling for 60 min. Filter through cheesecloth. Reserve filtrate.

Preparation of Medium: Add components to distilled/deionized water and bring volume to 1.0L. Mix thoroughly. Gently heat and bring to boiling. Distribute into tubes or flasks. Autoclave for 15 min at 15 psi pressure–121°C. Pour into sterile Petri dishes or leave in tubes.

Use: For the cultivation of yeasts and molds from dairy products and other foods.

Potato Dextrose Agar and Yeast Medium (PDA and Yeast Medium) (ATCC Medium 104)

Composition per liter:

Glucose	20.0g
Agar	15.0g
KH_2PO_4	2.0g
Yeast extract	1.5g
$MgSO_4 \cdot 7H_2O$	0.5g
Potato infusion	500.0mL

Potato Infusion:

Composition per 500.0mL:

Potatoes	300.0g

Preparation of Potato Infusion: Peel and dice potatoes. Add 500.0mL of distilled/deionized water. Gently heat and bring to boiling. Continue boiling for 30 min. Filter through cheesecloth. Reserve filtrate.

Preparation of Medium: Add components to distilled/deionized water and bring volume to 1.0L. Mix thoroughly. Gently heat and bring to boiling. Distribute into tubes or flasks. Autoclave for 15 min at 15 psi pressure–121°C. Pour into sterile Petri dishes or leave in tubes.

Use: For the cultivation and maintenance of *Pseudomonas fluorescens, Streptoverticillium baldaccii,* and *Xanthomonas campestris.*

Potato Dextrose Agar with Thiamine (ATCC Medium 2391)

Composition per liter:

Glucose	20.0g
Agar	15.0g
Thiamine·HCl	2.0mg
Potatoes, infusion from	500.0mL

pH 5.6 ± 0.2 at 25°C

Potato Infusion:

Composition per 500.0mL:

Potatoes	300.0g

Preparation of Potato Infusion: Peel and dice potatoes. Add 500.0mL of distilled/deionized water. Gently heat and bring to boiling. Continue boiling for 60 min. Filter through cheesecloth. Bring volume to 500.0mL with distilled/deionized water.Reserve filtrate.

Preparation of Medium: Add components to distilled/deionized water and bring volume to 1.0L. Mix thoroughly. Gently heat and bring to boiling. Distribute into tubes or flasks. Autoclave for 15 min at 15 psi pressure–121°C. Pour into sterile Petri dishes or leave in tubes.

Use: For the cultivation of yeasts and molds.

Potato Dextrose Broth

Composition per liter:

Potatoes, infusion from	200.0g
Glucose	20.0g

pH 5.1 ± 0.2 at 25°C

Source: This medium is available as a premixed powder from BD Diagnostic Systems.

Potatoes, Infusion From:

Composition per 500.0mL:

Potatoes	300.0g

Preparation of Potatoes, Infusion From: Peel and dice potatoes. Add 500.0mL of distilled/deionized water. Gently heat and bring to boiling. Continue boiling for 30 min. Filter through cheesecloth.

Preparation of Medium: Add components to distilled/deionized water and bring volume to 1.0L. Mix thoroughly. Distribute into tubes or flasks. Autoclave for 15 min at 15 psi pressure–121°C.

Use: For the cultivation of a wide variety of yeasts and molds.

Potato Dextrose Broth with Yeast Extract

Composition per liter:

Potato, peeled and cut	200.0g
Glucose	10.0g
Yeast extract	3.0g

pH 7.2 ± 0.2 at 25°C

Preparation of Medium: Peel and cut potatoes. Add potatoes to 500.0mL of water. Boil potatoes for 20 min. Filter through cheesecloth. Add glucose and yeast extract to filtrate. Bring volume to 1.0L with distilled/deionized water . Mix thoroughly. Adjust pH to 7.2. Distribute into tubes or flasks. Autoclave for 15 min at 15 psi pressure–121°C.

Use: For the cultivation of a variety of fungi.

Potato Dextrose L-IsoleucineAgar (ATCC Medium 2205)

Composition per liter:

Glucose	20.0g
Agar	15.0g
Potatoes, infusion from	500.0mL
L-Isoleucine solution	50.0mL

pH 5.6 ± 0.2 at 25°C

Potato Infusion:

Composition per 500.0mL:

Potatoes	300.0g

Preparation of Potato Infusion: Peel and dice potatoes. Add 500.0mL of distilled/deionized water. Gently heat and bring to boiling. Continue boiling for 30 min. Filter through cheesecloth. Reserve filtrate.

L-Isoleucine Solution:

Composition per 50.0mL:

L-Isoleucine	0.13g

Preparation of L-Isoleucine Solution: Add L-isoleucine to distilled/deionized water and bring volume to 50.0mL. Mix thoroughly. Filter sterilize.

Preparation of Medium: Add components, except L-isoleucine solution, to distilled/deionized water and bring volume to 950.0L. Mix thoroughly. Gently heat and bring to boiling. Autoclave for 15 min at 15 psi pressure–121°C. Cool to 45°–50°C. Aseptically add 50.0mL of sterile L-isoleucine solution. Mix thoroughly. Pour into sterile Petri dishes or distribute to sterile tubes.

Use: For the cultivation of yeasts and molds.

Potato Dextrose Salt Agar (BAM M127)

Composition per liter:

NaCl	75.0g
Agar	20.0g
Glucose	20.0g
Potato infusion	1.0L

pH 5.6 ± 0.2 at 25°C

Potato Infusion:

Composition per liter:

Potatoes, unpeeled and sliced	200.0g

Preparation of Potato Infusion: Add unpeeled potato slices to 1.0L of distilled/deionized water. Gently heat and bring to boiling. Continue boiling for 30 min. Filter through cheesecloth. Reserve filtrate. Bring volume to 1.0L with distilled/deionized water.

Preparation of Medium: Add agar and glucose to 1.0L potato infusion. Mix thoroughly. Gently heat and bring to boiling. Distribute into tubes or flasks. Autoclave for 15 min at 15 psi pressure–121°C. Pour into sterile Petri dishes or leave in tubes.

Use: For the cultivation and enumeration of holophilic yeasts and halophilic filamentous fungi (molds) from foods.

Potato Dextrose Yeast Agar (PDY Agar)

Composition per liter:

Glucose	20.0g
Agar	15.0g
Yeast extract	5.0g
Potato infusion	500.0mL

pH 5.6 ± 0.2 at 25°C

Potato Infusion:

Composition per 500.0mL:

Potatoes	300.0g

Preparation of Potato Infusion: Peel and dice potatoes. Add 500.0mL of distilled/deionized water. Gently heat and bring to boiling. Continue boiling for 30 min. Filter through cheesecloth. Reserve filtrate.

Preparation of Medium: Add components to distilled/deionized water and bring volume to 1.0L. Mix thoroughly. Gently heat and bring to boiling. Distribute into tubes or flasks. Autoclave for 15 min at 15 psi pressure–121°C. Pour into sterile Petri dishes or leave in tubes.

Use: For the cultivation and maintenance of *Bacillus* species and fungi; also used to induce sporulation in many fungi.

Potato Extract Agar

Composition per liter:

Agar	15.0g
Peptone	5.0g
NaCl	5.0g
Yeast extract	2.0g
Beef extract powder	1.0g
Potato extract	20.0mL

pH 7.4 ± 0.2 at 25°C

Potato Extract:

Composition per liter:

Potatoes	300.0g

Preparation of Potato Extract: Peel and dice potatoes. Add 500.0mL of distilled/deionized water. Gently heat and bring to boiling. Continue boiling for 30 min. Filter through cheesecloth.

Use: For the cultivation of a wide variety of yeasts and molds.

Potato Flakes Agar

Composition per liter:

Potato flakes	20.0g
Agar	15.0g
Glucose	10.0g

Preparation of Medium: Add components to distilled/deionized water and bring volume to 1.0L. Mix thoroughly. Gently heat and bring to boiling. Distribute into tubes or flasks. Autoclave for 15 min at 15 psi pressure–121°C. Pour into sterile Petri dishes or leave in tubes.

Use: For the cultivation and induction of sporulation in all fungi.

Potato Glucose Agar

Composition per liter:

Potato, infusion from	500.0g
Glucose	20.0g
Agar	15.0g

Preparation of Medium: Peel and slice potatoes thinly. Add 800.0mL of distilled/deionized water immediately to potatoes to prevent oxidation. Gently heat and bring to 60°C. Maintain at 60°C for 60 min. Filter through cheesecloth. Adjust volume of filtrate to 1.0L with distilled/deionized water. Add agar. Gently heat and bring to boiling. Add glucose. Mix thoroughly. Distribute into tubes or flasks. Autoclave for 20 min at 10 psi pressure–115°C. Pour into sterile Petri dishes or leave in tubes.

Use: For the cultivation and maintenance of *Nocardia asteroides, Pseudomonas caryophylli, Pseudomonas syringae, Rhodococcus* species, *Streptomyces nobilis, Streptomyces prasinosporus,* and *Streptomyces* species.

Potato Infusion Agar

Composition per liter:

Potatoes, infusion from	200.0g
Agar	15.0g
Glucose	10.0g
Proteose peptone	10.0g
Beef extract	5.0g
NaCl	5.0g
Glycerol	20.0mL

pH 6.8 ± 0.2 at 25°C

Source: This medium is available as a premixed powder from BD Diagnostic Systems.

Potatoes, Infusion From:

Composition per 500.0mL:

Potatoes	300.0g

Preparation of Potatoes, Infusion From: Peel and dice potatoes. Add 500.0mL of distilled/deionized water. Gently heat and bring to boiling. Continue boiling for 30 min. Filter through cheesecloth.

Preparation of Medium: Add glycerol to 500.0mL of distilled/deionized water. Add remaining components. Mix thoroughly. Gently heat and bring to boiling. Distribute into tubes or flasks. Autoclave for 15 min at 15 psi pressure–121°C. Pour into sterile Petri dishes or leave in tubes.

Use: For the isolation of *Brucella abortus.*

Potato Infusion Agar (ATCC Medium 421)

Composition per liter:

Potato	200.0g
Agar	15.0g

Preparation of Medium: Peel and finely dice potatoes. Add to 500.0mL of distilled/deionized water. Gently heat and bring to boiling. Continue boiling for 20 min. Filter through cheesecloth. Bring volume of filtrate to 1.0L with distilled/deionized water. Add agar. Gently heat and bring to boiling. Distribute into tubes or flasks. Autoclave for 15 min at 15 psi pressure–121°C. Pour into sterile Petri dishes or leave in tubes.

Use: For the cultivation and maintenance of *Streptomyces fradiae.*

Potato Infusion HiVeg Agar

Composition per liter:

Potato, infusion from	200.0g
Agar	15.0g
Glucose	10.0g
Plant peptone	10.0g
Plant extract	5.0g
NaCl	5.0g

pH 6.8 ± 0.2 at 25°C

Source: This medium is available as a premixed powder from HiMedia.

Preparation of Medium: Add components to distilled/deionized water and bring volume to 1.0L. Mix thoroughly. Gently heat and bring to boiling. Distribute into tubes or flasks. Autoclave for 15 min at 15 psi pressure–121°C. Pour into sterile Petri dishes or leave in tubes.

Use: For the cultivation and maintenance of fungi and other aciduric microorganisms.

Potato Infusion with Inorganic Salts

Composition per liter:

Potato	200.0g
Agar	15.0g
K_2HPO_4	0.5g
$MgSO_4 \cdot 7H_2O$	0.4g
$CaCl_2 \cdot 2H_2O$	0.1g
$ZnSO_4 \cdot 7H_2O$	0.03g

$MnSO_4 \cdot 5H_2O$ 0.02g
$CuSO_4 \cdot 5H_2O$ 0.01g
$FeSO_4 \cdot 7H_2O$ 0.01g

Preparation of Medium: Peel and dice potatoes. Add 500.0mL of distilled/deionized water. Gently heat and bring to boiling. Continue boiling for 30 min. Filter through cheesecloth. Bring volume of filtrate to 1.0L. Add agar. Mix thoroughly. Gently heat and bring to boiling. Add remaining components. Mix thoroughly. Distribute into tubes or flasks. Autoclave for 15 min at 15 psi pressure–121°C. Pour into sterile Petri dishes or leave in tubes.

Use: For the cultivation and maintenance of *Bacillus macquariensis*.

Potato Malt Agar

Composition per liter:

Potatoes, infusion from 200.0g
Sucrose 60.0g
Agar 20.0g
Malt extract 20.0g
Peptone 1.0g

pH 7.4 ± 0.2 at 25°C

Source: This medium is available as a premixed powder from BD Diagnostic Systems.

Potatoes, Infusion From:

Composition per 500.0mL:

Potatoes 300.0g

Preparation of Potatoes, Infusion From: Peel and dice potatoes. Add 500.0mL of distilled/deionized water. Gently heat and bring to boiling. Continue boiling for 30 min. Filter through cheesecloth.

Preparation of Medium: Add components to distilled/deionized water and bring volume to 1.0L. Mix thoroughly. Gently heat and bring to boiling. Distribute into tubes or flasks. Autoclave for 15 min at 15 psi pressure–121°C. Pour into sterile Petri dishes or leave in tubes.

Use: For the cultivation and maintenance of fungi and other aciduric microorganisms.

Potato Malt Agar with Filter Paper

Composition per liter:

Potatoes 240.0g
Agar 15.0g
Malt extract 5.0g
Filter paper .... variable

Preparation of Medium: Wash and peel potatoes. Dice potatoes and place in 1.0L of tap water. Gently heat and bring to boiling. Boil for 30 min. Filter through cheesecloth. Bring volume of filtrate to 1.0L with tap water. Add 15.0g of agar and 5.0g of malt extract. Gently heat and bring to boiling. Distribute into tubes. Add a strip of white filter paper to each tube. Autoclave for 20 min at 15 psi pressure–121°C.

Use: For the cultivation and maintenance of *Chaetomium cochliodes* and *Chaetomium globosum.*

Potato Malt HiVeg Agar

Composition per liter:

Potatoes, infusion from 200.0g
Sucrose 60.0g
Agar 20.0g
Malt extract 20.0g
Plant peptone 1.0g

pH 7.4 ± 0.2 at 25°C

Source: This medium is available as a premixed powder from HiMedia.

Preparation of Medium: Add components to distilled/deionized water and bring volume to 1.0L. Mix thoroughly. Gently heat while stirring and bring to boiling. Distribute into tubes or flasks. Autoclave for 15 min at 15 psi pressure–121°C. Pour into sterile Petri dishes or leave in tubes.

Use: For the cultivation of yeasts and molds. For the cultivation and maintenance of smut fungi and other phytopathogenic fungi.

Potato Medium

Composition per liter:

Potato 60.0g
Agar 15.0g
Glucose 10.0g
Peptone 10.0g
Yeast extract 5.0g
$CaCO_3$ 1.0g

Preparation of Medium: Peel and dice potato. Homogenize in a blender. Add potato and remaining components to distilled/deionized water and bring volume to 1.0L. Mix thoroughly. Gently heat and bring to boiling. Distribute into tubes or flasks. Autoclave for 15 min at 15 psi pressure–121°C. Pour into sterile Petri dishes or leave in tubes.

Use: For the cultivation and maintenance of *Clostridium laniganii.*

Potato P-YE *Thermus* Medium

Composition per liter:

Agar 20.0g
Peptone 5.0g
Yeast extract 0.2g
Potatoes, infusion from 200.0mL

pH 7.8 ± 0.2 at 25°C

Potatoes, Infusion From:

Composition per 500.0mL:

Potatoes 300.0g

Preparation of Medium: Add components to distilled/deionized water and bring volume to 1.0L. Mix thoroughly. Gently heat and bring to boiling. Distribute into tubes or flasks. Autoclave for 15 min at 15 psi pressure–121°C. Pour into sterile Petri dishes or leave in tubes.

Use: For the cultivation and maintenance of *Thermus ruber*.

Potato Sucrose Agar

Composition per liter:

Potato extract 200.0g
Sucrose 20.0g
Agar 20.0g

pH 6.5 ± 0.2 at 25°C

Preparation of Medium: Wash and peel potatoes. Dice potatoes and place in a muslin bag. Suspend bag in 500.0mL of tap water. Gently heat and bring to boiling. Boil for 10 min. Remove the muslin bag with the potatoes. Bring volume of potato extract to 1.0L with tap water. Add 20.0g of agar and 5.0g of malt extract. Mix thoroughly. Adjust pH to 6.5 with $CaCO_3$. Gently heat and bring to boiling. Distribute into tubes or flasks. Autoclave for 20 min at 15 psi pressure–121°C. Pour into sterile Petri dishes or leave in tubes.

Use: For the cultivation and maintenance of *Candida famata, Colletotrichum capsici, Colletotrichum coccodes, Colletotrichum crassipes, Colletotrichum dematium, Colletotrichum gloesporioides, Microdochium nivale,* and numerous *Fusarium* species.

Potato Yeast Agar

Composition per liter:

Diced potatoes	300.0g
Glucose	20.0g
Agar	15.0g
Yeast extract	5.0g

Preparation of Medium: Dice potatoes and place in 500.0mL of boiling water for 30 min. Strain through cheesecloth. Adjust volume to 1.0L with distilled/deionized water. Mix thoroughly. Add agar. Gently heat and bring to boiling. Add 20.0g of glucose. Mix thoroughly. Distribute into tubes or flasks. Autoclave for 15 min at 15 psi pressure–121°C. Pour into sterile Petri dishes or leave in tubes.

Use: For the cultivation and maintenance of *Paecilomyces fumosoroseus.*

Powell and Errington's Medium

Composition per 1060.0mL:

Solution 1	50.0mL
Solution 3	50.0mL
Solution 4	10.0mL
Solution 2	5.0mL

pH 7.0 ± 0.2 at 25°C

Solution 1:

Composition per liter:

$(NH_4)_2HPO_4$	238.0g
K_2SO_4	70.0g
$NaH_2PO_4 \cdot 2H_2O$	31.0g

Preparation of Solution 1: Add components to distilled/deionized water and bring volume to 1.0L. Mix thoroughly.

Solution 2:

Composition per liter:

MgO	10.0g
$FeCl_3 \cdot 6H_2O$	5.4g
$CaCO_3$	2.0g
$ZnSO_4 \cdot 7H_2O$	1.44g
$MnSO_4 \cdot 4H_2O$	1.11g
$Na_2MoO_4 \cdot 2H_2O$	0.49g
$CoSO_4 \cdot 7H_2O$	0.28g
$CuSO_4 \cdot 5H_2O$	0.25g
H_3BO_4	0.062g
HCl, concentrated	50.0mL

Preparation of Solution 2: Add components to distilled/deionized water and bring volume to 1.0L. Mix thoroughly.

Solution 3:

Composition per 50.0mL:

Citric acid	4.2g
Glucose	3.6g
L-Glutamic acid	2.94g
Succinic acid	1.18g

Preparation of Solution 3: Add components to distilled/deionized water and bring volume to 50.0mL. Mix thoroughly. Filter sterilize.

Solution 4:

Composition per 10.0mL:

$Na_2S_2O_3 \cdot 5H_2O$	1.24g

Preparation of Solution 4: Add $Na_2S_2O_3 \cdot 5H_2O$ to distilled/deionized water and bring volume to 10.0mL. Mix thoroughly. Filter sterilize.

Preparation of Medium: Add 50.0mL of solution 1 and 5.0mL of solution 2. Mix thoroughly. Bring volume to 1.0L with distilled/deionized water. Autoclave for 15 min at 15 psi pressure–121°C. Cool to 25°C. Adjust pH to 7.0 with sterile NaOH. Aseptically add 50.0mL of solution 3 and 10.0mL of sterile solution 4. Mix thoroughly. Aseptically distribute into sterile tubes or flasks.

Use: For the cultivation of a variety of heterotrophic microorganisms.

PP Agar

Composition per liter:

Agar	15.0g
Polypeptone™	10.0g
Yeast extract	2.0g
$MgSO_4 \cdot 7H_2O$	1.0g

pH 6.8 ± 0.2 at 25°C

Preparation of Medium: Add components to distilled/deionized water and bring volume to 1.0L. Mix thoroughly. Gently heat and bring to boiling. Distribute into tubes or flasks. Autoclave for 15 min at 15 psi pressure–121°C. Pour into sterile Petri dishes or leave in tubes.

Use: For the cultivation and maintenance of *Bacillus sphaericus.*

PP Medium

Composition per liter:

Proteose peptone	10.0g
Pancreatic digest of peptone	10.0g
Ribonucleic acid from *Torula* yeast	1.0g
Asolectin	0.2g
Artificial seawater	167.0mL
Vitamin solution	2.0mL

Artificial Seawater:

Composition per 167.0mL:

Aqua-Marin sea salts	6.95g

Source: Aqua-Marin sea salts are available from Aquatrol, Inc., Anaheim, CA.

Preparation of Artificial Seawater: Add Aqua-Marin sea salts to distilled/deionized water and bring volume to 167.0mL. Mix thoroughly. Filter sterilize.

Vitamin Solution:

Composition per 100.0mL:

Thiamine·HCl	150.0mg
Calcium D-(+)-pantothenate	100.0mg
Folic acid	50.0mg
Nicotinamide	50.0mg
Pyridoxal·HCl	50.0mg
Riboflavin	50.0mg
DL-6 Thioctic acid	1.0mg
Biotin solution	10.0mL

Preparation of Vitamin Solution: Add components to distilled/deionized water and bring volume to 100.0mL. Mix thoroughly. Filter sterilize. For long-term storage, preserve under nitrogen at –20°C.

Biotin Solution:

Composition per 10.0mL:

Biotin 0.01mg

Preparation of Biotin Solution: Add biotin to 10.0mL of absolute ethanol. Mix thoroughly.

Preparation of Medium: Add ascolectin to 500.0mL of distilled/deionized water. Gently heat to 80°C. Mix thoroughly. Add other components, except artificial seawater and vitamin solution, to distilled/deionized water and bring volume to 831.0mL. Mix thoroughly. Adjust pH to 7.2. Autoclave for 15 min at 15 psi pressure–121°C. Aseptically add 167.0mL of sterile artificial seawater and 2.0mL of sterile vitamin solution. Mix thoroughly. Aseptically distribute into sterile tubes or flasks.

Use: For the cultivation of *Potomacus pottsi*.

PP Starch Medium

Composition per liter:

Polypeptone™ 10.0g
Soluble starch 10.0g
K_2HPO_4 .... 3.0g
$MgSO_4 \cdot 7H_2O$ 1.0g

Preparation of Medium: Add components to distilled/deionized water and bring volume to 1.0L. Mix thoroughly. Gently heat while stirring and bring to boiling. Distribute into tubes or flasks. Autoclave for 15 min at 15 psi pressure–121°C.

Use: For the cultivation and maintenance of *Bacillus mycoides*.

PPB, Modified Caldwell and Bryant

Composition per liter:

Pancreatic digest of casein 2.0g
Yeast extract 2.0g
Cellobiose 1.0g
Glucose 1.0g
Maltose 1.0g
Starch 1.0g
Resazurin 1.0mg
Rumen fluid, clarified 150.0mL
Mineral solution I 100.0mL
Mineral solution II 100.0mL
Na_2CO_3 solution 50.0mL
Hemin solution 10.0mL
L-Cysteine·HCl solution 10.0mL
Volatile fatty acid mixture 3.1mL

pH 6.8 ± 0.2 at 25°C

Mineral Solution I:

Composition per 100.0mL:

K_2HPO_4 .... 0.2g

Preparation of Mineral Solution I: Add K_2HPO_4 to distilled/deionized water and bring volume to 100.0mL. Mix thoroughly.

Mineral Solution II:

Composition per 100.0mL:

NaCl 0.4g
$(NH_4)_2SO_4$ 0.4g
KH_2PO_4 .... 0.3g
$MgSO_4 \cdot 7H_2O$ 0.09g
$CaCl_2$ 0.05g

Preparation of Mineral Solution II: Add components to distilled/deionized water and bring volume to 100.0mL. Mix thoroughly.

Na_2CO_3 Solution:

Composition per 100.0mL:

Na_2CO_3 .... 8.0g

Preparation of Na_2CO_3 Solution: Add Na_2CO_3 to distilled/deionized water and bring volume to 100.0mL. Mix thoroughly. Sparge with 100% CO_2. Autoclave for 15 min at 15 psi pressure–121°C.

Hemin Solution:

Composition per 100.0mL:

Hemin 1.0g
NaOH (1*N* solution) 10.0mL

Preparation of Hemin Solution: Add components to 100.0mL of distilled/deionized water. Mix thoroughly.

L-Cysteine·HCl Solution:

Composition per 10.0mL:

L-Cysteine·HCl 0.25g

Preparation of L-Cysteine·HCl Solution: Add L-cysteine·HCl to distilled/deionized water and bring volume to 10.0mL. Mix thoroughly. Autoclave under 100% N_2 for 15 min at 15 psi pressure–121°C.

Volatile Fatty Acid Mixture:

Composition per 31.0mL:

Acetic acid 17.0mL
Propionic acid 6.0mL
Butyric acid 4.0mL
DL–α-Methylbutyric acid 1.0mL
Isobutyric acid 1.0mL
Isovaleric acid 1.0mL
n-Valeric acid 1.0mL

Preparation of Volatile Fatty Acid Mixture: Combine components. Mix thoroughly. Store under 100% N_2.

Preparation of Medium: Prepare and dispense medium under 100% CO_2. Add components, except L-cysteine·HCl solution and Na_2CO_3 solution, to distilled/deionized water and bring volume to 930.0mL. Mix thoroughly. Sparge with 100% CO_2. Adjust pH to 6.8 with 1*N* NaOH. Distribute anaerobically 9.3mL volumes into Hungate tubes. Autoclave for 15 min at 15 psi pressure–121°C. Aseptically and anaerobically add 0.2mL of sterile L-cysteine·HCl solution and 0.5mL of sterile Na_2CO_3 solution. Check that final pH is 6.8. (Note: if not properly gassed with 100% CO_2, the medium pH can be as high as 9.5.)

Use: For the cultivation of *Anaerovibrio lipolytica, Bacteroides* species, *Butyrivibrio crossotus, Butyrivibrio fibrisolvens, Eubacterium cellulosolvans, Eubacterium ruminantium, Fibrobacter succinogenes, Lachnospira multiparus, Megasphaera elsdenii, Rhodococcus torques, Ruminobacter amylophilus, Ruminococcus albus, Ruminococcus bromii, Ruminococcus flavifaciens, Selenomonas ruminantium, Succinomonus amylolytica, Succinovibrio dextrinisolvens,* and *Veillonella parvula.*

PPES-II Agar Medium (DSMZ Medium 1075)

Composition per liter:

Agar 15.0g
Peptone 2.0g
Proteose peptone No. 3 .... 1.0g
Soytone 1.0g
Yeast extract 1.0g

Fe(III)-EDTA 0.1g
Artificial seawater 1.0L

pH 7.8 ± 0.2 at 25°C

Artificial Seawater:
Composition per 100.0mL:

$NaCl$ 30.0g
$MgCl_2 \cdot 6H_2O$ 10.8g
$MgSO_4 \cdot 7H_2O$ 5.4g
$CaCl_2 \cdot 2H_2O$ 1.0g
KCl 0.7g

Preparation of Artificial Seawater: Add components to distilled/deionized water and bring volume to 1.0L. Mix thoroughly.

Preparation of Medium: Add components to the artificial seawater and bring volume to 1.0L. Mix thoroughly. Adjust pH to 7.8. Distribute into tubes or flasks. Gently heat while stirring and bring to boiling. Mix thoroughly. Autoclave for 15 min at 15 psi pressure–121°C. Pour into Petri dishes or leave in tubes.

Use: For the cultivation of *Roseivivax halotolerans.*

PPES II Medium

Composition per liter:

Agar 15.0g
Polypeptone™ 2.0g
Yeast extract 1.0g
Papaic digest of soybean meal 1.0g
Proteose peptone No. 3 .... 1.0g
Ferric phosphate, soluble 0.1g
Marine mud extract 100.0mL

pH 7.6 ± 0.2 at 25°C

Preparation of Medium: Add components to distilled/deionized water and bring volume to 1.0L. Mix thoroughly. Gently heat and bring to boiling. Distribute into tubes or flasks. Autoclave for 15 min at 15 psi pressure–121°C. Pour into sterile Petri dishes or leave in tubes.

Use: For the cultivation of *Erythrobacter longus, Haloferax mediterranei,* and *Roseobacter denitrificans.*

PPGA Medium

Composition per liter:

Agar 18.0g
Glucose 5.0g
Peptone 5.0g
NaCl 3.0g
Na_2HPO_4 .... 1.2g
KH_2PO_4 .... 0.5g
Potato decoction 1.0L

pH 7.0 ± 0.2 at 25°C

Potato Decoction:
Composition per liter:

Potatoes 200.0g

Preparation of Potato Decoction: Peel and dice potatoes. Add 1.0L of distilled/deionized water. Gently heat and bring to boiling. Continue boiling for 20 min. Filter through two layers of cheesecloth. Bring volume of filtrate to 1.0L with distilled/deionized water.

Preparation of Medium: Combine components. Gently heat and bring to boiling. Adjust pH to 7.0. Distribute into tubes or flasks. Autoclave for 15 min at 15 psi pressure–121°C. Pour into sterile Petri dishes or leave in tubes.

Use: For the cultivation of *Burkholderia glumae* and *Burkholderia plantarii.*

PPLO Agar

Composition per liter:

Agar 14.0g
Beef heart, infusion from 50g 6.0g
NaCl 5.0g
Mycoploasma supplement solution 300.0mL

pH 7.8 ± 0.2 at 25°C

Source: This medium is available as a premixed powder from BD Diagnostic Systems.

***Mycoplasma* Supplement Solution:**
Composition per 300.0mL:

Horse serum, desiccated 16.0g
Yeast extract 0.1g

Preparation of *Mycoplasma* Supplement Solution: Add components to distilled/deionized water and bring volume to 300.0mL. Mix thoroughly. Filter sterilize.

Preparation of Medium: Add components, except *Mycoplasma* supplement solution, to distilled/deionized water and bring volume to 700.0mL. Mix thoroughly. Gently heat and bring to boiling. Boil for 1 min. Autoclave for 15 min at 15 psi pressure–121°C. Cool to 45°–50°C. Aseptically add sterile*Mycoplasma* supplement solution. Mix thoroughly. Pour into sterile Petri dishes or distribute into sterile tubes.

Use: For the cultivation of *Mycoplasma* species.

PPLO Agar

Composition per liter:

Agar 14.0g
Pancreatic digest of casein 7.0g
NaCl 5.0g
Beef extract 3.0g
Yeast extract 3.0g
Beef heart, solids from infusion 2.0g
Mycoploasma supplement solution 300.0mL

pH 7.8 ± 0.2 at 25°C

Source: This medium is available as a premixed powder from BD Diagnostic Systems.

***Mycoplasma* Supplement Solution:**
Composition per 300.0mL:

Horse serum 200.0mL
Yeast extract (fresh autolysate) 100.0mL
Thallium acetate .... 50.0 mg

Preparation of *Mycoplasma* Supplement Solution: Combine components. Mix thoroughly. Filter sterilize

Preparation of Medium: Add components, except *Mycoplasma* supplement solution, to distilled/deionized water and bring volume to 700.0mL. Mix thoroughly. Gently heat and bring to boiling. Boil for 1 min. Autoclave for 15 min at 15 psi pressure–121°C. Cool to 45°–50°C. Aseptically add sterile *Mycoplasma* supplement solution. Mix thoroughly. Pour into sterile Petri dishes or distribute into sterile tubes.

Use: For the cultivation of *Mycoplasma* species.

PPLO Agar

Composition per liter:

Beef heart, infusion from	50.0g
Agar	14.0g
Peptone	10.0g
NaCl	5.0g
Bovine serum	100.0mL

pH 7.8 ± 0.2 at 25°C

Preparation of Medium: Add components, except bovine serum, to distilled/deionized water and bring volume to 900.0mL. Mix thoroughly. Gently heat and bring to boiling. Autoclave for 15 min at 15 psi pressure–121°C. Cool to 45°–50°C. Aseptically add sterile bovine serum. Mix thoroughly. Pour into sterile Petri dishes or distribute into sterile tubes.

Use: For the isolation and cultivation of *Mycoplasma* species (pleuropneumonia-like organisms).

PPLO Agar

Composition per liter:

Beef heart, infusion from	50.0g
Agar	14.0g
Peptone	10.0g
NaCl	5.0g
Ascitic fluid	250.0mL

pH 7.8 ± 0.2 at 25°C

Preparation of Medium: Add components, except ascitic fluid, to distilled/deionized water and bring volume to 750.0mL. Mix thoroughly. Gently heat and bring to boiling. Autoclave for 15 min at 15 psi pressure–121°C. Cool to 45°–50°C. Aseptically add sterile ascitic fluid. Mix thoroughly. Pour into sterile Petri dishes or distribute into sterile tubes.

Use: For the isolation and cultivation of *Mycoplasma* species (pleuropneumonia-like organisms).

PPLO Agar Base

See: ***Mycoplasma*** **Agar Base**

PPLO Agar with Additives for *Mycoplasma*

Composition per 1010.0mL:

Agar	15.0g
Arginine	1.74g
Glutamine	1.46g
Phenol Red	0.02g
PPLO broth	700.0mL
Horse serum, not inactivated	200.0mL
Yeast extract solution, fresh	100.0mL
Vitamins in Eagle's medium, 100X	10.0mL

pH 7.1 ± 0.2 at 25°C

PPLO Broth:

Composition per liter:

Beef heart, infusion from	50.0g
Peptone	10.0g
NaCl	5.0g

Preparation of PPLO Broth: Add components to distilled/deionized water and bring volume to 1.0L. Mix thoroughly. Gently heat and bring to boiling.

Yeast Extract Solution:

Composition per 300.0mL:

Baker's yeast, live, pressed, starch-free	75.0g

Preparation of Yeast Extract Solution: Add the live Baker's yeast to 300.0mL of distilled/deionized water. Autoclave for 90 min at 15 psi pressure–121°C. Allow to stand. Remove supernatant solution. Adjust pH to 6.6–6.8. Filter sterilize.

Vitamins in Eagle's Medium, 100X:

Composition per liter:

Inositol	2.0mg
Calcium pantothenate	1.0mg
Choline chloride	1.0mg
Folic acid	1.0mg
Nicotinamide	1.0mg
Pyridoxal	1.0mg
Thiamine·HCl	1.0mg
Riboflavin	0.1mg

Preparation of Vitamins in Eagle's Medium, 100X: Add components to distilled/deionized water and bring volume to 1.0L. Mix thoroughly. Filter sterilize.

Preparation of Medium: Add components—except fresh yeast extract solution, horse serum, and vitamins in Eagle's medium, 100X—to distilled/deionized water and bring volume to 690.0mL. Mix thoroughly. Gently heat and bring to boiling. Autoclave for 15 min at 15 psi pressure–121°C. Cool to 45°–50°C. Aseptically add sterile fresh yeast extract solution, horse serum, and vitamins in Eagle's medium, 100X. Mix thoroughly. Pour into sterile Petri dishes or distribute into sterile tubes.

Use: For the cultivation and maintenance of *Mycoplasma arginini* and *Spiroplasma apis*.

PPLO Agar, pH 7.6 with Additives for *Mycoplasma*

Composition per liter:

Agar	15.0g
L-Cysteine·HCl·H_2O	1.0g
PPLO broth	700.0mL
Horse serum, not inactivated	200.0mL
Yeast extract solution, fresh	100.0mL

pH 7.6 ± 0.2 at 25°C

PPLO Broth:

Composition per liter:

Beef heart, infusion from	50.0g
Peptone	10.0g
NaCl	5.0g

Preparation of PPLO Broth: Add components to distilled/deionized water and bring volume to 1.0L. Mix thoroughly. Gently heat and bring to boiling.

Yeast Extract Solution:

Composition per 300.0mL:

Baker's yeast, live, pressed, starch-free	75.0g

Preparation of Yeast Extract Solution: Add the live Baker's yeast to 300.0mL of distilled/deionized water. Autoclave for 90 min at 15 psi pressure–121°C. Allow to stand. Remove supernatant solution. Adjust pH to 6.6–6.8. Filter sterilize.

Preparation of Medium: Add components, except fresh yeast extract solution and horse serum, to distilled/deionized water and bring volume to 700.0mL. Mix thoroughly. Gently heat and bring to boiling.

Autoclave for 15 min at 15 psi pressure–121°C. Cool to 45°–50°C. Aseptically add sterile fresh yeast extract solution and horse serum. Mix thoroughly. Pour into sterile Petri dishes or distribute into sterile tubes.

Use: For the cultivation and maintenance of *Mycoplasma faucium*.

PPLO Broth

Composition per liter:

Beef heart, infusion from 50g	6.0g
NaCl	5.0g
Mycoploasma supplement solution	300.0mL

pH 7.8 ± 0.2 at 25°C

Source: This medium is available as a premixed powder from BD Diagnostic Systems.

***Mycoplasma* Supplement Solution:**

Composition per 300.0mL:

Horse serum, desiccated	16.0g
Yeast extract	0.1g

Preparation of *Mycoplasma* Supplement Solution: Add components to distilled/deionized water and bring volume to 300.0mL. Mix thoroughly. Filter sterilize

Preparation of Medium: Add components, except *Mycoplasma* supplement solution, to distilled/deionized water and bring volume to 700.0mL. Mix thoroughly. Gently heat and bring to boiling. Boil for 1 min. Autoclave for 15 min at 15 psi pressure–121°C. Cool to 45°–50°C. Aseptically add sterile *Mycoplasma* supplement solution. Mix thoroughly.

Use: For the cultivation of *Mycoplasma* species.

PPLO Broth

Composition per liter:

Pancreatic digest of casein	7.0g
NaCl	5.0g
Beef extract	3.0g
Yeast extract	3.0g
Beef heart, solids from infusion	2.0g
Mycoploasma supplement solution	300.0mL

pH 7.8 ± 0.2 at 25°C

Source: This medium is available as a premixed powder from BD Diagnostic Systems.

***Mycoplasma* Supplement Solution:**

Composition per 300.0mL:

Horse serum	200.0mL
Yeast extract (fresh autolysate)	100.0mL
Thallium acetate	50.0 mg

Preparation of *Mycoplasma* Supplement Solution: Combine components. Mix thoroughly. Filter sterilize

Preparation of Medium: Add components, except *Mycoplasma* supplement solution, to distilled/deionized water and bring volume to 700.0mL. Mix thoroughly. Gently heat and bring to boiling. Boil for 1 min. Autoclave for 15 min at 15 psi pressure–121°C. Cool to 45°–50°C. Aseptically add sterile *Mycoplasma* supplement solution. Mix thoroughly.

Use: For the cultivation of *Mycoplasma* species.

PPLO Broth, pH 7.6 with Additives for *Mycoplasma*

Composition per liter:

L-Cysteine·HCl·H_2O	1.0g
PPLO broth	700.0mL
Horse serum, not inactivated	200.0mL
Yeast extract solution, fresh	100.0mL

pH 7.6 ± 0.2 at 25°C

PPLO Broth:

Composition per liter:

Beef heart, infusion from	50.0g
Peptone	10.0g
NaCl	5.0g

Preparation of PPLO Broth: Add components to distilled/deionized water and bring volume to 1.0L. Mix thoroughly. Gently heat and bring to boiling.

Yeast Extract Solution:

Composition per 300.0mL:

Baker's yeast, live, pressed, starch-free	75.0g

Preparation of Yeast Extract Solution: Add the live Baker's yeast to 300.0mL of distilled/deionized water. Autoclave for 90 min at 15 psi pressure–121°C. Allow to stand. Remove supernatant solution. Adjust pH to 6.6–6.8. Filter sterilize.

Preparation of Medium: Add components, except fresh yeast extract solution and horse serum, to distilled/deionized water and bring volume to 700.0mL. Mix thoroughly. Gently heat and bring to boiling. Autoclave for 15 min at 15 psi pressure–121°C. Cool to 45°–50°C. Aseptically add sterile fresh yeast extract solution and horse serum. Mix thoroughly. Aseptically distribute into sterile tubes or flasks.

Use: For the cultivation and maintenance of *Mycoplasma faucium*.

PPLO Broth with Additives for *Mycoplasma*

Composition per 1010.0mL:

Arginine	1.74g
Glutamine	1.46g
Phenol Red	0.02g
PPLO broth	700.0mL
Horse serum, not inactivated	200.0mL
Yeast extract solution, fresh	100.0mL
Vitamins in Eagle's medium, 100X	10.0mL

pH 7.1 ± 0.2 at 25°C

PPLO Broth:

Composition per liter:

Beef heart, infusion from	50.0g
Peptone	10.0g
NaCl	5.0g

Preparation of PPLO Broth: Add components to distilled/deionized water and bring volume to 1.0L. Mix thoroughly. Gently heat and bring to boiling.

Yeast Extract Solution:

Composition per 300.0mL:

Baker's yeast, live, pressed, starch-free	75.0g

Preparation of Yeast Extract Solution: Add the live Baker's yeast to 300.0mL of distilled/deionized water. Autoclave for 90 min at 15 psi pressure–121°C. Allow to stand. Remove supernatant solution. Adjust pH to 6.6–6.8. Filter sterilize.

Vitamins in Eagle's Medium, 100X:

Composition per liter:

Inositol	2.0mg
Calcium pantothenate	1.0mg
Choline chloride	1.0mg
Folic acid	1.0mg
Nicotinamide	1.0mg
Pyridoxal	1.0mg
Thiamine·HCl	1.0mg
Riboflavin	0.1mg

Preparation of Vitamins in Eagle's Medium, 100X: Add components to distilled/deionized water and bring volume to 1.0L. Mix thoroughly. Filter sterilize.

Preparation of Medium: Add components—except fresh yeast extract solution, horse serum, and vitamins in Eagle's medium, 100X—to distilled/deionized water and bring volume to 690.0mL. Mix thoroughly. Gently heat and bring to boiling. Autoclave for 15 min at 15 psi pressure–121°C. Cool to 45°–50°C. Aseptically add sterile fresh yeast extract solution, horse serum, and vitamins in Eagle's medium, 100X. Mix thoroughly. Aseptically distribute into sterile tubes or flasks.

Use: For the cultivation of *Mycoplasma arginini* and *Spiroplasma apis*.

PPLO Broth with Bovine Serum

Composition per liter:

Beef heart, infusion from	50.0g
Peptone	10.0g
NaCl	5.0g
Phenol Red (1% solution)	2.0mL
Yeast extract solution, fresh	100.0mL
Glucose solution	25.0mL
Bovine serum, filter sterilized	10.0mL

pH 7.5 ± 0.2 at 25°C

Yeast Extract Solution:

Composition per 300.0mL:

Baker's yeast, live, pressed, starch-free	75.0g

Preparation of Yeast Extract Solution: Add the live Baker's yeast to 300.0mL of distilled/deionized water. Autoclave for 90 min at 15 psi pressure–121°C. Allow to stand. Remove supernatant solution. Adjust pH to 6.6–6.8. Filter sterilize.

Glucose Solution:

Composition per 100.0mL:

D-Glucose	20.0g

Preparation of Glucose Solution: Add glucose to distilled/deionized water and bring volume to 100.0mL. Mix thoroughly. Filter sterilize.

Preparation of Medium: Add components—except fresh yeast extract solution, glucose solution, and bovine serum—to distilled/deionized water and bring volume to 865.0mL. Mix thoroughly. Gently heat and bring to boiling. Adjust pH to 7.5. Autoclave for 15 min at 15 psi pressure–121°C. Cool to 45°–50°C. Aseptically add sterile fresh yeast extract solution, glucose solution, and bovine serum. Mix thoroughly. Aseptically distribute into sterile tubes or flasks.

Use: For the cultivation and maintenance of *Acholeplasma morum*.

PPLO Broth with Crystal Violet

Composition per liter:

Beef heart, infusion from	50.0g
Peptone	10.0g
NaCl	5.0g
Crystal Violet	0.01g
Ascitic fluid	250.0mL
Chapman tellurite solution	2.85mL

pH 7.8 ± 0.2 at 25°C

Source: This medium is available as a premixed powder from BD Diagnostic Systems.

Chapman Tellurite Solution:

Composition per 100.0mL:

K_2TeO_3	1.0g

Preparation of Chapman Tellurite Solution: Add K_2TeO_3 to distilled/deionized water and bring volume to 100.0mL. Mix thoroughly. Filter sterilize.

Caution: Potassium tellurite is toxic.

Preparation of Medium: Add components, except ascitic fluid and Chapman tellurite solution, to distilled/deionized water and bring volume to 747.15mL. Mix thoroughly. Autoclave for 15 min at 15 psi pressure–121°C. Cool to less than 37°C. Aseptically add sterile ascitic fluid and 2.85mL of Chapman tellurite solution. Mix thoroughly. Aseptically distribute into sterile tubes or flasks.

Use: For the isolation of *Mycoplasma* species from clinical specimens.

PPLO Broth without Crystal Violet

Composition per liter:

Beef heart, infusion from	50.0g
Peptone	10.0g
NaCl	5.0g
Thallium acetate (optional)	0.5g
Penicillin (optional)	100,000U
Ascitic fluid	250.0mL

pH 7.8 ± 0.2 at 25°C

Source: This medium is available as a premixed powder from BD Diagnostic Systems.

Preparation of Medium: Add components, except ascitic fluid, to distilled/deionized water and bring volume to 750.0mL. Mix thoroughly. Autoclave for 15 min at 15 psi pressure–121°C. Cool to less than 37°C. Aseptically add sterile ascitic fluid. If desired, 0.5g of thallium acetate or 100,000U of penicillin may be added for a more selective medium. Mix thoroughly. Aseptically distribute into sterile tubes or flasks.

Use: For the enrichment of pleuro-pneumonia-like organisms (PPLOs) and *Mycoplasma* species from clinical specimens.

PPLO Broth without Crystal Violet with Calf Serum, Fresh Yeast Extract, and Sodium Acetate

Composition per liter:

Beef heart, infusion from	50.0g
Peptone	10.0g
Sodium acetate	9.0g
NaCl	5.0g

Yeast extract solution, fresh....................250.0mL
Calf serum....................100.0mL

pH 7.8 ± 0.2 at 25°C

Yeast Extract Solution:

Composition per 300.0mL:

Baker's yeast, live, pressed, starch-free....................75.0g

Preparation of Yeast Extract Solution: Add the live Baker's yeast to 300.0mL of distilled/deionized water. Autoclave for 90 min at 15 psi pressure–121°C. Allow to stand. Remove supernatant solution. Adjust pH to 6.6–6.8. Filter sterilize.

Preparation of Medium: Add components, except fresh yeast extract solution and calf serum, to distilled/deionized water and bring volume to 550.0mL. Mix thoroughly. Autoclave for 15 min at 15 psi pressure–121°C. Cool to 45°–50°C. Aseptically add sterile fresh yeast extract solution and calf serum. Mix thoroughly. Aseptically distribute into sterile tubes or flasks.

Use: For the cultivation and maintenance of *Mycoplasma* species.

PPLO Broth without Crystal Violet with Horse Serum

Composition per liter:

Beef heart, infusion from....................50.0g
Peptone....................10.0g
NaCl....................5.0g
Horse serum, inactivated....................200.0mL

pH 7.8 ± 0.2 at 25°C

Preparation of Medium: Add components, except horse serum, to distilled/deionized water and bring volume to 800.0mL. Mix thoroughly. Autoclave for 15 min at 15 psi pressure–121°C. Cool to 45°–50°C. Aseptically add sterile horse serum. Mix thoroughly. Aseptically distribute into sterile tubes or flasks.

Use: For the cultivation and maintenance of *Acholeplasma* species and *Mycoplasma* species.

PPLO Broth without Crystal Violet with Horse Serum and Fresh Yeast Extract

Composition per liter:

Beef heart, solids from infusion....................50.0g
Peptone....................10.0g
NaCl....................5.0g
Yeast extract solution, fresh....................250.0mL
Horse serum....................200.0mL

pH 7.8 ± 0.2 at 25°C

Yeast Extract Solution:

Composition per 300.0mL:

Baker's yeast, live, pressed, starch-free....................75.0g

Preparation of Yeast Extract Solution: Add the live Baker's yeast to 300.0mL of distilled/deionized water. Autoclave for 90 min at 15 psi pressure–121°C. Allow to stand. Remove supernatant solution. Adjust pH to 6.6–6.8. Filter sterilize.

Preparation of Medium: Add components, except fresh yeast extract solution and horse serum, to distilled/deionized water and bring volume to 550.0mL. Mix thoroughly. Autoclave for 15 min at 15 psi pressure–121°C. Cool to 45°–50°C. Aseptically add sterile fresh yeast extract solution and horse serum. Mix thoroughly. Aseptically distribute into sterile tubes or flasks.

Use: For the cultivation and maintenance of *Mycoplasma putrefaciens.*

PPLO Broth without Crystal Violet with Horse Serum, Glucose, and Fresh Yeast Extract

Composition per liter:

Beef heart, infusion from....................50.0g
Peptone....................10.0g
Glucose....................5.0g
NaCl....................5.0g
Yeast extract solution, fresh....................250.0mL
Horse serum....................200.0mL

pH 7.8 ± 0.2 at 25°C

Yeast Extract Solution:

Composition per 300.0mL:

Baker's yeast, live, pressed, starch-free....................75.0g

Preparation of Yeast Extract Solution: Add the live Baker's yeast to 300.0mL of distilled/deionized water. Autoclave for 90 min at 15 psi pressure–121°C. Allow to stand. Remove supernatant solution. Adjust pH to 6.6–6.8. Filter sterilize.

Preparation of Medium: Add components, except fresh yeast extract solution and horse serum, to distilled/deionized water and bring volume to 550.0mL. Mix thoroughly. Autoclave for 15 min at 15 psi pressure–121°C. Cool to 45°–50°C. Aseptically add sterile fresh yeast extract solution and horse serum. Mix thoroughly. Aseptically distribute into sterile tubes or flasks.

Use: For the cultivation and maintenance of *Mycoplasma putrefaciens, Mycoplasma collis,* and *Mycoplasma cricetuli.*

PPLO Broth without Crystal Violet with Sodium Acetate, Fresh Yeast Extract, and Calf Serum (ATCC Medium 843)

Composition per liter:

Beef heart, infusion from....................50.0g
Peptone....................10.0g
NaCl....................5.0g
Sodium acetate....................1.0g
Calf serum....................100.0mL
Yeast extract solution, fresh....................50.0mL

pH 7.8 ± 0.2 at 25°C

Yeast Extract Solution:

Composition per 300.0mL:

Baker's yeast, live, pressed, starch-free....................75.0g

Preparation of Yeast Extract Solution: Add the live Baker's yeast to 300.0mL of distilled/deionized water. Autoclave for 90 min at 15 psi pressure–121°C. Allow to stand. Remove supernatant solution. Adjust pH to 6.6–6.8. Filter sterilize.

Preparation of Medium: Add components, except fresh yeast extract solution and calf serum, to distilled/deionized water and bring volume to 850.0mL. Mix thoroughly. Autoclave for 15 min at 15 psi pressure–121°C. Cool to 25°C. Aseptically add sterile fresh yeast extract solution and calf serum. Mix thoroughly. Aseptically distribute into sterile tubes or flasks.

Use: For the cultivation and maintenance of *Acholeplasma laidlawii.*

PPLO Broth with Penicillin

Composition per 1010.0mL:

Pancreatic digest of casein....................7.0g
NaCl....................5.0g
Beef extract....................3.0g
Yeast extract....................3.0g

Beef heart, solids from infusion 2.0g
Mycoploasma supplement solution 300.0mL
Penicillin solution 10.0mL

pH 7.8 ± 0.2 at 25°C

Source: This medium is available as a premixed powder from BD Diagnostic Systems.

***Mycoplasma* Supplement Solution:**
Composition per 300.0mL:

Horse serum 200.0mL
Yeast extract (fresh autolysate) 100.0mL
Thallium acetate .... 50.0 mg

Preparation of *Mycoplasma* Supplement Solution: Combine components. Mix thoroughly. Filter sterilize.

Penicllin Solution:
Composition per 10.0mL:

Penicillin 500,000U

Preparation of Peniclin Solution: Add penicillin to distilled/deionized water and bring volume to 10.0mL. Mix thoroughly. Filter sterilize.

Preparation of Medium: Add components, except *Mycoplasma* supplement solution and peniclin solution, to distilled/deionized water and bring volume to 700.0mL. Mix thoroughly. Gently heat and bring to boiling. Boil for 1 min. Autoclave for 15 min at 15 psi pressure–121°C. Cool to 45°–50°C. Aseptically add sterile *Mycoplasma* supplement solution and penicllin solution. Mix thoroughly.

Use: For the selective cultivation of *Mycoplasma* species.

PPNG Selective Medium (Penicillinase-Producing *Neisseria gonorrhoeae* Medium)

Composition per plate:

Quadrant I 10.0mL
Quadrant II 10.0mL

Quadrant I:
Composition per 10.0mL:

Martin Lewis agar 10.0mL

Quadrant II:
Composition per 10.0mL:

Martin Lewis agar, enriched 10.0mL

Use: For the differentiation and presumptive identification of penicillinase-producing strains of *Neisseria gonorrhoeae*. The PPNG selective medium is a two-sectored plate, each containing a different medium. See Martin-Lewis agars for additional information.

PPT Agar, 1*M*

Composition per liter:

NaCl 58.4g
Agar 18.0g
Proteose peptone No. 3 .... 10.0g
Pancreatic digest of casein 10.0g

Preparation of Medium: Add components to distilled/deionized water and bring volume to 1.0L. Mix thoroughly. Gently heat and bring to boiling. Distribute into tubes or flasks. Autoclave for 15 min at 15 psi pressure–121°C. Pour into sterile Petri dishes or leave in tubes.

Use: For the cultivation of *Pseudomonas* species.

PPYG Medium

Composition per liter:

Agar 15.0g
Glucose 5.0g
Peptone 5.0g
NaCl 1.5g
$Na_2HPO_4 \cdot 12H_2O$ 1.5g
Yeast extract 1.5g
$MgCl_2 \cdot 6H_2O$ 0.1g
Na_2CO_3 solution 50.0mL
Glucose solution 50.0mL

pH 10.5–11.0 at 25°C

Na_2CO_3 Solution:
Composition per 50.0mL:

Na_2CO_3 .... 5.03g

Preparation of Na_2CO_3 Solution: Add Na_2CO_3 to distilled/deionized water and bring volume to 50.0mL. Mix thoroughly. Filter sterilize.

Glucose Solution:
Composition per 50.0mL:

D-Glucose 5.0g

Preparation of Glucose Solution: Add glucose to distilled/deionized water and bring volume to 50.0mL. Mix thoroughly. Filter sterilize.

Preparation of Medium: Add components, except Na_2CO_3 solution and glucose solution, to distilled/deionized water and bring volume to 900.0mL. Mix thoroughly. Gently heat and bring to boiling. Autoclave for 15 min at 15 psi pressure–121°C. Cool to 45°–50°C. Aseptically add sterile Na_2CO_3 solution and glucose solution. Mix thoroughly. Pour into sterile Petri dishes or distribute into sterile tubes.

Use: For the cultivation and maintenance of *Exiguobacterium aurantiacum*.

PRAS-PYG with Tween™ 80

Composition per 1054.25mL:

Glucose 10.0g
Yeast extract 10.0g
Peptone 5.0g
Pancreatic digest of casein 5.0g
L-Cysteine·HCl·H_2O 0.5g
Salts solution 40.0mL
Hemin solution 10.0mL
Resazurin 0.025% 4.0mL
Tween™ 80 .... 0.25mL
Vitamin K_1 solution 0.2mL

pH 7.0 ± 0.2 at 25°C

Salts Solution:
Composition per liter:

$NaHCO_3$ 10.0g
NaCl 2.0g
K_2HPO_4 .... 1.0g
KH_2PO_4 .... 1.0g
$CaCl_2$ (anhydrous) 0.2g
$MgSO_4$ 0.2g

Preparation of Salts Solution: Dissolve $CaCl_2$ and $MgSO_4$ in 300.0mL of distilled water. Add 500.0mL of water, and add the remaining salts while swirling slowly. Add 200.0mL of distilled water, mix, and store at 4°C.

Hemin Solution:

Composition per 100.0mL:

Hemin........50.0mg
NaOH (1*N* solution)........1.0mL

Preparation of Hemin Solution: Dissolve hemin in 1.0mL of 1*N* NaOH solution. Bring volume to 100.0mL with distilled/deionized water. Mix thoroughly. Autoclave for 15 min at 15 psi pressure–121°C.

Vitamin K_1 Solution:

Composition per 30.15mL:

Ethanol (95% solution)........30.0mL
Vitamin K_1........0.15mL

Preparation of Vitamin K_1 Solution: Combine components. Mix thoroughly. Store at 4°C in the dark. Discard solution after 1 month.

Preparation of Medium: Prepare and dispense medium under 97% N_2 + 3% H_2. Add components, except glucose, Tween™ 80, and L-cysteine·HCl·H_2O, to distilled/deionized water and bring volume to 1.0L. Mix thoroughly. Gently heat and bring to boiling. Continue boiling for 3 min. Cool to room temperature while sparging with 97% N_2 + 3% H_2. Add glucose, Tween™ 80, and L-cysteine·HCl·H_2O. Mix thoroughly. Anaerobically distribute into sterile tubes or flasks. Autoclave for 15 min at 15 psi pressure–121°C.

Use: For the cultivation of *Clostridium puniceum.*

Pre-Enrichment HiVeg Broth Base with Magnesium Sulfate and Calcium Chloride

Composition per liter:

Yeast extract........20.0g
Plant special peptone........10.0g
Na_2HPO_4........7.1g
KCl........1.0g
NaCl........1.0g
Magnesium sulfate solution........10.0mL
Calcium chloride soltuion........10.0mL

pH 8.3 ± 0.2 at 25°C

Source: This medium, without magnesium sulfate and calcium chloride solutions, is available as a premixed powder from HiMedia.

Magnesium Sulfate Solution:

Composition per 10.0mL:

$MgSO_4$........0.01g

Preparation of Magnesium Sulfate Solution: Add $MgSO_4$ to distilled/deionized water and bring volume to 10.0mL. Mix thoroughly. Filter sterilize.

Calcium Chloride Solution:

Composition per 10.0mL:

$CaCl_2$........0.01g

Preparation of Calcium Chloride Solution: Add $CaCl_2$ to distilled/deionized water and bring volume to 10.0mL. Mix thoroughly. Filter sterilize.

Preparation of Medium: Add components, except magnesium chloride solution and calcium chloride solution, to distilled/deionized water and bring volume to 980.0mL. Mix thoroughly. Gently heat while stirring and bring to boiling. Distribute into tubes or flasks. Autoclave for 15 min at 15 psi pressure–121°C. Cool to 45°–50°C. Aseptically add 10.0mL of sterile magnesium sulfate solution and 10.0mL of sterile calcium chloride solution. Mix thoroughly. Aseptically distribute to sterile tubes or flasks.

Use: For the isolation, enrichment, and cultivation of *Yersinia enterocolitica* from foods.

Preenrichment Medium (PEM)

Composition per liter:

Yeast extract........20.0g
Special peptone........10.0g
Na_2HPO_4........7.1g
NaCl........1.0g
KCl........1.0g
$MgSO_4·7H_2O$ solution........10.0mL
$CaCl_2·2H_2O$ solution........10.0mL

pH 8.3 ± 0.2 at 25°C

$MgSO_4·7H_2O$ Solution:

Composition per 10.0mL:

$MgSO_4·7H_2O$........0.01g

Preparation of $MgSO_4·7H_2O$ Solution: Add $MgSO_4·7H_2O$ to distilled/deionized water and bring volume to 10.0mL. Mix thoroughly. Filter sterilize.

$CaCl_2·2H_2O$ Solution:

$CaCl_2·2H_2O$........0.01g

Preparation of $CaCl_2·2H_2O$ Solution: Add the $CaCl_2·2H_2O$ to distilled/deionized water and bring volume to 10.0mL. Mix thoroughly. Filter sterilize.

Preparation of Medium: Add components, except $MgSO_4·7H_2O$ solution and $CaCl_2·2H_2O$ solution, to distilled/deionized water and bring volume to 980.0mL. Mix thoroughly. Adjust pH to 8.3. Gently heat and bring to boiling. Autoclave for 15 min at 15 psi pressure–121°C. Cool to 45°–50°C. Aseptically add sterile $MgSO_4·7H_2O$ solution and $CaCl_2·2H_2O$ solution. Mix thoroughly. Aseptically distribute into sterile tubes.

Use: For the isolation and enrichment of *Yersinia enterocolitica* from foods.

Preferred Medium

Composition per liter:

Peptone........20.0g
Glucose........20.0g
Agar........15.0g
Casamino acids........10.0g
Yeast extract........10.0g

Preparation of Medium: Add components to distilled/deionized water and bring volume to 1.0L. Mix thoroughly. Gently heat and bring to boiling. Distribute into tubes or flasks. Autoclave for 15 min at 15 psi pressure–121°C. Pour into sterile Petri dishes or leave in tubes.

Use: For the cultivation and maintenance of *Kluyveromyces lactis.*

Presence-Absence Broth (P-A Broth)

Composition per liter:

Pancreatic digest of casein........10.0g
Lactose........7.5g
Pancreatic digest of gelatin........5.0g
Beef extract........3.0g
NaCl........2.5g
K_2HPO_4........1.375g
KH_2PO_4........1.375g

Sodium lauryl sulfate	0.05g
Bromcresol Purple	8.5mg

pH 6.8 ± 0.2 at 25°C

Source: This medium is available as a premixed powder from BD Diagnostic Systems.

Preparation of Medium: Add components to distilled/deionized water and bring volume to 333.0mL. Mix thoroughly. Distribute into screw-capped 250.0mL milk dilution bottles in 50.0mL volumes. Autoclave for 15 min at 15 psi pressure–121°C.

Use: For the detection of coliform bacteria in water from treatment plants or distribution systems using the presence-absence coliform test.

Preston Blood-Free Medium
See: ***Campylobacter*** **Charcoal Differential Agar**

Preston Enrichment Broth

Composition per liter:

Beef extract	10.0g
Peptone	10.0g
NaCl	5.0g
Horse blood, lysed	50.0mL
Antibiotic solution	10.0mL

pH 7.5 ± 0.2 at 25°C

Antibiotic Solution:

Composition per 10.0mL:

Cycloheximide	0.1g
Rifampicin	0.01g
Trimethoprim lactate	0.01g
Polymyxin B	5000U

Preparation of Antibiotic Solution: Add components to distilled/deionized water and bring volume to 10.0mL. Mix thoroughly. Filter sterilize.

Caution: Cycloheximide is toxic. Avoid skin contact or aerosol formation and inhalation.

Preparation of Medium: Add components, except horse blood and antibiotic solution, to distilled/deionized water and bring volume to 940.0mL. Mix thoroughly. Gently heat and bring to boiling. Autoclave for 15 min at 15 psi pressure–121°C. Cool to 45°–50°C. Aseptically add sterile horse blood and antibiotic solution. Mix thoroughly. Aseptically distribute into sterile tubes or flasks.

Use: For the isolation and enrichment of *Campylobacter* species from foods.

Preston HiVeg Agar Base with Horse Blood and Antibiotics

Composition per liter:

Agar	12.0g
Plant extract	10.0g
Plant peptone	10.0g
NaCl	5.0g
Horse blood, lysed	50.0mL
Antibiotic solution	10.0mL

pH 7.5 ± 0.2 at 25°C

Source: This medium, without horse blood and antibiotics, is available as a premixed powder from HiMedia.

Antibiotic Solution:

Composition per 10.0mL:

Cycloheximide	0.1g
Rifampicin	0.01g
Trimethoprim lactate	0.01g
Polymyxin B	5000U

Preparation of Antibiotic Solution: Add components to distilled/deionized water and bring volume to 10.0mL. Mix thoroughly. Filter sterilize.

Caution: Cycloheximide is toxic. Avoid skin contact or aerosol formation and inhalation.

Preparation of Medium: Add components, except horse blood and antibiotic solution, to distilled/deionized water and bring volume to 940.0mL. Mix thoroughly. Gently heat and bring to boiling. Autoclave for 15 min at 15 psi pressure–121°C. Cool to 45°–50°C. Aseptically add sterile horse blood and antibiotic solution. Mix thoroughly. Aseptically distribute into sterile tubes or flasks.

Use: For the isolation and enrichment of *Campylobacter* species from foods.

Preston's *Campylobacter* Medium
See: ***Campylobacter*** **Selective Medium, Preston's**

Presumpto Media

Composition per plate:

Quadrant I	5.0mL
Quadrant II	5.0mL
Quadrant III	5.0mL
Quadrant IV	5.0mL

Quadrant I:

Composition per 5.0mL:

Lombard-Dowell agar	5.0mL

Quadrant II:

Composition per 5.0mL:

Lombard-Dowell bile agar	5.0mL

Quadrant III:

Composition per 5.0mL:

Lombard-Dowell egg yolk agar	5.0mL

Quadrant IV:

Composition per 5.0mL:

Lombard-Dowell esculin agar	5.0mL

Preparation of Quadrant Media: Sterilize Lombard-Dowell Agar by autocalving for 15 min at 15 psi pressure–121°C. Cool to 45°–50°C. Add additional components as filter sterilized solutions. Mix and distribute as 5.0mL aliquots into quadrants.

Use: For the differentiation and presumptive identification of anaerobic bacteria. The Presumpto media is a four-sectored plate each containing a different medium.

Prey Seawater Broth (DSMZ Medium 1013)

Composition per liter:

Artificial seawater	700.0mL
Trace elements solution	10.0mL

pH 7.7 ± 0.3 at 25°C

Artificial Seawater:

Composition per liter:

NaCl	27.7g
$MgSO_4 \cdot 7H_2O$	7.0g
$MgCl_2 \cdot 6H_2O$	5.5g
$CaCl_2 \cdot 2H_2O$	2.25g
KCl	0.65g
NaBr	0.1g
H_3BO_3	30.0mg
$SrCl_2 \cdot 6H_2O$	15.0mg
Citric acid	10.0mg
KI	0.05mg

Preparation of Artificial Seawater: Add components to distilled/deionized water and bring volume to 1.0L. Mix thoroughly.

Trace Elements Solution:

Composition per liter:

$MgSO_4 \cdot 7H_2O$	3.0g
Nitrilotriacetic acid	1.5g
NaCl	1.0g
$MnSO_4 \cdot 2H_2O$	0.5g
$CoSO_4 \cdot 7H_2O$	0.18g
$ZnSO_4 \cdot 7H_2O$	0.18g
$CaCl_2 \cdot 2H_2O$	0.1g
$FeSO_4 \cdot 7H_2O$	0.1g
$KAl(SO_4)_2 \cdot 12H_2O$	0.02g
$CuSO_4 \cdot 5H_2O$	0.01g
H_3BO_3	0.01g
$Na_2MoO_4 \cdot 2H_2O$	0.01g
$NiCl_2 \cdot 6H_2O$	0.025g
$Na_2SeO_3 \cdot 5H_2O$	0.3mg

Preparation of Trace Elements Solution: Add nitrilotriacetic acid to 500.0mL of distilled/deionized water. Adjust pH to 6.5 with KOH. Add remaining components. Add distilled/deionized water to 1.0L.

Preparation of Medium: Add components to distilled/deionized water and bring volume to 1.0L. Mix thoroughly. Adjust pH to 7.7. Autoclave for 15 min at 15 psi pressure–121°C. Cool to 25°C. Grow prey bacterium *E. coli* DSM 15416) on agar plates using marine agar 2216. For 50mL prey seawater broth use one 24 h old agar plate and wash off grown cells in 3.0–5.0mL artificial seawater. The prey seawater broth should contain approx. 10^7 to 10^9 cells. Inoculate the medium immediately after preparation and incubate the suspension with shaking until a decrease in turbidity is visible.

Use: For the cultivation of *Bacteriovorax litoralis.*

Pril Xylose Ampicillin Agar (PXA Agar)

Composition per liter:

Agar	15.0g
Xylose	10.0g
Pancreatic digest of gelatin	5.0g
Beef extract	3.0g
Pril	0.2g
Ampicillin	0.03g
Phenol Red	0.025g

pH 6.8 ± 0.2 at 25°C

Note: Pril is a quaternary ammonium detergent composed of a mixture of primary alkyl sulfate, alkyl-benzyl sulfonate, and salts. It is available from Böhme Fettchemie GmbH, Düsseldorf, Germany.

Preparation of Medium: Add components to distilled/deionized water and bring volume to 1.0L. Mix thoroughly. Gently heat and bring to boiling. Distribute into tubes or flasks. Autoclave for 15 min at 15 psi pressure–121°C. Pour into sterile Petri dishes or leave in tubes.

Use: For the selective isolation and cultivation of *Aeromonas hydrophila.*

Pringsheim's Medium

Composition per liter:

KNO_3	0.2g
$(NH_4)_2HPO_4$	0.02g
$MgSO_4 \cdot 7H_2O$	0.01g
$CaCl_2 \cdot 2H_2O$	0.005g
$FeCl_2$	0.5mg

Source: This medium is available from HiMedia.

Preparation of Medium: Add components to distilled/deionized water and bring volume to 1.0L. Mix thoroughly. Gently heat and bring to boiling. Distribute into tubes or flasks. Autoclave for 15 min at 15 psi pressure–121°C. Pour into sterile Petri dishes or leave in tubes.

Use: For the cultivation of cyanobacteria.

Propionibacterium Agar

Composition per liter:

Agar	15.0g
Casein peptone, tryptic digest	10.0g
Sodium lactate	10.0g
Yeast extract	5.0g

pH 7.0–7.2 at 25°C

Preparation of Medium: Add components to distilled/deionized water and bring volume to 1.0L. Mix thoroughly. Adjust pH to 7.0–7.2. Gently heat and bring to boiling. Distribute into tubes or flasks. Autoclave for 15 min at 15 psi pressure–121°C. Pour into sterile Petri dishes or leave in tubes.

Use: For the cultivation and maintenance of *Propionibacterium acidipropionici* and *Propionibacterium thoenii.*

Propionibacterium Agar

Composition per liter:

Agar	15.0g
Pancreatic digest of casein	10.0g
Sodium lactate	10.0g
Yeast extract	5.0g

pH 7.0 ± 0.2 at 25°C

Preparation of Medium: Add components to distilled/deionized water and bring volume to 1.0L. Mix thoroughly. Gently heat and bring to boiling. Distribute into tubes or flasks. Autoclave for 15 min at 15 psi pressure–121°C. Pour into sterile Petri dishes or leave in tubes.

Use: For the cultivation and maintenance of *Aerococcus viridans, Arthrobacter globiformis, Arthrobacter* species, *Brachybacterium faecium, Brochothrix campestris, Carnobacterium divergens, Carnobacterium gallinarum, Carnobacterium mobile, Carnobacterium piscicola, Clavibacter michiganensis, Clavibacter* species, *Corynebacterium amycolatum, Corynebacterium cystitidis, Corynebacterium jeikeium, Corynebacterium matruchotii, Corynebacterium mycetoides, Corynebacterium pilosum, Corynebacterium* species, *Corynebacterium urealyticum, Corynebacterium xerosis, Deinococcus radiophilus, Dermabacter hominus, Enterococcus avium, Enterococcus casseliflavus, Enterococcus cecorum, Enterococcus dispar, Enterococcus* *durans, Enterococcus faecalis, Enterococcus faecium, Enterococcus gallinarum, Enterococcus malodoratus, Enterococcus mundtii, Enterococcus pseudoavium, Enterococcus raffinosus, Enterococcus saccharolyticus, Enterococcus seriolicida, Enterococcus solitarius, Enterococcus sulfureus, Gluconobacter oxydans, Kurthia sibirica, Lactobacillus maltaromicus, Lactococcus garvieae, Lactococcus lactis, Lactococcus piscium, Lactococcus plantarum, Lactococcus raffinolactis, Microbacterium arborescens, Micrococcus luteus, Pediococcus urinaeequi, Pseudomonas putida, Rhodococcus equi, Rhodococcus maris, Rhodococcus* species, *Staphylococcus aureus, Staphylococcus capitis, Staphylococcus cohnii, Staphylococcus delphini, Staphylococcus lentus, Staphylococcus lugdunensis, Staphylococcus muscae, Staphylococcus schleiferi, Staphylococcus simulans, Staphylococcus xylosus, Streptococcus acidominimus, Streptococcus alactolyticus, Streptococcus anginosus, Streptococcus canis, Streptococcus cricetus, Streptococcus downei, Streptococcus dysgalactiae, Streptococcus equi, Streptococcus ferus, Streptococcus gordonii, Streptococcus hyointestinalis, Streptococcus macacae, Streptococcus mitis, Streptococcus mutans, Streptococcus parasanguis, Streptococcus parauberis, Streptococcus pneumoniae, Streptococcus porcinus, Streptococcus pyogenes, Streptococcus rattus, Streptococcus salivarius, Streptococcus sanguis, Streptococcus sobrinus, Streptococcus* species, *Streptococcus uberis, Streptococcus vestibularis, Thermoactinomyces intermedius, Vagococcus fluvialis,* and *Vagococcus salmoninarum.*

Propionibacterium Medium

Composition per liter:

Yeast extract	10.0g
$Na_2HPO_4 \cdot 2H_2O$	3.0g
KH_2PO_4	1.0g
Lactate solution	40.0mL

pH 7.0 ± 0.2 at 25°C

Lactate Solution:

Composition per 100.0mL:

Sodium lactate	70.0g

Preparation of Lactate Solution: Add 70.0g of sodium lactate to distilled/deionized water and bring volume to 100.0mL. Mix thoroughly.

Preparation of Medium: Add components, except lactate solution, to distilled/deionized water and bring volume to 960.0mL. Mix thoroughly. Add 40.0mL of lactate solution. Mix thoroughly. Adjust pH to 7.0. Distribute into screw-capped tubes. Autoclave for 15 min at 15 psi pressure–121°C.

Use: For the cultivation of *Propionibacterium acidipropionici, Propionibacterium freudenreichii, Propionibacterium jensenii,* and *Propionibacterium thoenii.*

Propionigenium maris Medium

Composition per 1014.0mL:

Solution A	940.0mL
Solution E ($NaHCO_3$ solution)	50.0mL
Solution F (Substrate solution)	10.0mL
Solution G ($Na_2S \cdot 9H_2O$ solution)	10.0mL
Solution B (Trace elements solution SL-10)	2.0mL
Solution C (Seven vitamin solution)	1.0mL
Solution D (Selenite-tungstate solution)	1.0mL

pH 7.2–7.4 at 25°C

Solution A:

Composition per 940.0mL:

NaCl	1.0g
Yeast extract	0.5g
KCl	0.5g
$MgCl_2 \cdot 6H_2O$	0.4g
NH_4Cl	0.25g
KH_2PO_4	0.2g
$CaCl_2 \cdot 2H_2O$	0.15g
Resazurin	0.5mg

Preparation of Solution A: Prepare and dispense under 80% N_2 + 20% CO_2. Add components to distilled/deionized water and bring volume to 940.0mL. Mix thoroughly. Autoclave for 15 min at 15 psi pressure–121°C.

Solution B (Trace Elements Solution SL-10):

Composition per liter:

$FeCl_2 \cdot 4H_2O$	1.5g
$CoCl_2 \cdot 6H_2O$	190.0mg
$MnCl_2 \cdot 4H_2O$	100.0mg
$ZnCl_2$	70.0mg
$Na_2MoO_4 \cdot 2H_2O$	36.0mg
$NiCl_2 \cdot 6H_2O$	24.0mg
H_3BO_3	6.0mg
$CuCl_2 \cdot 2H_2O$	2.0mg
HCl (25% solution)	10.0mL

Preparation of Solution B (Trace Elements Solution SL-10): Prepare and dispense under 100% N_2. Add $FeCl_2 \cdot 4H_2O$ to 10.0mL of HCl solution. Mix thoroughly. Add distilled/deionized water and bring volume to 1.0L. Add remaining components. Mix thoroughly. Autoclave for 15 min at 15 psi pressure–121°C.

Solution C (Seven Vitamin Solution):

Composition per liter:

Pyridoxine·HCl	300.0mg
Nicotinic acid	200.0mg
Thiamine·HCl	200.0mg
Calcium pantothenate	100.0mg
Cyanocobalamine	100.0mg
p-Aminobenzoic acid	80.0mg
D(+)-Biotin	20.0mg

Preparation of Solution C (Seven Vitamin Solution): Add components to distilled/deionized water and bring volume to 1.0L. Mix thoroughly. Filter sterilize. Sparge with 100% N_2.

Solution D (Selenite-Tungstate Solution):

Composition per liter:

NaOH	0.5g
$Na_2WO_4 \cdot 2H_2O$	4.0mg
$Na_2SeO_3 \cdot 5H_2O$	3.0mg

Preparation of Solution D (Selenite Tungstate Solution): Add components to distilled/deionized water and bring volume to 1.0L. Mix thoroughly. Filter sterilize. Sparge with 100% N_2.

$Na_2S \cdot 9H_2O$ Solution:

Composition per 10.0mL:

$Na_2S \cdot 9H_2O$	0.6g

Preparation of $Na_2S \cdot 9H_2O$ Solution: Add $Na_2S \cdot 9H_2O$ to distilled/deionized water and bring volume to 10.0mL. Mix thoroughly. Sparge with 100% N_2. Autoclave for 15 min at 15 psi pressure–121°C.

Solution E ($NaHCO_3$ Solution):

Composition per 50.0mL:

$NaHCO_3$ 2.5g

Preparation of Solution E ($NaHCO_3$ Solution): Add $NaHCO_3$ to distilled/deionized water and bring volume to 50.0mL. Mix thoroughly. Sparge with 100% N_2. Autoclave for 15 min at 15 psi pressure–121°C.

Solution F (Substrate Solution):

Composition per 10.0mL:

Disodium succinate 2.5g

Preparation of Solution F (Substrate Solution): Add disodium succinate to distilled/deionized water and bring volume to 10.0mL. Mix thoroughly. Sparge with 100% N_2. Autoclave for 15 min at 15 psi pressure–121°C.

Solution G ($Na_2S \cdot 9H_2O$ Solution):

Composition per 10.0mL:

$Na_2S \cdot 9H_2O$ 0.3g

Preparation of Solution G ($Na_2S \cdot 9H_2O$ Solution): Add $Na_2S \cdot 9H_2O$ to distilled/deionized water and bring volume to 10.0mL. Mix thoroughly. Sparge with 100% N_2. Autoclave for 15 min at 15 psi pressure–121°C.

Preparation of Medium: To 940.0mL of sterile solution A, aseptically and anaerobically add 1.0mL of sterile solution B, 1.0mL of sterile solution C, 1.0mL of sterile solution D, 50.0mL of sterile solution E, 10.0mL of sterile solution F, and 10.0mL of sterile solution G. Mix thoroughly. Aseptically and anaerobically distribute into sterile tubes or flasks.

Use: For the cultivation of *Propionigenium maris*.

Propionigenium modestum Medium (DSMZ Medium 293)

Composition per liter:

NaCl	20.0g
$MgCl_2 \cdot 6H_2O$	3.0g
KCl	0.5g
NH_4Cl	0.25g
KH_2PO_4	0.2g
$CaCl_2 \cdot 2H_2O$	0.15g
Resazurin	1.0mg
$NaHCO_3$ solution	10.0mL
Na_2-succinate solution	10.0mL
$Na_2S \cdot 9H_2O$ solution	10.0mL
Trace elements solution SL-10	1.0mL

pH 7.2 ± 0.2 at 25°C

$Na_2S \cdot 9H_2O$ Solution:

Composition per 10.0mL:

$Na_2S \cdot 9H_2O$ 0.36g

Preparation of $Na_2S \cdot 9H_2O$ Solution: Add $Na_2S \cdot 9H_2O$ to distilled/deionized water and bring volume to 10.0mL. Mix thoroughly. Autoclave under 100% N_2 for 15 min at 15 psi pressure–121°C. Cool to room temperature.

$NaHCO_3$ Solution:

Composition per 10.0mL:

$NaHCO_3$ 2.5g

Preparation of $NaHCO_3$ Solution: Add $NaHCO_3$ to distilled/deionized water and bring volume to 10.0mL. Mix thoroughly. Sparge with 80% N_2 + 20% CO_2. Filter sterilize.

Na_2-succinate Solution:

Composition per 10.0mL:

Na_2-succinate 3.25g

Preparation of Na_2-succinate Solution: Add Na_2-succinate to distilled/deionized water and bring volume to 10.0mL. Mix thoroughly. Sparge with 100% N_2. Filter sterilize.

Trace Elements Solution SL-10:

Composition per liter:

$FeCl_2 \cdot 4H_2O$	1.5g
$CoCl_2 \cdot 6H_2O$	190.0mg
$MnCl_2 \cdot 4H_2O$	100.0mg
$ZnCl_2$	70.0mg
$Na_2MoO_4 \cdot 2H_2O$	36.0mg
$NiCl_2 \cdot 6H_2O$	24.0mg
H_3BO_3	6.0mg
$CuCl_2 \cdot 2H_2O$	2.0mg
HCl (25% solution)	10.0mL

Preparation of Trace Elements Solution SL-10: Add $FeCl_2 \cdot 4H_2O$ to 10.0mL of HCl solution. Mix thoroughly. Add distilled/deionized water and bring volume to 1.0L. Add remaining components. Mix thoroughly. Sparge with 80% N_2 + 20% CO_2. Filter sterilize.

Preparation of Medium: Prepare and dispense medium under 80% N_2 + 20% CO_2 gas atmosphere. Add components, except $NaHCO_3$ solution, Na_2-succinate solution, $Na_2S \cdot 9H_2O$ solution, and trace elements solution SL-10, to distilled/deionized water and bring volume to 969.0mL. Mix thoroughly. Adjust pH to 7.2. Sparge with 80% N_2 + 20% CO_2. Autoclave for 15 min at 15 psi pressure–121°C. Aseptically and anaerobically add 10.0mL $NaHCO_3$ solution, 10.0mL Na_2-succinate solution, 10.0mL $Na_2S \cdot 9H_2O$ solution, and 1.0mL trace elements solution SL-10. Mix thoroughly. Aseptically and anaerobically distribute into sterile tubes or bottles. After inoculation, flush and repressurize the gas head space of culture bottles with sterile 80% N_2 + 20% CO_2 to 1 bar overpressure.

Use: For the cultivation of *Propionigenium modestum.*

Propionigenium modestum Medium (DSMZ Medium 293)

Composition per liter:

NaCl	20.0g
$MgCl_2 \cdot 6H_2O$	3.0g
KCl	0.5g
NH_4Cl	0.25g
KH_2PO_4	0.2g
$CaCl_2 \cdot 2H_2O$	0.15g
Resazurin	1.0mg
$NaHCO_3$ solution	10.0mL
Na_2-succinate solution	10.0mL
$Na_2S \cdot 9H_2O$ solution	10.0mL
Yeast extract solution	10.0mL
Trace elements solution SL-10	1.0mL

pH 7.2 ± 0.2 at 25°C

$Na_2S \cdot 9H_2O$ Solution:

Composition per 10.0mL:

$Na_2S \cdot 9H_2O$ 0.36g

Preparation of $Na_2S \cdot 9H_2O$ Solution: Add $Na_2S \cdot 9H_2O$ to distilled/deionized water and bring volume to 10.0mL. Mix thoroughly. Autoclave under 100% N_2 for 15 min at 15 psi pressure–121°C. Cool to room temperature.

$NaHCO_3$ Solution:
Composition per 10.0mL:
$NaHCO_3$ 2.5g

Preparation of $NaHCO_3$ Solution: Add $NaHCO_3$ to distilled/deionized water and bring volume to 10.0mL. Mix thoroughly. Sparge with 80% N_2 + 20% CO_2. Filter sterilize.

Na_2-succinate Solution:
Composition per 10.0mL:
Na_2-succinate 3.25g

Preparation of Na_2-succinate Solution: Add Na_2-succinate to distilled/deionized water and bring volume to 10.0mL. Mix thoroughly. Sparge with 100% N_2. Filter sterilize.

Yeast Extract Solution:
Composition per 10.0mL:
Yeast extract 1.0g

Preparation of Yeast Extract Solution: Add yeast extract to distilled/deionized water and bring volume to 10.0mL. Mix thoroughly. Sparge with 100% N_2. Autoclave under 100% N_2 for 15 min at 15 psi pressure–121°C. Cool to room temperature.

Trace Elements Solution SL-10:
Composition per liter:
$FeCl_2 \cdot 4H_2O$ 1.5g
$CoCl_2 \cdot 6H_2O$ 190.0mg
$MnCl_2 \cdot 4H_2O$ 100.0mg
$ZnCl_2$ 70.0mg
$Na_2MoO_4 \cdot 2H_2O$ 36.0mg
$NiCl_2 \cdot 6H_2O$ 24.0mg
H_3BO_3 .... 6.0mg
$CuCl_2 \cdot 2H_2O$ 2.0mg
HCl (25% solution) 10.0mL

Preparation of Trace Elements Solution SL-10: Add $FeCl_2 \cdot 4H_2O$ to 10.0mL of HCl solution. Mix thoroughly. Add distilled/deionized water and bring volume to 1.0L. Add remaining components. Mix thoroughly. Sparge with 80% N_2 + 20% CO_2. Filter sterilize.

Preparation of Medium: Prepare and dispense medium under 80% N_2 + 20% CO_2 gas atmosphere. Add components, except $NaHCO_3$ solution, Na_2-succinate solution, $Na_2S \cdot 9H_2O$ solution, yeast extract solution, and trace elements solution SL-10, to distilled/deionized water and bring volume to 959.0mL. Mix thoroughly. Adjust pH to 7.2. Sparge with 80% N_2 + 20% CO_2. Autoclave for 15 min at 15 psi pressure–121°C. Aseptically and anaerobically add 10.0mL $NaHCO_3$ solution, 10.0mL Na_2-succinate solution, 10.0mL $Na_2S \cdot 9H_2O$ solution, 10.0mL yeast extract solution, and 1.0mL trace elements solution SL-10. Mix thoroughly. Aseptically and anaerobically distribute into sterile tubes or bottles. After inoculation, flush and repressurize the gas head space of culture bottles with sterile 80% N_2 + 20% CO_2 to 1 bar overpressure.

Use: For the cultivation of *Propionigenium modestum* DSM 2376.

Propionigenium modestum Medium

Composition per 1001.0mL:
NaCl 20.0g
$MgCl_2 \cdot 6H_2O$ 3.0g
KCl 0.5g
NH_4Cl 0.25g
KH_2PO_4 .... 0.2g
$CaCl_2 \cdot 2H_2O$ 0.15g
Resazurin 1.0mg
$NaHCO_3$ solution 10.0mL
$Na_2S \cdot 9H_2O$ solution 10.0mL
Disodium succinate solution 10.0mL
Trace elements solution SL-10 .... 1.0mL
pH 7.2 ± 0.2 at 25°C

$NaHCO_3$ Solution:
Composition per 10.0mL:
$NaHCO_3$ 2.5g

Preparation of $NaHCO_3$ Solution: Add $NaHCO_3$ to distilled/deionized water and bring volume to 10.0mL. Mix thoroughly. Filter sterilize. Gas under 80% N_2 + 20% CO_2.

$Na_2S \cdot 9H_2O$ Solution:
Composition per 10.0mL:
$Na_2S \cdot 9H_2O$ 0.36g

Preparation of $Na_2S \cdot 9H_2O$ Solution: Add $Na_2S \cdot 9H_2O$ to distilled/deionized water and bring volume to 10.0mL. Mix thoroughly. Gas under 100% N_2. Autoclave for 15 min at 15 psi pressure–121°C.

Disodium Succinate Solution:
Composition per 10.0mL:
Disodium succinate 3.25g

Preparation of Disodium Succinate Solution: Add disodium succinate to distilled/deionized water and bring volume to 10.0mL. Mix thoroughly. Filter sterilize. Gas under 80% N_2 + 20% CO_2.

Trace Elements Solution SL-10:
Composition per liter:
$FeCl_2 \cdot 4H_2O$ 1.5g
$CoCl_2 \cdot 6H_2O$ 190.0mg
$MnCl_2 \cdot 4H_2O$ 100.0mg
$ZnCl_2$ 70.0mg
$Na_2MoO_4 \cdot 2H_2O$ 36.0mg
$NiCl_2 \cdot 6H_2O$ 24.0mg
H_3BO_3 .... 6.0mg
$CuCl_2 \cdot 2H_2O$ 2.0mg
HCl (25% solution) 10.0mL

Preparation of Trace Elements Solution SL-10: Prepare and dispense under 80% N_2 + 20% CO_2. Add $FeCl_2 \cdot 4H_2O$ to 10.0mL of HCl solution. Mix thoroughly. Add distilled/deionized water and bring volume to 1.0L. Add remaining components. Mix thoroughly. Gas under 80% N_2 + 20% CO_2. Autoclave for 15 min at 15 psi pressure–121°C.

Preparation of Medium: Prepare medium and dispense under 80% N_2 + 20% CO_2. Add components, except $NaHCO_3$ solution, $Na_2S \cdot 9H_2O$ solution, disodium succinate solution, and trace elements solution SL-10, to distilled/deionized water and bring volume to 969.0mL. Mix thoroughly. Sparge with 80% N_2 + 20% CO_2. Autoclave for 15 min at 15 psi pressure–121°C. Aseptically and anaerobically add 10.0mL of sterile $NaHCO_3$ solution, 10.0mL of sterile $Na_2S \cdot 9H_2O$ solution, 10.0mL of sterile disodium succinate solution, and 1.0mL of sterile trace elements solution SL-10. Mix thoroughly. Aseptically and anaerobically distribute into sterile tubes or flasks.

Use: For the cultivation and maintenance of *Propionigenium modestum.*

Propionispira Medium

Composition per liter:
Sodium lactate 4.0g
Yeast extract 1.0g
Mineral solution 2 .... 50.0mL
Sodium carbonate solution 50.0mL

Mineral solution 1 .... 25.0mL
L-Cysteine-sulfide reducing agent 20.0mL
Wolfe's mineral solution 10.0mL
Wolfe's Vitamin solution 10.0mL
Resazurin (0.025% solution) 4.0mL

pH 7.2 ± 0.2 at 25°C

Mineral Solution 1:

Composition per liter:

K_2HPO_4 .... 6.0g

Preparation of Mineral Solution 1: Add K_2HPO_4 to distilled/deionized water and bring volume to 1.0L. Mix thoroughly.

Mineral Solution 2:

Composition per liter:

NaCl 12.0g
KH_2PO_4 .... 6.0g
$(NH_4)_2SO_4$ 6.0g
$MgSO_4 \cdot 7H_2O$ 2.4g
$CaCl_2 \cdot 2H_2O$ 1.6g

Preparation of Mineral Solution 2: Add components to distilled/deionized water and bring volume to 1.0L. Mix thoroughly.

Sodium Carbonate Solution:

Composition per 100.0mL:

Na_2CO_3 .... 8.0g

Preparation of Sodium Carbonate Solution: Add Na_2CO_3 to distilled/deionized water and bring volume to 100.0mL Mix thoroughly.

L-Cysteine-Sulfide Reducing Agent:

Composition per 20.0mL:

L-Cysteine·HCl·H_2O 300.0mg
$Na_2S \cdot 9H_2O$ 300.0mg

Preparation of L-Cysteine-Sulfide Reducing Agent: Add L-cysteine·HCl·H_2O to 10.0mL of distilled/deionized water. Mix thoroughly. In a separate tube, add $Na_2S \cdot 9H_2O$ to 10.0mL of distilled/deionized water. Mix thoroughly. Gas both solutions with 100% N_2 and cap tubes. Autoclave both solutions for 15 min at 15 psi pressure–121°C using fast exhaust. Cool to 50°C. Aseptically combine the two solutions under 100% N_2.

Wolfe's Mineral Solution:

Composition per liter

$MgSO_4 \cdot 7H_2O$ 3.0g
Nitrilotriacetic acid 1.5g
NaCl 1.0g
$MnSO_4 \cdot H_2O$ 0.5g
$FeSO_4 \cdot 7H_2O$ 0.1g
$CoCl_2 \cdot 6H_2O$ 0.1g
$CaCl_2$ 0.1g
$ZnSO_4 \cdot 7H_2O$ 0.1g
$CuSO_4 \cdot 5H_2O$ 0.01g
$AlK(SO_4)_2 \cdot 12H_2O$ 0.01g
H_3BO_3 .... 0.01g
$Na_2MoO_4 \cdot 2H_2O$ 0.01g

Preparation of Wolfe's Mineral Solution: Add nitrilotriacetic acid to 500.0mL of distilled/deionized water. Dissolve by adjusting pH to 6.5 with KOH. Add remaining components. Add distilled/deionized water to 1.0L.

Wolfe's Vitamin Solution:

Composition per liter:

Pyridoxine·HCl 10.0mg
Thiamine·HCl 5.0mg
Riboflavin 5.0mg
Nicotinic acid 5.0mg
Calcium pantothenate 5.0mg
p-Aminobenzoic acid 5.0mg
Thioctic acid 5.0mg
Biotin 2.0mg
Folic acid 2.0mg
Cyanocobalamin 100.0μg

Preparation of Wolfe's Vitamin Solution: Add components to distilled/deionized water and bring volume to 1.0L. Mix thoroughly. Filter sterilize.

Preparation of Medium: Add components, except Wolfe's vitamin solution and L-cysteine-sulfide reducing agent, to distilled/deionized water and bring volume to 970.0mL. Mix thoroughly. Autoclave for 15 min at 15 psi pressure–121°C. Cool under 80% N_2 + 20% CO_2. Aseptically add the sterile Wolfe's vitamin solution and then the sterile L-cysteine-sulfide reducing agent. Adjust the pH to 7.2. Distribute aseptically and anaerobically into sterile tubes.

Use: For the cultivation and maintenance of *Propionispira arboris*.

Propionispora Medium (DSMZ Medium 503c)

Composition per 1010.0mL:

Solution A 940.0mL
Solution B 50.0mL
Solution C 10.0mL
Solution D 10.0mL

pH 7.1 ± 0.2 at 25°C

Solution A:

Composition per 940.0mL:

Yeast extract 1.0g
NaCl 1.0g
KCl 0.5g
$MgCl_2 \cdot 6H_2O$ 0.4g
KH_2PO_4 .... 0.2g
NH_4Cl 0.25g
$CaCl_2 \cdot 2H_2O$ 0.15g
Resazurin 0.5mg
Trace elements solution SL-10 .... 1.0mL

Preparation of Solution A: Prepare under 80% N_2 + 20% CO_2 gas atmosphere. Add components to distilled/deionized water and bring volume to 940.0mL. Mix thoroughly. Adjust pH to 7.2. Sparge with 80% N_2 + 20% CO_2. Autoclave for 15 min at 15 psi pressure–121°C. Cool to 25°C.

Trace Elements Solution SL-10:

Composition per liter:

$FeCl_2 \cdot 4H_2O$ 1.5g
$CoCl_2 \cdot 6H_2O$ 190.0mg
$MnCl_2 \cdot 4H_2O$ 100.0mg
$ZnCl_2$ 70.0mg
$Na_2MoO_4 \cdot 2H_2O$ 36.0mg
$NiCl_2 \cdot 6H_2O$ 24.0mg
H_3BO_3 .... 6.0mg
$CuCl_2 \cdot 2H_2O$ 2.0mg
HCl (25% solution) 10.0mL

Preparation of Trace Elements Solution SL-10: Add $FeCl_2 \cdot 4H_2O$ to 10.0mL of HCl solution. Mix thoroughly. Add distilled/deionized water and bring volume to 1.0L. Add remaining components. Mix thoroughly. Sparge with 80% N_2 + 20% CO_2. Autoclave for 15 min at 15 psi pressure–121°C.

Solution B:

Composition per 100.0mL:

$NaHCO_3$	5.0g

Preparation of Solution B: Add $NaHCO_3$ to distilled/deionized water and bring volume to 100.0mL. Mix thoroughly. Sparge with 100% N_2 gas mixture. Autoclave for 15 min at 15 psi pressure–121°C. Cool to 25°C.

Solution C:

Composition per 100.0mL:

Fructose	5.0g

Preparation of Solution C: Add fructose to distilled/deionized water and bring volume to 100.0mL. Mix thoroughly. Sparge with 100% N_2 gas mixture. Autoclave for 15 min at 15 psi pressure–121°C. Cool to 25°C.

Solution D:

Composition per 10.0mL:

$Na_2S \cdot 9H_2O$	0.125g

Preparation of Solution D: Add $Na_2S \cdot 9H_2O$ to distilled/deionized water and bring volume to 10.0mL. Mix thoroughly. Autoclave under 100% N_2 for 15 min at 15 psi pressure–121°C. Cool to 25°C.

Preparation of Medium: Prepare and dispense medium under 80% N_2 + 20% CO_2 gas atmosphere. Sequentially add 50.0mL solution B, 10.0mL solution C, and 10.0mL solution D, to 940.0mL solution A. Distribute anaerobically under 80% N_2 + 20% CO_2 into appropriate vessels. The pH should be 7.0–7.2.

Use: For the cultivation of *Propionispora vibrioides.*

Propionivibrio/Acetivibrio/Formivibrio Medium

Composition per 1012.0mL:

Solution A	950.0mL
Solution E	30.0mL
Solution D (Vitamin solution)	10.0mL
Solution F	10.0mL
Solution G	10.0mL
Solution B (Trace elements solution SL-10)	1.0mL
Solution C (Selenite-tungstate solution)	1.0mL

pH 6.7 ± 0.2 at 25°C

Solution A:

Composition per 950.0mL:

KH_2PO_4	1.4g
NH_4Cl	0.5g
$MgCl_2 \cdot 6H_2O$	0.2g
$CaCl_2 \cdot 2H_2O$	0.15g
Resazurin	1.0mg

Preparation of Solution A: Add components to distilled/deionized water and bring volume to 950.0mL. Mix thoroughly. Sparge with 80% N_2 + 20% CO_2. Autoclave for 15 min at 15 psi pressure–121°C.

Solution B (Trace Elements Solution SL-10):

Composition per liter:

$FeCl_2 \cdot 4H_2O$	1.5g
$CoCl_2 \cdot 6H_2O$	190.0mg
$MnCl_2 \cdot 4H_2O$	100.0mg
$ZnCl_2$	70.0mg
$Na_2MoO_4 \cdot 2H_2O$	36.0mg
$NiCl_2 \cdot 6H_2O$	24.0mg
H_3BO_3	6.0mg
$CuCl_2 \cdot 2H_2O$	2.0mg
HCl (25% solution)	10.0mL

Preparation of Solution B (Trace Elements Solution SL-10): Add $FeCl_2 \cdot 4H_2O$ to 10.0mL of HCl solution. Mix thoroughly. Add distilled/deionized water and bring volume to 1.0L. Add remaining components. Mix thoroughly. Sparge with 80% N_2 + 20% CO_2. Autoclave for 15 min at 15 psi pressure–121°C.

Solution C (Selenite-Tungstate Solution):

Composition per liter:

NaOH	0.5g
$Na_2WO_4 \cdot 2H_2O$	4.0mg
$Na_2SeO_3 \cdot 5H_2O$	3.0mg

Preparation of Solution C (Selenite-Tungstate Solution): Add components to distilled/deionized water and bring volume to 1.0L. Mix thoroughly. Sparge with 80% N_2 + 20% CO_2. Autoclave for 15 min at 15 psi pressure–121°C.

Solution D (Vitamin Solution):

Composition per liter:

Pyridoxine·HCl	10.0mg
Calcium DL-pantothenate	5.0mg
Lipoic acid	5.0mg
Nicotinic acid	5.0mg
p-Aminobenzoic acid	5.0mg
Riboflavin	5.0mg
Thiamine·HCl	5.0mg
Biotin	2.0mg
Folic acid	2.0mg
Vitamin B_{12}	0.1mg

Preparation of Solution D (Vitamin Solution): Add components to distilled/deionized water and bring volume to 1.0L. Mix thoroughly. Sparge with 80% N_2 + 20% CO_2. Filter sterilize.

Solution E:

Composition per 30.0mL:

$NaHCO_3$	1.5g

Preparation of Solution E: Add $NaHCO_3$ to distilled/deionized water and bring volume to 30.0mL. Mix thoroughly. Filter sterilize. Sparge with 80% N_2 + 20% CO_2.

Solution F:

Composition per 10.0mL:

Disodium maleate	1.6g

Preparation of Solution F: Add disodium maleate to distilled/deionized water and bring volume to 10.0mL. Mix thoroughly. Sparge with 80% N_2 + 20% CO_2. Autoclave for 15 min at 15 psi pressure–121°C.

Solution G:

Composition per 10.0mL:

$Na_2S \cdot 9H_2O$	0.25g

Preparation of Solution G: Add $Na_2S \cdot 9H_2O$ to distilled/deionized water and bring volume to 10.0mL. Mix thoroughly. Gas under 100% N_2. Autoclave for 15 min at 15 psi pressure–121°C.

Preparation of Medium: Aseptically and anaerobically combine 950.0mL of sterile solution A with 1.0mL of sterile solution B, 1.0mL of sterile solution C, 10.0mL of sterile solution D, 30.0mL of sterile solution E, 10.0mL of sterile solution F, and 10.0mL of sterile solution G, in that order. Mix thoroughly. Final pH should be 6.7–6.8. Aseptically and anaerobically distribute into sterile tubes or flasks under 80% N_2 + 20% CO_2.

Use: For the cultivation and maintenance of *Propionivibrio dicarboxylicus*.

Propionivibrio/Acetivibrio/Formivibrio Medium

Composition per 1012.0mL:

Solution A	950.0mL
Solution E	30.0mL
Solution D (Vitamin solution)	10.0mL
Solution F	10.0mL
Solution G	10.0mL
Solution B (Trace elements solution SL-10)	1.0mL
Solution C (Selenite-tungstate solution)	1.0mL

pH 7.7–7.9 at 25°C

Solution A:

Composition per 950.0mL:

KH_2PO_4	1.4g
NH_4Cl	0.5g
$MgCl_2 \cdot 6H_2O$	0.2g
$CaCl_2 \cdot 2H_2O$	0.15g
Yeast extract	50.0mg
Resazurin	1.0mg

Preparation of Solution A: Add components to distilled/deionized water and bring volume to 950.0mL. Mix thoroughly. Sparge with 100% N_2. Autoclave for 15 min at 15 psi pressure–121°C.

Solution B (Trace Elements Solution SL-10):

Composition per liter:

$FeCl_2 \cdot 4H_2O$	1.5g
$CoCl_2 \cdot 6H_2O$	190.0mg
$MnCl_2 \cdot 4H_2O$	100.0mg
$ZnCl_2$	70.0mg
$Na_2MoO_4 \cdot 2H_2O$	36.0mg
$NiCl_2 \cdot 6H_2O$	24.0mg
H_3BO_3	6.0mg
$CuCl_2 \cdot 2H_2O$	2.0mg
HCl (25% solution)	10.0mL

Preparation of Solution B (Trace Elements Solution SL-10): Add $FeCl_2 \cdot 4H_2O$ to 10.0mL of HCl solution. Mix thoroughly. Add distilled/deionized water and bring volume to 1.0L. Add remaining components. Mix thoroughly. Sparge with 100% N_2. Autoclave for 15 min at 15 psi pressure–121°C.

Solution C (Selenite-Tungstate Solution):

Composition per liter:

NaOH	0.5g
$Na_2WO_4 \cdot 2H_2O$	4.0mg
$Na_2SeO_3 \cdot 5H_2O$	3.0mg

Preparation of Solution C (Selenite-Tungstate Solution): Add components to distilled/deionized water and bring volume to 1.0L. Mix thoroughly. Sparge with 100% N_2. Autoclave for 15 min at 15 psi pressure–121°C.

Solution D (Vitamin Solution):

Composition per liter:

Pyridoxine·HCl	10.0mg
Calcium DL-pantothenate	5.0mg
Lipoic acid	5.0mg
Nicotinic acid	5.0mg
p-Aminobenzoic acid	5.0mg
Riboflavin	5.0mg
Thiamine·HCl	5.0mg
Biotin	2.0mg
Folic acid	2.0mg
Vitamin B_{12}	0.1mg

Preparation of Solution D (Vitamin Solution): Add components to distilled/deionized water and bring volume to 1.0L. Mix thoroughly. Sparge with 100% N_2. Filter sterilize.

Solution E:

Composition per 30.0mL:

$NaHCO_3$	1.5g

Preparation of Solution E: Add $NaHCO_3$ to distilled/deionized water and bring volume to 30.0mL. Mix thoroughly. Filter sterilize. Sparge with 100% N_2.

Solution F:

Composition per 10.0mL:

Cinnamic acid	1.6g
NaOH (1*N* solution)	variable

Preparation of Solution F: Add cinnamic acid to distilled/deionized water and bring volume to 8.0mL. Mix thoroughly. Add sufficient quantity of 1*N* NaOH solution to bring pH to 7.8. Sparge with 100% N_2. Autoclave for 15 min at 15 psi pressure–121°C.

Solution G:

Composition per 10.0mL:

$Na_2S \cdot 9H_2O$	0.25g

Preparation of Solution G: Add $Na_2S \cdot 9H_2O$ to distilled/deionized water and bring volume to 10.0mL. Mix thoroughly. Sparge with 100% N_2. Autoclave for 15 min at 15 psi pressure–121°C.

Preparation of Medium: Aseptically and anaerobically under 100% N_2 combine 950.0mL of sterile solution A with 1.0mL of sterile solution B, 1.0mL of sterile solution C, 10.0mL of sterile solution D, 30.0mL of sterile solution E, 10.0mL of sterile solution F, and 10.0mL of sterile solution G, in that order. Mix thoroughly. Final pH should be 7.7–7.9. If necessary, add about 15.0mL of sterile anaerobic 5% Na_2CO_3 solution to 1.0L of medium to adjust pH. Aseptically and anaerobically distribute into sterile tubes or flasks under 100% N_2.

Use: For the cultivation and maintenance of *Acetivibrio multivorans*.

Propionivibrio/Acetivibrio/Formivibrio Medium

Composition per 1012.0mL:

Solution A	950.0mL
Solution E	30.0mL
Solution D (Vitamin solution)	10.0mL
Solution F	10.0mL
Solution G	10.0mL
Solution B (Trace elements solution SL-10)	1.0mL
Solution C (Selenite-tungstate solution)	1.0mL

pH 7.5–7.7 at 25°C

Solution A:

Composition per 950.0mL:

KH_2PO_4	1.4g
NH_4Cl	0.5g
$MgCl_2 \cdot 6H_2O$	0.2g
$CaCl_2 \cdot 2H_2O$	0.15g
Yeast extract	50.0mg
Resazurin	1.0mg

Preparation of Solution A: Add components to distilled/deionized water and bring volume to 950.0mL. Mix thoroughly. Sparge with 100% N_2. Autoclave for 15 min at 15 psi pressure–121°C.

Solution B (Trace Elements Solution SL-10):

Composition per liter:

$FeCl_2 \cdot 4H_2O$	1.5g
$CoCl_2 \cdot 6H_2O$	190.0mg
$MnCl_2 \cdot 4H_2O$	100.0mg
$ZnCl_2$	70.0mg
$Na_2MoO_4 \cdot 2H_2O$	36.0mg
$NiCl_2 \cdot 6H_2O$	24.0mg
H_3BO_3	6.0mg
$CuCl_2 \cdot 2H_2O$	2.0mg
HCl (25% solution)	10.0mL

Preparation of Solution B (Trace Elements Solution SL-10): Add $FeCl_2 \cdot 4H_2O$ to 10.0mL of HCl solution. Mix thoroughly. Add distilled/deionized water and bring volume to 1.0L. Add remaining components. Mix thoroughly. Sparge with 100% N_2. Autoclave for 15 min at 15 psi pressure–121°C.

Solution C (Selenite-Tungstate Solution):

Composition per liter:

NaOH	0.5g
$Na_2WO_4 \cdot 2H_2O$	4.0mg
$Na_2SeO_3 \cdot 5H_2O$	3.0mg

Preparation of Solution C (Selenite-Tungstate Solution): Add components to distilled/deionized water and bring volume to 1.0L. Mix thoroughly. Sparge with 100% N_2. Autoclave for 15 min at 15 psi pressure–121°C.

Solution D (Vitamin Solution):

Composition per liter:

Pyridoxine·HCl	10.0mg
Calcium DL-pantothenate	5.0mg
Lipoic acid	5.0mg
Nicotinic acid	5.0mg
p-Aminobenzoic acid	5.0mg
Riboflavin	5.0mg
Thiamine·HCl	5.0mg
Biotin	2.0mg
Folic acid	2.0mg
Vitamin B_{12}	0.1mg

Preparation of Solution D (Vitamin Solution): Add components to distilled/deionized water and bring volume to 1.0L. Mix thoroughly. Sparge with 100% N_2. Filter sterilize.

Solution E:

Composition per 30.0mL:

$NaHCO_3$	1.5g

Preparation of Solution E: Add $NaHCO_3$ to distilled/deionized water and bring volume to 30.0mL. Mix thoroughly. Filter sterilize. Sparge with 100% N_2.

Solution F:

Composition per 10.0mL:

Trisodium citrate	2.94g

Preparation of Solution F: Add trisodium citrate to distilled/deionized water and bring volume to 10.0mL. Mix thoroughly. Sparge with 100% N_2. Autoclave for 15 min at 15 psi pressure–121°C.

Solution G:

Composition per 10.0mL:

$Na_2S \cdot 9H_2O$	0.25g

Preparation of Solution G: Add $Na_2S \cdot 9H_2O$ to distilled/deionized water and bring volume to 10.0mL. Mix thoroughly. Sparge with 100% N_2. Autoclave for 15 min at 15 psi pressure–121°C.

Preparation of Medium: Aseptically and anaerobically under 100% N_2 combine 950.0mL of sterile solution A with 1.0mL of sterile solution B, 1.0mL of sterile solution C, 10.0mL of sterile solution D, 30.0mL of sterile solution E, 10.0mL of sterile solution F, and 10.0mL of sterile solution G, in that order. Mix thoroughly. Final pH should be 7.5–7.7. If necessary, add sterile anaerobic 5% Na_2CO_3 solution to adjust pH. Aseptically and anaerobically distribute into sterile tubes or flasks under 100% N_2.

Use: For the cultivation and maintenance of *Formivibrio citricus*.

Proskauer-Beck Medium for *Mycobacterium*

Composition per liter:

Asparagine	5.0g
KH_2PO_4	5.0g
Magnesium citrate	2.5g
$MgSO_4 \cdot 7H_2O$	0.6g
Glycerol	20.0mL

pH 7.4 ± 0.2 at 25°C

Preparation of Medium: Add components, one at a time, to distilled/deionized water and bring volume to 1.0L. Mix thoroughly. Make sure one salt is totally dissolved before the next one is added. Adjust pH to 7.8 with 40% NaOH. Autoclave for 15 min at 15 psi pressure–121°C. The pH of the medium after autoclaving should be 7.4. Filter through Whatman #1 filter paper to remove any precipitate. Distribute into tubes or flasks. Autoclave again for 15 min at 15 psi pressure–121°C.

Use: For the cultivation and maintenance of *Mycobacterium tuberculosis*.

Prosthecobacter Medium

Composition per liter:

Glucose	0.25g
$(NH_4)_2SO_4$	0.25g
Na_2HPO_4	0.071g
Hutner's mineral base	20.0mL

Hutner's Mineral Base:

Composition per liter:

$MgSO_4 \cdot 7H_2O$	29.7g
Nitrilotriacetic acid	10.0g
$CaCl_2 \cdot 2H_2O$	3.34g
$FeSO_4 \cdot 7H_2O$	0.1g
$(NH_4)_2MoO_4$	9.25mg
Metals "44"	50.0mL

Preparation of Hutner's Mineral Base: Add nitrilotriacetic acid to 500.0mL of distilled/deionized water. Dissolve by adjusting pH to 6.5 with KOH. Add remaining components. Readjust pH to 7.2 with H_2SO_4 or KOH. Add distilled/deionized water to 1.0L. Store at 5°C.

Metals "44":

Composition per 100.0mL:

$ZnSO_4 \cdot 7H_2O$	1.1g
$FeSO_4 \cdot 7H_2O$	0.5g
EDTA	0.25g

$MnSO_4 \cdot 7H_2O$ 0.154g
$CuSO_4 \cdot 5H_2O$ 0.04g
$Co(NO_3)_2 \cdot 6H_2O$ 0.025g
$Na_2B_4O_7 \cdot 10H_2O$ 0.018g

Preparation of Metals "44": Add a few drops of H_2SO_4 to distilled/deionized water to inhibit precipitate formation. Add components to acidified distilled/deionized water and bring volume to 100.0mL. Mix thoroughly.

Preparation of Medium: Add components to distilled/deionized water and bring volume to 1.0L. Mix thoroughly. Distribute into tubes or flasks. Autoclave for 15 min at 15 psi pressure–121°C.

Use: For the cultivation of *Prosthecobacter fusiformis*.

Prosthecomicrobium and *Ancalomicrobium* Medium

Composition per 1030.0mL:

Ammonium sulfate 0.25g
Glucose 0.25g
Na_2HPO_4 .... 71.0mg
Modified Hutner's basal salts 20.0mL
Vitamin solution 10.0mL

Modified Hutner's Basal Salts:

Composition per liter:

$MgSO_4 \cdot 7H_2O$ 29.7g
Nitrilotriacetic acid 10.0g
$CaCl_2 \cdot 2H_2O$ 3.34g
$FeSO_4 \cdot 7H_2O$ 99.0mg
Ammonium molybdate 9.25mg
Metals "44" 50.0mL

Preparation of Modified Hutner's Basal Salts: Dissolve the nitrilotracetic acid first and neutralize the solution with KOH. Add other components and adjust the pH to 7.2 with KOH or H_2SO_4. There may be a slight precipitate. Store at 5°C.

Metals "44"

Composition per liter:

$ZnSO_4 \cdot 7H_2O$ 1.1g
$FeSO_4 \cdot 7H_2O$ 0.5g
$CuSO_4 \cdot 5H_2O$ 0.04g
EDTA 0.25g
$MnSO_4 \cdot 7H_2O$ 0.154g
$Co(NO_3)_2 \cdot 6H_2O$ 0.025g
$Na_2B_4O_7 \cdot 10H_2O$ 0.018g

Preparation of Metals "44": Add components to distilled/deionized water and bring volume to 100.0mL. Mix thoroughly. Autoclave for 15 min at 15 psi pressure–121°C. Add aseptically to sterile modified Hutner's basal salts solution.

Vitamin Solution:

Composition per liter:

Cyanocobalamin 10.0mg
Pyridoxine·HCl 10.0mg
Thiamine·HCl 10.0mg
Calcium DL-pantothenate 5.0mg
Nicotinamide 5.0mg
Biotin 2.0mg
Folic acid 2.0mg

Preparation of Vitamin Solution: Add components to distilled/deionized water and bring volume to 1.0L. Mix thoroughly. Filter sterilize.

Preparation of Medium: Add components, except modified Hutner's basal salts solution and vitamin solution, to distilled/deionized water and bring volume to 970.0mL. Mix thoroughly. Autoclave for 15 min at 15 psi pressure–121°C. Aseptically add 20.0mL of modified Hutner's basal salts solution and 10.0mL of sterile vitamin solution. Mix thoroughly. Aseptically distribute into sterile tubes or flasks.

Use: For the cultivation of *Prosthecomicrobium enhydrum* and *Prosthecomicrobium pneumaticum*.

Proteose Agar

Composition per liter:

Agar 15.0g
Proteose peptone No. 3 .... 15.0g
Yeast extract 7.5g
Casamino acids 5.0g
K_2HPO_4 .... 5.0g
$(NH_4)_2SO_4$ 1.5g
Starch, soluble 1.0g

pH 9.0 ± 0.2 at 25°C

Preparation of Medium: Add components to distilled/deionized water and bring volume to 1.0L. Mix thoroughly. Gently heat and bring to boiling. Distribute into tubes in 10.0mL volumes. Autoclave for 15 min at 15 psi pressure–121°C. Allow tubes to cool in a slanted position.

Use: For the cultivation of *Vibrio* species from foods.

Proteose HiVeg Agar

Composition per liter:

Agar 15.0g
Plant peptone No. 3 .... 15.0g
K_2HPO_4 .... 5.0g
Plant acid hydrolysate 5.0g
Yeast extract 7.5g
$(NH_4)_2SO_4$ 1.5g
Starch, soluble 1.0g

pH 9.0 ± 0.2 at 25°C

Source: This medium is available as a premixed powder from HiMedia.

Preparation of Medium: Add components to distilled/deionized water and bring volume to 1.0L. Mix thoroughly. Gently heat while stirring and bring to boiling. Distribute into tubes or flasks. Autoclave for 10 min at 15 psi pressure–121°C. Pour into sterile Petri dishes or leave in tubes.

Use: For the cultivation and maintenance of *Vibrio* species from foods.

Proteose No. 3 Agar (ATCC Medium 50)

Composition per liter:

Proteose peptone No. 3 .... 20.0g
Agar 15.0g
Na_2HPO_4 .... 5.0g
NaCl 5.0g
Glucose 0.5g

pH 7.3 ± 0.2 at 25°C

Source: This medium is available as a premixed powder from BD Diagnostic Systems.

Preparation of Medium: Add components to distilled/deionized water and bring volume to 1.0L. Mix thoroughly. Gently heat while stirring and bring to boiling. Distribute into tubes or flasks. Autoclave for 10 min at 15 psi pressure–121°C. Pour into sterile Petri dishes or leave in tubes.

Use: For the cultivation and maintenance of *Escherichia coli* and other bacteria.

Proteose No. 3 Agar

Composition per 1010.0mL:

Proteose peptone No. 3	20.0g
Agar	15.0g
Na_2HPO_4	5.0g
NaCl	5.0g
Glucose	0.5g
Hemoglobin solution	500.0mL
Supplement A	10.0mL

pH 7.3 ± 0.2 at 25°C

Source: This medium is available as a premixed powder from BD Diagnostic Systems.

Hemoglobin Solution:

Composition per 500.0mL:

Hemoglobin	10.0g

Preparation of Hemoglobin Solution: Add hemoglobin to distilled/deionized water and bring volume to 500.0mL. Mix thoroughly. Autoclave for 15 min at 15 psi pressure–121°C. Cool to 45°–50°C.

Supplement A:

Composition per 10.0.mL:

Supplement A contains yeast concentrate with Crystal Violet.

Preparation of Supplement A: Add components to distilled/deionized water and bring volume to 10.0mL. Mix thoroughly. Filter sterilize.

Preparation of Medium: Add components, except hemoglobin solution and supplement A, to distilled/deionized water and bring volume to 500.0mL. Mix thoroughly. Gently heat and bring to boiling. Autoclave for 15 min at 15 psi pressure–121°C. Cool to 50°–60°C. Aseptically add 500.0mL of sterile hemoglobin solution and 10.0mL of sterile supplement A. Mix thoroughly. Pour into sterile Petri dishes or distribute into sterile tubes.

Use: For the isolation and cultivation of *Neisseria* species, *Hemophilus* species, and other fastidious bacteria. For the cultivation and maintenance of *Escherichia coli.*

Proteose No. 3 Agar

Composition per 1010.0mL:

Proteose peptone No. 3	20.0g
Agar	15.0g
Na_2HPO_4	5.0g
NaCl	5.0g
Glucose	0.5g
Hemoglobin solution	500.0mL
Supplement B	10.0mL

pH 7.3 ± 0.2 at 25°C

Source: This medium is available as a premixed powder from BD Diagnostic Systems.

Hemoglobin Solution:

Composition per 500.0mL:

Hemoglobin	10.0g

Preparation of Hemoglobin Solution: Add hemoglobin to distilled/deionized water and bring volume to 500.0mL. Mix thoroughly. Autoclave for 15 min at 15 psi pressure–121°C. Cool to 45°–50°C.

Supplement B:

Composition per 10.0mL:

Supplement B contains yeast concentrate, glutamine, coenzyme, cocarboxylase, hematin, and growth factors.

Preparation of Supplement B: Add components to distilled/deionized water and bring volume to 10.0mL. Mix thoroughly. Filter sterilize.

Preparation of Medium: Add components, except hemoglobin solution and supplement B, to distilled/deionized water and bring volume to 500.0mL. Mix thoroughly. Gently heat and bring to boiling. Autoclave for 15 min at 15 psi pressure–121°C. Cool to 50°–60°C. Aseptically add 500.0mL of sterile hemoglobin solution and 10.0mL of sterile supplement B. Mix thoroughly. Pour into sterile Petri dishes or distribute into sterile tubes.

Use: For the isolation and cultivation of *Neisseria* species, *Hemophilus* species, and other fastidious bacteria. For the cultivation and maintenance of *Escherichia coli.*

Proteose No. 3 Agar

Composition per 1010.0mL:

Proteose peptone No. 3	20.0g
Agar	15.0g
Na_2HPO_4	5.0g
NaCl	5.0g
Glucose	0.5g
Hemoglobin solution	500.0mL
Supplement VX	10.0mL

pH 7.3 ± 0.2 at 25°C

Source: This medium is available as a premixed powder from BD Diagnostic Systems.

Hemoglobin Solution:

Composition per 500.0mL:

Hemoglobin	10.0g

Preparation of Hemoglobin Solution: Add hemoglobin to distilled/deionized water and bring volume to 500.0mL. Mix thoroughly. Autoclave for 15 min at 15 psi pressure–121°C. Cool to 45°–50°C.

Supplement VX:

Composition per 10.0mL:

Supplement B contains essential growth factors.

Preparation of Supplement VX: Add components to distilled/deionized water and bring volume to 10.0mL. Mix thoroughly. Filter sterilize.

Preparation of Medium: Add components, except hemoglobin solution and supplement VX, to distilled/deionized water and bring volume to 500.0mL. Mix thoroughly. Gently heat and bring to boiling. Autoclave for 15 min at 15 psi pressure–121°C. Cool to 50°–60°C. Aseptically add 500.0mL of sterile hemoglobin solution and 10.0mL of sterile supplement VX. Mix thoroughly. Pour into sterile Petri dishes or distribute into sterile tubes.

Use: For the isolation and cultivation of *Neisseria* species, *Hemophilus* species, and other fastidious bacteria. For the cultivation and maintenance of *Escherichia coli.*

Proteose Yeast Extract Medium

Composition per liter:

Proteose peptone	20.0g
Glucose	10.0g
Yeast extract	5.0g

Preparation of Medium: Add components to distilled/deionized water and bring volume to 1.0L. Mix thoroughly. Distribute into tubes or flasks. Autoclave for 15 min at 15 psi pressure–121°C.

Use: For the cultivation of a variety of bacteria.

Prototheca Isolation Agar (PIM)

Composition per liter:

Agar	20.0g
Glucose	10.0g
Potassium hydrogen phthalate	10.0g
NaOH	0.9g
NH_4Cl	0.3g
5-Fluorocytosine	0.25g
KH_2PO_4	0.2g
$MgSO_4$	0.1g
Thiamine·HCl	0.001g

pH 5.1 ± 0.1 at 25°C

Preparation of Medium: Add components to distilled/deionized water and bring volume to 1.0L. Mix thoroughly. Adjust pH to 5.1. Gently heat and bring to boiling. Distribute into tubes or flasks. Autoclave for 15 min at 15 psi pressure–121°C. Pour into sterile Petri dishes or leave in tubes.

Use: For the cultivation of *Prototheca moriformis* and *Prototheca ulmea.*

Provasoli Medium

Composition per liter:

NaCl	11.75g
$MgCl_2·6H_2O$	5.35g
Na_2SO_4	2.0g
$CaCl_2·2H_2O$	0.75g
Tris(hydroxymethyl)aminomethane	0.5g
KCl	0.35g
Na_2HPO_4	0.05g

pH 7.6 ± 0.2 at 25°C

Preparation of Medium: Add components to distilled/deionized water and bring volume to 1.0L. Mix thoroughly. Distribute into tubes or flasks. Autoclave for 15 min at 15 psi pressure–121°C.

Use: For the isolation and cultivation of *Leucothrix* species from marine habitats.

Prune Agar

Composition per liter:

Agar	17.0g
Lactose	5.0g
Prunes	5.0g
Yeast extract	1.0g

pH 5.8–6.0 at 25°C

Preparation of Medium: Add prunes to 1.0L of distilled/deionized water. Gently heat and bring to boiling. Continue boiling for 30 min. Pass the prunes and liquid through a sieve to include as much of the prune pulp as possible. Add agar. Gently heat and bring to boiling. Remove from heat and add the lactose and yeast extract. Mix thoroughly. Adjust pH to 5.8–6.0. Distribute into tubes or flasks. Autoclave for 20 min at 20 psi pressure–126°C. Pour into sterile Petri dishes or leave in tubes.

Use: For the cultivation and maintenance of *Pyricularia oryzae.*

PRYES Agar

See: **Pentachloronitrobenzene Rose Bengal Yeast Extract Sucrose Agar**

PSA (Potato Sucrose Agar)

Composition per liter:

Potatoes	200.0g
Agar	20.0g
Sucrose	20.0g

Preparation of Medium: Slice potatoes with skin. Place 200.0g of potatoes in 1.0L of distilled/deionized water. Gently heat and bring to boiling. Allow to boil for 20 min. Filter through Whatman filter paper. Add agar and sucrose to filtrate. Bring volume to 1.0L with distilled/deionized water. Mix thoroughly. Gently heat and bring to boiling. Distribute into tubes or flasks. Autoclave for 15 min at 15 psi pressure–121°C. Pour into sterile Petri dishes or leave in tubes.

Use: For the cultivation of *Aspergillus oryzae, Cylindrodendrum album, Nectria ipomoeae, Nectria haematococca,* and other fungi.

PSD Agar

See: **Peptone Starch Dextrose Agar**

PSE Agar

See: **Pfizer Selective *Enterococcus* Agar**

Pseudoalteromonas spiralis Medium (DSMZ Medium 1072)

Composition per liter:

NaCl	15.0g
Na-acetate	1.0g
Yeast extract	0.5g
Casamino acids	0.5g
$MgSO_4·7H_2O$	0.5g
KH_2PO_4	0.3g
NH_4Cl	0.3g
KCl	0.3g
$CaCl_2·2H_2O$	0.05g
Trace elements solution	2.0mL
Vitamin solution	2.0mL

pH 7.8 ± 0.2 at 25°C

Trace Elements Solution:

Composition per liter:

$MnCl_2·4H_2O$	0.3g
$FeSO_4·7H_2O$	3.0mg

Preparation of Trace Elements Solution: Add components to distilled/deionized water and bring volume to 1.0L. Mix thoroughly.

Vitamin Solution:

Composition per liter:

Thiamine-HCl·$2H_2O$	400.0mg
Nicotinic acid	400.0mg
Biotin	80.0mg
Vitamin B_{12}	20.0mg

Preparation of Vitamin Solution: Add components to distilled/deionized water and bring volume to 1.0L. Mix thoroughly. Filter sterilize.

Preparation of Medium: Add components, except vitamin solution, to distilled/deionized water and bring volume to 1.0L. Mix thoroughly. Adjust pH to 5.9. Distribute into tubes or flasks. Gently heat while stirring and bring to boiling. Mix thoroughly. Autoclave for 15 min at 15 psi pressure–121°C. Cool to room termperature. Aseptically add vitamin solution. Mix thorougly. Adjust to pH 7.8 with $NaHCO_3$. Aseptically distribute into culture vessels.

Use: For the cultivation of *Pseudoalteromonas spiralis.*

Pseudoamycolata halophobica Medium

Composition per liter:

Glucose	5.0g
Peptone	5.0g
Yeast extract	3.0g
K_2HPO_4	0.2g

pH 6.8 ± 0.2 at 25°C

Preparation of Medium: Add components to distilled/deionized water and bring volume to 1.0L. Mix thoroughly. Adjust pH to 6.8. Distribute into tubes or flasks. Autoclave for 15 min at 15 psi pressure–121°C.

Use: For the cultivation of *Pseudoamycolata halophobica.*

Pseudobutyrivibrio Medium

Composition per liter:

Disodium succinate	5.0g
Yeast extract	5.0g
NaCl	0.45g
$(NH_4)_2SO_4$	0.45g
K_2HPO_4	0.225g
KH_2PO_4	0.225g
$MgSO_4·7H_2O$	0.09g
$CaCl_2·2H_2O$	0.06g
Indigocarmine	5.0mg
Rumen fluid, clarified	400.0mL
Glucose solution	20.0mL
$NaHCO_3$ solution	10.0mL
L-Cysteine·HCl·H_2O solution	10.0mL
$Na_2S·9H_2O$ solution	10.0mL

pH 6.6–6.8 at 25°C

Glucose Solution:

Composition per 20.0mL:

D-Glucose	5.0g

Preparation of Glucose Solution: Add glucose to distilled/deionized water and bring volume to 20.0mL. Mix thoroughly. Sparge with 100% N_2. Autoclave for 15 min at 15 psi pressure–121°C.

$NaHCO_3$ Solution:

Composition per 10.0mL:

$NaHCO_3$	6.4g

Preparation of $NaHCO_3$ Solution: Add $NaHCO_3$ to distilled/deionized water and bring volume to 10.0mL. Mix thoroughly. Sparge with 100% N_2. Autoclave for 15 min at 15 psi pressure–121°C.

L-Cysteine·HCl·H_2O Solution:

Composition per 10.0mL:

L-Cysteine·HCl·H_2O	0.3g

Preparation of L-Cysteine·HCl·H_2O Solution: Add L-cysteine·HCl·H_2O to distilled/deionized water and bring volume to 10.0mL. Mix thoroughly. Sparge with 100% N_2. Autoclave for 15 min at 15 psi pressure–121°C.

$Na_2S·9H_2O$ Solution:

Composition per 10.0mL:

$Na_2S·9H_2O$	0.3g

Preparation of $Na_2S·9H_2O$ Solution: Add $Na_2S·9H_2O$ to distilled/deionized water and bring volume to 10.0mL. Mix thoroughly. Sparge with 100% N_2. Autoclave for 15 min at 15 psi pressure–121°C. Before use, neutralize to pH 7.0 with sterile HCl.

Preparation of Medium: Add components, except glucose solution, L-cysteine·HCl·H_2O solution, $Na_2S·9H_2O$ solution, and $NaHCO_3$ solution, to distilled/deionized water and bring volume to 960.0mL. Mix thoroughly. Gently heat and bring to boiling. Cool to room temperature while sparging with 100% CO_2. Autoclave for 15 min at 15 psi pressure–121°C. Aseptically and anaerobically add 10.0mL of sterile glucose solution, 10.0mL of sterile L-cysteine·HCl solution, 10.0mL of sterile $Na_2S·9H_2O$ solution, and 10.0mL of sterile $NaHCO_3$ solution. Mix thoroughly.

Use: For the cultivation of *Pseudobutyrivibrio ruminis.*

Pseudomonas aeruginosa Agar (*PA* Agar) (m–PA Agar) (m-*Pseudomonas aeruginosa* Agar)

Composition per liter:

Agar	15.0g
$Na_2S_2O_3$	6.8g
L-Lysine·HCl	5.0g
NaCl	5.0g
Xylose	2.5g
Yeast extract	2.0g
Lactose	1.25g
Sucrose	1.25g
Ferric ammonium citrate	0.8g
Sulfapyridine	0.176g
Cycloheximide	0.15g
Phenol Red	0.08g
Nalidixic acid	0.037g
Kanamycin	8.5mg

pH 7.1 ± 0.2 at 25°C

Caution: Cycloheximide is toxic. Avoid skin contact or aerosol formation and inhalation.

Preparation of Medium: Add components—except sulfapyridine, cycloheximide, nalidixic acid, and kanamycin—to distilled/deionized water and bring volume to 1.0L. Mix thoroughly. Adjust pH to 6.5. Autoclave for 15 min at 15 psi pressure–121°C. Cool to 55°–60°C. Readjust pH to 7.1. Aseptically add the sulfapyridine, cycloheximide, nalidixic acid, and kanamycin. Mix thoroughly. Pour into 50mm × 12mm Petri dishes in 3.0mL volumes.

Use: For the cultivation and estimation of numbers of *Pseudomonas aeruginosa* in water by the membrane filter method.

Pseudomonas Agar F (DSMZ Medium 907)

Composition per liter:

Agar	15.0g
Tryptone	10.0g
Proteose peptone	10.0g
Glycerol	10.0g
K_2HPO_4	1.5g
$MgSO_4$	1.5g

pH 7.1 ± 0.2 at 25°C

Preparation of Medium: Add components to distilled/deionized water and bring volume to 1.0L. Mix thoroughly. Gently heat and bring to boiling. Distribute into tubes or flasks. Autoclave for 15 min at 15 psi pressure–121°C. Pour into sterile Petri dishes or leave in tubes.

Use: For the cultivation and maintenance of *Schineria larvae.*

Pseudomonas Agar F

Composition per liter:

Proteose peptone No. 3	20.0g
Agar	15.0g
Glycerol	10.0g
Pancreatic digest of casein	10.0g
K_2HPO_4	1.5g
$MgSO_4 \cdot 7H_2O$	0.73g

pH 7.0 ± 0.2 at 25°C

Preparation of Medium: Add components to distilled/deionized water and bring volume to 1.0L. Mix thoroughly. Gently heat and bring to boiling. Distribute into tubes or flasks. Autoclave for 15 min at 15 psi pressure–121°C. Pour into sterile Petri dishes or leave in tubes.

Use: For the cultivation and observation of fluorescein production in *Pseudomonas* species.

Pseudomonas Agar F

Composition per liter:

Agar	15.0g
Glycerol	10.0g
Proteose peptone No. 3	10.0g
Pancreatic digest of casein	10.0g
K_2HPO_4	1.5g
$MgSO_4 \cdot 7H_2O$	1.5g

pH 7.0 ± 0.2 at 25°C

Source: This medium is available as a premixed powder from BD Diagnostic Systems.

Preparation of Medium: Add components to distilled/deionized water and bring volume to 1.0L. Mix thoroughly. Gently heat and bring to boiling. Distribute into tubes or flasks. Autoclave for 15 min at 15 psi pressure–121°C. Pour into sterile Petri dishes or leave in tubes.

Use: For the isolation, cultivation, and differentiation of *Pseudomonas aeruginosa* on the basis of pigment production.

Pseudomonas Agar (for Fluorescein), HiVeg

Composition per liter:

Agar	15.0g
Plant hydrolysate	10.0g
Plant peptone No. 3	10.0g
$MgSO_4$	1.5g
K_2HPO_4	1.5g
Glycerol	10.0mL

pH 7.0 ± 0.2 at 25°C

Source: This medium, without glycerol, is available as a premixed powder from HiMedia.

Preparation of Medium: Add components to distilled/deionized water and bring volume to 1.0L. Mix thoroughly. Gently heat and bring to boiling. Distribute into tubes or flasks. Autoclave for 15 min at 15 psi pressure–121°C. Pour into sterile Petri dishes or leave in tubes.

Use: For the cultivation and observation of fluorescein production in *Pseudomonas* species.

Pseudomonas Agar (for Pyocyanin), HiVeg

Composition per liter:

Plant special peptone	20.0g
Agar	15.0g
K_2SO_4	10.0g
$MnCl_2$	1.4g
Glycerol	10.0mL

pH 7.0 ± 0.2 at 25°C

Source: This medium, without glycerol, is available as a premixed powder from HiMedia.

Preparation of Medium: Add components to distilled/deionized water and bring volume to 1.0L. Mix thoroughly. Gently heat and bring to boiling. Distribute into tubes or flasks. Autoclave for 15 min at 15 psi pressure–121°C. Pour into sterile Petri dishes or leave in tubes.

Use: For the cultivation and observation of pyocyanin production in *Pseudomonas* species.

Pseudomonas Agar P

Composition per liter:

Proteose peptone No. 3	20.0g
Agar	15.0g
Glycerol	10.0g
K_2HPO_4	10.0g
$MgCl_2 \cdot 6H_2O$	1.4g

pH 7.0 ± 0.2 at 25°C

Source: This medium is available as a premixed powder from BD Diagnostic Systems.

Preparation of Medium: Add components to distilled/deionized water and bring volume to 1.0L. Mix thoroughly. Gently heat and bring to boiling. Distribute into tubes or flasks. Autoclave for 15 min at 15 psi pressure–121°C. Pour into sterile Petri dishes or leave in tubes.

Use: For the isolation, cultivation, and differentiation of *Pseudomonas aeruginosa* on the basis of pigment production.

Pseudomonas Asparagine Broth

Composition per liter:

DL-Asparagine	3.0g
K_2HPO_4	1.0g
$MgSO_4 \cdot 7H_2O$	0.5g

pH 7.1 ± 0.2 at 25°C

Source: This medium is available from HiMedia.

Preparation of Medium: Add components to distilled/deionized water and bring volume to 1.0L. Mix thoroughly. Gently heat and bring to boiling. Distribute into tubes or flasks. Autoclave for 15 min at 15 psi pressure–121°C.

Use: For the presumptive identification of *Pseudomonas aeruginosa* from recreational waters.

Pseudomonas Basal Mineral Medium

Composition per liter:

K_2HPO_4	12.5g
KH_2PO_4	3.8g
$(NH_4)_2SO_4$	1.0g
$MgSO_4 \cdot 7H_2O$	0.1g
Carbon source (0.8*M* solution)	100.0mL
Trace elements solution	5.0mL

pH 7.2 ± 0.2 at 25°C

Trace Elements Solution:

Composition per liter:

H_3BO_3	0.232g
$ZnSO_4 \cdot 7H_2O$	0.174g
$FeSO_4(NH_4)_2SO_4 \cdot 6H_2O$	0.116g
$CoSO_4 \cdot 7H_2O$	0.096g
$(NH_4)_6Mo_7O_{24} \cdot 4H_2O$	0.022g
$CuSO_4 \cdot 5H_2O$	8.0mg
$MnSO_4 \cdot 4H_2O$	8.0mg

Preparation of Trace Elements Solution: Add components to distilled/deionized water and bring volume to 1.0L. Mix thoroughly.

Carbon Source:

Composition per 100.0mL:

Glucose	14.4g

Preparation of Carbon Source: Add glucose to distilled/deionized water and bring volume to 100.0mL. Mix thoroughly. Filter sterilize. Other carbon sources may replace glucose. Prepare 0.8*M* carbon source solution.

Preparation of Medium: Add components, except carbon source, to distilled/deionized water and bring volume to 900.0mL. Mix thoroughly. Gently heat and bring to boiling. Autoclave for 15 min at 15 psi pressure–121°C. Cool to 45°–50°C. Aseptically add 100.0mL of sterile carbon source. Mix thoroughly. Aseptically distribute into sterile tubes or flasks.

Use: For the cultivation and differentiation of *Pseudomonas* species based on their ability to grow on different carbon sources.

Pseudomonas bathycetes Medium

Composition per liter:

NaCl	24.0g
Proteose peptone	10.0g
$MgSO_4 \cdot 7H_2O$	7.0g
$MgCl_2$	5.3g
Yeast extract	3.0g
KCl	0.7g

pH 7.2–7.4 at 25°C

Preparation of Medium: Add components to distilled/deionized water and bring volume to 1.0L. Mix thoroughly. Distribute into tubes or flasks. Autoclave for 15 min at 15 psi pressure–121°C.

Use: For the cultivation and maintenance of *Alteromonas haloplanktis, Alteromonas nigrifaciens, Pseudomonas bathycetes,* and *Pseudomonas elongata.*

Pseudomonas CFC Agar

Composition per liter:

Pancreatic digest of gelatin	16.0g
Agar	11.0g
Pancreatic digest of casein	10.0g
K_2SO_4	10.0g
$MgCl_2 \cdot 6H_2O$	1.4g
CFC selective supplement	10.0mL
Glycerol	10.0mL

pH 7.1 ± 0.2 at 25°C

Source: This medium is available as a premixed powder from Oxoid Unipath.

CFC Selective Supplement:

Composition per 10.0mL:

Cephaloridine	0.05g
Fucidin	0.01g
Cetrimide	0.01g

Preparation of CFC Selective Supplement: Add components to distilled/deionized water and bring volume to 10.0mL. Mix thoroughly. Filter sterilize.

Preparation of Medium: Add components, except CFC selective supplement, to distilled/deionized water and bring volume to 990.0mL. Mix thoroughly. Gently heat and bring to boiling. Autoclave for 15 min at 15 psi pressure–121°C. Cool to 45°–50°C. Aseptically add sterile CFC selective supplement. Mix thoroughly. Pour into sterile Petri dishes or distribute into sterile tubes.

Use: For the selective isolation and cultivation of *Pseudomonas* species.

Pseudomonas chloritidismutans Medium (DSMZ Medium 944)

Composition per 1001.0mL:

Solution A	900.0mL
Solution B	50.0mL
Solution C	50.0mL
Vitamin solution	1.0mL

pH 9.0 ± 0.2 at 25°C

Solution A:

Composition per 900mL:

$Na\text{-acetate} \cdot 3H_2O$	2.72g
$NaClO_3$	1.06g
KH_2PO_4	0.41g
Na_2HPO_4	0.53g
Resazurin	0.5mg
Selenite-tungstate solution	4.0mL
Trace elements solution SL-10	1.0mL

Trace Elements Solution SL-10:

Composition per liter:

$FeCl_2 \cdot 4H_2O$	1.5g
$CoCl_2 \cdot 6H_2O$	190.0mg
$MnCl_2 \cdot 4H_2O$	100.0mg
$ZnCl_2$	70.0mg
$Na_2MoO_4 \cdot 2H_2O$	36.0mg
$NiCl_2 \cdot 6H_2O$	24.0mg
H_3BO_3	6.0mg
$CuCl_2 \cdot 2H_2O$	2.0mg
HCl (25% solution)	10.0mL

Preparation of Trace Elements Solution SL-10: Add $FeCl_2 \cdot 4H_2O$ to 10.0mL of HCl solution. Mix thoroughly. Add distilled/deionized water and bring volume to 1.0L. Add remaining components. Mix thoroughly. Sparge with 80% N_2 + 20% CO_2.

Selenite-Tungstate Solution

Composition per liter:

NaOH	0.5g
$Na_2WO_4 \cdot 2H_2O$	4.0mg
$Na_2SeO_3 \cdot 5H_2O$	3.0mg

Preparation of Selenite-Tungstate Solution: Add components to distilled/deionized water and bring volume to 1.0L. Mix thoroughly. Sparge with 100% N_2.

Preparation of Solution A: Add components to distilled/deionized water and bring volume to 900.0mL. Mix thoroughly. Sparge with 100% N_2. Autoclave for 15 min at 15 psi pressure–121°C. Cool to room temperature.

Solution B:

Composition per 50.0mL:

$CaCl_2$	0.11g
$MgCl_2$	0.10g

Preparation of Solution B: Add components to distilled/deionized water and bring volume to 50.0mL. Mix thoroughly. Sparge with 100% N_2. Autoclave for 15 min at 15 psi pressure–121°C. Cool to room temperature.

Solution C:

Composition per 50.0mL:

$NaHCO_3$	3.73g
$Na_2S \cdot 9H_2O$	0.5g
NH_4HCO_3	0.44g

Preparation of Solution C: Add components to distilled/deionized water and bring volume to 50.0mL. Mix thoroughly. Sparge with 100% N_2. Autoclave for 15 min at 15 psi pressure–121°C. Cool to room temperature.

Vitamin Solution:

Composition per liter:

Vitamin B_{12}	50.0mg
Pantothenic acid	50.0mg
Riboflavin	50.0mg
Alpha-lipoic acid	50.0mg
p-Aminobenzoic acid	50.0mg
Thiamine-HCl·$2H_2O$	50.0mg
Nicotinic acid	25.0mg
Nicotine amide	25.0mg
Biotin	20.0mg
Folic acid	20.0mg
Pyridoxamine-HCl	10.0mg

Preparation of Vitamin Solution: Add components to distilled/deionized water and bring volume to 1.0L. Mix thoroughly. Filter sterilize.

Preparation of Medium: Prepare and dispense medium under 100% N_2 gas. Aseptically and anaerobically combine 900.0mL sterile solution A, 50.0mL sterile solution B, 50.0mL sterile solution C, and 1.0mL sterile vitamin solution. Mix thoroughly. The pH should be 9.0. Aseptically and anaerobically distribute into sterile tubes or flasks.

Use: For the cultivation of *Pseudomonas chloritidismutans (Pseudomonas stutzeri).*

Pseudomonas CN Agar

Composition per liter:

Pancreatic digest of gelatin	16.0g
Agar	11.0g
Pancreatic digest of casein	10.0g
K_2SO_4	10.0g
$MgCl_2 \cdot 6H_2O$	1.4g
CN selective supplement	10.0mL
Glycerol	10.0mL

pH 7.1 ± 0.2 at 25°C

Source: This medium is available as a premixed powder from Oxoid Unipath.

CN Selective Supplement:

Composition per 10.0mL:

Cetrimide	0.1g
Sodium nalidixate	7.5mg

Preparation of CN Selective Supplement: Add components to distilled/deionized water and bring volume to 10.0mL. Mix thoroughly. Filter sterilize.

Preparation of Medium: Add components, except CN selective supplement, to distilled/deionized water and bring volume to 990.0mL. Mix thoroughly. Gently heat and bring to boiling. Autoclave for 15 min at 15 psi pressure–121°C. Cool to 45°–50°C. Aseptically add sterile CN selective supplement. Mix thoroughly. Pour into sterile Petri dishes or distribute into sterile tubes.

Use: For the selective isolation and cultivation of *Pseudomonas* species.

Pseudomonas denitrificans Medium (LMG 153)

Composition per liter:

Agar	15.0g
Glucose	10.0g
Yeast extract	5.0g
$FeCl_3$ solution	20.0mL

$FeCl_3$ Solution:

Composition per 100.0mL:

$FeCl_3$	0.03g

Preparation of $FeCl_3$ Solution: Add $FeCl_3$ to distilled/deionized water and bring volume to 100.0mL. Mix thoroughly. Filter sterilize.

Preparation of Medium: Add components, except $FeCl_3$ solution, to distilled/deionized water and bring volume to 980.0mL. Mix thoroughly. Gently heat and bring to boiling. Autoclave for 15 min at 15 psi pressure–121°C. Cool to 45°–50°C. Aseptically add 20.0mL of sterile $FeCl_3$ solution. Mix thoroughly. Pour into sterile Petri dishes or distribute into sterile tubes.

Use: For the cultivation and maintenance of *Pseudomonas* species.

Pseudomonas Denitrification Medium

Composition per liter:

Glycerol	10.0g
KNO_3	10.0g
Yeast extract	3.0g
$(NH_4)_2SO_4$	1.5g
Agar	1.0g
$K_2HPO_4 \cdot 3H_2O$	0.8g
$MgSO_4 \cdot 7H_2O$	0.5g

KH_2PO_4 .... 0.2g
$CaCl_2$ 0.1g

pH 7.2 ± 0.2 at 25°C

Preparation of Medium: Add components to distilled/deionized water and bring volume to 1.0L. Mix thoroughly. Distribute into tubes in 10.0mL volumes. Autoclave for 15 min at 15 psi pressure–121°C.

Use: For the cultivation and differentiation of *Pseudomonas* species based on their ability to produce pyocin and other fluorescent pigments during denitrification.

Pseudomonas halophila Medium

Composition per liter:

Solution A 890.0mL
Solution B 100.0mL
Vitamin solution 10.0mL

pH 7.0 ± 0.2 at 25°C

Solution A:

Composition per 890.0mL:

NaCl 46.8g
$MgSO_4·7H_2O$ 39.4g
Glycerol 5.0g
NH_4Cl 1.0g
Trace elements solution SL-10 .... 1.0mL

Trace Elements Solution SL-10:

Composition per liter:

$FeCl_2·4H_2O$ 1.5g
$CoCl_2·6H_2O$ 0.19g
$MnCl_2·4H_2O$ 100.0mg
$ZnCl_2$ 70.0mg
$Na_2MoO_4·2H_2O$ 36.0mg
$NiCl_2·6H_2O$ 24.0mg
H_3BO_3 .... 6.0mg
$CuCl_2·2H_2O$ 2.0mg
HCl (25% solution) 10.0mL

Preparation of Trace Elements Solution SL-10: Add $FeCl_2·4H_2O$ to 10.0mL of HCl solution. Mix thoroughly. Add distilled/deionized water and bring volume to 1.0L. Add remaining components. Mix thoroughly.

Preparation of Solution A: Add components to distilled/deionized water and bring volume to 890.0mL. Mix thoroughly. Autoclave for 15 min at 15 psi pressure–121°C. Cool to room temperature.

Solution B:

Composition per 100.0mL:

KH_2PO_4 .... 1.0g

Preparation of Solution B: Add KH_2PO_4 to distilled/deionized water and bring volume to 100.0mL. Mix thoroughly. Autoclave for 15 min at 15 psi pressure–121°C. Cool to room temperature.

Vitamin Solution:

Composition per liter:

Pyridoxine·HCl 10.0mg
Calcium pantothenate 5.0mg
Nicotinic acid 5.0mg
Robiflavin 5.0mg
Thiamine·HCl 5.0mg
Biotin 2.0mg
Folic acid 2.0mg
p-Aminobenzoic acid 1.0mg
Cyanocobalamin 0.01mg

Preparation of Vitamin Solution: Add components to distilled/deionized water and bring volume to 1.0L. Mix thoroughly. Filter sterilize.

Preparation of Medium: Aseptically combine 890.0mL of cooled sterile solution A, 100.0mL of cooled sterile solution B, and 10.0mL of sterile vitamin solution. Mix thoroughly. Adjust pH to 7.0 with sterile 6*N* NaOH or HCl solution.

Use: For the cultivation and maintenance of *Pseudomonas halophila*.

Pseudomonas HiVeg Agar Base with Glycerol and Selective Supplement

Composition per liter:

Plant peptone No. 2 .... 16.0g
Agar 11.0g
Plant hydrolysate 10.0g
K_2SO_4 .... 10.0g
$MnCl_2$, anhydrous 1.4g
Glycerol 10.0mL
Selective supplement 10.0mL

pH 7.0 ± 0.2 at 25°C

Source: This medium, without glycerol or selective supplement, is available as a premixed powder from HiMedia.

Selective Supplement Solution:

Composition per 10.0mL:

Cetrimide 0.2g
Nalidixic acid 15.0mg

Preparation of Selective Supplement Solution: Add components to distilled/deionized water and bring volume to 10.0mL. Mix thoroughly. Filter sterilize.

Preparation of Medium: Add glycerol and then other components, except selective supplement, to distilled/deionized water and bring volume to 990.0mL. Mix thoroughly. Gently heat and bring to boiling. Distribute into tubes or flasks. Autoclave for 15 min at 15 psi pressure–121°C. Cool to 45°–50°C. Aseptically add 10.0mL selective supplement. Mix thoroughly. Pour into sterile Petri dishes or leave in tubes.

Use: For the selective isolation of *Pseudomonas* species.

Pseudomonas indigofera Agar

Composition per liter:

Agar 12.0g
Yeast extract 10.0g
Glucose 5.0g
Sodium acetate 0.5g

pH 7.0 ± 0.2 at 25°C

Preparation of Medium: Add components to distilled/deionized water and bring volume to 1.0L. Mix thoroughly. Gently heat and bring to boiling. Distribute into tubes or flasks. Autoclave for 15 min at 15 psi pressure–121°C. Pour into sterile Petri dishes or leave in tubes.

Use: For the cultivation and maintenance of *Pseudomonas indigofera*.

Pseudomonas Isolation Agar

Composition per liter:

Peptone 20.0g
Agar 13.6g
K_2SO_4 .... 10.0g
$MgCl_2·6H_2O$ 1.4g

Irgasan® (triclosan) 0.025g
Glycerol 20.0mL

pH 7.0 ± 0.2 at 25°C

Source: This medium is available as a premixed powder from BD Diagnostic Systems.

Preparation of Medium: Add components to distilled/deionized water and bring volume to 1.0L. Mix thoroughly. Gently heat and bring to boiling. Distribute into tubes or flasks. Autoclave for 15 min at 15 psi pressure–121°C. Pour into sterile Petri dishes or leave in tubes.

Use: For the isolation and cultivation of *Pseudomonas* species.

Pseudomonas Isolation HiVeg Agar Base with Glycerol

Composition per liter:

Plant peptone 20.0g
Agar 13.6g
K_2SO_4 .... 10.0g
$MnCl_2$ 1.4g
Triclosan (Irgasan®) 0.025g
Glycerol 10.0mL

pH 7.0 ± 0.2 at 25°C

Source: This medium, without glycerol, is available as a premixed powder from HiMedia.

Preparation of Medium: Add glycerol and then other components to distilled/deionized water and bring volume to 1.0L. Mix thoroughly. Gently heat and bring to boiling. Distribute into tubes or flasks. Autoclave for 15 min at 15 psi pressure–121°C. Pour into sterile Petri dishes or leave in tubes.

Use: For the selective isolation and identification of *Pseudomonas aeruginosa* from clinical and nonclinical specimens.

Pseudomonas lemoignei Agar

Composition per liter:

Agar 20.0g
Sodium pyruvate 4.0g
$MgSO_4 \cdot 7H_2O$ 0.2g
KH_2PO_4 .... 0.15g
K_2HPO_4 .... 0.05g
Salt solution 15.0mL

pH 7.0 ± 0.2 at 25°C

Salt Solution:

Composition per 100.0mL:

Ferric ammonium citrate 1.0g
$CaCl_2$ 0.1g

Preparation of Salt Solution: Add components to distilled/deionized water and bring volume to 100.0mL. Mix thoroughly.

Preparation of Medium: Add components to distilled/deionized water and bring volume to 1.0L. Mix thoroughly. Adjust pH to 7.0. Gently heat and bring to boiling. Distribute into tubes or flasks. Autoclave for 15 min at 15 psi pressure–121°C. Pour into sterile Petri dishes or leave in tubes.

Use: For the cultivation and maintenance of *Pseudomonas lemoignei.*

Pseudomonas Medium (ATCC Medium 59)

Composition per liter:

K_2HPO_4 .... 1.15g
NH_4NO_3 .... 1.0g
Yeast extract 1.0g
KH_2PO_4 .... 0.625g
$MgSO_4 \cdot 7H_2O$ 0.02g
Pyrrolidine 4.0mL

pH 7.0 ± 0.2 at 25°C

Preparation of Medium: Add pyrrolidine to 500.0mL of distilled/deionized water. Mix thoroughly. Adjust pH to 7.0. Add remaining components. Bring volume to 1.0L with distilled/deionized water. Distribute into tubes or flasks. Autoclave for 15 min at 15 psi pressure–121°C.

Use: For the cultivation and maintenance of *Pseudomonas fluorescens.*

Pseudomonas Medium (ATCC Medium 179)

Composition per 1020.0mL:

Solution 1 .... 1.0L
Solution 3 .... 15.0mL
Solution 2 .... 5.0mL

pH 6.8 ± 0.2 at 25°C

Solution 1:

Composition per liter:

Agar 20.0g
K_2HPO_4 .... 2.56g
KH_2PO_4 .... 2.08g
NH_4Cl 1.0g
$MgSO_4 \cdot 7H_2O$ 0.5g

Preparation of Solution 1: Add components to distilled/deionized water and bring volume to 1.0L. Mix thoroughly. Gently heat and bring to boiling. Adjust pH to 6.8. Distribute into tubes or flasks. Autoclave for 15 min at 15 psi pressure–121°C. Cool to 45°–50°C.

Solution 2:

Composition per 100.0mL:

Ferric ammonium citrate 1.0g
$CaCl_2$ 0.1g

Preparation of Solution 2: Add components to distilled/deionized water and bring volume to 100.0mL. Mix thoroughly. Filter sterilize.

Solution 3:

Composition per 100.0mL:

Succinic acid 11.8g

Preparation of Solution 3: Add succinic acid to distilled/deionized water and bring volume to 100.0mL. Mix thoroughly. Adjust pH to 6.0 with NaOH. Filter sterilize.

Preparation of Medium: To 1.0L of cooled sterile solution 1, aseptically add 5.0mL of sterile solution 2 and 15.0mL of sterile solution 3. Mix thoroughly. Pour into sterile Petri dishes or distribute into sterile tubes.

Use: For the cultivation and maintenance of *Pseudomonas lemoignei* and *Pseudomonas putida.*

Pseudomonas Medium (ATCC Medium 186)

Composition per liter:

Marine salts mix 19.0g
Glucose 10.0g
Yeast extract 10.0g
K_2HPO_4 .... 1.0g

Peptone	1.0g
KH_2PO_4	0.5g
NH_4Cl	0.5g
$CaCl_2$	0.1g
$Na2SO_3$	0.1g
$NaHCO_3$	0.1g
$FeCl_3$	0.05g

Preparation of Medium: Add components to distilled/deionized water and bring volume to 1.0L. Mix thoroughly. Distribute into tubes or flasks. Autoclave for 15 min at 15 psi pressure–121°C.

Use: For the cultivation and maintenance of *Bacillus sphaericus*.

Pseudomonas Medium (ATCC Medium 226)

Composition per liter:

Agar	20.0g
Yeast extract	10.0g
Glucose	5.0g
Sodium acetate	0.5g

Preparation of Medium: Add components to distilled/deionized water and bring volume to 1.0L. Mix thoroughly. Gently heat and bring to boiling. Distribute into tubes or flasks. Autoclave for 15 min at 15 psi pressure–121°C. Pour into sterile Petri dishes or leave in tubes.

Use: For the cultivation and maintenance of *Pseudomonas indigofera*.

Pseudomonas Medium (ATCC Medium 609)

Composition per liter:

Agar	15.0g
K_2HPO_4	8.71g
Nitrilotriacetic acid	1.91g
Na_2SO_4	0.57g
$MgSO_4$	0.25g
$FeSO_4$	0.5mg
$Ca(NO_3)_2$	0.5mg

pH 6.5 ± 0.2 at 25°C

Preparation of Medium: Add nitrilotriacetic acid to approximately 500.0mL of distilled/deionized water. Mix thoroughly. Adjust pH to 6.5. Add remaining components. Bring volume to 1.0L with distilled/deionized water. Gently heat and bring to boiling. Distribute into tubes or flasks. Autoclave for 15 min at 15 psi pressure–121°C. Pour into sterile Petri dishes or leave in tubes.

Use: For the cultivation and maintenance of *Pseudomonas* species.

Pseudomonas Medium (ATCC Medium 775)

Composition per liter:

NH_4Cl	5.0g
K_2HPO_4	1.5g
L-Tryptophan	1.0g
KH_2PO_4	0.5g
Yeast extract	0.5g
$MgSO_4$	0.2g

Preparation of Medium: Add components to distilled/deionized water and bring volume to 1.0L. Mix thoroughly. Distribute into tubes or flasks. Autoclave for 15 min at 15 psi pressure–121°C.

Use: For the cultivation and maintenance of *Comamonas acidovorans*.

Pseudomonas Medium A

Composition per liter:

Peptone	20.0g
Agar	15.0g
Glycerol	10.0g
K_2SO_4	10.0g
$MgCl_2$	1.4g

pH 7.2 ± 0.2 at 25°C

Preparation of Medium: Add components to distilled/deionized water and bring volume to 1.0L. Mix thoroughly. Gently heat and bring to boiling. Distribute into tubes or flasks. Autoclave for 10 min at 10 psi pressure–115°C. Pour into sterile Petri dishes or leave in tubes.

Use: For the cultivation and production of pyocyanin by *Pseudomonas* species.

Pseudomonas Medium B

Composition per liter:

Peptone	20.0g
Agar	15.0g
Glycerol	10.0g
$MgSO_4 \cdot 7H_2O$	1.5g
K_2HPO_4 solution	100.0mL

pH 7.2 ± 0.2 at 25°C

K_2HPO_4 Solution:

Composition per 100.0mL:

K_2HPO_4	1.5g

Preparation of K_2HPO_4 Solution: Add K_2HPO_4 to distilled/deionized water and bring volume to 100.0mL. Mix thoroughly. Autoclave for 15 min at 15 psi pressure–121°C. Cool to 45°–50°C.

Preparation of Medium: Add components, except K_2HPO_4 solution, to distilled/deionized water and bring volume to 900.0mL. Mix thoroughly. Gently heat and bring to boiling. Autoclave for 15 min at 15 psi pressure–121°C. Cool to 45°–50°C. Aseptically add 100.0mL of sterile K_2HPO_4 solution. Mix thoroughly. Pour into sterile Petri dishes or distribute into sterile tubes.

Use: For the cultivation and observation of fluorescin production by *Pseudomonas* species.

Pseudomonas Medium I

Composition per liter:

Pancreatic digest of casein	10.0g
Yeast extract	10.0g
Glucose	5.0g
K_2HPO_4	5.0g
Salts solution	5.0mL

Salts Solution:

Composition per 100.0mL:

$MgSO_4 \cdot 4H_2O$	4.0g
$FeSO_4$	0.2g
$MnSO_4 \cdot 4H_2O$	0.2g
NaCl	0.2g

Preparation of Salts Solution: Add components to distilled/deionized water and bring volume to 100.0mL. Mix thoroughly.

Preparation of Medium: Add components to distilled/deionized water and bring volume to 1.0L. Mix thoroughly. Distribute into tubes or flasks. Autoclave for 15 min at 15 psi pressure–121°C.

Use: For the cultivation and maintenance of *Bacillus sphaericus*.

Pseudomonas Medium No. 2

Composition per liter:

Agar	15.0g
$Na_2HPO_4 \cdot 12H_2O$	6.0g
Succinic acid	5.0g
KH_2PO_4	2.4g
NH_4Cl	1.0g
$MgSO_4 \cdot 7H_2O$	0.5g
$CaCl_2 \cdot 6H_2O$	0.01g
$FeCl_3 \cdot 6H_2O$	0.01g

pH 6.8 ± 0.2 at 25°C

Preparation of Medium: Add components to distilled/deionized water and bring volume to 1.0L. Mix thoroughly. Adjust pH to 6.8. Gently heat and bring to boiling. Distribute into tubes or flasks. Autoclave for 15 min at 15 psi pressure–121°C. Pour into sterile Petri dishes or leave in tubes.

Use: For the cultivation of *Pseudomonas* species and *Psychrobacter immobilis*.

Pseudomonas Phage Medium

Composition per liter:

Agar	15.0g
Nutrient broth	10.0g
K_2HPO_4	1.11g
Glucose	1.0g
KH_2PO_4	0.49g

pH 7.0 ± 0.2 at 25°C

Preparation of Medium: Add components to distilled/deionized water and bring volume to 1.0L. Mix thoroughly. Gently heat and bring to boiling. Distribute into tubes or flasks. Autoclave for 15 min at 15 psi pressure–121°C. Pour into sterile Petri dishes or leave in tubes.

Use: For the cultivation and maintenance of *Pseudomonas fluorescens*.

Pseudomonas pickettii Medium

Composition per liter:

Agar	15.0g
Peptone	5.0g
NaCl	5.0g
$Na_2HPO_4 \cdot 12H_2O$	2.39g
Yeast extract	2.0g
Beef extract	1.0g
K_2HPO_4	0.45g

pH 6.8 ± 0.2 at 25°C

Preparation of Medium: Add components to distilled/deionized water and bring volume to 1.0L. Mix thoroughly. Gently heat and bring to boiling. Distribute into tubes or flasks. Autoclave for 15 min at 15 psi pressure–121°C. Pour into sterile Petri dishes or leave in tubes.

Use: For the cultivation of *Burkholderia pickettii*.

Pseudomonas saccharophila Medium

Composition per 1015.0mL:

Agar	20.0g
Na_2HPO_4	4.8g
KH_2PO_4	4.4g
NH_4Cl	1.0g
$MgSO_4 \cdot 7H_2O$	0.5g
Solution A	5.0mL
Solution B	10.0mL

Solution A:

Composition per 100.0mL:

Ferric ammonium citrate	1.0g
$CaCl_2$	0.1g

Preparation of Solution A: Add components to distilled/deionized water and bring volume to 100.0mL. Mix thoroughly. Filter sterilize.

Solution B:

Composition per 100.0mL:

Sucrose	10.0g

Preparation of Solution B: Add sucrose to distilled/deionized water and bring volume to 100.0mL. Mix thoroughly. Filter sterilize.

Preparation of Medium: Add components, except solution A and solution B, to distilled/deionized water and bring volume to 1.0L. Mix thoroughly. Gently heat and bring to boiling. Autoclave for 15 min at 15 psi pressure–121°C. Cool to 45°–50°C. Aseptically add sterile solution A and sterile solution B. Mix thoroughly. Pour into sterile Petri dishes or distribute into sterile tubes.

Use: For the cultivation and maintenance of *Pseudomonas saccharophila* and other *Pseudomonas* species.

Pseudomonas solanacearum Medium

Composition per liter:

Agar	17.0g
Peptone	10.0g
Glucose	5.0g
Pancreatic digest of casein	1.0g

Preparation of Medium: Add components to distilled/deionized water and bring volume to 1.0L. Mix thoroughly. Gently heat and bring to boiling. Distribute into tubes or flasks. Autoclave for 15 min at 15 psi pressure–121°C. Pour into sterile Petri dishes or leave in tubes.

Use: For the cultivation and maintenance of *Pseudomonas solanacearum*.

Pseudomonas syngii Medium

Composition per liter:

Agar	15.0g
Acid casein hydrolysate	7.5g
Sucrose	2.0g
$MgSO_4 \cdot 7H_2O$	250.0mg
K_2HPO_4	500.0mg
Ammonium ferricitrate solution	20.0mL

Ammonium Ferricitrate Solution:

Composition per 20.0mL:

Ammonium ferricitrate	0.25g

Preparation of Ammonium Ferricitrate Solution: Add ammonium ferricitrate to distilled/deionized water and bring volume to 20.0mL. Mix thoroughly. Filter sterilize.

Preparation of Medium: Add components, except ammonium ferricitrate solution, to distilled/deionized water and bring volume to 980.0mL. Mix thoroughly. Gently heat and bring to boiling. Autoclave for 15 min at 15 psi pressure–121°C. Cool to 50°–55°C. Aseptically add 20.0mL of sterile ammonium ferricitrate solution. Mix thoroughly. Pour into sterile Petri dishes or distribute into sterile tubes.

Use: For the cultivation and maintenance of *Pseudomonas synygii*.

Pseudomonas syringae Selective Medium

Composition per liter:

Agar	15.0g
L-Proline	5.0g
$MgSO_4 \cdot 7H_2O$	0.2g
K_2HPO_4	0.08g
KH_2PO_4	0.02g
$MnSO_4 \cdot 4H_2O$ solution	10.0mL

pH 6.8 ± 0.2 at 25°C

$MnSO_4 \cdot 4H_2O$ Solution:

Composition per 10.0mL:

$MnSO_4 \cdot 4H_2O$	2.1g

Preparation of $MnSO_4 \cdot 4H_2O$ Solution: Add $MnSO_4 \cdot 4H_2O$ to distilled/deionized water and bring volume to 10.0mL. Mix thoroughly. Autoclave for 15 min at 15 psi pressure–121°C.

Preparation of Medium: Add components, except $MnSO_4 \cdot 4H_2O$ solution, to distilled/deionized water and bring volume to 990.0mL. Mix thoroughly. Gently heat and bring to boiling. Adjust pH to 6.8. Autoclave for 10 min at 10 psi pressure–115°C. Cool to 45°–50°C. Aseptically add sterile $MnSO_4 \cdot 4H_2O$ solution. Mix thoroughly. Pour into sterile Petri dishes.

Use: For the selective isolation and cultivation of *Pseudomonas syringae*.

Pseudosel™ Agar

See: **Cetrimide Agar, USP**

PSS Broth

See: **Peptone Succinate Salts Broth**

PSS Medium

See: **Peptone Succinate Salts Medium**

PSTA Enrichment HiVeg Broth Base

Composition per liter:

Tris hydroxymethyl aminomethane	3.0g
Plant peptone	1.0g
Sucrose	1.0g
NaN_3	0.192g
Brilliant Green	0.0125g

pH 7.0 ± 0.2 at 25°C

Source: This medium is available as a premixed powder from HiMedia.

Preparation of Medium: Add components to distilled/deionized water and bring volume to 1.0L. Mix thoroughly. Gently heat and bring to boiling. Distribute into tubes or flasks. Autoclave for 15 min at 15 psi pressure–121°C. Pour into sterile Petri dishes or leave in tubes.

Use: For the secondary enrichment of *Yersinia enterocolitica* from foods.

PT Agar

Composition per liter:

Agar	15.0g
Pancreatic digest of casein	4.0g
Yeast extract	4.0g
$MgSO_4 \cdot 7H_2O$	2.0g
$CaCl_2 \cdot 2H_2O$	1.0g

pH 7.2 ± 0.2 at 25°C

Preparation of Medium: Add components to distilled/deionized water and bring volume to 1.0L. Mix thoroughly. Gently heat and bring to boiling. Distribute into tubes or flasks. Autoclave for 15 min at 15 psi pressure–121°C. Pour into sterile Petri dishes or leave in tubes.

Use: For the cultivation of myxobacteria.

PTYG Medium (LMG Medium 238)

Composition per liter:

Agar	15.0g
Peptone	5.0g
Tryptone	5.0g
Yeast extract	5.0g
Glucose	5.0g

pH 7.2 ± 0.2 at 25°C

Preparation of Medium: Add components to distilled/deionized water and bring volume to 1.0L. Mix thoroughly. Gently heat and bring to boiling. Distribute into tubes or flasks. Autoclave for 15 min at 15 psi pressure–121°C. Pour into sterile Petri dishes or leave in tubes.

Use: For the cultivation and maintenance of *Sphingobium herbicidovorans* and *Sphingomonas pruni.*

PTYG Medium (DSMZ Medium 914)

Composition per liter:

Glucose	10.0g
Peptone	5.0g
Tryptone	5.0g
Yeast extract	5.0g
$MgSO_4 \cdot 7H_2O$	0.6g
$CaCl_2$	0.06g

pH 7.0 ± 0.2 at 25°C

Preparation of Medium: Add components to distilled/deionized water and bring volume to 1.0L. Mix thoroughly. Distribute into tubes or flasks. Autoclave for 15 min at 15 psi pressure–121°C.

Use: For the cultivation of *Kineococcus radiotolerans.*

Purple Agar

Composition per liter:

Agar	15.0g
Proteose peptone No. 3	10.0g
NaCl	5.0g
Beef extract	1.0g
Bromcresol Purple	0.02g
Carbohydrate solution	20.0mL

pH 6.8 ± 0.2 at 25°C

Source: This medium is available as a premixed powder from BD Diagnostic Systems.

Carbohydrate Solution:

Composition per 20.0mL:

Carbohydrate	10.0g

Preparation of Carbohydrate Solution: Add carbohydrate to distilled/deionized water and bring volume to 20.0mL. For expensive carbohydrates, 5.0g may be used instead of 10.0g. Mix thoroughly. Filter sterilize.

Preparation of Medium: Add components, except carbohydrate solution, to distilled/deionized water and bring volume to 980.0mL. Mix thoroughly. Gently heat and bring to boiling. Distribute into tubes in 9.8mL volumes. Autoclave for 15 min at 15 psi pressure–121°C. Cool to 45°–50°C. Aseptically add 0.2mL of sterile carbohydrate solution to each tube. Mix thoroughly. Allow tubes to cool in a slanted position.

Use: For the preparation of carbohydrate media used in fermentation studies for the identification of bacteria, especially members of the Enterobacteriaceae. Bacteria that can ferment the carbohydrate turn the medium yellow.

Purple Broth (Purple Carbohydrate Broth)

Composition per liter:

Proteose peptone No. 3	10.0g
NaCl	5.0g
Beef extract	1.0g
Bromcresol Purple	0.015g
Carbohydrate solution	20.0mL

pH 6.8 ± 0.2 at 25°C

Source: This medium is available as a premixed powder from BD Diagnostic Systems.

Carbohydrate Solution:
Composition per 20.0mL:

Carbohydrate	10.0g

Preparation of Carbohydrate Solution: Add carbohydrate to distilled/deionized water and bring volume to 20.0mL. For expensive carbohydrates, 5.0g may be used instead of 10.0g. Mix thoroughly. Filter sterilize.

Preparation of Medium: Add components, except carbohydrate solution, to distilled/deionized water and bring volume to 980.0mL. Mix thoroughly. Gently heat and bring to boiling. Distribute into tubes in 9.8mL volumes. Autoclave for 15 min at 15 psi pressure–121°C. Cool to 25°C. Aseptically add 0.2mL of sterile carbohydrate solution to each tube. Mix thoroughly.

Use: For the preparation of carbohydrate media used in fermentation studies for the identification of bacteria, especially members of the Enterobacteriaceae. Bacteria that can ferment the carbohydrate turn the medium yellow.

Purple Broth

Composition per liter:

Pancreatic digest of gelatin	10.0g
NaCl	5.0g
Bromcresol Purple	0.02g
Carbohydrate solution	20.0mL

pH 6.8 ± 0.2 at 25°C

Source: This medium is available as a premixed powder from BD Diagnostic Systems.

Carbohydrate Solution:
Composition per 20.0mL:

Carbohydrate	10.0g

Preparation of Carbohydrate Solution: Add carbohydrate to distilled/deionized water and bring volume to 20.0mL. For expensive carbohydrates, 5.0g may be used instead of 10.0g. Mix thoroughly. Filter sterilize.

Preparation of Medium: Add components, except carbohydrate solution, to distilled/deionized water and bring volume to 980.0mL. Mix thoroughly. Adjust pH to 7.4 if necessary. Distribute into tubes containing an inverted Durham tube. Fill each tube with 9.8mL of medium. Autoclave for 15 min at 15 psi pressure–121°C. Aseptically add 0.2mL of sterile carbohydrate solution to each tube.

Use: For the preparation of liquid fermentation media. Bacteria that can ferment the carbohydrate turn the medium yellow.

Purple Broth with Sodium Chloride (Purple Carbohydrate Broth with NaCl) (BAM M130)

Composition per liter:

Proteose peptone No. 3	10.0g
NaCl	5.0g
Beef extract	1.0g
Bromcresol Purple	0.02g
Carbohydrate solution	20.0mL

pH 6.8 ± 0.2 at 25°C

Source: This medium is available as a premixed powder from BD Diagnostic Systems.

Carbohydrate Solution:
Composition per 20.0mL:

Carbohydrate	10.0g

Preparation of Carbohydrate Solution: Add carbohydrate to distilled/deionized water and bring volume to 20.0mL. For expensive carbohydrates, 5.0g may be used instead of 10.0g. Mix thoroughly. Filter sterilize.

Preparation of Medium: Add components, except carbohydrate solution, to distilled/deionized water and bring volume to 980.0mL. Mix thoroughly. Gently heat and bring to boiling. Distribute into tubes in 9.8mL volumes. Autoclave for 15 min at 15 psi pressure–121°C. Cool to 25°C. Aseptically add 0.2mL of sterile carbohydrate solution to each tube. Mix thoroughly.

Use: For the preparation of carbohydrate media used in fermentation studies for the identification of bacteria, especially members of the Enterobacteriaceae. Bacteria that can ferment the carbohydrate turn the medium yellow.

Purple Carbohydrate Broth
See: **Purple Broth**

Purple Carbohydrate Fermentation Broth Base (BAM M130a)

Composition per liter:

Pancreatic digest of gelatin	10.0g
NaCl	5.0g
Bromcresol Purple	0.02g
Carbohydrate solution	100.0mL

pH 6.8 ± 0.2 at 25°C

Source: This medium is available as a premixed powder from BD Diagnostics.

Carbohydrate Solution:
Composition per 100.0mL:

Carbohydrate	5.0g

Preparation of Carbohydrate Solution: Add carbohydrate to distilled/deionized water and bring volume to 100.0mL. Mix thoroughly. Filter sterilize.

Preparation of Medium: Add components, except carbohydrate solution, to distilled/deionized water and bring volume to 900.0mL.

Mix thoroughly. Adjust pH to 7.4 if necessary. Distribute into tubes containing an inverted Durham tube. Fill each tube with 9.0mL of medium. Autoclave for 15 min at 15 psi pressure–121°C. Aseptically add 1.0mL of sterile carbohydrate solution to each tube.

Use: For the preparation of liquid fermentation media. Bacteria that can ferment the carbohydrate turn the medium yellow.

Purple Carbohydrate Fermentation Broth Base with Esculin (BAM M130a)

Composition per liter:

Pancreatic digest of gelatin	10.0g
NaCl	5.0g
Esculin	5.0g
Bromcresol Purple	0.02g
Carbohydrate solution	10.0mL

pH 6.8 ± 0.2 at 25°C

Source: This medium, without esculin, is available as a premixed powder from BD Diagnostics.

Carbohydrate Solution:

Composition per 100.0mL:

Carbohydrate	5.0g

Preparation of Carbohydrate Solution: Add carbohydrate to distilled/deionized water and bring volume to 100.0mL. Mix thoroughly. Filter sterilize.

Preparation of Medium: Add components, except carbohydrate solution and esculin, to distilled/deionized water and bring volume to 900.0mL. Mix thoroughly. Adjust pH to 7.4 if necessary. Distribute into tubes containing an inverted Durham tube. Fill each tube with 9.0mL of medium. Add 0.05g esculin to each tube. Autoclave for 15 min at 10 psi pressure–115°C. Aseptically add 1.0mL of sterile carbohydrate solution to each tube.

Use: For the preparation of liquid fermentation media, e.g., for *Listeria* spp. and *Enterococcus* spp. Bacteria that can ferment the carbohydrate turn the medium yellow.

Purple HiVeg Agar Base with Carbohydrate

Composition per liter:

Agar	15.0g
Plant special peptone	10.0g
NaCl	5.0g
Plant extract	1.0g
Bromcresol Purple	0.02g
Carbohydrate solution	20.0mL

pH 6.8 ± 0.2 at 25°C

Source: This medium, without carbohydrate, is available as a premixed powder from HiMedia.

Carbohydrate Solution:

Composition per 20.0mL:

Carbohydrate	10.0g

Preparation of Carbohydrate Solution: Add carbohydrate to distilled/deionized water and bring volume to 20.0mL. For expensive carbohydrates, 5.0g may be used instead of 10.0g. Mix thoroughly. Filter sterilize.

Preparation of Medium: Add components, except carbohydrate solution, to distilled/deionized water and bring volume to 980.0mL. Mix thoroughly. Gently heat and bring to boiling. Distribute into tubes in 9.8mL volumes. Autoclave for 15 min at 15 psi pressure–121°C. Cool to 45°–50°C. Aseptically add 0.2mL of sterile carbohydrate solution to each tube. Mix thoroughly. Allow tubes to cool in a slanted position.

Use: For the preparation of carbohydrate media used in fermentation studies for the identification of bacteria, especially members of the Enterobacteriaceae. Bacteria that can ferment the carbohydrate turn the medium yellow.

Purple HiVeg Broth Base with Carbohydrate

Composition per liter:

Plant special peptone	10.0g
NaCl	5.0g
Bromcresol Purple	0.02g
Carbohydrate solution	20.0mL

pH 6.8 ± 0.2 at 25°C

Source: This medium, without carbohydrate, is available as a premixed powder from HiMedia.

Carbohydrate Solution:

Composition per 20.0mL:

Carbohydrate	10.0g

Preparation of Carbohydrate Solution: Add carbohydrate to distilled/deionized water and bring volume to 20.0mL. For expensive carbohydrates, 5.0g may be used instead of 10.0g. Mix thoroughly. Filter sterilize.

Preparation of Medium: Add components, except carbohydrate solution, to distilled/deionized water and bring volume to 980.0mL. Mix thoroughly. Gently heat and bring to boiling. Distribute into tubes in 9.8mL volumes. Autoclave for 15 min at 15 psi pressure–121°C. Cool to 25°C. Aseptically add 0.2mL of sterile carbohydrate solution to each tube. Mix thoroughly.

Use: For the preparation of carbohydrate media used in fermentation studies for the identification of bacteria, especially members of the Enterobacteriaceae. Bacteria that can ferment the carbohydrate turn the medium yellow.

Purple Lactose Agar

Composition per liter:

Agar	10.0g
Lactose	10.0g
Peptone	5.0g
Beef extract	3.0g
Bromcresol Purple	0.025g

pH 6.8 ± 0.1 at 25°C

Source: This medium is available as a premixed powder from BD Diagnostic Systems.

Preparation of Medium: Add components to distilled/deionized water and bring volume to 1.0L. Mix thoroughly. Gently heat and bring to boiling. Distribute into tubes or flasks. Autoclave for 15 min at 15 psi pressure–121°C. Pour into sterile Petri dishes or leave in tubes. Allow tubes to cool in a slanted position.

Use: For the detection and differentiation of members of the Enterobacteriaceae. Bacteria that can ferment lactose turn the medium yellow.

Purple Serum Agar Base

Composition per liter:

Agar	20.0g
Lactose	20.0g

Peptone	20.0g
NaCl	5.0g
Bromcresol Purple	0.03g
Phenol Red	0.024g

pH 7.6 ± 0.2 at 25°C

Preparation of Medium: Add components to distilled/deionized water and bring volume to 1.0L. Mix thoroughly. Gently heat and bring to boiling. Distribute into tubes or flasks. Autoclave for 15 min at 15 psi pressure–121°C. Pour into sterile Petri dishes or leave in tubes.

Use: For the cultivation and differentiation of Gram-negative bacteria isolated from the urinary tract. Bacteria that can ferment lactose turn the medium yellow.

PV Blood Agar (Paromomycin Vancomycin Blood Agar)

Composition per liter:

Agar	20.0g
Pancreatic digest of casein	15.0g
NaCl	5.0g
Papaic digest of soybean meal	5.0g
Yeast extract	5.0g
L-Cystine	0.4g
Paromomycin	0.1g
Vancomycin	7.5mg
Hemin	5.0mg
Sheep blood, defibrinated	50.0mL
Vitamin K_1 solution	10.0mL

pH 7.5 ± 0.2 at 25°C

Vitamin K_1 Solution:

Composition per 10.0mL:

Vitamin K_1	0.01g
Ethanol	10.0mL

Preparation of Vitamin K_1 Solution: Add vitamin K_1 to 10.0mL of absolute ethanol. Mix thoroughly. Filter sterilize.

Preparation of Medium: Add components—except vitamin K_1 solution, sheep blood, paromomycin, and vancomycin—to distilled/deionized water and bring volume to 940.0mL. Mix thoroughly. Gently heat and bring to boiling for 1 min. Autoclave for 15 min at 15 psi pressure–121°C. Cool to 50°–55°C. Aseptically add the sterile vitamin K_1 solution, sheep blood, vancomycin, and paromomycin. Mix thoroughly. Pour into sterile Petri dishes.

Use: For the selective cultivation of fastidious anaerobic bacteria.

PW Medium (LMG 182)

Composition per 1100.0mL:

Agar	12.0g
Papaic digest of soybean meal	4.0g
KH_2PO_4	1.2g
K_2HPO_4	1.0g
Trypticase™ peptone	1.0g
$MgSO_4 \cdot 7H_2O$	0.4g
Glutamine solution	50.0mL
Bovine serum albumin solution	30.0mL
Phenol Red (0.2% solution)	10.0mL
Solution A	10.0mL

Glutamine Solution:

Composition per 50.0mL:

L-Glutamine	4.0g

Preparation of Glutamine Solution: Add glutamine to distilled/deionized water and bring volume to 50.0ml. Mix thoroughly. Filter sterilize.

Bovine Serum Albumin Solution:

Composition per 50.0mL:

Bovine serum albumin, fraction V	10.0g

Preparation of Bovine Serum Albumin Solution: Add bovine serum albumin to distilled/deionized water and bring volume to 50.0mL. Mix thoroughly. Filter sterilize.

Solution A:

Composition per 101.0mL:

NaOH (0.05*N* solution)	100.0mL
Hemin chloride	0.1g

Preparation of Medium: Add components, except glutamine solution and bovine serum albumin solution, to distilled/deionized water and bring volume to 920.0mL. Mix thoroughly. Gently heat and bring to boiling. Autoclave for 15 min at 15 psi pressure–121°C. Cool to 50°–55°C. Aseptically add 50.0mL of sterile glutamine solution and 10.0mL of sterile bovine serum albumin solution. Mix thoroughly. Pour into sterile Petri dishes or distribute into sterile tubes.

Use: For the cultivation and maintenance of *Xylella fastidiosa.*

PXA Agar

See: **Pril Xylose Ampicillin Agar**

PY Basal Medium

Composition per 104.0mL:

Yeast extract	1.0g
L-Cysteine·HCl·H_2O	0.5g
Peptone	0.5g
Pancreatic digest of casein	0.5g
Resazurin	0.16mg
Salts solution	4.0mL

Salts Solution:

Composition per liter:

$NaHCO_3$	10.0g
NaCl	2.0g
K_2HPO_4	1.0g
KH_2PO_4	1.0g
$CaCl_2$	0.2g
$MgSO_4$	0.2g

Preparation of Salts Solution: Add components to distilled/deionized water and bring volume to 1.0L. Mix thoroughly.

Preparation of Medium: Add components to distilled/deionized water and bring volume to 104.0mL. Mix thoroughly. Distribute into tubes or flasks. Autoclave for 15 min at 15 psi pressure–121°C.

Use: For the identification of treponemes.

PY Broth (DSMZ Medium 1071)

Composition per liter:

NaCl	30.0g
Agar	15.0g
Polypepton™	2.0g
Yeast extract	0.5g
$MgCl_2 \cdot 6H_2O$	0.5g
$FeCl_3 \cdot 6H_2O$	6.0mg

$CaCl_2 \cdot 2H_2O$ 5.0mg
$Na_2MoO_4 \cdot 2H_2O$ 5.0mg
$CuCl_2 \cdot 2H_2O$ 4.0mg
pH 8.0 ± 0.2 at 25°C

Preparation of Medium: Add components to distilled/deionized water and bring volume to 1.0L. Mix thoroughly. Adjust pH to 8.0. Distribute into tubes or flasks. Gently heat while stirring and bring to boiling. Mix thoroughly. Autoclave for 15 min at 15 psi pressure–121°C. Cool to room termperature.

Use: For the cultivation of *Desulfosporosinus orientis*.

PY Carbohydrate Medium

Composition per 110.0mL:
Polypeptone™ 1.0g
Yeast extract 1.0g
Glucose 0.4g
Maltose 0.4g
Ribose 0.4g
Starch, soluble 0.4g
L-Cysteine·HCl·H_2O 0.09g
Resazurin 0.16mg
Serum VFAH supplement 10.0mL
$NaHCO_3$ solution 10.0mL
Salts solution 4.0mL
pH 7.2–7.5 at 25°C

Serum VFAH Supplement:
Composition per 107.5mL:
Rabbit serum 100.0mL
Heme solution 5.0mL
VFA solution 1.5mL
Thiamine pyrophosphate solution 1.0mL

Preparation of Serum VFAH: Combine the four solutions. Mix thoroughly. Filter sterilize.

Thiamine Pyrophosphate Solution:
Composition per 10.0mL:
Thiamine pyrophosphate 0.05g

Preparation of Thiamine Pyrophosphate Solution: Add thiamine pyrophosphate to distilled/deionized water and bring volume to 10.0mL. Mix thoroughly. Filter sterilize.

Heme Solution:
Composition per 100.0mL:
Hemin 0.5g
NaOH (1*N* solution) 1.0mL

Preparation of Heme Solution: Add hemin to NaOH solution. Mix thoroughly. Bring volume to 100.0mL with distilled/deionized water.

VFA Solution:
Composition per 100.0mL:
Acetic acid, glacial 5.0mL
n-Butyric acid 4.0mL
n-Valeric acid 1.0mL
Isobutyric acid 1.0mL
Isovaleric acid 1.0mL
n-Butyric acid 1.0mL

Preparation of VFA Solution: Add components to distilled/deionized water and bring volume to 100.0mL. Mix thoroughly. Adjust pH to 7.0.

$NaHCO_3$ Solution:
Composition per 10.0mL:
$NaHCO_3$ 0.5g

Preparation of $NaHCO_3$ Solution: Add the $NaHCO_3$ to distilled/deionized water and bring volume to 10.0mL. Mix thoroughly. Filter sterilize.

Salts Solution:
Composition per 400.0mL:
K_2HPO_4 .... 0.9g
NaCl 0.8g
KH_2PO_4 .... 0.4g
$MnCl_2 \cdot 4H_2O$ 0.16g
$MgSO_4$ 0.08g

Preparation of Salts Solution: Add components to distilled/deionized water and bring volume to 400.0mL. Mix thoroughly.

Preparation of Medium: Add components, except serum VFAH supplement and $NaHCO_3$ solution, to distilled/deionized water and bring volume to 90.0mL. Mix thoroughly. Gently heat and bring to boiling. Continue boiling until resazurin turns colorless, indicating reduction. Autoclave for 15 min at 15 psi pressure–121°C. Cool to 25°C. Aseptically and anaerobically add 10.0mL of sterile serum VFAH supplement and 10.0mL of sterile $NaHCO_3$ solution. Anaerobically distribute into sterile tubes or flasks under 100% N_2.

Use: For the cultivation of oral treponemes.

PY CMC Medium (Peptone Yeast Extract Carboxymethyl Cellulose Medium)

Composition per liter:
Agar 15.0g
Carboxymethyl cellulose 10.0g
NaCl 5.0g
Polypeptone™ 5.0g
Yeast extract 5.0g
$MgSO_4 \cdot 7H_2O$ 2.0g
KH_2PO_4 .... 1.0g
Na_2CO_3 solution 100.0mL
pH 9.5 ± 0.2 at 25°C

Na_2CO_3 Solution:
Composition per 100.0mL:
Na_2CO_3 .... 10.0g

Preparation of Na_2CO_3 Solution: Add Na_2CO_3 to distilled/deionized water and bring volume to 100.0mL. Mix thoroughly. Autoclave for 15 min at 15 psi pressure–121°C. Cool to 45°–50°C.

Preparation of Medium: Add components, except Na_2CO_3 solution, to distilled/deionized water and bring volume to 900.0mL. Mix thoroughly. Gently heat and bring to boiling. Autoclave for 15 min at 15 psi pressure–121°C. Cool to 45°–50°C. Aseptically add sterile Na_2CO_3 solution. Mix thoroughly. Adjust pH to 9.5 if necessary. Pour into sterile Petri dishes or distribute into sterile tubes.

Use: For the cultivation and maintenance of alkalophilic *Bacillus* species.

PY Inositol Medium (Peptone Yeast Extract Inositol Medium)

Composition per liter:
i-Inositol 10.0g
Yeast extract 10.0g

Peptone 5.0g
Pancreatic digest of casein 5.0g
L-Cysteine·HCl·H_2O 0.5g
Salts solution 40.0mL
Hemin solution 10.0mL
Resazurin solution 4.0mL
Vitamin K_1 solution 0.2mL

pH 6.9 ± 0.2 at 25°C

Salts Solution:

Composition per liter:

$NaHCO_3$ 10.0g
NaCl 2.0g
K_2HPO_4 .... 1.0g
KH_2PO_4 .... 1.0g
$CaCl_2$, anhydrous 0.2g
$MgSO_4$ 0.2g

Preparation of Salts Solution: Add $CaCl_2$ and $MgSO_4$ to 300.0mL of distilled/deionized water. Mix thoroughly until dissolved. Bring volume to 800.0mL with distilled/deionized water. Add remaining components while stirring. Bring volume to 1.0L. Mix thoroughly. Store at 4°C.

Hemin Solution:

Composition per 100.0mL:

Hemin 0.05g
NaOH (1*N* solution) 1.0mL

Preparation of Hemin Solution: Add hemin to NaOH solution. Mix thoroughly. Adjust volume to 100.0mL with distilled/deionized water. Autoclave for 15 min at 15 psi pressure–121°C. Cool to 45°–50°C.

Resazurin Solution:

Composition per 44.0mL:

Resazurin 0.044g

Preparation of Resazurin Solution: Add resazurin to distilled/deionized water and bring volume to 44.0mL. Mix thoroughly.

Vitamin K_1 Solution:

Composition per 30.0mL:

Vitamin K_1 .... 0.15g
Ethanol (95% solution) 30.0mL

Preparation of Vitamin K_1 Solution: Add vitamin K_1 to ethanol. Mix thoroughly. Store in a brown bottle and keep under refrigeration. Discard after 1 month.

Preparation of Medium: Add components—except hemin solution, L-cysteine·HCl·H_2O, and vitamin K_1 solution—to distilled/deionized water and bring volume to 989.8mL. Mix thoroughly. Gently heat and bring to boiling under 80% N_2 + 10% CO_2 + 10% H_2. Continue boiling until resazurin turns colorless, indicating reduction. Cool to 50°C. Add the L-cysteine·HCl·H_2O, hemin solution, and vitamin K_1 solution. Adjust pH to 6.9 if necessary. Anaerobically distribute into tubes under 80% N_2 + 10% CO_2 + 10% H_2. Cap the tubes with rubber stoppers. Place tubes in a press. Autoclave for 15 min at 15 psi pressure–121°C with fast exhaust.

Use: For the cultivation and maintenance of *Eubacterium desmolans*.

PY Medium

See: **Peptone Yeast Extract Medium**

PY 1% Medium
(Peptone Yeast Extract 1% Medium)

Composition per liter:

Peptone 10.0g
Yeast extract 10.0g
NaCl 5.0g

Preparation of Medium: Add components to distilled/deionized water and bring volume to 1.0L. Mix thoroughly. Distribute into tubes or flasks. Autoclave for 15 min at 15 psi pressure–121°C.

Use: For the cultivation and maintenance of *Brevibacterium lactofermentum* and *Corynebacterium glutamicum*.

PY Medium with Fructose
(Peptone Yeast Extract Medium with Fructose)

Composition per liter:

Fructose 10.0g
Yeast extract 10.0g
Peptone 5.0g
Pancreatic digest of casein 5.0g
L-Cysteine·HCl·H_2O 0.5g
Salts solution 40.0mL
Hemin solution 10.0mL
Resazurin solution 4.0mL
Vitamin K_1 solution 0.2mL

pH 6.9 ± 0.2 at 25°C

Salts Solution:

Composition per liter:

$NaHCO_3$ 10.0g
NaCl 2.0g
K_2HPO_4 .... 1.0g
KH_2PO_4 .... 1.0g
$CaCl_2$, anhydrous 0.2g
$MgSO_4$ 0.2g

Preparation of Salts Solution: Add $CaCl_2$ and $MgSO_4$ to 300.0mL of distilled/deionized water. Mix thoroughly until dissolved. Bring volume to 800.0mL with distilled/deionized water. Add remaining components while stirring. Bring volume to 1.0L. Mix thoroughly. Store at 4°C.

Hemin Solution:

Composition per 100.0mL:

Hemin 0.05g
NaOH (1*N* solution) 1.0mL

Preparation of Hemin Solution: Add hemin to NaOH solution. Mix thoroughly. Adjust volume to 100.0mL with distilled/deionized water. Autoclave for 15 min at 15 psi pressure–121°C. Cool to 45°–50°C.

Resazurin Solution:

Composition per 44.0mL:

Resazurin 0.044g

Preparation of Resazurin Solution: Add resazurin to distilled/deionized water and bring volume to 44.0mL. Mix thoroughly.

Vitamin K_1 Solution:

Composition per 30.0mL:

Vitamin K_1 .... 0.15g
Ethanol (95% solution) 30.0mL

Preparation of Vitamin K_1 Solution: Add vitamin K_1 to ethanol. Mix thoroughly. Store in a brown bottle and keep under refrigeration. Discard after 1 month.

Preparation of Medium: Add components—except hemin solution, L-cysteine·HCl·H_2O, and vitamin K_1 solution—to distilled/deionized water and bring volume to 989.8mL. Mix thoroughly. Gently heat and bring to boiling under 80% N_2 + 10% CO_2 + 10% H_2. Continue boiling until resazurin turns colorless, indicating reduction. Cool to 45°–50°C. Add the L-cysteine·HCl·H_2O, hemin solution, and vitamin K_1 solution. Adjust pH to 6.9 if necessary. Anaerobically distribute into tubes under 80% N_2 + 10% CO_2 + 10% H_2. Cap the tubes with rubber stoppers. Place tubes in a press. Autoclave for 15 min at 15 psi pressure–121°C with fast exhaust.

Use: For the cultivation and maintenance of *Megasphaera cerevisiae*.

PY Medium with Glucose (Peptone Yeast Extract Medium with Glucose) (PYG Medium)

Composition per liter:

Glucose	10.0g
Yeast extract	10.0g
Peptone	5.0g
Pancreatic digest of casein	5.0g
L-cysteine·HCl·H_2O	0.5g
Salts solution	40.0mL
Hemin solution	10.0mL
Resazurin solution	4.0mL
Vitamin K_1 solution	0.2mL

pH 6.9 ± 0.2 at 25°C

Salts Solution:

Composition per liter:

$NaHCO_3$	10.0g
NaCl	2.0g
K_2HPO_4	1.0g
KH_2PO_4	1.0g
$CaCl_2$, anhydrous	0.2g
$MgSO_4$	0.2g

Preparation of Salts Solution: Add $CaCl_2$ and $MgSO_4$ to 300.0mL of distilled/deionized water. Mix thoroughly until dissolved. Bring volume to 800.0mL with distilled/deionized water. Add remaining components while stirring. Bring volume to 1.0L. Mix thoroughly. Store at 4°C.

Hemin Solution:

Composition per 100.0mL:

Hemin	0.05g
NaOH (1*N* solution)	1.0mL

Preparation of Hemin Solution: Add hemin to NaOH solution. Mix thoroughly. Adjust volume to 100.0mL with distilled/deionized water. Autoclave for 15 min at 15 psi pressure–121°C. Cool to 45°–50°C.

Resazurin Solution:

Composition per 44.0mL:

Resazurin	0.044g

Preparation of Resazurin Solution: Add resazurin to distilled/deionized water and bring volume to 44.0mL. Mix thoroughly.

Vitamin K_1 Solution:

Composition per 30.0mL:

Vitamin K_1	0.15g
Ethanol (95% solution)	30.0mL

Preparation of Vitamin K_1 Solution: Add vitamin K_1 to ethanol. Mix thoroughly. Store in a brown bottle and keep under refrigeration. Discard after 1 month.

Preparation of Medium: Add components—except hemin solution, L-cysteine·HCl·H_2O, and vitamin K_1 solution—to distilled/deionized water and bring volume to 989.8mL. Mix thoroughly. Gently heat and bring to boiling under 80% N_2 + 10% CO_2 + 10% H_2. Continue boiling until resazurin turns colorless, indicating reduction. Cool to 45°–50°C. Add the L-cysteine·HCl·H_2O, hemin solution, and vitamin K_1 solution. Adjust pH to 6.9 if necessary. Anaerobically distribute into tubes under 80% N_2 + 10% CO_2 + 10% H_2. Cap the tubes with rubber stoppers. Place tubes in a press. Autoclave for 15 min at 15 psi pressure–121°C with fast exhaust.

Use: For the cultivation and maintenance of *Clostridium* species.

PY Medium with Serum-Cocarboxylase

Composition per liter:

Yeast extract	10.0g
Peptone	5.0g
Pancreatic digest of casein	5.0g
L-Cysteine·HCl·H_2O	0.5g
Serum-cocarboxylase solution	100.0mL
Salts solution	40.0mL
Hemin solution	10.0mL
Resazurin solution	4.0mL
Vitamin K_1 solution	0.2mL

pH 6.9 ± 0.2 at 25°C

Salts Solution:

Composition per liter:

$NaHCO_3$	10.0g
NaCl	2.0g
K_2HPO_4	1.0g
KH_2PO_4	1.0g
$CaCl_2$, anhydrous	0.2g
$MgSO_4$	0.2g

Preparation of Salts Solution: Add $CaCl_2$ and $MgSO_4$ to 300.0mL of distilled/deionized water. Mix thoroughly until dissolved. Bring volume to 800.0mL with distilled/deionized water. Add remaining components while stirring. Bring volume to 1.0L. Mix thoroughly. Store at 4°C.

Hemin Solution:

Composition per 100.0mL:

Hemin	0.050g
NaOH (1*N* solution)	1.0mL

Preparation of Hemin Solution: Add hemin to NaOH solution. Mix thoroughly. Adjust volume to 100.0mL with distilled/deionized water. Autoclave for 15 min at 15 psi pressure–121°C. Cool to 45°–50°C.

Resazurin Solution:

Composition per 44.0mL:

Resazurin	0.044g

Preparation of Resazurin Solution: Add resazurin to distilled/deionized water and bring volume to 44.0mL. Mix thoroughly.

Vitamin K_1 Solution:

Composition per 30.0mL:

Vitamin K_1	0.15g
Ethanol (95% solution)	30.0mL

Preparation of Vitamin K_1 Solution: Add vitamin K_1 to ethanol. Mix thoroughly. Store in a brown bottle and keep under refrigeration. Discard after 1 month.

Serum-Cocarboxylase Solution:

Composition per 101.0mL:

Rabbit serum, heat inactivated....100.0mL
Cocarboxylase solution....1.0mL

Cocarboxylase Solution:

Composition per 100.0mL:

Cocarboxylase....5mg

Preparation of Cocarboxylase Solution: Add cocarboxylase to distilled/deionized water and bring volume to 100.0mL. Mix thoroughly. Filter sterilize.

Preparation of Serum-Cocarboxylase Solution: Aseptically combine 100.0mL of sterile, heat-inactivated rabbit serum with 1.0mL of sterile cocarboxylase solution. Mix thoroughly.

Preparation of Medium: Add components, except serum-cocarboxylase solution, L-cysteine·HCl·H_2O, hemin solution, and vitamin K_1 solution, to distilled/deionized water and bring volume to 889.8mL. Mix thoroughly. Gently heat and bring to boiling under 80% N_2 + 10% CO_2 + 10% H_2. Continue boiling until resazurin turns colorless, indicating reduction. Cool to 45°–50°C. Add the serum-cocarboxylase solution, L-cysteine·HCl·H_2O, hemin solution, and vitamin K_1 solution. Adjust pH to 6.9, if necessary. Anaerobically distribute into tubes under 80% N_2 + 10% CO_2 + 10% H_2. Cap the tubes with rubber stoppers. Place tubes in a press. Autoclave for 15 min at 15 psi pressure–121°C with fast exhaust.

Use: For the cultivation of *Treponema phagedenis.*

PY Salt Medium (Peptone Yeast Extract Salt Medium)

Composition per liter:

Peptone....9.0g
NaCl....5.0g
Yeast extract....5.0g

pH 7.2 ± 0.2 at 25°C

Preparation of Medium: Add components to distilled/deionized water and bring volume to 1.0L. Mix thoroughly. Distribute into tubes or flasks. Autoclave for 15 min at 15 psi pressure–121°C.

Use: For the cultivation and maintenance of *Bacillus brevis.*

PYb Agar

Composition per liter:

Agar....20.0g
Proteose peptone....1.0g
Yeast extract....1.0g
KH_2PO_4 solution....32.0mL
Na_2HPO_4 solution....8.0mL
$CaCl_2$ solution....4.0mL
$MgSO_4·7H_2O$ solution....2.5mL

pH 6.5 ± 0.5 at 25°C

$CaCl_2$ Solution:

Composition per 100.0mL:

$CaCl_2$....0.75g

Preparation of $CaCl_2$ Solution: Add $CaCl_2$ to distilled/deionized water and bring volume to 100.0mL. Mix thoroughly. Adjust pH to 6.5. Autoclave for 25 min at 15 psi pressure–121°C. Cool to 50°–55°C.

$MgSO_4·7H_2O$ Solution:

Composition per 100.0mL:

$MgSO_4·7H_2O$....9.8g

Preparation of $MgSO_4·7H_2O$ Solution: Add $MgSO_4·7H_2O$ to distilled/deionized water and bring volume to 100.0mL. Mix thoroughly. Adjust pH to 6.5. Autoclave for 25 min at 15 psi pressure–121°C. Cool to 50°–55°C.

Na_2HPO_4 Solution:

Composition per 100.0mL:

Na_2HPO_4....6.7g

Preparation of Na_2HPO_4 Solution: Add Na_2HPO_4 to distilled/deionized water and bring volume to 100.0mL. Mix thoroughly. Adjust pH to 6.5. Autoclave for 25 min at 15 psi pressure–121°C. Cool to 50°–55°C.

KH_2PO_4 Solution:

Composition per 100.0mL:

KH_2PO_4....3.4g

Preparation of KH_2PO_4 Solution: Add KH_2PO_4 to distilled/deionized water and bring volume to 100.0mL. Mix thoroughly. Adjust pH to 6.5. Autoclave for 25 min at 15 psi pressure–121°C. Cool to 50°–55°C.

Preparation of Medium: Add components, except KH_2PO_4 solution, Na_2HPO_4 solution, $CaCl_2$ solution, and $MgSO_4·7H_2O$ solution, to distilled/deionized water and bring volume to 953.5mL. Mix thoroughly. Gently heat and bring to boiling. Autoclave for 15 min at 15 psi pressure–121°C. Cool to 50°–55°C. Aseptically add 32.0mL of sterile KH_2PO_4 solution, 8.0mL of sterile Na_2HPO_4 solution, 4.0mL of sterile $CaCl_2$ solution, and 2.5mL of sterile $MgSO_4·7H_2O$ solution. Mix thoroughly. Pour into sterile Petri dishes or distribute into sterile tubes.

Use: For the cultivation of *Acanthamoeba hatchetti, Acanthamoeba jacobsi, Acanthamoeba polyphaga, Echinamoeba exundans, Naegleria gruberi, Paratetramitus jugosus, Rhizamoeba* species, *Tetramitus rostratus*, and *Vahlkampfia lobospinosa.*

PYCS Medium

Composition per liter:

KH_2PO_4....1.0g
$(NH_4)_2SO_4$....1.0g
$MgCl_2·6H_2O$....0.2g
NaCl....0.2g
$CaCl_2·2H_2O$....45.0mg
Fructose solution....50.0mL
$NaHCO_3$ solution....10.0mL
Wolfe's mineral solution....10.0mL

pH 7.2 ± 0.2 at 25°C

Fructose Solution:

Composition per 50.0mL:

D-Fructose....5.0g

Preparation of Fructose Solution: Add D-fructose to distilled/deionized water and bring volume to 50.0mL. Mix thoroughly. Filter sterilize.

$NaHCO_3$ Solution:

Composition per 10.0mL:

$NaHCO_3$....1.68g

Preparation of $NaHCO_3$ Solution: Add $NaHCO_3$ to distilled/deionized water and bring volume to 10.0mL. Mix thoroughly. Filter sterilize.

Wolfe's Mineral Solution:

Composition per liter:

$MgSO_4 \cdot 7H_2O$	3.0g
Nitrilotriacetic acid	1.5g
NaCl	1.0g
$MnSO_4 \cdot H_2O$	0.5g
$CaCl_2$	0.1g
$CoCl_2 \cdot 6H_2O$	0.1g
$FeSO_4 \cdot 7H_2O$	0.1g
$ZnSO_4 \cdot 7H_2O$	0.1g
$AlK(SO_4)_2 \cdot 12H_2O$	0.01g
$CuSO_4 \cdot 5H_2O$	0.01g
H_3BO_3	0.01g
$Na_2MoO_4 \cdot 2H_2O$	0.01g

Preparation of Wolfe's Mineral Solution: Add nitrilotriacetic acid to approximately 500.0mL of water and adjust to pH 6.5 with KOH to dissolve the compound. Bring volume to 1.0L with remaining water and add remaining components one at a time.

Preparation of Medium: Add components, except fructose solution and $NaHCO_3$ solution, to distilled/deionized water and bring volume to 940.0mL. Mix thoroughly. Autoclave for 15 min at 15 psi pressure–121°C. Aseptically add 50.0mL of sterile fructose solution and 10.0mL of sterile $NaHCO_3$ solution. Mix thoroughly. Adjust pH to 7.2. Aseptically distribute into sterile tubes or flasks. Fill containers to capacity.

Use: For the cultivation of *Rhodoferax fermentans* and *Thiocapsa halophila*.

PYE Medium

Composition per liter:

Yeast extract	4.0g
Sodium pyruvate	2.2g
K_2HPO_4	1.0g
$(NH_4)_2SO_4$	1.0g
$MgSO_4 \cdot 7H_2O$	0.2g
$Na_2S_2O_3 \cdot 5H_2O$	0.2g
$CaCl_2 \cdot 2H_2O$	0.02g
Trace elements solution SL-6	1.0mL

Trace Elements Solution SL-6:

Composition per liter:

$MnCl_2 \cdot 4H_2O$	0.5g
H_3BO_3	0.3g
$CoCl_2 \cdot 6H_2O$	0.2g
$ZnSO_4 \cdot 7H_2O$	0.1g
$Na_2MoO_4 \cdot 2H_2O$	0.03g
$NiCl_2 \cdot 6H_2O$	0.02g
$CuCl_2 \cdot 2H_2O$	0.01g

Preparation of Trace Elements Solution SL-6: Add components to distilled/deionized water and bring volume to 1.0L. Mix thoroughly.

Preparation of Medium: Add components to distilled/deionized water and bring volume to 1.0L. Mix thoroughly. Distribute into tubes or flasks. Autoclave for 15 min at 15 psi pressure–121°C.

Use: For the cultivation of *Heliobacterium modestocaldum*.

PYEA Agar

Composition per liter:

Agar	15.0g
Peptone	10.0g
Yeast extract	10.0g
NaCl	5.0g

pH 7.2 ± 0.2 at 25°C

Preparation of Medium: Add components to distilled/deionized water and bring volume to 1.0L. Mix thoroughly. Gently heat and bring to boiling. Distribute into tubes or flasks. Autoclave for 15 min at 15 psi pressure–121°C. Pour into sterile Petri dishes or leave in tubes.

Use: For the cultivation and maintenance of *Blastobacter natatorius* and *Deinobacter grandis*.

PYEM Medium (DSMZ Medium 1157)

Composition per liter:

Peptone	2.0g
Yeast extract	2.0g
NH_4Cl	0.5g
Riboflavin solution	5.0mL
Glucose solution	2.0mL
Magnesium sulfate solution	1.0mL
Calcium chloride solution	1.0mL

pH 7.0 ± 0.2 at 25°C

Glucose Solution:

Composition per 10.0mL:

Glucose	5.0g

Preparation of Glucose Solution: Add glucose to distilled/deionized water and bring volume to 10.0mL. Mix thoroughly. Filter sterilize.

Riboflavin Solution:

Composition per 10.0ml:

Riboflavin	2.0mg

Preparation of Riboflavin Solution: Add riboflavin to distilled/deionized water and bring volume to 10.0mL. Mix thoroughly. Filter sterilize.

Calcium Chloride Solution:

Composition per 10.0mL:

$CaCl_2 \cdot 2H_2O$	1.0g

Preparation of Calcium Chloride Solution: Add $CaCl_2 \cdot 2H_2O$ to distilled/deionized water and bring volume to 10.0mL. Mix thoroughly. Sparge with N_2. Autoclave for 15 min at 15 psi pressure–121°C. Cool to room temperature.

Magnesium Sulfate Solution:

Composition per 10.0mL:

$MgSO_4 \cdot 7H_2O$	2.0g

Preparation of Magnesium Sulfate Solution: Add $MgSO_4 \cdot 7H_2O$ to distilled/deionized water and bring volume to 10.0mL. Mix thoroughly. Autoclave for 15 min at 15 psi pressure–121°C. Cool to room temperature.

Preparation of Medium: Add components, except riboflavin, glucose, magnesium sulfate, and calcium chloride solutions, to distilled/deionized water and bring volume to 991.0mL. Mix thoroughly. Adjust pH to 7.0. Autoclave for 15 min at 15 psi pressure–121°C. Cool to 25°C. Aseptically add riboflavin, glucose, magnesium sulfate, and calcium chloride solutions. Mix thoroughly. Aseptically distribute into culture vessels.

Use: For the cultivation of *Phenylobacterium conjunctum*.

Pyes Medium (DSMZ Medium 937)

Composition per liter:

Peptone from casein 3.0g
Yeast extract 3.0g
Sodium succinate 2.3g

pH 7.2 ± 0.2 at 25°C

Preparation of Medium: Add components to distilled/deionized water and bring volume to 1.0L. Mix thoroughly. Distribute into tubes or flasks. Autoclave for 15 min at 15 psi pressure–121°C.

Use: For the cultivation of *Hymenobacter aerophilus* and *Hymenobacter* sp.

PYEX Glucose Salt Medium (Peptone Yeast Extract Glucose Salt Medium)

Composition per liter:

Peptone 10.0g
NaCl 5.0g
Yeast extract 5.0g
Glucose 1.0g

Preparation of Medium: Add components to distilled/deionized water and bring volume to 1.0L. Mix thoroughly. Distribute into tubes or flasks. Autoclave for 15 min at 15 psi pressure–121°C.

Use: For the cultivation and maintenance of *Micrococcus luteus*.

PYF Medium

Composition per liter:

Yeast extract 10.0g
Fructose 5.0g
Peptone 5.0g
Pancreatic digest of casein 5.0g
Na_2HPO_4 .... 2.0g
L-Cysteine·HCl 1.0mL
Tween™ 80 .... 1.0mL

pH 7.0 ± 0.2 at 25°C

Preparation of Medium: Add components to distilled/deionized water and bring volume to 1.0L. Mix thoroughly. Adjust pH to 7.0. Distribute into tubes or flasks. Autoclave for 15 min at 15 psi pressure–121°C.

Use: For the cultivation of *Clostridium acetobutylicum, Megasphaera cerevisiae, Megasphaera elsdenii, Pectinatus cerevisiiphilus, Pectinatus frisingensis, Selenomonas lacticifex, Zymophilus paucivorans*, and *Zymophilus raffinosivorans*.

PYG Agar

Composition per liter:

Glucose 5.0g
Peptone 5.0g
Yeast extract 5.0g

pH 7.0 ± 0.2 at 25°C

Preparation of Medium: Add components to distilled/deionized water and bring volume to 1.0L. Mix thoroughly. Gently heat and bring to boiling. Distribute into tubes or flasks. Autoclave for 15 min at 15 psi pressure–121°C. Pour into sterile Petri dishes or leave in tubes.

Use: For the cultivation and maintenance of *Mycoplana ramosa* and *Mycoplana segnis*.

PYG Agar

Composition per liter:

Agar 20.0g
Proteose peptone 20.0g
Yeast extract 1.0g
Glucose solution 50.0mL
Sodium citrate solution 34.0mL
Ferric ammonium sulfate 10.0mL
KH_2PO_4 solution 10.0mL
$MgSO_4 \cdot 7H_2O$ solution 10.0mL
Na_2HPO_4 solution 10.0mL
$CaCl_2$ solution 8.0mL

pH 6.5 ± 0.5 at 25°C

Glucose Solution:

Composition per 100.0mL:

Glucose 36.0g

Preparation of Glucose Solution: Add glucose to distilled/deionized water and bring volume to 100.0mL. Mix thoroughly. Filter sterilize. Warm to 55°C.

$MgSO_4 \cdot 7H_2O$ Solution:

Composition per 100.0mL:

$MgSO_4 \cdot 7H_2O$ 9.8g

Preparation of $MgSO_4 \cdot 7H_2O$ Solution: Add $MgSO_4 \cdot 7H_2O$ to distilled/deionized water and bring volume to 100.0mL. Mix thoroughly. Adjust pH to 6.5. Autoclave for 25 min at 15 psi pressure–121°C. Cool to 50°–55°C.

Ferric Ammonium Sulfate Solution:

Composition per 100.0mL:

$Fe(NH_4)_2(SO_4)_2 \cdot 6H_2O$ 0.135g

Preparation of Ferric Ammonium Sulfate Solution: Add $Fe(NH_4)_2(SO_4)_2 \cdot 6H_2O$ to distilled/deionized water and bring volume to 100.0mL. Mix thoroughly. Adjust pH to 6.5. Autoclave for 25 min at 15 psi pressure–121°C. Cool to 50°–55°C.

Na_2HPO_4 Solution:

Composition per 100.0mL:

Na_2HPO_4 .... 6.7g

Preparation of Na_2HPO_4 Solution: Add Na_2HPO_4 to distilled/deionized water and bring volume to 100.0mL. Mix thoroughly. Adjust pH to 6.5. Autoclave for 25 min at 15 psi pressure–121°C. Cool to 50°–55°C.

Sodium Citrate Solution:

Composition per 100.0mL:

Sodium citrate·$2H_2O$ 2.9g

Preparation of Sodium Citrate Solution: Add 2.9g of sodium citrate·$2H_2O$ to distilled/deionized water and bring volume to 100.0mL. Mix thoroughly. Adjust pH to 6.5. Autoclave for 25 min at 15 psi pressure–121°C. Cool to 50°–55°C.

KH_2PO_4 Solution:

Composition per 100.0mL:

KH_2PO_4 .... 3.4g

Preparation of KH_2PO_4 Solution: Add KH_2PO_4 to distilled/deionized water and bring volume to 100.0mL. Mix thoroughly. Adjust pH to 6.5. Autoclave for 25 min at 15 psi pressure–121°C. Cool to 50°–55°C.

$CaCl_2$ Solution:

Composition per 100.0mL:

$CaCl_2$ 0.75g

Preparation of $CaCl_2$ Solution: Add $CaCl_2$ to distilled/deionized water and bring volume to 100.0mL. Mix thoroughly. Adjust pH to 6.5. Autoclave for 25 min at 15 psi pressure–121°C. Cool to 50°–55°C.

Preparation of Medium: Add components, except glucose solution, sodium citrate solution, ferric ammonium sulfate solution, KH_2PO_4 solution, Na_2HPO_4 solution, $CaCl_2$ solution, and $MgSO_4·7H_2O$ solution, to distilled/deionized water and bring volume to 868.0mL. Mix thoroughly. Gently heat and bring to boiling. Autoclave for 15 min at 15 psi pressure–121°C. Cool to 50°–55°C. Aseptically add 50.0mL of sterile glucose solution, 34.0mL of sterile sodium citrate solution, 10.0mL of sterile ferric ammonium sulfate solution, 10.0mL of sterile KH_2PO_4 solution, 10.0mL of sterile Na_2HPO_4 solution, 8.0mL of sterile $CaCl_2$ solution, and 10.0mL of sterile $MgSO_4·7H_2O$ solution. Mix thoroughly. Pour into sterile Petri dishes or distribute into sterile tubes.

Use: For the cultivation of *Acanthamoeba astronyxis, Acanthamoeba castellanii, Acanthamoeba comandoni, Acanthamoeba culbertsoni, Acanthamoeba divionensis, Acanthamoeba healyi, Acanthamoeba quina, Acanthamoeba lenticulata, Acanthamoeba lugdunensis, Acanthamoeba mauritaniensis, Acanthamoeba palestinensis, Acanthamoeba pearcei, Acanthamoeba polyphaga, Acanthamoeba pustulosa, Acanthamoeba rhysodes, Acanthamoeba royreba, Acanthamoeba* species, *Acanthamoeba terricola*, and *Acanthamoeba triangularis*.

PYG Broth (Peptone Yeast Extract Glucose Broth)

Composition per liter:

Peptone 20.0g
D-Glucose 10.0g
Yeast extract 10.0g
L-cysteine·HCl·H_2O 0.5g
VPI salt solution 40.0mL
Resazurin solution 4.0mL

pH 7.2 ± 0.2 at 25°C

Resazurin Solution:

Composition per 44.0mL:

Resazurin 0.044g

Preparation of Resazurin Solution: Add resazurin to distilled/deionized water and bring volume to 44.0mL. Mix thoroughly.

VPI Salt Solution:

Composition per 40.0mL:

$CaCl_2$ 0.2g
$MgSO_4$ 0.2g
K_2HPO_4 .... 1.0g
KH_2PO_4 .... 1.0g

Preparation of VPI Salt Solution: Add $CaCl_2$ and $MgSO_4$ to 300.0mL of distilled/deionized water. Mix thoroughly until dissolved. Bring volume to 800.0mL with distilled/deionized water. Add remaining components while stirring. Bring volume to 1.0L. Mix thoroughly. Store at 4°C.

Preparation of Medium: Add components to distilled/deionized water and bring volume to 1.0L. Mix thoroughly. Distribute into screw-capped tubes in 7.0mL volumes. Autoclave for 15 min at 15 psi pressure–121°C. Cool to 45°–50°C under 100% N_2.

Use: For the cultivation of a wide variety of anaerobic bacteria.

PYG Medium

See: **PY Medium with Glucose**

PYG Medium (Peptone Yeast Extract Glucose Medium)

Composition per liter:

Proteose peptone 20.0g
Glucose 18.0g
Yeast extract 2.0g
Sodium citrate·$2H_2O$ 1.0g
$MgSO_4·7H_2O$ 0.98g
$Na_2HPO_4·7H_2O$ 0.355g
KH_2PO_4 .... 0.34g
$CaCl_2$ 0.059g
$Fe(NH_4)_2(SO_4)_2·6H_2O$ 0.02g

pH 6.5 ± 0.2 at 25°C

Preparation of Medium: Add components, except $CaCl_2$, to distilled/deionized water and bring volume to 900.0mL. Mix thoroughly until dissolved. Add $CaCl_2$. Mix thoroughly. Bring volume to 1.0L with distilled/deionized water. Distribute into screw-capped tubes in 5.0mL volumes. Autoclave for 15 min at 15 psi pressure–121°C.

Use: For the cultivation of *Acanthamoeba* species.

PYG Medium (Peptone Yeast Extract Glucose Medium)

Composition per liter:

Agar 15.0g
Glucose 0.25g
Peptone 0.25g
Yeast extract 0.25g
Hutner's modified salt solution 20.0mL
Vitamin solution 10.0mL

Hutner's Modified Salts Solution:

Composition per liter:

$MgSO_4·7H_2O$ 29.7g
Nitrilotriacetic acid 10.0g
$CaCl_2·2H_2O$ 3.34g
$FeSO_4·7H_2O$ 0.1g
Metals "44" 50.0mL

Preparation of Hutner's Modified Salts Solution: Add nitrilotriacetic acid to 500.0mL of distilled/deionized water. Dissolve by adjusting pH to 6.5 with KOH. Add remaining components. Add distilled/deionized water to 1.0L.

Metals "44":

Composition per 100.0mL:

$ZnSO_4·7H_2O$ 1.1g
$FeSO_4·7H_2O$ 0.5g
EDTA 0.25g
$MnSO_4·7H_2O$ 0.154g
$CuSO_4·5H_2O$ 0.04g
$Co(NO_3)_2·6H_2O$ 0.025g
$Na_2B_4O_7·10H_2O$ 0.018g

Preparation of Metals "44": Add components to distilled/deionized water and bring volume to 100.0mL. Mix thoroughly.

Vitamin Solution:

Composition per liter:

Pyridoxine·HCl 0.01g
Calcium pantothenate 5.0mg

Nicotinamide	5.0mg
Riboflavin	5.0mg
Thiamine·HCl	5.0mg
Biotin	2.0mg
Folic acid	2.0mg
Vitamin B_{12}	0.1mg

Preparation of Vitamin Solution: Add components to distilled/deionized water and bring volume to 1.0L. Mix thoroughly. Filter sterilize.

Preparation of Medium: Add components, except vitamin solution, to distilled/deionized water and bring volume to 990.0mL. Mix thoroughly. Gently heat and bring to boiling. Autoclave for 15 min at 15 psi pressure–121°C. Cool to 45°–50°C. Aseptically add 10.0mL of sterile vitamin solution. Mix thoroughly. Pour into sterile Petri dishes or distribute into sterile tubes.

Use: For the isolation and cultivation of *Pasteuria ramosa.*

PYG Medium (Peptone Yeast Extract Glucose Medium) (ATCC Medium 663)

Composition per liter:

Agar	20.0g
Glucose	3.0g
Peptone	1.25g
Yeast extract	1.25g

Preparation of Medium: Add components to distilled/deionized water and bring volume to 1.0L. Mix thoroughly. Gently heat and bring to boiling. Distribute into tubes or flasks. Autoclave for 15 min at 15 psi pressure–121°C. Pour into sterile Petri dishes or leave in tubes.

Use: For the cultivation of *Eikenella corrodens.*

PYG Medium (B) (DSMZ Medium 1139)

Composition per liter:

Yeast extract	10.0g
Glucose	10.0g
Polypeptone	5.0g
Tryptone	5.0g
Salts solution	40.0mL

pH 7.2 ± 0.2 at 25°C

Salts Solution:

Composition per 10.0mL:

$NaHCO_3$	10.0g
NaCl	2.0g
K_2HPO_4	1.0g
KH_2PO_4	1.0g
$MgSO_4·7H_2O$	0.4g
$CaCl_2·2H_2O$	0.2g

Preparation of Salts Solution: Add components to distilled/deionized water and bring volume to 1.0L. Mix thoroughly.

Preparation of Medium: Add components to distilled/deionized water and bring volume to 1.0L. Mix thoroughly. Adjust pH to 7.2. Distribute into tubes or flasks. Autoclave for 15 min at 15 psi pressure–121°C. Cool to 25°C.

Use: For the cultivation of *Flavobacterium omnivorum.*

PYG Medium (E) (DSMZ Medium 1140)

Composition per liter:

Agar	15.0g
Peptone	5.0g
Beef extract	3.0g
$MgSO_4·7H_2O$	1.5g
Yeast extract	0.2g
Glucose	5.0g
NaCl	0.5g

pH 7.0 ± 0.2 at 25°C

Preparation of Medium: Add components to distilled/deionized water and bring volume to 1.0L. Mix thoroughly. Adjust pH to 7.0. Distribute into tubes or flasks. Autoclave for 15 min at 15 psi pressure–121°C. Cool to 25°C. Pour into Petri dishes or leave in tubes.

Use: For the cultivation of *Flavobacterium glaciei.*

PYG Medium, Modified

Composition per 961.0mL:

Yeast extract	10.0g
Beef extract	5.0g
Glucose	5.0g
Peptone	5.0g
Trypticase™	5.0g
K_2HPO_4	2.0g
L-Cysteine·HCl	0.5g
Resazurin	1.0mg
Salt solution	40.0mL
Hemin solution	10.0mL
Tween™ 80	1.0mL
Vitamin K_1 solution	0.2mL

pH 7.2 ± 0.2 at 25°C

Salt Solution:

Composition per liter:

$NaHCO_3$	10.0g
NaCl	2.0g
K_2HPO_4	1.0g
KH_2PO_4	1.0g
$MgSO_4·7H_2O$	0.5g
$CaCl_2·2H_2O$	0.25g

Preparation of Salt Solution: Add components to distilled/deionized water and bring volume to 1.0L. Mix thoroughly.

Hemin Solution:

Composition per 100.0mL:

Hemin	50.0mg
NaOH (1*N* solution)	1.0mL

Preparation of Hemin Solution: Add hemin to 1.0mL of NaOH solution. Mix thoroughly until dissolved. Bring volume to 100.0mL with distilled/deionized water. Mix thoroughly. Store at 4°C.

Vitamin K_1 Solution:

Composition per 20.0mL:

Vitamin K_1	0.1g
Ethanol (95% solution)	20.0mL

Preparation of Vitamin K_1 Solution: Add vitamin K_1 to 95% ethanol and bring volume to 20.0mL. Mix thoroughly. Filter sterilize. Store in a brown glass bottle at 4°C.

Preparation of Medium: Add components, except L-cysteine·HCl, vitamin K_1 solution, and hemin solution, to distilled/deionized water and bring volume to 1.0L. Mix thoroughly. Gently heat and bring to boiling. Continue boiling for 5 min. Cool to room temperature while sparging with 100% CO_2. Add L-cysteine·HCl, vitamin K_1 solution, and hemin solution. Mix thoroughly. Adjust pH to 7.2 using 8*N* NaOH while continuing to sparge with 100% CO_2. After pH has been attained sparge with 100% N_2. Anaerobically distribute into tubes or flasks. Autoclave for 15 min at 15 psi pressure–121°C.

Use: For the cultivation of *Acetomicrobium flavidum, Actinomyces denticolens, Actinomyces hordeovulneris, Actinomyces meyeri, Actinomyces pyogenes, Anaerobiospirillum succiniciproducens, Arcanobacterium haemolyticum, Atopobium minutum, Atopobium parvulum, Atopobium rimae, Bacteroides eggerthii, Bacteroides helcogenes, Bacteroides pyogenes, Bacteroides* species, *Bacteroides suis, Bacteroides uniformis, Bifidobacterium gallinarum, Clostridium beijerinckii, Clostridium butyricum, Clostridium* species, *Clostridium xylanolyticum, Coriobacterium glomerans, Eubacterium combesii, Eubacterium limosum, Eubacterium multiforme, Eubacterium nitritogenes, Eubacterium tenue, Fusobacterium naviforme, Fusobacterium necrophorum, Fusobacterium nucleatum, Gemella morbillorum, Lactobacillus catenaformis, Lactobacillus crispatus, Megamonas hypermegas, Megasphaera cerevisiae, Megasphaera elsdenii, Mitsuokella multiacidus, Peptostreptococcus anaerobius, Peptostreptococcus indolicus, Peptostreptococcus magnus, Peptostreptococcus prevotii, Peptostreptococcus productus, Peptostreptococcus* species, *Peptostreptococcus tetradius, Prevotella bivia, Prevotella buccae, Prevotella buccalis, Prevotella denticola, Prevotella disiens, Prevotella intermedia, Propionibacterium acnes, Propionibacterium avidum, Propionibacterium freudenreichii, Propionibacterium granulosum, Propionibacterium lymphophilum, Selenomonas sputigena, Staphylococcus saccharolyticus, Streptococcus constellatus, Streptococcus hansenii, Streptococcus intermedius,* and *Streptococcus pleomorphus.*

PYG Medium for *Spirillum* (Peptone Yeast Extract Glucose Medium for *Spirillum*)

Composition per liter:

Agar	15.0g
Peptone	10.0g
Yeast extract	5.0g
Glucose solution	10.0mL

pH 7.2 ± 0.2 at 25°C

Glucose Solution:

Composition per 10.0mL:

D-Glucose	3.0g

Preparation of Glucose Solution: Add glucose to distilled/deionized water and bring volume to 10.0mL. Mix thoroughly. Filter sterilize.

Preparation of Medium: Add components, except glucose solution, to distilled/deionized water and bring volume to 990.0mL. Mix thoroughly. Gently heat and bring to boiling. Autoclave for 15 min at 15 psi pressure–121°C. Cool to 45°–50°C. Aseptically add sterile glucose solution. Mix thoroughly. Pour into sterile Petri dishes or distribute into sterile tubes.

Use: For the cultivation and maintenance of *Spirillum pleomorphum.*

PYG Medium with Volatile Fatty Acids

Composition per 1003.3mL:

Yeast extract	10.0g
Beef extract	5.0g
Glucose	5.0g
Trypticase™	5.0g
L-Cysteine·HCl	0.5g
$(NH_4)_2SO_4$	0.5g
Hemin	5.0mg
Resazurin	1.0mg
Mineral solution	40.0mL
Fatty acid mixture	3.1mL
NaOH (1*N* solution)	0.5mL
Vitamin K_1 solution	0.2mL

pH 6.9 ± 0.2 at 25°C

Mineral Solution:

Composition per liter:

$NaHCO_3$	10.0g
NaCl	2.0g
K_2HPO_4	1.0g
KH_2PO_4	1.0g
$MgSO_4·7H_2O$	0.48g
$CaCl_2·2H_2O$	0.3g

Preparation of Mineral Solution: Add components to distilled/deionized water and bring volume to 1.0L. Mix thoroughly.

Fatty Acid Mixture:

Composition per 31.0mL:

Acetic acid	17.0mL
Propionic acid	6.0mL
n-Butyric acid	4.0mL
DL-2-Methylbutyric acid	1.0mL
iso-Butyric acid	1.0mL
iso-Valeric acid	1.0mL
n-Valeric acid	1.0mL

Preparation of Fatty Acid Mixture: Combine components.

Vitamin K_1 Solution:

Composition per 10.0mL:

Vitamin K_1	0.05g
Ethanol (95% solution)	10.0mL

Preparation of Vitamin K_1 Solution: Add vitamin K_1 to 95% ethanol and bring volume to 10.0mL. Mix thoroughly. Filter sterilize. Store in a brown glass bottle at 4°C.

Preparation of Medium: Dissolve hemin in 0.5mL of 1*N* NaOH. Add remaining components, except L-cysteine·HCl, and bring volume to 1.0L with distilled/deionized water. Mix thoroughly. Gently heat and bring to boiling. Continue boiling for 5 min. Cool to room temperature while sparging with 100% CO_2. Add L-cysteine·HCl. Mix thoroughly. Adjust pH to 6.0 using 8*N* NaOH while continuing to sparge with 100% CO_2. After pH has been attained, sparge with 100% N_2. Anaerobically distribute into tubes or flasks. Autoclave for 15 min at 15 psi pressure–121°C.

Use: For the cultivation and maintenance of *Acetivibrio ethanolgignens* and *Bacteroides xylanolyticus.*

PYG with 0.1% Tween™ 80

Composition per 1004.2mL:

Glucose	10.0g
Yeast extract	10.0g
Peptone	5.0g

Pancreatic digest of casein	5.0g
Tween™ 80	1.0g
L-Cysteine·HCl·H_2O	0.5g
Salts solution	40.0mL
Hemin solution	10.0mL
Resazurin (0.025% solution)	4.0mL
Vitamin K_1 solution	0.2mL

pH 7.0 ± 0.2 at 25°C

Salts Solution:

Composition per liter:

$NaHCO_3$	10.0g
NaCl	2.0g
K_2HPO_4	1.0g
KH_2PO_4	1.0g
$CaCl_2$, anhydrous	0.2g
$MgSO_4$	0.2g

Preparation of Salts Solution: Add $CaCl_2$ and $MgSO_4$ to distilled/deionized water and bring volume to 300.0mL. Add 500.0mL of distilled/deionized water. Add the remaining components while swirling slowly. Add 200.0mL of distilled/deionized water. Mix thoroughly. Store at 4°C.

Hemin Solution:

Composition per 100.0mL:

Hemin	50.0mg
NaOH (1*N* solution)	1.0mL

Preparation of Hemin Solution: Dissolve hemin in 1.0mL of 1*N* NaOH solution. Bring volume to 100.0mL with distilled/deionized water. Mix thoroughly. Autoclave for 15 min at 15 psi pressure–121°C.

Vitamin K_1 Solution:

Composition per 30.15mL:

Ethanol (95% solution)	30.0mL
Vitamin K_1	0.15mL

Preparation of Vitamin K_1 Solution: Combine components. Mix thoroughly. Store at 4°C in the dark. Discard solution after 1 month.

Preparation of Medium: Add components, except the 0.5g of L-cysteine·HCl·H_2O, to distilled/deionized water and bring volume to 1.0L. Mix thoroughly. Adjust pH to 7.0. Gently heat and bring to boiling. Add L-cysteine·HCl·H_2O. Mix thoroughly. Distribute into tubes or flasks. Autoclave for 15 min at 15 psi pressure–121°C.

Use: For the cultivation of *Lactobacillus uli* and *Atopobium rimae*.

Pygeye Agar

Composition per liter:

Agar	15.0g
Glucose	10.0g
Neopeptone	5.0g
Yeast extract	0.5g
Egg yolk emulsion, 50%	200.0mL

Egg Yolk Emulsion, 50%:

Composition per 200mL:

Chicken egg yolks	22
Whole chicken egg	2
NaCl (0.9% solution)	100.0mL

Preparation of Egg Yolk Emulsion, 50%: Soak eggs with 1:100 dilution of saturated mercuric chloride solution for 1 min. Crack eggs and separate yolks from whites. Mix egg yolks with 2 chicken eggs. Beat to form emulsion. Measure 100.0mL of egg yolk emulsion and add to 100.0mL of 0.9% NaCl solution. Mix thoroughly. Filter sterilize. Warm to 45°–50°C.

Preparation of Medium: Add components, except egg yolk emulsion, 50%, to distilled/deionized water and bring volume to 800.0mL. Mix thoroughly. Gently heat while stirring and bring to boiling. Autoclave for 15 min at 15 psi pressure–121°C. Cool to 50°–55°C. Aseptically add 200.0mL of sterile egg yolk emulsion, 50%. Mix thoroughly. Pour into sterile Petri dishes in 15.0mL volumes.

Use: For the cultivation and maintenance of *Conidiobolus apiculatus, Conidiobolus obscurus, Dactylella oviparasitica, Entomophthora aphidis, Entomophthora ignobilis, Entomophthora sphaerosperma,* and *Erynia blunckii.*

PYGHS Medium

Composition per 60.0mL:

Beef heart, solids from infusion	10.0g
Glucose	1.0g
Polypeptone™	1.0g
Yeast extract	1.0g
L-Cysteine·HCl·H_2O	0.9g
Ionagar No. 2	0.72g
$NaHCO_3$	0.5g
Tryptose	0.2g
NaCl	0.1g
Resazurin	0.16mg
Rabbit serum, inactivated	5.0mL
Salts solution	5.0mL

pH 7.2–7.5 at 25°C

Salts Solution:

Composition per 400.0mL:

K_2HPO_4	0.9g
NaCl	0.8g
KH_2PO_4	0.4g
$MnCl_2·4H_2O$	0.16g
$MgSO_4$	0.08g

Preparation of Salts Solution: Add components to distilled/deionized water and bring volume to 400.0L. Mix thoroughly.

Preparation of Medium: Add components, except rabbit serum, to distilled/deionized water and bring volume to 55.0mL. Mix thoroughly. Gently heat and bring to boiling. Autoclave for 15 min at 15 psi pressure–121°C. Cool to 45°–50°C. Aseptically add 5.0mL of sterile rabbit serum. Mix thoroughly.

Use: For the cultivation of treponemes.

PYGM Medium (Peptone Yeast Extract Glucose Maltose Medium)

Composition per 261.0mL:

Peptone	5.0g
Yeast extract	2.5g
Glucose	1.25g
Maltose	1.25g
L-Cysteine·HCl·H_2O	0.125g
Salts solution	10.0mL
Resazurin (0.025% solution)	1.0mL

Salts Solution:

Composition per 100.0mL:

$NaHCO_3$	1.0g
NaCl	0.2g
K_2HPO_4	0.1g

KH_2PO_4 .... 0.1g
$CaCl_2$, anhydrous 0.02g
$MgSO_4$ 0.02g
H_2SO_4 (50% solution) 0.3mL
$Na_2MoO_4 \cdot 2H_2O$ 1.0μg
$CoCl_2 \cdot 6H_2O$ 1.0μg

Preparation of Salts Solution: Add components to distilled/deionized water and bring volume to 100.0mL. Mix thoroughly.

Preparation of Medium: Add components to distilled/deionized water and bring volume to 261.0mL. Mix thoroughly. Distribute into tubes or flasks. Autoclave for 15 min at 15 psi pressure–121°C.

Use: For the cultivation of *Bacteroides praeacutus, Eubacterium nitritogenes, Lactobacillus ruminis,* and *Tissierella praeacuta.*

PYGS Agar (Peptone Yeast Glucose Seawater Agar) (ATCC Medium 1973)

Composition per liter:
Agar 15.0g
Glucose 3.0g
Peptone 1.25g
Yeast extract 1.25g
Seawater 25.0mL

pH 7.3 ± 0.2 at 25°C

Preparation of Medium: Add components to cold distilled/deionized water and bring volume to 1.0L. Mix thoroughly. Gently heat and bring to boiling. Distribute into tubes or flasks. Autoclave for 15 min at 15 psi pressure–121°C. Pour into sterile Petri dishes or leave in tubes.

Use: For the cultivation of a variety of marine bacteria.

PYGV Agar

Composition per liter:
Agar 15.0g
Peptone 0.25g
Yeast extract 0.25g
Hutner's basal salts solution 20.0mL
Glucose solution 10.0mL
Vitamin solution 2X 5.0mL

pH 7.5 ± 0.2 at 25°C

Hutner's Basal Salts Solution:
Composition per liter:
$MgSO_4 \cdot 7H_2O$ 29.7g
Nitrilotriacetic acid 10.0g
$CaCl_2 \cdot 2H_2O$ 3.335g
$FeSO_4 \cdot 7H_2O$ 99.0mg
$(NH_4)_6MoO_7O_{24} \cdot 4H_2O$ 9.25mg
"Metals 44" 50.0mL

"Metals 44":
Composition per 100.0mL:
$ZnSO_4 \cdot 7H_2O$ 1.095g
$FeSO_4 \cdot 7H_2O$ 0.5g
Sodium EDTA 0.25g
$MnSO_4 \cdot H2O$ 0.154g
$CuSO_4 \cdot 5H_2O$ 39.2mg
$Co(NO_3)_2 \cdot 6H_2O$ 24.8mg
$Na_2B_4O_7 \cdot 10H_2O$ 17.7mg

Preparation of "Metals 44": Add sodium EDTA to distilled/deionized water and bring volume to 90.0mL. Mix thoroughly. Add a few drops of concentrated H_2SO_4 to retard precipitation of heavy metal ions. Add remaining components. Mix thoroughly. Bring volume to 100.0mL with distilled/deionized water.

Preparation of Hutner's Basal Salts Solution: Add nitrilotriacetic acid to 500.0mL of distilled/deionized water. Adjust pH to 6.5 with KOH. Add remaining components. Add distilled/deionized water to 1.0L. Adjust pH to 6.8.

Glucose Solution:
Composition per 10.0mL:
D-Glucose 0.25g

Preparation of Glucose Solution: Add glucose to distilled/deionized water and bring volume to 10.0mL. Mix thoroughly. Filter sterilize.

Vitamin Solution 2X:
Composition per liter:
Pyridoxine·HCl 20.0mg
p-Aminobenzoic acid 10.0mg
Calcium DL-pantothenate 10.0mg
Nicotinamide 10.0mg
Riboflavin 10.0mg
Thiamine·HCl 10.0mg
Biotin 4.0mg
Folic acid 4.0mg
Vitamin B_{12} 0.2mg

Preparation of Vitamin Solution 2X: Add components to distilled/deionized water and bring volume to 1.0L. Mix thoroughly. Filter sterilize. Store in the dark at 5°C.

Preparation of Medium: Add components, except glucose solution and vitamin solution 2X, to distilled/deionized water and bring volume to 985.0mL. Mix thoroughly. Gently heat and bring to boiling. Adjust pH to 7.5 with 6*N* KOH (approximately 6 drops). Autoclave for 20 min at 15 psi pressure–121°C. Cool to 50°–55°C. Aseptically add 10.0mL of sterile glucose solution and 5.0mL of sterile vitamin solution 2X. Mix thoroughly. Pour into sterile Petri dishes or distribute into sterile tubes.

Use: For the cultivation of *Blastobacter aggregatus, Blastobacter capsulatus, Blastobacter denitrificans, Planctomyces limnophilus,* and *Gemmobacter aquatilis.*

PYGV Marine Medium (Peptone Yeast Extract Glucose Vitamin Marine Medium)

Composition per liter:
Agar 15.0g
Peptone 0.25g
Yeast extract 0.25g
Mineral salt solution 20.0mL
Glucose solution 10.0mL
Vitamin solution 5.0mL

pH 7.5 ± 0.2 at 25°C

Mineral Salt Solution:
Composition per liter:
$MgSO_4 \cdot 7H_2O$ 29.7g
Nitrilotriacetic acid 10.0g
$CaCl_2 \cdot 2H_2O$ 3.34g
$FeSO_4 \cdot 7H_2O$ 0.099g

$Na_2MoO_4·2H_2O$ 0.013g
Metals "44" 50.0mL

Preparation of Mineral Salt Solution: Add nitrilotriacetic acid to 500.0mL of distilled/deionized water. Dissolve by adjusting pH to 6.5 with KOH. Add remaining components. Add distilled/deionized water to 1.0L. Readjust pH to 7.2.

Metals "44":
Composition per 100.0mL:

$ZnSO_4·7H_2O$ 1.1g
$FeSO_4·7H_2O$ 0.5g
EDTA 0.25g
$MnSO_4·7H_2O$ 0.154g
$CuSO_4·5H_2O$ 0.04g
$Co(NO_3)_2·6H_2O$ 0.025g
$Na_2B_4O_7·10H_2O$ 0.018g

Preparation of Metals "44": Add a few drops of H_2SO_4 to distilled/deionized water to inhibit precipitate formation. Add components to acidified distilled/deionized water and bring volume to 100.0mL. Mix thoroughly.

Glucose Solution:
Composition per 100.0mL:

D-Glucose 2.5g

Preparation of Glucose Solution: Add glucose to distilled/deionized water and bring volume to 100.0mL. Mix thoroughly. Filter sterilize.

Vitamin Solution:
Composition per liter:

Pyridoxine·HCl 0.02g
p-Aminobenzoic acid 0.01g
Calcium D-pantothenate 0.01g
Nicotinamide 0.01g
Riboflavin 0.01g
Thiamine·HCl 0.01g
Biotin 4.0mg
Folic acid 4.0mg
Cyanocobalamin 0.2mg

Preparation of Vitamin Solution: Add components to distilled/deionized water and bring volume to 1.0L. Mix thoroughly. Filter sterilize.

Preparation of Medium: Add components, except glucose solution and vitamin solution, to seawater and bring volume to 985.0mL. Mix thoroughly. Gently heat and bring to boiling. Autoclave for 15 min at 15 psi pressure–121°C. Cool to 45°–50°C. Aseptically add 10.0mL of sterile glucose solution and 5.0mL of sterile vitamin solution. Mix thoroughly. Adjust pH to 7.5 with sterile KOH if necessary. Pour into sterile Petri dishes or distribute into sterile tubes.

Use: For the cultivation and maintenance of *Planctomyces brasiliensis*.

PYGV Medium
(Peptone Yeast Extract Glucose Vitamin Medium)

Composition per liter:

Agar 15.0g
Peptone 0.25g
Yeast extract 0.25g
Mineral salt solution 20.0mL
Glucose solution 10.0mL
Vitamin solution 5.0mL

pH 7.5 ± 0.2 at 25°C

Mineral Salt Solution:
Composition per liter:

$MgSO_4.7H_2O$ 29.7g
Nitrilotriacetic acid 10.0g
$CaCl_2·2H_2O$ 3.34g
$FeSO_4·7H_2O$ 99.0mg
$Na_2MoO_4·2H_2O$ 12.67mg
Metals "44" 50.0mL

Preparation of Mineral Salt Solution: Add nitrilotriacetic acid to 500.0mL of distilled/deionized water. Dissolve by adjusting pH to 6.5 with KOH. Add remaining components. Add distilled/deionized water to 1.0L. Readjust pH to 7.2.

Metals "44":
Composition per 100.0mL:

$ZnSO_4·7H_2O$ 1.1g
$FeSO_4·7H_2O$ 0.5g
EDTA 0.25g
$MnSO_4·7H_2O$ 0.154g
$CuSO_4·5H_2O$ 0.04g
$Co(NO_3)_2·6H_2O$ 0.025g
$Na_2B_4O_7·10H_2O$ 0.018g

Preparation of Metals "44": Add a few drops of H_2SO_4 to distilled/deionized water to inhibit precipitate formation. Add components to acidified distilled/deionized water and bring volume to 100.0mL. Mix thoroughly.

Glucose Solution:
Composition per 100.0mL:

D-Glucose 2.5g

Preparation of Glucose Solution: Add glucose to distilled/deionized water and bring volume to 100.0mL. Mix thoroughly. Filter sterilize.

Vitamin Solution:
Composition per liter:

Pyridoxine·HCl 0.02g
p-Aminobenzoic acid 0.01g
Calcium D-pantothenate 0.01g
Nicotinamide 0.01g
Riboflavin 0.01g
Thiamine·HCl 0.01g
Biotin 4.0mg
Folic acid 4.0mg
Cyanocobalamin 0.2mg

Preparation of Vitamin Solution: Add components to distilled/deionized water and bring volume to 1.0L. Mix thoroughly. Filter sterilize.

Preparation of Medium: Add components, except glucose solution and vitamin solution, to distilled/deionized water and bring volume to 985.0mL. Mix thoroughly. Gently heat and bring to boiling. Autoclave for 15 min at 15 psi pressure–121°C. Cool to 45°–50°C. Aseptically add 10.0mL of sterile glucose solution and 5.0mL of sterile vitamin solution. Mix thoroughly. Adjust pH to 7.5 with sterile KOH if necessary. Pour into sterile Petri dishes or distribute into sterile tubes.

Use: For the cultivation and maintenance of *Blastobacter aggregatus, Blastobacter capsulatus, Blastobacter denitrificans,* and *Planctomyces limnophilus.*

PYGV Medium

Composition per liter:

Agar	15.0g
Peptone	0.25g
Yeast extract	0.25g
Mineral solution	20.0mL
Glucose solution	10.0mL
Vitamin solution	5.0mL

pH 7.5 ± 0.2 at 25°C

Mineral Solution:

Composition per liter:

$MgSO_4 \cdot 7H_20$	29.7g
$NaMoO_4 \cdot 2H_2O$	12.67g
Nitrilotriacetic acid	10.0g
$CaCl_2 \cdot 2H_2O$	3.34g
$FeSO_4 \cdot 7H_2O$	0.1g
Metals "44" solution	50.0mL

Preparation of Mineral Solution: Add nitrilotriacetic acid to 500.0mL of distilled/deionized water. Dissolve by adjusting pH to 6.5 with KOH. Add remaining components. Readjust pH to 7.2 with H_2SO_4 or KOH. Add distilled/deionized water to 1.0L. Store at 5°C.

Metals "44" Solution:

Composition per 100.0mL:

$ZnSO_4 \cdot 7H_2O$	1.1g
$FeSO_4 \cdot 7H_2O$	0.5g
EDTA	0.25g
$MnSO_4 \cdot 7H_2O$	0.154g
$CuSO_4 \cdot 5H_2O$	0.04g
$Co(NO_3)_2 \cdot 6H_2O$	0.025g
$Na_2B_4O_7 \cdot 10H_2O$	0.018g

Preparation of Metals "44" Solution: Add components to distilled/deionized water and bring volume to 100.0mL. Mix thoroughly.

Glucose Solution:

Composition per 100.0mL:

D-Glucose	2.5g

Preparation of Glucose Solution: Add D-glucose to distilled/deionized water and bring volume to 100.0mL. Mix thoroughly. Filter sterilize.

Vitamin Solution:

Composition per liter:

Pyridoxin·HCl	0.02g
p-Aminobenzoic acid	0.01g
Ca-panthothenate	0.01g
Nicotinamide	0.01g
Riboflavin	0.01g
Thiamine·HCl	0.01g
Biotin	4.0mg
Folic acid	4.0mg
Vitamin B_{12}	0.2mg

Preparation of Vitamin Solution: Add components to distilled/deionized water and bring volume to 1.0L.

Preparation of Medium: Add components, except glucose solution and vitamin solution, to distilled/deionized water and bring volume to 985.0mL. Mix thoroughly. Gently heat and bring to boiling. Autoclave for 20 min at 15 psi pressure–121°C. Cool to 60°C. Aseptically add 10.0mL of sterile glucose solution and 5.0mL of sterile vitamin solution. Mix thoroughly. Pour into sterile Petri dishes or distribute into sterile tubes.

Use: For the enrichment of *Stella* species from polluted waters.

Pyrazinamidase Agar (Pyrazinamide Medium)

Composition per liter:

Pancreatic digest of casein	15.0g
Agar	15.0g
Papaic digest of soybean meal	5.0g
NaCl	5.0g
Yeast extract	3.0g
Pyrazinecarboxamide	1.0g
Tris(hydroxymethyl)amino- methane maleate buffer (0.2*M*, pH 6.0)	1.0L

pH 6.0 ± 0.2 at 25°C

Preparation of Medium: Combine components. Mix thoroughly. Gently heat and bring to boiling. Distribute into tubes in 5.0mL volumes. Autoclave for 15 min at 15 psi pressure–121°C. Allow tubes to cool in a slanted position.

Use: For the cultivation, differentiation, and maintenance of pathogenic *Yersinia* species. Bacteria that produce pyrazinamidase turn the medium pink.

Pyrazinamidase Agar (BAM M131)

Composition per liter:

Pancreatic digest of casein	11.25g
Agar	11.25g
Papaic digest of soybean meal	3.75g
NaCl	3.75g
Yeast extract	3.0g
Pyrazine-carboxamide	1.0g
Tris maleate, 0.2*M*, pH6.0	1.0L

pH 6.0 ± 0.2 at 25°C

Preparation of Medium: Add components to 0.2*M* tris maleate and bring volume to 1.0L. Mix thoroughly. Gently heat and bring to boiling. Distribute into tubes or flasks. Autoclave for 15 min at 15 psi pressure–121°C. Pour into sterile Petri dishes or leave in tubes. For slants allow tubes to cool in an inclined position.

Use: For the cultivation of *Yersinia* spp.

Pyrazinamide Medium

Composition per liter:

Agar	15.0g
Na_2HPO_4	2.5g
Sodium pyruvate	2.0g
L-Asparagine	2.0g
KH_2PO_4	1.0g
Pancreatic digest of casein	0.5g
Tween™ 80	0.2g
Pyrazinamide	0.1g
$CaCl_2 \cdot 2H_2O$	0.5mg
$CuSO_4 \cdot 5H_2O$	0.1mg
$ZnSO_4 \cdot 7H_2O$	0.1mg
Ferric ammonium citrate	0.05g
$MgSO_4 \cdot 7H_2O$	0.01g

pH 6.6 ± 0.2 at 25°C

Preparation of Medium: Combine components. Mix thoroughly. Gently heat and bring to boiling. Distribute into tubes in 5.0mL volumes. Autoclave for 15 min at 15 psi pressure–121°C. Allow tubes to cool in a slanted position.

Use: For the cultivation and differentiation of *Corynebacterium* species and related organisms. Bacteria that produce pyrazinamidase turn the medium pink.

Pyridine Medium

Composition per 1001.0mL:

K_2HPO_4	0.61g
KH_2PO_4	0.39g
KCl	0.25g
Yeast extract	0.15g
$MgSO_4 \cdot 7H_2O$	0.13g
Pyridine	1.0mL
Trace elements solution	1.0mL

Trace Elements Solution:

Composition per liter:

$FeSO_4 \cdot 7H_2O$	40.0mg
$MnSO_4 \cdot 4H_2O$	40.0mg
$ZnSO_4 \cdot 7H_2O$	20.0mg
$CuSO_4 \cdot 5H_2O$	5.0mg
$Na_2MoO_4 \cdot 2H_2O$	5.0mg
$CoCl_2 \cdot 6H_2O$	4.0mg
$CaCl_2 \cdot 2H_2O$	0.4mg
NaCl	1.0g

Preparation of Trace Elements Solution: Add components to distilled/deionized water and bring volume to 1.0L. Mix thoroughly.

Preparation of Medium: Add components, except pyridine, to distilled/deionized water and bring volume to 1.0L. Mix thoroughly. Autoclave for 15 min at 15 psi pressure–121°C. Cool to room temperature. In a fume hood, aseptically add 1.0mL of pyridine. Mix thoroughly. Aseptically distribute into sterile tubes or flasks. Use polyurethane foam closures to eliminate odors caused by volatilization of pyridine.

Use: For the cultivation of *Micrococcus luteus.*

Pyridoxine Assay Medium

Composition per liter:

Sucrose	30.0g
Ammonium tartrate	10.0g
KH_2PO_4	5.0g
Sodium dihydrogen citrate	4.0g
$MgSO_4 \cdot 7H_2O$	1.0g
$CaCl_2 \cdot 2H_2O$	0.2g
NaCl	0.2g
Choline chloride	0.01g
$FeCl_3$	0.01g
Thiamine·HCl	0.01g
$ZnSO_4 \cdot 7H_2O$	4.0mg
Nicotinic acid	2.0mg
Calcium pantothenate	1.0mg
Riboflavin	1.0mg
p-Aminobenzoic acid	200μg
Biotin	8μg

pH 4.5 ± 0.2 at 25°C

Source: This medium is available as a prepared medium from BD Diagnostic Systems.

Preparation of Medium: Add components to distilled/deionized water and bring volume to 1.0L. Mix thoroughly. Gently heat and bring to boiling. Continue boiling for 2–3 min. Distribute into tubes in 5.0mL volumes. Add standard solutions and test solutions to each tube. Bring volume of each tube to 10.0mL with distilled/deionized water. Autoclave for 15 min at 15 psi pressure–121°C.

Use: For the microbiological assay of pyridoxine using *Neurospora sitophila* as the test organism.

Pyridoxine Y Medium

Composition per liter:

Glucose	40.0g
$(NH_4)_2SO_4$	4.0g
L-Asparagine	4.0g
KH_2PO_4	3.0g
$MgSO_4 \cdot 7H_2O$	1.0g
$CaCl_2 \cdot 2H_2O$	0.49g
DL-Isoleucine	0.04g
DL-Methionine	0.04g
DL-Tryptophan	0.04g
DL-Valine	0.04g
L-Histidine·HCl	0.02g
Riboflavin	0.02g
Biotin salt	8.0mg
Inositol	5.0mg
$FeSO_4 \cdot 7H_2O$	0.5mg
Calcium pantothenate	0.4mg
Nicotinic acid	0.4mg
Thiamine·HCl	0.4mg
H_3BO_3	0.2mg
KI	0.2mg
$CuSO_4 \cdot 5H_2O$	0.09mg
$MnSO_4 \cdot H_2O$	0.08mg
$ZnSO_4 \cdot 7H_2O$	0.08mg
$(NH_4)_2MoO_4$	0.04mg

pH 4.4 ± 0.2 at 25°C

Source: This medium is available as a prepared medium from BD Diagnostic Systems.

Preparation of Medium: Add components to distilled/deionized water and bring volume to 1.0L. Mix thoroughly. Gently heat and bring to boiling. Continue boiling for 2–3 min. Distribute into tubes in 5.0mL volumes. Add standard solutions and test solutions to each tube. Bring volume of each tube to 10.0mL with distilled/deionized water. Autoclave for 15 min at 15 psi pressure–121°C.

Use: For the microbiological assay of pyridoxine using *Saccharomyces uvarum* as the test organism.

Pyrobaculum calidifontis Medium (DSMZ Medium 1090)

Composition per liter:

Tryptone	10.0g
$Na_2S_2O_3$	3.0g
Yeast extract	1.0g

pH 7.0 ± 0.2 at 25°C

Preparation of Medium: Add components to distilled/deionized water and bring volume to 1.0L. Mix thoroughly. Adjust pH to 7.0 with NaOH. Distribute into tubes or flasks. Autoclave for 15 min at 15 psi pressure–121°C.

Use: For the cultivation of *Pyrobaculum calidifontis.*

Pyrobaculum Medium

Composition per liter:

$Na_2S_2O_3 \cdot 5H_2O$	2.0g
$(NH_4)_2SO_4$	1.3g
Peptone	0.5g
$Na_2S \cdot 9H_2O$	0.5g
KH_2PO_4	0.28g
$MgSO_4 \cdot 7H_2O$	0.25g
Yeast extract	0.2g
$CaCl_2 \cdot 2H_2O$	0.07g
$FeCl_3 \cdot 6H_2O$	0.02g
Resazurin	1.0mg
$MnCl_2 \cdot 4H_2O$	1.8mg
$Na_2B_4 \cdot 10H_2O$	4.5mg
$ZnSO_4 \cdot 7H_2O$	0.22mg
$CuCl_2 \cdot 2H_2O$	0.05mg
$Na_2MoO_4 \cdot 2H_2O$	0.03mg
$VOSO_4 \cdot 2H_2O$	0.03mg
$CoSO_4$	0.01mg

Preparation of Medium: Add components, except peptone, yeast extract, and $Na_2S \cdot 9H_2O$, to distilled/deionized water and bring volume to 1.0L. Mix thoroughly. Bring pH to 6.0 using 8*N* NaOH. Sparge with 100% N_2 for 30 min. Add peptone, yeast extract, and $Na_2S \cdot 9H_2O$. Bring pH back to 6.0 using 10*N* H_2SO_4. Anaerobically distribute into sterile tubes or flasks under 100% N_2. Do not autoclave medium. If not used immediately, heat the medium to 90°C for 60 min on each of 3 consecutive days.

Use: For the cultivation and maintenance of *Pyrobaculum islandicum*.

Pyrobaculum Medium

Composition per liter:

Sulfur, powdered	20.0g
$(NH_4)_2SO_4$	1.3g
Peptone	0.5g
$Na_2S \cdot 9H_2O$	0.5g
KH_2PO_4	0.28g
$MgSO_4 \cdot 7H_2O$	0.25g
Yeast extract	0.2g
$CaCl_2 \cdot 2H_2O$	0.07g
$FeCl_3 \cdot 6H_2O$	0.02g
Resazurin	1.0mg
$MnCl_2 \cdot 4H_2O$	1.8mg
$Na_2B_4 \cdot 10H_2O$	4.5mg
$ZnSO_4 \cdot 7H_2O$	0.22mg
$CuCl_2 \cdot 2H_2O$	0.05mg
$Na_2MoO_4 \cdot 2H_2O$	0.03mg
$VOSO_4 \cdot 2H_2O$	0.03mg
$CoSO_4$	0.01mg

Preparation of Medium: Add components, except peptone, yeast extract, and $Na_2S \cdot 9H_2O$, to distilled/deionized water and bring volume to 1.0L. Mix thoroughly. Bring pH to 6.0 using 8*N* NaOH. Sparge with 100% N_2 for 30 min. Add peptone, yeast extract, and $Na_2S \cdot 9H_2O$. Bring pH back to 6.0 using 10*N* H_2SO_4. Anaerobically distribute into sterile tubes or flasks under 100% N_2. Do not autoclave medium. If not used immediately, heat the medium to 90°C for 60 min on each of three consecutive days.

Use: For the cultivation and maintenance of *Pyrobaculum organotrophum*.

Pyrococcus endeavori Medium ES4

Composition per 3.0L:

Solution A	1.0L
Solution B	1.0L
Solution C	1.0L

Solution A:

Composition per liter:

NaCl	47.8g
Na_2SO_4	8.0g
KCl	1.4g
$NaHCO_3$	0.4g
KBr	0.2g
H_3BO_3	0.06g

Preparation of Solution A: Add components to distilled/deionized water and bring volume to 1.0L. Mix thoroughly. Autoclave for 15 min at 15 psi pressure–121°C.

Solution B:

Composition per liter:

$MgCl_2 \cdot 6H_2O$	21.6g
$CaCl_2 \cdot 2H_2O$	3.0g
$SrCl_2 \cdot 6H_2O$	0.05g

Preparation of Solution B: Add components to distilled/deionized water and bring volume to 1.0L. Mix thoroughly. Autoclave for 15 min at 15 psi pressure–121°C.

Solution C:

Composition per liter:

Sodium acetate	50.0g
NH_4Cl	12.5g
K_2HPO_4	7.0g

Preparation of Solution C: Add components to distilled/deionized water and bring volume to 1.0L. Mix thoroughly. Autoclave for 15 min at 15 psi pressure–121°C.

Preparation of Medium: Aseptically combine 1.0L of sterile solution A with 1.0L of sterile solution B and 1.0L of sterile solution C. Mix thoroughly. Aseptically distribute into sterile tubes or flasks.

Use: For the cultivation of *Pyrococcus endeavori*.

Pyrococcus furiosus Medium

Composition per liter:

NaCl	13.8g
Pancreatic digest of casein	5.0g
Yeast extract	5.0g
Maltose	5.0g
$MgSO_4$	3.5g
$MgCl_2$	2.75g
KH_2PO_4	0.5g
$CaCl_2$	0.75g
KCl	0.325g
NaBr	50.0mg
KI	50.0mg
H_3BO_3	15.0mg
$SrCl_2$	7.5mg
Citric acid	5.0mg
Resazurin	2.5mg
Mineral solution	10.0mL

pH 6.8 ± 0.2 at 25°C

Mineral Solution:
Composition per liter:

Nitrilotriacetic acid	1.0g
$MnSO_4$	0.5g
$FeCl_3·6H_2O$	1.1g
$Na_2WO_4·2H_2O$	0.3g
EDTA	0.292g
$NiCl_2·6H_2O$	0.2g
$CoSO_4·7H_2O$	0.1g
$ZnSO_4·7H_2O$	0.1g
$CuSO_4·5H_2O$	0.01g
$Na_2MoO_4·2H_2O$	0.01g

Preparation of Mineral Solution: Add nitrilotriacetic acid to 500.0mL of distilled/deionized water. Adjust pH to 6.5 with KOH. Add remaining components. Mix thoroughly. Add distilled/deionized water to 1.0L. Adjust pH to 6.8.

Preparation of Medium: Add components to distilled/deionized water and bring volume to 1.0L. Mix thoroughly. Adjust pH to 6.8. Distribute into tubes or flasks. Autoclave for 15 min at 15 psi pressure–121°C.

Use: For the cultivation of high cell concentrations of *Pyrococcus furiosus*.

Pyrococcus Medium

Composition per liter:

Sulfur	30.0g
NaCl	13.85g
Peptone	5.0g
$MgSO_4·7H_2O$	3.5g
$MgCl_2·6H_2O$	2.75g
Yeast extract	1.0g
$CaCl_2$	0.75g
KH_2PO_4	0.5g
KCl	0.325g
NaBr	0.05g
H_3BO_3	15.0mg
$SrCl_2·6H_2O$	7.5mg
Citric acid	5.0mg
$(NH_4)_2Ni(SO_4)_2$	2.0mg
Resazurin	1.0mg
KI	0.05mg
Trace minerals solution	10.0mL
$Na_2S·9H_2O$ solution	10.0mL

Trace Minerals Solution:
Compostion per liter:

$MgSO_4·7H_2O$	3.0g
Nitrilotritracetic acid	1.5g
NaCl	1.0g
$MnSO_4·H_2O$	0.5g
$CaCl_2·2H_2O$	0.1g
$CoSO_4$ (or $CoCl_2$)	0.1g
$FeSO_4·7H_2O$	0.1g
$ZnSO_4$	0.1g
$AIK(SO_4)_2$	0.01g
$CuSO_4·5H_2O$	0.01g
H_3BO_3	0.01g
$Na_2MoSO_42H_2O$	0.01g

Preparation of Trace Minerals Solution: Add nitrilotriacetic acid to 500.0mL of distilled/deionized water. Adjust pH to 6.5 with KOH. Add remaining components. Add distilled/deionized water to 1.0L. Adjust pH to 7.0.

$Na_2S·9H_2O$ Solution:
Composition per 10.0mL:

$Na_2S·9H_2O$	0.3g

Preparation of $Na_2S·9H_2O$ Solution: Add $Na_2S·9H_2O$ to distilled/deionized water and bring volume to 10.0mL. Mix thoroughly. Sparge with 100% N_2. Autoclave for 15 min at 15 psi pressure–121°C.

Preparation of Medium: Add components, except $Na_2S·9H_2O$ solution, to distilled/deionized water and bring volume to 1.0L. Mix thoroughly. Adjust pH to 6.5 with H_2SO_4. Do not autoclave. Sterilize by steaming at 100°C for 30 min on 3 consecutive days. Before inoculation, add 10.0mL of sterile $Na_2S·9H_2O$ solution. Mix thoroughly.

Use: For the cultivation and maintenance of *Pyrococcus furiosus* and *Pyrococcus woesei*.

Pyrococcus/Staphylothermus Medium

Composition per 1010.0mL:

Sulfur, powdered	30.0g
NaCl	13.85g
Peptone	5.0g
$MgSO_4·7H_2O$	3.5g
$MgCl_2·6H_2O$	2.75g
$NiCl_2·6H_2O$	2.0g
Yeast extract	1.0g
$CaCl_2·2H_2O$	0.75g
KH_2PO_4	0.5g
KCl	0.325g
NaBr	0.05g
H_3BO_3	0.015g
$(NH_4)_2SO_4$	10.0mg
$SrCl_2·6H_2O$	7.5mg
Citric acid	5.0mg
Resazurin	1.0mg
KI	0.05mg
Trace elements solution	10.0mL
$Na_2S·9H_2O$ solution	10.0mL

pH 6.5 ± 0.2 at 25°C

Trace Elements Solution:
Composition per liter:

$MgSO_4·7H_2O$	3.0g
Nitrilotriacetic acid	1.5g
NaCl	1.0g
$MnSO_4·2H_2O$	0.5g
$CoSO_4·7H_2O$	0.18g
$ZnSO_4·7H_2O$	0.18g
$CaCl_2·2H_2O$	0.1g
$FeSO_4·7H_2O$	0.1g
$NiCl_2·6H_2O$	0.025g
$KAl(SO_4)_2·12H_2O$	0.02g
$CuSO_4·5H_2O$	0.01g
H_3BO_3	0.01g
$Na_2MoO_4·2H_2O$	0.01g
$Na_2SeO_3·5H_2O$	0.3mg

Preparation of Trace Elements Solution: Add nitrilotriacetic acid to 500.0mL of distilled/deionized water. Adjust pH to 6.5 with KOH. Add remaining components. Adjust pH to 7.0. Add distilled/deionized water to 1.0L.

$Na_2S \cdot 9H_2O$ Solution:
Composition per 10.0mL:

$Na_2S \cdot 9H_2O$	0.5g

Preparation of $Na_2S \cdot 9H_2O$ Solution: Add $Na_2S \cdot 9H_2O$ to distilled/deionized water and bring volume to 10.0mL. Mix thoroughly. Sparge with 100% N_2. Autoclave for 15 min at 15 psi pressure–121°C.

Preparation of Medium: Prepare and dispense medium under 100% N_2. Add components, except $Na_2S \cdot 9H_2O$ solution, to distilled/deionized water and bring volume to 1.0L. Mix thoroughly. Gently heat and bring to boiling. Continue boiling for 5 min. Cool to room temperature while sparging with 100% N_2. Bring pH to 6.5 using 10*N* H_2SO_4. Anaerobically distribute into tubes or flasks. Autoclave for 15 min at 15 psi pressure–121°C. Immediately prior to inoculation, add 0.1mL of sterile $Na_2S \cdot 9H_2O$ solution to each 10.0mL of medium. Check that final pH is 6.5.

Use: For the cultivation and maintenance of *Pyrococcus furiosus, Pyrococcus woesei,* and *Staphylothermus marinus.*

Pyrodictium abyssi Medium

Composition per liter:

Sulfur, powdered	30.0g
NaCl	13.85g
$MgSO_4 \cdot 7H_2O$	3.5g
$MgCl_2 \cdot 6H_2O$	2.75g
$CaCl_2 \cdot 2H_2O$	0.75g
$Na_2S \cdot 9H_2O$	0.5g
KH_2PO_4	0.5g
Yeast extract	0.5g
KCl	0.325g
NaBr	0.05g
H_3BO_3	0.015g
$(NH_4)_2SO_4$	10.0mg
$SrCl_2 \cdot 6H_2O$	7.5mg
$NiCl_2 \cdot 6H_2O$	2.0mg
Resazurin	1.0mg
$Na_2WO_4 \cdot 2H_2O$	0.1mg
KI	0.05mg
Trace elements solution	10.0mL

pH 5.5–6.0 at 25°C

Trace Elements Solution:
Composition per liter:

$MgSO_4 \cdot 7H_2O$	3.0g
Nitrilotriacetic acid	1.5g
NaCl	1.0g
$MnSO_4 \cdot 2H_2O$	0.5g
$CoSO_4 \cdot 7H_2O$	0.18g
$ZnSO_4 \cdot 7H_2O$	0.18g
$CaCl_2 \cdot 2H_2O$	0.1g
$FeSO_4 \cdot 7H_2O$	0.1g
$NiCl_2 \cdot 6H_2O$	0.025g
$KAl(SO_4)_2 \cdot 12H_2O$	0.02g
$CuSO_4 \cdot 5H_2O$	0.01g
H_3BO_3	0.01g
$Na_2MoO_4 \cdot 2H_2O$	0.01g
$Na_2SeO_3 \cdot 5H_2O$	0.3mg

Preparation of Trace Elements Solution: Add nitrilotriacetic acid to 500.0mL of distilled/deionized water. Adjust pH to 6.5 with KOH. Add remaining components. Adjust pH to 7.0. Add distilled/deionized water to 1.0L.

Preparation of Medium: Prepare and dispense medium under 80% H_2 + 20% CO_2. Add components, except $Na_2S \cdot 9H_2O$, to distilled/deionized water and bring volume to 1.0L. Mix thoroughly. Gently heat and bring to boiling. Continue boiling for 5 min. Cool to room temperature while sparging with 80% H_2 + 20% CO_2. Add $Na_2S \cdot 9H_2O$. Mix thoroughly. Bring pH to 5.5 using 10*N* H_2SO_4. Anaerobically distribute into tubes or flasks making sure to evenly distribute sulfur. Do not autoclave. For immediate use, heat the medium in a boiling water bath for 60 min prior to inoculation. For storage of medium, heat medium in a boiling water bath for 60 min on 3 consecutive days. Store at room temperature.

Use: For the cultivation and maintenance of *Pyrodictium abyssi.*

Pyrodictium Medium

Composition per liter:

Sulfur, powdered	30.0g
NaCl	13.85g
$MgSO_4 \cdot 7H_2O$	3.5g
$MgCl_2 \cdot 6H_2O$	2.75g
Yeast extract	2.0g
$CaCl_2 \cdot 2H_2O$	0.75g
$Na_2S \cdot 9H_2O$	0.5g
KH_2PO_4	0.5g
KCl	0.325g
NaBr	0.05g
H_3BO_3	0.015g
$(NH_4)_2SO_4$	10.0mg
$SrCl_2 \cdot 6H_2O$	7.5mg
Citric acid	5.0mg
$NiCl_2 \cdot 6H_2O$	2.0mg
Resazurin	1.0mg
KI	0.05mg
Trace elements solution	10.0mL

pH 5.5 ± 0.2 at 25°C

Trace Elements Solution:
Composition per liter:

$MgSO_4 \cdot 7H_2O$	3.0g
Nitrilotriacetic acid	1.5g
NaCl	1.0g
$MnSO_4 \cdot 2H_2O$	0.5g
$CoSO_4 \cdot 7H_2O$	0.18g
$ZnSO_4 \cdot 7H_2O$	0.18g
$CaCl_2 \cdot 2H_2O$	0.1g
$FeSO_4 \cdot 7H_2O$	0.1g
$NiCl_2 \cdot 6H_2O$	0.025g
$KAl(SO_4)_2 \cdot 12H_2O$	0.02g
$CuSO_4 \cdot 5H_2O$	0.01g
H_3BO_3	0.01g
$Na_2MoO_4 \cdot 2H_2O$	0.01g
$Na_2SeO_3 \cdot 5H_2O$	0.3mg

Preparation of Trace Elements Solution: Add nitrilotriacetic acid to 500.0mL of distilled/deionized water. Adjust pH to 6.5 with KOH. Add remaining components. Adjust pH to 7.0. Add distilled/deionized water to 1.0L.

Preparation of Medium: Prepare and dispense medium under 80% H_2 + 20% CO_2. Add components, except $Na_2S \cdot 9H_2O$, to distilled/deionized water and bring volume to 1.0L. Mix thoroughly. Gently heat and bring to boiling. Continue boiling for 5 min. Cool to room temperature while sparging with 80% H_2 + 20% CO_2. Add $Na_2S \cdot 9H_2O$. Mix thoroughly. Bring pH to 5.5 using 10*N* H_2SO_4. Anaerobically distribute into tubes or flasks, making sure to evenly distribute sulfur. Do not autoclave. For immediate use, heat the medium in a boiling water bath for 60 min prior to inoculation. For storage of medium, heat medium in a boiling water bath for 60 min on 3 consecutive days. Store at room temperature.

Use: For the cultivation and maintenance of *Pyrodictium brockii, Pyrodictium occultum*, and a consortium consisting of *Lactobacillus brevis, Streptococcus lactis*, and *Saccharomyces cerevisiae.*

Pyrolobus fumarii Medium (DSMZ Medium 792)

Composition per liter:

NaCl	13.850g
$MgSO_4 \cdot 7H_2O$	3.5g
$MgCl_2 \cdot 6H_2O$	2.75g
KNO_3	1.0g
KH_2PO_4	0.5g
$CaCl_2 \cdot 2H_2O$	0.375g
KCl	0.325g
NaBr	0.05g
H_3BO_3	0.015g
$SrCl_2 \cdot 6H_2O$	7.5mg
Resazurin	1.0mg
Trace elements solution	10.0mL
$Na_2S \cdot 9H_2O$ solution	10.0mL
KI solution	0.05mL

pH 5.5 ± 0.2 at 25°C

Trace Elements Solution:

$MgSO_4 \cdot 7H_2O$	3.0g
NaCl	1.0g
$MnSO_4 \cdot 2H_2O$	0.5g
$ZnSO_4 \cdot 7H_2O$	0.18g
$CoSO_4 \cdot 7H_2O$	0.18g
$FeSO_4 \cdot 7H_2O$	0.1g
$CaCl_2 \cdot 2H_2O$	0.1g
$NiCl_2 \cdot 6H_2O$	0.025g
$KAl(SO_4)_2 \cdot 12H_2O$	0.02g
$CuSO_4 \cdot 5H_2O$	0.01g
H_3BO_3	0.01g
$Na_2MoO_4 \cdot 4H_2O$	0.01g
$Na_2WO_4 \cdot 2H_2O$	0.01g
$Na_2SeO_3 \cdot 5H_2O$	0.30mg

Preparation of Trace Elements Solution: Add components to distilled/deionized water and bring volume to 1.0L. Mix thoroughly. Adjust pH to 1.0 with H_2SO_4.

$Na_2S \cdot 9H_2O$ Solution:
Composition per 10.0mL:

$Na_2S \cdot 9H_2O$	0.3g

Preparation of $Na_2S \cdot 9H_2O$ Solution: Add $Na_2S \cdot 9H_2O$ to distilled/deionized water and bring volume to 10.0mL. Mix thoroughly. Autoclave under 100% N_2 for 15 min at 15 psi pressure–121°C. Cool to room temperature.

KI Solution:
Composition per 10.0mL:

KI	5.0mg

Preparation of KI Solution: Add KI to distilled/deionized water and bring volume to 10.0mL. Mix thoroughly. Autoclave under 100% N_2 for 15 min at 15 psi pressure–121°C. Cool to room temperature.

Preparation of Medium: Add components, except $Na_2S \cdot 9H_2O$ solution, to distilled/deionized water and bring volume to 990.0mL. Mix thoroughly. Gently heat and bring to boiling. Cool to room temperature while sparging with 80% H_2 + 20% CO_2. Distribute into serum bottles under 80% H_2 + 20% CO_2, e.g., 20mL into 120mL serum bottles. Autoclave for 15 min at 15 psi pressure–121°C. Cool to 25°C. Aseptically inject $Na_2S \cdot 9H_2O$ solution, 0.2mL per 20mL medium. Mix thoroughly. Adjust pH to 5.5. After inoculation pressurize vials to 2 bar overpressure with 80% H_2 + 20% CO_2 gas mixture.

Use: For the cultivation of *Pyrolobus fumarii.*

Pyrrolidone Agar

Composition per liter:

Noble agar	21.0g
K_2HPO_4	5.65g
KH_2PO_4	2.95g
$MgSO_4 \cdot 7H_2O$	1.0g
Pyrrolidone carboxylic acid solution	30.0mL
NaOH solution	30.0mL
Trace metals	6.3mL

Pyrrolidone Carboxylic Acid Solution:
Composition per 300.0mL:

Pyrrolidone carboxylic acid	50.0g

Preparation of Pyrrolidone Carboxylic Acid Solution: Add pyrrolidone carboxylic acid to distilled/deionized water and bring volume to 300.0mL. Mix thoroughly. Filter sterilize.

NaOH Solution:
Composition per 100.0mL:

NaOH	5.0g

Preparation of NaOH Solution: Add NaOH to distilled/deionized water and bring volume to 100.0mL. Mix thoroughly. Filter sterilize.

Trace Metals:
Composition per 100.0mL:

$FeSO_4 \cdot 7H_2O$	0.18g
$MnCl_2 \cdot 2H_2O$	0.13g
$CuSO_4 \cdot 5H_2O$	0.1g
$ZnSO_4 \cdot 7H_2O$	0.02g

Preparation of Trace Metals: Add a few drops of H_2SO_4 to distilled/deionized water to inhibit precipitate formation. Add components to acidified distilled/deionized water and bring volume to 100.0mL. Mix thoroughly.

Preparation of Medium: Add components, except pyrrolidone carboxylic acid solution and NaOH solution, to distilled/deionized water and bring volume to 940.0mL. Mix thoroughly. Gently heat and bring to boiling. Autoclave for 15 min at 15 psi pressure–121°C. Cool to 45°–50°C. Aseptically add 30.0mL of sterile pyrrolidone carboxylic acid solution and 30.0mL of sterile NaOH solution. Mix thoroughly. Pour into sterile Petri dishes or distribute into sterile tubes.

Use: For the cultivation and maintenance of *Pseudomonas fluorescens.*

Pyruvate Utilization Medium

Composition per liter:

Sodium pyruvate	10.0g
Pancreatic digest of casein	10.0g
K_2HPO_4	5.0g
NaCl	5.0g

Yeast extract	5.0g
Bromthymol Blue	0.1g

pH 7.1–7.4 at 25°C

Preparation of Medium: Add components to distilled/deionized water and bring volume to 1.0L. Mix thoroughly. Gently heat and bring to boiling. Adjust pH to 7.1–7.4. Distribute into tubes in 5.0mL volumes. Autoclave for 15 min at 15 psi pressure–121°C.

Use: For the cultivation of bacteria that can metabolize pyruvate. Bacteria that can utilize pyruvate turn the medium yellow.

Pyruvic Acid Egg Medium

Composition per 1640.0mL:

KH_2PO_4	11.4g
D-Glucose	10.0g
Na_2HPO_4	6.0g
Pyruvic acid	3.0g
$MgSO_4 \cdot 7H_2O$	0.3g
Malachite Green	0.125g
Egg, homogenized whole	1.0L
Penicillin solution	10.0mL

Source: This medium is available as a prepared medium from Oxoid Unipath.

Penicillin Solution:

Composition per 10.0mL:

Penicillin G	100,000U

Preparation of Penicillin Solution: Add penicillin G to distilled/deionized water and bring volume to 10.0mL. Mix thoroughly. Filter sterilize.

Homogenized Whole Egg:

Composition per liter:

Whole eggs	18–24

Preparation of Homogenized Whole Egg: Use fresh eggs, less than 1 week old. Scrub the shells with soap. Let stand in a soap solution for 30 min. Rinse in running water. Soak eggs in 70% ethanol for 15 min. Break the eggs into a sterile container. Homogenize by shaking. Filter through four layers of sterile cheesecloth into a sterile graduated cylinder. Measure out 1.0L.

Preparation of Medium: Add components, except homogenized whole egg and penicillin solution, to distilled/deionized water and bring volume to 630.0mL. Mix thoroughly. Autoclave for 15 min at 15 psi pressure–121°C. Cool to 45°–50°C. Aseptically add homogenized whole egg and penicillin solution to cooled sterile basal medium. Mix thoroughly. Aseptically distribute into sterile tubes. Inspissate at 85°–90°C (moist heat) for 45 min.

Use: For the isolation and cultivation of *Mycobacterium* species, especially ones that are drug resistant and difficult to grow.

PYS Agar (Peptone Yeast Extract Salt Agar)

Composition per liter:

Agar	15.0g
Peptone	15.0g
NaCl	5.0g
Yeast extract	5.0g

pH 7.2–7.4 at 25°C

Preparation of Medium: Add components to tap water and bring volume to 1.0L. Mix thoroughly. Gently heat and bring to boiling. Distribute into tubes or flasks. Autoclave for 15 min at 15 psi pressure–121°C. Pour into sterile Petri dishes or leave in tubes.

Use: For the cultivation and maintenance of *Actinomadura madurae.*

PYS Medium (DSMZ Medium 1117)

Composition per liter:

Peptone	8.0g
Yeast extract	4.0g
NaCl	2.0g

pH 7.2 ± 0.2 at 25°C

Preparation of Medium: Add components to tap water and bring volume to 1.0L. Mix thoroughly. Adjust pH to 7.2. Distribute into tubes or flasks. Autoclave for 15 min at 15 psi pressure–121°C.

Use: For the cultivation of *Geobacillus toebii* subsp. *decanicus.*

PYSE Medium (DSMZ Medium 1120)

Composition per liter:

NaCl	15.0g
$MgCl_2 \cdot 6H_2O$	5.4g
Peptone	8.0g
Yeast extract	3.0g
$MgSO_4 \cdot 7H_2O$	2.65g
$CaSO_4 \cdot 2H_2O$	0.65g
KCl	0.35g

pH 7.5 ± 0.2 at 25°C

Preparation of Medium: Add components to tap water and bring volume to 1.0L. Mix thoroughly. Adjust pH to 7.5. Distribute into tubes or flasks. Autoclave for 15 min at 15 psi pressure–121°C.

Use: For the cultivation of *Colwellia maris.*

Quinoline Medium

Composition per 1000.2mL:

K_2HPO_4	0.61g
KH_2PO_4	0.39g
KCl	0.25g
Yeast extract	0.1g
Wolfe's mineral solution	10.0mL
Quinoline	0.2mL

Wolfe's Mineral Solution:

Composition per liter:

$MgSO_4 \cdot 7H_2O$	3.0g
Nitrilotriacetic acid	1.5g
NaCl	1.0g
$MnSO_4 \cdot 2H_2O$	0.5g
$CoCl_2 \cdot 6H_2O$	0.1g
$ZnSO_4 \cdot 7H_2O$	0.1g
$CaCl_2 \cdot 2H_2O$	0.1g
$FeSO_4 \cdot 7H_2O$	0.1g
$NiCl_2 \cdot 6H_2O$	0.025g
$KAl(SO_4)_2 \cdot 12H_2O$	0.02g
$CuSO_4 \cdot 5H_2O$	0.01g
H_3BO_3	0.01g
$Na_2MoO_4 \cdot 2H_2O$	0.01g
$Na_2SeO_3 \cdot 5H_2O$	0.3mg

Preparation of Wolfe's Mineral Solution: Add nitrilotriacetic acid to 500.0mL of distilled/deionized water. Adjust pH to 6.5 with KOH. Add remaining components one at a time. Add distilled/deionized water to 1.0L. Adjust pH to 6.8.

Preparation of Medium: Add components, except quinoline, to distilled/deionized water and bring volume to 1.0L. Mix thoroughly. Autoclave for 15 min at 15 psi pressure–121°C. Cool to room temperature. In a fume hood, aseptically add 0.2mL of quinoline. Mix thoroughly. Aseptically distribute into sterile tubes or flasks. Use polyurethane foam closures to eliminate odors caused by volatilization of quinoline.

Use: For the cultivation of *Rhodococcus* species.

Quinolinic Acid Medium

Composition per liter:

Quinolinic acid	1.5g
K_2HPO_4	1.1g
NH_4NO_3	1.0g
KH_2PO_4	0.5g
$MgSO_4 \cdot 7H_2O$	0.25g

Preparation: Add quinolinic acid to distilled/deionized water and bring volume to 900.0mL. Mix thoroughly. Bring pH to 7.0 with NaOH. Add other components. Bring volume to 1.0L. Mix thoroughly.

Distribute into tubes or flasks. Autoclave for 15 min at 15 psi pressure–121°C.

Use: For the cultivation of microorganisms that can utilize quinolinic acid as sole carbon source.

R Agar

Composition per liter:

Agar	20.0g
Peptone	10.0g
Casamino acids	5.0g
Malt extract	5.0g
Yeast extract	5.0g
Beef extract	2.0g
Glycerol	2.0g
$MgSO_4 \cdot 7H_2O$	1.0g
Tween™ 80	50.0mg

pH 7.2 ± 0.2 at 25°C

Preparation of Medium: Add components to distilled/deionized water and bring volume to 1.0L. Mix thoroughly. Gently heat and bring to boiling. Distribute into tubes or flasks. Autoclave for 15 min at 15 psi pressure–121°C. Pour into sterile Petri dishes or leave in tubes.

Use: For the cultivation and maintenance of *Arthrobacter polychromogenes, Arthrobacter protophormaie, Aureobacterium* species, *Brevibacterium acetylicum, Brevibacterium linens, Cellulomonas cartae, Clavibacter michiganense, Corynebacterium* species, *Curtobacterium citreum, Curtobacterium albidum, Curtobacterium flaccumfaciens, Curtobacterium insectiphilium, Curtobacterium luteum, Curtobacterium pusillum, Curtobacterium* species, *Dermabacter hominus, Microbacterium arborescens, Microbacterium imperiale, Microbacterium lacticum, Mycobacterium acapulcensis, Mycobacterium* species, *Nocardioides fastidiosa, Rhizomonas suberifaciens, Rhodococcus luteus, Rhodococcus maris, Staphylococcus carnosus, Staphylococcus* species, *Terrabacter tumescens,* and *Tsukamurella paurometabolum.*

R Agar for Phage Lysates

Composition per liter:

Agar	12.0g
Pancreatic digest of casein	10.0g
NaCl	8.0g
Yeast extract	1.0g
Glucose solution	5.0mL
$CaCl_2 \cdot 2H_2O$ solution	2.0mL

pH 6.8 ± 0.2 at 25°C

Glucose Solution:

Composition per 10.0mL:

D-Glucose	2.0g

Preparation of Glucose Solution: Add glucose to distilled/deionized water and bring volume to 10.0mL. Mix thoroughly. Filter sterilize.

$CaCl_2 \cdot 2H_2O$ Solution:

Composition per 10.0mL:

$CaCl_2 \cdot 2H_2O$	1.47g

Preparation of $CaCl_2 \cdot 2H_2O$ Solution: Add the $CaCl_2 \cdot 2H_2O$ to distilled/deionized water and bring volume to 10.0mL. Mix thoroughly. Filter sterilize.

Preparation of Medium: Add components, except glucose solution and $CaCl_2 \cdot 2H_2O$ solution, to distilled/deionized water and bring volume to 993.0mL. Mix thoroughly. Gently heat and bring to boiling. Autoclave for 15 min at 15 psi pressure–121°C. Cool to 45°–50°C. Aseptically add 5.0mL of glucose solution and 2.0mL of $CaCl_2 \cdot 2H_2O$ solution. Mix thoroughly. Pour into sterile Petri dishes.

Use: For the cultivation of bacterial host cells in the production of bacteriophage lysates.

R Agar with Catalase

Composition per liter:

Agar	20.0g
Peptone	10.0g
Casamino acids	5.0g
Malt extract	5.0g
Yeast extract	5.0g
Beef extract	2.0g
Glycerol	2.0g
$MgSO_4 \cdot 7H_2O$	1.0g
Catalase	60.0mg
Tween™ 80	50.0mg

pH 7.2 ± 0.2 at 25°C

Preparation of Medium: Add components to distilled/deionized water and bring volume to 1.0L. Mix thoroughly. Gently heat and bring to boiling. Distribute into tubes or flasks. Autoclave for 15 min at 15 psi pressure–121°C. Pour into sterile Petri dishes or leave in tubes.

Use: For the cultivation and maintenance of *Rarobacter faecitabidus.*

R Agar with 3% Sodium Chloride

Composition per liter:

NaCl	30.0g
Agar	20.0g
Peptone	10.0g
Casamino acids	5.0g
Malt extract	5.0g
Yeast extract	5.0g
Beef extract	2.0g

Glycerol 2.0g
$MgSO_4 \cdot 7H_2O$ 1.0g
Tween™ 80 .... 50.0mg

pH 7.2 ± 0.2 at 25°C

Preparation of Medium: Add components to distilled/deionized water and bring volume to 1.0L. Mix thoroughly. Gently heat and bring to boiling. Distribute into tubes or flasks. Autoclave for 15 min at 15 psi pressure–121°C. Pour into sterile Petri dishes or leave in tubes.

Use: For the cultivation and maintenance of *Rhodococcus marinonascens.*

R Agar with 5% Sodium Chloride

Composition per liter:

NaCl	50.0g
Agar	20.0g
Peptone	10.0g
Casamino acids	5.0g
Malt extract	5.0g
Yeast extract	5.0g
Beef extract	2.0g
Glycerol	2.0g
$MgSO_4 \cdot 7H_2O$	1.0g
Tween™ 80	50.0mg

pH 7.2 ± 0.2 at 25°C

Preparation of Medium: Add components to distilled/deionized water and bring volume to 1.0L. Mix thoroughly. Gently heat and bring to boiling. Distribute into tubes or flasks. Autoclave for 15 min at 15 psi pressure–121°C. Pour into sterile Petri dishes or leave in tubes.

Use: For the cultivation and maintenance of *Marinococcus albus, Marinococcus halophilus,* and other *Marinococcus* species.

R Broth for Phage Lysates

Composition per liter:

Pancreatic digest of casein	10.0g
NaCl	8.0g
Yeast extract	1.0g
Glucose solution	5.0mL
$CaCl_2 \cdot 2H_2O$ solution	2.0mL

pH 6.8 ± 0.2 at 25°C

Glucose Solution:

Composition per 10.0mL:

D-Glucose 2.0g

Preparation of Glucose Solution: Add glucose to distilled/deionized water and bring volume to 10.0mL. Mix thoroughly. Filter sterilize.

$CaCl_2 \cdot 2H_2O$ Solution:

Composition per 10.0mL:

$CaCl_2 \cdot 2H_2O$ 1.47g

Preparation of $CaCl_2 \cdot 2H_2O$ Solution: Add the $CaCl_2 \cdot 2H_2O$ to distilled/deionized water and bring volume to 10.0mL. Mix thoroughly. Filter sterilize.

Preparation of Medium: Add components, except glucose solution and $CaCl_2 \cdot 2H_2O$ solution, to distilled/deionized water and bring volume to 993.0mL. Mix thoroughly. Gently heat and bring to boiling. Autoclave for 15 min at 15 psi pressure–121°C. Cool to 45°–50°C. Aseptically add 5.0mL of glucose solution and 2.0mL of $CaCl_2 \cdot 2H_2O$ solution. Mix thoroughly. Aseptically distribute into sterile tubes or flasks.

Use: For the cultivation of bacterial host cells in the production of bacteriophage lysates.

R Medium

Composition per liter:

Agar	30.0g
$NaHCO_3$	8.0g
K_2HPO_4	3.0g
Glucose	2.5g
Glutamine	0.61g
Serine	0.24g
Leucine	0.23g
Lysine	0.23g
Asparagine	0.18g
Valine	0.17g
Isoleucine	0.17g
Tyrosine	0.14g
Arginine·HCl	0.125g
Phenylalanine	0.125g
Threonine	0.12g
Methionine	0.073g
Glycine	0.065g
Histidine·HCl	0.055g
Proline	0.043g
Tryptophan	0.035g
L-Cystine	0.025g
$MgSO_4 \cdot H_2O$	9.9mg
$CaCl_2 \cdot 2H_2O$	7.4mg
Adenine sulfate	2.1mg
Uracil	1.4mg
Thiamine·HCl	1.0mg
$MnSO_4 \cdot H_2O$	0.9mg

pH 8.0 ± 0.2 at 25°C

Preparation of Medium: Add components, except agar, to distilled/deionized water and bring volume to 500.0mL. Mix thoroughly. Filter sterilize. Warm to 45°–50°C. Add agar to distilled/deionized water and bring volume to 500.0mL. Mix thoroughly. Autoclave for 15 min at 15 psi pressure–121°C. Cool to 45°–50°C. Aseptically combine both solutions. Mix thoroughly. Pour into sterile Petri dishes or distribute into sterile tubes.

Use: For the cultivation of *Bacillus anthracis*, especially for the production of toxins.

R-Top Agar

Composition per liter:

Pancreatic digest of casein	10.0g
NaCl	8.0g
Agar	5.0g
K_2HPO_4	2.3g
Yeast extract	1.0g
KH_2PO_4	0.67g
$(NH_4)_2SO_4$	0.33g
Glucose	0.33g
Sodium citrate	0.17g
$MgSO_4 \cdot 7H_2O$	0.03g
Glucose solution	5.0mL
$CaCl_2 \cdot 2H_2O$ solution	2.0mL

pH 7.0 ± 0.2 at 25°C

Glucose Solution:
Composition per 10.0mL:
D-Glucose 2.0g

Preparation of Glucose Solution: Add glucose to distilled/deionized water and bring volume to 10.0mL. Mix thoroughly. Filter sterilize.

$CaCl_2 \cdot 2H_2O$ Solution:
Composition per 10.0mL:
$CaCl_2 \cdot 2H_2O$ 1.47g

Preparation of $CaCl_2 \cdot 2H_2O$ Solution: Add the $CaCl_2 \cdot 2H_2O$ to distilled/deionized water and bring volume to 10.0mL. Mix thoroughly. Filter sterilize.

Preparation of Medium: Add components, except glucose solution and $CaCl_2 \cdot 2H_2O$ solution, to distilled/deionized water and bring volume to 993.0mL. Mix thoroughly. Gently heat and bring to boiling. Autoclave for 15 min at 15 psi pressure–121°C. Cool to 45°–50°C. Aseptically add 5.0mL of glucose solution and 2.0mL of $CaCl_2 \cdot 2H_2O$ solution. Mix thoroughly. Pour into sterile Petri dishes.

Use: For use as a top agar in the cultivation of bacterial host cells for the production of bacteriophage lysates.

R2 Broth (ATCC Medium 1795)

Composition per liter:
Tryptone 20.0g
Yeast extract 10.0g
NaCl 10.0g

pH7.2 ± 0.2 at 25°C

Preparation of Medium: Add components to distilled/deionized water and bring volume to 1.0L. Mix thoroughly. Distribute into tubes or flasks. Autoclave for 15 min at 15 psi pressure–121°C.

Use: For the cultivation of *Spiroplasma citri, S. floricola, S. apis, S. melliferum, and Mycoplasma iowae.*

R2 Broth

Composition per liter:
Peptones 1.0g
Yeast extract 0.5g
Glucose 0.5g
Starch, soluble 0.5g
KH_2PO_4 .... 0.3g
Sodium pyruvate 0.3g
$MgSO_4 \cdot 7H_2O$ 0.024g

pH 7.0 ± 0.2 at 25°C

Preparation of Medium: Add components to distilled/deionized water and bring volume to 1.0L. Mix thoroughly. Distribute into tubes or flasks. Autoclave for 15 min at 15 psi pressure–121°C.

Use: For the cultivation of *Bacillus* spp. and *Deinococcus* spp.

RA (Raulin Neutral of Dierckx)

Composition per liter:
Solution A 50.0mL
Solution B 900.0mL

Solution A:
Composition per 50.0mL:
Tartaric acid 0.47g
$MgCO_3$ 0.265g

Preparation of Solution A: Add components to distilled/deionized water and bring volume to 100.0mL. Mix thoroughly.

Solution B:
Composition per 900.0mL:
Sucrose 46.60g
NH_4NO_3 .... 2.66g
K_2CO_3 .... 0.40g
$(NH_4)_3PO_4$ 0.40g
$(NH_4)_2SO_4$ 0.16g
$FeSO_4$ 0.04g
$ZnSO_4$ 0.04g

Preparation of Solution B: Add components to distilled/deionized water and bring volume to 900.0mL. Mix thoroughly.

Preparation of Medium: Add 50.0mL of solution A and 900.0mL of solution B to distilled/deionized water and bring volume to 1.0L. Mix thoroughly. Distribute into tubes or flasks. Autoclave for 15 min at 15 psi pressure–121°C.

Use: For the cultivation and maintenance of *Penicillium* species.

R2A Agar

Composition per liter:
Agar 15.0g
Yeast extract 0.5g
Acid hydrolysate of casein 0.5g
Glucose 0.5g
Soluble starch 0.5g
K_2HPO_4 .... 0.3g
Sodium pyruvate 0.3g
Pancreatic digest of casein 0.25g
Peptic digest of animal tissue 0.25g
$MgSO_4$, anhydrous 0.024g

pH 7.2 ± 0.2 at 25°C

Source: This medium is available as a premixed powder from BD Diagnostic Systems.

Preparation of Medium: Add components to distilled/deionized water and bring volume to 1.0L. Mix thoroughly. Gently heat with mixing and bring to boiling. Distribute into tubes or flasks. Autoclave for 15 min at 15 psi pressure–121°C. Do not overheat. Pour into sterile Petri dishes or leave in tubes.

Use: For use in standard methods for pour plate, spread plate, and membrane filter analysis to enumerate heterotrophic bacteria from waters.

R2A Agar

Composition per liter:
Agar 15.0g
Yeast extract 0.5g
Acid hydrolysate of casein 0.5g
Glucose 0.5g
Soluble starch 0.5g
K_2HPO_4 .... 0.3g
Sodium pyruvate 0.3g
Pancreatic digest of casein 0.25g
Peptic digest of animal tissue 0.25g
$MgSO_4$, anhydrous 0.024g

pH 7.2 ± 0.2 at 25°C

Source: This medium is available as a premixed powder from BD Diagnostic Systems.

Preparation of Medium: Add components to distilled/deionized water and bring volume to 1.0L. Mix thoroughly. Gently heat with mixing and bring to boiling. Distribute into tubes or flasks. Autoclave for 15 min at 15 psi pressure–121°C. Do not overheat. Pour into sterile Petri dishes or leave in tubes.

Use: For use in standard methods for pour plate, spread plate, and membrane filter analysis to enumerate heterotrophic bacteria from potable waters.

R2A Agar, Modified

Composition per liter:

Agar	15.0g
NH_4Cl	0.8g
KNO_3	0.505g
Casamino acids	0.5g
Glucose	0.5g
Peptone	0.5g
Sodium pyruvate	0.5g
Starch, soluble	0.5g
Yeast extract	0.5g
K_2HPO_4	0.4g
KH_2PO_4	0.25g
$MgCl_2·6H_2O$	20.0mg
$CaCl_2·2H_2O$	15.0mg
$FeSO_4·7H_2O$	7.0mg
$MnCl_2·4H_2O$	5.0mg
Na_2SO_4	5.0mg
$CoCl_2·6H_2O$	0.5mg
H_3BO_3	0.5mg
$NiSO_4·6H_2O$	0.5mg
$ZnCl_2$	0.5mg
$CuCl_2·2H_2O$	0.3mg
$Na_2MoO_4·2H_2O$	10.0μg

pH 7.0 ± 0.2 at 25°C

Preparation of Medium: Add components to distilled/deionized water and bring volume to 1.0L. Mix thoroughly. Adjust pH to 7.0. Gently heat and bring to boiling. Distribute into tubes or flasks. Autoclave for 15 min at 15 psi pressure–121°C. Pour into sterile Petri dishes or leave in tubes.

Use: For the cultivation of *Azoarcus tolulyticus*.

R 2A HiVeg Agar

Composition per liter:

Agar	15.0g
Glucose	0.5g
Plant peptone No. 3	0.5g
Starch, soluble	0.5g
Plant acid hydrolysate	0.5g
Yeast extract	0.5g
K_2HPO_4	0.3g
Sodium pyruvate	0.3g
$MgSO_4$	0.024g

pH 7.2 ± 0.2 at 25°C

Source: This medium is available as a premixed powder from HiMedia.

Preparation of Medium: Add components to distilled/deionized water and bring volume to 1.0L. Mix thoroughly. Gently heat with mixing and bring to boiling. Distribute into tubes or flasks. Autoclave for 15 min at 15 psi pressure–121°C. Do not overheat. Pour into sterile Petri dishes or leave in tubes.

Use: For use in standard methods for pour plate, spread plate, and membrane filter analysis to enumerate heterotrophic bacteria from potable waters.

R3 Medium (DSMZ Medium 966)

Composition per liter:

Yeast extract	1.0g
Proteose peptone No.3	1.0g
Casamino acids	1.0g
Glucose	1.0g
K_2HPO_4	0.6g
$MgSO_4·7H_2O$	0.1g
Na-pyruvate	0.05g

pH 7.8 ± 0.2 at 25°C

Preparation of Medium: Add components to distilled/deionized water and bring volume to 1.0L. Mix thoroughly. Distribute into tubes or flasks. Autoclave for 15 min at 15 psi pressure–121°C.

Use: For the cultivation of *Rubritepida flocculans*.

R3A Agar

Composition per liter:

Agar	15.0g
Yeast extract	1.0g
Acid hydrolysate of casein	1.0g
Glucose	1.0g
Soluble starch	1.0g
K_2HPO_4	0.6g
Sodium pyruvate	0.6g
Pancreatic digest of casein	0.5g
Peptic digest of animal tissue	0.5g
$MgSO_4$, anhydrous	0.048g

pH 7.2 ± 0.2 at 25°C

Source: This medium is available as a premixed powder from BD Diagnostic Systems.

Preparation of Medium: Add components to distilled/deionized water and bring volume to 1.0L. Mix thoroughly. Gently heat with mixing and bring to boiling. Distribute into tubes or flasks. Autoclave for 15 min at 15 psi pressure–121°C. Do not overheat. Pour into sterile Petri dishes or leave in tubes.

Use: For the cultivation and maintenance of heterotrophic bacteria from potable waters.

R8 Medium (DSMZ Medium 912)

Composition per liter:

$NaHCO_3$	2.52g
MOPS	2.1g
NaCl	1.0g
Glucose	0.9g
$MgCl_2·6H_2O$	0.5g
$Na_2S·9H_2O$	0.6g
Cysteine-HCl	0.4g
NH_4Cl	0.3g
KCl	0.3g
K_2HPO_4	0.25g

KH_2PO_4 .... 0.2g
Yeast extract 0.19g
Peptone 0.19g
$CaCl_2 \cdot 2H_2O$ 0.015g
Resazurin 0.5mg
Rumen fluid 50.0mL
Na-pantothenate solution 10.0mL
Trace elements solution SL-10 .... 1.0mL

pH 7.2 ± 0.2 at 25°C

Trace Elements Solution SL-10:

Composition per liter:

$FeCl_2 \cdot 4H_2O$ 1.5g
$CoCl_2 \cdot 6H_2O$ 190.0mg
$MnCl_2 \cdot 4H_2O$ 100.0mg
$ZnCl_2$ 70.0mg
$Na_2MoO_4 \cdot 2H_2O$ 36.0mg
$NiCl_2 \cdot 6H_2O$ 24.0mg
H_3BO_3 .... 6.0mg
$CuCl_2 \cdot 2H_2O$ 2.0mg
HCl (25% solution) 10.0mL

Preparation of Trace Elements Solution SL-10: Add $FeCl_2 \cdot 4H_2O$ to 10.0mL of HCl solution. Mix thoroughly. Add distilled/deionized water and bring volume to 1.0L. Add remaining components. Mix thoroughly. Sparge with 100% N_2.

Rumen Fluid:

Composition per 50.0mL:

Rumen fluid, clarified 50.0mL

Preparation of Rumen Fluid: Sparge clarified rumen fluid with 100% N_2. Autoclave for 15 min at 15 psi pressure–121°C.

Na-Pantothenate Solution:

Composition per 10.0mL:

Na-pantothenate 0.1g

Preparation of Na-Pantothenate Solution: Add Na-pantothenate to distilled/deionized water and bring volume to 10.0mL. Mix thoroughly. Sparge with 100% N_2. Filter sterilize.

Preparation of Medium: Prepare and dispense medium under 80% N_2 + 20% CO_2 gas atmosphere. Add components, except $NaHCO_3$, pantothenate solution, $Na_2S \cdot 9H_2O$, cysteine-HCl, and rumen fluid, to distilled/deionized water and bring volume to 940.0mL. Mix thoroughly. Gently heat and bring to boiling. Boil for 3 min. Cool to room temperature while sparging with 80% N_2 + 20% CO_2. Add solid bicarbonate, sodium sulfide, and cysteine-HCl. Adjust pH to 7.2. Distribute under 80% N_2 + 20% CO_2 atmosphere into anaerobe tubes or bottles. Autoclave for 15 min at 15 psi pressure–121°C. Aseptically and anaerobically add, per liter of medium, 50.0mL rumen fluid and 10.0mL Na-pantothenate solution. Mix thoroughly. The final pH of the medium should be 7.2.

Use: For the cultivation of *Spirochaeta* spp.

R70-2 Agar, Modified with Fructose

Composition per liter:

Fructose 20.0g
Agar 15.0g
Yeast extract 5.0g
Dimethyl glutaric acid 4.01g
$(NH_4)_2SO_4$ 3.3g
Trisodium citrate·$2H_2O$ 1.18g
KH_2PO_4 .... 1.0g
$MgSO_4 \cdot 7H_2O$ 0.25g
Wolfe's vitamin solution 10.0mL
100X modified salts 10.0mL

pH 5.0 ± 0.2 at 25°C

Wolfe's Vitamin Solution:

Composition per liter:

Pyridoxine·HCl 0.01g
Thiamine·HCl 5.0mg
Riboflavin 5.0mg
Nicotinic acid 5.0mg
Calcium pantothenate 5.0mg
p-Aminobenzoic acid 5.0mg
Thioctic acid 5.0mg
Biotin 2.0mg
Folic acid 2.0mg
Cyanocobalamin 0.1mg

Preparation of Wolfe's Vitamin Solution: Add components to distilled/deionized water and bring volume to 1.0L. Mix thoroughly. Filter sterilize.

100X Modified Salts:

Composition per liter:

$CaCl_2 \cdot 2H_2O$ 1.47g
$FeCl_3 \cdot 6H_2O$ 0.27g
$ZnSO_4 \cdot 7H_2O$ 0.144g
$MnSO_4 \cdot H_2O$ 0.085g
$CoCl_2 \cdot 6H_2O$ 0.024g
$NiCl_2 \cdot 6H_2O$ 0.024g
$Na_2MoO_4 \cdot 2H_2O$ 0.024g
$CuSO_4 \cdot 5H_2O$ 0.016g
HCl, concentrated 4.1mL

Preparation of 100X Modified Salts: Add components to distilled/deionized water and bring volume to 1.0L. Mix thoroughly. Filter sterilize.

Preparation of Medium: Add components, except Wolfe's vitamin solution and 100X modified salts, to distilled/deionized water and bring volume to 980.0mL. Mix thoroughly. Adjust pH to 5.0. Gently heat and bring to boiling. Autoclave for 15 min at 15 psi pressure–121°C. Cool to 45°–50°C. Aseptically add sterile Wolfe's vitamin solution and 100X modified salts. Mix thoroughly. Pour into sterile Petri dishes or distribute into sterile tubes.

Use: For the cultivation of *Acetobacter xylinum.*

R70-2 Agar, Modified with Glucose

Composition per liter:

Glucose 20.0g
Agar 15.0g
Yeast extract 5.0g
Dimethyl glutaric acid 4.01g
$(NH_4)_2SO_4$ 3.3g
Trisodium citrate·$2H_2O$ 1.18g
KH_2PO_4 .... 1.0g
$MgSO_4 \cdot 7H_2O$ 0.25g
Wolfe's vitamin solution 10.0mL
100X modified salts 10.0mL

pH 5.0 ± 0.2 at 25°C

Wolfe's Vitamin Solution:

Composition per liter:

Pyridoxine·HCl 0.01g
Thiamine·HCl 5.0mg
Riboflavin 5.0mg

Nicotinic acid	5.0mg
Calcium pantothenate	5.0mg
p-Aminobenzoic acid	5.0mg
Thioctic acid	5.0mg
Biotin	2.0mg
Folic acid	2.0mg
Cyanocobalamin	0.1mg

Preparation of Wolfe's Vitamin Solution: Add components to distilled/deionized water and bring volume to 1.0L. Mix thoroughly. Filter sterilize.

100X Modified Salts:

Composition per liter:

$CaCl_2 \cdot 2H_2O$	1.47g
$FeCl_3 \cdot 6H_2O$	0.27g
$ZnSO_4 \cdot 7H_2O$	0.144g
$MnSO_4 \cdot H_2O$	0.085g
$CoCl_2 \cdot 6H_2O$	0.024g
$NiCl_2 \cdot 6H_2O$	0.024g
$Na_2MoO_4 \cdot 2H_2O$	0.024g
$CuSO_4 \cdot 5H_2O$	0.016g
HCl, concentrated	4.1mL

Preparation of 100X Modified Salts: Add components to distilled/deionized water and bring volume to 1.0L. Mix thoroughly. Filter sterilize.

Preparation of Medium: Add components, except Wolfe's vitamin solution and 100X modified salts, to distilled/deionized water and bring volume to 980.0mL. Mix thoroughly. Adjust pH to 5.0. Gently heat and bring to boiling. Autoclave for 15 min at 15 psi pressure–121°C. Cool to 45°–50°C. Aseptically add sterile Wolfe's vitamin solution and 100X modified salts. Mix thoroughly. Pour into sterile Petri dishes or distribute into sterile tubes.

Use: For the cultivation of *Acetobacter xylinum.*

R70-2 Broth, Modified with Fructose

Composition per liter:

Fructose	30.0g
Yeast extract	5.0g
Dimethyl glutaric acid	4.01g
$(NH_4)_2SO_4$	3.3g
Trisodium citrate·$2H_2O$	1.18g
KH_2PO_4	1.0g
$MgSO_4 \cdot 7H_2O$	0.25g
Wolfe's vitamin solution	10.0mL
100X modified salts	10.0mL

pH 5.0 ± 0.2 at 25°C

Wolfe's Vitamin Solution:

Composition per liter:

Pyridoxine·HCl	0.01g
Thiamine·HCl	5.0mg
Riboflavin	5.0mg
Nicotinic acid	5.0mg
Calcium pantothenate	5.0mg
p-Aminobenzoic acid	5.0mg
Thioctic acid	5.0mg
Biotin	2.0mg
Folic acid	2.0mg
Cyanocobalamin	0.1mg

Preparation of Wolfe's Vitamin Solution: Add components to distilled/deionized water and bring volume to 1.0L. Mix thoroughly. Filter sterilize.

100X Modified Salts:

Composition per liter:

$CaCl_2 \cdot 2H_2O$	1.47g
$FeCl_3 \cdot 6H_2O$	0.27g
$ZnSO_4 \cdot 7H_2O$	0.144g
$MnSO_4 \cdot H_2O$	0.085g
$CoCl_2 \cdot 6H_2O$	0.024g
$NiCl_2 \cdot 6H_2O$	0.024g
$Na_2MoO_4 \cdot 2H_2O$	0.024g
$CuSO_4 \cdot 5H_2O$	0.016g
HCl, concentrated	4.1mL

Preparation of 100X Modified Salts: Add components to distilled/deionized water and bring volume to 1.0L. Mix thoroughly. Filter sterilize.

Preparation of Medium: Add components, except Wolfe's vitamin solution and 100X modified salts, to distilled/deionized water and bring volume to 980.0mL. Mix thoroughly. Autoclave for 15 min at 15 psi pressure–121°C. Cool to 45°–50°C. Aseptically add sterile Wolfe's vitamin solution and 100X modified salts. Mix thoroughly. Aseptically distribute into sterile tubes or flasks.

Use: For the cultivation of *Acetobacter xylinum.*

R70-2 Broth, Modified with Glucose

Composition per liter:

Fructose	30.0g
Yeast extract	5.0g
Dimethyl glutaric acid	4.01g
$(NH_4)_2SO_4$	3.3g
Trisodium citrate·$2H_2O$	1.18g
KH_2PO_4	1.0g
$MgSO_4 \cdot 7H_2O$	0.25g
Wolfe's vitamin solution	10.0mL
100X modified salts	10.0mL

pH 5.0 ± 0.2 at 25°C

Wolfe's Vitamin Solution:

Composition per liter:

Pyridoxine·HCl	0.01g
Thiamine·HCl	5.0mg
Riboflavin	5.0mg
Nicotinic acid	5.0mg
Calcium pantothenate	5.0mg
p-Aminobenzoic acid	5.0mg
Thioctic acid	5.0mg
Biotin	2.0mg
Folic acid	2.0mg
Cyanocobalamin	0.1mg

Preparation of Wolfe's Vitamin Solution: Add components to distilled/deionized water and bring volume to 1.0L. Mix thoroughly. Filter sterilize.

100X Modified Salts:

Composition per liter:

$CaCl_2 \cdot 2H_2O$	1.47g
$FeCl_3 \cdot 6H_2O$	0.27g
$ZnSO_4 \cdot 7H_2O$	0.144g
$MnSO_4 \cdot H_2O$	0.085g

$CoCl_2 \cdot 6H_2O$ 0.024g
$NiCl_2 \cdot 6H_2O$ 0.024g
$Na_2MoO_4 \cdot 2H_2O$ 0.024g
$CuSO_4 \cdot 5H_2O$ 0.016g
HCl, concentrated 4.1mL

Preparation of 100X Modified Salts: Add components to distilled/deionized water and bring volume to 1.0L. Mix thoroughly. Filter sterilize.

Preparation of Medium: Add components, except Wolfe's vitamin solution and 100X modified salts, to distilled/deionized water and bring volume to 980.0mL. Mix thoroughly. Autoclave for 15 min at 15 psi pressure–121°C. Cool to 45°–50°C. Aseptically add sterile Wolfe's vitamin solution and 100X modified salts. Mix thoroughly. Aseptically distribute into sterile tubes or flasks.

Use: For the cultivation of *Acetobacter xylinum.*

Rabbit Blood Agar

Composition per 1250.0mL:

Pancreatic digest of casein 16.0g
Agar 13.5g
Brain heart, solids from infusion 8.0g
Peptic digest of animal tissue 5.0g
NaCl 5.0g
Glucose 2.0g
Na_2HPO_4 .... 2.5g
Rabbit blood, defibrinated 250.0mL

pH 7.4 ± 0.2 at 25°C

Preparation of Medium: Add components, except rabbit blood, to distilled/deionized water and bring volume to 1.0L. Mix thoroughly. Autoclave for 15 min at 15 psi–121°C. Aseptically add sterile rabbit blood. Pour into sterile Petri dishes or aseptically distribute into sterile tubes or flasks while shaking.

Use: For the cultivation and maintenance of *Corynebacterium diphtheriae, Haemophilus ducreyi,* and *Actinobacillus lignieresii.*

Rabbit Dung Agar

Composition per liter:

Rabbit dung 20.0g
Agar 15.0g

pH 7.2 ± 0.2 at 25°C

Preparation of Medium: Add rabbit dung to 1.0L of distilled/deionized water. Gently heat and bring to boiling. Continue boiling for 20 min. Filter through Whatman #1 filter paper. Bring volume of filtrate to 1.0L with distilled/deionized water. Add agar. Adjust pH to 7.2. Distribute into tubes or flasks. Autoclave for 15 min at 15 psi pressure–121°C. Pour into sterile Petri dishes or leave in tubes.

Use: For the cultivation of myxobacteria.

Rabbit Dung Agar

Composition per test tube:

Rabbit pellets, presterilized .... 3 or 4
Agar solution 4.0mL

Agar Solution:

Composition per 100.0mL:

Agar 1.5g

Preparation of Agar Solution: Add agar to distilled/deionized water and bring volume to 100.0mL. Mix thoroughly. Gently heat and bring to boiling.

Preparation of Medium: Place 3 or 4 presterilized rabbit pellets in each test tube. Dispense 4.0mL of 1.5% agar solution into the tubes. Autoclave for 15 min at 15 psi pressure–121°C. Allow tubes to cool in a slanted position so that rabbit pellets extend above agar surface. Inoculate microorganism on pellets.

Use: For the cultivation and maintenance of *Chaetomium adinocladium, Lophotrichus incarnatus, Mycoarachis inversa, Nigrosabulum globosum,* many *Pilobolus* species, *Spiromyces minutus,* and *Wardomyces simplex.*

Rabbit Food Agar

Composition per liter:

Rabbit food, commercial pellets 25.0g
Agar 15.0g

Preparation of Medium: Add rabbit food pellets to 1.0L of distilled/deionized water. Gently heat and bring to boiling. Let steep for 30 min. Filter solids through cheesecloth. Add agar to filtrate. Mix thoroughly. Bring volume to 1.0L with distilled/deionized water. Gently heat and bring to boiling. Distribute into tubes or flasks. Autoclave for 15 min at 15 psi pressure–121°C. Pour into sterile Petri dishes or leave in tubes.

Use: For the cultivation and maintenance of the yeasts *Candida glaebosa, Chaetomium virescens, Curvularia lunata, Filobasidium floriforme,* and numerous filamentous fungi.

Rabbit Heart Infusion Agar

Composition per liter:

Beef heart, infusion from 500.0g
Agar 15.0g
Tryptose 10.0g
NaCl 5.0g
Rabbit blood, defibrinated 50.0mL

pH 7.4 ± 0.2 at 25°C

Preparation of Medium: Add components, except rabbit blood, to distilled/deionized water and bring volume to 950.0mL. Mix thoroughly. Gently heat and bring to boiling. Autoclave for 15 min at 15 psi pressure–121°C. Cool to 50°–55°C. Aseptically add 50.0mL of sterile rabbit blood. Mix thoroughly. Pour into sterile Petri dishes or distribute into sterile tubes.

Use: For the culture and maintenance of *Bartonella quintana.*

Rabbit Laked Blood Agar

Composition per liter:

Agar 15.0g
Pancreatic digest of casein 10.0g
Peptic digest of animal tissue 10.0g
NaCl 5.0g
Yeast extract 2.0g
Glucose 1.0g
$NaHSO_3$ 0.1g
Rabbit blood, laked 50.0mL
Hemin solution 1.0mL
Vitamin K_1 solution 1.0mL

pH 7.0 ± 0.2 at 25°C

Hemin Solution:

Composition per 10.0mL:

Hemin 0.5g
NaOH (1*N* solution) 10.0mL

Preparation of Hemin Solution: Add hemin to 10.0mL of NaOH solution. Mix thoroughly.

Vitamin K_1 Solution:

Composition per 20.0mL:

Vitamin K_1 (phytomenadione)	0.2g
Ethanol (95% solution)	20.0mL

Preparation of Vitamin K_1 Solution: Add vitamin K_1 to 20.0mL of ethanol. Mix thoroughly.

Preparation of Medium: Add components, except vitamin K_1 solution and laked rabbit blood, to distilled/deionized water and bring volume to 849.0mL. Mix thoroughly. Gently heat and bring to boiling. Autoclave for 15 min at 15 psi pressure–121°C. Cool to 45°–50°C. Aseptically add 1.0mL of sterile vitamin K_1 solution and 50.0mL of sterile laked rabbit blood. Laked blood is prepared by freezing whole blood overnight and thawing to room temperature. Mix thoroughly. Pour into sterile Petri dishes or distribute into sterile tubes.

Use: For the cultivation and enhancement of pigment production of a variety of anaerobic bacteria.

Rabbit Serum Medium (Rabbit Serum Bovine Serum Albumin Tween™ 80 Medium) (Rabbit Serum BSA Tween™ 80 Medium)

Composition per liter:

Basal medium	900.0mL
Rabbit serum with supplements	100.0mL

pH 7.4 ± 0.2 at 25°C

Basal Medium:

Composition per 900.0mL:

Na_2HPO_4	1.0g
NaCl	1.0g
KH_2PO_4	0.3g
Glycerol (10% solution)	1.0mL
NH_4Cl (25% solution)	1.0mL
Sodium pyruvate (10% solution)	1.0mL
Thiamine (0.5% solution)	1.0mL

Preparation of Basal Medium: Add components to distilled/deionized water and bring volume to 900.0mL. Mix thoroughly. Gently heat and bring to boiling. Autoclave for 20 min at 15 psi pressure–121°C. Cool to 25°C.

Rabbit Serum with Supplements:

Composition per 106.0mL:

Rabbit serum	100.0mL
L-Asparagine (3% solution)	5.0mL
$MgCl_2$-$CaCl_2$ solution	1.0mL

Preparation of Rabbit Serum with Supplements: Combine the three solutions. Mix thoroughly. Filter sterilize.

$MgCl_2$-$CaCl_2$ Solution:

Composition per 100.0mL:

$CaCl_2 \cdot 2H_2O$	1.5g
$MgCl_2 \cdot 6H_2O$	1.5g

Preparation of $MgCl_2$-$CaCl_2$ Solution: Add components to distilled/deionized water and bring volume to 100.0mL. Mix thoroughly.

Preparation of Medium: Aseptically combine 900.0mL of cooled sterile basal medium and 100.0mL of sterile rabbit serum with supplements. Mix thoroughly. Aseptically distribute into sterile tubes or flasks.

Use: For the cultivation of *Leptospira* species.

RAE Medium

See: **Reinforced AE Medium**

R8AH Medium (DSMZ Medium 651)

Composition per liter:

Malic acid	2.5g
$(NH_4)_2SO_4$	1.25g
Yeast extract	1.0g
K_2HPO_4	0.9g
KH_2PO_4	0.6g
$MgSO_4 \cdot 7H_2O$	0.2g
$CaCl_2 \cdot 2H_2O$	0.07g
EDTA	0.02g
Ferric citrate	0.01g
Vitamin solution	7.5mL
Trace elements solution	1.0mL

pH 6.9 ± 0.2 at 25°C

Trace Elements Solution:

Composition per 100.0mL:

Ferric citrate	0.3g
EDTA	0.05g
$CaCl_2 \cdot 2H_2O$	0.02g
$MnSO_4 \cdot H_2O$	2.0mg
$(NH_4)_6Mo_7O_{24} \cdot 4H_2O$	2.0mg
H_3BO_3	1.0mg
$CuSO_4 \cdot 5H_2O$	1.0mg
$ZnSO_4$	1.0mg

Preparation of Trace Elements Solution: Add components to distilled/deionized water and bring volume to 100.0mL. Mix thoroughly.

Vitamin Solution:

Composition per liter:

Thiamine·HCl	0.4g
Nicotinic acid	0.2g
Nicotinamide	0.2g
Biotin	8.0mg

Preparation of Vitamin Solution: Add components to distilled/deionized water and bring volume to 1.0L. Mix thoroughly.

Preparation of Medium: Add malic acid to 500.0mL of distilled/deionized water. Adjust pH to 6.9 with NaOH. Add remaining components. Bring volume to 1.0L with distilled/deionized water. Mix thoroughly. Adjust pH to 6.9. Distribute into tubes or flasks. Autoclave for 15 min at 15 psi pressure–121°C.

Use: For the cultivation and maintenance of *Rhodopseudomonas palustris, Rhodobacter sphaeroides, Rhodocyclus tenuis, Rhodopseudomonas rutila, Rhodospirillum photometricum,* and *Rhodospirillum rubrum.*

Rainbow Agar O157

Composition per liter:

Proprietary.

Source: This medium is available as a premixed powder from Biolog Inc.

Preparation: Suspend 60.0g of the proprietary mixture in distilled/deionized water and bring volume to 1.0L. Mix thoroughly. Gently heat and bring to boiling. Autoclave for 15 min at 15 psi pressure–

121°C. Cool to 45°C–50°C. Mix thoroughly. Pour into sterile Petri dishes or distribute into sterile tubes. The final medium should be clear and virtually colorless. No pH adjustment is needed. The final pH should be pH 7.9–8.3. To increase the selectivity of the medium, a sterile solution containing 0.8mg potassium tellurite and 10mg novobiocin can be added. Caution must be used because tellurite is toxic.

Use: For the detection, isolation, and presumptive identification of verotoxin-producing strains of *Escherichia coli*, particularly serotype O157:H7. The medium contains chromogenic substrates that are specific for two *E. coli*-associated enzymes: β-galactosidase (a blue-black chromogenic substrate) and β-glucuronidase (a red chromogenic substrate). The distinctive black or gray coloration of *E. coli* O157:H7 colonies is easily viewed by laying the Petri plate against a white background. When O157 is surrounded by pink or magenta non-toxigenic colonies, it may have a bluish hue. The addition of selective agents improves performance. *E. coli* O157:H7 colony coloration will be slightly bluer with these selective agents added. Tellurite is highly selective for *E. coli* O157:H7 and can reduce background flora considerably. Novobiocin inhibits *Proteus* swarming and the growth of tellurite-reducing bacteria. Rare strains of O157:H7 are tellurite sensitive.

Rainbow Agar *Salmonella*

Composition per liter:

Proprietary

pH 7.2–7.6 at 25°C

Source: This medium is available from Biolog.

Preparation of Medium: Add components to distilled/deionized water and bring volume to 990.0mL. Mix thoroughly. Add 10.0mL 35% glycerol. Stir until components are evenly dispersed. Gently heat and boil. Autoclave for 10 min at 15 psi pressure–121°C. Do not exceed 10 min. Cool agar to 45°C–50°C before pouring plates.

Use: As a selective, chromogenic medium to aid in the detection and isolation of H_2S-producing *Salmonella* species. Black colonies are formed by even weak H_2S-producing strains.

RajHans Medium (HiCrome™ *Salmonella* Medium, Modified)

Composition per liter:

Agar	12.0 g
Casein enzymic hydrolysate	8.0 g
Yeast extract	5.0 g
NaCl	5.0 g
Chromogenic mixture	4.32 g
Peptic digest of animal tissue	4.0 g
Lactose	3.0 g
Sodium deoxycholate	1.0 g
Neutral Red	0.02 g

pH 7.3 ± 0.2 at 25°C

Source: This medium is available as a premixed powder from HiMedia.

Preparation of Medium: Add components to distilled/deionized water and bring volume to 1.0L. Mix thoroughly. Gently heat while stirring and bring to boiling. Mix to completely dissolve components. Do not autoclave. Cool to 50°C. Mix thoroughly. Pour into sterile Petri dishes.

Use: A selective chromgenic medium used for the isolation and differentiation of *Salmonella* species from the members of Enterobacteriaceae, especially *Proteus* species.

RajHans Medium, HiVeg (HiCrome™ *Salmonella* Medium, Modified)

Composition per liter:

Agar	12.0g
Propylene glycol	10.0g
Plant special peptone	8.0g
Yeast extract	2.0g
B.C. indicator	2.0g
Sodium deoxycholate	1.0g

pH 7.3 ± 0.2 at 25°C

Source: This medium is available as a premixed powder from HiMedia.

Preparation of Medium: Add components to distilled/deionized water and bring volume to 1.0L. Mix thoroughly. Gently heat while stirring and bring to boiling. Mix to completely dissolve components. Do not autoclave. Cool to 50°C. Mix thoroughly. Pour into sterile Petri dishes.

Use: A selective chromgenic medium used for the isolation and differentiation of *Salmonella* species from the members of Enterobacteriaceae, especially *Proteus* species.

Raka-Ray Agar

Composition per liter:

Pancreatic digest of casein	20.0g
Agar	17.0g
Maltose	10.0g
Fructose	5.0g
Glucose	5.0g
Yeast extract	5.0g
2-Phenylethanol	3.0g
Potassium aspartate	2.5g
Potassium glutamate	2.5g
Betaine·HCL	2.0g
Diammonium hydrogen citrate	2.0g
$MgSO_4 \cdot 7H_2O$	2.0g
KH_2PO_4	2.0g
Liver concentrate	1.0g
$MnSO_4 \cdot H_2O$	0.66g
N-Acetylglucosamine	0.5g
Cycloheximide	7.0mg
Sorbitan monooleate	10.0mL

pH 5.4 ± 0.2 at 25°C

Source: This medium is available as a premixed powder from Oxoid Unipath.

Caution: Cycloheximide is toxic. Avoid skin contact or aerosol formation and inhalation.

Preparation of Medium: Add components, except phenylethanol, to distilled/deionized water and bring volume to 1.0L. Mix thoroughly. Gently heat and bring to boiling. Autoclave for 15 min at 15 psi pressure–121°C. Cool to 45°–50°C. Aseptically add 3.0g of 2-phenylethanol. Mix thoroughly. Pour into sterile Petri dishes or distribute into sterile tubes.

Use: For the isolation of lactic acid bacteria in beer and brewing processes.

Raka-Ray No. 3 Medium (DSMZ Medium 1047)

Composition per liter:

Tryptone	20.0g
Agar	16.0g
Maltose	10.0g
Fructose	5.0g
Glucose	5.0g
Yeast extract	5.0g
Potassium aspartate	2.5g
Potassium glutamate	2.5g
Betaine·HCL	2.0g
Diammonium hydrogen citrate	2.0g
KH_2PO_4	2.0g
Liver concentrate	1.0g
$MgSO_4·7H_2O$	0.98g
$MnSO_4·H_2O$	0.66g
N-Acetylglucosamine	0.5g

pH 5.4 ± 0.2 at 25°C

Preparation of Medium: Add components to distilled/deionized water and bring volume to 1.0L. Mix thoroughly. Gently heat and bring to boiling. Autoclave for 15 min at 15 psi pressure–121°C. Pour into sterile Petri dishes or distribute into sterile tubes.

Use: For the cultivation of "*Lactobacillus backii.*"

Rambach® Agar

Composition per liter:

Agar	15.0g
Polypropylene glycol	10.5g
Peptone	8.0g
NaCl	5.0g
Chromogenic mix	1.5g
Na-desoxycholate	1.0g

pH 7.4 ± 0.2 at 25°C

Source: Rambach Agar is available from CHROMagar Microbiology and Merck.

Preparation of Medium: Add components to distilled/deionized water and bring volume to 1.0L. Mix thoroughly. Heat in a boiling water bath or in a current of steam, while shaking from time to time. The medium is totally suspended, if no visual particles stick to the glass wall. The medium should not be heat treated further. Complete dissolution with shaking in 5-min sequences is approximately 35–40 minutes. Do not autoclave. Do not overheat. Cool as fast as possible to 45°–50°C while gently shaking from time to time. Pour into sterile Petri dishes. To prevent any precipitate or clotting of the chromogenic mix in the plates, place Petri dishes during pouring procedure on a cool (max. 25°C) surface. The plates are opaque and pink.

Use: For the detection of enteric bacteria, including coliforms and *Salmonella* spp. Sodium desoxycholate inhibits the accompanying Gram-positive flora. This medium enables *Salmonella* spp. to be differentiated unambiguously from other bacteria. *Salmonella* spp. form a characteristic red color. In order to differentiate coliforms from Salmonellae, the medium contains a chromogene indicating the presence of β-galactosidase splitting, a characteristic of coliforms. Coliform microorganisms grow as blue-green or blue-violet colonies. Other Enterobacteriaceae and Gram-negative bacteria, such as *Proteus, Pseudomonas, Shigella, S. typhi, and S. parathyphi* A, grow as colorless-yellow colonies.

Rap Broth, Modified

See: **Rappaport Broth, Modified**

Raper *Achyla* Medium No. 1

Composition per liter:

Agar	20.0g
Lentil (hot water extract)	10.0g
Starch, soluble	3.0g
Peptone	1.0g
$CaCl_2$	1.0μg
$FeCl_3$	1.0μg
KH_2PO_4	1.0μg
$MgSO_4$	1.0μg
$ZnSO_4$	1.0μg

Preparation of Medium: Add components to distilled/deionized water and bring volume to 1.0L. Mix thoroughly. Gently heat and bring to boiling. Distribute into tubes or flasks. Autoclave for 15 min at 15 psi pressure–121°C. Pour into sterile Petri dishes or leave in tubes.

Use: For the cultivation of *Achyla* species.

Raper *Achyla* Medium No. 2

Composition per liter:

Agar	20.0g
Starch, soluble	3.0g
Inositol	1.0g
Peptone	1.0g

Preparation of Medium: Add components to distilled/deionized water and bring volume to 1.0L. Mix thoroughly. Gently heat and bring to boiling. Distribute into tubes or flasks. Autoclave for 15 min at 15 psi pressure–121°C. Pour into sterile Petri dishes or leave in tubes.

Use: For the cultivation of *Achyla* species.

RAPID′*E. coli* 2 Agar

Composition per liter:

Proprietary

Source: This medium is available from Biorad.

Use: For the direct enumeration of *E. coli* and coliforms in foods. Selectivity and electivity are based on the detection of glucuronidase and galactosidase activities. Hydrolysis of chromogenic substrate results in purple to pink *E. coli* colonies (gluc+/gal+) and blue-green coliform colonies (gluc-/gal+). RAPID′*E. coli* 2 agar is AFNOR validated according to ISO 16140 protocol to enumerate *E. coli* and coliforms on the same plate at 37°C, without any further confirmation of characteristic colonies.

RAPID′*Enterococcus* Agar

Composition per liter:

Proprietary

Source: This medium is available from Biorad.

Use: A selective chromogenic culture medium for the direct enumeration, without confirmation, of enterococci in water and in food products. The cleavage of the chromogenic substrate by glucosidase activity of Enterococci leads to specific blue colonies. RAPID′ *Enterococcus* totally inhibits growth of Gram-negative flora and that of practically all Gram-positive bacteria other than Enterococci, due to the combined action of temperature and selective media.

Rapid Fermentation Medium

Composition per liter:

Pancreatic digest of casein	20.0g
NaCl	5.0g
Agar	3.5g
L-Cystine	0.5g
Na_2SO_3	0.5g
Phenol Red	0.017g

pH 7.3 ± 0.2 at 25°C

Source: This medium is available as a premixed powder from BD Diagnostic Systems.

Preparation of Medium: Add components to distilled/deionized water and bring volume to 1.0L. Mix thoroughly. Distribute into tubes or flasks. Autoclave for 15 min at 15 psi pressure–121°C.

Use: For the differentiation of *Neisseria* species isolated from clinical specimens.

Rapid HiColiform Agar

Composition per liter:

Agar	15.0g
Peptone, special	5.0g
NaCl	5.0g
K_2HPO_4	2.7g
KH_2PO_4	2.0g
Sorbitol	1.0g
Sodium lauryl sulfate	0.1g
1-Isopropyl- ß-D-1-thiogalactopyranoside	0.1g
Chromogenic mixture	0.08g
Fluorogenic mixture	0.05g

pH 7.2 ± 0.2 at 25°C

Source: This medium is available as a premixed powder from HiMedia.

Preparation of Medium: Add components to distilled/deionized water and bring volume to 1.0L. Mix thoroughly. Gently heat and bring to boiling. Distribute into tubes or flasks. Autoclave for 15 min at 15 psi pressure–121°C. Pour into sterile Petri dishes or leave in tubes.

Use: For the detection and confirmation of *Escherichia coli* and total coliforms on the basis of enzyme substrate reaction from water samples, using a combination of chromogenic and fluorogenic substrates.

Rapid HiColiform Broth

Composition per liter:

Peptone, special	5.0g
NaCl	5.0g
K_2HPO_4	2.7g
KH_2PO_4	2.0g
Sorbitol	1.0g
Sodium lauryl sulfate	0.1g
IPTG	0.1g
Chromogenic substrate	0.08g
Fluorogenic substrate	0.05g

pH 6.8 ± 0.2 at 25°C

Source: This medium is available from HiMedia.

Preparation of Medium: Add components to distilled/deionized water and bring volume to 1.0L. Mix thoroughly. Gently heat and bring to boiling. Distribute into tubes or flasks. Autoclave for 15 min at 15 psi pressure–121°C.

Use: For the detection and confirmation of *Escherichia coli* and total coliforms from water samples, using a combination of chromogenic and fluorogenic substrates.

Rapid HiColiform HiVeg Agar

Composition per liter:

Agar	15.0g
Plant special peptone	5.0g
NaCl	5.0g
K_2HPO_4	2.7g
KH_2PO_4	2.0g
Sorbitol	1.0g
1-Isopropyl-ß-D-1-thiogalactopyranoside	0.1g
Sodium lauryl sulfate	0.1g
Chromogenic mixture	0.08g
Fluorogenic mixture	0.05g

pH 7.2 ± 0.2 at 25°C

Source: This medium is available as a premixed powder from HiMedia.

Preparation of Medium: Add components to distilled/deionized water and bring volume to 1.0L. Mix thoroughly. Gently heat and bring to boiling. Distribute into tubes or flasks. Autoclave for 15 min at 15 psi pressure–121°C. Pour into sterile Petri dishes or leave in tubes.

Use: For the detection and confirmation of *Escherichia coli* and total coliforms on the basis of enzyme substrate reaction from water samples, using a combination of chromogenic and fluorogenic substrates.

Rapid HiColiform HiVeg Agar

Composition per liter:

Agar	15.0g
Plant special peptone	5.0g
NaCl	5.0g
K_2HPO_4	2.7g
KH_2PO_4	2.0g
Sorbitol	1.0g
IPTG (Isopropyl-β-D-thiogalactopyranoside)	0.1g
Sodium lauryl sulfate	0.1g
Chromogenic substrate	0.08g
Fluorogenic substrate	0.05g

pH 7.2 ± 0.2 at 25°C

Source: This medium is available as a premixed powder from HiMedia.

Preparation of Medium: Add components to distilled/deionized water and bring volume to 1.0L. Mix thoroughly. Gently heat and bring to boiling. Distribute into tubes or flasks. Autoclave for 15 min at 15 psi pressure–121°C. Pour into sterile Petri dishes or leave in tubes.

Use: For the detection and confirmation of *Escherichia coli* and total coliforms on the basis of enzyme substrate reaction from water samples, using a combination of chromogenic and fluorogenic substrates.

Rapid HiColiform HiVeg Broth

Composition per liter:

Plant special peptone	5.0g
NaCl	5.0g
K_2HPO_4	2.7g
KH_2PO_4	2.0g
Sorbitol	1.0g
IPTG (Isopropyl-β-D-thiogalactopyranoside)	0.1g

Sodium lauryl sulfate 0.1g
Chromogenic substrate 0.08g
Fluorogenic substrate 0.05g
pH 7.2 ± 0.2 at 25°C

Source: This medium is available as a premixed powder from HiMedia.

Preparation of Medium: Add components to distilled/deionized water and bring volume to 1.0L. Mix thoroughly. Gently heat and bring to boiling. Distribute into tubes or flasks. Autoclave for 15 min at 15 psi pressure–121°C.

Use: For the detection and confirmation of *Escherichia coli* and total coliforms on the basis of enzyme substrate reaction from water samples, using a combination of chromogenic and fluorogenic substrates.

Rapid HiEnterococci Agar

Composition per liter:
Agar 15.0g
Peptone, special 10.0g
NaCl 5.0g
Polysorbate 80 .... 2.0g
NaN_3 .... 0.3g
Na_2HPO_4 .... 1.25g
Chromogenic mixture 0.06g
pH 7.5 ± 0.2 at 25°C

Source: This medium is available from HiMedia.

Caution: Sodium azide has a tendency to form explosive metal azides with plumbing materials. It is advisable to use enough water to flush off the disposables.

Preparation of Medium: Add components to distilled/deionized water and bring volume to 1.0L. Mix thoroughly. Gently heat and bring to boiling. Distribute into tubes or flasks. Autoclave for 15 min at 15 psi pressure–121°C. Pour into sterile Petri dishes or leave in tubes.

Use: For the rapid and easy identification and differentiation of enterococci from water samples.

RAPID′*L. mono* Medium (BAM M131a)

Composition per liter:
Peptones 30.0g
Agar B, proprietary 13.0g
D-Xylose 10.0g
LiCl 9.0g
Meat extract 5.0g
Yeast extract 1.0g
Phenol Red 0.12g
Selective supplement, proprietary 20.0g
Chromogenic substrate, proprietary 1.0mL
pH 7.3 ± 0.1 at 25°C

Source: This medium is available from Biorad.

Use: A selective chromogenic culture medium for the detection and differentiation of *Listeria* spp., including *L. ivanovii* and *L. monocytogenes*.

Rappaport Broth, Modified (Rap Broth, Modified)

Composition per 250.2mL:
Solution A 155.0mL
Solution C 53.0mL
Solution B 40.0mL
Solution D 1.6mL
Solution E 0.6mL

Solution A:
Composition per liter:
Pancreatic digest of casein 10.0g

Preparation of Solution A: Add pancreatic digest of casein to distilled/deionized water and bring volume to 1.0L. Mix thoroughly.

Solution B:
Composition per liter:
Na_2HPO_4 .... 9.5g

Preparation of Solution B: Add Na_2HPO_4 to distilled/deionized water and bring volume to 1.0L. Mix thoroughly.

Solution C:
Composition per 100.0mL:
$MgCl_2 \cdot 6H_2O$ 40.0g

Preparation of Solution C: Add $MgCl_2 \cdot 6H_2O$ to distilled/deionized water and bring volume to 100.0mL. Mix thoroughly. Autoclave for 15 min at 15 psi pressure–121°C. Cool to 25°C.

Solution D:
Composition per 100.0mL:
Malachite Green 0.2g

Preparation of Solution D: Add Malachite Green to sterile distilled/deionized water and bring volume to 100.0mL. Mix thoroughly. Do not sterilize.

Solution E:
Composition per 10.0mL:
Carbenicillin 0.01g

Preparation of Solution E: Add carbenicillin to distilled/deionized water and bring volume to 10.0mL. Mix thoroughly. Filter sterilize.

Preparation of Medium: Combine 155.0mL of solution A and 40.0mL of solution B. Mix thoroughly. Autoclave for 15 min at 15 psi pressure–121°C. Cool to 45°–50°C. Aseptically add 53.0mL of sterile solution C, 1.6mL of solution D, and 0.6mL of sterile solution E. Mix thoroughly. Aseptically distribute into sterile tubes or flasks.

Use: For the isolation and cultivation of *Yersinia enterocolitica* from foods.

Rappaport-Vassiliadis Enrichment Broth (RV Enrichment Broth)

Composition per liter:
NaCl 8.0g
Papaic digest of soybean meal 5.0g
KH_2PO_4 .... 1.6g
Magnesium chloride solution 100.0mL
Malachite Green solution 10.0mL
pH 5.2 ± 0.2 at 25°C

Source: This medium is available as a premixed powder from Oxoid Unipath.

Magnesium Chloride Solution:
Composition per 100.0mL:
$MgCl_2 \cdot 6H_2O$ 40.0g

Preparation of Magnesium Chloride Solution: Add 40.0g of $MgCl_2 \cdot 6H_2O$ to distilled/deionized water and bring volume to 100.0mL. Mix thoroughly. Autoclave for 15 min at 15 psi pressure–121°C. Cool to 45°–50°C.

Malachite Green Solution:

Composition per 10.0mL:

Malachite Green oxalate	0.04g

Preparation of Malachite Green Solution: Add Malachite Green to distilled/deionized water and bring volume to 10.0mL. Mix thoroughly. Autoclave for 15 min at 15 psi pressure–121°C. Cool to 45°–50°C.

Preparation of Medium: Add components to distilled/deionized water and bring volume to 1.0L. Mix thoroughly. Distribute into tubes in 10.0mL volumes. Autoclave for 15 min at 10 psi pressure–115°C.

Use: For the isolation and cultivation of *Salmonella* species from food and environmental specimens.

Rappaport-Vassiliadis R10 Broth

Composition per liter:

$MgCl_2$, anhydrous	13.4g
NaCl	7.2g
Papaic digest of soybean meal	4.54g
KH_2PO_4	1.45g
Malachite Green oxalate	0.036g

pH 5.1 ± 0.2 at 25°C

Preparation of Medium: Add components to distilled/deionized water and bring volume to 1.0L. Mix thoroughly. Distribute into screw-capped tubes in 10.0mL volumes. Autoclave for 15 min at 10 psi pressure–116°C.

Use: For the isolation and cultivation of *Salmonella* species from food and environmental specimens.

Rappaport-Vassiliadis Medium Semisolid, Modified with Novobiocin (MSRV)

Composition per liter:

$MgCl_2$	10.93g
NaCl	7.34g
Enzymatic digest of casein	4.59g
Casein acid hydrolysate	4.59g
Agar	2.7g
KH_2PO_4	1.4 g
Malachite Green oxalate	0.037g
Novobiocin solution	1.0mL

pH 5.6 ± 0.2 at 25°C

Novobiocin Solution:

Composition per 10.0mL:

Novobiocin	0.2g

Preparation of Novobiocin Solution: Add novobiocin to distilled/deionized water and bring volume to 10.0mL. Mix thoroughly. Filter sterilize.

Preparation of Medium: Add components, except novobiocin solutiont, to distilled/deionized water and bring volume to 990.0mL. Mix thoroughly. Gently heat and bring to boiling. Distribute into tubes or flasks. Do not autoclave. Do not overheat. Cool to 50°C. Aseptically add 1.0mL novobiocin solution. Pour into sterile Petri dishes or leave in tubes.

Use: For the rapid and sensitive isolation of motile *Salmonella* spp. from food products following pre-enrichment or selective enrichment. The semisolid medium allows motility to be detected as halos of growth around the original point of inoculation. Recommended by the European Chocolate Manufacturer's Association. For the isolation of f *Salmonella* spp. (other than *S. typhi* and *S. partyphi* type A) from stool specimens with high sensitivity and specificity.

Rappaport-Vassiliadis Soy Peptone Broth (RVS Broth)

Composition per liter:

$MgCl_2$, anhydrous	13.58g
NaCl	7.2g
Papaic digest of soybean meal	4.5g
KH_2PO_4	1.26g
K_2HPO_4	0.18g
Malachite Green	0.036g

pH 5.2 ± 0.2 at 25°C

Source: This medium is available as a premixed powder from Oxoid Unipath.

Preparation of Medium: Add components to distilled/deionized water and bring volume to 1.0L. Mix thoroughly. Distribute into screw-capped tubes in 10.0mL volumes. Autoclave for 15 min at 10 psi pressure–115°C.

Use: For the isolation and cultivation of *Salmonella* species from food and environmental specimens.

Raymond's Medium

Composition per liter:

Na_2HPO_4	3.0g
NaCl	3.0g
NH_4NO_3	2.0g
KH_2PO_4	2.0g
$MgSO_4$	0.2g
Na_2CO_3	0.1g
$MnSO_4$	0.02g
$CaCl_2$	0.01g
$FeSO_4$	0.01g
n-Hexadecane	10.0mL

pH 6.8–7.0 at 25°C

Preparation of Medium: Add components to distilled/deionized water and bring volume to 1.0L. Mix thoroughly. Distribute into tubes or flasks. Autoclave for 20 min at 15 psi pressure–121°C.

Use: For the cultivation of *Rhodococcus erythropolis, Rhodococcus luteus, Rhodococcus maris*, and other bacteria that can utilize *n*-hexadecane as a carbon source.

Razi's Medium

Composition per liter:

Beef extract	10.0g
Peptic digest of animal tissue	10.0g
Glucose	5.0g
NaCl	5.0g
Sodium acetate	3.0g
Yeast extract	3.0g
Potassium aspartate	2.0g
Starch, soluble	1.0g
Agar	0.5g
Cysteine hydrochloride	0.5g

pH 7.2 ± 0.2 at 25°C

Source: This medium is available as a premixed powder from HiMedia.

Preparation of Medium: Add components to distilled/deionized water and bring volume to 1.0L. Mix thoroughly. Gently heat and bring to boiling. Distribute into tubes or flasks. Autoclave for 15 min at 15 psi pressure–121°C.

Use: For the maintenance of cultures of *Campylobacter* spp.

RBA
See: **Diazotrophic Medium**

RB-1/RB-9 Medium

Composition per liter:

Glucose or starch	10.0g
$Na_2\ HPO_4 \cdot 12H_2O$	4.2g
KH_2PO_4	2.64g
Yeast extract	2.0g
Na_2SO_4	1.0g
NH_4Cl	0.5g
L-Cysteine·HCl	0.5g
$Na_2S \cdot 9H_2O$	0.5g
$MgCl_2 \cdot 6H_2O$	0.36g
Resazurin	1.0mg
Trace elements solution	10.0mL
Vitamin solution	10.0mL

pH 5.0–6.0 at 25°C

Vitamin Solution:

Composition per liter:

Pyridoxine·HCl	10.0mg
Calcium DL-pantothenate	5.0mg
Lipoic acid	5.0mg
Nicotinic acid	5.0mg
p-Aminobenzoic acid	5.0mg
Riboflavin	5.0mg
Thiamine·HCl	5.0mg
Biotin	2.0mg
Folic acid	2.0mg
Vitamin B_{12}	0.1mg

Preparation of Vitamin Solution: Add components to distilled/deionized water and bring volume to 1.0L. Mix thoroughly. Sparge with 100% N_2. Autoclave for 15 min at 15 psi pressure–121°C.

Trace Elements Solution:

Composition per liter:

$MgSO_4 \cdot 7H_2O$	3.0g
Nitrilotriacetic acid	1.5g
NaCl	1.0g
$MnSO_4 \cdot 2H_2O$	0.5g
$CoSO_4 \cdot 7H_2O$	0.18g
$ZnSO_4 \cdot 7H_2O$	0.18g
$CaCl_2 \cdot 2H_2O$	0.1g
$FeSO_4 \cdot 7H_2O$	0.1g
$NiCl_2 \cdot 6H_2O$	0.025g
$KAl(SO_4)_2 \cdot 12H_2O$	0.02g
$CuSO_4 \cdot 5H_2O$	0.01g
H_3BO_3	0.01g
$Na_2MoO_4 \cdot 2H_2O$	0.01g
$Na_2SeO_3 \cdot 5H_2O$	0.3mg

Preparation of Trace Elements Solution: Add nitrilotriacetic acid to 500.0mL of distilled/deionized water. Adjust pH to 6.5 with KOH. Add remaining components. Add distilled/deionized water to 1.0L.

Preparation of Medium: Prepare and dispense medium under 100% N_2. Add components to distilled/deionized water and bring volume to 1.0L. Mix thoroughly. Sparge with 100% N_2. Adjust pH to 5.0–6.0. Anaerobically distribute into tubes or flasks. Autoclave for 20 min at 15 psi pressure–121°C.

Use: For the cultivation of *Clostridium* species.

RC Agar
See: **Rippey-Cabelli Agar**

RCM Medium
See: **Reinforced Clostridial Agar**

RCM Medium, Modified

Composition per liter:

Casamino acids	15.0g
Tryptose	10.0g
Beef extract	10.0g
Glucose	5.0g
NaCl	5.0g
Yeast extract	3.0g
Sodium acetate	3.0g
Soluble starch	1.0g
L-Cysteine·HCl·H_2O	0.5g
Agar	0.5g

pH 6.8 ± 0.2 at 25°C

Source: Reinforced clostridial medium without casamino acids is available as a premixed powder from BD Diagnostic Systems and Oxoid Unipath.

Preparation of Medium: Add components to distilled/deionized water and bring volume to 1.0L. Mix thoroughly. Gently heat and bring to boiling. Distribute into tubes or flasks. Autoclave for 15 min at 10 psi pressure–115°C. Pour into sterile Petri dishes or leave in tubes.

Use: For the cultivation of *Clostridium acetireducens* and *Clostridium aminophilum.*

R-CW Medium (DSMZ Medium 775)

Composition per 1001.0mL:

Trypticase™	5.0g
Tryptone	5.0g
Yeast extract	5.0g
KH_2PO_4	5.0g
Na-acetate	5.0g
$(NH_4)_2$ citrate	2.0g
$MgSO_4 \cdot 7H_2O$	0.5g
$MnSO_4 \cdot 2H_2O$	0.2g
Cheese whey	1000.0mL
Tween™ 80	1.0mL

pH 5.5 ± 0.2 at 25°C

Preparation of Medium: Add components to 1.0L cheese whey. Mix thoroughly. Adjust pH to 5.5. Distribute into tubes or flasks. Autoclave for 15 min at 15 psi pressure–121°C.

Use: For the cultivation of *Lactobacillus kefirgranum* and *Lactobacillus parakefiri.*

R-CW Medium (LMG Medium 265)

Composition per liter:

Trypticase™	5.0g
Tryptone	5.0g

Yeast extract	5.0g
KH_2PO_4	5.0g
Na-acetate	5.0g
$(NH_4)_2$citrate	2.0g
$MgSO_4 \cdot 7H_2O$	0.5g
$MnSO_4 \cdot 5H_2O$	0.2g
Cheese whey	1.0L
Tween™ 80	1.0mL

pH 5.5 ± 0.2 at 25°C

Preparation of Medium: Add components to 1.0L cheese whey. Mix thoroughly. Distribute into tubes or flasks. Autoclave for 15 min at 15 psi pressure–121°C.

Use: For cultivation and maintenance of *Lactobacillus kefirgranum* and *Lactobacillus parakefiri.*

RE-101 Medium (DSMZ Medium 1130)

Composition per liter:

NaCl	175.0g
$MgCl_2 \cdot 6H_2O$	20.0g
K_2SO_4	5.0g
Yeast extract	5.0g
$CaCl_2 \cdot 2H_2O$	0.1g

pH 7.2 ± 0.2 at 25°C

Preparation of Medium: Add components to distilled/deionized water and bring volume to 1.0L. Mix thoroughly. Adjust pH to 7.0. Gently heat and bring to boiling. Autoclave for 15 min at 15 psi pressure–121°C. Pour into sterile Petri dishes or distribute into sterile tubes.

Use: For the cultivation of *Haloplanus natans.*

Reactivation with Liquid Medium 246 (DSMZ Medium 246a)

Composition per liter:

Peptone	10.0g
Yeast extract	10.0g
Artificial seawater, filtered	750.0mL

pH 7.3 ± 0.2 at 25°C

Artificial Seawater:

Composition per liter:

NaCl	28.13g
$MgCl_2 \cdot 6H_2O$	4.8g
$MgSO_4 \cdot 7H_2O$	3.5g
$CaCl_2 \cdot 2H_2O$	1.6g
KCl	0.77g
$NaHCO_3$	0.11g

Preparation of Artificial Seawater: Add components to distilled/deionized water and bring volume to 1.0L. Mix thoroughly. Filter sterilize.

Preparation of Medium: Add peptone and yeast extract to 250.0mL tap water. Mix thoroughly. Adjust pH to 7.8. Boil for 5 min. Filter and readjust the pH to 7.3. Distribute into tubes or flasks. Autoclave for 15 min at 15 psi pressure–121°C. Cool to 25°C. Add 750.0mL filter sterilized seawater. (Note: Natural seawater is stored in the dark for at least 3 weeks to age. If natural seawater is not available use artificial seawater.) Mix thoroughly.

Use: For the rehydration and cultivation of *Photorhabdus luminescens=Xenorhabdus luminescens* and marine bacteria from lyophilized ampules of stock cultures.

Reactivation with Tryptone Soya Broth (DSMZ Medium 220a)

Composition per liter:

Peptone from casein	17.0g
Peptone from soymeal	3.0g
NaCl	5.0g
D(+)-Glucose	2.5g
K_2HPO_4	2.5g

pH 7.3 ± 0.2 at 25°C

Preparation of Medium: Add components to distilled/deionized water and bring volume to 1.0L. Mix thoroughly. Gently heat and bring to boiling. Distribute into tubes or flasks. Autoclave for 15 min at 15 psi pressure–121°C.

Use: For the reactivation and cultivation of *Aeromonas* spp., *Pseudomonas* spp., *Burkholderia* spp., *Burkholderia phenazinium (Pseudomonas phenazinium), Plesiomonas shigelloides,* and *Sphingobium chlorophenolicum=Sphingomonas chlorophenolica.*

Reddy's Differential Agar, Modified (Lactic Streak HiVeg Agar)

Composition per liter:

Agar	15.0g
Sodium carboxymethylcellulose	10.0g
Calcium citrate	10.0g
Plant peptone	5.0g
Papaic digest of soybean meal	5.0g
Yeast extract	5.0g
Plant extract	5.0g
Lactose	1.5g
L-Arginine·HCl	1.5g
Bromcresol Purple	2.0mg

pH 6.0 ± 0.2 at 25°C

Source: This medium is available as a premixed powder from HiMedia.

Preparation of Medium: Add components to distilled/deionized water and bring volume to 1.0L. Mix thoroughly. Gently heat and bring to boiling. Distribute into tubes or flasks. Autoclave for 10 min at 10 psi pressure–115°C. Pour into sterile Petri dishes or leave in tubes.

Use: For the qualitative and quantitative differentiation of lactic streptococci.

Reduced Salt Solution Medium (RSS Medium)

Composition per liter:

$CaCl_2 \cdot H_2O$	20.0g
$NaHCO_3$	10.0g
Dithiothreitol	2.0g
$MgSO_4 \cdot 7H_2O$	2.0g
K_2HPO_4	1.0g
KH_2PO_4	1.0g
NaCl	0.2g

pH 9.2 ± 0.2 at 25°C

Preparation of Medium: Add components to distilled/deionized water and bring volume to 1.0L. Mix thoroughly. Distribute into screw-capped tubes or flasks. Autoclave for 15 min at 15 psi pressure–121°C.

Use: For the transport and isolation of bacteria from dental plaque, especially *Streptococcus mutans, Streptococcus sanguis*, and *Lactobacillus* species.

Reduced Transport Fluid

Composition per liter:

$(NH_4)_2SO_4$	9.0g
NaCl	9.0g
K_2HPO_4	4.5g
KH_2PO_4	4.5g
Na_2CO_3	4.0g
EDTA (ethylenediamine tetraacetic acid)	3.8g
Dithiothreitol	2.0g
$MgSO_4 \cdot 7H_2O$	1.8g

pH 8.0 ± 0.2 at 25°C

Preparation of Medium: Add components to distilled/deionized water and bring volume to 1.0L. Mix thoroughly. Filter sterilize. Aseptically distribute into sterile tubes with rubber stoppers.

Use: For the transport and isolation of bacteria from dental plaque, especially *Streptococcus mutans* and *Streptococcus sanguis*. Also used for the cultivation of a variety of Gram-positive bacteria from the oral cavity, especially streptococci, actinomycetes, lactobacilli, clostridia, *Bacteroides* species, *Fusobacterium* species, and *Veillonella* species.

Reduced Transport Fluid

Composition per liter:

Stock mineral salt solution No. 1	75.0mL
Stock mineral salt solution No. 2	75.0mL
Dithiothreitol (1% solution)	20.0mL
Ethylenediamine tetraacetic acid (1*M* solution)	10.0mL
Na_2CO_3 (8% solution)	5.0mL
Resazurin (0.1% solution)	1.0mL

pH 8.0 ± 0.2 at 25°C

Stock Mineral Salt Solution No. 1:
Composition per 100.0mL:

K_2HPO_4	0.6g

Preparation of Stock Mineral Salt Solution No. 1: Add K_2HPO_4 to distilled/deionized water and bring volume to 100.0mL. Mix thoroughly.

Stock Mineral Salt Solution No. 2:
Composition per 100.0mL:

NaCl	1.2g
$(NH_4)_2SO_4$	1.2g
K_2HPO_4	0.6g
$MgSO_4 \cdot 7H_2O$	0.25g

Preparation of Stock Mineral Salt Solution No. 2: Add components to distilled/deionized water and bring volume to 100.0mL. Mix thoroughly.

Preparation of Medium: Add components to distilled/deionized water and bring volume to 1.0L. Mix thoroughly. Filter sterilize. Aseptically distribute into sterile tubes with rubber stoppers.

Use: For the transport and isolation of bacteria from dental plaque, especially *Streptococcus mutans* and *Streptococcus sanguis*. Also used for the cultivation of a variety of Gram-positive bacteria from the oral cavity, especially streptococci, actinomycetes, lactobacilli, clostrida, *Bacteroides*, Fusobacteria, and *Veillonela*.

Regan-Lowe Charcoal Agar (Regan-Lowe Medium)

Composition per liter:

Agar	12.0g
Beef extract	10.0g
Pancreatic digest of gelatin	10.0g
Soluble starch	10.0g
NaCl	5.0g
Charcoal	4.0g
Niacin	0.01g
Horse blood, defibrinated	100.0mL
Cephalexin solution	10.0mL

pH 7.4 ± 0.2 at 25°C

Source: This medium is available as a premixed powder from BD Diagnostic Systems.

Cephalexin Solution:
Composition per 10.0mL:

Cephalexin	0.04g

Preparation of Cephalexin Solution: Add cephalexin to distilled/deionized water and bring volume to 10.0mL. Mix thoroughly. Filter sterilize.

Preparation of Medium: Add components, except horse blood and cephalexin solution, to distilled/deionized water and bring volume to 890.0mL. Mix thoroughly. Gently heat and bring to boiling. Autoclave for 15 min at 15 psi pressure–121°C. Cool to 45°–50°C. Aseptically add sterile horse blood and sterile cephalexin solution. Mix thoroughly. Pour into sterile Petri dishes or distribute into sterile tubes. Swirl medium while dispensing to keep charcoal in suspension.

Use: For the selective isolation and cultivation of *Bordetella pertussis* and *Bordetella parapertussis* from clinical specimens.

Regan-Lowe Semisolid Transport Medium

Composition per liter:

Agar	6.0g
Beef extract	5.0g
Pancreatic digest of gelatin	5.0g
Soluble starch	5.0g
NaCl	2.5g
Charcoal	2.0g
Niacin	0.01g
Horse blood, defibrinated	100.0mL
Cephalexin solution	10.0mL

pH 7.4 ± 0.2 at 25°C

Cephalexin Solution:
Composition per 10.0mL:

Cephalexin	0.04g

Preparation of Cephalexin Solution: Add cephalexin to distilled/deionized water and bring volume to 10.0mL. Mix thoroughly. Filter sterilize.

Preparation of Medium: Add components, except horse blood and cephalexin solution, to distilled/deionized water and bring volume to 890.0mL. Mix thoroughly. Gently heat and bring to boiling. Autoclave for 15 min at 15 psi pressure–121°C. Cool to 45°–50°C. Aseptically add sterile horse blood and sterile cephalexin solution. Mix thoroughly. Aseptically distribute into small, sterile, screw-capped tubes. Fill tubes half-full. Swirl medium while dispensing to keep charcoal in suspension.

Use: For the transport of *Bordetella pertussis* and *Bordetella parapertussis* isolated from clinical specimens.

Reinforced AE Medium (RAE Medium) (LMG Medium 239)

Composition per liter:

Base medium	500.0mL
Growth medium	500.0mL

pH 5.0 ± 0.2 at 25°C

Base Medium:

Composition per liter:

Agar	10.0g

Preparation of Base Medium: Add agar to distilled/deionized water and bring volume to 1.0L. Mix thoroughly. Gently heat and bring to boiling. Autoclave for 15 min at 15 psi pressure–121°C. Pour as a base layer into sterile Petri dishes.

Growth Medium:

Composition per liter:

Glucose	40.0g
Agar	20.0g
Yeast extract	10.0g
Peptone	10.0g
$Na_2HPO_4 \cdot 2H_2O$	3.38g
Citric acid·$2H_2O$	1.5g
Ethanol	20.0mL
Acetic acid	10.0mL

Preparation of Growth Medium: Add components, except ethanol and acetic acid, to distilled/deionized water and bring volume to 970.0mL. Mix thoroughly. Gently heat and bring to boiling. Autoclave for 15 min at 15 psi pressure–121°C. Cool to 45°–50°C. Aseptically add 20.0mL filter sterilized ethanol and 10.0mL filter sterilized acetic acid. Mix thoroughly.

Preparation of Growth Medium: This medium is used as a double layer. Pour as a layer of base medium into sterile Petri dishes. Allow to solidify. Pour a thin layer of growth medium over the solid base medium. Allow to solidify.

Use: For the isolation and cultivation of *Gluconacetobacter* spp. and *Acetobacter pomorum.*

Reinforced Clostridial Agar

Composition per liter:

Agar	13.5g
Beef extract	10.0g
Pancreatic digest of casein	10.0g
NaCl	5.0g
Glucose	5.0g
Yeast extract	3.0g
Sodium acetate	3.0g
Soluble starch	1.0g
L-Cysteine·HCl·H_2O	0.5g

pH 6.8 ± 0.2 at 25°C

Source: This medium is available as a premixed powder from BD Diagnostic Systems and Oxoid Unipath.

Preparation of Medium: Add components to distilled/deionized water and bring volume to 1.0L. Mix thoroughly. Gently heat and bring to boiling. Distribute into tubes or flasks. Autoclave for 15 min at 10 psi pressure–115°C. Pour into sterile Petri dishes or leave in tubes.

Use: For the cultivation and enumeration of *Clostridium* species, *Bifidobacterium* species, other anaerobes (e.g., lactobacilli), and facultative organisms from clinical specimens and foods.

Reinforced Clostridial Agar with Tween™ (LMG Medium 146)

Composition per liter:

Agar	13.5g
Beef extract	10.0g
Pancreatic digest of casein	10.0g
NaCl	5.0g
Glucose	5.0g
Yeast extract	3.0g
Sodium acetate	3.0g
Tween™ 80	1.0g
Soluble starch	1.0g
L-Cysteine·HCl·H_2O	0.5g

pH 6.8 ± 0.2 at 25°C

Preparation of Medium: Add components to distilled/deionized water and bring volume to 1.0L. Mix thoroughly. Gently heat and bring to boiling. Distribute into tubes or flasks. Autoclave for 15 min at 10 psi pressure–115°C. Pour into sterile Petri dishes or leave in tubes.

Use: For the cultivation and maintenance of *Bifidobacterium merycicum.*

Reinforced Clostridial HiVeg Agar

Composition per liter:

Agar	13.5g
Plant extract	10.0g
Plant hydrolysate	10.0g
Glucose	5.0g
NaCl	5.0g
Yeast extract	3.0g
Sodium acetate	3.0g
Starch, soluble	1.0g
L-Cysteine·HCl	0.5g

pH 6.8 ± 0.2 at 25°C

Source: This medium is available as a premixed powder from Hi-Media.

Preparation of Medium: Add components to distilled/deionized water and bring volume to 1.0L. Mix thoroughly. Gently heat and bring to boiling. Distribute into tubes or flasks. Autoclave for 15 min at 10 psi pressure–115°C. Pour into sterile Petri dishes or leave in tubes.

Use: For the cultivation and enumeration of clostridia and other anaerobes.

Reinforced Clostridial HiVeg Broth

Composition per liter:

Plant extract	10.0g
Plant hydrolysate	10.0g
Glucose	5.0g
NaCl	5.0g
Sodium acetate	3.0g
Yeast extract	3.0g

Starch, soluble	1.0g
Agar	0.5g
L-Cysteine·HCl	0.5g

pH 6.8 ± 0.2 at 25°C

Source: This medium is available as a premixed powder from HiMedia.

Preparation of Medium: Add components to distilled/deionized water and bring volume to 1.0L. Mix thoroughly. Gently heat and bring to boiling. Distribute into tubes or flasks. Autoclave for 15 min at 10 psi pressure–115°C.

Use: For the cultivation and enumeration of clostridia and other anaerobes.

Reinforced Clostridial Medium

Composition per liter:

Tryptose	10.0g
Beef extract	10.0g
Glucose	5.0g
NaCl	5.0g
Yeast extract	3.0g
Sodium acetate	3.0g
Soluble starch	1.0g
L-Cysteine·HCl·H_2O	0.5g
Agar	0.5g

pH 6.8 ± 0.2 at 25°C

Source: This medium is available as a premixed powder from BD Diagnostic Systems and Oxoid Unipath.

Preparation of Medium: Add components to distilled/deionized water and bring volume to 1.0L. Mix thoroughly. Gently heat and bring to boiling. Distribute into tubes or flasks. Autoclave for 15 min at 10 psi pressure–115°C. Pour into sterile Petri dishes or leave in tubes.

Use: For the nonselective cultivation and enumeration of *Clostridium* species, other anaerobes such as lactobacilli, and facultative organisms from clinical specimens and foods.

Reinforced Clostridial Medium with Casamino Acids

Composition per liter:

Casamino acids	15.0g
Agar	13.5g
Beef extract	10.0g
Pancreatic digest of casein	10.0g
NaCl	5.0g
Glucose	5.0g
Yeast extract	3.0g
Sodium acetate	3.0g
Soluble starch	1.0g
L-Cysteine·HCl·H_2O	0.5g

pH 6.8 ± 0.2 at 25°C

Preparation of Medium: Add components to distilled/deionized water and bring volume to 1.0L. Mix thoroughly. Distribute into tubes or flasks. Autoclave for 15 min at 15 psi pressure–121°C.

Use: For the cultivation of *Clostridium aminophilum.*

Reinforced Clostridial Medium with Glycerol

Composition per liter:

Agar	13.5g
Beef extract	10.0g
Pancreatic digest of casein	10.0g
NaCl	5.0g
Glucose	5.0g
Glycerol	5.0g
Yeast extract	3.0g
Sodium acetate	3.0g
Soluble starch	1.0g
L-Cysteine·HCl·H_2O	0.5g

pH 6.8 ± 0.2 at 25°C

Preparation of Medium: Add components to distilled/deionized water and bring volume to 1.0L. Mix thoroughly. Gently heat and bring to boiling. Distribute into tubes or flasks. Autoclave for 15 min at 10 psi pressure–115°C. Pour into sterile Petri dishes or leave in tubes.

Use: For the cultivation of *Anaerovibrio glycerini.*

Reinforced Clostridial Medium, Modified (ATCC Medium 2107)

Composition per liter:

Tryptose	10.0g
Beef extract	10.0g
Glucose	5.0g
NaCl	5.0g
Yeast extract	3.0g
Sodium acetate	3.0g
Soluble starch	1.0g
L-Cysteine·HCl·H_2O	0.5g

pH 6.8 ± 0.2 at 25°C

Preparation of Medium: Add components to distilled/deionized water and bring volume to 1.0L. Mix thoroughly. Gently heat and bring to boiling. Distribute into tubes or flasks. Autoclave for 15 min at 10 psi pressure–115°C. Pour into sterile Petri dishes or leave in tubes.

Use: For the cultivation of *Clostridium saccharobutylicum, Clostridium frigidicarnis,* and *Mitsuokella jalaludinii.*

Reinforced Clostridial Medium with Sodium Lactate

Composition per liter:

Tryptose	10.0g
Beef extract	10.0g
Glucose	5.0g
NaCl	5.0g
Yeast extract	3.0g
Sodium acetate	3.0g
Soluble starch	1.0g
L-Cysteine·HCl·H_2O	0.5g
Agar	0.5g
Sodium lactate (60% solution)	15.0mL

pH 6.8 ± 0.2 at 25°C

Preparation of Medium: Add components to distilled/deionized water and bring volume to 1.0L. Mix thoroughly. Gently heat and bring to boiling. Distribute into tubes or flasks. Autoclave for 15 min at 10 psi pressure–115°C. Pour into sterile Petri dishes or leave in tubes.

Use: For the nonselective cultivation and enumeration of *Clostridium* species, other anaerobes such as lactobacilli, and facultative organisms from clinical specimens and foods.

Reinforced Clostridial Medium with Uric Acid

Composition per liter:

Agar	13.5g
Beef extract	10.0g
Pancreatic digest of casein	10.0g
NaCl	5.0g

Glucose	5.0g
Uric acid	3.0g
Yeast extract	3.0g
Sodium acetate	3.0g
Soluble starch	1.0g
L-Cysteine·HCl·H_2O	0.5g

pH 6.8 ± 0.2 at 25°C

Preparation of Medium: Add components to distilled/deionized water and bring volume to 1.0L. Mix thoroughly. Distribute into tubes or flasks. Autoclave for 15 min at 15 psi pressure–121°C.

Use: For the cultivation of *Clostridium acidurici.*

Renibacterium KDM2 Medium

Composition per liter:

Agar	15.0g
Peptone	10.0g
L-Cysteine·HCl·H_2O	1.0g
Yeast extract	0.5g
Fetal calf serum	200.0mL

pH 6.5 ± 0.2 at 25°C

Preparation of Medium: Add components, except fetal calf serum and agar, to distilled/deionized water and bring volume to 800.0mL. Mix thoroughly. Adjust pH to 6.5 with NaOH. Add agar. Gently heat while stirring and bring to boiling. Autoclave for 15 min at 15 psi pressure–121°C. Cool to 45°–50°C. Aseptically add fetal calf serum. Mix thoroughly. Pour into sterile Petri dishes or distribute into sterile tubes.

Use: For the cultivation and maintenance of *Renibacterium salmoninarum.*

Reuters Sorbic Acid Agar Base

Composition per liter:

D-Glucose	20.0g
Agar	16.0g
Casein enzymic hydrolysate	10.0g
Meat extract	10.0g
Yeast extract	5.0g
Sodium acetate	5.0g
Sodium citrate	3.0g
Tween 80	1.0g
$MgSO_4·7H_2O$	0.2g
$MnSO_4·2H_2O$	0.05g
Selective supplement solution	10.0mL

pH 5.0 ± 0.2 at 25°C

Source: This medium is available from HiMedia.

Selective Supplement Solution:
Composition per 10.0mL:

Sorbic acid	0.4g

Preparation of Selective Supplement Solution: Add sorbic acid to distilled/deionized water and bring volume to 10.0mL. Mix thoroughly. Filter sterilize.

Preparation of Medium: Add components, except selective supplement solution, to distilled/deionized water and bring volume to 990.0mL. Mix thoroughly. Gently heat and bring to boiling. Mix to dissolve components completely. Cool to 50°C. Aseptically add selective supplement solution. Mix thoroughly. Sterilize for 30 min at 0 psi pressure–100°C. Pour into Petri dishes or aseptically distribute into sterile tubes.

Use: For the isolation and differentiation of lactobacilli from foodstuffs, feces, etc.

RF Medium

Composition per liter:

Yeast extract	0.05g
Peptone	0.05g
$(NH_4)_2SO_4$	0.05g
L-Cysteine·HCl·H_2O	0.05g
Salt solution	50.0mL
Rumen fluid, clarified	30.0mL
Resazurin (1% solution)	0.1mL

pH 7.4 ± 0.2 at 25°C

Salt Solution:
Composition per liter:

$NaHCO_3$	10.0g
NaCl	2.0g
K_2HPO_4	1.0g
KH_2PO_4	1.0g
$CaCl_2$, anhydrous	0.2g
$MgSO_4$	0.2g

Preparation of Salts Solution: Add $CaCl_2$ and $MgSO_4$ to 300.0mL of distilled/deionized water. Mix thoroughly until dissolved. Bring volume to 800.0mL with distilled/deionized water. Add remaining components while stirring. Bring volume to 1.0L. Mix thoroughly. Store at 4°C.

Preparation of Medium: Add components to distilled/deionized water and bring volume to 1.0L. Mix thoroughly. Adjust pH to 6.2–6.3 with 4*N* HCl. Gently heat and bring to boiling under 100% N_2. Anaerobically distribute into tubes in 7.0mL volumes. Cap with rubber stoppers. Place tubes in a press. Autoclave for 20 min at 15 psi pressure–121°C with fast exhaust. The pH of the medium should be 7.4 after autoclaving.

Use: For the cultivation and maintenance of *Treponema bryantii.*

RFC Agar

See: **Rumen Fluid Cellobiose Agar**

RGCA Medium (Rumen Fluid Glucose Cellobiose Agar)

Composition per 300.3mL:

Rumen fluid	120.0mL
Solution IV	65.0mL
Mineral solution I	45.0mL
Mineral solution II	45.0mL
Na_2CO_3 solution	20.0mL
L-Cysteine·HCl·H_2O solution	5.0mL
Solution III	0.3mL

pH 6.6 ± 0.2 at 25°C

Mineral Solution I:
Composition per 100.0mL:

K_2HPO_4	0.3g

Preparation of Mineral Solution I: Add K_2HPO_4 to distilled/deionized water and bring volume to 100.0mL. Mix thoroughly.

Mineral Solution II:
Composition per 100.0mL:

$(NH_4)_2SO_4$	0.6g
NaCl	0.6g

KH_2PO_4 .... 0.3g
$MgSO_4$ 0.06g
$CaCl_2$ 0.06g

Preparation of Mineral Solution II: Add components to distilled/deionized water and bring volume to 100.0mL. Mix thoroughly.

Solution III:

Composition per 10.0mL:

Resazurin 0.01g

Preparation of Solution III: Add resazurin to 10.0mL of distilled/deionized water. Mix thoroughly.

Solution IV:

Composition per 65.0mL:

Agar 4.5g
Glucose 0.6g
Cellobiose 0.6g

Preparation of Solution IV: Add components to distilled/deionized water and bring volume to 65.0mL. Mix thoroughly.

L-Cysteine·HCl·H_2O Solution:

Composition per 100.0mL:

L-Cysteine·HCl·H_2O 3.0g

Preparation of L-Cysteine·HCl·H_2O Solution: Add L-cysteine·HCl·H_2O to distilled/deionized water and bring volume to 100.0mL. Mix thoroughly. Filter sterilize.

Na_2CO_3 Solution:

Composition per 100.0mL:

Na_2CO_3 .... 6.0g

Preparation of Na_2CO_3 Solution: Add Na_2CO_3 to distilled/deionized water and bring volume to 100.0mL. Mix thoroughly. Filter sterilize.

Rumen Fluid:

Composition per 120.0mL:

Rumen fluid 120.0mL

Preparation of Rumen Fluid: Filter rumen contents, obtained from a cow on an alfalfa-hay concentrate ration, through two layers of cheesecloth to remove larger particles. Store under CO_2 in quart milk bottles in the refrigerator. Much of the particulate matter settles out. Use the supernatant fluid.

Preparation of Medium: Combine 45.0mL of mineral solution I, 45.0mL of mineral solution II, 0.3mL of solution III, and 65.0mL of solution IV in a 500.0mL flask. Gently heat and bring to boiling. Add 120.0mL of rumen fluid. Gently heat and bring to boiling under 100% CO_2. Cap with a rubber stopper and wire the stopper secure. Autoclave for 20 min at 15 psi pressure–121°C. Cool to 45°–50°C. Remove stopper and gas with 100% CO_2 to eliminate O_2. Aseptically add 5.0mL of sterile L-cysteine·HCl·H_2O solution and 20.0mL of sterile Na_2CO_3 solution. Mix thoroughly. Aseptically and anaerobically distribute into tubes under 100% CO_2 in 6.0mL volumes. Cap with rubber stoppers.

Use: For the cultivation and maintenance of *Ruminococcus albus, Ruminococcus flavifaciens,* and *Succinimonas amylolytica.*

Rhamnose Salts Medium

Composition per liter:

Rhamnose 10.0g
Yeast extract 3.0g
K_2HPO_4 .... 2.9g
KH_2PO_4 .... 2.1g
$NH_4 \cdot Cl$ 2.0g
$MgSO_4 \cdot 7H_2O$ 0.4g
NaCl 30.0mg
$CaCl_2$ 3.0mg
$FeSO_4 \cdot 7H_2O$ 1.0mg

pH 7.0 ± 0.2 at 25°C

Preparation of Medium: Add components to distilled/deionized water and bring volume to 1.0L. Mix thoroughly. Distribute into tubes or flasks. Autoclave for 15 min at 15 psi pressure–121°C.

Use: For the cultivation of *Rhodococcus chlorophenolicus.*

Rhizobium Agar (LMG 201)

Composition per liter:

Agar 20.0g
Mannitol 10.0g
Yeast extract 1.0g
Sodium glutamate 0.5g
KH_2PO_4 .... 0.5g
$MgSO_4 \cdot 7H_2O$ 0.1g
$CaCl_2 \cdot 2H_2O$ 40.0mg
$FeCl_3$ 4.0mg

pH 6.8 ± 0.2 at 25°C

Preparation of Medium: Add components to distilled/deionized water and bring volume to 1.0L. Mix thoroughly. Adjust pH to 6.8. Gently heat and bring to boiling. Distribute into tubes or flasks. Autoclave for 15 min at 15 psi pressure–121°C. Pour into sterile Petri dishes or leave in tubes.

Use: For the cultivation and maintenance of *Rhizobium fredii, Rhizobium galegae, Rhizobium leguminosarum, Rhizobium loti, Rhizobium meliloti,* and *Rhizobium tropici.*

Rhizobium BIII Defined Agar

Composition per liter:

Agar 13.0g
Mannitol 10.0g
Sodium glutamate 1.1g
K_2HPO_4 .... 0.23g
$MgSO_4 \cdot 7H_2O$ 0.1g
Trace elements stock 1.0mL
Vitamin stock 1.0mL

pH 7.0 ± 0.2 at 25°C

Trace Elements Stock:

Composition per liter:

Nitrilotriacetic acid 7.0g
$CaCl_2 \cdot 2H_2O$ 6.62g
H_3BO_3 .... 0.145g
$FeSO_4 \cdot 7H_2O$ 0.125g
Na_2MoO_4 .... 0.125g
$ZnSO_4 \cdot 7H_2O$ 0.108g
$CoSO_4 \cdot 7H_2O$ 0.07g
$CuSO_4 \cdot 5H_2O$ 5.0mg
$MnCl_2 \cdot 4H_2O$ 4.3mg

Preparation of Trace Elements Stock: Add components to 500.0mL of distilled/deionized water in the order: $CaCl_2 \cdot 2H_2O$, H_3BO_3, $FeSO_4 \cdot 7H_2O$, $CoSO_4 \cdot 7H_2O$, $CuSO_4 \cdot 5H_2O$, $MnCl_2 \cdot 4H_2O$, $ZnSO_4 \cdot 7H_2O$, and Na_2MoO_4. Adjust pH to 5.0. Add nitrilotriacetic acid. Bring volume to 1.0L with distilled/deionized water.

Vitamin Stock:

Composition per liter:

Inositol	0.12g
p-Aminobenzoic acid	0.02g
Biotin	0.02g
Calcium pantothenate	0.02g
Nicotinic acid	0.02g
Pyridoxine·HCl	0.02g
Riboflavin	0.02g
Thiamine·HCl	0.02g
Sodium phosphate buffer (50.0m*M* solution, pH 7.0)	1.0L

Preparation of Vitamin Stock: Combine components. Mix thoroughly. Filter sterilize. Store at 4°C in the dark.

Preparation of Medium: Add components, except vitamin stock, to distilled/deionized water and bring volume to 999.0mL. Mix thoroughly. Gently heat and bring to boiling. Autoclave for 15 min at 15 psi pressure–121°C. Cool to 45°–50°C. Aseptically add 1.0mL of sterile vitamin stock. Mix thoroughly. Pour into sterile Petri dishes or distribute into sterile tubes.

Use: For the isolation and cultivation of *Rhizobium* species from root nodules.

Rhizobium BIII Defined Broth

Composition per liter:

Mannitol	10.0g
Sodium glutamate	1.1g
K_2HPO_4	0.23g
$MgSO_4 \cdot 7H_2O$	0.1g
Trace elements stock	1.0mL
Vitamin stock	1.0mL

pH 7.0 ± 0.2 at 25°C

Trace Elements Stock:

Composition per liter:

Nitrilotriacetic acid	7.0g
$CaCl_2 \cdot 2H_2O$	6.62g
H_3BO_3	0.145g
$FeSO_4 \cdot 7H_2O$	0.125g
Na_2MoO_4	0.125g
$ZnSO_4 \cdot 7H_2O$	0.108g
$CoSO_4 \cdot 7H_2O$	0.07g
$CuSO_4 \cdot 5H_2O$	5.0mg
$MnCl_2 \cdot 4H_2O$	4.3mg

Preparation of Trace Elements Stock: Add components, except nitrilotriacetic acid, to 500.0mL of distilled/deionized water in the order listed. Adjust pH to 5.0. Add nitrilotriacetic acid. Bring volume to 1.0L with distilled/deionized water.

Vitamin Stock:

Composition per liter:

Inositol	0.12g
p-Aminobenzoic acid	0.02g
Biotin	0.02g
Calcium pantothenate	0.02g
Nicotinic acid	0.02g
Pyridoxine·HCl	0.02g
Riboflavin	0.02g
Thiamine·HCl	0.02g
Sodium phosphate buffer (50.0m*M* solution, pH 7.0)	1.0L

Preparation of Vitamin Stock: Combine components. Mix thoroughly. Filter sterilize. Store at 4°C in the dark.

Preparation of Medium: Add components, except vitamin stock, to distilled/deionized water and bring volume to 999.0mL. Mix thoroughly. Autoclave for 15 min at 15 psi pressure–121°C. Cool to 25°C. Aseptically add 1.0mL of sterile vitamin stock. Mix thoroughly. Aseptically distribute into sterile tubes or flasks.

Use: For the isolation and cultivation of *Rhizobium* species.

Rhizobium japonicum Agar

Composition per liter:

Agar	15.0g
Mannitol	10.0g
Yeast extract	1.0g
Soil extract	200.0mL

Soil Extract:

Composition per liter:

African Violet soil	77.0g
Na_2CO_3	0.2g

Preparation of Soil Extract: Add components to 1.0L of tap water. Autoclave for 15 min at 15 psi pressure–121°C. Filter through Whatman filter paper. Bring volume to 1.0L with tap water.

Preparation of Medium: Add components to distilled/deionized water and bring volume to 1.0L. Mix thoroughly. Gently heat and bring to boiling. Distribute into tubes or flasks. Autoclave for 15 min at 15 psi pressure–121°C. Pour into sterile Petri dishes or leave in tubes.

Use: For the cultivation and maintenance of *Bradyrhizobium japonicum.*

Rhizobium Medium 1

Composition per liter:

Agar	15.0g
Yeast extract	10.0g
K_2HPO_4	0.5g
$MgSO_4 \cdot 7H_2O$	0.2g
NaCl	0.2g
$FeCl_3 \cdot 6H_2O$	0.002g

pH 7.2 ± 0.2 at 25°C

Preparation of Medium: Add components, except agar, to distilled/deionized water and bring volume to 1.0L. Mix thoroughly. Adjust pH to 7.2. Add agar. Gently heat and bring to boiling. Distribute into tubes or flasks. Autoclave for 15 min at 15 psi pressure–121°C. Pour into sterile Petri dishes or leave in tubes.

Use: For the cultivation of members of the Rhizobiaceae.

Rhizobium Medium 2

Composition per liter:

Agar	15.0g
Glycerol	4.6g
$CaSO_4$	1.3g
K_2HPO_4	1.0g
L-Arabinose	1.0g
Yeast extract	1.0g
KNO_3	0.7g
$MgSO_4 \cdot 7H_2O$	0.36g
$FeCl_3 \cdot 6H_2O$	4.0mg

pH 7.2 ± 0.2 at 25°C

Preparation of Medium: Add components, except agar, to distilled/deionized water and bring volume to 1.0L. Mix thoroughly. Adjust pH to 7.2. Add agar. Gently heat and bring to boiling. Distribute into tubes or flasks. Autoclave for 15 min at 15 psi pressure–121°C. Pour into sterile Petri dishes or leave in tubes.

Use: For the cultivation of members of the Rhizobiaceae.

Rhizobium X Medium

Composition per liter:

Agar	15.0g
Mannitol	10.0g
Yeast extract	1.0g
Soil extract	200.0mL

pH 7.2 ± 0.2 at 25°C

Soil Extract:

Composition per 200.0mL:

African Violet soil	77.0g
Na_2CO_3	0.2g

Preparation of Soil Extract: Add components to tap water and bring volume to 200.0mL. Autoclave for 60 min at 15 psi pressure–121°C. Filter through Whatman #1 filter paper.

Preparation of Medium: Add components to distilled/deionized water and bring volume to 1.0L. Mix thoroughly. Gently heat and bring to boiling. Distribute into tubes or flasks. Autoclave for 15 min at 15 psi pressure–121°C. Pour into sterile Petri dishes or leave in tubes.

Use: For the cultivation and maintenance of *Bradyrhizobium japonicum, Rhizobium* species, and *Sinorhizobium xinjiangensis*.

Rhizobium X Medium with Thiram

Composition per liter:

Agar	15.0g
Mannitol	10.0g
Yeast extract	1.0g
Soil extract	200.0mL
Thiram solution	10.0mL

pH 7.2 ± 0.2 at 25°C

Thiram Solution:

Composition per 10.0mL:

Thiram	1.0mg
Ethanol, absolute	10.0mL

Preparation of Thiram Solution: Add thiram to 10.0mL of absolute ethanol. Mix thoroughly. Filter sterilize.

Soil Extract:

Composition per 200.0mL:

African Violet soil	77.0g
Na_2CO_3	0.2g

Preparation of Soil Extract: Add components to tap water and bring volume to 200.0mL.

Preparation of Medium: Add components, except thiram solution, to distilled/deionized water and bring volume to 990.0mL. Mix thoroughly. Gently heat and bring to boiling. Autoclave for 15 min at 15 psi pressure–121°C. Cool to 50°C. Aseptically add 10.0mL of sterile thiram solution. Pour into sterile Petri dishes or distribute into sterile tubes.

Use: For the cultivation and maintenance of *Bradyrhizobium japonicum, Rhizobium* species, and *Sinorhizobium xinjiangensis*.

Rhizoctonia Isolation Medium

Composition per liter:

Agar	20.0g
K_2HPO_4	1.0g
KCl	0.5g
$MgSO_4 \cdot 7H_2O$	0.5g
$NaNO_2$	0.2g
$FeSO_4 \cdot 7H_2O$	0.01g
Dexon® solution	10.0mL
Antibiotic solution	10.0mL
Gallic acid solution	10.0mL

Antibiotic Solution:

Composition per 10.0mL:

Chloramphenicol	0.05g
Streptomycin	0.05g

Preparation of Antibiotic Solution: Add components to distilled/deionized water and bring volume to 10.0mL. Mix thoroughly. Filter sterilize.

Dexon® Solution:

Composition per 10.0mL:

Dexon® (Chemagro®) wettable powder	0.09g

Preparation of Dexon® Solution: Add Dexon® to distilled/deionized water and bring volume to 10.0mL. Mix thoroughly. Filter sterilize.

Gallic Acid Solution:

Composition per 10.0mL:

Gallic acid	0.4g

Preparation of Gallic Acid Solution: Add gallic acid to distilled/deionized water and bring volume to 10.0mL. Mix thoroughly. Filter sterilize.

Preparation of Medium: Add components—except Dexon® solution, antibiotic solution, and gallic acid solution—to distilled/deionized water and bring volume to 970.0mL. Mix thoroughly. Gently heat and bring to boiling. Autoclave for 15 min at 15 psi pressure–121°C. Cool to 45°–50°C. Aseptically add sterile Dexon® solution, sterile antibiotic solution, and sterile gallic acid solution. Mix thoroughly. Pour into sterile Petri dishes or distribute into sterile tubes.

Use: For the isolation and cultivation of *Rhizoctonia* species.

Rhizomonas Medium

Composition per liter:

Noble agar	11.0g
Pancreatic digest of casein	5.0g
Glucose	2.5g
K_2HPO_4	1.0g
$MgSO_4 \cdot 7H_2O$	0.5g
KNO_3	0.5g
$Ca(NO_3)_2 \cdot 4H_2O$	0.06g

pH 7.2 ± 0.2 at 25°C

Preparation of Medium: Add components to distilled/deionized water and bring volume to 1.0L. Mix thoroughly. Gently heat and bring to boiling. Adjust pH to 7.2. Distribute into tubes or flasks. Autoclave for 15 min at 15 psi pressure–121°C. Pour into sterile Petri dishes or leave in tubes.

Use: For the cultivation and maintenance of *Rhizomonas suberifaciens*.

Rhizomonas suberifaciens Medium

Composition per liter:

Pancreatic digest of casein	5.0g
$K_2HPO_4 \cdot 3H_2O$	1.3g
Noble agar	1.1g
KNO_3	0.5g
$MgSO_4 \cdot 7H_2O$	0.5g
$Ca(NO_3)_2 \cdot 4H_2O$	60.0mg

pH 7.2 ± 0.2 at 25°C

Preparation of Medium: Add components to distilled/deionized water and bring volume to 1.0L. Mix thoroughly. Gently heat and bring to boiling. Adjust pH to 7.2. Distribute into tubes or flasks. Autoclave for 15 min at 15 psi pressure–121°C. Pour into sterile Petri dishes or leave in tubes.

Use: For the cultivation of *Rhizomonas suberifaciens.*

Rhodobacter adriaticus Medium

Composition per 1001.0mL:

NaCl	25.0g
$NaHCO_3$	3.0g
K_2HPO_4	1.0g
NH_4Cl	1.0g
$MgCl_2 \cdot 6H_2O$	0.5g
Sodium ascorbate	0.5g
$CaCl_2 \cdot 2H_2O$	0.1g
Trace elements solution SLA	1.0mL
Vitamin solution	1.0mL

pH 7.0 ± 0.2 at 25°C

Trace Elements Solution SLA:

Composition per liter:

$CuCl_2 \cdot 2H_2O$	10.0g
$FeCl_2 \cdot 4H_2O$	1.8g
H_3BO_3	0.5g
$CoCl_2 \cdot 6H_2O$	0.25g
$ZnCl_2$	0.1g
$MnCl_2 \cdot 4H_2O$	70.0mg
$Na_2MoO_4 \cdot 2H_2O$	30.0mg
$Na_2SeO_3 \cdot 5H_2O$	10.0mg
$NiCl_2 \cdot 6H_2O$	10.0mg

Preparation of Trace Elements Solution SLA: Add components to distilled/deionized water and bring volume to 1.0L. Mix thoroughly. Bring pH to 2.0–3.0.

Vitamin Solution:

Composition per liter:

Nicotinamide	35.0mg
Thiamine·HCl	30.0mg
p-Aminobenzoic acid	20.0mg
Pyridoxal·HCl	10.0mg
Calcium DL-pantothenate	10.0mg
Biotin	10.0mg
Vitamin B_{12}	5.0mg

Preparation of Vitamin Solution: Add components to distilled/deionized water and bring volume to 1.0L. Mix thoroughly. Filter sterilize.

Preparation of Medium: Add components, except vitamin solution, to distilled/deionized water and bring volume to 1.0L. Mix thoroughly. Autoclave for 15 min at 15 psi pressure–121°C. Aseptically add 1.0mL of sterile vitamin solution. Mix thoroughly. Aseptically distribute into sterile tubes or flasks.

Use: For the cultivation and maintenance of *Rhodobacter adraiticus*.

Rhodobacter changlensis Medium (DSMZ Medium 1197)

Composition per 1001.0mL:

Yeast extract	0.4g
Sodium pyruvate	3.0g
NH_4Cl	0.6g
$MgCl_2 \cdot 6H_2O$	0.5g
KH_2PO_4	0.5g
NaCl	0.4g
NH_4Cl	0.6g
$CaCl_2 \cdot 2H_2O$	0.05g
Trace elements solution SL-7	1.0mL
Vitamin solution	1.0mL

pH 7.2 ± 0.2 at 25°C

Trace Elements Solution SL-7:

Composition per liter:

$CoCl_2 \cdot 6H_2O$	200.0mg
$MnCl_2 \cdot 4H_2O$	100.0mg
$ZnCl_2$	70.0mg
H_3BO_3	60.0mg
$Na_2MoO_4 \cdot 2H_2O$	40.0mg
$CuCl_2 \cdot 2H_2O$	20.0mg
$NiCl_2 \cdot 6H_2O$	20.0mg
HCl (25%)	1.0mL

Preparation of Trace Elements Solution SL-7: Add components to distilled/deionized water and bring volume to 1.0L. Mix thoroughly.

Vitamin Solution:

Composition per 100.0mL:

Vitamin B_{12}	20.0mg

Preparation of Vitamin Solution: Add vitamin B_{12} to distilled/deionized water and bring volume to 100.0mL. Mix thoroughly. Filter sterilize.

Preparation of Medium: Add components, except vitamin solution, to distilled/deionized water and bring volume to 1.0L. Mix thoroughly. Adjust pH to 7.2. Gently heat and bring to boiling. Autoclave for 15 min at 15 psi pressure–121°C. Cool to room temperature. Aseptically add vitamin solution. Mix thoroughly. Aseptically distribute into culture vessels.

Use: For the cultivation of *Rhodobacter changlensis.*

Rhodobacter Medium (LMG Medium 80)

Composition per liter:

Yeast extract	1.0g
Disodium succinate	1.0g
KH_2PO_4	0.5g
$MgSO_4 \cdot 7H2O$	0.4g
NaCl	0.4g
NH_4Cl	0.4g
$CaCl_2 \cdot 2H_2O$	50.0mg
Ferric citrate solution	5.0mL
Trace elements solution	1.0mL
Ethanol	0.5mL

pH 5.8 ± 0.2 at 25°C

Ferric Citrate Solution :
Composition per 100.0mL:

Ferric citrate 0.1g

Preparation of Ferric Citrate Solution: Add ferric citrate to distilled/deionized water and bring volume to 100.0mL. Mix thoroughly.

Trace Elements Solution:
Composition per liter:

H_3BO_3 .... 0.3g
$CoCl_2 \cdot 6H_2O$ 0.2g
$ZnSO_4 \cdot 7H_2O$ 0.1g
$Na_2MoO_4 \cdot 2H_2O$ 30.0mg
$MnCl_2 \cdot 4H_2O$ 30.0mg
$NiCl_2 \cdot 6H_2O$ 20.0mg
$CuCl_2 \cdot 2H_2O$ 10.0mg

Preparation of Trace Elements Solution: Add components to distilled/deionized water and bring volume to 1.0L. Mix thoroughly.

Preparation of Medium: Add components to distilled/deionized water and bring volume to 1.0L. Mix thoroughly. Distribute 40 mL medium into 50 mL screw-capped bottles. Flush each bottle for 1 to 2 min with nitrogen gas and then close immediately with rubber septa and screw caps. Autoclave for 15 min at 15 psi pressure–121°C. Sterile syringes are used to inoculate and remove the samples. Incubate in light using a tungsten lamp.

Use: For the cultivation of *Rhodobacter capsulatus, Rhodobacter sphaeroides,* and *Rhodospirillum rubrum.*

Rhodobacter veldkampii Medium (DSMZ Medium 867)

Composition per 2780.0mL:

Solution 1 .... 1540.0mL
Solution 3 .... 1000.0mL
Solution 4 .... 120.0mL
Solution 5 .... 120.0mL

pH 4.0 ± 0.1 at 25°C

Solution 1:
Composition per 2500.0mL:

$CaCl_2$ 2.0g

Preparation of Solution 1: Add $CaCl_2$ to distilled/deionized water and bring volume to 2.5L. Mix thoroughly.

Solution 3:
Composition per liter:

$NaHCO_3$ 4.5g
Solution 2 .... 100.0mL

Preparation of Solution 3: Add $NaHCO_3$ to distilled/deionized water and bring volume to 900.0mL. Mix thoroughly. Sparge with gaseous CO_2 for at least 30 min. Add 100.0mL solution 2. Immediately filter sterilize using CO_2 pressure to push the liquid through (no suction).

Solution 2:
Composition per 100.0mL:

Sodium ascorbate 2.4g
KH_2PO_4 .... 1.0g
KCl 1.0g
NH_4Cl 0.8g
$MgCl_2 \cdot 6H_2O$ 0.8g
Heavy metal solution 50.0mL
Vitamin solution 15.0mL
Vitamin B_{12} solution 3.0mL

Preparation of Solution 2: Add components to distilled/deionized water and bring volume to 100.0mL. Mix thoroughly.

Heavy Metal Solution:
Composition per liter:

EDTA 1.50g
$FeSO_4 \cdot 7H_2O$ 0.2g
$ZnSO_4 \cdot 7H_2O$ 0.1g
$MnCl_2 \cdot 7H_2O$ 0.02g
Modified Hoagland trace elements solution 6.0mL

Preparation of Heavy Metal Solution: Add components to distilled/deionized water and bring volume to 1.0L. Mix thoroughly.

Modified Hoagland Trace Elements Solution:
Composition per 3.6L:

H_3BO_3 .... 11.0g
$MnCl_2 \cdot 4H_2O$ 7.0g
$ZnCl_2$ 1.0g
$CuCl_2$ 1.0g
$NiCl_2$ 1.0g
$CoCl_2$ 1.0g
$AlCl_3$ 1.0g
KI 1.0g
KBr 0.5g
LiCl 0.5g
$SnCl_2 \cdot 2H_2O$ 0.5g
$BaCl_2$ 0.5g
Na_2MoO_4 .... 0.5g
Na_2SeO_3 .... 0.5g
$NaVO_3 \cdot H_2O$ 0.1g

Preparation of Modified Hoagland Trace Elements Solution: Add components sequentially to distilled/deionized water and bring final volume to 3.6L. Mix thoroughly after adding each component until dissolved. Adjust pH to just below 7.0. Adjust the final pH to 3–4. The flaky yellow precipitate which is formed after mixing transforms after standing for one or a few days into a very fine white precipitate. Mix thoroughly before use.

Vitamin B_{12} Solution:
Composition per 3.0mL:

Vitamin B_{12} (cyanocobalamine) 2.0mg

Preparation of Vitamin B_{12} Solution: Add vitamin B_{12} to distilled/deionized water and bring volume to 3.0mL. Mix thoroughly.

Vitamin Solution:
Composition per 100.0mL:

Pyridoxamine·2HCl 5.0mg
Nicotinic acid 2.0mg
Thiamine 1.0mg
Pantothenic acid 0.5mg
Biotin 0.2mg
p-Aminobenzoic acid 0.1mg

Preparation of Vitamin Solution: Add components to distilled/deionized water and bring volume to 100.0mL. Mix thoroughly. Sparge with 100% CO_2 for 30 min.

Solution 4:
Composition per 200.0mL:

$Na_2S \cdot 9H_2O$ 3.0g

Preparation of Solution 4: Add $Na_2S \cdot 9H_2O$ to distilled/deionized water in a flask with a magnetic stirrer and bring volume to 200.0mL. Mix thoroughly. Autoclave under 100% N_2 for 15 min at 15 psi pressure–121°C. Cool to room temperature. Partially neutralize the sterilized solution by adding, on a magnetic stirrer, drop by drop, 2.0mL sterile 2*M* H_2SO_4.

Solution 5:

Composition per 200.0mL:

Na-acetate 6.0g

Preparation of Solution 5: Add Na-acetate to distilled/deionized water and bring volume to 200.0mL. Mix thoroughly. Sparge with 100% N_2 for 5 min. Autoclave under 100% N_2 for 15 min at 15 psi pressure–121°C. Cool to room temperature.

Preparation of Medium: Distribute solution 1 in 77.0mL amounts into 20 127mL screw-capped Boston round bottles. Autoclave for 15 min at 15 psi pressure–121°C. Cool to room temperature. Aseptically add 50.0mL sterile solution 3 to each of the 20 127mL bottles containing 77.0mL of sterile solution 1 so that the solution completely fills the bottle. Mix thoroughly. Remove 6.0mL of the medium from the completely filled bottles. Add 6.0mL of neutralized solution 4 so that the bottles are again completely filled. Mix thoroughly. Remove 6.0mL of the medium from the completely filled bottles. Add 6.0mL of solution 5 so that the bottles are again completely filled. Mix thoroughly. Adjust pH to 4.0. Allow the bottles to stand overnight to develop a hazy, white precipitate before inoculating. Mix the solution thoroughly before use. To inoculate, remove 6.0mL of completed medium and replace it with an equal volume of inoculum. Grow cultures under tungsten light.

Use: For the cultivation of *Rhodobacter veldkampii.*

Rhodobacter veldkampii Medium

Composition per 127.0mL:

Solution 1 .... 76.2mL
Solution 2 + Solution 3 .... 44.8mL
Solution 4 .... 6.0mL

Solution 1:

Composition per 2.5L:

$CaCl_2$ 2.0g

Preparation of Solution 1: Add $CaCl_2$ to distilled/deionized water and bring volume to 2.5L. Distribute in 80.0mL volumes into 127.0mL screw-capped bottles. Autoclave for 15 min at 15 psi pressure–121°C.

Solution 2:

Composition per 100.0mL:

Sodium ascorbate 2.4g
Sodium acetate 1.0g
KC1 .... 1.0g
KH_2PO_4 .... 1.0g
$MgCl_2·6H_2O$ 0.8g
NH_4Cl 0.8g
Heavy metal solution 50.0mL
Vitamin solution 15.0mL
Vitamin B_{12} solution 3.0mL

Preparation of Solution 2: Add components to distilled/deionized water and bring volume to 100.0mL. Mix thoroughly.

Heavy Metal Solution:

Composition per liter:

Ethylenediamine tetraacetate (EDTA) 1.5g
$FeSO_4·7H_2O$ 0.2g
$ZnSO_4·7H_2O$ 0.1g
$MnCl_2·4H_2O$ 0.02g
Modified Hoagland trace elements solution 6.0mL

Preparation of Heavy Metal Solution: Dissolve EDTA in 800.0mL of distilled/deionized water. Add remaining components. Bring volume to 1.0L with distilled/deionized water. Mix thoroughly.

Modified Hoagland Trace Elements Solution:

Composition per 3.6L:

H_3BO_3 .... 11.0g
$MnCl_2·4H_2O$ 7.0g
$AlCl_3$ 1.0g
$CoCl_2$ 1.0g
$CuCl_2$ 1.0g
KI 1.0g
$NiCl_2$ 1.0g
$ZnCl_2$ 1.0g
$BaCl_2$ 0.5g
KBr 0.5g
LiCl 0.5g
Na_2MoO_4 .... 0.5g
$SeCl_4$ 0.5g
$SnCl_2·2H_2O$ 0.5g
$NaVO_3·H_2O$ 0.1g

Preparation of Modified Hoagland Trace Elements Solution: Prepare each component as a separate solution. Dissolve each salt in approximately 100.0mL of distilled/deionized water. Adjust the pH of each solution to below 7.0. Combine all the salt solutions and bring the volume to 3.6L with distilled/deionized water. Adjust the pH to 3–4. A yellow precipitate may form after mixing. After a few days, it will turn into a fine white precipitate. Mix the solution thoroughly before using.

Vitamin Solution:

Composition per 100.0mL:

Pyridoxamine·2HCl 5.0mg
Nicotinic acid 2.0mg
Thiamine 1.0mg
Pantothenic acid 0.5mg
Biotin 0.2mg
p-Aminobenzoic acid 0.1mg

Preparation of Vitamin Solution: Add components to distilled/deionized water and bring volume to 100.0mL. Mix thoroughly.

Vitamin B_{12} Solution:

Composition per 100.0mL:

Vitamin B_{12} (cyanocobalamin) 2.0mg

Preparation of Vitamin B_{12} Solution: Add vitamin B_{12} to distilled/deionized water and bring volume to 100.0mL. Mix thoroughly.

Solution 3:

Composition per 900.0mL:

$NaHCO_3$ 4.5g

Preparation of Solution 3: Add $NaHCO_3$ to distilled/deionized water and bring volume to 900.0mL. Mix thoroughly. Bubble 100% CO_2 through the solution for 30 min. After CO_2 saturation of solution 3, add solution 2 and immediately filter the mixture through a Seitz filter (or a Millipore) using positive CO_2 pressure to push the liquid through.

Solution 4:

Composition per 200.0mL:

$Na_2S·9H_2O$ 3.0g

Preparation of Solution 4: Add $Na_2S·9H_2O$ to distilled/deionized water and bring volume to 200.0mL. Add a magnetic stir bar to the flask. Autoclave for 15 min at 15 psi pressure–121°C. On a magnetic stirrer, slowly add 2.0mL of sterile 2*M* H_2SO_4. This partially neutralizes the solution. The solution should turn yellow. H_2S gas will be liberated—neutralization and distribution of the solution should be done as rapidly as possible under adequate ventilation.

Preparation of Medium: To the 80.0mL of sterile solution 1 in screw-capped bottles, add combined solutions 2 and 3 immediately after filtration and fill bottles to capacity. Mix thoroughly. Aseptically remove 6.0mL of the medium from the bottles and replace it with 6.0mL of neutralized solution 4. Let stand for 24 hr. The medium should form a fine white precipitate before using. To inoculate, remove 6.0mL of the completed medium from the bottles and replace it with 6.0mL of inoculum.

Use: For the cultivation and maintenance of *Rhodobacter veldkampii*.

Rhodobium gokurnum Medium (DSMZ Medium 1129)

Composition per liter:

NaCl	20.0g
KH_2PO_4	0.5g
$MgCl_2·6H_2O$	1.0g
NH_4Cl	0.6g
$CaCl_2·2H_2O$	0.15g
Sorbitol solution	10.0mL
Yeast extract solution	10.0mL
Sodium pyruvate solution	10.0mL
Ferric citrate solution	5.0mL
Trace elements solution SL-7	1.0mL

pH 6.5 ± 0.2 at 25°C

Ferric Citrate Solution:

Composition per 10.0mL:

Ferric citrate	0.01g

Preparation of Ferric Citrate Solution: Add ferric citrate to distilled/deionized water and bring volume to 10.0mL. Mix thoroughly. Filter sterilize.

Trace Elements Solution SL-7:

Composition per liter:

$CoCl_2·6H_2O$	200.0mg
$MnCl_2·4H_2O$	100.0mg
$ZnCl_2$	70.0mg
H_3BO_3	60.0mg
$Na_2MoO_4·2H_2O$	40.0mg
$CuCl_2·2H_2O$	20.0mg
$NiCl_2·6H_2O$	20.0mg
HCl (25%)	1.0mL

Preparation of Trace Elements Solution SL-7: Add components to distilled/deionized water and bring volume to 1.0L. Mix thoroughly.

Sorbitol Solution:

Composition per 10.0mL:

Sorbitol	3.0g

Preparation of Sorbitol Solution: Add sorbitol to distilled/deionized water and bring volume to 10.0mL. Mix thoroughly. Filter sterilize.

Sodium Pyruvate Solution:

Composition per 10.0mL:

Sodium pyruvate	3.0g

Preparation of Sodium Pyruvate Solution: Add sodium pyruvate to distilled/deionized water and bring volume to 10.0mL. Mix thoroughly. Filter sterilize.

Yeast Extract Solution:

Composition per 10.0mL:

Yeast extract	0.4g

Preparation of Yeast Extract Solution: Add yeast extract to distilled/deionized water and bring volume to 10.0mL. Mix thoroughly. Filter sterilize.

Preparation of Medium: Add components, except ferric citrate, trace elements, sorbitol, sodium pyruvate, and yeast extract solutions, to distilled/deionized water and bring volume to 964.0mL. Mix thoroughly. Adjust pH to 6.5. Gently heat and bring to boiling. Autoclave for 15 min at 15 psi pressure–121°C.Cool to room temperature. Aseptically add ferric citrate, trace elements, sorbitol, sodium pyruvate, and yeast extract solutions. Mix thoroughly. Aseptically distribute into culture vessels.

Use: For the cultivation of *Rhodobium gokarnense* (*Rhodobium gokurnum*).

Rhodobium gokurnum Medium (DSMZ Medium 1129)

Composition per liter:

NaCl	20.0g
$MgCl_2·6H_2O$	1.0g
NH_4Cl	0.68g
KH_2PO_4	0.5g
$CaCl_2·2H_2O$	0.15g
Sulfide solution	10.0mL
Yeast extract solution	10.0mL
Sodium pyruvate solution	10.0mL
Ferric citrate solution	5.0mL
Trace elements solution SL-7	1.0mL
Vitamin solution	1.0mL

pH 7.2 ± 0.2 at 25°C

Ferric Citrate Solution:

Composition per 10.0mL:

Ferric citrate	0.01g

Preparation of Ferric Citrate Solution: Add ferric citrate to distilled/deionized water and bring volume to 10.0mL. Mix thoroughly. Filter sterilize.

Trace Elements Solution SL-7:

Composition per liter:

$CoCl_2·6H_2O$	200.0mg
$MnCl_2·4H_2O$	100.0mg
$ZnCl_2$	70.0mg
H_3BO_3	60.0mg
$Na_2MoO_4·2H_2O$	40.0mg
$CuCl_2·2H_2O$	20.0mg
$NiCl_2·6H_2O$	20.0mg
HCl (25%)	1.0mL

Preparation of Trace Elements Solution SL-7: Add components to distilled/deionized water and bring volume to 1.0L. Mix thoroughly.

Vitamin Solution:

Composition per liter:

Vitamin B_{12}	2.0mg

Preparation of Vitamin Solution: Add vitamin B_{12} to distilled/deionized water and bring volume to 1.0L. Mix thoroughly. Filter sterilize.

Sodium Pyruvate Solution:

Composition per 10.0mL:

Sodium pyruvate ... 3.0g

Preparation of Sodium Pyruvate Solution: Add sodium pyruvat to distilled/deionized water and bring volume to 10.0mL. Mix thoroughly. Filter sterilize.

Yeast Extract Solution:

Composition per 10.0mL:

Yeast extract ... 0.4g

Preparation of Yeast Extract Solution: Add yeast extract to distilled/deionized water and bring volume to 10.0mL. Mix thoroughly. Filter sterilize.

Sulfide Solution:

Composition per 10.0mL:

$Na_2S\cdot9H_2O$ ... 0.25g

Preparation of Sulfide Solution: Add $Na_2S\cdot9H_2O$ to distilled/deionized water and bring volume to 10.0mL. Mix thoroughly. Autoclave under 100% N_2 for 15 min at 15 psi pressure–121°C. Cool to room temperature.

Preparation of Medium: Add components, except ferric citrate, trace elements, sulfide, vitamin, sodium pyruvate, and yeast extract solutions, to distilled/deionized water and bring volume to 963.0mL. Mix thoroughly. Adjust pH to 7.2. Gently heat and bring to boiling. Autoclave for 15 min at 15 psi pressure–121°C. Cool to room temperature. Aseptically add ferric citrate, trace elements, vitamin, sulfide, sodium pyruvate, and yeast extract solutions. Mix thoroughly. Aseptically distribute into culture vessels.

Use: For the cultivation of *Marichromatium bheemlicum.*

Rhodobium gokurnum Medium (DSMZ Medium 1129)

Composition per liter:

NaCl ... 20.0g
$MgCl_2\cdot6H_2O$ ... 1.0g
NH_4Cl ... 0.6g
KH_2PO_4 ... 0.5g
$CaCl_2\cdot2H_2O$ ... 0.15g
Sulfide solution ... 10.0mL
Yeast extract solution ... 10.0mL
Sodium pyruvate solution ... 10.0mL
Ferric citrate solution ... 5.0mL
Trace elements solution SL-7 ... 1.0mL
Vitamin solution ... 1.0mL

pH 7.2 ± 0.2 at 25°C

Ferric Citrate Solution:

Composition per 10.0mL:

Ferric citrate ... 0.01g

Preparation of Ferric Citrate Solution: Add ferric citrate to distilled/deionized water and bring volume to 10.0mL. Mix thoroughly. Filter sterilize.

Trace Elements Solution SL-7:

Composition per 1001.0mL:

$CoCl_2\cdot6H_2O$ ... 200.0mg
$MnCl_2\cdot4H_2O$ ... 100.0mg
$ZnCl_2$ ... 70.0mg
H_3BO_3 ... 60.0mg
$Na_2MoO_4\cdot2H_2O$ ... 40.0mg
$CuCl_2\cdot2H_2O$ ... 20.0mg
$NiCl_2\cdot6H_2O$ ... 20.0mg
HCl (25%) ... 1.0mL

Preparation of Trace Elements Solution SL-7: Add components to distilled/deionized water and bring volume to 1.0L. Mix thoroughly.

Vitamin Solution:

Composition per liter:

Vitamin B_{12} ... 2.0mg

Preparation of Vitamin Solution: Add vitamin B_{12} to distilled/deionized water and bring volume to 1.0L. Mix thoroughly. Filter sterilize.

Sodium Pyruvate Solution:

Composition per 10.0mL:

Sodium pyruvate ... 3.0g

Preparation of Sodium Pyruvate Solution: Add sodium pyruvate to distilled/deionized water and bring volume to 10.0mL. Mix thoroughly. Filter sterilize.

Yeast Extract Solution:

Composition per 10.0mL:

Yeast extract ... 0.4g

Preparation of Yeast Extract Solution: Add yeast extract to distilled/deionized water and bring volume to 10.0mL. Mix thoroughly. Filter sterilize.

Sulfide Solution:

Composition per 10.0mL:

$Na_2S\cdot9H_2O$ ... 0.25g

Preparation of Sulfide Solution: Add $Na_2S\cdot9H_2O$ to distilled/deionized water and bring volume to 10.0mL. Mix thoroughly. Autoclave under 100% N_2 for 15 min at 15 psi pressure–121°C. Cool to room temperature.

Preparation of Medium: Add components, except ferric citrate, trace elements, sulfide, vitamin, sodium pyruvate, and yeast extract solutions, to distilled/deionized water and bring volume to 963.0mL. Mix thoroughly. Adjust pH to 7.2. Gently heat and bring to boiling. Autoclave for 15 min at 15 psi pressure–121°C. Cool to room temperature. Aseptically add ferric citrate, trace elements, vitamin, sulfide, sodium pyruvate, and yeast extract solutions. Mix thoroughly. Aseptically distribute into culture vessels.

Use: For the cultivation of *Thiorhodococcus bheemlicus.*

Rhodobium gokurnum Medium (DSMZ Medium 1129)

Composition per liter:

NaCl ... 20.0g
$MgCl_2\cdot6H_2O$ ... 1.0g
KH_2PO_4 ... 0.5g
NH_4Cl ... 0.34g
$CaCl_2\cdot2H_2O$ ... 0.15g
Sulfide solution ... 10.0mL
Thiosulfate solution ... 10.0mL
Bicarbonate solution ... 10.0mL
Yeast extract solution ... 10.0mL
Sodium pyruvate solution ... 10.0mL
Ferric citrate solution ... 5.0mL
Trace elements solution SL-7 ... 1.0mL
Vitamin solution ... 1.0mL

pH 7.2 ± 0.2 at 25°C

Ferric Citrate Solution:
Composition per 10.0mL:

Ferric citrate 0.01g

Preparation of Ferric Citrate Solution: Add ferric citrate to distilled/deionized water and bring volume to 10.0mL. Mix thoroughly. Filter sterilize.

Trace Elements Solution SL-7:
Composition per 1001.0mL:

$CoCl_2 \cdot 6H_2O$ 200.0mg
$MnCl_2 \cdot 4H_2O$ 100.0mg
$ZnCl_2$ 70.0mg
H_3BO_3 .... 60.0mg
$Na_2MoO_4 \cdot 2H_2O$ 40.0mg
$CuCl_2 \cdot 2H_2O$ 20.0mg
$NiCl_2 \cdot 6H_2O$ 20.0mg
HCl (25%) 1.0mL

Preparation of Trace Elements Solution SL-7: Add components to distilled/deionized water and bring volume to 1.0L. Mix thoroughly.

Vitamin Solution:
Composition per liter:

Vitamin B_{12} 2.0mg

Preparation of Vitamin Solution: Add vitamin B_{12} to distilled/deionized water and bring volume to 1.0L. Mix thoroughly. Filter sterilize.

Sodium Pyruvate Solution:
Composition per 10.0mL:

Sodium pyruvate 3.0g

Preparation of Sodium Pyruvate Solution: Add sodium pyruvate to distilled/deionized water and bring volume to 10.0mL. Mix thoroughly. Filter sterilize.

Yeast Extract Solution:
Composition per 10.0mL:

Yeast extract 0.4g

Preparation of Yeast Extract Solution: Add yeast extract to distilled/deionized water and bring volume to 10.0mL. Mix thoroughly. Filter sterilize.

Sulfide Solution:
Composition per 10.0mL:

$Na_2S \cdot 9H_2O$ 0.25g

Preparation of Sulfide Solution: Add $Na_2S \cdot 9H_2O$ to distilled/deionized water and bring volume to 10.0mL. Mix thoroughly. Autoclave under 100% N_2 for 15 min at 15 psi pressure–121°C. Cool to room temperature.

Thiosulfate Solution:
Composition per 10.0mL:

$Na_2S_2O_3 \cdot 5H_2O$ 1.6g

Preparation of Thiosulfate Solution: Add $Na_2S_2O_3 \cdot 5H_2O$ to distilled/deionized water and bring volume to 10.0mL. Mix thoroughly. Sparge with 100% N_2. Adjust to pH 10.0. Filter sterilize.

Bicarbonate Solution:
Composition per 100.0mL:

$NaHCO_3$ 10.0g

Preparation of Bicarbonate Solution: Add $NaHCO_3$ to distilled/deionized water and bring volume to 100.0mL. Mix thoroughly. Adjust pH to 6.5. Seal in bottles (half full) under an atmosphere of CO_2. Autoclave for 15 min at 15 psi pressure–121°C. Cool to room temperature.

Preparation of Medium: Add components, except ferric citrate, trace elements, sulfide, thiosulfate, bicarbonate, vitamin, sodium pyruvate, and yeast extract solutions, to distilled/deionized water and bring volume to 943.0mL. Mix thoroughly. Adjust pH to 7.2. Gently heat and bring to boiling. Autoclave for 15 min at 15 psi pressure–121°C. Cool to room temperature. Aseptically add ferric citrate, trace elements, sulfide, thiosulfate, bicarbonate, vitamin, sodium pyruvate, and yeast extract solutions. Mix thoroughly. Aseptically distribute into culture vessels.

Use: For the cultivation of *Allochromatium renukae* and *Allochromatium roseum.*

Rhodobium gokurnum Medium (DSMZ Medium 1129)

Composition per liter:

NaCl 20.0g
KH_2PO_4 .... 0.5g
$MgCl_2 \cdot 6H_2O$ 1.0g
NH_4Cl 0.6g
$CaCl_2 \cdot 2H_2O$ 0.15g
Sulfide solution 10.0mL
Thiosulfate solution 10.0mL
Bicarbonate solution 10.0mL
Yeast extract solution 10.0mL
Sodium pyruvate solution 10.0mL
Ferric citrate solution 5.0mL
Trace elements solution SL-7 .... 1.0mL
Vitamin solution 1.0mL

pH 6.9 ± 0.2 at 25°C

Ferric Citrate Solution:
Composition per 10.0mL:

Ferric citrate 0.01g

Preparation of Ferric Citrate Solution: Add ferric citrate to distilled/deionized water and bring volume to 10.0mL. Mix thoroughly. Filter sterilize.

Trace Elements Solution SL-7:
Composition per liter:

$CoCl_2 \cdot 6H_2O$ 200.0mg
$MnCl_2 \cdot 4H_2O$ 100.0mg
$ZnCl_2$ 70.0mg
H_3BO_3 .... 60.0mg
$Na_2MoO_4 \cdot 2H_2O$ 40.0mg
$CuCl_2 \cdot 2H_2O$ 20.0mg
$NiCl_2 \cdot 6H_2O$ 20.0mg
HCl (25%) 1.0mL

Preparation of Trace Elements Solution SL-7: Add components to distilled/deionized water and bring volume to 1.0L. Mix thoroughly.

Vitamin Solution:
Composition per liter:

Vitamin B_{12} 2.0mg

Preparation of Vitamin Solution: Add vitamin B_{12} to distilled/deionized water and bring volume to 1.0L. Mix thoroughly. Filter sterilize.

Sodium Pyruvate Solution:
Composition per 10.0mL:

Sodium pyruvate 3.0g

Preparation of Sodium Pyruvate Solution: Add sodium pyruvate to distilled/deionized water and bring volume to 10.0mL. Mix thoroughly. Filter sterilize.

Yeast Extract Solution:

Composition per 10.0mL:

Yeast extract	0.4g

Preparation of Yeast Extract Solution: Add yeast extract to distilled/deionized water and bring volume to 10.0mL. Mix thoroughly. Filter sterilize.

Sulfide Solution:

Composition per 10.0mL:

$Na_2S{\cdot}9H_2O$	0.25g

Preparation of Sulfide Solution: Add $Na_2S{\cdot}9H_2O$ to distilled/deionized water and bring volume to 10.0mL. Mix thoroughly. Autoclave under 100% N_2 for 15 min at 15 psi pressure–121°C. Cool to room temperature.

Thiosulfate Solution:

Composition per 10.0mL:

$Na_2S_2O_3{\cdot}5H_2O$	1.6g

Preparation of Thiosulfate Solution: Add $Na_2S_2O_3{\cdot}5H_2O$ to distilled/deionized water and bring volume to 10.0mL. Mix thoroughly. Sparge with 100% N_2. Adjust to pH 10.0. Filter sterilize.

Bicarbonate Solution:

Composition per 100.0mL:

$NaHCO_3$	10.0g

Preparation of Bicarbonate Solution: Add $NaHCO_3$ to distilled/deionized water and bring volume to 100.0mL. Mix thoroughly. Adjust pH to 6.5. Seal in bottles (half full) under an atmosphere of CO_2. Autoclave for 15 min at 15 psi pressure–121°C. Cool to room temperature.

Preparation of Medium: Add components, except ferric citrate, trace elements, sulfide, thiosulfate, bicarbonate, vitamin, sodium pyruvate, and yeast extract solutions, to distilled/deionized water and bring volume to 943.0mL. Mix thoroughly. Adjust pH to 6.9. Gently heat and bring to boiling. Autoclave for 15 min at 15 psi pressure–121°C.Cool to room temperature. Aseptically add ferric citrate, trace elements, sulfide, thiosulfate, bicarbonate, vitamin, sodium pyruvate, and yeast extract solutions. Mix thoroughly. Aseptically distribute into culture vessels.

Use: For the cultivation of *Roseospira visakhapatnamensis.*

Rhodobium gokurnum Medium (DSMZ Medium 1129)

Composition per liter:

NaCl	20.0g
$MgSO_4{\cdot}7H_2O$	1.5g
NH_4Cl	0.6g
KH_2PO_4	0.5g
$CaCl_2{\cdot}2H_2O$	0.15g
Sulfide solution	10.0mL
Thiosulfate solution	10.0mL
Bicarbonate solution	10.0mL
Yeast extract solution	10.0mL
Sodium pyruvate solution	10.0mL
Ferric citrate solution	5.0mL
Trace elements solution SL-7	1.0mL
Vitamin solution	1.0mL

pH 6.8 ± 0.2 at 25°C

Ferric Citrate Solution:

Composition per 10.0mL:

Ferric citrate	0.01g

Preparation of Ferric Citrate Solution: Add ferric citrate to distilled/deionized water and bring volume to 10.0mL. Mix thoroughly. Filter sterilize.

Trace Elements Solution SL-7:

Composition per liter:

$CoCl_2{\cdot}6H_2O$	200.0mg
$MnCl_2{\cdot}4H_2O$	100.0mg
$ZnCl_2$	70.0mg
H_3BO_3	60.0mg
$Na_2MoO_4{\cdot}2H_2O$	40.0mg
$CuCl_2{\cdot}2H_2O$	20.0mg
$NiCl_2{\cdot}6H_2O$	20.0mg
HCl (25%)	1.0mL

Preparation of Trace Elements Solution SL-7: Add components to distilled/deionized water and bring volume to 1.0L. Mix thoroughly.

Vitamin Solution:

Composition per liter:

Vitamin B_{12}	2.0mg

Preparation of Vitamin Solution: Add vitamin B_{12} to distilled/deionized water and bring volume to 1.0L. Mix thoroughly. Filter sterilize.

Sodium Pyruvate Solution:

Composition per 10.0mL:

Sodium pyruvate	3.0g

Preparation of Sodium Pyruvate Solution: Add sodium pyruvate to distilled/deionized water and bring volume to 10.0mL. Mix thoroughly. Filter sterilize.

Yeast Extract Solution:

Composition per 10.0mL:

Yeast extract	0.4g

Preparation of Yeast Extract Solution: Add yeast extract to distilled/deionized water and bring volume to 10.0mL. Mix thoroughly. Filter sterilize.

Sulfide Solution:

Composition per 10.0mL:

$Na_2S{\cdot}9H_2O$	0.25g

Preparation of Sulfide Solution: Add $Na_2S{\cdot}9H_2O$ to distilled/deionized water and bring volume to 10.0mL. Mix thoroughly. Autoclave under 100% N_2 for 15 min at 15 psi pressure–121°C. Cool to room temperature.

Thiosulfate Solution:

Composition per 10.0mL:

$Na_2S_2O_3{\cdot}5H_2O$	1.6g

Preparation of Thiosulfate Solution: Add $Na_2S_2O_3{\cdot}5H_2O$ to distilled/deionized water and bring volume to 10.0mL. Mix thoroughly. Sparge with 100% N_2. Adjust to pH 10.0. Filter sterilize.

Bicarbonate Solution:

Composition per 100.0mL:

$NaHCO_3$	10.0g

Preparation of Bicarbonate Solution: Add $NaHCO_3$ to distilled/deionized water and bring volume to 100.0mL. Mix thoroughly. Adjust pH to 6.5. Seal in bottles (half full) under an atmosphere of CO_2. Autoclave for 15 min at 15 psi pressure–121°C. Cool to room temperature.

Preparation of Medium: Add components, except ferric citrate, trace elements, sulfide, thiosulfate, bicarbonate, vitamin, sodium pyruvate, and yeast extract solutions, to distilled/deionized water and bring volume to 943.0mL. Mix thoroughly. Adjust pH to 6.8. Gently heat and bring to boiling. Autoclave for 15 min at 15 psi pressure–121°C. Cool to room temperature. Aseptically add ferric citrate, trace elements, sulfide, thiosulfate, bicarbonate, vitamin, sodium pyruvate, and yeast extract solutions. Mix thoroughly. Aseptically distribute into culture vessels.

Use: For the cultivation of *Roseospira goensis.*

Rhodobium Medium (DSMZ Medium 745)

Composition per liter:

NaCl	50.2g
Na-DL-malate	3.6g
Yeast extract	1.0g
KH_2PO_4	1.0g
$(NH_4)_2SO_4$	1.0g
$MgCl_2 \cdot 6H_2O$	0.2g
$CaCl_2 \cdot 2H_2O$	0.05g
$Na_2S \cdot 9H_2O$ solution	10.0mL
Trace elements solution SL-8	1.0mL

pH 6.8 ± 0.2 at 25°C

Trace Elements Solution SL-8:
Composition per liter:

Na_2-EDTA	5.2g
$FeCl_2 \cdot 4H_2O$	1.5g
$CoCl_2 \cdot 6H_2O$	190.0mg
$MnCl_2 \cdot 4H_2O$	100.0mg
$ZnCl_2$	70.0mg
H_3BO_3	62.0mg
$Na_2MoSO_4 \cdot 2H_2O$	36.0mg
$NiCl_2 \cdot 6H_2O$	24.0mg
$CuCl_2 \cdot 2H_2O$	17.0mg

Preparation of Trace Elements Solution SL-8: Add components to distilled/deionized water and bring volume to 1.0L. Mix thoroughly. Sparge with 100% N_2.

$Na_2S \cdot 9H_2O$ Solution:
Composition per 10.0mL:

$Na_2S \cdot 9H_2O$	0.5g

Preparation of $Na_2S \cdot 9H_2O$ Solution: Add $Na_2S \cdot 9H_2O$ to distilled/deionized water and bring volume to 10.0mL. Mix thoroughly. Adjust pH to 6.8. Autoclave under 100% N_2 for 15 min at 15 psi pressure–121°C. Cool to room temperature.

Preparation of Medium: Add components, except $Na_2S \cdot 9H_2O$ solution, to distilled/deionized water and bring volume to 1.0L. Mix thoroughly. Adjust to pH 6.8. Gently heat and bring to boiling. Cool to room temperature while sparging with 100% N_2. Autoclave for 15 min at 15 psi pressure–121°C. Cool to room temperature. Add 10.0mL sterile $Na_2S \cdot 9H_2O$ solution. Mix thoroughly. Aseptically and anaerobically under 100% N_2 distribute into sterile screw-cap tubes or flasks.

Use: For the cultivation of *Rhodobium orientis (Rhodovulum orientis).*

Rhodoblastus Medium (DSMZ Medium 1142)

Composition per liter:

Yeast extract	0.1g
Na_2-succinate	1.0g
KH_2PO_4	0.5g
$MgSO_4 \cdot 7H_2O$	0.4g
NaCl	0.4g
NH_4Cl	0.4g
$CaCl_2 \cdot 2H_2O$	0.05g
Trace elements solution SL-6	1.0mL

pH 5.7 ± 0.2 at 25°C

Trace Elements Solution SL-6:
Composition per liter:

$MnCl_2 \cdot 4H_2O$	0.5g
H_3BO_3	0.3g
$CoCl_2 \cdot 6H_2O$	0.2g
$ZnSO_4 \cdot 7H_2O$	0.1g
$Na_2MoO_4 \cdot 2H_2O$	0.03g
$NiCl_2 \cdot 6H_2O$	0.02g
$CuCl_2 \cdot 2H_2O$	0.01g

Preparation of Trace Elements Solution SL-6: Add components to distilled/deionized water and bring volume to 1.0L. Mix thoroughly. Autoclave for 15 min at 15 psi pressure–121°C.

Preparation of Medium: Add components to distilled/deionized water and bring volume to 1.0L. Mix thoroughly. Adjust pH to 5.7. Distribute into tubes or flasks. Gently heat and bring to boiling. Autoclave for 15 min at 15 psi pressure–121°C.

Use: For the cultivation of *Rhodoblastus* spp.

Rhodocyclus Medium (LMG Medium 82)

Composition per liter:

Yeast extract	1.0g
Ammonium acetate	0.5g
KH_2PO_4	0.5g
$MgSO_4 \cdot 7H2O$	0.4g
NaCl	0.4g
NH_4Cl	0.4g
$CaCl_2 \cdot 2H_2O$	50.0mg
Ferric citrate solution	5.0mL
Trace elements solution	1.0mL
Vitamin solution	1.0mL
Ethanol	0.5mL

pH 5.8 ± 0.2 at 25°C

Ferric Citrate Solution :
Composition per 100.0mL:

Ferric citrate	0.1g

Preparation of Ferric Citrate Solution: Add ferric citrate to distilled/deionized water and bring volume to 100.0mL. Mix thoroughly.

Trace Elements Solution:
Composition per liter:

H_3BO_3	0.3g
$CoCl_2 \cdot 6H_2O$	0.2g
$ZnSO_4 \cdot 7H_2O$	0.1g
$Na_2MoO_4 \cdot 2H_2O$	30.0mg
$MnCl_2 \cdot 4H_2O$	30.0mg

$NiCl_2 \cdot 6H_2O$ 20.0mg
$CuCl_2 \cdot 2H_2O$ 10.0mg

Preparation of Trace Elements Solution: Add components to distilled/deionized water and bring volume to 1.0L. Mix thoroughly.

Vitamin Solution:

Composition per 100.0mL:

Vitamin B_{12} 2.0mg

Preparation of Vitamin Solution: Add vitamin B_{12} to 100.0mL of distilled/deionized water. Mix thoroughly.

Preparation of Medium: Add components to distilled/deionized water and bring volume to 1.0L. Mix thoroughly. Distribute 40.0mL medium into 50 mL screw-capped bottles. Flush each bottle for 1 to 2 minutes with nitrogen gas and then close immediately with rubber septa and screw caps. Autoclave for 15 min at 15 psi pressure–121°C. Sterile syringes are used to inoculate and remove the samples. Incubate in light using a tungsten lamp.

Use: For the cultivation of *Rhodocyclus purpureus.*

Rhodocyclus purpureus Medium (DSMZ Medium 44)

Composition per 1056.9 mL:

Na_2-succinate 1.0g
(NH_4)-acetate 0.5g
KH_2PO_4 .... 0.5g
$MgSO_4 \cdot 7H_2O$ 0.4g
NaCl 0.4g
NH_4Cl 0.4g
Yeast extract 0.3g
L-Cysteine 0.3g
$CaCl_2 \cdot 2H_2O$ 0.05g
Ferric citrate solution 5.0mL
Trace elements solution SL-6 .... 1.0mL
Ethanol 0.5mL
Vitamin B_{12} solution 0.4mL

pH 6.8 ± 0.2 at 25°C

Trace Elements Solution SL-6:

Composition per liter:

$MnCl_2 \cdot 4H_2O$ 0.5g
H_3BO_3 .... 0.3g
$CoCl_2 \cdot 6H_2O$ 0.2g
$ZnSO_4 \cdot 7H_2O$ 0.1g
$Na_2MoO_4 \cdot 2H_2O$ 0.03g
$NiCl_2 \cdot 6H_2O$ 0.02g
$CuCl_2 \cdot 2H_2O$ 0.01g

Preparation of Trace Elements Solution SL-6: Add components to distilled/deionized water and bring volume to 1.0L. Mix thoroughly.

Vitamin B_{12} Solution:

Composition per 100.0mL:

Vitamin B_{12} 10.0mg

Preparation of Vitamin B_{12} Solution: Add vitamin B_{12} to distilled/deionized water and bring volume to 100.0mL. Mix thoroughly. Sparge under 100% N_2 gas for 3 min.

Ferric Citrate Solution:

Composition per 10.0mL:

Ferric citrate 10.0mg

Preparation of Ferric Citrate Solution: Add ferric citrate to distilled/deionized water and bring volume to 10.0mL. Mix thoroughly. Sparge under 100% N_2 gas for 3 min.

Preparation of Medium: Add components to 1050.0mL distilled/deionized water. Mix thoroughly. Gently heat and bring to boiling. Boil for 3–4 min under a stream of 100% N_2. Distribute 45.0mL of the prepared medium into 50.0mL screw-capped tubes that have been flushed with 100% N_2. Autoclave for 15 min at 15 psi pressure–121°C. Cool to room temperature. Sterile syringes are used to inoculate and remove samples. Incubate in the light using a tungsten lamp.

Use: For the cultivation and maintenance of brown and other oxygen sensitive Rhodospirillaceae.

Rhodocyclus purpureus Medium (DSMZ Medium 44)

Composition per 1056.9 mL:

Na_2-succinate 1.0g
(NH_4)-acetate 0.5g
KH_2PO_4 .... 0.5g
$MgSO_4 \cdot 7H_2O$ 0.4g
NaCl 0.4g
NH_4Cl 0.4g
Yeast extract 0.3g
$CaCl_2 \cdot 2H_2O$ 0.05g
Ferric citrate solution 5.0mL
Trace elements solution SL-6 .... 1.0mL
Ethanol 0.5mL
Vitamin B_{12} solution 0.4mL
Neutralized sulfide solution .... variable

pH 6.8 ± 0.2 at 25°C

Trace Elements Solution SL-6:

Composition per liter:

$MnCl_2 \cdot 4H_2O$ 0.5g
H_3BO_3 .... 0.3g
$CoCl_2 \cdot 6H_2O$ 0.2g
$ZnSO_4 \cdot 7H_2O$ 0.1g
$Na_2MoO_4 \cdot 2H_2O$ 0.03g
$NiCl_2 \cdot 6H_2O$ 0.02g
$CuCl_2 \cdot 2H_2O$ 0.01g

Preparation of Trace Elements Solution SL-6: Add components to distilled/deionized water and bring volume to 1.0L. Mix thoroughly.

Vitamin B_{12} Solution:

Composition per 100.0mL:

Vitamin B_{12} 10.0mg

Preparation of Vitamin B_{12} Solution: Add vitamin B_{12} to distilled/deionized water and bring volume to 100.0mL. Mix thoroughly. Sparge under 100% N_2 gas for 3 min.

Ferric Citrate Solution:

Composition per 10.0mL:

Ferric citrate 10.0mg

Preparation of Ferric Citrate Solution: Add ferric citrate to distilled/deionized water and bring volume to 10.0mL. Mix thoroughly. Sparge under 100% N_2 gas for 3 min.

Neutralized Sulfide Solution:

Composition per 100.0mL:

$Na_2S \cdot 9H_2O$ 1.5g

Preparation of Neutralized Sulfide Solution: Add $Na_2S \cdot 9H_2O$ to distilled/deionized water in a 250.0mL screw-capped bottle fitted with a butyl rubber septum and bring volume to 100.0mL. Add a magnetic stir bar. Mix thoroughly. Sparge under 100% N_2 gas for 3 min. Autoclave for 15 min at 15 psi pressure–121°C. Cool to room temperature. Adjust pH to about 7.0 with sterile 2*M* H_2SO_4. Do not open the bottle to add H_2SO_4; use a sterile syringe. Stir the solution continuously to avoid precipitation of elemental sulfur. The final solution should be clear and yellow in color.

Preparation of Medium: Add components, except neutralized sulfide solution, to 1050.0mL distilled/deionized water. Mix thoroughly. Gently heat and bring to boiling. Boil for 3–4 min under a stream of 100% N_2. Distribute 45.0mL of the prepared medium into 50.0mL screw-capped tubes that have been flushed with 100% N_2. Autoclave for 15 min at 15 psi pressure–121°C. Cool to room temperature. Before inoculation, aseptically and anaerobically add 0.25–0.50mL of neutralized sulfide solution. Sterile syringes are used to inoculate and remove samples. Incubate in the light using a tungsten lamp.

Use: For the cultivation and maintenance of brown and other oxygen-sensitive Rhodospirillaceae.

Rhodocyclus purpureus Medium (DSMZ Medium 44)

Composition per 1056.9 mL:

Na_2-succinate	1.0g
(NH_4)-acetate	0.5g
KH_2PO_4	0.5g
$MgSO_4 \cdot 7H_2O$	0.4g
NaCl	0.4g
NH_4Cl	0.4g
Yeast extract	0.3g
$CaCl_2 \cdot 2H_2O$	0.05g
Ferric citrate solution	5.0mL
Trace elements solution SL-6	1.0mL
Ethanol	0.5mL
Vitamin B_{12} solution	0.4mL

pH 6.8 ± 0.2 at 25°C

Trace Elements Solution SL-6:

Composition per liter:

$MnCl_2 \cdot 4H_2O$	0.5g
H_3BO_3	0.3g
$CoCl_2 \cdot 6H_2O$	0.2g
$ZnSO_4 \cdot 7H_2O$	0.1g
$Na_2MoO_4 \cdot 2H_2O$	0.03g
$NiCl_2 \cdot 6H_2O$	0.02g
$CuCl_2 \cdot 2H_2O$	0.01g

Preparation of Trace Elements Solution SL-6: Add components to distilled/deionized water and bring volume to 1.0L. Mix thoroughly.

Vitamin B_{12} Solution:

Composition per 100.0mL:

Vitamin B_{12}	10.0mg

Preparation of Vitamin B_{12} Solution: Add vitamin B_{12} to distilled/deionized water and bring volume to 100.0mL. Mix thoroughly. Sparge under 100% N_2 gas for 3 min.

Ferric Citrate Solution:

Composition per 10.0mL:

Ferric citrate	10.0mg

Preparation of Ferric Citrate Solution: Add ferric citrate to distilled/deionized water and bring volume to 10.0mL. Mix thoroughly. Sparge under 100% N_2 gas for 3 min.

Preparation of Medium: Add components to 1050.0mL distilled/deionized water. Mix thoroughly. Gently heat and bring to boiling. Boil for 3–4 min under a stream of 100% N_2. Distribute 45.0mL of the prepared medium into 50.0mL screw-capped tubes that have been flushed with 100% N_2. Autoclave for 15 min at 15 psi pressure–121°C. Cool to room temperature. Sterile syringes are used to inoculate and remove samples. Incubate in the light using a tungsten lamp.

Use: For the cultivation and maintenance of *Rhodocyclus purpureus.*

Rhodomicrobium Medium (LMG Medium 79)

Composition per liter:

Di-odium succinate	1.0g
KH_2PO_4	0.5g
$MgSO_4 \cdot 7H_2O$	0.4g
NaCl	0.4g
NH_4Cl	0.4g
Yeast extract	0.2g
$CaCl_2 \cdot 2H_2O$	50.0mg
Ferric citrate solution	5.0mL
Trace elements solution	1.0mL

pH 5.7 ± 0.2 at 25°C

Ferric Citrate Solution :

Composition per 100.0mL:

Ferric citrate	0.1g

Preparation of Ferric Citrate Solution: Add ferric citrate to distilled/deionized water and bring volume to 100.0mL. Mix thoroughly.

Trace Elements Solution:

Composition per liter:

H_3BO_3	0.3g
$CoCl_2 \cdot 6H_2O$	0.2g
$ZnSO_4 \cdot 7H_2O$	0.1g
$Na_2MoO_4 \cdot 2H_2O$	30.0mg
$MnCl_2 \cdot 4H_2O$	30.0mg
$NiCl_2 \cdot 6H_2O$	20.0mg
$CuCl_2 \cdot 2H_2O$	10.0mg

Preparation of Trace Elements Solution: Add components to distilled/deionized water and bring volume to 1.0L. Mix thoroughly.

Preparation of Medium: Add components to distilled/deionized water and bring volume to 1.0L. Mix thoroughly. Distribute 40.0mL medium into 50mL screw-capped bottles. Flush each bottle for 1 to 2 min with nitrogen gas and then close immediately with rubber septa and screw-caps. Autoclave for 15 min at 15 psi pressure–121°C. Sterile syringes are used to inoculate and remove the samples. Incubate in light using a tungsten lamp.

Use: For the cultivation of *Rhodomicrobium vannielii* and *Rhodoblastus acidophilus.*

Rhodopila globiformis Medium

Composition per liter:

Mannitol	1.5g
Sodium gluconate	0.56g
KH_2PO_4	0.4g
NaCl	0.4g
$MgCl_2 \cdot 6H_2O$	0.4g

NH_4Cl 0.4g
$Na_2S_2O_3 \cdot 5H_2O$ 0.2g
$CaCl_2 \cdot 2H_2O$ 0.05g
Ferric citrate 5.0mg
VA vitamins 1.0mL
Trace elements solution SL-6 .... 1.0mL
pH 4.9 ± 0.2 at 25°C

VA Vitamins:
Composition per 500.0mL:
Nicotinamide 0.175g
Thiamine·HCl 0.15g
p-Aminobenzoic acid 0.1g
Biotin 0.05g
Pyridoxine·2HCl 0.05g
Calcium pantothenate 0.05g
Cyanocobalamin 0.025g

Preparation of VA Vitamins: Add components to distilled/deionized water and bring volume to 500.0mL. Mix thoroughly.

Trace Elements Solution SL-6:
Composition per liter:
H_3BO_3 .... 0.3g
$CoCl_2 \cdot 6H_2O$ 0.2g
$ZnSO_4 \cdot 7H_2O$ 0.1g
$MnCl_2 \cdot 4H_2O$ 0.03g
$Na_2MoO_4 \cdot H_2O$ 0.03g
$NiCl_2 \cdot 6H_2O$ 0.02g
$CuCl_2 \cdot 2H_2O$ 0.01g

Preparation of Trace Elements Solution SL-6: Add components to distilled/deionized water and bring volume to 1.0L. Mix thoroughly. Adjust pH to 3.4.

Preparation of Medium: Add components to distilled/deionized water and bring volume to 1.0L. Mix thoroughly. Adjust pH to 4.9. Distribute into tubes or flasks. Autoclave for 15 min at 15 psi pressure–121°C.

Use: For the cultivation and maintenance of *Rhodopila globiformis.*

Rhodopila globiformis Medium (LMG Medium 81)

Composition per 1008.0mL:
Mannitol 1.5g
Sodium gluconate 0.5g
KH_2PO_4 .... 0.5g
$MgSO_4 \cdot 7H_2O$ 0.4g
NH_4Cl 0.4g
NaCl 0.4g
Yeast extract 0.25g
$CaCl_2 \cdot 2H_2O$ 50.0mg
Ferric citrate solution 5.0mL
Na_2SO_3 solution 2.0mL
Trace elements solution 1.0mL
Biotin solution 1.0mL
p-Aminobenzoic acid solution 1.0mL
pH 4.9 ± 0.2 at 25°C

Ferric Citrate Solution:
Composition per 100.0mL:
Ferric citrate 0.1g

Preparation of Ferric Citrate Solution: Add ferric citrate to distilled/deionized water and bring volume to 100.0mL. Mix thoroughly.

Trace Elements Solution:
Composition per liter:
H_3BO_3 .... 0.3g
$CoCl_2 \cdot 6H_2O$ 0.2g
$ZnSO_4 \cdot 7H_2O$ 0.1g
$Na_2MoO_4 \cdot 2H_2O$ 30.0mg
$MnCl_2 \cdot 4H_2O$ 30.0mg
$NiCl_2 \cdot 6H_2O$ 20.0mg
$CuCl_2 \cdot 2H_2O$ 10.0mg

Preparation of Trace Elements Solution: Add components to distilled/deionized water and bring volume to 1.0L. Mix thoroughly.

Biotin Solution:
Composition per 100.0mL:
Biotin 2.0mg

Preparation of Biotin Solution: Add biotin to distilled/deionized water and bring volume to 100.0mL. Mix thoroughly.

***p*-Aminobenzoic Acid Solution:**
Composition per 100.0mL:
p-Aminobenzoic Acid 10.0mg

Preparation of *p*-Aminobenzoic Acid Solution: Add *p*-aminobenzoic acid to distilled/deionized water and bring volume to 100.0mL. Mix thoroughly.

$Na_2S_2O_3$ Solution:
Composition per 10.0mL:
$Na_2S_2O_3$ 1.0g

Preparation of $Na_2S_2O_3$ Solution: Add $Na_2S_2O_3$ to distilled/deionized water and bring volume to 10.0mL. Mix thoroughly. Autoclave for 15 min at 15 psi pressure–121°C.

Preparation of Medium: Add components, except $Na_2S_2O_3$ solution, ferric citrate solution, biotin solution, and *p*-aminobenzoic acid solution, to distilled deionized water and bring volume to 1.0mL. Mix thoroughly. Adjust pH to 4.9. Add 5.0mL of ferric citrate solution, 1.0mL of biotin solution, and 1.0mL of *p*-aminobenzoic acid solution. Mix thoroughly. Distribute into screw-capped tubes or bottles. Autoclave for 15 min at 15 psi pressure–121°C. Allow to cool to room temperature. Aseptically add 0.2mL of sterile $Na_2S_2O_3$ solution to each 100.0mL of medium. Mix thoroughly.

Use: For the cultivation and maintenance of *Rhodopila globiformis* spp.

Rhodopseudomonas blastica Medium

Composition per liter:
Sodium pyruvate 1.5g
Sodium hydrogen malate 1.5g
Yeast extract 1.0g
NH_4Cl 0.5g
$MgSO_4 \cdot 7H_2O$ 0.4g
NaCl 0.4g
$CaCl_2 \cdot 2H_2O$ 0.05g
Sodium pyryuvate solution 50.0mL
Sodium hydrogen malate solution 50.0mL
Sodium phosphate buffer (0.1*M*, pH 6.8) 50.0mL
pH 6.8 ± 0.2 at 25°C

Sodium Pyruvate Solution :
Composition per 50.0mL:
Sodium pyruvate 1.5g

Preparation of Sodium Pyruvate Solution: Add sodium pyruvate to distilled/deionized water and bring volume to 50.0mL. Mix thoroughly. Filter sterilize.

Sodium Hydrogen Malate Solution :
Composition per 50.0mL:

Sodium hydrogen malate	1.5g

Preparation of Sodium Hydrogen Malate Solution: Add sodium hydrogen malate to distilled/deionized water and bring volume to 50.0mL. Mix thoroughly. Filter sterilize.

Preparation of Medium: Add components—except sodium pyruvate solution, sodium hydrogen malate solution, and sodium phosphate buffer—to distilled/deionized water and bring volume to 850.0mL. Mix thoroughly. Gently heat and bring to boiling. Adjust pH to 6.8 with KOH. Autoclave for 15 min at 15 psi pressure–121°C. Cool to 45°–50°C. Filter sterilize the sodium pyruvate solution, sodium hydrogen malate solution, and sodium phosphate buffer. Aseptically add 50.0mL of sterile sodium pyruvate solution, 50.0mL of sodium hydrogen malate solution, and 50.0mL of sodium phosphate buffer to cooled basal medium. Mix thoroughly. Pour into sterile Petri dishes or distribute into sterile tubes.

Use: For the cultivation and maintenance of *Rhodopseudomonas blastica* and other *Rhodopseudomonas* species.

Rhodopseudomonas globiformis Medium

Composition per 1002.0mL:

Mannitol	1.5g
KH_2PO_4	0.5g
Sodium gluconate	0.5g
$MgSO_4 \cdot 7H_2O$	0.4g
NaCl	0.4g
NH_4Cl	0.4g
Yeast extract	0.25g
$CaCl_2 \cdot 2H_2O$	0.05g
Ferric citrate solution	5.0mL
$Na_2S_2O_3$ solution	2.0mL
Biotin solution	1.0mL
p-Aminobenzoic acid solution	1.0mL
Trace elements solution SL-6	1.0mL

pH 4.9 ± 0.2 at 25°C

Ferric Citrate Solution:
Composition per 10.0mL:

Ferric citrate	0.01g

Preparation of Ferric Citrate Solution: Add ferric citrate to distilled/deionized water and bring volume to 10.0mL. Mix thoroughly. Autoclave for 15 min at 15 psi pressure–121°C.

$Na_2S_2O_3$ Solution :
Composition per 10.0mL:

$Na_2S_2O_3$	1.0g

Preparation of $Na_2S_2O_3$ Solution: Add $Na_2S_2O_3$ to distilled/deionized water and bring volume to 10.0mL. Mix thoroughly. Autoclave for 15 min at 15 psi pressure–121°C.

Biotin Solution:
Composition per 10.0mL:

Biotin	0.2mg

Preparation of Biotin Solution: Add biotin to distilled/deionized water and bring volume to 10.0mL. Mix thoroughly. Autoclave for 15 min at 15 psi pressure–121°C.

***p*-Aminobenzoic Acid Solution:**
Composition per 10.0mL:

p-Aminobenzoic acid	1.0mg

Preparation of *p*-Aminobenzoic Acid Solution: Add *p*-aminobenzoic acid to distilled/deionized water and bring volume to 10.0mL. Mix thoroughly. Autoclave for 15 min at 15 psi pressure–121°C.

Trace Elements Solution SL-6:
Composition per liter:

$MnCl_2 \cdot 4H_2O$	0.5g
H_3BO_3	0.3g
$CoCl_2 \cdot 6H_2O$	0.2g
$ZnSO_4 \cdot 7H_2O$	0.1g
$Na_2MoO_4 \cdot 2H_2O$	0.03g
$NiCl_2 \cdot 6H_2O$	0.02g
$CuCl_2 \cdot 2H_2O$	0.01g

Preparation of Trace Elements Solution SL-6: Add components to distilled/deionized water and bring volume to 1.0L. Mix thoroughly.

Preparation of Medium: Add components, except $Na_2S_2O_3$ solution, ferric citrate solution, biotin solution, and *p*-aminobenzoic acid solution, to distilled deionized water and bring volume to 993.0mL. Mix thoroughly. Adjust pH to 4.9. Add 5.0mL of ferric citrate solution, 1.0mL of biotin solution, and 1.0mL of *p*-aminobenzoic acid solution. Mix thoroughly. Distribute into screw-capped tubes or bottles. Autoclave for 15 min at 15 psi pressure–121°C. Allow to cool to room temperature. Aseptically add 0.2mL of sterile $Na_2S_2O_3$ solution to each 100.0mL of medium. Mix thoroughly.

Use: For the cultivation and maintenance of *Rhodopila globiformis*.

Rhodopseudomonas julia Medium

Composition per liter:

$NaHCO_3$	3.0g
NaCl	1.0g
KH_2PO_4	1.0g
NH_4Cl	1.0g
Sodium acetate	1.0g
Na_2SO_4	0.7g
$MgCl_2 \cdot 6H_2O$	0.5g
Sodium ascorbate	0.5g
$CaCl_2 \cdot 2H_2O$	0.1g
Yeast extract	0.1g
$Na_2S \cdot 9H_2O$ solution	10.0mL
SLA trace elements solution	1.0mL
VA vitamin solution	1.0mL

pH 6.9–7.0 at 25°C

$Na_2S \cdot 9H_2O$ Solution:
Composition per 10.0mL:

$Na_2S \cdot 9H_2O$	0.156g

Preparation of $Na_2S \cdot 9H_2O$ Solution: Add $Na_2S \cdot 9H_2O$ to distilled/deionized water and bring volume to 10.0mL. Mix thoroughly. Autoclave for 15 min at 15 psi pressure–121°C. Before use, neutralize to pH 7.0 with sterile HCl.

SLA Trace Elements Solution:
Composition per liter:

$FeCl_2 \cdot 4H_2O$	1.8g
H_3BO_3	0.5g
$CoCl_2 \cdot 6H_2O$	0.25g
$ZnCl_2$	0.1g

$MnCl_2 \cdot 4H_2O$ 70.0mg
$Na_2MoO_4 \cdot 2H_2O$ 30.0mg
$CuCl_2 \cdot 2H_2O$ 10.0mg
$Na_2SeO_3 \cdot 5H_2O$ 10.0mg
$NiCl_2 \cdot 6H_2O$ 10.0mg

Preparation of SLA Trace Elements Solution: Add components to distilled/deionized water and bring volume to 1.0L. Mix thoroughly. Adjust pH to 2.0–3.0.

VA Vitamin Solution:
Composition per liter:
Nicotinamide 0.175g
Thiamine·HCl 0.15g
p-Aminobenzoic acid 0.1g
Biotin 50.0mg
Calcium D-(+)-pantothenate 50.0mg
Pyridoxine·2HCl 50.0mg
Cyanocobalamin 25.0mg

Preparation of VA Vitamin Solution: Add components to distilled/deionized water and bring volume to 1.0L. Mix thoroughly. Filter sterilize.

Preparation of Medium: Add components, except $Na_2S \cdot 9H_2O$ solution, to distilled/deionized water and bring volume to 990.0mL. Mix thoroughly. Adjust pH to 6.9–7.0. Autoclave for 15 min at 15 psi pressure–121°C. Aseptically add 10.0mL of sterile $Na_2S \cdot 9H_2O$ solution. Mix thoroughly. Aseptically distribute into sterile tubes or flasks.

Use: For the cultivation of *Rhodopseudomonas julia.*

Rhodopseudomonas Medium (ATCC Medium 543)

Composition per liter:
Sodium succinate 2.5g
$(NH_4)_2SO_4$ 1.25g
K_2HPO_4 .... 0.9g
KH_2PO_4 .... 0.6g
Yeast extract 0.5g
$MgSO_4 \cdot 7H_2O$ 0.2g
$CaCl_2$ 0.07g
Ferric citrate 3.0mg
Ethylenediamine tetraacetate (EDTA) 2.0mg
pH 7.0 ± 0.2 at 25°C

Preparation of Medium: Add components to distilled/deionized water and bring volume to 1.0L. Mix thoroughly. Distribute into tubes or flasks. Autoclave for 15 min at 15 psi pressure–121°C.

Use: For the cultivation and maintenance of *Rhodopseudomonas* species.

Rhodopseudomonas Medium (ATCC Medium 650)

Composition per liter:
Sodium succinate 1.5g
KH_2PO_4 .... 1.0g
NH_4Cl 0.5g
$MgSO_4 \cdot 7H_2O$ 0.4g
NaCl 0.4g
$CaCl_2 \cdot 2H_2O$ 0.05g
Trace metals solution 10.0mL
pH 5.6–6.0 at 25°C

Trace Metals Solution:
Composition per 100.0mL:
Ferric citrate 0.3g
Ethylenediamine tetraacetic acid (EDTA) 0.05g
$CaCl_2 \cdot 2H_2O$ 0.02g
$MnSO_4 \cdot H_2O$ 0.002g
$(NH_4)_6Mo_7O_{24} \cdot 4H_2O$ 0.002g
H_3BO_3 .... 0.001g
$CuSO_4 \cdot 5H_2O$ 0.001g
$ZnSO_4$ 0.001g

Preparation of Trace Metals Solution: Add components to distilled/deionized water and bring volume to 100.0mL. Mix thoroughly. Filter sterilize.

Preparation of Medium: Add components, except trace metals solution, to distilled/deionized water and bring volume to 990.0mL. Mix thoroughly. Autoclave for 15 min at 15 psi pressure–121°C. Cool to 25°C. Aseptically add 10.0mL of trace metals solution. Mix thoroughly. Aseptically distribute into sterile tubes or flasks.

Use: For the cultivation and maintenance of *Rhodopseudomonas viridis, Rhodopseudomonas acidophila*, and other *Rhodopseudomonas* species.

Rhodopseudomonas rutila Medium

Composition per liter:
Sodium glutamate 2.0g
Sodium L-malate 2.0g
Yeast extract 2.0g
KH_2PO_4 .... 1.0g
$NaHCO_3$ 0.5g
$MgSO_4 \cdot 7H_2O$ 0.2g
$CaCl_2 \cdot 2H_2O$ 0.1g
$MnSO_4 \cdot H_2O$ 2.0mg
Biotin 1.0mg
Nicotinic acid 1.0mg
Thiamine·HCl 1.0mg
$CoCl_2 \cdot 6H_2O$ 0.5mg
$FeSO_4 \cdot 7H_2O$ 0.5mg

Preparation of Medium: Add components to distilled/deionized water and bring volume to 1.0L. Mix thoroughly. Distribute into tubes or flasks. Autoclave for 15 min at 15 psi pressure–121°C.

Use: For the cultivation of *Rhodopseudomonas palustris* and *Rhodopseudomonas rutila.*

Rhodopseudomonas sulfoviridis Medium

Composition per 1050.0mL:
Ammonium acetate 1.5g
Sodium malate 1.0g
KH_2PO_4 .... 0.5g
$Na_2S_2O_3$ 0.5g
$MgSO_4 \cdot 7H_2O$ 0.4g
NaCl 0.4g
NH_4Cl 0.4g
Yeast extract 0.3g
Disodium succinate 0.25g
$CaCl_2 \cdot 2H_2O$ 0.05g
Ferric citrate solution 5.0mL
Trace elements solution SL-6 .... 1.0mL
Ethanol 0.5mL

Vitamin B_{12} solution 0.4mL
Neutralized sulfide solution .... variable

pH 6.8 ± 0.2 at 25°C

Ferric Citrate Solution:
Composition per 10.0mL:
Ferric citrate 10.0mg

Preparation of Ferric Citrate Solution: Add ferric citrate to distilled/deionized water and bring volume to 10.0mL. Mix thoroughly. Sparge under 100% N_2 for 3 min. Autoclave for 15 min at 15 psi pressure–121°C. Store under N_2.

Trace Elements Solution SL-6:
Composition per liter:
$MnCl_2·4H_2O$ 0.5g
H_3BO_3 .... 0.3g
$CoCl_2·6H_2O$ 0.2g
$ZnSO_4·7H_2O$ 0.1g
$Na_2MoO_4·2H_2O$ 0.03g
$NiCl_2·6H_2O$ 0.02g
$CuCl_2·2H_2O$ 0.01g

Preparation of Trace Elements Solution SL-6: Add components to distilled/deionized water and bring volume to 1.0L. Mix thoroughly.

Vitamin B_{12} Solution:
Composition per 100.0mL:
Vitamin B_{12} 10.0mg

Preparation of Vitamin B_{12} Solution: Add vitamin B_{12} to distilled/deionized water and bring volume to 100.0mL. Mix thoroughly. Sparge under 100% N_2 for 3 min. Autoclave for 15 min at 15 psi pressure–121°C. Store under N_2.

Neutralized Sulfide Solution:
Composition per 100.0mL:
$Na_2S·9H_2O$ 1.5g

Preparation of Neutralized Sulfide Solution: Add $Na_2S·9H_2O$ to distilled/deionized water in a 250mL screw-capped bottle fitted with a butyl rubber septum and bring volume to 100.0mL. Add a magnetic stir bar. Mix thoroughly. Sparge under 100% N_2 for 3 min. Autoclave for 15 min at 15 psi pressure–121°C. Cool to room temperature. Adjust pH to about 7.3 with sterile 2*M* H_2SO_4. Do not open the bottle to add H_2SO_4; use a sterile syringe. Stir the solution continuously to avoid precipitation of elemental sulfur. The final solution should be clear and yellow in color.

Preparation of Medium: Add components, except neutralized sulfide solution, to distilled/deionized water and bring volume to 1050.0mL. Mix thoroughly. Gently heat and bring to boiling. Boil for 3–4 min under a stream of 100% N_2. Distribute 45.0mL of the prepared medium into 50.0mL screw-capped tubes that have been flushed with 100% N_2. Autoclave for 15 min at 15 psi pressure–121°C. Cool to room temperature. Before inoculation, aseptically and anaerobically add 0.25–0.50mL of neutralized sulfide solution to each tube.

Use: For the cultivation and maintenance of *Rhodopseudomonas sulfoviridis*.

Rhodospirillaceae Enrichment Medium

Composition per liter:
Dicarboxylic acid substrate 1.0g
KH_2PO_4 .... 0.5g
NaCl 0.4g
NH_4Cl 0.4g
$MgSO_4·7H_2O$ 0.2g
Yeast extract 0.2g
$CaCl_2·2H_2O$ 0.05g
Ferric citrate solution 5.0mL
Trace elements solution SL-7 .... 1.0mL
Vitamin B_{12} solution 1.0mL

pH 6.8 ± 0.2 at 25°C

Ferric Citrate Solution:
Composition per 100.0mL:
Ferric citrate 0.1g

Preparation of Ferric Citrate Solution: Add ferric citrate to distilled/deionized water and bring volume to 100.0mL. Mix thoroughly.

Trace Elements Solution SL-7:
Composition per liter:
$CoCl_2·6H_2O$ 0.2g
$MnCl_2·4H_2O$ 0.1g
$ZnCl_2$ 0.07g
H_3BO_3 .... 0.06g
$NaMoO_4·2H_2O$ 0.04g
$CuCl_2·2H_2O$ 0.02g
$NiCl_2·6H_2O$ 0.02g
HCl (25% solution) 1.0mL

Preparation of Trace Elements Solution SL-7: Add components to distilled/deionized water and bring volume to 1.0L. Mix thoroughly.

Vitamin B_{12} Solution:
Composition per 100.0mL:
Vitamin B_{12} 1.0mg

Preparation of Vitamin B_{12} Solution: Add vitamin B_{12} to distilled/deionized water and bring volume to 100.0mL. Mix thoroughly.

Preparation of Medium: Add components to distilled/deionized water and bring volume to 1.0L. Succinic acid or glutaric acid may be used for the dicarboxylic acid substrate. Mix thoroughly. Adjust pH to 6.8. Distribute into tubes or flasks. Autoclave for 15 min at 15 psi pressure–121°C.

Use: For the enrichment and isolation of members of the Rhodospirillaceae.

Rhodospirillaceae Enrichment Medium

Composition per liter:
Fatty acid substrate 1.0g
KH_2PO_4 .... 0.5g
NaCl 0.4g
NH_4Cl 0.4g
$MgSO_4·7H_2O$ 0.2g
Yeast extract 0.2g
$CaCl_2·2H_2O$ 0.05g
$NaHCO_3$ solution 40.0mL
Ferric citrate solution 5.0mL
Trace elements solution SL-7 .... 1.0mL
Vitamin B_{12} solution 1.0mL

pH 7.3 ± 0.2 at 25°C

Ferric Citrate Solution:
Composition per 100.0mL:
Ferric citrate 0.1g

Preparation of Ferric Citrate Solution: Add ferric citrate to distilled/deionized water and bring volume to 100.0mL. Mix thoroughly.

Trace Elements Solution SL-7:

Composition per liter:

$CoCl_2 \cdot 6H_2O$	0.2g
$MnCl_2 \cdot 4H_2O$	0.1g
$ZnCl_2$	0.07g
H_3BO_3	0.06g
$NaMoO_4 \cdot 2H_2O$	0.04g
$CuCl_2 \cdot 2H_2O$	0.02g
$NiCl_2 \cdot 6H_2O$	0.02g
HCl (25% solution)	1.0mL

Preparation of Trace Elements Solution SL-7: Add components to distilled/deionized water and bring volume to 1.0L. Mix thoroughly.

Vitamin B_{12} Solution:

Composition per 100.0mL:

Vitamin B_{12}	1.0mg

Preparation of Vitamin B_{12} Solution: Add vitamin B_{12} to distilled/deionized water and bring volume to 100.0mL. Mix thoroughly.

$NaHCO_3$ Solution:

Composition per 100.0mL:

$NaHCO_3$	5.0g

Preparation of $NaHCO_3$ Solution: Add $NaHCO_3$ to distilled/deionized water and bring volume to 100.0mL. Mix thoroughly. Filter sterilize.

Preparation of Medium: Add components, except $NaHCO_3$ solution, to distilled/deionized water and bring volume to 1.0L. Acetate, propionate, or butyrate salts may be used for the fatty acid substrate. Mix thoroughly. Adjust pH to 7.3. Distribute into flasks in 50.0mL volumes. Autoclave for 15 min at 15 psi pressure–121°C. Cool to 25°C. Immediately prior to inoculation, aseptically add 2.0mL of sterile $NaHCO_3$ solution to each flask containing 50.0mL of medium.

Use: For the enrichment and isolation of members of the Rhodospirillaceae.

Rhodospirillaceae Enrichment Medium

Composition per liter:

Fatty acid or dicarboxylic acid substrate	1.0g
KH_2PO_4	0.5g
NaCl	0.4g
NH_4Cl	0.4g
$MgSO_4 \cdot 7H_2O$	0.2g
Yeast extract	0.2g
$CaCl_2 \cdot 2H_2O$	0.05g
Ferric citrate solution	5.0mL
Trace elements solution SL-7	1.0mL
Vitamin B_{12} solution	1.0mL

pH 5.2–5.5 at 25°C

Ferric Citrate Solution:

Composition per 100.0mL:

Ferric citrate	0.1g

Preparation of Ferric Citrate Solution: Add ferric citrate to distilled/deionized water and bring volume to 100.0mL. Mix thoroughly.

Trace Elements Solution SL-7:

Composition per liter:

$CoCl_2 \cdot 6H_2O$	0.2g
$MnCl_2 \cdot 4H_2O$	0.1g
$ZnCl_2$	0.07g
H_3BO_3	0.06g
$NaMoO_4 \cdot 2H_2O$	0.04g
$CuCl_2 \cdot 2H_2O$	0.02g
$NiCl_2 \cdot 6H_2O$	0.02g
HCl (25% solution)	1.0mL

Preparation of Trace Elements Solution SL-7: Add components to distilled/deionized water and bring volume to 1.0L. Mix thoroughly.

Vitamin B_{12} Solution:

Composition per 100.0mL:

Vitamin B_{12}	1.0mg

Preparation of Vitamin B_{12} Solution: Add vitamin B_{12} to distilled/deionized water and bring volume to 100.0mL. Mix thoroughly.

Preparation of Medium: Add components to distilled/deionized water and bring volume to 1.0L. Acetate, propionate, or butyrate salts may be used for the fatty acid substrate. Succinic acid or glutaric acid may be used for the dicarboxylic acid substrate. Lactate or ethanol may be used as an alternate substrate. Mix thoroughly. Adjust pH to 5.2–5.5. Distribute into tubes or flasks. Autoclave for 15 min at 15 psi pressure–121°C.

Use: For the enrichment and isolation of *Rhodopseudomonas acidophila* and *Rhodomicrobium vannielii.*

Rhodospirillaceae Medium

Composition per liter:

Succinic acid	1.0g
KH_2PO_4	0.5g
NaCl	0.4g
NH_4Cl	0.4g
$MgSO_4 \cdot 7H_2O$	0.2g
Yeast extract	0.2g
$CaCl_2 \cdot 2H_2O$	0.05g
Ferric citrate solution	5.0mL
Vitamin B_{12} solution	1.0mL
Trace elements solution SL7	1.0mL

pH 6.8 ± 0.2 at 25°C

Ferric Citrate Solution:

Composition per 100.0mL:

Ferric citrate	0.1g

Preparation of Ferric Citrate Solution: Add ferric citrate to distilled/deionized water and bring volume to 100.0mL. Mix thoroughly. Filter sterilize.

Vitamin B_{12} Solution:

Composition per 100.0mL:

Vitamin B_{12}	1.0mg

Preparation of Vitamin B_{12} Solution: Add vitamin B_{12} to distilled/deionized water and bring volume to 100.0mL. Mix thoroughly. Filter sterilize.

Trace Elements Solution SL-7:

Composition per liter:

$MnCl_2 \cdot 4H_2O$	100.0mg
ZnCl	70.0mg
H_3Bo_3	60.0mg
$NaMoO_4 \cdot 2H_2O$	40.0mg
$CoCl_2 \cdot 2H_2O$	20.0mg
$CuCl_2 \cdot 2H_2O$	20.0mg

$NiCl_2 \cdot 6H_2O$ 20.0mg
HCl (25%) 1.0mL

Preparation of Trace Elements Solution SL7: Add 100mg of $MnCl_2 \cdot 4H_2O$ to 1.0mL of HCl solution. Mix thoroughly. Add distilled/deionized water and bring volume to 1.0L. Add remaining components. Mix thoroughly. Sparge with 100% N_2. Autoclave for 15 min at 15 psi pressure–121°C.

Preparation of Medium: Add components, except ferric -citrate solution, vitamin B_{12} solution, and trace elements solution SL-7, to distilled/deionized water and bring volume to 993.0mL. Mix thoroughly. Adjust pH to 6.8. Autoclave for 30 min at 15 psi pressure–121°C. Aseptically add 5.0mL of sterile ferric citrate solution, 1.0mL of sterile vitamin B_{12} solution, and 1.0mL of sterile trace elements solution SL-7. Mix thoroughly. Aseptically distribute into sterile screw-capped tubes under anaerobic conditions. Tighten screw caps.

Use: For the cultivation of *Rhodospirillum* species and other members of the family Rhodospirillaceae.

Rhodospirillaceae Medium, Modified

Composition per 1050.0mL:

Ammonium acetate 1.5g
KH_2PO_4 .... 0.5g
$MgSO_4.7H_2O$ 0.4g
NaCl 0.4g
NH_4Cl 0.4g
Yeast extract 0.3g
Disodium succinate 0.25g
$CaCl_2 \cdot 2H_2O$ 0.05g
Ferric citrate solution 5.0mL
Trace elements solution SL-6 .... 1.0mL
Ethanol 0.5mL
Vitamin B_{12} solution 0.4mL
Neutralized sulfide solution .... variable

pH 6.8 ± 0.2 at 25°C

Ferric Citrate Solution:
Composition per 10.0mL:

Ferric citrate 10.0mg

Preparation of Ferric Citrate Solution: Add ferric citrate to distilled/deionized water and bring volume to 10.0mL. Mix thoroughly. Sparge under 100% N_2 for 3 min. Autoclave for 15 min at 15 psi pressure–121°C. Store under N_2 gas.

Trace Elements Solution SL-6:
Composition per liter:

$MnCl_2 \cdot 4H_2O$ 0.5g
H_3BO_3 .... 0.3g
$CoCl_2 \cdot 6H_2O$ 0.2g
$ZnSO_4 \cdot 7H_2O$ 0.1g
$Na_2MoO_4 \cdot 2H_2O$ 0.03g
$NiCl_2 \cdot 6H_2O$ 0.02g
$CuCl_2 \cdot 2H_2O$ 0.01g

Preparation of Trace Elements Solution SL-6: Add components to distilled/deionized water and bring volume to 1.0L. Mix thoroughly.

Vitamin B_{12} Solution:
Composition per 100.0mL:

Vitamin B_{12} 10.0mg

Preparation of Vitamin B_{12} Solution: Add vitamin B_{12} to distilled/deionized water and bring volume to 100.0mL. Mix thoroughly. Sparge under 100% N_2 gas for 3 min. Autoclave for 15 min at 15 psi pressure–121°C. Store under N_2 gas.

Neutralized Sulfide Solution:
Composition per 100.0mL:

$Na_2S \cdot 9H_2O$ 1.5g

Preparation of Neutralized Sulfide Solution: Add $Na_2S \cdot 9H_2O$ to distilled/deionized water in a 250mL screw-capped bottle fitted with a butyl rubber septum and bring volume to 100.0mL. Add a magnetic stir bar. Mix thoroughly. Sparge under 100% N_2 gas for 3 min. Autoclave for 15 min at 15 psi pressure–121°C. Cool to room temperature. Adjust pH to about 7.3 with sterile 2*M* H_2SO_4. Do not open the bottle to add H_2SO_4; use a sterile syringe. Stir the solution continuously to avoid precipitation of elemental sulfur. The final solution should be clear and yellow in color.

Preparation of Medium: Add components, except neutralized sulfide solution, to distilled/deionized water and bring volume to 1050.0mL. Mix thoroughly. Gently heat and bring to boiling. Boil for 3–4 min under a stream of 100% N_2. Distribute 45.0mL of the prepared medium into 50.0mL screw-capped tubes that have been flushed with 100% N_2. Autoclave for 15 min at 15 psi pressure–121°C. Cool to room temperature. Before inoculation, aseptically and anaerobically add 0.25–0.50mL of neutralized sulfide solution to each tube.

Use: For the cultivation and maintenance of *Ectothiorhodospira marismortui, Rhodobacter capsulatus, Rhodobacter sphaeroides, Rhodobacter sulfidophilus, Rhodocyclus tenuis, Rhodopseudomonas blastica, Rhodopseudomonas marina, Rhodopseudomonas palustris, Rhodopseudomonas rosea, Rhodopseudomonas viridis, Rhodospirillum fulvum, Rhodospirillum molischianum, Rhodospirillum photometricum, Rhodospirillum rubrum, Rhodospirillum salexigens,* and *Rubrivivax gelatinosus.*

Rhodospirillum Medium

Composition per 1006.5mL:

Disodium succinate 1.0g
Yeast extract 0.8g
KH_2PO_4 .... 0.5g
$MgSO_4 \cdot 7H_2O$ 0.4g
NaCl 0.4g
NH_4Cl 0.4g
Yeast extract 0.2g
$CaCl_2 \cdot H_2O$ 50.0mg
Ferric citrate solution 5.0mL
Trace elements solution 1.0mL
Ethanol 0.5mL

pH 6.8 ± 0.2 at 25°C

Ferric Citrate Solution:
Composition per 10.0mL:

Ferric citrate 0.01g

Preparation of Ferric Citrate Solution: Add ferric citrate to distilled/deionized water and bring volume to 10.0mL. Mix thoroughly. Filter sterilize.

Trace Elements Solution:
Composition per liter:

H_3BO_3 .... 0.3g
$CoCl_2 \cdot 6H_2O$ 0.2g
$ZnSO_4 \cdot 7H_2O$ 0.1g
$MnCl_2 \cdot 4H_2O$ 0.03g
$Na_2MoO_4 \cdot 2H_2O$ 0.03g
$NiCl_2 \cdot 6H_2O$ 0.02g
$CuCl_2 \cdot 2H_2O$ 0.01g

Preparation of Trace Elements Solution: Add components to distilled/deionized water and bring volume to 1.0L. Mix thoroughly. Filter sterilize.

Preparation of Medium: Add components, except ferric citrate solution, trace elements solution, and ethanol, to distilled/deionized water and bring volume to 1.0L. Mix thoroughly. Autoclave for 15 min at 15 psi pressure–121°C. Aseptically add 5.0mL of sterile ferric citrate solution, 1.0mL of sterile trace elements solution, and 0.5mL of filter-sterilized ethanol. Mix thoroughly. Aseptically distribute into sterile tubes or flasks.

Use: For the cultivation of *Rhodospirillum fulvum.*

Rhodospirillum Medium (ATCC Medium 1308)

Composition per liter:

Yeast extract	1.0g
Disodium succinate	1.0g
KH_2PO_4	0.5g
Sodium ascorbate	0.5g
$MgSO_4 \cdot 7H_2O$	0.4g
NaCl	0.4g
NH_4Cl	0.4g
$CaCl_2 \cdot 2H_2O$	0.05g
Ferric citrate (0.1% solution)	5.0mL
Trace elements solution SL-6	1.0mL
Ethanol	0.5mL

pH 6.0 ± 0.2 at 25°C

Trace Elements Solution SL-6:

Composition per liter:

H_3BO_3	0.3g
$CoCl_2 \cdot 6H_2O$	0.2g
$ZnSO_4 \cdot 7H_2O$	0.1g
$MnCl_2 \cdot 4H_2O$	0.03g
$Na_2MoO_4 \cdot H_2O$	0.03g
$NiCl_2 \cdot 6H_2O$	0.02g
$CuCl_2 \cdot 2H_2O$	0.01g

Preparation of Trace Elements Solution SL-6: Add components to distilled/deionized water and bring volume to 1.0L. Mix thoroughly. Adjust pH to 3.4.

Preparation of Medium: Add components to distilled/deionized water and bring volume to 1.0L. Mix thoroughly. Adjust pH to 6.0. Distribute into tubes or flasks. Autoclave for 15 min at 15 psi pressure–121°C.

Use: For the cultivation and maintenance of *Rhodospirillum* species.

Rhodospirillum Medium (ATCC Medium 1408)

Composition per liter:

NaCl	100.0g
$MgCl_2 \cdot 6H_2O$	3.5g
Yeast extract	1.5g
Peptone	1.5g
Sodium malate	1.4g
KH_2PO_4	0.3g
SLA trace elements	1.0mL

pH 7.0 ± 0.2 at 25°C

SLA Trace Elements:

Composition per liter:

$FeCl_2 \cdot 4H_2O$	1.8g
H_3BO_3	0.5g
$CoCl_2 \cdot 6H_2O$	0.25g
$ZnCl_2$	0.1g
$MnCl_2 \cdot 4H_2O$	0.07g
$Na_2MoO_4 \cdot 2H_2O$	0.03g
$NiCl_2 \cdot 6H_2O$	0.01g
$CuCl_2 \cdot 2H_2O$	0.01g
$Na_2SeO_3 \cdot 5H_2O$	0.01g

Preparation of SLA Trace Elements: Add components to distilled/deionized water and bring volume to 1.0L. Mix thoroughly. Adjust pH to 2–3.

Preparation of Medium: Add components to distilled/deionized water and bring volume to 1.0L. Mix thoroughly. Adjust medium to final pH of 7.0. Sterilize by autoclaving at 121°C for 15 min.

Use: For the cultivation of *Rhodospirillum* species.

Rhodospirillum Medium, Modified I

Composition per liter:

Disodium succinate	1.0g
KH_2PO_4	0.5g
$MgSO_4 \cdot 7H_2O$	0.4g
NaCl	0.4g
NH_4Cl	0.4g
Yeast extract	0.2g
$CaCl_2 \cdot 2H_2O$	0.05g
Ferric citrate (0.1% solution)	5.0mL
Trace elements solution SL-6	1.0mL

pH 5.7 ± 0.2 at 25°C

Trace Elements Solution SL-6:

Composition per liter:

H_3BO_3	0.3g
$CoCl_2 \cdot 6H_2O$	0.2g
$ZnSO_4 \cdot 7H_2O$	0.10g
$MnCl_2 \cdot 4H_2O$	0.03g
$Na_2MoO_4 \cdot H_2O$	0.03g
$NiCl_2 \cdot 6H_2O$	0.02g
$CuCl_2 \cdot 2H_2O$	0.01g

Preparation of Trace Elements Solution SL-6: Add components to distilled/deionized water and bring volume to 1.0L. Mix thoroughly. Adjust pH to 3.4.

Preparation of Medium: Add components to distilled/deionized water and bring volume to 1.0L. Mix thoroughly. Adjust pH to 5.7. Distribute 40.0mL volumes into tubes or bottles. Sparge with 100% N_2 for 1–2 min. Autoclave for 15 min at 15 psi pressure–121°C.

Use: For the cultivation and maintenance of *Rhodomicrobium vannielii, Rhodopseudomonas acidophila,* and *Rhodobacter capsulatus.*

Rhodospirillum Medium, Modified II

Composition per liter:

Yeast extract	1.0g
Disodium succinate	1.0g
KH_2PO_4	0.5g
$MgSO_4 \cdot 7H_2O$	0.4g
NaCl	0.4g
NH_4Cl	0.4g

$CaCl_2 \cdot 2H_2O$ 0.05g
Ferric citrate (0.1% solution) 5.0mL
Trace elements solution SL-6 .... 1.0mL
Ethanol 0.5mL

pH 6.8 ± 0.2 at 25°C

Trace Elements Solution SL-6:

Composition per liter:

H_3BO_3 .... 0.3g
$CoCl_2 \cdot 6H_2O$ 0.2g
$ZnSO_4 \cdot 7H_2O$ 0.10g
$MnCl_2 \cdot 4H_2O$ 0.03g
$Na_2MoO_4 \cdot H_2O$ 0.03g
$NiCl_2 \cdot 6H_2O$ 0.02g
$CuCl_2 \cdot 2H_2O$ 0.01g

Preparation of Trace Elements Solution SL-6: Add components to distilled/deionized water and bring volume to 1.0L. Mix thoroughly. Adjust pH to 3.4.

Preparation of Medium: Add components to distilled/deionized water and bring volume to 1.0L. Mix thoroughly. Adjust pH to 6.8. Distribute 40.0mL volumes into tubes or bottles. Sparge with 100% N_2 for 1–2 min. Autoclave for 15 min at 15 psi pressure–121°C.

Use: For the cultivation and maintenance of *Rhodobacter sphaeroides, Rhodocyclus tenuis, Rhodopseudomonas blastica, Rhodopseudomonas palustris, Rhodopseudomonas viridis, Rhodospirillum fulvum, Rhodospirillum molischianum, Rhodospirillum photometricum, Rhodospirillum rubrum,* and *Rubrivivax gelatinosus.*

Rhodospirillum Medium, Modified III

Composition per liter:

Yeast extract 1.0g
Ammonium acetate 0.5g
KH_2PO_4 .... 0.5g
$MgSO_4 \cdot 7H_2O$ 0.4g
NaCl 0.4g
NH_4Cl 0.4g
$CaCl_2 \cdot 2H_2O$ 0.05g
Vitamin B_{12} 20.0mg
Ferric citrate (0.1% solution) 5.0mL
Trace elements solution SL-6 .... 1.0mL
Ethanol 0.5mL

pH 6.8 ± 0.2 at 25°C

Trace Elements Solution SL-6:

Composition per liter:

H_3BO_3 .... 0.3g
$CoCl_2 \cdot 6H_2O$ 0.2g
$ZnSO_4 \cdot 7H_2O$ 0.1g
$MnCl_2 \cdot 4H_2O$ 0.03g
$Na_2MoO_4 \cdot H_2O$ 0.03g
$NiCl_2 \cdot 6H_2O$ 0.02g
$CuCl_2 \cdot 2H_2O$ 0.01g

Preparation of Trace Elements Solution SL-6: Add components to distilled/deionized water and bring volume to 1.0L. Mix thoroughly. Adjust pH to 3.4.

Preparation of Medium: Add components to distilled/deionized water and bring volume to 1.0L. Mix thoroughly. Adjust pH to 6.8. Distribute 40.0mL volumes into tubes or bottles. Sparge with 100% N_2 for 1–2 min. Autoclave for 15 min at 15 psi pressure–121°C.

Use: For the cultivation and maintenance of *Rhodocyclus purpureus.*

Rhodospirillum Medium, Modified IV

Composition per liter:

NaCl 25.0g
Yeast extract 1.0g
Disodium succinate 1.0g
KH_2PO_4 .... 0.5g
$MgSO_4 \cdot 7H_2O$ 0.4g
NaCl 0.4g
NH_4Cl 0.4g
$CaCl_2 \cdot 2H_2O$ 0.05g
Ferric citrate (0.1% solution) 5.0mL
Trace elements solution SL-6 .... 1.0mL
Ethanol 0.5mL

pH 6.8 ± 0.2 at 25°C

Trace Elements Solution SL-6:

Composition per liter:

H_3BO_3 .... 0.3g
$CoCl_2 \cdot 6H_2O$ 0.2g
$ZnSO_4 \cdot 7H_2O$ 0.1g
$MnCl_2 \cdot 4H_2O$ 0.03g
$Na_2MoO_4 \cdot H_2O$ 0.03g
$NiCl_2 \cdot 6H_2O$ 0.02g
$CuCl_2 \cdot 2H_2O$ 0.01g

Preparation of Trace Elements Solution SL-6: Add components to distilled/deionized water and bring volume to 1.0L. Mix thoroughly. Adjust pH to 3.4.

Preparation of Medium: Add components to distilled/deionized water and bring volume to 1.0L. Mix thoroughly. Adjust pH to 6.8. Distribute 40.0mL volumes into tubes or bottles. Sparge with 100% N_2 for 1–2 min. Autoclave for 15 min at 15 psi pressure–121°C.

Use: For the cultivation and maintenance of *Rhodobacter sulfidophilus.*

Rhodospirillum salinarum Medium

Composition per liter:

NaCl 100.0g
KCl 5.0g
$MgCl_2 \cdot 6H_2O$ 5.0g
$MgSO_4 \cdot 7H_2O$ 5.0g
$NH_4 \cdot Cl$ 5.0g
Peptone solution 30.0mL
Yeast extract solution 30.0mL
Ferric citrate solution 10.0mL
Trace elements solution 5.0mL

Peptone Solution:

Composition per 100.0mL:

Peptone 15.0g

Preparation of Peptone Solution: Add peptone to distilled/deionized water and bring volume to 100.0mL. Mix thoroughly. Filter sterilize.

Yeast Extract Solution:

Composition per 10.0mL:

Yeast extract 15.0g

Preparation of Yeast Extract Solution: Add yeast extract to distilled/deionized water and bring volume to 100.0mL. Mix thoroughly. Filter sterilize.

Ferric Citrate Solution:

Composition per 10.0mL:

Ferric citrate 0.1g

Preparation of Ferric Citrate Solution: Add ferric citrate to distilled/deionized water and bring volume to 10.0mL. Mix thoroughly. Filter sterilize.

Trace Elements Solution:
Composition per liter:

$CoCl_2 \cdot 6H_2O$	10.0g
$ZnSO_4 \cdot 7H_2O$	220.0mg
$MgCl_2 \cdot 4H_2O$	180.0mg
$Na_2MoO_4 \cdot 2H_2O$	6.3mg
$CuSO_4 \cdot 5H_2O$	1.0mg

Preparation of Medium: Add components, except peptone solution, yeast extract solution, ferric citrate solution, and trace elements solution to distilled/deionized water and bring volume to 925.0mL. Mix thoroughly. Autoclave for 15 min at 15 psi pressure–121°C. Aseptically add 30.0mL of sterile peptone solution, 30.0mL of sterile yeast extract solution, 10.0mL of sterile ferric citrate solution, and 5.0mL of sterile trace elements solution. Mix thoroughly. Aseptically distribute into sterile tubes or flasks.

Use: For the cultivation of *Rhodospirillum salinarum*.

Rhodovulum iodosum Rhodovulum robiginosum Medium (DSMZ Medium 929)

Composition per liter:

NaCl	26.4g
$MgSO_4 \cdot 7H_2O$	6.8g
$MgCl_2 \cdot 6H_2O$	5.7g
$CaCl_2 \cdot 2H_2O$	1.5g
KCl	0.66g
KBr	0.09g
$NaHCO_3$ solution	30.0mL
Phosphate solution	10.0mL
Sodium acetate solution	10.0mL
Iron sulfate solution	10.0mL
NH_4Cl solution	1.0mL
Thiosulfate solution	1.0mL
Selenite-tungstate solution	1.0mL
Trace elements solution	1.0mL
Vitamin solution	1.0mL
Vitamin B_{12} solution	1.0mL
Vitamin B_1 solution	1.0mL

pH 6.8 ± 0.2 at 25°C

Sodium Acetate Solution:
Composition per 100.0mL:

Na-acetate	4.1g

Preparation of Sodium Acetate Solution: Add sodium acetate to distilled/deionized water and bring volume to 100.0mL. Mix thoroughly. Sparge with 100% N_2. Filter sterilize.

Trace Elements Solution:
Composition per liter:

$CoCl_2 \cdot 6H_2O$	190.0mg
$ZnSO_4 \cdot 7H_2O$	144.0mg
$MnCl_2 \cdot 4H_2O$	100.0mg
$Na_2MoO_4 \cdot 4H_2O$	36.0mg
H_3BO_3	30.0mg
$NiCl_2 \cdot 6H_2O$	24.0mg
Na_2EDTA	5.2mg
$FeSO_4 \cdot 7H_2O$	2.1mg
$CuCl_2 \cdot 2H_2O$	2.0mg

Preparation of Trace Elements Solution: Add components to distilled/deionized water and bring volume to 1.0L. Mix thoroughly. Adjust pH to 6.0. Sparge with 100% N_2. Autoclave for 15 min at 15 psi pressure–121°C. Cool to room temperature.

Vitamin Solution:
Composition per 100.0mL:

Pyridoxine	15.0mg
Nicotinate	10.0mg
Pantothenate	5.0mg
Para-aminobenzoic acid	4.0mg
Biotin	1.0mg

Preparation of Vitamin Solution: Add components to distilled/deionized water and bring volume to 100.0mL. Mix thoroughly. Sparge with 100% N_2. Filter sterilize.

Vitamin B_{12} Solution:
Composition per 100.0mL:

Cyanocobalamine	5.0mg

Preparation of Vitamin B_{12} Solution: Add cyanocobalamine to distilled/deionized water and bring volume to 100.0mL. Mix thoroughly. Sparge with 100% N_2. Filter sterilize.

Vitamin B_1 Solution:
Composition per 100.0mL:

Thiamine	10.0mg

Preparation of Vitamin B_1 Solution: Add thiamine to distilled/deionized water and bring volume to 100.0mL. Mix thoroughly. Sparge with 100% N_2. Filter sterilize.

NH_4Cl Solution:
Composition per 10.0mL:

NH_4Cl	2.5g

Preparation of NH_4Cl Solution: Add NH_4Cl to distilled/deionized water and bring volume to 10.0mL. Mix thoroughly. Sparge with 100% N_2. Autoclave for 15 min at 15 psi pressure–121°C. Cool to room temperature.

Thiosulfate Solution:
Composition per 10.0mL:

$Na_2S_2O_3 \cdot 5H_2O$	1.24g

Preparation of Thiosulfate Solution: Add $Na_2S_2O_3 \cdot 5H_2O$ to distilled/deionized water and bring volume to 10.0mL. Mix thoroughly. Sparge with 100% N_2. Autoclave for 15 min at 15 psi pressure–121°C. Cool to room temperature.

$NaHCO_3$ Solution:
Composition per 100.0mL:

$NaHCO_3$	8.4g

Preparation of $NaHCO_3$ Solution: Add $NaHCO_3$ to distilled/deionized water and bring volume to 100.0mL. Mix thoroughly. Sparge with 80% N_2 + 20% CO_2. Autoclave for 15 min at 15 psi pressure–121°C under an atmosphere of CO_2. Cool to room temperature.

Selenite-Tungstate Solution
Composition per liter:

NaOH	0.5g
$Na_2WO_4 \cdot 2H_2O$	4.0mg
$Na_2SeO_3 \cdot 5H_2O$	3.0mg

Preparation of Selenite-Tungstate Solution: Add components to distilled/deionized water and bring volume to 1.0L. Mix thoroughly. Sparge with 100% N_2. Autoclave for 15 min at 15 psi pressure–121°C. Cool to room temperature.

Phosphate Solution:

Composition per 10.0mL:

KH_2PO_4	0.4g

Preparation of Phosphate Solution: Add KH_2PO_4 to distilled/deionized water and bring volume to 10.0mL. Mix thoroughly. Sparge with 100% N_2. Autoclave for 15 min at 15 psi pressure–121°C. Cool to room temperature.

Iron Sulfate Solution:

Composition per 10.0mL:

$FeSO_4$	1.52g

Preparation of Iron Sulfate Solution: Add $FeSO_4$ to distilled/deionized water and bring volume to 10.0mL. Mix thoroughly. Sparge with 100% N_2. Autoclave for 15 min at 15 psi pressure–121°C. Cool to room temperature.

Preparation of Medium: Prepare and dispense medium under 90% N_2 + 10% CO_2 gas mixture. Add components, except $NaHCO_3$ solution, phosphate solution, sodium acetate solution, iron sulfate solution, NH_4Cl solution, thiosulfate solution, selenite-tungstate solution, trace elements solution, vitamin solution, vitamin B_{12} solution, and vitamin B_1 solution, to distilled/deionized water and bring volume to 933.0mL. Mix thoroughly. Sparge with 90% N_2 + 10% CO_2 gas mixture. Autoclave for 15 min at 15 psi pressure–121°C. Aseptically and anaerobically add 30.0mL $NaHCO_3$ solution, 10.0mL phosphate solution, 10.0mL sodium acetate solution, 10.0mL iron sulfate solution, 1.0mL NH_4Cl solution, 1.0mL thiosulfate solution, 1.0mL selenite-tungstate solution, 1.0mL trace elements solution, 1.0mL vitamin solution, 1.0mL vitamin B_{12} solution, and 1.0mL vitamin B_1 solution. When the iron is added a white precipitate may form. Adjust pH to 6.8. Aseptically and anaerobically distribute to tubes or bottles.

Use: For the cultivation of *Rhodovulum iodosum* and *Rhodovulum robiginosum.*

Rhodovulum kholense Medium (DSMZ Medium 1136)

Composition per liter:

NaCl	20.0g
Sodium pyruvate	3.0g
$MgSO_4·7H_2O$	2.0g
NH_4Cl	0.64g
KH_2PO_4	0.5g
Yeast extract	0.5g
$CaCl_2·2H_2O$	0.12g
Trace elements solution SL-12	1.0mL

pH 5.7 ± 0.2 at 25°C

Trace Elements Solution SL-12:

Composition per liter:

$FeSO_4·7H_2O$	1.1g
H_3BO_3	0.3g
$CoCl_2·6H_2O$	0.19g
$MnCl_2·2H_2O$	0.05g
$ZnCl_2$	42.0mg
$NiCl_2·6H_2O$	24.0mg
$Na_2MoO_4·4H_2O$	18.0mg
$CuCl_2·2H_2O$	2.0mg

Preparation of Trace Elements Solution SL-12: Add components to distilled/deionized water and bring volume to 1.0L. Mix thoroughly. Filter sterilize.

Preparation of Medium: Add components to distilled/deionized water and bring volume to 1.0L. Mix thoroughly. Adjust pH to 7.0. Distribute into tubes or flasks. Gently heat and bring to boiling. Autoclave for 15 min at 15 psi pressure–121°C.

Use: For the cultivation of *Rhodoblastus* spp.

Rhodovulum strictum Medium (DSMZ Medium 746)

Composition per liter:

Solution 1	500.0mL
Solution 2	500.0mL

pH 7.8 ± 0.2 at 25°C

Solution 1:

Composition per 500.0mL:

NaCl	8.2g
Na-DL-malate	3.6g
Yeast extract	1.0g
$(NH_4)_2SO_4$	1.0g
$Na2_S2O_3·5H_2O$	0.5g
$MgCl_2·6H_2O$	0.2g
$CaCl_2·2H_2O$	0.05g
Trace elements solution SL-8	1.0mL

Trace Elements Solution SL-8:

Composition per liter:

Na_2-EDTA	5.2g
$FeCl_2·4H_2O$	1.5g
$CoCl_2·6H_2O$	190.0mg
$MnCl_2·4H_2O$	100.0mg
$ZnCl_2$	70.0mg
H_3BO_3	62.0mg
$Na_2MoSO_4·2H_2O$	36.0mg
$NiCl_2·6H_2O$	24.0mg
$CuCl_2·2H_2O$	17.0mg

Preparation of Trace Elements Solution SL-8: Add components to distilled/deionized water and bring volume to 1.0L. Mix thoroughly. Sparge with 100% N_2.

Preparation of Solution 1: Prepare and dispense medium under an oxygen-free 100% N_2. Add components to distilled/deionized water and bring volume to 500.0L. Mix thoroughly. Sparge with 100% N_2. Autoclave for 15 min at 15 psi pressure–121°C. Cool to 25°C.

Solution 2:

Composition per 500mL:

K_2HPO_4	1.35g
KH_2PO_4	0.35g

Preparation of Solution 2: Add components to distilled/deionized water and bring volume to 500.0mL. Mix thoroughly. Adjust pH to 7.8. Autoclave for 15 min at 15 psi pressure–121°C. Cool to room temperature.

Preparation of Medium: Aseptically and anaerobically combine 500.0mL solution 1 and 500.0mL solution 2 under N_2. Mix thoroughly. Adjust pH to 7.8. Aseptically and anaerobically distribute into sterile screw-cap tubes or flasks.

Use: For the cultivation of *Rhodovulum strictum.*

Rhodovulum sulfidophilum Medium (LMG Medium 84)

Composition per liter:

NaCl	25.0g
Yeast extract	1.0g
Disodium succinate	1.0g
KH_2PO_4	0.5g
$MgSO_4$·7H2O	0.4g
NaCl	0.4g
NH_4Cl	0.4g
$CaCl_2·2H_2O$	50.0mg
Ferric citrate solution	5.0mL
Trace elements solution	1.0mL
Ethanol	0.5mL

pH 5.8 ± 0.2 at 25°C

Ferric Citrate Solution :
Composition per 100.0mL:

Ferric citrate	0.1g

Preparation of Ferric Citrate Solution: Add ferric citrate to distilled/deionized water and bring volume to 100.0mL. Mix thoroughly.

Trace Elements Solution:
Composition per liter:

H_3BO_3	0.3g
$CoCl_2·6H_2O$	0.2g
$ZnSO_4·7H_2O$	0.1g
$Na_2MoO_4·2H_2O$	30.0mg
$MnCl_2·4H_2O$	30.0mg
$NiCl_2·6H_2O$	20.0mg
$CuCl_2·2H_2O$	10.0mg

Preparation of Trace Elements Solution: Add components to distilled/deionized water and bring volume to 1.0L. Mix thoroughly.

Preparation of Medium: Add components to distilled/deionized water and bring volume to 1.0L. Mix thoroughly. Distribute 40.0mL medium into 50mL screw-capped bottles. Flush each bottle for 1 to 2 min with nitrogen gas and then close immediately with rubber septa and screw caps. Autoclave for 15 min at 15 psi pressure–121°C. Sterile syringes are used to inoculate and remove the samples. Incubate in light using a tungsten lamp.

Use: For the cultivation of *Rhodovulum sulfidophilum.*

Rhodovulum visakhum Medium (DSMZ Medium 1128)

Composition per liter:

NaCl	20.0g
$MgCl_2·6H_2O$	1.0g
NH_4Cl	0.6g
KH_2PO_4	0.5g
Sorbitol	3.0g
Sodium pyruvate	3.0g
Yeast extract	0.4g
$CaCl_2·2H_2O$	0.15g
Trace elements solution SL-8	1.0mL
Vitamin solution	1.0mL

pH 6.5 ± 0.2 at 25°C

Trace Elements Solution SL-8:
Composition per liter:

Disodium EDTA	5.2g
$FeCl_2·4H_2O$	1.5g
$CoCl_2·6H_2O$	0.19g
$MnCl_2·4H_2O$	0.1g
$ZnCl_2$	0.07g
H_3BO_3	0.06g
$NaMoO_4·2H_2O$	0.04g
$CuCl_2·2H_2O$	0.02g
$NiCl_2·6H_20$	0.02g

Preparation of Trace Elements Solution SL-8: Add components to distilled/deionized water and bring volume to 1.0L. Mix thoroughly. Filter sterilize.

Vitamin Solution:
Composition per 10.0mL:

Vitamin B_{12}	0.2mg

Preparation of Vitamin Solution: Add components to distilled/deionized water and bring volume to 10.0mL. Mix thoroughly. Filter sterilize.

Preparation of Medium: Add components, except trace elements SL-8 and vitamin solutions, to distilled/deionized water and bring volume to 1.0L. Mix thoroughly. Adjust pH to 6.5. Distribute into tubes or flasks. Gently heat and bring to boiling. Autoclave for 15 min at 15 psi pressure–121°C. Cool to room temperature. Aseptically add trace elements SL-8 and vitamin solutions. Mix thoroughly. Adjust pH to 8.3. Aseptically distribute into culture vessels.

Use: For the cultivation of *Rhodovulum visakhum.*

Riboflavin Assay Medium

Composition per liter:

Peptone, photolyzed	22.0g
Glucose	20.0g
Yeast supplement	2.0g
Sodium acetate	1.8g
K_2HPO_4	1.0g
KH_2PO_4	1.0g
$MgSO_4·7H_2O$	0.4g
L-Cystine	0.2g
$FeSO_4·7H_2O$	20.0mg
$MnSO_4·H_2O$	20.0mg
NaCl	20.0mg

pH 6.8 ± 0.2 at 25°C

Source: This medium is available as a premixed powder from BD Diagnostic Systems.

Preparation of Medium: Add components to distilled/deionized water and bring volume to 1.0L. Mix thoroughly. Distribute into tubes in 5.0mL volumes. Add standard solutions and test solutions to each tube. Bring volume of each tube to 10.0mL. Autoclave for 10 min at 15 psi pressure–121°C.

Use: For the microbiological assaying of riboflavin using *Lactobacillus casei* as the test organism.

Riboflavin Medium

Composition per liter:

Glucose	20.0g
Yeast extract	10.0g
Peptone	10.0g
Riboflavin	0.02g

Preparation of Medium: Add components to distilled/deionized water and bring volume to 1.0L. Mix thoroughly. Gently heat and bring

to boiling. Distribute into tubes or flasks. Autoclave for 15 min at 15 psi pressure–121°C.

Use: For the cultivation and maintenance of *Saccharomyces cerevisiae.*

Ribose Production Medium

Composition per liter:

D-Glucose	150.0g
$CaCO_3$	20.0g
Dried yeast	10.0g
$(NH_4)_2SO_4$	5.0g
Tryptophan	0.05g
Tyrosine	0.05g
Phenylalanine	0.05g

Preparation of Medium: Add components to distilled/deionized water and bring volume to 1.0L. Mix thoroughly. Distribute into tubes or flasks. Autoclave for 15 min at 15 psi pressure–121°C.

Use: For the cultivation and maintenance of *Bacillus subtilis.*

Rice Extract Agar

Composition per liter:

Agar	20.0g
White rice, solids from extract	5.0g
Polysorbate 80	10.0mL

pH 6.6 ± 0.2 at 25°C

Source: This medium is available as a premixed powder from BD Diagnostic Systems.

Preparation of Medium: Add components, except polysorbate 80, to distilled/deionized water and bring volume to 990.0mL. Mix thoroughly. Gently heat and bring to boiling. Add polysorbate 80. Mix thoroughly. Distribute into tubes or flasks. Autoclave for 15 min at 15 psi pressure–121°C. Pour into sterile Petri dishes.

Use: For the cultivation and differentiation of *Candida albicans* and *Candida stellatoidea* from other *Candida* species based on chlamydospore formation.

Rice Extract Agar

Composition per liter:

Agar	20.0g
White rice, solids from extract	20.0g
Polysorbate 80	10.0mL

pH 7.1 ± 0.2 at 25°C

Source: This medium is available as a premixed powder from BD Diagnostic Systems.

Preparation of Medium: Add components, except polysorbate 80, to distilled/deionized water and bring volume to 990.0mL. Mix thoroughly. Gently heat and bring to boiling. Add polysorbate 80. Mix thoroughly. Distribute into tubes or flasks. Autoclave for 15 min at 15 psi pressure–121°C. Pour into sterile Petri dishes.

Use: For the cultivation and differentiation of *Candida albicans* and *Candida stellatoidea* from other *Candida* species based on chlamydospore formation.

Rice Grain Medium

Composition per 25.0mL:

White rice, polished and without added vitamins	8.0g

Preparation of Medium: Add 8.0g of white rice to 25.0mL of distilled/deionized water. Mix thoroughly. Autoclave for 15 min at 15 psi pressure–121°C. Pour into sterile Petri dishes.

Use: For the identification of *Microsporum audovini* (no growth) from other *Microsporum* species (growth and sporulation).

Rice Infusion Oxgall Tween™ 80 Agar

See: **RIOT Agar**

Rich Medium (DSMZ Medium 736)

Composition per liter:

Agar	20.0g
Peptone	10.0g
Yeast extract	5.0g
Casamino acids	5.0g
Malt extract	5.0g
Meat extract	2.0g
Glycerol	2.0g
$MgSO_4 \cdot 7H_2O$	1.0g
Tween™ 80	0.05g

pH 7.2 ± 0.2 at 25°C

Preparation of Medium: Add components to distilled/deionized water and bring volume to 1.0L. Mix thoroughly. Gently heat and bring to boiling. Distribute into tubes or flasks. Autoclave for 15 min at 15 psi pressure–121°C. Pour into sterile Petri dishes or leave in tubes.

Use: For the cultivation and maintenance of *Cryobacterium psychrophilum=Curtobacterium psychrophilum, Demetria terragena, Knoellia sinensis, Knoellia subterranea, Beutenbergia cavernae, Ornithinicoccus hortensis, Ornithinimicrobium humiphilum, Janibacter terrae,* and *Tetrasphaera elongata.*

Rifampicin Luria Agar

Composition per liter:

Agar	15.0g
Pancreatic digest of casein	10.0g
Yeast extract	5.0g
NaCl	0.5g
Glucose solution	20.0mL
Rifampicin solution	10.0mL

Glucose Solution:

Composition per 50.0mL:

Glucose	5.0g

Preparation of Glucose Solution: Add glucose to distilled/deionized water and bring volume to 50.0mL. Mix thoroughly. Filter sterilize.

Rifampicin Solution:

Composition per 10.0mL:

Rifampicin	30mg

Preparation of Rifampicin Solution: Add rifampicin to distilled/deionized water and bring volume to 10.0mL. Mix thoroughly. Filter sterilize.

Preparation of Medium: Add components, except glucose solution and rifampicin solution, to distilled/deionized water and bring volume to 970.0mL. Mix thoroughly. Autoclave for 15 min at 15 psi pressure–121°C. Aseptically add 20.0mL of sterile glucose solution and 10.0mL of sterile rifampicin solution. Mix thoroughly. Aseptically distribute into sterile tubes or flasks.

Use: For the cultivation of *Escherichia coli.*

Rila Marine Medium

Composition per liter:

Agar	15.0g
Peptone	0.5g
Yeast extract	0.5g
Pancreatic digest of casein	0.5g
Marine salts mixture	800.0mL

pH 7.6–8.0 at 25°C

Preparation of Medium: Add components to distilled/deionized water and bring volume to 1.0L. Mix thoroughly. Gently heat and bring to boiling. Autoclave for 15 min at 15 psi pressure–121°C. Adjust pH to 7.6–8.0. Pour into sterile Petri dishes or distribute into sterile tubes.

Use: For the cultivation and maintenance of *Alteromonas denitrificans*.

Rimler-Shotts Medium (RS Medium)

Composition per liter:

Agar	13.5g
$Na_2S_2O_3 \cdot 5H_2O$	6.8g
L-Ornithine·HCl	6.5g
NaCl	5.0g
L-Lysine·HCl	5.0g
Maltose	3.5g
Yeast extract	3.0g
Sodium deoxycholate	1.0g
Ferric ammonium citrate	0.8g
L-Cysteine·HCl	0.3g
Bromthymol Blue	0.03g
Novobiocin solution	10.0mL

pH 7.0 ± 0.2 at 25°C

Novobiocin Solution:

Composition per 10.0mL:

Novobiocin	5.0mg

Preparation of Novobiocin Solution: Add novobiocin to distilled/deionized water and bring volume to 10.0mL. Mix thoroughly. Filter sterilize.

Preparation of Medium: Add components, except novobiocin solution, to distilled/deionized water and bring volume to 990.0mL. Mix thoroughly. Gently heat and bring to boiling. Autoclave for 15 min at 15 psi pressure–121°C. Cool to 45°–50°C. Aseptically add sterile novobiocin solution. Mix thoroughly. Pour into sterile Petri dishes or distribute into sterile tubes.

Use: For the selective isolation, cultivation, and presumptive identification of *Aeromonas hydrophila* and other Gram-negative bacteria based on their ability to decarboxylate lysine and ornithine, ferment maltose, and produce H_2S. Maltose-fermenting bacteria appear as yellow colonies. Bacteria that produce lysine or ornithine decarboxylase turn the medium greenish-yellow to yellow. Bacteria that produce H_2S appear as colonies with black centers.

RIOT Agar (Rice Infusion Oxgall Tween™ 80 Agar)

Composition per 1010.0mL:

Agar	10.0g
Oxgall	10.0g
Rice extract	1.0L
Tween™ 80	10.0mL

pH 7.3 ± 0.2 at 25°C

Rice Extract:

Composition per liter:

Cream of rice cereal	10.0g

Preparation of Rice Extract: Add cream of rice cereal to 1.0L of boiling tap water. Mix thoroughly. Filter quickly through cheesecloth. Bring volume of filtrate to 1.0L with tap water.

Preparation of Medium: Combine components. Mix thoroughly. Gently heat and bring to boiling. Distribute into tubes or flasks. Autoclave for 15 min at 15 psi pressure–121°C. Pour into sterile Petri dishes or leave in tubes.

Use: For the cultivation and differentiation of *Candida albicans* and *Candida stellatoidea* from other *Candida* species based on chlamydospore formation.

Rippey-Cabelli Agar (RC Agar)

Composition per liter:

Agar	15.0g
Meat peptone	5.0g
Trehalose	5.0g
NaCl	3.0g
KCl	2.0g
Yeast extract	2.0g
Bromthymol Blue	0.44g
$MgSO_4 \cdot 7H_2O$	0.2g
$FeCl_3 \cdot 6H_2O$	0.1g
Sodium deoxycholate	0.1g
Ampicillin solution	10.0mL
Ethanol	10.0mL

pH 8.0 ± 0.2 at 25°C

Ampicillin Solution:

Composition per 10.0mL:

Ampicillin	0.02g

Preparation of Ampicillin Solution: Add ampicillin to distilled/deionized water and bring volume to 10.0mL. Mix thoroughly. Filter sterilize.

Preparation of Medium: Add components—except sodium deoxycholate, ampicillin solution, and ethanol—to distilled/deionized water and bring volume to 980.0mL. Mix thoroughly. Gently heat and bring to boiling. Autoclave for 15 min at 15 psi pressure–121°C. Cool to 45°–50°C. Aseptically add sodium deoxycholate, 10.0mL of sterile ampicillin solution, and 10.0mL of ethanol. Mix thoroughly. Pour into sterile Petri dishes or distribute into sterile tubes.

Use: For the isolation, cultivation, and differentiation of *Aeromonas* species and *Plesiomonas* species from water samples using the membrane filter method. This medium differentiates bacteria on the basis of trehalose fermentation. Bacteria that ferment trehalose turn the medium yellow.

Rippey-Cabelli HiVeg Agar Base with Ethanol and Ampicillin

Composition per liter:

Agar	15.0g
Plant hydrolysate No. 1	5.0g
Trehalose	5.0g

NaCl 3.0g
KCl 2.0g
Yeast extract 2.0g
$MgSO_4$ 0.2g
Iron (III) chloride 0.1g
Bromthymol Blue 0.04g
Ampicillin solution 10.0mL
Ethanol 10.0mL

pH 8.0 ± 0.2 at 25°C

Source: This medium, without ampicillin solution and ethanol, is available as a premixed powder from HiMedia.

Ampicillin Solution:
Composition per 10.0mL:
Ampicillin 0.02g

Preparation of Ampicillin Solution: Add ampicillin to distilled/deionized water and bring volume to 10.0mL. Mix thoroughly. Filter sterilize.

Preparation of Medium: Add components—except ampicillin solution and ethanol—to distilled/deionized water and bring volume to 990.0mL. Mix thoroughly. Gently heat and bring to boiling. Autoclave for 15 min at 15 psi pressure–121°C. Cool to 45°–50°C. Aseptically add 10.0mL of sterile ampicillin solution, and 10.0mL of ethanol. Mix thoroughly. Pour into sterile Petri dishes or distribute into sterile tubes.

Use: For the isolation, cultivation, and differentiation of *Aeromonas* species and *Plesiomonas* species from water samples using the membrane filter method. For the differential and selective isolation of *Aeromonas hydrophila* species from water samples

RM Medium

Composition per liter:
Glucose 20.0g
Agar 15.0g
Yeast extract 10.0g
KH_2PO_4 .... 2.0g
Solution 1 .... 250.0mL
Solution 2 .... 250.0mL
Solution 3 .... 250.0mL
Solution 4 .... 250.0mL

pH 6.0 ± 0.2 at 25°C

Solution 1:
Composition per 250.0mL:
Glucose 20.0g

Preparation of Solution 1: Add glucose to distilled/deionized water and bring volume to 250.0mL. Mix thoroughly. Autoclave for 15 min at 15 psi pressure–121°C. Cool to 45°–50°C.

Solution 2:
Composition per 250.0mL:
Agar 15.0g

Preparation of Solution 2: Add agar to distilled/deionized water and bring volume to 250.0mL. Mix thoroughly. Autoclave for 15 min at 15 psi pressure–121°C. Cool to 45°–50°C.

Solution 3:
Composition per 250.0mL:
Yeast extract 10.0g

Preparation of Solution 3: Add yeast extract to distilled/deionized water and bring volume to 250.0mL. Mix thoroughly. Autoclave for 15 min at 15 psi pressure–121°C. Cool to 45°–50°C.

Solution 4:
Composition per 250.0mL:
KH_2PO_4 .... 2.0g

Preparation of Solution 4: Add KH_2PO_4 to distilled/deionized water and bring volume to 250.0mL. Mix thoroughly. Autoclave for 15 min at 15 psi pressure–121°C. Cool to 45°–50°C.

Preparation of Medium: Aseptically combine the four sterile solutions. Mix thoroughly. Adjust pH to 6.0. Pour into sterile Petri dishes or distribute into sterile tubes.

Use: For the cultivation and maintenance of *Zymomonas mobilis*.

Rogosa Agar

Composition per liter:
Sodium acetate 25.0g
Agar 20.0g
Glucose 20.0g
Pancreatic digest of casein 10.0g
KH_2PO_4 .... 6.0g
Yeast extract 5.0g
Ammonium citrate 2.0g
Sorbitan monooleate 1.0g
$MgSO_4 \cdot 7H_2O$ 0.575g
$MnSO_4 \cdot H_2O$ 0.12g
$FeSO_4 \cdot 7H_2O$ 0.4mg
Acetic acid, glacial 1.32mL

pH 5.4 ± 0.2 at 25°C

Source: This medium is available as a premixed powder from Oxoid Unipath.

Preparation of Medium: Add components, except acetic acid, to distilled/deionized water and bring volume to 998.7mL. Mix thoroughly. Gently heat and bring to boiling. Add glacial acetic acid. Mix thoroughly. Gently heat while stirring and bring to 90°–100°C for 2–3 min. Do not autoclave. Pour into sterile Petri dishes or distribute into sterile tubes.

Use: For the isolation, cultivation, and enumeration of lactobacilli, especially from feces, saliva, vaginal specimens, and dairy products.

Rogosa Broth, Modified

Composition per 1005.0mL:
Glucose 20.0g
Trypticase™ 10.0g
Yeast extract 5.0g
K_2HPO_4 .... 3.0g
KH_2PO_4 .... 3.0g
Tryptose 3.0g
Ammonium citrate 2.0g
Sodium acetate 1.0g
Tween™ 80 .... 1.0g
L-Cysteine 0.2g
Salt solution 5.0mL

pH 6.8 ± 0.2 at 25°C

Salt Solution:
Composition per 100.0mL:
$MgSO_4 \cdot 7H_2O$ 11.5g
$MnSO_4$ 2.4g
$FeSO_4 \cdot 7H_2O$ 1.68g

Preparation of Medium: Add components to distilled/deionized water and bring volume to 100.0mL. Mix thoroughly. Distribute into tubes or flasks. Autoclave for 15 min at 15 psi pressure–121°C.

Preparation of Medium: Add components, except salt solution, to distilled/deionized water and bring volume to 995.0mL. Mix thoroughly. Autoclave for 15 min at 15 psi pressure–121°C. Aseptically add 5.0mL of sterile salt solution. Adjust pH to 6.8. Mix thoroughly. Aseptically distribute into sterile tubes or flasks.

Use: For the cultivation of *Lactobacillus* species.

Rogosa SL Agar (Rogosa Selective *Lactobacillus* Agar)

Composition per liter:

Agar	15.0g
Sodium acetate	15.0g
Glucose	10.0g
Pancreatic digest of casein	10.0g
K_2HPO_4	6.0g
Yeast extract	5.0g
Arabinose	5.0g
Sucrose	5.0g
Ammonium citrate	2.0g
Sorbitan monooleate	1.0g
$MgSO_4 \cdot 7H_2O$	0.57g
$MnSO_4 \cdot 7H_2O$	0.12g
$FeSO_4 \cdot H_2O$	0.03g
Acetic acid, glacial	1.32mL

pH 5.4 ± 0.2 at 25°C

Source: This medium is available as a premixed powder from BD Diagnostic Systems.

Preparation of Medium: Add components, except glacial acetic acid, to distilled/deionized water and bring volume to 998.7mL. Mix thoroughly. Gently heat and bring to boiling. Add glacial acetic acid. Mix thoroughly. Gently heat while stirring and bring to 90°–100°C for 2–3 min. Do not autoclave. Pour into sterile Petri dishes or distribute into sterile tubes.

Use: For the isolation, cultivation, and enumeration of lactobacilli, especially from feces, saliva, vaginal specimens, and dairy products.

Rogosa SL Broth (Rogosa Selective *Lactobacillus* Broth)

Composition per liter:

Sodium acetate	15.0g
Glucose	10.0g
Pancreatic digest of casein	10.0g
K_2HPO_4	6.0g
Yeast extract	5.0g
Arabinose	5.0g
Sucrose	5.0g
Ammonium citrate	2.0g
Sorbitan monooleate	1.0g
$MgSO_4 \cdot 7H_2O$	0.57g
$MnSO_4 \cdot 7H_2O$	0.12g
$FeSO_4 \cdot H_2O$	0.03g
Acetic acid, glacial	1.32mL

pH 5.4 ± 0.2 at 25°C

Source: This medium is available as a premixed powder from BD Diagnostic Systems.

Preparation of Medium: Add components, except glacial acetic acid, to distilled/deionized water and bring volume to 998.7mL. Mix thoroughly. Gently heat and bring to boiling. Add glacial acetic acid. Mix thoroughly. Gently heat while stirring and bring to 90°–100°C for 2–3 min. Do not autoclave. Aseptically distribute into sterile tubes.

Use: For the isolation, cultivation, and enumeration of lactobacilli, especially from feces, saliva, vaginal specimens, and dairy products.

Rogosa SL HiVeg Agar

Composition per liter:

Agar	15.0g
Sodium acetate	15.0g
Glucose	10.0g
Plant hydrolysate No. 1	10.0g
KH_2PO_4	6.0g
Arabinose	5.0g
Saccharose	5.0g
Yeast extract	5.0g
Ammonium citrate	2.0g
$MgSO_4$	0.57g
$MnSO_4$	0.12g
$FeSO_4$	0.03g
Acetic acid, glacial	1.32mL
Polysorbate 80	1.0mL

pH 5.4 ± 0.2 at 25°C

Source: This medium, without acetic acid or polysorbate 80, is available as a premixed powder from HiMedia.

Preparation of Medium: Add components to distilled/deionized water and bring volume to 1.0L. Mix thoroughly. Gently heat and bring to boiling. Distribute into tubes or flasks. Gently heat to 90–100°C. Hold at temperature for 2–3 min. Do not autoclave. Cool to 50°C. Pour into sterile Petri dishes or leave in tubes.

Use: For the isolation, cultivation, and enumeration of lactobacilli, especially from feces, saliva, vaginal specimens, and dairy products. For the cultivation of oral and fecal lactobacilli.

Rogosa SL HiVeg Broth

Composition per liter:

Sodium acetate	15.0g
Glucose	10.0g
Plant hydrolysate No. 1	10.0g
KH_2PO_4	6.0g
Arabinose	5.0g
Saccharose	5.0g
Yeast extract	5.0g
Ammonium citrate	2.0g
$MgSO_4$	0.57g
$MnSO_4$	0.12g
$FeSO_4$	0.03g
Acetic acid, glacial	1.32mL
Polysorbate 80	1.0mL

pH 5.4 ± 0.2 at 25°C

Source: This medium, without acetic acid or polysorbate 80, is available as a premixed powder from HiMedia.

Preparation of Medium: Add components to distilled/deionized water and bring volume to 1.0L. Mix thoroughly. Gently heat and bring to boiling. Distribute into tubes or flasks. Gently heat to 90–100°C. Hold at temperature for 2–3 min. Do not autoclave.

Use: For the isolation, cultivation, and enumeration of lactobacilli, especially from feces, saliva, vaginal specimens, and dairy products.

Rolled Oats Mineral Medium (DSMZ Medium 84)

Composition per 1001.0mL:

Rolled oats	20.0g
Agar	12.0g
Trace elements solution	1.0mL

pH 7.2 ± 0.2 at 25°C

Trace Elements Solution:

Composition per 100.0mL:

$FeSO_4 \cdot 7H_2O$	0.1g
$MnCl_2 \cdot 4H_2O$	0.1g
$ZnSO_4 \cdot 7H_2O$	0.1g

Preparation of Trace Elements Solution: Add components to distilled/deionized water and bring volume to 100.0mL. Mix thoroughly.

Preparation of Medium: Add rolled oats to 500.0mL distilled/deionized water. Gently heat and bring to boiling. Boil for 20 min. Filter. Add agar to the filtrate and bring volume to 1.0L with distilled/deionized water. Mix thoroughly. Add 1.0mL trace elements solution. Gently heat and bring to boiling. Distribute into tubes or flasks. Autoclave for 15 min at 15 psi pressure–121°C. Pour into sterile Petri dishes or leave in tubes.

Use: For the cultivation and maintenance of *Actinomadura rubrobrunea, Streptomyces chartreusis, Streptomyces aculeolatus, Streptomyces thermodiastaticus, Microbispora rosea, Micromonospora coerulea, Thermoactinomyces vulgaris, Thermoactinomyces sacchari, Streptosporangium album,* and *Planobispora rosea.*

Rose Bengal Chloramphenicol Agar

Composition per liter:

Agar	15.0g
Glucose	10.0g
Papaic digest of soybean meal	5.0g
KH_2PO_4	1.0g
$MgSO_4 \cdot 7H_2O$	0.5g
Rose Bengal	0.05g
Chloramphenicol solution	10.0mL

pH 7.0 ± 0.2 at 25°C

Source: This medium is available as a premixed powder from BD Diagnostic Systems and Oxoid Unipath.

Chloramphenicol Solution:

Composition per 10.0mL:

Chloramphenicol	0.1g

Preparation of Chloramphenicol Solution: Add chloramphenicol to distilled/deionized water and bring volume to 10.0mL. Mix thoroughly. Filter sterilize.

Preparation of Medium: Add components, except chloramphenicol solution, to distilled/deionized water and bring volume to 990.0mL. Mix thoroughly. Gently heat and bring to boiling. Autoclave for 15 min at 15 psi pressure–121°C. Cool to 45°C. Aseptically add sterile chloramphenicol solution. Mix thoroughly. Pour into sterile Petri dishes or distribute into sterile tubes.

Use: For the selective isolation, cultivation, and enumeration of yeasts and molds from environmental specimens and foods.

Rose Bengal Chloramphenicol HiVeg Agar

Composition per liter:

Agar	15.5g
Glucose	10.0g
Plant peptone No. 4	5.0g
KH_2PO_4	1.0g
$MgSO_4$	0.5g
Rose Bengal	0.05g
Chloramphenicol solution	10.0mL

pH 7.2 ± 0.2 at 25°C

Source: This medium, without chloramphenicol, is available as a premixed powder from HiMedia.

Chloramphenicol Solution:

Composition per 10.0mL:

Chloramphenicol	0.1g

Preparation of Chloramphenicol Solution: Add chloramphenicol to distilled/deionized water and bring volume to 10.0mL. Mix thoroughly. Filter sterilize.

Preparation of Medium: Add components, except chloramphenicol solution, to distilled/deionized water and bring volume to 990.0mL. Mix thoroughly. Gently heat and bring to boiling. Autoclave for 15 min at 15 psi pressure–121°C. Cool to 45°C. Aseptically add sterile chloramphenicol solution. Mix thoroughly. Pour into sterile Petri dishes or distribute into sterile tubes.

Use: For the selective isolation, cultivation, and enumeration of yeasts and molds from environmental specimens and foods.

Roseicyclus Medium (DSMZ Medium 1183)

Composition per liter:

Na_2SO_4	15.0g
$MgSO_4$	2.0g
Malic acid	1.0g
Na acetate	1.0g
Yeast extract	1.0g
Peptone	0.5g
KCl	0.3g
NH_4Cl	0.3g
Bicarbonate solution	5.0mL
Phosphate solution	3.0mL
Vitamin solution	2.0mL
Trace elements solution	2.0mL
Calcium chloride solution	0.5mL

pH 8.3 ± 0.2 at 25°C

Phosphate Solution:

Composition per 10.0mL:

K_2HPO_4	1.0g

Preparation of Phosphate Solution: Add K_2HPO_4 to distilled/deionized water and bring volume to 10.0mL. Mix thoroughly. Filter sterilize.

Calcium Chloride Solution:

Composition per 10.0mL:

$CaCl_2 \cdot 2H_2O$	1.0g

Preparation of Calcium Chloride Solution: Add $CaCl_2 \cdot 2H_2O$ to distilled/deionized water and bring volume to 10.0mL. Mix thoroughly. Autoclave for 15 min at 15 psi pressure–121°C. Cool to room temperature.

Bicarbonate Solution:
Composition per 10.0mL:

$NaHCO_3$	1.0g

Preparation of Bicarbonate Solution: Add $NaHCO_3$ to distilled/deionized water and bring volume to 10.0mL. Mix thoroughly. Adjust pH to 6.5. Filter sterilize.

Vitamin Solution:
Composition per liter:

Nicotinic acid	0.4g
Thiamine-HCl·$2H_2O$	0.4g
Biotin	80.0mg
Vitamin B_{12}	0.5mg

Preparation of Vitamin Solution: Add components to distilled/deionized water and bring volume to 1.0L. Mix thoroughly. Filter sterilize.

Trace Elements Solution:
Composition per liter:

$FeSO_4·7H_2O$	0.3g
$ZnSO_4·7H_2O$	5.0mg
$MnCl_2·4H_2O$	3.0mg
$NiCl_2·6H_2O$	3.0mg
$Na_2MoO_4·4H_2O$	3.0mg
H_3BO_3	2.0mg
$CuCl_2·2H_2O$	2.0mg
$CoCl_2·6H_2O$	1.0mg

Preparation of Trace Elements Solution: Add components to distilled/deionized water and bring volume to 1.0L. Mix thoroughly. Adjust pH to 3.0. Filter sterize.

Preparation of Medium: Add components, except phosphate, bicarbonate, calcium chloride, trace elements, and vitamin solutions, to distilled/deionized water and bring volume to 990.0mL. Mix thoroughly. Adjust pH to 5.95. Distribute into tubes or flasks. Gently heat and bring to boiling. Autoclave for 15 min at 15 psi pressure–121°C. Cool to room temperature. Aseptically add phosphate, bicarbonate, calcium chloride, trace elements, and vitamin solutions. Mix thoroughly. Aseptically distribute into culture vessels.

Use: For the cultivation of *Roseicyclus* spp.

Roseinatronobacter Agar (DSMZ Medium 928)

Composition per liter:

K_2HPO_4	25.0g
Na_2CO_3	11.0g
$NaHCO_3$	4.0g
NaCl	2.5g
Sodium acetate	0.8g
Yeast extract	0.5g
Peptone	0.5g
KNO_3	0.25g
Agar solution	500.0mL

pH 10.0 ± 0.2 at 25°C

Agar Solution:
Composition per 500.0mL:

Agar	20.0g

Preparation of Agar Solution: Add agar to distilled/deionized water and bring volume to 500.0mL. Mix thoroughly. Gently heat and bring to boiling. Autoclave for 15 min at 15 psi pressure–121°C. Cool to 55°C.

Preparation of Medium: Add components, except agar solution, to distilled/deionized water and bring volume to 500.0mL. Mix thoroughly. Autoclave for 15 min at 15 psi pressure–121°C. Cool to 55°C. Add 500.0mL sterile warm agar solution. Mix thoroughly. Pour into Petri dishes or distribute to sterile tubes.

Use: For the cultivation of *Roseinatronobacter thiooxidans.*

Roseinatronobacter Medium (DSMZ Medium 928)

Composition per liter:

K_2HPO_4	25.0g
Na_2CO_3	11.0g
$NaHCO_3$	4.0g
NaCl	2.5g
Sodium acetate	0.8g
Yeast extract	0.5g
Peptone	0.5g
KNO_3	0.25g

pH 10.0 ± 0.2 at 25°C

Preparation of Medium: Add components to distilled/deionized water and bring volume to 1.0L. Mix thoroughly. Distribute into tubes or flasks. Autoclave for 15 min at 15 psi pressure–121°C.

Use: For the cultivation of *Roseinatronobacter thiooxidans.*

Rouf's Medium (DSMZ Medium 1019)

Composition per liter:

Yeast extract	5.0g
Peptone	5.0g
$MgSO_4·7H_2O$	0.2g
Fe(NH3)citrate	0.15g
$CaCl_2·2H_2O$	0.05g
$MnSO_4·H_2O$	0.05g
$FeCl_3·6H_2O$	0.01g
Vitamin solution	10.0mL
Trace elements solution	1.0mL

pH 7.1 ± 0.2 at 25°C

Vitamin Solution:
Composition per liter:

Pyridoxine-HCl	10.0mg
Thiamine-HCl·$2H_2O$	5.0mg
Riboflavin	5.0mg
Nicotinic acid	5.0mg
D-Ca-pantothenate	5.0mg
p-Aminobenzoic acid	5.0mg
Lipoic acid	5.0mg
Biotin	2.0mg
Folic acid	2.0mg
Vitamin B_{12}	0.1mg

Preparation of Vitamin Solution: Add components to distilled/deionized water and bring volume to 1.0L. Mix thoroughly. Filter sterilize.

Trace Elements Solution:
Composition per liter:

Nitrilotriacetic acid	12.8g
$CoCl_2·6H_2O$	0.17g
$CaCl_2·2H_2O$	0.1g
$FeSO_4·7H_2O$	0.1g

$MnCl_2·4H_2O$	0.1g
NaCl	0.1g
$Na_2MoO_4·2H_2O$	0.1g

Preparation of Trace Elements Solution: Add nitrilotriacetic acid to 500.0mL of distilled/deionized water. Dissolve by adjusting pH to 6.5 with KOH. Add remaining components. Add distilled/deionized water to 1.0L. Mix thoroughly. Adjust pH to 7.0.

Preparation of Medium: Add components, except trace elements and vitamin solutions, to distilled/deionized water and bring volume to 989.0mL. Mix thoroughly. Adjust pH to 7.1. Distribute into tubes or flasks. Gently heat and bring to boiling. Autoclave for 15 min at 15 psi pressure–121°C. Cool to room temperature. Aseptically add trace elements and vitamin solutions. Mix thoroughly. Adjust pH to 7.1. Aseptically distribute into culture vessels.

Use: For the cultivation of *Phenylobacterium lituiforme.*

RP Medium

Composition per liter:

Solution 1	960.0mL
Solution 2	40.0mL

Solution 1:

RPMI 1640 solution	900.0mL
HEPES solution	60.0mL

RPMI 1640 Medium:

Composition per liter:

Inorganic salt solution	400.0mL
Other component solution	400.0mL
Amino acid solution	100.0mL
Vitamin solution	100.0mL

Inorganic Salt Solution:

Composition per 400.0mL:

NaCl	6.0g
$NaH_2PO_4·H_2O$	0.8g
KCl	0.4g
$Ca(NO_3)_2·4H_2O$	0.1g
$MgSO_4$ (anhydrous)	48.84mg

Preparation of Inorganic Salt Solution: Add components to distilled/deionized water and bring volume to 400.0mL. Mix thoroughly. Autoclave for 15 min at 15 psi pressure–121°C. Cool to 25°C.

Other Component Solution:

Composition per 400.0mL:

D-Glucose	2.0g
Phenol Red	5.0mg
Glutathione, reduced	1.0mg

Preparation of Other Component Solution: Add components to distilled/deionized water and bring volume to 400.0mL. Mix thoroughly. Autoclave for 15 min at 15 psi pressure–121°C. Cool to 25°C.

Amino Acid Solution:

Composition per 100.0mL:

L-Glutamine	300.0mg
L-Arginine	200.0mg
L-Cysteine·2HCl	65.0mg
L-Asparagine (free base)	50.0mg
L-Isoleucine	50.0mg
L-Leucine	50.0mg
L-Lysine·HCl	40.0mg
L-Serine	30.0mg
L-Tyrosine·2Na·$2H_2O$	29.0mg
L-Aspartic acid	20.0mg
L-Glutamic acid	20.0mg
L-Hydroxyproline	20.0mg
L-Proline	20.0mg
L-Threonine	20.0mg
L-Valine	20.0mg
L-Histidine (free base)	15.0mg
L-Methionine	15.0mg
L-Phenylalanine	15.0mg
Glycine	10.0mg
L-Tryptophan	5.0mg

Preparation of Amino Acid Solution: Add components to distilled/deionized water and bring volume to 100.0mL. Mix thoroughly. Filter sterilize.

Vitamin Solution:

Composition per 100.0mL:

i-Inositol	35.0mg
Folic acid	1.0mg
Niacinamide	1.0mg
Para-aminobenzoic acid	1.0mg
Pyridoxal·HCl	1.0mg
Pyridoxine·HCl	1.0mg
Thiamine·HCl	1.0mg
Choline chloride	3.0mg
D-Ca pantothenate	0.25mg
Biotin	0.2mg
Riboflavin	0.2mg
Vitamin B_{12}	0.005mg

Preparation of Vitamin Solution: Add components to distilled/deionized water and bring volume to 100.0mL. Mix thoroughly. Filter sterilize.

Preparation of RPMI 1640 Medium: Aseptically combine 400.0mL of sterile inorganic salt solution, 400.0mL of sterile other component solution, 100.0mL of sterile amino acid solution, and 100.0mL of sterile vitamin solution.

HEPES Solution:

Composition per 60.0mL:

HEPES	5.94g

Preparation of HEPES Solution: Add HEPES to distilled/deionized water and bring volume to 60.0mL. Mix thoroughly. Filter sterilize.

Preparation of Solution 1: Aseptically combine 960.0mL of sterile RPMI-1640 solution and 60.0mL sterile HEPES solution.

Solution 2:

Composition per 100.0mL:

$NaHCO_3$	5.0g

Preparation of Solution 2: Add $NaHCO_3$ to distilled/deionized water and bring volume to 100.0mL. Mix thoroughly. Filter sterilize.

Preparation of Medium: Aseptically combine 960.0mL of sterile solution 1 with 40.0mL of sterile solution 2. Aseptically distribute into sterile tubes or flasks.

Use: For the cultivation of *Balamuthia mandrillaris* and *Plasmodium falciparum.*

RPMI 1640 Medium with L-Glutamine

Composition per liter:

NaCl	6.0g
$NaHCO_3$	2.0g
D-Glucose	2.0g
$Na_2HPO_4 \cdot 7H_2O$	1.5g
KCl	0.4g
L-Glutamine	0.3g
L-Arginine	0.2g
$Ca(NO_3)_2 \cdot 4H_2O$	0.1g
$MgSO_4 \cdot 7H_2O$	0.1g
L-Asparagine	0.05g
L-Cystine	0.05g
L-Isoleucine, allo free	0.05g
L-Leucine, methionine free	0.05g
L-Lysine·HCl	0.04g
i-Inositol	0.035g
L-Serine	0.03g
L-Aspartic acid	0.02g
L-Glutamic acid	0.02g
L-Hydroxyproline	0.02g
L-Proline, hydroxy-L-proline free	0.02g
L-Threonine, allo free	0.02g
L-Tyrosine	0.02g
L-Valine	0.02g
L-Histidine, free base	0.015g
L-Methionine	0.015g
L-Phenylalanine	0.015g
Glycine	0.01g
L-Tryptophan	5.0mg
Phenol Red	5.0mg
Choline chloride	3.0mg
Glutathione, reduced	1.0mg
p-Aminobenzoic acid	1.0mg
Folic acid	1.0mg
Nicotinamide	1.0mg
Pyridoxine·HCl	1.0mg
Thiamine·HCl	1.0mg
D-Calcium pantothenate	0.25mg
Biotin	0.2mg
Riboflavin	0.2mg
Vitamin B_{12}	5.0μg

pH 7.3 ± 0.2 at 25°C

Preparation of Medium: Add components to distilled/deionized water and bring volume to 1.0L. Adjust pH to 7.3 with 1*N* HCl or 1*N* NaOH. Filter sterilize. Aseptically distribute into sterile tubes or flasks.

Use: For the cultivation of mammalian cells in tissue culture. Culture media for human immunodeficiency viruses.

RS Medium

See: **Rimler-Shotts Medium**

RS HiVeg Medium Base with Novobiocin (Rimler-Shotts Medium)

Composition per liter:

Agar	13.5g
$Na_2S_2O_3$	6.8g
L-Ornithine hydrochloride	6.5g
L-Lysine hydrochloride	5.0g
NaCl	5.0g
Maltose	3.5g
Yeast extract	3.0g
Synthetic detergent No. III	1.0g
Ferric ammonium citrate	0.8g
L-Cysteine·HCl	0.3g
Bromthymol Blue	0.03g
Novobiocin solution	10.0mL

pH 7.0 ± 0.2 at 25°C

Source: This medium, without novobiocin, is available as a premixed powder from HiMedia.

Novobiocin Solution:

Composition per 10.0mL:

Novobiocin	5.0mg

Preparation of Novobiocin Solution: Add novobiocin to distilled/deionized water and bring volume to 10.0mL. Mix thoroughly. Filter sterilize.

Preparation of Medium: Add components, except novobiocin solution, to distilled/deionized water and bring volume to 990.0mL. Mix thoroughly. Gently heat and bring to boiling. Autoclave for 15 min at 15 psi pressure–121°C. Cool to 45°–50°C. Aseptically add sterile novobiocin solution. Mix thoroughly. Pour into sterile Petri dishes or distribute into sterile tubes.

Use: For the selective isolation, cultivation, and presumptive identification of *Aeromonas hydrophila* and other Gram-negative bacteria based on their ability to decarboxylate lysine and ornithine, ferment maltose, and produce H_2S. Maltose-fermenting bacteria appear as yellow colonies. Bacteria that produce lysine or ornithine decarboxylase turn the medium greenish-yellow to yellow. Bacteria that produce H_2S appear as colonies with black centers.

RSS Medium

See: **Reduced Salt Solution Medium**

Rubitelea Medium (DSMZ Medium 1177)

Composition per liter:

Starch	10.0g
Yeast extract	4.0g
Peptone	2.0g
Seawater	1.0L

pH 7.5 ± 0.2 at 25°C

Preparation of Medium: Add components, except bicarbonate solution, to seawater and bring volume to 1.0L. Mix thoroughly. Adjust pH to 7.5. Distribute into tubes or flasks. Gently heat and bring to boiling. Autoclave for 15 min at 15 psi pressure–121°C.

Use: For the cultivation of *Rubitelea* spp.

Rubritalea Medium (DSMZ Medium 1137)

Composition per liter:

Peptone	1.5g
Yeast extract	1.5g
Glucose	1.5g
Concentrated artificial seawater	325.0mL
Hutner's basal salts solution	20.0mL
Bicarbonate solution	10.0mL
Tris-HCl (1*M*, pH 7.5)	5.0mL

pH 7.5 ± 0.2 at 25°C

Hutner's Basal Salts Solution:

Composition per liter:

$MgSO_4 \cdot 7H_2O$	29.7g
Nitrilotriacetic acid	10.0g
$CaCl_2 \cdot 2H_2O$	3.335g
$FeSO_4 \cdot 7H_2O$	99.0mg
$(NH_4)_6MoO_7O_{24} \cdot 4H_2O$	9.25mg
"Metals 44"	50.0mL

"Metals 44":

Composition per 100.0mL:

$ZnSO_4 \cdot 7H_2O$	1.095g
$FeSO_4 \cdot 7H_2O$	0.5g
Sodium EDTA	0.25g
$MnSO_4 \cdot H2O$	0.154g
$CuSO_4 \cdot 5H_2O$	39.2mg
$Co(NO_3)_2 \cdot 6H_2O$	24.8mg
$Na_2B_4O_7 \cdot 10H_2O$	17.7mg

Preparation of "Metals 44": Add sodium EDTA to distilled/deionized water and bring volume to 90.0mL. Mix thoroughly. Add a few drops of concentrated H_2SO_4 to retard precipitation of heavy metal ions. Add remaining components. Mix thoroughly. Bring volume to 100.0mL with distilled/deionized water.

Preparation of Hutner's Basal Salts Solution: Add nitrilotriacetic acid to 500.0mL of distilled/deionized water. Adjust pH to 6.5 with KOH. Add remaining components. Add distilled/deionized water to 1.0L. Adjust pH to 6.8. Autoclave for 15 min at 15 psi pressure–121°C. Cool to 25°C.

Concentrated Artificial Seawater:

Composition per liter:

NaCl	70.43g
Na_2SO_4	11.75g
$MgCl_2 \cdot 6H_2O$	31.86g
$CaCl_2 \cdot 2H_2O$	4.35g
KCl	1.99g
KBr	0.29g
H_3BO_3	0.08g

Preparation of Concentrated Artificial Seawater: Add components to distilled/deionized water and bring volume to 1.0L. Mix thoroughly.

Bicarbonate Solution:

Composition per 10.0mL:

$NaHCO_3$	2.88g

Preparation of Bicarbonate Solution: Add $NaHCO_3$ to distilled/deionized water and bring volume to 10.0mL. Mix thoroughly. Adjust pH to 6.5. Filter sterilize.

Preparation of Medium: Add components, except bicarbonate solution, to distilled/deionized water and bring volume to 990.0mL. Mix thoroughly. Adjust pH to 7.5. Gently heat and bring to boiling. Autoclave for 15 min at 15 psi pressure–121°C. Cool to room temperature. Aseptically add bicarbonate solution. Mix thoroughly. Adjust pH to 7.5. Aseptically distribute into culture vessels.

Use: For the cultivation of *Rubritalea* spp.

Rumen Bacteria Medium

Composition per 1001.0mL:

Na_2CO_3	4.0g
Trypticase™	2.0g
Yeast extract	0.5g
K_2HPO_4	0.3g
Hemin	1.0mg
Resazurin	1.0mg
Minerals solution	38.0mL
Carbohydrate solution	20.0mL
L-Cysteine·HCl·H_2O solution	10.0mL
$Na_2S \cdot 9H_2O$ solution	10.0mL
Volatile fatty acid mixture	3.1mL

pH 6.7 ± 0.2 at 25°C

Minerals Solution:

Composition per liter:

NaCl	12.0g
KH_2PO_4	6.0g
$(NH_4)_2SO_4$	6.0g
$MgSO_4 \cdot 7H_2O$	2.5g
$CaCl_2 \cdot 2H_2O$	1.6g

Preparation of Minerals Solution: Add components to distilled/deionized water and bring volume to 1.0L. Mix thoroughly.

L-Cysteine·HCl·H_2O Solution:

Composition per 10.0mL:

L-Cysteine·HCl·H_2O	0.25g

Preparation of L-Cysteine·HCl·H2O Solution: Add L-cysteine·HCl·H_2O to distilled/deionized water and bring volume to 10.0mL. Mix thoroughly. Sparge with 100% CO_2. Autoclave for 15 min at 15 psi pressure–121°C.

$Na_2S \cdot 9H_2O$ Solution:

Composition per 10.0mL:

$Na_2S \cdot 9H_2O$	0.25g

Preparation of $Na_2S \cdot 9H_2O$ Solution: Add $Na_2S \cdot 9H_2O$ to distilled/deionized water and bring volume to 10.0mL. Mix thoroughly. Sparge with 100% CO_2. Autoclave for 15 min at 15 psi pressure–121°C.

Carbohydrate Solution:

Composition per 20.0mL:

Glucose	0.5g
Cellobiose	0.5g
Glycerol	0.5g
Maltose	0.5g
Starch, soluble	0.5g

Preparation of Carbohydrate Solution: Add components to distilled/deionized water and bring volume to 20.0mL. Mix thoroughly. Sparge under 100% CO_2. Autoclave for 15 min at 15 psi pressure–121°C.

Volatile Fatty Acid Mixture:

Composition per 7.75mL:

Acetic acid	4.25mL
Propionic acid	1.50mL
Butyric acid	1.0mL
DL-2-Methyl butyric acid	0.25mL
iso-Butyric acid	0.25mL
iso-Valeric acid	0.25mL
n-Valeric acid	0.25mL

Preparation of Volatile Fatty Acid Mixture: Combine components. Mix thoroughly.

Preparation of Medium: Prepare and dispense medium under 100% CO_2. Add components, except carbohydrate solution, Na_2CO_3, L-cysteine·HCl·H_2O solution, and $Na_2S \cdot 9H_2O$ solution, to distilled/deionized water and bring volume to 960.0mL Mix thoroughly. Gently heat and bring to boiling. Continue boiling for 5 min. Cool to room temperature while sparging with 100% CO_2. Add Na_2CO_3. Continue sparging with 100% CO_2 until pH reaches 6.8. Distribute into rubber-stoppered tubes under 100% CO_2. Autoclave for 15 min at 15 psi pressure–121°C. Aseptically and anaerobically add 20.0mL of sterile carbohydrate solution, 10.0mL of sterile L-cysteine·HCl·H_2O solution, and 10.0mL of sterile $Na_2S·9H_2O$ solution or, using a syringe, inject the appropriate amount of sterile carbohydrate solution, sterile $Na_2S·9H_2O$ solution, and sterile L-cysteine·HCl·H_2O solution into individual tubes containing medium.

Use: For the cultivation and maintenance of *Anaerovibrio glycerini, Anaerovibrio lipolytica, Butyrivibrio fibrisolvens, Lachnospira multiparus, Succinimonas amylolytica,* and *Succinivibrio dextrinosolvens*.

Rumen Fluid Cellobiose Agar (RFC Agar)

Composition per 10.0mL:

Rumen fluid cellobiose base medium	8.9mL
$NaHCO_3$-rifampin solution	1.0mL
Cellobiose solution	0.1mL

Rumen Fluid Cellobiose Base Medium:

Composition per 89.0mL:

Noble agar	0.7g
Cysteine·HCl·H_2O	0.1g
Clarified rumen fluid	30.0mL
Salts solution A	20.0mL
Salts solution B	20.0mL
Resazurin (0.1% solution)	0.1mL

pH 6.7–7.0 at 25°C

Preparation of Rumen Fluid Cellobiose Base Medium: Add components to distilled/deionized water and bring volume to 89.0mL. Mix thoroughly. Gently heat and bring to boiling. Continue boiling until resazurin turns colorless, indicating reduction. Anaerobically distribute into tubes in 8.9mL volumes under 100% CO_2. Cap tubes with rubber stoppers. Autoclave for 15 min at 15 psi pressure–121°C. Cool to 25°C.

Salts Solution A:

Composition per liter:

$CaCl_2$	0.45g
$MgSO_4$	0.45g

Preparation of Salts Solution A: Add components to distilled/deionized water and bring volume to 1.0L. Mix thoroughly.

Salts Solution B:

Composition per liter:

NaCl	4.5g
$(NH_4)_2SO_4$	4.5g
KH_2PO_4	2.25g
K_2HPO_4	2.25g

Preparation of Salts Solution B: Add components to distilled/deionized water and bring volume to 1.0L. Mix thoroughly.

$NaHCO_3$-Rifampin Solution:

Composition per 10.0mL:

$NaHCO_3$	0.5g
Rifampin	0.1mg

Preparation of $NaHCO_3$-Rifampin Solution: Add components to distilled/deionized water and bring volume to 10.0mL. Mix thoroughly. Filter sterilize.

Cellobiose Solution:

Composition per 10.0mL:

Cellobiose	1.0g

Preparation of Cellobiose Solution: Add cellobiose to distilled/deionized water and bring volume to 10.0mL. Mix thoroughly. Filter sterilize.

Preparation of Medium: To each tube containing 8.9mL of sterile rumen fluid cellobiose base medium, aseptically add 1.0mL of sterile $NaHCO_3$-rifampin solution and 0.1mL of sterile cellobiose solution. Mix thoroughly.

Use: For the selective isolation of rumen treponemes.

Ruminobacter amylophilus Medium

Composition per liter:

Pancreatic digest of casein	10.0g
$NaHCO_3$	6.0g
Starch, soluble	5.0g
NaCl	0.9g
$(NH_4)_2SO_4$	0.9g
L-Cysteine·HCl	0.5g
K_2HPO_4	0.45g
KH_2PO_4	0.45g
$MgSO_4·7H_2O$	0.18g
$CaCl_2·2H_2O$	0.12g
Resazurin	1.0mg

pH 7.0 ± 0.2 at 25°C

Preparation of Medium: Prepare and dispense under 100% CO_2. Add components to distilled/deionized water and bring volume to 1.0L. Mix thoroughly. Sparge with 100% CO_2. Anaerobically distribute into tubes or flasks. Autoclave for 15 min at 15 psi pressure–121°C.

Use: For the cultivation and maintenance of *Ruminobacter amylophilus*.

Ruminococcus albus Medium

Composition per 1001.0mL:

Pancreatic digest of casein	5.0g
Na_2CO_3	4.0g
Glucose	3.0g
Cellobiose	2.0g
Yeast extract	2.0g
L-Cysteine·HCl	0.5g
Resazurin	1.0mg
Mineral solution 1	40.0mL
Mineral solution 2	40.0mL
Fatty acid mixture	1.0mL

Mineral Solution 1:

Composition per 100.0mL:

K_2HPO_4	0.6g

Preparation of Mineral Solution 1: Add K_2HPO_4 to distilled/deionized water and bring volume to 100.0mL. Mix thoroughly.

Mineral Solution 2:

Composition per 100.0mL:

$(NH_4)_2SO_4$	2.0g
NaCl	2.0g
KH_2PO_4	0.6g
$MgSO_4·7H_2O$	0.25g
$CaCl_2·7H_2O$	0.16g

Preparation of Mineral Solution 2: Add components to distilled/deionized water and bring volume to 100.0mL. Mix thoroughly.

Fatty Acid Mixture:
Composition per 100.0mL:

Isobutyric acid	10.0mL
Isovaleric acid	10.0mL
2-Methylbutyric acid	10.0mL

Preparation of Fatty Acid Mixture: Add components to distilled/deionized water and bring volume to 100.0mL. Sparge with 100% CO_2.

Preparation of Medium: Add components, except Na_2CO_3, L-cysteine·HCl, and fatty acid mixture, to distilled/deionized water and bring volume to 1.0L. Mix thoroughly. Gently heat and bring to boiling. Continue boiling for 5 min. Cool to room temperature while sparging with 100% CO_2. Add Na_2CO_3, L-cysteine·HCl, and fatty acid mixture. Adjust pH to 7.0. Anaerobically distribute into tubes or flasks under 100% N_2. Autoclave for 15 min at 15 psi pressure–121°C.

Use: For the cultivation and maintenance of *Ruminococcus albus.*

Ruminococcus pasteurii Medium

Composition per liter:

$NaHCO_3$	2.5g
Sodium tartrate	2.0g
NaCl	1.0g
KCl	0.5g
$MgCl_2·6H_2O$	0.4g
$Na_2S·9H_2O$	0.36g
NH_4Cl	0.25g
KH_2PO_4	0.2g
$CaCl_2·2H_2O$	0.15g
Resazurin	1.0mg
$NaHCO_3$ solution	10.0mL
$Na_2S·9H_2O$ solution	10.0mL
Trace elements solution SL-7	1.0mL

pH 7.2 ± 0.2 at 25°C

Trace Elements Solution SL-7:
Composition per liter:

$FeCl_2·4H_2O$	1.5g
$CoCl_2·6H_2O$	0.19g
$MnCl_2·4H_2O$	0.1g
$ZnCl_2$	0.07g
H_3BO_3	0.062g
$Na_2MoO_4·2H_2O$	0.036g
$NiCl_2·6H_2O$	0.024g
$CuCl_2·2H_2O$	0.017g
HCl (25% solution)	10.0mL

Preparation of Trace Elements Solution SL-7: Add the $FeCl_2·4H_2O$ to the HCl. Add distilled/deionized water and bring volume to 1.0L. Add remaining components. Mix thoroughly. Autoclave for 15 min at 15 psi pressure–121°C under 100% N_2. Cool to room temperature.

$NaHCO_3$ Solution:
Composition per 10.0mL:

$NaHCO_3$	2.5g

Preparation of $NaHCO_3$ Solution: Add the $NaHCO_3$ to distilled/deionized water and bring volume to 10.0mL. Mix thoroughly. Filter sterilize.

$Na_2S·9H_2O$ Solution:
Composition per 10.0mL:

$Na_2S·9H_2O$	0.36g

Preparation of $Na_2S·9H_2O$ Solution: Add $Na_2S·9H_2O$ to distilled/deionized water and bring volume to 10.0mL. Mix thoroughly. Autoclave for 15 min at 15 psi pressure–121°C under 100% N_2.

Preparation of Medium: Add components—except $NaHCO_3$ solution, $Na_2S·9H_2O$ solution, and trace elements solution SL-7—to distilled/deionized water and bring volume to 999.0mL. Mix thoroughly. Adjust pH to 7.2. Gently heat and bring to boiling under 80% N_2 + 20% CO_2. Distribute into tubes in 9.8mL volumes under 80% N_2 + 20% CO_2. Cool to 25°C. Aseptically add 0.1mL of sterile $NaHCO_3$ solution and 0.01mL of sterile trace elements solution SL-7 to each tube. Mix thoroughly. Immediately prior to inoculation, aseptically add 0.1mL of sterile $Na_2S·9H_2O$ solution to each tube.

Use: For the cultivation and maintenance of *Ruminococcus pasteurii.*

Ruminococcus pasteurii Medium

Composition per liter:

NaCl	1.0g
KCl	0.5g
$MgCl_2·6H_2O$	0.4g
NH_4Cl	0.25g
KH_2PO_4	0.2g
$CaCl_2·2H_2O$	0.15g
Resazurin	1.0mg
Biotin	0.4mg
2,3-Butanediol solution	50.0mL
$NaHCO_3$ solution	20.0mL
$Na_2S·9H_2O$ solution	10.0mL
Trace elements solution SL-10	1.0mL

pH 7.2 ± 0.2 at 25°C

Trace Elements Solution SL-10:
Composition per liter:

$FeCl_2·4H_2O$	1.5g
$CoCl_2·6H_2O$	190.0mg
$MnCl_2·4H_2O$	100.0mg
$ZnCl_2$	70.0mg
$Na_2MoO_4·2H_2O$	36.0mg
$NiCl_2·6H_2O$	24.0mg
H_3BO_3	6.0mg
$CuCl_2·2H_2O$	2.0mg
HCl (25% solution)	10.0mL

Preparation of Trace Elements Solution SL-10: Prepare and dispense under 80% N_2 + 20% CO_2. Add $FeCl_2·4H_2O$ to 10.0mL of HCl solution. Mix thoroughly. Add distilled/deionized water and bring volume to 1.0L. Add remaining components. Mix thoroughly. Filter sterilize. Gas under 80% N_2 + 20% CO_2.

2,3-Butanediol Solution:
Composition per 50.0mL:

2,3-Butanediol	0.9g

Preparation of 2,3-Butanediol Solution: Add 2,3-butanediol to distilled/deionized water and bring volume to 50.0mL. Mix thoroughly. Filter sterilize. Gas under 80% N_2 + 20% CO_2.

$NaHCO_3$ Solution:
Composition per 20.0mL:

$NaHCO_3$	2.5g

Preparation of $NaHCO_3$ Solution: Add $NaHCO_3$ to distilled/deionized water and bring volume to 20.0mL. Mix thoroughly. Filter sterilize. Gas under 80% N_2 + 20% CO_2.

$Na_2S{\cdot}9H_2O$ Solution:

Composition per 10.0mL:

$Na_2S{\cdot}9H_2O$	0.36g

Preparation of $Na_2S{\cdot}9H_2O$ Solution: Add $Na_2S{\cdot}9H_2O$ to distilled/deionized water and bring volume to 10.0mL. Mix thoroughly. Gas under 100% N_2. Autoclave for 15 min at 15 psi pressure–121°C.

Preparation of Medium: Prepare medium and dispense under 80% N_2 + 20% CO_2. Add components, except 2,3-butanediol solution, $NaHCO_3$ solution, and $Na_2S{\cdot}9H_2O$ solution, to distilled/deionized water and bring volume to 970.0mL. Mix thoroughly. Sparge with 80% N_2 + 20% CO_2. Autoclave for 15 min at 15 psi pressure–121°C. Aseptically and anaerobically add 50.0mL of sterile 2,3-butanediol solution, 20.0mL of sterile $NaHCO_3$ solution, and 10.0mL of sterile $Na_2S{\cdot}9H_2O$ solution. Mix thoroughly. Aseptically and anaerobically distribute into sterile tubes or flasks.

Use: For the cultivation and maintenance of *Ruminococcus pasteurii.*

Runella slithyformis Medium

Composition per liter:

Agar	10.0g
Peptone	2.0g
Yeast extract	1.0g
$MgSO_4{\cdot}7H_2O$	0.2g

Preparation of Medium: Add components to tap water and bring volume to 1.0L. Mix thoroughly. Gently heat and bring to boiling. Distribute into tubes or flasks. Autoclave for 15 min at 15 psi pressure–121°C. Pour into sterile Petri dishes or leave in tubes.

Use: For the cultivation and maintenance of *Runella slithyformis.*

Russell Double-Sugar Agar

Composition per liter:

Agar	15.0g
Proteose peptone No. 3	12.0g
Lactose	10.0g
NaCl	5.0g
Beef extract	1.0g
Glucose	1.0g
Phenol Red	0.025g

pH 7.5 ± 0.2 at 25°C

Preparation of Medium: Add components to distilled/deionized water and bring volume to 1.0L. Mix thoroughly. Gently heat and bring to boiling. Distribute into tubes. Autoclave for 15 min at 15 psi pressure–121°C. Allow tubes to cool in a slanted position.

Use: For the identification of Gram-negative enteric bacilli based on their fermentation of glucose and lactose. Bacteria that ferment both glucose and lactose produce a yellow slant and yellow butt. Bacteria that ferment glucose but do not ferment lactose produce a red slant and a yellow butt. Bacteria that ferment neither glucose nor lactose produce an unchanged pink-orange color.

RV Enrichment Broth

See: **Rappaport-Vassiliadis Enrichment Broth**

RVS Broth

See: **Rappaport-Vassiliadis Soy Peptone Broth**

RV5 Medium (DSMZ Medium 1147)

Composition per liter:

NaCl	15.0g
DL-malic acid	4.0g
BICINE [N,N-bis(2-hydroxyethyl)glycine] buffer	1.63g
K_2HPO_4	1.0g
NH_4Cl	0.52g
Yeast extract	0.5g
$MgSO_4{\cdot}7H_2O$	0.2g
EDTA	10.0mg
$CaCl_2{\cdot}2H_2O$	7.5gm
Vitamin solution	10.0mL
$NaHCO_3$ solution	10.0mL
Trace elements solution	1.0mL

pH 7.5 ± 0.2 at 25°C

Vitamin Solution:

Composition per 10.0mL:

Thiamine-HCl·$2H_2O$	1.0mg
Biotin	0.03mg
Vitamin B_{12}	0.02mg

Preparation of Vitamin Solution: Add components to distilled/deionized water and bring volume to 10.0mL. Mix thoroughly. Filter sterilize.

$NaHCO_3$ Solution:

Composition per 20.0mL:

$NaHCO_3$	2.5g

Preparation of $NaHCO_3$ Solution: Add $NaHCO_3$ to distilled/deionized water and bring volume to 20.0mL. Mix thoroughly. Filter sterilize. Gas under 80% N_2 + 20% CO_2.

Trace Elements Solution:

Composition per liter:

EDTA	5.2g
$FeCl_2{\cdot}4H_2O$	1.5g
$CoCl_2{\cdot}6H_2O$	190.0mg
$Na_2MoO_4{\cdot}2H_2O$	188.0mg
$MnCl_2{\cdot}4H_2O$	100.0mg
$ZnCl_2$	70.0mg
$VoSO_4{\cdot}2H_2O$	30.0mg
$Ni_2Cl_2{\cdot}6H_2O$	25.0mg
$CuCl_2{\cdot}2H_2O$	17.0mg
H_3BO_3	6.0mg
$Na_2WO_4{\cdot}2H_2O$	2.0mg
$NaHSeO_3$	2.0mg

Preparation of Trace Elements Solution: Add components to distilled/deionized water and bring volume to 1.0L. Adjust pH to 3.0. Mix thoroughly.

Preparation of Medium: Add components, except bicarbonate solution and vitamin solution, to distilled/deionized water and bring volume to 980.0mL. Mix thoroughly. Adjust pH to 9.5. Gently heat and bring to boiling. Autoclave for 15 min at 15 psi pressure–121°C. Cool to room temperature. Aseptically add vitamin solution. Mix thoroughly. Adjust pH to 9.5. Aseptically distribute into culture vessels. Before inoculation aseptically add bicarbonate solution.

Use: For the cultivation of *Rhodobaca bogoriensis.*

Ryan's *Aeromonas* Medium

See: ***Aeromonas*** **Medium**

R2YE Medium

Composition per 1062.2mL:

Thiostrepton	50.0mg
Basal solution	800.0mL
TES (*N*-tris[hydroxymethyl]methyl-2-amino–ethane-sulfonic acid) buffer	100.0mL
$CaCl_2 \cdot 2H_2O$ solution	80.2mL
Yeast extract solution	50.0mL
L-Proline solution	15.0mL
KH_2PO_4 solution	10.0mL
NaOH solution	5.0mL
Trace elements solution	2.0mL

pH 7.2 ± 0.2 at 25°C

Basal Solution:

Composition per 800.0mL:

Sucrose	103.0g
$MgCl_2 \cdot 6H_2O$	10.12g
D-Glucose	10.0g
K_2SO_4	0.25g
Casamino acids	0.1g

Preparation of Basal Solution: Add components to distilled/deionized water and bring volume to 800.0mL. Mix thoroughly. Autoclave for 15 min at 15 psi pressure–121°C. Cool to 25°C.

TES Buffer:

Composition per liter:

TES (*N*-tris[hydroxymethyl] methyl-2-amino–ethane-sulfonic acid) buffer	57.3g

Preparation of TES Buffer: Add TES to distilled/deionized water and bring volume to 1.0L. Mix thoroughly. Adjust pH to 7.2. Filter sterilize.

$CaCl_2 \cdot 2H_2O$ Solution:

Composition per 100.0mL:

$CaCl_2 \cdot 2H_2O$	3.68g

Preparation of $CaCl_2 \cdot 2H_2O$ Solution: Add the $CaCl_2 \cdot 2H_2O$ to distilled/deionized water and bring volume to 100.0mL. Mix thoroughly. Filter sterilize.

Yeast Extract Solution:

Composition per 100.0mL:

Yeast extract	10.0g

Preparation of Yeast Extract Solution: Add yeast extract to distilled/deionized water and bring volume to 100.0mL. Mix thoroughly. Filter sterilize.

L-Proline Solution:

Composition per 100.0mL:

L-Proline	20.0g

Preparation of L-Proline Solution: Add the proline to distilled/deionized water and bring volume to 100.0mL. Mix thoroughly. Filter sterilize.

KH_2PO_4 Solution:

Composition per 100.0mL:

KH_2PO_4	0.5g

Preparation of KH_2PO_4 Solution Solution: Add the KH_2PO_4 to distilled/deionized water and bring volume to 100.0mL. Mix thoroughly. Filter sterilize.

NaOH Solution:

Composition per 100.0mL:

NaOH	40.0g

Preparation of NaOH Solution: Add the NaOH to distilled/deionized water and bring volume to 100.0mL. Mix thoroughly. Filter sterilize

Trace Elements Solution:

Composition per liter:

$FeCl_3 \cdot 6H_2O$	0.2g
$ZnCl_2$	0.04g
$CuCl_2 \cdot 2H_2O$	0.01g
$MnCl_2 \cdot 4H_2O$	0.01g
$Na_2B_4O_7 \cdot 10H_2O$	0.01g
$(NH_4)_6Mo_7O_{24} \cdot 4H_2O$	0.01g

Preparation of Trace Elements Solution: Add components to distilled/deionized water and bring volume to 1.0L. Mix thoroughly. Filter sterilize.

Preparation of Medium: To 800.0mL of cooled, sterile basal solution, aseptically add the remaining components. Mix thoroughly. Aseptically distribute into sterile tubes or flasks.

Use: For the cultivation and maintenance of *Streptomyces lividans*.

S. aureus ID

Composition per liter:

Proprietary

Source: This medium is available from bioMérieux.

Use: For the direct identification of *Staphylococcus aureus* and the selective isolation of staphylococci. Direct identification of *S. aureus* is based on the spontaneous green coloration of α-glucosidase-producing colonies.

S Broth

Composition per liter:

Peptone	10.0g
Meat extract	2.4g
NaCl	2.0g

pH 7.0 ± 0.2 at 25°C

Preparation of Medium: Add components to distilled/deionized water and bring volume to 1.0L. Mix thoroughly. Distribute into tubes or flasks. Autoclave for 15 min at 15 psi pressure–121°C.

Use: For the cultivation of *Bacillus cereus*.

S Medium

Composition per liter:

Glycogen	3.0g
$MgSO_4 \cdot 7H_2O$	2.0g
L-Leucine	1.0g
L-Tyrosine	0.6g
L-Asparagine	0.5g
L-Isoleucine	0.5g
L-Proline	0.5g
L-Lysine	0.25g
KH_2PO_4	0.13g

Djenkolic acid 0.1g
L-Arginine 0.1g
L-Serine 0.1g
L-Threonine 0.1g
L-Valine 0.1g
L-Alanine 0.05g
L-Glycine 0.05g
L-Histidine 0.05g
L-Methionine 0.05g
L-Tryptophan 0.05g

pH 7.6 ± 0.2 at 25°C

Preparation of Medium: Add components to distilled/deionized water and bring volume to 1.0L. Mix thoroughly. Filter sterilize. Aseptically distribute into tubes or flasks.

Use: For the cultivation of *Myxococcus xanthus*.

S Salts

Composition per liter:

Dibenzothiophene 5.0g
NH_4Cl 0.5g
KH_2PO_4 .... 0.25g
$MgCl_2 \cdot 6H_2O$ 0.25g

pH 6.5–7.0 at 25°C

Preparation of Medium: Add components to distilled/deionized water and bring volume to 1.0L. Mix thoroughly. Adjust pH to 6.5–7.0 with KOH. Distribute into tubes or flasks. Autoclave for 15 min at 15 psi pressure–121°C.

Use: For the cultivation of *Bacillus sulfasportare*.

S6 Medium for Thiobacilli

Composition per liter:

Agar 15.0g
$Na_2S_2O_3$ 10.0g
KH_2PO_4 .... 11.8g
Na_2HPO_4 .... 1.2g
$MgSO_4 \cdot 7H_2O$ 0.1g
$(NH_4)_2SO_4$ 0.1g
$CaCl_2$ 0.03g
$FeCl_3$ 0.02g
$MnSO_4$ 0.02g

Preparation of Medium: Add components to distilled/deionized water and bring volume to 1.0L. Mix thoroughly. Gently heat and bring to boiling. Distribute into tubes or flasks. Autoclave for 15 min at 15 psi pressure–121°C. Pour into sterile Petri dishes or leave in tubes.

Use: For the cultivation and maintenance of *Thiobacillus denitrificans* and *Thiobacillus thioparus*.

S8 Medium for Thiobacilli

Composition per liter:

Agar 15.0g
KH_2PO_4 .... 11.8g
$Na_2S_2O_3$ 10.0g
KNO_3 .... 5.0g
Na_2HPO_4 .... 1.2g
$NaHCO_3$ 0.5g
$MgSO_4 \cdot 7H_2O$ 0.1g
$(NH_4)_2SO_4$ 0.1g
$CaCl_2$ 0.03g
$FeCl_3$ 0.02g
$MnSO_4$ 0.02g

Preparation of Medium: Add components to distilled/deionized water and bring volume to 1.0L. Mix thoroughly. Gently heat and bring to boiling. Distribute into tubes or flasks. Autoclave for 15 min at 15 psi pressure–121°C. Pour into sterile Petri dishes or leave in tubes.

Use: For the cultivation and maintenance of *Thiobacillus neapolitanus*.

SA Agar

Composition per liter:

Agar 15.0g
Pancreatic digest of casein 10.0g
NaCl 5.0g
Starch, soluble 1.0g
Ampicillin 0.01g
Phenol Red 0.018g

pH 7.4 ± 0.2 at 25°C

Preparation of Medium: Add components, except ampicillin, to distilled/deionized water and bring volume to 1.0L. Mix thoroughly. Gently heat and bring to boiling. Autoclave for 15 min at 15 psi pressure–121°C. Cool to 45°–50°C. Aseptically add ampicillin. Mix thoroughly. Pour into sterile Petri dishes.

Use: For the isolation, cultivation, and differentiation, based on starch hydrolysis, of *Aeromonas hydrophila* from foods. After inoculation of plates and growth of cultures, starch hydrolysis is determined by flooding each plate with 5.0mL of Lugol's iodine solution.

SA Agar, Modified (Lachica's Medium)

Composition per liter:

Beef heart, solids from infusion 500.0g
Agar 15.0g
Tryptose 10.0g
NaCl 5.0g
Amylose Azure 3.0g
Ampicillin 0.01mg

pH 7.4 ± 0.2 at 25°C

Preparation of Medium: Add components to distilled/deionized water and bring volume to 1.0L. Mix thoroughly. Gently heat and bring to boiling. Distribute into tubes or flasks. Autoclave for 15 min at 15 psi pressure–121°C. Pour into sterile Petri dishes.

Use: For the isolation and cultivation of *Aeromonas hydrophila* from foods. *Aeromonas hydrophila* appears as colonies surrounded by a light halo on a light blue background.

SA HiVeg Agar Base with Ampicillin

Composition per liter:

Agar 15.0g
Plant hydrolysate 10.0g
Starch, soluble 10.0g
NaCl 5.0g
Phenol Red 0.025g
Ampicillin solution 10.0mL

pH 7.4 ± 0.2 at 25°C

Source: This medium, without ampicillin, is available as a premixed powder from HiMedia.

Ampicillin Solution:

Composition per 10.0mL:

Ampicillin	0.01g

Preparation of Ampicillin Solution: Add ampicillin to distilled/deionized water and bring volume to 10.0mL. Mix thoroughly. Filter sterilize.

Preparation of Medium: Add components, except ampicillin solution, to distilled/deionized water and bring volume to 990.0mL. Mix thoroughly. Gently heat and bring to boiling. Autoclave for 15 min at 15 psi pressure–121°C. Cool to 45°C. Aseptically add sterile ampicillin solution. Mix thoroughly. Pour into sterile Petri dishes or distribute into sterile tubes.

Use: For the isolation, cultivation, and differentiation of *Aeromonas hydrophilia* from foods based on starch hydrolysis.

SABHI Agar (Sabouraud Glucose and Brain Heart Infusion Agar)

Composition per liter:

Glucose	21.0g
Agar	15.0g
Pancreatic digest of casein	10.5g
Peptic digest of animal tissue	5.0g
Brain heart, solids from infusion	4.0g
NaCl	2.5g
Na_2HPO_4	1.25g

pH 6.8 ± 0.2 at 25°C

Source: This medium is available as a premixed powder from BD Diagnostic Systems.

Preparation of Medium: Add components to distilled/deionized water and bring volume to 1.0L. Mix thoroughly. Gently heat and bring to boiling. Distribute into tubes or flasks. Autoclave for 15 min at 15 psi pressure–121°C. Pour into sterile Petri dishes in 20.0mL volumes or leave in tubes.

Use: For the cultivation of dermatophytes and other pathogenic and nonpathogenic fungi from clinical and nonclinical specimens.

SABHI Agar

Composition per liter:

Beef heart, infusion from	125.0g
Calf brains, infusion from	100.0g
Glucose	21.0g
Agar	15.0g
Neopeptone	5.0g
Proteose peptone	5.0g
NaCl	2.5g
Na_2HPO_4	1.25g
Chloromycetin solution	1.0mL

pH 7.0 ± 0.2 at 25°C

Source: This medium is available as a premixed powder from BD Diagnostic Systems.

Chloromycetin Solution:

Composition per 10.0mL:

Chloromycetin	1.0g

Preparation of Chloromycetin Solution: Add chloromycetin to distilled/deionized water and bring volume to 10.0mL. Mix thoroughly. Filter sterilize.

Preparation of Medium: Add components, except chloromycetin solution, to distilled/deionized water and bring volume to 999.0mL. Mix thoroughly. Gently heat and bring to boiling. Autoclave for 15 min at 15 psi pressure–121°C. Cool to 45°–50°C. Aseptically add 1.0mL of sterile chloromycetin solution. Mix thoroughly. Aseptically distribute into sterile tubes in 5.0mL volumes.

Use: For the cultivation of dermatophytes and other pathogenic and nonpathogenic fungi from clinical and nonclinical specimens.

SABHI Agar, Modified

Composition per liter:

Beef heart, infusion from	62.5g
Calf brain, infusion from	50.0g
Glucose	20.5g
Brain heart infusion broth	18.6g
Agar	7.5g
Neopeptone	5.0g
Pancreatic digest of gelatin	2.5g
NaCl	1.25g
Na_2HPO_4	0.625g

pH 6.8 ± 0.2 at 25°C

Preparation of Medium: Dissolve, then autoclave at 121°C for 15 min. Cool to 50°C and add 1.0mL of sterile chloramphenicol solution (100.0mg/mL). Mix well and dispense into sterile tubes. Slant and allow to harden. Refrigerate until needed.

Use: For the cultivation of dermatophytes and other pathogenic and nonpathogenic fungi from clinical and nonclinical specimens.

SABHI Blood Agar

Composition per liter:

Beef heart, infusion from	125.0g
Calf brains, infusion from	100.0g
Glucose	21.0g
Agar	15.0g
Neopeptone	5.0g
Proteose peptone	5.0g
NaCl	2.5g
Na_2HPO_4	1.25g
Blood	100.0mL
Chloromycetin solution	1.0mL

pH 7.0 ± 0.2 at 25°C

Source: This medium is available as a premixed powder from BD Diagnostic Systems.

Chloromycetin Solution:

Composition per 10.0mL:

Chloromycetin	1.0g

Preparation of Chloromycetin Solution: Add chloromycetin to distilled/deionized water and bring volume to 10.0mL. Mix thoroughly. Filter sterilize.

Preparation of Medium: Add components, except blood and chloromycetin solution, to distilled/deionized water and bring volume to 899.0mL. Mix thoroughly. Gently heat and bring to boiling. Autoclave for 15 min at 15 psi pressure–121°C. Cool to 45°–50°C. Aseptically add 100.0mL of sterile blood and 1.0mL of sterile chloromycetin solution. Sheep blood or human blood may be used. Mix thoroughly. Aseptically distribute into sterile tubes in 5.0mL volumes.

Use: For the cultivation of dermatophytes and other pathogenic and nonpathogenic fungi from clinical and nonclinical specimens. Blood enhances the recovery of *Blastomyces dermatitidis* and *Histoplasma capsulatum* and their conversion to the yeast phase.

SABHI HiVeg Agar Base with Chloramphenicol

Composition per liter:

Glucose	21.0g
Agar	15.0g
Plant infusion	5.14g
Plant peptone No. 3	5.0g
Plant special peptone	5.0g
Plant special infusion	4.11g
NaCl	2.5g
Na_2HPO_4	1.25g
Chloramphenicol solution	10.0mL

pH 7.0 ± 0.2 at 25°C

Source: This medium, without chloramphenicol, is available as a premixed powder from HiMedia.

Chloramphenicol Solution:

Composition per 10.0mL:

Chloramphenicol	0.1g

Preparation of Chloramphenicol Solution: Add chloramphenicol to distilled/deionized water and bring volume to 10.0mL. Mix thoroughly. Filter sterilize.

Preparation of Medium: Add components, except chloramphenicol solution, to distilled/deionized water and bring volume to 990.0mL. Mix thoroughly. Gently heat and bring to boiling. Autoclave for 15 min at 15 psi pressure–121°C. Cool to 45°–50°C. Aseptically add 10.0mL of sterile chloramphenicol solution. Mix thoroughly. Pour into Petri dishes or aseptically distribute into sterile tubes.

Use: For the cultivation of dermatophytes and other pathogenic and nonpathogenic fungi from clinical and nonclinical specimens.

SABHI HiVeg Agar Base with Chloromycetin

Composition per liter:

Glucose	21.0g
Agar	15.0g
Plant infusion	5.14g
Plant peptone No. 3	5.0g
Plant special peptone	5.0g
Plant special infusion	4.11g
NaCl	2.5g
Na_2HPO_4	1.25g
Chloromycetin solution	1.0mL

pH 7.0 ± 0.2 at 25°C

Source: This medium, without chloromycetin, is available as a premixed powder from HiMedia.

Chloromycetin Solution:

Composition per 10.0mL:

Chloromycetin	1.0g

Preparation of Chloromycetin Solution: Add chloromycetin to distilled/deionized water and bring volume to 10.0mL. Mix thoroughly. Filter sterilize.

Preparation of Medium: Add components, except chloromycetin solution, to distilled/deionized water and bring volume to 999.0mL. Mix thoroughly. Gently heat and bring to boiling. Autoclave for 15 min at 15 psi pressure–121°C. Cool to 45°–50°C. Aseptically add 1.0mL of sterile chloromycetin solution. Mix thoroughly. Aseptically distribute into sterile tubes in 5.0mL volumes.

Use: For the cultivation of dermatophytes and other pathogenic and nonpathogenic fungi from clinical and nonclinical specimens.

SABHI HiVeg Agar Base with Blood and Chloromycetin

Composition per liter:

Glucose	21.0g
Agar	15.0g
Plant infusion	5.14g
Plant peptone No. 3	5.0g
Plant special peptone	5.0g
Plant special infusion	4.11g
NaCl	2.5g
Na_2HPO_4	1.25g
Blood	100.0mL
Chloromycetin solution	1.0mL

pH 7.0 ± 0.2 at 25°C

Source: This medium, without blood and chloromycetin, is available as a premixed powder from HiMedia.

Chloromycetin Solution:

Composition per 10.0mL:

Chloromycetin	1.0g

Preparation of Chloromycetin Solution: Add chloromycetin to distilled/deionized water and bring volume to 10.0mL. Mix thoroughly. Filter sterilize.

Preparation of Medium: Add components, except blood and chloromycetin solution, to distilled/deionized water and bring volume to 899.0mL. Mix thoroughly. Gently heat and bring to boiling. Autoclave for 15 min at 15 psi pressure–121°C. Cool to 45°–50°C. Aseptically add 100.0mL of sterile blood and 1.0mL of sterile chloromycetin solution. Sheep blood or human blood may be used. Mix thoroughly. Aseptically distribute into sterile tubes in 5.0mL volumes.

Use: For the cultivation of dermatophytes and other pathogenic and nonpathogenic fungi from clinical and nonclinical specimens. Blood enhances the recovery of *Blastomyces dermatitidis* and *Histoplasma capsulatum* and their conversion to the yeast phase.

Sabouraud Agar

Composition per liter:

Neopeptone	30.0g
Agar	20.0g

pH 6.8 ± 0.2 at 25°C

Preparation of Medium: Add components to tap water and bring volume to 1.0L. Mix thoroughly. Gently heat and bring to boiling. Distribute into tubes or flasks. Autoclave for 15 min at 15 psi pressure–121°C. Pour into sterile Petri dishes or leave in tubes.

Use: For the cultivation of yeasts and molds.

Sabouraud Agar with CCG and 3% Sodium Chloride

Composition per 3031.5mL:

Glucose	120.0g
NaCl	90.0g
Agar	45.0g
Peptone	30.0g
Chloramphenicol solution	15.0mL

Cycloheximide solution	15.0mL
Gentamicin solution	1.5mL

Chloramphenicol Solution:

Composition per 15.0mL:

Chloramphenicol	0.15g

Preparation of Chloramphenicol Solution: Add chloramphenicol to distilled/deionized water and bring volume to 15.0mL. Mix thoroughly. Filter sterilize.

Cycloheximide Solution:

Composition per 15.0mL:

Cycloheximide	0.3g

Preparation of Cycloheximide Solution: Add cycloheximide to distilled/deionized water and bring volume to 15.0mL. Mix thoroughly. Filter sterilize.

Caution: Cycloheximide is toxic. Avoid skin contact or aerosol formation and inhalation.

Gentamicin Solution:

Composition per 10.0mL:

Gentamicin	0.4g

Preparation of Gentamicin Solution: Add gentamicin to distilled/deionized water and bring volume to 10.0mL. Mix thoroughly. Filter sterilize.

Preparation of Medium: Add components—except chloramphenicol solution, cycloheximide solution, and gentamicin solution—to distilled/deionized water and bring volume to 3.0L. Mix thoroughly. Gently heat and bring to boiling. Autoclave for 15 min at 15 psi pressure–121°C. Cool to 45°–50°C. Aseptically add 15.0mL of sterile chloramphenicol solution, 15.0mL of sterile cycloheximide solution, and 1.5mL of sterile gentamicin solution. Mix thoroughly. Aseptically distribute into sterile tubes. Allow tubes to cool in a slanted position.

Use: For the selective isolation and cultivation of fungi from specimens with a mixed flora.

Sabouraud Agar with CCG and 5% Sodium Chloride

Composition per 3031.5mL:

NaCl	150.0g
Glucose	120.0g
Agar	45.0g
Peptone	30.0g
Chloramphenicol solution	15.0mL
Cycloheximide solution	15.0mL
Gentamicin solution	1.5mL

Chloramphenicol Solution:

Composition per 15.0mL:

Chloramphenicol	0.15g

Preparation of Chloramphenicol Solution: Add chloramphenicol to distilled/deionized water and bring volume to 15.0mL. Mix thoroughly. Filter sterilize.

Cycloheximide Solution:

Composition per 15.0mL:

Cycloheximide	0.3g

Preparation of Cycloheximide Solution: Add cycloheximide to distilled/deionized water and bring volume to 15.0mL. Mix thoroughly. Filter sterilize.

Caution: Cycloheximide is toxic. Avoid skin contact or aerosol formation and inhalation.

Gentamicin Solution:

Composition per 10.0mL:

Gentamicin	0.4g

Preparation of Gentamicin Solution: Add gentamicin to distilled/deionized water and bring volume to 10.0mL. Mix thoroughly. Filter sterilize.

Preparation of Medium: Add components—except chloramphenicol solution, cycloheximide solution, and gentamicin solution—to distilled/deionized water and bring volume to 3.0L. Mix thoroughly. Gently heat and bring to boiling. Autoclave for 15 min at 15 psi pressure–121°C. Cool to 45°–50°C. Aseptically add 15.0mL of sterile chloramphenicol solution, 15.0mL of sterile cycloheximide solution, and 1.5mL of sterile gentamicin solution. Mix thoroughly. Aseptically distribute into sterile tubes. Allow tubes to cool in a slanted position.

Use: For the selective isolation and cultivation of fungi from specimens with a mixed flora.

Sabouraud Agar, Diluted 1/10

Composition per liter:

Agar	20.0g
Glucose	4.0g
KH_2PO_4	1.5g
$MgSO_4 \cdot 7H_2O$	1.0g
$NaNO_3$	1.0g
Peptone	1.0g

pH 6.8 ± 0.2 at 25°C

Preparation of Medium: Add components to distilled/deionized water and bring volume to 1.0L. Mix thoroughly. Adjust pH to 6.8. Gently heat and bring to boiling. Distribute into tubes or flasks. Autoclave for 15 min at 15 psi pressure–121°C. Pour into sterile Petri dishes or leave in tubes.

Use: For the cultivation and maintenance of a variety of fungi and heterotrophic bacteria.

Sabouraud Agar, Diluted 1/10 with Salt

Composition per liter:

Agar	15.0g
Glucose	2.0g
KH_2PO_4	1.0g
$MgSO_4 \cdot 7H_2O$	1.0g
Mycological peptone	1.0g

pH 6.8–7.0 at 25°C

Preparation of Medium: Add components to distilled/deionized water and bring volume to 1.0L. Mix thoroughly. Adjust pH to 6.8–7.0. Gently heat and bring to boiling. Distribute into tubes or flasks. Autoclave for 15 min at 15 psi pressure–121°C. Pour into sterile Petri dishes or leave in tubes.

Use: For the cultivation and maintenance of *Arthroderma benhamiae, Arthroderma vanbreuseghemii, Microsporum canis,* and *Trichophyton mentagrophytes.*

Sabouraud Agar, Modified

Composition per liter:

Agar	20.0g
Glucose	20.0g
Neopeptone	10.0g

pH 7.0 ± 0.2 at 25°C

Source: This medium is available as a premixed powder from BD Diagnostic Systems.

Preparation of Medium: Add components to distilled/deionized water and bring volume to 1.0L. Mix thoroughly. Gently heat and bring to boiling. Distribute into tubes or flasks. Autoclave for 15 min at 15 psi pressure–121°C. Pour into sterile Petri dishes or leave in tubes.

Use: For the cultivation of yeasts, molds, and aciduric bacteria.

Sabouraud Chloramphenicol HiVeg Agar

Composition per liter:

Glucose	40.0g
Agar	15.0g
Plant hydrolysate	5.0g
Plant peptone	5.0g
Chloramphenicol	0.05g
Chloramphenicol solution	10.0mL

pH 5.6 ± 0.2 at 25°C

Source: This medium is available as a premixed powder from HiMedia.

Preparation of Medium: Add components to distilled/deionized water and bring volume to 1.0L. Mix thoroughly. Gently heat and bring to boiling. Distribute into tubes or flasks. Autoclave for 15 min at 15 psi pressure–121°C. Pour into sterile Petri dishes or leave in tubes.

Use: For the cultivation and identification of yeasts.

Sabouraud Cycloheximide Chloramphenicol HiVeg Agar

Composition per liter:

Glucose	20.0g
Agar	15.0g
Plant peptone	10.0g
Cycloheximide	0.5g
Chloramphenicol	0.04g

pH 6.8 ± 0.2 at 25°C

Source: This medium is available as a premixed powder from HiMedia.

Caution: Cycloheximide is very toxic. Avoid skin contact or aerosol formation and inhalation.

Preparation of Medium: Add components to distilled/deionized water and bring volume to 1.0L. Mix thoroughly. Gently heat and bring to boiling. Distribute into tubes or flasks. Autoclave for 15 min at 15 psi pressure–121°C. Pour into sterile Petri dishes or leave in tubes.

Use: For the isolation and cultivation of pathogenic fungi.

Sabouraud Dextrose Agar (BAM M133)

Composition per liter:

Neopeptone	30.0g
Agar	20.0g

pH 5.6 ± 0.2 at 25°C

Preparation of Medium: Add components to tap water and bring volume to 1.0L. Mix thoroughly. Gently heat and bring to boiling. Distribute into tubes or flasks. Autoclave for 15 min at 15 psi pressure–121°C. Pour into sterile Petri dishes or leave in tubes.

Use: For the cultivation of yeasts and molds.

Sabouraud Dextrose Agar pH 5.6

Composition per liter:

Glucose	40.0g
Agar	15.0g
Mycological peptone	10.0 g

pH 5.6 ± 0.2 at 25°C

Source: This medium is available as a premixed powder from Oxoid Unipath.

Preparation of Medium: Add components to distilled/deionized water and bring volume to 1.0L. Mix thoroughly. Gently heat while stirring and bring to boiling. Autoclave for 15 min at 15 psi pressure–121°C. Cool to 50°C. Mix thoroughly. Pour into sterile Petri dishes or leave in tubes.

Use: For the cultivation of yeasts and other fungi. For use in qualitative and quantitative procedures for yeasts and fungi.

Sabouraud Glucose Agar (Saboraud Dextrose Agar) (SabDex, 2%)

Composition per liter:

Glucose	20.0g
Agar	15.0g
Pancreatic digest of casein	5.0g
Peptic digest of animal tissue	5.0g

pH 5.6 ± 0.2 at 25°C

Preparation of Medium: Add components to distilled/deionized water and bring volume to 1.0L. Mix thoroughly. Gently heat and bring to boiling. Distribute into tubes or flasks. Autoclave for 15 min at 15 psi pressure–121°C. Pour into sterile Petri dishes or leave in tubes.

Use: For the cultivation of yeast and filamentous fungi. For the cultivation of pathogenic and nonpathogenic fungi, especially dermatophytes. The medium may be made more selective for fungi by the addition of chloramphenicol. Fluconozole (final concentration of 8–16mg per mL) may also be added to test for antibiotic sensitivity.

Sabouraud Glucose Agar (Saboraud Dextrose Agar) (SabDex, 4%)

Composition per liter:

Glucose	40.0g
Agar	15.0g
Pancreatic digest of casein	5.0g
Peptic digest of animal tissue	5.0g

pH 5.6 ± 0.2 at 25°C

Source: This medium is available as a premixed powder from BD Diagnostic Systems and Oxoid Unipath.

Preparation of Medium: Add components to distilled/deionized water and bring volume to 1.0L. Mix thoroughly. Gently heat and bring to boiling. Distribute into tubes or flasks. Autoclave for 15 min at 15 psi pressure–121°C. Pour into sterile Petri dishes or leave in tubes.

Use: For the cultivation of yeast and filamentous fungi. For the cultivation of pathogenic and nonpathogenic fungi, especially dermatophytes. The medium may be made more selective for fungi by the addition of chloramphenicol. Fluconozole (final concentration of 8–16mg per mL) may also be added to test for antibiotic sensitivity.

Sabouraud Glucose Agar, Emmons

Composition per liter:

Glucose 20.0g
Agar 17.0g
Pancreatic digest of casein 5.0g
Peptic digest of animal tissue 5.0g

pH 6.9 ± 0.2 at 25°C

Source: This medium is available as a premixed powder from BD Diagnostic Systems.

Preparation of Medium: Add components to distilled/deionized water and bring volume to 1.0L. Mix thoroughly. Gently heat and bring to boiling. Distribute into tubes or flasks. Autoclave for 15 min at 13 psi pressure–118°C. Pour into sterile Petri dishes or leave in tubes.

Use: For the cultivation of dermatophytes and other pathogenic and nonpathogenic fungi from clinical and nonclinical specimens. For the cultivation of yeast and filamentous fungi.

Sabouraud Glucose Agar with Chloramphenicol and Cycloheximide

Composition per liter:

Glucose 40.0g
Agar 15.0g
Pancreatic digest of casein 5.0g
Peptic digest of animal tissue 5.0g
Cycloheximide solution 10.0mL
Chloramphenicol solution 10.0mL

pH 5.6 ± 0.2 at 25°C

Cycloheximide Solution:

Composition per 10.0mL:

Cycloheximide 0.5g
Acetone 10.0mL

Preparation of Cycloheximide Solution: Add cycloheximide to acetone. Mix thoroughly.

Caution: Cycloheximide is toxic. Avoid skin contact or aerosol formation and inhalation.

Chloramphenicol Solution:

Composition per 10.0mL:

Chloramphenicol 0.05g
Ethanol (95% solution) 10.0mL

Preparation of Chloramphenicol Solution: Add chloramphenicol to ethanol. Mix thoroughly.

Preparation of Medium: Add components, except cycloheximide solution and chloramphenicol solution, to distilled/deionized water and bring volume to 980.0mL. Mix thoroughly. Gently heat and bring to boiling. Add the cycloheximide solution and chloramphenicol solution. Mix thoroughly. Distribute into tubes or flasks. Autoclave for 15 min at 15 psi pressure–121°C. Pour into sterile Petri dishes or leave in tubes.

Use: For the cultivation and identification of yeasts.

Sabouraud Glucose Agar, HiVeg

Composition per liter:

Glucose 40.0g
Agar 15.0g
Plant peptone No. 4 .... 10.0g
Selective supplement 10.0mL

pH 6.9 ± 0.2 at 25°C

Source: This medium is available as a premixed powder from HiMedia.

Selective Supplement:

Composition per 10.0mL:

Cycloheximide 0.5g
Chloramphenicol 0.04g

Preparation of Selective Supplement: Add components to distilled/deionized water and bring volume to 10.0mL. Mix thoroughly. Filter sterilize.

Caution: Cycloheximide is very toxic. Avoid skin contact or aerosol formation and inhalation.

Preparation of Medium: Add components, except selective supplement, to distilled/deionized water and bring volume to 990.0mL. Mix thoroughly. Gently heat and bring to boiling. Distribute into tubes or flasks. Autoclave for 15 min at 15 psi pressure–121°C. Cool to 50°C. Aseptically add 10.0mL sterile selective supplement. Mix thoroughly. Pour into sterile Petri dishes or aseptically distribute into tubes.

Use: For the cultivation of dermatophytes and other pathogenic and nonpathogenic fungi from clinical and nonclinical specimens. For the cultivation of yeast and filamentous fungi.

Sabouraud Glucose Agar with Olive Oil

Composition per liter:

Glucose 40.0g
Agar 15.0g
Pancreatic digest of casein 5.0g
Peptic digest of animal tissue 5.0g
Olive oil 20.0mL
Tween™ 80 .... 2.0mL

pH 5.6 ± 0.2 at 25°C

Preparation of Medium: Add components to distilled/deionized water and bring volume to 1.0L. Mix thoroughly. Gently heat and bring to boiling. Distribute into tubes or flasks. Autoclave for 15 min at 15 psi pressure–121°C. Allow tubes to cool in a slanted position.

Use: For the cultivation and maintenance of *Malassezia* species.

Sabouraud Glucose and Brain Heart Infusion Agar
See: **SABHI Agar**

Sabouraud Glucose Broth

Composition per liter:

Glucose 20.0g
Neopeptone 10.0g

pH 5.6 ± 0.2 at 25°C

Source: This medium is available as a premixed powder from BD Diagnostic Systems.

Preparation of Medium: Add components to distilled/deionized water and bring volume to 1.0L. Mix thoroughly. Distribute into tubes or flasks. Autoclave for 15 min at 15 psi pressure–121°C. Avoid overheating.

Use: For the cultivation of pathogenic and nonpathogenic fungi, especially dermatophytes. The medium may be made more selective for fungi by the addition of chloramphenicol.

Sabouraud Glucose HiVeg Agar Base, Modified with Cycloheximide and Chloramphenicol (Glucose HiVeg Agar Base, Emmons)

Composition per liter:

Glucose	20.0g
Agar	17.0g
Plant special peptone	10.0g

pH 7.0 ± 0.2 at 25°C

Source: This medium is available as a premixed powder from HiMedia.

Preparation of Medium: Add components to distilled/deionized water and bring volume to 1.0L. Mix thoroughly. Gently heat and bring to boiling. Distribute into tubes or flasks. Autoclave for 15 min at 15 psi pressure–121°C. Pour into sterile Petri dishes or leave in tubes.

Use: For the cultivation of fungi.

Sabouraud Glucose HiVeg Broth (Sabouraud Liquid HiVeg Medium)

Composition per liter:

Glucose	20.0g
Plant special peptone	10.0g

pH 5.6 ± 0.2 at 25°C

Source: This medium is available as a premixed powder from HiMedia.

Preparation of Medium: Add components to distilled/deionized water and bring volume to 1.0L. Mix thoroughly. Distribute into tubes or flasks. Autoclave for 15 min at 15 psi pressure–121°C.

Use: For the cultivation of pathogenic and nonpathogenic fungi, especially dermatophytes. The medium may be made more selective for fungi by the addition of chloramphenicol.

Sabouraud Glucose Maltose HiVeg Agar

Composition per liter:

Agar	15.0g
Glucose	10.0g
Maltose	10.0g
Plant hydrolysate	5.0g
Plant peptone	5.0g

pH 5.6 ± 0.2 at 25°C

Source: This medium is available as a premixed powder from HiMedia.

Preparation of Medium: Add components to distilled/deionized water and bring volume to 1.0L. Mix thoroughly. Gently heat and bring to boiling. Distribute into tubes or flasks. Autoclave for 15 min at 15 psi pressure–121°C. Avoid overheating. Pour into sterile Petri dishes or leave in tubes.

Use: For the cultivation and maintenance of a variety of fungi.

Sabouraud Liquid Broth, Modified

See: **Antibiotic Medium 13**

Sabouraud Maltose Agar

Composition per liter:

Maltose	40.0g
Agar	15.0g
Pancreatic digest of casein	5.0g
Peptic digest of animal tissue	5.0g

pH 5.6 ± 0.2 at 25°C

Source: This medium is available as a premixed powder from BD Diagnostic Systems and Oxoid Unipath.

Preparation of Medium: Add components to distilled/deionized water and bring volume to 1.0L. Mix thoroughly. Gently heat and bring to boiling. Distribute into tubes or flasks. Autoclave for 15 min at 15 psi pressure–121°C. Avoid overheating. Pour into sterile Petri dishes or leave in tubes.

Use: For the cultivation and maintenance of a variety of fungi.

Sabouraud Maltose Broth

Composition per liter:

Maltose	40.0g
Neopeptone	10.0g

pH 5.6 ± 0.2 at 25°C

Source: This medium is available as a premixed powder from BD Diagnostic Systems.

Preparation of Medium: Add components to distilled/deionized water and bring volume to 1.0L. Mix thoroughly. Distribute into tubes or flasks. Autoclave for 15 min at 15 psi pressure–121°C. Avoid overheating.

Use: For the cultivation of a variety of fungi.

Sabouraud Maltose HiVeg Broth

Composition per liter:

Maltose	40.0g
Plant peptone No. 4	10.0g

pH 5.6 ± 0.2 at 25°C

Source: This medium is available as a premixed powder from HiMedia.

Preparation of Medium: Add components to distilled/deionized water and bring volume to 1.0L. Mix thoroughly. Gently heat and bring to boiling. Distribute into tubes or flasks. Autoclave for 15 min at 15 psi pressure–121°C. Avoid overheating.

Use: For the cultivation and maintenance of a variety of fungi.

Sabouraud Medium, Emmons Modification

Composition per liter:

Glucose	20.0g
Agar	15.0g
Peptone	10.0g

pH 6.8–7.0 at 25°C

Preparation of Medium: Add components to distilled/deionized water and bring volume to 1.0L. Mix thoroughly. Adjust pH to 6.8–7.0. Gently heat and bring to boiling. Distribute into tubes or flasks. Autoclave for 15 min at 15 psi pressure–121°C. Pour into sterile Petri dishes or leave in tubes.

Use: For the cultivation and maintenance of *Arthroderma benhamiae, Arthroderma vanbreuseghemii, Aureobasidium pullulans, Epidermophyton floccosum, Microsporum canis, Sporothrix schenckii, Trichophyton mentagrophytes,* and *Trichophyton rubrum.*

Sabouraud Medium, Fluid

Composition per liter:

Glucose	20.0g
Pancreatic digest of casein	5.0g
Peptamin	5.0g

pH 5.7 ± 0.2 at 25°C

Source: This medium is available as a premixed powder from BD Diagnostic Systems and Oxoid Unipath.

Preparation of Medium: Add components to distilled/deionized water and bring volume to 1.0L. Mix thoroughly. Distribute into tubes or flasks. Autoclave for 15 min at 15 psi pressure–121°C. Avoid overheating.

Use: For the isolation and cultivation of yeasts and molds.

Sabouraud Medium, Fluid, HiVeg (Fluid Sabouraud HiVeg Medium)

Composition per liter:

Glucose	20.0g
Plant hydrolysate	5.0g
Plant peptone	5.0g

pH 5.7 ± 0.2 at 25°C

Source: This medium is available as a premixed powder from HiMedia.

Preparation of Medium: Add components to distilled/deionized water and bring volume to 1.0L. Mix thoroughly. Gently heat and bring to boiling. Distribute into tubes or flasks. Autoclave for 15 min at 15 psi pressure–121°C.

Use: For the cultivation and maintenance of a variety of fungi. For the sterility testing of pharmaceutical preparations for the presence of molds and bacteria.

Saccharococcus Agar

Composition per liter:

Agar	20.0g
Beef extract	5.0g
Sucrose	5.0g
Pancreatic digest of casein	3.0g
Glucose	1.0g

pH 6.8 ± 0.2 at 25°C

Preparation of Medium: Add components to distilled/deionized water and bring volume to 1.0L. Mix thoroughly. Gently heat and bring to boiling. Distribute into tubes or flasks. Autoclave for 15 min at 15 psi pressure–121°C. Pour into sterile Petri dishes or leave in tubes.

Use: For the cultivation and maintenance of *Saccharococcus thermophilus.*

Saccharolytic Clostridia Medium

Composition per liter:

Pancreatic digest of casein	10.0g
Yeast extract	6.0g
Sodium thioglycolate	0.5g
Carbohydrate solution	100.0mL
Potassium phosphate (1*M* solution, pH 7.5)	30.0mL
$MgSO_4$ (1*M* solution)	1.0mL
Solution M	0.5mL
$FeSO_4$ solution	0.2mL

pH 7.0–7.2 at 25°C

Carbohydrate Solution:

Composition per 100.0mL:

Glucose or sucrose	20.0g

Preparation of Carbohydrate Solution: Add carbohydrate to distilled/deionized water and bring volume to 100.0mL. Mix thoroughly. Filter sterilize.

Solution M:

Composition per liter:

$CaCl_2$	3.33g
$MnCl_2·4H_2O$	1.98g
$Na_2MoO_4·2H_2O$	1.21g
$CoCl_2·6H_2O$	1.19g

Preparation of Solution M: Add components to distilled/deionized water and bring volume to 1.0L. Mix thoroughly.

$FeSO_4·7H_2O$ Solution:

Composition per 10.0mL:

$FeSO_4·7H_2O$	55.6g
H_2SO_4 (0.1*M* solution)	10.0mL

Preparation of $FeSO_4·7H_2O$ Solution: Add the $FeSO_4·7H_2O$ to 10.0mL of H_2SO_4 solution. Mix thoroughly. Filter sterilize.

Preparation of Medium: Add components, except carbohydrate solution, to distilled/deionized water and bring volume to 900.0mL. Mix thoroughly. Gently heat and bring to boiling. Autoclave for 20 min at 15 psi pressure–121°C. Cool to 45°–50°C. Aseptically add sterile carbohydrate solution. Mix thoroughly. Aseptically distribute into sterile tubes.

Use: For the cultivation of saccharolytic *Clostridium* species.

Saccharolytic Clostridia Medium

Composition per liter:

Sodium thioglycolate	1.0g
K_2HPO_4	0.8g
KH_2PO_4	0.2g
$MgSO_4·7H_2O$	0.2g
NaCl	0.2g
$Na_2MoO_4·2H_2O$	0.025g
Yeast extract	0.01g
$FeSO_4·7H_2O$	0.01g
$MnSO_4·4H_2O$	0.01g
CaCl2	0.01g
Carbohydrate solution	100.0mL
Soil extract	10.0mL
Trace elements solution	1.0mL

pH 7.2 ± 0.2 at 25°C

Carbohydrate Solution:

Composition per 100.0mL:

Glucose or sucrose	10.0g

Preparation of Carbohydrate Solution: Add glucose or sucrose to distilled/deionized water and bring volume to 100.0mL. Mix thoroughly. Filter sterilize.

Soil Extract:

Composition per 200.0mL:

Garden soil, neutral	100.0g

Preparation of Soil Extract: Add garden soil to 100.0mL of tap water. Gently heat and bring to 130°C for 60 min. Cool to 45°C. Filter through Whatman #1 filter paper. Autoclave for 15 min at 15 psi pressure–121°C. Cool to 45°–50°C.

Trace Elements Solution:

Composition per liter:

$Na_2B_4O_7 \cdot 10H_2O$	0.05g
$CoNO_3 \cdot 6H_2O$	0.05g
$CdSO_4 \cdot 2H_2O$	0.05g
$CuSO_4 \cdot 5H_2O$	0.05g
$ZnSO_4 \cdot 7H_2O$	0.05g
$MnSO_4 \cdot H_2O$	0.05g

Preparation of Trace Elements Solution: Add components to distilled/deionized water and bring volume to 1.0L. Mix thoroughly.

Preparation of Medium: Add components, except sodium thioglycolate and carbohydrate solution, to distilled/deionized water and bring volume to 900.0mL. Mix thoroughly. Gently heat and bring to boiling. Add sodium thioglycolate. Mix thoroughly. Distribute 9.5mL into test tubes that contain inverted Durham tubes. Autoclave for 15 min at 15 psi pressure–121°C. Cool to 45°–50°C. Aseptically add 0.5mL of sterile carbohydrate solution to each tube. Mix thoroughly.

Use: For the isolation of N_2-fixing, saccharolytic *Clostridium* species.

Saccharomyces Medium

Composition per liter:

Glucose	20.0g
Peptone	20.0g
Agar	15.0g
Yeast extract	10.0g
Yeast nitrogen base without amino acids	6.7g
$(NH_4)_2SO_4$	5.0g
Proline	4.0mg

Preparation of Medium: Add glucose, yeast nitrogen base without amino acids, and proline to distilled/deionized water and bring volume to 200.0mL. Mix thoroughly. Filter sterilize. Add peptone, yeast extract, $(NH_4)_2SO_4$, and agar to distilled/deionized water and bring volume to 800.0mL. Mix thoroughly. Gently heat and bring to boiling. Autoclave for 15 min at 15 psi pressure–121°C. Cool to 50°C. Aseptically combine the two sterile solutions. Mix thoroughly. Pour into sterile Petri dishes or aseptically distribute into sterile tubes.

Use: For the cultivation and maintenance of *Saccharomyces cerevisiae.*

Saccharomyces rouxii Medium

Composition per liter:

Glucose	400.0g
Agar	20.0g
Peptone	20.0g
Yeast extract	10.0g

Preparation of Medium: Add components to distilled/deionized water and bring volume to 1.0L. Mix thoroughly. Gently heat and bring to boiling. Distribute into tubes or flasks. Autoclave for 15 min at 15 psi pressure–121°C. Pour into sterile Petri dishes or leave in tubes.

Use: For the cultivation of *Zygosaccharomyces rouxii.*

Use: For the cultivation of *Monodictys austrina.*

Sakazakii DHL Agar

Composition per liter:

Agar	15.0g
Casein enzymic hydrolysate	10.0g
Meat peptone	10.0g
Lactose	10.0g
Sucrose	10.0g
Meat extract	3.0g
$Na_2S_2O_3$	2.0g
Sodium deoxycholate	1.5g
Sodium citrate	1.0g
Ammonium iron (III) citrate	1.0g
L-Cysteine·HCl·H_2O	0.2g
Neutral Red	0.03g

pH 7.2 ± 0.2 at 25°C

Source: This medium is available from HiMedia.

Preparation of Medium: Add components to distilled/deionized water and bring volume to 1.0L. Mix thoroughly. Gently heat and bring to boiling. Distribute into tubes or flasks. Autoclave for 15 min at 15 psi pressure–121°C. Pour into sterile Petri dishes or leave in tubes.

Use: For the detection and isolation of pathogenic Enterobacteriaceae from all types of specimens.

Saline Czapek Agar (SCZA)

Composition per liter:

Sucrose	30.0g
NaCl	25.0g
Agar	15.0g
$NaNO_3$	3.0g
K_2HPO_4	1.0g
KCl	0.5g
$MgSO_4 \cdot 7H_2O$	0.5g
$FeSO_4 \cdot 7H_2O$	0.01g

pH 7.3 ± 0.2 at 25°C

Preparation of Medium: Add components, except sucrose, to distilled/deionized water and bring volume to 900.0mL. Mix thoroughly. Distribute into tubes or flasks. In a separate flask, add sucrose to distilled/deionized water and bring volume to 100.0mL. Mix thoroughly. Autoclave both solutions separately for 15 min at 15 psi pressure–121°C. Cool to 50°C. Combine the sterile solutions. Mix thoroughly. Pour into sterile Petri dishes or distribute into sterile tubes.

Salinibacter ruber Agar (DSMZ Medium 936)

Composition per liter:

NaCl	195.0g
$MgSO_4 \cdot 7H_2O$	49.5g
$MgCl_2 \cdot 6H_2O$	34.6g
Agar	20.0g
KCl	5.0g
$CaCl_2 \cdot 2H_2O$	1.25g
Yeast extract	1.0g
NaBr	0.625g
$NaHCO_3$	0.25g

pH 7.2 ± 0.2 at 25°C

Preparation of Medium: Add components to distilled/deionized water and bring volume to 1.0L. Mix thoroughly. Gently heat and bring to boiling. Distribute into tubes or flasks. Autoclave for 15 min at 15 psi pressure–121°C. Pour into sterile Petri dishes or leave in tubes.

Use: For the cultivation and maintenance of *Salinibacter ruber.*

Salinibacter ruber Medium (DSMZ Medium 936)

Composition per liter:

NaCl	195.0g
$MgSO_4 \cdot 7H_2O$	49.5g
$MgCl_2 \cdot 6H_2O$	34.6g
KCl	5.0g
$CaCl_2 \cdot 2H_2O$	1.25g
Yeast extract	1.0g
NaBr	0.625g
$NaHCO_3$	0.25g

pH 7.2 ± 0.2 at 25°C

Preparation of Medium: Add components to distilled/deionized water and bring volume to 1.0L. Mix thoroughly. Distribute into tubes or flasks. Autoclave for 15 min at 15 psi pressure–121°C.

Use: For the cultivation of *Salinibacter ruber.*

Salinivibrio costicola Subspecies *vallismortis* Medium (DSMZ Medium 597)

Composition per liter:

NaCl	25.0g
$MgSO_4 \cdot 7H_2O$	9.6g
$MgCl_2 \cdot 6H_2O$	7.0g
Glucose	5.0g
KCl	3.8g
Yeast extract	1.0g
$CaCl_2 \cdot 2H_2O$	0.5g
$K_2HPO_4 \cdot 3H_2O$	0.4g
$NaHCO_3$ solution	100.0mL

pH 7.0 ± 0.2 at 25°C

$NaHCO_3$ Solution:

Composition per 100.0mL:

$NaHCO_3$	3.0g

Preparation of $NaHCO_3$ Solution: Add $NaHCO_3$ to distilled/deionized water and bring volume to 100.0mL. Mix thoroughly. Filter sterilize.

Preparation of Medium: Add components, except $NaHCO_3$ solution, to distilled/deionized water and bring volume to 900.0mL. Mix thoroughly. Autoclave for 15 min at 15 psi pressure–121°C. Cool to 25°C. Aseptically add 100.0mL $NaHCO_3$ solution. Mix thoroughly. Aseptically distribute into sterile tubes or flasks.

Use: For the cultivation of *Salinivibrio costicola* subsp. *vallismortis.*

Salmonella Chromogen Agar (Rambach Equivalent Agar)

Composition per liter:

Agar	15.0g
Peptone	5.0g
NaCl	5.0g
Yeast extract	2.0g
Meat extract	1.0g
Sodium deoxycholate	1.0g

pH 7.3 ± 0.2 at 25°C

Source: This medium is available from Fluka, Sigma-Aldrich.

Preparation of Medium: Add components to distilled/deionized water and bring volume to 1.0L. Mix thoroughly. Gently heat while stirring and bring to boiling. Autoclave for 15 min at 15 psi pressure–121°C. Pour into sterile Petri dishes.

Use: A differential diagnostic agar for the detection of *Salmonella* in food, including the isolation and enumeration of *Salmonella* from bivalves.

Salmonella Chromogenic Agar (OSCM)

Composition per liter:

Chromogenic mix	28.0g
Agar	12.0g
Special peptone	10.0g
Selective supplement solution	10.0mL

pH 7.2 ± 0.2 at 25°C

Source: This medium is available as a premixed powder from Oxoid Unipath.

Selective Supplement Solution:

Composition per 10.0mL:

Cefsulodin	12.0mg
Novobiocin	5.0mg

Preparation of Selective Supplement Solution: Add components to distilled/deionized water and bring volume to 10.0mL. Mix thoroughly. Filter sterilize.

Preparation of Medium: Add components to distilled/deionized water and bring volume to 1.0mL Mix thoroughly. Gently heat while stirring and bring to boiling. Do not autoclave. Cool quickly to 50°C. Mix thoroughly. Pour into sterile Petri dishes.

Use: For the identification of *Salmonella* species and differentiation of *Salmonella* spp. other organisms in the family Enterobacteriaceae. This medium combines two chromogens for the detection of *Salmonella* spp., 5-bromo-6-chloro-3-indolyl caprylate (Magenta-caprylate) and 5-bromo-4-chloro-3-indolyl-β-D galactopyranoside (X-gal). X-gal is a substrate for the enzyme β-D-galactosidase. Hydrolysis of the chromogen, Magenta-caprylate, by lactose-negative *Salmonella* species results in magenta colonies. The addition of the selective supplement solution increases the selectivity of the medium. Novobiocin inhibits *Proteus* growth and cefsulodin inhibits growth of pseudomonads.

Salmonella Differential Agar (RajHans Medium)

Composition per liter:

Agar	12.0g
Propylene glycol	10.0g
Peptone, special	8.0g
B.C. indicator	2.0g
Yeast extract	2.0g
Sodium deoxycholate	1.0g

pH 7.3 ± 0.2 at 25°C

Source: This medium is available as a premixed powder from HiMedia.

Preparation of Medium: Add components to distilled/deionized water and bring volume to 1.0L. Mix thoroughly. Gently heat while stirring and bring to boiling. Mix to completely dissolve components. Do not autoclave. Cool to 50°C. Mix thoroughly. Pour into sterile Petri dishes.

Use: A selective chromgenic medium used for the isolation and differentiation of *Salmonella* species from the members of Enterobacteriaceae, especially *Proteus* species.

Salmonella Differential HiVeg Agar, Modified (*Salmonella* Differential Agar, Modified, HiVeg)

Composition per liter:

Propylene glycol	10.0g
Plant special peptone	8.0g
NaCl	5.0g
Yeast extract	3.0g
B.C. indicator	2.0g
Synthetic detergent	1.0g

pH 7.3 ± 0.2 at 25°C

Source: This medium is available as a premixed powder from HiMedia.

Preparation of Medium: Add components to distilled/deionized water and bring volume to 1.0L. Mix thoroughly. Gently heat while stirring and bring to boiling. Mix to completely dissolve components. Do not autoclave. Cool to 50°C. Mix thoroughly. Pour into sterile Petri dishes.

Use: A selective chromogenic medium used for the isolation and differentiation of *Salmonella* species from the members of Enterobacteriaceae, especially *Proteus* species.

Salmonella HiVeg Agar, ONOZ

Composition per liter:

Agar	15.0g
Sucrose	13.0g
Lactose	11.5g
Na_3-citrate·$5H_2O$	9.3g
Plant peptone	8.625g
Plant extract No. 1	6.0g
L-Phenylalanine	5.0g
$Na_2S_2O_3 \cdot 5H_2O$	4.25g
Yeast extract	3.0g
Synthetic detergent No. 1	2.0g
$Na_2HPO_4 \cdot 2H_2O$	1.0g
Ferric citrate	0.5g
Metachrome Yellow	0.47g
$MgSO_4$	0.4g
Aniline Blue	0.25g
Neutral Red	0.022g
Brilliant Green	1.66mg

pH 7.4 ± 0.2 at 25°C

Preparation of Medium: Add components to distilled/deionized water and bring volume to 1.0L. Mix thoroughly. Distribute into tubes or flasks. Gently heat and bring to boiling. Do not autoclave. Cool to 50°C. Pour into sterile Petri dishes or leave in tubes.

Use: For the cultivation and maintenance *Salmonella* spp. from clinical specimens.

Salmonella Medium

Composition per liter:

Pancreatic digest of casein	10.0g
NaCl	5.0g

pH 7.4 ± 0.2 at 25°C

Preparation of Medium: Add components to distilled/deionized water and bring volume to 1.0L. Mix thoroughly. Distribute into tubes or flasks. Autoclave for 15 min at 15 psi pressure–121°C.

Use: For the cultivation and maintenance of *Escherichia coli* and *Salmonella choleraesuis*.

Salmonella Rapid Test Elective Medium

Composition per liter:

Tryptone	10.0g
Na_2HPO_4	9.0g
Sodium chloride	5.0g
Casein	5.0g
KH_2PO_4	1.5g
Malachite Green	0.0025g

pH 6.5 ± 0.2 at 25°C

Preparation of Medium: Add components to distilled/deionized water and bring volume to 1.0L. Mix thoroughly. Autoclave for 15 min at 15 psi pressure–121°C.

Use: For the Oxoid Salmonella Rapid Test which is for the presumptive detection of motile *Salmonella* in foods and environmental samples.

Salmonella Rapid Test Elective Medium, 2X

Composition per liter:

Tryptone	20.0g
Na_2HPO_4	18.0g
Sodium chloride	10.0g
KH_2PO_4	3.0g
Casein	10.0g
Malachite Green	0.005g

pH 6.5 ± 0.2 at 25°C

Preparation of Medium: Add components to distilled/deionized water and bring volume to 1.0L. Mix thoroughly. Autoclave for 15 min at 15 psi pressure–121°C.

Use: Use as described in the Oxoid Salmonella Rapid Test which is for the presumptive detection of motile *Salmonella* in foods and environmental samples.

Salmonella Shigella Agar (SS Agar)

Composition per liter:

Agar	13.5g
Lactose	10.0g
Bile salts	8.5g
$Na_2S_2O_3$	8.5g
Sodium citrate	8.5g
Beef extract	5.0g
Pancreatic digest of casein	2.5g
Peptic digest of animal tissue	2.5g
Ferric citrate	1.0g
Neutral Red	0.025g
Brilliant Green	0.33mg

pH 7.0 ± 0.2 at 25°C

Source: This medium is available as a premixed powder from BD Diagnostic Systems and Oxoid Unipath.

Preparation of Medium: Add components to distilled/deionized water and bring volume to 1.0L. Mix thoroughly. Gently heat while stirring and bring to boiling. Do not autoclave. Cool to 45°–50°C. Pour into sterile Petri dishes in 20.0mL volumes. Allow the surface of the plates to dry before inoculation.

Use: For the selective isolation and differentiation of pathogenic enteric bacilli, especially those belonging to the genus *Salmonella.* This medium is not recommended for the primary isolation of *Shigella* species. Lactose-fermenting bacteria such as *Escherichia coli* or *Klebsiella pneumoniae* appear as small pink or red colonies. Lactose-nonfermenting bacteria—such as *Salmonella* species, *Proteus* species, and *Shigella* species—appear as colorless colonies. Production of H_2S by *Salmonella* species turns the center of the colonies black.

Salmonella Shigella Agar, Modified (SS Agar, Modified)

Composition per liter:

Agar	12.0g
Lactose	10.0g
Sodium citrate	10.0g
$Na_2S_2O_3$	8.5g
Bile salts	5.5g
Beef extract	5.0g
Peptone	5.0g
Ferric citrate	1.0g
Neutral Red	0.025g
Brilliant Green	0.33mg

pH 7.3 ± 0.2 at 25°C

Source: This medium is available as a premixed powder from Oxoid Unipath.

Preparation of Medium: Add components to distilled/deionized water and bring volume to 1.0L. Mix thoroughly. Gently heat while stirring and bring to boiling. Do not autoclave. Cool to 45°–50°C. Pour into sterile Petri dishes in 20.0mL volumes. Allow the surface of the plates to dry before inoculation.

Use: For the selective isolation and differentiation of pathogenic enteric bacilli, especially those belonging to the genus *Salmonella.* This medium provides better growth of *Shigella* species. Lactose-fermenting bacteria such as *Escherichia coli* or *Klebsiella pneumoniae* appear as small pink or red colonies. Lactose-nonfermenting bacteria—such as *Salmonella* species, *Proteus* species, and *Shigella* species—appear as colorless colonies. Production of H_2S by *Salmonella* species turns the center of the colonies black.

Salmonella Shigella Deoxycholate Agar

See: **SS Deoxycholate Agar**

Salt Agar

Composition per liter:

NaCl	58.4g
Agar	15.0g
Proteose peptone	5.0g
Pancreatic digest of casein	5.0g

pH 6.9 ± 0.2 at 25°C

Preparation of Medium: Add components to distilled/deionized water and bring volume to 1.0L. Mix thoroughly. Gently heat and bring to boiling. Distribute into tubes or flasks. Autoclave for 15 min at 15 psi pressure–121°C. Pour into sterile Petri dishes or leave in tubes.

Use: For the cultivation and maintenance of *Marinococcus halophilus.*

Salt Broth, Modified

Composition per liter:

NaCl	65.0g
Enzymatic digest of animal tissue	10.0g
Heart digest	10.0g
Glucose	1.0g
Bromcresol Purple	0.016g

pH 7.2 ± 0.2 at 25°C

Source: Available as a prepared medium from BD Diagnostic Systems.

Preparation of Medium: Add components to distilled/deionized water and bring volume to 1.0L. Mix thoroughly. Distribute into tubes or flasks. Autoclave for 15 min at 15 psi pressure–121°C.

Use: For the cultivation and differentiation of the enterococcal group D streptococci from nonenterococcal group D streptococci based on salt tolerance.

Salt Colistin Broth

Composition per liter:

NaCl	20.0g
Peptone	10.0g
Yeast extract	3.0g
Colistin solution	10.0mL

pH 7.4 ± 0.2 at 25°C

Colistin Solution:

Composition per 10.0mL:

Colistin methane sulfonate	500,000U

Preparation of Colistin Solution: Add colistin methane sulfonate to distilled/deionized water and bring volume to 10.0mL. Mix thoroughly. Filter sterilize.

Preparation of Medium: Add components, except colistin solution, to distilled/deionized water and bring volume to 990.0mL. Mix thoroughly. Gently heat until dissolved. Autoclave for 15 min at 15 psi pressure–121°C. Cool to 25°C. Aseptically add sterile colistin solution. Mix thoroughly. Aseptically distribute into sterile tubes or flasks.

Use: For the cultivation of halophilic *Vibrio* species.

Salt Malt Agar (SMA1)

Composition per liter:

NaCl	25.0g
Agar	15.0g
Malt extract	10.0g

Preparation of Medium: Add components to distilled/deionized water and bring volume to 1.0L. Mix thoroughly. Gently heat and bring to boiling. Distribute into tubes or flasks. Autoclave for 15 min at 15 psi pressure–121°C. Pour into sterile Petri dishes or leave in tubes.

Use: For the cultivation of *Dendryphiella salina* and *Dendryphiella arenaria.*

Salt Meat Broth

Composition per liter:

NaCl	100.0g
Neutral ox-heart tissue	30.0g
Beef extract	10.0g
Peptone	10.0g

pH 7.6 ± 0.2 at 25°C

Source: This medium is available as tablets from Oxoid Unipath.

Preparation of Medium: Add components to distilled/deionized water and bring volume to 1.0L. Mix thoroughly. Distribute into tubes or flasks. Autoclave for 15 min at 15 psi pressure–121°C.

Use: For the isolation and cultivation of staphylococci from specimens with a mixed flora such as fecal specimens, especially during the investigation of staphylococcal food poisoning.

Salt Medium

Composition per liter:

NaCl	58.4g
Proteose peptone	5.0g
Pancreatic digest of casein	5.0g

pH 4.9 ± 0.2 at 25°C

Preparation of Medium: Add components to distilled/deionized water and bring volume to 1.0L. Mix thoroughly. Distribute into tubes or flasks. Autoclave for 15 min at 15 psi pressure–121°C.

Use: For the cultivation and maintenance of *Marinococcus halophilus*.

Salt Nutrient Agar

Composition per liter:

NaCl	25.0g
Agar	15.0g
Peptone	5.0g
Yeast extract	2.0g
Beef extract	1.0g

Preparation of Medium: Add components to distilled/deionized water and bring volume to 1.0L. Mix thoroughly. Gently heat and bring to boiling. Distribute into tubes or flasks. Autoclave for 15 min at 15 psi pressure–121°C. Pour into sterile Petri dishes or leave in tubes.

Use: For the cultivation of *Bacillus* species, *Pseudomonas beijerinckii*, and other *Pseudomonas* species.

Salt Polymyxin Broth (SPB)

Composition per liter:

NaCl	20.0g
Pancreatic digest of casein	10.0g
Yeast extract	3.0g
Polymyxin B	250,000U

pH 8.8 ± 0.2 at 25°C

Preparation of Medium: Add components to distilled/deionized water and bring volume to 1.0L. Mix thoroughly. Adjust pH to 8.8. Distribute into tubes or flasks. Autoclave for 10 min at 10 psi pressure–115°C.

Use: For the isolation and cultivation of *Vibrio* species from foods.

Salt Polymyxin HiVeg Broth Base with Polymyxin

Composition per liter:

NaCl	20.0g
Plant hydrolysate	10.0g
Yeast extract	3.0g
Selective supplement	10.0mL

pH 8.8 ± 0.2 at 25°C

Source: This medium, without selective supplement, is available as a premixed powder from HiMedia.

Selective Supplement:

Composition per 10.0mL:

Polymyxin B sulfate	100,000U

Preparation of Selective Supplement: Add polymycin B sulfate to distilled/deionized water and bring volume to 10.0mL. Mix thoroughly. Filter sterilize.

Preparation of Medium: Add components, except selective supplement, to distilled/deionized water and bring volume to 990.0mL. Mix thoroughly. Gently heat and bring to boiling. Distribute into tubes or flasks. Mix thoroughly. Adjust pH to 8.8. Distribute into tubes or flasks. Autoclave for 10 min at 10 psi pressure–115°C. Cool to 50°C. Aseptically add 10.0mL sterile selective supplement. Mix thoroughly. Pour into sterile Petri dishes or aseptically distribute into tubes.

Use: For the isolation and cultivation of *Vibrio* species from foods.

Salt Tolerance Medium

Composition per liter:

Beef heart, infusion from	500.0g
NaCl	65.0g
Tryptose	10.0g
Glucose	1.0g
Indicator solution	1.0mL

pH 7.4 ± 0.2 at 25°C

Indicator Solution:

Composition per 100.0mL:

Bromcresol Purple	1.6g
Ethanol (95% solution)	100.0mL

Preparation of Indicator Solution: Add Bromcresol Purple to ethanol. Mix thoroughly.

Preparation of Medium: Add components to distilled/deionized water and bring volume to 1.0L. Mix thoroughly. Distribute into tubes or flasks. Autoclave for 15 min at 15 psi pressure–121°C.

Use: For the cultivation of salt-tolerant *Streptococcus* species and other salt-tolerant Gram-positive cocci. For the differentiation of Gram-positive cocci based on salt tolerance.

Salt Tolerance Medium

Composition per liter:

NaCl	60.0g
Peptone	5.0g
Yeast extract	2.0g
Beef extract	1.0g

pH 7.4 ± 0.2 at 25°C

Preparation of Medium: Add components to distilled/deionized water and bring volume to 1.0L. Mix thoroughly. Distribute into tubes or flasks. Autoclave for 15 min at 15 psi pressure–121°C.

Use: For the cultivation and differentiation of *Aeromonas* and *Plesiomonas* species based on salt tolerance.

Salt Tolerance Medium

Composition per liter:

Beef heart, solids from infusion	500.0g
NaCl	65.0g
Tryptose	10.0g

pH 7.4 ± 0.2 at 25°C

Preparation of Medium: Add components to distilled/deionized water and bring volume to 1.0L. Mix thoroughly. Distribute into tubes or flasks. Autoclave for 15 min at 15 psi pressure–121°C.

Use: For testing the salt tolerance of a variety of microorganisms.

Salt Tolerance Medium, Gilardi

Composition per liter:

NaCl	65.0g
Pancreatic digest of casein	15.0g
Agar	15.0g
Papaic digest of soybean meal	5.0g

pH 7.3 ± 0.2 at 25°C

Preparation of Medium: Add components to distilled/deionized water and bring volume to 1.0L. Mix thoroughly. Gently heat and bring to boiling. Distribute into tubes or flasks. Autoclave for 15 min at 15 psi pressure–121°C. Do not overheat. Pour into sterile Petri dishes or leave in tubes.

Use: For the cultivation and maintenance of salt-tolerant, nonfermenting Gram-negative bacteria. For the differentiation of nonfermenting Gram-negative bacteria based on salt tolerance.

Salt Tolerance Medium, Tatum

Composition per liter:

NaCl	65.0g
Peptone	5.0g
Yeast extract	2.0g
Beef extract	1.0g

pH 7.4 ± 0.2 at 25°C

Preparation of Medium: Add components to distilled/deionized water and bring volume to 1.0L. Mix thoroughly. Distribute into tubes or flasks. Autoclave for 15 min at 15 psi pressure–121°C.

Use: For the cultivation of salt-tolerant, nonfermenting Gram-negative bacteria. For the differentiation of nonfermenting Gram-negative bacteria based on salt tolerance.

Sanfrancisco Medium

Composition per liter:

Rye bran or wheat bran	50.0g
Fresh baker's yeast	21.0g
Pancreatic digest of casein	10.0g
Fructose	7.0g
Glucose	7.0g
Maltose	7.0g
Yeast extract	7.0g
Diammonium citrate	5.0g
Sodium acetate·$3H_2O$	5.0g
KH_2PO_4·$3H_2O$	2.5g
Meat extract	2.0g
Sodium gluconate	2.0g
Tween™ 80	1.0g
L-Cysteine·HCl	0.5g
$MgSO_4$·$7H_2O$	0.2g
$MnSO_4$·$4H_2O$	0.05g
$FeSO_4$·$7H_2O$	0.01g

pH 5.4 ± 0.2 at 25°C

Preparation of Medium: Add components to distilled/deionized water and bring volume to 1.0L. Mix thoroughly. Adjust pH to 5.4. Distribute into tubes or flasks. Autoclave for 15 min at 15 psi pressure–121°C.

Use: For the cultivation of *Lactobacillus sanfrancisco.*

SAP 1 Agar

Composition per liter:

Agar	15.0g
Pancreatic digest of casein	5.0g
Yeast extract	5.0g
Artificial seawater	1.0L

pH 7.2 ± 0.2 at 25°C

Artificial Seawater:

Composition per liter:

NaCl	24.7g
$MgSO_4$·$7H_2O$	6.3g
$MgCl_2$·$6H_2O$	4.6g
$CaCl_2$	1.0g
KCl	0.7g
$NaHCO_3$	0.2g

Preparation of Artificial Seawater: Add components to distilled/deionized water and bring volume to 1.0L. Mix thoroughly.

Preparation of Medium: Add solid components to 1.0L of artificial seawater. Mix thoroughly. Gently heat and bring to boiling. Distribute into tubes or flasks. Autoclave for 15 min at 15 psi pressure–121°C. Pour into sterile Petri dishes or leave in tubes.

Use: For the isolation and cultivation of *Cytophaga* species, *Herpetosiphon* species, *Saprospira* species, and *Flexithrix* species.

SAP 2 Agar

Composition per liter:

Agar	15.0g
Pancreatic digest of casein	1.0g
Yeast extract	1.0g
Artificial seawater	1.0L

pH 7.2 ± 0.2 at 25°C

Artificial Seawater:

Composition per liter:

NaCl	24.7g
$MgSO_4$·$7H_2O$	6.3g
$MgCl_2$·$6H_2O$	4.6g
$CaCl_2$	1.0g
KCl	0.7g
$NaHCO_3$	0.2g

Preparation of Artificial Seawater: Add components to distilled/deionized water and bring volume to 1.0L. Mix thoroughly.

Preparation of Medium: Add solid components to 1.0L of artificial seawater. Mix thoroughly. Gently heat and bring to boiling. Distribute into tubes or flasks. Autoclave for 15 min at 15 psi pressure–121°C. Pour into sterile Petri dishes or leave in tubes.

Use: For the isolation and cultivation of *Cytophaga* species, *Herpetosiphon* species, *Saprospira* species, and *Flexithrix* species.

Saprospira grandis Medium

Composition per 1010.0mL:

Pancreatic digest of casein	5.0g
Yeast extract	5.0g
$Ca(NO_3)_2$·$4H_2O$	0.1g
K_2HPO_4	0.02g
Seawater, filtered	1.0L
Trace elements	10.0mL

pH 7.0 ± 0.2 at 25°C

Trace Elements:
Composition per liter:

$FeSO_4·7H_2O$	0.5mg
$ZnSO_4·7H_2O$	0.3mg
H_3BO_3	0.1mg
$CoCl_2·6H_2O$	0.1mg
$CuSO_4·5H_2O$	0.1mg
$MnSO_4·4H_2O$	0.1mg
$Na_2MoO_4·2H_2O$	0.1mg

Preparation of Trace Elements: Add components to distilled/deionized water and bring volume to 1.0L. Mix thoroughly.

Preparation of Medium: Combine components. Mix thoroughly. Adjust pH to 7.0. Filter sterilize.

Use: For the cultivation of *Saprospira grandis*.

Sarcina maxima Medium

Composition per liter:

Glucose	10.0g
Peptone	10.0g
Yeast extract	5.0g
L-Cysteine·HCl solution	10.0mL

pH 6.0 ± 0.2 at 25°C

L-Cysteine·HCl Solution:
Composition per 10.0mL:

L-Cysteine·HCl	0.5g

Preparation of L-Cysteine·HCl Solution: Add L-cysteine·HCl to distilled/deionized water and bring volume to 10.0mL. Mix thoroughly. Filter sterilize.

Preparation of Medium: Add components, except L-cysteine·HCl solution, to distilled/deionized water and bring volume to 990.0mL. Mix thoroughly. Gently heat and bring to boiling. Autoclave for 15 min at 15 psi pressure–121°C. Cool to 25°C. Aseptically add sterile L-cysteine·HCl solution. Mix thoroughly. Aseptically distribute into sterile tubes or flasks.

Use: For the cultivation of *Sarcina maxima*.

Sarcina Medium (DSMZ Medium 21)

Composition per liter:

Glucose	30.0g
Peptone	5.0g
Yeast extract	5.0g
Distilled water	1000.0mL

pH 6.0 ± 0.2 at 25°C

Preparation of Medium: Add components to distilled/deionized water and bring volume to 1.0L. Mix thoroughly. Adjust pH to 6.0. Gently heat and bring to boiling. Distribute into tubes or flasks. Autoclave for 15 min at 15 psi pressure–121°C.

Use: For the cultivation and maintenance of *Sarcina ventriculi* and *Sarcina maxima.*

Sarcina ventriculi Growth Medium

Composition per liter:

Glucose	30.0g
Peptone	5.0g
Yeast extract	5.0g

pH 6.0 ± 0.2 at 25°C

Preparation of Medium: Add components to distilled/deionized water and bring volume to 1.0L. Mix thoroughly. Distribute into tubes in 10.0mL volumes. Autoclave for 20 min at 15 psi pressure–121°C.

Use: For the cultivation and maintenance of *Sarcina maxima* and *Sarcina ventriculi.*

Sauton's Fluid Medium Base

Composition per liter:

L-Asparagine	1.33g
Polysorbate 80	0.833g
Citric acid	0.66g
K_2HPO_4	0.177g
NaH_2PO_4	0.056g
NaCl	0.035g
Ferric ammonium citrate (brown)	0.0167g
Glycerol	20.0mL

pH 7.2 ± 0.2 at 25°C

Source: This medium is available from HiMedia.

Preparation of Medium: Add glycerol to distilled/deionized water and bring volume to 1.0L. Mix thoroughly. Add remaining components. Mix thoroughly. Gently heat and bring to boiling. Distribute into tubes or flasks. Autoclave for 15 min at 15 psi pressure–121°C.

Use: For the enumeration of mycobacteria.

Sauton's Medium

Composition per liter:

L-Asparagine	4.0g
Citric acid	2.0g
K_2HPO_4	0.5g
$MgSO_4$	0.5g
Triton® WR 1339	0.25g
Ferric ammonium citrate	0.05g
Glycerol	40.0mL

Preparation of Medium: Add components to distilled/deionized water and bring volume to 1.0L. Mix thoroughly. Distribute into tubes or flasks. Autoclave for 15 min at 15 psi pressure–121°C.

Use: For the cultivation of *Mycobacterium tuberculosis* strain Bacille Calmette-Guèrin (BCG) for vaccine production.

SBG Enrichment Broth (Selenite Brilliant Green Enrichment Broth)

Composition per liter:

D-Mannitol	5.0g
Peptone	5.0g
Yeast extract	5.0g
$Na_2SeO_3·5H_2O$	4.0g
K_2HPO_4	2.65g
KH_2PO_4	1.02g
Sodium taurocholate	1.0g
Brilliant Green	5.0mg

pH 7.2 ± 0.2 at 25°C

Source: This medium is available as a premixed powder from BD Diagnostic Systems.

Preparation of Medium: Add components to distilled/deionized water and bring volume to 1.0L. Mix thoroughly. Gently heat and bring

to boiling. Continue boiling for 5–10 min. Do not autoclave. Distribute into sterile tubes or flasks.

Use: For the selective isolation of *Salmonella* species, especially from eggs and egg products.

SBG Sulfa Enrichment

Composition per liter:

D-Mannitol	5.0g
Peptone	5.0g
Yeast extract	5.0g
$Na_2SeO_3 \cdot 5H_2O$	4.0g
K_2HPO_4	2.65g
KH_2PO_4	1.02g
Sodium taurocholate	1.0g
Sodium sulfapyridine	0.5g
Brilliant Green	5.0mg

pH 7.2 ± 0.2 at 25°C

Source: This medium is available as a premixed powder from BD Diagnostic Systems.

Preparation of Medium: Add components to distilled/deionized water and bring volume to 1.0L. Mix thoroughly. Gently heat and bring to boiling. Continue boiling for 5–10 min. Do not autoclave. Distribute into sterile tubes or flasks.

Use: For the selective isolation of *Salmonella* species, especially from eggs and egg products.

SB/SW Medium

Composition per liter:

NaCl	1.0g
KCl	0.5g
$MgCl_2 \cdot 6H_2O$	0.4g
NH_4Cl	0.25g
KH_2PO_4	0.2g
$CaCl_2 \cdot 2H_2O$	0.15g
Resazurin	1.0mg
Sodium crotonate solution	50.0mL
$NaHCO_3$ solution	20.0mL
$Na_2S \cdot 9H_2O$ solution	10.0mL
Seven vitamin solution	10.0mL
Sodium dithionite solution	10.0mL
Trace elements solution SL-10	1.0mL

pH 7.2 ± 0.2 at 25°C

Trace Elements Solution SL-10:
Composition per liter:

$FeCl_2 \cdot 4H_2O$	1.5g
$CoCl_2 \cdot 6H_2O$	190.0mg
$MnCl_2 \cdot 4H_2O$	100.0mg
$ZnCl_2$	70.0mg
$Na_2MoO_4 \cdot 2H_2O$	36.0mg
$NiCl_2 \cdot 6H_2O$	24.0mg
H_3BO_3	6.0mg
$CuCl_2 \cdot 2H_2O$	2.0mg
HCl (25% solution)	10.0mL

Preparation of Trace Elements Solution SL-10: Prepare and dispense under 80% N_2 + 20% CO_2. Add $FeCl_2 \cdot 4H_2O$ to 10.0mL of HCl solution. Mix thoroughly. Add distilled/deionized water and bring volume to 1.0L. Add remaining components. Mix thoroughly. Sparge with 80% N_2 + 20% CO_2. Autoclave for 15 min at 15 psi pressure–121°C.

Sodium Crotonate Solution:
Composition per 50.0mL:

Sodium crotonate	1.1g

Preparation of Sodium Crotonate Solution: Add sodium crotonate to distilled/deionized water and bring volume to 50.0mL. Mix thoroughly. Filter sterilize. Sparge with 80% N_2 + 20% CO_2.

$NaHCO_3$ Solution:
Composition per 20.0mL:

$NaHCO_3$	2.5g

Preparation of $NaHCO_3$ Solution: Add $NaHCO_3$ to distilled/deionized water and bring volume to 20.0mL. Mix thoroughly. Filter sterilize. Sparge with 80% N_2 + 20% CO_2.

$Na_2S \cdot 9H_2O$ Solution:
Composition per 10.0mL:

$Na_2S \cdot 9H_2O$	0.36g

Preparation of $Na_2S \cdot 9H_2O$ Solution: Add $Na_2S \cdot 9H_2O$ to distilled/deionized water and bring volume to 10.0mL. Mix thoroughly. Sparge with 100% N_2. Autoclave for 15 min at 15 psi pressure–121°C.

Seven Vitamin Solution:
Composition per liter:

Pyridoxine·HCl	0.3g
Thiamine·HCl	0.2g
Nicotinic acid	0.2g
Calcium DL-pantothenate	0.1g
Vitamin B_{12}	0.1g
p-Aminobenzoic acid	80.0mg
Biotin	20.0mg

Preparation of Seven Vitamin Solution: Add components to distilled/deionized water and bring volume to 1.0L. Mix thoroughly.

Sodium Dithionite Solution:
Composition per 10.0mL:

Sodium dithioninium	0.2g

Preparation of Sodium Dithionite Solution: Add sodium dithioninium to distilled/deionized water and bring volume to 10.0mL. Mix thoroughly. Sparge with 100% N_2. Autoclave for 15 min at 15 psi pressure–121°C.

Preparation of Medium: Prepare medium and dispense under 80% N_2 + 20% CO_2. Add components, except sodium crotonate solution, seven vitamin solution, sodium dithionite solution, $NaHCO_3$ solution, and $Na_2S \cdot 9H_2O$ solution, to distilled/deionized water and bring volume to 910.0mL. Mix thoroughly. Sparge with 80% N_2 + 20% CO_2. Autoclave for 15 min at 15 psi pressure–121°C. Aseptically and anaerobically add 50.0mL of sterile sodium crotonate solution, 20.0mL of sterile $NaHCO_3$ solution, 10.0mL of seven vitamin solution, and 10.0mL of sterile $Na_2S \cdot 9H_2O$ solution. Mix thoroughly. Aseptically and anaerobically distribute into sterile tubes or flasks. After inoculation, add 0.1mL of sodium dithionite solution per 10.0mL of medium.

Use: For the cultivation of *Syntrophobacter buswellii.*

SB/SW Medium

Composition per liter:

NaCl	1.0g
KCl	0.5g
$MgCl_2 \cdot 6H_2O$	0.4g
NH_4Cl	0.25g
KH_2PO_4	0.2g
$CaCl_2 \cdot 2H_2O$	0.15g

Resazurin	1.0mg
Sodium pyruvate solution	50.0mL
$NaHCO_3$ solution	20.0mL
$Na_2S{\cdot}9H_2O$ solution	10.0mL
Seven vitamin solution	10.0mL
Sodium dithionite solution	10.0mL
Trace elements solution SL-10	1.0mL

pH 7.2 ± 0.2 at 25°C

Trace Elements Solution SL-10:
Composition per liter:

$FeCl_2{\cdot}4H_2O$	1.5g
$CoCl_2{\cdot}6H_2O$	190.0mg
$MnCl_2{\cdot}4H_2O$	100.0mg
$ZnCl_2$	70.0mg
$Na_2MoO_4{\cdot}2H_2O$	36.0mg
$NiCl_2{\cdot}6H_2O$	24.0mg
H_3BO_3	6.0mg
$CuCl_2{\cdot}2H_2O$	2.0mg
HCl (25% solution)	10.0mL

Preparation of Trace Elements Solution SL-10: Prepare and dispense under 80% N_2 + 20% CO_2. Add $FeCl_2{\cdot}4H_2O$ to 10.0mL of HCl solution. Mix thoroughly. Add distilled/deionized water and bring volume to 1.0L. Add remaining components. Mix thoroughly. Sparge with 80% N_2 + 20% CO_2. Autoclave for 15 min at 15 psi pressure–121°C.

Sodium Pyruvate Solution:
Composition per 50.0mL:

Sodium pyruvate	1.25g

Preparation of Sodium Pyruvate Solution: Add sodium pyruvate to distilled/deionized water and bring volume to 50.0mL. Mix thoroughly. Filter sterilize. Sparge with 80% N_2 + 20% CO_2.

$NaHCO_3$ Solution:
Composition per 20.0mL:

$NaHCO_3$	2.5g

Preparation of $NaHCO_3$ Solution: Add $NaHCO_3$ to distilled/deionized water and bring volume to 20.0mL. Mix thoroughly. Filter sterilize. Sparge with 80% N_2 + 20% CO_2.

$Na_2S{\cdot}9H_2O$ Solution:
Composition per 10.0mL:

$Na_2S{\cdot}9H_2O$	0.36g

Preparation of $Na_2S{\cdot}9H_2O$ Solution: Add $Na_2S{\cdot}9H_2O$ to distilled/deionized water and bring volume to 10.0mL. Mix thoroughly. Sparge with 100% N_2. Autoclave for 15 min at 15 psi pressure–121°C.

Seven Vitamin Solution:
Composition per liter:

Pyridoxine·HCl	0.3g
Thiamine·HCl	0.2g
Nicotinic acid	0.2g
Calcium DL-pantothenate	0.1g
Vitamin B_{12}	0.1g
p-Aminobenzoic acid	80.0mg
Biotin	20.0mg

Preparation of Seven Vitamin Solution: Add components to distilled/deionized water and bring volume to 1.0L. Mix thoroughly.

Sodium Dithionite Solution:
Composition per 10.0mL:

Sodium dithioninium	0.2g

Preparation of Sodium Dithionite Solution: Add sodium dithioninium to distilled/deionized water and bring volume to 10.0mL. Mix thoroughly. Sparge with 100% N_2. Autoclave for 15 min at 15 psi pressure–121°C.

Preparation of Medium: Prepare medium and dispense under 80% N_2 + 20% CO_2. Add components, except sodium pyruvate solution, seven vitamin solution, sodium dithionite solution, $NaHCO_3$ solution, and $Na_2S{\cdot}9H_2O$ solution, to distilled/deionized water and bring volume to 910.0mL. Mix thoroughly. Sparge with 80% N_2 + 20% CO_2. Autoclave for 15 min at 15 psi pressure–121°C. Aseptically and anaerobically add 50.0mL of sterile sodium pyruvate solution, 20.0mL of sterile $NaHCO_3$ solution, 10.0mL of seven vitamin solution, and 10.0mL of sterile $Na_2S{\cdot}9H_2O$ solution. Mix thoroughly. Aseptically and anaerobically distribute into sterile tubes or flasks. After inoculation, add 0.1mL of sodium dithionite solution per 10.0mL of medium.

Use: For the cultivation of *Syntrophobacter wolinii*.

SC Agar
(DSMZ Medium 751)

Composition per liter:

Agar	15.0g
Peptone from soy meal	8.0g
Corn meal, solids from infusion	2.0g
K_2HPO_4	1.0g
KH_2PO_4	1.0g
$MgSO_4{\cdot}7H_2O$	0.2g
Hemin solution	15.0mL
Serum albumin solution	10.0mL
Cysteine solution	10.0mL
Glucose solution	5.0mL

pH 6.6 ± 0.2 at 25°C

Hemin Solution:
Composition per 100.0mL:

Hemin	0.1g

Preparation of Hemin Solution: Add hemin to 100.0mL 0.05*N* NaOH. Mix thoroughly.

Serum Albumin Solution:
Composition per 10.0mL:

Bovine serum albumin, fraction V	2.0g

Preparation of Serum Albumin Solution: Add bovine serum albumin, fraction V to distilled/deionized water and bring volume to 10.0mL. Mix thoroughly. Filter sterilize.

Glucose Solution:
Composition per 10.0mL:

D-Glucose	1.0g

Preparation of Glucose Solution: Add glucose to distilled/deionized water and bring volume to 10.0mL. Mix thoroughly. Filter sterilize.

Cysteine Solution:
Composition per 10.0mL:

L-Cysteine·HCl·H_2O	1.0g

Preparation of Cysteine Solution: Add L-cysteine·HCl·H_2O to distilled/deionized water and bring volume to 10.0mL. Mix thoroughly. Filter sterilize.

Preparation of Medium: Add components, except glucose solution, serum albumin solution, and cysteine solution, to distilled/deionized water and bring volume to 975.0mL. Mix thoroughly. Gently heat and bring to boiling. Autoclave for 15 min at 15 psi pressure–121°C. Cool to 50°C. Aseptically add 10.0mL sterile cysteine solution, 10.0mL sterile glucose solution, and 10.0mL sterile serum albumin solution. Mix thoroughly. Adjust pH to 6.6. Pour into sterile Petri dishes or distribute into sterile tubes.

Use: For the cultivation and maintenance of *Leifsonia xyli* subsp. *cynodontis=Clavibacter xyli* subsp. *cynodonti.*

SC Agar

Composition per liter:

Agar	15.0g
Papaic digest of soybean meal	8.0g
Corn meal (solids from infusion)	2.0g
K_2HPO_4	1.0g
KH_2PO_4	1.0g
$MgSO_4·7H_2O$	0.2g
Hemin solution	15.0mL
Bovine serum albumin, fraction V solution	10.0mL
L-Cysteine·H_2O solution	10.0mL
Glucose solution	1.0mL

pH 6.6 ± 0.2 at 25°C

Hemin Solution:

Composition per 100.0mL:

Hemin	0.1g
NaOH (0.05*N* solution)	100.0mL

Preparation of Hemin Solution: Add hemin to NaOH solution. Mix thoroughly.

Bovine Serum Albumin, Fraction V Solution:

Composition per 10.0mL:

Bovine serum albumin, fraction V	2.0g

Preparation of Bovine Serum Albumin, Fraction V Solution: Add bovine serum albumin to distilled/deionized water and bring volume to 10.0mL. Mix thoroughly. Filter sterilize.

L-Cysteine·H_2O Solution:

Composition per 10.0mL:

L-Cysteine·H_2O	1.0g

Preparation of L-Cysteine·H_2O Solution: Add L-cysteine·H_2O to distilled/deionized water and bring volume to 10.0mL. Mix thoroughly. Filter sterilize.

Glucose Solution:

Composition per 10.0mL:

D-Glucose	5.0g

Preparation of Glucose Solution: Add glucose to distilled/deionized water and bring volume to 10.0mL. Mix thoroughly. Filter sterilize.

Preparation of Medium: Add components—except bovine serum albumin solution, L-cysteine·H_2O solution, and glucose solution—to distilled/deionized water and bring volume to 979.0mL. Mix thoroughly. Adjust pH to 6.6 with NaOH. Gently heat and bring to boiling. Autoclave for 15 min at 15 psi pressure–121°C. Cool to 45°–50°C. Aseptically add 10.0mL of sterile bovine serum albumin solution, 10.0mL of sterile L-cysteine·H_2O solution, and 1.0mL of sterile glucose solution. Mix thoroughly. Pour into sterile Petri dishes or distribute into sterile tubes.

Use: For the cultivation and maintenance of *Clavibacter xyli.*

SC Agar

Composition per liter:

Agar	20.0g
$(NH_4)_2SO_4$	5.0g
KH_2PO_4	1.0g
$MgSO_4·7H_2O$	0.5g
NaCl	0.1g
$CaCl_2·2H_2O$	0.1g
Inositol	2.0mg
KI	1.0mg
H_3BO_3	0.5mg
$ZnSO_4·7H_2O$	0.4mg
$MnSO_4·4H_2O$	0.4mg
Thiamine·HCl	0.4mg
Pyroxidine·HCl	0.4mg
Niacin	0.4mg
Calcium pantothenate	0.4mg
p-Aminobenzoic acid	0.2mg
Riboflavin	0.2mg
$FeCl_3$	0.2mg
$Na_2MoO_4·4H_2O$	0.2mg
$CuSO_4·5H_2O$	0.04mg
Folic acid	2.0μg
Biotin	2.0μg
Glucose solution	50.0mL
Synthetic stock solution	30.0mL

Glucose Solution:

Composition per 100.0mL:

Glucose	40.0g

Preparation of Glucose Solution: Add glucose to distilled/deionized water and bring volume to 100.0mL. Mix thoroughly. Filter sterilize.

Synthetic Stock Solution:

Composition per 300.0mL:

L-Isoleucine	5.25g
L-Arginine	3.48g
L-Aspartic acid	2.66g
L-Leucine	2.62g
L-Methionine	1.49g
L-Threonine	1.19g
L-Valine	1.17g
L-Serine	1.05g
L-Lysine	0.91g
L-Phenylalanine	0.83g
L-Tryptophan	0.82g
L-Histidine	0.58g
m-Inositol	0.36g
Uracil	0.22g
L-Tyrosine	0.18g
Adenine	0.135g

Preparation of Synthetic Stock Solution: Add components to distilled/deionized water and bring volume to 300.0mL. Mix thoroughly. Filter sterilize. Store in the dark at 25°C.

Preparation of Medium: Add components, except synthetic stock solution and glucose solution, to distilled/deionized water and bring volume to 920.0mL. Mix thoroughly. Gently heat and bring to boiling. Autoclave for 15 min at 15 psi pressure–121°C. Cool to 50°–55°C. Aseptically add 30.0mL of sterile synthetic stock solution and 50.0mL of sterile glucose solution. Mix thoroughly. Pour into sterile Petri dishes or distribute into sterile tubes.

Use: For the cultivation and maintenance of *Saccharomyces cerevisiae*.

SC Agar without Histidine

Composition per liter:

Agar	20.0g
$(NH_4)_2SO_4$	5.0g
KH_2PO_4	1.0g
$MgSO_4 \cdot 7H_2O$	0.5g
NaCl	0.1g
$CaCl_2 \cdot 2H_2O$	0.1g
Inositol	2.0mg
KI	1.0mg
H_3BO_3	0.5mg
$ZnSO_4 \cdot 7H_2O$	0.4mg
$MnSO_4 \cdot 4H_2O$	0.4mg
Thiamine·HCl	0.4mg
Pyroxidine·HCl	0.4mg
Niacin	0.4mg
Calcium pantothenate	0.4mg
p-Aminobenzoic acid	0.2mg
Riboflavin	0.2mg
$FeCl_3$	0.2mg
$Na_2MoO_4 \cdot 4H_2O$	0.2mg
$CuSO_4 \cdot 5H_2O$	0.04mg
Folic acid	2.0μg
Biotin	2.0μg
Glucose solution	50.0mL
Synthetic stock solution	30.0mL

Glucose Solution:

Composition per 100.0mL:

Glucose	40.0g

Preparation of Glucose Solution: Add glucose to distilled/deionized water and bring volume to 100.0mL. Mix thoroughly. Filter sterilize.

Synthetic Stock Solution:

Composition per 300.0mL:

L-Isoleucine	5.25g
L-Arginine	3.48g
L-Aspartic acid	2.66g
L-Leucine	2.62g
L-Methionine	1.49g
L-Threonine	1.19g
L-Valine	1.17g
L-Serine	1.05g
L-Lysine	0.91g
L-Phenylalanine	0.83g
L-Tryptophan	0.82g
Myo-inositol	0.36g
Uracil	0.22g
L-Tyrosine	0.18g
Adenine	0.135g

Preparation of Synthetic Stock Solution: Add components to distilled/deionized water and bring volume to 300.0mL. Mix thoroughly. Filter sterilize. Store in the dark at 25°C.

Preparation of Medium: Add components, except synthetic stock solution and glucose solution, to distilled/deionized water and bring volume to 920.0mL. Mix thoroughly. Gently heat and bring to boiling. Autoclave for 15 min at 15 psi pressure–121°C. Cool to 50°–55°C. Aseptically add 30.0mL of sterile synthetic stock solution and 50.0mL of sterile glucose solution. Mix thoroughly. Pour into sterile Petri dishes or distribute into sterile tubes.

Use: For the cultivation and maintenance of strains of *Saccharomyces cerevisiae* that do not require histidine.

SC Agar without Leucine and Tryptophan

Composition per liter:

Agar	20.0g
$(NH_4)_2SO_4$	5.0g
KH_2PO_4	1.0g
$MgSO_4 \cdot 7H_2O$	0.5g
NaCl	0.1g
$CaCl_2 \cdot 2H_2O$	0.1g
Inositol	2.0mg
KI	1.0mg
H_3BO_3	0.5mg
$ZnSO_4 \cdot 7H_2O$	0.4mg
$MnSO_4 \cdot 4H_2O$	0.4mg
Thiamine·HCl	0.4mg
Pyroxidine·HCl	0.4mg
Niacin	0.4mg
Calcium pantothenate	0.4mg
p-Aminobenzoic acid	0.2mg
Riboflavin	0.2mg
$FeCl_3$	0.2mg
$Na_2MoO_4 \cdot 4H_2O$	0.2mg
$CuSO_4 \cdot 5H_2O$	0.04mg
Folic acid	2.0μg
Biotin	2.0μg
Glucose solution	50.0mL
Synthetic stock solution	30.0mL

Glucose Solution:

Composition per 100.0mL:

Glucose	40.0g

Preparation of Glucose Solution: Add glucose to distilled/deionized water and bring volume to 100.0mL. Mix thoroughly. Filter sterilize.

Synthetic Stock Solution:

Composition per 300.0mL:

L-Isoleucine	5.25g
L-Arginine	3.48g
L-Aspartic acid	2.66g
L-Methionine	1.49g
L-Threonine	1.19g
L-Valine	1.17g
L-Serine	1.05g
Lysine	0.91g
L-Phenylalanine	0.83g
L-Histidine	0.58g
Myo-inositol	0.36g
Uracil	0.22g
Adenine	0.135g
L-Tyrosine	0.18g

Preparation of Synthetic Stock Solution: Add components to distilled/deionized water and bring volume to 300.0mL. Mix thoroughly. Filter sterilize. Store in the dark at 25°C.

Preparation of Medium: Add components, except synthetic stock solution and glucose solution, to distilled/deionized water and bring volume to 920.0mL. Mix thoroughly. Gently heat and bring to boiling. Autoclave for 15 min at 15 psi pressure–121°C. Cool to 50°–55°C. Aseptically add 30.0mL of sterile synthetic stock solution and 50.0mL of sterile glucose solution. Mix thoroughly. Pour into sterile Petri dishes or distribute into sterile tubes.

Use: For the cultivation and maintenance of strains of *Saccharomyces cerevisiae* that do not require leucine and tryptophan.

SC Agar without Uracil

Composition per liter:

Agar	20.0g
$(NH_4)_2SO_4$	5.0g
KH_2PO_4	1.0g
$MgSO_4·7H_2O$	0.5g
NaCl	0.1g
$CaCl_2·2H_2O$	0.1g
Inositol	2.0mg
KI	1.0mg
H_3BO_3	0.5mg
$ZnSO_4·7H_2O$	0.4mg
$MnSO_4·4H_2O$	0.4mg
Thiamine·HCl	0.4mg
Pyroxidine·HCl	0.4mg
Niacin	0.4mg
Calcium pantothenate	0.4mg
p-Aminobenzoic acid	0.2mg
Riboflavin	0.2mg
$FeCl_3$	0.2mg
$Na_2MoO_4·4H_2O$	0.2mg
$CuSO_4·5H_2O$	0.04mg
Folic acid	2.0μg
Biotin	2.0μg
Glucose solution	50.0mL
Synthetic stock solution	30.0mL

Glucose Solution:

Composition per 100.0mL:

Glucose	40.0g

Preparation of Glucose Solution: Add glucose to distilled/deionized water and bring volume to 100.0mL. Mix thoroughly. Filter sterilize.

Synthetic Stock Solution:

Composition per 300.0mL:

L-Isoleucine	5.25g
L-Arginine	3.48g
L-Aspartic acid	2.66g
L-Leucine	2.62g
L-Methionine	1.49g
L-Threonine	1.19g
L-Valine	1.17g
L-Serine	1.05g
L-Lysine	0.91g
L-Phenylalanine	0.83g
L-Tryptophan	0.82g
L-Histidine	0.58g
Myo-inositol	0.36g
L-Tyrosine	0.18g
Adenine	0.135g

Preparation of Synthetic Stock Solution: Add components to distilled/deionized water and bring volume to 300.0mL. Mix thoroughly. Filter sterilize.

Preparation of Medium: Add components, except synthetic stock solution and glucose solution, to distilled/deionized water and bring volume to 920.0mL. Mix thoroughly. Gently heat and bring to boiling. Autoclave for 15 min at 15 psi pressure–121°C. Cool to 50°–55°C. Aseptically add 30.0mL of sterile synthetic stock solution and 50.0mL of sterile glucose solution. Mix thoroughly. Pour into sterile Petri dishes or distribute into sterile tubes.

Use: For the cultivation and maintenance of strains of *Saccharomyces cerevisiae* that do not require uracil.

SC Broth

Composition per liter:

Papaic digest of soybean meal	8.0g
KH_2PO_4	1.5g
K_2HPO_4	0.5g
$MgSO_4·7H_2O$	0.2g
Hemin solution	15.0mL
Bovine serum albumin, fraction V solution	10.0mL
L-Cysteine·H_2O solution	10.0mL
Glucose solution	1.0mL

pH 6.6 ± 0.2 at 25°C

Hemin Solution:

Composition per 100.0mL:

Hemin	0.1g
NaOH (0.05*N* solution)	100.0mL

Preparation of Hemin Solution: Add hemin to NaOH solution. Mix thoroughly.

Bovine Serum Albumin, Fraction V Solution:

Composition per 10.0mL:

Bovine serum albumin, fraction V	2.0g

Preparation of Bovine Serum Albumin, Fraction V Solution: Add bovine serum albumin to distilled/deionized water and bring volume to 10.0mL. Mix thoroughly. Filter sterilize.

L-Cysteine·H_2O Solution:

Composition per 10.0mL:

L-Cysteine·H_2O	1.0g

Preparation of L-Cysteine·H_2O Solution: Add L-cysteine·H_2O to distilled/deionized water and bring volume to 10.0mL. Mix thoroughly. Filter sterilize.

Glucose Solution:

Composition per 10.0mL:

D-Glucose	5.0g

Preparation of Glucose Solution: Add glucose to distilled/deionized water and bring volume to 10.0mL. Mix thoroughly. Filter sterilize.

Preparation of Medium: Add components—except bovine serum albumin solution, L-cysteine·H_2O solution, and glucose solution—to distilled/deionized water and bring volume to 979.0mL. Mix thoroughly. Adjust pH to 6.6 with NaOH. Gently heat and bring to boiling. Autoclave for 15 min at 15 psi pressure–121°C. Cool to 45°–50°C. Aseptically add 10.0mL of sterile bovine serum albumin solution, 10.0mL of sterile L-cysteine·H_2O solution, and 1.0mL of sterile glucose solution. Mix thoroughly. Aseptically distribute into sterile tubes or flasks.

Use: For the cultivation of *Clavibacter xyli*.

SC Broth

Composition per liter:

$(NH_4)_2SO_4$	5.0g
KH_2PO_4	1.0g
$MgSO_4 \cdot 7H_2O$	0.5g
NaCl	0.1g
$CaCl_2 \cdot 2H_2O$	0.1g
Inositol	2.0mg
KI	1.0mg
H_3BO_3	0.5mg
$ZnSO_4 \cdot 7H_2O$	0.4mg
$MnSO_4 \cdot 4H_2O$	0.4mg
Thiamine·HCl	0.4mg
Pyroxidine·HCl	0.4mg
Niacin	0.4mg
Calcium pantothenate	0.4mg
p-Aminobenzoic acid	0.2mg
Riboflavin	0.2mg
$FeCl_3$	0.2mg
$Na_2MoO_4 \cdot 4H_2O$	0.2mg
$CuSO_4 \cdot 5H_2O$	0.04mg
Folic acid	2.0μg
Biotin	2.0μg
Glucose solution	50.0mL
Synthetic stock solution	30.0mL

Glucose Solution:

Composition per 100.0mL:

Glucose	40.0g

Preparation of Glucose Solution: Add glucose to distilled/deionized water and bring volume to 100.0mL. Mix thoroughly. Filter sterilize.

Synthetic Stock Solution:

Composition per 300.0mL:

L-Isoleucine	5.25g
L-Arginine	3.48g
L-Aspartic acid	2.66g
L-Leucine	2.62g
L-Methionine	1.49g
L-Threonine	1.19g
L-Valine	1.17g
L-Serine	1.05g
L-Lysine	0.91g
L-Phenylalanine	0.83g
L-Tryptophan	0.82g
L-Histidine	0.58g
m-inositol	0.36g
Uracil	0.22g
L-Tyrosine	0.18g
Adenine	0.135g

Preparation of Synthetic Stock Solution: Add components to distilled/deionized water and bring volume to 300.0mL. Mix thoroughly. Filter sterilize. Store in the dark at 25°C.

Preparation of Medium: Add components, except synthetic stock solution and glucose solution, to distilled/deionized water and bring volume to 920.0mL. Mix thoroughly. Autoclave for 15 min at 15 psi pressure–121°C. Aseptically add 30.0mL of sterile synthetic stock solution and 50.0mL of sterile glucose solution. Mix thoroughly. Aseptically distribute into sterile tubes or flasks.

Use: For the cultivation of *Saccharomyces cerevisiae*.

SC Broth without Histidine

Composition per liter:

$(NH_4)_2SO_4$	5.0g
KH_2PO_4	1.0g
$MgSO_4 \cdot 7H_2O$	0.5g
NaCl	0.1g
$CaCl_2 \cdot 2H_2O$	0.1g
Inositol	2.0mg
KI	1.0mg
H_3BO_3	0.5mg
$ZnSO_4 \cdot 7H_2O$	0.4mg
$MnSO_4 \cdot 4H_2O$	0.4mg
Thiamine·HCl	0.4mg
Pyroxidine·HCl	0.4mg
Niacin	0.4mg
Calcium pantothenate	0.4mg
p-Aminobenzoic acid	0.2mg
Riboflavin	0.2mg
$FeCl_3$	0.2mg
$Na_2MoO_4 \cdot 4H_2O$	0.2mg
$CuSO_4 \cdot 5H_2O$	0.04mg
Folic acid	2.0μg
Biotin	2.0μg
Glucose solution	50.0mL
Synthetic stock solution	30.0mL

Glucose Solution:

Composition per 100.0mL:

Glucose	40.0g

Preparation of Glucose Solution: Add glucose to distilled/deionized water and bring volume to 100.0mL. Mix thoroughly. Filter sterilize.

Synthetic Stock Solution:

Composition per 300.0mL:

L-Isoleucine	5.25g
L-Arginine	3.48g
L-Aspartic acid	2.66g
L-Leucine	2.62g
L-Methionine	1.49g
L-Threonine	1.19g
L-Valine	1.17g
L-Serine	1.05g
L-Lysine	0.91g
L-Phenylalanine	0.83g
L-Tryptophan	0.82g
Myo-inositol	0.36g
Uracil	0.22g
L-Tyrosine	0.18g
Adenine	0.135g

Preparation of Synthetic Stock Solution: Add components to distilled/deionized water and bring volume to 300.0mL. Mix thoroughly. Filter sterilize. Store in the dark at 25°C.

Preparation of Medium: Add components, except synthetic stock solution and glucose solution, to distilled/deionized water and bring volume to 920.0mL. Mix thoroughly. Gently heat and bring to boiling. Autoclave for 15 min at 15 psi pressure–121°C. Cool to 50°–55°C.

Aseptically add 30.0mL of sterile synthetic stock solution and 50.0mL of sterile glucose solution. Mix thoroughly. Aseptically distribute into sterile tubes or flasks.

Use: For the cultivation of strains of *Saccharomyces cerevisiae* that do not require histidine.

SC Broth without Leucine and Tryptophan

Composition per liter:

$(NH_4)_2SO_4$	5.0g
KH_2PO_4	1.0g
$MgSO_4 \cdot 7H_2O$	0.5g
NaCl	0.1g
$CaCl_2 \cdot 2H_2O$	0.1g
Inositol	2.0mg
KI	1.0mg
H_3BO_3	0.5mg
$ZnSO_4 \cdot 7H_2O$	0.4mg
$MnSO_4 \cdot 4H_2O$	0.4mg
Thiamine·HCl	0.4mg
Pyroxidine·HCl	0.4mg
Niacin	0.4mg
Calcium pantothenate	0.4mg
p-Aminobenzoic acid	0.2mg
Riboflavin	0.2mg
$FeCl_3$	0.2mg
$Na_2MoO_4 \cdot 4H_2O$	0.2mg
$CuSO_4 \cdot 5H_2O$	0.04mg
Folic acid	2.0μg
Biotin	2.0μg
Glucose solution	50.0mL
Synthetic stock solution	30.0mL

Glucose Solution:

Composition per 100.0mL:

Glucose	40.0g

Preparation of Glucose Solution: Add glucose to distilled/deionized water and bring volume to 100.0mL. Mix thoroughly. Filter sterilize.

Synthetic Stock Solution:

Composition per 300.0mL:

L-Isoleucine	5.25g
L-Arginine	3.48g
L-Aspartic acid	2.66g
L-Methionine	1.49g
L-Threonine	1.19g
L-Valine	1.17g
L-Serine	1.05g
Lysine	0.91g
L-Phenylalanine	0.83g
L-Histidine	0.58g
Myo-inositol	0.36g
Uracil	0.22g
L-Tyrosine	0.18g
Adenine	0.135g

Preparation of Synthetic Stock Solution: Add components to distilled/deionized water and bring volume to 300.0mL. Mix thoroughly. Filter sterilize. Store in the dark at 25°C.

Preparation of Medium: Add components, except synthetic stock solution and glucose solution, to distilled/deionized water and bring volume to 920.0mL. Mix thoroughly. Gently heat and bring to boiling. Autoclave for 15 min at 15 psi pressure–121°C. Cool to 50°–55°C. Aseptically add 30.0mL of sterile synthetic stock solution and 50.0mL of sterile glucose solution. Mix thoroughly. Aseptically distribute into sterile tubes or flasks.

Use: For the cultivation of strains of *Saccharomyces cerevisiae* that do not require leucine and tryptophan.

SC Broth without Uracil

Composition per liter:

$(NH_4)_2SO_4$	5.0g
KH_2PO_4	1.0g
$MgSO_4 \cdot 7H_2O$	0.5g
NaCl	0.1g
$CaCl_2 \cdot 2H_2O$	0.1g
Inositol	2.0mg
KI	1.0mg
H_3BO_3	0.5mg
$ZnSO_4 \cdot 7H_2O$	0.4mg
$MnSO_4 \cdot 4H_2O$	0.4mg
Thiamine·HCl	0.4mg
Pyroxidine·HCl	0.4mg
Niacin	0.4mg
Calcium pantothenate	0.4mg
p-Aminobenzoic acid	0.2mg
Riboflavin	0.2mg
$FeCl_3$	0.2mg
$Na_2MoO_4 \cdot 4H_2O$	0.2mg
$CuSO_4 \cdot 5H_2O$	0.04mg
Folic acid	2.0μg
Biotin	2.0μg
Glucose solution	50.0mL
Synthetic stock solution	30.0mL

Glucose Solution:

Composition per 100.0mL:

Glucose	40.0g

Preparation of Glucose Solution: Add glucose to distilled/deionized water and bring volume to 100.0mL. Mix thoroughly. Filter sterilize.

Synthetic Stock Solution:

Composition per 300.0mL:

L-Isoleucine	5.25g
L-Arginine	3.48g
L-Aspartic acid	2.66g
L-Leucine	2.62g
L-Methionine	1.49g
L-Threonine	1.19g
L-Valine	1.17g
L-Serine	1.05g
L-Lysine	0.91g
L-Phenylalanine	0.83g
L-Tryptophan	0.82g
L-Histidine	0.58g
Myo-inositol	0.36g
L-Tyrosine	0.18g
Adenine	0.135g

Preparation of Synthetic Stock Solution: Add components to distilled/deionized water and bring volume to 300.0mL. Mix thoroughly. Filter sterilize.

Preparation of Medium: Add components, except synthetic stock solution and glucose solution, to distilled/deionized water and bring volume to 920.0mL. Mix thoroughly. Gently heat and bring to boiling. Autoclave for 15 min at 15 psi pressure–121°C. Cool to 50°–55°C. Aseptically add 30.0mL of sterile synthetic stock solution and 50.0mL of sterile glucose solution. Mix thoroughly. Aseptically distribute into sterile tubes or flasks.

Use: For the cultivation of strains of *Saccharomyces cerevisiae* that do not require uracil.

SC Medium (DSMZ Medium 751)

Composition per liter:

Peptone from soy meal	8.0g
KH_2PO_4	1.5g
K_2HPO_4	0.5g
$MgSO_4·7H_2O$	0.2g
Hemin solution	15.0mL
Serum albumin solution	10.0mL
Cysteine solution	10.0mL
Glucose solution	5.0mL

pH 6.6 ± 0.2 at 25°C

Hemin Solution:
Composition per 100.0mL:

Hemin	0.1g

Preparation of Hemin Solution: Add hemin to 100mL 0.05*N* NaOH. Mix thoroughly.

Serum Albumin Solution:
Composition per 10.0mL:

Bovine serum albumin, fraction V	2.0g

Preparation of Serum Albumin Solution: Add bovine serum albumin, fraction V to distilled/deionized water and bring volume to 10.0mL. Mix thoroughly. Filter sterilize.

Glucose Solution:
Composition per 10.0mL:

D-Glucose	1.0g

Preparation of Glucose Solution: Add glucose to distilled/deionized water and bring volume to 10.0mL. Mix thoroughly. Filter sterilize.

Cysteine Solution:
Composition per 10.0mL:

L-Cysteine·HCl·H_2O	1.0g

Preparation of Cysteine Solution: Add L-cysteine·HCl·H_2O to distilled/deionized water and bring volume to 10.0mL. Mix thoroughly. Filter sterilize.

Preparation of Medium: Add components, except glucose solution, serum albumin solution, and cysteine solution, to distilled/deionized water and bring volume to 970.0mL. Mix thoroughly. Gently heat and bring to boiling. Autoclave for 15 min at 15 psi pressure–121°C. Cool to 50°C. Aseptically add 10.0mL sterile cysteine solution, 10.0mL sterile glucose solution, and 10.0mL sterile serum albumin solution. Mix thoroughly. Adjust pH to 6.6. Pour into sterile Petri dishes or distribute into sterile tubes.

Use: For the cultivation of *Leifsonia xyli* subsp. *cynodontis*.

SC Medium

Composition per 1021.0mL:

Agar	15.0g
Papaic digest of soybean meal	8.0g
Corn meal, solids from infusion	2.0g
Tween™ 80	1.0g
K_2HPO_4	1.0g
KH_2PO_4	1.0g
$MgSO_4·7H_2O$	0.2g
Hemin chloride solution	15.0mL
Bovine serum albumin solution	10.0mL
L-Cysteine solution	10.0mL
Glucose solution	1.0mL

pH 6.6 at 25°C

Hemin Chloride Solution:
Composition per 100.0mL:

Hemin chloride	0.1g
NaOH (0.05*N* solution)	100.0mL

Preparation of Hemin Chloride Solution: Add hemin chloride to 100.0mL of NaOH solution. Mix thoroughly.

Bovine Serum Albumin Solution:
Composition per 10.0mL:

Bovine serum albumin	2.0g

Preparation of Bovine Serum Albumin Solution: Add bovine serum albumin to distilled/deionized water and bring volume to 10.0mL. Mix thoroughly. Filter sterilize.

L-Cysteine Solution:
Composition per 10.0mL:

L-Cysteine, free base	1.0g

Preparation of L-Cysteine Solution: Add L-cysteine to distilled/deionized water and bring volume to 10.0mL. Mix thoroughly. Filter sterilize.

Glucose Solution:
Composition per 10.0mL:

Glucose	5.0g

Preparation of Glucose Solution: Add glucose to distilled/deionized water and bring volume to 10.0mL. Mix thoroughly. Autoclave for 15 min at 15 psi pressure–121°C. Cool to 25°C.

Preparation of Medium: Add components—except bovine serum albumin solution, L-cysteine solution, and glucose solution—to distilled/deionized water and bring volume to 1.0L. Mix thoroughly. Gently heat and bring to boiling. Autoclave for 15 min at 15 psi pressure–121°C. Cool to 45°–50°C. Aseptically add 10.0mL of sterile bovine serum albumin solution, 10.0mL of sterile L-cysteine solution, and 1.0mL of sterile glucose solution. Mix thoroughly. Pour into sterile Petri dishes or distribute into sterile tubes.

Use: For the isolation and cultivation of coryneform bacteria that cause ratoon stunting disease of sugarcane.

SCGYEM Medium

Composition per liter:

Casein, isoelectric	10.0g
Yeast extract	5.0g
Glucose	2.5g
Na_2HPO_4	1.325g
KH_2PO_4	0.8g
Calf serum, heat inactivated	100.0mL

Preparation of Medium: Add components, except calf serum, to distilled/deionized water and bring volume to 900.0mL. Mix thoroughly. Autoclave for 30 min at 15 psi pressure–121°C. Cool to 25°C. Aseptically add 100.0mL of sterile, heat-inactivated calf serum. Mix thoroughly. Aseptically distribute into sterile, screw-capped tubes or flasks.

Use: For the cultivation of *Naegleria australiensis* and *Naegleria fowleri.*

Schaedler Agar (Schaedler Anaerobic Agar)

Composition per liter:

Agar	13.5g
Glucose	5.83g
Pancreatic digest of casein	5.7g
Proteose peptone No. 3	5.0g
Yeast extract	5.0g
Tris(hydroxymethyl)aminomethane buffer	3.0g
NaCl	1.65g
Papaic digest of soybean meal	1.0g
K_2HPO_4	0.83g
L-Cystine	0.4g
Hemin	0.01g

pH 7.6 ± 0.2 at 25°C

Source: This medium is available as a premixed powder from BD Diagnostic Systems and Oxoid Unipath.

Preparation of Medium: Add components to distilled/deionized water and bring volume to 1.0L. Mix thoroughly. Gently heat and bring to boiling. Distribute into tubes or flasks. Autoclave for 15 min at 15 psi pressure–121°C. Pour into sterile Petri dishes or leave in tubes.

Use: For the isolation, cultivation, and enumeration of anaerobic and aerobic microorganisms.

Schaedler Agar

Composition per liter:

Agar	13.5g
Pancreatic digest of casein	8.2g
Glucose	5.8g
Yeast extract	5.0g
Tris(hydroxymethyl)aminomethane buffer	3.0g
Peptic digest of animal tissue	2.5g
NaCl	1.7g
Papaic digest of soybean meal	1.0g
K_2HPO_4	0.8g
L-Cystine	0.4g
Hemin	0.01g

pH 7.6 ± 0.2 at 25°C

Source: This medium is available as a premixed powder from BD Diagnostic Systems.

Preparation of Medium: Add components to distilled/deionized water and bring volume to 1.0L. Mix thoroughly. Gently heat and bring to boiling. Distribute into tubes or flasks. Autoclave for 15 min at 15 psi pressure–121°C. Pour into sterile Petri dishes or leave in tubes.

Use: For the isolation, cultivation, and enumeration of anaerobic and aerobic microorganisms.

Schaedler Agar with Vitamin K_1 and Sheep Blood

Composition per liter:

Agar	13.5g
Pancreatic digest of casein	8.2g
Glucose	5.8g
Yeast extract	5.0g
Tris(hydroxymethyl)aminomethane buffer	3.0g
Peptic digest of animal tissue	2.5g
Papaic digest of soybean meal	1.0g
NaCl	1.7g
K_2HPO_4	0.8g
L-Cystine	0.4g
Hemin	0.01g
Sheep blood, defibrinated	50.0mL
Vitamin K_1 solution	1.0mL

pH 7.6 ± 0.2 at 25°C

Vitamin K_1 Solution:
Composition per 10.0mL:

Vitamin K_1	5.0g
Ethanol, absolute	10.0mL

Preparation of Vitamin K_1 Solution: Add vitamin K_1 to ethanol. Mix thoroughly.

Preparation of Medium: Add components, except sheep blood, to distilled/deionized water and bring volume to 950.0mL. Mix thoroughly. Gently heat and bring to boiling. Autoclave for 15 min at 15 psi pressure–121°C. Cool to 45°–50°C. Aseptically add sterile sheep blood. Mix thoroughly. Pour into sterile Petri dishes or distribute into sterile tubes.

Use: For the recovery of fastidious anaerobic bacteria such as *Bacteroides* species.

Schaedler Anaerobic Agar
See: **Schaedler Agar**

Schaedler Anaerobic Broth
See: **Schaedler Broth**

Schaedler Broth (Schaedler Anaerobic Broth)

Composition per liter:

Pancreatic digest of casein	8.2g
Glucose	5.8g
Yeast extract	5.0g
Tris(hydroxymethyl)aminomethane buffer	3.0g
Peptic digest of animal tissue	2.5g
NaCl	1.7g
Papaic digest of soybean meal	1.0g
K_2HPO_4	0.8g
L-Cystine	0.4g
Hemin	0.01g

pH 7.6 ± 0.2 at 25°C

Source: This medium is available as a premixed powder from BD Diagnostic Systems and Oxoid Unipath.

Preparation of Medium: Add components to distilled/deionized water and bring volume to 1.0L. Mix thoroughly. Distribute into tubes or flasks. Autoclave for 15 min at 15 psi pressure–121°C.

Use: For the cultivation and maintenance of *Eubacterium combesii, Eubacterium contortum*, and a variety of other anaerobic bacteria.

Schaedler CNA Agar with Vitamin K_1 and Sheep Blood

Composition per liter:

Agar	13.5g
Pancreatic digest of casein	8.2g
Glucose	5.8g
Yeast extract	5.0g
Tris(hydroxymethyl)aminomethane buffer	3.0g
Peptic digest of animal tissue	2.5g
Papaic digest of soybean meal	1.0g
NaCl	1.7g
K_2HPO_4	0.8g
L-Cystine	0.4g
Hemin	0.01g
Colistin	0.01g
Nalidixic acid	0.01g
Sheep blood, defibrinated	50.0mL
Vitamin K_1 solution	1.0mL

pH 7.6 ± 0.2 at 25°C

Vitamin K_1 Solution:

Composition per 10.0mL:

Vitamin K_1	5.0g
Ethanol, absolute	10.0mL

Preparation of Vitamin K_1 Solution: Add vitamin K_1 to ethanol. Mix thoroughly.

Preparation of Medium: Add components, except sheep blood, to distilled/deionized water and bring volume to 950.0mL. Mix thoroughly. Gently heat and bring to boiling. Autoclave for 15 min at 15 psi pressure–121°C. Cool to 45°–50°C. Aseptically add sterile sheep blood. Mix thoroughly. Pour into sterile Petri dishes or distribute into sterile tubes.

Use: For the selective isolation of anaerobic, Gram-positive cocci, especially *Peptococcus* species and *Peptostreptococcus* species.

Schaedler HiVeg Agar with Blood

Composition per liter:

Agar	15.0g
Glucose	5.83g
Plant hydrolysate	5.67g
Plant peptone No. 3	5.0g
Yeast extract	5.0g
Tris(hydroxymethyl)aminomethane buffer	3.0g
NaCl	1.67g
Papaic digest of soybean meal	1.0g
K_2HPO_4	0.83g
L-Cystine	0.4g
$Fe_4(P_2O_7)_3 \cdot H_2O$	0.01g
Sheep blood, defibrinated	50.0mL

pH 7.6 ± 0.2 at 25°C

Source: This medium, without blood, is available as a premixed powder from HiMedia.

Preparation of Medium: Add components, except sheep blood, to distilled/deionized water and bring volume to 950.0mL. Mix thoroughly. Gently heat and bring to boiling. Autoclave for 15 min at 15 psi pressure–121°C. Cool to 45°–50°C. Aseptically add 50.0mL of sterile sheep blood. Mix thoroughly.

Use: For the enumeration of various aerobic and anaerobic bacterial species present in the gastrointestinal tract.

Schaedler HiVeg Broth

Composition per liter:

Glucose	5.83g
Plant hydrolysate	5.67g
Plant peptone No. 3	5.0g
Tris(hydroxymethyl)aminomethane buffer	3.0g
Yeast extract	5.0g
NaCl	1.67g
Papaic digest of soybean meal	1.0g
K_2HPO_4	0.83g
L-Cystine	0.4g
$Fe_4(P_2O_7)_3 \cdot H_2O$	0.01g
Sheep blood, defibrinated	50.0mL

pH 7.6 ± 0.2 at 25°C

Source: This medium, without blood, is available as a premixed powder from HiMedia.

Preparation of Medium: Add components, except sheep blood, to distilled/deionized water and bring volume to 950.0mL. Mix thoroughly. Gently heat and bring to boiling. Autoclave for 15 min at 15 psi pressure–121°C. Cool to 45°–50°C. Aseptically add 50.0mL of sterile sheep blood. Mix thoroughly. Aseptically distribute to tubes or flasks.

Use: For the enumeration of various aerobic and anaerobic bacterial species present in the gastrointestinal tract.

Schaedler KV Agar with Vitamin K_1 and Sheep Blood

Composition per liter:

Agar	13.5g
Pancreatic digest of casein	8.2g
Glucose	5.8g
Yeast extract	5.0g
Tris(hydroxymethyl)aminomethane buffer	3.0g
Peptic digest of animal tissue	2.5g
NaCl	1.7g
Papaic digest of soybean meal	1.0g
K_2HPO_4	0.8g
L-Cystine	0.4g
Hemin	0.01g
Kanamycin	0.01g
Vancomycin	7.5mg
Sheep blood, defibrinated	50.0 mL
Vitamin K_1 solution	1.0mL

pH 7.6 ± 0.2 at 25°C

Vitamin K_1 Solution:

Composition per 10.0mL:

Vitamin K_1	5.0g
Ethanol, absolute	10.0mL

Preparation of Vitamin K_1 Solution: Add vitamin K_1 to ethanol. Mix thoroughly.

Preparation of Medium: Add components, except sheep blood, to distilled/deionized water and bring volume to 950.0mL. Mix thoroughly. Gently heat and bring to boiling. Autoclave for 15 min at 15 psi pressure–121°C. Cool to 45°–50°C. Aseptically add sterile sheep blood. Mix thoroughly. Pour into sterile Petri dishes or distribute into sterile tubes.

Use: For the selective isolation of Gram-negative anaerobic bacteria.

Schiemmann's CIN Agar
See: Yersinia Selective (CIN) Agar

Schizophyllum Medium

Composition per liter:

Agar	20.0g
Glucose	20.0g
Peptone	2.0g
Yeast extract	2.0g
K_2HPO_4	1.0g
$MgSO_4 \cdot 7H_2O$	0.5g
KH_2PO_4	0.46g

Preparation of Medium: Add components to distilled/deionized water and bring volume to 1.0L. Mix thoroughly. Gently heat and bring to boiling. Distribute into tubes or flasks. Autoclave for 15 min at 15 psi pressure–121°C. Pour into sterile Petri dishes or leave in tubes.

Use: For the cultivation and maintenance of *Schizophyllum* species.

Schizosaccharomyces Malate Medium

Composition per liter:

Agar	30.0g
Glucose	10.0g
Malic acid	10.0g
Peptone	5.0g
Yeast extract	3.0g
Malt extract	3.0g

Preparation of Medium: Add components to distilled/deionized water and bring volume to 1.0L. Mix thoroughly. Gently heat and bring to boiling. Adjust pH to 4.8 with KOH. Distribute into tubes or flasks. Autoclave for 15 min at 15 psi pressure–121°C. Pour into sterile Petri dishes or leave in tubes.

Use: For the cultivation and maintenance of *Schizosaccharomyces pombe.*

Schleifer-Krämer Agar (SK Agar)

Composition per liter:

Agar	13.0g
Glycerol	10.0g
Sodium pyruvate	10.0g
Pancreatic digest of casein	10.0g
Beef extract	5.0g
Yeast extract	3.0g
Potassium isothiocyanate	2.25g
LiCl	2.0g
$Na_2HPO_4 \cdot 2H_2O$	0.9g
$NaH_2PO_4 \cdot H_2O$	0.6g
Glycine	0.5g
NaN_3 solution	10.0mL

pH 7.2 ± 0.2 at 25°C

NaN_3 Solution:

Composition per 10.0mL:

NaN_3	0.045g

Preparation of NaN_3 Solution: Add NaN_3 to distilled/deionized water and bring volume to 10.0mL. Mix thoroughly. Filter sterilize.

Preparation of Medium: Add components, except NaN_3 solution, to distilled/deionized water and bring volume to 990.0mL. Mix thoroughly. Adjust pH to 7.2. Gently heat and bring to boiling. Autoclave for 15 min at 15 psi pressure–121°C. Cool to 45°–50°C. Aseptically add sterile NaN_3 solution. Mix thoroughly. Pour into sterile Petri dishes or distribute into sterile tubes.

Use: For the isolation and cultivation of *Staphylococcus* species.

Schmitthenner's Agar
See: StA

Schneider's *Drosophila* Medium

Composition per liter:

$MgSO_4 \cdot 7H_2O$	3.7g
NaCl	2.1g
Yeast extract	2.0g
Trehalose	2.0g
D-Glucose	2.0g
L-Glutamine	1.8g
L-Lysine·HCl	1.7g
L-Proline	1.7g
KCl	1.6g
$Na_2HPO_4 \cdot 7H_2O$	1.3g
L-Glutamic acid	0.8g
L-Methionine	0.8g
$CaCl_2$, anhydrous	0.6g
KH_2PO_4	0.5g
β-Alanine	0.5g
L-Tyrosine	0.5g
L-Arginine	0.4g
L-Aspartic acid	0.4g
L-Histidine	0.4g
L-Threonine	0.4g
$NaHCO_3$	0.4g
Glycine	0.3g
L-Serine	0.3g
L-Valine	0.3g
L-Isoleucine	0.2g
L-Leucine	0.2g
L-Phenylalanine	0.2g
α-Ketoglutaric acid	0.2g
Fumaric acid	0.1g
Malic acid	0.1g
Succinic acid	0.1g
L-Cystine	0.1g
L-Tryptophan	0.1g
L-Cysteine	0.06g

Preparation of Medium: Add components to distilled/deionized water and bring volume to 1.0L. Mix thoroughly. Filter sterilize.

Use: For the cultivation of *Drosophila* and other insect species. Also used as a chemically defined supplement for the cultivation of fastidious microorganisms.

Schuberts Arginine Broth

Composition per liter:

Casein enzymic hydrolysate	17.0g
L-Arginine monohydrochloride	10.0g
NaCl	5.0g
Papaic digest of soabean meal	3.0g
D-Glucose	0.5g
Cresol Red	0.01g

Bromothymol Blue 0.0075g
Brilliant Green 0.00038g

pH 7.0 ± 0.2 at 25°C

Source: This medium is available from HiMedia.

Preparation of Medium: Add components to distilled/deionized water and bring volume to 1.0L. Mix thoroughly. Gently heat and bring to boiling. Distribute into tubes or flasks. Autoclave for 15 min at 15 psi pressure–121°C. pH 7.2 ± 0.2 at 25°C

Use: For the isolation of chlorine-damaged *Pseudomonas aeruginosa* from swimming pool water.

Schuster's Axenic *Naegleria* Medium

Composition per liter:

Peptone 2.5g
Yeast extract 2.5g
Fetal bovine serum, inactivated 100.0mL
Liver desiccate solution 100.0mL
Na_2HPO_4 (0.05*M*) 60.0mL
KH_2PO_4 (0.05*M*) 40.0mL

Liver Desiccate Solution:
Composition per 100.0mL:

Liver, desiccated 5.0g

Preparation of Liver Desiccate Solution: Add desiccated liver to distilled/deionized water and bring volume to 100.0mL. Mix thoroughly. Filter through Whatman no. 1 filter paper to remove sediment. Autoclave filtrate for 15 min at 15 psi pressure–121°C.

Preparation of Medium: Add components, except fetal bovine serum, to distilled/deionized water and bring volume to 900.0mL. Mix thoroughly. Autoclave for 15 min at 15 psi pressure–121°C. Immediately prior to use, aseptically add 100.0mL of sterile fetal bovine serum. Mix thoroughly. Aseptically distribute into sterile tubes or flasks.

Use: For the cultivation of *Naegleria gruberi.*

Schwartz Differential HiVeg Medium

Composition per liter:

Agar 20.0g
Glucose 10.0g
Plant peptone 5.0g
Malt extract 3.0g
Yeast extract 3.0g
Na_2SO_3 .... 2.92g
Basic Fuchsin 0.47g
Dextrin 0.11g

pH 6.9 ± 0.2 at 25°C

Source: This medium is available as a premixed powder from HiMedia.

Preparation of Medium: Add components to distilled/deionized water and bring volume to 1.0L. Mix thoroughly. Gently heat and bring to boiling. Do not autoclave. Cool to 45°C. Pour into sterile Petri dishes.

Use: For the differentiation of brewing yeasts from wild yeasts.

SCY Medium

Composition per liter:

Sucrose 1.0g
Pancreatic digest of casein 0.9g
NaCl 0.05g
Papaic digest of soybean meal 0.03g
K_2HPO_4 .... 0.025g
Vitamin solution 10.0mL

pH 7.1 ± 0.2 at 25°C

Vitamin Solution:
Composition per 10.0mL:

Thiamine 0.4mg
Vitamin B_{12} 0.01mg

Preparation of Vitamin Solution: Add components to distilled/deionized water and bring volume to 10.0mL. Mix thoroughly. Filter sterilize.

Preparation of Medium: Add components, except vitamin solution, to distilled/deionized water and bring volume to 990.0mL. Mix thoroughly. Adjust pH to 7.1. Autoclave for 15 min at 15 psi pressure–121°C. Aseptically add 10.0mL of sterile vitamin solution. Mix thoroughly. Aseptically distribute into sterile tubes or flasks.

Use: For the cultivation and maintenance of *Flavobacterium* species.

SCY Medium (Maintenance SCY Medium)

Composition per liter:

Agar 10.0g
Sucrose 1.0g
Pancreatic digest of casein 0.92g
Yeast extract 0.25g
NaCl 0.05g
Papaic digest of soybean meal 0.03g
K_2HPO_4 .... 0.025g
Thiamine 0.4mg
Cyanocobalamin 0.01mg

pH 7.3 ± 0.2 at 25°C

Preparation of Medium: Add components to distilled/deionized water and bring volume to 1.0L. Mix thoroughly. Filter sterilize.

Use: For the cultivation and maintenance of iron and sulfur bacteria.

SCY Medium (Maintenance SCY Medium)

Composition per liter:

Solution A 1.0L
Solution B 200.0mL

Solution A:
Composition per liter:

Agar 10.0g
Pancreatic digest of casein 0.92g
NaCl 0.05g
Papaic digest of soybean meal 0.03g
K_2HPO_4 .... 0.025g

pH 7.0 ± 0.2 at 25°C

Preparation of Solution A: Add components to distilled/deionized water and bring volume to 1.0L. Mix thoroughly. Gently heat and bring to boiling. Distribute into tubes in 10.0mL volumes. Autoclave for 15 min at 15 psi pressure–121°C. Allow tubes to cool in a slanted position.

Solution B:
Composition per 200.0mL:

Sucrose 2.0g
Yeast extract 0.5g

Thiamine 0.8mg
Vitamin B_{12} 0.02mg

pH 8.5 ± 0.2 at 25°C

Preparation of Solution B: Add components to slightly alkaline tap water, pH 8.5, and bring volume to 200.0mL. Mix thoroughly. Filter sterilize.

Preparation of Medium: Inoculate bacteria onto prepared slants of solution A. After inoculation of tubes, aseptically add 2.0mL of sterile solution B on top of each slant.

Use: For the cultivation and maintenance of iron bacteria. For the cultivation and maintenance of *Haliscomenobacter hydrossis*.

SCZA
See: **Saline Czapek Agar**

SD Medium
See: ***Serratia* Differential Medium**

SDO Medium (DSMZ Medium 1151)

Composition per liter:

NaCl 100.0g
$MgSO_4 \cdot 7H_2O$ 3.45g
$MgCl_2 \cdot 6H_2O$ 3.00g
$NaHCO_3$ 1.0g
NH_4Cl 0.25g
$CaSO_4 \cdot 2H_2O$ 0.25g
KH_2PO_4 .... 0.14g
$CaCl_2 \cdot 2H_2O$ 0.14g
Yeast extract 0.1g
Trace elements solution SL-6 .... 1.0mL
Vitamin solution 1.0mL

pH 6.0 ± 0.2 at 25°C

Vitamin Solution:
Composition per liter:

Pyridoxine-HCl 10.0mg
Thiamine-HCl·$2H_2O$ 5.0mg
Riboflavin 5.0mg
Nicotinic acid 5.0mg
D-Ca-pantothenate 5.0mg
p-Aminobenzoic acid 5.0mg
Lipoic acid 5.0mg
Biotin 2.0mg
Folic acid 2.0mg
Vitamin B_{12} 0.1mg

Preparation of Vitamin Solution: Add components to distilled/deionized water and bring volume to 1.0L. Mix thoroughly. Filter sterilize.

Trace Elements Solution SL-6:
Composition per liter:

$MnCl_2 \cdot 4H_2O$ 0.5g
H_3BO_3 .... 0.3g
$CoCl_2 \cdot 6H_2O$ 0.2g
$ZnSO_4 \cdot 7H_2O$ 0.1g
$Na_2MoO_4 \cdot 2H_2O$ 0.03g
$NiCl_2 \cdot 6H_2O$ 0.02g
$CuCl_2 \cdot 2H_2O$ 0.01g

Preparation of Trace Elements Solution SL-6: Add components to distilled/deionized water and bring volume to 1.0L. Mix thoroughly. Autoclave for 15 min at 15 psi pressure–121°C.

Preparation of Medium: Add components to distilled/deionized water and bring volume to 1.0L. Mix thoroughly. Adjust pH to 6.0. Autoclave for 15 min at 15 psi pressure–121°C.

Use: For the cultivation of *Salinisphaera shabanensis*.

SDS HiVeg Agar with Polymyxin B (Sodium Dodecyl Sulfate Polymyxin Sucrose HiVeg Agar)

Composition per liter:

NaCl 20.0g
Agar 15.0g
Sucrose 15.0g
Plant peptone No. 3 .... 10.0g
Plant extract 5.0g
Sodium dodecyl sulfate 1.0g
Bromthymol Blue 0.04g
Cresol Red 0.04g
Polymyxin Bsolution 1.0mL

pH 7.6 ± 0.2 at 25°C

Source: This medium, without polymyxin B solution, is available as a premixed powder from HiMedia.

Polymyxin B Solution:
Composition per 1.0mL:

Polymyxin B 1.0mg

Preparation of Polymyxin B Solution: Add polymyxin B to distilled/deionized water and bring volume to 1.0mL. Mix thoroughly. Filter sterilize.

Preparation of Medium: Add components, except polymyxin B solution, to distilled/deionized water and bring volume to 1.0L. Mix thoroughly. Gently heat and bring to boiling. Distribute into tubes or flasks. Autoclave for 15 min at 15 psi pressure–121°C. Cool to 45°–50°C. Aseptically add 1.0mL of polymyxin B solution. Mix thoroughly. Pour into sterile Petri dishes.

Use: For the enrichment, isolation, and enumeration of *Vibrio vulnificus* from seafood.

Seawater Agar (SWA)

Composition per liter:

Agar 15.0g
Peptone 5.0g
Yeast extract 5.0g
Beef extract 3.0g
Seawater, synthetic 1.0L

pH 7.5 ± 0.2 at 25°C

Seawater, Synthetic:
Composition per liter:

NaCl 27.0g
$MgSO_4 \cdot 7H_2O$ 7.0g
Tris(hydroxymethyl)aminomethane buffer 2.0g
KCl 0.6g
$CaCl_2$ 0.3g

Preparation of Seawater, Synthetic: Add components to distilled/deionized water and bring volume to 1.0L. Mix thoroughly.

Preparation of Medium: Combine components. Mix thoroughly. Gently heat and bring to boiling. Distribute into tubes or flasks. Autoclave for 15 min at 15 psi pressure–121°C. Pour into sterile Petri dishes or leave in tubes.

Use: For the isolation and cultivation of halophilic microorganisms from foods, such as *Pseudomonas* species and *Vibrio* species from fish.

Seawater Agar

Composition per liter:

Agar	20.0g
Beef extract	10.0g
Peptone	10.0g
Seawater	750.0mL

pH 7.2 ± 0.2 at 25°C

Preparation of Medium: Add components to tap water and bring volume to 1.0L. Mix thoroughly. Gently heat and bring to boiling. Distribute into tubes or flasks. Autoclave for 15 min at 15 psi pressure–121°C. Pour into sterile Petri dishes or leave in tubes.

Use: For the selective isolation and cultivation of *Planococcus* species.

Seawater Agar (SWA)

Composition per liter:

Agar	15.0g
Peptone	5.0g
Yeast extract	5.0g
Beef extract	3.0g
Seawater, synthetic	1.0L

pH 7.5 ± 0.2 at 25°C

Seawater, Synthetic:

Composition per liter:

NaCl	24.0g
$MgSO_4 \cdot 7H_2O$	7.0g
$MgCl_2 \cdot 6H_2O$	5.3g
KCl	0.7g
$CaCl_2$	0.1g

Preparation of Seawater, Synthetic: Add components to distilled/deionized water and bring volume to 1.0L. Mix thoroughly. Adjust pH to 7.5.

Preparation of Medium: Combine components. Mix thoroughly. Gently heat and bring to boiling. Distribute into tubes or flasks. Autoclave for 15 min at 15 psi pressure–121°C. Pour into sterile Petri dishes or leave in tubes.

Use: For the isolation and cultivation of halophilic microorganisms from foods, such as *Pseudomonas* species and *Vibrio* species from fish.

Seawater Agar

Composition per liter:

Agar	20.0g
Beef extract	10.0g
Peptone	10.0g
Artificial seawater	750.0mL

Artificial Seawater

Composition per liter:

NaCl	28.13g
$MgSO_4 \cdot 7H_2O$	3.5g
$MgCl_2$	2.55g
$CaCl_2$	1.2g
KCl	0.77g
$NaHCO_3$	0.11g

pH 7.3 ± 0.2 at 25°C

Preparation of Artificial Seawater: Natural seawater is stored in the dark for at least 3 weeks to "age." If natural seawater is not available, use artificial seawater. To prepare artificial seawater, add components to distilled/deionized water and bring volume to 1.0L. Mix thoroughly.

Preparation of Medium: Add beef extract and peptone to distilled/deionized water and bring volume to 250.0mL. Mix thoroughly. Adjust pH to 7.8. Gently heat and bring to boiling. Boil for 10 min. Add 750.0mL of natural or artificial seawater. Mix thoroughly. Adjust pH to 7.3. Add 20.0g of agar. Mix thoroughly. Gently heat and bring to boiling. Distribute into tubes or flasks. Autoclave for 15 min at 15 psi pressure–121°C. Pour into sterile Petri dishes or leave in tubes.

Use: For the cultivation and maintenance of *Alteromonas rubra, Brevibacterium stationis, Chromohalobacter marismortui, Flectobacillus marinus, Marinococcus albus, Marinococcus halophilus, Pasteurella piscicida, Photobacterium phosphoreum, Planococcus citreus, Planococcus kocurii, Vibrio adaptatus, Vibrio campbellii, Vibrio costicola, Vibrio harveyi, Vibrio mediterranei, Vibrio natriegens,* and other *Vibrio* species.

Seawater Agar with Fetal Calf Serum

Composition per liter:

Agar	20.0g
Beef extract	10.0g
Peptone	10.0g
Seawater	750.0mL
Fetal calf serum	100.0mL

pH 7.2 ± 0.2 at 25°C

Preparation of Medium: Add components, except fetal calf serum, to tap water and bring volume to 900.0mL. Mix thoroughly. Gently heat and bring to boiling. Autoclave for 15 min at 15 psi pressure–121°C. Cool to 50°–55°C. Aseptically add 100.0mL of sterile fetal calf serum warmed to 50°–55°C. Mix thoroughly. Pour into sterile Petri dishes or distribute into sterile tubes.

Use: For the cultivation of *Aeromonas* species and *Vibrio salmonicida*.

Seawater Agar with Horse Blood

Composition per liter:

Agar	20.0g
Beef extract	10.0g
Peptone	10.0g
Seawater	750.0mL
Horse blood	100.0mL

pH 7.2 ± 0.2 at 25°C

Preparation of Medium: Add components, except horse blood, to tap water and bring volume to 900.0mL. Mix thoroughly. Gently heat and bring to boiling. Autoclave for 15 min at 15 psi pressure–121°C. Cool to 50°–55°C. Aseptically add 100.0mL of sterile horse blood warmed to 50°–55°C. Mix thoroughly. Pour into sterile Petri dishes or distribute into sterile tubes.

Use: For the cultivation of *Aeromonas* species.

Seawater Agar Medium

Composition per liter:

Agar	15.0g
Beef extract	10.0g
Peptone	10.0g
Seawater, aged	750.0mL

pH 7.2–7.3 at 25°C

Preparation of Medium: Add components to distilled/deionized water and bring volume to 1.0L. Mix thoroughly. Gently heat and bring to boiling. Distribute into tubes or flasks. Autoclave for 15 min at 15 psi pressure–121°C. Pour into sterile Petri dishes or leave in tubes.

Use: For the isolation and cultivation of marine *Flavobacterium* species.

Seawater Agar Modified (DSMZ Medium 917)

Composition per liter:

NaCl	17.7g
Agar	15.0g
$MgSO_4 \cdot 7H_2O$	4.46g
$MgCl_2 \cdot 6H_2O$	3.4g
Peptone	2.5g
Hexadecane	2.0g
Yeast extract	1.5g
KCl	0.48g
Calcium chloride solution	10.0mL

pH 7.2 ± 0.2 at 25°C

Calcium Chloride Solution:

Composition per 10.0mL:

$CaCl_2 \cdot 2H_2O$	0.98g

Preparation of Calcium Chloride Solution: Add $CaCl_2 \cdot 2H_2O$ to distilled/deionized water and bring volume to 10.0mL. Mix thoroughly. Autoclave for 15 min at 15 psi pressure–121°C.

Preparation of Medium: Add components, except agar and calcium chloride solution, to distilled/deionized water and bring volume to 990.0mL. Adjust pH to 7.2. Mix thoroughly. Add 15.0g agar. Gently heat and bring to boiling. Autoclave for 15 min at 15 psi pressure–121°C. Cool to 50°C. Aseptically add 10.0mL sterile calcium chloride solution. Mix thoroughly. Pour into sterile Petri dishes or distribute to sterile tubes.

Use: For the cultivation and maintenance of *Muricauda ruestringensis.*

Seawater Agar with 1% Serum

Composition per liter:

Agar	12.0g
Seawater	990.0mL
Horse serum	10.0mL

Preparation of Medium: Add agar to 990.0mL of seawater. Mix thoroughly. Gently heat and bring to boiling. Autoclave for 15 min at 15 psi pressure–121°C. Cool to 45°–50°C. Aseptically add 10.0mL of sterile horse serum. Mix thoroughly. Pour into sterile Petri dishes or distribute into sterile tubes.

Use: For the cultivation and maintenance of *Basipetospora halophila, Halosphaeria retorquens, Thraustochytrium striatum,* and *Lagenidium callinectes.*

Seawater Basal Medium

Composition per liter:

$NH_4 \cdot Cl$	10.0g
Lactate	2.0g
Tris (hydroxymethyl) aminomethane	0.6g
$K_2HPO_4 \cdot 3H_2O$	75.0mg
$FeSO_4 \cdot 7H_2O$	29.0mg
Artificial seawater	500.0mL

Artificial Seawater:

Composition per liter:

$MgSO_4 \cdot 7H_2O$	24.65g
NaCl	23.37g
$CaCl_2 \cdot 2H_2O$	2.94g
KCl	1.49g

pH 7.3 ± 0.2 at 25°C

Preparation of Medium: Add components to distilled/deionized water and bring volume to 1.0L. Mix thoroughly. Distribute into tubes or flasks. Autoclave for 15 min at 15 psi pressure–121°C.

Use: For the cultivation of marine bacteria.

Seawater Complete Medium

Composition per liter:

Pancreatic digest of casein	5.0g
Yeast extract	3.0g
Seawater	750.0mL
Glycerol	3.0mL

Preparation of Medium: Add components to distilled/deionized water and bring volume to 1.0L. Mix thoroughly. Distribute into tubes or flasks. Autoclave for 15 min at 15 psi pressure–121°C.

Use: For the cultivation and maintenance of *Vibrio fischeri.*

Seawater Lemco Agar

Composition per liter:

Agar	15.0g
Beef extract	10.0g
Peptone	10.0g
Seawater, filtered aged	750.0mL

pH 7.3 ± 0.2 at 25°C

Preparation of Medium: Add components, except agar, to distilled/deionized water and bring volume to 1.0L. Mix thoroughly. Adjust pH to 7.8. Gently heat and bring to boiling. Boil for 3–5 min. Filter through Whatman filter paper. Adjust pH to 7.3. Add agar. Mix thoroughly. Gently heat and bring to boiling. Distribute into tubes or flasks. Autoclave for 15 min at 15 psi pressure–121°C. Pour into sterile Petri dishes or leave in tubes.

Use: For the cultivation of *Halococcus nondenitrificans.*

Seawater Lemco Broth

Composition per liter:

Beef extract	10.0g
Peptone	10.0g
Seawater, filtered aged	750.0mL

pH 7.3 ± 0.2 at 25°C

Preparation of Medium: Add components to distilled/deionized water and bring volume to 1.0L. Mix thoroughly. Adjust pH to 7.8. Gently heat and bring to boiling. Boil for 3–5 min. Filter through Whatman filter paper. Adjust pH to 7.3. Distribute into tubes or flasks. Autoclave for 15 min at 15 psi pressure–121°C.

Use: For the cultivation of *Halococcus nondenitrificans.*

Seawater Medium

Composition per liter:

Agar	15.0g
Peptone	5.0g
Beef extract	2.0g
KNO_3	0.5g
Seawater, aged	1.0L

pH 7.8 ± 0.2 at 25°C

Preparation of Medium: Combine components. Mix thoroughly. Gently heat and bring to boiling. Distribute into tubes or flasks. Autoclave for 15 min at 15 psi pressure–121°C. Pour into sterile Petri dishes or leave in tubes.

Use: For the cultivation of halophilic bacteria.

Seawater Medium

Composition per liter:

Agar	15.0g
Peptone	5.0g
Yeast extract	1.0g
$FeSO_4$	0.2g
Seawater, aged	1.0L

Preparation of Medium: Add components to 1.0L of seawater. Mix thoroughly. Gently heat and bring to boiling. Distribute into tubes or flasks. Autoclave for 15 min at 15 psi pressure–121°C. Pour into sterile Petri dishes or leave in tubes.

Use: For the cultivation and maintenance of *Cyclobacterium marinus.*

Seawater 802 Medium

Composition per liter:

Solution A	500.0mL
Solution B	500.0mL

Solution A:

Composition per 500.0mL:

NaCl	27.5g
$MgCl_2 \cdot 6H_2O$	5.38g
$MgSO_4 \cdot 7H_2O$	6.78g
KCl	0.72g
$NaHCO_3$	0.2g
$CaCL_2 \cdot 2H_2O$	1.4g

Preparation of Solution A: Add components to distilled/deionized water and bring volume to 500.0mL. Mix thoroughly. Filter sterilize.

Solution B:

Composition per liter:

Rye grass cerophyll	5.0g

Preparation of Solution B: Add cerophyll to distilled/deionized water and bring volume to 1.0L. Mix thoroughly. Gently heat and bring to a boil. Boil for 5 min. Filter through Whatman #1 filter paper. Add 0.5g of Na_2HPO_4. Bring volume to 1.0L with distilled/deionized water. Mix thoroughly. Distribute 10.0mL volumes into tubes. Autoclave for 15 min at 15 psi pressure–121°C. Cool to 25°C.

Source: Cerophyll can be obtained from Ward's Natural Science Establishment, Inc. Dairy Goat Nutrition distributes Grass Media Culture, which is equivalent. Cereal Leaf Product from Sigma Chemical is similar to cerophyll.

Preparation of Medium: Aseptically add 500.0mL of sterile solution A and 500.0mL of sterile solution B. Mix thoroughly. Aseptically distribute into sterile tubes or flasks.

Use: For the cultivation of *Amastigomonas bermudensis, Ancyromonas sigmoides, Bodo curvifilus, Bodo saliens, Cafeteria roenbergensis, Cafeteria minuta, Ciliophrys infusionum, Cruzella marina, Glauconema bermudense, Helkesimastix faecicola, Jakoba libera, Massisteria marina, Monosiga brevicollis, Percolomonas cosmopolitus, Pteridomonas danica,* and *Trimyema shoalsia.*

Seawater 802 Medium, Half-Strength

Composition per liter:

Solution A	500.0mL
Solution B	500.0mL

Solution A:

Composition per 500.0mL:

NaCl	27.5g
$MgCl_2 \cdot 6H_2O$	5.38g
$MgSO_4 \cdot 7H_2O$	6.78g
$CaCl_2 \cdot 2H_2O$	1.4g
KCl	0.72g
$NaHCO_3$	0.2g

Preparation of Solution A: Add components to distilled/deionized water and bring volume to 500.0mL. Mix thoroughly. Filter sterilize.

Solution B:

Composition per liter:

Rye grass cerophyll	2.5g
NaH_2PO_4	0.5g

Preparation of Solution B: Add cerophyll to distilled/deionized water and bring volume to 1.0L. Mix thoroughly. Gently heat and bring to a boil. Boil for 5 min. Filter through Whatman #1 filter paper. Add 0.5g of Na_2HPO_4. Bring volume to 1.0L with distilled/deionized water. Mix thoroughly. Distribute 10.0mL volumes into tubes. Autoclave for 15 min at 15 psi pressure–121°C. Cool to 25°C.

Source: Cerophyll can be obtained from Ward's Natural Science Establishment, Inc. Dairy Goat Nutrition distributes Grass Media Culture, which is equivalent. Cereal Leaf Product from Sigma Chemical is similar to cerophyll.

Preparation of Medium: Aseptically add 500.0mL of sterile solution A and 500.0mL of sterile solution B. Mix thoroughly. Aseptically distribute into sterile tubes or flasks.

Use: For the cultivation of *Euplotes harpa.*

Seawater *Nitrosomonas* Medium

Composition per 1003.3mL:

HEPES (*N*-[2-hydroxyethyl]piperazine-*N′*-2-ethanesulfonic acid) buffer	4.76g
$(NH_4)_2SO_4$	1.5g
$MgSO_4 \cdot 7H_2O$	0.2g
$CaCl_2 \cdot 2H_2O$	20.0mg
K_2HPO_4	15.0mg
Artificial seawater	1.0L
K_2CO_3 (5% solution)	2.0mL
Trace elements solution	1.0mL
Phenol Red (0.04% solution)	0.3mL

pH 7.8 ± 0.2 at 25°C

Trace Elements Solution:

Composition per liter:

EDTA	2.06g
$FeSO_4 \cdot 7H_2O$	1.54g
$MnCl_2 \cdot 4H_2O$	0.2g
$Na_2MoO_4 \cdot 2H_2O$	0.1g
$ZnSO_4 \cdot 7H_2O$	0.1g
$CuSO_4 \cdot 5H_2O$	20.0mg
$CoCl_2 \cdot 6H_2O$	2.0mg
HCl, concentrated	83.0mL

Preparation of Trace Elements: Add HCl and EDTA to distilled/deionized water and and bring volume to 900.0mL. Mix thoroughly. Add remaining components. Mix thoroughly.

Artificial Seawater:

Composition per liter:

NaCl	27.5g
$MgSO_4 \cdot 7H_2O$	6.78g
$MgCl_2 \cdot 6H_2O$	5.38g
$CaCl_2 \cdot 2H_2O$	1.4g
KCl	0.72g
$NaHCO_3$	0.2g

Preparation of Artificial Seawater: Add components to distilled/deionized water and bring volume to 1.0L. Mix thoroughly.

Preparation of Medium: Combine components. Mix thoroughly. Adjust pH to 7.8. Distribute into tubes or flasks. Autoclave for 15 min at 15 psi pressure–121°C.

Use: For the cultivation of *Nitrosomonas cryotolerans*.

Seawater Nutrient Agar (SNA) (ATCC Medium 2205)

Composition per liter:

Nutrient agar, 2X	500.0mL
Synthetic seawater, 2X	500.0mL

pH 7.3 ± 0.2 at 25°C

Nutrient Agar, 2X

Composition per 500.0mL:

Agar	15.0g
Peptone	5.0g
NaCl	5.0g
Yeast extract	2.0g
Beef extract	1.0g

Preparation of Nutrient Agar, 2X: Add components to distilled/deionized water and bring volume to 500.0mL. Mix thoroughly. Gently heat and bring to boiling. Autoclave for 15 min at 15 psi pressure–121°C. Cool to 45°–50°C.

Seawater, Synthetic, 2X:

Composition per 500.0mL:

NaCl	24.0g
$MgSO_4 \cdot 7H_2O$	7.0g
$MgCl_2 \cdot 6H_2O$	5.3g
KCl	0.7g
$CaCl_2$	0.1g

Preparation of Seawater, Synthetic, 2X: Add components to distilled/deionized water and bring volume to 500.0mL. Mix thoroughly. Adjust pH to 7.5. Filter sterilize.

Preparation of Medium: Warm synthetic seawater to 50°C. Aseptically combine sterile nutrient agar and sterile synthetic seawater. Mix thoroughly. Pour into sterile Petri dishes or distribute into sterile tubes.

Use: For the isolation and cultivation of marine bacteria.

Seawater *Spirillum* Medium

Composition per liter:

Calcium lactate	10.0g
Peptone	5.0g
Beef extract	3.0g
Yeast extract	3.0g
Seawater	750.0mL

pH 7.0 ± 0.2 at 25°C

Preparation of Medium: Add components to distilled/deionized water and bring volume to 1.0L. Mix thoroughly. Adjust pH to 7.0. Distribute into tubes or flasks. Autoclave for 20 min at 10 psi pressure–115°C. A precipitate will form during autoclaving.

Use: For the cultivation of marine *Spirillum* species.

Seawater with Serum

Composition per liter:

Agar	10.0g
Seawater	1.0L
Bovine serum, sterile	100.0mL

Preparation of Medium: Add agar to 1.0L of seawater. Mix thoroughly. Gently heat and bring to boiling. Autoclave for 15 min at 15 psi pressure–121°C. Cool to 45–50°C. Aseptically add 100.0mL of sterile liquid beef serum. Mix thoroughly. Pour into sterile Petri dishes or distribute into sterile tubes.

Use: For the cultivation and maintenance of *Basipetospora halophila, Halosphaeria retorquens, Lagenidium callinectes,* and *Thraustochytrium striatum.*

Seawater Yeast Extract Agar

Composition per liter:

Marine salts mix	37.9g
Agar	15.0g
Proteose peptone	10.0g
Yeast extract	3.0g

pH 7.2–7.4 at 25°C

Preparation of Medium: Add components to distilled/deionized water and bring volume to 1.0L. Mix thoroughly. Gently heat and bring to boiling. Distribute into tubes or flasks. Autoclave for 15 min at 15 psi pressure–121°C. Pour into sterile Petri dishes or leave in tubes.

Use: For the cultivation and maintenance of *Alteromonas* species, *Caulobacter halobacteroides*, *Caulobacter maris*, *Cytophaga marinoflava*, and *Cytophaga salmonicolor*.

Seawater Yeast Extract Broth, Modified

Composition per liter:

NaCl	23.4g
$MgSO_4 \cdot 7H_2O$	6.9g
Peptone	1.0g
Yeast extract	1.0g
KCl	0.75g

Preparation of Medium: Add components to distilled/deionized water and bring volume to 1.0L. Mix thoroughly. Distribute into tubes or flasks. Autoclave for 15 min at 15 psi pressure–121°C.

Use: For the cultivation and maintenance of *Proteus* species and *Vibrio* species.

Seawater Yeast Extract Peptone Medium

Composition per liter:

Agar	15.0g
Peptone	5.0g
Yeast extract	3.0g
Seawater, aged and filtered	750.0mL

pH 7.3 ± 0.2 at 25°C

Preparation of Medium: Add components, except agar, to distilled/deionized water and bring volume to 1.0L. Mix thoroughly. Adjust pH to 7.8. Gently heat and bring to boiling. Continue boiling for 3–5 min. Filter through Whatman filter paper. Adjust pH to 7.3. Add agar. Gently heat and bring to boiling. Distribute into tubes or flasks. Autoclave for 15 min at 15 psi pressure–121°C. Pour into sterile Petri dishes or leave in tubes.

Use: For the cultivation of *Planococcus kocurii.*

Seawater Yeast Peptone Agar (DSMZ Medium 243)

Composition per liter:

Agar	12.0g
Peptone	5.0g
Yeast extract	3.0g
Artificial seawater	750.0mL

pH 7.3 ± 0.2 at 25°C

Artificial Seawater:

NaCl	28.13g
$MgCl_2 \cdot 6H_2O$	4.8g
$MgSO_4 \cdot 7H_2O$	3.5g
$CaCl_2 \cdot 2H_2O$	1.6g
KCl	0.77g
$NaHCO_3$	0.11g

Preparation of Artificial Sea Water: Add components to distilled/deionized water and bring volume to 1.0L. Mix thoroughly. Filter sterilize.

Preparation of Medium: Add peptone and yeast extract to 250.0mL distilled/deionized water. Mix thoroughly. Adjust pH to 7.8. Boil for 5 min. Filter and readjust the pH to 7.3. Add agar. Gently heat and bring to boiling. Distribute into tubes or flasks. Autoclave for 15 min at 15 psi pressure–121°C. Cool to 45°C–50°C. Add 750.0mL filter-sterilized seawater that has been heated to 50°C. (Note: Natural seawater is stored in the dark for at least 3 weeks to age. If natural seawater is not available use artificial seawater.) Mix thorougly. Pour into sterile Petri dishes or leave in tubes.

Use: For the cultivation and maintenance of *Zobellia uliginosa=Cellulophaga uliginosa=Cytophaga uliginosa, Marinobacterium jannaschii=Oceanospirillum jannaschii, Vibrio harveyi=Lucibacterium harveyi, Halomonas* sp., *Pseudoalteromonas espejiana=Alteromonas espejiana.*

Seawater Yeast Peptone Medium (DSMZ Medium 949)

Composition per liter:

Peptone	5.0g
Yeast extract	3.0g
Seawater, filtered and aged	750.0mL

pH 7.3 ± 0.2 at 25°C

Preparation of Medium: Add components to distilled/deionized water and bring volume to 1.0L. Mix thoroughly. Adjust pH to 7.3. Distribute into tubes or flasks. Autoclave for 15 min at 15 psi pressure–121°C.

Use: For the cultivation of *Shewanella algae.*

Seawater YPG (DSMZ Medium 1169)

Composition per liter:

NaCl	20.0g
Peptone	10.0g
Glucose	10.0g
Yeast extract	5.0g

pH 7.4 ± 0.2 at 25°C

Preparation of Medium: Add components to distilled/deionized water and bring volume to 1.0L. Mix thoroughly. Adjust pH to 7.4. Autoclave for 15 min at 15 psi pressure–121°C.

Use: For the cultivation of *Altererythrobacter indicus.*

Seed Agar

See: **Antibiotic Medium 1**

Selection Agar I

Composition per liter:

Agar	25.0g
Yeast extract	1.0g
$CuSO_4$	0.5mg
Wort solution	1.0L

Wort Solution:

Composition per liter:

Malt extract	110.0g

Preparation of Wort Solution: Add malt extract to distilled/deionized water and bring volume to 1.0L. Mix thoroughly.

Preparation of Medium: Combine components. Mix thoroughly. Gently heat and bring to boiling. Distribute into tubes or flasks. Autoclave for 15 min at 15 psi pressure–121°C. Pour into sterile Petri dishes or leave in tubes.

Use: For the selective cultivation of *Saccharomyces cerevisiae.*

Selection Agar II

Composition per liter:

Sodium acetate	40.0g
Agar	25.0g
$(NH_4)_2SO_4$	5.0g
Maltose	2.5g
KH_2PO_4	1.0g
$MgSO_4 \cdot 7H_2O$	0.5g
NaCl	0.1g
$CaCl_2 \cdot 2H_2O$	0.1g
Inositol	2.0mg
KI	1.0mg
H_3BO_3	0.5mg
$ZnSO_4 \cdot 7H_2O$	0.4mg
$MnSO_4 \cdot 4H_2O$	0.4mg
Thiamine·HCl	0.4mg
Pyroxidine·HCl	0.4mg
Niacin	0.4mg
Calcium pantothenate	0.4mg

p-Aminobenzoic acid	0.2mg
Riboflavin	0.2mg
$FeCl_3$	0.2mg
$Na_2MoO_4 \cdot 4H_2O$	0.2mg
$CuSO_4 \cdot 5H_2O$	0.04mg
Folic acid	2.0μg
Biotin	2.0μg

Preparation of Medium: Add components to distilled/deionized water and bring volume to 1.0L. Mix thoroughly. Gently heat and bring to boiling. Distribute into tubes or flasks. Autoclave for 15 min at 15 psi pressure–121°C. Pour into sterile Petri dishes or leave in tubes.

Use: For the selective cultivation of *Saccharomyces cerevisiae*.

Selective 7H11 Agar
See: **Seven H11 Agar**

Selenite Brilliant Green Enrichment Broth
See: **SBG Enrichment Broth**

Selenite Broth (Selenite Broth, Lactose) (Selenite F Enrichment Medium) (Sodium Biselenite Medium) (Sodium Hydrogen Selenite Medium)

Composition per liter:

Na_2HPO_4	10.0g
Pancreatic digest of casein	5.0g
Lactose	4.0g
$NaHSeO_3 \cdot 5H_2O$	4.0g

pH 7.0 ± 0.2 at 25°C

Source: This medium is available as a premixed powder from BD Diagnostic Systems and a prepared medium from Oxoid Unipath.

Caution: Sodium biselenite is toxic and a potential teratogen and care must be taken to avoid inhalation of the powdered dye, contact with the skin, or ingestion, especially in pregnant laboratory workers.

Preparation of Medium: Add components to distilled/deionized water and bring volume to 1.0L. Mix thoroughly. Gently heat and bring to boiling. Do not autoclave. Distribute into sterile tubes in 10.0mL volumes.

Use: For the isolation and enrichment of *Salmonella* species from clinical specimens and food products.

Selenite Broth Base, Mannitol

Composition per liter:

Na_2HPO_4	10.0g
Peptone	5.0g
Mannitol	4.0g
$NaHSeO_3 \cdot 5H_2O$	4.0g

pH 7.1 ± 0.2 at 25°C

Source: This medium is available as a premixed powder from Oxoid Unipath.

Caution: Sodium selenite is toxic and a potential teratogen and care must be taken to avoid inhalation of the powdered dye, contact with the skin, or ingestion, especially in pregnant laboratory workers.

Preparation of Medium: Add components to distilled/deionized water and bring volume to 1.0L. Mix thoroughly. Gently heat. Do not autoclave. Distribute into sterile tubes in 10.0mL volumes. Sterilize for 10 min at 0 psi pressure–100°C.

Use: For the isolation and cultivation of *Salmonella typhi* and *Salmonella paratyphi*.

Selenite Cystine Broth

Composition per liter:

Na_2HPO_4	10.0g
Pancreatic digest of casein	5.0g
Lactose	4.0g
$Na_2SeO_3 \cdot 5H_2O$	4.0g
L-Cystine	0.02g

pH 7.0 ± 0.2 at 25°C

Source: This medium is available as a premixed powder from BD Diagnostic Systems and Oxoid Unipath.

Caution: Sodium selenite is toxic and a potential teratogen and care must be taken to avoid inhalation of the powdered dye, contact with the skin, or ingestion, especially in pregnant laboratory workers.

Preparation of Medium: Add components to distilled/deionized water and bring volume to 1.0L. Mix thoroughly. Gently heat. Do not autoclave. Distribute into sterile tubes in 10.0mL volumes. Sterilize for 15 min at 0 psi pressure–100°C.

Use: For the isolation and cultivation of *Salmonella* species from feces, dairy products, and other specimens.

Selenite Cystine Broth (BAM M134)

Composition per liter:

Na_2HPO_4	5.5g
Polypeptone	5.0g
KH_2PO_4	4.5g
Lactose	4.0g
$Na_2SeO_3 \cdot 5H_2O$	4.0g
L-Cystine	0.01g

pH 7.0 ± 0.2 at 25°C

Caution: Sodium selenite is toxic and a potential teratogen and care must be taken to avoid inhalation of the powdered dye, contact with the skin, or ingestion, especially in pregnant laboratory workers.

Preparation of Medium: Add components to distilled/deionized water and bring volume to 1.0L. Mix thoroughly. Gently heat. Do not autoclave. Distribute into sterile tubes in 10.0mL volumes. Sterilize for 10 min at 0 psi pressure–100°C in flowing steam.

Use: For the isolation and cultivation of *Salmonella* species from feces, dairy products, and other specimens.

Selenite F Broth

Composition per liter:

KH_2PO_4	7.0g
Pancreatic digest of casein	5.0g
Lactose	4.0g
$Na_2SeO_3 \cdot 5H_2O$	4.0g
Na_2HPO_4	3.0g

pH 7.0 ± 0.2 at 25°C

Source: This medium is available as a premixed powder from BD Diagnostic Systems.

Caution: Sodium selenite is toxic and a potential teratogen and care must be taken to avoid inhalation of the powdered dye, contact with the skin, or ingestion, especially in pregnant laboratory workers.

Preparation of Medium: Add components to distilled/deionized water and bring volume to 1.0L. Mix thoroughly. Gently heat. Do not autoclave. Distribute into sterile tubes in 10.0mL volumes. Sterilize for 30 min at 0 psi pressure–100°C.

Use: For the isolation and cultivation of *Salmonella* species from feces, dairy products, and other specimens.

Selenite F Enrichment Medium

See: **Selenite Broth**

Selenate Reducer Medium (DSMZ Medium 1049)

Composition per 1004mL:

Solution A	940.0mL
Solution B	9.4mL
Solution C	23.5mL
Solution D	12.2mL
Solution E	0.94mL
Solution F (vitamin solution)	4.7mL
Solution G (trace elements solution SL-10)	9.4mL
Solution H	4.7mL

pH 6.9 ± 0.2 at 25°C

Solution A:

Composition per 940.0mL:

NaCl	23.0g
$MgCl_2 \cdot 6H_2O$	3.0g
KCl	1.30g
DL-Na-lactate	1.12g
NH4Cl	0.5g
KH2PO4	0.2g
$CaCl_2 \cdot 2H_2O$	0.1g
Resazurin	0.5mg

Preparation of Solution A: Add components to distilled/deionized water and bring volume to 940.0mL. Mix thoroughly. Distribute to anaerobic cultivation vessels. Gently heat and bring to boiling. Cool to room temperature while sparging with a gas mixture of 80% N_2 + 20% CO_2. Seal vessels. Autoclave for 15 min at 15 psi pressure–121°C. Cool to room temperature.

Solution B:

Composition per 10.0mL:

Na_2SeO_4	0.019g

Preparation of Solution B: Add Na_2SeO_4 to distilled/deionized water and bring volume to 10.0mL. Mix thoroughly. Gently heat and bring to boiling. Cool to room temperature while sparging with a gas mixture of 80% N_2 + 20% CO_2. Autoclave for 15 min at 15 psi pressure–121°C. Cool to room temperature.

Solution C:

Composition per 10.0mL:

$NaHCO_3$	0.5g

Preparation of Solution C: Add $NaHCO_3$ to distilled/deionized water and bring volume to 10.0mL. Mix thoroughly. Gently heat and bring to boiling. Cool to room temperature while sparging with a gas mixture of 80% H_2 + 20% CO_2. Autoclave for 15 min at 15 psi pressure–121°C. Cool to room temperature.

Solution D:

Composition per 10.0mL:

$Na_2S \cdot 9H_2O$	0.3g

Preparation of Solution D: Add $Na_2S \cdot 9H_2O$ to distilled/deionized water and bring volume to 10.0mL. Mix thoroughly. Autoclave under 100% N_2 for 15 min at 15 psi pressure–121°C. Cool to room temperature.

Solution E:

Composition per 10.0mL:

Na_2WO_4	8.0mg
$Na_2S_2O_3$	3.0mg

Preparation of Solution E: Add components to distilled/deionized water and bring volume to 10.0mL. Mix thoroughly. Sparge with 100% N_2. Filter sterilize.

Solution F (Vitamin Solution):

Composition per liter:

Folic acid	20.0g
α-Lipoic acid	50.0mg
p-Aminobenzoic acid	50.0mg
Pantothenic acid	50.0mg
Riboflavin	50.0mg
Thamine·HCl	50.0mg
Vitamin B_{12}	50.0mg
Nicotine amide	25.0mg
Biotin	20.0mg
Nicotinic acid	20.0mg
Pyridoxamine·HCl	10.0mg

Preparation of Solution F (Vitamin Solution): Add components to distilled/deionized water and bring volume to 1.0L. Sparge with 100% N_2. Mix thoroughly. Stir for 2 hr. Filter sterilize.

Solution G (Trace Elements Solution SL-10):

Composition per liter:

$MgSO_4 \cdot 7H_2O$	3.0g
Nitrilotriacetic acid	1.5g
NaCl	1.0g
$MnSO_4 \cdot 2H_2O$	0.5g
$CoSO_4 \cdot 7H_2O$	0.18g
$ZnSO_4 \cdot 7H_2O$	0.18g
$CaCl_2 \cdot 2H_2O$	0.1g
$FeSO_4 \cdot 7H_2O$	0.1g
$NiCl_2 \cdot 6H_2O$	0.025g
$KAl(SO_4)_2 \cdot 12H_2O$	0.02g
H_3BO_3	0.01g
$Na_2MoO_4 \cdot 4H_2O$	0.01g
$CuSO_4 \cdot 5H_2O$	0.01g
$Na_2SeO_3 \cdot 5H_2O$	0.3mg

Preparation of Solution G (Trace Elements Solution SL-10): Add nitrilotriacetic acid to 500.0mL of distilled/deionized water. Dissolve by adjusting pH to 6.5 with KOH. Add remaining components. Add distilled/deionized water to 1.0L. Mix thoroughly. Adjust pH to 7.0. Sparge with 100% N_2. Filter sterilize.

Solution H:

Composition per 10.0mL:

1,4-Naphthaquinone	0.4mg
Hemin	0.1mg

Preparation of Solution H: Add components to distilled/deionized water and bring volume to 10.0mL. Mix thoroughly. Sparge with 100% N_2. Filter sterilize.

Preparation of Medium: Add the following to 10.0mL of solution A: 0.1mL solution B, 0.25mL solution C, 0.13mL solution D, 0.01mL solution E, 0.05mL solution F, 0.1mL solution G, and 0.05mL solution H.

Use: For the cultivation of *Sedimenticola selenatireducens.*

Selenomonas acidaminophila Medium

Composition per liter:

Disodium β-glycerophosphate	19.0g
Beef extract	5.0g
Lactose	5.0g
Papaic digest of soybean meal	5.0g
Sodium glutamate	3.4g
Pancreatic digest of casein	2.5g
Peptic digest of animal tissue	2.5g
Yeast extract	2.5g
Ascorbic acid	0.5g
$MgSO_4 \cdot 7H_2O$	0.25g

pH 7.15 ± 0.05 at 25°C

Preparation of Medium: Add components to distilled/deionized water and bring volume to 1.0L. Mix thoroughly. Distribute into tubes or flasks. Autoclave for 15 min at 15 psi pressure–121°C.

Use: For the cultivation and maintenance of *Selenomonas acidaminophila.*

Selenomonas ruminantium Medium

Composition per liter:

Pancreatic digest of casein	5.0g
Na_2CO_3	4.0g
Sodium acetate	4.0g
Yeast extract	2.0g
Glucose	1.0g
KH_2PO_4	1.0g
L-Cysteine·HCl	0.5g
Resazurin	1.0mg
n-Valeric acid	0.1mL

pH 7.0 ± 0.2 at 25°C

Preparation of Medium: Prepare and dispense medium under 100% CO_2. Add components to distilled/deionized water and bring volume to 1.0L. Mix thoroughly. Sparge with 100% CO_2. Adjust pH to 7.0. Anaerobically distribute into tubes or flasks. Autoclave for 20 min at 15 psi pressure–121°C.

Use: For the cultivation and maintenance of *Selenomonas ruminantium* and *Selenomonas* species.

Selenomonas Selective Medium (SS Medium)

Composition per 100.0mL:

Pancreatic digest of casein	0.5g
Mannitol	0.2g
$FeSO_4 \cdot 7H_2O$	0.1g
Sodium acetate	0.1g
Yeast extract	0.1g
L-Cysteine·HCl	0.08g
Mineral solution S	4.0mL
Sodium carbonate (8% solution)	2.5mL
n-Valeric acid	0.05mL

pH 5.9–6.1 at 25°C

Mineral Solution S:

Composition per liter:

KH_2PO_4	12.0g
NaCl	12.0g
$(NH_4)_2SO_4$	6.0g
$MgSO_4 \cdot 7H_2O$	2.5g
$CaCl_2 \cdot 2H_2O$	1.6g

Preparation of Mineral Solution S: Add components to distilled/deionized water and bring volume to 1.0L. Mix thoroughly.

Preparation of Medium: Add components to distilled/deionized water and bring volume to 100.0mL. Mix thoroughly. Filter sterilize. Aseptically distribute into sterile tubes or flasks.

Use: For the isolation and cultivation of *Selenomonas* species.

Sellers Agar (Sellers Differential Agar)

Composition per 1015.0mL:

Pancreatic digest of gelatin	20.0g
Agar	13.5g
D-Mannitol	2.0g
NaCl	2.0g
$MgSO_4 \cdot 7H_2O$	1.5g
K_2HPO_4	1.0g
L-Arginine	1.0g
$NaNO_3$	1.0g
Yeast extract	1.0g
$NaNO_3$	0.35g
Bromthymol Blue	0.04g
Phenol Red	8.0mg
Glucose solution	15.0mL

pH 6.7 ± 0.2 at 25°C

Source: This medium is available as a premixed powder from BD Diagnostic Systems.

Glucose Solution:

Composition per 10.0mL:

D-Glucose	5.0g

Preparation of Glucose Solution: Add D-glucose to distilled/deionized water and bring volume to 10.0mL. Mix thoroughly. Filter sterilize.

Preparation of Medium: Add components, except glucose solution, to distilled/deionized water and bring volume to 1.0L. Mix thoroughly. Gently heat and bring to boiling. Distribute into tubes in 10.0mL volumes. Autoclave for 15 min at 15 psi pressure–121°C. Allow tubes to cool in a slanted position to form a 3-inch slant with a 1.5-inch butt. Immediately prior to inoculation, aseptically add 0.15mL of sterile glucose solution to each tube. Let the glucose solution run down the side of the tube opposite the slant.

Use: For the cultivation and differentiation of nonfermentative Gram-negative bacilli, especially *Pseudomonas aeruginosa, Herellea vaginicola (Acinetobacter calcoaceticus), Mima polymorpha (Acinetobacter lwoffii), Alcaligenes faecalis,* and *Bacterium anitratum (Acinetobacter calcoaceticus).*

Sellers Differential HiVeg Agar

Composition per liter:

Plant peptone	20.0g
Agar	15.0g
D-Mannitol	2.0g

NaCl	2.0g
$MgSO_4 \cdot 7H_2O$	1.5g
K_2HPO_4	1.0g
L-Arginine	1.0g
Sodium nitrate	1.0g
Yeast extract	1.0g
Sodium nitrite	0.35g
Bromthymol Blue	0.04g
Phenol Red	8.0mg
Glucose solution	15.0mL

pH 6.7 ± 0.2 at 25°C

Source: This medium, without glucose solution, is available as a premixed powder from HiMedia.

Glucose Solution:
Composition per 20.0mL:

Glucose	10.0g

Preparation of Glucose Solution: Add glucose to distilled/deionized water and bring volume to 20.0mL. Mix thoroughly. Filter sterilize.

Preparation of Medium: Add components, except glucose solution, to distilled/deionized water and bring volume to 1.0L. Mix thoroughly. Gently heat and bring to boiling. Distribute into tubes in 10.0mL volumes. Autoclave for 15 min at 15 psi pressure–121°C. Allow tubes to cool in a slanted position to form a 3-inch slant with a 1.5-inch butt. Just before inoculation add 0.15mL sterile glucose solution to each slant.

Use: For the cultivation and differentiation of nonfermentative Gram-negative bacilli.

Semiselective Medium for *Legionella pneumophila*
See: **BMPA-α Medium**

Semisolid *Brucella* Broth

Composition per liter:

Peptone	10.0g
Pancreatic digest of casein	10.0g
NaCl	5.0g
Yeast extract	2.0g
Agar	1.6g
D-Glucose	1.0g

Preparation of Medium: Add components to distilled/deionized water and bring volume to 1.0L. Mix thoroughly. Distribute into tubes or flasks. Autoclave for 15 min at 15 psi pressure–121°C.

Use: For the cultivation of *Campylobacter* species.

Semisolid BSA Tween™ 80 Medium
See: **Bovine Serum Albumin Tween™ 80 Soft Agar**

Semisolid IMRV HiVeg Medium Base with Novobiocin

Composition per liter:

Plant hydrolysate	13.5g
Plant peptone	13.5g
$MnCl_2$	10.91g
Saccharose	7.5g
Agar	2.7g
KH_2PO_4	1.47g
$Na_2S_2O_3$	0.8g
Lactose	0.5g
$FeNH_4(SO_4)_2 \cdot 12H_2O$	0.2g
Bromcresol Purple	0.08g
Malachite Green	0.037g
Novobiocin solution	1.0mL

pH 5.6 ± 0.2 at 25°C

Source: This medium, without novobiocin, is available as a premixed powder from HiMedia.

Novobiocin Solution:
Composition per 10.0mL:

Novobiocin	0.1g

Novobiocin Solution: Add novobiocin to distilled/deionized water and bring volume to 10.0mL. Mix thoroughly. Filter sterilize.

Preparation of Medium: Add components, except novobiocin solutiont, to distilled/deionized water and bring volume to 999.0mL. Mix thoroughly. Gently heat and bring to boiling. Distribute into tubes or flasks. Do not autoclave. Do not overheat. Cool to 50°C. Aseptically add 1.0mL novobiocin solution. Pour into sterile Petri dishes or leave in tubes.

Use: For the isolation of motile *Salmonella* spp. from foods

Semiolid Medium, Modified (BAM M30c)

Composition per liter:

Peptone	10.0g
Lactalbumin hydrolysate	5.0g
Yeast extract	5.0g
NaCl	5.0g
Agar	1.8g
α-Ketoglutaric acid	1.0g
Na-pyruvate	0.5g
$Na_2S_2O_5$	0.5g
NaCO3	0.6g
Na-citrate	0.1g
Hemin	0.01g

pH 7.4 ± 0.2 at 25°C

Preparation of Medium: Add components to distilled/deionized water and bring volume to 1.0L. Mix thoroughly. Gently heat and bring to boiling. Cool to 50°C. Adjust pH to 7.4. Distribute into screw-capped tubes in 10.0mL volumes. Autoclave for 15 min at 15 psi pressure–121°C.

Use: For the storage and preservation of *Campylobacter* spp.

Semisolid Medium, Modified with Cysteine and Neutral Red (BAM M30d)

Composition per liter:

Peptone	10.0g
Lactalbumin hydrolysate	5.0g
Yeast extract	5.0g
NaCl	5.0g
Agar	1.8g
α-Ketoglutaric acid	1.0g
$NaCO_3$	0.6g
Na-pyruvate	0.5g
$Na_2S_2O_5$	0.5g
L-Cysteine·HCl·H_2O	0.2g

Hemin....0.01g
Neutral Red solution....10.0mL

pH 7.4 ± 0.2 at 25°C

Neutral Red Solution:
Composition per 100.0mL:
Neutral Red....0.2g
Ethanol....10.0mL

Preparation of Neutral Red Solution: Add 0.2g Neutral Red to 10.0mL ethanol. Mix thoroughly. Bring volume to 100.0mL with distilled/deionized water. Mix thoroughly.

Preparation of Medium: Add components to distilled/deionized water and bring volume to 1.0L. Mix thoroughly. Gently heat and bring to boiling. Cool to 50°C. Adjust pH to 7.4. Distribute into screw-capped tubes in 10.0mL volumes. Autoclave for 15 min at 15 psi pressure–121°C.

Use: For the biochemical identification of *Campylobacter* spp.

Semisolid Medium, Modified with Nitrate and without Neutral Red (BAM M30d)

Composition per liter:
KNO_3....10.0g
Peptone....10.0g
Lactalbumin hydrolysate....5.0g
Yeast extract....5.0g
NaCl....5.0g
Agar....1.8g
α-Ketoglutaric acid....1.0g
$NaCO_3$....0.6g
Na-pyruvate....0.5g
$Na_2S_2O_5$....0.5g
Hemin....0.01g

pH 7.4 ± 0.2 at 25°C

Preparation of Medium: Add components to distilled/deionized water and bring volume to 1.0L. Mix thoroughly. Gently heat and bring to boiling. Cool to 50°C. Adjust pH to 7.4. Distribute into screw-capped tubes in 10.0mL volumes. Autoclave for 15 min at 15 psi pressure–121°C.

Use: For the biochemical identification of *Campylobacter* spp.

Semisolid Medium, Modified with Sodium Chloride and Neutral Red (BAM M30d)

Composition per liter:
NaCl....35.0g
Peptone....10.0g
Lactalbumin hydrolysate....5.0g
Yeast extract....5.0g
Agar....1.8g
α-Ketoglutaric acid....1.0g
$NaCO_3$....0.6g
Na-pyruvate....0.5g
$Na_2S_2O_5$....0.5g
Hemin....0.01g
Neutral Red solution....10.0mL

pH 7.4 ± 0.2 at 25°C

Neutral Red Solution:
Composition per 100.0mL:
Neutral Red....0.2g
Ethanol....10.0mL

Preparation of Neutral Red Solution: Add 0.2g Neutral Red to 10.0mL ethanol. Mix thoroughly. Bring volume to 100.0mL with distilled/deionized water. Mix thoroughly.

Preparation of Medium: Add components to distilled/deionized water and bring volume to 1.0L. Mix thoroughly. Gently heat and bring to boiling. Cool to 50°C. Adjust pH to 7.4. Distribute into screw-capped tubes in 10.0mL volumes. Autoclave for 15 min at 15 psi pressure–121°C.

Use: For the biochemical identification of *Campylobacter* spp.

Semisolid Medium for Motility

Composition per liter:
Biosate....5.0g
Polypeptone™....5.0g
NaCl....5.0g
Agar....4.0g
Myosate....1.5g
Triphenyltetrazolium chloride solution....2.5mL

pH 6.9–7.1at 25°C

Triphenyltetrazolium Chloride Solution:
Composition per 10.0mL:
Triphenyltetrazolium chloride....0.1g
Ethanol (95% solution)....10.0mL

Preparation of Triphenyltetrazolium Chloride Solution: Add triphenyltetrazolium chloride to 10.0mL of ethanol. Mix thoroughly.

Preparation of Medium: Add components, except triphenyltetrazolium chloride solution, to distilled/deionized water and bring volume to 997.5mL. Mix thoroughly. Gently heat and bring to boiling. Add 2.5mL of triphenyltetrazolium chloride solution. Mix thoroughly. Distribute into tubes in 10.0mL volumes. Autoclave for 15 min at 15 psi pressure–121°C.

Use: For the differentiation of bacteria based on motility.

Semisolid Pectin Agar

Composition per liter:
Pectin....30.0g
Yeast extract....5.0g
Agar....3.0g
Bromthymol Blue (0.1% solution)....1.0mL
$CaCl_2 \cdot 2H_2O$ (10% solution)....0.6mL

pH 7.3 ± 0.2 at 25°C

Preparation of Medium: Add approximately 100.0mL of distilled/deionized water to a 2.0L flask. Place on a magnetic stirrer without heat. Slowly add the $CaCl_2 \cdot 2H_2O$ solution, Bromthymol Blue solution, yeast extract, and pectin while stirring. Mix thoroughly to ensure uniform wetting of the particles. Add agar. Gently heat and bring to almost boiling. Adjust pH to 7.3 with 1*N* NaOH if necessary. Do not overshoot pH above 7.3. Distribute into tubes or flasks. Autoclave for 15 min at 15 psi pressure–121°C.

Use: For the isolation and cultivation of bacteria such as *Erwinia* species and some *Klebsiella* species based on their ability to degrade pectin.

Semisolid RV HiVeg Medium Base with Novobiocin

Composition per liter:

$MnCl_2$, anhydrous	10.93g
NaCl	7.34g
Plant hydrolysate	4.6g
Plant hydrolysate No. 1	4.6g
Agar	2.7g
Malachite Green	0.037g
Novobiocin solution	1.0mL

pH 5.4 ± 0.2 at 25°C

Source: This medium, without novobiocin, is available as a premixed powder from HiMedia.

Novobiocin Solution:

Composition per 10.0mL:

Novobiocin	0.1g

Preparation of Novobiocin Solution: Add novobiocin to distilled/deionized water and bring volume to 10.0mL. Mix thoroughly. Filter sterilize.

Preparation of Medium: Add components, except novobiocin solution, to distilled/deionized water and bring volume to 999.0mL. Mix thoroughly. Gently heat and bring to boiling. Distribute into tubes or flasks. Do not autoclave. Do not overheat. Cool to 50°C. Aseptically add 1.0mL novobiocin solution. Pour into sterile Petri dishes or leave in tubes.

Use: For the isolation of motile *Salmonella* spp. from foods

Sensitest Agar

Composition per liter:

Pancreatic digest of casein	11.0g
Agar	8.0g
Buffer salts	3.3g
Peptone	3.0g
NaCl	3.0g
Glucose	2.0g
Starch	1.0g
Nucleoside bases	0.02g
Thiamine	0.02mg

pH 7.4 ± 0.2 at 25°C

Source: This medium is available as a premixed powder from Oxoid Unipath.

Preparation of Medium: Add components to distilled/deionized water and bring volume to 1.0L. Mix thoroughly. Gently heat and bring to boiling. Distribute into tubes or flasks. Autoclave for 15 min at 15 psi pressure–121°C. Pour into sterile Petri dishes.

Use: For the performance of antibiotic sensitivity assays.

Sensitivity Test HiVeg Medium with Blood Serum

Composition per liter:

Agar	15.0g
Glucose	10.0g
Veal, infusion from	10.0g
Plant peptone No. 3	10.0g
NaCl	3.0g
Na_2HPO_4	2.0g
Sodium acetate	1.0g
Guanine	0.01g
Uracil	0.01g
Xanthine	0.01g
Adenine sulfate	0.01g
Sterile bovine or sheep blood	50.0mL

pH 7.3 ± 0.2 at 25°C

Source: This medium, without blood, is available as a premixed powder from HiMedia.

Preparation of Medium: Add components, except blood, to distilled/deionized water and bring volume to 950.0mL. Mix thoroughly. Gently heat and bring to boiling. Autoclave for 15 min at 15 psi pressure–121°C. Cool to 45°–50°C. Aseptically add 50.0mL of sterile bovine blood or sheep blood. Mix thoroughly. Pour into sterile Petri dishes.

Use: For antimicroibal sensitivity testing of sulfonamides and other antimicrobics.

Serratia Differential Medium (SD Medium)

Composition per 102.0mL:

Solution A	92.0mL
Solution B	10.0mL

pH 6.7 ± 0.2 at 25°C

Solution A:

Composition per 92.0mL:

Yeast extract	1.0g
L-Ornithine	1.0g
NaCl	0.5g
Agar	0.4g
Irgasan inhibitor	1.0mL
Indicator solution	1.0mL

Preparation of Solution A: Add components to distilled/deionized water and bring volume to 92.0mL. Mix thoroughly. Adjust pH to 6.7 with 1*N* NaOH.

Irgasan Inhibitor:

Composition per 100.0mL:

Irgasan-DP-300 (4,2′, 4′-trichloro-2-hydroxydiphenylether)	0.1g
NaOH (1*N* solution)	10.0mL

Preparation of Irgasan Inhibitor: Add irgasan to 10.0mL of NaOH solution. Mix thoroughly. Gently heat to dissolve. Bring volume to 100.0mL with distilled/deionized water.

Indicator Solution:

Composition per 100.0mL:

Bromthymol Blue	0.2g
Phenol Red	0.1g

Preparation of Indicator Solution: Add components to 50.0mL of distilled/deionized water. Mix thoroughly for 1 hr. Bring volume to 100.0mL with distilled/deionized water.

Solution B:

Composition per 10.0mL:

L-Arabinose	1.0g

Preparation of Solution B: Add L-arabinose to distilled/deionized water and bring volume to 10.0mL. Mix thoroughly.

Preparation of Medium: Combine 92.0mL of solution A with 10.0mL of solution B. Mix thoroughly. Distribute into tubes. Autoclave for 15 min at 15 psi pressure–121°C. Allow tubes to cool in an upright position.

Use: For the cultivation and differentiation of *Serratia* species based on the fermentation of arabinose and production of ornithine decarboxylase. *Serratia marcescens* changes the medium to purple throughout the tube. *Serratia liquefaciens* changes the medium to a band of purple at the top of the tube with a green/yellow butt. *Serratia rubidaea* changes the medium to yellow throughout the tube.

Serratia Hd-MHr

Composition per liter:

Agar	15.0g
K_2HPO_4	7.0g
Glucose	5.0g
KH_2PO_4	3.0g
2-Methyl-DL-histidine·2HCl	1.0g
$(NH_4)_2SO_4$	1.0g
$MgSO_4 \cdot 7H_2O$	0.5g

Preparation of Medium: Add components to distilled/deionized water and bring volume to 1.0L. Mix thoroughly. Gently heat and bring to boiling. Distribute into tubes or flasks. Autoclave for 15 min at 15 psi pressure–121°C. Pour into sterile Petri dishes or leave in tubes.

Use: For the cultivation and maintenance of *Serratia marcescens*.

Serratia Medium (ATCC Medium 181)

Composition per liter:

Agar	20.0g
Pancreatic digest of casein	5.0g
Yeast extract	5.0g
Glucose	1.0g
K_2HPO_4	1.0g

pH 7.0 ± 0.2 at 25°C

Preparation of Medium: Add components to distilled/deionized water and bring volume to 1.0L. Mix thoroughly. Gently heat and bring to boiling. Distribute into tubes or flasks. Autoclave for 15 min at 15 psi pressure–121°C. Pour into sterile Petri dishes or leave in tubes.

Use: For the cultivation and maintenance of *Serratia marcescens*.

Serratia Medium (ATCC Medium 1399)

Composition per liter:

Agar	15.0g
K_2HPO_4	7.0g
Glucose	5.0g
KH_2PO_4	3.0g
Casein hydrolysate	1.0g
$(NH_4)_2SO_4$	1.0g
Yeast extract	1.0g
$MgSO_4 \cdot 7H_2O$	0.1g

pH 7.0 ± 0.2 at 25°C

Preparation of Medium: Add components to distilled/deionized water and bring volume to 1.0L. Mix thoroughly. Gently heat and bring to boiling. Distribute into tubes or flasks. Autoclave for 15 min at 15 psi pressure–121°C. Pour into sterile Petri dishes or leave in tubes.

Use: For the cultivation and maintenance of *Serratia marcescens*.

Serum Glucose Agar (Serum Dextrose Agar) (ATCC Medium 287)

Composition per 1060.0mL:

Agar	15.0g
Peptone	10.0g
Beef extract	5.0g
NaCl	5.0g
Serum-glucose solution	60.0mL

pH 7.3 ± 0.2 at 25°C

Serum-Glucose Solution:

Composition per 60.0mL:

D-Glucose	10.0g
Serum (inactivated at 56°C, 30 min)	50.0mL

Preparation of Serum-Glucose Solution: Add glucose to 50.0mL of heat-inactivated serum. Horse serum or ox serum may be used. Mix thoroughly. Filter sterilize.

Preparation of Medium: Add components, except serum-glucose solution, to distilled/deionized water and bring volume to 1.0L. Mix thoroughly. Gently heat and bring to boiling. Autoclave for 15 min at 10 psi pressure–115°C. Cool to 50°C. Aseptically add 60.0mL of sterile serum-glucose solution. Mix thoroughly. Pour into sterile Petri dishes or distribute into sterile tubes. Allow tubes to cool in a slanted position.

Use: For the cultivation and maintenance of *Brucella* species.

Serum Glucose Agar, Farrell Modified

Composition per 1086.9mL:

Agar	15.0g
Peptone	10.0g
Beef extract	5.0g
NaCl	5.0g
Serum-glucose solution	60.0mL
Bacitracin solution	12.5mL
Cycloheximide solution	10.0mL
Nystatin solution	2.0mL
Polymyxin B solution	1.0mL
Nalidixic acid solution	1.0mL
Vancomycin solution	0.4mL

pH 7.3 ± 0.2 at 25°C

Serum-Glucose Solution:

Composition per 60.0mL:

D-Glucose	10.0g
Serum (inactivated at 56°C, 30 min)	50.0mL

Preparation of Serum-Glucose Solution: Add glucose to 50.0mL of heat-inactivated serum. Horse serum or ox serum may be used. Mix thoroughly. Filter sterilize.

Bacitracin Solution:

Composition per 12.5mL:

Bacitracin	25,000U

Preparation of Bacitracin Solution: Add Bacitracin to distilled/deionized water and bring volume to 12.5mL. Mix thoroughly. Filter sterilize.

Cycloheximide Solution:

Composition per 100.0mL:

Cycloheximide	1.0g
Acetone	5.0mL

Preparation of Cycloheximide Solution: Add cycloheximide to 5.0mL of acetone. Mix thoroughly. Bring volume to 100.0mL with distilled/deionized water. Mix thoroughly. Filter sterilize.

Nystatin Solution:

Composition per 5.0mL:

Nystatin 250,000U

Preparation of Nystatin Solution: Add nystatin to distilled/deionized water and bring volume to 5.0mL. Mix thoroughly. Filter sterilize.

Polymyxin B Solution:

Composition per 2.0mL:

Polymyxin B 10,000U

Preparation of Polymyxin B Solution: Add polymyxin B to distilled/deionized water and bring volume to 2.0mL. Mix thoroughly. Filter sterilize.

Nalidixic Acid Solution:

Composition per 2.0mL:

Nalidixic acid 0.1g
NaOH (0.5*N* solution) 2.0mL

Preparation of Nalidixic Acid Solution: Add nalidixic acid to 2.0mL of NaOH solution. Mix thoroughly. Immediately before use, add 1.0mL of this stock solution to 9.0mL of distilled/deionized water. Mix thoroughly. Filter sterilize.

Vancomycin Solution:

Composition per 1.0mL:

Vancomycin 0.05g

Preparation of Vancomycin Solution: Add vancomycin to distilled/deionized water and bring volume to 1.0mL. Mix thoroughly. Filter sterilize.

Preparation of Medium: Add components—except serum-glucose solution, bacitracin solution, cycloheximide solution, nystatin solution, polymyxin B solution, nalidixic acid solution, and vancomycin solution—to distilled/deionized water and bring volume to 1.0L. Mix thoroughly. Gently heat and bring to boiling. Autoclave for 15 min at 10 psi pressure–115°C. Cool to 50°C. Aseptically add 60.0mL of sterile serum-glucose solution, 12.5mL of sterile bacitracin solution, 10.0mL of sterile cycloheximide solution, 2.0mL of sterile nystatin solution, 1.0mL of sterile polymyxin B solution, 1.0mL of sterile nalidixic acid solution, and 0.4mL of sterile vancomycin solution. Mix thoroughly. Pour into sterile Petri dishes or distribute into sterile tubes. Allow tubes to cool in a slanted position.

Use: For the selective isolation and cultivation of *Brucella* species.

Serum Potato Infusion Agar

Composition per 1120.0mL:

Agar 15.0g
Peptone 10.0g
Meat extract 5.0g
NaCl 5.0g
Potato infusion 1.0L
Horse serum, heat inactivated 100.0mL
Glycerol 20.0mL

pH 6.8 ± 0.2 at 25°C

Potato Infusion:

Composition per 10.0mL:

Potatoes 250.0g

Preparation of Potato Infusion: Add peeled, thinly sliced potatoes to 1.0L of distilled/deionized water. Infuse overnight at 60°C. Filter through Whatman #1 filter paper. Bring volume to 1.0L with distilled/deionized water.

Preparation of Medium: Combine components, except horse serum. Mix thoroughly. Gently heat and bring to boiling. Autoclave for 15 min at 15 psi pressure–121°C. Cool to 45°–50°C. Aseptically add 100.0mL of sterile horse serum. Mix thoroughly. Pour into sterile Petri dishes or distribute into sterile tubes.

Use: For the cultivation of *Brucella* species.

Serum Tellurite Agar

Composition per liter:

Agar 20.0g
Pancreatic digest of casein 10.0g
Peptic digest of animal tissue 10.0g
NaCl 5.0g
Glucose 2.0g
Lamb serum 50.0mL
Chapman tellurite solution 10.0mL

pH 7.5 ± 0.2 at 25°C

Source: This medium is available as a premixed powder from BD Diagnostic Systems.

Chapman Tellurite Solution:

Composition per 100.0mL:

K_2TeO_3 .... 1.0g

Preparation of Chapman Tellurite Solution: Add K_2TeO_3 to distilled/deionized water and bring volume to 100.0mL. Mix thoroughly. Filter sterilize.

Preparation of Medium: Add components, except lamb serum and Chapman tellurite solution, to distilled/deionized water and bring volume to 940.0mL. Mix thoroughly. Gently heat and bring to boiling. Autoclave for 15 min at 15 psi pressure–121°C. Cool to 45°–50°C. Aseptically add sterile lamb serum and 10.0mL of sterile Chapman tellurite solution. Mix thoroughly. Pour into sterile Petri dishes or distribute into sterile tubes.

Use: For the isolation and cultivation of *Corynebacterium* species, especially in the laboratory diagnosis of diphtheria.

Seven H11 Agar (Selective 7H11 Agar)

Composition per 1010.0mL:

Agar 13.5g
KH_2PO_4 .... 1.5g
Na_2HPO_4 .... 1.5g
Pancreatic digest of casein 1.0g
NaCl 0.85g
Monosodium glutamate 0.5g
$(NH_4)_2SO_4$ 0.5g
Sodium citrate 0.4g
$MgSO_4 \cdot 7H_2O$ 0.05g
Ferric ammonium citrate 0.04g
$CuSO_4 \cdot 5H_2O$ 1.0mg
Pyridoxine 1.0mg
$ZnSO_4 \cdot 7H_2O$ 1.0mg
Biotin 0.5mg
$CaCl_2 \cdot 2H_2O$ 0.5mg
Malachite Green 0.25mg

Middlebrook OADC enrichment 100.0mL
Antibiotic inhibitor 10.0mL
Glycerol 5.0mL

pH 6.6 ± 0.2 at 25°C

Source: This medium is available as a prepared medium from BD Diagnostic Systems.

Middlebrook OADC Enrichment:

Composition per liter:

Bovine albumin fraction V 5.0g
Glucose 2.0g
NaCl 0.85g
Catalase..3.0mg
Oleic acid 0.06mL

Preparation of Middlebrook OADC Enrichment: Add components to distilled/deionized water and bring volume to 100.0mL. Mix thoroughly. Filter sterilize.

Antibiotic Inhibitor:

Composition per 10.0mL:

Carbenicillin..0.05g
Trimethoprim lactate...0.02g
Amphotericin B..0.01g
Polymyxin B 200,000U

Preparation of Antibiotic Inhibitor: Add components to distilled/deionized water and bring volume to 10.0mL. Mix thoroughly. Filter sterilize.

Preparation of Medium: Add glycerol to 900.0mL of distilled/deionized water. Mix thoroughly. Add remaining components, except Middlebrook OADC enrichment and antibiotic inhibitor. Mix thoroughly. Gently heat. Do not boil. Autoclave for 10 min at 15 psi pressure–121°C. Cool to 50°–55°C. Aseptically add 100.0mL of sterile Middlebrook OADC enrichment and 10.0mL of sterile antibiotic solution. Mix thoroughly. Pour into sterile Petri dishes or distribute into sterile tubes.

Use: For the isolation and cultivation of *Mycobacterium* species from specimens with a mixed flora.

Seven-Hour Fecal Coliform Agar (Seven-Hour FC Agar) (m-Seven-Hour Fecal Coliform Agar)

Composition per liter:

Agar 15.0g
Lactose 10.0g
NaCl 7.5g
D-Mannitol...5.0g
Proteose peptone No. 3 ....5.0g
Yeast extract...3.0g
Bromcresol Purple 0.35g
Phenol Red 0.3g
Sodium lauryl sulfate 0.2g
Sodium deoxycholate...0.1g

pH 7.3 ± 0.1 at 25°C

Preparation of Medium: Add components to distilled/deionized water and bring volume to 1.0L. Mix thoroughly. Gently heat and bring to boiling. Continue boiling for 5 min. Cool to 55°–60°C. Adjust pH to 7.3 with 0.1*N* NaOH. Cool to 45°–50°C. Pour into sterile Petri dishes with tight-fitting lids in 5.0mL volumes. Store at 2°–10°C.

Use: For the rapid estimation of the bacteriological quality of water using the membrane filter method.

SF Broth (*Streptococcus faecalis* Broth)

Composition per liter:

Pancreatic digest of casein..20.0g
Glucose 5.0g
NaCl...5.0g
K_2HPO_4...4.0g
KH_2PO_4...1.5g
NaN_3 ....0.5g
Bromcresol Purple 0.032g

pH 6.9 ± 0.2 at 25°C

Source: This medium is available as a premixed powder from BD Diagnostic Systems.

Preparation of Medium: Add components to distilled/deionized water and bring volume to 1.0L. Mix thoroughly. Distribute into tubes or flasks. Autoclave for 15 min at 15 psi pressure–121°C.

Use: For the cultivation and differentiation of group D enterococci (*Streptococcus faecalis* and *Streptococcus faecium*) from group D nonenterococci and from other *Streptococcus* species. Group D enterococci turn the medium turbid and yellow-brown.

SF HiVeg Broth

Composition per liter:

Plant hydrolysate 20.0g
Glucose 5.0g
NaCl...5.0g
K_2HPO_4...4.0g
KH_2PO_4...1.5g
NaN_3 ....0.5g
Bromcresol Purple 0.032

pH 6.9 ± 0.2 at 25°C

Source: This medium is available as a premixed powder from HiMedia.

Preparation of Medium: Add components to distilled/deionized water and bring volume to 1.0L. Mix thoroughly. Distribute into tubes or flasks. Autoclave for 15 min at 15 psi pressure–121°C.

Use: For the cultivation and differentiation of group D enterococci (*Streptococcus faecalis* and *Streptococcus faecium*) from group D nonenterococci and from other *Streptococcus* species. Group D enterococci turn the medium turbid and yellow-brown.

SF1 Medium

Composition per liter:

NaCl...120.0g
$MgCl_2·6H_2O$ 7.0g
$MgSO_4·7H_2O$..6.0g
KCl...3.8g
Pancreatic digest of casein..2.0g
Yeast extract...2.0g
NH_4Cl 1.0g
$CaCl_2·2H_2O$ 0.5g
L-Cysteine·HCl 0.5g
$K_2HPO_4·3H_2O$ 0.4g
Resazurin 1.0mg
$Na_2SeO_3·5H_2O$..75.0μg
Na_2CO_3 solution 20.0mL

Trimethylamine·HCl solution 20.0mL
$Na_2S·9H_2O$ solution 10.0mL
Trace elements solution SL-10 ....1.0mL
NaOH (10*M* solution) 0.6mL

pH 7.3 ± 0.2 at 25°C

Na_2CO_3 Solution:
Composition per 20.0mL:
Na_2CO_3 ....10.0mg

Preparation of Na_2CO_3 Solution: Add Na_2CO_3 to distilled/deionized water and bring volume to 20.0mL. Mix thoroughly. Sparge under 100% N_2. Autoclave for 15 min at 15 psi pressure–121°C. Store under N_2.

Trimethylamine·HCl Solution:
Composition per 20.0mL:
Trimethylamine·HCl 10.0mg

Preparation of Trimethylamine·HCl Solution: Add trimethylamine·HCl to distilled/deionized water and bring volume to 20.0mL. Mix thoroughly. Sparge under 100% N_2. Autoclave for 15 min at 15 psi pressure–121°C. Store under N_2.

$Na_2S·9H_2O$ Solution:
Composition per 10.0mL:
$Na_2S·9H_2O$ 10.0mg

Preparation of $Na_2S·9H_2O$ Solution: Add $Na_2S·9H_2O$ to distilled/deionized water and bring volume to 10.0mL. Mix thoroughly. Sparge under 100% N_2. Autoclave for 15 min at 15 psi pressure–121°C. Store under N_2.

Trace Elements Solution SL-10:
Composition per liter:
$FeCl_2·4H_2O$ 1.5g
$CoCl_2·6H_2O$ 190.0mg
$MnCl_2·4H_2O$ 100.0mg
$ZnCl_2$ 70.0mg
$Na_2MoO_4·2H_2O$ 36.0mg
$NiCl_2·6H_2O$ 24.0mg
H_3BO_3 ....6.0mg
$CuCl_2·2H_2O$ 2.0mg
HCl (25% solution) 10.0mL

Preparation of Trace Elements Solution SL-10: Add $FeCl_2·4H_2O$ to 10.0mL of HCl solution. Mix thoroughly. Add distilled/deionized water and bring volume to 1.0L. Add remaining components. Mix thoroughly. Sparge with 80% N_2 + 20% CO_2. Autoclave for 15 min at 15 psi pressure–121°C.

Preparation of Medium: Prepare and dispense medium under 80% N_2 + 20% CO_2. Add components, except L-cysteine·HCl, NaOH, Na_2CO_3 solution, trimethylamine·HCl solution, and $Na_2S·9H_2O$ solution, to distilled/deionized water and bring volume to 950.0mL. Mix thoroughly. Gently heat and bring to boiling. Continue boiling for 5 min. Cool to room temperature while sparging with 80% N_2 + 20% CO_2. Add L-cysteine·HCl and NaOH while contiuning to sparge with 80% N_2 + 20% CO_2. Adjust pH to 6.7. Anaerobically distribute into tubes or bottles. Autoclave for 15 min at 15 psi pressure–121°C. Aseptically and anaerobically add 20.0mL of sterile trimethylamine·HCl solution, 20.0mL of sterile Na_2CO_3 solution, and 10.0mL of sterile $Na_2S·9H_2O$ solution per 950.0mL of medium. Check that final pH is 6.7.

Use: For the cultivation and maintenance of *Methanohalophilus* species.

SFP Agar
(Shahidi-Ferguson Perfringens Agar)

Composition per 2020.0mL:
Basal layer 1010.0mL
Cover layer 1010.0mL

Source: This medium is available as a premixed powder from BD Diagnostic Systems and Oxoid Unipath.

Basal Layer:
Composition per 1010.0mL:
Agar 20.0g
Tryptose 15.0g
Papaic digest of soybean meal 5.0g
Yeast extract 5.0g
Ferric ammonium citrate 1.0g
$NaHSO_3$ 1.0g
Egg yolk emulsion, 50% 100.0mL
Antibiotic inhibitor 10.0mL

pH 7.6 ± 0.2 at 25°C

Egg Yolk Emulsion, 50%:
Composition per 100.0mL:
Chicken egg yolks 11
Whole chicken egg 1
NaCl (0.9% solution) 50.0mL

Preparation of Egg Yolk Emulsion, 50%: Soak eggs with 1:100 dilution of saturated mercuric chloride solution for 1 min. Crack eggs and separate yolks from whites. Mix egg yolks with 1 chicken egg. Beat to form emulsion. Measure 50.0mL of egg yolk emulsion and add to 50.0mL of 0.9% NaCl solution. Mix thoroughly. Filter sterilize. Warm to 45°–50°C.

Antibiotic Inhibitor:
Composition per 10.0mL:
Kanamycin 0.012g
Polymyxin B sulfate 30,000U

Preparation of Antibiotic Inhibitor: Add components to distilled/deionized water and bring volume to 10.0mL. Mix thoroughly. Filter sterilize.

Preparation of Basal Layer: Add components—except egg yolk emulsion, 50%, and antibiotic inhibitor—to distilled/deionized water and bring volume to 990.0mL. Mix thoroughly. Gently heat and bring to boiling. Autoclave for 15 min at 15 psi pressure–121°C. Cool to 45°–50°C. Aseptically add sterile egg yolk emulsion, 50%, and antibiotic inhibitor. Mix thoroughly. Pour into sterile Petri dishes in 10.0mL volumes.

Cover Layer:
Composition per 1010.0mL:
Agar 20.0g
Tryptose 15.0g
Papaic digest of soybean meal 5.0g
Yeast extract 5.0g
Ferric ammonium citrate 1.0g
$NaHSO_3$ 1.0g
Antibiotic inhibitor 10.0mL

pH 7.6 ± 0.2 at 25°C

Preparation of Cover Layer: Add components—except antibiotic inhibitor—to distilled/deionized water and bring volume to 1.0L. Mix thoroughly. Gently heat and bring to boiling. Autoclave for 15 min at 15 psi pressure–121°C. Cool to 45°–50°C. Aseptically add sterile antibiotic inhibitor. Mix thoroughly.

Preparation of Medium: Prepare and dispense basal layer into sterile Petri dishes in 10.0mL volumes. Incubate overnight to dry plates and test for sterility. Inoculate plates using 0.1mL volume. Spread inoculum over suface of agar. Aseptically add 10.0mL of cover layer to each plate. Incubate at 37°C under 90% N_2 + 10% CO_2.

Use: For the isolation and enumeration of *Clostridium perfringens* from foods. *Clostridium perfringens* appears as black colonies surrounded by a precipitate.

S.F.P. HiVeg Agar Base with Egg Yolk and Antibiotics

Composition per liter:

Basal layer	500.0mL
Cover layer	500.0mL

Basal Layer:

Composition per liter:

Agar	20.0g
Plant hydrolysate No. 1	15.0g
Yeast extract	5.0g
Papaic digest of soybean meal	5.0g
$NaHSO_3$	1.0g
Ferric ammonium citrate	1.0g
Egg yolk emulsion, 50%	100.0mL
Antibiotic inhibitor	10.0mL

pH 7.6 ± 0.2 at 25°C

Source: This medium, without egg yolk emulsion and antibiotic inhibito, is available as a premixed powder from HiMedia.

Egg Yolk Emulsion, 50%:

Composition per 100.0mL:

Chicken egg yolks	11
Whole chicken egg	1
NaCl (0.9% solution)	50.0mL

Preparation of Egg Yolk Emulsion, 50%: Soak eggs with 1:100 dilution of saturated mercuric chloride solution for 1 min. Crack eggs and separate yolks from whites. Mix egg yolks with 1 chicken egg. Beat to form emulsion. Measure 50.0mL of egg yolk emulsion and add to 50.0mL of 0.9% NaCl solution. Mix thoroughly. Filter sterilize. Warm to 45°–50°C.

Antibiotic Inhibitor:

Composition per 10.0mL:

Kanamycin	0.012g
Polymyxin B sulfate	30,000U

Preparation of Antibiotic Inhibitor: Add components to distilled/deionized water and bring volume to 10.0mL. Mix thoroughly. Filter sterilize.

Preparation of Basal Layer: Add components—except egg yolk emulsion, 50%, and antibiotic inhibitor—to distilled/deionized water and bring volume to 990.0mL. Mix thoroughly. Gently heat and bring to boiling. Autoclave for 15 min at 15 psi pressure–121°C. Cool to 45°–50°C. Aseptically add sterile egg yolk emulsion, 50%, and antibiotic inhibitor. Mix thoroughly. Pour into sterile Petri dishes in 10.0mL volumes.

Cover Layer:

Composition per 1010.0mL:

Agar	20.0g
Tryptose	15.0g
Papaic digest of soybean meal	5.0g
Yeast extract	5.0g
Ferric ammonium citrate	1.0g
$NaHSO_3$	1.0g
Antibiotic inhibitor	10.0mL

pH 7.6 ± 0.2 at 25°C

Antibiotic Inhibitor:

Composition per 10.0mL:

Kanamycin	0.012g
Polymyxin B sulfate	30,000U

Preparation of Antibiotic Inhibitor: Add components to distilled/deionized water and bring volume to 10.0mL. Mix thoroughly. Filter sterilize.

Preparation of Cover Layer: Add components—except antibiotic inhibitor—to distilled/deionized water and bring volume to 1.0L. Mix thoroughly. Gently heat and bring to boiling. Autoclave for 15 min at 15 psi pressure–121°C. Cool to 45°–50°C. Aseptically add sterile antibiotic inhibitor. Mix thoroughly.

Preparation of Medium: Prepare and dispense basal layer into sterile Petri dishes in 10.0mL volumes. Incubate overnight to dry plates and test for sterility. Inoculate plates using 0.1mL volume. Spread inoculum over suface of agar. Aseptically add 10.0mL of cover layer to each plate. Incubate at 37°C under 90% N_2 + 10% CO_2.

Use: For the isolation and enumeration of *Clostridium perfringens* from foods. *Clostridium perfringens* appears as black colonies surrounded by a precipitate.

SG Agar

Composition per liter:

Agar	15.0g
Pancreatic digest of casein	15.0g
$CaCl_2 \cdot 2H_2O$	2.0g
$MgSO_4 \cdot 7H_2O$	1.0g

pH 7.0 ± 0.2 at 25°C

Preparation of Medium: Add components to distilled/deionized water and bring volume to 1.0L. Mix thoroughly. Gently heat and bring to boiling. Distribute into tubes or flasks. Autoclave for 15 min at 15 psi pressure–121°C. Pour into sterile Petri dishes or leave in tubes.

Use: For the cultivation of myxobacteria.

Shapton HiVeg Medium

Composition per liter:

Agar	15.0g
Plant peptone	5.0g
Plant extract	3.0g
Plant hydrolysate	2.5g
Yeast extract	1.0g
Glucose	1.0g
Bromcresol Purple	0.025g

pH 7.4 ± 0.2 at 25°C

Source: This medium is available as a premixed powder from HiMedia.

Preparation of Medium: Add components to distilled/deionized water and bring volume to 1.0L. Mix thoroughly. Gently heat while stirring and bring to boiling. Autoclave for 15 min at 15 psi pressure–121°C. Mix thoroughly. Pour into sterile Petri dishes.

Use: For the cultivation of aciduric and thermophilic flat sour sporeformers. For the enumeration of spores of *Bacillus stearothermophilus* which cause flat sour spoilage in canned foods with pH more than 4.5.

Shapton Medium

Composition per liter:

Agar	15.0g
Peptic digest of animal tissue	5.0g
Beef extract	3.0g
Casein enzymic hydrolysate	2.5g
Glucosec	1.0g
Yeast extract	1.0g
Bromo Cresol Purple	0.025g

pH 6.0 ± 0.2 at 25°C

Source: This medium is available as a premixed powder from HiMedia.

Preparation of Medium: Add components to distilled/deionized water and bring volume to 1.0L. Mix thoroughly. Gently heat while stirring and bring to boiling. Autoclave for 15 min at 15 psi pressure–121°C. Mix thoroughly. Pour into sterile Petri dishes.

Use: For the cultivation of aciduric and thermophilic flat sour sporeformers. For the enumeration of spores of *Bacillus stearothermophilus* which cause flat sour spoilage in canned foods with pH more than 4.5.

Sheep Blood Agar (BAM M135)

Composition per liter:

Proteose peptone	15.0g
Agar	12.0g
Liver digest	2.5g
Yeast extract	5.0g
NaCl	5.0g
Sheep blood, defibrinated	50.0mL

pH 7.3 ± 0.2 at 25°C

Source: This medium is available as a premixed powder from Oxoid Unipath.

Preparation of Medium: Add components, except sheep blood, to distilled/deionized water and bring volume to 950.0mL. Mix thoroughly. Heat with frequent agitation and boil for 1 min to completely dissolve. Autoclave for 15 min at 15 psi pressure–121°C. Cool to 45°–46°C. Aseptically add 50.0mL of sterile, defibrinated sheep blood. Mix thoroughly and pour into sterile Petri dishes.

Use: For the isolation, cultivation, and detection of hemolytic activity of streptococci and other fastidious microorganisms.

Sheep Blood Agar

Composition per liter:

Tryptone	14.0g
Agar	12.0g
NaCl	5.0g
Peptone, neutralized	4.5g
Yeast extract	4.5g
Sheep blood, defibrinated	70.0mL

pH 7.4 ± 0.2 at 25°C

Source: This medium is available as a premixed powder from Oxoid Unipath.

Preparation of Medium: Add components, except sheep blood, to distilled/deionized water and bring volume to 930.0mL. Mix thoroughly. Heat with frequent agitation and boil for 1 min to completely dissolve. Autoclave for 15 min at 15 psi pressure–121°C. Cool to 50°C. Aseptically add 70.0mL of sterile, defibrinated sheep blood. Mix thoroughly and pour into sterile Petri dishes.

Use: For the isolation, cultivation, and detection of hemolytic activity of streptococci and other fastidious microorganisms. Specifically formulated to give maximum recovery and improved hemolytic reactions with sheep blood.

Shepard's Differential Agar

See: **A7 Agar**

Shepard's M10 Medium

See: **Standard Fluid Medium 10B**

Shigella Broth

Composition per liter:

Pancreatic digest of casein	20.0g
NaCl	5.0g
K_2HPO_4	2.0g
KH_2PO_4	2.0g
Glucose	1.0g
Novobiocin solution	11.1mL
Tween™ 80	1.5mL

pH 7.0 ± 0.2 at 25°C

Novobiocin Solution:

Composition per liter:

Novobiocin	0.05g

Preparation of Novobiocin Solution: Add novobiocin to distilled/deionized water and bring volume to 1.0L. Mix thoroughly. Filter sterilize.

Preparation of Medium: Add components, except novobiocin solution, to distilled/deionized water and bring volume to 988.9mL. Mix thoroughly. Gently heat and bring to boiling. Autoclave for 15 min at 15 psi pressure–121°C. Cool to 45°–50°C. Aseptically add sterile novobiocin solution. Mix thoroughly. Aseptically distribute into sterile tubes.

Use: For the isolation and cultivation of *Shigella* species from food.

Shigella HiVeg Broth Base with Novobiocin

Composition per liter:

Plant hydrolysate	20.0g
NaCl	5.0g
K_2HPO_4	2.0g
KH_2PO_4	2.0g
Glucose	1.0g
Polysorbate 80	1.5 ml
Novobiocin solution	11.1mL

pH 7.0 ± 0.2 at 25°C

Source: This medium, without novobiocin, is available as a premixed powder from HiMedia.

Novobiocin Solution:

Composition per liter:

Novobiocin	0.05g

Preparation of Novobiocin Solution: Add novobiocin to distilled/deionized water and bring volume to 1.0L. Mix thoroughly. Filter sterilize.

Preparation of Medium: Add components, except novobiocin solution, to distilled/deionized water and bring volume to 988.9mL. Mix thoroughly. Gently heat and bring to boiling. Autoclave for 15 min at 15 psi pressure–121°C. Cool to 45°–50°C. Aseptically add sterile novobiocin solution. Mix thoroughly. Aseptically distribute into sterile tubes.

Use: For the isolation and cultivation of *Shigella* species from food.

Shiitake Agar

Composition per liter:

Malt extract	20.0g
Agar	15.0g
Yeast extract	1.0g

pH 7.0 ± 0.2 at 25°C

Preparation of Medium: Add components to distilled/deionized water and bring volume to 1.0L. Mix thoroughly. Adjust pH to 7.0. Gently heat and bring to boiling. Distribute into tubes or flasks. Autoclave for 15 min at 15 psi pressure–121°C. Pour into sterile Petri dishes or leave in tubes.

Use: For the cultivation and maintenance of *Lentinula edodes, Ochrobactrum anthropi*, and *Pseudomonas* species.

SI Agar

Composition per liter:

Peptone	15.6g
Agar	12.0g
NaCl	5.6g
Yeast extract	2.8g
D-Glucose	1.0g

pH 7.5 ± 0.2 at 25°C

Preparation of Medium: Add components to distilled/deionized water and bring volume to 1.0L. Mix thoroughly. Gently heat and bring to boiling. Distribute into tubes or flasks. Autoclave for 15 min at 15 psi pressure–121°C. Pour into sterile Petri dishes or leave in tubes.

Use: For the cultivation and maintenance of *Escherichia coli*.

Siderophore Mineral Medium

Composition per liter:

KH_2PO_4	8.2g
NaOH	1.6g
NH_4Cl	1.0g
KCl	0.5g
$CaSO_4 \cdot 2H_2O$	0.5mg
$CuSO_4 \cdot 5H_2O$	0.5mg
$FeCl_3 \cdot 6H_2O$	0.5mg
$ZnSO_4 \cdot 7H_2O$	0.5mg
Deferrioxamine B solution	10.0mL
$MgSO_4 \cdot 7H_2O$ solution	10.0mL
Wolfe's vitamin solution	5.0mL

Deferrioxamine B Solution:

Composition per 10.0mL:

Deferrioxamine B	1.0g

Preparation of Deferrioxamine B Solution: Add deferrioxamine B to distilled/deionized water and bring volume to 10.0mL. Mix thoroughly. Filter sterilize.

$MgSO_4 \cdot 7H_2O$ Solution:

Composition per 10.0mL:

$MgSO_4 \cdot 7H_2O$	0.5g

Preparation of $MgSO_4 \cdot 7H_2O$ Solution: Add $MgSO_4 \cdot 7H_2O$ to distilled/deionized water and bring volume to 10.0mL. Mix thoroughly. Filter sterilize.

Wolfe's Vitamin Solution:

Composition per liter:

Pyridoxine·HCl	0.01g
Thiamine·HCl	5.0mg
Riboflavin	5.0mg
Nicotinic acid	5.0mg
Calcium pantothenate	5.0mg
p-Aminobenzoic acid	5.0mg
Thioctic acid	5.0mg
Biotin	2.0mg
Folic acid	2.0mg
Cyanocobalamin	0.1mg

Preparation of Wolfe's Vitamin Solution: Add components to distilled/deionized water and bring volume to 1.0L. Mix thoroughly. Filter sterilize.

Preparation of Medium: Add components—except deferrioxamine B solution, $MgSO_4 \cdot 7H_2O$ solution, and Wolfe's vitamin solution—to distilled/deionized water and bring volume to 975.0mL. Mix thoroughly. Gently heat and bring to boiling. Autoclave for 15 min at 15 psi pressure–121°C. Cool to 45°–50°C. Aseptically add 10.0mL of sterile deferrioxamine B solution, 10.0mL of sterile $MgSO_4 \cdot 7H_2O$ solution, and 5.0mL of sterile Wolfe's vitamin solution. Mix thoroughly. Pour into sterile Petri dishes or distribute into sterile tubes.

Use: For the cultivation of ATCC strain 49538.

Sierra Medium

Composition per liter:

Agar	15.0g
Peptone	10.0g
NaCl	5.0g
$CaCl_2 \cdot H_2O$	0.1g
Tween™ 80	10.0mL

pH 7.4 ± 0.2 at 25°C

Preparation of Medium: Add components, except Tween™ 80, to distilled/deionized water and bring volume to 990.0mL. Mix thoroughly. Gently heat and bring to boiling. Autoclave for 15 min at 15 psi pressure–121°C. Cool to 45°–50°C. Separately autoclave Tween™ 80 for 15 min at 15 psi pressure–121°C. Cool to 45°–50°C. Aseptically add 10.0mL of sterile Tween™ 80. Mix thoroughly. Pour into sterile Petri dishes.

Use: For the differentiation of bacteria based on lipase activity. Bacteria with lipase activity form colonies surrounded by a white precipitate.

SIM HiVeg Medium

Composition per liter:

Plant peptone	30.0g
Agar	3.0g
Plant extract	3.0g
HiVeg peptonized iron	0.2g
$Na_2S_2O_3$	0.025g

pH 7.3 ± 0.2 at 25°C

Source: This medium is available as a premixed powder from HiMedia.

Preparation of Medium: Add components to distilled/deionized water and bring volume to 1.0L. Mix thoroughly. Gently heat and bring to boiling. Distribute into tubes in 15.0mL volumes. Autoclave for 15 min at 15 psi pressure–121°C. Allow tubes to cool in an upright position.

Use: For the determination of hydrogen sulfide production, indole formation, and motility of enteric bacilli.

SIM Medium

Composition per liter:

Peptone	30.0g
Agar	3.0g
Beef extract	3.0g
Peptonized iron	0.2g
$Na_2S_2O_3 \cdot 5H_2O$	0.025g

pH 7.3 ± 0.2 at 25°C

Source: This medium is available as a premixed powder from BD Diagnostic Systems.

Preparation of Medium: Add components to distilled/deionized water and bring volume to 1.0L. Mix thoroughly. Gently heat and bring to boiling. Distribute into tubes in 15.0mL volumes. Autoclave for 15 min at 15 psi pressure–121°C. Allow tubes to cool in an upright position.

Use: For the differentiation of members of the Enterobacteriaceae based on H_2S production, indole production, and motility.

SIM Medium

Composition per liter:

Pancreatic digest of casein	20.0g
Peptic digest of animal tissue	6.1g
Agar	3.5g
$Fe(NH_4)_2(SO_4)_2 \cdot 6H_2O$	0.2g
$Na_2S_2O_3 \cdot 5H_2O$	0.2g

pH 7.3 ± 0.2 at 25°C

Source: This medium is available as a premixed powder from BD Diagnostic Systems and Oxoid Unipath.

Preparation of Medium: Add components to distilled/deionized water and bring volume to 1.0L. Mix thoroughly. Gently heat and bring to boiling. Distribute into tubes in 15.0mL volumes. Autoclave for 15 min at 15 psi pressure–121°C. Allow tubes to cool in an upright position.

Use: For the differentiation of members of the Enterobacteriaceae based on H_2S production, indole production, and motility.

SIM Motility Medium (BAM M137)

Composition per liter:

Pancreatic digest of casein	20.0g
Peptic digest of animal tissue	6.1g
Agar	3.5g
$Fe(NH_4)_2(SO_4)_2 \cdot 6H_2O$	0.2g
$Na_2S_2O_3 \cdot 5H_2O$	0.2g

pH 7.3 ± 0.2 at 25°C

Source: This medium is available as a premixed powder from BD Diagnostic Systems and Oxoid Unipath.

Preparation of Medium: Add components to distilled/deionized water and bring volume to 1.0L. Mix thoroughly. Gently heat and bring to boiling. Distribute into tubes in 15.0mL volumes. Autoclave for 15 min at 15 psi pressure–121°C. Allow tubes to cool in an upright position.

Use: For the differentiation of members of the Enterobacteriaceae based on H_2S production, indole production, and motility.

Simmons' Citrate Agar (Citrate Agar)

Composition per liter:

Agar	15.0g
NaCl	5.0g
Sodium citrate	2.0g
K_2HPO_4	1.0g
$(NH_4)H_2PO_4$	1.0g
$MgSO_4 \cdot 7H_2O$	0.2g
Bromthymol Blue	0.08g

pH 6.9 ± 0.2 at 25°C

Source: This medium is available as a premixed powder from BD Diagnostic Systems and Oxoid Unipath.

Preparation of Medium: Add components to distilled/deionized water and bring volume to 1.0L. Mix thoroughly. Gently heat while stirring and bring to boiling. Distribute into tubes or flasks. Autoclave for 15 min at 15 psi pressure–121°C. Pour into sterile Petri dishes or leave in tubes.

Use: For the differentiation of Gram-negative bacteria on the basis of citrate utilization. Bacteria that can utilize citrate as sole carbon source turn the medium blue.

Simmons' Citrate Agar, Modified
See: **Acetate Differential Agar**

Simonsiella Agar (LMG Medium 31)

Composition per liter:

Tryptone	17.0g
Agar	15.0g
NaCl	5.0g
Yeast extract	4.0g
Soy peptone	3.0g
K_2HPO_4	2.5g
Bovine serum	100.0mL

pH 7.2 ± 0.2 at 25°C

Preparation of Medium: Add components, except bovine serum, to distilled/deionized water and bring volume to 900.0mL. Mix thoroughly. Gently heat and bring to boiling. Autoclave for 15 min at 15 psi pressure–121°C. Cool to 45°–50°C. Aseptically add 100.0mL sterile bovine serum. Mix thoroughly. Pour into sterile Petri dishes or distribute into sterile tubes.

Use: For the cultivation of *Simonsiella* spp.

Simonsiella Agar

Composition per liter:

Pancreatic digest of casein	17.0g
Agar	15.0g
NaCl	5.0g
Yeast extract	4.0g
Papaic digest of soybean meal	3.0g
K_2HPO_4	2.5g
Horse serum	100.0mL

Preparation of Medium: Add components, except horse serum, to distilled/deionized water and bring volume to 900.0mL. Mix thoroughly. Gently heat and bring to boiling. Autoclave for 15 min at 15 psi pressure–121°C. Cool to 50°–55°C. Aseptically add 100.0mL of sterile horse serum warmed to 50°–55°C. Mix thoroughly. Pour into sterile Petri dishes or distribute into sterile tubes.

Use: For the cultivation and maintenance of *Simonsiella muelleri* and *Simonsiella steedae*.

Simonsiella Broth

Composition per liter:

Pancreatic digest of casein	17.0g
NaCl	5.0g
Yeast extract	4.0g
Papaic digest of soybean meal	3.0g
K_2HPO_4	2.5g
Horse serum	100.0mL

Preparation of Medium: Add components, except horse serum, to distilled/deionized water and bring volume to 900.0mL. Mix thoroughly. Autoclave for 15 min at 15 psi pressure–121°C. Aseptically add 100.0mL of sterile bovine serum. Mix thoroughly. Aseptically distribute into sterile tubes or flasks.

Use: For the cultivation and maintenance of *Simonsiella muelleri* and *Simonsiella steedae*.

Simulated Grape Juice Medium

Composition per liter:

Glucose	16.0g
Tartaric acid	0.5g

pH 6.6 ± 0.2 at 25°C

Source: This medium is available as a premixed powder from HiMedia.

Preparation of Medium: Add components to distilled/deionized water and bring volume to 1.0L. Mix thoroughly. Distribute into tubes or flasks. Autoclave for 15 min at 15 psi pressure–121°C.

Use: For the cultivation of microorganisms associated with winemaking.

Singh's Medium, Modified

Composition per liter:

NaCl	8.75g
Lactalbumin hydrolysate	8.13g
Yeast extract	6.25g
D-Glucose	5.0g
$CaCl_2·2H_2O$	0.25g
KCl	0.25g
$NaH_2PO_4·H_2O$	0.25g
$NaHCO_3$	0.15g
$MgCl_2·6H_2O$	0.13g
Phenol Red	0.01g
Fetal bovine serum (heat inactivated at 56°C, 30 min)	200.0mL

pH 7.0 ± 0.2 at 25°C

Preparation of Medium: Add components to distilled/deionized water and bring volume to 1.0L. Mix thoroughly. Adjust pH to 7.0 with NaOH if necessary. Filter sterilize. Aseptically distribute into sterile tubes or flasks.

Use: For the cultivation and maintenance of *Spiroplasma* species.

Single-Layer Agar

Composition per 1050.0mL:

Tributyrin substrate	50.0g
Basal medium	1.0L

pH 6.8 ± 0.2 at 25°C

Basal Medium:
Composition per liter:

Agar	15.0g
Pancreatic digest of gelatin	5.0g
Beef extract	3.0g

Preparation of Basal Medium: Add components to 1.0L of distilled/deionized water. Autoclave for 15 min at 15 psi pressure–121°C. Cool to 50°C.

Tributyrin Substrate:
Composition:

Tributyrin substrate	50.0g

Preparation of Tributyrin Substrate: Remove free fatty acids in the tributyrin substrate by dissolving 50.0g in 500.0mL of petroleum ether. Pass the solution through an activated alumina column. Remove the petroleum ether by evaporation on a steam table under 100% N_2. Autoclave for 30 min at 15 psi pressure–121°C. Cool to 50°C.

Preparation of Medium: Aseptically combine 1.0L of sterile basal medium with 50.0g of sterile tributyrin substrate in a warm, sterile blender container. Blend for 1 min until homogenized. Rapidly pour into sterile Petri dishes in 7.0mL volumes. Dry the surface of the plates by partially opening the lids in a laminar flow hood for 15 min.

Use: For the isolation, cultivation, and identification of lipolytic microorganisms from food.

Single-Layer Agar

Composition per 1050.0mL:

Fat substrate	50.0g
Basal medium	1.0L

pH 6.8 ± 0.2 at 25°C

Basal Medium:
Composition per liter:

Agar	15.0g
Pancreatic digest of gelatin	5.0g
Beef extract	3.0g
Victoria Blue B solution	200.0mL

Preparation of Basal Medium: Add agar to 800.0mL of distilled/deionized water. Autoclave for 15 min at 15 psi pressure–121°C. Cool to 50°C. Aseptically add 200.0mL of Victoria Blue B solution. Mix thoroughly.

Victoria Blue B Solution:
Composition per 200.0mL:

Victoria Blue B	0.12g

Preparation of Victoria Blue B Solution: Add the Victoria Blue B to 200.0mL of distilled/deionized water. Mix thoroughly. Filter sterilize. Warm to 50°C.

Fat Substrate:
Composition:

Fat substrate	50.0g

Preparation of Fat Substrate: Corn oil, soybean oil, any cooking oil, lard, tallow, or triglycerides that do not contain antioxidants or other inhibitory substances may be used. Remove free fatty acids in the fat substrate by dissolving 50.0g of fat substrate in 500.0mL of petroleum ether. Pass the solution through an activated alumina column. Remove the petroleum ether by evaporation on a steam table under 100% N_2. Autoclave for 30 min at 15 psi pressure–121°C. Cool to 50°C.

Preparation of Medium: Aseptically combine 1.0L of sterile basal medium with 50.0g of sterile fat substrate in a warm, sterile blender container. Blend for 1 min until homogenized. Rapidly pour into sterile Petri dishes in 7.0mL volumes. Dry the surface of the plates by partially opening the lids in a laminar flow hood for 15 min.

Use: For the isolation, cultivation, and identification of lipolytic microorganisms from food.

Singulosphaera Medium (DSMZ Medium 1144)

Composition per liter:

Agar-agar	18.0g
N-acetylglucosamine	1.0g
KH_2PO_4	0.1g
Peptone	0.1g
Yeast extract	0.1g
Hutner's basal salts solution	20.0mL
Ampicillin soluiton	10.0mL

pH 5.8 ± 0.2 at 25°C

Hutner's Basal Salts Solution:
Composition per liter:

$MgSO_4 \cdot 7H_2O$	29.7g
Nitrilotriacetic acid	10.0g
$CaCl_2 \cdot 2H_2O$	3.335g
$FeSO_4 \cdot 7H_2O$	99.0mg
$(NH_4)_6Mo_7O_{24} \cdot 4H_2O$	9.25mg
"Metals 44"	50.0mL

"Metals 44":
Composition per 100.0mL:

$ZnSO_4 \cdot 7H_2O$	1.095g
$FeSO_4 \cdot 7H_2O$	0.5g
Sodium EDTA	0.25g
$MnSO_4 \cdot H2O$	0.154g
$CuSO_4 \cdot 5H_2O$	39.2mg
$Co(NO_3)_2 \cdot 6H_2O$	24.8mg
$Na_2B_4O_7 \cdot 10H_2O$	17.7mg

Preparation of "Metals 44": Add sodium EDTA to distilled/deionized water and bring volume to 90.0mL. Mix thoroughly. Add a few drops of concentrated H_2SO_4 to retard precipitation of heavy metal ions. Add remaining components. Mix thoroughly. Bring volume to 100.0mL with distilled/deionized water.

Preparation of Hutner's Basal Salts Solution: Add nitrilotriacetic acid to 500.0mL of distilled/deionized water. Adjust pH to 6.5 with KOH. Add remaining components. Add distilled/deionized water to 1.0L. Adjust pH to 6.8. Autoclave for 15 min at 15 psi pressure–121°C. Cool to 50°C.

Ampicillin Solution:
Composition per 10.0mL:

Na-ampicillin	0.2g

Preparation of Ampicillin Solution: Add Na-ampicillin to distilled/deionized water and bring volume to 10.0mL. Mix thoroughly. Filter sterilize.

Preparation of Medium: Add components, except Hutner's basal salts solution and ampicillin solution, to distilled/deionized water and bring volume to 970.0L. Mix thoroughly. Gently heat and bring to boiling. Autoclave for 15 min at 15 psi pressure–121°C. Cool to 50°C. Aseptically add Hutner's basal salts solution and ampicillin solution. Mix thoroughly. Adjust pH to 5.8. Pour into sterile Petri dishes or distribute into sterile tubes.

Use: For the cultivation of *Singulosphaera* spp.

Six B Agar (6 B Agar)

Composition per liter:

Glycerol	50.0g
Soluble starch	20.0g
Agar	15.0g
Glucose	10.0g
Yeast extract	5.0g
N-Z amine, type A	5.0g
$CaCO_3$	1.0g

Preparation of Medium: Add components to distilled/deionized water and bring volume to 1.0L. Mix thoroughly. Gently heat and bring to boiling. Distribute into tubes or flasks. Autoclave for 15 min at 15 psi pressure–121°C. Pour into sterile Petri dishes or leave in tubes.

Use: For the cultivation and maintenance of *Geodermatophilus obscurus, Micromonospora* species, *Nocardia carnea, Nocardia otitidiscaviarum,* and *Nocardia seriolae.*

SJ Agar

Composition per liter:

Agar	15.0g
K_2HPO_4	1.0g
KCl	0.5g
$MgSO_4 \cdot 7H_2O$	0.5g
$NaNO_3$	0.5g
$FeSO_4 \cdot 7H_2O$	0.01g
Glucose solution	100.0mL

pH 7.2 ± 0.2 at 25°C

Glucose Solution:
Composition per 100.0mL:

D-Glucose	1.0g

Preparation of Glucose Solution: Add D-glucose to distilled/deionized water and bring volume to 100.0mL. Mix thoroughly. Autoclave for 15 min at 15 psi pressure–121°C. Cool to 25°C.

Preparation of Medium: Add components, except glucose solution, to distilled/deionized water and bring volume to 900.0mL. Mix thoroughly. Gently heat and bring to boiling. Autoclave for 15 min at 15 psi pressure–121°C. Cool to 45°–50°C. Aseptically add sterile glucose solution. Mix thoroughly. Pour into sterile Petri dishes or distribute into sterile tubes.

Use: For the isolation and cultivation of *Cytophaga* species, *Herpetosiphon* species, *Saprospira* species, and *Flexithrix* species.

SK Agar

See: **Schleifer-Krämer Agar**

Skim Milk Acetate Medium

Composition per liter:

Agar	15.0g
Skim milk powder	5.0g

Yeast extract ... 0.5g
Sodium acetate ... 0.2g

Preparation of Medium: Add components to distilled/deionized water and bring volume to 1.0L. Mix thoroughly. Gently heat and bring to boiling. Distribute into tubes or flasks. Autoclave for 15 min at 15 psi pressure–121°C. Pour into sterile Petri dishes or leave in tubes.

Use: For the cultivation and maintenance of *Cytophaga johnsonae.*

Skim Milk Agar (Milk Agar) (ATCC Medium 377)

Composition per liter:
Agar ... 15.0g
Skim milk ... 8.0g

Preparation of Medium: Add components to distilled/deionized water and bring volume to 1.0L. Mix thoroughly. Gently heat and bring to boiling. Distribute into tubes or flasks. Autoclave for 15 min at 15 psi pressure–121°C. Pour into sterile Petri dishes or leave in tubes.

Use: For the cultivation and maintenance of *Herpetosiphon aurantiacus*.

Skim Milk Agar

Composition per 1100.0mL:
Agar ... 15.0g
Pancretic digest of casein ... 5.0g
Yeast extract ... 2.5g
Glucose ... 1.0g
Skim milk solution ... 100.0mL
pH 7.0 ± 0.1 at 25°C

Preparation of Skim Milk Solution: Add skim milk solids to distilled/deionized water and bring volume to 100.0mL. Mix thoroughly. Autoclave for 15 min at 15 psi pressure–121°C. Cool to 45°–50°C.

Preparation of Medium: Add components, except skim milk solution, to distilled/deionized water and bring volume to 1.0L. Mix thoroughly. Gently heat and bring to boiling. Distribute into tubes or flasks. Autoclave for 15 min at 15 psi pressure–121°C. Cool to 45°–50°C. Aseptically add 100.0mL of cooled, sterile skim milk solution. Mix thoroughly. Pour into sterile Petri dishes or aseptically distribute into sterile tubes.

Use: For the cultivation and differentiation of bacteria based on proteolytic activity.

Skim Milk Agar, Half Strength

Composition per liter:
Solution A ... 500.0mL
Solution B ... 500.0mL

Solution A:
Composition per 500.0mL:
Skim milk ... 25.0g

Preparation of Solution A: Add 25.0g of skim milk to distilled/deionized water and bring volume to 500.0mL. Mix thoroughly. Adjust pH to 7.0. Autoclave for 15 min at 15 psi pressure–121°C. Cool to 50°–55°C.

Solution B:
Composition per 500.0mL:
Agar ... 20.0g

Preparation of Solution B: Add 20.0g of agar to distilled/deionized water and bring volume to 500.0mL. Mix thoroughly. Gently heat and bring to boiling. Autoclave for 15 min at 15 psi pressure–121°C. Cool to 50°–55°C.

Preparation of Medium: Aseptically mix 500.0mL of solution A with 500.0mL of solution B. Pour into sterile Petri dishes or leave in tubes.

Use: For the cultivation and maintenance of *Pilimelia anulata.*

Skim Milk Glucose Agar (DSMZ Medium 810)

Composition per liter:
Solution A ... 500.0mL
Solution B ... 500.0mL
pH 6.8 ± 0.2 at 25°C

Solution A:
Composition per 500.0mL:
Skim milk ... 40.0g

Preparation of Solution A: Add skim milk to distilled/deionized water and bring volume to 500.0mL. Mix thoroughly. Autoclave for 15 min at 15 psi pressure–121°C.

Solution B:
Composition per 500.0mL:
Glucose ... 20.0g
Agar ... 20.0g

Preparation of Solution B: Add components to distilled/deionized water and bring volume to 500.0mL. Mix thoroughly. Autoclave for 15 min at 15 psi pressure–121°C.

Preparation of Medium: Aseptically mix 500.0mL sterile solution A and 500.0mL sterile solution B. Mix thoroughly. Adjust pH to 6.8. Aseptically distribute into sterile tubes or flasks.

Use: For the cultivation of *Trichophyton* spp.

Skim Milk HiVeg Agar

Composition per liter:
Skim milk powder ... 28.0g
Agar ... 15.0g
Plant hydrolysate ... 5.0g
Yeast extract ... 2.5g
Glucose ... 1.0g
pH 7.0 ± 0.2 at 25°C

Source: This medium is available as a premixed powder from HiMedia.

Preparation of Medium: Add components to distilled/deionized water and bring volume to 1.0L. Mix thoroughly. Gently heat and bring to boiling. Distribute into tubes or flasks. Autoclave for 15 min at 15 psi pressure–121°C. Pour into sterile Petri dishes or leave in tubes.

Use: For the cultivation and enumeration of microorganisms encountered in the dairy industry.

Skim Milk Horn Meal Mineral Agar

Composition per liter:
Agar ... 15.0g
Bovine horn meal ... 10.0g
$MgSO_4 \cdot 7H_2O$ ... 0.7g
$Ca(NO_3)_2 \cdot 4H_2O$ ... 0.5g

$NaHCO_3$ 0.2g
$FeCl_3$ 5.0mg
K_2HPO_4 .... 5.0mg
Skim milk solution 10.0mL

Skim Milk Solution:
Composition per 10.0mL:
Skim milk, powdered 2.5g

Preparation of Skim Milk Solution: Add powdered skim milk to distilled/deionized water and bring volume to 10.0mL. Mix thoroughly. Autoclave for 15 min at 15 psi pressure–121°C.

Preparation of Medium: Add components, except skim milk solution, to distilled/deionized water and bring volume to 990.0mL. Mix thoroughly. Gently heat and bring to boiling. Autoclave for 15 min at 15 psi pressure–121°C. Cool to 60°C. Aseptically add 10.0mL of sterile skim milk solution. Mix thoroughly. Pour into sterile Petri dishes or distribute into sterile tubes.

Use: For the cultivation and maintenance of *Capnocytophaga gingivalis, Capnocytophaga ochracea,* and *Capnocytophaga sputigena.*

Skirrow *Brucella* Medium

Composition per liter:
Blood agar base No. 2 .... 940.0mL
Horse blood, lysed defibrinated 50.0mL
Antibiotic solution 10.0mL

pH 7.4 ± 0.2 at 25°C

Blood Agar Base No. 2
Composition per 940.0mL:
Proteose peptone 15.0g
Agar 12.0g
NaCl 5.0g
Yeast extract 5.0g
Liver digest 2.5g

pH 7.4 ± 0.2 at 25°C

Preparation of Blood Agar Base No. 2: Add components to distilled/deionized water and bring volume to 940.0mL. Mix thoroughly. Gently heat while stirring and bring to boiling. Autoclave for 15 min at 15 psi pressure–121°C. Cool to 45°–50°C.

Antibiotic Solution:
Composition per 10.0mL:
Vancomycin 0.01g
Trimethoprim 5.0mg
Polymyxin B 2500U

Preparation of Antibiotic Solution: Add components to distilled/deionized water and bring volume to 10.0mL. Mix thoroughly. Filter sterilize.

Preparation of Medium: To 940.0mL of sterile cooled blood agar base No. 2, aseptically add 50.0mL of sterile, lysed defibrinated horse blood and 10.0mL of sterile antibiotic solution. Pour into sterile Petri dishes or distribute into sterile tubes.

Use: For the selective isolation and cultivation of *Campylobacter* species.

Skirrow's *Campylobacter* Agar

See: ***Campylobacter*** **Agar, Skirrow's**

SL Medium (DSMZ Medium 959)

Composition per liter:
NaCl 15.0g
PIPES 3.4g
Na-acetate·$3H_2O$ 2.72g
Yeast extract 2.0g
Trypticase™ 2.0g
NH_4Cl 1.0g
$MgCl_2·6H_2O$ 0.5g
K_2HPO_4 .... 0.35g
KH_2PO_4 .... 0.35g
KCl 0.2g
$CaCl_2·2H_2O$ 0.1g
Resazurin 0.5mg
Maltose solution 50.0mL
$Na_2S·9H_2O$ solution 10.0mL

pH 7.5 ± 0.2 at 25°C

Maltose Solution:
Composition per 100.0mL:
Maltose 3.5g

Preparation of Maltose Solution: Add maltose to distilled/deionized water and bring volume to 100.0mL. Mix thoroughly. Sparge with N_2. Autoclave for 15 min at 15 psi pressure–121°C. Cool to 25°C. Store anaerobically.

$Na_2S·9H_2O$ Solution:
Composition per 50.0mL:
$Na_2S·9H_2O$ 5.0g

Preparation of $Na_2S·9H_2O$ Solution: Add $Na_2S·9H_2O$ to distilled/deionized water and bring volume to 50.0mL. Sparge with N_2. Autoclave for 15 min at 15 psi pressure–121°C. Cool to 25°C. Store anaerobically.

Preparation of Medium: Add components, except maltose solution and $Na_2S·9H_2$ solution, to distilled/deionized water and bring volume to 940.0mL. Mix thoroughly. Gently heat and bring to boiling. Sparge with N_2. Adjust pH to 7.0. Autoclave for 15 min at 15 psi pressure–121°C. Cool to 25°C. Aseptically add 10.0mL sterile $Na_2S·9H_2O$ solution and 50.0mL sterile maltose solution. Mix thoroughly. Aspticaly and anaerobically distribute into sterile tubes or flasks under N_2. The final pH should be 7.5.

Use: For the cultivation of *Petrotoga olearia* and *Petrotoga sibirica.*

SL Medium

Composition per liter:
Agar 15.0g
Pancreatic digest of casein 10.0g
Glucose 10.0g
KH_2PO_4 .... 6.0g
Arabinose 5.0g
Sucrose 5.0g
Yeast extract 5.0g
Sodium acetate·$3H_2O$ 2.5g
Diammonium citrate 2.0g
Tween™ 80 .... 1.0g
$MgSO_4·7H_2O$ 0.58g
$MnSO_4·4H_2O$ 0.28g

pH 5.4 ± 0.2 at 25°C

Preparation of Medium: Add agar to 500.0mL of distilled/deionized water. Gently heat and bring to boiling. In a separate flask, add the remaining components, except sodium acetate, to distilled/deionized water and bring volume to 300.0mL. Mix thoroughly. Combine the two solutions. In a separate flask, add sodium acetate to distilled/deionized water and bring volume to100.0mL. Mix thoroughly. Adjust pH to 5.4 with glacial acetic acid. Add this solution to the agar solution. Mix thoroughly. Adjust pH to 5.4 with additional glacial acetic acid. Bring volume to 1.0L with distilled/deionized water. Do not autoclave. Aseptically distribute into sterile screw-capped tubes or flasks.

Use: For the selective isolation and cultivation of *Lactobacillus* species.

Slad Medium

Composition per liter:

Solution B	900.0mL
Solution A	100.0mL

Solution A:

Composition per 100.0mL:

Raffinose	20.0g
KH_2PO_4	1.0g
$MgSO_4 \cdot 7H_2O$	0.5g
NaCl	0.1g
$CaCl_2 \cdot 2H_2O$	0.1g
Ammonium sulfate	7.0mg
Inositol	2.0mg
KI	1.0mg
H_3BO_3	0.5mg
$ZnSO_4 \cdot 7H_2O$	0.4mg
$MnSO_4 \cdot 4H_2O$	0.4mg
Thiamine·HCl	0.4mg
Pyroxidine·HCl	0.4mg
Niacin	0.4mg
Calcium pantothenate	0.4mg
p-Aminobenzoic acid	0.2mg
Riboflavin	0.2mg
$FeCl_3$	0.2mg
$Na_2MoO_4 \cdot 4H_2O$	0.2mg
$CuSO_4 \cdot 5H_2O$	0.04mg
Folic acid	2.0μg
Biotin	2.0μg

Preparation of Solution A: Add components to distilled/deionized water and bring volume to 100.0mL. Mix thoroughly. Filter sterilize. Warm to 50°–55°C.

Solution B:

Composition per 900.0mL:

Agar	40.0g

Preparation of Solution B: Add agar to distilled/deionized water and bring volume to 900.0mL. Mix thoroughly. Gently heat and bring to boiling. Autoclave for 15 min at 15 psi pressure–121°C. Cool to 50°–55°C.

Preparation of Medium: Aseptically combine 100.0mL of sterile solution A and 900.0mL of sterile solution B. Mix thoroughly. Pour into sterile Petri dishes or distribute into sterile tubes.

Use: For the cultivation of *Saccharomyces cerevisiae.*

Slanetz and Bartley, HiVeg (Slanetz and Bartley HiVeg Medium)

Composition per liter:

Plant hydrolysate No. 1	20.0g
Agar	15.0g
Yeast extract	5.0g
Na_2HPO_4	4.0g
Glucose	2.0g
NaN_3	0.4g
2,3,5-Triphenyltetrazolium chloride	0.1g

pH 7.2 ± 0.2 at 25°C

Source: This medium is available as a premixed powder from HiMedia.

Caution: Sodium azide is toxic. Azides also react with metals and disposal must be highly diluted.

Preparation of Medium: Add components to distilled/deionized water and bring volume to 1.0L. Mix thoroughly. Gently heat and bring to boiling to dissolve the medium completely. Excessive heating should be avoided. Do not autoclave. Pour into sterile Petri dishes or leave in tubes.

Use: For the detection and enumeration of fecal streptococci by the membrane filter technique.

Slanetz and Bartley Medium

Composition per liter:

Tryptose	20.0g
Agar	10.0g
Yeast extract	5.0g
$Na_2HPO_4\ 2H_2O$	4.0g
Glucose	2.0g
NaN_3	0.4g
Tetrazolium chloride	0.1g

pH 7.2 ± 0.2 at 25°C

Source: This medium is available as a premixed powder from Oxoid Unipath.

Preparation of Medium: Add components to distilled/deionized water and bring volume to 1.0L. Mix thoroughly. Gently heat and bring to boiling. Distribute into tubes or flasks. Autoclave for 15 min at 15 psi pressure–121°C. Pour into sterile Petri dishes.

Use: For the detection and enumeration of enterococci by the membrane filter method.

Sludge Medium for Methanobacteria

Composition per liter:

$NaHCO_3$	4.0g
Sodium formate	2.0g
Sodium acetate	1.0g
Yeast extract	1.0g
L-Cysteine·HCl·H_2O	0.5g
KH_2PO_4	0.5g
$Na_2S \cdot 9H_2O$	0.5g
$MgSO_4 \cdot 7H_2O$	0.4g
NaCl	0.4g
NH_4Cl	0.4g
$CaCl_2 \cdot 2H_2O$	0.05g
$FeSO_4 \cdot 7H_2O$	2.0mg
Resazurin	1.0mg
Sludge fluid	50.0mL

Fatty acid mixture 20.0mL
Trace elements sollution SL-6 ...... 1.0mL
pH 6.7-7.0 at 25°C

Sludge Fluid:
Composition per 100.0mL:
Yeast extract 0.4g
Sludge 100.0mL

Preparation of Sludge Fluid: Add yeast extract to a concentration of 0.4% to sludge taken from an anaerobic digester. Gas with 100% N_2 for 5 min. Incubate at 37°C for 24 hr. Centrifuge at 13,000 × g. Remove the supernatant fluid. Gas with 100% N_2 for 5 min. Autoclave for 15 min at 15 psi pressure–121°C. Store at 25°C protected from light.

Fatty Acid Mixture:
Composition per 20.0mL:
α-Methylbutyric acid 0.5g
Isobutyric acid 0.5g
Isovaleric acid 0.5g
Valeric acid 0.5g

Preparation of Fatty Acid Mixture: Add components to distilled/deionized water and bring volume to 20.0mL. Mix thoroughly. Adjust pH to 7.5 with concentrated NaOH.

Trace Elements Solution SL-6:
Composition per liter:
H_3BO_3 ...... 0.3g
$CoCl_2·6H_2O$ 0.2g
$ZnSO_4·7H_2O$ 0.1g
$MnCl_2·4H_2O$ 0.03g
$Na_2MoO_4·H_2O$ 0.03g
$NiCl_2·6H_2O$ 0.02g
$CuCl_2·2H_2O$ 0.01g

Preparation of Trace Elements Solution SL-6: Add components to distilled/deionized water and bring volume to 1.0L. Mix thoroughly. Adjust pH to 3.4.

Preparation of Medium: Prepare and dispense medium under 80% N_2 + 20% CO_2. Add components to distilled/deionized water and bring volume to 1.0L. Mix thoroughly. Adjust pH to 6.7–7.0. Distribute anaerobically into tubes or bottles with aluminum seals. Autoclave for 15 min at 15 psi pressure–121°C with fast exhaust.

Use: For the cultivation of *Methanobacterium uliginosum* and *Methanobrevibacter ruminantium.*

Sludge Medium for Methanobacteria, pH 7.9

Composition per liter:
$NaHCO_3$ 4.0g
Sodium formate 2.0g
Sodium acetate 1.0g
Yeast extract 1.0g
L-Cysteine·HCl·H_2O 0.5g
KH_2PO_4 ...... 0.5g
$Na_2S·9H_2O$ 0.5g
$MgSO_4·7H_2O$ 0.4g
NaCl 0.4g
NH_4Cl 0.4g
$CaCl_2·2H_2O$ 0.05g
$FeSO_4·7H_2O$ 2.0mg
Resazurin 1.0mg
Sludge fluid 50.0mL
Fatty acid mixture 20.0mL
Trace elements solution SL-6 ...... 1.0mL
pH 7.9 ± 0.2 at 25°C

Sludge Fluid:
Composition per 100.0mL:
Yeast extract 0.4g
Sludge 100.0mL

Preparation of Sludge Fluid: Add yeast extract to a concentration of 0.4% to sludge taken from an anaerobic digester. Gas with 100% N_2 for 5 min. Incubate at 37°C for 24 hr. Centrifuge at 13,000 × g. Remove the supernatant fluid. Gas with 100% N_2 for 5 min. Autoclave for 15 min at 15 psi pressure–121°C. Store at 25°C protected from light.

Fatty Acid Mixture:
Composition per 20.0mL:
α-Methylbutyric acid 0.5g
Isobutyric acid 0.5g
Isovaleric acid 0.5g
Valeric acid 0.5g

Preparation of Fatty Acid Mixture: Add components to distilled/deionized water and bring volume to 20.0mL. Mix thoroughly. Adjust pH to 7.5 with concentrated NaOH.

Trace Elements Solution SL-6:
Composition per liter:
H_3BO_3 ...... 0.3g
$CoCl_2·6H_2O$ 0.2g
$ZnSO_4·7H_2O$ 0.1g
$MnCl_2·4H_2O$ 0.03g
$Na_2MoO_4·H_2O$ 0.03g
$NiCl_2·6H_2O$ 0.02g
$CuCl_2.2H_2O$ 0.01g

Preparation of Trace Elements Solution SL-6: Add components to distilled/deionized water and bring volume to 1.0L. Mix thoroughly. Adjust pH to 3.4.

Preparation of Medium: Prepare and dispense medium under 80% N_2 + 20% CO_2. Add components to distilled/deionized water and bring volume to 1.0L. Adjust pH to 7.9. Distribute anaerobically into tubes or bottles with aluminum seals. Autoclave for 15 min at 15 psi pressure–121°C with fast exhaust.

Use: For the cultivation of *Methanobacterium alcaliphilum* and *Methanobacterium thermoalcalip.*

SM Basal Salts Medium

Composition per liter:
Na_2HPO_4 ...... 4.5g
KH_2PO_4 ...... 1.5g
NH_4Cl 0.3g
$MgSO_4·7H_2O$ 0.1g
Trace metals solution 5.0mL
pH 6.0 ± 0.2 at 25°C

Trace Metals Solution:
Composition per liter:
Ethylenediamine tetraacetate 50.0g
$ZnSO_4·7H_2O$ 22.0g
$CaCl_2$ 5.54g
$MnCl_2·4H_2O$ 5.06g
$FeSO_4·7H_2O$ 4.99g
$CoCl_2·H_2O$ 1.61g

$CuSO_4 \cdot 5H_2O$ 1.57g
$(NH_4)_6Mo_7O_{24} \cdot 4H_2O$ 1.1g

Preparation of Trace Elements Solution: Add components to distilled/deionized water and bring volume to 1.0L. Mix thoroughly. Adjust pH to 6.0.

Preparation of Medium: Add components to distilled/deionized water and bring volume to 1.0L. Mix thoroughly. Adjust pH to 6.0. Distribute into tubes or flasks. Autoclave for 15 min at 15 psi pressure–121°C.

Use: For the cultivation of *Thiobacillus delicatus*.

SM ID2 Agar

Composition per liter:
Proprietary

Source: This medium is available from bioMérieux

Use: A new chromogenic medium for the selective isolation and detection of *Salmonella. S. typhi*, and *S. paratyphi* and most lactose(+) *Salmonella* present pale pink to mauve colonies. Other organisms are either inhibited, colorless, or pale blue in appearance.

SM Medium

Composition per liter:
Na_2HPO_4 .... 4.5g
KH_2PO_4 .... 1.5g
NH_4Cl 0.3g
$MgSO_4 \cdot 7H_2O$ 0.1g
$Na_2S_2O_3$ solution 100.0mL
Trace metals solution 5.0mL

pH 7.5 ± 0.2 at 25°C

$Na_2S_2O_3$ Solution
Composition per 100.0mL:
$Na_2S_2O_3$ solution 10.0g

Preparation of $Na_2S_2O_3$ Solution: Add 10.0g of $Na_2S_2O_3$ to distilled/deionized water and bring volume to 100.0mL. Mix thoroughly. Filter sterilize.

Trace Metals Solution:
Composition per liter:
Ethylenediaminetetraacetic acid 50.0g
$ZnSO_4 \cdot 7H_2O$ 22.0g
$CaCl_2$ 5.54g
$MnCl_2 \cdot 4H_2O$ 5.06g
$FeSO_4 \cdot 7H_2O$ 4.99g
$CoCl_2 \cdot 6H_2O$ 1.61g
$CuSO_4 \cdot 5H_2O$ 1.57g
$(NH_4)_6Mo_7O_{24} \cdot H_2O$ 1.10g

Preparation of Trace Metals Solution: Add components to distilled/deionized water and bring volume to 1.0L. Mix thoroughly. Adjust pH to 6.0 using KOH. Filter sterilize.

Preparation of Medium: Add components, except $Na_2S_2O_3$ solution and trace metals solution, to distilled/deionized water and bring volume to 895.0mL. Mix thoroughly. Autoclave for 15 min at 15 psi pressure–121°C. Aseptically add 100.0mL of sterile $Na_2S_2O_3$ solution and 5.0mL of sterile trace metals solution. Mix thoroughly. Aseptically distribute into sterile tubes or flasks.

Use: For the cultivation of *Thiobacillus thioparus*.

SM Selective Medium

Composition per liter:
Agar 15.0g
Mannitol 2.5g
L-Glutamic acid 1.0g
$MgSO_4 \cdot 7H_2O$ 0.16g
Triphenyltetrazolium·chloride solution 10.0mL
Antibiotic solution 10.0mL
KH_2PO_4 (0.2m*M* solution) 1.0mL
Metals solution 0.05mL

pH 7.2 ± 0.2 at 25°C

Triphenyltetrazolium Chloride Solution:
Composition per 10.0mL:
2,3,5-Triphenyltetrazolium chloride 0.1g
Ethanol (95% solution) 10.0mL

Preparation of Triphenyltetrazolium Chloride Solution: Add 2,3,5-triphenyltetrazolium chloride to 10.0mL of ethanol. Mix thoroughly.

Antibiotic Solution:
Composition per 10.0mL:
Bacitracin 0.05g
Cycloheximide 0.05g
Tyrothricin 0.02g
Captan 0.01g
Vancomycin 0.01g
Chloromycetin 5.0μg
Penicillin G 1.0μg

Preparation of Antibiotic Solution: Add components to distilled/deionized water and bring volume to 10.0mL. Mix thoroughly. Filter sterilize.

Metals Solution:
Composition per 100.0mL:
$ZnSO_4 \cdot 7H_2O$ 1.1g
$MnSO_4 \cdot H_2O$ 0.62g
$Fe(NH_4)_2(SO_4)_2 \cdot 6H_2O$ 0.18g
$CuSO_4 \cdot 5H_2O$ 0.029g
$CaSO_4 \cdot 5H_2O$ 0.029g
H_3PO_3 .... 0.011g
KI 0.013mg

Preparation of Metals Solution: Add components to distilled/deionized water and bring volume to 100.0mL. Mix thoroughly.

Preparation of Medium: Add components, except antibiotic solution, to distilled/deionized water and bring volume to 990.0mL. Mix thoroughly. Gently heat and bring to boiling. Adjust pH to 7.2 with KOH. Autoclave for 15 min at 15 psi pressure–121°C. Cool to 45°–50°C. Aseptically add sterile antibiotic solution. Mix thoroughly. Pour into sterile Petri dishes or distribute into sterile tubes.

Use: For the isolation and cultivation of *Pseudomonas solanacearum*.

SMA1

See: **Salt Malt Agar**

SMB Medium (DSMZ Medium 1185)

Composition per liter:
NaCl 50.0g
Tryptone 5.0g
Na_2HPO_4 .... 3.0g

KH_2PO_4	2.0g
Yeast extract	2.0g
NH_4NO_3	1.0g
$MgSO_4 \cdot 7H_2O$	0.2g
Na_2CO_3	0.1g
$CaCl_2 \cdot 2H_2O$	0.01g
Manganese sulfate solution	2.0mL
Iron sulfate solution	1.0mL

pH 6.8–7.1 at 25°C

Manganese Sulfate Solution:
Composition per 10.0mL:

$MnSO_4 \cdot 2H_2O$	0.1g

Preparation of Manganese Sulfate Solution: Add $MnSO_4 \cdot 2H_2O$ to distilled/deionized water and bring volume to 10.0mL. Mix thoroughly. Filter sterilize.

Iron Sulfate Solution:
Composition per 10.0mL:

$FeSO_4 \cdot 7H_2O$	0.1g

Preparation of Iron Sulfate Solution: Add $FeSO_4 \cdot 7H_2O$ to distilled/deionized water and bring volume to 10.0mL. Mix thoroughly. Filter sterilize.

Preparation of Medium: Add components to distilled/deionized water and bring volume to 1.0L. Mix thoroughly. Adjust pH to 6.8–7.1. Distribute into tubes or flasks. Autoclave for 15 min at 15 psi pressure–121°C.

Use: For the cultivation of *Salinicola socius.*

SMC Medium

Composition per liter:

Sorbitol	70.0g
Pancreatic digest of casein	17.0g
NaCl	5.0g
Beef extract	3.0g
Yeast extract	3.0g
Beef heart, solids from infusion	2.0g
Horse serum	200.0mL
Yeast extract solution	100.0mL
Phenol Red solution	20.0mL
Sucrose solution	20.0mL
L-Arginine·HCl solution	10.0mL
Fructose solution	2.0mL
Glucose solution	2.0mL

pH 7.5 ± 0.2 at 25°C

Yeast Extract Solution:
Composition per 100.0mL:

Yeast extract	25.0g

Preparation of Yeast Extract Solution: Add yeast extract to distilled/deionized water and bring volume to 100.0mL. Mix thoroughly. Autoclave for 15 min at 15 psi pressure–121°C. Cool to 45°–50°C.

Phenol Red Solution:
Composition per 100.0mL:

Phenol Red	0.01g

Preparation of Phenol Red Solution: Add Phenol Red to distilled/deionized water and bring volume to 100.0mL. Mix thoroughly. Autoclave for 15 min at 15 psi pressure–121°C. Cool to 45°–50°C.

Sucrose Solution:
Composition per 20.0mL:

Sucrose	10.0g

Preparation of Sucrose Solution: Add sucrose to distilled/deionized water and bring volume to 20.0mL. Mix thoroughly. Autoclave for 15 min at 15 psi pressure–121°C. Cool to 45°–50°C.

L-Arginine·HCl Solution:
Composition per 10.0mL:

L-Arginine·HCl	4.2g

Preparation of L-Arginine·HCl Solution: Add L-arginine·HCl to distilled/deionized water and bring volume to 10.0mL. Mix thoroughly. Autoclave for 15 min at 15 psi pressure–121°C. Cool to 45°–50°C.

Fructose Solution:
Composition per 10.0mL:

Fructose	5.0g

Preparation of Fructose Solution: Add fructose to distilled/deionized water and bring volume to 10.0mL. Mix thoroughly. Autoclave for 15 min at 15 psi pressure–121°C. Cool to 45°–50°C.

Glucose Solution:
Composition per 10.0mL:

Glucose	5.0g

Preparation of Glucose Solution: Add glucose to distilled/deionized water and bring volume to 10.0mL. Mix thoroughly. Autoclave for 15 min at 15 psi pressure–121°C. Cool to 45°–50°C.

Preparation of Medium: Add components—except horse serum, yeast extract solution, Phenol Red solution, sucrose solution, L-arginine·HCl solution, fructose solution, and glucose solution—to distilled/deionized water and bring volume to 646.0mL. Mix thoroughly. Gently heat and bring to boiling. Autoclave for 15 min at 15 psi pressure–121°C. Cool to 45°–50°C. Aseptically add 200.0mL of sterile horse serum, 100.0mL of sterile yeast extract solution, 20.0mL of sterile Phenol Red solution, 20.0mL of sterile sucrose solution, 10.0mL of sterile L-arginine·HCl solution, 2.0mL of sterile fructose solution, and 2.0mL of sterile glucose solution. Mix thoroughly. Aseptically distribute into sterile tubes or flasks.

Use: For the cultivation and maintenance of *Spiroplasma citri.*

SMC, Modified

Composition per liter:

Sorbitol	70.0g
Pancreatic digest of casein	17.0g
NaCl	5.0g
Beef extract	3.0g
Yeast extract	3.0g
Beef heart, solids from infusion	2.0g
Solution 1	100.0mL
Solution 3	20.0mL
Solution 2	10.0mL
NaOH (1*N* solution)	6.0mL

pH 7.7–7.8 at 25°C

Solution 1:
Composition per 100.0mL:

Sucrose	10.0g
Yeast extract	2.0g
Fructose	1.0g
Glucose	1.0g
Phenol Red	0.02g

Preparation of Solution 1: Add components to distilled/deionized water and bring volume to 100.0mL. Mix thoroughly. Filter sterilize.

Solution 2:
Composition per 10.0mL:
Bovine serum albumin, fraction V 0.1g

Preparation of Solution 2: Add bovine serum albumin to distilled/deionized water and bring volume to 10.0mL. Mix thoroughly. Filter sterilize.

Solution 3:
Composition per 20.0mL:
Horse serum 20.0mL

Preparation of Solution 3: Inactivate horse serum at 56°C for 30 min. Filter sterilize.

Preparation of Medium: Add components—except solution 1, solution 2, and solution 3—to distilled/deionized water and bring volume to 870.0mL. Autoclave for 15 min at 15 psi pressure–121°C. Cool to 45°–50°C. Aseptically add 100.0mL of sterile solution 1, 20.0mL of sterile solution 3, and 10.0mL of sterile solution 2. Mix thoroughly. Adjust pH to 7.7–7.8. Aseptically distribute into sterile tubes or flasks.

Use: For the cultivation and maintenance of *Spiroplasma citri*.

SME Agar (ATCC Medium 2345)

Composition per 1010.0mL:
NaCl 27.7g
Agar 18.0g
$MgSO_4 \cdot 7H_2O$ 7.0g
$MgCl_2 \cdot 6H_2O$ 5.5g
$CaCl_2 \cdot 2H_2O$ 1.5g
KCl 0.65g
NaBr 0.1g
H_3BO_3 .... 30.0mg
$SrCl \cdot 6H_2O$ 15.0mg
KI 0.05mg
Tris·HCl buffer (1.0*M* solution, pH 7.0) 25.0mL

Preparation of Medium: Add components to distilled/deionized water and bring volume to 1.0L. Mix thoroughly. Gently heat and bring to boiling. Distribute into tubes or flasks. Autoclave for 15 min at 15 psi pressure–121°C. Pour into sterile Petri dishes or leave in tubes.

Use: For the cultivation and maintenance of *Aquifex pyrophilus*.

SME Medium (DSMZ Medium 1155)

Composition per liter:
NaCl 27.7g
Sulfur, elemental 10.0g
$MgSO_4 \cdot 7H_2O$ 7.0g
$MgCl_2 \cdot 6H_2O$ 5.5g
KH_2PO_4 .... 1.0g
$CaCl_2 \cdot 2H_2O$ 0.75g
KCl 0.66g
$(NH_4)_2SO_4$ 0.5g
NaBr 0.10g
H_3BO_3 .... 30.0mg
$SrCl_2 \cdot 6H_2O$ 15.0mg
Sulfide solution 20.0mL
Yeast extract solution 10.0mL
KJ solution 0.05mL
pH 5.5 ± 0.2 at 25°C

KJ Solution:
Composition per 10.0mL:
KJ 10.0mg

Preparation of KJ Solution: Add KJ to distilled/deionized water and bring volume to 10.0mL. Mix thoroughly.

Sulfide Solution:
Composition per 10.0mL:
$Na_2S \cdot 9H_2O$ 0.25g

Preparation of Sulfide Solution: Add $Na_2S \cdot 9H_2O$ to distilled/deionized water and bring volume to 10.0mL. Mix thoroughly. Autoclave under 100% N_2 for 15 min at 15 psi pressure–121°C. Cool to room temperature.

Yeast Extract Solution:
Composition per 10.0mL:
Yeast extract 0.4g

Preparation of Yeast Extract Solution: Add yeast extract to distilled/deionized water and bring volume to 10.0mL. Mix thoroughly. Filter sterilize.

Preparation of Medium: Add components, except sulfide and yeast extract solutions, to distilled/deionized water and bring volume to 970.0mL. Mix thoroughly. Gently heat and bring to boiling. Sparge with a gas mixture of 80% H_2 + 20% CO_2. Autoclave for 15 min at 15 psi pressure–121°C. Cool to room temperature. Aseptically add sulfide solution. Mix thoroughly. Adjust pH to 5.5 with sulfuric acid. Aseptically add yeast extract solution. Mix thoroughly. Sparge with a gas mixture of 80% H_2 + 20% CO_2. Aseptically distribute into culture vessels.

Use: For the cultivation of *Thermodiscus maritimus*.

SME Medium, Modified

Composition per 1010.0mL:
NaCl 30.0g
$MgSO_4 \cdot 7H_2O$ 7.0g
$NaHCO_3$ 2.0g
KCl 0.65g
$CaCl_2 \cdot 2H_2O$ 0.5g
K_2HPO_4 .... 0.15g
NH_4Cl 0.15g
NaBr 0.1g
$MgCl_2 \cdot 6H_2O$ 5.5mg
Trace elements solution 10.0mL
pH 6.5 ± 0.2 at 25°C

Trace Elements Solution:
Composition per liter:
$MgSO_4 \cdot 7H_2O$ 3.0g
$(NH_4)_2Ni(SO_4)_2$ 2.0g
Nitrilotriacetic acid 1.5g
NaCl 1.0g
$MnSO_4 \cdot 2H_2O$ 0.5g
$CaCl_2 \cdot 2H_2O$ 0.1g
$FeSO_4 \cdot 7H_2O$ 0.1g
$NiCl_2 \cdot 6H_2O$ 0.025g
$KAl(SO_4)_2 \cdot 12H_2O$ 0.02g
$CoSO_4 \cdot 7H_2O$ 0.18g
$ZnSO_4 \cdot 7H_2O$ 0.18g
$CuSO_4 \cdot 5H_2O$ 0.01g
H_3BO_3 .... 0.01g
$Na_2MoO_4 \cdot 2H_2O$ 0.01g

Na_2WO_4	10.0mg
Na_2SeO_4	10.0mg
$Na_2SeO_3 \cdot 5H_2O$	0.3mg

Preparation of Trace Elements Solution: Add nitrilotriacetic acid to approximately 500.0mL of distilled/deionized water. Dissolve by adding KOH and adjust pH to 6.5. Add remaining components. Bring volume to 1.0L with additional distilled/deionized water. Adjust pH to 7.0 with KOH.

Preparation of Medium: Add components to distilled/deionized water and bring volume to 1.0L.Mix thoroughly. Sparge with 100% N_2 for 20 min. Adjust pH to 6.5–6.8 with H_2SO_4. Distribute 10.0mL volumes into 120.0mL serum bottles while gassing under 100% CO_2. Stopper each serum bottle tightly. Exchange the gas phase in each serum bottle with 79.75% H_2 + 19.75% CO_2 + 0.5% O_2 and bring pressure to 300KPa. Autoclave for 15 min at 15 psi pressure–121°C.

Use: For the cultivation and maintenance of *Aquifex pyrophilus.*

SME Medium Modified for IS7 (DSMZ Medium 891)

Composition per liter:

NaCl	30.0g
$NaHCO_3$	2.0g
$MgSO_4 \cdot 7H_2O$	0.7g
KCl	0.65g
$MgCl_2 \cdot 6H_2O$	0.55g
$CaCl_2 \cdot 2H_2O$	0.5g
Sulfur, powdered	0.5g
NH_4Cl	0.15g
K_2HPO_4	0.15g
NaBr	0.1g
Trace elements solution	10.0mL

pH 7.0 ± 0.2 at 25°C

Trace Elements Solution:

Composition per liter:

$MgSO_4 \cdot 7H_2O$	3.0g
Nitrilotriacetic acid	1.5g
NaCl	1.0g
$MnSO_4 \cdot 2H_2O$	0.5g
$CoSO_4 \cdot 7H_2O$	0.18g
$ZnSO_4 \cdot 7H_2O$	0.18g
$CaCl_2 \cdot 2H_2O$	0.1g
$FeSO_4 \cdot 7H_2O$	0.1g
$NiCl_2 \cdot 6H_2O$	0.025g
$KAl(SO_4)_2 \cdot 12H_2O$	0.02g
H_3BO_3	0.01g
$Na_2MoO_4 \cdot 4H_2O$	0.01g
$CuSO_4 \cdot 5H_2O$	0.01g
$Na_2SeO_3 \cdot 5H_2O$	0.3mg

Preparation of Trace Elements Solution: Add nitrilotriacetic acid to 500.0mL of distilled/deionized water. Dissolve by adjusting pH to 6.5 with KOH. Add remaining components. Add distilled/deionized water to 1.0L. Mix thoroughly.

Preparation of Medium: Add components to distilled/deionized water and bring volume to 1.0L. Mix thoroughly. Adjust pH to 7.0 using H_2SO_4. Distribute into anaerobe tubes or bottles. Autoclave for 15 min at 15 psi pressure–121°C. Provide an atmosphere of 78% H_2 + 20% CO_2 + 2% O_2.

Use: For the cultivation of unclassified bacterium DSM 12047.

Smibert's Semisolid *Brucella* Medium

Composition per liter:

Casein enzymic hydrolysate	10.0g
Peptic digest of animal tissue	10.0g
NaCl	5.0g
Yeast extract	2.0g
Agar	1.6g
Glucose	1.0g
$NaHSO_3$	0.1g
Neutral Red	0.02g

pH 7.2 ± 0.2 at 25°C

Source: This medium is available as a premixed powder from Hi-Media.

Preparation of Medium: Add components to distilled/deionized water and bring volume to 1.0L. Mix thoroughly. Gently heat and bring to boiling. Autoclave for 15 min at 15 psi pressure–121°C. Mix thoroughly. Aseptically distribute into sterile tubes.

Use: For the cultivation of *Campylobacter* species.

Smithella Medium (DSMZ Medium 1030)

Composition per liter:

Yeast extract	2.0g
Trypticase peptone	2.0g
NH_4Cl	1.0g
$MgCl_2 \cdot 6H_2O$	1.0g
$CaCl_2 \cdot 2H_2O$	0.4g
KH_2PO_4	0.4g
Wolfe's mineral elixir	1.0mL
Resazurin	0.5mg
Crotonate solution	10.0mL
Vitamin solution	10.0mL
Bicarbonate solution	10.0mL
Sulfide solution	10.0mL
Coenzyme M solution	10.0mL

pH 7.1 ± 0.2 at 25°C

Crotonate Solution:

Composition per 10.0mL:

Na-crotonate	1.7g

Preparation of Crotonate Solution: Add Na-crotonate to distilled/deionized water and bring volume to 10.0mL. Mix thoroughly. Sparge with 100% N_2. Filter sterilize.

Coenzyme M Solution:

Composition per 10.0mL:

Mercaptoethanesulfonic acid (coenzyme M)	0.5g

Preparation of Coenzyme M: Add coenzyme M to distilled/deionized water and bring volume to 10.0mL. Mix thoroughly. Sparge with 100% N_2. Filter sterilize.

Bicarbonate Solution:

Composition per 10.0mL:

$NaHCO_3$	2.0g

Preparation of Bicarbonate Solution: Add $NaHCO_3$ to distilled/deionized water and bring volume to 10.0mL. Mix thoroughly. Sparge with 20% CO_2 + 80% H_2. Autoclave for 15 min at 15 psi pressure–121°C. Cool to room temperature.

Sulfide Solution:

Composition per 10.0mL:

$Na_2S·9H_2O$	0.3g

Preparation of Sulfide Solution: Add $Na_2S·9H_2O$ to distilled/deionized water and bring volume to 10.0mL. Mix thoroughly. Autoclave under 100% N_2 for 15 min at 15 psi pressure–121°C. Cool to room temperature.

Vitamin Solution:

Composition per liter:

Pyridoxine-HCl	10.0mg
Thiamine-HCl·$2H_2O$	5.0mg
Riboflavin	5.0mg
Nicotinic acid	5.0mg
D-Ca-pantothenate	5.0mg
p-Aminobenzoic acid	5.0mg
Lipoic acid	5.0mg
Biotin	2.0mg
Folic acid	2.0mg
Vitamin B_{12}	0.1mg

Preparation of Vitamin Solution: Add components to distilled/deionized water and bring volume to 1.0L. Mix thoroughly. Sparge with 80% H_2 + 20% CO_2. Filter sterilize.

Wolfe's Mineral Elixir:

Composition per liter:

$MgSO_4·7H_2O$	30.0g
NaCl	10.0g
$MnSO_4·2H_2O$	5.0g
$(NH_4)_2NiSO_4·6H_2O$	2.8g
$CoCl_2·6H_2O$	1.8g
$ZnSO_4·7H_2O$	1.8g
$FeSO_4·7H_2O$	1.0g
$CaCl_2·2H_2O$	1.0g
$KAl(SO_4)_2·12H_2O$	0.18g
$CuSO_4·5H_2O$	0.1g
H_3BO_3	0.1g
$Na_2MoO_4·2H_2O$	0.1g
Na_2SeO_4	0.1g
$Na_2WO_4·2H_2O$	0.1g

Preparation of Wolfe's Mineral Elixir: Adjust pH of 1.0L of distilled/deionized water to 1.0 with dilute H_2SO_4. Add remaining components one at a time. Mix thoroughly to dissolve.

Preparation of Medium: Add components, except bicarbonate, vitamin, crotonate, coenzyme M, and sulfide solutions, to distilled/deionized water and bring volume to 950.0mL. Gently heat and bring to boiling. Boil for 1 min. Mix thoroughly. Cool to room temperature while sparging with 80% N_2 + 20% CO_2. Dispense into culture vessels under an atmosphere of 80% N_2 + 20% CO_2 (e.g., Balch tubes or serum vials). Autoclave for 15 min at 15 psi pressure–121°C. Aseptically and anoxically add bicarbonate, vitamin, crotonate, coenzyme M, and sulfide solutions. Adjust pH to 7.1.

Use: For the cultivation of *Smithella* spp.

Snyder Agar

Composition per liter:

Glucose	20.0g
Agar	16.0g
Pancreatic digest of casein	13.5g
Yeast extract	6.5g
NaCl	5.0g
Bromcresol Green	0.02g

pH 4.8 ± 0.2 at 25°C

Source: This medium is available as a premixed powder from BD Diagnostic Systems.

Preparation of Medium: Add components to distilled/deionized water and bring volume to 1.0L. Mix thoroughly. Gently heat and bring to boiling. Distribute into tubes in 10.0mL volumes. Autoclave for 15 min at 13 psi pressure–118°C. Do not overheat. Pour into sterile Petri dishes or leave in tubes.

Use: For the cultivation and enumeration of lactobacilli in saliva and indication of dental caries activity.

Snyder Test Agar

Composition per liter:

Agar	20.0g
Glucose	20.0g
Tryptose	20.0g
NaCl	5.0g
Bromcresol Green	0.02g

pH 4.8 ± 0.2 at 25°C

Source: This medium is available as a premixed powder from BD Diagnostic Systems.

Preparation of Medium: Add components to distilled/deionized water and bring volume to 1.0L. Mix thoroughly. Gently heat and bring to boiling. Distribute into tubes in 10.0mL volumes. Autoclave for 15 min at 13 psi pressure–118°C. Do not overheat. Pour into sterile Petri dishes or leave in tubes.

Use: For the cultivation and enumeration of lactobacilli in saliva and indication of dental caries activity.

Snyder Test HiVeg Agar (BCG-Glucose Agar, HiVeg)

Composition per liter:

Agar	20.0g
Glucose	20.0g
Plant peptone	20.0g
NaCl	5.0g
Bromcresol Green	0.02

pH 4.8 ± 0.2 at 25°C

Source: This medium is available as a premixed powder from HiMedia.

Preparation of Medium: Add components to distilled/deionized water and bring volume to 1.0L. Mix thoroughly. Gently heat and bring to boiling. Distribute into tubes in 10.0mL volumes. Autoclave for 15 min at 13 psi pressure–118°C. Do not overheat. Pour into sterile Petri dishes or leave in tubes.

Use: For the cultivation and enumeration of lactobacilli in saliva and indication of dental caries activity.

Soap Agar

Composition per 315.0mL:

Stearic acid	30.0g
NaCl	1.5g
Peptone	1.5g
Yeast extract	0.6g
Agar	0.33g

Beef extract 0.3g
NaOH (7*M*) 15.0mL

pH 8.5 ± 1.0 at 25°C

Preparation of Medium: Add components to distilled/deionized water and bring volume to 315.0mL. Mix thoroughly. Gently heat and bring to boiling. Distribute into tubes or flasks. Autoclave for 15 min at 15 psi pressure–121°C. Pour into sterile Petri dishes or leave in tubes.

Use: For the cultivation of *Arthrobacter* species.

Sodalis glossinidius Medium (DSMZ Medium 1021)

Composition per liter:

Agar 17.0g
NaCl 8.7g
Casein hydrolysate 8.1g
Yeast extract 6.2g
D-Glucose 5.0g
KCl 0.25g
$CaCl_2 \cdot 2H_2O$ 0.25g
$NaHCO_3$ 0.15g
$MgCl_2 \cdot 6H_2O$ 0.12g
Phosphate solution 100.0mL

pH 8.0 ± 0.2 at 25°C

Phosphate Solution:
Composition per 100.0mL:

NaH_2PO_4 .... 0.28g

Preparation of Phosphate Solution: Add NaH_2PO_4 to distilled/deionized water and bring volume to 100.0mL. Mix thoroughly. Autoclave or 20 min at 15 psi pressure–121°C. Cool to 50°C.

Preparation of Medium: Add components, except phosphate solution, to distilled/deionized water and bring volume to 900.0L. Mix thoroughly. Gently heat and bring to boiling. Autoclave for 20 min at 15 psi pressure–121°C. Cool to 50°C. Aseptically add phosphate solution. Mix thoroughly. Adjust pH to 8.0. Pour into sterile Petri dishes or distribute into sterile tubes.

Use: For the cultivation of *Sodalis glossinidius.*

Sodium Acetate Agar
See: **Acetate Differential Agar**

Sodium Acetate Medium I

Composition per liter:

Sodium acetate 33.0g
Agar 15.0g
Glucose 10.0g
Peptone 10.0g
K_2HPO_4 .... 5.0g
NaCl 5.0g
Yeast extract 3.0g

pH 7.0 ± 0.2 at 25°C

Preparation of Medium: Add components to distilled/deionized water and bring volume to 1.0L. Mix thoroughly. Gently heat and bring to boiling. Adjust pH to 7.0. Distribute into tubes or flasks. Autoclave for 10 min at 15 psi pressure–121°C. Pour into sterile Petri dishes or leave in tubes.

Use: For the cultivation and maintenance of *Pediococcus halophilus* and *Staphylococcus simulans.*

Sodium Biselenite Medium
See: **Selenite Broth**

Sodium Caseinate Agar

Composition per liter:

Agar 15.0g
Sodium caseinate 2.0g
K_2HPO_4 .... 0.5g
Peptone 0.5g
Yeast extract 0.5g

pH 7.4 ± 0.2 at 25°C

Preparation of Medium: Add components to distilled/deionized water and bring volume to 1.0L. Mix thoroughly. Gently heat and bring to boiling. Distribute into tubes or flasks. Autoclave for 15 min at 15 psi pressure–121°C. Pour into sterile Petri dishes or leave in tubes.

Use: For the cultivation of actinomycetes.

Sodium Chloride Broth, 6.5%

Composition per liter:

Beef heart, solids from infusion 500.0g
NaCl 65.0g
Tryptose 10.0g

pH 7.4 ± 0.2 at 25°C

Preparation of Medium: Add components to distilled/deionized water and bring volume to 1.0L. Mix thoroughly. Distribute into tubes or flasks. Autoclave for 15 min at 15 psi pressure–121°C.

Use: For the cultivation of enterococci and other salt-tolerant organisms. For the differentiation of microorganisms based on salt tolerance.

Sodium Chloride Sucrose Medium 900 (Sodium Chloride SUC Medium 900)

Composition per liter:

Sucrose 97.3g
Pancreatic digest of gelatin 14.5g
NaCl 14.3g
Agar 13.3g
Brain heart, solids from infusion 6.0g
Peptic digest of animal tissue 6.0g
Yeast extract 5.0g
Glucose 3.0g
Na_2HPO_4 .... 2.5g
$MgSO_4$ 0.25g
Horse serum (γ-globulin free, inactivated 30 min at 56°C) 100.0mL
Carbenicillin solution 10.0mL

pH 7.4 ± 0.2 at 25°C

Carbenicillin Solution:
Composition per 10.0mL:

Carbenicillin 5.0g

Preparation of Carbenicillin Solution: Add carbenicillin to distilled/deionized water and bring volume to 10.0mL. Mix thoroughly. Filter sterilize.

Preparation of Medium: Add components, except carbenicillin solution and horse serum, to distilled/deionized water and bring volume to 890.0mL. Mix thoroughly. Gently heat and bring to boiling. Autoclave for 15 min at 15 psi pressure–121°C. Cool to 45°–50°C. Aseptically add carbenicillin solution and horse serum. Mix thoroughly. Pour into sterile Petri dishes or distribute into sterile tubes.

Use: For the cultivation of *Pseudomonas aeruginosa.*

Sodium Chloride Sucrose Medium 900 with Penicillin G

Composition per liter:

Sucrose	97.3g
Pancreatic digest of gelatin	14.5g
NaCl	14.3g
Agar	13.3g
Brain heart, solids from infusion	6.0g
Peptic digest of animal tissue	6.0g
Yeast extract	5.0g
Glucose	3.0g
Na_2HPO_4	2.5g
$MgSO_4$	0.25g
Horse serum (γ-globulin free, inactivated 30 min at 56°C)	100.0mL
Penicillin solution	10.0mL

pH 7.4 ± 0.2 at 25°C

Penicillin Solution:

Composition per 10.0mL:

Penicillin G	500,000U

Preparation of Penicillin Solution: Add penicillin G to distilled/deionized water and bring volume to 10.0mL. Mix thoroughly. Filter sterilize.

Preparation of Medium: Add components, except penicillin solution and horse serum, to distilled/deionized water and bring volume to 900.0mL. Mix thoroughly. Gently heat and bring to boiling. Autoclave for 15 min at 15 psi pressure–121°C. Cool to 45°–50°C. Aseptically add penicillin solution and horse serum. Mix thoroughly. Pour into sterile Petri dishes or distribute into sterile tubes.

Use: For the cultivation of *Pseudomonas aeruginosa.*

Sodium Dodecyl Sulfate Polymyxin Sucrose Agar

Composition per liter:

NaCl	20.0g
Agar	15.0g
Sucrose	15.0g
Proteose peptone	10.0g
Beef extract	5.0g
Sodium lauryl sulfate	1.0g
Bromthymol Blue	0.04g
Cresol Red	0.04g
Polymyxin B solution	10.0mL

pH 7.6 ± 0.2 at 25°C

Polymyxin B Solution:

Composition per 10.0mL:

Polymyxin B sulfate	100,000U

Preparation of Polymyxin B Solution: Add polymyxin B sulfate to distilled/deionized water and bring volume to 10.0mL. Mix thoroughly. Filter sterilize.

Preparation of Medium: Add components, except polymyxin B solution, to distilled/deionized water and bring volume to 990.0mL. Mix thoroughly. Gently heat and bring to boiling. Autoclave for 15 min at 15 psi pressure–121°C. Cool to 45°–50°C. Aseptically add sterile polymyxin B solution. Mix thoroughly. Pour immediately into sterile Petri dishes or distribute into sterile tubes.

Use: For the isolation and cultivation of *Vibrio* species from foods.

Sodium Hippurate Broth (Hippurate Broth)

Composition per liter:

Beef heart, solids from infusion	500.0g
Tryptose	10.0g
Sodium hippurate	10.0g
NaCl	5.0g

pH 7.4 ± 0.2 at 25°C

Source: Heart infusion broth is available as a premixed powder from BD Diagnostic Systems.

Preparation of Medium: Add components to distilled/deionized water and bring volume to 1.0L. Mix thoroughly. Gently heat and bring to boiling. Distribute into screw-capped tubes or flasks. Autoclave for 15 min at 15 psi pressure–121°C. Tighten caps to prevent drying.

Use: For the identification and differentiation of β-hemolytic streptococci based on hippurate hydrolysis. After inoculation and incubation, tubes are treated with $FeCl_3$ reagent. A heavy precipitate remaining after 10–15 min indicates that hippurate has been hydrolyzed.

Sodium Hydrogen Selenite Medium

See: **Selenite Broth**

Sodium Lactate Agar

Composition per liter:

Agar	15.0g
Pancreatic digest of casein	10.0g
Sodium lactate	10.0g
Yeast extract	10.0g
K_2HPO_4	0.25g

pH 7.0 ± 0.2 at 25°C

Preparation of Medium: Add components to distilled/deionized water and bring volume to 1.0L. Mix thoroughly. Gently heat and bring to boiling. Distribute into tubes or flasks. Adjust pH to 7.0. Autoclave for 15 min at 15 psi pressure–121°C. Pour into sterile Petri dishes or leave in tubes.

Use: For the cultivation and isolation of *Propionibacterium* species from foods. For the isolation and cultivation of propionic acid-producing bacteria from cheese.

Sodium Lactate Agar, Modified

Composition per liter:

Agar	15.0g
Yeast extract	10.0g
Pancreatic digest of casein	5.1g
NaCl	1.5g
Papaic digest of soybean meal	0.9g
K_2HPO_4	0.75g
Glucose	0.75g
Sodium lactate, 60% syrup	20.0mL

pH 7.0 ± 0.2 at 25°C

Preparation of Medium: Add components to distilled/deionized water and bring volume to 1.0L. Mix thoroughly. Gently heat and bring to boiling. Adjust pH to 7.0. Distribute into tubes or flasks. Autoclave for 15 min at 15 psi pressure–121°C. Pour into sterile Petri dishes or leave in tubes.

Use: For the cultivation and isolation of *Propionibacterium* species from foods. For the isolation and cultivation of propionic acid-producing bacteria from cheese.

Soft Agar Gelatin Overlay

Composition per plate:

Base agar....15.0mL
Soft agar gelatin overlay....2.5mL
pH 7.0 ± 0.2 at 25°C

Base Agar:
Composition per liter:

Agar....15.0g
Peptone....5.0g
NaCl....5.0g
Beef extract....3.0g
$MnSO_4 \cdot H_2O$....0.05g

Preparation of Base Agar: Add components to distilled/deionized water and bring volume to 1.0L. Mix thoroughly. Gently heat and bring to boiling. Autoclave for 15 min at 15 psi pressure–121°C. Cool to 45°–50°C.

Soft Agar Gelatin Overlay:
Composition per liter:

Gelatin....15.0g
Agar....8.0g
Peptone....5.0g
NaCl....5.0g
Beef extract....3.0g
$MnSO_4 \cdot H_2O$....0.05g

Preparation of Soft Agar Gelatin Overlay: Add components to distilled/deionized water and bring volume to 1.0L. Mix thoroughly. Gently heat and bring to boiling. Autoclave for 15 min at 15 psi pressure–121°C. Cool to 45°–50°C.

Preparation of Medium: Aseptically pour cooled, sterile base agar into sterile Petri dishes in 15.0mL volumes. Allow agar to solidify. Inoculate plates with samples. Overlay each plate with 2.5mL of soft agar gelatin overlay.

Use: For the cultivation and differentiation of microorganisms based on proteolytic activity.

Soil Extract

Composition per 200.0mL:

African Violet soil....77.0g
Na_2CO_3....0.2g

Preparation of Medium: Add components to 200.0mL of distilled/deionized water. Mix thoroughly. Autoclave for 60 min at 15 psi pressure–121°C. Filter through paper and reserve filtrate.

Use: Used as a growth factor additive for the cultivation of soil bacteria and fungi.

Soil Extract Agar

Composition per liter:

Agar....20.0g
Glucose....1.0g
K_2HPO_4....1.0g
Peptone....1.0g
Yeast extract....1.0g
Soil extract....400.0mL
Cycloheximide solution....10.0mL
pH 6.6 ± 0.2 at 25°C

Soil Extract:
Composition per liter:

Garden soil, neutral....1.0Kg

Preparation of Soil Extract: Add garden soil to 1.0L of tap water. Autoclave for 20 min at 15 psi pressure–121°C. Filter through Whatman filter paper. Bring volume to 1.0L with tap water.

Cycloheximide Solution:
Composition per 10.0mL:

Cycloheximide....0.04g

Preparation of Cycloheximide Solution: Add cycloheximide to distilled/deionized water and bring volume to 10.0mL. Mix thoroughly. Filter sterilize.

Preparation of Medium: Add components, except cycloheximide solution, to distilled/deionized water and bring volume to 990.0mL. Mix thoroughly. Gently heat and bring to boiling. Autoclave for 15 min at 15 psi pressure–121°C. Cool to 45°–50°C. Aseptically add sterile cycloheximide solution. Mix thoroughly. Pour into sterile Petri dishes or distribute into sterile tubes.

Use: For the isolation and cultivation of *Arthrobacter* species.

Soil Extract Agar

Composition per liter:

Soil....500.0g
Agar....15.0g
Glucose....2.0g
Yeast extract....1.0g
KH_2PO_4....0.5g

Preparation of Medium: Add 500.0g of garden soil to 1.0L of tap water. Autoclave for 3 hr at 15 psi pressure–121°C. Filter through Whatman #2 filter paper. Add remaining components to filtrate. Bring volume to 1.0L with tap water. Gently heat and bring to boiling. Distribute into tubes in 7.0mL volumes. Autoclave for 15 min at 15 psi pressure–121°C. Allow tubes to cool in a slanted position.

Use: For the cultivation and identification of *Histoplasma capsulatum*, *Blastomyces dermatitidis*, and *Bacillus* species based on the formation of typical conidia.

Soil Extract Agar

Composition per liter:

Agar....15.0g
Soil extract....1.0L
pH 6.8 ± 0.2 at 25°C

Soil Extract:
Composition per liter:

Soil....400.0g

Preparation of Soil Extract: Air dry garden soil with a high content of organic matter and pass through a sieve. Weigh out 400.0g and add to 960.0mL of tap water. Autoclave for 60 min at 15 psi pressure–121°C. Cool to room temperature and allow soil to settle out. Decant supernatant solution. Filter through paper. Bring volume to 1.0L with distilled/deionized water.

Preparation of Medium: Add agar to 1.0L soil extract. Gently heat and bring to boiling. Autoclave for 15 min at 15 psi pressure–121°C. Pour into sterile Petri dishes or distribute into sterile tubes.

Use: For the cultivation of *Aureobacterium flavescens* and *Bacillus* species.

Soil Extract Glucose Yeast Extract Agar

Composition per liter:

Agar	15.0g
Glucose	2.0g
Yeast extract	1.0g
Soil extract	250.0mL

pH 6.8 ± 0.2 at 25°C

Soil Extract:

Composition per liter:

Garden soil	500.0g

Preparation of Soil Extract: Add 500.0g of garden soil to 1.0L of tap water. Autoclave for 1 hr at 15 psi pressure–121°C. Filter through Whatman #2 filter paper.

Preparation of Medium: Add components to distilled/deionized water and bring volume to 1.0L. Mix thoroughly. Gently heat and bring to boiling. Distribute into tubes or flasks. Autoclave for 15 min at 15 psi pressure–121°C. Pour into sterile Petri dishes or leave in tubes.

Use: For the cultivation and maintenance of *Streptomyces rectus*.

Soil Extract Glycerol Medium

Composition per liter:

Glycerol	20.0g
Peptone	5.0g
Beef extract	3.0g
Soil extract	150.0mL

pH 7.0 ± 0.2 at 25°C

Soil Extract:

Composition per liter:

Soil	400.0g

Preparation of Soil Extract: Allow garden soil to air dry. Add 400.0g of the air-dried garden soil to 960.0mL of tap water. Autoclave for 60 min at 15 psi pressure–121°C. Allow to cool to 25°C. Let stand until settling ceases. Decant the liquid through Whatman filter paper. Autoclave for 30 min at 15 psi pressure–121°C. Cool to room temeprature. Let stand until settling ceases.

Preparation of Medium: Add components to tap water and bring volume to 1.0L. Mix thoroughly. Adjust pH to 7.0. Distribute into tubes or flasks. Autoclave for 15 min at 15 psi pressure–121°C.

Use: For the cultivation and maintenance of *Mycobacterium terrae*.

Soil Extract Medium

Composition per liter:

Agar	15.0g
Pancreatic digest of gelatin	5.0g
Beef extract	3.0g
Soil extract	250.0mL

pH 6.8 ± 0.2 at 25°C

Soil Extract:

Composition per liter:

Garden soil	500.0g

Preparation of Soil Extract: Add 500.0g of garden soil to 1.0L of tap water. Autoclave for 1 hr at 15 psi pressure–121°C. Filter through Whatman #2 filter paper.

Preparation of Medium: Add components to distilled/deionized water and bring volume to 1.0L. Mix thoroughly. Gently heat and bring to boiling. Distribute into tubes or flasks. Autoclave for 15 min at 15 psi pressure–121°C. Pour into sterile Petri dishes or leave in tubes.

Use: For the cultivation and maintenance of *Streptomyces rectus*.

Soil Extract Medium

Composition per liter:

Soil	400.0g
Agar	15.0g

pH 6.8 ± 0.2 at 25°C

Preparation of Medium: Sieve air-dried garden soil with a high content of organic matter. To 400.0g of soil add 1.0L of distilled/deionized water. Autoclave for 60 min at 15 psi pressure–121°C. Cool to room temperature. Allow solids to sediment for a few hours. Carefully decant the supernatant solution. Centrifuge the supernatant solution at 10,000 × g for 15 min. Decant supernatant solution. Add agar (15.0g per liter of supernatant). Gently heat and bring to boiling. Distribute into tubes or flasks. Autoclave for 15 min at 15 psi pressure–121°C. Pour into sterile Petri dishes or leave in tubes.

Use: For the cultivation of *Arthrobacter* species, *Aureobacterium flavescens, Aureobacterium terregens,* and *Bacillus thiaminolyticus*.

Soil Extract Peptone Beef Extract Medium

Composition per liter:

Agar	15.0g
Peptone	5.0g
Beef extract	3.0g
Soil extract	1.0L

pH 7.0 ± 0.2 at 25°C

Soil Extract:

Composition per liter:

Garden soil	400.0g

Preparation of Soil Extract: Add garden soil to 1.0L of tap water. Autoclave for 1 hr at 15 psi pressure–121°C. Filter through cheesecloth and Whatman #2 filter paper. Autoclave filtrate again for 20 min at 15 psi pressure–121°C. Filter through Whatman #2 filter paper.

Preparation of Medium: Add agar, peptone, and beef extract to 1.0L of soil extract. Mix thoroughly. Gently heat and bring to boiling. Distribute into tubes or flasks. Autoclave for 15 min at 15 psi pressure–121°C. Pour into sterile Petri dishes or leave in tubes.

Use: For the cultivation and maintenance of *Oerskovia turbata* and *Oerskovia xanthineolytica*.

Soil Extract Potato Extract Medium

Composition per 510.0mL:

Malt extract	10.0g
Yeast extract	4.0g
Potato extract	250.0mL
Soil extract	250.0mL

pH 7.0 ± 0.2 at 25°C

Soil Extract:

Composition per liter:

Garden soil	400.0g

Preparation of Soil Extract: Add garden soil to 1.0L of tap water. Autoclave for 45 min at 15 psi pressure–121°C. Filter through cheesecloth.

Potato Extract:

Composition per liter:

Potatoes 400.0g

Preparation of Potato Extract: Peel and dice potatoes. Add 500.0mL of distilled/deionized water. Gently heat and bring to boiling. Continue boiling for 15 min. Filter through cheesecloth. Bring volume to 1.0L with distilled/deionized water.

Preparation of Medium: Combine components. Mix thoroughly. Distribute into tubes or flasks. Autoclave for 15 min at 15 psi pressure–121°C.

Use: For the cultivation and maintenance of *Saccharopolyspora rectivirgula.*

Soil Extract Salts Medium

Composition per liter:

Soil extract stock 625.0mL
Solution 1 .... 125.0mL
Solution 2 .... 125.0mL
Solution 3 .... 125.0mL

Soil Extract Stock:

Composition per liter:

Soil, air dried and sieved 333.3g

Preparation of Soil Extract Stock: Add soil to distilled/deionized water and bring volume to 1.0L. Mix thoroughly. Adjust pH to 8.0 with 1*N* NaOH or HCl. Autoclave for 30 min at 15 psi pressure–121°C. Allow soil to settle out. Carefully pour off supernatant. Filter through two layers of cheesecloth.

Solution 1:

Composition per liter:

K_2HPO_4 .... 1.0g

Preparation of Solution 1: Add K_2HPO_4 to distilled/deionized water and bring volume to 1.0L. Mix thoroughly.

Solution 2:

Composition per liter:

$MgSO_4 \cdot 7H_2O$ 1.0g

Preparation of Solution 2: Add $MgSO_4 \cdot 7H_2O$ to distilled/deionized water and bring volume to 1.0L. Mix thoroughly.

Solution 3:

Composition per liter:

KNO_3 .... 10.0g

Preparation of Solution 3: Add KNO_3 to distilled/deionized water and bring volume to 1.0L. Mix thoroughly.

Preparation of Medium: Combine components. Mix thoroughly. Filter sterilize. Aseptically distribute into sterile tubes or flasks.

Use: For the cultivation of *Chlamydomonas applanata, Polytoma uvella, Polytoma mirum, Polytoma ellipticum, Polytoma difficile,* and *Polytoma anomale.*

Soil Seawater Medium for Algae

Composition per liter:

HESNW medium solution 750.0mL
Soil extracts salts medium solution 250.0mL

HESNW Medium Solution:

Composition per 1011.0mL:

Natural seawater 1.0L
Enrichment solution 10.0mL
Vitamin solution 1.0mL

Preparation of HESNW Medium Solution: Allow natural seawater to age for 2 months. Filter sterilize. Aseptically add 10.0mL of sterile vitamin solution and 1.0mL enrichment solution. Mix thoroughly.

Enrichment Solution:

Composition per liter:

$NaNO_3$ 4.667g
$Na_2SiO_3 \cdot 9H_2O$ 3.000g
Sodium glycerophosphate 0.667g
$EDTA \cdot 2H_2O$ 0.553g
H_3BO_3 .... 0.380g
$Fe(NH_4)_2(SO_4)_2 \cdot 6H_2O$ 0.234g
$MnSO_4 \cdot 4H_2O$ 0.054g
$FeCl_3 \cdot 6H_2O$ 0.016g
$ZnSO_4 \cdot 7H_2O$ 7.3mg
$CoSO_4 \cdot 7H_2O$ 1.6mg

Preparation of Enrichment Solution: Add $Na_2SiO_3 \cdot 9H_2O$ to distilled/deionized water. Mix thoroughly. Neutralize $Na_2SiO_3 \cdot 9H_2O$ with 1*N* HCl. Add 500.0mL of distilled/deionized water. Mix thoroughly. Add remaining components and bring volume to 1.0L with distilled/deionized water. Mix thoroughly. Filter sterilize.

Vitamin Solution:

Composition per liter:

Thiamine 0.1g
Vitamin B_{12} 2.0mg
Biotin 1.0mg

Preparation of Vitamin Solution: Add components to distilled/deionized water and bring volume to 1.0L. Mix thoroughly. Filter sterilize.

Soil Extracts Salts Medium Solution:

Composition per liter:

Soil extract stock 625.0mL
Solution 1 .... 125.0mL
Solution 2 .... 125.0mL
Solution 3 .... 125.0mL

Soil Extract Stock:

Composition per liter:

Soil, air dried and sieved 333.3g

Preparation of Soil Extract Stock: Air dry soil. Sieve through fine-mesh screen. Add soil to distilled/deionized water and bring volume to 1.0L. Adjust pH to 8.0 with 1*N* NaOH or 1*N* HCl. Autoclave for 30 min at 15 psi pressure–121°C. Allow soil to settle. Decant liquid. Filter through cheesecloth.

Solution 1:

Composition per liter:

K_2HPO_4 .... 1.0g

Preparation of Solution 1: Add K_2HPO_4 to distilled/deionized water and bring volume to 1.0L. Mix thoroughly.

Solution 2:

Composition per liter:

$MgSO_4 \cdot 7H_2O$ 1.0g

Preparation of Solution 2: Add $MgSO_4 \cdot 7H_2O$ to distilled/deionized water and bring volume to 1.0L. Mix thoroughly.

Solution 3:

Composition per liter:

KNO_3	10.0g

Preparation of Solution 3: Add KNO_3 to distilled/deionized water and bring volume to 1.0L. Mix thoroughly.

Preparation of Soil Extracts Salts Medium Solution: Combine 625.0mL of soil extract stock, 125.0mL of solution 1, 125.0mL of solution 2, and 125.0mL of solution 3. Filter sterilize.

Preparation of Medium: Aseptically combine 750.0mL of sterile HESNW medium solution and 250.0mL of sterile soil extracts salts medium solution. Aseptically distribute into sterile tubes or flasks.

Use: For the cultivation of *Amphora roettgeri.*

Sonneborn's *Paramecium* Medium

Composition per liter:

Solution 1	1.0L
Klebsiella pneumoniae cultured on solution 2	variable

Solution 1:

Composition per liter:

Rye grass cerophyll	2.5g
Na_2HPO_4	0.5g

Preparation of Solution 1: Add cerophyll to distilled/deionized water and bring volume to 1.0L. Mix thoroughly. Gently heat and bring to boiling. Boil for 5 min. Filter through Whatman #1 filter paper. Add 0.5g of Na_2HPO_4. Bring volume to 1.0L with distilled/deionized water. Mix thoroughly. Distribute 10.0mL volumes into tubes. Autoclave for 15 min at 15 psi pressure–121°C.

Source: Cerophyll can be obtained from Ward's Natural Science Establishment, Inc. Dairy Goat Nutrition distributes Grass Media Culture, which is equivalent. Cereal Leaf Product from Sigma Chemical is similar to cerophyll.

Solution 2:

Composition per liter:

Agar	20.0g
Yeast extract	4.0g
Glucose	0.16g

Preparation of Solution 2: Add components to distilled/deionized water and bring volume to 1.0L. Mix thoroughly. Gently heat and bring to boiling. Distribute 5.0mL volumes into tubes. Autoclave for 15 min at 15 psi pressure–121°C. Allow tubes to cool in a slanted position.

Preparation of Medium: Inoculate the surface of the agar slants of solution 2 with a culture of *Klebsiella pneumoniae*. Incubate at 37°C for 24–48 hr. Scrape cells from the surface of the agar slants and add to 10.0mL of solution 1. Incubate at 30°C for 24 hr. Inoculate with protozoa.

Use: For the cultivation of a variety of *Paramecium* species, *Bodo* species, *Cercomonas* species, *Chlamydophrys* species, *Cyclidium* species, *Dimastigella* species, *Euplotes aediculatus*, *Glaucoma chattoni*, *Reclinomonas americana*, *Rhabdostyla* species, *Rhynchomonas nasuta*, *Stachyamoeba* species, *Thaumatomastix* species, and *Trichamoeba* species.

Sorangium Medium

Composition per liter:

Agar	10.0g
KNO_3	1.0g
K_2HPO_4	1.0g
$MgSO_4$	0.2g
$CaCl_2$	0.1g
$FeCl_3$	0.02g

Preparation of Medium: Add components to tap water and bring volume to 1.0L. Mix thoroughly. Gently heat and bring to boiling. Distribute into tubes or flasks. Autoclave for 15 min at 15 psi pressure–121°C. Pour into sterile Petri dishes or leave in tubes. Allow tubes to cool in a slanted position. Aseptically add a sterile strip (4.5cm × 1.0cm) of Whatman #1 filter paper to the surface of each slant or 4–6 sterile strips of filter paper to the surface of each agar plate.

Use: For the cultivation and maintenance of *Polyangium cellulosum.*

Sorbitol Agar

Composition per liter:

Agar	20.0g
Peptone	10.0g
NaCl	2.0g
Yeast extract	2.0g
Sorbitol solution	50.0mL

pH 7.0 ± 0.2 at 25°C

Sorbitol Solution:

Composition per 50.0mL:

Sorbitol	5.0g

Preparation of Sorbitol Solution: Add sorbitol to distilled/deionized water and bring volume to 50.0mL. Mix thoroughly. Filter sterilize.

Preparation of Medium: Add components, except sorbitol solution, to distilled/deionized water and bring volume to 950.0mL. Mix thoroughly. Adjust pH to 7.0. Gently heat and bring to boiling. Autoclave for 15 min at 15 psi pressure–121°C. Cool to 45°–50°C. Aseptically add sterile sorbitol solution. Mix thoroughly. Pour into sterile Petri dishes or distribute into sterile tubes.

Use: For the cultivation and maintenance of *Pseudomonas* species.

Sorbitol HiVeg Agar (Sorbitol MacConkey HiVeg Agar)

Composition per liter:

Plant peptone	17.0g
Agar	13.5g
D-Sorbitol	10.0g
Plant peptone No. 3	3.0g
Synthetic detergent	1.5g
NaCl	5.0g
Neutral Red	0.03g
Crystal Violet	1.0mg

pH 7.1 ± 0.2 at 25°C

Preparation of Medium: Add components to distilled/deionized water and bring volume to 1.0L. Mix thoroughly. Gently heat while stirring and bring to boiling. Autoclave for 15 min at 15 psi pressure–121°C. Avoid overheating. Mix thoroughly. Pour into sterile Petri dishes.

Use: For the isolation and identification of enteropathogenic *Escherichia coli* strains associated with infant diarrhea.

Sorbitol HiVeg Agar with Blood (Sorbitol MacConkey HiVeg Agar)

Composition per liter:

Plant peptone	17.0g
Agar	13.5g

D-Sorbitol	10.0g
Plant peptone No. 3	3.0g
Synthetic detergent	1.5g
NaCl	5.0g
Neutral Red	0.03g
Crystal Violet	1.0mg
Sterile bovine or sheep blood	50.0mL

pH 7.3 ± 0.2 at 25°C

Source: This medium, without blood, is available as a premixed powder from HiMedia.

Preparation of Medium: Add components, except blood, to distilled/deionized water and bring volume to 950.0mL. Mix thoroughly. Gently heat and bring to boiling. Autoclave for 15 min at 15 psi pressure–121°C. Cool to 45°–50°C. Aseptically add 50.0mL of sterile bovine blood or sheep blood. Mix thoroughly. Pour into sterile Petri dishes.

Use: For the isolation and identification of a wide variety of pathogenic bacteria.

Sorbitol Iron HiVeg Agar

Composition per liter:

Agar	20.0g
Plant peptone No. 3	15.0g
NaCl	5.0g
Plant extract	3.0g
D-Sorbitol	2.0g
Ferric ammonium citrate	0.5g
$Na_2S_2O_3$	0.5g
Phenol Red	0.03g

pH 7.6 ± 0.2 at 25°C

Source: This medium is available as a premixed powder from HiMedia.

Preparation of Medium: Add components to distilled/deionized water and bring volume to 1.0L. Mix thoroughly. Gently heat and bring to boiling. Distribute into tubes or flasks. Autoclave for 15 min at 15 psi pressure–121°C. Pour into sterile Petri dishes or leave in tubes.

Use: For the isolation and cultivation of coliform bacteria. For the identification and differentiation of enteropathogenic strains of *Escherichia coli* that do not ferment sorbitol.

Sorbitol MacConkey Agar (MacConkey Agar with Sorbitol)

Composition per liter:

Peptone	20.0g
Agar	15.0g
Sorbitol	10.0g
NaCl	5.0g
Bile salts No. 3	1.5g
Neutral Red	0.03g
Crystal Violet	1.0mg

pH 7.1 ± 0.2 at 25°C

Source: This medium is available as a premixed powder from BD Diagnostic Systems and Oxoid Unipath.

Preparation of Medium: Add components to distilled/deionized water and bring volume to 1.0L. Mix thoroughly. Gently heat and bring to boiling. Distribute into tubes or flasks. Autoclave for 15 min at 15 psi pressure–121°C. Pour into sterile Petri dishes or leave in tubes.

Use: For the isolation and cultivation of pathogenic *Escherichia coli.*

Sorbitol MacConkey Agar (MacConkey Agar with Sorbitol) (BAM M139)

Composition per liter:

Peptone or gelysate	17.0g
Agar	13.5g
Sorbitol	10.0g
NaCl	5.0g
Proteose peptone No. 3 or polypeptone	3.0g
Bile salts, purified	1.5g
Neutral Red	0.03g
Crystal Violet	1.0mg

pH 7.1 ± 0.2 at 25°C

Preparation of Medium: Add components to distilled/deionized water and bring volume to 1.0L. Mix thoroughly. Gently heat and bring to boiling. Distribute into tubes or flasks. Autoclave for 15 min at 15 psi pressure–121°C. Pour into sterile Petri dishes or leave in tubes.

Use: For the isolation and cultivation of pathogenic *Escherichia coli.*

Sorbitol MacConkey Agar with BCIG (SMAC with BCIG)

Composition per liter:

Peptone	20.0g
Agar	15.0g
Sorbitol	10.0g
NaCl	5.0g
Proteose peptone	3.0g
Bile salts mixture	1.5g
5-Bromo-4-chloro-3-indolyl-β-D-glucuronide sodium salt	0.1g
Neutral Red	0.03g

pH 7.1 ± 0.2 at 25°C

Source: This medium is available from Oxoid Unipath.

Preparation of Medium: Add components to distilled/deionized water and bring volume to 1.0L. Mix thoroughly. Gently heat while stirring and bring to boiling. Autoclave for 15 min at 15 psi pressure–121°C. Pour into sterile Petri dishes.

Use: A selective and differential medium for the detection of *Escherichia coli* O157 incorporating the chromogen 5-bromo-4-chloro-3-indolyl-β-D-glucuronide (BCIG). The medium combines two different screening mechanisms for the detection of *E. coli* O157, the failure to ferment sorbitol and the absence of β-glucuronidase activity. The non-sorbitol-fermenting and β-glucuronidase-negative *E. coli* O157 will appear as straw-colored colonies. Organisms with β-glucuronidase activity will cleave the substrate, leading to a distinct blue-green coloration of the colonies. The intestinal tract of ruminants is the prime reservoir of *E. coli* O157 and other enterohemorrhagic *E. coli* (EHEC) strains; therefore meats derived from cattle, sheep, goat, and deer can be expected to be contaminated. Foods implicated in human illness related to *E. coli* O157 include meats, dairy products, vegetables, salads, apple juice, and water.

Sorbitol Medium

Composition per liter:

Sorbitol	50.0g
Agar	15.0g
Peptone	10.0g
Yeast extract	10.0g

pH 6.0 ± 0.2 at 25°C

Preparation of Medium: Add components to distilled/deionized water and bring volume to 1.0L. Mix thoroughly. Adjust pH to 6.0 with HCl. Gently heat and bring to boiling. Distribute into tubes or flasks. Autoclave for 15 min at 15 psi pressure–121°C. Pour into sterile Petri dishes or leave in tubes.

Use: For the cultivation and maintenance of *Gluconobacter asaii* and *Gluconobacter frateurii*.

Sorbitol Medium, 5%

Composition per liter:

D-Sorbitol	50.0g
Agar	15.0g
Peptone	10.0g
Yeast extract	10.0g

pH 6.0 ± 0.2 at 25°C

Preparation of Medium: Add components to distilled/deionized water and bring volume to 1.0L. Mix thoroughly. Adjust to pH 6.0 with HCl. Gently heat and bring to boiling. Distribute into tubes or flasks. Autoclave for 15 min at 15 psi pressure–121°C. Pour into sterile Petri dishes or leave in tubes.

Use: For the cultivation of *Gluconobacter frateurii* and *Gluconobacter asaii*.

Sorogena Medium

Composition per 2000.0mL:

HI Agar	1.0L
HI/LV Broth	1.0L

HI Agar:

Composition per liter:

Agar	15.0g
Hay infusion broth	1.0L

pH 6.5 ± 0.2 at 25°C

Hay Infusion Broth:

Composition per liter:

Hay	2.5g

Preparation of Hay Infusion Broth: Add hay to distilled/deionized water and bring volume to 1.0L. Mix thoroughly. Gently heat and bring to boiling. Boil for 30 min. Filter through Whatman #1 filter paper. Bring volume to 1.0L with distilled/deionized water.

Preparation of HI Agar: Add agar to hay infusion broth and bring volume to 1.0L. Mix thoroughly. Adjust pH to 7.0 with 5% lactic acid or 1*N* NaOH. Gently heat and bring to boiling. Autoclave for 20 min at 15 psi pressure–121°C. Pour into sterile Petri dishes.

HI/LY Broth:

Composition per liter:

Lactose	0.2g
Yeast extract	0.1g
Hay infusion broth	1.0L

pH 6.0 ± 0.2 at 25°C

Hay Infusion Broth:

Composition per liter:

Hay	2.5g

Preparation of Hay Infusion Broth: Add lactose and yeast extract to distilled/deionized water and bring volume to 1.0L. Mix thoroughly. Gently heat and bring to boiling. Boil for 30 min. Filter through Whatman #1 filter paper. Bring volume to 1.0L with distilled/deionized water.

Preparation of HI/LY Broth: Add components to 1.0L hay infusion broth L. Mix thoroughly. Adjust pH to 6.0 with 5% lactic acid. Autoclave for 15 min at 15 psi pressure–121°C. Cool to 25°C.

Preparation of Medium: Aseptically add 15.0mL of HI/LY broth as an overlay over the surface of the HI agar plates.

Use: For the cultivation of *Sorogena stoianovitchae*.

SOT Medium

Composition per liter:

$NaHCO_3$	16.8g
$NaNO_3$	2.5g
K_2SO_4	1.0g
NaCl	1.0g
K_2HPO_4	0.5g
$MgSO_4 \cdot 7H_2O$	0.2g
Disodium EDTA·$2H_2O$	0.08g
$CaCl_2 \cdot 2H_2O$	0.04g
$FeSO_4 \cdot 7H_2O$	0.01g
Trace metals mix A5	1.0mL
Trace metals mix B6, modified	1.0mL

pH 9.0 ± 0.2 at 25°C

Trace Metals Mix A5:

Composition per liter:

H_3BO_3	2.86g
$MnCl_2 \cdot 4H_2O$	1.81g
$Na_2MoO_4 \cdot 2H_2O$	0.39g
$ZnSO_4 \cdot 7H_2O$	0.222g
$CuSO_4 \cdot 5H_2O$	0.079g
$Co(NO_3)_2 \cdot 6H_2O$	0.049g

Preparation of Trace Metals Mix A5: Add components to distilled/deionized water and bring volume to 1.0L. Mix thoroughly.

Trace Metals Mix B6, Modified:

Composition per liter:

NH_4NO_3	0.23g
$K_2Cr_2(SO_4)_4 \cdot 24H_2O$	0.096g
$NiSO_4 \cdot 7H_2O$	0.048g
$Ti_2(SO_4)_3$	0.04g
$Na_2WO_4 \cdot 2H_2O$	0.018g

Preparation of Trace Metals Mix B6, Modified: Add components to distilled/deionized water and bring volume to 1.0L. Mix thoroughly.

Preparation of Medium: Add components to distilled/deionized water and bring volume to 1.0L. Mix thoroughly. Adjust pH to 9.0. Distribute into tubes or flasks. Autoclave for 15 min at 15 psi pressure–121°C.

Use: For the cultivation and maintenance of *Spirulina maxima* and *Spirulina platensis*.

Sour Dough Medium

Composition per liter:

Maltose	20.0g
Pancreatic digest of casein	6.0g
Yeast extract	3.0g
Tween™ 80	0.3g
Fresh yeast extract solution	15.0mL

pH 5.6 ± 0.2 at 25°C

Fresh Yeast Extract Solution:

Composition per 100.0mL:

Baker's yeast, live, pressed, starch-free, ... 25.0g

Preparation of Fresh Yeast Extract Solution: Add the live Baker's yeast to 100.0mL of distilled/deionized water. Autoclave for 90 min at 15 psi pressure–121°C. Allow to stand. Remove supernatant solution. Adjust pH to 6.6–6.8.

Preparation of Medium: Add components to distilled/deionized water and bring volume to 1.0L. Mix thoroughly. Adjust pH to 5.6 with 20% lactic acid or 6*N* HCl. Distribute into tubes or flasks. Autoclave for 15 min at 15 psi pressure–121°C.

Use: For the cultivation and maintenance of *Lactobacillus sanfrancisco*.

Soy Peptone Broth

Composition per liter:

Papaic digest of soybean meal 20.0g
NaCl 5.0g
Phenol Red (2% solution) 1.0mL

pH 7.3 ± 0.2 at 25°C

Preparation of Medium: Add components to distilled/deionized water and bring volume to 1.0L. Mix thoroughly. Adjust pH to 7.3. Distribute into tubes or flasks. Autoclave for 15 min at 15 psi pressure–121°C.

Use: For the isolation and cultivation of *Mycoplasma* species and *Ureaplasma* species.

Soybean Agar

Composition per liter:

White soybeans 100.0g
Agar 15.0g

Preparation of Medium: Add soybeans to 1.0L of distilled/deionized water. Soak overnight. Autoclave for 60 min at 15 psi pressure–121°C. Filter through cheesecloth. Measure volume of filtrate. Add agar to a concentration of 1.5%. Gently heat and bring to boiling. Distribute into tubes or flasks. Autoclave for 15 min at 15 psi pressure–121°C. Pour into sterile Petri dishes or leave in tubes.

Use: For the cultivation and maintenance of *Bacillus subtilis* and *Pseudomonas syringae.*

Soybean Casein Digest Agar
See: **Trypticase™ Soy Agar**

Soybean Casein Digest Agar, HiVeg (Tryptone Soy Agar, HiVeg)

Composition per liter:

Agar 15.0g
Plant hydrolysate 15.0g
Papaic digest of soybean meal 5.0g
NaCl 5.0g

pH 7.2 ± 0.2 at 25°C

Source: This medium is available as a premixed powder from HiMedia.

Preparation of Medium: Add components to distilled/deionized water and bring volume to 1.0L. Mix thoroughly. Gently heat and bring to boiling. Distribute into tubes or flasks. Autoclave for 15 min at 15 psi pressure–121°C. Pour into sterile Petri dishes or leave in tubes.

Use: For the cultivation of a wide variety of bacteria.

Soybean Casein Digest Broth, USP
See: **Trypticase™ Soy Broth**

Soybean Casein Digest Medium, HiVeg (Tryptone Soy Broth, HiVeg)

Composition per liter:

Plant hydrolysate 17.0g
NaCl 5.0g
Papaic digest of soybean meal 3.0g
Glucose 2.5g
KH_2PO_4 .... 2.5g

pH 7.2 ± 0.2 at 25°C

Source: This medium is available as a premixed powder from HiMedia.

Preparation of Medium: Add components to distilled/deionized water and bring volume to 1.0L. Mix thoroughly. Gently heat and bring to boiling. Distribute into tubes or flasks. Autoclave for 15 min at 15 psi pressure–121°C. Store in a cool dark place preferably below 25°C.

Use: For the cultivation of a wide variety of bacteria.

Soybean Extract, M-1

Composition per liter:

Soybeans 50.0g
Soluble starch 15.0g
$(NH_4)_2HPO_3$ 10.0g
KCl 0.2g
$MgSO_4 \cdot 7H_2O$ 0.2g

pH 7.0 ± 0.2 at 25°C

Preparation of Medium: Add soybeans to 1.0L of distilled/deionized water. Soak overnight. Add 2.0g of NaOH. Adjust pH to 7.0 with HCl. Autoclave for 60 min at 0 psi pressure–100°C. Filter through cheesecloth. Bring volume of filtrate to 1.0L with distilled/deionized water. Add remaining components. Mix thoroughly. Distribute into tubes or flasks. Autoclave for 15 min at 15 psi pressure–121°C.

Use: For the cultivation and maintenance of *Bacillus subtilis*.

Soybean HiVeg Broth Base with Novobiocin

Composition per liter:

Plant hydrolysate 17.0g
NaCl 5.0g
K_2HPO_4 .... 4.0g
Papaic digest of soybean meal 3.0g
Glucose 2.5g
Synthetic detergent No. I 1.12g
Selective solution 10.0mL

pH 7.3 ± 0.2 at 25°C

Source: This medium, without selective novobiocin supplement, is available as a premixed powder from HiMedia.

Selective Solution:

Composition per 10.0mL:

Novobiocin 0.1g

Preparation of Selective Solution: Add novobiocin to distilled/deionized water and bring volume to 10.0mL. Mix thoroughly. Filter sterilize.

Preparation of Medium: Add components, except selective solution, to distilled/deionized water and bring volume to 990.0mL. Mix thoroughly. Gently heat and bring to boiling. Mix thoroughly. Autoclave for 15 min at 15 psi pressure–121°C. Cool to 45°–50°C. Aseptically add 10.0mL selective solution. Mix thoroughly. Distribute into tubes or flasks. Autoclave for 5 min at 15 psi pressure–121°C. Pour into sterile Petri dishes.

Use: For the enrichment and isolation of *Escherichia coli* O157:H7 from foods.

Soybean HiVeg Medium (Tryptone Soy HiVeg Broth without Glucose)

Composition per liter:

Plant hydrolysate	17.0g
NaCl	5.0g
Papaic digest of soybean meal	3.0g
K_2HPO_4	2.5g

pH 7.3 ± 0.2 at 25°C

Source: This medium is available as a premixed powder from HiMedia.

Preparation of Medium: Add components to distilled/deionized water and bring volume to 1.0L. Mix thoroughly. Gently heat and bring to boiling. Distribute into tubes or flasks. Autoclave for 15 min at 15 psi pressure–121°C.

Use: For the cultivation of a wide variety of microorganisms. For sterility testing.

Soybean HiVeg Medium with 0.1% Agar (Tryptone Soy HiVeg Broth with 0.1% Agar)

Composition per liter:

Plant hydrolysate	17.0g
NaCl	5.0g
Papaic digest of soybean meal	3.0g
K_2HPO_4	2.5g
Agar	1.0g

pH 7.3 ± 0.2 at 25°C

Source: This medium is available as a premixed powder from HiMedia.

Preparation of Medium: Add components to distilled/deionized water and bring volume to 1.0L. Mix thoroughly. Gently heat and bring to boiling. Distribute into tubes or flasks. Autoclave for 15 min at 15 psi pressure–121°C. Pour into sterile Petri dishes or leave in tubes.

Use: For the cultivation of anaerobes from root canals, blood, and other clinical specimens.

Soybean HiVeg Medium with 0.1% Agar with Glucose (Tryptone Soy HiVeg Broth with 0.1% Agar)

Composition per liter:

Plant hydrolysate	17.0g
NaCl	5.0g
Papaic digest of soybean meal	3.0g
Glucose	2.5g
K_2HPO_4	2.5g
Agar	1.0g

pH 7.3 ± 0.2 at 25°C

Source: This medium is available as a premixed powder from HiMedia.

Preparation of Medium: Add components to distilled/deionized water and bring volume to 1.0L. Mix thoroughly. Gently heat and bring to boiling. Distribute into tubes or flasks. Autoclave for 15 min at 15 psi pressure–121°C.

Use: For the cultivation of anaerobes from root canals, blood, and other clinical specimens and for determining glucose fermentation.

Soybean HiVeg Medium with Yeast Extract and Ferric Pyrophosphate

Composition per liter:

Plant hydrolysate	17.0g
NaCl	5.0g
Yeast extract	5.0g
Papaic digest of soybean meal	3.0g
K_2HPO_4	2.5g
$Fe_4(P_2O_7)_3 \cdot H_2O$	0.02g

pH 7.3 ± 0.2 at 25°C

Source: This medium is available as a premixed powder from HiMedia.

Preparation of Medium: Add components to distilled/deionized water and bring volume to 1.0L. Mix thoroughly. Gently heat and bring to boiling. Distribute into tubes or flasks. Autoclave for 15 min at 15 psi pressure–121°C.

Use: For the cultivation of fastidious bacteria.

SP Agar

Composition per liter:

Agar	15.0g
Pancreatic digest of casein	2.5g
Galactose	1.0g
Raffinose	1.0g
Sucrose	1.0g
$MgSO_4 \cdot 7H_2O$	0.5g
K_2HPO_4	0.25g
Vitamin solution	2.5mL

Vitamin Solution:

Composition per liter:

Inositol	1.0g
Calcium pantothenate	0.2g
Choline hydrochloride	0.2g
Thiamine	0.1g
Nicotinamide	0.75g
Pyridoxin	0.75g
Riboflavin	0.75g
p-Aminobenzoic acid	5.0mg
Folic acid	1.0mg
Biotin	0.05mg
Vitamin B_{12}	0.05mg
Ethanol	1.0L

Preparation of Vitamin Solution: Add solid components to 1.0L of ethanol. Mix thoroughly.

Preparation of Medium: Add components to distilled/deionized water and bring volume to 1.0L. Mix thoroughly. Gently heat and bring to boiling. Distribute into tubes or flasks. Autoclave for 15 min at 15 psi pressure–121°C. Pour into sterile Petri dishes or leave in tubes.

Use: For the cultivation of myxobacteria.

SP 2 Agar

Composition per liter:

Agar	15.0g
Pancreatic digest of casein	3.0g
Yeast extract	1.0g
Sodium acetate	0.02g
Artificial seawater	1.0L

pH 7.2 ± 0.2 at 25°C

Artificial Seawater:
Composition per liter:

NaCl	24.7g
$MgSO_4 \cdot 7H_2O$	6.3g
$MgCl_2 \cdot 6H_2O$	4.6g
$CaCl_2$	1.0g
KCl	0.7g
$NaHCO_3$	0.2g

Preparation of Artificial Seawater: Add components to distilled/deionized water and bring volume to 1.0L. Mix thoroughly.

Preparation of Medium: Add solid components to 1.0L of artificial seawater. Mix thoroughly. Gently heat and bring to boiling. Distribute into tubes or flasks. Autoclave for 15 min at 15 psi pressure–121°C. Pour into sterile Petri dishes or leave in tubes.

Use: For the isolation and cultivation of *Cytophaga* species, *Herpetosiphon* species, *Saprospira* species, and *Flexithrix* species.

SP Medium

Composition per liter:

Agar	15.0g
Soluble starch	5.0g
Pancreatic digest of casein	2.5g
Galactose	1.0g
Raffinose	1.0g
Sucrose	1.0g
$MgSO_4 \cdot 7H_2O$	0.5g
K_2HPO_4	0.25g

Preparation of Medium: Add components to distilled/deionized water and bring volume to 1.0L. Mix thoroughly. Gently heat and bring to boiling. Distribute into tubes or flasks. Autoclave for 15 min at 15 psi pressure–121°C. Pour into sterile Petri dishes or leave in tubes.

Use: For the cultivation and maintenance of *Archanigium gephyra*, *Cystobacter fuscus*, *Melittangium lichenicola*, *Myxococcus* species, *Polyangium brachysporum*, *Stigmatella aurantiaca*, and *Stigmatella erecta*.

SP 2 Agar

Composition per liter:

Agar	15.0g
Pancreatic digest of casein	3.0g
Yeast extract	1.0g
Sodium acetate	0.02g
Artificial seawater	1.0L

pH 7.2 ± 0.2 at 25°C

Artificial Seawater:
Composition per liter:

NaCl	24.7g
$MgSO_4 \cdot 7H_2O$	6.3g
$MgCl_2 \cdot 6H_2O$	4.6g
$CaCl_2$	1.0g
KCl	0.7g
$NaHCO_3$	0.2g

Preparation of Artificial Seawater: Add components to distilled/deionized water and bring volume to 1.0L. Mix thoroughly.

Preparation of Medium: Add solid components to 1.0L of artificial seawater. Mix thoroughly. Gently heat and bring to boiling. Distribute into tubes or flasks. Autoclave for 15 min at 15 psi pressure–121°C. Pour into sterile Petri dishes or leave in tubes.

Use: For the isolation and cultivation of *Cytophaga* species, *Herpetosiphon* species, *Saprospira* species, and *Flexithrix* species.

SP 4 Medium

Composition per liter:

Base solution	615.0mL
Fetal calf serum (inactivated at 56°C, 1 hr)	170.0mL
Yeast extract (2% solution)	100.0mL
CMRL 1066, 10X with glutamine	50.0mL
Fresh yeast extract solution	35.0mL
Phenol Red (0.1% solution)	20.0mL
Penicillin solution	10.0mL

pH 7.0–7.4 ± 0.2 at 25°C

Base Solution:
Composition per 615.0mL:

Pancreatic digest of casein	11.2g
Noble agar	8.0g
Pancreatic digest of gelatin	5.3g
Glucose	5.0g
NaCl	0.875g
Beef extract	0.525g
Yeast extract	0.525g
Beef heart, solids from infusion	0.35g

Preparation of Base Solution: Add components to distilled/deionized water and bring volume to 615.0mL. Mix thoroughly. Adjust pH to 7.5. Gently heat and bring to boiling. Autoclave for 15 min at 15 psi pressure–121°C. Cool to 45°–50°C.

CMRL 1066, 10X with Glutamine:
Composition per liter:

NaCl	6.8g
$NaHCO_3$	2.2g
D-Glucose	1.0g
KCl	0.4g
L-Cysteine·HCl·H_2O	0.26g
$CaCl_2$, anhydrous	0.2g
$MgSO_4 \cdot 7H_2O$	0.2g
$NaH_2PO_4 \cdot H_2O$	0.14g
L-Glutamine	0.1g
Sodium acetate·$3H_2O$	0.083g
L-Glutamic acid	0.075g
L-Arginine·HCl	0.07g
L-Lysine·HCl	0.07g
L-Leucine	0.06g
Glycine	0.05g
Ascorbic acid	0.05g
L-Proline	0.04g
L-Tyrosine	0.04g
L-Aspartic acid	0.03g
L-Threonine	0.03g
L-Alanine	0.025g
L-Phenylalanine	0.025g

L-Serine	0.025g
L-Valine	0.025g
L-Cystine	0.02g
L-Histidine·HCl·H_2O	0.02g
L-Isoleucine	0.02g
Phenol Red	0.02g
L-Methionine	0.015g
Deoxyadenosine	0.01g
Deoxycytidine	0.01g
Deoxyguanosine	0.01g
Glutathione, reduced	0.01g
Thymidine	0.01g
Hydroxy-L-proline	0.01g
L-Tryptophan	0.01g
Nicotinamide adenine dinucleotide	7.0mg
Tween™ 80	5.0mg
Sodium glucoronate·H_2O	4.2mg
Coenzyme A	2.5mg
Cocarboxylase	1.0mg
Flavin adenine dinucleotide	1.0mg
Nicotinamide adenine dinucleotide phosphate	1.0mg
Uridine triphosphate	1.0mg
Choline chloride	0.5mg
Cholesterol	0.2mg
5-Methyldeoxycytidine	0.1mg
Inositol	0.05mg
p-Aminobenzoic acid	0.05mg
Niacin	0.025mg
Niacinamide	0.025mg
Pyridoxine	0.025mg
Pyridoxal·HCl	0.025mg
Biotin	0.01mg
D-Calcium pantothenate	0.01mg
Folic acid	0.01mg
Riboflavin	0.01mg
Thiamine·HCl	0.01mg

Preparation of CMRL 1066, 10X with Glutamine: Add components to distilled/deionized water and bring volume to 1.0L. Mix thoroughly. Adjust pH to 7.2. Filter sterilize.

Fresh Yeast Extract Solution:
Composition per 100.0mL:

Baker's yeast, live, pressed, starch-free,	25.0g

Preparation of Fresh Yeast Extract Solution: Add the live Baker's yeast to 100.0mL of distilled/deionized water. Autoclave for 90 min at 15 psi pressure–121°C. Allow to stand. Remove supernatant solution. Adjust pH to 6.6–6.8.

Penicillin Solution:
Composition per 10.0mL:

Penicillin G	1,000,000U

Preparation of Penicillin Solution: Add penicillin G to distilled/deionized water and bring volume to 10.0mL. Mix thoroughly. Filter sterilize.

Preparation of Medium: To 615.0mL of cooled sterile base solution, aseptically add 170.0mL of sterile inactivated fetal calf serum, 100.0mL of sterile yeast extract, 50.0mL of sterile CMRL 1066, 10X with glutamine, 35.0mL of sterile fresh yeast extract solution, 20.0mL of Phenol Red solution, and 10.0mL of sterile penicillin solution. Mix thoroughly. Aseptically distribute into sterile tubes. Allow tubes to cool in a slanted position.

Use: For the cultivation of tick-derived *Mycoplasma* (*Spiroplasma*). Used for the enhanced recovery of *Mycoplasma pneumoniae, Mycoplasma alvi*, and *Mycoplasma hyopneumoniae*.

SP 4 Medium (DSMZ Medium 1076)

Composition per 510.2mL:

Tryptone	5.0g
Peptone	3.3g
NaCl	0.5g
Beef extract	0.3g
Yeast extract	0.3g
Beef heart, solids from infusion	0.2g
Fetal bovine serum (inactivated at 56°C, 1 hr)	90.0mL
CMRL 1066, 10X with glutamine	25.0mL
Yeast extract solution	17.5mL
Yeastolate solution	5.0mL
Glutamine solution	1.7mL
Phenol Red solution	1.0mL

pH 7.4 ± 0.2 at 25°C

CMRL 1066, 10X with Glutamine:
Composition per liter:

NaCl	6.8g
$NaHCO_3$	2.2g
D-Glucose	1.0g
KCl	0.4g
L-Cysteine·HCl·H_2O	0.26g
$CaCl_2$, anhydrous	0.2g
$MgSO_4 \cdot 7H_2O$	0.2g
$NaH_2PO_4 \cdot H_2O$	0.14g
L-Glutamine	0.1g
Sodium acetate·$3H_2O$	0.083g
L-Glutamic acid	0.075g
L-Arginine·HCl	0.07g
L-Lysine·HCl	0.07g
L-Leucine	0.06g
Glycine	0.05g
Ascorbic acid	0.05g
L-Proline	0.04g
L-Tyrosine	0.04g
L-Aspartic acid	0.03g
L-Threonine	0.03g
L-Alanine	0.025g
L-Phenylalanine	0.025g
L-Serine	0.025g
L-Valine	0.025g
L-Cystine	0.02g
L-Histidine·HCl·H_2O	0.02g
L-Isoleucine	0.02g
Phenol Red	0.02g
L-Methionine	0.015g
Deoxyadenosine	0.01g
Deoxycytidine	0.01g
Deoxyguanosine	0.01g
Glutathione, reduced	0.01g
Thymidine	0.01g
Hydroxy-L-proline	0.01g
L-Tryptophan	0.01g
Nicotinamide adenine dinucleotide	7.0mg
Tween™ 80	5.0mg
Sodium glucoronate·H_2O	4.2mg

Coenzyme A	2.5mg
Cocarboxylase	1.0mg
Flavin adenine dinucleotide	1.0mg
Nicotinamide adenine dinucleotide phosphate	1.0mg
Uridine triphosphate	1.0mg
Choline chloride	0.5mg
Cholesterol	0.2mg
5-Methyldeoxycytidine	0.1mg
Inositol	0.05mg
p-Aminobenzoic acid	0.05mg
Niacin	0.025mg
Niacinamide	0.025mg
Pyridoxine	0.025mg
Pyridoxal·HCl	0.025mg
Biotin	0.01mg
D-Calcium pantothenate	0.01mg
Folic acid	0.01mg
Riboflavin	0.01mg
Thiamine·HCl	0.01mg

Preparation of CMRL 1066, 10X with Glutamine: Add components to distilled/deionized water and bring volume to 1.0L. Mix thoroughly. Adjust pH to 7.2. Filter sterilize.

Yeast Extract Solution:

Composition per 10.0mL:

Yeast extract	2.0g

Preparation of Yeast Extract Solution: Add yesat extract to distilled/deionized water and bring volume to 10.0mL. Mix thoroughly. Autoclave for 15 min at 15 psi pressure–121°C.

Yeastolate Solution:

Composition per 10.0mL:

Yeastolate	2.0g

Preparation of Yeastolate Solution: Add yeastolate to distilled/deionized water and bring volume to 10.0mL. Mix thoroughly. Autoclave for 15 min at 15 psi pressure–121°C.

Phenol Red Solution:

Composition per 100.0mL:

Phenol Red	1.0g

Preparation of Phenol Red Solution: Add 1.0g of Phenol Red to distilled/deionized water and bring volume to 100.0mL. Mix thoroughly. Adjust pH to 7.0. Filter sterilize.

Glutamine Solution:

Composition per 10.0mL:

L-Glutamine	1.5g

Preparation of Glutamine Solution: Add 1.5g of L-glutamine to distilled/deionized water and bring volume to 10.0mL. Mix thoroughly. Filter sterilize.

Preparation of Medium: Add components, except fetal bovine serum, CMRL, yeast extract solution, yeastolate solution, Phenol Red solution, and glutamine solution, to distilled/deionized water and bring volume to 375.0mL. Mix thoroughly. Adjust pH to 7.4. Gently heat and bring to boiling. Autoclave for 15 min at 15 psi pressure–121°C. Cool to room temperature. Mix thoroughly. Aseptically add fetal bovine serum, CMRL, yeast extract solution, yeastolate solution, Phenol Red solution, and glutamine solution. Mix thoroughly.

Use: For the cultivation of *Mycoplasma fermentans.*

SP 4 Medium with Glucose (DSMZ Medium 1076a)

Composition per 515.4mL:

Tryptone	5.0g
Peptone	3.3g
NaCl	0.5g
Beef extract	0.3g
Yeast extract	0.3g
Beef heart, solids from infusion	0.2g
Fetal bovine serum (inactivated at 56°C, 1 hr)	90.0mL
CMRL 1066, 10X with glutamine	25.0mL
Yeast extract solution	17.5mL
Glucose solution	5.2mL
Yeastolate solution	5.0mL
Glutamine solution	1.7mL
Phenol Red solution	1.0mL

pH 7.4 ± 0.2 at 25°C

CMRL 1066, 10X with Glutamine:

Composition per liter:

NaCl	6.8g
$NaHCO_3$	2.2g
D-Glucose	1.0g
KCl	0.4g
L-Cysteine·HCl·H_2O	0.26g
$CaCl_2$, anhydrous	0.2g
$MgSO_4·7H_2O$	0.2g
$NaH_2PO_4·H_2O$	0.14g
L-Glutamine	0.1g
Sodium acetate·$3H_2O$	0.083g
L-Glutamic acid	0.075g
L-Arginine·HCl	0.07g
L-Lysine·HCl	0.07g
L-Leucine	0.06g
Glycine	0.05g
Ascorbic acid	0.05g
L-Proline	0.04g
L-Tyrosine	0.04g
L-Aspartic acid	0.03g
L-Threonine	0.03g
L-Alanine	0.025g
L-Phenylalanine	0.025g
L-Serine	0.025g
L-Valine	0.025g
L-Cystine	0.02g
L-Histidine·HCl·H_2O	0.02g
L-Isoleucine	0.02g
Phenol Red	0.02g
L-Methionine	0.015g
Deoxyadenosine	0.01g
Deoxycytidine	0.01g
Deoxyguanosine	0.01g
Glutathione, reduced	0.01g
Thymidine	0.01g
Hydroxy-L-proline	0.01g
L-Tryptophan	0.01g
Nicotinamide adenine dinucleotide	7.0mg
Tween™ 80	5.0mg
Sodium glucoronate·H_2O	4.2mg
Coenzyme A	2.5mg
Cocarboxylase	1.0mg
Flavin adenine dinucleotide	1.0mg

Nicotinamide adenine dinucleotide phosphate 1.0mg
Uridine triphosphate 1.0mg
Choline chloride 0.5mg
Cholesterol 0.2mg
5-Methyldeoxycytidine 0.1mg
Inositol 0.05mg
p-Aminobenzoic acid 0.05mg
Niacin 0.025mg
Niacinamide 0.025mg
Pyridoxine 0.025mg
Pyridoxal·HCl 0.025mg
Biotin 0.01mg
D-Calcium pantothenate 0.01mg
Folic acid 0.01mg
Riboflavin 0.01mg
Thiamine·HCl 0.01mg

Preparation of CMRL 1066, 10X with Glutamine: Add components to distilled/deionized water and bring volume to 1.0L. Mix thoroughly. Adjust pH to 7.2. Filter sterilize.

Yeast Extract Solution:
Composition per 10.0mL:
Yeast extract 2.0g

Preparation of Yeast Extract Solution: Add yesat extract to distilled/deionized water and bring volume to 10.0mL. Mix thoroughly. Autoclave for 15 min at 15 psi pressure–121°C.

Yeastolate Solution:
Composition per 10.0mL:
Yeastolate 2.0g

Preparation of Yeastolate Solution: Add yeastolate to distilled/deionized water and bring volume to 10.0mL. Mix thoroughly. Autoclave for 15 min at 15 psi pressure–121°C.

Phenol Red Solution:
Composition per 100.0mL:
Phenol Red 1.0g

Preparation of Phenol Red Solution: Add 1.0g of Phenol Red to distilled/deionized water and bring volume to 100.0mL. Mix thoroughly. Adjust pH to 7.0. Filter sterilize.

Glutamine Solution:
Composition per 10.0mL:
L-Glutamine 1.5g

Preparation of Glutamine Solution: Add 1.5g of L-glutamine to distilled/deionized water and bring volume to 10.0mL. Mix thoroughly. Filter sterilize.

Glucose Solution:
Composition per 10.0mL:
D-Glucose 1.0g

Preparation of Glucose Solution: Add D-glucose to distilled/deionized water and bring volume to 10.0mL. Mix thoroughly. Filter sterilize.

Preparation of Medium: Add components, except fetal bovine serum, CMRL, yeast extract solution, yeastolate solution, Phenol Red solution, and glutamine solution, to distilled/deionized water and bring volume to 375.0mL. Mix thoroughly. Adjust pH to 7.4. Gently heat and bring to boiling. Autoclave for 15 min at 15 psi pressure–121°C. Cool to room temperature. Mix thoroughly. Aseptically add fetal bovine serum, CMRL, yeast extract solution, yeastolate solution, glucose solution, Phenol Red solution, and glutamine solution. Mix thoroughly.

Use: For the cultivation of *Mycoplasma genitalium.*

SP 5 Broth

Composition per liter:
Pancreatic digest of casein 9.0g
Yeast extract 1.0g
Artificial seawater 1.0L
pH 7.2 ± 0.2 at 25°C

Artificial Seawater:
Composition per liter:
NaCl 24.7g
$MgSO_4 \cdot 7H_2O$ 6.3g
$MgCl_2 \cdot 6H_2O$ 4.6g
$CaCl_2$ 1.0g
KCl 0.7g
$NaHCO_3$ 0.2g

Preparation of Artificial Seawater: Add components to distilled/deionized water and bring volume to 1.0L. Mix thoroughly.

Preparation of Medium: Add solid components to 1.0L of artificial seawater. Mix thoroughly. Gently heat and bring to boiling. Distribute into tubes or flasks. Autoclave for 15 min at 15 psi pressure–121°C.

Use: For the isolation and cultivation of *Cytophaga* species, *Herpetosiphon* species, *Saprospira* species, and *Flexithrix* species.

SP 6 Agar

Composition per liter:
Agar 15.0g
Pancreatic digest of casein 3.0g
Yeast extract 1.0g
Artificial seawater 1.0L
pH 7.2 ± 0.2 at 25°C

Artificial Seawater:
Composition per liter:
NaCl 24.7g
$MgSO_4 \cdot 7H_2O$ 6.3g
$MgCl_2 \cdot 6H_2O$ 4.6g
$CaCl_2$ 1.0g
KCl 0.7g
$NaHCO_3$ 0.2g

Preparation of Artificial Seawater: Add components to distilled/deionized water and bring volume to 1.0L. Mix thoroughly.

Preparation of Medium: Add solid components to 1.0L of artificial seawater. Mix thoroughly. Gently heat and bring to boiling. Distribute into tubes or flasks. Autoclave for 15 min at 15 psi pressure–121°C. Pour into sterile Petri dishes or leave in tubes.

Use: For the isolation and cultivation of *Cytophaga* species, *Herpetosiphon* species, *Saprospira* species, and *Flexithrix* species.

SPB

See: **Salt Polymyxin Broth**

Special Infusion Agar, HiVeg with Blood

Composition per liter:
Agar 15.0g
Plant infusion 10.0g
Plant peptone No. 3 .... 10.0g
Plant special infusion 7.5g

NaCl	5.0g
Na_2HPO_4	2.5g
Glucose	2.0g
Blood, defibrinated	50.0mL

pH 7.4 ± 0.2 at 25°C

Source: This medium is available as a premixed powder from HiMedia.

Preparation of Medium: Add components, except blood, to distilled/deionized water and bring volume to 950.0L. Mix thoroughly. Gently heat until boiling. Autoclave for 15 min at 15 psi pressure–121°C. Cool to 50°C. Add 50.0mL sterile defibrinated blood. Mix thoroughly. Pour into sterile Petri dishes or distribute into sterile tubes.

Use: For the cultivation of a variety of fastidious and nonfastidious aerobic and anaerobic microorganisms, including streptococci, yeasts, and molds.

Special Infusion Broth, HiVeg with Blood

Composition per liter:

Plant infusion	10.0g
Plant peptone No. 3	10.0g
Plant special infusion	7.5g
NaCl	5.0g
Na_2HPO_4	2.5g
Glucose	2.0g
Blood, defibrinated	50.0mL

pH 7.4 ± 0.2 at 25°C

Source: This medium is available as a premixed powder from HiMedia.

Preparation of Medium: Add components, except blood, to distilled/deionized water and bring volume to 950.0L. Mix thoroughly. Gently heat until boiling. Autoclave for 15 min at 15 psi pressure–121°C. Cool to 50°C. Add 50.0mL sterile defibrinated blood. Mix thoroughly. Aseptically distribute into sterile tubes.

Use: For the cultivation of a variety of fastidious and nonfastidious aerobic and anaerobic microorganisms, including streptococci. For the propagation of pathogenic cocci and other fastidious organisms associated with blood culture work and allied pathological investigations.

Specimen Preservative Medium

Composition per liter:

NaCl	5.0g
Sodium citrate·$2H_2O$	5.0g
$(NH_4)_2HPO_4$	4.0g
KH_2PO_4	2.0g
Yeast extract	1.0g
Sodium deoxycholate	0.5g
$MgSO_4 \cdot 7H_2O$	0.4g
Glycerol	300.0mL

pH 7.0 ± 0.2 at 25°C

Preparation of Medium: Add components, except glycerol, to distilled/deionized water and bring volume to 700.0mL. Mix thoroughly. Gently heat and bring to boiling. Add 300.0mL of glycerol. Mix thoroughly. Distribute into tubes or flasks. Autoclave for 10 min at 11 psi pressure–116°C.

Use: For the preservation of viable microorganisms in stool specimens. For the transport of fecal material.

Sphaericus Spore Medium

Composition per liter:

Agar	15.0g
Pancreatic digest of gelatin	5.0g
Beef extract	3.0g
Yeast extract	0.5g
$MgCl_2$	0.095g
$CaCl_2$	0.078g
$MnCl_2$	6.0mg

Preparation of Medium: Add components to distilled/deionized water and bring volume to 1.0L. Mix thoroughly. Gently heat and bring to boiling. Distribute into tubes or flasks. Autoclave for 15 min at 15 psi pressure–121°C. Pour into sterile Petri dishes or leave in tubes.

Use: For the cultivation and maintenance of *Bacillus sphaericus*.

Sphaerotilus Agar (DSMZ Medium 51)

Composition per liter:

Agar	15.0g
Beef extract, Lab Lemco	5.0g

pH 7.1 ± 0.2 at 25°C

Preparation of Medium: Add components to distilled/deionized water and bring volume to 1.0L. Mix thoroughly. Adjust pH to 7.0. Gently heat and bring to boiling. Distribute into tubes. Autoclave for 15 min at 15 psi pressure–121°C. Cool in a sloping position to form slants. Cover solid slants with 2mL sterile tap water. Inoculate into the covering tap water and incubate at 20°C–25°C.

Use: For the cultivation and maintenance of *Sphaerotilus natans*.

Sphaerotilus CGYA Medium

Composition per liter:

Glycerol	10.0g
Pancreatic digest of casein	5.0g
Yeast extract	1.0g

Preparation of Medium: Add components to distilled/deionized water and bring volume to 1.0L. Mix thoroughly. Distribute into tubes or flasks. Autoclave for 15 min at 15 psi pressure–121°C.

Use: For the cultivation and maintenance of *Sphaerotilus natans* and *Sphaerotilus* species.

Sphaerotilus Defined Medium

Composition per liter:

Agar	15.0g
Glycerol	5.0g
Glutamic acid	0.9g
$FeSO_4 \cdot 7H_2O$	0.5g
$MgSO_4 \cdot 7H_2O$	0.1g
$CaCl_2 \cdot 2H_2O$	0.03g
$ZnSO_4 \cdot 7H_2O$	0.03g
Phosphate solution	100.0mL

pH 7.0 ± 0.2 at 25°C

Phosphate Solution:

Composition per 500.0mL:

K_2HPO_4	5.7g
KH_2PO_4	2.3g

Preparation of Phosphate Solution: Add components to distilled/deionized water and bring volume to 500.0mL. Mix thoroughly. Gently heat until dissolved. Autoclave for 15 min at 15 psi pressure–121°C.

Preparation of Medium: Add components, except phosphate solution, to distilled/deionized water and bring volume to 900.0mL. Mix thoroughly. Gently heat and bring to boiling. Autoclave for 10 min at 15 psi pressure–121°C. Cool to 45°–50°C. Aseptically add 100.0mL of sterile phosphate solution. Mix thoroughly. Pour into sterile Petri dishes or distribute into sterile tubes.

Use: For the cultivation of *Sphaerotilus* species.

Sphaerotilus discophorus Medium

Composition per liter:

Agar	12.0g
Peptone	5.0g
$MgSO_4 \cdot 7H_2O$	0.2g
$CaCl_2$	0.05g
$MnSO_4 \cdot H_2O$	0.05g
Ferric solution	100.0mL

pH 7.0 ± 0.2 at 25°C

Ferric Solution:

Composition per 100.0mL:

Ferric ammonium citrate	0.5g
$FeCl_3 \cdot 6H_2O$	0.01g

Preparation of Ferric Solution: Add components to distilled/deionized water and bring volume to 100.0mL. Mix thoroughly. Filter sterilize.

Preparation of Medium: Add components, except ferric solution, to tap water and bring volume to 900.0mL. Mix thoroughly. Gently heat and bring to boiling. Autoclave for 15 min at 15 psi pressure–121°C. Cool to 45°–50°C. Aseptically add sterile ferric solution. Mix thoroughly. Pour into sterile Petri dishes or distribute into sterile tubes.

Use: For the cultivation of *Sphaerotilus discophorus*.

Sphaerotilus Isolation Medium

Composition per liter:

Agar	15.0g
Glycerol	10.0g
Pancreatic digest of casein	5.0g
Yeast extract	1.0g

pH 7.0 ± 0.2 at 25°C

Preparation of Medium: Add components to distilled/deionized water and bring volume to 1.0L. Mix thoroughly. Gently heat and bring to boiling. Distribute into tubes or flasks. Autoclave for 15 min at 15 psi pressure–121°C. Pour into sterile Petri dishes or leave in tubes.

Use: For the isolation and cultivation of *Sphaerotilus* species.

Sphaerotilus/Leptothrix Agar

Composition per liter:

Agar	20.0g
Peptone	1.5g
Yeast extract	1.0g
Ferric ammonium citrate	0.5g
$MgSO_4 \cdot 7H_2O$	0.2g
$CaCl_2$	0.05g
$MnSO_4 \cdot H_2O$	0.05g
$FeCl_3 \cdot 6H_2O$	0.01g

pH 7.1 ± 0.2 at 25°C

Preparation of Medium: Add components to distilled/deionized water and bring volume to 1.0L. Mix thoroughly. Adjust pH to 7.1. Gently heat and bring to boiling. Distribute into tubes or flasks. Autoclave for 15 min at 15 psi pressure–121°C. Pour into sterile Petri dishes or leave in tubes.

Use: For the cultivation and maintenance of *Leptothrix cholodnii, Leptothrix* species, and *Sphaerotilus natans.*

Sphaerotilus and *Leptothrix* Enrichment Medium

Composition per liter:

Glucose	1.0g
Peptone	1.0g
$MgSO_4 \cdot 7H_2O$	0.2g
$FeCl_3 \cdot 6H_2O$	0.1g
$CaCl_2 \cdot 2H_2O$	0.05g

pH 7.0 ± 0.2 at 25°C

Preparation of Medium: Add components to distilled/deionized water and bring volume to 1.0L. Mix thoroughly. Distribute into tubes or flasks. Autoclave for 15 min at 15 psi pressure–121°C.

Use: For the enrichment and cultivation of *Sphaerotilus* species and *Leptothrix* species.

Sphaerotilus Leptothrix Medium (DSMZ Medium 803)

Composition per liter:

Agar	20.0g
Peptone	1.5g
Yeast extract	1.0g
Ferric ammonium citrate	0.5g
$MgSO_4 \cdot 7H_2O$	0.2g
$CaCl_2$	0.05g
$MnSO_4 \cdot 2H_2O$	0.05g
$FeCl_3 \cdot 6H_2O$	0.01g

pH 7.1 ± 0.2 at 25°C

Preparation of Medium: Add components to tap water and bring volume to 1.0L. Mix thoroughly. Gently heat and bring to boiling. Distribute into tubes or flasks. Autoclave for 15 min at 15 psi pressure–121°C. Pour into sterile Petri dishes or leave in tubes.

Use: For the cultivation and maintenance of *Leptothrix mobilis.*

Sphaerotilus Medium (DSMZ Medium 51)

Composition per liter:

Beef extract, Lab Lemco	5.0g

Preparation of Medium: Add beef extract to distilled/deionized water and bring volume to 1.0L. Mix thoroughly. Adjust pH to 7.0. Distribute into tubes or flasks. Autoclave for 15 min at 15 psi pressure–121°C.

Use: For the cultivation and maintenance of *Sphaerotilus natans.*

Sphaerotilus Medium

Composition per liter:

Agar	15.0g
Lab-Lemco powder	5.0g

pH 7.0 ± 0.2 at 25°C

Source: Lab-Lemco powder is available from Oxoid Unipath.

Preparation of Medium: Add components to distilled/deionized water and bring volume to 1.0L. Mix thoroughly. Gently heat and bring to boiling. Distribute into tubes or flasks. Autoclave for 15 min at 15 psi pressure–121°C. Pour into sterile Petri dishes or leave in tubes.

Use: For the cultivation of *Sphaerotilus natans*.

Sphaerotilus natans Enrichment Medium

Composition per liter:

Sodium lactate	0.1g
$Na_2HPO_4 \cdot 7H_2O$	0.034g
$CaCl_2$	0.027g
$MgSO_4 \cdot 7H_2O$	0.023g
K_2HPO_4	0.022g
KH_2PO_4	8.5mg
NH_4Cl	1.7mg
$FeCl_3 \cdot 6H_2O$	0.25mg

pH 7.1–7.2 at 25°C

Preparation of Medium: Add components to distilled/deionized water and bring volume to 1.0L. Mix thoroughly. Distribute into tubes or flasks. Autoclave for 15 min at 15 psi pressure–121°C.

Use: For the enrichment and cultivation of *Sphaerotilus natans*.

Sphaerotilus natans Isolation Agar

Composition per liter:

Agar	15.0g
Meat extract	0.5g

Preparation of Medium: Add components to tap water and bring volume to 1.0L. Mix thoroughly. Gently heat and bring to boiling. Distribute into tubes or flasks. Autoclave for 15 min at 15 psi pressure–121°C. Pour into sterile Petri dishes or leave in tubes.

Use: For the isolation and cultivation of *Sphaerotilus natans*.

Sphaerotilus natans Isolation Agar

Composition per liter:

Agar	15.0g
Casein hydrolysate	1.5g

Preparation of Medium: Add components to tap water and bring volume to 1.0L. Mix thoroughly. Gently heat and bring to boiling. Distribute into tubes or flasks. Autoclave for 15 min at 15 psi pressure–121°C. Pour into sterile Petri dishes or leave in tubes.

Use: For the isolation and cultivation of *Sphaerotilus natans*.

Sphaerotilus natans Medium (LMG Medium 33)

Composition per liter:

Yeast extract	10.0g
Peptone	5.0g
Casitone	5.0g
Glucose	5.0g
$(NH_4)_2SO_4$	0.5g
L-Cysteine·HCl	0.5g
Resazurin	1.0mg
Mineral solution	40.0mL
Fatty acid mixture	3.1mL
Hemin solution	0.5mL
Vitamin K_1	0.2mL

pH 6.9 ± 0.2 at 25°C

Mineral Solution:

Composition per liter:

$NaHCO_3$	10.0g
NaCl	2.0g
K_2HPO_4	1.0g
KH_2PO_4	1.0g
$MgSO_4 \cdot 7H_2O$	0.48g
$CaCl_2 \cdot 2H_2O$	0.3g

Preparation of Mineral Solution: Add components to distilled/deionized water and bring volume to 1.0L. Mix thoroughly.

Fatty Acid Mixture:

Composition per 31.0mL:

Acetic acid	17.0mL
Propionic acid	6.0mL
n-Butyric acid	4.0mL
n-Valeric acid	1.0mL
iso-Valeric acid	1.0mL
iso-Butyric acid	1.0mL
DL-2-Methylbutyric acid	1.0mL

Preparation of Fatty Acid Mixture: Combine components. Mix thoroughly. Adjust pH to 7.5 with concentrated NaOH.

Hemin Solution:

Composition per 1.0mL:

Hemin	5.0mg
NaOH (1*N* solution)	1.0mL

Preparation of Hemin Solution: Add hemin to 1.0mL of NaOH solution. Mix thoroughly.

Preparation of Medium: Add components, except L-cysteine·HCl, hemin solution, and fatty acid mixture, to distilled/deionized water and bring volume to 1.0L. Mix thoroughly. Gently heat and bring to boiling. Continue boiling for 5 min. Cool to room temperature while sparging with 100% CO_2. Add L-cysteine·HCl, hemin solution, and fatty acid mixture. Adjust pH to 6.9 with 8*N* NaOH while continuing to sparge with 100% CO_2. After pH has been reached, sparge with 100% N_2. Anaerobically distribute into tubes or flasks. Autoclave for 15 min at 15 psi pressure–121°C.

Use: For the cultivation and maintenance of *Sphaerotilus natans*.

Sphingobacterium Medium

Composition per liter:

Pancreatic digest of casein	10.0g
NaCl	5.0g
Yeast extract	3.0g

Preparation of Medium: Add components to distilled/deionized water and bring volume to 1.0L. Mix thoroughly. Distribute into tubes or flasks. Autoclave for 15 min at 15 psi pressure–121°C.

Use: For the cultivation and maintenance of *Sphingobacterium mizutae, Sphingobacterium multivorum,* and *Sphingobacterium spiritivorum.*

Spirillum gracile Agar

Composition per liter:

Agar	15.0g
Peptone	5.0g
Yeast extract	0.5g

K_2HPO_4 .... 0.1g
Tween™ 80 .... 0.02g
Tap water 1.0L

pH 7.2 ± 0.2 at 25°C

Preparation of Medium: Add components to tap water and bring volume to 1.0L. Mix thoroughly. Gently heat and bring to boiling. Distribute into tubes or flasks. Autoclave for 15 min at 15 psi pressure–121°C. Pour into sterile Petri dishes or leave in tubes.

Use: For the cultivation and maintenance of *Spirillum gracile.*

Spirillum gracile Broth

Composition per liter:

Peptone 5.0g
Yeast extract 0.5g
K_2HPO_4 .... 0.1g
Tween™ 80 .... 0.02g
Tap water 1.0L

pH 7.2 ± 0.2 at 25°C

Preparation of Medium: Add components to distilled/deionized water and bring volume to 1.0L. Mix thoroughly. Distribute into tubes or flasks. Autoclave for 15 min at 15 psi pressure–121°C.

Use: For the cultivation and maintenance of *Spirillum gracile.*

Spirillum gracile Medium

Composition per liter:

Agar 15.0g
Peptone 5.0g
Yeast extract 0.5g
K_2HPO_4 .... 0.1g
Tween™ 80 .... 0.02g

pH 7.2 ± 0.2 at 25°C

Preparation of Medium: Add components to tap water and bring volume to 1.0L. Mix thoroughly. Gently heat and bring to boiling. Distribute into tubes or flasks. Autoclave for 15 min at 15 psi pressure–121°C. Pour into sterile Petri dishes or leave in tubes.

Use: For the cultivation and maintenance of *Aquaspirillum gracile.*

Spirillum lipoferum Medium

Composition per liter:

Sodium malate 5.0g
Agar 3.5g
KH_2PO_4 .... 0.4g
$MgSO_4 \cdot 7H_2O$ 0.2g
K_2HPO_4 .... 0.1g
NaCl 0.1g
$CaCl_2$ 0.02g
$FeCl_3$ 0.01g
$NaMoO_4 \cdot 2H_2O$ 2.0mg
Bromthymol Blue solution 5.0mL

pH 6.8 ± 0.2 at 25°C

Bromthymol Blue Solution:
Composition per 10.0mL:

Bromthymol Blue 0.5g
Ethanol 10.0mL

Preparation of Bromthymol Blue Solution: Add Bromthymol Blue to 10.0mL of ethanol. Mix thoroughly.

Preparation of Medium: Add components to distilled/deionized water and bring volume to 1.0L. Mix thoroughly. Gently heat and bring to boiling. Distribute into tubes or flasks. Autoclave for 15 min at 15 psi pressure–121°C.

Use: For the isolation and cultivation of *Spirillum leptoferum.*

Spirillum lipoferum Medium

Composition per liter:

Malic acid 5.0g
NaOH 4.7g
Agar 1.75g
KH_2PO_4 .... 0.4g
$MgSO_4 \cdot 7H_2O$ 0.2g
K_2HPO_4 .... 0.1g
NaCl 0.1g
$CaCl_2$ 0.02g
$FeCl_3$ 0.01g
$NaMoO_4 \cdot 2H_2O$ 2.0mg
Bromthymol Blue solution 5.0mL

pH 6.8 ± 0.2 at 25°C

Bromthymol Blue Solution:
Composition per 10.0mL:

Bromthymol Blue 0.5g
Ethanol 10.0mL

Preparation of Bromthymol Blue Solution: Add Bromthymol Blue to 10.0mL of ethanol. Mix thoroughly.

Preparation of Medium: Add components to distilled/deionized water and bring volume to 1.0L. Mix thoroughly. Gently heat and bring to boiling. Distribute into tubes or flasks. Autoclave for 15 min at 15 psi pressure–121°C.

Use: For the isolation and cultivation of *Spirillum leptoferum.*

Spirillum lipoferum Medium

Composition per liter:

Malic acid 5.0g
KOH 4.0g
Agar 1.75g
$FeSO_4 \cdot 7H_2O$ 0.5g
K_2HPO_4 .... 0.5g
$MgSO_4 \cdot 7H_2O$ 0.2g
NaCl 0.1g
$CaCl_2$ 0.02g
$MnSO_4 \cdot H_2O$ 0.01g
$NaMoO_4 \cdot 2H_2O$ 2.0mg
Bromthymol Blue solution 5.0mL

pH 6.8 ± 0.2 at 25°C

Bromthymol Blue Solution:
Composition per 10.0mL:

Bromthymol Blue 0.5g
Ethanol 10.0mL

Preparation of Bromthymol Blue Solution: Add Bromthymol Blue to 10.0mL of ethanol. Mix thoroughly.

Preparation of Medium: Add components to distilled/deionized water and bring volume to 1.0L. Mix thoroughly. Gently heat and bring to boiling. Distribute into tubes or flasks. Autoclave for 15 min at 15 psi pressure–121°C.

Use: For the isolation and cultivation of *Spirillum leptoferum.*

Spirillum Medium

Composition per liter:

Calcium lactate	10.0g
Peptone	5.0g
Beef extract	3.0g
Yeast extract	3.0g

pH 7.0 ± 0.2 at 25°C

Preparation of Medium: Add components to distilled/deionized water and bring volume to 1.0L. Mix thoroughly. Adjust pH to 7.0. Distribute into tubes or flasks. Autoclave for 20 min at 11 psi pressure–116°C. A precipitate will form during autoclaving.

Use: For the cultivation of *Spirillum* species.

Spirillum Medium

Composition per liter:

Peptone	10.0g
$MgSO_4 \cdot 7H_2O$	1.0g
$(NH_4)_2SO_4$	1.0g
Succinic acid	1.0g
$FeCl_3 \cdot 6H_2O$	2.0mg
$MnSO_4 \cdot H_2O$	2.0mg

pH 6.8 ± 0.2 at 25°C

Preparation of Medium: Add components to distilled/deionized water and bring volume to 1.0L. Mix thoroughly. Distribute into tubes or flasks. Autoclave for 15 min at 15 psi pressure–121°C.

Use: For the cultivation of *Aquaspirillum autotrophicum, Aquaspirillum dispar, Aquaspirillum peregrinum,* and *Aquaspirillum serpens.*

Spirillum Nitrogen-Fixing Medium

Composition per liter:

Sodium malate	5.0g
KH_2PO_4	0.4g
$MgSO_4 \cdot 7H_2O$	0.2g
K_2HPO_4	0.1g
NaCl	0.1g
Yeast extract	0.05g
$CaCl_2$	0.02g
$FeCl_3$	0.01g
$NaMoO_4 \cdot 2H_2O$	2.0mg

pH 7.2-7.4 ± 0.2 at 25°C

Preparation of Medium: Add components to distilled/deionized water and bring volume to 1.0L. Mix thoroughly. Distribute into tubes or flasks. Autoclave for 15 min at 15 psi pressure–121°C.

Use: For the cultivation and maintenance of *Azospirillum brasilense, Azospirillum lipoferum*, and *Herbaspirillum seropedicae*.

Spirillum volutans Defined Medium

Composition per liter:

BES (*N,N*-bis[2-hydroxyethyl]-2-aminoethane sulfonic acid) buffer	1.07g
$MgSO_4 \cdot 7H_2O$	1.0g
$(NH_4)_2SO_4$	1.0g
Succinic acid	1.0g
L-Histidine	0.2g
L-Isoleucine	0.2g
L-Methionine	0.2g
L-Threonine	0.2g
NaCl	0.085g
L-Cystine	0.025g
K_2HPO_4	0.02g
$FeCl_3 \cdot 6H_2O$	3.0mg
DL-Norepinephrine	2.0mg
$MnSO_4 \cdot H_2O$	2.0mg
$CaCO_3$	1.0mg
$ZnSO_4 \cdot 7H_2O$	0.72mg
$Na_2MoO_4 \cdot 2H_2O$	0.245mg
$CoSO_4 \cdot 7H_2O$	0.14mg
$CuSO_4 \cdot 5H_2O$	0.13mg
H_2BO_3	0.031mg

pH 6.8 ± 0.2 at 25°C

Preparation of Medium: Add components to distilled/deionized water and bring volume to 1.0L. Mix thoroughly. Adjust pH to 6.8. Distribute into tubes or flasks. Autoclave for 15 min at 15 psi pressure–121°C.

Use: For the cultivation of *Spirillum volutans.*

Spirit Blue Agar

Composition per liter:

Agar	20.0g
Pancreatic digest of casein	10.0g
Yeast extract	5.0g
Spirit Blue	0.15g
Lipoidal emulsion	30.0mL

pH 6.8 ± 0.2 at 25°C

Lipoidal Emulsion:
Composition per 500.0mL:

Tween™ 80	1.0mL
Cottonseed oil or olive oil	100.0mL

Preparation of Lipoidal Emulsion: Add Tween™ 80 to 400.0mL of warm distilled/deionized water. Mix thoroughly. Add 100.0mL of cottonseed or olive oil. Emulsify in a blender. Autoclave for 15 min at 15 psi pressure–121°C. Cool to 45°–50°C.

Preparation of Medium: Add components, except lipoidal emulsion, to distilled/deionized water and bring volume to 970.0mL. Mix thoroughly. Gently heat and bring to boiling. Autoclave for 15 min at 15 psi pressure–121°C. Cool to 45°–50°C. Aseptically add 30.0mL of sterile lipoidal emulsion. Mix thoroughly. Pour into sterile Petri dishes while shaking flask to keep emulsion dispersed.

Use: For the detection, enumeration, and study of lipolytic microorganisms.

Spirit Blue HiVeg Agar

Composition per liter:

Agar	17.0g
Plant hydrolysate	10.0g
Yeast extract	5.0g
Spirit Blue	0.15g
Lipoidal emulsion	30.0mL

pH 6.8 ± 0.2 at 25°C

Source: This medium, without lipoidal emulsion, is available as a premixed powder from HiMedia.

Lipoidal Emulsion:
Composition per 500.0mL:

Tween™ 80	1.0mL
Cottonseed oil or olive oil	100.0mL

Preparation of Lipoidal Emulsion: Add Tween™ 80 to 400.0mL of warm distilled/deionized water. Mix thoroughly. Add 100.0mL of cottonseed or olive oil. Emulsify in a blender. Autoclave for 15 min at 15 psi pressure–121°C. Cool to 45°–50°C.

Preparation of Medium: Add components, except lipoidal emulsion, to distilled/deionized water and bring volume to 970.0mL. Mix thoroughly. Gently heat and bring to boiling. Autoclave for 15 min at 15 psi pressure–121°C. Cool to 45°–50°C. Aseptically add 30.0mL of sterile lipoidal emulsion. Mix thoroughly. Pour into sterile Petri dishes while shaking flask to keep emulsion dispersed.

Use: For the detection, enumeration, and study of lipolytic microorganisms.

Spirochaeta americana Medium (DSMZ Medium 1165)

Composition per liter:

NaCl	30.0g
$NaHCO_3$	24.0g
Na_2CO_3	2.76g
NH_4Cl	1.0g
KCl	0.2g
K_2HPO_4	0.2g
$MgCl_2 \cdot 6H_2O$	0.1g
Resazurin	1.0mg
Sulfide solution	10.0mL
Glucose solution	10.0mL
Yeast extract solution	10.0mL
Vitamin solution	2.0mL
Trace elements solution	1.0mL

pH 9.4 ± 0.2 at 25°C

Yeast Extract Solution:

Composition per 10.0mL:

Yeast extract	0.5g

Preparation of Yeast Extract Solution: Add yeast extract to distilled/deionized water and bring volume to 10.0mL. Mix thoroughly. Autoclave for 15 min at 15 psi pressure–121°C.

Glucose Solution:

Composition per 10.0mL:

D-Glucose	5.0g

Preparation of Glucose Solution: Add D-glucose to distilled/deionized water and bring volume to 10.0mL. Mix thoroughly. Filter sterilize.

Sulfide Solution:

Composition per 10.0mL:

$Na_2S \cdot 9H_2O$	0.4g

Preparation of Sulfide Solution: Add $Na_2S \cdot 9H_2O$ to distilled/deionized water and bring volume to 10.0mL. Mix thoroughly. Autoclave under 100% N_2 for 15 min at 15 psi pressure–121°C. Cool to room temperature.

Vitamin Solution:

Composition per liter:

Pyridoxine-HCl	10.0mg
Thiamine-HCl·$2H_2O$	5.0mg
Riboflavin	5.0mg
Nicotinic acid	5.0mg
D-Ca-pantothenate	5.0mg
p-Aminobenzoic acid	5.0mg
Lipoic acid	5.0mg
Biotin	2.0mg
Folic acid	2.0mg
Vitamin B_{12}	0.1mg

Preparation of Vitamin Solution: Add components to distilled/deionized water and bring volume to 1.0L. Mix thoroughly. Sparge with 80% H_2 + 20% CO_2. Filter sterilize.

Trace Elements Solution:

Composition per 200.0mL:

$MnCl_2 \cdot 4H_2O$	0.72g
$Fe(NH_4)_2(SO_4)_2 \cdot 6H_2O$	0.4g
$FeSO_4 \cdot 7H_2O$	0.2g
$CoCl_2 \cdot 6H_2O$	0.2g
$ZnSO_4 \cdot 7H_2O$	0.2g
$NiCl_2 \cdot 6H_2O$	0.1g
$Na_2MoO_4 \cdot 2H_2O$	0.02g
$CuSO_4 \cdot 5H_2O$	0.02g
H_3BO_3	0.02g
$KAl(SO_4)_2 \cdot 12H_2O$	0.02g
HCl	5.0mL

Preparation of Trace Elements Solution: Add components to distilled/deionized water and bring volume to 200.0mL. Mix thoroughly. Autoclave for 15 min at 15 psi pressure–121°C.

Preparation of Medium: Add components, except carbonate, bicarbonate, sulfide solution, yeast extract solution, glucose solution, and vitamin solution, to distilled/deionized water and bring volume to 968.0mL. Mix thoroughly. Gently heat and bring to boiling. Cool to room temperature while sparging with 100% N_2 gas. Add the Na_2CO_3 and $NaHCO_3$. Mix thoroughly while sparging with 100% N_2 gas. Adjust pH to 9.4. Dispense into culture vessels (Hungate tubes or serum bottles). Autoclave for 15 min at 15 psi pressure–121°C. Cool to room temperature. Aseptically add sulfide solution, yeast extract solution, glucose solution, and vitamin solution. Mix thoroughly.

Use: For the cultivation of *Spirochaeta americana.*

Spirochaeta aurantia Agar

Composition per 1010.0mL:

Solution A	1.0L
Solution B	10.0mL

pH 7.0 ± 0.2 at 25°C

Solution A:

Composition per liter:

Agar	10.0g
Pancreatic digest of casein	5.0g
Glucose	2.0g
Yeast extract	2.0g

Preparation of Solution A: Add components to distilled/deionized water and bring volume to 1.0L. Mix thoroughly. Adjust pH to 7.5 with KOH. Gently heat and bring to boiling. Autoclave for 15 min at 15 psi pressure–121°C. Cool to 50°–55°.

Solution B:

Composition per 200.0mL:

K_2HPO_4	21.25g
KH_2PO_4	10.62g

Preparation of Solution B: Add components to distilled/deionized water and bring volume to 200.0mL. Mix thoroughly. Adjust pH to 7.0. Autoclave for 15 min at 15 psi pressure–121°C. Cool to 50°–55°C.

Preparation of Medium: Combine 1.0L of sterile solution A with 10.0mL of sterile solution B. Mix thoroughly. Aseptically pour into sterile Petri dishes or distribute into sterile tubes.

Use: For the cultivation and maintenance of *Spirochaeta aurantia.*

Spirochaeta aurantia Growth Medium

Composition per liter:

Yeast extract	4.0g
Maltose	2.0g
Peptone	2.0g
Potassium phosphate buffer (0.1*M* solution, pH 7.0)	100.0mL

pH 7.2 ± 0.2 at 25°C

Preparation of Medium: Filter sterilize potassium phosphate buffer. Add components, except potassium phosphate buffer, to distilled/deionized water and bring volume to 900.0mL. Mix thoroughly. Gently heat and bring to boiling. Autoclave for 15 min at 15 psi pressure–121°C. Cool to 45°–50°C. Aseptically add sterile potassium phosphate buffer. Mix thoroughly. Adjust pH to 7.2. Aseptically distribute into sterile tubes or flasks.

Use: For the cultivation of *Spirochaeta aurantia.*

Spirochaeta aurantia Isolation Medium

Composition per liter:

Peptone	1.0g
Yeast extract	1.0g
Hay extract	500.0mL

pH 6.5 ± 0.2 at 25°C

Hay Extract:

Composition per liter:

Hay, dried	5.0g

Preparation of Hay Extract: Add hay to distilled/deionized water and bring volume to 1.0L. Mix thoroughly. Gently heat and bring to boiling. Continue boiling for 10 min. Filter through Whatman #1 filter paper.

Preparation of Medium: Add components to distilled/deionized water and bring volume to 1.0L. Mix thoroughly. Distribute into tubes or flasks. Autoclave for 15 min at 15 psi pressure–121°C.

Use: For the isolation and cultivation of *Spirochaeta aurantia.*

Spirochaeta caldaria Medium

Composition per liter:

Glucose	2.0g
Pancreatic digest of casein	2.0g
L-Cysteine	0.25g
Resazurin	0.5mg
Wolfe's mineral solution	25.0mL
Trace elements solution SL-10	10.0mL
Wolfe's vitamin solution	10.0mL

pH 7.0 ± 0.2 at 25°C

Wolfe's Mineral Solution:

Composition per liter:

$MgSO_4 \cdot 7H_2O$	3.0g
Nitrilotriacetic acid	1.5g
NaCl	1.0g
$MnSO_4 \cdot 2H_2O$	0.5g
$CoCl_2 \cdot 6H_2O$	0.1g
$ZnSO_4 \cdot 7H_2O$	0.1g
$CaCl_2 \cdot 2H_2O$	0.1g
$FeSO_4 \cdot 7H_2O$	0.1g
$NiCl_2 \cdot 6H_2O$	0.025g
$KAl(SO_4)_2 \cdot 12H_2O$	0.02g
$CuSO_4 \cdot 5H_2O$	0.01g
H_3BO_3	0.01g
$Na_2MoO_4 \cdot 2H_2O$	0.01g
$Na_2SeO_3 \cdot 5H_2O$	0.3mg

Preparation of Wolfe's Mineral Solution: Add nitrilotriacetic acid to 500.0mL of distilled/deionized water. Adjust pH to 6.5 with KOH. Add remaining components. Add distilled/deionized water to 1.0L. Adjust pH to 6.8. Filter sterilize. Sparge with 100% N_2.

Trace Elements Solution SL-10:

Composition per liter:

$FeCl_2 \cdot 4H_2O$	1.5g
$CoCl_2 \cdot 6H_2O$	190.0mg
$MnCl_2 \cdot 4H_2O$	100.0mg
$ZnCl_2$	70.0mg
$Na_2MoO_4 \cdot 2H_2O$	36.0mg
$NiCl_2 \cdot 6H_2O$	24.0mg
H_3BO_3	6.0mg
$CuCl_2 \cdot 2H_2O$	2.0mg
HCl (25% solution)	10.0mL

Preparation of Trace Elements Solution SL-10: Add $FeCl_2 \cdot 4H_2O$ to 10.0mL of HCl solution. Mix thoroughly. Add distilled/deionized water and bring volume to 1.0L. Add remaining components. Mix thoroughly. Filter sterilize. Sparge with 100% N_2.

Wolfe's Vitamin Solution:

Composition per liter:

Pyridoxine·HCl	10.0mg
p-Aminobenzoic acid	5.0mg
Lipoic acid	5.0mg
Nicotinic acid	5.0mg
Riboflavin	5.0mg
Thiamine·HCl	5.0mg
Calcium DL-pantothenate	5.0mg
Biotin	2.0mg
Folic acid	2.0mg
Vitamin B_{12}	0.1mg

Preparation of Wolfe's Vitamin Solution: Add components to distilled/deionized water and bring volume to 1.0L. Mix thoroughly. Filter sterilize. Sparge with 100% N_2.

Preparation of Medium: Prepare and dispense medium under 100% N_2. Add components, except Wolfe's mineral solution, trace elements SL-10 solution, and Wolfe's vitamin solution, to distilled/deionized water and bring volume to 955.0mL. Mix thoroughly. Sparge with 100% N_2. Autoclave for 15 min at 15 psi pressure–121°C. Aseptically and anaerobically add 25.0mL of sterile Wolfe's mineral solution, 10.0mL of sterile trace elements SL-10 solution, and 10.0mL of sterile Wolfe's vitamin solution. Mix thoroughly. Aseptically and anaerobically distribute into sterile tubes or flasks.Adjust pH to 7.0 with sterile NaOH.

Use: For the cultivation of *Spirochaeta caldaria.*

Spirochaeta halophila Medium

Composition per liter:

Glucose salts solution	970.0mL
Yeast extract peptone solution	30.0mL

Glucose Salts Solution:

Composition per liter:

NaCl	49.3g
$MgSO_4 \cdot 7H_2O$	49.2g
$CaCl_2 \cdot 2H_2O$	5.9g
Glucose solution	100.0mL
Sulfide solution	100.0mL

Preparation of Glucose Salts Solution: Add components, except glucose solution and sulfide solution, to distilled/deionized water and bring volume to 800.0mL. Mix thoroughly. Autoclave for 15 min at 15 psi pressure–121°C. Cool to 25°C. Aseptically add sterile glucose solution and sulfide solution. Mix thoroughly.

Sulfide Solution:

Composition per 100.0mL:

$Na_2S \cdot 9H_2O$	0.5g

Preparation of Sulfide Solution: Add $Na_2S \cdot 9H_2O$ to distilled/deionized water and bring volume to 100.0mL. Mix thoroughly. Autoclave for 15 min at 15 psi pressure–121°C. Cool to 25°C.

Glucose Solution:

Composition per 100.0mL:

Glucose	5.0g

Preparation of Glucose Solution: Add glucose to distilled/deionized water and bring volume to 100.0mL. Mix thoroughly. Filter sterilize.

Yeast Extract Peptone Solution:

Composition per 30.0mL:

Yeast extract	4.0g
Peptone	2.0g

Preparation of Yeast Extract Peptone Solution: Add components to distilled/deionized water and bring volume to 30.0mL. Mix thoroughly. Autoclave for 15 min at 15 psi pressure–121°C. Cool to 25°C.

Preparation of Medium: Aseptically combine 30.0mL of yeast extract peptone solution with 970.0mL of glucose salts solution. Mix thoroughly. Aseptically distribute into sterile tubes or flasks.

Use: For the isolation and cultivation of *Spirochaeta halophila*.

Spirochaeta isovalerica Medium

Composition per liter:

Glucose	2.0g
Pancreatic digest of casein	1.0g
Yeast extract	0.5g
L-Cysteine·HCl	0.05g
Resazurin	0.001g
Seawater	750.0mL
Tris·HCl buffer (0.2*M* solution, pH 7.5)	250.0mL

pH 7.5 ± 0.2 at 25°C

Preparation of Medium: Prepare and dispense medium under 100% N_2. Combine components. Mix thoroughly. Sparge with 100% N_2. Adjust pH to 7.5 while continuing to sparge with 100% N_2. Anaerobically distribute into tubes or flasks. Autoclave for 15 min at 15 psi pressure–121°C.

Use: For the cultivation and maintenance of *Spirochaeta isovalerica*.

Spirochaeta litoralis Medium

Composition per liter:

Pancreatic digest of casein	3.0g
NaCl	2.0g
Yeast extract	0.5g
Glucose solution	2.0mL
Potasssium phosphate buffer (1*M*, pH 7.4)	2.0mL
Sulfide solution	0.5mL
Salts solution	0.2mL

pH 7.3 ± 0.2 at 25°C

Glucose Solution:

Composition per 100.0mL:

D-Glucose	25.0g

Preparation of Glucose Solution: Add D-glucose to distilled/deionized water and bring volume to 100.0mL. Mix thoroughly. Filter sterilize.

Sulfide Solution:

Composition per 100.0mL:

$Na_2S \cdot 9H_2O$	10.0g

Preparation of Sulfide Solution: Add $Na_2S \cdot 9H_2O$ to distilled/deionized water and bring volume to 100.0mL. Autoclave for 15 min at 15 psi pressure–121°C. Cool to 25°C.

Salts Solution:

Composition per 100.0mL:

$MgSO_4 \cdot 7H_2O$	12.5g
$CaCl_2 \cdot 2H_2O$	3.75g
EDTA	1.0g
$FeSO_4 \cdot 7H_2O$	0.5g
Trace elements solution	25.0mL

Preparation of Salts Solution: Add components to distilled/deionized water and bring volume to 100.0mL. Mix thoroughly.

Trace Elements Solution:

Composition per 1800.0mL:

H_3BO_3	5.5g
$MnCl_2 \cdot 4H_2O$	3.5g
$AlCl_3 \cdot 6H_2O$	0.5g
$CoCl_2 \cdot 6H_2O$	0.5g
$CuCl_2 \cdot 2H_2O$	0.5g
$NiCl_2 \cdot 6H_2O$	0.5g
$ZnCl_2$	0.5g
KI	0.25g
LiCl	0.25g
$Na_2MoO_4 \cdot 2H_2O$	0.25g
$BaCl_2 \cdot 2H_2O$	0.15g
$SnCl_2 \cdot 2H_2O$	0.15g
$NaVO_3$	0.05g

Preparation of Trace Elements Solution: Add components to distilled/deionized water and bring volume to 1800.0mL. Mix thoroughly. Adjust pH to 3–4 with HCl.

Preparation of Medium: Add components, except glucose solution and sulfide solution, to distilled/deionized water and bring volume to 997.5mL. Mix thoroughly. Gently heat and bring to boiling. Autoclave for 15 min at 15 psi pressure–121°C. Cool to 45°–50°C. Aseptically add sterile glucose solution and sulfide solution. Mix thoroughly. Aseptically distribute into sterile tubes or flasks.

Use: For the isolation of *Spirochaeta litoralis* from marine habitats.

Spirochaeta litoralis Medium

Composition per liter:

Glucose	2.0g
Pancreatic digest of casein	1.0g
Yeast extract	1.0g
L-Cysteine	0.5g
Resazurin	1.0mg
Seawater	750.0mL
Tris·HCl buffer (1.0*M* solution, pH 7.5)	50.0mL

pH 7.2 ± 0.2 at 25°C

Preparation of Medium: Prepare and dispense medium under 100% N_2. Add components to distilled/deionized water and bring volume to 1.0L. Mix thoroughly. Sparge with 100% N_2. Adjust pH to 7.5 while continuing to sparge with 100% N_2. Anaerobically distribute into tubes or flasks. Autoclave for 15 min at 15 psi pressure–121°C. Final pH of medium after autoclaving should be 7.2.

Use: For the cultivation and maintenance of *Spirochaeta litoralis*.

Spirochaeta smaragdinae Medium (DSMZ Medium 819)

Composition per liter:

NaCl	50.0g
Yeast extract	5.0g
NH_4Cl	1.0g
Cysteine-HCl·H_2O	0.5g
K_2HPO_4	0.3g
KH_2PO_4	0.3g
$MgCl_2·6H_2O$	0.2g
KCl	0.2g
$CaCl_2·2H_2O$	0.1g
Resazurin	0.5mg
$NaHCO_3$ solution	80.0mL
$Na_2S·9H_2O$ solution	10.0mL
Trace elements solution	10.0mL

pH 7.8 ± 0.2 at 25°C

$Na_2S·9H_2O$ Solution:

Composition per 10.0mL:

$Na_2S·9H_2O$	0.2g

Preparation of $Na_2S·9H_2O$ Solution: Add $Na_2S·9H_2O$ to distilled/deionized water and bring volume to 10.0mL. Mix thoroughly. Sparge with 100% N_2. Autoclave for 15 min at 15 psi pressure–121°C. Neutralize to pH 7.0 with sterile HCl.

$NaHCO_3$ Solution:

Composition per 100.0mL:

$NaHCO_3$	5.0g

Preparation of $NaHCO_3$ Solution: Add $NaHCO_3$ to distilled/deionized water and bring volume to 100.0mL. Mix thoroughly. Sparge with 80% N_2 + 20% CO_2. Filter sterilize.

Trace Elements Solution:

Composition per liter:

$MgSO_4·7H_2O$	3.0g
Nitrilotriacetic acid	1.5g
NaCl	1.0g
$MnSO_4·2H_2O$	0.5g
$CoSO_4·7H_2O$	0.18g
$ZnSO_4·7H_2O$	0.18g
$CaCl_2·2H_2O$	0.1g
$FeSO_4·7H_2O$	0.1g
$NiCl_2·6H_2O$	0.025g
$KAl(SO_4)_2·12H_2O$	0.02g
H_3BO_3	0.01g
$Na_2MoO_4·4H_2O$	0.01g
$CuSO_4·5H_2O$	0.01g
$Na_2SeO_3·5H_2O$	0.3mg

Preparation of Trace Elements Solution: Add nitrilotriacetic acid to 500.0mL of distilled/deionized water. Dissolve by adjusting pH to 6.5 with KOH. Add remaining components. Add distilled/deionized water to 1.0L. Mix thoroughly.

Preparation of Medium: Prepare and dispense medium under 80% N_2 + 20% CO_2 gas atmosphere. Add components, except $NaHCO_3$ solution and $Na_2S·9H_2O$ solution, to distilled/deionized water and bring volume to 910.0mL. Mix thoroughly. Adjust pH to 7.8. Gently heat and bring to boiling. Cool while sparging with 80% N_2 + 20% CO_2. Autoclave for 15 min at 15 psi pressure–121°C. Cool to room temperature. Aseptically and anaerobically add 80.0mL $NaHCO_3$ solution and 10.0mL $Na_2S·9H_2O$ solution. Aseptically and anaerobically distribute into sterile tubes or bottles.

Use: For the cultivation of *Spirochaeta smaragdinae*.

Spirochaeta stenostrepta Medium

Composition per liter:

Glucose	5.0g
Peptone	2.0g
Yeast extract	0.3g
Vitamin B_{12}	0.01mg
Salts solution	100.0mL
Phosphate solution	15.0mL
Sulfide solution	10.0mL

pH 7.0 ± 0.2 at 25°C

Phosphate Solution:

Composition per liter:

KH_2PO_4	30.0g
K_2HPO_4	70.0g

Preparation of Phosphate Solution: Add components to distilled/deionized water and bring volume to 1.0L. Mix thoroughly. Filter sterilize.

Salts Solution:

Composition per liter:

$MgSO_4·7H_2O$	2.0g
$CaCl_2·2H_2O$	0.75g
EDTA	0.2g
$FeSO_4·7H_2O$	0.1g
Trace elements solution	5.0mL

Preparation of Salts Solution: Add EDTA to approximately 800.0mL of distilled/deionized water. Gently heat until dissolved. Adjust pH to 7.0 with 2.5% KOH. Add the remaining components. Mix thoroughly. Bring volume to 1.0L with distilled/deionized water.

Trace Elements Solution:

Composition per 1800.0mL:

H_3BO_3	5.5g
$MnCl_2·4H_2O$	3.5g
$AlCl_3·6H_2O$	0.5g
$CoCl_2·6H_2O$	0.5g
$CuCl_2·2H_2O$	0.5g
$NiCl_2·6H_2O$	0.5g
$ZnCl_2$	0.5g

KI	0.25g
LiCl	0.25g
$Na_2MoO_4 \cdot 2H_2O$	0.25g
$BaCl_2 \cdot 2H_2O$	0.15g
$SnCl_2 \cdot 2H_2O$	0.15g
$NaVO_3$	0.05g

Preparation of Trace Elements Solution: Add components to distilled/deionized water and bring volume to 1800.0mL. Mix thoroughly. Adjust pH to 3–4 with HCl.

Sulfide Solution:

Composition per 100.0mL:

$Na_2S \cdot 9H_2O$	2.0g

Preparation of Sulfide Solution: Add $Na_2S \cdot 9H_2O$ to distilled/deionized water and bring volume to 100.0mL. Autoclave for 15 min at 15 psi pressure–121°C. Cool to 25°C. Prepare solution freshly.

Preparation of Medium: Add components, except sulfide solution, to distilled/deionized water and bring volume to 990.0mL. Mix thoroughly. Gently heat and bring to boiling. Autoclave for 15 min at 15 psi pressure–121°C. Cool to 45°–50°C. Aseptically add 10.0mL of sterile sulfide solution. Mix thoroughly. Aseptically distribute into sterile tubes or flasks.

Use: For the isolation of *Spirochaeta stenostrepta.*

Spirochaeta stenostrepta Medium

Composition per liter:

Glucose	5.0g
Peptone	2.0g
Yeast extract	2.0g
L-Cysteine	0.5g
Resazurin	1.0mg

pH 7.3–7.6 at 25°C

Preparation of Medium: Prepare and dispense medium under 100% N_2. Add components to distilled/deionized water and bring volume to 1.0L. Mix thoroughly. Sparge with 100% N_2. Adjust pH to 7.3–7.6 while continuing to sparge with 100% N_2. Anaerobically distribute into tubes or flasks. Autoclave for 15 min at 15 psi pressure–121°C.

Use: For the cultivation and maintenance of *Spirochaeta stenostrepta.*

Spirochaeta zuelzerae Medium

Composition per liter:

Solution 1	480.0mL
Solution 2	480.0mL
Solution 3	20.0mL
Solution 4	20.0mL

pH 7.2 ± 0.2 at 25°C

Solution 1:

Composition per 480.0mL:

KH_2PO_4	0.75g
L-Cysteine·HCl·H_2O	0.5g
$NaH_2PO_4 \cdot H_2O$	0.25g

Preparation of Solution 1: Add components to distilled/deionized water and bring volume to 480.0mL. Mix thoroughly. Adjust pH to 7.2 with 5.0% KOH. Autoclave for 15 min at 15 psi pressure–121°C. Cool to 45°–50°C.

Solution 2:

Composition per 480.0mL:

NH_4Cl	1.0g
$MgSO_4 \cdot 7H_2O$	0.5g
Yeast extract	0.2g
$CaCl_2$	0.02g
Resazurin	1.0mg
$FeCl_3 \cdot 6H_2O$ solution (0.25g/L)	10.0mL
Trace elements solution	2.0mL

Preparation of Solution 2: Add components to distilled/deionized water and bring volume to 480.0mL. Mix thoroughly. Autoclave for 15 min at 15 psi pressure–121°C. Cool to 45°–50°C.

Trace Elements Solution:

Composition per 100.0mL:

$Na_2MoO_4 \cdot 2H_2O$	0.075g
H_3BO_3	0.056g
$ZnSO_4 \cdot 7H_2O$	0.044g
$CoCl_2 \cdot 6H_2O$	0.02g
$CuSO_4 \cdot 5H_2O$	2.0mg
$MnCl_2$	2.0mg

Preparation of Trace Elements Solution: Add components to distilled/deionized water and bring volume to 100.0mL. Mix thoroughly.

Solution 3:

Composition per 20.0mL:

$NaHCO_3$	1.0g

Preparation of Solution 3: Add $NaHCO_3$ to distilled/deionized water and bring volume to 20.0mL. Mix thoroughly. Filter sterilize under pressure.

Solution 4:

Composition per 20.0mL:

Glucose	2.0g

Preparation of Solution 4: Add glucose to distilled/deionized water and bring volume to 20.0mL. Mix thoroughly. Filter sterilize.

Preparation of Medium: Aseptically add 480.0 mL of sterile solution 1 to 480.0mL of sterile solution 2 under 80% N_2 + 20% CO_2. While gassing, add 20.0mL of sterile solution 3 and 20.0mL of sterile solution 4. Mix thoroughly. Adjust pH to 7.2. Aseptically and anaerobically distribute into tubes. Cap with rubber stoppers.

Use: For the cultivation and maintenance of *Spirochaeta zuelzerae.*

Spirochete Enrichment Medium

Composition per liter:

Agar	10.0g
Beef extract	1.0g
Peptone	1.0g
Yeast extract	1.0g
Seawater	500.0mL

Preparation of Medium: Add components to distilled/deionized water and bring volume to 1.0L. Mix thoroughly. Gently heat and bring to boiling. Distribute into tubes or flasks. Autoclave for 15 min at 15 psi pressure–121°C. Pour into sterile Petri dishes or leave in tubes.

Use: For the isolation of spirochetes from muds. A well is cut into the agar plate and filled with mud samples. Spirochetes migrate out of the mud into the agar surrounding the well.

Spirochete Medium (ATCC Medium 164)

Composition per liter:

Agar	15.0g
KH_2PO_4	1.0g

NH_4Cl	1.0g
Yeast extract	1.0g
$MgSO_4$	0.5g
$CaCl_2$	0.04g
$FeCl_3 \cdot 6H_2O$	1.25mg
$NaHCO_3$ solution	20.0mL
Glucose solution	10.0mL
$Na_2S \cdot 9H_2O$ solution	5.0mL

Glucose Solution:

Composition per 100.0mL:

Glucose	10.0g

Preparation of Glucose Solution: Add glucose to distilled/deionized water and bring volume to 100.0mL. Mix thoroughly. Filter sterilize.

$NaHCO_3$ Solution:

Composition per 100.0mL:

$NaHCO_3$	5.0g

Preparation of $NaHCO_3$ Solution: Add the $NaHCO_3$ to distilled/deionized water and bring volume to 100.0mL. Mix thoroughly. Filter sterilize.

$Na_2S \cdot 9H_2O$ Solution:

Composition per 100.0mL:

$Na_2S \cdot 9H_2O$	10.0g

Preparation of $Na_2S \cdot 9H_2O$ Solution: Add $Na_2S \cdot 9H_2O$ to distilled/deionized water and bring volume to 100.0mL. Mix thoroughly. Autoclave for 15 min at 15 psi pressure–121°C.

Preparation of Medium: Add components—except $NaHCO_3$ solution, glucose solution, and $Na_2S \cdot 9H_2O$ solution—to distilled/deionized water and bring volume to 965.0mL. Autoclave for 15 min at 15 psi pressure–121°C. Cool to 50°C. Aseptically add 20.0mL of sterile $NaHCO_3$ solution, 10.0mL of sterile glucose solution, and 5.0mL of sterile $Na_2S \cdot 9H_2O$ solution. Mix thoroughly. Pour into sterile Petri dishes or distribute into sterile tubes.

Use: For the cultivation of spirochetes.

Spirochete Medium (ATCC Medium 1712)

Composition per liter:

Tris(hydroxymethyl)aminomethane buffer	7.52g
Pancreatic digest of casein	1.0g
Yeast extract	1.0g
L-Cysteine·HCl·$2H_2O$	0.5g
Resazurin	1.0mg
Seawater	750.0mL
Glucose solution	20.0mL

pH 7.2 ± 0.2 at 25°C

Glucose Solution:

Composition per 20.0mL:

Glucose	2.0g

Preparation of Glucose Solution: Add glucose to distilled/deionized water and bring volume to 20.0mL. Mix thoroughly. Filter sterilize.

Preparation of Medium: Prepare and dispense medium under 100% N_2. Add components, except glucose solution, to distilled/deionized water and bring volume to 980.0mL. Mix thoroughly. Adjust pH to 7.5. Autoclave for 15 min at 15 psi pressure–121°C. Cool to 50°C. Aseptically add sterile glucose solution. Mix thoroughly. Aseptically distribute into sterile tubes or flasks.

Use: For the cultivation and maintenance of *Spirochaeta litoralis*.

Spirochete Thermophile Medium (DSMZ Medium 509)

Composition per 1012.0mL:

Solution A	920.0mL
Solution D	50.0mL
Solution E	20.0mL
Solution F	10.0mL
Solution G	10.0mL
Solution B	1.0mL
Solution C	1.0mL

pH 7.0 ± 0.2 at 25°C

Solution A:

Composition per 920.0mL:

NaCl	4.0g
$MgCl_2 \cdot 6H_2O$	0.8g
KCl	0.5g
NH_4Cl	0.3g
KH_2PO_4	0.2g
$CaCl_2 \cdot 2H_2O$	0.03g
Resazurin	1.0mg

Preparation of Solution A: Add components to 920.0mL distilled/deionized water. Mix thoroughly. Bring to boiling for a few minutes. Cool to room temperature while gassing with 80% N_2 + 20% CO_2 gas. Adjust pH to 6.0. Immediately distribute under N_2 into anaerobic tubes. Autoclave for 15 min at 15 psi pressure–121°C. Cool to 25°C.

Solution B:

Composition per liter:

$FeCl_2 \cdot 4H_2O$	1.5g
$CoCl_2 \cdot 6H_2O$	190.0mg
$MnCl_2 \cdot 4H_2O$	100.0mg
$ZnCl_2$	70.0mg
$Na_2MoO_4 \cdot 2H_2O$	36.0mg
$NiCl_2 \cdot 6H_2O$	24.0mg
H_3BO_3	6.0mg
$CuCl_2 \cdot 2H_2O$	2.0mg
HCl (25% solution)	10.0mL

Preparation of Solution B: Add $FeCl_2 \cdot 4H_2O$ to 10.0mL of HCl solution. Mix thoroughly. Add distilled/deionized water and bring volume to 1.0L. Add remaining components. Mix thoroughly. Sparge with 80% N_2 + 20% CO_2. Autoclave for 15 min at 15 psi pressure–121°C. Cool to 25°C.

Solution C:

Composition per liter:

NaOH	0.5g
$Na_2WO_4 \cdot 2H_2O$	4.0mg
$Na_2SeO_3 \cdot 5H_2O$	3.0mg

Preparation of Solution C: Add components to distilled/deionized water and bring volume to 1.0L. Mix thoroughly. Sparge with 100% N_2. Filter sterilize.

Solution D:

Composition per 100.0mL:

$NaHCO_3$	5.0g

Preparation of Solution D: Add $NaHCO_3$ to distilled/deionized water and bring volume to 100.0mL. Mix thoroughly. Sparge with

100% N_2 gas mixture. Autoclave for 15 min at 15 psi pressure–121°C. Cool to 25°C.

Solution E:

Composition per liter:

Pyridoxine-HCl	10.0mg
Thiamine-HCl·$2H_2O$	5.0mg
Riboflavin	5.0mg
Nicotinic acid	5.0mg
D-Ca-pantothenate	5.0mg
p-Aminobenzoic acid	5.0mg
Lipoic acid	5.0mg
Biotin	2.0mg
Folic acid	2.0mg
Vitamin B_{12}	0.1mg

Preparation of Solution E: Add components to distilled/deionized water and bring volume to 1.0L. Mix thoroughly. Sparge with 80% N_2 + 20% CO_2. Filter sterilize.

Solution F:

Composition per 10.0mL:

Starch	1.0g

Preparation of Solution F: Add starch to distilled/deionized water and bring volume to 10.0mL. Mix thoroughly. Sparge with 100% N_2 gas mixture. Autoclave for 15 min at 15 psi pressure–121°C. Cool to 25°C.

Solution G:

Composition per 10.0mL:

$Na_2S·9H_2O$	0.3g

Preparation of Solution G: Add $Na_2S·9H_2O$ to distilled/deionized water and bring volume to 10.0mL. Mix thoroughly. Autoclave under 100% N_2 for 15 min at 15 psi pressure–121°C. Cool to 25°C.

Preparation of Medium: Solution A is distributed into anaerobic tubes with rubber stoppers prior to autoclaving. Using aseptic and anaerobic conditions and syringes, appropriate volumes of sterile solutions B–G are injected into each tube to yield the specified concentrations.

Use: For the cultivation of *Spirochaeta thermophila.*

Spirolate Broth

Composition per liter:

Pancreatic digest of casein	15.0g
Glucose	5.0g
Yeast extract	5.0g
NaCl	2.5g
L-Cysteine·HCl·H_2O	1.0g
Sodium thioglycolate	0.5g
Palmitic acid	0.05g
Stearic acid	0.05g
Oleic acid	0.05g
Linoleic acid	0.05g
Serum	100.0mL

pH 7.1 ± 0.2 at 25°C

Source: This medium is available as a premixed powder from BD Diagnostic Systems.

Preparation of Medium: Add components, except serum, to distilled/deionized water and bring volume to 900.0mL. Mix thoroughly. Distribute into screw-capped tubes in 20.0mL volumes. Autoclave for 15 min at 15 psi pressure–121°C. Cool to 25°C. Aseptically add 2.0mL of serum to each tube. Heat-inactivated sheep, rabbit, or bovine serum may be used. Tighten caps. Mix thoroughly.

Use: For the cultivation of *Treponema phagedenis* and other spirochetes.

Spirolate HiVeg Broth, OMATA with Serum

Composition per liter:

Plant hydrolysate	15.0g
Glucose	5.0g
Yeast extract	5.0g
NaCl	2.5g
L-Cysteine·HCl	1.0g
Na-thioglycolate	0.5g
Sheep or rabbit blood serum, inactivated 10%	100.0mL

pH 7.1 ± 0.2 at 25°C

Source: This medium, without serum, is available as a premixed powder from HiMedia.

Preparation of Medium: Add components, except serum, to distilled/deionized water and bring volume to 900.0mL. Mix thoroughly. Distribute into screw-capped tubes in 20.0mL volumes. Autoclave for 15 min at 15 psi pressure–121°C. Cool to 25°C. Aseptically add 2.0mL of serum to each tube. Heat-inactivated sheep, rabbit, or bovine serum may be used. Tighten caps. Mix thoroughly.

Use: For the cultivation of *Treponema phagedenis* and other spirochetes.

Spiroplasma Agar MID

Composition per 291.2mL:

Schneider's *Drosophila* medium	160.0mL
Solution 1	80.0mL
Fetal calf serum	50.0mL
Phenol Red (0.5% solution)	1.2mL

pH 7.4 ± 0.2 at 25°C

Schneider's *Drosophila* Medium:

Composition per liter:

$MgSO_4·7H_2O$	3.7g
NaCl	2.1g
Yeast extract	2.0g
Trehalose	2.0g
D-Glucose	2.0g
L-Glutamine	1.8g
L-Lysine·HCl	1.7g
L-Proline	1.7g
KCl	1.6g
$Na_2HPO_4·7H_2O$	1.3g
L-Glutamic acid	0.8g
L-Methionine	0.8g
$CaCl_2$, anhydrous	0.6g
KH_2PO_4	0.5g
β-Alanine	0.5g
L-Tyrosine	0.5g
L-Arginine	0.4g
L-Aspartic acid	0.4g
L-Histidine	0.4g
L-Threonine	0.4g
$NaHCO_3$	0.4g
Glycine	0.3g
L-Serine	0.3g
L-Valine	0.3g

L-Isoleucine	0.2g
L-Leucine	0.2g
L-Phenylalanine	0.2g
α-Ketoglutaric acid	0.2g
Fumaric acid	0.1g
Malic acid	0.1g
Succinic acid	0.1g
L-Cystine	0.1g
L-Tryptophan	0.1g
L-Cysteine	0.06g

Preparation of Schneider's *Drosophila* Medium: Add components to 1.0L of distilled/deionized water. Mix thoroughly. Filter sterilize.

Solution 1:

Composition per 80.0mL:

Sorbitol	7.0g
Noble agar	5.0g
Beef heart, solids from infusion	5.0g
Peptone	1.8g
Sucrose	1.0g
Pancreatic digest of casein	1.0g
NaCl	0.5g
D-Fructose	0.1g
D-Glucose	0.1g

Preparation of Solution 1: Add components to distilled/deionized water and bring volume to 80.0mL. Mix thoroughly. Adjust pH to 7.8 with 1*N* NaOH. Autoclave for 15 min at 15 psi pressure–121°C. Cool to 50°C.

Preparation of Medium: Bring fetal calf serum and Phenol Red solution to 56°C. Rapidly bring Schneider's *Drosophila* medium to 37°C. Rapidly combine the components. Mix thoroughly. Pour into sterile Petri dishes or distribute into sterile tubes or flasks.

Use: For the cultivation and maintenance of *Spiroplasma kunkelii* and *Spiroplasma* species.

Spiroplasma Broth MID

Composition per 291.2mL:

Schneider's *Drosophila* medium	160.0mL
Solution 1	80.0mL
Fetal calf serum	50.0mL
Phenol Red (0.5% solution)	1.2mL

pH 7.4 ± 0.2 at 25°C

Schneider's Drosophila Medium:

Composition per liter:

$MgSO_4 \cdot 7H_2O$	3.7g
NaCl	2.1g
Yeast extract	2.0g
Trehalose	2.0g
D-Glucose	2.0g
L-Glutamine	1.8g
L-Lysine·HCl	1.7g
L-Proline	1.7g
KCl	1.6g
$Na_2HPO_4 \cdot 7H_2O$	1.3g
L-Glutamic acid	0.8g
L-Methionine	0.8g
$CaCl_2$, anhydrous	0.6g
KH_2PO_4	0.5g
β-Alanine	0.5g
L-Tyrosine	0.5g
L-Arginine	0.4g
L-Aspartic acid	0.4g
L-Histidine	0.4g
L-Threonine	0.4g
$NaHCO_3$	0.4g
Glycine	0.3g
L-Serine	0.3g
L-Valine	0.3g
L-Isoleucine	0.2g
L-Leucine	0.2g
L-Phenylalanine	0.2g
α-Ketoglutaric acid	0.2g
Fumaric acid	0.1g
Malic acid	0.1g
Succinic acid	0.1g
L-Cystine	0.1g
L-Tryptophan	0.1g
L-Cysteine	0.06g

Preparation of Schneider's *Drosophila* Medium: Add components to distilled/deionized water and bring volume to 1.0L. Mix thoroughly. Filter sterilize.

Solution 1:

Composition per 80.0mL:

Sorbitol	7.0g
Beef heart, solids from infusion	5.0g
Peptone	1.8g
Sucrose	1.0g
Pancreatic digest of casein	1.0g
NaCl	0.5g
D-Fructose	0.1g
D-Glucose	0.1g

Preparation of Solution 1: Add components to distilled/deionized water and bring volume to 80.0mL. Mix thoroughly. Adjust pH to 7.8 with 1*N* NaOH. Autoclave for 15 min at 15 psi pressure–121°C. Cool to 25°C.

Preparation of Medium: Bring fetal calf serum and Phenol Red solution to 56°C. Rapidly bring Schneider's *Drosophila* medium to 37°C. Rapidly combine the components. Mix thoroughly. Aseptically distribute into sterile tubes or flasks.

Use: For the cultivation and maintenance of *Spiroplasma kunkelii* and *Spiroplasma* species.

Spiroplasma Medium

Composition per liter:

Sucrose	80.0g
Beef heart, solids from infusion	34.7g
Peptone	6.9g
NaCl	3.5g
Horse serum, heat inactivated	100.0mL

pH 7.2 ± 0.2 at 25°C

Preparation of Medium: Add components, except horse serum, to distilled/deionized water and bring volume to 900.0mL. Mix thoroughly. Autoclave for 15 min at 15 psi pressure–121°C. Cool to 25°C. Aseptically add horse serum. Mix thoroughly. Aseptically distribute into sterile tubes or flasks.

Use: For the cultivation and maintenance of *Spiroplasma* species.

Spiroplasma Medium

Composition per liter:

Sorbitol	70.0g
Pancreatic digest of casein	7.0g
Yeast extract	5.0g
NaCl	5.0g
Beef extract	3.0g
Yeast extract	3.0g
Beef heart, solids from infusion	2.0g
Fructose	1.0g
Glucose	1.0g
Phenol Red	20.0mg
Horse serum	100.0mL

pH 7.8 ± 0.2 at 25°C

Preparation of Medium: Add components to distilled/deionized water and bring volume to 900.0mL. Mix thoroughly. Gently heat and bring to boiling. Autoclave for 15 min at 15 psi pressure–121°C. Cool to 50°–55°C. Aseptically add 100.0mL of sterile horse serum. Mix thoroughly.

Use: For the cultivation of *Spiroplasma citri.*

Spiroplasma Medium with 25 mg/L of Phenol Red

Composition per liter:

Sucrose	80.0g
Beef heart, solids from infusion	34.7g
Peptone	6.9g
NaCl	3.5g
Phenol Red	25.0mg
Horse serum, heat inactivated	100.0mL

pH 7.2 ± 0.2 at 25°C

Preparation of Medium: Add components, except horse serum, to distilled/deionized water and bring volume to 900.0mL. Mix thoroughly. Gently heat and bring to boiling. Autoclave for 15 min at 15 psi pressure–121°C. Cool to 45°–50°C. Aseptically add heat-inactivated horse serum. Mix thoroughly. Aseptically distribute into sterile tubes or flasks.

Use: For the cultivation and maintenance of *Spiroplasma floricola.*

Spizizen Potato Agar

Composition per liter:

Potatoes	200.0g
Agar	15.0g
$MnSO_4$	5.0mg

pH 6.8 ± 0.2 at 25°C

Preparation of Medium: Peel and dice potatoes. Add potatoes to 1.0L of tap water. Gently heat and bring to boiling. Continue boiling for 30 min. Filter through cheesecloth. Add $MnSO_4$ to filtrate and bring volume to 1.0L with tap water. Mix thoroughly. Adjust pH to 6.8. Add agar. Gently heat and bring to boiling. Distribute into tubes or flasks. Autoclave for 15 min at 15 psi pressure–121°C. Pour into sterile Petri dishes or leave in tubes.

Use: For the cultivation and maintenance of *Bacillus amyloliquefaciens*.

SPMA

See: **Mineral Salt Peptonized Milk Agar**

Spore Strip Broth

Composition per liter:

Spore strip broth	9.0g

Preparation of Medium: Add 9.0g of spore strip broth powder (a mixture of glucose, buffer salts, growth factors, and Bromthymol Blue) to distilled/deionized water and bring volume to 1.0L. Mix thoroughly. Distribute into tubes or flasks. Autoclave for 15 min at 15 psi pressure–121°C.

Use: For the recovery of spores of *Bacillus stearothermophilus* on spore strips used to determine the sterilization efficiency of autoclaves.

Sporobacter Medium (DSMZ Medium 711)

Composition per liter:

NH_4Cl	1.0g
NaCl	0.6g
Na-acetate·$3H_2O$	0.5g
Cysteine-HCl·H_2O	0.5g
K_2HPO_4	0.3g
KH_2PO_4	0.3g
Yeast extract	0.2g
$MgCl_2·6H_2O$	0.2g
$CaCl_2·2H_2O$	0.1g
KCl	0.1g
Resazurin	0.5mg
$NaHCO_3$ solution	40.0mL
Trimethoxycinnamate solution	10.0mL
$Na_2S·9H_2O$ solution	10.0mL
Trace elements solution SL-10	1.5mL

pH 7.1 ± 0.2 at 25°C

Trace Elements Solution SL-10:

Composition per liter:

$FeCl_2·4H_2O$	1.5g
$CoCl_2·6H_2O$	190.0mg
$MnCl_2·4H_2O$	100.0mg
$ZnCl_2$	70.0mg
$Na_2MoO_4·2H_2O$	36.0mg
$NiCl_2·6H_2O$	24.0mg
H_3BO_3	6.0mg
$CuCl_2·2H_2O$	2.0mg
HCl (25% solution)	10.0mL

Preparation of Trace Elements Solution SL-10: Add $FeCl_2·4H_2O$ to 10.0mL of HCl solution. Mix thoroughly. Add distilled/deionized water and bring volume to 1.0L. Add remaining components. Mix thoroughly. Sparge with 80% N_2 + 20% CO_2. Autoclave for 15 min at 15 psi pressure–121°C.

$Na_2S·9H_2O$ Solution:

Composition per 10.0mL:

$Na_2S·9H_2O$	0.3g

Preparation of $Na_2S·9H_2O$ Solution: Add $Na_2S·9H_2O$ to distilled/deionized water and bring volume to 10.0mL. Mix thoroughly. Autoclave under 100% N_2 for 15 min at 15 psi pressure–121°C. Cool to room temperature.

$NaHCO_3$ Solution:

Composition per 100.0mL:

$NaHCO_3$	10.0g

Preparation of $NaHCO_3$ Solution: Add $NaHCO_3$ to distilled/deionized water and bring volume to 100.0mL. Mix thoroughly. Sparge with 80% N_2 + 20% CO_2. Filter sterilize.

Trimethoxycinnamate Solution:
Composition per 10.0mL:

Trans-3,4,5-trimethoxycinnamate	1.2g

Preparation of Trimethoxycinnamate Solution: Add trans-3,4,5-trimethoxycinnamate to distilled/deionized water and bring volume to 10.0mL. Mix thoroughly. Neutralize with NaOH. Sparge with 100% N_2. Autoclave for 15 min at 15 psi pressure–121°C. Cool to room temperature.

Preparation of Medium: Prepare and dispense medium under 80% N_2 + 20% CO_2 gas atmosphere. Add components, except $NaHCO_3$ solution, $Na_2S{\cdot}9H_2O$ solution, trimethoxycinnamate solution, and trace elements solution SL-10, to distilled/deionized water and bring volume to 938.5mL. Mix thoroughly. Adjust pH to 7.1. Sparge with 80% N_2 + 20% CO_2. Autoclave for 15 min at 15 psi pressure–121°C. Aseptically and anaerobically add 40.0mL $NaHCO_3$ solution, 10.0mL $Na_2S{\cdot}9H_2O$ solution, 10.0mL trimethoxycinnamate solution, and 1.5mL trace elements solution SL-10. Mix thoroughly. Aseptically and anaerobically distribute into sterile tubes or bottles.

Use: For the cultivation of *Sporobacter termitidis.*

Sporocytophaga Medium

Composition per liter:

$NaNO_3$	2.0g
K_2HPO_4	1.2g
$MgSO_4{\cdot}7H_2O$	1.0g
KCl	0.5g
KH_2PO_4	0.14g
Yeast extract	0.02g
$FeSO_4{\cdot}7H_2O$	6.0mg
Filter paper strips	variable

pH 7.2 ± 0.2 at 25°C

Preparation of Medium: Add components, except filter paper strips, to distilled/deionized water and bring volume to 1.0L. Mix thoroughly. Distribute into tubes in 5.0mL volumes. Autoclave for 15 min at 15 psi pressure–121°C. Aseptically add a sterile filter paper strip to each tube so that 1.0–2.0cm of the strip extends above the medium.

Use: For the cultivation and maintenance of *Sporocytophaga myxococcoides*.

Sporohalobacter lortetii Agar

Composition per liter:

NaCl	105.0g
L-Glutamic acid	4.0g
Agar	20.0g
$CaCO_2$	5.0g
Soluble starch	2.0g
Casamino acids	2.0g
Nutrient broth	2.0g
Yeast extract	2.0g
KCl	0.75g
L-Cysteine	0.5g
$FeSO_4{\cdot}7H_2O$	0.002g
Resazurin	1.0mg
$MgCl_2{\cdot}6H_2O$ solution	40.0mL
$CaCl_2{\cdot}2H_2O$ solution	10.0mL
Trace elements solution	10.0mL
Vitamin solution	10.0mL

pH 6.5 ± 0.2 at 25°C

$MgCl_2{\cdot}6H_2O$ Solution:
Composition per 40.0mL:

$MgCl_2{\cdot}6H_2O$	0.01g

Preparation of $MgCl_2{\cdot}6H_2O$ Solution: Add $MgCl_2{\cdot}6H_2O$ to distilled/deionized water and bring volume to 40.0mL. Mix thoroughly. Sparge with 100% N_2. Autoclave for 15 min at 15 psi pressure–121°C.

$CaCl_2{\cdot}2H_2O$ Solution:
Composition per 40.0mL:

$CaCl_2{\cdot}2H_2O$	0.01g

Preparation of $CaCl_2{\cdot}2H_2O$ Solution: Add $CaCl_2{\cdot}2H_2O$ to distilled/deionized water and bring volume to 40.0mL. Mix thoroughly. Sparge with 100% N_2. Autoclave for 15 min at 15 psi pressure–121°C.

Trace Elements Solution:
Composition per liter:

$MgSO_4{\cdot}7H_2O$	3.0g
Nitrilotriacetic acid	1.5g
NaCl	1.0g
$MnSO_4{\cdot}2H_2O$	0.5g
$CoSO_4{\cdot}7H_2O$	0.18g
$ZnSO_4{\cdot}7H_2O$	0.18g
$CaCl_2{\cdot}2H_2O$	0.1g
$FeSO_4{\cdot}7H_2O$	0.1g
$NiCl_2{\cdot}6H_2O$	0.025g
$KAl(SO_4)_2{\cdot}12H_2O$	0.02g
$CuSO_4{\cdot}5H_2O$	0.01g
H_3BO_3	0.01g
$Na_2MoO_4{\cdot}2H_2O$	0.01g
$Na_2SeO_3{\cdot}5H_2O$	0.3mg

Preparation of Trace Elements Solution: Add nitrilotriacetic acid to 500.0mL of distilled/deionized water. Adjust pH to 6.5 with KOH. Add remaining components. Add distilled/deionized water to 1.0L.

Vitamin Solution:
Composition per liter:

Pyridoxine·HCl	10.0mg
Calcium DL-pantothenate	5.0mg
Lipoic acid	5.0mg
Nicotinic acid	5.0mg
p-Aminobenzoic acid	5.0mg
Riboflavin	5.0mg
Thiamine·HCl	5.0mg
Biotin	2.0mg
Folic acid	2.0mg
Vitamin B_{12}	0.1mg

Preparation of Vitamin Solution: Add components to distilled/deionized water and bring volume to 1.0L. Mix thoroughly.

Preparation of Medium: Prepare and dispense medium under 100% N_2. Add components, except $MgCl_2{\cdot}6H_2O$ and $CaCl_2{\cdot}2H_2O$ solutions, to distilled/deionized water and bring volume to 950.0mL. Mix thoroughly. Gently heat and bring to boiling. Sparge with 100% N_2. Autoclave for 15 min at 15 psi pressure–121°C. Aseptically add 40.0mL of sterile $MgCl_2{\cdot}6H_2O$ solution and 10.0mL of sterile $CaCl_2{\cdot}2H_2O$ solution. Mix thoroughly. Aseptically and anaerobically pour into sterile Petri dishes or distribute into sterile tubes.

Use: For the cultivation and maintenance of *Sporohalobacter lortetii.*

Sporohalobacter lortetii Broth

Composition per liter:

NaCl	105.0g
L-Glutamic acid	4.0g
Casamino acids	2.0g
Nutrient broth	2.0g
Yeast extract	2.0g
KCl	0.75g
L-Cysteine	0.5g
$FeSO_4·7H_2O$	0.002g
Resazurin	1.0mg
$MgCl_2·6H_2O$ solution	40.0mL
$CaCl_2·2H_2O$ solution	10.0mL
Trace elements solution	10.0mL
Vitamin solution	10.0mL

pH 6.5 ± 0.2 at 25°C

$MgCl_2·6H_2O$ Solution:

Composition per 40.0mL:

$MgCl_2·6H_2O$	0.01g

Preparation of $MgCl_2·6H_2O$ Solution: Add $MgCl_2·6H_2O$ to distilled/deionized water and bring volume to 40.0mL. Mix thoroughly. Sparge with 100% N_2. Autoclave for 15 min at 15 psi pressure–121°C.

$CaCl_2·2H_2O$ Solution:

Composition per 40.0mL:

$CaCl_2·2H_2O$	0.01g

Preparation of $CaCl_2·2H_2O$ Solution: Add $CaCl_2·2H_2O$ to distilled/deionized water and bring volume to 40.0mL. Mix thoroughly. Sparge with 100% N_2. Autoclave for 15 min at 15 psi pressure–121°C.

Trace Elements Solution:

Composition per liter:

$MgSO_4·7H_2O$	3.0g
Nitrilotriacetic acid	1.5g
NaCl	1.0g
$MnSO_4·2H_2O$	0.5g
$CoSO_4·7H_2O$	0.18g
$ZnSO_4·7H_2O$	0.18g
$CaCl_2·2H_2O$	0.1g
$FeSO_4·7H_2O$	0.1g
$NiCl_2·6H_2O$	0.025g
$KAl(SO_4)_2·12H_2O$	0.02g
$CuSO_4·5H_2O$	0.01g
H_3BO_3	0.01g
$Na_2MoO_4·2H_2O$	0.01g
$Na_2SeO_3·5H_2O$	0.3mg

Preparation of Trace Elements Solution: Add nitrilotriacetic acid to 500.0mL of distilled/deionized water. Adjust pH to 6.5 with KOH. Add remaining components. Add distilled/deionized water to 1.0L.

Vitamin Solution:

Composition per liter:

Pyridoxine·HCl	10.0mg
Calcium DL-pantothenate	5.0mg
Lipoic acid	5.0mg
Nicotinic acid	5.0mg
p-Aminobenzoic acid	5.0mg
Riboflavin	5.0mg
Thiamine·HCl	5.0mg
Biotin	2.0mg
Folic acid	2.0mg
Vitamin B_{12}	0.1mg

Preparation of Vitamin Solution: Add components to distilled/deionized water and bring volume to 1.0L. Mix thoroughly.

Preparation of Medium: Prepare and dispense medium under 100% N_2. Add components, except $MgCl_2·6H_2O$ and $CaCl_2·2H_2O$ solutions, to distilled/deionized water and bring volume to 950.0mL. Mix thoroughly. Sparge with 100% N_2. Autoclave for 15 min at 15 psi pressure–121°C. Aseptically add 40.0mL of sterile $MgCl_2·6H_2O$ solution and 10.0mL of sterile $CaCl_2·2H_2O$ solution. Mix thoroughly. Aseptically and anaerobically distribute into sterile tubes or flasks.

Use: For the cultivation and maintenance of *Sporohalobacter lortetii*.

Sporomusa Medium

Composition per 877.0mL:

NaCl	2.25g
Pancreatic digest of casein	2.0g
Yeast extract	2.0g
$MgSO_4·7H_2O$	0.5g
NH_4Cl	0.5g
K_2HPO_4	0.35g
KH_2PO_4	0.23g
$CaCl_2·2H_2O$	0.025g
$FeSO_4·7H_2O$	2.0mg
Resazurin	1.0mg
$NaHSeO_3$	15.0μg
$NaHCO_3$ solution	50.0mL
Glycine betaine solution	50.0mL
Wolfe's vitamin solution	10.0mL
L-Cysteine·HCl·H_2O solution	10.0mL
Trace elements solution SL-6	3.0mL

pH 7.0–7.2 at 25°C

$NaHCO_3$ Solution:

Composition per 50.0mL:

$NaHCO_3$	4.0g

Preparation of $NaHCO_3$ Solution: Add the $NaHCO_3$ to distilled/deionized water and bring volume to 50.0mL. Mix thoroughly. Gas under 80% N_2 + 20% CO_2 for 20 min.

Glycine Betaine Solution:

Composition per 50.0mL:

Glycine betaine	5.0g

Preparation of Glycine Betaine Solution: Add the glycine betaine to distilled/deionized water and bring volume to 50.0mL. Mix thoroughly. Filter sterilize. Aseptically gas under 80% N_2 + 20% CO_2.

Wolfe's Vitamin Solution:

Composition per liter:

Pyridoxine·HCl	0.01g
Thiamine·HCl	5.0mg
Riboflavin	5.0mg
Nicotinic acid	5.0mg
Calcium pantothenate	5.0mg
p-Aminobenzoic acid	5.0mg
Thioctic acid	5.0mg
Biotin	2.0mg
Folic acid	2.0mg
Cyanocobalamin	0.1mg

Preparation of Wolfe's Vitamin Solution: Add components to distilled/deionized water and bring volume to 1.0L. Mix thoroughly.

L-Cysteine·HCl·H_2O Solution:

Composition per 10.0mL:

L-Cysteine·HCl·H_2O	0.3g

Preparation of L-Cysteine·HCl·H_2O Solution: Add L-cysteine·HCl·H_2O to distilled/deionized water and bring volume to 10.0mL. Mix thoroughly. Filter sterilize. Aseptically gas under 100% N_2.

Trace Elements Solution SL-6:

Composition per liter:

H_3BO_3	0.3g
$CoCl_2·6H_2O$	0.2g
$ZnSO_4·7H_2O$	0.1g
$MnCl_2·4H_2O$	0.03g
$Na_2MoO_4·H_2O$	0.03g
$NiCl_2·6H_2O$	0.02g
$CuCl_2.2H_2O$	0.01g

Preparation of Trace Elements Solution SL-6: Add components to distilled/deionized water and bring volume to 1.0L. Mix thoroughly. Adjust pH to 3.4.

Preparation of Medium: Add components—except $NaHCO_3$ solution, glycine betaine solution, and L-cysteine·HCl·H_2O solution—to distilled/deionized water and bring volume to 890.0mL. Mix thoroughly. Gently heat and bring to boiling. Continue boiling for 5 min. Cool rapidly to 25°C under 80% N_2 + 20% CO_2. Add 50.0mL of $NaHCO_3$ solution. Mix thoroughly. Autoclave anaerobically for 15 min at 15 psi pressure–121°C. Cool to 25°C. Aseptically add sterile glycine betaine solution. Mix thoroughly. Immediately prior to inoculation, aseptically and anaerobically add L-cysteine·HCl·H_2O solution.

Use: For the cultivation and maintenance of *Sporomusa ovata* and *Sporomusa sphaeroides*.

Sporomusa Medium

Composition per 1010.0mL:

Betaine·H_2O	6.7g
$NaHCO_3$	4.0g
NaCl	2.25g
Pancreatic digest of casein	2.0g
Yeast extract	2.0g
$MgSO_4·7H_2O$	0.5g
NH_4Cl	0.5g
K_2HPO_4	0.348g
$CaCl_2·2H_2O$	0.25g
KH_2PO_4	0.227g
$FeSO_4·7H_2O$	2.0mg
Resazurin	1.0mg
$NaHSeO_3$	26.3μg
Vitamin solution	10.0mL
Reducing agent solution	10.0mL
Trace elements solution SL-10	1.0mL

pH 7.0 ± 0.2 at 25°C

Vitamin Solution:

Composition per liter:

Pyridoxine·HCl	10.0mg
Calcium DL-pantothenate	5.0mg
Lipoic acid	5.0mg
Nicotinic acid	5.0mg
p-Aminobenzoic acid	5.0mg
Riboflavin	5.0mg
Thiamine·HCl	5.0mg
Biotin	2.0mg
Folic acid	2.0mg
Vitamin B_{12}	0.1mg

Preparation of Vitamin Solution: Add components to distilled/deionized water and bring volume to 1.0L. Mix thoroughly.

Reducing Agent Solution:

Composition per 10.0mL:

L-Cysteine·HCl·H_2O	0.3g
$Na_2S·9H_2O$	0.3g

Preparation of Reducing Agent Solution: Add 10.0mL of distilled/deionized water to a flask. Gently heat and bring to boiling. Continue to boil for 1 min while sparging with 100% N_2. Cool to room temperature. Add L-cysteine·HCl·H_2O. Mix thoroughly. Adjust pH to 9 with 5*N* NaOH. Add $Na_2S·9H_2O$. Mix thoroughly. Autoclave for 10 min at 15 psi pressure–121°C.

Trace Elements Solution SL-10:

Composition per liter:

$FeCl_2·4H_2O$	1.5g
$CoCl_2·6H_2O$	190.0mg
$MnCl_2·4H_2O$	100.0mg
$ZnCl_2$	70.0mg
$Na_2MoO_4·2H_2O$	36.0mg
$NiCl_2·6H_2O$	24.0mg
H_3BO_3	6.0mg
$CuCl_2·2H_2O$	2.0mg
HCl (25% solution)	10.0mL

Preparation of Trace Elements Solution SL-10: Add $FeCl_2·4H_2O$ to 10.0mL of HCl solution. Mix thoroughly. Add distilled/deionized water and bring volume to 1.0L. Add remaining components. Mix thoroughly.

Preparation of Medium: Prepare and dispense medium under 80% N_2 + 20% CO_2. Add components, except reducing agent solution, to distilled/deionized water and bring volume to 1.0L. Mix thoroughly. Sparge with 80% N_2 + 20% CO_2. Anaerobically distribute into tubes or flasks. Autoclave for 15 min at 15 psi pressure–121°C. Using a syringe, aseptically and anaerobically add sterile reducing agent to each tube (10.0mL per liter of medium).

Use: For the cultivation and maintenance of *Sporomusa* species.

Sporomusa Medium, Modified

Composition per liter:

$NaHCO_3$	4.0g
NaCl	2.25g
Pancreatic digest of casein	2.0g
Yeast extract	2.0g
$MgSO_4·7H_2O$	0.5g
NH_4Cl	0.5g
K_2HPO_4	0.348g
$CaCl_2·2H_2O$	0.25g
KH_2PO_4	0.227g
$FeSO_4·7H_2O$	2.0mg
Resazurin	1.0mg
$NaHSeO_3$	26.3μg
Fructose solution	50.0mL
Reducing agent solution	10.0mL
Vitamin solution	10.0mL
Trace elements solution SL-10	1.0mL

pH 7.0 ± 0.2 at 25°C

Fructose Solution:

Composition per 50.0mL:

Fructose	10.0g

Preparation of Fructose Solution: Add fructose to distilled/deionized water and bring volume to 50.0mL. Mix thoroughly. Sparge under 100% N_2 gas for 3 min. Filter sterilize. Store under N_2 gas.

Vitamin Solution:

Composition per liter:

Pyridoxine·HCl	10.0mg
Calcium DL-pantothenate	5.0mg
Lipoic acid	5.0mg
Nicotinic acid	5.0mg
p-Aminobenzoic acid	5.0mg
Riboflavin	5.0mg
Thiamine·HCl	5.0mg
Biotin	2.0mg
Folic acid	2.0mg
Vitamin B_{12}	0.1mg

Preparation of Vitamin Solution: Add components to distilled/deionized water and bring volume to 1.0L. Mix thoroughly.

Reducing Agent Solution:

Composition per 10.0mL:

L-Cysteine·HCl·H_2O	0.3g
$Na_2S·9H_2O$	0.3g

Preparation of Reducing Agent Solution: Add 10.0mL of distilled/deionized water to a flask. Boil under N_2 gas for 1 min. Cool to room temperature. Add L-cysteine·HCl·H_2O and dissolve. Adjust to pH 9 with 5*N* NaOH. Add washed $Na_2S·9H_2O$ and dissolve. Mix thoroughly. Autoclave for 10 min at 15 psi pressure–121°C.

Trace Elements Solution SL-10:

Composition per liter:

$FeCl_2·4H_2O$	1.5g
$CoCl_2·6H_2O$	190.0mg
$MnCl_2·4H_2O$	100.0mg
$ZnCl_2$	70.0mg
$Na_2MoO_4·2H_2O$	36.0mg
$NiCl_2·6H_2O$	24.0mg
H_3BO_3	6.0mg
$CuCl_2·2H_2O$	2.0mg
HCl (25% solution)	10.0mL

Preparation of Trace Elements Solution SL-10: Add $FeCl_2·4H_2O$ to 10.0mL of HCl solution. Mix thoroughly. Add distilled/deionized water and bring volume to 1.0L. Add remaining components. Mix thoroughly.

Preparation of Medium: Add components, except $NaHCO_3$ and reducing agent solution, to distilled/deionized water and bring volume to 940.0mL. Gently heat and bring to boiling. Continue boiling for 3 min. Cool to room temperature under 80% N_2 + 20% CO_2. Add solid $NaHCO_3$ and bring pH to 7.0 by gassing. Distribute anaerobically under 80% N_2 + 20% CO_2 into tubes or flasks. Autoclave for 15 min at 15 psi pressure–121°C. Prior to inoculation of cultures, aseptically and anaerobically add 0.1mL of sterile reducing agent solution and 0.5mL of sterile fructose solution to each tube containing 9.4mL of sterile basal medium.

Use: For the cultivation and maintenance of *Sporomusa acidovorans*.

Sporomusa Medium, Modified

Composition per liter:

$NaHCO_3$	4.0g
NaCl	2.25g
Pancreatic digest of casein	2.0g
Yeast extract	2.0g
Betaine	1.35g
$MgSO_4·7H_2O$	0.5g
NH_4Cl	0.5g
K_2HPO_4	0.348g
$CaCl_2·2H_2O$	0.25g
KH_2PO_4	0.227g
$FeSO_4·7H_2O$	2.0mg
Resazurin	1.0mg
$NaHSeO_3$	26.3μg
Fructose solution	50.0mL
Reducing agent solution	10.0mL
Vitamin solution	10.0mL
Trace elements solution SL-10	1.0mL

pH 7.0 ± 0.2 at 25°C

Fructose Solution:

Composition per 50.0mL:

Fructose	10.0g

Preparation of Fructose Solution: Add fructose to distilled/deionized water and bring volume to 50.0mL. Mix thoroughly. Sparge under 100% N_2 gas for 3 min. Filter sterilize. Store under N_2 gas.

Vitamin Solution:

Composition per liter:

Pyridoxine·HCl	10.0mg
Calcium DL-pantothenate	5.0mg
Lipoic acid	5.0mg
Nicotinic acid	5.0mg
p-Aminobenzoic acid	5.0mg
Riboflavin	5.0mg
Thiamine·HCl	5.0mg
Biotin	2.0mg
Folic acid	2.0mg
Vitamin B_{12}	0.1mg

Preparation of Vitamin Solution: Add components to distilled/deionized water and bring volume to 1.0L. Mix thoroughly.

Reducing Agent Solution:

Composition per 10.0mL:

Dithiothreitol	0.154g

Preparation of Reducing Agent Solution: Add 10.0mL of distilled/deionized water to a flask. Boil under N_2 gas for 1 min. Cool to room temperature. Add dithiothreitol and dissolve. Autoclave for 10 min at 15 psi pressure–121°C.

Trace Elements Solution SL-10:

Composition per liter:

$FeCl_2·4H_2O$	1.5g
$CoCl_2·6H_2O$	190.0mg
$MnCl_2·4H_2O$	100.0mg
$ZnCl_2$	70.0mg
$Na_2MoO_4·2H_2O$	36.0mg
$NiCl_2·6H_2O$	24.0mg
H_3BO_3	6.0mg
$CuCl_2·2H_2O$	2.0mg
HCl (25% solution)	10.0mL

Preparation of Trace Elements Solution SL-10: Add $FeCl_2 \cdot 4H_2O$ to 10.0mL of HCl solution. Mix thoroughly. Add distilled/deionized water and bring volume to 1.0L. Add remaining components. Mix thoroughly.

Preparation of Medium: Add components, except $NaHCO_3$ and reducing agent solution, to distilled/deionized water and bring volume to 940.0mL. Gently heat and bring to boiling. Continue boiling for 3 min. Cool to room temperature under 80% N_2 + 20% CO_2. Add solid $NaHCO_3$ and bring pH to 7.0 by gassing. Distribute anaerobically under 80% N_2 + 20% CO_2 into tubes or flasks. Autoclave for 15 min at 15 psi pressure–121°C. Prior to inoculation of cultures, aseptically and anaerobically add 0.1mL of sterile reducing agent solution and 0.5mL of sterile fructose solution to each tube containing 9.4mL of sterile basal medium.

Use: For the cultivation and maintenance of *Sporomusa termitida.*

Sporomusa silvacetica Medium (DSMZ Medium 777)

Composition per liter:

NaCl	2.25g
Yeast extract	1.0g
Casitone	1.0g
NH_4Cl	0.5g
$MgSO_4 \cdot 7H_2O$	0.5g
K_2HPO_4	0.348g
KH_2PO_4	0.227g
$CaCl_2 \cdot 2H_2O$	0.25g
$FeSO_4 \cdot 7H_2O$	0.002g
Resazurin	1.0mg
$NaHSeO_3$	15.1μg
Cysteine solution	10.0mL
$NaHCO_3$ solution	10.0mL
$Na_2S \cdot 9H_2O$ solution	10.0mL
Fructose solution	10.0mL
Vitamin solution	10.0mL
Trace elements solution SL-10	1.0mL

pH 6.6 ± 0.2 at 25°C

Cysteine Solution:
Composition per 10.0mL:

L-Cysteine-HCl·H_2O	0.3g

Preparation of Cysteine Solution: Add L-cysteine·HCl·H_2O to distilled/deionized water and bring volume to 10.0mL. Mix thoroughly. Sparge with 100% N_2. Autoclave for 15 min at 15 psi pressure–121°C.

$Na_2S \cdot 9H_2O$ Solution:
Composition per 10.0mL:

$Na_2S \cdot 9H_2O$	0.3g

Preparation of $Na_2S \cdot 9H_2O$ Solution: Add $Na_2S \cdot 9H_2O$ to distilled/deionized water and bring volume to 10.0mL. Mix thoroughly. Autoclave under 100% N_2 for 15 min at 15 psi pressure–121°C. Cool to room temperature.

$NaHCO_3$ Solution:
Composition per 10.0mL:

$NaHCO_3$	1.5g

Preparation of $NaHCO_3$ Solution: Add $NaHCO_3$ to distilled/deionized water and bring volume to 10.0mL. Mix thoroughly. Sparge with 80% N_2 + 20% CO_2. Filter sterilize.

Fructose Solution:
Composition per 10.0mL:

Fructose	5.0g

Preparation of Fructose Solution: Add fructose to distilled/deionized water and bring volume to 10.0mL. Mix thoroughly. Sparge with 100% N_2. Autoclave for 15 min at 15 psi pressure–121°C. Cool to room temperature.

Trace Elements Solution SL-10:
Composition per liter:

$FeCl_2 \cdot 4H_2O$	1.5g
$CoCl_2 \cdot 6H_2O$	190.0mg
$MnCl_2 \cdot 4H_2O$	100.0mg
$ZnCl_2$	70.0mg
$Na_2MoO_4 \cdot 2H_2O$	36.0mg
$NiCl_2 \cdot 6H_2O$	24.0mg
H_3BO_3	6.0mg
$CuCl_2 \cdot 2H_2O$	2.0mg
HCl (25% solution)	10.0mL

Preparation of Trace Elements Solution SL-10: Add $FeCl_2 \cdot 4H_2O$ to 10.0mL of HCl solution. Mix thoroughly. Add distilled/deionized water and bring volume to 1.0L. Add remaining components. Mix thoroughly. Sparge with 80% N_2 + 20% CO_2. Autoclave for 15 min at 15 psi pressure–121°C.

Vitamin Solution:
Composition per liter:

Pyridoxine-HCl	10.0mg
Thiamine-HCl·$2H_2O$	5.0mg
Riboflavin	5.0mg
Nicotinic acid	5.0mg
D-Ca-pantothenate	5.0mg
p-Aminobenzoic acid	5.0mg
Lipoic acid	5.0mg
Biotin	2.0mg
Folic acid	2.0mg
Vitamin B_{12}	0.1mg

Preparation of Vitamin Solution: Add components to distilled/deionized water and bring volume to 1.0L. Mix thoroughly. Sparge with 80% H_2 + 20% CO_2. Filter sterilize.

Preparation of Medium: Prepare and dispense medium under 80% N_2 + 20% CO_2 gas atmosphere. Add components, except $NaHCO_3$ solution, fructose solution, cysteine solution, vitamin solution, and $Na_2S \cdot 9H_2O$ solution, to distilled/deionized water and bring volume to 950.0mL. Mix thoroughly. Adjust pH to 6.5–6.7. Autoclave for 15 min at 15 psi pressure–121°C. Cool to room temperature. Aseptically and anaerobically add 10.0mL $NaHCO_3$ solution, 10.0mL fructose solution, 10.0mL cysteine solution, 10.0mL vitamin solution, and 10.0mL $Na_2S \cdot 9H_2O$ solution. Mix thoroughly. Aseptically and anaerobically distribute into sterile tubes or bottles.

Use: For the cultivation of *Sporomusa silvacetica, Thermicanus aegyptius, Moorella thermoacetica,* and *Moorella mulderi.*

Sporosarcina halophila Agar

Composition per liter:

NaCl	30.0g
Agar	20.0g
$MgCl_2 \cdot 6H_2O$	5.0g
Peptone	5.0g
NaCl	5.0g

Yeast extract 2.0g
Beef extract 1.0g

pH 7.2 ± 0.2 at 25°C

Preparation of Medium: Add components to distilled/deionized water and bring volume to 1.0L. Mix thoroughly. Gently heat and bring to boiling. Distribute into tubes or flasks. Autoclave for 15 min at 15 psi pressure–121°C. Pour into sterile Petri dishes or leave in tubes.

Use: For the cultivation and maintenance of *Halomonas elongata, Halomonas halmophila, Listonella anguillara, Salinicoccus roseus, Sporosarcina halophila, Vibrio fluvialis, Vibrio furnissii, Vibrio hollisae, Vibrio ordalii,* and *Vibrio vulnificus.*

Sporosarcina ureae Medium

Composition per liter:

L-Asparagine·H_2O or L-glutamine 30.0g
KCl 3.4g
NaCl 2.92g
K_2HPO_4 .... 0.25g
$(NH_4)_2SO_4$ 0.2g
$MgSO_4 \cdot 7H_2O$ 0.05g
$FeSO_4 \cdot 7H_2O$ 2.5mg
$MnCl_2 \cdot 4H_2O$ 0.25mg
Biotin solution 10.0mL
L-Cysteine solution 10.0mL
$(NH_4)_2SO_4$ solution 10.0mL

pH 8.7 ± 0.2 at 25°C

Biotin Solution:

Composition per 10.0mL:

D-Biotin 1.0mg

Preparation of Biotin Solution: Add biotin to distilled/deionized water and bring volume to 10.0mL. Mix thoroughly. Filter sterilize.

L-Cysteine Solution:

Composition per 10.0mL:

L-Cysteine 0.04g

Preparation of L-Cysteine Solution: Add L-cysteine to distilled/deionized water and bring volume to 10.0mL. Mix thoroughly. Filter sterilize.

$(NH_4)_2SO_4$ Solution:

Composition per 10.0mL:

$(NH_4)_2SO_4$ 0.2g

Preparation of $(NH_4)_2SO_4$ Solution: Add $(NH_4)_2SO_4$ to distilled/deionized water and bring volume to 10.0mL. Mix thoroughly. Filter sterilize.

Preparation of Medium: Add components—except biotin solution, L-cysteine solution, and $(NH_4)_2SO_4$ solution—to distilled/deionized water and bring volume to 970.0mL. Mix thoroughly. Adjust pH to 8.7 with 1*N* NaOH. Autoclave for 15 min at 15 psi pressure–121°C. Cool to 45°–50°C. Aseptically add sterile biotin solution, L-cysteine solution, and $(NH_4)_2SO_4$ solution. Mix thoroughly. Aseptically distribute into sterile tubes.

Use: For the cultivation of *Sporosarcina ureae.*

Sporosarcina ureae Medium

Composition per liter:

Agar 30.0g
Glucose 4.0g
$(NH_4)_2SO_4$ 4.0g
Malt extract 3.0g
Peptone 3.0g
Yeast extract 2.0g
K_2HPO_4 .... 1.0g
$MgSO_4$ 0.8g
$CaCl_2$ 0.1g
$MnSO_4 \cdot H_2O$ 0.1g
$CuSO_4 \cdot 5H_2O$ 0.01g
$ZnSO_4$ 0.01g
$FeSO_4 \cdot 7H_2O$ 1.0mg

Preparation of Medium: Add components to 1.0L of distilled/deionized water. Mix thoroughly. Gently heat and bring to boiling. Distribute into tubes or flasks. Autoclave for 15 min at 15 psi pressure–121°C. Pour into sterile Petri dishes or leave in tubes.

Use: For the cultivation and induction of sporulation of *Sporosarcina ureae.*

Sporulating Agar

See: **AK Agar No. 2**

Sporulation Agar

Composition per liter:

Agar 15.0g
Glucose 10.0g
Tryptose 2.0g
Beef extract 1.0g
Yeast extract 1.0g
$FeSO_4$ 0.1mg

pH 7.2 ± 0.2 at 25°C

Preparation of Medium: Add components to distilled/deionized water and bring volume to 1.0L. Mix thoroughly. Adjust pH to 7.2. Gently heat and bring to boiling. Distribute into tubes or flasks. Autoclave for 15 min at 15 psi pressure–121°C. Pour into sterile Petri dishes or leave in tubes.

Use: For the cultivation of *Chlorella pyrenoidosa.*

Sporulation Agar

Composition per 1001.0mL:

Agar 10.0g
Pancreatic digest of gelatin 3.75
Beef extract 2.25
Trypticase™ 2.0g
Yeast extract 1.0g
Marine broth 2216 .... 0.5g
$MnSO_4 \cdot H_2O$ 0.5mg
Inosine (5m*M*) solution 1.0mL

Source: Marine broth 2216 is available from BD Diagnostic Systems.

Preparation of Medium: Add components to distilled/deionized water and bring volume to 1.0L. Mix thoroughly. Gently heat and bring to boiling. Distribute into tubes or flasks. Autoclave for 15 min at 15 psi pressure–121°C. Pour into sterile Petri dishes or leave in tubes.

Use: For the cultivation and maintenance of *Bacillus aminovorans.*

Sporulation Agar (m-Sporulation Agar)

Composition per liter:

Agar 15.0g
Glucose 10.0g

Tryptose	2.0g
Beef extract	1.0g
Yeast extract	1.0g
$FeSO_4$	1.0μg

pH 7.2 ± 0.2 at 25°C

Preparation of Medium: Add components to distilled/deionized water and bring volume to 1.0L. Mix thoroughly. Gently heat and bring to boiling. Distribute into tubes or flasks. Autoclave for 15 min at 15 psi pressure–121°C. Pour into sterile Petri dishes or leave in tubes.

Use: For the cultivation and sporulation of *Streptomyces*, *Streptoverticillium*, and *Thermoactinomyces* species. For the identification of sporulating bacteria by the membrane filter method.

Sporulation Broth

Composition per liter:

Polypeptone™	15.0g
Na_2HPO_4	11.0g
Starch, soluble	3.0g
Yeast extract	3.0g
Sodium thioglycolate	1.0g
$MgSO_4$, anhydrous	0.1g

pH 7.8 ± 0.1 at 25°C

Preparation of Medium: Add components to distilled/deionized water and bring volume to 1.0L. Mix thoroughly. Distribute into tubes in 15.0mL volumes. Autoclave for 15 min at 15 psi pressure–121°C.

Use: For the cultivation and observation of sporulation of *Clostridium perfringens.*

Sporulation Broth

Composition per liter:

Glucose	3.3g
Tryptose	0.66g
Beef extract	0.33g
Yeast extract	0.33g
$FeSO_4$	0.33μg

pH 7.2 ± 0.2 at 25°C

Preparation of Medium: Add components to distilled/deionized water and bring volume to 1.0L. Mix thoroughly. Distribute into tubes or flasks. Autoclave for 15 min at 15 psi pressure–121°C.

Use: For the cultivation and sporulation of *Streptomyces*, *Streptoverticillium*, and *Thermoactinomyces* species.

Sporulation Broth

Composition per liter:

Glucose	3.3g
Tryptose	0.7g
Beef extract	0.3g
Yeast extract	0.3g
$FeSO_4$	0.1mg

pH 7.2 ± 0.2 at 25°C

Preparation of Medium: Add components to distilled/deionized water and bring volume to 1.0L. Mix thoroughly. Adjust pH to 7.2. Distribute into tubes or flasks. Autoclave for 15 min at 15 psi pressure–121°C.

Use: For the cultivation of *Chlorella pyrenoidosa.*

Sporulation Medium, Modified

See: **Duncan-Strong Sporulation Medium, Modified**

Spray's Fermentation Medium

Composition per 1100.0mL:

Neopeptone	10.0g
Pancreatic digest of casein	10.0g
Agar	2.0g
Sodium thioglycolate	0.025g
Carbohydrate solution	110.0mL

pH 7.4 ± 0.1 at 25°C

Carbohydrate Solution:

Composition per 200.0mL:

Carbohydrate	20.0g

Preparation of Carbohydrate Solution: Add carbohydrate to distilled/deionized water and bring volume to 200.0mL. Glucose or glycerol may be used. Mix thoroughly. Filter sterilize.

Preparation of Medium: Add components, except agar and carbohydrate solution, to distilled/deionized water and bring volume to 990.0mL. Mix thoroughly. Adjust pH to 7.4. Add agar. Gently heat and bring to boiling. Distribute into tubes in 9.0mL volumes. Autoclave for 15 min at 15 psi pressure–121°C. Cool to 25°C. Immediately prior to use heat tubes in a boiling water bath for 10 min. Cool to 45°C. Aseptically add 1.0mL of sterile carbohydrate solution. Mix thoroughly.

Use: For the cultivation and differentiation of *Clostridium perfringens* based on carbohydrate fermentation patterns.

SPS Agar (Sulfite Polymyxin Sulfadiazine Agar)

Composition per liter:

Pancreatic digest of casein	15.0g
Agar	13.9g
Yeast extract	10.0g
Ferric citrate	0.5g
Na_2SO_3	0.5g
Sulfadiazine	0.12g
Polymyxin sulfate	0.01g

pH 7.0 ± 0.2 at 25°C

Source: This medium is available as a premixed powder from BD Diagnostic Systems.

Preparation of Medium: Add components to distilled/deionized water and bring volume to 1.0L. Mix thoroughly. Gently heat while stirring and bring to boiling. Distribute into tubes or flasks. Autoclave for 15 min at 13 psi pressure–118°C. Pour into sterile Petri dishes or leave in tubes.

Use: For the isolation and detection of *Clostridium perfringens* and *Clostridium botulinum* in foods and other materials.

SPS HiVeg Agar

Composition per liter:

Plant hydrolysate	15.0g
Agar	13.9g
Yeast extract	10.0g
Polysorbate 80	0.5g
Na_2SO_3	0.5g
Sulphadiazine	0.12g
Polymyxin B sulfate	0.01g

pH 7.0 ± 0.2 at 25°C

Source: This medium is available as a premixed powder from HiMedia.

Preparation of Medium: Add components to distilled/deionized water and bring volume to 1.0L. Mix thoroughly. Gently heat while stirring and bring to boiling. Distribute into tubes or flasks. Autoclave for 15 min at 13 psi pressure–118°C. Pour into sterile Petri dishes or leave in tubes.

Use: For the isolation and detection of *Clostridium perfringens* and *Clostridium botulinum* in foods and other materials.

SPS HiVeg Agar, Modified

Composition per liter:

Agar	15.0g
Plant hydrolysate	15.0g
Yeast extract	10.0g
Na_2SO_3	0.5g
Ferric citrate	0.5g
Sulphadiazine	0.12g
Na-thioglycollate	0.1g
Sorbitan monooleate	0.05g
Polymyxin B sulfate	0.01g

pH 7.0 ± 0.2 at 25°C

Source: This medium is available as a premixed powder from HiMedia.

Preparation of Medium: Add components to distilled/deionized water and bring volume to 1.0L. Mix thoroughly. Gently heat while stirring and bring to boiling. Distribute into tubes or flasks. Autoclave for 15 min at 13 psi pressure–118°C. Pour into sterile Petri dishes or leave in tubes.

Use: For the isolation and detection of *Clostridium perfringens* and *Clostridium botulinum* in foods and other materials.

SPYE Medium (LMG Medium 262)

Composition per liter:

Sea salts	30.0g
Agar No. 1	12.0g
Peptone	2.0g
Yeast extract	1.0g
Vitamin solution	5.0mL
Glucose solution	2.0mL

pH 7.3 ± 0.2 at 25°C

Vitamin Solution:
Composition per 10.0mL:

Riboflavin	2.0mg

Preparation of Vitamin Solution: Add riboflavin to distilled/deionized water and bring volume to 10.0mL. Mix thoroughly. Filter sterilize.

Glucose Solution:
Composition per 10.0mL:

Glucose	5.0g

Preparation of Glucose Solution: Add glucose to distilled/deionized water and bring volume to 100.0mL. Mix thoroughly. Filter sterilize.

Preparation of Medium: Add components, except vitamin solution and glucose solution, to 993.0mL distilled/deionized water. Mix thoroughly. Gently heat and bring to boiling. Autoclave for 15 min at 15 psi pressure–121°C. Cool to 45°–50°C. Aseptically add 5.0mL of sterile vitamin solution and 2.0mL of sterile glucose solution. Mix thoroughly. Pour into sterile Petri dishes or distribute into sterile tubes.

Use: For the cultivation of *Maricaulis* spp.

SP4-Z Medium with Arginine and DNA (DSMZ Medium 1076b)

Composition per 499.0mL:

Tryptone	5.0g
Peptone	3.3g
NaCl	0.5g
Beef extract	0.3g
Yeast extract	0.3g
Beef heart, solids from infusion	0.2g
DNA, herring sperm	0.04g
Fetal bovine serum (inactivated at 56°C, 1 hr)	90.0mL
CMRL 1066, 10X with glutamine	25.0mL
Yeastolate solution	5.0mL
Arginine solution	2.0mL
Glutamine solution	1.7mL
Phenol Red solution	1.0mL

pH 7.4 ± 0.2 at 25°C

CMRL 1066, 10X with Glutamine:
Composition per liter:

NaCl	6.8g
$NaHCO_3$	2.2g
D-Glucose	1.0g
KCl	0.4g
L-Cysteine·HCl·H_2O	0.26g
$CaCl_2$, anhydrous	0.2g
$MgSO_4\cdot7H_2O$	0.2g
$NaH_2PO_4\cdot H_2O$	0.14g
L-Glutamine	0.1g
Sodium acetate·$3H_2O$	0.083g
L-Glutamic acid	0.075g
L-Arginine·HCl	0.07g
L-Lysine·HCl	0.07g
L-Leucine	0.06g
Glycine	0.05g
Ascorbic acid	0.05g
L-Proline	0.04g
L-Tyrosine	0.04g
L-Aspartic acid	0.03g
L-Threonine	0.03g
L-Alanine	0.025g
L-Phenylalanine	0.025g
L-Serine	0.025g
L-Valine	0.025g
L-Cystine	0.02g
L-Histidine·HCl·H_2O	0.02g
L-Isoleucine	0.02g
Phenol Red	0.02g
L-Methionine	0.015g
Deoxyadenosine	0.01g
Deoxycytidine	0.01g
Deoxyguanosine	0.01g
Glutathione, reduced	0.01g
Thymidine	0.01g
Hydroxy-L-proline	0.01g
L-Tryptophan	0.01g
Nicotinamide adenine dinucleotide	7.0mg

Tween™ 80	5.0mg
Sodium glucoronate·H_2O	4.2mg
Coenzyme A	2.5mg
Cocarboxylase	1.0mg
Flavin adenine dinucleotide	1.0mg
Nicotinamide adenine dinucleotide phosphate	1.0mg
Uridine triphosphate	1.0mg
Choline chloride	0.5mg
Cholesterol	0.2mg
5-Methyldeoxycytidine	0.1mg
Inositol	0.05mg
p-Aminobenzoic acid	0.05mg
Niacin	0.025mg
Niacinamide	0.025mg
Pyridoxine	0.025mg
Pyridoxal·HCl	0.025mg
Biotin	0.01mg
D-Calcium pantothenate	0.01mg
Folic acid	0.01mg
Riboflavin	0.01mg
Thiamine·HCl	0.01mg

Preparation of CMRL 1066, 10X with Glutamine: Add components to distilled/deionized water and bring volume to 1.0L. Mix thoroughly. Adjust pH to 7.2. Filter sterilize.

Yeastolate Solution:
Composition per 10.0mL:

Yeastolate	2.0g

Preparation of Yeastolate Solution: Add yeastolate to distilled/deionized water and bring volume to 10.0mL. Mix thoroughly. Autoclave for 15 min at 15 psi pressure–121°C.

Phenol Red Solution:
Composition per 100.0mL:

Phenol Red	1.0g

Preparation of Phenol Red Solution: Add Phenol Red to distilled/deionized water and bring volume to 100.0mL. Mix thoroughly. Adjust pH to 7.0. Filter sterilize.

Glutamine Solution:
Composition per 10.0mL:

L-Glutamine	1.5g

Preparation of Glutamine Solution: Add L-glutamine to 10.0mL of distilled/deionized water.. Mix thoroughly. Filter sterilize.

Arginine Solution:
Composition per 10.0mL:

L-Arginine·HCl	5.0g

Preparation of Arginine Solution: Add L-arginine·HCl to distilled/deionized water and bring volume to 10.0mL. Mix thoroughly. Filter sterilize.

Preparation of Medium: Add components, except fetal bovine serum, CMRL, yeast extract solution, yeastolate solutuion, Phenol Red solution, arginine solution, and glutamine solution, to distilled/deionized water and bring volume to 375.0mL. Mix thoroughly. Adjust pH to 7.4. Gently heat and bring to boiling. Autoclave for 15 min at 15 psi pressure–121°C. Cool to room temperature. Mix thoroughly. Aseptically add fetal bovine serum, CMRL, yeast extract solution, yeastolate solution, arginine solution Phenol Red solution, and glutamine solution. Mix thoroughly.

Use: For the cultivation of *Mycoplasma* spp. *and Spiroplasma* spp., *including* the cultivation of *Mycoplasma genitalium, Mycoplasma gallinarum, Spiroplasma culicicola, Spiroplasma chinense, Spiroplasma corruscae, Spiroplasma alleghenense,* and *Spiroplasma clarkii.*

SP4-Z Medium with Glucose and DNA (DSMZ Medium 1076b)

Composition per 499.0mL:

Tryptone	5.0g
Peptone	3.3g
NaCl	0.5g
Beef extract	0.3g
Yeast extract	0.3g
Beef heart, solids from infusion	0.2g
DNA, herring sperm	0.04g
Fetal bovine serum (inactivated at 56°C, 1 hr)	90.0mL
CMRL 1066, 10X with glutamine	25.0mL
Yeastolate solution	5.0mL
Glucose solution	2.0mL
Glutamine solution	1.7mL
Phenol Red solution	1.0mL

pH 7.4 ± 0.2 at 25°C

CMRL 1066, 10X with Glutamine:
Composition per liter:

NaCl	6.8g
$NaHCO_3$	2.2g
D-Glucose	1.0g
KCl	0.4g
L-Cysteine·HCl·H_2O	0.26g
$CaCl_2$, anhydrous	0.2g
$MgSO_4 \cdot 7H_2O$	0.2g
$NaH_2PO_4 \cdot H_2O$	0.14g
L-Glutamine	0.1g
Sodium acetate·$3H_2O$	0.083g
L-Glutamic acid	0.075g
L-Arginine·HCl	0.07g
L-Lysine·HCl	0.07g
L-Leucine	0.06g
Glycine	0.05g
Ascorbic acid	0.05g
L-Proline	0.04g
L-Tyrosine	0.04g
L-Aspartic acid	0.03g
L-Threonine	0.03g
L-Alanine	0.025g
L-Phenylalanine	0.025g
L-Serine	0.025g
L-Valine	0.025g
L-Cystine	0.02g
L-Histidine·HCl·H_2O	0.02g
L-Isoleucine	0.02g
Phenol Red	0.02g
L-Methionine	0.015g
Deoxyadenosine	0.01g
Deoxycytidine	0.01g
Deoxyguanosine	0.01g
Glutathione, reduced	0.01g
Thymidine	0.01g
Hydroxy-L-proline	0.01g
L-Tryptophan	0.01g
Nicotinamide adenine dinucleotide	7.0mg
Tween™ 80	5.0mg
Sodium glucoronate·H_2O	4.2mg

Coenzyme A 2.5mg
Cocarboxylase 1.0mg
Flavin adenine dinucleotide 1.0mg
Nicotinamide adenine dinucleotide phosphate 1.0mg
Uridine triphosphate 1.0mg
Choline chloride 0.5mg
Cholesterol 0.2mg
5-Methyldeoxycytidine 0.1mg
Inositol 0.05mg
p-Aminobenzoic acid 0.05mg
Niacin 0.025mg
Niacinamide 0.025mg
Pyridoxine 0.025mg
Pyridoxal·HCl 0.025mg
Biotin 0.01mg
D-Calcium pantothenate 0.01mg
Folic acid 0.01mg
Riboflavin 0.01mg
Thiamine·HCl 0.01mg

Preparation of CMRL 1066, 10X with Glutamine: Add components to distilled/deionized water and bring volume to 1.0L. Mix thoroughly. Adjust pH to 7.2. Filter sterilize.

Yeastolate Solution:
Composition per 10.0mL:
Yeastolate 2.0g

Preparation of Yeastolate Solution: Add yeastolate to distilled/deionized water and bring volume to 10.0mL. Mix thoroughly. Autoclave for 15 min at 15 psi pressure–121°C.

Phenol Red Solution:
Composition per 100.0mL:
Phenol Red 1.0g

Preparation of Phenol Red Solution: Add Phenol Red to distilled/deionized water and bring volume to 100.0mL. Mix thoroughly. Adjust pH to 7.0. Filter sterilize.

Glutamine Solution:
Composition per 10.0mL:
L-Glutamine 1.5g

Preparation of Glutamine Solution: Add L-glutamine to distilled/deionized water and bring volume to 10.0mL. Mix thoroughly. Filter sterilize.

Glucose Solution:
Composition per 10.0mL:
D-Glucose 5.0g

Preparation of Glucose Solution: Add D-glucose to distilled/deionized water and bring volume to 10.0mL. Mix thoroughly. Filter sterilize.

Preparation of Medium: Add components, except fetal bovine serum, CMRL, yeast extract solution, yeastolate solution, Phenol Red solution, and glutamine solution, to distilled/deionized water and bring volume to 375.0mL. Mix thoroughly. Adjust pH to 7.4. Gently heat and bring to boiling. Autoclave for 15 min at 15 psi pressure–121°C. Cool to room temperature. Mix thoroughly. Aseptically add fetal bovine serum, CMRL, yeast extract solution, yeastolate solution, glucose solution Phenol Red solution, and glutamine solution. Mix thoroughly.

Use: For the cultivation of *Mycoplasma* spp. *and Spiroplasma* spp., *including* the cultivation of *Mycoplasma genitalium, Mycoplasma gallinarum, Spiroplasma culicicola, Spiroplasma chinense, Spiroplasma corruscae, Spiroplasma alleghenense,* and *Spiroplasma clarkii.*

SRB-Psychrophile Medium (DSMZ Medium 861)

Composition per 1158mL:
NaCl 20.0g
Na_2SO_4 .... 4.0g
$MgCl_2·6H_2O$ 3.0g
$CaCl_2·2H_2O$ 0.15g
KBr 0.09g
KCl 0.5g
Resazurin 0.5mg
NH_4Cl solution 49.5mL
KH_2PO_4 solution 49.5mL
$NaHCO_3$ solution 29.7mL
Vitamin solution 9.9mL
Substrate solution 9.9mL
$Na_2S·9H_2O$ solution 9.9mL
Dithionite solution 1.0mL
Trace elements solution SL-10 .... 1.0mL
Selenite-tungstate solution 1.0mL
pH 7.2 ± 0.2 at 25°C

Dithionite Solution
Composition per 10.0mL:
Na-dithionite 0.25g

Preparation of Dithionite Solution: Add Na-dithionite to distilled/deionized water and bring volume to 10.0mL. Mix thoroughly. Sparge with 100% N_2. Filter sterilize.

$Na_2S·9H_2O$ Solution:
Composition per 10.0mL:
$Na_2S·9H_2O$ 0.5g

Preparation of $Na_2S·9H_2O$ Solution: Add $Na_2S·9H_2O$ to distilled/deionized water and bring volume to 10.0mL. Mix thoroughly. Autoclave under 100% N_2 for 15 min at 15 psi pressure–121°C. Cool to room temperature.

Vitamin Solution:
Composition per liter:
Pyridoxine-HCl 10.0mg
Thiamine-HCl·$2H_2O$ 5.0mg
Riboflavin 5.0mg
Nicotinic acid 5.0mg
D-Ca-pantothenate 5.0mg
p-Aminobenzoic acid 5.0mg
Lipoic acid 5.0mg
Biotin 2.0mg
Folic acid 2.0mg
Vitamin B_{12} 0.1mg

Preparation of Vitamin Solution: Add components to distilled/deionized water and bring volume to 1.0L. Mix thoroughly. Sparge with 80% H_2 + 20% CO_2. Filter sterilize.

Selenite-Tungstate Solution
Composition per liter:
NaOH 0.5g
$Na_2WO_4·2H_2O$ 4.0mg
$Na_2SeO_3·5H_2O$ 3.0mg

Preparation of Selenite-Tungstate Solution: Add components to distilled/deionized water and bring volume to 1.0L. Mix thoroughly. Sparge with 100% N_2. Filter sterilize.

Trace Elements Solution SL-10:

Composition per liter:

$FeCl_2 \cdot 4H_2O$	1.5g
$CoCl_2 \cdot 6H_2O$	190.0mg
$MnCl_2 \cdot 4H_2O$	100.0mg
$ZnCl_2$	70.0mg
$Na_2MoO_4 \cdot 2H_2O$	36.0mg
$NiCl_2 \cdot 6H_2O$	24.0mg
H_3BO_3	6.0mg
$CuCl_2 \cdot 2H_2O$	2.0mg
HCl (25% solution)	10.0mL

Preparation of Trace Elements Solution SL-10: Add $FeCl_2 \cdot 4H_2O$ to 10.0mL of HCl solution. Mix thoroughly. Add distilled/deionized water and bring volume to 1.0L. Add remaining components. Mix thoroughly. Sparge with 80% N_2 + 20% CO_2. Autoclave for 15 min at 15 psi pressure–121°C.

$NaHCO_3$ Solution:

Composition per 100.0mL:

$NaHCO_3$	10.0g

Preparation of $NaHCO_3$ Solution: Add $NaHCO_3$ to distilled/deionized water and bring volume to 100.0mL. Mix thoroughly. Sparge with 80% N_2 + 20% CO_2. Filter sterilize.

NH_4Cl Solution:

Composition per 100.0mL:

NH_4Cl	0.5g

Preparation of NH_4Cl Solution: Add NH_4Cl to distilled/deionized water and bring volume to 100.0mL. Mix thoroughly. Sparge with 100% N_2. Filter sterilize.

KH_2PO_4 Solution:

Composition per 100.0mL:

KH_2PO_4	0.4g

Preparation of KH_2PO_4 Solution: Add KH_2PO_4 to distilled/deionized water and bring volume to 100.0mL. Mix thoroughly. Sparge with 100% N_2. Filter sterilize.

Substrate Solution:

Composition per 10.0mL:

Na-acetate	1.5g

Preparation of Substrate Solution: Add Na-acetate to distilled/deionized water and bring volume to 10.0mL. Sparge with N_2. Filter sterilize.

Preparation of Medium: Prepare and dispense medium under 80% N_2 + 20% CO_2. Add components, except $NaHCO_3$ solution, $Na_2S \cdot 9H_2O$ solution, substrate solution, selenite-tungstate solution, NH_4Cl solution, KH_2PO_4 solution, vitamin solution, dithionite solution, and trace elements solution SL-10. Distribute 30.0mL aliquots into 50mL serum bottles. Autoclave for 15 min at 15 psi pressure–121°C. Cool to room temperature. Aseptically and anaerobically for each 30.0mL medium add 1.5mL NH_4Cl solution, 1.5mL KH_2PO_4 solution, 0.9mL $NaHCO_3$ solution, and 0.3mL each of $Na_2S \cdot 9H_2O$, substrate, selenite-tungstate, and vitamin solutiona, and 0.03mL each of dithionite solutionand trace elements solution SL-10. Final pH is 7.2.

Use: For the cultivation of *Desulfofrigus oceanense* DSM 12341.

SRB-Psychrophile Medium (DSMZ Medium 861)

Composition per 1158.0mL:

NaCl	20.0g
Na_2SO_4	4.0g
$MgCl_2 \cdot 6H_2O$	3.0g
KCl	0.5g
$CaCl_2 \cdot 2H_2O$	0.15g
KBr	0.09g
Resazurin	0.5mg
NH_4Cl solution	49.5mL
KH_2PO_4 solution	49.5mL
$NaHCO_3$ solution	29.7mL
Vitamin solution	9.9mL
Substrate solution	9.9mL
$Na_2S \cdot 9H_2O$ solution	9.9mL
Dithionite solution	1.0mL
Trace elements solution SL-10	1.0mL
Selenite-tungstate solution	1.0mL

pH 7.2 ± 0.2 at 25°C

Dithionite Solution

Composition per 10.0mL:

Na-dithionite	0.25g

Preparation of Dithionite Solution: Add Na-dithionite to distilled/deionized water and bring volume to 10.0mL. Mix thoroughly. Sparge with 100% N_2. Filter sterilize.

$Na_2S \cdot 9H_2O$ Solution:

Composition per 10.0mL:

$Na_2S \cdot 9H_2O$	0.5g

Preparation of $Na_2S \cdot 9H_2O$ Solution: Add $Na_2S \cdot 9H_2O$ to distilled/deionized water and bring volume to 10.0mL. Mix thoroughly. Autoclave under 100% N_2 for 15 min at 15 psi pressure–121°C. Cool to room temperature.

Vitamin Solution:

Composition per liter:

Pyridoxine-HCl	10.0mg
Thiamine-HCl·$2H_2O$	5.0mg
Riboflavin	5.0mg
Nicotinic acid	5.0mg
D-Ca-pantothenate	5.0mg
p-Aminobenzoic acid	5.0mg
Lipoic acid	5.0mg
Biotin	2.0mg
Folic acid	2.0mg
Vitamin B_{12}	0.1mg

Preparation of Vitamin Solution: Add components to distilled/deionized water and bring volume to 1.0L. Mix thoroughly. Sparge with 80% H_2 + 20% CO_2. Filter sterilize.

Selenite-Tungstate Solution

Composition per liter:

NaOH	0.5g
$Na_2WO_4 \cdot 2H_2O$	4.0mg
$Na_2SeO_3 \cdot 5H_2O$	3.0mg

Preparation of Selenite-Tungstate Solution: Add components to distilled/deionized water and bring volume to 1.0L. Mix thoroughly. Sparge with 100% N_2. Filter sterilize.

Trace Elements Solution SL-10:

Composition per liter:

$FeCl_2 \cdot 4H_2O$	1.5g
$CoCl_2 \cdot 6H_2O$	190.0mg
$MnCl_2 \cdot 4H_2O$	100.0mg
$ZnCl_2$	70.0mg
$Na_2MoO_4 \cdot 2H_2O$	36.0mg

$NiCl_2 \cdot 6H_2O$ 24.0mg
H_3BO_3 .... 6.0mg
$CuCl_2 \cdot 2H_2O$ 2.0mg
HCl (25% solution) 10.0mL

Preparation of Trace Elements Solution SL-10: Add $FeCl_2 \cdot 4H_2O$ to 10.0mL of HCl solution. Mix thoroughly. Add distilled/deionized water and bring volume to 1.0L. Add remaining components. Mix thoroughly. Sparge with 80% N_2 + 20% CO_2. Autoclave for 15 min at 15 psi pressure–121°C.

$NaHCO_3$ Solution:
Composition per 100.0mL:
$NaHCO_3$ 10.0g

Preparation of $NaHCO_3$ Solution: Add $NaHCO_3$ to distilled/deionized water and bring volume to 100.0mL. Mix thoroughly. Sparge with 80% N_2 + 20% CO_2. Filter sterilize.

NH_4Cl Solution:
Composition per 100.0mL:
NH_4Cl 0.5g

Preparation of NH_4Cl Solution: Add NH_4Cl to distilled/deionized water and bring volume to 100.0mL. Mix thoroughly. Sparge with 100% N_2. Filter sterilize.

KH_2PO_4 Solution:
Composition per 100.0mL:
KH_2PO_4 .... 0.4g

Preparation of KH_2PO_4 Solution: Add KH_2PO_4 to distilled/deionized water and bring volume to 100.0mL. Mix thoroughly. Sparge with 100% N_2. Filter sterilize.

Substrate Solution:
Composition per 10.0mL:
Na-lactate 2.5g

Preparation of Substrate Solution: Add Na-lactate to distilled/deionized water and bring volume to 10.0mL. Sparge with N_2. Filter sterilize.

Preparation of Medium: Prepare and dispense medium under 80% N_2 + 20% CO_2. Add components, except $NaHCO_3$ solution, $Na_2S \cdot 9H_2O$ solution, substrate solution, selenite-tungstate solution, NH_4Cl solution, KH_2PO_4 solution, vitamin solution, dithionite solution, and trace elements solution SL-10. Distribute 30.0mL aliquots into 50mL serum bottles. Autoclave for 15 min at 15 psi pressure–121°C. Cool to room temperature. Aseptically and anaerobically for each 30.0mL medium add 0.9mL $NaHCO_3$ solution, 0.3mL $Na_2S \cdot 9H_2O$ solution, 0.3mL substrate solution, 0.03mL selenite-tungstate solution, 1.5mL NH_4Cl solution, 1.5mL KH_2PO_4 solution, 0.3mL vitamin solution, 0.03mL dithionite solution, and 0.03mL trace elements solution SL-10. Final pH is 7.2.

Use: For the cultivation of *Desulfotalea arctica* DSM 12342 and *Desulfofrigus fragile* DSM 12345.

SRB-Psychrophile Medium (DSMZ Medium 861)

Composition per 1158.0mL:
NaCl 10.0g
Na_2SO_4 .... 4.0g
$MgCl_2 \cdot 6H_2O$ 3.0g
KCl 0.5g
$CaCl_2 \cdot 2H_2O$ 0.15g
KBr 0.09g
Resazurin 0.5mg
NH_4Cl solution 49.5mL
KH_2PO_4 solution 49.5mL
$NaHCO_3$ solution 29.7mL
Vitamin solution 9.9mL
Substrate solution 9.9mL
$Na_2S \cdot 9H_2O$ solution 9.9mL
Dithionite solution 1.0mL
Trace elements solution SL-10 .... 1.0mL
Selenite-tungstate solution 1.0mL
pH 7.2 ± 0.2 at 25°C

Dithionite Solution
Composition per 10.0mL:
Na-dithionite 0.25g

Preparation of Dithionite Solution: Add Na-dithionite to distilled/deionized water and bring volume to 10.0mL. Mix thoroughly. Sparge with 100% N_2. Filter sterilize.

$Na_2S \cdot 9H_2O$ Solution:
Composition per 10.0mL:
$Na_2S \cdot 9H_2O$ 0.5g

Preparation of $Na_2S \cdot 9H_2O$ Solution: Add $Na_2S \cdot 9H_2O$ to distilled/deionized water and bring volume to 10.0mL. Mix thoroughly. Autoclave under 100% N_2 for 15 min at 15 psi pressure–121°C. Cool to room temperature.

Vitamin Solution:
Composition per liter:
Pyridoxine-HCl 10.0mg
Thiamine-HCl·$2H_2O$ 5.0mg
Riboflavin 5.0mg
Nicotinic acid 5.0mg
D-Ca-pantothenate 5.0mg
p-Aminobenzoic acid 5.0mg
Lipoic acid 5.0mg
Biotin 2.0mg
Folic acid 2.0mg
Vitamin B_{12} 0.1mg

Preparation of Vitamin Solution: Add components to distilled/deionized water and bring volume to 1.0L. Mix thoroughly. Sparge with 80% H_2 + 20% CO_2. Filter sterilize.

Selenite-Tungstate Solution
Composition per liter:
NaOH 0.5g
$Na_2WO_4 \cdot 2H_2O$ 4.0mg
$Na_2SeO_3 \cdot 5H_2O$ 3.0mg

Preparation of Selenite-Tungstate Solution: Add components to distilled/deionized water and bring volume to 1.0L. Mix thoroughly. Sparge with 100% N_2. Filter sterilize.

Trace Elements Solution SL-10:
Composition per liter:
$FeCl_2 \cdot 4H_2O$ 1.5g
$CoCl_2 \cdot 6H_2O$ 190.0mg
$MnCl_2 \cdot 4H_2O$ 100.0mg
$ZnCl_2$ 70.0mg
$Na_2MoO_4 \cdot 2H_2O$ 36.0mg
$NiCl_2 \cdot 6H_2O$ 24.0mg
H_3BO_3 .... 6.0mg
$CuCl_2 \cdot 2H_2O$ 2.0mg
HCl (25% solution) 10.0mL

Preparation of Trace Elements Solution SL-10: Add $FeCl_2·4H_2O$ to 10.0mL of HCl solution. Mix thoroughly. Add distilled/deionized water and bring volume to 1.0L. Add remaining components. Mix thoroughly. Sparge with 80% N_2 + 20% CO_2. Autoclave for 15 min at 15 psi pressure–121°C.

$NaHCO_3$ Solution:
Composition per 100.0mL:
$NaHCO_3$ 10.0g

Preparation of $NaHCO_3$ Solution: Add $NaHCO_3$ to distilled/deionized water and bring volume to 100.0mL. Mix thoroughly. Sparge with 80% N_2 + 20% CO_2. Filter sterilize.

NH_4Cl Solution:
Composition per 100.0mL:
NH_4Cl 0.5g

Preparation of NH_4Cl Solution: Add NH_4Cl to distilled/deionized water and bring volume to 100.0mL. Mix thoroughly. Sparge with 100% N_2. Filter sterilize.

KH_2PO_4 Solution:
Composition per 100.0mL:
KH_2PO_4 .... 0.4g

Preparation of KH_2PO_4 Solution: Add KH_2PO_4 to distilled/deionized water and bring volume to 100.0mL. Mix thoroughly. Sparge with 100% N_2. Filter sterilize.

Substrate Solution:
Composition per 10.0mL:
Na-lactate 2.5g

Preparation of Substrate Solution: Add Na-lactate to distilled/deionized water and bring volume to 10.0mL. Sparge with N_2. Filter sterilize.

Preparation of Medium: Prepare and dispense medium under 80% N_2 + 20% CO_2. Add components, except $NaHCO_3$ solution, $Na_2S·9H_2O$ solution, substrate solution, selenite-tungstate solution, NH_4Cl solution, KH_2PO_4 solution, vitamin solution, dithionite solution, and trace elements solution SL-10. Distribute 30.0mL aliquots into 50mL serum bottles. Autoclave for 15 min at 15 psi pressure–121°C. Cool to room temperature. Aseptically and anaerobically for each 30.0mL medium add 0.9mL $NaHCO_3$ solution, 0.3mL $Na_2S·9H_2O$ solution, 0.3mL substrate solution, 0.03mL selenite-tungstate solution, 1.5mL NH_4Cl solution, 1.5mL KH_2PO_4 solution, 0.3mL vitamin solution, 0.03mL dithionite solution, and 0.03mL trace elements solution SL-10. Final pH is 7.2.

Use: For the cultivation of *Desulfotalea psychrophila* DSM 12343.

SRB-Psychrophile Medium (DSMZ Medium 861)

Composition per 1158.0mL:
NaCl 20.0g
Na_2SO_4 .... 4.0g
$MgCl_2·6H_2O$ 3.0g
$CaCl_2·2H_2O$ 0.15g
KBr 0.09g
KCl 0.5g
Resazurin 0.5mg
NH_4Cl solution 49.5mL
KH_2PO_4 solution 49.5mL
$NaHCO_3$ solution 29.7mL
Vitamin solution 9.9mL
Substrate solution 9.9mL
$Na_2S·9H_2O$ solution 9.9mL
Dithionite solution 1.0mL
Trace elements solution SL-10 .... 1.0mL
Selenite-tungstate solution 1.0mL
pH 7.2 ± 0.2 at 25°C

Dithionite Solution
Composition per 10.0mL:
Na-dithionite 0.25g

Preparation of Dithionite Solution: Add Na-dithionite to distilled/deionized water and bring volume to 10.0mL. Mix thoroughly. Sparge with 100% N_2. Filter sterilize.

$Na_2S·9H_2O$ Solution:
Composition per 10.0mL:
$Na_2S·9H_2O$ 0.5g

Preparation of $Na_2S·9H_2O$ Solution: Add $Na_2S·9H_2O$ to distilled/deionized water and bring volume to 10.0mL. Mix thoroughly. Autoclave under 100% N_2 for 15 min at 15 psi pressure–121°C. Cool to room temperature.

Vitamin Solution:
Composition per liter:
Pyridoxine-HCl 10.0mg
Thiamine-HCl·$2H_2O$ 5.0mg
Riboflavin 5.0mg
Nicotinic acid 5.0mg
D-Ca-pantothenate 5.0mg
p-Aminobenzoic acid 5.0mg
Lipoic acid 5.0mg
Biotin 2.0mg
Folic acid 2.0mg
Vitamin B_{12} 0.1mg

Preparation of Vitamin Solution: Add components to distilled/deionized water and bring volume to 1.0L. Mix thoroughly. Sparge with 80% H_2 + 20% CO_2. Filter sterilize.

Selenite-Tungstate Solution
Composition per liter:
NaOH 0.5g
$Na_2WO_4·2H_2O$ 4.0mg
$Na_2SeO_3·5H_2O$ 3.0mg

Preparation of Selenite-Tungstate Solution: Add components to distilled/deionized water and bring volume to 1.0L. Mix thoroughly. Sparge with 100% N_2. Filter sterilize.

Trace Elements Solution SL-10:
Composition per liter:
$FeCl_2·4H_2O$ 1.5g
$CoCl_2·6H_2O$ 190.0mg
$MnCl_2·4H_2O$ 100.0mg
$ZnCl_2$ 70.0mg
$Na_2MoO_4·2H_2O$ 36.0mg
$NiCl_2·6H_2O$ 24.0mg
H_3BO_3 .... 6.0mg
$CuCl_2·2H_2O$ 2.0mg
HCl (25% solution) 10.0mL

Preparation of Trace Elements Solution SL-10: Add $FeCl_2·4H_2O$ to 10.0mL of HCl solution. Mix thoroughly. Add distilled/deionized water and bring volume to 1.0L. Add remaining components. Mix thoroughly. Sparge with 80% N_2 + 20% CO_2. Autoclave for 15 min at 15 psi pressure–121°C.

$NaHCO_3$ Solution:

Composition per 100.0mL:

$NaHCO_3$	10.0g

Preparation of $NaHCO_3$ Solution: Add $NaHCO_3$ to distilled/deionized water and bring volume to 100.0mL. Mix thoroughly. Sparge with 80% N_2 + 20% CO_2. Filter sterilize.

NH_4Cl Solution:

Composition per 100.0mL:

NH_4Cl	0.5g

Preparation of NH_4Cl Solution: Add NH_4Cl to distilled/deionized water and bring volume to 100.0mL. Mix thoroughly. Sparge with 100% N_2. Filter sterilize.

KH_2PO_4 Solution:

Composition per 100.0mL:

KH_2PO_4	0.4g

Preparation of KH_2PO_4 Solution: Add KH_2PO_4 to distilled/deionized water and bring volume to 100.0mL. Mix thoroughly. Sparge with 100% N_2. Filter sterilize.

Substrate Solution:

Composition per 10.0mL:

Na-propionate	1.5g

Preparation of Substrate Solution: Add Na-propionate to distilled/deionized water and bring volume to 10.0mL. Sparge with N_2. Filter sterilize.

Preparation of Medium: Prepare and dispense medium under 80% N_2 + 20% CO_2. Add components, except $NaHCO_3$ solution, $Na_2S{\cdot}9H_2O$ solution, substrate solution, selenite-tungstate solution, NH_4Cl solution, KH_2PO_4 solution, vitamin solution, dithionite solution, and trace elements solution SL-10. Distribute 30.0mL aliquots into 50mL serum bottles. Autoclave for 15 min at 15 psi pressure–121°C. Cool to room temperature. Aseptically and anaerobically for each 30.0mL medium add 0.9mL $NaHCO_3$ solution, 0.3mL $Na_2S{\cdot}9H_2O$ solution, 0.3mL substrate solution, 0.03mL selenite-tungstate solution, 1.5mL NH_4Cl solution, 1.5mL KH_2PO_4 solution, 0.3mL vitamin solution, 0.03mL dithionite solution, and 0.03mL trace elements solution SL-10. Final pH is 7.2.

Use: For the cultivation of *Desulfofaba gelida* DSM 12344.

SS Agar

See: ***Salmonella Shigella*** **Agar**

SS Agar, HiVeg (*Salmonella Shigella* Agar, HiVeg)

Composition per liter:

Agar	15.0g
Plant peptone	11.5g
Lactose	10.0g
Sodium citrate	10.0g
$Na_2S_2O_3$	8.5g
Plant extract	5.0g
Synthetic detergent No. I	2.0g
Ferric citrate	1.0g
Neutral Red	0.025g
Brilliant Green	0.33mg

pH 7.0 ± 0.2 at 25°C

Source: This medium is available as a premixed powder from HiMedia.

Preparation of Medium: Add components to distilled/deionized water and bring volume to 1.0L. Mix thoroughly. Gently heat while stirring and bring to boiling. Do not autoclave. Cool to 45°–50°C. Pour into sterile Petri dishes in 20.0mL volumes. Allow the surface of the plates to dry before inoculation.

Use: For the selective isolation and differentiation of pathogenic enteric bacilli, especially those belonging to the genus *Salmonella.* This medium is not recommended for the primary isolation of *Shigella* species. Lactose-fermenting bacteria such as *Escherichia coli* or *Klebsiella pneumoniae* appear as small pink or red colonies. Lactose-nonfermenting bacteria—such as *Salmonella* species, *Proteus* species, and *Shigella* species—appear as colorless colonies. Production of H_2S by *Salmonella* species turns the center of the colonies black.

SS Agar, Modified

See: ***Salmonella Shigella*** **Agar, Modified**

SS Deoxycholate Agar (*Salmonella Shigella* Deoxycholate Agar) (SSDC)

Composition per liter:

Agar	13.5g
Lactose	10.0g
Sodium deoxycholate	10.0g
Bile salts	8.5g
$Na_2S_2O_3$	8.5g
Sodium citrate	8.5g
Beef extract	5.0g
Pancreatic digest of casein	2.5g
Peptic digest of animal tissue	2.5g
$CaCl_2{\cdot}2H_2O$	1.0g
Ferric citrate	1.0g
Neutral Red	0.025g
Brilliant Green	0.33mg

pH 7.0 ± 0.2 at 25°C

Preparation of Medium: Add components to distilled/deionized water and bring volume to 1.0L. Mix thoroughly. Gently heat while stirring and bring to boiling. Do not autoclave. Cool to 45°–50°C. Pour into sterile Petri dishes in 20.0mL volumes. Allow the surface of the plates to dry before inoculation.

Use: For the isolation and cultivation of *Yersinia enterocolitica* from foods.

SSL Agar

Composition per liter:

Agar	2.5g
$CaCl_2{\cdot}2H_2O$	1.0g
Gelatin	1.0g
KNO_3	1.0g
$MgSO_4{\cdot}7H_2O$	1.0g
NaCl	1.0g
Pancreatic digest of casein	1.0g
Yeast extract	1.0g
Sodium glycerophosphate	0.1g
Cyanocobalamin	1.0μg
Trace elements solution	1.0mL

pH 7.5 ± 0.2 at 25°C

Trace Elements Solution:

Composition per liter:

Disodium EDTA	8.0g
$MnCl_2 \cdot 4H_2O$	0.1g
$CoCl_2 \cdot 6H_2O$	0.02g
KBr	0.02g
KI	0.02g
$ZnCl_2$	0.02g
$CuSO_4$	0.01g
H_3BO_3	0.01g
$Na_2MoO_4 \cdot 2H_2O$	0.01g
LiCl	5.0mg
$SnCl_2 \cdot 2H_2O$	5.0mg

Preparation of Trace Elements Solution: Add components to distilled/deionized water and bring volume to 1.0L. Mix thoroughly.

Preparation of Medium: Add components to distilled/deionized water and bring volume to 1.0L. Mix thoroughly. Gently heat and bring to boiling. Distribute into tubes or flasks. Autoclave for 15 min at 15 psi pressure–121°C. Pour into sterile Petri dishes or leave in tubes.

Use: For the isolation and cultivation of *Cytophaga* species, *Herpetosiphon* species, *Saprospira* species, and *Flexithrix* species.

SSM Agar

Composition per liter:

Solution A	850.0mL
Solution B	50.0mL
Solution C	100.0mL

Solution A:

Composition per 850.0mL:

Agar, noble	15.0g
Na_2HPO_4	6.0g
KH_2PO_4	3.0g
$MgSO_4 \cdot 7H_2O$	0.2g

Preparation of Solution A: Add components to distilled/deionized water and bring volume to 850.0mL. Mix thoroughly. Autoclave for 15 min at 15 psi pressure–121°C. Cool to 50°–55°C.

Solution B:

Composition per 50.0mL:

Glucose	5.0g

Preparation of Solution B: Add glucose to distilled/deionized water and bring volume to 50.0mL. Mix thoroughly. Autoclave for 15 min at 15 psi pressure–121°C. Cool to 50°–55°C.

Solution C:

Composition per 100.0mL:

Casein hydrolysate	20.0g

Preparation of Solution C: Add casein hydrolysate to distilled/deionized water and bring volume to 100.0mL. Mix thoroughly. Filter through Whatman #1 paper. Autoclave for 15 min at 15 psi pressure–121°C. Cool to 50°–55°C.

Preparation of Medium: Aseptically combine 850.0mL of sterile solution A, 50.0mL of sterile solution B, and 100.0mL of sterile solution C. Mix thoroughly. Pour into sterile Petri dishes or aseptically distribute into sterile tubes.

Use: For the cultivation and maintenance of a variety of bacteria.

ST Agar

See: ***Streptococcus thermophilus*** **Agar**

ST Holding Medium (m-ST Holding Medium)

Composition per liter:

KH_2PO_4	3.0g
Tris(hydroxymethyl)aminomethane buffer	3.0g
Sulfanilamide	1.5g
$NaH_2PO_4 \cdot H_2O$	0.1g
Ethanol (95% solution)	10.0mL

pH 8.6 ± 0.2 at 25°C

Preparation of Medium: Dissolve the sulfanilamide in the ethanol. Add all components to distilled/deionized water and bring volume to 1.0L. Mix thoroughly. Autoclave for 15 min at 15 psi pressure–121°C. Distribute in 1.8mL volumes to sterile Petri dishes with tight-fitting lids and an absorbent filter.

Use: For the cultivation and enumeration of coliform bacteria by the delayed-incubation total coliform procedure. For use as a holding or transport medium to keep coliform bacteria viable between sampling and laboratory culture.

StA (Schmitthenner's Agar)

Composition per liter:

Agar	20.0g
Sucrose	2.5g
Asparagine	0.27g
KH_2PO_4	0.15g
K_2HPO_4	0.15g
$MgSO_4$	0.1g
Sitosterol	0.01g

Preparation of Medium: Add components, except agar, to distilled/deionized water and bring volume to 1.0L. Mix thoroughly. Add agar. Swirl. Gently heat and bring to boiling. Distribute into tubes or flasks. Autoclave for 15 min at 15 psi pressure–121°C. Pour into sterile Petri dishes or leave in tubes.

Use: For the cultivation and maintenance of *Pythium aphanidermatum, Pythium graminicola, Pythium myriotylum, Pythiumsylvaticum,* and *Pythium ultimum.*

StA (Schmitthenner's Agar)

Composition per liter:

Sucrose	2.75g
$NaNO_3$	0.35g
KH_2PO_4	0.15g
K_2HPO_4	0.15g
$MgSO_4$	0.1g
Sitosterol	0.01g
Agar	20.0g

Preparation of Medium: Add components, in the order listed, to distilled/deionized water and bring volume to 1.0L. Mix thoroughly. Gently heat and bring to boiling. Distribute into tubes or flasks. Autoclave for 15 min at 15 psi pressure–121°C. Pour into sterile Petri dishes or leave in tubes.

Use: For the cultivation and maintenance of *Pythium aphanidermatum, Pythium graminicola, Pythium myriotylum, Pythiumsylvaticum,* and *Pythium ultimum.*

STAA Agar Base

Composition per liter:

Peptone	20.0g
Agar	13.0g
Glycerol	7.5g
Yeast extract	2.0g
K_2HPO_4	1.0g
$MgSO_4 \cdot 7H_2O$	1.0g
STAA selective supplement	10.0mL

pH 7.0 ± 0.1 at 25°C

Source: This medium is available as a premixed powder from Oxoid Unipath.

STAA Selective Supplement:

Composition per 10.0mL:

Streptomycin sulfate	0.5g
Cycloheximide	0.05g
Thallous acetate	0.05g

Preparation of STAA Selective Supplement: Add components to distilled/deionized water and bring volume to 10.0mL. Mix thoroughly. Filter sterilize.

Preparation of Medium: Add components, except glycerol and STAA selective supplement, to distilled/deionized water and bring volume to 985.0mL. Mix thoroughly. Gently heat and bring to boiling. Add glycerol. Mix thoroughly. Autoclave for 15 min at 15 psi pressure–121°C. Cool to 45°–50°C. Aseptically add sterile STAA selective supplement. Mix thoroughly. Pour into sterile Petri dishes or distribute into sterile tubes.

Use: For the isolation of *Brochothrix thermosphacta* from meat products.

Stab Agar

Composition per liter:

Nutrient broth	10.0g
Agar	6.0g
NaCl	5.0g
L-Cysteine·HCl·H_2O	0.01g
Thymine	0.01g

pH 7.0 ± 0.2 at 25°C

Preparation of Medium: Add components to distilled/deionized water and bring volume to 1.0L. Mix thoroughly. Gently heat and bring to boiling. Adjust pH to 7.0. Distribute into tubes. Autoclave for 15 min at 15 psi pressure–121°C.

Use: For the maintenance of bacterial strains, especially *Escherichia coli.*

Staib Agar

See: **Birdseed Agar**

Staley's Maintenance Agar

Composition per liter:

Agar	15.0g
Peptone	5.0g
Yeast extract	0.5g
Hutner's basal salts solution	20.0mL
Vitamin solution	10.0mL

Hutner's Basal Salts Solution:

Composition per liter:

$MgSO_4 \cdot 7H_2O$	29.7g
Nitrilotriacetic acid	10.0g
$CaCl_2 \cdot 2H_2O$	3.335g
$FeSO_4 \cdot 7H_2O$	99.0mg
$(NH_4)_6Mo_7O_{24} \cdot 4H_2O$	9.25mg
"Metals 44"	50.0mL

"Metals 44":

Composition per 100.0mL:

$ZnSO_4 \cdot 7H_2O$	1.095g
$FeSO_4 \cdot 7H_2O$	0.5g
Sodium EDTA	0.25g
$MnSO_4 \cdot H2O$	0.154g
$CuSO_4 \cdot 5H_2O$	39.2mg
$Co(NO_3)_2 \cdot 6H_2O$	24.8mg
$Na_2B_4O_7 \cdot 10H_2O$	17.7mg

Preparation of "Metals 44": Add sodium EDTA to distilled/deionized water and bring volume to 90.0mL. Mix thoroughly. Add a few drops of concentrated H_2SO_4 to retard precipitation of heavy metal ions. Add remaining components. Mix thoroughly. Bring volume to 100.0mL with distilled/deionized water.

Preparation of Hutner's Basal Salts Solution: Add nitrilotriacetic acid to 500.0mL of distilled/deionized water. Adjust pH to 6.5 with KOH. Add remaining components. Add distilled/deionized water to 1.0L. Adjust pH to 6.8. Autoclave for 15 min at 15 psi pressure–121°C. Cool to 50°C.

Vitamin Solution:

Composition per liter:

Nicotinamide	9.0mg
Calcium DL-pantothenate	5.0mg
Riboflavin	5.0mg
Thiamine·HCl	5.0mg
Biotin	2.0mg
Folic acid	2.0mg
Cyanocobalamin	0.1mg

Preparation of Vitamin Solution: Add components to distilled/deionized water and bring volume to 1.0L. Mix thoroughly. Filter sterilize.

Preparation of Medium: Add components, except Hutner's basal salts solution and vitamin solution, to distilled/deionized water and bring volume to 970.0mL. Mix thoroughly. Gently heat and bring to boiling. Autoclave for 15 min at 15 psi pressure–121°C. Cool to 50°–55°C. Aseptically add 20.0 mL of sterile Hutner's basal salts solution and 10.0mL of sterile vitamin solution. Pour into sterile Petri dishes or distribute into sterile tubes.

Use: For the cultivation of *Gemmata obscuriglobus*.

Staley's Maintenance Broth

Composition per liter:

Peptone	5.0g
Yeast extract	0.5g
Hutner's basal salts solution	20.0mL
Vitamin solution	10.0mL

Hutner's Basal Salts Solution:

Composition per liter:

$MgSO_4 \cdot 7H_2O$	29.7g
Nitrilotriacetic acid	10.0g
$CaCl_2 \cdot 2H_2O$	3.335g
$FeSO_4 \cdot 7H_2O$	99.0mg
$(NH_4)_6Mo_7O_{24} \cdot 4H_2O$	9.25mg
"Metals 44"	50.0mL

"Metals 44":

Composition per 100.0mL:

$ZnSO_4 \cdot 7H_2O$	1.095g
$FeSO_4 \cdot 7H_2O$	0.5g
Sodium EDTA	0.25g
$MnSO_4 \cdot H2O$	0.154g
$CuSO_4 \cdot 5H_2O$	39.2mg
$Co(NO_3)_2 \cdot 6H_2O$	24.8mg
$Na_2B_4O_7 \cdot 10H_2O$	17.7mg

Preparation of "Metals 44": Add sodium EDTA to distilled/deionized water and bring volume to 90.0mL. Mix thoroughly. Add a few drops of concentrated H_2SO_4 to retard precipitation of heavy metal ions. Add remaining components. Mix thoroughly. Bring volume to 100.0mL with distilled/deionized water.

Preparation of Hutner's Basal Salts Solution: Add nitrilotriacetic acid to 500.0mL of distilled/deionized water. Adjust pH to 6.5 with KOH. Add remaining components. Add distilled/deionized water to 1.0L. Adjust pH to 6.8. Autoclave for 15 min at 15 psi pressure–121°C. Cool to 25°C.

Vitamin Solution:

Composition per liter:

Nicotinamide	9.0mg
Calcium DL-pantothenate	5.0mg
Riboflavin	5.0mg
Thiamine·HCl	5.0mg
Biotin	2.0mg
Folic acid	2.0mg
Cyanocobalamin	0.1mg

Preparation of Vitamin Solution: Add components to distilled/deionized water and bring volume to 1.0L. Mix thoroughly. Filter sterilize.

Preparation of Medium: Add components, except Hutner's basal salts solution and vitamin solution, to distilled/deionized water and bring volume to 970.0mL. Mix thoroughly. Autoclave for 15 min at 15 psi pressure–121°C. Cool to 25°C. Aseptically add 20.0 mL of sterile Hutner's basal salts solution and 10.0mL of sterile vitamin solution. Mix thoroughly. Aseptically distribute into sterile tubes or flasks.

Use: For the cultivation of *Gemmata obscuriglobus*.

Stan 4 Agar

Composition per liter:

Solution B	650.0mL
Solution A	350.0mL

Solution A:

Composition per 350.0mL:

$CaCl_2 \cdot 2H_2O$	1.0g
KNO_3	1.0g
$MgSO_4 \cdot 7H_2O$	1.0g
Trace elements solution	1.0mL

Preparation of Solution A: Add components to distilled/deionized water and bring volume to 350.0mL. Mix thoroughly. Gently heat and bring to boiling. Autoclave for 15 min at 15 psi pressure–121°C. Cool to 45°–50°C.

Trace Elements Solution:

Composition per liter:

EDTA	8.0g
$MnCl_2 \cdot 4H_2O$	0.1g
$CoCl_2$	0.02g
KBr	0.02g
$ZnCl_2$	0.02g
$CuSO_4$	0.01g
H_3BO_3	0.01g
$NaMoO_4 \cdot 2H_2O$	0.01g
$BaCl_2$	5.0mg
LiCl	5.0mg
$SnCl_2 \cdot 2H_2O$	5.0mg

Preparation of Trace Elements Solution: Add components to distilled/deionized water and bring volume to 1.0L. Mix thoroughly.

Solution B:

Composition per 650.0mL:

Agar	10.0g
K_2HPO_4	1.0g

Preparation of Solution B: Add components to distilled/deionized water and bring volume to 650.0mL. Mix thoroughly. Gently heat and bring to boiling. Autoclave for 15 min at 15 psi pressure–121°C. Cool to 45°–50°C.

Preparation of Medium: Aseptically combine 350.0mL of cooled, sterile solution A and 650.0mL of cooled, sterile solution B. Mix thoroughly. Pour into sterile Petri dishes or distribute into sterile tubes.

Use: For the cultivation of myxobacteria.

Stan 5 Agar

Composition per liter:

Solution B	650.0mL
Solution A	350.0mL

Solution A:

Composition per 350.0mL:

$CaCl_2 \cdot 2H_2O$	1.0g
$(NH_4)_2SO_4$	1.0g
$MgSO_4 \cdot 7H_2O$	1.0g
Trace elements solution	1.0mL

Preparation of Solution A: Add components to distilled/deionized water and bring volume to 350.0mL. Mix thoroughly. Gently heat and bring to boiling. Autoclave for 15 min at 15 psi pressure–121°C. Cool to 45°–50°C.

Trace Elements Solution:

Composition per liter:

EDTA	8.0g
$MnCl_2 \cdot 4H_2O$	0.1g
$CoCl_2$	0.02g
KBr	0.02g
$ZnCl_2$	0.02g
$CuSO_4$	0.01g
H_3BO_3	0.01g
$NaMoO_4 \cdot 2H_2O$	0.01g
$BaCl_2$	5.0mg
LiCl	5.0mg
$SnCl_2 \cdot 2H_2O$	5.0mg

Preparation of Trace Elements Solution: Add components to distilled/deionized water and bring volume to 1.0L. Mix thoroughly.

Solution B:

Composition per 650.0mL:

Agar	10.0g
K_2HPO_4	1.0g

Preparation of Solution B: Add components to distilled/deionized water and bring volume to 650.0mL. Mix thoroughly. Gently heat and bring to boiling. Autoclave for 15 min at 15 psi pressure–121°C. Cool to 45°–50°C.

Preparation of Medium: Aseptically combine 350.0mL of cooled, sterile solution A and 650.0mL of cooled, sterile solution B. Mix thoroughly. Pour into sterile Petri dishes or distribute into sterile tubes.

Use: For the cultivation of myxobacteria.

Stan 6 Agar

Composition per liter:

Agar	10.0g
$CaCl_2 \cdot 2H_2O$	1.0g
K_2HPO_4	1.0g
$MgSO_4 \cdot 7H_2O$	1.0g
$(NH_4)_2SO_4$	1.0g
$FeCl_3$	0.2g
$MnSO_4 \cdot 7H_2O$	0.1g
Yeast extract	0.02g
Trace elements solution	1.0mL

Trace Elements Solution:

Composition per liter:

Disodium EDTA	8.0g
$MnCl_2 \cdot 4H_2O$	0.1g
$CoCl_2 \cdot 6H_2O$	0.02g
KBr	0.02g
KI	0.02g
$ZnCl_2$	0.02g
$CuSO_4$	0.01g
H_3BO_3	0.01g
$Na_2MoO_4 \cdot 2H_2O$	0.01g
LiCl	5.0mg
$SnCl_2 \cdot 2H_2O$	5.0mg

Preparation of Trace Elements Solution: Add components to distilled/deionized water and bring volume to 1.0L. Mix thoroughly.

Preparation of Medium: Add components to distilled/deionized water and bring volume to 1.0L. Mix thoroughly. Gently heat and bring to boiling. Distribute into tubes or flasks. Autoclave for 15 min at 15 psi pressure–121°C. Pour into sterile Petri dishes or leave in tubes.

Use: For the isolation and cultivation of *Cytophaga* species, *Herpetosiphon* species, *Saprospira* species, and *Flexithrix* species.

Stan 5 Mineral Medium

Composition per liter:

K_2HPO_4	1.0g
$(NH_4)_2SO_4$	1.0g
$MgSO_4 \cdot 7H_2O$	0.2g
$CaCl_2 \cdot 2H_2O$	0.1g
$FeCl_3$	0.02g

pH 7.0–7.5 at 25°C

Preparation of Medium: Add components to distilled/deionized water and bring volume to 1.0L. Mix thoroughly. Distribute into tubes or flasks. Autoclave for 15 min at 15 psi pressure–121°C.

Use: For the isolation and cultivation of *Cytophaga* species, *Herpetosiphon* species, *Saprospira* species, and *Flexithrix* species.

Standard Agar with Methanol and Yeast Extract

Composition per liter:

Base solution	982.0mL
Methanol	10.0mL
Solution B	5.0mL
Solution A	1.0mL
Solution C	1.0mL
Solution D	1.0mL

pH 4.0–4.5 at 25°C

Base Solution:

Composition per 982.0mL:

Agar	20.0g
Yeast extract	0.5g

Preparation of Base Solution: Add components to distilled/deionized water and bring volume to 982.0mL. Mix thoroughly. Gently heat and bring to boiling. Autoclave for 15 min at 15 psi pressure–121°C. Cool to 45°–50°C.

Methanol:

Composition per 10.0mL:

Methanol	10.0mL

Preparation of Methanol: Filter sterilize.

Solution A:

Composition per liter:

K_2HPO_4	87.09g
KH_2PO_4	68.05g

Preparation of Solution A: Add components to distilled/deionized water and bring volume to 1.0L. Mix thoroughly. Filter sterilize.

Solution B:

Composition per liter:

NH_4Cl	152.28g

Preparation of Solution B: Add NH_4Cl to distilled/deionized water and bring volume to 1.0L. Mix thoroughly. Filter sterilize.

Solution C:

Composition per liter:

$CaCl_2 \cdot 6H_2O$	5.47g

Preparation of Solution C: Add $CaCl_2 \cdot 6H_2O$ to distilled/deionized water and bring volume to 1.0L. Mix thoroughly. Filter sterilize.

Solution D:

Composition per liter:

$MgSO_4 \cdot 7H_2O$	71.2g
$FeSO_4 \cdot 7H_2O$	5.0g
$MnSO_4 \cdot 4H_2O$	0.81g
$CuSO_4 \cdot 5H_2O$	0.79g
$ZnSO_4 \cdot 7H_2O$	0.44g
$Na_2MoO_4 \cdot 2H_2O$	0.25g

Preparation of Solution D: Add components to distilled/deionized water and bring volume to 1.0L. Mix thoroughly. Filter sterilize.

Preparation of Medium: To 982.0mL of cooled, sterile base solution, aseptically add 10.0mL of sterile methanol, 1.0mL of sterile solution A, 5.0mL of sterile solution B, 1.0mL of sterile solution C, and 1.0mL of sterile solution D. Mix thoroughly. Adjust pH to 4.0–4.5 if necessary. Pour into sterile Petri dishes or distribute into sterile tubes.

Use: For the cultivation and maintenance of *Acetobacter methanolicus*.

Standard Fluid Medium 10B (Shepard's M10 Medium)

Composition per 102.5mL:

Base solution	70.0mL
Horse serum, unheated	20.0mL
Fresh yeast extract solution	10.0mL
Penicillin solution	1.0mL
CVA enrichment	0.5mL
L-Cysteine·HCl·H_2O solution	0.5mL
Urea solution	0.4mL
Phenol Red solution	0.1mL

pH 6.0 ± 0.2 at 25°C

Base Solution:

Composition per 70.0mL:

Beef heart, solids from infusion	5.0g
Peptone	1.0g
NaCl	0.5g

Preparation of Base Solution: Add components to distilled/deionized water and bring volume to 70.0mL. Mix thoroughly. Adjust pH to 5.5 with 2*N* HCl. Autoclave for 15 min at 15 psi pressure–121°C. Cool to 45°–50°C.

Fresh Yeast Extract Solution:

Composition per 100.0mL:

Baker's yeast, live, pressed, starch-free,	25.0g

Preparation of Fresh Yeast Extract Solution: Add the live Baker's yeast to 100.0mL of distilled/deionized water. Autoclave for 90 min at 15 psi pressure–121°C. Allow to stand. Remove supernatant solution. Adjust pH to 6.6–6.8.

Penicillin Solution:

Composition per 10.0mL:

Penicillin G	1,000,000U

Preparation of Penicillin Solution: Add penicillin to distilled/deionized water and bring volume to 10.0mL. Mix thoroughly. Filter sterilize.

CVA Enrichment:

Composition per liter:

Glucose	100.0g
L-Cysteine·HCl·H_2O	25.9g
L-Glutamine	10.0g
L-Cystine·2HCl	1.0g
Adenine	1.0g
Nicotinamide adenine dinucleotide	0.25g
Cocarboxylase	0.1g
Guanine·HCl	0.03g
$Fe(NO_3)_3$	0.02g
p-Aminobenzoic acid	0.013g
Vitamin B_{12}	0.01g
Thiamine·HCl	3.0mg

Preparation of CVA Enrichment: Add components to distilled/deionized water and bring volume to 1.0L. Mix thoroughly. Filter sterilize.

L-Cysteine·HCl·H_2O Solution:

Composition per 10.0mL:

L-Cysteine·HCl·H_2O	0.2g

Preparation of L-Cysteine·HCl·H_2O Solution: Add L-cysteine·HCl·H_2O to distilled/deionized water and bring volume to 10.0mL. Mix thoroughly. Filter sterilize.

Urea Solution:

Composition per 10.0mL:

Urea	1.0g

Preparation of Urea Solution: Add urea to distilled/deionized water and bring volume to 10.0mL. Mix thoroughly. Filter sterilize.

Phenol Red Solution:

Composition per 10.0mL:

Phenol Red	0.1g

Preparation of Phenol Red Solution: Add Phenol Red to distilled/deionized water and bring volume to 10.0mL. Mix thoroughly. Autoclave for 15 min at 15 psi pressure–121°C.

Preparation of Medium: To 70.0mL of cooled, sterile base solution, aseptically add 20.0mL of sterile horse serum, 10.0mL of sterile fresh yeast extract solution, 1.0mL of sterile penicillin solution, 0.5mL of sterile CVA enrichment, 0.5mL of sterile L-cysteine·HCl·H_2O solution, 0.4mL of sterile urea solution, and 0.1mL of sterile Phenol Red solution. Mix thoroughly. Aseptically distribute into sterile tubes or flasks.

Use: For the isolation and cultivation of *Ureaplasma urealyticum* from clinical specimens.

Standard I with Malt Agar

Composition per liter:

Agar	12.0g
Malt extract	10.0g
Peptone from meat	7.8g
Pancreatic digest of casein	7.8g
NaCl	5.6g
Yeast extract	2.8g
$CaCO_3$	2.0g
D(+)-Glucose	1.0g

pH 7.2 ± 0.2 at 25°C

Preparation of Medium: Add components to distilled/deionized water and bring volume to 1.0L. Mix thoroughly. Adjust pH to 7.2. Gently heat and bring to boiling. Distribute into tubes or flasks. Autoclave for 15 min at 15 psi pressure–121°C. Pour into sterile Petri dishes or leave in tubes.

Use: For the cultivation of *Thermocrispum municipale* and *Thermocrispum agreste*.

Standard I Medium (DSMZ Medium 453)

Composition per liter:

Peptone from meat	7.8g
Peptone from casein	7.8g
NaCl	5.6g
Yeast extract	2.8g
D(+)-Glucose	1.0g

pH 7.5 ± 0.2 at 25°C

Source: This medium is available from Merck.

Preparation of Medium: Add components to distilled/deionized water and bring volume to 1.0L. Mix thoroughly. Gently heat and bring to boiling. Distribute into tubes or flasks. Autoclave for 15 min at 15 psi pressure–121°C.

Use: For the cultivation of *Brevibacillus* spp., *Bacillus coagulans,* and *Thermoactinomyces vulgaris (Thermoactinomyces candidus).*

Standard Methods Agar (Tryptone Glucose Yeast Agar) (Plate Count Agar)

Composition per liter:

Agar	15.0g
Pancretic digest of casein	5.0g
Yeast extract	2.5g
Glucose	1.0g

pH 7.0 ± 0.1 at 25°C

Source: Available as a premixed powder from BD Diagnostic Systems.

Preparation of Medium: Add components to distilled/deionized water and bring volume to 1.0L. Mix thoroughly. Gently heat and bring to boiling. Distribute into tubes or flasks. Autoclave for 15 min at 15 psi pressure–121°C. Pour into sterile Petri dishes or leave in tubes.

Use: For the cultivation and enumeration by microbial plate counts of microorganisms isolated from milk and dairy products, foods, water, and other specimens.

Standard Methods Agar, HiVeg

Composition per liter:

Agar	15.0g
Plant peptone	5.0g
Yeast extract	2.5g
Glucose	1.0g

pH 7.0 ± 0.1 at 25°C

Source: This medium is available as a premixed powder from HiMedia.

Preparation of Medium: Add components to distilled/deionized water and bring volume to 1.0L. Mix thoroughly. Gently heat and bring to boiling. Distribute into tubes or flasks. Autoclave for 15 min at 15 psi pressure–121°C. Pour into sterile Petri dishes or leave in tubes.

Use: For the cultivation and enumeration by microbial plate counts of microorganisms isolated from milk and dairy products, foods, water, and other specimens.

Standard Methods Agar with Lecithin and Polysorbate 80

Composition per liter:

Agar	15.0g
Pancreatic digest of casein	5.0g
Polysorbate 80	5.0g
Yeast extract	2.5g
Glucose	1.0g
Lecithin	0.7g

pH 7.0 ± 0.2 at 25°C

Source: This medium is available as a premixed powder from BD Diagnostic Systems.

Preparation of Medium: Add components to distilled/deionized water and bring volume to 1.0L. Mix thoroughly. Gently heat and bring to boiling. Distribute into tubes or flasks. Autoclave for 15 min at 15 psi pressure–121°C. Pour into sterile Petri dishes or leave in tubes.

Use: For determination of the sterility of surfaces.

Standard Methods Broth (m-Standard Methods Broth) (Tryptone Glucose Yeast Broth) (m-Plate Count Broth)

Composition per liter:

Pancreatic digest of casein	10.0g
Yeast extract	5.0g
Glucose	2.0g

pH 7.0 ± 0.2 at 25°C

Source: Available as a premixed powder from BD Diagnostic Systems.

Preparation of Medium: Add components to distilled/deionized water and bring volume to 1.0L. Mix thoroughly. Distribute into tubes or flasks. Autoclave for 15 min at 15 psi pressure–121°C.

Use: For enumeration of the total number of microorganisms by the membrane filter method.

Standard Methods Caseinate Agar

Composition per liter:

Agar	15.0g
Sodium caseinate	10.0g
Pancreatic digest of casein	5.0g
Yeast extract	2.5g
Glucose	1.0g
Trisodium citrate solution	1.0L
$CaCl_2 \cdot 2H_2O$ (1*M* solution)	20.0mL

Trisodium Citrate Solution:

Composition per liter:

Trisodium citrate·$2H_2O$	4.41g

Preparation of Trisodium Citrate Solution: Add trisodium citrate·$2H_2O$ to distilled/deionized water and bring volume to 1.0L. Mix thoroughly.

Preparation of Medium: Add yeast extract, pancreatic digest of casein, glucose, and agar to 500.0mL of the trisodium citrate solution. Mix thoroughly. Gently heat until boiling. Add the remaining components. Bring volume to 1.0L with trisodium citrate solution. Autoclave for 15 min at 15 psi pressure–121°C. Cool to 45°–50°C. Pour into sterile Petri dishes.

Use: For the cultivation of proteolytic bacteria from foods.

Standard Methods Caseinate HiVeg Agar

Composition per liter:

Agar	15.0g
Sodium caseinate	10.0g
Plant hydrolysate	5.0g
Na_3-citrate	4.41g
Yeast extract	2.5g
$CaCl_2$	2.22g
Glucose	1.0g

pH 7.2 ± 0.2 at 25°C

Source: This medium is available as a premixed powder from HiMedia.

Preparation of Medium: Add components to distilled/deionized water and bring volume to 1.0L. Mix thoroughly. Gently heat while stirring and bring to boiling. Autoclave for 15 min at 15 psi pressure–121°C. Cool to 45°–50°C. Pour into sterile Petri dishes in 20.0mL volumes. Allow the surface of the plates to dry before inoculation.

Use: For the cultivation of proteolytic bacteria from foods.

Standard Methods HiVeg Agar with Tween 80 and Lecithin

Composition per liter:

Agar	15.0g
Polysorbate 80 (Tween 80)	5.0g
Plant hydrolysate	5.0g
Yeast extract	2.5g
Glucose	1.0g
Lecithin	0.7g

pH 7.0 ± 0.2 at 25°C

Source: This medium is available as a premixed powder from HiMedia.

Preparation of Medium: Add components to distilled/deionized water and bring volume to 1.0L. Mix thoroughly. Gently heat and bring to boiling. Distribute into tubes or flasks. Autoclave for 15 min at 15 psi pressure–121°C. Pour into sterile Petri dishes or leave in tubes.

Use: For determination of the sterility of surfaces.

Standard II Nutrient Agar

Composition per liter:

Agar	13.0g
Tryptose	7.0g
NaCl	5.0g

pH 7.5 ± 0.2 at 25°C

Source: This medium is available as a premixed powder from BD Diagnostic Systems.

Preparation of Medium: Add components to distilled/deionized water and bring volume to 1.0L. Mix thoroughly. Gently heat and bring to boiling. Distribute into tubes or flasks. Autoclave for 15 min at 15 psi pressure–121°C. Pour into sterile Petri dishes or leave in tubes.

Use: For the cultivation of nonfastidious microorganisms. For the maintenance of cultures of a wide variety of nonfastidious bacteria. May also be used as a base for blood and other enrichments for the cultivation of fastidious microorganisms. May be used to determine indole production.

Standard Nutrient HiVeg Agar

Composition per liter:

Agar	20.0g
Plant extract	10.0g
Plant peptone	10.0g
NaCl	5.0g

pH 7.6 ± 0.2 at 25°C

Source: This medium is available as a premixed powder from HiMedia.

Preparation of Medium: Add components to distilled/deionized water and bring volume to 1.0L. Mix thoroughly. Gently heat and bring to boiling. Distribute into tubes or flasks. Autoclave for 15 min at 15 psi pressure–121°C. Pour into sterile Petri dishes or leave in tubes.

Use: For the cultivation and enumeration of not particularly fastidious microorganisms.

Standard Nutrient HiVeg Broth (H.S. Vaccine HiVeg Medium)

Composition per liter:

Plant extract	10.0g
Plant peptone	10.0g
NaCl	5.0g

pH 7.6 ± 0.2 at 25°C

Source: This medium is available as a premixed powder from HiMedia.

Preparation of Medium: Add components to distilled/deionized water and bring volume to 1.0L. Mix thoroughly. Gently heat and bring to boiling. Distribute into tubes or flasks. Autoclave for 15 min at 15 psi pressure–121°C.

Use: For the cultivation and enumeration of not particularly fastidious microorganisms. For the culture of bacteria for vaccine production.

Standard *Staphylococcus* HiVeg Broth

Composition per liter:

Plant peptone	10.0g
NaCl	5.0g
Yeast extract	5.0g

pH 7.0 ± 0.2 at 25°C

Source: This medium is available as a premixed powder from HiMedia.

Preparation of Medium: Add components to distilled/deionized water and bring volume to 1.0L. Mix thoroughly. Gently heat and bring to boiling. Distribute into tubes or flasks. Autoclave for 15 min at 15 psi pressure–121°C.

Use: For the cultivation of *Staphylococcus* spp.

Stanier's Basal Medium with Pyridoxine and Yeast Extract

Composition per liter:

KH_2PO_4	2.78g
Na_2HPO_4	2.78g
$(NH_4)_2SO_4$	1.0g
Yeast extract	0.2g
Hutner's mineral base	20.0mL
Pyridoxine solution	10.0mL

pH 6.8 ± 0.2 at 25°C

Pyridoxine Solution:

Composition per 10.0mL:

Pyridoxine	2.0g

Preparation of Pyridoxine Solution: Add pyridoxine to distilled/deionized water and bring volume to 10.0mL. Mix thoroughly. Filter sterilize.

Hutner's Mineral Base:

Composition per liter:

$MgSO_4 \cdot 7H_2O$	29.7g
Nitrilotriacetic acid	10.0g
$CaCl_2 \cdot 2H_2O$	3.34g
$FeSO_4 \cdot 7H_2O$	99.0mg
$(NH_4)_2MoO_4$	9.25mg
Metals "44"	50.0mL

Preparation of Hutner's Mineral Base: Add nitrilotriacetic acid to 500.0mL of distilled/deionized water. Dissolve by adjusting pH to 6.5

with KOH. Add remaining components. Readjust pH to 7.2 with H_2SO_4 or KOH. Add distilled/deionized water to 1.0L.

Metals "44":

Composition per 100.0mL:

$ZnSO_4 \cdot 7H_2O$	1.1g
$FeSO_4 \cdot 7H_2O$	0.5g
EDTA	0.25g
$MnSO_4 \cdot 7H_2O$	0.154g
$CuSO_4 \cdot 5H_2O$	0.04g
$Co(NO_3)_2 \cdot 6H_2O$	0.025g
$Na_2B_4O_7 \cdot 10H_2O$	0.018g

Preparation of Metals "44": Add a few drops of H_2SO_4 to distilled/deionized water to inhibit precipitate formation. Add components to acidified distilled/deionized water and bring volume to 100.0mL. Mix thoroughly.

Preparation of Medium: Add components, except pyridoxine solution, to distilled/deionized water and bring volume to 990.0mL. Mix thoroughly. Adjust pH to 6.8 with 1*N* KOH. Autoclave for 15 min at 15 psi pressure–121°C. Cool to 45°–50°C. Aseptically add sterile pyridoxine solution. Mix thoroughly. Aseptically distribute into sterile tubes or flasks.

Use: For the cultivation and maintenance of *Pseudomonas* species.

Stanier's Basal Medium with Trichlorophenoxyacetate

Composition per liter:

KH_2PO_4	2.78g
Na_2HPO_4	2.78g
$(NH_4)_2SO_4$	1.0g
2,4,5-Trichlorophenoxyacetate	1.0g
Hutner's mineral base	20.0mL

Hutner's Mineral Base:

Composition per liter:

$MgSO_4 \cdot 7H_2O$	29.7g
Nitrilotriacetic acid	10.0g
$CaCl_2 \cdot 2H_2O$	3.34g
$FeSO_4 \cdot 7H_2O$	0.1g
$(NH_4)_2MoO_4$	9.25mg
Metals "44"	50.0mL

Preparation of Hutner's Mineral Base: Add nitrilotriacetic acid to 500.0mL of distilled/deionized water. Dissolve by adjusting pH to 6.5 with KOH. Add remaining components. Readjust pH to 7.2 with H_2SO_4 or KOH. Add distilled/deionized water to 1.0L. Store at 5°C.

Metals "44":

Composition per 100.0mL:

$ZnSO_4 \cdot 7H_2O$	1.1g
$FeSO_4 \cdot 7H_2O$	0.5g
EDTA	0.25g
$MnSO_4 \cdot 7H_2O$	0.154g
$CuSO_4 \cdot 5H_2O$	0.04g
$Co(NO_3)_2 \cdot 6H_2O$	0.025g
$Na_2B_4O_7 \cdot 10H_2O$	0.018g

Preparation of Metals "44": Add a few drops of H_2SO_4 to distilled/deionized water to inhibit precipitate formation. Add components to acidified distilled/deionized water and bring volume to 100.0mL. Mix thoroughly.

Preparation of Medium: Add components to distilled/deionized water and bring volume to 1.0L. Mix thoroughly. Distribute into tubes or flasks. Autoclave for 15 min at 15 psi pressure–121°C.

Use: For the cultivation and maintenance of *Pseudomonas cepacia*.

Staphylococcus Agar No. 110

Composition per liter:

NaCl	75.0g
Gelatin	30.0g
Agar	15.0g
D-Mannitol	10.0g
Pancreatic digest of casein	10.0g
K_2HPO_4	5.0g
Yeast extract	2.5g
Lactose	2.0g

pH 7.0 ± 0.2 at 25°C

Source: This medium is available as a premixed powder from BD Diagnostic Systems and Oxoid Unipath.

Preparation of Medium: Add components to distilled/deionized water and bring volume to 1.0L. Mix thoroughly. Gently heat and bring to boiling. Distribute into tubes or flasks. Autoclave for 15 min at 15 psi pressure–121°C. Pour into sterile Petri dishes or leave in tubes. Swirl flask while pouring plates to disperse precipitate.

Use: For the isolation and enumeration of staphylococci from clinical and nonclinical specimens.

Staphylococcus aureus Enrichment HiVeg Broth

Composition per liter:

Plant peptone	10.0g
Plant infusion	5.0g
NaCl	5.0g

pH 6.8 ± 0.2 at 25°C

Source: This medium is available as a premixed powder from HiMedia.

Preparation of Medium: Add components to distilled/deionized water and bring volume to 1.0L. Mix thoroughly. Gently heat and bring to boiling. Distribute into tubes or flasks. Autoclave for 15 min at 15 psi pressure–121°C.

Use: For the cultivation of *Staphylococcus aureus*.

Staphylococcus Broth (m-*Staphylococcus* Broth)

Composition per liter:

NaCl	75.0g
Mannitol	10.0g
Pancreatic digest of casein	10.0g
K_2HPO_4	5.0g
Yeast extract	2.5g
Lactose	2.0g

pH 7.0 ± 0.2 at 25°C

Source: This medium is available as a premixed powder from BD Diagnostic Systems.

Preparation of Medium: Add components to distilled/deionized water and bring volume to 1.0L. Mix thoroughly. Distribute into tubes or flasks. Autoclave for 15 min at 15 psi pressure–121°C.

Use: For the cultivation and enumeration of pathogenic and enterotoxigenic staphylococci by the membrane filter method. Also, when used in conjunction with Lipovitellin-salt-mannitol agar, for the detection of *Staphylococcus aureus* in swimming pool water.

Staphylococcus Medium

Composition per liter:

Agar	15.0g
Peptone	6.0g
Pancreatic digest of casein	4.0g
Yeast extract	3.0g
Beef extract	1.5g
Glucose	1.0g

pH 6.6 ± 0.2 at 25°C

Preparation of Medium: Add components to distilled/deionized water and bring volume to 1.0L. Mix thoroughly. Gently heat and bring to boiling. Distribute into tubes or flasks. Autoclave for 15 min at 15 psi pressure–121°C. Pour into sterile Petri dishes or leave in tubes.

Use: For the cultivation and maintenance of *Staphylococcus aureus*. For the enumeration of pathogenic and enterotoxigenic staphylococci by the membrane filter method.

Staphylococcus/Streptococcus Selective Medium

Composition per 1060.0mL:

Columbia blood agar base	1.0L
Horse blood, defibrinated	50.0mL
Antibiotic inhibitor	10.0mL

pH 7.3 ± 0.2 at 25°C

Columbia Blood Agar Base:

Composition per liter:

Special peptone	23.0g
Agar	10.0g
NaCl	5.0g
Starch	1.0g

Source: Columbia blood agar base is available as a premixed powder from Oxoid Unipath.

Preparation of Columbia Blood Agar Base: Add components to distilled/deionized water and bring volume to 1.0L. Mix thoroughly. Gently heat and bring to boiling. Autoclave for 15 min at 15 psi pressure–121°C. Cool to 45°–50°C.

Antibiotic Inhibitor:

Composition per 10.0mL:

Nalidixic acid	0.015g
Colistin sulfate	0.01g
Ethanol (95% solution)	10.0mL

Preparation of Antibiotic Inhibitor: Add components to 10.0mL ethanol. Mix thoroughly. Filter sterilize.

Preparation of Medium: To 1.0L of cooled sterile Columbia blood agar base, aseptically add sterile horse blood and sterile antibiotic inhibitor. Mix thoroughly. Pour into sterile Petri dishes or distribute into sterile tubes.

Use: For the selective isolation of *Staphylococcus aureus* and streptococci from clinical specimens or foods.

Starch Agar

Composition per liter:

Starch, soluble	20.0g
Agar	10.0g
$NaNO_3$	2.5g
K_2HPO_4	1.0g
$MgSO_4 \cdot 7H_2O$	0.6g
$CaCl_2 \cdot 2H_2O$	0.1g
NaCl	0.1g
$FeCl_3$	1mg

pH 7.2 ± 0.2 at 25°C

Preparation of Medium: Add components to distilled/deionized water and bring volume to 1.0L. Mix thoroughly. Gently heat and bring to boiling. Distribute into tubes or flasks. Autoclave for 15 min at 15 psi pressure–121°C. Pour into sterile Petri dishes or leave in tubes.

Use: For the cultivation of myxobacteria.

Starch Agar

Composition per liter:

Agar	15.0g
Potato starch	10.0g
Pancreatic digest of gelatin	5.0g
Beef extract	3.0g

pH 6.8 ± 0.2 at 25°C

Preparation of Medium: Add components, except potato starch, to distilled/deionized water and bring volume to 500.0mL. Mix thoroughly. Gently heat and bring to boiling. Add potato starch to distilled/deionized water and bring volume to 250.0mL. Gently heat and bring to boiling. Combine the two solutions and bring the volume to 1.0L with distilled/deionized water. Autoclave for 15 min at 15 psi pressure–121°C. Pour into sterile Petri dishes.

Use: For the cultivation and differentiation of aerobic *Actinomyces* species based on amylase production. After incubation, starch hydrolysis is determined by the addition of Gram's or Lugol's iodine solution. Organisms that produce amylase appear as colonies surrounded by a clear zone.

Starch Agar

Composition per liter:

Agar	12.0g
Soluble starch	10.0g
Beef extract	3.0g

pH 7.5 ± 0.2 at 25°C

Source: This medium is available as a premixed powder from BD Diagnostic Systems.

Preparation of Medium: Add components to distilled/deionized water and bring volume to 1.0L. Mix thoroughly. Gently heat and bring to boiling. Distribute into tubes or flasks. Autoclave for 15 min at 15 psi pressure–121°C. Pour into sterile Petri dishes.

Use: For the cultivation and differentiation of a variety of microorganisms based on amylase production. After incubation, starch hydrolysis is determined by the addition of Gram's or Lugol's iodine solution. Organisms that produce amylase appear as colonies surrounded by a clear zone.

Starch Agar

Composition per liter:

Agar	25.0g
Starch, soluble	2.0g
Mineral solution	200.0mL

Mineral Solution:

Composition per liter:

KH_2PO_4	2.0g
$(NH_4)_2SO_4$	1.4g

$MgSO_4 \cdot 7H_2O$ 0.3g
Calcium chloride solution 10.0mL
Trace elements solution 10.0mL

Calcium Chloride Solution:
Composition per 100.0mL:
$CaCl_2 \cdot 2H_2O$ 3.0g

Preparation of Calcium Chloride Solution: Add $CaCl_2 \cdot 2H_2O$ to distilled/deionized water and bring volume to 100.0mL. Mix thoroughly.

Trace Elements Solution:
Composition per liter:
$FeSO_4 \cdot 7H_2O$ 0.5g
$CoCl_2$ 0.2g
$MnSO_4 \cdot H_2O$ 0.16g
$ZnSO_4 \cdot 7H_2O$ 0.14g

Preparation of Trace Elements Solution: Add components to distilled/deionized water and bring volume to 1.0L. Mix thoroughly.

Preparation of Mineral Solution: Add components to distilled/deionized water and bring volume to 1.0L. Mix thoroughly.

Preparation of Medium: Add components to distilled/deionized water and bring volume to 1.0L. Mix thoroughly. Gently heat and bring to boiling. Distribute into tubes or flasks. Autoclave for 15 min at 15 psi pressure–121°C. Pour into sterile Petri dishes or leave in tubes.

Use: For the cultivation and maintenance of *Aspergillus awamori*.

Starch Agar with Bromcresol Purple

Composition per liter:
Agar 15.0g
Cornstarch 10.0g
Meat peptone 10.0g
Bromcresol Purple solution 1.2mL
pH 6.8 ± 0.2 at 25°C

Bromcresol Purple Solution:
Composition per 10.0mL:
Bromcresol Purple 0.16g
Ethanol (95% solution) 10.0mL

Preparation of Bromcresol Purple Solution: Add Bromcresol Purple to 10.0mL of 95% ethanol. Mix thoroughly.

Preparation of Medium: Add components to distilled/deionized water and bring volume to 1.0L. Mix thoroughly. Gently heat and bring to boiling. Distribute into tubes or flasks. Autoclave for 15 min at 15 psi pressure–121°C. Pour into sterile Petri dishes or leave in tubes.

Use: For the differentiation of *Gardnerella vaginalis (Haemophilus vaginalis, Corynebacterium vaginale)* from other microorganisms found in the genitourinary tract, with the exception of some strains of *Streptococcus* and *Lactobacillus*. Differentiation is based on starch hydrolysis. Bacteria that can hydrolyze starch appear as colonies surrounded by a yellow zone.

Starch Agar with Bromcresol Purple

Composition per liter:
Solution 1 .... 200.0mL
Solution 2 .... 20.0mL
pH 7.8 ± 0.2 at 25°C

Solution 1:
Composition per 200.0mL:
Heart infusion agar 5.0mL
Bromcresol Purple solution 0.2mL

Preparation of Solution 1: Add components to distilled/deionized water and bring volume to 200.0mL. Mix thoroughly. Gently heat while stirring and bring to boiling.

Heart Infusion Agar:
Composition per liter:
Beef heart, solids from infusion 500.0g
Agar 15.0g
Tryptose 10.0g
NaCl 5.0g

Preparation of Heart Infusion Agar: Add components to distilled/deionized water and bring volume to 1.0L. Mix thoroughly. Gently heat and bring to boiling.

Bromcresol Purple Solution:
Composition per 10.0mL:
Bromcresol Purple 0.16g
Ethanol (95% solution) 10.0mL

Preparation of Bromcresol Purple Solution: Add Bromcresol Purple to 10.0mL of ethanol. Mix thoroughly.

Solution 2:
Composition per 20.0mL:
Starch 0.4g

Preparation of Solution 2: Add starch to distilled/deionized water and bring volume to 20.0mL. Mix thoroughly. Gently heat while stirring and bring to boiling.

Preparation of Medium: Combine solution 1 and solution 2. Mix thoroughly. Autoclave for 15 min at 15 psi pressure–121°C. Pour into sterile Petri dishes or distribute into sterile tubes.

Use: For the differentiation of *Gardnerella vaginalis (Haemophilus vaginalis, Corynebacterium vaginale)* from other microorganisms found in the genitourinary tract, with the exception of some strains of *Streptococcus* and *Lactobacillus*. Differentiation is based on starch hydrolysis. Bacteria that can hydrolyze starch appear as colonies surrounded by a yellow zone.

Starch Agar Medium for *Pseudomonas*

Composition per liter:
Agar 15.0g
Peptone 5.0g
Yeast extract 5.0g
Soluble starch 3.0g
pH 7.0 ± 0.2 at 25°C

Preparation of Medium: Add components to distilled/deionized water and bring volume to 1.0L. Mix thoroughly. Gently heat and bring to boiling. Distribute into tubes or flasks. Autoclave for 15 min at 15 psi pressure–121°C. Pour into sterile Petri dishes or leave in tubes.

Use: For the cultivation and maintenance of *Pseudomonas* species and *Erwinia herbicola*.

Starch Casein Agar

Composition per liter:
Agar 15.0g
Soluble starch 10.0g
K_2HPO_4 .... 2.0g

KNO_3 .... 2.0g
NaCl 2.0g
Casein 0.3g
$MgSO_4 \cdot 7H_2O$ 0.05g
$CaCO_3$ 0.02g
$FeSO_4 \cdot 7H_2O$ 0.01g

Preparation of Medium: Add components to distilled/deionized water and bring volume to 1.0L. Mix thoroughly. Gently heat and bring to boiling. Distribute into tubes or flasks. For bottom layers, distribute into tubes in 15.0mL volumes. For top layers, distribute into tubes in 17.0mL volumes. Autoclave for 15 min at 15 psi pressure–121°C. Pour into sterile Petri dishes or leave in tubes.

Use: For the cultivation and enumeration of *Actinomycetes* species from water and soil samples by the double-layer agar technique.

Starch Casein Potassium Nitrate Agar

Composition per liter:

Agar 18.0g
Starch 10.0g
KNO_3 .... 2.0g
K_2HPO_4 .... 2.0g
NaCl 2.0g
Casein 0.3g
$MgSO_4 \cdot 7H_2O$ 0.05g
$CaCO_3$ 0.02g
$FeSO_4 \cdot 7H_2O$ 0.01g

Preparation of Medium: Add components to distilled/deionized water and bring volume to 1.0L. Mix thoroughly. Gently heat and bring to boiling. Distribute into tubes or flasks. Autoclave for 15 min at 15 psi pressure–121°C. Pour into sterile Petri dishes or leave in tubes.

Use: For the selective isolation and cultivation of streptomycetes.

Starch Fermentation Broth

Composition per 225.2mL:

Starch solution 20.0mL
Heart infusion broth 5.0mL
Bromcresol Purple solution 0.2mL

pH 7.8 ± 0.2 at 25°C

Heart Infusion Broth:

Composition per liter:

Beef heart, infusion from 500.0g
Tryptose 10.0g
NaCl 5.0g

pH 7.4 ± 0.2 at 25°C

Source: Heart infusion broth is available as a premixed powder from BD Diagnostic Systems.

Preparation of Heart Infusion Broth: Add components to distilled/deionized water and bring volume to 1.0L. Mix thoroughly. Distribute into tubes or flasks. Autoclave for 15 min at 15 psi pressure–121°C.

Bromcresol Purple Solution:

Composition per 10.0mL:

Bromcresol Purple 0.1g
Ethanol (95% solution) 10.0mL

Preparation of Bromcresol Purple Solution: Add Bromcresol Purple to 10.0mL ethanol. Mix thoroughly.

Starch Solution:

Composition per 20.0mL:

Starch 0.4g

Preparation of Starch Solution: Add starch to distilled/deionized water and bring volume to 20.0mL. Mix thoroughly. Gently heat while stirring and bring to boiling.

Preparation of Medium: Combine 5.0mL of heart infusion broth, 0.2mL of Bromcresol Purple solution, 200.0mL of distilled/deionized water, and 20.0mL of starch solution. Mix thoroughly. Distribute into tubes or flasks. Autoclave for 15 min at 15 psi pressure–121°C. Pour into sterile Petri dishes or leave in tubes.

Use: For the cultivation of *Corynebacterium* species.

Starch HiVeg Agar

Composition per liter:

Agar 15.0g
Plant peptone 5.0g
NaCl 5.0g
Starch, soluble 2.0g
Yeast extract 1.5g
Plant extract 1.5g

pH 7.2 ± 0.2 at 25°C

Source: This medium is available as a premixed powder from HiMedia.

Preparation of Medium: Add components to distilled/deionized water and bring volume to 1.0L. Mix thoroughly. Gently heat and bring to boiling. Distribute into tubes or flasks. Autoclave for 15 min at 15 psi pressure–121°C. Pour into sterile Petri dishes or leave in tubes.

Use: For the cultivation and differentiation of a variety of microorganisms based on amylase production. After incubation, starch hydrolysis is determined by the addition of Gram's or Lugol's iodine solution. Organisms that produce amylase appear as colonies surrounded by a clear zone.

Starch Hydrolysis Agar

Composition per liter:

Beef heart, infusion from 500.0g
Soluble starch 20.0g
Agar 15.0g
Tryptose 10.0g
NaCl 5.0g

pH 7.4 ± 0.2 at 25°C

Preparation of Medium: Add components to distilled/deionized water and bring volume to 1.0L. Mix thoroughly. Gently heat and bring to boiling. Distribute into tubes or flasks. Autoclave for 15 min at 15 psi pressure–121°C. Pour into sterile Petri dishes or leave in tubes.

Use: For the cultivation and differentiation of a variety of microorganisms based on amylase production. After incubation, starch hydrolysis is determined by the addition of Gram's or Lugol's iodine solution. Organisms that produce amylase appear as colonies surrounded by a clear zone.

Starch Hydrolysis Agar

Composition per liter:

Agar 12.0g
Soluble starch 10.0g
Beef exract 3.0g

pH 7.5 ± 0.2 at 25°C

Preparation of Medium: Add components to distilled/deionized water and bring volume to 1.0L. Mix thoroughly. Gently heat and bring to boiling. Distribute into tubes or flasks. Autoclave for 15 min at 15 psi pressure–121°C. Pour into sterile Petri dishes or leave in tubes.

Use: For the cultivation and differentiation of a variety of microorganisms based on amylase production. For the differentiation of bacteria, e.g. *Neisseria* sp. based upon starch hydrolysis. After incubation, starch hydrolysis is determined by the addition of Gram's or Lugol's iodine solution. Organisms that produce amylase appear as colonies surrounded by a clear zone.

Starch Medium

Composition per liter:

Agar	15.0g
Soluble starch	10.0g
Yeast extract	3.0g
$MgSO_4 \cdot 7H_2O$	0.25g

Preparation of Medium: Add components to distilled/deionized water and bring volume to 1.0L. Mix thoroughly. Gently heat and bring to boiling. Distribute into tubes or flasks. Autoclave for 15 min at 15 psi pressure–121°C. Pour into sterile Petri dishes or leave in tubes.

Use: For the cultivation and maintenance of *Guignardia laricina.*

Starch Mineral Salt Agar

Composition per liter:

Agar	12.0g
Starch, soluble	10.0g
$CaCO_3$	2.0g
$(NH_4)_2SO_4$	2.0g
K_2HPO_4	1.0g
$MgSO_4 \cdot 7H_2O$	1.0g
NaCl	1.0g

pH 7.2 ± 0.2 at 25°C

Preparation of Medium: Add components to distilled/deionized water and bring volume to 1.0L. Mix thoroughly. Gently heat and bring to boiling. Distribute into tubes or flasks. Autoclave for 15 min at 15 psi pressure–121°C. Pour into sterile Petri dishes or leave in tubes.

Use: For the cultivation and maintenance of *Streptoverticillium* species and *Thermomonospora formosensis.*

Starch Nitrate Medium (DSMZ Medium 856)

Composition per liter:

NaCl	100.0g
Agar	20.0g
Starch	20.0g
$CaCO_3$	3.0g
KNO_3	2.0g
K_2HPO_4	1.0g
$MgSO_4$	0.5g
Trace salts solution	1.0mL

pH 7.1 ± 0.2 at 25°C

Trace Salts Solution:

Composition per 100.0mL:

$FeSO_4 \cdot 7H_2O$	0.1g
$MnCl_2 \cdot 4H_2O$	0.1g
$ZnSO_4 \cdot 7H_2O$	0.1g

Preparation of Trace Salts Solution: Add components to distilled/deionized water and bring volume to 100.0mL. Mix thoroughly.

Preparation of Medium: Add components to distilled/deionized water and bring volume to 1.0L. Mix thoroughly. Gently heat and bring to boiling. Distribute into tubes or flasks. Autoclave for 15 min at 15 psi pressure–121°C. Pour into sterile Petri dishes or leave in tubes.

Use: For the cultivation and maintenance of *Saccharomonospora halophila (Microbispora* sp.*).*

Starch Salts Agar

Composition per liter:

Agar	20.0g
Solution A	500.0mL
Solution B	500.0mL

Solution A:

Composition per 500.0mL:

$CaCO_3$	2.0g
$(NH_4)_2SO_4$	2.0g
K_2HPO_4	1.0g
$MgSO_4 \cdot 7H_2O$	1.0g
NaCl	1.0g
Trace salts	1.0mL

Preparation of Solution A: Add components to distilled/deionized water and bring volume to 500.0mL. Mix thoroughly. Autoclave for 15 min at 15 psi pressure–121°C.

Trace Salts:

Composition per 100.0mL:

$FeSO_4 \cdot 7H_2O$	0.1g
$MnCl_2 \cdot 4H_2O$	0.1g
$ZnSO_4 \cdot 7H_2O$	0.1g

Preparation of Trace Salts: Add components to distilled/deionized water and bring volume to 100.0mL. Mix thoroughly. Autoclave for 15 min at 15 psi pressure–121°C.

Solution B:

Composition per 500.0mL:

Soluble starch	10.0g

Preparation of Solution B: Make a paste of the starch with a small volume of distilled/deionized water and then gradually add the starch to distilled/deionized water and bring volume to 500.0mL. Mix thoroughly. Autoclave for 15 min at 15 psi pressure–121°C.

Preparation of Medium: Thoroughly mix 500.0mL of solution A and 500.0mL of solution B. Adjust the pH to 7.2. Add 20.0g of agar and gently heat and bring to boiling. Autoclave for 15 min at 15 psi pressure–121°C. Cool to 50°C. Thoroughly mix to evenly distribute the precipitate. Pour into sterile Petri dishes or distribute into sterile tubes.

Use: For the cultivation of *Amycolatopsis rugosa, Nocardia lucida, Streptomyces canus, Streptomyces cyanogriseus, Streptomyces fradiae, Streptomyces hiroshimensis, Streptomyces kuwaitiensis, Streptomyces rubroverrucosus, Streptomyces spinoverrucosus, Streptomyces viridoverrucosus,* and *Thermoactinomyces dichotomicus.*

Starkey's Medium C, Modified

Composition per liter:

Sodium lactate	3.5g
$MgSO_4 \cdot 7H_2O$	2.0g
Na_2SO_4	1.0g
NH_4Cl	1.0g

Yeast extract ... 1.0g
KH_2PO_4 ... 0.5g
$CaCl_2 \cdot 2H_2O$ ... 0.1g
Ferrous ammonium sulfate solution ... 50.0mL
L-Cysteine·HCl·H_2O solution ... 10.0mL
pH 7.5 ± 0.2 at 25°C

Ferrous Ammonium Sulfate Solution:
Composition per 100.0mL:
$Fe(NH_4)_2(SO_4)_2 \cdot 6H_2O$ ... 1.0g

Preparation of Ferrous Ammonium Sulfate Solution: Add $Fe(NH_4)_2(SO_4)_2 \cdot 6H_2O$ to distilled/deionized water and bring volume to 100.0mL. Mix thoroughly. Filter sterilize.

L-Cysteine·HCl·H_2O Solution:
Composition per 10.0mL:
L-Cysteine·HCl·H_2O ... 0.75g

Preparation of L-Cysteine·HCl·H_2O Solution: Add L-cysteine·HCl·H_2O to distilled/deionized water and bring volume to 10.0mL. Mix thoroughly. Filter sterilize.

Preparation of Medium: Add components, except ferrous ammonium sulfate solution and L-cysteine·HCl·H_2O solution, to tap water and bring volume to 940.0mL. Mix thoroughly. Gently heat and bring to boiling. Autoclave for 15 min at 15 psi pressure–121°C. Cool to 45°–50°C. Aseptically add 50.0mL of sterile ferrous ammonium sulfate solution and 10.0mL of sterile L-cysteine·HCl·H_2O solution. Mix thoroughly. Adjust pH to 7.5 with filter-sterilized 2*N* NaOH. Pour into sterile Petri dishes or distribute into sterile tubes.

Use: For the cultivation and maintenance of *Desulfotomaculum* species and *Desulfovibrio* species.

Starkey's Medium C, Modified with Salt

Composition per liter:
NaCl ... 25.0g
Sodium lactate ... 3.5g
$MgSO_4 \cdot 7H_2O$ ... 2.0g
Na_2SO_4 ... 1.0g
NH_4Cl ... 1.0g
Yeast extract ... 1.0g
KH_2PO_4 ... 0.5g
$CaCl_2 \cdot 2H_2O$ ... 0.1g
Ferrous ammonium sulfate solution ... 50.0mL
L-Cysteine·HCl·H_2O solution ... 10.0mL
pH 7.5 ± 0.2 at 25°C

Ferrous Ammonium Sulfate Solution:
Composition per 100.0mL:
$Fe(NH_4)_2(SO_4)_2 \cdot 6H_2O$ ... 1.0g

Preparation of Ferrous Ammonium Sulfate Solution: Add $Fe(NH_4)_2(SO_4)_2 \cdot 6H_2O$ to distilled/deionized water and bring volume to 100.0mL. Mix thoroughly. Filter sterilize.

L-Cysteine·HCl·H_2O Solution:
Composition per 10.0mL:
L-Cysteine·HCl·H_2O ... 0.75g

Preparation of L-Cysteine·HCl·H_2O Solution: Add L-cysteine·HCl·H_2O to distilled/deionized water and bring volume to 10.0mL. Mix thoroughly. Filter sterilize.

Preparation of Medium: Add components, except ferrous ammonium sulfate solution and L-cysteine·HCl·H_2O solution, to tap water and bring volume to 940.0mL. Mix thoroughly. Gently heat and bring to boiling. Autoclave for 15 min at 15 psi pressure–121°C. Cool to 45°–50°C. Aseptically add 50.0mL of sterile ferrous ammonium sulfate solution and 10.0mL of sterile L-cysteine·HCl·H_2O solution. Mix thoroughly. Adjust pH to 7.5 with filter-sterilized 2*N* NaOH. Pour into sterile Petri dishes or distribute into sterile tubes.

Use: For the cultivation and maintenance of halophilic *Desulfovibrio* species.

Steenken and Smith Agar (Hohn's Medium, Modified)

Composition per 2065.0mL:
Homogenized whole egg ... 1500.0mL
Stock salts solution ... 500.0mL
Lacmoid solution ... 25.0mL
HCl (1*N* solution) ... 40.0mL
pH 6.6 ± 0.2 at 25°C

Stock Salts Solution:
Composition per 500.0mL:
KH_2PO_4 ... 2.0g
Asparagine ... 1.5g
Magnesium citrate ... 1.25 g
Na_2HPO_4, anhydrous ... 1.2 g
$MgSO_4$ ... 0.3g
Glycerol ... 60.0mL

Preparation of Stock Salts Solution: Add components, except glycerol, to distilled/deionized water that has been warmed to 80°C. Bring volume to 440.0mL. Mix thoroughly. Add 60.0mL of glycerol. Mix thoroughly. Autoclave for 20 min at 10 psi pressure–115°C. Cool to 25°C. Aseptically divide the solution into two 250.0mL parts.

Lacmoid Solution:
Composition per 100.0mL:
Lacmoid ... 1.0g
Ethanol (50% solution) ... 100.0mL

Preparation of Lacmoid Solution: Add lacmoid to 100.0mL of ethanol solution. Mix thoroughly.

Preparation of Medium: To one 250.0mL part of sterile stock salts solution, add 25.0mL of lacmoid solution. Mix thoroughly. To the other 250.0mL part of sterile stock salts solution, add 40.0mL of HCl solution. Mix thoroughly. Soak eggs in 70% ethanol for 10 min. Dry between sterile towels. Break eggs into a sterile container. Aseptically homogenize the whole eggs with a sterile glass rod. Add both stock salt solutions to the homogenized whole egg mixture. Mix thoroughly. Filter through sterile cheesecloth. Aseptically distribute into sterile tubes. Inspissate medium at 85°C (moist heat) for 90 min on 2 consecutive days.

Use: For the cultivation and maintenance of *Mycobacterium microti*.

Sterility Test Broth (USP Alternative Thioglycolate Medium)

Composition per liter:
Pancreatic digest of casein ... 15.0g
Glucose ... 5.0g
Yeast extract ... 5.0g
NaCl ... 2.5g
L-Cystine ... 0.5g
Sodium thioglycolate ... 0.5g
pH 7.1 ± 0.2 at 25°C

Source: This medium is available as a premixed powder from BD Diagnostic Systems.

Preparation of Medium: Add components to distilled/deionized water and bring volume to 1.0L. Mix thoroughly. Gently heat and bring to boiling. Distribute into tubes or flasks. Autoclave for 15 min at 15 psi pressure–121°C. Cool to 25°C. If not used immediately, prior to inoculation heat tubes in a boiling water bath for 5–10 min. Cool to 25°C.

Use: As an alternate medium, instead of fluid thioglycolate broth, for testing the sterility of a variety of specimens.

Stetteria Medium (DSMZ Medium 795)

Composition per liter:

Sulfur, powdered	10.0g
Peptone	2.0g
Yeast extract	1.0g
KH_2PO_4	0.5g
$NaHCO_3$	0.16g
$NiCl_2 \cdot 6H_2O$	3.0mg
Resazurin	0.75mg
Synthetic seawater, concentrated	500.0mL
Trace elements solution	15.0mL
$Na_2S \cdot 9H_2O$ solution	10.0mL
Selenite-tungstate solution	1.5mL

pH 7.2 ± 0.2 at 25°C

$Na_2S \cdot 9H_2O$ Solution:

Composition per 10.0mL:

$Na_2S \cdot 9H_2O$	0.5g

Preparation of $Na_2S \cdot 9H_2O$ Solution: Add $Na_2S \cdot 9H_2O$ to distilled/deionized water and bring volume to 10.0mL. Mix thoroughly. Autoclave under 100% N_2 for 15 min at 15 psi pressure–121°C. Cool to room temperature.

Selenite-Tungstate Solution

Composition per liter:

NaOH	0.5g
$Na_2WO_4 \cdot 2H_2O$	4.0mg
$Na_2SeO_3 \cdot 5H_2O$	3.0mg

Preparation of Selenite-Tungstate Solution: Add components to distilled/deionized water and bring volume to 1.0L. Mix thoroughly. Sparge with 100% N_2.

Synthetic Seawater, Concentrated:

Composition per liter:

NaCl	55.4g
$MgSO_4 \cdot 7H_2O$	14.0g
$MgCl_2 \cdot 6H_2O$	11.0g
$CaCl_2 \cdot 2H_2O$	1.5g
KCl	1.3g
NaBr	0.2g
H_3BO_3	0.06g
$SrCl_2 \cdot 6H_2O$	0.03g
Na_3-citrate	20.0mg
KI	0.1mg

Preparation of Synthetic Seawater, Concentrated: Add components to distilled/deionized water and bring volume to 1.0L. Mix thoroughly. Filter sterilize.

Trace Elements Solution:

Composition per liter:

$MgSO_4 \cdot 7H_2O$	3.0g
Nitrilotriacetic acid	1.5g
NaCl	1.0g
$MnSO_4 \cdot 2H_2O$	0.5g
$CoSO_4 \cdot 7H_2O$	0.18g
$ZnSO_4 \cdot 7H_2O$	0.18g
$CaCl_2 \cdot 2H_2O$	0.1g
$FeSO_4 \cdot 7H_2O$	0.1g
$NiCl_2 \cdot 6H_2O$	0.025g
$KAl(SO_4)_2 \cdot 12H_2O$	0.02g
H_3BO_3	0.01g
$Na_2MoO_4 \cdot 4H_2O$	0.01g
$CuSO_4 \cdot 5H_2O$	0.01g
$Na_2SeO_3 \cdot 5H_2O$	0.3mg

Preparation of Trace Elements Solution: Add nitrilotriacetic acid to 500.0mL of distilled/deionized water. Dissolve by adjusting pH to 6.5 with KOH. Add remaining components. Add distilled/deionized water to 1.0L. Mix thoroughly.

Preparation of Medium: Prepare and dispense medium under 80% N_2 + 20% CO_2 gas mixture. Add components, except synthetic seawater and $Na_2S \cdot 9H_2O$ solution, to 490.0mL distilled/deionized water. Mix thoroughly. Sparge with 80% N_2 + 20% CO_2. Autoclave for 15 min at 15 psi pressure–121°C. Cool to 25°C. Aseptically add 500.0mL filter-sterilized concentrated seawater. Flush with 80% N_2 + 20% CO_2 gas mixture for 20 min. Aseptically add 10.0mL $Na_2S \cdot 9H_2O$ solution. Adjust pH to 6.0 with H_2SO_4. Mix thoroughly. Aseptically and anaerobically distribute 20mL aliquots into sterile 100mL serum bottles. Pressurize bottles to 2 bar gas overpressure with 80% N_2 + 20% CO_2. Heat at 100°C for 1.5 h. Before use check that the medium pH is 6.0.

Use: For the cultivation of *Stetteria hydrogenophila* and *Staphylothermus hellenicus*.

STL Broth

Composition per liter:

Casamino acids	1.0g
Glucose	1.0g
Sodium glutamate	1.0g
$CaCl_2 \cdot 2H_2O$	0.1g
KNO_3	0.1g
$MgSO_4 \cdot 7H_2O$	0.1g
Sodium glycerophosphate	0.1g
Thiamine	1.0mg
Vitamin B_{12}	1.0μg
Trace elements solution	1.0mL

pH 7.5 ± 0.2 at 25°C

Trace Elements Solution:

Composition per liter:

Disodium EDTA	8.0g
$MnCl_2 \cdot 4H_2O$	0.1g
$CoCl_2 \cdot 6H_2O$	0.02g
KBr	0.02g
KI	0.02g
$ZnCl_2$	0.02g
$CuSO_4$	0.01g
H_3BO_3	0.01g
$Na_2MoO_4 \cdot 2H_2O$	0.01g
LiCl	5.0mg
$SnCl_2 \cdot 2H_2O$	5.0mg

Preparation of Trace Elements Solution: Add components to distilled/deionized water and bring volume to 1.0L. Mix thoroughly.

Preparation of Medium: Add components to distilled/deionized water and bring volume to 1.0L. Mix thoroughly. Gently heat and bring

to boiling. Distribute into tubes or flasks. Autoclave for 15 min at 15 psi pressure–121°C. Pour into sterile Petri dishes or leave in tubes.

Use: For the isolation and cultivation of *Cytophaga* species, *Herpetosiphon* species, *Saprospira* species, and *Flexithrix* species.

Stock Culture Agar

Composition per liter:

Beef heart infusion	500.0g
Gelatin	10.0g
Proteose peptone	10.0g
Agar	7.5g
Casein	5.0g
Na_2HPO_4	4.0g
Sodium citrate	3.0g
Glucose	0.5g

pH 7.5 ± 0.2 at 25°C

Source: This medium is available as a premixed powder from BD Diagnostic Systems.

Preparation of Medium: Add components to cold distilled/deionized water and bring volume to 1.0L. Mix thoroughly. Gently heat while stirring and bring to boiling. Distribute into tubes or flasks. Autoclave for 15 min at 15 psi pressure–121°C. Pour into sterile Petri dishes or leave in tubes.

Use: For the maintenance of pathogenic and nonpathogenic bacteria, especially streptococci.

Stock Culture Agar with L-Asparagine

Composition per liter:

Beef heart infusion	500.0g
Gelatin	10.0g
Proteose peptone	10.0g
Agar	7.5g
Casein	5.0g
Na_2HPO_4	4.0g
Sodium citrate	3.0g
L-Asparagine	1.0g
Glucose	0.5g

pH 7.5 ± 0.2 at 25°C

Preparation of Medium: Add components to cold distilled/deionized water and bring volume to 1.0L. Mix thoroughly. Gently heat while stirring and bring to boiling. Distribute into tubes or flasks. Autoclave for 15 min at 15 psi pressure–121°C. Pour into sterile Petri dishes or leave in tubes.

Use: For the maintenance of pathogenic and nonpathogenic bacteria, especially streptococci.

Stokes Agar

Composition per liter:

Agar	12.5g
Glucose	1.0g
Peptone	1.0g
$MgSO_4·7H_2O$	0.2g
$CaCl_2$	0.05g
$FeCl_3·6H_2O$	0.01g

Preparation of Medium: Add components to tap water and bring volume to 1.0L. Mix thoroughly. Gently heat and bring to boiling. Distribute into tubes or flasks. Autoclave for 15 min at 15 psi pressure–121°C. Pour into sterile Petri dishes or leave in tubes.

Use: For the isolation and cultivation of *Sphaerotilus natans*.

Stonebrink's Medium

Composition per 3040.0mL:

Homogenized whole egg	2.0L
Mineral salts solution	1.0L
Malachite Green solution	40.0mL

Mineral Salts Solution:

Composition per liter:

Na-pyruvate	12.5g
KH_2PO_4	7.0g
$Na_2HPO_4·7H_2O$	4.0g

Preparation of Mineral Salts Solution: Add components to distilled/deionized water and bring volume to 1.0L. Mix thoroughly. Autoclave for 15 min at 15 psi pressure–121°C. Cool to 50°C.

Malachite Green Solution:

Composition per 100.0mL:

Malachite Green	2.0g

Preparation of Malchite Green Solution: Add Malachite Green to distilled/deionized water and bring volume to 100.0mL. Mix thoroughly. Autoclave for 15 min at 15 psi pressure–121°C. Cool to 50°C.

Homogenized Whole Egg:

Composition per 2.0L:

Whole eggs	36–48

Preparation of Homogenized Whole Egg: Use fresh eggs, less than 1 week old. Scrub the shells with soap. Let stand in a soap solution for 30 min. Rinse in running water. Soak eggs in 70% ethanol for 15 min. Break the eggs into a sterile container. Homogenize by shaking. Filter through four layers of sterile cheesecloth into a sterile graduated cylinder. Measure out 2.0L.

Preparation of Medium: Aseptically add 40.0mL sterile Malachite Green solution to 1.0L of sterile mineral salts solution. Mix thoroughly. Aseptically add 2.0L of homogenized whole egg. Mix thoroughly. Distribute into sterile screw-capped tubes. Place tubes in a slanted position. Inspissate at 85°C (moist heat) for 45 min.

Use: For the cultivation of *Mycobacterium* species. For the isolation of *Mycobacterium bovis*.

Straw DYAA

Composition per liter:

Agar	20.0g
Glucose	10.0g
Yeast extract	1.0g
Asparagine	0.5g
$K_2HPO_4·3H_2O$	0.5g
$MgSO_4·7H_2O$	0.25g
$FeCl_3$ solution	0.5mL
Straw	variable

$FeCl_3$ Solution:

Composition per 10.0mL:

$FeCl_3$	1.0g

Preparation of $FeCl_3$ Solution: Add $FeCl_3$ to distilled/deionized water and bring volume to 10.0mL. Mix thoroughly.

Preparation of Medium: Add components, except straw, to distilled/deionized water and bring volume to 1.0L. Gently heat and bring to boiling. Distribute into tubes or flasks. Autoclave for 15 min at 15 psi pressure–121°C. Pour into sterile Petri dishes or leave in tubes. Autoclave straw for 15 min at 15 psi pressure–121°C. Aseptically add straw to the solidified agar.

Use: For the cultivation of *Cochliobolus sativus.*

Straw Malt Agar

Composition per liter:

Agar	15.0g
Malt extract	10.0g
Straw	variable

Preparation of Medium: Add components, except straw, to distilled/deionized water and bring volume to 1.0L. Mix thoroughly. Gently heat and bring to boiling. Distribute into tubes or flasks. Autoclave for 15 min at 15 psi pressure–121°C. Pour into sterile Petri dishes or leave in tubes. Autoclave straw for 15 min at 15 psi pressure–121°C. Aseptically add some straw to the solidified agar.

Use: For the cultivation of *Cladosporium vignae, Cochliobolus sativus,* and *Cochliobolus victoriae.*

StrepB Carrot Broth™

Composition per liter:

Proteose peptone No. 3	25.0g
Soluble sStarch	20.0g
Selective agents	12.2g
Morpholinepropanesulfonic aAcid (MOPS)	11.0g
Na_2HPO_4	8.5g
Glucose	2.5g
Sodium pyruvate	1.0g
$MgSO_4$	20.0g
StrepB carrot broth tiles with growth promoting factors	variable

pH 7.4 ± 0.1 at 25°C

Source: This medium is available from Hardy Diagnostics.

Preparation: This medium is supplied as a prepared broth in tubes. The StrepB carrot broth tile is added to a tube just prior to inoculation with a vaginal swab. The tile must remain submerged in the broth.

Use: For detecting the presence of Group B *Streptococcus* infections in pregnant women. This new screening test is an improvement over conventional methods, by increasing sensitivity, decreasing turn around time, while lowering overall cost. Tubes show an orange to red color change, typical of group B streptococci. The production of orange, red, or brick red pigment is a unique characteristic of hemolytic Group B streptococci due to reaction with substrates such as starch, proteose peptone, serum, and folate pathway inhibitors.

Strep ID Quad Plate

Composition per liter:

Quadrant I	5.0mL
Quadrant II	5.0mL
Quadrant III	5.0mL
Quadrant IV	5.0mL

Source: Available as a prepared medium from BD Diagnostic Systems.

Quadrant I:

Composition per 5.0mL:

Bacitracin	0.5mg
TSA II agar	5.0mL

Quadrant II:

Composition per 5.0mL:

TSA II agar	5.0mL
Sheep blood, defibrinated	0.25mL

Quadrant III:

Composition per 5.0mL:

Bile esculin agar	5.0mL

Quadrant IV:

Composition per 5.0mL:

Blood agar base with 6.5% NaCl	5.0mL

Preparation of Quadrant Media: Sterilize agars by autoclaving for 15 min at 15 psi pressure–121°C. Cool to 45°–50°C. Add additional components as filter sterilized solutions. Mix and distribute as 5.0mL aliquots into quadrants.

Use: For the differentiation and presumptive identification of streptococci. The Strep (*Streptococcus)* ID (Identification) Quad Plate is a four-sectored plate, each containing a different medium.

Streptococcal Growth Medium

Composition per liter:

Beef heart, solids from infusion	500.0g
Tryptose	10.0g
NaCl	5.0g
Glucose	1.0g
Bromcresol Purple solution	1.0mL

pH 7.4 ± 0.2 at 25°C

Bromcresol Purple Solution:

Composition per 10.0mL:

Bromcresol Purple	0.16g
Ethanol (95% solution)	10.0mL

Preparation of Bromcresol Purple Solution: Add Bromcresol Purple to 10.0mL of ethanol. Mix thoroughly.

Preparation of Medium: Add components to distilled/deionized water and bring volume to 1.0L. Mix thoroughly. Distribute into tubes in 5.0mL volumes. Autoclave for 15 min at 15 psi pressure–121°C.

Use: For the cultivation of *Streptococcus* species and other Gram-positive cocci. Growth in this medium turns the indicator yellow and the solution turbid.

Streptococcus agalactiae Selective HiVeg Agar Base with Blood and *Staphylococcus* B toxin

Composition per liter:

Agar	13.0g
Plant peptone	10.0g
NaCl	5.0g
Plant extract No. 1	5.0g
Esculin	1.0g
Thallous sulfate	0.333g
Crystal Violet	1.3g
Sheep blood, defibrinated	50.0mL
Staphylococcus B toxin	25.0mL

pH 7.0 ± 0.2 at 25°C

Source: This medium, without blood or toxin, is available as a premixed powder from HiMedia.

Preparation of Medium: Add components, except sheep blood and staphylococcal toxin, to distilled/deionized water and bring volume to 925.0mL. Mix thoroughly. Gently heat and bring to boiling. Autoclave for 15 min at 15 psi pressure–121°C. Cool to 45°–50°C. Aseptically add 50.0mL of sterile sheep blood and 25.0mL of staphylococcal toxin. Mix thoroughly. Pour into sterile Petri dishes or leave in tubes.

Use: For the selective cultivation of *Streptococcus agalactiae.*

Streptococcus Agar

Composition per liter:

Glucose	20.0g
Pancreatic digest of casein	20.0g
Agar	15.0g
K_2HPO_4	2.0g
$MgSO_4 \cdot 7H_2O$	0.1g

pH 6.8 ± 0.2 at 25°C

Preparation of Medium: Add components to distilled/deionized water and bring volume to 1.0L. Mix thoroughly. Gently heat and bring to boiling. Distribute into tubes or flasks. Autoclave for 15 min at 15 psi pressure–121°C. Pour into sterile Petri dishes or leave in tubes.

Use: For the cultivation and maintenance of *Streptococcus* species.

Streptococcus Blood Agar, Selective

Composition per liter:

Agar	15.0g
Pancreatic digest of casein	10.0g
Beef extract	6.7g
Nucleic acid	6.0g
NaCl	5.0g
Sheep blood, defibrinated	50.0mL
Maltose solution	10.0mL
Antibiotic inhibitor	10.0mL

pH 7.3 ± 0.2 at 25°C

Maltose Solution:

Composition per 10.0mL:

Maltose	0.25–5.0g

Preparation of Maltose Solution: Add maltose to distilled/deionized water and bring volume to 10.0mL. Mix thoroughly. Filter sterilize.

Antibiotic Inhibitor:

Composition per 10.0mL:

Polymyxin B sulfate	0.02g
Neomycin sulfate	0.01g

Preparation of Antibiotic Inhibitor: Add components to distilled/deionized water and bring volume to 10.0mL. Mix thoroughly. Filter sterilize.

Preparation of Medium: Add components—except sheep blood, maltose solution, and antibiotic inhibitor—to distilled/deionized water and bring volume to 930.0mL. Mix thoroughly. Gently heat and bring to boiling. Autoclave for 15 min at 15 psi pressure–121°C. Cool to 45°–50°C. Aseptically add sterile sheep blood, sterile maltose solution, and sterile antibiotic inhibitor. Mix thoroughly. Pour into sterile Petri dishes or distribute into sterile tubes.

Use: For the isolation and cultivation of group A hemolytic *Streptococcus* species from the human respiratory tract.

Streptococcus Enrichment HiVeg Broth (SE HiVeg Broth)

Composition per liter:

Plant hydrolysate	26.0g
Yeast extract	6.0g
NaCl	5.0g
Synthetic detergent	3.0g
Esculin	1.0g
Sodium citrate	1.0g
Ferric ammonium citrate	0.5g
NaN_3	0.25g

pH 7.0 ± 0.2 at 25°C

Source: This medium is available as a premixed powder from HiMedia.

Caution: Sodium azide is toxic. Azides also react with metals and disposal must be highly diluted.

Preparation of Medium: Add components to distilled/deionized water and bring volume to 1.0L. Mix thoroughly. Gently heat and bring to boiling. Distribute into tubes or flasks. Autoclave for 15 min at 15 psi pressure–121°C.

Use: For the selective isolation, cultivation, and enumeration of streptococci from specimens containing a mixed flora.

Streptococcus faecalis Broth

See: **SF Broth**

Streptococcus lactis Differential HiVeg Agar Base with Potassium Ferricyanide and Citrate

Composition per liter:

Agar	15.0g
Skim milk (nonfat milk)	10.0g
Glucose	5.0g
Plant hydrolysate No. 3	2.5g
Potassium ferricyanide solution	10.0mL
Citrate solution	10.0mL

pH 6.6 ± 0.2 at 25°C

Source: This medium is available as a premixed powder from HiMedia.

Citrate Solution:

Composition per 10.0mL:

Ferric citrate	0.25g
Sodium citrate	0.25g

Preparation of Citrate Solution: Add components to distilled/deionized water and bring volume to 10.0mL. Mix thoroughly. Sterilize using flowing steam for 30 min.

Potassium Ferricyanide Solution:

Composition per 10.0mL:

$K_3[Fe(CN)_6]$	1.0g

Preparation of Potassium Ferricyanide Solution: Add 1.0g of $K_3[Fe(CN)_6]$ to distilled/deionized water and bring volume to 10.0mL. Mix thoroughly. Sterilize using flowing steam for 30 min.

Preparation of Medium: Add components, except potassium ferricyanide and citrate solution, to distilled/deionized water and bring volume to 980.0L. Mix thoroughly. Gently heat until boiling. Autoclave for 12 min at 10 psi pressure–115°C. Cool to 50°C. Add 10.0mL sterile potassium ferricyanide solution and 10.0mL citrate solution. Mix thoroughly. Pour into sterile Petri dishes or aseptically distribute into sterile tubes.

Use: For the differentiation of citrate-utilizing lactic streptococci—*Lactobacillus lactis* (*Streptococcus lactis*) subspecies *diacetylactis*—from citrate-nonutilizing *Lactobacillus lactis* (*Streptococcus lactis*) and *Lactobacillus lactis* (*Streptococcus lactis*) subspecies *cremoris*.

Streptococcus Medium

Composition per liter:

Agar	15.0g
Glucose	4.0g
K_2HPO_4	3.8g
Pancreatic digest of casein	2.5g
Yeast extract	2.5g

pH 7.6 ± 0.2 at 25°C

Preparation of Medium: Add components to distilled/deionized water and bring volume to 1.0L. Mix thoroughly. Gently heat and bring to boiling. Distribute into tubes or flasks. Autoclave for 15 min at 15 psi pressure–121°C. Pour into sterile Petri dishes or leave in tubes.

Use: For the cultivation and maintenance of *Enterococcus faecalis*.

Streptococcus mutans Medium

Composition per 100.0mL:

Pancreatic digest of casein	2.0g
Mannitol	0.5g
NaCl	0.25g
Lactoalbumin	0.25g
Agar	0.075g
L-Cystine	0.05g
Sodium thioglycolate	0.05g
Thallium acetate	0.025g
Crystal Violet	0.1mg
Bromcresol Purple (0.04% solution)	15.0mL

pH 7.1 ± 0.2 at 25°C

Caution: Thallium salts are toxic.

Preparation of Medium: Add components to distilled/deionized water and bring volume to 1.0L. Mix thoroughly. Gently heat and bring to boiling. Distribute into tubes in 5.0mL volumes. Autoclave for 15 min at 15 psi pressure–121°C.

Use: For the selective isolation and cultivation of *Streptococcus mutans*. Bacteria that turn the medium yellow are presumptive for *Streptococcus mutans*.

Streptococcus pneumoniae Medium

Composition per liter:

Pancreatic digest of casein	17.0g
Glucose	10.0g
NaCl	5.0g
Papaic digest of soybean meal	3.0g
Yeast extract	3.0g
K_2HPO_4	2.5g

pH 7.2 ± 0.2 at 25°C

Preparation of Medium: Add components to distilled/deionized water and bring volume to 1.0L. Mix thoroughly. Adjust pH to 7.2. Distribute into tubes or flasks. Autoclave for 15 min at 15 psi pressure–121°C.

Use: For the cultivation of *Streptococcus pneumoniae*.

Streptococcus Selection HiVeg Agar

Composition per liter:

Agar	15.0g
Plant hydrolysate	15.0g
Glucose	5.0g
Papaic digest of soybean meal	5.0g
NaCl	4.0g
Sodium citrate	1.0g
L-Cystine	0.2g
NaN_3	0.2g
Na_2SO_3	0.2g
Crystal Violet	0.2mg

pH 7.4 ± 0.2 at 25°C

Source: This medium is available as a premixed powder from HiMedia.

Preparation of Medium: Add components to distilled/deionized water and bring volume to 1.0L. Mix thoroughly. Gently heat and bring to boiling. Distribute into tubes or flasks. Autoclave for 15 min at 15 psi pressure–121°C.

Use: For the selective isolation and enumeration of all types of streptococci including group A beta hemolytic strains.

Streptococcus Selection HiVeg Agar with Cycloheximide

Composition per liter:

Agar	15.0g
Plant hydrolysate	15.0g
Glucose	5.0g
Papaic digest of soybean meal	5.0g
NaCl	4.0g
Sodium citrate	1.0g
NaN_3	0.2g
Na_2SO_3	0.2g
L-Cystine	0.2g
Crystal Violet	0.2g
Cycloheximide solution	10.0mL

pH 7.4 ± 0.2 at 25°C

Source: This medium is available as a premixed powder from HiMedia.

Cycloheximide Solution:

Composition per 10.0mL:

Cycloheximide	0.01g

Preparation of Cycloheximide Solution: Add cycloheximide to distilled/deionized water and bring volume to 10.0mL. Mix thoroughly. Filter sterilize.

Preparation of Medium: Add components, except cycloheximide solution, to distilled/deionized water and bring volume to 990.0mL. Mix thoroughly. Gently heat and bring to boiling. Autoclave for 15 min at 15 psi pressure–121°C. Cool to 50°C. Asptically add 10.0mL cycloheximide solution. Mix thoroughly. Pour into sterile Petri dishes or aseptically distribute into tubes.

Use: For the selective isolation and enumeration of all types of streptococci including group A beta hemolytic strains.

Streptococcus Selection HiVeg Broth

Composition per liter:

Plant hydrolysate	15.0g
Glucose	5.0g
Papaic digest of soybean meal	5.0g
NaCl	4.0g
Sodium citrate	1.0g
L-Cystine	0.2g
NaN_3	0.2g

Na_2SO_3 .... 0.2g
Crystal Violet 0.2mg
pH 7.4 ± 0.2 at 25°C

Source: This medium is available as a premixed powder from HiMedia.

Preparation of Medium: Add components to distilled/deionized water and bring volume to 1.0L. Mix thoroughly. Gently heat and bring to boiling. Distribute into tubes or flasks. Autoclave for 15 min at 15 psi pressure–121°C.

Use: For the selective cultivation of streptococci including group A beta hemolytic strains.

Streptococcus Selective Medium

Composition per liter:
Special peptone 23.0g
Agar 10.0g
NaCl 5.0g
Starch 1.0g
Horse blood, defibrinated 50.0mL
Antibiotic inhibitor 10.0mL
pH 7.3± 0.2 at 25°C

Source: This medium is available as a premixed powder from Oxoid Unipath.

Antibiotic Inhibitor:
Composition per 10.0mL:
Colistin sulfate 10.0mg
Oxolinic acid 5.0mg

Preparation of Antibiotic Inhibitor: Add components to distilled/deionized water and bring volume to 10.0mL. Mix thoroughly. Filter sterilize.

Preparation of Medium: Add components, except horse blood and antibiotic inhibitor, to distilled/deionized water and bring volume to 940.0mL. Mix thoroughly. Gently heat and bring to boiling. Autoclave for 15 min at 15 psi pressure–121°C. Cool to 45°–50°C. Aseptically add sterile horse blood and sterile antibiotic inhibitor. Mix thoroughly. Pour into sterile Petri dishes or distribute into sterile tubes.

Use: For the selective isolation of streptococci from clinical specimens or foodstuffs.

Streptococcus suis Medium

Composition per liter:
Peptone 10.0g
Meat extract 8.0g
Glucose 5.0g
Lactose 5.0g
Yeast extract 3.0g
K_2HPO_4 .... 2.5g
KH_2PO_4 .... 2.5g
L-Cysteine·HCl 0.5g
$MgSO_4 \cdot 7H_2O$ 0.2g
$MnSO_4 \cdot 4H_2O$ 0.05g
Bovine serum 50.0mL
pH 6.8 ± 0.2 at 25°C

Preparation of Medium: Add components, except bovine serum, to distilled/deionized water and bring volume to 950.0mL. Mix thoroughly. Autoclave for 15 min at 15 psi pressure–121°C. Aseptically add 50.0mL of sterile bovine serum. Mix thoroughly. Aseptically distribute into sterile tubes or flasks.

Use: For the cultivation of *Streptococcus suis.*

Streptococcus thermophilus Agar (ST Agar)

Composition per liter:
Agar 15.0g
Sucrose 10.0g
Pancreatic digest of casein 10.0g
Yeast extract 5.0g
K_2HPO_4 .... 2.0g
pH 6.8 ± 0.2 at 25°C

Preparation of Medium: Add components to distilled/deionized water and bring volume to 1.0L. Mix thoroughly. Gently heat and bring to boiling. Distribute into tubes or flasks. Autoclave for 15 min at 15 psi pressure–121°C. Pour into sterile Petri dishes or leave in tubes.

Use: For the isolation and cultivation of *Streptococcus thermophilus* from dairy products.

Streptococcus thermophilus Isolation HiVeg Agar

Composition per liter:
Agar 15.0g
Plant hydrolysate 10.0g
Sucrose 10.0g
Yeast extract 5.0g
K_2HPO_4 .... 2.0g
pH 6.8 ± 0.2 at 25°C

Source: This medium is available as a premixed powder from HiMedia.

Preparation of Medium: Add components to distilled/deionized water and bring volume to 1.0L. Mix thoroughly. Gently heat and bring to boiling. Distribute into tubes or flasks. Autoclave for 15 min at 15 psi pressure–121°C. Pour into sterile Petri dishes or leave in tubes.

Use: For the selective isolation and cultivation of *Streptococcus thermophilus*.

Streptococcus uberis Broth

Composition per liter:
Peptone 0.5g
Yeast extract 0.5g
Skimmed milk 1.0L

Preparation of Medium: Add components to distilled/deionized water and bring volume to 1.0L. Mix thoroughly. Gently heat and bring to boiling. Distribute into tubes or flasks. Autoclave for 15 min at 6.2 psi pressure–110°C. Pour into sterile Petri dishes or leave in tubes.

Use: For the cultivation and maintenance of *Streptococcus uberis.*

Streptomyces Agar (LMG Medium 93)

Composition per liter:
Agar 20.0g
L-Asparagine 1.0g
K_2HPO_4 .... 1.0g
Glycerol 10.0mL
Trace salts solution 1.0mL
pH 7.0–7.4

Trace Salts Solution:
Composition per 100.0mL:

$FeSO_4 \cdot 7H_2O$	0.1g
$MnCl_2 \cdot 4H_2O$	0.1g
$ZnSO_4 \cdot 7H_2O$	0.1g

Preparation of Trace Salts Solution: Add components to distilled/deionized water and bring volume to 100.0mL. Mix thoroughly.

Preparation of Medium: Add components to distilled/deionized water and bring volume to 1.0L. Mix thoroughly. Gently heat and bring to boiling. Distribute into tubes or flasks. Autoclave for 15 min at 15 psi pressure–121°C. Pour into sterile Petri dishes or leave in tubes.

Use: For the cultivation of *Streptomyces* spp.

Streptomyces Agar

Composition per liter:

Agar	20.0g
Glucose	10.0g
Beef extract	4.0g
Peptone	4.0g
NaCl	2.5g
Yeast extract	1.0g

pH 7.2 ± 0.2 at 25°C

Preparation of Medium: Add components to distilled/deionized water and bring volume to 1.0L. Mix thoroughly. Gently heat and bring to boiling. Distribute into tubes or flasks. Autoclave for 15 min at 15 psi pressure–121°C. Pour into sterile Petri dishes or leave in tubes.

Use: For the cultivation and maintenance of *Actinomadura ferruginea, Actinomadura libanotica, Actinomadura madurae, Actinomadura pelletieri, Actinomadura roseoviolacea, Actinomadura spiralis, Actinomadura verrucosospora, Micropolyspora fascifera, Nocardioides albus, Nocardiopsis albus, Nocardiopsis dassonvillei, Saccharopolyspora rectivirgula, Streptococcus pluton,* and *Streptomyces griseus.*

Streptomyces Agar

Composition per liter:

Agar	12.0g
Malt extract	10.0g
Glucose	4.0g
Yeast extract	4.0g
$CaCO_3$	2.0g

Preparation of Medium: Add components to distilled/deionized water and bring volume to 1.0L. Mix thoroughly. Gently heat and bring to boiling. Distribute into tubes or flasks. Autoclave for 15 min at 15 psi pressure–121°C. Pour into sterile Petri dishes or leave in tubes.

Use: For the cultivation and maintenance of *Actinoplanes missouriensis, Saccharopolyspora rectivirgula, Streptococcus albogriseolus, Streptomyces badius, Streptomyces bikiniensis,* and *Thermomonospora mesophila.*

Streptomyces Agar

Composition per liter:

Agar	12.0g
Malt extract	10.0g
Glucose	4.0g
Yeast extract	4.0g
$CaCO_3$	2.0g

pH 7.2 ± 0.2 at 25°C

Preparation of Medium: Add components, except agar, to distilled/deionized water and bring volume to 1.0L. Mix thoroughly. Adjust pH (use pH indicator paper) to 7.2 using KOH. Add agar. Gently heat and bring to boiling. Distribute into tubes or flasks. Autoclave for 15 min at 15 psi pressure–121°C. Pour into sterile Petri dishes or leave in tubes.

Use: For the cultivation and maintenance of *Stygiolobus azoricus*.

Streptomyces Medium

Composition per liter:

Agar	25.0g
Glucose	5.0g
L-Glutamic acid	4.0g
KH_2PO_4	1.0g
NaCl	1.0g
$MgSO_4 \cdot 7H_2O$	0.7g
$FeSO_4 \cdot 7H_2O$	3.0mg

pH 7.0 ± 0.2 at 25°C

Preparation of Medium: Add components to distilled/deionized water and bring volume to 1.0L. Mix thoroughly. Gently heat and bring to boiling. Distribute into tubes or flasks. Autoclave for 15 min at 15 psi pressure–121°C. Pour into sterile Petri dishes or leave in tubes.

Use: For the cultivation and maintenance of *Streptomyces kanamyceticus*.

Streptomycete Antibiotic Activity Inoculum Medium

Composition per liter:

Pancreatic digest of casein	10.0g
D-Glucose	5.0g
Yeast extract	5.0g
K_2HPO_4	1.0g
Liver extract	100.0mL

pH 6.9 ± 0.2 at 25°C

Preparation of Medium: Add components to distilled/deionized water and bring volume to 1.0L. Mix thoroughly. Distribute into tubes or flasks. Autoclave for 15 min at 15 psi pressure–121°C.

Use: For the cultivation of *Streptomyces* species to be used in the antibiotic activity assay.

Streptomycete Antibiotic Activity Medium

Composition per liter:

Agar	15.0g
D-Glucose	15.0g
Soybean meal	15.0g
NaCl	5.0g
Yeast extract	1.0g
$CaCO_3$	1.0g
Glycerol	2.5mL

pH 6.8 ± 0.2 at 25°C

Preparation of Medium: Add components to distilled/deionized water and bring volume to 1.0L. Mix thoroughly. Gently heat and bring to boiling. Distribute into tubes or flasks. Autoclave for 15 min at 15 psi pressure–121°C. Pour into sterile Petri dishes.

Use: For the the cultivation and determination of antibiotic production of *Streptomyces* species by the streak method. *Bacillus subtilis* NRRL B-765, *Sarcina lutea* NRRL B-1018, *Escherichia coli* NRRL B-766, *Saccharomyces pasteurianus* NRRL Y-139, *Candida albicans* NRRL

Y-477, and *Mucor ramannianus* NRRL 1839 are used as test organisms.

Streptomycete Medium

Composition per liter:

Solution B	500.0mL
Solution A	400.0mL
Solution C	100.0mL

Solution A:

Composition per 400.0mL:

Glucose	20.0g
Agar	4.0g
Yeast extract	1.2g
$MgSO_4 \cdot 7H_2O$	0.25g
Bromcresol Purple	0.012g

Preparation of Solution A: Add components to distilled/deionized water and bring volume to 400.0mL. Mix thoroughly. Autoclave for 15 min at 15 psi pressure–121°C. Cool to 45°–50°C.

Solution B:

Composition per 500.0mL:

$Na_2HPO_4 \cdot 2H_2O$	534.0mg
KH_2PO_4	272.0mg

Preparation of Solution B: Add components to distilled/deionized water and bring volume to 500.0mL. Mix thoroughly. Autoclave for 15 min at 15 psi pressure–121°C. Cool to 45°–50°C.

Solution C:

Composition per 100.0mL:

$CaCO_3$	1.0g

Preparation of Solution C: Add $CaCO_3$ to distilled/deionized water and bring volume to 100.0mL. Mix thoroughly.

Preparation of Medium: Distribute solution C into test tubes in 0.2mL volumes. Autoclave for 15 min at 15 psi pressure–121°C. Cool to 45°–50°C. Combine cooled, sterile solution A and cooled, sterile solution B. Mix thoroughly. Add 1.8mL of solution A-B to each test tube containing sterile solution C. Mix thoroughly to distribute the $CaCO_3$. Cool tubes rapidly in an ice-water bath.

Use: For the cultivation and differentiation of streptomycetes based on their formation of organic acids. Bacteria that form organic acids turn the medium yellow and dissolve the $CaCO_3$.

Streptomycete Medium

Composition per liter:

Glycerol	5.0g
Agar	4.0g
NaCl	2.0g
KNO_3	1.0g
$Na_2HPO_4 \cdot 2H_2O$	0.534g
$MgSO_4 \cdot 7H_2O$	0.5g
KH_2PO_4	0.272g
Trace elements solution	1.0mL

pH 6.8 ± 0.2 at 25°C

Trace Elements Solution:

Composition per 100.0mL:

$FeSO_4 \cdot 7H_2O$	0.1g
$MnCl_2 \cdot 4H_2O$	0.1g
$ZnSO_4 \cdot 7H_2O$	0.1g

Preparation of Trace Elements Solution: Add components to distilled/deionized water and bring volume to 100.0mL. Mix thoroughly.

Preparation of Medium: Add components to distilled/deionized water and bring volume to 1.0L. Mix thoroughly. Gently heat and bring to boiling. Distribute into tubes in 1.0mL volumes. Autoclave for 15 min at 15 psi pressure–121°C.

Use: For the cultivation and differentiation of streptomycetes based on their reduction of nitrate to nitrite. Bacteria that reduce nitrate to nitrite form a red color after the addition of Griess-Ilosvay reagent.

Streptomycete Medium

Composition per liter:

Agar	12.0g
NaCl	5.0g
$Na_2HPO_4 \cdot 2H_2O$	1.98g
KH_2PO_4	1.51g
Glucose	1.0g
Pancreatic digest of casein	1.0g
$MgSO_4 \cdot 7H_2O$	0.5g
Phenol Red	0.012g
Urea solution	100.0mL

pH 6.8 ± 0.2 at 25°C

Urea Solution:

Composition per 100.0mL:

Urea	20.0g

Preparation of Urea Solution: Add urea to distilled/deionized water and bring volume to 100.0mL. Mix thoroughly. Filter sterilize.

Preparation of Medium: Add components, except urea solution, to distilled/deionized water and bring volume to 900.0mL. Mix thoroughly. Gently heat and bring to boiling. Autoclave for 15 min at 15 psi pressure–121°C. Cool to 45°–50°C. Aseptically add 100.0mL of sterile urea solution. Mix thoroughly. Aseptically distribute into sterile tubes. Allow tubes to cool in a slanted position.

Use: For the cultivation and differentiation of streptomycetes based on their ability to produce urease.

Streptomycete Medium

Composition per liter:

Sodium hippurate	10.0g
Na_2HPO_4	5.0g
Glucose	2.0g
Meat extract	2.0g
Peptone	2.0g
Yeast extract	2.0g

pH 7.0 ± 0.2 at 25°C

Preparation of Medium: Add components to distilled/deionized water and bring volume to 1.0L. Mix thoroughly. Distribute into tubes in 3.0mL volumes. Autoclave for 15 min at 15 psi pressure–121°C.

Use: For the cultivation and differentiation of streptomycetes based on their ability to hydrolyze hippurate.

Streptomycin Assay Agar with Yeast Extract

See: **Antibiotic Medium 5**

Streptomycin L Broth Medium

Composition per liter:

Pancreatic digest of casein	10.0g
NaCl	5.0g

Yeast extract	5.0g
Glucose	1.0g
Streptomycin solution	10.0mL

pH 7.0 ± 0.2 at 25°C

Streptomycin Solution:

Composition per 10.0mL:

Streptomycin sulfate	25.0mg

Preparation of Streptomycin Solution: Add streptomycin sulfate to distilled/deionized water and bring volume to 10.0mL. Mix thoroughly. Filter sterilize.

Preparation of Medium: Add components, except streptomycin solution, to distilled/deionized water and bring volume to 990.0mL. Mix thoroughly. Bring pH to 7.0. Autoclave for 15 min at 15 psi pressure–121°C. Aseptically add 10.0mL of sterile streptomycin solution. Mix thoroughly. Aseptically distribute into sterile tubes or flasks.

Use: For the cultivation of *Escherichia coli*.

Streptomycin Nutrient Agar

Composition per liter:

Agar	15.0g
Peptone	5.0g
Meat extract	3.0g
Streptomycin sulfate	40.0mg

Preparation of Medium: Add components to distilled/deionized water and bring volume to 1.0L. Mix thoroughly. Gently heat and bring to boiling. Distribute into tubes or flasks. Autoclave for 15 min at 15 psi pressure–121°C. Pour into sterile Petri dishes or leave in tubes.

Use: For the cultivation and maintenance of *Escherichia coli*.

Streptomycin Nutrient Agar No. 2

Composition per liter:

Agar	15.0g
Peptone	5.0g
NaCl	5.0g
Yeast extract	2.0g
Beef extract	1.0g
Streptomycin solution	10.0mL

pH 7.0 ± 0.2 at 25°C

Streptomycin Solution:

Composition per 10.0mL:

Streptomycin sulfate	125.0mg

Preparation of Streptomycin Solution: Add streptomycin sulfate to distilled/deionized water and bring volume to 10.0mL. Mix thoroughly. Filter sterilize.

Preparation of Medium: Add components, except streptomycin solution, to distilled/deionized water and bring volume to 990.0mL. Mix thoroughly. Bring pH to 7.0. Autoclave for 15 min at 15 psi pressure–121°C. Aseptically add 10.0mL of sterile streptomycin solution. Mix thoroughly. Pour into sterile Petri dishes or distribute into sterile tubes.

Use: For the cultivation of *Micrococcus luteus*.

Streptomycin Nutrient Agar No. 3

Composition per liter:

Agar	15.0g
Peptone	5.0g
NaCl	5.0g
Yeast extract	2.0g
Beef extract	1.0g
Streptomycin solution	10.0mL

pH 7.0 ± 0.2 at 25°C

Streptomycin Solution:

Composition per 10.0mL:

Streptomycin sulfate	500.0mg

Preparation of Streptomycin Solution: Add streptomycin sulfate to distilled/deionized water and bring volume to 10.0mL. Mix thoroughly. Filter sterilize.

Preparation of Medium: Add components, except streptomycin solution, to distilled/deionized water and bring volume to 990.0mL. Mix thoroughly. Bring pH to 7.0. Autoclave for 15 min at 15 psi pressure–121°C. Aseptically add 10.0mL of sterile streptomycin solution. Mix thoroughly. Pour into sterile Petri dishes or distribute into sterile tubes.

Use: For the cultivation of *Micrococcus luteus*.

Streptomycin Nutrient Agar No. 4

Composition per liter:

Agar	15.0g
Peptone	5.0g
NaCl	5.0g
Yeast extract	2.0g
Beef extract	1.0g
Streptomycin solution	10.0mL

pH 7.0 ± 0.2 at 25°C

Streptomycin Solution:

Composition per 10.0mL:

Streptomycin sulfate	80.0mg

Preparation of Streptomycin Solution: Add streptomycin sulfate to distilled/deionized water and bring volume to 10.0mL. Mix thoroughly. Filter sterilize.

Preparation of Medium: Add components, except streptomycin solution, to distilled/deionized water and bring volume to 990.0mL. Mix thoroughly. Bring pH to 7.0. Autoclave for 15 min at 15 psi pressure–121°C. Aseptically add 10.0mL of sterile streptomycin solution. Mix thoroughly. Pour into sterile Petri dishes or distribute into sterile tubes.

Use: For the cultivation of *Corynebacterium* species.

Streptomycin Terramycin® Malt Extract Agar

Composition per liter:

Malt extract	30.0g
Agar	15.0g
Peptone	5.0g
Streptomycin solution	100.0mL
Terramycin solution	100.0mL

pH 5.4 ± 0.2 at 25°C

Streptomycin Solution:

Composition per 100.0mL:

Streptomycin	0.07g

Preparation of Streptomycin Solution: Add streptomycin to distilled/deionized water and bring volume to 100.0mL. Mix thoroughly. Filter sterilize.

Terramycin Solution:
Composition per 100.0mL:

Terramycin	0.07g

Preparation of Terramycin Solution: Add terramycin to distilled/deionized water and bring volume to 100.0mL. Mix thoroughly. Filter sterilize.

Preparation of Medium: Add components, except streptomycin solution and terramycin solution, to distilled/deionized water and bring volume to 800.0mL. Mix thoroughly. Gently heat and bring to boiling. Autoclave for 15 min at 15 psi pressure–121°C. Cool to 45°–50°C. Aseptically add 100.0mL of sterile streptomycin solution and 100.0mL of sterile terramycin solution. Mix thoroughly. Pour into sterile Petri dishes in 20.0mL volumes.

Use: For the cultivation and enumeration of fungi isolated from sewage and polluted waters.

Streptosel™ Agar

Composition per liter:

Pancreatic digest of casein	15.0g
Agar	12.0g
Glucose	5.0g
Papaic digest of soybean meal	5.0g
NaCl	4.0g
Sodium citrate	1.0g
L-Cystine	0.2g
NaN_3	0.2g
Na_2SO_3	0.2g
Crystal Violet	0.2mg

pH 7.4 ± 0.2 at 25°C

Source: This medium is available as a premixed powder from BD Diagnostic Systems.

Caution: Sodium azide is toxic. Azides also react with metals and disposal must be highly diluted.

Preparation of Medium: Add components to distilled/deionized water and bring volume to 1.0L. Mix thoroughly. Gently heat and bring to boiling. Distribute into tubes or flasks. If medium is used the same day, do not autoclave. Pour into sterile Petri dishes or leave in tubes. If medium is to be stored, autoclave for 15 min at 13 psi pressure–118°C. Pour into sterile Petri dishes or leave in tubes.

Use: For the selective isolation, cultivation, and enumeration of streptococci from specimens containing a mixed flora.

Streptosel™ Broth

Composition per liter:

Pancreatic digest of casein	15.0g
Glucose	5.0g
Papaic digest of soybean meal	5.0g
NaCl	4.0g
Sodium citrate	1.0g
L-Cystine	0.2g
Na_2SO_3	0.2g
NaN_3	0.2g
Crystal Violet	0.2mg

pH 7.4 ± 0.2 at 25°C

Source: This medium is available as a premixed powder from BD Diagnostic Systems.

Caution: Sodium azide is toxic. Azides also react with metals and disposal must be highly diluted.

Preparation of Medium: Add components to distilled/deionized water and bring volume to 1.0L. Mix thoroughly. Distribute into tubes or flasks. Autoclave for 15 min at 13 psi pressure–118°C.

Use: For the selective isolation and cultivation of streptococci from specimens containing a mixed flora.

Steroidobacter Medium with Testosterone (DSMZ Medium 1116)

Composition per 1040.3mL:

NaCl	1.0g
KCl	0.5g
$NaNO_3$	0.42g
$MgCl_2 \cdot 6H_2O$	0.4g
NH_4Cl	0.25g
KH_2PO_4	0.2g
$CaCl_2 \cdot 2H_2O$	0.15g
Na_2SO_4	0.07g
Bicarbonate solution	30.0mL
Testosterone solution	10.0mL
Trace elements solution SL-10	0.1mL
Selenite-tungstate solution	0.1mL
Vitamin solution	0.1mL

pH 7.2 ± 0.2 at 25°C

Testosterone Solution:
Composition per 10.0mL:

Testosterone	0.2g
Acetone	10.0mL

Preparation of Testosterone Solution: Add tesosterone to acetone adn bring volume to 10.0mL. Mix thoroughly.

Selenite-Tungstate Solution:
Composition per liter:

NaOH	0.5g
$Na_2WO_4 \cdot 2H_2O$	4.0mg
$Na_2SeO_3 \cdot 5H_2O$	3.0mg

Preparation of Selenite-Tungstate Solution: Add components to distilled/deionized water and bring volume to 970.0mL. Mix thoroughly. Sparge with 100% N_2. Autoclave for 15 min at 15 psi pressure–121°C.

Trace Elements Solution SL-10:
Composition per liter:

$MgSO_4 \cdot 7H_2O$	3.0g
Nitrilotriacetic acid	1.5g
NaCl	1.0g
$MnSO_4 \cdot 2H_2O$	0.5g
$CoSO_4 \cdot 7H_2O$	0.18g
$ZnSO_4 \cdot 7H_2O$	0.18g
$CaCl_2 \cdot 2H_2O$	0.1g
$FeSO_4 \cdot 7H_2O$	0.1g
$NiCl_2 \cdot 6H_2O$	0.025g
$KAl(SO_4)_2 \cdot 12H_2O$	0.02g
H_3BO_3	0.01g
$Na_2MoO_4 \cdot 4H_2O$	0.01g
$CuSO_4 \cdot 5H_2O$	0.01g
$Na_2SeO_3 \cdot 5H_2O$	0.3mg

Preparation of Trace Elements Solution SL-10: Add nitrilotriacetic acid to 500.0mL of distilled/deionized water. Dissolve by adjusting pH to 6.5 with KOH. Add remaining components. Add distilled/deionized water to 1.0L. Mix thoroughly. Adjust pH to 7.0.

Vitamin Solution:
Composition per liter:

Vitamin B_{12}	50.0mg
Thiamine-HCl·$2H_2O$	50.0mg
Riboflavin	50.0mg
D-Ca-pantothenate	50.0mg
p-Aminobenzoic acid	50.0mg
Lipoic acid	50.0mg
Nicotinic acid	25.0mg
Nicotine amide	25.0mg
Biotin	20.0mg
Folic acid	20.0mg
Pyridoxine-HCl	10.0mg

Preparation of Vitamin Solution: Add components to distilled/deionized water and bring volume to 1.0L. Mix thoroughly for several hours. Filter sterilize.

Bicarbonate Solution:
Composition per 100.0mL:

$NaHCO_3$	8.4g

Preparation of Bicarbonate Solution: Add components to distilled/deionized water and bring volume to 100.0mL. Mix thoroughly. Sparge with 20% CO_2 + 80% H_2. Autoclave for 15 min at 15 psi pressure–121°C. Cool to room temperature.

Preparation of Medium: Dispense testosterone solution into anaerobic culture vessels. Allow solvent to evaporate to dryness. Add remaining components, except bicarbonate solution, vitamin solution, trace elements solution SL-10, and selenite-tungstate solution, to distilled/deionized water and bring volume to 1.0L. Mix thoroughly. Dispense 10.0mL portions of the medium to the culture vessels. Sparge with a gas mixture of 80% N_2 + 20% CO_2. Close the culture vessels. Autoclave for 15 min at 15 psi pressure–121°C. Cool to room temperature. For every 10.0mL of medium, add 0.3mL bicarbonate solution, 0.001mL trace elements solution SL-10, 0.001mL selenite-tungstate solution, and 0.001mL vitamin solution. Adjust pH to 7.2. Treat the vessels in an ultrasonic bath to detach and suspend the testosterone.

Use: For the cultivation of *Steroidobacter denitrificans*.

Steroidobacter Medium with Heptanoate (DSMZ Medium 1116)

Composition per 1090.3mL:

NaCl	1.0g
KCl	0.5g
$NaNO_3$	0.42g
$MgCl_2$·$6H_2O$	0.4g
NH_4Cl	0.25g
KH_2PO_4	0.2g
$CaCl_2$·$2H_2O$	0.15g
Na_2SO_4	0.07g
Heptanoate solution	50.0mL
Bicarbonate solution	30.0mL
Trace elements soltuion SL-10	0.1mL
Selenite-tungstate solution	0.1mL
Vitamin solution	0.1mL

pH 7.2 ± 0.2 at 25°C

Heptanoate Solution:
Composition per 100.0mL:

Heptanoate	0.85g

Preparation of Heptanoate Solution: Add components to distilled/deionized water and bring volume to 100.0mL. Mix thoroughly for several hours. Filter sterilize.

Selenite-Tungstate Solution:
Composition per liter:

NaOH	0.5g
Na_2WO_4·$2H_2O$	4.0mg
Na_2SeO_3·$5H_2O$	3.0mg

Preparation of Selenite-Tungstate Solution: Add components to distilled/deionized water and bring volume to 970.0mL. Mix thoroughly. Sparge with 100% N_2. Autoclave for 15 min at 15 psi pressure–121°C.

Trace Elements Solution SL-10:
Composition per liter:

$MgSO_4$·$7H_2O$	3.0g
Nitrilotriacetic acid	1.5g
NaCl	1.0g
$MnSO_4$·$2H_2O$	0.5g
$CoSO_4$·$7H_2O$	0.18g
$ZnSO_4$·$7H_2O$	0.18g
$CaCl_2$·$2H_2O$	0.1g
$FeSO_4$·$7H_2O$	0.1g
$NiCl_2$·$6H_2O$	0.025g
$KAl(SO_4)_2$·$12H_2O$	0.02g
H_3BO_3	0.01g
Na_2MoO_4·$4H_2O$	0.01g
$CuSO_4$·$5H_2O$	0.01g
Na_2SeO_3·$5H_2O$	0.3mg

Preparation of Trace Elements Solution SL-10: Add nitrilotriacetic acid to 500.0mL of distilled/deionized water. Dissolve by adjusting pH to 6.5 with KOH. Add remaining components. Add distilled/deionized water to 1.0L. Mix thoroughly. Adjust pH to 7.0.

Vitamin Solution:
Composition per liter:

Vitamin B_{12}	50.0mg
Thiamine-HCl·$2H_2O$	50.0mg
Riboflavin	50.0mg
D-Ca-pantothenate	50.0mg
p-Aminobenzoic acid	50.0mg
Lipoic acid	50.0mg
Nicotinic acid	25.0mg
Nicotine amide	25.0mg
Biotin	20.0mg
Folic acid	20.0mg
Pyridoxine-HCl	10.0mg

Preparation of Vitamin Solution: Add components to distilled/deionized water and bring volume to 1.0L. Mix thoroughly for several hours. Filter sterilize.

Bicarbonate Solution:
Composition per 100.0mL:

$NaHCO_3$	8.4g

Preparation of Bicarbonate Solution: Add $NaHCO_3$ to distilled/deionized water and bring volume to 100.0mL. Mix thoroughly. Sparge with 20% CO_2 + 80% N_2. Autoclave for 15 min at 15 psi pressure–121°C. Cool to room temperature.

Preparation of Medium: Add components, except heptanoate solution, bicarbonate solution, vitamin solution, trace elements solution, and selenite-tungstate solution, to distilled/deionized water and bring volume to 1.0L. Mix thoroughly. Dispense 10.0mL portions of the medium to the culture vessels. Sparge with a gas mixture of 80% N_2 + 20% CO_2. Close the culture vessels. Autoclave for 15 min at 15 psi pressure–121°C. Cool to room temperature. For every 10.0mL of medium, add 0.5mL heptanoate solution, 0.3mL bicarbonate solution, 0.001mL trace elements solution SL-10, 0.001mL selenite-tungstate solution, and 0.001mL vitamin solution. Adjust pH to 7.2. Treat the vessels in an ultrasonic bath to detach and suspend the heptanoate.

Use: For the cultivation of *Steroidobacter denitrificans*.

STT Agar

See: **Sucrose Teepol Tellurite Agar**

STTA Medium

Composition per liter:

Peptone	20.0g
Glycerol	15.0g
Agar	13.0g
Yeast extract	2.0g
K_2HPO_4	1.0g
$MgSO_4 \cdot 4H_2O$	1.0g
Antibiotic solution	10.0mL
Thallous acetate solution	10.0mL

pH 7.0 ± 0.2 at 25°C

Antibiotic Solution:

Composition per 10.0mL:

Streptomycin sulfate	0.5g
Cycloheximide	0.05g

Preparation of Antibiotic Solution: Add cycloheximide and streptomycin sulfate to distilled/deionized water and bring volume to 10.0mL. Mix thoroughly. Filter sterilize.

Thallous Acetate Solution:

Composition per 10.0mL:

Thallous acetate	0.05g

Preparation of Thallous Acetate Solution: Add thallous acetate to distilled/deionized water and bring volume to 10.0mL. Mix thoroughly. Filter sterilize.

Preparation of Medium: Add components, except antibiotic solution and thallous acetate solution, to distilled/deionized water and bring volume to 980.0mL. Mix thoroughly. Gently heat and bring to boiling. Autoclave for 15 min at 15 psi pressure–121°C. Cool to 45°–50°C. Aseptically add sterile antibiotic solution and thallous acetate solution. Mix thoroughly. Pour into sterile Petri dishes or distribute into sterile tubes.

Use: For the selective isolation and cultivation of *Brochothrix thermosphacta*.

Stuart *Leptospira* Broth, Modified

Composition per liter:

NaCl	1.93g
Na_2HPO_4	0.66g
NH_4Cl	0.34g
$MgCl_2 \cdot 6H_2O$	0.19g
L-Asparagine	0.13g
KH_2PO_4	0.08g
Glycerol	5.0mL
Rabbit serum, inactivated at 56°C, 30 min	100.0mL

pH 7.4 ± 0.2 at 25°C

Preparation of Medium: Add each component, except rabbit serum, to distilled/deionized water in separate flasks and bring each volume to 100.0mL. Mix thoroughly. Combine the seven solutions, except the rabbit serum. Mix thoroughly. Gently heat and bring to boiling. Autoclave for 15 min at 15 psi pressure–121°C. Cool to 45°–50°C. Aseptically add sterile rabbit serum. Mix thoroughly. Aseptically distribute into sterile tubes or flasks.

Use: For the cultivation of *Leptospira* species.

Stuart Medium Base

Composition per 1100.0mL:

NaCl	1.8g
Na_2HPO_4	0.67g
$MgCl_2 \cdot 6H_2O$	0.41g
NH_4Cl	0.27g
Asparagine	0.13g
KH_2PO_4	0.09g
Phenol Red	0.01g
Glycerol	5.0mL
Leptospira enrichment	100.0mL

pH 7.6 ± 0.2 at 25°C

Source: This medium is available as a premixed powder from BD Diagnostic Systems. *Leptospira* enrichment contains rabbit serum and hemoglobin and is available from BD Diagnostic Systems.

Preparation of Medium: Add components, except glycerol and *Leptospira* enrichment, to distilled/deionized water and bring volume to 995.0mL. Mix thoroughly. Add glycerol. Mix thoroughly. Autoclave for 15 min at 15 psi pressure–121°C. Cool to 45°–50°C. Aseptically add *Leptospira* enrichment. Mix thoroughly. Aseptically distribute into sterile screw-capped tubes in 10.0mL volumes.

Use: For the cultivation of *Leptospira* species.

Stuart Transport Medium

Composition per liter:

Sodium glycerophosphate	10.0g
Sodium thioglycolate	1.0g
$CaCl_2 \cdot 2H_2O$	0.1g
Methylene Blue	2.0mg

pH 7.4 ± 0.2 at 25°C

Preparation of Medium: Add components to distilled/deionized water and bring volume to 1.0L. Mix thoroughly. Gently heat and bring to boiling. Distribute into 7.0mL screw-capped tubes. Fill tubes to capacity. Autoclave for 15 min at 15 psi pressure–121°C.

Use: For the preservation of *Neisseria* species and other fastidious organisms during their transport from clinic to laboratory.

Stuart Transport Medium, Modified

Composition per liter:

Sodium glycerophosphate	10.0g
Agar	5.0g
L-Cysteine·HCl·H_2O	0.5g
Sodium thioglycolate	0.5g
$CaCl_2 \cdot 2H_2O$	0.1g
Methylene Blue	1.0mg

pH 7.4 ± 0.2 at 25°C

Source: This medium is available as a premixed powder from Oxoid Unipath.

Preparation of Medium: Add components to distilled/deionized water and bring volume to 1.0L. Mix thoroughly. Gently heat and bring to boiling. Distribute into 7.0mL screw-capped tubes. Fill tubes to capacity. Autoclave for 15 min at 15 psi pressure–121°C.

Use: For the preservation of *Neisseria* species and other fastidious organisms during their transport from clinic to laboratory.

Stygiolobus Medium

Composition per liter:

Sulfur flowers	5.0g
$(NH_4)_2SO_4$	1.3g
KH_2PO_4	0.28g
$MgSO_4 \cdot 7H_2O$	0.25g
Yeast extract	0.2g
$CaCl_2 \cdot 2H_2O$	0.07g
$FeCl_3 \cdot 6H_2O$	0.02g
$Na_2B_4O_7 \cdot 10H_2O$	4.5mg
$MnCl_2 \cdot 4H_2O$	1.8mg
Resazurin	0.5mg
$ZnSO_4 \cdot 7H_2O$	0.22mg
$CuCl_2 \cdot 2H_2O$	0.05mg
$Na_2MoO_4 \cdot 2H_2O$	0.03mg
$VOSO_4 \cdot 2H_2O$	0.03mg
$CoSO_4$	0.01mg

pH 2.5 ± 0.2 at 25°C

Preparation of Medium: Add components to distilled/deionized water and bring volume to 1.0L. Mix thoroughly. Adjust pH to 2.5 with 10*N* H_2SO_4. Distribute 20.0mL of medium into 100.0mL flasks or serum bottles. Autoclave for 15 min at 15 psi pressure–121°C. After inoculation, pressurize bottles to 200KPa with 80%H_2 + 20% CO_2.

Use: For the cultivation and maintenance of *Stygiolobus azoricus*.

Styrene Mineral Salts Agar

Composition per liter:

Agar	20.0g
$(NH_4)_2SO_4$	2.0g
K_2HPO_4	1.55g
$NaH_2PO \cdot 2H_2O$	0.85g
$MgCl_2 \cdot 6H_2O$	0.1g
EDTA	10.0mg
$FeSO_4 \cdot 7H_2O$	5.0mg
$ZnSO_4$	2.0mg
$CaCl_2 \cdot 2H_2O$	1.0mg
$MnCl_2 \cdot 2H_2O$	1.0mg
$CoCl_2 \cdot 6H_2O$	0.4mg
$CuSO_4 \cdot 5H_2O$	0.2mg
$Na_2MoO_4 \cdot 2H_2O$	0.2mg

Preparation of Medium: Add components to distilled/deionized water and bring volume to 1.0L. Mix thoroughly. Gently heat and bring to boiling. Distribute into tubes or flasks. Autoclave for 15 min at 15 psi pressure–121°C. Pour into sterile Petri dishes or leave in tubes. Place plates in a desiccator. Add to the desiccator an open bottle containing 10.0mL of dibutyl phthalate and 200.0μl of styrene.

Use: For the cultivation of styrene-utilizing microorganisms.

Styrene Mineral Salts Broth

Composition per liter:

$(NH_4)_2SO_4$	2.0g
K_2HPO_4	1.55g
$NaH_2PO \cdot 2H_2O$	0.85g
$MgCl_2 \cdot 6H_2O$	0.1g
EDTA	10.0mg
$FeSO_4 \cdot 7H_2O$	5.0mg
$ZnSO_4$	2.0mg
$CaCl_2 \cdot 2H_2O$	1.0mg
$MnCl_2 \cdot 2H_2O$	1.0mg
$CoCl_2 \cdot 6H_2O$	0.4mg
$CuSO_4 \cdot 5H_2O$	0.2mg
$Na_2MoO_4 \cdot 2H_2O$	0.2mg
Styrene	0.1mL

Preparation of Medium: Add components, except styrene, to distilled/deionized water and bring volume to 1.0L. Mix thoroughly. Distribute into tubes or flasks. Autoclave for 15 min at 15 psi pressure–121°C. Under a fume hood, aseptically add styrene to screw-capped tubes or flasks (5.0μl of styrene into 50.0mL of sterile mineral salts solution). Seal caps tightly.

Use: For the cultivation and maintenance of styrene-utilizing microorganisms.

Succinate Mineral Medium

Composition per 1001.0mL:

$MnSO_4 \cdot 7H_2O$	62.0g
Succinate	2.7g
$K_2HPO_4 \cdot 3H_2O$	1.63g
NH_4Cl	1.02g
KH_2PO_4	0.39g
Trace elements solution	1.0mL

pH 6.8 ± 0.2 at 25°C

Trace Elements Solution:

Composition per 100.0mL:

Solution A	50.0mL
Solution B	50.0mL

Preparation of Trace Elements Solution: Aseptically combine 50.0mL of sterile solution A with 50.0mL of sterile solution B.

Solution A:

Composition per liter:

Disodium EDTA	0.25g
$CaCl_2 \cdot 2H_2O$	0.14g
$FeCl_2 \cdot 4H_2O$	0.14g

Preparation of Solution A: Add components to distilled/deionized water and bring volume to 1.0L. Mix thoroughly. Autoclave for 15 min at 15 psi pressure–121°C.

Solution B:

Composition per liter:

Disodium EDTA	0.25g
H_3BO_3	30.0mg
$CoCl_2 \cdot 2H_2O$	28.0mg
$ZnCl_2$	5.0mg
$MnCl_2 \cdot 4H_2O$	4.0mg
$NaMoO_4 \cdot 2H_2O$	3.0mg
$NiCl_2 \cdot 6H_2O$	1.65mg
$CuCl_2 \cdot 2H_2O$	0.7mg

Preparation of Solution B: Add components to distilled/deionized water and bring volume to 1.0L. Mix thoroughly. Autoclave for 15 min at 15 psi pressure–121°C.

Preparation of Medium: Add components, except trace elements solution, to distilled/deionized water and bring volume to 1.0L. Mix thoroughly. Autoclave for 15 min at 15 psi pressure–121°C. Aseptically add 1.0mL of sterile trace elements solution. Mix thoroughly. Aseptically distribute into sterile tubes or flasks.

Use: For the cultivation of bacteria that can utilize succinate as a carbon source.

Succiniclasticum Medium

Composition per liter:

Disodium succinate	5.0g
Yeast extract	5.0g
NaCl	0.45g
$(NH_4)_2SO_4$	0.45g
K_2HPO_4	0.225g
KH_2PO_4	0.225g
$MgSO_4·7H_2O$	0.09g
$CaCl_2·2H_2O$	0.06g
Indigocarmine	5.0mg
Rumen fluid, clarified	400.0mL
$NaHCO_3$ solution	10.0mL
L-Cysteine·HCl·H_2O solution	10.0mL
$Na_2S·9H_2O$ solution	10.0mL

pH: 6.6–6.8

$NaHCO_3$ Solution:
Composition per 10.0mL:

$NaHCO_3$	6.4g

Preparation of $NaHCO_3$ Solution: Add $NaHCO_3$ to distilled/deionized water and bring volume to 10.0mL. Mix thoroughly. Sparge with 100% N_2. Autoclave for 15 min at 15 psi pressure–121°C.

L-Cysteine·HCl·H_2O Solution:
Composition per 10.0mL:

L-Cysteine·HCl·H_2O	0.3g

Preparation of L-Cysteine·HCl·H_2O Solution: Add L-cysteine·HCl·H_2O to distilled/deionized water and bring volume to 10.0mL. Mix thoroughly. Sparge with 100% N_2. Autoclave for 15 min at 15 psi pressure–121°C.

$Na_2S·9H_2O$ Solution:
Composition per 10.0mL:

$Na_2S·9H_2O$	0.3g

Preparation of $Na_2S·9H_2O$ Solution: Add $Na_2S·9H_2O$ to distilled/deionized water and bring volume to 10.0mL. Mix thoroughly. Sparge with 100% N_2. Autoclave for 15 min at 15 psi pressure–121°C. Before use, neutralize to pH 7.0 with sterile HCl.

Preparation of Medium: Add components, except $Na_2S·9H_2O$ solution, L-cysteine·HCl·H_2O solution, and $NaHCO_3$ solution, to distilled/deionized water and bring volume to 970.0mL. Mix thoroughly. Gently heat and bring to boiling. Cool to room temperature while sparging with 100% CO_2. Autoclave for 15 min at 15 psi pressure–121°C. Aseptically and anaerobically add 10.0mL of sterile L-cysteine·HCl·H_2O solution, 10.0mL of sterile $Na_2S·9H_2O$ solution, and 10.0mL of sterile $NaHCO_3$ solution to each tube. Mix thoroughly.

Use: For the cultivation of *Succiniclasticum ruminis.*

Sucrose Agar

Composition per liter:

Sucrose	50.0g
Agar	20.0g
Peptone	10.0g
NaCl	5.0g
Yeast extract	5.0g
$CaCO_3$	3.0g
Phenylethyl alcohol	3.0g
$MgSO_4·7H_2O$	0.5g
$MnSO_4·4H_2O$	0.5g
Tween™ 80	0.1g
Bromcresol Green	20.0mg
Cycloheximide solution	10.0mL

pH 6.2 ± 0.2 at 25°C

Cycloheximide Solution:
Composition per 10.0mL:

Cycloheximide	0.01g

Preparation of Cycloheximide Solution: Add cycloheximide to distilled/deionized water and bring volume to 10.0mL. Mix thoroughly. Filter sterilize.

Preparation of Medium: Add agar to distilled/deionized water and bring volume to 500.0mL. Mix thoroughly. Gently heat and bring to boiling. Add remaining components, except cycloheximide solution and phenylethyl alcohol, and bring volume to 990.0mL with distilled/deionized water. Adjust pH to 6.2. Autoclave for 15 min at 15 psi pressure–121°C. Cool to 45°–50°C. Aseptically add 10.0mL of sterile cycloheximide solution and 3.0g of phenylethyl alcohol. Mix thoroughly. Pour into sterile Petri dishes or distribute into sterile tubes.

Use: For the isolation and cultivation of *Lactobacillus* species from brewery isolates.

Sucrose Agar

Composition per liter:

Beef heart, solids from infusion	500.0g
Sucrose	50.0g
Agar	15.0g
Tryptose	10.0g
NaCl	5.0g

pH 7.4 ± 0.2 at 25°C

Preparation of Medium: Add components to distilled/deionized water and bring volume to 1.0L. Mix thoroughly. Gently heat and bring to boiling. Distribute into tubes or flasks. Autoclave for 15 min at 15 psi pressure–121°C. Pour into sterile Petri dishes or leave in tubes.

Use: For the differentiation of bacteria based on their ability to produce glucan. Dextran production, typical of *Streptococcus sanguis* and *Streptococcus mutans*, results in highly refractile-adherent or dry-adherent colonies. Levan production, typical of *Streptococcus salivarius*, results in opaque, gummy, nonadherent colonies. Colonies of *Streptococcus bovis* and *Leuconostoc mesenteroides* are similar to those of *Streptococcus salivarius* but are somewhat less gummy and rarely adhere to the medium. Large or small colonies that are mucoidal and nonadherent are considered negative or have no extracellular polysaccharide production.

Sucrose Broth

Composition per liter:

Solution A	500.0mL
Solution B	500.0mL

pH 7.1 ± 0.2 at 25°C

Solution A:
Composition per 500.0mL:

Pancreatic digest of casein	15.0g
Sodium acetate	12.0g

K_2HPO_4	10.0g
Glucose	5.5g
Yeast extract	5.0g
NaCl	2.5g
L-Cystine	0.5g
Sodium thioglycolate	0.5g

Preparation of Solution A: Add components to distilled/deionized water and bring volume to 500.0mL. Mix thoroughly. Autoclave for 15 min at 15 psi pressure–121°C. Cool to 45°–50°C.

Solution B:

Composition per 500.0mL:

Sucrose	50.0g

Preparation of Solution B: Add sucrose to distilled/deionized water and bring volume to 500.0mL. Mix thoroughly. Autoclave for 15 min at 15 psi pressure–121°C. Cool to 45°–50°C.

Preparation of Medium: Combine sterile solution A with sterile solution B. Mix thoroughly. Aseptically distribute into sterile tubes or flasks.

Use: For the differentiation of bacteria based on their ability to produce glucan. Production of glucan is indicated when the broth is partially or completely gelled—typical of *Streptococcus sanguis*—or when gelatinous, adherent deposits form on the bottom and walls of the tube—typical of *Streptococcus mutans*. An increase in the viscosity indicates the production of slime (unknown polysaccharide)—typical of *Streptococcus bovis*.

Sucrose HiVeg Agar for Brewery Isolates

Composition per liter:

Sucrose	50.0g
Agar	15.0g
Plant hydrolysate	10.0g
$(NH_4)_3PO_4$	5.0g
Yeast extract	5.0g
K_2HPO_4	5.0g
Cycloheximide solution	10.0mL

pH 6.2 ± 0.2 at 25°C

Source: This medium, without cycloheximide, is available as a premixed powder from HiMedia.

Cycloheximide Solution:

Composition per 10.0mL:

Cycloheximide	0.01g

Preparation of Cycloheximide Solution: Add cycloheximide to distilled/deionized water and bring volume to 10.0mL. Mix thoroughly. Filter sterilize.

Preparation of Medium: Add agar to distilled/deionized water and bring volume to 500.0mL. Mix thoroughly. Gently heat and bring to boiling. Add remaining components, except cycloheximide solution and phenylethyl alcohol, and bring volume to 990.0mL with distilled/deionized water. Adjust pH to 6.2. Autoclave for 15 min at 15 psi pressure–121°C. Cool to 45°–50°C. Aseptically add 10.0mL of sterile cycloheximide solution and 3.0g of phenylethyl alcohol. Mix thoroughly. Pour into sterile Petri dishes or distribute into sterile tubes.

Use: For the isolation and cultivation of *Lactobacillus* species, especially brewery isolates.

Sucrose Peptone Agar

Composition per liter:

Sucrose	20.0g
Agar	12.0g
Peptone	5.0g
K_2HPO_4	0.5g
$MgSO_4 \cdot 7H_2O$	0.25g

pH 7.2–7.4 at 25°C

Preparation of Medium: Add components to distilled/deionized water and bring volume to 1.0L. Mix thoroughly. Gently heat and bring to boiling. Adjust pH to 7.2–7.4. Distribute into tubes or flasks. Autoclave for 15 min at 15 psi pressure–121°C. Pour into sterile Petri dishes or leave in tubes.

Use: For the cultivation and maintenance of *Pseudomonas solanacearum* and *Xanthomonas albilineans*.

Sucrose Peptone Medium

Composition per liter:

Peptone	20.0g
Sucrose	20.0g

Preparation of Medium: Add components to distilled/deionized water and bring volume to 1.0L. Mix thoroughly. Distribute into tubes or flasks. Autoclave for 15 min at 15 psi pressure–121°C.

Use: For the cultivation of *Pseudomonas* species.

Sucrose Phosphate Glutamate Transport Medium

Composition per liter:

Sucrose	75.0g
Na_2HPO_4	1.22g
Glutamic acid	0.72g
K_2HPO_4	0.52g
Bovine serum	50.0mL
Antibiotic inhibitor	10.0mL

pH 7.4–7.6 at 25°C

Antibiotic Inhibitor:

Composition per 10.0mL:

Vancomycin	0.1g
Streptomycin	0.05g
Nystatin	25000U

Preparation of Antibiotic Inhibitor: Add components to distilled/deionized water and bring volume to 10.0mL. Mix thoroughly. Filter sterilize.

Preparation of Medium: Add components, except bovine serum and antibiotic inhibitor, to distilled/deionized water and bring volume to 940.0mL. Mix thoroughly. Gently heat and bring to boiling. Adjust pH to 7.4–7.6. Autoclave for 15 min at 15 psi pressure–121°C. Cool to 45°–50°C. Aseptically add sterile bovine serum and sterile antibiotic inhibitor. Mix thoroughly. Aseptically distribute into sterile tubes or flasks.

Use: For the maintenance of *Chlamydia* species during transport.

Sucrose Phosphate Transport Medium

Composition per liter:

Sucrose	68.5g
K_2HPO_4	2.1g
KH_2PO_4	1.1g

Bovine serum	50.0mL
Antibiotic inhibitor	10.0mL

pH 7.0 ± 0.2 at 25°C

Antibiotic Inhibitor:

Composition per 10.0mL:

Vancomycin	0.1g
Streptomycin	0.05g
Nystatin	25,000U

Preparation of Antibiotic Inhibitor: Add components to distilled/deionized water and bring volume to 10.0mL. Mix thoroughly. Filter sterilize.

Preparation of Medium: Add components, except bovine serum and antibiotic inhibitor, to distilled/deionized water and bring volume to 940.0mL. Mix thoroughly. Gently heat and bring to boiling. Adjust pH to 7.0. Autoclave for 15 min at 15 psi pressure–121°C. Cool to 45°–50°C. Aseptically add sterile bovine serum and sterile antibiotic inhibitor. Mix thoroughly. Aseptically distribute into sterile tubes or flasks.

Use: For the maintenance of *Chlamydia* species during transport.

Sucrose Teepol Tellurite Agar (STT Agar)

Composition per liter:

Agar	20.0g
Beef extract	1.0g
Peptone	1.0g
Sucrose	1.0g
NaCl	0.5g
Bromthymol Blue (0.2% solution)	2.5mL
Tellurite solution	2.5mL
Sodium lauryl sulfate (Teepol, 0.1% solution)	0.2mL

pH 8.0 ± 0.2 at 25°C

Tellurite Solution:

Composition per 100.0mL:

K_2TeO_3	0.05g

Preparation of Tellurite Solution: Add the K_2TeO_3 to distilled/deionized water and bring the volume to 100.0mL. Mix thoroughly. Filter sterilize. Use freshly prepared solution.

Caution: Potassium tellurite is toxic.

Preparation of Medium: Add components to distilled/deionized water and bring volume to 1.0L. Mix thoroughly. Gently heat and bring to boiling. Do not autoclave. Pour into sterile Petri dishes.

Use: For the selective isolation, cultivation, and differentiation of *Vibrio* species based on their ability to ferment sucrose. *Vibrio cholerae* appears as flat yellow colonies. *Vibrio parahaemolyticus* appears as elevated green-yellow mucoid colonies.

Sucrose Yeast Extract Medium

Composition per liter:

Sucrose	20.0g
Yeast extract	4.0g
K_2HPO_4	2.5g
$MgSO_4 \cdot 7H_2O$	1.0g
Trace elements solution	4.0mL

pH 7.0 ± 0.2 at 25°C

Trace Elements Solution:

Composition per liter:

$ZnSO_4 \cdot 7H_2O$	18.0g
$FeSO_4 \cdot 7H_2O$	9.0g
$MnSO_4 \cdot 4H_2O$	3.0g
$CoCl_2 \cdot 6H_2O$	0.9g
$CuSO_4 \cdot 5H_2O$	0.8g
Conc. H_2SO_4	5.0mL

Preparation of Trace Elements Solution: Add components to distilled/deionized water and bring volume to 1.0L. Mix thoroughly.

Preparation of Medium: Add components to distilled/deionized water and bring volume to 1.0L. Mix thoroughly. Distribute into tubes or flasks. Autoclave for 15 min at 15 psi pressure–121°C.

Use: For the cultivation of a variety of bacteria that can utilize sucrose as a carbon source.

Sulfate API Broth

Composition per liter:

NaCl	10.0g
Sodium lactate	5.2g
Yeast extract	1.0g
$MgSO_4 \cdot 7H_2O$	0.2g
Ascorbic acid	0.1g
$Fe(NH_4)_2(SO_4)_2 \cdot 6H_2O$	0.1g
K_2HPO_4	0.01g

pH 7.5 ± 0.2 at 25°C

Source: This medium is available as a premixed powder from BD Diagnostic Systems.

Preparation of Medium: Add the components to distilled/deionized water and bring volume to 1.0L. Mix thoroughly until dissolved. Distribute into tubes in 9.0mL volumes. Autoclave for 10 min at 15 psi pressure–121°C.

Use: For the detection, differentiation, and estimation of sulfate-reducing bacteria.

Sulfate-Reducing Bacteria Enrichment Medium

Composition per 1018.0mL:

Solution 1	970.0mL
Solution 4	30.0mL
Solution 6A, 6B, 6C, 6D, or 6E	10.0mL
Solution 5	3.0mL
Solution 2	1.0mL
Solution 3	1.0mL
Solution 7	1.0mL
Solution 8	1.0mL
Solution 9	1.0mL

pH 7.2 ± 0.2 at 25°C

Solution 1:

Composition per 970.0mL:

Na_2SO_4	3.0g
NaCl	1.2g
$MgCl_2 \cdot 6H_2O$	0.4g
KCl	0.3g
NH_4Cl	0.3g
KH_2PO_4	0.2g
$CaCl_2 \cdot 2H_2O$	0.15g

Preparation of Solution 1: Add components to distilled/deionized water and bring volume to 970.0mL. Mix thoroughly. Autoclave for 30 min at 15 psi pressure–121°C. Cool to 25°C under 90% N_2 + 10% CO_2.

Solution 2:

Composition per liter:

$FeCl_2 \cdot 4H_2O$	1.5g
$CoCl_2 \cdot 6H_2O$	0.12g
$MnCl_2 \cdot 4H_2O$	0.1g
$ZnCl_2$	0.07g
H_3BO_3	0.06g
$Na_2MoO_4 \cdot 2H_2O$	0.025g
$NiCl_2 \cdot 6H_2O$	0.025g
$CuCl_2 \cdot 2H_2O$	0.015g
HCl (25% solution)	6.5mL

Preparation of Solution 2: Add components to distilled/deionized water and bring volume to 1.0L. Mix thoroughly. Autoclave for 15 min at 15 psi pressure–121°C. Cool to 25°C.

Solution 3:

Composition per liter:

NaOH	0.5g
Na_2SeO_3	3.0mg

Preparation of Solution 3: Add components to distilled/deionized water and bring volume to 1.0L. Mix thoroughly. Autoclave for 15 min at 15 psi pressure–121°C. Cool to 25°C.

Solution 4:

Composition per 100.0mL:

$NaHCO_3$	8.5g

Preparation of Solution 4: Add $NaHCO_3$ to distilled/deionized water and bring volume to 100.0mL. Mix thoroughly. Saturate with 100% CO_2. Filter sterilize. Aseptically add solution to sterile, gas-tight, screw-capped bottles.

Solution 5:

Composition per 100.0mL:

$Na_2S \cdot 9H_2O$	12.0g

Preparation of Solution 5: Add $Na_2S \cdot 9H_2O$ to distilled/deionized water and bring volume to 100.0mL. Mix thoroughly. Add solution to gas-tight, screw-capped bottles. Gas under 100% N_2 for 20 min. Close caps tightly. Autoclave for 15 min at 15 psi pressure–121°C. Cool to 25°C.

Solution 6A:

Composition per 100.0mL:

Sodium acetate·$3H_2O$	20.0g

Preparation of Solution 6A: Add sodium acetate·$3H_2O$ to distilled/deionized water and bring volume to 100.0mL. Autoclave for 15 min at 15 psi pressure–121°C. Cool to 25°C.

Solution 6B:

Composition per 100.0mL:

n-Butyric acid	8.0g

Preparation of Solution 6B: Add *n*-butyric acid to distilled/deionized water and bring volume to 100.0mL. Adjust pH to 9.0 with NaOH. Autoclave for 15 min at 15 psi pressure–121°C. Cool to 25°C.

Solution 6C:

Composition per 100.0mL:

Propionic acid	7.0g

Preparation of Solution 6C: Add propionic acid to 100.0mL of distilled/deionized water. Adjust pH to 9.0 with NaOH. Autoclave for 15 min at 15 psi pressure–121°C. Cool to 25°C.

Solution 6D:

Composition per 100.0mL:

Benzoic acid	5.0g

Preparation of Solution 6D: Add benzoic acid to distilled/deionized water and bring volume to 100.0mL. Adjust pH to 9.0 with NaOH. Autoclave for 15 min at 15 psi pressure–121°C. Cool to 25°C.

Solution 6E:

Composition per 100.0mL:

n-Palmitic acid	5.0g
NaOH	0.78g

Preparation of Solution 6E: Add *n*-palmitic acid and NaOH to distilled/deionized water and bring volume to 100.0mL. Heat in a water bath until clear. Autoclave for 15 min at 15 psi pressure–121°C. Cool to 25°C.

Solution 7:

Composition per 100.0mL:

Thiamine	0.01g
p-Aminobenzoic acid	5.0mg
Vitamin B_{12}	5.0mg
Biotin	1.0mg

Preparation of Solution 7: Add components to distilled/deionized water and bring volume to 100.0mL. Mix thoroughly. Filter sterilize.

Solution 8:

Composition per liter:

Succinic acid	0.6g
Isobutyric acid	0.5g
2-Methylbutyric acid	0.5g
3-Methylbutyric acid	0.5g
Valeric acid	0.5g
Caproic acid	0.2g

Preparation of Solution 8: Add components to distilled/deionized water and bring volume to 100.0mL. Mix thoroughly. Adjust pH to 9.0 with NaOH. Autoclave for 15 min at 15 psi pressure–121°C. Cool to 25°C.

Solution 9:

Composition per 100.0mL:

$Na_2S_2O_4$	3.0g

Preparation of Solution 9: Add $Na_2S_2O_4$ to 100.0mL of O_2-free distilled/deionized water. Mix thoroughly. Anaerobically filter sterilize.

Preparation Medium: To 970.0mL of cooled, sterile solution 1, aseptically and anaerobically add 1.0mL of sterile solution 2, 1.0mL of sterile solution 3, 30.0mL of sterile solution 4, and 3.0mL of sterile solution 5. Mix thoroughly. Adjust pH to 7.2 with sterile HCl solution or sterile Na_2CO_3 solution. Aseptically and anaerobically distribute into sterile screw-capped bottles in 100.0mL volumes. Add 1.0mL of solution 6A, 6B, 6C, 6D, or 6E to each bottle containing 100.0mL of basal medium. Add 0.1mL of solution 7, 0.1mL of solution 8, and 0.1mL of solution 9 to each bottle containing 100.0mL of basal medium. Mix thoroughly.

Use: For the isolation, cultivation, and enrichment of sulfate-reducing bacteria.

Sulfate-Reducing Bacteria Medium

Composition per 1008.0mL:

Solution A	850.0mL
Solution C	100.0mL
Solution G	20.0mL

Solution D	10.0mL
Solution E (Wolfe's vitamin solution)	10.0mL
Solution H	10.0mL
Solution F	6.6mL
Solution B (Trace elements solution SL-10)	1.0mL
Solution I	0.4mL

pH 7.6 ± 0.2 at 25°C

Solution A:

Composition per 920.0mL:

Na_2SO_4	3.0g
NaCl	1.0g
KCl	0.5g
$MgCl_2 \cdot 6H_2O$	0.4g
NH_4Cl	0.3g
KH_2PO_4	0.2g
$CaCl_2 \cdot 2H_2O$	0.15g
Resazurin	0.5mg

Preparation of Solution A: Prepare and dispense solution anaerobically under 80% N_2 + 20% CO_2. Add components to distilled/deionized water and bring volume to 920.0mL. Mix thoroughly. Gently heat and bring to boiling. Continue boiling until resazurin turns colorless, indicating reduction, and a pH of 6.0 is reached. Cap with rubber stoppers. Autoclave for 15 min at 15 psi pressure–121°C. Cool to 25°C.

Solution B (Trace Elements Solution SL-10):

Composition per liter:

$FeCl_2 \cdot 4H_2O$	1.5g
$CoCl_2 \cdot 6H_2O$	0.19g
$MnCl_2 \cdot 4H_2O$	0.10g
$ZnCl_2$	0.070g
$Na_2MoO_4 \cdot 2H_2O$	0.036g
$NiCl_2 \cdot 6H_2O$	0.024g
H_3BO_3	6.0mg
$CuCl_2 \cdot 2H_2O$	2.0mg
HCl (25% solution)	10.0mL

Preparation of Solution B (Trace Elements Solution SL-10): Add the $FeCl_2 \cdot 4H_2O$ to 10.0mL of HCl solution. Mix thoroughly. Bring volume to approximately 900.0mL with distilled/deionized water. Mix thoroughly. Adjust pH to 6.0 with NaOH. Bring volume to 1.0L with distilled/deionized water. Filter sterilize. Aseptically gas under 100% N_2 for 20 min.

Solution C:

Composition per 100.0mL:

$NaHCO_3$	5.0g

Preparation of Solution C: Add $NaHCO_3$ to distilled/deionized water and bring volume to 100.0mL. Mix thoroughly. Filter sterilize. Aseptically gas under 80% N_2 + 20% CO_2 for 20 min.

Solution D:

Composition per 10.0mL:

Sodium propionate	1.5g

Preparation of Solution D: Prepare and dispense solution anaerobically under 80% N_2 + 20% CO_2. Add sodium propionate to distilled/deionized water and bring volume to 10.0mL. Mix thoroughly. Cap with a rubber stopper. Autoclave for 15 min at 15 psi pressure–121°C. Cool to 25°C.

Solution E (Wolfe's Vitamin Solution):

Composition per liter:

Pyridoxine·HCl	0.01g
Thiamine·HCl	5.0mg
Riboflavin	5.0mg
Nicotinic acid	5.0mg
Calcium pantothenate	5.0mg
p-Aminobenzoic acid	5.0mg
Thioctic acid	5.0mg
Biotin	2.0mg
Folic acid	2.0mg
Cyanocobalamin	0.1mg

Preparation of Solution E (Wolfe's Vitamin Solution): Add components to distilled/deionized water and bring volume to 1.0L. Mix thoroughly. Filter sterilize. Aseptically gas under 100% N_2 for 20 min.

Solution F:

Composition per 6.6mL:

$AlCl_3 \cdot 6H_2O$ (4.9% solution)	5.0mL
Na_2CO_3 (10.6% solution)	1.6mL

Preparation of Solution F: Combine both solutions. Mix thoroughly. Gas with 100% N_2. Cap with a rubber stopper. Autoclave for 15 min at 15 psi pressure–121°C. Cool to 25°C.

Solution G:

Composition per 10.0mL:

Rumen fluid, clarified	20.0mL

Preparation of Solution G: Gas rumen fluid under 100% N_2 for 20 min. Cap with a rubber stopper. Autoclave for 15 min at 15 psi pressure–121°C. Cool to 25°C.

Solution H:

Composition per 10.0mL:

$Na_2S \cdot 9H_2O$	0.4g

Preparation of Solution H: Add $Na_2S \cdot 9H_2O$ to distilled/deionized water and bring volume to 10.0mL. Gas under 100% N_2 for 20 min. Cap with a rubber stopper. Autoclave for 15 min at 15 psi pressure–121°C. Cool to 25°C.

Solution I:

Composition per 10.0mL:

$Na_2S_2O_4$	0.5g

Preparation of Solution I: Add $Na_2S_2O_4$ to distilled/deionized water and bring volume to 10.0mL. Mix thoroughly. Filter sterilize. Aseptically gas under 100% N_2 for 20 min. Prepare solution freshly.

Preparation of Medium: To 850.0mL of cooled, sterile solution A, aseptically and anaerobically add in the following order: 1.0mL of sterile solution B, 100.0mL of sterile solution C, 10.0mL of sterile solution D, 10.0mL of sterile solution E, 6.6mL of sterile solution F, 20.0mL of sterile solution G, and 10.0mL of sterile solution H. Mix thoroughly. Immediately prior to inoculation, aseptically and anaerobically add 0.4mL of sterile solution I. Mix thoroughly. Aseptically and anaerobically distribute into sterile tubes or flasks.

Use: For the cultivation and maintenance of a variety of sulfate-reducing bacteria.

Sulfate-Reducing Bacteria Medium with Lactate

Composition per liter:

Solution 1	980.0mL
Solution 2	10.0mL
Solution 3	10.0mL

pH 7.4 ± 0.2 at 25°C

Solution 1:

Composition per 980.0mL:

Sodium lactate (70% solution)	3.5g
$MgSO_4 \cdot 7H_2O$	2.0g
NH_4Cl	1.0g
Na_2SO_4	1.0g
Yeast extract	1.0g
K_2HPO_4	0.5g
$CaCl_2 \cdot 2H_2O$	0.1g

Preparation of Solution 1: Add components to distilled/deionized water and bring volume to 980.0mL. Mix thoroughly. Autoclave for 15 min at 15 psi pressure–121°C. Cool to 50°C.

Solution 2:

Composition per 10.0mL:

$FeSO_4 \cdot 7H_2O$	0.5g

Preparation of Solution 2: Add $FeSO_4 \cdot 7H_2O$ to distilled/deionized water and bring volume to 10.0mL. Mix thoroughly. Autoclave for 15 min at 15 psi pressure–121°C. Cool to 50°C.

Solution 3:

Composition per 10.0mL:

Ascorbic acid	0.1g
Sodium thioglycolate	0.1g

Preparation of Solution 3: Add components to distilled/deionized water and bring volume to 10.0mL. Mix thoroughly. Autoclave for 15 min at 15 psi pressure–121°C. Cool to 50°C.

Preparation of Medium: Aseptically combine 980.0mL of cooled, sterile solution 1, 10.0mL of cooled, sterile solution 2, and 10.0mL of cooled, sterile solution 3. Mix thoroughly. Aseptically distribute into sterile tubes or flasks.

Use: For the enrichment and isolation of sulfate-reducing bacteria.

Sulfate-Reducing HiVeg Medium

Composition per liter:

Part A	890.0mL
Part B	100.0mL
Part C	10.0mL

pH 7.5 ± 0.2 at 25°C

Source: This medium is available as a premixed powder from HiMedia.

Part A:

Composition per 890.0mL:

Plant peptone	2.0g
$MgSO_4 \cdot 7H_2O$	2.0g
Na_2SO_4	1.5g
Plant extract	1.0g
K_2HPO_4	0.5
$CaCl_2$	0.1g

Preparation of Part A: Add components to distilled/deionized water and bring volume to 890.0mL. Gently heat and bring to boiling. Mix thoroughly. Autoclave for 15 min at 15 psi pressure–121°C.

Part B:

Composition per 100.0mL:

Ferrous $(NH_4)_2SO_4$	0.392g
Sodium ascorbate	0.1g

Preparation of Part B: Add components to distilled/deionized water and bring volume to 100.0mL. Gently heat and bring to boiling. Mix thoroughly. Filter sterilize. Prepare on day of use.

Part C:

Composition per 100.0mL:

Sodium lactate	3.5g

Preparation of Part C: Add sodium lactate to distilled/deionized water and bring volume to 100.0mL. Gently heat and bring to boiling. Mix thoroughly. Autoclave for 15 min at 15 psi pressure–121°C.

Preparation of Medium: Aseptically combine 890.0mL part A, 100.0mL part B, and 10.0mL part C. Mix thoroughly. Distribute into screw-cap tubes. Completely fill the tubes.

Use: For the enumeration of sulfate reducing bacteria in water samples.

Sulfate-Reducing Medium

Composition per liter:

Sodium lactate	3.5g
$MgSO_4 \cdot 7H_2O$	2.0g
Peptone	2.0g
Na_2SO_4	1.5g
Beef extract	1.0g
K_2HPO_4	0.5g
$CaCl_2$	0.1g
$Fe(NH_4)_2(SO_4)_2 \cdot 6H_2O$ solution	10.0mL
Sodium ascorbate solution	10.0mL

pH 7.5 ± 0.3 at 25°C

$Fe(NH_4)_2(SO_4)_2 \cdot 6H_2O$ Solution:

Composition per 100.0mL:

$Fe(NH_4)_2(SO_4)_2 \cdot 6H_2O$	3.92g

Preparation of $Fe(NH_4)_2(SO_4)_2 \cdot 6H_2O$ Solution: Add 3.92g of $Fe(NH_4)_2(SO_4)_2 \cdot 6H_2O$ to distilled/deionized water and bring volume to 100.0mL. Mix thoroughly. Filter sterilize. Use freshly prepared solution.

Sodium Ascorbate Solution:

Composition per 100.0mL:

Sodium ascorbate	0.05g

Preparation of Sodium Ascorbate Solution: Add sodium ascorbate to distilled/deionized water and bring volume to 100.0mL. Mix thoroughly. Filter sterilize. Use freshly prepared solution.

Preparation of Medium: Add components, except sodium ascorbate solution and $Fe(NH_4)_2(SO_4)_2 \cdot 6H_2O$ solution , to distilled/deionized water and bring volume to 980.0mL. Mix thoroughly. Distribute into screw-capped tubes in 10.0mL volumes. Autoclave for 15 min at 15 psi pressure–121°C. Tubes must be filled to capacity after inoculation, so prepare extra medium and sterilize in a screw-capped flask or bottle. Prior to inoculation, aseptically add 0.1mL of freshly prepared sterile $Fe(NH_4)_2(SO_4)_2 \cdot 6H_2O$ solution for each 10.0mL of medium in the tubes. Also aseptically add 0.1mL of freshly prepared sterile sodium ascorbate solution for each 10.0mL of medium in the tubes. Inoculate tubes. Fill tubes to capacity with extra sterile medium. Screw caps tight.

Use: For the isolation, cultivation, and enumeration of iron and sulfur bacteria.

Sulfite Agar

Composition per liter:

Agar	20.0g
Pancreatic digest of casein	10.0g

Na_2SO_3 .... 1.0g
Iron nails 66

pH 7.6 ± 0.2 at 25°C

Source: This medium, without iron nails, is available as a premixed powder from BD Diagnostic Systems.

Preparation of Medium: Add components to distilled/deionized water and bring volume to 1.0L. Mix thoroughly. Gently heat and bring to boiling. Distribute into screw-capped tubes in 15.0mL volumes. Add a clean iron nail to each tube. Autoclave for 15 min at 15 psi pressure–121°C. Cool to 45°–50°C until ready to inoculate.

Use: For the detection and cultivation of thermophilic anaerobes that can produce H_2S from sulfite. Sulfite reduction appears as a blackening of the medium.

Sulfite HiVeg Agar with Iron

Composition per liter:
Agar 20.0g
Plant hydrolysate 10.0g
Na_2SO_3 .... 1.0g
Ferric citrate solution 10.0mL

pH 7.4 ± 0.2 at 25°C

Source: This medium is available as a premixed powder from Hi-Media.

Ferric Citrate Solution:
Composition per 10.0mL:
Ferric citrate 0.5g

Preparation of Ferric Citrate Solution: Add ferric citrate to distilled/deionized water and bring volume to 10.0mL. Mix thoroughly. Filter sterilize.

Preparation of Medium: Add components to distilled/deionized water and bring volume to 1.0L. Mix thoroughly. Gently heat and bring to boiling. Distribute into tubes or flasks. Autoclave for 15 min at 15 psi pressure–121°C. Pour into sterile Petri dishes or leave in tubes.

Use: For the detection of thermophilic sulfide-producing anaerobes.

Sulfite Polymyxin Sulfadiazine Agar
See: **SPS Agar**

Sulfitobacter pontiacus Medium (DSMZ Medium 733)

Composition per liter:
Basal salts solution 959.0mL
Biotin solution 10.0mL
Na-acetate solution 10.0mL
Yeast extract peptone solution 10.0mL
Magnesium calcium solution 10.0mL
Trace elements solution 1.0mL

pH 7.6 ± 0.2 at 25°C

Trace Elements Solution:
Composition per liter:
$MgSO_4 \cdot 7H_2O$ 3.0g
Nitrilotriacetic acid 1.5g
NaCl 1.0g
$MnSO_4 \cdot 2H_2O$ 0.5g
$CoSO_4 \cdot 7H_2O$ 0.18g
$ZnSO_4 \cdot 7H_2O$ 0.18g
$CaCl_2 \cdot 2H_2O$ 0.1g
$FeSO_4 \cdot 7H_2O$ 0.1g
$NiCl_2 \cdot 6H_2O$ 0.025g
$KAl(SO_4)_2 \cdot 12H_2O$ 0.02g
H_3BO_3 .... 0.01g
$Na_2MoO_4 \cdot 4H_2O$ 0.01g
$CuSO_4 \cdot 5H_2O$ 0.01g
$Na_2SeO_3 \cdot 5H_2O$ 0.3mg

Preparation of Trace Elements Solution: Add nitrilotriacetic acid to 500.0mL of distilled/deionized water. Dissolve by adjusting pH to 6.5 with KOH. Add remaining components. Add distilled/deionized water to 1.0L. Mix thoroughly. Filter sterilize.

Basal Salts Solution:
Composition per liter:
HEPES 8.0g
K_2HPO_4 .... 1.0g
NH_4Cl 0.5g
NaCl 15.0g

Preparation of Basal Salts Solution: Add components to 1.0L of distilled/deionized water. Adjust pH to 7.5–7.8 with NaOH. Mix thoroughly. Autoclave for 15 min at 15 psi pressure–121°C. Cool to 25°C.

Biotin Solution:
Composition per 10.0mL:
Biotin 0.1g

Preparation of Biotin Solution: Add biotin to 10.0mL of distilled/deionized water. Mix thoroughly. Filter sterilize.

Magnesium Calcium Solution:
Composition per 10.0mL:
$MgSO_4 \cdot 7H_2O$ 1.0g
$CaCl_2 \cdot 2H_2O$ 0.05g

Preparation of Magnesium Calcium Solution: Add components to 10.0mL of distilled/deionized water. Mix thoroughly. Filter sterilize.

Na-Acetate Solution:
Composition per 10.0mL:
Na-acetate 1.6g

Preparation of Na-Acetate Solution: Add Na-acetate to 10.0mL of distilled/deionized water. Mix thoroughly. Filter sterilize.

Yeast Extract Peptone Solution:
Composition per 10.0mL:
Yeast extract 1.0g
Peptone 0.5g

Preparation of Yeast Extract Peptone Solution: Add components to 10.0mL of distilled/deionized water. Mix thoroughly. Autoclave for 15 min at 15 psi pressure–121°C. Cool to 25°C.

Preparation of Medium: Aseptically combine 959.0mL basal salts solution, 1.0mL trace elements solution, 10.0mL Na-acetate solution, 10.0mL magnesium calcium solution, 10.0mL yeast extract peptone solution, and 10.0mL biotin solution. Mix thoroughly. Aseptically distribute to sterile tubes or flasks.

Use: For the cultivation of *Sulfitobacter pontiacus.*

Sulfobacillus disulfidooxidans Medium (DSMZ Medium 812)

Composition per liter:
$(NH_4)_2SO_4$ 3.0g
KH_2PO_4 .... 0.5g

$MgSO_4 \cdot 7H_2O$ 0.5g
KCl 0.1g
$Ca(NO_3)_2 \cdot 4H_2O$ 0.1g
Yeast extract 0.1g
Glutathione solution 10.0mL

pH 2.25 ± 0.1 at 25°C

Glutathione Solution:

Composition per 10.0mL:

Glutathione 1.0g

Preparation of Glutathione Solution: Add glutathione to distilled/deionized water and bring volume to 10.0mL. Mix thoroughly. Filter sterilize.

Preparation of Medium: Add components, except glutathione solution, to distilled/deionized water and bring volume to 990.0mL. Mix thoroughly. Adjust pH to 2.25. Autoclave for 15 min at 15 psi pressure–121°C. Cool to room temperature. Aseptically add 10.0mL sterile glutathione solution. Mix thoroughly. Aseptically distribute into sterile tubes or bottles.

Use: For the cultivation of *Sulfobacillus disulfidooxidans.*

Sulfobacillus Medium

Composition per 1020.0mL:

Solution A 700.0mL
Solution B 300.0mL
Solution C 20.0mL

pH 1.9–2.4 at 25°C

Solution A:

Composition per 700.0mL:

$(NH_4)_2SO_4$ 3.0g
KCl 0.1g
K_2HPO_4 .... 0.5g
$MgSO_4 \cdot 7H_2O$ 0.5g
$Ca(NO_3)_2$ 0.01g

Preparation of Solution A: Add components to distilled/deionized water and bring volume to 700.0mL. Mix thoroughly. Adjust pH to 2.0–2.2 with sulfuric acid. Autoclave for 15 min at 15 psi pressure–121°C.

Solution B:

Composition per 300.0mL:

$FeSO_4 \cdot 7H_2O$ 44.2g
H_2SO_4, 10*N* solution 1.0mL

Preparation of Solution B: Add components to distilled/deionized water and bring volume to 300.0mL. Mix thoroughly. Autoclave for 15 min at 15 psi pressure–121°C.

Solution C:

Composition per 20.0mL:

Yeast extract 0.2g

Preparation of Solution C: Add yeast extract to distilled/deionized water and bring volume to 20.0mL. Mix thoroughly. Autoclave for 15 min at 15 psi pressure–121°C.

Preparation of Medium: Aseptically combine 700.0mL of solution A, 300.0mL of solution B, and 20.0mL of solution C. Aseptically adjust pH to 1.9–2.4

Use: For the cultivation of *Sulfobacillus thermosulfidooxidans.*

Sulfolobus acidocaldarius Simplified Basal Medium

Composition per liter:

K_2SO_4 .... 6.0g
NaH_2PO_4 .... 1.0g
$MgSO_4 \cdot 7H_2O$ 0.6g
$CaCl_2 \cdot 7H_2O$ 0.2g
Trace minerals solution 0.04mL

pH 3.5 ± 0.2 at 25°C

Trace Minerals Solution:

Composition per 100.0mL:

$FeCl_3 \cdot 6H_2O$ 5.0g
$CuCl_2 \cdot 2H_2O$ 0.5g
$CoCl_2 \cdot 6H_2O$ 0.5g
$MnCl_2 \cdot 2H_2O$ 0.5g
$ZnCl_2$ 0.5g
HCl (1*N* solution) 100.0ml

Preparation of Trace Minerals Solution: Combine components. Mix thoroughly.

Preparation of Medium: Add components to distilled/deionized water and bring volume to 1.0L. Mix thoroughly. Adjust pH to 3.5 with H_2SO_4. Distribute into tubes or flasks. Autoclave for 15 min at 15 psi pressure–121°C.

Use: For the cultivation of *Sulfolobus acidocaldarius.*

Sulfolobus brierleyi Medium

Composition per liter:

Sulfur flowers 10.0g
$(NH_4)_2SO_4$ 3.0g
$K_2HPO_4 \cdot 3H_2O$ 0.5g
$MgSO_4 \cdot 7H_2O$ 0.5g
KCl 0.1g
$Ca(NO_3)$ 0.01g
Yeast extract solution 20.0mL

pH 1.5–2.5 at 25°C

Preparation of Sulfur: Autoclave sulfur at 8 psi pressure–112°C for 15 min.

Yeast Extract Solution:

Composition per 20.0mL:

Yeast extract 0.2g

Preparation of Yeast Extract Solution: Add yeast extract to distilled/deionized water and bring volume to 20.0mL. Mix thoroughly. Autoclave for 15 min at 15 psi pressure–121°C.

Preparation of Medium: Add components, except yeast extract solution and sulfur, to distilled/deionized water and bring volume to 980.0mL. Mix thoroughly. Adjust pH with 6*N* H_2SO_4 to 1.5–2.5. Autoclave for 15 min at 15 psi pressure–121°C. Aseptically add 20.0mL of sterile yeast extract solution and 10.0g of sterile sulfur. Mix thoroughly. Aseptically distribute into sterile tubes or flasks.

Use: For the cultivation of *Acidianus brierleyi.*

Sulfolobus Broth

Composition per liter:

Sucrose yeast solution 500.0mL
$CaCl_2 \cdot 2H_2O$ solution 250.0mL
Trace elements solution 250.0mL

pH 3.0–3.5 at 25°C

Sucrose Yeast Solution:
Composition per 500.0mL:

Sucrose ... 2.0g
Yeast extract ... 1.0g

Preparation of Sucrose Yeast Solution: Add components to distilled/deionized water and bring volume to 500.0mL. Mix thoroughly. Autoclave for 15 min at 15 psi pressure–121°C.

$CaCl_2 \cdot 2H_2O$ Solution:
Composition per 250.0mL:

$CaCl_2 \cdot 2H_2O$ ... 2.0g

Preparation of $CaCl_2 \cdot 2H_2O$ Solution: Add $CaCl_2 \cdot 2H_2O$ to distilled/deionized water and bring volume to 250.0mL. Mix thoroughly. Autoclave for 15 min at 15 psi pressure–121°C.

Trace Elements Solution:
Composition per 250.0mL:

$(NH_4)_2SO_4$ ... 1.3g
KH_2PO_4 ... 0.28g
$MgSO_4 \cdot 7H_2O$ ... 0.25g
$FeSO_4 \cdot 7H_2O$ ... 28.0mg
$Na_2B_4O_7 \cdot 10H_2O$ ... 4.5mg
$MnCl_2 \cdot 7H_2O$ ... 1.8mg
$ZnSO_4 \cdot 7H_2O$ ... 0.22mg
$CuCl_2 \cdot 2H_2O$ ... 0.05mg
$NaMoO_4 \cdot 2H_2O$ ... 0.03mg
$VOSO_4 \cdot 2H_2O$ ... 0.03mg
$CoSO_4 \cdot 2H_2O$ ... 0.01mg

Preparation of Trace Elements Solution: Add components to distilled/deionized water and bring volume to 250.0mL. Mix thoroughly. Autoclave for 15 min at 15 psi pressure–121°C.

Preparation of Medium: Aseptically combine 500.0mL of sterile sucrose yeast solution with 250.0mL of sterile $CaCl_2 \cdot 2H_2O$ solution and 250.0mL of sterile trace elements solution. Mix thoroughly. Adjust pH to 3.0–3.5 with sterile H_2SO_4. Aseptically distribute into sterile tubes or flasks.

Use: For the cultivation of *Sulfolobus* species.

Sulfolobus Medium

Composition per liter:

$(NH_4)_2SO_4$ ... 1.3g
Yeast extract ... 1.0g
KH_2PO_4 ... 0.28g
$MgSO_4 \cdot 7H_2O$ ... 0.25g
$CaCl_2 \cdot 2H_2O$ ... 0.07g
$FeCl_3 \cdot 6H_2O$ ... 0.02g
$Na_2B_4O_7 \cdot 10H_2O$ ... 4.5mg
$MnCl_2 \cdot 4H_2O$ ... 1.8mg
$ZnSO_4 \cdot 7H_2O$ ... 0.22mg
$CuCl_2 \cdot 2H_2O$ ... 0.05mg
$Na_2MoO_4 \cdot 2H_2O$ ... 0.03mg
$VOSO_4 \cdot 2H_2O$ ... 0.03mg
$CoSO_4$ ... 0.01mg

pH 2.0 ± 0.2 at 25°C

Preparation of Medium: Add components to distilled/deionized water and bring volume to 1.0L. Mix thoroughly. Adjust pH at 25°C to 2.0 with 10*N* H_2SO_4. Filter sterilize. Aseptically distribute into tubes or flasks.

Use: For the cultivation and maintenance of *Sulfolobus acidocaldarius*.

Sulfolobus Medium

Composition per liter:

Gellan sucrose yeast solution ... 500.0mL
$CaCl_2 \cdot 2H_2O/MgCl_2 \cdot 6H_2O$ solution ... 250.0mL
Trace elements solution ... 250.0mL

pH 3.0–3.5 at 25°C

Gellan Sucrose Yeast Solution:
Composition per 500.0mL:

Gellan gum ... 6.5g
Sucrose ... 2.0g
Yeast extract ... 1.0g

Source: Gellan gum is available from Kelco.

Preparation of Gellan Sucrose Yeast Solution: Add components to distilled/deionized water and bring volume to 500.0mL. Mix thoroughly. Gently heat and bring to boiling. Autoclave for 15 min at 15 psi pressure–121°C. Cool to 60°C.

$CaCl_2 \cdot 2H_2O/MgCl_2 \cdot 6H_2O$ Solution:
Composition per 250.0mL:

$CaCl_2 \cdot 2H_2O$ ... 2.44g
$MgCl_2 \cdot 6H_2O$ ... 2.0g

Preparation of $CaCl_2 \cdot 2H_2O/MgCl_2 \cdot 6H_2O$ Solution: Add components to distilled/deionized water and bring volume to 250.0mL. Mix thoroughly. Autoclave for 15 min at 15 psi pressure–121°C. Cool to 60°C.

Trace Elements Solution:
Composition per 250.0mL:

$(NH_4)_2SO_4$ ... 1.3g
KH_2PO_4 ... 0.28g
$MgSO_4 \cdot 7H_2O$ ... 0.25g
$FeSO_4 \cdot 7H_2O$ ... 28.0mg
$Na_2B_4O_7 \cdot 10H_2O$ ... 4.5mg
$ZnSO_4 \cdot 7H_2O$ ... 0.22mg
$CuCl_2 \cdot 2H_2O$ ... 0.05mg
$NaMoO_4 \cdot 2H_2O$ ... 0.03mg
$VOSO_4 \cdot 2H_2O$ ... 0.03mg
$CoSO_4 \cdot 2H_2O$ ... 0.01mg

Preparation of Trace Elements Solution: Add components to distilled/deionized water and bring volume to 250.0mL. Mix thoroughly. Autoclave for 15 min at 15 psi pressure–121°C. Cool to 60°C.

Preparation of Medium: Aseptically combine 500.0mL of sterile gellan sucrose yeast solution with 250.0mL of sterile $CaCl_2 \cdot 2H_2O/MgCl_2 \cdot 6H_2O$ solution and 250.0mL of sterile trace elements solution. Mix thoroughly. Adjust pH to 3.0–3.5 with sterile H_2SO_4. Pour into sterile Petri dishes or distribute into sterile tubes.

Use: For the cultivation of *Sulfolobus* species.

Sulfolobus Medium, Revised

Composition per liter:

$(NH_4)_2SO_4$ ... 1.3g
Tryptone ... 1.0g
KH_2PO_4 ... 0.28g
$MgSO_4 \cdot 7H_2O$ ... 0.25g
$CaCl_2 \cdot 2H_2O$ ... 0.07g
Yeast extract ... 0.05g
$FeCl_3 \cdot 6H_2O$ ... 0.02g
$Na_2B_4O_7$ ... 4.5mg
$MnCl_2 \cdot 4H_2O$ ... 1.8mg
$ZnSO_4 \cdot 7H_2O$ ... 0.22mg

$CuCl_2 \cdot H_2O$	0.05mg
$Na_2MoO_4 \cdot H_2O$	0.03mg
$VOSO_4 \cdot 2H_2O$	0.03mg
$CoSO_4$	0.01mg

pH 3.0 ± 0.2 at 25°C

Preparation of Medium: Add components to distilled/deionized water and bring volume to 1.0L. Mix thoroughly. Adjust pH at 25°C to 3.0 with 10*N* H_2SO_4. Filter sterilize. Aseptically distribute into tubes or flasks.

Use: For the cultivation and maintenance of *Sulfolobus* species.

Sulfolobus shibatae Medium

Composition per liter:

$(NH_4)_2SO_4$	1.3g
Yeast extract	1.0g
KH_2PO_4	0.28g
$MgSO_4 \cdot 7H_2O$	0.25g
$CaCl_2 \cdot 2H_2O$	0.07g
$FeCl_3 \cdot 6H_2O$	0.02g
$Na_2B_4O_7 \cdot 10H_2O$	4.5mg
$MnCl_2 \cdot 4H_2O$	1.8mg
$ZnSO_4 \cdot 7H_2O$	0.22mg
$CuCl_2 \cdot 2H_2O$	0.05mg
$Na_2MoO_4 \cdot 2H_2O$	0.03mg
$VOSO_4 \cdot 2H_2O$	0.03mg
$CoSO_4$	0.01mg

pH 3.5 ± 0.2 at 25°C

Preparation of Medium: Add components to distilled/deionized water and bring volume to 1.0L. Mix thoroughly. Adjust pH to 3.5 with 10*N* H_2SO_4. Filter sterilize. Aseptically distribute into tubes or flasks.

Use: For the cultivation of *Sulfolobus shibatae.*

Sulfolobus solfataricus Medium

Composition per liter:

KH_2PO_4	3.1g
$(NH_4)_2SO_4$	2.5g
Casamino acids	1.0g
Yeast extract	1.0g
$CaCl_2 \cdot 2H_2O$	0.25g
$MgSO_4 \cdot 7H_2O$	0.2g
$Na_2B_4O_7 \cdot 10H_2O$	4.5mg
$MnCl_2 \cdot 4H_2O$	1.8mg
$ZnSO_4 \cdot 7H_2O$	0.22mg
$CuCl_2 \cdot 2H_2O$	0.05mg
$Na_2MoO_4 \cdot 2H_2O$	0.03mg
$VOSO_4 \cdot 2H_2O$	0.03mg
$CoSO_4 \cdot 7H_2O$	0.01mg

pH 4.0–4.2 at 25°C

Preparation of Medium: Add components to distilled/deionized water and bring volume to 1.0L. Mix thoroughly. Adjust pH at 25°C to 4.0–4.2 with 10*N* H_2SO_4. Filter sterilize. Aseptically distribute into tubes or flasks.

Use: For the cultivation and maintenance of *Sulfolobus solfataricus*.

Sulfophobococcus zilligii Medium (DSMZ Medium 770)

Composition per 1035.0mL:

Glycine	1.5g
Na_2CO_3	230.0mg
$CaCl_2 \cdot 2H_2O$	66.0mg
Na_2-EDTA	32.0mg
$MgSO_4 \cdot 7H_2O$	31.0mg
KCl	31.0mg
$MnSO_4 \cdot 2H_2O$	2.3mg
$ZnCl_2$	2.1mg
$Na_2B_4O_7 \cdot 10H_2O$	1.8mg
Resazurin	0.5mg
Serum albumin solution	10.0mL
Dithiothreitol solution	10.0mL
Yeast extract solution	10.0mL
Iron sulfate solution	5.0mL

pH 7.6 ± 0.2 at 25°C

Dithiothreitol Solution:

Composition per 10.0mL:

Dithiothreitol	1.54mg

Preparation of Dithiothreitol Solution: Add dithiothreitol to distilled/deionized water and bring volume to 10.0mL. Mix thoroughly. Filter sterilize.

Iron Sulfate Solution:

Composition per 10.0mL:

$FeSO_4 \cdot 7H_2O$	0.1g

Preparation of Iron Sulfate Solution: Add $FeSO_4 \cdot 7H_2O$ to distilled/deionized water and bring volume to 10.0mL. Mix thoroughly. Filter sterilize.

Serum Albumin Solution:

Composition per 10.0mL:

Bovine serum albumin, fraction V	1.0g

Preparation of Serum Albumin Solution: Add bovine serum albumin, fraction V to distilled/deionized water and bring volume to 10.0mL. Mix thoroughly. Filter sterilize.

Yeast Extract Solution:

Composition per 10.0mL:

Yeast extract	1.0g

Preparation of Yeast Extract Solution: Add yeast extract to distilled/deionized water and bring volume to 10.0mL. Mix thoroughly. Sparge with 100% N_2. Autoclave under 100% N_2 for 15 min at 15 psi pressure–121°C. Cool to room temperature.

Preparation of Medium: Prepare medium anaerobically under 100% N_2 gas. Add components, except iron sulfate solution, serum albumin solution, dithiothreitol solution, and yeast extract solution, to distilled/deionized water and bring volume to 1.0L. Mix thoroughly. Gently heat and bring to boiling. Cool to 80–90°C. Adjust pH to 7.6. Cool to room temperature. Dispense 30.0mL aliquots into serum bottles. Autoclave for 15 min at 15 psi pressure–121°C. Cool to room temperature. Aseptically inject per each 30.0mL the following solutions: 0.3mL sterile yeast extract solution, 0.3mL sterile dithiothreitol solution, 0.15mL sterile iron sulfate solution, and 10.0mL sterile serum albumin solution. Final pH should be 7.6.

Use: For the cultivation of *Sulfophobococcus zilligii.*

Sulforhabdus Medium (DSMZ Medium 386a)

Composition per 1002.0mL:

Solution A	920.0mL
Solution C	50.0mL
Solution D	10.0mL

Solution E	10.0mL
Solution F	10.0mL
Solution B (Trace elements solution SL-10B)	1.0mL

pH 7.2–7.5 at 25°C

Solution A:

Composition per 920.0mL:

NaCl	1.0g
KCl	0.5g
$MgCl_2 \cdot 6H_2O$	0.4g
KH_2PO_4	0.2g
NH_4Cl	0.3g
$CaCl_2 \cdot 2H_2O$	0.15g

Preparation of Solution A: Add components to distilled/deionized water and bring volume to 920.0mL. Mix thoroughly. Sparge with 80% N_2 + 20% CO_2 gas until saturated. Autoclave for 15 min at 15 psi pressure–121°C. Cool to 25°C.

Solution B (Trace Elements Solution SL-10B):

Composition per liter:

$FeCl_2 \cdot 4H_2O$	1.5g
H_3BO_3	300.0mg
$CoCl_2 \cdot 6H_2O$	190.0mg
$MnCl_2 \cdot 4H_2O$	100.0mg
$ZnCl_2$	70.0mg
$Na_2MoO_4 \cdot 2H_2O$	36.0mg
$NiCl_2 \cdot 6H_2O$	24.0mg
$CuCl_2 \cdot 2H_2O$	2.0mg
HCl (25% solution)	7.7mL

Preparation of Solution B (Trace Elements Solution SL-10B): Add $FeCl_2 \cdot 4H_2O$ to 10.0mL of HCl solution. Mix thoroughly. Add distilled/deionized water and bring volume to 1.0L. Add remaining components. Mix thoroughly. Sparge with 100% N_2. Autoclave for 15 min at 15 psi pressure–121°C.

Solution C:

Composition per 100.0mL:

$NaHCO_3$	5.0g

Preparation of Solution C: Add $NaHCO_3$ to distilled/deionized water and bring volume to 100.0mL. Mix thoroughly. Sparge with 80% N_2 + 20% CO_2 gas until saturated, approximately 20 min. Filter sterilize under 100% CO_2 into a sterile, gas-tight 100.0mL screw-capped bottle.

Solution D:

Composition per 10.0mL:

Na_2-acetate·$3H_2O$	0.3g

Preparation of Solution D: Add Na_2-acetate to distilled/deionized water and bring volume to 10.0mL. Sparge with N_2. Filter sterilize. Store anaerobically.

Solution E:

Composition per 10.0mL:

$Na_2S \cdot 9H_2O$	0.4g

Preparation of Solution E: Add $Na_2S \cdot 9H_2O$ to distilled/deionized water and bring volume to 10.0mL. Sparge with N_2. Autoclave for 15 min at 15 psi pressure–121°C. Cool to 25°C. Store anaerobically.

Solution F:

Composition per 10.0mL:

Na-thiosulfate	2.5g

Preparation of Solution F: Add Na-thiosulfate to distilled/deionized water and bring volume to 10.0mL. Mix thoroughly. Filter sterilize. Flush with 80% N_2 + 20% CO_2 to remove dissolved oxygen.

Preparation of Medium: Add solution B, solution C, solution D, solution E, and solution F to solution A in that order under 80% N_2 + 20% CO_2 gas. Adjust the pH to 7.2–7.5.

Use: For the cultivation of *Desulfocapsa thiozymogenes.*

Sulfur Medium

Composition per liter:

Sulfur, elemental	10.0g
KH_2PO_4	3.0g
$MgSO_4 \cdot 7H_2O$	0.5g
$(NH_4)_2SO_4$	0.3g
$CaCl_2 \cdot 2H_2O$	0.25g
$FeCl_3 \cdot 6H_2O$	0.02g

pH 4.8 ± 0.2 at 25°C

Preparation of Medium: Add components, except sulfur, to distilled/deionized water and bring volume to 1.0L. Mix thoroughly. Add 1.0g of sulfur to each of ten 250.0mL flasks. Add 100.0mL of medium to each flask. Autoclave for 30 min at 0 psi pressure–100°C on 3 consecutive days.

Use: For the isolation, cultivation, and enumeration of iron and sulfur bacteria.

Sulfurimonas paralvinella Medium (DSMZ Medium 1053)

Composition per liter:

NaCl	20.0g
Sulfur, elemental	10.0g
$MgSO_4 \cdot 7H_2O$	4.0g
$MgCl_2 \cdot 6H_2O$	3.0g
$Na_2S_2O_3 \cdot 5H_2O$	1.0g
$NaNO_3$	1.0g
$CaCl_2 \cdot 2H_2O$	0.8g
KCl	0.33g
NH_4Cl	0.25g
K_2HPO_4	0.09g
KH_2PO_4	0.07g
$Fe_2(SO_4)_3 \cdot H_2O$	0.01g
Resazurin	0.5mg
Trace elements solution	10.0mL
Bicarbonate solution	10.0mL
Vitamin solution	10.0mL
Selenite-tungstate solution	1.0mL

pH 6.8 ± 0.2 at 25°C

Trace Elements Solution:

Composition per liter:

$MgSO_4 \cdot 7H_2O$	3.0g
Nitrilotriacetic acid	1.5g
NaCl	1.0g
$MnSO_4 \cdot 2H_2O$	0.5g
$CoSO_4 \cdot 7H_2O$	0.18g
$ZnSO_4 \cdot 7H_2O$	0.18g
$CaCl_2 \cdot 2H_2O$	0.1g
$FeSO_4 \cdot 7H_2O$	0.1g
$NiCl_2 \cdot 6H_2O$	0.025g
$KAl(SO_4)_2 \cdot 12H_2O$	0.02g
H_3BO_3	0.01g

$Na_2MoO_4 \cdot 4H_2O$	0.01g
$CuSO_4 \cdot 5H_2O$	0.01g
$Na_2SeO_3 \cdot 5H_2O$	0.3mg

Preparation of Trace Elements Solution: Add nitrilotriacetic acid to 500.0mL of distilled/deionized water. Dissolve by adjusting pH to 6.5 with KOH. Add remaining components. Add distilled/deionized water to 1.0L. Mix thoroughly.

Vitamin Solution:

Composition per liter:

Pyridoxine-HCl	10.0mg
Thiamine-HCl·$2H_2O$	5.0mg
Riboflavin	5.0mg
Nicotinic acid	5.0mg
D-Ca-pantothenate	5.0mg
p-Aminobenzoic acid	5.0mg
Lipoic acid	5.0mg
Biotin	2.0mg
Folic acid	2.0mg
Vitamin B_{12}	0.1mg

Preparation of Vitamin Solution: Add components to distilled/deionized water and bring volume to 1.0L. Mix thoroughly. Sparge with 80% H_2 + 20% CO_2. Filter sterilize.

Bicarbonate Solution:

Composition per 10.0mL:

$NaHCO_3$	1.0g

Preparation of Bicarbonate Solution: Add $NaHCO_3$ to distilled/deionized water and bring volume to 10.0mL. Mix thoroughly. Sparge with 20% CO_2 + 80% N_2. Autoclave for 15 min at 15 psi pressure–121°C. Cool to room temperature.

Selenite-Tungstate Solution:

Composition per liter:

NaOH	0.5g
$Na_2WO_4 \cdot 2H_2O$	4.0mg
$Na_2SeO_3 \cdot 5H_2O$	3.0mg

Preparation of Selenite-Tungstate Solution: Add components to distilled/deionized water and bring volume to 1.0L. Mix thoroughly. Sparge with 100% N_2. Autoclave for 15 min at 15 psi pressure–121°C.

Preparation of Medium: Steam sulfur for 3 hr on each of 3 successive days. Add the sulfur to the culture vessels.Add components, except vitamin solution, sulfur, and bicarbonate solution, to distilled/deionized water and bring volume to 975.0mL. Mix thoroughly. Gently heat and bring to boiling. Boil for 3 min. Cool to room temperature under 80% H_2 + 20% CO_2. Adjust pH to 6.8 with NaOH. Dispense under same gas atmosphere into the culture vessels containing the sulfur (up to volume of 20%). Autoclave for 20 min at 6 psi pressure–110°C. Aseptically add vitamin and bicarbonate solutions. Adjust the pH to 6.5. After inoculation pressurize vessels to 2 bar overpressure with 80% H_2 + 20% CO_2 gas mixture.

Use: For the cultivation of *Sulfurimonas paralvinella.*

Sulfurospirillum Medium

Composition per 1004.0mL:

Solution A	900.0mL
Solution C	80.0mL
Solution D	20.0mL
Solution B (Trace elements solution SL-10)	2.0mL
Solution E	2.0mL

pH 7.2 ± 0.2 at 25°C

Solution A:

Composition per 900.0mL:

KH_2PO_4	1.36g
$MgSO_4 \cdot 7H_2O$	0.37g
NH_4Cl	0.27g
$CaCl_2 \cdot 2H_2O$	0.1g

Preparation of Solution A: Add components to distilled/deionized water and bring volume to 900.0mL. Mix thoroughly. Sparge with 80% N_2 + 20% CO_2. Anaerobically distribute into tubes or flasks. Autoclave for 15 min at 15 psi pressure–121°C.

Solution B (Trace Elements Solution SL-10):

Composition per liter:

$FeCl_2 \cdot 4H_2O$	1.5g
$CoCl_2 \cdot 6H_2O$	190.0mg
$MnCl_2 \cdot 4H_2O$	100.0mg
$ZnCl_2$	70.0mg
$Na_2MoO_4 \cdot 2H_2O$	36.0mg
$NiCl_2 \cdot 6H_2O$	24.0mg
H_3BO_3	6.0mg
$CuCl_2 \cdot 2H_2O$	2.0mg
HCl (25% solution)	10.0mL

Preparation of Solution B (Trace Elements Solution SL-10): Add $FeCl_2 \cdot 4H_2O$ to 10.0mL of HCl solution. Mix thoroughly. Add distilled/deionized water and bring volume to 1.0L. Add remaining components. Mix thoroughly. Sparge with 100% N_2. Autoclave for 15 min at 15 psi pressure–121°C.

Solution C:

Composition per 80.0mL:

$NaHCO_3$	4.0g

Preparation of Solution C: Add $NaHCO_3$ to distilled/deionized water and bring volume to 80.0mL. Mix thoroughly. Sparge with 80% N_2 + 20% CO_2. Autoclave for 15 min at 15 psi pressure–121°C.

Solution D:

Composition per 20.0mL:

Sodium fumarate	4.0g

Preparation of Solution D: Add sodium fumarate to distilled/deionized water and bring volume to 20.0mL. Mix thoroughly. Sparge with 100% N_2. Autoclave for 15 min at 15 psi pressure–121°C.

Solution E:

Composition per 2.0mL:

L-Cysteine·HCl	0.063g

Preparation of Solution E: Add L-cysteine·HCl to distilled/deionized water and bring volume to 2.0mL. Mix thoroughly. Sparge with 100% N_2. Autoclave for 15 min at 15 psi pressure–121°C.

Preparation of Medium: To sterile solution A in tubes or flasks, add, using a syringe, appropriate volumes of sterile solution B, solution C, solution D, and solution E. Mix thoroughly.

Use: For the cultivation of *Sulfurospirillum deleyianum.*

Sulfurospirillum Medium

Composition per liter:

Solution A	953.0mL
Solution B	42.0mL
Solution C	5.0mL

pH 7.2 ± 0.2 at 25°C

Solution A:
Composition per 953.0mL:

KH_2PO_4	1.36g
$MgSO_4 \cdot 7H_2O$	0.37g
NH_4Cl	0.27g
$CaCl_2 \cdot 2H_2O$	0.10g

Preparation of Solution A: Add components to distilled/deionized water and bring volume to 953.0mL. Mix thoroughly. Autoclave for 15 min at 15 psi pressure–121°C. Cool to room temperature while sparging with 90% N_2 + 10% CO_2.

Solution B:
Composition per 42.0mL:

Sodium fumarate	4.0g
$NaHCO_3$	2.0g
Trace elements solution SL-10	2.0mL

Trace Elements Solution SL-10:
Composition per liter:

$FeCl_2 \cdot 4H_2O$	1.5g
$CoCl_2 \cdot 6H_2O$	190.0mg
$MnCl_2 \cdot 4H_2O$	100.0mg
$ZnCl_2$	70.0mg
$Na_2MoO_4 \cdot 2H_2O$	36.0mg
$NiCl_2 \cdot 6H_2O$	24.0mg
H_3BO_3	6.0mg
$CuCl_2 \cdot 2H_2O$	2.0mg
HCl (25% solution)	10.0mL

Preparation of Trace Elements Solution SL-10: Add $FeCl_2 \cdot 4H_2O$ to 10.0mL of HCl solution. Mix thoroughly. Add distilled/deionized water and bring volume to 1.0L. Add remaining components. Mix thoroughly.

Preparation of Solution B: Add components to distilled/deionized water and bring volume to 42.0mL. Mix thoroughly. Filter sterilize.

Solution C:
Composition per 5.0mL:

L-Cysteine·HCl	0.063g

Preparation of Solution C: Add L-cysteine·HCl to distilled/deionized water and bring volume to 5.0mL. Mix thoroughly. Filter sterilize.

Preparation of Medium: Aseptically combine 953.0mL of sterile solution A with 42.0mL of sterile solution B and 5.0mL of sterile solution C. Prepare freshly. Adjust pH to 7.2 with sterile 2*M* Na_2CO_3 solution or sterile 2*N* HCl solution.

Use: For the cultivation of *Sulfurospirillum deleyianum.*

Sulfurospirillum II Medium (DSMZ Medium 771)

Composition per 1080.0mL:

Yeast extract	1.0g
NaCl	460.0mg
K_2HPO_4	225.0mg
KH_2PO_4	225.0mg
$(NH_4)_2SO_4$	225.0mg
$MgSO_4 \cdot 7H_2O$	117.0mg
Resazurin	0.5mg
$NaHCO_3$ solution	30.0mL
$NaNO_3$ solution	10.0mL
$Na_2S \cdot 9H_2O$ solution	10.0mL
Na-lactate solution	10.0mL
Vitamin solution	10.0mL
Cysteine solution	10.0mL
Trace elements solution SL-10	1.0mL
Selenite-tungstate solution	1.0mL

pH 7.3 ± 0.2 at 25°C

Cysteine Solution:
Composition per 10.0mL:

L-Cysteine-HCl·H_2O	0.15g

Preparation of Cysteine Solution: Add L-cysteine·HCl·H_2O to distilled/deionized water and bring volume to 10.0mL. Mix thoroughly. Sparge with 100% N_2. Autoclave for 15 min at 15 psi pressure–121°C.

Selenite-Tungstate Solution
Composition per liter:

NaOH	0.5g
$Na_2WO_4 \cdot 2H_2O$	4.0mg
$Na_2SeO_3 \cdot 5H_2O$	3.0mg

Preparation of Selenite-Tungstate Solution: Add components to distilled/deionized water and bring volume to 1.0L. Mix thoroughly. Sparge with 100% N_2. Filter sterilize.

Na-lactate Solution:
Composition per 10.0mL:

Na-lactate	2.25g

Preparation of Na-lactate Solution: Add Na-lactate to distilled/deionized water and bring volume to 10.0mL. Mix thoroughly. Sparge with 100% N_2. Filter sterilize.

$NaNO_3$ Solution:
Composition per 10.0mL:

$NaNO_3$	1.7g

Preparation of $NaNO_3$ Solution: Add $NaNO_3$ to distilled/deionized water and bring volume to 10.0mL. Mix thoroughly. Sparge with 80% N_2 + 20% CO_2. Filter sterilize.

$Na_2S \cdot 9H_2O$ Solution:
Composition per 10.0mL:

$Na_2S \cdot 9H_2O$	0.1g

Preparation of $Na_2S \cdot 9H_2O$ Solution: Add $Na_2S \cdot 9H_2O$ to distilled/deionized water and bring volume to 10.0mL. Mix thoroughly. Autoclave under 100% N_2 for 15 min at 15 psi pressure–121°C. Cool to room temperature.

$NaHCO_3$ Solution:
Composition per 30.0mL:

$NaHCO_3$	4.2g

Preparation of $NaHCO_3$ Solution: Add $NaHCO_3$ to distilled/deionized water and bring volume to 30.0mL. Mix thoroughly. Sparge with 80% N_2 + 20% CO_2. Filter sterilize.

Trace Elements Solution SL-10:
Composition per liter:

$FeCl_2 \cdot 4H_2O$	1.5g
$CoCl_2 \cdot 6H_2O$	190.0mg
$MnCl_2 \cdot 4H_2O$	100.0mg
$ZnCl_2$	70.0mg
$Na_2MoO_4 \cdot 2H_2O$	36.0mg
$NiCl_2 \cdot 6H_2O$	24.0mg
H_3BO_3	6.0mg
$CuCl_2 \cdot 2H_2O$	2.0mg
HCl (25% solution)	10.0mL

Preparation of Trace Elements Solution SL-10: Add $FeCl_2·4H_2O$ to 10.0mL of HCl solution. Mix thoroughly. Add distilled/deionized water and bring volume to 1.0L. Add remaining components. Mix thoroughly. Sparge with 80% N_2 + 20% CO_2.

Vitamin Solution:

Composition per liter:

Pyridoxine-HCl	10.0mg
Thiamine-HCl·$2H_2O$	5.0mg
Riboflavin	5.0mg
Nicotinic acid	5.0mg
D-Ca-pantothenate	5.0mg
p-Aminobenzoic acid	5.0mg
Lipoic acid	5.0mg
Biotin	2.0mg
Folic acid	2.0mg
Vitamin B_{12}	0.1mg

Preparation of Vitamin Solution: Add components to distilled/deionized water and bring volume to 1.0L. Mix thoroughly. Sparge with 80% H_2 + 20% CO_2. Filter sterilize.

Preparation of Medium: Prepare and dispense medium under 80% N_2 + 20% CO_2. Add components, except $NaHCO_3$ solution, $Na_2S·9H_2O$ solution, cysteine solution, $NaNO_3$ solution, Na-lactate solution, and vitamin solution, to distilled/deionized water and bring volume to 1.0mL. Mix thoroughly. Sparge with 80% N_2 + 20% CO_2. Adjust pH to 7.3. Dispense either 10.0mL aliquots into 15mL Hungate tubes or 50.0mL aliquots into 100mL Hungate bottles. Autoclave for 15 min at 15 psi pressure–121°C. Aseptically and anaerobically inject from sterile stock solutions $NaHCO_3$ solution, $Na_2S·9H_2O$ solution, cysteine solution, vitamin solution, sodium nitrate solution, and sodium lactate solution. Final pH should be 7.3.

Use: For the cultivation of *Sulfurospirillum arsenophilum (Geospirillum* sp.*)* and *Sulfurospirillum barnesii.*

Sulfurospirillum MV Medium (DSMZ Medium 1097)

Composition per liter:

NaCl	20.0g
$MgCl_2·6H_2O$	3.0g
$CaCl_2·2H_2O$	0.8g
KCl	0.7g
NH_4Cl	0.2g
KH_2PO_4	0.2g
Resazurin	0.5mg
Vitamin solution	10.0mL
Sulfide solution	10.0mL
Fumarate solution	10.0mL
Bicarbonate solution	10.0mL
Trace elements solution SL-10	2.0mL
Selenite-tungstate solution	1.0mL

pH 7.2 ± 0.2 at 25°C

Bicarbonate Solution:

Composition per 10.0mL:

$NaHCO_3$	2.5g

Preparation of Bicarbonate Solution: Add $NaHCO_3$ to distilled/deionized water and bring volume to 10.0mL. Mix thoroughly. Sparge with 20% CO_2 + 80% N_2. Autoclave for 15 min at 15 psi pressure–121°C. Cool to room temperature.

Vitamin Solution:

Composition per liter:

Pyridoxine-HCl	10.0mg
Thiamine-HCl·$2H_2O$	5.0mg
Riboflavin	5.0mg
Nicotinic acid	5.0mg
D-Ca-pantothenate	5.0mg
p-Aminobenzoic acid	5.0mg
Lipoic acid	5.0mg
Biotin	2.0mg
Folic acid	2.0mg
Vitamin B_{12}	0.1mg

Preparation of Vitamin Solution: Add components to distilled/deionized water and bring volume to 1.0L. Mix thoroughly. Sparge with 100% N_2. Filter sterilize.

Fumarate Solution:

Composition per 10.0mL:

Na_2-fumarate	1.6g

Preparation of Fumarate Solution: Add Na_2-fumarate to distilled/deionized water and bring volume to 10.0mL. Mix thoroughly. Sparge with 100% N_2. Autoclave for 15 min at 15 psi pressure–121°C. Cool to room temperature.

Trace Elements Solution SL-10:

Composition per liter:

$MgSO_4·7H_2O$	3.0g
Nitrilotriacetic acid	1.5g
NaCl	1.0g
$MnSO_4·2H_2O$	0.5g
$CoSO_4·7H_2O$	0.18g
$ZnSO_4·7H_2O$	0.18g
$CaCl_2·2H_2O$	0.1g
$FeSO_4·7H_2O$	0.1g
$NiCl_2·6H_2O$	0.025g
$KAl(SO_4)_2·12H_2O$	0.02g
H_3BO_3	0.01g
$Na_2MoO_4·4H_2O$	0.01g
$CuSO_4·5H_2O$	0.01g
$Na_2SeO_3·5H_2O$	0.3mg

Preparation of Trace Elements Solution SL-10: Add nitrilotriacetic acid to 500.0mL of distilled/deionized water. Dissolve by adjusting pH to 6.5 with KOH. Add remaining components. Add distilled/deionized water to 1.0L. Mix thoroughly. Adjust pH to 7.0.

Selenite-Tungstate Solution:

Composition per liter:

NaOH	0.5g
$Na_2WO_4·2H_2O$	4.0mg
$Na_2SeO_3·5H_2O$	3.0mg

Preparation of Selenite-Tungstate Solution: Add components to distilled/deionized water and bring volume to 1.0L. Mix thoroughly. Sparge with 100% N_2. Autoclave for 15 min at 15 psi pressure–121°C.

Sulfide Solution:

Composition per 10.0mL:

$Na_2S·9H_2O$	0.3g

Preparation of Sulfide Solution: Add $Na_2S·9H_2O$ to distilled/deionized water and bring volume to 10.0mL. Mix thoroughly. Autoclave under 100% N_2 for 15 min at 15 psi pressure–121°C. Cool to room temperature.

Preparation of Medium: Add components, except vitamin solution, bicarbonate solution, fumarate solution, and sulfide solution, to distilled/deionized water and bring volume to 960.0mL. Mix thoroughly. Gently heat and bring to boiling. Boil for 1 min. Cool to room temperature under 80% N_2 + 20% CO_2. Dispense under same gas atmosphere into culture vessels. Autoclave for 15 min at 15 psi pressure–121°C. Aseptically add fumarate, sulfide, vitamin, and bicarbonate solutions. Adjust the pH to 7.2.

Use: For the cultivation of *Sulfurospirillum* spp.

Super HiVeg Broth

Composition per liter:

Plant hydrolysate	35.0g
Yeast extract	20.0g
NaCl	5.0g

pH 7.4 ± 0.2 at 25°C

Source: This medium is available as a premixed powder from HiMedia.

Preparation of Medium: Add components to distilled/deionized water and bring volume to 1.0L. Mix thoroughly. Distribute into tubes or flasks. Autoclave for 15 min at 15 psi pressure–121°C.

Use: For the cultivation of *Escherichia coli*.

Super MMB Medium (LMG Medium 188)

Yeast extract	1.0g
KH_2PO_4	1.0g
Peptone	0.4g
Sodium succinate	0.4g
NH_4Cl	0.2g
NaCl	0.2g
$MgSO_4·7H_2O$	0.2g
$CaCl_2·2H_2O$	10.0mg
Ferric citrate	5.0mg
Vitamin solution	20.0mL
Trace elements solution SL-6	1.0mL

pH 7.0 ± 0.2 at 25°C

Trace Elements Solution SL-6 :
Composition per liter:

H_3BO_3	0.3g
$CoCl_2·6H_2O$	0.2g
$ZnSO_4·7H_2O$	0.1g
$MnCl_2·4H_2O$	0.03g
$Na_2MoO_4·H_2O$	0.03g
$NiCl_2·6H_2O$	0.02g
$CuCl_2·2H_2O$	0.01g

Preparation ofTrace Elements Solution SL-6 : Add components to distilled/deionized water and bring volume to 1.0L. Mix thoroughly. Adjust pH to 3.4.

Vitamin Solution:
Composition per liter:

Calcium DL-pantothenate	5.0mg
Riboflavin	5.0mg
Thiamine·HCl	5.0mg
Biotin	2.0mg
Folic acid	2.0mg
Vitamin B_{12}	0.1mg

Preparation of Vitamin Solution: Add components to distilled/deionized water and bring volume to 1.0L. Mix thoroughly. Filter sterilize.

Preparation of Medium: Add components, except vitamin solution, to 980.0mL distilled/deionized water. Mix thoroughly. Autoclave for 15 min at 15 psi pressure–121°C. Cool to 25°C. Aseptically add 20.0mL sterile vitamin solution. Mix thoroughly. Aseptically distribute to sterile tubes or flasks.

Use: For the cultivation of *Aquabacter* spp.

Superbroth

Composition per liter:

Pancreatic digest of casein	32.0g
Yeast extract	20.0g
NaCl	5.0g
NaOH (1*N* solution)	5.0mL

Preparation of Medium: Add components to distilled/deionized water and bring volume to 1.0L. Mix thoroughly. Distribute into tubes or flasks. Autoclave for 15 min at 15 psi pressure–121°C.

Use: For the cultivation of *Escherichia coli*.

Supplemented (Arginine) M9 Medium
See: **M9 Medium with Arginine**

Supplemented *Aspergillus* Minimal Agar

Composition per liter:

Agar	20.0g
Glucose	20.0g
$NaNO_3$	6.0g
$(NH_4)_2SO_4$	5.0g
Casamino acids	2.0g
K_2HPO_4	1.52g
KH_2PO_4	1.0g
KCl	0.52g
$MgSO_4·7H_2O$	0.52g
NaCl	0.1g
$CaCl_2·2H_2O$	0.1g
Inositol	2.0mg
KI	1.0mg
H_3BO_3	0.5mg
$ZnSO_4·7H_2O$	0.4mg
$MnSO_4·4H_2O$	0.4mg
Thiamine·HCl	0.4mg
Pyroxidine·HCl	0.4mg
Niacin	0.4mg
Calcium pantothenate	0.4mg
p-Aminobenzoic acid	0.2mg
Riboflavin	0.2mg
$Na_2MoO_4·4H_2O$	0.2mg
$CuSO_4·5H_2O$	0.04mg
Folic acid	2.0μg
Biotin	2.0μg
$FeCl_3·6H_2O$	1.0μg
$ZnSO_4·7H_2O$	1.0μg

pH 6.5 ± 0.2 at 25°C

Preparation of Medium: Add components to distilled/deionized water and bring volume to 1.0L. Mix thoroughly. Adjust pH to 6.5. Gently heat and bring to boiling. Distribute into tubes or flasks. Autoclave for 15 min at 15 psi pressure–121°C. Pour into sterile Petri dishes or leave in tubes.

Use: For the cultivation and maintenance of *Aspergillus* species.

Supplemented *Aspergillus* Minimal Broth

Composition per liter:

Glucose	20.0g
$NaNO_3$	6.0g
$(NH_4)_2SO_4$	5.0g
Casamino acids	2.0g
K_2HPO_4	1.52g
KH_2PO_4	1.0g
KCl	0.52g
$MgSO_4 \cdot 7H_2O$	0.52g
NaCl	0.1g
$CaCl_2 \cdot 2H_2O$	0.1g
Inositol	2.0mg
KI	1.0mg
H_3BO_3	0.5mg
$ZnSO_4 \cdot 7H_2O$	0.4mg
$MnSO_4 \cdot 4H_2O$	0.4mg
Thiamine·HCl	0.4mg
Pyroxidine·HCl	0.4mg
Niacin	0.4mg
Calcium pantothenate	0.4mg
p-Aminobenzoic acid	0.2mg
Riboflavin	0.2mg
$Na_2MoO_4 \cdot 4H_2O$	0.2mg
$CuSO_4 \cdot 5H_2O$	0.04mg
Folic acid	2.0μg
Biotin	2.0μg
$FeCl_3 \cdot 6H_2O$	1.0μg
$ZnSO_4 \cdot 7H_2O$	1.0μg

pH 6.5 ± 0.2 at 25°C

Preparation of Medium: Add components to distilled/deionized water and bring volume to 1.0L. Mix thoroughly. Adjust pH to 6.5. Distribute into tubes or flasks. Autoclave for 15 min at 15 psi pressure–121°C.

Use: For the cultivation of *Aspergillus* species.

Supplemented (Tryptophan) M9 Medium

See: **M9 Medium with Tryptophan**

SW 2 Agar

Composition per liter:

Agar	15.0g
NH_4Cl	1.0g
Sodium acetate	0.02g
Artificial seawater	1.0L

Artificial Seawater:

Composition per liter:

NaCl	24.7g
$MgSO_4 \cdot 7H_2O$	6.3g
$MgCl_2 \cdot 6H_2O$	4.6g
$CaCl_2$	1.0g
KCl	0.7g
$NaHCO_3$	0.2g

Preparation of Artificial Seawater: Add components to distilled/deionized water and bring volume to 1.0L. Mix thoroughly.

Preparation of Medium: Add solid components to 1.0L of artificial seawater. Mix thoroughly. Gently heat and bring to boiling. Distribute into tubes or flasks. Autoclave for 15 min at 15 psi pressure–121°C. Pour into sterile Petri dishes or leave in tubes.

Use: For the isolation and cultivation of *Cytophaga* species, *Herpetosiphon* species, *Saprospira* species, and *Flexithrix* species.

SWA

See: **Seawater Agar**

Swampy Medium

Composition per liter:

Agar	10.0g
$CaCO_3$	10.0g
Peptone	0.5g
Yeast extract	0.5g

Preparation of Medium: Add components to seawater and bring volume to 1.0L. Mix thoroughly. Gently heat and bring to boiling. Distribute into tubes or flasks. Autoclave for 15 min at 15 psi pressure–121°C. Pour into sterile Petri dishes or leave in tubes.

Use: For the cultivation and maintenance of *Vibrio liquefaciens*.

Sweet E Broth for Anaerobes

Composition per 100.0mL:

Gelatin	0.3g
Cellobiose	0.1g
Fructose	0.1g
Glucose	0.1g
L-Arabinose	0.1g
Maltose	0.1g
Starch	0.1g
Agar	0.075g
Peptone	0.05g
L-Cysteine·HCl·H_2O	0.05g
$(NH_4)_2SO_4$	0.05g
Yeast extract	0.05g
Salts solution	50.0mL
Rumen fluid	30.0mL
Resazurin solution	0.4mL
Pyruvic acid	0.01mL

pH 6.5 ± 0.2 at 25°C

Salts Solution:

Composition per liter:

$NaHCO_3$	10.0g
NaCl	2.0g
K_2HPO_4	1.0g
KH_2PO_4	1.0g
$CaCl_2$, anhydrous	0.2g
$MgSO_4 \cdot 7H_2O$	0.2g

Preparation of Salts Solution: Add $CaCl_2$ and $MgSO_4 \cdot 7H_2O$ to distilled/deionized water and bring volume to 300.0mL. Mix thoroughly. Bring volume to 800.0mL with distilled/deionized water. Add remaining components while stirring. Bring volume to 1.0L with distilled/deionized water. Mix thoroughly. Store at 4°C.

Resazurin Solution:

Composition per 44.0mL:

Resazurin	0.011g

Preparation of Resazurin Solution: Add resazurin to distilled/deionized water and bring volume to 44.0mL. Mix thoroughly.

Preparation of Medium: Add components to distilled/deionized water and bring volume to 100.0mL. Mix thoroughly. Gently heat and bring to boiling under O_2-free 97% N_2 + 3% H_2. Adjust the pH to 6.5 if necessary. Continue boiling until the medium turns yellow. Distribute into tubes or flasks under O_2-free 97% N_2 + 3% H_2. Cap tubes with rubber stoppers. Place tubes in a press. Autoclave for 15 min at 15 psi pressure–121°C with fast exhaust.

Use: For the cultivation and maintenance of *Clostridium cocleatum* and *Clostridium spiroforme*.

SWM Medium

Composition per 1014.0mL:

Solution A	940.0mL
Solution E ($NaHCO_3$ solution)	50.0mL
Solution F (Substrate solution)	10.0mL
Solution G ($Na_2S{\cdot}9H_2O$ solution)	10.0mL
Solution B (Trace elements solution SL-10)	2.0mL
Solution C (Seven vitamin solution)	1.0mL
Solution D (Selenite-tungstate solution)	1.0mL

pH 7.2–7.4 at 25°C

Solution A:
Composition per 940.0mL:

NaCl	1.0g
KCl	0.5g
$MgCl_2{\cdot}6H_2O$	0.4g
NH_4Cl	0.25g
KH_2PO_4	0.2g
$CaCl_2{\cdot}2H_2O$	0.15g
Resazurin	0.5mg

Preparation of Solution A: Prepare and dispense under 80% N_2 + 20% CO_2. Add components to distilled/deionized water and bring volume to 940.0mL. Mix thoroughly. Autoclave for 15 min at 15 psi pressure–121°C.

Solution B (Trace Elements Solution SL-10):
Composition per liter:

$FeCl_2{\cdot}4H_2O$	1.5g
$CoCl_2{\cdot}6H_2O$	190.0mg
$MnCl_2{\cdot}4H_2O$	100.0mg
$ZnCl_2$	70.0mg
$Na_2MoO_4{\cdot}2H_2O$	36.0mg
$NiCl_2{\cdot}6H_2O$	24.0mg
H_3BO_3	6.0mg
$CuCl_2{\cdot}2H_2O$	2.0mg
HCl (25% solution)	10.0mL

Preparation of Solution B (Trace Elements Solution SL-10): Prepare and dispense under 100% N_2. Add $FeCl_2{\cdot}4H_2O$ to 10.0mL of HCl solution. Mix thoroughly. Add distilled/deionized water and bring volume to 1.0L. Add remaining components. Mix thoroughly. Autoclave for 15 min at 15 psi pressure–121°C.

Solution C (Seven Vitamin Solution):
Composition per liter:

Pyridoxine·HCl	300.0mg
Nicotinic acid	200.0mg
Thiamine·HCl	200.0mg
Calcium pantothenate	100.0mg
Cyanocobalamine	100.0mg
p-Aminobenzoic acid	80.0mg
D(+)-Biotin	20.0mg

Preparation of Solution C (Seven Vitamin Solution): Add components to distilled/deionized water and bring volume to 1.0L. Mix thoroughly. Filter sterilize. Sparge with 100% N_2.

Solution D (Selenite-Tungstate Solution):
Composition per liter:

NaOH	0.5g
$Na_2WO_4{\cdot}2H_2O$	4.0mg
$Na_2SeO_3{\cdot}5H_2O$	3.0mg

Preparation of Solution D (Selenite-Tungstate Solution): Add components to distilled/deionized water and bring volume to 1.0L. Mix thoroughly. Filter sterilize. Sparge with 100% N_2.

Solution E ($NaHCO_3$ Solution):
Composition per 50.0mL:

$NaHCO_3$	2.5g

Preparation of Solution E ($NaHCO_3$ Solution): Add $NaHCO_3$ to distilled/deionized water and bring volume to 50.0mL. Mix thoroughly. Gas under 100% N_2. Autoclave for 15 min at 15 psi pressure–121°C.

Solution F (Substrate Solution):
Composition per 10.0mL:

Pyrogallol	0.5g

Preparation of Solution F (Substrate Solution): Add pyrogallol to distilled/deionized water and bring volume to 10.0mL. Mix thoroughly. Gas under 100% N_2. Autoclave for 15 min at 15 psi pressure–121°C.

Solution G ($Na_2S{\cdot}9H_2O$ Solution):
Composition per 10.0mL:

$Na_2S{\cdot}9H_2O$	0.3g

Preparation of Solution G ($Na_2S{\cdot}9H_2O$ Solution): Add $Na_2S{\cdot}9H_2O$ to distilled/deionized water and bring volume to 10.0mL. Mix thoroughly. Gas under 100% N_2. Autoclave for 15 min at 15 psi pressure–121°C.

Preparation of Medium: To 940.0mL of sterile solution A, aseptically and anaerobically add 1.0mL of sterile solution B, 1.0mL of sterile solution C, 1.0mL of sterile solution D, 50.0mL of sterile solution E, 10.0mL of sterile solution F, and 10.0mL of sterile solution G. Mix thoroughly. Aseptically and anaerobically distribute into sterile tubes or flasks.

Use: For the cultivation and maintenance of *Pelobacter massiliensis*.

SWMTY Marine Medium

Composition per liter:

Marine salts mix	38.0g
Agar	15.0g
Pancreatic digest of casein	2.0g
Yeast extract	2.0g
Tris(hydroxymethyl)aminomethane buffer	1.0g
KNO_3	0.5g
Sodium glycerophosphate	0.1g
Trace elements solution HO-LE	1.0mL

pH 7.0 ± 0.2 at 25°C

Trace Elements Solution HO-LE:
Composition per liter:

H_3BO_3	2.85g
$MnCl_2{\cdot}4H_2O$	1.8g

Sodium tartrate 1.77g
$FeSO_4 \cdot 7H_2O$ 1.36g
$CoCl_2 \cdot 6H_2O$ 0.04g
$CuCl_2.2H_2O$ 0.027g
$Na_2MoO_4 \cdot 2H_2O$ 0.025g
$ZnCl_2$ 0.02g

Preparation of Trace Elements Solution HO-LE: Add components to distilled/deionized water and bring volume to 1.0L. Mix thoroughly. Filter sterilize.

Preparation of Medium: Add components to distilled/deionized water and bring volume to 1.0L. Mix thoroughly. Gently heat and bring to boiling. Distribute into tubes or flasks. Autoclave for 15 min at 15 psi pressure–121°C. Pour into sterile Petri dishes or leave in tubes.

Use: For the cultivation and maintenance of a variety of heterotrophic marine bacterial species.

SXT Blood Agar

Composition per liter:
Pancreatic digest of casein 14.5g
Agar 14.0g
NaCl 5.0g
Papaic digest of soybean meal 5.0g
Growth factor, BBL 1.5g
Sulfamethoxazole 0.024g
Trimethoprim 1.25mg
Sheep blood, defibrinated 50.0mL
pH 7.3 ± 0.2 at 25°C

Source: This medium is available as a premixed powder from BD Diagnostic Systems.

Preparation of Medium: Add components, except defibrinated sheep blood, to distilled/deionized water and bring volume to 950.0mL. Mix thoroughly. Gently heat and bring to boiling. Autoclave for 15 min at 15 psi pressure–121°C. Cool to 45°–50°C. Aseptically add 50.0mL of defibrinated sheep blood. Mix thoroughly. Pour into sterile Petri dishes or distribute into sterile tubes.

Use: For the selective isolation of Lancefield group A and group B streptococci from throat cultures and other clinical specimens.

SY Broth

Composition per liter:
$(NH_4)_2SO_4$ 2.0g
$Na_2HPO_4 \cdot 2H_2O$ 1.4g
KH_2PO_4 .... 0.7g
$MgSO_4 \cdot 7H_2O$ 0.2g
$FeSO_4$ 5.0mg
$MnSO_4$ 5.0mg
Glucose solution 100.0mL

Glucose Solution:
Composition per 100.0mL:
D-Glucose 10.0g

Preparation of Glucose Solution: Add D-glucose to distilled/deionized water and bring volume to 100.0mL. Mix thoroughly. Autoclave for 15 min at 15 psi pressure–121°C. Cool to 25°C.

Preparation of Medium: Add components, except glucose solution, to distilled/deionized water and bring volume to 900.0mL. Mix thoroughly. Gently heat and bring to boiling. Autoclave for 15 min at 15 psi pressure–121°C. Cool to 45°–50°C. Aseptically add sterile glucose solution. Mix thoroughly. Aseptically distribute into sterile tubes or flasks.

Use: For the isolation and cultivation of *Cytophaga* species, *Herpetosiphon* species, *Saprospira* species, and *Flexithrix* species.

SYA Medium

Composition per liter:
Agar 20.0g
Soluble starch 10.0g
Yeast extract 2.0g

Preparation of Medium: Add components to tap water and bring volume to 1.0L. Mix thoroughly. Gently heat and bring to boiling. Distribute into tubes or flasks. Autoclave for 15 min at 15 psi pressure–121°C. Pour into sterile Petri dishes or leave in tubes.

Use: For the cultivation and maintenance of *Streptomyces chartreusis*.

SYC Medium

Composition per liter:
Sucrose 10.0g
Pancreatic digest of casein 8.0g
Yeast extract 4.0g
K_2HPO_4 .... 3.0g
$MgSO_4 \cdot 7H_2O$ 0.3g
pH 7.0 ± 0.2 at 25°C

Preparation of Medium: Add components to distilled/deionized water and bring volume to 1.0L. Mix thoroughly. Distribute into tubes or flasks. Autoclave for 15 min at 15 psi pressure–121°C.

Use: For the cultivation and maintenance of *Agrobacterium tumefaciens*.

SYFAC Medium (DSMZ Medium 1041)

Composition per liter:
Sea salts, Sigma 35.0g
Yeast extract 1.0g
Resazurin .... 0.5 mg
Vitamin solution 10.0mL
Fumarate solution 10.0mL
Bicarbonate solution 10.0mL
Acetate solution 10.0mL
Sulfide solution 10.0mL
Wolfe's mineral elixir 1.0mL
pH 7.1 ± 0.1 at 25°C

Sulfide Solution:
Composition per 10.0mL:
$Na_2S \cdot 9H_2O$ 0.3g

Preparation of Sulfide Solution: Add $Na_2S \cdot 9H_2O$ to distilled/deionized water and bring volume to 10.0mL. Mix thoroughly. Autoclave under 100% N_2 for 15 min at 15 psi pressure–121°C. Cool to room temperature.

Bicarbonate Solution:
Composition per 10.0mL:
$NaHCO_3$ 2.0g

Preparation of Bicarbonate Solution: Add $NaHCO_3$ to distilled/deionized water and bring volume to 10.0mL. Mix thoroughly. Sparge with 20% CO_2 + 80% N_2. Autoclave for 15 min at 15 psi pressure–121°C. Cool to room temperature.

Acetate Solution:

Composition per 10.0mL:

Na-acetate 1.6g

Preparation of Acetate Solution: Add Na-acetate to distilled/deionized water and bring volume to 10.0mL. Mix thoroughly. Sparge with 100% N_2. Autoclave for 15 min at 15 psi pressure–121°C. Cool to room temperature.

Fumarate Solution:

Composition per 10.0mL:

Na_2-fumarate 3.2g

Preparation of Fumarate Solution: Add Na_2-fumarate to distilled/deionized water and bring volume to 10.0mL. Mix thoroughly. Sparge with 100% N_2. Autoclave for 15 min at 15 psi pressure–121°C. Cool to room temperature.

Wolfe's Mineral Elixir:

Composition per liter:

$MgSO_4 \cdot 7H_2O$ 30.0g
NaCl 10.0g
$MnSO_4 \cdot 2H_2O$ 5.0g
$(NH_4)_2NiSO_4 \cdot 6H_2O$ 2.8g
$CoCl_2 \cdot 6H_2O$ 1.8g
$ZnSO_4 \cdot 7H_2O$ 1.8g
$FeSO_4 \cdot 7H_2O$ 1.0g
$CaCl_2 \cdot 2H_2O$ 1.0g
$KAl(SO_4)_2 \cdot 12H_2O$ 0.18g
$CuSO_4 \cdot 5H_2O$ 0.1g
H_3BO_3 .... 0.1g
$Na_2MoO_4 \cdot 2H_2O$ 0.1g
Na_2SeO_4 .... 0.1g
$Na_2WO_4 \cdot 2H_2O$ 0.1g

Preparation of Wolfe's Mineral Elixir: Adjust pH of 1.0L of distilled/deionized water to 1.0 with dilute H_2SO_4. Add remaining components one at a time. Mix thoroughly to dissolve.

Vitamin Solution:

Composition per liter:

Pyridoxine-HCl 10.0mg
Thiamine-HCl·$2H_2O$ 5.0mg
Riboflavin 5.0mg
Nicotinic acid 5.0mg
D-Ca-pantothenate 5.0mg
p-Aminobenzoic acid 5.0mg
Lipoic acid 5.0mg
Biotin 2.0mg
Folic acid 2.0mg
Vitamin B_{12} 0.1mg

Preparation of Vitamin Solution: Add components to distilled/deionized water and bring volume to 1.0L. Mix thoroughly. Sparge with 80% H_2 + 20% CO_2. Filter sterilize.

Preparation of Medium: Add components, except sulfide, acetate, fumarate, bicarbonate, and vitamin solutions, to distilled/deionized water and bring volume to 950.0mL. Mix thoroughly. Sparge with 80% N_2 + 20% CO_2. Filter sterilize. Gently heat and bring to boiling. Boil for 1 min. Cool to room temperature while sparging with 80% N_2 + 20% CO_2 gas mixture. Dispense under same gas atmosphere in culture vessels. Autoclave for 15 min at 15 psi pressure–121°C. Cool to room temperature. Aspetically add acetate, bicarbonate, fumarate, vitamin, and sulfide solutions from sterile, anoxic solutions. Mix thoroughly. Adjust the pH to 7.0–7.2.

Use: For the cultivation of *Desulfuromusa ferrireducens.*

SYLC Medium (DSMZ Medium 1040)

Composition per liter:

Sea salts, Sigma 35.0g
Yeast extract 1.0g
Resazurin .... 0.5 mg
Vitamin solution 10.0mL
Bicarbonate solution 10.0mL
Lactate solution 10.0mL
Sulfide solution 10.0mL
Wolfe's mineral elixir 1.0mL

pH 7.1 ± 0.1 at 25°C

Sulfide Solution:

Composition per 10.0mL:

$Na_2S \cdot 9H_2O$ 0.3g

Preparation of Sulfide Solution: Add $Na_2S \cdot 9H_2O$ to distilled/deionized water and bring volume to 10.0mL. Mix thoroughly. Autoclave under 100% N_2 for 15 min at 15 psi pressure–121°C. Cool to room temperature.

Bicarbonate Solution:

Composition per 10.0mL:

$NaHCO_3$ 2.0g

Preparation of Bicarbonate Solution: Add $NaHCO_3$ to distilled/deionized water and bring volume to 10.0mL. Mix thoroughly. Sparge with 20% CO_2 + 80% N_2. Autoclave for 15 min at 15 psi pressure–121°C. Cool to room temperature.

Lactate Solution:

Composition per 10.0mL:

Na-lactate 2.5g

Preparation of Lactate Solution: Add Na-lactate to distilled/deionized water and bring volume to 10.0mL. Mix thoroughly. Sparge with 100% N_2. Autoclave for 15 min at 15 psi pressure–121°C. Cool to room temperature.

Wolfe's Mineral Elixir:

Composition per liter:

$MgSO_4 \cdot 7H_2O$ 30.0g
NaCl 10.0g
$MnSO_4 \cdot 2H_2O$ 5.0g
$(NH_4)_2NiSO_4 \cdot 6H_2O$ 2.8g
$CoCl_2 \cdot 6H_2O$ 1.8g
$ZnSO_4 \cdot 7H_2O$ 1.8g
$FeSO_4 \cdot 7H_2O$ 1.0g
$CaCl_2 \cdot 2H_2O$ 1.0g
$KAl(SO_4)_2 \cdot 12H_2O$ 0.18g
$CuSO_4 \cdot 5H_2O$ 0.1g
H_3BO_3 .... 0.1g
$Na_2MoO_4 \cdot 2H_2O$ 0.1g
Na_2SeO_4 .... 0.1g
$Na_2WO_4 \cdot 2H_2O$ 0.1g

Preparation of Wolfe's Mineral Elixir: Adjust pH of 1.0L of distilled/deionized water to 1.0 with dilute H_2SO_4. Add remaining components one at a time. Mix throughly to dissolve.

Vitamin Solution:

Composition per liter:

Pyridoxine-HCl 10.0mg
Thiamine-HCl·$2H_2O$ 5.0mg

Riboflavin	5.0mg
Nicotinic acid	5.0mg
D-Ca-pantothenate	5.0mg
p-Aminobenzoic acid	5.0mg
Lipoic acid	5.0mg
Biotin	2.0mg
Folic acid	2.0mg
Vitamin B_{12}	0.1mg

Preparation of Vitamin Solution: Add components to distilled/deionized water and bring volume to 1.0L. Mix thoroughly. Sparge with 80% H_2 + 20% CO_2. Filter sterilize.

Preparation of Medium: Add components, except sulfide, lactate, bicarbonate, and vitamin solutions, to distilled/deionized water and bring volume to 960.0mL. Mix thoroughly. Sparge with 80% N_2 + 20% CO_2. Gently heat and bring to boiling. Boil for 1 min. Cool to room temperature while sparging with 80% N_2 + 20% CO_2 gas mixture. Dispense under same gas atmosphere in culture vessels. Autoclave for 15 min at 15 psi pressure–121°C. Cool to room temperature. Aspetically add lactate, bicarbonate, vitamin, and sulfide solutions from sterile, anoxic solutions. Mix thoroughly. Adjust the pH to 7.0–7.2.

Use: For the cultivation of *Desulfovibrio ferrireducens.*

Syncase Broth

Composition per liter:

Casamino acids	20.0g
K_2HPO_4	8.71g
Yeast extract	6.0g
NaCl	2.5g

pH 8.5 ± 0.2 at 25°C

Preparation of Medium: Add components to distilled/deionized water and bring volume to 1.0L. Mix thoroughly. Adjust pH to 8.5. Distribute into tubes or flasks. Autoclave for 15 min at 15 psi pressure–121°C.

Use: For the cultivation of heat-labile, toxin-producing *Escherichia coli* from foods.

Synthetic Broth, AOAC (Synthetic Broth, Association of Official Analytical Chemists)

Composition per liter:

Na_2HPO_4	4.0g
NaCl	3.0g
K_2HPO_4	1.5g
L-Glutamic acid	1.3g
DL-Valine	1.0g
L-Lysine	0.85g
L-Leucine	0.8g
DL-Serine	0.61g
DL-Threonine	0.5g
L-Aspartic acid	0.45g
DL-Isoleucine	0.44g
DL-Alanine	0.43g
L-Arginine	0.4g
DL-Methionine	0.37g
DL-Histidine	0.3g
DL-Phenylalanine	0.26g
L-Tyrosine	0.21g
KCl	0.2g
Aminoacetic acid	0.06g
L-Cystine	0.05g
$MgSO_4$	0.05g
L-Proline	0.05g
DL-Tryptophan	0.05g
Nicotinamide	0.01g
Thiamine·HCl	0.01g

pH 7.1 ± 0.1 at 25°C

Source: This medium is available as a premixed powder from BD Diagnostic Systems.

Preparation of Medium: Add components to distilled/deionized water and bring volume to 1.0L. Mix thoroughly. Gently heat and bring to boiling. Distribute into tubes or flasks. Autoclave for 20 min at 15 psi pressure–121°C.

Use: For the determination of phenol coefficients of disinfectants.

Synthetic Complete Medium

Composition per liter:

Solution A	600.0mL
Solution B	400.0mL

Solution A:

Composition per 600.0mL:

Agar	20.0g

Preparation of Solution A: Add agar to distilled/deionized water and bring volume to 600.0mL. Mix thoroughly. Gently heat and bring to boiling. Autoclave for 15 min at 15 psi pressure–121°C. Cool to 45–50°C.

Solution B:

Composition per 400.0mL:

Glucose	20.0g
Yeast nitrogen base without amino acids	6.7g
L-Leucine	0.18g
L-Alanine	90.0mg
L-Arginine	90.0mg
L-Asparagine	90.0mg
L-Aspartic acid	90.0mg
L-Cysteine	90.0mg
L-Glutamine	90.0mg
L-Glutamic acid	90.0mg
Glycine	90.0mg
L-Histidine	90.0mg
i-Inositol	90.0mg
L-Isoleucine	90.0mg
L-Lysine	90.0mg
L-Methionine	90.0mg
L-Phenylalanine	90.0mg
L-Proline	90.0mg
L-Serine	90.0mg
L-Threonine	90.0mg
L-Tryptophan	90.0mg
L-Tyrosine	90.0mg
L-Valine	90.0mg
Adenine	23.0mg
p-Aminobenzoic acid	9.0mg

Preparation of Solution B: Add components to distilled/deionized water and bring volume to 400.0mL. Mix thoroughly. Filter sterilize. Warm to 45°–50°C prior to preparation of medium.

Preparation of Medium: Aseptically combine 600.0mL of sterile solution A with 400.0mL of sterile solution B. Mix thoroughly. Pour into sterile Petri dishes or distribute into sterile tubes.

Use: For the cultivation and maintenance of *Saccharomyces cerevisiae.*

Synthetic Malate Medium with 0.25% Sodium Chloride

Composition per 1010.0mL:

DL-Malic acid	5.0g
KOH	4.5g
NaCl	2.5g
KH_2PO_4	0.6g
NH_4Cl	0.5g
K_2HPO_4	0.4g
$MgSO_4 \cdot 7H_2O$	0.2g
Yeast extract	0.1g
$CaCl_2$	0.02g
$MnSO_4 \cdot H_2O$	0.01g
$Na_2MoO_4 \cdot 2H_2O$	0.002g
Biotin	0.1mg
Ferric EDTA solution	10.0mL

pH 7.2 ± 0.2 at 25°C

Ferric EDTA Solution:

Composition per 100.0mL:

Ferric EDTA	0.66g

Preparation of Ferric EDTA Solution: Add ferric EDTA to distilled/deionized water and bring volume to 100.0mL. Mix thoroughly.

Preparation of Medium: Add components to distilled/deionized water and bring volume to 1.0L. Mix thoroughly. Adjust pH to 7.2. Distribute into tubes or flasks. Autoclave for 15 min at 15 psi pressure–121°C.

Use: For the cultivation of *Azospirillum halopraeferens.*

Synthetic *Mucor* Agar

Composition per liter:

Glucose	40.0g
Agar	15.0g
Asparagine	2.0g
KH_2PO_4	0.5g
$MgSO_4$	0.25g
Thiamine·HCl	0.5mg

Preparation of Medium: Add components to distilled/deionized water and bring volume to 1.0L. Mix thoroughly. Gently heat and bring to boiling. Distribute into tubes or flasks. Autoclave for 15 min at 15 psi pressure–121°C. Pour into sterile Petri dishes or leave in tubes.

Use: For the cultivation and maintenance of *Mucor* species.

Synthetic Sea Salt

Composition per liter:

NaCl	14.9g
$MgSO_4 \cdot 7H_2O$	3.80g
$MnCl_2 \cdot 6H_2O$	2.94g
KCl	0.435g
$NaHCO_3$	0.1515g
Borax	3.0mg
$SrCl_2 \cdot 6H_2O$	0.7mg

pH 7.3 ± 0.2 at 25°C

Preparation of Medium: Add components to distilled/deionized water and bring volume to 1.0L. Mix thoroughly. Adjust pH to 7.3. Distribute into tubes or flasks. Autoclave for 15 min at 15 psi pressure–121°C.

Use: For the cultivation of marine bacteria.

Synthetic Seawater Medium

Composition per liter:

NaCl	27.0g
$MgSO_4 \cdot 7H_2O$	7.0g
Monosodium glutamate	5.0g
Tris(hydroxymethyl)aminomethane buffer	2.0g
Glucose	1.0g
KCl	0.6g
$CaCl_2$	0.3g
Sodium glycerophosphate	0.2g
Vitamin B_{12}	1.0μg

pH 7.5 ± 0.2 at 25°C

Preparation of Medium: Add components to distilled/deionized water and bring volume to 1.0L. Mix thoroughly. Adjust pH to 7.5. Distribute into tubes or flasks. Autoclave for 15 min at 15 psi pressure–121°C.

Use: For the cultivation and maintenance of *Leucothrix mucor.*

Syntrophobacter pfennigii Medium

Composition per liter:

NaCl	1.0g
Na_2SO_4	0.7g
KCl	0.5g
$MgCl_2 \cdot 6H_2O$	0.4g
NH_4Cl	0.25g
KH_2PO_4	0.2g
$CaCl_2 \cdot 2H_2O$	0.15g
Resazurin	1.0mg
Sodium propionate solution	50.0mL
$Na_2S \cdot 9H_2O$ solution	10.0mL
$Na_2S_2O_4$ solution	10.0mL
Trace elements solution SL-10	1.0mL
Seven vitamin solution	1.0mL
$NaHCO_3$ solution	variable

pH 7.2–7.4 at 25°C

Sodium Propionate Solution:

Composition per 50.0mL:

Sodium propionate	1.5g

Preparation of Sodium Propionate Solution: Add sodium propionate to distilled/deionized water and bring volume to 50.0mL. Mix thoroughly. Filter sterilize. Sparge with 80% N_2 + 20% CO_2.

$NaHCO_3$ Solution:

Composition per 20.0mL:

$NaHCO_3$	2.5g

Preparation of $NaHCO_3$ Solution: Add $NaHCO_3$ to distilled/deionized water and bring volume to 20.0mL. Mix thoroughly. Filter sterilize. Sparge with 80% N_2 + 20% CO_2.

$Na_2S \cdot 9H_2O$ Solution:

Composition per 10.0mL:

$Na_2S \cdot 9H_2O$	0.36g

Preparation of $Na_2S·9H_2O$ Solution: Add $Na_2S·9H_2O$ to distilled/deionized water and bring volume to 10.0mL. Mix thoroughly. Sparge with 100% N_2. Autoclave for 15 min at 15 psi pressure–121°C.

$Na_2S_2O_4$ Solution:

Composition per 10.0mL:

$Na_2S_2O_4·5H_2O$	2.0g

Preparation of $Na_2S_2O_4$ Solution: Add $Na_2S_2O_4·5H_2O$ to distilled/deionized water and bring volume to 10.0mL. Mix thoroughly. Sparge with 100% N_2. Autoclave for 15 min at 15 psi pressure–121°C.

Trace Elements Solution SL-10:

Composition per liter:

$FeCl_2·4H_2O$	1.5g
$CoCl_2·6H_2O$	190.0mg
$MnCl_2·4H_2O$	100.0mg
$ZnCl_2$	70.0mg
$Na_2MoO_4·2H_2O$	36.0mg
$NiCl_2·6H_2O$	24.0mg
H_3BO_3	6.0mg
$CuCl_2·2H_2O$	2.0mg
HCl (25% solution)	10.0mL

Preparation of Trace Elements Solution SL-10: Prepare and dispense under 80% N_2 + 20% CO_2. Add $FeCl_2·4H_2O$ to 10.0mL of HCl solution. Mix thoroughly. Add distilled/deionized water and bring volume to 1.0L. Add remaining components. Mix thoroughly. Sparge with 80% N_2 + 20% CO_2. Autoclave for 15 min at 15 psi pressure–121°C.

Seven Vitamin Solution:

Composition per liter:

Pyridoxine·HCl	0.3g
Thiamine·HCl	0.2g
Nicotinic acid	0.2g
Calcium DL-pantothenate	0.1g
Vitamin B_{12}	0.1g
p-Aminobenzoic acid	80.0mg
Biotin	20.0mg

Preparation of Seven Vitamin Solution: Add components to distilled/deionized water and bring volume to 1.0L. Mix thoroughly.

Preparation of Medium: Prepare medium and dispense under 80% N_2 + 20% CO_2. Add components, except sodium propionate solution, $NaHCO_3$ solution, $Na_2S·9H_2O$ solution, and $Na_2S_2O_3$ solution, to distilled/deionized water and bring volume to 930.0mL. Mix thoroughly. Sparge with 80% N_2 + 20% CO_2. Autoclave for 15 min at 15 psi pressure–121°C. Aseptically and anaerobically add 50.0mL of sterile sodium propionate solution, 10.0mL of sterile $Na_2S_2O_3$ solution, and 10.0mL of sterile $Na_2S·9H_2O$ solution. Mix thoroughly. Aseptically and anaerobically add sufficient volume of sterile $NaHCO_3$ solution to bring pH to 7.2–7.4. Aseptically and anaerobically distribute into sterile tubes or flasks.

Use: For the cultivation of *Syntrophobacter pfennigii.*

Syntrophobacter wolinii Medium

Solution A	916.0mL
Solution B	70.0mL
Solution C	10.0mL
Solution D	10.0mL

pH 7.2 ± 0.2 at 25°C

Solution A:

Composition per 916.0mL:

Na_2SO_4	2.8g
Sodium propionate	1.5g
Pancreatic digest of casein	1.0g
Resazurin	1.0mg
Mineral solution	50.0mL
Rumen fluid, clarified	50.0mL
Vitamin solution	5.0mL
Trace elements solution SL-10	1.0mL

Preparation of Solution A: Add components to distilled/deionized water and bring volume to 916.0mL. Adjust pH to 7.2. Gently heat and bring to boiling. Continue boiling for a few minutes. Allow to cool to room temperature under 80% N_2 + 20% CO_2. Distribute into bottles under 80% N_2 + 20% CO_2. Autoclave for 15 min at 15 psi pressure–121°C.

Mineral Solution:

Composition per liter:

Nitrilotriacetic acid	12.5g
NaCl	1.0g
$FeCl_3·4H_2O$	0.2g
$MnCl_2·4H_2O$	0.1g
$CaCl_2·2H_2O$	0.1g
$ZnCl_2$	0.1g
$CuCl_2$	0.02g
Na_2SeO_3	0.02g
$CoCl_2·6H_2O$	0.017g
H_3BO_3	0.01g
$Na_2MoO_4·2H_2O$	0.01g

Preparation of Mineral Solution: Add nitrilotriacetic acid to 500.0mL of distilled/deionized water. Adjust pH to 6.5 with KOH. Add remaining components. Add distilled/deionized water to 1.0L. Mix thoroughly.

Vitamin Solution:

Composition per liter:

Biotin	0.25mg
Nicotinic acid	2.5mg
Thiamine·HCl	1.25mg
p-Aminobenzoic acid	1.25mg
Pantothenic acid	0.62mg
Pyridoxine·HCl	6.2mg

Preparation of Vitamin Solution: Add components to distilled/deionized water and bring volume to 1.0L. Mix thoroughly.

Trace Elements Solution SL-10:

Composition per liter:

$FeCl_2·4H_2O$	1.5g
$CoCl_2·6H_2O$	190.0mg
$MnCl_2·4H_2O$	100.0mg
$ZnCl_2$	70.0mg
$Na_2MoO_4·2H_2O$	36.0mg
$NiCl_2·6H_2O$	24.0mg
H_3BO_3	6.0mg
$CuCl_2·2H_2O$	2.0mg
HCl (25% solution)	10.0mL

Preparation of Trace Elements Solution SL-10: Add $FeCl_2·4H_2O$ to 10.0mL of HCl solution. Mix thoroughly. Add distilled/deionized water and bring volume to 1.0L. Add remaining components. Mix thoroughly.

Solution B:

Composition per 70.0mL:

$NaHCO_3$ 3.5g

Preparation of Solution B: Add $NaHCO_3$ to distilled/deionized water and bring volume to 70.0mL. Mix thoroughly. Filter sterilize. Sparge with 80% N_2 + 20% CO_2 for 15 min.

Solution C:

Composition per 10.0mL:

L-Cysteine·HCl 0.3g

Preparation of Solution C: Add L-cysteine·HCl to distilled/deionized water and bring volume to 10.0mL. Mix thoroughly. Sparge with 100% N_2 for 3–4 min. Autoclave under 100% N_2 for 15 min at 15 psi pressure–121°C.

Solution D:

Composition per 10.0mL:

$Na_2S{\cdot}9H_2O$ 0.3g

Preparation of Solution D: Add $Na_2S{\cdot}9H_2O$ to distilled/deionized water and bring volume to 10.0mL. Mix thoroughly. Sparge with 100% N_2 for 3–4 min. Autoclave under 100% N_2 for 15 min at 15 psi pressure–121°C.

Preparation of Medium: To 916.0mL of sterile solution A, add 70.0mL of sterile solution B, 10.0mL of sterile solution C, and 10.0mL of sterile solution D. Mix thoroughly.

Use: For the cultivation and maintenance of *Syntrophobacter wolinii*.

Syntrophococcus sucromutans Medium

Composition per 1002.0mL:

Solution A 916.0mL
Solution C 50.0mL
Solution B 25.0mL
Solution D 10.0mL
Solution E 1.0mL

pH 7.2–7.4 at 25°C

Solution A:

Composition per 916.0mL:

Sodium formate 0.6g
Resazurin 1.0mg
Rumen fluid, clarified 300.0mL
Mineral solution 50.0mL
Vitamin solution 5.0mL
Trace elements solution SL-10 .... 1.0mL

Preparation of Solution A: Add components to distilled/deionized water and bring volume to 916.0mL. Mix thoroughly. Adjust pH to 6.4. Autoclave for 15 min at 15 psi pressure–121°C.

Mineral Solution:

Composition per liter:

KH_2PO_4 .... 10.0g
NaCl 8.0g
NH_4Cl 8.0g
$MgCl_2{\cdot}6H_2O$ 6.6g
$CaCl_2{\cdot}2H_2O$ 1.0g

Preparation of Mineral Solution: Add components to distilled/deionized water and bring volume to 1.0L. Mix thoroughly.

Vitamin Solution:

Composition per liter:

Pyridoxine·HCl 6.2mg
Nicotinic acid 2.5mg
p-Aminobenzoic acid 1.25mg
Thiamine·HCl 1.25mg
Pantothenic acid 0.62mg
Biotin 0.25mg

Preparation of Vitamin Solution: Add components to distilled/deionized water and bring volume to 1.0L. Mix thoroughly.

Trace Elements Solution SL-10:

Composition per liter:

$FeCl_2{\cdot}4H_2O$ 1.5g
$CoCl_2{\cdot}6H_2O$ 190.0mg
$MnCl_2{\cdot}4H_2O$ 100.0mg
$ZnCl_2$ 70.0mg
$Na_2MoO_4{\cdot}2H_2O$ 36.0mg
$NiCl_2{\cdot}6H_2O$ 24.0mg
H_3BO_3 .... 6.0mg
$CuCl_2{\cdot}2H_2O$ 2.0mg
HCl (25% solution) 10.0mL

Preparation of Trace Elements Solution SL-10: Add $FeCl_2{\cdot}4H_2O$ to 10.0mL of HCl solution. Mix thoroughly. Add distilled/deionized water and bring volume to 1.0L. Add remaining components. Mix thoroughly. Autoclave for 15 min at 15 psi pressure–121°C.

Solution B:

Composition per 25.0mL:

Lactose 5.0g

Preparation of Solution B: Add lactose to distilled/deionized water and bring volume to 25.0mL. Mix thoroughly. Filter sterilize.

Solution C:

Composition per 50.0mL:

$NaHCO_3$ 2.5g

Preparation of Solution C: Add $NaHCO_3$ to distilled/deionized water and bring volume to 50.0mL. Mix thoroughly. Autoclave for 15 min at 15 psi pressure–121°C.

Solution D:

Composition per 10.0mL:

L-Cysteine 0.24g

Preparation of Solution D: Add L-cysteine to distilled/deionized water and bring volume to 10.0mL. Mix thoroughly. Autoclave for 15 min at 15 psi pressure–121°C.

Solution E:

Composition per 1.0mL:

$Na_2S{\cdot}9H_2O$ 78.0mg

Preparation of Solution E: Add $Na_2S{\cdot}9H_2O$ to distilled/deionized water and bring volume to 1.0mL. Mix thoroughly. Autoclave for 15 min at 15 psi pressure–121°C.

Preparation of Medium: Prepare and dispense medium under H_2-free 80% N_2 + 20% CO_2. Aseptically and anaerobically combine 916.0mL of sterile solution A with 25.0mL of sterile solution B, 50.0mL of sterile solution C, 10.0mL of sterile solution D, and 1.0mL of sterile solution E. Mix thoroughly. Final pH should be 6.4–6.8

Use: For the cultivation and maintenance of *Syntrophococcus sucromutans*.

Syntrophomonas bryantii Medium

Composition per 1026.0mL:

Solution A 916.0mL
Solution B 70.0mL

Solution C	10.0mL
Solution D	10.0mL
Sodium laurate solution	10.0mL
$CaCl_2·2H_2O$ solution	10.0mL

pH 7.2 ± 0.2 at 25°C

Solution A:

Composition per 916.0mL:

PIPES (piperazine-*N,N'*-bis [2-ethanesulfonic acid]) buffer	15.12g
Na_2SO_4	2.8g
Butyric acid	1.7g
Pancreatic digest of casein	1.0g
Resazurin	1.0mg
Mineral solution	50.0mL
Rumen fluid, clarified	50.0mL
Vitamin solution	5.0mL
Trace elements solution SL-10	1.0mL

Mineral Solution:

Composition per liter:

Nitrilotriacetic acid	12.5g
NaCl	1.0g
$FeCl_3·4H_2O$	0.2g
$MnCl_2·4H_2O$	0.1g
$CaCl_2·2H_2O$	0.1g
$ZnCl_2$	0.1g
$CuCl_2$	0.02g
Na_2SeO_3	0.02g
$CoCl_2·6H_2O$	0.017g
H_3BO_3	0.01g
$Na_2MoO_4·2H_2O$	0.01g

Preparation of Mineral Solution: Add nitrilotriacetic acid to 500.0mL of distilled/deionized water. Adjust pH to 6.5 with KOH. Add remaining components. Add distilled/deionized water to 1.0L. Mix thoroughly.

Vitamin Solution:

Composition per liter:

Pyridoxine·HCl	6.2mg
Nicotinic acid	2.5mg
Thiamine·HCl	1.25mg
p-Aminobenzoic acid	1.25mg
Pantothenic acid	0.62mg
Biotin	0.25mg

Preparation of Vitamin Solution: Add components to distilled/deionized water and bring volume to 1.0L. Mix thoroughly.

Trace Elements Solution SL-10:

Composition per liter:

$FeCl_2·4H_2O$	1.5g
$CoCl_2·6H_2O$	190.0mg
$MnCl_2·4H_2O$	100.0mg
$ZnCl_2$	70.0mg
$Na_2MoO_4·2H_2O$	36.0mg
$NiCl_2·6H_2O$	24.0mg
H_3BO_3	6.0mg
$CuCl_2·2H_2O$	2.0mg
HCl (25% solution)	10.0mL

Preparation of Trace Elements Solution SL-10: Add $FeCl_2·4H_2O$ to 10.0mL of HCl solution. Mix thoroughly. Add distilled/deionized water and bring volume to 1.0L. Add remaining components. Mix thoroughly.

Preparation of Solution A: Add components to distilled/deionized water and bring volume to 916.0mL. Adjust pH to 7.2. Gently heat and bring to boiling. Continue boiling for a few minutes. Allow to cool to room temperature under 80% N_2 + 20% CO_2. Distribute into bottles under 80% N_2 + 20% CO_2. Autoclave for 15 min at 15 psi pressure–121°C.

Solution B:

Composition per 70.0mL:

$NaHCO_3$	3.5g

Preparation of Solution B: Add $NaHCO_3$ to distilled/deionized water and bring volume to 70.0mL. Mix thoroughly. Filter sterilize. Sparge with 80% N_2 + 20% CO_2 for 15 min.

Solution C:

Composition per 10.0mL:

L-Cysteine·HCl	0.3g

Preparation of Solution C: Add L-cysteine·HCl to distilled/deionized water and bring volume to 10.0mL. Mix thoroughly. Sparge with 100% N_2 for 3–4 min. Autoclave under 100% N_2 for 15 min at 15 psi pressure–121°C.

Solution D:

Composition per 10.0mL:

$Na_2S·9H_2O$	0.3g

Preparation of Solution D: Add $Na_2S·9H_2O$ to distilled/deionized water and bring volume to 10.0mL. Mix thoroughly. Sparge with 100% N_2 for 3–4 min. Autoclave under 100% N_2 for 15 min at 15 psi pressure–121°C.

Sodium Laurate Solution:

Composition per 10.0mL:

Sodium laurate	2.78g

Preparation of Sodium Laurate Solution: Add sodium laurate to distilled/deionized water and bring volume to 10.0mL. Mix thoroughly. Sparge with 100% N_2. Autoclave for 15 min at 15 psi pressure–121°C.

$CaCl_2·2H_2O$ Solution:

Composition per 40.0mL:

$CaCl_2·2H_2O$	1.84g

Preparation of $CaCl_2·2H_2O$ Solution: Add $CaCl_2·2H_2O$ to distilled/deionized water and bring volume to 40.0mL. Mix thoroughly. Sparge with 100% N_2. Autoclave for 15 min at 15 psi pressure–121°C.

Preparation of Medium: To 916.0mL of sterile solution A, add 70.0mL of sterile solution B, 10.0mL of sterile solution C, and 10.0mL of sterile solution D. Mix thoroughly. Prior to inoculation, aseptically add 10.0mL of sterile sodium laurate solution and 10.0mL of sterile $CaCl_2·2H_2O$ solution. Mix thoroughly.

Use: For the cultivation and maintenance of *Syntrophomonas sapovorans*.

Syntrophomonas Medium

Composition per 1006.0mL:

Solution A	916.0mL
Solution B	70.0mL
Solution C	10.0mL
Solution D	10.0mL

pH 7.2 ± 0.2 at 25°C

Solution A:

Composition per 916.0mL:

Na_2SO_4	2.8g
Butyric acid	1.7g
Pancreatic digest of casein	1.0g
Resazurin	1.0mg
Mineral solution	50.0mL
Rumen fluid, clarified	50.0mL
Vitamin solution	5.0mL
Trace elements solution SL-10	1.0mL

Mineral Solution:

Composition per liter:

Nitrilotriacetic acid	12.5g
NaCl	1.0g
$FeCl_3 \cdot 4H_2O$	0.2g
$MnCl_2 \cdot 4H_2O$	0.1g
$CaCl_2 \cdot 2H_2O$	0.1g
$ZnCl_2$	0.1g
$CuCl_2$	0.02g
Na_2SeO_3	0.02g
$CoCl_2 \cdot 6H_2O$	0.017g
H_3BO_3	0.01g
$Na_2MoO_4 \cdot 2H_2O$	0.01g

Preparation of Mineral Solution: Add nitrilotriacetic acid to 500.0mL of distilled/deionized water. Adjust pH to 6.5 with KOH. Add remaining components. Add distilled/deionized water to 1.0L. Mix thoroughly.

Vitamin Solution:

Composition per liter:

Pyridoxine·HCl	6.2mg
Nicotinic acid	2.5mg
Thiamine·HCl	1.25mg
p-Aminobenzoic acid	1.25mg
Pantothenic acid	0.62mg
Biotin	0.25mg

Preparation of Vitamin Solution: Add components to distilled/deionized water and bring volume to 1.0L. Mix thoroughly.

Trace Elements Solution SL-10:

Composition per liter:

$FeCl_2 \cdot 4H_2O$	1.5g
$CoCl_2 \cdot 6H_2O$	190.0mg
$MnCl_2 \cdot 4H_2O$	100.0mg
$ZnCl_2$	70.0mg
$Na_2MoO_4 \cdot 2H_2O$	36.0mg
$NiCl_2 \cdot 6H_2O$	24.0mg
H_3BO_3	6.0mg
$CuCl_2 \cdot 2H_2O$	2.0mg
HCl (25% solution)	10.0mL

Preparation of Trace Elements Solution SL-10: Add $FeCl_2 \cdot 4H_2O$ to 10.0mL of HCl solution. Mix thoroughly. Add distilled/deionized water and bring volume to 1.0L. Add remaining components. Mix thoroughly.

Preparation of Solution A: Add components to distilled/deionized water and bring volume to 916.0mL. Adjust pH to 7.2. Gently heat and bring to boiling. Continue boiling for a few minutes. Allow to cool to room temperature under 80% N_2 + 20% CO_2. Distribute into bottles under 80% N_2 + 20% CO_2. Autoclave for 15 min at 15 psi pressure–121°C.

Solution B:

Composition per 70.0mL:

$NaHCO_3$	3.5g

Preparation of Solution B: Add $NaHCO_3$ to distilled/deionized water and bring volume to 70.0mL. Mix thoroughly. Filter sterilize. Sparge with 80% N_2 + 20% CO_2 for 15 min.

Solution C:

Composition per 10.0mL:

L-Cysteine·HCl	0.3g

Preparation of Solution C: Add L-cysteine·HCl to distilled/deionized water and bring volume to 10.0mL. Mix thoroughly. Sparge with 100% N_2 for 3–4 min. Autoclave under 100% N_2 for 15 min at 15 psi pressure–121°C.

Solution D:

Composition per 10.0mL:

$Na_2S \cdot 9H_2O$	0.3g

Preparation of Solution D: Add $Na_2S \cdot 9H_2O$ to distilled/deionized water and bring volume to 10.0mL. Mix thoroughly. Sparge with 100% N_2 for 3–4 min. Autoclave under 100% N_2 for 15 min at 15 psi pressure–121°C.

Preparation of Medium: To 916.0mL of sterile solution A, add 70.0mL of sterile solution B, 10.0mL of sterile solution C, and 10.0mL of sterile solution D. Mix thoroughly.

Use: For the cultivation of *Syntrophomonas* species.

Syntrophomonas Medium, Sulfate-Free

Composition per 1006.0mL:

Solution A	916.0mL
Solution B	70.0mL
Solution C	10.0mL
Solution D	10.0mL

pH 7.2 ± 0.2 at 25°C

Solution A:

Composition per 916.0mL:

Butyric acid	1.7g
Pancreatic digest of casein	1.0g
Resazurin	1.0mg
Mineral solution	50.0mL
Rumen fluid, clarified	50.0mL
Vitamin solution	5.0mL
Trace elements solution SL-10	1.0mL

Mineral Solution:

Composition per liter:

Nitrilotriacetic acid	12.5g
NaCl	1.0g
$FeCl_3 \cdot 4H_2O$	0.2g
$MnCl_2 \cdot 4H_2O$	0.1g
$CaCl_2 \cdot 2H_2O$	0.1g
$ZnCl_2$	0.1g
$CuCl_2$	0.02g
Na_2SeO_3	0.02g
$CoCl_2 \cdot 6H_2O$	0.017g
H_3BO_3	0.01g
$Na_2MoO_4 \cdot 2H_2O$	0.01g

Preparation of Mineral Solution: Add nitrilotriacetic acid to 500.0mL of distilled/deionized water. Adjust pH to 6.5 with KOH. Add remaining components. Add distilled/deionized water to 1.0L. Mix thoroughly.

Vitamin Solution:
Composition per liter:

Pyridoxine·HCl	6.2mg
Nicotinic acid	2.5mg
Thiamine·HCl	1.25mg
p-Aminobenzoic acid	1.25mg
Pantothenic acid	0.62mg
Biotin	0.25mg

Preparation of Vitamin Solution: Add components to distilled/deionized water and bring volume to 1.0L. Mix thoroughly.

Trace Elements Solution SL-10:
Composition per liter:

$FeCl_2·4H_2O$	1.5g
$CoCl_2·6H_2O$	190.0mg
$MnCl_2·4H_2O$	100.0mg
$ZnCl_2$	70.0mg
$Na_2MoO_4·2H_2O$	36.0mg
$NiCl_2·6H_2O$	24.0mg
H_3BO_3	6.0mg
$CuCl_2·2H_2O$	2.0mg
HCl (25% solution)	10.0mL

Preparation of Trace Elements Solution SL-10: Add $FeCl_2·4H_2O$ to 10.0mL of HCl solution. Mix thoroughly. Add distilled/deionized water and bring volume to 1.0L. Add remaining components. Mix thoroughly.

Preparation of Solution A: Add components to distilled/deionized water and bring volume to 916.0mL. Adjust pH to 7.2. Gently heat and bring to boiling. Continue boiling for a few minutes. Allow to cool to room temperature under 80% N_2 + 20% CO_2. Distribute into bottles under 80% N_2 + 20% CO_2. Autoclave for 15 min at 15 psi pressure–121°C.

Solution B:
Composition per 70.0mL:

$NaHCO_3$	3.5g

Preparation of Solution B: Add $NaHCO_3$ to distilled/deionized water and bring volume to 70.0mL. Mix thoroughly. Filter sterilize. Sparge with 80% N_2 + 20% CO_2 for 15 min.

Solution C:
Composition per 10.0mL:

L-Cysteine·HCl	0.3g

Preparation of Solution C: Add L-cysteine·HCl to distilled/deionized water and bring volume to 10.0mL. Mix thoroughly. Sparge with 100% N_2 for 3–4 min. Autoclave under 100% N_2 for 15 min at 15 psi pressure–121°C.

Solution D:
Composition per 10.0mL:

$Na_2S·9H_2O$	0.3g

Preparation of Solution D: Add $Na_2S·9H_2O$ to distilled/deionized water and bring volume to 10.0mL. Mix thoroughly. Sparge with 100% N_2 for 3–4 min. Autoclave under 100% N_2 for 15 min at 15 psi pressure–121°C.

Preparation of Medium: To 916.0mL of sterile solution A, add 70.0mL of sterile solution B, 10.0mL of sterile solution C, and 10.0mL of sterile solution D. Mix thoroughly.

Use: For the cultivation of *Syntrophus buswelii*.

Syntrophomonas species Medium

Composition per 1006.0mL:

Solution A	916.0mL
Solution B	70.0mL
Solution C	10.0mL
Solution D	10.0mL

pH 7.2 ± 0.2 at 25°C

Solution A:
Composition per 916.0mL:

Na_2SO_4	2.8g
Pancreatic digest of casein	1.0g
Sodium stearate	0.61g
Resazurin	1.0mg
Mineral solution	50.0mL
Rumen fluid, clarified	50.0mL
Vitamin solution	5.0mL
Trace elements solution SL-10	1.0mL

Preparation of Solution A: Add components to distilled/deionized water and bring volume to 916.0mL. Adjust pH to 7.2. Gently heat and bring to boiling. Continue boiling for a few minutes. Allow to cool to room temperature under 80% N_2 + 20% CO_2. Distribute into bottles under 80% N_2 + 20% CO_2. Autoclave for 15 min at 15 psi pressure–121°C.

Mineral Solution:
Composition per liter:

Nitrilotriacetic acid	12.5g
NaCl	1.0g
$FeCl_3·4H_2O$	0.2g
$MnCl_2·4H_2O$	0.1g
$CaCl_2·2H_2O$	0.1g
$ZnCl_2$	0.1g
$CuCl_2$	0.02g
Na_2SeO_3	0.02g
$CoCl_2·6H_2O$	0.017g
H_3BO_3	0.01g
$Na_2MoO_4·2H_2O$	0.01g

Preparation of Mineral Solution: Add nitrilotriacetic acid to 500.0mL of distilled/deionized water. Adjust pH to 6.5 with KOH. Add remaining components. Add distilled/deionized water to 1.0L. Mix thoroughly.

Vitamin Solution:
Composition per liter:

Pyridoxine·HCl	6.2mg
Nicotinic acid	2.5mg
Thiamine·HCl	1.25mg
p-Aminobenzoic acid	1.25mg
Pantothenic acid	0.62mg
Biotin	0.25mg

Preparation of Vitamin Solution: Add components to distilled/deionized water and bring volume to 1.0L. Mix thoroughly.

Trace Elements Solution SL-10:
Composition per liter:

$FeCl_2·4H_2O$	1.5g
$CoCl_2·6H_2O$	190.0mg
$MnCl_2·4H_2O$	100.0mg
$ZnCl_2$	70.0mg
$Na_2MoO_4·2H_2O$	36.0mg
$NiCl_2·6H_2O$	24.0mg
H_3BO_3	6.0mg

$CuCl_2 \cdot 2H_2O$ 2.0mg
HCl (25% solution) 10.0mL

Preparation of Trace Elements Solution SL-10: Add $FeCl_2 \cdot 4H_2O$ to 10.0mL of HCl solution. Mix thoroughly. Add distilled/deionized water and bring volume to 1.0L. Add remaining components. Mix thoroughly.

Solution B:
Composition per 70.0mL:
$NaHCO_3$ 3.5g

Preparation of Solution B: Add $NaHCO_3$ to distilled/deionized water and bring volume to 70.0mL. Mix thoroughly. Filter sterilize. Sparge with 80% N_2 + 20% CO_2 for 15 min.

Solution C:
Composition per 10.0mL:
L-Cysteine·HCl 0.3g

Preparation of Solution C: Add L-cysteine·HCl to distilled/deionized water and bring volume to 10.0mL. Mix thoroughly. Sparge with 100% N_2 for 3–4 min. Autoclave under 100% N_2 for 15 min at 15 psi pressure–121°C.

Solution D:
Composition per 10.0mL:
$Na_2S \cdot 9H_2O$ 0.3g

Preparation of Solution D: Add $Na_2S \cdot 9H_2O$ to distilled/deionized water and bring volume to 10.0mL. Mix thoroughly. Sparge with 100% N_2 for 3–4 min. Autoclave under 100% N_2 for 15 min at 15 psi pressure–121°C.

Preparation of Medium: To 916.0mL of sterile solution A, add 70.0mL of sterile solution B, 10.0mL of sterile solution C, and 10.0mL of sterile solution D. Mix thoroughly.

Use: For the cultivation of *Syntrophomonas wolfei.*

Syntrophothermus Medium (DSMZ Medium 870)

Composition per liter:
$NaHCO_3$ 2.5g
NH_4Cl 0.54g
$MgCl_2 \cdot 6H_2O$ 0.2g
$CaCl_2 \cdot 2H_2O$ 0.15g
KH_2PO_4 .... 0.14g
Resazurin 0.5mg
$Na_2S \cdot 9H_2O$ solution 10.0mL
Cysteine solution 10.0mL
Vitamin solution 10.0mL
Substrate solution 10.0mL
Trace elements solution 1.0mL
Selenite-tungstate solution 1.0mL
pH 7.0 ± 0.2 at 25°C

Substrate Solution:
Composition per 10.0mL:
Na-crotonate 0.86g

Preparation of Substrate Solution: AddNa-crotonate to distilled/deionized water and bring volume to 10.0mL. Mix thoroughly. Sparge with 100% N_2. Filter sterilize.

$Na_2S \cdot 9H_2O$ Solution:
Composition per 10.0mL:
$Na_2S \cdot 9H_2O$ 0.3g

Preparation of $Na_2S \cdot 9H_2O$ Solution: Add $Na_2S \cdot 9H_2O$ to distilled/deionized water and bring volume to 10.0mL. Mix thoroughly. Autoclave under 100% N_2 for 15 min at 15 psi pressure–121°C. Cool to room temperature.

Cysteine Solution:
Composition per 10.0mL:
L-Cysteine·HCl·H_2O 0.3g

Preparation of Cysteine Solution: Add L-cysteine·HCl·H_2O to distilled/deionized water and bring volume to 10.0mL. Mix thoroughly. Sparge with 100% N_2. Autoclave for 15 min at 15 psi pressure–121°C.

Selenite-Tungstate Solution
Composition per liter:
NaOH 0.5g
$Na_2WO_4 \cdot 2H_2O$ 4.0mg
$Na_2SeO_3 \cdot 5H_2O$ 3.0mg

Preparation of Selenite-Tungstate Solution: Add components to distilled/deionized water and bring volume to 1.0L. Mix thoroughly. Sparge with 100% N_2. Filter sterilize.

Trace Elements Solution:
Composition per liter:
$MgSO_4 \cdot 7H_2O$ 3.0g
Nitrilotriacetic acid 1.5g
NaCl 1.0g
$MnSO_4 \cdot 2H_2O$ 0.5g
$CoSO_4 \cdot 7H_2O$ 0.18g
$ZnSO_4 \cdot 7H_2O$ 0.18g
$CaCl_2 \cdot 2H_2O$ 0.1g
$FeSO_4 \cdot 7H_2O$ 0.1g
$NiCl_2 \cdot 6H_2O$ 0.025g
$KAl(SO_4)_2 \cdot 12H_2O$ 0.02g
H_3BO_3 .... 0.01g
$Na_2MoO_4 \cdot 4H_2O$ 0.01g
$CuSO_4 \cdot 5H_2O$ 0.01g
$Na_2SeO_3 \cdot 5H_2O$ 0.3mg

Preparation of Trace Elements Solution: Add nitrilotriacetic acid to 500.0mL of distilled/deionized water. Dissolve by adjusting pH to 6.5 with KOH. Add remaining components. Add distilled/deionized water to 1.0L. Mix thoroughly.

Vitamin Solution:
Composition per liter:
Pyridoxine-HCl 10.0mg
Thiamine-HCl·$2H_2O$ 5.0mg
Riboflavin 5.0mg
Nicotinic acid 5.0mg
D-Ca-pantothenate 5.0mg
p-Aminobenzoic acid 5.0mg
Lipoic acid 5.0mg
Biotin 2.0mg
Folic acid 2.0mg
Vitamin B_{12} 0.1mg

Preparation of Vitamin Solution: Add components to distilled/deionized water and bring volume to 1.0L. Mix thoroughly. Sparge with 80% H_2 + 20% CO_2. Filter sterilize.

Preparation of Medium: Prepare and dispense medium under 80% N_2 + 20% CO_2 gas atmosphere. Add components, except cysteine solution, $Na_2S \cdot 9H_2O$ solution, and substrate solution, to distilled/deionized water and bring volume to 970.0mL. Mix thoroughly. Adjust pH to 7.0. Sparge with

80% N_2 + 20% CO_2 for 30 min. Distribute into Hungate tubes or serum bottles. Autoclave for 15 min at 15 psi pressure–121°C. For each 10.0mL medium, aseptically and anaerobically add 0.1mL cysteine solution, 0.1mL $Na_2S{\cdot}9H_2O$ solution, and 0.1mL substrate solution. Mix thoroughly.

Use: For the cultivation of *Syntrophothermus lipocalidus* DSM 12680.

Syntrophothermus Medium (DSMZ Medium 870)

Composition per liter:

$NaHCO_3$	2.5g
NH_4Cl	0.54g
$MgCl_2{\cdot}6H_2O$	0.2g
$CaCl_2{\cdot}2H_2O$	0.15g
KH_2PO_4	0.14g
Resazurin	0.5mg
$Na_2S{\cdot}9H_2O$ solution	10.0mL
Cysteine solution	10.0mL
Vitamin solution	10.0mL
Substrate solution	10.0mL
Trace elements solution	1.0mL
Selenite-tungstate solution	1.0mL

pH 7.0 ± 0.2 at 25°C

Substrate Solution:

Composition per 10.0mL:

Na-butyrate	2.2g

Preparation of Substrate Solution: Add Na-butyrate to distilled/deionized water and bring volume to 10.0mL. Mix thoroughly. Sparge with 100% N_2. Filter sterilize.

$Na_2S{\cdot}9H_2O$ Solution:

Composition per 10.0mL:

$Na_2S{\cdot}9H_2O$	0.3g

Preparation of $Na_2S{\cdot}9H_2O$ Solution: Add $Na_2S{\cdot}9H_2O$ to distilled/deionized water and bring volume to 10.0mL. Mix thoroughly. Autoclave under 100% N_2 for 15 min at 15 psi pressure–121°C. Cool to room temperature.

Cysteine Solution:

Composition per 10.0mL:

L-Cysteine·HCl·H_2O	0.3g

Preparation of Cysteine Solution: Add L-cysteine·HCl·H_2O to distilled/deionized water and bring volume to 10.0mL. Mix thoroughly. Sparge with 100% N_2. Autoclave for 15 min at 15 psi pressure–121°C.

Selenite-Tungstate Solution

Composition per liter:

NaOH	0.5g
$Na_2WO_4{\cdot}2H_2O$	4.0mg
$Na_2SeO_3{\cdot}5H_2O$	3.0mg

Preparation of Selenite-Tungstate Solution: Add components to distilled/deionized water and bring volume to 1.0L. Mix thoroughly. Sparge with 100% N_2. Filter sterilize.

Trace Elements Solution:

Composition per liter:

$MgSO_4{\cdot}7H_2O$	3.0g
Nitrilotriacetic acid	1.5g
NaCl	1.0g
$MnSO_4{\cdot}2H_2O$	0.5g
$CoSO_4{\cdot}7H_2O$	0.18g
$ZnSO_4{\cdot}7H_2O$	0.18g
$CaCl_2{\cdot}2H_2O$	0.1g
$FeSO_4{\cdot}7H_2O$	0.1g
$NiCl_2{\cdot}6H_2O$	0.025g
$KAl(SO_4)_2{\cdot}12H_2O$	0.02g
H_3BO_3	0.01g
$Na_2MoO_4{\cdot}4H_2O$	0.01g
$CuSO_4{\cdot}5H_2O$	0.01g
$Na_2SeO_3{\cdot}5H_2O$	0.3mg

Preparation of Trace Elements Solution: Add nitrilotriacetic acid to 500.0mL of distilled/deionized water. Dissolve by adjusting pH to 6.5 with KOH. Add remaining components. Add distilled/deionized water to 1.0L. Mix thoroughly.

Vitamin Solution:

Composition per liter:

Pyridoxine-HCl	10.0mg
Thiamine-HCl·$2H_2O$	5.0mg
Riboflavin	5.0mg
Nicotinic acid	5.0mg
D-Ca-pantothenate	5.0mg
p-Aminobenzoic acid	5.0mg
Lipoic acid	5.0mg
Biotin	2.0mg
Folic acid	2.0mg
Vitamin B_{12}	0.1mg

Preparation of Vitamin Solution: Add components to distilled/deionized water and bring volume to 1.0L. Mix thoroughly. Sparge with 80% H_2 + 20% CO_2. Filter sterilize.

Preparation of Medium: Prepare and dispense medium under 80% N_2 + 20% CO_2 gas atmosphere. Add components, except cysteine solution, $Na_2S{\cdot}9H_2O$ solution, and substrate solution, to distilled/deionized water and bring volume to 970.0mL. Mix thoroughly. Adjust pH to 7.0. Sparge with 80% N_2 + 20% CO_2 for 30 min. Distribute into Hungate tubes or serum bottles. Autoclave for 15 min at 15 psi pressure–121°C. For each 10.0mL medium, aseptically and anaerobically add 0.1mL cysteine solution, 0.1mL $Na_2S{\cdot}9H_2O$ solution, and 0.1mL substrate solution. Mix thoroughly.

Use: For the cultivation of *Syntrophothermus lipocalidus* DSM 12681.

Syntrophus buswellii II Medium

Composition per 1001.0mL:

Solution A	870.0mL
Solution C	100.0mL
Solution D	10.0mL
Solution E (Vitamin solution)	10.0mL
Solution F	10.0mL
Solution B (Trace elements solution SL-10)	1.0mL

pH 7.1–7.4 at 25°C

Solution A:

Composition per 870.0mL:

Na_2SO_4	3.0g
NaCl	1.0g
KCl	0.5g
$MgCl_2{\cdot}6H_2O$	0.4g
NH_4Cl	0.3g
KH_2PO_4	0.2g
$CaCl_2{\cdot}2H_2O$	0.15g
Resazurin	1.0mg

Preparation of Solution A: Add components to distilled/deionized water and bring volume to 870.0mL. Mix thoroughly. Gently heat and bring to boiling. Continue boiling for 3-4 min. Allow to cool to room temperature while gassing under 80% N_2 + 20% CO_2. Continue gassing until pH reaches below 6.0. Seal the flask under 80% N_2 + 20% CO_2. Autoclave for 15 min at 15 psi pressure–121°C.

Solution B (Trace Elements Solution SL-10):
Composition per liter:

$FeCl_2 \cdot 4H_2O$	1.5g
$CoCl_2 \cdot 6H_2O$	190.0mg
$MnCl_2 \cdot 4H_2O$	100.0mg
$ZnCl_2$	70.0mg
$Na_2MoO_4 \cdot 2H_2O$	36.0mg
$NiCl_2 \cdot 6H_2O$	24.0mg
H_3BO_3	6.0mg
$CuCl_2 \cdot 2H_2O$	2.0mg
HCl (25% solution)	10.0mL

Preparation of Solution B (Trace Elements Solution SL-10): Add $FeCl_2 \cdot 4H_2O$ to 10.0mL of HCl solution. Mix thoroughly. Add distilled/deionized water and bring volume to 1.0L. Add remaining components. Mix thoroughly. Gas under 100% N_2. Autoclave for 15 min at 15 psi pressure–121°C.

Solution C:
Composition per 100.0mL:

$NaHCO_3$	5.0g

Preparation of Solution C: Add $NaHCO_3$ to distilled/deionized water and bring volume to 100.0mL. Mix thoroughly. Filter sterilize. Gas under 80% N_2 + 20% CO_2.

Solution D:
Composition per 10.0mL:

Sodium benzoate	3.0g
Sodium acetate	1.0g

Preparation of Solution D: Add components to distilled/deionized water and bring volume to 10.0mL. Mix thoroughly. Gas under 100% N_2. Autoclave for 15 min at 15 psi pressure–121°C.

Solution E (Vitamin Solution):
Composition per liter:

Pyridoxine·HCl	10.0mg
Calcium DL-pantothenate	5.0mg
Lipoic acid	5.0mg
Nicotinic acid	5.0mg
p-Aminobenzoic acid	5.0mg
Riboflavin	5.0mg
Thiamine·HCl	5.0mg
Biotin	2.0mg
Folic acid	2.0mg
Vitamin B_{12}	0.1mg

Preparation of Solution E (Vitamin Solution): Add components to distilled/deionized water and bring volume to 1.0L. Mix thoroughly. Gas under 100% N_2. Autoclave for 15 min at 15 psi pressure–121°C.

Solution F:
Composition per 10.0mL:

$Na_2S \cdot 9H_2O$	0.4g

Preparation of Solution F: Add $Na_2S \cdot 9H_2O$ to distilled/deionized water and bring volume to 10.0mL. Mix thoroughly. Gas under 100% N_2. Autoclave for 15 min at 15 psi pressure–121°C.

Preparation of Medium: Aseptically and anaerobically combine 870.0mL of sterile solution A with 1.0mL of sterile solution B, 100.0mL of sterile solution C, 10.0mL of sterile solution D, 10.0mL of sterile solution E, and 10.0mL of sterile solution F, in that order. Mix thoroughly. Anaerobically distribute into sterile tubes or flasks under 100% N_2.

Use: For the cultivation and maintenance of *Syntrophus buswelii.*

Syntrophus Medium

Composition per 1006.0mL:

Solution A	916.0mL
Solution B	70.0mL
Solution C	10.0mL
Solution D	10.0mL

pH 7.2 ± 0.2 at 25°C

Solution A:
Composition per 916.0mL:

Na_2SO_4	2.8g
Sodium benzoate	2.0g
Pancreatic digest of casein	1.0g
Resazurin	1.0mg
Mineral solution	50.0mL
Rumen fluid, clarified	50.0mL
Vitamin solution	5.0mL
Trace elements solution SL-10	1.0mL

Mineral Solution:
Composition per liter:

Nitrilotriacetic acid	12.5g
NaCl	1.0g
$FeCl_3 \cdot 4H_2O$	0.2g
$MnCl_2 \cdot 4H_2O$	0.1g
$CaCl_2 \cdot 2H_2O$	0.1g
$ZnCl_2$	0.1g
$CuCl_2$	0.02g
Na_2SeO_3	0.02g
$CoCl_2 \cdot 6H_2O$	0.017g
H_3BO_3	0.01g
$Na_2MoO_4 \cdot 2H_2O$	0.01g

Preparation of Mineral Solution: Add nitrilotriacetic acid to 500.0mL of distilled/deionized water. Adjust pH to 6.5 with KOH. Add remaining components. Add distilled/deionized water to 1.0L. Mix thoroughly.

Vitamin Solution:
Composition per liter:

Pyridoxine·HCl	6.2mg
Nicotinic acid	2.5mg
Thiamine·HCl	1.25mg
p-Aminobenzoic acid	1.25mg
Pantothenic acid	0.62mg
Biotin	0.25mg

Preparation of Vitamin Solution: Add components to distilled/deionized water and bring volume to 1.0L. Mix thoroughly.

Trace Elements Solution SL-10:
Composition per liter:

$FeCl_2 \cdot 4H_2O$	1.5g
$CoCl_2 \cdot 6H_2O$	190.0mg
$MnCl_2 \cdot 4H_2O$	100.0mg
$ZnCl_2$	70.0mg
$Na_2MoO_4 \cdot 2H_2O$	36.0mg

$NiCl_2 \cdot 6H_2O$ 24.0mg
H_3BO_3 .... 6.0mg
$CuCl_2 \cdot 2H_2O$ 2.0mg
HCl (25% solution) 10.0mL

Preparation of Trace Elements Solution SL-10: Add $FeCl_2 \cdot 4H_2O$ to 10.0mL of HCl solution. Mix thoroughly. Add distilled/deionized water and bring volume to 1.0L. Add remaining components. Mix thoroughly.

Preparation of Solution A: Add components to distilled/deionized water and bring volume to 916.0mL. Adjust pH to 7.2. Gently heat and bring to boiling. Continue boiling for a few minutes. Allow to cool to room temperature under 80% N_2 + 20% CO_2. Distribute into bottles under 80% N_2 + 20% CO_2. Autoclave for 15 min at 15 psi pressure–121°C.

Solution B:
Composition per 70.0mL:
$NaHCO_3$ 3.5g

Preparation of Solution B: Add $NaHCO_3$ to distilled/deionized water and bring volume to 70.0mL. Mix thoroughly. Filter sterilize. Sparge with 80% N_2 + 20% CO_2 for 15 min.

Solution C:
Composition per 10.0mL:
L-Cysteine·HCl 0.3g

Preparation of Solution C: Add L-cysteine·HCl to distilled/deionized water and bring volume to 10.0mL. Mix thoroughly. Sparge with 100% N_2 for 3–4 min. Autoclave under 100% N_2 for 15 min at 15 psi pressure–121°C.

Solution D:
Composition per 10.0mL:
$Na_2S \cdot 9H_2O$ 0.3g

Preparation of Solution D: Add $Na_2S \cdot 9H_2O$ to distilled/deionized water and bring volume to 10.0mL. Mix thoroughly. Sparge with 100% N_2 for 3–4 min. Autoclave under 100% N_2 for 15 min at 15 psi pressure–121°C.

Preparation of Medium: To 916.0mL of sterile solution A, add 70.0mL of sterile solution B, 10.0mL of sterile solution C, and 10.0mL of sterile solution D. Mix thoroughly.

Use: For the cultivation and maintenance of *Syntrophus buswelii*.

Syntrophus Medium, Sulfate-Free

Composition per 1006.0mL:
Solution A 916.0mL
Solution B 70.0mL
Solution C 10.0mL
Solution D 10.0mL

pH 7.2 ± 0.2 at 25°C

Solution A:
Composition per 916.0mL:
Sodium benzoate 2.0g
Pancreatic digest of casein 1.0g
Resazurin 1.0mg
Mineral solution 50.0mL
Rumen fluid, clarified 50.0mL
Vitamin solution 5.0mL
Trace elements solution SL-10 .... 1.0mL

Mineral Solution:
Composition per liter:
Nitrilotriacetic acid 12.5g
NaCl 1.0g
$FeCl_3 \cdot 4H_2O$ 0.2g
$MnCl_2 \cdot 4H_2O$ 0.1g
$CaCl_2 \cdot 2H_2O$ 0.1g
$ZnCl_2$ 0.1g
$CuCl_2$ 0.02g
Na_2SeO_3 .... 0.02g
$CoCl_2 \cdot 6H_2O$ 0.017g
H_3BO_3 .... 0.01g
$Na_2MoO_4 \cdot 2H_2O$ 0.01g

Preparation of Mineral Solution: Add nitrilotriacetic acid to 500.0mL of distilled/deionized water. Adjust pH to 6.5 with KOH. Add remaining components. Add distilled/deionized water to 1.0L. Mix thoroughly.

Vitamin Solution:
Composition per liter:
Pyridoxine·HCl 6.2mg
Nicotinic acid 2.5mg
Thiamine·HCl 1.25mg
p-Aminobenzoic acid 1.25mg
Pantothenic acid 0.62mg
Biotin 0.25mg

Preparation of Vitamin Solution: Add components to distilled/deionized water and bring volume to 1.0L. Mix thoroughly.

Trace Elements Solution SL-10:
Composition per liter:
$FeCl_2 \cdot 4H_2O$ 1.5g
$CoCl_2 \cdot 6H_2O$ 190.0mg
$MnCl_2 \cdot 4H_2O$ 100.0mg
$ZnCl_2$ 70.0mg
$Na_2MoO_4 \cdot 2H_2O$ 36.0mg
$NiCl_2 \cdot 6H_2O$ 24.0mg
H_3BO_3 .... 6.0mg
$CuCl_2 \cdot 2H_2O$ 2.0mg
HCl (25% solution) 10.0mL

Preparation of Trace Elements Solution SL-10: Add $FeCl_2 \cdot 4H_2O$ to 10.0mL of HCl solution. Mix thoroughly. Add distilled/deionized water and bring volume to 1.0L. Add remaining components. Mix thoroughly.

Preparation of Solution A: Add components to distilled/deionized water and bring volume to 916.0mL. Adjust pH to 7.2. Gently heat and bring to boiling. Continue boiling for a few minutes. Allow to cool to room temperature under 80% N_2 + 20% CO_2. Distribute into bottles under 80% N_2 + 20% CO_2. Autoclave for 15 min at 15 psi pressure–121°C.

Solution B:
Composition per 70.0mL:
$NaHCO_3$ 3.5g

Preparation of Solution B: Add $NaHCO_3$ to distilled/deionized water and bring volume to 70.0mL. Mix thoroughly. Filter sterilize. Sparge with 80% N_2 + 20% CO_2 for 15 min.

Solution C:
Composition per 10.0mL:
L-Cysteine·HCl 0.3g

Preparation of Solution C: Add L-cysteine·HCl to distilled/deionized water and bring volume to 10.0mL. Mix thoroughly. Sparge with 100% N_2 for 3–4 min. Autoclave under 100% N_2 for 15 min at 15 psi pressure–121°C.

Solution D:

Composition per 10.0mL:

$Na_2S·9H_2O$	0.3g

Preparation of Solution D: Add $Na_2S·9H_2O$ to distilled/deionized water and bring volume to 10.0mL. Mix thoroughly. Sparge with 100% N_2 for 3–4 min. Autoclave under 100% N_2 for 15 min at 15 psi pressure–121°C.

Preparation of Medium: To 916.0mL of sterile solution A, add 70.0mL of sterile solution B, 10.0mL of sterile solution C, and 10.0mL of sterile solution D. Mix thoroughly.

Use: For the cultivation and maintenance of *Syntrophus buswelii*.

SYPC Medium (DSMZ Medium 1188)

Composition per liter:

Sea salts, Sigma	35.0g
Yeast extract	1.0g
Trypticase polypeptone	0.5g
Lactate solution	10.0mL
Wolfe's mineral elixir	1.0mL

pH 7.5 ± 0.2 at 25°C

Cellobiose Solution:

Composition per 10.0mL:

Cellobiose	2.0g

Preparation of Cellobiose Solution: Add cellobiose to distilled/deionized water and bring volume to 10.0mL. Mix thoroughly. Sparge with 100% N_2. Filter sterilize.

Wolfe's Mineral Elixir:

Composition per liter:

$MgSO_4·7H_2O$	30.0g
NaCl	10.0g
$MnSO_4·2H_2O$	5.0g
$(NH_4)_2NiSO_4·6H_2O$	2.8g
$CoCl_2·6H_2O$	1.8g
$ZnSO_4·7H_2O$	1.8g
$FeSO_4·7H_2O$	1.0g
$CaCl_2·2H_2O$	1.0g
$KAl(SO_4)_2·12H_2O$	0.18g
$CuSO_4·5H_2O$	0.1g
H_3BO_3	0.1g
$Na_2MoO_4·2H_2O$	0.1g
Na_2SeO_4	0.1g
$Na_2WO_4·2H_2O$	0.1g

T 7Agar Base (m-T7 Agar Base)

Composition per liter:

Lactose	20.0g
Agar	15.0g
Polyoxyethylene ether W-1	5.0g
Yeast extract	3.0g
Pancreatic digest of casein	2.5g
Peptic digest of animal tissue	2.5g
Sodium heptadecyl sulfate	0.1g
Bromthymol Blue	0.1g
Bromcresol Purple	0.1g

pH 7.4 ± 0.2 at 25°C

Source: This medium is available as a premixed powder from BD Diagnostic Systems.

Preparation of Medium: Add components to distilled/deionized water and bring volume to 1.0L. Mix thoroughly. Gently heat while stirring and bring to boiling. Distribute into tubes or flasks. Autoclave for 15 min at 15 psi pressure–121°C. Cool to 45°–50°C. The medium may be made more selective by adding 1.0mg of penicillin G per liter. Pour into sterile Petri dishes or leave in tubes.

Use: For the selective recovery and differential identification of injured coliform microorganisms from chlorinated water by the membrane filter method. For rapid estimation of the bacteriological quality of water using the membrane filter method.

T-ASW Medium

Composition per 1003.0mL:

NaCl	25.0g
$Na_2S_2O_3·5H_2O$	2.5g
$MgSO_4·7H_2O$	1.5g
$(NH_4)_2SO_4$	1.0g
KH_2PO_4	0.4g
$CaCl_2·2H_2O$	0.3g
$NaHCO_3$	0.2g
Tris·HCl buffer, 0.1*M*, pH 7.5	200.0mL
Phenol Red (0.5% solution)	2.0mL
Trace elements solution	1.0mL

pH 7.5 ± 0.2 at 25°C

Trace Elements Solution:

Composition per liter:

Disodium EDTA	50.0g
$CaCl_2·2H_2O$	5.5g
$MnCl_2·4H_2O$	5.1g
$FeSO_4·7H_2O$	5.0g
$ZnSO_4·7H_2O$	2.2g
$CoCl_2·6H_2O$	1.6g
$CuSO_4·5H_2O$	1.6g
$(NH_4)_6Mo_7O_{24}·4H_2O$	1.1g

Preparation of Trace Elements Solution: Add components to distilled/deionized water and bring volume to 1.0L. Mix thoroughly. Adjust pH to 6.0 with KOH. Autoclave for 15 min at 15 psi pressure–121°C.

Preparation of Medium: Add components to distilled/deionized water and bring volume to 1.0L. Mix thoroughly. Adjust pH to 7.5. Filter sterilize.

Use: For the cultivation and maintenance of *Thiobacillus hydrothermalis*.

T2 Medium for *Thiobacillus*

Composition per liter:

Solution A	250.0mL
Solution B	250.0mL
Solution C	250.0mL
Solution D	250.0mL

pH 7.0 ± 0.2 at 25°C

Solution A:

Composition per 250.0mL:

$Na_2S_2O_3 \cdot 5H_2O$	5.0g
KNO_3	2.0g
NH_4Cl	1.0g

Preparation of Solution A: Add components to distilled/deionized water and bring volume to 250.0mL. Mix thoroughly. Filter sterilize.

Solution B:

Composition per 250.0mL

KH_2PO_4	2.0g

Preparation of Solution B: Add KH_2PO_4 to distilled/deionized water and bring volume to 250.0mL. Mix thoroughly. Filter sterilize.

Solution C:

Composition per 250.0mL

$NaHCO_3$	2.0g

Preparation of Solution C: Add $NaHCO_3$ to distilled/deionized water and bring volume to 250.0mL. Mix thoroughly. Filter sterilize.

Solution D:

Composition per 250.0mL

$MgSO_4 \cdot 7H_2O$	0.8g
$FeSO_4 \cdot 7H_2O$ (2%, w/v, in 1*N* HCl)	1.0mL
Trace metal solution	1.0mL

Preparation of Solution D: Add components to distilled/deionized water and bring volume to 250.0mL. Mix thoroughly. Filter sterilize.

$FeSO_4 \cdot 7H_2O$ Solution:

Composition per 100.0mL

$FeSO_4 \cdot 7H_2O$	2.0g
HCl (1*N* solution)	100.0mL

Preparation of $FeSO_4 \cdot 7H_2O$ Solution: Add the $FeSO_4 \cdot 7H_2O$ to the HCl solution. Mix thoroughly.

Trace Metals Solution:

Composition per liter:

EDTA	50.0g
$ZnSO_4$	22.0g
$CaCl_2$	5.54g
$MnCl_2$	5.06g
$FeSO_4 \cdot 7H_2O$	4.99g
$CoCl_2$	1.61g
$CuSO_4$	1.57g
$(NH_4)_2MoO_4$	1.1g

Preparation of Trace Metals Solution: Add components to distilled/deionized water and bring volume to 1.0L. Mix thoroughly. Adjust pH to 6.0 with KOH.

Preparation of Medium: Aseptically combine the four sterile solutions: solution A, solution B, solution C, and solution D. Adjust the pH to 7.0. Aseptically distribute into sterile tubes or flasks.

Use: For the cultivation and maintenance of *Thiobacillus denitrificans* and other thiobacilli.

Tap Water Agar

Composition per liter:

Agar	15.0g
Tap water	1.0L

Preparation of Medium: Add agar to 1.0L of tap water. Mix thoroughly. Gently heat and bring to boiling. Autoclave for 15 min at 15 psi pressure–121°C. Pour into sterile Petri dishes.

Use: For the cultivation and differentiation of fungi and aerobic actinomycetes based on filament and aerial hyphae morphology.

Tarshis Blood Agar

Composition per 1050.0mL:

Beef heart infusion	500.0g
Agar	15.0g
Meat peptone	10.0g
NaCl	5.0g
Penicillin G, sterile	100,000U
Sheep blood, sterile	300.0mL
Glycerol	10.0mL

pH 6.6 ± 0.2 at 25°C

Preparation of Medium: Add components, except sheep blood and penicillin G, to distilled/deionized water and bring volume to 750.0mL. Mix thoroughly. Gently heat and bring to boiling. Autoclave for 15 min at 15 psi pressure–121°C. Cool to 45°–50°C. Aseptically add sterile sheep blood and sterile penicillin G. Mix thoroughly. Pour into sterile Petri dishes or distribute into sterile tubes.

Use: For the isolation and cultivation of *Mycobacterium tuberculosis.*

Tartoff-Hobbs HiVeg Broth with Glycerol (Terrific HiVeg Broth)

Composition per liter:

Yeast extract	24.0g
Plant hydrolysate	12.0g
KH_2PO_4	9.4g
K_2HPO_4	2.2g
Glycerol	4.0mL

pH 7.2 ± 0.2 at 25°C

Source: This medium, without glycerol, is available as a premixed powder from HiMedia.

Preparation of Medium: Add components to distilled/deionized water and bring volume to 1.0L. Mix thoroughly. Distribute into tubes or flasks. Autoclave for 15 min at 15 psi pressure–121°C.

Use: For the cultivation of plasmid-bearing strains of *Escherichia coli.*

TAT Broth Base (Trypticase™ Azolectin Tween™ Broth Base)

Composition per liter:

Pancreatic digest of casein	20.0g
Lecithin	5.0g
Polysorbate 20 (Tween™ 20)	40.0mL

pH 7.2 ± 0.2 at 25°C

Source: This medium is available as a premixed powder from BD Diagnostic Systems.

Preparation of Medium: Add pancreatic digest of casein and lecithin to distilled/deionized water and bring volume to 960.0mL. Add the Tween™ 20. Mix thoroughly. Gently heat and bring to 48°–50°C for 30 min. Distribute into tubes or flasks. Autoclave for 15 min at 15 psi pressure–121°C.

Use: For the isolation of Gram-negative organisms from topical drugs and cosmetics.

TAT HiVeg Broth Base with Polysorbate

Composition per liter:

Plant hydrolysate....................20.0g
Azolectin....................5.0g
Polysorbate 20 (Tween™ 20)....................40.0mL

pH 7.2 ± 0.2 at 25°C

Source: This medium, without polysorbate, is available as a premixed powder from HiMedia.

Preparation of Medium: Add plant hydrolysate and azolectin to distilled/deionized water and bring volume to 960.0mL. Add 40.0mL Tween™ 20. Mix thoroughly. Gently heat and bring to 48°–50°C for 30 min. Distribute into tubes or flasks. Autoclave for 15 min at 15 psi pressure–121°C.

Use: For the isolation of Gram-negative organisms from topical drugs and cosmetics.

Tatum Motility Test and Maintenance Medium

See: **Motility Test and Maintenance Medium, Tatum**

Taurocholate Tellurite Gelatin Agar

See: **Monsur Agar**

TB Broth Base

Composition per liter:

Proteose peptone....................4.0g
Na_2HPO_4....................2.5g
Yeast extract....................2.0g
Sodium citrate....................1.5g
KH_2PO_4....................1.0g
$MgSO_4 \cdot 7H_2O$....................0.6g
Polysorbate 80....................0.5g
Bovine albumin solution....................50.0mL
Glucose solution....................10.0mL
Glycerol....................5.0mL

pH 7.0 ± 0.2 at 25°C

Source: This medium is available from HiMedia.

Glucose Solution:

Composition per 10.0mL:

Glucose....................5.0g

Preparation of Glucose Solution: Add glucose to distilled/deionized water and bring volume to 10.0mL. Mix thoroughly. Filter sterilize.

Bovine Albumin Solution:

Composition per 50.0mL:

Bovine serum albumin....................5.0g

Preparation of Bovine Albumin Solution: Add bovine serum albuin to distilled/deionized water and bring volume to 50.0mL. Mix thoroughly. Filter sterilize.

Preparation of Medium: Add components, except bovine albumin and glucose solution, to distilled/deionized water and bring volume to 960.0mL. Mix thoroughly. Autoclave for 15 min at 15 psi pressure–121°C. Cool to 50°C. Aseptically add glucose solution and bovine albumin solution. Mix thoroughly. Aseptically distribute into sterile tubes or flasks.

Use: For the cultivation of *Mycobacterium tuberculosis.*

TB Broth Base without Polysorbate 80

Composition per liter:

Proteose peptone....................4.0g
Na_2HPO_4....................2.5g
Yeast extract....................2.0g
Sodium citrate....................1.5g
KH_2PO_4....................1.0g
$MgSO_4 \cdot 7H_2O$....................0.6g
Bovine albumin solution....................50.0mL
Glucose solution....................10.0mL

pH 7.0 ± 0.2 at 25°C

Source: This medium is available from HiMedia.

Glucose Solution:

Composition per 10.0mL:

Glucose....................5.0g

Preparation of Glucose Solution: Add glucose to distilled/deionized water and bring volume to 10.0mL. Mix thoroughly. Filter sterilize.

Bovine Albumin Solution:

Composition per 50.0mL:

Bovine serum albumin....................5.0g

Preparation of Bovine Albumin Solution: Add bovine serum albumin to distilled/deionized water and bring volume to 50.0mL. Mix thoroughly. Filter sterilize.

Preparation of Medium: Add glycerol to distilled/deionized water and bring volume to 955.0mL. Mix thoroughly. Add remaining components, except bovine albumin and glucose solutions, Autoclave for 15 min at 15 psi pressure–121°C. Cool to 50°C. Aseptically add glucose solution and bovine albumin solution. Mix thoroughly. Aseptically distribute into sterile tubes or flasks.

Use: For the cultivation of *Mycobacterium tuberculosis.*

TB HiVeg Broth Base with Bovine Albumin and Glucose

Composition per liter:

Plant peptone No. 3....................4.0g
Na_2HPO_4....................2.5g
Yeast extract....................2.0g
Sodium citrate....................1.5g
KH_2PO_4....................1.0g
$MgSO_4$....................0.6g
Polysorbate 80....................0.5g
Glucose solution....................50.0mL
Bovine albumin solution....................50.0mL

pH 7.0 ± 0.2 at 25°C

Source: This medium, without glucose or bovine albumin, is available as a premixed powder from HiMedia.

Glucose Solution:

Composition per 100.0mL:

Glucose....................10.0g

Preparation of Glucose Solution: Add glucose to distilled/deionized water and bring volume to 100.0mL. Mix thoroughly. Filter sterilize.

Bovine Albumin Solution:

Composition per 100.0mL:

Bovine albumin fraction V....................10.0g

Preparation of Bovine Albumin Solution: Add bovine albumin to distilled/deionized water and bring volume to 100.0mL. Mix thoroughly. Adjust pH to 7.8 with NaOH. Store at –20°C.

Preparation of Medium: Add components, except bovine albumin solution and glucose solution, to distilled/deionized water and bring volume to 900.0L. Mix thoroughly. Gently heat until boiling. Autoclave for 15 min at 15 psi pressure–121°C. Cool to 50°C. Add 50.0mL sterile glucose solution and 50.0mL bovine albumin solution. Mix thoroughly. Aseptically distribute into sterile tubes.

Use: For the cultivation of *Mycobacterium tuberculosis*.

TB HiVeg Broth Base with Bovine Serum and Glucose

Composition per liter:

Plant peptone No. 3	4.0g
Na_2HPO_4	2.5g
Yeast extract	2.0g
Sodium citrate	1.5g
KH_2PO_4	1.0g
$MgSO_4$	0.6g
Polysorbate 80	0.5g
Glucose solution	50.0mL
Bovine serum	50.0mL

pH 7.0 ± 0.2 at 25°C

Source: This medium, without glucose or bovine albumin, is available as a premixed powder from HiMedia.

Glucose Solution:
Composition per 100.0mL:

Glucose	10.0g

Preparation of Glucose Solution: Add glucose to distilled/deionized water and bring volume to 100.0mL. Mix thoroughly. Filter sterilize.

Preparation of Medium: Add components, except bovine serum and glucose solution, to distilled/deionized water and bring volume to 900.0L. Mix thoroughly. Gently heat until boiling. Autoclave for 15 min at 15 psi pressure–121°C. Cool to 50°C. Add 50.0mL sterile glucose solution and 50.0mL bovine serum. Mix thoroughly. Aseptically distribute into sterile tubes.

Use: For the cultivation of *Mycobacterium tuberculosis*.

TB HiVeg Broth Base without Tween™ 80 with Bovine Albumin and Glucose

Composition per liter:

Plant peptone No. 3	4.0g
Na_2HPO_4	2.5g
Yeast extract	2.0g
Sodium citrate	1.5g
KH_2PO_4	1.0g
$MgSO_4$	0.6g
Glucose solution	50.0mL
Bovine albumin solution	50.0mL

pH 7.0 ± 0.2 at 25°C

Source: This medium, without glucose or bovine albumin, is available as a premixed powder from HiMedia.

Glucose Solution:
Composition per 100.0mL:

Glucose	10.0g

Preparation of Glucose Solution: Add glucose to distilled/deionized water and bring volume to 100.0mL. Mix thoroughly. Filter sterilize.

Bovine Albumin Solution:
Composition per 100.0mL:

Bovine albumin fraction V	10.0g

Preparation of Bovine Albumin Solution: Add bovine albumin to distilled/deionized water and bring volume to 100.0mL. Mix thoroughly. Adjust pH to 7.8 with NaOH. Store at –20°C.

Preparation of Medium: Add components, except bovine albumin solution and glucose solution, to distilled/deionized water and bring volume to 900.0L. Mix thoroughly. Gently heat until boiling. Autoclave for 15 min at 15 psi pressure–121°C. Cool to 50°C. Add 50.0mL sterile glucose solution and 50.0mL bovine albumin solution. Mix thoroughly. Aseptically distribute into sterile tubes.

Use: For the cultivation of mycobacteria when the presence of oleic acid is undesirable. For the cultivation of *Mycobacterium tuberculosis*.

TB HiVeg Broth Base without Tween™ 80 with Bovine Serum and Glucose

Composition per liter:

Plant peptone No. 3	4.0g
Na_2HPO_4	2.5g
Yeast extract	2.0g
Sodium citrate	1.5g
KH_2PO_4	1.0g
$MgSO_4$	0.6g
Glucose solution	50.0mL
Bovine serum	50.0mL

pH 7.0 ± 0.2 at 25°C

Source: This medium is available as a premixed powder from HiMedia.

Glucose Solution:
Composition per 100.0mL:

Glucose	10.0g

Preparation of Glucose Solution: Add glucose to distilled/deionized water and bring volume to 100.0mL. Mix thoroughly. Filter sterilize.

Preparation of Medium: Add components, except bovine serum and glucose solution, to distilled/deionized water and bring volume to 900.0L. Mix thoroughly. Gently heat until boiling. Autoclave for 15 min at 15 psi pressure–121°C. Cool to 50°C. Add 50.0mL sterile glucose solution and 50.0mL bovine serum. Mix thoroughly. Aseptically distribute into sterile tubes.

Use: For the cultivation of mycobacteria when the presence of oleic acid is undesirable. For the cultivation of *Mycobacterium tuberculosis*.

TB Nitrate Reduction Broth

Composition per 100.0mL:

$Na_2HPO_4 \cdot 12H_2O$	0.485g
KH_2PO_4	0.117g
$NaNO_3$	0.085g

pH 7.0 ± 0.2 at 25°C

Preparation of Medium: Add components to distilled/deionized water and bring volume to 100.0mL. Mix thoroughly. Distribute into tubes or flasks. Autoclave for 15 min at 15 psi pressure–121°C.

Use: For the differentiation of *Mycobacterium* species based on nitrate reduction. After growth of cells in appropriate medium, nitrate reduction is determined by making a suspension of cells in TB nitrate reduction broth and adding hydrochloric acid, sulfanilamide, and *N*-naphylenendiamine. Nitrate reduction turns the medium pink. *Mycobacterium tuberculosis* reduces nitrate and turns the medium deep pink within 1 min. *Mycobacterium bovis* does not reduce nitrate and does not change the medium.

TBAB 298 Medium
(Tryptose 298 Blood Agar Base Medium)

Tryptose Blood Agar Base:

Composition per 409.6mL:

Agar	15.0g
Tryptose	10.0g
NaCl	5.0g
Beef extract	3.0g
Glucose solution	4.0mL
Thymine solution	2.0mL
D-Alanine solution	2.0mL
Streptomycin sulfate solution	1.6mL

pH 7.2 ± 0.2 at 25°C

Glucose Solution:

Composition per 100.0mL:

Glucose	50.0g

Preparation of Glucose Solution: Add glucose to distilled/deionized water and bring volume to 100.0mL. Mix thoroughly. Filter sterilize.

Thymine Solution:

Composition per 10.0mL:

Thymine	0.1g

Preparation of Thymine Solution: Add thymine to distilled/deionized water and bring volume to 10.0mL. Mix thoroughly. Filter sterilize.

D-Alanine Solution:

Composition per 10.0mL:

D-Alanine	0.1g

Preparation of D-Alanine Solution: Add D-alanine to distilled/deionized water and bring volume to 10.0mL. Mix thoroughly. Filter sterilize.

Streptomycin Sulfate Solution:

Composition per 10.0mL

Streptomycin sulfate	2.5g

Preparation of Streptomycin Sulfate Solution: Add streptomycin sulfate to distilled/deionized water and bring volume to 10.0mL. Mix thoroughly. Filter sterilize.

Preparation of Medium: Add agar, tryptose, NaCl, and beef extract to distilled/deionized water and bring volume to 400.0mL. Mix thoroughly. Autoclave for 15 min at 15 psi pressure–121°C. Cool to 50°C. Aseptically add 4.0mL of the sterile glucose solution, 2.0mL of the sterile thymine solution, 2.0mL of the sterile alanine solution, and 1.6mL of the sterile streptomycin sulfate solution. Mix thoroughly. Aseptically distribute into sterile tubes or flasks.

Use: For the cultivation of *Bacillus subtilis*.

TBX Agar
(Tryptone Bile X-glucuronide Agar)

Composition per liter:

Peptone	20.0g
Agar	15.0g
Bile salts	1.5g
X-β-D-glucuronide	0.075g

pH 7.2 ± 0.2 at 25°C

Source: This medium is available from Fluka, Sigma-Aldrich.

Preparation of Medium: Add components to distilled/deionized water and bring volume to 1.0L. Mix thoroughly. Gently heat while stirring and bring to boiling. Autoclave for 15 min at 15 psi pressure–121°C. Pour into sterile Petri dishes.

Use: For the detection and enumeration of *E. coli* in foodstuffs, animal food, and water without further confirmation. *E. coli* colonies are colored blue-green. The presence of the enzyme β-D-glucuronidase differentiates most *E. coli* sp. from other coliforms. *E. coli* absorbs the chromogenic substrate 5-bromo-4-chloro-3-indolyl-β-D-glucuronide. The enzyme β-glucuronidase splits the bond between the chromophore 5-bromo-4-chloro-3-indolyl and the β-D-glucuronide. Growth of accompanying Gram-positive flora is largely inhibited by the use of bile salts.

TBYA Agar
(Tryptone Beef Yeast Extract Acetate Agar)

Composition per liter:

Agar	15.0g
Pancreatic digest of casein	2.0g
Beef extract	0.5g
Yeast extract	0.5g
Sodium actuate	0.2g

pH 7.2–7.4 at 25°C

Preparation of Medium: Add components to distilled/deionized water and bring volume to 1.0L. Mix thoroughly. Adjust pH to 7.2–7.4. Gently heat and bring to boiling. Distribute into tubes or flasks. Autoclave for 15 min at 15 psi pressure–121°C. Pour into sterile Petri dishes or leave in tubes.

Use: For the cultivation and maintenance of *Leuconostoc* species.

TC Amino Acids, HeLa 100X
See: **Tissue Culture Amino Acids, HeLa 100X**

TC Amino Acids, Minimal Eagle 50X
See: **Tissue Culture Amino Acids, Minimal Eagle 50X**

TC Dulbecco Solution
See: **Tissue Culture Dulbecco Solution**

TC Earle Solution
See: **Tissue Culture Earle Solution**

TC Hanks Solution
See: **Tissue Culture Hanks Solution**

TC Medium 199
See: **Tissue Culture Medium 199**

TC Medium Eagle, HeLa
See: **Tissue Culture Medium Eagle, HeLa**

TC Medium Eagle with Earle BSS
See: **Tissue Culture Medium Eagle with Earle Balanced Salt Solution**

TC Medium Eagle with Hanks BSS
See: **Tissue Culture Medium Eagle with Hank's Balanced Salt Solution**

TC Medium Ham F10
See: **Tissue Culture Medium Ham F10**

TC Medium NCTC 109
See: **Tissue Culture Medium NCTC 109**

TC Medium RPMI #1640
See: **Tissue Culture Medium RPMI #1640**

TC Minimal Medium Eagle Spinner Modified MEM-S
See: **Tissue Culture Minimal Medium Eagle Spinner Modified**

TC Minimal Medium Eagle with Earle BSS
See: **Tissue Culture Minimal Medium Eagle with Earle Balanced Salts Solution**

TC Tyrode Solution
See: **Tissue Culture Tyrode Solution**

TC Vitamins Minimal Eagle, 100X
See: **Tissue Culture Vitamins Minimal Eagle, 100X**

TCBS Agar (Thiosulfate Citrate Bile Salt Sucrose Agar)

Composition per liter:

Sucrose	20.0g
Agar	14.0g
NaCl	10.0g
Sodium citrate	10.0g
$Na_2S_2O_3$	10.0g
Yeast extract	5.0g
Pancreatic digest of casein	5.0g
Peptic digest of animal tissue	5.0g
Oxgall	5.0g
Sodium cholate	3.0g
Ferric citrate	1.0g
Thymol Blue	0.04g
Bromthymol Blue	0.04g

pH 8.6 ± 0.2 at 25°C

Source: This medium is available as a premixed powder from BD Diagnostic Systems.

Preparation of Medium: Add components to distilled/deionized water and bring volume to 1.0L. Mix thoroughly. Gently heat while stirring and bring to boiling. Do not autoclave. Cool to 45°–50°C. Pour into sterile Petri dishes or distribute into sterile tubes.

Use: For the selective isolation of *Vibrio cholerae* and *Vibrio parahaemolyticus* from a variety of clinical and nonclinical specimens.

TCBS HiVeg Agar

Composition per liter:

Sucrose	20.0g
Agar	15.0g
Plant peptone No. 3	15.0g
NaCl	10.0g
Sodium citrate	10.0g
$Na_2S_2O_3$	10.0g
Yeast extract	6.0g
Synthetic detergent No. II	2.0g
Ferric citrate	1.0g
Thymol Blue	0.04g
Bromthymol Blue	0.04g

pH 8.6 ± 0.2 at 25°C

Source: This medium is available as a premixed powder from HiMedia.

Preparation of Medium: Add components to distilled/deionized water and bring volume to 1.0L. Mix thoroughly. Gently heat while stirring and bring to boiling. Do not autoclave. Cool to 45°–50°C. Pour into sterile Petri dishes or distribute into sterile tubes.

Use: For the selective isolation of *Vibrio cholerae* and *Vibrio parahaemolyticus* from a variety of clinical and nonclinical specimens. For the cultivation of enteropathogenic vibrios causing food poisoning.

TCBS HiVeg Agar (Selective)

Composition per liter:

Sucrose	20.0g
Agar	15.0g
Plant special peptone	14.5g
NaCl	10.0g
Sodium citrate	10.0g
$Na_2S_2O_3$	10.0g
Yeast extract	5.0g
Synthetic detergent No. II	2.0g
Synthetic detergent No. IV	1.5g
Ferric citrate	1.0g
BromthymolBlue	0.04g
Thymol Blue	0.04g

pH 8.6 ± 0.2 at 25°C

Source: This medium is available as a premixed powder from HiMedia.

Preparation of Medium: Add components to distilled/deionized water and bring volume to 1.0L. Mix thoroughly. Gently heat while stirring and bring to boiling. Do not autoclave. Cool to 45°–50°C. Pour into sterile Petri dishes or distribute into sterile tubes.

Use: For the selective isolation of *Vibrio cholerae* and other enteropathogenic vibrios.

TCG Medium (DSMZ Medium 1009)

Composition per liter:

Casitone	5.0g
Glucose	4.0g
Tryptone	3.0g
Artificial seawater	1.0L

pH 7.5 ± 0.2 at 25°C

Artificial Seawater:

Composition per 10.0mL:

Sea salts, Instant Ocean	22.0g

Preparation of Artificial Seawater: Add Instant Ocean sea salts to distilled/deionized water and bring volume to 1.0L. Mix thoroughly.

Preparation of Medium: Add components to artificial seawater and bring volume to 1.0L. Mix thoroughly. Distribute to tubes or flasks. Autoclave for 15 min at 15 psi pressure–121°C.

Use: For the cultivation of *Mechercharimyces mesophilus.*

TCH Medium
(Thiophene 2 Carboxylic Acid Hydrazide Medium)

Composition per 1105.0mL:

Thiophene-2-carboxylic acid hydrazide	1.1mg
Middlebrook 7H10 agar base	1.0L
OADC enrichment	100.0mL
Glycerol	5.0mL

pH 6.6 ± 0.2 at 25°C

Middlebrook 7H10 Agar Base:
Composition per liter:

Agar	15.0g
Na_2HPO_4	1.5g
KH_2PO_4	1.5g
$(NH_4)_2SO_4$	0.5g
L-Glutamic acid	0.5g
Sodium citrate	0.4g
Ferric ammonium citrate	0.04g
$MgSO_4 \cdot 7H_2O$	0.025g
$ZnSO_4 \cdot 7H_2O$	1.0mg
$CuSO_4 \cdot 5H_2O$	1.0mg
Pyridoxine	1.0mg
Biotin	0.5mg
$CaCl_2 \cdot 2H_2O$	0.5mg
Malachite Green	0.25mg

Preparation of Middlebrook 7H10 Agar Base: Add glycerol to 900.0mL of distilled/deionized water and add remaining components. Mix thoroughly. Gently heat and bring to boiling.

Middlebrook OADC Enrichment:
Composition per 100.0mL:

Bovine albumin fraction V	5.0g
Glucose	2.0g
NaCl	0.85g
Oleic acid	0.05g
Catalase	4.0mg

Source: This enrichment is available as a prepared enrichment from BD Diagnostic Systems.

Preparation of Middlebrook OADC Enrichment: Add components to distilled/deionized water and bring volume to 100.0mL. Mix thoroughly. Filter sterilize.

Preparation for Medium: Combine components. Mix thoroughly. Distribute into tubes or flasks. Autoclave for 15 min at 15 psi pressure–121°C. Pour into sterile Petri dishes or leave in tubes.

Use: For the differentiation of *Mycobacterium* species based on sensitivity to TCH. *Mycobacterium bovis* is inhibited by TCH. *Mycobacterium tuberculosis* and other mycobacteria are generally resistant to low concentrations of TCH. This distinguishes *Mycobacterium bovis* from other nonchromogenic, slow-growing mycobacteria.

TCY Agar

Composition per liter:

NaCl	31.3g
Agar	15.0g
$MgCl_2 \cdot 6H_2O$	10.8g
$CaCl_2 \cdot 2H_2O$	1.0g
Casamino acids	1.0g
Tryptone	1.0g
KCl	0.7g
Yeast extract	0.2g

pH 7.2 ± 0.2 at 25°C

Preparation of Medium: Add components to distilled/deionized water and bring volume to 1.0L. Mix thoroughly. Gently heat and bring to boiling. Distribute into tubes or flasks. Autoclave for 15 min at 15 psi pressure–121°C. Pour into sterile Petri dishes or leave in tubes.

Use: For the cultivation and maintenance of *Flexibacter maritimus.*

TCY Broth

Composition per liter:

NaCl	31.3g
Agar	15.0g
$MgCl_2 \cdot 6H_2O$	10.8g
$CaCl_2 \cdot 2H_2O$	1.0g
Casamino acids	1.0g
Tryptone	1.0g
KCl	0.7g
Yeast extract	0.2g

pH 7.2 ± 0.2 at 25°C

Preparation of Medium: Add components to distilled/deionized water and bring volume to 1.0L. Mix thoroughly. Distribute into tubes or flasks. Autoclave for 15 min at 15 psi pressure–121°C.

Use: For the cultivation of *Flexibacter maritimus.*

TD3 Medium
(DSMZ Medium 876)

Composition per 1078.0mL:

NaCl	20.0g
$MgCl_2 \cdot 6H_2O$	9.8g
Na_2SO_4	4.0g
KCl	0.5g
NH_4Cl	0.25g
KH_2PO_4	0.2g
$CaCl_2 \cdot 2H_2O$	0.1g
Resazurin	0.5mg
$NaHCO_3$ solution	50.0mL
$Na_2S \cdot 9H_2O$ solution	13.0mL
Sodium caproate solution	10.0mL
Selenite-tungstate solution	2.0mL
Seven vitamin solution	1.0mL
Vitamin solution	1.0mL
Trace elements solution SL-10	1.0mL

pH 6.8 ± 0.2 at 25°C

$Na_2S \cdot 9H_2O$ Solution:
Composition per 20.0mL:

$Na_2S \cdot 9H_2O$	0.6g

Preparation of $Na_2S \cdot 9H_2O$ Solution: Add $Na_2S \cdot 9H_2O$ to distilled/deionized water and bring volume to 20.0mL. Mix thoroughly. Autoclave under 100% N_2 for 15 min at 15 psi pressure–121°C. Cool to room temperature.

$NaHCO_3$ Solution:
Composition per 10.0mL:

$NaHCO_3$	5.0g

Preparation of $NaHCO_3$ Solution: Add $NaHCO_3$ to distilled/deionized water and bring volume to 10.0mL. Mix thoroughly. Sparge with 80% N_2 + 20% CO_2. Filter sterilize.

Sodium Caproate Solution:

Composition per 10.0mL:

Sodium caproate	0.5g

Preparation of Sodium Caproate Solution: Add sodium caproate to distilled/deionized water and bring volume to 10.0mL. Mix thoroughly. Sparge with 100% N_2. Filter sterilize.

Seven Vitamin Solution:

Composition per liter:

Pyridoxine hydrochloride	300.0mg
Thiamine-HCl·$2H_2O$	200.0mg
Nicotinic acid	200.0mg
Vitamin B_{12}	100.0mg
Calcium pantothenate	100.0mg
p-Aminobenzoic acid	80.0mg
D(+)-Biotin	20.0mg

Preparation of Seven Vitamin Solution: Add components to distilled/deionized water and bring volume to 1.0L. Sparge with 100% N_2. Mix thoroughly. Filter sterilize.

Vitamin Solution:

Composition per 100.0mL:

Pyridoxine-HCl	10.0mg
Thiamine-HCl·$2H_2O$	5.0mg
Riboflavin	5.0mg
Nicotinic acid	5.0mg
D-Ca-pantothenate	5.0mg
p-Aminobenzoic acid	5.0mg
Lipoic acid	5.0mg
Biotin	2.0mg
Folic acid	2.0mg
Vitamin B_{12}	0.1mg

Preparation of Vitamin Solution: Add components to distilled/deionized water and bring volume to 100.0mL. Mix thoroughly. Sparge with 80% H_2 + 20% CO_2. Filter sterilize.

Trace Elements Solution SL-10:

Composition per liter:

$FeCl_2$·$4H_2O$	1.5g
$CoCl_2$·$6H_2O$	190.0mg
$MnCl_2$·$4H_2O$	100.0mg
$ZnCl_2$	70.0mg
Na_2MoO_4·$2H_2O$	36.0mg
$NiCl_2$·$6H_2O$	24.0mg
H_3BO_3	6.0mg
$CuCl_2$·$2H_2O$	2.0mg
HCl (25% solution)	10.0mL

Preparation of Trace Elements Solution SL-10: Add $FeCl_2$·$4H_2O$ to 10.0mL of HCl solution. Mix thoroughly. Add distilled/deionized water and bring volume to 1.0L. Add remaining components. Mix thoroughly. Sparge with 80% N_2 + 20% CO_2. Autoclave for 15 min at 15 psi pressure–121°C.

Selenite-Tungstate Solution

Composition per liter:

NaOH	0.5g
Na_2WO_4·$2H_2O$	4.0mg
Na_2SeO_3·$5H_2O$	3.0mg

Preparation of Selenite-Tungstate Solution: Add components to distilled/deionized water and bring volume to 1.0L. Mix thoroughly. Sparge with 100% N_2. Filter sterilize.

Optional Supplemental Fatty Acid Mixture:

Composition per 20.0mL:

Valeric acid	0.5g
Isovaleric acid	0.5g
Alpha-Methylbutyric acid	0.5g
Isobutyric acid	0.5g

Preparation of Optional Supplemental Fatty Acid Mixture: Add components to 20.0mL distilled/deionized water. Mix thoroughly. Sparge with 100% N_2. Filter sterilize.

Preparation of Medium: Prepare and dispense medium under 80% N_2 + 20% CO_2 gas atmosphere. Add components, except $NaHCO_3$ solution, sodium caproate solution, Na_2S·$9H_2O$ solution, vitamin solution, seven vitamin solution, selenite-tungstate solution, and trace elements solution SL-10, to distilled/deionized water and bring volume to 1.0L. Mix thoroughly. Adjust pH to 7.2. Sparge with 80% N_2 + 20% CO_2. Autoclave for 15 min at 15 psi pressure–121°C. Aseptically and anaerobically add 50.0mL $NaHCO_3$ solution, 10.0mL sodium caproate solution, 10.0mL Na_2S·$9H_2O$ solution, 1.0mL vitamin solution, 1.0mL seven vitamin solution, 2.0mL selenite-tungstate solution, and 1.0mL trace elements solution SL-10. Mix thoroughly. Aseptically and anaerobically distribute into sterile tubes or bottles. Growth can be stimulated by addition of clarified rumen fluid or 10–20mL of a mixture of fatty acids. Final pH should be 6.7–6.9.

Use: For the cultivation of unclassified bacterium DSM 13418.

TDC Medium

Composition per liter:

Agar	20.0g
$CaCO_3$	10.0g
Glucose	5.0g
K_2HPO_4	1.0g
$MgSO_4$	1.0g

Preparation of Medium: Add components to tap water and bring volume to 1.0L. Mix thoroughly. Gently heat and bring to boiling. Distribute into tubes or flasks. Autoclave for 15 min at 15 psi pressure–121°C. Pour into sterile Petri dishes or leave in tubes.

Use: For the cultivation and maintenance of *Azotobacter beijerinckii* and other *Azotobacter* species.

TDN Broth (DSMZ Medium 1012)

Composition per liter:

Pancreatic digest of gelatin	0.5g
Casamino acids	0.5g
Beef extract	0.3g
Yeast extract	0.1g
Magnesium chloride solution	10.0mL
Calcium chloride solution	10.0mL

pH 7.2 ± 0.2 at 25°C

Calcium Chloride Solution:

Composition per 10.0mL:

$CaCl_2$·$2H_2O$	0.3g

Preparation of Calcium Chloride Solution: Add $CaCl_2$·$2H_2O$ to distilled/deionized water and bring volume to 10.0mL. Mix thoroughly. Autoclave for 15 min at 15 psi pressure–121°C. Cool to room temperature.

Magnesium Chloride Solution:

Composition per 10.0mL:

$MgCl_2 \cdot 6H_2O$	0.6g

Preparation of Magnesium Chloride Solution: Add $MgCl_2 \cdot 6H_2O$ to distilled/deionized water and bring volume to 10.0mL. Mix thoroughly. Autoclave for 15 min at 15 psi pressure–121°C. Cool to room temperature.

Preparation of Medium: Add components, except calcium chloride and magnesium chloride solutions, to distilled/deionized water and bring volume to 980.0mL. Mix thoroughly. Adjust the pH to 7.2 with NaOH. Gently heat and bring to boiling. Autoclave for 15 min at 15 psi pressure–121°C. Cool to room temperature. Aspetically add calcium chloride and magnesium chloride solutions. Mix thoroughly.

Use: For the cultivation of *Bdellovibrio bacteriovorus.*

Tea Fungus Medium (DSMZ Medium 268)

Composition per liter:

Sucrose	50.0g
Tea leaves, black	5.0g

Add tea and sucrose to a flask. Add 1.0L freshly boiled tap water. Allow to stand for 15 min. Filter and cool to room temperature. Place a small disc of a cork stopper that has been steamed for 15 min on 2 successive days on the surface of the tea. Place the inoculum onto the cork. Cover the beaker with aluminum foil and incubate.

Use: For the cultivation of tea fungus.

TEC Agar (m-TEC Agar)

Composition per liter:

Agar	15.0g
Lactose	10.0g
NaCl	7.5g
Proteose peptone	5.0g
K_2HPO_4	3.3g
Yeast extract	3.0g
KH_2PO_4	1.0g
Sodium lauryl sulfate	0.2g
Sodium deoxycholate	0.1g
Bromcresol Purple	0.08g
Bromphenol Red	0.08g

pH 5.0 ± 0.2 at 25°C

Source: This medium is available as a premixed powder from BD Diagnostic Systems.

Preparation of Medium: Add components to distilled/deionized water and bring volume to 1.0L. Mix thoroughly. Gently heat and bring to boiling. Adjust pH to 5.0. Sterilization is unnecessary. Pour into sterile Petri dishes or distribute into sterile tubes or flasks. Store at 2°–8°C. Use within 1 week.

Use: For detection of *Escherichia coli* in recreational waters by the membrane filter method. This agar is used in conjunction with a urea substrate to detect urease production. After addition of the urea substrate, *Escherichia coli* appears as yellow-yellow/brown colonies when viewed under a fluorescent lamp.

Tech Agar

Composition per liter:

Pancreatic digest of gelatin	20.0g
Agar	13.6g
$K_2SO_4 \cdot 7H_2O$	10.0g
$MgCl_2 \cdot 6H_2O$	1.4g
Glycerol	10.0mL

pH 7.2 ± 0.2 at 25°C

Source: This medium is available as a premixed powder from BD Diagnostic Systems.

Preparation of Medium: Add components, except glycerol, to distilled/deionized water and bring volume to 990.0mL. Mix thoroughly. Add glycerol. Gently heat and bring to boiling. Distribute into tubes or flasks. Autoclave for 15 min at 15 psi pressure–121°C. Pour into sterile Petri dishes or leave in tubes.

Use: For the production of pyocyanin pigment by *Pseudomonas* species.

Teepol Broth, Enriched (m-Teepol Broth, Enriched)

Composition per liter:

Peptone	40.0g
Lactose	30.0g
Yeast extract	6.0g
Phenol Red	0.2g
Sodium lauryl sulfate (Teepol, 0.1% solution)	4.0mL

pH 7.4 ± 0.2 at 25°C

Preparation of Medium: Add components to distilled/deionized water and bring volume to 1.0L. Mix thoroughly. Distribute into tubes or flasks. Autoclave for 15 min at 15 psi pressure–121°C.

Use: For the enumeration of coliform organisms and *Escherichia coli* in water by the membrane filter method.

Teepol HiVeg Broth

Composition per liter:

Plant peptone	20.0g
Lactose	10.0g
NaCl	5.0g
Teepol	1.0g
Phenol Red	0.02g

pH 7.6 ± 0.2 at 25°C

Source: This medium is available as a premixed powder from HiMedia.

Preparation of Medium: Add components to distilled/deionized water and bring volume to 1.0L. Mix thoroughly. Distribute into tubes or flasks. Autoclave for 15 min at 15 psi pressure–121°C.

Use: For the enumeration of coliform organisms and *Escherichia coli* in water by the membrane filter method. For the selective isolation and identification of enteric, lactose-fermenting bacteria.

Tellurite Blood Agar

See: **Chocolate Tellurite Agar**

Tellurite Glycine Agar

Composition per liter:

Agar	17.5g
Pancreatic digest of casein	10.0g
Glycine	10.0g

Yeast extract	6.5g
D-Mannitol	5.0g
K_2HPO_4	5.0g
LiCl	5.0g
Enzymatic hydrolysate of soybean meal	3.5g
Chapman tellurite solution	10.0mL

pH 7.2 ± 0.2 at 25°C

Source: This medium is available as a premixed powder from BD Diagnostic Systems.

Caution: Lithium chloride is harmful. Avoid bodily contact and inhalation of vapors. On contact with skin wash with plenty of water immediately.

Chapman Tellurite Solution:

Composition per 100.0mL:

K_2TeO_3	1.0g

Preparation of Chapman Tellurite Solution: Add K_2TeO_3 to distilled/deionized water and bring volume to 100.0mL. Mix thoroughly. Filter sterilize.

Caution: Potassium tellurite is toxic.

Preparation of Medium: Add components, except Chapman tellurite solution, to distilled/deionized water and bring volume to 990.0mL. Mix thoroughly. Gently heat and bring to boiling. Autoclave for 15 min at 15 psi pressure–121°C. Cool to 50°–55°C. Aseptically add 10.0mL of sterile Chapman tellurite solution. Mix thoroughly. Pour into sterile Petri dishes or distribute into sterile tubes. Allow the surface of the plates to dry before inoculating.

Use: For the isolation and cultivation of coagulase-positive staphylococci.

Tellurite Glycine Agar

Composition per liter:

Agar	16.0g
Pancreatic digest of casein	10.0g
Glycine	10.0g
Yeast extract	5.0g
D-Mannitol	5.0g
K_2HPO_4	5.0g
LiCl	5.0g
Chapman tellurite solution	20.0mL

pH 7.2 ± 0.2 at 25°C

Source: This medium is available as a premixed powder from BD Diagnostic Systems.

Chapman Tellurite Solution:

Composition per 100.0mL:

K_2TeO_3	1.0g

Preparation of Chapman Tellurite Solution: Add K_2TeO_3 to distilled/deionized water and bring volume to 100.0mL. Mix thoroughly. Filter sterilize.

Caution: Potassium tellurite is toxic.

Preparation of Medium: Add components, except Chapman tellurite solution, to distilled/deionized water and bring volume to 980.0mL. Mix thoroughly. Gently heat and bring to boiling. Autoclave for 15 min at 15 psi pressure–121°C. Cool to 50°–55°C. Aseptically add 20.0mL of sterile Chapman tellurite solution. Mix thoroughly. Pour into sterile Petri dishes or distribute into sterile tubes. Allow the surface of the plates to dry before inoculating.

Use: For the quantitative detection of coagulase-positive staphylococci from foods and other sources.

Tellurite Polymyxin Egg Yolk Agar

See: **TPEY Agar**

TEP Uric Acid Medium

Composition per liter:

Agar	20.0g
$Na_2HPO_4 \cdot 12H_2O$	9.0g
Uric acid	4.0g
Pancreatic digest of casein	1.7g
KH_2PO_4	1.5g
NaCl	0.5g
Papaic digest of soybean meal	0.3g
K_2HPO_4	0.25g
Glucose	0.25g
$MgSO_4 \cdot 7H_2O$	0.2g
$CaCl_2$	0.02g
Ferric ammonium citrate	1.2mg
$MnCl_2 \cdot 4H_2O$	1.0mg

pH 7.2 ± 0.2 at 25°C

Preparation of Medium: Add components to distilled/deionized water and bring volume to 1.0L. Mix thoroughly. Gently heat and bring to boiling. Distribute into tubes or flasks. Autoclave for 15 min at 15 psi pressure–121°C. Pour into sterile Petri dishes or leave in tubes.

Use: For the cultivation and maintenance of *Bacillus fastidiosus* and other microorganisms that can utilize uric acid as a carbon source.

Tepidanaerobacter Medium (DSMZ Medium 1051)

Composition per liter:

Yeast extract	2.3g
NH_4Cl	0.54g
$MgCl_2 \cdot 6H_2O$	0.2g
$CaCl_2 \cdot 2H_2O$	0.15g
KH_2PO_4	0.14g
Resazurin	0.5mg
Bicarbonate solution	10.0mL
Glucose solution	10.0mL
Sulfide solution	10.0mL
Cysteine solution	10.0mL
Trace elements solution	1.0mL
Vitamin solution	2.0mL

pH 7.1 ± 0.2 at 25°C

Trace Elements Solution:

Composition per liter:

Nitrilotriacetic acid	12.8g
$FeCl_3 \cdot 6H_2O$	1.35g
NaCl	1.0g
$NiCl_2 \cdot 6H_2O$	0.12g
$MnCl_2 \cdot 4H_2O$	0.1g
$CaCl_2 \cdot 2H_2O$	0.1g
$ZnCl_2$	0.1g
$Na_2SeO_3 \cdot 5H_2O$	0.026g
$CuCl_2 \cdot 2H_2O$	0.025g
$CoCl_2 \cdot 6H_2O$	0.024g
$Na_2MoO_4 \cdot 2H_2O$	0.024g
H_3BO_3	0.01g

Preparation of Trace Elements Solution: Add nitrilotriacetic acid to approximately 500.0mL of distilled/deionized water. Dissolve by adding KOH and adjust pH to 6.5. Add remaining components. Bring volume to 1.0L with additional distilled/deionized water. Adjust pH to 7.0 with KOH.

Glucose Solution:

Composition per 10.0mL:

D-Glucose	2.2g

Preparation of Glucose Solution: Add glucose to distilled/deionized water and bring volume to 10.0mL. Mix thoroughly. Sparge with 100% N_2. Filter sterilize.

Sulfide Solution:

Composition per 10.0mL:

$Na_2S \cdot 9H_2O$	0.3g

Preparation of Sulfide Solution: Add $Na_2S \cdot 9H_2O$ to distilled/deionized water and bring volume to 10.0mL. Mix thoroughly. Autoclave under 100% N_2 for 15 min at 15 psi pressure–121°C. Cool to room temperature.

Bicarbonate Solution:

Composition per 10.0mL:

$NaHCO_3$	2.5g

Preparation of Bicarbonate Solution: Add $NaHCO_3$ to distilled/deionized water and bring volume to 10.0mL. Mix thoroughly. Sparge with a gas mixture of 80% N_2 + 20% CO_2. Filter sterilize.

Cysteine Solution:

Composition per 10.0mL:

L-Cysteine-HCl·$2H_2O$	0.3g

Preparation of Cysteine Solution: Add L-cysteine to distilled/deionized water and bring volume to 10.0mL. Mix thoroughly. Sparge with 100% N_2. Filter sterilize.

Vitamin Solution:

Composition per liter:

Pyridoxine-HCl	10.0mg
Thiamine-HCl·$2H_2O$	5.0mg
Riboflavin	5.0mg
Nicotinic acid	5.0mg
D-Ca-pantothenate	5.0mg
p-Aminobenzoic acid	5.0mg
Lipoic acid	5.0mg
Biotin	2.0mg
Folic acid	2.0mg
Vitamin B_{12}	0.1mg

Preparation of Vitamin Solution: Add components to distilled/deionized water and bring volume to 1.0L. Mix thoroughly. Sparge with 100% N_2. Filter sterilize.

Preparation of Medium: Add components, except vitamin, bicarbonate, glucose, cysteine, and sulfide solutions, to distilled/deionized water and bring volume to 958.0mL. Mix thoroughly. Gently heat and bring to boiling. Boil for 1 min. Cool to room temperature while sparging with a gas mixture of 80% N_2 + 20% CO_2. Dispense into culture vessels under an atmosphere of 80% N_2 + 20% CO_2. Autoclave for 15 min at 15 psi pressure–121°C. Cool to room temperature. Aspetically add vitamin, bicarbonate, glucose, cysteine, and sulfide solutions. Mix thoroughly. Adjust the pH to 7.0–7.2.

Use: For the cultivation of *Tepidanaerobacter syntrophicus*.

Tepidibacter Medium (DSMZ Medium 985)

Composition per liter:

NaCl	18.0g
Casein	10.0g
$MgCl_2 \cdot 6H_2O$	4.0g
KCl	0.34g
NH_4Cl	0.25g
Yeast extract	0.2g
KH_2PO_4	0.18g
$CaCl_2 \cdot 2H_2O$	0.11g
$Fe(NH_4)_2(SO_4)_2 \cdot 6H_2O$	0.02g
Resazurin	1.0mg
Bicarbonate solution	10.0mL
Vitamin solution	10.0mL
Sulfide solution	10.0mL
Trace elements solution SL-10	1.0mL

pH 6.8 ± 0.2 at 25°C

Sulfide Solution:

Composition per 10.0mL:

$Na_2S \cdot 9H_2O$	0.5g

Preparation of Sulfide Solution: Add $Na_2S \cdot 9H_2O$ to distilled/deionized water and bring volume to 10.0mL. Mix thoroughly. Autoclave under 100% N_2 for 15 min at 15 psi pressure–121°C. Cool to room temperature.

Bicarbonate Solution:

Composition per 10.0mL:

$NaHCO_3$	5.0g

Preparation of Bicarbonate Solution: Add $NaHCO_3$ to distilled/deionized water and bring volume to 10.0mL. Mix thoroughly. Sparge with gas mixture of 80% N_2 + 20% CO_2. Filter sterilize.

Vitamin Solution:

Composition per liter:

Pyridoxine-HCl	10.0mg
Thiamine-HCl·$2H_2O$	5.0mg
Riboflavin	5.0mg
Nicotinic acid	5.0mg
D-Ca-pantothenate	5.0mg
p-Aminobenzoic acid	5.0mg
Lipoic acid	5.0mg
Biotin	2.0mg
Folic acid	2.0mg
Vitamin B_{12}	0.1mg

Preparation of Vitamin Solution: Add components to distilled/deionized water and bring volume to 1.0L. Mix thoroughly. Sparge with 80% H_2 + 20% CO_2. Filter sterilize.

Trace Elements Solution SL-10:

Composition per liter:

$FeCl_2 \cdot 4H_2O$	1.5g
$CoCl_2 \cdot 6H_2O$	190.0mg
$MnCl_2 \cdot 4H_2O$	100.0mg
$ZnCl_2$	70.0mg
$Na_2MoO_4 \cdot 2H_2O$	36.0mg
$NiCl_2 \cdot 6H_2O$	24.0mg
H_3BO_3	6.0mg
$CuCl_2 \cdot 2H_2O$	2.0mg
HCl (25% solution)	10.0mL

Preparation of Trace Elements Solution SL-10: Add $FeCl_2 \cdot 4H_2O$ to 10.0mL of HCl solution. Mix thoroughly. Add distilled/deionized water and bring volume to 1.0L. Add remaining components. Mix thoroughly. Sparge with 100% N_2. Autoclave for 15 min at 15 psi pressure–121°C.

Preparation of Medium: Add components, except vitamin, bicarbonate, and sulfide solutions, to distilled/deionized water and bring volume to 970.0mL. Mix thoroughly. Gently heat and bring to boiling. Boil for 1 min. Cool to room temperature while sparging with a gas mixture of 80% N_2 + 20% CO_2. Dispense into culture vessels under an atmosphere of 80% N_2 + 20% CO_2. Autoclave for 15 min at 15 psi pressure–121°C. Cool to room temperature. Aspetically add vitamin, bicarbonate, and sulfide solutions. Mix thoroughly. Adjust the pH to 6.6–7.0.

Use: For the cultivation of *Tepidibacter thalassicus* and *Deferribacter autotrophicus.*

Tepidibacter Medium with Peptone (DSMZ Medium 985)

Composition per liter:

NaCl	18.0g
Peptone	10.0g
$MgCl_2 \cdot 6H_2O$	4.0g
KCl	0.34g
NH_4Cl	0.25g
Yeast extract	0.2g
KH_2PO_4	0.18g
$CaCl_2 \cdot 2H_2O$	0.11g
$Fe(NH_4)_2(SO_4)_2 \cdot 6H_2O$	0.02g
Resazurin	1.0mg
Bicarbonate solution	10.0mL
Vitamin solution	10.0mL
Sulfide solution	10.0mL
Trace elements solution SL-10	1.0mL
Wolfe's mineral elixir	1.0mL

pH 6.8 ± 0.2 at 25°C

Sulfide Solution:

Composition per 10.0mL:

$Na_2S \cdot 9H_2O$	0.5g

Preparation of Sulfide Solution: Add $Na_2S \cdot 9H_2O$ to distilled/deionized water and bring volume to 10.0mL. Mix thoroughly. Autoclave under 100% N_2 for 15 min at 15 psi pressure–121°C. Cool to room temperature.

Bicarbonate Solution:

Composition per 10.0mL:

$NaHCO_3$	5.0g

Preparation of Bicarbonate Solution: Add $NaHCO_3$ to distilled/deionized water and bring volume to 10.0mL. Mix thoroughly. Sparge with a gas mixture of 80% N_2 + 20% CO_2. Filter sterilize.

Vitamin Solution:

Composition per liter:

Pyridoxine-HCl	10.0mg
Thiamine-HCl·$2H_2O$	5.0mg
Riboflavin	5.0mg
Nicotinic acid	5.0mg
D-Ca-pantothenate	5.0mg
p-Aminobenzoic acid	5.0mg
Lipoic acid	5.0mg
Biotin	2.0mg
Folic acid	2.0mg
Vitamin B_{12}	0.1mg

Preparation of Vitamin Solution: Add components to distilled/deionized water and bring volume to 1.0L. Mix thoroughly. Sparge with 80% H_2 + 20% CO_2. Filter sterilize.

Trace Elements Solution SL-10:

Composition per liter:

$FeCl_2 \cdot 4H_2O$	1.5g
$CoCl_2 \cdot 6H_2O$	190.0mg
$MnCl_2 \cdot 4H_2O$	100.0mg
$ZnCl_2$	70.0mg
$Na_2MoO_4 \cdot 2H_2O$	36.0mg
$NiCl_2 \cdot 6H_2O$	24.0mg
H_3BO_3	6.0mg
$CuCl_2 \cdot 2H_2O$	2.0mg
HCl (25% solution)	10.0mL

Preparation of Trace Elements Solution SL-10: Add $FeCl_2 \cdot 4H_2O$ to 10.0mL of HCl solution. Mix thoroughly. Add distilled/deionized water and bring volume to 1.0L. Add remaining components. Mix thoroughly. Sparge with 100% N_2. Autoclave for 15 min at 15 psi pressure–121°C.

Wolfe's Mineral Elixir:

Composition per liter:

$MgSO_4 \cdot 7H_2O$	30.0g
NaCl	10.0g
$MnSO_4 \cdot 2H_2O$	5.0g
$(NH_4)_2NiSO_4 \cdot 6H_2O$	2.8g
$CoCl_2 \cdot 6H_2O$	1.8g
$ZnSO_4 \cdot 7H_2O$	1.8g
$FeSO_4 \cdot 7H_2O$	1.0g
$CaCl_2 \cdot 2H_2O$	1.0g
$KAl(SO_4)_2 \cdot 12H_2O$	0.18g
$CuSO_4 \cdot 5H_2O$	0.1g
H_3BO_3	0.1g
$Na_2MoO_4 \cdot 2H_2O$	0.1g
Na_2SeO_4	0.1g
$Na_2WO_4 \cdot 2H_2O$	0.1g

Preparation of Wolfe's Mineral Elixir: Adjust pH of 1.0L of distilled/deionized water to 1.0 with dilute H_2SO_4. Add remaining components one at a time. Mix throughly to dissolve.

Preparation of Medium: Add components, except vitamin, bicarbonate, and sulfide solutions, to distilled/deionized water and bring volume to 970.0mL. Mix thoroughly. Gently heat and bring to boiling. Boil for 1 min. Cool to room temperature while sparging with a gas mixture of 80% N_2 + 20% CO_2. Dispense into culture vessels under an atmosphere of 80% N_2 + 20% CO_2. Autoclave for 15 min at 15 psi pressure–121°C. Cool to room temperature. Aspetically add vitamin, bicarbonate, and sulfide solutions. Mix thoroughly. Adjust the pH to 6.6–7.0.

Use: For the cultivation of *Clostridium tepidiprofundi.*

Teredinobacter Medium

Composition 1010.0mL:

Solution A	50.0mL
Solution C	750.0mL
Solution B	200.0mL
Solution D	10.0mL

Solution A:

Composition per 50.0mL:

K_2HPO_4	20.0mg
Na_2CO_3	20.0mg
Ferric ammonium citrate	6.0mg
EDTA	1.0mg
Trace metal mix A5	1.0mL

Trace Metal Mix A5:

Composition per liter:

H_3BO_3	2.86g
$MnCl_2 \cdot 4H_2O$	1.81g
$ZnSO_4 \cdot 7H_2O$	0.222g
$Na_2MoO_4 \cdot 2H_2O$	0.39g
$CuSO_4 \cdot 5H_2O$	0.079g
$Co(NO_3)_2 \cdot 6H_2O$	49.4mg

Preparation of Trace Metal Mix A5: Add components to distilled/deionized water and bring volume to 1.0L. Mix thoroughly.

Preparation of Solution A: Add components to distilled/deionized water and bring volume to 50.0mL. Mix thoroughly. Autoclave for 15 min at 15 psi pressure–121°C. Cool to 50°–55°C.

Solution B:

Composition per 200.0mL:

Agar, noble	2.0g

Preparation of Solution B: Add agar to distilled/deionized water and bring volume to 200.0mL. Mix thoroughly. Gently heat and bring to boiling. Autoclave for 15 min at 15 psi pressure–121°C. Cool to 50°–55°C.

Solution C:

Composition per 750.0mL:

Seawater	750.0mL

Preparation of Solution C: Autoclave for 15 min at 15 psi pressure–121°C. Cool to 50°–55°C.

Solution D:

Composition per 10.0mL:

Cellulose, Sigmacell Type 101	1.0g

Preparation of Solution D: Add cellulose to distilled/deionized water and bring volume to 10.0mL. Mix thoroughly. Autoclave for 15 min at 15 psi pressure–121°C. Cool to 50°–55°C.

Preparation of Medium: Aseptically combine 50.0mL of sterile solution A, 200.0mL of sterile solution B, 750.0mL of sterile solution C, and 10.0mL of sterile solution D. Mix thoroughly. Aseptically distribute into sterile tubes.

Use: For the cultivation of unidentified bacterium ATCC 39867.

Tergitol 7 Agar

Composition per liter:

Lactose	20.0g
Agar	13.0g
Peptone	10.0g
Yeast extract	6.0g
Meat extract	5.0g
Tergitol-7	0.1g
Bromthymol Blue	0.05g
TTC solution	5.0mL

pH 7.2 ± 0.2 at 25°C

Source: This medium is available as a premixed powder from Oxoid Unipath.

TTC Solution:

Composition per 100.0mL:

Triphenyltetrazolium chloride	0.05g

Preparation of TTC Solution: Add triphenyltetrazolium chloride to distilled/deionized water and bring volume to 100.0mL. Mix thoroughly. Filter sterilize.

Preparation of Medium: Add components to distilled/deionized water and bring volume to 995.0mL. Mix thoroughly. Gently heat and bring to boiling. Autoclave for 15 min at 15 psi pressure–121°C. Cool to 50°C. Aseptically add 5.0mL of sterile TTC solution. Mix thoroughly. Pour into sterile Petri dishes or distribute into sterile tubes.

Use: For the detection and enumeration of coliforms. Lactose-fermenting bacteria appear as yellow colonies. Lactose-nonfermenting bacteria appear as blue colonies.

Tergitol 7 Agar

Composition per liter:

Agar	15.0g
Lactose	10.0g
Yeast extract	3.0g
Pancreatic digest of casein	2.5g
Peptic digest of animal tissue	2.5g
Tergitol 7	0.1g
Bromthymol Blue	25.0mg
TTC solution	3.0mL

pH 6.9 ± 0.2 at 25°C

Source: This medium is available as a premixed powder from BD Diagnostic Systems.

TTC Solution:

Composition per 100.0mL:

Triphenyltetrazolium chloride	1.0g

Preparation of TTC Solution: Add triphenyltetrazolium chloride to distilled/deionized water and bring volume to 100.0mL. Mix thoroughly. Filter sterilize.

Preparation of Medium: Add components to distilled/deionized water and bring volume to 997.0mL. Mix thoroughly. Gently heat and bring to boiling. Autoclave for 15 min at 15 psi pressure–121°C. Cool to 50°C. Aseptically add 3.0mL of sterile TTC solution. Mix thoroughly. Pour into sterile Petri dishes or distribute into sterile tubes.

Use: For the selective isolation and differentiation of coliform bacteria based on lactose fermentation. Lactose-fermenting bacteria appear as yellow colonies. Lactose-nonfermenting bacteria appear as blue colonies.

Tergitol 7 Agar H

Composition per liter:

Agar	15.0g
Lactose	10.0g
Yeast extract	3.0g
Pancreatic digest of casein	2.5g
Peptic digest of animal tissue	2.5g
Ferric ammonium citrate	0.5g
$Na_2S_2O_3$	0.5g
Tergitol 7	0.1g
Bromthymol Blue	0.025g

pH 7.2 ± 0.2

Preparation of Medium: Add components to distilled/deionized water and bring volume to 1.0L. Mix thoroughly. Gently heat and bring

to boiling. Distribute into tubes or flasks. Autoclave for 15 min at 15 psi pressure–121°C. Pour into sterile Petri dishes or leave in tubes.

Use: For the selective isolation and differentiation of enteric bacteria from urine.

Tergitol 7 Broth

Composition per liter:

Lactose	10.0g
Yeast extract	3.0g
Pancreatic digest of casein	2.5g
Peptic digest of animal tissue	2.5g
Tergitol 7	0.1g
Bromthymol Blue	25.0mg
TTC solution	3.0mL

pH 6.9 ± 0.2 at 25°C

Source: This medium is available as a premixed powder from BD Diagnostic Systems.

TTC Solution:

Composition per 100.0mL:

Triphenyltetrazolium chloride	1.0g

Preparation of TTC Solution: Add triphenyltetrazolium chloride to distilled/deionized water and bring volume to 100.0mL. Mix thoroughly. Filter sterilize.

Preparation of Medium: Add components to distilled/deionized water and bring volume to 997.0mL. Mix thoroughly. Gently heat while stirring and bring to boiling. Autoclave for 15 min at 15 psi pressure–121°C. Cool to 25°C. Aseptically add 3.0mL of sterile TTC solution. Mix thoroughly.

Use: For the isolation and cultivation of coliforms, *Salmonella,* and other enteric bacteria.

Tergitol 7 HiVeg Agar Base with TTC

Composition per liter:

Agar	15.0g
Lactose	10.0g
Plant peptone No. 3	5.0g
Yeast extract	3.0g
Sodium heptadecyl sulfate	0.1g
Bromthymol Blue	0.025g
TTC solution	5.0mL

pH 7.2 ± 0.2 at 25°C

Source: This medium, without TTC solution, is available as a premixed powder from HiMedia.

TTC Solution:

Composition per 100.0mL:

Triphenyltetrazolium chloride	0.05g

Preparation of TTC Solution: Add triphenyltetrazolium chloride to distilled/deionized water and bring volume to 100.0mL. Mix thoroughly. Filter sterilize.

Preparation of Medium: Add components to distilled/deionized water and bring volume to 995.0mL. Mix thoroughly. Gently heat and bring to boiling. Autoclave for 15 min at 15 psi pressure–121°C. Cool to 50°C. Aseptically add 5.0mL of sterile TTC solution. Mix thoroughly. Pour into sterile Petri dishes or distribute into sterile tubes.

Use: For the detection and enumeration of coliforms. Lactose-fermenting bacteria appear as yellow colonies. Lactose-nonfermenting bacteria appear as blue colonies. For the selective enumeration and identification of coliform organisms, as per Indian Standard published by BIS.

Tergitol 7 HiVeg Agar H

Composition per liter:

Agar	15.0g
Lactose	10.0g
Plant peptone No. 3	5.0g
Yeast extract	3.0g
Ferric ammonium citrate	0.5g
$Na_2S_2O_3$	0.5g
Tergitol 7	0.1g
Bromthymol Blue	0.025g

pH 7.2 ± 0.2

Preparation of Medium: Add components to distilled/deionized water and bring volume to 1.0L. Mix thoroughly. Gently heat and bring to boiling. Distribute into tubes or flasks. Autoclave for 15 min at 15 psi pressure–121°C. Pour into sterile Petri dishes or leave in tubes.

Use: For the selective isolation and differentiation of enteric bacteria from urine.

Tergitol-7 HiVeg Broth

Composition per liter:

Lactose	10.0g
Plant peptone No. 3	5.0g
Yeast extract	3.0g
Tergitol 7	0.1g
Bromthymol Blue	0.025g

pH 6.9 ± 0.2

Source: This medium is available as a premixed powder from HiMedia.

Preparation of Medium: Add components to distilled/deionized water and bring volume to 1.0L. Mix thoroughly. Gently heat and bring to boiling. Distribute into tubes or flasks. Autoclave for 15 min at 15 psi pressure–121°C.

Use: For the selective and differential medium for detection and enumeration of coliforms.

Termitobacter Medium

Composition per 1060.0mL

NH_4Cl	1.0g
NaCl	0.6g
Sodium acetate·$3H_2O$	0.5g
L-Cysteine·HCl	0.5g
K_2HPO_4	0.3g
KH_2PO_4	0.3g
$MgCl_2·6H_2O$	0.2g
$CaCl_2·2H_2O$	0.1g
KCl	0.1g
Yeast extract	0.2g
Resazurin	0.5mg
$NaHCO_3$ solution	40.0mL
$Na_2S·9H_2O$ solution	10.0mL
Trans-3,4,5-trimethoxycinnamate solution	10.0mL
Trace elements solution SL-10	1.5mL

$NaHCO_3$ Solution:

Composition per 10.0mL:

$NaHCO_3$	1.0g

Preparation of $NaHCO_3$ Solution: Add $NaHCO_3$ to distilled/deionized water and bring volume to 10.0mL. Mix thoroughly. Sparge with 80% N_2 + 20% CO_2. Autoclave for 15 min at 15 psi pressure–121°C.

$Na_2S{\cdot}9H_2O$ Solution:
Composition per 10.0mL:
$Na_2S{\cdot}9H_2O$ 0.3g

Preparation of $Na_2S{\cdot}9H_2O$ Solution: Add $Na_2S{\cdot}9H_2O$ to distilled/deionized water and bring volume to 10.0mL. Mix thoroughly. Sparge with 100% N_2. Autoclave for 15 min at 15 psi pressure–121°C. Before use, neutralize to pH 7.0 with sterile HCl.

Trans-3,4,5-Trimethoxycinnamate Solution:
Composition per 10.0mL:
Trans-3,4,5-trimethoxycinnamate 1.19g

Preparation of Trans-3,4,5-Trimethoxycinnamate Solution: Add trans-3,4,5-trimethoxycinnamate to distilled/deionized water and bring volume to 10.0mL. Mix thoroughly. Sparge with 100% N_2. Neutralize with NaOH. Autoclave for 15 min at 15 psi pressure–121°C.

Preparation of Medium: Prepare and dispense medium under 80% N_2 + 20% CO_2 gas mixture. Add components, except $NaHCO_3$ solution, trans-3,4,5-trimethoxycinnamate solution, and $Na_2S{\cdot}9H_2O$ solution, to distilled/deionized water and bring volume to 1.0L. Mix thoroughly. Sparge with 80% N_2 + 20% CO_2 gas mixture. Autoclave for 15 min at 15 psi pressure–121°C. Aseptically and anaerobically add 40.0mL of sterile $NaHCO_3$ solution, 10.0mL of sterile trans-3,4,5-trimethoxycinnamate solution, and 10.0mL of sterile $Na_2S{\cdot}9H_2O$ solution. Mix thoroughly. Aseptically and anaerobically distribute into sterile tubes or bottles.

Use: For the cultivation of *Sporobacter termitidis.*

Terrific Broth

Composition per liter:
Yeast extract 24.0g
Tryptone 12.0g
KH_2PO_4 .... 9.4g
K_2HPO_4 .... 2.2g
pH 7.2 ± 0.2 at 25°C

Preparation of Medium: Add components to distilled/deionized water and bring volume to 1.0L. Mix thoroughly. Distribute into tubes or flasks. Autoclave for 15 min at 15 psi pressure–121°C.

Use: For the cultivation of plasmid-bearing strains of *Escherichia coli.*

Terrific Broth with 50 μg/ml Ampicillin (ATCC Medium 2140)

Composition per liter:
Yeast extract 24.0g
Tryptone 12.0g
Glycerol 12.0g
Buffer solution 100.0mL
Ampicillin solution 10.0mL
pH 7.2 ± 0.2 at 25°C

Buffer Solution:
Composition per 100.0mL:
KH_2PO_4 .... 2.31g
K_2HPO_4 .... 12.54g

Preparation of Buffer Solution: Add components to 100.0mL distilled/deionized water. Mix thoroughly. Autoclave for 15 min at 15 psi pressure–121°C. Cool to 25°C.

Ampicillin Solution:
Composition per 10.0mL:
Ampicillin 50μg

Preparation of Ampicillin Solution: Add ampicillin to distilled/deionized water and bring volume to 10.0mL. Mix thoroughly. Filter sterilize.

Preparation of Medium: Add components, except buffer and ampicillin solutions, to 890.0mL distilled/deionized water. Mix thoroughly. Autoclave for 15 min at 15 psi pressure–121°C. Cool to 25°C. Aseptically add the buffer and ampicillin solutions. Mix thoroughly. Aseptically distribute to tubes or flasks.

Use: For the cultivation of plasmid-bearing strains of *Escherichia coli.*

Terrific Broth with 100 μg/ml Ampicillin (ATCC Medium 1946)

Composition per liter:
Yeast extract 24.0g
Tryptone 12.0g
Glycerol 12.0g
Buffer solution 100.0mL
Ampicillin solution 10.0mL
pH 7.2 ± 0.2 at 25°C

Buffer Solution:
Composition per 100.0mL:
KH_2PO_4 .... 2.31g
K_2HPO_4 .... 12.54g

Preparation of Buffer Solution: Add components to 100.0mL distilled/deionized water. Mix thoroughly. Autoclave for 15 min at 15 psi pressure–121°C. Cool to 25°C.

Ampicillin Solution:
Composition per 10.0mL:
Ampicillin 100μg

Preparation of Ampicillin Solution: Add ampicillin to distilled/deionized water and bring volume to 10.0mL. Mix thoroughly. Filter sterilize.

Preparation of Medium: Add components, except buffer and ampicillin solutions, to 890.0mL distilled/deionized water. Mix thoroughly. Autoclave for 15 min at 15 psi pressure–121°C. Cool to 25°C. Aseptically add the buffer and ampicillin solutions. Mix thoroughly. Aseptically distribute to tubes or flasks.

Use: For the cultivation of plasmid-bearing strains of *Escherichia coli.* For the cloning of *p3PK* plasmid for studying intracellular protein targeting, *pSH* plasmid which is a strong mammalian expression vector, and *pOT182* plasmid for the construction and use of a self-cloning promoter probe vector for Gram-negative bacteria.

Terrific Broth with 200 μg/ml Ampicillin (ATCC Medium 1945)

Composition per liter:
Yeast extract 24.0g
Tryptone 12.0g
Glycerol 12.0g

Buffer solution 100.0mL
Ampicillin solution 10.0mL

pH 7.2 ± 0.2 at 25°C

Buffer Solution:
Composition per 100.0mL:
KH_2PO_4 .... 2.31g
K_2HPO_4 .... 12.54g

Preparation of Buffer Solution: Add components to 100.0mL distilled/deionized water. Mix thoroughly. Autoclave for 15 min at 15 psi pressure–121°C. Cool to 25°C.

Ampicillin Solution:
Composition per 10.0mL:
Ampicillin 200.0μg

Preparation of Ampicillin Solution: Add ampicillin to distilled/deionized water and bring volume to 10.0mL. Mix thoroughly. Filter sterilize.

Preparation of Medium: Add components, except buffer and ampicillin solutions, to 890.0mL distilled/deionized water. Mix thoroughly. Autoclave for 15 min at 15 psi pressure–121°C. Cool to 25°C. Aseptically add the buffer and ampicillin solutions. Mix thoroughly. Aseptically distribute to tubes or flasks.

Use: For the cultivation of plasmid-bearing strains of *Escherichia coli.* For the cloning of *pSVbeta, pCMVbeta, pTKbeta, pADbeta, pNASS-beta, and pBS-hTOP2* plasmids in *E. coli.*

Tetracycline L Broth Medium

Composition per liter:
Agar 15.0g
Pancreatic digest of casein 10.0g
NaCl 5.0g
Yeast extract 5.0g
Peptone 5.0g
NaCl 5.0g
Yeast extract 2.0g
Beef extract 1.0g
Glucose 1.0g
Tetracycline solution 10.0mL

pH 7.0 ± 0.2 at 25°C

Tetracycline Solution:
Composition per 10.0mL:
Tetracycline 12.5mg

Preparation of Tetracycline Solution: Add tetracycline to distilled/deionized water and bring volume to 10.0mL. Mix thoroughly. Filter sterilize.

Preparation of Medium: Add components, except tetracycline solution, to distilled/deionized water and bring volume to 990.0mL. Mix thoroughly. Bring pH to 7.0. Autoclave for 15 min at 15 psi pressure–121°C. Aseptically add 10.0mL of sterile tetracycline solution. Mix thoroughly. Aseptically distribute into sterile tubes or flasks.

Use: For the cultivation of *Escherichia coli.*

Tetracycline Luria Agar No. 1

Composition per liter:
Agar 15.0g
Pancreatic digest of casein 10.0g
Yeast extract 5.0g
NaCl 0.5g
Glucose solution 20.0mL
Tetracycline solution 10.0mL

pH 7.0 ± 0.2 at 25°C

Glucose Solution:
Composition per 50.0mL:
Glucose 5.0g

Preparation of Glucose Solution: Add glucose to distilled/deionized water and bring volume to 50.0mL. Mix thoroughly. Filter sterilize.

Tetracycline Solution:
Composition per 10.0mL:
Tetracycline 12.0mg

Preparation of Tetracycline Solution: Add tetracycline to distilled/deionized water and bring volume to 10.0mL. Mix thoroughly. Filter sterilize.

Preparation of Medium: Add components, except tetracycline solution and glucose solution, to distilled/deionized water and bring volume to 970.0mL. Mix thoroughly. Bring pH to 7.0. Gently heat and bring to boiling. Autoclave for 15 min at 15 psi pressure–121°C. Cool to 50°–55°C. Aseptically add 10.0mL of sterile tetracycline solution and 20.0mL of sterile glucose solution. Mix thoroughly. Pour into sterile Petri dishes or distribute into sterile tubes.

Use: For the cultivation of *Escherichia coli.*

Tetracycline Luria Agar No. 2

Composition per liter:
Agar 15.0g
Pancreatic digest of casein 10.0g
Yeast extract 5.0g
NaCl 0.5g
Glucose solution 20.0mL
Tetracycline solution 10.0mL

pH 7.0 ± 0.2 at 25°C

Glucose Solution:
Composition per 50.0mL:
Glucose 5.0g

Preparation of Glucose Solution: Add glucose to distilled/deionized water and bring volume to 50.0mL. Mix thoroughly. Filter sterilize.

Tetracycline Solution:
Composition per 10.0mL:
Tetracycline 10.0mg

Preparation of Tetracycline Solution: Add tetracycline to distilled/deionized water and bring volume to 10.0mL. Mix thoroughly. Filter sterilize.

Preparation of Medium: Add components, except tetracycline solution and glucose solution, to distilled/deionized water and bring volume to 970.0mL. Mix thoroughly. Bring pH to 7.0. Gently heat and bring to boiling. Autoclave for 15 min at 15 psi pressure–121°C. Cool to 50°–55°C. Aseptically add 10.0mL of sterile tetracycline solution and 20.0mL of sterile glucose solution. Mix thoroughly. Pour into sterile Petri dishes or distribute into sterile tubes.

Use: For the cultivation of *Escherichia coli.*

Tetracycline Luria Agar No. 3

Composition per liter:

Agar	15.0g
Pancreatic digest of casein	10.0g
Yeast extract	5.0g
NaCl	0.5g
Glucose solution	20.0mL
Tetracycline solution	10.0mL

pH 7.0 ± 0.2 at 25°C

Glucose Solution:

Composition per 50.0mL:

Glucose	5.0g

Preparation of Glucose Solution: Add glucose to distilled/deionized water and bring volume to 50.0mL. Mix thoroughly. Filter sterilize.

Tetracycline Solution:

Composition per 10.0mL:

Tetracycline	20.0mg

Preparation of Tetracycline Solution: Add tetracycline to distilled/deionized water and bring volume to 10.0mL. Mix thoroughly. Filter sterilize.

Preparation of Medium: Add components, except tetracycline solution and glucose solution, to distilled/deionized water and bring volume to 970.0mL. Mix thoroughly. Bring pH to 7.0. Gently heat and bring to boiling. Autoclave for 15 min at 15 psi pressure–121°C. Cool to 50°–55°C. Aseptically add 10.0mL of sterile tetracycline solution and 20.0mL of sterile glucose solution. Mix thoroughly. Pour into sterile Petri dishes or distribute into sterile tubes.

Use: For the cultivation of *Escherichia coli*.

Tetracycline TY Salt Medium

Composition per liter:

$NaCl_2$	10.0g
Pancreatic digest of casein	10.0g
Yeast extract	5.0g
Tetracycline solution	10.0mL

pH 7.0 ± 0.2 at 25°C

Tetracycline Solution:

Composition per 10.0mL:

Tetracycline	12.5mg

Preparation of Tetracycline Solution: Add tetracycline to distilled/deionized water and bring volume to 10.0mL. Mix thoroughly. Filter sterilize.

Preparation of Medium: Add components, except tetracycline solution, to distilled/deionized water and bring volume to 990.0mL. Mix thoroughly. Bring pH to 7.0. Gently heat and bring to boiling. Autoclave for 15 min at 15 psi pressure–121°C. Aseptically add 10.0mL of sterile tetracycline solution. Mix thoroughly. Aseptically distribute into sterile tubes or flasks.

Use: For the cultivation of *Escherichia coli*.

Tetrahymena Medium

Composition per liter:

Pancreatic digest of casein	5.0g
Proteose peptone	5.0g
K_2HPO_4	0.2g

pH 7.2 ± 0.2 at 25°C

Preparation of Medium: Add components to distilled/deionized water and bring volume to 1.0L. Mix thoroughly. Adjust pH to 7.2. Distribute into screw-capped tubes or flasks. Autoclave for 15 min at 15 psi pressure–121°C.

Use: For the cultivation of *Tetrahymena americanis, Tetrahymena asiatica, Tetrahymena australis, Tetrahymena borealis, Tetrahymena canadensis, Tetrahymena capricornis, Tetrahymena caudata, Tetrahymena corlissi, Tetrahymena elliotti, Tetrahymena furgasoni, Tetrahymena hegewischi, Tetrahymena hyperangularis, Tetrahymena malaccensis, Tetrahymena mimbres, Tetrahymena nanney, Tetrahymena nipissingi, Tetrahymena pigmentosa, Tetrahymena pyriformis, Tetrahymena silvana, Tetrahymena sonneborni, Tetrahymena thermophila,* and *Tetrahymena tropicalis.*

Tetramethyl Ammonium Chloride Agar

Composition per liter:

Agar	15.0g
Tetramethyl ammonium chloride	1.0g
Thiamine·HCl	0.5g
Standard mineral base	1.0L

pH 6.5 ± 0.2 at 25°C

Standard Mineral Base:

Composition per liter:

$(NH_4)_2SO_4$	1.0g
Phosphate buffer (1*M* solution, pH 6.8)	40.0mL
Huntner's vitamin-free mineral base	20.0mL

Huntner's Vitamin-Free Mineral Base:

Composition per liter:

$MgSO_4 \cdot 7H_2O$	14.45g
Nitrilotriacetic acid	10.0g
$CaCl_2 \cdot 2H_2O$	3.335g
$FeSO_4 \cdot 7H_2O$	99.0mg
Metals "44"	50.0mg
$(NH_4)_6Mo_7O_{24} \cdot 4H_2O$	9.25mg

Preparation of Huntner's Vitamin Free Mineral Base: Add nitrilotriacetic acid to 500.0mL of distilled/deionized water. Adjust pH to 6.5 with KOH. Add remaining components. Add distilled/deionized water to 1.0L. Adjust pH to 7.0.

Metals "44":

Composition per 100.0mL:

$ZnSO_4 \cdot 7H_2O$	1095.0mg
$FeSO_4 \cdot 7H_2O$	500.0mg
Ethylenediaminetetraacetic acid	250.0mg
$MnSO_4 \cdot H_2O$	154.0mg
$CuSO_4 \cdot 5H_2O$	39.2mg
$Co(NO_3)_2 \cdot 6H_2O$	24.8mg
$Na_2B_4O_7 \cdot 10H_2O$	17.7mg

Preparation of Metals "44": Add components to distilled/deionized water and bring volume to 1.0L. Mix thoroughly.

Preparation of Medium: Add components to distilled/deionized water and bring volume to 1.0L. Mix thoroughly. Gently heat and bring to boiling. Distribute into tubes or flasks. Autoclave for 15 min at 15 psi pressure–121°C. Pour into sterile Petri dishes or leave in tubes.

Use: For the cultivation and maintenance of *Paracoccus kocurii.*

Tetramitus rostratus Flagellate Medium, PY/2

Composition per liter:

Na_2HPO_4	3.55g
KH_2PO_4	3.40g

$MgSO_4 \cdot 7H_2O$ 2.47g
$CaCl_2$, anhydrous 1.12g
Proteose peptone 0.5g
Yeast extract 0.5g

pH 6.5 ± 0.2 at 25°C

Preparation of Medium: Add components to distilled/deionized water and bring volume to 1.0L. Mix thoroughly. Adjust pH to 6.5. Distribute into screw-capped tubes or flasks. Autoclave for 15 min at 15 psi pressure–121°C.

Use: For the cultivation of *Tetramitus rostratus*.

Tetrathionate Brilliant Green HiVeg Broth

Composition per liter:

$CaCO_3$ 20.0g
$K_2S_4O_6$ 20.0g
Plant peptone 8.6g
Synthetic detergent No. II 8.0g
NaCl 6.4g
Brilliant Green 0.07g

pH 7.0 ± 0.2 at 25°C

Source: This medium is available as a premixed powder from HiMedia.

Preparation of Medium: Add components to distilled/deionized water and bring volume to 1.0L. Mix thoroughly. Gently heat and bring to boiling. A slight precipitate will remain. Do not autoclave. Cool to 25°C. Store at 4°C.

Use: For the enrichment and isolation of salmonellae.

Tetrathionate Broth

Composition per liter:

$Na_2S_2O_3$ 40.7g
$CaCO_3$ 25.0g
NaCl 4.5g
Peptone 4.5g
Yeast extract 1.8g
Beef extract 0.9g
Iodine solution 20.0mL

Iodine Solution:

Composition per 20.0mL:

Iodine 6.0g
KI 5.0g

Preparation of Iodine Solution: Add iodine and KI to distilled/deionized water and bring volume to 20.0mL. Mix thoroughly.

Preparation of Medium: Add components, except iodine solution, to distilled/deionized water and bring volume to 980.0mL. Mix thoroughly. Gently heat and bring to boiling. Do not autoclave. Cool to 40°C. Add 20.0mL of iodine solution. Mix thoroughly. Distribute into tubes in 10.0mL volumes. Use medium the same day it is prepared.

Use: For the selective isolation and enrichment of *Salmonella typhi* and other salmonellae from fecal specimens, sewage, and other specimens.

Tetrathionate Broth (FDA M145)

Composition per 1030.0mL:

Tetrathionate broth base 1.0L
Iodine-potassium iodide solution 20.0mL
Brilliant Green solution 10.0mL

pH 8.4 ± 0.2 at 25°C

Tetrathionate Broth Base:

Composition per liter:

$Na_2S_2O_3 \cdot 5H_2O$ 30.0g
$CaCO_3$ 10.0g
Polypeptone™ 5.0g
Bile salts 1.0g

Preparation of Tetrathionate Broth Base: Add components to distilled/deionized water and bring volume to 1.0L. Mix thoroughly. Gently heat and bring to boiling. A slight precipitate will remain. Do not autoclave. Cool to 25°C. Store at 4°C.

Iodine-Potassium Iodide Solution:

Composition per 20.0mL:

Iodine, resublimed 6.0g
KI 5.0g

Preparation of Iodine-Potassium Iodide Solution: Add KI to 5.0mL of sterile distilled/deionized water. Mix thoroughly. Add iodine. Mix thoroughly. Bring volume to 20.0mL with sterile distilled/deionized water.

Brilliant Green Solution:

Composition per 100.0mL:

Brilliant Green 0.1g

Preparation of Brilliant Green Solution: Add Brilliant Green to sterile distilled/deionized water and bring volume to 100.0mL. Mix thoroughly.

Preparation of Medium: Combine 1.0L of tetrathionate broth base, 20.0mL of iodine-potassium iodide solution, and 10.0mL of Brilliant Green solution. Mix thoroughly. Aseptically distribute into tubes in 10.0mL volumes. Do not heat medium after it has been mixed.

Use: For the selective isolation and cultivation of *Salmonella* species from foods.

Tetrathionate Broth (TT Broth)

Composition per liter:

$Na_2S_2O_3$ 30.0g
$CaCO_3$ 10.0g
Proteose peptone 5.0g
Bile salts 1.0g
Iodine solution 20.0mL

pH 8.4± 0.2 at 25°C

Source: This medium is available as a premixed powder from BD Diagnostic Systems.

Iodine Solution:

Composition per 20.0mL:

Iodine 6.0g
KI 5.0g

Preparation of Iodine Solution: Add iodine and KI to distilled/deionized water and bring volume to 20.0mL. Mix thoroughly.

Preparation of Medium: Add components, except iodine solution, to distilled/deionized water and bring volume to 980.0mL. Mix thoroughly. Gently heat and bring to boiling. Do not autoclave. Cool to 40°C. Add 20.0mL of iodine solution. Mix thoroughly. Distribute into tubes in 10.0mL volumes. Use medium the same day it is prepared.

Use: For the selective isolation and enrichment of *Salmonella typhi* and other salmonellae from infectious material.

Tetrathionate Broth (m-Tetrathionate Broth) (m-TT Broth)

Composition per liter:

$Na_2S_2O_3$	30.0g
$CaCO_3$	10.0g
Pancreatic digest of casein	2.5g
Peptic digest of animal tissue	2.5g
Iodine–iodide solution	20.0mL

pH 8.0 ± 0.2 at 25°C

Iodine-Iodide Solution:

Composition per 20.0mL:

Iodine	6.0g
KI	5.0g

Preparation of Iodine-Iodide Solution: Add iodine and KI to distilled/deionized water and bring volume to 20.0mL. Mix thoroughly.

Preparation of Medium: Add components, except iodine-iodide solution, to distilled/deionized water and bring volume to 980.0mL. Mix thoroughly. Gently heat and bring to boiling. Do not autoclave. Cool to 40°C. Add 20.0mL of iodine-iodide solution. Mix thoroughly. Distribute into tubes in 10.0mL volumes. Use medium the same day it is prepared.

Use: For the selective isolation in the membrane filter method of *Salmonella* species from feces, urine, foods, and other specimens of sanitary importance.

Tetrathionate Broth (m-Tetrathionate Broth)

Composition per liter:

$Na_2S_2O_3$	30.0g
Proteose peptone	5.0g
Bile salts	1.0g
Iodine solution	20.0mL

pH 8.0 ± 0.2 at 25°C

Source: This medium is available as a premixed powder from BD Diagnostic Systems.

Iodine Solution:

Composition per 20.0mL:

Iodine	6.0g
KI	5.0g

Preparation of Iodine Solution: Add iodine and KI to distilled/deionized water and bring volume to 20.0mL. Mix thoroughly.

Preparation of Medium: Add components, except iodine solution, to distilled/deionized water and bring volume to 980.0mL. Mix thoroughly. Gently heat and bring to boiling. Do not autoclave. Cool to 40°C. Add 20.0mL of iodine solution. Mix thoroughly. Use medium the same day it is prepared.

Use: For the enrichment of *Salmonella* species in the membrane filter method prior to placing the filter on selective media such as Brilliant Green broth.

Tetrathionate Broth, Hajna (TT Broth, Hajna)

Composition per liter:

$Na_2S_2O_3$	38.0g
$CaCO_3$	25.0g
Casein/meat peptone (50/50)	18.0g
NaCl	5.0g
D-Mannitol	2.5g
Yeast extract	2.0g
Glucose	0.5g
Sodium deoxycholate	0.5g
Brilliant Green	0.01g
Iodine solution	40.0mL

pH 7.5–7.8 at 25°C

Source: This medium is available as a premixed powder from BD Diagnostic Systems.

Iodine Solution:

Composition per 40.0mL:

KI	8.0g
Iodine	5.0g

Preparation of Iodine Solution: Add iodine and KI to distilled/deionized water and bring volume to 40.0mL. Mix thoroughly.

Preparation of Medium: Add components, except iodine solution, to distilled/deionized water and bring volume to 960.0mL. Mix thoroughly. Gently heat and bring to boiling. Do not autoclave. Cool to 40°C. Add 40.0mL of iodine solution. Mix thoroughly. Distribute into tubes in 10.0mL volumes. Use medium the same day it is prepared.

Use: For the isolation of *Salmonella* species, except *Salmonella typhi*, and *Arizona* species from fecal specimens, urine, food samples, and other specimens of sanitary significance.

Tetrathionate Broth with Novobiocin

Composition per liter:

$Na_2S_2O_3$	38.0g
$CaCO_3$	25.0g
Casein/meat peptone (50/50)	18.0g
NaCl	5.0g
Yeast extract	2.0g
D-Mannitol	0.5g
Glucose	0.5g
Sodium deoxycholate	0.5g
Brilliant Green	0.01g
Novobiocin	4.0mg
Iodine solution	40.0mL

pH 7.5–7.8 at 25°C

Iodine Solution:

Composition per 40.0mL:

KI	8.0g
Iodine	5.0g

Preparation of Iodine Solution: Add iodine and KI to distilled/deionized water and bring volume to 40.0mL. Mix thoroughly.

Preparation of Medium: Add components, except iodine solution, to distilled/deionized water and bring volume to 960.0mL. Mix thoroughly. Gently heat and bring to boiling. Do not autoclave. Cool to 40°C. Add 40.0mL of iodine solution. Mix thoroughly. Distribute into tubes in 10.0mL volumes. Use medium the same day it is prepared.

Use: For the isolation of *Salmonella* species, except *Salmonella typhi*, and *Arizona* species from fecal specimens and other specimens of sanitary importance. Novobiocin suppresses the growth of *Proteus* species.

Tetrathionate Broth, USA (TT Broth, USA)

$Na_2S_2O_3$	30.0g
$CaCO_3$	10.0g

Casein peptone 2.5g
Meat peptone 2.5g
Bile salts 1.0g
Iodine-iodide solution 20.0mL

Source: This medium is available as a premixed powder from Oxoid Unipath.

Iodine–Iodide Solution:
Composition per 20.0mL:
Iodine 6.0g
KI 5.0g

Preparation of Iodine–Iodide Solution: Add iodine and KI to distilled/deionized water and bring volume to 20.0mL. Mix thoroughly.

Preparation of Medium: Add components, except iodine solution, to distilled/deionized water and bring volume to 980.0mL. Mix thoroughly. Gently heat and bring to boiling. Do not autoclave. Cool to 40°C. Add 20.0mL of iodine solution. Mix thoroughly. Distribute into tubes in 10.0mL volumes. Use medium the same day it is prepared.

Use: For the selective enrichment of *Salmonella* species from feces, urine, foods, and other specimens of sanitary importance.

Tetrathionate Crystal Violet Enhancement Broth

Composition per liter:
Potassium tetrathionate 20.0g
Casein/meat peptone (50/50) 8.6g
NaCl 6.4g
Crystal Violet 0.005g
pH 6.5 ± 0.2 at 25°C

Preparation of Medium: Add components to distilled/deionized water and bring volume to 1.0L. Mix thoroughly. Distribute into tubes or flasks. Autoclave for 15 min at 15 psi pressure–121°C.

Use: For the isolation of *Salmonella* species, except *Salmonella typhi*, and *Arizona* species from fecal specimens, urine, food samples, and other specimens of sanitary significance.

Tetrathionate HiVeg Broth Base, Hajna with Iodine (TT HiVeg Broth Base)

Composition per liter:
$Na_2S_2O_3$ 38.0g
$CaCO_3$ 25.0g
Plant special peptone 18.0g
NaCl 5.0g
D-Mannitol 2.5g
Yeast extract 2.0g
Glucose 0.5g
Synthetic detergent No. III 0.5g
Brilliant Green 0.01g
Iodine solution 40.0mL
pH 7.6 ± 0.2 at 25°C

Source: This medium, without iodine, is available as a premixed powder from HiMedia.

Iodine Solution:
Composition per 40.0mL:
KI 8.0g
Iodine 5.0g

Preparation of Iodine Solution: Add iodine and KI to distilled/deionized water and bring volume to 40.0mL. Mix thoroughly.

Preparation of Medium: Add components, except iodine solution, to distilled/deionized water and bring volume to 960.0mL. Mix thoroughly. Gently heat and bring to boiling. Do not autoclave. Cool to 40°C. Add 40.0mL of iodine solution. Mix thoroughly. Distribute into tubes in 10.0mL volumes. Use medium the same day it is prepared.

Use: For the isolation of *Salmonella* species, except *Salmonella typhi*, and *Arizona* species from fecal specimens, urine, food samples, and other specimens of sanitary significance.

Tetrathionate Reductase Medium

Composition per tube:
Solution I 10.0mL
Solution III 0.2mL
Solution II 0.1mL
Solution IV 0.1mL

Solution I:
Composition per liter:
$Na_2HPO_4 \cdot 12H_2O$ 3.6g
KH_2PO_4 .... 1.0g
NH_4Cl 0.5g
Peptone 0.25g
Yeast extract 0.25g
$MgSO_4 \cdot 7H_2O$ 0.03g

Preparation of Solution I: Add components to distilled/deionized water and bring volume to 1.0L. Mix thoroughly. Gently heat and bring to boiling. Distribute into tubes in 10.0mL volumes. Autoclave for 15 min at 15 psi pressure–121°C. Cool to 25°C.

Solution II:
Composition per 100.0mL:
$CaCl_2 \cdot 2H_2O$ 0.1g
Ferric ammonium citrate 0.05g

Preparation of Solution II: Add components to distilled/deionized water and bring volume to 100.0mL. Mix thoroughly. Gently heat and bring to boiling. Autoclave for 15 min at 15 psi pressure–121°C. Cool to 25°C.

Solution III:
Composition per 100.0mL:
Sodium succinate 15.0g

Preparation of Solution III: Add sodium succinate to distilled/deionized water and bring volume to 100.0mL. Mix thoroughly. Gently heat until dissolved. Autoclave for 15 min at 15 psi pressure–121°C. Cool to 25°C.

Solution IV:
Composition per 100.0mL:
$Na_2S_4O_6 \cdot 2H_2O$ 10.0g

Preparation of Solution IV: Add $Na_2S_4O_6 \cdot 2H_2O$ to distilled/deionized water and bring volume to 100.0mL. Mix thoroughly. Sterilize by filtration. Store at 4°C.

Preparation of Medium: To each tube containing 10.0mL of sterile solution I, aseptically add 0.1mL of sterile solution II, 0.2mL of sterile solution III, and 0.1mL of sterile solution IV. Mix thoroughly. Use immediately.

Use: For the cultivation and differentiation of hydrogen-oxidizing bacteria based on their production of tetrathionate reductase.

Tetrathionate Reductase Test Medium

Composition per 1025.0mL:

$K_2S_4O_6$	5.0g
Peptone water	1.0L
Bromthymol Blue (0.2% solution)	25.0mL

pH 7.4 ± 0.2 at 25°C

Peptone Water:

Composition per liter:

Peptone	10.0g
NaCl	5.0g

Preparation of Peptone Water: Add components to distilled/deionized water and bring volume to 1.0L. Mix thoroughly.

Preparation of Medium: Combine components. Mix thoroughly. Adjust pH to 7.4. Filter sterilize. Dispense into tubes in 1.0mL volumes or into wells of sterile microculture plates for replica inoculation.

Use: For the cultivation and identification of *Serratia* species based on their ability to reduce tetrathionate. Bacteria that reduce tetrathionate turn the medium yellow.

Tetrazolium Thallium Glucose Agar

Composition per liter:

Agar	14.0g
Beef extract	10.0g
Peptone	10.0g
Glucose solution	100.0mL
2,3,5-Triphenyltetrazolium·HCl solution	10.0mL
Thallous acetate solution	10.0mL

2,3,5-Triphenyltetrazolium·HCl Solution:

Composition per 10.0mL:

2,3,5-Triphenyltetrazolium·HCl	0.1g

Preparation of 2,3,5-Triphenyltetrazolium·HCl Solution: Add 2,3,5-triphenyltetrazolium·HCl to distilled/deionized water and bring volume to 10.0mL. Mix thoroughly. Autoclave for 7 min at 15 psi pressure–121°C.

Glucose Solution:

Composition per 100.0mL:

Glucose	10.0g

Preparation of Glucose Solution: Add glucose to distilled/deionized water and bring volume to 100.0mL. Mix thoroughly. Filter sterilize.

Thallous Acetate Solution:

Composition per 10.0mL:

Thallous acetate	1.0g

Preparation of Thallous Acetate Solution: Add thallous acetate to distilled/deionized water and bring volume to 10.0mL. Mix thoroughly. Filter sterilize.

Preparation of Medium: Add components—except glucose solution, 2,3,5-triphenyltetrazolium·HCl solution, and thallous acetate solution—to distilled/deionized water and bring volume to 880.0mL. Mix thoroughly. Gently heat and bring to boiling. Autoclave for 15 min at 15 psi pressure–121°C. Cool to 45°–50°C. Aseptically add sterile glucose solution, 2,3,5-triphenyltetrazolium·HCl solution, and thallous acetate solution. Mix thoroughly. Pour into sterile Petri dishes or distribute into sterile tubes.

Use: For the cultivation of *Streptococcus* species.

Tetrazolium Tolerance Agar (TTC Agar)

Composition per liter:

Pancreatic digest of casein	15.0g
Agar	15.0g
Triphenyltetrazolium chloride	10.0g
Papaic digest of soybean meal	5.0g
NaCl	5.0g

pH 7.3 ± 0.2 at 25°C

Preparation of Medium: Add components to distilled/deionized water and bring volume to 1.0L. Mix thoroughly. Gently heat and bring to boiling. Distribute into tubes or flasks. Autoclave for 15 min at 15 psi pressure–121°C. Do not overheat. Pour into sterile Petri dishes or leave in tubes.

Use: For the differentiation of bacteria based upon the ability to tolerate and grow in the presence of tetrazolium. *Streptococcus faecalis* (enterococci) rapidly reduces tetrazolium.

TF Medium

Composition per liter:

NaCl	7.0g
Pancreatic digest of casein	2.0g
Yeast extract	2.0g
$MgSO_4 \cdot 7H_2O$	1.8g
K_2HPO_4	1.6g
$MgCl_2 \cdot 6H_2O$	1.4g
$Na_2HPO_4 \cdot H_2O$	1.0g
NH_4Cl	0.5g
$Na_2S \cdot 9H_2O$	0.3g
$MgSO_4 \cdot 7H_2O$	0.16g
Resazurin	0.5mg
Wolfe's mineral solution	10.0mL
Wolfe's vitamin solution	10.0mL
Glucose solution	10.0mL
$CaCl_2 \cdot 2H_2O$ solution	10.0mL
L-Cysteine·HCl·H_2O solution	10.0mL

pH 6.8–7.0 at 25°C

Wolfe's Mineral Solution:

Composition per liter:

$MgSO_4 \cdot 7H_2O$	3.0g
Nitrilotriacetic acid	1.5g
NaCl	1.0g
$MnSO_4 \cdot 2H_2O$	0.5g
$CoCl_2 \cdot 6H_2O$	0.1g
$ZnSO_4 \cdot 7H_2O$	0.1g
$CaCl_2 \cdot 2H_2O$	0.1g
$FeSO_4 \cdot 7H_2O$	0.1g
$NiCl_2 \cdot 6H_2O$	0.025g
$KAl(SO_4)_2 \cdot 12H_2O$	0.02g
$CuSO_4 \cdot 5H_2O$	0.01g
H_3BO_3	0.01g
$Na_2MoO_4 \cdot 2H_2O$	0.01g
$Na_2SeO_3 \cdot 5H_2O$	0.3mg

Preparation of Wolfe's Mineral Solution: Add nitrilotriacetic acid to 500.0mL of distilled/deionized water. Adjust pH to 6.5 with KOH. Add remaining components. Add distilled/deionized water to 1.0L. Adjust pH to 6.8.

Wolfe's Vitamin Solution:
Composition per liter:

Pyridoxine·HCl	10.0mg
p-Aminobenzoic acid	5.0mg
Lipoic acid	5.0mg
Nicotinic acid	5.0mg
Riboflavin	5.0mg
Thiamine·HCl	5.0mg
Calcium DL-pantothenate	5.0mg
Biotin	2.0mg
Folic acid	2.0mg
Vitamin B_{12}	0.1mg

Preparation of Wolfe's Vitamin Solution: Add components to distilled/deionized water and bring volume to 1.0L. Mix thoroughly. Filter sterilize.

Glucose Solution:
Composition per 10.0mL:

D-Glucose	3.0g

Preparation of Glucose Solution: Add glucose to distilled/deionized water and bring volume to 10.0mL. Mix thoroughly. Sparge with 100% N_2. Autoclave for 15 min at 15 psi pressure–121°C.

$CaCl_2·2H_2O$ Solution:
Composition per 10.0mL:

$CaCl_2·2H_2O$	0.06g

Preparation of $CaCl_2·2H_2O$ Solution: Add $CaCl_2·2H_2O$ to distilled/deionized water and bring volume to 10.0mL. Mix thoroughly. Sparge with 100% N_2. Autoclave for 15 min at 15 psi pressure–121°C.

L-Cysteine·HCl·H_2O Solution:
Composition per 10.0mL:

L-Cysteine·HCl·H_2O	0.3g

Preparation of L-Cysteine·HCl·H_2O Solution: Add L-cysteine·HCl·H_2O to distilled/deionized water and bring volume to 10.0mL. Mix thoroughly. Sparge with 100% N_2. Autoclave for 15 min at 15 psi pressure–121°C.

Preparation of Medium: Prepare and dispense medium under 100% N_2. Add components, except glucose solution, $CaCl_2·2H_2O$ solution, and L-cysteine·HCl·H_2O solution, to distilled/deionized water and bring volume to 970.0mL. Mix thoroughly. Adjust pH to 6.8–7.0. Sparge with 100% N_2. Autoclave for 15 min at 15 psi pressure–121°C. Aseptically and anaerobically add 10.0mL of sterile glucose solution, 10.0mL of sterile $CaCl_2·2H_2O$ solution, and 10.0mL of sterile L-cysteine·HCl·H_2O solution. Mix thoroughly. Final pH should be 6.8–7.0.

Use: For the cultivation of *Thermosipho* species.

TF(A) Medium (DSMZ Medium 740)

Composition per liter:

K_2HPO_4	1.6g
Yeast extract	2.0g
Trypticase™	2.0g
$Na_2HPO_4·H_2O$	1.0g
NH_4Cl	0.5g
$MgSO_4·7H_2O$	0.16g
Resazurin	0.5mg
Calcium chloride solution	10.0mL
Glucose solution	10.0mL
Cysteine solution	10.0mL
$Na_2S·9H_2O$ solution	10.0mL
Trace elements solution	10.0mL
Vitamin solution	10.0mL

pH 6.8 ± 0.2 at 25°C

Calcium Chloride Solution:
Composition per 10.0mL:

$CaCl_2·2H_2O$	0.06g

Preparation of Calcium Chloride Solution: Add $CaCl_2·2H_2O$ to distilled/deionized water and bring volume to 10.0mL. Mix thoroughly. Sparge with 100% N_2. Autoclave for 15 min at 15 psi pressure–121°C. Cool to 25°C.

Cysteine Solution:
Composition per 10.0mL:

L-Cysteine·HCl·H_2O	0.3g

Preparation of Cysteine Solution: Add L-cysteine·HCl·H_2O to distilled/deionized water and bring volume to 10.0mL. Mix thoroughly. Sparge with 100% N_2. Autoclave for 15 min at 15 psi pressure–121°C. Cool to 25°C.

Glucose Solution:
Composition per 10.0mL:

Glucose	3.0g

Preparation of Glucose Solution: Add glucose to distilled/deionized water and bring volume to 10.0mL. Mix thoroughly. Sparge with 100% N_2. Filter sterilize.

$Na_2S·9H_2O$ Solution:
Composition per 10.0mL:

$Na_2S·9H_2O$	0.3g

Preparation of $Na_2S·9H_2O$ Solution: Add $Na_2S·9H_2O$ to distilled/deionized water and bring volume to 10.0mL. Sparge with N_2. Autoclave for 15 min at 15 psi pressure–121°C. Cool to 25°C. Store anaerobically.

Trace Elements Solution:
Composition per liter:

$MgSO_4·7H_2O$	3.0g
Nitrilotriacetic acid	1.5g
NaCl	1.0g
$MnSO_4·2H_2O$	0.5g
$CoSO_4·7H_2O$	0.18g
$ZnSO_4·7H_2O$	0.18g
$CaCl_2·2H_2O$	0.1g
$FeSO_4·7H_2O$	0.1g
$NiCl_2·6H_2O$	0.025g
$KAl(SO_4)_2·12H_2O$	0.02g
H_3BO_3	0.01g
$Na_2MoO_4·4H_2O$	0.01g
$CuSO_4·5H_2O$	0.01g
$Na_2SeO_3·5H_2O$	0.3mg

Preparation of Trace Elements Solution: Add nitrilotriacetic acid to 500.0mL of distilled/deionized water. Dissolve by adjusting pH to 6.5 with KOH. Add remaining components. Add distilled/deionized water to 1.0L. Mix thoroughly. Filter sterilize.

Vitamin Solution:
Composition per liter:

Pyridoxine-HCl	10.0mg
Thiamine-HCl·$2H_2O$	5.0mg
Riboflavin	5.0mg
Nicotinic acid	5.0mg
D-Ca-pantothenate	5.0mg

p-Aminobenzoic acid 5.0mg
Lipoic acid 5.0mg
Biotin 2.0mg
Folic acid 2.0mg
Vitamin B_{12} 0.1mg

Preparation of Vitamin Solution: Add components to distilled/deionized water and bring volume to 1.0L. Mix thoroughly. Sparge with 100% N_2. Filter sterilize.

Preparation of Medium: Prepare and dispense medium under an oxygen-free 100% N_2. Add components, except vitamin solution, cysteine solution, calcium chloride solution, glucose solution, trace elements solution, and $Na_2S{\cdot}9H_2O$ solution, to distilled/deionized water and bring volume to 940.0mL. Mix thoroughly. Sparge with 100% N_2. Autoclave for 15 min at 15 psi pressure–121°C. Cool to 25°C. Aseptically and anaerobically add 10.0mL sterile vitamin solution, 10.0mL of sterile cysteine solution, 10.0mL sterile glucose solution, 10.0mL sterile calcium chloride solution, 10.0mL sterile trace elements solution, and 10.0mL of sterile $Na_2S{\cdot}9H_2O$ solution. Mix thoroughly. Adjust pH to 6.8. Aseptically and anaerobically distribute into sterile tubes or flasks.

Use: For the cultivation of *Fervidobacterium pennivorans, Fervidobacterium gondwanense,* and *Fervidobacterium* sp.

TGA Medium

Composition per liter:

Pancreatic digest of casein 10.0g
NaCl 5.0g
Glucose 2.0g

pH 7.0 ± 0.2 at 25°C

Preparation of Medium: Add components to distilled/deionized water and bring volume to 1.0L. Mix thoroughly. Adjust pH to 7.0. Distribute into tubes or flasks. Autoclave for 15 min at 15 psi pressure–121°C.

Use: For the cultivation of *Salmonella typhimurium.*

TGB HiVeg Agar

Composition per liter:

Agar 15.0g
Plant hydrolysate 5.0g
Plant extract 3.0g
Glucose 1.0g

pH 7.0 ± 0.2 at 25°C

Source: This medium is available as a premixed powder from HiMedia.

Preparation of Medium: Add components to distilled/deionized water and bring volume to 1.0L. Mix thoroughly. Distribute into tubes or flasks. Autoclave for 15 min at 15 psi pressure–121°C. Pour into sterile Petri dishes.

Use: For the enumeration of bacteria in water, air, milk, and other dairy products.

TGE Broth

Composition per liter:

Pancreatic digest of casein 10.0g
Beef extract 6.0g
Glucose 2.0g

pH 7.0 ± 0.2 at 25°C

Source: This medium is available as a premixed powder from BD Diagnostic Systems.

Preparation of Medium: Add components to distilled/deionized water and bring volume to 1.0L. Mix thoroughly. Distribute into tubes or flasks. Autoclave for 15 min at 15 psi pressure–121°C.

Use: For the enumeration of bacteria by the membrane filter method.

TGMM-2 Medium

Composition per liter:

Saccharomyces cerevisiae
type II cells, dried 3.0g
Soy lecithin penterythritol mix 0.1g
Skim milk powder 1.0mg

Source: Dried *Saccharomyces cerevisiae* type II cells and soy lecithin (type II-S L-α-phosphatidyl choline) are available from Sigma Chemical Co.

Soy Lecithin Penterythritol Mix:
Composition per 0.11g:

Penterythritol 0.1g
Soy lecithin (type II-S
L-α-phosphatidyl choline) 10.0mg

Preparation of Soy Lecithin Penterythritol Mix: Mix soy lecithin with the inert solid diluent penterythritol as a 1:10 mixture and grind to a fine powder.

Preparation of Medium: Add components to distilled/deionized water and bring volume to 1.0L. Mix thoroughly. Gently heat to 85°C while vigorously mixing for 10 min. Distribute 10.0mL volumes into tubes. Autoclave for 15 min at 15 psi pressure–121°C.

Use: For the cultivation of *Tetrahymena limacis, Tetrahymena patula, Tetrahymena rostrata,* and *Tetrahymena setosa.*

TGY Medium
(Tryptone Glucose Yeast Extract Medium)

Composition per liter of tap water:

Agar 20.0g
Pancreatic digest of casein 5.0g
Yeast extract 5.0g
Glucose 1.0g
K_2HPO_4 .... 1.0g

pH 7.0 ± 0.2 at 25°C

Preparation of Medium: Add components to distilled/deionized water and bring volume to 1.0L. Mix thoroughly. Gently heat and bring to boiling. Distribute into tubes or flasks. Autoclave for 15 min at 15 psi pressure–121°C. Pour into sterile Petri dishes or leave in tubes.

Use: For the cultivation and maintenance of a variety of bacteria, including *Bacillus* species, *Corynebacterium* species, *Enterococcus* species, and *Pseudomonas* species.

TGYM Medium
(Tryptone Glucose Yeast Extract Methionine Medium)

Composition per liter:

Pancreatic digest of casein 5.0g
Yeast extract 3.0g
Glucose 1.0g
DL-Methionine 0.5g

Preparation of Medium: Add components to distilled/deionized water and bring volume to 1.0L. Mix thoroughly. Distribute into tubes or flasks. Autoclave for 15 min at 15 psi pressure–121°C.

Use: For the cultivation and maintenance of *Deinococcus* species.

Thauera aromatica AR-1 Medium (DSMZ Medium 855)

Composition per 992.0mL:

Solution A	870.0mL
Solution C	100.0mL
Solution D	10.0mL
Solution E (Vitamin solution)	10.0mL
Solution B (Trace elements solution SL-10)	1.0mL
Selenite-tungstate solution	1.0mL

pH 7.2 ± 0.2 at 25°C

Solution A:

Composition per 870.0mL:

Na_2SO_4	3.0g
NaCl	1.0g
KNO_3	0.6g
KCl	0.5g
$MgCl_2·6H_2O$	0.4g
NH_4Cl	0.3g
KH_2PO_4	0.2g
$CaCl_2·2H_2O$	0.15g
Resazurin	1.0mg

Preparation of Solution A: Add components to distilled/deionized water and bring volume to 870.0mL. Mix thoroughly.

Solution B (Trace Elements Solution SL-10):

Composition per liter:

$FeCl_2·4H_2O$	1.5g
$CoCl_2·6H_2O$	190.0mg
$MnCl_2·4H_2O$	100.0mg
$ZnCl_2$	70.0mg
$Na_2MoO_4·2H_2O$	36.0mg
$NiCl_2·6H_2O$	24.0mg
H_3BO_3	6.0mg
$CuCl_2·2H_2O$	2.0mg
HCl (25% solution)	10.0mL

Preparation of Solution B (Trace Elements Solution SL-10): Add $FeCl_2·4H_2O$ to 10.0mL of HCl solution. Mix thoroughly. Add distilled/deionized water and bring volume to 1.0L. Add remaining components. Mix thoroughly. Sparge with 100% N_2. Autoclave for 15 min at 15 psi pressure–121°C.

Solution C:

Composition per 100.0mL:

$NaHCO_3$	5.0g

Preparation of Solution C: Add $NaHCO_3$ to distilled/deionized water and bring volume to 100.0mL. Mix thoroughly. Filter sterilize. Flush with 80% N_2 + 20% CO_2 to remove dissolved oxygen.

Solution D:

Composition per 10.0mL:

Na-benzoate	0.7g

Preparation of Solution D: Add Na-benzoate to distilled/deionized water and bring volume to 10.0mL. Mix thoroughly. Sparge with 100% N_2. Autoclave for 15 min at 15 psi pressure–121°C.

Solution E (Vitamin Solution):

Composition per liter:

Pyridoxine-HCl	10.0mg
Thiamine-HCl·$2H_2O$	5.0mg
Riboflavin	5.0mg
Nicotinic acid	5.0mg
D-Ca-pantothenate	5.0mg
p-Aminobenzoic acid	5.0mg
Lipoic acid	5.0mg
Biotin	2.0mg
Folic acid	2.0mg
Vitamin B_{12}	0.10mg

Solution E (Vitamin Solution): Add components to distilled/deionized water and bring volume to 1.0L. Mix thoroughly. Sparge with 100% N_2. Autoclave for 15 min at 15 psi pressure–121°C.

Selenite-Tungstate Solution

Composition per liter:

NaOH	0.5g
$Na_2WO_4·2H_2O$	4.0mg
$Na_2SeO_3·5H_2O$	3.0mg

Preparation of Selenite-Tungstate Solution: Add components to distilled/deionized water and bring volume to 1.0L. Mix thoroughly. Sparge with 100% N_2. Filter sterilize.

Preparation of Medium: Gently heat solution A and bring to boiling. Boil solution A for a few minutes. Cool to room temperature. Gas with 80% N_2 + 20% CO_2 gas mixture to reach a pH below 6. Autoclave for 15 min at 15 psi pressure–121°C. Cool to room temperature. Sequentially add 1.0mL solution B, 100.0mL solution C, 10.0mL solution D, 10.0mL solution E, and 1.0mL sterile selenite-tungstate solution. Distribute anaerobically under 80% N_2 + 20% CO_2 into appropriate vessels. Addition of 10–20mg sodium dithionite per liter from a 5% (w/v) solution, freshly prepared under N_2 and filter-sterilized, may stimulate growth.

Use: For the anaerobic cultivation of *Thauera aromatica.*

Thauera aromatica AR-1 Medium (DSMZ Medium 855)

Composition per 892.0mL:

Solution A	870.0mL
Solution C	10.0mL
Solution D (Vitamin solution)	10.0mL
Solution B (Trace elements solution SL-10)	1.0mL
Selenite-tungstate solution	1.0mL

pH 7.2 ± 0.2 at 25°C

Solution A:

Composition per 870.0mL:

NaCl	1.0g
Na_2SO_4	3.0g
$MgCl_2·6H_2O$	0.4g
KH_2PO_4	0.2g
NH_4Cl	0.3g
KCl	0.5g
$CaCl_2·2H_2O$	0.15g
Resazurin	1.0mg

Preparation of Solution A: Add components to distilled/deionized water and bring volume to 870.0mL Mix thoroughly.

Solution B (Trace Elements Solution SL-10):

Composition per liter:

$FeCl_2·4H_2O$	1.5g
$CoCl_2·6H_2O$	190.0mg
$MnCl_2·4H_2O$	100.0mg
$ZnCl_2$	70.0mg
$Na_2MoO_4·2H_2O$	36.0mg
$NiCl_2·6H_2O$	24.0mg
H_3BO_3	6.0mg
$CuCl_2·2H_2O$	2.0mg
HCl (25% solution)	10.0mL

Preparation of Solution B (Trace Elements Solution SL-10): Add $FeCl_2·4H_2O$ to 10.0mL of HCl solution. Mix thoroughly. Add distilled/deionized water and bring volume to 1.0L. Add remaining components. Mix thoroughly. Sparge with 100% N_2. Autoclave for 15 min at 15 psi pressure–121°C.

Solution C:

Composition per 10.0mL:

Na-benzoate	0.7g

Preparation of Solution C: Add Na-benzoate to distilled/deionized water and bring volume to 10.0mL. Mix thoroughly. Sparge with 100% N_2. Autoclave for 15 min at 15 psi pressure–121°C.

Solution D (Vitamin Solution):

Composition per liter:

Pyridoxine-HCl	10.0mg
Thiamine-HCl·$2H_2O$	5.0mg
Riboflavin	5.0mg
Nicotinic acid	5.0mg
D-Ca-pantothenate	5.0mg
p-Aminobenzoic acid	5.0mg
Lipoic acid	5.0mg
Biotin	2.0mg
Folic acid	2.0mg
Vitamin B_{12}	0.10mg

Solution D (Vitamin Solution): Add components to distilled/deionized water and bring volume to 1.0L. Mix thoroughly. Sparge with 100% N_2. Autoclave for 15 min at 15 psi pressure–121°C.

Selenite-Tungstate Solution

Composition per liter:

NaOH	0.5g
$Na_2WO_4·2H_2O$	4.0mg
$Na_2SeO_3·5H_2O$	3.0mg

Preparation of Selenite-Tungstate Solution: Add components to distilled/deionized water and bring volume to 1.0L. Mix thoroughly. Sparge with 100% N_2. Filter sterilize.

Preparation of Medium: Adjust pH of solution A to 7.0. Autoclave for 15 min at 15 psi pressure–121°C. Cool to room temperature. Sequentially add 1.0mL solution B, 10.0mL solution C, 10.0mL solution D, and 1.0mL sterile selenite-tungstate solution. Adjust pH to 7.2. Aseptically distribute into tubes or flasks.

Use: For the aerobic cultivation of *Thauera aromatica.*

Thauera aromatica Medium

Composition 1015.0mL:

Solution A	500.0mL
Solution B	500.0mL
Trace elements solution SL-10	10.0mL
Vitamin solution	5.0mL

pH 7.5 ± 0.2 at 25°C

Solution A:

Composition per 500.0mL:

K_2HPO_4	5.92g
KH_2PO_4	0.816g

Preparation of Solution A: Add components to distilled/deionized water and bring volume to 500.0mL. Mix thoroughly. Adjust pH to 7.5. Autoclave for 15 min at 15 psi pressure–121°C. Cool to room temperature.

Solution B:

Composition per 500.0mL:

KNO_3	2.0g
Sodium benzoate	0.72g
NH_4Cl	0.267g
$MgSO_4·7H_2O$	0.197g
$CaCl_2·2H_2O$	0.025g

Preparation of Solution B: Add components to distilled/deionized water and bring volume to 500.0mL. Mix thoroughly. Adjust pH to 7.5. Autoclave for 15 min at 15 psi pressure–121°C. Cool to room temperature.

Trace Elements Solution SL-10:

Composition per liter:

$FeCl_2·4H_2O$	1.5g
$CoCl_2·6H_2O$	190.0mg
$MnCl_2·4H_2O$	100.0mg
$ZnCl_2$	70.0mg
$Na_2MoO_4·2H_2O$	36.0mg
$NiCl_2·6H_2O$	24.0mg
H_3BO_3	6.0mg
$CuCl_2·2H_2O$	2.0mg
HCl (25% solution)	10.0mL

Preparation of Trace Elements Solution SL-10: Add $FeCl_2·4H_2O$ to 10.0mL of HCl solution. Mix thoroughly. Add distilled/deionized water and bring volume to 1.0L. Add remaining components. Mix thoroughly. Sparge with 100% N_2. Autoclave for 15 min at 15 psi pressure–121°C. Cool to room temperature.

Vitamin Solution:

Composition per liter:

Pyridoxine·HCl	10.0mg
p-Aminobenzoic acid	5.0mg
Lipoic acid	5.0mg
Nicotinic acid	5.0mg
Riboflavin	5.0mg
Thiamine·HCl	5.0mg
Calcium DL-pantothenate	5.0mg
Biotin	2.0mg
Folic acid	2.0mg
Vitamin B_{12}	0.1mg

Preparation of Vitamin Solution: Add components to distilled/deionized water and bring volume to 1.0L. Mix thoroughly. Filter sterilize.

Preparation of Medium: Aseptically combine 500.0mL of sterile solution A, 500.0mL of sterile solution B, 10.0 mL of sterile trace elements solution SL-10, and 5.0 mL of sterile vitamin solution. Mix thoroughly. Aseptically distribute into sterile tubes or flasks.

Use: For the cultivation of *Thauera aromatica.*

Thauera mechernichi Medium (DSMZ Medium 918)

Composition per liter:

Na_2HPO_4	4.20g
Na-acetate	2.93g
KH_2PO_4	1.5g
NH_4Cl	0.3g
$MgSO_4 \cdot 7H_2O$	0.1g
Trace elements solution	2.0mL

pH 7.1 ± 0.2 at 25°C

Trace Elements Solution:

Composition per liter:

Na_2-EDTA	50.0g
$CaCl_2 \cdot 2H_2O$	5.5g
$MnCl_2 \cdot 4H_2O$	5.06g
$FeSO_4 \cdot 7H_2O$	5.0g
$ZnSO_4 \cdot 7H_2O$	2.0g
$CoCl_2 \cdot 6H_2O$	1.61g
$CuSO_4 \cdot 5H_2O$	1.57g
$(NH_4)_6Mo_7O_{24} \cdot 4H_2O$	1.1g

Preparation of Trace Elements Solution: Add components to distilled/deionized water and bring volume to 1.0L. Mix thoroughly. Adjust pH to 6.0.

Preparation of Medium: Add components to distilled/deionized water and bring volume to 1.0L. Mix thoroughly. Distribute into tubes or flasks. Autoclave for 15 min at 15 psi pressure–121°C.

Use: For the cultivation of *Thauera mechernichensis*.

Thayer-Martin Agar, Modified (MTM II) (Modified Thayer-Martin Agar)

Composition per liter:

Agar	12.0g
Hemoglobin	10.0g
Pancreatic digest of casein	7.5g
Selected meat peptone	7.5g
NaCl	5.0g
K_2HPO_4	4.0g
Cornstarch	1.0g
KH_2PO_4	1.0g
CNVT inhibitor	10.0mL
Supplement solution	10.0mL

pH 7.2 ± 0.2 at 25°C

CNVT Inhibitor:

Composition per 10.0mL:

Colistin sulfate	7.5mg
Trimethoprim lactate	5.0mg
Vancomycin	3.0mg
Nystatin	12,500U

Preparation of CNVT Inhibitor: Add components to distilled/deionized water and bring volume to 10.0mL. Mix thoroughly. Filter sterilize.

Supplement Solution:

Composition per liter:

Glucose	100.0g
L-Cysteine·HCl	25.9g
L-Glutamine	10.0g
L-Cystine	1.1g
Adenine	1.0g
Nicotinamide adenine dinucleotide	0.25g
Vitamin B_{12}	0.1g
Thiamine pyrophosphate	0.1g
Guanine·HCl	0.03g
$Fe(NO_3)_3 \cdot 6H_2O$	0.02g
p-Aminobenzoic acid	0.013g
Thiamine·HCl	3.0mg

Source: The supplement solution IsoVitaleX® enrichment is available from BD Diagnostic Systems. This enrichment may be replaced by supplement VX from BD Diagnostic Systems.

Preparation of Supplement Solution: Add components to distilled/deionized water and bring volume to 1.0L. Mix thoroughly. Filter sterilize.

Preparation of Medium: Add components, except CNVT inhibitor and supplement solution, to distilled/deionized water and bring volume to 990.0mL. Mix thoroughly. Gently heat and bring to boiling. Distribute into tubes or flasks. Autoclave for 15 min at 15 psi pressure–121°C. Cool to 45°–50°C. Aseptically add 10.0mL of sterile CNVT inhibitor and 10.0mL of sterile supplement solution. Mix thoroughly. Pour into sterile Petri dishes or distribute into sterile tubes.

Use: For the isolation of *Neisseria* species from specimens containing mixed flora of bacteria and fungi.

Thayer-Martin HiVeg Medium Base with Hemoglobin and Vitox Supplement

Composition per liter:

Plant special peptone	23.0g
Agar	13.0g
NaCl	5.0g
Starch	1.0g
Hemoglobin solution	250.0mL
Vitox supplement	10.0mL

pH 7.0 ± 0.2 at 25°C

Source: This medium, without hemoglobin or Vitox supplement, is available as a premixed powder from HiMedia.

Hemoglobin Solution:

Composition per 250.0mL:

Hemoglobin	5.0g

Preparation of Hemoglobin Solution: Add hemoglobin to distilled/deionized water and bring volume to 250.0mL. Mix thoroughly. Autoclave for 15 min at 15 psi pressure–121°C. Cool to 45°–50°C.

Vitox Supplement:

Composition per 10.0mL:

Glucose	2.0g
L-Cysteine·HCl	0.518g
L-Glutamine	0.2g
L-Cystine	0.022g
Adenine sulfate	0.01g
Nicotinamide adenine dinucleotide	5.0mg
Cocarboxylase	2.0mg
Guanine·HCl	0.6mg
$Fe(NO_3)_3 \cdot 6H_2O$	0.4mg
p-Aminobenzoic acid	0.26mg
Vitamin B_{12}	0.2mg
Thiamine·HCl	0.06mg

Preparation of Vitox Supplement: Add components to distilled/deionized water and bring volume to 10.0mL. Mix thoroughly. Filter sterilize.

Preparation of Medium: Add components, except hemoblobin and Vitox supplement, to distilled/deionized water and bring volume to 740.0mL. Mix thoroughly. Gently heat until boiling. Autoclave for 15 min at 15 psi pressure–121°C. Cool to 45°–50°C. Aseptically add 250.0mL of sterile hemoglobin solution and 10.0mL of sterile Vitox supplement. Mix thoroughly. Pour into sterile Petri dishes or distribute into sterile tubes.

Use: For the isolation and cultivation of fastidious microorganisms, especially *Neisseria* species. For the selective isolation of gonococci from pathological specimens.

Thayer-Martin Medium

Composition per liter:

GC agar base	740.0mL
Hemoglobin solution	250.0mL
Vitox supplement	10.0mL

pH 7.3 ± 0.2 at 25°C

GC Agar Base:

Composition per 740.0mL:

Special peptone	15.0g
Agar	10.0g
NaCl	5.0g
K_2HPO_4	4.0g
Cornstarch	1.0g
KH_2PO_4	1.0g

pH 7.2 ± 0.2 at 25°C

Preparation of GC Agar Base: Add components of GC medium base and the hemoglobin to distilled/deionized water and bring volume to 740.0mL. Mix thoroughly. Gently heat until boiling. Autoclave for 15 min at 15 psi pressure–121°C. Cool to 45°–50°C.

Hemoglobin Solution:

Composition per 250.0mL:

Hemoglobin	5.0g

Preparation of Hemoglobin Solution: Add hemoglobin to distilled/deionized water and bring volume to 250.0mL. Mix thoroughly. Autoclave for 15 min at 15 psi pressure–121°C. Cool to 45°–50°C.

Vitox Supplement:

Composition per 10.0mL:

Glucose	2.0g
L-Cysteine·HCl	0.518g
L-Glutamine	0.2g
L-Cystine	0.022g
Adenine sulfate	0.01g
Nicotinamide adenine dinucleotide	5.0mg
Cocarboxylase	2.0mg
Guanine·HCl	0.6mg
$Fe(NO_3)_3 \cdot 6H_2O$	0.4mg
p-Aminobenzoic acid	0.26mg
Vitamin B_{12}	0.2mg
Thiamine·HCl	0.06mg

Preparation of Vitox Supplement: Add components to distilled/deionized water and bring volume to 10.0mL. Mix thoroughly. Filter sterilize.

Preparation of Medium: To 740.0mL of cooled sterile GC agar base, aseptically add 250.0mL of sterile hemoglobin solution and 10.0mL of sterile Vitox supplement. Mix thoroughly. Pour into sterile Petri dishes or distribute into sterile tubes.

Use: For the isolation and cultivation of fastidious microorganisms, especially *Neisseria* species.

Thayer-Martin Medium

Composition per liter:

Hemoglobin	10.0g
GC medium base	980.0mL
CNVT inhibitor	10.0mL
Supplement B	10.0mL

pH 7.3 ± 0.2 at 25°C

Source: This medium is available as a prepared medium in tubes from BD Diagnostic Systems.

GC Medium Base:

Composition per 980.0mL:

Proteose peptone No. 3	15.0g
Agar	10.0g
NaCl	5.0g
K_2HPO_4	4.0g
Cornstarch	1.0g
KH_2PO_4	1.0g

pH 7.2 ± 0.2 at 25°C

Preparation of GC Medium Base: Add components of GC medium base and the hemoglobin to distilled/deionized water and bring volume to 1.0L. Mix thoroughly. Gently heat until boiling. Autoclave for 15 min at 15 psi pressure–121°C. Cool to 45°–50°C.

CNVT Inhibitor:

Composition per 10.0mL:

Colistin sulfate	7.5mg
Trimethoprim lactate	5.0mg
Vancomycin	3.0mg
Nystatin	12,500U

Preparation of CNVT Inhibitor: Add components to distilled/deionized water and bring volume to 10.0mL. Mix thoroughly. Filter sterilize.

Preparation of Medium: To 980.0mL of cooled sterile GC medium base, aseptically add 10.0mL of sterile CNVT inhibitor and 10.0mL of sterile supplement B. Mix thoroughly. Pour into sterile Petri dishes or distribute into sterile tubes.

Use: For the isolation and cultivation of fastidious microorganisms, especially *Neisseria* species.

Thayer-Martin Medium, Modified (Modified Thayer-Martin Agar)

Composition per liter:

GC agar base	720.0mL
Hemoglobin solution	250.0mL
GC supplement	30.0mL

pH 7.3 ± 0.2 at 25°C

GC Agar Base:

Composition per 720.0mL:

Special peptone	15.0g
Agar	10.0g
NaCl	5.0g
K_2HPO_4	4.0g

Cornstarch	1.0g
KH_2PO_4	1.0g

pH 7.2 ± 0.2 at 25°C

Preparation of GC Agar Base: Add components of GC medium base to distilled/deionized water and bring volume to 720.0mL. Mix thoroughly. Gently heat until boiling. Autoclave for 15 min at 15 psi pressure–121°C. Cool to 45°–50°C.

Hemoglobin Solution:

Composition per 250.0mL:

Hemoglobin	5.0g

Preparation of Hemoglobin Solution: Add hemoglobin to distilled/deionized water and bring volume to 250.0mL. Mix thoroughly. Autoclave for 15 min at 15 psi pressure–121°C. Cool to 45°–50°C.

GC Supplement:

Composition per 30.0mL:

Yeast autolysate	10.0g
Glucose	1.5g
$NaHCO_3$	0.15g
Colistin sulfate	7.5mg
Trimethoprim lactate	5.0mg
Vancomycin	3.0mg
Nystatin	12,500U

Preparation of GC Supplement: Add components to distilled/deionized water and bring volume to 30.0mL. Mix thoroughly. Filter sterilize.

Preparation of Medium: To 720.0mL of cooled sterile GC agar base, aseptically add 250.0mL of sterile hemoglobin solution and 30.0mL of sterile GC supplement. Mix thoroughly. Pour into sterile Petri dishes or distribute into sterile tubes.

Use: For the selective isolation and cultivation of fastidious microorganisms, especially *Neisseria* species.

Thayer-Martin Medium, Modified (Modified Thayer-Martin Agar)

Composition per liter:

GC agar base	730.0mL
Hemoglobin solution	250.0mL
Vitox supplement	10.0mL
VCNT antibiotic solution	10.0mL

pH 7.3 ± 0.2 at 25°C

GC Agar Base:

Composition per 730.0mL:

Special peptone	15.0g
Agar	10.0g
NaCl	5.0g
K_2HPO_4	4.0g
Cornstarch	1.0g
KH_2PO_4	1.0g

pH 7.2 ± 0.2 at 25°C

Preparation of GC Agar Base: Add components of GC medium base in to distilled/deionized water and bring volume to 730.0mL. Mix thoroughly. Gently heat until boiling. Autoclave for 15 min at 15 psi pressure–121°C. Cool to 45°–50°C.

Hemoglobin Solution:

Composition per 250.0mL:

Hemoglobin	5.0g

Preparation of Hemoglobin Solution: Add hemoglobin to distilled/deionized water and bring volume to 250.0mL. Mix thoroughly. Autoclave for 15 min at 15 psi pressure–121°C. Cool to 45°–50°C.

Vitox Supplement:

Composition per 10.0mL:

Glucose	2.0g
L-Cysteine·HCl	0.518g
L-Glutamine	0.2g
L-Cystine	0.022g
Adenine sulfate	0.01g
Nicotinamide adenine dinucleotide	5.0mg
Cocarboxylase	2.0mg
Guanine·HCl	0.6mg
$Fe(NO_3)_3 \cdot 6H_2O$	0.4mg
p-Aminobenzoic acid	0.26mg
Vitamin B_{12}	0.2mg
Thiamine·HCl	0.06mg

Preparation of Vitox Supplement: Add components to distilled/deionized water and bring volume to 10.0mL. Mix thoroughly. Filter sterilize.

VCNT Antibiotic Solution:

Composition per 10.0mL:

Colistin methane sulfonate	7.5mg
Trimethoprim lactate	5.0mg
Vancomycin	3.0mg
Nystatin	12,500U

Preparation of VCNT Antibiotic Solution: Add components to distilled/deionized water and bring volume to 10.0mL. Mix thoroughly. Filter sterilize.

Preparation of Medium: To 730.0mL of cooled, sterile GC agar base, aseptically add 250.0mL of sterile hemoglobin solution, 10.0mL of sterile Vitox supplement, and 10.0mL of VCNT antibiotic solution. Mix thoroughly. Pour into sterile Petri dishes or distribute into sterile tubes.

Use: For the selective isolation and cultivation of fastidious microorganisms, especially *Neisseria* species.

Thayer-Martin Medium, Selective

Composition per liter:

GC agar base	730.0mL
Hemoglobin solution	250.0mL
Vitox supplement	10.0mL
VCN antibiotic solution	10.0mL

pH 7.3 ± 0.2 at 25°C

GC Agar Base:

Composition per 730.0mL:

Special peptone	15.0g
Agar	10.0g
NaCl	5.0g
K_2HPO_4	4.0g
Cornstarch	1.0g
KH_2PO_4	1.0g

pH 7.2 ± 0.2 at 25°C

Preparation of GC Agar Base: Add components of GC medium base to distilled/deionized water and bring volume to 730.0mL. Mix thoroughly. Gently heat until boiling. Autoclave for 15 min at 15 psi pressure–121°C. Cool to 45°–50°C.

Hemoglobin Solution:
Composition per 250.0mL:

Hemoglobin	5.0g

Preparation of Hemoglobin Solution: Add hemoglobin to distilled/deionized water and bring volume to 250.0mL. Mix thoroughly. Autoclave for 15 min at 15 psi pressure–121°C. Cool to 45°–50°C.

Vitox Supplement:
Composition per 10.0mL:

Glucose	2.0g
L-Cysteine·HCl	0.518g
L-Glutamine	0.2g
L-Cystine	0.022g
Adenine sulfate	0.01g
Nicotinamide adenine dinucleotide	5.0mg
Cocarboxylase	2.0mg
Guanine·HCl	0.6mg
$Fe(NO_3)_3 \cdot 6H_2O$	0.4mg
p-Aminobenzoic acid	0.26mg
Vitamin B_{12}	0.2mg
Thiamine·HCl	0.06mg

Preparation of Vitox Supplement: Add components to distilled/deionized water and bring volume to 10.0mL. Mix thoroughly. Filter sterilize.

VCN Antibiotic Solution:
Composition per 10.0mL:

Colistin methane sulfonate	7.5mg
Vancomycin	3.0mg
Nystatin	12,500U

Preparation of VCN Antibiotic Solution: Add components to distilled/deionized water and bring volume to 10.0mL. Mix thoroughly. Filter sterilize.

Preparation of Medium: To 730.0mL of cooled, sterile GC agar base, aseptically add 250.0mL of sterile hemoglobin solution, 10.0mL of sterile Vitox supplement, and 10.0mL of VCN antibiotic solution. Mix thoroughly. Pour into sterile Petri dishes or distribute into sterile tubes.

Use: For the selective isolation and cultivation of fastidious microorganisms, especially *Neisseria* species.

Thayer-Martin Selective Agar

Composition per liter:

Agar	12.0g
Hemoglobin	10.0g
Pancreatic digest of casein	7.5g
Selected meat peptone	7.5g
NaCl	5.0g
K_2HPO_4	4.0g
Cornstarch	1.0g
KH_2PO_4	1.0g
Supplement solution	10.0mL
VCN inhibitor	10.0mL

pH 7.2 ± 0.2 at 25°C

Source: This medium is available as a premixed powder from BD Diagnostic Systems.

Supplement Solution:
Composition per liter:

Glucose	100.0g
L-Cysteine·HCl	25.9g
L-Glutamine	10.0g
L-Cystine	1.1g
Adenine	1.0g
Nicotinamide adenine dinucleotide	0.25g
Vitamin B_{12}	0.1g
Thiamine pyrophosphate	0.1g
Guanine·HCl	0.03g
$Fe(NO_3)_3 \cdot 6H_2O$	0.02g
p-Aminobenzoic acid	0.013g
Thiamine·HCl	3.0mg

Source: The supplement solution IsoVitaleX® enrichment is available from BD Diagnostic Systems. This enrichment may be replaced by supplement VX from BD Diagnostic Systems.

Preparation of Supplement Solution: Add components to distilled/deionized water and bring volume to 1.0L. Mix thoroughly. Filter sterilize.

VCN Inhibitor:
Composition per 10.0mL:

Colistin	7.5mg
Vancomycin	3.0mg
Nystatin	12,500U

Preparation of VCN Inhibitor: Add components to distilled/deionized water and bring volume to 10.0mL. Mix thoroughly. Filter sterilize.

Preparation of Medium: Add components, except supplement solution and VCN inhibitor, to distilled/deionized water and bring volume to 980.0mL. Mix thoroughly. Gently heat and bring to boiling. Autoclave for 15 min at 15 psi pressure–121°C. Cool to 45°–50°C. Aseptically add sterile VCN inhibitor and sterile supplement solution. Mix thoroughly. Pour into sterile Petri dishes or distribute into sterile tubes.

Use: For the selective isolation of *Neisseria gonorrhoeae* and *Neisseria meningitidis* from specimens containing mixed flora of bacteria and fungi.

Thermincola Medium (DSMZ Medium 1028)

Composition per liter:

NH_4Cl	1.0g
$MgCl_2 \cdot 6H_2O$	0.33g
KH_2PO_4	0.5g
KCl	0.33g
Na-acetate	0.2g
$CaCl_2 \cdot 2H_2O$	0.1g
Resazurin	0.5mg
Vitamin solution	20.0mL
Bicarbonate solution	10.0mL
Carbonate solution	10.0mL
Sulfide solution	10.0mL
Yeast extract solution	10.0mL
Wolfe's mineral elixir	1.0mL

pH 8.0 ± 0.2 at 25°C

Sulfide Solution:
Composition per 10.0mL:

$Na_2S \cdot 9H_2O$	1.0g

Preparation of Sulfide Solution: Add $Na_2S \cdot 9H_2O$ to distilled/deionized water and bring volume to 10.0mL. Mix thoroughly. Autoclave under 100% N_2 for 15 min at 15 psi pressure–121°C. Cool to room temperature.

Bicarbonate Solution:

Composition per 10.0mL:

$NaHCO_3$	0.5g

Preparation of Bicarbonate Solution: Add $NaHCO_3$ to distilled/deionized water and bring volume to 10.0mL. Mix thoroughly. Sparge with gas mixture of 80% N_2 + 20% CO_2. Filter sterilize.

Carbonate Solution:

Composition per 10.0mL:

Na_2CO_3	0.5g

Preparation of Carbonate Solution: Add Na_2CO_3 to distilled/deionized water and bring volume to 10.0mL. Mix thoroughly. Sparge with gas mixture of 80% N_2 + 20% CO_2. Filter sterilize.

Yeast Extract Solution:

Composition per 10.0mL:

Yeast extract	0.2g

Preparation of Yeast Extract Solution: Add yeast extract to distilled/deionized water and bring volume to 10.0mL. Mix thoroughly. Sparge with 100% N_2 gas. Autoclave for 15 min at 15 psi pressure–121°C.

Wolfe's Mineral Elixir:

Composition per liter:

$MgSO_4 \cdot 7H_2O$	30.0g
NaCl	10.0g
$MnSO_4 \cdot 2H_2O$	5.0g
$(NH_4)_2NiSO_4 \cdot 6H_2O$	2.8g
$CoCl_2 \cdot 6H_2O$	1.8g
$ZnSO_4 \cdot 7H_2O$	1.8g
$FeSO_4 \cdot 7H_2O$	1.0g
$CaCl_2 \cdot 2H_2O$	1.0g
$KAl(SO_4)_2 \cdot 12H_2O$	0.18g
$CuSO_4 \cdot 5H_2O$	0.1g
H_3BO_3	0.1g
$Na_2MoO_4 \cdot 2H_2O$	0.1g
Na_2SeO_4	0.1g
$Na_2WO_4 \cdot 2H_2O$	0.1g

Preparation of Wolfe's Mineral Elixir: Adjust pH of 1.0L of distilled/deionized water to 1.0 with dilute H_2SO_4. Add components one at a time. Mix thoroughly to dissolve.

Vitamin Solution:

Composition per liter:

Pyridoxine-HCl	10.0mg
Thiamine-HCl·$2H_2O$	5.0mg
Riboflavin	5.0mg
Nicotinic acid	5.0mg
D-Ca-pantothenate	5.0mg
p-Aminobenzoic acid	5.0mg
Lipoic acid	5.0mg
Biotin	2.0mg
Folic acid	2.0mg
Vitamin B_{12}	0.1mg

Preparation of Vitamin Solution: Add components to distilled/deionized water and bring volume to 1.0L. Mix thoroughly. Sparge with 80% H_2 + 20% CO_2. Filter sterilize.

Preparation of Medium: Add components, except vitamin solution, to distilled/deionized water and bring volume to 950.0mL. Mix thoroughly. Gently heat and bring to boiling. Boil for 1 min. Cool to room temperature while sparging with 100% N_2 gas. Dispense into culture vessels under an atmostphere of 100% CO (carbon monoxide) . Culture vessels should be filled to approximately 20%. Autoclave for 15 min at 15 psi pressure–121°C. Cool to room temperature. Aseptically add carbonate, bicarbonate, vitamin, and sulfide solutions. Mix thoroughly. Adjust the pH to 8.0 with sterile, anoxic 1*N* HCl. Prior to inoculation aseptically add yeast extract solution.

Use: For the cultivation of *Thermincola carboxydiphila*.

Thermoacetogenium phaeum Medium (DSMZ Medium 880)

Composition per liter:

$KHCO_3$	3.5g
NH_4Cl	1.0g
NaCl	0.6g
KH_2PO_4	0.3g
$MgCl_2 \cdot 6H_2O$	0.1g
$CaCl_2 \cdot 2H_2O$	0.08g
Resazurin	0.5mg
Sodium pyruvate solution	50.0mL
Vitamin solution	10.0mL
$Na_2S \cdot 9H_2O$ solution	10.0mL
Cysteine solution	10.0mL
Trace elements solution	1.0mL
Selenite-tungstate solution	1.0mL

pH 7.0–7.1 at 25°C

Sodium Pyruvate Solution:

Composition per 50.0mL:

Sodium pyruvate	5.0g

Preparation of Sodium Pyruvate Solution: Add sodium pyruvate to distilled/deionized water and bring volume to 50.0mL. Mix thoroughly. Sparge with 100% N_2. Filter sterilize.

Selenite-Tungstate Solution

Composition per liter:

NaOH	0.5g
$Na_2WO_4 \cdot 2H_2O$	4.0mg
$Na_2SeO_3 \cdot 5H_2O$	3.0mg

Preparation of Selenite-Tungstate Solution: Add components to distilled/deionized water and bring volume to 1.0L. Mix thoroughly. Sparge with 100% N_2. Filter sterilize.

$Na_2S \cdot 9H_2O$ Solution:

Composition per 10.0mL:

$Na_2S \cdot 9H_2O$	0.3g

Preparation of $Na_2S \cdot 9H_2O$ Solution: Add $Na_2S \cdot 9H_2O$ to distilled/deionized water and bring volume to 10.0mL. Mix thoroughly. Sparge with 100% N_2. Autoclave for 15 min at 15 psi pressure–121°C.

Cysteine Solution:

Composition per 10.0mL:

L-Cysteine·HCl·H_2O	0.3g

Preparation of Cysteine Solution: Add L-cysteine·HCl·H_2O to distilled/deionized water and bring volume to 10.0mL. Mix thoroughly. Sparge with 100% N_2. Autoclave for 15 min at 15 psi pressure–121°C.

Trace Elements Solution SL-10:

Composition per liter:

$FeCl_2 \cdot 4H_2O$	1.5g
Na_2-EDTA	0.5g
$CoCl_2 \cdot 6H_2O$	190.0mg
$MnCl_2 \cdot 4H_2O$	100.0mg

$ZnCl_2$ 70.0mg
$Na_2MoO_4 \cdot 2H_2O$ 36.0mg
$NiCl_2 \cdot 6H_2O$ 24.0mg
H_3BO_3 ....6.0mg
$CuCl_2 \cdot 2H_2O$ 2.0mg
HCl (25% solution) 10.0mL

Preparation of Trace Elements Solution SL-10: Add $FeCl_2 \cdot 4H_2O$ to 10.0mL of HCl solution. Mix thoroughly. Add distilled/deionized water and bring volume to 1.0L. Add remaining components. Mix thoroughly. Adjust pH to 6.5. Sparge with 100% N_2. Autoclave for 15 min at 15 psi pressure–121°C.

Vitamin Solution:

Composition per liter:
Pyridoxine-HCl 10.0mg
Thiamine-HCl·$2H_2O$ 5.0mg
Riboflavin 5.0mg
Nicotinic acid 5.0mg
D-Ca-pantothenate 5.0mg
p-Aminobenzoic acid 5.0mg
Lipoic acid 5.0mg
Biotin 2.0mg
Folic acid 2.0mg
Vitamin B_{12} 0.1mg

Preparation of Vitamin Solution: Add components to distilled/deionized water and bring volume to 1.0L. Mix thoroughly. Sparge with 80% H_2 + 20% CO_2. Filter sterilize.

Preparation of Medium: Prepare and dispense medium under 80% N_2 + 20% CO_2 gas atmosphere. Add components, except $KHCO_3$, sodium pyruvate solution, cysteine solution, and $Na_2S \cdot 9H_2O$ solution, to distilled/deionized water and bring volume to 930.0mL. Mix thoroughly. Gently heat and bring to boiling. Boil for 10 min. Cool to room temperature while sparging with 80% N_2 + 20% CO_2. Add 3.5g $KHCO_3$. Mix thoroughly while sparging with 80% N_2 + 20% CO_2 gas atmosphere. Autoclave for 15 min at 15 psi pressure–121°C. Aseptically and anaerobically add 50.0mL sodium pyruvate solution, 10.0mL cysteine solution, and 10.0mL $Na_2S \cdot 9H_2O$ solution. Mix thoroughly. Final pH is 7.0–7.1. Aseptically and anaerobically distribute into sterile tubes or bottles.

Use: For the cultivation of *Thermacetogenium phaeum.*

Thermoacidurans Agar

Composition per liter:
Agar 20.0g
Yeast extract 5.0g
Proteose peptone 5.0g
Glucose 5.0g
K_2HPO_4 ....4.0g

pH 5.0 ± 0.2 at 25°C

Source: This medium is available as a premixed powder from BD Diagnostic Systems.

Preparation of Medium: Add components to distilled/deionized water and bring volume to 1.0L. Mix thoroughly. Gently heat and bring to boiling. Distribute into tubes or flasks. Autoclave for 15 min at 15 psi pressure–121°C. Do not overheat. Pour into sterile Petri dishes or leave in tubes.

Use: For the isolation and cultivation of *Bacillus thermoacidurans* from food products.

Thermoacidurans HiVeg Agar

Composition per liter:
Agar 20.0g
Glucose 5.0g
Plant peptone No. 3 ....5.0g
Yeast extract 5.0g
K_2HPO_4 ....4.0g

pH 5.0 ± 0.2 at 25°C

Source: This medium is available as a premixed powder from HiMedia.

Preparation of Medium: Add components to distilled/deionized water and bring volume to 1.0L. Mix thoroughly. Gently heat and bring to boiling. Distribute into tubes or flasks. Autoclave for 15 min at 12 psi pressure–118°C. Do not overheat. Pour into sterile Petri dishes or leave in tubes.

Use: For the isolation and cultivation of *Bacillus thermoacidurans* from food products.

Thermoactinomyces dichotomicus Medium

Composition per liter:
Maize, split 50.0g
Agar 20.0g
Starch 10.0g
NaCl 5.0g
Peptone 5.0g
$CaCl_2$ 0.5g

pH 7.2 ± 0.2 at 25°C

Preparation of Medium: Add split maize (crushed corn) to 1.0L of boiling water. Steam for 30 min. Filter through Whatman #1 filter paper. Add remaining components to maize filtrate. Gently heat and bring to boiling. Distribute into tubes or flasks. Autoclave for 15 min at 15 psi pressure–121°C. Pour into sterile Petri dishes or leave in tubes.

Use: For the cultivation of *Thermoactinomyces dichotomicus.*

Thermoactinomyces Medium (DSMZ Medium 978)

Composition per liter:
Polypeptone™ 30.0g
Agar 15.0g
Glycerol 2.0g
L-Asparagine 1.0g
K_2HPO_4 ....1.0g
Vitamin B solution 10.0mL
Trace salts solution 1.0mL

pH 7.2 ± 0.2 at 25°C

Trace Salts Solution:

Composition per 100.0mL:
$FeSO_4 \cdot 7H_2O$ 0.1g
$MnCl_2 \cdot 4H_2O$ 0.1g
$ZnSO_4 \cdot 7H_2O$ 0.1g

Preparation of Trace Salts Solution: Add components to distilled/deionized water and bring volume to 100.0mL. Mix thoroughly.

Vitamin B Solution:

Composition per 200.0mL:
Thiamine-HCl 10.0mg
Riboflavin 10.0mg
Nicotinate 10.0mg
Pyridoxine-HCl 10.0mg

Inositol	10.0mg
Calcium pantothenate	10.0mg
p-Aminobenzoate	10.0mg
D-Biotin	5.0mg

Preparation of Vitamin B Solution: Add components to distilled/deionized water and bring volume to 200.0mL. Mix thoroughly. Filter sterilize.

Preparation of Medium: Add components, except vitamin B solution, to distilled/deionized water and bring volume to 900.0mL. Mix thoroughly. Autoclave for 15 min at 15 psi pressure–121°C. Cool to room temperature. Aseptically add 100.0mL vitamin B solution. Aseptically distribute into sterile tubes or flasks.

Use: For the cultivation of *Thermoactinomyces peptonophilus.*

Thermoactinomyces Medium

Composition per liter:

Agar	20.0g
Malt extract	10.0g
Yeast extract	4.0g
Glucose	4.0g

pH 7.3 ± 0.2 at 25°C

Preparation of Medium: Add components to distilled/deionized water and bring volume to 1.0L. Mix thoroughly. Gently heat and bring to boiling. Distribute into tubes or flasks. Autoclave for 15 min at 15 psi pressure–121°C. Pour into sterile Petri dishes or leave in tubes.

Use: For the cultivation and maintenance of *Thermoactinomyces sacchari.*

Thermoactinomyces Medium (DSMZ Medium 978)

Composition per liter:

Polypeptone	30.0g
Agar	15.0g
Glycerol	2.0g
L-Asparagine	1.0g
KH_2PO_4	1.0g
Vitamin solution	10.0mL
Trace elements solution	1.0mL

pH 7.2 ± 0.2 at 25°C

Vitamin Solution:

Composition per 200.0mL:

Pyridoxine-HCl	10.0mg
Thiamine-HCl·$2H_2O$	10.0mg
Riboflavin	10.0mg
Nicotinic acid	10.0mg
D-Ca-pantothenate	10.0mg
Inositol	10.0mg
p-Aminobenzoic acid	10.0mg
Biotin	5.0mg

Preparation of Vitamin Solution: Add components to distilled/deionized water and bring volume to 200.0mL. Mix thoroughly. Filter sterilize.

Trace Elements Solution:

Composition per 100.0mL:

$FeSO_4 \cdot 7H_2O$	0.1g
$MnCl_2 \cdot 4H_2O$	0.1g
$ZnSO_4 \cdot 7H_2O$	0.1g

Preparation of Trace Elements Solution: Add components to distilled/deionized water and bring volume to 100.0mL. Mix thoroughly.

Preparation of Medium: Add components, except vitamin solution, to distilled/deionized water and bring volume to 990.0mL. Mix thoroughly. Adjust the pH to 7.2. Autoclave for 15 min at 15 psi pressure–121°C. Cool to room temperature. Aseptically add vitamin solution. Mix thoroughly. Aseptically dispense into culture vessels.

Use: For the cultivation of *Thermoactinomyces* spp.

Thermoactinopolyspora Medium

Composition per liter:

Maltose	20.0g
Agar	15.0g
Papaic digest of soybean meal	15.0g
Yeast extract	2.0g

pH 7.2 ± 0.2 at 25°C

Preparation of Medium: Add components to tap water and bring volume to 1.0L. Mix thoroughly. Gently heat and bring to boiling. Distribute into tubes or flasks. Autoclave for 15 min at 15 psi pressure–121°C. Pour into sterile Petri dishes or leave in tubes.

Use: For the cultivation and maintenance of *Thermoactinomyces* and *Thermoactinopolyspora* species.

Thermoanaerobacter ethanolicus Medium

Composition per liter:

Glucose	8.0g
$Na_2HPO_4 \cdot 12H_2O$	4.2g
Yeast extract	2.0g
KH_2PO_4	1.5g
NH_4Cl	0.5g
$MgCl_2 \cdot 6H_2O$	0.18g
Reducing solution	40.0mL
Wolfe's modified mineral solution	5.0mL
Resazurin (0.1% solution)	1.0mL
Vitamin solution	0.5mL

Caution: This medium contains Na_2S, and H_2S production will occur, especially upon prolonged boiling. H_2S is hazardous and preparation of this medium should be done in a chemical fume hood.

Reducing Solution:

Composition per 200.0mL:

Cysteine·HCl·H_2O	2.5g
$Na_2S \cdot 9H_2O$	2.5g
NaOH (0.2*N* solution)	200.0mL

Preparation of Reducing Solution: Gently heat the NaOH solution and bring to boiling. Gas with 95% N_2 + 5% H_2. Cool to room temperature. Add the cysteine·HCl·H_2O and $Na_2S \cdot 9H_2O$. Anaerobically distribute into tubes. Cap with rubber stoppers. Autoclave for 15 min at 15 psi pressure–121°C.

Vitamin Solution:

Composition per 500.0mL:

Pyridoxine·HCl	0.1g
p-Aminobenzoic acid	0.05g
Calcium pantothenate	0.05g
Nicotinic acid	0.05g
Thioctic acid	0.05g
Biotin	0.02g
Folic acid	0.02g
Riboflavin	5.0mg

Thiamine·HCl 5.0mg
Vitamin B_{12} 1.0mg

Preparation of Vitamin Solution: Add components to distilled/deionized water and bring volume to 500.0mL. Mix thoroughly. Store solution in the dark at −10°C.

Wolfe's Modified Mineral Solution:
Composition per liter:

$MgSO_4 \cdot 7H_2O$	3.0g
Nitrilotriacetic acid	1.5g
NaCl	1.0g
$MnSO_4 \cdot H_2O$	0.5g
$CaCl_2$ (anhydrous)	0.1g
$Co(NO_3)_2 \cdot 6H_2O$	0.1g
$FeSO_4 \cdot 7H_2O$	0.1g
$ZnSO_4 \cdot 7H_2O$	0.1g
$AlK(SO_4)_2$ (anhydrous)	0.01g
$CuSO_4 \cdot 5H_2O$	0.01g
H_3BO_3	0.01g
$Na_2MoO_4 \cdot 2H_2O$	0.01g
Na_2SeO_3 (anhydrous)	1.0mg

Preparation of Wolfe's Modified Mineral Solution: Add nitrilotriacetic acid to 500.0mL of distilled/deionized water. Dissolve by adjusting pH to 6.5 with KOH. Add remaining components. Add distilled/deionized water to 1.0L.

Preparation of Medium: Add components, except reducing solution, to distilled/deionized water and bring volume to 1.0L. Gently heat and bring to boiling under 95% N_2 + 5% H_2. Continue boiling until color changes from blue to pink. Add the reducing solution. The pink color will disappear, indicating that the solution has been reduced. Distribute into tubes or flasks under 95% N_2 + 5% H_2 using anerobic techniques. Cap tubes with rubber stoppers. Autoclave for 15 min at 15 psi pressure–121°C.

Use: For the cultivation and maintenance of thermophilic anaerobes such as *Thermoanaerobacter* species and some *Clostridium* species.

Thermoanaerobacter subterraneus Medium (DSMZ Medium 899)

Composition per liter:

Yeast extract	2.0g
$MgCl_2 \cdot 6H_2O$	1.0g
NH_4Cl	1.0g
NaCl	0.6g
Cysteine-HCl·H_2O	0.5g
K_2HPO_4	0.3g
KH_2PO_4	0.3g
KCl	0.2g
$CaCl_2 \cdot 2H_2O$	0.1g
Resazurin	0.5mg
D-Glucose solution	30.0mL
$NaHCO_3$ solution	20.0mL
Trace mineral solution	10.0mL
$Na_2S_2O_3$ solution	10.0mL
$Na_2S \cdot 9H_2O$ solution	10.0mL

pH 7.0 ± 0.2 at 25°C

$Na_2S \cdot 9H_2O$ Solution:
Composition per 10.0mL:
$Na_2S \cdot 9H_2O$ 0.45g

Preparation of $Na_2S \cdot 9H_2O$ Solution: Add $Na_2S \cdot 9H_2O$ to distilled/deionized water and bring volume to 10.0mL. Mix thoroughly. Autoclave under 100% N_2 for 15 min at 15 psi pressure–121°C. Cool to room temperature.

$NaHCO_3$ Solution:
Composition per 20.0mL:
$NaHCO_3$ 4.0g

Preparation of $NaHCO_3$ Solution: Add $NaHCO_3$ to distilled/deionized water and bring volume to 20.0mL. Mix thoroughly. Sparge with 80% N_2 + 20% CO_2. Filter sterilize.

Glucose Solution:
Composition per 30.0mL:
D-Glucose 4.0g

Preparation of Glucose Solution: Add D-glucose to distilled/deionized water and bring volume to 30.0mL. Mix thoroughly. Sparge with 100% N_2. Filter sterilize.

$Na_2S_2O_3$ Solution:
Composition per 10.0mL:
$Na_2S_2O_3 \cdot 5H_2O$ 2.5g

Preparation of $Na_2S_2O_3$ Solution: Add $Na_2S_2O_3 \cdot 5H_2O$ to distilled/deionized water and bring volume to 10.0mL. Mix thoroughly. Autoclave under 100% N_2 for 15 min at 15 psi pressure–121°C. Cool to room temperature.

Trace Elements Solution:
Composition per liter:

$MgSO_4 \cdot 7H_2O$	3.0g
Nitrilotriacetic acid	1.5g
NaCl	1.0g
$MnSO_4 \cdot 2H_2O$	0.5g
$CoSO_4 \cdot 7H_2O$	0.18g
$ZnSO_4 \cdot 7H_2O$	0.18g
$CaCl_2 \cdot 2H_2O$	0.1g
$FeSO_4 \cdot 7H_2O$	0.1g
$NiCl_2 \cdot 6H_2O$	0.025g
$KAl(SO_4)_2 \cdot 12H_2O$	0.02g
H_3BO_3	0.01g
$Na_2MoO_4 \cdot 4H_2O$	0.01g
$CuSO_4 \cdot 5H_2O$	0.01g
$Na_2SeO_3 \cdot 5H_2O$	0.3mg

Preparation of Trace Elements Solution: Add nitrilotriacetic acid to 500.0mL of distilled/deionized water. Dissolve by adjusting pH to 6.5 with KOH. Add remaining components. Add distilled/deionized water to 1.0L. Mix thoroughly.

Preparation of Medium: Prepare and dispense medium under 80% N_2 + 20% CO_2 gas atmosphere. Add components, except $NaHCO_3$ solution, glucose solution, $Na_2S \cdot 9H_2O$ solution, and $Na_2S_2O_3$ solution, to distilled/deionized water and bring volume to 930.0mL. Mix thoroughly. Sparge with 80% N_2 + 20% CO_2. Distribute into sterile tubes or bottles. Autoclave for 15 min at 15 psi pressure–121°C. Aseptically and anaerobically per 1.0L of medium add 20.0mL $NaHCO_3$ solution, 30.0mL glucose solution, 10.0mL $Na_2S \cdot 9H_2O$ solution, and 10.0mL $Na_2S_2O_3$ solution. Mix thoroughly. The final pH should be 7.0.

Use: For the cultivation of *Thermoanaerobacter subterraneus.*

Thermoanaerobacter sulfurophilus Medium (DSMZ Medium 827)

Composition per 1055.0mL:

Sulfur, powdered	10.0g
NH_4Cl	0.33g

KCl....................0.33g
KH_2PO_4....................0.33g
$MgCl_2·6H_2O$....................0.33g
$CaCl_2·2H_2O$....................0.33g
Resazurin....................0.5mg
Glucose solution....................25.0mL
$Na_2S·9H_2O$ solution....................15.0mL
Vitamin solution....................10.0mL
Yeast extract solution....................5.0mL
Trace elements solution SL-10....................1.0mL

pH 7.0 ± 0.2 at 25°C

Vitamin Solution:
Composition per liter:

Pyridoxine-HCl....................10.0mg
Thiamine-HCl·$2H_2O$....................5.0mg
Riboflavin....................5.0mg
Nicotinic acid....................5.0mg
D-Ca-pantothenate....................5.0mg
p-Aminobenzoic acid....................5.0mg
Lipoic acid....................5.0mg
Biotin....................2.0mg
Folic acid....................2.0mg
Vitamin B_{12}....................0.1mg

Preparation of Vitamin Solution: Add components to distilled/deionized water and bring volume to 1.0L. Mix thoroughly. Sparge with 80% H_2 + 20% CO_2. Filter sterilize.

Glucose Solution:
Composition per 50.0mL:

Glucose....................10.0g

Preparation of Glucose Solution: Add glucose to distilled/deionized water and bring volume to 50.0mL. Mix thoroughly. Filter sterilize.

Yeast Extract Solution:
Composition per 10.0mL:

Yeast extract....................1.0g

Preparation of Yeast Extract Solution: Add yeast extract to distilled/deionized water and bring volume to 10.0mL. Mix thoroughly. Autoclave under 100% N_2 for 15 min at 15 psi pressure–121°C. Cool to room temperature.

Trace Elements Solution SL-10:
Composition per liter:

$FeCl_2·4H_2O$....................1.5g
$CoCl_2·6H_2O$....................190.0mg
$MnCl_2·4H_2O$....................100.0mg
$ZnCl_2$....................70.0mg
$Na_2MoO_4·2H_2O$....................36.0mg
$NiCl_2·6H_2O$....................24.0mg
H_3BO_3....................6.0mg
$CuCl_2·2H_2O$....................2.0mg
HCl (25% solution)....................10.0mL

Preparation of Trace Elements Solution SL-10: Add $FeCl_2·4H_2O$ to 10.0mL of HCl solution. Mix thoroughly. Add distilled/deionized water and bring volume to 1.0L. Add remaining components. Mix thoroughly. Sparge with 80% N_2 + 20% CO_2. Autoclave for 15 min at 15 psi pressure–121°C.

$Na_2S·9H_2O$ Solution:
Composition per 20.0mL:

$Na_2S·9H_2O$....................0.6g

Preparation of $Na_2S·9H_2O$ Solution: Add $Na_2S·9H_2O$ to distilled/deionized water and bring volume to 20.0mL. Mix thoroughly. Autoclave under 100% N_2 for 15 min at 15 psi pressure–121°C. Cool to room temperature.

Preparation of Medium: Prepare and dispense medium under 80% N_2 + 20% CO_2 gas mixture. Add components, except $Na_2S·9H_2O$ solution, glucose solution, vitamin solution, and yeast extract solution, to distilled/deionized water and bring volume to 1.0L. Mix thoroughly. Heat to 90°C on each of 3 successive days. Aseptically and anaerobically add 25.0mL sterile glucose solution, 15.0mL sterile $Na_2S·9H_2O$ solution, 10.0mL sterile vitamin solution, and 5.0mL sterile yeast extract solution. Mix thoroughly. Aseptically and anaerobically distribute to tubes or bottles. The pH should be 7.0.

Use: For the cultivation of *Thermoanaerobacter sulfurophilus.*

Thermoanaerobacter tengcongensis Medium (DSMZ Medium 965)

Composition per liter:

Soluble starch....................10.0g
NaCl....................2.0g
Tryptone....................2.0g
NH_4Cl....................1.0g
Yeast extract....................1.0g
$MgCl_2·6H_2O$....................0.5g
Cysteine-HCl·H_2O....................0.5g
K_2HPO_4....................0.3g
KH_2PO_4....................0.3g
KCl....................0.2g
$CaCl_2·2H_2O$....................0.05g
Resazurin....................0.5mg
Thiosulfate solution....................50.0mL
Trace elements solution....................10.0mL

pH 7.5 ± 0.2 at 25°C

Thiosulfate Solution:
Composition per 50.0mL:

$Na_2S_2O_3·5H_2O$....................5.0g

Preparation of Thiosulfate Solution: Add $Na_2S_2O_3·5H_2O$ to distilled/deionized water and bring volume to 50.0mL. Mix thoroughly. Autoclave for 15 min at 15 psi pressure–121°C. Cool to room temperature.

Trace Elements Solution:
Composition per liter:

$MgSO_4·7H_2O$....................3.0g
Nitrilotriacetic acid....................1.5g
NaCl....................1.0g
$MnSO_4·2H_2O$....................0.5g
$CoSO_4·7H_2O$....................0.18g
$ZnSO_4·7H_2O$....................0.18g
$CaCl_2·2H_2O$....................0.1g
$FeSO_4·7H_2O$....................0.1g
$NiCl_2·6H_2O$....................0.025g
$KAl(SO_4)_2·12H_2O$....................0.02g
H_3BO_3....................0.01g
$Na_2MoO_4·4H_2O$....................0.01g
$CuSO_4·5H_2O$....................0.01g
$Na_2SeO_3·5H_2O$....................0.3mg

Preparation of Trace Elements Solution: Add nitrilotriacetic acid to 500.0mL of distilled/deionized water. Dissolve by adjusting pH to 6.5 with KOH. Add remaining components. Add distilled/deionized water to 1.0L. Mix thoroughly.

Preparation of Medium: Add components, except thiosulfate solution, to distilled/deionized water and bring volume to 950.0mL. Mix thoroughly. Gently heat and bring to boiling. Boil for 3 min. Cool to room temperature while sparging with 100% N_2 gas. Adjust pH to 7.5. Distribute into tubes or bottles under 100% N_2 gas. Autoclave for 15 min at 15 psi pressure–121°C. Cool to room temperature. Aseptically and anaerobically under 100% N_2 gas, add 50.0mL sterile thiosulfate solution per liter of medium.

Use: For the cultivation of *Thermoanaerobacter tengcongensis.*

Thermoanaerobacterium Medium (DSMZ Medium 903)

Composition per 1030.0mL:

KH_2PO_4	0.5g
NaCl	0.4g
$MgCl_2 \cdot 6H_2O$	0.33g
Trypticase™	0.25g
$CaCl_2 \cdot 2H_2O$	0.05g
Resazurin	0.5mg
Sucrose solution	50.0mL
$NaHCO_3$ solution	20.0mL
$Na_2S \cdot 9H_2O$ solution	10.0mL
L-Cysteine solution	10.0mL
Selenite solution	1.0mL
Seven vitamin solution	1.0mL
Trace elements solution SL-10	1.0mL

pH 7.0 ± 0.2 at 25°C

Sucrose Solution:

Composition per 50.0mL:

Sucrose	5.0g

Preparation of Sucrose Solution: Add sucrose to distilled/deionized water and bring volume to 50.0mL. Mix thoroughly. Sparge with 100% N_2. Autoclave for 15 min at 15 psi pressure–121°C.

L-Cysteine Solution:

Composition per 10.0mL:

L-Cysteine·HCl·H_2O	0.3g

Preparation of L-Cysteine Solution: Add L-cysteine·HCl·H_2O to distilled/deionized water and bring volume to 10.0mL. Mix thoroughly. Sparge with 100% N_2. Autoclave for 15 min at 15 psi pressure–121°C.

$Na_2S \cdot 9H_2O$ Solution:

Composition per 10.0mL:

$Na_2S \cdot 9H_2O$	0.3g

Preparation of $Na_2S \cdot 9H_2O$ Solution: Add $Na_2S \cdot 9H_2O$ to distilled/deionized water and bring volume to 10.0mL. Mix thoroughly. Sparge with 100% N_2. Autoclave for 15 min at 15 psi pressure–121°C.

$NaHCO_3$ Solution:

Composition per 20.0mL:

$NaHCO_3$	2.5g

Preparation of $NaHCO_3$ Solution: Add $NaHCO_3$ to distilled/deionized water and bring volume to 20.0mL. Mix thoroughly. Sparge with 80% N_2 + 20% CO_2. Autoclave for 15 min at 15 psi pressure–121°C. Cool to 25°C. Must be prepared freshly.

Selenite Solution:

Composition per liter:

NaOH	0.5g
$Na_2SeO_3 \cdot 5H_2O$	3.0mg

Preparation of Selenite Solution: Add components to distilled/deionized water and bring volume to 1.0L. Mix thoroughly. Sparge with 100% N_2. Filter sterilize.

Seven Vitamin Solution:

Composition per liter:

Pyridoxine hydrochloride	300.0mg
Thiamine-HCl·$2H_2O$	200.0mg
Nicotinic acid	200.0mg
Vitamin B_{12}	100.0mg
Calcium pantothenate	100.0mg
p-Aminobenzoic acid	80.0mg
D(+)-Biotin	20.0mg

Preparation of Seven Vitamin Solution: Add components to distilled/deionized water and bring volume to 1.0L. Sparge with 100% N_2. Mix thoroughly. Filter sterilize.

Trace Elements Solution SL-10:

Composition per liter:

$FeCl_2 \cdot 4H_2O$	1.5g
$CoCl_2 \cdot 6H_2O$	190.0mg
$MnCl_2 \cdot 4H_2O$	100.0mg
$ZnCl_2$	70.0mg
$Na_2MoO_4 \cdot 2H_2O$	36.0mg
$NiCl_2 \cdot 6H_2O$	24.0mg
H_3BO_3	6.0mg
$CuCl_2 \cdot 2H_2O$	2.0mg
HCl (25% solution)	10.0mL

Preparation of Trace Elements Solution SL-10: Add $FeCl_2 \cdot 4H_2O$ to 10.0mL of HCl solution. Mix thoroughly. Add distilled/deionized water and bring volume to 1.0L. Add remaining components. Mix thoroughly. Sparge with 80% N_2 + 20% CO_2. Autoclave for 15 min at 15 psi pressure–121°C.

Preparation of Medium: Prepare and dispense medium under 80% N_2 + 20% CO_2. Add components, except seven vitamin solution, $NaHCO_3$ solution, sucrose solution, L-cysteine-HCl·H_2O solution, and $Na_2S \cdot 9H_2O$ solution, to distilled/deionized water and bring volume to 940.0mL. Mix thoroughly. Adjust pH to 7.0. Distribute into anaerobe tubes or bottles. Autoclave for 15 min at 15 psi pressure–121°C. Aseptically and anaerobically add, per liter, 10.0mL seven vitamin solution, 20.0mL $NaHCO_3$ solution, 50.0mL sucrose solution, 10.0mL L-cysteine-HCl·H_2O solution, and 10.0mL $Na_2S \cdot 9H_2O$ solution. Mix thoroughly. The final pH should be 7.0.

Use: For the cultivation of *Thermoanaerobacterium polysaccharolyticum* and *Thermoanaerobacterium zeae.*

Thermoanaerobium brockii Medium

Composition per liter:

Pancreatic digest of casein	10.0g
Yeast extract	3.0g
K_2HPO_4	1.5g
NH_4Cl	0.9g
NaCl	0.9g
KH_2PO_4	0.75g
$MgCl_2 \cdot 6H_2O$	0.2g
Glucose solution	25.0mL
$Na_2S \cdot 9H_2O$ (10% solution)	10.0mL

Trace elements solution	9.0mL
Wolfe's Vitamin solution	5.0mL
Resazurin (0.025% solution)	4.0mL
$FeSO_4 \cdot 7H_2O$ (10% solution)	0.03mL

pH 7.3 ± 0.2 at 25°C

Glucose Solution:

Composition per 100.0mL:

Glucose	20.0g

Preparation of Glucose Solution: Add glucose to distilled/deionized water and bring volume to 100.0mL. Mix thoroughly. Filter sterilize.

Trace Elements Solution:

Composition per liter:

Nitrilotriacetic acid	12.5g
NaCl	1.0g
$FeCl_3 \cdot 4H_2O$	0.2g
$MnCl_2 \cdot 4H_2O$	0.1g
$CaCl_2 \cdot 2H_2O$	0.1g
$ZnCl_2$	0.1g
$CuCl_2$	0.02g
Na_2SeO_3	0.02g
$CoCl_2 \cdot 6H_2O$	0.017g
H_3BO_3	0.01g
$Na_2MoO_4 \cdot 2H_2O$	0.01g

Preparation of Trace Elements Solution: Add nitrilotriacetic acid to 500.0mL of distilled/deionized water. Adjust pH to 6.5 with KOH. Add remaining components. Add distilled/deionized water to 1.0L.

Wolfe's Vitamin Solution:

Composition per liter:

Pyridoxine·HCl	10.0mg
Thiamine·HCl	5.0mg
Riboflavin	5.0mg
Nicotinic acid	5.0mg
Calcium pantothenate	5.0mg
p-Aminobenzoic acid	5.0mg
Thioctic acid	5.0mg
Biotin	2.0mg
Folic acid	2.0mg
Cyanocobalamin	100.0μg

Preparation of Wolfe's Vitamin Solution: Add components to distilled/deionized water and bring volume to 1.0L. Mix thoroughly.

Preparation of Medium: Add components, except glucose solution, to distilled/deionized water and bring volume to 975.0mL. Mix thoroughly. Autoclave for 15 min at 15 psi pressure–121°C. While still hot, aseptically add 25.0mL of the sterile glucose solution under 97% N_2 + 3% H_2. Adjust pH to 7.3 if necessary. Aseptically and anaerobically distribute into tubes. Cap with rubber stoppers.

Use: For the cultivation and maintenance of *Thermoanaerobium brockii.*

Thermoanaeromonas Medium (DSMZ Medium 963)

Composition per liter:

$NaHCO_3$	5.0g
K_2HPO_4	0.78g
KH_2PO_4	0.75g
NH_4Cl	0.5g
Glucose	0.5g
Yeast extract	0.5g
$MgSO_4 \cdot 7H_2O$	0.25g
NaCl	0.2g
Na_3-EDTA	0.04g
$CaCl_2 \cdot 2H_2O$	0.03g
$FeSO_4 \cdot 7H_2O$	0.01g
Resazurin	0.5mg
Thiosulfate solution	20.0mL
Vitamin solution	10.0mL
Trace elements solution	10.0mL
Cysteine solution	10.0mL

pH 6.5 ± 0.2 at 25°C

Vitamin Solution:

Composition per liter:

Pyridoxine-HCl	10.0mg
Thiamine-HCl·$2H_2O$	5.0mg
Riboflavin	5.0mg
Nicotinic acid	5.0mg
D-Ca-pantothenate	5.0mg
p-Aminobenzoic acid	5.0mg
Lipoic acid	5.0mg
Biotin	2.0mg
Folic acid	2.0mg
Vitamin B_{12}	0.1mg

Preparation of Vitamin Solution: Add components to distilled/deionized water and bring volume to 1.0L. Mix thoroughly. Sparge with 80% H_2 + 20% CO_2. Filter sterilize.

Trace Elements Solution:

Composition per liter:

$MgSO_4 \cdot 7H_2O$	3.0g
Nitrilotriacetic acid	1.5g
NaCl	1.0g
$MnSO_4 \cdot 2H_2O$	0.5g
$CoSO_4 \cdot 7H_2O$	0.18g
$ZnSO_4 \cdot 7H_2O$	0.18g
$CaCl_2 \cdot 2H_2O$	0.1g
$FeSO_4 \cdot 7H_2O$	0.1g
$NiCl_2 \cdot 6H_2O$	0.025g
$KAl(SO_4)_2 \cdot 12H_2O$	0.02g
H_3BO_3	0.01g
$Na_2MoO_4 \cdot 4H_2O$	0.01g
$CuSO_4 \cdot 5H_2O$	0.01g
$Na_2SeO_3 \cdot 5H_2O$	0.3mg

Preparation of Trace Elements Solution: Add nitrilotriacetic acid to 500.0mL of distilled/deionized water. Dissolve by adjusting pH to 6.5 with KOH. Add remaining components. Add distilled/deionized water to 1.0L. Mix thoroughly.

Cysteine Solution:

Composition per 10.0mL:

L-Cysteine·HCl·H_2O	0.25g

Preparation of Cysteine Solution: Add L-cysteine·HCl·H_2O to distilled/deionized water and bring volume to 10.0mL. Mix thoroughly. Sparge with 100% N_2. Autoclave for 15 min at 15 psi pressure–121°C.

Thiosulfate Solution:

Composition per 20.0mL:

$Na_2S_2O_3 \cdot 5H_2O$	1.24g

Preparation of Thiosulfate Solution: Add $Na_2S_2O_3 \cdot 5H_2O$ to distilled/deionized water and bring volume to 20.0mL. Mix thoroughly. Autoclave for 15 min at 15 psi pressure–121°C. Cool to room temperature.

Preparation of Medium: Add components, except $NaHCO_3$, vitamin solution, thiosulfate solution, and cysteine solution, to distilled/deionized water and bring volume to 960.0mL. Mix thoroughly. Gently heat and bring to boiling. Boil for 3 min. Cool to 25°C while sparging with 80% N_2 + 20% CO_2. Add solid $NaHCO_3$. Adjust pH to 6.8–7.0. Distribute to anaerobe tubes or bottles under 80% N_2 + 20% CO_2. Autoclave for 15 min at 15 psi pressure–121°C. Aseptically and anaerobically add, per liter of medium, 10.0mL sterile vitamin solution, 10.0mL sterile cysteine solution, and 20.0mL sterile thiosulfate solution. Mix thoroughly. The final pH should be 6.5.

Use: For the cultivation of *Thermanaeromonas toyohensis.*

Thermoanaerovibrio Medium (DSMZ Medium 873)

Composition per liter:

$NaHCO_3$	0.8g
NH_4Cl	0.33g
KH_2PO_4	0.33g
$MgCl_2 \cdot 6H_2O$	0.33g
$CaCl_2 \cdot 2H_2O$	0.22g
KCl	0.33g
Yeast extract	0.25g
Peptone	0.25g
Resazurin	0.5mg
$NaHCO_3$ solution	20.0mL
$Na_2S \cdot 9H_2O$ solution	10.0mL
Vitamin solution	10.0mL
Glucose solution	10.0mL
Calcium chloride solution	10.0mL
Magnesium chloride solution	10.0mL
Trace elements solution SL-10	1.0mL

pH 7.0–7.3 at 25°C

$NaHCO_3$ Solution:

Composition per 20.0mL:

$NaHCO_3$	2.0g

Preparation of $NaHCO_3$ Solution: Add $NaHCO_3$ to distilled/deionized water and bring volume to 20.0mL. Mix thoroughly. Sparge with 80% N_2 + 20% CO_2. Filter sterilize.

Magnesium Chloride Solution:

Composition per 10.0mL:

$MgCl_2 \cdot 6H_2O$	0.33g

Preparation of Magnesium Chloride Solution: Add 0.33g of $MgCl_2 \cdot 6H_2O$ to distilled/deionized water and bring volume to 10.0mL. Mix thoroughly. Sparge with 100% N_2. Filter sterilize.

Calcium Chloride Solution:

Composition per 10.0mL:

$CaCl_2 \cdot 2H_2O$	0.22g

Preparation of Calcium Chloride Solution: Add $CaCl_2 \cdot 2H_2O$ to distilled/deionized water and bring volume to 10.0mL. Mix thoroughly. Sparge with 100% N_2. Filter sterilize.

Glucose Solution:

Composition per 10.0mL:

Glucose	3.0g

Preparation of Glucose Solution: Add glucose to distilled/deionized water and bring volume to 10.0mL. Mix thoroughly. Sparge with 100% N_2. Filter sterilize.

$Na_2S \cdot 9H_2O$ Solution:

Composition per 10.0mL:

$Na_2S \cdot 9H_2O$	0.5g

Preparation of $Na_2S \cdot 9H_2O$ Solution: Add $Na_2S \cdot 9H_2O$ to distilled/deionized water and bring volume to 10.0mL. Mix thoroughly. Sparge with 100% N_2. Autoclave for 15 min at 15 psi pressure–121°C. Before use, neutralize to pH 7.0 with sterile HCl.

Vitamin Solution:

Composition per liter:

Pyridoxine-HCl	10.0mg
Thiamine-HCl·$2H_2O$	5.0mg
Riboflavin	5.0mg
Nicotinic acid	5.0mg
D-Ca-pantothenate	5.0mg
p-Aminobenzoic acid	5.0mg
Lipoic acid	5.0mg
Biotin	2.0mg
Folic acid	2.0mg
Vitamin B_{12}	0.1mg

Preparation of Vitamin Solution: Add components to distilled/deionized water and bring volume to 1.0L. Mix thoroughly. Sparge with 80% H_2 + 20% CO_2. Filter sterilize.

Trace Elements Solution SL-10:

Composition per liter:

$FeCl_2 \cdot 4H_2O$	1.5g
$CoCl_2 \cdot 6H_2O$	190.0mg
$MnCl_2 \cdot 4H_2O$	100.0mg
$ZnCl_2$	70.0mg
$Na_2MoO_4 \cdot 2H_2O$	36.0mg
$NiCl_2 \cdot 6H_2O$	24.0mg
H_3BO_3	6.0mg
$CuCl_2 \cdot 2H_2O$	2.0mg
HCl (25% solution)	10.0mL

Preparation of Trace Elements Solution SL-10: Add $FeCl_2 \cdot 4H_2O$ to 10.0mL of HCl solution. Mix thoroughly. Add distilled/deionized water and bring volume to 1.0L. Add remaining components. Mix thoroughly. Sparge with 80% N_2 + 20% CO_2. Autoclave for 15 min at 15 psi pressure–121°C.

Preparation of Medium: Prepare and dispense medium under 80% N_2 + 20% CO_2 gas atmosphere. Add components, except $NaHCO_3$ solution, glucose solution, calcium chloride solution, magnesium chloride solution, $Na_2S \cdot 9H_2O$ solution, vitamin solution, and trace elements solution SL-10, to distilled/deionized water and bring volume to 929.0mL. Mix thoroughly. Sparge with 80% N_2 + 20% CO_2. Autoclave for 15 min at 15 psi pressure–121°C. Aseptically and anaerobically add 10.0mL glucose solution, 10.0mL $Na_2S \cdot 9H_2O$ solution, 10.0mL magnesium chloride solution, 10.0ml calcium chloride solution, 10.0mL vitamin solution, and 1.0mL trace elements solution SL-10. Mix thoroughly. Adjust pH to 7.0–7.3 with 10.0mL $NaHCO_3$ solution. Aseptically and anaerobically distribute into sterile tubes or bottles.

Use: For the cultivation of *Thermanaerovibrio velox (Thermosinus velox).*

Thermobacterium Medium

Composition per liter:

Agar	20.0g
$(NH_4)_2SO_4$	1.3g
Yeast extract	1.0g
Pancreatic digest of casein	1.0g
KH_2PO_4	0.28g
$MgSO_4 \cdot 7H_2O$	0.247g
$CaCl_2 \cdot 2H_2O$	0.074g
$FeCl_3 .6H_2O$	0.019g
Salt solution	1.0mL

pH 8.5 + 0.2 at 25°C

Salt Solution:

Composition per liter:

$Na_2B_4O_7 \cdot 10H_2O$	4.4g
$MnCl_2 \cdot 4H_2O$	1.8g
$ZnSO_4 \cdot 7H_2O$	0.22g
$CuCl_2 \cdot H_2O$	0.05g
$Na_2MoO_4 .2H_2O$	0.03g
$VOSO_4 \cdot 2H_2O$	0.03g

Preparation of Salt Solution: Add components to distilled/deionized water and bring volume to 1.0L. Mix thoroughly. Adjust pH to 2.0 with H_2SO_4.

Preparation of Medium: Add components to distilled/deionized water and bring volume to 1.0L. Mix thoroughly. Gently heat and bring to boiling. Distribute into tubes in 11.0–12.0mL volumes. Autoclave for 15 min at 15 psi pressure–121°C. Allow tubes to solidify in a slanted position.

Use: For the cultivation and maintenance of *Thermomicrobium roseum*.

Thermobacteroides leptospartum Medium

Composition per 1168.1mL:

Yeast extract	2.0g
Trypticase™	2.0g
NaOH solution	1.0L
Glucose solution	113.0mL
Na_2S solution	22.6mL
Solution A	10.0mL
Mineral salts solution	10.0mL
Solution B	2.0mL
Resazurin solution	0.5mL

NaOH Solution:

Composition per liter:

NaOH	4.0g

Preparation of NaOH Solution: Add NaOH to distilled/deionized water and bring volume to 1.0L. Mix thoroughly.

Glucose Solution:

Composition per 100.0mL:

D-Glucose	5.0g

Preparation of Glucose Solution: Add glucose to distilled/deionized water and bring volume to 100.0mL. Mix thoroughly. Sparge with 100% N_2 for 15 min. Autoclave for 15 min at 15 psi pressure–121°C.

Na_2S Solution:

Na_2S	2.5g

Preparation of Na_2S Solution: Gently heat 100.0mL of distilled/deionized water to 100°C. Boil for 5 min. Sparge with 100% N_2 for 15 min. Add the Na_2S. Mix thoroughly. Sparge with 100% N_2 for 10 min. Autoclave for 15 min at 15 psi pressure–121°C.

Solution A:

Composition per liter:

NH_4Cl	100.0g
$MgCl_2 \cdot H_2O$	100.0g
$CaCl_2 \cdot 2H_2O$	40.0g

Preparation of Solution A: Add components to distilled/deionized water and bring volume to 1.0L. Mix thoroughly. Adjust pH to 4 with HCl.

Mineral Salts Solution:

Composition per liter:

$EDTA \cdot 2H_2O$	0.5g
$CoCl_2 \cdot H_2O$	0.15g
$MnCl_2 \cdot 4H_2O$	0.1g
$FeSO_4 \cdot 7H_2O$	0.1g
$ZnCl_2$	0.1g
$AlCl_3 \cdot H_2O$	40.0mg
$Na_2WO_4 \cdot 2H_2O$	30.0mg
$CuCl_2 \cdot 2H_2O$	20.0mg
$NiSO_4 \cdot H_2O$	20.0mg
H_2SeO_3	10.0mg
H_3BO_4	10.0mg
$NaMoO_4 \cdot 2H_2O$	10.0mg

Preparation of Mineral Salts Solution: Add components to distilled/deionized water and bring volume to 1.0L. Mix thoroughly. Adjust pH to 3 with HCl.

Solution B:

Composition per liter:

$K_2HPO_4 \cdot 3H_2O$	200.0g

Preparation of Solution B: Add $K_2HPO_4 \cdot 3H_2O$ to distilled/deionized water and bring volume to 1.0L. Mix thoroughly.

Resazurin Solution:

Composition per 100.0mL:

Resazurin	0.2g

Preparation of Resazurin Solution: Add resazurin to distilled/deionized water and bring volume to 100.0mL. Mix thoroughly.

Preparation of Medium: Sparge 1.0L of NaOH solution with 100% CO_2 for 30 min. Add 2.0g of yeast extract and 2.0g of Trypticase™. Mix thoroughly. Add 10.0mL of solution A, 2.0mL of solution B, 0.5mL of resazurin solution, and 10.0mL of mineral salts solution with pipets which have been flushed a few times with 100% N_2. Mix thoroughly. Anaerobically distribute 9.0mL volumes into anaerobic tubes fitted with butyl rubber stoppers. Autoclave for 15 min at 15 psi pressure–121°C. One hour prior to inoculation, add 1.0mL of sterile glucose solution and 0.2mL of sterile Na_2S solution to each 9.0mL of medium.

Use: For the cultivation of *Thermobacteroides leptospartum*.

Thermobacteroides proteolyticus Medium

Composition per 1010.0mL:

$NaHCO_3$	5.0g
Pancreatic digest of casein	2.0g
Yeast extract	2.0g
$MgCl_2 \cdot 6H_2O$	1.0g
NH_4Cl	1.0g
$CaCl_2 \cdot 2H_2O$	0.4g

K_2HPO_4 .... 0.4g
$Na_2S{\cdot}9H_2O$ 0.3g
Resazurin 1.0mg
Trace elements solution 10.0mL

pH 7.0 ± 0.2 at 25°C

Trace Elements Solution:

Composition per liter:

$MgSO_4{\cdot}7H_2O$ 3.0g
Nitrilotriacetic acid 1.5g
NaCl 1.0g
$MnSO_4{\cdot}2H_2O$ 0.5g
$CoSO_4{\cdot}7H_2O$ 0.18g
$ZnSO_4{\cdot}7H_2O$ 0.18g
$CaCl_2{\cdot}2H_2O$ 0.1g
$FeSO_4{\cdot}7H_2O$ 0.1g
$KAl(SO_4)_2{\cdot}12H_2O$ 0.02g
$CuSO_4{\cdot}5H_2O$ 0.01g
H_3BO_3 .... 0.01g
$Na_2MoO_4{\cdot}2H_2O$ 0.01g
$NiCl_2{\cdot}6H_2O$ 0.025g
$Na_2SeO_3{\cdot}5H_2O$ 0.3mg

Preparation of Trace Elements Solution: Add nitrilotriacetic acid to 500.0mL of distilled/deionized water. Adjust pH to 6.5 with KOH. Add remaining components. Add distilled/deionized water to 1.0L.

Preparation of Medium: Prepare and dispense medium under 80% N_2 + 20% CO_2. Add components to distilled/deionized water and bring volume to 1.0L. Mix thoroughly. Distribute into tubes or flasks. Autoclave for 15 min at 15 psi pressure–121°C.

Use: For the cultivation and maintenance of *Thermobacteroides proteolyticus*.

Thermococcus celer Medium

Composition per liter:

NaCl 40.0g
Sulfur 10.0g
Yeast extract 2.0g
$(NH_4)_2SO_4$ 1.3g
KH_2PO_4 .... 0.28g
$MgSO_4{\cdot}7H_2O$ 0.25g
$CaCl_2{\cdot}2H_2O$ 0.07g
$FeCl_2{\cdot}2H_2O$ 0.02g
$NaB_4O{\cdot}10H_2O$ 4.5mg
$MnCl_2{\cdot}4H_2O$ 1.8mg
Resazurin 1.0mg
$ZnSO_4{\cdot}7H_2O$ 0.22mg
$CuCl_2{\cdot}2H_2O$ 0.05mg
$NaMoO_4{\cdot}2H_2O$ 0.03mg
$VOSO_4{\cdot}2H_2O$ 0.03mg
$CoSO_4$ 0.01mg
$Na_2S{\cdot}9H_2O$ solution 10.0mL

pH 5.8 ± 0.2 at 25°C

$Na_2S{\cdot}9H_2O$ Solution:

Composition per 10.0mL:

$Na_2S{\cdot}9H_2O$ 0.3g

Preparation of $Na_2S{\cdot}9H_2O$ Solution: Add $Na_2S{\cdot}9H_2O$ to distilled/deionized water and bring volume to 10.0mL. Mix thoroughly. Sparge with 100% N_2. Autoclave for 15 min at 15 psi pressure–121°C.

Preparation of Medium: Add components, except $Na_2S{\cdot}9H_2O$ solution, to distilled/deionized water and bring volume to 1.0L. Mix thoroughly. Adjust pH to 5.8. Do not autoclave. Sterilize by steaming at 100°C for 30 min on 3 consecutive days. Before inoculation, add 10.0mL of sterile $Na_2S{\cdot}9H_2O$ solution. Mix thoroughly.

Use: For the cultivation of *Thermococcus celer.*

Thermococcus celer Medium

Composition per liter:

NaCl 40.0g
Sulfur, powdered 5.0g
$(NH_4)_2SO_4$ 1.3g
KH_2PO_4 .... 0.28g
$MgSO_4{\cdot}7H_2O$ 0.25g
$CaCl_2{\cdot}2H_2O$ 0.07g
$CuCl_2{\cdot}2H_2O$ 0.05g
$FeCl_3{\cdot}6H_2O$ 0.02g
$Na_2B_4O_7{\cdot}10H_2O$ 4.5mg
$MnCl_2{\cdot}4H_2O$ 1.8mg
Resazurin 1.0mg
$ZnSO_4{\cdot}7H_2O$ 0.22mg
$Na_2MoO_4{\cdot}2H_2O$ 0.03mg
$VOSO_4{\cdot}2H_2O$ 0.03mg
$CoSO_4{\cdot}7H_2O$ 0.01mg
Yeast extract solution 2.0g
$Na_2S{\cdot}9H_2O$ solution 10.0mL

pH 5.8 ± 0.2 at 25°C

Preparation of Sulfur: Add 10.0g of powdered sulfur to a flask and sterilize by steaming for 3 hr on 3 consecutive days.

Yeast Extract Solution:

Composition per 20.0mL:

Yeast extract 2.0g

Preparation of Yeast Extract Solution: Add yeast extract to distilled/deionized water and bring volume to 20.0mL. Mix thoroughly. Gently heat and bring to boiling. Boil for a few minutes. Do not autoclave.

$Na_2S{\cdot}9H_2O$ Solution:

Composition per 10.0mL:

$Na_2S{\cdot}9H_2O$ 0.3g

Preparation of $Na_2S{\cdot}9H_2O$ Solution: Add $Na_2S{\cdot}9H_2O$ to distilled/deionized water and bring volume to 10.0mL. Mix thoroughly. Sparge with 100% N_2. Autoclave for 15 min at 15 psi pressure–121°C.

Preparation of Medium: Prepare and dispense medium under 100% N_2. Add components, except sulfur, yeast extract solution, and $Na_2S{\cdot}9H_2O$ solution, to distilled/deionized water and bring volume to 980.0mL. Mix thoroughly. Sparge with 100% N_2. Autoclave for 15 min at 15 psi pressure–121°C. Aseptically add 5.0g of sterile sulfur, 10.0mL of sterile yeast extract solution, and 10.0mL of sterile $Na_2S{\cdot}9H_2O$ solution. Mix thoroughly. Aseptically and anaerobically distribute into sterile tubes or flasks.

Use: For the cultivation of *Thermococcus celer*.

Thermococcus chitinophagus Medium (DSMZ Medium 766)

Composition per liter:

Chitin, purified 4.0g
$(NH_4)_2SO_4$ 0.5g
KH_2PO_4 .... 0.5g

Resazurin 1.0mg
$(NH_4)_2Ni(SO_4)_2$ 0.3mg
$Na_2WO_4 \cdot 2H_2O$ 0.15mg
Na_2SeO_4 .... 0.15mg
Synthetic seawater 485.0mL
Trace elements solution SL-6 .... 15.0mL
$Na_2S \cdot 9H_2O$ solution 10.0mL
$NaHCO_3$ solution 10.0mL

pH 6.7 ± 0.2 at 25°C

$Na_2S \cdot 9H_2O$ Solution:

Composition per 10.0mL:

$Na_2S \cdot 9H_2O$ 0.5g

Preparation of $Na_2S \cdot 9H_2O$ Solution: Add $Na_2S \cdot 9H_2O$ to distilled/deionized water and bring volume to 10.0mL. Mix thoroughly. Autoclave under 100% N_2 for 15 min at 15 psi pressure–121°C. Cool to room temperature.

$NaHCO_3$ Solution:

Composition per 10.0mL:

$NaHCO_3$ 0.2g

Preparation of $NaHCO_3$ Solution: Add $NaHCO_3$ to distilled/deionized water and bring volume to 10.0mL. Mix thoroughly. Sparge with 80% N_2 + 20% CO_2. Filter sterilize.

Trace Elements Solution SL-6:

Composition per liter:

$MnCl_2 \cdot 4H_2O$ 0.5g
H_3BO_3 .... 0.3g
$CoCl_2 \cdot 6H_2O$ 0.2g
$ZnSO_4 \cdot 7H_2O$ 0.1g
$Na_2MoO_4 \cdot 2H_2O$ 0.03g
$NiCl_2 \cdot 6H_2O$ 0.02g
$CuCl_2 \cdot 2H_2O$ 0.01g

Preparation of Trace Elements Solution SL-6: Add components to distilled/deionized water and bring volume to 1.0L. Mix thoroughly. Autoclave for 15 min at 15 psi pressure–121°C.

Synthetic Seawater:

Composition per liter:

NaCl 23.477g
$MgCl_2 \cdot 6H_2O$ 4.981g
Na_2SO_4 .... 3.917g
$CaCl_2$ 1.12g
KCl 664.0mg
$NaHCO_3$ 192.0mg
H_3BO_3 .... 26.0mg
$SrCl_2$ 24.0mg
KBr 6.0mg
NaF 3.0mg

Preparation of Synthetic Seawater: Add components to distilled/deionized water and bring volume to 1.0L. Mix thoroughly. Filter sterilize.

Preparation of Purified Chitin: Cool 200.0mL of 37% HCl to 4°C. Add 20.0g chitin (practical grade from crab shells) to the cooled HCl. Mix thoroughly. Stir for 60 min at 4°C. Pour the suspension into 1 liter of distilled water, pre-cooled to 4°C. Filter through filter paper. Wash the residue five times with 500.0mL distilled water. Resuspend in 1.0L of distilled water. Neutralize the suspension with 10.0mL of 5*M* KOH to achieve a final pH of 6.5. Filter and wash with 3.0L of distilled water to remove KCl. Allow to air dry.

Preparation of Medium: Add components, except $NaHCO_3$ solution and $Na_2S \cdot 9H_2O$ solution, to distilled/deionized water and bring volume to 980.0mL. Gently heat and bring to boiling. Boil for 5 min. Cool to 25°C while sparging with 100% N_2. Add 10.0mL $NaHCO_3$ solution. Adjust pH to 6.7. Distribute the medium into Hungate tubes under an atmosphere of 100% N_2. Autoclave for 15 min at 15 psi pressure–121°C. Cool to room temperature. Reduce the medium by adding 10.0mL $Na_2S \cdot 9H_2O$ solution. Aseptically and anaerobically distribute to sterile tubes or bottles.

Use: For the cultivation of *Thermococcus chitinophagus.*

Thermococcus litoralis Medium

Composition per liter:

NaCl 25.0g
Sulfur 10.0g
$(NH_4)_2SO_4$ 1.3g
Yeast extract 1.0g
Peptone 0.5g
KH_2PO_4 .... 0.28g
$MgSO_4 \cdot 7H_2O$ 0.25g
$CaCl_2 \cdot 2H_2O$ 0.07g
$FeCl_2 \cdot 2H_2O$ 0.02g
$NaB_4O \cdot 10H_2O$ 4.5mg
$MnCl_2 \cdot 4H_2O$ 1.8mg
Resazurin 1.0mg
$ZnSO_4 \cdot 7H_2O$ 0.22mg
$CuCl_2 \cdot 2H_2O$ 0.05mg
$NaMoO_4 \cdot 2H_2O$ 0.03mg
$VOSO_4 \cdot 2H_2O$ 0.03mg
$CoSO_4$ 0.01mg
$Na_2S \cdot 9H_2O$ solution 10.0mL

pH 7.2 ± 0.2 at 25°C

$Na_2S \cdot 9H_2O$ Solution:

Composition per 10.0mL:

$Na_2S \cdot 9H_2O$ 0.3g

Preparation of $Na_2S \cdot 9H_2O$ Solution: Add $Na_2S \cdot 9H_2O$ to distilled/deionized water and bring volume to 10.0mL. Mix thoroughly. Sparge with 100% N_2. Autoclave for 15 min at 15 psi pressure–121°C.

Preparation of Medium: Add components, except $Na_2S \cdot 9H_2O$ solution, to distilled/deionized water and bring volume to 1.0L. Mix thoroughly. Adjust pH to 7.2. Do not autoclave. Sterilize by steaming at 100°C for 30 min on 3 consecutive days. Before inoculation, add 10.0mL of sterile $Na_2S \cdot 9H_2O$ solution. Mix thoroughly.

Use: For the cultivation of *Thermococcus litoralis*.

Thermococcus litoralis Medium

Composition per liter:

NaCl 19.45g
$MgCl_2 \cdot 6H_2O$ 12.6g
Sulfur, powdered 10.0g
Peptone 5.0g
Na_2SO_4 .... 3.42g
$CaCl_2 \cdot 2H_2O$ 2.38g
Yeast extract 1.0g
Na_2CO_3 .... 0.61g
KCl 0.55g
Resazurin 0.1g
KBr 0.08g
$SrCl_2 \cdot 6H_2O$ 0.057g

H_3BO_3 .... 0.022g
Na_2HPO_4 .... 0.01g
$Na_2SiO_3 \cdot 9H_2O$ 4.0mg
NaF 2.4mg
KNO_3 .... 1.6mg
$Na_2S \cdot 9H_2O$ solution 10.0mL

pH 6.5 ± 0.2 at 25°C

$Na_2S \cdot 9H_2O$ Solution:
Composition per 10.0mL:
$Na_2S \cdot 9H_2O$ 0.5g

Preparation of $Na_2S \cdot 9H_2O$ Solution: Add $Na_2S \cdot 9H_2O$ to distilled/deionized water and bring volume to 10.0mL. Mix thoroughly. Sparge with 100% N_2. Autoclave for 15 min at 15 psi pressure–121°C.

Preparation of Medium: Prepare and dispense medium under 100% N_2. Add components, except $Na_2S \cdot 9H_2O$ solution, to distilled/deionized water and bring volume to 990.0mL. Mix thoroughly. Adjust pH to 6.5. Sparge with 100% N_2. Do not autoclave. Sterilize by steaming for 3 hr at 100°C on 3 consecutive days. Prior to inoculation, aseptically and anaerobically add 10.0mL of sterile $Na_2S \cdot 9H_2O$ solution.

Use: For the cultivation of *Thermococcus litoralis.*

Thermococcus Medium (DSMZ Medium 806)

Composition per liter:
NaCl 18.0g
Sulfur 5.0g
$MgSO_4 \cdot 7H_2O$ 3.4g
$MgCl_2 \cdot 2H_2O$ 2.7g
Yeast extract 1.0g
Trypticase™ 1.0g
$NaHCO_3$ 1.0g
KCl 0.33g
NH_4Cl 0.25g
$CaCl_2 \cdot 2H_2O$ 0.14g
K_2HPO_4 .... 0.14g
Resazurin 0.001g
Na_2SeO_3 .... 0.001mg
$NiCl_2 \cdot 6H_2O$ 0.001mg
Trace elements solution 10.0mL
Vitamin solution 10.0mL
Cysteine solution 10.0mL
$Na_2S \cdot 9H_2O$ solution 10.0mL

pH 7.2 ± 0.2 at 25°C

$Na_2S \cdot 9H_2O$ Solution:
Composition per 10.0mL:
$Na_2S \cdot 9H_2O$ 0.5g

Preparation of $Na_2S \cdot 9H_2O$ Solution: Add $Na_2S \cdot 9H_2O$ to distilled/deionized water and bring volume to 10.0mL. Mix thoroughly. Autoclave under 100% N_2 for 15 min at 15 psi pressure–121°C. Cool to room temperature. Adjust pH to 7.0.

Cysteine Solution:
Composition per 10.0mL:
L-Cysteine·HCl·H_2O 0.3g

Preparation of Cysteine Solution: Add L-cysteine·HCl·H_2O to distilled/deionized water and bring volume to 10.0mL. Mix thoroughly. Sparge with 100% N_2. Autoclave for 15 min at 15 psi pressure–121°C. Cool to room temperature.

Trace Elements Solution:
Composition per liter:
$MgSO_4 \cdot 7H_2O$ 3.0g
Nitrilotriacetic acid 1.5g
NaCl 1.0g
$MnSO_4 \cdot 2H_2O$ 0.5g
$CoSO_4 \cdot 7H_2O$ 0.18g
$ZnSO_4 \cdot 7H_2O$ 0.18g
$CaCl_2 \cdot 2H_2O$ 0.1g
$FeSO_4 \cdot 7H_2O$ 0.1g
$NiCl_2 \cdot 6H_2O$ 0.025g
$KAl(SO_4)_2 \cdot 12H_2O$ 0.02g
H_3BO_3 .... 0.01g
$Na_2MoO_4 \cdot 4H_2O$ 0.01g
$CuSO_4 \cdot 5H_2O$ 0.01g
$Na_2SeO_3 \cdot 5H_2O$ 0.3mg

Preparation of Trace Elements Solution: Add nitrilotriacetic acid to 500.0mL of distilled/deionized water. Dissolve by adjusting pH to 6.5 with KOH. Add remaining components. Add distilled/deionized water to 1.0L. Mix thoroughly.

Vitamin Solution:
Composition per liter:
Pyridoxine-HCl 10.0mg
Thiamine-HCl·$2H_2O$ 5.0mg
Riboflavin 5.0mg
Nicotinic acid 5.0mg
D-Ca-pantothenate 5.0mg
p-Aminobenzoic acid 5.0mg
Lipoic acid 5.0mg
Biotin 2.0mg
Folic acid 2.0mg
Vitamin B_{12} 0.1mg

Preparation of Vitamin Solution: Add components to distilled/deionized water and bring volume to 1.0L. Mix thoroughly. Sparge with 100% N_2. Filter sterilize.

Preparation of Medium: Add components, except Fildes enrichment solution, $NaHCO_3$, vitamin solution, cysteine solution, and $Na_2S \cdot 9H_2O$ solution, to distilled/deionized water and bring volume to 970.0mL. Mix thoroughly. Gently heat and bring to boiling. Boil for 5 min. Cool to room temperature under 100% N_2. Add 1.0g solid $NaHCO_3$. Adjust pH to 7.2. Sterilize at 100°C for 3 hr on 3 consecutive days. Aseptically and anaerobically add 10.0mL vitamin solution, 10.0mL cysteine solution, and 10.0mL $Na_2S \cdot 9H_2O$ solution. Mix thoroughly. Adjust pH to 7.2. Aseptically and anaerobically distribute to sterile tubes or bottles.

Use: For the cultivation of *Thermococcus* spp.

Thermococcus profundus Medium

Composition per 1010.0mL:
NaCl 25.0g
Sulfur 10.0g
Peptone 5.0g
Yeast extract 1.0g
Resazurin 1.0mg
Salt base solution 1.0L
$Na_2S \cdot 9H_2O$ solution 10.0mL

pH 7.2 ± 0.2 at 25°C

Preparation of Sulfur: Sterilize powdered elemental sulfur by steaming for 3 hr at 0 psi pressure–100°C on 3 successive days.

$Na_2S \cdot 9H_2O$ Solution:
Composition per 10.0mL:

$Na_2S \cdot 9H_2O$	0.5g

Preparation of $Na_2S \cdot 9H_2O$ Solution: Add $Na_2S \cdot 9H_2O$ to distilled/deionized water and bring volume to 10.0mL. Mix thoroughly. Sparge with 100% N_2. Autoclave for 15 min at 15 psi pressure–121°C. Before use, neutralize to pH 7.0 with sterile HCl.

Salt Base Solution:
Composition per liter:

$(NH_4)_2SO_4$	1.3g
KH_2PO_4	0.28g
$MgSO_4 \cdot 7H_2O$	0.25g
$CaCl_2 \cdot 2H_2O$	0.07g
$FeCl_3 \cdot 6H_2O$	0.02g
$Na_2B_4O_7 \cdot 10H_2O$	4.5mg
$MnCl_2 \cdot 4H_2O$	1.8mg
$ZnSO_4 \cdot 7H_2O$	0.22mg
$CuCl_2 \cdot 2H_2O$	0.05mg
$Na_2MoO_4 \cdot 2H_2O$	0.03mg
$VOSO_4$	0.03mg
$CoSO_4 \cdot 7H_2O$	0.02mg

Preparation of Salt Base Solution: Add components to distilled/deionized water and bring volume to 1.0L. Mix thoroughly. Sparge with 100% N_2.

Preparation of Medium: Prepare and dispense medium under 100% N_2. Add components, except $Na_2S \cdot 9H_2O$ solution, to salt base solution and bring volume to 1.0L. Mix thoroughly. Adjust medium pH to 7.2 with H_2SO_4. Autoclave for 15 min at 15 psi pressure–121°C. Immediately prior to use, aseptically and anaerobically add 10.0mL of sterile $Na_2S \cdot 9H_2O$ solution. Mix thoroughly. Aseptically and anaerobically distribute into sterile tubes or bottles.

Use: For the cultivation of *Thermococcus profundus.*

Thermococcus stetteri Medium

Composition per liter:

NaCl	25.0g
Sulfur	10.0g
Yeast extract	3.0g
$(NH_4)_2SO_4$	1.3g
Peptone	0.5g
KH_2PO_4	0.28g
$MgSO_4 \cdot 7H_2O$	0.25g
$CaCl_2 \cdot 2H_2O$	0.07g
$FeCl_2 \cdot 2H_2O$	0.02g
$NaB_4O \cdot 10H_2O$	4.5mg
$MnCl_2 \cdot 4H_2O$	1.8mg
Resazurin	1.0mg
$ZnSO_4 \cdot 7H_2O$	0.22mg
$CuCl_2 \cdot 2H_2O$	0.05mg
$NaMoO_4 \cdot 2H_2O$	0.03mg
$VOSO_4 \cdot 2H_2O$	0.03mg
$CoSO_4$	0.01mg
$Na_2S \cdot 9H_2O$ solution	10.0mL

pH 6.5 ± 0.2 at 25°C

$Na_2S \cdot 9H_2O$ Solution:
Composition per 10.0mL:

$Na_2S \cdot 9H_2O$	0.3g

Preparation of $Na_2S \cdot 9H_2O$ Solution: Add $Na_2S \cdot 9H_2O$ to distilled/deionized water and bring volume to 10.0mL. Mix thoroughly. Sparge with 100% N_2. Autoclave for 15 min at 15 psi pressure–121°C.

Preparation of Medium: Add components, except $Na_2S \cdot 9H_2O$ solution, to distilled/deionized water and bring volume to 1.0L. Mix thoroughly. Adjust pH to 6.5. Do not autoclave. Sterilize by steaming at 100°C for 30 min on 3 consecutive days. Before inoculation, add 10.0mL of sterile $Na_2S \cdot 9H_2O$ solution. Mix thoroughly.

Use: For the cultivation of *Thermococcus stetteri*.

Thermodesulfobacterium Medium

Composition per liter:

Na_2SO_4	3.0g
$Na_2HPO_4 \cdot 12H_2O$	2.0g
NH_4Cl	1.0g
KH_2PO_4	0.3g
$MgCl_2 \cdot 6H_2O$	0.2g
$FeSO_4 \cdot 7H_2O$	1.5mg
Resazurin	1.0mg
Sodium lactate solution	20.0mL
Trace elements solution	10.0mL
Yeast extract solution	10.0mL
$Na_2S \cdot 9H_2O$ solution	10.0mL
Vitamin solution	5.0mL

pH 6.8 ± 0.2 at 25°C

Sodium Lactate Solution:
Composition per 20.0mL:

Sodium lactate	4.0g

Preparation of Sodium Lactate Solution: Add sodium lactate to distilled/deionized water and bring volume to 20.0mL. Mix thoroughly. Sparge with 100% N_2. Autoclave for 15 min at 15 psi pressure–121°C.

Trace Elements Solution:
Composition per liter:

Nitrilotriacetic acid	12.8g
NaCl	1.0g
$FeCl \cdot 4H_2O$	0.20
$CoCl_2 \cdot 6H_2O$	0.17g
$CaCl_2 \cdot 2H_2O$	0.1g
$MnCl_2 \cdot 4H_2O$	0.1g
$ZnCl_2$	0.1g
$CuCl_2$	0.02g
H_3BO_3	0.01g
$Na_2MoO_4 \cdot 2H_2O$	0.01g
$NiCl_2 \cdot 6H_2O$	0.026g
$Na_2SeO_3 \cdot 5H_2O$	0.02g

Preparation of Trace Elements Solution: Add nitrilotriacetic acid to 500.0mL of distilled/deionized water. Adjust pH to 6.5 with KOH. Add remaining components. Add distilled/deionized water to 1.0L.

Yeast Extract Solution:
Composition per 10.0mL:

Yeast extract	1.0g

Preparation of Yeast Extract Solution: Add yeast extract to distilled/deionized water and bring volume to 10.0mL. Mix thoroughly. Sparge with 100% N_2. Autoclave for 15 min at 15 psi pressure–121°C.

$Na_2S{\cdot}9H_2O$ Solution:

Composition per 10.0mL:

$Na_2S{\cdot}9H_2O$ 0.5g

Preparation of $Na_2S{\cdot}9H_2O$ Solution: Add $Na_2S{\cdot}9H_2O$ to distilled/deionized water and bring volume to 10.0mL. Mix thoroughly. Sparge with 100% N_2. Autoclave for 15 min at 15 psi pressure–121°C. Prior to use, neutralize solution by dropwise addition of sterile 1*N* HCl.

Vitamin Solution:

Composition per liter:

Pyridoxine·HCl 10.0mg
Calcium DL-pantothenate 5.0mg
Lipoic acid 5.0mg
Nicotinic acid 5.0mg
p-Aminobenzoic acid 5.0mg
Riboflavin 5.0mg
Thiamine·HCl 5.0mg
Biotin 2.0mg
Folic acid 2.0mg
Vitamin B_{12} 0.1mg

Preparation of Vitamin Solution: Add components to distilled/deionized water and bring volume to 1.0L. Mix thoroughly.

Preparation of Medium: Prepare and dispense medium under 100% N_2. Add components, except sodium lactate solution, yeast extract solution, and $Na_2S{\cdot}9H_2O$ solution, to distilled/deionized water and bring volume to 960.0mL. Mix thoroughly. Sparge with 100% N_2. Anaerobically distribute into tubes or bottles. Autoclave for 15 min at 15 psi pressure–121°C. Aseptically and anaerobically add to 1.0L of medium 20.0mL of sterile sodium lactate solution, 10.0mL of sterile yeast extract solution, and 10.0mL of sterile $Na_2S{\cdot}9H_2O$ solution. Mix thoroughly.

Use: For the cultivation and maintenance of *Thermodesulfobacterium commune*.

Thermodesulfobium Medium (DSMZ Medium 1005)

Composition per liter:

Na_2SO_4 .... 2.8g
KH_2PO_4 .... 0.78g
K_2HPO_4 .... 0.75g
NH_4Cl 0.5g
$MgSO_4{\cdot}7H_2O$ 0.25g
NaCl 0.2g
Na_3-EDTA 0.04g
$CaCl_2{\cdot}2H_2O$ 0.03g
Na-acetate 0.15g
$FeSO_4{\cdot}7H_2O$ 0.01g
Vitamin solution 10.0mL
Trace elements solution SL-9 .... 10.0mL
Cysteine solution 10.0mL

pH 5.7 ± 0.2 at 25°C

Cysteine Solution:

Composition per 10.0mL:

L-Cysteine-HCl·$2H_2O$ 0.25g

Preparation of Cysteine Solution: Add L-cysteine to to distilled/deionized water and bring volume to 10.0mL. Mix thoroughly. Sparge with 100% N_2. Filter sterilize.

Trace Elements Solution SL-9:

Composition per liter:

$MgSO_4{\cdot}7H_2O$ 3.0g
Nitrilotriacetic acid 1.5g
NaCl 1.0g
$MnSO_4{\cdot}2H_2O$ 0.5g
$CoSO_4{\cdot}7H_2O$ 0.18g
$ZnSO_4{\cdot}7H_2O$ 0.18g
$CaCl_2{\cdot}2H_2O$ 0.1g
$FeSO_4{\cdot}7H_2O$ 0.1g
$NiCl_2{\cdot}6H_2O$ 0.025g
$KAl(SO_4)_2{\cdot}12H_2O$ 0.02g
H_3BO_3 .... 0.01g
$Na_2MoO_4{\cdot}4H_2O$ 0.01g
$CuSO_4{\cdot}5H_2O$ 0.01g
$Na_2SeO_3{\cdot}5H_2O$ 0.3mg

Preparation of Trace Elements Solution SL-9: Add nitrilotriacetic acid to 500.0mL of distilled/deionized water. Dissolve by adjusting pH to 6.5 with KOH. Add remaining components. Add distilled/deionized water to 1.0L. Mix thoroughly.

Vitamin Solution:

Composition per liter:

Pyridoxine-HCl 10.0mg
Thiamine-HCl·$2H_2O$ 5.0mg
Riboflavin 5.0mg
Nicotinic acid 5.0mg
D-Ca-pantothenate 5.0mg
p-Aminobenzoic acid 5.0mg
Lipoic acid 5.0mg
Biotin 2.0mg
Folic acid 2.0mg
Vitamin B_{12} 0.1mg

Preparation of Vitamin Solution: Add components to distilled/deionized water and bring volume to 1.0L. Mix thoroughly. Sparge with 80% H_2 + 20% CO_2. Filter sterilize.

Preparation of Medium: Add components, except vitamin and cysteine solutions, to distilled/deionized water and bring volume to 980.0mL. Mix thoroughly. Gently heat and bring to boiling. Boil for 3 min. Cool to room temperature while sparging with a gas mixture of 80% N_2 + 20% CO_2. Adjust the pH to 5.5 with 10*N* H_2SO_4. Dispense under the same gas atmosphere in culture vessels (e.g., 20.0mL of the medium into 50mL serum bottles). Autoclave for 15 min at 15 psi pressure–121°C. Cool to room temperature. Aspetically add vitamin solution. Mix thoroughly. Prior to inoculation change atmosphere to 80% H_2 and 20% CO_2 gas mixture. Add the cysteine solution. Adjust pH to 5.5–6.0 if necessary. After inoculation pressurize vials to 0.5 bar overpressure with 80% H_2 and 20% CO_2 gas mixture.

Use: For the cultivation of *Thermodesulfobium narugense*.

Thermodesulforhabdus Medium

Composition per liter:

NaCl 10.0g
Na_2SO_4 .... 7.0g
Sodium acetate·$3H_2O$ 6.2g
$MgCl_2{\cdot}6H_2O$ 3.0g
KH_2PO_4 .... 1.0g
NH_4Cl 0.25g
$CaCl_2{\cdot}2H_2O$ 0.15g
Resazurin 0.5mg

$Na_2S{\cdot}9H_2O$ solution 10.0mL
$Na_2S_2O_4$ solution 10.0mL
Trace elements solution SL-10 .... 1.0mL
$NaHCO_3$ solution .... variable

pH 6.8 ± 0.2 at 25°C

$NaHCO_3$ Solution:

Composition per 20.0mL:

$NaHCO_3$ 1.0g

Preparation of $NaHCO_3$ Solution: Add $NaHCO_3$ to distilled/deionized water and bring volume to 20.0mL. Mix thoroughly. Sparge with 80% N_2 + 20% CO_2. Autoclave for 15 min at 15 psi pressure–121°C.

$Na_2S{\cdot}9H_2O$ Solution:

Composition per 10.0mL:

$Na_2S{\cdot}9H_2O$ 0.15g

Preparation of $Na_2S{\cdot}9H_2O$ Solution: Add $Na_2S{\cdot}9H_2O$ to distilled/deionized water and bring volume to 10.0mL. Mix thoroughly. Sparge with 100% N_2. Autoclave for 15 min at 15 psi pressure–121°C.

$Na_2S_2O_3$ Solution:

Composition per 10.0mL:

$Na_2S_2O_3{\cdot}5H_2O$ 2.0g

Preparation of $Na_2S_2O_3$ Solution: Add $Na_2S_2O_3{\cdot}5H_2O$ to distilled/deionized water and bring volume to 10.0mL. Mix thoroughly. Sparge with 100% N_2. Autoclave for 15 min at 15 psi pressure–121°C.

Trace Elements Solution SL-10:

Composition per liter:

$FeCl_2{\cdot}4H_2O$ 1.5g
$CoCl_2{\cdot}6H_2O$ 190.0mg
$MnCl_2{\cdot}4H_2O$ 100.0mg
$ZnCl_2$ 70.0mg
$Na_2MoO_4{\cdot}2H_2O$ 36.0mg
$NiCl_2{\cdot}6H_2O$ 24.0mg
H_3BO_3 .... 6.0mg
$CuCl_2{\cdot}2H_2O$ 2.0mg
HCl (25% solution) 10.0mL

Preparation of Trace Elements Solution SL-10: Add $FeCl_2{\cdot}4H_2O$ to 10.0mL of HCl solution. Mix thoroughly. Add distilled/deionized water and bring volume to 1.0L. Add remaining components. Mix thoroughly.

Preparation of Medium: Prepare and dispense medium under 100% N_2. Add components, except $Na_2S{\cdot}9H_2O$ solution, $Na_2S_2O_3$ solution, and $NaHCO_3$ solution, to distilled/deionized water and bring volume to 980.0mL. Mix thoroughly. Gently heat and bring to boiling. Cool to room temperature while sparging with 100% N_2. Anaerobically distribute 9.8mL volumes into tubes. Autoclave for 15 min at 15 psi pressure–121°C. Aseptically and anaerobically add 0.1mL of sterile $Na_2S{\cdot}9H_2O$ solution and 0.1mL of sterile $Na_2S_2O_4$ solution to each tube. Aseptically and anaerobically add a sufficient volume of sterile $NaHCO_3$ solution to each tube to bring the pH to 6.8.

Use: For the cultivation of *Thermodesulforhabdus norvegicus.*

Thermodesulfotobacterium Agar

Composition per liter:

Na_2SO_4 .... 30.0g
Agar 20.0g
Sodium lactate 4.0g
Yeast extract 1.0g
Mineral solution 2 .... 50.0mL
Na_2CO_3 solution 50.0mL
Mineral solution 1 .... 25.0mL
Cysteine-sulfide reducing agent 20.0mL
Wolfe's mineral solution 10.0mL
Wolfe's vitamin solution 10.0mL
Resazurin (0.025% solution) 4.0mL

pH 7.2 ± 0.2 at 25°C

Mineral Solution 1:

Composition per liter:

K_2HPO_4 .... 6.0g

Preparation of Mineral Solution 1: Add K_2HPO_4 to distilled/deionized water and bring volume to 1.0L. Mix thoroughly.

Mineral Solution 2:

Composition per liter:

NaCl 12.0g
KH_2PO_4 .... 6.0g
$(NH_4)_2SO_4$ 6.0g
$MgSO_4{\cdot}7H_2O$ 2.4g
$CaCl_2{\cdot}2H_2O$ 1.6g

Preparation of Mineral Solution 2: Add components to distilled/deionized water and bring volume to 1.0L. Mix thoroughly.

Na_2CO_3 Solution:

Composition per 100.0mL:

Na_2CO_3 .... 8.0g

Preparation of Na_2CO_3 Solution: Add Na_2CO_3 to distilled/deionized water and bring volume to 100.0mL. Mix thoroughly.

Cysteine-Sulfide Reducing Agent:

Composition per 20.0mL:

L-Cysteine·HCl·H_2O 300.0mg
$Na_2S{\cdot}9H_2O$ 300.0mg

Preparation of Cysteine-Sulfide Reducing Agent: Add L-cysteine·HCl·H_2O to 10.0mL of distilled/deionized water. Mix thoroughly. In a separate tube, add $Na_2S{\cdot}9H_2O$ to 10.0mL of distilled/deionized water. Mix thoroughly. Gas both solutions with 100% N_2 and cap tubes. Autoclave both solutions for 15 min at 15 psi pressure–121°C using fast exhaust. Cool to 50°C. Aseptically combine the two solutions under 100% N_2.

Wolfe's Mineral Solution:

Composition per liter:

$MgSO_4{\cdot}7H_2O$ 3.0g
Nitrilotriacetic acid 1.5g
NaCl 1.0g
$MnSO_4{\cdot}H_2O$ 0.5g
$FeSO_4{\cdot}7H_2O$ 0.1g
$CoCl_2{\cdot}6H_2O$ 0.1g
$CaCl_2$ 0.1g
$ZnSO_4{\cdot}7H_2O$ 0.1g
$CuSO_4{\cdot}5H_2O$ 0.01g
$AlK(SO_4)_2{\cdot}12H_2O$ 0.01g
H_3BO_3 .... 0.01g
$Na_2MoO_4{\cdot}2H_2O$ 0.01g

Preparation of Wolfe's Mineral Solution: Add nitrilotriacetic acid to 500.0mL of distilled/deionized water. Dissolve by adjusting pH to 6.5 with KOH. Add remaining components. Add distilled/deionized water to 1.0L.

Wolfe's Vitamin Solution:

Composition per liter:

Pyridoxine·HCl	10.0mg
Thiamine·HCl	5.0mg
Riboflavin	5.0mg
Nicotinic acid	5.0mg
Calcium pantothenate	5.0mg
p-Aminobenzoic acid	5.0mg
Thioctic acid	5.0mg
Biotin	2.0mg
Folic acid	2.0mg
Cyanocobalamin	100.0μg

Preparation of Wolfe's Vitamin Solution: Add components to distilled/deionized water and bring volume to 1.0L. Mix thoroughly. Filter sterilize.

Preparation of Medium: Add components, except vitamin solution and cysteine-sulfide reducing agent, to distilled/deionized water and bring volume to 970.0mL. Mix thoroughly. Gently heat and bring to boiling. Autoclave for 15 min at 15 psi pressure–121°C. Cool to 50°–55°C under 80% N_2 + 20% CO_2. Aseptically add the sterile vitamin solution and then the sterile cysteine-sulfide reducing agent. Adjust the pH to 7.2. Distribute aseptically and anaerobically into sterile tubes.

Use: For the cultivation and maintenance of *Thermodesulfobacterium commune* and other *Thermodesulfobacterium* species.

Thermodesulfotobacterium Broth

Composition per liter:

Na_2SO_4	30.0g
Sodium lactate	4.0g
Yeast extract	1.0g
Mineral solution 2	50.0mL
Na_2CO_3 solution	50.0mL
Mineral solution 1	25.0mL
Cysteine-sulfide reducing agent	20.0mL
Wolfe's mineral solution	10.0mL
Wolfe's vitamin solution	10.0mL
Resazurin (0.025% solution)	4.0mL

pH 7.2 ± 0.2 at 25°C

Mineral Solution 1:

Composition per liter:

K_2HPO_4	6.0g

Preparation of Mineral Solution 1: Add K_2HPO_4 to distilled/deionized water and bring volume to 1.0L. Mix thoroughly.

Mineral Solution 2:

Composition per liter:

NaCl	12.0g
KH_2PO_4	6.0g
$(NH_4)_2SO_4$	6.0g
$MgSO_4·7H_2O$	2.4g
$CaCl_2·2H_2O$	1.6g

Preparation of Mineral Solution 2: Add components to distilled/deionized water and bring volume to 1.0L. Mix thoroughly.

Na_2CO_3 Solution:

Composition per 100.0mL:

Na_2CO_3	8.0g

Preparation of Na_2CO_3 Solution: Add Na_2CO_3 to distilled/deionized water and bring volume to 100.0mL Mix thoroughly.

Cysteine-Sulfide Reducing Agent:

Composition per 20.0mL:

L-Cysteine·HCl·H_2O	300.0mg
$Na_2S·9H_2O$	300.0mg

Preparation of Cysteine-Sulfide Reducing Agent: Add L-cysteine·HCl·H_2O to 10.0mL of distilled/deionized water. Mix thoroughly. In a separate tube, add $Na_2S·9H_2O$ to 10.0mL of distilled/deionized water. Mix thoroughly. Gas both solutions with 100% N_2 and cap tubes. Autoclave both solutions for 15 min at 15 psi pressure–121°C using fast exhaust. Cool to 50°C. Aseptically combine the two solutions under 100% N_2.

Wolfe's Mineral Solution:

Composition per liter:

$MgSO_4·7H_2O$	3.0g
Nitrilotriacetic acid	1.5g
NaCl	1.0g
$MnSO_4·H_2O$	0.5g
$FeSO_4·7H_2O$	0.1g
$CoCl_2·6H_2O$	0.1g
$CaCl_2$	0.1g
$ZnSO_4·7H_2O$	0.1g
$CuSO_4·5H_2O$	0.01g
$AlK(SO_4)_2·12H_2O$	0.01g
H_3BO_3	0.01g
$Na_2MoO_4·2H_2O$	0.01g

Preparation of Wolfe's Mineral Solution: Add nitrilotriacetic acid to 500.0mL of distilled/deionized water. Dissolve by adjusting pH to 6.5 with KOH. Add remaining components. Add distilled/deionized water to 1.0L.

Wolfe's Vitamin Solution:

Composition per liter:

Pyridoxine·HCl	10.0mg
Thiamine·HCl	5.0mg
Riboflavin	5.0mg
Nicotinic acid	5.0mg
Calcium pantothenate	5.0mg
p-Aminobenzoic acid	5.0mg
Thioctic acid	5.0mg
Biotin	2.0mg
Folic acid	2.0mg
Cyanocobalamin	100.0μg

Preparation of Wolfe's Vitamin Solution: Add components to distilled/deionized water and bring volume to 1.0L. Mix thoroughly. Filter sterilize.

Preparation of Medium: Add components, except vitamin solution and cysteine-sulfide reducing agent, to distilled/deionized water and bring volume to 970.0mL. Mix thoroughly. Autoclave for 15 min at 15 psi pressure–121°C. Cool under 80% N_2 + 20% CO_2. Aseptically add the sterile vitamin solution and then the sterile cysteine-sulfide reducing agent. Adjust the pH to 7.2. Distribute aseptically and anaerobically into sterile tubes.

Use: For the cultivation and maintenance of *Thermodesulfobacterium commune* and other *Thermodesulfobacterium* species.

Thermodesulfovibrio yellowstonii Medium (DSMZ Medium 749)

Composition per liter:

Na_2SO_4	4.0g
Na-lactate	2.5g

$NaHCO_3$	1.3g
KCl	0.5g
Yeast extract	0.5g
$MgCl_2·6H_2O$	0.4g
NH_4Cl	0.25g
Na_2HPO_4	0.2g
Na-thioglycolate	0.2g
L-Ascorbic acid	0.2g
$CaCl_2·2H_2O$	0.15g
Resazurin	0.5mg
Trace elements solution	10.0mL
Vitamin solution	10.0mL

pH 7.5 ± 0.2 at 25°C

Trace Elements Solution:

Composition per liter:

$MgSO_4·7H_2O$	3.0g
Nitrilotriacetic acid	1.5g
NaCl	1.0g
$MnSO_4·2H_2O$	0.5g
$CoSO_4·7H_2O$	0.18g
$ZnSO_4·7H_2O$	0.18g
$CaCl_2·2H_2O$	0.1g
$FeSO_4·7H_2O$	0.1g
$NiCl_2·6H_2O$	0.025g
$KAl(SO_4)_2·12H_2O$	0.02g
H_3BO_3	0.01g
$Na_2MoO_4·4H_2O$	0.01g
$CuSO_4·5H_2O$	0.01g
$Na_2SeO_3·5H_2O$	0.3mg

Preparation of Trace Elements Solution: Add nitrilotriacetic acid to 500.0mL of distilled/deionized water. Dissolve by adjusting pH to 6.5 with KOH. Add remaining components. Add distilled/deionized water to 1.0L. Mix thoroughly.

Vitamin Solution:

Composition per liter:

Pyridoxine-HCl	10.0mg
Thiamine-HCl·$2H_2O$	5.0mg
Riboflavin	5.0mg
Nicotinic acid	5.0mg
D-Ca-pantothenate	5.0mg
p-Aminobenzoic acid	5.0mg
Lipoic acid	5.0mg
Biotin	2.0mg
Folic acid	2.0mg
Vitamin B_{12}	0.1mg

Preparation of Vitamin Solution: Add components to distilled/deionized water and bring volume to 1.0L. Mix thoroughly. Sparge with 80% H_2 + 20% CO_2. Filter sterilize.

Preparation of Medium: Prepare and dispense medium under an oxygen-free 100% N_2. Add components, except vitamin solution, Na-thioglycolate, $NaHCO_3$, and L-Ascorbic acid, to distilled/deionized water and bring volume to 990.0L. Mix thoroughly. Gently heat and bring to boiling. Cool while sparging with 100% N_2. Add 0.2g Na-thioglycolate, 1.3g $NaHCO_3$, and 0.2g L-ascorbic acid. Mix thoroughly. Adjust pH to 7.5. Autoclave for 15 min at 15 psi pressure–121°C. Cool to 25°C. Aseptically and anaerobically add 10.0mL sterile vitamin solution. Mix thoroughly. Aseptically and anaerobically distribute into sterile tubes or flasks.

Use: For the cultivation of *Thermodesulfovibrio yellowstonii.*

Thermofilum pendens Medium

Composition per liter:

Sulfur, powdered	10.0g
$(NH_4)_2SO_4$	1.3g
KH_2PO_4	0.28g
$MgSO_4·7H_2O$	0.25g
$CaCl_2·2H_2O$	0.07g
$Na_2S·9H_2O$	0.3g
$FeCl_3·6H_2O$	0.02g
$Na_2B_4O_7·10H_2O$	4.5mg
$MnCl_2·4H_2O$	1.8mg
$ZnSO_4·7H_2O$	0.22mg
$CuCl_2·2H_2O$	0.05mg
$Na_2MoO_4·2H_2O$	0.03mg
$VOSO_4·2H_2O$	0.03mg
$CoSO_4·7H_2O$	0.01mg
Yeast extract solution	20.0mL
Sucrose solution	20.0mL
Polar lipid fraction	6.0–12.0mL

pH 5.2 ± 0.2 at 25°C

Yeast Extract Solution:

Composition per 20.0mL:

Yeast extract	2.0g

Preparation of Yeast Extract Solution: Add yeast extract to distilled/deionized water and bring volume to 20.0mL. Mix thoroughly. Gently heat and bring to boiling. Boil for a few minutes. Sparge with 100% N_2. Do not autoclave.

Sucrose Solution:

Composition per 20.0mL:

Sucrose	2.0g

Preparation of Sucrose Solution: Add sucrose to distilled/deionized water and bring volume to 20.0mL. Mix thoroughly. Filter sterilize. Sparge with 100% N_2.

Preparation of Sulfur: Add 10.0g of powdered sulfur to a flask and sterilize by steaming for 3 hr on 3 consecutive days.

Polar Lipid Fraction:

Composition per 20.0mL:

Thermoproteus tenax cells (wet weight)	10.0g
Chloroform	500.0mL
Acetone	500.0mL
Methanol	500.0mL
TA buffer solution	80.0mL
Chloroform/methanol 1:1 (v/v)	20.0mL

TA Buffer Solution:

Composition per 100.0mL:

Tris·HCl	0.79g
β-mercaptoethanol	0.78g
NH_4Cl	0.118g
EDTA	0.029g

Preparation of TA Buffer Solution: Add components to distilled/deionized water and bring volume to 100.0mL. Mix thoroughly.

Preparation of Polar Lipid Fraction: Add 10.0g (wet weight) of *Thermoproteus tenax* cells to 20.0mL of TA buffer solution. Mix thoroughly. Sonicate for 2 min. Centrifuge at 20,000 rpm for 20 min. Resuspend pellet in 20.0mL of fresh TA buffer solution. Recentrifuge at 20,000 rpm for 20 min. Again resuspend pellet in 20.0mL of fresh TA buffer solution. Recentrifuge at 20,000 rpm for 20 min. Resuspend pellet in 20.0mL of fresh TA buffer solution. Centrifuge at 5,000 rpm for

5 min. Decant the supernatant solution and discard the pellet. Extract the supernatant solution twice with 20.0mL of chloroform/methanol (1:1) each time. Chromatograph the extract on a SIL-LC (325 mesh) silicic acid column (20cm × 2cm) using 500.0mL of chloroform, followed by 500.0mL of acetone, followed by 500.0mL of methanol. The methanol fraction is further purified by DEAE chromatography using chloroform/methanol 1:1 and methanol.

Preparation of Medium: Prepare and dispense medium under 100% N_2. Add components, except yeast extract solution, sucrose solution, sulfur, and polar lipid fraction, to distilled/deionized water and bring volume to 950.0mL. Mix thoroughly. Sparge with 100% N_2. Anaerobically distribute into tubes or bottles. Autoclave for 15 min at 15 psi pressure–121°C. Aseptically and anaerobically add to 1.0L of medium 20.0mL of sterile yeast extract solution, 20.0mL of sterile sucrose solution, 10.0g of sterile sulfur, and 6.0–12.0mL of sterile polar lipid fraction. Mix thoroughly.

Use: For the cultivation and maintenance of *Thermofilum pendens*.

Thermogymnomonas Medium (DSMZ Medium 1141)

Composition per liter:

KH_2PO_4	3.0g
$(NH_4)_2SO_4$	0.2g
$MgSO_4·7H_2O$	0.5g
$CaCl_2·2H_2O$	.025g
Glucose solution	50.0mL
Yeast extract solution	50.0mL

pH 3.0 ± 0.2 at 25°C

Glucose Solution:

Composition per 50.0mL:

Glucose	10.0g

Preparation of Glucose Solution: Add glucose to distilled/deionized water and bring volume to 50.0mL. Mix thoroughly. Filter sterilize.

Yeast Extract Solution:

Composition per 50.0mL:

Yeast extract	1.0g

Preparation of Yeast Extract Solution: Add yeast extract to distilled/deionized water and bring volume to 50.0mL. Mix thoroughly. Filter sterilize.

Preparation of Medium: Add components, except glucose and yeast extract solutions, to distilled/deionized water and bring volume to 900.0mL. Mix thoroughly. Autoclave for 15 min at 15 psi pressure–121°C. Cool to room temperature. Aspetically add glucose and yeast extract solutions. Mix thoroughly. Adjust pH to 3.0 with sterile 10*N* H_2SO_4.

Use: For the cultivation of *Thermogymnomonas* spp.

Thermoleophilum Medium

Composition per liter:

$NaNO_2$	2.0g
Na_2HPO_4	0.21g
$MgSO_4·7H_2O$	0.2g
KCl	0.04g
NaH_2PO_4	90.0mg
$CaCl_2$	15.0mg
$FeSO_4·7H_2O$	1.0mg
$ZnSO_4·7H_2O$	70.0μg
H_3BO_3	10.0μg
$MnSO_4·5H_2O$	10.0μg
MoO_3	10.0μg
$CuSO_4·5H_2O$	5.0μg
n-Heptadecane	1.0mL

pH 7.0 ± 0.2 at 25°C

Preparation of Medium: Add components, except *n*-heptadecane, to distilled/deionized water and bring volume to 1.0L. Mix thoroughly. Autoclave for 15 min at 15 psi pressure–121°C. Aseptically add 1.0mL of *n*-heptadecane. Mix thoroughly. Aseptically distribute into sterile tubes or flasks.

Use: For the cultivation of *Thermoleophilum album* and *Thermoleophilum minutum.*

Thermomicrobium fosteri Agar

Composition per liter:

Agar	20.0g
NH_4Cl	2.0g
Na_2HPO_4	0.21g
$MgSO_4·7H_2O$	0.2g
NaH_2PO_4	0.09g
KCl	0.04g
$CaCl_2$	0.015g
$ZnSO_4·7H_2O$	70.0μg
H_3BO_3	10.0μg
$MnSO_4·5H_2O$	10.0μg
MoO_3	10.0μg
$CuSO_4·5H_2O$	5.0μg
$FeSO_4·7H_2O$	1.0mg
Heptadecane, filter sterilized	20.0mL

pH 7.2 ± 0.2 at 25°C

Preparation of Medium: Add components, except heptadecane, to distilled/deionized water and bring volume to 980.0mL. Mix thoroughly. Gently heat and bring to boiling. Autoclave for 15 min at 15 psi pressure–121°C. Cool to 50°–55°C. Aseptically add 20.0mL of sterile heptadecane. Mix thoroughly. Aseptically distribute into sterile tubes. Cool tubes rapidly in a slanted position.

Use: For the cultivation and maintenance of *Thermomicrobium fosteri.*

Thermomicrobium fosteri Broth

Composition per liter:

NH_4Cl	2.0g
Na_2HPO_4	0.21g
$MgSO_4·7H_2O$	0.2g
NaH_2PO_4	0.09g
KCl	0.04g
$CaCl_2$	0.015g
$ZnSO_4·7H_2O$	70.0μg
H_3BO_3	10.0μg
$MnSO_4·5H_2O$	10.0μg
MoO_3	10.0μg
$CuSO_4·5H_2O$	5.0μg
$FeSO_4·7H_2O$	1.0mg
Heptadecane, filter sterilized	20.0mL

pH 7.2 ± 0.2 at 25°C

Preparation of Medium: Add components, except heptadecane, to distilled/deionized water and bring volume to 980.0mL. Mix thoroughly. Autoclave for 15 min at 15 psi pressure–121°C. Aseptically add

20.0mL of sterile heptadecane. Mix thoroughly. Aseptically distribute into sterile tubes or flasks.

Use: For the cultivation of *Thermomicrobium fosteri.*

Thermomicrobium roseum Agar (DSMZ Medium 592)

Composition per liter:

Agar	20.0g
$(NH_4)_2SO_4$ (sublimed)	1.3g
Yeast extract	1.0g
Tryptone	1.0g
$MgSO_4 \cdot 7H_2O$	0.247g
KH_2PO_4	0.280g
$CaCl_2 \cdot 2H_2O$	0.074g
$FeCl_3 \cdot 6H_2O$	0.019g
Salt solution	1.0mL

pH 8.5 ± 0.2 at 25°C

Salt Solution:

Composition per liter:

$Na_2B_4O_7 \cdot 10H_2O$	4.4g
$MnCl_2 \cdot 4H_2O$	1.8g
$ZnSO_4 \cdot 7H_2O$	0.22g
$CuCl_2 \cdot H_2O$	0.05g
$Na_2MoO_4 \cdot 4H_2O$	0.03g
$VOSO_4 \cdot 2H_2O$	0.03g

Preparation of Salt Solution: Add components to distilled/deionized water and bring volume to 100.0mL. Adjust pH to 2.0. Mix thoroughly.

Preparation of Medium: Add components to distilled/deionized water and bring volume to 1.0L. Mix thoroughly. Adjust pH to 8.5. Gently heat and bring to boiling. Autoclave for 15 min at 15 psi pressure–121°C. Pour into Petri dishes or pour short slants with a long butt in screw-capped tubes.

Use: For the cultivation and maintenance of *Thermomicrobium roseum.*

Thermomicrobium roseum Medium (DSMZ Medium 592)

Composition per liter:

$(NH_4)_2SO_4$ (sublimed)	1.3g
Yeast extract	1.0g
Tryptone	1.0g
$MgSO_4 \cdot 7H_2O$	0.247g
KH_2PO_4	0.280g
$CaCl_2 \cdot 2H_2O$	0.074g
$FeCl_3 \cdot 6H_2O$	0.019g
Salt solution	1.0mL

pH 8.5 ± 0.2 at 25°C

Salt Solution:

Composition per liter:

$Na_2B_4O_7 \cdot 10H_2O$	4.4g
$MnCl_2 \cdot 4H_2O$	1.8g
$ZnSO_4 \cdot 7H_2O$	0.22g
$CuCl_2 \cdot H_2O$	0.05g
$Na_2MoO_4 \cdot 4H_2O$	0.03g
$VOSO_4 \cdot 2H_2O$	0.03g

Preparation of Salt Solution: Add components to distilled/deionized water and bring volume to 100.0mL. Adjust pH to 2.0. Mix thoroughly.

Preparation of Medium: Add components to distilled/deionized water and bring volume to 1.0L. Mix thoroughly. Adjust pH to 8.5. Distribute into tubes or flasks. Autoclave for 15 min at 15 psi pressure–121°C.

Use: For the cultivation of *Thermomicrobium roseum.*

Thermomonospora Medium

Composition per liter:

Sucrose	30.0g
Agar	15.0g
Casamino acids	6.0g
$NaNO_3$	3.0g
Yeast extract	2.0g
K_2HPO_4	1.0g
$MgSO_4 \cdot 7H_2O$	0.5g
KCl	0.5g
$FeSO_4 \cdot 7H_2O$	0.01g

pH 8.0 ± 0.2 at 25°C

Preparation of Medium: Add components to distilled/deionized water and bring volume to 1.0L. Mix thoroughly. Distribute into tubes or flasks. Autoclave for 15 min at 15 psi pressure–121°C. Pour into sterile Petri dishes or leave in tubes.

Use: For the cultivation of *Thermomonospora alba* and *Thermomonospora mesophila.*

Thermophilic *Bacillus* Medium

Composition per liter:

Peptone	8.0g
Yeast extract	4.0g
NaCl	3.0g

pH 7.5 ± 0.2 at 25°C

Preparation of Medium: Add components to distilled/deionized water and bring volume to 1.0L. Mix thoroughly. Distribute into tubes or flasks. Autoclave for 15 min at 15 psi pressure–121°C.

Use: For the cultivation and maintenance of a variety of thermophilic *Bacillus* species.

Thermophilic Hydrogen-Bacteria Medium

Composition per 1000.5mL:

$Na_2HPO_4 \cdot 12H_2O$	4.5g
KH_2PO_4	1.5g
NaCl	1.0g
NH_4NO_3	1.0g
$MgSO_4 \cdot 7H_2O$	0.2g
$CaCl_2 \cdot 2H_2O$	10.0mg
$FeSO_4 \cdot 7H_2O$	10.0mg
Trace elements solution	0.5mL

pH 7.0 ± 0.2 at 25°C

Trace Elements Solution:

Composition per liter:

$ZnSO_4 \cdot 7H_2O$	28.0mg
$CoCl_2 \cdot 6H_2O$	4.0mg
H_3BO_3	4.0mg
$MnSO_4 \cdot 5H_2O$	4.0mg
MoO_3	4.0mg
$CuSO_4 \cdot 5H_2O$	2.0mg

Preparation of Trace Elements Solution: Add components to distilled/deionized water and bring volume to 1.0L. Mix thoroughly.

Preparation of Medium: Add components to distilled/deionized water and bring volume to 1.0L. Mix thoroughly. Distribute into tubes or flasks. Autoclave for 15 min at 15 psi pressure–121°C. Incubate cultures in 5% O_2 + 80% H_2 + 10% CO_2.

Use: For the cultivation and maintenance of *Hydrogenobacter thermophilus* and *Pseudomonas* species.

Thermophilic Maintenance Medium

Composition per liter:

$NaHCO_3$	3.0g
Yeast extract	1.0g
NH_4Cl	1.0g
KH_2PO_4	0.4g
K_2HPO_4	0.4g
$MgSO_4 \cdot 7H_2O$	0.1g
Cysteine-sulfide reducing solution	40.0mL
Fructose solution	25.0mL
Wolfe's vitamin solution	10.0mL
Wolfe's mineral solution	10.0mL
Resazurin (0.01% solution)	1.0mL

pH 5.6 ± 0.2 at 25°C

Cysteine-Sulfide Reducing Solution:

Composition per 100.0mL:

L-Cysteine·HCl·H_2O	1.25g
$Na_2S \cdot 9H_2O$	1.25g

Preparation of Cysteine-Sulfide Reducing Solution: Add L-cysteine·HCl·H_2O and $Na_2S \cdot 9H_2O$ to distilled/deionized water and bring volume to 100.0mL. Mix thoroughly.

Fructose Solution:

Composition per 100.0mL:

Fructose	20.0g

Preparation of Fructose Solution: Add fructose to distilled/deionized water and bring volume to 100.0mL. Mix thoroughly. Filter sterilize.

Wolfe's Vitamin Solution:

Composition per liter:

Pyridoxine·HCl	0.01g
Thiamine·HCl	5.0mg
Riboflavin	5.0mg
Nicotinic acid	5.0mg
Calcium pantothenate	5.0mg
p-Aminobenzoic acid	5.0mg
Thioctic acid	5.0mg
Biotin	2.0mg
Folic acid	2.0mg
Cyanocobalamin	100.0μg

Preparation of Wolfe's Vitamin Solution: Add components to distilled/deionized water and bring volume to 1.0L. Mix thoroughly.

Wolfe's Mineral Solution:

Composition per liter:

$MgSO_4 \cdot 7H_2O$	3.0g
Nitrilotriacetic acid	1.5g
NaCl	1.0g
$MnSO_4 \cdot H_2O$	0.5g
$FeSO_4 \cdot 7H_2O$	0.1g
$CoCl_2 \cdot 6H_2O$	0.1g
$CaCl_2$	0.1g
$ZnSO_4 \cdot 7H_2O$	0.1g
$CuSO_4 \cdot 5H_2O$	0.01g
$AlK(SO_4)_2 \cdot 12H_2O$	0.01g
H_3BO_3	0.01g
$Na_2MoO_4 \cdot 2H_2O$	0.01g

Preparation of Wolfe's Mineral Solution: Add nitrilotriacetic acid to 500.0mL of distilled/deionized water. Dissolve by adjusting pH to 6.5 with KOH. Add remaining components. Add distilled/deionized water to 1.0L.

Preparation of Medium: Add components, except cysteine-sulfide reducing solution and fructose solution, to distilled/deionized water and bring volume to 935.0mL. Mix thoroughly. Gently heat and bring to boiling. Continue boiling until resazurin turns colorless, indicating reduction. Add 40.0mL of the cysteine-sulfide reducing solution. Autoclave for 15 min at 15 psi pressure–121°C. Cool to 50°C under O_2-free 90% N_2 + 10% CO_2. Add 25.0mL of the sterile fructose solution. Adjust the pH to 5.6 if necessary. Aseptically and anaerobically distribute into sterile tubes. Cap with rubber stoppers.

Use: For the cultivation and maintenance of a variety of thermophilic anaerobes, including *Clostridium thermoautotrophicum*.

Thermophilic *Methanosarcina* Medium

Composition per 1021.0mL:

NaCl	2.25g
Pancreatic digest of casein	2.0g
Yeast extract	2.0g
$NaHCO_3$	0.85g
$MgSO_4 \cdot 7H_2O$	0.5g
NH_4Cl	0.5g
K_2HPO_4	0.348g
$CaCl_2 \cdot 2H_2O$	0.25g
KH_2PO_4	0.227g
$FeSO_4 \cdot 7H_2O$	2.0mg
Resazurin	1.0mg
Rumen fluid, clarified	50.0mL
Methanol solution	10.0mL
Vitamin solution	10.0mL
L-Cysteine·HCl·H_2O solution	10.0mL
$Na_2S \cdot 9H_2O$ solution	10.0mL
Trace elements solution SL-10	1.0mL

pH 6.5–6.8 at 25°C

Methanol Solution:

Composition per 10.0mL:

Methanol	5.0mL

Preparation of Methanol Solution: Add methanol to distilled/deionized water and bring volume to 10.0mL. Sparge with 100% N_2. Autoclave for 15 min at 15 psi pressure–121°C.

Vitamin Solution:

Composition per liter:

Pyridoxine·HCl	10.0mg
Calcium DL-pantothenate	5.0mg
Lipoic acid	5.0mg
Nicotinic acid	5.0mg
p-Aminobenzoic acid	5.0mg
Riboflavin	5.0mg
Thiamine·HCl	5.0mg
Biotin	2.0mg
Folic acid	2.0mg
Vitamin B_{12}	0.1mg

Preparation of Vitamin Solution: Add components to distilled/deionized water and bring volume to 1.0L. Mix thoroughly. Filter sterilize. Sparge with 80% N_2 + 20% CO_2.

L-Cysteine·HCl·H_2O Solution:
Composition per 10.0mL:

L-Cysteine·HCl·H_2O	0.3g

Preparation of L-Cysteine·HCl·H_2O Solution: Add L-cysteine·HCl·H_2O to distilled/deionized water and bring volume to 10.0mL. Mix thoroughly. Autoclave under 100% N_2 for 15 min at 15 psi pressure–121°C.

Na_2S·9H_2O Solution:
Composition per 10.0mL:

Na_2S·9H_2O	0.3g

Preparation of Na_2S·9H_2O Solution: Add Na_2S·9H_2O to distilled/deionized water and bring volume to 10.0mL. Mix thoroughly. Sparge with 100% N_2. Autoclave for 15 min at 15 psi pressure–121°C.

Trace Elements Solution SL-10:
Composition per liter:

$FeCl_2$·4H_2O	1.5g
$CoCl_2$·6H_2O	190.0mg
$MnCl_2$·4H_2O	100.0mg
$ZnCl_2$	70.0mg
Na_2MoO_4·2H_2O	36.0mg
$NiCl_2$·6H_2O	24.0mg
H_3BO_3	6.0mg
$CuCl_2$·2H_2O	2.0mg
HCl (25% solution)	10.0mL

Preparation of Trace Elements Solution SL-10: Add $FeCl_2$·4H_2O to 10.0mL of HCl solution. Mix thoroughly. Add distilled/deionized water and bring volume to 1.0L. Add remaining components. Mix thoroughly. Sparge with 100% N_2.

Preparation of Medium: Add components, except methanol solution, L-cysteine·HCl·H_2O solution, and Na_2S·9H_2O solution, to distilled/deionized water and bring volume to 960.0mL. Mix thoroughly. Sparge under 80% N_2 + 20% CO_2 for 3–4 min. Autoclave for 15 min at 15 psi pressure–121°C. Aseptically and anaerobically add 20.0mL of sterile vitamin solution, 10.0mL of sterile L-cysteine·HCl·H_2O solution, and 10.0mL of sterile Na_2S·9H_2O solution. Mix thoroughly. Aseptically and anaerobically distribute into sterile screw-capped bottles under 80% N_2 + 20% CO_2.

Use: For the cultivation and maintenance of *Methanosarcina thermophila.*

Thermophilic *Methanothrix* Medium

Composition per liter:

NH_4Cl	0.5g
K_2HPO_4	0.4g
$MgCl_2$·6H_2O	0.1g
Resazurin	1.0mg
$NaHCO_3$ solution	20.0mL
Trace elements solution	10.0mL
$CaCl_2$·2H_2O solution	10.0mL
Sodium acetate solution	10.0mL
Vitamin solution	10.0mL
Coenzyme M solution	10.0mL
Na_2S·9H_2O solution	5.0mL

pH 6.5 ± 0.2 at 25°C

$NaHCO_3$ Solution:
Composition per 20.0mL:

$NaHCO_3$	1.0g

Preparation of $NaHCO_3$ Solution: Add $NaHCO_3$ to distilled/deionized water and bring volume to 20.0mL. Mix thoroughly. Sparge with 80% N_2 + 20% CO_2 for 15 min. Autoclave for 15 min at 15 psi pressure–121°C.

Trace Elements Solution:
Composition per liter:

$MgSO_4$·7H_2O	3.0g
Nitrilotriacetic acid	1.5g
NaCl	1.0g
$MnSO_4$·2H_2O	0.5g
$CoSO_4$·7H_2O	0.18g
$ZnSO_4$·7H_2O	0.18g
$CaCl_2$·2H_2O	0.1g
$FeSO_4$·7H_2O	0.1g
$NiCl_2$·6H_2O	0.025g
$KAl(SO_4)_2$·12H_2O	0.02g
$CuSO_4$·5H_2O	0.01g
H_3BO_3	0.01g
Na_2MoO_4·2H_2O	0.01g
Na_2SeO_3·5H_2O	0.3mg

Preparation of Trace Elements Solution: Add nitrilotriacetic acid to approximately 500.0mL of distilled/deionized water. Dissolve by adding KOH and adjust pH to 6.5. Add remaining components. Bring volume to 1.0L with additional distilled/deionized water. Adjust pH to 7.0 with KOH.

$CaCl_2$·2H_2O Solution:
Composition per 10.0mL:

$CaCl_2$·2H_2O	0.1g

Preparation of $CaCl_2$·2H_2O Solution: Add $CaCl_2$·2H_2O to distilled/deionized water and bring volume to 10.0mL. Mix thoroughly. Sparge with 100% N_2. Autoclave for 15 min at 15 psi pressure–121°C.

Sodium Acetate Solution:
Composition per 10.0mL:

Sodium acetate	3.3g

Preparation of Sodium Acetate Solution: Add sodium acetate to distilled/deionized water and bring volume to 10.0mL. Mix thoroughly. Sparge with 100% N_2. Autoclave for 15 min at 15 psi pressure–121°C.

Vitamin Solution:
Composition per liter:

Pyridoxine·HCl	10.0mg
Calcium DL-pantothenate	5.0mg
Lipoic acid	5.0mg
Nicotinic acid	5.0mg
p-Aminobenzoic acid	5.0mg
Riboflavin	5.0mg
Thiamine·HCl	5.0mg
Biotin	2.0mg
Folic acid	2.0mg
Vitamin B_{12}	0.1mg

Preparation of Vitamin Solution: Add components to distilled/deionized water and bring volume to 1.0L. Mix thoroughly. Sparge with 80% N_2 + 20% CO_2.

Coenzyme M Solution:
Composition per 10.0mL:

Coenzyme M	0.142g

Preparation of Coenzyme M Solution: Add coenzyme M to distilled/deionized water and bring volume to 10.0mL. Mix thoroughly. Sparge with 100% N_2. Autoclave for 15 min at 15 psi pressure–121°C.

$Na_2S·9H_2O$ Solution:
Composition per 10.0mL:

$Na_2S·9H_2O$	0.5g

Preparation of $Na_2S·9H_2O$ Solution: Add $Na_2S·9H_2O$ to distilled/deionized water and bring volume to 10.0mL. Mix thoroughly. Sparge with 80% N_2 + 20% CO_2. Autoclave for 15 min at 15 psi pressure–121°C.

Preparation of Medium: Add components, except $NaHCO_3$ solution, $CaCl_2·2H_2O$ solution, sodium acetate solution, vitamin solution, coenzyme M solution, and $Na_2S·9H_2O$ solution, to distilled/deionized water and bring volume to 935.0mL. Gently heat and bring to boiling. Continue boiling for 10 min. Cool to room temperature while sparging with 80% N_2 + 20% CO_2. Gas the medium until the pH reaches 5.8. Anaerobically distribute the medium into serum bottles. Autoclave for 15 min at 15 psi pressure–121°C. Aseptically and anaerobically add 20.0mL of sterile $NaHCO_3$ solution, 10.0mL of sterile $CaCl_2·2H_2O$ solution, 10.0mL of sterile sodium acetate solution, 10.0mL of sterile vitamin solution, 10.0mL of sterile coenzyme M solution, and 5.0mL of sterile $Na_2S·9H_2O$ solution. Bring gas atmosphere in each bottle to 30% CO_2.

Use: For the cultivation and maintenance of *Methanothrix thermophila.*

Thermophilic Spirochete Medium

Composition per 1012.0mL:

Solution A	920.0mL
Solution D	50.0mL
Solution E (Vitamin solution)	20.0mL
Solution F	10.0mL
Solution G	10.0mL
Solution B (Trace elements solution SL-10)	1.0mL
Solution C (Selenite-tungstate solution)	1.0mL

pH 6.9 ± 0.2 at 25°C

Solution A:
Composition per 920.0mL:

NaCl	4.0g
$MgCl_2·6H_2O$	0.8g
KCl	0.5g
NH_4Cl	0.3g
KH_2PO_4	0.2g
$CaCl_2·2H_2O$	0.03g
Resazurin	1.0mg

Preparation of Solution A: Prepare and dispense under 80% N_2 + 20% CO_2. Add components to distilled/deionized water and bring volume to 920.0mL. Mix thoroughly. Gently heat and bring to boiling. Continue boiling for a few minutes. Cool to room temperature while sparging with 80% N_2 + 20% CO_2. Anaerobically distribute into tubes or bottles. Autoclave for 15 min at 15 psi pressure–121°C.

Solution B (Trace Elements Solution SL-10):
Composition per liter:

$FeCl_2·4H_2O$	1.5g
$CoCl_2·6H_2O$	190.0mg
$MnCl_2·4H_2O$	100.0mg
$ZnCl_2$	70.0mg
$Na_2MoO_4·2H_2O$	36.0mg
$NiCl_2·6H_2O$	24.0mg
H_3BO_3	6.0mg
$CuCl_2·2H_2O$	2.0mg
HCl (25% solution)	10.0mL

Preparation of Solution B (Trace Elements Solution SL-10): Add $FeCl_2·4H_2O$ to 10.0mL of HCl solution. Mix thoroughly. Add distilled/deionized water and bring volume to 1.0L. Add remaining components. Mix thoroughly. Sparge with 100% N_2. Autoclave for 15 min at 15 psi pressure–121°C.

Solution C (Selenite-Tungstate Solution):
Composition per liter:

NaOH	0.5g
$Na_2SeO_3·5H_2O$	3.0mg
$Na_2WO_4·2H_2O$	4.0mg

Preparation of Solution C (Selenite-Tungstate Solution): Add components to distilled/deionized water and bring volume to 1.0L. Mix thoroughly. Sparge with 100% N_2. Autoclave for 15 min at 15 psi pressure–121°C.

Solution D:
Composition per 50.0mL:

$NaHCO_3$	3.0mg

Preparation of Solution D: Add $NaHCO_3$ to distilled/deionized water and bring volume to 50.0mL. Mix thoroughly. Sparge with 80% N_2 + 20% CO_2. Autoclave for 15 min at 15 psi pressure–121°C.

Solution E (Vitamin Solution):
Composition per liter:

Pyridoxine·HCl	10.0mg
Calcium DL-pantothenate	5.0mg
Lipoic acid	5.0mg
Nicotinic acid	5.0mg
p-Aminobenzoic acid	5.0mg
Riboflavin	5.0mg
Thiamine·HCl	5.0mg
Biotin	2.0mg
Folic acid	2.0mg
Vitamin B_{12}	0.1mg

Preparation of Solution E (Vitamin Solution): Add components to distilled/deionized water and bring volume to 1.0L. Mix thoroughly. Sparge with 100% N_2. Autoclave for 15 min at 15 psi pressure–121°C.

Solution F:
Composition per 10.0mL:

Starch, soluble	1.0g

Preparation of Solution F: Add starch to distilled/deionized water and bring volume to 10.0mL. Mix thoroughly. Sparge with 100% N_2. Autoclave for 15 min at 15 psi pressure–121°C.

Solution G:
Composition per 10.0mL:

$Na_2S·9H_2O$	0.3g

Preparation of Solution G: Add $Na_2S·9H_2O$ to distilled/deionized water and bring volume to 10.0mL. Mix thoroughly. Sparge with 100% N_2. Autoclave for 15 min at 15 psi pressure–121°C.

Preparation of Medium: Prepare and dispense medium under 80% N_2 + 20% CO_2. To 920.0mL of sterile solution A, aseptically and anaerobically add 1.0mL of sterile solution B, 1.0mL of sterile solution C, 50.0mL of sterile solution D, 20.0mL of sterile solution E, 10.0mL of sterile solution F, and 10.0mL of sterile solution G in that order. Mix thoroughly.

Use: For the cultivation and maintenance of *Spirochaeta thermophila.*

Thermophilic Streptomycete Medium

Composition per liter:

Agar	20.0g
Maltose	20.0g
Soybean meal	5.0g
Yeast extract	2.0g

pH 6.5 ± 0.2 at 25°C

Preparation of Medium: Add components to tap water and bring volume to 1.0L. Mix thoroughly. Gently heat and bring to boiling. Distribute into tubes or flasks. Autoclave for 15 min at 15 psi pressure–121°C. Pour into sterile Petri dishes or leave in tubes.

Use: For the isolation and cultivation of thermophilic streptomycetes.

Thermophilic Streptomycete Medium

Composition per liter:

Soybean oil meal	20.0g
Glucose	10.0g
NaCl	10.0g
Pancreatic digest of casein	10.0g
Silica solution (Ludox)	500.0mL

Preparation of Medium: Add components, except silica solution, to distilled/deionized water and bring volume to 500.0mL. Mix thoroughly. Gently heat until dissolved. Autoclave this solution and the 500.0mL of silica solution separately for 15 min at 15 psi pressure–121°C. Cool to 25°C. Adjust the pH of both solutions to 7.0. Aseptically combine the two sterile solutions. Mix thoroughly. Pour into sterile Petri dishes in 40.0mL volumes.

Use: For the isolation and cultivation of thermophilic streptomycetes.

Thermophilic Streptomycete Medium IA

Composition per liter:

Agar	20.0g
Sucrose	5.0g
Pancreatic digest of casein	5.0g
Yeast extract	3.0g
$MgSO_4 \cdot 7H_2O$	0.5g
$FeSO_4 \cdot 7H_2O$	0.01g
Dung extract	5.0mL
Molasses	5.0mL
Trace elements solution	1.0mL

pH 7.2 ± 0.2 at 25°C

Dung Extract:

Composition per 100.0mL:

Sheep manure, dried	25.0g

Preparation of Dung Extract: Add dried sheep manure to 100.0mL of tap water. Mix thoroughly. Autoclave for 30 min at 15 psi pressure–121°C. Filter through Whatman #1 filter paper. Store at 4°C under toluene.

Trace Elements Solution:

Composition per 100.0mL:

$Fe(NH_4)_2SO_4$	0.1g
$ZnSO_4$	0.1g
$MnSO_4$	0.05g
$CoSO_4$	0.01g
H_3BO_3	0.01g
$CuSO_4$	8.0mg

Preparation of Trace Elements Solution: Add components to distilled/deionized water and bring volume to 100.0mL. Mix thoroughly.

Preparation of Medium: Add components to distilled/deionized water and bring volume to 1.0L. Mix thoroughly. Gently heat and bring to boiling. Distribute into tubes or flasks. Autoclave for 15 min at 15 psi pressure–121°C. Pour into sterile Petri dishes or leave in tubes.

Use: For the isolation and cultivation of thermophilic streptomycetes.

Thermoplasma acidophilum Growth Medium 7B

Composition per liter:

Sucrose	17.0g
$(NH_4)_2SO_4$	.6.8g
KOH	1.22g
Yeast extract	1.0g
$MgSO_4$	0.5g
$CaCl_2 \cdot 2H_2O$	0.25g
H_3PO_4 solution	1.5mL
Antifoam A	10.0μL

pH 1.60 ± 0.2 at 25°C

H_3PO_4 Solution:

Composition per 100.0mL:

H_3PO_4	85.0g

Preparation of H_3PO_4 Solution: Add H_3PO_4 to distilled/deionized water and bring volume to 100.0mL. Mix thoroughly.

Preparation of Medium: Add components, except Antifoam A, to distilled/deionized water and bring volume to 1.0L. Mix thoroughly. Adjust pH to 1.60 with 50% H_2SO_4. Distribute into tubes or flasks. Sterilize by heating to 100°C for 30 min. Allow to stand at room temperature for 24 hr. Add 10.0μL of Antifoam A per liter.

Use: For the cultivation of *Thermococcus acidophilum.*

Thermoplasma acidophilum Medium

Composition per liter:

$(NH_4)_2SO_4$	1.32g
Yeast extract solution	1.0g
KH_2PO_4	0.372g
$MgSO_4 \cdot 7H_2O$	0.247g
$CaCl_2 \cdot 2H_2O$	0.074g
Glucose solution	20.0mL
Yeast extract solution	10.0mL
Trace elements solution	10.0mL

pH 1.0–2.0 at 25°C

Glucose Solution:

Composition per 20.0mL:

Glucose	10.0g

Preparation of Glucose Solution: Add glucose to distilled/deionized water and bring volume to 20.0mL. Mix thoroughly. Autoclave for 15 min at 15 psi pressure–121°C.

Yeast Extract Solution:

Composition per 10.0mL:

Yeast extract	1.0g

Preparation of Yeast Extract Solution: Add yeast extract to distilled/deionized water and bring volume to 10.0mL. Mix thoroughly. Autoclave for 15 min at 15 psi pressure–121°C.

Trace Elements Solution:

Composition per liter:

$FeCl_3 \cdot 6H_2O$	1.93g
$Na_2B_4O_7 \cdot 10H_2O$	0.45g
$MnCl_2 \cdot 4H_2O$	0.18g
$ZnSO_4 \cdot 7H_2O$	22.0mg
$CuCl_2 \cdot 2H_2O$	5.0mg
$VOSO_4 \cdot 5H_2O$	3.8mg
$Na_2MoO_4 \cdot 2H_2O$	3.0mg
$CoSO_4 \cdot 7H_2O$	2.0mg

Preparation of Trace Elements Solution: Add components to distilled/deionized water and bring volume to 1.0L. Mix thoroughly.

Preparation of Medium: Add components, except glucose solution and yeast extract solution, to distilled/deionized water and bring volume to 970.0mL. Mix thoroughly. Adjust pH to 1.0–2.0 with 10*N* H_2SO_4. Autoclave for 15 min at 15 psi pressure–121°C. Aseptically add 20.0mL of sterile glucose solution and 10.0mL of sterile yeast extract solution. Mix thoroughly. Aseptically distribute into sterile tubes or flasks.

Use: For the cultivation and maintenance of *Thermoplasma acidophilum.*

Thermoplasma acidophilum Medium 7A

Composition per liter:

Glucose	10.0g
$(NH_4)_2SO_4$	6.8g
KH_2PO_4	3.0g
Yeast extract	1.0g
$MgSO_4$	0.5g
$CaCl_2 \cdot 2H_2O$	0.25g

pH 1.65 ± 0.2 at 25°C

Preparation of Medium: Add components to distilled/deionized water and bring volume to 1.0L. Mix thoroughly. Adjust pH to 1.65 with 50% H_2SO_4. Distribute into tubes or flasks. Sterilize by heating to 100°C for 30 min.

Use: For the cultivation of *Thermococcus acidophilum.*

Thermoplasma Agar

Composition per liter:

Basal solution	450.0mL
Solution B	450.0mL
Solution C	100.0mL

pH 2.0 ± 0.2 at 25°C

Basal Solution:

Composition per 500.0mL:

KH_2PO_4	3.0g
Yeast extract	1.0g
$MgSO_4 \cdot 7H_2O$	0.5g
$CaCl_2 \cdot 2H_2O$	0.25g
$(NH_4)_2SO_4$	0.2g

Preparation of Basal Solution: Add components to distilled/deionized water and bring volume to 500.0mL. Mix thoroughly. Adjust pH to 2.0 with 10*N* H_2SO_4. Autoclave for 15 min at 15 psi pressure–121°C. Cool to 55°C.

Solution B:

Composition per 450.0mL:

Noble agar	12.0g

Preparation of Solution B: Add agar to distilled/deionized water and bring volume to 450.0mL. Mix thoroughly. Gently heat and bring to boiling. Autoclave for 15 min at 15 psi pressure–121°C. Cool to 55°C.

Solution C:

Composition per 100.0mL:

Glucose	10.0g

Preparation of Solution C: Add glucose to distilled/deionized water and bring volume to 100.0mL. Mix thoroughly. Filter sterilize.

Preparation of Medium: Aseptically combine the cooled, sterile basal medium with sterile solution B and sterile solution C. Mix thoroughly. Pour into sterile Petri dishes or distribute into sterile tubes.

Use: For the cultivation and maintenance of *Thermoplasma acidophilum* and other *Thermoplasma* species.

Thermoplasma Broth

Composition per liter:

Basal solution	500.0mL
Solution C	100.0mL

pH 2.0 ± 0.2 at 25°C

Basal Solution:

Composition per 500.0mL:

KH_2PO_4	3.0g
Yeast extract	1.0g
$MgSO_4 \cdot 7H_2O$	0.5g
$CaCl_2 \cdot 2H_2O$	0.25g
$(NH_4)_2SO_4$	0.2g

Preparation of Basal Solution: Add components to distilled/deionized water and bring volume to 500.0mL. Mix thoroughly. Adjust pH to 2.0 with 10*N* H_2SO_4.

Solution C:

Composition per 100.0mL:

Glucose	10.0g

Preparation of Solution C: Add glucose to distilled/deionized water and bring volume to 100.0mL. Mix thoroughly. Filter sterilize.

Preparation of Medium: Add 500.0mL of basal solution to 400.0mL of distilled/deionized water. Autoclave for 15 min at 15 psi pressure–121°C. Cool to 55°C. Aseptically add 100.0mL of sterile glucose solution. Mix thoroughly. Aseptically distribute into sterile tubes.

Use: For the cultivation and maintenance of *Thermoplasma acidophilum* and other *Thermoplasma* species.

Thermoplasma volcanium Medium

Composition per liter:

KH_2PO_4	3.0g
$MgSO_4 \cdot 7H_2O$	1.0g
$CaCl_2 \cdot 2H_2O$	0.25g
$(NH_4)_2SO_4$	0.2g
Glucose solution	10.0mL
Yeast extract solution	10.0mL

pH 6.5 ± 0.2 at 25°C

Glucose Solution:

Composition per 10.0mL:

Glucose	5.0g

Preparation of Glucose Solution: Add glucose to distilled/deionized water and bring volume to 10.0mL. Mix thoroughly. Filter sterilize.

Yeast Extract Solution:

Composition per 10.0mL:

Yeast extract	1.0g

Preparation of Yeast Extract Solution: Add yeast extract to distilled/deionized water and bring volume to 10.0mL. Mix thoroughly. Filter sterilize.

Preparation of Medium: Add components, except glucose solution and yeast extract solution, to distilled/deionized water and bring volume to 980.0mL. Mix thoroughly. Adjust pH to 2.0 with 10*N* H_2SO_4. Autoclave for 15 min at 15 psi pressure–121°C. Aseptically add 10.0mL of sterile glucose solution and 10.0mL of sterile yeast extract solution. Mix thoroughly. Aseptically distribute into sterile tubes or flasks. Final pH should be 2.0–3.0.

Use: For the aerobic cultivation and maintenance of *Thermoplasma volcanium.*

Thermoproteus Medium

Composition per liter:

Solution A	500.0mL
Solution B	450.0mL
Solution C	50.0mL

pH 4.8–5.6 at 25°C

Solution A:

Composition per 500.0mL:

Glucose	10.0g
$FeSO_4 \cdot 7H_2O$	0.556g
$MgSO_4 \cdot 7H_2O$	0.492g
$CaSO_4 \cdot 2H_2O$	0.344g
$(NH_4)_2SO_4$	0.264g
Yeast extract	0.2g
KH_2PO_4	0.014g
Resazurin	1.0mg
Trace elements	10.0mL

Preparation of Solution A: Add components to distilled/deionized water and bring volume to 500.0mL. Mix thoroughly. Immediately filter sterilize.

Trace Elements:

Composition per liter:

$Na_2B_4O_7 \cdot 10H_2O$	0.45g
$MnCl_2 \cdot 4H_2O$	0.18g
$ZnSO_4 \cdot 7H_2O$	0.022g
$CuCl_2 \cdot 2H_2O$	5.0mg
$Na_2MoO_4 \cdot 2H_2O$	3.6mg
$VOSO_4 \cdot 5H_2O$	3.6mg
$CoSO_4 \cdot 7H_2O$	1.2mg

Preparation of Trace Elements: Add components to distilled/deionized water and bring volume to 1.0L. Mix thoroughly. Adjust pH to 3.0 with H_2SO_4 to retard precipitation.

Solution B:

Composition per 450.0mL:

Sulfur	10.0g

Preparation of Solution B: Add sulfur to 450.0mL of distilled/deionized water. Autoclave for 30 min at 0 psi pressure–100°C on 3 consecutive days.

Solution C:

Composition per 50.0mL:

$Na_2S \cdot 9H_2O$	0.85g

Preparation of Solution C: Add $Na_2S \cdot 9H_2O$ to distilled/deionized water and bring volume to 50.0mL. Mix thoroughly. Autoclave for 15 min at 15 psi pressure–121°C.

Preparation of Medium: Aseptically combine solutions A, B, and C under 97% N_2 + 3% H_2. Adjust pH to 4.8–5.6 with H_2SO_4. Aseptically and anaerobically distribute into sterile tubes or flasks under 97% N_2 + 3% H_2.

Use: For the cultivation and maintenance of *Thermoproteus tenax* and other *Thermoproteus* species.

Thermoproteus neutrophilus Medium

Composition per liter:

Sulfur, powdered	8.0g
$(NH_4)_2SO_4$	1.3g
$NaHCO_3$	0.85g
KH_2PO_4	0.28g
$MgSO_4 \cdot 7H_2O$	0.25g
$CaCl_2 \cdot 2H_2O$	0.07g
$FeCl_3 \cdot 6H_2O$	0.02g
$Na_2B_4 \cdot 10H_2O$	4.5mg
$MnCl_2 \cdot 4H_2O$	1.8mg
Resazurin	0.4mg
$ZnSO_4 \cdot 7H_2O$	0.22mg
$CuCl_2 \cdot 2H_2O$	0.05mg
$Na_2MoO_4 \cdot 2H_2O$	0.03mg
$VOSO_4 \cdot 2H_2O$	0.03mg
$CoSO_4$	0.01mg
Dithionite solution	1.0mL

pH 6.5 ± 0.2 at 25°C

Dithionite Solution:

Composition per 10.0mL:

$Na_2S_2O_4$	0.25g

Preparation of Dithionite Solution: Add $Na_2S_2O_4$ to distilled/deionized water and bring volume to 10.0mL. Mix thoroughly. Filter sterilize.

Preparation of Medium: Add components, except sulfur, $NaHCO_3$, resazurin, and dithionite solution, to distilled/deionized water and bring volume to 1.0L. Mix thoroughly. Add sulfur and resazurin. Adjust pH to 6.5 with NaOH. Gently heat and bring to boiling. Continue boiling for 5 min. Cool to room temperature while sparging with 80% H_2 + 20% CO_2. Add $NaHCO_3$. Mix thoroughly. Continue to sparge with 80% H_2 + 20% CO_2 until pH reaches 6.5. Distribute anaerobically 10.0mL of medium into 30.0mL serum bottles. Sterilize medium by heating at 85°C for 1 hr on 3 consecutive days. Prior to inoculation, add 10.0μL of sterile dithionite solution to each bottle.

Use: For the aerobic cultivation and maintenance of *Thermoproteus neutrophilus*.

Thermosipho africanus Medium

Composition per liter:

$NaHCO_3$	4.0g
Sodium acetate	4.0g

Sodium formate	2.0g
Yeast extract	1.0g
L-Cysteine·HCl	0.5g
KH_2PO_4	0.5g
$Na_2S·9H_2O$	0.5g
$MgSO_4·7H_2O$	0.4g
NaCl	0.4g
NH_4Cl	0.4g
$CaCl_2·2H_2O$	0.05g
$NiCl_2·6H_2O$	24.0mg
$FeSO_4·7H_2O$	2.0mg
Resazurin	1.0mg
Sludge fluid	50.0mL
Fatty acid mixture	20.0mL
Trace elements solution SL-10	1.0mL

pH 6.7 ± 0.2 at 25°C

Sludge Fluid:

Composition per 100.0mL:

Sludge	100.0mL
Yeast extract	0.4g

Preparation of Sludge Fluid: To 100.0mL of sludge from an anaerobic digester, add 0.4g of yeast extract. Sparge with 100% N_2 for a few minutes. Incubate at 37°C for 24 hr. Centrifuge the sludge at 13,000 × g for 15 min. Decant the clear supernatant solution. Sparge with 100% N_2 for a few minutes. Store in screw-capped bottles at room temperature in the dark.

Fatty Acid Mixture:

Composition per 20.0mL:

α-Methylbutyric acid	0.5g
Isobutyric acid	0.5g
Isovaleric acid	0.5g
Valeric acid	0.5g

Preparation of Fatty Acid Mixture: Add components to distilled/deionized water and bring volume to 20.0mL. Mix thoroughly. Adjust pH to 7.5 with concentrated NaOH.

Trace Elements Solution SL-10:

Composition per liter:

$FeCl_2·4H_2O$	1.5g
$CoCl_2·6H_2O$	190.0mg
$MnCl_2·4H_2O$	100.0mg
$ZnCl_2$	70.0mg
$Na_2MoO_4·2H_2O$	36.0mg
$NiCl_2·6H_2O$	24.0mg
H_3BO_3	6.0mg
$CuCl_2·2H_2O$	2.0mg
HCl (25% solution)	10.0mL

Preparation of Trace Elements Solution SL-10: Add $FeCl_2·4H_2O$ to 10.0mL of HCl solution. Mix thoroughly. Add distilled/deionized water and bring volume to 1.0L. Add remaining components. Mix thoroughly. Sparge with 100% N_2. Autoclave for 15 min at 15 psi pressure–121°C.

Preparation of Medium: Prepare and dispense medium anaerobically under 80% N_2 + 20% CO_2. Add components to distilled/deionized water and bring volume to 1.0L. Mix thoroughly. Sparge with 80% N_2 + 20% CO_2. Autoclave for 15 min at 15 psi pressure–121°C.

Use: For the cultivation and maintenance of *Thermosipho africanus*.

Thermosphaera Medium (DSMZ Medium 817)

Composition per liter:

$MgCl_2·6H_2O$	2.2g
Yeast extract	1.0g
Peptone	1.0g
NaCl	0.9g
KCl	17.0mg
NH_4Cl	12.5mg
$CaCl_2·2H_2O$	7.0mg
$K_2HPO_4·3H_2O$	7.0mg
Resazurin	0.4mg
$FeCl_3$	0.05mg
Vitamin solution	10.0mL
$Na_2S·9H_2O$ solution	10.0mL
$NaHCO_3$ solution	10.0mL

pH 6.5 ± 0.2 at 25°C

$NaHCO_3$ Solution:

Composition per 10.0mL:

$NaHCO_3$	1.0g

Preparation of $NaHCO_3$ Solution: Add $NaHCO_3$ to distilled/deionized water and bring volume to 10.0mL. Mix thoroughly. Sparge with 100% N_2. Autoclave for 15 min at 15 psi pressure–121°C. Cool to 25°C. Must be prepared freshly.

$Na_2S·9H_2O$ Solution:

Composition per 10.0mL:

$Na_2S·9H_2O$	0.5g

Preparation of $Na_2S·9H_2O$ Solution: Add $Na_2S·9H_2O$ to distilled/deionized water and bring volume to 10.0mL. Sparge with N_2. Autoclave for 15 min at 15 psi pressure–121°C. Cool to 25°C. Store anaerobically.

Vitamin Solution:

Composition per liter:

Pyridoxine-HCl	10.0mg
Thiamine-HCl·2H$_2$O	5.0mg
Riboflavin	5.0mg
Nicotinic acid	5.0mg
D-Ca-pantothenate	5.0mg
p-Aminobenzoic acid	5.0mg
Lipoic acid	5.0mg
Biotin	2.0mg
Folic acid	2.0mg
Vitamin B_{12}	0.1mg

Preparation of Vitamin Solution: Add components to distilled/deionized water and bring volume to 1.0L. Mix thoroughly. Sparge with 80% H_2 + 20% CO_2. Filter sterilize.

Preparation of Medium: Prepare and dispense medium under 80% N_2 + 20% CO_2. Add components, except $Na_2S·9H_2O$ solution, vitamin solution, and $NaHCO_3$ solution, to distilled/deionized water and bring volume to 970.0mL. Mix thoroughly. Flush medium with 80% N_2 + 20% CO_2. for 5 min. Adjust medium pH to 6.53. Autoclave for 15 min at 15 psi pressure–121°C. Cool to 25°C. Aseptically and anaerobically add 10.0mL sterile $Na_2S·9H_2O$ solution, 10.0mL sterile vitamin solution, and 10.0mL sterile $NaHCO_3$ solution. Mix thoroughly. Aseptically and anaerobically distribute into sterile thick-walled tubes or thick-walled bottles. Pressurize with 2 bar atmosphere of 80% N_2 + 20% CO_2.

Use: For the cultivation of *Thermosphaera aggregans*.

Thermosulfidibacter Medium (DSMZ Medium 1092)

Composition per liter:

Sea salts, Sigma	30.0g
Sulfur, elemental	3.0g
Resazurin	0.5mg
Vitamin solution	20.0mL
Sulfide solution	10.0mL
Yeast extract solution	10.0mL
Bicarbonate solution	10.0mL
Wolfe's mineral elixir	1.0mL

pH 7.1 ± 0.2 at 25°C

Sulfide Solution:

Composition per 10.0mL:

$Na_2S{\cdot}9H_2O$	0.5g

Preparation of Sulfide Solution: Add $Na_2S{\cdot}9H_2O$ to distilled/deionized water and bring volume to 10.0mL. Mix thoroughly. Autoclave under 100% N_2 for 15 min at 15 psi pressure–121°C. Cool to room temperature.

Wolfe's Mineral Elixir:

Composition per liter:

$MgSO_4{\cdot}7H_2O$	30.0g
NaCl	10.0g
$MnSO_4{\cdot}2H_2O$	5.0g
$(NH_4)_2NiSO_4{\cdot}6H_2O$	2.8g
$CoCl_2{\cdot}6H_2O$	1.8g
$ZnSO_4{\cdot}7H_2O$	1.8g
$FeSO_4{\cdot}7H_2O$	1.0g
$CaCl_2{\cdot}2H_2O$	1.0g
$KAl(SO_4)_2{\cdot}12H_2O$	0.18g
$CuSO_4{\cdot}5H_2O$	0.1g
H_3BO_3	0.1g
$Na_2MoO_4{\cdot}2H_2O$	0.1g
Na_2SeO_4	0.1g
$Na_2WO_4{\cdot}2H_2O$	0.1g

Preparation of Wolfe's Mineral Elixir: Adjust pH of 1.0L of distilled/deionized water to 1.0 with dilute H_2SO_4. Add remaining components one at a time. Mix throughly to dissolve.

Vitamin Solution:

Composition per liter:

Pyridoxine-HCl	10.0mg
Thiamine-HCl·$2H_2O$	5.0mg
Riboflavin	5.0mg
Nicotinic acid	5.0mg
D-Ca-pantothenate	5.0mg
p-Aminobenzoic acid	5.0mg
Lipoic acid	5.0mg
Biotin	2.0mg
Folic acid	2.0mg
Vitamin B_{12}	0.1mg

Preparation of Vitamin Solution: Add components to distilled/deionized water and bring volume to 1.0L. Mix thoroughly. Sparge with 80% H_2 + 20% CO_2. Filter sterilize.

Bicarbonate Solution:

Composition per 10.0mL:

$NaHCO_3$	2.5g

Preparation of Bicarbonate Solution: Add $NaHCO_3$ to distilled/deionized water and bring volume to 10.0mL. Mix thoroughly. Sparge with gas mixture of 80% N_2 + 20% CO_2. Filter sterilize.

Yeast Extract Solution:

Composition per 10.0mL:

Yeast extract	2.0g

Preparation of Yeast Extract Solution: Add yeast extract to distilled/deionized water and bring volume to 10.0mL. Mix thoroughly. Sparge with 100% N_2 gas. Autoclave for 15 min at 15 psi pressure–121°C.

Preparation of Medium: Add components, except vitamin, bicarbonate, sulfide, and yeast extract solutions, to distilled/deionized water and bring volume to 980.0mL. Mix thoroughly. Gently heat and bring to boiling. Boil for 1 min. Cool to room temperature while sparging with a gas mixture of 80% H_2 + 20% CO_2. Adjust the pH to 5.5 with 10*N* H_2SO_4. Dispense under the same gas atmosphere in culture vessels (balch tubes). Autoclave for 20 min at 6 psi pressure–110°C. Cool to room temperature. Aspetically add vitamin, bicarbonate, sulfide, and yeast extract solutions. Mix thoroughly. Prior to inoculation change atmosphere to 80% H_2 and 20% CO_2 gas mixture. Adjust pH to 7.0–7.2. After inoculation pressurize vials to 2.0 bar overpressure with 80% H_2 and 20% CO_2 gas mixture.

Use: For the cultivation of *Thermosulfidibacter* spp.

Thermosyntropha Medium (DSMZ Medium 731)

Composition per liter:

Yeast extract	10.0g
Na_2CO_3	3.0g
NH_4Cl	1.0g
NaCl	0.5g
K_2HPO_4	0.3g
KCl	0.3g
$MgCl_2{\cdot}6H_2O$	0.3g
$CaCl_2{\cdot}2H_2O$	0.05g
Resazurin	0.5mg
Vitamin solution	40.0mL
Trace elements solution	10.0mL
Cysteine solution	10.0mL
$Na_2S{\cdot}9H_2O$ solution	10.0mL
$NaHCO_3$ solution	10.0mL
Selenite-tungstate solution	1.0mL

pH 8.5 ± 0.2 at 25°C

Selenite-Tungstate Solution

Composition per liter:

NaOH	0.5g
$Na_2WO_4{\cdot}2H_2O$	4.0mg
$Na_2SeO_3{\cdot}5H_2O$	3.0mg

Preparation of Selenite-Tungstate Solution: Add components to distilled/deionized water and bring volume to 1.0L. Mix thoroughly. Sparge with 100% N_2. Filter sterilize.

$NaHCO_3$ Solution:

Composition per 10.0mL:

$NaHCO_3$	3.0g

Preparation of $NaHCO_3$ Solution: Add $NaHCO_3$ to distilled/deionized water and bring volume to 10.0mL. Mix thoroughly. Autoclave for 15 min at 15 psi pressure–121°C. Cool to 25°C. Must be prepared freshly.

Cysteine Solution:

Composition per 10.0mL:

L-Cysteine-HCl·H_2O	0.15g

Preparation of Cysteine Solution: Add L-cysteine·HCl·H_2O to distilled/deionized water and bring volume to 10.0mL. Mix thoroughly. Sparge with 100% N_2. Autoclave for 15 min at 15 psi pressure–121°C.

Trace Elements Solution:

Composition per liter:

$MgSO_4·7H_2O$	3.0g
Nitrilotriacetic acid	1.5g
NaCl	1.0g
$MnSO_4·2H_2O$	0.5g
$CoSO_4·7H_2O$	0.18g
$ZnSO_4·7H_2O$	0.18g
$CaCl_2·2H_2O$	0.1g
$FeSO_4·7H_2O$	0.1g
$NiCl_2·6H_2O$	0.025g
$KAl(SO_4)_2·12H_2O$	0.02g
H_3BO_3	0.01g
$Na_2MoO_4·4H_2O$	0.01g
$CuSO_4·5H_2O$	0.01g
$Na_2SeO_3·5H_2O$	0.3mg

Preparation of Trace Elements Solution: Add nitrilotriacetic acid to 500.0mL of distilled/deionized water. Dissolve by adjusting pH to 6.5 with KOH. Add remaining components. Add distilled/deionized water to 1.0L. Mix thoroughly.

Vitamin Solution:

Composition per liter:

Pyridoxine-HCl	10.0mg
Thiamine-HCl·$2H_2O$	5.0mg
Riboflavin	5.0mg
Nicotinic acid	5.0mg
D-Ca-pantothenate	5.0mg
p-Aminobenzoic acid	5.0mg
Lipoic acid	5.0mg
Biotin	2.0mg
Folic acid	2.0mg
Vitamin B_{12}	0.1mg

Preparation of Vitamin Solution: Add components to distilled/deionized water and bring volume to 1.0L. Mix thoroughly. Sparge with 80% H_2 + 20% CO_2. Filter sterilize.

$Na_2S·9H_2O$ Solution:

Composition per 10.0mL:

$Na_2S·9H_2O$	0.5g

Preparation of $Na_2S·9H_2O$ Solution: Add $Na_2S·9H_2O$ to distilled/deionized water and bring volume to 10.0mL. Sparge with N_2. Autoclave for 15 min at 15 psi pressure–121°C. Cool to 25°C. Store anaerobically.

Preparation of Medium: Prepare and dispense medium under 100% N_2. Add components, except $NaHCO_3$ solution, $Na_2S·9H_2O$ solution, cysteine solution, vitamin solution, selenite-tungstate solution, and trace elements solution, to distilled/deionized water and bring volume to 919.0mL. Mix thoroughly. Sparge with 100% N_2. Adjust pH to 8.5 with HCl. Autoclave for 15 min at 15 psi pressure–121°C. Aseptically and anaerobically add 10.0mL $NaHCO_3$ solution, 10.0mL $Na_2S·9H_2O$ solution, 10.0mL cysteine solution, 40.0mL vitamin solution, 1.0mL selenite-tungstate solution, and 10.0mL trace elements solution. Mix thoroughly. Aseptically and anaerobically distribute into sterile tubes or bottles.

Use: For the cultivation of *Thermosyntropha lipolytica.*

Thermoterrabacterium Medium (DSMZ Medium 778)

Composition per liter:

$NaHCO_3$	10.0g
Na_2-9,10-anthraquinone-2,6-disulfonate	8.25g
Yeast extract	1.0g
KH_2PO_4	0.33g
NH_4Cl	0.33g
KCl	0.33g
$MgCl_2·6H_2O$	0.33g
$NiCl_2·6H_2O$	200.0μg
$Na_2SeO_3·5H_2O$	120.0μg
$Na_2WO_4·2H_2O$	30.0μg
Calcium chloride solution	10.0mL
Vitamin solution	10.0mL
Glycerol (87%)	3.0mL
Trace elements solution SL-10	1.0mL

pH 6.8 ± 0.2 at 25°C

Calcium Chloride Solution:

Composition per 10.0mL:

$CaCl_2·2H_2O$	0.33g

Preparation of Calcium Chloride Solution: Add $CaCl_2·2H_2O$ to 10.0mL of distilled/deionized water. Mix thoroughly. Sparge with 100% N_2. Autoclave for 15 min at 15 psi pressure–121°C. Cool to room temperature.

Trace Elements Solution SL-10:

Composition per liter:

$FeCl_2·4H_2O$	1.5g
$CoCl_2·6H_2O$	190.0mg
$MnCl_2·4H_2O$	100.0mg
$ZnCl_2$	70.0mg
$Na_2MoO_4·2H_2O$	36.0mg
$NiCl_2·6H_2O$	24.0mg
H_3BO_3	6.0mg
$CuCl_2·2H_2O$	2.0mg
HCl (25% solution)	10.0mL

Preparation of Trace Elements Solution SL-10: Add $FeCl_2·4H_2O$ to 10.0mL of HCl solution. Mix thoroughly. Add distilled/deionized water and bring volume to 1.0L. Add remaining components. Mix thoroughly. Sparge with 80% N_2 + 20% CO_2. Autoclave for 15 min at 15 psi pressure–121°C.

Vitamin Solution:

Composition per liter:

Pyridoxine-HCl	10.0mg
Thiamine-HCl·$2H_2O$	5.0mg
Riboflavin	5.0mg
Nicotinic acid	5.0mg
D-Ca-pantothenate	5.0mg
p-Aminobenzoic acid	5.0mg
Lipoic acid	5.0mg
Biotin	2.0mg
Folic acid	2.0mg
Vitamin B_{12}	0.1mg

Preparation of Vitamin Solution: Add components to distilled/deionized water and bring volume to 1.0L. Mix thoroughly. Sparge with 80% H_2 + 20% CO_2. Filter sterilize.

Preparation of Medium: Add components, except $NaHCO_3$, calcium chloride solution, and vitamin solution, to distilled/deionized water and bring volume to 980.0mL. Mix thoroughly. Gently heat and bring to boiling. Boil for several minutes to dissolve the antraquinone. Cool to room temperature under an atmosphere of 100% CO_2. Add solid $NaHCO_3$. Adjust pH to 6.8 wtih NaOH. Dispense medium under CO_2 into tubes or bottles. Autoclave for 15 min at 15 psi pressure–121°C. Cool to room temperature. Aseptically and anaerobically distribute into sterile tubes or bottles. Before use, add calcium chloride and vitamin solutions from anaerobic, sterile stock solution.

Use: For the cultivation of *Thermoterrabacterium ferrireducens.*

Thermotoga elfii Medium

Composition per liter:

NaCl	10.0g
Pancreatic digest of casein	2.0g
Yeast extract	2.0g
NH_4Cl	1.0g
K_2HPO_4	0.3g
KH_2PO_4	0.3g
$MgCl_2 \cdot 6H_2O$	0.2g
$CaCl_2 \cdot 2H_2O$	0.1g
KCl	0.1g
L-Cysteine·HCl·H_2O	0.5g
Sodium acetate	0.5g
Resazurin	1.0mg
Na_2CO_3 solution	20.0mL
$Na_2S \cdot 9H_2O$ solution	20.0mL
Modified Wolfe's mineral solution	10.0mL

pH 8.0 ± 0.2 at 25°C

$NaHCO_3$ Solution:

Composition per 20.0mL:

$NaHCO_3$	2.0g

Preparation of $NaHCO_3$ Solution: Add $NaHCO_3$ to distilled/deionized water and bring volume to 20.0mL. Mix thoroughly. Sparge with 80% N_2 + 20% CO_2. Autoclave for 15 min at 15 psi pressure–121°C.

$Na_2S \cdot 9H_2O$ Solution:

Composition per 20.0mL:

$Na_2S \cdot 9H_2O$	0.4g

Preparation of $Na_2S \cdot 9H_2O$ Solution: Add $Na_2S \cdot 9H_2O$ to distilled/deionized water and bring volume to 20.0mL. Mix thoroughly. Sparge with 100% N_2. Autoclave for 15 min at 15 psi pressure–121°C. Before use, neutralize to pH 7.0 with sterile HCl.

Modified Wolfe's Mineral Solution:

Composition per liter:

$MgSO_4 \cdot 7H_2O$	3.0g
Nitrilotriacetic acid	1.5g
NaCl	1.0g
$MnSO_4 \cdot H_2O$	0.5g
$CaCl_2$	0.1g
$CoCl_2 \cdot 6H_2O$	0.1g
$FeSO_4 \cdot 7H_2O$	0.1g
$ZnSO_4 \cdot 7H_2O$	0.1g
$AlK(SO_4)_2 \cdot 12H_2O$	0.01g
$CuSO_4 \cdot 5H_2O$	0.01g
H_3BO_3	0.01g
$Na_2MoO_4 \cdot 2H_2O$	0.01g
Na_2SeO_3	0.01g
$NaWO_4 \cdot 2H_2O$	0.01g
$NiCl_2 \cdot 6H_2O$	0.01g

Preparation of Modified Wolfe's Mineral Solution: Add nitrilotriacetic acid to 500.0mL of distilled/deionized water. Adjust pH to 6.5 with KOH. Add remaining components one at a time. Add distilled/deionized water to 1.0L. Adjust pH to 6.8.

Preparation of Medium: Prepare and dispense medium under 100% N_2. Add components, except Na_2CO_3 solution and $Na_2S \cdot 9H_2O$ solution, to distilled/deionized water and bring volume to 960.0mL. Mix thoroughly. Adjust pH to 8.0 with 10*M* KOH. Gently heat and bring to boiling. Cool to room temperature while sparging with 100% N_2. Anaerobically dispense into tubes in 5.0mL aliquots under an atmosphere of 80% N_2 + 20% CO_2. Autoclave for 45 min at 6 psi pressure–110°C. Just prior to use, aseptically and anaerobically add 0.1mL of sterile Na_2CO_3 solution and 0.1mL of sterile $Na_2S \cdot 9H_2O$ solution to each tube.

Use: For the cultivation of *Thermotoga elfii.*

Thermotoga elfii Medium

Composition per liter:

NaCl	10.0g
Sodium thiosulfate·$5H_2O$	5.0g
Pancreatic digest of casein	5.0g
Yeast extract	5.0g
NH_4Cl	1.0g
K_2HPO_4	0.3g
KH_2PO_4	0.3g
$MgCl_2 \cdot 6H_2O$	0.2g
$CaCl_2 \cdot 2H_2O$	0.1g
KCl	0.1g
Sodium acetate	0.5g
Resazurin	0.5mg
Glucose solution	20.0mL
L-Cysteine solution	10.0mL
Na_2CO_3 solution	10.0mL
$Na_2S \cdot 9H_2O$ solution	10.0mL
Wolfe's mineral solution	10.0mL

Glucose Solution:

Composition per 20.0mL:

Glucose	4.0g

Preparation of Glucose Solution: Add glucose to distilled/deionized water and bring volume to 20.0mL. Mix thoroughly. Sparge with 100% N_2. Autoclave for 15 min at 15 psi pressure–121°C.

L-Cysteine Solution:

Composition per 10.0mL:

L-Cysteine·HCl·H_2O	0.5g

Preparation of L-Cysteine Solution: Add L-cysteine·HCl·H_2O to distilled/deionized water and bring volume to 10.0mL. Mix thoroughly. Sparge with 100% N_2. Autoclave for 15 min at 15 psi pressure–121°C.

Na_2CO_3 Solution:

Composition per 10.0mL:

Na_2CO_3	2.0g

Preparation of Na_2CO_3 Solution: Add Na_2CO_3 to distilled/deionized water and bring volume to 10.0mL. Mix thoroughly. Sparge with 80% N_2 + 20% CO_2. Autoclave for 15 min at 15 psi pressure–121°C.

$Na_2S·9H_2O$ Solution:

Composition per 10.0mL:

$Na_2S·9H_2O$	0.5g

Preparation of $Na_2S·9H_2O$ Solution: Add $Na_2S·9H_2O$ to distilled/deionized water and bring volume to 10.0mL. Mix thoroughly. Sparge with 100% N_2. Autoclave for 15 min at 15 psi pressure–121°C. Before use, neutralize to pH 7.0 with sterile HCl.

Wolfe's Mineral Solution:

Composition per liter:

$MgSO_4·7H_2O$	3.0g
Nitrilotriacetic acid	1.5g
NaCl	1.0g
$MnSO_4·2H_2O$	0.5g
$CoCl_2·6H_2O$	0.1g
$ZnSO_4·7H_2O$	0.1g
$CaCl_2·2H_2O$	0.1g
$FeSO_4·7H_2O$	0.1g
$NiCl_2·6H_2O$	0.025g
$KAl(SO_4)_2·12H_2O$	0.02g
$CuSO_4·5H_2O$	0.01g
H_3BO_3	0.01g
$Na_2MoO_4·2H_2O$	0.01g
$Na_2SeO_3·5H_2O$	0.3mg

Preparation of Wolfe's Mineral Solution: Add nitrilotriacetic acid to 500.0mL of distilled/deionized water. Adjust pH to 6.5 with KOH. Add remaining components. Add distilled/deionized water to 1.0L. Adjust pH to 6.8.

Preparation of Medium: Prepare and dispense medium under 80% N_2 + 20% CO_2 gas atmosphere. Add components, except glucose solution, Na_2CO_3 solution, L-cysteine solution, and $Na_2S·9H_2O$ solution, to distilled/deionized water and bring volume to 950.0mL. Mix thoroughly. Adjust pH to 6.8–7.0. Sparge with 80% N_2 + 20% CO_2. Autoclave for 15 min at 15 psi pressure–121°C. Aseptically and anaerobically add 20.0mL of sterile glucose solution, 10.0mL of sterile Na_2CO_3 solution, 10.0mL of sterlie L-cysteine solution, and 10.0mL of sterile $Na_2S·9H_2O$ solution. Mix thoroughly. Aseptically and anaerobically distribute into sterile tubes or bottles.

Use: For the cultivation of *Thermotoga elfii.*

Thermotoga hypogea Medium (DSMZ Medium 794)

Composition per liter:

NaCl	10.0g
Yeast extract	2.0g
Trypticase™	2.0g
NH_4Cl	1.0g
Na-acetate	0.5g
L-Cysteine	0.5g
K_2HPO_4	0.3g
KH_2PO_4	0.3g
$MgCl_2·6H_2O$	0.2g
$CaCl_2·2H_2O$	0.1g
KCl	0.1g
Resazurin	0.5mg
Na-thiosulfate solution	20.0mL
$NaHCO_3$ solution	20.0mL
Xylose solution	20.0mL
$Na_2S·9H_2O$ solution	14.0mL
Trace elements solution	10.0mL

pH 7.3 ± 0.2 at 25°C

Xylose Solution:

Composition per 20.0mL:

Xylose	3.0g

Preparation of Xylose Solution: Add xylose to distilled/deionized water and bring volume to 20.0mL. Mix thoroughly. Sparge with 100% N_2. Filter sterilize.

$NaHCO_3$ Solution:

Composition per 20.0mL:

$NaHCO_3$	2.0g

Preparation of $NaHCO_3$ Solution: Add $NaHCO_3$ to distilled/deionized water and bring volume to 20.0mL. Mix thoroughly. Autoclave for 15 min at 15 psi pressure–121°C. Cool to 25°C. Must be prepared freshly.

Na-thiosulfate Solution:

Composition per 20.0mL:

$Na_2S_2O_3·5H_2O$	5.0g

Preparation of Na-thiosulfate Solution: Add $Na_2S_2O_3·5H_2O$ to distilled/deionized water and bring volume to 20.0mL. Mix thoroughly. Sparge with 100% N_2. Filter sterilize.

$Na_2S·9H_2O$ Solution:

Composition per 20.0mL:

$Na_2S·9H_2O$	0.6g

Preparation of $Na_2S·9H_2O$ Solution: Add $Na_2S·9H_2O$ to distilled/deionized water and bring volume to 20.0mL. Sparge with N_2. Autoclave for 15 min at 15 psi pressure–121°C. Cool to 25°C. Store anaerobically.

Trace Elements Solution:

Composition per liter:

$MgSO_4·7H_2O$	3.0g
Nitrilotriacetic acid	1.5g
NaCl	1.0g
$MnSO_4·2H_2O$	0.5g
$CoSO_4·7H_2O$	0.18g
$ZnSO_4·7H_2O$	0.18g
$CaCl_2·2H_2O$	0.1g
$FeSO_4·7H_2O$	0.1g
$NiCl_2·6H_2O$	0.025g
$KAl(SO_4)_2·12H_2O$	0.02g
H_3BO_3	0.01g
$Na_2MoO_4·4H_2O$	0.01g
$CuSO_4·5H_2O$	0.01g
$Na_2SeO_3·5H_2O$	0.3mg

Preparation of Trace Elements Solution: Add nitrilotriacetic acid to 500.0mL of distilled/deionized water. Dissolve by adjusting pH to 6.5 with KOH. Add remaining components. Add distilled/deionized water to 1.0L. Mix thoroughly.

Preparation of Medium: Prepare and dispense medium under 80% N_2 + 20% CO_2 gas mixture. Add components, except xylose solution, $Na_2S·9H_2O$ solution, Na-thiosulfate solution, and $NaHCO_3$ solution, to 926.0mL distilled/deionized water. Mix thoroughly. Sparge with 80% N_2 + 20% CO_2. Adjust pH to 7.2–7.4. Autoclave for 15 min at 15 psi pressure–121°C. Cool to 25°C while sparging with 80% N_2 + 20% CO_2. Aseptically and anaerobically add 20.0mL sterile xylose solution, 14.0mL $Na_2S·9H_2O$ solution, 20.0mL $NaHCO_3$ solution, and 20.0mL

Na-thiosulfate solution, . Mix thoroughly. Aseptically and anaerobically distribute into sterile tubes or flasks.

Use: For the cultivation of *Thermotoga hypogea.*

Thermotoga Medium

Composition per 1017.0mL:

NaCl	20.0g
Starch, soluble	5.0g
$NiCl_2 \cdot 6H_2O$	2.0g
KH_2PO_4	0.5g
$Na_2S \cdot 9H_2O$	0.5g
Yeast extract	0.5g
Resazurin	1.0mg
Artificial seawater	250.0mL
Trace elements solution	15.0mL

pH 6.5 ± 0.2 at 25°C

Artificial Seawater:

Composition per liter:

NaCl	27.7g
$MgSO_4 \cdot 7H_2O$	7.0g
$MgCl_2 \cdot 6H_2O$	5.5g
$CaCl_2 \cdot 2H_2O$	2.25g
KCl	0.65g
NaBr	0.1g
H_3BO_3	30.0mg
$SrCl_2 \cdot 6H_2O$	15.0mg
Citric acid	10.0mg
KI	0.05mg

Preparation of Artificial Seawater: Add components to distilled/deionized water and bring volume to 1.0L. Mix thoroughly.

Trace Elements Solution:

Composition per liter:

$MgSO_4 \cdot 7H_2O$	3.0g
Nitrilotriacetic acid	1.5g
NaCl	1.0g
$MnSO_4 \cdot 2H_2O$	0.5g
$CoSO_4 \cdot 7H_2O$	0.18g
$ZnSO_4 \cdot 7H_2O$	0.18g
$CaCl_2 \cdot 2H_2O$	0.1g
$FeSO_4 \cdot 7H_2O$	0.1g
$KAl(SO_4)_2 \cdot 12H_2O$	0.02g
$CuSO_4 \cdot 5H_2O$	0.01g
H_3BO_3	0.01g
$Na_2MoO_4 \cdot 2H_2O$	0.01g
$NiCl_2 \cdot 6H_2O$	0.025g
$Na_2SeO_3 \cdot 5H_2O$	0.3mg

Preparation of Trace Elements Solution: Add nitrilotriacetic acid to 500.0mL of distilled/deionized water. Adjust pH to 6.5 with KOH. Add remaining components. Add distilled/deionized water to 1.0L.

Preparation of Medium: Prepare and dispense medium under 100% N_2. Add components to distilled/deionized water and bring volume to 1.0L. Mix thoroughly. Sparge with 100% N_2. Distribute into tubes or flasks. Autoclave for 15 min at 15 psi pressure–121°C.

Use: For the cultivation and maintenance of *Thermotoga maritima* and *Thermotoga neapolitana.*

Thermotoga 2 Medium

Composition per 1015.0mL:

Starch, soluble	5.0g
NaCl	3.46g
$MgSO_4 \cdot 7H_2O$	0.88g
$EDTA \cdot Na_2$	0.768g
$MgCl_2 \cdot 6H_2O$	0.69g
KH_2PO_4	0.5g
$Na_2S \cdot 9H_2O$	0.5g
Yeast extract	0.5g
$CaCl_2 \cdot 2H_2O$	0.14g
KCl	0.08g
NaBr	12.5mg
H_3BO_3	3.75mg
$(NH_4)_2Ni(SO_4)_2$	3.0mg
$SrCl_2 \cdot 6H_2O$	1.9mg
Resazurin	1.0mg
KI	0.006mg
Trace elements solution	15.0mL

pH 7.0 ± 0.2 at 25°C

Trace Elements Solution:

Composition per liter:

$MgSO_4 \cdot 7H_2O$	3.0g
Nitrilotriacetic acid	1.5g
NaCl	1.0g
$MnSO_4 \cdot 2H_2O$	0.5g
$CoSO_4 \cdot 7H_2O$	0.18g
$ZnSO_4 \cdot 7H_2O$	0.18g
$CaCl_2 \cdot 2H_2O$	0.1g
$FeSO_4 \cdot 7H_2O$	0.1g
$KAl(SO_4)_2 \cdot 12H_2O$	0.02g
$CuSO_4 \cdot 5H_2O$	0.01g
H_3BO_3	0.01g
$Na_2MoO_4 \cdot 2H_2O$	0.01g
$NiCl_2 \cdot 6H_2O$	0.025g
$Na_2SeO_3 \cdot 5H_2O$	0.3mg

Preparation of Trace Elements Solution: Add nitrilotriacetic acid to 500.0mL of distilled/deionized water. Adjust pH to 6.5 with KOH. Add remaining components. Add distilled/deionized water to 1.0L.

Preparation of Medium: Prepare and dispense medium under 100% N_2. Add components to distilled/deionized water and bring volume to 1.0L. Mix thoroughly. Sparge with 100% N_2. Distribute into tubes or flasks. Autoclave for 15 min at 15 psi pressure–121°C.

Use: For the cultivation and maintenance of *Thermotoga thermarum.*

Thermotoga petrophila Medium (DSMZ Medium 913)

Composition per liter:

MOPS	5.0g
Yeast extract	2.0g
$(NH_4)_2SO_4$	1.0g
KH_2PO_4	0.5g
Resazurin	0.5mg
Artificial seawater	750.0mL
Trace elements solution	10.0mL
Vitamin solution	10.0mL
$Na_2S \cdot 9H_2O$ solution	10.0mL

pH 7.0 ± 0.2 at 25°C

Artificial Seawater:
Composition per liter:

NaCl	27.7g
$MgSO_4 \cdot 7H_2O$	7.0g
$MgCl_2 \cdot 6H_2O$	5.5g
KCl	0.65g
NaBr	0.1g
H_3BO_3	30.0mg
$SrCl_2 \cdot 6H_2O$	15.0mg
Citric acid	10.0mg
KI	0.05mg
$CaCl_2 \cdot 2H_2O$	2.25 g

Preparation of Artificial Seawater: Add components to distilled/deionized water and bring volume to 1.0L. Mix thoroughly.

Trace Elements Solution:
Composition per liter:

$MgSO_4 \cdot 7H_2O$	3.0g
Nitrilotriacetic acid	1.5g
NaCl	1.0g
$MnSO_4 \cdot 2H_2O$	0.5g
$CoSO_4 \cdot 7H_2O$	0.18g
$ZnSO_4 \cdot 7H_2O$	0.18g
$CaCl_2 \cdot 2H_2O$	0.1g
$FeSO_4 \cdot 7H_2O$	0.1g
$NiCl_2 \cdot 6H_2O$	0.025g
$KAl(SO_4)_2 \cdot 12H_2O$	0.02g
H_3BO_3	0.01g
$Na_2MoO_4 \cdot 4H_2O$	0.01g
$CuSO_4 \cdot 5H_2O$	0.01g
$Na_2SeO_3 \cdot 5H_2O$	0.3mg

Preparation of Trace Elements Solution: Add nitrilotriacetic acid to 500.0mL of distilled/deionized water. Dissolve by adjusting pH to 6.5 with KOH. Add remaining components. Add distilled/deionized water to 1.0L. Mix thoroughly.

Vitamin Solution:
Composition per liter:

Pyridoxine-HCl	10.0mg
Thiamine-HCl·$2H_2O$	5.0mg
Riboflavin	5.0mg
Nicotinic acid	5.0mg
D-Ca-pantothenate	5.0mg
p-Aminobenzoic acid	5.0mg
Lipoic acid	5.0mg
Biotin	2.0mg
Folic acid	2.0mg
Vitamin B_{12}	0.1mg

Preparation of Vitamin Solution: Add components to distilled/deionized water and bring volume to 1.0L. Mix thoroughly. Sparge with 80% H_2 + 20% CO_2. Filter sterilize.

$Na_2S \cdot 9H_2O$ Solution:
Composition per 10.0mL:

$Na_2S \cdot 9H_2O$	0.3g

Preparation of $Na_2S \cdot 9H_2O$ Solution: Add $Na_2S \cdot 9H_2O$ to distilled/deionized water and bring volume to 10.0mL. Sparge with N_2. Autoclave for 15 min at 15 psi pressure–121°C. Cool to 25°C. Store anaerobically.

Preparation of Medium: Add components, except vitamin solution and $Na_2S \cdot 9H_2O$ solution, to 750.0mL artificial seawater. Bring volume to 980.0mL with distilled/deionized water. Mix thoroughly. Gently heat and bring to boiling. Boil for 3 min. Cool to room temperature while sparging with 100% N_2. Distribute under 100% N_2 into anaerobe tubes or bottles. Autoclave for 15 min at 15 psi pressure–121°C. Aseptically and anaerobically add, per liter of medium, 10.0mL vitamin solution and 10.0mL $Na_2S \cdot 9H_2O$ solution. Mix thoroughly. The final pH of the medium should be 7.0.

Use: For the cultivation of *Thermotoga petrophila* and *Thermotoga naphthophila.*

Thermotoga subterranea Medium

Composition per liter:

NaCl	12.0g
$MgSO_4 \cdot 7H_2O$	0.5g
PIPES	3.4g
KCl	2.0g
Peptone	1.0g
$Na_2S \cdot 9H_2O$	0.5g
Yeast extract	0.5g
NH_4Cl	0.1g
$CaCl_2 \cdot 2H_2O$	0.025g
K_2HPO_4	0.02g
Resazurin	0.5mg
Wolfe's mineral solution	10.0mL
Wolfe's vitamin solution	10.0mL

pH 7.0 ± 0.2 at 25°C

Wolfe's Vitamin Solution:
Composition per liter:

Pyridoxine·HCl	10.0mg
p-Aminobenzoic acid	5.0mg
Lipoic acid	5.0mg
Nicotinic acid	5.0mg
Riboflavin	5.0mg
Thiamine·HCl	5.0mg
Calcium DL-pantothenate	5.0mg
Biotin	2.0mg
Folic acid	2.0mg
Vitamin B_{12}	0.1mg

Preparation of Wolfe's Vitamin Solution: Add components to distilled/deionized water and bring volume to 1.0L. Mix thoroughly. Filter sterilize.

Wolfe's Mineral Solution:
Composition per liter:

$MgSO_4 \cdot 7H_2O$	3.0g
Nitrilotriacetic acid	1.5g
NaCl	1.0g
$MnSO_4 \cdot 2H_2O$	0.5g
$CoCl_2 \cdot 6H_2O$	0.1g
$ZnSO_4 \cdot 7H_2O$	0.1g
$CaCl_2 \cdot 2H_2O$	0.1g
$FeSO_4 \cdot 7H_2O$	0.1g
$NiCl_2 \cdot 6H_2O$	0.025g
$KAl(SO_4)_2 \cdot 12H_2O$	0.02g
$CuSO_4 \cdot 5H_2O$	0.01g
H_3BO_3	0.01g
$Na_2MoO_4 \cdot 2H_2O$	0.01g
$Na_2SeO_3 \cdot 5H_2O$	0.3mg

Preparation of Wolfe's Mineral Solution: Add nitrilotriacetic acid to 500.0mL of distilled/deionized water. Adjust pH to 6.5 with

KOH. Add remaining components. Add distilled/deionized water to 1.0L. Adjust pH to 6.8.

Preparation of Medium: Prepare and dispense medium under 100% N_2. Add components, except Wolfe's vitamin solution, to distilled/deionized water and bring volume to 990.0mL. Mix thoroughly. Adjust pH to 7.0. Sparge with 100% N_2. Autoclave for 15 min at 15 psi pressure–121°C. Aseptically and anaerobically add 10.0mL of sterile Wolfe's vitamin solution. Mix thoroughly. Adjust medium pH to 7.5 by adding sterile anaerobic 1*N* NaOH. Aseptically and anaerobically distribute into sterile tubes or bottles.

Use: For the cultivation of *Thermotoga subterranea.*

Thermovenabulum Medium (DSMZ Medium 962)

Composition per liter:

$NaHCO_3$	0.7g
NH_4Cl	0.33g
KH_2PO_4	0.33g
$MgCl_2·6H_2O$	0.33g
$CaCl_2·2H_2O$	0.33g
KCl	0.33g
Yeast extract	0.05g
Meat extract solution	50.0mL
Sodium fumarate solution	50.0mL
Vitamin solution	10.0mL
Trace elements solution	10.0mL
Selenite-tungstate solution	1.0mL

pH 7.0 ± 0.2 at 25°C

Selenite-Tungstate Solution

Composition per liter:

NaOH	0.5g
$Na_2WO_4·2H_2O$	4.0mg
$Na_2SeO_3·5H_2O$	3.0mg

Preparation of Selenite-Tungstate Solution: Add components to distilled/deionized water and bring volume to 1.0L. Mix thoroughly. Sparge with 100% N_2. Filter sterilize.

Trace Elements Solution:

Composition per liter:

$MgSO_4·7H_2O$	3.0g
Nitrilotriacetic acid	1.5g
NaCl	1.0g
$MnSO_4·2H_2O$	0.5g
$CoSO_4·7H_2O$	0.18g
$ZnSO_4·7H_2O$	0.18g
$CaCl_2·2H_2O$	0.1g
$FeSO_4·7H_2O$	0.1g
$NiCl_2·6H_2O$	0.025g
$KAl(SO_4)_2·12H_2O$	0.02g
H_3BO_3	0.01g
$Na_2MoO_4·4H_2O$	0.01g
$CuSO_4·5H_2O$	0.01g
$Na_2SeO_3·5H_2O$	0.3mg

Preparation of Trace Elements Solution: Add nitrilotriacetic acid to 500.0mL of distilled/deionized water. Dissolve by adjusting pH to 6.5 with KOH. Add remaining components. Add distilled/deionized water to 1.0L. Mix thoroughly.

Vitamin Solution:

Composition per liter:

Pyridoxine-HCl	10.0mg
Thiamine-HCl·$2H_2O$	5.0mg
Riboflavin	5.0mg
Nicotinic acid	5.0mg
D-Ca-pantothenate	5.0mg
p-Aminobenzoic acid	5.0mg
Lipoic acid	5.0mg
Biotin	2.0mg
Folic acid	2.0mg
Vitamin B_{12}	0.1mg

Preparation of Vitamin Solution: Add components to distilled/deionized water and bring volume to 1.0L. Mix thoroughly. Sparge with 80% H_2 + 20% CO_2. Filter sterilize.

Sodium Fumarate Solution:

Composition per 50.0mL:

Na_2-fumarate	3.20g

Preparation of Sodium Fumarate Solution: Add Na_2-fumarate to distilled/deionized water and bring volume to 50.0mL. Sparge with N_2. Autoclave for 15 min at 15 psi pressure–121°C. Cool to 25°C.

Meat Extract Solution:

Composition per 50.0mL:

Meat extract	3.0g

Preparation of Meat Extract Solution: Add meat extract to distilled/deionized water and bring volume to 50.0mL. Sparge with N_2. Autoclave for 15 min at 15 psi pressure–121°C. Cool to 25°C.

Preparation of Medium: Add components, except $NaHCO_3$, vitamin solution, sodium fumarate solution, and meat extract solution, to distilled/deionized water and bring volume to 890.0mL. Mix thoroughly. Gently heat and bring to boiling. Boil for 3 min. Cool to 25°C while sparging with 80% N_2 + 20% CO_2. Add solid $NaHCO_3$. Adjust pH to 6.8–7.0. Distribute to anaerobe tubes or bottles under 80% N_2 + 20% CO_2. Autoclave for 15 min at 15 psi pressure–121°C. Aseptically and anaerobically add, per liter of medium, 10.0mL sterile vitamin solution, 50.0mL sterile meat extract solution, and 50.0mL sterile sodium fumarate solution. Mix thoroughly. The final pH should be 6.5.

Use: For the cultivation of *Thermovenabulum ferriorganovorum.*

Thermovibrio Medium (DSMZ Medium 961)

Composition per liter:

Synthetic seawater	970.0mL
Potassium nitrate solution	10.0mL
Yeast extract solution	10.0mL
$Na_2S·9H_2O$ solution	10.0mL

pH 6.5 ± 0.2 at 25°C

Synthetic Seawater:

Composition per liter:

NaCl	27.7g
$MgSO_4·7H_2O$	7.0g
$MgCl_2·6H_2O$	5.5g
$CaCl_2·2H_2O$	0.75g
KCl	0.65g
KH_2PO_4	0.5g
$(NH_4)_2SO_4$	0.25g
NaBr	0.1g
H_3BO_3	0.03g

$SrCl_2{\cdot}6H_2O$	15.0mg
Resazurin	1.0mg
KI	0.05mg

Preparation of Synthetic Seawater: Add components to distilled water and bring volume to 1.0L. Mix thoroughly.

$Na_2S{\cdot}9H_2O$ Solution:

Composition per 10.0mL:

$Na_2S{\cdot}9H_2O$	0.5g

Preparation of $Na_2S{\cdot}9H_2O$ Solution: Add $Na_2S{\cdot}9H_2O$ to distilled/deionized water and bring volume to 10.0mL. Mix thoroughly. Autoclave under 100% N_2 for 15 min at 15 psi pressure–121°C. Cool to room temperature.

Yeast Extract Solution:

Composition per 10.0mL:

Yeast extract	1.0g

Preparation of Yeast Extract Solution: Add yeast extract to distilled/deionized water and bring volume to 10.0mL. Mix thoroughly. Autoclave under 100% N_2 for 15 min at 15 psi pressure–121°C. Cool to room temperature.

Potassium Nitrate Solution:

Composition per 10.0mL:

KNO_3	0.33g

Preparation of Potassium Nitrate Solution: Add KNO_3 to distilled/deionized water and bring volume to 10.0mL. Mix thoroughly. Autoclave for 15 min at 15 psi pressure–121°C. Cool to room temperature.

Preparation of Medium: Gently heat and bring 1.0L synthetic seawater to boiling. Boil for 3 min. Cool to 25°C while sparging with 80% H_2 + 20% CO_2. Adjust pH to 6.8–7.0. Distribute 20.0mL aliquots into 100mL serum bottles under 80% H_2 + 20% CO_2. Autoclave for 15 min at 15 psi pressure–121°C. Aseptically and anaerobically add per 20.0mL of medium, 0.2mL sterile potassium nitrate solution, 0.2mL sterile yeast extract solution, and 0.2mL sterile $Na_2S{\cdot}9H_2O$ solution. Mix thoroughly. Adjusted pH should be 6.5.

Use: For the cultivation of *Thermovibrio ruber.*

Thermus Agar

Composition per liter:

Agar	28.0g
Pancreatic digest of casein	2.5g
Yeast extract	2.5g
$Na_2HPO_4{\cdot}12H_2O$	0.43g
$MgCl_2{\cdot}6H_2O$	0.2g
Nitrilotriacetic acid	0.1g
KH_2PO_4	54.0mg
$CaSO_4{\cdot}2H_2O$	40.0mg
Micronutrient solution	1.0mL
Fe-citrate solution	0.5mL

pH 7.2 ± 0.2 at 25°C

Micronutrient Solution:

Composition per liter:

$MnSO_4{\cdot}H_2O$	2.28g
H_3BO_3	0.5g
$ZnSO_4{\cdot}7H_2O$	0.5g
$CoCl_2{\cdot}6H_2O$	45.0mg
$CuSO_4{\cdot}5H_2O$	25.0mg
$Na_2MoO_4{\cdot}2H_2O$	25.0mg
Concentrated H_2SO_4	0.5mL

Preparation of Micronutrient Solution: Add components to distilled/deionized water and bring volume to 1.0L. Mix thoroughly.

Fe-Citrate Solution:

Composition per liter:

Fe-citrate	24.5mg

Preparation of Fe-Citrate Solution: Add Fe-citrate to distilled/deionized water and bring volume to 1.0L. Mix thoroughly.

Preparation of Medium: Add nitrilotriacetic acid to 100.0mL of distilled/deionized water. Mix thoroughly. Adjust pH to 6.5 with KOH. Add remaining components and bring volume to 1.0L with distilled/deionized water. Mix thoroughly. Adjust pH to 7.2. Gently heat and bring to boiling. Distribute into tubes or flasks. Autoclave for 15 min at 15 psi pressure–121°C. Pour into sterile Petri dishes or leave in tubes.

Use: For the cultivation and maintenance of *Thermus* species.

Thermus BP Medium (*Thermus* Beef Extract Polypeptone™ Medium)

Composition per liter:

Agar	25.0g
Beef extract	4.0g
Polypeptone™	4.0g
K_2HPO_4	3.0g
KH_2PO_4	1.0g

pH 7.0 ± 0.2 at 25°C

Preparation of Medium: Add components to distilled/deionized water and bring volume to 1.0L. Mix thoroughly. Gently heat and bring to boiling. Distribute into tubes or flasks. Autoclave for 15 min at 15 psi pressure–121°C. Pour into sterile Petri dishes or leave in tubes.

Use: For the cultivation and maintenance of *Thermus aquaticus* and other *Thermus* species.

Thermus brockii Medium

Composition per liter:

Pancreatic digest of casein	1.0g
Yeast extract	1.0g
Salts solution	100.0mL

pH 7.6 ± 0.2 at 25°C

Salts Solution:

Composition per liter:

$NaNO_3$	6.89g
KNO_3	1.03g
$MgSO_4{\cdot}7H_2O$	1.0g
Nitrilotriacetic acid	1.0g
$CaSO_4{\cdot}2H_2O$	0.6g
NaCl	80.0mg
$FeCl_3$ solution	10.0mL
Trace elements solution	10.0mL

Preparation of Salts Solution: Add nitrilotriacetic acid to 500.0mL of distilled/deionized water. Adjust pH to 8.2 with 1*M* NaOH. Add remaining components. Add distilled/deionized water to 1.0L.

$FeCl_3$ Solution:

Composition per 100.0mL:

$FeCl_3$	28.0mg

Preparation of $FeCl_3$ Solution: Add $FeCl_3$ to distilled/deionized water and bring volume to 100.0mL. Mix thoroughly.

Trace Elements Solution:

Composition per liter:

$MnSO_4 \cdot H_2O$	2.2g
H_3BO_3	0.5g
$ZnSO_4 \cdot 7H_2O$	0.5g
$CoCl_2 \cdot 6H_2O$	46.0mg
$Na_2MoO_4 \cdot 2H_2O$	25.0mg
$CuSO_4$	16.0mg
H_2SO_4	0.5mL

Preparation of Trace Elements Solution: Add components to distilled/deionized water and bring volume to 1.0L. Mix thoroughly.

Preparation of Medium: Add components to distilled/deionized water and bring volume to 1.0L. Mix thoroughly. Distribute into tubes or flasks. Autoclave for 15 min at 15 psi pressure–121°C.

Use: For the cultivation of *Thermus brockii*.

Thermus Broth

Composition per liter:

Pancreatic digest of casein	2.5g
Yeast extract	2.5g
$Na_2HPO_4 \cdot 12H_2O$	0.43g
$MgCl_2 \cdot 6H_2O$	0.2g
Nitrilotriacetic acid	0.1g
KH_2PO_4	54.0mg
$CaSO_4 \cdot 2H_2O$	40.0mg
Micronutrient solution	1.0mL
Fe-citrate solution	0.5mL

pH 7.2 ± 0.2 at 25°C

Micronutrient Solution:

Composition per liter:

$MnSO_4 \cdot H_2O$	2.28g
H_3BO_3	0.5g
$ZnSO_4 \cdot 7H_2O$	0.5g
$CoCl_2 \cdot 6H_2O$	45.0mg
$CuSO_4 \cdot 5H_2O$	25.0mg
$Na_2MoO_4 \cdot 2H_2O$	25.0mg
H_2SO_4, concentrated	0.5mL

Preparation of Micronutrient Solution: Add components to distilled/deionized water and bring volume to 1.0L. Mix thoroughly.

Fe-Citrate Solution:

Composition per liter:

Fe-citrate	24.5mg

Preparation of Fe-Citrate Solution: Add Fe-citrate to distilled/deionized water and bring volume to 1.0L. Mix thoroughly.

Preparation of Medium: Add nitrilotriacetic acid to 100.0mL of distilled/deionized water. Mix thoroughly. Adjust pH to 6.5 with KOH. Add remaining components and bring volume to 1.0L with distilled/deionized water. Mix thoroughly. Adjust pH to 7.2. Distribute into tubes or flasks. Autoclave for 15 min at 15 psi pressure–121°C.

Use: For the cultivation and maintenance of *Thermus* species.

Thermus Enhanced Agar

Composition per liter:

Agar	28.0g
Pancreatic digest of casein	2.5g
Yeast extract	2.5g
$MgCl_2 \cdot 6H_2O$	0.2g
Nitrilotriacetic acid	0.1g
$CaSO_4 \cdot 2H_2O$	40.0mg
Phosphate buffer solution	100.0mL
Ferric citrate solution	0.5mL
Trace elements solution	0.5mL

Phosphate Buffer Solution:

Composition per liter:

$Na_2HPO_4 \cdot 12H_2O$	43.0g
KH_2PO_4	5.44g

Preparation of Phosphate Buffer Solution: Add components to distilled/deionized water and bring volume to 1.0L. Mix thoroughly. Adjust pH to 7.2. Autoclave for 15 min at 15 psi pressure–121°C. Cool to 50°–55°C.

Ferric Citrate Solution:

Composition per 10.0mL:

Ferric citrate	24.5mg

Preparation of Ferric Citrate Solution: Add ferric citrate to distilled/deionized water and bring volume to 10.0mL. Mix thoroughly.

Trace Elements Solution:

Composition per liter:

Nitrilotriacetic acid	12.8g
$FeCl_2 \cdot 4H_2O$	1.0g
$MnCl_2 \cdot 4H_2O$	0.5g
$CoCl_2 \cdot 6H_2O$	0.3g
$CuCl_2 \cdot 2H_2O$	50.0mg
$Na_2MoO_4 \cdot 2H_2O$	50.0mg
H_3BO_3	20.0mg
$NiCl_2 \cdot 6H_2O$	20.0mg

Preparation of Trace Elements Solution: Add nitrilotriacetic acid to 500.0mL of distilled/deionized water. Adjust pH to 6.5 with KOH. Add remaining components. Add distilled/deionized water to 1.0L.

Preparation of Medium: Add components, except phosphate buffer solution, to distilled/deionized water and bring volume to 900.0mL. Mix thoroughly. Gently heat and bring to boiling. Adjust pH to 7.2 with NaOH. Autoclave for 15 min at 15 psi pressure–121°C. Cool to 50°–55°C. Aseptically add 100.0mL of sterile phosphate buffer solution. Mix thoroughly. Pour into sterile Petri dishes or distribute into sterile tubes.

Use: For the cultivation of *Thermus* species.

Thermus Enhanced Agar with 1% Sodium Chloride

Composition per liter:

Agar	28.0g
NaCl	10.0g
Pancreatic digest of casein	2.5g
Yeast extract	2.5g
$MgCl_2 \cdot 6H_2O$	0.2g
Nitrilotriacetic acid	0.1g
$CaSO_4 \cdot 2H_2O$	40.0mg
Phosphate buffer solution	100.0mL
Ferric citrate solution	0.5mL
Trace elements solution	0.5mL

Phosphate Buffer Solution:

Composition per liter:

$Na_2HPO_4 \cdot 12H_2O$	43.0g
KH_2PO_4	5.44g

Preparation of Phosphate Buffer Solution: Add components to distilled/deionized water and bring volume to 1.0L. Mix thoroughly. Adjust pH to 7.2. Autoclave for 15 min at 15 psi pressure–121°C. Cool to 50–55°C.

Ferric Citrate Solution:

Composition per 10.0mL:

Ferric citrate	24.5mg

Preparation of Ferric Citrate Solution: Add ferric citrate to distilled/deionized water and bring volume to 10.0mL. Mix thoroughly.

Trace Elements Solution:

Composition per liter:

Nitrilotriacetic acid	12.8g
$FeCl_2 \cdot 4H_2O$	1.0g
$MnCl_2 \cdot 4H_2O$	0.5g
$CoCl_2 \cdot 6H_2O$	0.3g
$CuCl_2 \cdot 2H_2O$	50.0mg
$Na_2MoO_4 \cdot 2H_2O$	50.0mg
H_3BO_3	20.0mg
$NiCl_2 \cdot 6H_2O$	20.0mg

Preparation of Trace Elements Solution: Add nitrilotriacetic acid to 500.0mL of distilled/deionized water. Adjust pH to 6.5 with KOH. Add remaining components. Add distilled/deionized water to 1.0L.

Preparation of Medium: Add components, except phosphate buffer solution, to distilled/deionized water and bring volume to 900.0mL. Mix thoroughly. Gently heat and bring to boiling. Adjust pH to 7.2 with NaOH. Autoclave for 15 min at 15 psi pressure–121°C. Cool to 50°–55°C. Aseptically add 100.0mL of sterile phosphate buffer solution. Mix thoroughly. Pour into sterile Petri dishes or distribute into sterile tubes.

Use: For the cultivation of *Thermus* species.

Thermus Medium

Composition per liter:

Agar	30.0g
Polypeptone™	8.0g
Yeast extract	4.0g
NaCl	2.0g

pH 7.5 ± 0.2 at 25°C

Preparation of Medium: Add components to distilled/deionized water and bring volume to 1.0L. Mix thoroughly. Gently heat and bring to boiling. Distribute into tubes or flasks. Autoclave for 15 min at 15 psi pressure–121°C. Pour into sterile Petri dishes or leave in tubes.

Use: For the cultivation and maintenance of *Thermus aquaticus* and other *Thermus* species.

Thermus Medium (DSMZ Medium 1033)

Composition per liter:

Tryptone	1.0g
Yeast extract	1.0g
Mineral solution 1	100.0mL
Mineral solution 2	10.0mL
Ferric chloride solution	10.0mL

pH 8.2 ± 0.2 at 25°C

Mineral Solution 1:

Composition per liter:

$NaNO_3$	6.9g
KNO_3	1.8g
Na_2HPO_4	1.11g
$MgSO_4 \cdot 7H_2O$	1.0g
Nitrilotriacetic acid	1.0g
$CaSO_4 \cdot 2H_2O$	0.6g
NaCl	0.08g

Preparation of Mineral Solution 1: Add nitrilotriacetic acid to 500.0mL of distilled/deionized water. Dissolve by adjusting pH to 6.5 with KOH. Add remaining components. Add distilled/deionized water to and bring volume to 1.0L. Mix thoroughly.

Mineral Solution 2:

Composition per liter:

$MnSO_4 \cdot 2H_2O$	0.22g
$ZnSO_4 \cdot 7H_2O$	0.05g
H_3BO_3	0.05g
$CuSO_4 \cdot 5H_2O$	2.5mg

Preparation of Mineral Solution 2: Add components to distilled/deionized water to 1.0L. Mix thoroughly.

Ferric Chloride Solution:

Composition per 100.0mL:

$FeCl_3 \cdot 6H_2O$	4.6mg

Preparation of Ferric Chloride Solution: Add $FeCl_3 \cdot 6H_2O$ to distilled/deionized water and bring volume to 100.0mL. Mix thoroughly.

Preparation of Medium: Add components to distilled/deionized water and bring volume to 1.0L. Mix thoroughly. Adjust pH to 8.2 with 1*M* NaOH. Dispense into tubes or flasks. Autoclave for 15 min at 15 psi pressure–121°C.

Use: For the cultivation of *Thermus* spp.

Thermus 162 Medium (DSMZ Medium 878)

Composition per liter:

Agar	28.0g
Yeast extract	1.0g
Tryptone	1.0g
$MgCl_2 \cdot 6H_2O$	200.0mg
Nitrilotriacetic acid	100.0mg
$CaSO_4 \cdot 2H_2O$	40.0mg
Phosphate buffer	100.0mL
Ferric citrate solution	0.5mL
Trace elements solution	0.5mL

pH 7.2 ± 0.2 at 25°C

Ferric Citrate Solution:

Composition per 10.0mL:

Ferric citrate	24.5mg

Preparation of Ferric Citrate Solution: Add ferric citrate to distilled/deionized water and bring volume to 10.0mL. Mix thoroughly.

Trace Elements Solution:

Composition per liter:

$CoCl_2 \cdot 6H_2O$	45.0g
$CuSO_4 \cdot 5H_2O$	25.0g
$Na_2MoO_4 \cdot 4H_2O$	25.0g
$MnSO_4 \cdot 2H_2O$	2.28g

$ZnSO_4 \cdot 7H_2O$ 0.5g
H_3BO_3 .... 0.5g
H_2SO_4 .... 0.5mL

Preparation of Trace Elements Solution: Add components to distilled/deionized water and bring volume to 1.0L. Mix thoroughly.

Phosphate Buffer:

Composition per liter:

$Na_2HPO_4 \cdot 12H_2O$ 43.0g
KH_2PO_4 .... 5.44g

Preparation of Phosphate Buffer: Add components to distilled/deionized water and bring volume to 1.0L. Mix thoroughly. Adjust pH to 7.2. Autoclave for 15 min at 15 psi pressure–121°C. Cool to 50°C.

Preparation of Medium: Add components, except phosphate buffer, to distilled/deionized water and bring volume to 900.0mL. Mix thoroughly. Gently heat and bring to boiling. Autoclave for 15 min at 15 psi pressure–121°C. Cool to 50°C. Add 100.0mL warm phosphate buffer. Mix thoroughly. Pour into Petri dishes or distribute into sterile tubes.

Use: For the cultivation and maintenance of *Thermus* spp., *Rubrobacter xylanophilus, Thermonema rossianum, Deinococcus geothermalis, Deinococcus murrayi,* and *Tepidimonas ignava.*

Thermus 162 Medium

Composition per 1010.0mL:

Agar 28.0g
Tryptone 2.5g
Yeast extract 2.5g
$MgCl_2 \cdot 6H_2O$ 0.2g
Nitrilotriacetic acid 0.1g
$CaSO_4 \cdot 2H_2O$ 40.0mg
Phosphate buffer solution 100.0mL
Ferric citrate solution 0.5mL
Trace elements solution 0.5mL

pH 7.2 ± 0.2 at 25°C

Phosphate Buffer Solution:

Composition per liter:

$Na_2HPO_4 \cdot 12H_2O$ 43.0g
KH_2PO_4 .... 5.44g

Preparation of Phosphate Buffer Solution: Add components to distilled/deionized water and bring volume to 1.0L. Mix thoroughly. Adjust pH to 7.2. Autoclave for 15 min at 15 psi pressure–121°C.

Ferric Citrate Solution:

Composition per 10.0mL:

Ferric citrate 24.5mg

Preparation of Ferric Citrate Solution: Add ferric citrate to distilled/deionized water and bring volume to 10.0mL. Mix thoroughly.

Trace Elements Solution:

Composition per liter:

Nitrilotriacetic acid 12.8g
$FeCl_2 \cdot 4H_2O$ 1.0g
$MnCl_2 \cdot 4H_2O$ 0.5g
$CoCl_2 \cdot 4H_2O$ 0.3g
$CuCl_2 \cdot 2H_2O$ 50.0mg
$Na_2MoO_4 \cdot 2H_2O$ 50.0mg
H_3BO_3 .... 20.0mg
$NiCl_2 \cdot 6H_2O$ 20.0mg

Preparation of Trace Elements Solution: Add nitrilotriacetic acid to 500.0mL of distilled/deionized water. Adjust pH to 6.5 with KOH. Add remaining components. Add distilled/deionized water to 1.0L. Adjust pH to 7.0.

Preparation of Medium: Add components, except phosphate buffer solution, to distilled/deionized water and bring volume to 900.0mL. Mix thoroughly. Gently heat and bring to boiling. Adjust pH to 7.2 with NaOH. Autoclave for 15 min at 15 psi pressure–121°C. Cool to 50°–55°C. Aseptically add 100.0mL of sterile phosphate buffer solution. Mix thoroughly. Pour into sterile Petri dishes or distribute into sterile tubes.

Use: For the cultivation of *Thermus* species.

Thermus Medium Enhanced

Composition per liter:

Agar 28.0g
Yeast extract 2.5g
Tryptone 2.5g
$MgCl_2 \cdot 6H_2O$ 0.2g
Nitrilotriacetic acid 0.1g
$CaSO_4 \cdot 2H_2O$ 0.04g
Phosphate buffer solution 100.0mL
Ferric citrate solution 0.5mL
Trace elements solution 0.5mL

Ferric Citrate Solution:

Composition per 100.0mL:

Ferric citrate 0.24g

Preparation of Ferric Citrate Solution: Add ferric citrate to distilled/deionized water and bring volume to 100.0mL. Mix thoroughly.

Phosphate Buffer Solution:

Composition per liter:

$Na_2HPO_4 \cdot 12H_2O$ 43.0g
KH_2PO_4 .... 5.44g

Preparation of Phosphate Buffer Solution: Add components to distilled/deionized water and bring volume to 1.0L. Mix thoroughly. Adjust pH to 7.2. Autoclave for 15 min at 15 psi pressure–121°C.

Trace Elements Solution:

Composition per liter:

Nitrilotriacetic acid 12.8g
$FeCl_3 \cdot 4H_2O$ 1.0g
$MnCl_2 \cdot 4H_2O$ 0.5g
$CoCl_2 \cdot 6H_2O$ 0.3g
$CuCl_2 \cdot 2H_2O$ 0.05g
$Na_2MoO_4 \cdot 2H_2O$ 0.05g
H_3BO_3 .... 0.02g
$NiCl_2 \cdot 6H_2O$ 0.02g

Preparation of Trace Elements Solution: Add components to distilled/deionized water and bring volume to 1.0L. Mix thoroughly.

Preparation of Medium: Add components, except phosphate buffer solution, to distilled/deionized water and bring volume to 990.0mL. Mix thoroughly. Gently heat and bring to boiling. Autoclave for 15 min at 15 psi pressure–121°C. Aseptically add 10.0mL of sterile phosphate buffer solution. Mix thoroughly. Pour into sterile Petri dishes or distribute into sterile tubes.

Use: For the cultivation and maintenance of *Thermus* species.

Thermus Medium Enhanced with 1% NaCl

Composition per liter:

Agar	28.0g
NaCl	10.0g
Yeast extract	2.5g
Tryptone	2.5g
$MgCl_2 \cdot 6H_2O$	0.2g
Nitrilotriacetic acid	0.1g
$CaSO_4 \cdot 2H_2O$	0.04g
Phosphate buffer solution	100.0mL
Ferric citrate solution	0.5mL
Trace elements solution	0.5mL

Ferric Citrate Solution:

Composition per 100.0mL:

Ferric citrate	0.24g

Preparation of Ferric Citrate Solution: Add ferric citrate to distilled/deionized water and bring volume to 100.0mL. Mix thoroughly.

Phosphate Buffer Solution:

Composition per liter:

$Na_2HPO_4 \cdot 12H_2O$	43.0g
KH_2PO_4	5.44g

Preparation of Phosphate Buffer Solution: Add components to distilled/deionized water and bring volume to 1.0L. Mix thoroughly. Adjust pH to 7.2. Autoclave for 15 min at 15 psi pressure–121°C.

Trace Elements Solution:

Composition per liter:

Nitrilotriacetic acid	12.8g
$FeCl_3 \cdot 4H_2O$	1.0g
$MnCl_2 \cdot 4H_2O$	0.5g
$CoCl_2 \cdot 6H_2O$	0.3g
$CuCl_2 \cdot 2H_2O$	0.05g
$Na_2MoO_4 \cdot 2H_2O$	0.05g
H_3BO_3	0.02g
$NiCl_2 \cdot 6H_2O$	0.02g

Preparation of Trace Elements Solution: Add components to distilled/deionized water and bring volume to 1.0L. Mix thoroughly.

Preparation of Medium: Add components, except phosphate buffer solution, to distilled/deionized water and bring volume to 990.0mL. Mix thoroughly. Gently heat and bring to boiling. Autoclave for 15 min at 15 psi pressure–121°C. Aseptically add 10.0mL of sterile phosphate buffer solution. Mix thoroughly. Pour into sterile Petri dishes or distribute into sterile tubes.

Use: For the cultivation and maintenance of *Rhodothermus marinus.*

Thermus Peptone Meat Extract Yeast Extract Agar

See: ***Thermus*** **PMY Agar**

Thermus Peptone Meat Extract Yeast Extract Broth

See: ***Thermus*** **PMY Broth**

Thermus PMY Agar (*Thermus* Peptone Meat Extract Yeast Extract Agar)

Composition per liter:

Agar	15.0g
Peptone	5.0g
Meat extract	3.5g
Yeast extract	1.5g
NaCl	1.5g

pH 7.0 ± 0.2 at 25°C

Preparation of Medium: Add components to distilled/deionized water and bring volume to 1.0L. Mix thoroughly. Gently heat and bring to boiling. Distribute into tubes or flasks. Autoclave for 15 min at 15 psi pressure–121°C. Pour into sterile Petri dishes or leave in tubes.

Use: For the cultivation and maintenance of *Thermus aquaticus* and other *Thermus* species.

Thermus PMY Broth (*Thermus* Peptone Meat Extract Yeast Extract Broth)

Composition per liter:

Peptone	5.0g
Meat extract	3.5g
Agar	3.0g
Yeast extract	1.5g
NaCl	1.5g

pH 7.0 ± 0.2 at 25°C

Preparation of Medium: Add components to distilled/deionized water and bring volume to 1.0L. Mix thoroughly. Distribute into tubes or flasks. Autoclave for 15 min at 15 psi pressure–121°C.

Use: For the cultivation and maintenance of *Thermus aquaticus* and other *Thermus* species.

Thermus ruber Medium

Composition per liter:

Agar	12.0g
Universal peptone	5.0g
Starch, soluble	1.0g
Yeast extract	1.0g

pH 8.0 ± 0.2 at 25°C

Source: Universal peptone is available from Merck, Sharpe, and Dohme.

Preparation of Medium: Add components to distilled/deionized water and bring volume to 1.0L. Mix thoroughly. Gently heat and bring to boiling. Distribute into tubes or flasks. Autoclave for 15 min at 15 psi pressure–121°C. Pour into sterile Petri dishes or leave in tubes.

Use: For the cultivation and maintenance of *Thermus ruber.*

Thermus sp. Medium (DSMZ Medium 1045)

Composition per liter:

Peptone	8.0g
Yeast extract	4.0g
NaCl	2.0g

pH 7.0 ± 0.2 at 25°C

Preparation of Medium: Add components to distilled/deionized water and bring volume to 1.0L. Mix thoroughly. Adjust pH to 7.0. Distribute into tubes or flasks. Autoclave for 15 min at 15 psi pressure–121°C.

Use: For the cultivation of *Thermus* spp.

Thermus thermophilus Medium (DSMZ Medium 74)

Polypeptone™	8.0g
Yeast extract	4.0g
NaCl	2.0g

pH 7.0 ± 0.2 at 25°C

Preparation of Medium: Add components to distilled/deionized water and bring volume to 1.0L. Mix thoroughly. Adjust pH to 7.0. Distribute into tubes or flasks. Autoclave for 15 min at 15 psi pressure–121°C.

Use: For the cultivation and maintenance of *Thermus thermophilus.*

Thiamine Assay Medium

Composition per liter:

Glucose	40.0g
Peptone	22.0g
Sodium acetate	15.0g
Vitamin assay casamino acids	5.0g
K_2HPO_4	1.0g
KH_2PO_4	1.0g
$MgSO_4 \cdot 7H_2O$	0.4g
Adenine sulfate	0.02g
$FeSO_4 \cdot 7H_2O$	0.02g
Guanine·HCl	0.02g
$MnSO_4 \cdot 5H_2O$	0.02g
NaCl	0.02g
Uracil	0.02g
L-Cystine	0.2mg
p-Aminobenzoic acid	0.2mg
Calcium pantothenate	0.2mg
Niacin	0.2mg
Pyridoxine·HCl	0.2mg
Riboflavin	0.2mg
Folic acid	0.5μg
Biotin	0.8μg

pH 6.5 ± 0.2 at 25°C

Source: This medium is available as a premixed powder from BD Diagnostic Systems.

Preparation of Medium: Add components to distilled/deionized water and bring volume to 1.0L. Mix thoroughly. Gently heat and bring to boiling. Continue boiling for 2–3 min. Distribute into tubes or flasks in 5.0mL volumes while swirling the flask to disperse the precipitate. Add standard solutions or test solutions and bring volume of each tube to 10.0mL with distilled/deionized water. Autoclave for 15 min at 15 psi pressure–121°C.

Use: For the microbiological assaying of thiamine using *Lactobacillus fermentum* as the test organism.

Thiamine Assay Medium LV

Composition per liter:

Glucose	20.0g
Pancreatic digest of casein	20.0g
K_2HPO_4	10.0g
NaCl	10.0g
Sodium citrate	10.0g
Yeast extract, thiamine-free	10.0g
Sorbitan monooleate complex	2.0g
$MgSO_4 \cdot 7H_2O$	1.6g
$MnCl_2 \cdot 4H_2O$	0.28g
$FeSO_4 \cdot 7H_2O$	0.08g

pH 6.0 ± 0.2 at 25°C

Source: This medium is available as a premixed powder from BD Diagnostic Systems.

Preparation of Medium: Add components to distilled/deionized water and bring volume to 1.0L. Mix thoroughly. Gently heat and bring to boiling. Continue boiling for 2–3 min. Distribute into tubes or flasks in 5.0mL volumes while swirling the flask to disperse the precipitate. Add standard solutions or test solutions and bring volume of each tube to 10.0mL with distilled/deionized water. Autoclave for 15 min at 15 psi pressure–121°C.

Use: For the microbiological assaying of thiamine using *Lactobacillus viridescens* as the test organism.

Thiamine Salts Medium

Composition per liter:

KH_2PO_4	1.0g
$FeSO_4 \cdot 7H_2O$	0.05g
$MgSO_4 \cdot 7H_2O$	0.02g
$CaCl_2$	0.02g
$MnCl_2 \cdot 4H_2O$	1.0mg
$Na_2MoO_4 \cdot 2H_2O$	1.0mg
Thiamine·HCl solution	10.0mL

pH 7.0 ± 0.2 at 25°C

Thiamine·HCl Solution:

Composition per 10.0mL:

Thiamine·HCl	3.0g

Preparation of Thiamine·HCl Solution: Add thiamine·HCl to distilled/deionized water and bring volume to 10.0mL. Mix thoroughly. Filter sterilize.

Preparation of Medium: Add components, except thiamine·HCl solution, to distilled/deionized water and bring volume to 990.0mL. Mix thoroughly. Adjust pH to 7.0 with KOH. Autoclave for 20 min at 15 psi pressure–121°C. Cool to 45°–50°C. Aseptically add the sterile thiamine·HCl solution. Mix thoroughly. Distribute into sterile tubes or flasks.

Use: For the cultivation of ATCC strain 25589.

Thiazole Medium

Composition per 110.0mL:

Solution A	100.0mL
Solution B	5.0mL
Solution C	5.0mL

pH 7.0 ± 0.2 at 25°C

Solution A:

Composition per 100.0mL:

Benzothiazole	30.0mg
NaOH (10% solution)	90.0mL

Preparation of Solution A: Combine components. Mix thoroughly. Adjust pH to 7.0 with concentrated HCl. Autoclave for 15 min at 15 psi pressure–121°C.

Solution B:

Composition per 50.0mL:

KH_2PO_4	0.3g

Preparation of Solution B: Add KH_2PO_4 to distilled/deionized water and bring volume to 50.0mL. Mix thoroughly. Autoclave for 15 min at 15 psi pressure–121°C.

Solution C:

Composition per 50.0mL:

$MgSO_4 \cdot 7H_2O$	50.0mg
$FeCl_3$	10.0mg
$CaCl_2 \cdot 2H_2O$	6.0mg

Preparation of Solution C: Add components to distilled/deionized water and bring volume to 50.0mL. Mix thoroughly. Autoclave for 15 min at 15 psi pressure–121°C.

Preparation of Medium: Aseptically combine 100.0mL of sterile solution A with 5.0mL of sterile solution B and 5.0mL of sterile solution C. Mix thoroughly.

Use: For the cultivation of unidentified bacteria DSMZ 8993 and DSMZ 8994.

THIO Medium

See: **Thioglycolate Medium, Enriched**

THIO + Bile Medium

See: **Thioglycolate Medium with 20% Bile**

Thioalkalivibrio halophilus Medium (DSMZ Medium 1014)

Composition per liter:

NaCl	175.0g
K_2HPO_4	1.5g
NH_4Cl	0.5g
Magnesium chloride solution	10.0mL
Magnesium sulfate solution	10.0mL
Thiosulfate solution	10.0mL
Bicarbonate solution	10.0mL
Trace element solution SL-4	1.0mL

pH 8.2 ± 0.2 at 25°C

Magnesium Sulfate Solution:

Composition per 10.0mL:

$MgSO_4 \cdot 7H_2O$	0.25g

Preparation of Magnesium Sulfate Solution: Add $MgSO_4 \cdot 7H_2O$ to distilled/deionized water and bring volume to 10.0mL. Mix thoroughly. Autoclave for 15 min at 15 psi pressure–121°C. Cool to room temperature.

Magnesium Chloride Solution:

Composition per 10.0mL:

$MgCl_2 \cdot 6H_2O$	0.2g

Preparation of Magnesium Chloride Solution: Add $MgCl_2 \cdot 6H_2O$ to distilled/deionized water and bring volume to 10.0mL. Mix thoroughly. Autoclave for 15 min at 15 psi pressure–121°C. Cool to room temperature.

Thiosulfate Solution:

Composition per 10.0mL:

$Na_2S_2O_3 \cdot 5H_2O$	5.0g

Preparation of Thiosulfate Solution: Add $Na_2S_2O_3 \cdot 5H_2O$ to distilled/deionized water and bring volume to 10.0mL. Mix thoroughly. Sparge with 100% N_2. Adjust to pH 10.0. Filter sterilize.

Bicarbonate Solution:

Composition per 10.0mL:

$NaHCO_3$	4.0g

Preparation of Bicarbonate Solution: Add $NaHCO_3$ to distilled/deionized water and bring volume to 10.0mL. Mix thoroughly. Filter sterilize.

Trace Elements Solution SL-4:

Composition per liter:

EDTA	0.5g
$FeSO_4 \cdot 7H_2O$	0.2g
Trace elements solution SL-6	100.0

Preparation of Trace Elements Solution SL-4: Add components to distilled/deionized water and bring volume to 1.0L. Mix thoroughly.

Trace Elements Solution SL-6:

Composition per liter:

H_3BO_3	0.3g
$CoCl_2 \cdot 6H_2O$	0.2g
$ZnSO_4 \cdot 7H_2O$	0.1g
$MnCl_2 \cdot 4H_2O$	0.03g
$Na_2MoO_4 \cdot H_2O$	0.03g
$NiCl_2 \cdot 6H_2O$	0.02g
$CuCl_2 \cdot 2H_2O$	0.01g

Preparation of Trace Elements Solution SL-6: Add components to distilled/deionized water and bring volume to 1.0L. Mix thoroughly. Adjust pH to 3.4.

Preparation of Medium: Add components, except magnesium chloride, magnesium sulfate, thiosulfate, and bicarbonate solutions, to distilled/deionized water and bring volume to 960.0mL. Mix thoroughly. Adjust pH to 8.0–8.5. Dispense into screw-capped Erlenmeyer flasks (fill to 1/10 volume). Autoclave for 30 min at 6 psi pressure–110°C. Cool to room temperature. Aseptically add magnesium chloride, magnesium sulfate, thiosulfate, and bicarbonate solutions. Mix thoroughly.

Use: For the cultivation of *Thioalkalivibrio halophilus.*

Thiobacillus A2 Agar (T3 Agar)

Composition per 1100.0mL:

Solution B	1.0L
Solution A	100.0mL

pH 8.5 ± 0.2 at 25°C

Solution A:

Composition per 100.0mL:

$Na_2S_2O_3 \cdot 5H_2O$	5.0g
Na_2HPO_4	4.2g
KH_2PO_4	1.5g
NH_4Cl	1.0g
Phenol Red (0.2% solution)	1.0mL

Preparation of Solution A: Add components to distilled/deionized water and bring volume to 100.0mL. Mix thoroughly. Adjust pH to 9.0. Autoclave for 15 min at 15 psi pressure–121°C. Cool to 45°–50°C.

Solution B:

Composition per liter:

Agar	15.0g
$MgSO_4 \cdot 7H_2O$	0.1g
Trace metal solution	5.0mL

Preparation of Solution B: Add components to distilled/deionized water and bring volume to 1.0mL. Mix thoroughly. Autoclave for 15 min at 15 psi pressure–121°C. Cool to 45°–50°C.

Trace Metal Solution:

Composition per liter:

EDTA	50.0g
$ZnSO_4$	22.0g
$CaCl_2$	5.54g
$MnCl_2$	5.06g
$FeSO_4 \cdot 7H_2O$	4.99g
$CoCl_2$	1.61g
$CuSO_4$	1.57g
$(NH_4)_2MoO_4 \cdot 4H_2O$	1.1g

Preparation of Trace Metal Solution: Add components to distilled/deionized water and bring volume to 1.0L. Mix thoroughly. Adjust pH to 6.0 with KOH.

Preparation of Medium: Aseptically add 100.0mL of sterile solution A to 1.0L of sterile solution B. Mix thoroughly. Adjust pH to 8.5 if necessary. Pour into sterile Petri dishes or distribute into sterile tubes.

Use: For the cultivation and maintenance of *Thiobacillus versutus* and other *Thiobacillus* species.

Thiobacillus A2 Broth (T3 Broth)

Composition per 1100.0mL:

Solution B	1.0L
Solution A	100.0mL

pH 8.5 ± 0.2 at 25°C

Solution A:

Composition per 100.0mL:

$Na_2S_2O_3 \cdot 5H_2O$	5.0g
Na_2HPO_4	4.2g
KH_2PO_4	1.5g
NH_4Cl	1.0g
Phenol Red (0.2% solution)	1.0mL

Preparation of Solution A: Add components to distilled/deionized water and bring volume to 100.0mL. Mix thoroughly. Adjust pH to 9.0. Autoclave for 15 min at 15 psi pressure–121°C. Cool to 45°–50°C.

Solution B:

Composition per liter:

$MgSO_4 \cdot 7H_2O$	0.1g
Trace metal solution	5.0mL

Preparation of Solution B: Add components to distilled/deionized water and bring volume to 1.0mL. Mix thoroughly. Autoclave for 15 min at 15 psi pressure–121°C. Cool to 45°–50°C.

Trace Metal Solution:

Composition per liter:

EDTA	50.0g
$ZnSO_4$	22.0g
$CaCl_2$	5.54g
$MnCl_2$	5.06g
$FeSO_4 \cdot 7H_2O$	4.99g
$CoCl_2$	1.61g
$CuSO_4$	1.57g
$(NH_4)_2MoO_4 \cdot 4H_2O$	1.1g

Preparation of Trace Metal Solution: Add components to distilled/deionized water and bring volume to 1.0mL. Mix thoroughly. Adjust pH to 6.0 with KOH.

Preparation of Medium: Aseptically add 100.0mL of sterile solution A to 1.0L of sterile solution B. Mix thoroughly. Adjust pH to 8.5 if necessary. Distribute into sterile tubes or flasks.

Use: For the cultivation and maintenance of *Thiobacillus versutus* and other *Thiobacillus* species.

Thiobacillus acidophilus Agar

Composition per liter:

Agar	15.0g
$(NH_4)_2SO_4$	3.0g
$MgSO_4 \cdot 7H_2O$	1.0g
KH_2PO_4	0.5g
KCl	0.1g
$Ca(NO_3)_2 \cdot 4H_2O$	18.0mg
$FeSO_4 \cdot 7H_2O$	0.01mg
Glucose solution	20.0mL

pH 4.5 ± 0.2 at 25°C

Glucose Solution:

Composition per 20.0mL:

Glucose	10.0g

Preparation of Glucose Solution: Add glucose to distilled/deionized water and bring volume to 20.0mL. Mix thoroughly. Autoclave for 15 min at 15 psi pressure–121°C.

Preparation of Medium: Add components, except glucose solution, to distilled/deionized water and bring volume to 980.0mL. Mix thoroughly. Gently heat and bring to boiling. Autoclave for 15 min at 15 psi pressure–121°C. Cool to 50°–55°C. Adjust pH to 4.5 with H_2SO_4. Aseptically add 20.0mL of sterile glucose solution. Mix thoroughly. Pour into sterile Petri dishes or distribute into sterile tubes.

Use: For the cultivation and maintenance of *Thiobacillus acidophilus*.

Thiobacillus acidophilus Broth

Composition per liter:

Glucose	10.0g
$(NH_4)_2SO_4$	3.0g
$MgSO_4 \cdot 7H_2O$	1.0g
KH_2PO_4	0.5g
KCl	0.1g
$Ca(NO_3)_2 \cdot 4H_2O$	18.0mg
$FeSO_4 \cdot 7H_2O$	0.01mg

pH 3.5 ± 0.2 at 25°C

Preparation of Medium: Add components to distilled/deionized water and bring volume to 1.0L. Mix thoroughly. Adjust pH to 3.5 with H_2SO_4. Distribute into tubes or flasks. Autoclave for 15 min at 15 psi pressure–121°C.

Use: For the cultivation of *Thiobacillus acidophilus*.

Thiobacillus acidophilus Medium (DSMZ Medium 108)

Composition per liter:

Agar	15.0g
$(NH_4)_2SO_4$	3.0g
$MgSO_4 \cdot 7H_2O$	1.0g
KH_2PO_4	0.5g
KCl	0.1g
$Ca(NO_3)_2 \cdot 4H_2O$	18.0mg
$FeSO_4 \cdot 7H_2O$	0.01mg
Glucose solution	50.0mL

pH 4.5 ± 0.2 at 25°C

Glucose Solution:

Composition per 100.0mL:

D-Glucose	10.0g

Preparation of Glucose Solution: Add glucose to distilled/deionized water and bring volume to 100.0mL. Mix thoroughly. Filter sterilize.

Preparation of Medium: Add components, except glucose solution, to distilled/deionized water and bring volume to 950.0mL. Mix thoroughly. Gently heat and bring to boiling. Autoclave for 15 min at 15 psi pressure–121°C. Cool to 45°–50°C. Aseptically add 50.0mL sterile glucose solution. Mix thoroughly. Adjust pH to 4.5 with H_2SO_4. Pour into sterile Petri dishes or distribute into sterile tubes.

Use: For the cultivation and maintenance of *Acidiphilium acidophilum.*

Thiobacillus acidophilus Medium (DSMZ Medium 108)

Composition per liter:

$(NH_4)_2SO_4$	3.0g
$MgSO_4 \cdot 7H_2O$	1.0g
KH_2PO_4	0.5g
KCl	0.1g
$Ca(NO_3)_2 \cdot 4H_2O$	18.0mg
$FeSO_4 \cdot 7H_2O$	0.01mg
Glucose solution	50.0mL

pH 3.5 ± 0.2 at 25°C

Glucose Solution:

Composition per 100.0mL:

D-Glucose	10.0g

Preparation of Glucose Solution: Add glucose to distilled/deionized water and bring volume to 100.0mL. Mix thoroughly. Filter sterilize.

Preparation of Medium: Add components, except glucose solution, to distilled/deionized water and bring volume to 950.0mL. Mix thoroughly. Gently heat and bring to boiling. Autoclave for 15 min at 15 psi pressure–121°C. Cool to 25°C. Aseptically add 50.0mL sterile glucose solution. Mix thoroughly. Adjust pH to 3.5 with H_2SO_4. Aseptically distribute into sterile tubes or flasks.

Use: For the cultivation of *Acidiphilium acidophilum.*

Thiobacillus Agar

Composition per liter:

Agar, noble	20.0g
$Na_2S_2O_3 \cdot 5H_2O$	5.0g
KH_2PO_4	3.0g
$CaCl_2$	0.1g
$MgCl_2 \cdot 6H_2O$	0.1g
$(NH_4)_2 \cdot Cl$	0.1g

pH 4.2 ± 0.2 at 25°C

Preparation of Medium: Add components, except agar, to distilled/deionized water and bring volume to 500.0mL. Mix thoroughly. Adjust pH to 4.2. Autoclave for 15 min at 15 psi pressure–121°C. Cool to 50°–55°C. Add agar to distilled/deionized water and bring volume to 500.0mL. Mix thoroughly. Gently heat and bring to boiling. Autoclave for 15 min at 15 psi pressure–121°C. Cool to 50°–55°C. Aseptically combine 500.0mL of the sterile basal medium with 500.0mL of the sterile agar solution. Pour into sterile Petri dishes or distribute into sterile tubes.

Use: For the cultivation of aciduric *Thiobacillus* species such as *Thiobacillus concretivorus* and *Thiobacillus intermedius*.

Thiobacillus Agar

Composition per liter:

Ionagar No. 2	12.0g
$Na_2S_2O_3 \cdot 5H_2O$	10.0g
K_2HPO_4	4.0g
KH_2PO_4	4.0g
$(NH_4)_2 \cdot SO_4$	0.1g
$MgSO_4 \cdot 7H_2O$	0.1g
$CaCl_2$	0.1g
$FeCl_3 \cdot 6H_2O$	2.0mg
$MnSO_4 \cdot 4H_2O$	2.0mg

pH 6.6 ± 0.2 at 25°C

Source: Ionagar No. 2 is available from Oxoid Unipath.

Preparation of Medium: Add components, except the agar, to distilled/deionized water and bring volume to 950.0mL. Mix thoroughly. Adjust the pH to 6.6. Add the agar. Bring volume to 1.0L with distilled/deionized water. Gently heat and bring to boiling. Distribute into tubes or flasks. Autoclave for 20 min at 15 psi pressure–121°C. Pour into sterile Petri dishes or leave in tubes.

Use: For the cultivation of nonaciduric *Thiobacillus* species such as *Thiobacillus neapolitanus, Thiobacillus novellus,* and *Thiobacillus thioparus*.

Thiobacillus Agar I for Acidophilic *Thiobacillus*

Composition per liter:

Solution A	500.0mL
Solution B	500.0mL

pH 4.2 ± 0.2 at 25°C

Solution A:

Composition per 500.0mL:

$Na_2S_2O_3$	5.0g
KH_2PO_4	3.0g
$CaCl_2$	0.1g
$MgCl_2 \cdot 6H_2O$	0.1g
NH_4Cl	0.1g

Preparation of Solution A: Add components to distilled/deionized water and bring volume to 500.0mL. Adjust pH to 4.2 with 1*N* HCl. Mix thoroughly. Autoclave for 15 min at 15 psi pressure–121°C. Cool to 50°C.

Solution B:

Composition per 500.0mL:

Agar, noble	20.0g

Preparation of Solution A: Add agar to distilled/deionized water and bring volume to 500.0mL. Autoclave for 15 min at 15 psi pressure–121°C. Cool to 50°C.

Preparation of Medium: Aseptically combine 500.0mL of sterile solution A and 500.0mL of sterile solution B. Combine the two solutions while still hot. Mix thoroughly. Pour into sterile Petri dishes or distribute into sterile tubes.

Use: For the cultivation of *Thiobacillus thiooxidans*.

Thiobacillus albertis Agar

Composition per 500.0mL:

Solution A	250.0mL
Solution B	250.0mL

pH 4.0 ± 0.2 at 25°C

Solution A:

Composition per 250.0mL:

$Na_2S_2O_3 \cdot 5H_2O$	5.0g
KH_2PO_4	3.0g
$MgSO_4 \cdot 7H_2O$	0.5g
$(NH_4)_2SO_4$	0.4g
$CaCl_2 \cdot 2H_2O$	0.25g
$FeSO_4 \cdot 7H_2O$	10.0mg

Preparation of Solution A: Add components to distilled/deionized water and bring volume to 250.0mL. Mix thoroughly. Adjust pH to 4.0. Autoclave for 15 min at 15 psi pressure–121°C. Cool to 50°–55°C.

Solution B:

Composition per 250.0mL:

Agar	15.0g

Preparation of Solution B: Add agar to distilled/deionized water and bring volume to 250.0mL. Mix thoroughly. Gently heat and bring to boiling. Autoclave for 15 min at 15 psi pressure–121°C. Cool to 50°–55°C.

Preparation of Medium: Aseptically combine 250.0mL of sterile solution A with 250.0mL of sterile solution B. Mix thoroughly. Pour into sterile Petri dishes or distribute into sterile tubes.

Use: For the cultivation of *Thiobacillus albertis*.

Thiobacillus albertis Broth

Composition per liter:

$Na_2S_2O_3 \cdot 5H_2O$	5.0g
KH_2PO_4	3.0g
$MgSO_4 \cdot 7H_2O$	0.5g
$(NH_4)_2SO_4$	0.4g
$CaCl_2 \cdot 2H_2O$	0.25g
$FeSO_4 \cdot 7H_2O$	10.0mg

pH 4.5 ± 0.2 at 25°C

Preparation of Medium: Add components to distilled/deionized water and bring volume to 1.0L. Mix thoroughly. Adjust pH to 4.5. Distribute into tubes or flasks. Autoclave for 15 min at 15 psi pressure–121°C.

Use: For the cultivation of *Thiobacillus albertis*.

Thiobacillus aquaesulis Agar

Composition per liter:

Solution B	900.0mL
Solution A	100.0mL

pH 7.4 ± 0.2 at 25°C

Solution A:

Composition per 100.0mL:

$Na_2HPO_42H_2O$	7.9g
KH_2PO_4	1.5g

Preparation of Solution A: Add components to distilled/deionized water and bring volume to 100.0mL. Mix thoroughly. Adjust pH to 7.4. Autoclave for 15 min at 15 psi pressure–121°C. Cool to 50°–55°C.

Solution B:

Composition per 900.0mL:

Agar	15.0g
$Na_2S_2O_3$	5.0g
NH_4Cl	0.4g
$MgSO_4 \cdot 7H_2O$	0.1g
Phenol Red	3.0mg
Trace elements solution	10.0mL

Trace Elements Solution:

Composition per liter:

Disodium EDTA	50.0g
NaOH	11.0g
$ZnSO_4 \cdot 7H_2O$	11.0g
$CaCl_2 \cdot 2H_2O$	7.34g
$FeSO_4 \cdot 7H_2O$	5.0g
$MnCl_2 \cdot 4H_2O$	2.5g
$CoCl_2 \cdot 6H_2O$	0.5g
$(NH_4)_6MoO_{24} \cdot 4H_2O$	0.5g
$CuSO_4 \cdot 5H_2O$	0.2g

Preparation of Trace Elements Solution: Add components to distilled/deionized water and bring volume to 1.0L. Mix thoroughly. Adjust pH to 6.0.

Preparation of Solution B: Add components to distilled/deionized water and bring volume to 900.0mL. Mix thoroughly. Gently heat and bring to boiling. Autoclave for 15 min at 15 psi pressure–121°C. Cool to 50°–55°C.

Preparation of Medium: Aseptically combine 100.0mL of cooled, sterile solution A with 900.0mL of cooled, sterile solution B. Mix thoroughly. Pour into sterile Petri dishes or distribute into sterile tubes.

Use: For the cultivation and maintenance of *Thiobacillus aquaesulis*.

Thiobacillus aquaesulis Broth

Composition per liter:

Solution B	900.0mL
Solution A	100.0mL

pH 7.4 ± 0.2 at 25°C

Solution A:

Composition per 100.0mL:

$Na_2HPO_42H_2O$	7.9g
KH_2PO_4	1.5g

Preparation of Solution A: Add components to distilled/deionized water and bring volume to 100.0mL. Mix thoroughly. Adjust pH to 7.4. Autoclave for 15 min at 15 psi pressure–121°C.

Solution B:

Composition per 900.0mL:

$Na_2S_2O_3$	5.0g
NH_4Cl	0.4g
$MgSO_4 \cdot 7H_2O$	0.1g
Phenol Red	3.0mg
Trace elements solution	10.0mL

Trace Elements Solution:

Composition per liter:

Disodium EDTA	50.0g
NaOH	11.0g
$ZnSO_4 \cdot 7H_2O$	11.0g
$CaCl_2 \cdot 2H_2O$	7.34g
$FeSO_4 \cdot 7H_2O$	5.0g
$MnCl_2 \cdot 4H_2O$	2.5g
$CoCl_2 \cdot 6H_2O$	0.5g
$(NH_4)_6MoO_{24} \cdot 4H_2O$	0.5g
$CuSO_4 \cdot 5H_2O$	0.2g

Preparation of Trace Elements Solution: Add components to distilled/deionized water and bring volume to 1.0L. Mix thoroughly. Adjust pH to 6.0.

Preparation of Solution B: Add components to distilled/deionized water and bring the volume to 900.0mL. Mix thoroughly. Autoclave for 15 min at 15 psi pressure–121°C.

Preparation of Medium: Aseptically combine 100.0mL of cooled, sterile solution A with 900.0mL of cooled, sterile solution B. Mix thoroughly. Aseptically distribute into sterile tubes or flasks.

Use: For the cultivation and maintenance of *Thiobacillus aquaesulis*.

Thiobacillus aquaesulis Medium

Composition per liter:

Solution B	900.0mL
Solution A	100.0mL

Solution A:

Composition per 100.0mL:

$Na_2HPO_4 \cdot 2H_2O$	7.9g
KH_2HPO_4	1.5g

Preparation of Solution A: Add components to distilled/deionized water and bring volume to 100.0mL. Mix thoroughly. Adjust pH to 7.6. Autoclave for 15 min at 15 psi pressure–121°C. Cool to 50°–55°C.

Solution B:

Composition per 900.0mL:

Agar	15.0g
$Na_2S_2O_3$	5.0g
NH_4Cl	0.4g
$MgSO_4 \cdot 7H_2O$	0.1g
Phenol Red	3.0mg
Trace metals	10.0mL

Preparation of Solution B: Add components to distilled/deionized water and bring volume to 900.0mL. Mix thoroughly. Gently heat and bring to boiling. Autoclave for 15 min at 15 psi pressure–121°C. Cool to 50°–55°C.

Trace Metals:

Composition per liter:

EDTA	50.0g
NaOH	11.0g
$ZnSO_4 \cdot 7H_2O$	11.0g
$CaCl_2 \cdot 2H_2O$	7.34g
$FeSO_4 \cdot 7H_2O$	5.0g
$MnCl_2 \cdot 4H_2O$	2.5g
$CoCl_2 \cdot 6H_2O$	0.5g
$(NH_4)_2Mo_2O_{27}$	0.5g
$CuSO_4 \cdot 5H_2O$	0.2g

Preparation of Trace Metals: Add components to distilled/deionized water and bring volume to 1.0L. Mix thoroughly. Adjust pH to 6.0 with H_2SO_4.

Preparation of Medium: Aseptically combine 100.0mL of sterile solution A with 900.0mL of sterile solution B. Pour into sterile Petri dishes or distribute into sterile tubes.

Use: For the cultivation of *Thiobacillus aquaesulis.*

Thiobacillus Broth I for Acidophilic *Thiobacillus*

Composition per liter:

$Na_2S_2O_3$	5.0g
KH_2PO_4	3.0g
$CaCl_2$	0.1g
$MgCl_2 \cdot 6H_2O$	0.1g
NH_4Cl	0.1g

pH 4.2 ± 0.2 at 25°C

Preparation of Medium: Add components to distilled/deionized water and bring volume to 1.0L. Adjust pH to 4.2 with 1*N* HCl. Mix thoroughly. Distribute into screw-capped tubes or flasks. Autoclave for 15 min at 15 psi pressure–121°C.

Use: For the cultivation of *Thiobacillus thiooxidans*.

Thiobacillus caldus Agar

Composition per liter:

Solution E	500.0mL
Solution A	460.0mL
Solution D	20.0mL
Solution B	10.0mL
Solution C	10.0mL

pH 2.5 ± 0.2 at 25°C

Solution A:

Composition per 460.0mL:

$Na_2SO_4 \cdot 10H_2O$	3.2g
$(NH_4)_2SO_4$	3.0g
$MgSO_4 \cdot 7H_2O$	0.5g
KCl	0.1g
K_2HPO_4	50.0mg

Preparation of Solution A: Add components to distilled/deionized water and bring volume to 460.0mL. Mix thoroughly. Adjust pH to 1.75 with H_2SO_4. Autoclave for 15 min at 15 psi pressure–121°C. Cool and maintain above 60°C.

Solution B:

Composition per 10.0mL:

$FeCl_3 \cdot 6H_2O$	11.0mg
$Ca(NO_3)_2 \cdot 4H_2O$	10.0mg
H_3BO_3	2.0mg
$MnSO_4 \cdot H_2O$	2.0mg
$ZnSO_4 \cdot 7H_2O$	0.9mg
$Na_2MoO_4 \cdot 2H_2O$	0.8mg
$CoCl_2 \cdot 6H_2O$	0.6mg
$CuSO_4 \cdot 5H_2O$	0.5mg

Preparation of Solution B: Add components to distilled/deionized water and bring volume to 10.0mL. Mix thoroughly. Filter sterilize.

Solution C:

Composition per 10.0mL:

Glucose	0.45g

Preparation of Solution C: Add glucose to distilled/deionized water and bring volume to 10.0mL. Mix thoroughly. Filter sterilize.

Solution D:

Composition per 20.0mL:

$Na_2S_4O_6$	0.77g

Preparation of Solution D: Add $Na_2S_4O_6$ to distilled/deionized water and bring volume to 20.0mL. Mix thoroughly. Filter sterilize.

Solution E:

Composition per 500.0mL:

Phytagel™ (Gellan gum; available from Sigma Chemical Co.)	15.0g

Preparation of Solution E: Add phytagel to distilled/deionized water and bring volume to 500.0mL. Mix thoroughly. Autoclave for 15 min at 15 psi pressure–121°C. Cool and maintain above 60°C.

Preparation of Medium: Maintain solutions A and E above 60°C to prevent rapid gelling of medium. Aseptically combine 460.0mL of sterile solution A with 10.0mL of sterile solution B, 10.0mL of sterile solution C, 20.0mL of sterile solution D, and 500.0mL of sterile solution E. Mix throughly. Pour into sterile Petri dishes or distribute into sterile tubes.

Use: For the heterotrophic cultivation and maintenance of *Thiobacillus caldus.*

Thiobacillus caldus Broth

Composition per liter:

Solution A	960.0mL
Solution D	20.0mL
Solution B	10.0mL
Solution C	10.0mL

pH 2.5 ± 0.2 at 25°C

Solution A:

Composition per 960.0mL:

$Na_2SO_4 \cdot 10H_2O$	3.2g
$(NH_4)_2SO_4$	3.0g
$MgSO_4 \cdot 7H_2O$	0.5g
KCl	0.1g
K_2HPO_4	50.0mg

Preparation of Solution A: Add components to distilled/deionized water and bring volume to 960.0mL. Mix thoroughly. Adjust pH to 1.75 with H_2SO_4. Autoclave for 15 min at 15 psi pressure–121°C.

Solution B:

Composition per 10.0mL:

$FeCl_3 \cdot 6H_2O$	11.0mg
$Ca(NO_3)_2 \cdot 4H_2O$	10.0mg
H_3BO_3	2.0mg
$MnSO_4 \cdot H_2O$	2.0mg
$ZnSO_4 \cdot 7H_2O$	0.9mg
$Na_2MoO_4 \cdot 2H_2O$	0.8mg
$CoCl_2 \cdot 6H_2O$	0.6mg
$CuSO_4 \cdot 5H_2O$	0.5mg

Preparation of Solution B: Add components to distilled/deionized water and bring volume to 10.0mL. Mix thoroughly. Filter sterilize.

Solution C:

Composition per 10.0mL:

Glucose	0.45g

Preparation of Solution C: Add glucose to distilled/deionized water and bring volume to 10.0mL. Mix thoroughly. Filter sterilize.

Solution D:

Composition per 20.0mL:

$Na_2S_4O_6$	0.77g

Preparation of Solution D: Add $Na_2S_4O_6$ to distilled/deionized water and bring volume to 20.0mL. Mix thoroughly. Filter sterilize.

Preparation of Medium: Aseptically combine 960.0mL of sterile solution A with 10.0mL of sterile solution B, 10.0mL of sterile solution C, and 20.0mL of sterile solution D. Mix thoroughly. Aseptically distribute into sterile tubes or flasks.

Use: For the heterotrophic cultivation and maintenance of *Thiobacillus caldus.*

Thiobacillus caldus Broth

Composition per liter:

Solution A	970.0mL
Solution C	20.0mL
Solution B	10.0mL

pH 2.5 ± 0.2 at 25°C

Solution A:

Composition per 970.0mL:

$Na_2SO_4 \cdot 10H_2O$	3.2g
$(NH_4)_2SO_4$	3.0g
$MgSO_4 \cdot 7H_2O$	0.5g
KCl	0.1g
K_2HPO_4	50.0mg

Preparation of Solution A: Add components to distilled/deionized water and bring volume to 970.0mL. Mix thoroughly. Adjust pH to 1.75 with H_2SO_4. Autoclave for 15 min at 15 psi pressure–121°C.

Solution B:

Composition per 10.0mL:

$FeCl_3 \cdot 6H_2O$	11.0mg
$Ca(NO_3)_2 \cdot 4H_2O$	10.0mg
H_3BO_3	2.0mg
$MnSO_4 \cdot H_2O$	2.0mg
$ZnSO_4 \cdot 7H_2O$	0.9mg
$Na_2MoO_4 \cdot 2H_2O$	0.8mg
$CoCl_2 \cdot 6H_2O$	0.6mg
$CuSO_4 \cdot 5H_2O$	0.5mg

Preparation of Solution B: Add components to distilled/deionized water and bring volume to 10.0mL. Mix thoroughly. Filter sterilize.

Solution C:

Composition per 20.0mL:

$Na_2S_4O_6$	0.77g

Preparation of Solution C: Add $Na_2S_4O_6$ to distilled/deionized water and bring volume to 20.0mL. Mix thoroughly. Filter sterilize.

Preparation of Medium: Aseptically combine 970.0mL of sterile solution A with 10.0mL of sterile solution B and 20.0mL of sterile solution C. Mix thoroughly. Aseptically distribute into sterile tubes or flasks.

Use: For the chemolithotrophic cultivation of *Thiobacillus caldus.*

Thiobacillus caldus Medium

Composition per liter:

Solution E	500.0mL
Solution A	460.0mL
Solution D	20.0mL
Solution B	10.0mL
Solution C	10.0mL

pH 2.5 ± 0.2 at 25°C

Solution A:

Composition per 460.0mL:

$Na_2SO_4 \cdot 10H_2O$	3.2g
$(NH_4)_2SO_4$	3.0g
$MgSO_4 \cdot 7H_2O$	0.5g
KCl	0.1g
K_2HPO_4	50.0mg

Preparation of Solution A: Add components to distilled/deionized water and bring volume to 460.0mL. Mix thoroughly. Adjust pH to 1.75 with H_2SO_4. Autoclave for 15 min at 15 psi pressure–121°C. Cool to 60°–65°C.

Solution B:

Composition per 10.0mL:

$FeCl_3 \cdot 6H_2O$	11.0mg
$Ca(NO_3)_2 \cdot 4H_2O$	10.0mg
H_3BO_3	2.0mg
$MnSO_4 \cdot H_2O$	2.0mg
$ZnSO_4 \cdot 7H_2O$	0.9mg
$Na_2MoO_4 \cdot 2H_2O$	0.8mg
$CoCl_2 \cdot 6H_2O$	0.6mg
$CuSO_4 \cdot 5H_2O$	0.5mg

Preparation of Solution B: Add components to distilled/deionized water and bring volume to 10.0mL. Mix thoroughly. Filter sterilize. Warm to 60°C.

Solution C:

Composition per 10.0mL:

Glucose	0.45g

Preparation of Solution C: Add glucose to distilled/deionized water and bring volume to 10.0mL. Mix thoroughly. Filter sterilize. Warm to 60°C.

Solution D:

Composition per 20.0mL:

Sodium tetrathionate	0.77g

Preparation of Solution D: Add sodium tetrathionate to distilled/deionized water and bring volume to 20.0mL. Mix thoroughly. Filter sterilize. Warm to 60°C.

Solution E:

Composition per 500.0mL:

Phytagel	15.0g

Preparation of Solution E: Add phytagel to distilled/deionized water and bring volume to 500.0mL. Mix thoroughly. Autoclave for 15 min at 15 psi pressure–121°C. Cool to 60°–65°C.

Preparation of Medium: Aseptically combine 460.0mL of sterile solution A, 10.0mL of sterile solution B, 10.0mL of sterile solution C, 20.0mL of sterile solution D, and 500.0mL of sterile solution E. Mix thoroughly. Aseptically pour into sterile Petri dishes or distrbute into sterile tubes.

Use: For the cultivation of *Thiobacillus caldus*.

Thiobacillus cuprinus Medium

Composition per 1010.0mL:

$MgSO_4 \cdot 7H_2O$	3.45g
$MgCl_2 \cdot 6H_2O$	2.75g
NH_4Cl	1.25g
NaCl	0.5g
Sulfur, powdered	0.5g
KCl	0.33g
$CaCl_2 \cdot 2H_2O$	0.14g
K_2HPO_4	0.14g
$NiCl_2 \cdot 6H_2O$	2.0mg
Trace elements solution	10.0mL

pH 3.5 ± 0.2 at 25°C

Preparation of Sulfur: Sterilize 1.0g of powdered sulfur by steaming for 1 hr on three consecutive days.

Trace Elements Solution:

Composition per liter:

$MgSO_4 \cdot 7H_2O$	3.0g
Nitrilotriacetic acid	1.5g
NaCl	1.0g
$MnSO_4 \cdot 2H_2O$	0.5g
$CoSO_4 \cdot 7H_2O$	0.18g
$ZnSO_4 \cdot 7H_2O$	0.18g
$CaCl_2 \cdot 2H_2O$	0.1g
$FeSO_4 \cdot 7H_2O$	0.1g
$KAl(SO_4)_2 \cdot 12H_2O$	0.02g
$CuSO_4 \cdot 5H_2O$	0.01g
H_3BO_3	0.01g
$Na_2MoO_4 \cdot 2H_2O$	0.01g
$NiCl_2 \cdot 6H_2O$	0.025g
$Na_2SeO_3 \cdot 5H_2O$	0.3mg

Preparation of Trace Elements Solution: Add nitrilotriacetic acid to 500.0mL of distilled/deionized water. Adjust pH to 6.5 with KOH. Add remaining components. Add distilled/deionized water to 1.0L.

Preparation of Medium: Add components, except sulfur, to distilled/deionized water and bring volume to 1.0L. Mix thoroughly. Adjust pH to 3.5 with H_2SO_4. Autoclave for 15 min at 15 psi pressure–121°C. Aseptically add 0.5g of sterile sulfur. Mix thoroughly. Aseptically distribute into sterile tubes or flasks.

Use: For the cultivation and maintenance of *Thiobacillus cuprinus*.

Thiobacillus denitrificans Medium

Composition per liter:

KNO_3	5.0g
$Na_2S_2O_3 \cdot 5H_2O$	5.0g
$NaHCO_3$	1.0g
K_2HPO_4	0.2g
$MgCl_2$	0.1g

pH 7.0 ± 0.2 at 25°C

Preparation of Medium: Add components to distilled/deionized water and bring volume to 1.0L. Mix thoroughly. Distribute into tubes or flasks. Autoclave for 15 min at 15 psi pressure–121°C.

Use: For the cultivation of *Thiobacillus denitrificans*.

Thiobacillus denitrificans Medium

Composition per liter:

$Na_2S_2O_3 \cdot 5H_2O$	5.0g
KNO_3	2.0g
KH_2PO_4	2.0g
$NaHCO_3$	2.0g
NH_4Cl	1.0g
$MgSO_4 \cdot 7H_2O$	0.8g
Trace metals solution	1.0mL

pH 6.8–7.0 at 25°C

Trace Metals Solution:

Composition per liter:

Disodium EDTA	50.0g
NaOH	11.0g
$CaCl_2 \cdot 2H_2O$	7.34g
$FeSO_4 \cdot 7H_2O$	5.0g
$MnCl_2 \cdot 2H_2O$	2.5g
$ZnSO_4 \cdot 7H_2O$	2.2g
$CoCl_2 \cdot 6H_2O$	0.5g
$(NH_4)_6Mo_7O_{24} \cdot 4H_2O$	0.5g
$CuSO_4 \cdot 5H_2O$	0.2g

Preparation of Trace Metals Solution: Add EDTA to distilled/deionized water and bring volume to 500.0mL. Mix thoroughly. Adjust pH to 6.0 with NaOH. Add remaining components, one by one. Maintain the pH at 6.0. After dissolution of all the salts, adjust the pH to 4.0 with HCl. Store at 4°C.

Preparation of Medium: Add components to distilled/deionized water and bring volume to 1.0L. Mix thoroughly. Distribute into tubes or flasks. Autoclave for 15 min at 15 psi pressure–121°C.

Use: For the isolation and cultivation of *Thiobacillus denitrificans*.

Thiobacillus ferrooxidans Medium

Composition per liter:

$Al_2(SO_4)_3 \cdot 12H_2O$	1.4g
NaCl	1.0g
KH_2PO_4	0.4g
$MgSO_4 \cdot 7H_2O$	0.1g
$(NH_4)_2SO_4$	0.1g
$CaCl_2$	0.03g
$MnSO_4 \cdot 4H_2O$	0.02g
$FeSO_4 \cdot 7H_2O$ solution	100.0mL

$FeSO_4 \cdot 7H_2O$ Solution:

Composition per 100.0mL:

$FeSO_4 \cdot 7H_2O$	10.0g
H_2SO_4, concentrated	0.09mL

Preparation of $FeSO_4 \cdot 7H_2O$ Solution: Add $FeSO_4 \cdot 7H_2O$ and H_2SO_4 to distilled/deionized water and bring volume to 100.0mL. Mix thoroughly. Autoclave for 15 min at 15 psi pressure–121°C.

Preparation of Medium: Add components, except $FeSO_4 \cdot 7H_2O$ solution, to distilled/deionized water and bring volume to 900.0mL. Mix thoroughly. Gently heat and bring to boiling. Distribute into flasks in 90.0mL volumes. Autoclave for 15 min at 15 psi pressure–121°C. Cool to 25°C. Aseptically add 10.0mL of sterile $FeSO_4 \cdot 7H_2O$ solution to each flask. Mix thoroughly.

Use: For the cultivation of *Thiobacillus ferrooxidans*.

Thiobacillus ferrooxidans Medium

Composition per liter:

Solution I	400.0mL
Solution III	400.0mL
Solution II	200.0mL

Solution I:

Composition per 500.0mL:

K_2HPO_4	0.5g
$MgSO_4 \cdot 7H_2O$	0.5g
$(NH_4)_2SO_4$	0.5g
H_2SO_4 (1*N* solution)	5.0mL

Preparation of Solution I: Add components to distilled/deionized water and bring volume to 500.0mL. Mix thoroughly. Autoclave for 15 min at 15 psi pressure–121°C. Cool to 45°–50°C.

Solution II:

Composition per liter:

$FeSO_4 \cdot 7H_2O$	167.0g
1*N* H_2SO_4	50.0mL

Preparation of Solution II: Add components to distilled/deionized water and bring volume to 1.0L. Mix thoroughly. Filter sterilize. Warm to 45°–50°C.

Solution III:

Composition per liter:

Agar	10.0g

Preparation of Solution III: Add agar to distilled/deionized water and bring volume to 1.0L. Mix thoroughly.

Preparation of Medium: Aseptically combine 400.0mL of sterile solution I, 200.0mL of sterile solution II, and 400.0mL of sterile solution III. Mix thoroughly. Aseptically distribute into sterile tubes or flasks.

Use: For the isolation and cultivation of *Thiobacillus ferrooxidans*.

Thiobacillus ferrooxidans Medium, APH

Composition per liter:

$FeSO_4 \cdot 7H_2O$	40.0g
$(NH_4)_2SO_4$	2.0g
K_2HPO_4	0.5g
$MgSO_4 \cdot 7H_2O$	0.5g
KCl	0.1g
$Ca(NO_3)_2$	0.01g

pH 3.0 ± 0.2 at 25°C

Preparation of Medium: Add components to distilled/deionized water and bring volume to 1.0L. Mix thoroughly. Adjust pH to 3.0 with dilute H_2SO_4. Distribute into tubes or flasks. Filter sterilize. Aseptically distribute into sterile tubes or flasks.

Use: For the cultivation and maintenance of *Leptospirillum ferrooxidans* and *Thiobacillus ferrooxidans*.

Thiobacillus ferrooxidans Medium with Ferrous Sulfate

Composition per liter:

$FeSO_4 \cdot 7H_2O$	33.3g
KH_2PO_4	0.4g
$MgSO_4 \cdot 7H_2O$	0.4g
$(NH_4)_2SO_4$	0.4g
0.1*N* H_2SO_4	1000.0mL

pH 1.4 ± 0.2 at 25°C

Preparation of Medium: Add components to distilled/deionized water and bring volume to 1.0L. Mix thoroughly. Adjust pH to 1.4 with H_2SO_4. Distribute into tubes or flasks. Autoclave for 15 min at 15 psi pressure–121°C.

Use: For the cultivation and maintenance of *Clostridium acetobutylicum*.

Thiobacillus ferrooxidans Medium with Tetrathionate

Composition per liter:

$K_2S_4O_6$	5.0g
KH_2PO_4	3.0g
$(NH_4)_2SO_4$	3.0g
$MgSO_4 \cdot 7H_2O$	0.5g
$CaCl_2 \cdot 2H_2O$	0.25g

pH 4.4 ± 0.2 at 25°C

Preparation of Medium: Add components to distilled/deionized water and bring volume to 1.0L. Mix thoroughly. Adjust pH to 4.4 with H_2SO_4. Distribute into tubes or flasks. Autoclave for 15 min at 15 psi pressure–121°C.

Use: For the cultivation and maintenance of *Thiobacillus ferrooxidans*.

Thiobacillus ferrooxidans Medium with Thiosulfate

Composition per liter:

$Na_2S_2O_3 \cdot 5H_2O$	5.0g
KH_2PO_4	3.0g
$(NH_4)_2SO_4$	3.0g
$MgSO_4 \cdot 7H_2O$	0.5g
$CaCl_2 \cdot 2H_2O$	0.25g

pH 4.4 ± 0.2 at 25°C

Preparation of Medium: Add components to distilled/deionized water and bring volume to 1.0L. Mix thoroughly. Adjust pH to 4.4 with

H_2SO_4. Distribute into tubes or flasks. Autoclave for 15 min at 15 psi pressure–121°C.

Use: For the cultivation and maintenance of *Thiobacillus ferrooxidans* and *Thiobacillus thiooxidans*.

Thiobacillus halophilus Agar

Composition per liter:

NaCl	50.0g
Agar	15.0g
$Na_2HPO_4·2H_2O$	7.9g
$Na_2S_2O_3·5H_2O$	5.0g
KH_2PO_4	1.5g
NH_4Cl	0.4g
$MgSO_4·7H_2O$	0.1g
Phenol Red (saturated aqueous solution)	12.5mL
Trace metals solution	10.0mL

pH 7.3 ± 0.2 at 25°C

Trace Metals Solution:

Composition per liter:

Disodium EDTA	50.0g
NaOH	11.0g
$ZnSO_4·7H_2O$	11.0g
$CaCl_2·2H_2O$	7.34g
$FeSO_4·7H_2O$	5.0g
$MnCl_2·4H_2O$	2.5g
$CoCl_2·6H_2O$	0.5g
$(NH_4)_6Mo_7O_{24}·4H_2O$	0.5g
$CuSO_4·5H_2O$	0.2g

Preparation of Trace Metals Solution: Add components to distilled/deionized water and bring volume to 1.0L. Adjust pH to 6.0. Mix thoroughly.

Preparation of Medium: Add components to distilled/deionized water and bring volume to 1.0L. Mix thoroughly. Gently heat and bring to boiling. Distribute into tubes or flasks. Autoclave for 15 min at 15 psi pressure–121°C. Pour into sterile Petri dishes or leave in tubes.

Use: For the cultivation and maintenance of *Thiobacillus halophilus*.

Thiobacillus halophilus Broth

Composition per liter:

Solution A	900.0mL
Solution B	100.0mL

pH 7.3 ± 0.2 at 25°C

Solution A:

Composition per 900.0mL:

NaCl	50.0g
$Na_2S_2O_3·5H_2O$	5.0g
NH_4Cl	0.4g
$MgSO_4·7H_2O$	0.1g
Phenol Red (saturated aqueous solution)	12.5mL
Trace metals solution	10.0mL

Trace Metals Solution:

Composition per liter:

Disodium EDTA	50.0g
NaOH	11.0g
$ZnSO_4·7H_2O$	11.0g
$CaCl_2·2H_2O$	7.34g
$FeSO_4·7H_2O$	5.0g
$MnCl_2·4H_2O$	2.5g
$CoCl_2·6H_2O$	0.5g
$(NH_4)_6Mo_7O_{24}·4H_2O$	0.5g
$CuSO_4·5H_2O$	0.2g

Preparation of Trace Metals Solution: Add components to distilled/deionized water and bring volume to 1.0L. Adjust pH to 6.0. Mix thoroughly.

Preparation of Solution A: Add components to distilled/deionized water and bring volume to 900.0mL. Mix thoroughly. Autoclave for 15 min at 15 psi pressure–121°C.

Solution B:

Composition per 100.0mL:

$Na_2HPO_4·2H_2O$	7.9g
KH_2PO_4	1.5g

Preparation of Solution B: Add components to distilled/deionized water and bring volume to 100.0mL. Mix thoroughly. Autoclave for 15 min at 15 psi pressure–121°C.

Preparation of Medium: Aseptically combine 900.0mL of sterile solution A with 100.0mL of sterile solution B. Mix thoroughly. Aseptically distribute into sterile tubes or flasks.

Use: For the cultivation and maintenance of *Thiobacillus halophilus*.

Thiobacillus halophilus Medium

Composition per liter:

Solution B	900.0mL
Solution A	100.0mL

pH 7.3 ± 0.2 at 25°C

Solution B:

Composition per 900.0mL:

NaCl	50.0g
Agar, noble	15.0g
$Na_2S_2O_3·5H_2O$	5.0g
NH_4Cl	0.4g
Phenol Red	10.0mg
Hutner's basal salts solution	20.0mL

Preparation of Solution B: Add components to distilled/deionized water and bring volume to 900.0mL. Mix thoroughly. Gently heat and bring to boiling. Autoclave for 15 min at 15 psi pressure–121°C. Cool to 50°–55°C.

Hutner's Basal Salts Solution:

Composition per liter:

$MgSO_4·7H_2O$	29.7g
Nitrilotriacetic acid	10.0g
$CaCl_2·2H_2O$	3.335g
$FeSO_4·7H_2O$	99.0mg
$(NH_4)_6MoO_7O_{24}·4H_2O$	9.25mg
"Metals 44"	50.0mL

"Metals 44":

Composition per 100.0mL:

$ZnSO_4·7H_2O$	1.095g
$FeSO_4·7H_2O$	0.5g
Sodium EDTA	0.25g
$MnSO_4·H2O$	0.154g
$CuSO_4·5H_2O$	39.2mg
$Co(NO_3)_2·6H_2O$	24.8mg
$Na_2B_4O_7·10H_2O$	17.7mg

Preparation of "Metals 44": Add sodium EDTA to distilled/deionized water and bring volume to 90.0mL. Mix thoroughly. Add a few drops of concentrated H_2SO_4 to retard precipitation of heavy metal ions. Add remaining components. Mix thoroughly. Bring volume to 100.0mL with distilled/deionized water.

Preparation of Hutner's Basal Salts Solution: Add nitrilotriacetic acid to 500.0mL of distilled/deionized water. Adjust pH to 6.5 with KOH. Add remaining components. Add distilled/deionized water to 1.0L. Adjust pH to 6.8.

Solution A:

Composition per 100.0mL:

Na_2HPO_4	6.3g
KH_2PO_4	1.5g

Preparation of Solution A: Add components to distilled/deionized water and bring volume to 100.0mL. Mix thoroughly. Autoclave for 15 min at 15 psi pressure–121°C. Cool to 50°–55°C.

Preparation of Medium: Aseptically combine 900.0mL of sterile solution B with 100.0mL of sterile solution A. Mix thoroughly. Pour into sterile Petri dishes or distribute into sterile tubes.

Use: For the cultivation of *Thiobacillus halophilus.*

Thiobacillus Heterotrophic Medium

Composition per liter:

Glucose	5.0g
$MgSO_4 \cdot 7H_2O$	0.5g
$(NH_4)_2SO_4$	0.15g
KH_2PO_4	0.1g
KCl	0.05g
$Ca(NO_3)_2$	0.01g

pH 3.0 ± 0.2 at 25°C

Preparation of Medium: Add components to distilled/deionized water and bring volume to 1.0L. Mix thoroughly. Filter sterilize.

Use: For the cultivation and maintenance of *Thiobacillus organoparus* and other heterotrophic *Thiobacillus* species.

Thiobacillus intermedius Medium

Composition per 1010.0mL:

$Na_2S_2O_3 \cdot 5H_2O$	10.0g
Solution I	1.0L
Solution II	10.0mL

Solution I:

Composition per liter:

NH_4Cl	1.0g
K_2HPO_4	0.6g
$MgCl_2 \cdot 6H_2O$	0.5g
KH_2PO_4	0.4g
$MgSO_4$	0.3g
$CaCl_2 \cdot 2H_2O$	0.2g
$FeCl_3 \cdot 6H_2O$	0.02g

Preparation of Solution I: Add components to distilled/deionized water and bring volume to 1.0L. Mix thoroughly.

Solution II:

Composition per liter:

$CaCl_2 \cdot 2H_2O$	0.1g
$ZnSO_4 \cdot 7H_2O$	0.09g
$CuSO_4 \cdot 5H_2O$	0.04g
$MnSO_4$	0.02g
$Na_2B_4O_7$	0.01g
$(NH_4)_6Mo_7O_{24} \cdot 4H_2O$	5.0mg

Preparation of Solution II: Add components to distilled/deionized water and bring volume to 1.0L. Mix thoroughly.

Preparation of Medium: Combine solution I, 10.0mL of solution II, and 10.0g of $Na_2S_2O_3 \cdot 5H_2O$. Mix thoroughly. Filter sterilize. Aseptically distribute into sterile tubes or flasks.

Use: For the isolation and autotrophic cultivation of *Thiobacillus intermedius.*

Thiobacillus intermedius Medium

Composition per 1010.0mL:

Glucose	10.0g
$Na_2S_2O_3 \cdot 5H_2O$	10.0g
Solution I	1.0L
Solution II	10.0mL

Solution I:

Composition per liter:

NH_4Cl	1.0g
K_2HPO_4	0.6g
$MgCl_2 \cdot 6H_2O$	0.5g
KH_2PO_4	0.4g
$MgSO_4$	0.3g
$CaCl_2 \cdot 2H_2O$	0.2g
$FeCl_3 \cdot 6H_2O$	0.02g

Preparation of Solution I: Add components to distilled/deionized water and bring volume to 1.0L. Mix thoroughly.

Solution II:

Composition per liter:

$CaCl_2 \cdot 2H_2O$	0.1g
$ZnSO_4 \cdot 7H_2O$	0.09g
$CuSO_4 \cdot 5H_2O$	0.04g
$MnSO_4$	0.02g
$Na_2B_4O_7$	0.01g
$(NH_4)_6Mo_7O_{24} \cdot 4H_2O$	5.0mg

Preparation of Solution II: Add components to distilled/deionized water and bring volume to 1.0L. Mix thoroughly.

Preparation of Medium: Combine 1.0L of solution I, 10.0mL of solution II, 10.0g of glucose, and 10.0g of $Na_2S_2O_3 \cdot 5H_2O$. Mix thoroughly. Filter sterilize. Aseptically distribute into sterile tubes or flasks.

Use: For the isolation and mixotrophic cultivation of *Thiobacillus intermedius.*

Thiobacillus intermedius Medium

Composition per 1010.0mL:

Glucose	10.0g
Yeast extract	0.3g
Solution I	1.0L
Solution II	10.0mL

Solution I:

Composition per liter:

NH_4Cl	1.0g
K_2HPO_4	0.6g
$MgCl_2 \cdot 6H_2O$	0.5g
KH_2PO_4	0.4g
$MgSO_4$	0.3g
$CaCl_2 \cdot 2H_2O$	0.2g
$FeCl_3 \cdot 6H_2O$	0.02g

Preparation of Solution I: Add components to distilled/deionized water and bring volume to 1.0L. Mix thoroughly.

Solution II:

Composition per liter:

$CaCl_2 \cdot 2H_2O$	0.1g
$ZnSO_4 \cdot 7H_2O$	0.09g
$CuSO_4 \cdot 5H_2O$	0.04g
$MnSO_4$	0.02g
$Na_2B_4O_7$	0.01g
$(NH_4)_6Mo_7O_{24} \cdot 4H_2O$	5.0mg

Preparation of Solution II: Add components to distilled/deionized water and bring volume to 1.0L. Mix thoroughly.

Preparation of Medium: Combine 1.0L of solution I, 10.0mL of solution II, 10.0g of glucose, and 0.3g of yeast extract. Mix thoroughly. Filter sterilize. Aseptically distribute into sterile tubes or flasks.

Use: For the isolation and heterotrophic cultivation of *Thiobacillus intermedius*.

Thiobacillus intermedius Medium (LMG 113)

Composition per liter:

Agar	15.0g
$Na_2S_2O_3 \cdot 5H_2O$	10.0g
$MgCl_2 \cdot H_2O$	1.0g
NH_4Cl	1.0g
Yeast extract	1.0g
K_2HPO_4	0.6g
KH_2PO_4	0.4g
Chlorophenol Red	80.0mg
$FeCl_3 \cdot H_2O$	33.0mg

Preparation of Medium: Add components to distilled/deionized water and bring volume to 1.0L. Mix thoroughly. Gently heat and bring to boiling. Distribute into tubes or flasks. Autoclave for 15 min at 15 psi pressure–121°C. Pour into sterile Petri dishes or leave in tubes.

Use: For the cultivation of *Thiobacillus intermedius*.

Thiobacillus Medium

Composition per 100.0mL:

$Na_2S_2O_3 \cdot 5H_2O$	1.0g
KH_2PO_4	0.1g
NH_4Cl	0.1g
$MgCl_2 \cdot 7H_2O$	0.05g

pH 6.8 ± 0.2 at 25°C

Preparation of Medium: Add components to distilled/deionized water and bring volume to 1.0L. Mix thoroughly. Distribute into tubes or flasks. Autoclave for 15 min at 15 psi pressure–121°C.

Use: For the cultivation of *Thiobacillus thioparus* and *Thiobacillus thiooxidans*.

Thiobacillus Medium

Composition per liter:

$Na_2S_2O_3 \cdot 5H_2O$	10.0g
K_2HPO_4	4.0g
KH_2PO_4	4.0g
$CaCl_2$	0.1g
$MgSO_4 \cdot 7H_2O$	0.1g
$(NH_4)_2SO_4$	0.1g
$FeCl_3 \cdot 6H_2O$	0.02g
$MnSO_4 \cdot 4H_2O$	0.02g

pH 6.6 ± 0.2 at 25°C

Preparation of Medium: Add components to distilled/deionized water and bring volume to 1.0L. Mix thoroughly. Distribute into flasks in 100.0mL volumes. Autoclave for 60 min at 0 psi pressure–100°C on 3 consecutive days.

Use: For the cultivation of nonaciduric *Thiobacillus* species.

Thiobacillus Medium

Composition per liter:

$Na_2S_2O_3 \cdot 5H_2O$	10.0g
$Na_2HPO_4 \cdot 7H_2O$	7.9g
Sodium formate	6.8g
Glucose	3.6g
KNO_3	2.0g
KH_2PO_4	1.5g
NH_4Cl	0.3g
$MgSO_4 \cdot 7H_2O$	0.1g
Trace metals solution	5.0mL

pH 7.6–8.5 at 25°C

Trace Metals Solution:

Composition per liter:

Disodium EDTA	50.0g
NaOH	11.0g
$CaCl_2 \cdot 2H_2O$	7.34g
$FeSO_4 \cdot 7H_2O$	5.0g
$MnCl_2 \cdot 2H_2O$	2.5g
$ZnSO_4 \cdot 7H_2O$	2.2g
$CoCl_2 \cdot 6H_2O$	0.5g
$(NH_4)_6Mo_7O_{24} \cdot 4H_2O$	0.5g
$CuSO_4 \cdot 5H_2O$	0.2g

Preparation of Trace Metals Solution: Add EDTA to distilled/deionized water and bring volume to 500.0mL. Mix thoroughly. Adjust pH to 6.0 with NaOH. Add remaining components, one by one. Maintain the pH at 6.0. After dissolution of all the salts, adjust the pH to 4.0 with HCl. Store at 4°C.

Preparation of Medium: Add components to distilled/deionized water and bring volume to 1.0L. Mix thoroughly. Adjust pH to 7.6–8.5. Filter sterilize. Aseptically distribute into sterile tubes or flasks.

Use: For the isolation and anaerobic cultivation of *Thiobacillus* species.

Thiobacillus Medium

Composition per liter:

$Na_2S_2O_3 \cdot 5H_2O$	10.0g
$Na_2HPO_4 \cdot 7H_2O$	7.9g
Sodium formate	6.8g
Glucose	3.6g
KH_2PO_4	1.5g
NH_4Cl	0.3g
$MgSO_4 \cdot 7H_2O$	0.1g
Trace metals solution	5.0mL

pH 7.6–8.5 at 25°C

Trace Metals Solution:

Composition per liter:

Disodium EDTA	50.0g
NaOH	11.0g

$CaCl_2 \cdot 2H_2O$	7.34g
$FeSO_4 \cdot 7H_2O$	5.0g
$MnCl_2 \cdot 2H_2O$	2.5g
$ZnSO_4 \cdot 7H_2O$	2.2g
$CoCl_2 \cdot 6H_2O$	0.5g
$(NH_4)_6Mo_7O_{24} \cdot 4H_2O$	0.5g
$CuSO_4 \cdot 5H_2O$	0.2g

Preparation of Trace Metals Solution: Add EDTA to distilled/deionized water and bring volume to 500.0mL. Mix thoroughly. Adjust pH to 6.0 with NaOH. Add remaining components, one by one. Maintain the pH at 6.0. After dissolution of all the salts, adjust the pH to 4.0 with HCl. Store at 4°C.

Preparation of Medium: Add components to distilled/deionized water and bring volume to 1.0L. Mix thoroughly. Adjust pH to 7.6–8.5. Filter sterilize. Aseptically distribute into sterile tubes or flasks.

Use: For the isolation and aerobic cultivation of *Thiobacillus* species.

Thiobacillus Medium (ATCC Medium 64)

Composition per 500.0mL:

Solution A	400.0mL
Solution B	100.0mL

pH 2.8 ± 0.2 at 25°C

Solution A:

Composition per 400.0mL:

$(NH_4)_2SO_4$	0.4g
KH_2PO_4	0.2g
$MgSO_4 \cdot 7H_2O$	0.08g

Preparation of Solution A: Add components to distilled/deionized water and bring volume to 400.0mL. Mix thoroughly. Autoclave for 15 min at 15 psi pressure–121°C. Cool to 45°–50°C.

Solution B:

Composition per 100.0mL:

$FeSO_4 \cdot 7H_2O$	10.0g
H_2SO_4 (1*N* solution)	1.0mL

Preparation of Solution B: Add components to distilled/deionized water and bring volume to 100.0mL. Mix thoroughly. Autoclave for 15 min at 15 psi pressure–121°C. Cool to 45°–50°C.

Preparation of Medium: Aseptically add 100.0mL of cooled sterile solution B to 400.0mL of cooled sterile solution A. Mix thoroughly. Adjust pH to 2.8. Aseptically distribute into sterile tubes or flasks.

Use: For the cultivation and maintenance of a variety of *Thiobacillus* species.

Thiobacillus Medium (ATCC Medium 125)

Composition per liter:

Sulfur	10.0g
KH_2PO_4	3.0g
$MgSO_4 \cdot 7H_2O$	0.5g
$CaCl_2$	0.25g
$(NH_4)_2SO_4$	0.2g
$FeSO_4 \cdot 7H_2O$	5.0mg

Preparation of Medium: Add components, except sulfur, to tap water and bring volume to 1.0L. Mix thoroughly. Add 1.0g of sulfur to each of 10 flasks. Distribute the broth in 100.0mL volumes into the flasks. Pour the broth down the side of the flask so that the sulfur is not wetted. Autoclave for 30 min at 0 psi pressure–100°C on 3 consecutive days. Be sure that sulfur remains on the surface of the broth during the sterilization.

Use: For the cultivation and maintenance of a variety of *Thiobacillus* species.

Thiobacillus Medium (ATCC Medium 152)

Composition per liter:

Agar	15.0g
$Na_2S_2O_3 \cdot 5H_2O$	10.0g
NH_4Cl	1.0g
Yeast extract	1.0g
K_2HPO_4	0.6g
$MgCl_2$	0.5g
KH_2PO_4	0.4g
Chlorophenol Red	0.08g
$FeCl_3$	0.02g

Preparation of Medium: Add components to distilled/deionized water and bring volume to 1.0L. Mix thoroughly. Gently heat and bring to boiling. Distribute into tubes or flasks. Autoclave for 15 min at 15 psi pressure–121°C. Pour into sterile Petri dishes or leave in tubes.

Use: For the cultivation and maintenance of a variety of *Thiobacillus* species.

Thiobacillus Medium (ATCC Medium 426)

Composition per liter:

$Na_2S_2O_3 \cdot 5H_2O$	10.0g
$Na_2HPO_4 \cdot 7H_2O$	7.9g
KH_2PO_4	1.5g
NH_4Cl	0.3g
$MgSO_4 \cdot 7H_2O$	0.1g
Phenol Red	2.0mg
Trace metals solution	5.0mL

pH 8.5 ± 0.2 at 25°C

Trace Metals Solution:

Composition per liter:

EDTA	50.0g
$ZnSO_4$	22.0g
$CaCl_2$	5.54g
$MnCl_2$	5.06g
$FeSO_4 \cdot 7H_2O$	4.99g
$CoCl_2$	1.61g
$CuSO_4$	1.57g
$(NH_4)_2MoO_4 \cdot 4H_2O$	1.1g

Preparation of Trace Metals Solution: Add components to distilled/deionized water and bring volume to 1.0L. Mix thoroughly. Adjust pH to 6.0 with KOH.

Preparation of Medium: Add components to distilled/deionized water and bring volume to 1.0L. Mix thoroughly. Adjust pH to 8.5 with 10% Na_2CO_3. Distribute into tubes or flasks. Autoclave for 15 min at 15 psi pressure–121°C. Adjust pH to 8.5 with sterile 10% Na_2CO_3 if necessary. The broth should be pink.

Use: For the cultivation and maintenance of a variety of *Thiobacillus* species.

Thiobacillus Medium (ATCC Medium 528)

Composition per liter:

$Na_2S_2O_3$	10.0g
Yeast extract	5.0g
NH_4Cl	1.0g
K_2HPO_4	0.6g
$MgCl_2$	0.5g
KH_2PO_4	0.4g
$MgSO_4$	0.3g
Bromthymol Blue	0.03g
$FeCl_3$	0.02g
Heavy metal solution	30.0mL

pH 6.8 ± 0.2 at 25°C

Heavy Metal Solution:

Composition per liter:

Ethylenediamine tetraacetate	1.5g
$FeSO_4 \cdot 7H_2O$	0.2g
$ZnSO_4 \cdot 7H_2O$	0.1g
$MnCl_2 \cdot 4H_2O$	0.02g
Modified Hoagland trace elements solution	6.0mL

Preparation of Heavy Metal Solution: Add EDTA to approximately 900.0mL of distilled/deionized water. Dissolve by adjusting pH to 7.0 with NaOH. Bring volume to 1.0L with distilled/deionized water.

Modified Hoagland Trace Elements Solution:

Composition per 3.6L:

H_3BO_3	11.0g
$MnCl_2 \cdot 4H_2O$	7.0g
$AlCl_3$	1.0g
$CoCl_2$	1.0g
$CuCl_2$	1.0g
KI	1.0g
$NiCl_2$	1.0g
$ZnCl_2$	1.0g
$BaCl_2$	0.5g
KBr	0.5g
LiCl	0.5g
Na_2MoO_4	0.5g
$SeCl_4$	0.5g
$SnCl_2 \cdot 2H_2O$	0.5g
$NaVO_3 \cdot H_2O$	0.1g

Preparation of Modified Hoagland Trace Elements Solution: Prepare each component as a separate solution. Dissolve each salt in approximately 100.0mL of distilled/deionized water. Adjust the pH of each solution to below 7.0. Combine all the salt solutions and bring the volume to 3.6L with distilled/deionized water. Adjust the pH to 3–4. A yellow precipitate may form after mixing. After a few days, it will turn into a fine white precipitate. Mix the solution thoroughly .

Preparation of Medium: Add components to distilled/deionized water and bring volume to 1.0L. Mix thoroughly. Distribute into tubes or flasks. Autoclave for 15 min at 15 psi pressure–121°C.

Use: For the cultivation and maintenance of a variety of *Thiobacillus* species.

Thiobacillus Medium B

Composition per liter:

Noble agar	15.0g
$Na_2S_2O_3.5H_2O$	5.0g
KH_2PO_4	3.0g
NH_4Cl	0.1g
$MgCl_2$	0.1g
$CaCl_2$	0.1g

pH 4.2 ± 0.2 at 25°C

Preparation of Medium: Add components to distilled/deionized water and bring volume to 1.0L. Mix thoroughly. Gently heat and bring to boiling. Distribute into tubes or flasks. Autoclave for 15 min at 15 psi pressure–121°C. Pour into sterile Petri dishes or leave in tubes.

Use: For the cultivation and maintenance of *Thiobacillus thiooxidans* and *Streptomyces scabies*.

Thiobacillus neapolitanus Medium

Composition per 1002.0mL:

Solution I	1.0L
Solution II	2.0mL

pH 6.2–7.0 at 25°C

Solution I:

Composition per liter:

$Na_2S_2O_3 \cdot 5H_2O$	10.0g
KH_2PO_4	4.0g
K_2HPO_4	4.0g
$MgSO_4 \cdot 7H_2O$	0.8g
$KHCO_3$	0.7g
NH_4Cl	0.4g

Preparation of Solution I: Add components to distilled/deionized water and bring volume to 1.0L. Mix thoroughly.

Solution II:

Composition per liter:

Disodium EDTA	50.0g
NaOH	11.0g
$CaCl_2 \cdot 2H_2O$	7.34g
$FeSO_4 \cdot 7H_2O$	5.0g
$MnCl_2 \cdot 2H_2O$	2.5g
$ZnSO_4 \cdot 7H_2O$	2.2g
$CoCl_2 \cdot 6H_2O$	0.5g
$(NH_4)_6Mo_7O_{24} \cdot 4H_2O$	0.5g
$CuSO_4 \cdot 5H_2O$	0.2g

Preparation of Solution II: Add EDTA to distilled/deionized water and bring volume to 500.0mL. Mix thoroughly. Adjust pH to 6.0 with NaOH. Add remaining components, one by one. Maintain the pH at 6.0. After dissolution of all the salts, adjust the pH to 4.0 with HCl. Store at 4°C.

Preparation of Medium: Aseptically combine 1.0L of solution I and 2.0mL of solution II. Mix thoroughly. Adjust pH to 6.2–7.0. Distribute into tubes or flasks. Autoclave for 15 min at 15 psi pressure–121°C.

Use: For the isolation and cultivation of *Thiobacillus neapolitanus*.

Thiobacillus novellus Medium

Composition per liter:

$Na_2S_2O_3 \cdot 5H_2O$	10.0g
K_2HPO	4.0g
KH_2PO_4	1.5g
$MgSO_4 \cdot 7H_2O$	0.5g
$(NH_4)_2SO_4$	0.3g

Yeast extract	0.3g
Trace metals solution	10.0mL

pH 6.8–7.2 at 25°C

Trace Metals Solution:

Composition per liter:

Disodium EDTA	50.0g
NaOH	11.0g
$CaCl_2 \cdot 2H_2O$	7.34g
$FeSO_4 \cdot 7H_2O$	5.0g
$MnCl_2 \cdot 2H_2O$	2.5g
$ZnSO_4 \cdot 7H_2O$	2.2g
$CoCl_2 \cdot 6H_2O$	0.5g
$(NH_4)_6Mo_7O_{24} \cdot 4H_2O$	0.5g
$CuSO_4 \cdot 5H_2O$	0.2g

Preparation of Trace Metals Solution: Add EDTA to distilled/deionized water and bring volume to 500.0mL. Mix thoroughly. Adjust pH to 6.0 with NaOH. Add remaining components, one by one. Maintain the pH at 6.0. After dissolution of all the salts, adjust the pH to 4.0 with HCl. Store at 4°C.

Preparation of Medium: Add components to distilled/deionized water and bring volume to 1.0L. Mix thoroughly. Distribute into tubes or flasks. Autoclave for 15 min at 15 psi pressure–121°C.

Use: For the isolation and cultivation of *Thiobacillus novellus*.

Thiobacillus perometabolis Medium

Composition per liter:

$Na_2S_2O_3$	10.0g
Yeast extract	5.0g
NH_4Cl	1.0g
K_2HPO_4	0.6g
$MgCl_2$	0.5g
KH_2PO_4	0.4g
$MgSO_4$	0.3g
Bromthymol Blue	0.03g
$FeCl_3$	0.02g
Heavy metal solution	30.0mL

pH 3.0–4.0 at 25°C

Heavy Metal Solution:

Composition per liter:

Ethylenediamine tetraacetate	1.5g
$FeSO_4 \cdot 7H_2O$	0.2g
$ZnSO_4 \cdot H_2O$	0.1g
$MnCl_2 \cdot 4H_2O$	0.02g
Modified Hoagland trace elements solution	6.0mL

Preparation of Heavy Metal Solution: Add ethylenediamine tetraacetate to distilled/deionized water and bring volume to 500.0mL. Mix thoroughly. Adjust pH to 6.8 to dissolve EDTA. Add remaining components. Bring volume to 1.0L with distilled/deionized water. Mix thoroughly. Adjust pH to 6.8.

Modified Hoagland Trace Elements Solution:

Composition per liter:

H_3BO_3	11.0g
$MnCl_2 \cdot 4H_2O$	7.0g
$AlCl_3$	1.0g
$CoCl_2$	1.0g
$CuCl_2$	1.0g
KI	1.0g
$NiCl_2$	1.0g
$ZnCl_2$	1.0g
$BaCl_2$	0.5g
KBr	0.5g
LiCl	0.5g
Na_2MoO_4	0.5g
SeS_2	0.5g
$SnCl_2 \cdot 2H_2O$	0.5g
$NaVO_3 \cdot H_2O$	0.1g

Preparation of Modified Hoagland Trace Elements Solution: Add components seqentially to distilled/deionized water and bring volume to 1.0L. Mix thoroughly. Adjust pH to 6.8.

Preparation of Medium: Add components to distilled/deionized water and bring volume to 1.0L. Mix thoroughly. Adjust pH to 3–4. Distribute into tubes or flasks. Autoclave for 15 min at 15 psi pressure–121°C. A flaky yellow precipitate forms after mixing but will change in a few days to a fine, white precipitate. Mix the medium thoroughly before use.

Use: For the cultivation of *Thiobacillus perometabolis* and other bacteria that can utilize thiosulfate as an energy source.

Thiobacillus plumbophilus Medium

Composition per 1010.0mL:

$NiCl_2 \cdot 6H_2O$	5.0g
$MgSO_4 \cdot 7H_2O$	3.45g
$MgCl_2 \cdot 6H_2O$	2.75g
NH_4Cl	1.25g
NaCl	0.5g
KCl	0.3g
$CaCl_2 \cdot 2H_2O$	0.14g
K_2HPO_4	0.14g
$Na_2WO_4 \cdot 2H_2O$	0.5mg
Trace elements solution	10.0mL

pH 6.0 ± 0.2 at 25°C

Trace Elements Solution:

Composition per liter:

$MgSO_4 \cdot 7H_2O$	3.0g
Nitrilotriacetic acid	1.5g
NaCl	1.0g
$MnSO_4 \cdot 2H_2O$	0.5g
$CoSO_4 \cdot 7H_2O$	0.18g
$ZnSO_4 \cdot 7H_2O$	0.18g
$CaCl_2 \cdot 2H_2O$	0.1g
$FeSO_4 \cdot 7H_2O$	0.1g
$KAl(SO_4)_2 \cdot 12H_2O$	0.02g
$CuSO_4 \cdot 5H_2O$	0.01g
H_3BO_3	0.01g
$Na_2MoO_4 \cdot 2H_2O$	0.01g
$NiCl_2 \cdot 6H_2O$	0.025g
$Na_2SeO_3 \cdot 5H_2O$	0.3mg

Preparation of Trace Elements Solution: Add nitrilotriacetic acid to 500.0mL of distilled/deionized water. Adjust pH to 6.5 with KOH. Add remaining components. Add distilled/deionized water to 1.0L.

Preparation of Medium: Add components to distilled/deionized water and bring volume to 1.0L. Mix thoroughly. Adjust pH to 6.0 with H_2SO_4. Distribute 20.0mL volumes into 100.0mL serum bottles. Sparge with 80% H_2 + 20% CO_2. Autoclave for 15 min at 15 psi pressure–121°C. Check pH of medium after autoclaving and bring to 6.0 if necessary. After inoculation add 3% sterile air (9.0mL per serum bottle) with a syringe. Pressurize bottles to 2 bar with 80% H_2 + 20% CO_2.

Use: For the cultivation and maintenance of *Thiobacillus plumbophilus*.

Thiobacillus prosperus Medium

Composition per 1010.0mL:

$MgSO_4 \cdot 7H_2O$	3.45g
$MgCl_2 \cdot 6H_2O$	2.75g
NH_4Cl	1.25g
NaCl	0.5g
Sulfur, powdered	0.5g
KCl	0.33g
$CaCl_2 \cdot 2H_2O$	0.14g
K_2HPO_4	0.14g
KH_2PO_4	0.14g
$NiCl_2 \cdot 6H_2O$	2.0mg
Trace elements solution	10.0mL

pH 2.5 ± 0.2 at 25°C

Preparation of Sulfur: Add 10.0g of powdered sulfur to a flask and sterilize by steaming for 3 hr on 3 consecutive days.

Trace Elements Solution:

Composition per liter:

$MgSO_4 \cdot 7H_2O$	3.0g
Nitrilotriacetic acid	1.5g
NaCl	1.0g
$MnSO_4 \cdot 2H_2O$	0.5g
$CoSO_4 \cdot 7H_2O$	0.18g
$ZnSO_4 \cdot 7H_2O$	0.18g
$CaCl_2 \cdot 2H_2O$	0.1g
$FeSO_4 \cdot 7H_2O$	0.1g
$NiCl_2 \cdot 6H_2O$	0.025g
$KAl(SO_4)_2 \cdot 12H_2O$	0.02g
$CuSO_4 \cdot 5H_2O$	0.01g
H_3BO_3	0.01g
$Na_2MoO_4 \cdot 2H_2O$	0.01g
$Na_2SeO_3 \cdot 5H_2O$	0.3mg

Preparation of Trace Elements Solution: Add nitrilotriacetic acid to 500.0mL of distilled/deionized water. Adjust pH to 6.5 with KOH. Add remaining components. Add distilled/deionized water to 1.0L.

Preparation of Medium: Add components, except sulfur, to distilled/deionized water and bring volume to 1.0L. Mix thoroughly. Adjust pH to 2.5 with H_2SO_4. Autoclave for 15 min at 15 psi pressure–121°C. Aseptically add 0.5g of sterile sulfur. Mix thoroughly. Aseptically distribute into sterile tubes or flasks.

Use: For the cultivation and maintenance of *Thiobacillus prosperus*.

Thiobacillus tepidarius Medium

Composition per liter:

Agar	10.0g
$Na_2S_2O_3 \cdot 5H_2O$	4.96g
$MgSO_4 \cdot 7H_2O$	0.8g
NH_4Cl	0.4g
Phosphate solution	100.0mL
Bromcresol Purple, saturated solution	2.0mL
Trace metals A-5	1.0mL

Phosphate Solution:

Composition per 100.0mL:

KH_2PO_4	4.0g
K_2HPO_4	4.0g

Preparation of Phosphate Solution: Add components to distilled/deionized water and bring volume to 100.0mL. Mix thoroughly. Autoclave for 15 min at 15 psi pressure–121°C.

Trace Metals A-5:

Composition per liter:

H_3BO_3	2.86g
$MnCl_2 \cdot 4H_2O$	1.81g
$Na_2MoO_4 \cdot 2H_2O$	0.39g
$ZnSO_4 \cdot 7H_2O$	0.222g
$CuSO_4 \cdot 5H_2O$	0.079g
$Co(NO_3)_2 \cdot 6H_2O$	49.4mg

Preparation of Trace Metals A-5: Add components to distilled/deionized water and bring volume to 1.0L. Mix thoroughly.

Preparation of Medium: Add components, except phosphate solution, to distilled/deionized water and bring volume to 900.0mL. Autoclave for 15 min at 15 psi pressure–121°C. Aseptically add 100.0mL of the sterile phosphate solution. Mix thoroughly. Aseptically distribute into sterile tubes or flasks.

Use: For the cultivation and maintenance of *Thiobacillus tepidarius*.

Thiobacillus/Thermophilus Medium

Composition per liter:

$Na_2S_2O_3 \cdot 5H_2O$	5.0g
$NaHCO_3$	1.0g
Na_2HPO_4	0.2g
$MgCl_2$	0.1g
$NH_4 \cdot Cl$	0.1g

pH 7.0 ± 0.2 at 25°C

Preparation of Medium: Add components to distilled/deionized water and bring volume to 1.0L. Mix thoroughly. Distribute into tubes or flasks. Autoclave for 20 min at 6 psi pressure–109°C or filter sterilize.

Use: For the cultivation of *Thiobacillus* species and *Thermophilus* species.

Thiobacillus thiooxidans Medium

Composition per liter:

Sulfur, powdered	10.0g
KH_2PO_4	5.0g
$MgSO_4 \cdot 7H_2O$	0.5g
$CaCl_2$	0.25g
$(NH_4)_2SO_4$	0.2g
$FeSO_4$	0.01g

pH 7.0 ± 0.2 at 25°C

Preparation of Medium: Add components, except sulfur, to distilled/deionized water and bring volume to 1.0L. Mix thoroughly. Distribute into flasks in 100.0mL volumes. Add 1.0g of sulfur to each flask. Autoclave for 30 min at 0 psi pressure–100°C on 3 consecutive days.

Use: For the cultivation of *Thiobacillus thiooxidans*.

Thiobacillus thiooxidans Medium

Composition per liter:

Flowers of sulfur	5.0g
K_2HPO	3.5g
$MgSO_4 \cdot 7H_2O$	0.5g
$(NH_4)_2SO_4$	0.3g

$CaCl_2$ 0.25g
$FeSO_4 \cdot 7H_2O$ 0.02g

pH 4.5 ± 0.2 at 25°C

Preparation of Medium: Add components, except flowers of sulfur, to distilled/deionized water and bring volume to 1.0L. Mix thoroughly. Gently heat and bring to boiling. Distribute into flasks or bottles in 100.0mL volumes. Add 0.5g of flowers of sulfur to each flask or bottle. Autoclave for 15 min at 15 psi pressure–121°C.

Use: For the isolation and cultivation of *Thiobacillus thiooxidans*.

Thiobacillus thiooxidans Medium

Composition per liter:

Sulfur, powdered 10.0g
KH_2PO_4 .... 3.0g
$CaCl_2 \cdot 2H_2O$ 0.14g
$MgCl_2 \cdot 6H_2O$ 0.1g
NH_4Cl 0.1g

pH 4.2 ± 0.2 at 25°C

Preparation of Sulfur: Add 10.0g of powdered sulfur to a flask and sterilize by autoclaving for 15 min at 8 psi pressure–112°C.

Preparation of Medium: Add components, except sulfur, to distilled/deionized water and bring volume to 1.0L. Mix thoroughly. Adjust pH to 2.5 with H_2SO_4. Autoclave for 15 min at 15 psi pressure–121°C. Aseptically layer 0.5g of sterile sulfur onto the surface of the medium.

Use: For the cultivation and maintenance of *Thiobacillus* species.

Thiobacillus thioparus Agar

Composition per liter:

Agar, purified 12.0g
$Na_2S_2O_3 \cdot 5H_2O$ 10.0g
K_2HPO_4 .... 4.0g
KH_2PO_4 .... 4.0g
$MgSO_4 \cdot 7H_2O$ 0.1g
$(NH_4)_2SO_4$ 0.1g
$CaCl_2$ 0.1g
$FeCl_3 \cdot 6H_2O$ 0.02g
$MnSO_4 \cdot H_2O$ 0.02g

pH 6.6 ± 0.2 at 25°C

Preparation of Medium: Add components, except agar, to distilled/deionized water and bring volume to 1.0L. Mix thoroughly. Adjust pH to 6.6. Add agar. Gently heat and bring to boiling. Distribute into tubes or flasks. Autoclave for 20 min at 10 psi pressure–115°C. Pour into sterile Petri dishes or leave in tubes.

Use: For the cultivation and maintenance of *Thiobacillus thioparus*.

Thiobacillus thioparus Broth

Composition per liter:

$Na_2S_2O_3 \cdot 5H_2O$ 10.0g
K_2HPO_4 .... 4.0g
KH_2PO_4 .... 4.0g
$MgSO_4 \cdot 7H_2O$ 0.1g
$(NH_4)_2SO_4$ 0.1g
$CaCl_2$ 0.1g
$FeCl_3 \cdot 6H_2O$ 0.02g
$MnSO_4 \cdot H_2O$ 0.02g

pH 6.6 ± 0.2 at 25°C

Preparation of Medium: Add components to distilled/deionized water and bring volume to 1.0L. Mix thoroughly. Adjust pH to 6.6. Distribute into tubes or flasks. Autoclave for 20 min at 10 psi pressure–115°C.

Use: For the cultivation and maintenance of *Thiobacillus thioparus*.

Thiobacillus thioparus Medium

Composition per liter:

$Na_2S_2O_3 \cdot 5H_2O$ 5.0g
K_2HPO_4 .... 4.0g
$MgSO_4 \cdot 7H_2O$ 0.5g
$(NH_4)_2SO_4$ 0.4g
$CaCl_2$ 0.25g
$FeSO_4$ 0.01g

pH 7.0 ± 0.2 at 25°C

Preparation of Medium: Add components to distilled/deionized water and bring volume to 1.0L. Mix thoroughly. Distribute into tubes or flasks. Autoclave for 15 min at 15 psi pressure–121°C.

Use: For the cultivation of *Thiobacillus thioparus*.

Thiobacillus thioparus II Medium (DSMZ Medium 486)

Composition per 1003.0mL:

Solution A 900.0mL
Solution B 100.0mL
Vitamin solution 3.0mL

pH 7.4 ± 0.2 at 25°C

Solution A:

Composition per 900.0mL:

$Na_2S_2O_3 \cdot 5H_2O$ 5.0g
NH_4Cl 0.4g
Na_2CO_3 .... 0.4g
$MgCl_2 \cdot 6H_2O$ 0.2g
Bromcresol Purple (saturated solution) 2.0mL
Trace metals solution 1.0mL

Preparation of Solution A: Add components to distilled/deionized water and bring volume to 900.0mL. Mix thoroughly. Autoclave for 15 min at 15 psi pressure–121°C. Cool to room temperature.

Solution B:

Composition per 100.0mL:

KH_2PO_4 .... 2.0g
K_2HPO_4 .... 2.0g

Preparation of Solution B: Add components to distilled/deionized water and bring volume to 100.0mL. Mix thoroughly. Adjust pH to 7.6. Autoclave for 15 min at 15 psi pressure–121°C. Cool to room temperature.

Trace Elements Solution:

Composition per liter:

Na_2-EDTA 50.0g
NaOH 11.0g
$ZnSO_4 \cdot 7H_2O$ 11.0g
$CaCl_2 \cdot 2H_2O$ 7.34g
$FeSO_4 \cdot 7H_2O$ 5.0g
$MnCl_2 \cdot 4H_2O$ 2.5g
$CoCl_2 \cdot 6H_2O$ 0.5g
$(NH_4)_6Mo_7O_{24} \cdot 4H_2O$ 0.5g
$CuSO_4 \cdot 5H_2O$ 0.2g

Preparation of Trace Elements Solution: Add components to distilled/deionized water and bring volume to 1.0L. Mix thoroughly.

Vitamin Solution:

Composition per liter:

Riboflavin	20.0mg
Ca-pantothenate	20.0mg
Nicotinic acid	20.0mg
Pyridoxine-HCl	20.0mg
p-Aminobenzoic acid	10.0mg
Thiamine-HCl·$2H_2O$	10.0mg
Biotin	1.0mg
Vitamin B_{12}	1.0mg

Preparation of Vitamin Solution: Add components to distilled/deionized water and bring volume to 1.0L. Mix thoroughly. Filter sterilize.

Preparation of Medium: Aseptically combine 100.0mL solution A, 900.0mL solution B, and 3.0mL vitamin solution. Mix thoroughly. Aseptically distribute into sterile tubes or flasks.

Use: For the cultivation and maintenance of *Thiobacillus thioparus*.

Thiobacillus thioparusii Agar

Composition per 1003.0mL:

Solution B	900.0mL
Solution A	100.0mL
Vitamin solution	3.0mL

pH 7.1 ± 0.2 at 25°C

Solution A:

Composition per 100.0mL:

K_2HPO_4	2.0g
KH_2PO_4	2.0g

Preparation of Solution A: Add components to distilled/deionized water and bring volume to 100.0mL. Mix thoroughly. Autoclave for 15 min at 15 psi pressure–121°C. Cool to 50°–55°C.

Solution B:

Composition per 900.0mL:

Agar	15.0g
$Na_2S_2O_3 \cdot 5H_2O$	5.0g
Na_2CO_3	0.4g
NH_4Cl	0.4g
$MgCl_2 \cdot 6H_2O$	0.2g
Bromcresol Purple (saturated aqueous solution)	2.0mL
Trace elements solution	1.0mL

Trace Elements Solution:

Composition per liter:

Disodium EDTA	50.0g
NaOH	11.0g
$ZnSO_4 \cdot 7H_2O$	11.0g
$CaCl_2 \cdot 2H_2O$	7.34g
$FeSO_4 \cdot 7H_2O$	5.0g
$MnCl_2 \cdot 4H_2O$	2.5g
$CoCl_2 \cdot 6H_2O$	0.5g
$(NH_4)_6MoO_{24} \cdot 4H_2O$	0.5g
$CuSO_4 \cdot 5H_2O$	0.2g

Preparation of Trace Elements Solution: Add components to distilled/deionized water and bring volume to 1.0L. Mix thoroughly. Adjust pH to 6.0.

Preparation of Solution B: Add components to distilled/deionized water and bring volume to 900.0mL. Mix thoroughly. Gently heat and bring to boiling. Autoclave for 15 min at 15 psi pressure–121°C. Cool to 50°–55°C.

Vitamin Solution:

Composition per liter:

Calcium DL-pantothenate	20.0mg
Nicotinic acid	20.0mg
Pyridoxine·HCl	20.0mg
Riboflavin	20.0mg
p-Aminobenzoic acid	10.0mg
Thiamine·HCl	10.0mg
Biotin	1.0mg
Vitamin B_{12}	1.0mg

Preparation of Vitamin Solution: Add components to distilled/deionized water and bring volume to 1.0L. Mix thoroughly. Adjust pH to 7.0. Filter sterilize.

Preparation of Medium: Aseptically combine 100.0mL of sterile solution A with 900.0mL of sterile solution B and 3.0mL of sterile vitamin solution. Mix thoroughly. Pour into sterile Petri dishes or distribute into sterile tubes.

Use: For the cultivation and maintenance of *Thiobacillus thioparus*.

Thiobacillus thioparusii Broth

Composition per 1003.0mL:

Solution B	900.0mL
Solution A	100.0mL
Vitamin solution	3.0mL

pH 7.1 ± 0.2 at 25°C

Solution A:

Composition per 100.0mL:

K_2HPO_4	2.0g
KH_2PO_4	2.0g

Preparation of Solution A: Add components to distilled/deionized water and bring volume to 100.0mL. Mix thoroughly. Autoclave for 15 min at 15 psi pressure–121°C. Cool to room temperature.

Solution B:

Composition per 900.0mL:

$Na_2S_2O_3 \cdot 5H_2O$	5.0g
Na_2CO_3	0.4g
NH_4Cl	0.4g
$MgCl_2 \cdot 6H_2O$	0.2g
Bromcresol Purple (saturated aqueous solution)	2.0mL
Trace elements solution	1.0mL

Trace Elements Solution:

Composition per liter:

Disodium EDTA	50.0g
NaOH	11.0g
$ZnSO_4 \cdot 7H_2O$	11.0g
$CaCl_2 \cdot 2H_2O$	7.34g
$FeSO_4 \cdot 7H_2O$	5.0g
$MnCl_2 \cdot 4H_2O$	2.5g
$CoCl_2 \cdot 6H_2O$	0.5g
$(NH_4)_6MoO_{24} \cdot 4H_2O$	0.5g
$CuSO_4 \cdot 5H_2O$	0.2g

Preparation of Trace Elements Solution: Add components to distilled/deionized water and bring volume to 1.0L. Mix thoroughly. Adjust pH to 6.0.

Preparation of Solution B: Add components to distilled/deionized water and bring volume to 900.0mL. Mix thoroughly. Gently heat and bring to boiling. Autoclave for 15 min at 15 psi pressure–121°C. Cool to room temperature.

Vitamin Solution:

Composition per liter:

Calcium DL-pantothenate	20.0mg
Nicotinic acid	20.0mg
Pyridoxine·HCl	20.0mg
Riboflavin	20.0mg
p-Aminobenzoic acid	10.0mg
Thiamine·HCl	10.0mg
Biotin	1.0mg
Vitamin B_{12}	1.0mg

Preparation of Vitamin Solution: Add components to distilled/deionized water and bring volume to 1.0L. Mix thoroughly. Adjust pH to 7.0. Filter sterilize.

Preparation of Medium: Aseptically combine 100.0mL of sterile solution A with 900.0mL of sterile solution B and 3.0mL of sterile vitamin solution. Mix thoroughly. Aseptically distribute into sterile tubes or flasks.

Use: For the cultivation and maintenance of *Thiobacillus thioparus*.

Thiobacillus thyasiris Agar

Composition per 1010.0mL:

Solution A	100.0mL
Solution B	900.0mL
Vitamin solution	10.0mL

pH 7.3–7.6 at 25°C

Solution A:

Composition per 100.0mL:

$Na_2HPO_42H_2O$	7.9g
KH_2PO_4	1.5g

Preparation of Solution A: Add components to distilled/deionized water and bring volume to 100.0mL. Mix thoroughly. Adjust pH to 7.4. Autoclave for 15 min at 15 psi pressure–121°C. Cool to 50°–55°C.

Solution B:

Composition per 900.0mL:

NaCl	25.1g
Agar	15.0g
$Na_2S_2O_3$	5.0g
NH_4Cl	0.4g
$MgSO_4·7H_2O$	0.1g
Phenol Red	3.0mg
Trace elements solution	10.0mL

Trace Elements Solution:

Composition per liter:

Disodium EDTA	50.0g
NaOH	11.0g
$ZnSO_4·7H_2O$	11.0g
$CaCl_2·2H_2O$	7.34g
$FeSO_4·7H_2O$	5.0g
$MnCl_2·4H_2O$	2.5g
$CoCl_2·6H_2O$	0.5g
$(NH_4)6MoO_{24}·4H_2O$	0.5g
$CuSO_4·5H_2O$	0.2g

Preparation of Trace Elements Solution: Add components to distilled/deionized water and bring volume to 1.0L. Mix thoroughly. Adjust pH to 6.0.

Preparation of Solution B: Add components to distilled/deionized water and bring volume to 900.0mL. Mix thoroughly. Gently heat and bring to boiling. Autoclave for 15 min at 15 psi pressure–121°C. Cool to 50°–55°C.

Vitamin Solution:

Composition per liter:

Calcium DL-pantothenate	20.0mg
Nicotinic acid	20.0mg
Pyridoxine·HCl	20.0mg
Riboflavin	20.0mg
p-Aminobenzoic acid	10.0mg
Thiamine·HCl	10.0mg
Biotin	1.0mg
Vitamin B_{12}	1.0mg

Preparation of Vitamin Solution: Add components to distilled/deionized water and bring volume to 1.0L. Mix thoroughly. Adjust pH to 7.0. Filter sterilize.

Preparation of Medium: Aseptically combine 100.0mL of sterile solution A with 900.0mL of sterile solution B and 10.0mL of sterile vitamin solution. Mix thoroughly. Pour into sterile Petri dishes or distribute into sterile tubes.

Use: For the cultivation and maintenance of *Thiobacillus thyasiris*.

Thiobacillus thyasiris Broth

Composition per 1010.0mL:

Solution B	900.0mL
Solution A	100.0mL
Vitamin solution	10.0mL

pH 7.3–7.6 at 25°C

Solution A:

Composition per 100.0mL:

$Na_2HPO_42H_2O$	7.9g
KH_2PO_4	1.5g

Preparation of Solution A: Add components to distilled/deionized water and bring volume to 100.0mL. Mix thoroughly. Adjust pH to 7.4. Autoclave for 15 min at 15 psi pressure–121°C. Cool to room temperature.

Solution B:

Composition per 900.0mL:

NaCl	25.1g
$Na_2S_2O_3$	5.0g
NH_4Cl	0.4g
$MgSO_4·7H_2O$	0.1g
Phenol Red	3.0mg
Trace elements solution	10.0mL

Trace Elements Solution:

Composition per liter:

Disodium EDTA	50.0g
NaOH	11.0g
$ZnSO_4·7H_2O$	11.0g
$CaCl_2·2H_2O$	7.34g
$FeSO_4·7H_2O$	5.0g
$MnCl_2·4H_2O$	2.5g
$CoCl_2·6H_2O$	0.5g

$(NH_4)_6MoO_{24}\cdot4H_2O$ 0.5g
$CuSO_4\cdot5H_2O$ 0.2g

Preparation of Trace Elements Solution: Add components to distilled/deionized water and bring volume to 1.0L. Mix thoroughly. Adjust pH to 6.0.

Preparation of Solution B: Add components to distilled/deionized water and bring volume to 900.0mL. Mix thoroughly. Autoclave for 15 min at 15 psi pressure–121°C. Cool to room temperature.

Vitamin Solution:
Composition per liter:
Calcium DL-pantothenate 20.0mg
Nicotinic acid 20.0mg
Pyridoxine·HCl 20.0mg
Riboflavin 20.0mg
p-Aminobenzoic acid 10.0mg
Thiamine·HCl 10.0mg
Biotin 1.0mg
Vitamin B_{12} 1.0mg

Preparation of Vitamin Solution: Add components to distilled/deionized water and bring volume to 1.0L. Mix thoroughly. Adjust pH to 7.0. Filter sterilize.

Preparation of Medium: Aseptically combine 100.0mL of sterile solution A with 900.0mL of sterile solution B and 10.0mL of sterile vitamin solution. Mix thoroughly. Aseptically distribute into sterile tubes or flasks.

Use: For the cultivation and maintenance of *Thiobacillus thyasiris*.

Thiobacillus thyasiris/Thiobacillus halophilus Medium (DSMZ Medium 484)

Composition per 1010.0mL:
Solution B 900.0mL
Solution A 100.0mL
Vitamin solution 10.0mL
pH 7.3–7.6 at 25°C

Solution A:
Composition per 100.0mL:
$Na_2HPO_4\cdot2H_2O$ 7.9g
KH_2PO_4 .... 1.5g

Preparation of Solution A: Add components to distilled/deionized water and bring volume to 100.0mL. Mix thoroughly. Adjust pH to 7.6. Autoclave for 15 min at 15 psi pressure–121°C. Cool to room temperature.

Solution B:
Composition per 900.0mL:
NaCl 25.1g
$Na_2S_2O_3$ 5.0g
NH_4Cl 0.4g
$MgSO_4\cdot7H_2O$ 0.1g
Phenol Red 3.0mg
Trace elements solution 10.0mL

Preparation of Solution B: Add components to distilled/deionized water and bring volume to 900.0mL. Mix thoroughly. Autoclave for 15 min at 15 psi pressure–121°C. Cool to room temperature.

Trace Elements Solution:
Composition per liter:
Na_2-EDTA 50.0g
NaOH 11.0g
$ZnSO_4\cdot7H_2O$ 11.0g
$CaCl_2\cdot2H_2O$ 7.34g
$FeSO_4\cdot7H_2O$ 5.0g
$MnCl_2\cdot4H_2O$ 2.5g
$CoCl_2\cdot6H_2O$ 0.5g
$(NH_4)_6Mo_7O_{24}\cdot4H_2O$ 0.5g
$CuSO_4\cdot5H_2O$ 0.2g

Preparation of Trace Elements Solution: Add components to distilled/deionized water and bring volume to 1.0L. Mix thoroughly.

Vitamin Solution:
Composition per liter:
Riboflavin 20.0mg
Ca-pantothenate 20.0mg
Nicotinic acid 20.0mg
Pyridoxine-HCl 20.0mg
p-Aminobenzoic acid 10.0mg
Thiamine-HCl·$2H_2O$ 10.0mg
Biotin 1.0mg
Vitamin B_{12} 1.0mg

Preparation of Vitamin Solution: Add components to distilled/deionized water and bring volume to 1.0L. Mix thoroughly. Filter sterilize.

Preparation of Medium: Aseptically combine 100.0mL solution A, 900.0mL solution B, and 10.0mL vitamin solution. Mix thoroughly. Aseptically distribute into sterile tubes or flasks.

Use: For the cultivation and maintenance of *Thiobacillus thyasiris, Thiobacillus halophilus, Thiomicrospira thyasirae,* and *Halothiobacillus halophilus* DSM6132.

Thiocapsa halophila Medium

Composition per 1061.0mL:
NaCl 70.0g
$MgCl_2\cdot6H_2O$ 4.5g
$MgSO_4\cdot7H_2O$ 3.0g
NH_4Cl 0.5g
KH_2PO_4 .... 0.3g
$CaCl_2\cdot2H_2O$ 0.05g
$NaHCO_3$ solution 40.0mL
$Na_2S\cdot9H_2O$ solution 10.0mL
$Na_2S_2O_3$ solution 10.0mL
Vitamin B_{12} solution 1.0mL
Trace elements solution SL-12B 1.0mL
pH 7.2 ± 0.2 at 25°C

$NaHCO_3$ Solution:
Composition per 100.0mL:
$NaHCO_3$ 5.0g

Preparation of $NaHCO_3$ Solution: Add $NaHCO_3$ to distilled/deionized water and bring volume to 100.0mL. Mix thoroughly. Filter sterilize.

$Na_2S\cdot9H_2O$ Solution:
Composition per 10.0mL:
$Na_2S\cdot9H_2O$ 0.6g

Preparation of $Na_2S\cdot9H_2O$ Solution: Add $Na_2S\cdot9H_2O$ to distilled/deionized water and bring volume to 10.0mL. Mix thoroughly. Filter sterilize.

$Na_2S_2O_3$ Solution :

Composition per 10.0mL:

$Na_2S_2O_3 \cdot 5H_2O$	0.5g

Preparation of $Na_2S_2O_3$ Solution: Add $Na_2S_2O_3 \cdot 5H_2O$ to distilled/deionized water and bring volume to 10.0mL. Mix thoroughly. Filter sterilize.

Vitamin B_{12} Solution:

Composition per 10.0mL:

Vitamin B_{12}	0.2mg

Preparation of Vitamin B_{12} Solution: Add vitamin B_{12} to distilled/deionized water and bring volume to 10.0mL. Mix thoroughly. Filter sterilize.

Trace Elements Solution SL-12B:

Composition per liter:

$MnCl_2 \cdot 2H_2O$	50.0g
Disodium EDTA	3.0g
$FeSO_4 \cdot 7H_2O$	1.1g
H_3BO_3	0.3g
$CoCl_2 \cdot 6H_2O$	0.19g
$ZnCl_2$	42.0mg
$NiCl_2 \cdot 6H_2O$	24.0mg
$Na_2MoO_4 \cdot 2H_2O$	18.0mg
$CuCl_2 \cdot 2H_2O$	2.0mg

Preparation of Trace Elements Solution SL-12B: Add components to distilled/deionized water and bring volume to 1.0L. Mix thoroughly.

Preparation of Medium: Add components, except $NaHCO_3$ solution, $Na_2S \cdot 9H_2O$ solution, $Na_2S_2O_3$ solution, and vitamin B_{12} solution, to distilled/deionized water and bring volume to 1.0L. Mix thoroughly. Autoclave for 15 min at 15 psi pressure–121°C. Cool to room temperature while sparging with 90% N_2 + 10% CO_2. Aseptically and anaerobically add 40.0mL of sterile $NaHCO_3$ solution, 10.0mL of sterile $Na_2S \cdot 9H_2O$ solution, 10.0mL of sterile $Na_2S_2O_3$ solution, and 1.0mL of sterile vitamin B_{12} solution. Mix thoroughly. Aseptically and anaerobically distribute into sterile tubes or flasks.

Use: For the cultivation and maintenance of *Thiocapsa halophila.*

Thiocapsa Medium

Composition per 127.0mL:

Solution 1	76.2mL
Solution 2 + Solution 3	44.8mL
Solution 4	6.0mL

Solution 1:

Composition per 2.5L:

NaCl	39.68g
$CaCl_2$	2.0g

Preparation of Solution 1: Add components to distilled/deionized water and bring volume to 2.5L. Distribute in 80.0mL volumes into 127.0mL screw-capped bottles. Autoclave for 15 min at 15 psi pressure–121°C.

Solution 2:

Composition per 100.0mL:

Sodium ascorbate	2.4g
KCl	1.0g
KH_2PO_4	1.0g
$MgCl_2 \cdot 6H_2O$	0.8g
NH_4Cl	0.8g
Heavy metal solution	50.0mL
Vitamin solution	15.0mL
Vitamin B_{12} solution	3.0mL

Preparation of Solution 2: Add components to distilled/deionized water and bring volume to 100.0mL. Mix thoroughly.

Heavy Metal Solution:

Composition per liter:

Ethylenediamine tetraacetate (EDTA)	1.5g
$FeSO_4 \cdot 7H_2O$	0.2g
$ZnSO_4 \cdot 7H_2O$	0.1g
$MnCl_2 \cdot 4H_2O$	0.02g
Modified Hoagland trace elements solution	6.0mL

Preparation of Heavy Metal Solution: Dissolve EDTA in approximately 800.0mL of distilled/deionized water. Add remaining components. Bring volume to 1.0L with distilled/deionized water. Mix thoroughly.

Modified Hoagland Trace Elements Solution:

Composition per 3.6L:

H_3BO_3	11.0g
$MnCl_2 \cdot 4H_2O$	7.0g
$AlCl_3$	1.0g
$CoCl_2$	1.0g
$CuCl_2$	1.0g
KI	1.0g
$NiCl_2$	1.0g
$ZnCl_2$	1.0g
$BaCl_2$	0.5g
KBr	0.5g
LiCl	0.5g
Na_2MoO_4	0.5g
$SeCl_4$	0.5g
$SnCl_2 \cdot 2H_2O$	0.5g
$NaVO_3 \cdot H_2O$	0.1g

Preparation of Modified Hoagland Trace Elements Solution: Prepare each component as a separate solution. Dissolve each salt in approximately 100.0mL of distilled/deionized water. Adjust the pH of each solution to below 7.0. Combine all the salt solutions and bring the volume to 3.6L with distilled/deionized water. Adjust the pH to 3–4. A yellow precipitate may form after mixing. After a few days, it will turn into a fine white precipitate. Mix the solution thoroughly before using.

Vitamin Solution:

Composition per 100.0mL:

Pyridoxamine·2HCl	5.0mg
Nicotinic acid	2.0mg
Thiamine	1.0mg
Pantothenic acid	0.5mg
Biotin	0.2mg
p-Aminobenzoic acid	0.1mg

Preparation of Vitamin Solution: Add components to distilled/deionized water and bring volume to 100.0mL. Mix thoroughly.

Vitamin B_{12} Solution:

Composition per 100.0mL:

Vitamin B_{12} (cyanocobalamin)	2.0mg

Preparation of Vitamin B_{12} Solution: Add vitamin B_{12} to distilled/deionized water and bring volume to 100.0mL. Mix thoroughly.

Solution 3:

Composition per 900.0mL:

$NaHCO_3$ 4.5g

Preparation of Solution 3: Add $NaHCO_3$ to distilled/deionized water and bring volume to 900.0mL. Mix thoroughly. Bubble 100% CO_2 through the solution for 30 min. After CO_2 saturation of solution 3, add solution 2 and immediately filter the mixture through a Seitz filter (or a Millipore) using positive CO_2 pressure to push the liquid through.

Solution 4:

Composition per 200.0mL:

$Na_2S{\cdot}9H_2O$ 3.0g

Preparation of Solution 4: Add $Na_2S{\cdot}9H_2O$ to distilled/deionized water and bring volume to 200.0mL. Add a magnetic stir bar to the flask. Autoclave for 15 min at 15 psi pressure–121°C. On a magnetic stirrer, slowly add 2.0mL of sterile 2*M* H_2SO_4. This partially neutralizes the solution. The solution should turn yellow. H_2S gas will be liberated; neutralization and distribution of the solution should be done as rapidly as possible under adequate ventilation.

Preparation of Medium: To the 80.0mL of sterile solution 1 in screw-capped bottles, add combined solutions 2 and 3 immediately after filtration and fill bottles to capacity. Mix thoroughly. Aseptically remove 6.0mL of the medium from the bottles and replace it with 6.0mL of neutralized solution 4. Let stand for 24 hr. The medium should form a fine white precipitate before using. To inoculate, remove 6.0mL of the completed medium from the bottles and replace it with 6.0mL of inoculum.

Use: For the cultivation and maintenance of a variety of *Thiocapsa* species.

Thiocyanate Agar

Composition per liter:

Solution A 800.0mL
Solution B 100.0mL
Solution C 100.0mL

Solution A:

Composition per 800.0mL:

Agar, noble 30.0g
K_2HPO_4 .... 1.0g
KH_2PO_4 .... 1.0g
$MgSO_4{\cdot}7H_2O$ 0.2g
$CaCl_2$ 20.0mg
$FeCl_3{\cdot}6H_2O$ (60%) 0.1mL

Preparation of Solution A: Add components to distilled/deionized water and bring volume to 1.0L. Mix thoroughly. Gently heat and bring to boiling. Autoclave for 15 min at 15 psi pressure–121°C. Cool to 50°–55°C.

Solution B:

Composition per 100.0mL:

KCNS 3.6g

Preparation of Solution B: Add KCNS to distilled/deionized water and bring volume to 100.0mL. Mix thoroughly. Autoclave for 15 min at 15 psi pressure–121°C. Cool to 50°–55°C.

Solution C:

Composition per 100.0mL:

Disodium succinate 1.5g

Preparation of Solution C: Add disodium succinate to distilled/deionized water and bring volume to 100.0mL. Mix thoroughly. Autoclave for 15 min at 15 psi pressure–121°C. Cool to 50°–55°C.

Preparation of Medium: Aseptically combine 800.0mL of solution A with 100.0mL of solution B and 100.0mL of solution C. Mix thoroughly. Pour into sterile Petri dishes or aseptically distribute into sterile tubes.

Use: For the cultivation and maintenance of a variety of microorganisms that can utilize thiocyanante as sole source of nitrogen and sulfur.

Thiocyanate Utilization Medium

Composition per 1225.0mL:

Basal solution 1.0L
Solution C 200.0mL
Solution B 20.0mL
Solution A 5.0mL

Basal Solution:

Composition per liter:

Na_2HPO_4 .... 4.8g
KH_2PO_4 .... 4.4g
$MgSO_4{\cdot}7H_2O$ 0.5g

Preparation of Basal Solution: Add components to distilled/deionized water and bring volume to 1.0L. Mix thoroughly. Autoclave for 15 min at 15 psi pressure–121°C. Cool to 45°–50°C.

Solution A:

Composition per 100.0mL:

$FeCl_3{\cdot}6H_2O$ 1.0g
$CaCl_2$ 0.1g

Preparation of Solution A: Add components to distilled/deionized water and bring volume to 100.0mL. Mix thoroughly. Filter sterilize.

Solution B:

Composition per 100.0mL:

D-Glucose 10.0g

Preparation of Solution B: Add glucose to distilled/deionized water and bring volume to 100.0mL. Mix thoroughly. Filter sterilize.

Solution C:

Composition per 200.0mL:

NaSCN 1.0g

Preparation of Solution C: Add NaSCN to distilled/deionized water and bring volume to 200.0mL. Mix thoroughly. Filter sterilize.

Preparation of Medium: To 1.0L of cooled, sterile basal solution, aseptically add 5.0mL of sterile solution A, 20.0mL of sterile solution B, and 200.0mL of sterile solution C. Mix thoroughly. Aseptically distribute into sterile tubes or flasks.

Use: For the cultivation and maintenance of a variety of microorganisms that can utilize thiocyanate as sole source of nitrogen and sulfur.

Thiofaba Medium (DSMZ Medium 1114)

Composition per liter:

Solution A 960.0mL
Solution B 10.0mL
Solution C 10.0mL
Solution D 10.0mL
Solution E 10.0mL

pH 6.5 ± 0.2 at 25°C

Solution A:

Composition per 960.0mL:

$MgCl_2 \cdot 6H_2O$	0.75g
$CaCl_2 \cdot 2H_2O$	0.15g
NH_4Cl	0.54g
Trace elements solution	2.0mL

Trace Elements Solution:

Composition per liter:

$MgSO_4 \cdot 7H_2O$	3.0g
Nitrilotriacetic acid	1.5g
NaCl	1.0g
$MnSO_4 \cdot 2H_2O$	0.5g
$CoSO_4 \cdot 7H_2O$	0.18g
$ZnSO_4 \cdot 7H_2O$	0.18g
$CaCl_2 \cdot 2H_2O$	0.1g
$FeSO_4 \cdot 7H_2O$	0.1g
$NiCl_2 \cdot 6H_2O$	0.025g
$KAl(SO_4)_2 \cdot 12H_2O$	0.02g
H_3BO_3	0.01g
$Na_2MoO_4 \cdot 4H_2O$	0.01g
$CuSO_4 \cdot 5H_2O$	0.01g
$Na_2SeO_3 \cdot 5H_2O$	0.3mg

Preparation of Trace Elements Solution: Add nitrilotriacetic acid to 500.0mL of distilled/deionized water. Dissolve by adjusting pH to 6.5 with KOH. Add remaining components. Add distilled/deionized water to 1.0L. Mix thoroughly.

Preparation of Solution A: Add components to distilled/deionized water and bring volume to 960.0mL. Mix thoroughly. Sparge with a gas mixture of 80% N_2 + 20% CO_2. Autoclave for 15 min at 15 psi pressure–121°C. Cool to room temperature.

Solution B:

Composition per 10.0mL:

KH_2PO_4	1.19g
KH_2PO_4	0.21g

Preparation of Solution B: Add components to distilled/deionized water and bring volume to 10.0mL. Mix thoroughly. Sparge with 100% N_2. Autoclave or 15 min at 15 psi pressure–121°C. Cool to room temperature.

Solution C:

Composition per 10.0mL:

Na_2CO_3	4.0g

Preparation of Solution C: Add Na_2CO_3 to distilled/deionized water and bring volume to 10.0mL. Mix thoroughly. Sparge with a gas mixture of 80% N_2 + 20% CO_2. Filter sterilize.

Solution D:

Composition per 10.0mL:

$Na_2S_2O_3 \cdot 5H_2O$	1.25g

Preparation of Solution D: Add $Na_2S_2O_3 \cdot 5H_2O$ to distilled/deionized water and bring volume to 10.0mL. Mix thoroughly. Sparge with 100% N_2. Filter sterilize.

Solution E:

Composition per liter:

Pyridoxine-HCl	10.0mg
Thiamine-HCl·$2H_2O$	5.0mg
Riboflavin	5.0mg
Nicotinic acid	5.0mg
D-Ca-pantothenate	5.0mg
p-Aminobenzoic acid	5.0mg
Lipoic acid	5.0mg
Biotin	2.0mg
Folic acid	2.0mg
Vitamin B_{12}	0.1mg

Preparation of Solution E: Add components to distilled/deionized water and bring volume to 1.0L. Mix thoroughly. Sparge with 100% N_2. Filter sterilize.

Preparation of Medium: Dispense solution A under a gas atmosphere of 80% N_2 + 20% CO_2 into serum vials to 20% of volume. Autoclave for 15 min at 15 psi pressure–121°C. Cool to room temperature. Aspetically add sterile solutions B–E in that order. Adjust the pH to 6.5. After inoculation pressurize with an amount of sterile air that is equivalent to 30% of the volume of the cultivation vessel.

Use: For the cultivation of *Thiofaba* spp.

Thiogel® Medium

Composition per liter:

Gelatin	50.0g
Pancreatic digest of casein	17.0g
Glucose	6.0g
Papaic digest of soybean meal	3.0g
NaCl	2.5g
Sodium thioglycolate	0.5g
Agar	0.7g
Na_2SO_3	0.1g
L-Cystine	0.25g

pH 7.0 ± 0.2 at 25°C

Source: This medium is available as a premixed powder from BD Diagnostic Systems.

Preparation of Medium: Add components to distilled/deionized water preheated to 50°C and bring volume to 1.0L. Mix thoroughly. Let stand for 5 min. Gently heat while stirring and bring to boiling. Distribute into tubes, filling them half full. Autoclave for 15 min at 13 psi pressure–118°C. Pour into sterile Petri dishes or leave in tubes.

Use: For the differentiation of microorganisms based on their ability to liquefy gelatin.

Thioglycolate Bile Broth

Composition per 1050.0mL:

Pancreatic digest of casein	15.0g
Glucose	5.5g
Yeast extract	5.0g
NaCl	2.5g
Agar	0.75g
L-Cystine	0.5g
Sodium thioglycolate	0.5g
Bile solution	50.0mL

pH 7.1 ± 0.2 at 25°C

Bile Solution:

Composition per 100.0mL:

Oxgall	40.0g
Sodium deoxycholate	2.0g

Preparation of Bile Solution: Add components to distilled/deionized water and bring volume to 100.0mL. Mix thoroughly. Filter sterilize.

Preparation of Medium: Add components, except bile solution, to distilled/deionized water and bring volume to 1.0L. Mix thoroughly. Gently heat and bring to boiling. Distribute into tubes in 10.0mL volumes. Autoclave for 15 min at 15 psi pressure–121°C. Cool to 45°–50°C. Aseptically add 0.5mL of sterile bile solution to each tube. Mix thoroughly.

Use: For the cultivation of *Bacteroides fragilis* and *Clostridium perfringens* from clinical specimens.

Thioglycolate Broth USP, Alternative

Composition per liter:

Pancreatic digest of casein	15.0g
Glucose	5.5g
Yeast extract	5.0g
NaCl	2.5g
L-Cystine	0.5g
Sodium thioglycolate	0.5g

pH 7.1 ± 0.2 at 25°C

Source: This medium is available as a premixed powder from Oxoid Unipath.

Preparation of Medium: Add components to distilled/deionized water and bring volume to 1.0L. Mix thoroughly. Distribute into tubes or flasks. Autoclave for 15 min at 15 psi pressure–121°C. Prepare freshly or boil and cool the medium just before use.

Use: For the cultivation of both aerobic and anaerobic organisms in the performance of sterility tests of turbid or viscous specimens.

Thioglycolate Gelatin Medium

Composition per liter:

Gelatin	50.0g
Pancreatic digest of casein	15.0g
Yeast extract	5.0g
NaCl	2.5g
Glucose	2.0g
Agar	0.75g
L-Cystine	0.25g
Na_2SO_3	0.1g
Thioglycollic acid	0.3mL

pH 7.0 ± 0.2 at 25°C

Source: This medium is available as a premixed powder from BD Diagnostic Systems.

Preparation of Medium: Add components to distilled/deionized water and bring volume to 1.0L. Mix thoroughly. Gently heat and bring to 50°C. Let stand 5 min. Gently heat and bring to boiling. Distribute into tubes or flasks. Autoclave for 15 min at 15 psi pressure–121°C.

Use: For the determination of gelatin liquefaction by aerobes, microaerophiles, and anaerobes without special incubation.

Thioglycolate HiVeg Agar

Composition per liter:

Agar	20.0g
Plant hydrolysate	15.0g
Glucose	5.5g
Yeast extract	5.0g
NaCl	2.5g
L-Cystine	0.5g
Na-thioglycolate	0.5g
Resazurin	1.0mg

pH 7.1 ± 0.2 at 25°C

Source: This medium is available as a premixed powder from HiMedia.

Preparation of Medium: Add components to distilled/deionized water and bring volume to 1.0L. Mix thoroughly. Distribute into tubes or flasks. Autoclave for 15 min at 15 psi pressure–121°C. Prepare freshly or boil and cool the medium just before use. Pour into sterile Petri dishes or distribute into sterile tubes.

Use: For the cultivation of facultative and anaerobic organisms. For the performance of sterility tests of turbid or viscous specimens.

Thioglycolate HiVeg Medium without Indicator

Composition per liter:

Plant hydrolysate	17.0g
Glucose	6.0g
Papaic digest of soybean meal	3.0g
NaCl	2.5g
Agar	0.7g
Na-thioglycolate	0.5g
L-Cystine	0.25g
Na_2SO_3	0.1g

pH 7.1 ± 0.2 at 25°C

Source: This medium is available as a premixed powder from HiMedia.

Preparation of Medium: Add components to distilled/deionized water and bring volume to 1.0L. Mix thoroughly. Distribute into tubes or flasks. Autoclave for 15 min at 15 psi pressure–121°C. Prepare freshly or boil and cool the medium just before use.

Use: For the cultivation of facultative and anaerobic organisms. For the performance of sterility tests of turbid or viscous specimens.

Thioglycolate Medium (DSMZ Medium 530)

Composition per liter:

Agar	15.0g
Peptone from casein	5.0g
Meat extract	3.0g
Na-thioglycolate solution	100.0mL

pH 5.5 ± 0.2 at 25°C

Thioglycolate Solution :
Composition per 100.0mL:

Na-thioglycolate	0.5g

Preparation of Thioglycolate Solution: Add thioglycolate to distilled/deionized water and bring volume to 100.0mL. Mix thoroughly. Filter sterilize. Warm to 50°C.

Preparation of Medium: Add components, except thioglycolate solution, to distilled/deionized water and bring volume to 900.0mL. Mix thoroughly. Gently heat and bring to boiling. Autoclave for 15 min at 15 psi pressure–121°C. Cool to 50°C. Aseptically add 100.0mL warm thioglycolate solution. Adjust pH to 5.5. Pour into sterile Petri dishes or distribute into sterile tubes.

Use: For the cultivation and maintenance of various anaerobic bacteria.

Thioglycolate Medium with 20% Bile (THIO + Bile Medium)

Composition per liter:

Oxgall	20.0g
Thioglycolate medium without indicator	1.0L

Hemin solution 0.5mL
Vitamin K_1 solution 0.1mL

pH 7.0 ± 0.2 at 25°C

Thioglycolate Medium without Indicator:
Composition per liter:

Pancreatic digest of casein	17.0g
Glucose	6.0g
Papaic digest of soybean meal	3.0g
NaCl	2.5g
Agar	0.7g
Sodium thioglycolate	0.5g
L-Cystine	0.25g
Na_2SO_3	0.1g

Preparation of Thioglycolate Medium Without Indicator: Add components to distilled/deionized water and bring volume to 1.0L. Mix thoroughly.

Vitamin K_1 Solution:
Composition per 100.0mL:

Vitamin K_1	1.0g

Preparation of Vitamin K_1 Solution: Add vitamin K_1 to 99.0mL of absolute ethanol. Mix thoroughly.

Hemin Solution:
Composition per 100.0mL:

Hemin	1.0g
NaOH (1*N* solution)	20.0mL

Preparation of Hemin Solution: Add hemin to 20.0mL of 1*N* NaOH solution. Mix thoroughly. Bring volume to 100.0mL with distilled/deionized water.

Preparation of Medium: Add 0.5mL of hemin solution, 0.1mL of vitamin K_1 solution, and oxgall to 1.0L of thioglycolate medium without indicator. Mix thoroughly. Distribute into screw-capped tubes or flasks. Autoclave for 15 min at 15 psi pressure–121°C. Cool tubes or flasks under 85% N_2 + 10% H_2 + 5% CO_2. Tighten caps.

Use: For the isolation, cultivation, and identification of a variety of obligate anaerobic bile-tolerant bacteria.

Thioglycolate Medium, Brewer

Composition per liter:

Glucose	5.0g
Peptone	5.0g
NaCl	5.0g
Yeast extract	2.0g
Sodium thioglycolate	1.1g
Agar	1.0g
Beef extract	1.0g
Methylene Blue	2.0mg

pH 7.2 ± 0.2 at 25°C

Source: This medium is available as a premixed powder from Oxoid Unipath.

Preparation of Medium: Add components to distilled/deionized water and bring volume to 1.0L. Mix thoroughly. Gently heat and bring to boiling. Distribute into tubes or flasks. Autoclave for 15 min at 15 psi pressure–121°C.

Use: For determination of the sterility of solutions containing mercurial preservatives.

Thioglycolate Medium, Brewer Modified

Composition per liter:

Pancreatic digest of casein	17.5g
Glucose	10.0g
NaCl	5.0g
Papaic digest of soybean meal	2.5g
K_2HPO_4	2.0g
Sodium thioglycolate	1.0g
Agar	0.5g
Methylene Blue	0.002g

pH 7.2 ± 0.2 at 25°C

Source: This medium is available as a premixed powder from BD Diagnostic Systems.

Preparation of Medium: Add components to distilled/deionized water and bring volume to 1.0L. Mix thoroughly. Gently heat while stirring and bring to boiling. Distribute into tubes or flasks, filling them half full. Autoclave for 15 min at 15 psi pressure–121°C.

Use: For the cultivation of obligate anaerobes, microaerophiles, and facultative organisms.

Thioglycolate Medium, Enriched (THIO Medium) (Thioglycolate Medium with Vitamin K_1 and Hemin)

Composition per liter:

Thioglycolate medium without indicator	1.0L
Hemin solution	0.5mL
Vitamin K_1 solution	0.1mL

pH 7.0 ± 0.2 at 25°C

Thioglycolate Medium without Indicator:
Composition per liter:

Pancreatic digest of casein	17.0g
Glucose	6.0g
Papaic digest of soybean meal	3.0g
NaCl	2.5g
Agar	0.7g
Sodium thioglycolate	0.5g
L-Cystine	0.25g
Na_2SO_3	0.1g

Source: Thioglycolate medium without indicator is available as a premixed powder from Oxoid Unipath and BD Diagnostic Systems.

Preparation of Thioglycolate Medium without Indicator: Add components to distilled/deionized water and bring volume to 1.0L. Mix thoroughly.

Vitamin K_1 Solution:
Composition per 100.0mL:

Vitamin K_1	1.0g

Preparation of Vitamin K_1 Solution: Add vitamin K_1 to 99.0mL of absolute ethanol. Mix thoroughly.

Hemin Solution:
Composition per 100.0mL:

Hemin	1.0g
NaOH (1*N* solution)	20.0mL

Preparation of Hemin Solution: Add hemin to 20.0mL of 1*N* NaOH solution. Mix thoroughly. Bring volume to 100.0mL with distilled/deionized water.

Preparation of Medium: Add 0.5mL of hemin solution and 0.1mL of vitamin K_1 solution to 1.0L of thioglycolate medium without indicator. Mix thoroughly. Distribute into screw-capped tubes or flasks. Autoclave for 15 min at 15 psi pressure–121°C. Cool tubes or flasks under 85% N_2 + 10% H_2 + 5% CO_2. Tighten caps.

Use: For the isolation, cultivation, and identification of a wide variety of obligate anaerobic bacteria.

Thioglycolate Medium, Fluid (Fluid Thioglycolate Medium) (FTG) (BAM M146)

Composition per liter:

Pancreatic digest of casein	15.0g
Glucose	5.0g
Yeast extract	5.0g
NaCl	2.5g
Agar	0.75g
L-cystine	0.5g
Sodium thioglycolate	0.5g
Resazurin solution	1.0mL

pH 7.1 ± 0.2 at 25°C

Resazurin Solution:

Composition per 10.0mL:

Na-resazurin	10.0mg

Preparation of Resazurin Solution: Add Na-resazurin to 10.0mL of distilled/deionized water. Mix thoroughly. Prepare freshly.

Preparation of Medium: Add components, except sodium thioglycolate and resazurin solution, to distilled/deionized water and bring volume to 1.0L. Mix thoroughly. Gently heat and bring to boiling. Add 0.5g sodium thioglycolate. Mix thoroughly. Adjust pH to 7.1. Add 1.0mL resazurin solution. Mix thoroughly. Distribute into tubes or flasks. Autoclave for 20 min at 15 psi pressure–121°C.

Use: For the cultivation of both aerobic and anaerobic organisms in the performance of sterility tests.

Thioglycolate Medium without Glucose

Composition per liter:

Pancreatic digest of casein	15.0g
Yeast extract	5.0g
NaCl	2.5g
Agar	0.75g
L-Cystine	0.25g
Methylene Blue	2.0mg
Thioglycolic acid	0.3mL

pH 7.2 ± 0.2 at 25°C

Source: This medium is available as a premixed powder from BD Diagnostic Systems.

Preparation of Medium: Add components to distilled/deionized water and bring volume to 1.0L. Mix thoroughly. Gently heat and bring to boiling. Distribute into tubes or flasks. Autoclave for 15 min at 15 psi pressure–121°C. If medium becomes oxidized before use (Methylene Blue turns blue), heat in a boiling water bath to expel absorbed O_2. Cool to 25°C.

Use: For the cultivation of anaerobic, microaerophilic, and aerobic microorganisms. For use in sterility testing of a variety of specimens.

Thioglycolate Medium without Glucose

Composition per liter:

Pancreatic digest of casein	20.0g
NaCl	2.5g
K_2HPO_4	1.5g
Sodium thioglycolate	0.6g
Agar	0.5g
L-Cystine	0.4g
Na_2SO_3	0.2g
Methylene Blue	2.0mg

pH 7.2 ± 0.2 at 25°C

Source: This medium is available as a premixed powder from BD Diagnostic Systems.

Preparation of Medium: Add components to distilled/deionized water and bring volume to 1.0L. Mix thoroughly. Gently heat while stirring and bring to boiling. Distribute into tubes or flasks, filling them half full. Autoclave for 15 min at 15 psi pressure–121°C.

Use: Use as a base for fermentation studies of anaerobic bacteria and for the promotion of endospore formation.

Thioglycolate Medium without Glucose and Indicator

Composition per liter:

Pancreatic digest of casein	15.0g
Yeast extract	5.0g
NaCl	2.5g
Agar	0.75g
L-Cystine	0.25g
Thioglycolic acid	0.3mL

pH 7.2 ± 0.2 at 25°C

Source: This medium is available as a premixed powder from BD Diagnostic Systems.

Preparation of Medium: Add components to distilled/deionized water and bring volume to 1.0L. Mix thoroughly. Gently heat and bring to boiling. Distribute into tubes or flasks. Autoclave for 15 min at 15 psi pressure–121°C. If medium becomes oxidized before use, heat in a boiling water bath to expel absorbed O_2. Cool to 25°C.

Use: For the cultivation of anaerobic, microaerophilic, and aerobic microorganisms. For use in sterility testing of a variety of specimens.

Thioglycolate Medium without Indicator

Composition per liter:

Pancreatic digest of casein	15.0g
Yeast extract	5.0g
Glucose	5.0g
NaCl	2.5g
Agar	0.75g
Sodium thioglycolate	0.5g
L-Cystine	0.25g

pH 7.2 ± 0.2 at 25°C

Source: This medium is available as a premixed powder from BD Diagnostic Systems.

Preparation of Medium: Add components to distilled/deionized water and bring volume to 1.0L. Mix thoroughly. Gently heat and bring to boiling. Distribute into tubes or flasks. Autoclave for 15 min at 15 psi pressure–121°C. If medium becomes oxidized before use, heat in a boiling water bath to expel absorbed O_2. Cool to 25°C.

Use: For the cultivation of anaerobic, microaerophilic, and aerobic microorganisms. For use in sterility testing of a variety of specimens.

Thioglycolate Medium without Indicator

Composition per liter:

Pancreatic digest of casein	17.0g
Glucose	6.0g
Papaic digest of soybean meal	3.0g
NaCl	2.5g
Agar	0.7g
Sodium thioglycolate	0.5g
L-Cystine	0.25g
Na_2SO_3	0.1g

pH 7.0 ± 0.2 at 25°C

Source: This medium is available as a premixed powder from Oxoid Unipath.

Preparation of Medium: Add components to distilled/deionized water and bring volume to 1.0L. Mix thoroughly. Distribute into tubes or flasks. Autoclave for 15 min at 15 psi pressure–121°C. Prepare freshly or boil and cool the medium just before use.

Use: For the growth of aerobic and anaerobic microorganisms in diagnostic bacteriology.

Thioglycolate Medium without Indicator-135C

Composition per liter:

Pancreatic digest of casein	17.0g
Glucose	6.0g
Papaic digest of soybean meal	3.0g
NaCl	2.5g
Agar	0.7g
Sodium thioglycolate	0.5g
Na_2SO_3	0.1g
L-Cystine	0.25g

pH 7.0 ± 0.2 at 25°C

Source: This medium is available as a premixed powder from BD Diagnostic Systems.

Preparation of Medium: Add components to distilled/deionized water and bring volume to 1.0L. Mix thoroughly. Gently heat while stirring and bring to boiling. Distribute into tubes or flasks, filling them half full. For maintenance of cultures, a small quantity of $CaCO_3$ may be added to tubes before adding medium. Autoclave for 15 min at 13 psi pressure–118°C. Prepare freshly or boil and cool the medium just before use. Store prepared medium at 2°–8°C in the dark.

Use: For the isolation and cultivation of a wide variety of microorganisms, particularly obligate anaerobes, from clinical specimens and other materials.

Thioglycolate Medium without Indicator with Hemin (Thioglycolate Medium, Supplemented)

Composition per liter:

Pancreatic digest of casein	17.0g
$CaCO_3$, chips or powder	10.0g
Glucose	6.0g
Papaic digest of soybean meal	3.0g
NaCl	2.5g
Agar	0.7g
Sodium thioglycolate	0.5g
L-Cystine	0.25g
Na_2SO_3	0.1g
Hemin	5.0mg
Na_2CO_3 solution	10.0mL
Vitamin K_1 solution	10.0mL

pH 7.2 ± 0.2 at 25°C

Na_2CO_3 Solution:

Composition per 10.0mL:

Na_2CO_3	1.0g

Preparation of Na_2CO_3 Solution: Add Na_2CO_3 to distilled/deionized water and bring volume to 10.0mL. Mix thoroughly. Filter sterilize.

Vitamin K_1 Solution:

Composition per 100.0mL:

Vitamin K_1	1.0g
Ethanol, absolute	99.0mL

Preparation of Vitamin K_1 Solution: Add vitamin K_1 to 99.0mL of absolute ethanol. Mix thoroughly.

Preparation of Medium: Add components, except $CaCO_3$, Na_2CO_3 solution, and vitamin K_1 solution, to distilled/deionized water and bring volume to 990.0mL. Mix thoroughly. Gently heat and bring to boiling. Add 0.1g of $CaCO_3$ chips or powder to each of 100 test tubes. Distribute broth into the same tubes in 10.0mL volumes. Autoclave for 15 min at 15 psi pressure–121°C. Cool to 45°–50°C. Aseptically add 0.1mL of sterile Na_2CO_3 solution and 0.1mL of sterile vitamin K_1 solution to each tube. Mix thoroughly.

Use: For the cultivation of a wide variety of obligate anaerobes.

Thioglycolate Medium, Supplemented
See: **Thioglycolate Medium without Indicator with Hemin**

Thioglycolate Medium, USP

Composition per liter:

Pancreatic digest of casein	15.0g
Glucose	5.5g
Yeast extract	5.0g
NaCl	2.5g
Agar	0.5g
L-cystine	0.5g
Sodium thioglycolate	0.5g
Resazurin	1.0mg

pH 7.1 ± 0.2 at 25°C

Source: This medium is available as a premixed powder from Oxoid Unipath.

Preparation of Medium: Add components to distilled/deionized water and bring volume to 1.0L. Mix thoroughly. Gently heat and bring to boiling. Distribute into tubes or flasks. Autoclave for 15 min at 15 psi pressure–121°C.

Use: For the cultivation of both aerobic and anaerobic organisms in the performance of sterility tests.

Thioglycolate Medium with Vitamin K_1 and Hemin
See: **Thioglycolate Medium, Enriched**

Thioglycolate Peptone Glucose Yeast Extract Medium
See: **TPGY Medium**

Thioglycolate Potato Liver Medium
See: **TPL Medium**

Thiol Broth

Composition per liter:

Proteose peptone, No. 3	10.0g
Thiol complex	8.0g
Yeast extract	5.0g
NaCl	5.0g
Glucose	1.0g
p-Aminobenzoic acid	0.05g

pH 7.1 ± 0.2 at 25°C

Source: This medium is available as a premixed powder from BD Diagnostic Systems.

Preparation of Medium: Add components to distilled/deionized water and bring volume to 1.0L. Mix thoroughly. Gently heat and bring to boiling. Distribute into tubes or flasks. For neutralization of penicillin, distribute medium into tubes to a depth of 60.0mm. For neutralization of streptomycin, distribute medium into tubes in shallow layers. Autoclave for 15 min at 15 psi pressure–121°C.

Use: For the cultivation of bacteria from body fluids and other materials containing penicillin, streptomycin, or sulfonamides. Also used for the cultivation and maintenance of *Bifidobacterium* species.

Thiohalophilus Medium (DSMZ Medium 1058)

Composition per liter:

NaCl	120.0g
K_2HPO_4	1.5g
NH_4Cl	0.5g
Bicarbonate solution	10.0mL
Calcium chloride solution	10.0mL
Magnesium chloride solution	10.0mL
Seven vitamin solution	1.0mL
Trace elements solution SL-10 with EDTA	1.0mL
Carbonate solution	variable

pH 7.6 ± 0.2 at 25°C

Thiosulfate Solution:

Composition per 10.0mL:

$Na_2S_2O_3 \cdot 5H_2O$	5.0g

Preparation of Thiosulfate Solution: Add components to distilled/deionized water and bring volume to 10.0mL. Mix thoroughly. Sparge with 100% N_2. Filter sterilize.

Bicarbonate Solution:

Composition per 10.0mL:

$NaHCO_3$	5.0g

Preparation of Bicarbonate Solution: Add $NaHCO_3$ to distilled/deionized water and bring volume to 10.0mL. Mix thoroughly. Sparge with a gas mixture of 80% N_2 + 20% CO_2. Filter sterilize.

Carbonate Solution:

Composition per 10.0mL:

Na_2CO_3	2.0g

Preparation of Carbonate Solution: Add Na_2CO_3 to distilled/deionized water and bring volume to 10.0mL. Mix thoroughly. Sparge with a gas mixture of 80% N_2 + 20% CO_2. Filter sterilize.

Magnesium Chloride Solution:

Composition per 10.0mL:

$MgCl_2 \cdot 6H_2O$	0.4g

Preparation of Magnesium Chloride Solution: Add 0.4g of $MgCl_2 \cdot 6H_2O$ to distilled/deionized water and bring volume to 10.0mL. Mix thoroughly. Autoclave for 15 min at 15 psi pressure–121°C. Cool to room temperature. Sparge with 100% N_2.

Calcium Chloride Solution:

Composition per 10.0mL:

$CaCl_2 \cdot 2H_2O$	0.05g

Preparation of Calcium Chloride Solution: Add components to distilled/deionized water and bring volume to 10.0mL. Mix thoroughly. Autoclave for 15 min at 15 psi pressure–121°C. Cool to room temperature. Sparge with 100% N_2 gas.

Seven Vitamin Solution:

Composition per liter:

Pyridoxine·HCl	300.0mg
Nicotinic acid	200.0mg
Thiamine·HCl	200.0mg
Calcium DL-pantothenate	100.0mg
Cyanocobalamine	100.0mg
p-Aminobenzoic acid	80.0mg
D(+)-Biotin	20.0mg

Preparation of Seven Vitamin Solution: Add components to distilled/deionized water and bring volume to 1.0L. Mix thoroughly. Filter sterilize. Sparge with 100% N_2.

Trace Elements Solution SL-10 with EDTA:

Composition per liter:

$FeCl_2 \cdot 4H_2O$	1.5g
Na_2-EDTA	0.5g
$CoCl_2 \cdot 6H_2O$	190.0mg
$MnCl_2 \cdot 4H_2O$	100.0mg
$ZnCl_2$	70.0mg
$Na_2MoO_4 \cdot 2H_2O$	36.0mg
$NiCl_2 \cdot 6H_2O$	24.0mg
H_3BO_3	6.0mg
$CuCl_2 \cdot 2H_2O$	2.0mg
HCl (25% solution)	10.0mL

Preparation of Trace Elements Solution SL-10 with EDTA: Add $FeCl_2 \cdot 4H_2O$ to 10.0mL of HCl solution. Mix thoroughly. Add distilled/deionized water and bring volume to 1.0L. Add remaining components. Mix thoroughly. Sparge with 100% N_2. Autoclave for 15 min at 15 psi pressure–121°C. Adjust pH to 7.0.

Preparation of Medium: Add components, except magnesium chloride, thiosulfate, bicarbonate, and vitamin solutions, to distilled/deionized water and bring volume to 970.0mL. Mix thoroughly. Sparge with 80% N_2 + 20% CO_2 gas mixture for at least 30 min to remove dissolved oxygen and to saturate the solution with CO_2. Dispense into anaerobic culture vessels (e.g., Balch tubes) to 1/2 volume under air atmosphere. Close vials with butyl rubber septa to prevent free exchange of oxygen with the external atmosphere. Autoclave for 15 min at 15 psi pressure–121°C. Cool to room temperature. Aseptically add magnesium chloride, thiosulfate, bicarbonate, and vitamin solutions. Mix thoroughly. Adjust pH to 7.5–7.8 using a sterile solution of sodium carbonate (5% w/v).

Use: For the cultivation of *Thiohalomonas denitrificans, Thiohalophilus thiocyanatoxydans,* and *Thiomicrospira halophila.*

Thiohalophilus Medium (DSMZ Medium 1058)

Composition per liter:

NaCl	233.0g
K_2HPO_4	1.5g

NH_4Cl 0.5g
Bicarbonate solution 10.0mL
Calcium chloride solution 10.0mL
Magnesium chloride solution 10.0mL
Seven vitamins solution 1.0mL
Trace elements solution SL-10 with EDTA 1.0mL
Carbonate solution .... variable

pH 7.2 ± 0.2 at 25°C

Thiosulfate Solution:
Composition per 10.0mL:
$Na_2S_2O_3 \cdot 5H_2O$ 5.0g

Preparation of Thiosulfate Solution: Add $Na_2S_2O_3 \cdot 5H_2O$ to distilled/deionized water and bring volume to 10.0mL. Mix thoroughly. Sparge with 100% N_2. Filter sterilize.

Bicarbonate Solution:
Composition per 10.0mL:
$NaHCO_3$ 5.0g

Preparation of Bicarbonate Solution: Add $NaHCO_3$ to distilled/deionized water and bring volume to 10.0mL. Mix thoroughly. Sparge with a gas mixture of 80% N_2 + 20% CO_2. Filter sterilize.

Carbonate Solution:
Composition per 10.0mL:
Na_2CO_3 .... 2.0g

Preparation of Carbonate Solution: Add Na_2CO_3 to distilled/deionized water and bring volume to 10.0mL. Mix thoroughly. Sparge with a gas mixture of 80% N_2 + 20% CO_2. Filter sterilize.

Magnesium Chloride Solution:
Composition per 10.0mL:
$MgCl_2 \cdot 6H_2O$ 0.4g

Preparation of Magnesium Chloride Solution: Add 0.4g of $MgCl_2 \cdot 6H_2O$ to distilled/deionized water and bring volume to 10.0mL. Mix thoroughly. Autoclave for 15 min at 15 psi pressure–121°C. Cool to room temperature. Sparge with 100% N_2.

Calcium Chloride Solution:
Composition per 10.0mL:
$CaCl_2 \cdot 2H_2O$ 0.05g

Preparation of Calcium Chloride Solution: Add $CaCl_2 \cdot 2H_2O$ to distilled/deionized water and bring volume to 10.0mL. Mix thoroughly. Autoclave for 15 min at 15 psi pressure–121°C. Cool to room temperature. Sparge with 100% N_2.

Seven Vitamin Solution:
Composition per liter:
Pyridoxine·HCl 300.0mg
Nicotinic acid 200.0mg
Thiamine·HCl 200.0mg
Calcium DL-pantothenate 100.0mg
Cyanocobalamine 100.0mg
p-Aminobenzoic acid 80.0mg
D(+)-Biotin 20.0mg

Preparation of Seven Vitamin Solution: Add components to distilled/deionized water and bring volume to 1.0L. Mix thoroughly. Filter sterilize. Sparge with 100% N_2.

Trace Elements Solution SL-10 with EDTA:
Composition per liter:
$FeCl_2 \cdot 4H_2O$ 1.5g
Na_2-EDTA 0.5g
$CoCl_2 \cdot 6H_2O$ 190.0mg
$MnCl_2 \cdot 4H_2O$ 100.0mg
$ZnCl_2$ 70.0mg
$Na_2MoO_4 \cdot 2H_2O$ 36.0mg
$NiCl_2 \cdot 6H_2O$ 24.0mg
H_3BO_3 .... 6.0mg
$CuCl_2 \cdot 2H_2O$ 2.0mg
HCl (25% solution) 10.0mL

Preparation of Trace Elements Solution SL-10 with EDTA: Add $FeCl_2 \cdot 4H_2O$ to 10.0mL of HCl solution. Mix thoroughly. Add distilled/deionized water and bring volume to 1.0L. Add remaining components. Mix thoroughly. Sparge with 100% N_2. Autoclave for 15 min at 15 psi pressure–121°C. Adjust pH to 7.0.

Preparation of Medium: Add components, except magnesium chloride, thiosulfate, bicarbonate, and vitamin solutions, to distilled/deionized water and bring volume to 970.0mL. Mix thoroughly. Sparge with 80% N_2 + 20% CO_2 gas mixture for at least 30 min to remove dissolved oxygen and to saturate the solution with CO_2. Dispense into anaerobic culture vessels (e.g., Balch tubes) to 1/2 volume under air atmosphere. Close vials with butyl rubber septa to prevent free exchange of oxygen with the external atmosphere. Autoclave for 15 min at 15 psi pressure–121°C. Cool to room temperature. Aseptically add magnesium chloride, thiosulfate, bicarbonate, and vitamin solutions. Mix thoroughly. Adjust pH to 7.2 using a sterile solution of sodium carbonate (5% w/v).

Use: For the cultivation of *Thiomicrospira halophila* DSM 15071.

Thiohalophilus Medium (DSMZ Medium 1058)

Composition per liter:
NaCl 180.0g
K_2HPO_4 .... 1.5g
KNO_3 .... 1.0g
NH_4Cl 0.5g
Bicarbonate solution 10.0mL
Calcium chloride solution 10.0mL
Magnesium chloride solution 10.0mL
Seven vitamin solution 1.0mL
Trace elements solution SL-10 with EDTA 1.0mL
Carbonate solution .... Variable

pH 7.6 ± 0.2 at 25°C

Thiosulfate Solution:
Composition per 10.0mL:
$Na_2S_2O_3 \cdot 5H_2O$ 5.0g

Preparation of Thiosulfate Solution: Add $Na_2S_2O_3 \cdot 5H_2O$ to distilled/deionized water and bring volume to 10.0mL. Mix thoroughly. Sparge with 100% N_2. Filter sterilize.

Bicarbonate Solution:
Composition per 10.0mL:
$NaHCO_3$ 5.0g

Preparation of Bicarbonate Solution: Add $NaHCO_3$ to distilled/deionized water and bring volume to 10.0mL. Mix thoroughly. Sparge with a gas mixture of 80% N_2 + 20% CO_2. Filter sterilize.

Carbonate Solution:
Composition per 10.0mL:
Na_2CO_3 .... 2.0g

Preparation of Carbonate Solution: Add Na_2CO_3 to distilled/deionized water and bring volume to 10.0mL. Mix thoroughly. Sparge with a gas mixture of 80% N_2 + 20% CO_2. Filter sterilize.

Magnesium Chloride Solution:

Composition per 10.0mL:

$MgCl_2 \cdot 6H_2O$ 0.4g

Preparation of Magnesium Chloride Solution: Add 0.4g of $MgCl_2 \cdot 6H_2O$ to distilled/deionized water and bring volume to 10.0mL. Mix thoroughly. Autoclave for 15 min at 15 psi pressure–121°C. Cool to room temperature. Sparge with 100% N_2.

Calcium Chloride Solution:

Composition per 10.0mL:

$CaCl_2 \cdot 2H_2O$ 0.05g

Preparation of Calcium Chloride Solution: Add $CaCl_2 \cdot 2H_2O$ to distilled/deionized water and bring volume to 10.0mL. Mix thoroughly. Autoclave for 15 min at 15 psi pressure–121°C. Cool to room temperature. Sparge with 100% N_2.

Seven Vitamin Solution:

Composition per liter:

Pyridoxine·HCl 300.0mg
Nicotinic acid 200.0mg
Thiamine·HCl 200.0mg
Calcium DL-pantothenate 100.0mg
Cyanocobalamine 100.0mg
p-Aminobenzoic acid 80.0mg
D(+)-Biotin 20.0mg

Preparation of Seven Vitamin Solution: Add components to distilled/deionized water and bring volume to 1.0L. Mix thoroughly. Filter sterilize. Sparge with 100% N_2.

Trace Elements Solution SL-10 with EDTA:

Composition per liter:

$FeCl_2 \cdot 4H_2O$ 1.5g
Na_2-EDTA 0.5g
$CoCl_2 \cdot 6H_2O$ 190.0mg
$MnCl_2 \cdot 4H_2O$ 100.0mg
$ZnCl_2$ 70.0mg
$Na_2MoO_4 \cdot 2H_2O$ 36.0mg
$NiCl_2 \cdot 6H_2O$ 24.0mg
H_3BO_3 .... 6.0mg
$CuCl_2 \cdot 2H_2O$ 2.0mg
HCl (25% solution) 10.0mL

Preparation of Trace Elements Solution SL-10 with EDTA: Add $FeCl_2 \cdot 4H_2O$ to 10.0mL of HCl solution. Mix thoroughly. Add distilled/deionized water and bring volume to 1.0L. Add remaining components. Mix thoroughly. Sparge with 100% N_2. Autoclave for 15 min at 15 psi pressure–121°C. Adjust pH to 7.0.

Preparation of Medium: Add components, except magnesium chloride, thiosulfate, bicarbonate, and vitamin solutions, to distilled/deionized water and bring volume to 970.0mL. Mix thoroughly. Sparge with 80% N_2 + 20% CO_2 gas mixture for at least 30 min. to remove dissolved oxygen and to saturate the solution with CO_2. Dispense into anaerobic culture vessels (e.g., Balch tubes) to 1/2 volume under air atmosphere. Close vials with butyl rubber septa to prevent free exchange of oxygen with the external atmosphere. Autoclave for 15 min at 15 psi pressure–121°C. Cool to room temperature. Aseptically add magnesium chloride, thiosulfate, bicarbonate, and vitamin solutions. Mix thoroughly. Adjust pH to 7.5–7.8 using a sterile solution of sodium carbonate (5% w/v).

Use: For the cultivation of *Thiohalomonas denitrificans*.

Thiohalophilus Medium (DSMZ Medium 1058)

Composition per liter:

NaCl 180.0g
K_2HPO_4 .... 1.5g
KNO_3 .... 2.0g
NH_4Cl 0.5g
Bicarbonate solution 10.0mL
Calcium chloride solution 10.0mL
Magnesium chloride solution 10.0mL
Seven vitamin solution 1.0mL
Trace elements solution SL-10 with EDTA 1.0mL
Carbonate solution .... Variable

pH 7.6 ± 0.2 at 25°C

Thiosulfate Solution:

Composition per 10.0mL:

$Na_2S_2O_3 \cdot 5H_2O$ 5.0g

Preparation of Thiosulfate Solution: Add $Na_2S_2O_3 \cdot 5H_2O$ to distilled/deionized water and bring volume to 10.0mL. Mix thoroughly. Sparge with 100% N_2. Filter sterilize.

Bicarbonate Solution:

Composition per 10.0mL:

$NaHCO_3$ 5.0g

Preparation of Bicarbonate Solution: Add $NaHCO_3$ to distilled/deionized water and bring volume to 10.0mL. Mix thoroughly. Sparge with a gas mixture of 80% N_2 + 20% CO_2. Filter sterilize.

Carbonate Solution:

Composition per 10.0mL:

Na_2CO_3 .... 2.0g

Preparation of Carbonate Solution: Add Na_2CO_3 to distilled/deionized water and bring volume to 10.0mL. Mix thoroughly. Sparge with a gas mixture of 80% N_2 + 20% CO_2. Filter sterilize.

Magnesium Chloride Solution:

Composition per 10.0mL:

$MgCl_2 \cdot 6H_2O$ 0.4g

Preparation of Magnesium Chloride Solution: Add 0.4g of $MgCl_2 \cdot 6H_2O$ to distilled/deionized water and bring volume to 10.0mL. Mix thoroughly. Autoclave for 15 min at 15 psi pressure–121°C. Cool to room temperature. Sparge with 100% N_2.

Calcium Chloride Solution:

Composition per 10.0mL:

$CaCl_2 \cdot 2H_2O$ 0.05g

Preparation of Calcium Chloride Solution: Add $CaCl_2 \cdot 2H_2O$ to distilled/deionized water and bring volume to 10.0mL. Mix thoroughly. Autoclave for 15 min at 15 psi pressure–121°C. Cool to room temperature. Sparge with 100% N_2.

Seven Vitamin Solution:

Composition per liter:

Pyridoxine·HCl 300.0mg
Nicotinic acid 200.0mg
Thiamine·HCl 200.0mg
Calcium DL-pantothenate 100.0mg
Cyanocobalamine 100.0mg
p-Aminobenzoic acid 80.0mg
D(+)-Biotin 20.0mg

Preparation of Seven Vitamin Solution: Add components to distilled/deionized water and bring volume to 1.0L. Mix thoroughly. Filter sterilize. Sparge with 100% N_2.

Trace Elements Solution SL-10 with EDTA:

Composition per liter:

$FeCl_2 \cdot 4H_2O$	1.5g
Na_2-EDTA	0.5g
$CoCl_2 \cdot 6H_2O$	190.0mg
$MnCl_2 \cdot 4H_2O$	100.0mg
$ZnCl_2$	70.0mg
$Na_2MoO_4 \cdot 2H_2O$	36.0mg
$NiCl_2 \cdot 6H_2O$	24.0mg
H_3BO_3	6.0mg
$CuCl_2 \cdot 2H_2O$	2.0mg
HCl (25% solution)	10.0mL

Preparation of Trace Elements Solution SL-10 with EDTA: Add $FeCl_2 \cdot 4H_2O$ to 10.0mL of HCl solution. Mix thoroughly. Add distilled/deionized water and bring volume to 1.0L. Add remaining components. Mix thoroughly. Sparge with 100% N_2. Autoclave for 15 min at 15 psi pressure–121°C. Adjust pH to 7.0.

Preparation of Medium: Add components, except magnesium chloride, thiosulfate, bicarbonate, and vitamin solutions, to distilled/deionized water and bring volume to 970.0mL. Mix thoroughly. Sparge with 80% N_2 + 20% CO_2 gas mixture for at least 30 min. to remove dissolved oxygen and to saturate the solution with CO_2. Dispense into anaerobic culture vessels (e.g., Balch tubes) to 1/2 volume under air atmosphere. Close vials with butyl rubber septa to prevent free exchange of oxygen with the external atmosphere. Autoclave for 15 min at 15 psi pressure–121°C. Cool to room temperature. Aseptically add magnesium chloride, thiosulfate, bicarbonate, and vitamin solutions. Mix thoroughly. Adjust pH to 7.5–7.8 using a sterile solution of sodium carbonate (5% w/v).

Use: For the cultivation of *Thiohalomonas nitratireducens*.

Thiol HiVeg Broth

Composition per liter:

Plant peptone No. 3	10.0g
Thiol compound	8.0g
NaCl	5.0g
Yeast extract	5.0g
Glucose	1.0g
p-Aminobenzoic acid (PABA)	0.05g

pH 7.1 ± 0.2 at 25°C

Source: This medium is available as a premixed powder from HiMedia.

Preparation of Medium: Add components to distilled/deionized water and bring volume to 1.0L. Mix thoroughly. Gently heat and bring to boiling. Distribute into tubes or flasks. For neutralization of penicillin, distribute medium into tubes to a depth of 60.0mm. For neutralization of streptomycin, distribute medium into tubes in shallow layers. Autoclave for 15 min at 15 psi pressure–121°C.

Use: For the cultivation of bacteria from body fluids and other materials containing penicillin, streptomycin, or sulfonamides. Also used for the cultivation and maintenance of *Bifidobacterium* species.

Thiol HiVeg Medium

Composition per liter:

Plant peptone No. 3	10.0g
Thiol compound	8.0g
Yeast extract	5.0g
NaCl	5.0g
Agar	1.0g
Glucose	1.0g
p-Aminobenzoic acid (PABA)	0.05g

pH 7.1 ± 0.2 at 25°C

Source: This medium is available as a premixed powder from HiMedia.

Preparation of Medium: Add components to distilled/deionized water and bring volume to 1.0L. Mix thoroughly. Gently heat and bring to boiling. Distribute into tubes or flasks. For neutralization of penicillin, distribute medium into tubes to a depth of 60 mm. For neutralization of streptomycin, distribute medium into tubes in shallow layers. Autoclave for 15 min at 15 psi pressure–121°C.

Use: For the cultivation of bacteria from body fluids and other materials containing penicillin, streptomycin, or sulfonamides. Also used for the cultivation and maintenance of *Bifidobacterium* species.

Thiol Medium

Composition per liter:

Proteose peptone No. 3	10.0g
Thiol complex	8.0g
Yeast extract	5.0g
NaCl	5.0g
Glucose	1.0g
Agar	1.0g
p-Aminobenzoic acid	0.05g

pH 7.1 ± 0.2 at 25°C

Source: This medium is available as a premixed powder from BD Diagnostic Systems.

Preparation of Medium: Add components to distilled/deionized water and bring volume to 1.0L. Mix thoroughly. Gently heat and bring to boiling. Distribute into tubes or flasks. For neutralization of penicillin, distribute medium into tubes to a depth of 60mm. For neutralization of streptomycin, distribute medium into tubes in shallow layers. Autoclave for 15 min at 15 psi pressure–121°C.

Use: For the cultivation of bacteria from body fluids and other materials containing penicillin, streptomycin, or sulfonamides. Also used for the cultivation and maintenance of *Bifidobacterium* species.

Thiomicrospira denitrificans Agar

Composition per 1001.0mL:

Solution A	940.0mL
Solution B	40.0mL
Solution C	20.0mL
Solution D	1.0mL

pH 7.0 ± 0.2 at 25°C

Solution A:

Composition per 940.0mL:

Agar	15.0g
KH_2PO_4	2.0g
KNO_3	2.0g
NH_4Cl	1.0g
$MgSO_4 \cdot 7H_2O$	0.8g
Trace elements solution SL-4	2.0mL

Preparation of Solution A: Add components to distilled/deionized water and bring volume to 940.0mL. Mix thoroughly. Gently heat and bring to boiling. Adjust pH to 7.0 with NaOH. Autoclave for 15 min at 15 psi pressure–121°C. Cool to 45°–50°C.

Trace Elements Solution SL-4:

Composition per liter:

EDTA 0.5g
$FeSO_4 \cdot 7H_2O$ 0.2g
Trace elements solution SL-6 .... 100.0

Preparation of Trace Elements Solution SL-4: Add components to distilled/deionized water and bring volume to 1.0L. Mix thoroughly.

Trace Elements Solution SL-6:

Composition per liter:

H_3BO_3 .... 0.3g
$CoCl_2 \cdot 6H_2O$ 0.2g
$ZnSO_4 \cdot 7H_2O$ 0.1g
$MnCl_2 \cdot 4H_2O$ 0.03g
$Na_2MoO_4 \cdot H_2O$ 0.03g
$NiCl_2 \cdot 6H_2O$ 0.02g
$CuCl_2 \cdot 2H_2O$ 0.01g

Preparation of Trace Elements Solution SL-6: Add components to distilled/deionized water and bring volume to 1.0L. Mix thoroughly. Adjust pH to 3.4.

Solution B:

Composition per 40.0mL:

$Na_2S_2O_3 \cdot 5H_2O$ 5.0g

Preparation of Solution B: Add $Na_2S_2O_3 \cdot 5H_2O$ to distilled/deionized water and bring volume to 40.0mL. Mix thoroughly. Autoclave for 15 min at 15 psi pressure–121°C. Cool to 45°–50°C.

Solution C:

Composition per 20.0mL:

$NaHCO_3$ 1.0g

Preparation of Solution C: Add $NaHCO_3$ to distilled/deionized water and bring volume to 20.0mL. Mix thoroughly. Filter sterilize.

Solution D:

Composition per liter:

$FeSO_4 \cdot 7H_2O$ 2.0mg
H_2SO_4 (0.1*N* solution) 1.0mL

Preparation of Solution D: Add $FeSO_4 \cdot 7H_2O$ to 1.0mL of 0.1*N* H_2SO_4 solution. Mix thoroughly. Autoclave for 15 min at 15 psi pressure–121°C. Cool to 45°–50°C.

Preparation of Medium: Aseptically add 40.0mL of sterile solution B, 20.0mL of sterile solution C, and 1.0mL of sterile solution D to 940.0mL of sterile solution A. Mix thoroughly. Aseptically and anaerobically distribute into sterile tubes under 100% N_2.

Use: For the cultivation and maintenance of *Thiomicrospira denitrificans*.

Thiomicrospira denitrificans Broth

Composition per 1001.0mL:

Solution A 940.0mL
Solution B 40.0mL
Solution C 20.0mL
Solution D 1.0mL

pH 7.0 ± 0.2 at 25°C

Solution A:

Composition per 940.0mL:

KH_2PO_4 .... 2.0g
KNO_3 .... 2.0g
NH_4Cl 1.0g
$MgSO_4 \cdot 7H_2O$ 0.8g
Trace elements solution SL-4 .... 2.0mL

Preparation of Solution A: Add components to distilled/deionized water and bring volume to 940.0mL. Mix thoroughly. Adjust pH to 7.0 with NaOH. Autoclave for 15 min at 15 psi pressure–121°C. Cool to 45°–50°C.

Solution B:

Composition per 40.0mL:

$Na_2S_2O_3 \cdot 5H_2O$ 5.0g

Preparation of Solution B: Add $Na_2S_2O_3 \cdot 5H_2O$ to distilled/deionized water and bring volume to 40.0mL. Mix thoroughly. Autoclave for 15 min at 15 psi pressure–121°C. Cool to 45°–50°C.

Solution C:

Composition per 20.0mL:

$NaHCO_3$ 1.0g

Preparation of Solution C: Add $NaHCO_3$ to distilled/deionized water and bring volume to 20.0mL. Mix thoroughly. Filter sterilize.

Solution D:

Composition per liter:

$FeSO_4 \cdot 7H_2O$ 2.0mg
H_2SO_4 (0.1*N* solution) 1.0mL

Preparation of Solution D: Add $FeSO_4 \cdot 7H_2O$ to 1.0mL of 0.1*N* H_2SO_4 solution. Mix thoroughly. Autoclave for 15 min at 15 psi pressure–121°C. Cool to 45°–50°C.

Trace Elements Solution SL-4:

Composition per liter:

EDTA 0.5g
$FeSO_4 \cdot 7H_2O$ 0.2g
Trace elements solution SL-6 .... 100.0mL

Preparation of Trace Elements Solution SL-4: Add components to distilled/deionized water and bring volume to 1.0L. Mix thoroughly.

Trace Elements Solution SL-6:

Composition per liter:

H_3BO_3 .... 0.3g
$CoCl_2 \cdot 6H_2O$ 0.2g
$ZnSO_4 \cdot 7H_2O$ 0.1g
$MnCl_2 \cdot 4H_2O$ 0.03g
$Na_2MoO_4 \cdot H_2O$ 0.03g
$NiCl_2 \cdot 6H_2O$ 0.02g
$CuCl_2 . 2H_2O$ 0.01g

Preparation of Trace Elements Solution SL-6: Add components to distilled/deionized water and bring volume to 1.0L. Mix thoroughly. Adjust pH to 3.4.

Preparation of Medium: Aseptically add 40.0mL of sterile solution B, 20.0mL of sterile solution C, and 1.0mL of sterile solution D to 940.0mL of sterile solution A. Mix thoroughly. Aseptically and anaerobically distribute into sterile tubes under 100% N_2.

Use: For the cultivation and maintenance of *Thiomicrospira denitrificans*.

Thiomicrospira denitrificans Medium

Composition per liter:

Part I	500.0mL
Part II	500.0mL

pH 7.0–8.0 at 25°C

Part I:

Composition per liter:

NaCl	20.0g
KNO_3	4.0g
$(NH_4)_2SO_4$	2.0g
$MgSO_4·7H_2O$	1.5g
K_2HPO_4	0.6g
KH_2PO_4	0.4g
$FeSO_4$ solution	2.0mL
Trace metals solution	2.0mL
HCl, concentrated	1.0mL

Preparation of Part I: Add components to distilled/deionized water and bring volume to 1.0L. Mix thoroughly. Autoclave for 15 min at 15 psi pressure–121°C. Cool to 25°C.

$FeSO_4$ Solution:

Composition per 100.0mL:

$FeSO_4·7H_2O$	0.5g
HCl (1*N* solution)	100.0mL

Preparation of $FeSO_4$ Solution: Combine the $FeSO_4·7H_2O$ and 100.0mL of HCl solution. Mix thoroughly.

Trace Metals Solution:

Composition per liter:

Disodium EDTA	50.0g
NaOH	11.0g
$CaCl_2·2H_2O$	7.34g
$MnCl_2·2H_2O$	2.5g
$ZnSO_4·7H_2O$	2.2g
$CoCl_2·6H_2O$	0.5g
$(NH_4)_6Mo_7O_{24}·4H_2O$	0.5g
$CuSO_4·5H_2O$	0.2g

Preparation of Trace Metals Solution: Add EDTA to distilled/deionized water and bring volume to 500.0mL. Mix thoroughly. Adjust pH to 6.0 with NaOH. Add remaining components, one by one. Maintain the pH at 6.0. After dissolution of all the salts, adjust the pH to 4.0 with HCl. Store at 4°C.

Part II:

Composition per liter:

$Na_2S_2O_3·5H_2O$	10.0g
$NaHCO_3$	3.0g
NaOH	0.05g

Preparation of Part II: Add components to distilled/deionized water and bring volume to 1.0L. Mix thoroughly. Autoclave for 15 min at 15 psi pressure–121°C. Cool to 25°C.

Preparation of Medium: Aseptically combine 500.0mL of sterile part I and 500.0mL of sterile part II. Mix thoroughly. Aseptically distribute into sterile tubes or flasks.

Use: For the isolation and cultivation of *Thiomicrospira denitrificans.*

Thiomicrospira Medium (ATCC Medium 1036)

Composition per liter:

NaCl	25.0g
$Na_2S_2O_3·5H_2O$	8.0g
$MgSO_4·7H_2O$	1.5g
$(NH_4)_2SO_4$	1.0g
K_2HPO_4	0.5g
$CaCl_2$	0.3g
Vitamin B_{12}	15.0μg
Vishniac and Santer trace metals	0.2mL
Bromcresol Purple (0.05% solution)	0.1mL

pH 7.2 ± 0.2 at 25°C

Vishniac and Santer Trace Metals:

Composition per liter:

Ethylenediamine tetraacetic acid (EDTA)	50.0g
$ZnSO_4·7H_2O$	22.0g
$CaCl_2$	5.54g
$MnCl_2·4H_2O$	5.06g
$FeSO_4·7H_2O$	4.99g
$CoCl_2·6H_2O$	1.61g
$CuSO_4·5H_2O$	1.57g
$(NH_4)_6Mo_7O_{24}·4H_2O$	1.1g

Preparation of Vishniac and Santer Trace Metals: Add components to distilled/deionized water and bring volume to 1.0L. Adjust pH to 6.0 with KOH. Mix thoroughly.

Preparation of Medium: Add components to distilled/deionized water and bring volume to 1.0L. Mix thoroughly. Adjust pH to 7.2. Filter sterilize. Aseptically distribute into sterile tubes or flasks.

Use: For the cultivation and maintenance of *Thiomicrospira* species.

Thiomicrospira Medium (ATCC Medium 1422)

Composition per liter:

NaCl	25.1g
Tris·HCl	3.07g
$Na_2S_2O_3·5H_2O$	2.48g
$MgSO_4·7H_2O$	1.5g
$(NH_4)_2SO_4$	1.0g
KH_2PO_4	0.42g
$CaCl_2·2H_2O$	0.29g
$NaHCO_3$	0.2g
Phenol Red (0.5% solution)	1.0mL
Vishniac and Santer trace metals	0.2mL

pH 7.5 ± 0.2 at 25°C

Vishniac and Santer Trace Metals:

Composition per liter:

Ethylenediamine tetraacetic acid (EDTA)	50.0g
$ZnSO_4·7H_2O$	22.0g
$CaCl_2$	5.54g
$MnCl_2·4H_2O$	5.06g
$FeSO_4·7H_2O$	4.99g
$CoCl_2·6H_2O$	1.61g
$CuSO_4·5H_2O$	1.57g
$(NH_4)_6Mo_7O_{24}·4H_2O$	1.1g

Preparation of Vishniac and Santer Trace Metals: Add components to distilled/deionized water and bring volume to 1.0L. Adjust pH to 6.0 with KOH. Mix thoroughly.

Preparation of Medium: Add components to distilled/deionized water and bring volume to 1.0L. Mix thoroughly. Adjust pH to 7.5. Filter sterilize. Aseptically distribute into sterile tubes or flasks.

Use: For the cultivation and maintenance of *Thiomicrospira* species.

Thiomicrospira pelophila Medium

Composition per liter:

NaCl	25.0g
Agar	10.0g
$Na_2S_2O_3 \cdot 5H_2O$	5.0–8.0g
$MgSO_4 \cdot 7H_2O$	1.5g
$(NH_4)_2SO_4$	1.0g
K_2HPO_4	0.5g
$CaCl_2$	0.3g
Vitamin B_{12}	0.15mg
Trace metals solution	0.2mL

Trace Metals Solution:

Composition per liter:

Disodium EDTA	50.0g
NaOH	11.0g
$CaCl_2 \cdot 2H_2O$	7.34g
$FeSO_4 \cdot 7H_2O$	5.0g
$MnCl_2 \cdot 2H_2O$	2.5g
$ZnSO_4 \cdot 7H_2O$	2.2g
$CoCl_2 \cdot 6H_2O$	0.5g
$(NH_4)_6Mo_7O_{24} \cdot 4H_2O$	0.5g
$CuSO_4 \cdot 5H_2O$	0.2g

Preparation of Trace Metals Solution: Add EDTA to distilled/deionized water and bring volume to 500.0mL. Mix thoroughly. Adjust pH to 6.0 with NaOH. Add remaining components, one by one. Maintain the pH at 6.0. After dissolution of all the salts, adjust the pH to 4.0 with HCl. Store at 4°C.

Preparation of Medium: Add components to distilled/deionized water and bring volume to 1.0L. Mix thoroughly. Gently heat and bring to boiling. Distribute into tubes or flasks. Autoclave for 15 min at 15 psi pressure–121°C. Pour into sterile Petri dishes or leave in tubes.

Use: For the cultivation of *Thiomicrospira pelophila.*

Thiomicrospira pelophila Medium

Composition per 1000.2mL:

NaCl	25.0g
$MgSO_4 \cdot 7H_2O$	1.5g
$(NH_4)_2SO_4$	1.0g
$CaCl_2 \cdot 2H_2O$	0.42g
Bromthymol Blue	4.0mg
K_2HPO_4 solution	100.0mL
$Na_2S_2O_3 \cdot 5H_2O$ solution	100.0mL
Vitamin B_{12} solution	10.0mL
Trace elements solution	0.2mL
Na_2CO_3 solution	variable

pH 7.2 ± 0.2 at 25°C

K_2HPO_4 Solution:

Composition per 100.0mL:

K_2HPO_4	0.5g

Preparation of K_2HPO_4 Solution: Add K_2HPO_4 to distilled/deionized water and bring volume to 100.0mL. Mix thoroughly. Autoclave for 15 min at 15 psi pressure–121°C.

$Na_2S_2O_3$ Solution:

Composition per 10.0mL:

$Na_2S_2O_3 \cdot 5H_2O$	5.0g

Preparation of $Na_2S_2O_3$ Solution: Add $Na_2S_2O_3 \cdot 5H_2O$ to distilled/deionized water and bring volume to 10.0mL. Mix thoroughly. Autoclave for 15 min at 15 psi pressure–121°C.

Vitamin B_{12} Solution:

Composition per 10.0mL:

Vitamin B_{12}	15.0mg

Preparation of Vitamin B_{12} Solution: Add vitamin B_{12} to distilled/deionized water and bring volume to 10.0mL. Mix thoroughly. Filter sterilize.

Trace Elements Solution:

Composition per liter:

Disodium EDTA	50.0g
$ZnSO_4 \cdot 7H_2O$	22.0g
$CaCl_2 \cdot 2H_2O$	5.54g
$MnCl_2 \cdot 4H_2O$	5.06g
$FeSO_4 \cdot 7H_2O$	5.0g
$CoCl_2 \cdot 6H_2O$	1.61g
$CuSO_4 \cdot 5H_2O$	1.57g
$(NH_4)_6Mo_7O_{24} \cdot 4H_2O$	1.1g

Preparation of Trace Elements Solution: Add components to distilled/deionized water and bring volume to 1.0L. Mix thoroughly.

Na_2CO_3 Solution:

Composition per 100.0mL:

Na_2CO_3	0.4g

Preparation of Na_2CO_3 Solution: Add Na_2CO_3 to distilled/deionized water and bring volume to 100.0mL. Mix thoroughly. Autoclave for 15 min at 15 psi pressure–121°C.

Preparation of Medium: Add components, except K_2HPO_4 solution, $Na_2S_2O_3 \cdot 5H_2O$ solution, vitamin B_{12} solution, and Na_2CO_3 solution, to distilled/deionized water and bring volume to 790.0mL. Mix thoroughly. Adjust pH to 7.2. Autoclave for 15 min at 15 psi pressure–121°C. Aseptically add 100.0mL of sterile K_2HPO_4 solution, 100.0mL of sterile $Na_2S_2O_3 \cdot 5H_2O$ solution, and 10.0mL of sterile vitamin B_{12} solution. Mix thoroughly. Aseptically adjust pH to 7.2 with the appropriate volume of sterile Na_2CO_3 solution. Aseptically distribute into sterile tubes or flasks.

Use: For the cultivation and maintenance of *Thiomicrospira pelophila.*

Thiomonas delicata Medium (DSMZ Medium 1037)

Composition per liter:

$Na_2S_2O_3 \cdot 5H_2O$	5.0g
Na_2HPO_4	4.5g
K_2HPO_4	1.5g
Yeast extract	1.0g
Na-aspartate	1.0g
NH_4Cl	0.3g
$MgSO_4 \cdot 7H_2O$	0.1g

pH 6.0 ± 0.2 at 25°C

Preparation of Medium: Add components to distilled/deionized water and bring volume to 1.0L. Mix thoroughly. Adjust pH to 6.0. Dispense into culture vessels. Autoclave for 15 min at 15 psi pressure–121°C.

Use: For the cultivation of *Thiomonas delicata.*

Thiomonas delicata Medium (DSMZ Medium 1037)

Composition per liter:

$Na_2S_2O_3 \cdot 5H_2O$	5.0g
Na_2HPO_4	4.5g

K_2HPO_4	1.5g
Yeast extract	1.0g
NH_4Cl	0.3g
$MgSO_4·7H_2O$	0.1g

pH 6.0 ± 0.2 at 25°C

Preparation of Medium: Add components to distilled/deionized water and bring volume to 1.0L. Mix thoroughly. Adjust pH to 6.0. Dispense into culture vessels. Autoclave for 15 min at 15 psi pressure–121°C.

Use: For the cultivation of *Thiomonas perometabolis*.

Thiophaeococcus mangrovi Medium (DSMZ Medium 1162)

Composition per liter:

NaCl	20.0g
Sodium pyruvate	3.0g
$MgSO_4·7H_2O$	2.0g
NH_4Cl	0.64g
K_2HPO_4	0.5g
Yeast extract	0.4g
$CaCl_2·2H_2O$	0.15g
Bicarbonate solution	10.0mL
Sulfide solution	10.0mL
Thiosulfate solution	10.0mL
Trace element solution SL-8	1.0mL
Vitamin solution	1.0mL

pH 7.5 ± 0.2 at 25°C

Sulfide Solution:

Composition per 10.0mL:

$Na_2S·9H_2O$	1.0g

Preparation of Sulfide Solution: Add $Na_2S·9H_2O$ to distilled/deionized water and bring volume to 10.0mL. Mix thoroughly. Sparge with 100% N_2. Filter sterilize.

Thiosulfate Solution:

Composition per 10.0mL:

$Na_2S_2O_3·5H_2O$	1.5g

Preparation of Thiosulfate Solution: Add $Na_2S_2O_3·5H_2O$ to distilled/deionized water and bring volume to 10.0mL. Mix thoroughly. Sparge with 100% N_2. Filter sterilize.

Bicarbonate Solution:

Composition per 10.0mL:

$NaHCO_3$	1.0g

Preparation of Bicarbonate Solution: Add $NaHCO_3$ to distilled/deionized water and bring volume to 10.0mL. Mix thoroughly. Sparge with a gas mixture of 80% N_2 + 20% CO_2. Filter sterilize.

Vitamin Solution:

Composition per 10.0mL:

Vitamin B_{12}	2.0mg

Preparation of Vitamin Solution: Add Vitamin B_{12} to distilled/deionized water and bring volume to 10.0mL. Mix thoroughly. Filter sterilize.

Trace Elements Solution SL-8:

Composition per liter:

Disodium EDTA	5.2g
$FeCl_2·4H_2O$	1.5g
$CoCl_2·6H_2O$	0.19g
$MnCl_2·4H_2O$	0.1g
$ZnCl_2$	0.07g
H_3BO_3	0.06g
$NaMoO_4·2H_2O$	0.04g
$CuCl_2·2H_2O$	0.02g
$NiCl_2·6H_20$	0.02g

Preparation of Trace Elements Solution SL-8: Add components to distilled/deionized water and bring volume to 1.0L. Mix thoroughly.

Preparation of Medium: Add components, except sulfide, bicarbonate, thiosulfate, and vitamin solutions, to distilled/deionized water and bring volume to 970.0mL. Mix thoroughly. Sparge with 80% N_2 + 20% CO_2 gas mixture. Dispense into culture vessels. Autoclave for 15 min at 15 psi pressure–121°C. Cool to room temperature. Aseptically add sulfide, thiosulfate, bicarbonate, and vitamin solutions. Mix thoroughly. Adjust pH to 7.5.

Use: For the cultivation of *Thiophaeococcus mangrovi*.

Thiophene-2-Carboxylic Acid Hydrazide
See: **TCH Medium**

Thiorhodococcus Medium (DSMZ Medium 28a)

Composition per 5.0L:

Solution A	4.0L
Solution B	860.0mL
Solution E	100.0mL
Solution F	20.0mL
Solution C	5.0mL
Solution D	5.0mL

pH 7.3 at 25°C

Solution A:

Composition per 4.0L:

NaCl	100.0g
$MgCl_2·6H_2O$	5.0g
$MgSO_4$	2.5g
$Na_2S_2O_3·5H_2O$	2.5g
KH_2PO_4	1.7g
NH_4Cl	1.7g
KCl	1.7g
$CaCl_2·2H_2O$	1.25g

Preparation of Solution A: Add components to 4.0L distilled water. Mix thoroughly. Autoclave for 45 min at 15 psi pressure–121°C in a 5-liter special bottle or flask with four openings at the top, together with a teflon-coated magnetic bar. In this 5-liter bottle, two openings for tubes are in the central, silicon rubber stopper; one is a short, gas-inlet tube with a sterile cotton filter, and the other is an outlet tube for medium, which reaches the bottom of the vessel at one end and has, at the other end, a silicon rubber tube with a pinch cock and a bell for aseptic dispensing of the medium into bottles. The other two openings have gas-tight screw caps; one of these openings is for the addition of sterile solutions and the other serves as a gas outlet. After autoclaving, cool to room temperature under a 100% N_2 atmosphere with a positive pressure of 0.05– 0.1 atm (a manometer for low pressure will be required). Saturate the cold medium with CO_2 by magnetic stirring for 30 min under a CO_2 atmosphere of 0.05–0.1 atm.

Solution B:

Distilled water	860.0mL

Preparation of Solution B: Autoclave distilled water for 15 min at 15 psi pressure–121°C in a cotton-stoppered Erlenmeyer flask. Cool to room temperature under an atmosphere of N_2 in an anaerobic jar.

Solution C:

Composition per 100.0mL:

Vitamin B_{12} 2.0mg

Preparation of Solution C: Add vitamin B_{12} to distilled/deionized water and bring volume to 100.0mL. Mix thoroughly. Sparge under 100% N_2 gas for 3 min. Filter sterilize Store under N_2 gas.

Solution D:

Composition per liter:

Disodium ethylendiamine tetraacetate (Disodium EDTA) 3.0g
$FeSO_4 \cdot 7H_2O$ 1.1g
H_3BO_3 .... 0.3g
$CoCl_2 \cdot 6H_2O$ 0.19g
$MnCl_2 \cdot 2H_2O$ 50.0mg
$ZnCl_2$ 42.0mg
$NiCl_2 \cdot 6H_2O$ 24.0mg
$Na_2MoO_4 \cdot 2H_2O$ 18.0mg
$CuCl_2 \cdot 2H_2O$ 2.0mg

Preparation of Solution D: Add components to distilled/deionized water and bring volume to 1.0L. Mix thoroughly. Autoclave for 15 min at 15 psi pressure–121°C.

Solution E:

Composition per 100.0mL:

$NaHCO_3$ 7.5g

Preparation of Solution E: Add $NaHCO_3$ to distilled/deionized water and bring volume to 100.0mL. Mix thoroughly. Sparge with 100% CO_2 until saturated. Filter sterilize under 100% CO_2 into a sterile, gas-tight 100.0mL screw-capped bottle.

Solution F:

Composition per 100.0mL:

$Na_2S \cdot 9H_2O$ 10.0g

Preparation of Solution F: Add $Na_2S \cdot 9H_2O$ to distilled/deionized water in a 250.0mL screw-capped bottle fitted with a butyl rubber septum and bring volume to 100.0mL. Mix thoroughly. Sparge under 100% N_2 gas for 3 min. Autoclave for 15 min at 15 psi pressure–121°C. Cool to room temperature.

Neutralized Sulfide Solution:

Composition per 100.0mL:

$Na_2S \cdot 9H_2O$ 1.5g

Preparation of Neutralized Sulfide Solution: Add $Na_2S \cdot 9H_2O$ to distilled/deionized water in a 250.0mL screw-capped bottle fitted with a butyl rubber septum and bring volume to 100.0mL. Add a magnetic stir bar. Mix thoroughly. Sparge under 100% N_2 gas for 3 min. Autoclave for 15 min at 15 psi pressure–121°C. Cool to room temperature. Adjust pH to about 7.3 with sterile 2*M* H_2SO_4. Do not open the bottle to add H_2SO_4; use a sterile syringe. Stir the solution continuously to avoid precipitation of elemental sulfur. The final solution should be clear and yellow in color.

Preparation of Medium: Add solutions B, C, D, E, and F to solution A through one of the screw-cap openings against a stream of either N_2 gas or, better, a mixture of 95% N_2 and 5% CO_2 while the medium is magnetically stirred. Adjust the pH of the medium with sterile HCl or Na_2CO_3 solution (2*M* solutions) to pH 7.3. Distribute the medium aseptically through the medium outlet tube into sterile, 100mL bottles (with metal caps and autoclavable rubber seals) using the positive gas pressure (0.05–0.1 atm) of the N_2/CO_2 gas mixture. Leave a small air bubble in each bottle to meet possible pressure changes. The tightly sealed, screw-cap bottles can be stored for several weeks or months in the dark. During the first 24 hr, the iron of the medium precipitates in the form of black flocs. No other sediment should arise in the otherwise clear medium. Incubate in the light at 500–1,000 lux intensity. Feed periodically with neutralized solution of sodium sulfide to replenish sulfide and with other supplement solutions.

Use: For the cultivation of *Thiorhodococcus minor*.

Thiosphaera Agar

Composition per liter:

Agar 15.0g
Na_2HPO_4 .... 4.2g
KH_2PO_4 .... 1.5g
NH_4Cl 0.3g
$MgSO_4 \cdot 7H_2O$ 0.1g
KNO_3 .... 0.1g
Vishniac and Santer trace metals 2.0mL

pH 8.0–8.2 at 25°C

Vishniac and Santer Trace Metals:

Composition per liter:

Ethylenediamine tetraacetic acid (EDTA) 50.0g
$ZnSO_4 \cdot 7H_2O$ 22.0g
$CaCl_2$ 5.54g
$MnCl_2 \cdot 4H_2O$ 5.06g
$FeSO_4 \cdot 7H_2O$ 4.99g
$CoCl_2 \cdot 6H_2O$ 1.61g
$CuSO_4 \cdot 5H_2O$ 1.57g
$(NH_4)_6Mo_7O_{24} \cdot 4H_2O$ 1.1g

Preparation of Vishniac and Santer Trace Metals: Add components to distilled/deionized water and bring volume to 1.0L. Adjust pH to 6.0 with KOH. Mix thoroughly.

Preparation of Medium: Add components, except agar, to distilled/deionized water and bring volume to 500.0mL. Mix thoroughly. Adjust pH to 8.0–8.2. Filter sterilize. Warm to 45°–50°C. Add agar to distilled/deionized water and bring volume to 500.0mL. Mix thoroughly. Gently heat and bring to boiling. Autoclave for 15 min at 15 psi pressure–121°C. Cool to 45°–50°C. Aseptically combine the two sterile solutions. Mix thoroughly. Pour into sterile Petri dishes or distribute into sterile tubes.

Use: For the cultivation and maintenance of *Thiosphaera pantotropha*.

Thiosphaera Broth

Composition per liter:

Na_2HPO_4 .... 4.2g
KH_2PO_4 .... 1.5g
NH_4Cl 0.3g
$MgSO_4 \cdot 7H_2O$ 0.1g
KNO_3 .... 0.1g
Vishniac and Santer trace metals 2.0mL

pH 8.0–8.2 at 25°C

Vishniac and Santer Trace Metals:

Composition per liter:

Ethylenediamine tetraacetic acid (EDTA) 50.0g
$ZnSO_4 \cdot 7H_2O$ 22.0g
$CaCl_2$ 5.54g
$MnCl_2 \cdot 4H_2O$ 5.06g

$FeSO_4 \cdot 7H_2O$	4.99g
$CoCl_2 \cdot 6H_2O$	1.61g
$CuSO_4 \cdot 5H_2O$	1.57g
$(NH_4)_6Mo_7O_{24} \cdot 4H_2O$	1.1g

Preparation of Vishniac and Santer Trace Metals: Add components to distilled/deionized water and bring volume to 1.0L. Adjust pH to 6.0 with KOH. Mix thoroughly.

Preparation of Medium: Add components to distilled/deionized water and bring volume to 1.0L. Mix thoroughly. Adjust pH to 8.0–8.2. Filter sterilize. Aseptically distribute into sterile tubes or flasks.

Use: For the cultivation and maintenance of *Thiosphaera pantotropha.*

Thiosphaera pantotropha Medium

Composition per 1001.0mL:

Agar	20.0g
$Na_2HPO_4 \cdot 2H_2O$	7.9g
KH_2PO_4	1.5g
NH_4Cl	0.3g
$MgSO_4 \cdot 7H_2O$	0.1g
Yeast extract solution	10.0mL
Trace elements solution SL-10	1.0mL

pH 7.5 ± 0.2 at 25°C

Yeast Extract Solution:

Composition per 10.0mL:

Yeast extract	1.0g

Preparation of Yeast Extract Solution: Add yeast extract to distilled/deionized water and bring volume to 10.0mL. Mix thoroughly. Autoclave for 15 min at 15 psi pressure–121°C.

Trace Elements Solution SL-10:

Composition per liter:

$FeCl_2 \cdot 4H_2O$	1.5g
$CoCl_2 \cdot 6H_2O$	190.0mg
$MnCl_2 \cdot 4H_2O$	100.0mg
$ZnCl_2$	70.0mg
$Na_2MoO_4 \cdot 2H_2O$	36.0mg
$NiCl_2 \cdot 6H_2O$	24.0mg
H_3BO_3	6.0mg
$CuCl_2 \cdot 2H_2O$	2.0mg
HCl (25% solution)	10.0mL

Preparation of Trace Elements Solution SL-10: Add $FeCl_2 \cdot 4H_2O$ to 10.0mL of HCl solution. Mix thoroughly. Add distilled/deionized water and bring volume to 1.0L. Add remaining components. Mix thoroughly. Sparge with 100% N_2. Autoclave for 15 min at 15 psi pressure–121°C.

Preparation of Medium: Add components, except yeast extract solution, to distilled/deionized water and bring volume to 990.0mL. Mix thoroughly. Gently heat and bring to boiling. Autoclave for 15 min at 15 psi pressure–121°C. Cool to 50°–55°C. Aseptically add 10.0mL of sterile yeast extract solution. Mix thoroughly. Pour into sterile Petri dishes or distribute into sterile tubes.

Use: For the cultivation and maintenance of *Paracoccus denitrificans*.

Thiosulfate Citrate Bile Salts Sucrose Agar

See: **TCBS Agar**

Thiosulfate-Oxidizing Medium

Composition per liter:

K_2HPO_4	2.0g
$MgSO_4 \cdot 7H_2O$	0.1g
$CaCl_2 \cdot 2H_2O$	0.1g
$FeCl_3 \cdot 6H_2O$	0.02g
$(NH_4)_2SO_4$ solution	100.0mL
Thiosulfate solution	100.0mL

pH 7.8 ± 0.2 at 25°C

$(NH_4)_2SO_4$ Solution:

Composition per 100.0mL:

$(NH_4)_2SO_4$	0.1g

Preparation of $(NH_4)_2SO_4$ Solution: Add the $(NH_4)_2SO_4$ to distilled/deionized water and bring volume to 100.0mL. Mix thoroughly. Autoclave for 15 min at 15 psi pressure–121°C. Cool to 45°–50°C.

Thiosulfate Solution:

Composition per 100.0mL:

$Na_2S_2O_3 \cdot 5H_2O$	10.0g

Preparation of Thiosulfate Solution: Add the $Na_2S_2O_3 \cdot 5H_2O$ to distilled/deionized water and bring volume to 100.0mL. Mix thoroughly. Autoclave for 15 min at 15 psi pressure–121°C. Cool to 45°–50°C.

Preparation of Medium: Add components, except $(NH_4)_2SO_4$ solution and thiosulfate solution, to distilled/deionized water and bring volume to 800.0mL. Mix thoroughly. Autoclave for 15 min at 15 psi pressure–121°C. Cool to 45°–50°C. Aseptically add the sterile $(NH_4)_2SO_4$ solution and the sterile thiosulfate solution. Mix thoroughly. Adjust the pH to 7.8 if necessary. Aseptically distribute into sterile tubes or flasks.

Use: For the isolation and cultivation of iron and sulfur bacteria.

Thiosulfate Salts Broth

Composition per liter:

$Na_2S_2O_3 \cdot 5H_2O$	24.81g
$NH_4 \cdot Cl$	2.2g
KH_2PO_4	2.0g
Artificial seawater	500.0mL

Artificial Seawater:

Composition per liter:

NaCl	23.476g
$MgCl_2$	4.981g
Na_2SO_4	3.917g
$CaCl_2$	1.102g
KCl	0.664g
$NaHCO_3$	0.192g
KBr	0.096g
H_3BO_3	0.026g
$SrCl_3$	0.024g
NaF	3.0mg

pH 5.0 ± 0.2 at 25°C

Preparation of Artificial Seawater: Add components to distilled/deionized water and bring volume to 1.0L. Mix thoroughly.

Preparation of Medium: Add components to distilled/deionized water and bring volume to 1.0L. Mix thoroughly. Adjust pH to 5.0. Distribute into tubes or flasks. Autoclave for 15 min at 15 psi pressure–121°C.

Use: For the cultivation of *Thiobacillus* species.

Thiothrix Agar (DSMZ Medium 573)

Composition per liter:

Agar	12.0g
NH_4Cl	0.2g
Na-acetate	0.1g
K_2HPO_4	0.01g
$MgSO_4 \cdot 7H_2O$	0.01g
$CaSO_4$ (saturated solution)	20.0mL
$Na_2S \cdot 9H_2O$ solution	10.0mL
Trace elements solution	5.0mL

pH 7.5 ± 0.2 at 25°C

$Na_2S \cdot 9H_2O$ Solution:

Composition per 10.0mL:

$Na_2S \cdot 9H_2O$	0.3g

Preparation of $Na_2S \cdot 9H_2O$ Solution: Add $Na_2S \cdot 9H_2O$ to distilled/deionized water and bring volume to 10.0mL. Sparge with N_2. Autoclave for 15 min at 15 psi pressure–121°C. Cool to 25°C. Store anaerobically.

Trace Elements Solution:

Composition per liter:

$FeSO_4 \cdot 7H_2O$	0.7g
EDTA	0.2g
$ZnSO_4 \cdot 7H_2O$	10.0mg
H_3BO_3	10.0mg
$MnSO_4 \cdot 4H_2O$	2.0mg
$Co(NO_3)_2$	1.0mg
$Na_2MoO_4 \cdot 4H_2O$	1.0mg
$CuSO_4 \cdot 5H_2O$	5.0μg

Preparation of Trace Elements Solution: Add components to distilled/deionized water and bring volume to 1.0L. Mix thoroughly.

Preparation of Medium: Add components, except $Na_2S \cdot 9H_2O$ solution, to distilled/deionized water and bring volume to 990.0mL. Mix thoroughly. Gently heat and bring to boiling. Autoclave for 15 min at 15 psi pressure–121°C. Cool to 50°C. Aseptically add 10.0mL sterile $Na_2S \cdot 9H_2O$ solution. Mix thoroughly. Pour into Petri dishes or aseptically distribute into sterile tubes.

Use: For the cultivation and maintenance of *Thiothrix nivea.*

Thiothrix Agar

Composition per 1003.0mL:

Agar	12.0g
NH_4Cl	0.2g
Sodium acetate	0.1g
K_2HPO_4	0.01g
$MgSO_4 \cdot 7H_2O$	0.01g
$CaSO_4$ (saturated solution)	20.0mL
$Na_2S \cdot 9H_2O$ solution	3.0mL
Trace elements solution	5.0mL

pH 7.5 ± 0.2 at 25°C

$Na_2S \cdot 9H_2O$ Solution:

Composition per 10.0mL:

$Na_2S \cdot 9H_2O$	1.0g

Preparation of $Na_2S \cdot 9H_2O$ Solution: Add $Na_2S \cdot 9H_2O$ to distilled/deionized water and bring volume to 10.0mL. Mix thoroughly. Autoclave for 15 min at 15 psi pressure–121°C.

Trace Elements Solution:

Composition per liter:

$FeSO_4 \cdot 7H_2O$	0.7g
EDTA	0.2g
$ZnSO_4 \cdot 7H_2O$	0.01g
$MnSO_4 \cdot 4H_2O$	0.002g
H_3BO_3	10.0mg
$CO(NO_3)_2$	1.0mg
$Na_2MoO_4 \cdot 2H_2O$	1.0mg
$CuSO_4 \cdot 5H_2O$	5.0μg

Preparation of Trace Elements Solution: Add components to distilled/deionized water and bring volume to 1.0L. Mix thoroughly.

Preparation of Medium: Add components, except $Na_2S \cdot 9H_2O$ solution, to distilled/deionized water and bring volume to 1.0L. Mix thoroughly. Gently heat and bring to boiling. Autoclave for 15 min at 15 psi pressure–121°C. Cool to 50°–55°C. Aseptically add 3.0mL of sterile $Na_2S \cdot 9H_2O$ solution. Mix thoroughly. Pour into sterile Petri dishes or distribute into sterile tubes.

Use: For the cultivation and maintenance of *Thiothrix nivea.*

Thiothrix Medium (DSMZ Medium 573)

Composition per liter:

NH_4Cl	0.2g
Na-acetate	0.1g
K_2HPO_4	0.01g
$MgSO_4 \cdot 7H_2O$	0.01g
$CaSO_4$ (saturated solution)	20.0mL
$Na_2S \cdot 9H_2O$ solution	10.0mL
Trace elements solution	5.0mL

pH 7.5 ± 0.2 at 25°C

$Na_2S \cdot 9H_2O$ Solution:

Composition per 10.0mL:

$Na_2S \cdot 9H_2O$	0.3g

Preparation of $Na_2S \cdot 9H_2O$ Solution: Add $Na_2S \cdot 9H_2O$ to distilled/deionized water and bring volume to 10.0mL. Sparge with N_2. Autoclave for 15 min at 15 psi pressure–121°C. Cool to 25°C. Store anaerobically.

Trace Elements Solution:

Composition per liter:

$FeSO_4 \cdot 7H_2O$	0.7g
EDTA	0.2g
$ZnSO_4 \cdot 7H_2O$	10.0mg
H_3BO_3	10.0mg
$MnSO_4 \cdot 4H_2O$	2.0mg
$Co(NO_3)_2$	1.0mg
$Na_2MoO_4 \cdot 4H_2O$	1.0mg
$CuSO_4 \cdot 5H_2O$	5.0μg

Preparation of Trace Elements Solution: Add components to distilled/deionized water and bring volume to 1.0L. Mix thoroughly.

Preparation of Medium: Add components, except $Na_2S \cdot 9H_2O$ solution, to distilled/deionized water and bring volume to 990.0mL. Mix thoroughly. Autoclave for 15 min at 15 psi pressure–121°C. Aseptically add 10.0mL sterile $Na_2S \cdot 9H_2O$ solution. Mix thoroughly. Aseptically distribute into sterile tubes or flasks.

Use: For the cultivation and maintenance of *Thiothrix nivea.*

Thorne Medium, Modified

Composition per liter:

Glycerol	20.0g
L-Glutamic acid	4.0g
Citric acid	2.0g
$MgSO_4 \cdot 7H_2O$	1.0g
Ferric ammonium citrate	0.5g
K_2HPO_4	0.5g

pH 7.4 ± 0.2 at 25°C

Preparation of Medium: Add components to distilled/deionized water and bring volume to 1.0L. Mix thoroughly. Adjust pH using NH_4OH (not NaOH). Distribute into tubes or flasks. Autoclave for 15 min at 15 psi pressure–121°C.

Use: For the cultivation of *Bacillus licheniformis*.

Thymidine Auxotroph XPS Medium

See: **XPS Broth with Thymidine**

Tibi Medium

Composition per liter:

Sucrose	100.0–150.0g
Fig, dried, quartered	1
Lemon wedge (0.5cm segment)	1

Preparation of Medium: Add components to tap water and bring volume to 1.0L in a 1.0L Erlenmeyer flask fitted with a cotton stopper. Autoclave for 15 min at 15 psi pressure–121°C. Cool to room temperature. Inoculate with about 50.0mL of Tibi grains.

Use: For the cultivation of osmophilic bacteria and fungi from tibi grains.

Tieghemiomyces Medium

Composition per liter:

Agar	20.0g
Casein hydrolysate	10.0g
KH_2PO_4	1.0g
$MgSO_4 \cdot 7H_2O$	0.5g
$ZnSO_4 \cdot 7H_2O$	0.2mg
$FeSO_4$	0.2mg
$MnSO_4 \cdot H_2O$	0.2mg
Thiamine	0.2mg
Biotin	0.01mg
Glycerol	20.0mL

pH 6.0–6.5 at 25°C

Preparation of Medium: Add components to distilled/deionized water and bring volume to 1.0L. Mix thoroughly. Gently heat and bring to boiling. Adjust pH to 6.0–6.5. Distribute into tubes or flasks. Autoclave for 15 min at 15 psi pressure–121°C. Pour into sterile Petri dishes or leave in tubes.

Use: For the cultivation and maintenance of *Tieghemiomyces parasiticus*.

Tindallia Medium (DSMZ Medium 798)

Composition per liter:

NaCl	10.0g
Na_2CO_3	8.0g
Yeast extract	4.0g
NH_4Cl	0.5g
KCl	0.2g
K_2HPO_4	0.2g
$Na_2S \cdot 9H_2O$ solution	10.0mL
Trace elements solution SL-10	1.0mL
Vitamin solution	1.0mL

pH 9.0 ± 0.2 at 25°C

Vitamin Solution:

Composition per liter:

Pyridoxine-HCl	10.0mg
Thiamine-HCl·$2H_2O$	5.0mg
Riboflavin	5.0mg
Nicotinic acid	5.0mg
D-Ca-pantothenate	5.0mg
p-Aminobenzoic acid	5.0mg
Lipoic acid	5.0mg
Biotin	2.0mg
Folic acid	2.0mg
Vitamin B_{12}	0.1mg

Preparation of Vitamin Solution: Add components to distilled/deionized water and bring volume to 1.0L. Mix thoroughly. Sparge with 80% H_2 + 20% CO_2. Filter sterilize.

$Na_2S \cdot 9H_2O$ Solution:

Composition per 10.0mL:

$Na_2S \cdot 9H_2O$	0.5g

Preparation of $Na_2S \cdot 9H_2O$ Solution: Add $Na_2S \cdot 9H_2O$ to distilled/deionized water and bring volume to 10.0mL. Sparge with N_2. Autoclave for 15 min at 15 psi pressure–121°C. Cool to 25°C. Store anaerobically.

Preparation of Medium: Prepare and dispense medium under 100% N_2. Add components, except vitamin solution, to distilled/deionized water and bring volume to 999.0mL. Mix thoroughly. Gently heat and bring to boiling. Boil for 5 min. Cool to 25°C while sparging with 100% N_2. Adjust pH to 9.0. Autoclave for 15 min at 15 psi pressure–121°C. Cool to 25°C. Aseptically and anaerobically add 1.0mL vitamin solution. Mix thoroughly. Aseptically and anaerobically under 100% N_2 distribute into sterile tubes or bottles.

Use: For the cultivation of *Tindallia magadiensis.*

Tindallia Medium (DSMZ Medium 1148)

Composition per liter:

NaCl	30.0g
Peptone	5.0g
NH_4Cl	1.0g
Yeast extract	0.5g
KCl	0.2g
K_2HPO_4	0.2g
$MgCl_2 \cdot 6H_2O$	0.1g
Resazurin	1.0mg
Carbonate solution	10.0mL
Bicarbonate solution	10.0mL
Sulfide solution	10.0mL
Vitamin solution	2.0mL
Trace elements solution	1.0mL

pH 9.7 ± 0.2 at 25°C

Sulfide Solution:

Composition per 10.0mL:

$Na_2S \cdot 9H_2O$	0.25g

Preparation of Sulfide Solution: Add $Na_2S{\cdot}9H_2O$ to distilled/deionized water and bring volume to 10.0mL. Mix thoroughly. Autoclave under 100% N_2 for 15 min at 15 psi pressure–121°C. Cool to room temperature.

Bicarbonate Solution:

Composition per 10.0mL:

$NaHCO_3$	1.5g

Preparation of Bicarbonate Solution: Add components to distilled/deionized water and bring volume to 10.0mL. Mix thoroughly. Sparge with a gas mixture of 80% N_2 + 20% CO_2. Filter sterilize.

Carbonate Solution:

Composition per 10.0mL:

Na_2CO_3	2.76g

Preparation of Carbonate Solution: Add components to distilled/deionized water and bring volume to 10.0mL. Mix thoroughly. Sparge with a gas mixture of 80% N_2 + 20% CO_2. Filter sterilize.

Vitamin Solution:

Composition per liter:

Pyridoxine-HCl	10.0mg
Thiamine-HCl·$2H_2O$	5.0mg
Riboflavin	5.0mg
Nicotinic acid	5.0mg
D-Ca-pantothenate	5.0mg
p-Aminobenzoic acid	5.0mg
Lipoic acid	5.0mg
Biotin	2.0mg
Folic acid	2.0mg
Vitamin B_{12}	0.1mg

Preparation of Vitamin Solution: Add components to distilled/deionized water and bring volume to 1.0L. Mix thoroughly. Sparge with 80% H_2 + 20% CO_2. Filter sterilize.

Trace Elements Solution:

Composition per 200.0mL:

$MnCl_2{\cdot}4H_2O$	0.72g
$Fe(NH_4)_2(SO_4)_2{\cdot}6H_2O$	0.4g
$FeSO_4{\cdot}7H_2O$	0.2g
$CoCl_2{\cdot}6H_2O$	0.2g
$ZnSO_4{\cdot}7H_2O$	0.2g
$NiCl_2{\cdot}6H_2O$	0.1g
$Na_2MoO_4{\cdot}2H_2O$	0.02g
$CuSO_4{\cdot}5H_2O$	0.02g
H_3BO_3	0.02g
$KAl(SO_4)_2{\cdot}12H_2O$	0.02g
HCl	5.0mL

Preparation of Trace Elements: Add components to distilled/deionized water and bring volume to 200.0mL. Mix thoroughly. Autoclave for 15 min at 15 psi pressure–121°C.

Preparation of Medium: Add components, except sulfide, bicarbonate, carbonate, and vitamin solutions, to distilled/deionized water and bring volume to 970.0mL. Mix thoroughly. Sparge with 80% N_2 + 20% CO_2 gas mixture. Dispense into culture vessels. Autoclave for 15 min at 15 psi pressure–121°C. Cool to room temperature. Aseptically add sulfide, carbonate, bicarbonate, and vitamin solutions. Mix thoroughly. Adjust pH to 9.5.–10.0

Use: For the cultivation of *Tindallia* spp.

Tinsdale Agar

Composition per 1100.0mL:

Proteose peptone	20.0g
Agar	15.0g
NaCl	5.0g
Yeast extract	5.0g
L-Cystine	0.24g
Tinsdale supplement	150.0mL

pH 7.4 ± 0.2 at 25°C

Source: This medium is available as a premixed powder from BD Diagnostic Systems and Oxoid Unipath.

Tinsdale Supplement:

Composition per 100.0mL:

$Na_2S_2O_3$	0.43g
K_2TeO_3	0.35g
Serum	100.0mL

Caution: Potassium tellurite is toxic.

Preparation of Tinsdale Supplement: Add $Na_2S_2O_3$ and K_2TeO_3 to serum. Mix thoroughly. Filter sterilize.

Preparation of Medium: Add components, except Tinsdale supplement, to distilled/deionized water and bring volume to 1.0L. Mix thoroughly. Gently heat and bring to boiling. Autoclave for 15 min at 15 psi pressure–121°C. Cool to 50°–55°C. Aseptically add 100.0mL of sterile Tinsdale supplement. Mix thoroughly. Pour into sterile Petri dishes or distribute into sterile tubes.

Use: For the primary isolation and identification of *Corynebacterium diphtheriae*.

Tinsdale HiVeg Agar Base with Tinsdale Supplement

Composition per liter:

Plant peptone	20.0g
Agar	15.0g
NaCl	5.0g
$Na_2S_2O_3$	0.43g
L-Cystine	0.24g
Tinsdale supplement	150.0mL

pH 7.4 ± 0.2 at 25°C

Source: This medium is available as a premixed powder from HiMedia.

Tinsdale Supplement:

Composition per 100.0mL:

$Na_2S_2O_3$	0.43g
K_2TeO_3	0.35g
Serum	100.0mL

Caution: Potassium tellurite is toxic.

Preparation of Tinsdale Supplement: Add $Na_2S_2O_3$ and K_2TeO_3 to serum. Mix thoroughly. Filter sterilize.

Preparation of Medium: Add components, except Tinsdale supplement, to distilled/deionized water and bring volume to 1.0L. Mix thoroughly. Gently heat and bring to boiling. Autoclave for 15 min at 15 psi pressure–121°C. Cool to 50°–55°C. Aseptically add 100.0mL of sterile Tinsdale supplement. Mix thoroughly. Pour into sterile Petri dishes or distribute into sterile tubes.

Use: For the primary isolation and identification of *Corynebacterium diphtheriae*.

Tissierella creatinophila Medium (DSMZ Medium 824)

Composition per 1022.0mL:

Creatine	3.8g
Na-formate	2.72g
Yeast extract	2.0g
KCl	1.0g
$MgSO_4 \cdot 7H_2O$	0.5g
NaCl	0.5g
NH_4Cl	0.25g
KH_2PO_4	0.2g
$CaCl_2 \cdot 2H_2O$	0.15g
Resazurin	0.5mg
$Na_2SeO_3 \cdot 5H_2O$	0.26mg
$NaHCO_3$ solution	40.0mL
$Na_2S \cdot 9H_2O$ solution	20.0mL
Vitamin solution	10.0mL
Seven vitamin solution	1.0mL
Trace elements solution SL-10	1.0mL

pH 7.5–7.8 ± 0.2 at 25°C

$Na_2S \cdot 9H_2O$ Solution:
Composition per 20.0mL:

$Na_2S \cdot 9H_2O$	0.6g

Preparation of $Na_2S \cdot 9H_2O$ Solution: Add $Na_2S \cdot 9H_2O$ to distilled/deionized water and bring volume to 20.0mL. Mix thoroughly. Autoclave under 100% N_2 for 15 min at 15 psi pressure–121°C. Cool to room temperature.

$NaHCO_3$ Solution:
Composition per 100.0mL:

$NaHCO_3$	10.0g

Preparation of $NaHCO_3$ Solution: Add $NaHCO_3$ to distilled/deionized water and bring volume to 100.0mL. Mix thoroughly. Sparge with 80% N_2 + 20% CO_2. Filter sterilize.

Vitamin Solution:
Composition per liter:

Pyridoxine-HCl	10.0mg
Thiamine-HCl·$2H_2O$	5.0mg
Riboflavin	5.0mg
Nicotinic acid	5.0mg
D-Ca-pantothenate	5.0mg
p-Aminobenzoic acid	5.0mg
Lipoic acid	5.0mg
Biotin	2.0mg
Folic acid	2.0mg
Vitamin B_{12}	0.1mg

Preparation of Vitamin Solution: Add components to distilled/deionized water and bring volume to 1.0L. Mix thoroughly. Sparge with 100% N_2. Filter sterilize.

Seven Vitamin Solution:
Composition per liter:

Pyridoxine hydrochloride	300.0mg
Thiamine-HCl·$2H_2O$	200.0mg
Nicotinic acid	200.0mg
Vitamin B_{12}	100.0mg
Calcium pantothenate	100.0mg
p-Aminobenzoic acid	80.0mg
D(+)-Biotin	20.0mg

Preparation of Seven Vitamin Solution: Add components to distilled/deionized water and bring volume to 1.0L. Sparge with 100% N_2. Mix thoroughly. Filter sterilize.

Trace Elements Solution SL-10:
Composition per liter:

$FeCl_2 \cdot 4H_2O$	1.5g
$CoCl_2 \cdot 6H_2O$	190.0mg
$MnCl_2 \cdot 4H_2O$	100.0mg
$ZnCl_2$	70.0mg
$Na_2MoO_4 \cdot 2H_2O$	36.0mg
$NiCl_2 \cdot 6H_2O$	24.0mg
H_3BO_3	6.0mg
$CuCl_2 \cdot 2H_2O$	2.0mg
HCl (25% solution)	10.0mL

Preparation of Trace Elements Solution SL-10: Add $FeCl_2 \cdot 4H_2O$ to 10.0mL of HCl solution. Mix thoroughly. Add distilled/deionized water and bring volume to 1.0L. Add remaining components. Mix thoroughly.

Preparation of Medium: Prepare and dispense medium under 100% N_2. Add components, except $NaHCO_3$ solution, $Na_2S \cdot 9H_2O$ solution, seven vitamin solution, and vitamin solution, to 950.0mL distilled/deionized water. Mix thoroughly. Sparge with 100% N_2 + 20% CO_2. Adjust pH to 7.6. Dispense 10.0mL aliquots into bottles. Autoclave for 15 min at 15 psi pressure–121°C. For each 10.0mL medium aseptically and anaerobically inject from sterile stock solutions 0.4mL $NaHCO_3$ solution, 0.2mL $Na_2S \cdot 9H_2O$ solution, 0.1mL vitamin solution, and 0.01mL seven vitamin solution. Final pH should be 7.5–7.8.

Use: For the cultivation of *Tissierella creatinophila.*

Tissue Culture Amino Acids, HeLa 100X (TC Amino Acids, HeLa 100X)

Composition per liter:

L-Lysine	0.029g
L-Isoleucine	0.026g
L-Leucine	0.026g
L-Threonine	0.023g
L-Valine	0.023g
L-Tyrosine	0.018g
L-Arginine	0.017g
L-Phenylalanine	0.016g
L-Cystine	0.012g
L-Histidine	7.8mg
L-Methionine	7.5mg
L-Tryptophan	4.1mg

pH 7.2–7.4 at 25°C

Preparation of Tissue Culture Amino Acids, HeLa 100X: Add components to distilled/deionized water and bring volume to 1.0L. Mix thoroughly. Adjust pH to 7.2–7.4. Filter sterilize.

Use: For the preparation of Eagle HeLa medium for tissue culture procedures and virus studies.

Tissue Culture Amino Acids, Minimal Eagle 50X (TC Amino Acids, Minimal Eagle 50X)

Composition per liter:

L-Arginine	0.1g
L-Lysine	0.058g
L-Isoleucine	0.052g
L-Leucine	0.052g

L-Threonine	0.048g
L-Valine	0.046g
L-Tyrosine	0.036g
L-Phenylalanine	0.032g
L-Histidine	0.031g
L-Cystine	0.024g
L-Methionine	0.015g
L-Tryptophan	0.01g

pH 7.2–7.4 at 25°C

Preparation of Tissue Culture Amino Acids, Minimal Eagle 50X: Add components to distilled/deionized water and bring volume to 1.0L. Mix thoroughly. Adjust pH to 7.2–7.4. Filter sterilize.

Use: For the preparation of TC minimal medium Eagle for tissue culture procedures and virus studies.

Tissue Culture Dulbecco Solution (TC Dulbecco Solution)

Composition per liter:

NaCl	8.0g
Na_2HPO_4	1.15g
KH_2PO_4	0.2g
KCl	0.2g
$CaCl_2 \cdot 2H_2O$	0.1g
$MgCl_2 \cdot 6H_2O$	0.1g

pH 7.2–7.4 at 25°C

Preparation of Tissue Culture Dulbecco Solution: Add components to distilled/deionized water and bring volume to 1.0L. Mix thoroughly. Adjust pH to 7.2–7.4. Filter sterilize.

Use: For use in tissue culture and virus preparations.

Tissue Culture Earle Solution (TC Earle Solution)

Composition per 1002.0mL:

NaCl	6.8g
$NaHCO_3$	2.2g
Glucose	1.0g
KCl	0.4g
$CaCl_2 \cdot 2H_2O$	0.2g
NaH_2PO_4	0.125g
$MgSO_4 \cdot 7H_2O$	0.1g
Phenol Red (1% solution)	2.0mL

pH 7.2–7.4 at 25°C

Preparation of Tissue Culture Earle Solution: Add components, except Phenol Red, to distilled/deionized water and bring volume to 1.0L. Mix thoroughly. Add 2.0mL of Phenol Red solution. Adjust pH to 7.2–7.4. Filter sterilize.

Use: For use in tissue culture and virus preparations.

Tissue Culture Hanks Solution (TC Hanks Solution)

Composition per liter:

NaCl	8.0g
Glucose	1.0g
KCl	0.4g
$NaHCO_3$	0.35g
$CaCl_2 \cdot 2H_2O$	0.14g
$MgCl_2 \cdot 6H_2O$	0.1g
$MgSO_4 \cdot 7H_2O$	0.1g
Na_2HPO_4	0.06g
KH_2PO_4	0.06g
Phenol Red	0.02g

pH 7.2–7.4 at 25°C

Source: This medium is available as a premixed solution from BD Diagnostic Systems.

Preparation of Tissue Culture Hanks Solution: Add components to distilled/deionized water and bring volume to 1.0L. Mix thoroughly. Adjust pH to 7.2–7.4. Filter sterilize.

Use: For use in tissue culture procedures.

Tissue Culture Medium 199 (TC Medium 199)

Composition per 1050.0mL:

NaCl	8.0g
Glucose	1.0g
KCl	0.4g
$NaHCO_3$	0.35g
DL-Glutamic acid	0.15g
$CaCl_2 \cdot 2H_2O$	0.14g
DL-Leucine	0.12g
L-Glutamine	0.1g
$MgSO_4 \cdot 7H_2O$	0.1g
L-Arginine	0.07g
L-Lysine	0.07g
DL-Aspartic acid	0.06g
Na_2HPO_4	0.06g
KH_2PO_4	0.06g
DL-Threonine	0.06g
DL-Alanine	0.05g
Glycine	0.05g
DL-Phenylalanine	0.05g
DL-Serine	0.05g
Sodium acetate	0.05g
DL-Valine	0.05g
DL-Isoleucine	0.04g
L-Proline	0.04g
L-Tyrosine	0.04g
DL-Methionine	0.03g
L-Cystine	0.02g
L-Histidine	0.02g
Phenol Red	0.02g
DL-Tryptophan	0.02g
Adenine	0.01g
L-Hydroxyproline	0.01g
Tween™ 80	5.0mg
Adenosine triphosphate	1.0mg
Choline	0.5mg
Deoxyribose	0.5mg
Ribose	0.5mg
Guanine	0.3mg
Hypoxanthine	0.3mg
Thymine	0.3mg
Uracil	0.3mg
Xanthine	0.3mg
Adenylic acid	0.2mg
Cholesterol	0.2mg
Calciferol	0.1mg
$Fe(NO_3)_3 \cdot 9H_2O$	0.1mg
L-Cysteine	0.1mg

Vitamin A 0.1mg
α-Tocopherol phosphate 0.01mg
Biotin 0.01mg
Calcium pantothenate 0.01mg
Folic acid 0.01mg
Menadione 0.01mg
Riboflavin 0.01mg
Thiamine·HCl 0.01mg
p-Aminobenzoic acid 0.05mg
Ascorbic acid 0.05mg
Glutathione 0.05mg
Inositol 0.05mg
Niacin 0.025mg
Niacinamide 0.025mg
Pyridoxine·HCl 0.025mg
Pyridoxal·HCl 0.025mg
Serum 50.0–100.0mL

pH 7.2–7.4 at 25°C

Preparation of Medium: Add components, except serum, to distilled/deionized water and bring volume to 1.0L. Mix thoroughly. Adjust pH to 7.2–7.4 with 10% Na_2CO_3 solution. Filter sterilize. Aseptically add 50.0–100.0mL of sterile serum. Human serum, bovine serum, horse serum, or fetal calf serum may be used. Mix thoroughly. If desired, antibacterial inhibitors may be added. Aseptically add 500,000U of penicillin and 0.5g of streptomycin to 1050.0mL of the complete medium to increase selectivity.

Use: For the cultivation of a wide variety of cell lines in tissue culture. It is especially useful for the detection, titering, and identification of viruses in tissue culture cells.

Tissue Culture Medium Eagle with Earle Balanced Salt Solution (TC Medium Eagle with Earle BSS)

Composition per 1056.0mL:

NaCl 6.8g
$NaHCO_3$ 2.2g
Glucose 1.0g
KCl 0.4g
$CaCl_2 \cdot 2H_2O$ 0.2g
NaH_2PO_4 .... 0.125g
$MgSO_4 \cdot 7H_2O$ 0.1g
L-Isoleucine 0.026g
L-Leucine 0.026g
L-Lysine 0.026g
L-Threonine 0.024g
L-Valine 0.0235g
L-Tyrosine 0.018g
L-Arginine 0.0174g
L-Phenylalanine 0.0165g
L-Cystine 0.012g
L-Histidine 8.0mg
L-Methionine 7.5mg
Phenol Red 5.0mg
L-Tryptophan 4.0mg
Inositol 1.8mg
Biotin 1.0mg
Calcium pantothenate 1.0mg
Choline chloride 1.0mg
Folic acid 1.0mg
Nicotinamide 1.0mg
Pyridoxal·HCl 1.0mg
Thiamine·HCl 1.0mg
Riboflavin 0.1mg
Serum 50.0–100.0mL
Glutamine solution 6.0mL

pH 7.2–7.4 at 25°C

Glutamine Solution:

Composition per 100.0mL:

L-Glutamine 5.0g
NaCl (0.85% solution) 100.0mL

Preparation of Glutamine Solution: Add the glutamine to the 0.85% NaCl solution. Mix thoroughly. Filter sterilize.

Preparation of Medium: Add components, except glutamine and serum, to distilled/deionized water and bring volume to 1.0L. Mix thoroughly. Adjust pH to 7.2–7.4. Filter sterilize. Aseptically add 6.0mL of sterile glutamine solution and 50.0–100.0mL of sterile serum. Human serum, bovine serum, horse serum, or fetal calf serum may be used. Mix thoroughly.

Use: For the cultivation of HeLa, KB, and other tissue culture cell lines.

Tissue Culture Medium Eagle with Hanks Balanced Salt Solution (TC Medium Eagle with Hanks BSS)

Composition per 1056.0mL:

NaCl 8.0g
Glucose 1.0g
KCl 0.4g
$CaCl_2 \cdot 2H_2O$ 0.14g
$MgSO_4 \cdot 7H_2O$ 0.1g
KH_2PO_4 .... 0.06g
Na_2HPO_4 .... 0.05g
L-Isoleucine 0.026g
L-Leucine 0.026g
L-Lysine 0.026g
L-Threonine 0.024g
L-Valine 0.0235g
L-Tyrosine 0.018g
L-Arginine 0.0174g
L-Phenylalanine 0.0165g
L-Cystine 0.012g
L-Histidine 8.0mg
L-Methionine 7.5mg
Phenol Red 5.0mg
L-Tryptophan 4.0mg
Inositol 1.8mg
Biotin 1.0mg
Folic acid 1.0mg
Calcium pantothenate 1.0mg
Choline chloride 1.0mg
Nicotinamide 1.0mg
Pyridoxal·HCl 1.0mg
Thiamine·HCl 1.0mg
Riboflavin 0.1mg
Serum 50.0–100.0mL
Glutamine solution 6.0mL

pH 7.2–7.4 at 25°C

Glutamine Solution:

Composition per 100.0mL:

L-Glutamine	5.0g
NaCl (0.85% solution)	100.0mL

Preparation of Glutamine Solution: Add the glutamine to the 0.85% NaCl solution. Mix thoroughly. Filter sterilize.

Preparation of Medium: Add components, except glutamine and serum, to distilled/deionized water and bring volume to 1.0L. Mix thoroughly. Adjust pH to 7.2–7.4. Filter sterilize. Aseptically add 6.0mL of sterile glutamine solution and 50.0–100.0mL of sterile serum. Human serum, bovine serum, horse serum, or fetal calf serum may be used. Mix thoroughly.

Use: For use as a base in the preparation of liquid media used for the cultivation of tissue culture cell lines.

Tissue Culture Medium Eagle, HeLa (TC Medium Eagle, HeLa)

Composition per 1056.0mL:

NaCl	5.85g
$NaHCO_3$	1.68g
Glucose	0.9g
KCl	0.373g
NaH_2PO_4	0.138g
$MgCl_2 \cdot 6H_2O$	0.12g
$CaCl_2 \cdot 2H_2O$	0.11g
L-Lysine	0.0269g
L-Isoleucine	0.0262g
L-Leucine	0.0262g
L-Threonine	0.0238g
L-Valine	0.0234g
L-Tyrosine	0.0181g
L-Arginine	0.0174g
L-Phenylalanine	0.0165g
L-Cystine	0.012g
L-Histidine	7.8mg
L-Methionine	7.5mg
Phenol Red	5.0mg
L-Tryptophan	4.1mg
Folic acid	0.44mg
Thiamine·HCl	0.34mg
Biotin	0.24mg
Pantothenic acid	0.22mg
Pyridoxal·HCl	0.2mg
Choline chloride	0.14mg
Nicotinamide	0.12mg
Riboflavin	0.04mg
Serum	50.0mL–100.0mL
Glutamine solution	6.0mL

pH 7.2–7.4 at 25°C

Glutamine Solution:

Composition per 100.0mL:

L-Glutamine	5.0g
NaCl (0.85% solution)	100.0mL

Preparation of Glutamine Solution: Add the glutamine to the 0.85% NaCl solution. Mix thoroughly. Filter sterilize.

Preparation of Medium: Add components, except glutamine and serum, to distilled/deionized water and bring volume to 1.0L. Mix thoroughly. Adjust pH to 7.2–7.4. Filter sterilize. Aseptically add 6.0mL of sterile glutamine solution and 50.0–100.0mL of sterile serum. Human serum, bovine serum, horse serum, or fetal calf serum may be used. Mix thoroughly.

Use: For the cultivation and maintenance of HeLa and other cell lines in tissue culture, and for studying the cytopathogenicity of viral agents.

Tissue Culture Medium Ham F10 (TC Medium Ham F10)

Composition per 1050.0mL:

NaCl	7.4g
Glucose	1.1g
Na_2HPO_4	0.29g
KCl	0.285g
L-Arginine	0.211g
$MgSO_4 \cdot 7H_2O$	0.153g
L-Glutamine	0.1462g
Sodium pyruvate	0.11g
KH_2PO_4	0.083g
$CaCl_2 \cdot 2H_2O$	0.044g
L-Cystine	0.0315g
L-Lysine	0.0293g
L-Histidine	0.021g
L-Asparagine	0.015g
L-Glutamic acid	0.0147g
L-Aspartic acid	0.0133g
L-Leucine	0.0131g
L-Proline	0.0115g
L-Serine	0.0105g
L-Alanine	8.91mg
Glycine	7.51mg
L-Phenylalanine	4.96mg
L-Methionine	4.48mg
Hypoxanthine	4.0mg
L-Threonine	3.57mg
L-Valine	3.5mg
L-Isoleucine	2.6mg
L-Tyrosine	1.81mg
Cyanocobalamin	1.3mg
Folic acid	1.3mg
Phenol Red	1.2mg
Thiamine·HCl	1.0mg
$FeSO_4 \cdot 7H_2O$	0.83mg
Calcium pantothenate	0.7mg
Thymidine	0.7mg
Choline chloride	0.69mg
Niacinamide	0.6mg
L-Tryptophan	0.6mg
i-Inositol	0.54mg
Riboflavin	0.37mg
Lipoic acid	0.2mg
Pyridoxine·HCl	0.2mg
$ZnSO_4 \cdot 7H_2O$	0.028mg
Biotin	0.024mg
$CuSO_4 \cdot 5H_2O$	2.5μg
Fetal calf serum	50.0–100.0mL

pH 7.2–7.4 at 25°C

Preparation of Medium: Add components, except fetal calf serum, to distilled/deionized water and bring volume to 1.0L. Mix thoroughly. Adjust pH to 7.2–7.4 with 10% Na_2CO_3 solution. Filter sterilize. Aseptically add 50.0–100.0mL of sterile fetal calf serum. Mix thoroughly.

Use: For the cultivation of a wide variety of cell lines in tissue culture.

Tissue Culture Medium NCTC 109 (TC Medium NCTC 109)

Composition per 1050.0mL:

NaCl	6.8g
$NaHCO_3$	2.2g
Glucose	1.0g
KCl	0.4g
L-Cysteine	0.26g
$CaCl_2 \cdot 2H_2O$	0.2g
NaH_2PO_4	0.14g
L-Glutamine	0.14g
$MgSO_4 \cdot 7H_2O$	0.1g
Sodium acetate	0.05g
Ascorbic acid	0.05g
L-Alanine	0.03g
L-Lysine	0.03g
L-Arginine	0.026g
L-Valine	0.025g
L-Leucine	0.02g
Phenol Red	0.02g
L-Histidine	0.019g
L-Threonine	0.019g
L-Isoleucine	0.018g
L-Tryptophan	0.017.g
L-Phenylalanine	0.017g
L-Tyrosine	0.016g
Glycine	0.014g
Tween™ 80	0.012g
L-Serine	0.011g
L-Cystine	0.01g
Glutathione	0.01g
Cyanocobalamin	0.01g
Deoxycytidine	0.01g
Deoxyguanosine	0.01g
Deoxyadenosine	0.01g
Thymidine	0.01g
L-Aspartic acid	9.91mg
L-Glutamic acid	8.26mg
L-Arginine	8.09mg
L-Ornithine	7.38mg
Nicotinamide adenine dinucleotide	7.0mg
L-Proline	6.13mg
L-α-N-butyric acid	5.51mg
L-Methionine	4.44mg
L-Taurine	4.18mg
L-Hydroxyproline	4.09mg
D-Glucosamine	3.2mg
Coenzyme A	2.5mg
Glucuronolactone	1.8mg
Sodium glucuronate	1.8mg
Choline chloride	1.25mg
Cocarboxylase	1.0mg
Flavin adenine dinucleotide	1.0mg
Uridine triphosphate	1.0mg
Nicotinamide adenine dinucleotide phosphate	1.0mg
Vitamin A	0.25mg
Calciferol	0.25mg
i-Inositol	0.125mg
p-Aminobenzoic acid	0.125mg
5-Methylcytosine	0.1mg
Pyridoxine·HCl	0.0625mg
Pyridoxal·HCl	0.0625mg
Niacin	0.0625mg
Niacinamide	0.0625mg
Biotin	0.025mg
Folic acid	0.025mg
Menadione	0.025mg
Pantothenate	0.025mg
Riboflavin	0.025mg
Thiamine·HCl	0.025mg
α-Tocopherol phosphate	0.025mg
Serum	50.0–100.0mL

pH 7.2–7.4 at 25°C

Preparation of Medium: Add components, except serum, to distilled/deionized water and bring volume to 1.0L. Mix thoroughly. Adjust pH to 7.2–7.4 with 10% Na_2CO_3 solution. Filter sterilize. Aseptically add 50.0–100.0mL of sterile serum. Human serum, bovine serum, horse serum, or fetal calf serum may be used. Mix thoroughly.

Use: For the cultivation of a wide variety of cell lines in tissue culture.

Tissue Culture Medium RPMI No. 1640 (TC Medium RPMI #1640)

Composition per liter:

NaCl	6.46g
Glucose	2.0g
$NaHCO_3$	2.0g
NaH_2PO_4	1.512g
KCl	0.4g
L-Glutamine	0.3g
L-Arginine	0.2g
Calcium nitrate	0.1g
$MgSO_4 \cdot 7H_2O$	0.1g
L-Asparagine	0.05g
L-Cystine	0.05g
L-Isoleucine	0.05g
L-Leucine	0.05g
L-Lysine·HCl	0.04g
Inositol	0.035g
L-Serine	0.03g
Hydroxy-L-proline	0.02g
L-Aspartic acid	0.02g
L-Glutamic acid	0.02g
L-Proline	0.02g
L-Threonine	0.02g
L-Tyrosine	0.02g
L-Valine	0.02g
L-Histidine	0.015g
L-Methionine	0.015g
L-Phenylalanine	0.015g
Glycine	0.01g
L-Tryptophan	5.0mg
Phenol Red	5.0mg
Choline chloride	3.0mg
p-Aminobenzoic acid	1.0mg
Folic acid	1.0mg
Glutathione	1.0mg
Nicotinamide	1.0mg
Pyridoxine·HCl	1.0mg
Thiamine·HCl	1.0mg
Calcium pantothenate	0.25mg
Biotin	0.2mg

Riboflavin 0.2mg
Vitamin B_{12} 5.0mg
Serum 50.0–100.0mL

pH 7.2–7.4 at 25°C

Source: This medium is available as a premixed powder and solution from BD Diagnostic Systems.

Preparation of Medium: Add components, except serum, to distilled/deionized water and bring volume to 1.0L. Mix thoroughly. Adjust pH to 7.2–7.4 with 10% Na_2CO_3 solution. Filter sterilize. Aseptically add 50.0–100.0mL of sterile serum. Human serum, bovine serum, horse serum, or fetal calf serum may be used. Mix thoroughly.

Use: For the cultivation of a wide variety of cell lines in tissue culture.

Tissue Culture Minimal Medium Eagle

Composition per liter:
Sterile salt solution 944.0mL
TC amino acids, minimal Eagle 50X 20.0mL
TC $NaHCO_3$, 10% 20.0mL
TC vitamins, minimal Eagle 100X 10.0mL
TC glutamine, 5% 6.0mL

pH 7.2–7.4 at 25°C

Sterile Salt Solution:

Composition per 944.0mL:
NaCl 6.8g
Glucose 1.0g
KCl 0.4g
$CaCl_2$ 0.2g
$MgCl_2$ 0.2g
NaH_2PO_4 .... 0.15g

Preparation of Sterile Salt Solution: Add components to distilled/deionized water and bring volume to 944.0mL. Mix thoroughly. Filter sterilize.

TC Amino Acids, Minimal Eagle 50X:

Composition per liter:
L-Arginine 0.1g
L-Lysine 0.06g
L-Isoleucine 0.05g
L-Leucine 0.05g
L-Threonine 0.05g
L-Valine 0.05g
L-Tyrosine 0.04g
L-Phenylalanine 0.03g
L-Histidine 0.03g
L-Cystine 0.02g
L-Methionine 0.02g
L-Tryptophan 0.01g

Preparation of TC Amino Acids, Minimal Eagle 50X: Add components to distilled/deionized water and bring volume to 1.0L. Mix thoroughly. Adjust pH to 7.2–7.4. Filter sterilize.

TC $NaHCO_3$, 10%:

Composition per 100.0mL:
$NaHCO_3$ 10.0g

Preparation of TC $NaHCO_3$, 10%: Add $NaHCO_3$ to distilled/deionized water and bring volume to 100.0mL. Mix thoroughly. Filter sterilize.

TC Vitamins, Minimal Eagle 100X:

Composition per liter:
Inositol 2.0mg
Calcium pantothenate 1.0mg
Choline chloride 1.0mg
Folic acid 1.0mg
Nicotinamide 1.0mg
Pyridoxal 1.0mg
Thiamine·HCl 1.0mg
Riboflavin 0.1mg

Preparation of TC Vitamins, Minimal Eagle 100X: Add components to distilled/deionized water and bring volume to 1.0L. Mix thoroughly. Filter sterilize.

TC Glutamine, 5%:

Composition per 100.0mL:
L-Glutamine 5.0g
NaCl (0.85% solution) 100.0mL

Preparation of TC Glutamine, 5%: Add the glutamine to the 0.85% NaCl solution. Mix thoroughly. Filter sterilize.

Preparation of Medium: Aseptically combine 944.0mL of sterile salt solution, 20.0mL of sterile TC amino acids, minimal Eagle 50X, 20.0mL of sterile TC $NaHCO_3$, 10%, 10.0mL of sterile TC vitamins, minimal Eagle 100X, and 6.0mL of sterile TC glutamine, 5%. Mix thoroughly. Adjust pH to 7.2–7.4 if necessary.

Use: For the cultivation of mammalian cells in monolayer or suspension for tissue culture procedures and virus preparation.

Tissue Culture Minimal Medium Eagle with Earle Balanced Salts Solution (TC Minimal Medium Eagle with Earle BSS)

Composition per 1056.0mL:
NaCl 6.8g
Glucose 1.0g
KCl 0.4g
$CaCl_2 \cdot 2H_2O$ 0.2g
$MgCl_2 \cdot 6H_2O$ 0.2g
NaH_2PO_4 .... 0.15g
L-Arginine 0.1g
L-Lysine 0.06g
L-Isoleucine 0.05g
L-Leucine 0.05g
L-Threonine 0.05g
L-Valine 0.05g
L-Tyrosine 0.04g
L-Phenylalanine 0.03g
L-Histidine 0.03g
L-Cystine 0.02g
L-Methionine 0.02g
L-Tryptophan 0.01g
i-Inositol 2.0mg
Calcium pantothenate 1.0mg
Choline chloride 1.0mg
Folic acid 1.0mg
Nicotinamide 1.0mg
Pyridoxal 1.0mg
Thiamine·HCl 1.0mg
Riboflavin 0.1mg
Serum 50.0–100.0mL

Glutamine solution 6.0mL
$CaCl_2 \cdot 2H_2O$ solution (optional) 2.0mL
pH 7.2–7.4 at 25°C

Glutamine Solution:

Composition per 100.0mL:

L-Glutamine 5.0g
NaCl (0.85% solution) 100.0mL

Preparation of Glutamine Solution: Add the glutamine to the 0.85% NaCl solution. Mix thoroughly. Filter sterilize.

Preparation of Medium: Add components, except glutamine and serum, to distilled/deionized water and bring volume to 1.0L. Mix thoroughly. Adjust pH to 7.2–7.4 with 10% Na_2CO_3 solution. Filter sterilize. Aseptically add 6.0mL of sterile glutamine solution and 50.0–100.0mL of sterile serum. Human serum, bovine serum, horse serum, or fetal calf serum may be used. Mix thoroughly. To grow cells in a monolayer, aseptically add 2.0mL of a sterile 10% $CaCl_2 \cdot 2H_2O$ solution. To grow cells in suspension, omit the $CaCl_2 \cdot 2H_2O$ solution.

Use: For preparation of Eagle's minimal medium for the cultivation of cells in monolayer or suspension in tissue culture.

Tissue Culture Minimal Medium Eagle Spinner Modified (TC Minimal Medium Eagle Spinner Modified MEM-S)

Composition per 1056.0mL:

NaH_2PO_4 .... 1.35g
NaCl 6.8g
$NaHCO_3$ 2.2g
Glucose 1.0g
KCl 0.4g
$CaCl_2 \cdot 2H_2O$ 0.2g
NaH_2PO_4 .... 0.125g
$MgSO_4 \cdot 7H_2O$ 0.1g
L-Isoleucine 0.026g
L-Leucine 0.026g
L-Lysine 0.026g
L-Threonine 0.024g
L-Valine 0.0235g
L-Tyrosine 0.018g
L-Arginine 0.0174g
L-Phenylalanine 0.0165g
L-Cystine 0.012g
L-Histidine 8.0mg
L-Methionine 7.5mg
Phenol Red 5.0mg
L-Tryptophan 4.0mg
Inositol 1.8mg
Biotin 1.0mg
Calcium pantothenate 1.0mg
Choline chloride 1.0mg
Folic acid 1.0mg
Nicotinamide 1.0mg
Pyridoxal·HCl 1.0mg
Thiamine·HCl 1.0mg
Riboflavin 0.1mg
Serum 50.0mL–100.0mL
Glutamine solution 6.0mL
pH 7.2–7.4 at 25°C

Glutamine Solution:

Composition per 100.0mL:

L-Glutamine 5.0g
NaCl (0.85% solution) 100.0mL

Preparation of Glutamine Solution: Add the glutamine to the 0.85% NaCl solution. Mix thoroughly. Filter sterilize.

Preparation of Medium: Add components, except glutamine and serum, to distilled/deionized water and bring volume to 1.0L. Mix thoroughly. Adjust pH to 7.2–7.4 with 10% Na_2CO_3 solution. Filter sterilize. Aseptically add 6.0mL of sterile glutamine solution and 50.0–100.0mL of sterile serum. Human serum, bovine serum, horse serum, or fetal calf serum may be used. Mix thoroughly.

Use: For the cultivation of mammalian cells in suspension.

Tissue Culture Tyrode Solution (TC Tyrode Solution)

Composition per 1002.0mL:

NaCl 8.0g
Glucose 1.0g
$NaHCO_3$ 1.0g
$CaCl_2 \cdot 2H_2O$ 0.2g
KCl 0.2g
$MgCl_2 \cdot 6H_2O$ 0.1g
NaH_2PO_4 .... 0.05g
Phenol Red (1% solution) 2.0mL
pH 7.2–7.4 at 25°C

Preparation of Tissue Culture Tyrode Solution: Add components, except Phenol Red, to distilled/deionized water and bring volume to 1.0L. Mix thoroughly. Add 2.0mL of Phenol Red solution. Adjust pH to 7.2–7.4. Filter sterilize.

Use: For use in tissue culture procedures.

Tissue Culture Vitamins Minimal Eagle, 100X (TC Vitamins Minimal Eagle, 100X)

Composition per liter:

Inositol 2.0mg
Calcium pantothenate 1.0mg
Choline chloride 1.0mg
Folic acid 1.0mg
Nicotinamide 1.0mg
Pyridoxal 1.0mg
Thiamine·HCl 1.0mg
Riboflavin 0.1mg
pH 7.2–7.4 at 25°C

Preparation of TC Vitamins, Minimal Eagle 100X: Add components to distilled/deionized water and bring volume to 1.0L. Mix thoroughly. Filter sterilize.

Use: For the preparation of Tissue Culture minimal medium Eagle used in tissue culture procedures.

TMA Mineral Medium

Composition per liter:

Agar, noble 20.0g
KH_2PO_4 .... 2.78g
Na_2HPO_4 .... 2.78g
$(NH_4)_2SO_4$ 1.0g

Tetramethylammonium perchlorate solution 20.0mL
Hutner's basal salts solution 20.0mL

pH 6.8 ± 0.2 at 25°C

Tetramethylammonium Perchlorate Solution:
Composition per 20.0mL:

Tetramethylammonium perchlorate 1.0g

Preparation of Tetramethylammonium Perchlorate Solution: Add tetramethylammonium perchlorate to distilled/deionized water and bring volume to 20.0mL. Mix thoroughly. Filter sterilize.

Hutner's Basal Salts Solution:
Composition per liter:

$MgSO_4 \cdot 7H_2O$ 29.7g
Nitrilotriacetic acid 10.0g
$CaCl_2 \cdot 2H_2O$ 3.335g
$FeSO_4 \cdot 7H_2O$ 99.0mg
$(NH_4)_6MoO_7O_{24} \cdot 4H_2O$ 9.25mg
"Metals 44" 50.0mL

"Metals 44":
Composition per 100.0mL:

$ZnSO_4 \cdot 7H_2O$ 1.095g
$FeSO_4 \cdot 7H_2O$ 0.5g
Sodium EDTA 0.25g
$MnSO_4 \cdot H2O$ 0.154g
$CuSO_4 \cdot 5H_2O$ 39.2mg
$Co(NO_3)_2 \cdot 6H_2O$ 24.8mg
$Na_2B_4O_7 \cdot 10H_2O$ 17.7mg

Preparation of "Metals 44": Add sodium EDTA to distilled/deionized water and bring volume to 90.0mL. Mix thoroughly. Add a few drops of concentrated H_2SO_4 to retard precipitation of heavy metal ions. Add remaining components. Mix thoroughly. Bring volume to 100.0mL with distilled/deionized water.

Preparation of Hutner's Basal Salts Solution: Add nitrilotriacetic acid to 500.0mL of distilled/deionized water. Adjust pH to 6.5 with KOH. Add remaining components. Add distilled/deionized water to 1.0L. Adjust pH to 6.8.

Preparation of Medium: Add components, except tetramethylammonium perchlorate solution, to distilled/deionized water and bring volume to 980.0mL. Mix thoroughly. Gently heat and bring to boiling. Adjust pH to 6.8. Autoclave for 15 min at 15 psi pressure–121°C. Cool to 50°–55°C. Aseptically add 20.0mL of sterile tetramethylammonium perchlorate solution. Mix thoroughly. Pour into sterile Petri dishes or distribute into sterile tubes.

Use: For the cultivation of *Paracoccus kocurii.*

TMAO HiVeg Medium (Trimethylamine-*N*-Oxide HiVeg Medium)

Composition per liter:

Plant extract 10.0g
Plant peptone 10.0g
NaCl 5.0g
Agar 2.0g
Trimethylamine-*N*-oxide 1.0g
Yeast extract 1.0g

pH 7.5 ± 0.2 at 25°C

Source: This medium is available as a premixed powder from HiMedia.

Preparation of Medium: Add components to distilled/deionized water and bring volume to 1.0L. Mix thoroughly. Gently heat and bring to boiling. Distribute into screw-capped tubes in 4.0mL volumes. Autoclave for 15 min at 15 psi pressure–121°C. Allow tubes to cool in an upright position.

Use: For the cultivation and differentiation of *Campylobacter* species from foods. *Campylobacter jejuni* and *Campylobacter coli* will not grow.

TMAO Medium

See: **Trimethylamine *N*-Oxide Medium**

TMBS4 Medium (DSMZ Medium 559)

Composition per 1003.0mL:

Solution A 870.0mL
Solution C 100.0mL
Solution D 10.0mL
Solution E (Vitamin solution) 10.0mL
Solution F 10.0mL
Solution B (Trace elements solution SL-10) 1.0mL
Solution G 1.0mL
Solution H 1.0mL

pH 7.1–7.4 at 25°C

Solution A:
Composition per 870.0mL:

NaCl 1.0g
$MgCl_2 \cdot 6H_2O$ 0.4g
Na_2SO_4 .... 3.0g
KCl 0.5g
NH_4Cl 0.3g
KH_2PO_4 .... 0.2g
$CaCl_2 \cdot 2H_2O$ 0.15g
Resazurin 1.0mg

Preparation of Solution A: Add components to distilled/deionized water and bring volume to 870.0mL Mix thoroughly.

Solution B (Trace Elements Solution SL-10):
Composition per liter:

$FeCl_2 \cdot 4H_2O$ 1.5g
$CoCl_2 \cdot 6H_2O$ 190.0mg
$MnCl_2 \cdot 4H_2O$ 100.0mg
$ZnCl_2$ 70.0mg
$Na_2MoO_4 \cdot 2H_2O$ 36.0mg
$NiCl_2 \cdot 6H_2O$ 24.0mg
H_3BO_3 .... 6.0mg
$CuCl_2 \cdot 2H_2O$ 2.0mg
HCl (25% solution) 10.0mL

Preparation of Solution B (Trace Elements Solution SL-10): Add $FeCl_2 \cdot 4H_2O$ to 10.0mL of HCl solution. Mix thoroughly. Add distilled/deionized water and bring volume to 1.0L. Add remaining components. Mix thoroughly. Sparge with 100% N_2. Autoclave for 15 min at 15 psi pressure–121°C.

Solution C:
Composition per 100.0mL:

$NaHCO_3$ 5.0g

Preparation of Solution C: Add $NaHCO_3$ to distilled/deionized water and bring volume to 100.0mL Mix thoroughly. Filter sterilize. Flush with 80% N_2 + 20% CO_2 to remove dissolved oxygen.

Solution D:

Composition per 10.0mL:

Syringic acid 0.6g

Preparation of Solution D: Add syringic acid to distilled/deionized water and bring volume to 10.0mL. Adjust pH to 8.0 with NaOH. Mix thoroughly. Sparge with 100% N_2. Autoclave for 15 min at 15 psi pressure–121°C.

Solution E (Vitamin Solution):

Composition per liter:

Pyridoxine-HCl 10.0mg
Thiamine-HCl·$2H_2O$ 5.0mg
Riboflavin 5.0mg
Nicotinic acid 5.0mg
D-Ca-pantothenate 5.0mg
p-Aminobenzoic acid 5.0mg
Lipoic acid 5.0mg
Biotin 2.0mg
Folic acid 2.0mg
Vitamin B_{12} 0.10mg

Solution E (Vitamin Solution): Add components to distilled/deionized water and bring volume to 1.0L. Mix thoroughly. Sparge with 100% N_2. Autoclave for 15 min at 15 psi pressure–121°C.

Solution F:

Composition per 10.0mL:

$Na_2S \cdot 9H_2O$ 0.4g

Preparation of Solution F: Add $Na_2S \cdot 9H_2O$ to distilled/deionized water and bring volume to 10.0mL. Mix thoroughly. Sparge with 100% N_2. Autoclave for 15 min at 15 psi pressure–121°C.

Solution G:

Composition per 10.0mL:

Na-dithionite 0.25g

Preparation of Solution G: Add Na-dithionite to distilled/deionized water and bring volume to 10.0mL. Mix thoroughly. Sparge with 100% N_2. Autoclave for 15 min at 15 psi pressure–121°C.

Solution H:

Composition per liter:

NaOH 0.5g
$Na_2WO_4 \cdot 2H_2O$ 4.0mg
$Na_2SeO_3 \cdot 5H_2O$ 3.0mg

Preparation of Solution H: Add components to distilled/deionized water and bring volume to 1.0L. Mix thoroughly. Sparge with 100% N_2. Filter sterilize.

Preparation of Medium: Gently heat solution A and bring to boiling. Boil solution A for a few minutes. Cool to room temperature. Gas with 80% N_2 + 20% CO_2 gas mixture to reach a pH below 6. Autoclave for 15 min at 15 psi pressure–121°C. Cool to room temperature. Sequentially add 1.0mL solution B, 100.0mL solution C, 10.0mL solution D, 10.0mL solution E, 10.0mL solution F, 1.0mL solution G, and 1.0mL solution H. Distribute anaerobically under 80% N_2 + 20% CO_2 into appropriate vessels.

Use: For the cultivation of *Holophaga foetida.*

TMBS4 Medium

Composition per 1002.0mL:

Solution A 870.0mL
Solution C 100.0mL
Solution D 10.0mL
Solution E (Vitamin solution) 10.0mL
Solution F 10.0mL
Solution B (Trace elements solution SL-10) 1.0mL
Solution G 1.0mL

pH 7.1–7.4 at 25°C

Solution A:

Composition per 870.0mL:

Na_2SO_4 .... 3.0g
NaCl 1.0g
KCl 0.5g
$MgCl_2 \cdot 6H_2O$ 0.4g
NH_4Cl 0.3g
KH_2PO_4 .... 0.2g
$CaCl_2 \cdot 2H_2O$ 0.15g
Resazurin 1.0mg

Preparation of Solution A: Add components to distilled/deionized water and bring volume to 870.0mL. Mix thoroughly. Gently heat and bring to boiling. Continue boiling for 3–4 min. Allow to cool to room temperature while gassing under 80% N_2 + 20% CO_2. Continue gassing until pH reaches below 6.0. Seal the flask under 80% N_2 + 20% CO_2. Autoclave for 15 min at 15 psi pressure–121°C.

Solution B (Trace Elements Solution SL-10):

Composition per liter:

$FeCl_2 \cdot 4H_2O$ 1.5g
$CoCl_2 \cdot 6H_2O$ 190.0mg
$MnCl_2 \cdot 4H_2O$ 100.0mg
$ZnCl_2$ 70.0mg
$Na_2MoO_4 \cdot 2H_2O$ 36.0mg
$NiCl_2 \cdot 6H_2O$ 24.0mg
H_3BO_3 .... 6.0mg
$CuCl_2 \cdot 2H_2O$ 2.0mg
HCl (25% solution) 10.0mL

Preparation of Solution B (Trace Elements Solution SL-10): Add $FeCl_2 \cdot 4H_2O$ to 10.0mL of HCl solution. Mix thoroughly. Add distilled/deionized water and bring volume to 1.0L. Add remaining components. Mix thoroughly. Gas under 100% N_2. Autoclave for 15 min at 15 psi pressure–121°C.

Solution C:

Composition per 100.0mL:

$NaHCO_3$ 5.0g

Preparation of Solution C: Add $NaHCO_3$ to distilled/deionized water and bring volume to 100.0mL. Mix thoroughly. Filter sterilize. Gas under 80% N_2 + 20% CO_2.

Solution D:

Composition per 10.0mL:

Syringic acid 0.5g

Preparation of Solution D: Add syringic acid to distilled/deionized water and bring volume to 10.0mL. Mix thoroughly. Adjust pH to 8 with NaOH. Gas under 100% N_2. Autoclave for 15 min at 15 psi pressure–121°C.

Solution E (Vitamin Solution):

Composition per liter:

Pyridoxine·HCl 10.0mg
Calcium DL-pantothenate 5.0mg
Lipoic acid 5.0mg
Nicotinic acid 5.0mg
p-Aminobenzoic acid 5.0mg
Riboflavin 5.0mg

Thiamine·HCl 5.0mg
Biotin 2.0mg
Folic acid 2.0mg
Vitamin B_{12} 0.1mg

Preparation of Solution E (Vitamin Solution): Add components to distilled/deionized water and bring volume to 1.0L. Mix thoroughly. Gas under 100% N_2. Autoclave for 15 min at 15 psi pressure–121°C.

Solution F:

Composition per 10.0mL:

$Na_2S \cdot 9H_2O$ 0.25g
$Na_2S_2O_4$ 0.25g

Preparation of Solution F: Add components to distilled/deionized water and bring volume to 10.0mL. Mix thoroughly. Gas under 100% N_2. Autoclave for 15 min at 15 psi pressure–121°C.

Solution G:

Composition per liter:

NaOH 0.5g
$Na_2WO_4 \cdot 2H_2O$ 4.0mg
$Na_2SeO_3 \cdot 5H_2O$ 3.0mg

Preparation of Solution G: Add components to distilled/deionized water and bring volume to 1.0L. Mix thoroughly. Gas under 100% N_2. Autoclave for 15 min at 15 psi pressure–121°C.

Preparation of Medium: Aseptically and anaerobically combine solution A with solution B, solution C, solution D, solution E, solution F, and solution G, in that order. Mix thoroughly. Anaerobically distribute into sterile tubes or flasks under 80% N_2 + 20% CO_2.

Use: For the cultivation and maintenance of *Pelobacter* species.

TN Broth

See: **Trypticase™ Novobiocin Broth**

T_1N_0 Broth (Tryptone Broth)

Composition per liter:

Pancreatic digest of casein 10.0g

pH 7.1 ± 0.2 at 25°C

Preparation of Medium: Add pancreatic digest of casein to distilled/deionized water and bring volume to 1.0L. Mix thoroughly. Gently heat and bring to boiling. Distribute into tubes or flasks. Autoclave for 15 min at 15 psi pressure–121°C.

Use: For the cultivation of *Vibrio cholerae* and other *Vibrio* species.

T_1N_1 Agar (Tryptone Salt Agar)

Composition per liter:

Agar 20.0g
NaCl 10.0g
Pancreatic digest of casein 10.0g

pH 7.1 ± 0.2 at 25°C

Preparation of Medium: Add components to distilled/deionized water and bring volume to 1.0L. Mix thoroughly. Gently heat and bring to boiling. Distribute into tubes or flasks. Autoclave for 15 min at 15 psi pressure–121°C. Pour into sterile Petri dishes or leave in tubes. Allow tubes to cool in a slanted position.

Use: For the cultivation of *Vibrio cholerae* and other *Vibrio* species.

T_1N_1 Broth (Tryptone Salt Broth)

Composition per liter:

NaCl 10.0g
Pancreatic digest of casein 10.0g

pH 7.1 ± 0.2 at 25°C

Preparation of Medium: Add components to distilled/deionized water and bring volume to 1.0L. Mix thoroughly. Gently heat and bring to boiling. Distribute into tubes or flasks. Autoclave for 15 min at 15 psi pressure–121°C.

Use: For the cultivation of *Vibrio cholerae* and other *Vibrio* species.

T_1N_2 Agar (Tryptone Salt Agar)

Composition per liter:

Agar 20.0g
NaCl 20.0g
Pancreatic digest of casein 10.0g

pH 7.1 ± 0.2 at 25°C

Preparation of Medium: Add components to distilled/deionized water and bring volume to 1.0L. Mix thoroughly. Gently heat and bring to boiling. Distribute into tubes or flasks. Autoclave for 15 min at 15 psi pressure–121°C. Pour into sterile Petri dishes or leave in tubes. Allow tubes to cool in a slanted position.

Use: For the cultivation of *Vibrio cholerae* and other *Vibrio* species.

T_1N_3 Broth (Tryptone Salt Broth)

Composition per liter:

NaCl 30.0g
Pancreatic digest of casein 10.0g

pH 7.1 ± 0.2 at 25°C

Preparation of Medium: Add components to distilled/deionized water and bring volume to 1.0L. Mix thoroughly. Gently heat and bring to boiling. Distribute into tubes or flasks. Autoclave for 15 min at 15 psi pressure–121°C.

Use: For the cultivation of *Vibrio cholerae* and other *Vibrio* species.

T_1N_6 Broth (Tryptone Salt Broth)

Composition per liter:

NaCl 60.0g
Pancreatic digest of casein 10.0g

pH 7.1 ± 0.2 at 25°C

Preparation of Medium: Add components to distilled/deionized water and bring volume to 1.0L. Mix thoroughly. Gently heat and bring to boiling. Distribute into tubes or flasks. Autoclave for 15 min at 15 psi pressure–121°C.

Use: For the cultivation of *Vibrio cholerae* and other *Vibrio* species.

T_1N_8 Broth (Tryptone Salt Broth)

Composition per liter:

NaCl 80.0g
Pancreatic digest of casein 10.0g

pH 7.1 ± 0.2 at 25°C

Preparation of Medium: Add components to distilled/deionized water and bring volume to 1.0L. Mix thoroughly. Gently heat and bring to boiling. Distribute into tubes or flasks. Autoclave for 15 min at 15 psi pressure–121°C.

Use: For the cultivation of *Vibrio cholerae* and other *Vibrio* species.

T_1N_{10} Broth (Tryptone Salt Broth)

Composition per liter:

NaCl	100.0g
Pancreatic digest of casein	10.0g

pH 7.1 ± 0.2 at 25°C

Preparation of Medium: Add components to distilled/deionized water and bring volume to 1.0L. Mix thoroughly. Gently heat and bring to boiling. Distribute into tubes or flasks. Autoclave for 15 min at 15 psi pressure–121°C.

Use: For the cultivation of *Vibrio cholerae* and other *Vibrio* species.

TN HiVeg Agar

Composition per liter:

Agar	15.0g
Plant hydrolysate	10.0g
NaCl	10.0g

pH 7.2 ± 0.2 at 25°C

Source: This medium is available as a premixed powder from HiMedia.

Preparation of Medium: Add components to distilled/deionized water and bring volume to 1.0L. Mix thoroughly. Gently heat and bring to boiling. Distribute into tubes or flasks. Autoclave for 15 min at 15 psi pressure–121°C. Do not overheat. Pour into sterile Petri dishes or leave in tubes.

Use: For the isolation and cultivation of vibrios from food samples.

TNSA Agar

See: **Trypaflavin Nalidixic Acid Serum Agar**

TNT Medium (Tryptone NaCl Thiamine Medium)

Composition per liter:

Pancreatic digest of casein	10.0g
NaCl	5.0g
Thiamine·HCl	1.0mg

pH 7.3 ± 0.2 at 25°C

Preparation of Medium: Add components to distilled/deionized water and bring volume to 1.0L. Mix thoroughly. Distribute into tubes or flasks. Autoclave for 15 min at 15 psi pressure–121°C.

Use: For the cultivation and maintenance of *Escherichia coli.*

TOC Agar (Tween™ 80 Oxgall Caffeic Acid Agar)

Composition per liter:

Agar	20.0g
Oxgall	10.0g
Caffeic acid	0.3g
Tween™ 80	10.0mL

Source: This medium is available as a prepared medium from BD Diagnostic Systems.

Preparation of Medium: Add components to distilled/deionized water and bring volume to 1.0L. Mix thoroughly. Gently heat and bring to boiling. Autoclave for 15 min at 15 psi pressure–121°C. Pour into sterile Petri dishes.

Use: For the differentiation and identification of *Candida albicans* and *Cryptococcus neoformans. Cryptococcus albicans* produces germ tubes and chlamydospores when grown on this medium. *Cryptococcus neoformans* appears as tan to brown colonies.

Todd-Hewitt Broth

Composition per liter:

Beef heart, infusion from	500.0g
Neopeptone	20.0g
Na_2CO_3	2.5g
Glucose	2.0g
NaCl	2.0g
Na_2HPO_4	0.4g

pH 7.8 ± 0.2 at 25°C

Source: This medium is available as a premixed powder from BD Diagnostic Systems.

Preparation of Medium: Add components to distilled/deionized water and bring volume to 1.0L. Mix thoroughly. Distribute into tubes or flasks. Autoclave for 15 min at 15 psi pressure–121°C.

Use: For the cultivation of group A streptococci used in serological typing, and for the cultivation of a variety of pathogenic microorganisms.

Todd-Hewitt Broth

Composition per liter:

Pancreatic digest of casein	20.0g
Infusion from 450.0g fat-free minced meat	10.0g
Glucose	2.0g
$NaHCO_3$	2.0g
NaCl	2.0g
Na_2HPO_4	0.4g

pH 7.8 ± 0.2 at 25°C

Source: This medium is available as a premixed powder from Oxoid Unipath.

Preparation of Medium: Add components to distilled/deionized water and bring volume to 1.0L. Mix thoroughly. Distribute into tubes or flasks. Autoclave for 10 min at 10 psi pressure–115°C.

Use: For the cultivation of group A streptococci used in serological typing, and for the cultivation of a variety of pathogenic microorganisms.

Todd-Hewitt Broth (ATCC Medium 235)

Composition per liter:

Peptone	20.0g
Beef heart, solids from infusion	3.1g
Na_2CO_3	2.5g
Glucose	2.0g
NaCl	2.0g
Na_2HPO_4	0.4g

pH 7.8 ± 0.2 at 25°C

Source: This medium is available as a premixed powder from BD Diagnostic Systems.

Preparation of Medium: Add components to distilled/deionized water and bring volume to 1.0L. Mix thoroughly. Distribute into tubes or flasks. Autoclave for 15 min at 15 psi pressure–121°C.

Use: For the cultivation of group A streptococci used in serological typing, and for the cultivation of a variety of pathogenic microorganisms.

Todd-Hewitt Broth, Modified

Composition per liter:

Neopeptone	20.0g
Glucose	2.0g
$NaHCO_3$	2.0g
NaCl	2.0g
Na_2HPO_4	0.4g
Beef heart infusion	1.0L

pH 7.8 ± 0.2 at 25°C

Preparation of Medium: Add components to distilled/deionized water and bring volume to 1.0L. Mix thoroughly. Distribute into tubes or flasks. Autoclave for 10 min at 10 psi pressure–115°C.

Use: For the cultivation of streptococci for serological identification.

Todd-Hewitt HiVeg Broth

Composition per liter:

Plant peptone	20.0g
Plant special infusion	10.0g
Na_2CO_3	2.5g
NaCl	2.0g
Glucose	2.0g
Na_2HPO_4	0.4g

pH 7.8 ± 0.2 at 25°C

Source: This medium is available as a premixed powder from HiMedia.

Preparation of Medium: Add components to distilled/deionized water and bring volume to 1.0L. Mix thoroughly. Distribute into tubes or flasks. Autoclave for 15 min at 15 psi pressure–1°C.

Use: For the cultivation of group A streptococci used in serological typing, and for the cultivation of a variety of pathogenic microorganisms.

Todd-Hewitt Medium (DSMZ Medium 697)

Composition per liter:

Casein peptone	20.0g
Meat infusion	10.0g
Glucose	2.0g
$NaHCO_3$	2.0g
NaCl	2.0g
Na_2HPO_4	0.4g

pH 7.8 ± 0.2 at 25°C

Preparation of Medium: Add components to distilled/deionized water and bring volume to 1.0L. Mix thoroughly. Distribute into tubes or flasks. Autoclave for 15 min at 15 psi pressure–121°C.

Use: For the cultivation of *Globicatella sanguinis, Streptococcus cristatus,* and *Actinomyces hyovaginalis.*

Toluidine Blue DNA Agar

Composition per liter:

Agar	10.0g
NaCl	10.0g
Tris(hydroxymethyl)aminomethane buffer	6.1g
Deoxyribonucleic acid	0.3g
Toluidine Blue O	0.083g
$CaCl_2$, anhydrous	1.1mg

pH 9.0 ± 0.2 at 25°C

Preparation of Medium: Add tris(hydroxymethyl)aminomethane buffer to distilled/deionized water and bring volume to 1.0L. Mix thoroughly. Adjust pH to 9.0. Add the remaining components, except Toluidine Blue O. Mix thoroughly. Gently heat and bring to boiling. Add Toluidine Blue O. Mix thoroughly. If used the same day, sterilization is not necessary. Cool to 50°C. Pour into sterile Petri dishes or distribute into sterile tubes.

Use: For the cultivation and differentiation of *Staphylococcus aureus* from foods.

Toluidine Blue DNA Agar

Composition per liter:

Agar	10.0g
NaCl	10.0g
Tris(hydroxymethyl)aminomethane buffer	6.1g
Deoxyribonucleic acid (DNA)	0.3g
Toluidine Blue O	0.083g
$CaCl_2$, anhydrous	1.1mg

pH 7.3 ± 0.2 at 25°C

Preparation of Medium: Add components, except Toluidine Blue O, to distilled/deionized water and bring volume to 1.0L. Mix thoroughly. Gently heat and bring to boiling. Add Toluidine Blue O. Mix thoroughly. Medium does not have to be sterilized if used immediately. Pour into sterile Petri dishes or distribute into sterile tubes. Allow tubes to cool in a slanted position.

Use: For the cultivation and differentiation of bacteria based on their production of deoxyribonuclease (DNase). Bacteria that produce DNase turn the medium pink.

Tomato Dextrin Yeast Medium

Composition per liter:

Tomato paste	20.0g
Dextrin	20.0g
Agar	20.0g
Baker's yeast	10.0g
$CoCl_2 \cdot 6H_2O$	5.0mg

pH 7.2–7.4 at 25°C

Preparation of Medium: Add components, except agar, to distilled/deionized water and bring volume to 1.0L. Mix thoroughly. Adjust pH to 7.2–7.4. Add agar. Gently heat and bring to boiling. Distribute into tubes or flasks. Autoclave for 15 min at 15 psi pressure–121°C. Pour into sterile Petri dishes or leave in tubes.

Use: For the cultivation of *Streptomyces avermitilis*.

Tomato Juice Agar

Composition per liter:

Agar	12.0g
Pancreatic digest of casein	10.0g

Peptonized milk 10.0g
Tomato juice 400.0mL

pH 6.1 ± 0.2 at 25°C

Source: This medium is available as a premixed powder from BD Diagnostic Systems and Oxoid Unipath.

Preparation of Medium: Add components to distilled/deionized water and bring volume to 1.0L. Mix thoroughly. Gently heat and bring to boiling. Distribute into tubes or flasks. Autoclave for 15 min at 15 psi pressure–121°C. Pour into sterile Petri dishes or leave in tubes.

Use: For the cultivation of lactobacilli, especially *Lactobacillus acidophilus.*

Tomato Juice Agar (ATCC Medium 33)

Composition per liter:

Agar 11.0g
Pancreatic digest of casein 10.0g
Yeast extract 10.0g
Tomato juice, filtered 200.0mL

pH 7.2 ± 0.2 at 25°C

Preparation of Medium: Add components to distilled/deionized water and bring volume to 1.0L. Mix thoroughly. Adjust pH to 7.2. Gently heat and bring to boiling. Distribute into tubes or flasks. Autoclave for 15 min at 15 psi pressure–121°C. Pour into sterile Petri dishes or leave in tubes.

Use: For the cultivation and maintenance of a variety of bacteria including *Lactobacillus, Leuconostoc, Pediococcus,* and *Propionibacterium* species.

Tomato Juice Agar Special

Composition per liter:

Agar 20.0g
Pancreatic digest of casein 10.0g
Peptonized milk 10.0g
Tomato juice 400.0mL

pH 5.0 ± 0.2 at 25°C

Source: This medium is available as a premixed powder from BD Diagnostic Systems.

Preparation of Medium: Add components to distilled/deionized water and bring volume to 1.0L. Mix thoroughly. Gently heat and bring to boiling. Distribute into tubes or flasks. Autoclave for 15 min at 15 psi pressure–121°C. Avoid overheating—it results in a soft agar. Pour into sterile Petri dishes or leave in tubes.

Use: For the cultivation and enumeration of lactobacilli.

Tomato Juice Broth

Composition per liter:

Tomato juice, dessicated 20.0g
Glucose 10.0g
Yeast extract 10.0g
K_2HPO_4 .... 0.5g
KH_2PO_4 .... 0.5g
$MgSO_4 \cdot 7H_2O$ 0.2g
$FeSO_4 \cdot 7H_2O$ 0.01g
$MnSO_4 \cdot 7H_2O$ 0.01g
NaCl 0.01g

pH 6.7 ± 0.2 at 25°C

Source: This medium is available as a premixed powder from BD Diagnostic Systems.

Preparation of Medium: Add components to distilled/deionized water and bring volume to 1.0L. Mix thoroughly. Distribute into tubes or flasks. Autoclave for 15 min at 15 psi pressure–121°C.

Use: For the cultivation of yeast and other aciduric microorganisms.

Tomato Juice Broth (ATCC Medium 433)

Composition per liter:

Pancreatic digest of casein 10.0g
Yeast extract 10.0g
Tomato juice, filtered 200.0mL

pH 7.2 ± 0.2 at 25°C

Preparation of Medium: Add components to distilled/deionized water and bring volume to 1.0L. Mix thoroughly. Adjust pH to 7.2. Gently heat and bring to boiling. Distribute into tubes or flasks. Autoclave for 15 min at 15 psi pressure–121°C.

Use: For the cultivation and maintenance of a variety of fastidious bacteria that require complex growth factors, including *Lactobacillus, Aerococcus, Bifidobacterium,* and *Pediococcus* species.

Tomato Juice HiVeg Agar

Composition per liter:

Agar 11.0g
Plant hydrolysate 10.0g
Plant hydrolysate No. 3 .... 10.0g
Tomato juice 400.0mL

pH 5.0 ± 0.2 at 25°C

Source: This medium is available as a premixed powder from HiMedia.

Preparation of Medium: Add components to distilled/deionized water and bring volume to 1.0L. Mix thoroughly. Gently heat and bring to boiling. Distribute into tubes or flasks. Autoclave for 15 min at 15 psi pressure–121°C. Avoid overheating—it results in a soft agar. Pour into sterile Petri dishes or leave in tubes.

Use: For the cultivation and enumeration of lactobacilli.

Tomato Juice HiVeg Agar, Special

Composition per liter:

Agar 20.0g
Plant hydrolysate No. 3 .... 10.0g
Plant peptone 10.0g
Tomato juice 400.0mL

pH 5.0 ± 0.2 at 25°C

Source: This medium is available as a premixed powder from HiMedia.

Preparation of Medium: Add components to distilled/deionized water and bring volume to 1.0L. Mix thoroughly. Gently heat and bring to boiling. Distribute into tubes or flasks. Autoclave for 15 min at 15 psi pressure–121°C. Avoid overheating—it results in a soft agar. Pour into sterile Petri dishes or leave in tubes.

Use: For the cultivation and enumeration of lactobacilli.

Tomato Juice HiVeg Medium Base with Tomato Juice and Cycloheximide

Composition per liter:

Agar	15.0g
Glucose	10.0g
Tomato juice solids, from 150.0mL	7.5g
Yeast extract	5.0g
Plant special peptone	5.0g
KH_2PO_4	0.5g
$CaCl_2$	0.125g
$MgSO_4$	0.125g
KCl	0.125g
NaCl	0.125g
Bromcresol Green	0.03g
$MnSO_4$	0.03g
Tomato juice, canned	150.0mL
Cycloheximide solution	10.0mL

pH 5.0 ± 0.2 at 25°C

Cycloheximide Solution:

Composition per 10.0mL:

Cycloheximide	0.1g

Preparation of Cycloheximide Solution: Add cycloheximide to distilled/deionized water and bring volume to 10.0mL. Mix thoroughly. Filter sterilize.

Caution: Cycloheximide is toxic. Avoid skin contact or aerosol formation and inhalation.

Preparation of Medium: Add components, except cycloheximide solution, to distilled/deionized water and bring volume to 990.0mL. Mix thoroughly. Gently heat and bring to boiling. Autoclave for 15 min at 15 psi pressure–121°C. Cool to 45°–50°C. Aseptically add sterile cycloheximide solution. Mix thoroughly. Pour into sterile Petri dishes or distribute into sterile tubes.

Use: For the isolation and cultivation of lactobacilli.

Tomato Juice Medium

Composition per liter:

Agar	15.0g
Glucose	10.0g
Polypeptone™	5.0g
Yeast extract	5.0g
KH_2PO_4	0.5g
$CaCl_2 \cdot 2H_2O$	0.125g
KCl	0.125g
$MgSO_4 \cdot 7H_2O$	0.125g
NaCl	0.125g
Bromcresol Green	0.03g
$MnSO_4 \cdot 4H_2O$	3.0mg
Tomato juice, canned	150.0mL
Cycloheximide solution	10.0mL

pH 5.0 ± 0.2 at 25°C

Cycloheximide Solution:

Composition per 10.0mL:

Cycloheximide	0.1g

Preparation of Cycloheximide Solution: Add cycloheximide to distilled/deionized water and bring volume to 10.0mL. Mix thoroughly. Filter sterilize.

Caution: Cycloheximide is toxic. Avoid skin contact or aerosol formation and inhalation.

Preparation of Medium: Add components, except cycloheximide solution, to distilled/deionized water and bring volume to 990.0mL. Mix thoroughly. Gently heat and bring to boiling. Autoclave for 15 min at 15 psi pressure–121°C. Cool to 45°–50°C. Aseptically add sterile cycloheximide solution. Mix thoroughly. Pour into sterile Petri dishes or distribute into sterile tubes.

Use: For the isolation and cultivation of lactobacilli from wine.

Tomato Juice Medium

Composition per liter:

Tryptic digest of casein	10.0g
Yeast extract	10.0g
Tomato juice, filtered, pH 7.0	200.0mL
Tween™ 80	1.0mL

pH 6.5 ± 0.2 at 25°C

Preparation of Medium: Add components to distilled/deionized water and bring volume to 1.0L. Mix thoroughly. Distribute into tubes or flasks. Autoclave for 15 min at 15 psi pressure–121°C.

Use: For the cultivation and maintenance of *Lactobacillus collinoides*.

Tomato Juice Milk Agar (DSMZ Medium 353)

Composition per liter:

Skim milk	100.0g
Yeast extract	5.0g
Tomato juice	100.0mL

pH 7.0 ± 0.2 at 25°C

Preparation of Medium: Filter canned tomatoes through paper to produce tomato juice. Leave overnight at 10°C. Add components to distilled/deionized water and bring volume to 1.0L. Mix thoroughly. Gently heat and bring to boiling. Distribute into tubes or flasks. Autoclave for 15 min at 15 psi pressure–121°C.

Use: For the isolation and cultivation of *Enterococcus faecalis=Streptococcus faecalis.*

Tomato Juice Yeast Extract Medium

Composition per liter:

Skim milk	100.0g
Yeast extract	5.0g
Tomato juice, filtered	100.0mL

Preparation of Medium: Add components to distilled/deionized water and bring volume to 1.0L. Mix thoroughly. Distribute into tubes or flasks. Autoclave for 15 min at 15 psi pressure–121°C.

Use: For the cultivation of yeasts and fungi.

Tomato Juice Yeast Extract Milk Medium

Composition per liter:

Skim milk	100.0g
Yeast extract	5.0g
Tomato juice, filtered	100.0mL

pH 7.0 ± 0.2 at 25°C

Preparation of Medium: Filter canned tomato juice through paper. Let stand overnight at 10°C. Add remaining components and bring to 1.0L with distilled/deionized water. Mix thoroughly. Distribute into tubes or flasks. Autoclave for 15 min at 15 psi pressure–121°C.

Use: For the cultivation and maintenance of a variety of fastidious bacteria that require complex growth factors, including *Lactobacillus, Streptococcus,* and *Enterococcus* species.

Tomato Paste Oatmeal Agar

Composition per liter:

Oatmeal (dried baby food)	20.0g
Tomato paste	20.0g
Agar	15.0g

pH 7.0 ± 0.2 at 25°C

Preparation of Medium: Add components to distilled/deionized water and bring volume to 1.0L. Mix thoroughly. Melt agar by steaming for 20–30 min at 0 psi pressure–100°C. Distribute into tubes or flasks. Autoclave for 15 min at 15 psi pressure–121°C. Pour into sterile Petri dishes or leave in tubes.

Use: For the cultivation and maintenance of *Flexibacter* species.

Top Agarose

Composition per liter:

Pancreatic digest of casein	10.0g
NaCl	8.0g
Agarose	6.0g

pH 7.0 ± 0.2 at 25°C

Preparation of Medium: Add components to distilled/deionized water and bring volume to 1.0L. Mix thoroughly. Distribute into tubes or flasks. Autoclave for 25 min at 15 psi pressure–121°C.

Use: For the distribution of bacteriophage or bacterial cells, especially *Escherichia coli*, evenly in a thin layer over the surface of a plate.

Torulopsis Medium

Composition per liter:

Glucose	100.0g
Agar	20.0g
Casamino acids	4.0g
Yeast extract	1.0g
KH_2PO_4	1.0g
$MgSO_4·7H_2O$	0.5g
$CaCl_2·2H_2O$	0.1g
NaCl	0.1g

pH 5.6 ± 0.2 at 25°C

Preparation of Medium: Add components to distilled/deionized water and bring volume to 1.0L. Mix thoroughly. Gently heat and bring to boiling. Adjust pH to 5.6. Distribute into tubes or flasks. Autoclave for 15 min at 15 psi pressure–121°C. Pour into sterile Petri dishes or leave in tubes.

Use: For the cultivation and maintenance of *Candida versatilis.*

Toxoplasma Medium

Composition per liter:

NaCl	6.8g
$NaHCO_3$	2.2g
Glucose	1.0g
KCl	0.4g
$CaCl_2$	0.2g
$NaH_2PO_4·H_2O$	0.125g
Arginine	0.105g
$MgSO_4$	0.1g
L-Cystine	0.024g
Glutamine	0.292g
Histidine	0.031g
Lysine	0.058g
Isoleucine	0.052g
Leucine	0.052g
Phenol Red	0.050g
Threonine	0.048g
Valine	0.046g
Tyrosine	0.036g
Phenylalanine	0.032g
Methionine	0.015g
Tryptophan	0.010g
Inositol	2.0mg
Choline	1.0mg
Folic acid	1.0mg
Nicotinamide	1.0mg
Pantothenic acid	1.0mg
Pyridoxal·HCl	1.0mg
Thiamine·HCl	1.0mg
Riboflavin	0.1mg
Fetal bovine serum, heat inactivated	100.0mL

pH 7.2–7.4 at 25°C

Preparation of Medium: Add components, except fetal bovine serum, to distilled/deionized water and bring volume to 905.0mL. Mix thoroughly. Adjust pH to 7.2–7.4. Autoclave for 15 min at 15 psi pressure–121°C. Aseptically add 100.0mL of sterile, heat-inactivated fetal bovine serum. Mix thoroughly. Aseptically distribute into sterile tubes or flasks.

Use: For the cultivation of *Toxoplasma gondii.*

TPBY

See: **Tryptone Phosphate Brain Heart Infusion Yeast Extract Agar**

TPEY Agar (Tellurite Polymyxin Egg Yolk Agar)

Composition per liter:

NaCl	20.0g
Agar	15.5g
Pancreatic digest of casein	10.0g
Yeast extract	5.0g
D-Mannitol	5.0g
LiCl	2.0g
Egg yolk emulsion (30% solution)	100.0mL
Chapman tellurite solution	10.0mL
Polymyxin B solution	0.4mL

pH 7.1 ± 0.2 at 25°C

Source: This medium is available as a premixed powder from BD Diagnostic Systems.

Egg Yolk Emulsion (30% Solution):

Composition per 100.0mL:

NaCl	0.6g
Egg yolk	30.0mL

Preparation of Egg Yolk Emulsion (30% Solution): Add NaCl and egg yolk to distilled/deionized water and bring volume to 100.0mL. Mix thoroughly. Filter sterilize.

Chapman Tellurite Solution:

Composition per 100.0mL:

K_2TeO_3	1.0g

Preparation of Chapman Tellurite Solution: Add K_2TeO_3 to distilled/deionized water and bring volume to 100.0mL. Mix thoroughly. Filter sterilize.

Polymyxin B Solution:

Composition per 100.0mL:

Polymyxin B	1.0g

Preparation of Polymyxin B Solution: Add polymyxin B to distilled/deionized water and bring volume to 100.0mL. Mix thoroughly. Filter sterilize.

Caution: Potassium tellurite is toxic.

Preparation of Medium: Add components—except 30% egg yolk emulsion, Chapman tellurite solution, and polymyxin B solution—to distilled/deionized water and bring volume to 890.0mL. Mix thoroughly. Gently heat and bring to boiling. Autoclave for 15 min at 15 psi pressure–121°C. Cool to 45°–50°C. Aseptically add 100.0mL of sterile 30% egg yolk emulsion, 10.0mL of sterile Chapman tellurite solution, and 0.4mL of sterile polymyxin B solution. Mix thoroughly. Pour into sterile Petri dishes or distribute into sterile tubes.

Use: For the recovery of staphylococci from foods and other materials.

TPEY HiVeg Agar Base with Egg Yolk, Tellurite, and Polymyxin B

Composition per liter:

NaCl	20.0g
Agar	18.0g
Plant hydrolysate	10.0g
D-Mannitol	5.0g
Yeast extract	5.0g
LiCl	2.0g
Egg yolk emulsion (30% solution)	100.0mL
Chapman tellurite solution	10.0mL
Polymyxin B solution	0.4mL

pH 7.1 ± 0.2 at 25°C

Source: This medium, without egg yolk, tellurite, and polymyxin B, is available as a premixed powder from HiMedia.

Egg Yolk Emulsion (30% Solution):

Composition per 100.0mL:

NaCl	0.6g
Egg yolk	30.0mL

Preparation of Egg Yolk Emulsion (30% Solution): Add NaCl and egg yolk to distilled/deionized water and bring volume to 100.0mL. Mix thoroughly. Filter sterilize.

Chapman Tellurite Solution:

Composition per 100.0mL:

K_2TeO_3	1.0g

Preparation of Chapman Tellurite Solution: Add K_2TeO_3 to distilled/deionized water and bring volume to 100.0mL. Mix thoroughly. Filter sterilize.

Polymyxin B Solution:

Composition per 100.0mL:

Polymyxin B	1.0g

Preparation of Polymyxin B Solution: Add polymyxin B to distilled/deionized water and bring volume to 100.0mL. Mix thoroughly. Filter sterilize.

Caution: Potassium tellurite is toxic.

Preparation of Medium: Add components—except 30% egg yolk emulsion, Chapman tellurite solution, and polymyxin B solution—to distilled/deionized water and bring volume to 890.0mL. Mix thoroughly. Gently heat and bring to boiling. Autoclave for 15 min at 15 psi pressure–121°C. Cool to 45°–50°C. Aseptically add 100.0mL of sterile 30% egg yolk emulsion, 10.0mL of sterile Chapman tellurite solution, and 0.4mL of sterile polymyxin B solution. Mix thoroughly. Pour into sterile Petri dishes or distribute into sterile tubes.

Use: For the recovery of staphylococci from foods and other materials.

TPGY Broth

See: **Trypticase™ Peptone Glucose Yeast Extract Broth**

TPGY Medium (Thioglycolate Peptone Glucose Yeast Extract Medium)

Composition per liter:

Pancreatic digest of casein	50.0g
Peptone	5.0g
Yeast extract	5.0g
Glucose	1.0g
Sodium thioglycolate	1.0g

pH 7.1 ± 0.2 at 25°C

Preparation of Medium: Add components to distilled/deionized water and bring volume to 1.0L. Mix thoroughly. Distribute into tubes or flasks. Autoclave for 15 min at 15 psi pressure–121°C.

Use: For the cultivation of a variety of anaerobic bacteria.

TPGYT Broth

See: **Trypticase™ Peptone Glucose Yeast Extract Broth with Trypsin**

TPL Medium (Thioglycolate Potato Liver Medium)

Composition per liter:

Potato	200.0g
Yeast extract	31.0g
Liver	25.0g
Glycerol	15.0g
Agar	15.0g
Meat extract	5.5g
Glucose	7.5g
Peptone	2.5g
NaCl	2.5g
Sodium thioglycolate	0.5g
Methylene Blue	1.0mg

pH 7.0 ± 0.2 at 25°C

Preparation of Medium: Add peeled, sliced potato to approximately 500.0mL of distilled/deionized water. Gently heat and bring to boiling. Continue boiling for 30 min. Filter through cheesecloth. Cut up liver into small pieces and add to approximately 150.0mL of distilled/deionized water. Gently heat and bring to boiling. Continue boiling for 30 min. Filter through cheesecloth. Add boiled potato solids, boiled liver solids, and remaining components to distilled/deionized water and bring volume to 1.0L. Mix thoroughly. Gently heat and bring to boiling. Distribute into tubes or flasks. Make sure each of the tubes receives a few pieces of liver. Autoclave for 15 min at 15 psi pressure–121°C.

Use: For the cultivation and maintenance of *Pseudomonas* species.

TPT 18 Medium (DSMZ Medium 1127)

Composition per liter:

Glucose	0.5g
Yeast extract	0.1g
Pancreatic digest of casein	0.1g
$MgSO_4 \cdot 7H_2O$	0.05g
$CaCl_2 \cdot 2H_2O$	0.02g

pH 6.0 ± 0.2 at 25°C

Preparation of Medium: Add components to distilled/deionized water and bring volume to 1.0L. Mix thoroughly. Autoclave for 15 min at 15 psi pressure–121°C. Adjust pH to 6.0.

Use: For the cultivation of *Mucilaginibacter gracilis*.

TPY Medium

Composition per liter:

Pancreatic digest of casein	10.0g
Glucose	5.0g
Pancreatic digest of soybean meal	5.0g
Yeast extract	2.5g
K_2HPO_4	2.0g
Agar	1.5g
Cysteine·HCl	0.5g
$MgCl_2 \cdot 6H_2O$	0.5g
$ZnSO_4 \cdot 7H_2O$	0.25g
$CaCl_2$	0.15g
$FeCl_3$	1.0μg
Tween™ 80	1.0mL

pH 6.5 ± 0.2 at 25°C

Preparation of Medium: Add components to distilled/deionized water and bring volume to 1.0L. Mix thoroughly. Gently heat until dissolved. Distribute into tubes or flasks. Autoclave for 15 min at 15 psi pressure–121°C.

Use: For the isolation and cultivation of *Bifidobacterium* species.

TPYG Medium

Composition per liter:

Pancreatic digest of casein	10.0g
Peptone	5.0g
Yeast extract	5.0g
Glucose solution	50.0mL

Glucose Solution:

Composition per 100.0mL:

Glucose	20.0g

Preparation of Glucose Solution: Add glucose to distilled/deionized water and bring volume to 100.0mL. Mix thoroughly. Filter sterilize.

Preparation of Medium: Add components, except glucose solution, to distilled/deionized water and bring volume to 950.0mL. Mix thoroughly. Autoclave for 15 min at 15 psi pressure–121°C. Aseptically add 50.0mL of sterile glucose solution. Mix thoroughly. Aseptically distribute into sterile tubes or flasks.

Use: For the cultivation of *Clostridium felsineum*.

Trace Elements Solution HO-LE

Composition per liter:

H_3BO_3	2.85g
$MnCl_2 \cdot 4H_2O$	1.8g
Sodium tartrate	1.77g
$FeSO_4 \cdot 7H_2O$	1.36g
$CoCl_2 \cdot 6H_2O$	0.04g
$CuCl_2 . 2H_2O$	0.027g
$Na_2MoO_4 \cdot 2H_2O$	0.025g
$ZnCl_2$	0.02g

Preparation of Trace Elements Solution HO-LE: Add components to distilled/deionized water and bring volume to 1.0L. Mix thoroughly. Filter sterilize.

Use: For the enrichment of other media requiring added trace metals.

Transgrow Medium

Composition per liter:

GC agar base	730.0mL
Hemoglobin solution	250.0mL
Vitox supplement	10.0mL
VCN antibiotic solution	10.0mL

pH 7.3 ± 0.2 at 25°C

GC Agar Base:

Composition per 730.0mL:

Special peptone	15.0g
Agar	20.0g
NaCl	5.0g
K_2HPO_4	4.0g
Cornstarch	1.0g
KH_2PO_4	1.0g

pH 7.2 ± 0.2 at 25°C

Preparation of GC Agar Base: Add components of GC medium base and the hemoglobin to distilled/deionized water and bring volume to 730.0mL. Mix thoroughly. Gently heat until boiling. Autoclave for 15 min at 15 psi pressure–121°C. Cool to 45°–50°C.

Hemoglobin Solution:

Composition per 250.0mL:

Hemoglobin	5.0g

Preparation of Hemoglobin Solution: Add hemoglobin to distilled/deionized water and bring volume to 250.0mL. Mix thoroughly. Autoclave for 15 min at 15 psi pressure–121°C. Cool to 45°–50°C.

Vitox Supplement:

Composition per 10.0mL:

Glucose	2.0g
L-Cysteine·HCl	0.518g
L-Glutamine	0.2g
L-Cystine	0.022g
Adenine sulfate	0.01g
Nicotinamide adenine dinucleotide	5.0mg
Cocarboxylase	2.0mg
Guanine·HCl	0.6mg
$Fe(NO_3)_3 \cdot 6H_2O$	0.4mg
p-Aminobenzoic acid	0.26mg
Vitamin B_{12}	0.2mg
Thiamine·HCl	0.06mg

Preparation of Vitox Supplement: Add components to distilled/deionized water and bring volume to 10.0mL. Mix thoroughly. Filter sterilize.

VCN Antibiotic Solution:

Composition per 10.0mL:

Colistin methane sulfonate	7.5mg
Vancomycin	3.0mg
Nystatin	12,500U

Preparation of VCN Antibiotic Solution: Add components to distilled/deionized water and bring volume to 10.0mL. Mix thoroughly. Filter sterilize.

Preparation of Medium: To 730.0mL of cooled, sterile GC agar base, aseptically add 250.0mL of sterile hemoglobin solution, 10.0mL of sterile Vitox supplement, and 10.0mL of VCN antibiotic solution. Mix thoroughly. Pour into sterile Petri dishes or distribute into sterile tubes.

Use: For the cultivation and transport of fastidious microorganisms, especially *Neisseria* species.

Transgrow Medium

Composition per liter:

GC medium base	730.0mL
Hemoglobin solution	250.0mL
Supplement B	10.0mL
VCNT antibiotic solution	10.0mL

pH 7.3 ± 0.2 at 25°C

GC Medium Base:

Composition per 730.0mL:

Proteose peptone No. 3	15.0g
Agar	20.0g
NaCl	5.0g
K_2HPO_4	4.0g
Glucose	1.5g
Cornstarch	1.0g
KH_2PO_4	1.0g

pH 7.2 ± 0.2 at 25°C

Preparation of GC Medium Base: Add components to distilled/deionized water and bring volume to 730.0mL. Mix thoroughly. Gently heat until boiling. Autoclave for 15 min at 15 psi pressure–121°C. Cool to 45°–50°C.

Hemoglobin Solution:

Composition per 250.0mL:

Hemoglobin	10.0g

Preparation of Hemoglobin Solution: Add hemoglobin to distilled/deionized water and bring volume to 250.0mL. Mix thoroughly. Autoclave for 15 min at 15 psi pressure–121°C. Cool to 45°–50°C.

Supplement B:

Composition per 10.0mL:

Supplement B contains yeast concentrate, glutamine, coenzyme, cocarboxylase, hematin, and growth factors.

Preparation of Supplement B: Add components to distilled/deionized water and bring volume to 10.0mL. Mix thoroughly. Filter sterilize.

Source: Supplement B is available as a premixed powder from BD Diagnostic Systems.

VCNT Antibiotic Solution:

Composition per 10.0mL:

Colistin methane sulfonate	7.5mg
Trimethoprim lactate	5.0mg
Vancomycin	3.0mg
Nystatin	12,500U

Preparation of VCNT Antibiotic Solution: Add components to distilled/deionized water and bring volume to 10.0mL. Mix thoroughly. Filter sterilize.

Preparation of Medium: To 730.0mL of cooled, sterile GC medium base, aseptically add 250.0mL of sterile hemoglobin solution, 10.0mL of sterile supplement B, and 10.0mL of VCNT antibiotic solution. Mix thoroughly. Pour into sterile Petri dishes or distribute into sterile tubes.

Use: For the cultivation and transport of fastidious microorganisms, especially *Neisseria* species.

Transgrow Medium with Trimethoprim

Composition per liter:

Agar	20.0g
Hemoglobin	10.0g
Pancreatic digest of casein	7.5g
Selected meat peptone	7.5g
NaCl	5.0g
K_2HPO_4	4.0g
Glucose	1.5g
Cornstarch	1.0g
KH_2PO_4	1.0g
Supplement solution	10.0mL
VCNT inhibitor	10.0mL

pH 6.7 ± 0.2 at 25°C

Source: This medium is available as a prepared medium from BD Diagnostic Systems.

Supplement Solution:

Composition per liter:

Glucose	100.0g
L-Cysteine·HCl	25.9g
L-Glutamine	10.0g
L-Cystine	1.1g
Adenine	1.0g
Nicotinamide adenine dinucleotide	0.25g
Vitamin B_{12}	0.1g
Thiamine pyrophosphate	0.1g
Guanine·HCl	0.03g
$Fe(NO_3)_3 \cdot 6H_2O$	0.02g
p-Aminobenzoic acid	0.013g
Thiamine·HCl	3.0mg

Source: The supplement solution (IsoVitaleX® enrichment) is available from BD Diagnostic Systems. This enrichment may be replaced by supplement VX from BD Diagnostic Systems.

Preparation of Supplement Solution: Add components to distilled/deionized water and bring volume to 1.0L. Mix thoroughly. Filter sterilize.

VCNT Inhibitor:

Composition per 10.0mL:

Colistin	7.5mg
Trimethoprim lactate	5.0mg
Vancomycin	3.0mg
Nystatin	12,500U

Preparation of VCNT Inhibitor: Add components to distilled/deionized water and bring volume to 10.0mL. Mix thoroughly. Filter sterilize.

Preparation of Medium: Add components, except supplement solution and VCNT inhibitor, to distilled/deionized water and bring volume to 980.0mL. Mix thoroughly. Gently heat and bring to boiling. Autoclave for 15 min at 15 psi pressure–121°C. Cool to 45°–50°C under 5–30% CO_2. Aseptically add 10.0mL of sterile supplement solution and 10.0mL of sterile VCNT inhibitor. Mix thoroughly. Aseptically distribute under 5–30% CO_2 into sterile screw-capped tubes.

Use: For the transportation and recovery of pathogenic *Neisseria* species.

Transgrow Medium without Trimethoprim

Composition per liter:

Agar	20.0g
Hemoglobin	10.0g
Pancreatic digest of casein	7.5g
Selected meat peptone	7.5g
NaCl	5.0g
K_2HPO_4	4.0g
Glucose	1.5g
Cornstarch	1.0g
KH_2PO_4	1.0g
Supplement solution	10.0mL
VCN inhibitor	10.0mL

pH 6.7 ± 0.2 at 25°C

Source: This medium is available as a prepared medium from BD Diagnostic Systems.

Supplemement Solution:

Composition per liter:

Glucose	100.0g
L-Cysteine·HCl	25.9g
L-Glutamine	10.0g
L-Cystine	1.1g
Adenine	1.0g
Nicotinamide adenine dinucleotide	0.25g
Vitamin B_{12}	0.1g
Thiamine pyrophosphate	0.1g
Guanine·HCl	0.03g
$Fe(NO_3)_3·6H_2O$	0.02g
p-Aminobenzoic acid	0.013g
Thiamine·HCl	3.0mg

Source: The supplement solution IsoVitaleX® enrichment is available from BD Diagnostic Systems. This enrichment may be replaced by supplement VX from BD Diagnostic Systems.

Preparation of Supplement Solution: Add components to distilled/deionized water and bring volume to 1.0L. Mix thoroughly. Filter sterilize.

VCN Inhibitor:

Composition per 10.0mL:

Colistin	7.5mg
Vancomycin	3.0mg
Nystatin	12,500U

Preparation of VCN Inhibitor: Add components to distilled/deionized water and bring volume to 10.0mL. Mix thoroughly. Filter sterilize.

Preparation of Medium: Add components, except supplement solution and VCN inhibitor, to distilled/deionized water and bring volume to 980.0mL. Mix thoroughly. Gently heat and bring to boiling. Autoclave for 15 min at 15 psi pressure–121°C. Cool to 45°–50°C under 5–30% CO_2. Aseptically add 10.0mL of sterile supplement solution and 10.0mL of sterile VCN inhibitor. Mix thoroughly. Aseptically distribute under 5–30% CO_2 into sterile screw-capped tubes.

Use: For the transportation and recovery of pathogenic *Neisseria* species.

Transport Medium

Composition per liter:

Sodium glycerophosphate	10.0g
Agar	3.0g
Sodium thioglycolate	1.0g
$CaCl_2·2H_2O$	0.1g
Methylene Blue	2.0mg

pH 7.3 ± 0.2 at 25°C

Source: This medium is available as a premixed powder from BD Diagnostic Systems.

Preparation of Medium: Add components to distilled/deionized water and bring volume to 1.0L. Mix thoroughly. Gently heat while stirring and bring to boiling. Distribute into screw-capped tubes or vials. Fill tubes nearly to capacity. Leave only enough space so that when a small swab is introduced the tube does not overflow. Autoclave for 10 min at 15 psi pressure–121°C. Tighten caps on tubes.

Use: For the transportation of swab specimens for the recovery of a wide variety of microorganisms, including *Neisseria gonorrhoeae.*

Transport Medium Stuart

Composition per liter:

Sodium glycerophosphate	10.0g
Agar	3.0g
Sodium thioglycolate	0.9g
$CaCl_2·2H_2O$	0.1g
Methylene Blue	2.0mg

pH 7.3 ± 0.2 at 25°C

Source: This medium is available as a premixed powder from BD Diagnostic Systems.

Preparation of Medium: Add components to distilled/deionized water and bring volume to 1.0L. Mix thoroughly. Gently heat while stirring and bring to boiling. Distribute into screw-capped tubes or vials. Fill tubes nearly to capacity. Leave only enough space so that when a small swab is introduced the tube does not overflow. Autoclave for 10 min at 15 psi pressure–121°C. Tighten caps on tubes.

Use: For the transportation of swab specimens for the recovery of a wide variety of microorganisms, including *Neisseria gonorrhoeae.*

Trebouxia Agar

Composition per liter:

Bristol's solution	850.0mL
Soil extract	140.0mL
Glucose	20.0g
Agar	15.0g
Proteose peptone	10.0g

Bristol's Solution:

Composition per 1000.1mL:

$NaNO_3$ solution	10.0g
KH_2PO_4 solution	7.0g
K_2HPO_4 solution	3.0g
$MgSO_4·7H_2O$ solution	3.0g
$CaCl_2$ solution	1.0g

NaCl solution 1.0g
$FeCl_3$ solution 0.1mL

$NaNO_3$ Solution:
Composition per 400.0mL:
$NaNO_3$ 10.0g

Preparation of $NaNO_3$ Solution: Add $NaNO_3$ to distilled/deionized water and bring volume to 400.0mL. Mix thoroughly.

$CaCl_2$ Solution:
Composition per 400.0mL:
$CaCl_2$ 1.0g

Preparation of $CaCl_2$ Solution: Add $CaCl_2$ to distilled/deionized water and bring volume to 400.0mL. Mix thoroughly.

$MgSO_4 \cdot 7H_2O$ Solution:
Composition per 400.0mL:
$MgSO_4 \cdot 7H_2O$ 3.0g

Preparation of $MgSO_4 \cdot 7H_2O$ Solution: Add $MgSO_4 \cdot 7H_2O$ to distilled/deionized water and bring volume to 400.0mL. Mix thoroughly.

K_2HPO_4 Solution:
Composition per 400.0mL:
K_2HPO_4 .... 3.0g

Preparation of K_2HPO_4 Solution: Add K_2HPO_4 to distilled/deionized water and bring volume to 400.0mL. Mix thoroughly.

KH_2PO_4 Solution:
Composition per 400.0mL:
KH_2PO_4 .... 7.0g

Preparation of KH_2PO_4 Solution: Add KH_2PO_4 to distilled/deionized water and bring volume to 400.0mL. Mix thoroughly.

NaCl Solution:
Composition per 400.0mL:
NaCl 1.0g

Preparation of NaCl Solution: Add NaCl to distilled/deionized water and bring volume to 400.0mL. Mix thoroughly.

$FeCl_3$ Solution:
Composition per 100.0mL:
$FeCl_3$ 1.0g

Preparation of $FeCl_3$ Solution: Add $FeCl_3$ to distilled/deionized water and bring volume to 100.0mL. Mix thoroughly.

Preparation of Bristol's Solution: Add 10.0mL of $NaNO_3$ solution, 10.0mL of $CaCl_2$ solution, 10.0mL of $MgSO_4 \cdot 7H_2O$ solution, 10.0mL of $NaNO_3$ solution, 10.0mL of K_2HPO_4 solution, 10.0mL of KH_2PO_4 solution, and 10.0mL of NaCl solution to distilled/deionized water and bring volume to 1.0L. Mix thoroughly. Add 0.1mL of $FeCl_3$ solution. Mix thoroughly.

Soil Extract:
Composition per 200.0mL:
African Violet soil 77.0g
Na_2CO_3 .... 0.2g

Preparation of Soil Extract: Add components to distilled/deionized water and bring volume to 200.0mL. Mix thoroughly. Autoclave for 60 min at 15 psi pressure–121°C. Filter through Whatman #1 filter paper.

Preparation of Medium: Combine components. Mix thoroughly. Gently heat and bring to boiling. Distribute into tubes or flasks. Autoclave for 15 min at 15 psi pressure–121°C. Pour into sterile Petri dishes or leave in tubes.

Use: For the cultivation of *Brachiomonas submarina* and *Trebouxia magna.*

Treponema bryantii Medium

Composition per liter:
L-Cysteine·HCl 1.0g
NaCl 0.9g
$(NH_4)_2SO_4$ 0.9g
K_2HPO_4 .... 0.45g
KH_2PO_4 .... 0.45g
$MgSO_4 \cdot 7H_2O$ 0.18g
$CaCl_2 \cdot 2H_2O$ 0.12g
Resazurin 1.0mg
Rumen fluid, clarified 300.0mL
$NaHCO_3$ solution 100.0mL
Glucose solution (10% w/v) 20.0mL
pH 7.0 ± 0.2 at 25°C

$NaHCO_3$ Solution:
Composition per 100.0mL:
$NaHCO_3$ 5.0g

Preparation of $NaHCO_3$ Solution: Add $NaHCO_3$ to distilled/deionized water and bring volume to 100.0mL. Mix thoroughly. Filter sterilize. Sparge with 100% CO_2.

Glucose Solution:
Composition per 20.0mL:
Glucose 2.0g

Preparation of Glucose Solution: Add glucose to distilled/deionized water and bring volume to 20.0mL. Mix thoroughly. Filter sterilize. Sparge with 100% CO_2.

Preparation of Medium: Prepare and dispense medium under 100% CO_2. Add components, except rumen fluid, $NaHCO_3$ solution, and glucose solution, to distilled/deionized water and bring volume to 580.0mL. Mix thoroughly. Adjust pH to 7.0 with KOH. Sparge with 100% CO_2. Autoclave for 15 min at 15 psi pressure–121°C. Cool to room temperature. Aseptically add 300.0mL of sterile rumen fluid, 100.0mL of sterile $NaHCO_3$ solution, and 20.0mL of sterile glucose solution. Mix thoroughly. Aseptically distribute into sterile tubes or flasks.

Use: For the cultivation and maintenance of *Treponema bryantii.*

Treponema denticola Medium (DSMZ Medium 909)

Composition per 1045.0mL:
Solution A 1.0L
Solution B 45.0mL
pH 7.1 ± 0.2 at 25°C

Solution A:
Composition per liter:
Trypticase™ 10.0g
Yeast extract 2.5g
Agar 3.0g
Glucose 2.0g
L-Cysteine·HCl 1.0g
Na-thioglycolate 0.5g
L-Asparagine 0.25g
Brain heart infusion broth 450.0mL

Preparation of Solution A: Add components to distilled/deionized water and bring volume to 1.0L. Mix thoroughly. Gently heat and bring to boiling. Boil for 5 min. Cool to room temperature while sparging with 80% N_2 + 20% CO_2. Distribute to anaerobe tubes or bottles under 80% N_2 + 20% CO_2. Autoclave for 15 min at 15 psi pressure–121°C. Cool to room temperature.

Brain Heart Infusion Broth:
Composition per liter:

Pancreatic digest of gelatin	14.5g
Brain heart, solids from infusion	6.0g
Peptic digest of animal tissue	6.0g
NaCl	5.0g
Casein	5.0g
Glucose	3.0g
Na_2HPO_4	2.5g

Preparation of Brain Heart Infusion Broth: Add components to distilled/deionized water and bring volume to 1.0L. Mix thoroughly.

Solution B:
Composition per 65.0mL:

$NaHCO_3$	2.0g
Thiamine pyrophosphate	6.0mg
Volatile fatty acid solution	20.0mL
Rabbit serum	20.0mL

Volatile Fatty Acid Solution:
Composition per 102.0mL:

KOH, 0.1*N*	100.0mL
Isobutyric acid	0.5mL
D,L-2-methylbutyric acid	0.5mL
Isovaleric acid	0.5mL
Valeric acid	0.5mL

Preparation of Volatile Fatty Acid Solution: Combine components and mix thoroughly.

Preparation of Solution B: Add components, except rabbit serum, to distilled/deionized water and bring volume to 45.0mL. Mix thoroughly. Filter sterilize. Add 20.0mL rabbit serum.

Preparation of Medium: Aseptically and anaerobically add 0.45mL solution B and 10.0mL solution A to individual tubes. Pressurize the tubes with H_2 to 0.5 bar.

Use: For the cultivation of *Treponema denticola.*

Treponema Isolation Medium

Composition per liter:

Solution A	450.0mL
Spirolate broth	450.0mL
Rabbit serum, inactivated at 56°C for 30 min	100.0mL

pH 7.4 ± 0.2 at 25°C

Solution A:
Composition per 450.0mL:

Agar	8.0g
Asparagine	0.25g
Sodium thioglycolate	0.25g
Pancreatic digest of casein	0.25g
Brain heart infusion broth	450.0mL

Preparation of Solution A: Combine components. Mix thoroughly. Gently heat and bring to boiling. Autoclave for 15 min at 15 psi pressure–121°C. Cool to 45°–50°C.

Brain Heart Infusion Broth:
Composition per liter:

Pancreatic digest of gelatin	14.5g
Brain heart, solids from infusion	6.0g
Peptic digest of animal tissue	6.0g
NaCl	5.0g
Casein	5.0g
Glucose	3.0g
Na_2HPO_4	2.5g

Preparation of Brain Heart Infusion Broth: Add components to distilled/deionized water and bring volume to 1.0L. Mix thoroughly.

Spirolate Broth:
Composition per liter:

Pancreatic digest of casein	15.0g
Glucose	5.0g
Yeast extract	5.0g
NaCl	2.5g
L-Cysteine·HCl·H_2O	1.0g
Sodium thioglycolate	0.5g
Palmitic acid	0.05g
Stearic acid	0.05g
Oleic acid	0.05g
Linoleic acid	0.05g

Preparation of Spirolate Broth: Add components to distilled/deionized water and bring volume to 1.0L. Mix thoroughly. Autoclave for 15 min at 15 psi pressure–121°C. Cool to 25°C.

Preparation of Medium: Combine 450.0mL of sterile solution A, 450.0mL of sterile spirolate broth, and 100.0mL of rabbit serum. Mix thoroughly. Aseptically distribute into sterile tubes or flasks.

Use: For the isolation and cultivation of oral, genital, and fecal treponemes.

Treponema Isolation Medium

Composition per liter:

Beef heart, solids from infusion	20.0g
Ionagar No. 2	7.2g
K_2HPO_4	2.0g
Arabinose	0.8g
Glucose	0.8g
Maltose	0.8g
Polypeptone™	0.8g
Pyruvate	0.8g
Starch, soluble	0.8g
Sucrose	0.8g
Cysteine·HCl	0.68g
$(NH_4)_2SO_4$	0.6g
Serine	0.4g
Tryptose	0.4g
Yeast extract	0.4g
NaCl	0.2g
Rumen fluid	500.0mL
Rabbit serum-cocarboxylase solution	100.0mL

pH 7.2 ± 0.2 at 25°C

Rabbit Serum-Cocarboxylase Solution:
Composition per liter:

Rabbit serum, heat inactivated	100.0mL
Cocarboxylase solution	1.0mL

Preparation of Rabbit Serum-Cocarboxylase Solution: Heat rabbit serum at 56°C for 1 hr. Add 1.0mL of cocarboxylase solution. Mix thoroughly.

Cocarboxylase Solution:

Composition per 1.0mL:

Cocarboxylase	0.5g

Preparation of Cocarboxylase Solution: Add cocarboxylase to 1.0mL of distilled/deionized water. Mix thoroughly. Filter sterilize.

Preparation of Medium: Add components, except rumen fluid and rabbit serum-cocarboxylase solution, to distilled/deionized water and bring volume to 400.0mL. Mix thoroughly. Gently heat and bring to boiling. Autoclave for 15 min at 15 psi pressure–121°C. Cool to 45°–50°C. Aseptically add 500.0mL of sterile rumen fluid and 100.0mL of sterile rabbit serum-cocarboxylase solution. Mix thoroughly. Pour into sterile Petri dishes or distribute into sterile tubes.

Use: For the isolation of oral treponemes.

Treponema macrodentium Medium

Composition per liter:

Glucose	1.0g
Nicotinamide	0.4g
Spermine·4HCl	0.15g
Sodium isobutyrate	0.02g
Carboxylase	5.0mg
PPLO agar	900.0mL
Bovine serum	100.0mL

pH 7.0 ± 0.2 at 25°C

PPLO Agar:

Composition per 900.0mL:

Beef heart, infusion from	50.0g
Agar	14.0g
Peptone	10.0g
NaCl	5.0g

pH 7.8 ± 0.2 at 25°C

Preparation of PPLO Agar: Add components to distilled/deionized water and bring volume to 900.0mL. Mix thoroughly.

Preparation of Medium: Combine components, except bovine serum. Mix thoroughly. Autoclave for 15 min at 15 psi pressure–121°C. Cool to 45°–50°C. Aseptically add sterile bovine serum. Mix thoroughly. Aseptically distribute into sterile tubes or flasks.

Use: For the isolation and cultivation of *Treponema macrodentium*.

Treponema Medium

Composition per liter:

Pancreatic digest of casein	30.0g
Ionagar No. 2	8.0g
Glucose	5.0g
Yeast extract	5.0g
NaCl	2.5g
Cysteine·HCl	0.75g
Horse serum, inactivated	100.0mL

pH 7.4 ± 0.2 at 25°C

Preparation of Medium: Add components, except horse serum, to distilled/deionized water and bring volume to 900.0mL. Mix thoroughly. Gently heat and bring to boiling. Autoclave for 15 min at 15 psi pressure–121°C. Cool to 45°–50°C. Aseptically add 100.0mL of sterile horse serum. Mix thoroughly. Pour into sterile Petri dishes or distribute into sterile tubes.

Use: For the isolation and cultivation of oral treponemes.

Treponema Medium

Composition per liter:

Spirolate agar	900.0mL
Rabbit serum, inactivated at 56°C for 30 min	100.0mL

Spirolate Agar:

Composition per liter:

Pancreatic digest of casein	15.0g
Agar	14.0g
Glucose	5.0g
Yeast extract	5.0g
NaCl	2.5g
L-Cysteine·HCl·H_2O	1.0g
Sodium thioglycolate	0.5g
Palmitic acid	0.05g
Stearic acid	0.05g
Oleic acid	0.05g
Linoleic acid	0.05g

Preparation of Spirolate Agar: Add components to distilled/deionized water and bring volume to 1.0L. Mix thoroughly. Gently heat and bring to boiling. Autoclave for 15 min at 15 psi pressure–121°C. Cool to 45°–50°C.

Preparation of Medium: To 900.0mL of cooled, sterile spirolate agar, aseptically add 100.0mL of rabbit serum. Mix thoroughly. Aseptically distribute into sterile tubes or flasks.

Use: For the isolation of oral treponemes.

Treponema Medium

Composition per liter:

Spirolate agar	675.0mL
Brain heart infusion broth	225.0mL
Rabbit serum, inactivated at 56°C for 30 min	100.0mL

pH 7.0–7.2 ± 0.2 at 25°C

Spirolate Agar:

Composition per 675.0mL:

Pancreatic digest of casein	15.0g
Ionagar No. 2	8.0g
Glucose	5.0g
Yeast extract	5.0g
NaCl	2.5g
L-Cysteine·HCl·H_2O	1.0g
Sodium thioglycolate	0.5g
Palmitic acid	0.05g
Stearic acid	0.05g
Oleic acid	0.05g
Linoleic acid	0.05g

Preparation of Spirolate Agar: Add components to distilled/deionized water and bring volume to 675.0mL. Mix thoroughly. Autoclave for 15 min at 15 psi pressure–121°C. Cool to 45°–50°C.

Brain Heart Infusion Broth:

Composition per liter:

Pancreatic digest of gelatin	14.5g
Brain heart, solids from infusion	6.0g

Peptic digest of animal tissue	6.0g
NaCl	5.0g
Casein	5.0g
Glucose	3.0g
Na_2HPO_4	2.5g

Preparation of Brain Heart Infusion Broth: Add components to distilled/deionized water and bring volume to 1.0L. Mix thoroughly.

Preparation of Medium: Aseptically combine 675.0mL of cooled, sterile spirolate agar, 225.0mL of cooled, sterile brain heart infusion broth, and 100.0mL of rabbit serum. Mix thoroughly. Pour into sterile Petri dishes or distribute into sterile tubes.

Use: For the isolation of oral treponemes.

Treponema Medium

Composition per liter:

Solution A	440.0mL
Spirolate broth	440.0mL
Rabbit serum, inactivated at 56°C for 30 min	100.0mL
Mucin solution	20.0mL

pH 7.8 ± 0.2 at 25°C

Solution A:

Composition per 440.0mL:

Ionagar No. 2	8.0g
Brain heart infusion broth	440.0mL

Brain Heart Infusion Broth:

Composition per liter:

Pancreatic digest of gelatin	14.5g
Brain heart, solids from infusion	6.0g
Peptic digest of animal tissue	6.0g
NaCl	5.0g
Casein	5.0g
Glucose	3.0g
Na_2HPO_4	2.5g

Preparation of Brain Heart Infusion Broth: Add components to distilled/deionized water and bring volume to 1.0L. Mix thoroughly.

Preparation of Solution A: Add 8.0g of ionagar to 440.0mL of brain heart infusion broth. Mix thoroughly. Gently heat and bring to boiling. Autoclave for 15 min at 15 psi pressure–121°C. Cool to 45°–50°C.

Spirolate Broth:

Composition per liter:

Pancreatic digest of casein	15.0g
Glucose	5.0g
Yeast extract	5.0g
NaCl	2.5g
L-Cysteine·HCl·H_2O	1.0g
Sodium thioglycolate	0.5g
Palmitic acid	0.05g
Stearic acid	0.05g
Oleic acid	0.05g
Linoleic acid	0.05g

Preparation of Spirolate Broth: Add components to distilled/deionized water and bring volume to 1.0L. Mix thoroughly. Autoclave for 15 min at 15 psi pressure–121°C. Cool to 25°.

Mucin Solution:

Composition per 20.0mL:

Mucin	0.2g

Preparation of Mucin Solution: Add mucin to distilled/deionized water and bring volume to 20.0mL. Mix thoroughly. Filter sterilize.

Preparation of Medium: Aseptically combine 440.0mL of solution A, 440.0mL of spirolate broth, 100.0mL of rabbit serum, and 20.0mL of mucin solution. Mix thoroughly. Aseptically distribute into sterile tubes or flasks.

Use: For the isolation of intestinal treponemes.

Treponema Medium

Composition per liter:

Agar	13.0g
Glucose	1.4g
Cysteine·HCl	0.64g
$(NH_4)_2SO_4$	0.5g
Polypeptone™	0.5g
Starch, soluble	0.5g
Yeast extract	0.5g
Resazurin	1.6mg
Salts solution	500.0mL
Bovine rumen fluid	280.0mL

pH 7.2–7.5 at 25°C

Salts Solution:

Composition per liter:

$NaHCO_3$	10.0g
NaCl	2.0g
K_2HPO_4	1.0g
KH_2PO_4	1.0g
$CaCl_2$	0.2g
$MgSO_4$	0.2g
CoCl	3.4mg
$MnSO_4$	3.4mg
$NaMoO_4$	3.4mg

Preparation of Salts Solution: Add components to distilled/deionized water and bring volume to 1.0L. Mix thoroughly.

Preparation of Medium: Add components, except bovine rumen fluid, to distilled/deionized water and bring volume to 720.0mL. Mix thoroughly. Gently heat and bring to boiling. Autoclave for 15 min at 15 psi pressure–121°C. Cool to 45°–50°C. Aseptically add bovine rumen fluid. Mix thoroughly. Pour into sterile Petri dishes or distribute into sterile tubes.

Use: For the isolation of intestinal treponemes.

Treponema Medium

Composition per liter:

Cysteine·HCl·H_2O	1.0g
Glucose	1.0g
Nicotinamide	0.4g
Spermidine·4HCl	0.15g
Sodium isobutyrate	0.02g
Thiamine pyrophosphate	5.0mg
PPLO broth	900.0mL
Rabbit serum, inactivated	100.0mL

pH 7.8 ± 0.2 at 25°C

PPLO Broth:

Composition per 900.0mL:

Beef heart, infusion from solids	50.0g
Peptone	10.0g
NaCl	5.0g

Preparation of PPLO Broth: Add components to distilled/deionized water and bring volume to 900.0mL. Mix thoroughly.

Preparation of Medium: Combine components, except rabbit serum. Mix thoroughly. Filter sterilize. Aseptically add sterile rabbit serum. Mix thoroughly. Aseptically distribute into sterile tubes or flasks.

Use: For the cultivation of oral treponemes. For the cultivation of *Treponema denticola, Treponema macrodentium,* and *Treponema oralis*.

Treponema Medium

Composition per liter:

Spirolate broth	675.0mL
Brain heart infusion broth	225.0mL
Rabbit serum	100.0mL

Spirolate Broth:
Composition per liter:

Pancreatic digest of casein	15.0g
Glucose	5.0g
Yeast extract	5.0g
NaCl	2.5g
L-Cysteine·HCl·H_2O	1.0g
Sodium thioglycolate	0.5g
Palmitic acid	0.05g
Stearic acid	0.05g
Oleic acid	0.05g
Linoleic acid	0.05g

Preparation of Spirolate Broth: Add components to distilled/deionized water and bring volume to 1.0L. Mix thoroughly. Autoclave for 15 min at 15 psi pressure–121°C. Cool to 25°C.

Brain Heart Infusion Broth:
Composition per liter:

Pancreatic digest of gelatin	14.5g
Brain heart, solids from infusion	6.0g
Peptic digest of animal tissue	6.0g
NaCl	5.0g
Casein	5.0g
Glucose	3.0g
Na_2HPO_4	2.5g

Preparation of Brain Heart Infusion Broth: Add components to distilled/deionized water and bring volume to 1.0L. Mix thoroughly. Autoclave for 15 min at 15 psi pressure–121°C. Cool to 25°C.

Preparation of Medium: Aseptically combine 675.0mL of cooled, sterile spirolate broth, 225.0mL of cooled, sterile brain heart infusion broth, and 100.0mL of rabbit serum. Mix thoroughly.

Use: For the cultivation of oral treponemes.

Treponema Medium

Composition per liter:

Heart infusion broth, modified	450.0mL
Spirolate broth	450.0mL
Rabbit serum, inactivated	100.0mL

pH 7.4 ± 0.2 at 25°C

Heart Infusion Broth, Modified:
Composition per liter:

Beef heart, solids from infusion	500.0g
Tryptose	10.0g
NaCl	5.0g
Asparagine	2.5g
Sodium thioglycolate	2.5g
Pancreatic digest of casein	2.5g

Preparation of Heart Infusion Broth, Modified: Add components to distilled/deionized water and bring volume to 1.0L. Mix thoroughly. Gently heat and bring to boiling. Autoclave for 15 min at 15 psi pressure–121°C. Cool to 25°C.

Spirolate Broth:
Composition per liter:

Pancreatic digest of casein	15.0g
Glucose	5.0g
Yeast extract	5.0g
NaCl	2.5g
L-Cysteine·HCl·H_2O	1.0g
Sodium thioglycolate	0.5g
Palmitic acid	0.05g
Stearic acid	0.05g
Oleic acid	0.05g
Linoleic acid	0.05g

Preparation of Spirolate Broth: Add components to distilled/deionized water and bring volume to 1.0L. Mix thoroughly. Autoclave for 15 min at 15 psi pressure–121°C. Cool to 25°C.

Preparation of Medium: Aseptically combine 450.0mL of cooled, sterile spirolate broth, 450.0mL of cooled, sterile heart infusion broth, modified, and 100.0mL of rabbit serum. Mix thoroughly. Aseptically distribute into sterile tubes or flasks.

Use: For the cultivation of treponemes.

Treponema Medium

Composition per 500.0mL:

Beef heart, solids from infusion	250.0g
Sucrose	50.0g
Tryptose	5.0g
NaCl	2.5g
Yeast extract	2.5g
Agar	0.5g
Sodium thioglycolate	0.38g
$MgSO_4$	0.05g
Horse serum, inactivated	100.0mL

pH 7.4 ± 0.2 at 25°C

Preparation of Medium: Add components, except horse serum, to distilled/deionized water and bring volume to 400.0mL. Mix thoroughly. Adjust pH to 7.4. Gently heat and bring to boiling. Distribute into tubes in 4.0mL volumes. Autoclave for 15 min at 15 psi pressure–121°C. Cool to 25°C. Prior to inoculation, add 1.0mL sterile horse serum to each tube.

Use: For the cultivation and maintenance of *Treponema pallidum* and other *Treponema* species.

Treponema Medium 1

Composition per liter:

Thioglycolate agar USP, alternate	900.0mL
Normal calf serum	100.0mL

pH 7.1 ± 0.2 at 25°C

Thioglycolate Agar USP, Alternate:
Composition per 900.0mL:

Pancreatic digest of casein	15.0g
Ionagar No. 2	7.0g

Glucose	5.5g
Yeast extract	5.0g
NaCl	2.5g
L-Cystine	0.5g
Sodium thioglycolate	0.5g

Preparation of Thioglycolate Agar USP, Alternate: Add components to distilled/deionized water and bring volume to 900.0mL. Mix thoroughly. Autoclave for 15 min at 15 psi pressure–121°C. Cool to 45°–50°C.

Preparation of Medium: Aseptically combine 900.0mL of cooled sterile thioglycolate agar USP, alternate, and 100.0mL of calf serum. Mix thoroughly. Pour into sterile Petri dishes or distribute into sterile tubes.

Use: For the cultivation of treponemes.

Treponema Medium 2

Composition per liter:

Pancreatic digest of casein	30.0g
Ionagar No. 2	7.0g
Glucose	5.0g
Yeast extract	5.0g
NaCl	2.5g
L-Cysteine·HCl·H_2O	2.0g
Rabbit serum	100.0mL

pH 7.2 ± 0.2 at 25°C

Preparation of Medium: Add components, except rabbit serum, to distilled/deionized water and bring volume to 900.0mL. Mix thoroughly. Gently heat and bring to boiling. Autoclave for 15 min at 15 psi pressure–121°C. Cool to 45°–50°C. Aseptically add sterile rabbit serum. Mix thoroughly. Pour into sterile Petri dishes or distribute into sterile tubes.

Use: For the cultivation of treponemes.

Treponema Medium 3

Composition per liter:

Spirolate agar	675.0mL
Brain heart infusion broth	225.0mL
Rabbit serum	100.0mL

Spirolate Agar:

Composition per liter:

Pancreatic digest of casein	15.0g
Ionagar No. 2	7.0g
Glucose	5.0g
Yeast extract	5.0g
NaCl	2.5g
L-Cysteine·HCl·H_2O	1.0g
Sodium thioglycolate	0.5g
Palmitic acid	0.05g
Stearic acid	0.05g
Oleic acid	0.05g
Linoleic acid	0.05g

Preparation of Spirolate Agar: Add components to distilled/deionized water and bring volume to 1.0L. Mix thoroughly. Gently heat and bring to boiling. Autoclave for 15 min at 15 psi pressure–121°C. Cool to 25°C.

Brain Heart Infusion Broth:

Composition per liter:

Pancreatic digest of gelatin	14.5g
Brain heart, solids from infusion	6.0g
Peptic digest of animal tissue	6.0g
NaCl	5.0g
Casein	5.0g
Glucose	3.0g
Na_2HPO_4	2.5g

Preparation of Brain Heart Infusion Broth: Add components to distilled/deionized water and bring volume to 1.0L. Mix thoroughly. Gently heat and bring to boiling. Autoclave for 15 min at 15 psi pressure–121°C. Cool to 25°C.

Preparation of Medium: Aseptically combine 675.0mL of cooled, sterile spirolate broth, 225.0mL of cooled, sterile brain heart infusion broth, and 100.0mL of rabbit serum. Mix thoroughly.

Use: For the cultivation of treponemes.

Treponema Medium, Prereduced

Composition per liter:

Agar	1.6g
Glucose	1.4g
Cysteine·HCl·H_2O	0.64g
$(NH_4)_2SO_4$	0.5g
Polypeptone™	0.5g
Starch, soluble	0.5g
Yeast extract	0.5g
Resazurin	1.6mg
Salts solution	500.0mL
Bovine rumen fluid	280.0mL

pH 7.2–7.5 at 25°C

Salts Solution:

Composition per liter:

$NaHCO_3$	10.0g
NaCl	2.0g
K_2HPO_4	1.0g
KH_2PO_4	1.0g
$CaCl_2$	0.2g
$MgSO_4$	0.2g
CoCl	3.4mg
$MnSO_4$	3.4mg
$NaMoO_4$	3.4mg

Preparation of Salts Solution: Add components to distilled/deionized water and bring volume to 1.0L. Mix thoroughly.

Preparation of Medium: Add components, except bovine rumen fluid, to distilled/deionized water and bring volume to 720.0mL. Mix thoroughly. Gently heat and bring to boiling. Autoclave for 15 min at 15 psi pressure–121°C. Cool to 45°–50°C. Aseptically add 280.0mL of sterile bovine rumen fluid. Mix thoroughly. Aseptically and anaerobically distribute into sterile tubes or flasks under 100% N_2.

Use: For the cultivation of fecal and intestinal treponemes.

Treponema saccharophilum Medium

Composition per liter:

Pancreatic digest of casein	2.0g
Yeast extract	2.0g
L-Cysteine·HCl	1.0g
Resazurin	1.0mg
Salt solution A	200.0mL
Salt solution B	200.0mL
$NaHCO_3$ solution	100.0mL
Glucose solution	20.0mL

iso-Butyric acid......0.4mL
n-Butyric acid......0.4mL
DL-2-Methylbutyric acid......0.2mL
iso-Valeric acid......0.2mL
n-Valeric acid......0.2mL

pH 6.7–7.0 at 25°C

Salt Solution A:

Composition per liter:

$MgSO_4 \cdot 7H_2O$......0.96g
$CaCl_2 \cdot 2H_2O$......0.59g

Preparation of Salt Solution A: Add components to distilled/deionized water and bring volume to 1.0L. Mix thoroughly.

Salt Solution B:

Composition per liter:

K_2HPO_4......2.25g
KH_2PO_4......2.25g
NaCl......4.5g
$(NH_4)_2SO_4$......4.5g

Preparation of Salt Solution B: Add components to distilled/deionized water and bring volume to 1.0L. Mix thoroughly.

$NaHCO_3$ Solution:

Composition per 100.0mL:

$NaHCO_3$......5.0g

Preparation of $NaHCO_3$ Solution: Add $NaHCO_3$ to distilled/deionized water and bring volume to 100.0mL. Mix thoroughly. Filter sterilize. Sparge with 100% CO_2.

Glucose Solution:

Composition per 20.0mL:

Glucose......2.0g

Preparation of Glucose Solution: Add glucose to distilled/deionized water and bring volume to 20.0mL. Mix thoroughly. Filter sterilize. Sparge with 100% CO_2.

Preparation of Medium: Prepare and dispense medium under 100% CO_2. Add components, except $NaHCO_3$ solution and glucose solution, to distilled/deionized water and bring volume to 880.0mL. Mix thoroughly. Adjust pH to 7.0 with KOH. Gently heat and bring to boiling. Continue boiling for 5 min. Cool to room temperature while sparging with 100% CO_2. Anaerobically distribute 8.8mL into anaerobic culture tubes. Autoclave for 15 min at 15 psi pressure–121°C. Using a syringe, add 1.0mL of sterile $NaHCO_3$ solution and 0.2 mL of sterile glucose solution to each tube. Check that final pH is 6.7–7.0.

Use: For the cultivation and maintenance of *Treponema saccharophilum.*

Treponema succinifaciens Medium (DSMZ Medium 275)

Composition per liter:

Solution A......875.0mL
Solution B......50.0mL
Solution D......50.0mL
Solution C......25.0mL

pH 7.0 ± 0.2 at 25°C

Solution A:

Composition per 875.0mL:

NaCl......1.0g
K_2HPO_4......0.5g
Cysteine-HCl·H_2O......0.5g
KH_2PO_4......0.5g
Yeast extract......0.5g
Peptone......0.5g
$(NH_4)_2SO_4$......0.5g
$CaCl_2 \cdot 2H_2O$......0.1g
$MgSO_4 \cdot 7H_2O$......0.1g
Resazurin......0.001g
Rumen fluid, clarified......300.0mL

Preparation of Solution A: Add components to 575.0mL of distilled/deionized water. Mix thoroughly. Adjust pH to 6.2–6.3 with 4*N* HCl. Heat to boiling point. Cool to room temperature under 100% N_2. Autoclave for 15 min at 15 psi pressure–121°C.

Solution B:

Composition per 100.0mL:

Glucose......20.0g

Preparation of Solution B: Add glucose to 100.0mL of distilled/deionized water. Mix thoroughly. Sparge with 100% N_2. Autoclave for 15 min at 15 psi pressure–121°C.

Solution C:

Composition per 100.0mL:

KH_2PO_4......0.45g
$Na_2HPO_4 \cdot 2H_2O$......0.58g

Preparation of Solution C: Add components to distilled/deionized water and bring volume to 1.0L. Mix thoroughly. Adjust pH to 7.4. Sparge with 100% N_2. Autoclave for 15 min at 15 psi pressure–121°C.

Solution D:

Composition per 100.0mL:

$NaHCO_3$......5.0g

Preparation of Solution D: Add $NaHCO_3$ to distilled/deionized water and bring volume to 100.0mL. Mix thoroughly. Sparge with 80% N_2 + 20% CO_2. Filter sterilize.

Preparation of Medium: Distribute solution A under 100% N_2 into anaerobic tubes. Autoclave for 20 min at 15 psi pressure–121°C. Cool to 25°C. Aseptically and anaerobically add appropriate amounts of solutions B, C, and D to achieve final concentrations.

Use: For the cultivation of *Treponema succinifaciens.*

Tributyrin Agar

Composition per liter:

Agar......15.0g
Tributyrin (glyceryl tributyrate)......10.0g
Peptone......5.0g
Yeast extract......3.0g

pH 7.5 ± 0.2 at 25°C

Source: This medium is available as a prepared medium from Oxoid Unipath.

Preparation of Medium: Add components to distilled/deionized water and bring volume to 1.0L. Mix thoroughly. Gently heat and bring to boiling. Distribute into tubes or flasks. Autoclave for 15 min at 15 psi pressure–121°C. Pour into sterile Petri dishes.

Use: For the cultivation and enumeration of lipolytic fungi and bacteria, especially *Staphylococcus* species, *Flavobacterium* species, *Clostridium* species, and *Pseudomonas* species from butter. Lipolytic bacteria appear as colonies surrounded by a clear zone.

Tributyrin HiVeg Agar Base with Tributyrin

Composition per liter:

Agar	15.0g
Plant peptone	5.0g
Yeast extract	3.0g
Tributyrin (glyceryl tributyrate)	10.0mL

pH 7.5 ± 0.2 at 25°C

Source: This medium, without tributyrin, is available as a premixed powder from HiMedia.

Preparation of Medium: Add components to distilled/deionized water and bring volume to 990.0mL. Add 10.0mL of tributyrin. Mix thoroughly. Gently heat and bring to boiling. Distribute into tubes or flasks. Autoclave for 15 min at 15 psi pressure–121°C. Cool to 45°–50°C. Mix thoroughly. Pour into sterile Petri dishes or leave in tubes.

Use: For the detection of lipolytic microorganisms.

Trichlorophenol Medium

Composition per liter:

Pancreatic digest of casein	8.5g
NaCl	2.5g
Papaic digest of soybean meal	1.5g
K_2HPO_4	1.25g
Glucose	1.25g
2,4,6-Trichlorophenol	1.25g

pH 7.3 ± 0.2 at 25°C

Preparation of Medium: Add components to distilled/deionized water and bring volume to 1.0L. Mix thoroughly. Gently heat until dissolved. Distribute into tubes or flasks. Autoclave for 15 min at 15 psi pressure–121°C.

Use: For the cultivation and maintenance of *Arthrobacter* species and other microorganisms that can degrade chlorinated phenols.

Trichococcus Medium

Composition per liter:

Pancreatic digest of casein	10.0g
Na_2SO_4	4.0g
$MgCl_2 \cdot 6H_2O$	1.1g
KCl	0.7g
Glucose solution	30.0mL

pH 7.3 ± 0.2 at 25°C

Glucose Solution:

Composition per 30.0mL:

Glucose	3.0g

Preparation of Glucose Solution: Add glucose to distilled/deionized water and bring volume to 30.0mL. Mix thoroughly. Filter sterilize.

Preparation of Medium: Add components, except glucose solution, to distilled/deionized water and bring volume to 970.0mL. Mix thoroughly. Autoclave for 15 min at 15 psi pressure–121°C. Aseptically add 30.0mL of sterile glucose solution. Mix thoroughly. Aseptically distribute into sterile tubes or flasks.

Use: For the cultivation of *Trichococcus flocculiformis.*

Trichomonas HiVeg Agar Base with Serum

Composition per liter:

Plant extract No. 2	25.0g
NaCl	6.5g
Glucose	5.0g
Agar	1.0g
Horse serum	80.0mL

pH 6.4 ± 0.2 at 25°C

Source: This medium, without horse serum, is available as a premixed powder from HiMedia.

Horse Serum:

Composition per 80.0mL:

Horse serum	80.0mL

Preparation of Horse Serum: Gently heat sterile horse serum to 56°C for 30 min. Aseptically adjust pH to 6.0 with 0.1*N* HCl. Use immediately.

Preparation of Medium: Add components, except horse serum, to distilled/deionized water and bring volume to 920.0mL. Mix thoroughly. Gently heat and bring to boiling. Autoclave for 15 min at 15 psi pressure–121°C. Cool to 45°–50°C. Aseptically add 80.0mL of freshly prepared sterile horse serum. Mix thoroughly. Aseptically distribute into sterile tubes or flasks.

Use: For the cultivation of *Trichomonas vaginalis.*

Trichomonas HiVeg Agar Base with Serum and Selective Supplement

Composition per liter:

Plant extract No. 2	25.0g
NaCl	6.5g
Glucose	5.0g
Agar	1.0g
Horse serum	80.0mL
Selective supplement	10.0mL

pH 6.4 ± 0.2 at 25°C

Source: This medium, without horse serum or selective supplement, is available as a premixed powder from HiMedia.

Horse Serum:

Composition per 80.0mL:

Horse serum	80.0mL

Preparation of Horse Serum: Gently heat sterile horse serum to 56°C for 30 min. Aseptically adjust pH to 6.0 with 0.1*N* HCl. Use immediately.

Selective Supplement:

Composition per 10.0mL:

Streptomycin	0.5g
Penicllin	1,000,000U

Preparation of Selective Supplement: Add components to distilled/deionized water and bring volume to 10.0mL. Mix thoroughly. Filter sterilize.

Preparation of Medium: Add components, except horse serum and selective supplement, to distilled/deionized water and bring volume to 910.0mL. Mix thoroughly. Gently heat and bring to boiling. Autoclave for 15 min at 15 psi pressure–121°C. Cool to 45°–50°C. Aseptically add 80.0mL of freshly prepared sterile horse serum and 10.0mL sterile selective supplement. Mix thoroughly. Aseptically distribute into sterile tubes or flasks.

Use: For the cultivation of *Trichomonas vaginalis*.

Trichomonas Medium

Composition per liter:

Liver digest	25.0g
NaCl	6.5g
Glucose	5.0g
Agar	1.0g
Horse serum	80.0mL

pH 6.4 ± 0.2 at 25°C

Source: This medium is available as a premixed powder from Oxoid Unipath.

Horse Serum:

Composition per 80.0mL:

Horse serum	80.0mL

Preparation of Horse Serum: Gently heat sterile horse serum to 56°C for 30 min. Aseptically adjust pH to 6.0 with 0.1*N* HCl. Use immediately.

Preparation of Medium: Add components, except horse serum, to distilled/deionized water and bring volume to 920.0mL. Mix thoroughly. Gently heat and bring to boiling. Autoclave for 15 min at 15 psi pressure–121°C. Cool to 45°–50°C. Aseptically add 80.0mL of freshly prepared sterile horse serum. Mix thoroughly. Aseptically distribute into sterile tubes or flasks.

Use: For the cultivation of *Trichomonas vaginalis*.

Trichomonas Medium No. 2

Composition per liter:

Glucose	22.5g
Liver digest	18.0g
Pancreatic digest of casein	17.0g
NaCl	5.0g
Pancreatic digest of soybean meal	3.0g
K_2HPO_4	2.5g
Chloramphenicol	0.125g
Horse serum	250.0mL
Calcium pantothenate (0.5% solution)	1.0mL

pH 6.2 ± 0.2 at 25°C

Source: This medium is available as a prepared medium from Oxoid Unipath.

Preparation of Medium: Add components, except horse serum, to distilled/deionized water and bring volume to 750.0mL. Mix thoroughly. Autoclave for 15 min at 5 psi pressure–108°C. Cool to 45°–50°C. Aseptically add 250.0mL of sterile horse serum. Mix thoroughly. Aseptically distribute into sterile tubes or flasks.

Use: For the isolation of *Trichomonas vaginalis*.

Trichomonas Selective Medium

Composition per liter:

Liver digest	25.0g
NaCl	6.5g
Glucose	5.0g
Agar	1.0g
Horse serum	80.0mL
Antibiotic inhibitor	10.0mL

pH 6.4 ± 0.2 at 25°C

Source: This medium is available as a premixed powder from Oxoid Unipath.

Horse Serum:

Composition per 80.0mL:

Horse serum	80.0mL

Preparation of Horse Serum: Gently heat sterile horse serum to 56°C for 30 min. Aseptically adjust pH to 6.0 with 0.1*N* HCl. Use immediately.

Antibiotic Inhibitor:

Composition per 10.0mL:

Streptomycin	500.0mg
Penicillin G	1,000,000U

Preparation of Antibiotic Inhibitor: Add components to distilled/deionized water and bring volume to 10.0mL. Mix thoroughly. Filter sterilize.

Preparation of Medium: Add components, except horse serum and antibiotic inhibitor, to distilled/deionized water and bring volume to 910.0mL. Mix thoroughly. Gently heat and bring to boiling. Autoclave for 15 min at 15 psi pressure–121°C. Cool to 45°–50°C. Aseptically add 80.0mL of freshly prepared sterile horse serum and 10.0mL of sterile antibiotic inhibitor. Mix thoroughly. Aseptically distribute into sterile tubes or flasks.

Use: For the cultivation of *Trichomonas vaginalis* from specimens with a mixed bacterial flora.

Trichomonas Selective Medium

Composition per liter:

Liver digest	25.0g
NaCl	6.5g
Glucose	5.0g
Agar	1.0g
Horse serum	80.0mL
Antibiotic inhibitor	10.0mL

pH 6.4 ± 0.2 at 25°C

Horse Serum:

Composition per 80.0mL:

Horse serum	80.0mL

Preparation of Horse Serum: Gently heat sterile horse serum to 56°C for 30 min. Aseptically adjust pH to 6.0 with 0.1*N* HCl. Use immediately.

Antibiotic Inhibitor:

Composition per 10.0mL:

Chloramphenicol	100.0mg

Preparation of Antibiotic Inhibitor: Add chloramphenicol to distilled/deionized water and bring volume to 10.0mL. Mix thoroughly. Filter sterilize.

Preparation of Medium: Add components, except horse serum and antibiotic inhibitor, to distilled/deionized water and bring volume to 910.0mL. Mix thoroughly. Gently heat and bring to boiling. Autoclave for 15 min at 15 psi pressure–121°C. Cool to 45°–50°C. Aseptically add 80.0mL of freshly prepared sterile horse serum and 10.0mL of sterile antibiotic inhibitor. Mix thoroughly. Aseptically distribute into sterile tubes or flasks.

Use: For the cultivation of *Trichomonas vaginalis* from specimens with a mixed bacterial flora.

Trichophyton Agar 1

Composition per liter:

Glucose	40.0g
Agar	15.0g

Vitamin assay casamino acids 2.5g
KH_2PO_4 .... 1.8g
$MgSO_4 \cdot 7H_2O$ 0.1g
pH 6.8 ± 0.2 at 25°C

Source: This medium is available as a premixed powder from BD Diagnostic Systems.

Preparation of Medium: Add components to distilled/deionized water and bring volume to 1.0L. Mix thoroughly. Gently heat and bring to boiling. Distribute into tubes. Autoclave for 15 min at 15 psi pressure–121°C. Allow tubes to cool in a slanted position.

Use: For the differentiation of the *Trichophyton* species.

Trichophyton Agar 2

Composition per liter:
Glucose 40.0g
Agar 15.0g
Vitamin assay casamino acids 2.5g
KH_2PO_4 .... 1.8g
$MgSO_4 \cdot 7H_2O$ 0.1g
Inositol 50.0mg
pH 6.8 ± 0.2 at 25°C

Source: This medium is available as a premixed powder from BD Diagnostic Systems.

Preparation of Medium: Add components to distilled/deionized water and bring volume to 1.0L. Mix thoroughly. Gently heat and bring to boiling. Distribute into tubes. Autoclave for 15 min at 15 psi pressure–121°C. Allow tubes to cool in a slanted position.

Use: For the differentiation of the *Trichophyton* species.

Trichophyton Agar 3

Composition per liter:
Glucose 40.0g
Agar 15.0g
Vitamin assay casamino acids 2.5g
KH_2PO_4 .... 1.8g
$MgSO_4 \cdot 7H_2O$ 0.1g
Inositol 0.05g
Thiamine·HCl 0.2mg
pH 6.8 ± 0.2 at 25°C

Source: This medium is available as a premixed powder from BD Diagnostic Systems.

Preparation of Medium: Add components to distilled/deionized water and bring volume to 1.0L. Mix thoroughly. Gently heat and bring to boiling. Distribute into tubes. Autoclave for 15 min at 15 psi pressure–121°C. Allow tubes to cool in a slanted position.

Use: For the differentiation of the *Trichophyton* species.

Trichophyton Agar 4

Composition per liter:
Glucose 40.0g
Agar 15.0g
Vitamin assay casamino acids 2.5g
KH_2PO_4 .... 1.8g
$MgSO_4 \cdot 7H_2O$ 0.1g
Thiamine·HCl USP 200.0μg
pH 6.8 ± 0.2 at 25°C

Source: This medium is available as a premixed powder from BD Diagnostic Systems.

Preparation of Medium: Add components to distilled/deionized water and bring volume to 1.0L. Mix thoroughly. Gently heat and bring to boiling. Distribute into tubes. Autoclave for 15 min at 15 psi pressure–121°C. Allow tubes to cool in a slanted position.

Use: For the differentiation of the *Trichophyton* species.

Trichophyton Agar 5

Composition per liter:
Glucose 40.0g
Agar 15.0g
Vitamin assay casamino acids 2.5g
KH_2PO_4 .... 1.8g
$MgSO_4 \cdot 7H_2O$ 0.1g
Nicotinic acid 2.0mg
pH 6.8 ± 0.2 at 25°C

Source: This medium is available as a premixed powder from BD Diagnostic Systems.

Preparation of Medium: Add components to distilled/deionized water and bring volume to 1.0L. Mix thoroughly. Gently heat and bring to boiling. Distribute into tubes. Autoclave for 15 min at 15 psi pressure–121°C. Allow tubes to cool in a slanted position.

Use: For the differentiation of the *Trichophyton* species.

Trichophyton Agar 6

Composition per liter:
Glucose 40.0g
Agar 15.0g
KH_2PO_4 .... 1.8g
Ammonium nitrate 1.5g
$MgSO_4 \cdot 7H_2O$ 0.1g
pH 6.8 ± 0.2 at 25°C

Source: This medium is available as a premixed powder from BD Diagnostic Systems.

Preparation of Medium: Add components to distilled/deionized water and bring volume to 1.0L. Mix thoroughly. Gently heat and bring to boiling. Distribute into tubes. Autoclave for 15 min at 15 psi pressure–121°C. Allow tubes to cool in a slanted position.

Use: For the differentiation of the *Trichophyton* species.

Trichophyton Agar 7

Composition per liter:
Glucose 40.0g
Agar 15.0g
KH_2PO_4 .... 1.8g
Ammonium nitrate 1.5g
$MgSO_4 \cdot 7H_2O$ 0.1g
Histidine·HCl 0.03g
pH 6.8 ± 0.2 at 25°C

Source: This medium is available as a premixed powder from BD Diagnostic Systems.

Preparation of Medium: Add components to distilled/deionized water and bring volume to 1.0L. Mix thoroughly. Gently heat and bring to boiling. Distribute into tubes. Autoclave for 15 min at 15 psi pressure–121°C. Allow tubes to cool in a slanted position.

Use: For the differentiation of the *Trichophyton* species.

Trichophyton HiVeg Agar 1

Composition per liter:

Glucose	40.0g
Agar	15.0g
Vitamin-free casein enzymic hydrolysate	2.5g
KH_2PO_4	1.8g
$MgSO_4$	0.1g

pH 6.8 ± 0.2 at 25°C

Source: This medium is available as a premixed powder from HiMedia.

Preparation of Medium: Add components to distilled/deionized water and bring volume to 1.0L. Mix thoroughly. Gently heat and bring to boiling. Distribute into tubes. Autoclave for 15 min at 15 psi pressure–121°C. Allow tubes to cool in a slanted position.

Use: For the differentiation of the *Trichophyton* species.

Trichophyton HiVeg Agar 2

Composition per liter:

Glucose	40.0g
Agar	15.0g
Vitamin-free plant hydrolysate	2.5g
KH_2PO_4	1.8g
$MgSO_4$	0.1g
Inositol	5.0mg

pH 6.8 ± 0.2 at 25°C

Source: This medium is available as a premixed powder from HiMedia.

Preparation of Medium: Add components to distilled/deionized water and bring volume to 1.0L. Mix thoroughly. Gently heat and bring to boiling. Distribute into tubes. Autoclave for 15 min at 15 psi pressure–121°C. Allow tubes to cool in a slanted position.

Use: For the differentiation of the *Trichophyton* species.

Trichophyton HiVeg Agar 3

Composition per liter:

Glucose	40.0g
Agar	15.0g
Vitamin-free plant hydrolysate	2.5g
KH_2PO_4	1.8g
$MgSO_4$	0.1g
Inositol	5.0mg
Thiamine	5.0mg

pH 6.8 ± 0.2 at 25°C

Source: This medium is available as a premixed powder from HiMedia.

Preparation of Medium: Add components to distilled/deionized water and bring volume to 1.0L. Mix thoroughly. Gently heat and bring to boiling. Distribute into tubes. Autoclave for 15 min at 15 psi pressure–121°C. Allow tubes to cool in a slanted position.

Use: For the differentiation of the *Trichophyton* species.

Trichophyton HiVeg Agar 4

Composition per liter:

Glucose	40.0g
Agar	15.0g
KH_2PO_4	1.8g
$MgSO_4$	0.1g
Vitamin-free plant hydrolysate	2.5g
Thiamine hydrochloride	0.2mg

pH 6.8 ± 0.2 at 25°C

Source: This medium is available as a premixed powder from HiMedia.

Preparation of Medium: Add components to distilled/deionized water and bring volume to 1.0L. Mix thoroughly. Gently heat and bring to boiling. Distribute into tubes. Autoclave for 15 min at 15 psi pressure–121°C. Allow tubes to cool in a slanted position.

Use: For the differentiation of the *Trichophyton* species.

Trichophyton HiVeg Agar 5

Composition per liter:

Glucose	40.0g
Agar	15.0g
Vitamin-free plant hydrolysate	2.5g
KH_2PO_4	1.8g
$MgSO_4$	0.1g
Nicotinic acid	0.02g

pH 6.8 ± 0.2 at 25°C

Source: This medium is available as a premixed powder from BD Diagnostic Systems.

Preparation of Medium: Add components to distilled/deionized water and bring volume to 1.0L. Mix thoroughly. Gently heat and bring to boiling. Distribute into tubes. Autoclave for 15 min at 15 psi pressure–121°C. Allow tubes to cool in a slanted position.

Use: For the differentiation of the *Trichophyton* species.

Trichosel™ Broth, Modified

Composition per liter:

Pancreatic digest of casein	12.0g
Yeast extract	5.0g
Liver extract	2.0g
Maltose	2.0g
L-Cysteine·HCl	1.0g
Agar	1.0g
Chloramphenicol	0.1g
Methylene Blue	3.0mg
Horse serum	50.0mL

pH 6.0 ± 0.2 at 25°C

Source: This medium is available as a premixed powder from BD Diagnostic Systems.

Preparation of Medium: Add components, except horse serum, to distilled/deionized water and bring volume to 950.0mL. Mix thoroughly. Gently heat while stirring and bring to boiling. Autoclave for 15 min at 13 psi pressure–118°C. Cool to 45°–50°C. Aseptically add 50.0mL of sterile horse serum. Mix thoroughly. Aseptically distribute into sterile tubes or flasks.

Use: For the isolation and cultivation of *Trichomonas* species.

Trimethylamine *N*-Oxide Medium (TMAO Medium)

Composition per liter:

Beef extract	10.0g
Peptone	10.0g
NaCl	5.0g

Agar 2.0g
Trimethylamine *N*-oxide 1.0g
Yeast extract 1.0g

pH 7.5 ± 0.2 at 25°C

Source: This medium is available as a premixed powder from Oxoid Unipath.

Preparation of Medium: Add components to distilled/deionized water and bring volume to 1.0L. Mix thoroughly. Gently heat and bring to boiling. Distribute into screw-capped tubes in 4.0mL volumes. Autoclave for 15 min at 15 psi pressure–121°C. Allow tubes to cool in an upright position.

Use: For the cultivation and differentiation of *Campylobacter* species from foods. *Campylobacter jejuni* and *Campylobacter coli* will not grow.

Triple Sugar Iron Agar (TSI Agar)

Composition per liter:

Peptone 20.0g
Agar 12.0g
Lactose 10.0g
Sucrose 10.0g
NaCl 5.0g
Beef extract 3.0g
Yeast extract 3.0g
Glucose 1.0g
Ferric citrate 0.3g
$Na_2S_2O_3$ 0.3g
Phenol Red 0.025g

pH 7.4 ± 0.2 at 25°C

Source: This medium is available as a premixed powder from BD Diagnostic Systems and Oxoid Unipath.

Preparation of Medium: Add components to distilled/deionized water and bring volume to 1.0L. Mix thoroughly. Gently heat and bring to boiling. Distribute into tubes or flasks. Autoclave for 15 min at 15 psi pressure–121°C. Allow tubes to cool in a slanted position to form a 1.0-inch butt.

Use: For the differentiation of members of the Enterobacteriaceae based on their fermentation of lactose, sucrose, and glucose and the production of H_2S.

Triple Sugar Iron Agar (TSI Agar)

Composition per liter:

Agar 13.0g
Pancreatic digest of casein 10.0g
Peptic digest of animal tissue 10.0g
Lactose 10.0g
Sucrose 10.0g
NaCl 5.0g
Glucose 1.0g
$Fe(NH_4)_2(SO_4)_2{\cdot}6H_2O$ 0.2g
$Na_2S_2O_3$ 0.2g
Phenol Red 0.025g

pH 7.3 ± 0.2 at 25°C

Source: This medium is available as a premixed powder from BD Diagnostic Systems.

Preparation of Medium: Add components to distilled/deionized water and bring volume to 1.0L. Mix thoroughly. Gently heat and bring to boiling. Distribute into tubes or flasks. Autoclave for 15 min at 15 psi pressure–121°C. Allow tubes to cool in a slanted position to form a 1.0-inch butt.

Use: For the differentiation of members of the Enterobacteriaceae based on their fermentation of lactose, sucrose, and glucose and the production of H_2S.

Triple Sugar Iron Agar (TSI Agar) (BAM M149 Medium 2)

Composition per liter:

Peptone 15.0g
Agar 12.0g
Lactose 10.0g
Sucrose 10.0g
Proteose peptone 5.0g
NaCl 5.0g
Beef extract 3.0g
Yeast extract 3.0g
Glucose 1.0g
$Na_2S_2O_3$ 0.3g
$FeSO_4$ 0.2g
Phenol Red 0.024g

pH 7.4 ± 0.2 at 25°C

Preparation of Medium: Add components to distilled/deionized water and bring volume to 1.0L. Mix thoroughly. Gently heat and bring to boiling. Distribute into tubes or flasks. Autoclave for 15 min at 15 psi pressure–121°C. Allow tubes to cool in a slanted position to form a 1.0-inch butt.

Use: For the differentiation of members of the Enterobacteriaceae based on their fermentation of lactose, sucrose, and glucose and the production of H_2S.

Triple Sugar Iron Agar, HiVeg

Composition per liter:

Agar 12.0g
Plant hydrolysate 10.0g
Plant peptone 10.0g
Lactose 10.0g
Sucrose 10.0g
NaCl 5.0g
Plant extract 3.0g
Yeast extract 3.0g
Glucose 1.0g
$Na_2S_2O_3$ 0.3g
$FeSO_4$ 0.2g
Phenol Red 0.024g

pH 7.4 ± 0.2 at 25°C

Source: This medium is available as a premixed powder from HiMedia.

Preparation of Medium: Add components to distilled/deionized water and bring volume to 1.0L. Mix thoroughly. Gently heat and bring to boiling. Distribute into tubes or flasks. Autoclave for 15 min at 15 psi pressure–121°C. Allow tubes to cool in a slanted position to form a 1.0-inch butt.

Use: For the differentiation of members of the Enterobacteriaceae based on their fermentation of lactose, sucrose, and glucose and the production of H_2S.

Tris YP Agar (Tris Yeast Extract Peptone Agar)

Composition per liter:

Agar	19.0g
Yeast extract	3.0g
Glucose	1.0g
Peptone	0.6g
Tris buffer (0.05*M*, pH 7.5)	1.0L

pH 7.5 ± 0.2 at 25°C

Preparation of Medium: Add components to distilled/deionized water and bring volume to 1.0L. For top layer agar, add 6.0g of agar instead of 19.0g. Mix thoroughly. Gently heat and bring to boiling. Distribute into tubes or flasks. Autoclave for 15 min at 15 psi pressure–121°C. Pour into sterile Petri dishes.

Use: For the cultivation and maintenance of *Bdellovibrio* species.

Tris YP Broth (Tris Yeast Extract Peptone Broth)

Composition per liter:

Yeast extract	3.0g
Glucose	1.0g
Peptone	0.6g
Tris buffer (0.05*M*, pH 7.5)	1.0L

pH 7.5 ± 0.2 at 25°C

Preparation of Medium: Add components to distilled/deionized water and bring volume to 1.0L. Mix thoroughly. Distribute into tubes or flasks. Autoclave for 15 min at 15 psi pressure–121°C.

Use: For the cultivation and maintenance of *Bdellovibrio* species.

Trypaflavin Nalidixic Acid Serum Agar (TNSA Agar)

Composition per liter:

Ionagar No. 2	12.0g
Peptone	10.0g
Beef extract	3.0g
H_2O	926.5mL
Bovine serum, heat inactivated	50.0mL
Nalidixic acid solution	20.0mL
Trypaflavin solution	3.5mL

pH 7.2–7.4 at 25°C

Nalidixic Acid Solution:

Composition per 10.0mL:

Nalidixic acid	0.02g

Preparation of Nalidixic Acid Solution: Add nalidixic acid to distilled/deionized water and bring volume to 10.0mL. Mix thoroughly. Filter sterilize.

Trypaflavin Solution:

Composition per 10.0mL:

Trypaflavin	0.1g

Preparation of Trypaflavin Solution: Add trypaflavin to distilled/deionized water and bring volume to 10.0mL. Mix thoroughly. Filter sterilize.

Preparation of Medium: Add components—except bovine serum, nalidixic acid solution, and trypaflavin solution—to distilled/deionized water and bring volume to 926.5mL. Mix thoroughly. Gently heat and bring to boiling. Autoclave for 15 min at 15 psi pressure–121°C. Cool to 45°–50°C. Aseptically add 50.0mL of sterile bovine serum, 20.0mL of sterile nalidixic acid solution, and 3.5mL of trypaflavin solution. Mix thoroughly. Pour into sterile Petri dishes or distribute into sterile tubes.

Use: For the isolation and cultivation of *Listeria* species from preenriched specimens.

Trypanosome Medium

Composition per 1300.0mL:

Solid phase	1.0L
Liquid phase (Locke's solution)	300.0mL

pH 7.2–7.4 at 25°C

Solid Phase:

Composition per liter:

Agar	15.0g
NaCl	8.0g
Peptone	5.0g
Beef extract	3.0g
Rabbit blood, defibrinated	300.0mL

Preparation of Solid Phase: Add components, except rabbit blood, to distilled/deionized water and bring volume to 700.0mL. Mix thoroughly. Adjust pH to 7.2–7.4. Gently heat and bring to boiling. Autoclave for 15 min at 15 psi pressure–121°C. Cool to 50°–55°C. Aseptically add 300.0mL of sterile defibrinated rabbit blood. Mix thoroughly. Distribute 10.0mL aliquots into sterile screw-capped tubes. Allow to solidify in a slanted position.

Liquid Phase (Locke's Solution):

Composition per liter:

NaCl	8.0g
Glucose	2.5g
KH_2PO_4	0.3g
$CaCl_2$	0.2g
KCl	0.2g

Preparation of Liquid Phase (Locke's Solution): Add components to distilled/deionized water and bring volume to 1.0L. Mix thoroughly. Autoclave for 15 min at 15 psi pressure–121°C.

Preparation of Medium: Aseptically overlay agar slants (solid phase) with 3.0mL per tube of sterile liquid phase (Locke's solution).

Use: For the cultivation of *Leishmania donovani*, *Leishmania braziliensis*, *Trypanosoma gambiense*, and *Trypanosoma rhodesiense*.

Tryptic Digest Broth

Composition per liter:

Tryptic digest of beef heart	10.0g
NaCl	5.0g
Glucose	1.0g

pH 7.6 ± 0.2 at 25°C

Source: This medium is available as a premixed powder from BD Diagnostic Systems.

Preparation of Medium: Add components to distilled/deionized water and bring volume to 1.0L. Mix thoroughly. Distribute into tubes or flasks. Autoclave for 15 min at 15 psi pressure–121°C.

Use: For use as a base medium to which enrichments are added. For the cultivation of fastidious microorganisms.

Tryptic Nitrate Medium

Composition per liter:

Tryptose	20.0g
Na_2HPO_4	2.0g
Agar	1.0g
Glucose	1.0g
KNO_3	1.0g

pH 7.6 ± 0.2 at 25°C

Source: This medium is available as a premixed powder from BD Diagnostic Systems.

Preparation of Medium: Add components to distilled/deionized water and bring volume to 1.0L. Mix thoroughly. Gently heat and bring to boiling. Distribute into tubes in 10.0mL volumes. Autoclave for 15 min at 15 psi pressure–121°C.

Use: For the cultivation and differentiation of *Pseudomonas* and related genera. For the differentiation of bacteria based on their reduction of nitrate to nitrite. After incubation of the bacterium in tryptic nitrate medium for 18–24 hr, sulfanillic acid and α-naphthol reagents are added. Nitrate reduction is indicated by the development of a red to violet color.

Tryptic Soy Agar

See: **Trypticase™ Soy Agar**

Tryptic Soy Agar Blood Agar Base

See: **Trypticase™ Soy Agar with Sheep Blood**

Tryptic Soy Agar with Magnesium Ions

Composition per liter:

Pancreatic digest of casein	17.0g
Agar	15.0g
NaCl	5.0g
Pancreatic digest of soybean meal	3.0g
Glucose	2.5g
K_2HPO_4	2.5g
$MgCl_2$	0.95g

pH 7.3 ± 0.2 at 25°C

Preparation of Medium: Add components to tap water and bring volume to 1.0L. Mix thoroughly. Gently heat and bring to boiling. Distribute into tubes or flasks. Autoclave for 15 min at 15 psi pressure–121°C. Pour into sterile Petri dishes or leave in tubes.

Use: For the cultivation of *Escherichia coli* for bacteriophage production.

Tryptic Soy Agar with Magnesium Sulfate

Composition per liter:

Agar	15.0g
Pancreatic digest of casein	15.0g
NaCl	5.0g
Pancreatic digest of soybean meal	5.0g
$MgSO_4 \cdot 7H_2O$	1.5g

pH 7.3 ± 0.2 at 25°C

Preparation of Medium: Add components to distilled/deionized water and bring volume to 1.0L. Mix thoroughly. Gently heat and bring to boiling. Autoclave for 15 min at 15 psi pressure–121°C. Pour into sterile Petri dishes in 20.0mL volumes.

Use: For the cultivation of *Escherichia coli* from foods.

Tryptic Soy Agar with Magnesium Sulfate and Sodium Chloride

Composition per liter:

Pancreatic digest of casein	50.0g
NaCl	30.0g
Agar	15.0g
Pancreatic digest of soybean meal	5.0g
$MgSO_4 \cdot 7H_2O$	1.5g

pH 7.3 ± 0.2 at 25°C

Preparation of Medium: Add components to distilled/deionized water and bring volume to 1.0L. Mix thoroughly. Gently heat and bring to boiling. Autoclave for 15 min at 15 psi pressure–121°C. Pour into sterile Petri dishes in 20.0mL volumes.

Use: For the cultivation of *Vibrio* species from foods.

Tryptic Soy Agar with Sodium Chloride (ATCC Medium 2276)

Composition per liter:

Pancreatic digest of casein	50.0g
NaCl	20.0g
Agar	15.0g
Pancreatic digest of soybean meal	5.0g
$MgSO_4 \cdot 7H_2O$	1.5g

pH 7.3 ± 0.2 at 25°C

Preparation of Medium: Add components to distilled/deionized water and bring volume to 1.0L. Mix thoroughly. Gently heat and bring to boiling. Autoclave for 15 min at 15 psi pressure–121°C. Pour into sterile Petri dishes or leave in tubes.

Use: For the cultivation of *Vibrio* species.

Tryptic Soy Agar with 0.6% Yeast Extract

Composition per liter:

Agar	15.0g
Pancreatic digest of casein	15.0g
Yeast extract	6.0g
Pancreatic digest of soybean meal	5.0g
NaCl	5.0g

pH 7.0–7.5 at 25°C

Source: This medium is available as a premixed powder from BD Diagnostic Systems.

Preparation of Medium: Add components to distilled/deionized water and bring volume to 1.0L. Mix thoroughly. Gently heat and bring to boiling. Distribute into tubes or flasks. Autoclave for 15 min at 15 psi pressure–121°C. Pour into sterile Petri dishes or leave in tubes.

Use: For the isolation and cultivation of *Listeria monocytogenes* from foods.

Tryptic Soy Blood Agar

See: **Trypticase™ Soy Agar with Sheep Blood**

Tryptic Soy Blood Agar with VAN 4

See: **Trypticase™ Soy Agar with Sheep Blood and Vancomycin**

Tryptic Soy Broth

Composition per liter:

Pancreatic digest of casein	18.0g
Papaic digest of soybean meal	6.0g
NaCl	6.0g

pH 7.3 ± 0.2 at 25°C

Source: This medium is available as a premixed powder from BD Diagnostic Systems.

Preparation of Medium: Add components to distilled/deionized water and bring volume to 1.0L. Mix thoroughly. Gently heat and bring to boiling. Distribute into tubes or flasks. Autoclave for 15 min at 15 psi pressure–121°C. Mix thoroughly.

Use: For the isolation and cultivation of a wide variety of microorganisms.

Tryptic Soy Broth with 0.001*M* Calcium Chloride (ATCC Medium 1380)

Composition per liter:

Pancreatic digest of casein	18.0g
Papaic digest of soybean meal	6.0g
NaCl	6.0g
$CaCl_2$ solution	10.0mL

pH 7.3 ± 0.2 at 25°C

Source: This medium, without $CaCl_2$, is available as a premixed powder from BD Diagnostic Systems.

Calcium Chloride Solution:

Composition per 10.0mL:

$CaCl_2$	0.111g

Preparation of Calcium Chloride Solution: Add $CaCl_2$ to distilled/deionized water and bring volume to 10.0mL. Mix thoroughly. Filter sterilize.

Preparation of Medium: Add components, except $CaCl_2$ solution, to distilled/deionized water and bring volume to 990mL. Mix thoroughly. Gently heat and bring to boiling. Autoclave for 15 min at 15 psi pressure–121°C. Aseptically add 10.0mL $CaCl_2$ solution. Mix thoroughly. Distribute into tubes or flasks.

Use: For the isolation and cultivation of *Bronchothrix thermospacta.*

Tryptic Soy Broth with 0.1% Potassium Nitrate (ATCC Medium 1183)

Composition per liter:

Pancreatic digest of casein	15.0g
Agar	15.0g
Papaic digest of soybean meal	5.0g
NaCl	5.0g
KNO_3	1.0g

pH 7.3 ± 0.2 at 25°C

Source: This medium, without nitrate, is available as a premixed powder from BD Diagnostic Systems.

Preparation of Medium: Add components to distilled/deionized water and bring volume to 1.0L. Mix thoroughly. Gently heat and bring to boiling. Distribute into tubes or flasks. Autoclave for 15 min at 15 psi pressure–121°C. Do not overheat. Pour into sterile Petri dishes or leave in tubes.

Use: For the isolation and cultivation of *Flavobacterium* sp.

Tryptic Soy Broth with Magnesium Ions (ATCC Medium 1588)

Composition per liter:

Pancreatic digest of casein	17.0g
NaCl	5.0g
Pancreatic digest of soybean meal	3.0g
Glucose	2.5g
K_2HPO_4	2.5g
$MgCl_2$	0.95g

pH 7.3 ± 0.2 at 25°C

Preparation of Medium: Add components to distilled/deionized water and bring volume to 1.0L. Mix thoroughly. Distribute into flasks or tubes. Autoclave for 15 min at 15 psi pressure–121°C.

Use: For the cultivation of *Escherichia coli* for bacteriophage production.

Tryptic Soy Broth with 1*M* Potassium Chloride (ATCC Medium 2074)

Composition per liter:

KCl	74.5g
Pancreatic digest of casein	17.0g
NaCl	5.0g
Papaic digest of soybean meal	3.0g
K_2HPO_4	2.5g
Glucose	2.5g

pH 7.3 ± 0.2 at 25°C

Preparation of Medium: Add components to distilled/deionized water and bring volume to 1.0L. Mix thoroughly. Distribute into tubes or flasks. Autoclave for 15 min at 15 psi pressure–121°C.

Use: For the cultivation and maintenance of *Gracilibacillus dipsosauri.*

Tryptic Soy Fast Green Agar (TSFA)

Composition per liter:

Pancreatic digest of casein	17.0g
Agar	15.0g
NaCl	5.0g
Papaic digest of soybean meal	3.0g
K_2HPO_4	2.5g
Glucose	2.5g
Fast Green FCF	0.25g

pH 7.3 ± 0.2 at 25°C

Preparation of Medium: Add components to distilled/deionized water and bring volume to 1.0L. Mix thoroughly. Gently heat and bring to boiling. Distribute into tubes or flasks. Autoclave for 15 min at 15 psi pressure–121°C. Cool to 45°–50°C. Aseptically adjust pH to 7.3. Pour into sterile Petri dishes.

Use: For the isolation and cultivation of *Salmonella* species from foods.

Trypticase™ Agar Base

Composition per liter:

Pancreatic digest of casein	20.0g
Agar	3.5g
Phenol Red	0.02g

pH 7.4 ± 0.2 at 25°C

Source: This medium is available as a premixed powder from BD Diagnostic Systems and Oxoid Unipath.

Preparation of Medium: Add components to distilled/deionized water and bring volume to 1.0L. Mix thoroughly. Gently heat and bring to boiling. Distribute into tubes or flasks. Autoclave for 15 min at 15 psi pressure–121°C. Pour into sterile Petri dishes or leave in tubes.

Use: For the differentiation of microorganisms based on their motility.

Trypticase™ Agar Base with Carbohydrate

Composition per liter:

Pancreatic digest of casein 20.0g
Carbohydrate 5.0g
Agar 3.5g
Phenol Red 0.02g

pH 7.4 ± 0.2 at 25°C

Preparation of Medium: Add components to distilled/deionized water and bring volume to 1.0L. Mix thoroughly. Gently heat and bring to boiling. Distribute into tubes. Autoclave for 15 min at 13 psi pressure–118°C. Do not overheat. Pour into sterile Petri dishes or leave in tubes.

Use: For differentiation of microorganisms based on their motility and fermentation reactions. Fermentation of carbohydrate turns the medium yellow.

Trypticase™ Azolectin Tween™ Broth Base

See: **TAT Broth Base**

Trypticase™ Broth, Supplemented

Composition per liter:

Pancreatic digest of casein 20.0g
$MgSO_4 \cdot 7H_2O$ 0.015g
$FeCl_3$ 7.0mg

pH 7.2 ± 0.2 at 25°C

Preparation of Medium: Add components to distilled/deionized water and bring volume to 1.0L. Mix thoroughly. Distribute into tubes or flasks. Autoclave for 15 min at 15 psi pressure–121°C.

Use: For the cultivation of *Bacillus stearothermophilus*.

Trypticase™ Glucose Agar

Composition per liter:

Pancreatic digest of casein 17.0g
Agar 15.0g
NaCl 5.0g
Papaic digest of soybean meal 3.0g
Glucose 7.5g
Yeast extract 3.0g
KH_2PO_4 .... 2.5g

Preparation of Medium: Add components to distilled/deionized water and bring volume to 1.0L. Mix thoroughly. Gently heat and bring to boiling. Distribute into tubes or flasks. Autoclave for 15 min at 15 psi pressure–121°C. Pour into sterile Petri dishes or leave in tubes.

Use: For the cultivation and maintenance of *Lactococcus raffinolactis, Streptococcus equi, Streptococcus pneumoniae,* and *Streptococcus uberis.*

Trypticase™ Glucose Extract Agar

Composition per liter:

Agar 15.0g
Pancreatic digest of casein 5.0g
Beef extract 3.0g
Glucose 1.0g

pH 7.0 ± 0.2 at 25°C

Source: This medium is available as a premixed powder from BD Diagnostic Systems.

Preparation of Medium: Add components to distilled/deionized water and bring volume to 1.0L. Mix thoroughly. Gently heat and bring to boiling. Distribute into tubes or flasks. Autoclave for 15 min at 15 psi pressure–121°C. Pour into sterile Petri dishes or leave in tubes.

Use: For the enumeration of bacteria in water, milk, and other specimens.

Trypticase™ Novobiocin Broth (TN Broth)

Composition per liter:

Pancreatic digest of casein 17.0g
NaCl 5.0g
Papaic digest of soybean meal 3.0g
K_2HPO_4 .... 2.5g
Glucose 2.5g
Bile salts No. 3 .... 1.5g
K_2HPO_4 .... 1.5g
Novobiocin solution 10.0mL

pH 7.3 ± 0.2 at 25°C

Novobiocin Solution:

Composition per 10.0mL:

Novobiocin 0.02g

Preparation of Novobiocin Solution: Add novobiocin to distilled/deionized water and bring volume to 10.0mL. Mix thoroughly. Filter sterilize.

Preparation of Medium: Add components, except novobiocin solution, to distilled/deionized water and bring volume to 990.0mL. Mix thoroughly. Gently heat and bring to boiling. Autoclave for 15 min at 15 psi pressure–121°C. Cool to 45°–50°C. Aseptically add sterile novobiocin solution. Mix thoroughly. Pour into sterile Petri dishes or distribute into sterile tubes.

Use: For the cultivation of verotoxin-producing *Escherichia coli.*

Trypticase™ Peptone Glucose Yeast Extract Broth (TPGY Broth)

Composition per liter:

Pancreatic digest of casein 50.0g
Yeast extract 20.0g
Peptone 5.0g
Glucose 4.0g
Sodium thioglycolate 1.0g

pH 7.0 ± 0.2 at 25°C

Preparation of Medium: Add components to distilled/deionized water and bring volume to 1.0L. Mix thoroughly. Distribute into tubes in 15.0mL volumes. Autoclave for 10 min at 15 psi pressure–121°C.

Use: For the cultivation of *Clostridium botulinum.*

Trypticase™ Peptone Glucose Yeast Extract Broth, Buffered

Composition per liter:

Pancreatic digest of casein 50.0g
Yeast extract 20.0g

Na_2HPO_4	5.0g
Peptone	5.0g
Glucose	4.0g
Sodium thioglycolate	1.0g

pH 7.3 ± 0.2 at 25°C

Preparation of Medium: Add components to distilled/deionized water and bring volume to 1.0L. Mix thoroughly. Gently heat until dissolved. Adjust pH to 7.3. Distribute into tubes in 15.0mL volumes. Autoclave for 8 min at 15 psi pressure–121°C.

Use: For the isolation and cultivation of *Clostridium perfringens* from foods.

Trypticase™ Peptone Glucose Yeast Extract Broth with Trypsin (TPGYT Broth)

Composition per 1067.0mL:

Pancreatic digest of casein	50.0g
Yeast extract	20.0g
Peptone	5.0g
Glucose	4.0g
Sodium thioglycolate	1.0g
Trypsin solution	67.0mL

pH 7.0 ± 0.2 at 25°C

Trypsin Solution:

Composition per 100.0mL:

Trypsin	1.5g

Preparation of Trypsin Solution: Add trypsin to distilled/deionized water and bring volume to 100.0mL. Mix thoroughly. Filter sterilize.

Preparation of Medium: Add components, except trypsin solution, to distilled/deionized water and bring volume to 1.0L. Mix thoroughly. Gently heat and bring to boiling. Distribute into tubes in 15.0mL volumes. Autoclave for 10 min at 15 psi pressure–121°C. Immediately prior to use, aseptically add 1.0mL of sterile trypsin solution to each tube. Mix thoroughly.

Use: For the cultivation of *Clostridium botulinum.*

Trypticase™ Phytone™ Glucose Medium

Composition per liter:

Glucose	15.0g
Pancreatic digest of casein	10.0g
Agar	8.0g
Papaic digest of soybean meal	5.0g
Yeast extract	2.5g
K_2HPO_4	2.0g
L-Cysteine·HCl·H_2O	0.5g
$MgCl_2$	0.5g
$ZnSO_4 \cdot 7H_2O$	0.25g
$FeCl_3$	1.0mg

Preparation of Medium: Add $ZnSO_4$ to approximately 100.0mL of distilled/deionized water and dissolve. Add remaining components and bring volume to 1.0L with distilled/deionized water. Mix thoroughly. Distribute into tubes or flasks. Autoclave for 15 min at 15 psi pressure–121°C. Pour into sterile Petri dishes or leave in tubes.

Use: For the cultivation and maintenance of *Bifidobacterium* species.

Trypticase™ Phytone™ Glucose Medium with Tween™ 80

Composition per liter:

Glucose	15.0g
Pancreatic digest of casein	10.0g
Agar	8.0g
Papaic digest of soybean meal	5.0g
Yeast extract	2.5g
K_2HPO_4	2.0g
L-Cysteine·HCl·H_2O	0.5g
$MgCl_2$	0.5g
$ZnSO_4 \cdot 7H_2O$	0.25g
$FeCl_3$	1.0mg
Tween™ 80	2.0mL

Preparation of Medium: Add $ZnSO_4$ to approximately 100.0mL of distilled/deionized water and dissolve. Add remaining components and bring volume to 1.0L with distilled/deionized water. Mix thoroughly. Distribute into tubes or flasks. Autoclave for 15 min at 15 psi pressure–121°C. Pour into sterile Petri dishes or leave in tubes.

Use: For the cultivation and maintenance of *Bifidobacterium* species.

Trypticase™ Phytone Medium (DSMZ Medium 75)

Composition per liter:

Trypticase™ peptone	17.0g
Phytone™ peptone	3.0g
NaCl	5.0g
K_2HPO_4	2.5g
Glucose	2.5g
Distilled water	1000.0mL

pH 7.3 ± 0.2 at 25°C

Preparation of Medium: Add components to distilled/deionized water and bring volume to 1.0L. Mix thoroughly. Adjust pH to 7.3. Distribute into tubes or flasks. Autoclave for 15 min at 15 psi pressure–121°C.

Use: For the cultivation and maintenance of *Acinetobacter* spp.

Trypticase™ Phytone™ Medium

Composition per liter:

Pancreatic digest of casein	17.0g
NaCl	5.0g
Phytone™ peptone	3.0g
Glucose	2.5g
K_2HPO_4	2.5g

pH 7.3 ± 0.2 at 25°C

Preparation of Medium: Add components to distilled/deionized water and bring volume to 1.0L. Mix thoroughly. Distribute into tubes or flasks. Autoclave for 15 min at 15 psi pressure–121°C.

Use: For the cultivation of *Acinetobacter* species.

Trypticase™ Serum Seawater Agar (ATCC Medium 1359)

Composition per liter:

Agar	12.0g
Pancreatic digest of casein	1.0g
Seawater	990.0mL
Horse serum	10.0mL

pH 7.3 ± 0.2 at 25°C

Preparation of Medium: Add components, except horse serum, to 990.0mL seawater. Mix thoroughly. Gently heat and bring to boiling. Autoclave for 15 min at 15 psi pressure–121°C. Cool to 45°–50°C. Aseptically add sterile horse serum. Mix thoroughly. Pour into sterile Petri dishes or distribute into sterile tubes.

Use: For the cultivation and maintenance of fastidious marine bacterial species.

Trypticase™ Serum Seawater Agar

Composition per liter:

Agar	12.0g
Pancreatic digest of casein	1.0g
Seawater	990.0mL
Horse serum	10.0mL

Preparation of Medium: Add agar and pancreatic digest of casein to 990.0mL of seawater. Mix thoroughly. Gently heat and bring to boiling. Autoclave for 15 min at 15 psi pressure–121°C. Cool to 45°–50°C. Aseptically add 10.0mL of sterile horse serum. Mix thoroughly. Pour into sterile Petri dishes or distribute into sterile tubes.

Use: For the cultivation and maintenance of *Diplophyrs marina, Haliphthoros milfordensis,* and *Ostracoblabe implexa.*

Trypticase™ Soy Agar (ATCC Medium 18)

Composition per liter:

Pancreatic digest of casein	17.0g
Agar	15.0g
NaCl	5.0g
Papaic digest of soybean meal	3.0g
K_2HPO_4	2.5g
Glucose	2.5g

pH 7.3 ± 0.2 at 25°C

Preparation of Medium: Add components to distilled/deionized water and bring volume to 1.0L. Mix thoroughly. Gently heat and bring to boiling. Distribute into tubes or flasks. Autoclave for 15 min at 15 psi pressure–121°C. Pour into sterile Petri dishes or leave in tubes.

Use: For the cultivation and maintenance of a wide variety of heterotrophic microorganisms. For the cultivation of a wide variety of fastidious and nonfastidious microorganisms from clinical and nonclinical specimens. Also used for the rapid estimation of the bacteriological quality of water.

Trypticase™ Soy Agar (Tryptic Soy Agar) (Soybean Casein Digest Agar) (ATCC Medium 77)

Composition per liter:

Pancreatic digest of casein	15.0g
Agar	15.0g
Papaic digest of soybean meal	5.0g
NaCl	5.0g

pH 7.3 ± 0.2 at 25°C

Source: This medium is available as a premixed powder from BD Diagnostic Systems.

Preparation of Medium: Add components to distilled/deionized water and bring volume to 1.0L. Mix thoroughly. Gently heat and bring to boiling. Distribute into tubes or flasks. Autoclave for 15 min at 15 psi pressure–121°C. Do not overheat. Pour into sterile Petri dishes or leave in tubes.

Use: For the isolation and cultivation of a wide variety of fastidious as well as nonfastidious microorganisms.

Trypticase™ Soy Agar with Cefazidime (ATCC Medium 1997)

Composition per liter:

Pancreatic digest of casein	15.0g
Agar	15.0g
Papaic digest of soybean meal	5.0g
NaCl	5.0g
Ceftazidime solution	10.0mL

pH 7.3 ± 0.2 at 25°C

Ceftazidime Solution:

Composition per 10.0mL:

Ceftazidime	4.0mg

Preparation of Ceftazidime Solution: Add ceftazidime to distilled/deionized water and bring volume to 10.0mL. Mix thoroughly. Filter sterilize.

Preparation of Medium: Add components, except ceftazidime solution, to distilled/deionized water and bring volume to 990.0mL. Mix thoroughly. Gently heat and bring to boiling. Autoclave for 15 min at 15 psi pressure–121°C. Cool to 45°–50°C. Aseptically add 10.0mL of sterile ceftazidime solution. Mix thoroughly. Pour into sterile Petri dishes or distribute into sterile tubes.

Use: For the cultivation and maintenance of *Klebsiella oxytoca.*

Trypticase™ Soy Agar with 1% Glucose

Composition per liter:

Pancreatic digest of casein	15.0g
Agar	15.0g
Glucose	10.0g
Papaic digest of soybean meal	5.0g
NaCl	5.0g

pH 7.3 ± 0.2 at 25°C

Preparation of Medium: Add components to distilled/deionized water and bring volume to 1.0L. Mix thoroughly. Gently heat and bring to boiling. Distribute into tubes or flasks. Autoclave for 15 min at 15 psi pressure–121°C. Do not overheat. Pour into sterile Petri dishes or leave in tubes.

Use: For the cultivation and maintenance of *Saccharomonospora viridis.*

Trypticase™ Soy Agar with Glycerol

Composition per liter:

Pancreatic digest of casein	15.0g
Agar	15.0g
Papaic digest of soybean meal	5.0g
NaCl	5.0g
Glycerol	50.0mL

pH 7.3 ± 0.2 at 25°C

Preparation of Medium: Add components to distilled/deionized water and bring volume to 1.0L. Mix thoroughly. Gently heat while stirring and bring to boiling. Distribute into tubes or flasks. Autoclave for 15 min at 15 psi pressure–121°C. Do not overheat. Pour into sterile Petri dishes or leave in tubes. For blood plates, 50.0–100.0mL of sterile defibrinated sheep blood may be added to sterile medium that has been melted and cooled to 45°–50°C.

Use: For the cultivation and maintenance of *Acinetobacter calcoaceticus*.

Trypticase™ Soy Agar with Human Blood

Composition per liter:

Pancreatic digest of casein	15.0g
Agar	15.0g
Papaic digest of soybean meal	5.0g
NaCl	5.0g
Human blood, defibrinated	50.0mL

pH 7.3 ± 0.2 at 25°C

Preparation of Medium: Add components, except human blood, to distilled/deionized water and bring volume to 950.0mL. Mix thoroughly. Gently heat and bring to boiling. Autoclave for 15 min at 15 psi pressure–121°C. Cool to 45°–50°C. Aseptically add sterile human blood. Mix thoroughly. Pour into sterile Petri dishes in 17.0mL volumes or distribute into sterile tubes.

Use: For the cultivation of a wide variety of fastidious microorganisms. For the observation of hemolytic reactions of a variety of bacteria. May be used to perform the CAMP test for the presumptive identification of group B streptococci (*Streptococcus agalactiae*).

Trypticase™ Soy Agar with Lecithin and Polysorbate 80 (Microbial Content Test Agar)

Composition per liter:

Pancreatic digest of casein	15.0g
Agar	15.0g
Papaic digest of soybean meal	5.0g
NaCl	5.0g
Polysorbate 80 (Tween™ 80)	5.0g
Lecithin	0.7g

pH 7.3 ± 0.2 at 25°C

Source: This medium is available as a premixed powder from BD Diagnostic Systems.

Preparation of Medium: Add components to distilled/deionized water and bring volume to 1.0L. Mix thoroughly. Gently heat and bring to boiling. Distribute into tubes or flasks. Autoclave for 15 min at 13 psi pressure–118°C. Cool to 45°–50°C. Pour into sterile Petri dishes in 17.0mL volumes or leave in tubes.

Use: For the detection and enumeration of microorganisms in replicate plating techniques. Also used for the detection and enumeration of microorganisms present on surfaces of sanitary importance.

Trypticase™ Soy Agar-Magnesium Sulfate-Sodium Chloride Agar (TSAMS) (BAM M152a)

Composition per liter:

NaCl	20.0g
Trypticase peptone	15.0g
Agar	15.0g
Phytone™ peptone	5.0g
$MgSO_4 \cdot 7H_2O$	1.5g

pH 7.3 ± 0.2 at 25°C

Preparation of Medium: Add components to distilled/deionized water and bring volume to 1.0L. Mix thoroughly. Gently heat and bring to boiling. Autoclave for 15 min at 15 psi pressure–121°C. Pour into sterile Petri dishes in 20.0mL volumes.

Use: For the cultivation of *Vibrio* species from foods.

Trypticase™ Soy Agar, Modified

Composition per liter:

Pancreatic digest of casein	17.0g
Agar	15.0g
NaCl	5.0g
Yeast extract	4.0g
Papaic digest of soybean meal	3.0g
K_2HPO_4	2.5g

Preparation of Medium: Add components to distilled/deionized water and bring volume to 1.0L. Mix thoroughly. Gently heat and bring to boiling. Distribute into tubes or flasks. Autoclave for 15 min at 15 psi pressure–121°C. Pour into sterile Petri dishes or leave in tubes.

Use: For the cultivation and maintenance of the *Simonsiella* species.

Trypticase™ Soy Agar, Modified (ATCC Medium 1386)

Composition per liter:

Agar	18.0g
Pancreatic digest of casein	17.0g
NaCl	5.0g
Papaic digest of soybean meal	3.0g
K_2HPO_4	2.5g
Glucose	2.5g
Yeast extract	0.4g
NH_4OH, concentrated	0.035mL

pH 7.5 ± 0.2 at 25°C

Preparation of Medium: Add components, except NH_4OH, to distilled/deionized water and bring volume to 1.0L. Mix thoroughly. Gently heat and bring to boiling. Autoclave for 15 min at 15 psi pressure–121°C. Cool to 45°–50°C. Aseptically add NH_4OH. Mix thoroughly. Adjust pH to 7.5 if necessary. Pour into sterile Petri dishes or distribute into sterile tubes.

Use: For the cultivation of ATCC strain 31205.

Trypticase™ Soy Agar, Modified (ATCC Medium 1481)

Composition per liter:

Pancreatic digest of casein	17.0g
Agar	15.0g
NaCl	5.0g
Papaic digest of soybean meal	3.0g
K_2HPO_4	2.5g
Glucose	2.5g
L-Glutamine	10.0mL

pH 6.5 ± 0.2 at 25°C

Preparation of Medium: Add components, except glutamine, to distilled/deionized water and bring volume to 990.0mL. Mix thoroughly. Gently heat and bring to boiling. Autoclave for 15 min at 15 psi pressure–121°C. Cool to 45°–50°C. Aseptically add 10.0mL of sterile glutamine. Mix thoroughly. Adjust pH to 6.5. Pour into sterile Petri dishes or distribute into sterile tubes.

Use: Used as a base that is supplemented. For the cultivation of fastidious microorganisms. When supplemented with sheep blood, this medium is useful for the observation of hemolytic reactions of a variety of bacteria.

Trypticase™ Soy Agar, Modified (TSA II™)

Composition per liter:

Pancreatic digest of casein	14.5g
Agar	14.0g
Papaic digest of soybean meal	5.0g
NaCl	5.0g
Growth factors (BBL)	1.5g

pH 7.3 ± 0.2 at 25°C

Source: This medium is available as a premixed powder from BD Diagnostic Systems.

Preparation of Medium: Add components to distilled/deionized water and bring volume to 1.0L. Mix thoroughly. Gently heat while stirring and bring to boiling. Distribute into tubes or flasks. Autoclave for 15 min at 15 psi pressure–121°C. Do not overheat. Pour into sterile Petri dishes or leave in tubes. For blood plates, 50.0–100.0mL of sterile defibrinated sheep blood may be added to sterile medium that has been melted and cooled to 45°–50°C.

Use: Used as a base that is supplemented. For the cultivation of fastidious microorganisms. When supplemented with sheep blood, this medium is useful for the observation of hemolytic reactions of a variety of bacteria. It may be used to perform the CAMP test for the presumptive identification of group B streptococci (*Streptococcus agalactiae*).

Trypticase™ Soy Agar, Modified with Horse Serum

Composition per liter:

Pancreatic digest of casein	17.0g
Agar	15.0g
NaCl	5.0g
Yeast extract	4.0g
Papaic digest of soybean meal	3.0g
K_2HPO_4	2.5g
Horse serum	100.0mL

pH 7.3 ± 0.2 at 25°C

Preparation of Medium: Add components, except horse serum, to distilled/deionized water and bring volume to 900.0mL. Mix thoroughly. Gently heat and bring to boiling. Autoclave for 15 min at 15 psi pressure–121°C. Cool to 45°–50°C. Aseptically add sterile horse serum. Mix thoroughly. Pour into sterile Petri dishes or distribute into sterile tubes.

Use: For the cultivation and maintenance of *Simonsiella* species, *Alysiella* species, and *Moraxella* species.

Trypticase™ Soy Agar with Sheep Blood (Tryptic Soy Blood Agar) (TSA Blood Agar)

Composition per liter:

Pancreatic digest of casein	15.0g
Agar	15.0g
Papaic digest of soybean meal	5.0g
NaCl	5.0g
Sheep blood, defibrinated	50.0mL

pH 7.3 ± 0.2 at 25°C

Preparation of Medium: Add components, except sheep blood, to distilled/deionized water and bring volume to 950.0mL. Mix thoroughly. Gently heat and bring to boiling. Autoclave for 15 min at 15 psi pressure–121°C. Cool to 45°–50°C. Aseptically add sterile sheep blood. Mix thoroughly. Pour into sterile Petri dishes in 17.0mL volumes or distribute into sterile tubes.

Use: For the cultivation of a wide variety of fastidious microorganisms. For the observation of hemolytic reactions of a variety of bacteria. May be used to perform the CAMP test for the presumptive identification of group B streptococci (*Streptococcus agalactiae*).

Trypticase™ Soy Agar with Sheep Blood, Formate, and Fumarate

Composition per liter:

Pancreatic digest of casein	14.5g
Agar	14.0g
Papaic digest of soybean meal	5.0g
NaCl	5.0g
Sucrose	2.0g
Growth factors	1.5g
Sheep blood, defibrinated	50.0mL
Formate-fumarate solution	13.0mL

pH 7.3 ± 0.2 at 25°C

Source: Growth Factors are available as a premixed powder from BD Diagnostic Systems.

Formate-Fumarate Solution:

Composition per 100.0mL:

Sodium formate	6.0g
Fumaric acid	6.0g

Preparation of Formate-Fumarate Solution: Add components to distilled/deionized water and bring volume to 100.0mL. Mix thoroughly. Adjust pH to 7.0. Filter sterilize.

Preparation of Medium: Add components, except sheep blood, to distilled/deionized water and bring volume to 950.0mL. Mix thoroughly. Gently heat and bring to boiling. Autoclave for 15 min at 15 psi pressure–121°C. Cool to 45°–50°C. Aseptically add sterile sheep blood. Mix thoroughly. Pour into sterile Petri dishes. Prior to inoculation, aseptically spread 0.2mL of sterile formate-fumarate solution on each plate.

Use: For the isolation of *Streptococcus pneumoniae* from a variety of clinical specimens.

Trypticase™ Soy Agar with Sheep Blood and Gentamicin (TSA II™ with Sheep Blood and Gentamicin)

Composition per liter:

Pancreatic digest of casein	14.5g
Agar	14.0g
Papaic digest of soybean meal	5.0g
NaCl	5.0g
Growth factors	1.5g
Sheep blood, defibrinated	50.0mL
Gentamicin solution	10.0mL

pH 7.3 ± 0.2 at 25°C

Source: This medium is available as a premixed powder from BD Diagnostic Systems.

Gentamicin Solution:
Composition per 10.0mL:

Gentamicin	2.5mg

Preparation of Gentamicin Solution: Add gentamicin to distilled/deionized water and bring volume to 10.0mL. Mix thoroughly. Filter sterilize.

Preparation of Medium: Add components, except sheep blood and gentamicin solution, to distilled/deionized water and bring volume to 940.0mL. Mix thoroughly. Gently heat and bring to boiling. Autoclave for 15 min at 15 psi pressure–121°C. Cool to 45°–50°C. Aseptically add sterile sheep blood and sterile gentamicin solution. Mix thoroughly. Pour into sterile Petri dishes or distribute into sterile tubes.

Use: For the isolation of *Streptococcus pneumoniae* from a variety of clinical specimens.

Trypticase™ Soy Agar with Sheep Blood and Streptomycin (ATCC Medium 1810)

Composition per liter:

Pancreatic digest of casein	14.5g
Agar	14.0g
Papaic digest of soybean meal	5.0g
NaCl	5.0g
Growth factors	1.5g
Sheep blood, defibrinated	50.0mL
Streptomycin solution	10.0mL

pH 7.3 ± 0.2 at 25°C

Source: Growth factors are available as a premixed powder from BD Diagnostic Systems.

Streptomycin Solution:
Composition per 10.0mL:

Streptomycin	0.5g

Preparation of Streptomycin Solution: Add streptomycin to distilled/deionized water and bring volume to 10.0mL. Mix thoroughly. Filter sterilize.

Preparation of Medium: Add components, except sheep blood and streptomycin solution, to distilled/deionized water and bring volume to 940.0mL. Mix thoroughly. Gently heat and bring to boiling. Autoclave for 15 min at 15 psi pressure–121°C. Cool to 45°–50°C. Aseptically add sterile sheep blood and sterile streptomycin solution. Mix thoroughly. Pour into sterile Petri dishes or distribute into sterile tubes.

Use: For the cultivation of *Streptococcus gordonii.*

Trypticase™ Soy Agar with Sheep Blood, Sucrose, and Tetracycline

Composition per liter:

Pancreatic digest of casein	14.5g
Agar	14.0g
Papaic digest of soybean meal	5.0g
NaCl	5.0g
Sucrose	2.0g
Growth factors	1.5g
Sheep blood, defibrinated	50.0mL
Tetracycline solution	10.0mL

pH 7.3 ± 0.2 at 25°C

Source: Growth factors are available as a premixed powder from BD Diagnostic Systems.

Tetracycline Solution:
Composition per 10.0mL:

Tetracycline	0.5mg

Preparation of Tetracycline Solution: Add tetracycline to distilled/deionized water and bring volume to 10.0mL. Mix thoroughly. Filter sterilize.

Preparation of Medium: Add components, except sheep blood and tetracycline solution, to distilled/deionized water and bring volume to 940.0mL. Mix thoroughly. Gently heat and bring to boiling. Autoclave for 15 min at 15 psi pressure–121°C. Cool to 45°–50°C. Aseptically add sterile sheep blood and sterile tetracycline solution. Mix thoroughly. Pour into sterile Petri dishes or distribute into sterile tubes.

Use: For the isolation of *Streptococcus pneumoniae* from a variety of clinical specimens.

Trypticase™ Soy Agar with Sheep Blood and Tween™ 80 (ATCC Medium 1893)

Composition per liter:

Pancreatic digest of casein	15.0g
Agar	15.0g
Papaic digest of soybean meal	5.0g
NaCl	5.0g
Sheep blood, defibrinated	50.0mL
Tween™ 80	10.0mL

pH 7.3 ± 0.2 at 25°C

Preparation of Medium: Add components, except sheep blood, to distilled/deionized water and bring volume to 950.0mL. Mix thoroughly. Gently heat and bring to boiling. Autoclave for 15 min at 15 psi pressure–121°C. Cool to 45°–50°C. Aseptically add sterile sheep blood. Mix thoroughly. Pour into sterile Petri dishes in 17.0mL volumes or distribute into sterile tubes.

Use: For the cultivation of a wide variety of fastidious microorganisms.

Trypticase™ Soy Agar with Sheep Blood and Vancomycin (ATCC Medium 1976)

Composition per liter:

Pancreatic digest of casein	14.5g
Agar	14.0g
Papaic digest of soybean meal	5.0g
NaCl	5.0g
Growth factors	1.5g
Sheep blood, defibrinated	50.0mL
Vancomycin solution	10.0mL

pH 7.3 ± 0.2 at 25°C

Source: This medium is available as a premixed powder from BD Diagnostic Systems.

Vancomycin Solution:
Composition per 10.0mL:

Vancomycin	4.0mg

Preparation of Vancomycin Solution: Add vancomycin to distilled/deionized water and bring volume to 10.0mL. Mix thoroughly. Filter sterilize.

Preparation of Medium: Add components, except sheep blood and vancomycin solution, to distilled/deionized water and bring volume to 940.0mL. Mix thoroughly. Gently heat and bring to boiling. Autoclave for 15 min at 15 psi pressure–121°C. Cool to 50°–55°C. Aseptically add sterile sheep blood and sterile vancomycin solution. Mix thoroughly. Pour into sterile Petri dishes or distribute into sterile tubes.

Use: For the isolation of *Streptococcus and Enterococcus* spp. from a variety of clinical specimens.

Trypticase™ Soy Agar with Sodium Chloride (ATCC Medium 176)

Composition per liter:

NaCl	30.0g
Pancreatic digest of casein	15.0g
Agar	15.0g
Papaic digest of soybean meal	5.0g
Bile salts No. 3	1.0g

pH 7.3 ± 0.2 at 25°C

Preparation of Medium: Add components to distilled/deionized water and bring volume to 1.0L. Mix thoroughly. Gently heat and bring to boiling. Distribute into tubes or flasks. Autoclave for 15 min at 15 psi pressure–121°C. Pour into sterile Petri dishes or leave in tubes.

Use: For the cultivation and maintenance of *Vibrio alginolyticus.*

Trypticase™ Soy Agar with 3% Sodium Chloride (TSA NaCl)

Composition per liter:

NaCl	30.0g
Agar	15.0g
Pancreatic digest of casein	15.0g
Papaic digest of soybean meal	5.0g

pH 7.3 ± 0.2 at 25°C

Preparation of Medium: Add components to distilled/deionized water and bring volume to 1.0L. Mix thoroughly. Gently heat and bring to boiling. Distribute into tubes or flasks. Autoclave for 15 min at 15 psi pressure–121°C. Pour into sterile Petri dishes or leave in tubes.

Use: For the cultivation of halophilic microorganisms isolated from foods.

Trypticase™ Soy Agar with Sodium Chloride, Horse Serum, and Penicillin

Composition per liter:

NaCl	35.0g
Pancreatic digest of casein	15.0g
Agar	15.0g
Papaic digest of soybean meal	5.0g
Horse serum, inactivated	100.0mL
Penicillin solution	10.0mL

pH 7.3 ± 0.2 at 25°C

Penicillin Solution:

Composition per 10.0mL:

Penicillin	1,000,000U

Preparation of Penicillin Solution: Add penicillin to distilled/deionized water and bring volume to 10.0mL. Mix thoroughly. Filter sterilize.

Preparation of Medium: Add components, except horse serum and penicillin solution, to distilled/deionized water and bring volume to 890.0mL. Mix thoroughly. Gently heat and bring to boiling. Autoclave for 15 min at 15 psi pressure–121°C. Do not overheat. Cool to 50°C. Aseptically add 100.0mL of sterile horse serum and 10.0mL of sterile penicillin solution. Mix thoroughly. Pour into sterile Petri dishes or distribute into sterile tubes.

Use: For the isolation and cultivation of fungi.

Trypticase™ Soy Agar with Starch (ATCC Medium 1818)

Composition per liter:

Pancreatic digest of casein	15.0g
Agar	15.0g
Starch, soluble	10.0g
Papaic digest of soybean meal	5.0g
NaCl	5.0g

pH 5.5 ± 0.2 at 25°C

Preparation of Medium: Add components to distilled/deionized water and bring volume to 1.0L. Mix thoroughly. Adjust pH to 5.5. Gently heat and bring to boiling. Distribute into tubes or flasks. Autoclave for 15 min at 15 psi pressure–121°C. Pour into sterile Petri dishes or leave in tubes.

Use: For the cultivation of *Bacillus naganoensis.*

Trypticase™ Soy Agar with Tobramycin

Composition per liter:

Pancreatic digest of casein	17.0g
Agar	15.0g
NaCl	5.0g
Papaic digest of soybean meal	3.0g
K_2HPO_4	2.5g
Glucose	2.5g
Tobramycin solution	10.0mL

pH 7.3 ± 0.2 at 25°C

Tobromycin Solution:

Composition per 10.0mL:

Tobramycin	8.0mg

Preparation of Tobromycin Solution: Add tobramycin to distilled/deionized water and bring volume to 10.0mL. Mix thoroughly. Filter sterilize.

Preparation of Medium: Add components, except tobramycin solution, to distilled/deionized water and bring volume to 990.0mL. Mix thoroughly. Gently heat and bring to boiling. Autoclave for 15 min at 15 psi pressure–121°C. Cool to 45°–50°C. Aseptically add sterile tobramycin solution. Mix thoroughly. Pour into sterile Petri dishes or distribute into sterile tubes.

Use: For the cultivation and maintenance of *Serratia marcescens.*

Trypticase™ Soy Agar with Yeast Extract (TSAYE)

Composition per liter:

Pancreatic digest of casein	17.0g
Agar	15.0g
Yeast extract	6.0g
NaCl	5.0g
Papaic digest of soybean meal	3.0g

K_2HPO_4	2.5g
Glucose	2.5g

pH 7.3 ± 0.2 at 25°C

Preparation of Medium: Add components to distilled/deionized water and bring volume to 1.0L. Mix thoroughly. Gently heat and bring to boiling. Distribute into tubes or flasks. Autoclave for 15 min at 15 psi pressure–121°C. Pour into sterile Petri dishes or leave in tubes.

Use: For the cultivation and maintenance of a wide variety of heterotrophic microorganisms. For the isolation and cultivation of *Listeria monocytogenes* from foods.

Trypticase™ Soy Agar with Yeast Extract and Glucose

Composition per liter:

Pancreatic digest of casein	17.0g
Agar	15.0g
Glucose	7.5g
NaCl	5.0g
Yeast extract	3.0g
Papaic digest of soybean meal	3.0g
K_2HPO_4	2.5g

pH 7.0–7.2 at 25°C

Preparation of Medium: Add components to distilled/deionized water and bring volume to 1.0L. Mix thoroughly. Gently heat and bring to boiling. Distribute into tubes or flasks. Autoclave for 15 min at 15 psi pressure–121°C. Pour into sterile Petri dishes or leave in tubes.

Use: For the cultivation and maintenance of *Pediococcus urinaeequi.*

Trypticase™ Soy Broth (Soybean Casein Digest Broth, USP) (LMG Medium 185)

Composition per liter:

Pancreatic digest of casein	17.0g
NaCl	5.0g
Papaic digest of soybean meal	3.0g
K_2HPO_4	2.5g
Glucose	2.5g

pH 7.3 ± 0.2 at 25°C

Source: This medium is available as a premixed powder from BD Diagnostic Systems.

Preparation of Medium: Add components to distilled/deionized water and bring volume to 1.0L. Mix thoroughly. Distribute into tubes or flasks. Autoclave for 15 min at 15 psi pressure–121°C.

Use: For the cultivation and maintenance of *Bordetella* spp., *Capnocytophaga* spp., *Brachybacterium* spp., *Erwinia amylovora, Curtobacterium flaccumfaciens, Aeromonas* spp., *Ralstonia* spp., *Ornithobacterium rhinotracheale, Riemerella* spp., *Arthrobacter* spp., *Pedobacter* spp., *Burkholderia vietnamiensis, Myroides* spp., *Chryseobacterium* spp., *Microbacterium* spp., *Nocardioides* spp., *Cellulomonas* spp., Agromyces spp., *Pelistega europaea, Bergeyella zoohelcum, Bacillus* spp., *Burkholderia* spp., *Brevibacillus* spp., *Aneurinibacillus aneurinilyticus, Pandoraea* spp., and *Pseudomonas* spp. For the cultivation of a wide variety of fastidious and nonfastidious microorganisms from clinical and nonclinical specimens. Also used for the rapid estimation of the bacteriological quality of water.

Trypticase™ Soy Broth with 0.1% Agar

Composition per liter:

Pancreatic digest of casein	17.0g
NaCl	5.0g
Papaic digest of soybean meal	3.0g
K_2HPO_4	2.5g
Glucose	2.5g
Agar	1.0g

pH 7.3 ± 0.2 at 25°C

Source: This medium is available as a premixed powder from BD Diagnostic Systems.

Preparation of Medium: Add components to distilled/deionized water and bring volume to 1.0L. Mix thoroughly. Distribute into tubes or flasks. Autoclave for 15 min at 15 psi pressure–121°C.

Use: For the cultivation of anaerobic microorganisms. For the cultivation of microorganisms isolated from root canals and other clinical specimens.

Trypticase™ Soy Broth Agar Modified (DSMZ Medium 535a)

Composition per liter:

Agar	15.0g
Pancreatic digest of casein	17.0g
NaCl	5.0g
Pancreatic digest of soybean meal	3.0g
K_2HPO_4	2.5g
Glucose	2.5g
Horse blood, defibrinated	50.0mL

pH 7.3 ± 0.2 at 25°C

Preparation of Medium: Add components, except horse blood, to distilled/deionized water and bring volume to 950.0mL. Mix thoroughly. Heat with frequent agitation and boil for 1 min to completely dissolve. Autoclave for 15 min at 15 psi pressure–121°C. Cool to 45°–50°C. Aseptically add 50.0mL of sterile, defibrinated horse blood. Mix thoroughly and pour into sterile Petri dishes.

Use: For cultivation and maintenance of *Corynebacterium glucuronolyticum, Corynebacterium durum, Corynebacterium falsenii, Corynebacterium singulare,* and *Corynebacterium xerosis.*

Trypticase™ Soy Broth with Calcium Chloride

Composition per liter:

Pancreatic digest of casein	17.0g
NaCl	5.0g
Papaic digest of soybean meal	3.0g
K_2HPO_4	2.5g
Glucose	2.5g
$CaCl_2 \cdot 2H_2O$	0.15g

pH 7.3 ± 0.2 at 25°C

Preparation of Medium: Add components to distilled/deionized water and bring volume to 1.0L. Mix thoroughly. Distribute into tubes or flasks. Autoclave for 15 min at 15 psi pressure–121°C.

Use: For the cultivation and maintenance of *Brochothrix thermosphacta.*

Trypticase™ Soy Broth with Cefazidime

Composition per liter:

Pancreatic digest of casein	17.0g
NaCl	5.0g
Papaic digest of soybean meal	3.0g
K_2HPO_4	2.5g

Glucose 2.5g
Ceftazidime solution 10.0mL
pH 7.3 ± 0.2 at 25°C

Ceftazidime Solution:
Composition per 10.0mL:
Ceftazidime 4.0mg

Preparation of Ceftazidime Solution: Add ceftazidime to distilled/deionized water and bring volume to 10.0mL. Mix thoroughly. Filter sterilize.

Preparation of Medium: Add components, except ceftazidime solution, to distilled/deionized water and bring volume to 990.0mL. Mix thoroughly. Autoclave for 15 min at 15 psi pressure–121°C. Cool to 25°C. Aseptically add 10.0mL of sterile ceftazidime solution. Mix thoroughly. Aseptically distribute into sterile tubes or flasks.

Use: For the cultivaion of *Klebsiella oxytoca.*

Trypticase™ Soy Broth with Ferrous Sulfate (Tryptic Soy Broth with Ferrous Sulfate) (BAM M186)

Composition per liter:
Pancreatic digest of casein 17.0g
NaCl 5.0g
Papaic digest of soybean meal 3.0g
K_2HPO_4 .... 2.5g
Glucose 2.5g
$FeSO_4$ 35.0mg
pH 7.3 ± 0.2 at 25°C

Preparation of Medium: Add components to distilled/deionized water and bring volume to 1.0L. Mix thoroughly. Gently heat until dissolved. Adjust pH to 7.3. Distribute into tubes in 10.0mL volumes. Autoclave for 15 min at 15 psi pressure–121°C.

Use: For the detection of *Salmonella* spp. from foods.

Trypticase™ Soy Broth with 10m*M* Glucose

Composition per liter:
Pancreatic digest of casein 17.0g
NaCl 5.0g
Papaic digest of soybean meal 3.0g
K_2HPO_4 .... 2.5g
Glucose 1.8g
pH 7.3 ± 0.2 at 25°C

Preparation of Medium: Add components to distilled/deionized water and bring volume to 1.0L. Mix thoroughly. Distribute into tubes or flasks. Autoclave for 15 min at 15 psi pressure–121°C.

Use: For the cultivation of a variety of fastidious and nonfastidious microorganisms from clinical and nonclinical specimens.

Trypticase™ Soy Broth with Fetal Calf Serum

Composition per liter:
Pancreatic digest of casein 17.0g
NaCl 5.0g
Papaic digest of soybean meal 3.0g
K_2HPO_4 .... 2.5g
Glucose 2.5g
Fetal calf serum 100.0mL
pH 7.3 ± 0.2 at 25°C

Preparation of Medium: Add components, except fetal calf serum, to distilled/deionized water and bring volume to 900.0mL. Mix thoroughly. Autoclave for 15 min at 15 psi pressure–121°C. Cool to 25°C. Aseptically add sterile fetal calf serum. Mix thoroughly. Aseptically distribute into sterile tubes.

Use: For the cultivation and maintenance of *Serpula innocens* and *Brachyspira innocens.*

Trypticase™ Soy Broth with 10m*M* Glucose (ATCC Medium 1189)

Composition per liter:
Pancreatic digest of casein 18.0g
Papaic digest of soybean meal 6.0g
NaCl 6.0g
Glucose 1.0g
pH 7.3 ± 0.2 at 25°C

Source: This medium, without glucose, is available as a premixed powder from BD Diagnostic Systems.

Preparation of Medium: Add components to distilled/deionized water and bring volume to 1.0L. Mix thoroughly. Gently heat and bring to boiling. Distribute into tubes or flasks. Autoclave for 15 min at 15 psi pressure–121°C. Do not overheat. Pour into sterile Petri dishes or leave in tubes.

Use: For the isolation and cultivation of *Staphylococcus aureus* subsp. *aureus.*

Trypticase™ Soy Broth with Glycerol

Composition per liter:
Pancreatic digest of casein 17.0g
NaCl 15.0g
Papaic digest of soybean meal 3.0g
K_2HPO_4 .... 2.5g
Glycerol 240.0mL
pH 7.3 ± 0.2 at 25°C

Preparation of Medium: Add components to distilled/deionized water and bring volume to 1.0L. Mix thoroughly. Gently heat until dissolved. Adjust pH to 7.3. Distribute into tubes in 10.0mL volumes. Autoclave for 15 min at 15 psi pressure–121°C.

Use: For the cultivation and maintenance of a wide variety of microorganisms from foods.

Trypticase™ Soy Broth with Horse Serum

Composition per liter:
Pancreatic digest of casein 17.0g
NaCl 5.0g
Papaic digest of soybean meal 3.0g
K_2HPO_4 .... 2.5g
Glucose 2.5g
Horse serum, inactivated 100.0mL
pH 7.3 ± 0.2 at 25°C

Preparation of Medium: Add components, except horse serum, to distilled/deionized water and bring volume to 900.0mL. Mix thoroughly. Autoclave for 15 min at 15 psi pressure–121°C. Cool to 25°C. Aseptically add 100.0mL of sterile horse serum. Mix thoroughly. Aseptically distribute into sterile tubes.

Use: For the cultivation and maintenance of *Serpula innocens.*

Trypticase™ Soy Broth with Human Blood

Composition per liter:

Pancreatic digest of casein	17.0g
NaCl	5.0g
Papaic digest of soybean meal	3.0g
K_2HPO_4	2.5g
Glucose	2.5g
Human blood, defibrinated	50.0mL

pH 7.3 ± 0.2 at 25°C

Preparation of Medium: Add components, except human blood, to distilled/deionized water and bring volume to 950.0mL. Mix thoroughly. Autoclave for 15 min at 15 psi pressure–121°C. Cool to 25°C. Aseptically add 50.0mL of sterile human blood. Mix thoroughly. Aseptically distribute into sterile tubes.

Use: For the cultivation and maintenance of *Serpula innocens*.

Trypticase™ Soy Broth with 0.1% DL-7-Hydroxybutyric Acid (LMG Medium 266)

Composition per liter:

Pancreatic digest of casein	17.0g
NaCl	5.0g
Papaic digest of soybean meal	3.0g
K_2HPO_4	2.5g
Glucose	2.5g
DL-7-Hydroxybutyric acid	1.0g

pH 7.2 ± 0.2 at 25°C

Preparation of Medium: Add components to distilled/deionized water and bring volume to 1.0L. Mix thoroughly. Distribute into tubes or flasks. Gently warm with mixing until solution is complete. Autoclave for 15 min at 15 psi pressure–121°C.

Use: For cultivation and maintenance of *Paucimonas lemoignei*.

Trypticase™ Soy Broth, Modified

Composition per 1000.2mL:

Pancreatic digest of casein	17.0g
NaCl	15.0g
K_2HPO_4	4.0g
Papaic digest of soybean meal	3.0g
Glucose	2.5g
Bile salts No. 3	1.5g
Novobiocin solution	0.2mL

pH 7.3 ± 0.2 at 25°C

Novobiocin Solution:

Composition per liter:

Novobiocin	0.05g

Preparation of Novobiocin Solution: Add novobiocin to distilled/deionized water and bring volume to 1.0L. Mix thoroughly. Filter sterilize.

Preparation of Medium: Add components, except novobiocin solution, to distilled/deionized water and bring volume to 1.0L. Mix thoroughly. Gently heat and bring to boiling. Autoclave for 15 min at 15 psi pressure–121°C. Cool to 45°–50°C. Aseptically add sterile novobiocin solution. Mix thoroughly. Aseptically distribute into sterile tubes.

Use: For the isolation and cultivation of *Shigella* species from food.

Trypticase™ Soy Broth, Modified (mTSB) (BAM M156)

Composition per 1000.2mL:

Pancreatic digest of casein	17.0g
NaCl	5.0g
K_2HPO_4	4.0g
Papaic digest of soybean meal	3.0g
Glucose	2.5g
Bile salts No. 3	1.5g
Novobiocin solution	0.2mL

pH 7.3 ± 0.2 at 25°C

Novobiocin Solution:

Composition per liter:

Novobiocin	0.05g

Preparation of Novobiocin Solution: Add novobiocin to distilled/deionized water and bring volume to 1.0L. Mix thoroughly. Filter sterilize.

Preparation of Medium: Add components, except novobiocin solution, to distilled/deionized water and bring volume to 1.0L. Mix thoroughly. Gently heat and bring to boiling. Autoclave for 15 min at 15 psi pressure–121°C. Cool to 45°–50°C. Aseptically add sterile novobiocin solution. Mix thoroughly. Aseptically distribute into sterile tubes.

Use: For the isolation and cultivation of *Shigella* species from food.

Trypticase™ Soy Broth with Neomycin

Composition per liter:

Pancreatic digest of casein	17.0g
NaCl	5.0g
Papaic digest of soybean meal	3.0g
K_2HPO_4	2.5g
Glucose	2.5g
Agar	1.0g
Neomycin solution	10.0mL

pH 7.3 ± 0.2 at 25°C

Neomycin Solution:

Composition per 10.0mL:

Neomycin	5.0mg

Preparation of Neomycin Solution: Add neomycin to distilled/deionized water and bring volume to 10.0mL. Mix thoroughly. Filter sterilize.

Preparation of Medium: Add components, except neomycin solution, to distilled/deionized water and bring volume to 990.0mL. Mix thoroughly. Gently heat and bring to boiling. Autoclave for 15 min at 15 psi pressure–121°C. Cool to 45°–50°C. Aseptically add sterile neomycin solution. Mix thoroughly. Pour into sterile Petri dishes or distribute into sterile tubes.

Use: For the cultivation and maintenance of *Bacillus megaterium*.

Trypticase™ Soy Broth with 1*M* Potassium Chloride (LMG Medium 256)

Composition per liter:

KCl	74.5g
Pancreatic digest of casein	17.0g
NaCl	5.0g
Papaic digest of soybean meal	3.0g

K_2HPO_4	2.5g
Glucose	2.5g

pH 7.2 ± 0.2 at 25°C

Preparation of Medium: Add components to distilled/deionized water and bring volume to 1.0L. Mix thoroughly. Distribute into tubes or flasks. Autoclave for 15 min at 15 psi pressure–121°C.

Use: For the cultivation and maintenance of *Gracilibacillus dipsosauri.*

Trypticase™ Soy Broth with Sea Salts (LMG Medium 249)

Composition per liter:

Sea salts	20.0g
Pancreatic digest of casein	17.0g
NaCl	5.0g
Papaic digest of soybean meal	3.0g
K_2HPO_4	2.5g
Glucose	2.5g
Yeast extract	1.0g

pH 7.2 ± 0.2 at 25°C

Preparation of Medium: Add components to distilled/deionized water and bring volume to 1.0L. Mix thoroughly. Distribute into tubes or flasks. Autoclave for 15 min at 15 psi pressure–121°C.

Use: For the cultivation and maintenance of *Cellulophaga baltica* and *Cellulophaga fucicola.*

Trypticase™ Soy Broth with Sheep Blood (LMG Medium 189)

Composition per liter:

Pancreatic digest of casein	17.0g
NaCl	5.0g
Papaic digest of soybean meal	3.0g
K_2HPO_4	2.5g
Glucose	2.5g
Sheep blood, defibrinated	50.0mL

Preparation of Medium: Add components, except sheep blood, to distilled/deionized water and bring volume to 950.0mL. Mix thoroughly. Autoclave for 15 min at 15 psi pressure–121°C. Cool to 45°–50°C. Aseptically add 50.0mL of sterile, defibrinated sheep blood. Mix thoroughly. Aseptically distribute into sterile tubes or flasks.

Use: For the cultivation and maintenance of *Bartonella bacilliformis.*

Trypticase™ Soy Broth with 1.5% Sodium Chloride

Composition per liter:

Pancreatic digest of casein	17.0g
NaCl	15.0g
Papaic digest of soybean meal	3.0g
K_2HPO_4	2.5g
Glucose	2.5g

pH 7.3 ± 0.2 at 25°C

Preparation of Medium: Add components to distilled/deionized water and bring volume to 1.0L. Mix thoroughly. Distribute into tubes or flasks. Autoclave for 15 min at 15 psi pressure–121°C.

Use: For the cultivation and maintenance of *Pasteurella* species and *Listonella anguillarum.*

Trypticase™ Soy Broth with Sodium Chloride and Sodium Pyruvate

Composition per liter:

NaCl	100.0g
Pancreatic digest of casein	17.0g
Sodium pyruvate	10.0g
Papaic digest of soybean meal	3.0g
Glucose	2.5g
K_2HPO_4	2.5g

pH 7.3 ± 0.2 at 25°C

Preparation of Medium: Add components to distilled/deionized water and bring volume to 1.0L. Mix thoroughly. Gently heat until dissolved. Adjust pH to 7.3. Distribute into tubes in 10.0mL volumes. Autoclave for 15 min at 15 psi pressure–121°C.

Use: For the isolation and cultivation of *Staphylococcus aureus* from foods.

Trypticase™ Soy Broth with Starch (LMG Medium 232)

Composition per liter:

Pancreatic digest of casein	17.0g
Starch, soluble	10.0g
NaCl	5.0g
Papaic digest of soybean meal	3.0g
K_2HPO_4	2.5g
Glucose	2.5g

pH 7.2 ± 0.2 at 25°C

Preparation of Medium: Add components to distilled/deionized water and bring volume to 1.0L. Mix thoroughly. Distribute into tubes or flasks. Autoclave for 15 min at 15 psi pressure–121°C.

Use: For the cultivation and maintenance of *Bacillus naganoensis.*

Trypticase™ Soy Broth with Tobramycin

Composition per liter:

Pancreatic digest of casein	17.0g
NaCl	5.0g
Papaic digest of soybean meal	3.0g
K_2HPO_4	2.5g
Glucose	2.5g
Agar	1.0g
Tobramycin solution	10.0mL

pH 7.3 ± 0.2 at 25°C

Tobramycin Solution:

Composition per 10.0mL:

Tobramycin	8.0mg

Preparation of Tobramycin Solution: Add tobramycin to distilled/deionized water and bring volume to 10.0mL. Mix thoroughly. Filter sterilize.

Preparation of Medium: Add components, except tobramycin solution, to distilled/deionized water and bring volume to 990.0mL. Mix thoroughly. Autoclave for 15 min at 15 psi pressure–121°C. Cool to 45°–50°C. Aseptically add sterile tobramycin solution. Mix thoroughly. Aseptically distribute into sterile tubes.

Use: For the cultivation and maintenance of *Serratia marcescens.*

Trypticase™ Soy Broth with Tween™ 80

Composition per liter:

Pancreatic digest of casein	17.0g
NaCl	5.0g
Papaic digest of soybean meal	3.0g
K_2HPO_4	2.5g
Glucose	2.5g
Agar	1.0g
Tween™ 80	1.0mL

pH 7.3 ± 0.2 at 25°C

Preparation of Medium: Add components to distilled/deionized water and bring volume to 1.0L. Mix thoroughly. Distribute into tubes or flasks. Autoclave for 15 min at 15 psi pressure–121°C.

Use: For the cultivation and maintenance of *Corynebacterium genitalium*.

Trypticase™ Soy Broth with Yeast Extract

Composition per liter:

Pancreatic digest of casein	17.0g
Yeast extract	6.0g
NaCl	5.0g
Papaic digest of soybean meal	3.0g
K_2HPO_4	2.5g
Glucose	2.5g

pH 7.3 ± 0.2 at 25°C

Preparation of Medium: Add components to distilled/deionized water and bring volume to 1.0L. Mix thoroughly. Gently heat and bring to boiling. Distribute into tubes or flasks. Autoclave for 15 min at 15 psi pressure–121°C.

Use: For the cultivation and maintenance of a wide variety of heterotrophic microorganisms.

Trypticase™ Soy Broth with 0.1% Yeast Extract (LMG Medium 235)

Composition per liter:

Pancreatic digest of casein	17.0g
NaCl	5.0g
Papaic digest of soybean meal	3.0g
K_2HPO_4	2.5g
Glucose	2.5g
Yeast extract	1.0g

pH 7.2 ± 0.2 at 25°C

Preparation of Medium: Add components to distilled/deionized water and bring volume to 1.0L. Mix thoroughly. Distribute into tubes or flasks. Autoclave for 15 min at 15 psi pressure–121°C.

Use: For teh cultivation and maintenance of *Bacillus* spp. and other heterotrophic bacteria.

Trypticase Soy Broth without Glucose

Composition per liter:

Pancreatic digest of casein	17.0g
NaCl	5.0g
Papaic digest of soybean meal	3.0g
K_2HPO_4	2.5g

pH 7.3 ± 0.2 at 25°C

Source: This medium is available as a premixed powder from BD Diagnostic Systems.

Preparation of Medium: Add components to distilled/deionized water and bring volume to 1.0L. Mix thoroughly. Distribute into tubes or flasks. Autoclave for 15 min at 15 psi pressure–121°C.

Use: For the cultivation of a wide variety of microorganisms when the presence of carbohydrate is undesirable.

Trypticase™ Soy Broth with Yeast Extract (TSBYE)

Composition per liter:

Pancreatic digest of casein	17.0g
Yeast extract	6.0g
NaCl	5.0g
Papaic digest of soybean meal	3.0g
K_2HPO_4	2.5g
Glucose	2.5g

pH 7.3 ± 0.2 at 25°C

Preparation of Medium: Add components to distilled/deionized water and bring volume to 1.0L. Mix thoroughly. Gently heat and bring to boiling. Distribute into tubes or flasks. Autoclave for 15 min at 15 psi pressure–121°C.

Use: For the cultivation of *Listeria monocytogenes* from foods.

Trypticase™ Soy Glucose Medium

Composition per liter:

Glucose	50.0g
Pancreatic digest of casein	7.5g
Agar	7.5g
Papaic digest of soybean meal	2.5g
NaCl	2.5g

pH 7.3 ± 0.2 at 25°C

Preparation of Medium: Add components to distilled/deionized water and bring volume to 1.0L. Mix thoroughly. Gently heat and bring to boiling. Distribute into tubes or flasks. Autoclave for 15 min at 15 psi pressure–121°C.

Use: For the cultivation and maintenance of *Pseudomonas cepacia*.

Trypticase™ Soy Polymyxin Broth

Composition per 1006.67mL:

Pancreatic digest of casein	17.0g
NaCl	5.0g
Papaic digest of soybean meal	3.0g
K_2HPO_4	2.5g
Glucose	2.5g
Polymyxin B solution	6.67mL

pH 7.3 ± 0.2 at 25°C

Polymyxin B Solution:

Composition per 10.0mL:

Polymyxin B	0.015g

Preparation of Polymyxin B Solution: Add polymyxin B to distilled/deionized water and bring volume to 10.0mL. Mix thoroughly. Filter sterilize.

Preparation of Medium: Add components, except polymyxin B solution, to distilled/deionized water and bring volume to 1.0L. Mix thoroughly. Gently heat and bring to boiling. Distribute into tubes in 15.0mL volumes. Autoclave for 15 min at 15 psi pressure–121°C. Cool to 45°–50°C. Aseptically add 0.1mL of sterile polymyxin B solution to each tube. Mix thoroughly.

Use: For the isolation and cultivation of *Bacillus cereus* from foods.

Trypticase™ Soy Sheep Blood Agar (Tryptic Soy Blood Agar) (TSA Blood Agar) (BAM M159)

Composition per 1050.0mL:

Pancreatic digest of casein	15.0g
Agar	15.0g
Papaic digest of soybean meal	5.0g
NaCl	5.0g
Sheep blood, defibrinated	50.0mL

pH 7.3 ± 0.2 at 25°C

Preparation of Medium: Add components, except sheep blood, to distilled/deionized water and bring volume to 1.0L. Mix thoroughly. Gently heat and bring to boiling. Autoclave for 15 min at 15 psi pressure–121°C. Cool to 45°–50°C. Aseptically add 50.0mL sterile sheep blood. Mix thoroughly. Pour into sterile Petri dishes in 20.0mL volumes or distribute into sterile tubes.

Use: For the cultivation of a wide variety of fastidious microorganisms. For the observation of hemolytic reactions of a variety of bacteria. May be used to perform the CAMP test for the presumptive identification of group B streptococci (*Streptococcus agalactiae*).

Trypticase™ Soy Soil Extract

See: **TS Soil Extract**

Trypticase™ Soy Tryptose Broth

Composition per liter:

Pancreatic digest of casein	13.5g
Peptic digest of animal tissue	5.0g
NaCl	5.0g
Yeast extract	3.0g
Glucose	1.75g
Papaic digest of soybean meal	1.5g
K_2HPO_4	1.25g

pH 7.2 ± 0.2 at 25°C

Preparation of Medium: Add components to distilled/deionized water and bring volume to 1.0L. Mix thoroughly. Distribute into tubes in 5.0mL volumes. Autoclave for 15 min at 15 psi pressure–121°C.

Use: For the enrichment of *Salmonella* species from foods.

Trypticase™ Soy Yeast Extract Agar

Composition per liter:

Agar	28.0g
NaCl	10.0g
Pancreatic digest of casein	2.5g
Yeast extract	2.5g
Base medium	100.0mL
Phosphate buffer	100.0mL
Fe-citrate solution	5.0mL
Trace elements solution	5.0mL

pH 7.2 ± 0.2 at 25°C

Base Medium:

Composition per liter:

$MgCl_2 \cdot 6H_2O$	2.0g
Titriplex I	1.0g
$CaSO_4 \cdot 2H_2O$	0.4g
NaOH pellets	0.2g

Fe-Citrate Solution:

Composition per 100.0mL:

Fe-citrate	0.25g

Preparation of Fe-Citrate Solution: Add Fe-citrate to distilled/deionized water and bring volume to 100.0mL. Mix thoroughly.

Trace Elements Solution:

Composition per 5.0mL:

$CuCl_2 \cdot 2H_2O$	50.0mg
Titriplex I	12.8mg
$FeCl_2 \cdot 4H_2O$	1.0mg
$MnCl_2 \cdot 4H_2O$	0.5mg
$CoCl_2 \cdot 6H_2O$	0.3mg
$ZnCl_2$	0.2mg
$Na_2MoO_4 \cdot 2H_2O$	50.0mg
$NiCl_2 \cdot 6H_2O$	50.0mg
H_3BO_3	20.0mg

Preparation of Trace Elements Solution: Add components to distilled/deionized water and bring volume to 5.0mL. Mix thoroughly.

Preparation of Base Medium: Add components to distilled/deionized water and bring volume to 1.0L. Mix thoroughly.

Phosphate Buffer:

Composition per liter:

$Na_2HPO_4 \cdot 12H_2O$	43.0g
KH_2PO_4	5.44g

Preparation of Phosphate Buffer: Add components to distilled/deionized water and bring volume to 1.0L. Mix thoroughly. Autoclave for 15 min at 15 psi pressure–121°C. Cool to 50°C.

Preparation of Medium: Add components, except phosphate buffer, and bring volume to 900.0mL. Mix thoroughly. Gently heat and bring to boiling. Autoclave for 15 min at 15 psi pressure–121°C. Cool to 50°–55°C. Aseptically add 100.0mL of sterile phosphate buffer. Mix thoroughly. Pour into sterile Petri dishes or distribute into sterile tubes.

Use: For the cultivation and maintenance of *Lactococcus lactis* and *Lactococcus plantarum*.

Trypticase™ Soy Yeast Extract Agar

Composition per liter:

Pancreatic digest of casein	17.0g
Agar	15.0g
NaCl	5.0g
Yeast extract	3.0g
Papaic digest of soybean meal	3.0g
K_2HPO_4	2.5g
Glucose	2.5g

pH 7.1 ± 0.2 at 25°C

Preparation of Medium: Add components to distilled/deionized water and bring volume to 1.0L. Mix thoroughly. Gently heat and bring to boiling. Distribute into tubes or flasks. Autoclave for 15 min at 15 psi pressure–121°C. Pour into sterile Petri dishes or leave in tubes.

Use: For the cultivation of *Staphylococcus carnosus.*

Trypticase™ Soy Yeast Extract Medium

See: **TSY Medium**

Trypticase™ Soy Yeast Extract Medium (LMG Medium 217)

Composition per liter:

Pancreatic digest of casein	17.0g
NaCl	8.0g
Papaic digest of soybean meal	3.0g
K_2HPO_4	2.5g
Glucose	2.5g
Sheep blood, defibrinated	50.0mL

Preparation of Medium: Add components to distilled/deionized water and bring volume to 1.0L. Mix thoroughly. Distribute into tubes or flasks. Autoclave for 15 min at 15 psi pressure–121°C.

Use: For the cultivation and maintenance of *Staphylococcus carnosus* and *Brachybacterium* spp.

Trypticase™ Soy Yeast Extract Starch Medium

See: **TSYES Medium**

Trypticase™ Starch Agar (DSMZ Medium 56)

Composition per liter:

Pancreatic digest of casein	17.0g
Agar	15.0g
NaCl	5.0g
Papaic digest of soybean meal	3.0g
K_2HPO_4	2.5g
Glucose	2.5g
Yeast extract	2.0g
Soluble starch	1.0g

pH 7.3 ± 0.2 at 25°C

Preparation of Medium: Add components to distilled/deionized water and bring volume to 1.0L. Mix thoroughly. Gently heat and bring to boiling. Distribute into tubes or flasks. Autoclave for 15 min at 15 psi pressure–121°C. Pour into sterile Petri dishes or leave in tubes.

Use: For the cultivation and maintenance of *Bacillus* spp.

Trypticase™ Sulfite Neomycin Agar

See: **TSN Agar**

Trypticase™ Tellurite Agar Base

Composition per liter:

Agar	20.0g
Pancreatic digest of casein	10.0g
Peptic digest of animal tissue	10.0g
NaCl	5.0g
Glucose	2.0g
Serum	50.0mL
Chapman tellurite solution	10.0mL

pH 7.5 ± 0.2 at 25°C

Source: This medium is available as a premixed powder from BD Diagnostic Systems.

Chapman Tellurite Solution:

Composition per 100.0mL:

K_2TeO_3	1.0g

Preparation of Chapman Tellurite Solution: Add K_2TeO_3 to distilled/deionized water and bring volume to 100.0mL. Mix thoroughly. Filter sterilize.

Caution: Potassium tellurite is toxic.

Preparation of Medium: Add components, except serum and Chapman tellurite solution, to distilled/deionized water and bring volume to 940.0mL. Mix thoroughly. Gently heat and bring to boiling. Autoclave for 15 min at 15 psi pressure–121°C. Cool to 45°–50°C. Aseptically add sterile serum and sterile Chapman tellurite solution. Sheep serum, rabbit serum, or human serum may be used. Mix thoroughly. Pour into sterile Petri dishes.

Use: For the selective isolation of microorganisms from clinical specimens, especially from the nose, throat, and vagina.

Trypticase™ Yeast Extract Glucose Medium

Composition per liter:

Pancreatic digest of casein	10.0g
Glucose	10.0g
KH_2PO_4	6.8g
Yeast extract	5.0g
$NaHCO_3$	1.0g
Tween™ 80	0.5g
Sodium formaldehyde sulfoxalate	0.5g
$CaCl_2$	0.02g
$MgSO_4$	0.02g
NaCl	0.02g

pH 7.0 ± 0.2 at 25°C

Preparation of Medium: Add components to distilled/deionized water and bring volume to 1.0L. Mix thoroughly. Adjust pH to 7.0. Distribute into tubes or flasks. Autoclave for 15 min at 15 psi pressure–121°C.

Use: For the cultivation of *Propionibacterium* species.

Trypticase™ Yeast Extract Glucose Medium

See: **TYEG Medium**

Trypticase™ Yeast Extract Glucose Medium

See: **TYG Medium**

Tryptone Agar

Composition per liter:

Agar	15.0g
Pancreatic digest of casein	8.0g
NaCl	8.0g

pH 7.2 ± 0.2 at 25°C

Preparation of Medium: Add components to distilled/deionized water and bring volume to 1.0L. Mix thoroughly. Gently heat and bring to boiling. Distribute into tubes or flasks. Autoclave for 15 min at 15 psi pressure–121°C. Pour into sterile Petri dishes or leave in tubes.

Use: For the cultivation and maintenance of fastidious aerobic and facultative microorganisms such as *Escherichia coli* and *Pseudomonas* species.

Tryptone Agar Base, HiVeg

Composition per liter:

Plant hydrolysate	20.0g
Agar	3.5g
Phenol Red	0.02g

pH 7.2 ± 0.2 at 25°C

Source: This medium is available as a premixed powder from HiMedia.

Preparation of Medium: Add components to distilled/deionized water and bring volume to 1.0L. Mix thoroughly. Gently heat and bring

to boiling. Distribute into tubes or flasks. Autoclave for 15 min at 15 psi pressure–121°C. Pour into sterile Petri dishes or leave in tubes.

Use: For the cultivation and maintenance of fastidious aerobic and facultative microorganisms such as *Escherichia coli* and *Pseudomonas* species.

Tryptone Agar, HiVeg

Composition per liter:

Plant hydrolysate	20.0g
Agar	15.0g
Synthetic detergent No. 1	1.5g

pH 7.2 ± 0.2 at 25°C

Source: This medium is available as a premixed powder from HiMedia.

Preparation of Medium: Add components to distilled/deionized water and bring volume to 1.0L. Mix thoroughly. Gently heat and bring to boiling. Distribute into tubes or flasks. Autoclave for 15 min at 15 psi pressure–121°C. Pour into sterile Petri dishes or leave in tubes.

Use: For the cultivation and maintenance of fastidious aerobic and facultative microorganisms such as *Escherichia coli* and *Pseudomonas* species.

Tryptone Beef Yeast Extract Acetate Agar

See: **TBYA Agar**

Tryptone Bile Agar

Composition per liter:

Pancreatic digest of casein	20.0g
Agar	15.0g
Bile salts No. 3	1.5g

pH 7.2 ± 0.2 at 25°C

Source: This medium is available as a premixed powder from BD Diagnostic Systems and Oxoid Unipath.

Preparation of Medium: Add components to distilled/deionized water and bring volume to 1.0L. Mix thoroughly. Gently heat and bring to boiling. Distribute into tubes or flasks. Autoclave for 15 min at 15 psi pressure–121°C. Pour into sterile Petri dishes.

Use: For the isolation and enumeration of *Escherichia coli* from foods.

Tryptone Bile Glucuronide Agar, Harlequin (Harlequin TBGA)

Composition per liter:

Tryptone	20.0g
Agar	15.0g
Bile salts No. 3	1.5g
X-glucuronide	0.075g

pH 7.2 ± 0.2 at 25°C

Source: This medium is available from lab m.

Preparation of Medium: Add components to distilled/deionized water and bring volume to 1.0L. Mix thoroughly. Allow to soak for 10 min. Autoclave for 15 min at 15 psi pressure–121°C. Cool to 45°–50°C. Pour into sterile Petri dishes.

Use: For the simple enumeration of *E. coli* without the need for membranes or pre-incubation The medium has been modified by the addition of a chromogenic substrate to detect the β-glucuronidase, which is highly specific for *E. coli.* The advantage of the chromogenic substrate is that it requires no UV lamp to visualize the reaction, and it is concentrated within the colony, facilitating easier enumeration in the presence of other organisms, or when large numbers are present on the plate.

Tryptone Bile X-glucuronide Agar, Chromocult (Chromocult® TBX) (ChromocultTryptone Bile X-glucuronide Agar)

Composition per liter:

Peptone	20.0g
Agar	15.0g
Bile salts No. 3	1.5g
X-β-D-glucuronide	0.075g

pH 7.2 ± 0.2 at 25°C

Source: This medium is available from Merck.

Preparation of Medium: Add components to distilled/deionized water and bring volume to 1.0L. Mix thoroughly. Autoclave for 15 min at 15 psi pressure–121°C. Cool to 45°–50°C. Pour into sterile Petri dishes.

Use: For the differentiation of *E. coli* from other coliforms. The presence of the enzyme β-D-glucuronidase differentiates most *E. coli* spp. from other coliforms. *E. coli* absorbs the chromogenic substrate 5-bromo-4-chloro-3-indolyl-β-D-glucuronide (X-β-D-glucuronide). The enzyme β-glucuronidase splits the bond between the chromophore 5-bromo-4-chloro-3-indolyle and the β-D-glucuronide. *E. coli* colonies are colored blue-green. Growth of accompanying Gram-positive flora is largely inhibited by the use of bile salts. The prepared medium is clear and yellowish.

Tryptone Broth

Composition per liter:

Pancreatic digest of casein	10.0g
Glucose	5.0g
K_2HPO_4	1.25g
Yeast extract	1.0g
Bromcresol Purple solution	2.0mL

Bromcresol Purple Solution:

Composition per 100.0mL:

Bromcresol Purple	2.0g
Ethanol	10.0mL

Preparation of Bromcresol Purple Solution: Add Bromcresol Purple to 10.0mL of ethanol. Mix thoroughly. Bring volume to 100.0mL with distilled/deionized water.

Preparation of Medium: Add components to distilled/deionized water and bring volume to 1.0L. Mix thoroughly. Distribute into screw-capped tubes in 10.0mL volumes. Autoclave for 20 min at 15 psi pressure–121°C.

Use: For the cultivation of *Salmonella* species from foods.

Tryptone Broth

See: **T_1N_0 Broth**

Tryptone Broth

See: **Tryptone Water Broth**

Tryptone Broth (ATCC Medium 274)

Composition per liter:

Pancreatic digest of casein	10.0g

pH 7.2 ± 0.2 at 25°C

Preparation of Medium: Add pancreatic digest of casein to distilled/deionized water and bring volume to 1.0L. Mix thoroughly. Distribute into tubes or flasks. Autoclave for 15 min at 15 psi pressure–121°C.

Use: For the cultivation and maintenance of fastidious aerobic and facultative microorganisms such as *Escherichia coli* and *Pseudomonas* species.

Tryptone Broth, 1% (ATCC Medium 274)

Composition per liter:

Pancreatic digest of casein	1.0g

pH 7.2 ± 0.2 at 25°C

Preparation of Medium: Add pancreatic digest of casein to distilled/deionized water and bring volume to 1.0L. Mix thoroughly. Distribute into tubes or flasks. Autoclave for 15 min at 15 psi pressure–121°C.

Use: For the cultivation and maintenance of *Escherichia coli* and *Pseudomonas* species. For the identification and confirmation of *Vibrio cholerae* in foods.

Tryptone Broth with $CaCl_2$

Composition per liter:

Pancreatic digest of casein	10.0g
$CaCl_2$	5.5g

pH 7.2 ± 0.2 at 25°C

Preparation of Medium: Add components to distilled/deionized water and bring volume to 1.0L. Mix thoroughly. Distribute into tubes or flasks. Autoclave for 15 min at 15 psi pressure–121°C.

Use: For the cultivation and maintenance of fastidious aerobic and facultative microorganisms such as *Escherichia coli.*

Tryptone Broth, HiVeg (Tryptone Water, HiVeg)

Composition per liter:

Plant hydrolysate	10.0g
NaCl	5.0g

pH 7.5 ± 0.2 at 25°C

Source: This medium is available as a premixed powder from HiMedia.

Preparation of Medium: Add components to distilled/deionized water and bring volume to 1.0L. Mix thoroughly. Gently heat and bring to boiling. Distribute into tubes or flasks. Autoclave for 15 min at 15 psi pressure–121°C.

Use: For the detection of indole-producing microorganisms.

Tryptone Glucose Beef Extract Agar (Tryptone Glucose Extract Agar)

Composition per liter:

Agar	15.0g
Pancreatic digest of casein	5.0g
Beef extract	3.0g
Glucose	1.0g

pH 7.0 ± 0.2 at 25°C

Source: This medium is available as a premixed powder from BD Diagnostic Systems and Oxoid Unipath.

Preparation of Medium: Add components to distilled/deionized water and bring volume to 1.0L. Mix thoroughly. Gently heat and bring to boiling. Distribute into tubes or flasks. Autoclave for 15 min at 15 psi pressure–121°C. Cool to 45°–50°C. If the dilution of the specimen is greater than 1:10, add 10.0mL of sterile 10% skim milk solution. Mix thoroughly. Pour into sterile Petri dishes or leave in tubes.

Use: For the enumeration of bacteria by the standard plate count procedure. For the cultivation and enumeration of bacteria from milk and dairy products. For the detection of thermophilic organisms.

Tryptone Glucose Beef Extract Agar with Sucrose

Composition per liter:

Agar	15.0g
Pancreatic digest of casein	5.0g
Beef extract	3.0g
Glucose	1.0g
Sucrose	5.0g

pH 7.0 ± 0.2 at 25°C

Preparation of Medium: Add components to distilled/deionized water and bring volume to 1.0L. Mix thoroughly. Gently heat and bring to boiling. Distribute into tubes or flasks. Autoclave for 15 min at 15 psi pressure–121°C. Pour into sterile Petri dishes or leave in tubes.

Use: For the cultivation and maintenance of *Saccharococcus thermophilus*.

Tryptone Glucose Beef Extract Agar with Yeast Extract

Composition per liter:

Agar	15.0g
Pancreatic digest of casein	5.0g
Beef extract	3.0g
Glucose	1.0g
Yeast extract	1.0g

pH 7.0 ± 0.2 at 25°C

Preparation of Medium: Add components to distilled/deionized water and bring volume to 1.0L. Mix thoroughly. Gently heat and bring to boiling. Distribute into tubes or flasks. Autoclave for 15 min at 15 psi pressure–121°C. Pour into sterile Petri dishes or leave in tubes.

Use: For the cultivation and maintenance of *Ancylobacter aquaticus* and *Spirosoma linguale*.

Tryptone Glucose Extract Agar

See: **Tryptone Glucose Beef Extract Agar**

Tryptone Glucose Extract HiVeg Agar (Tryptone Glucose Yeast Extract HiVeg Agar)

Composition per liter:

Agar	15.0g
Plant hydrolysate	5.0g
Yeast extract	3.0g
Glucose	1.0g

pH 7.0 ± 0.2 at 25°C

Source: This medium is available as a premixed powder from HiMedia.

Preparation of Medium: Add components to distilled/deionized water and bring volume to 1.0L. Mix thoroughly. Gently heat and bring to boiling. Distribute into tubes or flasks. Autoclave for 15 min at 15 psi pressure–121°C.

Use: For the enumeration of bacteria by the standard plate count procedure. For the cultivation and enumeration of bacteria from milk and dairy products.

Tryptone Glucose HiVeg Agar

Composition per liter:

Plant hydrolysate	20.0g
Glucose	5.0g
Agar	3.5g
Bromthymol Blue	0.01g

pH 7.0 ± 0.2 at 25°C

Source: This medium is available as a premixed powder from HiMedia.

Preparation of Medium: Add components to distilled/deionized water and bring volume to 1.0L. Mix thoroughly. Gently heat and bring to boiling. Distribute into tubes or flasks. Autoclave for 15 min at 15 psi pressure–121°C.

Use: For the enumeration of bacteria by the standard plate count procedure. For the cultivation and enumeration of bacteria from milk and dairy products.

Tryptone Glucose Medium

Composition per liter:

Pancreatic digest of casein	20.0g
Glucose	5.0g
K_2HPO_4	0.35g
KH_2PO_4	0.2g
$MgCl_2·6H_2O$	0.1g
$CaCl_2·2H_2O$	0.07g
Vitamin solution	10.0mL

Vitamin Solution:

Composition per 10.0mL:

Thiamine	200.0μg
Biotin	50.0μg

Preparation of Vitamin Solution: Add components to distilled/deionized water and bring volume to 10.0mL. Mix thoroughly. Filter sterilize.

Preparation of Medium: Add components, except vitamin solution, to distilled/deionized water and bring volume to 990.0mL. Mix thoroughly. Gently heat and bring to boiling. Autoclave for 15 min at 15 psi pressure–121°C. Aseptically add 10.0mL of sterile vitamin solution. Mix thoroughly. Aseptically distribute into sterile flasks or tubes.

Use: For the cultivation and maintenance of *Amoebidium parasiticum, Capniomyces stellatus, Smittium culicis, Smittium culisetae, Smittium simulii,* and *Smittium* species.

Tryptone Glucose Yeast Extract HiVeg Broth

Composition per liter:

Plant hydrolysate	10.0g
Glucose	5.0g
K_2HPO_4	1.25g
Yeast extract	1.0g

pH 6.8 ± 0.2 at 25°C

Source: This medium is available as a premixed powder from HiMedia.

Preparation of Medium: Add components to distilled/deionized water and bring volume to 1.0L. Mix thoroughly. Gently heat and bring to boiling. Distribute into tubes or flasks. Autoclave for 15 min at 15 psi pressure–121°C.

Use: For the enumeration of bacteria by the standard plate count procedure. For the cultivation and enumeration of bacteria from milk and dairy products.

Tryptone with Sodium Chloride Broth

Composition per liter:

Pancreatic digest of casein	8.0g
NaCl	0.5g

Preparation of Medium: Add components to distilled/deionized water and bring volume to 1.0L. Mix thoroughly. Distribute into tubes or flasks. Autoclave for 15 min at 15 psi pressure–121°C.

Use: For the cultivation and maintenance of fastidious aerobic and facultative microorganisms such as *Escherichia coli* and *Pseudomonas* species.

Tryptone Glucose Yeast Agar
See: **Plate Count Agar**

Tryptone Glucose Yeast Agar
See: **Standard Methods Agar**

Tryptone Glucose Yeast Broth
See: **Standard Methods Broth**

Tryptone Glucose Yeast Extract Medium
See: **TGY Medium**

Tryptone Glucose Yeast Extract Methionine Medium
See: **TGYM Medium**

Tryptone Lactose Iron HiVeg Agar

Composition per liter:

Plant hydrolysate	20.0g
Lactose	10.0g
Agar	3.5g
Na_2SO_3	0.4g
$FeSO_4$	0.2g
$Na_2S_2O_3$	0.08g
Phenol Red	0.02g

pH 7.2 ± 0.2 at 25°C

Source: This medium is available as a premixed powder from HiMedia.

Preparation of Medium: Add components to distilled/deionized water and bring volume to 1.0L. Mix thoroughly. Gently heat and bring to boiling. Distribute into tubes or flasks. Autoclave for 15 min at 15 psi pressure–121°C. Pour into sterile Petri dishes or leave in tubes.

Use: For the cultivation and maintenance of fastidious aerobic and facultative microorganisms such as *Escherichia coli* and *Pseudomonas* species.

Tryptone NaCl Thiamine Medium
See: **TNT Medium**

Tryptone Nitrate HiVeg Medium (Indole Nitrate HiVeg Medium)

Composition per liter:

Plant peptone	20.0g
Na_2HPO_4	2.0g
Agar	1.0g
Glucose	1.0g
KNO_3	1.0g

pH 7.2 ± 0.2 at 25°C

Source: This medium is available as a premixed powder from BD Diagnostic Systems.

Preparation of Medium: Add components to distilled/deionized water and bring volume to 1.0L. Mix thoroughly. Gently heat and bring to boiling with frequent agitation. Distribute into tubes or flasks. Autoclave for 15 min at 15 psi pressure–121°C.

Use: For the identification of microorganisms by means of the nitrate reduction and indole tests.

Tryptone Peptone Glucose Yeast Extract HiVeg Broth Base without Trypsin

Composition per liter:

Plant hydrolysate	50.0g
Yeast extract	20.0g
Plant peptone	5.0g
Glucose	4.0g
Na-thioglycolate	1.0g

pH 7.0 ± 0.2 at 25°C

Source: This medium is available as a premixed powder from HiMedia.

Preparation of Medium: Add components to distilled/deionized water and bring volume to 1.0L. Mix thoroughly. Gently heat and bring to boiling. Distribute into tubes or flasks. Autoclave for 15 min at 15 psi pressure–121°C.

Use: For the enumeration of bacteria by the standard plate count procedure. For the cultivation and enumeration of bacteria from milk and dairy products. For the detection of thermophilic organisms.

Tryptone Phosphate Brain Heart Infusion Yeast Extract Agar (TPBY)

Composition per liter:

Pancreatic digest of casein	20.0g
Agar	15.0g
NaCl	5.14g
K_2HPO_4	2.0g
KH_2PO_4	2.0g
Yeast extract	1.0g
Oxgall	0.5g
Pancreatic digest of gelatin	0.4g
Brain heart, solids from infusion	0.16g
Peptic digest of animal tissue	0.16g
Glucose	0.08g
Na_2HPO_4	0.06g
Tween™ 80	1.5mL

pH 7.0 ± 0.2 at 25°C

Preparation of Medium: Add components to distilled/deionized water and bring volume to 1.0L. Mix thoroughly. Gently heat and bring to boiling. Adjust pH to 7.0. Distribute into tubes or flasks. Autoclave for 15 min at 15 psi pressure–121°C. Pour into sterile Petri dishes or leave in tubes.

Use: For the cultivation of coliform bacteria, such as *Escherichia coli*, from foods.

Tryptone Phosphate Broth

Composition per liter:

Pancreatic digest of casein	20.0g
NaCl	5.0g
K_2HPO_4	2.0g
KH_2PO_4	2.0g
Tween™ 80	15.0mL

pH 7.0 ± 0.2 at 25°C

Preparation of Medium: Add components to distilled/deionized water and bring volume to 1.0L. Mix thoroughly. Distribute into tubes or flasks. Autoclave for 15 min at 15 psi pressure–121°C.

Use: For the cultivation of enteropathogenic *Escherichia coli.*

Tryptone Phosphate Broth

Composition per liter:

Pancreatic digest of casein	20.0g
NaCl	5.0g
K_2HPO_4	2.0g
KH_2PO_4	2.0g
Tween™ 80	1.5mL

pH 7.0 ± 0.2 at 25°C

Preparation of Medium: Add components to distilled/deionized water and bring volume to 1.0L. Mix thoroughly. Distribute into tubes or flasks. Autoclave for 15 min at 15 psi pressure–121°C.

Use: For the cultivation of coliform bacteria, such as *Escherichia coli*, from foods.

Tryptone Phosphate HiVeg Broth

Composition per liter:

Plant hydrolysate	20.0g
NaCl	5.0g
K_2HPO_4	2.0g
KH_2PO_4	2.0g
Polysorbate 80	1.5g

pH 7.0 ± 0.2 at 25°C

Source: This medium is available as a premixed powder from HiMedia.

Preparation of Medium: Add components to distilled/deionized water and bring volume to 1.0L. Mix thoroughly. Distribute into tubes or flasks. Autoclave for 15 min at 15 psi pressure–121°C.

Use: For the cultivation of enteropathogenic *Escherichia coli.* For the enrichment and cultivation of enteropathogenic *Escherichia coli* from suspected food samples.

Tryptone Salt Agar
See: **T_1N_1 Agar**

Tryptone Salt Agar
See: **T_1N_2 Agar**

Tryptone Salt Broth
See: T_1N_1 Broth

Tryptone Salt Broth
See: T_1N_3 Broth

Tryptone Salt Broth
See: T_1N_6 Broth

Tryptone Salt Broth
See: T_1N_8 Broth

Tryptone Salt Broth
See: T_1N_{10} Broth

Tryptone in Seawater Agar

Composition per liter:

Agar 15.0g
Pancreatic digest of casein 1.0g

pH 7.0 ± 0.2 at 25°C

Preparation of Medium: Add components to seawater and bring volume to 1.0L. Mix thoroughly. Gently heat and bring to boiling. Distribute into tubes or flasks. Autoclave for 15 min at 15 psi pressure–121°C. Pour into sterile Petri dishes or leave in tubes.

Use: For the cultivation and maintenance of *Bacillus pacificus.*

Tryptone Soy Agar

Composition per liter:

Agar 15.0g
Pancreatic digest of casein 15.0g
NaCl 5.0g
Pancreatic digest of soybean meal 5.0g

pH 7.3 ± 0.2 at 25°C

Source: This medium is available as a premixed powder from Oxoid Unipath.

Preparation of Medium: Add components to distilled/deionized water and bring volume to 1.0L. Mix thoroughly. Gently heat and bring to boiling. Distribute into tubes or flasks. Autoclave for 15 min at 15 psi pressure–121°C. Pour into sterile Petri dishes or leave in tubes.

Use: For the cultivation and maintenance of a wide variety of microorganisms.

Tryptone Soy Agar pH 5.5 (LMG Medium 241)

Composition per liter:

Tryptone 15.0g
Agar 15.0g
Soy peptone 5.0g
NaCl 5.0g

pH 5.5 ± 0.2 at 25°C

Preparation of Medium: Add components to distilled/deionized water and bring volume to 1.0L. Mix thoroughly. Adjust pH to 5.5. Gently heat and bring to boiling. Distribute into tubes or flasks. Autoclave for 15 min at 15 psi pressure–121°C. Pour into sterile Petri dishes or leave in tubes.

Use: For the cultivation and maintenance of heterotrophic bacteria.

Tryptone Soy Agar pH 6.5 (LMG Medium 242)

Composition per liter:

Tryptone 15.0g
Agar 15.0g
Soy peptone 5.0g
NaCl 5.0g

pH 6.5 ± 0.2 at 25°C

Preparation of Medium: Add components to distilled/deionized water and bring volume to 1.0L. Mix thoroughly. Adjust pH to 6.5. Gently heat and bring to boiling. Distribute into tubes or flasks. Autoclave for 15 min at 15 psi pressure–121°C. Pour into sterile Petri dishes or leave in tubes.

Use: For the cultivation and maintenance of heterotrophic bacteria.

Tryptone Soy Agar, HiVeg (Antibiotic Assay Medium - J, HiVeg)

Composition per liter:

Agar 15.0g
Plant hydrolysate 15.0g
NaCl 5.0g
Papaic digest of soybean meal 5.0g

pH 7.3 ± 0.1 at 25°C

Source: This medium is available as a premixed powder from HiMedia.

Preparation of Medium: Add components to distilled/deionized water and bring volume to 1.0L. Mix thoroughly. Gently heat and bring to boiling. Distribute into tubes or flasks. Autoclave for 15 min at 15 psi pressure–121°C. Pour into sterile Petri dishes.

Use: A general purpose medium for cultivating a wide variety of fastidious microorganisms.

Tryptone Soy Agar with Horse Blood (LMG Medium 50)

Composition per liter:

Tryptone 15.0g
Agar 15.0g
Soy peptone 5.0g
NaCl 5.0g
Horse blood, sterile defibrinated 50.0mL

pH 7.3 ± 0.2 at 25°C

Preparation of Medium: Add components, except horse blood, to distilled/deionized water and bring volume to 950.0mL. Mix thoroughly. Gently heat and bring to boiling. Autoclave for 15 min at 15 psi pressure–121°C. Cool to 45°–50°C. Aseptically add 50.0mL sterile horse blood. Mix thoroughly. Pour into sterile Petri dishes or distribute into sterile tubes.

Use: For the cultivation and maintenance of *Paenibacillus lentimorbus.*

Tryptone Soy Agar with Horse Blood

Composition per liter:

Agar 15.0g
Pancreatic digest of casein 15.0g
NaCl 5.0g

Pancreatic digest of soybean meal 5.0g
Horse blood, defibrinated 50.0mL
pH 7.3 ± 0.2 at 25°C

Preparation of Medium: Add components, except horse blood, to distilled/deionized water and bring volume to 950.0mL. Mix thoroughly. Heat with frequent agitation and boil for 1 min to completely dissolve. Autoclave for 15 min at 15 psi pressure–121°C. Cool to 45°–50°C. Aseptically add 50.0mL of sterile, defibrinated horse blood. Mix thoroughly and pour into sterile Petri dishes.

Use: For cultivation and maintenance of *Bacillus lentimorbus*.

Tryptone Soy Agar with 1% Sodium Chloride

Composition per liter:
Agar 15.0g
Pancreatic digest of casein 15.0g
NaCl 15.0g
Pancreatic digest of soybean meal 5.0g
Cocarboxylase solution 100.0mL
pH 7.3 ± 0.2 at 25°C

Cocarboxylase Solution:
Composition per 100.0mL:
Cocarboxylase 2.0mg

Preparation of Cocarboxylase Solution: Add cocarboxylase to distilled/deionized water and bring volume to 100.0mL. Mix thoroughly. Filter sterilize.

Preparation of Medium: Add components, except cocarboxylase solution, to distilled/deionized water and bring volume to 900.0mL. Mix thoroughly. Gently heat and bring to boiling. Autoclave for 15 min at 15 psi pressure–121°C. Cool to 50°–55°C. Aseptically add 100.0mL of sterile cocarboxylase solution. Mix thoroughly. Pour into sterile Petri dishes or distribute into sterile tubes.

Use: For the cultivation and maintenance of *Haemophilus piscium*.

Tryptone Soy Agar with 4.5% Sodium Chloride

Composition per liter:
NaCl 45.0g
Pancreatic digest of casein 17.0g
Agar 12.0g
Papaic digest of soybean meal 3.0g
Glucose 2.5g
K_2HPO_4 .... 2.5g
pH 7.0 ± 0.2 at 25°C

Preparation of Medium: Add components to distilled/deionized water and bring volume to 1.0L. Mix thoroughly. Gently heat and bring to boiling. Distribute into tubes or flasks. Autoclave for 15 min at 15 psi pressure–121°C. Pour into sterile Petri dishes or leave in tubes.

Use: For the isolation and cultivation of *Brevibacterium linens*.

Tryptone Soy Agar with Sheep Blood (LMG Medium 175)

Composition per liter:
Agar 15.0g
Pancreatic digest of casein 15.0g
NaCl 5.0g
Pancreatic digest of soybean meal 5.0g
Sheep blood, defibrinated 50.0mL
pH 7.3 ± 0.2 at 25°C

Preparation of Medium: Add components, except sheep blood, to distilled/deionized water and bring volume to 950.0mL. Mix thoroughly. Heat with frequent agitation and boil for 1 min to completely dissolve. Autoclave for 15 min at 15 psi pressure–121°C. Cool to 45°–50°C. Aseptically add 50.0mL of sterile, defibrinated sheep blood. Mix thoroughly and pour into sterile Petri dishes.

Use: For the cultivation and maintenance of *Bacillus* spp.

Tryptone Soy Broth

Composition per liter:
Pancreatic digest of casein 17.0g
NaCl 5.0g
Pancreatic digest of soybean meal 3.0g
K_2HPO_4 .... 2.5g
Glucose 2.5g
pH 7.3 ± 0.2 at 25°C

Source: This medium is available as a premixed powder from Oxoid Unipath.

Preparation of Medium: Add components to distilled/deionized water and bring volume to 1.0L. Mix thoroughly. Distribute into tubes or flasks. Autoclave for 15 min at 15 psi pressure–121°C.

Use: For the cultivation of a wide variety of microorganisms.

Tryptone Soy Broth, HiVeg (Antibiotic Assay No. 37, HiVeg)

Composition per liter:
Plant hydrolysate 15.0g
NaCl 5.0g
Papaic digest of soybean meal 5.0g
pH 7.3 ± 0.1 at 25°C

Source: This medium is available as a premixed powder from HiMedia.

Preparation of Medium: Add components to distilled/deionized water and bring volume to 1.0L. Mix thoroughly. Distribute into tubes or flasks. Autoclave for 10 min at 15 psi pressure–121°C.

Use: A general purpose medium for cultivating a wide variety of fastidious microorganisms.

Tryptone Soy Broth with 0.1% Tween™ 80

Composition per liter:
Pancreatic digest of casein 15.0g
Papaic digest of soybean meal 5.0g
NaCl 5.0g
Tween™ 80 .... 1.0g

Preparation of Medium: Add components to distilled/deionized water and bring volume to 1.0L. Mix thoroughly. Distribute into tubes or flasks. Autoclave for 15 min at 15 psi pressure–121°C.

Use: For the cultivation of *Corynebacterium genitalium* and *Corynebacterium pseudogenitalium*.

Tryptone Soy Broth with Yeast Extract

Composition per liter:
Pancreatic digest of casein 17.0g
NaCl 5.0g
Yeast extract 3.0g
Pancreatic digest of soybean meal 3.0g

K_2HPO_4 .... 2.5g
Glucose 2.5g

pH 7.3 ± 0.2 at 25°C

Preparation of Medium: Add components to distilled/deionized water and bring volume to 1.0L. Mix thoroughly. Distribute into tubes or flasks. Autoclave for 15 min at 15 psi pressure–121°C.

Use: For the cultivation of *Carnobacterium piscicola.*

Tryptone Soy HiVeg Agar with Added Sodium Chloride and Cocarboxylase

Composition per liter:

Agar 15.0g
Plant hydrolysate 15.0g
NaCl 10.0g
Papaic digest of soybean meal 5.0g
Cocarboxylase solution 100.0mL

pH 7.3 ± 0.2 at 25°C

Source: This medium, without cocarboxylase, is available as a premixed powder from HiMedia.

Cocarboxylase Solution:

Composition per 100.0mL:

Cocarboxylase 2.0mg

Preparation of Cocarboxylase Solution: Add cocarboxylase to distilled/deionized water and bring volume to 100.0mL. Mix thoroughly. Filter sterilize.

Preparation of Medium: Add components, except cocarboxylase solution, to distilled/deionized water and bring volume to 900.0mL. Mix thoroughly. Gently heat and bring to boiling. Autoclave for 15 min at 15 psi pressure–121°C. Cool to 50°–55°C. Aseptically add 100.0mL of sterile cocarboxylase solution. Mix thoroughly. Pour into sterile Petri dishes or distribute into sterile tubes.

Use: For the cultivation and maintenance of *Haemophilus piscium.*

Tryptone Soy HiVeg Agar with Lecithin and Polysorbate 80

Composition per liter:

Agar 15.0g
Plant hydrolysate 15.0g
Papaic digest of soybean meal 5.0g
NaCl 5.0g
Lecithin 0.7g
Polysorbate 80 .... 5.0mL

pH 7.3 ± 0.2 at 25°C

Source: This medium, without polysorbate 80, is available as a premixed powder from HiMedia.

Preparation of Medium: Add components to tap water and bring volume to 1.0L. Mix thoroughly. Gently heat and bring to boiling. Distribute into tubes or flasks. Autoclave for 15 min at 15 psi pressure–121°C. Pour into sterile Petri dishes or leave in tubes.

Use: For determining the efficiency of sanitization of containers, surfaces, and water-miscible cosmetics.

Tryptone Soy HiVeg Agar with Magnesium Sulfate (TSAM HiVeg)

Composition per liter:

Agar 15.0g
Plant hydrolysate 15.0g
Plant peptone 5.0g
NaCl 5.0g
$MgSO_4$ 1.5g

pH 7.3 ± 0.2 at 25°C

Source: This medium is available as a premixed powder from HiMedia.

Preparation of Medium: Add components to tap water and bring volume to 1.0L. Mix thoroughly. Gently heat and bring to boiling. Distribute into tubes or flasks. Autoclave for 15 min at 15 psi pressure–121°C. Pour into sterile Petri dishes or leave in tubes.

Use: For the cultivation of *Escherichia coli* for bacteriophage production.

Tryptone Soy HiVeg Broth with 0.1% Agar (Soybean HiVeg Medium with 0.1% Agar)

Composition per liter:

Plant hydrolysate 17.0g
NaCl 5.0g
Papaic digest of soybean meal 3.0g
Glucose 2.5g
K_2HPO_4 .... 2.5g
Agar 1.0g

pH 7.3 ± 0.2 at 25°C

Source: This medium is available as a premixed powder from HiMedia.

Preparation of Medium: Add components to distilled/deionized water and bring volume to 1.0L. Mix thoroughly. Gently heat and bring to boiling. Distribute into tubes or flasks. Autoclave for 15 min at 15 psi pressure–121°C.

Use: For the cultivation of anaerobes from root canals, blood, and other clinical specimens and for determining glucose fermentation.

Tryptone SoyHiVeg Broth without Glucose (Soybean HiVeg Medium)

Composition per liter:

Plant hydrolysate 17.0g
NaCl 5.0g
Papaic digest of soybean meal 3.0g
K_2HPO_4 .... 2.5g

pH 7.3 ± 0.2 at 25°C

Source: This medium is available as a premixed powder from HiMedia.

Preparation of Medium: Add components to distilled/deionized water and bring volume to 1.0L. Mix thoroughly. Gently heat and bring to boiling. Distribute into tubes or flasks. Autoclave for 15 min at 15 psi pressure–121°C.

Use: For the cultivation of a wide variety of microorganisms. For sterility testing.

Tryptone Soy HiVeg Broth with 10% Sodium Chloride and 1% Sodium Pyruvate

Composition per liter:

NaCl 105.0g
Plant hydrolysate 17.0g
Sodium pyruvate 10.0g
Papaic digest of soybean meal 3.0g

Glucose 2.5g
K_2HPO_4.......... 2.5g

pH 7.3 ± 0.2 at 25°C

Source: This medium is available as a premixed powder from HiMedia.

Preparation of Medium: Add components to distilled/deionized water and bring volume to 1.0L. Mix thoroughly. Gently heat until dissolved. Adjust pH to 7.3. Distribute into tubes in 10.0mL volumes. Autoclave for 15 min at 15 psi pressure–121°C.

Use: For the isolation and cultivation of *Staphylococcus aureus* from foods.

Tryptone Soy Salt HiVeg Agar with Magnesium Sulfate

Composition per liter:

Plant hydrolysate.......... 50.0g
NaCl.......... 30.0g
Agar.......... 15.0g
Papaic digest of soybean meal.......... 5.0g
$MgSO_4$.......... 1.5g

pH 7.3 ± 0.2 at 25°C

Source: This medium is available as a premixed powder from HiMedia.

Preparation of Medium: Add components to tap water and bring volume to 1.0L. Mix thoroughly. Gently heat and bring to boiling. Distribute into tubes or flasks. Autoclave for 15 min at 15 psi pressure–121°C. Pour into sterile Petri dishes or leave in tubes.

Use: For the cultivation of *Escherichia coli* for bacteriophage production.

Tryptone Soy Yeast Extract HiVeg Agar

Composition per liter:

Plant hydrolysate.......... 17.0g
Agar.......... 15.0g
Yeast extract.......... 6.0g
NaCl.......... 5.0g
Papaic digest of soybean meal.......... 3.0g
Glucose.......... 2.5g
K_2HPO_4.......... 2.5g

pH 7.3 ± 0.2 at 25°C

Source: This medium is available as a premixed powder from HiMedia.

Preparation of Medium: Add components to distilled/deionized water and bring volume to 1.0L. Mix thoroughly. Gently heat and bring to boiling. Distribute into tubes or flasks. Autoclave for 15 min at 15 psi pressure–121°C. Pour into sterile Petri dishes or leave in tubes.

Use: For the cultivation and maintenance of a wide variety of heterotrophic microorganisms. For the confirmation of *Listeria* in Henry's light.

Tryptone Soy Yeast Extract HiVeg Broth

Composition per liter:

Plant hydrolysate.......... 17.0g
Yeast extract.......... 6.0g
NaCl.......... 5.0g
Papaic digest of soyabean meal.......... 3.0g
Glucose.......... 2.5g
K_2HPO_4.......... 2.5g

pH 7.3 ± 0.2 at 25°C

Source: This medium is available as a premixed powder from HiMedia.

Preparation of Medium: Add components to distilled/deionized water and bring volume to 1.0L. Mix thoroughly. Gently heat and bring to boiling. Distribute into tubes or flasks. Autoclave for 15 min at 15 psi pressure–121°C.

Use: For the cultivation and maintenance of a wide variety of heterotrophic microorganisms.For the confirmation of *Listeria* in Henry's light.

Tryptone Thioglycolate Medium (DSMZ Medium 48)

Composition per liter:

Yeast extract.......... 6.0g
K_2HPO_4.......... 5.45g
Peptone.......... 2.0g
Tryptone.......... 2.0g
KH_2PO_4.......... 1.2g
Na-thioglycolate.......... 0.5g
$MgSO_4 \cdot 7H_2O$.......... 0.025g
$CaCl_2 \cdot 2H_2O$.......... 0.015g
$FeSO_4 \cdot 7H_2O$.......... 0.01g
$CoCl_2 \cdot 6H_2O$.......... 2.5mg
$Na_2MoO_4 \cdot 2H_2O$.......... 2.5mg
$MnCl_2 \cdot 4H_2O$.......... 2.0mg
Glucose solution.......... 50.0mL

pH 7.5 ± 0.2 at 25°C

Glucose Solution:
Composition per 50.0mL:

D-Glucose.......... 20.0g

Preparation of Glucose Solution: Add glucose to distilled/deionized water and bring volume to 50.0mL. Mix thoroughly. Filter sterilize.

Preparation of Medium: Add components, except glucose solution, to distilled/deionized water and bring volume to 950.0mL. Mix thoroughly. Gently heat and bring to boiling. Autoclave for 15 min at 15 psi pressure–121°C. Cool to 25°C. Aseptically add 50.0mL of sterile glucose solution. Mix thoroughly. Aseptically distribute into sterile tubes or flasks.

Use: For the cultivation of *Clostridium beijerinckii.*

Tryptone Thioglycolate Medium

Composition per liter:

Yeast extract.......... 6.0g
K_2HPO_4.......... 5.45g
Peptone.......... 2.0g
Tryptone.......... 2.0g
KH_2PO_4.......... 1.25g
Sodium thioglycolate.......... 0.5g
$MgSO_4 \cdot 7H_2O$.......... 0.025g
$CaCl_2 \cdot 2H_2O$.......... 0.015g
$FeSO_4 \cdot 7H_2O$.......... 0.01g
$CoCl_2 \cdot 6H_2O$.......... 2.5mg
$Na_2MoO_4 \cdot 2H_2O$.......... 2.5mg

$MnCl_2 \cdot 4H_2O$ 2.0mg
Glucose solution 50.0mL

pH 7.5 ± 0.2 at 25°C

Glucose Solution:

Composition per 50.0mL:

Glucose 20.0g

Preparation of Glucose Solution: Add glucose to distilled/deionized water and bring volume to 50.0mL. Mix thoroughly. Filter sterilize.

Preparation of Medium: Add components, except glucose solution, to distilled/deionized water and bring volume to 950.0mL. Mix thoroughly. Autoclave for 15 min at 15 psi pressure–121°C. Aseptically add 50.0mL of sterile glucose solution. Mix thoroughly. Aseptically distribute into sterile tubes or flasks.

Use: For the cultivation and maintenance of *Clostridium butyricum* and *Clostridium roseum*.

Tryptone Water Broth (Tryptone Broth)

Composition per liter:

Pancreatic digest of casein 10.0g
NaCl 5.0g

pH 7.5 ± 0.2 at 25°C

Source: This medium is available as a premixed powder from Oxoid Unipath.

Preparation of Medium: Dissolve 15.0g in 1.0L of distilled water and distribute into final containers. Sterilize by autoclaving at 121°C for 15 min.

Use: For the cultivation of production of indole by microorganisms.

Tryptone Water, HiVeg (Tryptone Broth, HiVeg)

Composition per liter:

Plant hydrolysate 20.0g
NaCl 5.0g

pH 7.0 ± 0.2 at 25°C

Source: This medium is available as a premixed powder from HiMedia.

Preparation of Medium: Add components to tap water and bring volume to 1.0L. Mix thoroughly. Gently heat and bring to boiling. Distribute into tubes or flasks. Autoclave for 15 min at 15 psi pressure–121°C.

Use: For production of indole by microorganisms.

Tryptone Water Broth, HiVeg

Composition per liter:

Plant hydrolysate 10.0g
Glucose 5.0g
K_2HPO_4 .... 1.25g
Yeast extract 1.0g
Bromcresol Purple 0.04g

pH 7.3 ± 0.2 at 25°C

Source: This medium is available as a premixed powder from HiMedia.

Preparation of Medium: Add components to tap water and bring volume to 1.0L. Mix thoroughly. Gently heat and bring to boiling. Distribute into tubes or flasks. Autoclave for 15 min at 15 psi pressure–121°C.

Use: For the cultivation of *Salmonella* species from foods.

Tryptone with Sodium Chloride Broth

Composition per liter:

Pancreatic digest of casein 8.0g
NaCl 0.5g

Preparation of Medium: Add components to distilled/deionized water and bring volume to 1.0L. Mix thoroughly. Distribute into tubes or flasks. Autoclave for 15 min at 15 psi pressure–121°C.

Use: For the cultivation and maintenance of fastidious aerobic and facultative microorganisms such as *Escherichia coli* and *Pseudomonas* species.

Tryptone Yeast Extract Agar

Composition per liter:

Pancreatic digest of casein 10.0g
Agar 2.0g
Yeast extract 1.0g
Bromcresol Purple 0.04g
Carbohydrate solution 100.0mL

pH 7.0 ± 0.2 at 25°C

Carbohydrate Solution:

Composition per 100.0mL:

Carbohydrate 10.0g

Preparation of Carbohydrate Solution: Add carbohydrate to distilled/deionized water and bring volume to 100.0mL. Glucose or mannitol may be used. Mix thoroughly. Filter sterilize.

Preparation of Medium: Add components, except carbohydrate solution, to distilled/deionized water and bring volume to 900.0mL. Mix thoroughly. Adjust pH to 7.0. Gently heat and bring to boiling. Distribute into tubes in 13.5mL volumes. Autoclave for 20 min at 10 psi pressure–115°C. Cool to 45°–50°C. Aseptically add 1.5mL of carbohydrate solution to each tube. Mix thoroughly. Solidify agar quickly by placing tubes in ice water.

Use: For the cultivation and differentiation of *Staphylococcus aureus* based on glucose and mannitol fermentation. Bacteria that ferment the added carbohydrate turn the medium yellow.

Tryptone Yeast Extract Agar 1

Composition per liter:

Agar 20.0g
Pancreatic digest of casein 10.0g
Yeast extract 5.0g
K_2HPO_4 .... 4.4g
Glucose 2.0g
NaCl 2.0g

pH 7.2 ± 0.2 at 25°C

Preparation of Medium: Add components to distilled/deionized water and bring volume to 1.0L. Mix thoroughly. Adjust pH to 7.2. Gently heat and bring to boiling. Distribute into tubes or flasks. Autoclave for 15 min at 15 psi pressure–121°C. Pour into sterile Petri dishes or leave in tubes.

Use: For the cultivation and maintenance of a wide variety of heterotrophic bacteria.

Tryptone Yeast Extract Broth
See: **ISP Medium 1**

Tryptone Yeast Extract Glucose Medium
See: **TYG Medium**

Tryptone Yeast Extract Glucose Salt Medium
See: **TYGS Medium**

Tryptone Yeast Extract HiVeg Agar with Carbohydrate

Composition per liter:

Agar 12.0g
Plant hydrolysate........ 6.0g
Yeast extract powder........ 3.0g
Carbohydrate solution........ 100.0mL

pH 7.0 ± 0.2 at 25°C

Source: This medium, without carbohydrate solution, is available as a premixed powder from HiMedia.

Carbohydrate Solution:
Composition per 100.0mL:

Carbohydrate........ 10.0g

Preparation of Carbohydrate Solution: Add carbohydrate to distilled/deionized water and bring volume to 100.0mL. Glucose or mannitol may be used. Mix thoroughly. Filter sterilize.

Preparation of Medium: Add components, except carbohydrate solution, to distilled/deionized water and bring volume to 900.0mL. Mix thoroughly. Adjust pH to 7.0. Gently heat and bring to boiling. Distribute into tubes in 13.5mL volumes. Autoclave for 20 min at 10 psi pressure–115°C. Cool to 45°–50°C. Aseptically add 1.5mL of carbohydrate solution to each tube. Mix thoroughly. Solidify agar quickly by placing tubes in ice water.

Use: For the cultivation and differentiation of *Staphylococcus aureus* based on glucose and mannitol fermentation. Bacteria that ferment the added carbohydrate turn the medium yellow.

Tryptone Yeast Extract Medium

Composition per liter:

Pancreatic digest of casein........ 10.0g
Yeast extract........ 1.0g

Preparation of Medium: Add components to distilled/deionized water and bring volume to 1.0L. Mix thoroughly. Distribute into tubes or flasks. Autoclave for 15 min at 15 psi pressure–121°C.

Use: For the cultivation of *Haloferax volcanii.*

Tryptone Yeast Extract Mineral Medium

Composition per liter:

Pancreatic digest of casein........ 10.0g
Yeast extract........ 10.0g
$NaHCO_3$........ 6.0g
Arginine........ 3.0g
NaCl........ 1.0g
K_2HPO_4........ 0.5g
KH_2PO_4........ 0.5g
L-Cysteine·HCl........ 0.3g
$CaCl_2$........ 0.1g
$MgSO_4$........ 0.1g
Resazurin........ 0.1mg

pH 7.2 ± 0.2 at 25°C

Preparation of Medium: Add components to distilled/deionized water and bring volume to 1.0L. Mix thoroughly. Distribute into tubes or flasks. Autoclave for 15 min at 15 psi pressure–121°C.

Use: For the cultivation and maintenance of *Peptostreptococcus heliotrinreducens.*

Tryptone Yeast Extract Salt Medium
See: **TYES Medium**

Tryptone Yeast Extract Salt Medium

Composition per liter:

Solution A........ 500.0mL
Solution B........ 500.0mL

pH 6.8 ± 0.2 at 25°C

Solution A:
Composition per 500.0mL:

NaCl........ 125.0g
$MgCl_2·6H_2O$........ 50.0g
K_2SO_4........ 5.0g
$CaCl_2·6H_2O$........ 0.2g

Preparation of Solution A: Add components to distilled/deionized water and bring volume to 500.0mL. Mix thoroughly. Adjust pH to 6.8. Autoclave for 15 min at 15 psi pressure–121°C.

Solution B:
Composition per 500.0mL:

Pancreatic digest of casein........ 5.0g
Yeast extract........ 5.0g

Preparation of Solution B: Add components to distilled/deionized water and bring volume to 500.0mL. Mix thoroughly. Adjust pH to 6.8. Autoclave for 15 min at 15 psi pressure–121°C.

Preparation of Medium: Aseptically combine 500.0mL of solution A with 500.0mL of solution B. Mix thoroughly. Aseptically distribute into sterile tubes or flasks.

Use: For the cultivation of *Haloferax volcanii.*

Tryptophan Assay Medium

Composition per liter:

Glucose........ 40.0g
Sodium acetate........ 20.0g
Casamino acids........ 12.0g
K_2HPO_4........ 1.0g
KH_2PO_4........ 1.0g
$MgSO_4·7H_2O$........ 0.4g
L-Cystine........ 0.2g
Adenine sulfate........ 0.02g
$FeSO_4·7H_2O$........ 0.02g
Guanine·HCl........ 0.02g
$MnSO_4·7H_2O$........ 0.02g
NaCl........ 0.02g
Uracil........ 0.02g
Pyridoxine·HCl........ 0.4mg
Riboflavin........ 0.4mg
p-Aminobenzoic acid........ 0.2mg
Calcium pantothenate........ 0.2mg
Niacin........ 0.2mg

Thiamine·HCl 0.2mg
Biotin 0.8μg

pH 6.7 ± 0.2 at 25°C

Source: This medium is available as a premixed powder from BD Diagnostic Systems.

Preparation of Medium: Add components to distilled/deionized water and bring volume to 1.0L. Mix thoroughly. Distribute into tubes in 5.0mL volumes. Add standard solutions and test solutions to each tube. Bring volume of each tube to 10.0mL. Autoclave for 15 min at 15 psi pressure–121°C.

Use: For the assay of tryptophan using *Lactobacillus plantarum* as an indicator organism.

Tryptophan Broth

Composition per 100.0mL:

L-Tryptophan 0.5g
NaCl 0.5g
KH_2PO_4 .... 0.25g

pH 7.4 ± 0.2 at 25°C

Preparation of Medium: Add components to distilled/deionized water and bring volume to 100.0mL. Mix thoroughly. Adjust pH to 7.4. Filter sterilize. Aseptically distribute in 1.0mL volumes into sterile screw-capped tubes.

Use: For the cultivation of *Flavobacterium* species and a variety of other bacteria. Also used to differentiate bacteria based on indole production. Indole is determined by the addition of modified Kovacs reagent to cultures that have incubated for 18–24 hr. Formation of a red color in the upper layer indicates indole formation.

Tryptophan HiVeg Medium

Composition per liter:

Plant hydrolysate 10.0g
NaCl 5.0g
DL-Tryptophan 1.0g

pH 7.5 ± 0.2 at 25°C

Source: This medium is available as a premixed powder from HiMedia.

Preparation of Medium: Add components to distilled/deionized water and bring volume to 100.0mL. Mix thoroughly. Adjust pH to 7.4. Filter sterilize. Aseptically distribute in 1.0mL volumes into sterile screw-capped tubes.

Use: For the cultivation of *Flavobacterium* species and a variety of other bacteria. Also used to differentiate bacteria based on indole production. Indole is determined by the addition of modified Kovacs reagent to cultures that have incubated for 18–24 hr. Formation of a red color in the upper layer indicates indole formation.

Tryptophan 1% Solution (Trypticase™ 1% Solution) (Tryptone 1% Solution)

Composition per liter:

Pancreatic digest of casein 10.0g

pH 7.0 ± 0.2 at 25°C

Source: This medium is available as a premixed powder from BD Diagnostic Systems.

Preparation of Medium: Add pancreatic digest of casein to distilled/deionized water and bring volume to 1.0L. Mix thoroughly. Distribute into tubes or flasks. Autoclave for 15 min at 15 psi pressure–121°C.

Use: For the differentiation of bacteria, especially members of the Enterobacteriaceae, based on their production of indole.

Tryptose Agar

Composition per liter:

Agar 15.0g
Pancreatic digest of casein 10.0g
Peptic digest of animal tissue 10.0g
NaCl 5.0g
Glucose 1.0g

pH 7.2 ± 0.2 at 25°C

Source: This medium is available as a premixed powder from BD Diagnostic Systems.

Preparation of Medium: Add components to distilled/deionized water and bring volume to 1.0L. Mix thoroughly. Gently heat and bring to boiling. Distribute into tubes or flasks. Autoclave for 15 min at 15 psi pressure–121°C. Pour into sterile Petri dishes or leave in tubes.

Use: For the cultivation and maintenance of fastidious aerobic and facultative microorganisms.

Tryptose Agar (BAM M167)

Composition per liter:

Tryptose 20.0g
Agar 15.0g
NaCl 5.0g
Glucose 1.0g

pH 7.2 ± 0.2 at 25°C

Preparation of Medium: Add components to distilled/deionized water and bring volume to 1.0L. Mix thoroughly. Autoclave for 15 min at 15 psi pressure–121°C. Cool to 45°–50°C. Pour into sterile Petri dishes or leave in tubes. For slants allow tubes to cool in an inclined position.

Use: For the cultivation of a variety of bacteria for serology.

Tryptose Agar with Citrate

Composition per liter:

Agar 15.0g
Pancreatic digest of casein 10.0g
Peptic digest of animal tissue 10.0g
Sodium citrate 10.0g
NaCl 5.0g
Glucose 1.0g

pH 7.2 ± 0.2 at 25°C

Preparation of Medium: Add components to distilled/deionized water and bring volume to 1.0L. Mix thoroughly. Gently heat and bring to boiling. Distribute into tubes or flasks. Autoclave for 15 min at 15 psi pressure–121°C. Pour into sterile Petri dishes or leave in tubes.

Use: For the cultivation and maintenance of fastidious aerobic and facultative microorganisms, including *Brucella* species and streptococci.

Tryptose Agar, HiVeg

Composition per liter:

Plant hydrolysate No. 1	20.0g
Agar	15.0g
NaCl	5.0g
Glucose	1.0g

pH 7.2 ± 0.2 at 25°C

Source: This medium is available as a premixed powder from HiMedia.

Preparation of Medium: Add components to distilled/deionized water and bring volume to 1.0L. Mix thoroughly. Gently heat and bring to boiling. Distribute into tubes or flasks. Autoclave for 15 min at 15 psi pressure–121°C. Pour into sterile Petri dishes or leave in tubes.

Use: For the cultivation and maintenance of fastidious aerobic and facultative microorganisms. For the isolation, cultivation, and differentiation of *Brucella*, sreptococci, and pneumococci.

Tryptose Agar with Sheep Blood (ATCC Medium 546)

Composition per liter:

Agar	15.0g
Tryptose	10.0g
NaCl	5.0g
Beef extract	3.0g
Sheep blood, defibrinated	50.0–100.0mL

pH 7.3 ± 0.2 at 25°C

Preparation of Medium: Add components, except sheep blood, to distilled/deionized water and bring volume to 900–950mL. Mix thoroughly. Gently heat and bring to boiling. Distribute into tubes. Autoclave for 15 min at 15 psi pressure–121°C. Cool to 45°–50°C. Aseptically add sterile sheep blood. Mix thoroughly. Pour into sterile Petri dishes in 17.0mL volumes or distribute into sterile tubes.

Use: For the cultivation and maintenance of a wide variety of fastidious microorganisms.

Tryptose Agar with Thiamine

Composition per liter:

Agar	15.0g
Pancreatic digest of casein	10.0g
Peptic digest of animal tissue	10.0g
NaCl	5.0g
Glucose	1.0g
Thiamine·HCl	5.0mg

pH 7.2 ± 0.2 at 25°C

Preparation of Medium: Add components to distilled/deionized water and bring volume to 1.0L. Mix thoroughly. Gently heat and bring to boiling. Distribute into tubes or flasks. Autoclave for 15 min at 15 psi pressure–121°C. Pour into sterile Petri dishes or leave in tubes.

Use: For the cultivation and maintenance of fastidious aerobic and facultative microorganisms, including *Brucella* species and streptococci.

Tryptose Agar with Thiamine HCl, HiVeg

Composition per liter:

Agar	15.0g
Plant peptone	10.0g
Plant hydrolysate	10.0g
NaCl	5.0g
Glucose	1.0g
Thiamine·HCl	5.0mg

pH 7.2 ± 0.2 at 25°C

Preparation of Medium: Add components to distilled/deionized water and bring volume to 1.0L. Mix thoroughly. Gently heat and bring to boiling. Distribute into tubes or flasks. Autoclave for 15 min at 15 psi pressure–121°C. Pour into sterile Petri dishes or leave in tubes.

Use: For the cultivation and maintenance of fastidious aerobic and facultative microorganisms, including *Brucella* species and streptococci.

Tryptose Blood Agar

Composition per liter:

Agar	12.0g
Tryptose	10.0g
NaCl	5.0g
Beef extract	3.0g
Sheep blood, defibrinated	70.0mL

pH 7.2 ± 0.2 at 25°C

Source: This medium is available as a premixed powder from Oxoid Unipath.

Preparation of Medium: Add components, except sheep blood, to distilled/deionized water and bring volume to 930.0mL. Mix thoroughly. Gently heat and bring to boiling. Autoclave for 15 min at 15 psi pressure–121°C. Cool to 45°–50°C. Aseptically add sterile sheep blood. Mix thoroughly. Pour into sterile Petri dishes in 17.0mL volumes or distribute into sterile tubes.

Use: For the cultivation and maintenance of a wide variety of fastidious microorganisms.

Tryptose Blood Agar

Composition per liter:

Agar	20.0g
Proteose peptone No. 3	10.0g
Tryptose	10.0g
Beef extract	5.0g
NaCl	5.0g
Yeast extract	5.0g
Sheep blood	100.0mL
L-Cysteine·HCl solution	2.5mL

L-Cysteine·HCl Solution:

Composition per 10.0mL:

L-cysteine·HCl	1.0g

Preparation of L-Cysteine·HCl Solution: Dissolve 1.0g of L-cysteine·HCl in distilled/deionized water and bring volume to 10.0mL. Mix thoroughly. Filter sterilize. Warm to 50°C.

Preparation of Medium: Add components, except sheep blood and L-cysteine·HCl solution, to distilled/deionized water and bring volume to 887.5mL. Mix thoroughly. Gently heat and bring to boiling. Autoclave for 15 min at 15 psi pressure–121°C. Cool to 50°–55°C. Warm sheep blood to 50°C. Aseptically add 100.0mL of sterile sheep blood and 2.5mL of sterile L-cysteine·HCl solution. Mix thoroughly. Pour into sterile Petri dishes or distribute into sterile tubes.

Use: For the cultivation and maintenance of *Corynebacterium matruchotii, Propionibacterium propionicum,* and *Staphylococcus saccharolyticus.*

Tryptose Blood Agar Base

Composition per liter:

Agar	15.0g
Tryptose	10.0g
NaCl	5.0g
Beef extract	3.0g

Preparation of Medium: Add components to distilled/deionized water and bring volume to 1.0L. Mix thoroughly. Gently heat and bring to boiling. Distribute into tubes. Autoclave for 15 min at 15 psi pressure–121°C. Allow tubes to cool in a slanted position to obtain a 4–5.0cm slant and a 2–3.0cm butt.

Use: For the cultivation and enumeration of *Salmonella* species from foods.

Tryptose Blood Agar Base, HiVeg with Sheep Blood

Composition per liter:

Agar	15.0g
Plant hydrolysate No. 1	10.0g
NaCl	5.0g
Plant extract	3.0g
Sheep blood, defibrinated	70.0mL

pH 7.2 ± 0.2 at 25°C

Source: This medium, without blood, is available as a premixed powder from HiMedia.

Preparation of Medium: Add components, except sheep blood, to distilled/deionized water and bring volume to 930.0mL. Mix thoroughly. Gently heat and bring to boiling. Autoclave for 15 min at 15 psi pressure–121°C. Cool to 45°–50°C. Aseptically add sterile sheep blood. Mix thoroughly. Pour into sterile Petri dishes in 17.0mL volumes or distribute into sterile tubes.

Use: For the cultivation and maintenance of a wide variety of fastidious microorganisms. For the isolation of fastidious organisms and determining hemolytic reactions.

Tryptose Blood Agar Base with Yeast Extract

Composition per liter:

Agar	15.0g
Tryptose	10.0g
NaCl	5.0g
Beef extract	3.0g
Yeast extract	1.0g
Sheep blood, defibrinated	50.0mL

pH 7.3 ± 0.2 at 25°C

Source: This medium is available as a premixed powder from BD Diagnostic Systems.

Preparation of Medium: Add components, except sheep blood, to distilled/deionized water and bring volume to 950.0mL. Mix thoroughly. Gently heat and bring to boiling. Autoclave for 15 min at 15 psi pressure–121°C. Cool to 45°–50°C. Aseptically add sterile sheep blood. Mix thoroughly. Pour into sterile Petri dishes in 17.0mL volumes or distribute into sterile tubes.

Use: For the cultivation and maintenance of a wide variety of fastidious microorganisms.

Tryptose Blood Agar Base with Yeast Extract, HiVeg

Composition per liter:

Agar	15.0g
Plant hydrolysate No. 1	10.0g
NaCl	5.0g
Plant extract	3.0g
Yeast extract	1.0g
Sheep blood, defibrinated	70.0mL

pH 7.2 ± 0.2 at 25°C

Source: This medium, without blood, is available as a premixed powder from HiMedia.

Preparation of Medium: Add components, except sheep blood, to distilled/deionized water and bring volume to 930.0mL. Mix thoroughly. Gently heat and bring to boiling. Autoclave for 15 min at 15 psi pressure–121°C. Cool to 45°–50°C. Aseptically add sterile sheep blood. Mix thoroughly. Pour into sterile Petri dishes in 17.0mL volumes or distribute into sterile tubes.

Use: For the cultivation and maintenance of a wide variety of fastidious microorganisms. For the isolation of fastidious organisms and determining hemolytic reactions.

Tryptose Blood Agar Base 298 Medium

See: **TBAB 298 Medium**

Tryptose Broth

Composition per liter:

Pancreatic digest of casein	10.0g
Peptic digest of animal tissue	10.0g
NaCl	5.0g
Glucose	1.0g

pH 7.2 ± 0.2 at 25°C

Source: This medium is available as a premixed powder from BD Diagnostic Systems.

Preparation of Medium: Add components to distilled/deionized water and bring volume to 1.0L. Mix thoroughly. Distribute into tubes or flasks. Autoclave for 15 min at 15 psi pressure–121°C.

Use: For the cultivation of fastidious aerobic and facultative microorganisms, including streptococci. For the cultivation of fastidious aerobic and facultative microorganisms.

Tryptose Broth (BAM M167)

Composition per liter:

Tryptose	20.0g
NaCl	5.0g
Glucose	1.0g

pH 7.2 ± 0.2 at 25°C

Preparation of Medium: Add components to distilled/deionized water and bring volume to 1.0L. Mix thoroughly. Distribute into tubes or flasks. Autoclave for 15 min at 15 psi pressure–121°C.

Use: For the cultivation of a variety of bacteria for serology.

Tryptose Broth with Citrate

Composition per liter:

Pancreatic digest of casein	10.0g
Peptic digest of animal tissue	10.0g
Sodium citrate	10.0g
NaCl	5.0g
Glucose	1.0g
Thiamine·HCl	5.0mg

pH 7.2 ± 0.2 at 25°C

Preparation of Medium: Add components to distilled/deionized water and bring volume to 1.0L. Mix thoroughly. Distribute into tubes or flasks. Autoclave for 15 min at 15 psi pressure–121°C.

Use: For the isolation and cultivation of a variety of fastidious aerobic microorganisms, especially *Brucella* species, from clinical sources and dairy products.

Tryptose Broth, HiVeg

Composition per liter:

Plant hydrolysate No. 1	20.0g
NaCl	5.0g
Glucose	1.0g

pH 7.2 ± 0.2 at 25°C

Source: This medium is available as a premixed powder from HiMedia.

Preparation of Medium: Add components to distilled/deionized water and bring volume to 1.0L. Mix thoroughly. Distribute into tubes or flasks. Autoclave for 15 min at 15 psi pressure–121°C.

Use: For the cultivation of fastidious aerobic and facultative microorganisms, including streptococci.

Tryptose Cycloserine Glucose HiVeg Agar Base with Cycloserine

Composition per liter:

Agar	20.0g
Plant hydrolysate No. 1	15.0g
Papaic digest of soybean meal	5.0g
Yeast extract	5.0g
Ferric ammonium citrate	1.0g
Cycloserine solution	10.0mL

pH 7.6 ± 0.2 at 25°C

Source: This medium, without cycloserine, is available as a premixed powder from HiMedia.

Cycloserine Solution:

Composition per 10.0mL:

D-Cycloserine	0.4g

Preparation of Cycloserine Solution: Add D-cycloserine to distilled/deionized water and bring volume to 10.0mL. Mix thoroughly. Filter sterilize.

Preparation of Medium: Add components, except cycloserine solution, to distilled/deionized water and bring volume to 990.0mL. Mix thoroughly. Gently heat and bring to boiling. Autoclave for 15 min at 15 psi pressure–121°C. Cool to 45°–50°C. Aseptically add sterile cycloserine solution. Mix thoroughly. Pour into sterile Petri dishes or distribute into sterile tubes.

Use: For the isolation and cultivation of *Clostridium* species, especially *Clostridium botulinum*, from foods.

Tryptose Cycloserine Dextrose Agar

Composition per liter:

Agar	20.0g
Tryptose	15.0g
Pancreatic digest of soybean meal	5.0g
Yeast extract	5.0g
Ferric ammonium citrate	1.0g
Cycloserine solution	10.0mL

pH 7.6 ± 0.2 at 25°C

Cycloserine Solution:

Composition per 10.0mL:

D-Cycloserine	0.4g

Preparation of Cycloserine Solution: Add D-cycloserine to distilled/deionized water and bring volume to 10.0mL. Mix thoroughly. Filter sterilize.

Preparation of Medium: Add components, except cycloserine solution, to distilled/deionized water and bring volume to 990.0mL. Mix thoroughly. Gently heat and bring to boiling. Autoclave for 15 min at 15 psi pressure–121°C. Cool to 45°–50°C. Aseptically add sterile cycloserine solution. Mix thoroughly. Pour into sterile Petri dishes or distribute into sterile tubes.

Use: For the isolation and cultivation of *Clostridium* species, especially *Clostridium botulinum*, from foods.

Tryptose Phosphate Agar

Composition per liter:

Tryptose	20.0g
Agar	15.0g
NaCl	5.0g
Na_2HPO_4	2.5g
Glucose	2.0g

pH 7.3 ± 0.2 at 25°C

Preparation of Medium: Add components to distilled/deionized water and bring volume to 1.0L. Mix thoroughly. Gently heat and bring to boiling. Distribute into tubes or flasks. Autoclave for 15 min at 15 psi pressure–121°C. Pour into sterile Petri dishes or leave in tubes.

Use: For the cultivation and maintenance of *Erysipelothris tonsillarum.*

Tryptose Phosphate Broth

Composition per liter:

Tryptose	20.0g
NaCl	5.0g
Na_2HPO_4	2.5g
Glucose	2.0g

pH 7.3 ± 0.2 at 25°C

Source: This medium is available as a premixed powder from BD Diagnostic Systems and Oxoid Unipath.

Preparation of Medium: Add components to distilled/deionized water and bring volume to 1.0L. Mix thoroughly. Distribute into tubes or flasks. Autoclave for 15 min at 15 psi pressure–121°C. Prior to the inoculation of anaerobic microorganisms, place tubes of sterile medium in a 100°C bath for 15 min and cool undisturbed.

Use: For the cultivation of a variety of bacteria. For cell culture.

Tryptose Phosphate Broth, HiVeg

Composition per liter:

Plant hydrolysate No. 1	20.0g
NaCl	5.0g
Na_2HPO_4	2.5g
Glucose	2.0g

pH 7.3 ± 0.2 at 25°C

Source: This medium is available as a premixed powder from HiMedia.

Preparation of Medium: Add components to distilled/deionized water and bring volume to 1.0L. Mix thoroughly. Distribute into tubes or flasks. Autoclave for 15 min at 15 psi pressure–121°C. Prior to the inoculation of anaerobic microorganisms, place tubes of sterile medium in a 100°C bath for 15 min and cool undisturbed.

Use: For the cultivation of a variety of fastidious bacteria. For the cultivation of fastidious bacteria and as an adjuvant to tissue culture media.

Tryptose Phosphate Broth, Modified

Composition per liter:

Enzymatic digest of casein	20.0g
NaCl	5.0g
Na_2HPO_4	2.5g
Glucose	2.0g

pH 7.3 ± 0.2 at 25°C

Source: This medium is available as a premixed powder from BD Diagnostic Systems.

Preparation of Medium: Add components to distilled/deionized water and bring volume to 1.0L. Mix thoroughly. Distribute into tubes or flasks. Autoclave for 15 min at 15 psi pressure–121°C. Prior to the inoculation of anaerobic microorganisms, place tubes of sterile medium in a 100°C bath for 15 min and cool undisturbed.

Use: For the cultivation of a variety of fastidious microorganisms, including pneumococci, streptococci, and meningococci.

Tryptose Sulfite Cycloserine Agar (TSC Agar)

Composition per liter:

Tryptose	15.0g
Agar	14.0g
Pancreatic digest of soybean meal	5.0g
Yeast extract	5.0g
Ferric ammonium citrate	1.0g
$Na_2S_2O_5$	1.0g
Cycloserine solution	10.0mL

pH 7.6 ± 0.2 at 25°C

Cycloserine Solution:

Composition per 10.0mL:

D-Cycloserine	0.4g

Preparation of Cycloserine Solution: Add cycloserine to distilled/deionized water and bring volume to 10.0mL. Mix thoroughly. Filter sterilize.

Preparation of Medium: Add components, except cycloserine solution, to distilled/deionized water and bring volume to 990.0mL. Mix thoroughly. Gently heat and bring to boiling. Autoclave for 15 min at 15 psi pressure–121°C. Cool to 45°–50°C. Aseptically add sterile cycloserine solution. Mix thoroughly. Pour into sterile Petri dishes.

Use: For the presumptive identification and enumeration of *Clostridium perfringens*.

Tryptose Sulfite Cycloserine Agar (TSC Agar)

Composition per liter:

Tryptose	15.0g
Agar	14.0g
Beef extract	5.0g
Pancreatic digest of soybean meal	5.0g
Yeast extract	5.0g
Ferric ammonium citrate	1.0g
$Na_2S_2O_5$	1.0g
Egg yolk emulsion	50.0mL
Cycloserine solution	10.0mL

pH 7.6 ± 0.2 at 25°C

Source: This medium is available as a premixed powder from Oxoid Unipath.

Egg Yolk Emulsion:

Composition:

Chicken egg yolks	11
Whole chicken egg	1

Preparation of Egg Yolk Emulsion: Soak eggs with 1:100 dilution of saturated mercuric chloride solution for 1 min. Crack eggs and separate yolks from whites. Mix egg yolks with 1 chicken egg.

Cycloserine Solution:

Composition per 10.0mL:

D-Cycloserine	0.4g

Preparation of Cycloserine Solution: Add cycloserine to distilled/deionized water and bring volume to 10.0mL. Mix thoroughly. Filter sterilize.

Preparation of Medium: Add components, except cycloserine solution and egg yolk emulsion, to distilled/deionized water and bring volume to 940.0mL. Mix thoroughly. Gently heat and bring to boiling. Autoclave for 15 min at 15 psi pressure–121°C. Cool to 45°–50°C. Aseptically add sterile cycloserine solution and egg yolk emulsion. Mix thoroughly. Pour into sterile Petri dishes.

Use: For the presumptive identification and enumeration of *Clostridium perfringens*.

Tryptose Sulfite Cycloserine Agar with Polymyxin and Kanamycin

Composition per liter:

Tryptose	15.0g
Agar	14.0g
Beef extract	5.0g
Pancreatic digest of soybean meal	5.0g
Yeast extract	5.0g
Ferric ammonium citrate	1.0g
$Na_2S_2O_5$	1.0g
Antibiotic solution	10.0mL

pH 7.6 ± 0.2 at 25°C

Antibiotic Solution:

Composition per 10.0mL:

D-Cycloserine	0.4g
Polymyxin B sulfate	0.03g
Kanamycin sulfate	0.012g

Preparation of Antibiotic Solution: Add components to distilled/deionized water and bring volume to 10.0mL. Mix thoroughly. Filter sterilize.

Preparation of Medium: Add components, except antibiotic solution, to distilled/deionized water and bring volume to 990.0mL. Mix thoroughly. Gently heat and bring to boiling. Autoclave for 15 min at 15 psi pressure–121°C. Cool to 45°–50°C. Aseptically add sterile antibiotic solution. Mix thoroughly. Pour into sterile Petri dishes.

Use: For the isolation and enumeration of *Clostridium perfringens* from foods and clinical specimens.

Tryptose Sulfite Cycloserine Agar without Egg Yolk (TSC Agar without Egg Yolk)

Composition per liter:

Tryptose	15.0g
Agar	14.0g
Beef extract	5.0g
Pancreatic digest of soybean meal	5.0g
Yeast extract	5.0g
Ferric ammonium citrate	1.0g
$Na_2S_2O_5$	1.0g
Cycloserine solution	10.0mL

pH 7.6 ± 0.2 at 25°C

Cycloserine Solution:

Composition per 10.0mL:

D-Cycloserine	0.4g

Preparation of Cycloserine Solution: Add cycloserine to distilled/deionized water and bring volume to 10.0mL. Mix thoroughly. Filter sterilize.

Preparation of Medium: Add components, except cycloserine solution, to distilled/deionized water and bring volume to 990.0mL. Mix thoroughly. Gently heat and bring to boiling. Autoclave for 15 min at 15 psi pressure–121°C. Cool to 45°–50°C. Aseptically add sterile cycloserine solution. Mix thoroughly. Pour into sterile Petri dishes.

Use: For the presumptive identification and enumeration of *Clostridium perfringens*.

TS Agar (DSMZ Medium 893)

Composition per liter:

Agar	15.0g
Sucrose	5.0g
Tryptone	5.0g
Beef extract	3.0g
Glucose	1.0g

pH 7.0 ± 0.2 at 25°C

Preparation of Medium: Add components to distilled/deionized water and bring volume to 1.0L. Mix thoroughly. Gently heat and bring to boiling. Distribute into tubes or flasks. Autoclave for 15 min at 15 psi pressure–121°C. Pour into sterile Petri dishes or leave in tubes.

Use: For the cultivation and maintenance of *Saccharococcus thermophilus*.

TS Medium for *Spirochaeta caldaria*

Composition per liter:

Pancreatic digest of casein	2.0g
Cellobiose	1.0g
Maltose	1.0g
Yeast extract	1.0g
Dithiothreitol	0.15g
Resazurin	1.0mg

pH 7.0 ± 0.2 at 25°C

Preparation of Medium: Prepare and dispense medium under 100% N_2. Add components to distilled/deionized water and bring volume to 1.0L. Mix thoroughly. Gently heat and bring to boiling. Continue boiling for 3 min. Cool to room temperature while sparging with 100% N_2. Anaerobically distribute into anaerobic tubes. Autoclave for 15 min at 15 psi pressure–121°C. Adjust pH to 7.0

Use: For the cultivation of *Spirochaeta caldaria.*

TS Soil Extract (Trypticase™ Soy Soil Extract)

Composition per liter:

Pancreatic digest of casein	17.0g
Agar	15.0g
NaCl	5.0g
Papaic digest of soybean meal	3.0g
K_2HPO_4	2.5g
Glucose	2.5g
Soil extract	250.0mL

Soil Extract:

Composition per 400.0mL:

African Violet soil	154.0g
Na_2CO_3	0.4g

Preparation of Soil Extract: Add components to tap water and bring volume to 400.0mL. Autoclave for 60 min at 15 psi pressure–121°C. Filter through Whatman filter paper.

Preparation of Medium: Add components to tap water and bring volume to 1.0L. Mix thoroughly. Gently heat and bring to boiling. Distribute into tubes or flasks. Autoclave for 15 min at 15 psi pressure–121°C. Pour into sterile Petri dishes or leave in tubes.

Use: For the cultivation and maintenance of *Bacillus xerothermodurans*.

TSA 5400 Selective Isolation Medium

Composition per liter:

Pancreatic digest of casein	15.0g
Agar	15.0g
Papaic digest of soybean meal	5.0g
NaCl	5.0g
Bovine blood, citrated	100.0mL
Spectinomycin solution	10.0mL

pH 7.3 ± 0.2 at 25°C

Spectinomycin Solution:

Composition per 10.0mL:

Spectinomycin	0.4g

Preparation of Spectinomycin Solution: Add spectinomycin to distilled/deionized water and bring volume to 10.0mL. Mix thoroughly. Filter sterilize.

Preparation of Medium: Add components, except bovine blood and spectinomycin solution, to distilled/deionized water and bring volume to 890.0mL. Mix thoroughly. Gently heat and bring to boiling. Autoclave for 15 min at 15 psi pressure–121°C. Cool to 45°–50°C. Aseptically add sterile bovine blood and 10.0mL of sterile spectinomycin solution. Mix thoroughly. Pour into sterile Petri dishes or distribute into sterile tubes.

Use: For the isolation of *Treponema hyodysenteriae.*

TSA Blood Agar
See: **Trypticase™ Soy Agar with Sheep Blood**

TSA NaCl
See: **Trypticase™ Soy Agar with 3% NaCl**

TSA II™
See: **Trypticase™ Soy Agar, Modified**

TSA II™with Sheep Blood and Gentamicin
See: **Trypticase™ Soy Agar with Sheep Blood and Gentamicin**

TSBY Salt Medium

Composition per liter:

NaCl	18.0g
Pancreatic digest of casein	17.0g
$MgCl_2 \cdot 6H_2O$	4.0g
$MgSO_4 \cdot 7H_2O$	3.45g
Yeast extract	3.0g
Papaic digest of soybean meal	3.0g
K_2HPO_4	2.5g
Glucose	2.5g
KCl	0.34g
NH_4Cl	0.25g
$CaCl_2 \cdot 2H_2O$	0.14g

pH 7.2 ± 0.2 at 25°C

Preparation of Medium: Add components to distilled/deionized water and bring volume to 1.0L. Mix thoroughly. Distribute into tubes or flasks. Autoclave for 15 min at 15 psi pressure–121°C.

Use: For the cultivation and maintenance of *Bacillus subtilis, Carnobacterium alterfunditum,* and *Carnobacterium funditum.*

TSBY Salt Medium
See: ***Bacillus mascerans*** **Medium**

TSC Agar
See: **Tryptose Sulfite Cycloserine Agar**

TSC Agar, Fluorocult (Fluorocult TSC Agar) (Fluorocult Tryptose Sulfite Cycloserine Agar) (Tryptose Sulfite Cycloserine Agar, Fluorocult)

Composition per liter:

Agar	15.0g
Tryptose	15.0g
Peptone from soymeal	5.0g
Yeast extract	5.0g
$Na_2S_2O_5$	1.0g
Ammonium ferric citrate	1.0g
D-Cycloserine	0.2g
4-Methylumbelliferyl-phosphate disodium salt	50.0mg

Source: This medium is available from Merck.

Preparation of Medium: Add components to distilled/deionized water and bring volume to 1.0L. Mix thoroughly. Autoclave for 15 min at 15 psi pressure–121°C. Cool to 45°–50°C. Pour into sterile Petri dishes.

Use: For the isolation and enumeration of the vegetative and spore forms of *Clostridium perfringens* in foodstuffs. The culture medium complies with the recommendations of the International Organization for Standardization (ISO) (1978) and the DIN Norm 10165 for the examination of meat and meat products. It also conforms with the APHA recommendations for the examination of foods (1992). D-Cycloserine inhibits the accompanying bacterial flora and causes the colonies which develop to remain smaller. 4-Methylumbelliferyl-phosphate (MUP) is a fluorogenic substrate for the alkaline and acid phosphatase. The acid phosphatase is a highly specific indicator for *C. perfringens*. The acid phosphatase splits the fluorogenic substrate MUP forming 4-methylumbelliferone which can be identified as fluorescence in long-wave UV light. Thus a strong suggestion for the presence of *C. perfringens* can be obtained.

TSC Agar without Egg Yolk
See: **Tryptose Sulfite Cycloserine Agar without Egg Yolk**

TSFA
See: **Tryptic Soy Fast Green Agar**

TSI Agar
See: **Triple Sugar Iron Agar**

TSN Agar (Trypticase™ Sulfite Neomycin Agar)

Composition per liter:

Pancreatic digest of casein	15.0g
Agar	13.5g
Yeast extract	10.0g
Na_2SO_3	1.0g
Ferric citrate	0.5g
Neomycin sulfate	0.05g
Polymyxin sulfate	0.02g
Buffered thioglycolate solution	50.0mL

pH 7.2 ± 0.2 at 25°C

Source: This medium is available as a premixed powder from BD Diagnostic Systems.

Buffered Thioglycolate Solution:
Composition per 50.0mL:

Buffer solution	35.0mL
Sodium thioglycolate solution	15.0mL

Preparation of Buffered Thioglycolate Solution: Combine components. Mix thoroughly. Autoclave for 15 min at 15 psi pressure–121°C. Cool to 45°–50°C.

Buffer Solution:
Composition per 100.0mL:

Na_2CO_3	28.0g
K_2HPO_4	5.7g

Preparation of Buffer Solution: Add components to distilled/deionized water and bring volume to 100.0mL. Mix thoroughly.

Thioglycolate Solution:
Composition per 100.0mL:

Sodium thioglycolate	13.3g

Preparation of Thioglycolate Solution: Add sodium thioglycolate to distilled/deionized water and bring volume to 100.0mL. Mix thoroughly.

Preparation of Medium: Add components, except buffered thioglycolate solution, to distilled/deionized water and bring volume to 950.0mL. Mix thoroughly. Gently heat and bring to boiling. Autoclave for 12 min at 13 psi pressure–118°C. Do not overheat. Cool to 45°–50°C. Aseptically add buffered thioglycolate solution. Mix thoroughly. Pour into sterile Petri dishes or distribute into sterile tubes.

Use: For the selective isolation of *Clostridium perfringens.*

TSY Medium (Trypticase™ Soy Yeast Extract Medium)

Agar	20.0g
Pancreatic digest of casein	17.0g
Yeast extract	5.0g

NaCl	5.0g
Papaic digest of soybean meal	3.0g
K_2HPO_4	2.5g
Glucose	2.5g

pH 7.3 ± 0.2 at 25°C

Preparation of Medium: Add components to distilled/deionized water and bring volume to 1.0L. Mix thoroughly. Gently heat and bring to boiling. Distribute into tubes or flasks. Autoclave for 15 min at 15 psi pressure–121°C. Pour into sterile Petri dishes or leave in tubes.

Use: For the cultivation and maintenance of *Escherichia coli.*

TSYES Medium (Trypticase™ Soy Yeast Extract Starch Medium)

Composition per liter:

Pancreatic digest of casein	17.0g
Agar	15.0g
NaCl	5.0g
Papaic digest of soybean meal	3.0g
K_2HPO_4	2.5g
Glucose	2.5g
Yeast extract	2.0g
Soluble starch	1.0g

pH 7.3 ± 0.2 at 25°C

Preparation of Medium: Add components to distilled/deionized water and bring volume to 1.0L. Mix thoroughly. Gently heat while stirring and bring to boiling. Distribute into tubes or flasks. Autoclave for 15 min at 15 psi pressure–121°C. Pour into sterile Petri dishes or leave in tubes.

Use: For the cultivation and maintenance of *Bacillus* species.

TT Broth

See: **Tetrathionate Broth**

TT Broth, Hajna

See: **Tetrathionate Broth, Hajna**

TT Broth, USA

See: **Tetrathionate Broth, USA**

TTC Agar

See: **Tetrazolium Tolerance Agar**

TTD Medium (DSMZ Medium 480b)

Composition per liter:

NaCl (marine salts)	25.0g
Sulfur, powder	10.0g
Casitone	5.0g
NH_4Cl	0.33g
$CaCl_2 \cdot 2H_2O$	0.33g
$MgCl_2 \cdot 6H_2O$	0.33g
KCl	0.33g
KH_2PO_4	0.33g
Resazurin	1.0mg
$Na_2S \cdot 9H_2O$ solution	10.0mL
Vitamin solution	10.0mL
Trace elements solution SL-10	1.0mL

pH 6.9 ± 0.2 at 25°C

Trace Elements Solution SL-10:

Composition per liter:

$FeCl_2 \cdot 4H_2O$	1.5g
$CoCl_2 \cdot 6H_2O$	190.0mg
$MnCl_2 \cdot 4H_2O$	100.0mg
$ZnCl_2$	70.0mg
$Na_2MoO_4 \cdot 2H_2O$	36.0mg
$NiCl_2 \cdot 6H_2O$	24.0mg
H_3BO_3	6.0mg
$CuCl_2 \cdot 2H_2O$	2.0mg
HCl (25% solution)	10.0mL

Preparation of Trace Elements Solution SL-10: Add $FeCl_2 \cdot 4H_2O$ to 10.0mL of HCl solution. Mix thoroughly. Add distilled/deionized water and bring volume to 1.0L. Add remaining components. Mix thoroughly. Sparge with 100% N_2. Autoclave for 15 min at 15 psi pressure–121°C.

Vitamin Solution:

Composition per liter:

Pyridoxine-HCl	10.0mg
Thiamine-HCl·$2H_2O$	5.0mg
Riboflavin	5.0mg
Nicotinic acid	5.0mg
D-Ca-pantothenate	5.0mg
p-Aminobenzoic acid	5.0mg
Lipoic acid	5.0mg
Biotin	2.0mg
Folic acid	2.0mg
Vitamin B_{12}	0.1mg

Preparation of Vitamin Solution: Add components to distilled/deionized water and bring volume to 1.0L. Mix thoroughly. Sparge with 80% H_2 + 20% CO_2. Filter sterilize.

$Na_2S \cdot 9H_2O$ Solution:

Composition per 10.0mL:

$Na_2S \cdot 9H_2O$	0.5g

Preparation of $Na_2S \cdot 9H_2O$ Solution: Add $Na_2S \cdot 9H_2O$ to distilled/deionized water and bring volume to 10.0mL. Sparge with N_2. Autoclave for 15 min at 15 psi pressure–121°C. Cool to 25°C. Store anaerobically.

Preparation of Medium: Prepare and dispense medium under an oxygen-free 80% N_2 + 20% CO_2 gas mixture. Add components, except vitamin solution and $Na_2S \cdot 9H_2O$ solution, to 980.0mL distilled/deionized water. Mix thoroughly. Sparge with 80% N_2 + 20% CO_2. Adjust pH to 5.9 with concentrated NaOH. Sterilize medium by heating for 1 hr at 90°C–100°C on 3 subsequent days. Sparge with 80% N_2 + 20% CO_2. Before use, aseptically and anaerobically add 10.0mL sterile vitamin solution and 10.0mL sterile $Na_2S \cdot 9H_2O$ solution. Mix thoroughly. Aseptically and anaerobically distribute into sterile tubes or flasks.

Use: For the cultivation of *Thermococcus stetteri.*

TTYSH Medium

Composition per liter:

Pancreatic digest of casein	10.0g
Tryptose	10.0g
Yeast extract	10.0g
Glucose	5.0g
NaCl	5.0g
L-Cysteine·HCl	1.0g
K_2HPO_4	0.8g
KH_2PO_4	0.8g

Ascorbic acid 0.2g
Bovine serum, heat inactivated 50.0mL
Hemin solution 20.0mL

Hemin Solution:

Composition per 50.0mL:

Hemin 50.0mg

Preparation of Hemin Solution: Add 50.0mg of hemin to distilled/deionized water and bring volume 50.0mL. Adjust the pH to 10.5 with 1*N* NaOH. Mix thoroughly. Filter sterilize.

Preparation of Medium: Add components, except tryptose, hemin solution, and bovine serum, to distilled/deionized water and bring volume to 600.0mL. Mix thoroughly. Add tryptose. Mix thoroughly. Gently heat and bring to boiling. Cool to 25°C. Autoclave for 15 min at 15 psi pressure–121°C. Aseptically add 50.0mL of sterile, heat-inactivated bovine serum and 20.0mL of sterile hemin solution. Mix thoroughly. Aseptically distribute into sterile tubes or flasks.

Use: For the cultivation of *Crithidia fasciculata.*

Tween™ 80A Agar

Composition per liter:

Agar 15.0g
Pancreatic digest of casein 5.0g
Yeast extract 3.0g
$MgSO_4 \cdot 7H_2O$ 2.0g
Tween™ 80 solution 50.0mL

pH 7.2 ± 0.2 at 25°C

Tween™ 80 Solution:

Composition per 50.0mL:

Tween™ 80 .... 10.0g

Preparation of Tween™ 80 Solution: Add Tween™ 80 to distilled/deionized water and bring volume to 50.0mL. Mix thoroughly. Autoclave for 15 min at 15 psi pressure–121°C.

Preparation of Medium: Add components, except Tween™ 80 solution, to distilled/deionized water and bring volume to 950.0mL. Gently heat and bring to boiling. Adjust pH to 7.2. Autoclave for 15 min at 15 psi pressure–121°C. Cool to 50°–55°C. Aseptically add 50.0mL of sterile Tween™ 80 solution. Mix thoroughly. Pour into sterile Petri dishes or distribute into sterile tubes.

Use: For the cultivation of *Agitococcus lubricus.*

Tween™ 80 Agar

See: **Polysorbate 80 Agar**

Tween™ 80 Agar (DSMZ Medium 884)

Composition per liter:

Solution A 900.0mL
Solution B 100.0mL

Solution A:

Composition per 900.0mL:

Agar 15.0g
Peptone 10.0g
NaCl 5.0g
$CaCl_2 \cdot 2H_2O$ 0.1g

pH 7.1 ± 0.2 at 25°C

Preparation of Solution A: Add components to distilled/deionized water and bring volume to 900.0mL. Mix thoroughly. Gently heat and bring to boiling. Autoclave for 15 min at 15 psi pressure–121°C. Cool to 50°C.

Solution B:

Composition per 100.0mL:

Tween™ 80 .... 10.0g

Preparation of Solution B: Add Tween™ 80 to distilled/deionized water and bring volume to 100.0mL. Mix thoroughly. Autoclave for 15 min at 15 psi pressure–121°C. Cool to 50°C.

Preparation of Medium: Aseptically combine 900.0mL sterile solution A and 100.0mL sterile solution B. Mix thoroughly. Pour into sterile Petri dishes.

Use: For the cultivation of unclassified bacterium DSM 13023.

Tween™ 80A Broth

Composition per liter:

Pancreatic digest of casein 5.0g
Yeast extract 3.0g
$MgSO_4 \cdot 7H_2O$ 2.0g
Tween™ 80 solution 50.0mL

pH 7.2 ± 0.2 at 25°C

Tween™ 80 Solution:

Composition per 50.0mL:

Tween™ 80 .... 10.0g

Preparation of Tween™ 80 Solution: Add Tween™ 80 to distilled/deionized water and bring volume to 50.0mL. Mix thoroughly. Autoclave for 15 min at 15 psi pressure–121°C.

Preparation of Medium: Add components, except Tween™ 80 solution, to distilled/deionized water and bring volume to 950.0mL. Mix thoroughly. Adjust pH to 7.2. Autoclave for 15 min at 15 psi pressure–121°C. Aseptically add 50.0mL of sterile Tween™ 80 solution. Mix thoroughly. Aseptically distribute into sterile tubes or flasks.

Use: For the cultivation of *Agitococcus lubricus.*

Tween™ 80A Medium (DSMZ Medium 618)

Composition per liter:

Casitone 5.0g
Yeast extract 3.0g
$MgSO_4 \cdot 7H_2O$ 2.0g
Tween™ 80 solution 50.0mL

pH 7.2 ± 0.2 at 25°C

Tween™ 80 Solution:

Composition per 50.0mL:

Tween™ 80 .... 10.0g

Preparation of Tween™ 80 Solution: Add Tween™ 80 to distilled/deionized water and bring volume to 50.0mL. Mix thoroughly. Autoclave for 15 min at 15 psi pressure–121°C.

Preparation of Medium: Add components, except Tween™ 80 solution, to distilled/deionized water and bring volume to 950.0mL. Mix thoroughly. Adjust pH to 7.2. Autoclave for 15 min at 15 psi pressure–121°C. Aseptically add 50.0mL of sterile Tween™ 80 solution. Mix thoroughly. Aseptically distribute into sterile tubes or flasks.

Use: For the cultivation of *Agitococcus lubricus.*

Tween™ 80 Hydrolysis Broth

Composition per liter:

Na_2HPO_4 .... 5.79g
NaH_2PO_4 .... 3.53g
Neutral Red 0.02g
Tween™ 80 solution 5.0mL

pH 7.0 ± 0.2 at 25°C

Tween™ 80 Solution:

Composition per 50.0mL:

Tween™ 80 .... 10.0g

Preparation of Tween™ 80 Solution: Add Tween™ 80 to distilled/deionized water and bring volume to 50.0mL. Mix thoroughly. Autoclave for 15 min at 15 psi pressure–121°C.

Preparation of Medium: Add components, except Tween™ 80 solution, to distilled/deionized water and bring volume to 995.0mL. Mix thoroughly. Adjust pH to 7.2. Autoclave for 15 min at 15 psi pressure–121°C. Aseptically add 5.0mL of sterile Tween™ 80 solution. Mix thoroughly. Aseptically distribute into sterile tubes or flasks.

Use: For the differentiation of *Mycobacterium* species. Strains that hydrolyze Tween™ 80 within 5 days turn the medium pink to red.

Tween™ 80 Hydrolysis Broth

Composition per 125.0mL:

Neutral Red 0.1g
Solution 1 .... 38.9mL
Solution 2 .... 61.1mL
Tween™ 80 solution 25.0mL

pH 7.0 ± 0.2 at 25°C

Solution 1:

Composition per 400.0mL:

KH_2PO_4 .... 22.7g

Preparation of Solution 1: Add KH_2PO_4 to distilled/deionized water and bring volume to 400.0mL. Mix thoroughly.

Solution 2:

Composition per 400.0mL:

Na_2HPO_4 .... 23.8g

Preparation of Solution 2: Add Na_2HPO_4 to distilled/deionized water and bring volume to 400.0mL. Mix thoroughly.

Tween™ 80 Solution:

Composition per 50.0mL:

Tween™ 80 .... 10.0g

Preparation of Tween™ 80 Solution: Add Tween™ 80 to distilled/deionized water and bring volume to 50.0mL. Mix thoroughly. Autoclave for 15 min at 15 psi pressure–121°C.

Preparation of Medium: Add components, except Tween™ 80 solution, to distilled/deionized water and bring volume to 975.0mL. Mix thoroughly. Adjust pH to 7.2. Autoclave for 15 min at 15 psi pressure–121°C. Aseptically add 25.0mL of sterile Tween™ 80 solution. Mix thoroughly. Aseptically distribute into sterile tubes or flasks.

Use: For the differentiation of *Mycobacterium* species. Strains that hydrolyze Tween™ 80 within 5 days turn the medium pink to red.

Tween™ 80 Hydrolysis Broth

Composition per 102.5mL:

$NaHPO_4$ (0.066*M* solution) 61.1mL
KH_2PO_4 (0.066*M* solution) 38.9mL
Neutral Red (0.1% solution) 2.0mL
Tween™ 80 .... 0.5mL

pH 7.0 ± 0.2 at 25°C

Preparation of Medium: Combine components. Mix thoroughly. Distribute into tubes or flasks. Autoclave for 15 min at 15 psi pressure–121°C.

Use: For the differentiation of *Mycobacterium* species. Strains that hydrolyze Tween™ 80 within 5 days turn the medium pink to red.

Tween™ 80 Hydrolysis Medium

Composition per liter:

Agar 12.0g
Peptone 10.0g
Tween™ 80 .... 10.0g
NaCl 5.0g
$CaCl_2$ 0.1g

pH 7.2–7.4 at 25°C

Preparation of Medium: Add components to distilled/deionized water and bring volume to 1.0L. Mix thoroughly. Gently heat and bring to boiling. Distribute into tubes or flasks. Autoclave for 15 min at 15 psi pressure–121°C. Pour into sterile Petri dishes.

Use: For the cultivation and differentiation of *Pseudomonas* species based on their ability to hydrolyze Tween™ 80. Bacteria that hydrolyze Tween™ 80 appear as colonies surrounded by an opaque zone.

Tween™ 80 Oxgall Caffeic Acid Agar

See: **TOC Agar**

TY Medium

Composition per liter:

Agar 15.0g
Pancreatic digest of casein 5.0g
Yeast extract 3.0g
$CaCl_2 \cdot 6H_2O$ 1.3g

Preparation of Medium: Add components to distilled/deionized water and bring volume to 1.0L. Mix thoroughly. Gently heat and bring to boiling. Distribute into tubes or flasks. Autoclave for 15 min at 15 psi pressure–121°C. Pour into sterile Petri dishes or leave in tubes.

Use: For the cultivation of a wide variety of bacteria.

TY Medium (DSMZ Medium 1143)

Composition per liter:

Tryptone 5.0g
Yeast extract 3.0g
$CaCl_2 \cdot 2H_2O$ 0.9g

pH 6.8 ± 0.2 at 25°C

Preparation of Medium: Add components to distilled/deionized water and bring volume to 1.0L. Mix thoroughly. Autoclave for 15 min at 15 psi pressure–121°C. Adjust pH to 6.8.

Use: For the cultivation of *Azorhizobium doebereinerae.*

TY Medium, 2X

Composition per liter:

Pancreatic digest of casein 16.0g
Yeast extract 10.0g
NaCl 5.0g

pH 7.0 ± 0.2 at 25°C

Preparation of Medium: Add components to distilled/deionized water and bring volume to 1.0L. Mix thoroughly. Distribute into tubes or flasks. Autoclave for 25 min at 15 psi pressure–121°C.

Use: For the cultivation of *Escherichia coli.*

TY Salt Medium

Composition per liter:

NaCl	10.0g
Pancreatic digest of casein	10.0g
Yeast extract	5.0g

pH 7.0 ± 0.2 at 25°C

Preparation of Medium: Add components to distilled/deionized water and bring volume to 1.0L. Mix thoroughly. Distribute into tubes or flasks. Autoclave for 15 min at 15 psi pressure–121°C.

Use: For the cultivation of a wide variety of bacteria.

TY Salts Medium

Composition per liter:

Pancreatic digest of casein	1.0g
Yeast extract	1.0g
Salts solution	100.0mL

pH 8.2 ± 0.2 at 25°C

Salts Solution:

Composition per liter:

$NaNO_3$	6.89g
KNO_3	1.03g
$MgSO_4 \cdot 7H_2O$	1.0g
Nitrilotriacetic acid	1.0g
$CaSO_4 \cdot 2H_2O$	0.6g
NaCl	80.0mg
$FeCl_3$ solution	10.0mL
Trace elements solution	10.0mL

Preparation of Salts Solution: Add components to distilled/deionized water and bring volume to 1.0L. Mix thoroughly. Adjust pH to 8.2 with 1*M* NaOH. Autoclave for 15 min at 15 psi pressure–121°C.

$FeCl_3$ Solution:

Composition per 100.0mL:

$FeCl_3$	28.0mg

Preparation of $FeCl_3$ Solution: Add $FeCl_3$ to distilled/deionized water and bring volume to 100.0mL. Mix thoroughly.

Trace Elements Solution:

Composition per liter:

$MnSO_4 \cdot H_2O$	2.2g
H_3BO_3	0.5g
$ZnSO_4 \cdot 7H_2O$	0.5g
$CoCl_2 \cdot 6H_2O$	46.0mg
$Na_2MoO_4 \cdot 2H_2O$	25.0mg
$CuSO_4$	16.0mg
H_2SO_4	0.5mL

Preparation of Trace Elements Solution: Add components to distilled/deionized water and bring volume to 1.0L. Mix thoroughly.

Preparation of Medium: Add components, except salts solution, to distilled/deionized water and bring volume to 900.0mL. Mix thoroughly. Autoclave for 15 min at 15 psi pressure–121°C. Aseptically add 100.0mL of sterile salts solution. Mix thoroughly. Aseptically distribute into sterile tubes or flasks.

Use: For the cultivation of *Thermomonospora aquaticus, Thermus filiformis, Thermus flavus, Thermus ruber,* and other *Thermus* species.

TYE Broth Medium
(ATCC Medium 1972)

Composition per liter:

Tryptone	16.0g
Yeast extract	10.0g
NaCl	5.0g
Glucose	4.0g

pH 7.2 ± 0.2 at 25°C

Preparation of Medium: Add components to distilled/deionized water and bring volume to 1.0L. Mix thoroughly. Adjust pH to 7.2. Gently heat and bring to boiling. Distribute into tubes or flasks. Autoclave for 15 min at 15 psi pressure–121°C.

Use: For the cultivation and maintenance of *Aeromicrobium erythreum.*

TYE-CO

See: ***Clostridium thermoaceticum*** **Medium**

TYE HES Medium

Composition per 950.0mL:

NaCl	49.7g
$MgSO_4 \cdot 7H_2O$	49.3g
Noble agar	10.0g
Yeast extract	0.5g
Pancreatic digest of casein	0.5g
$CaCl_2 \cdot 2H_2O$ solution	50.0mL

pH 7.2 ± 0.2 at 25°C

$CaCl_2 \cdot 2H_2O$ Solution:

Composition per 100.0mL:

$CaCl_2 \cdot 2H_2O$	0.3g

Preparation of $CaCl_2 \cdot 2H_2O$ Solution: Add the $CaCl_2 \cdot 2H_2O$ to distilled/deionized water and bring volume to 100.0mL. Mix thoroughly. Autoclave for 15 min at 15 psi pressure–121°C. Cool to 45°–50°C.

Preparation of Medium: Add components, except $CaCl_2 \cdot 2H_2O$ solution, to distilled/deionized water and bring volume to 950.0mL. Mix thoroughly. Gently heat and bring to boiling. Autoclave for 15 min at 15 psi pressure–121°C. Cool to 45°–50°C. Aseptically add 50.0mL of sterile $CaCl_2 \cdot 2H_2O$ solution. Mix thoroughly. Adjust pH to 7.2. Pour into sterile Petri dishes or distribute into sterile tubes.

Use: For the cultivation of *Planococcus* species.

TYEG Medium
(Trypticase™ Yeast Extract Glucose Medium)

Composition per 1050.0mL:

NaCl	100.0g
Pancreatic digest of casein	10.0g
$Na_2HPO_4 \cdot 7H_2O$	2.1g
NH_4Cl	1.0g
KH_2PO_4	0.3g
$MgCl_2 \cdot 6H_2O$	0.2g
Glucose solution	50.0mL
$Na_2S \cdot 7H_2O$ solution	25.0mL
Trace minerals solution II	10.0mL
Wolfe's vitamin solution	10.0mL
Yeast extract solution	5.0mL

Resazurin (0.2% solution) 1.0mL
$FeSO_4 \cdot 9H_2O$ (2.5% solution) 25.0µl

pH 7.3 ± 0.1 at 25°C

Glucose Solution:

Composition per 100.0mL:

D-Glucose 10.0g

Preparation of Glucose Solution: Add glucose to distilled/deionized water and bring volume to 100.0mL. Mix thoroughly. Filter sterilize. Aseptically bubble with 90% N_2 + 10% CO_2 to reduce.

$Na_2S \cdot 7H_2O$ Solution:

Composition per 100.0mL:

$Na_2S \cdot 7H_2O$ 2.5g

Preparation of $Na_2S \cdot 7H_2O$ Solution: Add $Na_2S \cdot 7H_2O$ to distilled/deionized water and bring volume to 100.0mL. Mix thoroughly. Autoclave for 15 min at 15 psi pressure–121°C. Use freshly prepared solution.

Trace Minerals Solution II:

Composition per liter:

Nitrilotriacetic acid	12.8g
$CoCl_2 \cdot 6H_2O$	0.17g
$CaCl_2 \cdot 2H_2O$	0.1g
FeSO4·7H2O	0.1g
$MnCl_2 \cdot 4H_2O$	0.1g
NaCl	0.1g
$ZnCl_2$	0.1g
$NiSO_4 \cdot 6H_2O$	0.026g
$CuCl_2 \cdot 2H_2O$	0.02g
Na_2SeO_3	0.017g
H_3BO_3	0.01g
$Na_2MoO_4 \cdot 2H_2O$	0.01g

Preparation of Trace Minerals Solution II: Add nitrilotriacetic acid to 500.0mL of distilled/deionized water. Dissolve by adjusting pH to 6.5 with KOH. Add remaining components. Add distilled/deionized water to 1.0L. Filter through Whatman filter paper. Store under N_2.

Wolfe's Vitamin Solution:

Composition per liter:

Pyridoxine·HCl	10.0mg
Thiamine·HCl	5.0mg
Riboflavin	5.0mg
Nicotinic acid	5.0mg
Calcium pantothenate	5.0mg
p-Aminobenzoic acid	5.0mg
Thioctic acid	5.0mg
Biotin	2.0mg
Folic acid	2.0mg
Cyanocobalamin	100.0µg

Preparation of Wolfe's Vitamin Solution: Add components to distilled/deionized water and bring volume to 1.0L. Mix thoroughly. Filter sterilize. Aseptically bubble with 90% N_2 + 10% CO_2 to reduce.

Yeast Extract Solution:

Composition per 100.0mL:

Yeast extract 10.0g

Preparation of Yeast Extract Solution: Add yeast extract to distilled/deionized water and bring volume to 100.0mL. Mix thoroughly. Filter sterilize. Aseptically bubble with 90% N_2 + 10% CO_2 to reduce.

Preparation of Medium: Add components—except glucose solution, yeast extract solution, and Wolfe's vitamin solution—to distilled/deionized water and bring volume to 960.0mL. Mix thoroughly. Adjust pH to 7.3. Gently heat and bring to boiling under 90% N_2 + 10% CO_2. Autoclave for 15 min at 15 psi pressure–121°C. Cool to 45°–50°C. Aseptically and anaerobically add 50.0mL of sterile glucose solution, 10.0mL of sterile Wolfe's vitamin solution, and 5.0mL of sterile yeast extract solution. Aseptically and anaerobically distribute into sterile tubes in 5.0mL volumes. Immediately prior to inoculation, aseptically add 0.125mL of sterile $Na_2S \cdot 9H_2O$ solution per tube.

Use: For the cultivation and maintenenace of *Halobacteroides acetoethylicus*.

TYES Medium (Tryptone Yeast Extract Salt Medium)

Composition per liter:

Agar	15.0g
Pancreatic digest of casein	10.0g
NaCl	8.0g
Yeast extract	1.0g
$CaCl_2$	0.3g

pH 7.2 ± 0.2 at 25°C

Preparation of Medium: Add components to distilled/deionized water and bring volume to 1.0L. Mix thoroughly. Gently heat and bring to boiling. Distribute into tubes or flasks. Autoclave for 15 min at 15 psi pressure–121°C. Pour into sterile Petri dishes or leave in tubes.

Use: For the cultivation and maintenenace of *Escherichia coli*.

TYG Medium

Composition per liter:

Yeast extract	5.0g
K_2HPO_4	3.5g
Tryptone	1.0g
L-Cysteinium chloride·H_2O	0.5g
Na_2SO_4	0.2g
Biotin	10.0mg
p-Aminobenzoic acid	10.0mg
Sugar solution	5.0mL
Resazurin solution	4.0mL
Trace elements solution	1.0mL

Resazurin Solution:

Composition per 100.0mL:

Resazurin 25.0mg

Preparation of Resazurin Solution: Add resazurin to distilled/deionized water and bring volume to 100.0mL. Mix thoroughly.

Trace Elements Solution:

Composition per 100.0mL:

$FeCl_3 \cdot 6H_2O$	2.7g
$MgSO_4$	1.2g
$NaMoO_4 \cdot 2H_2O$	0.24g
$MnSO_4 \cdot 7H_2O$	0.17g
$CaCl_2 \cdot 2H_2O$	0.15g
$ZnSO_4 \cdot 7H_2O$	29.0mg
$CuSO_4 \cdot 5H_2O$	25.0mg
$CoCl_2 \cdot 6H_2O$	24.0mg
H_2SO_4	2.8mL

Preparation of Trace Elements Solution: Add components to distilled/deionized water and bring volume to 100.0mL. Mix thoroughly.

Sugar Solution:

Composition per 100.0mL:

Glucose 1.0g

Preparation of Sugar Solution: Add glucose to distilled/deionized water and bring volume to 100.0mL. Mix thoroughly. Filter sterilize.

Preparation of Medium: Add components, except L-cysteinium chloride, biotin, *p*-aminobenzoic acid, and sugar solution, to distilled/deionized water and bring volume to 995.0mL. Mix thoroughly. Adjust pH to 7.0. Gently heat but do not boil. Cook for 5–10 min so that color first turns red and then turns yellow. Cool on ice. Add L-cysteinium chloride, biotin, and *p*-aminobenzoic acid. Mix thoroughly. Autoclave for 15 min at 15 psi pressure–121°C. Aseptically add 5.0 mL of sterile sugar solution. Mix thoroughly.

Use: For the cultivation of *Clostridium* species.

TYG Medium

Composition per liter:

Pancreatic digest of casein	20.0g
Glucose	5.0g
KH_2PO_4	4.0g
Sodium thioglycolate	0.5g
Yeast extract	0.5g
$MgSO_4 \cdot 7H_2O$	0.2g
$FeSO_4 \cdot 7H_2O$	5.0mg
$MnSO_4 \cdot 4H_2O$	5.0mg
NH_4MoO_4	5.0mg

pH 7.4 ± 0.2 at 25°C

Preparation of Medium: Add components to distilled/deionized water and bring volume to 1.0L. Mix thoroughly. Adjust pH to 7.4 with NaOH. Distribute into tubes or flasks. Autoclave for 10 min at 11 psi pressure–116°C.

Use: For the cultivation of *Sprolactobacillus cellulosolvens*.

TYG Medium (Trypticase™ Yeast Extract Glucose Medium) (ATCC Medium 603)

Composition per liter:

Pancreatic digest of casein	10.0g
NaCl	8.0g
Yeast extract	1.0g
Glucose	1.0g
$CaCl_2 \cdot 2H_2O$	0.3g

Preparation of Medium: Add components to distilled/deionized water and bring volume to 1.0L. Mix thoroughly. Distribute into tubes or flasks. Autoclave for 15 min at 15 psi pressure–121°C.

Use: For the cultivation and maintenenace of *Escherichia coli*.

TYG Medium (Tryptone Yeast Extract Glucose Medium) (ATCC Medium 741)

Composition per liter:

Agar	20.0g
Pancreatic digest of casein	3.0g
Yeast extract	3.0g
Glucose	3.0g
K_2HPO_4	1.0g

pH 7.4 ± 0.2 at 25°C

Preparation of Medium: Add components to distilled/deionized water and bring volume to 1.0L. Mix thoroughly. Gently heat and bring to boiling. Distribute into tubes or flasks. Autoclave for 15 min at 15 psi pressure–121°C. Pour into sterile Petri dishes or leave in tubes.

Use: For the cultivation and maintenenace of *Thermomonospora fusca*.

TYGM-9 Medium

Composition per liter:

NaCl	7.5g
K_2HPO_4	2.8g
Casein digest	2.0g
Gastric mucin	2.0g
Yeast extract	1.0g
KH_2PO_4	0.4g
Bovine serum, heat inactivated	30.0mL
Rice starch solution	30.0mL
Tween™ solution	0.5mL

pH 7.4 ± 0.2 at 25°C

Tween™ Solution:

Composition per 100.0mL:

Tween™ 80 .... 10.0g

Preparation of Tween™ Solution: Add Tween™ 80 to absolute ethanol and bring volume to 100.0mL. Mix thoroughly. Filter sterilize.

Rice Starch Solution:

Composition per 100.0mL:

Rice starch	5.0g
Phosphate-buffered saline solution	100.0mL

Preparation of Rice Starch Solution: Heat sterilize rice starch at 150°C for 2 hr. Aseptically add 100.0mL of sterile phosphate-buffered saline solution. Mix thoroughly. Use immediately.

Phosphate-Buffered Saline Solution:

Composition per liter:

NaCl	9.0g
$Na_2HPO_4 \cdot 7H_2O$	0.795g
KH_2PO_4	0.114g

Preparation of Phosphate-Buffered Saline Solution: Add components to distilled/deionized water and bring volume to 1.0L. Mix thoroughly. Adjust pH to 7.4. Autoclave for 15 min at 15 psi pressure–121°C. Cool to 25°C.

Preparation of Medium: Add components, except rice starch solution, Tween™ solution, and bovine serum, to distilled/deionized water and bring volume to 939.5mL. Mix thoroughly. Autoclave for 15 min at 15 psi pressure–121°C. Cool to 25°C. Aseptically add 30.0mL of sterile bovine serum, 30.0mL of sterile rice starch solution, and 0.5mL of sterile Tween™ solution. Mix thoroughly. Aseptically distribute into sterile, screw-capped tubes or flasks.

Use: For the cultivation of *Dientamoeba fragilis, Ditrichomonas honigbergii, Entamoeba coli, Entamoeba dispar, Entamoeba gingivalis, Entamoeba insolita, Entamoeba histolytica, Entamoeba moshkovskii, Entamoeba polecki, Entamoeba ranarum, Pseudotrichomonas keilini*, and *Trepomonas agilis*.

TYGPN Medium

Composition per liter:

Pancreatic digest of casein	20.0g
KNO_3	10.0g
Yeast extract	10.0g

Na_2HPO_4 .... 5.0g
Glycerol (80% solution) 10.0mL

pH 7.0 ± 0.2 at 25°C

Preparation of Medium: Add components to distilled/deionized water and bring volume to 1.0L. Mix thoroughly. Distribute into tubes or flasks. Autoclave for 25 min at 15 psi pressure–121°C.

Use: For the cultivation of *Escherichia coli.*

TYGS Medium (Tryptone Yeast Extract Glucose Salt Medium)

Composition per liter:

Agar 15.0g
Pancreatic digest of casein 10.0g
NaCl 8.0g
Yeast extract 1.0g
$CaCl_2 \cdot 2H_2O$ solution 100.0mL
Glucose solution 100.0mL

$CaCl_2 \cdot 2H_2O$ Solution:
Composition per 100.0mL:

$CaCl_2 \cdot 2H_2O$ 0.3g

Preparation of $CaCl_2 \cdot 2H_2O$ Solution: Add the $CaCl_2 \cdot 2H_2O$ to distilled/deionized water and bring volume to 100.0mL. Mix thoroughly. Filter sterilize.

Glucose Solution:
Composition per 100.0mL:

D-Glucose 1.0g

Preparation of Glucose Solution: Add glucose to distilled/deionized water and bring volume to 100.0mL. Mix thoroughly. Filter sterilize.

Preparation of Medium: Add components, except $CaCl_2 \cdot 2H_2O$ solution and glucose solution, to distilled/deionized water and bring volume to 800.0mL. Mix thoroughly. Gently heat and bring to boiling. Autoclave for 15 min at 15 psi pressure–121°C. Cool to 45°–50°C. Aseptically add the sterile $CaCl_2 \cdot 2H_2O$ solution and sterile glucose solution. Mix thoroughly. Pour into sterile Petri dishes or distribute into sterile tubes.

Use: For the cultivation and maintenance of a variety of bacteria.

TYI-S-33 Medium

Composition per liter:

Pancreatic digest of casein 20.0g
Yeast extract 10.0g
NaCl 2.0g
L-Cysteine·HCl 1.0g
Ascorbic acid 0.2g
Ferric ammonium citrate 22.8mg
Bovine serum, heat inactivated 100.0mL
Special 107 vitamin mix 100.0mL
Buffer solution 50.0mL
Glucose solution 50.0mL

pH 6.8 ± 0.2 at 25°C

Glucose Solution:
Composition per 50.0mL:

Glucose 10.0g

Preparation of Glucose Solution: Add glucose to distilled/deionized water and bring volume to 50.0mL. Mix thoroughly. Filter sterilize.

Buffer Solution:
Composition per 50.0mL:

K_2HPO_4 .... 1.0g
KH_2PO_4 .... 0.6g

Preparation of Buffer Solution: Add components to distilled/deionized water and bring volume to 50.0mL. Mix thoroughly. Adjust pH to 6.8. Autoclave for 15 min at 15 psi pressure–121°C. Cool to 25°C.

Special 107 Vitamin Mix:
Composition per 120.0.0mL:

Solution 1 .... 1.2mL
Solution 2 .... 0.4mL
Solution 3 .... 0.4mL
Solution 4 vitamins 100.0mL

Preparation of Special 107 Vitamin Mix: Aseptically combine 1.2mL of sterile solution 1, 0.4mL of sterile solution 2, 0.4mL of sterile solution 3, and 100.0mL of sterile solution 4 vitamins. Bring volume to 120.0mL with sterile distilled/deionized water.

Solution 1:
Composition per 100.0mL:

Vitamin B_{12} 40.0mg

Preparation of Solution 1: Add vitamin B_{12} to distilled/deionized water and bring volume to 100.0mL. Mix thoroughly. Filter sterilize.

Solution 2:
Composition per 100.0.0mL:

DL-6,8-Thioctic acid, oxidized 100.0mg
Absolute ethanol 100.0mL

Preparation of Solution 2: Add DL-6,8-thioctic acid to absolute ethanol and bring volume to 100.0mL. Mix thoroughly. Filter sterilize.

Solution 3:
Composition per 100.0mL:

Tween™ 80 .... 50.0g
Absolute ethanol 100.0mL

Preparation of Solution 3: Add Tween™ 80 to absolute ethanol and bring volume to 100.0mL. Mix thoroughly. Filter sterilize.

Solution 4 Vitamins:
Composition per liter:

Vitamin B_{12} 10.0mg
Calcium D-(+)-pantothenate 2.3mg
Choline chloride 1.25mg
Riboflavin 0.7mg
Calciferol (vitamin D_2) 0.25mg
Vitamin A, crystallized alcohol 0.25mg
p-Aminobenzoic acid 0.125mg
i-Inositol 0.125mg
Biotin 0.025mg
α-Tocopherol phosphate, disodium salt 0.025mg
Folic acid 0.025mg
Menadione (vitamin K_3) 0.025mg
Thiamine·HCl 0.025mg
Niacin 0.0625mg
Niacinamide 0.0625mg
Pyridoxal·HCl 0.0625mg
Pyridoxine·HCl 0.0625mg

Preparation of Solution 4 Vitamins: Add components to distilled/deionized water and bring volume to 1.0L. Mix thoroughly. Filter sterilize.

Preparation of Medium: Add components, except heat-inactivated bovine serum, special 107 vitamin mix, glucose solution, and buffer solution, to distilled/deionized water and bring volume to 700.0mL. Mix thoroughly. Autoclave for 15 min at 15 psi pressure–121°C. Aseptically add 100.0mL of sterile, heat-inactivated bovine serum, 50.0mL of sterile glucose solution, 100.0mL of sterile special 107 vitamin mix, and 50.0mL of sterile buffer solution. Mix thoroughly. Aseptically distribute into sterile tubes or flasks.

Use: For the cultivation of *Entamoeba barreti, Entamoeba histolytica, Entamoeba insolita, Entamoeba invadens, Entamoeba moshkovskii, Entamoeba ranarum, Entamoeba terrapinae, Monocercomonas* species, *Phreatamoeba balamuthi, Spironucleus vortens,* and *Trichomonas tenax*.

TYM Basal Medium, Modified 1

Composition per liter:

Pancreatic digest of peptone	20.0g
Yeast extract	10.0g
Maltose	5.0g
L-Cysteine·HCl	1.0g
K_2HPO_4	0.8g
KH_2PO_4	0.8g
Agar, noble	0.5g
L-Ascorbic acid	0.2g

pH 7.8 ± 0.2 at 25°C

Preparation of Medium: Add components, except agar, to distilled/deionized water and bring volume to 1.0L. Mix thoroughly. Adjust pH to 7.8 with NaOH. Add the agar. Mix thoroughly. Gently heat and bring to boiling. Distribute into screw-capped tubes. Autoclave for 15 min at 15 psi pressure–121°C.

Use: For the cultivation of *Trichomonas* species.

TYM Basal Medium, Modified 1

Composition per liter:

Pancreatic digest of casein	20.0g
Yeast extract	10.0g
Maltose	5.0g
L-Cysteine·HCl	1.0g
K_2HPO_4	0.8g
KH_2PO_4	0.8g
Agar, noble	0.5g
L-Ascorbic acid	0.2g
Lamb serum, heat inactivated	300.0mL

pH 7.2 ± 0.2 at 25°C

Preparation of Medium: Add components, except agar and lamb serum, to distilled/deionized water and bring volume to 700.0mL. Mix thoroughly. Adjust to pH 7.2 with NaOH. Add the agar. Mix thoroughly. Gently heat and bring to boiling. Autoclave for 15 min at 15 psi pressure–121°C. Cool to 25°C. Aseptically add 300.0mL of sterile, heat-inactivated lamb serum. Mix thoroughly. Aseptically distribute into sterile screw-capped tubes. Use soon after preparation.

Use: For the cultivation of *Hypotrichomonas acosta, Tetratrichomonas gallinarum, Trichomitus batrachorum, Trichomonas gallinae, Trichomonas vaginalis, Tritrichomonas augusta,* and *Tritrichomonas suis.*

TYM Basal Medium, Modified 2

Composition per liter:

Pancreatic digest of casein	20.0g
Yeast extract	10.0g
Maltose	5.0g
L-Cysteine·HCl	1.0g
K_2HPO_4	0.8g
KH_2PO_4	0.8g
Agar, noble	0.5g
L-Ascorbic acid	0.2g
Horse serum, heat inactivated	300.0mL

pH 7.2 ± 0.2 at 25°C

Preparation of Medium: Add components, except agar and horse serum, to distilled/deionized water and bring volume to 700.0mL. Mix thoroughly. Adjust pH to 7.2. Add the agar. Mix thoroughly. Gently heat and bring to boiling. Autoclave for 15 min at 15 psi pressure–121°C. Cool to 50°–55°C. Aseptically add 300.0mL of heat-inactivated horse serum. Mix thoroughly. Aseptically distribute into screw-capped tubes.

Use: For the cultivation of *Trichomonas gallinae* and *Tritrichomonas foetus*.

TYM Basal Medium, Modified 2

Composition per liter:

Pancreatic digest of casein	20.0g
Yeast extract	10.0g
Maltose	5.0g
L-Cysteine·HCl	1.0g
K_2HPO_4	0.8g
KH_2PO_4	0.8g
Agar, noble	0.5g
L-Ascorbic acid	0.2g
Lamb serum, heat inactivated	300.0mL

pH 7.0 ± 0.2 at 25°C

Preparation of Medium: Add components, except agar and lamb serum, to distilled/deionized water and bring volume to 700.0mL. Mix thoroughly. Adjust to pH 7.0 with NaOH. Add the agar. Mix thoroughly. Gently heat and bring to boiling. Autoclave for 15 min at 15 psi pressure–121°C. Cool to 25°C. Aseptically add 300.0mL of sterile, heat-inactivated lamb serum. Mix thoroughly. Aseptically distribute into sterile screw-capped tubes. Use soon after preparation.

Use: For the cultivation of *Trichomonas vaginalis* and *Hypotrichomonas* species.

TYM Basal Medium, Modified 3

Composition per liter:

Pancreatic digest of casein	20.0g
Yeast extract	10.0g
Maltose	5.0g
L-Cysteine·HCl	1.0g
K_2HPO_4	0.8g
KH_2PO_4	0.8g
Agar, noble	0.5g
L-Ascorbic acid	0.2g
Horse serum, heat inactivated	300.0mL

pH 7.0 ± 0.2 at 25°C

Preparation of Medium: Add components, except agar and horse serum, to distilled/deionized water and bring volume to 700.0mL. Mix thoroughly. Adjust pH to 7.0. Add the agar. Mix thoroughly. Gently heat and bring to boiling. Autoclave for 15 min at 15 psi pressure–121°C. Cool to 50°–55°C. Aseptically add 300.0mL of heat-inactivated horse serum. Mix thoroughly. Aseptically distribute into screw-capped tubes.

Use: For the cultivation of *Pentatrichomonas hominis, Proteromonas lacertae,* and *Tritrichomonas foetus.*

TYM Basal Medium, Modified 3

Composition per liter:

Pancreatic digest of casein	20.0g
Yeast extract	10.0g
Maltose	5.0g
L-Cysteine·HCl	1.0g
K_2HPO_4	0.8g
KH_2PO_4	0.8g
Agar, noble	0.5g
L-Ascorbic acid	0.2g
Lamb serum, heat inactivated	200.0mL
Dubos medium serum	100.0mL

pH 6.0–6.5 at 25°C

Preparation of Medium: Add components, except agar lamb serum, and Dubos medium serum, to distilled/deionized water and bring volume to 700.0mL. Mix thoroughly. Adjust to pH to 6.0–6.5. Add the agar. Mix thoroughly. Gently heat and bring to boiling. Autoclave for 15 min at 15 psi pressure–121°C. Cool to 25°C. Aseptically add 200.0mL of sterile, heat-inactivated lamb serum and 100.0mL of sterile Dubos medium serum. Mix thoroughly. Aseptically distribute into sterile, screw-capped tubes. Use soon after preparation.

Use: For the cultivation of *Pentatrichomonas hominis*, *Trichomonas vaginalis*, *Tritrichomonas suis*, and *Tritrichomonas foetus*.

TYM Basal Medium, Modified 4

Composition per liter:

Pancreatic digest of casein	20.0g
Yeast extract	10.0g
Maltose	5.0g
L-Cysteine·HCl	1.0g
K_2HPO_4	0.8g
KH_2PO_4	0.8g
Agar, noble	0.5g
L-Ascorbic acid	0.2g
Horse serum, heat inactivated	300.0mL

pH 6.0 ± 0.2 at 25°C

Preparation of Medium: Add components, except agar and horse serum, to distilled/deionized water and bring volume to 700.0mL. Mix thoroughly. Adjust pH to 6.0. Add the agar. Mix thoroughly. Gently heat and bring to boiling. Autoclave for 15 min at 15 psi pressure–121°C. Cool to 50°–55°C. Aseptically add 300.0mL of heat-inactivated horse serum. Mix thoroughly. Aseptically distribute into screw-capped tubes.

Use: For the cultivation of *Tritrichomonas foetus* and *Trichomonas vaginalis*.

TYM Medium

Composition per liter:

Pancreatic digest of casein	10.0g
Yeast extract	10.0g
$NaHCO_3$	6.0g
NaCl	1.0g
K_2HPO_4	0.5g
KH_2PO_4	0.5g
L-Cysteine·HCl·H_2O	0.3g
$MgSO_4$	0.1g
$CaCl_2$	0.1g

pH 6.7–6.8 at 25°C

Preparation of Medium: Add components, except $NaHCO_3$ and L-cysteine·HCl·H_2O, to distilled/deionized water and bring volume to 1.0L. Mix thoroughly. Gently heat and bring to boiling. Cool to room temperature while sparging with 100% CO_2. Add $NaHCO_3$ and L-cysteine·HCl·H_2O. Mix thoroughly. Continue sparging with 100% CO_2 for 5–10 min. Anaerobically distribute into tubes or flasks. Autoclave for 15 min at 15 psi pressure–121°C. Final pH should be 6.7–6.8.

Use: For the cultivation of *Peptostreptococcus heliotrinreducens*.

TYN Medium

Composition per liter:

$Na_2S_2O_3$·$5H_2O$	10.0g
Pancreatic digest of casein	1.0g
Yeast extract	1.0g
Na_2SO_4	1.0g

Preparation of Medium: Add components to distilled/deionized water and bring volume to 1.0L. Mix thoroughly. Distribute into tubes or flasks. Autoclave for 15 min at 15 psi pressure–121°C.

Use: For the cultivation and maintenance of *Thiobacillus* species.

Tyrosine Agar

Composition per liter:

Solution 1	900.0mL
Solution 2	100.0mL

pH 7.0 ± 0.2 at 25°C

Solution 1:

Composition per 900.0mL:

Agar	15.0g
Pancreatic digest of gelatin	5.0g
Beef extract	3.0g

Preparation of Solution 1: Add components to distilled/deionized water and bring volume to 900.0mL. Mix thoroughly. Gently heat and bring to boiling.

Solution 2:

Composition per 100.0mL:

Tyrosine	5.0g

Preparation of Solution 2: Add tyrosine to distilled/deionized water and bring volume to 100.0mL. Mix thoroughly. Gently heat and bring to boiling.

Preparation of Medium: Combine solutions 1 and 2. Mix thoroughly. Distribute into tubes or flasks. Autoclave for 15 min at 15 psi pressure–121°C. Pour into sterile Petri dishes or leave in tubes.

Use: For the differentiation of aerobic *Actinomycete* species. Clearing around a colony indicates utilization of tyrosine. *Streptomyces* and *Actinomadura* species utilize tyrosine. *Nocardia asteroides, Nocardia caviae,* and *Mycobacterium fortuitum* do not utilize tyrosine.

Tyrosine Agar (International Streptomyces Project Medium 7) (ISP Medium 7) (ATCC Medium 1776)

Composition per liter:

Agar	20.0g
Glycerol	15.0g
L-Tyrosine	0.5g
L-Asparagine	1.0g
K_2HPO_4	0.5g

$MgSO_4 \cdot 7H_2O$ 0.5g
NaCl 0.5g
$FeSO_4 \cdot 7H_2O$ 0.01g
Trace elements solution HO-LE 1.0mL

pH 7.3 ± 0.1 at 25°C

Trace Elements Solution HO-LE:

Composition per liter:

H_3BO_3 .... 2.85g
$MnCl_2 \cdot 4H_2O$ 1.8g
Sodium tartrate 1.77g
$FeSO_4 \cdot 7H_2O$ 1.36g
$CoCl_2 \cdot 6H_2O$ 0.04g
$CuCl_2 . 2H_2O$ 0.027g
$Na_2MoO_4 \cdot 2H_2O$ 0.025g
$ZnCl_2$ 0.02g

Preparation of Trace Elements Solution HO-LE: Add components to distilled/deionized water and bring volume to 1.0L. Mix thoroughly. Filter sterilize.

Preparation of Medium: Add components to distilled/deionized water and bring volume to 1.0L. Mix thoroughly. Adjust pH to 7.3. Gently heat and bring to boiling. Distribute into tubes or flasks. Autoclave for 15 min at 15 psi pressure–121°C. Pour into sterile Petri dishes or leave in tubes.

Use: For the cultivation and maintenance of *Streptoalloteichus* species. For the differentiation of *Streptomyces* species based on melanine production.

Tyrosine Casein Nitrate Medium (TCN Medium)

Composition per liter:

Sodium caseinate 25.0g
Agar 15.0g
$NaNO_3$ 10.0g
L-Tyrosine 1.0g

Preparation of Medium: Add components to tap water and bring volume to 1.0L. Mix thoroughly. Gently heat and bring to boiling. Distribute into tubes or flasks. Autoclave for 15 min at 15 psi pressure–121°C. Pour into sterile Petri dishes or leave in tubes.

Use: For the isolation and cultivation of streptomycetes from infected plants.

TYX Medium

Composition per liter:

Yeast extract 5.0g
K_2HPO_4 .... 3.5g
Tryptone 1.0g
Na_2SO_4 .... 0.2g
L-cysteinium chloride·H_2O 0.5g
Biotin 10.0mg
p-Aminobenzoic acid 10.0mg
Sugar solution 5.0mL
Resazurin solution 4.0mL
Trace elements solution 1.0mL

Resazurin Solution:

Composition per 100.0mL:

Resazurin 25.0mg

Preparation of Resazurin Solution: Add resazurin to distilled/deionized water and bring volume to 100.0mL. Mix thoroughly.

Trace Elements Solution:

Composition per 100.0mL:

$FeCl_3 \cdot 6H_2O$ 2.7g
$MgSO_4$ 1.2g
$NaMoO_4 \cdot 2H_2O$ 0.24g
$MnSO_4 \cdot 7H_2O$ 0.17g
$CaCl_2 \cdot 2H_2O$ 0.15g
$ZnSO_4 \cdot 7H_2O$ 29.0mg
$CuSO_4 \cdot 5H_2O$ 25.0mg
$CoCl_2 \cdot 6H_2O$ 24.0mg
H_2SO_4 .... 2.8mL

Preparation of Trace Elements Solution: Add components to distilled/deionized water and bring volume to 100.0mL. Mix thoroughly.

Sugar Solution:

Composition per 100.0mL:

Xylose 1.0g

Preparation of Sugar Solution: Add xylose to distilled/deionized water and bring volume to 100.0mL. Mix thoroughly. Filter sterilize.

Preparation of Medium: Add components, except L-cysteinium chloride, biotin, *p*-aminobenzoic acid, and sugar solution, to distilled/deionized water and bring volume to 995.0mL. Mix thoroughly. Adjust pH to 7.0. Gently heat but do not boil. Cook for 5–10 min so that color first turns red and then turns yellow. Cool on ice. Add L-cysteinium chloride, biotin, and *p*-aminobenzoic acid. Mix thoroughly. Autoclave for 15 min at 15 psi pressure–121°C. Aseptically add 5.0 mL of sterile sugar solution. Mix thoroughly.

Use: For the cultivation of *Clostridium* species.

TZC Selective Medium

Composition per liter:

Agar 17.0g
Peptone 1.0g
Glucose 0.5g
Pancreatic digest of casein 0.1g
2,3,5-Triphenyltetrazolium·HCl solution 10.0mL

2,3,5-Triphenyltetrazolium·HCl Solution:

Composition per 10.0mL:

2,3,5-Triphenyltetrazolium·HCl 0.05g

Preparation of 2,3,5-Triphenyltetrazolium·HCl Solution: Add 2,3,5-triphenyltetrazolium·HCl to distilled/deionized water and bring volume to 10.0mL. Mix thoroughly. Autoclave for 15 min at 15 psi pressure–121°C.

Preparation of Medium: Add components, except 2,3,5-triphenyltetrazolium·HCl solution, to distilled/deionized water and bring volume to 990.0mL. Mix thoroughly. Gently heat and bring to boiling. Autoclave for 15 min at 15 psi pressure–121°C. Cool to 45°–50°C. Aseptically add 10.0mL of sterile 2,3,5-triphenyltetrazolium·HCl solution. Mix thoroughly. Pour into sterile Petri dishes.

Use: For the isolation, cultivation, and differentiation of *Pseudomonas solanacearum*. The virulent, wild-type strains appear as irregular to round white colonies with a pink center. Avirulent mutants, which readily occur in nature, appear as round, deep red colonies with a narrow blue border.

U Agar Plates
(*Ureaplasma* Agar Plates)
(MES Agar)

Composition per 100.2mL:

Base agar	65.0mL
Horse serum	20.0mL
Yeast dialysate	10.0mL
MES (2-*N*-morpholinoethane sulfonic acid) buffer solution	3.0mL
Penicillin solution	2.0mL
Urea solution	0.2mL

pH 5.5 ± 0.2 at 25°C

Base Agar:

Composition per liter:

Papaic digest of soybean meal	20.0g
Agarose	10.0g
NaCl	5.0g
Phenol Red (2% solution)	1.0mL

Preparation of Base Agar: Add components to distilled/deionized water and bring volume to 1.0L. Mix thoroughly. Gently heat and bring to boiling. Adjust pH to 7.3. Autoclave for 15 min at 15 psi pressure–121°C. Cool to 45°–50°C.

Yeast Dialysate:

Composition per 10.0mL:

Yeast, active dried	450.0g

Preparation of Yeast Dialysate: Add active, dried yeast to distilled/deionized water and bring volume to 1250.0mL. Gently heat and bring to 40°C. Autoclave for 15 min at 15 psi pressure–121°C. Put into dialysis tubing. Dialyze against 1.0L of distilled/deionized water for 2 days at 4°C. Discard tubing and its contents. Autoclave dialysate for 15 min at 15 psi pressure–121°C. Store at –20°C.

MES Buffer Solution:

Composition per 100.0mL:

MES (2-*N*-morpholinoethane sulfonic acid) buffer	19.52g

Preparation of MES Buffer Solution: Add MES buffer to distilled/deionized water and bring volume to 100.0mL. Mix thoroughly. Adjust pH to 5.5. Filter sterilize.

Penicillin Solution:

Composition per 10.0mL:

Penicillin	100,000U

Preparation of Penicillin Solution: Add penicillin to distilled/deionized water and bring volume to 10.0mL. Mix thoroughly. Filter sterilize.

Urea Solution:

Composition per 100.0mL:

Urea	6.0g

Preparation of Urea Solution: Add urea to distilled/deionized water and bring volume to 100.0mL. Mix thoroughly. Filter sterilize.

Preparation of Medium: To 65.0mL of cooled, sterile base agar, aseptically add 10.0mL of sterile yeast dialysate, 20.0mL of horse serum, 2.0mL of sterile penicillin solution, 3.0mL of sterile MES buffer solution, and 0.2mL of sterile urea solution. Mix thoroughly. Pour into 10mm × 35mm Petri dishes in 5.0mL volumes. Allow plates to stand overnight at 25°C to remove excess surface moisture.

Use: For the isolation and cultivation of *Ureaplasma* species.

U Broth
(*Ureaplasma* Broth)

Composition per 99.5mL:

Base agar	65.0mL
Horse serum	20.0mL
Yeast dialysate	10.0mL
Penicillin solution	2.0mL
MES (2-*N*-morpholinoethane sulfonic acid) buffer solution	1.0mL
Na_2SO_3 solution	1.0mL
Urea solution	0.5mL

pH 5.5 ± 0.2 at 25°C

Base Agar:

Composition per liter:

Papaic digest of soybean meal	20.0g
Agarose	10.0g
NaCl	5.0g
Phenol Red (2% solution)	1.0mL

Preparation of Base Agar: Add components to distilled/deionized water and bring volume to 1.0L. Mix thoroughly. Gently heat and bring to boiling. Adjust pH to 7.3. Autoclave for 15 min at 15 psi pressure–121°C. Cool to 45°–50°C.

Yeast Dialysate:

Composition per 10.0mL:

Yeast, active dried	450.0g

Preparation of Yeast Dialysate: Add active, dried yeast to distilled/deionized water and bring volume to 1250.0mL. Gently heat and bring to 40°C. Autoclave for 15 min at 15 psi pressure–121°C. Put into dialysis tubing. Dialyze against 1.0L of distilled/deionized water for 2 days at 4°C. Discard tubing and its contents. Autoclave dialysate for 15 min at 15 psi pressure–121°C. Store at –20°C.

Penicillin Solution:

Composition per 10.0mL:

Penicillin	100,000U

Preparation of Penicillin Solution: Add penicillin to distilled/deionized water and bring volume to 10.0mL. Mix thoroughly. Filter sterilize.

MES Buffer Solution:

Composition per 100.0mL:

MES (2-*N*-morpholinoethane sulfonic acid) buffer	19.52g

Preparation of MES Buffer Solution: Add MES buffer to distilled/deionized water and bring volume to 100.0mL. Mix thoroughly. Adjust pH to 5.5. Filter sterilize.

Na_2SO_3 Solution:

Composition per 10.0mL:

Na_2SO_3	0.126g

Preparation of Na_2SO_3 Solution: Add Na_2SO_3 to distilled/deionized water and bring volume to 10.0mL. Mix thoroughly. Filter sterilize.

Urea Solution:

Composition per 100.0mL:

Urea	6.0g

Preparation of Urea Solution: Add urea to distilled/deionized water and bring volume to 100.0mL. Mix thoroughly. Filter sterilize.

Preparation of Medium: To 65.0mL of cooled, sterile base agar, aseptically add 10.0mL of sterile yeast dialysate, 20.0mL of horse serum, 2.0mL of sterile penicillin solution, 1.0mL of sterile MES buffer solution, 1.0mL of sterile Na_2SO_3 solution, and 0.5mL of sterile urea solution. Mix thoroughly. Pour into 10mm × 35mm Petri dishes in 5.0mL volumes. Allow plates to stand overnight at 25°C to remove excess surface moisture. Use within 48 hr.

Use: For the isolation and cultivation of *Ureaplasma urealyticum.*

U9 Broth (Urease Color Test Medium)

Composition per 101.6mL:

U9 base	95.0mL
Horse serum, unheated	5.0mL
Penicillin G solution	1.0mL
Urea solution	0.5mL
Phenol Red solution	0.1mL

pH 6.0 ± 0.2 at 25°C

U9 Base:

Composition per 100.0mL:

NaCl	0.63g
Pancreatic digest of casein	0.425g
Papaic digest of soybean meal	0.075g
K_2HPO_4	0.063g
Glucose	0.063g
KH_2PO_4	0.02g

Preparation of U9 Base: Add components to distilled/deionized water and bring volume to 100.0mL. Mix thoroughly. Adjust pH to 5.5 with 1*N* HCl. Autoclave for 15 min at 15 psi pressure–121°C. Cool to 45°–50°C.

Penicillin G Solution:

Composition per 10.0mL:

Penicillin G	0.63g

Preparation of Penicillin G Solution: Add penicillin G to distilled/deionized water and bring volume to 10.0mL. Mix thoroughly. Filter sterilize.

Urea Solution:

Composition per 30.0mL:

Urea	3.0g

Preparation of Urea Solution: Add urea to distilled/deionized water and bring volume to 30.0mL. Mix thoroughly. Filter sterilize.

Phenol Red Solution:

Composition per 10.0mL:

Phenol Red	0.1g

Preparation of Phenol Red Solution: Add Phenol Red to distilled/deionized water and bring volume to 10.0mL. Mix thoroughly. Filter sterilize.

Preparation of Medium: To 95.0mL of cooled, sterile U9 base, aseptically add 5.0mL of sterile horse serum, 1.0mL of sterile penicillin G solution, 0.5mL of sterile urea solution, and 0.1mL of sterile Phenol Red solution. Mix thoroughly. Aseptically distribute into sterile tubes or flasks.

Use: For the isolation and identification of T-strain mycoplasmas from clinical specimens, especially *Ureaplasma urealyticum.* T-mycoplasmas are the only members of the *Mycoplasma* group known to contain urease. Bacteria with urease activity turn the medium dark pink.

U9 Broth with Amphotericin B

Composition per 101.6mL:

U9 base	95.0mL
Horse serum, unheated	5.0mL
Antibiotic solution	1.0mL
Urea solution	0.5mL
Phenol Red solution	0.1mL

pH 6.0 ± 0.2 at 25°C

U9 Base:

Composition per 100.0mL:

NaCl	0.63g
Pancreatic digest of casein	0.425g
Papaic digest of soybean meal	0.075g
K_2HPO_4	0.063g
Glucose	0.063g
KH_2PO_4	0.02g

Preparation of U9 Base: Add components to distilled/deionized water and bring volume to 100.0mL. Mix thoroughly. Adjust pH to 5.5 with 1*N* HCl. Autoclave for 15 min at 15 psi pressure–121°C. Cool to 45°–50°C.

Antibiotic Solution:

Composition per 10.0mL:

Penicillin G	0.63g
Amphotericin B	2.5mg

Preparation of Antibiotic Solution: Add penicillin G and amphotericin B to distilled/deionized water and bring volume to 10.0mL. Mix thoroughly. Filter sterilize.

Urea Solution:

Composition per 30.0mL:

Urea	3.0g

Preparation of Urea Solution: Add urea to distilled/deionized water and bring volume to 30.0mL. Mix thoroughly. Filter sterilize.

Phenol Red Solution:

Composition per 10.0mL:

Phenol Red	0.1g

Preparation of Phenol Red Solution: Add Phenol Red to distilled/deionized water and bring volume to 10.0mL. Mix thoroughly. Filter sterilize.

Preparation of Medium: To 95.0mL of cooled, sterile U9 base, aseptically add 5.0mL of sterile horse serum, 1.0mL of sterile antibiotic solution, 0.5mL of sterile urea solution, and 0.1mL of sterile Phenol Red solution. Mix thoroughly. Aseptically distribute into sterile tubes or flasks.

Use: For the isolation and identification of T-strain mycoplasmas from clinical specimens, especially *Ureaplasma urealyticum.* T-mycoplasmas are the only members of the *Mycoplasma* group known to contain urease. Bacteria with urease activity turn the medium dark pink.

U9B Broth

Composition per 102.1mL:

U9 base	95.0mL
Horse serum, unheated	5.0mL
Penicillin G solution	1.0mL
Urea solution	0.5mL
L-Cysteine·HCl·H_2O solution	0.5mL
Phenol Red solution	0.1mL

pH 6.0 ± 0.2 at 25°C

U9 Base:

Composition per 100.0mL:

NaCl	0.63g
Pancreatic digest of casein	0.425g
Papaic digest of soybean meal	0.075g
K_2HPO_4	0.063g
Glucose	0.063g
KH_2PO_4	0.02g

Preparation of U9 Base: Add components to distilled/deionized water and bring volume to 100.0mL. Mix thoroughly. Adjust pH to 5.5 with 1*N* HCl. Autoclave for 15 min at 15 psi pressure–121°C. Cool to 45°–50°C.

Penicillin G Solution:

Composition per 10.0mL:

Penicillin G	0.63 g

Preparation of Penicillin G Solution: Add penicillin G to distilled/deionized water and bring volume to 10.0mL. Mix thoroughly. Filter sterilize.

Urea Solution:

Composition per 30.0mL:

Urea	3.0g

Preparation of Urea Solution: Add urea to distilled/deionized water and bring volume to 30.0mL. Mix thoroughly. Filter sterilize.

L-Cysteine·HCl·H_2O Solution:

Composition per 50.0mL:

L-Cysteine·HCl·H_2O	1.0g

Preparation of L-Cysteine·HCl·H_2O Solution: Add L-cysteine·HCl·H_2O to distilled/deionized water and bring volume to 50.0mL. Mix thoroughly. Filter sterilize.

Phenol Red Solution:

Composition per 10.0mL:

Phenol Red	0.1g

Preparation of Phenol Red Solution: Add Phenol Red to distilled/deionized water and bring volume to 10.0mL. Mix thoroughly. Filter sterilize.

Preparation of Medium: To 95.0mL of cooled, sterile U9 base, aseptically add 5.0mL of sterile horse serum, 1.0mL of sterile penicillin G solution, 0.5mL of sterile urea solution, 0.5mL of sterile L-cysteine·HCl·H_2O solution, and 0.1mL of sterile Phenol Red solution. Mix thoroughly. Aseptically distribute into sterile tubes or flasks.

Use: For the isolation and identification of T-strain mycoplasmas from clinical specimens, especially *Ureaplasma urealyticum*. T-mycoplasmas are the only members of the *Mycoplasma* group known to contain urease. Bacteria with urease activity turn the medium dark pink.

U9C Broth

Composition per 102.0mL:

U9C base	90.0mL
Horse serum, unheated	10.0mL
Penicillin G solution	1.0mL
Urea solution	0.3mL
L-Cysteine·HCl·H_2O solution	0.5mL
GHL tripeptide solution	0.1mL
Phenol Red solution	0.1mL

pH 6.0 ± 0.2 at 25°C

U9C Base:

Composition per 100.0mL:

NaCl	0.85g
Pancreatic digest of casein	0.25g
Papaic digest of soybean meal	0.15g
K_2HPO_4	0.12g
Glucose	0.12g
$MgCl_2 \cdot 6H_2O$	0.2g
Yeast extract	0.1g
KH_2PO_4	0.02g

Preparation of U9C Base: Add components to distilled/deionized water and bring volume to 100.0mL. Mix thoroughly. Adjust pH to 5.5 with 2*N* HCl. Autoclave for 15 min at 15 psi pressure–121°C. Cool to 45°–50°C.

Penicillin G Solution:

Composition per 10.0mL:

Penicillin G	0.63g

Preparation of Penicillin G Solution: Add penicillin G to distilled/deionized water and bring volume to 10.0mL. Mix thoroughly. Filter sterilize.

Urea Solution:

Composition per 30.0mL:

Urea	3.0g

Preparation of Urea Solution: Add urea to distilled/deionized water and bring volume to 30.0mL. Mix thoroughly. Filter sterilize.

L-Cysteine·HCl·H_2O Solution:

Composition per 50.0mL:

L-Cysteine·HCl·H_2O	1.0g

Preparation of L-Cysteine·HCl·H_2O Solution: Add L-cysteine·HCl·H_2O to distilled/deionized water and bring volume to 50.0mL. Mix thoroughly. Filter sterilize.

GHL Tripeptide Solution:

Composition per 10.0mL:

GHL tripeptide	0.2mg

Preparation of GHL Tripeptide Solution: Add GHL tripeptide (glycyl-L-histidyl-L-lysine acetate) to distilled/deionized water and bring volume to 10.0mL. Mix thoroughly. Filter sterilize.

Phenol Red Solution:

Composition per 10.0mL:

Phenol Red	0.1g

Preparation of Phenol Red Solution: Add Phenol Red to distilled/deionized water and bring volume to 10.0mL. Mix thoroughly. Filter sterilize.

Preparation of Medium: To 90.0mL of cooled, sterile U9C base, aseptically add 10.0mL of sterile horse serum, 1.0mL of sterile penicillin G solution, 0.3mL of sterile urea solution, 0.5mL of sterile L-cysteine·HCl·H_2O solution, 0.1mL of sterile GHL tripeptide solution, and 0.1mL of sterile Phenol Red solution. Mix thoroughly. Aseptically distribute into sterile tubes or flasks.

Use: For the isolation and identification of T-strain mycoplasmas from clinical specimens, especially *Ureaplasma urealyticum*. T-mycoplasmas are the only members of the *Mycoplasma* group known to contain urease. Bacteria with urease activity turn the medium dark pink.

U4 Medium

Composition per 100.0mL:

Hartley's digest broth	20.0mL
Fetal calf serum	15.0mL

Fresh yeast extract solution....................10.0mL
Hanks' balanced salt solution, 10X....................4.0mL
$MgSO_4 \cdot 5H_2O$ (0.025% solution)....................1.0mL
Urea (20% solution)....................0.25mL
Phenol Red (1% solution)....................0.2mL

pH 6.0–6.2 at 25°C

Hartley's Digest Broth:

Composition per 10.0L:

Ox heart....................3000.0g
Pancreatin....................50.0g
Na_2CO_3, anhydrous (0.8% solution)....................5.0L
HCl, concentrated....................80.0mL

pH 7.5 ± 0.2 at 25°C

Preparation of Hartley's Digest Broth: Finely mince the ox heart. Add the meat to 5.0L of distilled/deionized water. Gently heat and bring to 80°C. Add the 5.0L of Na_2CO_3 solution. Cool to 45°C. Add pancreatin and maintain at 45°C for 4 hr while stirring. Add the HCl and steam at 100°C for 30 min. Cool to room temperature. Adjust pH to 8.0 with 1*N* NaOH. Gently heat and bring to boiling. Continue boiling for 25 min. Filter while hot. Cool to room temperature. Adjust pH to 7.5. Autoclave for 15 min at 15 psi pressure–121°C.

Fresh Yeast Extract Solution:

Composition per 100.0mL:

Baker's yeast, live, pressed, starch-free....................25.0g

Preparation of Fresh Yeast Extract Solution: Add the live Baker's yeast to 100.0mL of distilled/deionized water. Autoclave for 90 min at 15 psi pressure–121°C. Allow to stand. Remove supernatant solution. Adjust pH to 6.6–6.8.

Hanks' Balanced Salt Solution, 10X:

Composition per liter:

Na_2Cl....................80.0g
Glucose....................10.0g
KCl....................4.0g
$CaCl_2$....................1.4g
$MgCl_2 \cdot 6H_2O$....................1.0g
$MgSO_4 \cdot 7H_2O$....................1.0g
$Na_2HPO_4 \cdot 7H_2O$....................0.9g
KH_2PO_4....................0.6g

Preparation of Hanks' Balanced Salt Solution, 10X: Add components to distilled/deionized water and bring volume to 100.0mL. Mix thoroughly.

Preparation of Medium: Add components to distilled/deionized water and bring volume to 1.0L. Mix thoroughly. Adjust pH to 6.0–6.2 with HCl. Filter-sterilize medium. Aseptically distribute into sterile tubes or flasks.

Use: For the cultivation and maintenance of *Ureaplasma diversum*.

UBA Medium (Universal Beer Agar)

Composition per liter:

Glucose....................16.1g
Peptonized milk....................15.0g
Agar....................12.0g
Tomato juice, dessicated....................12.2g
Yeast extract....................6.1g
K_2HPO_4....................0.31g
KH_2PO_4....................0.31g
$MgSO_4 \cdot 7H_2O$....................0.12g
$FeSO_4$....................6.0mg
$MnSO_4 \cdot 5H_2O$....................6.0mg
NaCl....................6.0mg
Beer....................250.0mL

pH 6.3 ± 0.2 at 25°C

Source: This medium is available as a premixed powder from BD Diagnostic Systems and Oxoid Unipath.

Preparation of Medium: Add components, except beer, to distilled/deionized water and bring volume to 750.0mL. Mix thoroughly. Gently heat and bring to boiling. Add beer. Mix thoroughly. Distribute into tubes or flasks. Autoclave for 10 min at 15 psi pressure–121°C. Pour into sterile Petri dishes or leave in tubes.

Use: For the cultivation of microorganisms of significance in the brewing industry.

Universal Agar for Yeasts
See: **Yeast Malt Extract Agar**

Universal Beer Agar

Composition per liter:

Peptonized milk....................15.0g
Agar....................12.0g
Yeast extract....................10.0g
Glucose....................10.0g
Tomato juice solids....................7.0g
K_2HPO_4....................0.5g
KH_2PO_4....................0.5g
$MgSO_4 \cdot 5H_2O$....................0.2g
NaCl....................0.01g
$FeSO_4 \cdot 7H_2O$....................0.01g
$MnSO_4 \cdot H_2O$....................0.01g
Beer....................250.0mL

pH 6.3 ± 0.2 at 25°C

Source: This medium is available as a premixed powder from BD Diagnostic Systems.

Preparation of Medium: Add components, except beer, to distilled/deionized water and bring volume to 750.0mL. Mix thoroughly. Gently heat and bring to boiling. Add beer. Mix thoroughly. Distribute into tubes or flasks. Autoclave for 10 min at 15 psi pressure–121°C. Cool to 50°C. Aseptically add 250.0mL of beer without degassing. Pour into sterile Petri dishes or leave in tubes.

Use: For the enumeration of contaminating bacteria and yeasts encountered in wort and beer.

Universal Beer Agar
See: **UBA Medium**

Universal Beer HiVeg Agar with Beer (UB HiVeg Agar)

Composition per liter:

Glucose....................16.1g
Plant hydrolysate No. 4....................15.0g
Tomato juice....................12.2g
Agar....................12.0g
Yeast extract....................6.1g
K_2HPO_4....................0.31g
KH_2PO_4....................0.31g
$MgSO_4$....................0.12
$FeSO_4$....................6.0mg

$MnSO_4$	6.0mg
NaCl	6.0mg
Beer	250.0mL

pH 6.3 ± 0.2 at 25°C

Source: This medium, without beer, is available as a premixed powder from HiMedia.

Preparation of Medium: Add components, except beer, to distilled/deionized water and bring volume to 750.0mL. Mix thoroughly. Gently heat and bring to boiling. Add beer. Mix thoroughly. Distribute into tubes or flasks. Autoclave for 10 min at 15 psi pressure–121°C. Pour into sterile Petri dishes or leave in tubes.

Use: For cultivation of microorganisms of significance in the brewing industry.

Universal Medium for Yeasts (YM) (DSMZ Medium 186)

Composition per liter:

Agar	15.0g
Glucose	10.0g
Peptone	5.0g
Yeast extract	3.0g
Malt extract	3.0g

pH 6.8 ± 0.2 at 25°C

Preparation of Medium: Add components to distilled/deionized water and bring volume to 1.0L. Mix thoroughly. Gently heat and bring to boiling. Distribute into tubes or flasks. Autoclave for 15 min at 15 psi pressure–121°C. Pour into sterile Petri dishes or leave in tubes.

Use: For the cultivation and maintenance of *Saccharmomyces* spp. and other yeasts.

Universal Preenrichment Broth (BAM M188)

Composition per liter:

KH_2PO_4	15.0g
Na_2HPO_4	7.0g
Tryptone	5.0g
Proteose peptone	5.0g
NaCl	5.0g
Glucose	0.5g
$MgSO_4$	0.25g
Sodium pyruvate	0.2g
Ferric ammonium citrate	0.1g

pH 6.3 ± 0.2 at 25°C

Preparation of Medium: Add components to distilled/deionized water and bring volume to 1.0L. Mix thoroughly. Gently heat until dissolved. Adjust pH to 6.3. Distribute into tubes in 10.0mL volumes. Autoclave for 15 min at 15 psi pressure–121°C.

Use: For the enrichment of injured foodborne pathogens of different genera simultaneously in lieu of having to undergo separate simultaneous enrichment cultures for subsequent detection or isolation of each pathogen.

University of Vermont *Listeria* Enrichment Broth

See: **UVM *Listeria* Enrichment Broth**

University of Vermont I *Listeria* Primary Selective Enrichment Broth

See: ***Listeria* Enrichment Broth I, USDA FSIS**

University of Vermont II *Listeria* Primary Selective Enrichment Broth

See: ***Listeria* Enrichment Broth II, USDA FSIS**

University of Vermont Modified *Listeria* Enrichment Broth

See: **UVM Modified *Listeria* Enrichment Broth**

Urea Agar

Composition per liter:

Agar	15.0g
Urea	10.0g
$Na_2HPO_4 \cdot 12H_2O$	9.0g
NaCl	5.0g
KH_2PO_4	1.5g
Meat extract	1.0g
Yeast extract	1.0g
$MgSO_4 \cdot 7H_2O$	0.2g
$MnCl_2 \cdot 4H_2O$	20.0mg
$CaCl_2$	1.2mg
Glucose solution	100.0mL

Glucose Solution:

Composition per 100.0mL:

Glucose	5.0g

Preparation of Glucose Solution: Add glucose to distilled/deionized water and bring volume to 100.0mL. Mix thoroughly. Filter sterilize. Warm to 50°C.

Preparation of Medium: Add components, except glucose solution, to distilled/deionized water and bring volume to 900.0mL. Mix thoroughly. Gently heat and bring to boiling. Autoclave for 15 min at 15 psi pressure–121°C. Cool to 50°–55°C. Aseptically add 100.0mL of sterile glucose solution. Mix thoroughly. Pour into sterile Petri dishes or distribute into sterile tubes.

Use: For the cultivation and maintenance of *Corynebacterium glutamicum*.

Urea Agar

Composition per liter:

Agar	15.0g
$Na_2HPO_4 \cdot 12H_2O$	9.0g
NaCl	5.0g
KH_2PO_4	1.5g
Meat extract	1.0g
Yeast extract	1.0g
$MgSO_4 \cdot 7H_2O$	0.2g
$MnCl_2 \cdot 4H_2O$	20.0mg
$CaCl_2$	1.2mg
Glucose-urea solution	100.0mL

Glucose-Urea Solution:

Composition per 100.0mL:

Urea	10.0g
Glucose	5.0g

Preparation of Glucose-Urea Solution: Add components to distilled/deionized water and bring volume to 100.0mL. Mix thoroughly. Filter sterilize. Warm to 50°C.

Preparation of Medium: Add components, except glucose-urea solution, to distilled/deionized water and bring volume to 900.0mL. Mix thoroughly. Gently heat and bring to boiling. Autoclave for 15 min at 15 psi pressure–121°C. Cool to 50°–55°C. Aseptically add 100.0mL of sterile glucose-urea solution. Mix thoroughly. Pour into sterile Petri dishes or distribute into sterile tubes.

Use: For the cultivation and maintenance of *Corynebacterium glutamicum.*

Urea Agar (Urease Test Agar) (Urea Agar Base, Christensen)

Composition per liter:

Urea	20.0g
Agar	15.0g
NaCl	5.0g
KH_2PO_4	2.0g
Peptone	1.0g
Glucose	1.0g
Phenol Red	0.012g

pH 6.8 ± 0.2 at 25°C

Source: This medium is available as a premixed powder from BD Diagnostic Systems.

Preparation of Medium: Add components, except agar, to distilled/deionized water and bring volume to 100.0mL. Mix thoroughly. Filter sterilize. Add agar to distilled/deionized water and bring volume to 900.0mL. Mix thoroughly. Gently heat and bring to boiling. Autoclave for 15 min at 15 psi pressure–121°C. Cool to 50°C. Aseptically add the 100.0mL of sterile basal medium. Mix thoroughly. Distribute into sterile tubes. Allow tubes to solidify in a slanted position.

Use: For the differentiation of a variety of microorganisms, especially members of the Enterobacteriaceae, aerobic actinomycetes, streptococci, and nonfermenting Gram-negative bacteria, on the basis of urease production.

Urea Agar Base

Composition per liter:

Agar	15.0g
NaCl	5.0g
Na_2HPO_4	1.2g
Peptone	1.0g
Glucose	1.0g
KH_2PO_4	0.8g
Phenol Red	0.012g
Urea solution	50.0mL

pH 6.8 ± 0.2 at 25°C

Source: This medium is available as a premixed powder from Oxoid Unipath.

Urea Solution:

Composition per 100.0mL:

Urea	40.0g

Preparation of Urea Solution: Add urea to distilled/deionized water and bring volume to 100.0mL. Mix thoroughly. Filter sterilize.

Preparation of Medium: Add components, except urea solution, to distilled/deionized water and bring volume to 950.0mL. Mix thoroughly. Gently heat and bring to boiling. Autoclave for 20 min at 10 psi pressure–115°C. Cool to 50°C. Aseptically add 50.0mL of sterile urea solution. Mix thoroughly. Pour into sterile Petri dishes or distribute into sterile tubes. Allow tubes to solidify in a slanted position.

Use: For the detection of *Proteus* species based on rapid urease activity and the identification of other members of the Enterobacteriaceae based on urease activity. Urease-positive bacteria turn the medium pink.

Urea Agar Base, Christensen

See: **Urea Agar**

Urea Broth

Composition per liter:

$Na_2HPO_4 \cdot 12H_2O$	9.0g
NaCl	5.0g
KH_2PO_4	1.5g
Meat extract	1.0g
Yeast extract	1.0g
$MgSO_4 \cdot 7H_2O$	0.2g
$MnCl_2 \cdot 4H_2O$	20.0mg
$CaCl_2$	1.2mg
Glucose-urea solution	100.0mL

Glucose-Urea Solution:

Composition per 100.0mL:

Urea	10.0g
Glucose	5.0g

Preparation of Glucose-Urea Solution: Add components to distilled/deionized water and bring volume to 100.0mL. Mix thoroughly. Filter sterilize.

Preparation of Medium: Add components, except glucose-urea solution, to distilled/deionized water and bring volume to 900.0mL. Mix thoroughly. Autoclave for 15 min at 15 psi pressure–121°C. Aseptically add 100.0mL of sterile glucose-urea solution. Mix thoroughly. Aseptically distribute into sterile tubes or flasks.

Use: For the cultivation and maintenance of *Corynebacterium glutamicum.*

Urea Broth

See: **Urease Test Broth**

Urea Broth 10B for *Ureaplasma urealyticum*

Composition per 100.5mL:

PPLO broth without Crystal Violet	70.0mL
Horse serum, unheated	20.0mL
Fresh yeast extract solution	10.0mL
L-Cysteine·HCl·H_2O solution	0.5mL
CVA enrichment	0.5mL
Urea solution	0.4mL
Phenol Red	0.1mL

PPLO Broth without Crystal Violet:

Composition per 900.0mL:

Beef heart, solids from infusion	16.1g
Peptone	3.25g
NaCl	1.61g

Preparation of PPLO Broth without Crystal Violet: Add components to distilled/deionized water and bring volume to 900.0mL. Adjust pH to 5.5 with 2*N* HCl. Autoclave for 15 min at 15 psi pressure–121°C. Cool to 37°C.

Fresh Yeast Extract Solution:

Composition per 100.0mL:

Baker's yeast live, pressed, starch-free	25.0g

Preparation of Fresh Yeast Extract Solution: Add the live Baker's yeast to 100.0mL of distilled/deionized water. Autoclave for 90 min at 15 psi pressure–121°C. Allow to stand. Remove supernatant solution. Adjust pH to 6.6–6.8.

L-Cysteine·HCl·H_2O Solution:

Composition per 50.0mL:

L-Cysteine·HCl·H_2O	1.0g

Preparation of L-Cysteine·HCl·H_2O Solution: Add L-cysteine·HCl·H_2O to distilled/deionized water and bring volume to 50.0mL. Mix thoroughly. Filter sterilize.

CVA Enrichment:

Composition per liter:

Glucose	100.0g
L-Cysteine·HCl·H_2O	25.9g
L-Glutamine	10.0g
Adenine	1.0g
L-Cystine·2HCl	1.0g
Nicotinamide adenine dinucleotide	0.25g
Cocarboxylase	0.1g
Guanine·HCl	0.03g
$Fe(NO_3)_3$	0.02g
Vitamin B_{12}	0.01g
p-Aminobenzoic acid	0.013g
Thiamine·HCl	3.0mg

Preparation of CVA Enrichment: Add components to distilled/deionized water and bring volume to 1.0L. Mix thoroughly. Filter sterilize.

Urea Solution:

Composition per 30.0mL:

Urea	3.0g

Preparation of Urea Solution: Add urea to distilled/deionized water and bring volume to 30.0mL. Mix thoroughly. Filter sterilize.

Preparation of Medium: Aseptically combine the components, except the PPLO broth without Crystal Violet. Aseptically add this mixture to the cooled, sterile PPLO broth without Crystal Violet. Mix thoroughly. Aseptically distribute into sterile tubes or flasks.

Use: For the cultivation and maintenance of *Ureaplasma urealyticum* and other *Ureaplasma* species. Urease-positive bacteria turn the medium peach orange.

Urea Broth Base

Composition per liter:

NaCl	5.0g
Na_2HPO_4	1.2g
Peptone	1.0g
Glucose	1.0g
KH_2PO_4	0.8g
Phenol Red	0.012g
Urea solution	50.0mL

pH 6.8 ± 0.2 at 25°C

Source: This medium is available as a premixed powder from Oxoid Unipath.

Urea Solution:

Composition per 100.0mL:

Urea	40.0g

Preparation of Urea Solution: Add urea to distilled/deionized water and bring volume to 100.0mL. Mix thoroughly. Filter sterilize.

Preparation of Medium: Add components, except urea solution, to distilled/deionized water and bring volume to 950.0mL. Mix thoroughly. Autoclave for 20 min at 10 psi pressure–115°C. Cool to 50°C. Aseptically add 50.0mL of sterile urea solution. Mix thoroughly. Aseptically distribute into sterile tubes or flasks.

Use: For the differentiation of members of the Enterobacteriaceae based on urease production. Urease-positive bacteria turn the medium pink.

Urea HiVeg Agar Base Autoclavable with Urea (Christensen HiVeg Agar Autoclavable)

Composition per liter:

Agar	15.0g
NaCl	5.0g
Na_2HPO_4	1.2g
Glucose	1.0g
Plant peptone	1.0g
KH_2PO_4	0.8g
Phenol Red	0.012g
Urea solution	50.0mL

pH 6.8 ± 0.2 at 25°C

Source: This medium, without urea solution, is available as a premixed powder from HiMedia.

Urea Solution:

Composition per 100.0mL:

Urea	40.0g

Preparation of Urea Solution: Add urea to distilled/deionized water and bring volume to 100.0mL. Mix thoroughly. Filter sterilize.

Preparation of Medium: Add components, except urea solution, to distilled/deionized water and bring volume to 950.0mL. Mix thoroughly. Gently heat and bring to boiling. Autoclave for 20 min at 10 psi pressure–115°C. Cool to 50°C. Aseptically add 50.0mL of sterile urea solution. Mix thoroughly. Pour into sterile Petri dishes or distribute into sterile tubes. Allow tubes to solidify in a slanted position.

Use: For the detection of *Proteus* species based on rapid urease activity and the identification of other members of the Enterobacteriaceae based on urease activity. Urease-positive bacteria turn the medium pink. For the detection of urease production, particularly by *Proteus vulgaris*, micrococci and paracolon organisms.

Urea Nutrient Agar

Composition per 1050.0mL:

Agar	15.0g
Peptone	5.0g
NaCl	5.0g
Yeast extract	2.0g
Beef extract	1.0g
Urea solution	50.0mL

Urea Solution:

Composition per 100.0mL:

Urea	20.0g

Preparation of Urea Solution: Add urea to distilled/deionized water and bring volume to 100.0mL. Mix thoroughly. Filter sterilize.

Preparation of Medium: Add components, except urea solution, to distilled/deionized water and bring volume to 1.0L. Mix thoroughly. Gently heat and bring to boiling. Autoclave for 15 min at 15 psi pressure–121°C. Cool to 50°–55°C. Aseptically add 50.0mL of sterile urea solution. Mix thoroughly. Pour into sterile Petri dishes or distribute into sterile tubes.

Use: For the cultivation of *Bacillus pantothenticus*.

Urea R Broth (Urea Rapid Broth)

Composition per liter:

Urea	20.0g
Yeast extract	0.1g
Na_2HPO_4	0.095g
KH_2PO_4	0.091g
Phenol Red	0.01g

pH 6.9 ± 0.2 at 25°C

Source: This medium is available as a prepared medium from BD Diagnostic Systems.

Preparation of Medium: Add components to distilled/deionized water and bring volume to 1.0L. Mix thoroughly. Filter sterilize. Aseptically distribute into sterile tubes or flasks.

Use: For the differentiation of members of the Enterobacteriaceae based on the rapid detection of urease activity. Urease-positive bacteria turn the medium cerise.

Urea Semisolid Medium

Composition per 450.0:

Solution A	400.0mL
Solution B	50.0mL

Solution A:

Composition per 400.0mL:

Pancreatic digest of casein	6.0g
Yeast extract	2.0g
NaCl	1.0g
Yeast extract	0.8g
Agar	0.3g
L-Cystine	0.1g
Thioglycolic acid	0.12mL

pH 7.2 ± 0.2 at 25°C

Preparation of Solution A: Add components to distilled/deionized water and bring volume to 400.0mL. Mix thoroughly. Gently heat and bring to boiling. Autoclave for 15 min at 15 psi pressure–121°C. Cool to 60°C.

Solution B:

Composition per 50.0mL:

Urea	8.0g
Na_2HPO_4	3.8g
KH_2PO_4	3.64g
Yeast extract	0.04g
Phenol Red	4.0mg

Preparation of Solution B: Add components to distilled/deionized water and bring volume to 50.0mL. Mix thoroughly. Filter sterilize.

Preparation of Medium: Aseptically combine 400.0mL of sterile solution A and 50.0mL of sterile solution B. Mix thoroughly. Aseptically distribute into sterile screw-capped tubes in 7.0mL volumes. Pass the tubes into an anaerobic chamber containing 85% N_2 + 10% H_2 + 5% CO_2 for 60 min. Close screw caps tightly.

Use: For the cultivation and differentiation of anaerobic bacteria based on their production of urease. Bacteria that produce urease turn the medium bright red.

Urea Test Broth

Composition per liter:

Urea	20.0g
Na_2HPO_4	9.5g
KH_2PO_4	9.1g
Yeast extract	0.1g
Phenol Red	0.01g
Urea solution	100.0mL

Urea Solution:

Composition per 100.0mL:

Urea	20.0g

Preparation of Urea Solution: Add urea to distilled/deionized water and bring volume to 100.0mL. Mix thoroughly. Filter sterilize.

Preparation of Medium: Add components, except urea solution, to distilled/deionized water and bring volume to 900.0mL. Mix thoroughly. Autoclave for 15 min at 15 psi pressure–121°C. Cool to 45°–50°C. Aseptically add sterile urea solution. Mix thoroughly. Aseptically distribute into sterile tubes in 3.0mL volumes.

Use: For the cultivation and differentiation of members of the Enterobacteriaceae and aerobic actinomycetes based on their production of urease. Bacteria that produce urease turn the medium bright red.

Urea Test Broth

Composition per 99.6mL:

H broth base	85.0mL
Horse serum	10.0mL
Penicillin solution	2.0mL
MES (2-*N*-morpholinoethane sulfonic acid) buffer solution	1.0mL
Na_2SO_3 solution	1.0mL
Urea solution	0.5mL
Phenol Red solution	0.1mL

pH 7.2 ± 0.2 at 25°C

H Broth Base:

Composition per liter:

NaCl	5.0g
Pancreatic digest of casein	5.0g
Peptone	5.0g
Beef extract	3.0g
K_2HPO_4	2.5g
Glucose	1.0g

Preparation of H Broth Base: Add components to distilled/deionized water and bring volume to 1.0L. Mix thoroughly. Gently heat and bring to boiling. Distribute into tubes in 4.0mL volumes. Autoclave for 15 min at 10 psi pressure–115°C. Cool to 45°–50°C.

Penicillin Solution:

Composition per 10.0mL:

Penicillin	100,000U

Preparation of Penicillin Solution: Add penicillin to distilled/deionized water and bring volume to 10.0mL. Mix thoroughly. Filter sterilize.

MES Buffer Solution:
Composition per 100.0mL:

MES (2-*N*-morpholinoethane sulfonic acid) buffer	19.52g

Preparation of MES Buffer Solution: Add MES buffer to distilled/deionized water and bring volume to 100.0mL. Mix thoroughly. Adjust pH to 5.5. Filter sterilize.

Na_2SO_3 Solution:
Composition per 10.0mL:

Na_2SO_3	0.126g

Preparation of Na_2SO_3 Solution: Add Na_2SO_3 to distilled/deionized water and bring volume to 10.0mL. Mix thoroughly. Filter sterilize.

Urea Solution:
Composition per 100.0mL:

Urea	6.0g

Preparation of Urea Solution: Add urea to distilled/deionized water and bring volume to 100.0mL. Mix thoroughly. Filter sterilize.

Phenol Red Solution:
Composition per 10.0mL:

Phenol Red	0.1g

Preparation of Phenol Red Solution: Add Phenol Red to distilled/deionized water and bring volume to 10.0mL. Mix thoroughly. Filter sterilize.

Preparation of Medium: To 85.0mL of cooled sterile H broth base, aseptically add 10.0mL of sterile horse serum, 2.0mL of sterile penicillin solution, 1.0mL of MES buffer solution, 1.0mL of Na_2SO_3 solution, 0.5mL of urea solution, and 0.1 mL of sterile Phenol Red solution. Mix thoroughly. Aseptically distribute into test tubes in 3.0mL volumes.

Use: For the cultivation and differentiation of *Ureaplasma* species based on their production of urease.

Ureaplasma Medium (DSMZ Medium 1096)

Composition per 1005.0mL:

Solution A	700.0mL
Solution B	305.0mL

pH 6.0 ± 0.2 at 25°C

Solution A:
Composition per 700.0mL:

Pancreatic digest of casein	7.0g
NaCl	5.0g
Beef extract	3.0g
Yeast extract	3.0g
Beef heart, solids from infusion	2.0g
Urea	0.4g
L-Cysteine-HCl·$2H_2O$	0.1g
DNA, fish sperm	0.2g
Phenol Red	0.02g

Preparation of Solution A: Add components to distilled/deionized water and bring volume to 700.0mL. Mix thoroughly. Adjust pH to 6.0. Autoclave for 15 min at 15 psi pressure–121°C. Cool to room temperature.

Solution B:
Composition per 305.0mL:

Horse serum	200.0mL
Yeastolate solution	100.0mL
Isovitalex	5.0mL

Yeastolate Solution:
Composition per 100.0mL:

Yeastolate	20.0g

Preparation of Yeastolate Solution: Add yeastolate to distilled/deionized water and bring volume to 100.0mL. Mix thoroughly.

Preparation of Solution B: Combine components. Filter sterilize.

Preparation of Medium: Aseptically add 305.0mL solution B to 700.0mL solution A. Mix thoroughly.

Use: For the cultivation of *Ureaplasma* spp.

Urease Color Test Medium
See: **U9 Broth**

Urease Indole Test Broth
See: **F35M Hajna Broth**

Urease Test Agar
See: **Urea Agar**

Urease Test Broth (Urea Broth)

Composition per liter:

Urea	20.0g
Na_2HPO_4	9.5g
KH_2PO_4	9.1g
Yeast extract	0.1g
Phenol Red	0.01g

pH 6.8 ± 0.2 at 25°C

Source: This medium is available as a premixed powder from BD Diagnostic Systems.

Preparation of Medium: Add components to distilled/deionized water and bring volume to 1.0L. Mix thoroughly. Filter sterilize. Aseptically distribute into sterile tubes or flasks.

Use: For the differentiation of organisms, especially the Enterobacteriaceae, on the basis of urease production. Urease-positive bacteria turn the medium pink.

Uric Acid Agar

Composition per liter:

Agar	20.0g
Uric acid	10.0g
KH_2PO_4	0.5g

pH 7.0 ± 0.2 at 25°C

Preparation of Medium: Add components to distilled/deionized water and bring volume to 1.0L. Mix thoroughly. Adjust pH to 7.0. Gently heat and bring to boiling. Distribute into tubes or flasks. Autoclave for 15 min at 15 psi pressure–121°C. Pour into sterile Petri dishes or leave in tubes. Swirl flask while pouring medium.

Use: For the cultivation and maintenance of microorganisms, such as *Bacillus fastidiosus*, that can utilize uric acid as the sole source of carbon, nitrogen, and energy.

Uric Acid Agar

Composition per liter:

Agar	15.0g
$Na_2HPO_4 \cdot 12H_2O$	9.0g
NaCl	5.0g
KH_2PO_4	1.5g
Meat extract	1.0g
Yeast extract	1.0g
$MgSO_4 \cdot 7H_2O$	0.2g
$MnCl_2 \cdot 4H_2O$	20.0mg
$CaCl_2$	1.2mg
Glucose-uric acid solution	100.0mL

Glucose-Uric Acid Solution:

Composition per 100.0mL:

Glucose	5.0g
Uric acid	0.4g

Preparation of Glucose-Uric Acid Solution: Add components to distilled/deionized water and bring volume to 100.0mL. Mix thoroughly. Filter sterilize. Warm to 50°C.

Preparation of Medium: Add components, except glucose-uric acid solution, to distilled/deionized water and bring volume to 900.0mL. Mix thoroughly. Gently heat and bring to boiling. Autoclave for 15 min at 15 psi pressure–121°C. Cool to 50°–55°C. Aseptically add 100.0mL of sterile glucose-uric acid solution. Mix thoroughly.

Use: For the cultivation and maintenance of *Bacillus* species.

Uric Acid Agar for Clostridia

Composition per liter:

Agar	20.0g
K_2HPO_4	4.0g
Uric acid	3.0g
Yeast extract	1.0g
Sodium thioglycolate	0.5g
$MgSO_4 \cdot 7H_2O$	0.1g
$FeSO_4 \cdot 7H_2O$	5.0mg
Phenol Red (0.04% solution)	1.0mL

pH 7.6–8.0 at 25°C

Preparation of Medium: Add components, except uric acid, to approximately 900.0mL of distilled/deionized water. Mix thoroughly. Gently heat and bring to boiling. Adjust pH to 7.6 with 1*N* NaOH. Add the uric acid. Mix thoroughly. Adjust pH to 7.6. Distribute into tubes or flasks. Autoclave for 15 min at 15 psi pressure–121°C. Pour into sterile Petri dishes or leave in tubes.

Use: For the cultivation and maintenance of anaerobic bacteria, such as *Clostridium acidurici* and *Clostridium cylindrosporum*, that can utilize uric acid as the sole source of carbon and energy.

Uric Acid Broth

Composition per liter:

$Na_2HPO_4 \cdot 12H_2O$	9.0g
NaCl	5.0g
KH_2PO_4	1.5g
Meat extract	1.0g
Yeast extract	1.0g
$MgSO_4 \cdot 7H_2O$	0.2g
$MnCl_2 \cdot 4H_2O$	20.0mg
$CaCl_2$	1.2mg
Glucose-uric acid solution	100.0mL

Glucose-Uric Acid Solution:

Composition per 100.0mL:

Glucose	5.0g
Uric acid	0.4g

Preparation of Glucose-Uric Acid Solution: Add components to distilled/deionized water and bring volume to 100.0mL. Mix thoroughly. Filter sterilize.

Preparation of Medium: Add components, except glucose-uric acid solution, to distilled/deionized water and bring volume to 900.0mL. Mix thoroughly. Autoclave for 15 min at 15 psi pressure–121°C. Cool to 50°–55°C. Aseptically add 100.0mL of sterile glucose-uric acid solution. Mix thoroughly. Aseptically distribute into sterile tubes or flasks.

Use: For the cultivation and maintenance of *Bacillus* species.

Uric Acid Broth for Clostridia

Composition per liter:

K_2HPO_4	4.0g
Uric acid	3.0g
Yeast extract	1.0g
Sodium thioglycolate	0.5g
$MgSO_4 \cdot 7H_2O$	0.1g
$FeSO_4 \cdot 7H_2O$	5.0mg
Phenol Red (0.04% solution)	1.0mL

pH 7.6–8.0 at 25°C

Preparation of Medium: Add components, except uric acid, to approximately 900.0mL of distilled/deionized water. Mix thoroughly. Gently heat and bring to boiling. Adjust pH to 7.6 with 1*N* NaOH. Add the uric acid. Mix thoroughly. Adjust pH to 7.6. Distribute into tubes or flasks. Autoclave for 15 min at 15 psi pressure–121°C.

Use: For the cultivation and maintenance of anaerobic bacteria, such as *Clostridium acidurici* and *Clostridium cylindrosporum*, that can utilize uric acid as the sole source of carbon and energy.

Uric Acid Medium

Composition per liter:

Agar	15.0g
Peptone	5.0g
NaCl	5.0g
Yeast extract	2.0g
Beef extract	1.0g
Uric acid solution	150.0mL

Uric Acid Solution:

Composition per 150.0mL:

Uric acid	6.0g

Preparation of Uric Acid Solution: Add uric acid to distilled/deionized water and bring volume to 150.0mL. Mix thoroughly. Autoclave for 15 min at 15 psi pressure–121°C. Cool to 50°–55°C.

Preparation of Medium: Add components, except uric acid solution, to distilled/deionized water and bring volume to 750.0mL. Mix thoroughly. Gently heat and bring to boiling. Autoclave for 15 min at 15 psi pressure–121°C. Cool to 50°–55°C. Aseptically add 150.0mL of sterile uric acid solution. Mix thoroughly. Pour into sterile Petri dishes or distribute into sterile tubes while agitating.

Use: For the cultivation of *Bacillus fastidiosus*.

Uric Acid Semisolid Agar for Clostridia

Composition per liter:

K_2HPO_4	4.0g
Uric acid	3.0g
Agar	2.0g
Yeast extract	1.0g
Sodium thioglycolate	0.5g
$MgSO_4 \cdot 7H_2O$	0.1g
$FeSO_4 \cdot 7H_2O$	5.0mg
Phenol Red (0.04% solution)	1.0mL

pH 7.6–8.0 at 25°C

Preparation of Medium: Add components, except uric acid, to approximately 900.0mL of distilled/deionized water. Mix thoroughly. Gently heat and bring to boiling. Adjust pH to 7.6 with 1*N* NaOH. Add the uric acid. Mix thoroughly. Adjust pH to 7.6. Distribute into tubes or flasks. Autoclave for 15 min at 15 psi pressure–121°C.

Use: For the cultivation and maintenance of *Clostridium acidurici* and *Clostridium cylindrosporum* that can utilize uric acid.

Uric Acid Utilization Agar

Composition per liter:

Agar	15.0g
Uric acid	10.0g
K_2HPO_4	0.5g

pH 7.2 ± 0.2 at 25°C

Preparation of Medium: Add components to distilled/deionized water and bring volume to 1.0L. Mix thoroughly. Gently heat and bring to boiling. Distribute into tubes or flasks. Autoclave for 15 min at 15 psi pressure–121°C. Pour into sterile Petri dishes or leave in tubes.

Use: For the cultivation and maintenance of anaerobic bacteria, such as *Bacillus fastidiosus*, that can utilize uric acid as the sole source of carbon and energy.

Urogenital Mycoplasma Broth Base

Composition per liter:

Heart infusion powder	8.0g
Casein enzymatic hydrolysate	8.0g
Yeast extract	4.0g
NaCl	3.5g
Arginine hydrochloride	5.0g
Cysteine hydrochloride	0.1g
Phenol Red	0.05g
Horse serum	50.0ml
Urea solution	10.0mL
Vitamin solution	10.0mL
Selective supplement solution	10.0mL

pH 6.3 ± 0.2 at 25°C

Source: This medium is available from HiMedia.

Selective Supplement Solution:

Composition per 10.0mL:

Penicillin	5.0mg
Amphotericin B	1.0mg
Penicillin	100,000U

Preparation of Selective Supplement Solution: Add components to distilled/deionized water and bring volume to 10.0mL. Mix thoroughly. Filter sterilize.

Urea Solution:

Composition per 10.0mL:

Urea	0.5g

Preparation of Urea Solution: Add urea to distilled/deionized water and bring volume to 10.0mL. Mix thoroughly. Filter sterilize.

Vitamain Solution:

Composition per 10.0mL:

Glucose	2.0g
L-Cysteine·HCl	0.518g
L-Glutamine	0.2g
L-Cystine	0.022g
Adenine sulfate	0.02g
Nicotinamide adenine dinucleotide	5.0mg
Cocarboxylase	2.0mg
Guanine·HCl	0.6mg
$Fe(NO_3)_3 \cdot 6H_2O$	0.4mg
p-Aminobenzoic acid	0.26mg
Vitamin B_{12}	0.2mg
Thiamine·HCl	0.06mg

Preparation of Vitamin solution: Add components to distilled/deionized water and bring volume to 10.0mL. Mix thoroughly. Filter sterilize.

Preparation of Medium: Add components, except vitamin solution, urea solution, horse serum, and selective supplement solution, to distilled/deionized water and bring volume to 920.0mL. Mix thoroughly. Autoclave for 15 min at 15 psi pressure–121°C. Cool to 50°C. Aseptically add vitamin solution, urea solution, horse serum, and selective supplement solution. Mix thoroughly. Aseptically distribute into sterile tubes.

Use: For the selective culture of *Mycoplasma hominis* and *Ureaplasma urealyticum*.

USP Alternative Thioglycolate Medium

See: **Sterility Test Broth**

Ustilago Complete Agar II

Composition per liter:

Agar	20.0g
Glucose	10.0g
Hydrolyzed casein	2.5g
NH_4NO_3	1.5g
Yeast extract	1.0g
Salt solution	62.5mL
Vitamin solution	10.0mL
Hydrolyzed nucleic acids solution	5.0mL

pH 7.0 ± 0.2 at 25°C

Salt Solution:

Composition per liter:

KH_2PO_4	16.0g
KCl	8.0g
Na_2SO_4	4.0g
$MgSO_4$	2.0g
$CaCl_2$	1.0g
Trace elements solution	8.0mL

Trace Elements Solution:

Composition per liter:

$CuSO_4 \cdot 5H_2O$	0.4g
$ZnCl_2$	0.4g
$MnCl_2 \cdot 4H_2O$	0.14g
$FeCl_3 \cdot 6H_2O$	100.0mg
H_3BO_3	60.0mg
$Na_2MoO_4 \cdot 2H_2O$	40.0mg

Preparation of Trace Elements Solution: Add components to distilled/deionized water and bring volume to 1.0L. Mix thoroughly.

Preparation of Salt Solution: Add components to distilled/deionized water and bring volume to 1.0L. Mix thoroughly.

Vitamin Solution:

Composition per liter:

Inositol	1.0g
Calcium pantothenate	0.2g
Choline chloride	0.2g
Nicotinic acid	0.2g
Thiamine	100.0mg
p-Aminobenzoic acid	50.0mg
Pyridoxine	50.0mg
Riboflavin	50.0mg

Preparation of Vitamin Solution: Add components to distilled/deionized water and bring volume to 1.0L. Mix thoroughly.

Hydrolyzed Nucleic Acids Solution:

Composition per 80.0mL:

DNA, calf thymus	2.0g
RNA	2.0g
HCl (1*M* solution)	30.0mL
NaOH (1*M* solution)	30.0mL

Preparation of Hydrolyzed Nucleic Acids Solution: Add DNA to 30.0mL of 1*M* NaOH solution. Add RNA to 30.0mL of 1*M* HCl solution. Autoclave the two solutions separately for 10 min at 15 psi pressure–121°C. Mix the two solutions. Adjust the pH to 6.0. Centrifuge at 5000 × g for 10 min. Decant supernatant solution and filter. Bring volume to 80.0mL with distilled/deionized water. Store at –20°C.

Preparation of Medium: Add components to distilled/deionized water and bring volume to 1.0L. Mix thoroughly. Gently heat and bring to boiling. Adjust pH to 7.0. Distribute into tubes or flasks. Autoclave for 15 min at 15 psi pressure–121°C. Pour into sterile Petri dishes or leave in tubes.

Use: For the cultivation and maintenance of *Ustilago* species.

Ustilago Complete Broth II

Composition per liter:

Glucose	10.0g
Hydrolyzed casein	2.5g
NH_4NO_3	1.5g
Yeast extract	1.0g
Salt solution	62.5mL
Vitamin solution	10.0mL
Hydrolyzed nucleic acids solution	5.0mL

pH 7.0 ± 0.2 at 25°C

Salt Solution:

Composition per liter:

KH_2PO_4	16.0g
KCl	8.0g
Na_2SO_4	4.0g
$MgSO_4$	2.0g
$CaCl_2$	1.0g
Trace elements solution	8.0mL

Trace Elements Solution:

Composition per liter:

$CuSO_4 \cdot 5H_2O$	0.4g
$ZnCl_2$	0.4g
$MnCl_2 \cdot 4H_2O$	0.14g
$FeCl_3 \cdot 6H_2O$	100.0mg
H_3BO_3	60.0mg
$Na_2MoO_4 \cdot 2H_2O$	40.0mg

Preparation of Trace Elements Solution: Add components to distilled/deionized water and bring volume to 1.0L. Mix thoroughly.

Preparation of Salt Solution: Add components to distilled/deionized water and bring volume to 1.0L. Mix thoroughly.

Vitamin Solution:

Composition per liter:

Inositol	1.0g
Calcium pantothenate	0.2g
Choline chloride	0.2g
Nicotinic acid	0.2g
Thiamine	100.0mg
p-Aminobenzoic acid	50.0mg
Pyridoxine	50.0mg
Riboflavin	50.0mg

Preparation of Vitamin Solution: Add components to distilled/deionized water and bring volume to 1.0L. Mix thoroughly.

Hydrolyzed Nucleic Acids Solution:

Composition per 80.0mL:

DNA, calf thymus	2.0g
RNA	2.0g
HCl (1*M* solution)	30.0mL
NaOH (1*M* solution)	30.0mL

Preparation of Hydrolyzed Nucleic Acids Solution: Add DNA to 30.0mL of 1*M* NaOH solution. Add RNA to 30.0mL of 1*M* HCl solution. Autoclave the two solutions separately for 10 min at 15 psi pressure–121°C. Mix the two solutions. Adjust the pH to 6.0. Centrifuge at 5000 × g for 10 min. Decant supernatant solution and filter. Bring volume to 80.0mL with distilled/deionized water. Store at –20°C.

Preparation of Medium: Add components to distilled/deionized water and bring volume to 1.0L. Mix thoroughly. Adjust pH to 7.0. Distribute into tubes or flasks. Autoclave for 15 min at 15 psi pressure–121°C.

Use: For the cultivation of *Ustilago* species.

Ustilago Medium

Composition per liter:

Yeast extract	11.0g
Glucose	10.0g
NH_4NO_3	1.5g
Salt solution	62.5mL
Vitamin solution	10.0mL

Salt Solution:

Composition per liter:

KH_2PO_4	16.0g
KCl	8.0g
Na_2SO_4	4.0g
$MgSO_4 \cdot 7H_2O$	2.0g
$CaCl_2$	1.0g
Trace elements solution	8.0mL

Preparation of Salt Solution: Add components to distilled/deionized water and bring volume to 1.0L. Mix thoroughly.

Trace Elements Solution:

Composition per 500.0mL:

$CuSO_4 \cdot 5H_2O$	0.2g
$ZnCl_2$	0.2g

$MnCl_2 \cdot 4H_2O$	0.07g
$FeCl_3 \cdot 6H_2O$	0.05g
H_3BO_3	0.03g
$Na_2MoO_4 \cdot 2H_2O$	0.02g

Preparation of Trace Elements Solution: Add components to distilled/deionized water and bring volume to 500.0mL. Mix thoroughly.

Vitamin Solution:

Composition per liter:

Inositol	0.4g
Calcium pantothenate	0.2g
Choline chloride	0.2g
Nicotinic acid	0.2g
Thiamine	0.1g
Pyridoxine	0.05g
Riboflavin	0.05g

Preparation of Vitamin Solution: Add components to distilled/deionized water and bring volume to 1.0L. Mix thoroughly. Filter sterilize.

Preparation of Medium: Add components, except vitamin solution, to distilled/deionized water and bring volume to 990.0mL. Mix thoroughly. Gently heat and bring to boiling. Autoclave for 15 min at 15 psi pressure–121°C. Cool to 45°–50°C. Aseptically add 10.0mL of sterile vitamin solution. Mix thoroughly. Aseptically distribute into sterile tubes or flasks.

Use: For the cultivation of *Ustilago* species.

Ustilago Medium

Composition per liter:

Agar	20.0g
Glucose	10.0g
Peptone	10.0g
Malt extract	3.0g
Yeast extract	3.0g
Beef extract	1.0g

pH 5.7 ± 0.2 at 25°C

Preparation of Medium: Add components to distilled/deionized water and bring volume to 1.0L. Mix thoroughly. Gently heat and bring to boiling. Adjust pH to 5.7. Distribute into tubes or flasks. Autoclave for 15 min at 15 psi pressure–121°C. Pour into sterile Petri dishes or leave in tubes.

Use: For the cultivation and maintenance of *Ustilago* species.

Ustilago Minimal Medium

Composition per liter:

Glucose	10.0g
KNO_3	3.0g
Salt solution	62.5mL

Salt Solution:

Composition per liter:

KH_2PO_4	16.0g
KCl	8.0g
Na_2SO_4	4.0g
$MgSO_4 \cdot 7H_2O$	2.0g
$CaCl_2$	1.0g
Trace elements solution	8.0mL

Preparation of Salt Solution: Add components to distilled/deionized water and bring volume to 1.0L. Mix thoroughly.

Trace Elements Solution:

Composition per 500.0mL:

$CuSO_4 \cdot 5H_2O$	0.2g
$ZnCl_2$	0.2g
$MnCl_2 \cdot 4H_2O$	0.07g
$FeCl_3 \cdot 6H_2O$	0.05g
H_3BO_3	0.03g
$Na_2MoO_4 \cdot 2H_2O$	0.02g

Preparation of Trace Elements Solution: Add components to distilled/deionized water and bring volume to 500.0mL. Mix thoroughly.

Preparation of Medium: Add components to distilled/deionized water and bring volume to 1.0L. Mix thoroughly. Distribute into tubes or flasks. Autoclave for 15 min at 15 psi pressure–121°C.

Use: For the cultivation of *Ustilago* species.

UVM *Listeria* Enrichment Broth (University of Vermont *Listeria* Enrichment Broth)

Composition per liter:

NaCl	20.0g
Na_2HPO_4	9.6g
Pancreatic digest of casein	5.0g
Peptic digest of animal tissue	5.0g
Beef extract	5.0g
Yeast extract	5.0g
KH_2PO_4	1.35g
Esculin	1.0g
Nalidixic acid	40.0mg
Acriflavine·HCl	12.0mg

pH 7.2 ± 0.2 at 25°C

Preparation of Medium: Add components to distilled/deionized water and bring volume to 1.0L. Mix thoroughly. Distribute into tubes or flasks. Autoclave for 15 min at 15 psi pressure–121°C.

Use: For the selective isolation of *Listeria monocytogenes.*

UVM Modified *Listeria* Enrichment Broth (University of Vermont Modified *Listeria* Enrichment Broth)

Composition per liter:

NaCl	20.0g
Na_2HPO_4	9.6g
Pancreatic digest of casein	5.0g
Peptic digest of animal tissue	5.0g
Beef extract	5.0g
Yeast extract	5.0g
KH_2PO_4	1.35g
Esculin	1.0g
Nalidixic acid	20.0mg
Acriflavine·HCl	12.0mg

pH 7.2 ± 0.2 at 25°C

Source: This medium is available as a premixed powder from BD Diagnostic Systems.

Preparation of Medium: Add components to distilled/deionized water and bring volume to 1.0L. Mix thoroughly. Distribute into tubes or flasks. Autoclave for 15 min at 15 psi pressure–121°C.

Use: For the selective isolation of *Listeria monocytogenes.*

V Agar

Composition per liter:

Agar	13.5g
Pancreatic digest of casein	12.0g
Peptone	10.0g
Peptic digest of animal tissue	5.0g
NaCl	5.0g
Beef extract	3.0g
Yeast extract	3.0g
Cornstarch	1.0g
Human blood, anticoagulated	50.0mL

pH 7.4 ± 0.2 at 25°C

Source: This medium is available as a prepared medium from BD Diagnostic Systems.

Preparation of Medium: Add components, except human blood, to distilled/deionized water and bring volume to 950.0mL. Mix thoroughly. Gently heat and bring to boiling. Distribute into tubes or flasks. Autoclave for 15 min at 15 psi pressure–121°C. Cool to 50°C. Aseptically add 50.0mL of human blood. Mix thoroughly. Pour into sterile Petri dishes or leave in tubes.

Use: For the isolation and differentiation of *Gardnerella vaginalis* from clinical specimens. Plates are incubated under an atmosphere with 3–10% CO_2. *Gardnerella vaginalis* appears as small white colonies with diffuse β-hemolysis.

V-8™ Agar

Composition per liter:

Agar	20.0g
$CaCO_3$	4.0g
V-8™ canned vegetable juice	200.0mL

pH 7.3 ± 0.2 at 25°C

Preparation of Medium: Add components to distilled/deionized water and bring volume to 1.0L. Mix thoroughly. Gently heat and bring to boiling. Distribute into tubes or flasks. Autoclave for 15 min at 15 psi pressure–121°C. Pour into sterile Petri dishes or leave in tubes.

Use: For the isolation and cultivation of *Actinomadura* species, *Actinopolyspora* species, *Excellospora* species, and *Microspora* species.

V-8™ Agar

Composition per liter:

Agar	15.0g
$CaCO_3$	2.0g
V-8™ canned vegetable juice	200.0mL

Preparation of Medium: Add components to distilled/deionized water and bring volume to 1.0L. Mix thoroughly. Gently heat and bring to boiling. Distribute into tubes or flasks. Autoclave for 15 min at 15 psi pressure–121°C. Pour into sterile Petri dishes or leave in tubes.

Use: For the cultivation of numerous yeasts and filamentous fungi.

V-8™ Agar, Half-strength (ATCC Medium 2211)

Composition per liter:

Agar	15.0g
$CaCO_3$	1.5g
V-8™ canned vegetable juice	100.0mL

pH 7.2 ± 0.2 at 25°C

Preparation of Medium: Add components to tap water and bring volume to 1.0L. Mix thoroughly. Gently heat and bring to boiling. Distribute into tubes or flasks. Autoclave for 15 min at 15 psi pressure–121°C. Pour into sterile Petri dishes or leave in tubes.

Use: For the isolation and cultivation of *Actinomadura* species, *Actinopolyspora* species, *Excellospora* species, and *Microspora* species.

V-8™ Agar, Half-strength (ATCC Medium 2216)

Composition per liter:

Agar	15.0g
$CaCO_3$	0.75g
V-8™ canned vegetable juice	50.0mL

pH 7.2 ± 0.2 at 25°C

Preparation of Medium: Add components to 950.0mL distilled/deionized water. Mix thoroughly. Bring volume to 1.0L with tap water. Mix thoroughly. Adjust to pH 7.2 ± 0.2. Gently heat and bring to boiling. Distribute into tubes or flasks. Autoclave for 15 min at 15 psi pressure–121°C. Pour into sterile Petri dishes or leave in tubes.

Use: For the isolation and cultivation of *Actinomadura* species, *Actinopolyspora* species, *Excellospora* species, and *Microspora* species.

V-8™ Agar, pH 7.2

Composition per liter:

Agar	15.0g
$CaCO_3$	3.0g
V-8™ canned vegetable juice	200.0mL

pH 7.2 ± 0.2 at 25°C

Preparation of Medium: Add components to distilled/deionized water and bring volume to 1.0L. Mix thoroughly. Adjust pH to 7.2. Gently heat and bring to boiling. Distribute into tubes or flasks. Autoclave for 15 min at 15 psi pressure–121°C. Pour into sterile Petri dishes or leave in tubes.

Use: For the cultivation of *Bipolaris leersiae, Bipolaris micropus, Camposporium pellucidum, Cochliobolus spicifer, Cochliobolus bicolor, Cochliobolus miyabeanus, Cochliobolus australiensis, Cochliobolus sativus, Cochliobolus victoriae, Curvularia inaequalis, Drechslera catenaria, Drechslera panicimiliacei, Embellisia chlamydospora, Helicosporium pallidum, Helminthosporium papulosum, Phytophthora cinnamomi, Phytophthora cryptogea, Setosphaeria rostrata,* and other yeasts and filamentous fungi.

V-8™-0 Agar

Composition per liter:

Agar	15.0 g
$CaCO_3$	5.0 g
$CaCl_3$	100.0mg
β-Sitosterol	30.0mg
Tryptophan	20.0mg
Thiamine	1.0mg
V-8™ canned vegetable juice	354.0mL

Preparation of Medium: Add $CaCO_3$ to V-8. Centrifuge for 20 min at 4000 rpm. Decant the supernatant. Add the supernatant to distilled/deionized water and bring volume to 1.0L. Mix thoroughly. Add remaining components. Gently heat and bring to boiling. Distribute into tubes or flasks. Autoclave for 15 min at 15 psi pressure–121°C. Pour into sterile Petri dishes or leave in tubes.

Use: For the cultivation of *Phytophthora syringae, Phytophthora palmivora, Phytophthora nicotianae, Phytophthora erythroseptica, Phytophthora drechsleri,* and *Phytophthora citrophthora.*

V-8™ Juice Agar (ATCC Medium 343)

Composition per liter:

Agar	15.0g
$CaCO_3$	4.0g
V-8™ canned vegetable juice	200.0mL

pH 7.2 ± 0.2 at 25°C

Preparation of Medium: Add components to distilled/deionized water and bring volume to 1.0L. Mix thoroughly. Gently heat and bring to boiling. Adjust pH to 7.2. Distribute into tubes or flasks. Autoclave for 15 min at 15 psi pressure–121°C. Pour into sterile Petri dishes or leave in tubes.

Use: For the isolation and cultivation of *Filobasidiella depauperata* and *Leucosporidium scottii.*

V-8™ Juice Agar II (ATCC Medium 2040)

Composition per liter:

Agar	30.0g
V-8™ canned vegetable juice	300.0mL

pH 7.3± 0.2 at 25°C

Preparation of Medium: Add components to distilled/deionized water and bring volume to 1.0L. Mix thoroughly. Gently heat and bring to boiling. Distribute into tubes or flasks. Autoclave for 15 min at 15 psi pressure–121°C. Pour into sterile Petri dishes or leave in tubes.

Use: For the cultivation of various yeasts and filamentous fungi.

V-8™ Juice Seawater Agar

Composition per liter:

Agar	15.0g
$CaCO_3$	3.0g
Artificial seawater	800.0mL
V-8™ canned vegetable juice, unsalted	200.0mL

pH 7.0 ± 0.2 at 25°C

Artificial Seawater:

Composition per liter:

NaCl	20.0g
$MgCl_2 \cdot 6H_2O$	5.38g
$MgSO_4 \cdot 7H_2O$	6.78g
KCl	0.72g
$NaHCO_3$	0.2g
$CaCl_2 \cdot 2H_2O$	1.4g

Preparation of Artificial Seawater: Add components to distilled/deionized water and bring volume to 1.0L. Mix thoroughly.

Preparation of Medium: Combine components. Mix thoroughly. Adjust pH to 7.0. Gently heat and bring to boiling. Distribute into tubes or flasks. Autoclave for 15 min at 15 psi pressure–121°C. Pour into sterile Petri dishes or leave in tubes.

Use: For the cultivation of *Halophytophthora masteri* and *Halophytophthora tartarea.*

V-8™ Rye Agar

Composition per 1050.0mL:

Agar	20.0g
$CaCO_3$	0.2g
Rye broth	1.0L
V-8™ canned vegetable juice	50.0mL

Rye Broth:

Composition per liter:

Whole rye grains	50.0g

Preparation of Rye Broth: Soak the rye grains in 1.1L of distilled/deionized water for 24–36 hr at 24°C. Autoclave at 15 psi pressure–121°C for 30 min. Filter through four layers of cheesecloth. Bring the final filtrate volume to 1.0L with distilled/deionized water.

Preparation of Medium: Add components to distilled/deionized water and bring volume to 1.0L. Mix thoroughly. Gently heat and bring to boiling. Distribute into tubes or flasks. Autoclave for 15 min at 15 psi pressure–121°C. Pour into sterile Petri dishes or leave in tubes.

Use: For the cultivation and maintenance of *Phytophthora infestans.*

V24N Medium (DSMZ Medium 88b)

Composition per liter:

Starch, soluble	2.0g
$(NH_4)_2SO_4$	1.3g
Sulfur, powdered	1.0g
$Na_2S \cdot 9H_2O$	0.5g
KH_2PO_4	0.28g
$MgSO_4 \cdot 7H_2O$	0.25g
Yeast extract	0.2g
$CaCl_2 \cdot 2H_2O$	0.07g
$FeCl_3 \cdot 6H_2O$	0.02g
$Na_2B_4O_7 \cdot 10H_2O$	4.5mg
$MnCl_2 \cdot 4H_2O$	1.8mg
Resazurin	0.4mg
$ZnSO_4 \cdot 7H_2O$	0.22mg
$CuCl_2 \cdot 2H_2O$	0.05mg
$Na_2MoO_4 \cdot 2H_2O$	0.03mg
$VOSO_4 \cdot 2H_2O$	0.03mg
$CoSO_4$	0.01mg

pH 6.0 ± 0.2 at 25°C

Preparation of Medium: Add components to distilled/deionized water and bring volume to 1.0L. Mix thoroughly. Adjust the pH to 6.0 with H_2SO_4 (25% v/v). Sparge medium with 100% N_2 gas. Distribute anaerobically into rubber-stoppered tubes or bottles that have been presterilized by autoclaving. On 2 successive days heat the medium for 2 hr at 85 °C.

Use: For the growth and maintenance of *Thermofilum librum.*

Van Niel's Agar

Composition per liter:

Agar	20.0g
Yeast extract	10.0g
K_2HPO_4	1.0g
$MgSO_4$	0.5g

pH 7.0 ± 0.2 at 25°C

Preparation of Medium: Add components to distilled/deionized water and bring volume to 1.0L. Mix thoroughly. Gently heat and bring

to boiling. Distribute into tubes or flasks. Autoclave for 15 min at 15 psi pressure–121°C. Pour into sterile Petri dishes or leave in tubes.

Use: For the cultivation of *Rhodomicrobium vannielii*.

Van Niel's Medium, Modified

Composition per liter:

Yeast extract	10.0g
$MgSO_4$	0.1g
EDTA	2.0mg
Trace elements solution	10.0mL
K_2HPO_4 solution	2.5mL

pH 7.1 ± 0.2 at 25°C

Trace Elements Solution:

Composition per 100.0mL:

$CaCl_2 \cdot 2H_2O$	0.3g
Ferric ammonium citrate	0.2g

Preparation of Trace Elements Solution: Add components to distilled/deionized water and bring volume to 100.0mL. Mix thoroughly. Filter sterilize.

K_2HPO_4 Solution:

Composition per 100.0mL:

K_2HPO_4	4.0g

Preparation of K_2HPO_4 Solution: Add K_2HPO_4 to distilled/deionized water and bring volume to 100.0mL. Mix thoroughly. Filter sterilize.

Preparation of Medium: Add components, except K_2HPO_4 and trace elements solution, to distilled/deionized water and bring volume to 987.5mL. Mix thoroughly. Autoclave for 15 min at 15 psi pressure–121°C. Cool to 50°C. Aseptically add trace elements and K_2HPO_4 solutions. Mix thoroughly. Distribute into sterile tubes or flasks.

Use: For the cultivation and maintenance of *Rhodobacter sphaeroides*.

Van Niel's Yeast Agar

See: **Van Niel's Agar**

Van Niel's Yeast Agar with Glutamate (ATCC Medium 1370)

Composition per liter:

Agar	20.0g
Yeast extract	10.0g
K_2HPO_4	1.0g
Glutamate	0.7g
$MgSO_4$	0.5g

pH 7.1 ± 0.2 at 25°C

Preparation of Medium: Add components to tap water and bring volume to 1.0L. Mix thoroughly. Gently heat and bring to boiling. Distribute into tubes or flasks. Autoclave for 15 min at 15 psi pressure–121°C. Pour into sterile Petri dishes or leave in tubes.

Use: For the cultivation of *Rhodomicrobium* spp.

Van Niel's Yeast Agar with Sodium Chloride

Composition per liter:

NaCl	25.0g
Agar	20.0g
Yeast extract	10.0g
K_2HPO_4	1.0g
$MgSO_4$	0.5g

pH 7.1 ± 0.2 at 25°C

Preparation of Medium: Add components to tap water and bring volume to 1.0L. Mix thoroughly. Gently heat and bring to boiling. Distribute into tubes or flasks. Autoclave for 15 min at 15 psi pressure–121°C. Pour into sterile Petri dishes or leave in tubes.

Use: For the cultivation of halophilic *Rhodomicrobium* spp.

Van Niel's Yeast Agar with 25% Sodium Chloride (ATCC Medium 217)

Composition per liter:

NaCl	250.0g
Agar	20.0g
Yeast extract	10.0g
K_2HPO_4	1.0g
$MgSO_4$	0.5g

pH 7.1 ± 0.2 at 25°C

Preparation of Medium: Add components to tap water and bring volume to 1.0L. Mix thoroughly. Gently heat and bring to boiling. Distribute into tubes or flasks. Autoclave for 15 min at 15 psi pressure–121°C. Pour into sterile Petri dishes or leave in tubes.

Use: For the cultivation of halophilic *Rhodomicrobium* spp.

Van Niel's Yeast Agar with Succinate (ATCC Medium 1243)

Composition per liter:

Agar	20.0g
Yeast extract	10.0g
K_2HPO_4	1.0g
Succinate	0.6g
$MgSO_4$	0.5g

pH 7.1 ± 0.2 at 25°C

Preparation of Medium: Add components to tap water and bring volume to 1.0L. Mix thoroughly. Gently heat and bring to boiling. Distribute into tubes or flasks. Autoclave for 15 min at 15 psi pressure–121°C. Pour into sterile Petri dishes or leave in tubes.

Use: For the cultivation of *Rhodomicrobium* spp.

Van Niel's Yeast Medium with Pyruvate, Modified

Composition per liter:

Yeast extract	10.0g
$MgSO_4$	0.1g
EDTA	2.0mg
Sodium pyruvate solution	100.0mL
Trace elements solution	10.0mL
K_2HPO_4 solution	5.0mL
Trace metal A-5 solution	1.0mL

pH 7.1± 0.1 at 25°C

Sodium Pyruvate Solution:

Composition per 100.0mL:

Sodium pyruvate	1.1g

Preparation of Sodium Pyruvate Solution: Add sodium pyruvate to distilled/deionized water and bring volume to 100.0mL. Mix thoroughly. Filter sterilize.

Trace Elements Solution:

Composition per 100.0mL:

$CaCl_2 \cdot 2H_2O$	0.3g
Ferric ammonium citrate	0.2g

Preparation of Trace Elements Solution: Add components to distilled/deionized water and bring volume to 100.0mL. Mix thoroughly. Filter sterilize.

K_2HPO_4 Solution:

Composition per 100.0mL:

K_2HPO_4	4.0g

Preparation of K_2HPO_4 Solution: Add K_2HPO_4 to distilled/deionized water and bring volume to 100.0mL. Mix thoroughly. Filter sterilize.

Trace Metal A-5 Solution:

Composition per liter:

H_3BO_3	2.86g
$MnCl_2 \cdot 4H_2O$	1.81g
$Na_2MoO_4 \cdot 2H_2O$	0.39g
$ZnSO_4 \cdot 7H_2O$	0.222g
$CuSO_4 \cdot 5H_2O$	0.079g
$Co(NO_3)_2 \cdot 6H_2O$	0.049g

Preparation of Trace Metal A-5 Solution: Add components to distilled/deionized water and bring volume to 1.0L. Mix thoroughly. Filter sterilize.

Preparation of Medium: Add components, except sodium pyruvate, trace elements, K_2HPO_4, and trace metal A-5 solutions, to distilled/deionized water and bring volume to 884.0mL. Mix thoroughly. Autoclave for 15 min at 15 psi pressure–121°C. Cool to 25°C. Aseptically add 100.0mL of sterile sodium pyruvate solution, 10.0mL of sterile trace elements solution, 5.0mL of sterile K_2HPO_4 solution, and 1.0mL of sterile trace metal A-5 solution. Mix thoroughly. Adjust pH to 7.1 ± 0.1. Aseptically distribute into sterile tubes or flasks.

Use: For the cultivation and maintenance of photosynthetic bacteria, such as *Heliobacillus mobilis* and *Rhodopseudomonas palustris*.

Vanillate Medium

Composition per liter:

Agar	20.0g
$(NH_4)_2SO_4$	1.0g
KH_2PO_4	0.4g
Yeast extract	0.1g
$MgSO_4 \cdot 7H_2O$	0.01g
Trace elements solution	10.0mL
Vanillic acid solution	10.0mL

Trace Elements Solution:

Composition per liter:

$MnSO_4 \cdot 4H_2O$	0.4g
H_3BO_3	0.5mg
$ZnSO_4 \cdot 7H_2O$	0.4mg
$FeCl_3$	0.2mg
$(NH_4)_6Mo_7O_{24} \cdot 4H_2O$	0.2mg
KI	0.1mg
$CuSO_4 \cdot 5H_2O$	0.04mg

Preparation of Trace Elements Solution: Add components to distilled/deionized water and bring volume to 1.0L. Mix thoroughly. Filter sterilize.

Vanillic Acid Solution:

Composition per 10.0mL:

Vanillic acid, sodium salt	1.5g

Preparation of Vanillic Acid Solution: Add vanillic acid to distilled/deionized water and bring volume to 10.0mL. Mix thoroughly. Filter sterilize.

Preparation of Medium: Add components, except vanillic acid solution and trace elements solution, to distilled/deionized water and bring volume to 980.0mL. Mix thoroughly. Gently heat and bring to boiling. Autoclave for 15 min at 15 psi pressure–121°C. Cool to 50°–55°C. Warm the vanillic acid solution and the trace elements solution to 50°–55°C. Aseptically add 10.0mL of sterile vanillic acid solution and 10.0mL of sterile trace elements solution. Mix thoroughly. Pour into sterile Petri dishes or distribute into sterile tubes.

Use: For the cultivation of *Pseudomonas fluorescens*.

VCR Medium

Composition per 1001.0mL:

Agar, noble	15.0g
$NaNO_3$	2.0g
KH_2PO_4	1.5g
K_2HPO_4	1.2g
NH_4Cl	0.5g
$MgSO_4 \cdot 7H_2O$	0.2g
$FeCl_3$	0.01g
$CaCl_2 \cdot 2H_2O$	15.0mg
$CuSO_4 \cdot 5H_2O$	1.0mg
Vitamin B_{12} solution	1.0mL

pH 7.2 ± 0.2 at 25°C

Vitamin B_{12} Solution:

Composition per 10.0mL:

Vitamin B_{12}	10.0μg

Preparation of Vitamin B_{12} Solution: Add vitamin B_{12} to distilled/deionized water and bring volume to 10.0mL. Mix thoroughly. Filter sterilize.

Preparation of Medium: Add components, except vitamin B_{12} solution, to distilled/deionized water and bring volume to 1.0L. Mix thoroughly. Gently heat and bring to boiling. Autoclave for 15 min at 15 psi pressure–121°C. Cool to 50°–55°C. Aseptically add 1.0mL of sterile vitamin B_{12} solution. Mix thoroughly. Pour into sterile Petri dishes or distribute into sterile tubes.

Use: For the cultivation of *Sphaerobacter thermomethanica*.

Veal Infusion Agar

Composition per liter:

Agar	15.0g
Veal, infusion from	10.0g
Pancreatic digest of casein	5.0g
Peptic digest of animal tissue	5.0g
NaCl	5.0g

pH 7.4 ± 0.2 at 25°C

Source: This medium is available as a premixed powder from BD Diagnostic Systems.

Preparation of Medium: Add components to distilled/deionized water and bring volume to 1.0L. Mix thoroughly. Gently heat and bring to boiling. Distribute into tubes or flasks. Autoclave for 15 min at 15 psi pressure–121°C. Pour into sterile Petri dishes or leave in tubes.

Use: For the cultivation and maintenance of a variety of microorganisms. Can be used for the cultivation of fastidious microorganisms when enriched with blood or serum.

Veal Infusion Agar (ATCC Medium 521)

Composition per liter:

Veal, infusion from	500.0g
Agar	15.0g
Pancreatic digest of casein	5.0g
Peptic digest of animal tissue	5.0g
NaCl	5.0g

pH 7.4 ± 0.2 at 25°C

Source: This medium is available as a premixed powder from BD Diagnostic Systems.

Preparation of Medium: Add components to distilled/deionized water and bring volume to 1.0L. Mix thoroughly. Gently heat and bring to boiling. Distribute into tubes or flasks. Autoclave for 15 min at 15 psi pressure–121°C. Pour into sterile Petri dishes or leave in tubes.

Use: For the cultivation and maintenance of a variety of microorganisms. Can be used for the cultivation of fastidious microorganisms when enriched with blood or serum.

Veal Infusion Agar (BAM M173)

Composition per liter:

Veal, infusion from	500.0g
Agar	15.0g
Proteose peptone No. 3	10.0g
NaCl	5.0g

pH 7.3 ± 0.2 at 25°C

Source: This medium is available from BD Diagnostic Systems.

Preparation of Medium: Add components to distilled/deionized water and bring volume to 1.0L. Mix thoroughly. Autoclave for 15 min at 15 psi pressure–121°C. Cool to 45°–50°C. Pour into sterile Petri dishes or leave in tubes. For slants allow tubes to cool in an inclined position.

Use: For the cultivation of fastidious microorganisms.

Veal Infusion Broth

Composition per liter:

Veal, infusion from	10.0g
Pancreatic digest of casein	5.0g
Peptic digest of animal tissue	5.0g
NaCl	5.0g

pH 7.4 ± 0.2 at 25°C

Source: This medium is available from BD Diagnostic Systems.

Preparation of Medium: Add components to distilled/deionized water and bring volume to 1.0L. Mix thoroughly. Gently heat and bring to boiling. Distribute into tubes or flasks. Autoclave for 15 min at 15 psi pressure–121°C. Use freshly prepared solution.

Use: For the cultivation of streptococci and other microorganisms.

Veal Infusion Broth (ATCC Medium 521)

Composition per liter:

Veal, infusion from	500.0g
Pancreatic digest of casein	5.0g
Peptic digest of animal tissue	5.0g
NaCl	5.0g

pH 7.4 ± 0.2 at 25°C

Source: This medium is available as a premixed powder from BD Diagnostic Systems.

Preparation of Medium: Add components to distilled/deionized water and bring volume to 1.0L. Mix thoroughly. Distribute into tubes or flasks. Autoclave for 15 min at 15 psi pressure–121°C. Use freshly prepared solution.

Use: For the cultivation and maintenance of *Arthrobacter* species, streptococci, and other microorganisms.

Veal Infusion Broth (BAM M173)

Composition per liter:

Veal, infusion from	500.0g
Proteose peptone No.3	10.0g
NaCl	5.0g

pH 7.4 ± 0.2 at 25°C

Source: This medium is available from BD Diagnostic Systems.

Preparation of Medium: Add components to distilled/deionized water and bring volume to 1.0L. Mix thoroughly. Distribute into tubes or flasks. Autoclave for 15 min at 15 psi pressure–121°C.

Use: For the cultivation of fastidious microorganisms.

Veal Infusion Broth with Horse Serum

Composition per liter:

Veal, infusion from	500.0g
Pancreatic digest of casein	5.0g
Peptic digest of animal tissue	5.0g
NaCl	5.0g
Horse serum, heat inactivated	100.0mL

pH 7.4 ± 0.2 at 25°C

Preparation of Medium: Add components, except horse serum, to distilled/deionized water and bring volume to 900.0mL. Mix thoroughly. Gently heat and bring to boiling. Autoclave for 15 min at 15 psi pressure–121°C. Cool to 50°C. Aseptically add 100.0mL of horse serum. Mix thoroughly. Aseptically distribute into tubes or flasks. Use freshly prepared solution or boil without mixing prior to use.

Use: For the cultivation and maintenance of *Streptococcus pyogenes*.

Veal Infusion Broth with Rabbit Serum

Composition per liter:

Veal, infusion from	500.0g
Pancreatic digest of casein	5.0g
Peptic digest of animal tissue	5.0g
NaCl	5.0g
Rabbit serum, heat inactivated	150.0mL

pH 7.4 ± 0.2 at 25°C

Preparation of Medium: Add components, except rabbit serum, to distilled/deionized water and bring volume to 850.0mL. Mix thoroughly. Gently heat and bring to boiling. Autoclave for 15 min at 15 psi pressure–121°C. Cool to 50°C. Aseptically add 150.0mL of rabbit serum. Mix thoroughly. Aseptically distribute into sterile tubes or flasks. Boil without agitation prior to use. Use freshly prepared solution or boil without mixing prior to use.

Use: For the cultivation and maintenance of *Proteus mirabilis*.

Veal Yeast Extract Medium
See: **VY Medium**

Veillonella Agar

Composition per liter:

Agar	15.0g
Pancreatic digest of casein	5.0g
Yeast extract	3.0g
Sodium thioglycolate	0.75g
Vancomycin	7.5mg
Basic Fuchsin	2.0mg
Sodium lactate (60% solution)	21.0mL

pH 7.5± 0.2 at 25°C

Source: This medium is available as a premixed powder from BD Diagnostic Systems.

Caution: Basic Fuchsin is a potential carcinogen and care must be taken to avoid inhalation of the powdered dye and contact with the skin.

Preparation of Medium: Add components to distilled/deionized water and bring volume to 1.0L. Mix thoroughly. Gently heat and bring to boiling. Distribute into tubes or flasks. Autoclave for 15 min at 15 psi pressure–121°C. Pour into sterile Petri dishes or leave in tubes.

Use: For the isolation and cultivation of *Veillonella* species.

Veillonella HiVeg Agar Base with Lactate

Composition per liter:

Agar	15.0g
Plant hydrolysate	5.0g
Yeast extract	3.0g
Na-thioglycolate	0.75g
Basic Fuchsin	2.0mg
Sodium lactate (60% solution)	21.0mL

pH 7.5± 0.2 at 25°C

Source: This medium, without lactate solution, is available as a premixed powder from HiMedia.

Caution: Basic Fuchsin is a potential carcinogen and care must be taken to avoid inhalation of the powdered dye and contact with the skin.

Preparation of Medium: Add components to distilled/deionized water and bring volume to 1.0L. Mix thoroughly. Gently heat and bring to boiling. Distribute into tubes or flasks. Autoclave for 15 min at 15 psi pressure–121°C. Mix thoroughly. Pour into sterile Petri dishes or leave in tubes.

Use: For the selective isolation and cultivation of *Veillonella* species.

Veillonella HiVeg Agar Base with Lactate and Vancomycin

Composition per liter:

Agar	15.0g
Plant hydrolysate	5.0g
Yeast extract	3.0g
Na-thioglycolate	0.75g
Basic Fuchsin	2.0mg
Sodium lactate (60% solution)	21.0mL
Selective supplement solution	10.0mL

pH 7.5± 0.2 at 25°C

Source: This medium, without lactate solution, is available as a premixed powder from HiMedia.

Caution: Basic Fuchsin is a potential carcinogen and care must be taken to avoid inhalation of the powdered dye and contact with the skin.

Selective Supplement Solution:

Composition per 10.0mL:

Vancomycin	7.5mg

Preparation of Selective Supplement Solution: Add vancomycin to distilled/deionized water and bring volume to 10.0mL. Mix thoroughly. Filter sterilize.

Preparation of Medium: Add components, except selective supplement solution, to distilled/deionized water and bring volume to 990.0mL. Mix thoroughly. Gently heat and bring to boiling. Autoclave for 15 min at 15 psi pressure–121°C. Cool to 45°–50°C. Aseptically add 10.0mL sterile selective supplement solution. Mix thoroughly. Pour into sterile Petri dishes or aseptically distribute into sterile tubes or flasks.

Use: For the selective isolation and cultivation of *Veillonella* species.

Veillonella Medium

Composition per liter:

Pancreatic digest of casein	5.0g
Yeast extract	3.0g
Tween™ 80	1.0g
Glucose	1.0g
Sodium thioglycolate	0.75g
Sodium lactate (60% solution)	21.0mL

pH 7.5 ± 0.2 at 25°C

Preparation of Medium: Add components to distilled/deionized water and bring volume to 1.0L. Mix thoroughly. Adjust pH to 7.5 with K_2CO_3. Distribute into tubes or flasks. Autoclave for 15 min at 15 psi pressure–121°C.

Use: For the cultivation and maintenance of *Veillonella* species.

Veillonella Medium, DSM

Composition per liter:

Sodium lactate (60% solution)	7.5g
Pancreatic digest of casein	5.0g
Yeast extract	3.0g
Tween™ 80	1.0g
Glucose	1.0g
Sodium thioglycolate	0.75g
Putrescine	3.0mg
Resazurin	1.0mg

pH 7.5 ± 0.2 at 25°C

Preparation of Medium: Prepare and dispense medium anaerobically under 100% N_2. Add components to distilled/deionized water and bring volume to 1.0L. Mix thoroughly. Adjust pH to 7.5 with K_2CO_3. Anaerobically distribute into tubes or flasks. Autoclave for 15 min at 15 psi pressure–121°C.

Use: For the cultivation and maintenance of *Veillonella parvula* and other *Veillonella* species.

Veillonella Selective Medium

Composition per liter:

Pancreatic digest of casein	5.0g
Yeast extract	3.0g
Tween™ 80	1.0g
Sodium thioglycolate	0.75g

Sodium lactate (50% solution) 25.0mL
Streptomycin solution 10.0mL

pH 6.6 ± 0.2 at 25°C

Streptomycin Solution:
Composition per 10.0mL:
Streptomycin 5.0mg

Preparation of Streptomycin Solution: Add streptomycin to distilled/deionized water and bring volume to 10.0mL. Mix thoroughly. Filter sterilize.

Preparation of Medium: Add components, except streptomycin solution, to distilled/deionized water and bring volume to 990.0mL. Mix thoroughly. Gently heat and bring to boiling. Adjust pH to 6.6 with K_2CO_3. Autoclave for 15 min at 15 psi pressure–121°C. Cool to 45°–50°C. Aseptically add sterile streptomycin solution. Mix thoroughly. Aseptically distribute into sterile tubes or flasks.

Use: For the cultivation of *Veillonella* species.

VEN CHI2 Medium (DSMZ Medium 293a)

Composition per liter:
NaCl 20.0g
$MgCl_2 \cdot 6H_2O$ 3.0g
KCl 0.5g
NH_4Cl 0.25g
KH_2PO_4 .... 0.2g
$CaCl_2 \cdot 2H_2O$ 0.15g
Resazurin 1.0mg
$NaHCO_3$ solution 10.0mL
Na_2-succinate solution 10.0mL
$Na_2S \cdot 9H_2O$ solution 10.0mL
Vitamin solution 10.0mL
Quinic acid solution 10.0mL
Trace elements solution SL-10 .... 1.0mL

pH 7.2 ± 0.2 at 25°C

$Na_2S \cdot 9H_2O$ Solution:
Composition per 10.0mL:
$Na_2S \cdot 9H_2O$ 0.36g

Preparation of $Na_2S \cdot 9H_2O$ Solution: Add $Na_2S \cdot 9H_2O$ to distilled/deionized water and bring volume to 10.0mL. Mix thoroughly. Autoclave under 100% N_2 for 15 min at 15 psi pressure–121°C. Cool to room temperature.

$NaHCO_3$ Solution:
Composition per 10.0mL:
$NaHCO_3$ 2.5g

Preparation of $NaHCO_3$ Solution: Add $NaHCO_3$ to distilled/deionized water and bring volume to 10.0mL. Mix thoroughly. Sparge with 80% N_2 + 20% CO_2. Filter sterilize.

Na_2-succinate Solution:
Composition per 10.0mL:
Na_2-succinate 3.25g

Preparation of Na_2-succinate Solution: Add Na_2-succinate to distilled/deionized water and bring volume to 10.0mL. Mix thoroughly. Sparge with 100% N_2. Filter sterilize.

Vitamin Solution:
Composition per liter:
Pyridoxine-HCl 10.0mg
Thiamine-HCl·$2H_2O$ 5.0mg
Riboflavin 5.0mg
Nicotinic acid 5.0mg
D-Ca-pantothenate 5.0mg
p-Aminobenzoic acid 5.0mg
Lipoic acid 5.0mg
Biotin 2.0mg
Folic acid 2.0mg
Vitamin B_{12} 0.1mg

Preparation of Vitamin Solution: Add components to distilled/deionized water and bring volume to 1.0L. Mix thoroughly. Sparge with 80% H_2 + 20% CO_2. Filter sterilize.

Quinic Acid Solution:
Composition per 10.0mL:
Quinic acid 1.0g

Preparation of Quinic Acid Solution: Add quinic acid to distilled/deionized water and bring volume to 10.0mL. Mix thoroughly. Adjust pH to 7.0. Sparge with 80% N_2 + 20% CO_2. Filter sterilize.

Trace Elements Solution SL-10:
Composition per liter:
$FeCl_2 \cdot 4H_2O$ 1.5g
$CoCl_2 \cdot 6H_2O$ 190.0mg
$MnCl_2 \cdot 4H_2O$ 100.0mg
$ZnCl_2$ 70.0mg
$Na_2MoO_4 \cdot 2H_2O$ 36.0mg
$NiCl_2 \cdot 6H_2O$ 24.0mg
H_3BO_3 .... 6.0mg
$CuCl_2 \cdot 2H_2O$ 2.0mg
HCl (25% solution) 10.0mL

Preparation of Trace Elements Solution SL-10: Add $FeCl_2 \cdot 4H_2O$ to 10.0mL of HCl solution. Mix thoroughly. Add distilled/deionized water and bring volume to 1.0L. Add remaining components. Mix thoroughly. Sparge with 80% N_2 + 20% CO_2. Filter sterilize.

Preparation of Medium: Prepare and dispense medium under 80% N_2 + 20% CO_2 gas atmosphere. Add components, except $NaHCO_3$ solution, Na_2-succinate solution, $Na_2S \cdot 9H_2O$ solution, vitamin solution, quinic acid solution, and trace elements solution SL-10, to distilled/deionized water and bring volume to 949.0mL. Mix thoroughly. Adjust pH to 7.2. Sparge with 80% N_2 + 20% CO_2. Autoclave for 15 min at 15 psi pressure–121°C. Aseptically and anaerobically add 10.0mL $NaHCO_3$ solution, 10.0mL Na_2-succinate solution, 10.0mL $Na_2S \cdot 9H_2O$ solution, 10.0mL vitamin solution, 10.0mL quinic acid solution, and 1.0mL trace elements solution SL-10. Mix thoroughly. Aseptically and anaerobically distribute into sterile tubes or bottles. After inoculation, flush and repressurize the gas head space of culture bottles with sterile 80% N_2 + 20% CO_2 to 1 bar overpressure.

Use: For the cultivation of *Ilyobacter insuetus*.

Venenivibrio stagnispumantis Medium (DSMZ Medium 1146)

Composition per liter:
$MgSO_4 \cdot 7H_2O$ 7.0g
$Na_2S_2O_3 \cdot 5H_2O$ 2.0g
MES 1.95g
KCl 0.5g
$MgCl_2 \cdot 6H_2O$ 0.4g
$CaCl_2 \cdot 2H_2O$ 0.4g
NaOH 1.36g
NH_4Cl 0.2g

KH_2PO_4 .... 0.25g
Trace elements solution 10.0mL

pH 5.5 ± 0.2 at 25°C

Trace Elements Solution:

Composition per liter:

EDTA 5.0g
$CoCl_2 \cdot 6H_2O$ 150.0mg
$MnCl_2 \cdot 4H_2O$ 100.0mg
$ZnCl_2$ 100.0mg
$FeSO_4 \cdot 7H_2O$ 100.0mg
$AlCl_3 \cdot 6H_2O$ 40.0mg
$Na_2WO_4 \cdot 2H_2O$ 30.0mg
$CuCl_2 \cdot 2H_2O$ 20.0mg
$NiSO_4 \cdot 6H_2O$ 20.0mg
$NaHSeO_3$ 10.0mg
H_3BO_3 .... 10.0mg
$Na_2MoO_4 \cdot 2H_2O$ 10.0mg

Preparation of Trace Elements Solution: Add components to distilled/deionized water and bring volume to 1.0L. Adjust pH to 3.0 with HCl. Mix thoroughly.

Preparation of Medium: Prepare anaerobic distilled/deionized water by sparging with with 100% CO_2. Adjust pH to 5.5. Sparge with CO_2 for 15 min. Add components to the distilled/deionized anaerobic water and bring volume to 1.0L. Dispense into culture tubes. Stopper and seal tubes by crimping caps onto stoppers. Autoclave for 20 min at 15 psi pressure–121°C.

Use: For the cultivation of *Venenivibrio stagnispumantis.*

Viability-Preserving Microbiostatic Medium

See: **VMGII Medium**

Vibrio Agar

Composition per liter:

Sucrose 20.0g
Agar 15.0g
NaCl 10.0g
Sodium citrate·$2H_2O$ 10.0g
$Na_2S_2O_3 \cdot 5H_2O$ 6.5g
Oxgall 5.0g
Yeast extract 5.0g
Pancreatic digest of casein 4.0g
Proteose peptone 3.0g
Sodium deoxycholate 1.0g
Sodium lauryl sulfate 0.2g
Water Blue 0.2g
Cresol Red 0.02g

pH 8.5 ± 0.2 at 25°C

Preparation of Medium: Add components to distilled/deionized water and bring volume to 1.0L. Mix thoroughly. Adjust pH to 8.5. Gently heat and bring to boiling. Do not autoclave. Pour into sterile Petri dishes or distribute into sterile tubes.

Use: For the isolation and cultivation of the *Vibrio cholerae.*

Vibrio costicola Medium

Composition per liter:

NaCl 81.0g
$MgSO_4 \cdot 7H_2O$ 19.7g
Agar 15.0g
$MgCl \cdot H_2O$ 15.0g
Yeast extract 10.0g
Proteose peptone 5.0g
KCl 2.0g
Glucose 1.0g
$CaCl_2 \cdot 2H_2O$ 0.48g
$NaHCO_3$ 60.0mg
NaBr 26.0mg

Preparation of Medium: Add components to distilled/deionized water and bring volume to 1.0L. Mix thoroughly. Gently heat and bring to boiling. Distribute into tubes or flasks. Autoclave for 15 min at 15 psi pressure–121°C. Pour into sterile Petri dishes or leave in tubes.

Use: For the cultivation and maintenance of *Vibrio costicola.*

Vibrio HiVeg Agar

Composition per liter:

Sucrose 20.0g
Agar 15.0g
NaCl 10.0g
Sodium citrate·$2H_2O$ 10.0g
Plant hydrolysate 8.0g
$Na_2S_2O_3 \cdot 5H_2O$ 6.5g
Yeast extract 5.0g
Plant peptone No. 3 .... 3.0g
Synthetic detergent No. II 1.0g
Synthetic detergent No. III 1.0g
Sodium lauryl sulfate 0.2g
China Blue 0.2g
Cresol Red 0.2g

pH 8.5 ± 0.2 at 25°C

Source: This medium is available as a premixed powder from HiMedia.

Preparation of Medium: Add components to distilled/deionized water and bring volume to 1.0L. Mix thoroughly. Adjust pH to 8.5. Gently heat and bring to boiling. Do not autoclave. Pour into sterile Petri dishes or distribute into sterile tubes.

Use: For the isolation and cultivation of the *Vibrio cholerae.*

Vibrio Medium

Composition per liter:

NaCl 10.0g
Pancreatic digest of casein 10.0g
$MgCl_2 \cdot 6H_2O$ 4.0g
KCl 1.0g

pH 7.5 ± 0.2 at 25°C

Preparation of Medium: Add components to distilled/deionized water and bring volume to 1.0L. Mix thoroughly. Distribute into tubes or flasks. Autoclave for 15 min at 15 psi pressure–121°C.

Use: For the cultivation of *Vibrio diazotrophicus.*

Vibrio natriegens Medium

Composition per liter:

Urea 20.0g
NaCl 15.0g
Agar 15.0g
Peptone 5.0g
Meat extract 3.0g

pH 7.0 ± 0.2 at 25°C

Preparation of Medium: Add components to distilled/deionized water and bring volume to 1.0L. Mix thoroughly. Gently heat and bring to boiling. Adjust pH to 7.0. Distribute into tubes or flasks. Autoclave for 15 min at 15 psi pressure–121°C. Pour into sterile Petri dishes or leave in tubes.

Use: For the cultivation of *Vibrio natriegens.*

Vibrio parahaemolyticus Agar (VP Agar)

Composition per liter:

Agar	20.0g
NaCl	20.0g
Sucrose	20.0g
Sodium citrate	10.0g
$Na_2S_2O_3 \cdot 5H_2O$	10.0g
Peptone	10.0g
Sodium taurocholate	5.0g
Yeast extract	5.0g
Sodium lauryl sulfate	0.2g
Bromthymol Blue	0.04g
Thymol Blue	0.04g

pH 8.6 ± 0.2 at 25°C

Preparation of Medium: Add components to distilled/deionized water and bring volume to 1.0L. Mix thoroughly. Gently heat and bring to boiling. Do not autoclave. Pour into sterile Petri dishes.

Use: For the isolation, cultivation, enumeration, and presumptive identification of coliforms in milk, food, and other specimens of sanitary significance. For the enumeration of bacteria in cheese, especially *Pseudomonas fragi, Pseudomonas viscosa,* and *Alcaligenes metalcaligenes.* Sucrose-fermenting bacteria appear as yellow colonies with pale yellow peripheries. Sucrose-nonfermenting bacteria appear as mucoid, green colonies with a dark green center.

Vibrio parahaemolyticus Sucrose Agar (VPSA)

Composition per liter:

NaCl	30.0g
Agar	15.0g
Sucrose	10.0g
Yeast extract	7.0g
Tryptose	5.0g
Pancreatic digest of casein	5.0g
Bile salts No. 3	1.5g
Bromthymol Blue	0.025g

pH 8.6 ± 0.2 at 25°C

Preparation of Medium: Add components to distilled/deionized water and bring volume to 1.0L. Mix thoroughly. Gently heat and bring to boiling. Do not autoclave. Cool to 50°C. Pour into sterile Petri dishes in 20.0mL volumes. Allow plates to dry before using.

Use: For the isolation, cultivation, and differentiation of *Vibrio parahaemolyticus* from seafood. *Vibrio parahaemolyticus* and *Vibrio vulnificus* appear as blue to green colonies. Other *Vibrio* species appear as yellow colonies.

Vibrio parahaemolyticus Sucrose HiVeg Agar

Composition per liter:

NaCl	30.0g
Agar	15.0g
Sucrose	10.0g
Yeast extract	7.0g
Plant hydrolysate	5.0g
Plant hydrolysate No. 1	5.0g
Synthetic detergent No. I	1.5g
Bromthymol Blue	0.025g

pH 8.6 ± 0.2 at 25°C

Source: This medium is available as a premixed powder from HiMedia.

Preparation of Medium: Add components to distilled/deionized water and bring volume to 1.0L. Mix thoroughly. Gently heat and bring to boiling. Do not autoclave. Cool to 50°C. Pour into sterile Petri dishes in 20.0mL volumes. Allow plates to dry before using.

Use: For the isolation, cultivation, and differentiation of *Vibrio parahaemolyticus* from seafood. *Vibrio parahaemolyticus* and *Vibrio vulnificus* appear as blue to green colonies. Other *Vibrio* species appear as yellow colonies.

Vibrio vallismortis Medium

Composition per liter:

NaCl	25.0g
$MgSO_4 \cdot 7H_2O$	9.6g
$MgCl_2 \cdot 6H_2O$	7.0g
Glucose	5.0g
KCl	3.8g
Yeast extract	1.0g
$CaCl_2 \cdot 2H_2O$	0.5g
$K_2HPO_4 \cdot 3H_2O$	0.4g
$NaHCO_3$ solution	20.0mL

pH 7.0 ± 0.2 at 25°C

$NaHCO_3$ Solution:

Composition per 20.0mL:

$NaHCO_3$	3.0g

Preparation of $NaHCO_3$ Solution: Add $NaHCO_3$ to distilled/deionized water and bring volume to 20.0mL. Mix thoroughly. Filter sterilize.

Preparation of Medium: Add components, except $NaHCO_3$ solution, to distilled/deionized water and bring volume to 980.0mL. Mix thoroughly. Autoclave for 15 min at 15 psi pressure–121°C. Cool to room temperature. Aseptically add 20.0mL of sterile $NaHCO_3$ solution. Mix thoroughly. Aseptically distribute into sterile tubes or flasks.

Use: For the cultivation of *Vibrio vallismortis.*

Vibrio vulnificus Agar (VVA) (BAM M190)

Composition per liter:

NaCl	30.0g
Agar	25.0g
Peptone	20.0g
Cellobiose solution	100.0mL
Dye solution	10.0mL

pH 8.2 ± 0.2 at 25°C

Dye Solution:

Composition per 100.0mL:

Bromthymol Blue	0.6g
Ethanol, 70%	100.0mL

Preparation of Dye Solution: Add Bromthymol Blue to 100.0mL of 70% ethanol. Mix thoroughly.

Cellobiose Solution:

Composition per 100.0mL:

Cellobiose	10.0g

Preparation of Cellobiose Solution: Add cellobiose to distilled/deionized water and bring volume to 100.0mL. Mix thoroughly. Gently heat while mixing to dissolve the cellobiose. Cool. Filter sterilize.

Preparation of Medium: Add components, except cellobiose solution, to distilled/deionized water and bring volume to 900.0mL. Mix thoroughly. Gently heat until dissolved. Adjust pH to 8.2. Autoclave for 15 min at 15 psi pressure–121°C. Cool to 50°C. Aseptically add 100.0mL sterile cellobiose solution. Mix thoroughly and pour into sterile Petri dishes. Final color of medium should be light blue.

Use: For the detection of *Vibrio vulnificus* from seafoods.

Violet Peptone Bile Lactose Broth

Composition per liter:

Lactose	10.0g
Peptone	10.0g
Bile salts	5.0g
Gentian Violet	0.04g

pH 7.6 ± 0.2 at 25°C

Preparation of Medium: Add components to distilled/deionized water and bring volume to 1.0L. Mix thoroughly. Gently heat and bring to boiling. Distribute into tubes or flasks. Autoclave for 15 min at 15 psi pressure–121°C. Pour into sterile Petri dishes or leave in tubes.

Use: For the selective cultivation of members of the Enterobacteriaceae.

Violet Red Bile Agar

Composition per liter:

Agar	15.0g
Lactose	10.0g
Glucose	10.0g
Pancreatic digest of gelatin	7.0g
NaCl	5.0g
Yeast extract	3.0g
Bile salts	1.5g
Neutral Red	0.03g
Crystal Violet	2.0mg

pH 7.4 ± 0.2 at 25°C

Preparation of Medium: Add components to distilled/deionized water and bring volume to 1.0L. Mix thoroughly. Gently heat while stirring and bring to boiling. Distribute into tubes or flasks. Autoclave for 15 min at 15 psi pressure–121°C. Pour immediately into sterile Petri dishes or leave in tubes.

Use: For the isolation and cultivation of members of the Enterobacteriaceae from brined vegetables. For the enumeration of members of the Enterobacteriaceae from brined vegetables by the pour plate technique.

Violet Red Bile Agar (VRB Agar)

Composition per liter:

Agar	15.0g
Lactose	10.0g
Pancreatic digest of gelatin	7.0g
NaCl	5.0g
Yeast extract	3.0g
Bile salts	1.5g
Neutral Red	0.03g
Crystal Violet	2.0mg

pH 7.4 ± 0.2 at 25°C

Source: This medium is available as a premixed powder from BD Diagnostic Systems and Oxoid Unipath.

Preparation of Medium: Add components to distilled/deionized water and bring volume to 1.0L. Mix thoroughly. Gently heat while stirring and bring to boiling. Distribute into tubes or flasks. Autoclave for 15 min at 15 psi pressure–121°C. Pour immediately into sterile Petri dishes or leave in tubes.

Use: For the detection of coliform bacteria in water and food.

Violet Red Bile Agar, HiVeg

Composition per liter:

Agar	15.0g
Lactose	10.0g
Plant peptone	7.0g
NaCl	5.0g
Yeast extract	3.0g
Synthetic detergent No. I	1.5g
Neutral Red	0.03g
Crystal Violet	2.0mg

pH 7.4 ± 0.2 at 25°C

Source: This medium is available as a premixed powder from HiMedia.

Preparation of Medium: Add components to distilled/deionized water and bring volume to 1.0L. Mix thoroughly. Gently heat and bring to boiling. Distribute into tubes or flasks. Boil to dissolve components completely. Do not autoclave. Cool to 45°C. Pour into sterile Petri dishes or leave in tubes.

Use: For the detection of coliform bacteria in water and food.

Violet Red Bile Agar with MUG

Composition per liter:

Agar	15.0g
Lactose	10.0g
Pancreatic digest of gelatin	7.0g
NaCl	5.0g
Yeast extract	3.0g
Bile salts	1.5g
MUG (4-methylumbelliferyl-β-D-glucuronide)	0.1g
Neutral Red	0.03g
Crystal Violet	2.0mg

pH 7.4 ± 0.2 at 25°C

Source: This medium is available as a premixed powder from BD Diagnostic Systems.

Preparation of Medium: Add components to distilled/deionized water and bring volume to 1.0L. Mix thoroughly. Gently heat while stirring and bring to boiling. Distribute into tubes or flasks. Autoclave for 15 min at 15 psi pressure–121°C. Pour immediately into sterile Petri dishes or leave in tubes.

Use: For the differentiation of *Escherichia coli* from dairy products and other foods based on their ability to produce β-glucuronidase.

Violet Red Bile Glucose Agar

Composition per liter:

Agar	12.0g
Glucose	10.0g
Peptone	7.0g
NaCl	5.0g
Yeast extract	3.0g
Bile salts No. 3	1.5g
Neutral Red	0.03g
Crystal Violet	2.0mg

pH 7.4 ± 0.2 at 25°C

Source: This medium is available as a premixed powder from Oxoid Unipath.

Preparation of Medium: Add components to distilled/deionized water and bring volume to 1.0L. Mix thoroughly. Gently heat and bring to boiling. Do not autoclave. Pour into sterile Petri dishes or distribute into sterile tubes.

Use: For the detection and enumeration of Enterobacteriaceae from foods.

Violet Red Glucose HiVeg Agar with Lactose

Composition per liter:

Agar	15.0g
Lactose monohydrate	9.5g
Glucose monohydrate	9.09g
Plant peptone	7.0g
NaCl	5.0g
Yeast extract	3.0g
Synthetic detergent No. I	1.5g
Neutral Red	0.03g
Crystal Violet	2.0mg

pH 7.4 ± 0.2 at 25°C

Source: This medium is available as a premixed powder from HiMedia.

Preparation of Medium: Add components to distilled/deionized water and bring volume to 1.0L. Mix thoroughly. Gently heat and bring to boiling. Do not autoclave. Pour into sterile Petri dishes or distribute into sterile tubes.

Use: For the detection and enumeration of Enterobacteriaceae from raw foods.

Violet Red Glucose HiVeg Agar without Lactose

Composition per liter:

Agar	12.0g
Glucose	10.0g
Plant peptone	7.0g
NaCl	5.0g
Yeast extract	3.0g
Synthetic detergent No. I	1.5g
Neutral Red	0.03g
Crystal Violet	2.0mg

pH 7.4 ± 0.2 at 25°C

Source: This medium is available as a premixed powder from HiMedia.

Preparation of Medium: Add components to distilled/deionized water and bring volume to 1.0L. Mix thoroughly. Gently heat and bring to boiling. Do not autoclave. Pour into sterile Petri dishes or distribute into sterile tubes.

Use: For the detection and enumeration of Enterobacteriaceae from raw foods.

Violet Red HiVeg Agar

Composition per liter:

Agar	15.0g
Lactose	10.0g
Plant peptone	7.0g
NaCl	5.0g
Yeast extract	3.0g
Synthetic detergent No. I	1.5g
Neutral Red	0.03g
Crystal Violet	2.0mg

pH 7.4 ± 0.2 at 25°C

Source: This medium is available as a premixed powder from HiMedia.

Preparation of Medium: Add components to distilled/deionized water and bring volume to 1.0L. Mix thoroughly. Gently heat and bring to boiling. Do not autoclave. Pour into sterile Petri dishes or distribute into sterile tubes.

Use: For the detection and enumeration of Enterobacteriaceae from foods. Recommended by the ISO Committee for selective isolation and enumeration of coli-aerogenes bacteria in water. For the detection and enumeration of coliforms from water and food.

Violet Red HiVeg Agar (1.2%)

Composition per liter:

Agar	12.0g
Lactose	10.0g
Plant peptone	7.0g
NaCl	5.0g
Yeast extract	3.0g
Synthetic detergent No. I	1.5g
Neutral Red	0.03g
Crystal Violet	2.0mg

pH 7.4 ± 0.2 at 25°C

Source: This medium is available as a premixed powder from HiMedia.

Preparation of Medium: Add components to distilled/deionized water and bring volume to 1.0L. Mix thoroughly. Gently heat and bring to boiling. Do not autoclave. Pour into sterile Petri dishes or distribute into sterile tubes.

Use: For the detection and enumeration of Enterobacteriaceae from foods. Recommended by the ISO Committee for selective isolation and enumeration of coli-aerogenes bacteria in water. For the detection and enumeration of coliforms from water and food.

Violet Red HiVeg Broth

Composition per liter:

Plant peptone	7.0g
NaCl	5.0g
Yeast extract	3.0g
Lactose	1.5g
Synthetic detergent No. I	1.5g
Neutral Red	0.03g
Crystal Violet	2.0mg

pH 7.4 ± 0.2 at 25°C

Source: This medium is available as a premixed powder from HiMedia.

Preparation of Medium: Add components to distilled/deionized water and bring volume to 1.0L. Mix thoroughly. Gently heat and bring to boiling. Do not autoclave.

Use: For the isolation and detection of coliforms from water, milk, and other foods.

Viral Transport Medium (VTM)

Composition per 104.1mL:

Bovine serum albumin	0.5g
Veal infusion broth	100.0mL
Phenol Red	0.4mL
Amphotericin B solution	2.0mL
Gentamicin solution	1.0mL
Vancomycin solution	0.2mL

pH 7.4 ± 0.2 at 25°C

Veal Infusion Broth:

Composition per liter:

Veal, infusion from	500.0g
NaCl	5.0g
Pancreatic digest of casein	5.0g
Peptic digest of animal tissue	5.0g

Preparation of Veal Infusion Broth: Add components to distilled/deionized water and bring volume to 1.0L. Mix thoroughly. Distribute into tubes or flasks. Autoclave for 15 min at 15 psi pressure–121°C. Use freshly prepared solution.

Amphotericin B Solution:

Composition per 10.0mL:

Amphotericin B	2.5g

Preparation of Amphotericin B Solution: Add amphotericin B to distilled/deionized water and bring volume to 10.0mL. Mix thoroughly. Filter sterilize.

Gentamicin Solution:

Composition per 10.0mL:

Gentamicin	0.5g

Preparation of Gentamicin Solution: Add gentamicin to distilled/deionized water and bring volume to 10.0mL. Mix thoroughly. Filter sterilize.

Vancomycin Solution:

Composition per 10.0mL:

Vancomycin	0.5g

Preparation of Vancomycin Solution: Add vancomycin to distilled/deionized water and bring volume to 10.0mL. Mix thoroughly. Filter sterilize.

Preparation of Medium: To 100.0mL of sterile veal infusion broth, aseptically add bovine serum albumin, Phenol Red, amphotericin B solution, gentamicin solution, and vancomycin solution. Mix thoroughly. Dispense 2.0mL of medium into serum vials. Store at 4°C and use for up to 2 months.

Use: For the maintenance and transport of specimens suspected of being virally infected.

Vitamin B_6 Blood Agar (ATCC Medium 860)

Composition per liter:

Agar	15.0g
Pancreatic digest of casein	15.0g
Papaic digest of soybean meal	5.0g
NaCl	5.0g
Sheep blood, defibrinated	50.0mL

pH 7.3 ± 0.2 at 25°C

Preparation of Medium: Add components, except sheep blood, to distilled/deionized water and bring volume to 1.0L. Mix thoroughly. Gently heat and bring to boiling. Autoclave for 15 min at 15 psi pressure–121°C. Cool to 45°–50°C. Aseptically add sterile, defibrinated sheep blood. Pour into sterile Petri dishes or distribute into sterile tubes.

Use: For the cultivation and maintenance of fastidious microorganisms, especially *Streptococcus* species.

Vitamin B_6 Blood Agar with Pyridoxal-HCl (ATCC Medium 1511)

Composition per liter:

Agar	15.0g
Pancreatic digest of casein	15.0g
Papaic digest of soybean meal	5.0g
NaCl	5.0g
Pyridoxal·HCl	0.01g
Sheep blood, defibrinated	50.0mL

pH 7.3 ± 0.2 at 25°C

Preparation of Medium: Add components, except sheep blood, to distilled/deionized water and bring volume to 1.0L. Mix thoroughly. Gently heat and bring to boiling. Autoclave for 15 min at 15 psi pressure–121°C. Cool to 45°–50°C. Aseptically add sterile, defibrinated sheep blood. Pour into sterile Petri dishes or distribute into sterile tubes.

Use: For the cultivation and maintenance of fastidious microorganisms, especially *Streptococcus* species.

Vitamin B_{12} Assay Medium

Composition per liter:

Glucose	40.0g
Sodium acetate	20.0g
Vitamin assay casamino acids	12.0g
Sorbitan monooleate complex	2.0g
K_2HPO_4	1.0g
KH_2PO_4	1.0g
$MgSO_4 \cdot 7H_2O$	0.4g
DL-Tryptophan	0.2g
L-Cystine	0.2g
Adenine	0.02g
$FeSO_4$	0.02g
Guanine	0.02g
$MnSO_4 \cdot 5H_2O$	0.02g
NaCl	0.02g
Uracil	0.02g
Pyridoxine·HCl	4.0mg
Niacin	2.0mg
Riboflavin	2.0mg
Thiamine·HCl	2.0mg
Xanthine	1.0mg

Calcium pantothenate 200μg
p-Aminobenzoic acid 200μg
Folic acid 100μg
Biotin 10μg

pH 6.3 ± 0.2 at 25°C

Source: This medium is available as a premixed powder from BD Diagnostic Systems.

Preparation of Medium: Add components to distilled/deionized water and bring volume to 1.0L. Mix thoroughly. Gently heat and bring to boiling. Continue boiling for 2–3 min. Distribute into tubes in 5.0mL volumes. Add standard solution or test solutions to each tube. Adjust the volume of each tube to 10.0mL with distilled/deionized water. Autoclave for 15 min at 15 psi pressure–121°C.

Use: For the microbiological assaying of vitamin B_{12} using *Lactobacillus leichmannii* as the test organism.

Vitamin B_{12} Assay Medium with Colistin

Composition per liter:

Glucose 40.0g
Sodium acetate 20.0g
Vitamin assay casamino acids 12.0g
Sorbitan monooleate complex 2.0g
K_2HPO_4 .... 1.0g
KH_2PO_4 .... 1.0g
$MgSO_4 \cdot 7H_2O$ 0.4g
Colistin sulfate 0.5g
DL-Tryptophan 0.2g
L-Cystine 0.2g
Adenine 0.02g
$FeSO_4$ 0.02g
Guanine 0.02g
$MnSO_4 \cdot 5H_2O$ 0.02g
NaCl 0.02g
Uracil 0.02g
Pyridoxine·HCl 4.0mg
Niacin 2.0mg
Riboflavin 2.0mg
Thiamine·HCl 2.0mg
Xanthine 1.0mg
Calcium DL-pantothenate 200μg
p-Aminobenzoic acid 200μg
Folic acid 100μg
Biotin 10μg
Cyanocobalamin 250.0ng

pH 6.3 ± 0.2 at 25°C

Preparation of Medium: Add components to distilled/deionized water and bring volume to 1.0L. Mix thoroughly. Gently heat and bring to boiling. Continue boiling for 2–3 min. Distribute into tubes in 5.0mL volumes. Add standard solution or test solutions to each tube. Adjust the volume of each tube to 10.0mL with distilled/deionized water. Autoclave for 15 min at 15 psi pressure–121°C.

Use: For the cultivation of *Lactobacillus leichmannii*.

Vitamin B_{12} HiVeg Agar

Composition per liter:

Glucose 20.0g
K_2SO_4 .... 20.0g
Agar 15.0g
Sodium acetate 12.0g
Plant acid hydrolysate, vitamin free 10.0g
Soypeptone, vitamin free 5.0g
Na-thioglycollate 1.7g
K_2HPO_4 .... 1.0g
KH_2PO_4 .... 1.0g
Polysorbate 80 .... 1.0g
Ribonucleic acid 1.0g
$MgSO_4$ 0.4g
L-Cystine 0.2g
DL-Tryptophan 0.2g
NaCl 0.02g
$FeSO_4$ 0.02g
$MnSO_4$ 0.02g
Adenine sulfate 0.0176g
Guanine hydrochloride 0.0124g
Uracil 0.01g
Xanthine (sodium) 0.01g
Pyridoxal-5-phosphate 4.0mg
Pyridoxine hydrochloride 4.0mg
Calcium pantothenate 2.0mg
Niacin 2.0mg
Riboflavin 2.0mg
Thiamine hydrochloride 2.0mg
Folic acid 1.0mg
Biotin 1.0μg

pH 6.3 ± 0.2 at 25°C

Source: This medium is available as a premixed powder from HiMedia.

Preparation of Medium: Add components to distilled/deionized water and bring volume to 1.0L. Mix thoroughly. Gently heat and bring to boiling. Continue boiling for 2–3 min. Distribute into tubes in 5.0mL volumes. Add standard solution or test solutions to each tube. Adjust the volume of each tube to 10.0mL with distilled/deionized water. Autoclave for 15 min at 15 psi pressure–121°C.

Use: For the microbiological assaying of vitamin B_{12} using *Lactobacillus leichmannii* as the test organism.

Vitamin B_{12} Medium

See: **B_{12} Medium**

Vitamin B_{12} Medium

Composition per liter:

Agar 15.0g
Casein hydrolysate 6.0g
K_2HPO_4 .... 0.2g
$MgSO_4 \cdot 7H_2O$ 0.2g
Asparagine 0.15g
Vitamin B_{12} 40.0μg
$FeSO_4 \cdot 7H_2O$ 0.1μg
Glycerol 2.0mL

pH 7.0 ± 0.2 at 25°C

Preparation of Medium: Add components to distilled/deionized water and bring volume to 1.0L. Mix thoroughly. Gently heat and bring to boiling. Distribute into tubes or flasks. Autoclave for 15 min at 15 psi pressure–121°C. Pour into sterile Petri dishes or leave in tubes.

Use: For the cultivation of *Escherichia coli*.

Vitamin B_{12} Nutrient Agar

Composition per liter:

Agar	15.0g
Peptone	5.0g
NaCl	5.0g
Yeast extract	2.0g
Beef extract	1.0g
Vitamin B_{12}	0.4mg

Preparation of Medium: Add components to distilled/deionized water and bring volume to 1.0L. Mix thoroughly. Gently heat and bring to boiling. Distribute into tubes or flasks. Autoclave for 15 min at 15 psi pressure–121°C. Pour into sterile Petri dishes or leave in tubes.

Use: For the cultivation of *Escherichia coli.*

Vitamin Medium for *Microbacterium*

Composition per liter:

Casamino acids	10.0g
Glucose	10.0g
$(NH_4)_2SO_4$	5.0g
KH_2PO_4	5.0g
K_2HPO_4	5.0g
$MgSO_4 \cdot 7H_2O$	0.5g
Vitamin solution	4.0mL

pH 7.0 ± 0.2 at 25°C

Vitamin Solution:

Composition per 100.0mL:

Thiamine	0.05g
Riboflavin	0.05g
Pyridoxine·HCl	0.05g
Calcium pantothenate	0.05g
Nicotinic acid	0.01g
Biotin	0.01g
Folic acid	0.01g
p-Aminobenzoic acid	0.01g

Preparation of Vitamin Solution: Add components to distilled/deionized water and bring volume to 100.0mL. Mix thoroughly. Filter sterilize.

Preparation of Medium: Add components, except vitamin solution, to distilled/deionized water and bring volume to 996.0mL. Mix thoroughly. Autoclave for 10 min at 15 psi pressure–121°C. Cool to 45°–50°C. Aseptically add 4.0mL of sterile vitamin solution. Mix thoroughly. Distribute into sterile tubes or flasks.

Use: For the cultivation and maintenance of *Microbacterium* species.

VL Agar with Blood

Composition per liter:

Agar	20.0g
Tryptone	10.0g
NaCl	5.0g
Yeast extract	5.0g
Beef extract	2.0g
Glucose	2.0g
L-Cysteine·HCl	0.3g
Sheep blood or horse blood	100.0mL

pH 7.4 ± 0.2 at 25°C

Preparation of Medium: Add components, except sheep blood, to distilled/deionized water and bring volume to 900.0mL. Mix thoroughly. Adjust pH to 7.4. Gently heat and bring to boiling. Autoclave for 15 min at 15 psi pressure–121°C. Cool to 50°–55°C. Warm sheep blood to 50°C. Aseptically add 100.0mL of sterile sheep blood or 100.0mL of sterile horse blood. Mix thoroughly. Pour into sterile Petri dishes or distribute into sterile tubes.

Use: For the cultivation and maintenance of *Bacterionema helcogenes, Bacteroides nodosus, Bacteroides pyogenes, Bacteroides salivosus, Bacteroides suis, Bifidobacterium adolescentis, Bifidobacterium bifidum, Bifidobacterium breve, Bifidobacterium longum, Campylobacter coli, Campylobacter concisus, Campylobacter fetus, Campylobacter hyointestinalis, Campylobacter jejuni, Campylobacter lari, Campylobacter mucosalis, Campylobacter* species, *Campylobacter sputorum, Capnocytophaga gingivalis, Capnocytophaga ochracea, Capnocytophaga sputigena, Clostridium colinum, Clostridium difficile, Clostridium* species, *Clostridium spiroforme, Falcivibrio grandis, Falcivibrio vaginalis, Fusobacterium simiae, Gardnerella vaginalis, Leptotrichia buccalis, Pectinatus frisingensis, Peptostreptococcus anaerobius, Peptostreptococcus asaccharolyticus, Peptostreptococcus indolicus, Peptostreptococcus magnus, Peptostreptococcus micros, Peptostreptococcus prevotii, Peptostreptococcus tetradius, Propionibacterium acnes, Propionibacterium avidum, Propionibacterium granulosum, Propionibacterium lymphophilum,* and *Tonsillophilus suis.*

VL Medium

Composition per liter:

Pancreatic digest of casein	10.0g
Agar	6.0g
NaCl	5.0g
Yeast extract	5.0g
Meat extract	2.0g
Glucose	2.0g
L-Cysteine·HCl·H_2O	0.3g
Antibiotic solution	10.0mL

pH 7.4 ± 0.2 at 25°C

Antibiotic Solution:

Composition per 10.0mL:

Kanamycin	0.1g
Vancomycin	7.5mg

Preparation of Antibiotic Solution: Add components to distilled/deionized water and bring volume to 10.0mL. Mix thoroughly. Filter sterilize.

Preparation of Medium: Add components, except antibiotic solution, to distilled/deionized water and bring volume to 990.0mL. Mix thoroughly. Gently heat and bring to boiling. Autoclave for 15 min at 15 psi pressure–121°C. Cool to 45°–50°C. Aseptically add sterile antibiotic solution. Mix thoroughly. Aseptically distribute into sterile tubes or flasks.

Use: For the isolation and cultivation of *Bacteroides* species.

VL Medium

Composition per liter:

Pancreatic digest of casein	10.0g
Agar	6.0g
NaCl	5.0g
Yeast extract	5.0g
Meat extract	2.0g
Glucose	2.0g
L-Cysteine·HCl·H_2O	0.3g

NaN_3 .... 0.05g
Ethyl Violet 0.05g

pH 7.4 ± 0.2 at 25°C

Caution: Sodium azide is toxic. Azides also react with metals and disposal must be highly diluted.

Preparation of Medium: Add components to distilled/deionized water and bring volume to 1.0L. Mix thoroughly. Gently heat and bring to boiling. Distribute into tubes or flasks. Autoclave for 15 min at 15 psi pressure–121°C.

Use: For the isolation and cultivation of *Fusobacterium* species.

VM Medium

Composition per liter:

KOH 4.5g
Beef extract 3.0g
DL-Malic acid 2.5g
Agar 2.0g
NaCl 1.0g
Yeast extract 1.0g
KH_2PO_4 .... 0.6g
NH_4Cl 0.5g
K_2HPO_4 .... 0.4g
$MgSO_4 \cdot 7H_2O$ 0.2g
Ferric EDTA 66.0mg
$CaCl_2$ 20.0mg
$MnSO_4 \cdot H_2O$ 10.0mg
$Na_2MoO_4 \cdot 2H_2O$ 2.0mg
Biotin 0.1mg

pH 6.8 ± 0.2 at 25°C

Preparation of Medium: Add components to distilled/deionized water and bring volume to 1.0L. Mix thoroughly. Gently heat and bring to boiling. Adjust pH to 6.8. Distribute into tubes or flasks. Autoclave for 15 min at 15 psi pressure–121°C.

Use: For the cultivation of unidentified bacteria ATCC 51563 and ATCC 51564.

VM1 Medium (DSMZ Medium 890)

Composition per liter:

NaCl 20.0g
$MgCl_2 \cdot 6H_2O$ 12.6g
Na_2SO_4 .... 3.24g
$CaCl_2 \cdot 2H_2O$ 2.38g
KCl 0.56g
Sulfur, powdered 0.5g
NH_4Cl 0.3g
K_2HPO_4 .... 0.2g
$NaHCO_3$ 0.16g
Trace elements solution 10.0mL

pH 7.0 ± 0.2 at 25°C

Trace Elements Solution:
Composition per liter:

$MgSO_4 \cdot 7H_2O$ 3.0g
Nitrilotriacetic acid 1.5g
NaCl 1.0g
$MnSO_4 \cdot 2H_2O$ 0.5g
$CoSO_4 \cdot 7H_2O$ 0.18g
$ZnSO_4 \cdot 7H_2O$ 0.18g
$CaCl_2 \cdot 2H_2O$ 0.1g
$FeSO_4 \cdot 7H_2O$ 0.1g
$NiCl_2 \cdot 6H_2O$ 0.025g
$KAl(SO_4)_2 \cdot 12H_2O$ 0.02g
H_3BO_3 .... 0.01g
$Na_2MoO_4 \cdot 4H_2O$ 0.01g
$CuSO_4 \cdot 5H_2O$ 0.01g
$Na_2SeO_3 \cdot 5H_2O$ 0.3mg

Preparation of Trace Elements Solution: Add nitrilotriacetic acid to 500.0mL of distilled/deionized water. Dissolve by adjusting pH to 6.5 with KOH. Add remaining components. Add distilled/deionized water to 1.0L. Mix thoroughly.

Preparation of Medium: Add components to distilled/deionized water and bring volume to 1.0L. Mix thoroughly. Adjust pH to 7.0 using H_2SO_4. Fill 20.0mL medium into 100.0mL serum bottles and seal with a rubber stopper. Add atmosphere of 78% H_2 + 20% CO_2 + 2% O_2 with an overpressure. Autoclave for 15 min at 15 psi pressure–121°C.

Use: For the autotrophic cultivation of *Hydrogenothermus marinus*.

VM1 Medium (DSMZ Medium 890)

Composition per liter:

NaCl 20.0g
$MgCl_2 \cdot 6H_2O$ 12.6g
Peptone 5.0g
Na_2SO_4 .... 3.24g
$CaCl_2 \cdot 2H_2O$ 2.38g
Starch 2.0g
Yeast extract 1.0g
KCl 0.56g
Sulfur, powdered 0.5g
NH_4Cl 0.3g
K_2HPO_4 .... 0.2g
$NaHCO_3$ 0.16g
Trace elements solution 10.0mL

pH 7.0 ± 0.2 at 25°C

Trace Elements Solution:
Composition per liter:

$MgSO_4 \cdot 7H_2O$ 3.0g
Nitrilotriacetic acid 1.5g
NaCl 1.0g
$MnSO_4 \cdot 2H_2O$ 0.5g
$CoSO_4 \cdot 7H_2O$ 0.18g
$ZnSO_4 \cdot 7H_2O$ 0.18g
$CaCl_2 \cdot 2H_2O$ 0.1g
$FeSO_4 \cdot 7H_2O$ 0.1g
$NiCl_2 \cdot 6H_2O$ 0.025g
$KAl(SO_4)_2 \cdot 12H_2O$ 0.02g
H_3BO_3 .... 0.01g
$Na_2MoO_4 \cdot 4H_2O$ 0.01g
$CuSO_4 \cdot 5H_2O$ 0.01g
$Na_2SeO_3 \cdot 5H_2O$ 0.3mg

Preparation of Trace Elements Solution: Add nitrilotriacetic acid to 500.0mL of distilled/deionized water. Dissolve by adjusting pH to 6.5 with KOH. Add remaining components. Add distilled/deionized water to 1.0L. Mix thoroughly.

Preparation of Medium: Add components to distilled/deionized water and bring volume to 1.0L. Mix thoroughly. Adjust pH to 7.0 using H_2SO_4. Fill 20.0mL medium into 100.0mL serum bottles and seal with a rubber stopper. Add atmosphere of 80% N_2 + 20% air. Autoclave for 15 min at 15 psi pressure–121°C.

Use: For the heterotrophic cultivation of *Hydrogenothermus marinus.*

VMGII Medium (Viability-Preserving Microbiostatic Medium)

Composition per 1100.0mL:

Solution 1	900.0mL
Solution 2	100.0mL
Salt stock solution	100.0mL

Solution 1:

Composition per 900.0mL:

Noble agar	0.1g

Preparation of Solution 1: Add agar to distilled/deionized water and bring volume to 900.0mL. Mix thoroughly. Gently heat and bring to boiling. Cool to 45°–50°C.

Solution 2:

Composition per 100.0mL:

Charcoal, bacteriological	10.0g
Gelatin peptone	10.0g
Meat peptone	1.0g
Cysteine·HCl	0.5g
Thioglycolic acid	0.5mL

Preparation of Solution 2: Add components to distilled/deionized water and bring volume to 100.0mL. Mix thoroughly.

Stock Salt Solution:

Composition per liter:

Sodium glycerophosphate	100.0g
NaCl	10.0g
KCl	4.2g
$CaCl_2·6H_2O$	2.4g
$MgSO_4·7H_2O$	1.0g
Phenylmercuric acetate	0.03g

Preparation of Stock Salt Solution: Add phenylmercuric acetate to approximately 800.0mL of distilled/deionized water. Gently heat. Add remaining components. Bring volume to 1.0L with distilled/deionized water.

Preparation of Medium: To 900.0mL of cooled solution 1, add 100.0mL of solution 2 and 100.0mL of stock salt solution. Mix thoroughly. Distribute into screw-capped tubes. Autoclave for 15 min at 15 psi pressure–121°C.

Use: For the isolation and cultivation of oral streptococci, including *Streptococcus mutans* and *Streptococcus sanguis*, and nonspore-forming bacteria, including *Lactobacillus* species from human dental plaque.

Vogel-Bonner Minimal Medium

Composition per liter:

K_2HPO_4	10.0g
$NaNH_4HPO_4·H_2O$	3.5g
Citric acetate	2.0g
Glucose	2.0g
$MgSO_4·7H_2O$	200.0mg
Biotin solution	0.1mL

Biotin Solution:

Composition per 100.0mL:

Biotin	2.5g

Preparation of Biotin Solution: Add biotin to 50% ethanol and bring volume to 100.0mL. Mix thoroughly. Filter sterilize. Store at 5°C.

Preparation of Medium: Add components, except biotin solution, to distilled/deionized water and bring volume to 999.9mL. Mix thoroughly. Autoclave for 15 min at 15 psi pressure–121°C. Aseptically add 0.1mL of sterile biotin solution. Mix thoroughly. Aseptically distribute into sterile tubes or flasks.

Use: For the cultivation of *Neurospora* species.

Vogel and Johnson Agar

Composition per liter:

Agar	16.0g
Pancreatic digest of casein	10.0g
D-Mannitol	10.0g
Glycine	10.0g
Yeast extract	5.0g
K_2HPO_4	5.0g
LiCl	5.0g
Phenol Red	0.025g
K_2TeO_3 solution	20.0mL

pH 7.2 ± 0.2 at 25°C

Source: This medium is available as a premixed powder from BD Diagnostic Systems and Oxoid Unipath.

Caution: Lithium chloride is harmful. Avoid bodily contact and inhalation of vapours. On contact with skin wash with plenty of water immediately.

K_2TeO_3 Solution:

Composition per 100.0mL:

K_2TeO_3	1.0g

Preparation of K_2TeO_3 Solution: Add K_2TeO_3 to distilled/deionized water and bring volume to 100.0mL. Mix thoroughly. Filter sterilize.

Caution: Potassium tellurite is toxic.

Preparation of Medium: Add components, except K_2TeO_3 solution, to distilled/deionized water and bring volume to 980.0mL. Mix thoroughly. Gently heat and bring to boiling. Autoclave for 15 min at 15 psi pressure–121°C. Cool to 45°–50°C. Aseptically add 20.0mL of sterile K_2TeO_3 solution. Mix thoroughly. Pour into sterile Petri dishes or distribute into sterile tubes.

Use: For the detection of coagulase-positive *Staphylococcus aureus.*

Vogel-Johnson Agar Base, HiVeg

Composition per liter:

Agar	16.0g
K_2HPO_4	5.0g
Glycine	10.0g
Plant hydrolysate	10.0g
Mannitol	10.0g
LiCl	5.0g
Yeast extract	5.0g
Phenol Red	0.025g
K_2TeO_3 solution	20.0mL

pH 7.2 ± 0.2 at 25°C

Source: This medium, without tellurite, is available as a premixed powder from HiMedia.

Caution: Lithium chloride is harmful. Avoid bodily contact and inhalation of vapours. On contact with skin wash with plenty of water immediately.

K_2TeO_3 Solution:

Composition per 100.0mL:

K_2TeO_3	1.0g

Preparation of K_2TeO_3 Solution: Add K_2TeO_3 to distilled/deionized water and bring volume to 100.0mL. Mix thoroughly. Filter sterilize.

Caution: Potassium tellurite is toxic.

Preparation of Medium: Add components, except K_2TeO_3 solution, to distilled/deionized water and bring volume to 980.0mL. Mix thoroughly. Gently heat and bring to boiling. Autoclave for 15 min at 15 psi pressure–121°C. Cool to 45°–50°C. Aseptically add 20.0mL of sterile K_2TeO_3 solution. Mix thoroughly. Pour into sterile Petri dishes or distribute into sterile tubes.

Use: For the detection of coagulase-positive *Staphylococcus aureus.*

Vogel N Medium
See: **N DeVogel Medium**

Vogel S Medium for Slime-Like *Neurospora*

Yeast extract	0.75g
Pancreatic digest of gelatin	0.5g
Beef extract	0.3g
Vogel N 10X solution	10.0mL
Sorbitol	7.5mL

Vogel N 10X Solution:

Composition per 100.0mL:

Sucrose	15.0g
KH_2PO_4	5.0g
Trisodium phosphate	3.0g
$MgSO_4 \cdot 7H_2O$	0.2g
$CaCl_2 \cdot H_2O$ solution	20.0mL
Biotin solution	5.0mL
Trace elements solution	5.0mL

Preparation of Vogel N Solution: Add components, except biotin solution and trace elements solution, to distilled/deionized water and bring volume to 70.0mL. Mix thoroughly. Autoclave for 15 min at 15 psi pressure–121°C. Aseptically add 5.0 mL of sterile biotin solution and 5.0mL of sterile trace elements solution. Mix thoroughly. Aseptically distribute into sterile tubes or flasks.

$CaCl_2 \cdot H_2O$ Solution:

Composition per 20.0mL:

$CaCl_2 \cdot H_2O$	0.1g

Preparation of $CaCl_2 \cdot H_2O$ Solution: Add $CaCl_2 \cdot H_2O$ to distilled/deionized water and bring volume to 20.0mL. Mix thoroughly.

Biotin Solution:

Composition per 100.0mL:

Biotin	5.0mg

Preparation of Biotin Solution: Add biotin to 50% ethanol and bring volume to 100.0mL. Mix thoroughly. Filter sterilize. Store at 5°C.

Trace Elements Solution:

Composition per 100.0mL:

Citric acid·H_2O	5.0g
$ZnSO_4 \cdot 7H_2O$	5.0g
$Fe(NH_4)_2(SO_4)_2 \cdot 6H_2O$	1.0g
$CuSO_4 \cdot 5H_2O$	0.25g
H_3BO_3, anhydrous	0.05g
$MnSO_4 \cdot H_2O$	0.05g
$Na_2MoO_4 \cdot 2H_2O$	0.05g

Preparation of Trace Elements Solutions: Add components successively to distilled/deionized water and bring volume to 100.0mL. Mix thoroughly after addition of each component. Filter sterilize. Add 2–3mL of chloroform as a preservative. Store at 25°C.

Preparation of Medium: Add components, except Vogel N 10X solution, to distilled/deionized water and bring volume to 990.0mL. Mix thoroughly. Autoclave for 15 min at 15 psi pressure–121°C. Aseptically add 10.0mL of sterile Vogel N 10X solution. Mix thoroughly. Aseptically distribute into sterile tubes or flasks.

Use: For the cultivation of *Neurospora* species.

Voges-Proskauer Medium
See: **VP Medium**

Von Hofsten & Malmquist Medium B

Composition per liter:

Agar	15.0g
$NaNO_3$	2.0g
K_2HPO_4	0.5g
$CaCl_2 \cdot H_2O$	0.02g
$FeSO_4 \cdot 7H_2O$	0.02g
$MgSO_4 \cdot 7H_2O$	0.02g
$MnSO_4 \cdot H_2O$	0.02g

pH 7.5 ± 0.2 at 25°C

Preparation of Medium: Add components to distilled/deionized water and bring volume to 1.0L. Mix thoroughly. Gently heat and bring to boiling. Distribute into tubes or flasks. Autoclave for 15 min at 15 psi pressure–121°C. Pour into sterile Petri dishes or leave in tubes.

Use: For the cultivation of *Alteromonas* species and *Cytophaga saccharophila.*

VP Agar
See: ***Vibrio parahaemolyticus* Agar**

VP Broth, Modified, Smith, Gordon, and Clark

Composition per liter:

Proteose peptone	7.0g
Glucose	5.0g
NaCl	5.0g

Preparation of Medium: Add components to distilled/deionized water and bring volume to 1.0L. Mix thoroughly. Distribute into tubes in 5.0mL volumes. Autoclave for 15 min at 15 psi pressure–121°C.

Use: For the isolation and cultivation of *Bacillus cereus* from foods.

VP HiVeg Medium

Composition per liter:

Agar	20.0g
NaCl	20.0g
Sucrose	20.0g
Plant peptone	10.0g
Sodium citrate	10.0g
$Na_2S_2O_3$	10.0g

Synthetic detergent No. V 5.0g
Yeast extract 5.0g
Sodium lauryl sulfate 0.2g
Bromthymol Blue 0.04g
Thymol Blue 0.04g

pH 6.9 ± 0.2 at 25°C

Source: This medium is available as a premixed powder from HiMedia.

Preparation of Medium: Add components to distilled/deionized water and bring volume to 1.0L. Mix thoroughly. Adjust pH to 6.9. Distribute into tubes in 3.0mL volumes. Autoclave for 15 min at 15 psi pressure–121°C.

Use: For the cultivation and differentiation of bacteria based on their ability to produce acetoin.

VP Medium (Voges-Proskauer Medium)

Composition per liter:

Peptone 7.0g
K_2HPO_4 .... 5.0g
Glucose 5.0g

pH 6.9 ± 0.2 at 25°C

Preparation of Medium: Add components to distilled/deionized water and bring volume to 1.0L. Mix thoroughly. Adjust pH to 6.9. Distribute into tubes in 3.0mL volumes. Autoclave for 15 min at 15 psi pressure–121°C.

Use: For the cultivation and differentiation of bacteria based on their ability to produce acetoin.

VPSA

See: ***Vibrio parahaemolyticus*** **Sucrose Agar**

VRB Agar

See: **Violet Red Bile Agar**

VRB Agar, Fluorocult (Fluorocult VRB Agar)

Composition per liter:

Agar 13.0g
Lactose 10.0g
Peptone from meat 7.0g
NaCl 5.0g
Yeast extract 3.0g
Bile salts mixture 1.5g
4-Methylumbelliferyl-β-D-glucuronide 0.1g
Neutral Red 0.03g
Crystal Violet 0.002g

pH 7.4 ± 0.2 at 25°C

Source: This medium is available from Merck.

Preparation of Medium: Add components to distilled/deionized water and bring volume to 1.0L. Mix thoroughly. Heat in a boiling water bath or in free flowing steam with frequent stirring until completely dissolved. Do not boil for more than 2 min. Do not autoclave. Do not overheat. Pour into sterile Petri dishes. The plates are clear and dark red.

Use: For the detection and enumeration of coliform bacteria, in particular *E. coli*. Crystal Violet and bile salts largely inhibit the growth of Gram-positive accompanying bacterial flora. Lactose-positive colonies show a color change to red of the pH indicator. *E. coli* colonies show a fluorescence under UV light. Lactose-negative Enterobacteriaceae are colorless. Lactose-positive colonies are red and often surrounded by a turbid zone due to the precipitation of bile acids. Light blue fluorescing colonies denote *E. coli.*

VRB MUG Agar (Violet Red Bile Lactose MUG Agar)

Composition per liter:

Agar 13.0g
Lactose 10.0g
Meat peptone 7.0g
NaCl 5.0g
Yeast extract 3.0g
Bile salts mixture 1.5g
4-Methylumbelliferyl-β-D-glucuronide 0.1g
Neutral Red 0.03g
Crystal Violet 0.002g

pH 7.4 ± 0.2 at 37°C

Source: This medium is available from Fluka, Sigma-Aldrich.

Preparation of Medium: Add components to distilled/deionized water and bring volume to 1.0L. Mix thoroughly. Gently heat while stirring and bring to boiling. Autoclave for 15 min at 15 psi pressure–121°C. Cool to 50°C. Pour into sterile Petri dishes.

Use: For the detection and enumeration of coliform bacteria, in particular *E. coli.* Gram-positive accompanying flora are extensively inhibited by Crystal Violet and bile salts. A color change to red indicates lactose-positive colonies, within which *E. coli* can be demonstrated by fluorescence in the UV.

VRE Agar

Composition per 1004.0mL:

Tryptone 20.0g
Agar 10.0g
Yeast extract 5.0g
NaCl 5.0g
Sodium citrate 1.0g
Esculin 1.0g
Ferric ammonium citrate 0.5g
NaN_3 .... 0.15g
Selective supplement solution 4.0mL

pH 7.0 ± 0.2 at 25°C

Source: This medium is available as a premixed powder from Oxoid Unipath.

Selective Supplement Solution:

Composition per 4.0mL:

Meropenum 1.0mg
Vancomycin 6.0mg

Preparation of Selective Supplement Solution: Add components to distilled/deionized water and bring volume to 4.0mL. Mix thoroughly. Filter sterilize.

Preparation of Medium: Add components, except selective supplement solution, to distilled/deionized water and bring volume to 1.0L. Mix thoroughly. Gently heat while stirring and bring to boiling. Autoclave for 15 min at 15 psi pressure–121°C. Cool to 50°C. Aseptially add 4.0mL selective supplement solution. Mix thoroughly . Pour into sterile Petri dishes.

Use: For the isolation of vancomycin resistant enterococci (VRE) from clinical samples. Nonresistant enterococci containing the *Van* C genes will not grow on this medium.The selective supplement suppresses growth of Gram-negative bacteria and *E. gallinarum*. The medium contains an indicator system to detect the growth of esculin-hydrolyzing organisms. Enterococci produce black zones around the colonies from the formation of black iron phenolic compounds derived from esculin-hydrolyis products and ferrous iron.

VRE Agar

Composition per 1004.0mL:

Tryptone	20.0g
Agar	10.0g
Yeast extract	5.0g
NaCl	5.0g
Sodium citrate	1.0g
Esculin	1.0g
Ferric ammonium citrate	0.5g
NaN_3	0.15g
Selective supplement solution	4.0mL

pH 7.0 ± 0.2 at 25°C

Source: This medium is available as a premixed powder from Oxoid Unipath.

Selective Supplement Solution:

Composition per 4.0mL:

Gentamicin	512.0mg

Preparation of Selective Supplement Solution: Add gentamicin to distilled/deionized water and bring volume to 4.0mL. Mix thoroughly. Filter sterilize.

Preparation of Medium: Add components, except selective supplement solution, to distilled/deionized water and bring volume to 1.0L. Mix thoroughly. Gently heat while stirring and bring to boiling. Autoclave for 15 min at 15 psi pressure–121°C. Cool to 50°C. Aseptially add 4.0mL selective supplement solution. Mix thoroughly . Pour into sterile Petri dishes.

Use: For the isolation of high-level aminoglycoside-resistant enterococci (HLARE) from clinical samples. Nonresistant enterococci containing the *Van* C genes will not grow on this medium.The selective supplement suppresses growth of Gram-negative bacteria and *E. gallinarum*. The medium contains an indicator system to detect the growth of esculin-hydrolyzing organisms. Enterococci produce black zones around the colonies from the formation of black iron phenolic compounds derived from esculin hydrolyis products and ferrous iron.

VRE Broth

Composition per 1004.0mL:

Calf brain infusion solids	12.5g
Proteose peptone	10.0g
Beef heart infusion solids	5.0g
NaCl	5.0g
Na_2HPO_4	2.5g
Glucose	2.0g
Selective supplement solution	4.0mL

pH 7.4 ± 0.2 at 25°C

Source: This medium is available as a premixed powder from Oxoid Unipath.

Selective Supplement Solution:

Composition per 4.0mL:

Meropenum	2.0mg

Preparation of Selective Supplement Solution: Add meropenum to distilled/deionized water and bring volume to 4.0mL. Mix thoroughly. Filter sterilize.

Preparation of Medium: Add components, except selective supplement solution, to distilled/deionized water and bring volume to 1.0L. Mix thoroughly. Gently heat while stirring and bring to boiling. Autoclave for 15 min at 15 psi pressure–121°C. Cool to 50°C. Aseptially add 4.0mL selective supplement solution. Mix thoroughly . Aseptically distribute into sterile tubes.

Use: For the isolation of high-level aminoglycoside-resistant enterococci (HLARE) from clinical samples. Nonresistant enterococci will not grow on this medium. The selective supplement suppresses growth of Gram-negative bacteria and *E. gallinarum*.

VTM

See: **Viral Transport Medium**

Vulcanibacillus Medium (DSMZ Medium 1042)

Composition per liter:

Sea salts, Sigma	30.0g
$NaNO_3$	1.0g
Na-pyruvate	1.0g
Resazurin	0.5mg
$VoSO_4 \cdot 2H_2O$	0.05mg
Vitamin solution	20.0mL
Yeast extract solution	10.0mL
Bicarbonate solution	10.0mL
Iron sulfate solution	10.0mL
Wolfe's mineral elixir	2.0mL

pH 6.9 ± 0.2 at 25°C

Yeast Extract Solution:

Composition per 10.0mL:

Yeast extract	0.5g

Preparation of Yeast Extract Solution: Add yeast extract to distilled/deionized water and bring volume to 10.0mL. Mix thoroughly. Autoclave for 15 min at 15 psi pressure–121°C. Cool to room temperature. Sparge with 100% N_2.

Vitamin Solution:

Composition per liter:

Pyridoxine-HCl	10.0mg
Thiamine-HCl·$2H_2O$	5.0mg
Riboflavin	5.0mg
Nicotinic acid	5.0mg
D-Ca-pantothenate	5.0mg
p-Aminobenzoic acid	5.0mg
Lipoic acid	5.0mg
Biotin	2.0mg
Folic acid	2.0mg
Vitamin B_{12}	0.1mg

Preparation of Vitamin Solution: Add components to distilled/deionized water and bring volume to 1.0L. Mix thoroughly. Sparge with 100% N_2. Filter sterilize.

Wolfe's Mineral Elixir:

Composition per liter:

$MgSO_4\cdot 7H_2O$	30.0g
NaCl	10.0g
$MnSO_4\cdot 2H_2O$	5.0g
$(NH_4)_2NiSO_4\cdot 6H_2O$	2.8g
$CoCl_2\cdot 6H_2O$	1.8g
$ZnSO_4\cdot 7H_2O$	1.8g
$FeSO_4\cdot 7H_2O$	1.0g
$CaCl_2\cdot 2H_2O$	1.0g
$KAl(SO_4)_2\cdot 12H_2O$	0.18g
$CuSO_4\cdot 5H_2O$	0.1g
H_3BO_3	0.1g
$Na_2MoO_4\cdot 2H_2O$	0.1g
Na_2SeO_4	0.1g
$Na_2WO_4\cdot 2H_2O$	0.1g

Preparation of Wolfe's Mineral Elixir: Adjust pH of 1.0L of distilled/deionized water to 1.0 with dilute H_2SO_4. Add remaining components one at a time. Mix thoroughly to dissolve.

Bicarbonate Solution:

Composition per 10.0mL:

$NaHCO_3$	1.0g

Preparation of Bicarbonate Solution: Add $NaHCO_3$ to distilled/deionized water and bring volume to 10.0mL. Mix thoroughly. Sparge with a gas mixture of 80% N_2 + 20% CO_2. Filter sterilize.

Iron Sulfate Solution:

Composition per 10.0mL:

$FeSO_4\cdot 7H_2O$	0.1g

Preparation of Iron Sulfate Solution: Add $FeSO_4\cdot 7H_2O$ to 0.1*N* sulfuric acid and bring volume to 10.0mL. Mix thoroughly. Sparge with 100% N_2. Filter sterilize.

Preparation of Medium: Add components, except iron sulfate, bicarbonate, yeast extract, and vitamin solutions, to distilled/deionized water and bring volume to 940.0mL. Mix thoroughly. Adjust pH to 7.0. Gently heat and bring to boiling. Boil for 1 min. Cool to room temperature while sparging with 100% N_2. Dispense into culture vessels under an atmosphere of 100% N_2. Autoclave for 15 min at 15 psi pressure–121°C. Cool to room temperature. Aseptically add iron sulfate, bicarbonate, yeast extract, and vitamin solutions. Mix thoroughly. Adjust pH to 6.8–7.0. Aseptically dispense into culture vessels. It may be necessary to add 10–20mg sodium dithionite per liter (e.g., from a 5% (w/v) solution, freshly prepared under N_2 and filter-sterilized), if the medium is not completely reduced after inoculation.

Use: For the cultivation of *Vulcanibacillus modesticaldus.*

Vulcanibacillus Medium (DSMZ Medium 1042)

Composition per liter:

Sea salts, Sigma	30.0g
$NaNO_3$	1.0g
Na-pyruvate	1.0g
Resazurin	0.5mg
$VoSO_4\cdot 2H_2O$	0.05mg
Vitamin solution	20.0mL
Yeast extract solution	10.0mL
Bicarbonate solution	10.0mL
Iron sulfate solution	10.0mL
Wolfe's mineral elixir	2.0mL

pH 6.9 ± 0.2 at 25°C

Yeast Extract Solution:

Composition per 10.0mL:

Yeast extract	2.0g

Preparation of Yeast Extract Solution: Add yeast extract to distilled/deionized water and bring volume to 10.0mL. Mix thoroughly. Autoclave for 15 min at 15 psi pressure–121°C. Cool to room temperature. Sparge with 100% N_2.

Vitamin Solution:

Composition per liter:

Pyridoxine-HCl	10.0mg
Thiamine-HCl·$2H_2O$	5.0mg
Riboflavin	5.0mg
Nicotinic acid	5.0mg
D-Ca-pantothenate	5.0mg
p-Aminobenzoic acid	5.0mg
Lipoic acid	5.0mg
Biotin	2.0mg
Folic acid	2.0mg
Vitamin B_{12}	0.1mg

Preparation of Vitamin Solution: Add components to distilled/deionized water and bring volume to 1.0L. Mix thoroughly. Sparge with 100% N_2. Filter sterilize.

Wolfe's Mineral Elixir:

Composition per liter:

$MgSO_4\cdot 7H_2O$	30.0g
NaCl	10.0g
$MnSO_4\cdot 2H_2O$	5.0g
$(NH_4)_2NiSO_4\cdot 6H_2O$	2.8g
$CoCl_2\cdot 6H_2O$	1.8g
$ZnSO_4\cdot 7H_2O$	1.8g
$FeSO_4\cdot 7H_2O$	1.0g
$CaCl_2\cdot 2H_2O$	1.0g
$KAl(SO_4)_2\cdot 12H_2O$	0.18g
$CuSO_4\cdot 5H_2O$	0.1g
H_3BO_3	0.1g
$Na_2MoO_4\cdot 2H_2O$	0.1g
Na_2SeO_4	0.1g
$Na_2WO_4\cdot 2H_2O$	0.1g

Preparation of Wolfe's Mineral Elixir: Adjust pH of 1.0L of distilled/deionized water to 1.0 with dilute H_2SO_4. Add remaining components one at a time. Mix thoroughly to dissolve.

Bicarbonate Solution:

Composition per 10.0mL:

$NaHCO_3$	1.0g

Preparation of Bicarbonate Solution: Add $NaHCO_3$ to distilled/deionized water and bring volume to 10.0mL. Mix thoroughly. Sparge with a gas mixture of 80% N_2 + 20% CO_2. Filter sterilize.

Iron Sulfate Solution:

Composition per 10.0mL:

$FeSO_4\cdot 7H_2O$	0.1g

Preparation of Iron Sulfate Solution: Add $FeSO_4\cdot 7H_2O$ to 0.1*N* sulfuric acid and bring volume to 10.0mL. Mix thoroughly. Sparge with 100% N_2. Filter sterilize.

Preparation of Medium: Add components, except iron sulfate, bicarbonate, yeast extract, and vitamin solutions, to distilled/deionized water and bring volume to 940.0mL. Mix thoroughly. Adjust pH to 7.0. Gently heat and bring to boiling. Boil for 1 min. Cool to room temperature while sparging with 100% N_2. Dispense into culture vessels under an atmosphere of 100% N_2. Autoclave for 15 min at 15 psi pressure–121°C. Cool to room temperature. Aseptically add iron sulfate, bicarbonate, yeast extract, and vitamin solutions. Mix thoroughly. Adjust pH to 6.8–7.0. Aseptically dispense into culture vessels. It may be necessary to add 10–20mg sodium dithionite per liter (e.g., from a 5% (w/v) solution, freshly prepared under N_2 and filter-sterilized), if the medium is not completely reduced after inoculation.

Use: For the cultivation of *Clostridiisalibacter paucivoran.*

Vulcanithermus Medium (DSMZ Medium 977)

Composition per liter:

NaCl	25.0g
NH_4Cl	0.33g
KCl	0.33g
Calcium chloride solution	10.0mL
Magnesium chloride solution	10.0mL
Potassium nitrate solution	10.0mL
Tryptone solution	10.0mL
Sucrose solution	10.0mL
Yeast extract solution	10.0mL
PIPES solution	3.6mL
Vitamin solution	1.0mL
Trace elements solution	1.0mL

pH 6.8 ± 0.2 at 25°C

Trace Elements Solution:

Composition per liter:

$MgSO_4·7H_2O$	3.0g
Nitrilotriacetic acid	1.5g
NaCl	1.0g
$MnSO_4·2H_2O$	0.5g
$CoSO_4·7H_2O$	0.18g
$ZnSO_4·7H_2O$	0.18g
$CaCl_2·2H_2O$	0.1g
$FeSO_4·7H_2O$	0.1g
$NiCl_2·6H_2O$	0.025g
$KAl(SO_4)_2·12H_2O$	0.02g
H_3BO_3	0.01g
$Na_2MoO_4·4H_2O$	0.01g
$CuSO_4·5H_2O$	0.01g
$Na_2SeO_3·5H_2O$	0.3mg

Preparation of Trace Elements Solution: Add nitrilotriacetic acid to 500.0mL of distilled/deionized water. Dissolve by adjusting pH to 6.5 with KOH. Add remaining components. Add distilled/deionized water to 1.0L. Mix thoroughly.

Vitamin Solution:

Composition per liter:

Pyridoxine-HCl	10.0mg
Thiamine-HCl·$2H_2O$	5.0mg
Riboflavin	5.0mg
Nicotinic acid	5.0mg
D-Ca-pantothenate	5.0mg
p-Aminobenzoic acid	5.0mg
Lipoic acid	5.0mg
Biotin	2.0mg
Folic acid	2.0mg
Vitamin B_{12}	0.1mg

Preparation of Vitamin Solution: Add components to distilled/deionized water and bring volume to 1.0L. Mix thoroughly. Sparge with 80% H_2 + 20% CO_2. Filter sterilize.

PIPES Solution:

Composition per liter:

PIPES (Piperazine-N,N'-bis[2-ethane-sulfonic acid])	3.62

Preparation of PIPES Solution: Add PIPES to distilled/deionized water and bring volume to 1.0L. Mix thoroughly.

Yeast Extract Solution:

Composition per 10.0mL:

Yeast extract	0.5g

Preparation of Yeast Extract Solution: Add yeast extract to distilled/deionized water and bring volume to 10.0mL. Mix thoroughly. Autoclave for 15 min at 15 psi pressure–121°C. Cool to room temperature. Sparge with 100% N_2.

Sucrose Solution:

Composition per 10.0mL:

Sucrose	1.0g

Preparation of Sucrose Solution: Add sucrose to distilled/deionized water and bring volume to 10.0mL. Mix thoroughly. Autoclave for 15 min at 15 psi pressure–121°C. Cool to room temperature. Sparge with 100% N_2.

Tryptone Solution:

Composition per 10.0mL:

Tryptone	1.0g

Preparation of Tryptone Solution: Add tryptone to distilled/deionized water and bring volume to 10.0mL. Mix thoroughly. Autoclave for 15 min at 15 psi pressure–121°C. Cool to room temperature. Sparge with 100% N_2.

Postassium Nitrate Solution:

Composition per 10.0mL:

KNO_3	0.33g

Preparation of Potassium Nitrate Solution: Add KNO_3 to distilled/deionized water and bring volume to 10.0mL. Mix thoroughly. Autoclave for 15 min at 15 psi pressure–121°C. Cool to room temperature. Sparge with 100% N_2.

Magnesium Chloride Solution:

Composition per 10.0mL:

$MgCl_2·6H_2O$	0.33g

Preparation of Magnesium Chloride Solution: Add0.33g of $MgCl_2·6H_2O$ to distilled/deionized water and bring volume to 10.0mL. Mix thoroughly. Autoclave for 15 min at 15 psi pressure–121°C. Cool to room temperature. Sparge with 100% N_2.

Calcium Chloride Solution:

Composition per 10.0mL:

$CaCl_2·2H_2O$	0.33g

Preparation of Calcium Chloride Solution: Add $CaCl_2·2H_2O$ to distilled/deionized water and bring volume to 10.0mL. Mix thoroughly. Autoclave for 15 min at 15 psi pressure–121°C. Cool to room temperature. Sparge with 100% N_2.

Preparation of Medium: Add components, except calcium chloride, magnesium chloride, potassium nitrate, tryptone, yeast extract, vitamin, and sucrose solutions, to distilled/deionized water and bring volume to 940.0mL. Mix thoroughly. Sparge with 100% N_2. Adjust pH to 6.8. Dispense into vessels suitable for anaerobic growth (Hungate tubes or serum bottles) under an atmosphere of 100% N_2. Autoclave for 15 min at 15 psi pressure–121°C. Cool to room temperature. Add calcium chloride, magnesium chloride, potassium nitrate, tryptone, yeast extract, vitamin, and sucrose solutions. Mix thoroughly. Adjust pH to 6.8. Aseptically dispense into culture vessels.

Use: For the cultivation of *Vulcanithermus mediatlanticus*.

VY Agar

Composition per liter:

Agar	15.0g
Baker's yeast	10.0g
$CaCl_2 \cdot 2H_2O$	1.0g
Cyanocobalamin	5.0mg

pH 7.2 ± 0.2 at 25°C

Preparation of Medium: Add components to distilled/deionized water and bring volume to 1.0L. Mix thoroughly. Gently heat and bring to boiling. Distribute into tubes or flasks. Autoclave for 15 min at 15 psi pressure–121°C. Pour into sterile Petri dishes or leave in tubes.

Use: For the cultivation and maintenance of myxobacteria.

VY2 Agar

Composition per liter:

Agar	15.0g
Baker's yeast	5.0g
$CaCl_2 \cdot 2H_2O$	1.0g
Cyanocobalamin	5.0mg

pH 7.2 ± 0.2 at 25°C

Preparation of Medium: Add components to distilled/deionized water and bring volume to 1.0L. Mix thoroughly. Gently heat and bring to boiling. Distribute into tubes or flasks. Autoclave for 15 min at 15 psi pressure–121°C. Pour into sterile Petri dishes or leave in tubes.

Use: For the cultivation and maintenance of *Myxococcus amylovorans*.

VY5 Agar

Composition per liter:

Agar	15.0g
Baker's yeast	2.0g
$CaCl_2 \cdot 2H_2O$	1.0g
Cyanocobalamin	5.0mg

pH 7.2 ± 0.2 at 25°C

Preparation of Medium: Add components to distilled/deionized water and bring volume to 1.0L. Mix thoroughly. Gently heat and bring to boiling. Distribute into tubes or flasks. Autoclave for 15 min at 15 psi pressure–121°C. Pour into sterile Petri dishes or leave in tubes.

Use: For the cultivation and maintenance of myxobacteria.

VY Medium (Veal Yeast Extract Medium)

Composition per liter:

Veal, solids from infusion	10.0g
Pancreatic digest of casein	5.0g
Peptic digest of animal tissue	5.0g
NaCl	5.0g
Yeast extract	5.0g

Preparation of Medium: Add components to distilled/deionized water and bring volume to 1.0L. Mix thoroughly. Distribute into tubes or flasks. Autoclave for 15 min at 15 psi pressure–121°C.

Use: For the cultivation and maintenance of *Bacillus subtilis*.

VY/4-SWS Agar (DSMZ Medium 958)

Composition per 1001.0mL:

NaCl	20.0g
Agar	15.0g
Yeast cell paste (baker's yeast, washed in deionized water) wet weight	2.5g
Seawater salts solution	1.0L
Vitamin B_{12} solution	1.0mL

pH 7.5 ± 0.2 at 25°C

Seawater Salts Solution:

Composition per liter:

$MgSO_4 \cdot 7H_2O$	8.0g
$CaCl_2 \cdot 2H_2O$	1.0g
KCl	0.5g
$NaHCO_3$	0.16g
KBr	0.08g
$SrCl_2 \cdot 6H_2O$	0.03g
H_3BO_3	0.02g
Ferric citrate	0.01g
di-Na-ß-glycerophosphate	0.01g
Trace elements solution SL-4	1.0mL

Trace Elements Solution SL-4:

Composition per liter:

EDTA	0.5g
$FeSO_4 \cdot 7H_2O$	0.2g
Trace elements solution SL-6	100.0mL

Preparation of Trace Elements Solution SL-4: Add components to distilled/deionized water and bring volume to 1.0L. Mix thoroughly.

Trace Elements Solution SL-6:

Composition per liter:

H_3BO_3	0.3g
$CoCl_2 \cdot 6H_2O$	0.2g
$ZnSO_4 \cdot 7H_2O$	0.1g
$MnCl_2 \cdot 4H_2O$	0.03g
$Na_2MoO_4 \cdot H_2O$	0.03g
$NiCl_2 \cdot 6H_2O$	0.02g
$CuCl_2 \cdot 2H_2O$	0.01g

Preparation of Trace Elements Solution SL-6: Add components to distilled/deionized water and bring volume to 1.0L. Mix thoroughly. Adjust pH to 3.4.

Preparation of Seawater Salts Solution: Add components to distilled/deionized water and bring volume to 1.0L. Mix thoroughly.

Vitamin B_{12} Solution:

Composition per 10.0mL:

Cyanocobalamin	5.0mg

Preparation of Vitamin B_{12} Solution: Add cyanocobalamine to distilled/deionized water and bring volume to 10.0mL. Mix thoroughly. Sparge with 100% N_2. Filter sterilize.

Preparation of Medium: Add NaCl, agar, and yeast cell paste to 1.0L seawater salts solution. Mix thoroughly. Adjust pH to 7.5 with 1*M* NaOH. Gently heat and bring to boiling. Autoclave for 20 min at 15 psi pressure–121°C. Cool to 50°C. Aseptically add 1.0mL sterile vitamin

B_{12} solution. Mix thoroughly. Pour into sterile Petri dishes or leave in tubes.

Use: For the cultivation and maintenance of *Haliangium ochraceum, Haliangium tepidum, Plesiocystis pacifica,* and *Enhygromyxa salina (Thaxtera salina).*

W Medium

Composition per liter:

Sulfur	10.0g
KH_2PO_4	3.0g
$MgSO_4 \cdot 7H_2O$	0.5g
$CaCl_2 \cdot 2H_2O$	0.25g
$(NH_4)_2SO_4$	0.2g
$FeSO_4 \cdot 7H_2O$	10.0mg

pH 3.0 ± 0.2 at 25°C

Preparation of Sulfur: Sterilize by steaming at 100°C for 60 min on 3 consecutive days.

Preparation of Medium: Add components, except sulfur, to distilled/deionized water and bring volume to 990.0mL. Mix thoroughly. Autoclave for 15 min at 15 psi pressure–121°C. Aseptically add 10.0g of sterile sulfur. Mix thoroughly. Aseptically distribute into sterile tubes or flasks.

Use: For the cultivation of *Thiobacillus thiooxidans.*

Wadowsky-Yee Medium
See: **BCYE Selective Agar with PAV**

Wadowsky and Yee Medium, Modified
See: **MWY Medium**

Wagatsuma Agar

Composition per 1050.0mL:

NaCl	70.0g
Agar	15.0g
Mannitol	10.0g
Peptone	10.0g
K_2HPO_4	5.0g
Yeast extract	3.0g
Crystal Violet	1.0mg
Red blood cells	50.0mL

pH 8.0 ± 0.2 at 25°C

Red Blood Cells:

Composition per 100.0mL:

Blood, human or rabbit	100.0mL

Preparation of Red Blood Cells: Mix freshly drawn human or rabbit blood with anticoagulant and an equal volume of sterile 0.85% saline solution. Centrifuge cells at 4000 × g at 4°C for 15 min. Pour off saline and wash two more times with sterile saline. After last wash, pour off saline and resuspend cells to their original volume.

Preparation of Medium: Add components, except blood, to distilled/deionized water and bring volume to 1.0L. Mix thoroughly. Adjust pH to 8.0. Place in a steam bath for 30 min. Do not autoclave. Cool to 45°–50°C. Add 50.0mL of washed red blood cells. Mix thoroughly. Pour into sterile Petri dishes. Dry plates before using.

Use: For the cultivation and detection of thermostable hemolysin of *Vibrio parahaemolyticus* by the Kanagawa reaction.

Wagatsuma HiVeg Agar Base with Red Blood Cells

Composition per 1050.0mL:

NaCl	70.0g
Agar	15.0g
Plant peptone	10.0g
Mannitol	10.0g
K_2HPO_4	5.0g
Yeast extract	3.0g
Crystal Violet	1.0mg
Red blood cells	50.0mL

pH 8.0 ± 0.2 at 25°C

Source: This medium, without red blood cells, is available as a premixed powder from HiMedia.

Red Blood Cells:

Composition per 100.0mL:

Blood, human or rabbit	100.0mL

Preparation of Red Blood Cells: Mix freshly drawn human or rabbit blood with anticoagulant and an equal volume of sterile 0.85% saline solution. Centrifuge cells at 4000 × g at 4°C for 15 min. Pour off saline and wash two more times with sterile saline. After last wash, pour off saline and resuspend cells to their original volume.

Preparation of Medium: Add components, except blood, to distilled/deionized water and bring volume to 1.0L. Mix thoroughly. Adjust pH to 8.0. Place in a steam bath for 30 min. Do not autoclave. Cool to 45°–50°C. Add 50.0mL of washed red blood cells. Mix thoroughly. Pour into sterile Petri dishes. Dry plates before using.

Use: For the cultivation and detection of thermostable hemolysin of *Vibrio parahaemolyticus* by the Kanagawa reaction.

Wakimoto Medium, Modified

Composition per liter:

Agar	15.0g
Sucrose	15.0g
Peptone	5.0g
$Na_2HPO_4 \cdot 12\ H_2O$	2.0g
$Ca(NO_3)_2 \cdot 4H_2O$	0.5g
$FeSO_4 \cdot 7H_2O$	0.5g

Preparation of Medium: Add components to distilled/deionized water and bring volume to 1.0L. Mix thoroughly. Gently heat and bring to boiling. Distribute into tubes or flasks. Autoclave for 15 min at 15 psi pressure–121°C. Pour into sterile Petri dishes or leave in tubes.

Use: For the cultivation and maintenance of *Corynebacterium* species and *Pseudomonas* species.

Waksman's Glucose Agar

Composition per liter:

Agar	12.5g
Glucose	10.0g
Peptone	5.0g
Beef extract	5.0g
NaCl	5.0g

pH 7.4–7.6 at 25°C

Preparation of Medium: Add components to distilled/deionized water and bring volume to 1.0L. Mix thoroughly. Gently heat and bring to boiling. Distribute into tubes or flasks. Autoclave for 15 min at 15 psi pressure–121°C. Pour into sterile Petri dishes or leave in tubes.

Use: For the cultivation and maintenance of *Streptomyces* species.

Waksman's Sulfur Medium

Composition per liter:

KH_2PO_4	3.0g
$MgSO_4 \cdot 7H_2O$	0.5g
$(NH_4)_2SO_4$	0.2g
$CaCl_2 \cdot 2H_2O$	0.2g
$Fe_2(SO_4)_3$	0.1mg

Preparation of Medium: Add components to distilled/deionized water and bring volume to 1.0L. Mix thoroughly. It is not necessary to sterilize this medium. Distribute into sterile tubes or flasks.

Use: For the cultivation of sulfate-reducing microorganisms from soil.

Wall Defective Bacterial Medium

Composition per liter:

Sucrose	100.0g
Papaic digest of soybean meal	20.0g
Agarose	10.0g
Yeast extract	10.0g
NaCl	5.0g
$MgSO_4 \cdot 7H_2O$	2.5g
Horse serum	200.0mL
Cholesterol solution	10.0mL

pH 7.8 ± 0.2 at 25°C

Cholesterol Solution:

Composition per 10.0mL:

Cholesterol	0.04g
Ethanol (95% solution)	10.0mL

Preparation of Cholesterol Solution: Add cholesterol to 10.0mL of 95% ethanol. Mix thoroughly. Filter sterilize.

Preparation of Medium: Add components, except cholesterol solution and horse serum, to distilled/deionized water and bring volume to 790.0mL. Mix thoroughly. Adjust pH to 7.8. Add 10.0mL of cholesterol solution. Mix thoroughly. Gently heat and bring to boiling. Autoclave for 15 min at 15 psi pressure–121°C. Cool to 50°–55°C. Aseptically add 200.0mL of sterile horse serum. Mix thoroughly. Pour into sterile Petri dishes or distribute into sterile tubes.

Use: For the cultivation of cell wall-deficient bacteria, such as L forms, that depend on osmotic stabilization.

Wallenstein Medium

Composition per 4.225L:

Malachite Green	0.75g
Egg yolk emulsion	3.125L
Glycerol	100.0mL

pH 6.75 ± 0.2 at 25°C

Egg Yolk Emulsion:

Composition:

Chicken egg yolks	66
Whole chicken egg	6

Preparation of Egg Yolk Emulsion: Soak eggs with 1:100 dilution of saturated mercuric chloride solution for 1 min. Crack eggs and separate yolks from whites. Mix egg yolks with 6 chicken eggs.

Preparation of Medium: Add components to distilled/deionized water and bring volume to 1.0L. Mix thoroughly. Distribute into tubes. Autoclave for 15 min at 15 psi pressure–121°C.

Use: For the isolation of *Mycobacterium* species other than *Mycobacterium leprae.*

Wallerstein Laboratory Differential Agar

See: **WL Differential Agar**

Wallerstein Laboratory Differential Medium

See: **WL Differential Medium**

Wallerstein Laboratory Medium with Tomato Juice

See: **WL Medium with Tomato Juice**

Wallerstein Laboratory Nutrient Agar

See: **WL Nutrient Agar**

Wallerstein Laboratory Nutrient Broth

See: **WL Nutrient Broth**

Walsby Medium

Composition per liter:

$MgSO_4 \cdot 7H_2O$	0.075g
K_2HPO_4	0.039g
Na_2CO_3	0.02g
$CaCl_2 \cdot 2H_2O$	0.018g
H_3BO_3	2.8mg
$MnSO_4 \cdot 4H_2O$	2.0mg
$ZnSO_4$	0.22mg
MoO_3	0.18mg
$CuSO_4 \cdot 5H_2O$	0.08mg
$Co(NO_3)_2 \cdot 6H_2O$	0.05mg
Iron-EDTA solution	1.0mL

pH 8.5 ± 0.2 at 25°C

Iron-EDTA Solution:

Composition per liter:

EDTA	12.7g
$FeSO_4 \cdot 7H_2O$	4.98g

Preparation of Iron-EDTA Solution: Add components to distilled/deionized water and bring volume to 1.0L. Mix thoroughly.

Preparation of Medium: Add components to distilled/deionized water and bring volume to 1.0L. Mix thoroughly. Distribute into tubes or flasks. Autoclave for 15 min at 15 psi pressure–121°C.

Use: For the isolation and cultivation of planktonic gas-vacuolate cyanobacteria.

Wang's Semisolid HiVeg Medium with Blood

Composition per liter:

Plant hydrolysate	10.0g
Plant peptone	10.0g
NaCl	5.0g
Agar	4.0g
Yeast extract	2.0g
Glucose	1.0g
$NaHSO_3$	0.1g
Sheep blood, defibrinated	100.0mL

pH 7.0 ± 0.2 at 25°C

Source: This medium, without sheep blood, is available as a premixed powder from HiMedia.

Preparation of Medium: Add components, except sheep blood, to distilled/deionized water and bring volume to 900.0mL. Mix thoroughly. Gently heat and bring to boiling. Autoclave for 15 min at 15 psi pressure–121°C. Cool to 45°–50°C. Aseptically add sterile sheep blood. Sheep blood may be replaced by 50.0mL of horse blood and

50.0mL of sterile distilled/deionized water. Mix thoroughly. Aseptically distribute into sterile screw-capped tubes in 4.0mL volumes. Allow tubes to cool in an upright position.

Use: For the cultivation, transport, and maintenance of *Campylobacter* species from foods.

Wang's Transport Storage Medium

Composition per liter:

Pancreatic digest of casein	10.0g
Peptic digest of animal tissue	10.0g
NaCl	5.0g
Agar	4.0g
Yeast extract	2.0g
Glucose	1.0g
$NaHSO_3$	0.1g
Sheep blood, defibrinated	100.0mL

pH 7.0 ± 0.2 at 25°C

Preparation of Medium: Add components, except sheep blood, to distilled/deionized water and bring volume to 900.0mL. Mix thoroughly. Gently heat and bring to boiling. Autoclave for 15 min at 15 psi pressure–121°C. Cool to 45°–50°C. Aseptically add sterile sheep blood. Sheep blood may be replaced by 50.0mL of horse blood and 50.0mL of sterile distilled/deionized water. Mix thoroughly. Aseptically distribute into sterile screw-capped tubes in 4.0mL volumes. Allow tubes to cool in an upright position.

Use: For the cultivation, transport, and maintenance of *Campylobacter* species from foods.

Water Agar

Composition per liter:

Agar	20.0g

Preparation of Medium: Add agar to distilled/deionized water and bring volume to 1.0L. Mix thoroughly. Gently heat and bring to boiling. Distribute into tubes or flasks. Autoclave for 15 min at 15 psi pressure–121°C. Pour into sterile Petri dishes or leave in tubes.

Use: For the cultivation and observation of sporulation of some fungi.

Water Agar

Composition per liter:

Agar	15.0g
$CaCl_2 \cdot 2H_2O$	1.0g

pH 7.2 ± 0.2 at 25°C

Preparation of Medium: Add components to distilled/deionized water and bring volume to 1.0L. Mix thoroughly. Gently heat and bring to boiling. Distribute into tubes or flasks. Autoclave for 15 min at 15 psi pressure–121°C. Pour into sterile Petri dishes or leave in tubes.

Use: For the cultivation of myxobacteria.

Waxy Maize Starch Medium

Composition per liter:

Agar	20.0g
Waxy maize starch	5.0g
Pancreatic digest of casein	5.0g
Yeast extract	5.0g
$CoCl_2 \cdot 6H_2O$	0.1g
$CaCl_2 \cdot 2H_2O$	0.1g
Maltose solution	100.0mL

pH 6.7 ± 0.2 at 25°C

Maltose Solution:

Composition per 100.0mL:

Maltose	10.0g

Preparation of Maltose Solution: Add maltose to distilled/deionized water and bring volume to 100.0mL. Mix thoroughly. Filter sterilize.

Preparation of Medium: Add components, except maltose solution, to distilled/deionized water and bring volume to 900.0mL. Mix thoroughly. Gently heat and bring to boiling. Adjust pH to 6.7. Distribute into tubes or flasks. Autoclave for 15 min at 15 psi pressure–121°C. Aseptically add maltose solution. Pour into sterile Petri dishes or leave in tubes.

Use: For the cultivation and maintenance of *Bacillus* species.

Wayne Sulfatase Agar

See: **Arylsulfatase Agar**

WCX Agar

Composition per liter:

Agar	15.0g
$CaCl_2 \cdot 2H_2O$	1.0g
Cycloheximide solution	100.0mL

pH 7.2 ± 0.2 at 25°C

Cycloheximide Solution

Composition per 100.0mL:

Cycloheximide	2.5mg

Preparation of Cycloheximide Solution: Add cycloheximide to distilled/deionized water and bring volume to 100.0mL. Mix thoroughly. Filter sterilize.

Caution: Cycloheximide is toxic. Avoid skin contact or aerosol formation and inhalation.

Preparation of Medium: Add components, except cycloheximide solution, to distilled/deionized water and bring volume to 900.0mL. Mix thoroughly. Gently heat and bring to boiling. Autoclave for 15 min at 15 psi pressure–121°C. Cool to 45°–50°C. Aseptically add sterile cycloheximide solution. Mix thoroughly. Pour into sterile Petri dishes or distribute into sterile tubes.

Use: For the cultivation of myxobacteria.

Weitzman, Silva-Hutner Agar (WSH Agar) (ATCC Medium 2032)

Composition per liter:

Agar	20.0g
Alphacel	20.0g
Pablum Baby Oatmeal	10.0g
Hunt's Tomato Paste	10.0g
KH_2PO_4	1.5g
$MgSO_4 \cdot 7H_2O$	1.0g
$NaNO_3$	1.0g

Preparation of Medium: Add components to distilled/deionized water and bring volume to 1.0L. Mix thoroughly. Gently heat and bring to boiling. Distribute into tubes or flasks. Autoclave for 15 min at 15 psi pressure–121°C. Pour into sterile Petri dishes or leave in tubes.

Use: For the cultivation and maintenance of *Penicillium* spp. and other fungi.

Wesley Broth

Composition per liter:

Tryptose	20.0g
Bicine (*N,N*-bis-2-[hydroxyethyl]glycine) buffer	10.0g
NaCl	5.0g
Yeast extract	2.5g
Agar	1.0g
$FeSO_4$	0.25g
$Na_2S_2O_5$	0.25g
Sodium pyruvate	0.25g
Antibiotic solution	10.0mL
Alkaline hematin solution	6.25mL

Antibiotic Solution:

Composition per 10.0mL:

Rifampin	0.025g
Cefsulodin	6.25mg
Polymyxin B sulfate	20,000U

Preparation of Antibiotic Solution: Add components to distilled/deionized water and bring volume to 10.0mL. Mix thoroughly. Filter sterilize.

Alkaline Hematin Solution:

Composition per 10.0mL:

Hemin	0.032g
NaOH (0.15*N* solution)	10.0mL

Preparation of Hemin Solution: Add hemin to 10.0mL of NaOH solution. Mix thoroughly. Autoclave for 30 min at 5 psi pressure–108°C. Cool to 25°C.

Preparation of Medium: Add components, except antibiotic solution and alkaline hematin solution, to distilled/deionized water and bring volume to 983.75mL. Mix thoroughly. Gently heat and bring to boiling. Autoclave for 15 min at 15 psi pressure–121°C. Cool to 45°–50°C. Aseptically add 10.0mL of sterile antibiotic solution and 6.25mL of sterile alkaline hematin solution. Mix thoroughly. Aseptically distribute into sterile tubes. Use medium immediately or store overnight at 4°C.

Use: For the enrichment of *Campylobacter* species from foods.

Wesley Broth Base with Antibiotics and Hematin

Composition per liter:

Tryptose	20.0g
Bicine	10.0g
NaCl	5.0g
Yeast extract	2.5g
Agar	1.0g
$FeSO_4$	0.25g
$Na_2S_2O_3$	0.25g
Sodium pyruvate	0.25g
Antibiotic solution	10.0mL
Alkaline hematin solution	6.25mL

pH 8.0 ± 0.2 at 25°C

Source: This medium, without antibiotic and alkaline hematin solutions, is available as a premixed powder from HiMedia.

Antibiotic Solution:

Composition per 10.0mL:

Rifampin	0.025g
Cefsulodin	6.25mg
Polymyxin B sulfate	20,000U

Preparation of Antibiotic Solution: Add components to distilled/deionized water and bring volume to 10.0mL. Mix thoroughly. Filter sterilize.

Alkaline Hematin Solution:

Composition per 10.0mL:

Hemin	0.032g
NaOH (0.15*N* solution)	10.0mL

Preparation of Hemin Solution: Add hemin to 10.0mL of NaOH solution. Mix thoroughly. Autoclave for 30 min at 5 psi pressure–108°C. Cool to 25°C.

Preparation of Medium: Add components, except antibiotic solution and alkaline hematin solution, to distilled/deionized water and bring volume to 983.75mL. Mix thoroughly. Gently heat and bring to boiling. Autoclave for 15 min at 15 psi pressure–121°C. Cool to 45°–50°C. Aseptically add 10.0mL of sterile antibiotic solution and 6.25mL of sterile alkaline hematin solution. Mix thoroughly. Aseptically distribute into sterile tubes. Use medium immediately or store overnight at 4°C.

Use: For the enrichment of *Campylobacter* species from foods.

Wesley HiVeg Broth Base with Antibiotics and Hematin

Composition per liter:

Plant hydrolysate No. 1	20.0g
Bicine	10.0g
NaCl	5.0g
Yeast extract	2.5g
Agar	1.0g
$FeSO_4$	0.25g
$Na_2S_2O_3$	0.25g
Sodium pyruvate	0.25g
Antibiotic solution	10.0mL
Alkaline hematin solution	6.25mL

pH 8.0 ± 0.2 at 25°C

Source: This medium, without antibiotic and alkaline hematin solutions, is available as a premixed powder from HiMedia.

Antibiotic Solution:

Composition per 10.0mL:

Rifampin	0.025g
Cefsulodin	6.25mg
Polymyxin B sulfate	20,000U

Preparation of Antibiotic Solution: Add components to distilled/deionized water and bring volume to 10.0mL. Mix thoroughly. Filter sterilize.

Alkaline Hematin Solution:

Composition per 10.0mL:

Hemin	0.032g
NaOH (0.15*N* solution)	10.0mL

Preparation of Hemin Solution: Add hemin to 10.0mL of NaOH solution. Mix thoroughly. Autoclave for 30 min at 5 psi pressure–108°C. Cool to 25°C.

Preparation of Medium: Add components, except antibiotic solution and alkaline hematin solution, to distilled/deionized water and bring volume to 983.75mL. Mix thoroughly. Gently heat and bring to boiling. Autoclave for 15 min at 15 psi pressure–121°C. Cool to 45°–50°C. Aseptically add 10.0mL of sterile antibiotic solution and 6.25mL of sterile alkaline hematin solution. Mix thoroughly. Aseptically distribute into sterile tubes. Use medium immediately or store overnight at 4°C.

Use: For the enrichment of *Campylobacter* species from foods.

Wheat Peptone Agar

Composition per 750.0mL:

Agar	20.0g
Peptone	20.0g
Yeast extract	10.0g
Glycerol	10.0g
Wheat grain extract	500.0mL

pH 6.8–7.0 at 25°C

Wheat Grain Extract:

Composition per 500.0mL:

Whole wheat	30.0g

Preparation of Wheat Grain Extract: Add whole wheat to tap water and bring volume to 500.0mL. Mix thoroughly. Gently heat and bring to boiling. Continue boiling for 60 min. Filter through several thicknesses of cheesecloth padded with cotton.

Preparation of Medium: Add agar, peptone, yeast extract, and glycerol to whole grain extract filtrate. Mix thoroughly. Gently heat and bring to boiling. Autoclave for 15 min at 15 psi pressure–121°C. Pour into sterile Petri dishes or aseptically distribute into sterile tubes.

Use: For the cultivation and utilization of *Septoria nodorum.*

Whey Agar

Composition per liter:

Whey permeate	50.0g
Casein hydrolysate	20.0g
Agar	15.0g
Yeast extract	10.0g
Tween™ 80	1.0mL

pH 5.6 ± 0.2 at 25°C

Preparation of Medium: Add components to distilled/deionized water and bring volume to 1.0L. Mix thoroughly. Adjust pH to 5.6. Gently heat and bring to boiling. Distribute into tubes or flasks. Autoclave for 15 min at 15 psi pressure–121°C. Pour into sterile Petri dishes or leave in tubes.

Use: For the cultivation and maintenance of *Lactobacillus* species.

Whey Agar

Composition per liter:

Whey permeate	50.0g
Casein hydrolysate	20.0g
Agar	15.0g
Yeast extract	10.0g
Tween™ 80	1.0mL

pH 6.5 ± 0.2 at 25°C

Preparation of Medium: Add components to distilled/deionized water and bring volume to 1.0L. Mix thoroughly. Adjust pH to 6.5. Gently heat and bring to boiling. Distribute into tubes or flasks. Autoclave for 15 min at 15 psi pressure–121°C. Pour into sterile Petri dishes or leave in tubes.

Use: For the cultivation and maintenance of *Lactococcus* species.

Whey Broth

Composition per liter:

Whey permeate	50.0g
Casein hydrolysate	20.0g
Yeast extract	10.0g
Tween™ 80	1.0mL

pH 5.6 ± 0.2 at 25°C

Preparation of Medium: Add components to distilled/deionized water and bring volume to 1.0L. Mix thoroughly. Adjust pH to 5.6. Distribute into tubes or flasks. Autoclave for 15 min at 15 psi pressure–121°C.

Use: For the cultivation of *Lactobacillus* species.

Whey Broth

Composition per liter:

Whey permeate	50.0g
Casein hydrolysate	20.0g
Yeast extract	10.0g
Tween™ 80	1.0mL

pH 6.5 ± 0.2 at 25°C

Preparation of Medium: Add components to distilled/deionized water and bring volume to 1.0L. Mix thoroughly. Adjust pH to 6.5. Distribute into tubes or flasks. Autoclave for 15 min at 15 psi pressure–121°C.

Use: For the cultivation of *Lactococcus* species.

Wickerham Broth

Composition per 100.0mL:

Carbohydrate	10.0g
Yeast nitrogen base	100.0mL

Yeast Nitrogen Base, 10X:

Composition per liter:

Glucose	10.0g
KH_2PO_4	1.0g
$MgSO_4 \cdot 7H_2O$	0.5g
NaCl	0.1g
$CaCl_2 \cdot 2H_2O$	0.1g
DL-Methionine	0.02g
DL-Tryptophan	0.02g
L-Histidine·HCl	0.01g
Inositol	2.0mg
H_3BO_3	0.5mg
$ZnSO_4 \cdot 7H_2O$	0.4mg
$MnSO_4 \cdot 4H_2O$	0.4mg
Thiamine·HCl	0.4mg
Pyridoxine	0.4mg
Niacin	0.4mg
Calcium pantothenate	0.4mg
p-Aminobenzoic acid	0.2mg
Riboflavin	0.2mg
$FeCl_3$	0.2mg
$Na_2MoO_4 \cdot 4H_2O$	0.2mg
KI	0.1mg
$CuSO_4 \cdot 5H_2O$	0.04mg

Folic acid......2.0μg
Biotin......2.0μg

pH 4.5 ± 0.2 at 25°C

Preparation of Yeast Nitrogen Base: Add components to distilled/deionized water and bring volume to 1.0L. Mix thoroughly.

Preparation of Medium: To 100.0mL of yeast nitrogen base, add 10.0g of carbohydrate. Mix thoroughly. Filter sterilize. Aseptically distribute 0.5mL into tubes containing 4.5mL of sterile, distilled/deionized water.

Use: For the cultivation and differentiation of bacteria based on carbohydrate assimilation.

Wickerham Broth

Composition per 100.0mL:

KNO_3	0.78g
Yeast carbon base	100.0mL

pH 4.5 ± 0.2 at 25°C

Yeast Carbon Base:
Composition per liter:

Glucose	10.0g
KH_2PO_4	1.0g
$MgSO_4 \cdot 7H_2O$	0.5g
NaCl	0.1g
$CaCl_2 \cdot 2H_2O$	0.1g
DL-Methionine	0.02g
DL-Tryptophan	0.02g
L-Histidine·HCl	0.01g
Inositol	2.0mg
H_3BO_3	0.5mg
$ZnSO_4 \cdot 7H_2O$	0.4mg
$MnSO_4 \cdot 4H_2O$	0.4mg
Thiamine·HCl	0.4mg
Pyridoxine	0.4mg
Niacin	0.4mg
Calcium pantothenate	0.4mg
p-Aminobenzoic acid	0.2mg
Riboflavin	0.2mg
$FeCl_3$	0.2mg
$Na_2MoO_4 \cdot 4H_2O$	0.2mg
KI	0.1mg
$CuSO_4 \cdot 5H_2O$	0.04mg
Folic acid	2.0μg
Biotin	2.0μg

Preparation of Yeast Carbon Base: Add components to distilled/deionized water and bring volume to 1.0L. Mix thoroughly.

Preparation of Medium: To 100.0mL of yeast carbon base, add 0.78g of KNO_3 (or peptone). Mix thoroughly. Filter sterilize. Aseptically distribute 0.5mL into tubes containing 4.5mL of sterile distilled/deionized water.

Use: For the cultivation and differentiation of bacteria based on nitrate assimilation.

Wickerham Broth with Raffinose

Composition per 100.0mL:

Raffinose	20.0g
Yeast nitrogen base	100.0mL

Yeast Nitrogen Base, 10X:
Composition per liter:

Glucose	10.0g
KH_2PO_4	1.0g
$MgSO_4 \cdot 7H_2O$	0.5g
NaCl	0.1g
$CaCl_2 \cdot 2H_2O$	0.1g
DL-Methionine	0.02g
DL-Tryptophan	0.02g
L-Histidine·HCl	0.01g
Inositol	2.0mg
H_3BO_3	0.5mg
$ZnSO_4 \cdot 7H_2O$	0.4mg
$MnSO_4 \cdot 4H_2O$	0.4mg
Thiamine·HCl	0.4mg
Pyridoxine	0.4mg
Niacin	0.4mg
Calcium pantothenate	0.4mg
p-Aminobenzoic acid	0.2mg
Riboflavin	0.2mg
$FeCl_3$	0.2mg
$Na_2MoO_4 \cdot 4H_2O$	0.2mg
KI	0.1mg
$CuSO_4 \cdot 5H_2O$	0.04mg
Folic acid	2.0μg
Biotin	2.0μg

pH 4.5 ± 0.2 at 25°C

Preparation of Yeast Nitrogen Base: Add components to distilled/deionized water and bring volume to 1.0L. Mix thoroughly.

Preparation of Medium: To 100.0mL of yeast nitrogen base, add 20.0g of raffinose. Mix thoroughly. Filter sterilize. Aseptically distribute 0.5mL into tubes containing 4.5mL of sterile distilled/deionized water.

Use: For the cultivation and differentiation of bacteria based on carbohydrate assimilation.

Wickerham Carbon Base Broth

See: **Yeast Carbon Base, 10X**

Wilbrinck Agar for *Xanthomonas*

Composition per liter:

Sucrose	20.0g
Agar	12.0g
Peptone	5.0g
K_2HPO_4	0.5g
$MgSO_4 \cdot 7H_2O$	0.25g

pH 7.2 ± 0.2 at 25°C

Preparation of Medium: Add components to distilled/deionized water and bring volume to 1.0L. Mix thoroughly. Adjust pH to 7.2. Gently heat and bring to boiling. Distribute into tubes or flasks. Autoclave for 15 min at 15 psi pressure–121°C. Pour into sterile Petri dishes or leave in tubes.

Use: For the cultivation and maintenance of *Pseudomonas caryophylli, Xanthomonas albilineans,* and *Xanthomonas axonopodis.*

Wilbrinck Agar for *Xanthomonas albilineans*

Composition per liter:

Agar	20.0g
Sucrose	10.0g
Peptone	5.0g

K_2HPO_4 .... 0.5g
$MgSO_4 \cdot 7H_2O$ 0.25g
Na_2SO_3, anhydrous 0.05g
pH 7.2 ± 0.2 at 25°C

Preparation of Medium: Add components to distilled/deionized water and bring volume to 1.0L. Mix thoroughly. Gently heat and bring to boiling. Distribute into tubes or flasks. Autoclave for 15 min at 15 psi pressure–121°C. Pour into sterile Petri dishes or leave in tubes.

Use: For the cultivation and maintenance of *Xanthomonas albilineans* and other *Xanthomonas* species.

Wilkins-Chalgren Agar

Composition per liter:
Agar 15.0g
Gelatin peptone 10.0g
Pancreatic digest of casein 10.0g
NaCl 5.0g
Yeast extract 5.0g
Glucose 1.0g
L-Arginine 1.0g
Sodium pyruvate 1.0g
Hemin 5.0mg
Vitamin K_1 (menadione) 0.5mg
pH 7.1 ± 0.2 at 25°C

Source: This medium is available as a premixed powder from BD Diagnostic Systems.

Preparation of Medium: Add components to distilled/deionized water and bring volume to 1.0L. Mix thoroughly. Gently heat and bring to boiling. Distribute into tubes or flasks. Autoclave for 15 min at 15 psi pressure–121°C. Cool to 50°–55°C. Add antibiotic to be assayed; varying concentrations of antibiotics are used. Mix thoroughly. Pour into sterile Petri dishes or leave in tubes.

Use: For the cultivation and maintenance of anaerobic bacteria. For standardized antimicrobic susceptibility testing to determine the minimum inhibitory concentrations of antimicrobics for anaerobic bacteria.

Wilkins-Chalgren Anaerobe Agar

Composition per liter:
Agar 10.0g
Pancreatic digest of casein 10.0g
Gelatin peptone 10.0g
NaCl 5.0g
Yeast extract 5.0g
Glucose 1.0g
L-Arginine 1.0g
Sodium pyruvate 1.0g
Hemin 5.0mg
Menadione 0.5mg
Defibrinated blood 50.0mL
Tween™ 80 .... 1.0mL
pH 7.1 ± 0.2 at 25°C

Source: This medium is available as a premixed powder from Oxoid Unipath.

Preparation of Medium: Add components, except defibrinated blood, to distilled/deionized water and bring volume to 950.0mL. Mix thoroughly. Gently heat and bring to boiling. Distribute into tubes or flasks. Autoclave for 15 min at 15 psi pressure–121°C. Cool to 50°–55°C. Aseptically add 50.0mL of defibrinated blood. Mix thoroughly. Pour into sterile Petri dishes or leave in tubes.

Use: For the cultivation of nonsporulating anaerobes.

Wilkins-Chalgren Anaerobe Agar with Cysteine

Composition per liter:
Agar 10.0g
Pancreatic digest of casein 10.0g
Gelatin peptone 10.0g
NaCl 5.0g
Yeast extract 5.0g
Glucose 1.0g
L-Arginine 1.0g
Sodium pyruvate 1.0g
L-Cysteine·HCl 0.3g
Hemin 5.0mg
Menadione 0.5mg
Defibrinated blood 50.0mL
Tween™ 80 .... 1.0mL
pH 7.1 ± 0.2 at 25°C

Preparation of Medium: Add components, except defibrinated blood, to distilled/deionized water and bring volume to 950.0mL. Mix thoroughly. Gently heat and bring to boiling. Distribute into tubes or flasks. Autoclave for 15 min at 15 psi pressure–121°C. Cool to 50°–55°C. Aseptically add 50.0mL of defibrinated blood. Mix thoroughly. Pour into sterile Petri dishes or leave in tubes.

Use: For the cultivation of *Agromonas oligotrophica, Falcivibrio grandis, and Falcivibrio vaginalis*.

Wilkins-Chalgren Anaerobe Agar with GN Supplement

Composition per liter:
Agar 10.0g
Pancreatic digest of casein 10.0g
Gelatin peptone 10.0g
NaCl 5.0g
Yeast extract 5.0g
Glucose 1.0g
L-Arginine 1.0g
Sodium pyruvate 1.0g
Hemin 5.0mg
Menadione 0.5mg
Defibrinated blood 50.0mL
GN anaerobe selective supplement 20.0mL
pH 7.1 ± 0.2 at 25°C

Source: This medium is available as a premixed powder from Oxoid Unipath.

GN Anaerobe Selective Supplement
Composition per 20.0mL:
Nalidixic acid 10.0mg
Hemin 5.0mg
Sodium succinate 2.5mg
Vancomycin 2.5mg
Menadione 0.5mg

Preparation of GN Anaerobe Selective Supplement: Add components to distilled/deionized water and bring volume to 20.0mL. Mix thoroughly. Filter sterilize.

Preparation of Medium: Add components, except defibrinated blood and GN anaerobe selective supplement, to distilled/deionized water and bring volume to 900.0mL. Mix thoroughly. Gently heat and bring to boiling. Distribute into tubes or flasks. Autoclave for 15 min at 15 psi pressure–121°C. Cool to 50°–55°C. Aseptically add 20.0mL of GN anaerobe selective supplement and 50.0mL of defibrinated blood. Bring volume to 1.0L with distilled/deionized water. Mix thoroughly. Pour into sterile Petri dishes or leave in tubes.

Use: For the selective isolation of Gram-negative anaerobes.

Wilkins-Chalgren Anaerobe Agar with NS Supplement

Composition per liter:

Agar	10.0g
Pancreatic digest of casein	10.0g
Gelatin peptone	10.0g
NaCl	5.0g
Yeast extract	5.0g
Glucose	1.0g
L-Arginine	1.0g
Sodium pyruvate	1.0g
Hemin	5.0mg
Menadione	0.5mg
Defibrinated blood	50.0mL
NS anaerobe selective supplement	20.0 mL
Tween™ 80	1.0mL

pH 7.1 ± 0.2 at 25°C

Source: This medium is available as a premixed powder from Oxoid Unipath.

NS Anaerobe Selective Supplement:

Composition per 20.0mL:

Sodium pyruvate	1.0g
Nalidixic acid	0.01g
Hemin	5.0mg
Menadione	0.5mg

Preparation of NS Anaerobe Selective Supplement: Add components to distilled/deionized water and bring volume to 20.0mL. Mix thoroughly. Filter sterilize.

Preparation of Medium: Add components, except defibrinated blood and NS anaerobe selective supplement, to distilled/deionized water and bring volume to 900.0mL. Mix thoroughly. Gently heat and bring to boiling. Distribute into tubes or flasks. Autoclave for 15 min at 15 psi pressure–121°C. Cool to 50°–55°C. Aseptically add 20.0mL of NS anaerobe selective supplement and 50.0mL of defibrinated blood. Bring volume to 1.0L with distilled/deionized water. Mix thoroughly. Pour into sterile Petri dishes or leave in tubes.

Use: For the selective isolation of nonsporulating anaerobes.

Wilkins-Chalgren Anaerobe Broth (Anaerobe Broth, MIC)

Composition per liter:

Pancreatic digest of casein	10.0g
Gelatin peptone	10.0g
NaCl	5.0g
Yeast extract	5.0g
Glucose	1.0g
L-Arginine	1.0g
Sodium pyruvate	1.0g
Hemin	5.0mg
Menadione	0.5mg

pH 7.1 ± 0.2 at 25°C

Source: This medium is available as a premixed powder from BD Diagnostic Systems and Oxoid Unipath.

Preparation of Medium: Add components to distilled/deionized water and bring volume to 1.0L. Mix thoroughly. Distribute into tubes or flasks. Autoclave for 15 min at 15 psi pressure–121°C.

Use: For the cultivation and antimicrobial susceptibility (MIC) testing of anaerobic bacteria.

Wilkins-Chalgren Anaerobe Medium with Yeast Extract

Composition per liter:

Yeast extract	20.0g
Pancreatic digest of casein	10.0g
Gelatin peptone	10.0g
NaCl	5.0g
Yeast extract	5.0g
Glucose	1.0g
L-Arginine	1.0g
Sodium pyruvate	1.0g
Hemin	5.0mg
Menadione	0.5mg

pH 7.1 ± 0.2 at 25°C

Preparation of Medium: Add components to distilled/deionized water and bring volume to 1.0L. Mix thoroughly. Distribute into tubes or flasks. Autoclave for 15 min at 15 psi pressure–121°C.

Use: For the cultivation of *Selenomonas acidaminovorans.*

Wilkins-Chalgren Anaerobic HiVeg Agar Base with Blood

Composition per liter:

Agar	10.0g
Plant hydrolysate	10.0g
Plant peptone	10.0g
NaCl	5.0g
Yeast extract	5.0g
L-Arginine	1.0g
Glucose	1.0g
Sodium pyruvate	1.0g
$Fe_4(P_2O_7)_3 \cdot H_2O$	5.0mg
Menadione	0.5mg
Defibrinated blood	50.0mL

pH 7.1 ± 0.2 at 25°C

Source: This medium, without blood, is available as a premixed powder from HiMedia.

Preparation of Medium: Add components, except defibrinated blood, to distilled/deionized water and bring volume to 950.0mL. Mix thoroughly. Gently heat and bring to boiling. Distribute into tubes or flasks. Autoclave for 15 min at 15 psi pressure–121°C. Cool to 50°–55°C. Aseptically add 50.0mL of defibrinated blood. Mix thoroughly. Pour into sterile Petri dishes or leave in tubes.

Use: For the cultivation of nonsporulating anaerobes. For the cultivation and maintenance of anaerobic bacteria. For standardized antimicrobic susceptibility testing to determine the minimum inhibitory concentrations of antimicrobics for anaerobic bacteria.

Wilkins-Chalgren Anaerobic HiVeg Agar Base with Blood and Nonspore Anaerobic Supplement

Composition per liter:

Agar	10.0g
Plant hydrolysate	10.0g
Plant peptone	10.0g
NaCl	5.0g
Yeast extract	5.0g
L-Arginine	1.0g
Glucose	1.0g
Sodium pyruvate	1.0g
$Fe_4(P_2O_7)_3 \cdot H_2O$	5.0mg
Menadione	0.5mg
Defibrinated blood	50.0mL
Nonspore anaerobic supplement	

pH 7.1 ± 0.2 at 25°C

Source: This medium, without blood or nonspore anaerobic supplement, is available as a premixed powder from HiMedia.

Nonspore Anaerobic Supplement:
Composition per 10.0mL:

Sodium pyruvate	1.0g
Nalidixic acid	10.0mg
Ferric pyrophosphate, soluble	5.0mg
Menadione	0.5mg

Preparation of Nonspore Anaerobic Supplement: Add components to distilled/deionized water and bring volume to 10.0mL. Mix thoroughly. Filter sterilize.

Preparation of Medium: Add components, except defibrinated blood and nonspore anaerobic supplement, to distilled/deionized water and bring volume to 950.0mL. Mix thoroughly. Gently heat and bring to boiling. Distribute into tubes or flasks. Autoclave for 15 min at 15 psi pressure–121°C. Cool to 50°–55°C. Aseptically add 50.0mL of defibrinated blood and 10.0mL nonspore anaerobic supplement. Mix thoroughly. Pour into sterile Petri dishes or leave in tubes.

Use: For the cultivation of nonsporulating anaerobes. For the cultivation and maintenance of anaerobic bacteria. For standardized antimicrobic susceptibility testing to determine the minimum inhibitory concentrations of antimicrobics for anaerobic bacteria.

Wilkins-Chalgren Anaerobic HiVeg Broth Base with Blood

Composition per liter:

Plant hydrolysate	10.0g
Plant peptone	10.0g
Yeast extract	5.0g
NaCl	5.0g
Sodium pyruvate	1.0g
Glucose	1.0g
L-Arginine	1.0g
$Fe_4(P_2O_7)_3 \cdot H_2O$	5.0mg
Menadione	0.5mg
Defibrinated blood	50.0mL

pH 7.1 ± 0.2 at 25°C

Source: This medium, without blood, is available as a premixed powder from HiMedia.

Preparation of Medium: Add components, except defibrinated blood, to distilled/deionized water and bring volume to 950.0mL. Mix thoroughly. Gently heat and bring to boiling. Distribute into tubes or flasks. Autoclave for 15 min at 15 psi pressure–121°C. Cool to 50°–55°C. Aseptically add 50.0mL of defibrinated blood. Mix thoroughly. Aseptically distribute into tubes or leave in flasks.

Use: For the cultivation of nonsporulating anaerobes. For the cultivation and maintenance of anaerobic bacteria. For standardized antimicrobic susceptibility testing to determine the minimum inhibitory concentrations of antimicrobics for anaerobic bacteria.

Wilson and Blair's Medium

Composition per 165.0mL:

Nutrient agar solution	100.0mL
Solution B	45.0mL
Solution A	20.0mL

pH 6.8 ± 0.2 at 25°C

Nutrient Agar Solution:
Composition per 100.0mL:

Agar	1.95g
Pancreatic digest of gelatin	0.65g
Beef extract	0.39g

Preparation of Nutrient Agar Solution: Aseptically add components to distilled/deionized water and bring volume to 100.0mL. Mix thoroughly. Autoclave for 15 min at 15 psi pressure–121°C. Cool to 45°–50°C.

Solution A:
Composition per liter:

Na_2HPO_4	100.0g
Na_2SO_3	100.0g
Glucose	50.0g
Bismuth ammonium citrate	30.0g

Preparation of Solution A: Aseptically add 30.0g of bismuth ammonium citrate to 250.0mL of boiling distilled/deionized water. Aseptically add Na_2SO_3 to 500.0mL of boiling distilled/deionized water. Mix the two solutions. Add the Na_2HPO_4 to the boiling mixture. Cool to 45°C. In a separate flask, aseptically add glucose to 250.0mL of sterile distilled/deionized water. Mix thoroughly. Aseptically add the glucose solution to the other cooled solution.

Solution B:
Composition per 225.0mL:

Ferric citrate	2.0g
Brilliant Green	0.55g

Preparation of Solution B: Aseptically add 2.0g of ferric citrate to 200.0mL of sterile distilled/deionized water. Mix thoroughly. Aseptically add Brilliant Green to 25.0mL of sterile distilled/deionized water. Mix thoroughly. Aseptically combine the two solutions.

Preparation of Medium: Aseptically combine 100.0mL of nutrient agar solution, 20.0mL of solution A, and 45.0mL of solution B. Mix thoroughly. Pour into sterile Petri dishes or distribute into sterile tubes.

Use: For the cultivation of *Clostridium welchii.*

Wilson Blair Base

Composition per liter:

Agar	30.0g
Proteose peptone No. 3	10.0g
Glucose	10.0g
Beef extract	5.0g
NaCl	5.0g

Selective reagent....70.0mL
Brilliant Green (1% solution)....4.0mL
pH 7.3 ± 0.2 at 25°C

Selective Reagent:

Composition per 320.2mL:

Solution 1....100.0mL
Solution 2....100.0mL
Solution 3....100.0mL
Solution 4....20.2mL

Preparation of Selective Reagent: Combine 100.0mL of solution 1, 100.0mL of solution 2, 100.0mL of solution 3, and 20.2mL of solution 4. Mix thoroughly. Gently heat to boiling until a slate-grey color develops. Cool to 50°C.

Solution 1:

Composition per 100.0mL:

$NaHSO_3$....40.0g

Preparation of Solution 1: Add $NaHSO_3$ to 100.0mL of distilled/deionized water. Mix thoroughly.

Solution 2:

Composition per 100.0mL:

NaH_2PO_4....21.0g

Preparation of Solution 2: Add NaH_2PO_4 to 100.0mL of distilled/deionized water. Mix thoroughly.

Solution 3:

Composition per 100.0mL:

Bismuth ammonium citrate....12.5g

Preparation of Solution 3: Add bismuth ammonium citrate to 100.0mL of distilled/deionized water. Mix thoroughly.

Solution 4:

Composition per 20.2mL:

$FeSO_4$....0.96g

Preparation of Solution 4: Add $FeSO_4$ to 20.0mL of distilled/deionized water. Add 0.2mL of concentrated HCl. Mix thoroughly.

Preparation of Medium: Add components, except selective reagent and Brilliant Green solution, to distilled/deionized water and bring volume to 976.0mL. Mix thoroughly. Gently heat and bring to boiling. Distribute into tubes or flasks. Autoclave for 15 min at 15 psi pressure–121°C. Cool to 50°C. Aseptically add selective reagent and Brilliant Green solution. Mix thoroughly. Pour into sterile Petri dishes or leave in tubes.

Use: For the isolation and cultivation of *Salmonella*, especially *Salmonella typhi.*

Wilson Blair HiVeg Agar with Brilliant Green and Selective Reagent

Composition per liter:

Agar....20.0g
Plant peptone....10.0g
Bismuth sulfite indicator....8.0g
Glucose....5.0g
Plant extract....5.0g
Na_2HPO_4....4.0g
$FeSO_4$....0.3g
Brilliant Green....0.025g
Selective reagent....70.0mL
pH 7.3 ± 0.2 at 25°C

Source: This medium, without selective reagent, is available as a premixed powder from HiMedia.

Selective Reagent:

Composition per 320.2mL:

Solution 1....100.0mL
Solution 2....100.0mL
Solution 3....100.0mL
Solution 4....20.2mL

Preparation of Selective Reagent: Combine 100.0mL of solution 1, 100.0mL of solution 2, 100.0mL of solution 3, and 20.2mL of solution 4. Mix thoroughly. Gently heat to boiling until a slate-grey color develops. Cool to 50°C.

Solution 1:

Composition per 100.0mL:

$NaHSO_3$....40.0g

Preparation of Solution 1: Add $NaHSO_3$ to 100.0mL of distilled/deionized water. Mix thoroughly.

Solution 2:

Composition per 100.0mL:

NaH_2PO_4....21.0g

Preparation of Solution 2: Add NaH_2PO_4 to 100.0mL of distilled/deionized water. Mix thoroughly.

Solution 3:

Composition per 100.0mL:

Bismuth ammonium citrate....12.5g

Preparation of Solution 3: Add bismuth ammonium citrate to 100.0mL of distilled/deionized water. Mix thoroughly.

Solution 4:

Composition per 20.2mL:

$FeSO_4$....0.96g

Preparation of Solution 4: Add $FeSO_4$ to 20.0mL of distilled/deionized water. Add 0.2mL of concentrated HCl. Mix thoroughly.

Preparation of Medium: Add components, except selective reagent and Brilliant Green solution, to distilled/deionized water and bring volume to 976.0mL. Mix thoroughly. Gently heat and bring to boiling. Distribute into tubes or flasks. Autoclave for 15 min at 15 psi pressure–121°C. Cool to 50°C. Aseptically add selective reagent and Brilliant Green solution. Mix thoroughly. Pour into sterile Petri dishes or leave in tubes.

Use: For the isolation and cultivation of *Salmonella*, especially *Salmonella typhi.*

Wilson Blair HiVeg Agar Base with Selective Reagent and Brilliant Green

Composition per liter:

Agar....30.0g
Glucose....10.0g
Plant special peptone....10.0g
Plant extract....5.0g
NaCl....5.0g
Selective reagent....70.0mL
Brilliant Green (1% solution)....4.0mL
pH 7.3 ± 0.2 at 25°C

Source: This medium, without selective reagent or Brilliant Green, is available as a premixed powder from HiMedia.

Selective Reagent:
Composition per 100.0mL:

Solution 1	100.0mL
Solution 2	100.0mL
Solution 3	100.0mL
Solution 4	20.2mL

Preparation of Selective Reagent: Combine 100.0mL of solution 1, 100.0mL of solution 2, 100.0mL of solution 3, and 20.2mL of solution 4. Mix thoroughly. Gently heat to boiling until a slate-grey color develops. Cool to 50°C.

Solution 1:
Composition per 100.0mL:

$NaHSO_3$	40.0g

Preparation of Solution 1: Add $NaHSO_3$ to 100.0mL of distilled/deionized water. Mix thoroughly.

Solution 2:
Composition per 100.0mL:

NaH_2PO_4	21.0g

Preparation of Solution 2: Add NaH_2PO_4 to 100.0mL of distilled/deionized water. Mix thoroughly.

Solution 3:
Composition per 100.0mL:

Bismuth ammonium citrate	12.5g

Preparation of Solution 3: Add bismuth ammonium citrate to 100.0mL of distilled/deionized water. Mix thoroughly.

Solution 4:
Composition per 20.2mL:

$FeSO_4$	0.96g

Preparation of Solution 4: Add $FeSO_4$ to 20.0mL of distilled/deionized water. Add 0.2mL of concentrated HCl. Mix thoroughly.

Preparation of Medium: Add components, except selective reagent and Brilliant Green solution, to distilled/deionized water and bring volume to 976.0mL. Mix thoroughly. Gently heat and bring to boiling. Distribute into tubes or flasks. Autoclave for 15 min at 15 psi pressure–121°C. Cool to 50°C. Aseptically add selective reagent and Brilliant Green solution. Mix thoroughly. Pour into sterile Petri dishes or leave in tubes.

Use: For the isolation and cultivation of *Salmonella*, especially *Salmonella typhi*.

Winge Agar

Composition per liter:

Glucose	20.0g
Agar	15.0g
Yeast extract	3.0g

pH 7.2 ± 0.2 at 25°C

Preparation of Medium: Add components to distilled/deionized water and bring volume to 1.0L. Mix thoroughly. Adjust pH to 7.2. Gently heat and bring to boiling. Distribute into tubes or flasks. Autoclave for 15 min at 15 psi pressure–121°C. Pour into sterile Petri dishes or leave in tubes.

Use: For the cultivation and maintenance of *Candida albicans, Candida kefyr, Candida tropicalis, Citeromyces matritensis, Cryptococcus albidus, Neurospora crassa, Octosporomyces octosporus, Neurospora sitophila,* and *Saccharomyces* species.

Winge Melibiose Agar

Composition per liter:

Agar	15.0g
Melibiose	10.0g
Yeast extract	3.0g

pH 7.2 ± 0.2 at 25°C

Preparation of Medium: Add components to distilled/deionized water and bring volume to 1.0L. Mix thoroughly. Adjust pH to 7.2. Gently heat and bring to boiling. Distribute into tubes or flasks. Autoclave for 15 min at 15 psi pressure–121°C. Pour into sterile Petri dishes or leave in tubes.

Use: For the cultivation and maintenance of *Saccharomyces cerevisiae.*

Winogradsky's Medium, Modified

Composition per liter:

$CaCO_3$	5.0g
$(NH_4)_2SO_4$	1.0g
K_2HPO_4	1.0g
NaCl	1.0g
$MgSO_4 \cdot 7H_2O$	0.5g
$FeSO_4$	0.4g

Preparation of Medium: Add components to distilled/deionized water and bring volume to 1.0L. Mix thoroughly. Gently heat until dissolved. Do not autoclave. Distribute into tubes or flasks. Swirl flask while dispensing to suspend precipitate.

Use: For the cultivation of nitrifying bacteria.

Winogradsky's N-Free Medium

Composition per liter:

Agar	20.0g
$CaCO_3$	5.0mg
Sugar solution	100.0mL
Concentrated salt solution	5.0mL

pH 7.2 ± 0.2 at 25°C

Sugar Solution:
Composition per 100.0mL:

Sucrose or glucose	10.0g

Preparation of Sugar Solution: Add sucrose or glucose to 100.0mL of distilled/deionized water. Mix thoroughly. Autoclave for 10 min at 10 psi pressure–115°C. Cool to 50°C.

Concentrated Salt Solution:
Composition per liter:

KH_2PO_4	50.0g
$MgSO_4 \cdot 7H_2O$	25.0g
NaCl	25.0g
$FeSO_4 \cdot 7H_2O$	1.0g
$MnSO_4 \cdot 4H_2O$	1.0g
$Na_2MoO_4 \cdot 4H_2O$	1.0g

Preparation of Concentrated Salt Solution: Add components to tap water and bring volume to 1.0L. Mix thoroughly. Filter sterilize.

Preparation of Medium: Add components, except sugar solution, to distilled/deionized water and bring volume to 900.0mL. Mix thoroughly. Gently heat and bring to boiling. Distribute into tubes or flasks. Autoclave for 15 min at 15 psi pressure–121°C. Cool to 50°C. Aseptically add sugar solution. Adjust pH to 7.2. Mix thoroughly. Pour into sterile Petri dishes or leave in tubes.

Use: For the cultivation and maintenance of *Azomonas insignis.*

Winogradsky's Nitrite Medium

Composition per liter:

Agar	15.0g
$NaNO_2$	2.0g
Na_2CO_3, anhydrous	1.0g
K_2HPO_4	0.5g

Preparation of Medium: Add components to distilled/deionized water and bring volume to 1.0L. Mix thoroughly. Gently heat and bring to boiling. Distribute into tubes. Autoclave for 15 min at 15 psi pressure–121°C.

Use: For the selective isolation and cultivation of *Nocardia* species and *Rhodococcus* species.

WL Differential Agar (Wallerstein Laboratory Differential Agar)

Composition per liter:

Glucose	50.0g
Agar	20.0g
Pancreatic digest of casein	5.0g
Yeast extract	4.0g
KH_2PO_4	0.55g
KCl	0.425g
$CaCl_2 \cdot 2H_2O$	0.125g
$MgSO_4 \cdot 7H_2O$	0.125g
Bromcresol Green	0.022g
Actidione® (cycloheximide)	4.0mg
$FeCl_3$	2.5mg
$MnSO_4 \cdot 4H_2O$	2.5mg

pH 5.5 ± 0.2 at 25°C

Caution: Cycloheximide is toxic. Avoid skin contact or aerosol formation and inhalation.

Source: This medium is available as a premixed powder from BD Diagnostic Systems.

Preparation of Medium: Add components to distilled/deionized water and bring volume to 1.0L. Mix thoroughly. Gently heat with mixing and bring to boiling. Distribute into tubes or flasks. Autoclave for 15 min at 15 psi pressure–121°C. Pour into sterile Petri dishes or leave in tubes.

Use: For the differential cultivation of bacteria from industrial fermentation processes. Growth of yeasts and molds is inhibited.

WL Differential HiVeg Agar

Composition per liter:

Glucose	50.0g
Agar	20.0g
Plant hydrolysate	5.0g
Yeast extract	4.0g
KH_2PO_4	0.55g
KCl	0.425g
$MgSO_4$	0.125g
$CaCl_2$	0.125g
Bromcresol Green	0.022g
Cycloheximide	4.0mg
$FeCl_3$	2.5mg
$MnSO_4$	2.5mg

pH 5.5 ± 0.2 at 25°C

Source: This medium is available as a premixed powder from HiMedia.

Caution: Cycloheximide is toxic. Avoid skin contact or aerosol formation and inhalation.

Preparation of Medium: Add components to distilled/deionized water and bring volume to 1.0L. Mix thoroughly. Gently heat with mixing and bring to boiling. Distribute into tubes or flasks. Autoclave for 15 min at 15 psi pressure–121°C. Pour into sterile Petri dishes or leave in tubes.

Use: For the differential cultivation of bacteria from industrial fermentation processes. Growth of yeasts and molds is inhibited. For the selective isolation and enumeration of bacteria encountered in breweries and industrial fermentations.

WL Differential HiVeg Broth

Composition per liter:

Glucose	50.0g
Plant hydrolysate	5.0g
Yeast extract	4.0g
KH_2PO_4	0.55g
KCl	0.425g
$MgSO_4$	0.125g
$CaCl_2$	0.125g
Bromcresol Green	0.022g
Cycloheximide	4.0mg
$FeCl_3$	2.5mg
$MnSO_4$	2.5mg

pH 5.5 ± 0.2 at 25°C

Source: This medium is available as a premixed powder from HiMedia.

Caution: Cycloheximide is toxic. Avoid skin contact or aerosol formation and inhalation.

Preparation of Medium: Add components to distilled/deionized water and bring volume to 1.0L. Mix thoroughly. Gently heat with mixing and bring to boiling. Distribute into tubes or flasks. Autoclave for 15 min at 15 psi pressure–121°C.

Use: For the differential cultivation of bacteria from industrial fermentation processes. Growth of yeasts and molds is inhibited. For the selective isolation and enumeration of bacteria encountered in breweries and industrial fermentations.

WL Differential Medium (Wallerstein Laboratory Differential Medium)

Composition per liter:

Glucose	50.0g
Agar	20.0g
Pancreatic digest of casein	5.0g
Yeast extract	4.0g
KH_2PO_4	0.55g
KCl	0.425g
$CaCl_2 \cdot 2H_2O$	0.125g
$MgSO_4 \cdot 7H_2O$	0.125g
Bromcresol Green	0.022g
Actidione® (cycloheximide)	4.0mg
$FeCl_3$	2.5mg
$MnSO_4 \cdot 4H_2O$	2.5mg

pH 5.5 ± 0.2 at 25°C

Source: This medium is available as a premixed powder from BD Diagnostic Systems.

Caution: Cycloheximide is toxic. Avoid skin contact or aerosol formation and inhalation.

Preparation of Medium: Add components to distilled/deionized water and bring volume to 1.0L. Mix thoroughly. Gently heat with mixing and bring to boiling. Distribute into tubes or flasks. Autoclave for 15 min at 15 psi pressure–121°C. Pour into sterile Petri dishes or leave in tubes.

Use: For the differential cultivation of bacteria from industrial fermentation processes. Growth of yeasts and molds is inhibited.

WL Differential Medium (Wallerstein Laboratory Differential Medium)

Composition per liter:

Glucose	50.0g
Agar	20.0g
Pancreatic digest of casein	5.0g
Yeast extract	4.0g
KH_2PO_4	0.55g
KCl	0.425g
$CaCl_2 \cdot 2H_2O$	0.125g
$MgSO_4 \cdot 7H_2O$	0.125g
Bromcresol Green	0.022g
Actidione® (cycloheximide)	4.0mg
$FeCl_3$	2.5mg
$MnSO_4 \cdot 4H_2O$	2.5mg
Na_2CO_3 (1% solution)	30.0mL

pH 6.5 ± 0.2 at 25°C

Source: This medium is available as a premixed powder from BD Diagnostic Systems.

Caution: Cycloheximide is toxic. Avoid skin contact or aerosol formation and inhalation.

Preparation of Medium: Add components to distilled/deionized water and bring volume to 1.0L. Mix thoroughly. Gently heat with mixing and bring to boiling. Distribute into tubes or flasks. Autoclave for 15 min at 15 psi pressure–121°C. Pour into sterile Petri dishes or leave in tubes.

Use: For the differential cultivation of bacteria from industrial fermentation processes. Growth of yeasts and molds is inhibited.

WL Medium with Tomato Juice (Wallerstein Laboratory Medium with Tomato Juice)

Composition per liter:

Glucose	50.0g
Pancreatic digest of casein	5.0g
Yeast extract	4.0g
KH_2PO_4	0.55g
KCl	0.425g
$CaCl_2$	0.125g
$MgSO_4 \cdot 7H_2O$	0.125g
Bromcresol Green	0.022g
$FeCl_3$	2.5mg
$MnSO_4 \cdot 4H_2O$	2.5mg
Tomato juice, canned, clarified	400.0mL

pH 5.5 ± 0.2 at 25°C

Preparation of Medium: Add components to distilled/deionized water and bring volume to 1.0L. Mix thoroughly. Distribute into tubes or flasks. Autoclave for 15 min at 15 psi pressure–121°C.

Use: For the cultivation of microorganisms from alcoholic mash.

WL Nutrient Agar (Wallerstein Laboratory Nutrient Agar)

Composition per liter:

Glucose	50.0g
Agar	20.0g
Pancreatic digest of casein	5.0g
Yeast extract	4.0g
KH_2PO_4	0.55g
KCl	0.425g
$CaCl_2 \cdot 2H_2O$	0.125g
$MgSO_4 \cdot 7H_2O$	0.125g
Bromcresol Green	0.022g
$FeCl_3$	2.5mg
$MnSO_4 \cdot 4H_2O$	2.5mg

pH 5.5 ± 0.2 at 25°C

Source: This medium is available as a premixed powder from BD Diagnostic Systems and Oxoid Unipath.

Preparation of Medium: Add components to distilled/deionized water and bring volume to 1.0L. Mix thoroughly. Gently heat with mixing and bring to boiling. Distribute into tubes or flasks. Autoclave for 15 min at 15 psi pressure–121°C. Pour into sterile Petri dishes or leave in tubes.

Use: For the detection of bacteria and yeasts in industrial fermentation processes, particularly from beer processing.

WL Nutrient Agar (Wallerstein Laboratory Nutrient Agar)

Composition per liter:

Glucose	50.0g
Agar	20.0g
Pancreatic digest of casein	5.0g
Yeast extract	4.0g
KH_2PO_4	0.55g
KCl	0.425g
$CaCl_2 \cdot 2H_2O$	0.125g
$MgSO_4 \cdot 7H_2O$	0.125g
Bromcresol Green	0.022g
$FeCl_3$	2.5mg
$MnSO_4 \cdot 4H_2O$	2.5mg
Na_2CO_3 (1% solution)	30.0mL

pH 6.5 ± 0.2 at 25°C

Source: This medium is available as a premixed powder from BD Diagnostic Systems and Oxoid Unipath.

Preparation of Medium: Add components to distilled/deionized water and bring volume to 1.0L. Mix thoroughly. Gently heat with mixing and bring to boiling. Distribute into tubes or flasks. Autoclave for 15 min at 15 psi pressure–121°C. Pour into sterile Petri dishes or leave in tubes.

Use: For the detection of bacteria and yeasts from industrial fermentation processes, particularly from beer processing.

WL Nutrient Broth (Wallerstein Laboratory Nutrient Broth)

Composition per liter:

Glucose	50.0g
Pancreatic digest of casein	5.0g
Yeast extract	4.0g
KH_2PO_4	0.55g
KCl	0.425g
$CaCl_2$	0.125g
$MgSO_4 \cdot 7H_2O$	0.125g
Bromcresol Green	0.022g
$FeCl_3$	2.5mg
$MnSO_4 \cdot 4H_2O$	2.5mg

pH 5.5 ± 0.2 at 25°C

Source: This medium is available as a premixed powder from BD Diagnostic Systems and Oxoid Unipath.

Preparation of Medium: Add components to distilled/deionized water and bring volume to 1.0L. Mix thoroughly. Distribute into tubes or flasks. Autoclave for 15 min at 15 psi pressure–121°C.

Use: For the cultivation of yeasts, molds, and bacteria found in brewing and other industrial fermentation processes.

WL Nutrient Broth (Wallerstein Laboratory Nutrient Broth)

Composition per liter:

Glucose	50.0g
Pancreatic digest of casein	5.0g
Yeast extract	4.0g
KH_2PO_4	0.55g
KCl	0.425g
$CaCl_2$	0.125g
$MgSO_4 \cdot 7H_2O$	0.125g
Bromcresol Green	0.022g
$FeCl_3$	2.5mg
$MnSO_4 \cdot 4H_2O$	2.5mg
Na_2CO_3 (1% solution)	30.0mL

pH 6.5 ± 0.2 at 25°C

Source: This medium is available as a premixed powder from BD Diagnostic Systems and Oxoid Unipath.

Preparation of Medium: Add components to distilled/deionized water and bring volume to 1.0L. Mix thoroughly. Distribute into tubes or flasks. Autoclave for 15 min at 15 psi pressure–121°C.

Use: For the cultivation of yeasts, molds, and bacteria found in brewing and other industrial fermentation processes.

WL Nutrient Broth (Wallerstein Laboratory Nutrient Broth)

Composition per liter:

Glucose	50.0g
Pancreatic digest of casein	5.0g
Yeast extract	4.0g
$FeCl_3$	2.5g
$MnSO_4 \cdot 4H_2O$	2.5g
KH_2PO_4	0.55g
KCl	0.425g
$CaCl_2 \cdot 2H_2O$	0.125g
$MgSO_4 \cdot 7H_2O$	0.125g
Bromcresol Green	0.022g

pH 5.5 ± 0.2 at 25°C

Preparation of Medium: Add components to distilled/deionized water and bring volume to 1.0L. Mix thoroughly. Distribute into tubes or flasks. Autoclave for 15 min at 15 psi pressure–121°C.

Use: For the control of brewing and other fermentation processes.

WL Nutrient HiVeg Broth

Composition per liter:

Glucose	50.0g
Plant hydrolysate	5.0g
Yeast extract	4.0g
KH_2PO_4	0.55g
KCl	0.425g
$MgSO_4$	0.125g
$CaCl_2$	0.125g
Bromcresol green	0.022g
$FeCl_3$	2.5mg
$MnSO_4$	2.5mg

pH 5.5 ± 0.2 at 25°C

Source: This medium is available as a premixed powder from HiMedia.

Preparation of Medium: Add components to distilled/deionized water and bring volume to 1.0L. Mix thoroughly. Gently heat with mixing and bring to boiling. Distribute into tubes or flasks. Autoclave for 15 min at 15 psi pressure–121°C.

Use: For the differential cultivation of bacteria from industrial fermentation processes. Growth of yeasts and molds is inhibited. For the selective isolation and enumeration of bacteria encountered in breweries and industrial fermentations.

WL Nutrient HiVeg Medium

Composition per liter:

Glucose	50.0g
Agar	20.0g
Plant hydrolysate	5.0g
Yeast extract	4.0g
KH_2PO_4	0.55g
KCl	0.425g
$CaCl_2$	0.125g
$MgSO_4$	0.125g
Bromcresol green	0.022g
$FeCl_3$	2.5mg
$MnSO_4$	2.5mg

pH 5.5 ± 0.2 at 25°C

Source: This medium is available as a premixed powder from HiMedia.

Preparation of Medium: Add components to distilled/deionized water and bring volume to 1.0L. Mix thoroughly. Gently heat with mixing and bring to boiling. Distribute into tubes or flasks. Autoclave for 15 min at 15 psi pressure–121°C.

Use: For the differential cultivation of bacteria from industrial fermentation processes. Growth of yeasts and molds is inhibited. For the selective isolation and enumeration of bacteria encountered in breweries and industrial fermentations.

WMC Medium

Composition per 1010.0mL:

$NaHCO_3$	5.0g
Sodium acetate	1.0g
L-Cysteine	0.5g

Resazurin	1.0mg
Mineral salt solution 2xW	500.0mL
LIP solution	50.0mL
TYC solution	50.0mL
Trace elements solution	10.0mL
$Na_2S \cdot 9H_2O$ solution	10.0mL

pH 7.2 ± 0.2 at 25°C

Mineral Solution 2xW:

Composition per liter:

NaCl	40.0g
$MgCl_2 \cdot 6H_2O$	5.6g
$MgSO_4 \cdot 7H_2O$	0.7g
KCl	0.68g
NH_4Cl	0.5g
$CaCl_2 \cdot 2H_2O$	0.28g
K_2HPO_4	0.28g

Preparation of Mineral Solution 2xW: Add components to distilled/deionized water and bring volume to 1.0L. Mix thoroughly. Filter sterilize. Store at room temperature in the dark.

LIP Solution:

Composition per liter:

L-Isoleucine	10.0g
L-Leucine	5.0g
Pantothenate	0.1g

Preparation of LIP Solution: Add components to distilled/deionized water and bring volume to 1.0L. Mix thoroughly. May be stored unsterilized at –20°C.

TYC Solution:

Composition per liter:

Casamino acids	100.0g
Yeast extract	50.0g
L-Tryptophan	1.0g

Preparation of TYC Solution: Add components to distilled/deionized water and bring volume to 1.0L. Mix thoroughly. Sparge with 100% N_2. Autoclave for 15 min at 15 psi pressure–121°C.

Trace Elements Solution:

Composition per liter:

$MgSO_4 \cdot 7H_2O$	3.0g
Nitrilotriacetic acid	1.5g
NaCl	1.0g
$MnSO_4 \cdot 2H_2O$	0.5g
$CoSO_4 \cdot 7H_2O$	0.18g
$ZnSO_4 \cdot 7H_2O$	0.18g
$FeSO_4 \cdot 7H_2O$	0.1g
$CaCl_2 \cdot 2H_2O$	0.1g
$KAl(SO_4)_2 \cdot 12H_2O$	0.02g
$CuSO_4 \cdot 5H_2O$	0.01g
H_3BO_3	0.01g
$Na_2MoO_4 \cdot 2H_2O$	0.01g
$NiCl_2 \cdot 6H_2O$	0.025g
$Na_2SeO_3 \cdot 5H_2O$	0.3mg

Preparation of Trace Elements Solution: Add nitrilotriacetic acid to 500.0mL of distilled/deionized water. Adjust pH to 6.5 with KOH. Add remaining components. Add distilled/deionized water to 1.0L.

$Na_2S \cdot 9H_2O$ Solution:

Composition per 10.0mL:

$Na_2S \cdot 9H_2O$	0.5g

Preparation of $Na_2S \cdot 9H_2O$ Solution: Add $Na_2S \cdot 9H_2O$ to distilled/deionized water and bring volume to 10.0mL. Mix thoroughly. Sparge with 100% N_2. Autoclave for 15 min at 15 psi pressure–121°C.

Preparation of Medium: Prepare and dispense medium under 80% H_2 + 20% CO_2. Add components, except TYC solution and $Na_2S \cdot 9H_2O$ solution, to distilled/deionized water and bring volume to 940.0mL. Mix thoroughly. Sparge with 80% H_2 + 20% CO_2. Autoclave for 15 min at 15 psi pressure–121°C. Aseptically and anaerobically add 50.0mL of sterile TYC solution and 10.0mL of sterile $Na_2S \cdot 9H_2O$ solution. Mix thoroughly.

Use: For the cultivation and maintenance of *Methanococcus voltae*.

Wolin-Bevis Medium

Composition per liter:

Agar	20.0g
$(NH_4)_2SO_4$	1.0g
KH_2HPO_4	1.0g
Glucose	0.25g
L-Histidine·HCl	0.25g
Tween™ 80 (polysorbate 80)	3.0mL

pH 5.4 ± 0.2 at 25°C

Preparation of Medium: Add components to distilled/deionized water and bring volume to 1.0L. Mix thoroughly. Gently heat and bring to boiling. Distribute into tubes or flasks. Autoclave for 15 min at 15 psi pressure–121°C. Pour into sterile Petri dishes or leave in tubes.

Use: For the enhanced production of chlamydospores by *Candida albicans*.

Wolinella succinogenes Medium

Composition per liter:

K_2HPO_4	5.0g
Fumaric acid	3.0g
Sodium formate	3.0g
$(NH_4)_2SO_4$	1.0g
Yeast extract	1.0g
$MgCl_2 \cdot 6H_2O$	0.2g
$FeSO_4 \cdot 7H_2O$	0.02g
Resazurin	1.0mg
Sodium thioglycolate solution	10.0mL

pH 7.0 ± 0.2 at 25°C

Sodium Thioglycolate Solution:

Composition per 10.0mL:

Sodium thioglycolate	0.5g

Preparation of Sodium Thioglycolate Solution: Add sodium thioglycolate to distilled/deionized water and bring volume to 10.0mL. Mix thoroughly. Sparge with 100% N_2. Autoclave for 15 min at 15 psi pressure–121°C.

Preparation of Medium: Add components, except sodium thioglycolate solution, to distilled/deionized water and bring volume to 990.0mL. Mix thoroughly. Autoclave for 15 min at 15 psi pressure–121°C. Aseptically add 10.0mL of sterile sodium thioglycolate solution. Mix thoroughly. Aseptically distribute into sterile tubes or flasks.

Use: For the cultivation and maintenance of *Wolinella succinogenes*.

Woods and Welton Agar

Composition per liter:

NaCl	23.4g
Casein hydrolysate	17.0g
Agar	15.0g
Glycerol	10.0g
Pancreatic digest of gelatin	5.0g
Glucose	5.0g
Papaic digest of soybean meal	3.0g
Beef extract	3.0g
Yeast extract	2.0g
Casamino acids, vitamin free	0.5g
Pancreatic digest of casein	0.5g
Na_2SO_3	0.1g

pH 7.6 ± 0.2 at 25°C

Preparation of Medium: Add components to distilled/deionized water and bring volume to 1.0L. Mix thoroughly. Gently heat and bring to boiling. Adjust pH to 7.6. Distribute into tubes or flasks. Autoclave for 15 min at 15 psi pressure–121°C. Pour into sterile Petri dishes or leave in tubes.

Use: For the cultivation of *Vibrio alginolyticus*.

Worfel-Ferguson Agar

Composition per liter:

Sucrose	20.0g
Agar	15.0g
NaCl	2.0g
Yeast extract	2.0g
K_2SO_4	1.0g
$MgSO_4 \cdot 7H_2O$	0.25g

pH 6.5 ± 0.2 at 25°C

Preparation of Medium: Add components to distilled/deionized water and bring volume to 1.0L. Mix thoroughly. Gently heat and bring to boiling. Distribute into tubes or flasks. Autoclave for 15 min at 15 psi pressure–121°C. Pour into sterile Petri dishes or leave in tubes.

Use: For the detection of capsule production by *Klebsiella*. For serological detection of the Neufeld (Quellung) reaction.

Wort Agar

Composition per liter:

Agar	15.0g
Malt extract	15.0g
Maltose	12.75g
Dextrin	2.75g
Glycerol	2.35g
K_2HPO_4	1.0g
NH_4Cl	1.0g
Pancreatic digest of gelatin	0.78g

pH 4.8 ± 0.2 at 25°C

Source: This medium is available as a premixed powder from BD Diagnostic Systems and Oxoid Unipath.

Preparation of Medium: Add components to distilled/deionized water and bring volume to 1.0L. Mix thoroughly. Gently heat and bring to boiling. Boil for 1 min with mixing. Distribute into tubes or flasks. Autoclave for 15 min at 15 psi pressure–121°C. Do not overheat, as this will result in hydrolysis of the agar. An additional 5.0g of agar can be used to make a firmer agar. Pour into sterile Petri dishes or leave in tubes.

Use: For the cultivation and enumeration of yeasts. The low pH of the agar selectively inhibits bacterial growth.

Wort Agar

Composition per liter:

Agar	25.0g
Yeast extract	1.0g
Wort solution	1.0L

Wort Solution:

Composition per liter:

Malt extract	110.0g

Preparation of Wort Solution: Add malt extract to distilled/deionized water and bring volume to 1.0L. Mix thoroughly.

Preparation of Medium: Combine components. Mix thoroughly. Gently heat and bring to boiling. Distribute into tubes or flasks. Autoclave for 15 min at 15 psi pressure–121°C. Pour into sterile Petri dishes or leave in tubes.

Use: For the cultivation and mainteneance of a wide variety of yeasts and filamentous fungi.

Wort Agar 6°Brix

Composition per liter:

Malt	250.0g
Agar	20.0g

Preparation of Medium: Add 250.0g of ground malt to tap water and bring volume to 1.0L. Mix thoroughly. Gently heat to 55°–60°C. Maintain at 55°–60°C for 1.5–2.0 hr while mixing regularly. Elevate temperature to 80°C and maintain for 10 min with thorough mixing. Cool the wort to 25°C and filter through a linen bag. Adjust the concentration of sugars to 6° Brix (specific gravity of 1.024 at 24°C). Add 20.0g of agar. Gently heat and bring to boiling. Mix thoroughly. Distribute into tubes or flasks. Autoclave for 30 min at 3 psi pressure–105°C. Pour into sterile Petri dishes or leave in tubes.

Use: For the cultivation and maintenance of *Rhodococcus erythropolis.*

Wort Broth

Composition per liter:

Yeast extract	1.0g
Wort solution	1.0L

Wort Solution:

Composition per liter:

Malt extract	110.0g

Preparation of Wort Solution: Add malt extract to distilled/deionized water and bring volume to 1000.0mL. Mix thoroughly.

Preparation of Medium: Combine components. Mix thoroughly. Distribute into tubes or flasks. Autoclave for 15 min at 15 psi pressure–121°C.

Use: For the cultivation of a wide variety of yeasts and filamentous fungi.

Wort HiVeg Agar

Composition per liter:

Agar	15.0g
Malt extract	15.0g
Maltose	12.75g
Dextrin	2.75g

NH_4Cl 1.0g
K_2HPO_4 .... 1.0g
Plant peptone 0.78g

pH 4.8 ± 0.2 at 25°C

Source: This medium is available as a premixed powder from HiMedia.

Preparation of Medium: Add components to distilled/deionized water and bring volume to 1.0L. Mix thoroughly. Gently heat and bring to boiling. Boil for 1 min with mixing. Distribute into tubes or flasks. Autoclave for 15 min at 15 psi pressure–121°C. Do not overheat, as this will result in hydrolysis of the agar. An additional 5.0g of agar can be used to make a firmer agar. Pour into sterile Petri dishes or leave in tubes.

Use: For the cultivation and enumeration of yeasts. The low pH of the agar selectively inhibits bacterial growth.

Wort HiVeg Broth

Composition per liter:
Malt extract 15.0g
Maltose 12.75g
Dextrin 2.75g
NH_4Cl 1.0g
K_2HPO_4 .... 1.0g
Plant peptone 0.78g

pH 4.8 ± 0.2 at 25°C

Source: This medium is available as a premixed powder from HiMedia.

Preparation of Medium: Add components to distilled/deionized water and bring volume to 1.0L. Mix thoroughly. Gently heat and bring to boiling. Boil for 1 min with mixing. Distribute into tubes or flasks. Autoclave for 15 min at 15 psi pressure–121°C.

Use: For the cultivation and enumeration of yeasts. The low pH of the agar selectively inhibits bacterial growth.

Wort Sucrose Agar

Composition per liter:
Agar 25.0g
Yeast extract 1.0g
Wort solution 500.0mL
Sucrose solution 500.0mL

Wort Solution:
Composition per 500.0mL:
Malt extract 55.0g

Preparation of Wort Solution: Add malt extract to distilled/deionized water and bring volume to 500.0mL. Mix thoroughly.

Sucrose Solution:
Composition per 500.0mL:
Sucrose 50.0g

Preparation of Sucrose Solution: Add sucrose to distilled/deionized water and bring volume to 500.0mL. Mix thoroughly.

Preparation of Medium: Combine components. Mix thoroughly. Gently heat and bring to boiling. Distribute into tubes or flasks. Autoclave for 15 min at 15 psi pressure–121°C. Pour into sterile Petri dishes or distribute into sterile tubes.

Use: For the cultivation and maintenance of *Aspergillus oryzae.*

Wort Sucrose Broth

Composition per liter:
Yeast extract 1.0g
Wort solution 500.0mL
Sucrose solution 500.0mL

Wort Solution:
Composition per 500.0mL:
Malt extract 55.0g

Preparation of Wort Solution: Add malt extract to distilled/deionized water and bring volume to 500.0mL. Mix thoroughly.

Sucrose Solution:
Composition per 500.0mL:
Sucrose 50.0g

Preparation of Sucrose Solution: Add sucrose to distilled/deionized water and bring volume to 500.0mL. Mix thoroughly.

Preparation of Medium: Combine components. Mix thoroughly. Distribute into tubes or flasks. Autoclave for 15 min at 15 psi pressure–121°C.

Use: For the cultivation of *Aspergillus oryzae.*

Xanthine Agar

Composition per liter:
Solution 1 .... 900.0mL
Solution 2 .... 100.0mL

pH 7.0 ± 0.2 at 25°C

Solution 1:
Composition per 900.0mL:
Agar 15.0g
Pancreatic digest of gelatin 5.0g
Beef extract 3.0g

Preparation of Solution 1: Add components to distilled/deionized water and bring volume to 900.0mL. Mix thoroughly. Gently heat and bring to boiling.

Solution 2:
Composition per 100.0mL:
Xanthine 4.0g

Preparation of Solution 2: Add xanthine to distilled/deionized water and bring volume to 100.0mL. Mix thoroughly. Gently heat and bring to boiling.

Preparation of Medium: Combine solutions 1 and 2. Mix thoroughly. Distribute into tubes or flasks. Autoclave for 15 min at 15 psi pressure–121°C. Pour into sterile Petri dishes or leave in tubes.

Use: For the differentiation of aerobic *Actinomycete* species. Clearing around a colony indicates utilization of xanthine. *Streptomyces* species utilize xanthine; most *Nocardia* and *Actinomadura* species do not utilize xanthine.

Xanthobacter agilis Agar

Composition per 1100.0mL:
Solution A 1.0L
Solution B 100.0mL

Solution A:
Composition per liter:
Agar 15.0g
$NaH_2PO_4 \cdot 12H_2O$ 9.0g
KH_2PO_4 .... 1.5g

$NH_4{\cdot}Cl$ 1.0g
Sodium propionate or 3-hydroxybutyrate 1.0g
$MgSO_4{\cdot}7H_2O$ 0.2g
Trace elements solution 1.0mL

pH 7.0 ± 0.2 at 25°C

Trace Elements Solution:
Composition per 2.0mL:

H_3BO_4 .... 560.0μg
$ZnSO_4{\cdot}7H_2O$ 350.0μg
$NiCl_2{\cdot}H_2O$ 160.0μg
$Na_2MoO_4{\cdot}2H_2O$ 100.0μg
$CuSO_4{\cdot}5H_2O$ 16.0μg
$MnCl_2{\cdot}4H_2O$ 16.0μg

Preparation of Trace Elements Solution: Add components to distilled/deionized water and bring volume to 2.0mL. Mix thoroughly.

Preparation of Solution A: Add components to distilled/deionized water and bring volume to 1.0L. Mix thoroughly. Gently heat and bring to boiling. Autoclave for 15 min at 15 psi pressure–121°C. Cool to 50°–55°C.

Solution B:
Composition per 100.0mL:

Ferric ammonium citrate 50.0mg
$CaCl_2{\cdot}2H_2O$ 100.0mg

Preparation of Solution B: Add components to distilled/deionized water and bring volume to 100.0mL. Mix thoroughly. Autoclave for 15 min at 15 psi pressure–121°C. Cool to 50°–55°C.

Preparation of Medium: Aseptically combine 1.0L of sterile solution A with 100.0mL of sterile solution B. Mix thoroughly. Pour into sterile Petri dishes or distribute into sterile tubes.

Use: For the cultivation of *Xanthobacter agilis*.

Xanthomonas Agar

Composition per liter:

Agar 15.0g
Pancreatic digest of gelatin 10.0g
Sucrose 10.0g
Beef extract 6.0g

pH 6.8 ± 0.2 at 25°C

Preparation of Medium: Add components to distilled/deionized water and bring volume to 1.0L. Mix thoroughly. Gently heat and bring to boiling. Distribute into tubes or flasks. Autoclave for 15 min at 15 psi pressure–121°C. Pour into sterile Petri dishes or leave in tubes.

Use: For the cultivation and maintenance of *Xanthomonas* species.

Xanthomonas Agar

Composition per liter:

$CaCO_3$ 30.0g
Agar 15.0g
Glucose 10.0g
Yeast extract 5.0g

pH 7.0 ± 0.2 at 25°C

Preparation of Medium: Add components to distilled/deionized water and bring volume to 1.0L. Mix thoroughly. Gently heat and bring to boiling. Distribute into tubes or flasks. Autoclave for 15 min at 15 psi pressure–121°C. Pour into sterile Petri dishes or leave in tubes. Cool rapidly.

Use: For the cultivation and maintenance of *Alcaligenes latus, Erwinia tracheiphila, Pseudomonas amygdali, Xanthomonas albilineans, Xanthomonas axonopodis, Xanthomonas campestris, Xanthomonas fragariae, Xanthomonas maltophilia, Xanthomonas oryzae,* and *Xylophilus ampelinus.*

Xanthomonas albilineans Agar

Composition per liter:

Sucrose 20.0g
Agar 15.0g
Peptone 10.0g
Yeast extract 5.0g

pH 7.0 ± 0.2 at 25°C

Preparation of Medium: Add components to distilled/deionized water and bring volume to 1.0L. Mix thoroughly. Adjust pH to 7.0. Gently heat and bring to boiling. Distribute into tubes or flasks. Autoclave for 15 min at 15 psi pressure–121°C. Pour into sterile Petri dishes or leave in tubes.

Use: For the cultivation and maintenance of *Xanthomonas alblinieans*.

Xanthomonas maltophilia Medium

Composition per liter:

Trizma® (tris[hydroxymethyl] aminomethane) base 6.04g
Glucose 5.0g
KCl 1.0g
NaCl 1.0g
L-Phenylalanine 0.9g
$MgSO_4$ 0.2g
L-Arginine 0.1g
L-Methionine 0.1g
$(NH_4)_2SO_4$ 0.1g
NH_4Cl 0.1g
Glycerol 0.68g
L-Serine 0.22g
L-Alanine 0.18g

pH 7.2 ± 0.2 at 25°C

Preparation of Medium: Add components to distilled/deionized water and bring volume to 1.0L. Mix thoroughly. Adjust pH to 7.2. Filter sterilize. Aseptically distribute into sterile tubes or flasks.

Use: For the cultivation of *Stenotrophomonas maltophilia.*

Xanthomonas Medium

Composition per liter:

Pancreatic digest of gelatin 10.0g
Sucrose 10.0g
Beef extract 6.0g

pH 6.8 ± 0.2 at 25°C

Preparation of Medium: Add components to distilled/deionized water and bring volume to 1.0L. Mix thoroughly. Gently heat with mixing. Distribute into screw-capped test tubes. Autoclave for 15 min at 15 psi pressure–121°C.

Use: For the cultivation and maintenance of *Xanthomonas* species.

Xanthomonas TYG Agar (*Xanthomonas* Tryptone Yeast Extract Glucose Agar)

Composition per liter:

Agar 20.0g
Pancreatic digest of casein 5.0g

Glucose	5.0g
Yeast extract	3.0g
K_2HPO_4	0.7g
$MgSO_4·7H_2O$	0.25g

Preparation of Medium: Add components to distilled/deionized water and bring volume to 1.0L. Mix thoroughly. Gently heat and bring to boiling. Distribute into tubes or flasks. Autoclave for 15 min at 15 psi pressure–121°C. Pour into sterile Petri dishes or leave in tubes.

Use: For the cultivation and maintenance of *Xanthomonas* species.

XB45/XB90/PB90-2 Medium (DSMZ Medium 862)

Composition per 1069.2mL:

Solution A	940.0mL
Solution E	100.0mL
Solution D	10.0mL
Solution G	10.0mL
Solution F	7.2mL
Solution B	1.0mL
Solution C	1.0mL

pH 7.2 ± 0.2 at 25°C

Solution A:

Composition per 940.0mL:

NaCl	1.0g
KCl	0.5g
$MgCl_2·6H_2O$	0.4g
KH_2PO_4	0.2g
NH_4Cl	0.25g
$CaCl_2·2H_2O$	0.15g
Resazurin	0.5mg

Preparation of Solution A: Prepare under 80% N_2 + 20% CO_2 gas atmosphere. Add components to distilled/deionized water and bring volume to 940.0mL. Mix thoroughly. Adjust pH to 7.2. Sparge with 80% N_2 + 20% CO_2. Autoclave for 15 min at 15 psi pressure–121°C. Cool to 25°C.

Solution B:

Composition per liter:

$FeCl_2·4H_2O$	1.5g
$CoCl_2·6H_2O$	190.0mg
$MnCl_2·4H_2O$	100.0mg
$ZnCl_2$	70.0mg
$Na_2MoO_4·2H_2O$	36.0mg
$NiCl_2·6H_2O$	24.0mg
H_3BO_3	6.0mg
$CuCl_2·2H_2O$	2.0mg
HCl (25% solution)	10.0mL

Preparation of Solution B: Add $FeCl_2·4H_2O$ to 10.0mL of HCl solution. Mix thoroughly. Add distilled/deionized water and bring volume to 1.0L. Add remaining components. Mix thoroughly. Sparge with 80% N_2 + 20% CO_2. Autoclave for 15 min at 15 psi pressure–121°C. Cool to 25°C.

Solution C:

Composition per liter:

Pyridoxine hydrochloride	300.0mg
Thiamine-HCl·$2H_2O$	200.0mg
Nicotinic acid	200.0mg
Vitamin B_{12}	100.0mg
Calcium pantothenate	100.0mg
p-Aminobenzoic acid	80.0mg
D(+)-Biotin	20.0mg

Preparation of Solution C: Add components to distilled/deionized water and bring volume to 1.0L. Sparge with 100% N_2. Mix thoroughly. Filter sterilize.

Solution D:

Composition per liter:

Pyridoxine-HCl	10.0mg
Thiamine-HCl·$2H_2O$	5.0mg
Riboflavin	5.0mg
Nicotinic acid	5.0mg
D-Ca-pantothenate	5.0mg
p-Aminobenzoic acid	5.0mg
Lipoic acid	5.0mg
Biotin	2.0mg
Folic acid	2.0mg
Vitamin B_{12}	0.1mg

Preparation of Solution D: Add components to distilled/deionized water and bring volume to 1.0L. Mix thoroughly. Sparge with 100% N_2. Filter sterilize.

Solution E:

Composition per 100.0mL:

$NaHCO_3$	5.0g

Preparation of Solution E: Add $NaHCO_3$ to distilled/deionized water and bring volume to 100.0mL. Mix thoroughly. Sparge with 100% N_2. Autoclave for 15 min at 15 psi pressure–121°C. Cool to 25°C.

Solution F:

Composition per 10.0mL:

Glucose	1.0g

Preparation of Solution F: Add glucose to distilled/deionized water and bring volume to 10.0mL. Mix thoroughly. Sparge with 100% N_2. Autoclave for 15 min at 15 psi pressure–121°C. Cool to 25°C.

Solution G:

Composition per 10.0mL:

$Na_2S·9H_2O$	0.125g

Preparation of Solution G: Add $Na_2S·9H_2O$ to distilled/deionized water and bring volume to 10.0mL. Mix thoroughly. Autoclave under 100% N_2 for 15 min at 15 psi pressure–121°C. Cool to 25°C.

Preparation of Medium: Prepare and dispense medium under 80% N_2 + 20% CO_2 gas atmosphere. Sequentially add 1.0mL solution B, 1.0mL solution C, 10.0mL solution D, 100.0mL solution E, 7.2mL solution F, and 10.0mL solution G to 940.0mL solution A. Distribute anaerobically under 80% N_2 + 20% CO_2 into appropriate vessels. The pH should be 7.2.

Use: For the cultivation of unclassified bacteria DSM 12558, DSM 12559, and DSM 12595.

XED-AGAR (DSMZ Medium 1026)

Composition per liter:

Agar	18.0g
Xylan	7.0g
Yeast extract	3.0g

pH 7.0 ± 0.2 at 25°C

Preparation of Medium: Add components to distilled/deionized water and bring volume to 1.0L. Mix thoroughly. Adjust pH to 7.0. Distribute into tubes or flasks. Autoclave for 15 min at 15 psi pressure–121°C. Pour into sterile Petri dishes or leave in tubes.

Use: For the cultivation of *Microbacterium ulmi.*

Xenorhabdus Agar

Composition per liter:

Agar	15.0g
Peptone	10.0g
NaCl	5.0g
Yeast extract	5.0g

pH 7.2 ± 0.2 at 25°C

Preparation of Medium: Add components to distilled/deionized water and bring volume to 1.0L. Mix thoroughly. Gently heat and bring to boiling. Adjust pH to 7.2. Distribute into tubes or flasks. Autoclave for 20 min at 15 psi pressure–121°C. Pour into sterile Petri dishes or leave in tubes.

Use: For the cultivation and maintenance of *Bacteroides galacturonicus* and *Xenorhabdus nematophilus.*

Xenorhabdus Broth

Composition per liter:

Peptone	10.0g
NaCl	5.0g
Yeast extract	5.0g

pH 7.2 ± 0.2 at 25°C

Preparation of Medium: Add components to distilled/deionized water and bring volume to 1.0L. Mix thoroughly. Adjust pH to 7.2. Distribute into tubes or flasks. Autoclave for 20 min at 15 psi pressure–121°C.

Use: For the cultivation and maintenance of *Bacteroides galacturonicus* and *Xenorhabdus nematophilus.*

XL Agar Base (Xylose Lysine Agar Base)

Composition per liter:

Agar	13.5g
Lactose	7.5g
Sucrose	7.5g
L-Lysine	5.0g
NaCl	5.0g
Xylose	3.5g
Yeast extract	3.0g
Phenol Red	0.08g
Thiosulfate-citrate solution	20.0mL

pH 7.5 ± 0.2 at 25°C

Source: This medium is available as a premixed powder from BD Diagnostic Systems.

Thiosulfate-Citrate Solution:

Composition per 100.0mL:

$Na_2S_2O_3$	34.0g
Ferric ammonium citrate	4.0g

Preparation of Thiosulfate-Citrate Solution: Add components to distilled/deionized water and bring volume to 100.0mL. Mix thoroughly.

Preparation of Medium: Add components, except thiosulfate-citrate solution, to distilled/deionized water and bring volume to 980.0mL. Mix thoroughly. Gently heat while stirring and bring to boiling. Distribute into tubes or flasks. Autoclave for 10 min at 14 psi pressure–118°C. Cool to 55°C. Aseptically add 20.0 mL of the sterile thiosulfate-citrate solution. Mix thoroughly. Pour into sterile Petri dishes or leave in tubes.

Use: For the isolation, cultivation, and differentiation of enteric pathogens. Nonfermenting xylose/lactose/sucrose bacteria appear as red colonies. Xylose-fermenting, lysine-decarboxylating bacteria appear as red colonies. Xylose-fermenting, lysine-nondecarboxylating bacteria appear as opaque yellow colonies. Lactose- or sucrose-fermenting bacteria appear as yellow colonies.

XL Agar Base

Composition per liter:

Agar	15 g
Lactose	7.5g
Sucrose	7.5g
L-Lysine	5.0g
NaCl	5.0g
Xylose	3.75g
Yeast extract	3.0g
Phenol Red	0.08g
Thiosulfate-citrate solution	20.0mL

pH 7.4 ± 0.2 at 25°C

Source: This medium is available as a premixed powder from BD Diagnostic Systems.

Thiosulfate-Citrate Solution:

Composition per 100.0mL:

$Na_2S_2O_3$	34.0g
Ferric ammonium citrate	4.0g

Preparation of Thiosulfate-Citrate Solution: Add components to distilled/deionized water and bring volume to 100.0mL. Mix thoroughly.

Preparation of Medium: Add components, except thiosulfate-citrate solution, to distilled/deionized water and bring volume to 980.0mL. Mix thoroughly. Gently heat while stirring and bring to boiling. Distribute into tubes or flasks. Autoclave for 10 min at 14 psi pressure–118°C. Cool to 55°C. Aseptically add 20.0 mL of the sterile thiosulfate-citrate solution. Mix thoroughly. Pour into sterile Petri dishes or leave in tubes.

Use: For the isolation, cultivation, and differentiation of enteric pathogens. Nonfermenting xylose/lactose/sucrose bacteria appear as red colonies. Xylose-fermenting, lysine-decarboxylating bacteria appear as red colonies. Xylose-fermenting, lysine-nondecarboxylating bacteria appear as opaque yellow colonies. Lactose- or sucrose-fermenting bacteria appear as yellow colonies.

XLD Agar (Xylose Lysine Deoxycholate Agar)

Composition per liter:

Agar	13.5g
Lactose	7.5g
Sucrose	7.5g
$Na_2S_2O_3$	6.8g
L-Lysine	5.0g
NaCl	5.0g
Xylose	3.5g

Yeast extract 3.0g
Sodium desoxycholate 2.5g
Ferric ammonium citrate 0.8g
Phenol Red 0.08g

pH 7.5 ± 0.2 at 25°C

Source: This medium is available as a premixed powder from BD Diagnostic Systems and Oxoid Unipath.

Preparation of Medium: Add components to distilled/deionized water and bring volume to 1.0L. Mix thoroughly. Gently heat and bring to boiling. Do not overheat. Distribute into tubes or flasks. Autoclave for 15 min at 15 psi pressure–121°C. Pour into sterile Petri dishes or leave in tubes. Plates should be poured as soon as possible to avoid precipitation.

Use: For the isolation and differentiation of enteric pathogens, especially *Shigella* and *Providencia* species. Nonfermenting xylose/lactose/sucrose bacteria appear as red colonies. Xylose-fermenting, lysine-decarboxylating bacteria appear as red colonies. Xylose-fermenting, lysine-nondecarboxylating bacteria appear as opaque yellow colonies. Lactose- or sucrose-fermenting bacteria appear as yellow colonies.

XLD Agar (Xylose Lysine Deoxycholate Agar) (BAM M179)

Composition per liter:

Agar 15.0g
Lactose 7.5g
Sucrose 7.5g
$Na_2S_2O_3$ 6.8g
L-Lysine 5.0g
NaCl 5.0g
Xylose 3.75g
Yeast extract 3.0g
Sodium deoxycholate 2.5g
Ferric ammonium citrate 0.8g
Phenol Red 0.08g

pH 7.5 ± 0.2 at 25°C

Source: This medium is available as a premixed powder from BD Diagnostic Systems and Oxoid.

Preparation of Medium: Add components to distilled/deionized water and bring volume to 1.0L. Mix thoroughly. Gently heat and bring to boiling. Do not overheat. Distribute into tubes or flasks. Autoclave for 15 min at 15 psi pressure–121°C. Pour into sterile Petri dishes or leave in tubes. Plates should be poured as soon as possible to avoid precipitation.

Use: For the isolation and differentiation of enteric pathogens, especially *Shigella* and *Providencia* species. Nonfermenting xylose/lactose/sucrose bacteria appear as red colonies. Xylose-fermenting, lysine-decarboxylating bacteria appear as red colonies. Xylose-fermenting, lysine-nondecarboxylating bacteria appear as opaque yellow colonies. Lactose- or sucrose-fermenting bacteria appear as yellow colonies.

XLD Agar, HiVeg (Xylose Lysine Deoxycholate HiVeg Agar)

Composition per liter:

Agar 15.0g
Lactose 7.5g
Sucrose 7.5g
$Na_2S_2O_3$ 6.8g
L-Lysine 5.0g
NaCl 5.0g
Yeast extract 4.0g
Xylose 3.5g
Synthetic detergent No. III 1.5g
Ferric ammonium citrate 0.8g
Phenol Red 0.08g
Selective supplement solution 4.6mL

pH 7.4 ± 0.2 at 25°C

Source: This medium is available as a premixed powder from HiMedia.

Selective Supplement Solution:
Composition per 100.0mL:

Tergitol™ 4 .... Proprietary

Preparation of Selective Supplement Solution: Available as premixed solution.

Preparation of Medium: Add components to distilled/deionized water and bring volume to 1.0L. Mix thoroughly. Gently heat and bring to boiling. Do not overheat. Distribute into tubes or flasks. Do not autoclave. Pour into sterile Petri dishes or leave in tubes. Plates should be poured as soon as possible to avoid precipitation.

Use: For the isolation and differentiation of enteric pathogens, especially *Shigella* and *Providencia* species. Nonfermenting xylose/lactose/sucrose bacteria appear as red colonies. Xylose-fermenting, lysine-decarboxylating bacteria appear as red colonies. Xylose-fermenting, lysine-nondecarboxylating bacteria appear as opaque yellow colonies. Lactose- or sucrose-fermenting bacteria appear as yellow colonies.

XLT4 HiVeg Agar Base

Composition per liter:

Agar 18.0g
Lactose 7.5g
Saccharose 7.5g
$Na_2S_2O_3$ 6.8g
L-Lysine 5.0g
NaCl 5.0g
Xylose 3.75g
Yeast extract 3.0g
Plant peptone No. 3 .... 1.6g
Ferric ammonium citrate 0.8g
Phenol Red 0.08g
Selective supplement solution 4.6mL

pH 7.4 ± 0.2 at 25°C

Source: This medium, without selective supplement solution, is available as a premixed powder from HiMedia.

Selective Supplement Solution:
Composition per 100.0mL:

Tergitol™ 4 .... Proprietary

Preparation of Selective Supplement Solution: Available as premixed solution.

Preparation of Medium: Add components to distilled/deionized water and bring volume to 1.0L. Mix thoroughly. Gently heat and bring to boiling. Do not overheat. Distribute into tubes or flasks. Autoclave for 15 min at 15 psi pressure–121°C. Pour into sterile Petri dishes or leave in tubes. Plates should be poured as soon as possible to avoid precipitation.

Use: For the isolation and differentiation of enteric pathogens, especially *Shigella* and *Providencia* species.

XPS Agar

Composition per liter:

Solution A	500.0mL
Solution B	500.0mL

pH 5.1 ± 0.2 at 25°C

Solution A:

Composition per 500.0mL:

Potatoes, infusion from	40.0g
Sucrose	15.0g
Peptone	5.0g
Glucose	4.0g
Casamino acids	1.0g
Na_2HPO_4	0.79g
$Ca(NO_3)_2 \cdot 4H_2O$ solution	10.0mL

Potatoes, Infusion From:

Composition per 500.0mL:

Potatoes	4.0g

Preparation of Potatoes, Infusion From: Peel and dice potatoes. Add 400.0mL of distilled/deionized water. Gently heat and bring to boiling. Continue boiling for 30 min. Filter through cheesecloth.

$Ca(NO3)_2 \cdot 4H_2O$ Solution:

Composition per 10.0mL:

$Ca(NO_3)_2 \cdot 4H_2O$	0.5g

Preparation of $Ca(NO_3)_2 \cdot 4H_2O$ Solution: Add $Ca(NO_3)_2 \cdot 4H_2O$ to distilled/deionized water and bring volume to 10.0mL. Mix thoroughly. Filter sterilize.

Preparation of Solution A: Add components, except $Ca(NO_3)_2 \cdot 4H_2O$ solution, to distilled/deionized water and bring volume to 490.0mL. Mix thoroughly. Autoclave for 15 min at 15 psi pressure–121°C. Aseptically add 10.0mL of sterile $Ca(NO_3)_2 \cdot 4H_2O$ solution. Mix thoroughly. Cool to 50°–55°C.

Solution B:

Composition per 500.0mL:

Agar	20.0g

Preparation of Solution A: Add agar to distilled/deionized water and bring volume to 500.0mL. Mix thoroughly. Gently heat and bring to boiling. Autoclave for 15 min at 15 psi pressure–121°C. Cool to 50°–55°C.

Preparation of Medium: Aseptically combine 500.0mL of solution A with 500.0mL of solution B. Mix thoroughly. Aseptically pour into sterile Petri dishes or distribute into sterile tubes.

Use: For the cultivation of *Xanthomonas campestris.*

XPS Broth

Composition per liter:

Potatoes, infusion from	40.0g
Sucrose	15.0g
Peptone	5.0g
Glucose	4.0g
Casamino acids	1.0g
Na_2HPO_4	0.79g
$Ca(NO_3)_2 \cdot 4H_2O$ solution	10.0mL

pH 5.1 ± 0.2 at 25°C

Potatoes, Infusion From:

Composition per 500.0mL:

Potatoes	4.0g

Preparation of Potatoes, Infusion From: Peel and dice potatoes. Add 500.0mL of distilled/deionized water. Gently heat and bring to boiling. Continue boiling for 30 min. Filter through cheesecloth.

$Ca(NO3)_2 \cdot 4H_2O$ Solution:

Composition per 10.0mL:

$Ca(NO_3)_2 \cdot 4H_2O$	0.5g

Preparation of $Ca(NO_3)_2 \cdot 4H_2O$ Solution: Add $Ca(NO_3)_2 \cdot 4H_2O$ to distilled/deionized water and bring volume to 10.0mL. Mix thoroughly. Filter sterilize.

Preparation of Medium: Add components, except $Ca(NO_3)_2 \cdot 4H_2O$ solution, to distilled/deionized water and bring volume to 990.0mL. Mix thoroughly. Autoclave for 15 min at 15 psi pressure–121°C. Aseptically add 10.0mL of sterile $Ca(NO_3)_2 \cdot 4H_2O$ solution. Mix thoroughly. Aseptically distribute into sterile tubes or flasks.

Use: For the cultivation of *Xanthomonas campestris.*

XPS Broth with Thymidine (Thymidine Auxotroph XPS Medium)

Composition per liter:

Potatoes, infusion from	40.0g
Sucrose	15.0g
Peptone	5.0g
Glucose	4.0g
Casamino acids	1.0g
Na_2HPO_4	0.79g
Thymidine	10.0mg
$Ca(NO_3)_2 \cdot 4H_2O$ solution	10.0mL

pH 5.1 ± 0.2 at 25°C

Potatoes, Infusion From:

Composition per 500.0mL:

Potatoes	4.0g

Preparation of Potatoes, Infusion From: Peel and dice potatoes. Add 500.0mL of distilled/deionized water. Gently heat and bring to boiling. Continue boiling for 30 min. Filter through cheesecloth.

$Ca(NO3)_2 \cdot 4H_2O$ Solution:

Composition per 10.0mL:

$Ca(NO_3)_2 \cdot 4H_2O$	0.5g

Preparation of $Ca(NO_3)_2 \cdot 4H_2O$ Solution: Add $Ca(NO_3)_2 \cdot 4H_2O$ to distilled/deionized water and bring volume to 10.0mL. Mix thoroughly. Filter sterilize.

Preparation of Medium: Add components, except $Ca(NO_3)_2 \cdot 4H_2O$ solution, to distilled/deionized water and bring volume to 990.0mL. Mix thoroughly. Autoclave for 15 min at 15 psi pressure–121°C. Aseptically add 10.0mL of sterile $Ca(NO_3)_2 \cdot 4H_2O$ solution. Mix thoroughly. Aseptically distribute into sterile tubes or flasks.

Use: For the cultivation of *Xanthomonas oryzae.*

XSM Agar

Composition per liter:

Agar	15.0g
Glucose	5.0g
Sucrose	2.0g
Malt extract	1.0g

Yeast extract 1.0g
Liver extract concentrate 1.0g
Corn steep liquor 1.0g

pH 7.0 ± 0.2 at 25°C

Preparation of Medium: Add components to tap water and bring volume to 1.0L. Mix thoroughly. Gently heat and bring to boiling. Distribute into screw-capped test tubes. Autoclave for 15 min at 15 psi pressure–121°C. Pour into sterile Petri dishes or leave in tubes.

Use: For the cultivation and maintenance of *Streptomyces cinereus* and *Streptomyces flaveus.*

Xylan Medium

Composition per liter:

Xylan 30.0g
Agar 12.0g
Peptone 2.0g
Yeast extract 0.5g
L-Cysteine·HCl·H_2O 0.25g
$Na_2S \cdot 9H_2O$ 0.25g
Rumen fluid 400.0mL
$NaHCO_3$ solution 40.0mL
Mineral solution I 25.0mL
Mineral solution II 25.0mL
Wolfe's vitamin solution 10.0mL
VFA solution 10.0mL
Hemin solution 10.0mL
Trace elements solution SL-6 .... 1.0mL

pH 7.0 ± 0.2 at 25°C

$NaHCO_3$ Solution:

Composition per 100.0mL:

$NaHCO_3$ 3.96g

Preparation of $NaHCO_3$ Solution: Add $NaHCO_3$ to distilled/deionized water and bring volume to 100.0mL. Mix thoroughly. Gas with 100% CO_2.

Mineral Solution I:

Composition per liter:

K_2HPO_4 .... 3.0g

Preparation of Mineral Solution I: Add K_2HPO_4 to distilled/deionized water and bring volume to 1.0L. Mix thoroughly.

Mineral Solution II:

Composition per liter:

Sodium citrate 20.0g
NaCl 12.0g
KH_2PO_4 .... 6.0g
$MgCl_2 \cdot 6H_2O$ 2.0g
$CaCl_2$ 1.2g

Preparation of Mineral Solution II: Add components to distilled/deionized water and bring volume to 1.0L. Mix thoroughly.

Wolfe's Vitamin Solution:

Composition per liter:

Pyridoxine·HCl 10.0mg
Thiamine·HCl 5.0mg
Riboflavin 5.0mg
Nicotinic acid 5.0mg
Calcium pantothenate 5.0mg
p-Aminobenzoic acid 5.0mg
Thioctic acid 5.0mg
Biotin 2.0mg
Folic acid 2.0mg
Cyanocobalamin 100.0μg

Preparation of Wolfe's Vitamin Solution: Add components to distilled/deionized water and bring volume to 1.0L. Mix thoroughly.

VFA Solution:

Composition per liter:

Acetic acid 178.3mL
Propionic acid 59.6mL
n-Butyric acid 38.4mL
Isobutyric acid 9.5mL
n-Valeric acid 9.4mL
Isovaleric acid 9.3mL
DL-α-Methylbutyric acid 4.4mL

Preparation of VFA Solution: Add components to distilled/deionized water and bring volume to approximately 500.0mL. Adjust pH to 7.5 with NaOH. Mix thoroughly. Bring volume to 1.0L with distilled/deionized water.

Hemin Solution:

Composition per 100.0mL:

Hemin 0.01g

Preparation of Hemin Solution: Add hemin to 100.0mL of 0.01*N* NaOH. Mix thoroughly.

Trace Elements Solution SL-6:

Composition per liter:

H_3BO_3 .... 0.3g
$CoCl_2 \cdot 6H_2O$ 0.2g
$ZnSO_4 \cdot 7H_2O$ 0.1g
$MnCl_2 \cdot 4H_2O$ 0.03g
$Na_2MoO_4 \cdot H_2O$ 0.03g
$NiCl_2 \cdot 6H_2O$ 0.02g
$CuCl_2.2H_2O$ 0.01g

Preparation of Trace Elements Solution SL-6: Add components to distilled/deionized water and bring volume to 1.0L. Mix thoroughly. Adjust pH to 3.4.

Preparation of Medium: Add components, except $Na_2S \cdot 9H_2O$, $NaCHO_3$,and L-cysteine·HCl·H_2O solutions, to distilled/deionized water and bring volume to 1.0L. Mix thoroughly. Gently heat and bring to boiling. Cool under 80% N_2 + 20% CO_2. Add L-cysteine·HCl·H_2O and $Na_2S \cdot 9H_2O$. Add sufficient $NaCHO_3$ solution to bring pH to 7.2 under 80% N_2 + 20% CO_2. Anaerobically distribute into tubes under 80% N_2 + 20% CO_2. Autoclave for 15 min at 15 psi pressure–121°C.

Use: For the cultivation and maintenance of *Clostridium xylanolyticum* and other microorganisms that can utilize xylan as a carbon source.

Xylella Agar (LMG Medium 115)

Composition per liter:

Agar 17.0g
Yeast extract 10.0g
ACES 10.0g
Activated charcoal 2.0g
α-ketoglutarate 1.0g
KOH, 1*N* 40mL
L-Cysteine-iron solution 20.0mL

pH 6.9 ± 0.2 at 25°C

L-Cysteine-Iron Solution:

Composition per 20.0mL:

L-Cysteine·HCl	0.4g
$Fe_4(P_2O_7)_3$	0.25g

Preparation of L-Cysteine-Iron Solution: Add components to distilled/deionized water and bring volume to 20.0mL. Mix thoroughly. Filter sterilize.

Preparation of Medium: Add ACES to 500.0mL of distilled water at 50°C. Combine with a solution containing 40.0mL of 1*N* KOH in 440.0mL of distilled water. Add the other components except cysteine iron solution. Mix thoroughly. Gently heat and bring to boiling. Autoclave for 15 min at 15 psi pressure–121°C. Cool to 50°C. Aseptically add 20.0mL sterile cysteine iron solution. Pour into sterile Petri dishes or distribute into sterile tubes.

Use: For the cultivation of *Xylella* spp.

Xylella fastidiosa Medium (LMG 115)

Composition per liter:

Agar	17.0g
Yeast extract	10.0g
ACES buffer	10.0g
Activated charcoal	2.0g
L-Cysteine-iron solution	20.0mL

pH 6.9 ± 0.2 at 25°C

L-Cysteine-Iron Solution:

Composition per 20.0mL:

L-Cysteine·HCl	0.4g
$Fe_4(P_2O_7)_3$	0.25g

Preparation of L-Cysteine-Iron Solution: Add components to distilled/deionized water and bring volume to 20.0mL. Mix thoroughly. Filter sterilize.

Preparation of Medium: Add ACES to 500.0mL of distilled/deionized water at 50°C. Add a solution containing 40.0mL of 1*N* KOH in 440.0mL of distilled water. Mix thoroughly. Add the remaining components, except L-cysteine-iron solution. Mix thoroughly. Gently heat and bring to boiling. Adjust pH to 6.9 with KOH. Autoclave for 15 min at 15 psi pressure–121°C. Cool to 50°C. Aseptically add 20.0mL of sterile L-cysteine-iron solution. Mix thoroughly. Pour into sterile Petri dishes or distribute into sterile tubes.

Use: For the cultivation of *Xylella fastidiosa*.

Xylophilus Medium

Composition per liter:

$CaCO_3$	20.0g
Agar	15.0g
D-Galactose	10.0g
Yeast extract	10.0g
Ferric ammonium citrate solution	10.0mL

Ferric Ammonium Citrate Solution:

Composition per 10.0mL:

Ferric ammonium citrate	0.25g

Preparation of Ferric Ammonium Citrate Solution: Add ferric ammonium citrate to distilled/deionized water and bring volume to 10.0mL. Mix thoroughly. Autoclave for 15 min at 15 psi pressure–121°C.

Preparation of Medium: Add components, except ferric ammonium citrate solution, to distilled/deionized water and bring volume to 990.0mL. Mix thoroughly. Gently heat and bring to boiling. Autoclave for 15 min at 15 psi pressure–121°C. Cool to 50°–55°C. Aseptically add 10.0mL of sterile ferric ammonium citrate solution. Mix thoroughly. Pour into sterile Petri dishes or distribute into sterile tubes.

Use: For the cultivation and maintenance of *Xylophilus ampelina*.

Xylose Lactose Tergitol™ 4 (XLT-4)

Composition per 1004.6mL:

Agar	18.0g
Lactose	7.5g
Sucrose	7.5g
$Na_2S_2O_3$	6.8g
Lysine	5.0g
NaCl	5.0g
Xylose	3.75g
Yeast extract	3.0g
Proteose peptone	1.6g
Ferric ammonium citrate	0.8g
Phenol Red	0.08g
Selective supplement solution	4.6mL

pH 7.4 ± 0.2 at 25°C

Source: This medium is available as a premixed powder from Oxoid Unipath.

Selective Supplement Solution:

Composition per 100.0mL:

Tergitol™ 4	Proprietary

Preparation of Selective Supplement Solution: Available as a premixed solution.

Preparation of Medium: Add components, except selective supplement solution, to distilled/deionized water and bring volume to 1.0L. Mix thoroughly. Add 4.6mL of selective supplement solution. Mix thoroughly. Gently heat while stirring and bring to boiling. Do not autoclave. Cool to 50°C. Mix thoroughly. Pour into sterile Petri dishes.

Use: For the isolation and identification of salmonellae from clinical, environmental, and food samples. The presence of the selective agent, Tergitol™ 4, in this medium inhibits many organisms that can be problematic on other plating media. In addition, biochemical and pH changes within the medium allow *Salmonella* spp. (black colonies) to be differentiated from organisms such as *E. coli* (yellow colonies) and *Shigella* spp. (red colonies). The enhanced selectivity of XLT-4 Agar reduces the need for further identification procedures, saving time and money, and results in fewer false presumptive positive colonies when compared to other *Salmonella* plating media.

Xylose Lysine Agar Base
See: **XL Agar Base**

Xylose Lysine Desoxycholate Agar
See: **XLD Agar**

Xylose Sodium Deoxycholate Citrate Agar

Composition per liter:

Agar	12.0g
Xylose	10.0g
Sodium citrate	5.0g
$Na_2S_2O_3·5H_2O$	5.0g

Beef extract	5.0g
Peptone	5.0g
NaCl	2.5g
Sodium deoxycholate	2.5g
Ferric ammonium citrate	1.0g
Neutral Red (1% solution)	2.5mL

pH 7.5 ± 0.2 at 25°C

Preparation of Medium: Add components to distilled/deionized water and bring volume to 1.0L. Mix thoroughly. Gently heat and bring to boiling for 20 sec. Do not autoclave. Cool to 45°–50°C. Pour into sterile Petri dishes.

Use: For the cultivation of *Salmonella* species and some *Shigella* species.

Xylose YP Agar (Xylose Yeast Extract Peptone Agar)

Composition per liter:

$CaCO_3$	20.0g
Agar	15.0g
Xylose	10.0g
Yeast extract	10.0g
Peptone	10.0g
$MgSO_4 \cdot 7H_2O$	0.2g
$MnSO_4 \cdot 4H_2O$	0.01g
$FeSO_4 \cdot 7H_2O$	0.01g
NaCl	0.01g

pH 6.8 ± 0.2 at 25°C

Preparation of Medium: Add components to distilled/deionized water and bring volume to 1.0L. Mix thoroughly. Gently heat and bring to boiling. Distribute into screw-capped test tubes. Autoclave for 15 min at 15 psi pressure–121°C. Adjust pH to 6.8. Mix thoroughly. Pour into sterile Petri dishes or leave in tubes.

Use: For the cultivation and maintenance of *Lactobacillus vaccinostercus* and other microorganisms that utilize xylose as a carbon source.

Xylose YP Broth (Xylose Yeast Extract Peptone Broth)

Composition per liter:

Xylose	10.0g
Yeast extract	10.0g
Peptone	10.0g
$MgSO_4 \cdot 7H_2O$	0.2g
$MnSO_4 \cdot 4H_2O$	0.01g
$FeSO_4 \cdot 7H_2O$	0.01g
NaCl	0.01g

pH 6.8 ± 0.2 at 25°C

Preparation of Medium: Add components to distilled/deionized water and bring volume to 1.0L. Mix thoroughly. Distribute into screw-capped test tubes. Autoclave for 15 min at 15 psi pressure–121°C.

Use: For the cultivation of *Lactobacillus vaccinostercus* and other microorganisms that utilize xylose as a carbon source.

Y 1 Adrenal Cell Growth Medium

Composition per 101.0mL:

Ham's F-10 medium	90.0mL
Fetal bovine serum	10.0mL
Penicillin-streptomycin solution	1.0mL

pH 7.0 ± 0.2 at 25°C

Ham's F-10 Medium:

Composition per liter:

NaCl	7.4g
$NaHCO_3$	1.2g
Glucose	1.1g
$NaH_2PO_4 \cdot H_2O$	0.29g
KCl	0.28g
L-Arginine·HCl	0.21g
L-Glutamine	0.15g
$MgSO_4 \cdot 7H_2O$	0.15g
Sodium pyruvate	0.11g
KH_2PO_4	0.08g
$CaCl_2 \cdot 2H_2O$	0.04g
L-Cystine·2HCl	0.04g
L-Histidine·HCl·H_2O	0.02g
L-Lysine·HCl	0.02g
L-Asparagine-H_2O	0.01g
L-Aspartic Acid	0.01g
L-Glutamic acid	0.01g
L-Leucine	0.01g
L-Proline	0.01g
L-Serine	0.01g
L-Alanine	8.9mg
Glycine	7.5mg
D-Phenylalanine	5.0mg
L-Methionine	4.5mg
Hypoxanthine	4.1mg
L-Threonine	3.6mg
L-Valine	3.5mg
L-Isoleucine	2.6mg
L-Tyrosine	1.8mg
Vitamin B_{12}	1.4mg
Folic acid	1.3mg
Phenol Red	1.2mg
Thiamine·HCl	1.0mg
$FeSO_4 \cdot 7H_2O$	0.8mg
Choline chloride	0.7mg
D-Calcium pantothenate	0.7mg
Thymidine	0.7mg
Niacinamide	0.6mg
L-Tryptophan	0.6mg
Isoinositol	0.5mg
Riboflavin	0.4mg
Lipoic acid	0.2mg
Pyridoxine·HCl	0.2mg
$ZnSO_4 \cdot 7H_2O$	0.03mg
Biotin	0.02mg
$CuSO_4 \cdot 5H_2O$	3.0μg

Preparation of Ham's F-10 Medium: Add components to distilled/deionized water and bring volume to 1.0L. Mix thoroughly.

Penicillin-Streptomycin Solution:

Composition per 100.0mL:

Streptomycin	0.5g
Penicillin G	500,000U

Preparation of Penicillin-Streptomycin Solution: Add components to distilled/deionized water and bring volume to 100.0mL. Mix thoroughly. Filter sterilize.

Preparation of Medium: Aseptically combine components. Filter sterilize. Store at 4–5°C.

Use: For the cultivation of Y-1 mouse adrenal tissue culture cells used for the detection of heat-labile toxin (LT) produced by enterotoxigenic strains of *Escherichia coli.* LT causes the conversion of elongated fibroblast-like cells into round, refractile cells.

Y 1 Adrenal Cell Growth Medium

Composition per 580.0mL:

Ham's F-10 medium	500.0mL
Fetal bovine serum	75.0mL
Penicillin-streptomycin solution	5.0mL

pH 7.0 ± 0.2 at 25°C

Ham's F-10 Medium:
Composition per liter:

NaCl	7.4g
$NaHCO_3$	1.2g
Glucose	1.1g
$NaH_2PO_4 \cdot H_2O$	0.29g
KCl	0.28g
L-Arginine·HCl	0.21g
L-Glutamine	0.15g
$MgSO_4 \cdot 7H_2O$	0.15g
Sodium pyruvate	0.11g
KH_2PO_4	0.08g
$CaCl_2 \cdot 2H_2O$	0.04g
L-Cystine·2HCl	0.04g
L-Histidine·HCl·H_2O	0.02g
L-Lysine·HCl	0.02g
L-Asparagine-H_2O	0.01g
L-Aspartic Acid	0.01g
L-Glutamic acid	0.01g
L-Leucine	0.01g
L-Proline	0.01g
L-Serine	0.01g
L-Alanine	8.9mg
Glycine	7.5mg
D-Phenylalanine	5.0mg
L-Methionine	4.5mg
Hypoxanthine	4.1mg
L-Threonine	3.6mg
L-Valine	3.5mg
L-Isoleucine	2.6mg
L-Tyrosine	1.8mg
Vitamin B_{12}	1.4mg
Folic acid	1.3mg
Phenol Red	1.2mg
Thiamine·HCl	1.0mg
$FeSO_4 \cdot 7H_2O$	0.8mg
Choline chloride	0.7mg
D-Calcium pantothenate	0.7mg
Thymidine	0.7mg
Niacinamide	0.6mg
L-Tryptophan	0.6mg
Isoinositol	0.5mg
Riboflavin	0.4mg
Lipoic acid	0.2mg
Pyridoxine·HCl	0.2mg
$ZnSO_4 \cdot 7H_2O$	0.03mg
Biotin	0.02mg
$CuSO_4 \cdot 5H_2O$	3.0μg

Preparation of Ham's F-10 Medium: Add components to distilled/deionized water and bring volume to 1.0L. Mix thoroughly.

Penicillin-Streptomycin Solution:
Composition per 100.0mL:

Streptomycin	0.5g
Penicillin G	500,000U

Preparation of Penicillin-Streptomycin Solution: Add components to distilled/deionized water and bring volume to 100.0mL. Mix thoroughly. Filter sterilize.

Preparation of Medium: Aseptically combine components. Filter sterilize. Store at 4°–5°C.

Use: For the cultivation of Y-1 mouse adrenal tissue culture cells used for the detection of cholera enterotoxin (CT) produced by enterotoxigenic strains of *Vibrio cholerae* or *Vibrio mimicus.* CT causes the conversion of elongated fibroblast-like cells into round, refractile cells.

YA12

Composition per liter:

Agar	10.0g
Glucose	0.6g
NaCl	0.5g
Beef extract	0.2g
Yeast extract	0.02g

pH 6.5–6.7 at 25°C

Preparation of Medium: Add components to distilled/deionized water and bring volume to 1.0L. Mix thoroughly. Adjust pH to 6.5–6.7. Gently heat and bring to boiling. Distribute into tubes or flasks. Autoclave for 15 min at 15 psi pressure–121°C. Pour into sterile Petri dishes or leave in tubes.

Use: For the cultivation of *Adelphamoeba galeacystis.*

YA Halophile Medium

Composition per liter:

NaCl	100.0g
Agar	15.0g
Sodium acetate·$3H_2O$	10.0g
Na_2HPO_4	3.8g
KH_2PO_4	1.3g
$Mg(NO_3)_2 \cdot 6H_2O$	1.0g
$(NH_4)_2SO_4$	1.0g
Yeast extract	1.0g

pH 7.2 ± 0.2 at 25°C

Preparation of Medium: Add components, except magnesium nitrate, to tap water and bring volume to 1.0L. Mix thoroughly. Distribute into tubes or flasks. Autoclave for 15 min at 15 psi pressure–121°C. Aseptically add magnesium nitrate. Adjust pH 7.2 with sterile KOH. Pour into sterile Petri dishes or leave in tubes.

Use: For the cultivation and maintenance of halophilic microorganisms, including *Bacillus halodenitrificans.*

YB Medium (Yeast Extract Beef Extract Medium)

Composition per liter:

Agar	20.0g
Peptone	10.0g
Beef extract	7.0g
Yeast extract	5.0g
NaCl	3.0g
Thiourea	0.1g
Methanol	20.0mL

pH 7.2 ± 0.2 at 25°C

Preparation of Medium: Add components, except methanol, to distilled/deionized water and bring volume to 1.0L. Mix thoroughly. Distribute into tubes or flasks. Autoclave for 15 min at 15 psi pressure–121°C. Aseptically add filter-sterilized methanol. Pour into sterile Petri dishes or leave in tubes.

Use: For the cultivation and maintenance of bacteria that can utilize methanol as a carbon source, including *Achromobacter methanolophila, Methanomonas methylovora, Methylobacterium* species, and *Pseudomonas methanolica.*

YC Agar without Tryptophan

Composition per liter:

Solution A	500.0mL
Solution B	500.0mL

Solution A:

Composition per 500.0mL:

Glucose	20.0g
Succinic acid	10.0g
NaOH	6.0g
$(NH_4)_2SO_4$	5.0g
KH_2PO_4	0.6g
$MgSO_4·7H_2O$	0.3g
Adenine	0.1g
L-Arginine	0.1g
L-Cysteine	0.1g
L-Leucine	0.1g
L-Lysine	0.1g
L-Threonine	0.1g
Uracil	0.1g
L-Aspartic acid	0.05g
L-Histidine	0.05g
L-Isoleucine	0.05g
L-Methionine	0.05g
L-Phenylalanine	0.05g
L-Proline	0.05g
L-Serine	0.05g
L-Tyrosine	0.05g
L-Valine	0.05g
NaCl	0.05g
$CaCl_2·2H_2O$	0.05g
DL-Methionine	0.01g
L-Histidine·HCl	0.005g
Inositol	1.0mg
KI	0.5mg
H_3BO_3	0.3mg
$ZnSO_4·7H_2O$	0.2mg
$MnSO_4·4H_2O$	0.2mg
Thiamine·HCl	0.2mg
Pyroxidine·HCl	0.2mg
Niacin	0.2mg
Calcium pantothenate	0.2mg
p-Aminobenzoic acid	0.1mg
Riboflavin	0.1mg
$FeCl_3$	0.1mg
$Na_2MoO_4·4H_2O$	0.1mg
$CuSO_4·5H_2O$	0.02mg
Folic acid	1.0μg
Biotin	1.0μg

pH 5.7 ± 0.2 at 25°C

Preparation of Solution A: Add components to distilled/deionized water and bring volume to 500.0mL. Mix thoroughly. Adjust pH to 5.6–5.7. Filter sterilize. Warm to 50°–55°C.

Solution B:

Composition per 500.0mL:

Agar, noble	20.0g

Preparation of Solution B: Add agar to distilled/deionized water and bring volume to 500.0mL. Mix thoroughly. Gently heat and bring to boiling. Autoclave for 15 min at 15 psi pressure–121°C. Cool to 50°–55°C.

Preparation of Medium: Aseptically mix 500.0mL of solution A and 500.0mL of solution B. Mix thoroughly. Pour into sterile Petri dishes or distribute into sterile tubes.

Use: For the cultivation and maintenance of *Saccharomyces cerevisiae.*

YC Medium without Tryptophan

Composition per liter:

Solution A	500.0mL
Solution B	500.0mL

Solution A:

Composition per 500.0mL:

Glucose	20.0g
Succinic acid	10.0g
NaOH	6.0g
$(NH_4)_2SO_4$	5.0g
KH_2PO_4	1.0g
$MgSO_4·7H_2O$	0.5g
NaCl	0.1g
$CaCl_2·2H_2O$	0.1g
Adenine	0.1g
L-Arginine	0.1g
L-Cysteine	0.1g
L-Leucine	0.1g
L-Lysine	0.1g
L-Threonine	0.1g
Uracil	0.1g
L-Aspartic acid	0.05g
L-Histidine	0.05g
L-Isoleucine	0.05g
L-Methionine	0.05g
L-Phenylalanine	0.05g
L-Proline	0.05g
L-Serine	0.05g
L-Tyrosine	0.05g
L-Valine	0.05g
Inositol	2.0mg
KI	1.0mg
H_3BO_3	0.5mg
$ZnSO_4·7H_2O$	0.4mg
$MnSO_4·4H_2O$	0.4mg
Thiamine·HCl	0.4mg
Pyroxidine·HCl	0.4mg
Niacin	0.4mg
Calcium pantothenate	0.4mg
p-Aminobenzoic acid	0.2mg
Riboflavin	0.2mg
$FeCl_3$	0.2mg
$Na_2MoO_4·4H_2O$	0.2mg

$CuSO_4 \cdot 5H_2O$ 0.04mg
Folic acid 2.0μg
Biotin 2.0μg

Preparation of Solution A: Add components to distilled/deionized water and bring volume to 500.0mL. Adjust pH to 5.6–5.7. Filter sterilize. Warm to 50°–55°C.

Solution B:

Composition per 500.0mL:

Agar, noble 20.0g

Preparation of Solution B: Add agar to distilled/deionized water and bring volume to 500.0mL. Gently heat and bring to boiling. Autoclave for 15 min at 15 psi pressure–121°C. Cool to 50–55°C.

Preparation of Medium: Aseptically combine 500.0mL of solution A and 500.0mL of solution B. Pour into sterile Petri dishes or distribute into sterile tubes.

Use: For the cultivation of *Saccharomyces cerevisiae.*

YDC Agar (Yeast Extract Dextrose Calcium Carbonate Agar)

Composition per liter:

$CaCO_3$, finely divided 20.0g
Glucose 20.0g
Agar 15.0g
Yeast extract 10.0g

pH 7. 0 ± 0.2 at 25°C

Preparation of Medium: Add components to distilled/deionized water and bring volume to 1.0L. Mix thoroughly. Gently heat and bring to boiling. Distribute into tubes in 10.0mL volumes. Autoclave for 15 min at 15 psi pressure–121°C. Allow tubes to cool in a slanted position.

Use: For the cultivation and maintenance of *Pseudomonas* species on agar slants.

YDC Medium

Composition per liter:

$CaCO_3$ 20.0g
Glucose 20.0g
Agar 15.0g
Yeast extract 10.0g

pH 7.2 ± 0.2 at 25°C

Preparation of Medium: Add components to distilled/deionized water and bring volume to 1.0L. Mix thoroughly. Gently heat and bring to boiling. Distribute into tubes or flasks. Autoclave for 15 min at 15 psi pressure–121°C. Pour into sterile Petri dishes or leave in tubes.

Use: For the cultivation of *Bdellovibrio* species.

Yeast Agar, Van Niel's

Composition per liter:

Agar 20.0g
Yeast extract 10.0g
K_2HPO_4 .... 1.0g
$MgSO_4 \cdot 7H_2O$ 0.5g

pH 7.0–7.2 at 25°C

Preparation of Medium: Add components to tap water and bring volume to 1.0L. Mix thoroughly. Gently heat and bring to boiling. Distribute into tubes or flasks. Autoclave for 15 min at 15 psi pressure–121°C. Pour into sterile Petri dishes or leave in tubes.

Use: For the cultivation and maintenance of a variety of microorganisms, including *Cytophaga* species, *Heliobacterium chlorum, Rhodomicrobium vannielii, Lysobacter enzymogenes, Rhodobacter* species, *Rhodocyclus gelatinosus, Rhodopseudomonas palustris,* and *Rhodospirillum rubrum.*

Yeast Agar, Van Niel's with Glutamate

Composition per liter:

Agar 20.0g
Yeast extract 10.0g
K_2HPO_4 .... 1.0g
$MgSO_4$ 0.5g
Monosodium glutamate 0.85g

pH 7.0–7.2 at 25°C

Preparation of Medium: Add components to tap water and bring volume to 1.0L. Mix thoroughly. Gently heat and bring to boiling. Distribute into tubes or flasks. Autoclave for 15 min at 15 psi pressure–121°C. Pour into sterile Petri dishes or leave in tubes.

Use: For the cultivation and maintenance of a variety of bacteria, including *Bacillus firmus, Cytophaga johnsonae, Heliobacterium chlorum, Lysobacter enzymogenes, Rhodobacter capsulatus, Rhodomicrobium vannielii, Rhodobacter sphaeroides, Rhodocyclus gelatinosus*, *Rhodocyclus gelatinosus*, *Rhodo-pseudomonas palustris*, and *Rhodospirillum rubrum*.

Yeast Agar, Van Niel's with 2.5% Sodium Chloride (ATCC Medium 1370)

Composition per liter:

NaCl 25.0g
Agar 20.0g
Yeast extract 10.0g
K_2HPO_4 .... 1.0g
$MgSO_4$ 0.5g

pH 7.0–7.2 at 25°C

Preparation of Medium: Add components to tap water and bring volume to 1.0L. Mix thoroughly. Gently heat and bring to boiling. Distribute into tubes or flasks. Autoclave for 15 min at 15 psi pressure–121°C. Pour into sterile Petri dishes or leave in tubes.

Use: For the cultivation and maintenance of *Chromatium vinosum* and *Rhodopseudomonas* species.

Yeast Agar, Van Niel's with 25% Sodium Chloride

Composition per liter:

NaCl 250.0g
Agar 20.0g
Yeast extract 10.0g
K_2HPO_4 .... 1.0g
$MgSO_4$ 0.5g

pH 7.0–7.2 at 25°C

Preparation of Medium: Add components to tap water and bring volume to 1.0L. Mix thoroughly. Gently heat and bring to boiling. Distribute into tubes or flasks. Autoclave for 15 min at 15 psi pressure–121°C. Pour into sterile Petri dishes or leave in tubes.

Use: For the cultivation and maintenance of halophilic bacteria, including *Haloarcula vallismortis, Halococcus morrhuae,* and *Halobacterium* species.

Yeast Agar, Van Niel's with Succinate

Composition per liter:

Agar	20.0g
Yeast extract	10.0g
Sodium succinate	1.35g
K_2HPO_4	1.0g
$MgSO_4·7H_2O$	0.5g

pH 7.0–7.2 at 25°C

Preparation of Medium: Add components to tap water and bring volume to 1.0L. Mix thoroughly. Gently heat and bring to boiling. Distribute into tubes or flasks. Autoclave for 15 min at 15 psi pressure–121°C. Pour into sterile Petri dishes or leave in tubes.

Use: For the cultivation and maintenance of *Rhodobacter capsulatus*.

Yeast Ascospore Agar

Composition per liter:

Agar	30.0g
Potassium acetate	10.0g
Yeast extract	2.5g
Glucose	1.0g

Preparation of Medium: Add components to distilled/deionized water and bring volume to 1.0L. Mix thoroughly. Gently heat and bring to boiling. Distribute into tubes or flasks. Autoclave for 15 min at 15 psi pressure–121°C.

Use: For the cultivation and observation of ascospore formation of yeast.

Yeast Beef Agar

See: **Antibiotic Medium 4**

Yeast Carbon Base, 10X (Wickerham Carbon Base Broth)

Composition per liter:

Glucose	10.0g
KH_2PO_4	1.0g
$MgSO_4·7H_2O$	0.5g
NaCl	0.1g
$CaCl_2·2H_2O$	0.1g
DL-Methionine	0.02g
DL-Tryptophan	0.02g
L-Histidine·HCl	0.01g
Inositol	2.0mg
H_3BO_3	0.5mg
$ZnSO_4·7H_2O$	0.4mg
$MnSO_4·4H_2O$	0.4mg
Thiamine·HCl	0.4mg
Pyridoxine	0.4mg
Niacin	0.4mg
Calcium pantothenate	0.4mg
p-Aminobenzoic acid	0.2mg
Riboflavin	0.2mg
$FeCl_3$	0.2mg
$Na_2MoO_4·4H_2O$	0.2mg
KI	0.1mg
$CuSO_4·5H_2O$	0.04mg
Folic Acid	2.0μg
Biotin	2.0μg

pH 5.5 ± 0.2 at 25°C

Source: This medium is available as a premixed powder from BD Diagnostic Systems.

Preparation of Medium: Add components to distilled/deionized water and bring volume to 1.0L. Mix thoroughly. Filter sterilize.

Use: Used as a base to which different nitrogen sources may be added. For the cultivation and differentiation of bacteria based on their ability to utilize diverse added nitrogen sources.

Yeast Dextrose Agar

Composition per liter:

Agar	15.0g
Glucose	10.0g
Yeast extract	10.0g

pH 7.0 ± 0.2 at 25°C

Preparation of Medium: Add components to distilled/deionized water and bring volume to 1.0L. Mix thoroughly. Gently heat and bring to boiling. Adjust pH to 7.0. Distribute into tubes or flasks. Autoclave for 15 min at 15 psi pressure–121°C.

Use: For the cultivation of a variety of heterotrophic microorganisms.

Yeast Extract Agar

Composition per liter:

Agar	15.0g
Malt extract	10.0g
Glucose	4.0g
Yeast extract	4.0g

pH 7.0 ± 0.2 at 25°C

Preparation of Medium: Add components to distilled/deionized water and bring volume to 1.0L. Mix thoroughly. Gently heat and bring to boiling. Distribute into tubes or flasks. Autoclave for 15 min at 15 psi pressure–121°C. Pour into sterile Petri dishes or leave in tubes.

Use: For the isolation and cultivation of *Actinomadura* species, *Actinopolyspora* species, *Excellospora* species, and *Microspora* species.

Yeast Extract Agar

Composition per liter:

Agar	15.0g
Peptone	5.0g
Yeast extract	3.0g

pH 7.2 ± 0.2 at 25°C

Source: This medium is available as a premixed powder from Oxoid Unipath.

Preparation of Medium: Add components to distilled/deionized water and bring volume to 1.0L. Mix thoroughly. Gently heat and bring to boiling. Distribute into tubes or flasks. Autoclave for 15 min at 15 psi pressure–121°C. Pour into sterile Petri dishes or leave in tubes.

Use: For the enumeration of microorganisms in potable and fresh water samples.

Yeast Extract Agar

Composition per liter:

Agar	15.0g
Peptone	9.5g
Yeast extract	7.0g
Beef extract	5.0g
NaCl	5.0g

pH 7.0 ± 0.2 at 25°C

Preparation of Medium: Add components to distilled/deionized water and bring volume to 1.0L. Mix thoroughly. Gently heat and bring to boiling. Distribute into tubes or flasks. Autoclave for 15 min at 15 psi pressure–121°C. Pour into sterile Petri dishes or leave in tubes.

Use: For the cultivation of *Aeromonas salmonicida.*

Yeast Extract Agar

Composition per liter:

Agar	20.0g
Yeast extract	1.0g
Buffer solution	2.0mL

pH 6.0 ± 0.2 at 25°C

Buffer Solution:

Composition per 400.0mL:

KH_2PO_4	60.0g
Na_2HPO_4	40.0g

Preparation of Buffer Solution: Add 40.0g of Na_2HPO_4 to 300.0mL of distilled/deionized water. Mix thoroughly. Add 60.0g of KH_2PO_4. Mix thoroughly. Adjust pH to 6.0.

Preparation of Medium: Add components to distilled/deionized water and bring volume to 1.0L. Mix thoroughly. Autoclave for 15 min at 15 psi pressure–121°C. Pour into sterile Petri dishes.

Use: For the identification of *Histoplasma capsulatum, Blastomyces dermatitidis*, and *Coccidioides immitis*.

Yeast Extract Agar

Composition per liter:

Agar	15.0g
Proteose peptone	10.0g
NaCl	5.0g
Yeast extract	3.0g

Preparation of Medium: Add components to distilled/deionized water and bring volume to 1.0L. Mix thoroughly. Gently heat and bring to boiling. Distribute into tubes or flasks. Autoclave for 15 min at 15 psi pressure–121°C. Pour into sterile Petri dishes or leave in tubes.

Use: For the cultivation of a variety of heterotrophic microorganisms.

Yeast Extract Agar for *Schizosaccharomyces*

Composition per liter:

Glucose	30.0g
Agar	15.0g
Yeast extract	5.0g

Preparation of Medium: Add components to distilled/deionized water and bring volume to 1.0L. Mix thoroughly. Gently heat and bring to boiling. Distribute into tubes or flasks. Autoclave for 15 min at 15 psi pressure–121°C. Pour into sterile Petri dishes or leave in tubes.

Use: For the cultivation and maintenance of *Schizosaccharomyces japonicus* and *Schizosaccharomyces pombe.*

Yeast Extract Beef Extract Medium
See: **YB Medium**

Yeast Extract Dextrose Calcium Carbonate Agar
See: **YDC Agar**

Yeast Extract Glucose Agar

Composition per liter:

Agar	15.0g
Glucose	15.0g
K_2HPO_4	5.2g
KH_2PO_4	3.18g
NH_4Cl	0.54g
Yeast extract	0.5g
$MgSO_4$	0.12g
Trace elements solution	5.0mL

pH 7.0 ± 0.2 at 25°C

Trace Elements Solution:

Composition per liter:

$ZnSO_4 \cdot 7H_2O$	0.287g
$CuSO_4 \cdot 5H_2O$	0.249g
$MnSO_4 \cdot 4H_2O$	0.223g
$Na_2MoO_4 \cdot 2H_2O$	0.124g
$CaCl_2 \cdot 6H_2O$	0.118g
KJ	0.083g
H_3BO_3	0.03g

Preparation of Trace Elements Solution: Add components to distilled/deionized water and bring volume to 1.0L. Mix thoroughly.

Preparation of Medium: Add components to distilled/deionized water and bring volume to 1.0L. Mix thoroughly. Adjust pH to 7.0. Gently heat and bring to boiling. Distribute into tubes or flasks. Autoclave for 15 min at 15 psi pressure–121°C. Pour into sterile Petri dishes or leave in tubes.

Use: For the cultivation and maintenance of *Bacillus licheniformis, Bacillus* species, *Clavibacter michiganense, Flavobacterium indologenes, Hafnia alvei, Pseudomonas fluorescens*, and *Serratia marcescens.*

Yeast Extract Glucose Broth

Composition per liter:

Glucose	15.0g
K_2HPO_4	5.2g
KH_2PO_4	3.18g
NH_4Cl	0.54g
Yeast extract	0.5g
$MgSO_4$	0.12g
Trace elements solution	5.0mL

pH 7.0 ± 0.2 at 25°C

Trace Elements Solution:

Composition per liter:

$ZnSO_4 \cdot 7H_2O$	0.287g
$CuSO_4 \cdot 5H_2O$	0.249g
$MnSO_4 \cdot 4H_2O$	0.223g
$Na_2MoO_4 \cdot 2H_2O$	0.124g
$CaCl_2 \cdot 6H_2O$	0.118g
KJ	0.083g
H_3BO_3	0.03g

Preparation of Trace Elements Solution: Add components to distilled/deionized water and bring volume to 1.0L. Mix thoroughly.

Preparation of Medium: Add components to distilled/deionized water and bring volume to 1.0L. Mix thoroughly. Adjust pH to 7.0. Distribute into tubes or flasks. Autoclave for 15 min at 15 psi pressure–121°C.

Use: For the cultivation and maintenance of *Bacillus licheniformis, Bacillus* species, *Clavibacter michiganense, Flavobacterium indologenes, Hafnia alvei, Pseudomonas fluorescens*, and *Serratia marcescens.*

Yeast Extract Glucose Calcium Carbonate Agar

Composition per liter:

$CaCO_3$	20.0g
Glucose	20.0g
Agar	15.0g
Yeast extract	10.0g

Preparation of Medium: Add components to distilled/deionized water and bring volume to 1.0L. Mix thoroughly. Gently heat and bring to boiling. Distribute into tubes or flasks. Autoclave for 15 min at 15 psi pressure–121°C. Pour into sterile Petri dishes or leave in tubes.

Use: For the isolation and cultivation of *Erwinia* species.

Yeast Extract Glucose Carbonate Medium
See: **YGC Medium**

Yeast Extract Glucose Carbonate Peptone Medium
See: **YGCP Medium**

Yeast Extract Glucose Citrate Medium
See: **YGC Medium**

Yeast Extract Glucose Citrate Medium with Cysteine
See: **YGC Medium with Cysteine**

Yeast Extract Glucose Medium

Composition per liter:

Agar	15.0g
Yeast extract	10.0g
Glucose	10.0g

Preparation of Medium: Add components to tap water and bring volume to 1.0L. Mix thoroughly. Gently heat and bring to boiling. Distribute into tubes or flasks. Autoclave for 15 min at 15 psi pressure–121°C. Pour into sterile Petri dishes or leave in tubes.

Use: For the cultivation of a variety of bacteria, including *Streptomyces* species, *Rhodococcus* species, and others.

Yeast Extract Glycerol Medium

Composition per liter:

Agar	15.0g
Yeast extract	5.0g
Glycerol	50.0mL

Preparation of Medium: Add components to distilled/deionized water and bring volume to 1.0L. Mix thoroughly. Gently heat and bring to boiling. Distribute into tubes or flasks. Autoclave for 15 min at 15 psi pressure–121°C. Pour into sterile Petri dishes or leave in tubes.

Use: For the cultivation and maintenance of *Geodermatophilus obscurus* subspecies *utahensis*.

Yeast Extract HiVeg Agar

Composition per liter:

Agar	15.0g
Plant peptone	5.0g
Yeast extract	3.0g

pH 7.2 ± 0.2 at 25°C

Source: This medium is available as a premixed powder from HiMedia.

Preparation of Medium: Add components to distilled/deionized water and bring volume to 1.0L. Mix thoroughly. Gently heat and bring to boiling. Distribute into tubes or flasks. Autoclave for 15 min at 15 psi pressure–121°C. Pour into sterile Petri dishes or leave in tubes.

Use: For the enumeration of microorganisms in potable and freshwater samples. A highly nutritive medium recommended for plate count of microorganisms in water.

Yeast Extract Malt Extract Agar
See: **ISP Medium 2**

Yeast Extract Malt Extract Agar, Diluted 1/10

Composition per liter:

Agar	20.0g
Malt extract	1.0g
Yeast extract	0.4g
Glucose	0.4g

pH 7.3 ± 0.2 at 25°C

Preparation of Medium: Add components to distilled/deionized water and bring volume to 1.0L. Mix thoroughly. Adjust pH to 7.3. Gently heat and bring to boiling. Distribute into tubes or flasks. Autoclave for 15 min at 15 psi pressure–121°C. Pour into sterile Petri dishes or leave in tubes.

Use: For the cultivation and maintenance of *Actinomadura atramentaria, Microtetraspora africana, Parvopolyspora pallida,* and *Streptosporangium fragile.*

Yeast Extract Malt Extract Glucose Agar

Composition per liter:

Agar	20.0g
Glucose	10.0g
Neopeptone	5.0g
Malt extract	3.0g
Yeast extract	3.0g

Preparation of Medium: Add components to distilled/deionized water and bring volume to 1.0L. Mix thoroughly. Gently heat and bring to boiling. Distribute into tubes or flasks. Autoclave for 15 min at 15 psi pressure–121°C. Pour into sterile Petri dishes or leave in tubes.

Use: For the isolation and cultivation of yeasts.

Yeast Extract Malt Extract Glucose Agar

Composition per liter:

Agar	20.0g
Malt extract	10.0g
Glucose	4.0g
Yeast extract	4.0g

pH 7.2 ± 0.2 at 25°C

Preparation of Medium: Add components to distilled/deionized water and bring volume to 1.0L. Mix thoroughly. Gently heat and bring to boiling. Distribute into tubes or flasks. Autoclave for 15 min at 15 psi pressure–121°C. Pour into sterile Petri dishes or leave in tubes.

Use: For the cultivation and maintenance of *Nocardia asteroides, Nocardia farcinica, Streptomyces antibioticus, Streptomyces argenteolus, Streptomyces aureofaciens, Streptomyces bluensis, Streptomyces caelestis, Streptomyces cinnamoneus, Streptomyces echinatus, Streptomyces griseocarneus, Streptomyces griseus, Streptomyces hawaiien-*

sis, Streptomyces kanamyceticus, Streptomyces kentuckensis, Streptomyces murinus, Streptomyces netropsis, Streptomyces niveus, Streptomyces nogalater, Streptomyces nousei, Streptomyces paucisporogenes, Streptomyces rimosus, Streptomyces sparsogenes, Streptomyces spectabilis, Streptomyces tendae, Streptomyces tenebraruis, Streptomyces violaceoruber, Streptomyces viridifaciens, and *Vibrio salmonicida.*

Yeast Extract Mannitol Agar

Composition per liter:

Agar	15.0g
Mannitol	10.0g
$CaCO_3$	4.0g
K_2HPO_4	0.5g
Yeast extract	0.4g
$MgSO_4 \cdot 7H_2O$	0.2g
NaCl	0.1g

pH 6.8–7.0 ± 0.2 at 25°C

Preparation of Medium: Add components to distilled/deionized water and bring volume to 1.0L. Omit $CaCO_3$ if a clear solution is needed. Mix thoroughly. Gently heat and bring to boiling. Distribute into tubes or flasks. Autoclave for 15 min at 15 psi pressure–121°C. Pour into sterile Petri dishes or leave in tubes.

Use: For the cultivation of members of the rhizobiaceae.

Yeast Extract Medium

Composition per liter:

Yeast extract	10.0g

Preparation of Medium: Add yeast extract to distilled/deionized water and bring volume to 1.0L. Mix thoroughly. Distribute into tubes or flasks. Autoclave for 15 min at 15 psi pressure–121°C.

Use: For the cultivation of *Pseudomonas cepacia.*

Yeast Extract Medium with Sodium Sulfide

Composition per liter:

Yeast extract	10.0g
Na_2S	0.15g

Preparation of Medium: Add yeast extract to distilled/deionized water and bring volume to 1.0L. Mix thoroughly. Autoclave for 15 min at 15 psi pressure–121°C. Immediately prior to inoculation, add 0.15g of Na_2S. Mix thoroughly. Aseptically distribute into sterile tubes or flasks.

Use: For the cultivation of *Rhodospirillum molischianum.*

Yeast Extract Mineral Agar

Composition per liter:

Agar	15.0g
Yeast extract	4.0g
$NaHPO_4 \cdot 12H_2O$	3.5g
K_2HPO_4	1.0g
NH_4Cl	0.5g
$MgSO_4 \cdot 7H_2O$	0.03g

pH 7.0 ± 0.2 at 25°C

Preparation of Medium: Add components to distilled/deionized water and bring volume to 1.0L. Mix thoroughly. Gently heat and bring to boiling. Distribute into tubes or flasks. Autoclave for 15 min at 15 psi pressure–121°C. Pour into sterile Petri dishes or leave in tubes.

Use: For the cultivation of *Bacillus azotoformans.*

Yeast Extract Mineral Medium (DSMZ Medium 259)

Composition per liter:

Agar	15.0g
Yeast extract	4.0g
$Na_2HPO_4 \cdot 12H_2O$	3.5g
K_2HPO_4	1.0g
NH_4Cl	0.5g
$MgSO_4 \cdot 7H_2O$	0.03g

pH 7.1 ± 0.2 at 25°C

Preparation of Medium: Add components to distilled/deionized water and bring volume to 1.0L. Mix thoroughly. Gently heat and bring to boiling. Distribute into tubes or flasks. Autoclave for 15 min at 15 psi pressure–121°C. Pour into sterile Petri dishes or leave in tubes.

Use: For the cultivation and maintenance of *Bacillus azotoformans.*

Yeast Extract Nutrient Agar Medium

See: **YNA Medium**

Yeast Extract Nutrient Gelatin Medium

See: **YNG Medium**

Yeast Extract Peptone Beef Extract Medium

See: **YEPB Medium**

Yeast Extract Peptone Starch Agar

Composition per liter:

Agar	18.0g
Soluble starch	10.0g
Peptone	10.0g
$CaCO_3$	5.0g
Sodium acetate	5.0g
Yeast extract	3.0g
KH_2PO_4	0.5g
K_2HPO_4	0.5g
$MgSO_4 \cdot 7H_2O$	0.3g
Sodium citrate	0.027g
NaCl	0.01g
$MnSO_4 \cdot 5H_2O$	0.01g
$CuSO_4 \cdot 5H_2O$	1.0mg
$CoCl_2 \cdot 6H_2O$	1.0mg
$FeSO_4 \cdot 7H_2O$	1.0mg

Preparation of Medium: Add components to tap water and bring volume to 1.0L. Mix thoroughly. Gently heat and bring to boiling. Distribute into tubes or flasks. Autoclave for 15 min at 15 psi pressure–121°C. Pour into sterile Petri dishes or leave in tubes.

Use: For the cultivation and maintenance of *Bacillus* species that utilize starch as a carbon source.

Yeast Extract Peptone Sulfate Cysteine Medium

See: **YPSC Medium**

Yeast Extract Phosphate Agar (YEP Agar)

Composition per liter:

Agar	20.0g
Yeast extract	1.0g
KH_2PO_4	0.3g

Na_2HPO_4	0.2g
Phenol Red	1.0mg

Source: This medium is available as a premixed powder from BD Diagnostic Systems.

Preparation of Medium: Add components to distilled/deionized water and bring volume to 1.0L. Mix thoroughly. Gently heat and bring to boiling. Distribute into tubes or flasks. Autoclave for 15 min at 15 psi pressure–121°C. Pour into sterile Petri dishes or leave in tubes.

Use: For the isolation of dimorphic pathogenic fungi from clinical specimens.

Yeast Extract Proteose Peptone Medium
See: **YEPP Medium**

Yeast Extract Rose Bengal HiVeg Broth Base with Sorbose

Composition per liter:

Na_2HPO_4	17.25g
Yeast extract	5.0g
Synthetic detergent	2.0g
NaCl	1.0g
Sodium pyruvate	1.0g
Rose Bengal	0.04g
$MgSO_4$	0.01g
Sorbose solution	100.0mL

pH 7.9 ± 0.2 at 25°C

Source: This medium, without sorbose solution, is available as a premixed powder from HiMedia.

Sorbose Solution:
Composition per 100.0mL:

Sorbose	4.0g

Preparation of Sorbose Solution: Add sorbose to distilled/deionized water and bring volume to 100.0mL. Filter sterilize.

Preparation of Medium: Add components, except sorbose solution, to distilled/deionized water and bring volume to 900.0mL. Mix thoroughly. Gently heat and bring to boiling. Autoclave for 15 min at 15 psi pressure–121°C. Cool to 45–50°C. Aseptically add 100.0mL of sterile sorbose solution. Mix thoroughly. Pour into sterile Petri dishes or distribute into sterile tubes.

Use: For the cold enrichment for recovery of *Yersinia enterocolitica* and *Yersinia pseudotuberculosis* from foods.

Yeast Extract Skim Milk Agar

Composition per liter:

Agar	15.0g
Skim milk powder	10.0g
Yeast extract	1.0g

Preparation of Medium: Add components to distilled/deionized water and bring volume to 1.0L. Mix thoroughly. Gently heat and bring to boiling. Distribute into tubes or flasks. Autoclave for 15 min at 15 psi pressure–121°C. Pour into sterile Petri dishes or leave in tubes.

Use: For the cultivation of *Lysobacter enzymogenes*.

Yeast Extract Sodium Lactate Medium

Composition per liter:

Agar	15.0g
Pancreatic digest of casein	10.0g
Yeast extract	10.0g
Sodium lactate	10.0g
KH_2PO_4	2.5g
$MnSO_4$	5.0mg

pH 7.0 ± 0.2 at 25°C

Preparation of Medium: Add components to distilled/deionized water and bring volume to 1.0L. Mix thoroughly. Gently heat and bring to boiling. Distribute into tubes or flasks. Autoclave for 15 min at 15 psi pressure–121°C. Pour into sterile Petri dishes or leave in tubes.

Use: For the isolation, cultivation, and maintenance of *Propionibacterium* species.

Yeast Extract Sucrose Agar (YESA) (ATCC Medium 2125)

Composition per liter:

Sucrose	20.0g
Agar	15.0g
Yeast extract	.4.0g
KH_2PO_4	1.0g
$MgSO_4 \cdot 7H_2O$	0.5g

pH 6.2 ± 0.2 at 25°C

Preparation of Medium: Add components to distilled/deionized water and bring volume to 1.0L. Mix thoroughly. Adjust pH to 6.2. Gently heat and bring to boiling. Distribute into tubes or flasks. Autoclave for 15 min at 15 psi pressure–121°C. Pour into sterile Petri dishes or leave in tubes.

Use: For the cultivation and maintenance of various fungi.

Yeast Extract Sucrose Agar

Composition per liter:

Sucrose	150.0g
Agar	20.0g
Yeast extract	.20.0g
$MgSO_4 \cdot 7H_2O$	0.5g
Trace metals solution	1.0mL

pH 6.5 ± 0.2 at 25°C

Trace Metals Solution:
Composition per 100.0mL:

$ZnSO_4 \cdot 7H_2O$	1.0g
$CuSO_4 \cdot 5H_2O$	0.5g

Preparation of Trace Metals Solution: Add components to distilled/deionized water and bring volume to 100.0mL. Mix thoroughly.

Preparation of Medium: Add components to distilled/deionized water and bring volume to 1.0L. Mix thoroughly. Adjust pH to 6.5. Gently heat and bring to boiling. Distribute into tubes or flasks. Autoclave for 15 min at 15 psi pressure–121°C. Pour into sterile Petri dishes or leave in tubes.

Use: For the cultivation and maintenance of various fungi.

Yeast Extract Tryptone Medium
See: **YT Medium**

Yeast Extract Tryptone NaCl Medium
See: **YTN Medium**

Yeast Fermentation Broth

Composition per liter:

Carbohydrate	10.0g
Pancreatic digest of gelatin	7.5g
Yeast extract	5.5g
Bromcresol Purple	16.0mg

Source: This medium is available as a premixed powder from BD Diagnostic Systems.

Preparation of Medium: Add components to distilled/deionized water and bring volume to 1.0L. Mix thoroughly. Distribute into test tubes, each containing an inverted Durham tube. Autoclave for 15 min at 15 psi pressure–121°C.

Use: For fermentation tests of specific carbohydrates used in the characterization and identification of yeasts. Gas accumulation in the Durham tube and a color change of the medium to yellow indicates carbohydrate fermentation.

Yeast Fermentation HiVeg Broth Base with Carbohydrate (Bromcresol Purple HiVeg Broth Base)

Composition per liter:

Plant peptone	10.0g
NaCl	5.0g
Plant extract	3.0g
Bromcresol Purple	0.04g
Carbohydrate solution	50.0mL

pH 7.0 ± 0.2 at 25°C

Source: This medium, without carbohydrate solution, is available as a premixed powder from HiMedia.

Carbohydrate Solution:

Composition per 100.0mL:

Carbohydrate	10.0g

Preparation of Carbohydrate Solution: Add carbohydrate to distilled/deionized water and bring volume to 100.0mL. Adonitol, arabinose, cellobiose, glucose, dulcitol, fructose, galactose, inositol, lactose, maltose, mannitol, raffinose, rhamnose, salicin, sorbitol, sucrose, trehalose, xylose, or other carbohydrates may be used. Mix thoroughly. Filter sterilize.

Preparation of Medium: Add components, except carbohydrate solution, to distilled/deionized water and bring volume to 1.0L. Mix thoroughly. Gently heat and bring to boiling. Distribute in 10.0mL volumes into test tubes containing inverted Durham tubes. Autoclave for 15 min at 15 psi pressure–121°C. Cool to 25°C. Add 0.5mL of sterile carbohydrate solution to each tube.

Use: For the determination of carbohydrate fermentation reactions of microorganisms.

Yeast Fermentation Medium

Composition per liter:

Peptone	7.5g
Yeast extract	4.5g
Bromthymol Blue (1.6% solution)	1.0mL
Carbohydrate solution	1.0mL

Carbohydrate Solution:

Composition per 10.0mL:

Carbohydrate	0.6g

Preparation of Carbohydrate Solution: Add carbohydrate to distilled/deionized water and bring volume to 10.0mL. Glucose, maltose, lactose, galactose, or trehalose may be used. If raffinose is used, prepare a 12% solution. Mix thoroughly. Filter sterilize.

Preparation of Medium: Add components, except carbohydrate solution, to distilled/deionized water and bring volume to 1.0L. Mix thoroughly. Gently heat and bring to boiling. Distribute in 2.0mL volumes into test tubes that contain an inverted Durham tube. Autoclave for 15 min at 15 psi pressure–121°C. Cool to 45°–50°C. Aseptically add 1.0mL of sterile carbohydrate solution. Mix thoroughly.

Use: For the cultivation and differentiation of yeast based on carbohydrate fermentation patterns. Yeasts that can ferment a specific carbohydrate turn the medium yellow.

Yeast Glucose Agar

Composition per liter:

Agar	15.0g
Pancreatic digest of gelatin	7.75g
Beef extract	4.75g
Yeast extract	2.5g
K_2HPO_4	2.5g
Glucose	1.0g

pH 7.0 ± 0.2 at 25°C

Preparation of Medium: Add components to distilled/deionized water and bring volume to 1.0L. Mix thoroughly. Gently heat and bring to boiling. Distribute into tubes or flasks. Autoclave for 15 min at 15 psi pressure–121°C. Pour into sterile Petri dishes or leave in tubes.

Use: For the cultivation and maintenance of a wide variety of bacteria.

Yeast Glucose Agar

Composition per liter:

Glucose	20.0g
Agar	15.0g
Yeast extract	10.0g

pH 7.0 ± 0.2 at 25°C

Preparation of Medium: Add components to distilled/deionized water and bring volume to 1.0L. Mix thoroughly. Gently heat and bring to boiling. Distribute into tubes or flasks. Autoclave for 15 min at 15 psi pressure–121°C. Pour into sterile Petri dishes or leave in tubes.

Use: For the cultivation of *Acetobacter aceti, Acetobacter hansenii, Acetobacter liquefaciens, Acetobacter pasteurianus, Acetobacter* species, *Brevibacterium* species, *Dermabacter hominus, Clostridium saccharoperbutylacetonicum, Corynebacterium amycolatum, Gluconobacter asaii, Gluconobacter cerinus, Gluconobacter frateurii, Gluconobacter oxydans, Gluconobacter* species, *Lactococcus piscium,* and *Streptococcus mutans.*

Yeast Glucose Agar for *Acetobacter*

Composition per liter:

Glucose	100.0g
$CaCO_3$	20.0g
Agar	15.0g
Yeast extract	10.0g

pH 7.0 ± 0.2 at 25°C

Preparation of Medium: Add components to distilled/deionized water and bring volume to 1.0L. Mix thoroughly. Gently heat and bring to boiling. Distribute into tubes or flasks. Autoclave for 15 min at 15 psi pressure–121°C. Pour into sterile Petri dishes or leave in tubes.

Use: For the cultivation and maintenance of *Acetobacter aceti, Acetobacter pasteurianus,* and *Gluconobacter oxydans.*

Yeast Glucose Broth

Composition per liter:

Pancreatic digest of gelatin	7.75g
Beef extract	4.75g
Yeast extract	2.5g
K_2HPO_4	2.5g
Glucose	1.0g

pH 7.0 ± 0.2 at 25°C

Preparation of Medium: Add components to distilled/deionized water and bring volume to 1.0L. Mix thoroughly. Distribute into tubes or flasks. Autoclave for 15 min at 15 psi pressure–121°C.

Use: For the cultivation of *Staphylococcus caseolyticus.*

Yeast Glucose Broth

Composition per liter:

Glucose	20.0g
Yeast extract	10.0g

Preparation of Medium: Add components to distilled/deionized water and bring volume to 1.0L. Mix thoroughly. Distribute into tubes or flasks. Autoclave for 15 min at 15 psi pressure–121°C.

Use: For the cultivation of *Enterococcus mundtii, Enterococcus pseudoavium, Enterococcus raffinosus, Enterococcus solitarius, Enterococcus sulfureus, Lactococcus raffinolactis,* and *Vagococcus fluvialis.*

Yeast Glucose Broth (YGB)

Composition per liter:

Beef extract	10.0g
Peptone	10.0g
NaCl	5.0g
Glucose	5.0g
Yeast extract	3.0g

pH 6.8 ± 0.2 at 25°C

Preparation of Medium: Add components to distilled/deionized water and bring volume to 1.0L. Mix thoroughly. Adjust pH to 6.8. Distribute into tubes or flasks. Autoclave for 15 min at 15 psi pressure–121°C.

Use: For the cultivation of *Enterococcus faecalis, Streptococcus anginosus,* and *Rhodobacter sphaeroides.*

Yeast Glucose Litmus Milk

See: **YGLM**

Yeast Glucose Litmus Milk with Chalk

See: **YGLM with Chalk**

Yeast HiVeg Agar (Antibiotic HiVeg Assay Medium No. 4)

Composition per liter:

Agar	15.0g
Plant peptone	6.0g
Yeast extract	3.0g
Plant extract	1.5g
Glucose	1.0g

pH 6.6 ± 0.05 at 25°C

Source: This medium is available as a premixed powder from HiMedia.

Preparation of Medium: Add components to distilled/deionized water and bring volume to 1.0L. Mix thoroughly. Gently heat and bring to boiling. Distribute into tubes or flasks. Autoclave for 15 min at 15 psi pressure–121°C. Pour into sterile Petri dishes or leave in tubes.

Use: For antibiotic assay testing.

Yeast Malate Medium

Composition per liter:

Yeast extract	5.0g
Sodium malate	1.0g

pH 7.0 ± 0.2 at 25°C

Preparation of Medium: Add components to distilled/deionized water and bring volume to 1.0L. Mix thoroughly. Distribute into tubes or flasks. Autoclave for 15 min at 15 psi pressure–121°C.

Use: For the cultivation of *Rhodopseudomonas viridis.*

Yeast Malate Medium

Composition per liter:

$KH_2PO_4 \cdot 12H_2O$	1.0g
$NaHCO_3$	1.0g
Sodium malate	1.0g
$(NH_4)_2SO_4$	0.5g
Trace elements solution	1.0mL

pH 6.8 ± 0.2 at 25°C

Trace Elements Solution:

Composition per liter:

H_3BO_3	0.3g
$CoCl_2 \cdot 6H_2O$	0.2g
$ZnSO_4 \cdot 7H_2O$	0.1g
$MnCl_2 \cdot 4H_2O$	0.03g
$Na_2MoO_4 \cdot 2H_2O$	0.03g
$NiCl_2 \cdot 6H_2O$	0.02g
$CuCl_2 \cdot 2H_2O$	0.01g

Preparation of Trace Elements Solution: Add components to distilled/deionized water and bring volume to 1.0L. Mix thoroughly. Filter sterilize.

Preparation of Medium: Add components, except trace elements solution, to distilled/deionized water and bring volume to 999.0mL. Mix thoroughly. Autoclave for 15 min at 15 psi pressure–121°C. Aseptically add 1.0mL of sterile trace elements solution. Mix thoroughly. Aseptically distribute into sterile tubes or flasks.

Use: For the cultivation of *Rhodopseudomonas viridis.*

Yeast Malate Medium (LMG 176)

Composition per 1001.0mL:

$KH_2PO_4 \cdot 12H_2O$	1.0g
$NaHCO_3$	1.0g
Sodium malate	1.0g
Yeast extract	1.0g
$(NH_4)_2SO_4$	0.5g
Trace elements solution	1.0mL

pH 6.8–7.0 at 25°C

Trace Elements Solution:
Composition per liter:

H_3BO_3	0.3g
$CoCl_2 \cdot 6H_2O$	0.2g
$ZnSO_4 \cdot 7H_2O$	0.1g
$MnCl_2 \cdot 4H_2O$	0.03g
$Na_2MoO_4 \cdot 2H_2O$	0.03g
$NiCl_2 \cdot 6H_2O$	0.02g
$CuCl_2 \cdot 2H_2O$	0.01g

Preparation of Trace Elements Solution: Add components to distilled/deionized water and bring volume to 1.0L. Mix thoroughly.

Preparation of Medium: Add components to distilled/deionized water and bring volume to 1.0L. Mix thoroughly. Adjust pH to 6.8–7.0. Distribute into tubes or flasks. Autoclave for 15 min at 15 psi pressure–121°C.

Use: For the cultivation of *Rhodopseudomonas viridis*.

Yeast Malt Agar

Composition per liter:

Agar	20.0g
Malt extract	10.0g
Glucose	4.0g
Yeast extract	4.0g

pH 7.2 ± 0.2 at 25°C

Preparation of Medium: Add components to distilled/deionized water and bring volume to 1.0L. Mix thoroughly. Gently heat and bring to boiling. Adjust pH to 7.2. Distribute into tubes or flasks. Autoclave for 15 min at 15 psi pressure–121°C. Pour into sterile Petri dishes or leave in tubes.

Use: For the cultivation and maintenance of actinomycetes, yeasts, and fungi.

Yeast Malt Agar for *Arthrobacter viscosus*

Composition per liter:

NaCl	10.0g
Pancreatic digest of casein	10.0g
Yeast extract	5.0g

pH 7.0 ± 0.2 at 25°C

Preparation of Medium: Add components to distilled/deionized water and bring volume to 1.0L. Mix thoroughly. Distribute into tubes or flasks. Autoclave for 15 min at 15 psi pressure–121°C.

Use: For the cultivation of *Arthrobacter viscosus* and *Escherichia coli*.

Yeast Malt Extract Agar (YM Agar)

Composition per liter:

Agar	20.0g
Glucose	10.0g
Peptone	5.0g
Yeast extract	3.0g
Malt extract	3.0g

pH 6.2 ± 0.2 at 25°C

Source: This medium is available as a premixed powder from BD Diagnostic Systems.

Preparation of Medium: Add components to distilled/deionized water and bring volume to 1.0L. Mix thoroughly. Gently heat and bring to boiling. Distribute into tubes or flasks. Autoclave for 15 min at 15 psi pressure–121°C. The medium may be rendered selective by adjusting the pH to 3.0–4.0 at 45°–55°C or by the addition of antibiotics at 45°–50°C or below. Pour into sterile Petri dishes or leave in tubes.

Use: For the cultivation of fungi, including yeasts, and other aciduric microorganisms such as *Actinoplanes* species, *Streptomyces* species, *Streptoverticillium* species, and *Nocardia* species.

Yeast Malt Extract Broth (YM Broth)

Composition per liter:

Glucose	10.0g
Peptone	5.0g
Yeast extract	3.0g
Malt extract	3.0g

pH 6.2 ± 0.2 at 25°C

Source: This medium is available as a premixed powder from BD Diagnostic Systems.

Preparation of Medium: Add components to distilled/deionized water and bring volume to 1.0L. Mix thoroughly. Distribute into tubes or flasks. Autoclave for 15 min at 15 psi pressure–121°C. The medium may be rendered selective by adjusting the pH to 3.0–4.0 at 45°–55°C or by the addition of antibiotics at 45°–50°C or below.

Use: For the cultivation of yeasts, molds, and other aciduric microorganisms such as *Actinoplanes* species, *Streptomyces* species, *Streptoverticillium* species, and *Nocardia* species.

Yeast Malt Extract Broth with 0.5% Calcium Carbonate (YM Broth with 0.5% $CaCO_3$)

Composition per liter:

Glucose	10.0g
Peptone	5.0g
$CaCO_3$	5.0g
Yeast extract	3.0g
Malt extract	3.0g

pH 6.2 ± 0.2 at 25°C

Preparation of Medium: Add components to distilled/deionized water and bring volume to 1.0L. Mix thoroughly. Distribute into tubes or flasks. Autoclave for 15 min at 15 psi pressure–121°C. The medium may be rendered selective by adjusting the pH to 3.0–4.0 at 45–55°C or by the addition of antibiotics at 45–50°C or below.

Use: For the cultivation and maintenance of *Dekkera anomala, Dekkera bruxellensis, Dekkera claussenii,* and *Dekkera lambica.*

Yeast Malt Extract Broth with 2.0% Calcium Carbonate (YM Broth with 2.0% $CaCO_3$)

Composition per liter:

$CaCO_3$	20.0g
Glucose	10.0g
Peptone	5.0g
Yeast extract	3.0g
Malt extract	3.0g

pH 6.2 ± 0.2 at 25°C

Preparation of Medium: Add components to distilled/deionized water and bring volume to 1.0L. Mix thoroughly. Distribute into tubes or flasks. Autoclave for 15 min at 15 psi pressure–121°C. The medium may be rendered selective by adjusting the pH to 3.0–4.0 at 45°–55°C or by the addition of antibiotics at 45°–50°C or below.

Use: For the cultivation and maintenance of *Dekkera abstinens*.

Yeast Malt Extract Broth with Glucose (YM Broth with Glucose)

Composition per liter:

Glucose	290.0g
Peptone	5.0g
$CaCO_3$	5.0g
Yeast extract	3.0g
Malt extract	3.0g

pH 6.2 ± 0.2 at 25°C

Preparation of Medium: Add components to distilled/deionized water and bring volume to 1.0L. Mix thoroughly. Distribute into tubes or flasks. Autoclave for 15 min at 15 psi pressure–121°C. The medium may be rendered selective by adjusting the pH to 3.0–4.0 at 45°–55°C or by the addition of antibiotics .

Use: For the cultivation of yeasts and filamentous fungi.

Yeast Malt Extract Broth with 1.0% Methanol (YM Broth with 1.0% Methanol)

Composition per liter:

Glucose	10.0g
Peptone	5.0g
Yeast extract	3.0g
Malt extract	3.0g
Methanol	10.0mL

pH 6.2 ± 0.2 at 25°C

Preparation of Medium: Add components, except methanol, to distilled/deionized water and bring volume to 990.0mL. Mix thoroughly. Autoclave for 15 min at 15 psi pressure–121°C. Cool to room temperature. Filter sterilize the methanol. Aseptically add the sterile methanol to the cooled sterile basal medium. Mix thoroughly. Distribute into sterile flasks or tubes.

Use: For the cultivation and maintenance of *Pichia angusta* and *Wickerhamiella domercqiae*.

Yeast Malt Extract Broth with 18% Sodium Chloride (YM Broth with 18% NaCl)

Composition per liter:

NaCl	180.0g
Glucose	10.0g
Peptone	5.0g
Yeast extract	3.0g
Malt extract	3.0g

pH 6.2 ± 0.2 at 25°C

Preparation of Medium: Add components to distilled/deionized water and bring volume to 1.0L. Mix thoroughly. Distribute into tubes or flasks. Autoclave for 15 min at 15 psi pressure–121°C.

Use: For the cultivation and maintenance of *Zygosaccharomyces rouxii*.

Yeast Malt Extract Broth with 40.0% Sucrose (YM Broth with 40.0% Sucrose)

Composition per liter:

Sucrose	400.0g
Glucose	10.0g
Peptone	5.0g
Yeast extract	3.0g
Malt extract	3.0g

pH 6.2 ± 0.2 at 25°C

Preparation of Medium: Add components to distilled/deionized water and bring volume to 1.0L. Mix thoroughly. Distribute into flasks or tubes. Autoclave for 15 min at 15 psi pressure–121°C.

Use: For the cultivation of *Chrysosporium fastidium, Chrysosporium xerophilum, Curvularia pallescens, Eupenicillium molle,* and *Talaromyces ucrainicus.*

Yeast Malt Extract Broth with 70.0% Sucrose (YM Broth with 70.0% Sucrose)

Composition per liter:

Sucrose	700.0g
Glucose	10.0g
Peptone	5.0g
Yeast extract	3.0g
Malt extract	3.0g

pH 6.2 ± 0.2 at 25°C

Preparation of Medium: Add components to distilled/deionized water and bring volume to 1.0L. Mix thoroughly. Distribute into flasks or tubes. Autoclave for 15 min at 15 psi pressure–121°C.

Use: For the cultivation of *Aspergillus penicilloides.*

Yeast Malt Extract Catalase Agar (YM Catalase Agar)

Composition per liter:

Agar	15.0g
K_2HPO_4	5.74g
Malt extract	5.0g
Yeast extract	5.0g
$NH_4H_2PO_4$	1.15g
Magnesium sulfate solution	10.0mL
Catalase solution	10.0mL

Catalase Solution:

Composition per 10.0mL:

Catalase	60.0mg

Preparation of Catalase Solution: Add catalase to distilled/deionized water and bring volume to 10.0mL. Mix thoroughly. Filter sterilize.

Magnesium Sulfate Solution:

Composition per 10.0mL:

$MgSO_4 \cdot 7H_2O$	205.0mg

Preparation of Magnesium Sulfate Solution: Add $MgSO_4 \cdot 7H_2O$ to distilled/deionized water and bring volume to 10.0mL. Mix thoroughly. Filter sterilize.

Preparation of Medium: Add components, except catalase and magnesium sulfate solutions, to distilled/deionized water and bring volume to 980.0mL. Mix thoroughly. Gently heat and bring to boiling. Distribute into tubes or flasks. Autoclave for 15 min at 15 psi pressure–121°C. Cool to 50°C. Aseptically add filter-sterilized catalase and magnesium sulfate solutions.

Use: For the cultivation and maintenance of *Rarobacter faecitabidus*.

Yeast Malt HiVeg Agar (YM HiVeg Agar)

Composition per liter:

Agar	20.0g
Glucose	10.0g
Plant peptone	5.0g
Malt extract	3.0g
Yeast extract	3.0g

pH 7.2 ± 0.2 at 25°C

Source: This medium is available as a premixed powder from HiMedia.

Preparation of Medium: Add components to distilled/deionized water and bring volume to 1.0L. Mix thoroughly. Gently heat and bring to boiling. Adjust pH to 7.2. Distribute into tubes or flasks. Autoclave for 15 min at 15 psi pressure–121°C.

Use: For the cultivation and maintenance of actinomycetes, yeasts, and fungi.

Yeast Malt HiVeg Broth (YM HiVeg Broth)

Composition per liter:

Glucose	10.0g
Plant peptone	5.0g
Malt extract	3.0g
Yeast extract	3.0g

pH 7.2 ± 0.2 at 25°C

Source: This medium is available as a premixed powder from HiMedia.

Preparation of Medium: Add components to distilled/deionized water and bring volume to 1.0L. Mix thoroughly. Gently heat and bring to boiling. Adjust pH to 7.2. Distribute into tubes or flasks. Autoclave for 15 min at 15 psi pressure–121°C.

Use: For the cultivation and maintenance of actinomycetes, yeasts, and fungi.

Yeast Mannitol Agar

Composition per liter:

Agar	15.0g
Mannitol	10.0g
K_2HPO_4	0.5g
Yeast extract	0.4g
$MgSO_4 \cdot 7H_2O$	0.2g
NaCl	0.1g

pH 6.8 ± 0.2 at 25°C

Preparation of Medium: Add components to distilled/deionized water and bring volume to 1.0L. Mix thoroughly. Gently heat and bring to boiling. Distribute into tubes or flasks. Autoclave for 15 min at 15 psi pressure–121°C. Pour into sterile Petri dishes or leave in tubes.

Use: For the cultivation of *Rhizobium* and *Azorhizobium* species.

Yeast Mannitol Agar

Composition per liter:

Agar	20.0g
Mannitol	10.0g
$Na_2HPO_4 \cdot 12H_2O$	1.2g
KH_2PO_4	0.55g
$MgSO_4 \cdot 7H_2O$	0.25g
NaCl	0.25g
Yeast extract	0.25g
$CaSO_4 \cdot 2H_2O$	30.0mg
$FeSO_4 \cdot 7H_2O$	3.5mg
H_3BO_3	500.0μg
$MnSO_4 \cdot 4H_2O$	400.0μg
$ZnSO_4 \cdot 7H_2O$	160.0μg
$CuSO_4 \cdot 5H_2O$	80.0μg

pH 7.0 ± 0.2 at 25°C

Preparation of Medium: Add components to distilled/deionized water and bring volume to 1.0L. Mix thoroughly. Gently heat and bring to boiling. Distribute into tubes or flasks. Autoclave for 15 min at 15 psi pressure–121°C. Pour into sterile Petri dishes or leave in tubes.

Use: For the cultivation of *Bradyrhizobium japonicum* and *Rhizobium leguminosarum.*

Yeast Mannitol Agar, Modified

Composition per liter:

Agar	15.0g
Mannitol	10.0g
$CaCO_3$	4.0g
K_2HPO_4	0.5g
Yeast extract	0.4g
$MgSO_4 \cdot 7H_2O$	0.2g
NaCl	0.1g

pH 6.8 ± 0.2 at 25°C

Preparation of Medium: Add components to distilled/deionized water and bring volume to 1.0L. Mix thoroughly. Gently heat and bring to boiling. Distribute into tubes or flasks. Autoclave for 15 min at 15 psi pressure–121°C. Pour into sterile Petri dishes or leave in tubes.

Use: For the cultivation of *Amycolata autotrophica, Amycolatopsis orientalis, Nocardioides albus, Promicromonospora citrea, Rhizobium fredii, Rhizobium galegae, Rhizobium leguminosarum, Rhizobium loti, Rhizobium meliloti,* and *Rhizobium tropici.*

Yeast Milk Medium

Composition per liter:

Agar	15.0g
Skim milk	10.0g
Yeast extract	1.0g

Preparation of Medium: Add components to distilled/deionized water and bring volume to 1.0L. Mix thoroughly. Gently heat and bring to boiling. Distribute into tubes or flasks. Autoclave for 15 min at 15 psi pressure–121°C. Pour into sterile Petri dishes or leave in tubes.

Use: For the cultivation and maintenance of *Lysobacter enzymogenes.*

Yeast Morphology Agar

Composition per liter:

$(NH_4)_2SO_4$	32.5g
Agar	18.0g
Glucose	10.0g
Asparagine	1.5g
KH_2PO_4	1.0g
$MgSO_4 \cdot 7H_2O$	0.5g
NaCl	0.5g
$CaCl_2 \cdot 2H_2O$	0.1g
DL-Methionine	0.02g
DL-Tryptophan	0.02g

L-Histidine·HCl	0.01g
Inositol	2.0mg
H_3BO_3	0.5mg
Calcium pantothenate	0.4mg
$MgSO_4 \cdot 7H_2O$	0.4mg
Niacin	0.4mg
Pyridoxine·HCl	0.4mg
Thiamine·HCl	0.4mg
$ZnSO_4 \cdot 7H_2O$	0.4mg
p-Aminobenzoic acid	0.2mg
$FeCl_3$	0.2mg
Riboflavin	0.2mg
$Na_2MoO_4 \cdot 4H_2O$	0.2mg
KI	0.1mg
$CuSO_4 \cdot 5H_2O$	0.04mg
Biotin	2.0μg
Folic acid	2.0μg

pH 5.6 ± 0.2 at 25°C

Source: This medium is available as a premixed powder from BD Diagnostic Systems.

Preparation of Medium: Add components to distilled/deionized water and bring volume to 1.0L. Mix thoroughly. Gently heat and bring to boiling. Distribute into tubes or flasks. Autoclave for 15 min at 15 psi pressure–121°C.

Use: For agar dilution and diffusion disk testing with 5-flucytosine. For the microbiological assay of flucytosine using *Candida kefyr* ATCC 28838 or *Saccharomyces cerevisiae* ATCC 36375 as the indicator microorganism.

Yeast Nitrogen Base

Composition per liter:

$(NH_4)_2SO_4$	5.0g
KH_2PO_4	1.0g
$MgSO_4 \cdot 7H_2O$	0.5g
NaCl	0.1g
$CaCl_2 \cdot 2H_2O$	0.1g
DL-Methionine	0.02g
DL-Tryptophan	0.02g
L-Histidine·HCl	0.01g
Inositol	2.0mg
KI	1.0mg
H_3BO_3	0.5mg
$ZnSO_4 \cdot 7H_2O$	0.4mg
$MnSO_4 \cdot 4H_2O$	0.4mg
Thiamine·HCl	0.4mg
Pyroxidine·HCl	0.4mg
Niacin	0.4mg
Calcium pantothenate	0.4mg
p-Aminobenzoic acid	0.2mg
Riboflavin	0.2mg
$FeCl_3$	0.2mg
$Na_2MoO_4 \cdot 4H_2O$	0.2mg
$CuSO_4 \cdot 5H_2O$	0.04mg
Folic acid	2.0μg
Biotin	2.0μg

pH 5.5 ± 0.2 at 25°C

Source: This medium is available as a premixed powder from BD Diagnostic Systems.

Preparation of Medium: Add components to distilled/deionized water and bring volume to 1.0L. Mix thoroughly. Distribute into tubes or flasks. Autoclave for 15 min at 15 psi pressure–121°C. Alternately for carbon assimilation tests, prepare a 10X concentrated solution by adding components to distilled/deionized water and bringing volume to 100.0mL. Mix thoroughly. Distribute into tubes or flasks. Autoclave for 15 min at 15 psi pressure–121°C. Prepare a carbohydrate solution by adding 0.5g of carbohydrate to 90.0mL of distilled/deionized water. Mix thoroughly. Filter sterilize. Aseptically add 0.5mL of the 10X concentrated solution to 4.5mL of the filter-sterilized carbohydrate solution. Mix thoroughly.

Use: For carbohydrate assimilation tests in the characterization and identification of yeasts.

Yeast Nitrogen Base, 10X with Asparagine and Glucose

Composition per liter:

Glucose	10.0g
$(NH_4)_2SO_4$	5.0g
L-Asparagine	1.5g
KH_2PO_4	1.0g
$MgSO_4 \cdot 7H_2O$	0.5g
NaCl	0.1g
$CaCl_2 \cdot 2H_2O$	0.1g
DL-Methionine	0.02g
DL-Tryptophan	0.02g
L-Histidine·HCl	0.01g
Inositol	2.0mg
H_3BO_3	0.5mg
$ZnSO_4 \cdot 7H_2O$	0.4mg
$MnSO_4 \cdot 4H_2O$	0.4mg
Thiamine·HCl	0.4mg
Pyridoxine·HCl	0.4mg
Niacin	0.4mg
Calcium pantothenate	0.4mg
p-Aminobenzoic acid	0.2mg
Riboflavin	0.2mg
$FeCl_3$	0.2mg
$Na_2MoO_4 \cdot 4H_2O$	0.2mg
KI	0.1mg
$CuSO_4 \cdot 5H_2O$	0.04mg
Folic acid	2.0μg
Biotin	2.0μg

pH 5.6 ± 0.2 at 25°C

Preparation of Medium: Add components to distilled/deionized water and bring volume to 1.0L. Dilute 100.0mL of 10X medium with 900.0mL of distilled/deionized water. Mix thoroughly. Filter sterilize. Aseptically distribute into sterile tubes or flasks.

Use: For susceptibility tests with yeasts and fungi.

Yeast Nitrogen Base with Carbohydrate

Composition per liter:

Carbohydrate	5.0g
$(NH_4)_2SO_4$	5.0g
KH_2PO_4	1.0g
$MgSO_4 \cdot 7H_2O$	0.5g
NaCl	0.1g
$CaCl_2 \cdot 2H_2O$	0.1g
DL-Methionine	0.02g
DL-Tryptophan	0.02g
L-Histidine·HCl	0.01g
Inositol	2.0mg

KI	0.1mg
H_3BO_3	0.5mg
$ZnSO_4 \cdot 7H_2O$	0.4mg
$MnSO_4 \cdot 4H_2O$	0.4mg
Thiamine·HCl	0.4mg
Pyroxidine·HCl	0.4mg
Niacin	0.4mg
Calcium pantothenate	0.4mg
p-Aminobenzoic acid	0.2mg
Riboflavin	0.2mg
$FeCl_3$	0.2mg
$Na_2MoO_4 \cdot 4H_2O$	0.2mg
$CuSO_4 \cdot 5H_2O$	0.04mg
Folic acid	2.0μg
Biotin	2.0μg

pH 5.6 ± 0.2 at 25°C

Preparation of Medium: Add components to distilled/deionized water and bring volume to 1.0L. Mix thoroughly. Filter sterilize. Aseptically distribute into tubes or flasks.

Use: For carbohydrate assimilation tests in the characterization and identification of yeasts.

Yeast Nitrogen Base Glucose Broth

Composition per liter:

Yeast nitrogen base	25.0mL
Glucose solution	25.0mL

pH 5.6 ± 0.2 at 25°C

Yeast Nitrogen Base:

Composition per 500.0mL:

$(NH_4)_2SO_4$	5.0g
KH_2PO_4	1.0g
$MgSO_4 \cdot 7H_2O$	0.5g
NaCl	0.1g
$CaCl_2 \cdot 2H_2O$	0.1g
DL-Methionine	0.02g
DL-Tryptophan	0.02g
L-Histidine·HCl	0.01g
Inositol	2.0mg
H_3BO_3	0.5mg
$ZnSO_4 \cdot 7H_2O$	0.4mg
$MnSO_4 \cdot 4H_2O$	0.4mg
Thiamine·HCl	0.4mg
Pyroxidine·HCl	0.4mg
Niacin	0.4mg
Calcium pantothenate	0.4mg
p-Aminobenzoic acid	0.2mg
Riboflavin	0.2mg
$FeCl_3$	0.2mg
$Na_2MoO_4 \cdot 4H_2O$	0.2mg
KI	0.1mg
$CuSO_4 \cdot 5H_2O$	0.04mg
Folic acid	2.0μg
Biotin	2.0μg

Source: Yeast nitrogen base is available as a premixed powder from BD Diagnostic Systems.

Preparation of Yeast Nitrogen Base: Add components to distilled/deionized water and bring volume to 500.0mL. Mix thoroughly. Filter sterilize.

Glucose Solution:

Composition per 500.0mL:

Glucose	10.0g

Preparation of Glucose Solution: Add glucose to distilled/deionized water and bring volume to 500.0mL. Mix thoroughly. Filter sterilize.

Preparation of Medium: Aseptically combine 25.0mL of sterile yeast nitrogen base and 25.0mL of sterile glucose solution. Mix thoroughly.

Use: For the cultivation and enrichment of yeast from sewage and polluted waters.

Yeast Peptone Agar (ATCC Medium 1858)

Composition per liter:

Agar	15.0g
Yeast extract	2.5g
Peptone	2.5g

pH 7.0 ± 0.4 at 25°C

Preparation of Medium: Add components to distilled/deionized water and bring volume to 1.0L. Mix thoroughly. Gently heat and bring to boiling. Distribute into tubes or flasks. Autoclave for 15 min at 15 psi pressure–121°C. Pour into sterile Petri dishes or leave in tubes.

Use: For the cultivation of *Cytophaga lytica, Pseudomonas lanceolata, Rhodobacter capsulatus, Rhodobacter sphaeroides, Rhodopseudomonas blastica, Rhodopseudomonas palustris, Rhodospirillum rubrum, Rubrivivax gelatinosus, Sphaerotilus natans,* and *Zoogloea ramigera.*

Yeast Peptone Broth

Composition per liter:

Yeast extract	2.5g
Peptone	2.5g

pH 7.0 ± 0.2 at 25°C

Preparation of Medium: Add components to distilled/deionized water and bring volume to 1.0L. Mix thoroughly. Distribute into tubes or flasks. Autoclave for 15 min at 15 psi pressure–121°C.

Use: For the cultivation of *Rhodopseudomonas* species.

Yeast Peptone Salt Medium

Composition per liter:

Agar	15.0g
Peptone	2.5g
Yeast extract	2.5g
NaCl	1.25g

pH 7.0 ± 0.4 at 25°C

Preparation of Medium: Add components to distilled/deionized water and bring volume to 1.0L. Mix thoroughly. Gently heat and bring to boiling. Distribute into tubes or flasks. Autoclave for 15 min at 15 psi pressure–121°C. Pour into sterile Petri dishes or leave in tubes.

Use: For the cultivation of *Deinobacter grandis*.

Yeast Peptone Succinate Medium (DSMZ Medium 988)

Composition per liter:

Yeast extract	3.0g
Peptone	3.0g
Sodium succinate	2.3g

pH 7.2 ± 0.2 at 25°C

Preparation of Medium: Add components to distilled/deionized water and bring volume to 1.0L. Mix thoroughly. Adjust pH to 7.2. Distribute into tubes or flasks. Autoclave for 15 min at 15 psi pressure–121°C.

Use: For the cultivation of *Sphingomonas aurantiaca* and other *Sphingomonas* spp.

Yeast Starch Agar

Composition per liter:

Agar	15.0g
Starch, soluble	15.0g
Yeast extract	4.0g
K_2HPO_4	0.5g
$MgSO_4 \cdot 7H_2O$	0.5g

pH 7.4 ± 0.2 at 25°C

Preparation of Medium: Add components to distilled/deionized water and bring volume to 1.0L. Mix thoroughly. Adjust pH to 7.4. Gently heat and bring to boiling. Distribute into tubes or flasks. Autoclave for 15 min at 15 psi pressure–121°C. Pour into sterile Petri dishes or leave in tubes.

Use: For the cultivation of *Streptomyces albospinus, Streptomyces longwoodensis, Streptomyces cystargineus,* and *Promicromonospora sukumoe.*

Yeast Starch Agar, Diluted 1/10

Composition per liter:

Agar	15.0g
Soluble starch	1.0g
Yeast extract	0.2g

pH 7.3 ± 0.2 at 25°C

Preparation of Medium: Add components to distilled/deionized water and bring volume to 1.0L. Mix thoroughly. Adjust pH to 7.3. Gently heat and bring to boiling. Distribute into tubes or flasks. Autoclave for 15 min at 15 psi pressure–121°C. Pour into sterile Petri dishes or leave in tubes.

Use: For the cultivation and maintenance of *Catellatospora citrea, Spirillospora* species, and *Streptosporangium viridialbum.*

Yeast Starch Agar, Half Strength

Composition per liter:

Agar	15.0g
Soluble starch	5.0g
Yeast extract	1.0g

pH 7.3 ± 0.2 at 25°C

Preparation of Medium: Add components to distilled/deionized water and bring volume to 1.0L. Mix thoroughly. Gently heat and bring to boiling. Distribute into tubes or flasks. Autoclave for 15 min at 15 psi pressure–121°C. Pour into sterile Petri dishes or leave in tubes.

Use: For the cultivation and maintenance of *Microtetraspora ferruginea* and *Streptomyces sporoverrucocus.*

Yeast Starch Agar A (DSMZ Medium 1027)

Composition per liter:

Agar	15.0g
Starch, soluble	10.0g
Yeast extract	2.0g

pH 7.3 ± 0.2 at 25°C

Preparation of Medium: Add components to distilled/deionized water and bring volume to 1.0L. Mix thoroughly. Gently heat and bring to boiling. Distribute into tubes or flasks. Autoclave for 15 min at 15 psi pressure–121°C. Pour into sterile Petri dishes or leave in tubes.

Use: For the cultivation of *Streptomyces* spp. For the cultivation and maintenance of *Actinomadura livida,* numerous *Actinoplanes* species, *Actinosynnema mirum, Amycolata hydrocarbonoxydans, Amycolatopsis mediterranei, Amycolatopsis orientalis, Catellatospora ferruginea, Dactylosporangium aurantiacum, Microbispora chromogenes, Microbispora thermodiastatica, Micromonospora chalcea, Microtetraspora angiospora, Microtetraspora incanescens, Microtetraspora niveoalba, Microtetraspora spiralis, Nocardia nitrifians, Planobispora rosea, Planomonospora venezuelensis, Saccharothrix aerocolonigenes, Saccharothrix mutabilis, Sporichthya polymorpha,* most *Streptomyces* species, many *Streptosporangium* species, and *Trichotomospora caesia.*

Yeast Synthetic Medium with 5′-dTMP

Composition per liter:

Glucose	20.0g
Agar	15.0g
Yeast nitrogen base without amino acids	6.7g
Casamino acids without vitamins	2.0g
2′-Deoxythymidine-5′-monophosphate	7.2mg

Preparation of Medium: Add glucose, yeast nitrogen base without amino acids, casamino acids without vitamins, and 2′-deoxythymidine-5′-monophosphate to distilled/deionized water and bring volume to 200.0mL. Mix thoroughly. Filter sterilize. Add agar to distilled/deionized water and bring volume to 800.0mL. Mix thoroughly. Gently heat and bring to boiling. Autoclave for 15 min at 15 psi pressure–121°C. Cool to 50°C. Aseptically combine the two sterile solutions. Mix thoroughly. Pour into sterile Petri dishes or aseptically distribute into sterile tubes.

Use: For the cultivation and maintenance of *Saccharomyces cerevisiae.*

Yeast Synthetic Minimal Medium

Composition per liter:

D-Glucose	20.0g
Agar	15.0g
$(NH_4)_2SO_4$	5.0g
KH_2PO_4	1.0g
$MgSO_4 \cdot 7H_2O$	0.5g
NaCl	0.1g
$CaCl_2 \cdot 2H_2O$	0.1g
Inositol	2.0mg
H_3BO_3	0.5mg
$ZnSO_4 \cdot 7H_2O$	0.4mg
$MnSO_4 \cdot 4H_2O$	0.4mg
Thiamine·HCl	0.4mg
Pyridoxine·HCl	0.4mg
Niacin	0.4mg
Calcium pantothenate	0.4mg
p-Aminobenzoic acid	0.2mg
Riboflavin	0.2mg
$FeCl_3$	0.2mg
$Na_2MoO_4 \cdot 4H_2O$	0.2mg
KI	0.1mg
$CuSO_4 \cdot 5H_2O$	0.04mg

Folic acid 2.0μg
Biotin 2.0μg

pH 5.6 ± 0.2 at 25°C

Preparation of Medium: Add agar to 900.0mL of distilled/deionized water. Mix thoroughly. Gently heat and bring to boiling. Distribute into tubes or flasks. Autoclave for 15 min at 15 psi pressure–121°C. Cool to 45°–50°C. In a separate flask, add remaining components to 100.0mL of distilled/deionized water. Mix thoroughly. Filter sterilize. Aseptically combine the two sterile solutions. Mix thoroughly. Pour into sterile Petri dishes.

Use: For the cultivation of a wide variety of heterotrophic microorganisms.

Yeast Synthetic Minimal Medium

Composition per liter:

Glucose 20.0g
Agar 15.0g
Yeast nitrogen base without amino acids 6.7g

Preparation of Medium: Add glucose and yeast nitrogen base without amino acids to distilled/deionized water and bring volume to 100.0mL. Mix thoroughly. Filter sterilize. Add agar to distilled/deionized water and bring volume to 900.0mL. Mix thoroughly. Gently heat and bring to boiling. Autoclave for 15 min at 15 psi pressure–121°C. Cool to 50°C. Aseptically combine the two sterile solutions. Mix thoroughly. Pour into sterile Petri dishes or aseptically distribute into sterile tubes.

Use: For the cultivation and maintenance of *Pichia angusta* and *Saccharomyces cerevisiae.*

Yeast Tryptone Medium

Composition per liter:

Pancreatic digest of casein 10.0g
NaCl 10.0g
Yeast extract 5.0g

pH 7.0 ± 0.2 at 25°C

Preparation of Medium: Add components to distilled/deionized water and bring volume to 1.0L. Mix thoroughly. Distribute into tubes or flasks. Autoclave for 15 min at 15 psi pressure–121°C.

Use: For the cultivation of *Escherichia coli.*

Yeast Tryptone Medium with Streptomycin

Composition per liter:

Pancreatic digest of casein 10.0g
NaCl 10.0g
Yeast extract 5.0g
Streptomycin 0.2g

pH 7.0 ± 0.2 at 25°C

Preparation of Medium: Add components to distilled/deionized water and bring volume to 1.0L. Mix thoroughly. Distribute into tubes or flasks. Autoclave for 15 min at 15 psi pressure–121°C.

Use: For the cultivation of *Escherichia coli.*

Yeast Tryptone Starch Medium

Composition per liter:

Agar 15.0g
Soluble starch 10.0g
Yeast extract 5.0g
Pancreatic digest of casein 5.0g
KH_2PO_4 .... 2.0g
$CaCl_2 \cdot 2H_2O$ 0.5g
$MnCl_2 \cdot 4H_2O$ 0.5g

Preparation of Medium: Add components to distilled/deionized water and bring volume to 1.0L. Mix thoroughly. Gently heat and bring to boiling. Distribute into tubes or flasks. Autoclave for 15 min at 15 psi pressure–121°C. Pour into sterile Petri dishes or leave in tubes.

Use: For the cultivation and maintenance of *Bacillus circulans.*

Yeast Water Agar

Composition per liter:

Glucose 20.0g
Agar 15.0g
Casein hydrolysate 5.0g
Yeast extract 4.0g
KH_2PO_4 .... 0.55g
KCl 0.4g
$CaCl_2$ 0.13g
$MgCl_2 \cdot 7H_2O$ 0.13g
$FeCl_3 \cdot 6H_2O$ 2.5mg
$MnSO_4 \cdot 4H_2O$ 2.5mg
Bromcresol Green solution 1.0mL

Bromcresol Green Solution:

Composition per 10.0mL:

Bromcresol Green 0.22g
Ethanol 10.0mL

Preparation of Bromcresol Green Solution: Add Bromcresol Green to 10.0mL of ethanol. Mix thoroughly. Filter sterilize.

Preparation of Medium: Add components to distilled/deionized water and bring volume to 1.0L. Mix thoroughly. Gently heat and bring to boiling. Distribute into tubes or flasks. Autoclave for 15 min at 15 psi pressure–121°C. Pour into sterile Petri dishes or leave in tubes.

Use: For the cultivation of *Zymomonas* species.

Yeastrel Agar

Composition per liter:

Agar 15.0g
Peptone 9.5g
Yeastrel 7.0g
Lab-lemco (meat extract) 5.0g
NaCl 5.0g

pH 7.0 ± 0.2 at 25°C

Source: Lab-lemco is available from Oxoid Unipath. Yeastrel is produced by Mapleton's Foods Ltd., Moss Street, Liverpool and is available from health food shops.

Preparation of Medium: Add components to distilled/deionized water and bring volume to 1.0L. Mix thoroughly. Gently heat and bring to boiling. Distribute into tubes or flasks. Autoclave for 15 min at 15 psi pressure–121°C. Pour into sterile Petri dishes or leave in tubes.

Use: For the cultivation of *Aeromonas salmonicida.*

YED Medium, Salted (DSMZ Medium 1123)

Composition per liter:

Agar 20.0g
NaCl 10.0g

Glucose 7.0g
Yeast extract 4.0g

pH 7.0 ± 0.2 at 25°C

Preparation of Medium: Add components to distilled/deionized water and bring volume to 1.0L. Mix thoroughly. Adjust pH to 7.0. Distribute into tubes or flasks. Autoclave for 15 min at 15 psi pressure–121°C. Pour into sterile Petri dishes or leave in tubes.

Use: For the cultivation of *Martelella mediterranea.*

YEME Thiostrepton Medium

Composition per liter:

Sucrose 340.0g
Peptone 5.0g
Malt extract 3.0g
Yeast extract 3.0g
Thiostrepton solution 10.0mL
$MgCl_2 \cdot 6H_2O$ solution 2.0mL

Thiostrepton Solution:
Composition per 10.0mL:

Thiostrepton 10.0mg

Preparation of Thiostrepton Solution: Add thiostrepton to distilled/deionized water and bring volume to 10.0mL. Mix thoroughly. Filter sterilize.

$MgCl_2 \cdot 6H_2O$ Solution:
Composition per 100.0mL:

$MgCl_2 \cdot 6H_2O$ 50.8g

Preparation of $MgCl_2 \cdot 6H_2O$ Solution: Add $MgCl_2 \cdot 6H_2O$ to distilled/deionized water and bring volume to 100.0mL. Mix thoroughly. Autoclave for 15 min at 15 psi pressure–121°C.

Preparation of Medium: Add components, except thiostrepton solution and $MgCl_2 \cdot 6H_2O$ solution, to distilled/deionized water and bring volume to 990.0mL. Mix thoroughly. Autoclave for 15 min at 15 psi pressure–121°C. Cool to 50°C. Aseptically add 10.0mL of sterile thiostrepton solution and 2.0mL of sterile solution. Mix thoroughly. Aseptically distribute into sterile tubes or flasks.

Use: For the cultivation of *Streptomyces lividans.*

YEP Agar

See: **Yeast Extract Phosphate Agar**

YEP Galactose Agar

Composition per liter:

Agar 20.0g
Galactose 20.0g
Peptone 20.0g
Yeast extract 10.0g

Preparation of Medium: Add components to distilled/deionized water and bring volume to 1.0L. Mix thoroughly. Gently heat and bring to boiling. Distribute into tubes or flasks. Autoclave for 15 min at 15 psi pressure–121°C. Pour into sterile Petri dishes or leave in tubes.

Use: For the cultivation of a variety of heterotrophic microorganisms.

YEP Galactose Medium

Composition per liter:

Agar 20.0g
Galactose 20.0g
Peptone 20.0g
Yeast extract 10.0g

Preparation of Medium: Add components to distilled/deionized water and bring volume to 1.0L. Mix thoroughly. Gently heat and bring to boiling. Distribute into tubes or flasks. Autoclave for 15 min at 15 psi pressure–121°C. Pour into sterile Petri dishes or leave in tubes.

Use: For the cultivation and maintenance of *Saccharomyces cerevisiae* and *Yarrowia lipolytica.*

YEPB Medium (Yeast Extract Peptone Beef Extract Medium)

Composition per liter:

Beef extract 10.0g
Polypeptone™ 10.0g
Yeast extract 5.0g
NaCl 3.0g
$MnCl_2 \cdot 4H_2O$ 0.1g

pH 7.0 ± 0.2 at 25°C

Preparation of Medium: Add components to distilled/deionized water and bring volume to 1.0L. Mix thoroughly. Adjust pH to 7.0 with KOH. Distribute into tubes or flasks. Autoclave for 15 min at 15 psi pressure–121°C. Adjust pH to 7.0 with KOH.

Use: For the cultivation of *Microbacterium* species.

YEPD Agar

Composition per liter:

Glucose 20.0g
Mycological peptone 20.0g
Agar 15.0g
Yeast extract 10.0g

pH 5.5 ± 0.2 at 25°C

Preparation of Medium: Add components to distilled/deionized water and bring volume to 1.0L. Mix thoroughly. Adjust pH to 5.5. Gently heat and bring to boiling. Distribute into tubes or flasks. Autoclave for 15 min at 15 psi pressure–121°C. Pour into sterile Petri dishes or leave in tubes.

Use: For the cultivation and maintenance of a variety of yeasts including *Botryozyma nematodophila, Bullera crocea, Candida* species, *Cryptococcus neoformans, Hanseniaspora uvarum, Kluyveromyces lactis, Kluyveromyces marxianus, Metschnikowia pulcherrima, Phaffia rhodozyma, Pichia* species, *Rhodosporidium toruloides, Rhodotorula graminis, Saccharomyces* species, *Schizosaccharomyces pombe, Schwanniomyces occidentalis, Trichosporon* species, *Yamadazyma stipitis, Yarrowia lipolytica, Zygosaccharomyces bailii,* and *Zygosaccharomyces rouxii.*

YEPD-FA Medium

Composition per liter:

Agar 20.0g
Glucose 20.0g
Peptone 20.0g
Yeast extract 10.0g
Tween™ 40 .... 10.0g
KH_2PO_4 .... 5.0g
K_2HPO_4 .... 5.0g
Myristic acid 70.0mg
Palmitic acid 70.0mg
Stearic acid 70.0mg

Preparation of Medium: Add components to distilled/deionized water and bring volume to 1.0L. Mix thoroughly. Gently heat and bring to boiling. Distribute into tubes or flasks. Autoclave for 15 min at 15 psi pressure–121°C. Pour into sterile Petri dishes or leave in tubes.

Use: For the cultivation and maintenance of *Saccharomyces cerevisiae.*

YEPD Inositol Agar

Composition per liter:

Glucose	20.0g
Mycological peptone	20.0g
Agar	15.0g
Yeast extract	10.0g
Inositol	180.0mg

Preparation of Medium: Add components to distilled/deionized water and bring volume to 1.0L. Mix thoroughly. Gently heat and bring to boiling. Distribute into tubes or flasks. Autoclave for 15 min at 15 psi pressure–121°C. Pour into sterile Petri dishes or leave in tubes.

Use: For the cultivation and maintenance of *Saccharomyces cerevisiae.*

YEPD Medium

Composition per liter:

Agar	20.0g
Glucose	20.0g
Peptone	20.0g
Yeast extract	10.0g

Preparation of Medium: Add components to distilled/deionized water and bring volume to 1.0L. Mix thoroughly. Gently heat and bring to boiling. Distribute into tubes or flasks. Autoclave for 15 min at 15 psi pressure–121°C. Pour into sterile Petri dishes or leave in tubes.

Use: For the cultivation of a variety of heterotrophic microorganisms.

YEPD Medium

Composition per liter:

Peptone	20.0g
Yeast extract	10.0g

Preparation of Medium: Add components to distilled/deionized water and bring volume to 1.0L. Mix thoroughly. Gently heat and bring to boiling. Distribute into tubes or flasks. Autoclave for 15 min at 15 psi pressure–121°C.

Use: For the cultivation and maintenance of a variety of yeasts, including *Botryozyma nematodophila, Bullera crocea, Candida* species, *Cryptococcus neoformans, Hanseniaspora uvarum, Kluyveromyces lactis, Kluyveromyces marxianus, Metschnikowia pulcherrima, Phaffia rhodozyma, Pichia* species, *Rhodosporidium toruloides, Rhodotorula graminis, Saccharomyces* species, *Schizosaccharomyces pombe, Schwanniomyces occidentalis, Trichosporon* species, *Yamadazyma stipitis, Yarrowia lipolytica, Zygosaccharomyces bailii,* and *Zygosaccharomyces rouxii.*

YEPD Medium with Heme

Composition per liter:

Agar	20.0g
Glucose	20.0g
Peptone	20.0g
Yeast extract	10.0g
Hemin stock solution	20.0mL

Hemin Stock Solution:

Composition per 10.0mL:

NaOH	0.04g
Hemin chloride	65.0mg
Ethanol (50% solution)	100.0mL

Preparation of Hemin Stock Solution: Combine components and mix thoroughly.

Preparation of Medium: Add components to distilled/deionized water and bring volume to 1.0L. Mix thoroughly. Gently heat and bring to boiling. Distribute into tubes or flasks. Autoclave for 15 min at 15 psi pressure–121°C. Pour into sterile Petri dishes or leave in tubes.

Use: For the cultivation and maintenance of *Saccharomyces cerevisiae.*

YEPD Medium with 0.3*M* Sodium Chloride

Composition per liter:

Agar	20.0g
Glucose	20.0g
Peptone	20.0g
NaCl	17.53g
Yeast extract	10.0g

Preparation of Medium: Add components to distilled/deionized water and bring volume to 1.0L. Mix thoroughly. Gently heat and bring to boiling. Distribute into tubes or flasks. Autoclave for 15 min at 15 psi pressure–121°C. Pour into sterile Petri dishes or leave in tubes.

Use: For the cultivation and maintenance of *Debaryomyces hansenii.*

YEPG Medium

Composition per liter:

Glucose	1.0g
Polypeptone™	2.0g
NH_4NO_3	0.2g
Yeast extract	0.2g

pH 7.0 ± 0.2 at 25°C

Preparation of Medium: Add components to distilled/deionized water and bring volume to 1.0L. Mix thoroughly. Adjust pH to 7.0. Distribute into tubes or flasks. Autoclave for 15 min at 15 psi pressure–121°C.

Use: For the cultivation of *Pseudomonas putida* and *Pseudomonas fluorescens.*

YEPG with 0.5% $CaCO_3$

See: ***Bacillus racemilacticus* Agar**

YEPP Medium (Yeast Extract Proteose Peptone Medium)

Composition per liter:

Agar	15.0g
Proteose peptone	10.0g
NaCl	5.0g
Yeast extract	3.0g

pH 7.2–7.4 ± 0.2 at 25°C

Preparation of Medium: Add components to distilled/deionized water and bring volume to 1.0L. Mix thoroughly. Gently heat and bring to boiling. Distribute into tubes or flasks. Autoclave for 15 min at 15 psi pressure–121°C. Pour into sterile Petri dishes or leave in tubes.

Use: For the cultivation and maintenance of *Pseudomonas* species.

Yersinia Isolation Agar
See: **CAL Agar**

Yersinia Isolation HiVeg Agar

Composition per liter:

Agar	15.0g
Plant peptone	15.0g
Lactose	10.0g
Sodium citrate	10.0g
Plant extract No. 1	8.5g
$Na_2S_2O_3$	8.5g
Yeast extract	5.0g
Synthetic detergent No. II	3.0g
Synthetic detergent No. III	2.0g
$CaCl_2$	1.0g
Ferric citrate	1.0g
Neutral Red	0.025g
Brilliant Green	0.3mg

pH 7.3 ± 0.2 at 25°C

Preparation of Medium: Add components to distilled/deionized water and bring volume to 1.0L. Mix thoroughly. Heat to boiling. Do not autoclave. Pour into sterile Petri dishes.

Use: For the isolation and characterization of *Yersinia enterocolitica* from fecal specimens and enumeration of *Yersinia enterocolitica* from water and other liquid specimens.

Yersinia Selective Agar
See: **CIN Agar**

Yersinia Selective Agar Base

Composition per liter:

Mannitol	20.0g
Peptone	17.0g
Agar	12.5g
Proteose peptone	3.0g
Yeast extract	2.0g
Sodium pyruvate	2.0g
NaCl	1.0g
Sodium desoxycholate	0.5g
$MgSO_4 \cdot 7H_2O$	0.01g
Neutral Red	0.03g
Crystal Violet	1.0mg
Selective supplement	6.0mL

pH 7.4 ± 0.2 at 25°C

Source: This medium is available as a premixed powder from BD Diagnostic Systems and Oxoid Unipath.

Selective Supplement:
Composition per 6.0mL:

Cefsulodin	15.0mg
Irgasan	4.0mg
Novobiocin	2.5mg
Ethanol	2.0mL

Preparation of Selective Supplement: Aseptically add components to 4.0mL of distilled/deionized water and 2.0mL of ethanol. Mix thoroughly.

Preparation of Medium: Add components to distilled/deionized water and bring volume to 1.0L. Mix thoroughly. Gently heat and bring to boiling. Distribute into tubes or flasks. Autoclave for 15 min at 15 psi pressure–121°C. Cool to 50°C. Aseptically add selective supplement. Mix thoroughly. Pour into sterile Petri dishes or leave in tubes.

Use: For the isolation and enumeration of *Yersinia enterocolitica* from food and clinical specimens.

Yersinia Selective HiVeg Agar Base with Selective Supplement

Composition per liter:

Plant special peptone	20.0g
Mannitol	20.0g
Agar	12.5g
Sodium pyruvate	2.0g
Yeast extract	2.0g
NaCl	1.0g
Synthetic detergent No. III	0.5g
Neutral Red	0.03g
$MgSO_4$	0.01g
Crystal Violet	1.0mg
Selective supplement	6.0mL

pH 7.4 ± 0.2 at 25°C

Source: This medium, without selective supplement, is available as a premixed powder from HiMedia.

Selective Supplement:
Composition per 6.0mL:

Cefsulodin	15.0mg
Irgasan	4.0mg
Novobiocin	2.5mg
Ethanol	2.0mL

Preparation of Selective Supplement: Aseptically add components to 4.0mL of distilled/deionized water and 2.0mL of ethanol. Mix thoroughly.

Preparation of Medium: Add components to distilled/deionized water and bring volume to 1.0L. Mix thoroughly. Gently heat and bring to boiling. Distribute into tubes or flasks. Autoclave for 15 min at 15 psi pressure–121°C. Cool to 50°C. Aseptically add selective supplement. Mix thoroughly. Pour into sterile Petri dishes or leave in tubes.

Use: For the isolation and enumeration of *Yersinia enterocolitica* from food and clinical specimens.

YGB
See: **Yeast Glucose Broth**

YGC Medium (Yeast Extract Glucose Carbonate Medium) (ATCC Medium 73)

Composition per liter:

Agar	20.0g
Glucose	20.0g
$CaCO_3$	20.0g
Yeast extract	10.0g

Preparation of Medium: Add components to distilled/deionized water and bring volume to 1.0L. Mix thoroughly. Gently heat and bring to boiling. Distribute into tubes or flasks. Autoclave for 30 min at 10 psi pressure–115°C. Cool to 48°C. Mix thoroughly. Pour into sterile Petri dishes or leave in tubes.

Use: For the cultivation of *Xanthomonas* species, *Erwinia* species, *Kluyvera* species, *Rhodococcus* species, *Streptomyces* species, *Pseudomonas pseudoalcaligenes*, and *Xylophilus ampelinus*.

YGC Medium (Yeast Extract Glucose Carbonate Medium) (ATCC Medium 459)

Composition per liter:

Glucose	50.0g
Agar	15.0g
$CaCO_3$	12.5g
Yeast extract	5.0g

Preparation of Medium: Add components to distilled/deionized water and bring volume to 1.0L. Mix thoroughly. Gently heat and bring to boiling. Distribute into tubes or flasks. Autoclave for 15 min at 15 psi pressure–121°C. Cool to 50°C. Pour into sterile Petri dishes or leave in tubes.

Use: For the cultivation and maintenance of *Acetobacter* species.

YGC Medium (Yeast Extract Glucose Citrate Medium) (ATCC Medium 216)

Composition per liter:

Beef extract	10.0g
Glucose	10.0g
Peptone	10.0g
Ammonium citrate	5.0g
Yeast extract	5.0g
Sodium acetate	2.0g
Tween™ 80	1.0g
$MgSO_4 \cdot 7H_2O$	0.2g
$MnSO_4 \cdot 4H_2O$	0.05g

pH 6.5 ± 0.2 at 25°C

Preparation of Medium: Add components to distilled/deionized water and bring volume to 1.0L. Mix thoroughly. Distribute into tubes or flasks. Autoclave for 15 min at 15 psi pressure–121°C.

Use: For the isolation and cultivation of *Leuconostoc* species.

YGC Medium with Cysteine (Yeast Extract Glucose Citrate Medium with Cysteine)

Composition per liter:

Glucose	10.0g
Peptone	10.0g
Beef extract	10.0g
Yeast extract	5.0g
Ammonium citrate	5.0g
Sodium acetate	2.0g
Tween™ 80	1.0g
$MgSO_4 \cdot 7H_2O$	0.2g
$MnSO_4 \cdot 4H_2O$	0.05g
L-Cysteine·HCl·H_2O solution	10.0mL

pH 6.5 ± 0.2 at 25°C

L-Cysteine·HCl Solution:

Composition per 10.0mL:

L-Cysteine·HCl·H_2O	0.5g

Preparation of L-Cysteine·HCl Solution: Add L-cysteine·HCl·H_2O to distilled/deionized water and bring volume to 10.0mL. Mix thoroughly. Filter sterilize.

Preparation of Medium: Add components, except for the L-cysteine·HCl·H_2O, to distilled/deionized water and bring volume to 990.0mL. Mix thoroughly. Distribute into tubes or flasks. Autoclave for 15 min at 15 psi pressure–121°C. Aseptically add L-cysteine hydrochloride solution.

Use: For the cultivation and maintenance of *Leuconostoc mesenteroides*.

YGC Medium with Glutamic Acid (Yeast Extract Glucose Carbonate Medium with Glutamic Acid)

Composition per liter:

Agar	20.0g
Glucose	20.0g
$CaCO_3$	20.0g
Agar	20.0g
Yeast extract	10.0g
Glutamic acid	0.1g

Preparation of Medium: Add components to distilled/deionized water and bring volume to 1.0L. Mix thoroughly. Gently heat and bring to boiling. Distribute into tubes or flasks. Autoclave for 30 min at 10 psi pressure–115°C. Cool to 48°C. Mix thoroughly. Pour into sterile Petri dishes or leave in tubes.

Use: For the cultivation and maintenance of *Xanthomonas campestris*.

YGCB Salt Medium

Composition per liter:

NaCl	50.0g
Beef broth	10.0g
Glucose	10.0g
Peptone	10,0g
Triammonium citrate	5.0g
Yeast extract	5.0g
Sodium acetate	2.0g
$MgSO_4 \cdot 7H_2O$	0.2g
$MnSO_4 \cdot 4H_2O$	50.0mg
Tween™ 80	1.0mL

pH 6.7 ± 0.2 at 25°C

Preparation of Medium: Add components to distilled/deionized water and bring volume to 1.0L. Mix thoroughly. Distribute into tubes or flasks. Autoclave for 15 min at 15 psi pressure–121°C.

Use: For the cultivation of *Tetragenococcus halophila.*

YGCP Medium (Yeast Extract Glucose Carbonate Peptone Medium)

Composition per liter:

Glucose	20.0g
Agar	17.5g
$CaCO_3$	10.0g
Yeast extract	2.5g
Peptone	2.5g
NaCl	1.0g
K_2HPO_4	1.0g
$MgSO_4$	0.5g

Preparation of Medium: Add components, except calcium carbonate, to distilled/deionized water and bring volume to 1.0L. Mix thoroughly. Gently heat and bring to boiling. Adjust pH to 7.2. Add calcium carbonate. Mix thoroughly. Distribute into tubes or flasks. Autoclave for 15 min at 15 psi pressure–121°C. Pour into sterile Petri dishes or leave in tubes.

Use: For the cultivation and maintenance of *Xanthomonas campestris* and *Xanthomonas oryzae.*

YGLM (Yeast Glucose Litmus Milk)

Composition per liter:

Glucose	10.0g
Skim milk powder	8.0g
Yeast extract	3.0g
Litmus	0.75g

Preparation of Medium: Add components to distilled/deionized water and bring volume to 1.0L. Mix thoroughly for 15–20 min. Distribute into tubes or flasks. Autoclave for 10 min at 10 psi pressure–115°C. Incubate for 1 week at 30°C to check for sterility before use.

Use: For the cultivation of *Gemella morbillorum.*

YGLM with Chalk (Yeast Glucose Litmus Milk with Chalk)

Composition per liter:

$CaCO_3$	20.0g
Glucose	10.0g
Yeast extract	3.0g
Litmus	0.75g
Skim milk	100.0mL

Preparation of Medium: Add components, except skim milk, to distilled/deionized water and bring volume to 900.0mL. Mix thoroughly. Autoclave for 15 min at 15 psi pressure–121°C. Cool to 25°C. Add 10.0mL of skim milk. Distribute into tubes. Autoclave for 10 min at 10 psi pressure–115°C. Incubate for 1 week at 30°C to check for sterility before use.

Use: For the cultivation of *Gemella morbillorum.*

YGLPB Medium

Composition per liter:

Peptone	10.0g
Lab-Lemco	8.0g
Glucose	5.0g
Lactose	5.0g
Yeast extract	3.0g
KH_2PO_4	2.5g
K_2HPO_4	2.5g
$MgSO_4 \cdot 7H_2O$	0.2g
$MnSO_4 \cdot 4H_2O$	0.05g

Preparation of Medium: Add components to distilled/deionized water and bring volume to 1.0L. Mix thoroughly. Distribute into tubes or flasks. Autoclave for 20 min at 15 psi pressure–121°C.

Use: For the cultivation of *Lactobacillus delbrueckii.*

YGLPB Medium

Composition per liter:

Peptone	10.0g
Beef extract	8.0g
Glucose	5.0g
Lactose	5.0g
Yeast extract	3.0g
K_2HPO_4	2.5g
KH_2PO_4	2.5g
$MgSO_4 \cdot 7H_2O$	0.2g
$MnSO_4 \cdot 4H_2O$	0.05g

pH 6.8 ± 0.2 at 25°C

Preparation of Medium: Add components to distilled/deionized water and bring volume to 1.0L. Mix thoroughly. Adjust pH to 6.8. Distribute into tubes or flasks. Autoclave for 15 min at 15 psi pressure–121°C.

Use: For the cultivation of *Carnobacterium divergens, Carnobacterium piscicola, Lactobacillus delbrueckii, Lactobacillus fermentum, Lactococcus lactis, Streptococcus ferus,* and *Streptococcus sobrinus.*

YGLPB Medium

Composition per liter:

Peptone	1.0g
Lab Lemco (meat extract)	0.8g
Glucose	0.5g
Lactose	0.5g
Yeast extract	0.3g
K_2HPO_4	0.25g
KH_2PO_4	0.25g
$MgSO_4 \cdot 7H_2O$	0.02g
$MnSO_4 \cdot 4H_2O$	5.0mg

pH 6.8 ± 0.2 at 25°C

Preparation of Medium: Add components to distilled/deionized water and bring volume to 1.0L. Mix thoroughly. Distribute into tubes or flasks. Autoclave for 15 min at 15 psi pressure–121°C.

Use: For the cultivation of *Carnobacterium gallinarum, Carnobacterium mobile, Enterococcus dispar, Lactobacillus fructivorans, Leuconostoc carnosum, Leuconostoc gelidum,* and *Vagococcus salmoninarum.*

YGLPB Medium without Lactose

Composition per liter:

Peptone	10.0g
Lab Lemco	8.0g
Yeast extract	3.0g
Glucose	5.0g
KH_2PO_4	2.5g
K_2HPO_4	2.5g
$MgSO_4 \cdot 7H_2O$	0.2g
$MnSO_4 \cdot 4H_2O$	0.05g

Preparation of Medium: Add components to distilled/deionized water and bring volume to 1.0L. Mix thoroughly. Distribute into tubes or flasks. Autoclave for 20 min at 15 psi pressure–121°C.

Use: For the cultivation of *Vagococcus fluvialis* and *Vagococcus salmoninarum.*

YI-S Medium

Composition per liter:

YI broth	880.0mL
Bovine serum, heat inactivated	100.0mL
Vitamin mixture 18	20.0mL

Source: Vitamin mixture 18 is available from Bio-fluids, Inc., Rockville, MD.

YI Broth:

Composition per liter:

YI base stock	780.0mL
10X Glucose buffer stock	100.0mL

YI Base Stock:

Composition per 780.0mL:

Yeast extract	30.0g
L-Cysteine·HCl	1.0g
NaCl	1.0g
Ascorbic acid	0.2g
Ferric ammonium citrate	228.0mg

10X Glucose Buffer Stock:

Composition per 100.0.0mL:

Glucose	10.0g
K_2HPO_4	1.0g
KH_2PO_4	0.6g

Preparation of 10X Glucose Buffer Stock: Add components to distilled/deionized water and bring volume to 100.0mL. Mix thoroughly. Filter sterilize.

Preparation of YI Base Stock: Add components to 600.0mL of distilled/deionized water. Mix thoroughly. Bring volume to 780.0mL with distilled/deionized water. Adjust pH to 6.8 with 1*N* NaOH. Distribute in 78.0mL aliquots to 100.0mL screw-capped bottles. Autoclave for 15 min at 15 psi pressure–121°C. Cool to room temperature.

Preparation of YI Broth: Aseptically add 10.0mL of 10X glucose buffer stock to 78.0mL of cooled YI base stock. Adjust osmolarity with NaCl to 380.0milliosmols/kg.

Preparation of Medium: Aseptically add 2.0mL of vitamin mixture 18 and 10.0mL of heat-inactivated bovine serum to 88.0mL of YI broth. Distribute in 13.0mL aliquots to 16 x 125mm screw-capped test tubes. Store at 4°C in the dark with the caps screwed on tightly. Use within 96 hr.

Use: For the cultivation of *Entamoeba* species.

YM Agar

See: **Yeast Malt Extract Agar**

YM Broth

See: **Yeast Malt Extract Broth**

YM Broth with 0.5%CaCO$_3$

See: **Yeast Malt Extract Broth with 0.5% CaCO$_3$**

YM Broth with 2.0%CaCO$_3$

See: **Yeast Malt Extract Broth with 2.0% CaCO$_3$**

YM Broth with Glucose

See: **Yeast Malt Extract Broth with Glucose**

YM Broth with 1.0% Methanol

See: **Yeast Malt Extract Broth with 1.0% Methanol**

YM Broth with 18% NaCl

See: **Yeast Malt Extract Broth with 18% NaCl**

YM Broth with 40% Sucrose

See: **Yeast Malt Extract Broth with 40% Sucrose**

YM Broth with 70% Sucrose

See: **Yeast Malt Extract Broth with 70% Sucrose**

YM Catalase Agar

See: **Yeast Malt Extract Catalase Agar**

YM-1L Broth

Composition per liter:

Sodium lactate	30.0g
Mycological peptone	10.0g
Succinic acid	10.0g
NaOH	6.0g
Yeast extract	5.0g
Adenine	0.01g
Uracil	0.01g
Yeast nitrogen base solution	100.0mL

Yeast Nitrogen Base Solution:

Composition per 100.0mL:

$(NH_4)_2SO_4$	5.0g
KH_2PO_4	1.0g
$MgSO_4·7H_2O$	0.5g
NaCl	0.1g
$CaCl_2·2H_2O$	0.1g
DL-Methionine	0.02g
DL-Tryptophan	0.02g
L-Histidine·HCl	0.01g
Inositol	2.0mg
KI	0.1mg
H_3BO_3	0.5mg
$ZnSO_4·7H_2O$	0.4mg
$MnSO_4·4H_2O$	0.4mg
Thiamine·HCl	0.4mg
Pyroxidine·HCl	0.4mg
Niacin	0.4mg
Calcium pantothenate	0.4mg
p-Aminobenzoic acid	0.2mg
Riboflavin	0.2mg
$FeCl_3$	0.2mg
$Na_2MoO_4·4H_2O$	0.2mg
$CuSO_4·5H_2O$	0.04mg
Folic acid	2.0μg
Biotin	2.0μg

Preparation of Yeast Nitrogen Base Solution: Add components to distilled/deionized water and bring volume to 1.0L. Mix thoroughly. Filter sterilize.

Preparation of Medium: Add components, except yeast nitrogen base solution, to distilled/deionized water and bring volume to 900.0mL. Mix thoroughly. Autoclave for 15 min at 15 psi pressure–121°C. Aseptically add 1.0L of sterile yeast nitrogen base solution. Mix thoroughly. Aseptically distribute into sterile tubes or flasks.

Use: For the cultivation of *Saccharomyces cerevisiae.*

YM Medium

See: **Universal Agar for Yeasts**

YM Medium (DSMZ Medium 1070)

Composition per liter:

Mannitol	10.0g
Yeast extract	0.5g
K_2HPO_4	0.5g
NaCl	0.2g
$CaCl_2·2H_2O$	0.2g
$MgSO_4·7H_2O$	0.1g

pH 7.0 ± 0.2 at 25°C

Preparation of Medium: Add components to distilled/deionized water and bring volume to 1.0L. Mix thoroughly. Adjust pH to 7.0. Distribute into tubes or flasks. Autoclave for 15 min at 15 psi pressure–121°C. Pour into sterile Petri dishes or leave in tubes.

Use: For the cultivation of *Labrys miyagiensis*.

YM5 with 10% Sorbitol

Composition per liter:

Sorbitol	100.0g
Glucose	10.0g
Succinic acid	10.0g
NaOH	6.0g
Peptone	2.0g
Yeast extract	1.0g
Adenine	0.01g
Uracil	0.01g
Yeast nitrogen base solution	100.0mL

pH 5.8 ± 0.2 at 25°C

Yeast Nitrogen Base Solution:

Composition per 100.0mL:

$(NH_4)_2SO_4$	5.0g
KH_2PO_4	1.0g
$MgSO_4{\cdot}7H_2O$	0.5g
NaCl	0.1g
$CaCl_2{\cdot}2H_2O$	0.1g
DL-Methionine	0.02g
DL-Tryptophan	0.02g
L-Histidine·HCl	0.01g
Inositol	2.0mg
KI	0.1mg
H_3BO_3	0.5mg
$ZnSO_4{\cdot}7H_2O$	0.4mg
$MnSO_4{\cdot}4H_2O$	0.4mg
Thiamine·HCl	0.4mg
Pyroxidine·HCl	0.4mg
Niacin	0.4mg
Calcium pantothenate	0.4mg
p-Aminobenzoic acid	0.2mg
Riboflavin	0.2mg
$FeCl_3$	0.2mg
$Na_2MoO_4{\cdot}4H_2O$	0.2mg
$CuSO_4{\cdot}5H_2O$	0.04mg
Folic acid	2.0μg
Biotin	2.0μg

Source: Yeast nitrogen base is available as a premixed powder from BD Diagnostic Systems.

Preparation of Yeast Nitrogen Base Solution: Add components to distilled/deionized water and bring volume to 100.0mL. Mix thoroughly. Filter sterilize.

Preparation of Medium: Add components, except yeast nitrogen base solution, to distilled/deionized water and bring volume to 900.0mL. Mix thoroughly. Autoclave for 15 min at 15 psi pressure–121°C. Cool to room temperature. Aseptically add 100.0mL of sterile yeast nitrogen base solution to the cooled sterile basal medium. Mix thoroughly. Adjust final pH to 5.8. Distribute into sterile flasks or tubes.

Use: For the cultivation of a sorbitol-utilizing fungus.

YMA Agar

Composition per liter:

Agar	15.0g
Mannitol	10.0g
$CaCO_3$	4.0g
KH_2PO_4	0.5g
Yeast extract	0.4g
$MgSO_4{\cdot}7H_2O$	0.2g
NaCl	0.1g

pH 6.8 ± 0.2 at 25°C

Preparation of Medium: Add components to distilled/deionized water and bring volume to 1.0L. Mix thoroughly. Gently heat and bring to boiling. Distribute into tubes or flasks. Autoclave for 15 min at 15 psi pressure–121°C. Pour into sterile Petri dishes or leave in tubes.

Use: For the cultivation of *Rhizobium fredii, Rhizobium galegae, Rhizobium huakuii, Rhizobium leguminosarum, Rhizobium loti, Rhizobium meliloti, Rhizobium phaseoli,* and *Rhizobium trifolii.*

YMA Medium (DSMZ Medium 1031)

Composition per liter:

Agar	20.0g
Mannitol	10.0g
Yeast extract	0.3g
$MgSO_4{\cdot}7H_2O$	0.2g
K_2HPO_4	0.2g
NaCl	0.05g

pH 7.0 ± 0.2 at 25°C

Preparation of Medium: Add components to distilled/deionized water and bring volume to 1.0L. Mix thoroughly. Adjust pH to 7.0. Distribute into tubes or flasks. Autoclave for 15 min at 15 psi pressure–121°C. Pour into sterile Petri dishes or leave in tubes.

Use: For the cultivation of *Phyllobacterium trifolii.*

YMF Agar

Composition per liter:

Agar	20.0g
Peptone	5.0g
Sugar, brown	3.0g
Malt extract	3.0g
Yeast extract	3.0g

pH 6.2 ± 0.2 at 25°C

Preparation of Medium: Add components to distilled/deionized water and bring volume to 1.0L. Mix thoroughly. Adjust pH to 6.2. Gently heat and bring to boiling. Distribute into tubes or flasks. Autoclave for 15 min at 15 psi pressure–121°C. Pour into sterile Petri dishes or leave in tubes.

Use: For the cultivation of various fungi.

YMF Broth

Composition per liter:

Peptone	5.0g
Sugar, brown	3.0g
Malt extract	3.0g
Yeast extract	3.0g

pH 6.2 ± 0.2 at 25°C

Preparation of Medium: Add components to distilled/deionized water and bring volume to 1.0L. Mix thoroughly. Adjust pH to 6.2.

Distribute into tubes or flasks. Autoclave for 15 min at 15 psi pressure–121°C.

Use: For the cultivation of various fungi.

YNA Medium (Yeast Extract Nutrient Agar Medium)

Composition per liter:

Agar	15.0g
NaCl	5.0g
Peptone	5.0g
Meat extract	4.0g
Yeast extract	2.5g

pH 7.0 ± 0.2 at 25°C

Preparation of Medium: Add components to distilled/deionized water and bring volume to 1.0L. Mix thoroughly. Gently heat and bring to boiling. Distribute into tubes or flasks. Autoclave for 15 min at 15 psi pressure–121°C. Pour into sterile Petri dishes or leave in tubes.

Use: For the isolation and cultivation of *Kurthia* species according to the agar streak method.

YNG Medium (Yeast Extract Nutrient Gelatin Medium)

Composition per liter:

Gelatin	100.0g
NaCl	5.0g
Peptone	5.0g
Meat extract	4.0g
Yeast extract	2.5g

pH 7.0 ± 0.2 at 25°C

Preparation of Medium: Add components to distilled/deionized water and bring volume to 1.0L. Mix thoroughly. Gently heat until dissolved. Distribute into tubes or flasks. Autoclave for 30 min at 10 psi pressure–115°C.

Use: For the isolation and cultivation of *Kurthia* species using the gelatin streak method.

Yolk Milk Medium (YOM)

Composition per liter:

Egg yolk	500.0mL
Milk	500.0mL

Egg Yolk:

Composition per 500.0mL:

Chicken egg yolks	variable

Preparation of Egg Yolk: Soak eggs with 1:100 dilution of saturated mercuric chloride solution for 1 min. Crack eggs and separate yolks from whites. Add sufficient egg yolk to bring volume to 500.0mL. Mix thoroughly.

Preparation of Milk: Autoclave 500.0mL of milk for 20 min at 15 psi pressure–115°C.

Preparation of Medium: Combine 500.0mL of sterile egg yolk and 500.0mL of sterile milk. Mix thoroughly. Distribute into sterile tubes or flasks. Heat to 95°C for 20–25 min.

Use: For the cultivation of *Condiobolus obscurus*.

Yopp's Medium

Composition per liter:

NaCl	116.88g
$MgCl_2·6H_2O$	10.68g
$MgSO_4·7H_2O$	10.0g
KCl	2.0g
$CaNO_3·4H_2O$	1.0g
Glycyl-glycine buffer	0.5g
$K_2HPO_4·3H_2O$	0.065g
Ferric EDTA	5.0mg
Trace metals solution	1.0mL

pH 7.8 ± 0.2 at 25°C

Trace Metals Solution:

Composition per liter:

$MnCl_2·4H_2O$	2.0g
H_3BO_3	0.5g
$ZnNO_3·6H_2O$	0.5g
$Co(NO_3)_2·6H_2O$	0.025g
$CuCl_2·2H_2O$	0.025g
$Na_2MoO_4·2H_2O$	0.025g
$VOSO_4·6H_2O$	0.025g
HCl	3.0mL

Preparation of Trace Metals Solution: Add components to distilled/deionized water and bring volume to 1.0L. Mix thoroughly.

Preparation of Medium: Add components to distilled/deionized water and bring volume to 1.0L. Mix thoroughly. Gently heat and bring to boiling. Distribute into tubes or flasks. Autoclave for 15 min at 15 psi pressure–121°C.

Use: For the isolation and cultivation of halophilic cyanobacteria.

YP87 Medium

Composition per liter:

Na_2SO_4	4.0g
$NaHCO_3$	1.3g
KCl	0.5g
Yeast extract	0.5g
$MgCl_2·6H_2O$	0.4g
NH_4Cl	0.25g
L-Ascorbic acid	0.2g
Na2HPO4	0.2g
Sodium thioglycolate	0.2g
$CaCl_2·2H_2O$	0.15g
Resazurin	1.0mg
Modified Wolfe's mineral solution	10.0mL
Wolfe's vitamin solution	10.0mL
Sodium lactate, 60% syrup	3.0mL

pH 7.5 ± 0.2 at 25°C

Modified Wolfe's Mineral Solution:

Composition per liter:

$MgSO_4·7H_2O$	3.0g
Nitrilotriacetic acid	1.5g
NaCl	1.0g
$MnSO_4·H_2O$	0.5g
$CaCl_2$	0.1g
$CoCl_2·6H_2O$	0.1g
$FeSO_4·7H_2O$	0.1g
$ZnSO_4·7H_2O$	0.1g
$AlK(SO_4)_2·12H_2O$	0.01g
$CuSO_4·5H_2O$	0.01g

H_3BO_3 .... 0.01g
$Na_2MoO_4·2H_2O$ 0.01g
Na_2SeO_3 .... 0.01g
$NaWO_4·2H_2O$ 0.01g
$NiCl_2·6H_2O$ 0.01g

Preparation of Modified Wolfe's Mineral Solution: Add nitrilotriacetic acid to 500.0mL of distilled/deionized water. Adjust pH to 6.5 with KOH. Add remaining components one at a time. Add distilled/deionized water to 1.0L. Adjust pH to 6.8.

Wolfe's Vitamin Solution:
Composition per liter:
Pyridoxine·HCl 10.0mg
p-Aminobenzoic acid 5.0mg
Lipoic acid 5.0mg
Nicotinic acid 5.0mg
Riboflavin 5.0mg
Thiamine·HCl 5.0mg
Calcium DL-pantothenate 5.0mg
Biotin 2.0mg
Folic acid 2.0mg
Vitamin B_{12} 0.1mg

Preparation of Wolfe's Vitamin Solution: Add components to distilled/deionized water and bring volume to 1.0L. Mix thoroughly.

Preparation of Medium: Prepare and dispense medium under 100% N_2. Add components, except L-ascorbic acid, $NaHCO_3$, and sodium thioglycolate, to distilled/deionized water and bring volume to 1.0L. Mix thoroughly. Adjust pH to 7.5. Gently heat and bring to boiling. Cool while sparging with 100% N_2. Add L-ascorbic acid, $NaHCO_3$, and sodium thioglycolate. Mix thoroughly. Sparge with 100% N_2. Distribute into tubes or bottles. Autoclave for 15 min at 15 psi pressure–121°C.

Use: For the cultivation of *Thermodesulfovibrio yellowstonii.*

YPAD Medium for MAK Mutants of *Saccharomyces*

Composition per liter:
Peptone 20.0g
Glucose 20.0g
Agar 20.0g
Yeast extract 10.0g
Adenine sulfate 0.4g

Preparation of Medium: Add components to distilled/deionized water and bring volume to 1.0L. Mix thoroughly. Gently heat and bring to boiling. Distribute into tubes or flasks. Autoclave for 15 min at 15 psi pressure–121°C. Pour into sterile Petri dishes or leave in tubes.

Use: For the cultivation and maintenance of a variety of yeasts, including *Candida albicans, Candida boidinii, Candida pintolopesii, Saccharomyces cerevisiae,* and *Schizosaccharomyces pombe.*

YPC Medium

Composition per liter:
Agar 15.0g
Proteose peptone 15.0g
Yeast extract 5.0g
KH_2PO_4 .... 4.0g
Sucrose 2.5g
Glucose 2.0g
L-Cystine 0.5g
Na_2SO_3 .... 0.2g

pH 7.2 ± 0.2 at 25°C

Preparation of Medium: Add components to distilled/deionized water and bring volume to 1.0L. Mix thoroughly. Gently heat and bring to boiling. Distribute into tubes or flasks. Autoclave for 15 min at 15 psi pressure–121°C. Pour into sterile Petri dishes or leave in tubes.

Use: For the cultivation of *Pasteurella multocida.*

YPD Medium (DSMZ Medium 393)

Composition per liter:
Peptone 20.0g
Glucose 20.0g
Yeast extract 10.0g

pH 6.5 ± 0.2 at 25°C

Preparation of Medium: Add components to distilled/deionized water and bring volume to 1.0L. Mix thoroughly. Adjust pH to 6.5. Distribute into tubes or flasks. Autoclave for 15 min at 15 psi pressure–121°C.

Use: For the isolation and cultivation of *Yarrowia lipolytica (Candida lipolytica), Kluyveromyces* spp., *Saccharomyces* spp., *Pichia* spp., and *Candida* spp.

YPDA (Yeast Peptone Dextrose Agar)

Composition per liter:
Agar 20.0g
Glucose 20.0g
Peptone 20.0g
Yeast extract 10.0g
Adenine sulfate 0.4g

Preparation of Medium: Add components to distilled/deionized water and bring volume to 1.0L. Mix thoroughly. Gently heat and bring to boiling. Distribute into tubes or flasks. Autoclave for 15 min at 15 psi pressure–121°C. Pour into sterile Petri dishes or leave in tubes.

Use: For the cultivation of *Taphrina populina.*

YPDP Medium with 5′-TMP

Composition per liter:
Glucose 20.0g
Peptone 20.0g
Agar 15.0g
Yeast extract 10.0g
KH_2PO_4 .... 1.5g
Thymidine-5′-monophosphate solution 10.0mL

Thymidine-5′-Monophosphate Solution:
Composition per 10.0mL:
Thymidine-5′-monophosphate 100.0mg

Preparation of Thymidine-5′-Monophosphate Solution: Add thymidine-5′-monophosphate to distilled/deionized water and bring volume to 10.0mL. Mix thoroughly. Filter sterilize.

Preparation of Medium: Add components, except thymidine-5′-monophosphate solution, to distilled/deionized water and bring volume to 990.0mL. Mix thoroughly. Gently heat and bring to boiling. Autoclave for 15 min at 15 psi pressure–121°C. Cool to 50°C. Aseptically add 10.0mL of sterile thymidine-5′-monophosphate solution. Mix thoroughly. Pour into sterile Petri dishes or aseptically distribute into sterile tubes.

Use: For the cultivation and maintenance of *Saccharomyces cerevisiae.*

YPG Agar with 2% Sodium Chloride

Composition per liter:

Agar	20.0g
Glucose	20.0g
NaCl	20.0g
Peptone	10.0g
Yeast extract	10.0g

pH 5.2 ± 0.2 at 25°C

Preparation of Medium: Add components to distilled/deionized water and bring volume to 1.0L. Mix thoroughly. Adjust pH to 5.2. Gently heat and bring to boiling. Distribute into tubes or flasks. Autoclave for 15 min at 15 psi pressure–121°C. Pour into sterile Petri dishes or leave in tubes.

YPG Medium (DSMZ Medium 1017)

Composition per liter:

Glucose	70.0gl
Yeast extract	10.0g
Peptone	10.0g

pH 6.0 ± 0.2 at 25°C

Preparation of Medium: Add components to distilled/deionized water and bring volume to 1.0L. Mix thoroughly. Adjust pH to 6.0 with dilute HCl. Distribute into tubes or flasks. Autoclave for 15 min at 15 psi pressure–121°C.

Use: For the cultivation of *Saccharibacter floricola.*

YPG Medium (DSMZ Medium 1172)

Composition per liter:

Yeast extract	1.0g
Peptone	1.0g
Glucose	1.0g

pH 5.7 ± 0.2 at 25°C

Preparation of Medium: Add components to distilled/deionized water and bring volume to 1.0L. Mix thoroughly. Adjust pH to 5.6–6.0 with dilute HCl. Distribute into tubes or flasks. Autoclave for 15 min at 15 psi pressure–121°C.

Use: For the cultivation of *Asticcacaulis benevestitus.*

YPGA (DSMZ Medium 1015)

Composition per liter:

Agar	15.0g
Yeast extract	7.0g
Peptone	7.0g
Glucose	7.0g

pH 7.3 ± 0.2 at 25°C

Preparation of Medium: Add components to distilled/deionized water and bring volume to 1.0L. Mix thoroughly. Adjust pH to 7.3. Distribute into tubes or flasks. Autoclave for 15 min at 15 psi pressure–121°C. Pour into sterile Petri dishes or leave in tubes.

Use: For the cultivation of *Stenotrophomonas maltophilia.*

Use: For the cultivation of a variety of yeasts and other fungi.

YPM Agar

Composition per liter:

Mannitol	25.0g
Agar	12.0g
Yeast extract	5.0g
Peptone	3.0g

Preparation of Medium: Add components to distilled/deionized water and bring volume to 1.0L. Mix thoroughly. Gently heat and bring to boiling. Do not adjust pH. Distribute into tubes or flasks. Autoclave for 15 min at 15 psi pressure–121°C. Pour into sterile Petri dishes or leave in tubes.

Use: For the cultivation of *Acetobacter aceti, Acetobacter pasteurianus, Acetobacter xylinum, Frateuria aurantia,* and *Pseudomonas aeruginosa.*

YPNC Medium

Composition per liter:

Agar	18.0g
NaCl	2.92g
KH_2PO_4	0.596g
Yeast extract	0.5g
Sodium hydrogen glutamate (pH 6.0)	0.37g
K_2HPO_4	0.107g
NH_4Cl	0.107g
$MgSO_4 \cdot 7H_2O$	0.049g
Glucose solution	10.0mL
Trace metals solution	1.0mL

pH 6.0 ± 0.2 at 25°C

Glucose Solution:

Composition per 10.0mL:

Glucose	4.0g

Preparation of Glucose Solution: Add glucose to distilled/deionized water and bring volume to 10.0mL. Mix thoroughly. Filter sterilize.

Trace Metals Solution:

Composition per liter:

EDTA	50.0g
$ZnSO_4 \cdot 7H_2O$	22.0g
$CaCl_2$	5.54g
$MnCl_2 \cdot 4H_2O$	5.06g
$FeSO_4 \cdot 7H_2O$	4.99g
$(NH_4)_6Mo_7O_{14} \cdot H_2O$	1.10g
$CoSO_4 \cdot 5H_2O$	1.57g
$CoCl_2 \cdot 6H_2O$	1.61g

Preparation of Trace Metals Solution: Add components, one at a time, to distilled/deionized water and bring volume to 1.0L. Mix thoroughly. Filter sterilize.

Preparation of Medium: Add components, except glucose solution and trace metals solution, to distilled/deionized water and bring volume to 989.0mL. Mix thoroughly. Gently heat and bring to boiling. Adjust pH to 6.0 with KOH. Autoclave for 15 min at 15 psi pressure–121°C. Cool to 50°C. Aseptically add 10.0mL of sterile glucose solution and 1.0mL of sterile trace metals solution. Mix thoroughly. Pour into sterile Petri dishes or aseptically distribute into sterile tubes.

Use: For the cultivation and maintenance of a variety of *Cryptococcus* species.

YPS Medium (DSMZ Medium 990)

Composition per liter:

Sea salts, Sigma	35.0g
Sulfur, elemental	5.0g
PIPES (piperazine-N,N'-bis[2-ethane-sulfonic acid])	3.46g
NH_4Cl	0.5g
KH_2PO_4	0.35g
$CaCl_2·2H_2O$	0.2g
$FeCl_3·6H_2O$	6.7mg
Na_2WO_4	2.9mg
Resazurin	0.1mg
Yeast extract solution	10.0mL
Peptone solution	10.0mL
Sulfide solution	5.0mL

pH 6.8 ± 0.2 at 25°C

Sulfide Solution:

Composition per 10.0mL:

$Na_2S·9H_2O$	0.5g

Preparation of Sulfide Solution: Add $Na_2S·9H_2O$ to distilled/deionized water and bring volume to 10.0mL. Mix thoroughly. Autoclave under 100% N_2 for 15 min at 15 psi pressure–121°C. Cool to room temperature. Adjust pH to 7.0.

Yeast Extract Solution:

Composition per 10.0mL:

Yeast extract	1.0g

Preparation of Yeast Extract Solution: Add yeast extract to distilled/deionized water and bring volume to 10.0mL. Mix thoroughly. Autoclave for 15 min at 15 psi pressure–121°C. Cool to room temperature. Sparge with 100% N_2.

Peptone Solution:

Composition per 10.0mL:

Peptone	4.0g

Preparation of Peptone Solution: Add peptone to distilled/deionized water and bring volume to 10.0mL. Mix thoroughly. Autoclave for 15 min at 15 psi pressure–121°C. Cool to room temperature. Sparge with 100% N_2.

Preparation of Medium: Add components, except yeast extract, peptone, and sulfide solutions, to distilled/deionized water and bring volume to 975.0mL. Mix thoroughly. Gently heat and bring to boiling. Cool to room temperature while sparging with 100% N_2. Adjust pH to 6.8. Prepare the medium without the yeast extract, peptone, and sodium sulfide. Boil the medium and cool under nitrogen. Adjust the pH to 6.8. Dispense into Hungate tubes or serum bottles under a nitrogen atmosphere. Sterilize the medium at 100°C for 3 hr on 3 consecutive days. Aseptically add the peptone and yeast extract solutions. Adding the sterile, neutralized sulfide solution to an end concentration of 0.025%. Final pH should be 6.8.

Use: For the cultivation of *Thermococcus marinus*.

YPS Medium (DSMZ Medium 1168)

Composition per liter:

Sea salts, Sigma	25.0g
Yeast extract	4.0g
Peptone	2.0g

pH 7.5 ± 0.2 at 25°C

Preparation of Medium: Add components to distilled/deionized water and bring volume to 1.0L. Mix thoroughly. Adjust pH to 7.5. Distribute into tubes or flasks. Autoclave for 15 min at 15 psi pressure–121°C.

Use: For the cultivation of *Owenweeksia hongkongensis*.

YPSC Agar (Yeast Extract Peptone Sulfate Cysteine Agar)

Composition per liter:

Agar	15.0g
Yeast extract	1.0g
Peptone	1.0g
Sodium acetate·$3H_2O$	0.5g
$MgSO_4·7H_2O$	0.25g
$CaCl_2·2H_2O$	0.25g
L-Cysteine·HCl·H_2O	0.05g

pH 7.5 ± 0.2 at 25°C

Preparation of Medium: Add components to distilled/deionized water and bring volume to 1.0L. Mix thoroughly. Gently heat and bring to boiling. Distribute into tubes or flasks. Autoclave for 15 min at 15 psi pressure–121°C. Adjust pH to 7.5 with sterile 10*M* NaOH. Pour into sterile Petri dishes or leave in tubes.

Use: For the cultivation and maintenance of *Bdellovibrio* species.

YPSC Agar, Cation-Supplemented

Composition per liter:

Sodium acetate·$3H_2O$	50.0g
Agar	15.0g
Peptone	10.0g
Yeast extract	10.0g
$MgSO_4·7H_2O$	0.74g
$CaCl_2·2H_2O$	0.29g
L-Cysteine·HCl·H_2O	0.05g
Bacitracin solution	10.0mL

Bacitracin Solution:

Composition per 10.0mL:

Bacitracin	6,000U

Preparation of Bacitracin Solution: Add bacitracin to distilled/deionized water and bring volume to 10.0mL. Mix thoroughly. Filter sterilize.

Preparation of Medium: Add components, except bacitracin solution, to distilled/deionized water and bring volume to 990.0mL. Mix thoroughly. Gently heat and bring to boiling. Autoclave for 15 min at 15 psi pressure–121°C. Cool to 45°–50°C. Aseptically add sterile bacitracin solution. Mix thoroughly. Pour into sterile Petri dishes or distribute into sterile tubes.

Use: For the cultivation and enumeration of *Bdellovibrio* species.

YPSC Medium (Yeast Extract Peptone Sulfate Cysteine Medium)

Composition per liter:

Yeast extract	1.0g
Peptone	1.0g
Sodium acetate·$3H_2O$	0.5g
$MgSO_4·7H_2O$	0.25g
$CaCl_2·2H_2O$	0.25g
L-Cysteine·HCl·H_2O	0.05g

pH 7.5 ± 0.2 at 25°C

Preparation of Medium: Add components to distilled/deionized water and bring volume to 1.0L. Mix thoroughly. Distribute into tubes or flasks. Autoclave for 15 min at 15 psi pressure–121°C. Adjust pH to 7.5 with sterile 10*M* NaOH.

Use: For the cultivation and enumeration of *Bdellovibrio* species.

YPSC Soft Agar (Yeast Extract Peptone Sulfate Cysteine Soft Agar)

Composition per liter:

Agar	6.0g
Yeast extract	1.0g
Peptone	1.0g
Sodium acetate·$3H_2O$	0.5g
$MgSO_4 \cdot 7H_2O$	0.25g
$CaCl_2 \cdot 2H_2O$	0.25g
L-Cysteine·HCl·H_2O	0.05g

pH 7.5 ± 0.2 at 25°C

Preparation of Medium: Add components to distilled/deionized water and bring volume to 1.0L. Mix thoroughly. Gently heat and bring to boiling. Distribute into tubes or flasks. Autoclave for 15 min at 15 psi pressure–121°C. Adjust pH to 7.5 with sterile 10*M* NaOH. Pour into sterile Petri dishes or leave in tubes.

Use: For the cultivation and maintenance of *Bdellovibrio* species.

YpSs Agar

Composition per liter:

Agar	15.0g
Soluble starch	15.0g
Yeast extract	4.0g
K_2HPO_4	1.0g
$MgSO_4 \cdot 7H_2O$	0.5g

Preparation of Medium: Add components to distilled/deionized water and bring volume to 1.0L. Mix thoroughly. Gently heat and bring to boiling. Distribute into tubes or flasks. Autoclave for 15 min at 15 psi pressure–121°C. Pour into sterile Petri dishes or leave in tubes.

Use: For the cultivation of bacteria that can utilize starch as a carbon source.

YPSS, Emerson Agar

Composition per liter:

Agar	15.0g
Soluble starch	15.0g
Yeast extract	4.0g
K_2HPO_4	1.0g
$MgSO_4 \cdot 7H_2O$	0.5g

Preparation of Medium: Add components to distilled/deionized water and bring volume to 1.0L. Mix thoroughly. Gently heat and bring to boiling. Distribute into tubes or flasks. Autoclave for 15 min at 15 psi pressure–121°C. Pour into sterile Petri dishes or leave in tubes.

Use: For the cultivation and maintenance of *Allomyces javanicus, Melanospora tiffanii,* and *Sporothrix schenckii.*

YSP Agar

Composition per liter:

Sucrose	20.0g
Agar	12.0g
Peptone	10.0g
Yeast extract	5.0g

pH 7.0 ± 0.2 at 25°C

Preparation of Medium: Add components to distilled/deionized water and bring volume to 1.0L. Mix thoroughly. Gently heat and bring to boiling. Distribute into tubes or flasks. Autoclave for 15 min at 15 psi pressure–121°C. Pour into sterile Petri dishes or leave in tubes.

Use: For the cultivation of *Xanthomonas albilineans.*

YT HiVeg Broth

Composition per liter:

Plant hydrolysate	16.0g
Yeast extract	10.0g
NaCl	5.0g

pH 7.2 ± 0.2 at 25°C

Source: This medium is available as a premixed powder from HiMedia.

Preparation of Medium: Add components to distilled/deionized water and bring volume to 1.0L. Mix thoroughly. Distribute into tubes or flasks. Autoclave for 15 min at 15 psi pressure–121°C.

Use: For the cultivation of *Escherichia coli.*

YT Medium (Yeast Extract Tryptone Medium)

Composition per liter:

Pancreatic digest of casein	8.0g
Yeast extract	5.0g
NaCl	5.0g

pH 7.2 ± 0.2 at 25°C

Preparation of Medium: Add components to distilled/deionized water and bring volume to 1.0L. Mix thoroughly. Distribute into tubes or flasks. Autoclave for 15 min at 15 psi pressure–121°C.

Use: For the cultivation of *Escherichia coli.*

YTG Medium

Composition per liter:

Tryptone	10.0g
Na_2CO_3	5.3g
Yeast extract	5.0g
$Na_2HPO_4 \cdot 2H_2O$	0.356g
L-Cysteine·HCl	0.2g
$Na_2S \cdot 9H_2O$	0.2g
KCl	0.075g
Resazurin	1.0mg
Glucose solution	20.0mL

pH 10.1 ± 0.2 at 25°C

Glucose Solution:

Composition per 20.0mL:

D-Glucose	3.0g

Preparation of Glucose Solution: Add glucose to distilled/deionized water and bring volume to 20.0mL. Mix thoroughly. Sparge with 100% N_2. Autoclave for 15 min at 15 psi pressure–121°C.

Preparation of Medium: Prepare and dispense medium under 100% N_2. Add components, except glucose solution, to distilled/deionized water and bring volume to 980.0L. Mix thoroughly. Sparge with 100% N_2 for 30 min. Anaerobically distribute 9.8mL volumes into tubes. Autoclave for 15 min at 15 psi pressure–121°C. Aseptically and anaerobically add 0.2mL of sterile glucose solution to each tube. Adjust pH to 10.1 with sterile anaerobic 3*N* NaOH solution.

Use: For the cultivation of *Clostridium paradoxum* and *Clostridium thermoalcaliphilum.*

YTN Medium (Yeast Extract Tryptone NaCl Medium)

Composition per liter:

NaCl	30.0g
Agar	15.0g
Yeast extract	10.0g
Pancreatic digest of casein	10.0g
Glucose	1.0g
Trace elements solution	1.0mL

pH 7.2 ± 0.2 at 25°C

Trace Elements Solution:

Composition per liter:

H_3BO_3	2.85g
$MnCl_2 \cdot 4H_2O$	1.8g
Sodium tartrate	1.77g
$FeSO_4$	1.36g
$CoCl_2 \cdot 6H_2O$	0.04g
$CuCl_2.2H_2O$	0.027g
$Na_2MoO_4 \cdot 2H_2O$	0.025g
$ZnCl_2$	0.021g

pH 7.2 ± 0.2 at 25°C

Preparation of Trace Elements Solution: Add components to distilled/deionized water and bring volume to 1.0L. Mix thoroughly.

Preparation of Medium: Add components to tap water and bring volume to 1.0L. Mix thoroughly. Gently heat and bring to boiling. Distribute into tubes or flasks. Autoclave for 15 min at 15 psi pressure–121°C. Pour into sterile Petri dishes or leave in tubes.

Use: For the cultivation of ATCC strain 21588.

YTSS Medium, Half Strength (DSMZ Medium 974)

Composition per liter:

Sea salts	20.0g
Yeast extract	2.0g
Tryptone	1.25g

pH 7.0 ± 0.2 at 25°C

Preparation of Medium: Add components to distilled/deionized water and bring volume to 1.0L. Mix thoroughly. Distribute into tubes or flasks. Autoclave for 15 min at 15 psi pressure–121°C.

Use: For the cultivation of *Roseovarius nubinhibens* and *Silicibacter pomeroyi.*

Z Agar

Composition per liter:

Agar	16.0
K_2HPO_4	5.0g
K_2SO_4	2.0g
KH_2PO_4	1.0g
$MgSO_4 \cdot 7H_2O$	0.05g
Acetamide solution	40.0mL

pH 7.2 ± 0.2 at 25°C

Acetamide Solution:

Composition per 100.0mL:

Acetamide	10.0g

Preparation of Acetamide Solution: Add acetamide to distilled/deionized water and bring volume to 100.0mL. Mix thoroughly. Filter sterilize.

Preparation of Medium: Add components, except acetamide solution, to distilled/deionized water and bring volume to 960.0mL. Mix thoroughly. Autoclave for 20 min at 15 psi pressure–121°C. Cool to 70°C. Aseptically add 40.0mL sterile acetamide solution. Mix thoroughly. Pour into sterile Petri dishes. Dry plates at 37°C for 30 min.

Use: For the isolation of *Pseudomonas aeruginosa* from milk.

Z Broth

Composition per liter:

Acetamide	5.0g
K_2HPO_4	5.0g
KH_2PO_4	3.0g
KNO_3	1.0g
$K_2S_4O_6$	1.0g
$MgSO_4 \cdot 7H_2O$	0.05g

pH 7.2 ± 0.2 at 25°C

Preparation of Medium: Add components to distilled/deionized water and bring volume to 1.0L. Mix thoroughly. Gently heat and bring to boiling. Cool. Aseptically distribute 10.0mL volumes into test tubes containing inverted Durham tubes. Heat for 15 min at 0 psi pressure–100°C.

Use: For the cultivation of *Pseudomonas aeruginosa* from milk.

Z Medium

Composition per liter:

Casein hydrolysate	10.0g
NaCl	10.0g
Yeast extract	5.0g
Glucose	1.0g
$CaCl_2 \cdot 2H_2O$	0.367g

Preparation of Medium: Add components to distilled/deionized water and bring volume to 1.0L. Mix thoroughly. Distribute into tubes or flasks. Autoclave for 15 min at 15 psi pressure–121°C.

Use: For the cultivation of *Alcaligenes eutrophus.*

Zavarzinella formosa Medium (DSMZ Medium 1196)

Composition per liter:

N-acetylglucosamine	1.0g
Glucose	0.5g
KH_2PO_4	0.1g
Peptone	0.1g
Yeast extract	0.1g
$MgSO_4 \cdot 7H_2O$	0.1g
Casamino acids	0.1g
$CaCl_2 \cdot 2H_2O$	0.05g
NaCl	0.01g
"Metals 44"	1.0mL

pH 5.9 ± 0.2 at 25°C

"Metals 44":

Composition per 100.0mL:

$ZnSO_4 \cdot 7H_2O$	1.095g
$FeSO_4 \cdot 7H_2O$	0.5g

Sodium EDTA 0.25g
$MnSO_4·H2O$ 0.154g
$CuSO_4·5H_2O$ 39.2mg
$Co(NO_3)_2·6H_2O$ 24.8mg
$Na_2B_4O_7·10H_2O$ 17.7mg

Preparation of "Metals 44": Add sodium EDTA to distilled/deionized water and bring volume to 90.0mL. Mix thoroughly. Add a few drops of concentrated H_2SO_4 to retard precipitation of heavy metal ions. Add remaining components. Mix thoroughly. Bring volume to 100.0mL with distilled/deionized water.

Preparation of Medium: Add components to distilled/deionized water and bring volume to 1.0L. Mix thoroughly. Adjust pH to 5.8–6.0. Distribute into tubes or flasks. Autoclave for 15 min at 15 psi pressure–121°C.

Use: For the cultivation of *Zavarzinella formosa.*

ZF2 Medium (DSMZ Medium 943)

Composition per liter:
$NaHCO_3$ 3.8g
Yeast extract 3.0g
$(NH_4)HCO_3$ 0.45g
$MgSO_4·6H_2O$ 0.13g
$CaCl_2·2H_2O$ 0.12g
Resazurin 0.5mg
Phosphate buffer 10.0mL
Glycine solution 5.0mL
Arginine solution 5.0mL
$Na_2S·9H_2O$ solution 5.0mL
Dithionite solution 1.0mL
Seven vitamin solution 1.0mL
Trace elements solution SL-10 .... 1.0mL
Selenite-tungstate solution 1.0mL
pH 7.3 ± 0.2 at 25°C

Arginine Solution:
Composition per 10.0mL:
Arginine-HCl 3.5g

Preparation of Arginine Solution: Add arginine-HCl to distilled/deionized water and bring volume to 10.0mL. Mix thoroughly. Filter sterilize.

Glycine Solution:
Composition per 100.0mL:
Glycine 15.0g

Preparation of Glycine Solution: Add glycine to distilled/deionized water and bring volume to 100.0mL. Mix thoroughly. Sparge with 100% N_2. Autoclave for 15 min at 15 psi pressure–121°C. Cool to room temperature.

$Na_2S·9H_2O$ Solution:
Composition per 10.0mL:
$Na_2S·9H_2O$ 0.3g

Preparation of $Na_2S·9H_2O$ Solution: Add $Na_2S·9H_2O$ to distilled/deionized water and bring volume to 10.0mL. Mix thoroughly. Autoclave under 100% N_2 for 15 min at 15 psi pressure–121°C. Cool to room temperature.

Dithionite Solution:
Composition per 10.0mL:
Na_2-dithionite 0.25g

Preparation of Dithionite Solution: Add Na_2-dithionite to distilled/deionized water and bring volume to 10.0mL. Mix thoroughly. Autoclave under 100% N_2 for 15 min at 15 psi pressure–121°C. Cool to 25°C.

Seven Vitamin Solution:
Composition per liter:
Pyridoxine hydrochloride 300.0mg
Thiamine-HCl·$2H_2O$ 200.0mg
Nicotinic acid 200.0mg
Vitamin B_{12} 100.0mg
Calcium pantothenate 100.0mg
p-Aminobenzoic acid 80.0mg
D(+)-Biotin 20.0mg

Preparation of Seven Vitamin Solution: Add components to distilled/deionized water and bring volume to 1.0L. Sparge with 100% N_2. Mix thoroughly. Filter sterilize.

Phosphate Buffer:
Composition per liter:
$Na_2HPO_4·12H_2O$ 43.0g
KH_2PO_4 .... 5.44g

Preparation of Phosphate Buffer: Add components to distilled/deionized water and bring volume to 1.0L. Mix thoroughly. Adjust pH to 7.3. Autoclave for 15 min at 15 psi pressure–121°C. Cool to 50°C.

Trace Elements Solution SL-10:
Composition per liter:
$FeCl_2·4H_2O$ 1.5g
$CoCl_2·6H_2O$ 190.0mg
$MnCl_2·4H_2O$ 100.0mg
$ZnCl_2$ 70.0mg
$Na_2MoO_4·2H_2O$ 36.0mg
$NiCl_2·6H_2O$ 24.0mg
H_3BO_3 .... 6.0mg
$CuCl_2·2H_2O$ 2.0mg
HCl (25% solution) 10.0mL

Preparation of Trace Elements Solution SL-10: Add $FeCl_2·4H_2O$ to 10.0mL of HCl solution. Mix thoroughly. Add distilled/deionized water and bring volume to 1.0L. Add remaining components. Mix thoroughly. Sparge with 80% N_2 + 20% CO_2. Autoclave for 15 min at 15 psi pressure–121°C.

Selenite-Tungstate Solution
Composition per liter:
NaOH 0.5g
$Na_2WO_4·2H_2O$ 4.0mg
$Na_2SeO_3·5H_2O$ 3.0mg

Preparation of Selenite-Tungstate Solution: Add components to distilled/deionized water and bring volume to 1.0L. Mix thoroughly. Sparge with 100% N_2. Filter sterilize.

Preparation of Medium: Prepare and dispense medium under 80% N_2 + 20% CO_2 gas mixture. Add components, except phosphate buffer, glycine solution, arginine solution, dithionite solution, seven vitamin solution, and $Na_2S·9H_2O$ solution, to distilled/deionized water and bring volume to 973.0mL. Mix thoroughly. Equilibrate with 80% N_2 + 20% CO_2 to reach pH 7.3. Autoclave for 15 min at 15 psi pressure–121°C. Cool to 25°C. Aseptically and anaerobically add 10.0mL phosphate buffer, 5.0mL glycine solution, 5.0mL arginine solution, 1.0mL dithionite solution, 1.0mL seven vitamin solution, and 5.0mL $Na_2S·9H_2O$ solution. Mix thoroughly. Aseptically and anaerobically distribute into sterile tubes or flasks.

Use: For the cultivation of *Sedimentibacter saalensis.*

Zoogloea Medium

Composition per liter:

Agar	15.0g
Pancreatic digest of casein	5.0g
Glycerol	5.0g
Yeast autolysate	1.0g
Sodium lactate	0.5g

Preparation of Medium: Add components to distilled/deionized water and bring volume to 1.0L. Mix thoroughly. Gently heat and bring to boiling. Distribute into screw-cap test tubes. Autoclave for 15 min at 15 psi pressure–121°C. Pour into sterile Petri dishes or leave in tubes.

Use: For the cultivation and maintenance of *Zoogloea ramigera* and other *Zoogloea* species.

Zymobacterium Agar

Composition per liter:

Pancreatic digest of casein	20.0g
Agar	15.0g
Orotic acid	2.0g
$Na_3PO_4 \cdot 12H_2O$	1.5g
Sodium thioglycolate	1.0g
NaOH	0.5g
Riboflavin	0.015g

pH 7.9 ± 0.2 at 25°C

Preparation of Medium: Add components to distilled/deionized water in the following order: tryptone, orotic acid, sodium hydroxide, sodium phosphate, sodium thioglycolate, riboflavin, agar and bring volume to 1.0L . Mix thoroughly. Some orotic acid will remain undissolved. Adjust pH to 7.9 using NaH_2PO_4. Distribute into tubes or flasks. Autoclave for 15 min at 15 psi pressure–121°C. Mix thoroughly to dissolve orotic acid. Pour into sterile Petri dishes or leave in tubes.

Use: For the cultivation and maintenance of *Clostridium (Zymobacterium) oroticum.*

Zymobacterium Broth

Composition per liter:

Pancreatic digest of casein	20.0g
Orotic acid	2.0g
$Na_3PO_4 \cdot 12H_2O$	1.5g
Sodium thioglycolate	1.0g
NaOH	0.5g
Riboflavin	0.015g

pH 7.9 ± 0.2 at 25°C

Preparation of Medium: Add components to distilled/deionized water in the following order: tryptone, orotic acid, sodium hydroxide, sodium phosphate, sodium thioglycolate, riboflavin and bring volume to 1.0L. Mix thoroughly. Some orotic acid will remain undissolved. Adjust pH to 7.9 using NaH_2PO_4. Distribute into tubes or flasks. Autoclave for 15 min at 15 psi pressure–121°C. Mix again to dissolve orotic acid.

Use: For the cultivation and maintenance of *Clostridium (Zymobacterium) oroticum.*

Zymomonas Agar

Composition per liter:

Glucose	20.0g
Agar	15.0g
Yeast extract	5.0g

Preparation of Medium: Add components to distilled/deionized water and bring volume to 1.0L. Mix thoroughly. Gently heat and bring to boiling. Distribute into tubes or flasks. Autoclave for 15 min at 15 psi pressure–121°C. Pour into sterile Petri dishes or leave in tubes.

Use: For the cultivation and maintenance of *Zymomonas mobilis* and other *Zymomonas* species.

Zymomonas Medium

Composition per liter:

Glucose	20.0g
Agar	15.0g
Peptone	10.0g
Yeast extract	10.0g

pH 6.8 ± 0.2 at 25°C

Preparation of Medium: Add components to distilled/deionized water and bring volume to 1.0L. Mix thoroughly. Gently heat and bring to boiling. Distribute into tubes or flasks. Autoclave for 15 min at 15 psi pressure–121°C. Pour into sterile Petri dishes or leave in tubes.

Use: For the cultivation of *Zymomonas anaerobia.*

Zymomonas Medium

Composition per liter:

Glucose	10.0g
Yeast extract	10.0g

pH 4.8 ± 0.2 at 25°C

Preparation of Medium: Add components to tap water and bring volume to 1.0L. Mix thoroughly. Adjust pH to 4.8. Distribute into tubes or flasks. Autoclave for 20 min at 15 psi pressure–121°C. Boil the medium immediately before use.

Use: For the cultivation of *Zymomonas mobilis* and other *Zymomonas* species.

Zymomonas Medium (ATCC Medium 845) (ATCC Medium 948)

Composition per liter:

Glucose	20.0g
Yeast extract	5.0g

Preparation of Medium: Add components to distilled/deionized water and bring volume to 1.0L. Mix thoroughly. Autoclave for 15 min at 15 psi pressure–121°C. Pour into sterile Petri dishes or distribute into sterile tubes.

Use: For the cultivation of *Zymomonas mobilis* and other *Zymomonas* species.

Zymomonas Sucrose Medium

Composition per liter:

Sucrose	150.0g
Yeast extract	2.0g
Peptone	2.0g
KH_2PO_4	2.0g
$(NH_4)_2SO_4$	2.0g
$MgSO_4 \cdot 7H_2O$	2.0g

pH 7.0 ± 0.2 at 25°C

Preparation of Medium: Add components to distilled/deionized water and bring volume to 1.0L. Mix thoroughly. Adjust pH to 7.0. Distribute into tubes or flasks. Autoclave for 15 min at 15 psi pressure–121°C.

Use: For the cultivation of *Zymomonas mobilis.*

A

C

D

E

F

G

O

P

Z